AMERICAN
MEN AND WOMEN
OF SCIENCE

AMERICAN MEN AND WOMEN OF SCIENCE

13TH EDITION

Edited by the Jaques Cattell Press

Discipline and
Geographic Indexes

R. R. BOWKER COMPANY
A Xerox Publishing Company
New York & London, 1976

Contents

Advisory Committee

Dr. Dael L. Wolfle, Chairman
Graduate School of Public Affairs
University of Washington

Dr. Randolph W. Bromery
Chancellor,
University of Massachusetts

Dr. Janet W. Brown
Program Head,
Office of Opportunities
in Science
American Association for the
Advancement of Science

Dr. Robert W. Cairns
Executive Director,
American Chemical Society

Dr. S. D. Cornell
Assistant to the President
National Academy of Sciences

Dr. Ruth M. Davis
Director,
Institute for Computer Science
and Technology
National Bureau of Standards

Dr. Carl D. Douglass
Deputy Director,
Division of Research Grants
National Institutes of Health

Dr. Richard G. Folsom
President Emeritus,
Rensselaer Polytechnic Institute

Dr. Robert E. Henze
Director,
Membership Division
American Chemical Society

Dr. Eugene L. Hess
Executive Director,
Federation of American Societies
for Experimental Biology

Dr. William C. Kelly
Executive Director,
Commission on Human Resources
National Research Council

Dr. Kenneth B. Raper
Department of Bacteriology
University of Wisconsin

Dr. A. L. Schawlow
Department of Physics
Stanford University

Dr. John F. Sherman
Vice President,
Association of American Medical
Colleges

Dr. Matthias Stelly
Executive Vice President,
American Society of Agronomy

Dr. John R. Whinnery
Department of Electrical Engineering
and Computer Sciences
University of California, Berkeley

Preface

The observance of an anniversary prompts reflection on the past, and it is natural to note here that the astounding growth in the dimension and stature of this 200 year old American nation has been influenced immeasurably by the achievements of her scientists. 1976 also marks the 70th anniversary of AMERICAN MEN & WOMEN OF SCIENCE as a chronicle of the lives and professional activities of those men and women most instrumental in affecting the shape and quality of science in America. The explosion in scientific activity over the past seven decades is clearly evident when one compares the 1906 edition, a single volume containing 4,000 entries, with the present edition, six volumes profiling nearly 110,000 men and women of importance in their fields.

This edition of AMERICAN MEN & WOMEN OF SCIENCE is a landmark in biographic achievement. The information contained in all seven volumes has been gathered, edited and compiled by the Jaques Cattell Press in the space of ten months. This is a radical, and beneficial, departure from the production of the 12th edition, which took three years to publish in its entirety. The acceleration was made possible by the use of a computerized printing method and more efficient production procedures.

The editors have not sacrificed quality in the interest of speed, however. The criteria were stringently applied in the selection of new entrants, all nominated by former biographees. The criteria follow.

Achievement, by reason of experience and training, of a stature in scientific work equivalent to that associated with the doctoral degree, coupled with presently continued activity in such work;

or

Research activity of high quality in science as evidenced by publication in reputable scientific journals; or, for those whose work cannot be published because of governmental or industrial security, research activity of high quality in science as evidenced by the judgment of the individual's peers;

or

Attainment of a position of substantial responsibility requiring scientific training and experience to the extent described for (1) and (2).

Geographic and discipline indexes make up this seventh volume of the set. The discipline index has been rearranged and is organized with major subject headings, providing easier access to the information. The subheadings are com-

piled from disciplines requested by the biographees, and thirty percent of the names appear under more than one subject. In most cases the geographic index reflects the business location of the person listed.

Certain disciplines previously included in the directory are not represented in this edition. Engineering and economics will appear in separate directories with expanded criteria, to enable even broader coverage of these areas. Others, including sociology, psychology and political science were omitted because they are fully covered by membership directories in each field.

Great appreciation is expressed to the AMERICAN MEN & WOMEN OF SCIENCE Advisory Committee for their guidance in the planning of the 13th edition. Their efforts have contributed to an unusually good response to our requests for information and nominations, which has enhanced the value of the publication. Also to be thanked are the many scientific societies that provided membership lists for the use of our researchers or that published announcements in their bulletins and journals.

The staff of the Jaques Cattell Press deserves the highest accolade for their sustained interest, devotion and good will through the many hours of learning and implementing the new procedures necessary to the successful completion of this book. The overwhelming workload was shared by temporary employees who performed with the greatest diligence and responsibility. The job could not have been completed without their fine help. Everyone involved with this project gave outstanding service, but the contributions of Alice Smith, Pauline Stump, Joyce Howell and Ila Martin cannot go without mention. Special acknowledgement is given to Fred Scott, former general manager of the Jaques Cattell Press, who was responsible for initiating the formation of the advisory committee and for overseeing the planning and early stages of work on the 13th edition.

Comments and suggestions are invited and should be addressed to The Editors, American Men & Women of Science, Jaques Cattell Press, P.O. Box 25001, Tempe, Arizona, 85282.

Renee Lautenbach, Supervising Editor
Anne Rhodes, Administrative Managing Editor

JAQUES CATTELL PRESS

Desmond Reaney, Manager Book Editorial

R. R. BOWKER COMPANY

October, 1976

Discipline Index

1

ACOUSTICS

Vent, Robert Joseph
Walker, Richard Alden
White, Richard Wallace

AGRICULTURE

Agriculture

Aagan, Raymond John
Amerine, Maynard Andrew
Arkin, Gerald Franklin
Ball, Wilbur Perry
Bertrand, Forest
Bowers, Clarence C
Bowling, John Dalton
Boynton, Damon
Bristol, Benton Keith
Burke, Philip William
Butler, Karl Douglas, Sr
Call, Edward Prior
Capo, Bernardo Guillermo
Cerbulis, Janis
Cobb, Walter R
Conley, Janis
Cooper, Tommye
Coulombe, Louis Joseph
Czarnetzky, Edward John
Dillon, Roy Dean
Dowler, Lloyd
Eby, Robert L
Emmons, Douglas Byron
Ferguson, Carl E
Fifield, Willard Marvin
Fisher, Theodore Roosevelt
Freeman, Verne Crawford
Garman, Willard Hershel
Garrigus, Wesley Patterson
Glover, Earl Robert
Goto, Shisuke
Hamilton, Richard Airth
Hatchcock, Bobby Ray
Hopkins, Homer Thawley, Jr
Householder, William Allen
Hull, Jerome, Jr
Huntington, David Hans
Ideker, Richard Louis
James, David Winston
Kanemasu, Edward Tsukasa
Kearney, Philip C
Kuratle, Henry III
Larson, Vernon C
Little, Charles Oran
Lukosevicius, Petras Povilas
MacLeod, Guy Franklin
Marx, George Donald
McCloud, Darell Edison
McMurray, Birch Lee
Mercier, Ernest
Moseman, Albert Henry
Nittler, LeRoy Walter
Noffsinger, Terrell L
Oomens, Frederick Walter
Perumal, Alexander
Pisano, Rocci George
Rammer, Irwin Alden
Ritchie, Austin E
Ross, James George
Rousek, Edwin J
Sandhu, Shabeg Singh
Schowengerdt, George Carl
Sigafus, Roy Edward
Smith, Allan Edward
Smith, Vearl Robert
Snyder, Fred Calvin
Spahr, Sidney Louis
Stringam, Elwood Williams
Sutherland, William Neil
Switzer, Clayton Macfie
Tait, Robert Malcolm
Tan, Kim H
Transtrum, Lloyd G
Van Krey, Harry Peter
von Schmeling, Bogislav G
Walker, Courtney Emery
Whitlock, Gaylord Purcell
Whitney, Wendell Keith
Wiley, William Henry
Williams, John Simeon
Wilson, George Rodger
Wolff, Robert L
Wondering, Thomas Franklin
Young, Bruce Arthur

Agricultural Biochemistry

Bennett, Raymond Dudley
Bice, Claude Wesley
Boyd, John Edward
Chen, Steve Shih-Chieh
Clark, Burr, Jr
Corsini, Dennis Lee
Craig, Burton Mackay
Crane, Elliott Maurice
Cunningham, David Kenneth
Dirks, Brinton Mario
El-Negoumy, Abdul Monem
Evans, James William
Fan, Hsing Yun

Fischer, William Carl, Jr
Frear, Donald Stuart
Garner, George Bernard
Gilles, Kenneth Albert
Golub, Tomasz
Gordon, Arthur Leonard
Greenaway, Walter Thomas
Hamilton, John Kelvin
Hayman, Ernest Paul
Heots, James Peter
Hill, Robert D
Hope, Hugh Johnson
Hunter, George William
Jackson, John Edward
Jaffe, Lionel
Jenner, Edward Levant
Keating, Eugene Kneeland
Knoche, Herman William
Kuck, James Chester
Leng, Marguerite Lambert
Loeffler, Josef Ernst
Mahon, John Harold
Matsumoto, Hiromu
McLaren, George Aiken
Melnychyn, Paul
Mitchell, Howard Lee
Montoure, John Ernest
Olson, Oscar Edward
Oonnithan, Easwaran Sukumaran
Palmere, Raymond M
Prebluda, Harry Jacob
Prentice, Neville
Privett, Orville Samuel
Rogers, Charles Fletcher
Sayre, Robert Newton
Verma, Devi C
Vose, John Randal

Agricultural Chemistry

Abramitis, Walter William
Advani, Shyam Bhojraj
Anastassiadis, Phoebus A
Ballantine, Larry Gene
Barthel, William Frederick
Beestman, George Bernard
Bishop, John Russell
Bjork, Carl Kenneth, Sr
Brewer, Arthur David
Budde, Paul Bernard
Burton, Joe Covington
Cattani, Ray August
Chan, Hak-Foon
Chin, Wei Tsung
Common, Robert Haddon
Docks, Edward Leon
Duisberg, Peter Caspar
Dyck, Arnold Wolff Jan
Erickson, David R
Fallscheer, Herman O
Fang, Sheng Chung
Felton, Staley Lee
Fitzpatrick, Thomas Joseph
Gatterdam, Paul Esch
Gerwitz, David L
Greenwald, Bernard William
Halverson, Andrew Wayne
Hamaker, John Warren
Hardesty, John Oliver
Hess, Earl Hollinger
Ikeda, Robert Misuru
Inglett, George Everett
Johnson, Wayne Orrin
Khan, Shahamat Ullah
Khayat, Ali
Kilsheimer, John Robert
Knapp, Joseph Leonce, Jr
Kuhr, Ronald John
Labbe, Marcel D
L'Annunziata, Michael Frank
Lepage, Marius
Lichy, Charles Thorne
Loeffler, Erwin Stanley
Luckenbaugh, Raymond Wilson
Lutz, Albert Miriam
Mabrouk, Ahmed Fahmy
MacIntyre, Thomas Martin
Mackiney, Gordon
Mann, Robert Leslie
Maravetz, Lester L
Marks, Alfred Finlay
Martin, Edward Eugene
Masri, Merle Sid
May, Ralph Forrest
McInroy, Elmer Eastwood
Mosier, Arvin Ray
Mounts, Timothy Lee
Newsom, William S, Jr
Noznick, Peter Paul
Olney, Charles Edward
Owens, Daniel Kenyon
Parrill, Irwin Homer
Pearce, John Archibald
Pence, James William
Peterson, Richard Grant
Polen, Percy B
Reed, John Fielding

Rigby, F Lloyd
Rockland, Louis B
Rogers, Richard Brewer
Rogerson, Thomas Dean
Rosenfield, Christine Ann Culp
Rupp, Eldor Gustav
Salomon, Milton
Schaefer, Wilbur Carl
Schechter, Milton Seymour
Shasha, Baruch
Skapason, Joseph Bjorn
Smith, Joseph A
Spencer, Elvins Yuill
Stansbury, Mack Fulton
Stein, Robert George
Strohm, Paul F
Swift, Lyle James
Talley, Eugene Alton
Thorn, George Denis
Thurman, Duane Edward
Treichler, Ray
Treves, Gino Robert
Turner, Fred, Jr
Van Landingham, Audrey Howard
Von Runkle, Rosmarie
Walker, Howard George, Jr
Wallace, Volney
White, Jonathan Winborne, Jr
Witman, Eugene DeWald
Wommack, Joel Benjamin, Jr
Wrolstad, Ronald Earl
Yih, Roy Yangming
Young, Donald C
Young, Hong Yip

Agricultural Economics

Adams, Dale W
Anderson, James Richard
Baldwin, Eldon Dean
Beer, Charles
Bressler, Glenn Otto
Buddemeier, Wilbur Dahl
Burns, David Jerome
Cummings, Ralph Waldo, Jr
Daum, Richard J
Epp, Donald James
Erven, Bernard Lee
Forster, D Lynn
Found, William Charles
Fuller, Wayne Arthur
Glover, Loyd, Jr
Halter, Albert Nelson
Halvorson, Lloyd Chester
Hardy, Ernest Edward
Harris, Charles Lawrence
Henderson, Dennis Roger
Herdt, Robert William
Hess, Carroll V
Hesser, Leon Francis
Hicks, John W, III
Hieronymus, Thomas Applegate
Higbee, Edward
Hushak, Leroy J
Kelly, Bernard Wayne
Larson, Donald W
Lee, Warren Ford
Luby, Patrick Joseph
McCormick, Francis B
Meyer, Richard Lee
Miller, William Lloyd
Nelson, A Gene
Niles, James Alfred
Nixon, Donald Merwin
Plaxico, James Samuel
Purcell, Joseph Carroll
Rask, Norman
Rawlins, Nolan Omri
Seitz, Wesley Donald
Shaudys, Edgar T
Sheppard, Charles Campbell
Siebert, Jerome Bernard
Smith, Ralph Emerson
Stopper, William W
Vertrees, Robert Layman
Walker, Francis Edwin
Ward, Ronald Wayne
Whitson, Robert Edd
Wilson, Richard Heilbron
Workman, John Paul
Workman, William Glenn

Agricultural Education

Collins, Robert Matthew
Larson, Russell Edward
Pesson, Lynn E
Ritchie, Austin Frederick
Welton, Richard Frederick
Wilson, Richard Heilbron

Agricultural Geography

Blaut, James Morris
Chakravarti, Aninda Kumar
Hanson, Austin Moe
Hills, Theodore Lewis
Kollmorgen, Walter Martin
Malone, Philip Garin
Olmstead, Clarence Walter
Reeds, Lloyd George

Agricultural Meteorology

Baier, Wolfgang
Bark, Laurence Dean
Brown, Donald Murray
Campbell, Gaylon Sanford
Caprio, Joseph Michael
Gerber, John Francis
Gillespie, Terry James
Gillette, Dale Alan
Marsolf, J David
Pack, Albert Boyd
Robertson, George Wilber
Shaw, Robert Harold
Taylor, Sterling Elwynn
Wesley, Marvin Larry
Wright, James Louis

Agricultural Microbiology

Appleton, George Sanders
Baribo, Lester E
Barran, Leslie Rohit
Bell, Robert Graham
Chandra, Purna
Cheng, Kuo-Joan
Cook, Harold Andrew
Jones, Graham Alfred
Kennedy, Elhart James
Marsden, David Henry
Norstadt, Fred A
Steinke, Paul Karl Willi
Stevenson, Ian Lawrie
Stoller, Benjamin Boris
Volz, Michael George

Agricultural Statistics

Gates, Charles Edgar
Hills, F Jackson
Hurt, Paul Victor
Littell, Ramon Clarence
Munson, Arvid W

Agronomy

Ahlgren, Henry Lawrence
Ahring, Robert M
Akeson, Walter Roy
Albrecht, Herbert Richard
Aldrich, Samuel Roy
Alexander, Charles William
Allan, Robert Emerson
Allen, Freddie Lewis
Allen, Robert John, Jr
Allred, Keith Reid
Allred, Rodney Chase
Alvey, David Dale
Anderson, Melvern K
Anderson, Oscar Emmett
Anderson, Stanley Robert
Andrew, Robert Harry
Andrews, Cecil Hunter
Arceneaux, George
Archimovich, Alexander S
Asay, Kay Harris
Ashford, Ross
Ashley, Doyle Allen
Atkins, Irvin Milburn
Atkins, Richard Elton
Austenson, Herman Milton
Ayo, Donald Joseph
Bader, Kenneth H
Baker, Barton Scofield
Balasko, John Allan
Bandeen, John Drummond
Bandel, Vernon Allan
Barker, LeRoy N
Barnes, Carl Eldon
Barnes, Robert F.
Barnett, Ronald David
Barney, Archie Fay
Barrett, Thomas Wilson
Barta, Allan Lee
Bass, Garland Booker
Batchelder, Arthur Roland
Battle, Richard Pierce
Bauer, Warren Rich
Baumgardner, Marion F
Baylor, John E
Beaty, Elvis Roy
Beavers, Alvin Herman
Beinhart, Ernest George, Jr
Bell, Frank F
Bennett, William Frederick
Bennett, William Hunter
Bergeaux, Phillip James
Bertinuson, Torvald Arthur
Bertramson, Bertram Rodney
Bieber, Gene Lawrence
Bieberly, Frank Gearhart
Bishnoi, Udai Ram
Bitzer, Morris Jay
Blair, Byron Oliver
Blake, Carl Thomas
Blaser, Roy Emil
Boehle, John, Jr
Bohning, John William
Bond, Andrew

Borgeson, Carl
Bortner, Charles Eugene
Boswell, Fred Carlen
Bowman, Donald Houts
Boyd, Charles Curtis
Boyd, Frederick Tilghman
Bradfield, Richard
Bradford, Willis Warren
Brady, Nyle C
Breland, Herman Leroy
Brickbauer, Elwood Arthur
Briggs, Robert Eugene
Brooks, Stanley Nelson
Brown, Acton Richard
Brown, Ronald Harold
Bryant, Harry Talbot
Bryner, Clarence Sheldon
Bubar, John Stephen
Buckardt, Henry Lloyd
Bula, Raymond J
Bunch, Harry Dean
Burger, Ambrose William
Burger, Othmar Joseph
Burleson, Charles Albertis
Burnside, Orvin C
Burrows, William Chapel
Burt, Evert Oakley
Burt, Gordon Willis
Burton, Glenn Willard
Burzlaff, Donald Frederick
Caffey, Horace Rouse
Calhoun, Wheeler, Jr
Callahan, Lloyd Milton
Canode, Chester Lang
Cardwell, Vernon Bruce
Carlson, Richard Eugene
Carpenter, Paul Nathaniel
Carson, Eugene Watson, Jr
Carter, Neri Anthony
Carter, Jack Franklin
Carter, John Newton
Carter, Lark Poland
Casady, Alfred Jackson
Cavanah, Lloyd (Earl)
Chamblee, Douglas Scales
Chandler, Robert Flint, Jr
Chapman, Stephen R
Chapman, Willis Harleston
Chevrette, Joseph Edgar
Chilcote, David Owen
Chomchalow, Narong
Chowdhury, Ikbalur Rashid
Christmas, Ellsworth P
Clapp, John Garland, Jr
Clark, Benjamin Edward
Clark, Neri Anthony
Clift, Cecil William
Coffman, Charles Benjamin
Cole, Richard H
Coleman, Otto Harvey
Collier, Jesse Wilton
Collins, Henry A
Conard, Elverne Clyde
Cook, Elton Davis
Cook, Harry Lee
Cook, Maurice Gayle
Cooper, Clee S
Cords, Howard Paul
Cowett, Everett R
Craddock, Garnet Roy
Craig, William F
Craigmiles, Julian Pryor
Crane, Paul Levi
Crawford, Robert Field
Crumpacker, David Wilson
Cummins, David Gray
Cushing, Robert Leavitt
Dade, Philip Eugene
Dalton, Lonnie Gene
Daniel, William Hugh
Daugherty, LeRoy Arthur
Davidson, Steve Edwin
Davis, Charles Homer
Davis, Dick D
Davis, Johnny Henry
Davis, Richard Richardson
Deal, Elwyn Ernest
Decker, Alvin Morris, Jr
De Ment, Jack (Donovan)
Deming, John Miley
Dennis, Robert E
Dessureaux, Lionel
Dickenson, Donald Dwight
Ditterline, Raymond Lee
Dorschner, Kenneth Peter
Dotzenko, Alexander Daniel
Douglas, Alvin Gene
Douglass, Charles Francis
Dowler, Clyde Cecil
Drolsom, Paul Newell
Duble, Richard Lee
Dudeck, Albert Eugene
Duell, Robert William
Duich, Joseph M
Dunavin, Leonard Sypret, Jr
Eakin, James (Henry), Jr
Earley, Ernest Benton
Eastin, Jerry Dean
Eastin, John A

Edwards, Lewis Hiram
Egli, Dennis B
Elkins, Donald Marcum
Ensign, Ronald D
Ensminger, Leonard Elroy
Erickson, Lambert Cornelius
Eslick, Robert Freeman
Evans, David W
Evans, Raeford G
Everson, Leroy Everett
Ewing, John Arthur
Faix, James Jacob
Farnsworth, Raymond Bartlett
Faw, Wade Farris
Felch, Richard Elroy
Fendall, Roger K
Feuer, Reeshon
Fick, Gary Warren
Fike, William Thomas, Jr
Finfrock, Dwight Curtis
Fink, Dwayne Harold
Fink, Rodney James
Finkner, Ralph Eugene
Fisher, Warner Douglass
Flanagan, Theodore Ross
Foote, Wilson Hoover
Forbes, Ian
Ford, John Harlan
Forsberg, Robert Arnold
Fortmann, Henry Raymond
Foth, Henry Donald
Foutch, Harley Wayne
Frakes, Rodney Vance
Francis, Charles Andrew
Frans, Robert Earl
Freyman, Stanislaw
Fribourg, Henry August
Frolik, Elvin Frank
Fuess, Frederick William, III
Fulkerson, Robert Serpell
Furrer, John D
Galloway, Harry M
Gardner, Franklin Pierce
Gascho, Gary John
Gausman, Harold Wesley
Genter, Clarence Frederick
Gentry, Claude Edwin
Gerloff, Eldean D
Gervais, Paul
Gilbert, Norris W
Gilbert, William Best
Gill, William Robert
Gist, George Reinecker
Goetze, Norman Richard
Gooding, John Alan
Gorbet, Daniel Wayne
Goss, Roy Leon
Gossett, Dorsey McPeake
Graffis, Don Warren
Graham, James Carl
Graumann, Hugo Oswalt
Gray, Clarence Cornelius, III
Gray, James Robert
Green, Detroy Edward
Gregg, Cecil Manren
Greub, Louis John
Griffith, William Kirk
Grimes, Donald Wilburn
Gritton, Earl Thomas
Gross, Harry Douglass
Grossman, Robert Bruce
Guenthner, Harold Reinhold
Haghiri, Faz
Hagstrom, Gerow Richard
Haise, Howard Ross
Hall, Jon K
Hallock, Daniel Leroy
Hammond, James Jacob
Hammons, Jasper Glen
Hammons, Ray Otto
Hanna, Michael Ross
Hanrahan, Sanford Franklin
Hanson, Angus Alexander
Hanway, Donald Grant
Harlan, Jack Rodney
Harper, James Eugene
Harrington, Joseph Donald
Harris, Wallace Wayne
Harrison, Robert Louis
Hart, Richard Harold
Harvey, Clark
Harvey, Robert Gordon, Jr
Harwood, Richard Roland
Hatfield, William Charles
Havelka, Ulysses D
Hawk, Virgil Brown
Hawkes, George Rogers
Hay, Russell Earl, Jr
Haynes, James Lester
Hebert, Leo Placide
Hebert, Richard Henry
Heinrichs, David Henry
Henderlong, Paul Robert
Henderson, Merlin Theodore
Herr, Donald Edward
Hess, Delbert Coy

Hesse, Walter Herman
Hexem, Rodney Orlyn
Heyne, Elmer George
Hiatt, Andrew Jackson
Hill, Gideon D
Hill, Kenneth Wilford
Hills, F Jackson
Hilst, Arvin Rudolph
Hinish, Wilmer Wayne
Hinkle, Dale Albert
Hobbs, James Arthur
Hock, Arthur George
Hodgson, Harlow James
Hoeft, Robert Gene
Hoffbeck, Loren John
Hollowell, Eugene A
Holt, Donald Alexander
Holt, Ethan Cleddy
Hooks, James A
Hoover, Clifford Dale
Horrocks, Rodney Dwain
Horton, Maurice Lee
Hoveland, Carl Soren
Hueg, William Frederick, Jr
Huffine, Wayne Winfield
Hughes, Luther Bertram, Jr
Humburg, Neil Edward
Husted, Robert Forest
Hutchcroft, Charles Dennett
Hutcheson, Thomas Barksdale, Jr
Indyk, Henry Walter
Islam, Nurul
Isom, William Howard
Izuno, Takumi
Jackobs, Joseph Alden
Jackson, Ernest Baker
Jackson, Marion Leroy
Jackson, Thomas Lloyd
Jacobs, Hyde Spencer
James, Norman Ivan
Jansen, Ivan John
Jeffers, Daniel L
Jellum, Milton Delbert
Jensen, Edwin Harry
Jensen, Neal Frederick
Johannes, Russell Frederick
Johannsen, Christian Jakob
Johnson, Corwin McGillivray
Johnson, David Robert
Johnson, Jay Wolbert
Johnson, Jerry Wayne
Johnson, Malcolm Julius
Johnson, Melvin Walter, Jr
Johnson, Richard Ray
Johnson, Virgil Allen
Johnson, Walter Lee
Johnston, Taylor Jimmie
Johnston, Theodore Herron
Jolliff, Gary David
Jones, Edward Raymond
Jones, Jack Earl
Josephson, Leonard Melvin
Judd, Benjamin Ira
Jung, Gerald Alvin
Justin, James Robert
Jutras, Michel Wilfrid
Justus, Norman Edward
Kantack, Benjamin H
Kehr, William R
Kemper, William Doral
Kennedy, Wilbert Keith
Keogh, Joseph Lloyd
Kidder, Gerald
Kilcher, Mark R
Killinger, Gordon Beverly
Kimbrough, Everett Lamar
King, John William
Kinman, Murray Luther
Klinck, Harold Rutherford
Klingman, Dayton L
Klute, Arnold
Knake, Ellery Louis
Kneebone, William Robert
Knowles, Paulden Ford
Koch, David William
Kohn, Gaston G
Kohnke, Helmut
Kolp, Bernard J
Kramer, Nicholas William
Kronstad, Warren Ervind
Kroonje, Wybe
Krueger, William Arthur
Kuhn, Albin Owings
Kurtz, Lester Touby
Lambert, Jean William
Lambert, Royce Leone
Lancaster, James D
Lang, Robert Lee
Langford, Walter Robert
Larsen, Arnold Lewis
Laude, Horton Meyer
Law, Alvin George
Lawton, Kirkpatrick
Lebsock, Kenneth L
Lechtenberg, Victor Louis
Leffel, Robert Cecil
Lejeune, Andre Joseph
Lewis, Cornelius Crawford

Lewis, David Thomas
Lewis, Robert Donald
Leyden, Robert Fullerton
Linscott, Dean L
Litzenberger, Samuel Cameron
Loper, Gerald Milton
Lovvorn, Roy Lee
Lucey, Robert Francis
Lueschen, William Everett
Lund, Hartvig Roald
Lund, Steve
Lund, Zane Franklin
Lutrick, Monroe Cornealous
Lyetly, Paul Junior
MacLeod, LLoyd Beck
Maguire, James Dale
Malm, Norman R
Mannering, Jerry Vincent
Maranville, Jerry Wesley
Marble, Vern L
Marriott, Lawrence Frederick
Marten, Gordon C
Martin, David P
Martin, William C
Marx, Gerald Alvin
Massengale, Martin Andrew
Massey, Herbert Fane, Jr
Massey, John Hubert
Matches, Arthur Gerald
Matlock, Ralph S
Matthews, David Livingstone
Mayes, McKinley
McAlister, Dean Ferdinand
McAlister, DeVere Richard
McCloud, Darell Edison
McCreery, Robert Atkeson
McGill, David Park
McGinnis, Robert Cameron
McGlamery, Marshal Dean
McGuire, Charles Francis
McGuire, William Saxon
McKee, Claude Gibbons
McKee, Guy William
McKee, William Henry, Jr
McLaughlin, Foil William
McMurphy, Wilfred E
McVickar, John S
Meade, John Arthur
Meggitt, William Fredric
Melsted, Sigurd Walter
Merwine, Norman Charles
Metcalfe, Darrel Seymour
Meyer, Dwain Wilber
Mikkelsen, Duane Soren
Mikulcik, John D
Miller, Dwane Gene
Miller, Frederick Powell
Miller, Robert W
Miller, Russell Lee
Miner, Gordon Stanley
Mislevy, Paul
Mitchell, Roger L
Mitchell, William H
Mitchell, William Warren
Mixon, Aubrey Clifton
Mock, James Joseph
Moh, Carl Craig
Moline, Waldemar John
Monson, Warren Glenn
Moomaw, James Curtis
Moore, Robert Parker
Moore, Raymond A
Morey, Darrell Dorr
Morris, Harold Donald
Morrison, Kenneth Jess
Moser, Lowell E
Moss, Dale Nelson
Mott, Gerald O
Mulchi, Charles Lee
Mulkey, James Robert, Jr
Murphy, Larry S
Murray, Calvin Clyde
Murray, Glen A
Murray, Jay Clarence
Musgrave, Robert Burns
Myers, Harold Edwin
Nass, Hans George
Naylor, Gerald Wayne
Nelson, Curtis Jerome
Nelson, Lenis Alton
Nelson, Leyton Vincent
Nelson, Werner Lu Lind
Newell, Laurence Cutler
Newlin, Owen Jay
Newman, James Edward
Nichols, James T
Niehaus, Merle Hinson
Niffenegger, Daniel Arvid
Niles, George Alva
Nilson, Erick Bogseth
Norwood, Charles Arthur
Obendorf, Ralph Louis
Oelke, Ervin Albert
Offutt, Marion Samuel
Ogg, Alex Grant, Jr
Oldemeyer, Robert King
Olsen, Farrel John
Omid, Ahmad

O'Neal, Thomas Denny
Oplinger, Edward Scott
Osler, Robert Donald
Otto, Harley John
Overman, Allen Ray
Owens, Clarence Burgess
Owings, Addison Davis
Papendick, Robert I
Pardee, William Durley
Parker, Frank Wilson
Parkman, Sammie Bell
Parochetti, James V
Parsons, John Lawrence
Patrick, William H, Jr
Pawlisch, Paul E
Payne, Kenyon Thomas
Peacock, Hugh Anthony
Pearce, Robert Brent
Pedersen, Marion Walter
Pendleton, Johnny Wryas
Perkins, Alvin Thomas
Perry, Astor
Peters, Elroy John
Peterson, Clarence James, Jr
Peterson, John Booth
Peterson, Maurice Lewellen
Petr, Frank Charles
Phillips, Ronald Edward
Pierre, William Henry
Place, Gerald Alan
Plate, Henry
Plucknett, Donald L
Pope, Warren Kirkpatrick
Porter, Owen Archuel
Posler, Gerry Lynn
Potts, Richard Carmechial
Powell, Andrew Jackson
Powell, Jerrel B
Price, Philip B
Prine, Gordon Madison
Probst, Albert Henry
Quesenberry, Kenneth Hays
Raese, John Thomas
Rasmussen, Lowell W
Raspet, Mabel Wilson
Reeves, Dale Leslie
Rehm, George W
Reich, Vernon Henry
Reicosky, Donald Charles
Reid, Donald J
Reid, William Shaw
Reiss, William Max
Renney, Arthur James
Reynolds, John Horace
Richardson, Grant Lee
Ririe, David
Ritenour, Gary Lee
Robinson, Frank Ernest
Robinson, Joseph Lee
Robinson, Robert George
Robocker, Willard Charles
Rodgers, Earl Gilbert
Rogers, Homer Eugene
Rogers, John Sinclair
Rogerson, Asa Benjamin
Rohweder, Dwayne A
Ronningen, Thomas Spooner
Ross, William Max
Roth, Charles Barron
Rothman, Paul George
Rouquette, Francis Marion, Jr
Roussel, John S
Ruelke, Otto Charles
Rumburg, Charles Buddy
Russel, Darrell Arden
Russell, Morell Belote
Sanchez, Pedro Antonio
Sander, Donald Henry
Santelmann, Paul William
Sapperfield, William Paul
Savage, Robert Gilmore
Schaaf, Herbert Martin
Schaller, Charles William
Schilling, John Andrew, Jr
Schmid, Alois Rudolph
Schmidt, Bertie Louis
Schmidt, John Wesley
Schmidt, Richard Edward
Schmidt, Walter Harold
Schmitz, George William
Schnappinger, Melvin Gerhardt, Jr
Schneider, Bernard Arnold
Scholl, Jesse Myron
Schonhorst, Melvin Herman
Schooler, Arnold B
Schrader, Marvin Mandel
Schreiber, John Herman
Schweitzer, Edward Ray
Schweitzer, Leland Ray
Schwendiman, John Leo
Schwer, Joseph Francis
Scott, Walter O'Daniel
Seaman, Donald Edward
Searcy, Virgil Shell
Sedberry, Joseph E, Jr
Seely, Clarence Ivan

Segars, William Isaac
Shands, Hazel Lee
Shank, Daniel Boyd
Shaw, Warren Cleaton
Shenk, John Stoner
Sheth, Anil Amratlal
Shibles, Richard Marwood
Shipp, Raymond Francis
Shoop, George Jerome
Sieveking, William Earl
Simkins, Charles Abraham
Simmons, Charles Ferdinand
Simmons, Gerald Herman
Simonson, Allan Barnard
Sinclair, Thomas Russell
Sims, John Leonidas
Singh, Rabindar Nath
Singh, Raghbir
Skogley, Conrad Richard
Skogley, Earl Arthur
Skold, Laurence Nelson
Skrdla, Willis Howard
Slife, Fred Warren
Slinkard, Alfred Eugene
Small, Howard G, Jr
Smalley, Ralph Ray
Smeck, Neil Edward
Smeltzer, Dale Gardner
Smiley, Jones Hazelwood
Smith, Andrew Douglas
Smith, Dale
Smith, Don Wiley
Smith, Townsend Jackson
Smith, William Kenneth
Snyder, George Heft
Soltanpour, Parviz Neil
Sosuiski, Frank Walter
Soto, Gerardo H
Southwick, Lawrence
Spencer, Jack T
Spiers, James Monroe
Spooner, Arthur Elmon
Sprague, George Frederick
Sprague, Howard Bennett
Sprague, Milton Alan
Stangel, Harvey J
Stanford, John Pershing
Stanley, David William
Stanley, Robert Lee, Jr
Starostka, Raymond Walter
Staten, Raymond Dale
Steppler, Howard Alvey
Stickler, Fred Charles
Stobbe, Elmer Henry
Strand, Oliver Eric
Strauch, Fred
Streeter, John Gennnil
Stritzel, Joseph Andrew
Stritzke, Jimmy Franklin
Stroube, Edward W
Stute, Charles Auten
Sullivan, Edward Francis
Svec, Leroy Vernon
Swan, Dean George
Swearingin, Marvin Laverne
Szabo, Steve Stanley
Talbert, Ronald Edward
Taylor, Lincoln Homer
Taylor, Norman Linn
Taylor, Richard Melvin
Taylor, Roscoe L
Taylor, Timothy H
Teater, Robert Woodson
Templeton, William Cheicy, Jr
Terman, Gilbert Leroy
Tesar, Milo
Thien, Stephen John
Thomas, Gerald Waylett
Thomas, Johnny Ray
Thomas, Ronald Leslie
Thomas, Walter Ivan
Thomas, Winfred
Thompson, Donald Loraine
Thompson, Harvey E
Thompson, John R
Thompson, Roy Lloyd
Thompson, Wilfred Roland, Jr
Thorne, Mariowe Driggs
Thorup, James Tat
Thorup, Richard M
Timm, Herman
Tossell, William Elwood
Triplett, Glover Brown, Jr
Trouse, Albert Charles
Trujillo, Philip M
Tseng, Shu-Ten
Tucker, Billy Bob
Ullery, Charles Howard
Unger, Victor Herman
Vadhwa, Om Parkash
Van Bavel, Cornelius H M
Vanderford, Harvey Birch
Vanderlip, Richard L
Van Doren, David Miller, Jr
Van Eck, Willem Adolph
Van Keuren, Robert W
Van Riper, Gordon Everett
Vaughan, Charles Edwin
Veatch, Collins

Vengris, Jonas
Vitosh, Maurice Lee
Vittum, Morrill Thayer
Vomocil, James Arthur
Vorst, James J
Waddle, Bradford Avon
Wagner, Robert Earl
Wakefield, Robert Chester
Waldron, Acie Chandler
Walgenbach, David D
Wang, Maw Shiu
Ward, Coleman Younger
Warner, James Northrup
Warner, Robert Lewis
Warnes, Dennis Daniel
Warren, Francis Shirley
Washko, John Blasius
Washko, Walter William
Wasson, Clyde E
Watschke, Thomas Lee
Watson, Vance H
Wear, John Ingram
Weaver, James Bode, Jr
Webb, Burleigh C
Webb, John Raymond
Webster, Gilbert Theodore
Webster, Orrin John
Wedin, Walter F
Weeks, Martin Edward
Weibel, Dale Eldon
Weismiller, Richard A
Welch, Charles Darrel
Welch, Kenneth Lincoln
Wesenberg, Darrell
Wesley, William Keith
Wesley, Ray Edward
West, Sherlie Hill
Westbrook, Fred Emerson
Whigham, David Keith
White, George Albert
Whited, Dean Allen
Whitney, Arthur Sheldon
Whitty, Elmo Benjamin
Wilcox, James Raymond
Wilcox, Wesley Crain
Wilding, Lawrence Paul
Wilkinson, James Freeman
Wilkinson, Stanley W
Williams, Curtis
Williams, James Henry, Jr
Williams, William Arnold
Winter, Steven Ray
Wise, Louis Neal
Wofford, Irvin Mirle
Wolf, Dale Duane
Wolf, Dale E
Woltz, Willie Garland
Wood, Donald Roy
Wood, Glen Meredith
Wood, Reed Ralph
Wooding, Frank James
Woodruff, Clarence Merrill
Woolley, Donald Grant
Worker, George F, Jr
Worley, Ray Edward
Worley, Smith, Jr
Wright, James Louis
Wright, Madison Johnston
Young, Arthur Wesley
Young, Evie Fountain, Jr
Youngman, Vern E
Youngman, Victor Bernar
Younts, Sanford Eugene
Zech, Arthur Conrad
Zilke, Samuel
Zimdahl, Robert Lawrence
Zimmerman, Lester J
Zuber, Marcus Stanley

Animal Husbandry
Alexander, Robert Allen
Anthony, Wilson Brady
Baker, Bryan, Jr
Baker, Frank Sloan, Jr
Bedell, Thomas Donald
Bovard, Kenly Paul
Briggs, Hilton Marshall
Brown, Herbert
Buchanan, Marion Lynn
Buck, Charles Frank
Burris, Martin Joe
Carroll, Floyd Dale
Clark, Jack L
Cole, Clarence Lorraine
Cunha, Tony Joseph
Dahl, Billie Eugene
Daniel, O'Dell G
Day, Billy Neil
Deans, Robert Jack
de Alba Martinez Jorge
Dinuson, William Erling
Dowe, Thomas Whitfield
Embry, Lawrence Bryan
England, David Charles
Evans, Lee E
Fahmy, Mohamed Hamed
Gnaedinger, Richard H
Gobble, James Lawrence

Griffin, Sumner Albert
Grummer, Robert Henry
Guyer, Paul Quentin
Hathorn, Fred
Hedrick, Harold Burdette
Heidenrich, Charles John
Henderson, Hugh E
Hickey, Wayne C, Jr
Hodgson, Charles Worth
Hoefer, Jacob A
Holden, Palmer Joseph
Hollandbeck, Richard
Holt, Leroy Henry
Johnson, George Robert
Johnson, LaDon Jerome
Jones, James Robert
Jordan, Robert Manseau
Kelly, Robert Frank
Kennington, Mack Humpherys
Keyes, Everett A
Klosterman, Earle Wayne
Kropf, Donald Harris
Lewis, John Morgan
Lindley, Charles Edward
Litton, George Washington
Madsen, Milton Andrew
Mattingly, Steele F
Meiske, Jay C
Menzies, Carl Stephen
Miller, John Ivan
Newland, Herman William
Paules, Leon H
Pease, Lawrence Honeyman
Peters, John Burl
Ramsey, Clovis Boyd
Ray, Maurice L
Ritchie, Harlan
Rust, Joseph William
Sharma, Udhishtra Deva
Smith, Edgar Fitzhugh
Spalding, Robert Wilber
Squiers, Clifford Dale
Stockbridge, Robert R
Tait, Robert Malcolm
Taylor, Jack Crossman
Taysom, Elvin David
Thrasher, Donald Miller
Tillman, Allen Douglas
Van Dongen, Cornelis Godefridus
Varney, William York
Wakeman, Donald Lee
Warner, William Michael
Webb, Robert Johnson
Welch, James Alexander
Zinn, Dale Wendel

Apiculture
Berthold, Robert, Jr
Berthold, Lloyd Millard
Burke, Philip William
Caron, Dewey Maurice
Connor, Lawrence John
Dietz, Alfred
Gary, Norman Ervin
Harbo, John Russell
Kaufeld, Norbert M
Laidlaw, Harry Hyde, Jr
Levin, Marshall David
Moeller, Floyd Edward
Nelson, Eric V
Newton, David C
Nye, William Preston
Sanford, Malcolm Thomas
Schwartz, Edward Leo
Smith, Maurice Vernon
Stephen, William Archibald
Taber, Stephen, III
Thompson, Victor Carl
Townsend, Gordon Frederick
Tucker, Kenneth Wilburn
Waller, Gordon David

Crop Breeding
Assy, Kay Harris
Chi, Kuo-Ruey
Christie, Bertram Rodney
Dickenson, Donald Dwight
Domingo, Wayne Elvin
Duell, Robert William
Ensign, Ronald D
Evans, Marshall Pierson
Frazier, Floyd Wendell
Hamilton, Donald Gregory
Higgs, Roger L
Hittle, Carl Nelson
Ho, Keh Ming
Hyer, Angus Hillyard
Kalton, Robert Rankin
Keim, Wayne Franklin
Lucken, Karl Allen
Munger, Henry Martin
Oswalt, Dallas Leon
Oswalt, Robert Donald
Remley, Frank Morris
Ronningen, Thomas Spooner
Rotar, Peter P

Rumbaugh, Melvin Dale
Schillinger, John Andrew, Jr
Shih, Ching-Yuan
Smith, Stuart Newton
Stokes, Granville Woolman
Stoskopf, N C
Timmermann, Dan, Jr
Wernsman, Earl Allen
Widner, Jimmy Newton

Crop Physiology
Balasko, John Allan
Barta, Allan Lee
Bennett, James Peter
Boote, Kenneth Jay
Brown, Jarvis Howard
Brun, William Alexander
Buren, Lawrence Lamont
Buxton, Dwayne Revere
Cardwell, Vernon Bruce
Copeland, Lawrence O
Criswell, Jerome Glenn
Croy, Lavoy I
Davis, Larry Alan
Deckard, Edward Lee
Duff, Dale Thomas
Egli, Dennis B
Fick, Gary Warren
Flowerday, Albert Dale
Fowler, James Lowell
Frey, Nicholas Martin
Freyman, Stanislaw
Fullerton, Thomas M
Gasser, Heinz
Hall, Anthony Elmitt
Hopper, Norman Wayne
Hume, David John
Johnson, David Robert
Joliff, Gary David
Kaul, Rudolf
Knievel, Daniel Paul
Langille, Alan Ralph
Martin, Freddie Anthony
Massengale, Martin Andrew
Mitchell, Roger L
Mullen, Russell Edward
Murray, Glen A
Patterson, Robert Preston
Teare, Iwan Dale
Victor, Donald Melvin

Field Crops
Anderson, Laurel Ethan
Barnes, Carl Eldon
Beeks, John Charles
Elliott, Fred Craig
Freeman, Kelly Carey
Green, Victor Eugene, Jr
Justin, James Robert
LeBaron, Marshall John
Lee, William Orvid
Litzenberger, Samuel Cameron
Marble, Vern L
Peck, Raymond A
Reid, Donald J
Van Epps, Gordon Almon

Range Management
Bleak, Alvin Thomas
Box, Thadis Wayne
Burzlaff, Donald Frederick
Chapline, William Ridgely, (Jr)
Dahl, Billie Eugene
Drawe, D Lynn
Driscoll, Richard Stark
Dwyer, Don D
Ehrenreich, John Helmuth
Fisser, Herbert George
Gibbens, Robert Parker
Goebel, Carl Jerome
Gomm, Fred Bryant
Green, Lisle Royal
Grumbles, Jim Bob
Harris, Venoia M
Hazell, Don Bliss
Hickey, Wayne C, Jr
Johnsen, Thomas Norman, Jr
Kothmann, Merwyn Mortimer
Lewis, Clifford Eugene
McKell, Cyrus Milo
Nichols, James T
Payne, Gene F
Roche, Ben F, Jr
Schmutz, Ervin Marcell
Schuster, Joseph L
Sharp, Lee Ajax
Smith, Edwin Lamar, Jr
White, Larry Dale
Whitson, Robert Edd
Williams, Robert E
Workman, John Paul
Wright, Henry Albert

Range Science
Allen, Thomas J
Bailey, Arthur W
Bement, Robert Earl
Bjugstad, Ardell Jerome
Claveran, Ramon A
Cook, Charles Wayne
Donart, Gary B
Eckert, Richard Edgar, Jr
Evans, Raymond Arthur
Grelen, Harold Eugene
Haas, Robert Henry
Herbel, Carlton Homer
Hyder, Donald N
Jordan, Gilbert Leroy
Kothmann, Merwyn Mortimer
Krueger, William Clement
Lavin, Fred
Lewis, James Kelley
Malechek, John Charles
Marquiss, Robert W
McCully, Wayne Gunter
McGinnies, William Joseph
Morris, Melvin Solomon
Morrow, Larry Alan
Pearson, Henry Alexander
Pieper, Rex Delane
Powell, Jeff
Ries, Ronald Edward
Robertson, Joseph Henry
Scifres, Charles Joel
Severson, Kieth Edward
Sims, Philip Leon
Skovlin, Jon Matthew
Tisdale, Edwin William
Vallentine, John Franklin
Vogel, Willis Gene
White, Larry Melvin
Whitman, Warren Charles
Wright, Jerald Ross
Wood, Benjamin W

Enology
Adams, Angus Macaulay
Alley, Curtis J
Baldy, Marian Wendorf
Beelman, Robert B
Guymon, James Fuqua
Kunkee, Ralph Edward
Simpson, John Ernest
Singleton, Vernon Leroy
Toczek, Donald Richard
Webb, Albert Dinsmoor

Floriculture
Allen, Raymond Clayton
Beck, Gail Edwin
Carpenter, William John
Culbert, John Robert
Fischer, Charles Clayton
Gartner, John Bernard
Hanchey, Richard Howard
Kalin, Elwood Walter
Kofranek, Anton Miles
Koths, Jay Sanford
Langhans, Robert W
Larmie, Walter Esmond
Mastalerz, John W
McDowell, Theodore C
Morr, Robert G
Orr, Henry Porter
Parvin, Philip Eugene
Rogers, Martin Norbert
Sanderson, Kenneth Chapman
Schekel, Kurt Anthony
Seeley, John George
Shaw, Richard John
Smith, James Elmo, Jr
Smith, Vincent C
Stinson, Richard Floyd
Tayama, Harry K

Seed Physiology
Berkey, Dennis Alan
Wiesner, Loren Elwood

Vegetable Crops
Andrew, William Treleaven
Brecht, Patrick Ernest
Cantliffe, Daniel James
Chambliss, Oyette Lavaughn
Davis, David Warren
Eisa, Hamdy Mahmoud
Elle, George O
Everett, Paul Harrison
Ewing, Elmer Ellis
Feddema, Leonard William
Forbes, Richard Brainard
Gull, Dwain D
Hall, Charles Virdus
Hamson, Alvin Russell
Hepton, Anthony
Kemble, William Cary
Lambeth, Victor Neal
Lorenz, Oscar Anthony
McFerran, Joe
Mills, Harry Arvin
Minges, Philip Adams
Morris, Leonard Leslie
Nettles, Victor Fleetwood
Nishimoto, Roy Katsuto

Ozaki, Henry Yoshio
Pratt, Arthur John
Sandsted, Roger France
Schweitzer, Leland Ray
Sims, William Lynn
Smith, Paul Gordon
Sweet, Robert Dean
Thomas, Paul Clarence
Thompson, Buford Dale
Thomson, Cecil Lyman
Topoleski, Leonard Daniel
Vandemark, J S
Wilson, Lorenzo George
Workman, Ralph Burns

Viticulture
Alley, Curtis J
Brown, Gerald Richard
Olmo, Harold Paul
Shaulis, Nelson Jacob

Weed Science
Addink, Sylvan
Ahrens, John Frederick
Allen, Thomas J
Alley, Harold Pugmire
Anderson, Jay LaMar
Andres, Lloyd A
Appleby, Arnold Pierce
Ashley, Richard Allan
Ashton, Floyd Milton
Baker, Ralph Stanley
Behrens, Richard
Bingham, Samuel Wayne
Boehle, John, Jr
Bovey, Rodney William
Burnside, Orvin C
Caulder, Jerry Dale
Chang, Fa Yan
Coartney, James S
Cole, Avean Wayne
Comes, Richard Durward
Cooper, Raymond B
Corbin, Frederick Thomas
Cords, Howard Paul
Corns, William George
Coulter, Llewellyn Legrande
Crabtree, Garvin (Dudley)
Crafts, Alden Springer
Dawson, Jean Howard
Delorit, Richard John
Dickerson, Chester T, Jr
Doersch, Ronald Ernest
Domanski, John Joseph, Jr
Dowler, Clyde Cecil
Ellis, John Fletcher
Erickson, Lambert Cornelius
Evans, Raymond Arthur
Fawcett, Richard Steven
Fertig, Stanford Newton
Fink, Rodney James
Fisher, Charles E
Fletchall, Oscar Hale
Foret, James A
Frankton, Clarence
Fuller, Thomas Charles
Gantz, Ralph Lee
Goeden, Richard Dean
Gould, Walter Leonard
Grande, John Anthony
Greer, Howard A L
Gruenhagen, Richard Dale
Haderlie, Lloyd Conn
Haramaki, Chiko
Hardcastle, Willis Sanford
Harris, Peter
Hartwig, Nathan Leroy
Harvey, Robert Gordon, Jr
Hauser, Ellis W
Hay, James Robert
Ilnicki, Richard Demetry
Kerr, Harold Delbert
Kincade, Robert Tyrus
Kingman, Glenn Charles
Knake, Ellery Louis
Lavy, Terry Lee
Lee, Gary Albert
Lewis, William Mason
McCarty, Melvin Knight
McGlamery, Marshal Dean
McWhorter, Chester Gray
Miller, Gerald R
Miller, Stephen Douglas
Moreland, Donald Edwin
Morrow, Larry Alan
Nalewaja, John Dennis
Noll, Charles Joseph
Norris, Robert Francis
Orsenigo, Joseph Reuter
Osgood, Robert Vernon
Palmer, Rupert Dewitt
Parka, Stanley John
Patterson, David Thomas
Peabody, Dwight Van Dorn, Jr
Peabody, Dwight Van Dorn, Jr
Penner, Donald P
Pulido, Miguel

Radosevich, Steven Robert
Reese, Robert Lewis
Richards, Russell Fayette
Roberts, Donald Ray
Rodgers, Earl Gilbert
Roeth, Frederick Warren
Ross, Merrill Arthur, Jr
Saidak, Walter John
Santelmann, Paul William
Schwartzbeck, Richard Arthur
Scifres, Charles Joel
Sckerl, Max Michael
Scudder, Walter Tredwell
Seaman, Donald Edward
Skroch, Walter Arthur
Smith, Dudley Templeton
Smith, Roy Jefferson, Jr
Stobbe, Elmer Henry
Stoller, Edward W
Switzer, Clayton Macfie
Talbert, Ronald Edward
Truelove, Bryan
Turgeon, Alfred J
Upchurch, Robert Phillip
Vanden Born, William Henry
Warnes, Dennis Daniel
Warren, George Frederick
Watson, Andrew John
Welker, William V, Jr
Wiese, Allen F
Wilkinson, Robert Eugene
Williams, James Lovon, Jr
Woofter, Harvey Darrell
Worsham, Arch Douglas
Wright, Wayne Gordon
Wright, William Leland
Yamaguchi, Shogo
Zimdahl, Robert Lawrence

ANATOMY

Anatomy
Ackerman, Gustave Adolph, Jr
Adams, Andrew Borden
Adrian, Erle Keys, Jr
Ajemian, Martin
Albert, Ernest Narinder
Albright, Raymond Gerard
Alcala, Jose Ramon
Alden, Roland Herrick
Aldridge, William Gordon
Alger, Elizabeth A
Al-lami, Fadhil
Allan, Frank Duane
Allen, Emory Raworth
Allen, Lane
Allin, Edgar Francis
Allison, John Everett
Amenta, Peter Sebastian
Anderson, Frank David
Anderson, James Edward
Anderson, John Walberg
Andrew, Warren
Aplington, Henry Webster, Jr
April, Ernest W
Arluk, David Jay
Ashburn, Allen David
Ashworth, Harold David
Askew, Harold Cochran
Asling, Clarence Willet
Astruc, Juan A
Atnip, Robert Lee
Aulsebrook, Lucille Hagan
Avery, James Knuckey
Bachhuber, Edward A
Bacon, Robert Lewis
Baird, Irwin Lewis
Baker, Burton Lowell
Ball, Carroll Raybourne
Barr, Murray Llewellyn
Barrington, Burness Austin, Jr
Barrnett, Russell Joffree
Barron, Donald Henry
Bar-Sela, Mildred Elwers
Bartley, Murray Hill, Jr
Barton, John M
Basmajian, John V
Basom, Charles Ray
Bauer, Gustav Eric
Baumel, Julian Joseph
Beal, John Anthony
Beary, Dexter F
Beasley, Andrew Bowie
Bedford, John Michael
Belt, Warner Duane
Benjamin, Hiram Bernard
Bennett, Henry Stanley
Benoit, Peter Wells
Benton, Robert S
Berendsen, Peter Barney
Berger, Andrew John
Bergmann, Louis Lawrence
Berman, Irwin
Bernick, Sol
Bernstein, Maurice Harry

ANATOMY

Bernstorf, Earl Cranston
Bhatnagar, Kunwar Prasad
Bhussry, Baldev Raj
Billenstien, Dorothy Corinne
Binhammer, Robert T
Black, John B
Blaha, Gordon C
Blevins, Charles Edward
Bloch, Edward Henry
Blount, Raymond Frank
Bo, Walter John
Boccabella, Anthony Vincent
Bockman, Dale Edward
Bodley, Herbert Daniel, II
Bois, Pierre
Bok, P Dean
Bolender, David Leslie
Boler, Reginald Keith
Bondareff, William
Bongard, Steven J
Bothner, Richard Charles
Boucher, Louis Jack
Bourne, Geoffrey Howard
Boving, Bent Giede
Boyden, Edward Allen
Boyne, Philip John
Brandel, Bruce Reeves
Brandt, Philip Williams
Breazile, James E
Bridgman, Charles Floyd
Brody, Harold
Brown, Esther Marie
Brown, Herbert Ensign
Brown, Jerry William
Brown, Roger E
Browning, Henry (Charles)
Bruesch, Simon Rulin
Buchanan, George Dale
Buck, Robert Crawforth
Buell, Katherine Mayhew
Bulger, Ruth Ellen
Bunge, Richard Paul
Burke, Jack Denning
Burkel, William E
Burnside, Mary Beth
Burrows, Leslie Raymond
Calabrisi, Paul
Calatroni, Donald R
Callahan, William Paxton, III
Callas, Gerald
Cameron, Ivan Lee
Campbell, John Howland
Campbell, Robert Benoni
Carihers, Jeanine Rutherford
Carlson, Bruce Martin
Carpenter, Anna-Mary
Carpenter, Malcolm Breckenridge
Carr, David Harvey
Carr, Malcolm Wallace
Castelli, Walter Andrew
Cauna, Nikolajs
Cavazos, Lauro Fred
Cave, Mac Donald
Ceron, Gabriel
Chapman, Albert Lee
Chapman, David MacLean
Chester, Clarence Lucian
Chakulas, John James
Ching, Melvin Chung Hing
Christensen, Albert Kent
Christensen, John Bert
Chu, Chi Hsuin Ulli
Church, Lloyd Eugene
Clabough, Jeanne Whitaker
Clark, David Lee
Clark, Sam Lillard, Jr
Coalson, Robert Ellis
Cobb, William Montague
Coggeshall, Richard E
Colborn, Gene Louis
Colomier, Marc
Combs, Clarence Murphy
Conklin, James L
Connelly, Thomas George
Constantinides, Paris
Contu, Paolo
Cooper, Margaret Hardesty
Cooperstein, Sherwin Jerome
Copenhaver, Wilfred Monroe
Corliss, Clark Edward
Corner, George Washington
Cotter, William Bryan, Jr
Coulter, Herbert David, Jr
Crabill, Edward Vaughn
Crafts, Roger Conant
Crisp, Thomas Mitchell, Jr
Critchlow, Burtis Vaughn
Croley, Thomas Edgar
Crouse, Gail
Crow-Baste, Claudia Adkison
Culberson, James Lee
Dalgleish, Arthur E
Daron, Garman Harlow
Davies, Jack
Davis, Robert Wilson
Dearden, Lyle Conway
Decker, John D

DeFouw, David O
De Groot, Jack
de Laubenfels, David John
De Lorenzo, Anthony John
De Pace, Dennis Michael
DeProspo, Nicholas Dominick
Deuschle, Frederick Marion
Dewey, Maynard Merle
DiDio, Liberato John Alphonse
Dienhart, Charlotte Marie
Dietert, Scott Edward
Di Stefano, Henry Saverio
Dixit, Padmaker Kashinath
Dixon, Andrew Derart
Dodge, Alice Hribal
Domm, Lincoln Valentine
Donati, Edward Joseph
Dornfest, Burton S
Dougherty, Harry L
Dow, Robert Stone
DuBrul, E Lloyd
Duke, Kenneth Lindsay
Duncan, Donald
Dutta, Hiram Moyee
Dvornik, Julian Jonathan
Dyer, Robert Frank
Eagles, Jan
Earle, Alvin Mathews
Edmonds, Richard H
Edwards, Betty F
Egar, Margaret Wells
Eglitis, Irma
Eglitis, John Arnold
Elfman, Alice G
Elfman, Herbert (Oliver)
Elias, Joel Jesse
Ely, Charles A
Enlow, Donald Hugh
Erickson, George Emil
Ettinger, Anna Marie Conway
Evans, Francis Gaynor
Evans, James Spurgeon
Everett, John Wendell
Everett, Newton B
Everingham, John
Farrell, Edward Joseph
Fasano, Anthony Vincent
Faulkin, Leslie J, Jr
Faulkner, Kenneth Keith
Fawcett, Don Wayne
Feagans, William Marion
Feder, Harvey Herman
Fedinec, Alexander
Feinman, Max L
Feldcher, Carl M
Felts, William Joseph Lawrence
Fife, William Paul
Finch, Robert Allen
Finck, Henry
Finerty, John Charles
Firriolo, Domenic
Fish, Harold Somers
Fitch, Kenneth Leonard
Fix, James D
Flanigan, Norbert James
Flexner, Louis Barkhouse
Floyd, Alton David
Foltz, Floyd Mathew
Fontaine, Julia Clare
Foote, Florence Martindale
Forbes, Thomas Rogers
Ford, Donald Herbert
Fox, Marjorie Hopkins
Frandson, Rowen Dale
Frederickson, Richard Gordon
Friedman, Constance Livingstone
Fritz, Carl J
Frommer, Jack
Fruhman, George Joshua
Furry, Donald Edward
Fursman, Lawrence L
Fuson, Roger Baker
Fyfe, Forest William
Gagnon, Real
Gale, Thomas Francis
Gammal, Elias Bichara
Garcia, Alfredo Mariano
Gardner, Paul Jay
Gardner, William Ullman
Gasser, Raymond Frank
Gaughran, George Richard Lawrence
Gavan, James Anderson
Gennaro, Joseph Francis
Gibbs, Finley P
Gibson, Maurice Henry Lindsay
Gilmore, Shirley Ann
Glass, Laurel Ellen
Glonski, Chester Anthony
Goldsberry, Steve
Goss, Charles Mayo
Gowgiel, Joseph Michael
Grancy, Daniel O
Grasso, Joseph Anthony
Gray, Donald James
Gray, Stephen Wood
Green, James Arnold
Greenwald, Gilbert Saul

Greep, Roy Orval
Gregg, Robert Vincent
Grenell, Robert Gordon
Gresik, Edward William
Greulich, Richard Curtice
Gross, James Harrison
Grove, Alvin Russell, Jr
Haar, Jack Luther
Haft, Jay Stuart
Haidak, Gerald Lewis
Hall, James Lawrence
Halpern, Earl Gregory, Jr
Hamel, Benjamin Barnett
Hamilton, David Whitman
Hammond, James Bruce
Hammond, Warner Smith
Hampton, James C
Hand, Peter James
Hansell, Margaret Mary
Harding, Fann
Harn, Stanton Douglas
Harrison, Frank
Harrison, Frederick Williams
Harvey, Michael
Harvey, Joseph Eldon
Haun, Charles Kenneth
Hausberger, Franz X
Hay, Elizabeth Dexter
Hayashida, Tetsuo
Hayden, Jess, Jr
Hayes, Everett Russell
Hayes, Raymond L, Jr
Heath, Everett
Heggestad, Carl B
Hegre, Erling Stanford
Hegre, Orion Donald
Heidger, Paul McClay, Jr
Heim, Louise Mann
Heinrikson, Ray Charles
Henry, Joseph L
Henry, Raymond Leo
Henson, O'Dell Williams, Jr
Herbener, George Henry
Herring, Susan Weller
Herzberg, Fred
Hess, Arthur
Hewes, Cecil Gordon
Higginbotham, Arlyn Curtis
Higginbotham, Frances Heffrin
Highower, James Anderson
Hild, Walther
Hill, Marvin Francis
Hill, Robert Towner
Hilliard, Jessamine
Hilloowala, Rusi Ardeshir
Hinds, James Wadsworth
Hinke, Joseph Anthony Michael
Hirt, Bethold Joseph
Hofer, Helmut Otto
Hoffman, Henry Harland
Holland, Robert Campbell
Hollenberg, Martin James
Hollinshead, May B
Hollinshead, William Henry
Holmes, Ian
Holyoke, Edward Augustus
Horak, Karel
Horn, Eugene Harold
Hoshino, Kazumasa
Hosteller, Jeptha Ray
House, Earl Lawrence
Hovde, Christian Arneson
Huber, John Franklin
Huelke, Donald Fred
Humberson, Albert O, Jr
Hung, Kuen-Shan
Hungerford, Gerald Fred
Hunter, Robert L
Hutchinson, Robert Cranford
Hwang, Ung Kee
Ift, John Dempster
Imai, Hideshige
Ingersoll, Everett Harold
Inke, Gabor
Inman, Verne Thomson
Iorio, Robert John
Isaacson, Robert John
Ishag, Mohammed
Israel, Harry, III
Jacobs, Allen Wayne
Jacobs, Richard M
Jacobs, Virgil Leon
Jakway, Jacqueline Sinks
Jampel, Robert Steven
Jande, Sohan Singh
Jayne, Edgar Pleasant
Jee, Webster Shew Shun
Jenkins, Thomas William
Jensen, Richard Harvey
Jimenez-Marin, Edmund
Johnson, Elmer Marshall
Johnson, Robert Joseph
Jollie, William Pucette
Jones, David Smith
Jones, Edward George
Jones, Oliver Perry
Jordan, Robert Lawrence
Judd, Teresita

Julyan, Frederick John
Jump, Ellis Burnett
Kabisch, William Thomas
Kalt, Marvin Robert
Kamrin, Benjamin Barnett
Kanczak, Norbert M
Karlson, Ulf Lennart
Kashiwa, Herbert Koro
Kasirsky, Gilbert
Kasprow, Barbara Ann
Katzberg, Allan Alfred
Kaye, Gordon I
Kaylor, Cornelius Timpson
Kennedy, Joseph Patrick
Kent, Barbara Wynne
Kernis, Marten Murray
Kessel, Richard Glen
Kiely, Lawrence J
Kiely, Michael Lawrence
Kelly, Robert Edward
Keller, Leland Edward
King, Lloyd Elijah, Jr
King, John Edward
Kinzey, Warren Glenford
Kirgis, Homer Dale
Kirkman, Hadley
Klass, Alan Arnold
Klein, Albert William
Klein, Michael
Kleiss, Ekkehard
Klinworth, Gordon K
Knisely, William Hagerman
Koch, William Edward
Kochhar, Devendra M
Koenig, Marilyn Jean
Kozam, George
Krahl, Vernon Edward
Kramer, Theodore Christian
Krause, William John
Krehbiel, Robert Henry
Kronman, Joseph Henry
Krutzsch, Philip Henry
Ladman, Aaron Julius
Lake, Lorraine Frances
Lakshman, Akarasi Bhojaraj
Lambson, Roger O
Lansing, Albert Ingram
Larkin, Lynn Haydock
Lash, James (Jay) W
Lauer, Edward Willard
Laurenson, Rae Duncan
Lawton, Irene Elizabeth
Lay, Douglas M
Lebond, Charles Philippe
Lee, Joseph Ching-Yuen
Leeson, Charles Roland
Leeson, Thomas Sydney
Leinonen, Ellen A
Leppi, Theodore John
Levene, Cyril
Leveque, Theodore Francois
Levitan, Max
Liu, Chan Nao
Lillie, John Howard
Littlefield, Gayle
Lloyd, Ruth Smith
Lobl, Richard Tolstoi
Lockard, Isabel
Loevy, Hannelore Taschini
Long, John Arthur
Low, Frank Norman
Lowrance, Edward Walton
Lucas, Alfred Martin
Lucas, Edgar Arthur
Luckett, Winter Patrick
Lutt, Carl J
Macaluso, Mary Christelle
MacCallum, Donald Kenneth
Macdonald, Gordon J
Mack, Harry Patterson
MacRae, Edith Krugelis
Maibenco, Helen Craig
Malewitz, Thomas Donald
Marceau, Gilles
Marchand, E Roger
Marieb, Elaine Nicpon
Marks, Sandy Cole, Jr
Marovitz, William F
Martin, James Harold
Martin, William David
Martinek, John Joel
Marvin, Horace Newell
Mason, Karl Ernest
Massopust, Leo Carl, Jr
Masters, Edwin M
Mathews, Murray Albert
Matulionis, Daniel H
Maxwell, David Samuel
McCafferty, Robert Eugene
McCauley, William John
McClugage, Samuel Gardner, Jr
McCreight, Charles Edward
McCrum, Wilbur Ross
McCuskey, Robert Scott
McMahan, Uel Jackson, II
McNamara, James Alyn, Jr

McNary, William Francis, Jr
Meader, Ralph Gibson
Meader, Roland Darrell
Meineke, Howard Albert
Mennega, Aaldert
Menton, David Norman
Meyer, David Bernard
Miller, Inglis, Jr
Miller, James Albert, Jr
Miller, Malcolm Ray
Mixter, Russell Lowell
Mizeres, Nicholas James
Mohammed, Clive Imram
Mohn, Melvin P
Momsen, Richard Paul, Jr
Monie, Ian Whitelaw
Monroe, Barbara Samson Granger
Monsen, Harry
Montgomery, Royce Lee
Moore, Keith Leon
Moore, Richard Dana
Moosman, Darvan Albert
Morest, Donald Kent
Morgan, Carl Robert
Morrison, Edward Joseph
Mortensen, Otto Axel
Moscovici, Mauricio
Moskowitz, Norman
Moss, Melvin Lionel
Moss-Salentijn, Letty
Motzkin, Shirley M
Muir, William A
Mulroy, Michael Joseph
Mundell, Robert David
Munigle, Jo Anne
Munkacsi, Istvan
Murnane, Thomas William
Murphy, Henry D
Murphy, Robert Carl
Murray, Irwin MacKay
Murrell, Leonard Richard
Nabors, Charles J, Jr
Nadler, Norman Jacob
Nakajima, Yasuko
Napolitano, Leonard Michael
Nathaniel, Edward J H
Nauta, Walle J H
Neaves, William Barlow
Nelsen, Olin E
Nelson, Edward Mons
Nelson, Gayle Herbert
Nemeth, Andrew Martin
Newhall, Chester Albert
Newman, Bertha L
Nicolls, Ken E
Niewenhuis, Robert James
Nikitovitch-Winer, Miroslava B
Niles, Nelson Robinson
Noback, Charles Robert
Noe, Bryan Dale
Nokes, Richard Francis
O'Connell, Alice L
Odland, George Fisher
Odor, Dorothy Louise
O'Leary, James Lee
Ollerich, Dwayne A
Olsen, Edmund Severn, Jr
O'Morchoe, Charles Christopher C
Orr, Mary Faith
Osinchak, Joseph
Osmanski, C Paul
Osmond, Dennis Gordon
O'Steen, Wendall Keith
Outzen, Henry Clair, Jr
Overholser, Milton David
Owers, Noel Oscar
Padawer, Jacques
Pakurar, Alice Swope
Pansky, Ben
Papka, Raymond Edward
Parks, Harold Francis
Parnell, Jerome Patrick
Parshley, Mary Stearns
Patek, Paul R
Paule, Wendelin Joseph
Pauly, John Edward
Paynter, Kenneth Jack
Peebles, Edward McCrady
Pekarthy, James Maurice
Pennington, Raymond Carroll
Pepe, Frank Albert
Peppler, Richard Douglas
Pereira, Gerard P
Perkins, Lois Claire
Perry, John Harold
Persaud, Triveddi Vidhya Nandan
Pesetsky, Alan
Peters, Alan
Peterson, Roy Reed
Petras, James Minas
Pettersen, James Clark
Pfeiffer, Carroll Athey
Phillips, Dwight Edward
Phillips, Steven Jones
Piatt, Jean
Pietsch, Paul Andrew
Plagge, James Clarence
Pliske, Edward Carl

Poorman, Douglas Harold
Poteat, William Louis
Pourcho, Roberta Grace
Pratt, Neal Edwin
Price, Joseph Levering
Printz, Richard H
Prutkin, Lawrence
Pullen, Edwin Wesley
Quattropani, Steven L
Rafferty, Nancy S
Rafols, Jose Antonio
Raikow, Robert Jay
Ramsay, Frederick J
Rasweiler, John Jacob, IV
Ratzlaff, Marc Henry
Rauchwerger, Joel M
Redmond, Billy Lee
Reed, Adrian Faragher
Rehman, Irving
Reinhardt, William Oscar
Reith, Edward John
Reynolds, Samuel R M
Rhines, Ruth
Rhodin, Johannes Arne Gosta
Richardson, Elisha Roscoe
Richins, Calvin Alexander
Rieck, Norman Wilbur
Rieke, William Oliver
Riley, Danny Arthur
Ring, John Robert
Rinker, George Clark
Ripley, Robert Clarence
Roberts, Walter Herbert B
Robertson, Douglas Reed
Robertson, George Gordon
Robertson, James David
Rodin, Martha Kinscher
Rodriguez-Peralta, Lorenzo Alberto
Rogers, Nelson Floyd
Romero-Sierra, Cesar Aurelio
Ronstrom, George Nelson
Roofe, Paul Gibbons
Roos, Henry
Ross, Leonard Lester
Ross, Michael H
Rosse, Cornelius
Rothman, Richard Harrison
Royce, George James
Rozier, Carolyn K
Ruby, John Robert
Rutherford, John Garvey
Saccoman, Frank (Michael)
Sack, Wolfgang Otto
Safanie, Alvin H
Saland, Linda C
Samorajski, Thaddeus
Sampson, Herschel Wayne
Sandborn, Bertil Sigvard Edmund
Sauerland, Eberhardt Karl
Sawyer, Charles Henry
Saxon, James Glenn
Scapino, Robert Peter
Schapiro, Herbert
Scharrer, Berta Vogel
Schatz, Leo
Scheving, Lawrence Einar
Schmedje, John Frederick
Schneider, Lawrence Kruse
Schon, Miguel Antonio
Schooley, Robert
Schuit, Kenneth Edward
Schulter, Frances Pierce
Schultz, Robert Lowell
Schweisthal, Michael R
Scott, David Alexander
Scott, David Evans
Scott, Earl B
Scott, Elton Monroe
Scott, John Watts, Jr
Seale, Raymond Ulric
Searls, James Collier
Seefeldt, Vern Dennis
Seibel, Hugo Rudolf
Seliger, William George
Sensenig, Edgar Carl
Sesco, Jerry Anthony
Sether, Lowell Albert
Settles, Harry Emerson
Setty, Laurel Raymond
Severn, Charles B
Shackleford, John Murphy
Sharawy, Mohamed
Shea, John Raymond Michael, Jr
Shellhamer, Robert Howard
Sheridan, Michael N
Sherman, Burton Stuart
Sherman, Jerome Kalman
Showers, Mary Jane C
Shriver, Joyce Elizabeth
Simard, Therese Gabrielle
Singer, Marcus
Singer, Ronald
Singh, Inder Jit
Singh, Roderick Pataudi
Singleton, Mary Clyde
Sinha, Akhouri Achyutanand
Sippel, Theodore Otto

Sistek, Vladimir
Skalko, Richard Gallant
Skjonsby, Harold Samuel
Slavin, Bernard Geoffrey
Slonecker, Charles Edward
Smith, Catherine Agnes
Smith, John Chandler
Smith, Raymond Dale
Smith, Stephen D
Smith, Stuart Werner
Smyser, Gerald Stanley
Snell, Richard Saxon
Snider, Ray S
Snodgrasse, Richard Montgomery
Snyder, George Edward
Sodicoff, Marvin
Soloff, Bernard Leroy
Solomon, Gordon Charles
Soper, Edward Henry
Sorenson, Robert Lowell
Speed, Edwin Maurice
Sperber, Geoffrey Hilliard
Spira, Arthur William
Spofford, Walter Richardson
Sprague, James Mather
Srebnik, Herbert Harry
Stallard, Richard E
Stempak, Jerome G
Stevens, Walter
Stilwell, Donald Lonson
Strauss, Helen Lorna Puttkammer
Strauss, Elliott William
Stultz, John Joseph
Sundberg, Ruth Dorothy
Sundsten, John Wallin
Susi, Frank Robert
Suzuki, Howard Kazuro
Swan, Roy Craig, Jr
Swanson, Ernest Allen, Jr
Swartz, Frank Joseph
Swett, John Emery
Swinyard, Chester Allan
Szepsenwol, Josel
Talbert, George Brayton
Tarby, Theodore John
Taylor, Anna Newman
Taylor, John Joseph
Tebo, Heyl Gremmer
Tedford, Myron Duncan
Telford, Ira Rockwood
Templeton, McCormick
Ten Cate, Arnold Richard
Thaemert, Jona Carl
Thalmann, Robert H
Thompson, James Scott
Thurow, Gordon Ray
Titkemeyer, Charles William
Tobin, Charles Emil
Todd, Gordon Livingston
Todd, Mary Elizabeth
Townsend, Samuel Franklin
Troyer, John Robert
Truex, Raymond Carl
Truscott, Basil Lionel
Tunturi, Archie Robert
Tyler, Walter Steele
Uyeda, Carl Kaoru
Van Cleave, Charles Durward
Van Dyke, John Howard
Vaughn, James E, Jr
Vaupel, Martin Robert
Velardo, Joseph Thomas
Vernon, Mina Lee
Vethamany, Victor Gladstone
Vijayan, Vijaya Kumari
Wadsworth, Gladys Elizabeth
Walker, Bruce Edward
Walker, Donald Gregory
Walker, Leon Bryan, Jr
Ward, James William
Warfel, John Hiatt
Warner, Francis James
Warner, Louise
Watkins, Dudley T
Weary, Marlys E
Weathersby, Hal Thompson
Weaver, Morris Eugene
Webb, Sawney David
Webber, Richard Harry
Webster, Richard Milroy
Weiss, Peddrick
Weiss, Leon
Welband, Wilbur A
Welter, Alphonse Nicholas
Welter, Dave Allen
West, William T
Weymouth, Richard J
White, Raymond Petrie, Jr
Whitmore, Mary (Elizabeth) Rowe
Wilcox, Harry Hammond
Williams, Benjamin Hayden
Williams, Mary Louise Monica Fritts
Williams, Vick Franklin
Willson, William Lane
Willson, John Tucker
Wilson, Doris Burda
Wilson, Frank Joseph
Wilson, Jack Lowery

Winborn, William Burt
Windle, William Frederick
Wischnitzer, Saul
Wismar, Beth Louise
Witzel, Everet Wayne
Wolfe, Jack Morris
Wood, Joe George
Woodburne, Russell Thomas
Worthington, Ward Curtis, Jr
Wotton, Robert Moore
Wragg, Laurence Edward
Yakaitis-Surbis, Albina Ann
Yeager, Vernon LeRoy
Yntema, Chester Loomis
Yollick, Bernard Lawrence
Young, Joseph Marvin
Young, Margaret Claire
Young, Paul Andrew
Young, Richard Wain
Young, William Johnson, II
Youngstom, Karl Arden
Yu, Mang Chung
Zeit, Walter
Zimmerman, Emery Gilroy
Zimny, Marilyn Lucile
Zingeser, Maurice Roy
Zygmunt, Wendell

Anatomic Pathology

Baden, Ernest
Coye, Robert Dudley
Cruickshank, Bruce
Fischer, Craig Leland
Gertz, Samuel David
Goodale, Fairfield
Krumerman, Martin Saul
Leipold, Horst Wilhelm
Lohse, Carleton Leslie
Russi, Simon
Schwartz, Colin John
Taylor, D Dax
Victor, Leonard Baker
Wellmann, Klaus Frederich

Comparative Anatomy

Ankel-Simons, Friderun Annursel
Blair, Charles Barkley, Jr
Chiasson, Robert Breton
Eastman, Joseph Thornton
Evans, Howard Edward
Fierstine, Harry Lee
Forman, G Lawrence
Garon, Olivier
Ghoshal, Nani Gopal
Hofer, Helmut Otto
Holmes, Edward Bruce
Jenkins, Floyd Albert
Jollie, Malcolm Thomas
Lee, Donald Gifford
Metcalf, Isaac Stevens Halstead
Minkoff, Eli Cooperman
Putnam, Jerry L
Rylander, Michael Kent
Schreiweis, Donald Otto
Smeriglio, Alfred John
Smith, Byron Colman
Spence, Alexander Perkins
Stahl, Barbara Jaffe
Struthers, Robert Claflin
Szebenyi, Emil

Developmental Anatomy

Bressler, Robert S
Campbell, William (Aloysius)
Chamberlain, Jack G
Diboll, Alfred
Gona, Amos G
Goss, Richard Johnson
Gray, James Clarke
Haley, Samuel Randolph
Hein, Rosemary Ruth
Hoar, Richard Morgan
Kelly, Douglas Elliott
Langdon, Herbert Lincoln
Maderson, Paul F A
Maue-Dickson, Wilma
Meyer, Conrad Frederick
Minor, Ronald R
Struthers, Robert Claflin
Sucheston, Martha Elaine
Torrey, Theodore Willett
Towers, Bernard
White, Richard Alan

Gross Anatomy

Bachop, William Earl
Blanton, Patricia Louise
Dalley, Arthur Frederick, II
Gardner, Weston Deuain
Grand, Theodore I
Hale, Gerry Alwyn
Harkins, Rosemary Knighton
Heltne, Paul Gregory
Jenkins, David Bruce
Kendall, Michael Welt
Krupp, Patricia Powers
Langdon, Herbert Lincoln
Metcalf, Isaac Stevens Halstead

ANATOMY

Mizell, Sherwin
Seibel, Werner
Stromberg, Melvin Willard
Sturtevant, Ruthann Patterson
Van De Graaff, Kent Marshall

Human Anatomy
Aycock, Nancy Rae
Bell, Mary
Cannon, Marvin Samuel
Carmichael, Stephen Webb
Carnes, James Edgar
Case, Norman Mondell
Clark, Glenn R
Cralley, John Clement
Crelin, Edmund Slocum
DeSesso, John Michael
Donaldson, Donald Jay
Eastman, Joseph Thornton
Enders, Allen Coffin
Eroschenko, Victor Paul
Fischer, Theodore Vernon
Friedman, Morton Henry
Gerstner, Robert
Gonzalez, Paula
Goodge, William Russell
Gray, Edwin R K
Grimm, James K
Haller, Ann Cordwell
Hawkins, Isaac Kinney
Hilliard, Stephen Dale
Joy, Edward Albert
Kapan, Shakti Prakash
Kaplan, Stanley
Keis, Adelbert Ferdinand Richard
Kleinhenz, Margie Joyce
Lasker, Gabriel (Ward)
Magruder, Samuel Rossington
Maraspin, Lyno Evelino
Martinez, Margaret Yarnall
Metcalf, William Kenneth
Miller, Marion Paul
Mitchell, Ormond Glenn
Morse, Dennis Ervin
Munger, Bryce Leon
Nelson, Marita Lee
Neufeld, Daniel Arthur
Pallie, Wazir
Rana, Mohammed Waheeduz-Zaman
Reilly, Frank Daniel
Rink, Richard Donald
Romrell, Lynn John
Sansone, Frances Marie
Saunders, John Bertrand De Cusance Morant
Shafland, James L
Walker, Elizabeth Reed
Waterhouse, Joseph Stallard
Wilbon, Walter Harrison
Williams, Thomas Walley, Jr
Zakhary, Rizkalla
Zwaan, Johan

Microscopic Anatomy
Al-Saadi, Abdul A
Benes, Elinor Simson
Burns, Edward Robert
Clermont, Yves Wilfred
Cotton, William Robert
Dickson, Douglas Howard
Dietlein, Lawrence Frederick
Dugan, Kimiko Hatta
Elias, Michael Hans
Epling, Glenwood Pershing
Feleppa, Alfred E, Jr
Friedman, Morton Henry
Golarz-De Bourne, Maria Nelly
Gwinnett, A John
Hadek, Robert
Hooker, William Mead
Jersild, Ralph Alvin, Jr
Kelly, Douglas Elliott
King, Barry Frederick
LeBouton, Albert V
Luchtel, Daniel Lee
Matthews, James Lester
McCallister, Lawrence P
Richer, Claude-Lise
Roosen-Runge, Edward C
Rosenthal, Theodore Bernard
Sedar, Albert William
Shryock, Edwin Harold
Snodgrass, Michael Jens

Neuroanatomy
Abplanalp, Paul LeRoy
Adams Smith, William Nelson
Aker, Franklin David
Albernaz, Jose Geraldo
Anderson, William John
Angevine, Jay Bernard, Jr
Augustine, James Robert
Baton, Robert Ralph
Bayer, Shirley Ann
Beal, John Anthony
Bennett, Marvin Herbert
Beresford, William Anthony
Berman, Alvin Leonard
Bernstein, Jerald Jack
Bertram, Ewart George
Bos, Jane
Brightman, Milton Wilfred
Brizzee, Kenneth Raymond
Brownson, Robert Henry
Butler, Ann Benedict
Campbell, Carlos Boyd Godfrey
Chambers, Wilbert Franklin
Clark, Ronald Grey
Clemente, Carmine Domenic
Cole, Wilbur Vose
Cookson, Francis Bernard
Coppock, Henry Aaron
Courville, Jacques
Cowan, W Maxwell
Cowley, A Ronald
Coyle, Peter
Creps, Elaine Sue
Curtis, Robin Livingstone
Debacker, Hilda Spodheim
Demski, Leo Stanley
DeVito, June Logan
Diamond, Ivan
Ebbesson, Sven O E
Feldman, Martin Leonard
Felten, David L
Finger, Thomas Emanuel
Fox, Clement Alphonsine
Freedman, Steven Leslie
Frontera-Reichard, Jose Guillermo
Fuller, Peter McAfee
Garrett, Frederic Daugherty
Gfeller, Eduard
Ghoshal, Nani Gopal
Giolli, Roland A
Globus, Albert
Gomez, Daniel Guillermo
Goodman, Donald Charles
Goodfellow, Elsie F
Glover, Roy Andrew
Guillery, Rainer Walter
Guth, Lloyd
Hafner, Gary Stuart
Haines, Duane Edwin
Hard, Walter Leon
Hartmann, James Francis
Hunt, Guy Marion, Jr
Hurst, Edith Marie Maclennan
Hyde, John Baskerville
Jacobson, Stanley
Johnson, John Irwin, Jr
Johnson, Thomas Nick
Johnston, Naomi Lemkey
Karten, Harvey J
Keller, Jeffrey Thomas
Kichler, Ernest Earl, Jr
Kimmel, Donald Loraine
King, James S
Kolb, Helga Ellen Thor
Kriebel, Richard Marvin
Krieg, Wendell Jordan
Kruger, Lawrence
LaMotte, Carole Choate
Larson, Sanford J
LaVail, Jennifer Hart
Lavelle, George Arthur
Leichnetz, George Robert
Leonard, Christiana Morison
Loewy, Arthur DeCosta
Lu Qui, Ivan James
Magoun, Horace Winchell
Martin, George Franklin, Jr
Matzke, Howard Arthur
McClure, Theodore Dean
Mehler, William Raphael
Mensah, Patricia Lucas
Meszler, Richard M
Moore, Josephine Carroll
Morrison, Adrian Russel
Mugnaini, Enrico
Nolan, Michael Francis
Norman, Wesley P
Northcutt, Richard Glenn
Norvell, John Edmondson, III
Palay, Sanford Louis
Peele, Talmage Lee
Penny, Joe Edward
Potter, Henson David
Powell, Ervin William
Ralston, Henry James, III
Ramon-Moliner, Enrique
Rennels, Marshall L
Riss, Walter
Rothballer, Alan Burns
Rubinson, Kalman
Russell, Glenn Vinton
Saunders, Richard L
Schnitzlein, Harold Norman
Schroeder, Dolores Margaret
Schwartz, Ilsa Roslow
Sechrist, John William
Severin, Charles Matthew
Shanthaveerappa, Totada Ramaiah
Sjostrand, Fritiof S
Smith, Diane Elizabeth
Smithberg, Morris
Stensaas, Larry J
Stensaas, Suzanne Sperling
Sterling, Peter
Stromberg, Melvin Willard
Stromberg, Norman Lewis
Sullivan, James Michael
Sutin, Jerome
Swigart, Richard Hanawalt
Taslitz, Norman
Thomas, Carolyn Eyster
Tigges, Johannes
Trachtenberg, Michael Carl
Turner, Robert Stuart
Ulinski, Philip Steven
Vaughan, Deborah Whittaker
Votaw, Charles Lesley
Ward, James Wellington
Warr, William Bruce
Wells, Joseph
Werner, Joan Kathleen
White, Edward Lewis
Whitlock, David Graham
Williams, Terence Heaton
Winans, Sarah Schilling
Wong-Riley, Margaret Tze Tung
Wright, Charles Gary
Yoss, Robert Eugene
Young, M Wharton

Plant Anatomy
Becker, Steven Allan
Berlyn, Graeme Pierce
Boke, Norman Hill
Brennan, James Robert
Byrne, John Maxwell
Calvin, Clyde Lacey
Carlson, John Bernard
Cecich, Robert Allen
Davidson, Christopher
Decker, Jane M
Derr, William Frederick
England, Wayne H
Eyde, Richard Husted
Freeman, Thomas Patrick
Goodman, Victor Herke
Harrison, H Keith
Heimsch, Charles W
Hillson, Charles James
Ikenberry, Gifford John, Jr
Kuijt, Job
Kukachka, Bohumil Francis
Larisey, Mary Maxine
Lott, John Norman Arthur
Mahlberg, Paul Gordon
Markle, Carrolle Anderson
McGrath, James J
Meyer, Conrad Frederick
Mikesell, Jan Erwin
Miller, Regis Bolden
Morey, Philip Richard
Moseley, Maynard Fowle, Jr
Myers, Gerald Andy
Neubauer, Benedict Francis
Oliver, Jeanette Clements
Pizzolato, Thompson Demetrio
Reeve, Marian E
Reeve, Roger Mansfield
Richardson, Paul Ernest
Rost, Thomas Lowell
Saigo, Roy Hirofumi
Stern, William Louis
Susalla, Ann A
Tepper, Herbert Bernard
Thompson, Neal Philip
Tseng, Charles C
Walker, Dan B
Wheeler, George Edward
Zimmermann, Martin Huldrych

Veterinary Anatomy
Atkins, David Lynn
Bartlett, Lawrence Matthews
Bell, John Thomas, Jr
Bhatnagar, Mahesh Kumar
Bisaillon, Andre
Bratton, Gerald Roy
Chibuzo, Gregory Anenonu
Christensen, George Curtis
Cummings, John Francis
Czarnecki, Caroline Mary Anne
DeLahunta, Alexander
Diesem, Charles D
Fletcher, Thomas Francis
Habel, Robert Earl
Hare, William Currie Douglas
Hinsman, Edward James
Horowitz, Aaron
Hullinger, Ronald Loral
Julian, Logan M
Kitchell, Ralph Lloyd
Lovell, James Edgeley
Magilton, James Henry
McClure, Robert Charles
McCurdy, Jon Alan
McKibben, John Scott
Meyer, Hermann
Ownby, Charlotte Ledbetter
Parke, Wesley Wilkin
Pierard, Jean Arthur
Sis, Raymond Francis
Skold, Bernard Harold
Staley, Theodore Earnest Leon
Venable, John Howard
Venzke, Walter George
Westerfield, Clifford
Williams, Raymond Crawford
Worthman, Robert Paul

Vertebrate Anatomy
Alexander, A Allan
Bentley, Cleo L
Chantell, Charles J
Galton, Peter Malcolm
Hart, Nathan Hoult
Kallen, Frank Clements
Liberti, Alfred Vincent
Plakke, Ronald Keith
St. Clair, Lorenz Edward
Stump, John Edward
Waters, James Frederick
Whiting, Anne Margaret

ANTHROPOLOGY

Anthropology
Abbott, Susan
Abu-Zahra, Nadia
Ackerman, Robert Edwin
Adams, Richard Edward Wood
Adams, Richard Newbold
Adelman, Fred
Aginsky, Bernard Willard
Aginsky, Ethel G
Ahler, Stanley Albert
Aigner, Jean Stephanie
Aikens, Clyde Melvin
Allard, Ethel Mary
Alland, Alexander, Jr
Alschuler, Milton
Alverson, Hoyt Sutliff
Anastasio, Angelo
Anderson, Arthur James Outram
Anderson, Douglas Dorland
Anderson, Eugene N, Jr
Anderson, James Nelson
Archer, Allan Frost
Arewa, E Ojo
Armstrong, Robert Plant
Aschenbrenner, Joyce Cathryn
Ascher, Robert
Ayres, William Stanley
Baerreis, David Albert
Bailey, Frederick George
Bailey, Wilfrid Charles
Bamberger, Joan Thatcher
Banks, Eugene Pendleton
Barnes, Annie Shaw
Barnett, Homer Garner
Barnouw, Victor
Barrett, Richard Allan
Basehart, Harry Wetherald
Basso, Ellen Becker
Basso, Keith Hamilton
Battersby, Harold Ronald Eric
Baty, Roger M
Baumhoff, Martin A
Beals, Ralph Leon
Beatty, John Joseph
Beck, Brenda E F
Becker, Marshall Joseph
Beckerman, Stephen Joel
Bee, Robert L
Beeson, William Jean
Befu, Harumi
Bell, Robert Eugene
Belshaw, Cyril Shirley
Ben-Dor, Shmuel
Benfer, Robert Alfred
Berger, Rainer
Berleant-Schiller, Riva
Berlin, Brent
Bernard, Harvey Russell
Berreman, Gerald Duane
Bessac, Frank Bagnall
Biebuyck, Daniel P
Binford, Lewis R
Bird, Junius Bouton
Bishop, Charles Aldrich
Blakeman, Crawford Harris, Jr
Blanton, Richard Edward
Blau, Harold
Bleibtreu, Hermann Karl
Blu, Karen Isobell
Bluebond-Langner, Myra Honore
Bock, Philip Karl
Bodine, John James
Bock, Walter Erwin
Boggs, Stephen Taylor
Bohannan, Laura
Bohannan, Paul James
Boissevain, Ethel (Mrs Arthur Lesser, Jr)

Bond, George Clement
Bourguignon, Erika Eichhorn
Bourque, Bruce Joseph
Bowles, Gordon Townsend
Boyd, John Paul
Boyer, Ruth McDonald
Bradfield, Stillman
Braidwood, Robert J
Brain, James Lewton
Brandt, Elizabeth Anne
Brant, Charles Sanford
Breternitz, David Alan
Briggs, Jean Louise
Briggs, Lloyd Cabot
Brockington, Donald Leslie
Brodsky, Carroll M
Brown, Kathleen Older
Brown, Leonard Keith
Brown, Paula
Bruhns, Karen Olsen
Bryant, Alan Lyle
Bryant, Vaughn Motley
Buchler, Ira Richard
Burling, Robbins
Bushnell, John Hempstead
Buxbaum, Edwin Clarence
Cain, H Thomas
Calabrese, Francis Anthony
Caldwell, Warren W
Campbell, John Martin
Campbell, Thomas Nolan
Carlson, David Sten
Carlson, Gustav Gunnar
Carmack, Robert
Carneiro, Robert Leonard
Carrasco, Pedro
Carroll, Vern
Carter, Elizabeth Francis
Casagrande, Joseph Bartholomew
Case, Charles Calvin
Casselberry, Samuel Emerson
Chagnon, Napoleon Alphonseau
Chang, Kwang-Chih
Chapman, Carl Haley
Chapman, Jefferson
Chard, Chester Stevens
Chenhall, Robert Gene
Chilcott, John Henry
Christensen, James Boyd
Christman, Luther Parmalee
Clark, Geoffrey Anderson
Clune, Francis Joseph, Jr
Codere, Helen
Coe, Michael Douglas
Cohen, Lucy M
Cohen, Yehudi Aryeh
Colby, Benjamin N
Cole, Johnnetta B
Collier, Donald
Collins, Henry B
Collins, June McCormick
Collins, Lloyd Raymond
Conklin, Harold Colyer
Conrad, Geoffrey Wentworth
Conrad, Jack Randolph
Cook, Edwin Aubrey
Coon, Carleton Stevens
Cooper, Matthew Owen
Cove, John James
Cowgill, George L
Cox, Bruce Alden
Crabb, David Wendell
Craig, Alan Knowlton
Crain, Jay Bouton
Crapanzano, Vincent Bernard
Crawford, Michael H
Creider, Chester Arthur, III
Crumley, Carole Linda
Crumrine, Norman Ross, II
Culbert, Thomas Patrick
Dahlberg, Frances Murray
Daifuku, Hiroshi
D'Andrade, Roy G
Daniels, Gert L
Daniels, Robert Edward
Danson, Edward Bridge
Dark, Philip John C
Daugherty, Richard Deo
Davis, Edward Mott
Davis, Wilbur Arthur
Day, Gordon Malcolm
d'Azevedo, Warren Leonard
Deagan, Kathleen A
De'ath, Colin Edward
DeGarmo, Glen Dean
DeJarnette, David Lloyd
Delson, Eric
Denman, Clayton Charlton
Desmond, Gerald Raymond
Despres, Leo Arthur
DeVore, Boyd Irven
Diamond, Norma Joyce
Diaz, May Nordquist
Dibble, Charles Elliot
Dick, Herbert William
Dickeman, Mildred
Di Peso, Charles Corradino
Dittert, Alfred Edward, Jr

Dixon, Keith Alan
Dobyns, Henry Farmer
Dockstader, Frederick J
Donnan, Christopher B
Dorjahn, Vernon Robert
Dorwin, John T
Downs, Richard Erskine
Drews, Robin Arthur
Du Bois, Cora
Ducey, Paul Richard
Duff, Wilson
Duffield, Lathel Flay
Dumond, Don Edward
Dunn, John Asher
Dupertuis, Clarence Welsey
Durbin, Mridula Adenwala
du Toit, Brian Murray
Dutton, Bertha Pauline
Dyson, Robert Harris, Jr
Eames, Edwin
Eastman, Carol Mary
Eder, James Farnum, Jr
Edmonson, Munro Sterling
Eggan, Fred R
Ehrich, Robert William
Ehrich, Allen S
Eiseley, Loren Corey
Ekholm, Gordon Frederick
Ekvall, Robert Brainerd
Elsasser, Albert B
Emerick, Richard Gibbs
Emory, Kenneth Pike
Epstein, Jeremiah Fain
Erasmus, Charles John
Erickson, Edwin E
Erickson, Vincent Oliver
Essene, Frank J
Euler, Robert Clark
Evans, Clifford
Evans, David Kenneth
Ewald, Robert Harold
Ewers, John Canfield
Ezell, Paul Howard
Faris, James Chester
Farmer, Malcolm French
Faron, Louis Charles
Faulkner, Charles Herman
Faust, Richard Deane
Fenton, William Nelson
Ferdon, Edwin Nelson, Jr
Ferguson, Leland Greer
Fernandez, Frank
Fernandez, James William
Finney, Ben R
Fischer, John Lyle
Fisher, James F
Fitzhugh, William W
Flannery, Kent Vaughn
Flannery, Regina (Mrs Karl F Herzfeld)
Flores-Meiser, Enya P
Foner, Nancy
Fontana, Bernard Lee
Force, Roland Wynfield
Ford, Richard Irving
Ford, Virginia
Forman, Shepard L
Fortier, David Harry
Foster, Brian Lee
Foster, George McClelland, Jr
Fowler, Don D
Fowler, Melvin Leo
Fraker, Charles Oliver
Fraser, Thomas Mott, Jr
Freed, Stanley Arthur
Freedman, James M
Freeman, Joan Elizabeth
Freeman, Susan Tax
French, Kathrine Story
Frison, George Carr
Fritz, John Merwin
Frost, Everett Lloyd
Fruzzetti, Lina Maria
Fry, Christine L
Fry, Robert E
Furst, Peter
Gale, Hoyt Rodney
Gallagher, Orvoell Roger
Gallaher, Art, Jr
Garbarino, Merwyn Stephens
Gardner, Peter Michael
Gardner, William Milton
Garrison, Vivian Eva
Gavan, James Anderson
Gehrie, Mark Joshua
Genoves, T Santiago
Getty, Bert Alfred
Getty, Harry Thomas
Gibbs, James Lowell, Jr
Gibson, Gordon Davis
Gibson, Kathleen Rita
Gindhart, Patricia S
Glazer, Mark
Goldstein, Melvin C
Golla, Victor Karl
Goodale, Jane Carter
Goode, Judith Granich
Goodman, Felicitas Daniels
Goodwin, Stefan Cornelius

Grabert, Garland Frederick
Gradwohl, David Mayer
Granberry, Julian
Grange, Roger T, Jr
Granger, Joseph Edward
Granzberg, Gary Robert
Grayson, Donald Kenneth
Green, Ernestene Leverne
Green, Roger Curtis
Greenberg, Joseph Harold
Greengo, Robert Eugene
Greenway, John
Gregersen, Edgar Alstrup
Griffin, William Bedford
Griffith, James Bennett
Griffith, Charles Ray
Grollig, Francis Xavier
Gross, Daniel Russell
Groscup, Gordon Leonard
Grove, David Cliff
Gruber, Jacob William
Gruhn, Ruth
Guemple, Lee
Gulick, John
Gulliver, Philip Hugh
Gumerman, George John, III
Gunnerson, James Howard
Guthe, Alfred Kidder
Gutkind, Peter C W
Gwaltney, John L
Gwynne-Thomas, Eric Hubert
Haag, William George
Hahn, Paul Gene
Hall, Edward Twitchell
Hall, Robert Leonard
Hally, David Judson
Halpern, Katherine Spencer
Hammel, Eugene Alfred
Hammond, Peter (Boyd)
Hamp, Eric Pratt
Hansen, Asael Tanner
Hansen, Edward Charles
Hardesty, Donald Lynn
Harner, Michael James
Harrington, Charles Christopher
Harrison, Peter D'Arcy
Hart, Donn Vorhis
Haury, Emil Walter
Hay, Thomas Hamilton
Hays, Thomas Reese
Heath, Dwight Braley
Heider, Karl Gustav
Heizer, Robert Fleming
Helm, June
Hemmings, E Thomas
Hendel-Sebestyen, Giselle
Henderson, John Stanley
Hester, James J
Hester, Joseph Aaron, Jr
Hewes, Gordon Winant
Hibben, Frank Cummings
Hickerson, Harold
Hickman, John Marshall
Hicks, Frederic Noble
Hill, James Newlin
Hill, Jane Hassler
Hinton, Thomas Benjamin
Hirabayashi, James A
Ho, Ting-Jui
Hockett, Charles Francis
Hodge, William Howard
Hoebel, Edward Adamson
Hoffman, John Leslie
Hogg, Thomas Clark
Hollos, Marida Clara
Holmes, Lowell Don
Honea, Kenneth Howard
Honigmann, John Joseph
Horner, George Roland
Horr, David Agee
Hosley, Edward
Howell, Alvin Hercules
Howell, Richard Wesley
Hubbell, Linda Jean
Hudson, Alfred Bacon
Hudson, Dee Travis
Hughes, Jack Thomas
Hulse, Frederick Seymour
Humphrey, Robert Lee, Jr
Hunt, M Eva
Hunter, William Stuart
Hutterer, Karl Leopold
Icenogle, David William
Ikawa-Smith, Fumiko
Ingersoll, Jasper
Inke, Gabor
Irving, William Nathaniel
Irwin-Williams, Cynthia Cora
Ishino, Iwao
Jablow, Joseph
James, Bernard Joseph
Jelinek, Arthur J
Jenkins, Charles Rivington
Jennings, Calvin Hunt
Jennings, Jesse David
Johnson, Alfred Edwin
Jones, Grant Drummond
Joseph, Roger

Judkins, Russell Alan
Kaplan, David
Kaschube, Dorothea Vedral
Kay, Paul
Kehoe, Alice Beck
Kehoe, Thomas Francis
Kellar, James Harley
Kelley, John Charles
Kelly, Arthur Randolph
Kelly, Isabel Truesdell
Kennedy, William Jerald
Kent, Kate Peck
Kernan, Keith Thomas
Keslin, Richard Orville
Kessler, Evelyn Seinfeld
Keyes, Charles Fenton
Kickert, Robert Warren
Kidd, Kenneth E
Kidder, Alfred, II
Kiefer, Christie Weber
Kikuchi, William Kenji
Kimball, Solon T
King, Arden Ross
Kinsey, W Fred, III
Kirch, Patrick Vinton
Kiriazis, James William
Kiste, Robert Carl
Klein, Harriet Esther Manelis
Klein, Richard G
Knez, Eugene Irving
Knudson, Ruthann
Konitzky, Gustav Adolf
Kopytoff, Igor
Koss, Joan Dee
Krader, Lawrence
Kraft, Herbert Clemens
Krieger, Alex Dony
Kuper, Hilda Beemer
Kupferer, Harriet Jane
Kurjack, Edward Barna
La Barre, Weston
Lamberg-Karlovsky, Clifford Charles
Lambert, Bernd
Lambert, Marjorie Ferguson
Landes, Ruth
Landman, Ruth Hallo
Landy, David
Lang, Gottfried Otto
Langness, Lewis L
Lanning, Edward P
Lantis, Margaret Lydia
Larson, Lewis Henry, Jr
Lathrap, Donald Ward
Laughlin, Robert Moody
Lawless, Robert Dale
Layton, Thomas Nutter
Leach, Larry Lamont
Leacock, Seth
Leap, William L
Lebra, Takie Sugiyama
Lebra, William Philip
Leechman, Douglas
Leeds, Anthony
Lehr, Raymond Bruce
Leininger, Madeleine Monica
Leone, Mark Paul
Leons, Madeline Barbara
Lessa, William Armand
LeVine, Robert Alan
Levy, Robert Isaac
Lewin, Ellen
Lewis, Herbert Samuel
Lewis, Phillip H
Lewis, Ralph Kepler
Libby, Dorothy
Lieberman, Leonard
Linares, Olga Frances
Long, Joseph K
Longacre, William Atlas, II
Luebben, Ralph A
Lundsgaarde, Henry P
Lurie, Nancy Oestreich
Lynch, Thomas Francis
Lyon, Patricia Jean
Macklin, June
Madsen, William
Mahar, J Michael
Maher, Robert Francis
Malouf, Carling I
Mani, Srinivasa Balasubra
Mann, Charles E
Manners, Robert Alan
Maranda, Elli Kongas
Maranda, Pierre
Marcus, Joyce
Maretzki, Thomas Walter
Margolis, Maxine Luanna
Marks, Stuart A
Marquardt, William Harrison
Martin, Marilynn Kay
Maruyama, Magoroh
Mason, Ronald James
Maxwell, Moreau Sanford
Maxwell, Thomas James
Mayer-Oakes, William James
McAllester, David Park
McBryde, Felix Webster
McCall, Daniel Francis

9

ANTHROPOLOGY

McCartney, Allen Papin
McClellan, Catharine
McCorkle, Thomas
McCracken, Robert Dale
McEwen, William John
McFee, Malcolm
McHugh, William Paul
McKay, Ruth Blumenfeld
McKennan, Robert Addison
McKenzie, Douglas Hugh
McKusick, Marshall Bassford
McNett, Charles William, Jr
McPherron, Alan
Meggers, Betty Jane
Meighan, Clement Woodward
Mendez, Eugenio Fernandez
Merriam, Alan Parkhurst
Merriam, Willis Bungay
Meyers, Allan Richard
Michael, Henry N
Michaelson, Evalyn Jacobson
Michels, Joseph William
Michener, Bryan Paul
Middleton, John F M
Milanich, Jerald Thomas
Miller, Beatrice Diamond
Miller, Elmer S
Miller, Frank Charles
Miller, Robert James
Mines, Rene
Mines, Mattison
Mintz, Jerome Richard
Mitchel-Kernan, Claudia I
Mobley, Harris W
Mochon, Marion Johnson
Moholy-Nagy, Hattula
Montgomery, Christine Anne
Montgomery, George Edward
Moore, G Alexander, Jr
Moore, Harvey Cleaver
Moore, Joseph Graessle
Morrill, Warren Thomas
Morris, Edward Craig
Morris, Richard Knowles
Morse, Dan Franklin
Moss, Leonard Wallace
Mountjoy, Joseph Bode
Muller, Jon David
Mulloy, William
Munn, Nancy D
Murdock, George Peter
Murphy, Robert Francis
Murrill, Rupert Ivan
Myers, James Edward
Nader, Laura
Nag, Moni
Nash, June C
Nash, Manning
Neely, James Alan
Nesbitt, Paul Homer
Neville, Gwen Kennedy
Neville, Melvin K
Nowak, Michael
Nunez, Theron A, Jr
Nunez, Hugo Gino
O'Barr, William McAlston
Officer, James Eoff
Olien, Michael David
Olien, Alan Peter
Olson, Ronald Leroy
O'Nell, Carl William
Opler, Marvin Kaufmann
Opler, Morris Edward
Orenstein, Henry
Osborne, Douglas
Osborne, Richard Hazelet
Ottenheimer, Harriet Joseph
Ottenheimer, Martin
Otterbein, Keith Frederick
Paine, Robert Patrick Barten
Paredes, James Anthony
Pasternak, Burton
Patterson, Thomas Carl
Peacock, James Lowe
Pearson, Richard Joseph
Peebles, Christopher Spalding
Pelto, Pertti Juho
Petzel, John Campbell
Peterson, Frederick Alvin
Peterson, Jon Holbrook, Jr
Pierce, Joe Eugene
Piker, Steven J
Pilling, Arnold Remington
Pitkin, Donald Stevenson
Plattner, Stuart Mark
Plog, Fred T
Plotnicov, Leonard
Poggie, John Joseph, Jr

Pollock, Harry Evelyn Dorr
Pond, Alonzo William
Posinsky, Sollie Henry
Potter, Jack M
Pressel, Esther Joan
Proulx, Donald Allen
Provencher, Ronald
Provinzano, James
Prufer, Olaf
Puleston, Dennis Edward
Purdy, Barbara Ann
Purrington, Burton Lewin
Quintana, Bertha Beatrice
Race, George Justice
Rachlin, Carol King
Raemsch, Bruce Ellenwood
Rainey, Froelich Gladstone
Rands, Robert Lawrence
Rappaport, Roy A
Rathje, William Laurens
Ravicz, Robert S
Ray, Verne Frederick
Raymond, James Scott
Read, Dwight Webster
Redman, Charles Lincoln
Reed, Charles Allen
Reed, Erik Kellerman
Reinhart, Theodore Russell
Reinman, Fred M
Rice, David Gordon
Ridington, William Robin
Riegelhaupt, Joyce Firstenberg
Riesenberg, Saul Herbert
Riesenfeld, Alphonse
Riesman, Paul Hastings
Rigby, Bruce Joseph
Riley, Thomas Joseph
Ripley, Suzanne
Ritchie, William Augustus
Robbins, Michael C
Roberts, John Milton
Rohlen, Thomas Payne
Rohr, Arthur Henry, Jr
Rohrl, Vivian
Rolingson, Martha Ann
Romanucci-Ross, Lola
Romney, Antone Kimball
Romney, James G, Jr
Rosen, Lawrence
Ross, Harold Marion
Ross, William T
Rouse, Timothy Gerald
Rouse, Irving
Rowe, Chandler William
Rowe, John Howland
Rowlett, Ralph Morgan
Rubin, Joan
Rubin, Vera
Ruby, Jay W
Ruffini, Julio Lawrence
Ruppe, Reynold Joseph
Ruyle, Eugene Edward
Rynkiewich, Michael Allen
Sabloff, Jeremy Arac
Sahlins, Marshall David
Salwen, Bert
Salzman, Philip Carl
Sanders, William T
Sanger, David
Sapir, J David
Sarles, Harvey B
Sarma, Akkaraju V N
Sasaki, Tom Taketo
Schaefer, James Michael
Scheele, Daniel Joseph
Scheele, Harry George
Schensul, Stephen Lewis
Schildkrout, Enid
Schlegel, Stuart Allen
Schneider, Frederick Ewing
Scholte, Bob
Schorr, Thomas S
Schusky, Ernest Lester
Schwartz, Douglas Wright
Schwerin, Karl H
Seaford, Henry Wade, Jr
Sears, William Hulse
Sebeok, Thomas Albert
Sexton, James D
Shack, William Alfred
Shane, Orrin Clifton
Shankman, Paul Andrew
Sharrock, Floyd Wayne
Shenkel, J Richard
Shimony, Annemarie Anrod
Shiner, Joel Lewis
Shook, Edwin Martin
Shreve, Gregory Monroe
Shutler, Mary Elizabeth
Shutler, Richard, Jr
Sibley, Willis Elbridge
Siegel, James T
Silverberg, James Mark
Silverman, Sydel
Simmons, William Scranton
Simonds, Paul Emery

Simpson, Ruth DeEtte
Sinder, Leon
Singer, Milton
Sinoto, Yoshiko
Skinner, Elliott Percival
Slobodin, Richard
Slocum, Sally Virginia
Smith, Carlyle Shreeve
Smith, Derek George
Smith, Gerald Patrick
Smith, Philip Edward Lake
Smith, Watson
Smith, William Charles
Snow, Dean Richard
Snyder, Richard Gerald
Snyder, Sally
Snyder, Warren Arthur
Soday, Frank John
Solecki, Ralph Stefan
Solheim, Wilhelm Gerhard, II
Sordinas, Augustus
Sorenson, John Leon
Southall, Aidan William
Spain, David Howard
Spaulding, Albert Clanton
Spencer, Robert Francis
Speth, John David
Spier, Robert Forest Gayton
Spoehr, Alexander
Spores, Ronald
Spradley, James Phillip
Sprague, Roderick
Spuhler, James Norman
Stancliff, Merton Wesley
Stark, Barbara Louise
Stephenson, Robert Lloyd
Stern, Theodore
Stevenson, Robert Findlay
Stewart, Kenneth Malcolm
Stewart, Thomas Dale
Stipe, Claude Edwin
Stirling, James Heber
Stoltman, James Bernard
Stone, Doris Zemurray
Stone, Richard L
Story, Dee Ann
Struever, Stuart
Stuart, David Edward
Sturtevant, William Curtis
Sutherland, Donald Ralph
Suttles, Wayne Prescott
Sutton, Constance Rita
Swartz, Marc Jerome
Swauger, James Lee
Swindler, Daris Ray
Szwed, John Francis
Taylor, Dee C
Taylor, Herbert Cecil
Taylor, Royal Ervin, Jr
Taylor, Walter Willard
Thiene, Frederick Patton
Thomas, David Hurst
Thomas, David John
Thomas, Norman Dwight
Thompson, Donald Enrique
Thompson, Raymond Harris
Thurman, Melburn D
Tiffany, Sharon Weston
Tiffany, Walter Warren
Tillson, David Stanley
Todd, Harry Flynn, Jr
Tooker, Elisabeth (Jane)
Toups, Polly Anticich
Townsend, Joan B
Turnbull, Christopher John
Turnbull, Colin MacMillan
Turner, Billie Lee, II
Tuttle, Russell H
Tyler, Stephen Albert
Underwood, Frances Wenrich
Underwood, Jane Hainline
Useem, John
Vastokas, Joan M
Vatuk, Sylvia Dutra
Vaughan, James Herbert, Jr
Vayda, Andrew Peter
Vilakazi, Absolom
Vincze, Lajos
Viola, Herman Joseph
Vogelin, Charles Frederick
Vogt, Evon Zartman
Voigt, David Quentin
Von Mering, Otto O
Voorhies, Barbara
Vreeland, Herbert Harold, III
Waddell, Eric Wilson
Wagley, Charles W
Wagner, Richard Vernon
Wagner, Roy
Walker, Deward Edgar, Jr
Wallace, Dwight Tousch
Warner, Frederic William
Washburn, Sherwood Larned
Washburn, William
Watkins, Mark Hanna
Watrous, Blanche Greene

Watson, Gordon Dulmage
Watson, James Bennett
Wax, Rosalie H
Weakly, Ward Fredrick
Weaver, Sally Mae
Webster, David Lee
Weiss, Gerald
Wendorf, Fred
Werner, Oswald
Whallon, Robert, Jr
Wheat, Joe Ben
Wheeler, Margaret Cameron
Wheeler, Tamara Stech
White, Anta M
White, Douglas Richie
Whitney, Daniel DeWayne
Whitten, Norman Earl, Jr
Whitter, Herbert Lincoln
Wicke, Charles Robinson
Wilbert, Johannes
Willems, Emilio
Williams, Aubrey Willis, Jr
Williams, Herbert H
Williams, James Raymond
Williams, Melvin Donald
Williams, Stephen
Willis, William Shedrick, Jr
Wilmsen, Edwin Norman
Wilson, John Philip
Winans, Edgar Vincent
Winter, Edward H
Withers, Arnold Moore
Withoft, John
Witucki, Jeannette Renner
Wolfe, Alvin William
Woodbury, Richard Benjamin
Woods, Clyde M
Wright, Barton Allen
Wynn, Jack Thomas
Yarnell, Richard Asa
Yengoyan, Aram A
Yinger, John Milton
Young, Jon Nathan
Young, Ulysses Simpson
Zenner, Walter P
Zimmerman, Lorraine May

Anthropometrics

Adams, Robert McCormick
Bass, William Marvin, III
Brand, Donald Dilworth
Damas, David John
Gustav, Bonnie Lee
Herron, Robert Ernest
Kottak, Conrad Philip
Reynolds, Herbert McGaughey
Roche, Alexander F
Smith, Allan Hathorn
White, Robert Manson

Applied Anthropology

Barnett, Homer Garner
Brittan, Norman
Brownrigg, Leslie Ann
Carter, William Earl
Clifton, James Alfred
Corrigan, Samuel Walter
Cox, Bruce Alden
Denton, Trevor
Dillon, Wilton Sterling
Doughty, Paul Larrabee
Downs, James Francis
Early, John Drennan
Erickson, Charles John
Fuller, Charles E
Gallaher, Art, Jr
Graves, Nancy Beatrice
Halpern, Joel Martin
Halpern, Katherine Spencer
Hatfield, Colby Ray, Jr
Hickman, John Marshall
Hiebert, Paul Gordon
Hughes, Charles Campbell
Jacobs, Sue-Ellen
Kaplan, Bernice Antoville (Mrs Gabriel Lasker)
Kelly, William Henderson
Kushner, Gilbert
Lange, Charles Henry
Leacock, Eleanor Burke
Lindquist, Lawrence Willard
Mason, Leonard Edward
Mayers, Marvin Keene
McKnight, Robert Kellogg
Mencher, Joan Phyllis
Metraux, Rhoda
Mithun, Jacqueline Stearns
Moran, Emilio Federico
Nash, Philleo
Nickerson, Gifford Spruce
Niehoff, Arthur Herman
Padfield, Harland Irvine
Pattison, Edward Mansell
Pearson, Keith Laurence
Rauf, Mohammad A
Reid, Marlene Barnes
Safa, Helen Icken
Safier, Gwendolyn

Shimkin, Demitri Boris
Smith, Courtland Lester
Spicer, Edward Holland
Stewart, Omer Call
Tax, Sol
Thompson, Laura
Tiedke, Kenneth Earle Riordan
Van Willigen, John Gilbert
Walker, Willard Brewer
Wallace, Ben J
Weaver, Thomas
West, Stanley A

Behavioral Anthropology
Dillon, Wilton Sterling
Graves, Theodore Dumaine
Holzinger, Charles Henry
Leibowitz, Lila
Padfield, Harland Irvine
Sanday, Peggy R
Scheflen, Albert E
Seltzer, Michael Rogers
Silverberg, James Mark
Singer, Philip
Van Der Elst, Dirk H

Biological Anthropology
Armelagos, George John
Austin, Donald Mac
Bajema, Carl J
Baker, Paul Thornell
Barden, Howard Stavers
Bennett, Kenneth A
Bland, John Edward
Buikstra, Jane Ellen
De Pena, Joan Finkle
Dethlefsen, Edwin Stewart
Dyke, Bennett
Dyson-Hudson, V Rada
Eckhardt, Robert Barry
Freeman, Milton Malcolm Roland
Fry, Edward Irad
Garruto, Ralph Michael
Gibbons, Michael Francis, Jr
Harpending, Henry Cosad
Howell, Francis Clark
Klepinger, Linda Lehman
Kunstadter, Peter
Lamb, Neven P
Lasker, Gabriel (Ward)
Lieberman, Leslie Sue
Mazess, Richard B
McHenry, Henry Malcolm
Miller, Peter S
Murad, Turhon Allen
Novak, Ladislav Peter
Olsen, Stanley John
Otten, Charlotte Marie
Oxnard, Charles Ernest
Poirier, Frank Eugene
Reid, Russell Martin
Rightmire, George Philip
Sank, Diane
Saul, Frank Philip
Sciulli, Paul William
Shapiro, Harry Lionel
Steegmann, Albert Theodore, Jr
Swedlund, Alan Charles
Tiger, Lionel
Warren, Charles Preston
Weiss, Kenneth Monrad
Wetherington, Ronald K
Witt, Shirley Hill
Zegura, Stephen Luke

Cultural Anthropology
Adams, Richard Newbold
Adams, William Yewdale
Aginsky, Bernard Willard
Aguilera, Francisco Enrique
Alkire, William Henry
Alland, Alexander, Jr
Ames, David Wason
Anderson, Barbara Gallatin
Anderson, Robert
Anderson, Robert Thomas
Atherton, John Harvey
Ayres, Barbara Chartier
Barclay, Harold Barton
Beals, Alan Robin
Beals, Ralph Leon
Beardsley, Richard King
Benet, Sula
Bennett, John William
Bharati, Agehananda
Bruner, Edward M
Burger, Henry G
Butterworth, Douglas Stanley
Canby, Joel Shackelford
Chance, Norman Allee
Chinas, Beverly Newbold
Clark, Mary Margaret
Cleland, Charles Alfred
Clifton, James Alfred
Cole, John Wallace
Cornell, John B
Crane, Julia Gorham

Crapo, Richley H
Crocker, William Henry
Crowley, Daniel John
De Laguna (Lopez de Leo), Frederica (Annis)
Dole, Gertrude Evelyn
Downs, James Francis
Driver, Harold Edson
Drucker, Philip
Eberhart, Hal
Ebihara, May Mayko
Edgerton, Robert Breckenridge
Elick, John William
Ember, Carol Ruchlis
Epstein, David George
Escobar, M Gabriel
Foster, Joseph Frederick
Fox, Richard G
Freilich, Morris
Fried, Morton H
Friedl, Ernestine
Friedl, John
Frisbie, Charlotte Johnson
Galaska, Chester F
Gallin, Bernard
Garland, William
Geertz, Hildred
Gerlach, Luther Paul
Glasse, Robert Marshall
Goldkind, Victor
Goldman, Irving
Goldschmidt, Walter Rochs
Gonzalez, Nancie L
Goodenough, Ward Hunt
Gould, Harold A
Gould, Richard Allan
Graburn, Nelson Hayes Henry
Gravel, Pierre Bettez
Gray, Robert Fred
Greenfeld, Philip John
Greenwood, Davydd James
Grindal, Bruce Theodore
Gunn, Harold Dale
Halpern, Joel Martin
Harman, Robert Charles
Harris, George Lawrence
Harris, Howard Leroy
Harris, Marvin
Harwood, Alan
Hasan, Khwaja Arif
Hatcher, Evelyn Payne
Helms, Mary Wallace
Henney, Jeannette Hillman
Hicks, George Leon, Jr
Hiebert, Paul Gordon
Himes, Ronald Stewart
Hippler, Arthur Edwin
Hitchcock, John Thayer
Hoffman, Charles Andrew, Jr
Hoffmann, Hans
Hotchkiss, John Calvin
Howard, Alan
Hsu, Francis Lang-Kwang
Hughes, Daniel Thomas
Hunter, David Emanuel
Hutchinson, Harry William
Isaac, Barry LaMont
Jantzen, Carl Raymond
Johnson, Allen Willard
Jones, Delmos Jehu
Jordan, David K
Kelly, Anthony Henderson
Kirsch, Anthony Thomas
Klass, Morton
Kohl, Seena B
Kutsche, Paul
Lander, Patricia Slade
Lange, Charles Henry
LaRuffa, Anthony Louis
Leighton, Alexander Hamilton
Leis, Philip Edward
Leons, Madeline Barbara
Lesser, Alexander
Lex, Barbara Wendy
Lieban, Richard Warren
Lindquist, Lawrence Willard
Lister, Robert Hill
Littleton, C Scott
Lockwood, William Grover
Luomala, Katharine
Magnarella, Paul J
Maloney, Thomas J
Mani, Srinivasa Balasubra
Maquet, Jacques
McCone, Robert Clyde
McCurdy, David Whitwell
McIrvin, Ronald Ray
Mead, Margaret
Meeker, Michael Elliott
Messenger, John Cowan
Messing, Simon D
Metraux, Rhoda
Miller, Solomon
Mitchell, William Edward
Montagu, Ashley
Moore, Frank William
Moore, Harvey Cleaver
Morey, Robert V

Nance, Charles Roger
Naroll, Raoul
Nason, James Duane
Nelson, Cynthia
Nicholas, Ralph Wallace
Niehoff, Arthur Herman
Nimmo, Harry Arlo
Norbeck, Edward
Nurge, Ethel
O'Brien, Denise (Mrs Jay Ruby)
Oliver, Symmes Chadwick
Ortner, Sherry B
Oswalt, Wendell Hillman
Owen, Roger C
Parker, Seymour
Peterson, Edwin Loose
Pope, Polly Holman
Price, John Andrew
Rahman, Mushtaq-Ur
Reina, Ruben E
Richardson, Miles Edward
Ross, Hubert Barnes
Rowe, William Leal
Rubel, Arthur Joseph
Rubel, Paula G
Sanday, Peggy R
Schlegel, Alice Elizabeth
Schmidt, Nancy Jeanne
Schneider, David M
Schorger, William Davison
Segal, Edwin Stanley
Seltzer, Michael Rogers
Service, Elman Rogers
Shelton, Austin Jesse
Shiloh, Ailon
Silverman, Philip Samuel
Simmons, Donald C
Skinner, George William
Slater, Mariam Kreiselman
Smith, Courtland Lester
Smith, Hale Gilliam
Smith, M Estellie
Smith, Robert Jack
Smith, Robert John
Smith, Valene Lucy
Spielberg, Joseph
Spindler, Louise Schaubel
Spiro, Melford Elliot
Stanislawski, Michael Barr
Stauder, Jack
Stover, Leon Eugene
Stricken, Arnold
Sweet, Louise Elizabeth
Tanner, Clara Lee
Tax, Sol
Taylor, Robert Bartley
Tefft, Stanton Knight
Textor, Robert Bayard
Theodoratus, Robert James
Theuws, Jacques Antoine
Thomas, Prentice Marquet, Jr
Thompson, Laura
Townsend, Patricia Kathryn
Tuden, Arthur
Tweddell, Colin Ellidge
Van Der Elst, Dirk H
Voget, Fred W
Wallace, Anthony Francis Clarke
Ward, Martha Coonfield
Warren, Claude Nelson
Watson, Lawrence Craig
Watson, Linvill
Weinberg, Daniela
Weiss, Melford Stephen
Williams, Thomas Rhys
Willner, Dorothy
Winner, Irene Portis
Winzeler, Robert Lee
Wood, John Jackson
Woodbury, Nathalie Ferris Sampson
Woolfson, Arnold Peter

Dental Anthropology
Dahlberg, Albert A
Mayhall, John Tarkington
Sirianni, Joyce E

Ethnobotany
Bohrer, Vorsila Laurene
Krauss, Beatrice Hilmer
Lawrence, Donald Buermann
Towle, Margaret Ashley
Wasson, Robert Gordon
Whiting, Alfred Frank

Ethnography
Ackerman, Robert Edwin
Agar, Michael Henry
Andrews, David Henry
Beardsley, Richard King
Bishop, Charles Aldrich
Clark, John Desmond
Clinton, Charles Anthony
Conklin, Harold Colyer
Ekvall, Robert Brainerd
Freed, Ruth Shelley
Gulliver, Philip Hugh
Gwaltney, John L

Hatt, Robert Torrens
Hohenthal, William Dalton, Jr
Kehoe, Thomas Francis
Kelly, Isabel Truesdell
King, Mary Elizabeth
Kiriazis, James William
Kniffen, Fred Bowerman
Lewis, Herbert Samuel
Luomala, Katharine
Moerman, Michael
Ottenberg, Simon
Pang, Hidegard Elisabeth
Perdue, Charles L, Jr
Price, Richard Swee
Sabella, James C
Salzman, Philip Carl
Salzmann, Zdenek
Schildkrout, Enid
Shreve, Gregory Monroe
Smith, Margo Lane
Spradley, James Phillip
Van Willigen, John Gilbert
Williams, Thomas Rhys
Winzeler, Robert Lee
Woodbury, Nathalie Ferris Sampson

Ethnology
Aberle, David Friend
Abler, Thomas Struthers
Alkire, William Henry
Anderson, Jonathan Gary
Aumann, Glenn D
Bailey, Milton (Edward)
Barclay, Harold Barton
Barnouw, Victor
Barrett, Richard Allan
Bascom, William Russel
Basehart, Harry Wetherald
Bidney, David
Bittle, William Elmer
Boon, James Alexander
Bricker, Victoria Reifler
Brown, Leonard Keith
Canby, Joel Shackelford
Carter, William Earl
Chinas, Beverly Newbold
Collins, June McCormick
Conant, Francis Paine
Damas, David John
Dark, Philip John C
De Saint-Jean, Alain Y
Desmond, Gerald Raymond
Dole, Gertrude Evelyn
Driver, Harold Edson
Dupree, Louis Benjamin
Eames, Edwin
Ebihara, May Mayko
Ellis, Florence Hawley
Elmendorf, William Welcome
Ewers, John Canfield
Fay, George Emory
Fenton, William Nelson
Forbis, Richard George
Ford, Richard Irving
Frantz, Charles
Freed, Stanley Arthur
Freeman, Susan Tax
Frisbie, Charlotte Johnson
Fuller, Charles E
Galusha, Joseph G, Jr
Garnst, Frederick Charles
Handler, Jerome Sidney
Hanks, Jane Richardson
Hart, Donn Vorhis
Hatch, Elvin James
Hatfield, Colby Ray, Jr
Heath, Dwight Braley
Helm, June
Helms, Mary Wallace
Hoebel, Edward Adamson
Hohenthal, William Dalton, Jr
Howard, James Henri
Hurt, Wesley Robert
Jachowski, Richard Leo
Jones, Grant Drummond
Kickert, Robert Warren
Klass, Morton
Knez, Eugene Irving
Krader, Lawrence
Kuper, Hilda Beemer
Kurtz, Ronald Joseph
Lanham, Betty Bailey
Lewis, Phillip H
Lounsbury, Floyd Glenn
Maranda, Pierre
Marwitt, John Paul
McFeat, Tom Farrar Scott
Michaelson, Karen L
Miller, Verna Jean
Morey, Robert V
Morrill, Warren Thomas
Murra, John Victor
Newcomb, William Wilmon, Jr
Newman, Philip Lee
Nickerson, Gifford Spruce
Orona, Angelo Raymond
Plotnicov, Leonard
Price, Richard Swee

ANTHROPOLOGY

Quimby, George Irving
Rachlin, Carol King
Riesenberg, Saul Herbert
Rogers, Spencer Lee
Ross, Harold Marion
Rothstein, Frances Marilyn
Rouls, Timothy Gerald
Shalter, Michael David, e
Shelton, Austin Jesse
Sorenson, Marion W
Stewart, Kenneth Malcolm
Stout, John Frederick
Tavakolian, Bahram Mehdi
Tefft, Stanton Knight
Theodoratus, Robert James
Thomas, Norman Dwight
Tiedke, Kenneth Earle Riordan
Titiev, Mischa
Vogel, Fred W
Wagley, Charles W
Walker, Willard Brewer
Wallace, Ben J
Watson, Lawrence Craig
Weist, Katherine Morrett
Whiteford, Andrew Hunter
Winter, Edward H
Young, Philip D

History of Anthropology

Bidney, David
Burridge, Kenelm Oswald Lancelot
Ducey, Paul Richard
Faron, Louis Charles
Freed, Ruth Shelley
Hatch, Elvin James
Holmes, Lowell Don
Holtzman, Stephen Ford
Kirsch, Anthony Thomas

Medical Anthropology

Abernethy, Virginia Deane
Christman, Noel Judson
Hackenberg, Robert Allan
Hochstrasser, Donald Lee
Jordan, Brigitte
Kennedy, Donald Alexander
Leslie, Charles Miller
Mazzur, Scott Ruigh
Messing, Simon D
Nydegger, Corinne Nemetz
Pearsall, Marion
Simeon, George John
Sipes, Richard Grey
Snow, Loudell Fromme
Todd, Harry Flynn, Jr
Townsend, John Marshall
Treloar, Alan Edward
Wiese, Helen Jean Coleman
Zarrugh, Laura Hoffman

Physical Anthropology

Anderson, Jonathan Gary
Anderson, James Edward
Angel, John Lawrence
Bass, William Marvin, III
Bell, Robert Eugene
Benoist, Jean
Biggerstaff, Robert Huggins
Birdsell, Joseph Benjamin
Bleibtreu, Hermann Karl
Bodel, John Knox
Brace, C Loring
Bramblett, Claud Allen
Brooks, Sheilagh Thompson
Brues, Alice Mossie
Buettner-Janusch, John
Cadien, James David
Cartmill, Matt
Chandler, Kirby
Charney, Michael
Chevalier-Skolnikoff, Suzanne
Clabeaux, Marie Striegel
Coelho, Anthony Mendes, Jr
Cooper, Harold Eugene
Damon, Albert
Dolhinow, Phyllis Carol
Field, Henry
Finnegan, Michael
Fix, James D
Friedlaender, Jonathan Scott
Frisancho, Roberto
Gaherty, Geoffrey George
Garn, Stanley Marion
Giles, Eugene
Gill, George Wilhelm
Gillette, Charles Edgar
Goldstein, Marcus Solomon
Greene, David Lee
Gustav, Bonnie Lee
Hanna, Joel Michael
Heath, Barbara Honeyman
Helmuth, Herman Siegfried
Herzberg, Hans Theodore Edward
Hiloowala, Rusi Ardeshir
Holcomb, George Ruhle
Holloway, Ralph L, Jr
Holtzman, Stephen Ford

Howells, William White
Hughes, David Rees
Hunt, Edward Eyre
Johnston, Francis E
Jolly, Clifford J
Jurmain, Robert Douglas
Kelso, Alec John
Kennedy, Kenneth Adrian Raine
Kerley, Ellis R
Kinzey, Warren Glenford
Krogman, Wilton Marion
Laughlin, William Sceva
Lindburg, Donald Gilson
Little, Michael Alan
Livingstone, Frank Brown
Longyear, John Munro, III
Malina, Robert Marion
Mann, Alan Eugene
Maples, William Ross
McConvile, John Theodore
McCullough, John Martin
McWilliams, Kenneth Richard
Milan, Frederick Arthur
Montagu, Ashley
Moreno-Black, Geraldine S
Napton, Lewis Kyle
Newell, Laura
O'Connor, Brian Lee
Ortner, Donald John
Phelps, David Sutton
Pietruswsky, Michael, Jr
Pilbeam, David Roger
Pollitzer, William Sprott
Protsch, Reiner Robert Rudolf
Rathbun, Ted Allan
Reid, Russell Martin
Reynolds, Herbert McGaughey
Rightmire, George Philip
Robbins, Louise Marie
Rogers, Spencer Lee
Rose, Michael Dudley
Rosen, Stephen I
St Hoyme, Lucile Eleanor
Sarich, Vincent M
Schon, Miguel Antonio
Schulter, Frances Pierce
Seltzer, Carl Coleman
Siegel, Michael Ian
Simmons, Michael Patrick
Singer, Ronald
Singh, Ripu Daman
Siriamni, Joyce E
Smith, Charline Galloway
Snodgrasse, Richard Montgomery
Snow, Clyde Collins
Staley, Robert Newton
Steen, John Carl
Stini, William Arthur
Sublett, Audrey J
Tappen, Neil Campbell
Tattersall, Ian Michael
Tinsman, James Herbert, Jr
Trinkaus, Erik
Turner, Christy Gentry, II
Ubelaker, Douglas Henry
Weiss, Mark Lawrence
White, Robert Manson
Wienker, Curtis Wakefield
Williams, Bobby Joe
Wolpoff, Milford Howell
Wood, Corinne Shear
Zihlman, Adrienne Louella

ASTRONOMY

Abbas, Mian Mohammad
Abell, George Ogden
Ables, Harold Dwayne
Abt, Helmut Arthur
Adler, Stephen Miller
Aksnes, Kaare
Albers, Henry
Alexander, Robert Stanley
Aller, Margo Friedel
Anderson, Christopher Marlowe
Anderson, Claude M
Anderson, Jean Hackett
Annear, Paul Richard
Argo, Harold Virgil
Armstrong, John William
Arp, Halton Christian
Ashbrook, Joseph
Atkinson, Robert d'Escourt
Aveni, Anthony
Babcock, Horace Welcome
Baez, Silvio
Bahng, John Deuck Ryong
Baker, James Gilbert
Bakos, Gustav Alfons
Barry, Don Cary
Bartlett, James Jefferson
Batten, Alan Henry
Bauer, Carl August
Baum, William Alvin

Bautz, Laura Patricia
Beavers, Willet I
Beer, Reinhard
Belserene, Emilia Pisani
Belton, Michael J S
Benedict, George Frederick
Berendzen, Richard Earl
Berg, Richard Allen
Berman, Louis
Bidelman, William Pendry
Binnendijk, Leendert
Birney, Dion Scott, Jr
Blanco, Victor Manuel
Bless, Robert Charles
Blitzstein, William
Blom, Christian James
Boardman, William Jarvis
Boesgaard, Ann Merchant
Boggess, Nancy Weber
Bok, Bart Jan
Bolton, Charles Thomas
Bond, Howard Emerson
Bookmyer, Beverly Brandon
Bowyer, C Stuart
Boyce, Peter Bradford
Bracher, Katherine
Bradley, Charles Crane
Brandt, John Conrad
Branley, Franklin M
Brittin, Alan Henry
Brooks, Walter Lyda
Brown, Robert Lawrence
Brunk, William Edward
Burke, J Anthony
Burns, Joseph A
Burton, William Butler
Byard, Paul L
Cameron, Winifred Sawtell
Campbell, Donald Bruce
Campbell, Warren Adams
Capen, Charles Franklin, Jr
Capriotti, Eugene Raymond
Carlson, Eric Dungan
Carpenter, Martha Stahr
Carpenter, Roland LeRoy
Carr, Roger Byington
Carr, Thomas Deaderick
Chamberlain, Joseph Miles
Chamberlain, Joseph Wyan
Chambers, Robert J
Chandrasekhar, Subrahmanyan
Chapman, Clark Russell
Chartrand, Mark Ray, III
Chen, Kwan-Yu
Cherrington, Ernest Hurst, Jr
Clark, Barry Gillespie
Clark, John Fulmer
Clark, Thomas Alan
Code, Arthur Dodd
Cohen, Arthur Dodd
Cohen, Marshall Harris
Collagan, Robert Bruce
Collins, George W, II
Collins, Thomas Elbert
Counselman, Charles Claude, III
Cowley, Anne Pyne
Cowley, Charles Ramsay
Coyne, George Vincent
Crampton, David
Crawford, David Livingstone
Cudworth, Kyle McCabe
Cuffey, James
Culver, Roger Bruce
Dahn, Conard Curtis
Davies, Merton Edward
Davis, Morris Schuyler
Deeming, Terence James
Demers, Serge
Dennis, Tom Ross
De Vaucouleurs, Gerard Henri
Dickel, Helene Ramseyer
Dieke, Sally Harrison
Dolan, Joseph Francis
Donn, Bertram (David)
Dufour, Reginald James
Duke, Douglas
Duncombe, Raynor Lockwood
Dunham, David Waring
Dunkelman, Lawrence
Dunn, Richard B
Dupuy, David Lorraine
Dyce, Rolf Buchanan
Ebbighausen, Edwin G
Ebert, Paul J
Edmondson, Frank Kelley
Edwards, Terry Winslow
Eichhorn-von Wurmb, Heinrich Karl
Elliot, James Ludlow
Ellis, Fred E
Elmore, Robert E
Engle, Paul Randal
Epps, Harland Warren
Epstein, Eugene Ethan
Epstein, Isadore
Erickson, William Clarence
Evans, David Stanley
Evans, Nancy Remage
Everhart, Edgar

Faber, Sandra Moore
Fallon, Frederick Walter
Fanale, Fraser P
Fastie, William George
Febelman, Walter A
Feibelman, John Donald
Fernie, John Donald
Fiala, Alan Dale
Fisher, Philip Chapin
Fitch, Walter Stewart
Fleischer, Robert
Fliegel, Henry Frederick
Ford, Clinton Banker
Ford, Holland Cole
Frank, Louis Albert
Franklin, Fred Aldrich
Franklin, Kenneth Linn
Fredrick, Laurence William
Garfinkel, Boris
Garrison, Robert Frederick
Gasteyer, Charles Earl
Gatewood, George David
Gault, Donald E
Gehrels, Tom (Anton Marie Jacob)
Geliker, Charles Don
Gibson, James (Benjamin)
Giclas, Henry Lee
Gierasch, Peter Jay
Gill, Jocelyn Ruth
Gilvarry, John James
Gingerich, Owen (Jay)
Goedicke, Victor Alfred
Gold, Thomas
Gordon, Courtney Parks
Gordon, Kurtiss Jay
Greenberg, Richard Joseph
Gull, Theodore Raymond
Gursky, Herbert
Hagen, John Peter
Hall, Douglas Scott
Hall, Richard Chandler
Hansen, Richard (Thomas)
Hanson, Robert Bruce
Hapke, Bruce W
Hardie, Robert Howie
Harper, Doyal Alexander, Jr
Harrington, Robert Sutton
Harris, Alan Joel
Harris, William Edgar
Harrison, Marjorie Hall
Hartoog, Mark Richard
Hartung, Jack Burdair
Harwick, Frederick David Alfred
Harwit, Christopher Alvin
Haupt, Ralph Freeman
Hayes, Donald Scott
Heap, Sara Ridgway
Heckathorn, Harry Mervin, III
Heintz, Wulff Dieter
Heiser, Arnold M
Hemenway, Curtis Leland
Henize, Karl Gordon
Herbig, George Howard
Herget, Paul
Herr, Richard Baessler
Hershey, John Landis
Herzog, Emil Rudolph
Hett, John Henry
Hewitt, Anthony Victor
Heyden, Francis Joseph
Heymann, Dieter
Hildebrand, Roger Henry
Hill, Sarah Jeanette
Hintzen, Paul Michael
Hoag, Arthur Allen
Hobbs, Lewis Mankin
Hodge, Paul William
Hoffleit, Ellen Dorrit
Hoffmann, William Frederick
Hogg, David Edward
Hogg, Helen (Battles) Sawyer
Hohl, Frank
Holt, Herbert Barthold
Holzinger, Joseph Rose
Hooper, William John, Jr
Houk, Nancy (Mia)
Housley, Robert Melvin
Houston, Walter Scott
Howard, Robert Franklin
Howard, William Eager, III
Hubbard, William Bogel, Jr
Hube, Douglas Peter
Hughes, Victor A
Huguenin, George Richard
Hujer, Karel
Humphreys, Roberta Marie
Hunter, Donald Mount
Hunter, James Hardin, Jr
Hunziker, Rodney William
Hutchings, John Barrie
Ingrao, Hector Carlos
Irwin, John (Henry) Barrows
Jacchia, Luigi Giuseppe
Jackson, Edward Soth
Jaffe, Leonard David
Jaffe, Walter Joseph
Janiczek, Paul Michael

Janssen, Michael Allen
Jefferys, William H, III
Jenkins, Edward Felix
Jenner, David Charles
Jenzano, Anthony Francis
Johns, Roman Karol Chrzaszcz
Johnson, Harold Lester
Johnson, Hugh Mitchell
Johnson, Torrence Vaino
Johnston, Kenneth John
Jones, Florence Shirley (Patterson)
Kaler, James Bailey
Kalnajs, Agris Janis
Karpov, Boris George
Keenan, Philip Childs
Keller, Geoffrey
Kesteven, Michael
Kieffer, Hugh Hartman
Kiewiet De Jonge, Joost Herman Albert
King, Ivan Robert
Kinman, Thomas David
Kirshner, Robert Paul
Kissell, Kenneth Eugene
Kiemola, Arnold R
Klock, Benny LeRoy
Knacke, Roger Fritz
Knapp, Stephen Laurence
Knappenberger, Paul Henry, Jr
Koch, Robert Harry
Kondo, Yoji
Kopal, Zdenek
Kraft, Robert Paul
Kraus, John Daniel
Krause, Helmut G L
Krienke, Ora Karl, Jr
Kristian, Jerome
Kron, Gerald Edward
Krzywoblocki, Maria Zbigniew von
Kumar, Shiv Sharan
Kunkel, William Eckart
Landolt, Arlo Udell
Langebartel, Ray Gartner
Langerbeck, Mary Therese
Larson, Harold Phillip
Latham, David Winslow
Layzer, David
Leacock, Robert Jay
Lebofsky, Larry Allen
Leclaire, Roger
Lee, Paul D
Lee, Thomas Alan
Leibhardt, Edward
Leonard, Charles Grant
Leung, Kam-Ching
Levitt, Israel Monroe
Liller, Martha Hazen
Liller, William
Lilley, Arthur Edward
Limber, David Nelson
Lippincott, Sarah Lee
Livingston, William Charles
Lu, Phillip Ketwa
Lutz, Julie Haynes
Luyten, Willem Jacob
Lynds, Beverly T
Lynds, Clarence Roger
Macklin, Philip Alan
MacRae, Donald Alexander
Malitson, Harriet Hutzler
Margrave, Thomas Ewing, Jr
Markowitz, William
Marlborough, John Michael
Marsden, Brian Geoffrey
Mathews, Robert Thomas
Matson, Dennis Ludwig
Matsushima, Satoshi
Mattei, Janet Akyuz
Mavco, George Edward
Mayall, Margaret Walton
Mayall, Nicholas Ulrich
Mayer, Cornell Henry
McAlister, Harold Alister
McCallion, William James
McCarthy, Dennis Dean
McCarthy, Martin
McClure, Robert D
McCook, George Patrick
McCord, Thomas Bard
McCormick, Philip Thomas
McCracken, Curtis W
McCrosky, Richard Eugene
McCuskey, Sidney Wilcox
McKinney, William Mark
McNamara, Delbert Harold
Mead, Jaylee Montague
Meisel, David Dering
Meltzer, Alan Sidney
Mengel, John Geist
Mengel, Donald Howard
Merrill, John Ellsworth
Middlehurst, Barbara Mary
Milkey, Robert William
Miller, Freeman Devold
Miller, Richard William
Millikan, Allan G
Millis, Robert Lowell
Milone, Eugene Frank
Mitchell, Walter Edmund, Jr

Mohler, Orren (Cuthbert)
Mook, Delo Emerson, II
Morgan, Thomas Harlow
Morgan, William Wilson
Morrison, David Douglas
Morrison, Nancy Dunlap
Moses, Ray Napoleon, Jr
Mulholland, John Derral
Mumford, George Saltonstall, III
Munch, Guido
Murphy, Elias Smith, Jr
Murray, Bruce C
Musen, Peter
Nacozy, Paul E
Naqvi, Saiyid Ishrat Husain
Nather, Roy Edward
Neff, John S
Nelson, Burt
Nelson, Harry Ernest
Newburn, Ray Leon, Jr
Newsom, Gerald Higley
Nicholson, Thomas Dominic
Noonan, Thomas Wyatt
Novotny, Eva
O'Brien, James Edward
O'Dell, Charles Robert
Oesterwinter, Claus
Oke, John Beverly
O'Keefe, John Aloysius
O'Leary, Brian Todd
Oliver, John Parker
Opal, Chet Brian
Opik, Ernest Julius
Osmer, Patrick Stewart
Osterbrock, Donald Edward
Osvalds, Valfrids
Ovenden, Michael W
Owen, Frazer Nelson
Owen, Tobias Chant
Parker, Robert Allan Ridley
Partridge, Robert Bruce
Pasachoff, Jay Myron
Pascu, Dan
Pataki, Louis Peter, Jr
Peacock, Keith
Peery, Benjamin Franklin, Jr
Peltier, Leslie Copus
Percy, John Rees
Perry, Charles Lewis
Persson, Sven Eric
Pesch, Peter
Peters, Geraldine Joan
Peterson, Alan W
Pettengill, Gordon Hemenway
Pfeiffer, Raymond John
Philip, A G Davis
Pieters, Carle
Pilachowski, Catherine Anderson
Pilcher, Carl Bernard
Pillans, Helen Mead
Pituga, George E
Popper, Daniel Magnes
Poss, Howard Lionel
Prendergast, Kevin Henry
Protheroe, William Mansel
Prouse, Ervin Joseph
Purton, Christopher Roger
Pyper, Diane Marie
Racine, Rene
Rambauske, Werner
Rathmann, Franz Heinrich
Rea, Donald George
Reaves, Gibson
Reitmeyer, William L
Reynolds, Ray Thomas
Rhynsburger, Robert Whitman
Richer, Harvey Brian
Richstone, Douglas Orange
Riegel, Kurt Wetherhold
Rieke, Carol Anger
Riggs, Philip Shaefer
Roark, Terry P
Roberts, Leonidas Howard
Roberts, William Woodruff, Jr
Rochester, Michael Grant
Roeder, Robert Charles
Roemer, Elizabeth
Rohrer, William Glen
Roman, Nancy Grace
Rothe, Carl Frederick
Rubin, Vera Cooper
Russell, John Albert
Ryan, Donald Edwin
Rymer, Harry
Sagan, Carl
Salanave, Leon Edward
Sandage, Allan Rex
Sander, Nestor John
Sandmann, William Henry
Sanner, Frederick Charles
Savage, Blair DeWillis
Scarfe, Colin David
Scargle, Jeffrey D
Schmidt, Edward George
Schmidt, Maarten
Schmidt, John Leigh
Schombert, John Leonard
Schopp, John David

Schroeder, Daniel John
Schubert, Gerald
Schulte, Daniel Herman
Schwarzschild, Martin
Scott, Elizabeth Leonard
Scott, Roderic MacDonald
Seaquist, Ernest Raymond
Seeds, Michael August
Seidelmann, Paul Kenneth
Sekanina, Zdenek
Shane, Charles Donald
Shane, Mary Lea
Shapiro, Irwin Ira
Shapiro, Lee Tobey
Sharpless, Stewart Lane
Shaw, James Scott
Shilts, James Leonard
Shurman, Michael Mendelsohn
Sida, Derek William
Simkin, Susan Marguerite
Simon, Lee Will
Simon, Michal
Simonson, Simon Christian, III
Singh, Jag Jeet
Sinton, William Merz
Sitterly, Charlotte Moore
Slaucitajs, Sergejs
Smith, Bradford Adelbert
Smith, Clayton Albert, Jr
Smith, Elske Van Panhuys
Smith, Harlan J
Smith, Haywood Clark, Jr
Smith, William Hayden
Southworth, Richard Boynton
Spinrad, Hyron
Spitzer, Lyman, Jr
Sprague, Newton G
Standeford, Leo Vern
Standish, E Myles, Jr
Stanger, Philip Charles
Stauffer, George Franklin
Stecher, Theodore P
Stecker, Floyd William
Stephenson, Charles Bruce
Stern, Albert Victor
Stipe, John Gordon, Jr
Stoeckley, Thomas Robert
Stoeckly, Robert E
Strand, Kaj Aage
Strobel, Darrell Fred
Strom, Stephen
Strong, Ian B
Sturch, Conrad Ray
Sulentic, Jack William
Swope, Henrietta Hill
Szebehely, Victor
Taylor, Donald James
Taylor, Joseph Hooton, Jr
Teiger, Martin
Terzian, Yervant
Teske, Richard Glenn
Thomas, Clinton Edward
Thomas, Jon Charles
Tifft, William Grant
Tolbert, Charles Ray
Toomre, Alar
Trimble, Virginia Louise
Tull, Robert Gordon
Turnrose, Barry Edmund
Ulrich, Marie-Helene DeMoulin
Underhill, Glenn
Upgren, Arthur Reinhold, Jr
Usher, Peter Denis
van Altena, William F
van de Kamp, Peter
van den Bergh, Sidney
Vandervoort, Peter Oliver
Van Till, Howard Jay
Vasilevkis, Stanislaus
Vaughan, Arthur Harris, Jr
Vedder, James Forrest
Verschuur, Gerrit L
Veverka, Joseph
Vrba, Frederick John
Wade, Campbell Marion
Wagman, Nicholas Emory
Wagner, Raymond Lee
Walker, Arthur Bertram Cuthbert, Jr
Walker, Merle F
Wallerstein, George
Wampler, E Joseph
Ward, William Roger
Warner, Jeffrey
Warner, John Ward
Weaver, Harold Francis
Webster, William John, Jr
Wehinger, Peter Augustus
Wehlau, Amelia W
Wehlau, William Henry
Weiler, Edward John
Weistrop, Donna Etta
Westerhout, Gart
Westphal, James Adolph
Weymann, Ray J
Whipple, Fred Lawrence
Whitaker, Ewen A
White, Raymond E

White, William Charles
Whitehead, Andrew Bruce
Wildey, Robert Leroy
Wilkening, Laurel Lynn
Williams, Carol Ann
Williams, John Albert
Williamson, Ralph Elmore
Wilson, Lee Anne Mordy
Wilson, Christopher Paul
Wilson, Olin Chaddock
Wilson, Raymond Hiram, Jr
Wing, Robert Farquhar
Winkler, Louis
Wolf, George William
Wood, David Belden
Wood, Frank Bradshaw
Wood, Howard John
Wray, James David
Wright, Frances Woodworth
Wyatt, Stanley Porter, Jr
Wyckoff, Susan
Yeomans, Donald Keith
Yoss, Kenneth M
Young, Andrew Tipton
Young, Arthur
Young, Warren Melvin
Zabriskie, Franklin Robert
Zellner, Benjamin Holmes
Zinn, Robert James
Zirin, Harold
Zuckerman, Benjamin Michael

Astrogeology
Bunch, Theodore Eugene
Carr, Michael H
Clanton, Uel S, Jr
England, Anthony W
French, Bevan Meredith
Gault, Donald E
Howard, James Hatten, III
Lowman, Paul Daniel, Jr
Lucchitta, Baerbel Koesters
Malcuit, Robert Joseph
Masursky, Harold
McCauley, John Francis
Morris, Elliot Cobia
Schaber, Gerald Gene
Schultz, Peter Hewlett
Smith, Eugene Irwin
Strom, Robert Gregson
Swann, Gordon Alfred
Walters, Charles Philip
Weihaupt, John George

Cosmology
Baierlein, Ralph Frederick
Dyer, Charles Chester
Geller, Margaret Joan
Hogarth, Jacke Edwin
Jacobs, Kenneth Charles
Keeney, Joe
Krogdahl, Wasley Sven
Liang, Edison Park-Tak
Misner, Charles William
Motill, Ronald Allen
Nissim-Sabat, Charles
Parker, Leonard Emanuel
Rauscher, Elizabeth Ann
Ravindra, Ravi
Schwartz, Daniel Alan
Shepley, Lawrence Charles
Tinsley, Beatrice Muriel
Tryon, Edward Polk
Wickes, William Castles
Wilkinson, David Todd

Planetary Atmospheres
Fellows, Robert Francis
Fels, Stephen Brook
Fymat, Alain L
Gross, Stanley H
Hess, Seymour Lester
Hodges, Ralph Richard, Jr
Ingersoll, Andrew Perry
Irvine, William Michael
Kuhn, William R
Levine, Joel Stewart
Mange, Phillip Warren
Mickelson, Michael Eugene
Olsen, Edward Tait
Orton, Glenn Scott
Plass, Gilbert Norman
Potter, Andrew Elwin, Jr
Traub, Wesley Arthur
Varanasi, Prasad
Williams, Dudley

Radio Astronomy
Andrew, Bryan Haydn
Barrett, Alan H
Basart, John Philip
Bash, Frank Ness
Bracewell, Ronald Newbold
Buhl, David
Castelli, John P
Christiansen, Wayne Arthur
Clark, Thomas Arvid
Conklin, Edward Kirkham

Cordes, James Martin
Counselman, Charles Claude, III
Craft, Harold Dumont, Jr
Cudaback, David Dill
Davis, Michael Moore
De Jong, Marvin Lee
Dieter-Conklin, Nannielou
Donivan, Frank Forbes, Jr
Douglas, James Nathaniel
Drake, Frank Donald
Dulk, George A
Ewing, Martin Sipple
Fainberg, Joseph
Firor, John William
Goldstein, Samuel Joseph, (Jr)
Gordon, Mark A
Gottesman, Stephen T
Greisen, Eric Winslow
Guidice, Donald Anthony
Haddock, Frederick Theodore, Jr
Hankins, Timothy Hamilton
Hardebeck, Ellen Jean
Heeschen, David Surphin
Higgs, Lloyd Albert
Hollinger, James Pippert
Irvine, William Michael
Jansson, Michael Allen
Johnson, Donald Rex
Kaftan-Kassim, May A
Katzin, Joel Carl
Kellermann, Kenneth Irwin
Kerr, Frank John
Klein, Michael John
Knapp, Gillian Revill
Knowles, Stephen H
Lebo, George Robert
Legg, Thomas Harry
Lo, Kwok-Yung
Lovas, Francis John
McCutcheon, William Henry
Meeks, Marion Littleton
Moffet, Alan Theodore
Morris, Mark Root
Myers, Philip Cherdak
Niell, Arthur Edwin
Olsen, Edward Tait
Olsson, Carl Niels
Pacholczyk, Andrzej Grzegorz
Palmer, Patrick Edward
Penzias, Arno A
Philip, Kenelm Winslow
Rankin, John Metcalf
Read, Richard Bradley
Roberts, Morton Spitz
Rogers, Alan Ernest Exel
Sakurai, Kunitomo
Seielstad, George A
Shaffer, David Bruce
Shimabukuro, Fred Ichiro
Shuter, William Leslie Hazlewood
Slade, Martin Alphonse, III
Stanek, Richard Anthony
Staelin, David Hudson
Swanson, Paul N
Thompson, Anthony Richard
Turner, Barry Earl
Turner, Kenneth Clyde
Ulich, Bobby Lee
Wardle, John Francis Carleton
Warwick, James Walter
Weber, Richard Rand
Weinreb, Sander
Welch, William John
Wilson, Robert Woodrow
Wilson, William John

X-ray Astronomy

Alexandropoulos, Nikos G
Burek, Anthony John
Canizares, Claude Roger
Catura, Richard Clarence
Cruddace, Raymond Gibson
Davidsen, Arthur Falnes
Davis, John Moulton
Fishman, Gerald Jay
Fritz, Gilbert Geiger
Garmire, Gordon Paul
Hearn, David Russell
Hoffman, Jeffrey Alan
Joss, Paul Christopher
Kellogg, Edwin M
Lampton, Michael Logan
Meekins, John Fred
Murray, Stephen S
Parsignault, Daniel Raymond
Peterson, Laurence E
Schnopper, Herbert William
Schreier, Ethan Joshua
Schwartz, Daniel Alan
Share, Gerald Harvey
Shulman, Seth David
Stevens, John Charles
Toraskar, Jayashree Ravalnath
Ulmer, Melville Paul
Underwood, James Henry
VanSpeybroeck, Leon Paul
Wilson, Brian Graham
Wilson, Robert E

ASTROPHYSICS

Astrophysics

Adel, Arthur
Ahluwalia, Harjit Singh
Aikman, George Christopher Lawrence
Aitken, Donald W, Jr
Aller, Lawrence Hugh
Ames, Susan
Anderson, Roland Carl
Arny, Thomas Travis
Assousa, George Elias
Atwood, Bruce
Auer, Lawrence H
Auman, Jason Reid
Axford, William Ian
Bahcall, John Norris
Bahcall, Neta Assaf
Baldwin, Ralph Belknap
Banderman, Lothar W
Banks, Harvey Washington
Barnes, Charles Andrew
Barnes, Ronnie C
Barnothy, Jeno Michael
Beavers, Willet I
Beckers, Jacques Maurice
Bekenstein, Jacob David
Bell, Graydon Dee
Bell, Roger Alistair
Berkey, Gordon Bruce
Billings, Donald Earl
Bjorkholm, Paul J
Blake, J Bernard
Blake, Richard Edward
Bodenheimer, Peter Herman
Boggess, Albert, III
Bohlin, Ralph Charles
Boley, Forrest Irving
Bonsack, Walter Karl
Boynton, William Vandegrift
Brandt, John Conrad
Branch, David Reed
Browne, Robert Lamme
Brown, Robert Rex
Buffington, Andrew
Burke, J Anthony
Burkhead, Martin Samuel
Buscombe, William
Byard, Paul L
Cann, Julius Hofeller
Calame, Gerald Paul
Carleton, Nathaniel Phillips
Caroff, Lawrence John
Carovillano, Robert L
Carpenter, Roland LeRoy
Cartwright, Brian Grant
Cassinelli, Joseph Patrick
Castor, John I
Caughlan, Georgeanne Robertson
Cernuschi, Felix
Chan, Kwing Lam
Chandrasekhar, Subrahmanyan
Chapman, Robert DeWitt
Chiu, Hong-Yee
Chow, Tai-Low
Chung, Kuk Pyo
Clark, Patricia Ann Andre
Clayton, Donald Delbert
Climenhaga, John Leroy
Cloutman, Lawrence Dean
Code, Arthur Dodd
Cohen, Jeffrey M
Cohen, Judith Gamora
Cook, Allan Fairchild, II
Cox, Arthur Nelson
Crannell, Carol Jo Argus
Curott, David Richard
Curtis, George William
Dalgarno, Alexander
Danielson, Robert E
Daub, Clarence Theodore, Jr
Davids, Cary Nathan
Davidson, Arthur Falnes
Davidson, Kris
Davis, Marc
Davis, Robert James
Degen, Vladimir
Delsemme, Armand Hubert
Demarque, Pierre
DeNoyer, Linda Kay
Dicke, Robert Henry
Dinger, Ann St Clair
Doherty, Lowell Ralph
Dolan, Joseph Francis
Douglass, David Holmes
Dubin, Maurice
Duke, Douglas
Dulk, George A
Dunham, Theodore, Jr
Dupree, Andrea K
Dyal, Palmer
Eddy, John Allen
Edmonds, Frank Norman, Jr
Eggen, Olin Jeuck
Eilgroth, Peter George
Epstein, Isadore
Erickson, Edwin Francis

Evans, John C
Evans, John Wainwright, Jr
Evans, Neal John, II
Faller, James E
Fazio, Giovanni Gene
Feibelman, Walter A
Feinstein, Alejandro
Felten, James Edgar
Fichtel, Carl Edwin
Fischel, David
Fisher, Richard R
Fishman, Gerald Jay
Fitzgerald, Maurice Pim
Fix, John Dekle
Fontaine, Gilles Joseph
Fox, Kenneth
Frazier, Edward Nelson
Gallagher, John Sill
Galt, John (Alexander)
Garcia-Munoz, Moises
Garmire, Gordon Paul
Garrison, Robert Frederick
Garstang, Roy Henry
Gebbie, Katharine Blodgett
Gehrz, Robert Douglas
Gerasimenko, Michel
Giacconi, Riccardo
Gilman, John Richard, Jr
Goldberg, Leo
Goldreich, Peter
Goorvitch, David
Gorenstein, Paul
Gould, Robert Joseph
Green, Louis Craig
Greenberg, Jerome Mayo
Greene, Thomas Frederick
Greenstein, Jesse Leonard
Gregory, Stephen Albert
Greig, William Elliott
Grenchik, Raymond Thomas
Grindlay, Jonathan Ellis
Gross, Peter George
Grossman, Allen S
Groth, Edward John, III
Gunn, James Edward
Hackwell, John Arthur
Hall, Donald Eugene
Halliday, Ian
Hansen, Carl John
Hardorp, Johannes Christfried
Harrington, James Patrick
Harrison, Edward Robert
Harrison, Marjorie Hall
Hartle, Richard Eastham
Harvey, Gale Allen
Harvey, John Warren
Harwit, Martin Otto
Hawkins, Charles Edward
Haymes, Robert
Hegyi, Dennis
Helfer, Herman Lawrence
Henry, Joseph Patrick
Henry, Richard Conn
Hesser, James Edward
Hill, Henry Allen
Hill, Stephen James
Hillendahl, Richard Warren
Hiltner, William Albert
Hobbs, Lewis Mankin
Holt, Stephen S
Hoover, Paul Swegman
Houck, James Richard
House, Lewis Lundberg
Howard, William Michael
Huang, Su-Shu
Huebner, Walter F
Huffman, Donald Ray
Huguenin, George Richard
Hundhausen, Arthur James
Hynek, Joseph Allen
Iben, Icko, Jr
Innanen, Kimmo A
Jeffries, John Trevor
Jenkins, Alvin Wilkins, Jr
Jenkins, Edward Beynon
Jenkins, Thomas Llewellyn
Johnson, Hollis Ralph
Johnson, George Lawrence
Jokipii, Jack Randolph
Jura, Michael Alan
Kalkofen, Wolfgang
Kelch, Walter L
Kemp, James Chalmers
King, David Solomon
Kinzer, Robert Lee
Knacke, Roger Fritz
Kniffen, Donald Avery
Koga, Rokutaro
Kuhi, Leonard Vello
Kumar, Cidambi Krishna
Kurfess, James Daniel
Kwan, John Ying-Kuen
Lande, Kenneth
Landstreet, John Darlington
Larson, Richard Bondo
Lasker, Gordon (Jewett)
Lasker, Barry Michael
Lea, Susan Maureen

Lebovitz, Norman Ronald
Lecar, Myron
Leckrone, David Stanley
Lee, Paul D
Leibacher, John William
Leighton, Robert Benjamin
Lerche, Ian
Lesh, Janet Rountree
Leung, Kam-Ching
Leventhal, Marvin
Levine, Randolph Herbert
Liang, Edison Park-Tak
L'Heureux, Jacques (Jean)
Lichtenstein, Pearl Rubenstein
Lillie, Charles Frederick
Limber, David Nelson
Lind, Vance Gordon
Lingenfelter, Richard Emery
Linnell, Albert Paul
Little-Marenin, Irene Renate
Lo, Kwok-Yung
Locanthi, Dorothy Davis
Lucke, Robert Lancaster
Lutz, Barry Lafean
Macalpine, Gordon Madeira
Malville, John McKim
Mankin, William Gray
Mansfield, Victor Neil
Maran, Stephen Paul
Marks, Dennis William
Marks, James Wai-Kee
Marshak, Robert Eugene
Martins, Donald Henry
Mathis, John Samuel
Matsushima, Satoshi
Max, Claire Ellen
McCluskey, George E, Jr
McCray, Richard Alan
McKee, Christopher Fulton
McNamara, Delbert Harold
Meinel, Aden Baker
Meisel, David Dering
Merker, Milton
Merrill, Athel Lavelle
Meyerott, Roland Edward
Michaud, Georges Joseph
Michel, F Curtis
Mickey, Donald Lee
Mihalas, Dimitri
Miller, Richard Henry
Mitalas, Romas
Mitchell, Robert Curtis
Morgan, Henri Emmanuel
Morgan, Thomas Edward
Morgan, Thomas Harlow
Morton, Donald Charles
Motz, Lloyd
Mulders, Gerard Francis William
Mullan, Dermot Joseph
Mullen, Joseph Matthew
Muller, Dietrich
Munch, Guido
Mutschlecner, Joseph Paul
Nagy, Theresa Ann
Neff, John S
Nelson, Jerry Earl
Nelson, Mark Radford
Newkirk, Gordon Allen
Noerdlinger, Peter David
Noyes, Robert Wilson
Olsen, Kenneth Harold
Olson, Edward Cooper
Opal, Chet Brian
Opher, Reuven
Opp, Albert Geelmuyden
Oppenheimer, Michael
Orrall, Frank Quimby
Orth, Charles Douglas
Oster, Ludwig Friedrich
Ostriker, Jeremiah P
Ovenden, Michael W
Owen, Tobias Chant
Page, Thornton Leigh
Papagiannis, Michael D
Parker, Eugene Newman
Parkinson, William Hambleton
Pasachoff, Jay Myron
Peale, Stanton Jerrold
Peery, Benjamin Franklin, Jr
Pellerin, Charles James, Jr
Perry, Judith Joanna
Peterson, Vern Leroy
Pfeiffer, Raymond John
Pilachowski, Catherine Anderson
Plavec, Miroslav (Mirek) Josef
Pneuman, Gerald W
Poland, Arthur I
Pottasch, Stuart Robert
Price, Michael J
Price, Richard Henry
Ptak, Roger Leon
Ramaty, Reuven
Rao, Kandarpa Narahari
Rasband, S Neil
Rhee, John Williams
Rich, Arthur
Rich, John Charles

Roark, Terry P
Robbins, Ralph Robert
Roberts, David Llewellyn
Rodman, James Purcell
Rodney, William Stanley
Rogerson, John Bernard, Jr
Rood, Robert Thomas
Rose, William K
Ross, Dennis Kent
Rossner, Lawrence Franklin
Routly, Paul McRae
Rust, David Maurice
Sackmann, I Juliana
Sahade, Jorge
Salpeter, Edwin Ernest
Sampson, Douglas Howard
Sandberg, Vernon Dean
Sanders, Robert Hugh
Sargent, Wallace Leslie William
Savedoff, Malcolm Paul
Sawyer, Constance B
Schardt, Alois Wolfgang
Schindler, Stephen Michael
Schlesinger, Barry Michael
Schmalberger, Donald C
Schnopper, Herbert William
Schreier, Ethan Joshua
Schroeder, Leon William
Schwartz, Robert Alan
Schwarzschild, Martin
Sears, Richard Langley
Sekanina, Zdenek
Serkowski, Krzysztof
Shapiro, Maurice Mandel
Shapiro, Stuart Louis
Shen, Benjamin Shih-Ping
Shen, Chun-Shan
Shivanandan, Kandiah
Simon, George Warren
Simon, Norman Robert
Siquig, Richard Anthony
Sitterly, Charlotte Moore
Skadron, George
Skumanich, Andrew
Slettebak, Arne
Smalley, Larry L
Smith, Alexander Goudy
Smith, Dean Francis
Smith, Richard Lloyd
Smoot, George Fitzgerald, III
Snider, Joseph Lyons
Snyder, Lewis Emil
Sofia, Sabatino
Solomon, Philip M
Speiser, Theodore Wesley
Spergel, Martin Samuel
Spiegel, Edward A
Spitzer, Lyman, Jr
Sreenivasan, Sreenivasa Ranga
Steigman, Gary
Stein, Robert Foster
Stein, Wayne Alfred
Stokes, Sidney Norman
Stone, Sidney Norman
Straka, William Charles
Strom, Stephen
Sturrock, Peter Andrew
Sweeney, Mary Ann
Swihart, Thomas Lee
Tananbaum, Harvey Dale
Tapia, Santiago
Tarter, Curtis Bruce
Tassoul, Jean-Louis
Tebo, Edith Janssen
Thaddeus, Patrick
Theys, John C
Thomas, Richard Nelson
Thompson, David John
Thompson, Rodger Irwin
Thonnard, Norbert
Thorne, Kip Stephen
Thuan, Trinh Xuan
Torres-Peimbert, Silvia
Toton, Edward Thomas
Traub, Wesley Arthur
Travis, Larry Dean
Trimble, Virginia Louise
Truran, James Wellington, Jr
Tyson, J Anthony
Underhill, Anne Barbara
Van Horn, Hugh Moody
Varshni, Yatendra Pal
Venkatesan, Doraswamy
Vila, Samuel Campdettos
Waddington, Cecil Jacob
Wagner, William John
Walker, Gordon Arthur Hunter
Wark, David Quentin
Watson, William Douglas
Wayland, James Robert, Jr
Webber, William R
Weekes, Trevor Cecil
Wefel, John Paul
Weinberg, Jerry L
Wentzel, Donat Gotthard
Werner, Michael Wolock
West, Donald K
White, Robert Stephen

Whitford, Albert Edward
Whitney, Charles Allen
Wildey, Robert Leroy
Williamson, William, Jr
Wilson, Olin Chaddock
Wilson, Robert E
Winkler, Paul Frank
Winters, Ronald Ross
Withbroe, George Lund
Witt, Adolf Nicolaus
Witteborn, Fred Carl
Wolff, Richard Alan
Wolff, Sidney Carne
Wolstencroft, Ramon David
Woltjer, Lodewyk
Woodgate, Bruce Edward
Wright, Kenneth Osborne
York, Donald Gilbert
Youtcheff, John Sheldon
Zalubas, Romuald
Zechiel, Leon Norris
Zimmerman, R Erik
Zych, Allen Dale

Celestial Mechanics

Blitzer, Leon
Burns, Joseph A
Cunningham, Leland E
Danby, John Michael Anthony
Dunham, David Waring
Everhart, Edgar
Greenberg, Richard Joseph
Hertz, Hans Georg
Liu, Joseph Jeng-Fu
Marsden, Brian Geoffrey
McLaughlin, William Irving
Nacozy, Paul E
Oesterwinter, Claus
Palmore, Julian Ivanhoe, III
Pascu, Dan
Saari, Donald Fene
Seidelmann, Paul Kenneth
Slade, Martin Alphonse, III
Van Flandern, Thomas Charles
Vinh, Nguyen Xuan
Vinti, John Pascal
Williams, Carol Ann
Wolf, Henry

Cosmochemistry

Anders, Edward
Duke, Michael B
Grossman, Lawrence
Lipschutz, Michael Elazar
Marti, Kurt
Mueller, George
Schaeffer, Oliver Adam
Srinivasan, B
Wasson, John Taylor

Theoretical Astrophysics

Bardeen, James Maxwell
Black, David Charles
Brecher, Kenneth
Chanmugam, Ganesar
Christy, Robert Frederick
Clement, Maurice James
Cox, John Paul
DeYoung, David Spencer
Endal, Andrew Samson
Field, George Brooks
Geller, Margaret Joan
Hills, Jack Gilbert
Hummer, David Graybill
Ipser, James Reid
Jacobs, Kenneth Charles
Jones, Frank Culver
Jones, Thomas Walter
Joss, Paul Christopher
Kegeles, Lawrence Steven
Lightman, Alan Paige
Liu, Chuan Sheng
O'Connell, Robert F
Omidvar, Kazem
Pacholczyk, Andrzej Grzegorz
Pines, David
Roberts, David Hall
Rouse, Carl Albert
Sartori, Leo
Schramm, David N
Schutz, Bernard Frederick
Seguin, Fredrick Hampton
Starrfield, Sumner Grosby
Talbot, Raymond James, Jr
Teukolsky, Saul Arno
Tinsley, Beatrice Muriel
Wagner, Raymond Lee
Wagner, Robert Vernon, Jr
Wheeler, James Richard
Wheeler, John Craig
Will, Clifford Martin
Wilson, James R

BACTERIOLOGY

Bacteriology

Abbey, Anthony Alfred
Abeling, Edwin John
Adams, Stuart Lyle
Adler, Frank Leo
Adler, Howard Irving
Ainis, Herman
Albert, Oscar J
Alegnani, William Charles
Alexander-Jackson, Eleanor Gertrude
Allen, James Harrill
Allen, Oscar Nelson
Amos, Harold
Anderson, Arthur W
Anderson, Kenneth Ellsworth
Anderson, Theodore Gustave
Andreoli, Anthony Joseph
Anellis, Abe
Appleman, Maria Duarte
Ardrey, William Boyle
Barbiers, Arthur Robert, Jr
Barnes, Glover William
Baron, Louis Sol
Barry, Arthur Leland
Bartholomew, James William
Bass, Joseph Alonzo
Bates, Joseph H
Batlin, Alexander
Bauer, Henry
Bayliss, Berenice
Bayne, Henry Godwin
Beaulieu, J A E M
Beck, Sidney M
Becker, Maurice Edwin
Bell, John Frederick
Bennett, Edward Owen
Berkman, Sam
Berryhill, David Lee
Beskid, George
Best, Audrey Nance
Betz, John Vianney
Bigley, Nancy Jane
Birkeland, Jorgen Maurice
Bismanis, Jekabs Edwards
Blitz, Ruth R
Blouse, Louis E, Jr
Bogacz, John
Boley, Robert B
Borick, Paul Michael
Bornside, George Harry
Bowers, Landon Emanual
Boyd, Frank McCalla
Boyd, William Lee
Bozeman, Samuel Richmond
Bradford, William L
Bradner, William Turnbull
Brillhart, Russell Edward
Brodie, Arnold Frank
Bromel, Mary Cook
Brown, Lewis Raymond
Bruno, Charles Frank
Bryan, Frank Leon
Buck, John David
Buck, Charles Elon
Buggs, Charles Wesley
Bullock, Graham Lambert
Burchard, Robert P
Burdash, Nicholas Michael
Burrows, William
Caldwell, Mary Estill
Calkins, Harmon Eldred
Campbell, Jack James Ramsay
Cardullo, Maria Ann
Carls, Ralph A
Carlson, Harve J
Carpenter, Charles Patten
Carr, John Halden
Carter, Paul Bearnson
Caruthers, John Quincy
Casey, Helen Liles
Centifanto, Ysolina M
Chambers, Cecil William
Chang, Timothy Scott
Chanock, Robert Merritt
Charlton, David Berry
Cherry, William Bailey
Chisler, John Adam
Cinquina, Carmela Louise
Clancy, Carl Francis
Clark, Alvin John
Clark, William Arthur
Claybaugh, Glenn Alan
Clegg, Lawrence Frank Levey
Clemens, William Bryson
Cole, Basil Chambrus
Colingsworth, Donald Rudolph
Colmer, Arthur Russell
Conant, Norman Francis
Conner, Ray M
Cooke, Patricia M
Coriell, Lewis L
Cosenza, Benjamin John
Cravitz, Leo

Crisley, Francis Daniel
Cutchins, Ernest Charles
Davenport, Calvin Armstrong
Dawson, James Robertson, Jr
Deibel, Robert Howard
Deitz, William Harris
DeLong, Robert Francis
DeLorenzo, William F
Delwiche, Eugene Albert
Dennis, Emery Westervelt
De Repentigny, Jacques
Dettwiler, Herman Andrew
Deufel, Robert
Devine, Leonard Francis
DeWitt, Charles Wayne, Jr
Diena, Benito B
Dmochowski, Leon Ludomir
Dodd, Matthew Charles
Doede, Dorothy Ruth
Doetsch, Raymond Nicholas
Domingue, Gerald James
Donaldson, David Miller
Donovick, Richard
Dorrell, William Woodrow
Doty, Robert Bruce
Dowell, Vulus Raymond, Jr
Drake, Charles Hadley
Drew, Ruth Miriam
Driesens, Robert James
Duncan, Charles Lee
Dunlop, Stuart George
Durham, Norman Nevill
Eddy, Bernice Elaine
Efthymiou, Constantine John
Egermeier, Edward R
Eisenstark, Abraham
Eisler, Daniel M
Eisman, Leon Philip
Elkan, Gerald Hugh
Ellias, Loretta Christine
Ellinghausen, Herman Charles, Jr
Elliott, Robert A
Ellis, John George
Ellis, Robert J
Elvin-Lewis, Memory P F
Emerson, Robert L
Emery, Jerrell Bemis
English, Arthur Robert
Erlandson, Arvid Leonard
Ervin, Robert Francis
Etchells, John Lincoln
Evans, James Brainerd
Evans, John Edward
Farkas-Himsley, Hannah
Fay, Rimmon C
Feiner, Rose Resnick
Feinstone, Wolffe Harry
Feltz, Elmer T
Firman, Melvin Curtis
Fisher, Mike W
Fisher, Paul John
Flannery, William Louis
Foster, Edwin Michael
Foster, John Wallace
Fournelle, Harold John
Fowler, Elizabeth Haddock
Frances, Saul
Frappier, Armand
Frea, James Irving
Fred, Edwin Broun
Frey, James R
Friedman, Mischa Elliot
Frisch, Arthur Wain
Frian, Laura Brilliantine
Fulghum, Robert Schmidt
Fuller, Wallace Hamilton
Gaby, William Lawrence
Gall, Lorraine Sibley
Ganley, Oswald Harold
Garcia, Manuel Mariano
Gerencser, Mary Ann (Aiken)
Gerencser, Vincent Frederic
Gershenfeld, Louis
Gershman, Melvin
Gilmour, Campbell Morrison
Gilroy, James Joseph
Girard, Kenneth Francis
Goldin, Milton
Golub, Orville Joseph
Gordon, Ruth Evelyn
Gottshall, Russell Y
Graber, Charles David
Green, William Asa
Greenberg, Louis
Greenberg, Richard Aaron
Griffith, Lewis John
Griffiths, Francis Priday
Griner, Lynn Adel
Gross, Noel Harden
Gruenwald, Ruben
Grundy, Walton Earle
Gygi, Francis Richard
Haas, Felix Levere
Hackman, Abigail Salyers
Hajna, Anthony Alphonse
Hall, Elizabeth Rose
Halpern, Bernard
Halpern, Bernard

BACTERIOLOGY

Handy, Mostafa Kamal
Hampp, Edward Gottlieb
Hanks, John Harold
Harden, Virginia Pauline
Harrell, William Knox
Hartmann, Floyd Wellington
Hasenclever, Herbert Frederick
Hauschild, Andreas H W
Hayes, Sheldon P
Heck, Joseph Gerard
Hedgecock, Loyd Wilson
Heitz, Carl Louis
Heilman, Bernard
Heinemann, Bernard
Henderson, Dorothy Henderson
Henderson, Thomas Otis
Hendrickson, Adolph Alexander
Herbert, Michael
Herman, Lloyd George
Hibbs, Robert A
Highlands, Matthew Edward
Hill, James Carroll
Hiramoto, Raymond Natsuo
Hitzman, Donald Oliver
Hochstein, Lawrence I
Hodes, Horace Louis
Hogue, Ralph Stewart
Holdeman, Lillian Virginia
Hood, Marion Winifred
Hood, Mary Noka
Hopper, Samuel Hersey
Hoskisson, William A
Hottle, George Austin
Houston, Chester Warren
Howard, Ronald M
Hufham, James Birk
Humphries, James Charles
Hunter, John Earl
Hurd, Robert Charles
Hurwitz, Charles
Iannarone, Michael
ichida, Allan A
llavsky, Jan
Imaeda, Tamotsu
Insalata, Nino F
Jablon, James Martin
Jacobs, Nicholas Joseph
Janeff, Donka Grigorova
Janeff, Jan Dimitroff
Janes, Donald Wallace
Jarvis, Francis George
Jay, James Monroe
Jeter, Max Albert
Johnson, David Franklin
Johnson, George Thomas
Johnson, Mary Knettles
Johnson, Rother Rodenious
Johnstone, Donald Lee
Jourdonais, Leonard Francis
Kashket, Eva Ruth
Kennedy, Eugene Richard
Kenner, Bernard Alexander
Kershaw, Edythe Marie
Kerwin, Richard Martin
Kestenbaum, Richard Charles
Kester, Andrew Stephen
Kidd, John Graydon
Kimball, Grace Caroline
Kirchner, Carl Edward John
Kirshbaum, Amiel
Kite, Joseph Hiram, Jr
Klinger, Lawrence Edward
Knight, Ralph Amber
Knighton, Holmes Tutt
Koft, Bernard Waldemar
Konopka, Edward Alexander
Kopper, Paul Heinz
Kramer, Norman
Krampitz, Lester Orville
Krasner, Robert Irving
Kroll, Henry Michael
Kronenwett, Frederick Rudolph
Kruse, Conrad Edward
Kuehn, Audrey Olson
Kuehner, Richard Louis
Kull, Frederick Charles
LaBrec, Theodore Robert
Lackman, David Buell
Laffer, Norman Callender
Lambert, Reginald Max
Lancefield, Rebecca Craighill
Lang, Raymond W
Langan, M Regina
Lankford, Charles Ely
Larson, Carl Leonard
Larson, Nora Leona
Lattuada, Charles P
Lawrence, James Vantine
Layton, Herbert Wallace
Legator, Marvin (Seymour)
Lerke, Peter A
Levin, Simon Eugene
Levine, Hillel Benjamin

Levine, Pincus Philip
Lewandowski, Thaddeus
Lewis, Lewis James
Lin, Fu Hai
Lind, Howard Eric
Lindorfer, Robert Karl
Lindstrom, Eugene Shipman
Litsky, Warren
Lloyd, Prescott Rees
Lockhart, William Raymond
London, Jack P
Lozano, Edgardo A
Ludovici, Peter Paul
Lund, Arnold Jerome
Luria, Salvador Edward
Lyles, Sanders Truman
Lynch, John Edward
Lyons, Don Chalmers
MacDonald, Russell Earl
Mack, Walter Noel
Mackevie, Robin Maxwell
MacMahon, Harold Edward
Madore, Bernadette
Majors, Paul Alexander
Maki, Leroy Noel
Malzahn, Ronald C
Mansfield, Tom
Marinelarena, Rafael
Marshall, Rosemarie
Martin, Donald Stover
Matheson, Ballem Howard
Matney, Thomas Stull
Mattick, Joseph Francis
Mattman, Lida Holmes
McBride, Tom Joseph
McDowell, Margaret Ann
McKay, Kenneth Alexander
McKee, Albert Preston
McMahon, Kenneth James
McVeigh, Ilda
Meyer, Daniel L
Middaugh, Paul Richard
Minard, Edwin Lincoln
Miller, Anna Kathrine
Mitwer, Tod Edwin
Mizuba, Seth Setsuo
Mizuno, William George
Moe, Paul G
Mohler, Irvin C, Jr
Mongeau, J Denis
Moore, Thomas D
Mora, Emilio Chavez
Morigi, Eugene Mario Edmund
Morris, Joseph Anthony
Morrison, Marcus Eugene
Morton, Harry E
Moyed, Harris S
Muenterer, Donald Arthur
Mull, Leon Edmund
Munch, Theodore
Murray, Francis Joseph
Murray, Robert George Everitt
Nagy, Julius G
Nakata, Herbert Minoru
Naylor, Harry Brooks
Needham, Gerald Morton
Nemes, Marjorie M
Neter, Erwin
Nevin, Thomas Andrew
Newman, Jack Huff
Nicholes, Paul Scott
Nielsen, Peter Adams
Nolte, William Anthony
Noyes, Howard Ellis
O'Kane, Daniel Joseph
Olitzky, Irving
Olson, Joseph Carl, Jr
O'Neill, Richard Delos
Orcutt, Fred Scott, Sr
Osborne, William Wesley
Otero, Joseph Guillermo
Ott, John Lewis
Otto, Robert H
Owen, Cora Rust
Packchanian, Ardzrooni (Arthur)
Pannell, Lolita
Panos, Charles
Parker, Don Timothy
Parker, Richard Bennett
Parsons, Jesse Leroy
Patterson, Anne Louise
Patterson, Harry Robert
Peniston, Francis L
Perrozzi, Joseph Richard
Peterson, Arthur Carl
Peterson, Glen Ervin
Phelps, Allen Spencer
Phillips, Haskell C
Pickett, Morris John
Pierce, Leon L
Pike, Robert Merrett
Pine, Ellen Kann
Pine, Martin J
Pittillo, Robert Francis
Pittman, Margaret
Pivnick, Hilliard
Platt, Thomas Boyne
Pomales-Lebron, Americo

Porter, Frederic Edwin
Portwood, Lucile Mitchell
Pratt, Darrell Bradford
Pressman, Ralph
Quinn, Loyd Yost
Ramsey, Hal Harrison
Rasilewicz, Casimir E
Read, Ralston Baker, Jr
Reid, Roger Delbert
Reinbold, George W
Reynolds, John Theodore
Rhymer, Ione
Rice, Marion McBurney
Richardson, Robert Louis
Righsel, Wilton Adair
Rittenberg, Sydney Charles
Roberstad, Gordon Wesley
Robinson, John
Robinson, Roslyn Quinby
Rode, Leonard John
Rodgers, Nelson Earl
Rodriguez-Leiva, Manuel
Romeyn, James Augustus
Romig, William Robert
Rosenberg, Eugene
Rosenthal, Sol Roy
Rosenwald, Albert John
Rosenzweig, Abraham Leon
Ross, Richard Travis
Rossmoore, Harold W
Rout, Mohammed Abdur
Roy, Theodore Ernest
Rypka, Eugene Weston
Sames, Richard William
Sandine, William Ewald
Sbarra, Anthony J
Schacter, Julius
Schaefter, Warren Ira
Schenken, John Rudolph
Schlamm, Norbert Arnold
Schuchardt, Lee Frank
Schwab, John Harris
Schwab, Mark Adam
Seeley, Harry Wilbur
Seneca, Harry
Seo, John S
Serafini, Angela
Seraichekas, Helen Rose
Serota, Cornelia Ann Roach
Shay, Donald Emerson
Shipp-Watson, Mary Elizabeth
Shirk, Richard Jay
Shirling, Elwood Brent
Shively, Carl E
Shockley, Thomas E
Silverman, Myron Simeon
Silverman, Sidney Joseph
Simpson, Frederick James
Singer, Arnold Jack
Sistrunk, William Allen
Skopek, Jerry
Slocum, Glenn Gerald
Slotnick, Irving James
Smiley, Karl L
Smith, Elinor Van Dorn
Smith, James William
Smith, Paul Howard
Smith, Richard Scott
Smith, Winslow Whitney
Snudden, Birdell Harry
Sokatch, John Robert
Solomon, John Junior
Solotorovsky, Morris
Solowey, Mathilde
Spendlove, John Clifton
Splitstoesser, Clara Quinnell
Splitstoesser, Don Frederick
Staley, James Trotter
Stauch, John Edward
Stern, Ivan J
Stern, Robert Malcolm
Stevens, David Arthur
Stevenson, Isaac Glenn
Stewart, Sarah Elizabeth
Stirrat, James Hill
Stoenner, Herbert George
Storlazzi, Joseph Jordan
Strandberg, Gerald William
Straughn, William Ringgold, Jr
Stuy, Johan Harrie
Sugiyama, Hiroshi
Surdy, Ted E
Sweet, Charles Edward
Tankersley, Robert Walker, Jr
Tarver, Fred Russell, Jr
Taylor, Gerald C
Teodoro, Rosario Reyes
Teresa, George Washington
Thomassen, Paul R, Jr
Thurston, John Robert
Tikemeyer, Charles William
Tove, Shirley Ruth
Trezise, William Joseph
Trivett, Terrence Lynn
Trussell, Paul Chandos
Tsuchiya, Henry Mitsumasa
Tufte, Marilyn Jean

Ulrich, John August
Umbreit, Wayne William
Underdahl, Norman Russell
Uscavage, Joseph Peter
Van Eseltine, William Parker
Vicher, Edward Ernest
Vidaver, Anne Marie Kopecky
Wagner, Morris
Walls, Nancy Williams
Warren, Halleck Burkett, Jr
Watkins, Harry Mitchell Sherman
Watson, Barbara Kascenko
Watson, Stanley W
Webb, Arthur Harper
Weber, George Russell
Weber, William Alexander
Weinberg, Ralph
Weiser, Russell Shivley
Welker, George W
Welshimer, Herbert Jefferson
Wessman, Garner Elmer
West, John Leslie
Wheeler, Albert Harold
White, Alan George Castle
White, John David
White, Lendell Aaron
Whitehead, Howard Allan
Whitehill, Alvin Richard
Whitmire, Carrie Ella
Widner, William Richard
Wilkins, Judd Rice
Williams, Fred Devoe
Williams, Joy Elizabeth P
Willis, William Hillman
Wilson, Joe Bransford
Wilson, Perry William
Wilson, Robert James
Wilt, John Charles
Wise, Robert Irby
Wiseman, Gordon Marcy
Witlin, Bernard
Wittenberger, Charles Louis
Wolf, Mark Adam
Wood, Alex James
Wood, Norris Philip
Wood, Robert Charles
Woodward, John Morrill
Wright, Donald N
Wyckoff, Delaphine Grace Rosa
Wynick, Priscilla Blakeney
Yaphe, Wilfred
Yarbrough, Henry Floyd, Jr
York, Charles James
Youmans, Guy Parry
Young, William Donald, Jr
Zeya, Hasan Ismail
Zimmerman, Leonard Norman
Zimpfer, Paul (Ellsworth)

Bacterial Genetics

Austrian, Robert
Duggan, Dennis E
Elliott, Rosemary Eskridge
Ihler, Karin Ippen
Roth, John R
Sunshine, Melvin Gilbert

Dairy Bacteriology

Bennett, Frederick William
Deane, Darrell Dwight
Gibson, Douglas (Lorne)
Lawton, Wallace Clayton
Luedecke, Lloyd O
Morgan, Max Eugene
Olson, Harold Cecil
Overcast, Woodrow Webb
Schultze, Walter Donald
Thomas, William Robb
Wagenaar, Raphael Omer
Ward, Paul J
Wilkowske, Howard Hugo

Food Bacteriology

Appleman, Milo Don
Elliott, Robert Paul
Fishbein, Morris
Rice, Andrew Cyrus
Wilson, Charles Marshall

Industrial Bacteriology

Malaspina, Alex
Shapiro, Rebecca Lillian

Medical Bacteriology

Binn, Leonard Norman
Blair, John Hanna
Brewer, John Hanna
Briody, Bernard Aloysius
Bubel, Hans Curt
Calderone, Julius G
Deal, Samuel Joseph
Diamond, Ben Elkan
Eigelsbach, Henry Thomas
Eveland, Warren C
Foster, Robert Scott
Freter, Rolf Gustav
Gaines, Sidney
Gilardi, Gerald Leland

Gilmore, Eleanor La Verne
Going, Dora Henley
Goldenberg, Martin Irwin
Grainger, Thomas Hutcheson, Jr
Holtman, Darlington Frank
Hugh, Rudolph
Jellard, Charles H
Kramer, Henry Herman
Kubica, George P
Martineau, Bernard
McCarty, Maclyn
Michelson, Israel David
Milgrom, Felix
Moody, Max Dale
Moore, Harold Beveridge
Raffel, Sidney
Ransom, John Paul
Sarber, Raymond William
Shepard, Maurice Charles
Slack, John Madison
Smith, Peter Byrd
Smith, Philip Nixon
Speck, Reinhard Staniford
Stone, Joseph Louis
Syeklocha, Delfa
Toshach, Sheila
Walker, Richard V
Weaver, Robert Elwin
Willoughby, Donald S
Wilson, Raphael
Woolridge, Robert Leonard

Physiological Bacteriology
Barnekow, Russell George, Jr
Baugh, Clarence L
Boehms, Charles Nelson
Booth, James Samuel
Burchall, James J
Fisher, Robert John
Fraenkel, Dan Gabriel
Gibbons, Ronald J
Hageman, James Howard
Hug, Daniel Hartz
Iandolo, John Joseph
Juni, Elliot
Kory, Mitchell
Martinez, Rafael Juan
McIntosh, Elaine Nelson
Mickelson, Milo Norval
O'Brien, Robert Thomas
Perry, Dennis
Reeves, Henry Courtland
Stewart, James Edward
Thompson, Thomas Leo
Torriani Gorini, Annamaria
Traxler, Richard Warwick

Veterinary Bacteriology
Biberstein, Ernest Ludwig
Fales, William Harold
Kirkbride, Clyde Arnold
Langford, Edgar Verden
Morse, Guy Emery
Wills, Franklin Knight

BIOCHEMISTRY

Biochemistry
Abbott, Donald Clayton
Abbott, Lynn De Forrest, Jr
Abbott, Mitchel Theodore
Abbott, Okra Jones
Abeles, Robert Heinz
Abdel-Latif, Ata A
Abell, Creed Wills
Abell, Liese Lewis
Abood, Leo George
Abraham, Edatbara Chacko
Abraham, Samuel
Abrams, Adolph
Abrams, Richard
Abramsky, Tessa
Abrash, Henry I
Abshire, Claude James
Abul-Hajj, Jusuf J
Ackerman, Clemens John
Ackerman, Philip Gulick
Acs, George
Adair, Winston Lee, Jr
Adams, Alden Ross
Adams, Alfred Birk
Adams, Elijah
Adams, Ernest Clarence
Adams, James Miller
Adams, Philip Delmar
Adamson, Richard H
Addanki, Somasundaram
Adelman, Richard Charles
Adkins, Benjamin Jefferson
Adler, John Henry
Adler, Julius
Adman, Raymond Lance
Adrouny, George Adour (Kuyumjian)
Aftergood, Lilla
Agarwal, Kailash C
Agosin, Moises

Agranoff, Bernard William
Ahluwalia, Gurjit Singh
Ahmed, Khalil
Ahuja, Jagan N
Airee, Shakti Kumar
Aisen, Philip
Akeson, Walter Roy
Aktipis, Stelios
Alam, Ashraf Ul
Alaupovic, Petar
Albanese, Anthony August
Albaum, Harry Gregory
Alben, James O
Albert, Mary Roberts Forbes (Day)
Alberte, Randall Sheldon
Albright, Fred Ronald
Abro, Phillip William
Alburn, Harvey Eugene
Aldous, Duane Leo
Aldridge, Mary Hennen
Alexander, George Jay
Alexander, Herman Davis
Alexander, James Craig
Alexander, James King
Alexander, Nicholas Michael
Alexander, Renee R
Alfin-Slater, Roslyn Bernice (Mrs Grant G Slater)
Alivisatos, Spyridon Gerasimos Anastasios
Alkjaersig, Norma Kirstine (Mrs A P Fletcher)
Allan, Robert K
Allard, Claude
Allen, Arthur
Allen, Charles Eugene
Allen, Charles Freeman
Allen, Charles Marshall, Jr
Allen, Cheryl
Allen, John Rybolt
Allen, Julius Cadden
Allen, Robert Scott
Allen, Sydney Henry George
Allmann, David William
Allred, John B
Alper, Robert
Alpers, Joseph Benjamin
Alphin, Reevis Stancil
Alsmeyer, Richard Harvey
Alsmeyer, William Louis
Altman, Kurt Ison
Alvarado, Francisco
Alvares, Alvito Peter
Amer, Mohamed Samir
Aminoff, David
Amiraian, Kenneth
Amoore, John Ernest
Anand, Amarjit S
Anchel, Marjorie Wolff (Mrs Herbert Rackow)
Andersen, Niels Hjorth
Andersen, Roger Allen
Andersen, William Ralph
Anderson, Bruce Murray
Anderson, Byron
Anderson, Carl Elmore
Anderson, David Gordon
Anderson, Ethel Irene
Anderson, Harlan Dwight
Anderson, John Arthur
Anderson, John Seymour
Anderson, Julius Horne, Jr
Anderson, Larry Ernest
Anderson, Laurens
Anderson, Louise Eleanor
Anderson, Paul M
Anderson, Richard Lee
Anderson, Robert E
Anderson, Robert L
Anderson, Robert Lewis
Anderson, Theodore Edmund
Anderson, Thomas Alexander
Anderson, W French
Andreoli, Anthony Joseph
Andrews, Fred Albert
Andrews, John Stevens, Jr
Anfinsen, Christian Boehmer
Angel, Charles
Angel, Joseph Francis
Angstadt, Carol Newborg
Anker, Herbert S
Annett, Robert Gordon
Ansbacher, Stefan
Anthony, Luean Evangeline
Antoniades, Harry Nicholas
Anwar, Rashid Ahmad
Appleman, M Michael
Appleton, Martin David
Archibald, Reginald MacGregor
Arcos, Joseph (Charles)
Arfin, Stuart Michael
Argoudelis, Chris J
Argus, Mary Frances
Arion, William Joseph
Arlinghaus, Ralph B
Armbrecht, Bernard Henry
Armbruster, Frederick Carl
Armstrong, Frank Bradley, Jr
Armstrong, John Briggs

Armstrong, Marvin Douglas
Armstrong, Robert Lee
Armstrong, Wallace David
Arneson, Dora Williams
Arneson, Richard Michael
Arnold, Jeffrey
Arnold, Roy Gary
Arnold, Wilfred Niels
Arnon, Daniel Israel
Arnone, Arthur Richard
Aronoff, Samuel
Aronson, John Noel
Aronson, Nathan Ned, Jr
Aronson, Robert Bernard
Arora, Kasturi Lal
Arthur, Robert David
Arvan, Dean Andrew
Asakura, Toshio
Ascione, Richard
Ash, Kenneth Owen
Ashe, Warren (Kelly)
Ashmore, Charles Robert
Ashwell, G Gilbert
Asimov, Isaac
Askari, Amir
Asplund, Russell Owen
Astill, Bernard Douglas
Astrachan, Lazarus
Atassi, Zouhair
Atencio, Alonzo C
Atkinson, Burr Gervais
Atkinson, Daniel Edward
Attaway, David Henry
Auerbach, Victor Hugo
Augenlicht, Leonard Harold
Augustyn, Joan Mary
Auld, David Stuart
Aurand, Leonard William
Aust, Steven Douglas
Austern, Barry M
Autor, Anne Pomeroy
Avadhani, Narayan G
Avigad, Gad
Avigan, Joel
Awapara, Jorge
Awasthi, Yogesh C
Axelrod, Abraham Edward
Axelrod, Bernard
Ayengar, Padmasini (Mrs Frederick Aladjem)
Ayre, Charles A
Azari, Parviz
Babbitt, Jerry
Babior, Bernard M
Bacchi, Cyrus Joseph
Bach, Michael Klaus
Bachrach, Howard Lloyd
Bachur, Nicholas R, Sr
Bade, Maria Leipelt
Badenhop, Arthur Fredrick
Bagatell, Fillmore Kenneth
Bagchi, Sakti Prasad
Bagdasarian, Andranik
Baggett, Billy
Baggott, James Patrick
Baginski, Eugene S
Bagshaw, Joseph Charles
Baich, Annette
Baig, Mirza Mansoor
Bailey, David George
Bailey, David Tiffany
Bailey, Gordon Burgess
Bailey, John Martyn
Bailin, Gary
Baines, A D
Baisted, Derek John
Baker, Alan Paul
Baker, Charles Albert
Baker, Dwight Lynds
Baker, Graeme Levo
Baker, Harold Nordean
Baker, Nome
Baker, Thomas Irving
Baker, Wilber Winston
Bakke, Jerome E
Bakshy, Stanley
Balazs, Endre Alexander
Baldeschwieler, John Dickson
Baldridge, Robert Crary
Baldwin, Heber Ross
Baldwin, Robert Russel
Baldwin, Thomas Oakley
Baldwin, William Walter
Balis, Moses Earl
Balish, Edward
Ball, Eric Glendinning
Ballentine, Robert
Ballou, Clinton Edward
Ballou, David Penfield
Baltimore, David
Bamburg, James Robert
Banaszak, Leonard Jerome
Banasik, Orville James
Banay-Schwartz, Miriam
Banks, William Louis
Barak, Anthony Joseph
Barany, Michael
Barban, Stanley

Barbehenn, Elizabeth Kern
Barber, Eugene Douglas
Barber, George Alfred
Barber, George Winston
Barbour, Stephen D
Barcelo, Raymond
Barclay, Ralph Kinney
Barker, Horace Albert
Barker, Kenneth Leroy
Barker, Robert
Barlow, John Slaney
Barnes, Eugene Miller, Jr
Barnet, Harry Nathan
Barnhart, James William
Baron, Ronald L
Baronowsky, Paul E
Barr, Charles Richard
Barrett, Harold Whilbert
Barrows, Charles Harry, Jr
Barrueto, Richard Benigno
Barry, Roger Donald
Bartholomew, William Holden
Bartlett, Grant Rogers
Bartlett, Paul Devere
Bartnicki-Garcia, Salomon
Barton, Ambrose Donald
Basch, Jay Justin
Basford, Robert Eugene
Bashey, Reza Ismail
Bass, Abraham
Bassham, James Alan
Basu, Subhash Chandra
Bates, Margaret Westbrook
Bates, William K
Batra, Prem Parkash
Batt, Conrad William
Batt, William George
Battaile, Julian
Bauer, Clifford David
Bauer, Roger Duane
Bauernfeind, Jacob Christopher
Baugh, Charles M
Baum, Robert Harold
Bauman, John W, Jr
Baumann, Wolfgang Josef
Baumgarten, Werner
Baumgartner, Luther Leroy
Baumstark, John Spann
Baur, Joseph Ralph
Bauriedel, Wallace Robert
Bavetta, Lucien Andrew
Bavisotto, Vincent
Baxter, Claude Frederick
Baxter, John Edwards
Bayley, Henry Shaw
Bayley, Stanley Thomas
Bazzano, Gaetano
Beach, Eliot Frederick
Beall, Desmond
Beall, Robert Joseph
Bean, Ross Coleman
Beare-Rogers, Joyce Louise
Beasley, C A (Bud)
Beaton, George Hector
Beattie, Diana Scott
Beaudreau, George Stanley
Beaven, Michael Anthony
Beck, David Paul
Beck, Jay Vern
Beck, Jeanne Crawford
Beck, William Samson
Becker, Barbara
Becker, Benjamin
Becker, Charles Edward
Becker, Gerald Leonard
Becker, Joseph F
Becker, Milton J
Becker, Robert Richard
Becker, Wayne Marvin
Becking, George C
Bednekoff, Alexander G
Beeler, Donald A
Beers, Roland Frank, Jr
Beerstecher, Ernest, Jr
Begg, Robert William
Beher, Francis Joseph
Beher, William Tyers
Behnke, William David
Behrens, Otto Karl
Behrisch, Hans Werner
Behrman, Edward Joseph
Behrman, Harold R
Beinert, Helmut
Beining, Paul R
Bell, Fred E
Bell, John Perkins
Bell, Oliver E, Jr
Bell, Paul Hadley
Bell, Robert Gale
Bell, Robert Maurice
Bellamy, Winthrop Dexter
Bellhorn, Margaret Burns
Bellino, Francis Leonard
Belman, Leonard John
Belman, Sidney
Beltz, Richard Edward
BeMiller, James Noble
Bendich, Aaron

BIOCHEMISTRY

Benedict, James Harold
Benedict, Jean Davidson
Benedict, Robert Curtis
Benesch, Reinhold
Benesch, Ruth Erica
Benevenga, Norlin Jay
Benham, Graham Harvey
Benisek, William Frank
Benne, Erwin John
Bennett, Edward Leigh
Bennett, Emmett
Bennu, Alfredo
Benoit, George Julien, Jr
Benoiton, Normand Leo
Ben-Porat, Tamar
Benson, Andrew Alm
Bensusan, Howard Bernard
Bentley, J Peter
Bentley, Ronald
Benton, Allen William
Benton, Duane Allen
Berenbom, Max
Berezney, Ronald
Berg, Clarence Peter
Berg, Marie Hirsch
Berg, Paul
Bergdoll, Merlin Scott
Berger, Franklin Gordon
Berger, Julius
Berger, Shelby Louise
Berk, Richard Samuel
Berke, Harry L
Berkowitz, David B
Berkut, Michael Kalen
Berlin, Elliott
Berlin, Soll
Berman, Eleanor
Bernard, Patrick Spitaletta
Berner, David Leo
Bernfeld, Peter
Bernlohr, Mary Lilias Christian
Bernlohr, Robert William
Bernofsky, Carl
Berns, Kenneth
Bernstein, Elaine Katz
Bernstein, Eugene H
Bernstein, Isadore A
Berry, James Frederick
Berry, Robert Eddy
Bertani, Laura Marie
Bertrand, Helen Anne
Bessey, Otto Arthur
Bessman, Maurice Jules
Bessman, Samuel Paul
Best, Audrey Nance
Bethell, Joseph Jay
Bett, Hillyard Dobson
Betz, Robert F
Beuk, Jack Frank
Bever, Arley Tunis, (Jr)
Bever, Enid L
Beveridge, James MacDonald Richardson
Bevill, Rardon Dixon, III
Bewley, John Derek
Beychok, Sherman
Beyer, Robert Edward
Bezkorovainy, Anatoly
Bhagavan, Hennnige
Bhatnagar, Rajendra Sahai
Bhattacharyya, Maryka Horsting
Bhavnani, Bhagu R
Bhuvaneswaran, Chidambaram
Bhuyan, Bijoy Kumar
Biaglow, John E
Bieber, Allan Leroy
Bieber, Loran Lamoine
Bieber, Theodore Immanuel
Bieri, John Genther
Biggs, Homer Gates
Bigler, William Norman
Bihler, Ivan
Bikle, Daniel David
Bilimoria, Minoo Hormasji
Billiar, Reinhart Billie
Billingsley, Lawrence Winston
Bingham, Robert J
Binkley, Francis
Binkley, Stephen Bennett
Bird, John William Clyde
Birdsall, John J
Bishop, Muriel Boyd
Bishop, Stephen Hurst
Biswas, Donald Lynn
Biswas, Shib D
Bitman, Joel
Black, Arthur Leo
Black, Billy C, II
Black, Clanton Candler, Jr
Black, John Alexander
Black, John Larry

Black, William James
Blackmore, Robert Valentine
Blackwell, Richard Quentin
Blackwood, Carlton E
Blackwood, Chesley M
Blair, Alan Huntley
Blair, Barbara Ann
Blair, Donald George Ralph
Blair, James Bryan
Blair, Paul V
Blake, James L
Blake, Robert L
Blakley, Raymond L
Blanchaer, Marcel Corneille
Blaney, Donald John
Blankenship, James W
Blankley, Madison Van
Blatt, Joel Martin
Blatt, Stanley Parris
Blatz, Paul E
Blaydes, David Fairchild
Blazyk, Jack
Blecher, Melvin
Bleiweis, Arnold Sheldon
Blevins, Raymond Dean
Bloch, Alfred
Bloch, Eric
Bloch, Konrad Emil
Bloch, Walter David
Blohm, Thomas Robert
Blomquist, Charles Howard
Blomquist, Gary James
Bloomfield, Daniel Kermit
Blostein, Rhoda
Blout, Elkin Rogers
Blum, Stanley Walter
Blumenfeld, Olga O
Blumenstein, Michael
Blumenthal, Herbert
Boatman, Sandra
Bobb, Yvonne Dolores
Bobbi, Jesse Leroy
Bocage, Albert J
Bocchieri, Samuel Francis
Bock, Rose Mary
Bock, Fred G
Bodansky, Oscar
Bodley, James William
Boeker, Elizabeth Anne
Boezi, John A
Boggs, Dallas Ervin
Bogoch, Samuel
Boguslawski, George
Bohinski, Robert Clement
Bohonos, Nestor
Boldt, Roger Earl
Bole, Giles G, Jr
Bollinger, James Norman
Bollum, Frederick James
Boltralik, John Joseph
Bonar, Robert Addison
Bond, Andrew
Bond, Judith
Bond, Thomas Jackson
Bondy, Stephen Claude
Bonhorst, Carl W
Bonner, John Franklin, Jr
Bonney, Robert John
Bono, Vincent Horace, Jr
Bonsnes, Roy Walter
Bonting, Sjoerd Lieuwe
Boone, Charles Walter
Borchers, Raymond (Lester)
Borders, Charles LaMonte, Jr
Bordner, Jon D B
Borek, Blanche Ann
Borek, Ernest
Borenfreund, Ellen
Borgese, Thomas A
Borglum, Gerald Baltzer
Borkovec, Alexej B
Bornstein, Aleck
Bornstein, Paul
Boroughs, Howard
Borowska, Zofia Kurylo
Borsook, Henry
Borun, Thaddeus W
Bosch, Arthur James
Bosc, Roland Andrew
Boston, James D
Botan, Edward Allan
Botero, J M
Bottino, Nestor Rodolfo
Boulet, Marcel
Bourke, John Butts
Bousquet, William F
Bouthillier, Louis-Philippe
Boutwell, Roswell Knight
Bowden, Joe Allen
Bowen, Charles Allen
Bowen, Vaughan Tabor
Bowles, William Henry
Bowman, Donald Edwin
Bowman, Roger Holmes
Boyd, Donald Mitchell
Boyer, Alvin C
Boyer, Paul Delos

Boylen, Joyce Beatrice
Bozian, Robert C
Brachfeld, Norman
Brackett, Benjamin Gaylord
Bradber, Clive
Bradfield, Robert B
Bradford, Reagan Howard
Bradham, Laurence Stobo
Bradovold, Donald Keith
Bradshaw, Edwin D, Jr
Bradshaw, Herbert Leon
Bradshaw, Ralph Alden
Bradshaw, William S
Brady, Robert Nyle
Brady, Roscoe Owen
Bragg, Philip Dell
Brake, Jon Michael
Brand, Ludwig
Brand, Karl Garet
Brandt, Richard Bernard
Brandts, John Frederick
Brandvold, Donald Keith
Brandwein, Bernard Jay
Bransome, Edwin D, Jr
Braselton, Webb Emmett, Jr
Brattsten, Lena B
Braverman, George
Brazda, Fred George
Brecher, Arthur Seymour
Brecher, Peter I
Breen, Moira
Breene, William Michael
Breitman, Theodore Ronald
Brenn, Thomas Peter
Bresnick, Edward
Bretthauer, Roger K
Brewer, John Michael
Brezenski, Francis T
Bridger, William Aitken
Bridgham, Catherine Mitchell
Brieley, Gerald Philip
Briggs, Thomas
Brigham, M Prince
Brin, Myron
Brinck-Johnson, Truls
Brinigar, William Seymour, Jr
Brink, Robert Harold, Jr
Brinkman, Gail Lynn
Briscoe, Anne M
Briskey, Ernest J
Brockerhoff, Hans
Brockman, Robert W
Brodie, Angela (Hartley)
Brodie, Arnold Frank
Brody, Marcia
Brody, Stuart
Brohn, Frederick Herman
Bromer, William Wallis
Bronk, John Ramsey
Bronsky, Albert J
Brooker, Gary
Brooks, James Lee
Brooks, John Bill
Brooks, Samuel Carroll
Broomfield, Clarence A
Broquist, Harry Pearson
Brosemer, Ronald Webster
Brostrom, Charles Otto
Brostrom, Margaret Ann
Brot, Nathan
Browder, Henry Polk, Jr
Brown, Douglas Markham
Brown, Fountaine Christine
Brown, Frederick S
Brown, Gene Monte
Brown, George Willard, Jr
Brown, Harry Darrow
Brown, James Walker, Jr
Brown, Jerry L
Brown, John Clifford
Brown, John Haynes
Brown, John Wesley
Brown, Neal Curtis
Brown, Olen Ray
Brown, Phyllis R
Brown, Raymond Arthur
Brown, Raymond Russell
Brown, Robert Lee
Brown, Ross Duncan, Jr
Brown, Theodore Llewellyn
Brown, William Duane
Brown, William Everett
Brown, William Henry
Brownie, Alexander C
Broyles, Robert Herman
Bruckner, Benjamin Harry
Brummond, Dewey Otto
Brunelle, Thomas Lee
Brunngraber, Ellnor Flora
Brunngraber, Eric Gustav
Bruno, Charles Frank
Bruno, Gerald A
Bruns, Lester George
Brush, James S
Bryan, John Kent
Bryan, Sara E
Bryan, William Phelan
Brysk, Jay Clark
Brysk, Miriam Mason
Bublitz, Clark

Buchanan, Bob Branch
Buchanan, John Machlin
Buchanan-Davidson, Dorothy Jean
Buchanan, James Arthur
Buckley, James Thomas
Buckley, Patricia M
Buckley, Ramon D
Bucolo, Giovanni
Bucovaz, Edsel Tony
Bueding, Ernest
Buell, George Christopher
Bukowick, Peter Anthony
Bull, Richard C
Bulla, Lee Austin, Jr
Bumpus, Francis Merlin
Bunce, George Edwin
Bundy, Hallie Flowers
Bunge, Wylie D
Bunkfeldt, Rudolf
Bunting, John William
Buono, Frederick J
Burch, Helen Bulbrook
Burch, Robert Emmett
Burchfield, Harry P
Burck, Philip John
Burd, John Frederick
Burdick, Donald
Burge, Alan Walter
Burge, Wylie D
Burger, Richard Melton
Burgess, Benjamin Franklin, Jr
Burgess, Richard Ernest
Burgus, Roger Cecil
Burk, Dean
Burka, Edward Richard
Burke, William Thomas, Jr
Burkhall, Raymond Kenneth
Burkhart, Morris Paton, Jr
Burleigh, Bruce Daniel, Jr
Burnett, Jean Bullard
Burnham, Bruce Franklin
Burns, John J
Burns, Mary Grace
Burns, Moore J
Burns, Richard Charles
Burr, George Oswald
Burr, William Wesley, Jr
Burris, Robert Harza
Burstein, Shlomo
Burtis, Carl A, Jr
Burton, Albert Frederick
Burton, Alice Jean
Burton, David Norman
Burton, Robert McMahon
Burton, Robert Main
Burton, Sheril Dale
Busby, William Fisher, Jr
Busch, Harris
Bush, Ian
Bush, Karen Jean
Busse, Robert Franklyn
Bustos-Valdes, Sergio Enrique
Buyske, Donald Albert
Butcher, Fred Ray
Butcher, Henry Clay, IV
Butcher, Reginald William
Butler, Gordon Cecil
Butler, Larry G
Butler, Lillian Ida
Butler, William Thomas
Butow, Ronald A
Button, Don K
Butzow, James J
Buzard, James Albert
Buzzee, David H
Byard, James Leonard
Byerrum, Richard Uglow
Byers, Larry Douglas
Byers, Lawrence Wallace
Byers, Sanford Oscar
Byrne, William Lawrence
Cabib, Enrico
Cagan, Robert Howard
Cahill, Charles L
Cain, Robert Howard
Cairns, William Louis
Calandra, Joseph Carl
Calbert, Harold Edward
Caldwell, Daniel R
Caldwell, Roger Lee
Calhoun, David H
Callahan, John William
Calvanico, Nickolas Joseph
Camerino, Pat William
Cameron, Colin Robert
Cameron, Merrill Nelson
Camien, Merrill Nelson
Camnener, Gerald Walter
Cammarata, Peter S
Camp, Bennie Joe
Campbell, Alice Del Campillo
Campbell, Alfred Duncan
Campbell, Benedict James
Campbell, Bruce (Nelson), Jr
Campbell, Clyde Del
Campbell, Harold Alexander
Campbell, Hayward
Campbell, James
Campbell, James Nicoll
Campbell, Linzy Leon

Campbell, Thomas Colin
Campo, Robert D
Canady, William James
Cancio, Marta
Canellakis, Evangelo S
Cann, Malcolm Calvin
Cannon, Donald Joseph
Cannon, Jerry Wayne
Cantarow, Abraham
Cantero, Antonio
Cantor, Abraham
Cape, Ronald Elliot
Caplan, Arnold I
Caplis, Michael E
Caplow, Michael
Caponio, Joseph Francis
Capp, Grayson L
Caprioli, Richard Michael
Capstack, Ernest
Caraway, Wendell Thomas
Carbon, John Anthony
Cardinale, George Joseph
Carew, Lyndon Belmont, Jr
Carey, Paul L
Cargill, David Innes
Carlson, Gerald Lowell
Carlson, James Roy
Carlson, Mildred V
Carpenter, Carolyn Virus
Carpenter, Frederick Hiltman
Carpenter, Lawrence Edward
Carpenter, Mary Pitynski
Carr, Charles William
Carr, Daniel Oscar
Carr, Julius Jay
Carraway, Kermit Lee
Carrico, Robert Joseph
Carroll, Edward James, Jr
Carroll, Harold Wilson
Carroll, James Joseph
Carroll, Kenneth Kitchener
Carroll, William Robert
Carson, Frederick Wallace
Carsten, Mary E
Carter, Charles Edward
Carter, Herbert Edmund
Carubelli, Raoul
Carver, Michael Joseph
Casey, Martha L
Cash, William Davis
Cashel, Michael
Casillas, Edmund Rene
Casjens, Sherwood Reid
Caspe, Saul
Caspi, Eliahu
Cass, Carol E
Cassens, Robert G
Castanera, Esther Goossen
Castellino, Francis Joseph
Caston, J Douglas
Castro, Alberto
Cater, Carl Malcom
Catravas, George Nicholas
Caughey, Winslow Spaulding
Cauther, Sally Eugenia
Cayen, Mitchell Ness
Cederquist, Dena Caroline
Ceithaml, Joseph James
Celander, David Robert
Celander, Evelyn Faun
Cenedella, Richard J
Cerbulis, Janis
Ceruti, Peter A
Cevallos, William Hernan
Cha, Sungman
Chacko, George Kutty
Chader, Gerald Joseph
Chae, Kun
Chaffee, Rowand R J
Chakrabarti, Siba Gopal
Chalkley, G Roger
Chambers, Robert Warner
Chan, David Siupoon
Chan, Ming Sui Michael
Chan, Peter Sinchun
Chan, Samuel H P
Chan, Shung Kai
Chance, Britton
Chance, Ronald E
Chandan, Ramesh Chandra
Chandler, Albert Morrell
Chandler, Michael Lynn
Chandra, G Ram
Chaney, Stephen Gifford
Chang, Albert Yen
Chang, Jaw-Kang
Chang, John Wan-Yuin
Chang, Lucy Ming-Shih
Chang, Shen Chin
Chang, Yung-Feng
Chanley, Jacob David
Chao, Fu-chuan
Chapman, Astrid Gronneberg
Chappelle, Emmett W
Charalampous, Frixos C
Chargaff, Erwin
Charney, Jesse

Charnock, John S
Chasin, Mark
Chassy, Bruce Matthew
Chatterton, Robert Treat, Jr
Chaykin, Sterling
Chaykovsky, Michael
Chefurka, William
Chen, Chiadao
Chen, Chung-Ho
Chen, Harry Wu-Shiong
Chen, I-Wen
Chen, James Pai-Fun
Chen, John Heng
Chen, Linda Li-Yueh Huang
Chen, Shao Lin
Cheng, Frank Hsieh Fu
Cherayil, George Devassia
Cherkin, Arthur
Cherniak, Robert
Chernick, Sidney Samuel
Cherry, Joe H
Chervenka, Charles Henry
Chesley, Leon Carey
Chetsanga, Christopher J
Cheung, Wai Yiu
Chheda, Girish B
Ch'ih, John Juwei
Childress, Charles Curtis
Chin, Byong Han
Chinn, Herman Isaac
Chiou, George Chung-Yih
Choules, George Lew
Chipault, Jacques Robert
Chipman, Wilmon B
Chirigos, Michael Anthony
Chirikjian, Jack G
Chiu, Jen-Fu
Cho, Kon Ho
Choi, Yong Chun
Chornock, Francis William
Chow, Fu Ho Chen
Christensen, Glenn Marvin
Christensen, Halvor Niels
Christensen, John
Christian, Samuel Terry
Christiansen, James Brackney
Christiansen, Marjorie Miner
Christman, Judith Kershaw
Chu, Fun Sun
Chu, Luke Lo-Hwa
Chu, Tsann Ming
Chuang, Hanson Yii-Kuan
Chuang, Ronald Yan-Li
Chung, Albert Edward
Chung, Jiwhey
Chung, Kwok-Leung
Chytil, Frank
Ciaccio, Edward I
Ciereszko, Leon Stanley
Cilley, Jonathan Hubbard
Cirillo, Vincent Paul
Civen, Morton
Clagett, Carl Owen
Clare, Stewart
Clark, Brian Roger
Clark, C Elmer
Clark, Charles Christopher
Clark, Dale Allen
Clark, Eloise Elizabeth
Clark, Irwin
Clark, Jeffrey Lee
Clark, John Magruder, Jr
Clark, Julia Berg
Clark, Leland Charles, Jr
Clark, Mary Jane
Clark, William Gilbert
Clark, William Richmond
Clarke, Donald Dudley
Clarke, Donald Walter
Claybrook, James Russell
Claycomb, Cecil Keith
Claycomb, William Creighton
Clayton, Charles Curtis
Clayton, James Wallace
Clayton, Raymond Brazenor
Clayton, Robert Allen
Clegg, James S
Clegg, Robert Edward
Cleland, William Wallace
Clemmons, Jackson Joshua Walter
Clemson, Harry C
Clesceri, Lenore Stanke
Cleveland, Anne Stack
Clevenger, Richard Lee
Clewell, Don Bert
Cloutier, Paul Frederick
Coburn, Stephen Putnam
Cockrell, Ronald Spencer
Codington, John F
Coe, Elmon Lee
Coffey, Donald Straley
Coffey, John Joseph
Coffey, John William
Coffey, Ronald Gibson
Cohen, Arthur Isaac
Cohen, Harold P
Cohen, Herman

Cohen, Leonard Harvey
Cohen, Natalie Shulman
Cohen, Pinya
Cohen, Robert Jay
Cohen, Saul Louis
Cohen, Seymour Stanley
Cohen, Stanley
Cohen, William
Cohen, William David
Cohn, David Valor
Cohn, Mildred
Cohn, Waldo E
Coke, James Logan
Colas, Antonio E
Colbourn, Joseph Leason
Colburn, Nancy Hall
Colburn, Robert Warren
Colby, Robert William
Cole, Edmond Ray
Cole, Thomas A
Coleman, Joseph Emory
Coleman, Lester Lyman
Coleman, Peter Stephen
Coleman, Ronald Leon
Collier, Herbert Bruce
Collins, Delwood C
Collins, James Francis
Collins, James Malcolm
Collins, John W
Collins, Vernon Kirkpatrick
Colman, Brian
Colman, Roberta F
Colowick, Sidney Paul
Colvin, Harry Walter, Jr
Colvin, John Ross
Commerford, Spencer Lewis
Condliffe, Peter George
Coniglio, John Giglio
Conley, Cecil
Conn, Eric Edward
Connell, George Leonard
Connelly, Jerald Leonard
Conner, Robert Thomas
Conney, Allan Howard
Connolly, John W
Connors, William Matthew
Conover, Thomas Ellsworth
Conrad, Harry Edward
Conrad, Herbert M
Conrad, Joseph H
Considine, Judith Mayberry
Constantinides, Spiros Minas
Constantopoulos, George
Conway, Thomas William
Cook, David Edgar
Cook, Robert Merold
Cook, Robert Thomas
Cook, William Harrison
Coolidge, Thomas Buckingham
Coon, Minor J
Cooper, Cecil
Cooper, John Allen Dicks
Coots, Robert Herman
Copenhaver, John Harrison, Jr
Corbin, Jack David
Corcoran, John William
Cordes, Eugene H
Cori, Carl Ferdinand
Cormier, Milton Joseph
Cornatzer, William Eugene
Cornblath, Marvin
Cornell, Creighton N
Cornell, James S
Cornell, Neal William
Cornwell, David George
Corrigan, John Joseph
Corwin, Laurence Martin
Cory, Joseph G
Coryell, Margaret E
Coscarelli, Waldimero
Coscia, Carmine James
Cossins, Edwin Albert
Costlow, Richard Dale
Cote, Lucien Joseph
Cottam, Gene Larry
Couch, James Russell
Coulson, Roland Armstrong
Cousineau, Gilles H
Coutinho, Claude Bernard
Cowan, James W
Cowger, Marilyn L
Cowgill, Robert Warren
Cowman, Richard Ammon
Cox, Andrew Chadwick
Cox, George Stanley
Cox, Ray
Cozzarelli, Nicholas Robert
Crabtree, Gerald Winston
Cramer, John Wesley
Crandall, Dana Irving
Crane, Charles Russell
Crane, Frederick Loring
Cramer, Robert Kellogg
Crass, Maurice Frederick, III
Craven, Charles Waller
Crawford, Richard Bradway
Crawhall, John C

Creasey, William Alfred
Crecelius, Samuel Brown
Creger, Clarence R
Creveling, Cyrus Robbins
Criddle, Richard S
Criscuolo, Dominic
Cristofalo, Vincent Joseph
Cromartie, Thomas Houston
Cronan, John Emerson, Jr
Cronheim, Georg Erich
Cronin, John Read
Croom, Henrietta Brown
Croteau, Rodney
Crounse, Robert Griffith
Crowe, Arlene Joyce
Crowshaw, Keith
Crusberg, Theodore Clifford
Cuatrecasas, Pedro
Cullen, Marion Permilla
Culley, William James
Cunningham, Bruce Arthur
Cunningham, Bryce A
Cunningham, Dennis Dean
Cunningham, Earlene Brown
Cunningham, Franklin E
Cunningham, Glenn N
Cunningham, Leon William
Currie, William Deems
Curtis, Gary Lynn
Curtis, Jerry Leon
Cusanovich, Michael A
Cushman, Samuel Wright
Cymerman, Allen
Cynkin, Morris Abraham
Czech, Michael Paul
Czerlinski, George Heinrich
Dabbah, Roger
Da Costa, William A
Dacre, Jack Craven
Dagley, Stanley
Dahlberg, Albert Edward
Dahlberg, James Eric
Dahlstrom, Robert V
Dahm, Karl Heinz
Dahm, Paul Adolph
Dahms, Arthur Stephen, Jr
Dahmus, Michael E
Dailey, Robert Engle
Dain, Joel A
Dairman, Wallace
Dakshinamurti, Krishnamurti
Dalal, Fram Rustom
Dallam, Richard Duncan
Dalton, Harry P
Dalton, Jack L
Daly, Marie Maynard
Dam, Richard
Damadian, Raymond
Damaskus, Charles William
Damle, Suresh B
Dan, Teruki Clark
Dancis, Joseph
Daniel, Louise Jane
Daniels, William Fowler
Danishefsky, Isidore
Danner, Dean Jay
Danner, Jean
D'Aoust, Brian Gilbert
Daravingas, George Vassilios
Darby, Eleanor Muriel Kapp
Darby, William Jefferson
Darnall, Dennis W
Daron, Harlow Hoover
Darragh, Marvin
Darragh, Richard T
Dasgupta, Uttam
Dash, Harriman Harvey
Dasler, Waldemar
Datta, Padma Rag
Datta, Prasanta
Davenport, Guy Rodman
David, Jean
Davidian, Nancy McConnell
Davidson, Betty
Davidson, Eugene Abraham
Davidson, Harold Michael
Davidson, Norman Ralph
Davidson, Samuel James
Davie, Earl W
Davies, Helen Jean Conrad
Davies, Ronald Ernest
Davies, Ronald Edgar
Davies, Thomas Harrison
Davis, Alvie Lee
Davis, Bernard David
Davis, Carl Lee
Davis, Eldred Jack
Davis, Frank
Davis, Frank French
Davis, James Wendell
Davis, Lawrence Clark
Davis, Leodis
Davis, Neil Clifton
Davis, Norman Rodger
Davis, Norman Seymour
Davis, Peyton Nelson
Davis, Raymond Vincent
Davis, Richard Henry, Jr

Davis, Robert Leo
Davis, Robert Paul
Davis, Virginia Eischen
Davis, Ward B
Davison, John Philip
Dawson, Igor Bert
Dawson, Charles Reginald
Dawson, Earl B
Dawson, Glyn
Day, Paul Louis
Day, Richard Allen
Dayhoff, Margaret Oakley
Deasy, Clara Louise
De Balbian Verster, Floris
Deber, Charles Michael
Debich, Danica
Decker, Richard H
Decker, Walter Johns
de Duve, Christian Rene
Deeney, Anne O'Connell
Deese, Dawson Charles
Dekker, Charles Abram
Dekker, Eugene Earl
DeKloet, Siwo R
de la Haba, Gabriel Luis
de Lamirande, Gaston
de Freitas, Anthony S
de Fremery, Donald
De Haas, Herman
Deitrich, Richard Adam
Dekirmenjian, Haroutune
DeLange, Robert J
Delaney, Robert
Deluva, Adelaide Marie
Delmer, Deborah P
De Long, Chester Wallace
De Luca, Chester
De Luca, Donald Carl
De Luca, Hector Floyd
De Luca, Luigi M
DeLuca, Marlene
De Master, Eugene Glenn
DeMedicis, E M J A
DeMeio, Romano Humberto
Demetriou, James A
De Miranda, Paulo
Dempsey, Mary Elizabeth
Dempsey, Walter B
Denko, Charles W
Dennen, David W
Dennis, Don
Dennis, Edward A
Dennis, Joe
Denton, Arnold Eugene
Deodhar, Sharad Dinkar
DePinto, John A
Deranleau, David A
De Renzo, Edward Clarence
Derr, Robert Frederick
Desai, Indrajit Dayalji
de Salegui, Miriam
DeShazo, Mary Lynn Davison
Deshmukh, Diwakar Shankar
Desiderio, Dominic Morse, (Jr)
DeSombre, Eugene Robert
Dessauer, Herbert Clay
Deszyck, Edward John
Detroy, Robert William
Detwiler, Thomas C
Deutsch, Mike John
Deutscher, Murray Paul
Devartanian, Daniel Vartan
DeVenuto, Frank
Dever, John E, Jr
DeVincenzi, Donald Louis
Devlin, Thomas McKeown
Devor, Arthur William
Devor, Kenneth Arthur
Dewar, Norman Ellison
Dewey, Virginia Caroline
Dewey, William Leo
DeWitt, William
Deyoe, Charles W
De Zeeuw, John Robert
Dhyse, Frederick George
Diab, Ihsan M
Dicc, J Fred
Dickerman, Herbert W
Dickie, John Peter
Dickman, John Theodore
Dickman, Sherman Russell
Di Cuollo, C John
Dieckert, Julius Walter
Diehl, John Edwin
Dietz, Albert Arnold Clarence
Dietz, Albert J Jr
Dietz, George William, Jr
Di Ferrante, Nicola Mario
Diller, Erold Ray
Dilley, Richard Alan
Dilworth, Benjamin Conroy
DiMaggio, Anthony, III
Dion, Arnold Silva
D'Iorio, Antoine
Di Pietro, David Louis
Dirksen, Thomas Reed
Dittner, John Charles

Diven, Warren Field
DiVincenzo, George D
Dixon, Jack Edward
Dixon, Robert Louis
Dobrogosz, Walter Jerome
Dobson, Harold Lawrence
Doctor, Bhupendra Pannalal
Doctor, Vasant Manilal
Dodds, Alvin Franklin
Dodgen, Charles Lee
Dodson, Vernon N
Doebbler, Gerald Francis
Doellgast, George John
Doering, Charles Henry
Doherty, David George
Doig, Marion Tilton, III
Doisy, Edward Adelbert
Doisy, Edward Adelbert, Jr
Doisy, Richard Joseph
Dolin, Morton Irwin
Domanik, Richard Anthony
Dombro, Roy S
Dominguez, Oscar V
Donaldson, Wayne
Donelson, John Everett
Donisch, Valentine
Donoho, Alvin Leroy
Donovan, Gerald Alton
Donovan, Ross Grant
Doolittle, Russell F
Dorchester, John Edmund Carleton
Dorer, Frederic Edmund
Dorfman, Albert
Dorfman, Ralph Isadore
Dorman, Homer Lee
Dorman, Stephen Charles
Dorner, Robert Wilhelm
Dorrington, Keith John
Dorsey, Thomas Edward
Doscher, Marilyn Scott
Doskotch, Raymond Walter
Doty, Paul Mead
Doudney, Charles Owen
Dougall, Donald K
Doughty, Clyde Carl
Douglass, Jocelyn Fielding
Douglass, Carl Dean
Doukas, Harry Michael
Dounce, Alexander Latham
Downing, Donald Talbot
Downing, Mancour
Downs, Frederick Jon
Doyle, Darrell Joseph
Doyle, Miles Lawrence
Doyle, Richard Robert
Doyle, Thomas Harry
Drake, Billy Blandin
Draper, Roy Douglas
Dratz, Edward Alexander
Draus, Frank John
Dreiling, Charles Ernest
Drell, William
Dresden, Carlton F
Dresden, Marc Henri
Dressler, David
Drewes, Lester Richard
Dreyfuss, Jacques
Drucker, Harvey
Drummond, George I
Drummond, Margaret Crawford
Drury, Horace Featherstone
Drysdale, James Wallace
Du, Julie (Yi-Fang) Tsai
Dubbs, Clyde Andrew
Dubbs, Del Rose M
Dubin, Alvin
Dubin, Donald T
DuBrul, Ernest
Ducharme, Jacques R
Duck, William N, Jr
Duckworth, Donna Hardy
Dudock, Bernard S
Duell, Elizabeth Ann
Dugan, Patrick R
Duke, Philip S
Dukes, Peter Paul
Dulaney, John Thornton
Dumas, Lawrence Bernard
Duncan, Gordon Duke
Dunham, Linda Thompson
Dunkley, Colleen Rose
Dunn, Danny Leroy
Dunn, Doris Frankel
Dunn, Floyd Warren
Dunn, Michael F
Dupont, Claire Hammel
Dupuis, Gilles
Dupuy, Harold Paul
Dure, Leon S, III
Durell, Jack
Durham, William Fay
Durham, Emmett Leigh
Dursch, Harry Robert
Durst, Jack Rowland
Durzan, Donald John

Dus, Karl M
Duval, Anna Marie
Du Vigneaud, Vincent
Dvornik, Dushan Michael
Dyckes, Douglas Franz
Dyer, Denzel Leroy
Dyer, Randolph H
Dyer, William John
Dymsza, Henry A
Dyson, John Edgar
Dziewiatkowski, Dominic Donald
Eades, Charles Hubert, Jr
Eakin, Robert Edward
Easley, Eliza (Lila) Waller
Eaks, Irving Leslie
Eberhard, Anatol
Eberly, Robert C
Eble, John Nelson
Eble, Thomas Eugene
Ebner, Kurt E
Eckardt, Robert E
Eddington, Carl Lee
Eddy, Dennis Eugene
Edelman, Gerald Maurice
Edelman, Jerome
Edelstein, Stuart J
Edgell, Marshall Hall
Edmonds, Mary P
Edmondson, Dale Edward
Edmundowicz, John Michael
Edsall, John Tileston
Edstrom, Ronald Dwight
Edward, Deirdre Waldron
Edwards, Cecile Hoover
Edwards, Harold Herbert
Edwards, John R
Edwards, Lawrence Jay
Ehlig, Carl F
Ehrenpreis, Seymour
Ehrenthal, Irving
Ehrlich, Melanie
Eich, Stephen Joseph
Eichberg, Joseph
Eichel, Herbert Joseph
Eichel, Herman Joseph
Eichholz, Alexander
Eidels, Leon
Eilberg, Ralph G
Eiler, John Joseph
Eisenberg, Frank, Jr
Eisenberg, Max A
Eisenberg, Roselyn Jane
Eisenstadt, Jerome Melvin
El Attar, Tawfik Mohammed Ali
Elbein, Alan D
Eleftheriou, Basil E
Eley, James H
Elford, Howard Lee
Eliceiri, George Louis
Elion, Thomas Edward
Elman, George Leon
Elson, Elliot
Elwood, John Clint
Elwyn, David Hunter
Elzinga, Marshall
Emerson, Geraldine Mariellen
Emerson, Gladys Anderson
Emery, Arthur James, Jr
Emery, Roy Saltsman
Emery, Thomas Fred
Endahl, Gerald Leroy
Eng, Lawrence F
Engel, Lewis Libman
Engel, Ruben William
Engelman, Cecil
Engler, Merlin Duane
England, Sasha
Englund, Paul Theodore
Entenman, Cecil
Entner, Nathan
Eoff, Kay M
Epler, James L
Eppright, Margaret Anne
Epstein, Wolfgang
Erecinska, Maria
Ericson, Alfred (Theodore)
Erlanger, Bernard Ferdinand
Ernst-Fonberg, Marylou
Erwin, Virgil Gene
Esders, Theodore Walter
Esfahani, Mojtaba
Eskelson, Cleamond D
Essenberg, Margaret Kottke
Essenberg, Richard Charles
Esser, Alfred F
Estabrook, Ronald (Winfield)

Eudy, William Wayne
Evans, Audrey Elizabeth
Evans, Earl Alison, Jr
Evans, Gary William
Evans, Helen Harrington
Evans, Robert John
Evard, Rene
Everett, George Albert
Everett, Mark Reuben
Everhart, Donald Lee
Ewan, Richard Colin
Ewart, Mervyn H
Exton, John Howard
Exton, Lee Edward
Fabianek, John
Fagan, Robert V
Fahey, Robert C
Fahien, Leonard A
Fahrney, David Emory
Fain, John Nicholas
Fairley, James Lafayette, Jr
Fairley, Wells Eugene
Falb, Richard D
Fales, Frank Weck
Fanger, Michael Walter
Farber, Emmanuel
Farkas, Walter Robert
Farnsworth, Patricia Nordstrom
Farnsworth, Wells Eugene
Farrell, Harold Maron, Jr
Farren, Ann Louise
Fasman, Gerald David
Faulkner, Willard Riley
Fawaz, George
Fay, Rimmon C
Fayle, Harlan Downing
Fazekas, Arpad Gyula
Feaster, John Pipkin
Feather, Milton S
Feder, Joseph
Feeney, Robert Earl
Feigelson, Muriel
Feigelson, Philip
Feigenbaum, Abraham Samuel
Feild, Frank Joseph
Feinberg, Benjamin Allen
Feingold, David Sidney
Feinman, Richard David
Feinstein, Louis
Feinstein, Robert Norman
Feldberg, Ross Sheldon
Feldman, Fred
Feldman, Jacob Harold
Fels, Irving Gordon
Felsenfeld, Herbert William
Feltham, Lewellyn Allister Woodrow
Fenichel, Richard Lee
Fennimore, David Clarke
Fenton, John William, II
Fenton, Paul Fredric
Ferguson, James Joseph, Jr
Ferguson, Karen Anne
Fernandez, Alberto Antonio
Ferone, Robert
Ferrara, Louis W
Ferrari, Richard Alan
Ferraro, John J
Ferrell, William James
Ferren, Larry Gene
Ferretti, Joseph Jerome
Ferrier, Barbara May
Ferris, Robert Monsour
Fessenden-Raden, June Marion
Fevold, Harry Richard
Fielding, Christopher J
Filmer, David Lee
Filios, Louis Charles
Fine, Albert Samuel
Fine, Richard Eliot
Fine, Philip
Fine, William James
Fineberg, Richard Arnold
Finelli, Vincent Nicola
Fink, Kathryn Ferguson
Fink, Robert Morgan
Finkelstein, Frances
Finkelstein, James David
Finkelstein, Lawrence
Finkelstein, Paul
Finkenstaedt, John Turner
Finlayson, John Sylvester
Finley, Wayne House
Finn, Frances M
Finnerty, James William
Finnerty, James Lawrence
Fiore, Joseph Vincent
Firschein, Hillard E
Fischbach, Fritz Alvin
Fischer, Edmond H
Fischer, George A
Fischer, Mark Samuel
Fischer, Roland
Fish, Wayne William
Fishbein, Lawrence
Fishbein, William Nichols
Fisher, Harvey Franklin
Fisher, James Richard
Fisher, Ronald Richard
Fisher, Waldo Reynolds
Fishkin, Arthur Frederic
Fishman, Jack

Fishman, Louis
Fishman, Myer M
Fishman, Peter Harvey
Fishman, William Harold
Fitch, Coy Dean
Fitch, Walter M
Fitt, Peter Stanley
Fitzgerald, Glenna Gibbs
Flaks, Joel George
Flanagan, Thomas Leo
Flavin, Martin
Fleeker, James R
Fleischer, Becca Catherine
Fleischer, Sidney
Fleischmajer, Raul
Fleischman, Alan Isadore
Fletcher, Dean Charles
Fletcher, Lewis Arrowood
Fletcher, Martin J
Flick, Donald Franklin
Flock, Eunice Verna
Flora, Robert Montgomery
Florini, James Ralph
Floss, Heinz G
Flouret, George R
Fluck, Eugene Richards
Fluharty, Arvan Lawrence
Flynn, Thomas Geoffrey
Folch-Pi, Jordi
Folk, John Edard
Fonda, Margaret Lee
Fondy, Thomas Paul
Fong, Kuo-Lan
Fontaine, Thomas Davis
Fontenelle, Lydia Julia
Foote, Carlton Dan
Foote, Joel Lindsley
Foote, Murray Wilbur
Fopeano, John Vincent, Jr
Forchielli, Enrico Henry
Forist, Arlington Ardeane
Forman, Donald T
Formica, Joseph Victor
Forrest, Irene Stephanie
Forrest, Robert J
Forrey, Arden W
Forscher, Bernard Kronman
Forsdyke, Donald Roy
Foster, Donald Myers
Foster, George A, Jr
Foster, John McGaw
Foster, Joseph Franklin
Foster, Robert Joe
Foster, Thomas Salisbury
Fouche, Clarence Estes, Jr
Foulds, John Douglas
Fournier, Maurille Joseph, Jr
Fox, Allen Sander
Fox, Jack Jay
Fox, Jack Lawrence
Fox, Jay B, Jr
Foy, Robert Bastian
Fraenkel-Conrat, Jane E
Frajola, Walter Joseph
Francesconi, Ralph P
Francis, Faith Ellen
Francis, Marion David
Frank, Bruce Hill
Frank, David Stanley
Frank, Fred R
Frank, Leonard Harold
Frankel, Edwin N
Franklin, Arthur Edmund
Franklin, Robert Louis
Franklin, Samuel Gregg
Franson, Richard Carl
Frantz, Ivan DeRay, Jr
Franz, John Matthias
Franzblau, Carl
Franzblau, Daniel W
Frascella, Daniel W
Frazier, Patricia Dianne (Murphy)
Frea, James Irving
Fredrick, Jerome Frederick
Fredricks, Walter William
Free, Alfred Henry
Free, Charles Alfred
Freed, Myer
Freed, Simon
Freed, Virgil Haven
Freedman, Aaron David
Freedman, Murray H
Freeman, Karl Boruch
Freeman, Smith
Freer, Stephan T
Freisheim, James Harold
French, Dexter
French, Ian Wilfred
Frenkel, Rowland Barnes
Frenkel, Rene
Frerichs, Wayne Marvin
Fresco, Jacques Robert
Freund, Thomas Steven
Freundlich, Martin
Frey, Perry Allen
Fricke, Howard Henry
Fridovich, Irwin
Fried, Melvin

Fried, Rainer
Friedberg, Errol Clive
Friedberg, Felix
Frieden, Carl
Frieden, Earl
Frieden, Edward Hirsch
Friedkin, Morris Enton
Friedland, Joan Martha
Friedland, Melvyn
Friedman, Leonard
Friedman, Robert Bernard
Friedman, Selwyn Marvin
Friedmann, Herbert Claus
Fritsch, Carl Walter
Fritz, Herbert Ira
Fritz, Irving Bamdas
Fritz, Paul John
Friz, Carl T
Frobish, Lowell T
Froede, Harry Curt
Frohman, Charles Edward
Frolich, Per Keyser
Fromageot, Henri Pierre-Marcel
Fromm, Herbert Jerome
Fruton, Joseph Stewart
Fu, Shou-Cheng Joseph
Fuchs, James Allen
Fugate, Kearby Joe
Fujii, Akira
Fujimoto, George Iwao
Fujimura, Robert
Fukushima, David Kenzo
Fukuyama, Thomas T
Fulco, Armand J
Fuleki, Tibor
Fuller, Gerald Maxwell
Fuller, Ray W
Fuller, Rufus Clinton
Fuller, Wallace Hamilton
Fullington, J Garrin
Fullmer, Curtis Sheridan
Furano, Anthony
Furchgott, Robert Francis
Furfine, Charles Stuart
Furlong, Clement Eugene
Furlong, Norman Burr, Jr
Furth, John J
Futrell, Mary Feltner
Futterman, Sidney
Gaballah, Saeed S
Gabay, Sabit
Gabbay, Edmond J
Gabbert, Paul George
Gabler, Walter Louis
Gabriel, R Othmar
Gadsden, Richard Hamilton
Gaebler, Oliver Henry
Gaffney, Barbara Jean
Gagnon, Arthur
Gaines, Robert D
Gale, Nord Loran
Galivan, John H
Gall, William Einar
Gallagher, Thomas Francis
Gallo, Duane Gordon
Gallo, Robert C
Gallopo, Andrew Robert
Gally, Joseph Anthony
Galster, William Allen
Galsworthy, Peter Robert
Galton, Suzanne A
Gambal, David
Gamble, James Lawder, Jr
Gamble, Samuel James Reeves
Gamble, Wilbert
Gan, Jose Cajilig
Gander, John E
Gang, Henry
Ganguly, Rama
Ganis, Frank Michael Gangarosa
Ganschow, Roger Elmer
Garbers, David Lorn
Garbutt, John Thomas
Garcia, Eugene N
Gardner, David Arnold
Gardner, Harold Wayne
Gardner, Richard Lynn
Garg, Hari Gopal
Garibaldi, John Attilio
Garlich, Jimmy Dale
Garrett, Reginald Hooker
Garrick, Michael D
Garst, Josephine Burgis
Gates, Robert Leroy
Gates, Ronald Eugene
Gatewood, Dean Charles
Gaucher, George Maurice
Gaudette, Leo Edward
Gawienowski, Anthony Michael
Gawron, Oscar
Gaylor, James Leroy
Gear, Adrian R L
Gear, James Richard
Gefter, Malcolm Lawrence
Geiger, Edwin Otto
Geiger, Paul Jerome
Gelboin, Harry Victor
Geller, Arthur Michael

Geller, David Melville
Genaux, Charles Thomas
George, Elmer, Jr
George, Harvey
Gerber, Louis P
Gergely, John
Gerhart, John C
Gerin, John Louis
Gerloff, Eldean D
Gerlt, John Alan
Gerritsen, Theo
Gershbein, Leon Lee
Gershoff, Stanley Norton
Gershon, Herman
Gerwin, Brenda Isen
Geschwind, Irving I
Gesteland, Raymond Frederick
Getz, Godfrey S
Geyer, Robert Pershing
Ghering, Mary Virgil
Gholson, Robert Karl
Ghosh, Amal Kumar
Gianetto, Robert
Gibbons, Barbara Hollingsworth
Gibbons, Katherine Bond
Giblin, Frank Joseph
Gibson, Audrey Jane
Gibson, David Mark
Gidez, Lewis Irwin
Gigliotti, Helen Jean
Gilbertson, John R
Gilboe, Daniel Pierre
Gilboe, David Dougherty
Gilder, Helena
Giles, Ralph E
Gilgan, Michael Wilson
Gilham, Peter Thomas
Gill, David Michael
Gill, James Wallace
Gillchriest, William Clarence
Gillespie, Elizabeth
Gillett, James Warren
Gilpin, Richard William
Gilvarg, Charles
Giner-Sorolla, Alfreso
Gingell, Ralph
Gingery, Roy Evans
Ginoza, Herbert S
Ginsburg, Victor
Ginsler, Victor William
Giorgio, Anthony Joseph
Giotta, Gregory John
Girotti, Albert William
Gittes, Hyman Raphael
Givner, Morris Lincoln
Gizis, Evangelos John
Gjerstad, Gunnar
Gladner, Jules A
Glaid, Andrew Joseph, III
Glaser, Jay Arthur
Glaser, Charles
Glaser, Luis
Glasky, Alvin Jerald
Glass, Robert Louis
Glassman, Harold Nelson
Glaze, Robert P
Gledhill, Robert Hamor
Gledhill, William Emerson
Gleiter, Melvin Earl
Glenn, Joseph Leonard
Glick, David M
Glitz, Dohn George
Glockin, Vera Charlotte
Glogovsky, Robert L
Glomset, John A
Gluck, Louis
Goatley, James Leon
Gochman, Nathan
Godfrey, Paul Russell
Godin, Claude
Goebel, Walther Frederick
Goff, Sidney
Gold, Allen Morton
Gold, Alvin Hirsh
Gold, Norman Irving
Goldberg, Alfred L
Goldberg, Melvin Leonard
Goldberg, Morris H
Goldberg, Nelson D
Goldberger, Robert Frank
Goldbloom, David Ellis
Golder, Richard Harry
Goldfarb, Abraham Robert
Goldfine, Howard
Goldman, Dexter Stanley
Goldman, Peter
Goldsby, Richard Allen
Goldsmith, Dale Preston Joel
Goldstein, Allan L
Goldstein, Fred Bernard
Goldstein, Gilbert
Goldstein, Jack
Goldstein, Menek
Goldsworthy, Patrick Donovan
Goldthwait, David Atwater
Goldwasser, Eugene
Goll, Darrel Eugene

Golub, Ellis Eckstein
Golubow, Julius
Golumbic, Calvin
Gomatos, Peter John
Goodfellow, Robert David
Goodfriend, Theodore L
Goodgal, Sol Howard
Goodhue, Charles Thomas
Gooding, Ronald Harry
Goodman, DeWitt Stetten
Goodman, Howard Michael
Goodman, Irving
Goodman, Norman L
Goodridge, Alan G
Goor, Ronald Stephen
Gordon, Annette Waters
Gordon, Edgar Stillwell
Gordon, Harold Thomas
Gordon, Malcolm Wofsy
Gordon, Milton Paul
Gordon, Nathan
Gore, Edward Michael
Gorelic, Lester Sylvan
Goren, Howard Joseph
Gorman, Robert Roland
Gornall, Allan Godfrey
Gorsica, Henry Jan
Gorski, Jack
Gorski, Theodore William
Gortner, Ross Aiken, Jr
Gotterer, Gerald S
Gottlieb, A Arthur
Gotto, Antonio Marion, Jr
Gough, Patricia Marie
Gould, Bernard Sidney
Gould, Robert Gordon
Gounaris, Anne Demetra
Goyco, Daubon, Jose A
Goyings, Lloyd Samuel
Goz, Barry
Grabske, Robert Jerold
Grady, Joseph Edward
Graf, Lloyd Herbert
Grafius, Melba A
Graham, Lewis Texada, Jr
Graham, Walter Donald
Grainger, Robert Ball
Gralla, Jay Douglas
Gram, Theodore Edward
Gramera, Robert Eugene
Grandjean, Carter Jules
Granick, Sam
Grant, Donald R
Grant, Norman Howard
Grattan, Jerome Francis
Grauer, Amelie L
Graven, Stanley N
Graves, Charles Norman
Graves, Donald J
Gray, Clarke Thomas
Gray, Ernest David
Gray, Gary D
Gray, Gary M
Gray, Gary Ronald
Gray, Irving
Gray, Michael William
Gray, Peter Norman
Gray, Reed Alden
Gray, Robert Dee
Grebner, Eugene Ernest
Green, Beverley R
Green, David Ezra
Green, Harry
Green, John H
Green, Maurice
Green, Melvin Howard
Green, Saul
Greenawalt, John W
Greenbaum, Lowell Marvin
Greenberg, David Morris
Greenberg, Goodwin Robert
Greenberg, Jacob
Greenberg, Leonard Jason
Greenberg, Louis Donald
Greenberg, Samuel Mendel.
Greenblatt, Irving Jules
Greene, Daryle E
Greene, Frank Clemson
Greene, Lewis Joel
Greene, Ronald C
Greenfield, Leonard Julian
Greenfield, Robert Edman, Jr
Greenfield, Seymour
Greengard, Olga
Greengard, Paul
Greenhouse, Walter Van Vleck
Greenlee, Lorance Lisle
Greenwald, Isidor
Greenwell, Benjamin Elmer
Greenwood, Frederick C
Gregg, Charles Thornton
Gregory, Francis Joseph
Gregory, John Delafield
Gribbins, Myers Floyd
Griffin, Amos Clark
Griffin, Charles Campbell
Griffin, John Henry

BIOCHEMISTRY

Griffin, Martin John
Griffin, Travis Barton
Griffith, Owen Malcolm
Griffith, Robert Bell
Griffith, Thomas
Grigsby, William Redman
Grill, Herman A
Grinna, Lynn Sharon
Grinnell, Frederick
Grisham, Charles Milton
Grisolia, Santiago
Groce, John Wesley
Grodsky, Gerold Morton
Grooms, Thomas Albin
Groskopf, William R
Gross, Erhard
Gross, Michael Alan
Gross, Stephen Richard
Grossman, Lawrence
Grossman, Steven Harris
Groves, William Ernest
Gruemer, Hanns-Dieter
Gruenstein, Eric Ian
Grunbaum, Benjamin Wolf
Grundy, Scott Montgomery
Gruner, Rudolf Richard
Gryder, Rosa Meyersburg
Guarino, Armand John
Gubler, Clark Johnson
Guchhait, Ras Bihari
Guerguian, John Leo
Gunning, Barbara E
Gunsalus, Irwin Clyde
Gunsberg, Ephraim
Gunter, Claude Ray
Gunter, Donald Albert
Gunther, Jay Kenneth
Gupta, Naba K
Gurd, Frank Ross Newman
Gurin, Samuel
Gurley, Lawrence Ray
Gurpide, Erlio
Gustine, David Lawrence
Gucho, Sidney Jack
Guthrie, Christine
Guthrie, George Drake
Guthrie, John Daulton
Guthwin, Hyman
Gutmann, Helmut Rudolph
Gutman, Helene Augusta Nathan
Guyer, Kenneth Eugene, Jr
Haab, Walter
Haagen-Smit, Arie Jan
Haas, Erwin
Haas, Gerhard Julius
Haas, Ward John
Haber, Bernard
Hac, Lucile (R)
Hackett, Patricia Lou
Hadd, Harry Earle
Hadler, Herbert Isaac
Haft, David Edward
Hagan, James Joyce
Hageman, James Howard
Hagen, Paul Beo
Hagen, Richard Eugene
Hager, Lowell Paul
Hageman, Dwain Douglas
Haggard, James Herbert
Haggerty, James Francis
Hagopian, Miasnig
Haining, Joseph Leo
Hainline, Adrian, Jr
Hainski, Martha Barrionuevo
Hajra, Amiya Kumar
Hakomori, Sen-Itiroh
Haldar, Dipak
Haley, Boyd Eugene
Haley, Edward Everett
Halkerson, Ian D K
Hall, Frank Foy
Hall, Iris Beryl Haddon
Hall, Leo McAloon
Hall, Michael Oakley
Hall, Ross Hume
Hall, Walfred Sigmon
Haller, Walter Knowlton
Halpern, Ephraim Philip
Halpern, Bernard
Halpin, Kenneth M
Halsall, Hallen Brian
Hamar, Dwayne Walter
Hamilton, Ian Robert
Hamilton, James Guthrie
Hamilton, James Wilburn
Hamilton, Jean A
Hamilton, Pat Brooks
Hamilton, Paul Barnard
Hamilton, Robert Houston

Hamkalo, Barbara Ann
Hamm, Thomas Edward
Hammel, Jay Morris
Hammer, Frank E
Hammersted, Roy H
Hammes, Gordon G
Hammond, Ray Kenneth
Hamori, Charles Edward, Jr
Hampton, Alexander
Hanahan, Donald James
Hancock, Ronald Lee
Handler, Philip
Hanig, Ruth Belle Cohn
Hankes, Lawrence Valentine
Hanlon, Lester
Hanlon, David Paul
Hanly, Mary Sue
Hanly, W Carey
Hanson, John Norman
Hansen, Roger Gaurth
Hanson, Douglas MacArthur
Hanson, Kenneth Ralph
Hanson, Richard Steven
Hanson, Ronald W
Hanson, Ronald Lee
Hanson, Russell Floyd
Hanson, Thomas Lawrence
Hanstein, Walter Georg
Hao, Yu-Lee
Hapner, Kenneth D
Harary, Isaac
Hardy, James D
Hardy, Ralph Wilbur Frederick
Hargrave, Paul Allan
Harharan, Palghat Venkteswar
Harkness, Donald R
Harkins, John McLay
Harmon, Denham
Harmon, George Andrew
Harold, Franklin Marcel
Harold, Stephen
Harper, Alfred Edwin
Harper, Edwin T
Harper, Elvin
Harper, Harold Anthony
Harper, Robert Peter
Harrill, Inez Kemble
Harrington, Glenn William
Harris, Don Navarro
Harris, Lewis Philip
Harris, Robert Allison
Harris, Sheldon Richard
Harris, Wayne G
Harris, Edward David
Harris, Henry Earl
Harris, John Edward
Harris, Joseph
Harrison, Edward Merle
Harrison, Helen Coplan
Harrison, John Henry, IV
Harrison, William Henry
Harshman, Sidney
Hart, Kathleen Therese
Hart, Lewis Thomas
Hart, Richard Allen
Harte, Robert Adolph
Hartenstein, Roy
Hartline, Richard
Hartman, David Robert
Hartman, Frederick Cooper
Hartman, Iclal Sirel
Hartman, Ronald Earl
Hartman, Standish Chard
Hartnett, John (Conrad)
Hartsuck, John Ann
Hartzler, Eva Ruth
Harvey, Cecil Claude
Harvey, Richard Alexander
Hascall, Vincent Charles, Jr
Haschemeyer, Audrey Elizabeth Veazie
Hash, John H
Haskell, Albert Russell
Hass, Louis F
Hassell, John Robert
Hastings, Albert Baird
Hatanaka, Masakazu
Hatch, Frederick Tasker
Hatcher, Herbert John
Hatefi, Youssef
Hatfield, G Wesley
Haugaard, Ella Shwartzman
Haugaard, Niels
Haurowitz, Felix (Michael)
Hauser, George
Hausmann, Ernest
Haust, Heinz (Heinrich) Leonhard
Havir, Evelyn A
Havran, Robert Thomas

Hawker, Charles Davis
Hawkins, Richard Albert
Hawkinson, Stuart Winfield
Hawrylewicz, Ervin J
Hayashi, James Akira
Hayashi, Teru Terry
Hayden, George A
Hayes, Andrew Wallace
Hayes, Donald Charles
Hayes, Dora Kruse
Hayes, Edward Lee
Hayes, John William
Hayes, Joseph Edward, Jr
Haymovits, Asher
Hays, Edwin Everett
Hays, Ruth Lanier
Hazzard, DeWitt George
Heacock, Ronald A
Heady, Fred Clark
Heagy, George McNeice
Hearing, Vincent Joseph, Jr
Hearn, Walter Russell
Heath, Edward Charles
Heath, Robert Louis
Hecht, Sidney Michael
Heck, Nicole Begin
Hedges, Dorothea Huseby
Hedgcoth, Charlie, Jr
Hedrick, Jerry Leo
Hedin, Paul A
Heegaard, Erik Villhelm
Heer, Leonard J
Hefmann, Erich
Hegarty, Charles Paul
Hegeman, George D
Hegre, Carman Stanford
Hegsted, David Mark
Hegyeli, Andrew Francis
Heikkila, Richard Elmer
Heidelberger, Charles
Heidelberger, Margaret Louise
Heim, Harold Clifford
Heimberg, Murray
Heiner, Ralph
Heineke, Herbert Raymond
Heinrich, Milton Rollin
Heinrichs, W LeRoy
Heinstein, Peter
Heintz, Roger Lewis
Heisler, Charles Rankin
Heitz, James Robert
Helbert, James Raymond
Hele, Priscilla
Helinski, Donald Raymond
Hellener, Christopher Walter
Hellman, Kenneth P
Hellriegel, John Curtis, Jr
Helmreich, Ernst
Helwig, Harold Lavern
Heming, Arthur Edward
Hendershot, William Fred
Henderson, Ellen Jane
Henderson, Joseph Franklin
Henderson, Lavell Merl
Henderson, Louis E
Henderson, Ralph Joseph, Jr
Henderson, Rogene Faulkner
Henderson, Thomas Otis
Henderson, Thomas Richard
Hendler, Richard Wallace
Hendrickson, Herman Stewart, II
Hendry, Richard Allan
Heninger, Richard Wilford
Henkart, Pierre
Henning, Susan June
Henricks, Donald Maurice
Henry, Harry James
Henry, Richard Joseph
Henges, David John
Henze, Robert Edgerton
Herbert, Raymond
Herbert, Edward
Herbert, Jack Durnin
Herbert, Michael
Herbert, William Fitts
Herbst, Edward John
Herdon, John Francis
Herr, Earl Binkley, Jr
Herrett, Richard Allison
Herriott, Harold Keith
Herriott, Jon R
Herriott, Roger Moss
Herrmann, Klaus Manfred
Herrmann, Harvey R
Herscovics, Annette Antoinette
Herschman, Harvey Lawrence
Hersh, Louis Barry
Hershey, John William Baker
Herz, Fritz
Hess, Eugene Lyle
Hess, George Paul
Hess, Helen Hope
Hess, Sidney Marvin
Hess, Walter Cohen
Heymann, Hans
Heyliger, Peter George
Heywood, Stuart Mackenzie
Hibler, Charles Philip

Hickenbottom, John Powell
Hickey, Richard James
Hicks, Patricia Fain
Hicks, Sonja Elaine
Hieber, Thomas Eugene
Higgins, Edwin Stanley
Higgins, Joan Amy
Higgins, Harvey (M), Jr
Higgins, Michael Lee
High, Robert James
Hilborn, David Alan
Hildebrand, Frank Childs
Hildebrand, John Grant, III
Hiles, Richard Allen
Hilf, Russell
Hill, Doris M
Hill, Charles Horace, Jr
Hill, Charles Whitacre
Hill, Donald Lynch
Hill, Douglas Calvert
Hill, Doyle Eugene
Hill, Edward Jay
Hill, Franklin D
Hill, George Carver
Hill, John Mayes, Jr
Hill, Robert James
Hill, Robert Lee
Hill, Robert Mathew
Hill, Roberta Bieler
Hill, Walter Ensign
Hilleoat, Brian Leslie
Hillegass, Donald Victor
Hills, C Loran
Hinkle, David Currier
Hinkle, Peter Currier
Hinman, Jack Wiley
Hill-Samli, Marqueta
Himer, Paul Edward
Hitltbran, Robert Comegys
Hilton, Mary Anderson
Hinman, Norman Dean
Hinz, Charles F
Hirschberg, Carlos Benjamin
Hirschberg, Erich
Hirschman, Albert
Hirschman, Hans
Hitchings, George Herbert
Hixson, Susan Harvill
Hiyama, Tetsuo
Hnilica, Lubomir Sidonius
Ho, Chien
Ho, Peter Peck Koh
Hoagland, Vincent DeForest, Jr
Hoard, Donald Wayne
Hoard, Donald Ellsworth
Hoare, Robert
Hobbs, Donald Clifford
Hoberman, Henry Don
Hobkirk, Ronald
Hoch, Frederic Louis
Hoch, George Edward
Hoch, James Alfred
Hoch, Sallie O'Neil
Hochstadt, Joy
Hochstein, Paul Eugene
Hockett, Robert Casad
Hodgins, Daniel Stephen
Hoddon, John M
Hoeksema, Walter David
Hoerman, Kirk Conklin
Hof, Liselotte Bertha
Hof, John Frederick
Hoftee, Patricia Anne
Hoffman, Allan Jordan
Hoffman, David J
Hoffman, Philip
Hoffmann, Dietrich
Hoffmann, Hans-Peter Gerhard
Hoffmann, Klaus Heinrich
Hofmann, Theo
Hofstee, Barend Hendrik Jan
Hogan, John Mathew
Hogenkamp, Henricus Petrus C
Hogg, James Felter
Hohmann, Philip George
Hohnadel, David Charles
Hokin, Lowell Edward
Hokin-Neaverson, Mabel
Holbrook, David James, Jr
Holden, Joseph August
Holden, Joseph Thaddeus
Holland, Robert Joseph
Holleman, James William
Holleman, William H
Hollenberg, Paul Frederick
Hollett, Charlotte R
Hollett, Andrew
Hollocher, Thomas Clyde, Jr
Holm, Robert Eric
Holm, Ralph Theodore
Holman, David G

Holmes, Richard
Holmes, William Farrar
Holmes, William Leighton
Holmlund, Chester Eric
Holmquist, Barton
Holmquist, Richard
Holoubek, Viktor
Holstein, Arthur G
Holsten, Richard David
Holt, James Allen
Holten, Darold Duane
Holten, Virginia Zewe
Homann, H Robert
Homann, Peter H
Homer, George Mohn
Honeycutt, Richard Carl
Hoober, John Kenneth
Hood, Samuel Lowry
Hopfinger, Anton J
Hopkins, Thomas R
Hopper, Sarah Priestly
Horecker, Bernard Leonard
Hornemann, Ulfert
Horner, Alan Alfred
Horner, William Harry
Horowitz, Jack
Horowitz, Martin I
Horowitz, Myer George
Horowitz, Norman Harold
Horowitz, Paul Martin
Horton, Horace Robert
Horwitt, Benjamin Norman
Horwitt, Max Kenneth
Horwitz, Susan Band
Hosein, Esau Abbas
Hoskins, Francis Clifford George
Hoskins, Dale Douglas
Hosler, Charles Frederick, Jr
Hospelhorn, Verne D
Hossain, Afzal
Hotchkiss, Rollin Douglas
Hotta, Shoichi Steven
Hou, Ching-Tsang
Houchin, Ollie Boyd
Houk, Albert Edward Hennessee
House, Theodore Ching-Teh
Houston, Forrest Gish
Houston, L L
Howard, Barbara V
Howard, Bruce David
Howard, Charles Frank, Jr
Howard, James Bryant
Howard, John Charles
Howard, Robert Eugene
Howell, Ralph Rodney
Howland, John LaFollette
Howton, David Ronald
Hoyle, Merrill Cassius
Hrazdina, Geza
Hsia, Sung Lan
Hsiao, Theodore Ching-Teh
Hsu, Howard Huai Ta
Hsu, Jeng Mein
Hsu, Quei-Shiow
Hsu, Robert Ying
Hsu, Wen-Tah
Hu, Alfred Soy Lan
Huang, Charles T L
Huang, Ching-hsien
Huang, Laura Chi
Huang, Pien-Chien
Huang, Sylvia Lee
Huang, Yung-Chen
Hubbard, Colin D
Hubbard, Richard W
Hubbard, William Ralph
Huber, Rueben Eugene
Huber, Wolfgang
Hudson, Billy Gerald
Huennekens, Frank Matthew, Jr
Huff, Jesse William
Huffaker, Ray C
Huffman, George Wallen
Huffman, Max Niel
Hug, George
Huggins, Clyde Griffin
Hughes, Luther Bertram, Jr
Hughes, Walter Lee, Jr
Hugli, Tony Edward
Huijing, Frans
Huisman, Titus Hendrik Jan
Hulcher, Frank H
Hullinger, Clifford
Hultquist, Donald Elliott
Humes, John Leroy
Hummel, Brian Christopher W
Humphreys, Wallace F
Hung, Paul P
Hunsley, James Ray
Hunt, Sara McClanahan
Hunter, Francis Edmund, Jr
Hurd, Suzanne Sheldon
Hurlbert, Robert Boston
Hurley, William Charles
Hurst, Robert Osmond
Hussa, Robert Oscar
Hussain, Mehdi Hajiyani
Huston, Charles K

Huszar, Gabor
Hutcheson, Eldridge Tilmon, III
Hutchinson, Franklin
Hutsell, Thomas Carlyle
Hutton, John James, Jr
Hyde, Paul Martin
Hyman, Richard W
Hyndman, Lee Allen
Iacono, James M
Ibanez, Manuel Luis
Ibsen, Kenneth Howard
Idell-Wenger, Jane Arlene
Idler, David Richard
Iijima, Herbert
Ikeda, George J
Ikels, Kenneth G
Imondi, Anthony Rocco
Imsande, John
Inagami, Tadashi
Inamine, Edward Seiyu
Inchiosa, Mario Anthony, Jr
Infante, Anthony A
Ingle, Morton Blakeman
Ingledew, William Michael
Ingles, Charles James
Ingram, Jordan Miles
Ingram, Robert Lee
Ingram, Vernon Martin
Ingram, William Prentiss, Jr
Inscoe, Joseph Kenneth
Insull, William, Jr
Ionescu, Lavinel G
Iodice, Arthur Alfonso
Ip, Clement Cheung-Yung
Ip, Margot Morris
Irish, Don DeLance
Irreverre, Filadelfo
Irvin, Joseph Logan
Irving, Charles Clayton
Irwin, William Elliot
Ise, Charles Masao
Isenberg, Irvin
Isler, Henri Gustave
Isselbacher, Kurt Julius
Issenberg, Phillip
Itaba, Kibe
Ito, Takeru
Itschner, Kenneth Frank
Itter, Stuart
Ives, David Homer
Iyengar, Raja M
Jack, Robert Cecil Milton
Jackel, Simon Samuel
Jacks, Thomas Jerome
Jackson, Albert William
Jackson, Carlton Darnell
Jackson, Craig Merton
Jackson, James Albert
Jackson, Johnny
Jackson, Larry Lee
Jackson, Peter H
Jackson, Richard Lee
Jackson, Richard Lee
Jackson, Sanford (Hugh)
Jacob, Theodore August
Jacobs, Francis Albin
Jacobs, Richard L
Jacobs, Ross
Jacobsen, Donald Weldon
Jacobsohn, Gert Max
Jacobson, Ada Leah Hyne
Jacobson, Bernard Jerome
Jacobson, Bruce Shell
Jacobson, Don Richard
Jacobson, Elaine Louise
Jacobson, Gail M
Jacobson, Herbert (Irving)
Jacobson, Karl Bruce
Jacobson, Myron Kenneth
Jacobson, Ralph Allen
Jacobus, William Edward
Jacoli, Giulio Guido
Jaffe, Werner G
Jainchill, Jerome
Jakoby, William Bernard
James, Emory Albarte, Jr
James, Harold Lee
James, Jesse
Jamieson, Graham Archibald
Jandorf, Bernard Joseph
Janik, Borek
Jarett, Leonard
Jarvis, Francis George
Jass, Herman Earl
Jaworski, Ernest George
Jaworski, Jan Guy
Jean, Marcel
Jeanloz, Roger William
Jeffay, Henry
Jefferson, William Emmett, Jr
Jeffrey, John J
Jelinek, Bohdan
Jellinck, Peter Harry
Jen, Joseph Jwu-Shan
Jencks, William Platt
Jenifer, Franklyn G
Jen-Jacobson, Linda
Jenkins, Kenneth James William

Jenkins, Lloyd Theodore
Jenkins, Winborne Terry
Jenness, Robert
Jennings, Allen Lee
Jennings, Robert Kimmel
Jensen, Elwood Vernon
Jensen, Richard Grant
Jensen, Robert Gordon
Jensen, Roy A
Jerina, Donald M
Jesaitis, Margeris Adomas
Jewett, Sandra Lynne
Ji, Tae Hwa
Jiang, Nai-Siang
Jirgensons, Bruno
Jirku, Helmut
Joel, Cliffe David
Johns, David Garrett
Johns, Philip Timothy
Johnson, Alan J
Johnson, Archie Doyle
Johnson, B Connor
Johnson, Bobby Ray
Johnson, Carl Lynn
Johnson, Corinne Lessig
Johnson, David Freeman
Johnson, Dewey, Jr
Johnson, Donald Edgar
Johnson, Donovan Earl
Johnson, Garland A
Johnson, John Alexander
Johnson, John Arnold
Johnson, Julius Earl
Johnson, LaVell R
Johnson, Paul Hickok
Johnson, Ralph M, Jr
Johnson, Richard James
Johnson, Robert Michael
Johnson, Ronald Roy
Johnson, Sharon Leijoy
Johnson, Willard Jesse
Johnston, Carter Dupuy
Johnston, Harry Henry
Johnston, John Marshall
Johnston, Marilyn Frances Meyers
Johnston, Martha M
Johnston, Michael Adair
Johnston, Robert Benjamin
Johnstone, Rose M
Joklik, Wolfgang Karl
Jomain-Baum, Mireille
Jondorf, Werner Robert
Jones, Chase Breese
Jones, David Hartley
Jones, Don Carl
Jones, John Patrick
Jones, Evan Earl
Jones, George Henry
Jones, James Donald
Jones, Mary Ellen
Jones, Oliver William
Jones, Richard Theodore
Jones, Robin Richard
Jones, Theodore Charles
Jones, Theodore Harold Douglas
Jonsson, Haldor Turner, Jr
Joos, Richard W
Jordan, John Patrick
Jordan, Wayne Robert
Jorgensen, George Norman
Joshi, Jayant Gopal
Joshi, Vasudev Chhotalal
Jost, Jean-Pierre
Jourdian, George William
Juhasz, Roderick
Jukes, Thomas Hughes
Jungalwala, Firoze Barnanshaw
Jungas, Robert Leando
Jungman, Richard A
Justice, Parvin
Kaback, Howard Ronald
Kabat, David
Kabat, Elvin Abraham
Kachmar, John Frederick
Kadaba, Pankaja Kooveli
Kadin, Harold
Kadis, Barney Morris
Kadis, Vincent William
Kaeppner, Werner Martin
Kaesberg, Paul Joseph
Kagan, Herbert Marcus
Kagan, Jacques
Kahlenberg, Arthur
Kahn, A Clark
Kahn, Kenneth
Kaiser, Armin Dale
Kaiser, Ivan Irvin
Kaji, Akira
Kaji, Hideko (Katayama)
Kakari, Sophia
Kako, Kyohei Joe
Kalab, Miloslav
Kalckar, Herman Moritz
Kalf, George Frederick
Kallen, Roland Gilbert
Kalman, Burton Jay
Kalman, Sumner Myron
Kalnitsky, George

Kalra, Vijay Kumar
Kaltenbach, John Paul
Kambli, Vijaykant Bhagwan
Kamin, Henry
Kaminski, Edward Jozef
Kaminski, Laurence Samuel
Kammen, Harold Oscar
Kampschmidt, Ralph Fred
Kanabrocki, Eugene Ladislaus
Kandutsch, Andrew August
Kane, James Francis
Kane, John Power
Kaneda, Toshi
Kaneshiro, Edna Sayomi
Kanfer, Julian Norman
Kao, Kung-Ying Tang
Kao, Oranda Hai-Wen
Kaper, Jacobus M
Kaplan, Ann Esther
Kaplan, Barry Hubert
Kaplan, Emanuel
Kaplan, Harvey
Kaplan, Morris Aaron
Kaplan, Nathan Oram
Kaplan, Phyllis Deen
Kaplan-Koch, Dora Deborah
Kapsalis, John George
Karam, Jim Daniel
Karasek, Marvin A
Karavolas, Harry J
Kargl, Thomas E
Karkas, John D
Karnovsky, Manfred L
Karp, Warren B
Karpatkin, Simon
Karpel, Richard Leslie
Karu, Alexander Edwin
Kasbekar, Dinkar Kashinath
Kashket, Eva Ruth
Kashket, Shelby
Kasinsky, Harold Edward
Kaslow, Haven DeLoss
Kasper, Charles Boyer
Kass, Leon Richard
Kassell, Beatrice
Kassner, Richard J
Kasting, Robert
Katchen, Bernard
Kates, Joseph R
Katsoyannis, Panayotis G
Katz, Albert Barry
Katz, Edward
Katz, Joseph
Katz, Morris Howard
Katz, Sam
Katzen, Howard M
Kaufman, Philip Aaron
Kauffman, Frederick C
Kaufman, Bernard
Kaufman, Bernard Tobias
Kaufman, Seymour
Kaufmann, Anthony J
Kay, Cyril Max
Kay, Ernest Robert MacKenzie
Kay, Robert Eugene
Kayne, Fredrick Jay
Kazal, Louis Anthony
Kazarinoff, Michael N
Keele, Bernard B, Jr
Keeney, Dennis Raymond
Kefalides, Nicholas Alexander
Keirns, James Jeffery
Keith, Eaden Francis
Kelleher, Philip Conboy
Kelleher, William Joseph
Kelleher, Raymond Joseph, Jr
Keller, Elizabeth Beach
Keller, Frederick Albert, Jr
Keller, Patricia J
Keller, Stephen Jay
Kelley, George Greene
Kelley, John Francis, Jr
Kelley, Leon A
Kellogg, Thomas Floyd
Kelly, Jeffrey John
Kelly, Robert Frank
Kelsall, Margaret Aston
Kemmerer, Arthur Russel
Kemp, Anthony Lionel
Kemp, John Daniel
Kemp, Robert Grant
Kenkare, Divaker B
Kennedy, Eugene P
Kennedy, John Elmo, Jr
Kennedy, Maurice Venson
Kennelly, Bruce
Kenney, Francis T
Kensler, William Clark
Kensler, Charles Joseph
Kent, Claudia Marie
Kent, Stephen Brian Henry
Kenyon, Alan J
Keough, Kevin Michael William

23

Keresztes-Nagy, Steven
Kern, Harold L
Kern, Milton
Kerr, Sylvia Jean
Kersey, William Hewell
Kessel, David Harry
Kessler, Gerald
Kessler, Michael J
Ketchie, Delmer O
Ketring, Darold L
Kezdy, Ferenc J
Khalifah, Raja Gabriel
Khan, Abdul Waheed
Kidder, George Wallace
Kiefer, Helen Chilton
Kielley, William Wayne
Kieras, Fred John
Kies, Marian Wood
Kihara, Hayato
Kilgour, Gordon Leslie
Kim, Hyun Dju
Kim, Ki-Han
Kim, Sangduk
Kim, Yee Sik
Kim, Young Tai
Kimbrough, Theo Daniel, Jr
Kimelberg, Harold Keith
Kimmel, Joe Robert
Kimmich, George Arthur
Kind, Charles Albert
Kindel, Paul Kurt
Kindt, Thomas James
King, Charles Glen
King, Charles Miller
King, Jonathan Stanton
King, Robert Edward
King, Tsoo E
Kingdon, Henry Shannon
Kingsbury, David Wilson
Kinney, Roland Walter
Kinoshita, Jin Harold
Kirby, Edward Paul
Kirby, Kenneth William
Kirdani, Rashad Y
Kirkham, William R
Kirksey, Avanelle
Kirkwood, Samuel
Kirsch, Jack Frederick
Kirschenbaum, Donald (Monroe)
Kirschenbaum, Joel Jerome
Kirschman, John C
Kirshner, Norman
Kirtley, Mary Elizabeth
Kishimoto, Yasuo
Kisliuk, Roy Louis
Kistler, Wilson Stephen, Jr
Kit, Saul
Kita, Donald Albert
Kitabchi, Abbas E
Kitchen, Hyram
Kitos, Paul Alan
Kittinger, George William
Kitto, George Barrie
Kitzes, George
Kiyasu, John Yutaka
Kizer, Donald Earl
Klaas, Rosalind Amelia (Mrs G Weber Schimpff)
Klaassen, Dwight Homer
Klapper, Michael H
Klebanoff, Seymour J
Klebe, Robert John
Klee, Lucille Holljes
Klein, LeRoy
Klein, Peter Douglas
Klein, Sigrid Marta
Kleinhofs, Andris
Kleinman, Roberta Wilma
Kleinschmidt, Walter John
Kleinsmith, Lewis Joel
Kleyn, Dick Henry
Kliewer, Walter Mark
Kline, Edward Samuel
Kline, Irene Tabitha
Kline, Oral Lee
Klingman, Jack Dennis
Klopfenstein, William Elmer
Klopping, Hein Louis
Klosterman, Harold Joseph
Kluepfel, Dieter
Kluge, Harlan Lyle
Kluger, Ronald H
Kmetec, Emil Philip
Knaak, James Bruce
Knauff, Raymond Eugene
Knecht, Albert T
Knight, Katherine Lathrop
Knoblock, Edward C
Kobayashi, Yutaka
Koch, Robert B
Kochakian, Charles Daniel
Kochwa, Shaul
Kodras, Rudolph
Koechlin, Bernard Alphons

Koehn, Paul V
Koenig, Virgil Leroy
Koeppe, Owen John
Koeppe, Roger Erdman
Koerber, Walter Ludwig
Koerner, James Frederick
Kohler, Diona Heather
Kohl, David Martin
Kohlhaw, Gunter B
Koide, George Oscar
Koide, Samuel Saburo
Koizumi, Kiyomi
Kokatnur, Mohan Gundo
Kokesh, Fritz Carl
Kolbeck, Ralph Carl
Kollman, George
Kominek, Leo Aloysius
Konigsbacher, Kurt S
Konigsberg, William Henry
Kono, Tetsuro
Kontaxis, Nicholas E
Koobs, Dick Herman
Koppelman, Ray
Koppenheffer, Thomas Lynn
Koreman, Stanley G
Koritz, Seymour Benjamin
Korn, Edward David
Kornberg, Roger David
Kornberg, Arthur
Korngtuh, Steven E
Krakow, Joseph S
Krakauer, Henry
Kraemer, Louise Margaret
Kraml, Michael Joseph Anthony
Kramer, Fred Russell
Kramer, Elizabeth
Krall, Albert Raymond
Krane, Stephen Martin
Krasna, Alvin Isaac
Krasner, Joseph
Krasnow, Frances
Kratochvil, Clyde Harding
Krauel, Kathryn Kreamer Kopf
Kraus, Lorraine Marquardt
Krause, Leonard Anthony
Krause, Reginald Frederick
Kravitz, Edward Arthur
Kraybill, Herman Fink
Kream, Jacob
Krebs, Edwin Gerhard
Kredich, Nicholas M
Kreiser, Thomas Harry
Kremzner, Leon T
Kremitzky, Thomas Anthony
Kress, Lawrence Francis
Kretchmar, Arthur Lockwood
Kretchner, Norman
Krichevsky, Micah I
Krieg, Daniel R
Krieger, Carl Henry
Krinsky, Norman Irving
Krishna, Gopal X
Krisko, Istavan
Kroeker, Warren Dean
Kroll, Emanuel
Krueger, Keatha Kathrine
Krueger, Robert Carl
Kruger, Fred Albert
Krulwich, Terry Ann
Krumdieck, Carlos L
Krupka, Richard Morley
Ku, Edmond Chiu-Choon
Ku, Han San
Kubena, Leon Franklin
Kubin, Rosa
Kuby, Stephen A
Kuchinskas, Edward Joseph
Kuchmak, Myron
Kuck, John Frederick Read, Jr
Kuehl, LeRoy Robert
Kuehn, Glenn Dean
Kuehner, Carl Albert
Kuettner, Klaus
Kuiken, Kenneth (Alfred)
Kuksis, Arnis
Kull, Fredrick J
Kulwich, Roman
Kumar, Soma
Kumar, Sudhir
Kumar, Suriender
Kumaroo, Kuziyiethu Krishnan
Kun, Ernest
Kundu, Nakuleswar

Kunitz, Moses
Kunkel, Ralph Edward
Kunkel, Harriott Orren
Kuntzman, Ronald Grover
Kuo, Harng-Shen
Kuo, Mau H
Kuo, Jyh-Fa
Kupiecki, Floyd Peter
Kupke, Donald Walter
Kuramitsu, Howard Kikuo
Kurmick, Nathaniel Bertrand
Kurtzman, Ralph Harold, Jr
Kushinsky, Stanley
Kusiak, John Warren
Kusyk, Christine Johanna
Kutchai, Howard C
Kuwahara, Steven Sadao
Kyker, Granvil Charles
Kyte, Jack Ernst
Labbe, Robert Ferdinand
Labrie, Fernand
LaBrosse, Elwood Henry
Lack, Leon
Lacko, Andras Gyorgy
Lacks, Sanford
Lada, Arnold
Laden, Karl
Ladisch, Rolf Karl
Laine, Roger Allan
Lajtha, Abel
Lakshmanan, Florence Lazicki
Lakshminarayanan, Krishnaiyer
Lam, Kwok-Wai
Lamberg, Stanley Lawrence
Lambert, Glenn Frederick
Lambert, James LeBeau
Lamberts, Burton Lee
Lamborg, Marvin
Lamden, Merton Philip
Lamoureux, Gerald Lee
Lamport, Derek Thomas Anthony
Lampson, George Peter
Lamy, Francois
Lanchantin, Gerard Francis
Landau, Bernard Robert
Lande, Saul
Landel, Aurora Mamaug
Landmann, Wendell August
Landon, Erwin Jacob
Lands, William Edward Mitchell
Lane, Alexander Z
Lane, Byron George
Lane, Gary (Thomas)
Lane, Malcolm Daniel
Lang, Joseph Herman
Langan, Thomas Augustine
Langdon, Robert Godwin
Lange, Charles Ford, Jr
La Noue, Kathryn P
Lansford, Edwin Myers, Jr
Lanyi, Janos K
Lanyi, Andre
Lapidus, Milton
Lara-Braud, Carolyn Weathersbee
Lardy, Henry Arnold
Larner, Joseph
Larrabee, Allan Roger
Larsen, Earl George
Larsen, James Bouton
Larson, Bruce Linder
Larson, David L
Larson, Rachel Harris (Mrs John Watson Henry)
Larson, Russell L
Lartigue, Donald Joseph
LaRue, Thomas A
Lascelles, June
Laseter, John Luther
Laskowski, Michael
Laskowski, Michael Jr
Last, Jerold Alan
Laszlo, John
Lata, Gene Frederick
Latimer, Steve B
Laughlin, Alice
Laughlin, Ethelreda R
Lavik, Paul Sophus
Law, John Harold
Lawford, George Ross
Lawrence, John McCune
Lawrence, Paul J
Lawrence, Robert Howard, Jr
Lawson, William Burrows
Layne, Donald Sainteval
Layne, Ennis C
Layton, Laurence Laird
Lazen, Alvin Gordon
Lazier, Catherine Beatrice
Lazzari, Eugene Paul
Lea, Michael Anthony
Leach, Byron Elwood
Leach, Franklin Rollin
Leary, John Dennis
Lease, Elmer John
Leaver, Frederick Wilson

LeBaron, Francis Newton
LeBherz, Herbert G
Lebowitz, Jacob
Leboy, Phoebe Starfield
Leden, Robert
Leder, Irwin Gordon
Leduc, Gerard
Lee, Agnes C
Lee, Charlotte Elizabeth Outland
Lee, Chong Sung
Lee, Chuan-Pu
Lee, Henry C
Lee, Howard Augustus
Lee, Jean Chor-Yin
Lee, John Chung
Lee, Kai-Lin
Lee, Kyu Yawp
Lee, Lih-Syng
Lee, Martin Jerome
Lee, Melvin
Lee, Michael John
Lee, Nancy Zee-Nee Ma
Lee, Norman David
Lee, Shaw-Guang Lin
Lee, Si Duk
Lee, Sung Gue
Lee, Sylvan Burton
Lee, Tee-Ping
Lee, Teh Hsun
Lee, Virginia Ann
Lee, Wei-Li S
Lee, Ya Pin
Lee, Yuan Chuan
Lee, Yuen San
Lee-Huang, Sylvia
Leeling, Jerry L
Lees, Helen
Lees, Howard
Lees, Robert S
Lees, Thomas Masson
Legator, Marvin (Seymour)
Lehman, Ernest Dale
Lehman, Israel Robert
Lehninger, Albert Lester
Lehoux, Jean-Guy
Lehrer, Harris Irving
Lehrer, Sherwin Samuel
Leibach, Fredrick Hartmut
Leicester, Henry Marshall
Leighty, Edith Gardner
Leiter, Joseph
Leitzmann, Claus
Leive, Loretta
Lembach, Kenneth James
Lemonde, Andre
Lenard, John
Lengyel, Peter
Lenhoff, Howard Maer
Lennarz, William Joseph
Lennox, Edwin Samuel
Lentz, Kenneth Eugene
Leon, Shalom A
Leonard, Charles Brown, Jr
Leonards, Jack Ralph
Leong, Kam Choy
Leoschke, William Leroy
LePage, Gerald Alvin
Lepp, Albert
Lerner, Aaron Bunsen
Lerner, Joseph
Lerner, Leon Maurice
Lesko, Stephen Albert
Lessard, James Louis
Lester, David
Lester, Robert Leonard
Letarte, Jacques
Letarte-Muirhead, Michelle
Leung, Philip Min Bun
Levander, Orville Arvid
Levedahl, Blaine Hess
Leveille, Gilbert Antonio
Levenbook, Leo
Levin, Herman Westley
Levin, Judith Goldstein
Levin, Louis
Levin, Samuel Joseph
Levintow, Leon
Levitz, Mortimer
Levy, Daniel
Levy, Hans Richard
Levy, Harvey Merrill
Levy, Milton
Levy, Robert I
Levy, Robert
Levy, Robert Sigmund
Lewbart, Marvin Louis
Lewin, Lawrence M
Lewis, Bertha Ann
Lewis, Donald Everett
Lewis, Florence Scott
Lewis, James Clement
Lewis, Jasper Phelps
Lewis, Marc Simon
Lewis, Roger Allen
Lewis, Roger Wolcott
Lewis, Roscoe Warfield
L'Heureux, Maurice Victor

Li, Choh Hao
Li, Jeanne B
Li, Li-Hsieng
Li, Lu Ku
Li, Steven Shoei-Lung
Li, Ting Kai
Li, Yu-Teh
Lianides, Sylvia Panagos
Liao, Shutsung
Libby, Paul Robert
Liberatore, Frederick Anthony
Liddell, Robert William, Jr
Lieberman, Seymour
Lien, Eric Louis
Liener, Irvin Ernest
Lienhard, Gustav E
Light, Albert
Light, Robley Jasper
Lightbody, James James
Likins, Robert Campbell
Lillehoj, Eivind B
Lilevik, Hans Andreas
Lin, Ada Wen-Shung Ma
Lin, Chi-Wei
Lin, Edmund Chi Chien
Lin, Fu Hai
Lin, Jiann-Tsyh
Lin, Leu-Fen Hou
Lin, Reng-Lang
Lin, Shin
Lin, Tsau-Yen
Lindenbaum, Arthur
Lindenblad, Gordon Eric
Lindquist, Robert Nels
Lindsay, Raymond H
Ling, Chung-Mei
Lingrel, Jerry B
Link, Karl Paul
Linker, Alfred
Linn, Stuart Michael
Linn, Tracy Claud
Linsky, Cary Bruce
Lionetti, Fabian Joseph
Lipke, Herbert
Lipmann, Fritz (Albert)
Lippel, Kenneth
Lippincott, Elizabeth Lois
Lipsett, Marie Nieft
Lipsky, Seymour Richard
Lipton, Samuel Harry
Lis, Adam W
Lis, Elaine Walker
Liss, Maurice
Listowsky, Irving
Litchfield, Charles Carter
Litman, Burton Joseph
Litman, Gary William
Litt, Michael
Little, Henry Nelson
Little, John Ernest
Littlewood, Barbara Shaffer
Litwack, Gerald
Liu, Adam Edwin Chiap Henn
Liu, Su-Chin Chang
Liu, Teh-Yung
Liu, Wen Chih
Livermore, Arthur Hamilton
Livermore, James Leslie
Lizardi, Paul Modesto
Ljungdahl, Lars Gerhard
Lloyd, Kenneth Oliver
Lloyd, Norman Edward
Loach, Paul A
Locke, Krystyna Kopaczyk
Locke, Raymond Kenneth
Lockhart, Haines Boots
Lodish, Harvey Franklin
Loeb, Lawrence Arthur
Loeb, Marilyn Rosenthal
Loewus, Frank A
Loftfield, Robert Berner
Logan, David Mackenzie
Loh, Horace H
Lolley, Richard Newton
Lombardo, Michael E
Lonberg-Holm, Knud Karl
London, Jack P
London, Morris
Long, Calvin Lee
Long, Cedric William
Long, George Louis
Long, James William
Long, Lawrence William
Long, Mary Jean
Longenecker, Herbert Eugene
Longenecker, John Bender
Longmire, William Joseph
Longmuir, Ian Stewart
Loo, Yen Hoong
Loomis, Walter David
Loper, Gerald Milton
Lopez-Santolino, Alfredo
Lorand, Laszlo
Lorenzetti, Olfeo J
Lorincz, Andrew Endre
Losee, Fred Lester
Lospalluto, Joseph John

Lotlikar, Prabhakar Dattaram
Lou, Marjorie Feng
Loucas, Spiro P
Loudfoot, James Herbert
Loudon, Gordon Marcus
Louis, Lawrence Hua-Hsien
Lovelace, C James
Lovell, Richard Arlington
Lovenberg, Walter McKay
Low, Barbara Wharton
Lowe, Richie Howard
Lowenstein, John Martin
Lowenstein, Leah Miriam
Lowenthal, Julius
Lower, Gerald Malcolm, Jr
Lowrey, Robert S
Lowry, Oliver Howe
Lowy, Bertram Alan
Lu, Anthony Y H
Lucas, Colin Cameron
Luchsinger, Wayne Wesley
Luckey, Thomas Donnell
Ludowieg, Julio
Ludvigsen, Bernhard (Thoger) Frants (Josef)
Ludwig, Martha Louise
Luecke, Richard William
Luisada-Opper, Anita Victoria
Lukemeyer, Jack Warren
Lukens, Lewis Nelson
Lukin, Marvin
Lumeng, Lawrence
Luman, Kenneth Dale
Lundberg, Walter Oscar Paul
Lundblad, Roger Lauren
Lundeen, Carl Victor, Jr
Lurie, Aron Osher
Lushbough, Channing Harden
Lusk, Joan Edith
Lust, George
Lustig, Bernard
Lutz, Wilson Boyd
Lyford, Sidney John, Jr
Lygre, David Gerald
Lyman, Richard Lee
Lyon, John Blakeslee, Jr
Lyon, Richard Hale
Ma, Pang-Fai
Maag, Dale D
Maass, Alfred Roland
MacColl, Robert Joseph
MacDonald, Alex Bruce
MacDonald, Donald Laurie
MacDonald, James Cameron
MacDonald, Norman Scott
MacDonald, Roderick Patterson
Macdonald, Timothy
MacFadyen, Douglas Archibald
MacGee, Joseph
MacGregor, Ronal Roy
MacInnis, Austin J
Mack, Lawrence Lloyd
Mackal, Roy Paul
Mackenzie, Charles Westlake, III
Mackenzie, Cosmo Glenn
Mackenzie, Julia Buzz
MacLeod, Robert Douglas
MacLeod, Robert Meredith
Macmanus, John Patrick
MacNintch, John Edwin
Macpherson, Lloyd Bertram
MacRae, Herbert F
Maddaiah, Vaddanahally Thimmaiah
Maddy, Kenneth Hilton
Madera-Orsini, Frank
Madhosingh, Clarence
Madison, James Thomas
Madsen, David Christy
Madsen, Neil Bernard
Maga, Joseph Andrew
Magee, Paul Terry
Magee, Steve Carl
Magee, Wayne Edward
Magee, William Lovel
Mager, Milton
Magill, Jane Mary (Oakes)
Magliulo, Anthony Rudolph
Magnuson, James Andrew
Mahajan, Damodar K
Mahajan, Kishan Paul
Mahler, Henry Ralph
Mahlum, Daniel Dennis
Mahoney, Joan Munroe
Mahowald, Theodore Augustus
Maickel, Roger Philip
Maier, Roger Philip
Main, Alexander Russell
Maio, Joseph James
Maitra, Shyamal Kumar
Maitra, Umadas
Majchrowicz, Edward
Majer, Jaroslav
Makarem, Anis H
Makman, Maynard Harlan
Malacinksi, George M
Malbica, Joseph Orazio
Maley, Frank
Maley, Gladys Feldott

Malicki, Carol Ann
Malins, Donald Clive
Malkin, Harold Marshall
Malkin, Leonard Isadore
Malkin, Richard
Mallery, Charles Henry
Mallette, Manney Frank
Mallin, Morton Lewis
Mandel, Lewis Richard
Mandeles, Stanley
Mandelstam, Paul
Mandl, Ines
Mandula, Barbara Blumenstein
Manery, Jeanne Forest
Mangan, George Francis, Jr
Mangelson, Farrin Leon
Mangum, John Harvey
Mani, Rama I
Maniatis, George Marinos
Maniscalco, Ignatius Anthony
Manly, Richard Samuel
Mann, George Vernon
Mann, Kenneth Gerard
Manner, Georg Karl
Mannering, Gilbert James
Manning, James Matthew
Manning, Jerry Edsel
Manning, Maurice
Manning, Robert Thomas
Mans, Rusty Jay
Mansour, Agnes Mary
Mansour, Tag Eldin
Manyan, David Richard
Mao, James Chieh Hsia
Marchalonis, John Jacob
Marchesi, Vincent T
Marciani, Dante Juan
Marco, Gino Joseph
Marcus, Abraham
Marcus, Carol Joyce
Marek, Jerry William
Maretzki, Andrew
Marfey, Sviatopolk Peter
Margolis, Richard Urdangen
Margolis, Sam Aaron
Margolis, Simeon
Margulies, Maurice
Mariani, Henry A
Marinetti, Guido V
Marini, Mario Anthony
Markiw, Roman Teodor
Markland, Francis Swaby, Jr
Marko, Arthur Myroslaw
Markovetz, Allen John
Marks, Bernard Herman
Marks, Paul A
Marks, Richard Henry Lee
Markus, Gabor
Markus, Helene Babad
Marlborough, David Ian
Marmur, Julius
Marquardt, Ronald Ralph
Marquez, Ernest Domingo
Marquez, Norman A
Marquis, Norman Ronald
Marra, Michael Dominick
Marrs, Barry Lee
Marschke, Charles Keith
Marsh, Connell Leroy
Marsh, John MacClenahan
Marsh, Julian Bunsick
Marsh, Walton Howard
Marshall, Garland Ross
Marshall, James John
Marshall, Lawrence Marcellus
Marshall, Lynmor Beverly
Marshall, Richard
Marsters, Roger Westcott
Marten, James Frederick
Martin, Arlene Patricia
Martin, Charles J
Martin, David Lee
Martin, Horace F
Martin, Jack E
Martin, John Lee
Martin, Julia Mae
Martin, Margaret Eileen
Martin, Michael McCulloch
Martin, Robert O
Martin, Roy Joseph, Jr
Martin, Sarah Smith
Martin, Stanley Morris
Martin, William Gerald
Martin, William Gilbert
Martindale, William Earl
Martinez-Carrion, Marino
Martinez Nadal, Noemi G
Martonosi, Anthony
Maruyama, Hitoshi
Marvel, John Thomas
Marx, Walter
Marx, George A
Marzluf, George A
Marzluff, William Frank, Jr
Masat, Robert James
Mashburn, Louise Tull
Mashburn, Thompson Arthur, Jr
Masken, James Frederick
Mason, Howard Stanley

Mason, Merle
Mason, Michael E
Mason, Morton Freeman
Mason, Norman Ronald
Mason, Reginald G, Jr
Massaro, Edward Joseph
Massey, Vincent
Masters, Bettie Sue Siler
Matacic, Slavica Smit
Mather, Adaline Nicoles
Mather, Jane H
Matheson, Alastair Taylor
Mathews, Christopher King
Mathews, Francis Scott
Mathews, Martin B
Mathies, James Crosby
Mathieu, Leo Gilles
Matschiner, John Thomas
Matsumoto, Charles
Matsuo, Robert R
Matta, Michael Stanley
Mattenheimer, Hermann G W
Matteo, Martha R
Matthews, Kathleen Shive
Matthews, Rowena Green
Matthijssen, Charles
Mattick, Joseph Francis
Mattick, Leonard Robert
Mattoon, James Richard
Mattson, Fred Hugh
Maturo, Joseph Martin, III
Matz, John J, Jr
Mautner, Henry George
Mavis, Richard David
Mavrides, Charalampos
Max, Stephen Richard
Maxon, William Densmore
Maxwell, Elizabeth Starbuck
Maxwell, Richard Elmore
May, Hubert Eugene
May, Sheldon William
Mayer, Manfred Martin
Mayer, Marion Sidney
Mayer, Steven Edward
Mayes, Jary S
Maynard, Donald Earle
Maynard, William Rose, Jr
Mayo, Joseph William
Mayol, Robert Francis
Mayron, Lewis Walter
Mazel, Paul
Mazumder, Rajarshi
Mazur, Abraham
Mazzone, Horace M
McAleer, William Joseph
McAllister, Harmon Carlyle, Jr
McArthur, Charles Stewart
McCaa, Connie Smith
McCall, John Temple
McCall, Keith Bradley
McCalla, Dennis Robert
McCallum, Roderick Eugene
McCaman, Richard Eugene
McCann, Daisy S
McCarl, Richard Lawrence
McCarthy, John Lawrence, Jr
McCarthy, Joseph Edward
McCarthy, Timothy Edward
McCarthy, William John
McCarty, Kenneth Scott
McCarty, Richard Earl
McCay, Paul Baker
McChesney, Evan William
McClary, Joseph Edward
McClintock, David K
McCluer, Robert Hampton
McClure, Claude
McClure, Michael Edward
McClurg, James Edward
McColloch, Robert James
McConn, Rita
McConn, Bruce
McConnell, David Graham
McConnell, Kenneth Paul
McConnell, Wallace Beverly
McCord, Tommy Joe
McCord, William Mellen
McCormick, Donald Bruce
McCormick, J Robert D
McCorquodale, Donald James
McCoy, Donald W
McCoy, Lowell Eugene
McCoy, Richard Hugh
McCoy, Sue
McCoy, Thomas Aylesbury
McCune, Ronald William
McDonagh, Jan M
McDonald, Francis Guy
McDonald, George Gordon
McDonald, Hugh Joseph
McDonald, John Kennely
McDonald, Margaret Ritchie
McDonald, Ted Painter
McElhaney, Ronald Nelson
McElroy, William David
McEvoy, Donald
McEwen, William Kirk
McFadden, Bruce Alden
McFall, Elizabeth

BIOCHEMISTRY

McFarland, James Thomas
McFarlane, Ellen Sandra
McGarry, John Denis
McGaughey, Charles Gilbert
McGeachin, Robert Lorimer
McGeer, Patrick L
McGilvery, Robert Warren
McGinnis, Arthur James
McGraw, Gerald Wayne
McGregor, Robert Finley
McGuinness, Eugene T
McGuire, Robert Francis
McIntire, Floyd Cottam
McIntire, Junius Merlin
McIntyre, Russell Theodore
McKay, Robert Harvey
McKay, Robert Wendell
McKee, Ralph Wendell
McKeehan, Wallace Lee
McKenna, Herbert, Jr
McKerns, Kenneth (Wilshire)
McKibbin, John Mead
McKigney, John Ignatius
McLaughlin, Calvin Sturgis
McLean, John Robert
McLean, Katharine Weidman
McLees, Byron D
McLellan, William L, Jr
McLemore, Willie O
McLennan, Barry Dean
McMahon, Daniel Stanton
McManus, Ivy Rosabelle
McMaster, Marvin Clayton, Jr
McMaster, Paul D
McMenamy, Rapier Hayden
McMillan, Paul Junior
McMorris, Frederick Arthur
McMurray, Walter Joseph
McNairy, William Colin Campbell
McNairy, Sidney A, Jr
McNall, Earl George
McNary, Robert Reed
McNutt, Walter Scott
McPhail, Murchie Kilburn
McPherson, James C, Jr
McRoberts, Milton R
McKorie, Robert Anderson
McShan, William Hartford
Mealey, Edward H
Mead, James Franklyn
Means, Gary Edward
Means, James Austin
Mecca, Christyna Emma
Mechanic, Gerald
Medina, Miguel Angel
Medzihradsky, Fedor
Meezan, Elias
Mego, John L
Mehl, John Wilbur
Mehler, Alan Haskell
Mehrotra, Bam Deo
Mehta, Paul Martin, Jr
Meienhofer, Johannes Arnold
Meighen, Edward Arthur
Meikle, Richard William
Meinke, Wilmon William
Meins, Frederick, Jr
Meister, Alton
Mela, Leena Marja
Melachouris, Nicholas
Melbye, Susanne Warner
Melcher, Irving
Melchior, Jacklyn Butler
Melchior, Norten Cass
Melera, Peter William
Melius, Paul
Mell, Galen P
Mellin, Theodore Nelson
Mellors, Alan
Melnick, Daniel
Meltzer, Herbert Lewis
Melville, Donald Burton
Melvin, Robert Burrow
Mende, Thomas Julius
Mendicino, Joseph Frank
Mendoza, Celso Enriquez
Menon, Aravindakshan I
Menzies, Robert Allen
Merdinger, Emanuel
Merola, A John
Merriam, Esther Virginia
Merrick, Joseph M
Merrifield, Robert Bruce
Merriman, Charles Richard
Mersmann, Harry John
Mertz, Edwin Theodore
Mertz, Walter
Messier, Robert Louis
Messineo, Luigi
Messmer, Dennis A
Metrione, Robert M
Metzenberg, Robert Lee, Jr
Metzger, H Peter
Metzger, Henry
Metzger, Robert P

Metzler, David Everett
Meyer, Delbert Eugene
Meyer, Franz
Meyer, Haruko
Meyer, Karl
Meyer, Ralph Roger
Meyer, William Laros
Mezei, Catherine
Mgbodile, Marcel Ume
Mhatre, NageshShanmao
Michael, Harry Oscar
Michael, William R
Mickelsen, Olaf
Miech, Ralph Patrick
Miedema, Eddy
Mieyal, John Joseph
Migicovsky, Bert Baruch
Mihalyi, Elemer
Milch, Lawrence Jacques
Miles, James Lowell
Militzer, Walter Ernest
Miller, Bradford
Miller, Charles G
Miller, Edward Godfrey, Jr
Miller, Elizabeth Eshelman
Miller, Glenn Joseph
Miller, Herbert Kenneth
Miller, Herman T
Miller, James Eugene
Miller, Jon Philip
Miller, Joyce Mary
Miller, Kent D
Miller, Leon Lee
Miller, Lila
Miller, Lorraine Theresa
Miller, Lowell D
Miller, Oscar Neal
Miller, Richard Lee
Miller, Richard Pressly
Miller, Richard Wilson
Miller, Ronald Lee
Miller, Sanford Arthur
Miller, Terry Lee
Miller, Thomas Lee
Miller, Wilbur Hobart
Miller, William Louis
Millett, Francis Spencer
Millette, Robert Loomis
Milligan, Larry Patrick
Mills, Gordon Candee
Mills, John Blakely, III
Mills, John Norman
Milner, Alice N
Milo, George Edward
Milstein, Julie Block
Milstone, Jacob Haskell
Minard, Frederick Nelson
Minton, Martin Lloyd, Jr
Mirone, Leonora
Misra, Hara Prasad
Mistry, Sorab Pirozshah
Mitacek, Eugene Jaroslav
Mitchel, Ronald Edward John
Mitchell, Alexander Rebar
Mitchell, Earl Douglass, Jr
Mitchell, George Weston
Mitchell, Herschel Kenworthy
Mitchell, Jack Harris, Jr
Mitchell, Robert Alexander
Mitra, Sankar
Mitz, Milton Aaron
Mittag, Thomas Waldemar
Miwa, Thomas Kanji
Miyada, Don Shuso
Mize, Charles Edward
Mizuno, Nobuko Shimotori
Model, Peter
Moffia, David Joseph
Moffitt, Robert Allan
Mohr, Scott Chalmers
Mohrenweiser, Harvey Walter
Mokrasch, Lewis Carl
Moldave, Kivie
Molinary, Samuel Victor
Moline, Sheldon Walter
Molnar, Janos
Momparler, Richard Lewis
Monder, Carl
Moner, John George
Monroe, Pearle Arvel
Montalvo, Francisco Emilio
Montgomery, Morris William
Montgomery, Rex
Montie, Thomas C
Monty, Kenneth James
Mookerjea, Salen
Mooney, Paul David
Moore, Blake William
Moore, Cyril L
Moore, Erin Colleen
Moore, Frank Archer
Moore, Patricia Ann
Moore, Peter Bartlett
Moore, Richard Owen

Moore, Stanford
Moorehead, Wells Rufus
Moos, Carl
Mooz, Elizabeth Dodd
Morais, Rejean
Morck, Roland Anton
Morehouse, Alpha L
Morehouse, Margaret Gulick
Morell, Pierre
Morell, Samuel Allan
Morgan, Howard E
Morgan, Page Wesley
Morgan, Paul Harper
Morgan, Thomas Edward, Jr
Mori, Kanaka Fred
Morley, Colin Godfrey Dennis
Moro, D James
Morrill, Gene A
Morris, Allan J
Morris, David Robert
Morris, Harold Paul
Morris, John Emory
Morris, Manford D
Morris, Roy Owen
Morris, Sidney Machen, Jr
Morrisett, Joel David
Morrison, George Robert
Morrison, John Coulter
Morrison, Joseph Louis
Morrison, Martin
Morrissey, Bruce William
Morse, Lewis David
Morse, Lura Myra
Mortenson, Leonard Earl
Mortimore, Glenn Edward
Morton, Bruce Eldine
Morton, Helen Janet
Mosbach, Erwin Heinz
Moscarello, Mario Antonio
Moscatelli, Ezio Anthony
Moschera, John Anthony
Moses, Henry A
Mosesson, Michael W
Moss, Bernard
Moss, Claude Wayne
Moss, Lloyd Kent
Moss, Melvin Lane
Mosteller, Raymond Dee
Moulton, Bruce Carl
Moury, Daniel Norman
Mowles, Thomas Francis
Moxon, Alvin Lloyd
Moyed, Harris S
Moyer, Rudolph Henry
Mozersky, Samuel M
Mraz, Frank Rudolph
Muck, George A
Mudd, John Brian
Mudd, Stuart Harvey
Mueller, Arthur Jacob
Mueller, Gerald Conrad
Mueller, Helmut
Mueller, Joseph Robert
Muench, Karl Hugo
Muhler, Joseph Charles
Mui, Paul Ting-Kai
Muirhead, Merle E
Muldoon, Thomas George
Mule, Salvatore Joseph
Mulford, Dwight James
Mulhausen, Hedy Ann
Mullen, Joseph David
Muller-Eberhard, Hans Joachim
Muller-Eberhard, Ursula
Mullinix, Kathleen Patricia
Mumma, Ralph O
Munavalli, Somashekhar
Munk, Vladimir
Munns, Theodore Willard
Munro, Hamish N
Muntz, John Adolph
Murano, Genesio
Murayama, Makio
Murdock, Archie Lee
Murdock, Larry Lee
Murphey, William Howard
Murphy, Alexander James
Murphy, Beverley (Elaine) Pearson
Murphy, John F
Murphy, John Joseph
Murphy, Marjory Beth
Murphy, Patrick Joseph
Murphy, Terence Martin
Murray, Robert Edward
Murray, Robert Kincaid
Murty, Mahadi Raghavandrarao V
Murty, Veeraraghavan Krishna
Mushak, Paul
Mushinski, Joseph Frederic
Musser, Samuel John
Muus, Jytte Marie
Mycek, Mary J
Myers, Dirck V

Myers, Lyle Leslie
Myint, Than
Myoda, Toshio Timothy
Myron, Duane R
Nabb, Dale Preston
Nacarrato, William Frank
Nadler, Kenneth David
Nagel, Glenn M
Nager, Urs Felix
Nagy, Bela Ferenc
Naimark, George Modell
Nair, Padmanabhan Padmanabhan
Nakada, Daisuke
Nakada, Henry Isao
Nakamoto, Tokumasa
Nakashima, Tadayoshi
Nandan, Rajiva
Nandedkar, Arvindkumar Narhari
Naqvi, Sayid Mahmoodul Hasan
Narahara, Hiromichi Tsuda
Narrod, Stuart Allan
Nash, Harold Anthony
Nasir-Ud-Din, Nasir
Nayeh, Shihadeh Nasri
Neagle, Lyle H
Neal, Arthur Leslie
Neal, Robert A
Needleman, Saul Ben
Neelakantan, Lakshmanan
Neelin, James Michael
Neely, Brock Wesley
Neeman, Moshe
Neet, Kenneth Edward
Neff, Alven William
Neiderhiser, Dewey Harold
Neidhardt, Frederick Carl
Neidlemann, Saul L
Neims, Allen Howard
Nelsestuen, Gary Lee
Nes, William Robert
Nelson, Charles A
Nelson, Dennis Raymond
Nelson, Donald John
Nelson, Edward Blake
Nelson, Eldon Carl
Nelson, Gary Joe
Nelson, George Humphry
Nelson, Thomas Evar
Nelson, Walter Ludwig
Nemerson, Yale
Nemer, Martin Joseph
Nettleton, Donald Edward, Jr
Neufeld, Elizabeth Fondal
Neufeld, Harold Alex
Neuhaus, Francis Clemens
Neuhaus, Otto Wilhelm
Neuman, William Frederick
Neumann, George Joseph
Neumann, Norbert Paul
Neurath, Hans
Neve, Richard Anthony
Nevyas, Jacob
Newbrun, Ernest
Newburgh, Robert Warren
Newell, Jon Albert
Newman, David John
Newman, Howard Abraham Ira
Newman, Jack Huff
Newman, Robert Alwin
Newmark, Harold Leon
Newmark, Marjorie Zeiger
Newton, William Anson, Jr
Nichoalds, George Edward
Nichol, Charles Adam
Nichol, Christina Janet
Nicholas, Harold Joseph
Nicholls, Doris Margaret
Nichols, George, Jr
Nichols, Jack Loran
Nicholson, Larry Michael
Nickerson, Walter John
Nicolaides, Nicholas
Nicolson, Margery O'Neal
Niederman, Robert Aaron
Niedermeier, William
Niederpruem, Donald J
Niehaus, Walter G, Jr
Nielsen, Forrest Harold
Nielsen, Harald Christian
Nielsen, Larry Dennis
Nielsen, Eldon Denzel
Nigam, Vijai Nandan
Nikaido, Hiroshi
Nikiforuk, Gordon
Nisen-Hamilton, Marit
Nimni, Marcel Efraim
Nirenberg, Marshall Warren
Nishihara, Mutsuko
Nishikawa, Alfred Hiriooshi
Nishimura, Jonathan Sei
Nishizawa, Edward Eichi
Nisselbaum, Jerome Seymour
Nitecki, Danute Emilija
Niu, Mann Chiang
Noall, Matthew Wilcox
Noble, Ernest Pascal
Noble, Nancy Lee
Noda, Lafayette Hachiro

Nolan, Chris
Nolan, Linda Lee
Noland, Jerre Lancaster
Noller, Harry Francis, Jr
Noltmann, Ernst August
Nooden, Larry Donald
Noonan, Kenneth Daniel
Norcia, Leonard Nicholas
Nordby, Harold Edwin
Nordin, John Hoffman
Nordin, Philip
Nordlie, Robert Conrad
Nordschow, Carleton Deane
Nordyke, Ellis Larrimore
Norman, Anthony Westcott
Norman, Philip Sidney
Norris, Frank Arthur
Norris, Thomas Elfred
Northrop, John Howard
Northrop, Robert L
Norton, Scotty Jim
Nossal, Nancy
Notation, Albert David
Noteboom, William Duane
Notides, Angelo C
Notrica, Solomon
Noval, Joseph James
Novelli, Guerino David
Novick, William Joseph, Jr
Novikoff, Alex Benjamin
Novoa, William Brewster
Nowak, Thomas
Nuenke, Richard Harold
Nussbaum, Alexander Leopold
Nussbaum, Siegfried
Nutter, William Ermal
Nygaard, Oddvar Frithjof
Nyhan, William Leo
Ober, Robert Elwood
O'Brien, Paul J
O'Brien, Peter J
O'Brien, Thomas W
Ockerse, Ralph
Ockerse, Paul Ralph
O'Connell, Paul William
O'Connor, John Dennis
O'Connor, John Francis
O'Connor, Timothy Edmond
O'Dell, Boyd Lee
Odense, Paul Holger
O'Donnell, James Francis
O'Donnell, Vincent Joseph
O'Donovan, Gerard Anthony
Oelshlegel, Frederick James, Jr
Oesper, Peter
Oesterling, Myrna Jane
O'Farrell, Helen Krogull
Ofengand, Edward James
Ogilvie, James William, Jr
Ogle, James D
Ogur, Maurice
Ohe, Keiji
Ohnishi, Tsuyoshi
Oien, Helen Grossbeck
Okerholm, Richard Arthur
O'Leary, Gerard Paul, Jr
O'Leary, Marion Hugh
O'Leary, William Michael
Oleinick, Nancy Landy
Oleson, Jerome Jordan
Olins, Ada Levy
Oliphant, Donald Edward
Oliphant, Edward Eugene
Oliver, Denis Richard
Oliver, Eugene Joseph
Olsen, Kenneth Wayne
Olsen, Richard William
Olson, Anita Cora
Olson, Arthur Olaf
Olson, Mark Obed Jerome
Olson, Merle Stratte
Olson, Robert Eugene
Olson, William Arthur
Omata, Robert Rokuro
O'Neal, Charles Harold
O'Neill, John Joseph
Ong, Eng-Bee
Ontko, Joseph Andrew
Oppenheimer, Norman Joseph
Opperman, James Alex
Opperman, Robert Arthur
Oratz, Murray
Orcutt, Donald Adelbert
Orcutt, Fred Scott, Sr
Orengo, Antonio
Oreskes, Irwin
Orlando, Joseph Alexander
Ormsbee, Richard Armstrong
Oro, Juan
Oronsky, Arnold Lewis
Orr, James Cameron
Orten, James M
Ortwerth, Beryl John
Osbahr, Albert J, Jr
Osborn, Mary Jane
Oshiro, Yuki
Oster, Mark Otho

Ostwald, Rosemarie
Osuch, Mary Ann V
Oswald, Edward Odell
Otani, Theodore Toshiro
Otero, Raymond B
Ottke, Robert Crittenden
Ottoboni, Minna Alice
Ousterhout, Lawrence Elwyn
Outlaw, Henry Earl
Ove, Peter
Overby, Lacy Rasco
Overturf, Merrill L
Owades, Joseph Lawrence
Owen, Charles Archibald, Jr
Owen, Stanley Paul
Owens, Kenneth
Oxender, Dale LaVern
Oyama, Jiro
Oyama, Vance I
Oyler, James Russell
Pabst, Michael John
Pace, Norman R
Pace Asciak, Cecil
Pachman, Elliott A
Packer, Lester
Packett, Leonard Vasco
Packham, Marian Aitchison
Padron, Jorge Louis
Page, Edouard
Paik, Woon Ki
Paine, Clair Maynard
Painter, Robert Hilton
Pairent, Frederick William
Paisley, Nancy Sandelin
Pal, Bimal Chandra
Pall, Martin L
Pallansch, Michael J
Palmer, Frederick B St Clair
Palmer, Graham
Palmer, Grant H
Palmer, John Warren
Palmer, Keith Henry
Palmer, Winifred G
Palmes, Edward Dannelly
Pamer, Treva Louise
Pan, Samuel Cheng
Panagides, John
Pande, Shri Vardhan
Pandya, Mahendra Kodarlal
Pannbacker, Richard George
Panos, Charles
Papaconstantinou, John
Papadopoulos, Nicholas M
Papageorge, Evangeline (Thomas)
Papahadjopoulos, Panayotis Demetrios
Papaioannou, Stamatios E
Papas, Takis S
Papastephanou, Constantin
Papirmeister, Bruno
Papkoff, Harold
Pappenheimer, Alwin Max, Jr
Paquin, Roger Joseph Alfred
Paranchych, William
Parcells, Alan Jerome
Pardee, Arthur Beck
Pardini, Ronald Shields
Pargaonkar, Padmaker Shankar
Parikh, Indu
Park, Charles Rawlinson
Park, James Theodore
Park, Jane Harting
Park, Won Kil
Parke, Donna
Parker, Charles J, Jr
Parker, Frank S
Parker, Herbert Edmund
Parks, Robert Emmett, Jr
Parlow, Albert Francis
Parrish, Donald Baker
Parrish, Frederick Charles, Jr
Parson, William Wood
Passaniti, John Martin
Passonneau, Janet Vivian
Pasterczyk, William Robert
Pastuszyn, Andrzej
Patel, Mulchand Shambhubhai
Patel, Sharad
Paterson, Alan Robb Phillips
Pathak, Keshav Dattatray
Pathak, Madhukar
Patrick, Richard Allen
Patterson, Elizabeth Knight
Patterson, Ernest Leonard
Patterson, Manford Kenneth, Jr
Paul, Benoy Bhushan
Paul, Iain C
Paul, Jeddeo
Paulsen, Elizabeth Charlotte
Paulson, Gaylord D
Paulson, James Carsten
Paulsrud, John Reynold
Paulton, Richard John Laurance
Pavlos, John
Pawelek, John Mason
Payne, Anita H
Payne, Merle G
Paynter, Malcolm James Benjamin
Paz, Mercedes Aurora

Pazur, John Howard
Peabody, Richard Arthur
Peacock, Andrew Clinton
Peacock, Milton O
Peake, Clinton J
Peanasky, Robert Joseph
Pearlman, William Henry
Pearlmutter, Anne Frances
Pearson, John Richard
Pecci, Joseph
Peck, Ernest James, Jr
Peckham, William Dierolf
Peczon, Benigno David
Pedersen, Peter L
Pedersen, Svend
Peeler, Herbert Tremble
Peets, Edwin Arnold
Peifer, James J
Peisach, Jack
Pekarek, Robert Sidney
Penhoet, Edward Etienne
Pennell, Robert Brown
Penner, Peter Edwin
Penniall, Ralph
Penniston, John Thomas
Pensky, Jack
Pentz, Ella Irene
Peraino, Carl
Perdue, Henry Stafford
Perdue, James F
Pereira, Joseph
Perisho, Clarence R
Perkins, Harold Jackson
Perkins, John Phillip
Perkinson, Jesse Dean, Jr
Perlgut, Louis E
Perlish, Jerome Seymour
Perlman, David
Perlman, Robert
Perry, Billy Wayne
Perry, Malcolm Blythe
Perry, Thomas Lockwood
Persky, Harold
Person, Philip
Pessen, Helmut
Pestka, Sidney
Peter, James Bernard
Petering, Harold George
Peterkofsky, Alan
Peterkofsky, Beverly
Peters, James Milton
Peters, John Henry
Peters, Theodore, Jr
Peterson, C Denis
Peterson, Daniel Walter
Peterson, Dennis Randall
Peterson, Durey Harold
Peterson, Elbert Axel
Peterson, Glen Ervin
Peterson, Julian Arnold
Peterson, Merlin Henry
Peterson, Neal Alfred
Peterson, Rudolph Nicholas
Peticlerc, Claude Jean
Petrack, Barbara Kepes
Petsko, Gregory Anthony
Pettinga, Cornelius Wesley
Petzold, Edgar
Pfeffer, Morris
Pfleger, Gerard David
Pfleger, Raymond C
Phalen, William Edmund
Phansalkar, Sadashiv Vinayak
Phares, Cleveland Kirk
Pharo, Richard Levers
Pharriss, Bruce Bailey
Phelps, Richard A
Phillips, Allen Thurman
Phillips, David Richard
Phillips, John Howell, Jr
Phillips, Marshall
Phillips, William Armin
Philpot, Richard Michael
Piatigorsky, Joram Paul
Pickrell, John A
Pierce, Jack Vincent
Pierce, John Grissim
Pieringer, Ronald Arthur
Pierre, Leon L
Pietruszko, Regina
Piez, Karl Anton
Pigman, Ward
Pilgeram, Laurence Oscar
Pilsbury, Harold C
Pincus, Jack Howard
Pincus, Joseph B
Pine, Ellen Kann
Pine, Leo
Pine, Martin J
Pinto, John
Pinto, P Vincent C
Piper, Walter Nelson
Pipkin, George Erwin
Pisano, John Joseph
Piszkiewicz, Dennis
Pitesky, Isadore
Pitha, Paula Marie
Pitlick, Frances Ann

Pitot, Henry C, III
Pizer, Lewis Ivan
Piziak, Veronica Kelly
Plagemann, Peter Guenter Wilhelm
Plager, Clinton J
Plantz, Philip Edward
Plapp, Bryce Vernon
Plate, Janet Margaret
Plaut, Gerhard Wolfgang Eugen
Plechner, Sophie L
Plowman, Kent Milton
Plummer, Thomas H, Jr
Pocker, Anna
Podleski, Thomas Roger
Poet, Raymond B
Poffenbarger, Philip Lynn
Pogell, Burton M
Poincelot, Raymond Paul, Jr
Polan, Carl E
Polatnick, Jerome
Polis, Beryl David
Pollack, Robert Leon
Pollak, Otakar Jaroslav
Pollard, Donald Ray
Pollard, Harvey Bruce
Polley, John Richard
Pomerantz, Seymour Herbert
Pomeranz, Yeshajahu
Pon, Ning Gin
Poole, Brian Howard
Poole, Doris Theodore
Popenoe, Edwin Alonzo
Popjak, George Joseph
Poretz, Ronald David
Porter, Charles Jack
Porter, Curt Culwell
Porter, John Willard
Porter, William L
Portner, Allen
Portugal, Franklin H
Posner, Aaron Sidney
Posner, Herbert S
Possmayer, Fred
Potter, Richard Lyle
Potter, Van Rensselaer
Potts, Albert Mintz
Poulsen, Lawrence Leroy
Pour-El, Akiva
Powell, Gary Lee
Powers, John Clancey, Jr
Powers, Wendell Holmes
Pownall, Henry Joseph
Poyer, Joe Lee
Pradhan, Tapas Kumar
Prager, Morton David
Prahl, James William
Prairie, Richard Lane
Praissman, Melvin
Prather, Charles Wayne
Preiss, Benjamin
Preiss, Jack
Prejean, Joe David
Prescott, Benjamin
Prescott, John Mack
Prescott, Lansing M
Presnell, Alexander Koehne
Pressey, Russell
Pressman, Berton Charles
Prestayko, Archie William
Preston, James Faulkner, III
Price, Alan Roger
Price, Frederick William
Price, Howard Charles
Price, Jack D
Price, Paul Arms
Price, Ralph Lorin
Price, Vincent Edward
Price, Winston Harvey
Priest, David Gerard
Pringle, Ross Barton
Printz, Morton Philip
Prior, Ronald Leon
Pritchard, Ernest Thackeray
Probst, Gerald William
Prockop, Darwin J
Proctor, Charles Mahan
Proksch, Gary J
Prough, Russell Allen
Proulx, Pierre R
Pruett, Patricia Onderdonk
Pruitt, Kenneth M
Prusoff, William Herman
Pruzansky, Jacob Julius
Pryor, William Austin
Prytz, Bo
Pscheidt, Gordon Robert
Psychoyos, Stacy
Pudelkiewicz, Walter Joseph
Pudles, Julio
Puett, John David
Pullman, Maynard Edward
Punch, James Darrell
Purcell, Albert Ernest
Purdom, Martha Elda
Purdy, Richard Eugene
Purdy, Robert H
Purkhiser, E Dale
Purvis, John L

Puski, Gabor
Putman, Frank William
Putterman, Gerald Joseph
Pye, Edward Kendall
Pyke, Thomas Richard
Pyler, Richard Ernst
Pynadath, Thomas I
Pynes, Alvan Wesley
Pynes, Gene Dale
Quackenbush, Forrest Ward
Quaife, Mary Louise
Quarles, Richard Hudson
Quastel, Juda Hirsch
Quener, Sherry Fream
Quenelle, Alan Courtland, Jr
Quinn, Caroline Elisabeth
Quinn, James Gerard
Quiocho, Florante A
Quissno, George L
Quissell, David Olin
Quo, Sih-Gwan
Rabinovitz, Marco
Rabinovitz, Israel Nathan
Rabinovitz, Jesse Charles
Rabinowitz, Joseph Loshak
Rabinowitz, Murray
Rachele, Julian Richard
Rachinsky, Michael Richard
Rackis, Joseph John
Racker, Efraim
Radding, Charles Meyer
Radhakrishnamurthy, Bhandaru
Radin, Norman Samuel
Radke, Frederick Herbert
Radloff, Harold David
Radzialowski, Frederick M
Rafelson, Max Emanuel, Jr
Rafter, Gale William
Ralston, Douglas Edmund
Ragheb, Hussein S
Ram, Gerson Louis
Ram, J Sri
Ragland, James Benjamin
Ragland, William Lauman, III
Raha, Chitta Ranjan
Raheja, Manu Chaturmal
Rainey, John Marion, Jr
Raju, Pullarkat Krishnan
Rajagopalan, K V
Rake, Adrian Vaughan
Rall, Stanley Carlton, Jr
Ramachandran, Janakiraman
Ramachandran, Subramania
Ramaley, Robert Folk
Rambaut, Paul Christopher
Ramp, Warren Kibby
Rand, Phillip Gordon
Randerath, Kurt
Rao, Ananda G
Rao, Ghanta Nageswara
Rapaport, Eliezer
Rapp, Gustav William
Rapp, Waldean G
Rappaport, Harry P
Rapport, Maurice M
Rasco, Marilyn Arnott
Raskas, Heschel Joshua
Rasmussen, Edith Svoboda
Rasmussen, Oscar Gustav
Rathbun, William B
Rathbun, Premila
Ratliff, Charles Ray
Ratner, Sarah
Raud, Heinz Randar
Raun, Ned S
Rausch, David John
Raveed, Dan
Ravel, Joanne Macow
Rawalay, Surjan Singh
Rawat, Arun Kumar
Rawitch, Allen Barry
Ray, Lee Edmisten
Ray, Paul Dean
Ray, Paul H
Ray, Verne A
Razzell, Wilfred Edwin
Read, Kenneth Richard Hodgson
Read, Merrill Stafford
Read, Virginia Hall
Reagor, John Charles
Reazin, George Harvey, Jr
Reber, Elwood Frank
Reckel, Rudolph P
Reddy, Bandaru Sivarama
Reddy, Venkat N
Reddy, William John
Redlich, Dorothy Von
Redman, Colvin Manuel
Redwood, William Raymond
Reeck, Gerald Russell
Reed, Donald James
Reed, Donald Eugene
Reed, George Henry
Reed, Jura Kuttis
Reed, Lester James
Reed, Peter William
Reed, Warren Douglas
Reese, Floyd Ernest

Reese, Harni Darwin
Reese, Nathan Allan
Reeves, Henry Courtland
Reeves, Ogle Raymond
Reeves, William John, Jr
Regen, David Marvin
Register, Ulma Doyle
Regnier, Frederick Eugene
Reich, Melvin
Reichert, Leo E, Jr
Reichmann, Manfred Eliezer
Reid, Bobby Leroy
Reid, Brian Robert
Reid, Parlane John
Reid, Ted Warren
Reimann, Erwin M
Reimann, Jessica Elizabeth
Reinosa Fuller, Jose Angel
Reiser, Raymond
Reiser, Sheldon
Reiss, Oscar Kully
Reithel, Francis Joseph
Reiter, Ronald Charles
Reller, Herbert Henry
Remy, Charles Nicholas
Rendina, George
Rendon, Leandro
Renis, Harold E
Rennert, Owen M
Rennert, Sheldon Samson
Renold, Albert Ernst
Reporter, Minocher C
Resnick, Harold
Resnik, Robert Alan
Ressler, Newton
Ressler, Charlotte
Retsema, James Allan
Reyes, Philip
Reynolds, Jacqueline Ann
Reynolds, Orland Bruce
Reynolds, Paul Joseph
Reynolds, Robert David
Rho, Joon H
Rhoades, James Lawrence
Rhoades, William Denham
Rhoads, James B
Rhodes, Charlotte
Rhodes, William Gale
Ribbons, Douglas William
Rice, Eldon Emerson
Rice, Elmer Harold
Rice, Jerry Mercer
Rice, Leslie Irene
Rice, Richard W
Rice, Robert Vernon
Richards, Frank Frederick
Richards, James Frederick
Richards, John Hall
Richards, Oliver Christopher
Richards, William Reese
Richardson, Arlan Gilbert
Richardson, Charles Clifton
Richardson, John Paul
Richardson, Stephen H
Richmond, Jonas Edward
Richmond, Martha Ellis
Richmond, Virginia
Richter, Erwin (William)
Rickert, Frederic Lawrence
Riddell, William A
Riden, Joseph Robert, Jr
Ridgway, George Junior
Ridgway, Helen Jane
Rieder, Sidney Victor
Riehm, John P
Rieske, John Samuel
Riggs, Demetrios A
Riggs, Arthur Dale
Riggs, Austen Fox, II
Riggs, Thomas Rowland
Rigney, James Arthur
Riley, Michael Verity
Rilling, Hans Christopher
Rinderknecht, Heinrich
Rinehart, Keith Edward
Rinehart, Roy K
Ringler, Ira
Ringler, Robert L
Riordan, James F
Riordan, John Richard
Ripperton, Lyman Alonzo
Risser, Nancy M
Ritchie, Kim
Rittenbury, Max Sanford
Rittenhouse, Harry George
Ritter, Edmond Jean
Ritter, Preston Otto
Ritzert, Roger William
Rivett, Robert Wyman
Rizack, Martin A
Rizzuto, Anthony B
Roach, Mary Katherine
Roark, Dennis Edward
Robb, Leslie Allan
Robbins, Ernest Aleck
Robbins, John Edward
Robbins, Kenneth Carl
Robbins, Phillips Wesley
Roberge, Andree Groleau

Roberson, Robert H
Roberts, David Wilfred Alan
Roberts, Eugene
Roberts, Joseph
Roberts, Joseph Linton
Roberts, Kenneth David
Roberts, Martin
Roberts, Phyllis Silver
Roberts, Richard Norman
Roberts, Robert Michael
Roberts, Sidney
Roberts, Thomas L
Roberts, Walden Kay
Roberts, William Van Bogaert
Robertson, Donald Claus
Robertson, Hugh Elburn
Robertson, Trevor
Robinson, Richard Morris
Robinson, Robert James
Robinson, Neal Clark
Robinson, James Lawrence
Robinson, Jack Bert, Jr
Robinson, George Waller
Robins, Eli
Rochovansky, Olga Maria
Roche, Thomas Edward
Rochefort, Joseph Guy
Ro-Choi, Tae Suk
Rockenmacher, Morris
Roepke, Raymond Rollin
Rogan, Eleanor Groeniger
Rodbell, Martin
Roden, Carl Nils Lennart
Rodgers, Nelson Earl
Rodgers, Richard Michael
Rodgers, Charles Graham
Rodgey, Frederick Lee
Rodney, Gertrude
Rodriguez, Mario Santos
Rodwell, Victor William
Roeder, Martin
Roels, Oswald A
Rogers, Beverly Jane
Rogers, Stearns Walter
Rogers, Samuel John
Rogers, Quinton Ray
Rogers, Robert Larry
Rogers, Kenneth Scipio
Rogers, Lorene Lane
Rogers, Palmer, Jr
Rogers, Thomas D
Rogers, William Edward, Jr
Rogers, William Irvine
Rohlfing, Duane L
Roll, Paul Melvin
Rolleston, Francis Stopford
Romani, Roger Joseph
Romanowski, Robert David
Romans, John Richard
Rongone, Edward Laurel
Ronzio, Robert A
Roodman, Stanford Trent
Roon, Robert Jack
Rosa, Nestor
Roscoe, Henry George
Rose, Irwin Allan
Rose, Joseph Edward
Rose, William Cumming
Roseman, Saul
Rosen, Barry Philip
Rosen, Fred
Rosen, Jeffrey Mark
Rosen, John Friesner
Rosenberg, Abraham
Rosenberg, Barbara Hatch
Rosenberg, Eugene
Rosenberg, Harry
Rosenbloom, Joel
Rosenbloom, Terrone Lee
Rosenfeld, George
Rosenfeld, Leonard M
Rosenfeld, Louis
Rosenstein, Robert William
Rosenstein, Arthur Frederick.
Rosenthal, Gerald A
Rosenthal, Harold Leslie
Rosenthal, Nathan Raymond
Rosett, Theodore
Rosevear, John William
Roskoski, Robert, Jr
Rosner, Lawrence
Ross, Doris Laune
Ross, Richard Travis
Rossiter, Roger James
Rossmiller, John David
Roth, Jay Sanford
Roth, Jerome Allan
Rothberg, Simon
Rothblat, George H
Rotherham, Jean
Rothfield, Lawrence I
Rothhus, John Arden
Rothrock, John William

Rothstein, Morton
Rotstein, Jerome
Rothman, Fritz M
Roufa, Donald Jay
Rouser, George
Roush, Allan Herbert
Rouslin, William
Roveto, Michael Julien
Rowe, Arthur Wilson
Rowe, Mark J
Rowe, William Bruce
Rowley, George Richard
Rownd, Robert Harvey
Roxby, Robert
Roy, Arun K
Roy, Debdutta
Roy-Burman, Pradip
Royer, Garfield Paul
Roze, Uldis
Rozhin, Jurij
Rubenstein, Kenneth E
Rubin, Charles Stuart
Rubin, Martin Israel
Rubin, Max
Rubin, Robert Jay
Rubin, Saul Howard
Rubinstein, David
Rubulis, Albert
Rudney, Harry
Rudolph, Guilford George
Ruegamer, William Raymond
Rutishauser, Urs Stephen
Ruffin, Spaulding Merrick
Ruliffson, Willard Sloan
Rutman, Robert Jesse
Rupley, John Allen
Russell, Charlotte Sananes
Russell, Diane Haddock
Russell, Douglas William
Russell, George Keith
Russell, Paul Telford
Russell, Percy J
Russell, Thomas J, Jr
Rutherford, Charles
Rutter, William J
Rutter, Henry Alouis, Jr
Ryan, James Walter
Ryan, Kenneth John
Ryan, Michael T
Ryan, Wayne L
Rynbrandt, Donald Jay
Saad, Farida M
Sabatini, David Domingo
Sable, Henry Zodoc
Sabo, Dennis John
Sabol, Steven Layne
Sabry, Zakaria I
Sachs, George
Sachs, Howard
Sacks, William
Sacktor, Bertram
Sadoff, Harold Lloyd
Sadler, John
See, Andy S W
Saffran, Murray
Saffran, Judith
Sahyun, Melville
Saidel, Leo James
Saifer, Abraham
St Clair, Richard William
St John, Judith Brook
Sakami, Warwick
Saldarini, Ronald John
Salo, Wilmar Lawrence
Salomon, Lothar I
Salser, Josephine See
Saltman, Paul David
Salzberg, David Aaron
Salzman, Lois Ann
Samis, Harvey Voorhees, Jr
Sampson, Phyllis Marie
Sampugna, Joseph
Samuels, Leo Tolstoy
Sanadi, D Rao
Sanchez, Albert
Sander, Eugene G
Sanders, Benjamin Elbert
Sanders, Louis Lee
Sanders, Marilyn Magdanz
Sanders, Robert B
Sandler, Thomas Garrison
Sandman, Robert Paul
Sandor, Thomas
Sands, Howard
Sandza, Joseph Gerard
Sanfilippo, Francis Anthony
Sani, Brahma Porinchu
Sankar, D V Siva
Sansing, Norman Glenn
Sansone-Bazzano, Gail
Santi, Daniel V
Sapico, Virginia L.
Sapolsky, Asher Isadore
Saponara, Arthur G
Sarcione, Edward James
Sardesai, Vishwanath M
Sardinas, Joseph Louis
Sarett, Herbert Paul

Sarin, Prem S
Sarkar, Bibudhendra
Sarma, Dittakavi S R
Sarma, Raghupathy
Sarma, Ramaswamy Harihara
Saroff, Jack
Saslaw, Leonard David
Sassa, Shigeru
Sasse, Edward Alexander
Sato, Clifford Shinichi
Satoh, Paul Shigemi
Sauberlich, Howerde Edwin
Sauer, Leonard A
Saunders, Joseph Francis
Savage, Jane Ramsdell
Savard, Francis Gerald Kenneth
Saxena, Brij B
Sayre, Francis Warren
Saz, Howard Jay
Scaffer, Stephen Ward
Scala, James
Scallen, Terence
Scannell, James Parnell
Scanu, Angelo
Scarborough, Gene Allen
Scarpa, Antonio
Schachman, Howard Kapnek
Schachter, H
Schaefer, John F, Jr
Schafer, Thomas Wayne
Schaffer, Frederick Leland
Schaffer, Sheldon Arthur
Schaffner, Carl Paul
Schallhorn, Robert George
Schanbacher, Floyd Leon
Schantz, Edward Joseph
Schapira, Harriette Charlotte
Scherberg, Neal Harvey
Scheuer, James
Schiaffino, Silvio Stephen
Schiffman, Sandra
Schiffmann, Elliot
Schildkraut, Carl Louis
Schiller, Sara
Schimke, Robert T
Schlesinger, Milton J
Schimmel, Paul Reinhard
Schimmel, Steven David
Schliselfeld, Louis Harold
Schlom, Jeffrey
Schlueter, Robert John
Schmid, Harald Heinrich Otto
Schmid, Karl
Schmid, Peter
Schmid, Rudi
Schmidt, Gerhard
Schmidt, Gilbert Carl
Schmidt, Robert Reinhart
Schmitt, Robert W
Schmir, Gaston L
Schmukler, Morton
Schnaitman, Carl A
Schneider, Allan Stanford
Schneider, Donald Leonard
Schneider, Donald Louis
Schneider, Henry
Schneider, Howard Albert
Schneider, John H
Schneider, John Joseph
Schnell, Gene Wheeler
Schnell, Jerome Vincent
Schnoes, Heinrich Konstantin
Schoellmann, Guenther
Schoenfeld, Robert George
Scholar, Eric M
Scholefield, Peter Gordon
Schonbeck, Niels Daniel
Schooler, James M, Jr
Schor, Joseph Martin
Schottelius, Dorothy Dickey
Schotz, Michael C
Schraer, Rosemary
Schram, Alfred C
Schramm, Vern Lee
Schrecker, Anthony Wolfgang
Schreiber, Manuel
Schreiner, Heinz Rupert
Schroeder, Duane David

Schroeder, Hartmut Richard
Schroepfer, George John, Jr
Schrohenloher, Ralph Edward
Schrum, Mary Irene Knoller
Schubert, Edward Thomas
Schubert, Walter John
Schucher, Reuben
Schuder, John Claude
Schuel, Herbert
Schuler, Martin N
Schulman, Herbert Michael
Schulman, LaDonne Heaton
Schultz, Julius
Schultz, Richard Michael
Schultze, Max Otto
Schulz, Arthur R
Schulze, Irene Theresa
Schumacher, Gebhard Friederich B
Schumer, William
Schumm, Dorothy Elaine
Schutzbach, John Stephen
Schwabe, Christian
Schwartz, Arnold
Schwartz, Edith Richmond
Schwartz, Gerald Peter
Schwartz, Harold Leon
Schwartz, Ira
Schwartz, Martin
Schwartz, Morton K
Schwarz, Henry P
Schwarz, Klaus
Schwarz, Otto John
Schwenzer, Kathryn Sarah
Schwert, George William
Scocca, John Joseph
Scorpio, Ralph M
Scott, David Frederick
Scott, Don
Scott, Dwight Baker McNair
Scott, Jesse Friend
Scott, Ronald McLean
Scott, William Addison, III
Scraba, Douglas G
Scrimgeour, Kenneth Gray
Scrutton, Michael Christopher
Searle, Gilbert Leslie
Searls, Robert L
Seed, Randolph William
Seeds, Nicholas Warren
Seegers, Walter Henry
Seegmiller, Jarvis Edwin
Seeley, Robert Dudley
Seery, Virginia Lee
Segal, Alvin
Segal, Harold Lewis
Segal, Stanton
Segel, Irwin Harvey
Seib, Paul A
Seibert, Florence Barbara
Seibert, Mary Angelice
Seidel, John Charles
Seifter, Sam
Seligson, David
Sell, Harold Melvin
Sell, John Edward
Sellinger, Otto Zivko
Sells, Bruce Howard
Seltzer, Jo Louise
Senciall, Ian Robert
Senn, Vincent John
Senum, Tellef
Serat, William Felkner
Serif, George Samuel
Serve, Munson Paul
Seshachalam, Dutta
Sessa, Grazia L
Setlow, Peter
Settlemire, Carl Thomas
Settoon, Patrick Delano
Sevall, Jack Sanders
Sexton, Edwin Leon
Seyer, Jerome Michael
Sgoutas, Demetrios Spiros
Shacter, Bernard
Shaeffer, Joseph Robert
Shaffer, Patricia Marie
Shah, Shantilal Nathubhai
Shahani, Khem Motumal
Shalita, Alan Remi
Shamberger, Raymond J
Shannon, Charles Francis
Shantz, Edgar Moore
Shapira, Raymond
Shapiro, Bennett Michaels
Shapiro, David Jordan
Shapiro, David M
Shapiro, Herman Simon
Shapiro, Irving Meyer
Shapiro, Irwin Louis
Shapiro, Lucille
Shapiro, Ralph
Shapiro, Robert
Shapiro, Stanley Kallick
Shapiro, Stanley Seymour
Shappirio, David Gordon
Sharp, John Joseph
Sharpless, Nansie Sue
Shashoua, Victor E

Shatkin, Aaron Jeffrey
Shaw, Elliott Nathan
Shaw, Emil Gilbert
Shaw, James Headon
Shaw, Kenneth Noel Francis
Shaw, Norman Yon-Shong
Shaw, Paul Dale
Shaw, Walter Norman
Shearer, Thomas Robert
Sheek, Martha Reyburn
Shefer, Sarah
Sheid, Bertrum
Sheinin, Rose
Shellenberger, Thomas E
Shelton, Damon Charles
Shelton, Keith Ray
Shemin, David
Shen, Nai-Hsuan Chang
Shepherd, George Robbins
Shepherd, Herndon Guinn, Jr
Sheppard, Herbert
Sheppard, John Richard
Sherman, William Reese
Sherr, Stanley I
Shetlar, Marvin Roy
Shibko, Samuel Issac
Shichi, Hitoshi
Shideler, Robert Weaver
Shields, James Edwin
Shier, Wayne Thomas
Shigeura, Harold Takeo
Shih, Thomas Yutzong
Shiman, Ross
Shimizu, C Susan
Shiota, Tetsuo
Shively, Jessup MacLean
Shively, John Ernest
Shneour, Elie Alexis
Shockman, Gerald David
Shome, Basudev
Shonk, Carl Ellsworth
Shooter, Eric Manvers
Shore, Bernard
Shore, Virgie Guinn
Short, Everett C, Jr
Short, William Arthur
Showe, Michael Kent
Shrago, Earl
Shu, Ping
Shugarman, Peter Melvin
Shugart, Lee Raleigh
Shull, Gilbert Malcolm
Shull, Kenneth Henry
Shum, Wan-Kyng Liu
Shuster, Robert C
Siakotos, Aristotle N
Sie, Edward Hsien Choh
Sie, Hsien-Gieh
Siebert, Karl Joseph
Siegel, Frank Leonard
Siegel, Jack Morton
Siegel, Lewis Melvin
Siegel, Sanford Marvin
Siegman, Fred Stephen
Siehr, Donald Joseph
Siek, Theodore John
Siekevitz, Philip
Sievert, Herman William
Sigler, Paul Benjamin
Sigman, David Stephan
Silber, Robert L
Silbert, David Frederick
Silhacek, Donald Le Roy
Silver, Bernard Euric
Silver, Melvin Joel
Silverman, Donald A
Silverman, Morris
Silverman, Philip Michael
Silverstein, Emanuel
Silverstein, Ronald
Simkins, Ronald Allen
Simmonds, Sofia
Simmons, Jean Elizabeth Margaret
Simmons, John Robert
Simmons, Noel
Simon, Sanford Ralph
Simoni, Robert Dario
Simonsen, Donald Howard
Simpicio, Jon
Simpson, John Wayne
Simpson, Melvin Vernon
Simpson, Robert Todd
Sims, Ethan Allen Hitchcock
Sinclair, Ronald
Sinclair, Walton B
Sinex, Francis Marott
Singer, Leon
Singer, Maxine Frank
Singer, Sanford Sandy
Singer, Thomas Peter
Singh, Eric John
Singh, Harbhajan
Singh, Harwant
Singh, Jagat
Singh, Kartar
Singh, Pritam
Singhal, Ram Pratap
Sink, John Davis

Sinsheimer, Robert Louis
Sipe, Jerry Eugene
Siperstein, Marvin David
Sipos, Tibor
Siracusano, Vincent C
Sirbasku, David Andrew
Siteri, Pentti Kasper
Sitz, Thomas O
Siu, Patrick Mew Lum
Sivak, Andrew
Sizer, Irwin Whiting
Sjogren, Robert Erik
Skarnes, Robert C
Skarstedt, Mark Teofil
Skeggs, Leonard Tucker
Skelton, Marilyn Mae
Skidmore, Wesley Dean
Skinner, Charles Gordon
Skipper, Howard Earle
Skipski, Vladimir Pavlovich
Skogerson, Lawrence Eugene
Skye, George Eri, II
Sky-Peck, Howard H
Slagel, Donald E
Slapikoff, Saul Abraham
Slater, George P
Slater, Grant Gay
Slaunwhite, Wilson Roy, Jr
Sleeman, Harry Kenneth
Slein, Milton Wilbur
Sligar, Stephen Gary
Sloan, Donald Leroy, Jr
Sloane, Nathan Howard
Slodki, Morey Eli
Slonim, Arnold Robert
Slotta, Karl Heinrich
Small, Gary D
Smalley, Robert Lee
Smeby, Robert Rudolph
Smillie, Lawrence Bruce
Smith, Archie Lee
Smith, Benjamin Williams
Smith, Carl Clinton
Smith, Carroll Ward
Smith, Charles Giles
Smith, Charlotte Damron
Smith, Colleen Mary
Smith, David Burrard
Smith, David Joseph
Smith, Donald Edgar
Smith, Eddie Carol
Smith, Edgar Eugene
Smith, Edith Lucile
Smith, Edward Russell
Smith, Edwin Barkley, Jr
Smith, Ellen Marian Evans
Smith, Emil L
Smith, Eric Ernest
Smith, Eugene Joseph
Smith, Frank Houston
Smith, Frederick George
Smith, Gary E
Smith, Grant Newey
Smith, Harold Carter
Smith, Irvin Darrow
Smith, Jack Louis
Smith, James Cecil
Smith, John Milton
Smith, John Thurmond
Smith, Joseph Donald
Smith, Kendric Charles
Smith, Leland Leroy
Smith, Leonard Charles
Smith, Lester
Smith, Louis C
Smith, Marian Jose
Smith, Marion Edmonds
Smith, Michael
Smith, Milton Watkins
Smith, Olive Watkins
Smith, Quenton Terrill
Smith, Richard Alan
Smith, Robbin Peggy Piety
Smith, Robert C
Smith, Robert Lewis
Smith, Robert William
Smith, Roberts Angus
Smith, Samuel Cooper
Smith, Sharron Williams
Smith, Sidney R, Jr
Smith, Steven Joel
Smith, Susan T
Smith, Thomas Elijah
Smith, Thomas Marion
Smith, William Grady
Smith, William Lee
Smoake, James Alvin
Smouse, Thomas Hadley
Smucker, Arthur Allan
Smuckler, Edward Aaron
Smulson, Mark Elliott
Smyrniotis, Pauline Zoe
Smyth, Robert Daniel
Sneider, Thomas W
Snell, Esmond Emerson
Snell, Junius Fielding
Snoke, John Edward
Snyder, Fred Hugh

BIOCHEMISTRY

Snyder, Fred Leonard
Snyder, Harry E
Snyder, Jack Austin
Soares, Joseph Henry, Jr
Soares, Stephen Laurie
Sobel, Robert Edward
Soberon, Guillermo
Soderling, Thomas Richard
Soeiro, Ruy
Soffer, Richard Luber
Sogn, John Allen
Sohler, Arthur
Sokol, Frantisek
Sokoloff, Louis
Soldo, Anthony Thomas
Solis-Gaffar, Maria Corazon
Solomon, John Junior
Solomon, Samuel
Solomons, Clive (Charles)
Solomonson, Larry Paul
Somberg, Ethel Weiss
Somero, George Nicholls
Somers, Perrie Daniel
Somerville, Ronald Lamont
Sonenberg, Martin
Soodak, Morris
Soodsma, James Franklin
Sophianopoulos, Alkis John
Sordahl, Louis John
Sorensen, Leif Boge
Sorof, Sam
Soslau, Gerald
Soto, Aida R
Soulen, Thomas Kay
Soupart, Pierre
Sourkes, Theodore Lionel
Southard, Wendell Homer
Sowinski, Raymond
Spackman, Darrel H
Spain, James Dorris, Jr
Spallholz, Julian Ernest
Spangenberg, Dorothy Breslin
Spangler, Martin Ord Lee
Spano, Francis A
Spearing, Cecilia W
Speck, John Clarence, Jr
Spector, Abraham
Spector, Arthur Abraham
Spector, Leonard B
Spector, Thomas
Speer, Henry Lee
Speer, Vaughn C
Speidel, Edna W
Spell, William Hux, Jr
Spelsberg, Thomas Coonan
Spence, Kemet Dean
Spencer, Elaine
Spencer, Howard Carnac
Spencer, John Hedley
Spencer, Mary Stapleton
Spencer, Richard L
Spencer, Richard Paul
Sperber, William Henry
Spero, Leonard
Spicer, Daniel Shields
Spiegel, Herbert Eli
Spilman, Edra Lavergene
Spining, Arthur Milton, III
Spiro, Mary Jane
Spiro, Robert Gunter
Spitzer, Judy A
Spitzer, Robert Harry
Spitzizen, John
Splittstoesser, Walter E
Sporn, Michael Benjamin
Sprecher, Howard W
Spremulli, Gertrude H
Sprince, Herbert
Sprinson, David Benjamin
Spritz, Norton
Spudich, James Anthony
Sreebny, Leo Morris
Sreter, Paul Arnold
Sreter, Frank A
Sribney, Michael
Srinivasan, Sathanur Ramachandran
Srinivasan, Vadake Ram
Srinivasava, Bejai Inder Sahai
Srivastava, Satish Kumar
Stadman, Earl Reece
Stadtman, Thressa Campbell
Stageman, Paul Jerome
Stahl, William J
Stahl, William Louis
Stahmann, Mark Arnold
Stahnke, Dennis William
Stambaugh, Richard L
Stancel, George Michael
Stanacev, Nikola Ziva
Stanfield, Manie K
Stanford, John W
Stanley, John Pearson
Stanley, Richard W
Stanley, Wendell Meredith, Jr
Stanley, Philip Gerald

Starcher, Barry Chapin
Stark, George Robert
Stark, James Cornelius
Starr, Jason Leonard
Stasiw, Roman Orest
Statz, Herman
Staub, Herbert Warren
Stauffer, Clyde E
Stearns, Eugene Marion, Jr
Stearns, Thomas W
Steck, Theodore Lyle
Steele, Richard Harold
Steers, Edward, Jr
Stegink, Lewis D
Stein, Abraham Morton
Stein, Gary S
Stein, Herman H
Stein, T Peter
Stein, William Howard
Steinberg, Daniel
Steinberg, James
Steiner, Donald Frederick
Steiner, Marion Rothberg
Steinhardt, Jacinto
Steinman, Charles Robert
Steinraut, Larry King
Steitz, Joan Argetsinger
Steitz, Robert C
Stellwagen, Earle C
Stellwagen, Robert Harwood
Stempien, Martin F, Jr
Stenesh, Jochanan
Stephens, Marvin Wayne
Stephenson, Norman Robert
Sterling, Rex Elliott
Stern, Ivan J
Stern, Joel R
Stern, Michele Suchard
Stern, Robert
Sternglanz, Rolf
Stetten, Marjorie Roloff
Stevens, Ann Rebecca (Larkin)
Stevens, Audrey L
Stevens, Carl Mantle, II
Stevens, Charles David
Stevens, Frits Christiaan
Stevens, Sue Cassell
Stevens, Vernon Lewis
Stevens, Vincent Leroy
Stevenson, David
Stevenson, Irene Edmund, Jr
Stevenson, William Campbell
Stewart, Allan Greenwood
Stewart, C Gordon
Stewart, Charles Jack
Stewart, Harold Brown
Stewart, James Anthony
Stewart, John Morrow
Stewart, Kent Kallam
Stewart, Lynn Martin
Stewart, Mark Armstrong
Stickle, Gene P
Stidworthy, George H
Stifel, Frederick Benton
Stier, Frederick Benton
Still, Gerald G
Stiller, Mary Louise
Stiller, Richard L
Stillway, Lewis William
Stinson, Robert Anthony
Stith, William Joseph
Stjernholm, Rune Leonard
Stock, Werner
Stockell-Hartree, Anne
Stoffyn, Pierre Jules
Stohrer, Gerhard
Stohs, Sidney John
Stojanovic, Borislav Jovan
Stokstad, Evan Ludvig Robert
Stoloff, Leonard
Stone, Charles Dean
Stone, David
Stone, Edward John
Stone, Gilbert C H
Stone, Henry Otto, Jr
Stone, Irwin
Stone, Joseph
Stone, Stanley S
Stoolmiller, Allen Charles
Stoops, James King
Stopkie, Roger John
Storck, Roger Louis
Storm, Daniel Ralph
Storrs, Eleanor Emerett
Storvick, Waldemar O
Story, Jon Alan
Stotz, Elmer Henry
Stouffer, John Emerson
Stout, Ernest Ray
Stowe, Bruce Bernot
Stowell, Ellery Cory, (Jr)
Straat, Patricia Ann
Stracher, Alfred
Strasdine, George Alfred
Stratton, Lewis Palmer
Straus, David Bradley
Straus, Werner

Strauss, James Henry
Strauss, Robert N
Strehler, Bernard Louis
Strength, Delphin Ralph
Strickland, Kenneth Percy
Stritmatter, Cornelius Frederick
Stritmatter, Philip
Strmiste, Gary F
Strobel, Gary F
Strobel, Donald Roy
Strobel, Rudolf G K
Stroman, David Womack
Stromer, Marvin Henry
Strong, Frank Morgan
Strother, Allen
Struck, Jacob, Jr
Strumeyer, David Hyman
Struve, William George
Stryer, Lubert
Stubbings, Robert Lamb
Stubbs, John Dorton
Stucki, William Paul
Stulberg, Melvin Philip
Stull, James Travis
Stumpf, Paul Karl
Sturman, John Andrew
Sturz, Robert Eugene
Su, Kwei Lee
Su, Gian Chand
Suderman, Harold Julius
Sudo, Sara Zeece
Suhadolnik, Robert J
Sukowski, Eugene
Sullivan, Andrew Jackson
Sullivan, Ann Clare
Sullivan, Betty
Sullivan, James Bolling
Sullivan, Lloyd John
Sullivan, Thomas Wesley
Sullivan, William Daniel
Sulya, Louis Leon
Summers, George Kendrick
Summers, William Cofield
Sun, Albert Yung-Kwang
Sun, Alexander Shihkaung
Sun, Frank F
Sun, Ansel Parrish
Sundharadas, Gnanasigamoni
Sung, Cheng-Po
Surdy, Ted E
Surgenor, Douglas MacNevin
Sussman, Howard H
Sutherland, Donald James
Sutter, Richard P
Suttie, John Weston
Suzuki, Michio
Suzuki, Tsuneo
Svacha, Anna Johnson
Svokos, Steve George
Swan, Patricia B
Swank, Richard Tilghman
Swanson, Arnold Martin
Swanson, Arthur Martin
Swanson, Curtis James
Swanson, Jack Lee
Swanson, John Robert
Swanson, Ronald Frederick
Sweat, Floyd Walter
Sweat, Max Leroy
Sweeley, Charles Crawford
Sweet, Frederick
Sweetman, Lawrence
Swell, Leon
Swendseid, Marian Edna
Swensen, Albert Donald
Swick, Robert Winfield
Swift, Terrence James
Swislocki, Norbert Ira
Switzer, Boyd Ray
Switzer, Robert Lee
Swope, Fred C
Sy, Jose
Syner, Frank N
Sze, Paul Yi Ling
Szent-Gyorgyi, Albert
Szent-Gyorgyi, Andrew Gabriel
Szer, Wlodzimierz
Szuhaj, Bernard F
Szybalski, Waclaw
Szymanski, Chester Dominic
Tabachnick, Milton
Tabor, Elbert Cecil
Tabor, Herbert
Taborsky, George
Tacker, Martha McClelland
Takayama, Kuni
Taketa, Fumito
Tallan, Harris H
Tallent, William Hugh
Tamaoki, Taiki
Tamir, Hadassah
Tan, Celine G L
Tan, Ah-Ti Chu
Tan, Charles Hua-Min
Tanenbaum, Stuart William
Tanford, Charles
Tang, Jordan J N
Tannenbaum, Carl Martin
Tanzer, Marvin Lawrence

Tappan, Donald Vester
Tappel, Aloys Louis
Tarantino, Laura Mary
Tarczy, E K
Tarleton, Raymond Joseph
Tartof, Alvin R
Taunton-Rigby, Alison
Taurog, Alvin
Tausig, Andrew
Taylor, Alan Neil
Taylor, Harold Allison, Jr
Taylor, Harold Leland
Taylor, Iain Edgar Park
Taylor, James A, Jr
Taylor, John Fuller
Taylor, Julius David
Taylor, Kenneth Boivin
Taylor, Norman Fletcher
Taylor, Robert Thomas
Taylor, Russell James, Jr
Tchen, Tche Tsing
Tcholakian, Robert Kevork
Teaford, Margaret Elaine
Teckell, Roger Alton
Teeri, Arthur Eino
Teipel, John William
Teitel, Alvin Gilbert
Telfer, Alvin Gilbert
Tenenhouse, Alan M
Tener, Gordon Malcolm
Teng, Ching Sung
Teng, Christina Wei-Tien Tu
Tennent, David Maddux
Teply, Lester Joseph
Teresi, Joseph Dominic
Termine, John David
Terner, Charles
Terriere, Leon C
Tershakovec, George Andrew
Tesar, Charles
Testa, Raymond Thomas
Tester, Cecil Fred
Tewksbury, Duane Allan
Tews, Jean Kring
Thampi, Nagendran Sankaranarayanan
Thanassi, John Walter
Thau, Rosemarie B Zischka
Thayer, Donald Wayne
Theil, Elizabeth
Theiner, Micha
Thenen, Shirley Warnock
Therriault, Donald G
Thiele, Elizabeth Henriette
Thiessen, Reinhardt, Jr
Thoma, John Anthony
Thomas, Barry Holland
Thomas, Byron Henry
Thomas, Dudley Watson
Thomas, Edward Wilfrid
Thomas, James Arthur
Thomas, John Alva
Thomas, Joseph James
Thomas, Kurian K
Thomas, Patricia Zeis
Thomas, Robert E
Thompson, Chester Ray
Thompson, Guy A, Jr
Thompson, Jerry Nelson
Thompson, John Eveleigh
Thompson, Marvin P
Thompson, Richard Michael
Thompson, Roy Charles, Jr
Thompson, Thomas Edward
Thompson, William Raymond
Thompson, Wynelle Doggett
Thomson, John Ferguson
Thomson, William Alexander Brown
Thorp, Frank Kedzie
Thorp, Neal Owen
Thurlow, John Frank
Thurman, Ronald Glenn
Thysen, Benjamin
Tietze, Frank
Tilden, J Tyson
Tinker, David Owen
Tinsley, Ian James
Tint, Howard
Tipton, Carl Lee
Tipton, Henry C
Tischio, John Patrick
Titchener, Edward Bradford
Tocci, Paul M
Tocco, Dominick Joseph
Todd, Charles Wyvil
Todd, Wilbert Remington
Toji, Lorraine Hellenga
Tokes, Zoltan Andras
Tolbert, Bert Mills
Tolbert, Margaret Ellen Mayo
Tolbert, Nathan Edward
Tollefson, Charles Ivar
Tomarelli, Rudolph Michael
Tomasi, Gordon Ernest
Tomasi, Thomas B Jr
Tomasz, Alexander
Tomasz, Maria
Tombropoulos, Elias George

Tometsko, Andrew M
Tomisek, Arthur John
Tomizawa, Henry Hideo
Tomlinson, Geraldine Ann
Tomlinson, Raymond Valentine
Tompkins, Paul Carter
Tong, Winton
Tonks, David Bayard
Tookey, Harvey Llewellyn
Toom, Paul Marvin
Tooney, Nancy Marion
Topham, Richard Walton
Toporek, Milton
Topper, Yale Jerome
Toralballa, Gloria C
Tornqvist, Erik Gustav Markus
Toro-Goyco, Efrain
Totter, John Randolph
Totton, Ezra Lester
Touchstone, Joseph Cary
Tourtellotte, Charles Dee
Tourtellotte, Mark Eton
Touster, Oscar
Tove, Samuel B
Tove, Shirley Ruth
Towne, Jack C
Trainer, John Ezra, Jr.
Trakatellis, Anthony C
Tramell, Paul Richard
Trams, Eberhard Georg
Traugh, Jolinda Ann
Traut, Robert Rush
Trautman, Jack Carl
Traver, Janet Hope
Travers, John Joseph
Tritsch, George Leopold
Travis, James
Traylor, Patricia Shizuko
Treadwell, Carleton Raymond
Treadwell, George Edward, Jr
Treble, Donald Harold
Treece, Jack Milan
Tremblay, George Charles
Trenholm, Harold Locksley
Trenkle, Allen H
Trent, Dennis W
Trevithick, John Richard
Trew, John Allan
Trezise, Willard Joseph
Trimble, Robert Bogue
Trione, Edward John
Tripathi, Kamala Kant
Troll, Walter
Tropp, Burton E
Troy, Frederic Arthur
Trujillo, Ralph Eusebio
Trumpower, Bernard Lee
Trupin, Joel Sunrise
Tryfiates, George P
Tsai, Chia-Yin
Tsai, Chishiun S
Tsai, Ming-Jer
Tsai, Min-Shen Chen
Tschudy, Donald P
Tsen, Cho Ching
Tsiapalis, Chris Milton
Tsibris, John Constantine Michael
Tsolas, Orestes
Tsong, Yun Yen
Tsou, Kwan Chung
Tsuboi, Kenneth Kaz
Tsuji, Frederick Ichiro
Tsutsui, Ethel Ashworth
Tu, Anthony T
Tu, Chin Ming
Tu, Chingkuang
Tucci, Anthony Frederick
Tullis, James Lyman
Tuma, Dean J
Tumbleson, Myron Eugene
Tung, Fred Fu
Tunis, Marvin
Turinsky, Jiri
Turk, Donald Earle
Turkington, Roger W
Turner, Earl Wilbert
Turner, James E
Turner, Michael D
Turner, Ralph B
Turnipseed, Marvin Roy
Tustanoff, Eugene Reno
Tutas, Daniel Joseph
Tutwiler, Gene Floyd
Tweto, John Halvor
Tyce, Gertrude Mary
Tyler, Tipton Ransom
Tyrell, David
Tyrell, Alfred A
Udenfriend, Sidney
Ugarte, Eduardo
Uhlenhopp, Elliott Lee
Ulto, Jouni Jorma
Ulm, Edgar H
Ulsamer, Andrew George, Jr
Umberger, Ernest Joy
Umbreit, Wayne William
Underwood, Arthur Louis

Underwood, Barbara Ann
Ungar, Frank
Unrau, Abraham Martin
Unrau, David George
Unsworth, Brian Russell
Upadhyay, Jagdish M
Upton, G Virginia
Urbas, Branko
Urry, Dan Wesley
Usdin, Vera Rudin
Utter, Merton Franklin
Uyeda, Kosaku
Uziel, Mayo
Vagelos, P Roy
Valadares, Joseph R E
Valentine, Raymond Carlyle
Vallee, Bert L
Vallee, Richard Bert
Van Abele, Frederick Richard
van Aller, Robert Thomas
Vanaman, Thomas Clark
Van Bruggen, John Timothy
Van Buren, Jerome Paul
Van Campen, Darrell R
Vance, Hugh Gordon
Vandegrift, Vaughn
Van Denheuvel, Franz Aime
Vanderheiden, Bernardo S
Vanderhoek, Jack Yehudi
van der Hoeven, Theo A
Vander Jagt, David Lee
Vanderlinde, Raymond E
Vander Wende, Christina
Vanderzant, Erma Schumacher
Vandevoorde, Jacques Pierre
Vande Woude, George
Van Dreal, Paul Arthur
Van Dyke, Knox
Van Dyke, Russell Austin
Van Eys, Jan
Van Fossan, Donald Duane
Van Gelder, Nico Michel
Van Huystee, Robert Bernard
Van Kley, Harold
Vanko, Michael
Van Lier, Johannes Ernestinus
Van Loon, Edward John
Van Meter, John Connell
Van Middelem, Charles Henry
Van Niel, Cornelis Bernardus
Van Pilsum, John Franklin
Van Reen, Robert
Van Vunakis, Helen
Van Wagtendonk, Willem Johan
Van Winkle, Walton, Jr
Varanasi, Usha Suryam
Varandani, Partab T
Vardanis, Alexander
Varma, Rajendra
Varnell, Thomas Raymond
Varricchio, Frederick
Vasington, Frank D
Vassel, Bruno
Vasta, Bruno Morreale
Vaughan, Burton Eugene
Vaughan, James Roland
Vaughan, Martha
Vavich, Mitchell George
Veech, Richard L
Vegotsky, Allen
Veis, Arthur
Veitch, Fletcher Pearre
Velick, Sidney Frederick
Veliky, Ivan Alois
Vella, Francis
Venditti, John M
Veneziale, Carlo Marcello
Vennesland, Birgit
Vercellotti, John R
Verlangieri, Anthony Joseph
Verly, Walter G
Vermeulen, Carl William
Vernon, Leo Preston
Vester, John William
Vestling, Carl Swensson
Viale, Richard O
Vice, John Leonard
Viciedo, Eusebio
Vickers, David Hyle
Vickery, Hubert Bradford
Viebrock, Frederick William
Vidaver, George Alexander
Vignos, Paul Joseph, Jr
Villafranca, German Bait
Villanueva, German John
Villarejo, Merna
Villar-Palasi, Carlos
Ville, Claude Alvin, Jr
Villee, Dorothy
Villemez, Clarence Louis, Jr
Vincent, Joseph Francis
Vincent, Phillip G
Vinograd, Jerome
Vinogradov, Serge
Vinson, Leonard J
Visceli, Thomas Alfonse
Visser, Donald Willis
Viswanatha, Thammaiah

Vitale, Joseph John
Vocci, Frank Joseph
Voet, Donald Herman
Vogel, James John
Vogel, Wolfgang Hellmut
Vogt, Molly Thomas
Vohra, Pran Nath
Voigt, Walter
Volcani, Benjamin Elazari
Vold, Barbara Schneider
Volenec, Frank Jerry
Volker, Joseph Francis
Volkin, Elliot
Volpert, Eugene M
Von Korff, Richard Walter
Von Schuching, Susanne
Vorbeck, Marie L
Votaw, Robert Grimm
Voth, Orville Lester
Vreman, Hendrik Jan
Vroman, Hugh Egmont
Vygantas, Auste Marija
Wachsman, Joseph T
Wachtel, Louis William
Wachter, Ralph Franklin
Wachtl, Carl
Wade, Adelbert Elton
Wade, David Robert
Wadkins, Charles Leroy
Wadman, W Hugh
Waggenknecht, Austin Clayton
Waggoner, Terry Bill
Wagh, Premanand Vinayak
Wagner, Conrad
Wagner, Eugene Stephen
Wagner, Frederick William
Wagner, Martin James
Wagner, Robert H
Wagner, Thomas Edwards
Wagreich, Harry
Wainer, Arthur
Wainfan, Elsie
Wainio, Walter W
Waite, Moseley
Waitzman, Morton Benjamin
Wakil, Salih J
Walborg, Earl Fredrick, Jr
Wald, George
Waldern, Donald E
Waldroup, Park William
Walker, Brian Lawrence
Walker, Charles R
Walker, Glenn Anthony
Walker, Howard David
Walker, Ian Gardiner
Walker, James Benjamin
Wall, Joseph Sennen
Wall, Monroe Eliot
Wallace, Herbert William
Wallace, Robert Allan
Wallach, Donald P
Wallander, Jerome F
Wallcave, Lawrence
Waller, George Rozier, Jr
Waller, James R
Wallick, Earl Taylor
Walsh, Christopher Thomas
Walsh, Joseph Matthew
Walsh, Kenneth Andrew
Walter, Charles Frank
Walters, Ronald Arlen
Walz, Frederick George
Wampler, Donald Eugene
Wang, Ching Chung
Wang, Chi-Sun
Wang, Hwa Lih
Wang, Jerry Hsueh-Ching
Wang, Jui Hsin
Wang, Li Chuan
Wang, Su-Sun
Wang, Taitzer
Wang, Yeu-Ming Alexander
Wannemacher, Robert, Jr
Waravdekar, Vaman Shivram
Ward, Darrell N
Ward, David Christian
Ward, James B
Ward, Samuel
Wardi, Ahmad Hassan
Ware, Arnold Grassel
Wargel, Robert Joseph
Warme, Paul Kenneth
Warner, Carol Miller
Warner, Donald Theodore
Warner, Robert Collett
Warnock, Laken Guinn
Warren, James C
Warren, William A
Warrington, Terrell L
Wase, Arthur William
Waslien, Carol Irene
Wasserman, Aaron Reuben
Wasserman, Robert Harold
Watanabe, Kyoichi A
Watanabe, Mamoru
Watanabe, Ronald S

Watanabe, Shizuo
Waters, Larry Charles
Waters, Michael Dee
Watkins, Paul Donald
Watrel, Warren George
Watson, John Alfred
Watson, Stanley Arthur
Watt, Dean Day
Wawszkiewicz, Edward John
Wax, Harry
Waxdal, Myron John
Waygood, Ernest Roy
Waymouth, Charity
Weatherby, Gerald Duncan
Weaver, John Martin
Weaver, Robert F
Weaver, Robert Hinchman
Webb, Neil Broyles
Webb, Thomas Evan
Webber, William R
Weber, Annemarie
Weber, Bruce Howard
Weber, Charles Walter
Weber, Darrell J
Weber, George
Weber, Gregorio
Weber, Heather Ross
Weber, Janet Crosby
Weber, Morton M
Weber, Peter B
Weber, Thomas Byrnes
Webster, Dale Arroy
Webster, George Calvin
Webster, Marion Elizabeth
Wedding, Randolph Townsend
Wedemeyer, Gary Alvin
Wedler, Frederick Charles Oliver
Wedmid, George Yuri
Weeks, Gerald
Wegener, Warner Smith
Wegner, Marcus Immanuel
Weibel, Michael Kent
Weichselbaum, Theodore Edwin
Weiland, Glenn Statler
Weil-Malherbe, Hans
Weimberg, Ralph
Weinbach, Eugene Clayton
Weinbaum, George
Weiner, Henry
Weinfeld, Herbert
Weinhold, Paul Allen
Weinhouse, Sidney
Weinryb, Ira
Weinstein, Bernard Ira
Weinstein, Constance de Courcy
Weinstein, Hyman Gabriel
Weinstock, Irwin Morton
Weintraub, Robert Louis
Weisberg, Samuel Myer
Weisgraber, Karl Heinrich
Weisiger, James Richard
Weisman, Robert A
Weiss, Benjamin
Weiss, Gary Bruce
Weiss, Irma Tuck
Weiss, Richard Louis
Weiss, Samuel Bernard
Weiss, Sidney
Weissbach, Arthur
Weissbach, Herbert
Weissman, Sherman Morton
Weissmann, Bernard
Welch, Billy Edward
Welch, Richard Stanley
Welch, William Henry, Jr
Welker, Neil Ernest
Weller, David Lloyd
Weller, Lowell Ernest
Wellner, Daniel
Wellner, Vaira Pamiljans
Wells, George Sherman
Wells, Ibert Clifton
Wells, Michael Arthur
Wells, Robert Dale
Wells, Warren F
Wells, William Wood
Welsh, Richard Stanley
Weltman, Joel Kenneth
Wendel, William Bean
Wender, Simon Harold
Wenner, Charles Earl
Wentland, Stephen Henry
Wenzel, Frederick J
Weppelman, Roger Michael
Werbin, Harold
Wergedal, Jon E
Wermus, Gerald R
Wessel, Carl John
West, Bruce David
West, Charles Allen
West, Charles Donald
West, Edward Staunton
West, Harold D
Westall, Frederick Charles
Westcott, Wayne Leslie
Westerfeld, Wilfred Wiedey
Westfall, Robert Judson
Westhead, Edward William, Jr
Westley, John William

Westphal, Ulrich Friedrich
Wetlaufer, Donald Burton
Wetstone, Howard J
Wetter, Leslie Robert
Wharton, David Carrie
Wheat, Robert Wayne
Wheatley, Victor Richard
Wheeler, Glynn Pearce
Wheeler, Willis Boly
Whelan, James Donlan
Whelan, William Joseph
Whikehart, David Ralph
Whisler, Walter William
Whitaker, John Robert
White, Abraham
White, Arnold Allen
White, Bernard J
White, Booker Taliaferro
White, David Cleveland
White, Edward Austin
White, Florence Roy
White, Fredric Paul
White, Gordon Allan
White, Harold Birts, Jr
White, Harold Keith
White, Helen Lyng
White, James Rushton
White, Roseann Spicola
Whitenberg, David Calvin
Whitfield, Carol Faye
Whitfield, Carolyn Dickson
Whitfield, George Buckmaster, Jr
Whiting, Frank M
Whiting, James R
Whitlock, Gaylord Purcell
Whitney, Philip Lawrence
Whittle, John Antony
Whitty, John Samuel
Wickerhauser, Milan
Wickrema Sinha, Asoka J
Wicks, Wesley Doane
Widder, James Stone
Widmann, Carl
Widnell, Christopher Courtenay
Wiebelhaus, Virgil D
Wiech, Norbert Leonard
Wiegand, Ronald Gay
Wiener, Benjamin
Wierbicki, Eugen
Wiese, Alvin Carl
Wiesner, Rakoma
Wiest, Walter Gibson
Wigand, Jeffrey Stephen
Wiggans, Donald Sherman
Wigler, Paul William
Wikman-Coffelt, Joan
Wikoff, Helen Landman
Wilcox, Archer Carl
Wilcox, Henry G
Wilcox, Ronald Bruce
Wild, Gaynor (Clarke)
Wild, Gene Muriel
Wildasin, Harry Lewis
Wilder, Violet Myrtle
Wildfeuer, Marvin Emanuel
Wiley, William Rodney
Wilhelmi, Alfred Ellis
Wilk, Sherwin
Wilken, David Richard
Wilkening, Marvin C
Wilkes, John Stuart
Wilkinson, David Ian
Wilkoff, Lee Joseph
Willard, James Matthew
Williams, Charles Haddon, Jr
Williams, Charles Herbert
Williams, Clara Hinton
Williams, Curtis Alvin, Jr
Williams, Edward Foster, Jr
Williams, George Ronald
Williams, Harold Henderson
Williams, Jimmy Calvin
Williams, Joy Elizabeth P
Williams, Mary Ann
Williams, Phletus P
Williams, Roger John
Williams, Susan Catherine Frary
Williams, William Joseph
Williams, William Lawrence
Williams, William Thomas
Williams-Ashman, Howard Guy
Williamson, Denis George
Williamson, John Richard
Willingham, Allan King
Willis, Dawn Butler
Wills, Charles Ronald
Wilms, Allan Charles
Wilson, David Buckingham
Wilson, David F
Wilson, Donald Alan
Wilson, George Donald
Wilson, Glenn Rhodes
Wilson, Irwin B
Wilson, Jack Harold
Wilson, James Albert
Wilson, Jerry Lee
Wilson, John Eric

Wilson, John Howard
Wilson, Karl A
Wilson, Lowell D
Wilson, Perry William
Wilson, Richard Hansel
Wilson, Robert G
Wilson, Robert Paul
Windmueller, Herbert George
Windsor, Emanuel
Winer, Alfred D
Winfield, Arnold Francis
Wingo, William Jacob
Winkel, Cleve R
Winkler, Bruce Conrad
Winkler, Herbert H
Winkler, Norman Walter
Winn, Grant Saunders
Winner, Bernard Mark
Winstead, Jack Alan
Winsten, Seymour
Winston, Walter Abbott
Winter, Charles Gordon
Winter, Harry Clark
Winter, William Phillips
Winters, Harvey
Winters, Mary Ann
Wirth, John Christian
Wishner, Lawrence Arndt
Wishnick, Marcia M
Withrow, Alice Phillips
Witmer, Herman John
Witte, David L
Wittels, Benjamin
Wittenberg, Beatrice A
Wittenberg, Jonathan B
Wittliff, James Lawrence
Wittman, James Smythe, III
Witty, Ralph
Wixon, Robert Llewellyn
Woessner, Jacob Frederick, Jr
Wohlrab, Hartmut
Wolcott, Robert Michael
Wold, Finn
Wolf, Don Paul
Wolf, George
Wolf, Paul A
Wolf, Robert Lawrence
Wolf, Walter Alan
Wolf, Walter J
Wolfe, Bernard Martin
Wolfe, Leonhard Scott
Wolfe, Raymond Grover, Jr
Wolfenden, Richard Vance
Wolff, Donald John
Wolff, Jan
Wolff, John Shearer, III
Wong, David Taiwai
Wong, Jeffrey Tze-Fei
Wong, Keith Kam-Kin
Wong, Kin-Ping
Wong, Ming Dak
Wong, Patrick Yui-Kwong
Woo, Alex James
Wood, Don Clifton
Wood, Garnett Elmer
Wood, Harland G
Wood, Henry Nelson
Wood, Jeanie McMillin
Wood, John Lewis
Wood, John Martin
Wood, Peter Douglas
Wood, Randall Dudley
Wood, Thomas Ross
Wood, Willis A
Wood, William Barry, III
Woodard, Helen Quincy
Woodard, Marie Kelly
Woodard, Vernon Rich, Jr
Woodfin, Beulah Marie
Woodford, Vernon Rich, Jr
Woodley, Charles Leon
Woodman, Richard J
Woods, Wendell David
Woodside, Kenneth Hall
Woodstock, Lowell Willard
Woodward, Clare K
Woodward, Glenn Jones
Woodworth, Robert Cummings
Woody, Robert Wayne
Wool, Ira Goodwin
Wootton, John Francis
Workman, Erwin Franklin, Jr
Woronick, Charles Louis
Worthen, Howard George
Wortman, Bernard
Woslman, Walter Daniel
Wostmann, Bernard Stephan
Wotiz, Herbert Henry
Woychik, John Henry
Wray, Granville Wayne
Wray, Virginia Lee Pollan
Wright, Barbara Evelyn
Wright, Donald N
Wright, George Joseph
Wright, Lemuel Dary
Wriston, John Clarence, Jr
Wrogemann, Klaus
Wu, Cheng-Wen

Wu, Chung
Wu, Felicia Ying-Hsiueh
Wu, Henry Chi-Ping
Wu, Ming-Chi
Wu, Ray J
Wu, Roy Shih-Shyong
Wu, Shirley Shao-ning
Wu, Tai Wing
Wu, Ting-Chi
Wuthier, Roy Edward
Wyckoff, Harland Dewitt
Wye, Edwin James
Wyle, Robert Lee
Wyngaarden, James Barnes
Wynston, Leslie K
Wyrick, Ronald Earl
Wyse, Roger Earl
Yamaguchi, Masatoshi
Yamamoto, Masatoshi
Yamamoto, Richard Susumu
Yamamoto, Harry Y
Yamazaki, Hiroshi
Yanari, Sam Satomi
Yang, Chung Shu
Yang, David Chih-Hsin
Yang, Man-chiu
Yang, Wen-Kuang
Yankelov, John Allen, Jr
Yankell, Samuel L
Yarus, Michael J
Yasmineh, Walid Gabriel
Yasunobu, Kerry T
Yearick, Elisabeth Stelle
Yeh, Samuel D J
Yehle, Clifford Omer
Yellin, Tobias O
Yen, Terence Tsin Tsu
Yeoman, Lynn Chalmers
Yesair, David Wayne
Yguerabide, Juan
Yip, Cecil Cheung-Ching
Yip, George
Yip, Lily Chung
Yoch, Duane Charles
Yocum, Charles Fredrick
Yonetani, Takashi
Yong, Fook Choy
Yonuschot, Gene R
Yoshida, Akira
Younathan, Erzat Saad
Young, Delano Victor
Young, Franklin
Young, George Robert
Young, Harold
Young, Janis Dillaha
Young, Marvin Kendall, Jr
Young, Nancy Lizotte
Young, Nelson Forsaith
Young, Peter Chun Man
Young, Robert M
Young, Ronald Jerome
Young, Ruth Steuart
Young, Vernon Robert
Youngian, Edward Victor
Yount, Ralph Granville
Yphantis, David Andrew
Yu, Byung Pal
Yu, Robert Kuan-Jen
Yu, Shiu Yeh
Yunice, Andy Aniece
Yurkowski, Michael
Yushok, Wasley Donald
Yuwiler, Arthur
Zabin, Irving
Zaborsky, Oskar Rudolf
Zafiraron, Alejandro
Zahalsky, Arthur C
Zahler, Warren Leigh
Zahnd, Hugo
Zahnley, James Curry
Zakrzewski, Sigmund Felix
Zalik, Saul
Zalkin, Howard
Zamenhof, Stephen
Zaneveld, Lourens Jan Dirk
Zannoni, Vincent G
Zapisek, William Francis
Zapp, John Adam, Jr
Zaroslinski, John F
Zaugg, Waldo S
Zbarsky, Sidney Howard
Zealey, Marion Edward
Zech, Arthur Conrad
Zedeck, Morris Samuel
Zee, Paulus
Zeitlin, Israel
Zelson, Philip Richard
Zemlicka, Jiri
Zemp, John Workman
Ziboh, Vincent Azubike
Ziccardi, Robert John
Zichis, Joseph
Ziegler, Daniel
Ziegler, Frederick Dixon
Ziegler, Peter
Zielke, Horst Ronald
Zigman, Seymour
Zikakis, John Philip

Zile, Maija Helene
Zilinskas, Barbara Ann
Zilkey, Bryan Frederick
Zill, Leonard Peter
Ziltz, Melvin Leonard
Zimmerman, Daniel Hill
Zimmerman, Don Charles
Zimmerman, James Kenneth
Zimmerman, Morris
Zimmerman, Sarah E
Zimmerman, Steven B
Ziporin, Zigmund Zangwill
Zorzoli, Anita
Zuccarello, William A
Zull, James E
Zuravski, Vincent Richard, Jr
Zygmunt, Walter A

Agricultural Biochemistry

Bennett, Raymond Dudley
Bice, Claude Wesley
Boyd, John Edward
Chen, Steve Shih-Chieh
Clark, Burr, Jr
Corsini, Dennis Lee
Craig, Burton Mackay
Craine, Elliott Maurice
Cunningham, David Kenneth
Dirks, Brinton Marlo
El-Negoumy, Abdul Monem
Evans, James William
Fan, Hsing Yun
Fischer, William Carl, Jr
Frear, Donald Stuart
Garner, George Bernard
Gilles, Kenneth Albert
Golab, Tomasz
Gordon, Arthur Leonard
Greenaway, Walter Thomas
Hamilton, John Kelvin
Hayman, Ernest Paul
Heotis, James Peter
Hill, Robert D
Hope, Hugh Johnson
Hunter, George William
Jackson, John Edward
Jaffe, Lionel
Jenner, Edward Levant
Keating, Eugene Kneeland
Kneen, Eric
Knoche, Herman William
Kuck, James Chester
Leng, Marguerite Lambert
Loeffler, Josef Ernst
Mahon, John Harold
Matsumoto, Hiromu
McLaren, George Aiken
Melnychyn, Paul
Mitchell, Howard Lee
Mondy, Nell Irene
Nightingale, Harry Irving
Olson, Oscar Edward
Palmere, Raymond M
Prebluda, Harry Jacob
Prentice, Neville
Privett, Orville Samuel
Rogers, Charles Fletcher
Sayre, Robert Newton
Verma, Devi C
Vose, John Randal

Analytical Biochemistry

Behar, Marjam Gojchlerner
Bleidner, William Egidius
Bosshardt, David Kirn
Brooks, Marvin Alan
Davis, Bruce Allan
Dzidic, Ismet
Fitzpatrick, Francis Anthony
Frantz, Beryl May
Gaines, Tinsley Powell
Griffin, Richard Norman
Ke, Paul Jenn
Kohlhepp, Sue Joanne
Krzeminski, Leo Francis
Lindner, Elek
Machlowitz, Roy Alan
Mayberry, William Roy
McAnally, John Sackett
Noyes, Claudia Margaret
Ohms, Jack Ivan
Pelletier, Omer
Pellizzari, Edo Domenico
Pennington, Sammy Noel
Penton, Zelda Eve
Pfaffenberger, Carl Dale
Price, Byron Frederick
Rechnitz, Garry Arthur
Reinhold, Vernon Nye
Reynolds, Warren Dudley
Robinson, Ross Utley
Sadee, Wolfgang
Schwartz, Robert Saul
Seifert, William Edgar, Jr
Seitz, Larry Max
Skala, James Herbert
Throop, Lewis John

Wallace, Jack E
Wiebers, Joyce Adams
Yadav, Kamaleshwari Prasad
Young, Richard Lawrence

Biochemical Genetics
Ahmed, Saiyed I
Ames, Bruce Nathan
Baglioni, Corrado
Barnett, Donald Ray
Benson, Charles Everett
Benzinger, Rolf Hans
Bode, Vernon Cecil
Brill, Winston J
Brot, Frederick Elliot
Brown, Loretta Ann Port
Calvert, Allen Fisher
Chalmers, John Harvey, Jr
Champney, William Scott
Chasin, Lawrence Allen
Chen, Shi-Han
Colby, Clarence
Cortner, Jean A
Dalby, Arthur
Del Vecchio, Vito Gerard
Eberhart, Bruce Maclean
Farmer, James Lee
Ferrell, Robert Edward
Folk, William Robert
Fraser, Murray Judson
Fuchs, Morton S
Gall, Graham A E
Grimwood, Brian Gene
Hall, Benjamin Downs
Henke, Randolph Ray
Holmes, Edward Warren
Hudock, George Anthony
Kaczmarczyk, Walter J
Kadner, Robert Joseph
Koch, Elizabeth Anne
Lightfoot, Donald Richard
Ling, Hubert
Liu, Houng-Zung
Lucas, Myron Cran
Malcolm Alexander Russell
Matsushita, Tatsuo
Meade, Linda Celida
Nash, David
Paigen, Kenneth
Patterson, David
Rattazzi, Mario Cristiano
Rowley, Rodney Ray
Sargent, Malcolm Lee
Sass-Kortsak, Andrew
Schneider, Julius Edward
Sheppard, David E
Simonds, Josephine Abigail
Sorger, George Joseph
Summers, Wilma Poos
Tashian, Richard Earl
Tedesco, Thomas Albert
Tourian, Ara Yervant
Ulter, Fred Madison
Wenger, David Arthur
Weyter, Frederick William
Whitt, Gregory Sidney
Wilcox, Gary Lynn
Wong, Paul Wing-Kon

Biochemical Pharmacology
Agarwal, Ram Prakash
Axelrod, Julius
Bain, James Arthur
Barboriak, Joseph Jan
Barnard, Eric A
Bellward, Gail Dianne
Bennett, Leonard Lee, Jr
Besch, Paige Keith
Bhargava, Hemendra Nath
Bloch, Alexander
Block, Ronald Edward
Buhler, Donald Raymond
Byington, Keith H
Chang, Shaw Fai
Chang, Yi-Han
Chapman, George Herbert
Chello, Paul Larson
Cheng, Yung-Chi
Cicero, Theodore James
Collins, Allan Clifford
Conard, Gordon Joseph
Cooke, William David
Cushman, David Wayne
Dalton, Colin
Dannemburg, Warren Nathaniel
Deckert, Fred W
Diedrich, Donald Frank
Drach, John Charles
Duggan, Daniel Edward
Ecobichon, Donald John
Ellenbogen, Leon
Finger, Kenneth F
Fischer, Lawrence J
Freudenthal, Ralph Ira
Fuller, George Charles
Galbraith, William
Gentry, Glenn Aden
George, William Jacob

Gibb, James Wooley
Gillette, James Robert
Gordon, Harry William
Green, Donald Eugene
Green, Vernon Albert
Guarino, Anthony Michael
Hacker, Bruce
Hakala, Maire Tellervo
Harrison, Yvonne E
Hebborn, Peter
Ho, Beng Thong
Ho, Dah-Hsi
Irving, George Washington, Jr
Jacobson, Martin Michael
Judis, Joseph
Kalman, Thomas Ivan
Kamm, Jerome Jr
Kary, Christina Dolores
Khanna, Jatinder Mohan
Killinger, Joanne Marie
Kohn, Leonard David
Krell, Robert Donald
Kupfer, David
La Du, Bert Nichols, Jr
Langner, Ronald O
Lanman, Robert Charles
Leibman, Kenneth Charles
Levin, Jerome Allen
Lin, Chin-Chung
Lindenmayer, George Earl
Lippmann, Wilbur
Lockridge, Oksana Maslivec
Maragoudakis, Michael E
Matthews, Hazel Benton, Jr
Milman, Harry Abraham
Mitoma, Chozo
Muni, Indu A
Muschek, Lawrence David
Nadkarni, Moreshwar Vithal
Nahas, Aly
Neil, Gary Lawrence
Nelson, Donald J
Pantuck, Eugene Joel
Pereira, Michael Alan
Persico, Francis J
Petering, Harold George
Peters, John Henry
Peters, Marvin Arthur
Pinson, Rex, Jr
Randerath, Erika
Rikans, Lora Elizabeth
Roberts, DeWayne
Robison, George Alan
Roth, Robert Andrew, Jr
Sargent, Thornton William, III
Sartorelli, Alan Clayton
Saunders, Priscilla Prince
Schayer, Richard William
Schroeder, David Henry
Schroer, Richard Allen
Schulman, Martin Phillip
Seifried, Harold Edwin
Shamoian, Charles Anthony
Slaga, Thomas Joseph
Snyder, Robert
Somani, Satu M
Sommer, Kathleen Ruth
Steele, William John
Sterbenz, Francis Joseph
Symchowicz, Samson
Trevor, Anthony John
Tsong, Tian Yow
Vandor, Sandor Laszlo
Wang, Howard Hao
Watkins, Mary Louise
Weisburger, John Hans
Welch, Richard Martin
Wilkens, Hans J
Wiseman, Edward H
Wykes, Arthur Albert
Zimmerman, Thomas Paul

Bioinorganic Chemistry
Addison, Anthony William
Amma, Elmer Louis
Baum, Stuart J
Bereman, David Arthur
Bereman, Robert Deane
Birnbaum, Edward Robert
Boschmann, Erwin
Brisbin, Doreen A
Brown, David Henry
Brubaker, George Randell
Burness, James Hubert
Cannon, John Burns
Chasteen, Norman Dennis
Corden, Brian Joseph
Cummings, Sue Carol
Darnall, Dennis W
Dolphin, David Henry
Dukes, Gary Rinehart
Eichhorn, Gunther Louis
Farina, Robert Donald
Farrier, Noel John
Fife, Wilmer Krafft
Fleischer, Everly B
Flynn, Arthur
Francis, Dawn Elizabeth

Fuchsman, William Harvey
Fuentes, Ricardo, Jr
Goff, Harold Milton
Grimes, Carol Jane Galles
Herskovitz, Thomas
Hodges, Helen Leslie
Hodgson, Keith Owen
Hoeschele, James David
Holwerda, Robert Alan
Horrocks, William DeWitt, Jr
Hurst, James Kendall
Jack, Thomas Richard
Katzenellenbogen, John Albert
Koppa, Vasantha
Legg, John Ivan
Maher, George Garrison
Miller, William Theodore
Milne, David Bayard
Neilands, John Brian
Nyc, Joseph Frank
Plane, Robert Allen
Sawyer, Donald Turner, Jr
Schubert, Jack
Schugar, Harvey
Schwarz, Klaus
Sherry, Allan Dean
Scovell, William Martin
Skinner, Helen Catherine W
Stadtherr, Leon Gregory
Storm, Carlyle Bell
Stucky, Gary Lee
Sumner, Kenneth
Swinehart, James Herbert
Tobias, Russell Stuart
Ucko, David A
Valentine, Joan Selverstone
Verkade, John George
Watters, Kenneth Lynn
Weidig, Charles Ferdinand
Weirich, Gunter Friedrich
Wright, Jeffrey Lawson Cameron
Wright, John Ricken

Bio-organic Chemistry
Abernethy, John Leo
Abrell, John William
Acton, Edward McIntosh
Alworth, William Lee
Anderson, Charles Dean
Anderson, Laurens
Arsenault, Guy Pierre
Aszalos, Adorjan
Atwood, Linda
Ball, M Isabel
Bays, James Philip
Beardsley, George Peter
Bergren, William Raymond
Biersmith, Edward L
Bittman, Robert
Bland, Jeffrey S
Bodanszky, Miklos
Borowitz, Irving Julius
Bovard, Freeman Carroll
Broom, Arthur Davis
Brown, George Bosworth
Bruice, Thomas Charles
Burrows, Elizabeth Parker
Cane, David Earl
Celmer, Walter Daniel
Chapman, Toby Marshall
Chaturvedi, Rama Kant
Chipman, David Mayer
Christensen, Burton Grant
Cooper, David John
Coward, James Kenderdine
Dafforn, Geoffrey Alan
Daves, Glenn Doyle, Jr
Dawson, Marcia Ilton
De, Nimai C
Denning, George Smith, Jr
Dolphin, David Henry
Douglas, Kenneth Thomas
Dunn, Ben Monroe
Dvonch, William
Eagar, Robert Gouldman, Jr
Ehler, Kenneth Walter
Epstein, William Warren
Farina, Peter R
Feinberg, Robert Samuel
Felix, Arthur M
Fitch, William Lawrence
Fong, Conrad Tuck Onn
Franz, John Edward
Friedman, Orrie Max
Fujii, Atsushi
Galaway, Ronald Alvin
Garfinkel, Harmon M
Gerig, John Thomas
Giacobbe, Thomas Joseph
Gilleland, Martha Jane
Glaser, Robert
Gleim, Robert David
Glinski, Ronald P
Goldberg, Stanley Irwin
Gorenstein, David George
Gorman, Marvin
Gould, Steven James
Grattan, James Alex

Gruger, Edward H, Jr
Grunewald, Gary Lawrence
Guthrie, James Peter
Gutowski, Gerald Edward
Gutsche, Carl David
Haag, William George
Hall, Stan Stanley
Hamilton, Gordon Andrew
Hamsher, James J
Hanely, Wayne Stewart
Hanzlik, Robert Paul
Harpold, Michael Alan
Harris, Durward Smith
Hawley, Lewis Burton, Jr
Hogg, John Leslie
Horowitz, Sylvia Teich
Hruby, Victor J
Iacobucci, Guillermo Arturo
Idoux, John Paul
Jewett, Sandra Lynne
Jordan, Frank
Juneja, Prem S
Kayser, Robert Helmut
Kiefer, Hansruedi
Kokesh, Fritz Carl
Kovacs, Joseph
Langone, John Joseph
Larson, Richard Allen
Laursen, Richard Allan
Lawton, Richard G
Lee, Tzoong-Chyh
Lerman, Charles Lew
Lin, Yong Yeng
Lobo, Angelo Peter
Lohrmann, Rolf
Lotspeich, Frederick Jackson
Maccoss, Malcolm
Mandava, Nagabhushanam
Manning, David Treadway
Martin, Ned Harold
Mathewson, James H
May, Paul David
Mayers, George Louis
McChesney, James Dewey
McCormick, John Pauling
McKenna, Charles Edward
McQuate, Robert Samuel
Michel, Karl Heinz
Miller, Paul Scott
Minnemeyer, Harry Joseph
Mitscher, Lester Allan
Mold, James Davis
Moser, Robert E
Mosher, Carol Walker
Moyer, Walter Allen, Jr
Muggli, Robert Zeno
Mungall, William Stewart
Myers, Harvey Nathaniel
Nair, Vasu
Oehlschlager, Allan Cameron
Ogilvie, Kelvin Kenneth
Ortiz De Montellano, Paul R
Overman, Larry Eugene
Page, David Sanborn
Pages, Robert Alex
Parham, James Crowder, II
Parry, Ronald John
Peck, Merlin Larry
Pence, Leland Hadley
Pitha, Josef
Pollack, Norman Mark
Pollack, Ralph Martin
Ponnamperuma, Cyril Andrew
Ramirez, J Roberto
Rasmussen, Russell Lee
Reed, Roberta Gable
Rich, Daniel Hulbert
Richards, William Reese
Ringold, Howard Joseph
Roberson, Jill Sharon
Rodia, Jacob Stephen
Roeske, Roger William
Rosazza, John Paul
Roubal, William Theodore
Sanchez, Robert A
Satterthwait, Arnold Chase, Jr
Sayer, Jane McKinley
Schaffer, Robert
Schmid, Donald Emil, Jr
Schray, Keith James
Schuck, James Michael
Servis, Robert Eugene
Sevenair, John P
Siddall, John Brian
Sih, Charles John
Silver, Marc Stamm
Singer, Alan Granger
Smith, Gerald Floyd
Smith, Jerry Howard
Smith, Roy George
Spatola, Arno F
Steffens, James Jeffrey
Stephenson, Betty Ann
Taber, Richard Lawrence
Thornton, Edward Ralph
Tieckelmann, Howard
Tolkmith, Henry
Usher, David Anthony

BIOCHEMISTRY

Valenty, Vivian Briones
Van Lanen, Robert Jerome
Vellaccio, Frank
Villafranca, Joseph John
Vining, Leo Charles
Waddell, Thomas Groth
Wallen, Lowell Lawrence
Weinstein, Boris
Wendel, Samuel Reece
Williamson, Charles Elvin
Wilson, Gustavus Edwin, Jr
Wineman, Robert Judson
Wishinsky, Henry
Zeffren, Eugene

Biopharmaceutics

Adair, Dennis Wilton
Adair, Suzanne Frank
Ayres, James Walter
Benet, Leslie Z
Bennaman, Joseph David
Conklin, John Douglas
Dighe, Shrikant Vishwanath
Feldman, Stuart
Fink, Anthony Lawrence
Foster, Thomas Scott
Garrett, Edward Robert
Goettsch, Robert Wayne
Goldberg, Arthur H
Grassetti, Davide Riccardo
Grebow, Peter Eric
Heman-Ackah, Samuel Monie
Hynes, John Thomas
Kamath, Burde Laxminarayan
Kleber, John William
Kripalani, Kishin J
Lamy, Peter Paul
Ludden, Thomas Marcellus
Masih, Shabir Zahoor
Matthews, David Allan
Miller, Kenneth Wayne
Nightingale, Charles Henry
Patton, Thomas Floyd
Randinitis, Edward J
Ritschel, Wolfgang Adolf
Sawchuk, Ronald John
Shultz, Walter
Singhvi, Sampat Manakchand
Skelly, Jerome Philip, Sr
Smith, Harold Linwood
Vitti, Trieste Guido
Weintraub, Howard Steven
Whitney, Charles Candee, Jr
Wilken, Leon Otto, Jr
Wilkinson, Paul Kenneth
Wright, Walter Eugene

Biophysical Chemistry

Adams, Emory Temple, Jr
Adiarte, Arthur Lardizabal
Adler, Alice Joan
Albers, Robert Jay
Allerhand, Adam
Alletton, Samuel E
Andrews, Stephen Brian
Applequist, Jon Barr
Bahary, William S
Bahr, James Theodore
Barton, Janice Sweeny
Berkowitz, Steven Arlen
Berliner, Lawrence J
Bitonen, Rodney Lincoln
Blei, Ira
Bloomfield, Victor Alfred
Bobst, Albert M
Bock, Robert Manley
Bolton, James R
Bothner-By, Aksel Arnold
Breslow, Esther, M G
Brundage, Robert Scott
Bryant, Loren Conrad
Bryner, Robert George
Bush, C Allen
Butterfield, David Allan
Byrn, Stephen Robert
Campbell, Mary Kathryn
Cann, John Rusweiler
Cantor, Charles Robert
Carter, John Vernon
Cerny, Laurence Charles
Chan, Sunney Ignatius
Colen, Alan Hugh
Copeland, Edmund Sargent
Coulter, Charles L
Crippen, Gordon Marvin
Crothers, Donald M
Czeisler, Jeffrey Lance
Dahlquist, Frederick Willis
Deonier, Richard Charles
Dickinson, Leonard Charles
Diehn, Bodo
Donnelly, Thomas Henry
Douthart, Richard James
Eakin, Richard Timothy
Eanes, Edward David
Ehrlich, Julian
Ehrlich, Robert Stark

Eisenberg, David
Epand, Richard Mayer
Evans, Douglas Fennell
Everett, Wilbur Wayne
Fabry, Mary E Riepe
Fabry, Thomas Lester
Fink, Thomas Robert
Fossel, Eric Thor
Franzen, James
Friedman, Michael E
Fuhr, Irvin
Gaffney, Betty Jean
Gennis, Robert Bennett
Gent, Martin Paul Neville
George, Philip
Gibson, Quentin Howieson
Goldsack, Douglas E
Goss, Dixie J
Gray, Donald
Griffith, O Hayes
Grigsby, Ronald Davis
Haddad, Louis Charles
Halvorson, Herbert Russell
Harbour, John Richard
Harpst, Jerry Adams
Harrison, Stephen Coplan
Hartzell, Charles Ross, III
Hearst, John Eugene
Hendrickson, Constance McRight
Henkens, Robert William
Hiatt, Caspar Wistar, III
Hooker, Thomas M, Jr
Horton, Aaron Wesley
Ifft, James Brown
James, Thomas Larry
Jensen, Lyle Howard
Jensen, Ronald Harry
Johnson, Walter Curtis
Jordan, Frank
Jost, Patricia Cowan
Kahn, Leo David
Kalkwarf, Donald Riley
Kallenbach, Neville R
Kegeles, Gerson
Keltz, Alan
Kluetz, Michael David
Knopp, James A
Koltun, Walter Lang
Kooi, Earl Robert
Kotowycz, George
Kowalsky, Arthur
Kreishman, George Paul
Kropf, Allen
Krugh, Thomas Richard
Kuhlmann, Karl Frederick
Leslie, James
Liebe, Donald Charles
Lippard, Stephen J
London, Robert Elliot
Macfarlane, Ronald Duncan
Madison, Vincent Stewart
Maki, August Harold
Mandel, Frederic
Manning, Gerald Stuart
Matthews, Charles Robert
McConnell, Harden Marsden
Minton, Allen Paul
Noble, Robert Warren, Jr
Nozaki, Yasuhiko
O'Brien, David F
Olson, Wilma King
Pease, Lila Gierasch
Pezolet, Michel
Phillips, Leo Augustus
Plocke, Donald J
Polnaszek, Carl Francis
Rahn, Ronald Otto
Record, M Thomas, Jr
Roche, Rodney Sylvester
Rosenberg, Jerome Laib
Ross, Robert Talman
Rowlands, John Rhys
Rubin, John Ronald
Sass, Ronald L
Sauer, Kenneth
Scheraga, Harold Abraham
Schleyer, Walter Leo
Schmeltz, Irwin
Schmidt, Carl William
Schmidt, Paul Gardner
Schmitz, Kenneth Stanley
Schoor, W Peter
Schuster, Todd Mervyn
Schwartz, Albert Truman
Seewald, David Allan
Shafer, Richard Howard
Shaw, Barbara Ramsay
Sheetz, Michael Patrick
Simons, Elizabeth Reiman
Spivey, Howard Olin
Squire, Phil George
Starzak, Michael Edward
Stevens, Charles Le Roy
Strauss, George
Stuehr, John Edward
Sturtevant, Julian Munson
Sukow, Wayne William
Tallman, Dennis Earl

Thomas, Charles Allen, Jr
Tollin, Gordon
Troxell, Terry Charles
Ts'o, Paul On Pong
Tu, Shu-i
Turner, Douglas Hugh
Uhlenbeck, Olke Cornelis
Ulmans, Robert Scott
Underkofler, Leland Alfred
Verpoorte, Jacob A
Vickery, Larry Edward
von Hippel, Peter Hans
Waggoner, Alan Stuart
Walton, Alan George
Wang, James C
Warden, Joseph Tallman
Wartell, Roger Martin
Watson, Barry
Wee, Elizabeth Liu
Weinstein, Harel
Wetmur, James Gerard
Whitten, David G
Wickstrom, Eric
Wiechelman, Karen Janice
Winkler, Marvin Howard
Worley, John David
Yang, Jen Tsi
York, Sheldon Stafford
Yu, Nai-Teng
Zand, Robert
Zimm, Bruno Hasbrouck
Zipp, Adam Peter
Zobel, Carl Richard

Clinical Biochemistry

Albert, Jerry David
Alper, Carl
Applegarth, Derek A
Barton, Edward J
Bartos, Dagmar
Bartos, Frantisek
Bauer, Robert
Bhagavan, Nadhipuram V
Blattler, Delbert Paul
Bondar, Richard Jay Laurent
Boutwell, Joseph Haskell
Breckenridge, Carl
Brewster, Marjorie Ann
Brownstone, Yehoshua Shieky
Bryson, Melvin Joseph
Campbell, William Jackson
Clements, Robert Lawrence
Collins, Galen Franklin
Crane, Laura Jane
Dautlick, Joseph X
Davidow, Bernard
Dhople, Arvind Madhav
Dubowski, Kurt Max
Fischer, Imre A
Foster, Raymond Orrville
Giegel, Joseph Lester
Gilbertson, Terry Joel
Gold, Martin
Goodwin, Jesse Francis
Grady, Harold James
Grannis, George Franklin
Griswold, Kenneth Edwin, Jr
Guerrant, Ralph Eugene
Gyure, William Louis
Hisey, Alan
Hoag, Gordon Neil
Hutterer, Ferenc
Jgou, Donald Kaye
Kapur, Bhushan M
Karinattu, Joseph J
Kathan, Ralph Herman
Kim, Gino
Law, Amy Stauber
Lehmann, Hermann Peter
Lepp, Cyrus Andrew
Leutner, Frederick Stanley
Linke, Ernest George
Lobstein, Otto Ervin
Maggio, Edward Thomas
Malya, Govinda P A
Mann, Lewis Theodore, Jr
Marx, Joseph Vincent
Mather, Alan
Mayfield, Ernest Durward, Jr
Melville, Robert Seaman
Mendel, Julius Louis
Morales, Daniel Richard
Morin, Leo Gregory
Patton, Selma Hicks
Petticrec, Claude Jean
Pragay, Desider Alexander
Rice, Eugene Worthington
Robinson, James McOmber
Rosenfeld, Martin Herbert
Routh, Joseph Isaac
St Rose, John Ellison
Sandberg, Robert Gustave
Sanford, Karl John
Schneider, Richard S
Schwartz, Peter Larry
Seidah, Nabil George
Shrawder, Elsie June
Simon, John Antony

Smith, Elizabeth Knapp
Timberlake, Joseph William
Toews, Cornelius J
Varma, Ranbir S
Vratsanos, Spyros M
Whiteley, Norman McKee
Zak, Bennie

Comparative Biochemistry

Campbell, James Wayne
Decker, Joan Elise
Delwiche, Constant Collin
de Villafranca, George Warren
DeVillez, Edward Joseph
Hochachka, Peter William'
Hodgson, Ernest
Kobayashi, George S
Moses, William
Peterson, Gary Lee
Tabachnick, Joseph
Young, Roger Grierson

Enzymology

Aldous, John Gray
Baccanari, David Patrick
Bailey, David George
Ballou, David Penfield
Benson, Robert Leland
Bernfeld, Peter
Beuk, Jack Frank
Black, William James
Boger, Eliahu
Cabib, Enrico
Caplow, Michael
Cardenas, Mary Janet M
Carper, William Robert
Cayle, Theodore A
Chasin, Mark
Chipman, David Mayer
Clifton, Eugene Everett
Cohen, Philip Pacy
Darrow, Robert A
De Sa, Richard John
Doman, Elvira (Mrs John H Holder)
Drake, Billy Blandin
Dunlap, Robert Bruce
Dunn, Michael F
Dyson, Robert Duane
Eichel, Herbert Joseph
Elliott, Bernard Burton
Erenrich, Evelyn Schwartz
Evans, Ben Edward
Feinstein, Robert Norman
Ferrier, Leslie Kenneth
Flora, Robert Montgomery
Fridovich, Irwin
Frisell, Wilhelm Richard
Fyfe, James Arthur
Gelbard, Alan Stewart
Ghering, Mary Virgil
Gillard, Baiba Kunins
Goren, Howard Joseph
Gracy, Robert Wayne
Gregory, Eugene Michael
Grooms, Thomas Albin
Guha, Arabinda
Gumport, Richard I
Harris, Ben Gerald
Hayman, Selma
Hedges, Dorothea Huseby
Heppel, Leon Alma
Hollenberg, Paul Frederick
Holten, David Duane
Ives, David Homer
Jenkins, Winborne Terry
Jennings, Allen Lee
Joseph, Ramon R
Jorns, Marilyn Schuman
Kapoor, Manju
Kelly, Susan Jean
Kennedy, James A
Kiefer, Helen Chilton
Kimball, Aubrey Pierce
Kimura, Tokuji
Kitz, Richard J
Kolenbrander, Harold Mark
Krakow, Gladys
Kushner, Sidney Ralph
Lartigue, Donald Joseph
Levin, Herman Westley
Lieberman, Jack
Lin, Ada Wen-Shung Ma
Long, George Louis
Main, Alexander Russell
Malhotra, Ashwani
Markert, Clement Lawrence
McCarville, Michael Edward
Meagher, Richard Brian
Messing, Ralph Allan
Mhatre, Nageshshamrao
Mildvan, Albert S
Miller, Richard Wilson
Morris, James Albert
Mozersky, Samuel M
Nason, Alvin
Nicholls, Doris Margaret

Nicoli, Miriam Ziegler
Noltmann, Ernst August
Passananti, Gaetano Thomas
Peanasky, Robert Joseph
Phillips, Louise Lang
Priest, David Gerard
Rao, Girimaji J Sathyanarayana
Rosen, Ora Mendelsohn
Rosner, Anthony Leopold
Schleyer, Heinz
Schwimmer, Sigmund
Scouten, William Henry
Shore, Joseph D
Singer, Anita Larks
Smith, Nathan Lewis, III
Snoke, Roy Eugene
Soodsma, James Franklin
Spradlin, Joseph Edward
Stadtman, Thressa Campbell
Stevens, Evelyn Victoria
Stewart, Lynn Martin
Straat, Patricia Ann
Strittmatter, Philipp
Suelter, Clarence Henry
Szymanski, Edward Stanley
Taylor, Kenneth Boivin
Terminiello, Louis
Trenholm, Harold Locksley
Tsai, Chishiun S
Vnek, John
Von Korff, Richard Walter
Wainio, Walter W
Weibel, Michael Kent
Westheimer, Frank Henry
Westley, John Leonard
Patton, Stuart
Schwimmer, Sigmund
Swaisgood, Harold Everett
Youngquist, Rudolph William

Food Biochemistry
Betz, Norman Leo
Cherry, John Paul
Crawford, David Lee
Deatherage, Fred E
Dollar, Alexander M
Honig, David Herman
Ke, Paul Jenn
Moats, William Alden
Palmer, James Kenneth
Patton, Stuart

Microbial Biochemistry
Aaronson, Sheldon
Adams, Gordon Albert
Aleem, M I Hussian
Andres, William Wolcott
Baptist, James (Noel)
Blakley, Edwin Raymond
Brown, Robert George
Bungay, Henry Robert, III
Burg, Richard William
Cort, Winifred Mitchell
Dawson, Peter Stephen Shevyn
Ferguson, Donald Allen, Jr
Fitz-James, Philip Chester
Georgi, Carl Edward
Greasham, Randolph Louis
Gum, Ernest Kemp, Jr
Hanson, Barbara Ann
Hidy, Phil Harter
Ho, Richard I-Fu
Hook, Derek John
Hurley, Laurence Harold
Hutchings, Brian Lamar
Jiu, James
Johnson, Marvin Joyce
Keister, Donald Lee
Killick, Kathleen Ann
Klucas, Robert Vernon
Kushner, Donn Jean
Lieberman, Hillel
Lueking, Donald Robert
Merkel, Joseph Robert
Morris, James Albert
Murray, Edward Donald
Peruzzotti, George Peter
Repaske, Roy
Righthand, Vera Fay
Roland, John Francis
Salsbury, Robert Lawrence
Sandford, Paul A
Sansing, Gerald Allen
Schwartz, Jeffrey Lee
Seitz, Eugene W
Shaw, Derek Humphrey
Shedlarski, Joseph George, Jr
Singh, Akhand Pratap
Twarog, Robert
Wagman, Gerald Howard
Walter, Richard Webb, Jr
Wise, Edmund Merriman, Jr
Wodzinski, Rudy Joseph

Nutritional Biochemistry
Adamson, Lucile Frances
Alam, Syed Qamar

Allison, Richard Gall
Ambrose, John Augustine
Ames, Stanley Richard
Barker, Norval Glen
Baumann, Carl August
Beecher, Gary Richard
Beitz, Donald Clarence
Bentley, Orville George
Berdanier, Carolyn Dawson
Bert, Mark Henry
Carlotti, Ronald John
Castell, John Daniel
Catignani, George Louis
Chang, Hsing-Tze Ruan
Chang, Mei Ling (Wu)
Charkey, Lowell William
Chow, Ching Kuang
Cook, David Allan
Cooperman, Jack M
Cousins, Robert John
Crosby, Lon Owen
Day, Harry Gilbert
Dixit, Padmaker Kashinath
Easterling, William Ewart, Jr
Edwards, Hardy Malcolm, Jr
Ellis, William Wesley
Farrier, Noel John
Feland, Sarah Elizabeth
Fordham, Joseph Raymond
Fox, Mattie Rae Spivey
Frier, Henry Ira
Fritz, James Clarence
Grove, John Amos
Hartsook, Elmer William
Haskell, Betty Echternach
Hayes, Johnnie Ray
Herting, David Clair
Hill, Eldon G
Hodson, Adrian Zachariah
Hoekstra, William George
Holub, Bruce John
Hoobler, Icie Macy
Hopkins, Daniel T
Hornstein, Irwin
Hunt, Charles E
Khairallah, Edward A
Koldovsky, Otakar
Lall, Santosh Prakash
Lee, Donald Jack
Lopez, Hady
Machlin, Lawrence Judah
Madappally, Mathew Mathai
Madsen, Kenneth Olaf
McCaughey, William Frank
McClain, Philip Edwin
Mehta, Tara
Mercer, Leonard Preston, II
Milne, David Bayard
Mohammed, Kasheed
Narayan, Krishamurthi Ananth
Narins, Dorice Marie
O'Dell, James Allen
Olson, James Allen
Painter, Ruth Coburn Robbins
Pelletier, Omer
Pinto, John
Prosky, Leon
Pubols, Merton Harold
Pyke, Ralph Edward
Ranhotra, Gurbachan Singh
Rechcigl, Miloslav, Jr
Roberts, Willard Lewis
Rodriquez, Mildred Shepherd
Roland, David Alfred, Sr
Sachan, Dileep Singh
Sanslone, William Robert
Sass, Neil Leslie
Scholz, Richard W
Schubert, John Rockwell
Sheppard, Alan Jonathan
Sherman, William Cyrus
Smith, John Edgar
Sundaresan, Peruvemba Ramnathan
Webb, Ryland Edwin
Whanger, Philip Daniel

Organic Biochemistry
Anglin, J Hill, Jr
Baum, George
Berlinguet, Louis
Blomstrom, Dale Clifton
Bryson, Thomas Allan
Cipera, John Dominik
Costello, Catherine E
Daniels, Peter John Lovell
Dheer, Surendra Kumar
Emken, Edward Allen
England, James Donald
Erickson, Bruce Wayne
Fendler, Janos Hugo
Fisher, George Harold, Jr
Ford, Jared Hewes
Giza, Yueh-Hua Chen
Gortatowski, Melvin Jerome
Han, Jerry C-Y
Hornig, Lilli Schwenk
Jones, John Bryan
Leonard, Nelson Jordan

Lutz, John George
Morneweck, Samuel
Mortensen, Harley Eugene
Newirth, Terry L
O'Connor, Rod
Powers, James Cecil
Ramanujam, V M Sadagopa
Regna, Peter P
Schroeder, Robert Samuel
Sessa, David Joseph
Sims, Homer Jennings
Tan, Liat
Thedford, Roosevelt
Tuhy, Peter Mirko
Wan, Kwok Ming

Physical Biochemistry
Abramson, Morris Barnet
Ackers, Gary Keith
Anderson, David Gene
Anderson, Lewis L
Antholine, William E
Aune, Kirk Carl
Bailey, James Michael
Baldwin, Robert Lesh
Baptist, Victor Harry
Barlow, Grant Harold
Barnett, Lewis Brinkly
Bello, Jake
Bernardin, John Emile
Berube, Gene Roland
Bleich, Hermann Ewald
Bolen, David Wayne
Brown, Eleanor Moore
Bucci, Enrico
Cameron, Bruce Francis
Cavalieri, Liebe Frank
Chun, Paul W
Cohen, Jack Sidney
Conover, Woodrow Wilson
Cope, Freeman Widener
Cox, David Jackson
Deal, William Cecil, Jr
Decker, Rolan Van
Denues, Arthur Russell Taylor
Deslauriers, Roxanne Marie Lorraine
Duke, Joseph A
Dunford, Hugh Brian
Dunn, Ben Monroe
Eaton, William Allen
Eisenhardt, Rudolph Hermann
Erenrich, Evelyn Schwartz
Erman, James Edwin
Fee, James Arthur
Fuller, Robert Arthur
Gaber, Bruce Paul
Gaskin, Felicia
Gellert, Martin Frank
Glusker, Jenny Pickworth
Gorin, George
Gray, Horace Benton, Jr
Halsey, John Frederick
Hamilton, Mary Jane Gill
Harris, Daniel Charles
Haselkorn, Robert
Hensens, Otto Derk
Hersh, Leroy S
Hill, Terrell Leslie
Hopkins, Thomas A
Ingham, Kenneth Culver
Irvine, George Norman
Jacobson, Homer
Janson, Thomas Ralph
Kelly, Philip James
Killion, Philip James
Klotz, Irving Myron
Klotz, Lynn Charles
Konecny, Jan
Kraut, Joseph
Kresheck, Gordon C
Kroon, Paulus Arie
Landsberger, Frank Robbert
Liberti, Paul A
Loach, Paul A
Lohr, Dennis Evan
Lovrien, Rex Eugene
Lyndrup, Mark Leroy
Manuck, Barbara Ann
Markley, John Lute
Martin, Francis Hall
Mattice, Wayne Lee
May, Leopold
Millar, David Bosie-Seurs, III
Mueller, Delbert Dean
Myer, Yash Paul
Pace, Carlos Nick
Parker, Helen Meister
Peticolas, Warner Leland
Raval, Dilip N
Redfield, Alfred Guillou
Rees, Allan W
Rifkind, Joseph Moses
Rill, Randolph Lynn
Roberts, Ronald C
Rosenberg, Robert Melvin
Rowe, Elizabeth Snow
Russo, Salvatore Frank
Saxena, Vishv Prakash

Schullery, Stephen Edmund
Schumaker, Verne Norman
Seaman, Geoffrey Vincent F
Shiao, Daniel Da-Fong
Shore, Joseph D
Smith, Linda Lou
Steiner, Robert Frank
Stone, Deborah Bennett
Storey, Bayard Thayer
Szuchet, Sara
Takahashi, Mark T
Takashima, Shiro
Teller, David Chambers
Timasheff, Serge Nicholas
Townend, Robert Edward
Trefonas, Louis Marco
Ward, Raymond Leland
Wei, Guang-Jong Jason
Westley, John Leonard
Wilhoit, Randolph Carroll
Williams, John Warren
Williams, Robley Cook, Jr
Wisnieski, Bernadine Joann
Young, Jay Maitland

Plant Biochemistry
Abrahamson, Lila
Albersheim, Peter
Augustin, Jorg A L
Bandurski, Robert Stanley
Benedict, Chauncey
Biggins, John
Breidenbach, Rowland William
Brown, Stewart Anglin
Bryan, Ashley Monroe
Camper, Nyal Dwight
Canvin, David T
Carson, Eugene Watson, Jr
Chapman, David J
Chen, Chong Maw
Cheniae, George Maurice
Christianson, Donald Duane
Clark, Ralph B
Clevenger, Sarah
Daly, Joseph Michael
D'Aoust, Andre Lucien
Dennis, David Thomas
Dull, Gerald G
Durley, Richard Charles
Ellsworth, Robert King
Evans, Harold J
Finkle, Bernard Joseph
Fox, J Eugene
Freebairn, Hugh Taylor
French, Richard Collins
Gamborg, Oluf Lind
Haard, Norman F
Hess, John Lloyd
Humphreys, Thomas Elder
Hylin, John Walter
Ibrahim, Ragai Kamel
Johnson, Morris Alfred
Jones, John Dewi
Kadkade, Prakash Gopal
Kahn, Joseph Stephan
Khan, Anwar Ahmad
Koeppe, David Edward
Kosuge, Tsune
Kuc, Joseph
LeTourneau, Duane John
Linden, James Carl
Lipke, William G
Maass, Wolfgang Siegfried Gunther
Maclachlan, Gordon Alistair
Mangat, Bhupinder Singh
Marriage, Paul Bernard
Marsho, Thomas V
Massey, Louis Melville, Jr
Mazelis, Mendel
McCalla, Arthur Gilbert
McCollum, Robert Edmund
McMahon, Vern August
Meeuse, Bastiaan Jacob Dirk
Meheriuk, Michael
Miller, Gene Walker
Norman, Arthur Geoffrey
Nozzolillo, Constance
Olson, Lee Charles
Pattee, Harold Edward
Patterson, Glenn Wayne
Peverly, John Howard
Phan, Chon-Ton
Punnett, Thomas R
Putman, Edison Walker
Racusen, David
Rauser, Wilfried Ernest
Rebstock, Theodore Lynn
Richards, George Manning
Roy, Harry
Schrader, Lawrence Edwin
Scott, Eion George
Shannon, Leland Marion
Shargool, Peter Douglas
Shrimpton, Douglas Malcolm
Siegelman, Harold William
Stallknecht, Gilbert Franklin
Sweetser, Philip Bliss
Thompson, John Fanning

Thornber, James Philip
Tolbert, Robert John
Towers, George Hugh Neil
Underhill, Edward Wesley
Uribe, Ernest Gilbert
Wallace, James William, Jr
Wallace, Joan M
Walton, Daniel C
Wang, Dalton Ta Tung
Watkin, John Emrys
Webb, Bill D
Weete, John Donald
White, Fred G
Whitehead, Eugene Irving
Williams, Gene R
Williams, John Peter
Wilson, Curtis Marshall
Wilson, David George
Winget, Gary Douglas
Zachariua, Robert Marvin
Zitnak, Ambrose

Soil Biochemistry
Albritten, Herbert Graves
Clapp, Charles Edward
Engbrog, James Charles
Skujins, John Janis
Young, J Lowell

Bond, Ted P

Biology
Abbatiello, Michael James
Adler, Beatriz Raquel
Ahlberg, Henry David
Ahlstrom, Elbert Halvor
Akins, Virginia
Albright, John T
Alderman, Louis Cleveland, Jr
Allegre, Charles Frederick
Alleva, John J
Allewell, Norma Mary
Amarose, Anthony Philip
Anderson, Ernest Clifford
Anderson, George Cameron
Anderson, Herbert Godwin, Jr
Anderson, John Murray
Anderson, Roger Arthur
Ansevin, Krystyna D
Araki, George Shoichi
Arce, Gina
Archimovich, Alexander S
Arnold, Frederic G
Avault, James W, Jr
Avena, Remedios M
Awbrey, Frank Thomas
Axman, Mary Claudine
Babanian, Rostom
Bacharach, Martin Max
Bagby, John R, Jr
Bailey, Donald Etheridge
Bailey, Norman Sprague
Baker, Edgar Gates Stanley
Bamrick, John Francis
Bancroft, John Basil
Banks, Edwin Melvin
Bannister, Thomas Turpin
Barclay, Eugene Samuel
Barclay, Frank Hunt
Barnes, Adele
Barnes, Eugene Miller, Jr
Barnett, Eugene Victor
Barr, Harold Jay
Bartel, Harold William
Bashour, Fouad A
Baughn, Charles (Otto), (Jr)
Baum, Werner Christian
Bayless, Laurence Emery
Be, Allan Wie Hwa
Beary, Dexter F
Becker, Gweneth (Leslie)
Bell, John Perkins
Bellamy, Raymond Edward
Belmonte, Rocco George
Belsky, Melvin Myron
Benedict, Irvin J
Benoit, Jean Claude
Benson, Norman G
Benzing, David H
Berg, William Eugene
Bergstresser, Kenneth A
Berman, Rose Louise
Bernard, Marie (Witte)
Bhalla, Satish Chander
Biggers, Charles James
Bingham, Nelson Eldred
Birdsong, Ray Stuart
Birkenholz, Dale Eugene
Bishop, Everett Lassiter, Jr
Blank, Robert H
Blanton, Madison Van
Bloom, William Whiley
Boell, Edgar John
Boeng, Robert William
Bogenschutz, Robert Parks
Bonfiglio, Michael
Bongiovanni, Alfred Marius
Booth, Ernest Sheldon
Boozer, Reuben Bryan
Borenstein, Samuel R
Borysko, Emil
Boudreault, Armand
Boyarsky, Lila Harriet
Boyd, Ivan Louis
Brained, John Whiting
Bratz, Robert Davis
Breakey, Donald Ray
Brenowitz, A Harry
Bridgman, Anna Josephine
Briggs, Robert Wilbur
Britten, Bryan Terrence
Brock, Mary Anne
Brooks, Merle Eugene
Brown, Donald Frederick Mackenzie
Brown, Frank Arthur, Jr
Brown, Howard S
Brown, Robert Francis
Brown, Walter Creighton
Brown, William Louis
Brubaker, Kenton Kaylor
Bruce, Harold Asa
Bruton, Charles William
Bucy, LaVerne
Buffaloe, Neal Dollison
Buffett, Rita Frances
Burdick, Harold Charles
Burke, William Thomas, Jr
Burnett, Allison L
Burt, Michael David Brunskill
Bussey, Arthur Howard
Busted, Robert Charles
Butler, Earl Orlo
Butts, David
Byers, Thomas Jones
Campbell, James A
Campo, Robert D
Caraway, Prentice Alvin
Carson, Theophilus Roosevelt
Cerroni, Rose E
Chappel, Clifford
Chen, Shepley S
Cheng, Thomas Clement
Chesnut, Clarence, Jr
Chew, Frances Sze-Ling
Chin, Edward
Chirst, John Conrad
Christensen, Eric
Christian, Howard Harris
Churchill, Helen Mar
Cibula, Adam Burt
Claflin, Tom O
Clark, Ervil Delwyn
Clark, Charles Henry Douglas
Clay, Mary Ellen
Cohen, Alan Mathew
Cohn, Sidney Arthur
Coleman, Richard Walter
Collings, William Doyne
Commoner, Barry
Cook, Eugene Wilbur, Jr
Cooper, Cary Wayne
Cooper, Edwin Lowell
Cooper, Robert Woodrow
Copeland, Donald Eugene
Copeland, Frederick Cleveland
Coppinger, Raymond Parke
Coyle, Elizabeth Eleanor
Coyle, Frederick Alexander
Craik, Eva Lee
Crall, Howard William
Crissey, Walter Ford
Cummings, Jean Marie
Curtin, Charles Byron
Cushing, John (Eldridge), Jr
Daggy, Tom
Dalton, Albert Joseph
Daniel, Charles Pack
Daniel, Charles Walter
Danielli, James F
DaVanzo, John Paul
Day, Lore Rose
Day, Ivana Podvalova
DeBoer, Kenneth F
Deevey, Edward Smith, Jr
DeHaan, Robert Lawrence
DeLanney, Louis Edgerton
Delbruck, Max
Delfin, Eliseo Dais
Denning, Jack
DeShaw, James Richard
de Torok, Denes Gabor
DiPasquale, Gene
Dobbs, Harry Donald
Doell, Ruth Gertrude
Doermann, August Henry
Donaldson, Lauren Russell
Doty, Stephen Bruce
Dowdy, William Wallace
Dropp, John Jerome
Drum, Ryan William
Dudzinski, Diane Marie
Dufour, Didier
Dunn, Richard Hudson
DuPraw, Ernest Joseph
Durand, James Blanchard
Dutta, Hiram Moyee
Eberhardt, Lester Lee
Edney, Norris Allen
Ehrlich, Paul Ralph
Ellis, Derek V
Ellis, John Francis
Emmons, Lyman Randlett
Enesco, Mircea Aaron
Eschmeyer, Paul Henry
Eskew, Cletis Theodore
Everhart, Watson Harry
Fan, David P
Feinberg, Edwin Harold
Fender, Derek Henry
Ferm, Vergil Harkness
Fetner, Robert Henry
Finley, David Emanuel
Fischer, Richard Bernard
Fisler, George Frederick
Fontaine, Julia Clare
Foote, Kenneth Gerald
Ford, James
Fortner, Joseph Gerald
Fowler, Horatio Seymour
Fozdar, Birendra Singh
Freiberg, Samuel Robert
Fremount, Henry Neil
Frenzel, Louis Daniel, Jr
Friedland, Beatrice L
Froiland, Sven Gordon
Frye, Ozro Earle, Jr
Fryer, Holly Claire
Fuller, Dorothy Langford
Fulton, John Donaldson
Gabriel, Mordecai Lionel
Gambrell, Lydia Jahn
Gammon, James Robert
Garcia, Alfredo Mariano
Gardner, Gerard
Garth, Richard Edwin
Gary, Roland Thacher
Gatzy, John Thomas
Gautereaux, Ione
Gay, Helen
Geddes, David Darwin
Geis, Aelred Dean
Geisler, Grace
George, John Lothar
Gerber, Bernard Robert
Germann, Paul Julian
Ghidoni, John Julian
Giacometti, Luigi
Gibbs, Martin
Gibson, Walter William
Giese, Arthur Charles
Gilbert, Frank Albert
Gilbert, Margaret Shea
Giltz, Maurice Leroy
Glicksman, Arvin Sigmund
Goldman, Max
Goldsmith, Mary Helen Martin
Gonzales, Federico
Goodheart, Clyde Raymond
Goodnight, Marie Louise
Gorovsky, Martin A
Gorski, Leon John
Graef, Philip Edwin
Greeley, Frederick
Green, Marcus Herbert
Greenberg, Joseph
Griffin, Donald Redfield
Griffin, Gary J
Griffiths, Raymond Bert
Grillo, Ramon S
Grobstein, Clifford
Groner, Miriam Georgia
Gross, Jerome
Gross, Phyllis P
Gundy, Samuel Charles
Gunther, Waldemar Carl
Gusmano, Ernest Ambrose
Gutman, Burton Samuel
Hagmeier, Edwin Moyer
Halazon, George Christ
Hamann, Cecil Boyce
Hamburgh, Max
Hamilton, John Meacham
Hamre, Harold Thomas
Hancock, Kenneth Farrell
Handler, Shirley Wolz
Hanes, Deanne Meredith
Hanes, Harry Louis
Hansen, Keith Leyton
Hansen, Willis Dale
Hardin, Garrett (James)
Hargis, Betty Jean
Harmsen, Rudolf
Harris, Venoia M
Harris, John Wallace
Hartig, William John
Harvey, Harry Thomas
Harvey, William Royal
Hassan, Ikram Ul
Hathaway, Wilfred Bostock
Haubrich, Robert Rice
Hauser, Mary Martin
Hawbecker, Albert Claude
Hayat, M A
Hayes, John Thompson
Heine, Ursula Ingrid
Heinrich, Gerd H
Heise, John J
Hemphill, Donald Vincent
Hensill, John Samuel
Heyer, William Ronald
Heyn, Anton Nicolaas Johannes
Hexter, William Michael
Hew, Choy-Leong
Hess, Archie Davilla
Hess, Edwin A
Hill, Helene Zimmermann
Hillman, Ralph
Himes, Marion
Himmel, Keith LaVern
Hinkley, Robert Edwin, Jr
Hitchcock, Dorothy Jean
Ho, Andrew K S
Hofsetter, Adrian Marie
Holle, Paul August
Holm, Richard William
Hoogstraal, Harry
Hooper, George Bates
Howard, Robert Stearns
Hsu, Yu-Chih
Huang, Chester Chen-Chiu
Hubbard, Ruth
Hubby, John L
Huff, George Charles
Hungate, Robert Edward
Hurd, Paul DeHart
Hussey, Kathleen Louise
Huszar, Gabor
Hutchinson, Robert Lynn
Inoue, Shinya
Isenberg, George Raymond, Jr
Isseroff, Hadar
Iverson, Ray Mads
Jackson, Elizabeth Burger
Jacobs, George Joseph
Jacobsohn, Myra K
Jacques, Felix Anthony
Jahoda, William John
Jakus, Marie A A
Jeffers, Edmund E
Jenner, Charles Edwin
Johansson, Mildred P
Johnson, David Franklin
Johnson, Thomas A
Jones, Allan W
Jones, Kenneth La Mar
Joram, Philip Robert
Jordan, Chris Sullivan
Josephson, Betty Louise
Kahlon, Prem Singh
Kalberer, John Theodore, Jr
Kang, Yeou-Jan
Kapler, Joseph Edward
Karas, James Glynn
Karklins, Olgerts Longins
Karolyi, Elmer Joseph
Kaston, Benjamin Julian
Kastrinos, William, Jr
Katsh, Seymour
Kay, Elizabeth Alison
Keck, Konrad
Keefe, Thomas Leeven
Keen, Veryl F
Keeshan, Margaret M
Kelly, Richard Delmer
Kelner, Albert
Kernaghan, Roy Peter
Kethley, Thomas William
Kidder, George Wallace
Kimball, Richard Fuller
Kimeldorf, Donald Jerome
King, David Beeman
King, John Wesley
Kirk, Daniel Eddins
Kiser, Roy Stone
Klein, Jan
Klinge, Paul E
Klug, William Stephen
Koevenig, James L
Komarek, Edwin Vaclav
Konishi, Masakazu
Koppelman, Ray
Kouba, John Emanuel, Jr
Kramer, Julian
Kramer, Sol
Krisko, Istavan
Krogmann, David William
Kuehn, Harold Herman
Kuralle, Henry III
Kutsky, Roman Joseph
Lachance, Jean Paul
Lake, James Albert
Landt, James Frederick
Lapin, David Marvin
Laug, George Milton
Lawrence, (Roche), Mary Anna
Lawson, Chester Alvin
Lazaruk, William
Lea, Malcolm Sinclair

Leach, Berton Joe
Leavitt, Lewis A
Ledney, George David
Lee, Joseph Chuen Kwun
Leraas, Harold J
Levin, Simon Asher
Levine, Rhea Joy Cottler
Lewis, Edward B
Lightner, Jerry P
Lind, Owen Thomas
Lindstedt-Siva, K June
Linduska, Joseph Paul
Lisonbee, Lorenzo Kenneth
Liss, Maurice
Little, Ellis Beecher
Livezey, Robert Lee
Lockett, M Clodovia
Loefer, John B
Lofgren, Ruth
Loughton, Barry G
Loveland, Robert Edward
Lowry, Edward MacLean
Ludwig, Carl Edward
Lundy, Talmage E
Lyman, Charles Peirson
MacFadden, Donald Lee
MacFarlane, Constance Ida
Mackey, James P
MacLean, William Plannette, III
Madden, John William
Mahoney, Colette
Mahoney, Francis Joseph
Maloney, Mary Adelaide
Malsberger, Richard Griffith
Mandell, Alan
Mandl, Richard H
Manley, Lillian C
Mann, Stanley Joseph
Manner, Harold Wallace
Markovetz, Allen John
Martin, M Celine
Martin, Virginia Lorelle
Masteller, Edwin C
Mattingly, Mary Ellen
McClurkin, Iola Taylor
McClurkin, John Irving, Jr
McCluskey, Elwood Sturges
McCullough, Herbert Alfred
McCully, Margaret E
McCutchen, Charles Walter
McDiarmid, Roy Wallace
McDonald, Stuart
McElroy, William David
McGee-Russell, Samuel M
McKee, John W
McLaren, Arthur Douglas
McLaren, Ian Alexander
McManus, Margaret Ann (Mary Annunciata)
McMillan, Harlan L
McMorris, Frederick Arthur
Mecca, Christyna Emma
Medora, Rustem Sohrab
Meiselman, Newton
Meiss, Alfred Nelson
Meng, Heinz Karl
Merrill, Reynold Cluff
Mettler, Lawrence Eugene
Michael, William Earl
Miller, Charles William
Miller, Donald Elbert
Miller, William Henry
Milne, Lorus Johnson
Milne, Margery (Joan) (Greene)
Mintz, Beatrice
Mirand, Edwin Albert
Moloney, John Bromley
Money, Kenneth Eric
Moorehead, William Douglas
Mork, David Peter Sogn
Morowitz, Harold Joseph
Morton, Joseph James Pandozzi
Morton, Thomas Harlow
Mouser, Gilbert Warren
Muffley, Harry Chilton
Mullahy, John Henry
Mulrennan, Cecilia Agnes
Munro, David (Aird)
Murphey, Frank J
Murphy, Clifford Elyman
Murphy, Ted Daniel
Musgrave, Anthony John
Nafpaktitis, Basil G
Napolitano, Joseph J
Nash, Carroll Blue
Nasrallah, Mikhail Elia
Nauman, Charles Hartley
Neal, Louise Adelaide
Neff, Robert Jack
Neff, Stuart Edmund
Nelson, Arthur Hansen
Nelson, Clarence Herbert
Nelson, Gideon Edmund, Jr
Newell, John T, II
Nichols, Herbert Wayne
Nishizawa, Edward Eichi
Norris, Bill Eugene
Norris, Dale Melvin, Jr

Norstog, Knut Jonson
Northrop, John Howard
Novak, Joseph Donald
Nunnally, David Ambrose
Nutting, Leighton Adams
Nye, Warren Edward
Oberlander, Herbert
Odell, Lois Dorothea
Okonkwo, Augustine Ikechukwuka
Ondricek, Anatola
Oppenheimer, Jane Marion
Orgebin-Crist, Marie-Claire
Orr, Henry Clayton
Oshima, Eugene Akio
Ostapiak, Mykola
Osterman, George B
Ove, Peter
Overmire, Thomas Gordon
Owens, John Harold
Owers, Noel Oscar
Ozburn, George W
Ozer, Harvey Leon
Palincsar, Edward Emil
Pallotta, Dominick John
Palm, John Daniel
Parrish, Fred Kenneth
Patt, Donald Irving
Patten, John A
Pattie, Donald L
Pearson, Allen Mobley
Peightel, William Edgar
Pentney, Roberta Pierson
Pepinsky, Raymond
Perkins, David Dexter
Perrine, John W, Jr
Peterle, Tony J
Peters, Joseph John
Petry, Robert Kendrick
Pielou, William P
Pierro, Louis John
Pierson, Edgar Franklin
Pisano, Rocci George
Pittendrigh, Colin Stephenson
Polinger, Iris Sandra
Pollard, Joseph Page
Pollister, Priscilla Frew
Porter, Homer Clifford
Pottinger, M Aelred
Powell, Thomas Edward
Power, Richard B
Powers, William Thomas
Preble, Norman Alexander
Prestwidge, Kathleen Joyce
Prevec, Ludvik Anthony
Prince, Alton Ernest
Privitera, Carmelo Anthony
Proietto, Lillian Jacqueline
Proudfoot, Bernadette Agnes
Quinn, Cosmas Edward
Quintarelli, Giuliano
Racle, Fred Arnold
Rainbolt, Mary Louise
Ramsey, John S
Randall, John Frank
Rasch, Ellen M
Rathjen, Warren Francis
Raychaudhuri, Anilbaran
Reardon, John Joseph
Reddish, Paul Sigman
Redman, Charles Edwin
Reedy, John Joseph
Reese, William Dean
Regnery, David Cook
Remington, Charles Lee
Remsen, Charles C, III
Reynolds, William Wallace
Riches, Ralph Harvard
Riemer, William John
Riffey, Meribeth M
Rilett, Robert Omar
Riley, Charles Victor
Ripley, Thomas H
Roberts, Evan Paul
Robinson, Kent
Robinson, Radcliffe Franklin
Rodden, Robert Morris
Roedel, Philip Morgan
Roeder, Kenneth David
Rogers, Rodney Albert
Rollins, Wade Cuthbert
Romanoff, Anastasia
Romsdahl, Marvin Magnus
Rosenberg, Evelyn Kivy
Rosenberg, Marvin J
Rosi, David
Ross, Morris H
Roth, Robert Mark
Rothman, Alvin Harvey
Rovee, David Thomas
Rucker, Ellis Suttle
Runyon, Ernest Hocking
Rusch, Wilbert, Sr
Russell, Helen Ross
Russell, Norman Hudson, Jr
Ryan, Simeon P
Ryan, Thomas I
Sacher, George Alban, Jr
Sanders, Frank Kingsley

Sato, Gordon Hisashi
Scharf, Arthur Alfred
Scheving, Lawrence Einar
Schloemer, Clarence Louis
Schmid, Peter
Schmitt, Allen F
Schmitt, Francis Otto
Schwartz, Norman Louis
Schwartz, Norman Martin
Schwassman, Horst Otto
Schwendinger, Richard B
Scott, David Bytovetzski
Scudder, Harvey Israel
Seaman, Edna
Seeburger, George Harold
Selberg, Edith Marie
Selser, Will Lindsey
Shade, Elwood B
Shannon, Ira Lenwood
Shannon, Jerry A, Jr
Shawver, Murl Charles
Shear, William Albert
Shepard, David C
Sheridan, Richard P
Sherman, Frederick George
Sherman, Thomas Fairchild
Shields, Arthur Randolph
Shields, Lora Mangum
Shilling, Paul R
Shinn, Eugene Allen
Siegel, Barbara Zenz
Siegel, Charles David
Siemens, George John
Simmons, Eric Leslie
Simmons, Samuel William
Simon, Stephen Wistar
Simpson, Tracy L
Singer, Seymour Jonathan
Slavin, Ovid
Sledge, Eugene Bondurant
Sleeper, David Allanbrook
Sloan, William Cooper
Smeriglio, Alfred John
Smith, Allyn Goodwin
Smith, Arlo Irving
Smith, Charles G
Smith, Hugh Burnice
Smith, James Lee
Smith, Lawrie Booth
Smith, Norman Cutler
Smith, W John
Snitgen, Donald Albert
Snyder, Donald Benjamin
Snyder, Jack Willard
Sognnaes, Reidar Fauske
Sohal, Rajinder Singh
Solberger, Dwight Ellsworth
Solomon, Marvin David
Soltero, Raymond Arthur
Spann, Liza Agnes
Spear, Irwin
Spears, Joseph Faulconer
Spencer, Selden J
Spiegel, Melvin
Spiroff, Boris E N
Spivak, Monroe Leon
Sprague, Isabelle Baird
Sprague, John Booty
Springer, Martha Edith
Springer, Victor Gruschka
Spuhler, James Norman
Srivastava, Lalit Mohan
Staal, Gerardus Benardus
Stableford, Louis Tranter
Starling, James Holt
Stauffer, Robert Clinton
Stechschulte, Agnes Louise
Steen, James Southworth
Stegner, Robert W
Stere, Athleen Jacobs
Stevenson, Robert Thomas
Stewart, Herbert
Stewart, James Ray
Stock, David Allen
Stokes, Lee W
Stombaugh, Tom Atkins
Stone, Frederick Logan
Storlazzi, Joseph Jordan
Strand, Fleur Lillian
Straughan, Isdale (Dale) Margaret
Straus, Helen Lorna Putkammer
Streett, James Clark, Jr
Strehler, Bernard Louis
Strickland, John Claiborne
Strohecker, Henry Frederick
Styles, Twitty Junius
Sullivan, Patricia Ann Nagengast
Sumner, Lowell
Suter, M St Agatha
Sutton, Dallas Albert
Swallow, Richard Louis
Swan, Lawrence Wesley
Sweet, Merrill Henry, II
Swezey, William Weekley
Sziklai, Oscar
Tanner, Gary Dale
Tautvydas, Kestutis Jonas
Taylor, Albert Cecil

Taylor, Dorothy Jane
Taylor, James Herbert
Tegenkamp, Thomas Richard
Te Paske Everett Russell
Thomas, Dempsey Lee
Thomas, Martin Lewis Hall
Thomas, McCalip Joseph
Thompson, Henry Joseph
Thomson, Junius Richard
Thorington, Richard Wainwright, Jr
Thorne, Oakleigh II
Thornton, Robert Melvin
Threlkeld, Stephen Francis H
Thureson-Klein, Asa Kristina
Tobin, Thomas Vincent
Tokay, Elbert
Topoff, Howard Ronald
Towner, Howard Frost
Townsend, Samuel Franklin
Trapido, Harold
Trimble, Mary Ellen
Truscott, Basil Lionel
Tucker, Charles Eugene
Tucker, Marie
Tyndall, Jesse Parker
Uhlig, Hans Gerd
Uhlrich, Helen Marie
Underhill, Raymond Alden
Urban, John
Uricchio, William Andrew
Van Bavel, Cornelius H M
Van Denack, Julia Marie
Van Deventer, William Carl
Van Overbeek, Johannes
Vasu, Bangalore Seshachalam
Vergeer, Teunis
Verna, John E
Vethamany, Victor Gladstone
Vinje, Mary M
Vogt, Peter Klaus
Wagner, Frederic Hamilton
Wagoner, Dale E
Walker, Kenneth Merriam
Walker, Philip Caleb
Walsh, Robert A
Walton, Cyprian James
Waugh, David Floyd
Weaver, Charles R
Webber, Herbert H
Webber, Patrick John
Webster, Dwight Albert
Webster, John Malcolm
Weis, Jerry Samuel
Weiss, Paul Alfred
Wells, Patrick Harrington
Wells, Russell Frederick
Wenzel, Frederick J
Westfall, Minter Jackson, Jr
Westman, James Ross
Whaley, William Gordon
White, James Edwin
White, Winifred Sharlene
Whitney, Marion Isabelle
Whitney, Rae
Widdowson, David Carl
Wiebe, Harold T
Wilde, Charles Edward, Jr
Wilkes, Alfred
Wilkes, James C
Williams, Carroll Milton
Williams, Cecil R
Williams, Donald Benjamin
Williams, Frederick McGee
Willis, Robert Ellis
Willis, David Lee
Wills, Christopher J
Willson, Dan Leroy
Wilmoth, James Herdman
Wilson, Frances G
Wilson, Leslie
Winchester, Albert McCombs
Winfree, Arthur T
Winnail, Douglas Samuel
Winokur, Morris
Winsmann, Fred Rudolph
Winston, Paul Wolf
Wischnitzer, Saul
Witkus, Eleanor Ruth
Witters, Weldon L
Wittry, Esperance
Wolf, Frank E
Wolf, Leonard Nicholas
Woodard, James W
Woodard, John
Woodworth, Robert Hugo
Work, Telford Hindley
Wu, Ching Kuei
Wu, Chun Kwun
Wylie, Richard Michael
Wyllie, Gilbert Alexander
Yates, Harris Oliver
Ycas, Martynas
Yerganian, George
Young, Frank Nelson, Jr
Young, Joseph Hardie
Young, Robert M
Youngren, Newell A
Zavortink, Thomas James

37

Zikakis, John Philip
Zimmer, George P
Zubay, Geoffrey
Zucker, Robert Martin

Animal Parasitology
Allen, Rex Wayne
Amrein, Yost Ursus Lucius
Behlow, Robert Frank
Chobotar, Bill
Collins, Jeffery Allen
Cuckler, Ashton Clinton
Eckman, Michael Kent
Esch, Gerald Wisler
Hughins, Ernest Jay
Mead, Robert Warren
Nathan, Henry C
Schaefer, Frank William, III
Seidel, Michael Caspar
Vegors, Halsey Hugh
Williams, James Carl
Wilson, Grant Ivins

Aquatic Biology
Bartsch, Alfred Frank
Beal, Ernest O
Beyers, Robert John
Bishop, John Watson
Bisson, Peter Andre
Blake, Peter Andre
Bodola, Anthony
Boles, Robert Joe
Brehmer, Morris Leroy
Buikema, Arthur L, Jr
Chamberlain, William Maynard
Chapman, Gary Adair
Clother, William Delbert
Colby, Peter J
Davies, Harold William, (Jr)
Davis, William Jackson
Deonier, D L
Dickson, Kenneth Lynn
Dinsmore, Bruce Heasley
Duffer, William Riley
Garton, Ronald Ray
Geen, Glen Howard
Geiger, Robert Warren
Grantham, Billy Joe
Hall, Gordon Earl
Hamilton, Robert Duncan
Hart, Charles Willard, Jr
Hawxby, Keith William
Hayes, Helen Landau
Hedgpeth, Joel Walker
Heller, Thomas Robert, Jr
Henschele, Ann
Horne, Alexander John
Howell, Henry Howze
Jahn, Lawrence A
Kerr, John Polk
Kevern, Niles Russell
Knight, Luther Augustus, Jr
Mackenthun, Kenneth Marsh
Magnuson, John Joseph
McCosker, John E
Mette, Maurice Ferdinand
Moore, Walter Guy
Oglesby, Ray Thurmond
Oliver, Kelly Hoyet, Jr
Olive, John H
Quinn, Barry George
Rees, John Tonks
Riemer, Donald Neil
Roelofs, Eugene Woodrow
Rofen, Robert Rees
Rose, Frederick Louis, Jr
Saksena, Vishnu P
Sawell, James Joseph
Schumacher, Robert E
Smith, Alden Ernest
Smith, Stanford Dennis
Sparks, Richard Edward
Stewart, Rolland Keith
Straw, Thomas Eugene
Tucker, John Shepard
Verch, Richard Lee
Wallace, James Bruce
Walter, William T
Warner, Mark Clayson
Webber, Harold Haskell
Weis, Judith Shulman
Whitworth, Walter Richard
Wildman, Ruth Bowman
Windell, John Thomas

Behavioral Biology
Adkins, Elizabeth Kocher
Beeman, Elizabeth Ann
Blizard, David Arthur
Brady, Joseph Vincent
Brener, Frederic J
Brodie, Edmund Darrell, Jr
Carde, Ring Richard Tomlinson
Chapman, Loring Frederick
Davis, Roger (Edward)
Doty, Richard Leroy
Eberhard, William Granville
Ferguson, James Mecham

Forester, Donald Charles
Griswold, Joseph Garland
Halpern, Bruce Peter
Hamilton, Charles R
Hayes, Keith James
Hill, James Leslie
Holldobler, Berthold Karl
Johnson, Alan Kim
Kreithen, Melvin Louis
Kristal, Mark Bennett
Lent, Peter C
McKaye, Kenneth Robert
Mushinsky, Henry Richard
O'Brien, William Daniel, Jr
Rake, Adrian Vaughan
Regal, Philip Joe
Ryker, Lee Chester
Sachs, Benjamin David
Shaklee, Alfred Barral
Wilson, Edward Osborne
Wilson, James Russell

Bioacoustics
Baptista, Luis Felipe
Best, LaVar
Brown, Richard Dean
Hersey, John Brackett
Johnson, Daniel Leon
Kelly-Fry, Elizabeth
Krenkau, Frederick William
O'Brien, William Daniel, Jr
Ryker, Lee Chester
Stebbins, William Cooper
Von Gierke, Henning Edgar
Young, Ruth Steuart

Bioengineering
Abbrecht, Peter H
Ackerman, Roy Alan
Alfrey, Clarence P, Jr
Amstutz, Harlan Cabot
Asgar, Kamal
Atinger, Ernst Otto
Bacus, James William
Bahill, Andrew Terry
Barry, William Earl
Cain, Charles Alan
Chang, Peter Hon
Cox, Robert Harold
Downie, David Ernest
Drinker, Philip Aldrich
Dunn, Floyd
Ewing, Channing Lester
Feinstein, Robert
Fetzner, Victor H
Frankel, Victor H
Gardner, Reed McArthur
Goodman, Lester
Gruber, George J
Hall, Ernest Lenard
Hansmann, Douglas R
Hartrum, Thomas Charles
Heetderks, William John
Herzinger, George Arthur
Hughes, Everett C
Kahn, Alan Richard
Kleiman, Devra Gail
Krouskop, Thomas Alan
Lemp, John Frederick, Jr
Lewis, Edwin Reynolds
Liang, Tung
Lieberman, Emanuel Roy
Longini, Richard Leon
Maseidvaag, Frode
Meerbaum, Samuel
Miraldi, Floro D
Murphy, Donald Henry
Murphy, Terence W
Newkirk, John Burt
Nicoloff, Demetre M
Noordergraaf, Abraham
Nudelman, Harvey Banet
Oakley, Burks, II
Pauley, James Donald
Polhemus, John Thomas
Replogle, Clyde R
Rushmer, Robert Frazer
Sachs, Alvin Howard
Salkind, Alvin J
Saltzberg, Bernard
Schamadan, James Louis
Schmidt, Edward Matthews
Schuder, John Claude
Sheehan, Brian Talbot
Shapiro, Howard Maurice
Singer, Jerome Ralph
Smolen, Victor Frank
Stacy, Ralph Winston
Stark, Lawrence
Steadman, John William
Strabucker, Robert A
Su, Tah-Mun
Susskind, Charles
Szarka, Laszlo Joseph
Topham, William Sanford
Van Den Bosch, Frank Joseph Gerard
Wall, Conrad, III
Ware, Ray Wilsford
Werblin, Frank Simon

Whiffen, James Douglass
Wilkins, Ebtisam A M Seoudi
Yee, Sinclair Shee-Sing
Yellin, Edward L
Zelman, Allen
Zingg, Walter
Zweifach, Benjamin William

Biogeography
Andrle, Robert Francis
Bennett, Charles Franklin
Burrill, Robert Meredith
Cornell, Howard Vernon
Crosswhite, Frank Samuel
Fonaroff, Leonard Schuyler
Foster, Robert H
Fraser, Donald Alexander
Gade, Daniel Wayne
Gaines, John Franklin
Gordon, Burton LeRoy
Gressitt, Judson Linsley
Holzael, Christina Marie
Isaac, Erich
Johannessen, Carl L
McMillan, R Bruce
Mikesell, Marvin Wray
Powell, John Martin
Rees, John David
Rounds, Richard Clifford
Sauer, Jonathon Deninger
Schiel, Joseph Bernard, Jr
Schmid, James Addison
Street, John Addison
Thorne, Robert Folger
Van Royen, Pieter
Walter, Hartmut
Wilhelm, Eugene J, Jr

Biological Anthropology
Armelagos, George John
Austin, Donald Mac
Baker, Paul Thornell
Barden, Howard Stavers
Bennett, Kenneth A
Bland, John Edward
Buikstra, Jane Ellen
De Pena, Joan Finkle
Dethlefsen, Edwin Stewart
Dyke, Bennett
Dyson-Hudson, V Rada
Eckhardt, Robert Barry
Freeman, Milton Malcolm Roland
Fry, Edward Irad
Garruto, Ralph Michael
Gibbons, Michael Francis, Jr
Harpending, Henry Cosad
Howell, Francis Clark
Klepinger, Linda Lehman
Kunstadter, Peter
Lamb, Neven P
Lieberman, Leslie Sue
Mazess, Richard B
McHenry, Henry Malcolm
Miller, Peter S
Murad, Turhon Allen
Novak, Ladislav Peter
Olsen, Stanley John
Otten, Charlotte Marie
Oxnard, Charles Ernest
Poirier, Frank Eugene
Sank, Diane
Saul, Frank Philip
Sciulli, Paul William
Shapiro, Harry Lionel
Steegmann, Albert Theodore, Jr
Swedlund, Alan Charles
Tiger, Lionel
Warren, Charles Preston
Wetherington, Ronald K
Witt, Shirley Hill

Biological Oceanography
Abbott, Isabella Aiona
Alvarino de Leira, Angeles
Alverson, Dayton L
Anderson, Paul Sigfried, Jr
Aron, William Irwin
Ayers, John Carr
Backus, Richard Haven
Banner, Albert Henry
Banse, Karl
Beers, John R
Berner, Leo Devitte, Jr
Best, Edgar Allan
Bienfang, Paul Kenneth
Bigelow, Maurice Hubbard
Blanton, William George
Bliss, Chester Ittner
Blumenthal, Reuben R
Boden, Brian Peter
Boesch, Donald Friedrich
Borom, John Lee
Boyd, Carl M
Bradshaw, John Stratili
Bright, Thomas J
Brinton, Edward
Brooks, Albert Lane
Brown, Dail Woodward

Brunbaugh, Joe H
Bullis, Harvey Raymond, Jr
Butman, John D
Bush, Louise Fulton
Butler, Philip Alan
Carey, Andrew Galbraith, Jr
Carlisle, John Griffin, Jr
Carpenter, Edward J
Carter, John C H
Cheung, Paul James
Chew, Kenneth Kendall
Childress, James J
Collier, Albert Walker
Courtenay, Walter Rowe, Jr
Crabtree, David Melvin
Cummings, William Charles
Crandell, George Frank
Curl, Herbert (Charles), Jr
Cutler, Edward Bayler
Davis, Curtiss Owen
Dean, David
Deevey, Georgiana Baxter
Demond, Joan
de Sylva, Donald Perrin
Devol, Allan Houston
Dickie, Lloyd Merlin
Digby, Peter Saki Bassett
Di Girolamo, Rudolph Gerard
Dobkin, Sheldon
Dugdale, Richard Cooper
Dunstan, William Morgan
Eickstaedt, Lawrence Lee
Eisler, Ronald
Eldredge, Lucius G
English, Thomas Saunders
Epifanio, Charles Edward
Eppley, Richard Wayne
Esaias, Wayne Evor
Fast, Thomas Normand
Fay, Roger Richard
Feder, Howard Mitchell
Fehlmann, Herman Adair
Fischer, Eugene Charles
Gilmartin, Malvern
Glude, John Bryce
Gonor, Jefferson John
Gonzalez, Juan Gerardo
Gordon, Bernard Ludwig
Gordon, Malcolm Stephen
Graham, Jeffrey Brent
Granger, Edward Henry
Green, George C
Green, John M
Grinols, Richard Byron
Guillard, Robert Russell Louis
Gutknecht, John William
Haderlie, Eugene Clinton
Haffner, Rudolph Eric
Hamby, Robert Jay
Hands, James Elden
Hargis, William Jennings, Jr
Hargraves, Paul E
Harris, Alva H
Harrison, Florence Louise
Hart, Josephine Frances Lavinia
Hastings, John Woodland
Helfrich, Philip
Herman, Sidney Samuel
Herrkind, William Frank
Hildebrand, Henry H
Hillis-Colinvaux, Llewellya Williams
Hillman, Robert Edward
Hoese, Hinton Dickson
Holmes, Robert W
Holmes, Robert Lawrence
Holton, Robert Lawrence
Hopkins, Sewell Hepburn
Hopkins, Thomas Sterling
Horton, Donald Bion
Hulsemann, Kuni
Humm, Harold Judson
Hunter, John Roe
Hutcheson, Michael Scott
Ingham, Merton Charles
Ingle, Robert Maurice
Jachowski, Richard Leo
Jachowski, Bela Michael
James, Bela Michael
Jones, Everet Clyde
Jones, Galen Everts
Joseph, James
Jumars, Peter Alfred
Kamykowski, Daniel
Kohler, Carl
Kraeuter, John Norman
Lalli, Carol Marie
Larsen, Peter Foster
Layton, William Malloy, Jr

Leavitt, Benjamin Burton
Lewis, Alan Graham
Lewis, John Bradley
Lewis, Roger Wolcott
Lindner, Elek
Lippson, Robert Lloyd
Littlepage, Jack Leroy
Loesch, Harold Carl
Loesch, Joseph
Lynch, Maurice Patrick
MacKenzie, Clyde Leonard, Jr
Macurda, Donald Bradford, Jr
Manning, Raymond B
Mansfield, Arthur Walter
Marshall, Harold George
Manzer, James Ivan
Mariscal, Richard North
Marshall, Nelson
Maturo, Frank Juan Sarno, Jr
McCain, John Charles
McCarthy, Francis Davey
McCarthy, James Joseph
McCauley, James Elias
McConnaughey, Bayard Harlow
McGowan, John Arthur
McLean, Richard Bea
McRoy, C Peter
Menzel, Robert Winston
Merriman, Daniel
Mertens, Edward William
Miller, Charles Benedict
Mills, Eric Leonard
Mitsui, Akira
Moeller, Henry William
Mohr, John Luther
Moore, Richard Byron
Morgan, David William
Moser, H Geoffrey
Mullin, Michael Mahlon
Myrberg, Arthur August, Jr
Nace, Paul Foley
Nakamura, Eugene Leroy
Napora, Theodore Alexander
Needler, Alfred Walker Hollinshead
Newman, William Anderson
Nichols-Driscoll, Jean Ann
Nicol, Joseph Arthur Colin
Nuzzi, Robert
Orth, Robert Joseph
Otsu, Tamio
Otto, David Arthur
Overstreet, Robin Miles
Packard, Theodore Train
Pamatmat, Mario M
Pattabhiraman, Tammanur R
Pearse, John Stuart
Pequegnat, Linda Lee Haithcock
Perkins, Frank Overton
Perry, Mary Jane
Phelps, Donald Kenneth
Phillips, David William
Pieper, Richard Edward
Pilson, Michael Edward Quinton
Pratt, David Mariotti
Price, Kent Sparks, Jr
Pulley, Thomas Edward
Rae, Kenneth MacFarlane
Randall, John Ernest
Rankin, John Stewart, Jr
Ray, Sammy Mehedy
Raymond, Anne Frances
Renfro, William Charles
Richards, Thomas L
Rodriguez, Gilberto
Roper, Clyde Forrest Eugene
Ross, June Rosa Pitt
Roth, Ariel A
Rowe, Gilbert Thomas
Ruggieri, George D
Sastry, Akella N
Scarratt, David Johnson
Scheltema, Rudolf S
Schoener, Amy
Sears, Mary
Shabica, Stephen Vale
Sherbine, K Bruce
Sherman, Kenneth
Silver, Mary Wilcox
Sinderman, Carl James
Skud, Bernard Einar
Small, Lawrence Frederick
Smayda, Theodore John
Smith, Vann Elliott
Songdahl, John Harald
Soule, Dorothy (Fisher)
Sparks, Albert Kirk
Spies, Robert Bernard
Squires, Donald Fleming
Staiger, Jon Crawford
Stephens, John Stewart, Jr
Stevenson, James Cameron
Strathmann, Richard Ray
Swift, Dorothy Garrison
Swift, Elijah, V
Taft, Jay Leslie
Taylor, Frank John Rupert (Max)
Teal, John Moline
Templeman, Wilfred

Tenore, Kenneth Robert
Thomas, Lowell Phillip
Thomas, William Hewitt
Thompson, Rosemary Ann
Tietjen, John H
Trench, Robert Kent
True, Renate (Schlenz)
Van Engel, Willard Abraham
Vernberg, Frank John
Voss, Gilbert Lincoln
Wacasey, Jervis Winn
Wade, Richard Archer
Weaver, Sylvia Short
Welch, Walter Raynes
Wells, John Morgan, Jr
Westley, Ronald E
Wheeler, Ellsworth Haines, Jr
White, Colin
Wiles, Michael
Wilkie, Donald W
Wilson, William Buford
Winn, Howard Elliott
Yang, Won-Tack
Yentsch, Charles Samuel
Young, Richard Edward
Zaneveld, Jacques Simon
Zubkoff, Paul Leon

Biological Rhythms
Bennett, Miriam Frances
Binkley, Sue Ann
Brown, Judith Adele
Bruce, Victor Gardiner
Cardoso, Sergio Steiner
Edmunds, Leland Nicholas, Jr
Hayes, Carol J
Hirshon, Jordon Barry
Mitchell, John Laurin Amos
Natalini, John Joseph
Norton, David L
Ortoleva, Peter Joseph
Palmer, John Derry
Stetson, Milton H
Sturtevant, Ruthann Patterson
Sweeney, Beatrice Marcy
Underwood, Herbert Arthur, Jr
Winget, Charles M
Zucker, Irving

Biological Structure
Adman, Elinor Thomson
Baskin, Denis George
Berman, Helen Miriam
Cavey, Michael John
De Harven, Etienne
Dorset, Douglas Lewis
Ellis, Richard Akers
Flint, Franklin Ford
Gaddum-Rosse, Penelope
Gantt, Elisabeth
Hammerman, Ira Saul
Humphreys, Walter James
Jeffrey, Jackson Eugene
Jenkins, Jimmy Raymond
Lee, Byungkook
Marshall Anne (Corinne)
McPherson, Alexander
Owczarzak, Alfred
Prothero, John W
Richards, Thomas Charles
Rippon, William Barton
Robertson, Charles William
Sattler, Carol Ann
Severson, Arlen Raynold
Suddath, Fred Leroy, (Jr)
Weinreb, Eva Lurie
Weinstock, Melvyn

Biomaterials
Brown, Stanley Alfred
Cowsar, Donald Roy
Houge, James C
Karagianes, Manuel Tom
Kim, Sung Wan
Lautenschlager, Eugene Paul
Marshall, Grayson William, Jr
Rechtien, James Joseph
Refojo, Miguel Fernandez
Rijke, Arie Marie
Walker, Thomas Carl

Biomathematics
Altshuler, Bernard
Apter, Julia Tutelman
Bernard, Mones
Bernard, Selden Robert
Bertell, Rosalie
Blumenson, Leslie Eli
Bremermann, Hans J
Bright, Peter Bowman
Brown, Barry W
Cardus, David
Carpenter, Gail Alexandra
Caswell, Hal
Cerimele, Benito Joseph
Cronin, Jane Smiley (Mrs Joseph C Scanlon)
Deysach, Lawrence George

Dixon, Wilfrid Joseph
Evans, John W
Feldman, Marcus William
Francis, Robert Colgate
Freedman, Herbert Irving
Garfinkel, David
Gold, Harvey Joseph
Gridgeman, Norman Theodore
Hirschfeld, William Jacob
Howell, John Robert
Hutchison, Gerald Andrew
Inselberg, Alfred
Jansson, Birger
Jolicoeur, Pierre
Karreman, George
Katholi, Charles Robinson
Kempthorne, Oscar
Klipper, Robert William
Koong, Ling-Jung
Landahl, Herbert Daniel
Levins, Richard
Licko, Vojtech
Maraman, Grady Vancil
Marks, Louis Sheppard
McMorris, Fred Raymond
Miller, Donald Richard
Niklas, Karl Joseph
Okubo, Akira
Oster, George F
Paloheimo, Jyri Erkki
Parker, Rodger D
Reiner, John Maximilian
Rescigno, Aldo
Rosen, Robert
Rudman, Sanford Winton
Schimmel, Herbert
Siler, William MacDowell
Silvers, Abraham
Simberloff, Daniel S
Smeach, Stephen Charles
Somorjai, Rajmund Lewis
Spalding, Gary E
Tallarida, Ronald Joseph
Thames, Howard Davis, Jr
Thompson, Howard K, Jr
Turner, Malcolm Elijah, Jr
Voorhees, Burton Hamilton
Walter, Charles Frank
Warren, Peter
Whittemore, Alice S
Wilson, P David
Woodbury, Max Atkin
Yorke, James Alan
Zimmerman, Stuart O
Zuker, Michael

Biomechanics
Becker, Roland Frederick
Chandran, Krishnan Bala
Eickelberg, W Warren B
Hodgson, Voigt R
Kazarian, Leon Edward
Liu, Young King
Roberts, David
Rubinow, Sol Isaac
Scheuchenzuber, H Joseph
Stern, Jack Tuteur, Jr

Biometeorology
Kutschenreuter, Paul Herbert
McIntire, Kenneth Robert
Parton, William Julian, Jr
Ripley, Earl Allison

Biometrics
Bearman, Jacob Eleazer
Bowden, David Clark
Briese, Franklin Wagner
Carmer, Samuel Grant
Chapman, Douglas George
Clutter, Jerome Lee
Dahlberg, Michael Lee
Damon, Richard Alan, Jr
Flueck, John A
Forbes, William Frederick
Garber, Morris Joseph
Ghent, Arthur W
Gill, John Leslie
Goldman, Anne Ipsen
Gosslee, David Gilbert
Gould, A Lawrence
Hardin, Robert Toombs
Hayne, Don William
Heath, Robert Gardner
Homer, Louis David
Hsi, Bartholomew P
Jensen, Chester E
Jolicoeur, Pierre
Juillet, Jacques Andre
Lake, Robin Benjamin
Linton, Kenneth Jack
Little, Thomas Morton
Loewenson, Ruth Brandenburger
Lucas, Joseph James
Mackey, Bruce Ernest
Marcus, Leslie F
Mattson, Dale Edward
McCann, James Alwyn

McCaughran, Donald Alistair
McGuire, Judson Ulery, Jr
McHugh, Richard B
Meade, James Horace, Jr
Moon, Thomas Edward
Mosteller, Robert Cobb
Mumm, Robert Franklin
Myers, Max H
Ott, Walther Henry
Petersen, Roger Gene
Pielou, Evelyn C
Rawlings, John Oren
Reutzel, Lawrence Frederick
Robson, Douglas Sherman
Rohlf, F James
Rowe, Kenneth Eugene
Scott, David Paul
Seif, Robert Dale
Shah, Bhupendra K
Singh, Madho
Siniff, Donald Blair
Sollberger, Arne Rudolph
Solomon, Daniel Lester
Sullivan, Alfred Dewitt
Thomas, John M
Townsend, Edwin C
Urban, Willard Edward, Jr
Walker, Rufus Floyd, Jr
Weaver, Clyde Richard
Wright, Allen Kent
Wunder, William W
Zelen, Marvin
Zubin, Joseph

Bionics
Bass, J Carl
Borth, Rudi
Ko, Wen Hsiung
Leet, Duane Gary
Wiitanen, Wayne Alfred
Wong, Jacob Yau-Man

Bionucleonics
Angstadt, Carol Newborg
Born, Gordon Stuart
Bruno, Gerald A
Grunewald, Ralph
Harris, Wayne G
Heeg, Joel Francis
Kessler, Wayne Vincent
Landolt, Robert Raymond
Llewellyn, Gerald Cecil
Miller, Charles Ellsworth
Painter, Kent
Pipes, Gayle Woody
Rothschild, Henri Charles
Schermerhorn, John W
Shaw, Stanley Miner
Shysh, Alec
Spitznagle, Larry Allen
Staiff, Donald C
Stiver, James Frederick
Vacik, James P
Wiebe, Leonard Irving
Zimmer, Albert Michael

Biospeleology
Holsinger, John Robert
Mitchell, Robert W

Biostatistics
Abbey, Helen
Abernathy, James Ralph
Allaway, Norman C
Anello, Charles
Arrington, Wendell S
Assenzo, Joseph Robert
Auerbach, Harry
Badger, George Franklin
Banghart, Frank W
Bartsch, Glenn Emil
Benedetti, Jacqueline Kay
Biot, William James
Boen, James Robert
Brandt, Edward Newman, Jr
Brown, Byron William, Jr
Buncher, Charles Ralph
Carr, Raymond Niel
Chang, Potter Chien-Tien
Chase, Gerald Roy
Chase, Helen Christina (Matulic)
Chen, Edwin Hung-Teh
Chen, Ming-Fen Myra
Chiang, Chin Long
Chiazze, Leonard, Jr
Choi, Sung Chil
Clarkson, Quentin Deane
Cornell, Richard Garth
Crowley, John James
Davis, Kathryn Bullock
Deane, Margaret
Densen, Paul M
Downs, Thomas D
Dumbroff, Erwin Bernard
Dunn, Olive Jean
Dyer, Alan Richard
Edelman, David Anthony

BIOLOGY

Ederer, Fred
Edwards, Brenda Kay
Elkin, William Futter
Ellenberg, Jonas Harold
Elveback, Lillian Rose
Federer, Walter Theodore
Federspiel, Charles Foster
Feigl, Polly Catherine
Feinlein, Manning
Feldman, Joseph Gerald
Feltner, William Henry
Fertig, John William
Fisher, Lloyd D, Jr
Fisher, Pearl Davidowitz
Flora, Jairus Dale, (Jr)
Flora, Roger E
Forsythe, Alan Barry
Frazier, Todd Mearl
Free, Spencer Michael, Jr
Gaffey, William Robert
Gartside, Peter Stuart
Gaylor, David William
Gelber, Richard David
George, Stephen L
Givens, Samuel Virtue
Glasser, Jay Howard
Gleason, Ray Edward
Goldberg, Irving David
Goldstein, Hyman
Greenberg, Bernard George
Greenberg, Richard Alvin
Grizzle, James Ennis
Haenszel, William Manning
Hagans, James Albert
Hawkins, C Morton
Hebel, John Richard
Hoel, David Gerhard
Holford, Theodore Richard
Hopkins, Carl Edward
Horn, Susan Dadakis
Hurley, Frank Leo
Hutcheson, David Paul
Hutcheson, Kermit
Jablon, Seymour
Jacobs, David R, Jr
Jain, Anrudh Kumar
Jessup, Gordon L, Jr
Johnson, Eugene A
Johnson, Dennis Addington
Jones, Paul Kenneth
Jones, Richard Hunn
Keenan, Kathleen Margaret
Keller, Edward Clarence, Jr
Kilpatrick, S James, Jr
Kimball, Allyn Winthrop
Kitler, Mary Ellen
Kjelsberg, Marcus Olaf
Klotz, Jerome Hamilton
Knatterud, Genell Lavonne
Knoke, James Dean
Kraemer, Helena Chmura
Krall, John Morton
Kramer, Morton
Kronmal, Richard Aaron
Kuzma, Jan Waldemar
Lachenbruch, Peter Anthony
Lagakos, Stephen William
Lavin, Philip Todd
Levy, Paul Samuel
Leyton, Morley Kamler
Litt, Bertram D
Littell, Arthur Simpson
Loadholt, Claude Boyd
Lu, Kuo Hwa
Lurie, Dan
Lynch, Cornelius James
Manos, Nicholas Emmanuel
Mason, Thomas Joseph
McGinnis, John Thurlow
McHenry, Hugh Lansden
Meiner, Curtis Lynea
Mellis, E David
Menduke, Hyman
Merchant, Roland Samuel
Meter, Gerald Edward
Meydrech, Edward F
Mielowski, William Leonard
Miller, Millage Clinton, III
Mohberg, Noel Ross
Moore, Dan Houston, II
Moore, Felix E
Morrison, Robert Dean
Mosimann, James Emile
Mullooly, John P
Neely, Peter Munro
Norusis, Marija Jurate
Oates, Richard Patrick
Odoroff, Charles Lazar
O'Fallon, William M
Ordille, Carol Maria
Orlando, Anthony Michael
Pell, Sidney
Perrin, Edward Burton
Powers, Jean Hensel
Rafter, John Arthur
Rao, Mamidanna S
Reading, James Cardon

Reinke, William Andrew
Remington, Richard Delleraine
Rice, Frank J
Rider, Rowland Vance
Roberts-Marcus, Helen Miriam
Robinson, Harry
Rockette, Howard Earl, Jr
Rosenberg, Saul H
Royal, Richard Miles
Ryel, Lawrence Atwell
Schoenfeld, David Alan
Schor, Stanley
Schork, Michael Anthony
Sen, Pranab Kumar
Serfling, Robert Elton
Sharma, Ran S
Shonick, William
Siegel, Carole Ethel
Siguel, Eduardo Nestor
Singer, Arthur Chester
Slack, Nelson Hosking
Smoler, Sylvia Wassertheil
Stephens, Newman Lloyd
Swallow, William Hutchinson
Tan, Wai-Yuan
Tarter, Michael E
Taylor, William F
Teichman, Robert
Thompson, Gordon Merle
Thompson, Donovan Jerome
Tonascia, James A
Tsay, Jia-Yeong
Ullman, Betty M
Ury, Hans Konrad
Vander Zwaag, Roger
Varady, John Carl
Vaughn, William King
Wahl, Patricia Walker
Weckwerth, Vernon Ervin
Weinberg, Roger
Weinstein, Abbott Samson
Weiss, Edward Sebastian
Weiss, William
Wells, Henry Bradley
Wette, Reimut
Williams, George W
Wittes, Janet Turk
Wolff, Albert Eli
Woodbury, Lowell Angus
Wyshak, Grace
Zielezny, Maria Anna
Zippin, Calvin

Biosystematics

Anderson, Gregory Joseph
Arp, Gerald Kench
Beaudry, Jean Romuald
Bishop, Yvonne M
Cicchinelli, Alexander L
Fincher, Anthony Graham
Ganders, Fred Russell
Guthrie, Roland L
Haig, Janet
Henderson, Douglas Miles
Simpson, Beryl Brintnall
Small, Ernest
Sullivan, Victoria I
Vogt, George Britton
Wiens, Delbert

Cryobiology

Graham, Edmund F
Greiff, Donald
Huggins, Charles Edward
Karow, Armand Monfort, Jr
Leibo, Stanley Paul
Mazur, Peter
McGann, Locksley Earl
Menz, Leo Joseph
Meryman, Harold Thayer
Pribor, Donald B
Rapatz, Gabriel Louis
Robinson, David Mason
Rowe, Arthur Wilson
Sherman, Jerome Kalman
Simon, Ellen McMurtrie
Weaver, William Judson

Developmental Biology

Abbott, Joan
Allen, William Ross
Alienspach, Allan Leroy
Armstrong, Peter Brownell
Ashmore, Charles Robert
Atherton, Robert W
Atkinson, Burr Gervais
Auclair, Walter
Auerbach, Robert
Auferheide, Karl John
Babich, George Leon
Bashop, William Earl
Baker, Jeffrey John Wheeler
Baker, Peter C
Baker, Patricia Cooper
Baker, Aimee Hayes
Bakken, Aimee Hayes
Ball, William David

Baratz, Robert Sears
Barker, Jane Ellen
Barrett, Dennis
Battle, Helen Irene
Bell, Eugene
Berlinwood, Martin
Berliowitz, Laurence Jack
Berns, Michael W
Berrill, Norman John
Bhaskaran, Govindan
Bieber, Samuel
Bischoff, Eric Richard
Black, Jessie Kate
Black, Virginia H
Blackler, Antonie W C
Bonner, James (Fredrick)
Bonner, John Tyler
Boohar, Richard Kenneth
Boyer, Barbara Conta
Brackenbury, Robert William
Bradshaw, William S
Brammer, Jimmie Duane
Brandom, William Franklin
Branham, Joseph Morhart
Brinkley, Linda Lee
Brinton, Elias (Lyle) Patterson
Bromley, Stephen C
Brookbank, John Warren
Browder, Leon Wilfred
Brown, Donald D
Brown, Ronald David
Bryant, Susan Victoria
Bull, Alice Louise
Burgess, David Ray
Byrd, Earl William, Jr
Cairnie, Alan B
Campbell, Richard Dana
Caplan, Arnold I
Carroll, Edward James, Jr
Cattolico, Rose Ann
Chamberlain, John Paul
Chan, Lee-Nien Lillian
Chen, James Che Wen
Chepenik, Kenneth Paul
Cherbas, Lucy Fuchsman
Cherbas, Peter Thomas
Charodo, Andrew
Church, Robert Bertram
Cloney, Richard Alan
Clothier, Galen Edward
Cohen, Sanford Ned
Coleman, Nicholas
Collins, James Malcolm
Collins, James Russell
Compher, Marvin Keen, Jr
Condoulis, William V
Conrad, Gary Warren
Cook, Marie Mildred
Coward, Stuart Jess
Cox, Prentiss Gwendolyn
Craft, Thomas Jacob, Sr
Cromer, Jerry Hattwanger
Crosby, Gayle Marcella
Crowell, (Prince) Sears, (Jr)
Cutler, Leslie Stuart
Daniel, Jon Cameron
Daniel, Joseph Car, Jr
Davidson, Eric Harris
Davis, Bill David
Davis, Elizabeth Allaway
Davis, Marjorie
Dawid, Igor Bert
Decker, Robert Scott
del Cerro, Manuel (Perez)
Derby, Albert
DiBerardino, Marie Antoinette
Diehl, Fred A
Dingle, Allan Douglas
Dinsmore, Charles Earle
Dirksen, Ellen Roter
Ditmer, John Edward
Dixon, Gordon H
Donaldson, Donald Jay
Dorgan, William Joseph
Dresden, Marc Henri
DuBrul, Ernest
Duncan, James Thayer
Edds, Louise Luckenbill
Edds, Mac Vincent
Eddy, Edward Mitchell
Edwards, Benjamin Frank
Edwards, Kiah, III
Edwards, Nancy Claire
Elgaard, Erik G
Enochs, Nettie Jean
Epel, David
Epp, Leonard G
Erickson, Ralph O
Etheridge, Albert Louis
Fankhauser, Gerhard
Feit, Ira (Nathan)
Finnegan, Cyril Vincent
Fisher, Kenneth Robert Stanley
Fitzharris, Timothy Patrick
Flickinger, Reed Adams

Flower, Michael Joe
Foret, John Emil
Foster, Donald Bartley
Francis, David W
Franklin, Luther Edward
Fraser, Ronald Chester
Fry, Anne Evans
Frye, Billy Eugene
Fulilove, Susan Louise
Fulton, Chandler Montgomery
Furcelle, Robert Peel
Garber, Beatrice B
Garen, Alan
George, Robert Porter
Gerdy, James Robert
Glass, Laurel Ellen
Goertemiller, Clarence C, Jr
Goldhor, Susan
Goldie, Mark
Goldsmith, Eli David
Gorecki, Donna
Gorell, Thomas Andrew
Gottlieb, Frederick Jay
Gould-Somero, Meredith
Grant, Philip
Greenhouse, Gerald Alan
Gregg, James Henderson
Gressel, Jonathan Ben
Grey, Robert Dean
Griffith, Jeffrey Knowles
Griswold, Michael David
Grodner, Mary Laslie
Gross, Paul Randolph
Grubb, Randall Barth
Gulyas, Bela Janos
Haas, Hermann Josef
Halaban, Ruth
Hall, Brian Keith
Hamilton, Howard Laverne
Hampton, Suzanne Harvey
Hanson, James Charles
Hardy Falding, Margaret Hurlstone
Hasell, John Robert
Hauschka, Stephen D
Hay, Robert J
Hayashida, Kaye
Hayes, Bridget Ann
Healey, Patrick Leonard
Heath, Harrison Duane
Hein, Rosemary Ruth
Hennen, Sally
Herold, Richard Carl
Hilfer, Saul Robert
Hill, Rolla B, Jr
Hinegardner, Ralph
Hoffman, Daniel Lewis
Hoffman, David J
Hood, Ronald David
Hopper, James Ernest
Horenstein, Evelyn Anne
Hsiao, Sidney Chihi
Huebner, Erwin
Humphreys, Tom Daniel
Hunter, Roy, Jr
Jacobson, Antone Gardner
Jeffery, William Richard
Jenkins, Kenneth Dunning
Jensen, Lawrence Craig-Winston
Jeon, Kwang Wu
Joftes, David Lion
Johnson, Kurt Edward
Justus, Jerry T
Kafatos, Fotis C
Kahn, Albert
Kankel, Douglas Ray
Karasaki, Shuichi
Karfunkel, Perry
Kasinsky, Harold Edward
Katoh, Arthur
Keefe, John Richard
Kelley, Robert Otis
Kemp, Norman Everett
Kerr, Marilyn Sue
Kerr, Norman Story
Ketley, Jeanne Nelson
Kieffer, Barry Irwin
Kille, John William
Kimmel, Charles Brown
Kimmel, Donald Loraine, Jr
Kirk, David Livingstone
Kirk, Marilyn Chaloupka
Kocher, Lawrence O
Kohl, David Martin
Konigsberg, Irwin R
Kornberg, Thomas B
Kosin, Igor Leonid
Kosher, Robert Andrew
Krukowski, Marilyn
Kumaran, A Krishna
LaMarca, Michael James
Lambert, Charles Calvin
Landesman, Richard
Lang, Anton
Langridge, William Henry Russell
Larkin, Jeanne Holden
Lash, James (Jay) W
Laufer, Hans

Lavietes, Beverly Blatt
Lee, Harold Hon-Kwong
Lee, Hsin-Yi
Lehman, Harvey Eugene
Lemanski, Larry Fredrick
Leonard, Thomas Joseph
Lesh-Laurie, Georgia Elizabeth
Lesseps, Roland Joseph
Lindsay, David Taylor
Liversage, Richard Albert
Lonski, Joseph
Loomis, William Farnsworth, Jr
MacCabe, Jeffrey Allan
MacDonald, Eve Lapeyrouse
Mack, Clinton Olmsted
Mahowald, Anthony P
Malacinksi, George M
Manasek, Francis John
Manes, Cole
Mangan, Jerrome
Maroni, Gustavo Primo
Martin, Gordon Wyatt
Mascarenhas, Joseph Peter
Masui, Yoshio
Mayer, Thomas C
McCrady, Edward, III
McDevitt, David Stephen
McKinnell, Robert Gilmore
McMenamin, John William
McWhinnie, Dolores J
Meins, Frederick, Jr
Meints, Russel H
Meizel, Stanley
Merriam, Robert William
Metz, Charles B
Miller, Larry O'Dell
Miller, Richard Lee
Miller, Tony Jasper
Mitchell, John Taylor
Moment, Gairdner Bostwick
Moog, Florence
Moretti, Richard Leo
Morrill, Gene A
Morrill, John Barstow, Jr
Morris, John Edward
Moscona, Aron Arthur
Moser, Charles M
Motzkin, Shirley M
Moustafa, Laila Ahmed
Mueller, Nancy Schneider
Mulcare, Donald J
Mun, Alton M
Murison, Gerald Leonard
Murphy, Collin Grisseau
Nace, George William
Nameroff, Mark A
Newman, Stuart Alan
Nihei, Taiichi
Niu, Mann Chiang
Norman, Wesley P
Norr, Sigmund Carl
O'Day, Danton Harry
Oka, Takami
Oppenheimer, Steven Bernard
Overton, Jane Harper
Paoletti, Robert Anthony
Pardue, Mary Lou
Parker, Nancy Johanne Rentner
Paul, Miles Richard
Persaud, Trivedi Vidhya Nandan
Pfohl, Ronald John
Piatigorsky, Joram Paul
Platzer, Anna (Colville)
Polinger, Iris Sandra
Pollock, Edward G
Postlethwait, John Harvey
Powell, Jeanne Adele
Prahlad, Kadaba V
Pritchett, William Henry
Purko, John
Raisbeck, Barbara
Rakic, Pasko
Rao, H V Ramakrishna
Rauch, Nancy
Reams, Willie Mathews, Jr
Reeves, Ogle Raymond
Reporter, Minocher C
Reyer, Randall William
Rhodes, Rondell H
Rickenberg, Howard V
Robinson, Kenneth Ronald
Rollins, Earl Arthur
Rosales-Sharp, Maria Consolacion
Roth, Stephen (Allen)
Roth, Thomas Frederic
Ryan, Edward Parsons
Sabharwal, Pritam Singh
Sacks, Marie Luisetti
Samuel, Edmund William
Sanford, William Corbin
Sanger, Joseph William
Satter, Ruth
Sawhney, Vipen Kumar
Sawyer, Roger Holmes
Schneiderman, Howard Allen
Schultz, Gilbert Allan
Schultz, Phyllis W
Scott, William James, Jr

Sehe, Charles Theodore
Sherer, Glenn Keith
Sherman, Michael Ian
Shininger, Terry Lynn
Shivers, Charles Alex
Shoger, Richard Eugene
Shore, Richard Eugene
Shostak, Stanley
Silver, Alene Freudenheim
Simpson, Sidney Burgess, Jr
Siu, Chi-Hung
Sivasubramanian, Pakkirisamy
Smith, Allen Anderson
Smith, Gerald Nelson, Jr
Smith, Michael Joseph
Smuts, Mary Elizabeth
Sofer, William Howard
Soll, David Richard
Solter, Davor
Solursh, Michael
Sonneborn, David R
Sorensen, Ralph Albrecht
Spiegel, Evelyn Sclufer
Spiess, Luretta Davis
Spooner, Brian Sandford
Stabler, Timothy Allen
Steinberg, Malcolm Saul
Stevens, Leroy Carlton, Jr
Stevenson, Joseph Ross
Stockdale, Frank Edward
Sussman, Maurice
Swanson, Ronald Frederick
Sweeney, Robert Milton
Tassava, Roy A
Taylor, George Thomas
Taylor, John Dirk
Telfer, William Harrison
Terry, Robert James
Thaler, M Michael
Theil, Elizabeth
Thomas, Mary Beth
Thompson, Robert Poole
Thomson, Dale S
Timberlake, William Edward
Timourian, Hector
Tobin, Allan Joshua
Tokes, Zoltan Andras
Torbit, Charles Allen, Jr
Trinkaus, John Philip
Turner, Robert Scott, Jr
Vanable, Joseph William, Jr
Van Ummersen, Claire Ann
Varricchio, Frederick
Vopicka, Ellen Vandersee
Waelsch, Salome Gluecksohn
Waggoner, Phillip Ray
Waibot, Virginia Elizabeth
Walker, Charles Robert
Walters, David Royal
Warner, Alden Howard
Webb, Andrew Clive
Weis, Judith Shulman
Wenger, Byron Sylvester
Wenger, Eleanor Lerner
Wermuth, Jerome Francis
Wessells, Norman Keith
Weston, Charles Richard
Weston, James A
Wiley, Lynn M
Williams, Geneva Hyland
Williams, Leah Ann
Willis, Judith Horwitz
Wise, Benjamin Nathan
Wiseman, Lawrence Linden
Wolk, Coleman Peter
Wolsky, Alexander Albert
Wood, Pauline J
Woodruff, Richard Ira
Woodside, Gilbert Llewellyn
Wourms, John Barton
Wright, David Anthony
Wright, Mary Lou
Wyttenbach, Charles Richard
Yoho, Timothy Price
Young, Richard S
Yund, Mary Alice
Zalik, Sara E
Zapisek, William Francis
Zarcaro, Robert Michael
Zelenka, Peggy Sue
Zwaan, Johan

Economic Biology
Roia, Frank Costa, Jr
Vietmeyer, Noel Duncan

Evolution
Adams, Robert Phillip
Adler, Kraig (Kerr)
Arnheim, Norman
Bader, Robert Smith
Baker, Herbert George
Baldwin, Thomas Oakley
Banta, Benjamin Harrison
Belcher, Jane Colburn
Bennett, John Francis
Borowsky, Richard Lewis
Burns, John McLauren

Bush, Guy L
Callison, George
Cameron, David Glen
Carpenter, Frances Lynn
Carson, Hampton Lawrence
Chiscon, J Alfred
Choate, Jerry Ronald
Christiansen, Kenneth Allen
Cichocki, Frederick Paul
Cook, Paul Pakes, Jr
Cook, Stanton Arnold
Corbin, Kendall Wallace
Craddock, Elysse Margaret
Cruden, Robert William
Dawson, Wallace Douglas, Jr
Dillon, Lawrence Samuel
Dingman, Jane Van Zandt
Dodson, Edward O
Dronamraju, Krishna Rao
Easton, Thomas W
Edmunds, George Francis, Jr
Endler, John Arthur
Feduccia, John Alan
Fisher, Francis John Fulton
Fossland, Robert Gerard
Francoeur, Robert Thomas
Futch, David Gardner
Gillis, John Ericsen
Goodman, Major M
Gould, Stephen Jay
Hall, Barry Gordon
Hanson, Earl Dorchester
Heltne, Paul Gregory
Henry, Charles Stuart
Hinderstein, Barry
Hsu, Laura Hwei-Nien Ling
Jameson, David Lee
Janzen, Daniel Hunt
Jeffery, Duane Eldro
Justice, Keith Evans
Kaneshiro, Kenneth Yoshimitsu
Lee, Merlin Raymond
Levy, Morris
Linhart, Yan Bohumil
Lloyd, James Edward
Low, Bobbi Stiers
MacLeod, Ellis Gilmore
Maiorana, Virginia Catherine
Maly, Edward J
Margulis, Lynn
Mayr, Ernst
Mecham, John Stephen
Minkoff, Eli Cooperman
Moore, John Alexander
Mulcahy, David Louis
Nagle, James John
Nei, Masatoshi
Nelson, Craig Eugene
Orians, Gordon Howell
Palmblad, Ivan G
Paulson, Dennis Roy
Pease, Roger Waterman, Jr
Peck, Stewart Blaine
Platz, James Ernest
Power, Dennis Michael
Raacke, Ilse Dorothea
Rand, Austin Stanley
Rasmussen, David Irvin
Regal, Philip Joe
Reiskind, Jonathan
Rohlfing, Duane L
Salmon, Theodora Nussmann
Sathe, Stanley Norman
Schaffer, William Morris
Slobodchikoff, Constantine Nicholas
Smith, Christopher Carlisle
Smith, Michael Howard
Smith, Neal Griffith
Stahl, Barbara Jaffe
Stalker, Harrison Dailey
Stanley, Steven Mitchell
Stephens, Stanley George
Strickberger, Monroe Wolf
Thompson, Frederic Christian
Todd, Neil Bowman
Van Valen, Leigh
Vermeij, Geerat Jacobus
Vickery, Robert Kingston, Jr
Vrijenhoek, Robert Charles
Wake, David Burton
Walker, James Willard
Warburton, Frederick E
Wasserman, Marvin
Wassersug, Richard Joel
Watt, Ward Belfield
West Eberhard, Mary Jane
Williams, Mary Bearden
Willson, Mary Frances
Wilson, Edward Osborne
Wright, Sewall
Yoon, Jong Sik

Exobiology
Billingham, John
DeVincenzi, Donald Louis
Keosian, John
MacElroy, Robert David
Picciolo, Grace Lee

Schwartz, Alan William
Young, Richard S

Experimental Biology
Bolton, Laura Lee
Christensen, Thomas Gash
Clifton, Kelly Hardenbrook
Grad, Bernard
Hathaway, Ralph Robert
Humphrey, Rufus R
Johnson, Ralph Emil
Kammeraad, Adrian
Laurie, John Sewall
Oldstone, Michael Beaurguard Alan
Rosenbaum, Robert Morris
Trentin, John Joseph
Tyan, Marvin L
Whitacre, Francis Marion
Zahl, Paul Arthur

Fish Biology
Anagnostopoulos, Constantine E
Batts, Billy Stuart
Baxter, George T
Beardsley, Grant Lindley, Jr
Bevan, Donald Edward
Bond, Lyndon Herrick
Borgeson, David P
Brewer, Gary David
Broadhead, Gordon Clifford
Brown, Bradford E
Bulow, Frank Joseph
Burgner, Robert Louis
Cahn, Phyllis Hofstein
Caillouet, Charles W
Campbell, Charles J
Cleaver, Frederick Charles
Cole, Charles Franklyn
Collins, Richard Arlen
Cooper, Edwin Lavern
Cooper, Gerald Paul
Cope, Oliver Brewern
Davis, William Spencer
Dizon, Andrew Edward
Dovel, William Lawrence
Dowell, Virgil Eugene
Duncan, Thomas O
Ebert, Earl Ernest
Eicher, George J
Eipper, Alfred Ward
Elson, Paul Frederick
Flickinger, Stephen Albert
Fore, Paul Lewis
Funk, John Leon
Greene, George Nystrom
Griswold, Bernard Lee
Grosslein, Marvin Darrel
Grover, John Harris
Gunning, Gerald Eugene
Hagen, Harold Kolstoe
Hales, Donald Caleb
Hall, James Dane
Hamilton, James Arthur Roy
Hatch, Richard Wallace
Hauser, William Joseph
Haven, Dexter Stearns
Head, Eric James
Heard, William R
Henry, Kenneth Albin
Heyerdahl, Eugene Gerhardt
Hodder, Vincent MacKay
Hokanson, Kenneth Eric Fabian
Horton, Howard Franklin
Hunt, Robert L
Huntsman, Gene Raymond
Idyll, Clarence Purvis
Ihssen, Peter Edward
Iversen, Edwin Severin
Jenkins, Robert M
Johnson, James Edward
Jones, Bernard R
Juhl, Rolf
June, Fred C
Keleher, James J
Kennedy, William Alexander
Kilpatrick, Earl Buddy
Kinney, Jerry Bruce
Kinney, Edward Coyle, Jr
Klawe, Witold L
Koo, Ted Swei-Yen
Kuehn, Jerome H
Larimore, Richard Weldon
Latta, William Carl
Lennon, Robert Earl
Leon, Kenneth Allen
Lewis, Robert Minturn
Martin, William Robert
Mathur, Dilip
Maxfield, Galen Harry
May, Arthur William
May, Robert Carlyle
McCauley, Robert William
McLain, Albertson Lamson
Meehan, William Robert
Merrell, Theodore Reed, Jr
Merriner, John Vennor
Mitchell, Lawrence Gustave

BIOLOGY

Momo, Walter Thomas
Moss, Donovan Dean
Mount, Donald I
Murphy, Garth Ivor
Nakatani, Roy E
Neill, William Harold
Nelson, Philip R
Olla, Bori Liborio
Orcutt, Harold George
Pacheco, Anthony Louis
Patrarche, Mercer Harding
Power, Geoffrey
Purkett, Charles A, Jr
Rajagopal, P K
Raleigh, Robert Franklin
Reed, James Robert, Jr
Reed, Roger J
Reynolds, James Blair
Ricker, William Edwin
Ridenhour, Richard Lewis
Roessler, Martin A
Rogers, Donald Eugene
Rothschild, Brian James
Saila, Saul Bernhard
Scalet, Charles George
Scattergood, Leslie Wayne
Schmulbach, James C
Schoettger, Richard A
Seymour, Allyn H
Shaklee, James Brooker
Shell, Eddie Wayne
Shomura, Richard Sunao
Siliman, Ralph Parks
Skinner, John Eugene
Sliger, Wilburn Andrew
Smith, Stanford Henry
Smith, Wendell Eugene
Smitherman, Renford Oneal
Southward, Glen Morris
Stalnaker, Clair B
Stauffer, Thomas Miel
Stevenson, James Harold
Strasburg, Donald Wishart
Struhsaker, Jeannette Adair Whipple
Summerfelt, Robert C
Tack, Peter Isaac
Tait, James Simpson
Talbot, Gerald Byron
Thompson, Richard Baxter
Toetz, Dale W
Tuttle, Russell H
Verhoeven, Leon A
Wagner, Harry Henry
Walburg, Charles Herman
Walford, Lionel Albert
Ward, Fredrick James
Warren, Charles Edward
Wendler, Henry O
West, Jerry Lee
Whitney, Richard Ralph
Wisby, Warren Jensen
Wise, John P

Fresh Water Biology
Baker, M Michelle
Berg, Clifford Osburn
Bouchard, Raymond William
Brown, Jack Stanley
Duthie, Hamish
Fernando, Constantine Herbert
Flannagan, John Fullan
Gilbert, John Jouett
Harrison, Arthur Desmond
Ide, Frederick Palmer
Klemm, Donald J
Mercer, Edward King
Rosengren, John
Schindler, John Frederick
Simpson, Karl William
Wiggins, Glenn Blakely

History of Biology
Allen, Garland Edward, III
Dexter, Ralph Warren
Ewan, Joseph (Andorfer)
Farley, John
Necker, Walter Ludwig
Weiner, Charles
Yos, David Albert

Human Biology
Bresler, Jack Barry
Feinberg, Richard
Garone, John Edward
Hughes, David Rees
Kamien, Ethel N
Lovejoy, Owen
Marsh, Richard Riley
O'Connor, Brian Lee
Reuss, Ronald Merl
Siiri, William Arthur
Zegura, Stephen Luke

Immunobiology
Aden, David Paul
Agnew, Robert Morson
Arala-Chaves, Mario Passalaqua

Badger, Alison Mary
Baker, Joseph George
Baxter, William D
Bellone, Clifford John
Beychok, Sherman
Blakeslee, Dennis Lauren
Blazkovec, Andrew A
Borysenko, Myrin
Byers, Vera Steinberger
Carlson, Donald Eugene
Carter, Bettina Bush (Mrs Daniel F Jackson)
Carter, Brian Geoffrey
Cebra, John Joseph
Chiscon, Martha Oakley
Chorpenning, Frank Winslow
Claflin, Alice J
Click, Robert Edward
Craig, Susan Walker
Cudkowicz, Gustavo
Feldbush, Thomas Lee
Field, Arthur Kirk
Fraker, Pamela Jean
Franzl, Robert E
Freeman, Max James
Gale, Robert Peter
Hanna, Michael G, Jr
Harris, Nick Steven
Hirschhorn, Rochelle
Houston, William Eddie
Hsu, Konrad Chang
Jensen, Joerg
Johnston, Muriel Evelyn
Kent, Naim Hasan
Killion, Jerald Jay
Kornfeld, Lottie
Kripke, Margaret Louise (Cook)
Kuhn, Raymond Eugene
Kulangara, Abraham Chakko
Lausch, Robert Nagle
Loan, Raymond Wallace
Lopez, Carlos
Loughman, Barbara Ellen Evers
Lukasewycz, Omelan Alexander
Mandy, William John
Manson, Lionel Arnold
Maurer, Bruce Anthony
McBride, Raymond Andrew
McConnachie, Peter Ross
McElree, Helen
Miller, Gary William
Miller, John Johnston, III
Miller, Richard Graham
Miller, Ronald Kent
Mitchell, Kenneth Frank
Nash, Donald Robert
Nathenson, Stanley G
Pascal, Theresa A
Patterson, Ronald James
Pearson, David D
Pisano, Joseph Carmen
Ranney, David Francis
Rice, Thomas Kenneth
Rich, Robert Regier
Sabbadini, Edris Rinaldo
Sabet, Tawfik Younis
Schenkel, Robert H
Segre, Mariangela Bertani
Selin, Helen Gill
Smart, Keith Lorenzo
Smith, Kenneth Larry
Speirs, Robert Sisson
Szakal, Andras Kalman
Terres, Geronimo
Thomas, Paul Milton
Thorn, Richard Mark
Toben, Howard Ray
Vann, Douglas Carroll
Warner, Carol Miller
Willis, Judith Ione
Wilson, Darcy Benoit
Winter, Alexander J
Wojnar, Robert John
Yoshino, Timothy Phillip
Zighelboim, Jacob
Zwilling, Bruce Stephen

Medical Parasitology
Abadie, Stanley Herbert
Beck, Jacob Walter
Boyles, James McGregor
Bruckner, David Alan
Chen, David Hou-Chung
Chernin, Eli
Contacos, Peter George
Cross, John Henry
Daly, James Joseph
Ferguson, Malcolm Stuart
Harper, Kathleen Lucille
Headlee, William Hugh
Hunter, George William, III
Ivey, Michael Hamilton
Jacobs, Leon
Larsh, John E, Jr
Lesser, Elliott
Lewert, Robert Murdoch
Little, Maurice Dale
Luttermoser, George William

Malek, Emile Abdel
Mathews, Henry Mabbett
McConnell, Elliott
McQuay, Russell Michael, Jr
Melvin, Dorothy Mae
Miller, Joseph Henry
Miller, Max Joseph
Orihel, Thomas Charles
Palmer, Timothy Trow
Radke, Myron Glen
Ritchie, Lawrence Starr
Ritterson, Albert L
Schiller, Everett L
Sheffield, Harley George
Sirewalt, Margaret Amelia
Sulzer, Alexander Jackson
Swartzwelder, John Clyde
Thorson, Ralph Edward
Weatherly, Norman F
Woodruff, David Scott

Molecular Biology
Abelson, John Norman
Abou-Sabe, Morad A
Abrahamson, Edwin William
Adams, Gabrielle H M
Adler, Alan David
Alberts, Bruce M
Alexander, Renee R
Altman, Sidney
Anderson, Carl William
Anderson, William Alan
Angerer, Robert Clifford
Arnheim, Norman
Arnott, Struther
Atherly, Alan G
Axel, Richard
Bagshaw, Joseph Charles
Baker, Robert Frank
Bandman, Everett
Barnett, William Edgar
Bar-Zev, Asher
Bauerle, Ronald H
Baxter-Gabbard, Karen Lee
Bear, Richard Scott
Beckman, Lewis David
Beer, Michael
Behki, Ram M
Bell, Robert Maurice
Bellino, Francis Leonard
Berger, Franklin Gordon
Berkowitz, David B
Bernstein, Carol
Bigger, Cynthia Anita Hopwood
Bird, Robert Earl
Birnbaum, Linda Silber
Bishop, David Hugh Langler
Blackman, Carl F, Jr
Blatt, Jeremiah Lion
Blumenfeld, Martin
Boezi, John A
Bohannon, Randolph F
Bolla, Robert Irving
Bonaventura, Joseph
Bonner, Tom Ivan
Borisy, Gary Guy
Bose, Subir Kumar
Bothwell, Alfred Lester Meador
Bourque, Don Philippe
Bowman, Ray Douglas
Branscomb, Elbert Warren
Bredderman, Paul John
Breen, Gail Anne Marie
Bremer, Hans
Brenchley, Jean Elnora
Briehl, Robin Walt
Brockman, William Warner
Brunner, David Warren
Brutlag, Douglas Lee
Budd, Thomas Wayne
Bukhari, Ahmad Iqbal
Burger, Richard Melton
Burgess, Richard Ray
Burness, Alfred Thomas Henry
Calendar, Richard
Camerman, Norman
Candelas, Graciela C
Caro, Lucien D
Carroll, Dana
Celis, Teodoro F R
Champe, Sewell Preston
Chan, Samuel H P
Chase, John William
Chetsanga, Christopher J
Chiang, Kwen-Sheng
Chow, Louise Tsi
Christman, Judith Kershaw
Clark, Roger William
Cleary, Paul Patrick
Cohen, Paul Sidney
Cole, Ronald Sinclair
Coleman, William Warner
Collins, Carolyn Jane
Cousineau, Gilles H
Cox, Donald Cody
Cox, George Stanley
Cramer, Jane Harris

Creighton, Thomas Edwin
Cronan, John Emerson, Jr
D'Adamo, Amedeo Filiberto, Jr
Dahlberg, James Eric
Daily, Otis Patrick
Daniel, Ellen
Datta, Prasanta
Davidson, Eric Harris
Davis, Craig H
Davis, Lawrence Clark
Delihas, Nicholas
Denhardt, David Tilton
DeRosier, David J
Deuscher, Murray Paul
Dietz, George William, Jr
Dixon, Gordon H
Doniger, Jay
Doolittle, Warren Ford, III
Douthit, Harry Anderson, Jr
Dressler, David
Dreyer, William J
Drlica, Karl
Dubnau, David
Duell, Elizabeth Ann
Duerksen, Jacob Dietrich
Dumas, Lawrence Bernard
Dunn, John Patrick James
Earhart, Charles Franklin, Jr
Eberle, Helen I
Eckhart, Walter
Eilem, Kay Adrian Oswald
Engelhardt, Dean Lee
Erikson, Raymond Leo
Ernst, Susan Gwenn
Estrup, Faiza Fawaz
Evans, Thomas Edward
Farber, Florence Eileen
Farrand, Stephen Kendall
Faust, Charles Harry, Jr
Feese, Bennie Taylor
Fenna, Roger Edward
Fermi, Giulio
Fessler, John Hans
Fields, Kay Louise
Fisher, Harold Wilbur
Flower, Michael Joe
Fong, Peter
Forrest, Hugh Sommerville
Foulds, John Douglas
Fournier, Maurille Joseph, Jr
Fraenkel-Conrat, Heinz Ludwig
Frankel, Fred Robert
Fraser, William Dean
Freese, Ernst
Freifelder, David
Fresco, Jacques Robert
Freundlich, Martin
Friesen James Donald
Gaballah, Saeed S
Gallant, Jonathan A
Ganesan, Adayapalam T
Garfin, David Edward
Garrick, Laura Morris
Gefter, Malcolm Lawrence
Geiduschek, Ernest Peter
Gerard, Gary Floyd
Gerschenson, Lazaro E
Gesteland, Raymond Frederick
Gibbons, Ian Read
Gibson, Kenneth David
Gilbert, Walter
Gilchrist, William Clarence
Glaser, Donald Arthur
Glick, Jane Mills
Godchaux, Walter, III
Godson, Godfrey Nigel
Goff, Christopher Godfrey
Goldberg, Edward B
Goll, Darrel Eugene
Goodman, Richard E
Goor, Ronald Stephen
Graham, Dale Elliot
Greenberg, Jay R
Griffiths, Anthony J F
Gross, Paul Randolph
Guild, Walter Rufus
Gulati, Subhash Chander
Gurney, Theodore, Jr
Gussin, Gary Nathaniel
Guterman, Sonia Kosow
Guthrie, Christine
Guthrie, George Drake
Haff, Lawrence Allen
Hahn, Fred Ernst
Hahn, William Eugene
Haidle, Charles Walter
Hall, Barry Gordon
Hankalo, Barbara Ann
Hanawalt, Philip Courtland
Hanpel, Arnold E
Hansen, John Norman
Hanson, Douglas MacArthur
Hari, V
Harmon, Shirley Ann
Harrington, William Fields
Hartman, Karl August, Jr
Hatfield, Dolph Lee
Hauala, Judith Ann

Hedman, Stephen Clifford
Heere, Leonard J
Heffernan, Laurel Grace
Hendrix, Roger Walden
Henry, Timothy James
Herbert, Edward
Herriott, Jon R
Herrmann, Klaus Manfred
Hershberger, Charles Lee
Heywood, Stuart Mackenzie
Hoagland, Mahlon Bush
Hofmann, Theo
Holmes, George Edward
Holoubek, Viktor
Holt, Charles Edward
Hopkins, Johns Wilson
Hopkins, Thomas R
Hopper, Anita Klein
Horgen, Paul Arthur
Horiuchi, Kensuke
Hough, Paul Van Campen
Howard, Bruce David
Howard, Glenn Willard, Jr
Howard, Guy Allen
Hsu, Ming-Ta
Huang, Ru-Chih Chow
Huang, Wai Mun
Hutchison, Clyde Allen, III
Hutton, James Robert
Igarashi, Satomi J
Iglewski, Wallace
Ihler, Garret Martin
Iijima, Herbert
Imsande, John
Ingersoll, Ronald John
Ingles, Charles James
Ingwall, Joanne S
Inman, Ross
Inners, Lamar Daniel
Jackson, David Archer
Jackson, Ethel Noland
Jacobson, Lewis A
Jacoli, Giulio Guido
Janik, Borek
Jardetzky, Oleg
Jaworski, Ernest George
Jeffery, William Richard
Jesaitis, Margeris Adomas
Johnson, Christine Margaret
Johnson, Frank Harris
Johnson, Lee Frederick
Johnson, Paul Hickok
Jones, George Henry
Kado, Clarence Isao
Kantor, George Joseph
Kaper, Jacobus M
Kaplan, Jacob Gordin
Kapp, Leon Neal
Kataja, Eva I
Kato, Karen Friedman
Katz, Louis
Kaye, Alvin Maurice
Keller, Stephen Jay
Kelley, William Sheldon
Kemper, Jost Hansjosef Karifried
Kennell, David Epperson
Kerwar, Suresh
Kimball, Paul Clark
King, Jonathan Alan
Kirk, Ralph Gary
Klem, Edward Benson
Klevecz, Robert Raymond
Kligman, Lorraine H L
Kline, Bruce Clayton
Kline, Larry Keith
Knight, Claude Arthur
Knight, Robert Hallowell
Kolodner, Richard David
Konrad, Michael Warren
Korn, David
Kozloff, Lloyd M
Krakow, Joseph S
Kramer, Fred Russell
Krawiec, Steven Stack
Kretsinger, Robert
Kubinski, Henry A
Kuff, Edward Louis
Kung, Shain-Dow
Kutter, Elizabeth Martin
Ladner, Jane Ellen Crawford
Laipis, Philip James
Landau, Joseph Victor
Landy, Arthur H
Lang, Dimitrij Adolf
Langridge, Robert
Lanks, Karl William
Lawrence, William Chase
Lazda, Velta Abuls
Lederberg, Seymour
Lee-Huang, Sylvia
Lerman, Leonard Solomon
Lesiewicz, Jeanne Lee
Lewis, James Bryan
Lightfoot, Haideh Nezam
Littlewood, Roland Kay
Lockwood, Arthur H
Logan, David Mackenzie
Losick, Richard Marc

Lu, Ponzy
Lucas-Lenard, Jean Marian
Luck, Dennis Noel
Lusena, Charles V
Lusis, Aldons Jekabs
Lyman, Harvard
Maciag, Thomas Edward
MacKay, Vivian Louise
MacLeod, Michael Christopher
Maher, Veronica Mary
Maitra, Umadas
Malt, Ronald A
Mandel, Manley
Mandel, Morton
Maniatis, Thomas Peter
Mansour, A Maher
Marchin, George Leonard
Margoliash, Emanuel
Margolis, Sam Aaron
Marmur, Julius
Marotta, Charles Anthony
Marsh, Robert Cecil
Martin, Terence Edwin
Martinson, Harold Gerhard
Marvin, Donald Arthur
Masker, Warren Edward
Massar, Ann Roller
Massie, Harold Raymond
Matthews, Brian Wesley
Maxwell, Elizabeth Starbuck
Mayfield, John Eric
Mayor, Heather Donald
Mazur, Barbara Jean
McAllister, William Turner
McCarthy, Brian John
McCormick, J Justin
McMurtrey, Marion John
Meisler, Arnold Irwin
Melcher, Ulrich Karl
Mendelson, Robert Alexander, Jr
Menninger, John Robert
Merril, Carl R
Mertz, Janet Elaine
Meselson, Matthew Stanley
Michalski, Chester James
Miner, Norman Allen
Mitra, Sankar
Mizuno, Shigeki
Morgan, Antony Richard
Morimoto, Hideo
Morris, Daniel W
Morse, Philip Dexter, II
Mount, David William Alexander
Mulder, Carel
Nagle, William Arthur
Narang, Saran A
Nathans, Daniel
Neufeld, Berney Roy
Neville, David Michael, Jr
Newbold, John Edward
Niederman, Robert Aaron
Noll, Hans
Nonoyama, Meihan
Ofengand, Edward James
Ohlsson-Wilhelm, Betsy Mae
Oldfield, Eric
Olenick, John George
Pero, Janice Gay
O'Malley, Bert W
Osborn, Mary Jane
Outlaw, Henry Earl
Palmer, Jon (Carl)
Parsons, Donald Frederick
Pastan, Ira Harry
Patel, Gordhan
Pauling, Edward Grellin
Pazoles, Christopher James
Pearlman, Ronald E
Pedersen, Peter L
Peets, Edwin Arnold
Perry, Kenneth W
Pettijohn, David E
Phillips, Leo Augustus
Phillips, Stephen Lee
Philpott, Delbert E
Pieczenik, George
Pierucci, Olga
Pietsch, Paul Andrew
Pitha, Paula Marie
Pomato, Nicholas
Poyton, Robert Oliver
Price, Frederick William
Propst-Ricciuti, Catherine Lamb
Prashne, Mark
Purcell, William Paul
Purifoy, Dorothy Jane Martin
Rado, Thomas A
Ragin, James Frederick
Ramsey, James Carroll
Ray, Dan S
Reeder, Ronald Howard
Reif-Lehrer, Liane
Reissig, Jose Luis
Reusser, Fritz
Rice, Nancy Reed
Rich, Alexander
Riggsby, William Stuart
Roberts, Richard John

Robinson, William Sidney
Roess, William B
Rogerson, Allen Collingwood
Roozen, Kenneth James
Rosner, Judah Leon
Ross, Jeffrey
Rownd, Robert Harvey
Roy-Burman, Pradip
Rubenstein, Irwin
Sadler, John
Safer, Brian
Salser, Winston Albert
Samuel, Albert
Sanger, Joseph William
Sarma, Raghupathy
Saunders, Grady Franklin
Schachman, Howard Kapnek
Schaechter, Moselio
Scheele, Robert Blain
Scheffler, Immo Erich
Scherberg, Neal Harvey
Schleich, Thomas W
Schleif, Robert Ferber
Schlessinger, David
Schoenborn, Benno P
Schrader, William Thurber
Schulman, LaDonne Heaton
Scocca, Roy Albert, III
Scott, Roy Albert, III
Sena, Elissa Purnell
Seufert, Wolf D
Shaeffer, Joseph Robert
Shafritz, David Andrew
Shahn, Ezra
Shanmugam, Govindaswamy
Shapiro, Lucille
Sharp, Philip Allen
Shih, Thomas Yutzong
Showe, Michael Kent
Siegel, Pamela Jean
Signer, Ethan Royal
Silver, Simon David
Silverman, Philip Michael
Simard, Rene
Singer, Beatrice Adell
Sjostrand, Fritiof S
Skalka, Anna Marie
Skinner, Dorothy M
Smith, Cassandra Lynn
Smith, Charles Allen
Smith, David Allen
Smith, David Waldo Edward
Smith, Douglas Wemp
Smith, Gerald Ralph
Smith, Issar
Snyder, Loren Russell
Soll, Dieter Gerhard
Sollner-Webb, Barbara Thea
Sonenshein, Abraham Lincoln
Spanier, Bonnie Barbara
Srivastava, Bejai Inder Sahai
Stanley, Wendell Meredith, Jr
Steitz, Joan Argetsinger
Steitz, Thomas Arthur
Stent, Gunther Siegmund
Stephenson, Mary Louise
Sternglanz, Rolf
Stout, Ernest Ray
Straus, Neil Alexander
Strauss, Bernard S
Strniste, Gary F
Stuart, Sarah Elizabeth
Stubbs, John Dorton
Stukus, Philip Eugene
Summers, William Cofield
Sussman, Maurice
Sutter, Richard P
Swedes, Jean Susanne
Sweet, Robert Mahlon
Szepesi, Bela
Taber, Harry Warren
Tessman, Irwin
Tevethia, Mary Judith (Robinson)
Thedford, Roosevelt
Thompson, Edward Ivins Bradbridge
Thorsett, Grant Orel
Tipper, Donald John
Tobin, Allan Joshua
Touster, Oscar
Towle, David Walter
Trachewsky, Daniel
Trakatellis, Anthony C
Traugh, Jolinda Ann
Traut, Robert Rush
Truesdell, Susan Jane
Tsai, Ming-Jer
Tsolas, Orestes
Upholt, William Boyce
Verheyden, Julien P H
Verses, Christ James
Vinograd, Jerome
Vizard, Douglas Lincoln
Voumakis, John Nicholas
Wahl, Geoffrey Myles
Wall, Thomas Randolph
Wallace, Susan Scholes
Warner, Jonathan Robert
Warshel, Arieh

Watson, James Dewey
Webster, Robert Edward
Weeks, Donald Paul
Weiler, Eberhardt
Weissbach, Herbert
Weller, David Lloyd
Wells, Robert Dale
Welsch, Federico
Wettstein, Felix O
White, Elizabeth Lloyd
Whitfield, Harvey James, Jr
Wild, James Robert
Wilhelm, James Maurice
Williams, Luther Steward
Williams, Robley Cook
Willson, Clyde D
Wilson, David Louis
Wimmer, Eckard
Winicov, Ilga Butelis
Witkin, Steven S
Wohlpart, Alfred
Wolgamott, Gary
Wolstenholme, David Robert
Woods, Philip Sargent
Woodward, Dow Owen
Wright, Andrew
Wright, Christine Gerda
Wyman, Jeffries
Yamamoto, Keith Robert
Yanofsky, Charles
Yarmolinsky, Michael Bezalel
Yarus, Michael J
Yielding, K Lemone
Young, Elton Theodore
Young, Ronald Jerome
Yourno, Joseph Dominic
Zabin, Irving
Zimmerman, Burke Kisling
Zimmermann, Robert Alan
Zuccarelli, Anthony Joseph

Neurobiology
Alley, Keith Edward
Anderson, Margaret
Arch, Stephen William
Barker, David Lowell
Barker, Jeffery Lange
Barmack, Neal Herbert
Barondes, Samuel Herbert
Bastian, Joseph
Berry, Robert Wayne
Besso, Joseph Augustus, Jr
Bland, Brian Herbert
Bodian, David
Bownds, M Deric
Brandt, Bruce Losure
Browner, Robert Herman
Bullock, Theodore Holmes
Burgess, Paul Richards
Bystrom, Barbara Gillooly
Campenot, Robert Barry
Capranica, Robert R
Chappell, Richard Lee
Claude, Philippa
Cohen, Adolph Irvin
Coleman, Paul David
Dahl, Nancy Ann
Davis, Richard
Dennis, Michael Joseph
DeSantis, Mark Edward
Descarries, Laurent
Diamond, Marian C
Disterhoft, John Francis
Doolin, Paul F
Dowling, John Elliott
Dubin, Mark William
Dubner, Ronald
Faber, Donald S
Fambrough, Douglas McIntosh
Fentress, John Carroll
Fernald, Russell Dawson
Fisher, Leslie John
Fisher, Steven Kay
Frigyesi, Tamas L
Furshpan, Edwin Jean
Gainer, Harold
Geinisman, Yuri
Gerstein, George Leonard
Getting, Peter Alexander
Gibbs, James Gendron, Jr
Gladfelter, Wilbert Eugene
Glasser, Richard Lee
Glassman, Edward
Goldman, Leonard Jay
Grabowski, Sandra Raynolds
Gwilliam, Gilbert Franklin
Haight, John Richard
Hara, Toshiaki
Harris, Charles Leon
Helson, Lawrence
Henson, Anna Miriam (Morgan)
Hickey, Terry Lee
Hillman, Dean Elof
Hirsch, Helmut V B
Hubbard, Jack Edward, Jr
Ingram, Walter Robinson
Irwin, Louis Neal
Kaiserman-Abramof, Ita Rebeca

BIOLOGY

Kanmer, Ann Emma
Kandel, Eric Richard
Kankel, Douglas Ray
Karlin, Arthur
Kelly, James P
Kelly, Regis Baker
Kendig, Joan Johnston
Kennedy, Michael Craig
Konopka, Ronald J
Koenig, Edward
Landis, Story Cleland
Lang, Frederick
Larimer, James Lynn
Lasek, Raymond J
Lester, Henry Allen
Levitt, Melvin
Llinas, Rodolfo
Marks, Neville
Maynard, Edith Adele
McEwen, Bruce Sherman
McGaugh, James L
McKelvy, Jeffrey Forrester
McLardy, Turner
McLaughlin, Barbara Jean
McNamara, Mary Colleen
Mischis, Richard Robert
Mittenthal, Jay Edward
Moran, David Taylor
Murphey, Rodney Keith
Murray, Marion
Nadelhaft, Irving
Nicholson, Charles (Godfrey)
Niklowitz, Werner Johannes
Norden, Jeanette Jean
Nornes, Howard Onsgaard
Oakley, Bruce
Okun, Lawrence M
Olivo, Richard Francis
Orkand, Richard K
Pak, William Louis
Pappas, George Demetrios
Paul, Dorothy Hayman
Pearlman, Alan Lee
Peterson, Richard George
Potter, Lincoln Truslow
Prior, David James
Puszkin, Saul
Rash, John Edward
Redburn, Dianna Ammons
Robertson, Richard Thomas
Rubel, Edwin W
Russell, Richard Lawson
Salpeter, Miriam Mirl
Salzberg, Brian Matthew
Scalia, Frank
Schafer, Rollie R
Schlapfer, Werner T
Schezer, Jeri Alineu
Sharma, Sansar C
Shear, Charles Robert
Siegel, Allan
Sinha, Arabinda Kumar
Sisken, Betty Florio
Slagel, Donald E
Somers, Michael Eugene
Spanis, Curt William
Spirito, Carl Peter
Stefano, George Bogdon
Steinhardt, Richard Antony
Stell, William Kenyon
Stent, Gunther Siegmund
Stoffolano, John George, Jr
Stout, John Frederick
Strada, Samuel Joseph
Strumwasser, Felix
Stuart, Ann Elizabeth
Sze, Paul Yi Ling
Talamo, Barbara Lisann
Turner, James Eldridge
Tweedle, Charles David
Uzman, Betty Geren
Van Orden, Lucas Schuyler, III
Vernadakis, Antonia (Mrs H L Ockerman)
Voneida, Theodore J
Weinreich, Daniel
Wilkens, Lon Alan
Wood, John Grady
Woodson, Paul Bernard
Wurtman, Richard Jay
Yu, Riley Chaoping
Zagon, Ian Stuart
Zigmond, Richard Eric
Zomzely-Neurath, Claire Eleanore
Zottoli, Steven Jaynes

Lilienthal, Bernard
Mandel, Irwin D
McDonald, James Lee, Jr
Morawa, Arnold Peter
Newbrun, Ernest
Phillips, Carleton Jaffrey
Pollock, Robert J, Jr
Redman, Robert Shelton
Richelle, Leon Joseph
Robinovitch, Murray R
Ruben, Morris P
Sandham, Herbert James
Shahrik, H Arto
Siegel, Ivens Aaron
Smith, Quenton Terrill
Smith, Daniel James
Strachan, Donald Stewart
Stringham, Reed Millington, Jr
Tamarin, Arnold
Taubman, Martin Arnold
Wolf, Robert Oliver
Zaki, Abd El-Moneim Emam

Paleobiology

Armstrong, Augustus K
Barghoorn, Elso Sterrenberg
Bell, Bruce McConnell
Berry, William Benjamin Newell
Blake, Daniel Bryan
Cooper, Gustav Arthur
Cvancara, Alan Milton
Dodson, Peter
Eldredge, Niles
Fisk, Lanny Herbert
Glenister, Brian Frederick
Hickman, Carole Stentz
Kaufman, Erle Galen
Maglio, Vincent Joseph
McAlester, Arcie Lee, Jr
Pirozynski, Krzysztof Andrzej
Ross, Charles Alexander
Schopf, James William
Schopf, Thomas Joseph Morton
Simmonds, Robert T
Stanley, Edward Alex
Sutherland, Patrick Kennedy
Thompson, Ida
West, Ronald Robert

Parasitology

Abram, James Baker, Jr
Agosin, Moises
Alderman, Louis Cleveland, Jr
Aldrich, Lewis Eugene, Jr
Alicata, Joseph Everett
Aliff, John Vincent
Andersen, Ferron Lee
Anderson, Cyrus Vincent
Anderson, David Robert
Anderson, Gary
Anderson, Robert I
Anderson, Roy Clayton
Angelopoulos, Edith W
Anthony, James Douglas
Arai, Hisao Philip
Armstrong, Howard Wayne
Ash, Lawrence Robert
Auerbach, Earl
Babero, Bert Bell
Bacchi, Cyrus Joseph
Bacha, William Joseph, Jr
Balamuth, William
Ball, Gordon Harold
Ball, William
Ballard, Neil Brian
Baron, Robert Richard
Barriga, Omar Oscar
Barrow, James Howell, Jr
Basch, Paul Frederick
Bawden, Monte Paul
Beaver, Paul Chester
Beilfuss, Erwin Roland
Bemrick, William Joseph
Berger, Harold
Berry, Jewell Edward
Bettencourt, Joseph S, Jr
Blankespoor, Harvey Dale
Boertje, Stanley
Bogitsh, Burton Jerome
Bourns, Thomas Kenneth Richard
Bowen, Richard Eli
Box, Edith Darrow
Bradley, Richard E
Branch, John Curtis
Bridgman, John Francis
Briggs, Norman Theodore
Brooke, Marion Murphy
Brown, Harold William
Bruce, John Irvin, Jr
Bullock, Wilbur Lewis
Burns, William Chandler
Burrows, Robert Beck
Burt, Michael David Brunskill
Buscher, Henry Neil
Cain, George D
Carter, Richard Thomas
Chandran, Satish Raman
Chang, Kwang-Poo
Chao, Jowett
Chi, Lois Wong
Ching, Hilda
Chitwood, May Belle Hutson
Ciordia, Honorico
Clark, David Thurmond
Clark, Glen W
Clarkson, Allen Boykin, Jr
Clem, Judy Roberta
Cmejla, Howard Edward
Coffey, James Cecil, Jr
Coil, William Herschell
Coleman, Robert Marshall
Colley, Fredrick Christensen
Collins, Richard Cornelius
Concannon, Joseph N
Connell, Frank Herman
Connor, Robert Sherman
Corkum, Kenneth C
Coulston, Frederick
Cramer, Ardis Lahann
Crandall, Richard B
Cressey, Roger F
Crofton, Neil Argo
Crook, James Richard
D'Alessandro, Philip Anthony
D'Alessandro-Bacigalupo, Antonio
Damian, Raymond T
De Giusti, Dominic Lawrence
Demaree, Richard Spottswood, Jr
Dennis, Emmet Adolphus
Desowitz, Robert
Despommier, Dickson
Desser, Sherwin S
Dixon, Carl Franklin
Dollahon, Norman Richard
Donaldson, Alan Weston
Doran, David James
Douvres, Frank William
Dowell, Frank Herbert
Dronen, Norman Obert, Jr
Durio, Walter O'Neal
Dusanic, Donald G
Duszynski, Donald Walter
Dwyer, Dennis Michael
Egerton, John Richard
Elliott, Alice
Erickson, Duane Gordon
Ernst, John Verlon
Ernst, Sue Carlisle
Esslinger, Jack Houston
Etges, Frank Joseph
Eure, Herman Edward
Ewert, Adam
Fallis, Albert Murray
Farmer, John Neville
Farr, Marion Margaret
Federowicz, Rose Ann
Fernando, Constantine Herbert
Fischthal, Jacob Henry
Fisher, Elwood
Fisher, Frank M, Jr
Fisk, LeRoy (Henry)
Fitzgerald, Paul Ray
Font, William Francis
Foor, W Eugene
Frandsen, John Christian
Freele, Hugh W
Freeman, Reino Samuel
French, Frank Elwood, Jr
Fried, Bernard
Furman, Deane Philip
Galvin, Thomas Joseph
Garcia, Richard
George, Charles Redgenal
Garoian, George
Gibbs, Harold Cuthbert
Gibson, Colvin Lee
Gilbertson, Donald Edmund
Gleason, Larry Neil
Goldman, Morris
Goodman, John David
Goulson, Hilton Thomas
Graham, Charles Lee
Grassmick, Robert Alan
Greene, Nathan Doyle
Greichus, Algirdas
Griffo, James Vincent, Jr
Grundmann, Albert Wendell
Guilford, Harry Garrett
Gutierrez, Jose
Haley, Albert James
Hall, John Edgar
Halberg, Carl William
Hampton, Carolyn Hutchins
Hanson, Merle Frederick
Hare, Gerard Murdock
Harkema, Reinard
Harley, John Paul
Harmon, Wallace Morrow
Harris, Clarence Eugene
Harrington, Glenn William
Harris, Alva H
Harris, Antonio Efthiemios
Harrison, Robert J
Hass, D Kendall
Hathaway, Ronald Philip
Healy, George Richard
Heck, Oscar Benjamin
Heckmann, Richard Anderson
Held, Joe R
Hendricks, James Richard
Henson, Richard Nelson
Herman, Carlton Martin
Herman, Robert
Hewitt, Redginal Irving
Heyneman, Donald
Hibbert, Larry Eugene
Hill, George Carver
Hillyer, George Vanzandt
Hinck, Lawrence Wilson
Ho, Ju-Shey
Hoffman, Glenn Lyle
Hoffman, Rhodes Burns
Holloway, Harry Lee, Jr
Hood, Marion Winifred
Howes, Harold, Jr
Hsu, Hsi Fan
Hsu, Shu Ying Li
Huff, Dennis Karl
Huffman, David George
Huizinga, Harry William
Hunes, Arthur Grover
Hurley, Francis Joseph
Husain, Ansar
Hutchison, William Forrest
Hwang, Joseph Cen
Hyland, Kerwin Ellsworth, Jr
Jachowski, Leo Albert, Jr
Jackson, George John
James, Hugo A
Jaskoski, Benedict Jacob
Jensen, Emron Alfred
Johnson, Arthur Albin
Johnson, David Thomas
Johnson, John Ronald
Johnson, John Christopher, Jr
Jones, Arthur Wynne
Jones, Ira
Justus, David Eldon
Kabata, Zbigniew
Kagan, Irving George
Kalkofen, Ulrich Paul
Kantor, Sidney
Kaplan, Eugene Herbert
Kates, Kenneth Casper, Sr
Katz, Frank Fred
Kemp, Walter Michael
Kenney, Michael
Keppner, Edwin James
Kerr, Kathel Bedortha
Kim, Charles Wesley
Kingston, Newton
Kinsella, John Michael
Knight, Robert Arthur
Kniskern, Verne Burton
Knuckles, Joseph Lewis
Krassner, Stuart M
Krupa, Paul L
Krupp, Iris M
Kuntz, Robert Elroy
Laird, Marshall
Lang, Bruce Z
Larson, Ingemar W
Larson, Omer P
Lautenschlager, Edward Walter
Lautter, Felix H
Lawrence, James Lester
LeFlore, William B
Leigh, Walter Henry
Leland, Stanley Edward, Jr
Levin, Norman Lewis
Levine, Donald Martin
Levine, Harvey Robert
Levine, Norman Dion
Levy, Michael Green
Lichtenfels, James Ralph
Lincicome, David Richard
Litchford, Robert Gary
Loomis, Edmond Charles
Lumsden, Richard
Lund, Everett Eugene
Lynch, John Edward
Lyons, Eugene T
Lysenko, Michael George
MacInnis, Austin J
Mackiewicz, John Stanley
Macy, Ralph William
Maddison, Shirley Eunice
Mahrt, Jerome L
Mankau, Sarojam Kurudamanil
Maples, William Paul
Margolis, Leo
Mark, Daniel Lee
Markell, Edward Kingsmill
Marquardt, William Charles
Martin, Gordon Wyatt
Martin, Virginia Loretta
Martin, Walter Edwin
Marty, Wayne George
Mayberry, Lillian Faye
McCarthy, Vincent Cormac
McCowen, Max Creager

Oral Biology

Berman, Kenneth Sidney
Fitzgerald, Robert James
Gay, Thomas John
Greulich, Richard Curtice
Hunt, Lindsay McLaurin, Jr
Kleinberg, Israel
Kollar, Edward James
Koulourides, Theodore I
Levine, Philip Theodore

McCoy, Oliver Rufus
McCue, John Francis
McDaniel, James Scott
McDermott, John Joseph
McDonald, Malcolm Edwin
McDougald, Larry Robert
McDowell, John Willis
McGavock, Walter Donald
McGhee, Robert Barclay
McGraw, James Carmichael
McGraw, John Leon, Jr
McKnight, Thomas John
McLoughlin, Donald Keith
McManus, Edward Clayton
Meade, Thomas Gerald
Meerovitch, Eugene
Mehra, Krishna Nandan
Meszoely, Charles Aladar Maria
Mettrick, David Francis
Meyer, Fred Paul
Miller, Joseph Nelson
Miller, Louis Howard
Mitchell, Joseph Christopher
Moore, Donald Vincent
Morris, Gerald Patrick
Morrison, Eston Odell
Muller, Miklos
Munson, Donald Albert
Murrell, Kenneth Darwin
Myers, Betty June
Nahhas, Fuad Michael
Najarian, Haig Hagop
Nash, Reginald George
Neghme, Amador
Neilson, John Taylor McLaren
Nelson, Elvin Clifford
Newton, Walter Lloyd
Nickel, Phillip Arnold
Nickol, Brent Bonner
Nigrelli, Ross Franco
Noble, Elmer Ray
Noble, Glenn Arthur
Noda, Kaoru
Nolf, Luther Owen
Nollen, Paul Marion
Norris, Donald Earl, Jr
Norris, Mark Gilbert, Jr
Nusser, Wilford Lee
Nydegger, LeRoy B
Oaks, John Adams
Odlaug, Theron Oswald
Oliver, Victor L
Oliver-Gonzales, Jose
Olson, Andrew Clarence, Jr
Olson, Robert Eldon
Ostlind, Dan A
Otero, Joseph Guillermo
Ott, Karen Jacobs
Overstreet, Robin Miles
Page, Clayton R, III
Panitz, Eric
Pankavich, John Anthony
Pappas, Peter William
Parker, John Clarence
Patillo, Walter Hugh, Jr
Patton, Curtis Leverne
Paulson, Carlton
Peebles, Charles Robert
Penn, James H
Penner, Lawrence Raymond
Perkins, Kenneth Warren
Peters, Lewis
Peters, Paul James
Petri, Leo Henry
Pfefferkorn, Elmer Roy, Jr
Phares, Cleveland Kirk
Phifer, Kenneth Oscar
Pitts, Thomas Dennis
Platzer, Edward George
Poeschel, Gordon Paul
Porter, Clarence A
Porter, Dale Albert
Porter, Richard Janvier
Powders, Vernon Neil
Prince, Buford Earl
Pritchard, Mary (Louise) Hanson
Prudhon, Rolland A, Jr
Rabalais, Francis Cleo
Raghunathan, Lalitha
Reid, Willard Malcolm
Reissig, Magdalena
Ridgeway, Bill Tom
Rigby, Donald W
Risby, Edward Louis
Roberson, Edward Lee
Roberts, Irwin Herbert
Roberts, Larry Spurgeon
Robinson, Edwin James, Jr
Roche, Marcel
Rodrick, Gary Eugene
Rogers, Steffen Harold
Rogers, William Edwin
Rossan, Richard Norman
Rouse, Thomas C
Rowland, May Eloise
Ruff, Michael David
Sadler, Clarence Reagan

Sadun, Elvio Herbert
Samuel, William Morris
Sanchez, Gilbert
Sargent, Roger Gary
Savard, Edward Victor
Schacher, John Fredrick
Schad, Gerhard Adam
Schell, Stewart Claude
Schenkel, Robert H
Schmidt, Gerald D
Schneider, Curt Richard
Seed, John Richard
Senft, Alfred Walter
Sengbusch, Howard George
Shaver, Robert John
Shepperson, Jacqueline Ruth
Sherman, Irwin William
Shields, David Allen
Shoemaker, Jon Philip
Short, Robert Brown
Shumard, Raymond Fred
Siddiqui, Wasim A
Sillman, Emmanuel I
Silverman, Paul Hyman
Simmons, John Everette, Jr
Simpson, Myron Lee
Singer, Ira
Slagle, Wayne Grey
Slocombe, Joseph Owen Douglas
Slonka, Gerald Francis
Smith, Barnett Frissell
Smith, Philip Edward
Smith, Thomas Marion
Sogandares-Bernal, Franklin
Solomon, Gene Barry
Soulsby, Ernest Jackson Lawson
Spitalny, George Leonard
Spuller, Robert L
Stahl, Walter Bernard
Standifer, Lonnie Nathaniel
Steen, Edwin Benzel
Stewart, Thomas Bonner
Stockton, Jack Jenks
Stone, William Morgan
Strout, Richard Goold
Stueben, Edmund Bruno
Styles, Twitty Junius
Sudds, Richard Huyette, Jr
Summers, William Allen, Sr
Swartz, Leslie Gerard
Tan, Bian Djoen
Tanner, Charles E
Tarshis, Irvin Barry
Thatcher, Vernon Everett
Thomas, Leo Alvon
Thompson, John Harold, Jr
Threlfall, William
Tipton, Vernon John
Trager, William
Train, Carl T
Tromba, Francis Gabriel
Tuff, Donald Wray
Tulloch, George Sherlock
Turco, Charles Paul
Turner, Henry Ford
Turner, James Henry
Twohy, Donald Wilfred
Ubelaker, John E
Uglem, Gary Lee
Unden, Albert Harold
Vande Vusse, Frederick John
Van Zandt, Paul Doyle
Vatne, Robert Dahlmeier
Vaughn, Charles Melvin
Velasquez, Carmen C
Vetterling, John Martin
Voge, Marietta
von Zellen, Bruce Walfred
Wagner, Edward D
Wallace, Franklin Gerhard
Walls, Kenneth W
Walton, Bryce Calvin
Wang, Ching Chung
Wang, Guang Tsan
Ward, Helen Lavina
Ward, James Wellington
Ward, James Francis
Warnock, Robert G
Warren, Lionel Gustave
Watt, John Yin Chieh
Webster, Jackson Dan
Webster, John Malcolm
Weinmann, David, II
Weinmann, Clarence Jacob
Weinstein, Paul P
Welker, George W
Weppelman, Roger Michael
Wessenberg, Harry Sanders
West, Arthur James, II
Westervelt, Clinton Albert, Jr
White, Francis Michael
White, Jesse Steven
Whittaker, Frederick Horace
Wiles, Michael
Wilhelm, Walter Eugene
Wilkes, Stanley Northrup
Williams, Jeffrey F
Williams, Russell Raymond

Wilson, William D
Wittner, Murray
Wong, Ming Ming
Wood, Irwin Boyden
Wood, Raymond Arthur
Wootton, Donald Merchant
Worley, David Eugene
Wright, Kenneth A
Wyatt, Ellis Junior
Wykoff, Dale Emerson
Yaeger, Robert George
Yarinsky, Allen
Yoeli, Meir
Yogore, Mariano G, Jr
Yoshino, Timothy Phillip
Young, Martin Dunaway
Yunker, Conrad Erhardt
Zahalsky, Arthur C
Zam, Stephen G, III
Zimmermann, William John
Zischke, James Albert

Pathobiology
Bang, Frederik Barry
Christian, John Jermyn
Collier, Robert John
Couch, John Alexander
Damon, Edward George
Foley, David Allen
Harshbarger, John Carl, Jr
Jakowska, Sophie (Mrs C L Jeannopoulos)
Rifkin, Erik
Schultz, Warren Walter
Smith, Albert Carl
Taylor, Donald Fulton
Vernick, Sanford H
Yang, Tsu-Ju (Thomas)

Perinatal Biology
Alleva, Frederic Remo
Harding, Paul George Richard
Hasselmeyer, Eileen Grace
Hemberger, Judith Ann
Kelman, Bruce Jerry
Rosso, Pedro
Stern, Leo

Photobiology
Bellin, Judith Schryver
Bjorkman, Olle
Bonaventura, Celia Jean
Brand, Jerry Jay
Brody, Marcia
Brown, Jeanette Snyder
Cairns, William Louis
Chappelle, Emmett W
Cripps, Derek J
Diner, Bruce Aaron
Gaffron, Hans
Girsch, Stephen John
Goodman, Lionel
Hall, Robert Lindsay
Harner, Carol Frances Hodgson
Hewitt, Roger R
Jagger, John
Job, Donald Dexter
Knaff, David Barry
Lamola, Angelo Anthony
Lin, Lily
Myers, Jack Edgar
Nachtwey, David Stuart
Nair, Sreedhar
Olson, John Melvin
Pon, Ning Gin
Pooler, John Preston
Schneider, Michael J
Seibert, Michael
Seliger, Howard Harold
Smith, Kendric Charles
Stern, Arthur Irving
Straight, Richard Coleman
Turek, Fred William
Withrow, Alice Phillips
Worrest, Robert Charles

Physical Biology
Curby, William Adolph
Fullmer, Curtis Sheridan
Lengemann, Frederick William
Macklin, Martin
Olson, Sidney Andrew
Thompson, John C, Jr

Polar Biology
McWhinnie, Mary Alice

Pollution Biology
Anderson, Bertil Gottfrid
Bender, Michael E
Benedict, Harris Miller
Brungs, William Aloysius, Jr
Clark, Donald Ray, Jr
Feder, William Adolph
Fetterolf, Carlos De La Mesa, Jr
Jackson, Herbert William
LeGore, Richard Stephen
Leone, Ida Alba

Martin, Michael
Merkley, Wayne Bingham
Reese, Weldon Harold
Rice, Stanley Donald
Smith, Lloyd Lyman
Spacie, Anne
Stendell, Rey Carl
Strand, John A, III
Teeri, James Arthur

Population Biology
Balsano, Joseph Silvio
Berger, Beverly Jane
Boyer, John Frederick
Brussard, Peter Frans
Cameron, Guy Neil
Campbell, Howard Wallace
Collier, Boyd David
Connell, Joseph H
Costantino, Robert Francis
Dawson, Peter Sanford
Emmel, Thomas C
Endler, John Arthur
Futuyma, Douglas Joel
Gaines, Michael Stephen
Giesel, James Theodore
Gill, Ayesha Elenin
Graham, Joseph James
Grant, Michael Clarence
Hacker, Carl Sidney
Hamrick, James Lewis
Horn, Edward Gustav
Hummon, William Dale
Hunt, Henry William
Jain, Subodh K
King, Charles Everett
LaBar, Martin
Levins, Richard
Lewontin, Richard Charles
Lidicker, William Zander, Jr
Lister, Bradford Carlton
Makielski, Sally Kimball
Mason, Larry Gordon
McKinney, Charles Oran
Mearns, Alan John
Metcalf, Robert Alan
Mitchell, Rodger (David)
Moore, William Samuel
Murdoch, William W
Nabi, Isidore
Ode, Philip E
Pienaar, Leon Visser
Plank, Stephen J
Platt, Austin Pickard
Prager, Denis Jules
Rabb, Robert Lamar
Ramsey, Paul Roger
Raven, Peter Hamilton
Roughgarden, Jonathan David
Schaal, Barbara Anna
Schennum, Wayne Edward
Schultz, Roland Jack
Shapiro, Arthur Maurice
Slakin, Montgomery Wilson
Snyder, John Crayton
Sokal, Robert Reuven
Soule, Michael E
Speidel, John Joseph
Stiven, Alan Ernest
Straw, Richard Myron
Taylor, Norman Burwell George
Terborgh, John J
Thompson, Charles Frederick
Tilley, Stephen George
Tordoff, Walter, III
Udovic, Joseph Daniel
Vaillant, Henry Winchester
Vandermeer, John H
Watson, Maxine Amanda
Weiss, Kenneth Monrad

Psychobiology
Buerger, Alfred Arthur
Burghardt, Gordon Martin
Cheal, MaryLou
Dews, Peter Booth
Ellingson, Robert James
Farel, Paul Bertrand
Gazzaniga, Michael Saunders
Goldschmidt, Leontine
Gottlieb, Gilbert
Grossman, Sebastian Peter
Hauty, George Thomas
Hunsaker, Don, II
Kakolewski, Jan Wiktor
Kluver, Heinrich
Knight, Walter Rea
Komisaruk, Barry Richard
Maxson, Stephen C
Moorcroft, William Herbert
Myers, Robert Durant
Oppenheim, Ronald William
Richter, Curt Paul
Teng, Evelyn Lee
Wallace, Roger B
Walsh, Roger Nugent
Weiler, Ivan-Jeanne Mayfield
Werboff, Jack

Williams, Theodore P
Zornetzer, Steven Frank

Radiobiology
Ainsworth, Earl John
Alpen, Edward Lewis
Arnols, Howard Ira
Anderson, Donald Rex
Bagshaw, Malcolm A
Bair, William J
Baker, Max Leslie
Baldwin, William F
Banerjee, Satyendra Nath
Barranco, San Christopher
Bateman, John Laurens
Baum, Siegmund Jacob
Berdjis, Charles Choaib
Boecker, Bruce Bernard
Bonham, Kelshaw
Brace, Kirkland
Brennan, James Thomas
Brown, J Martin
Bruce, Alan Kenneth
Bruce, Austin M
Burba, John Vyautas
Burger, Charles L
Bustad, Leo Kenneth
Cairnie, Alan B
Carlson, Donald Eugene
Carlson, James Gordon
Carpenter, Russell Le Grand
Carsten, Arland L
Casarett, Alison Provoost
Cember, Herman
Chavin, Walter
Chen, I-Wen
Clapp, Neal K
Clark, Gordon Murray
Clement, Jacob James
Cohen, Alan B
Cohn, Stanton Harry
Cole, Ronald Sinclair
Concannon, Joseph N
Conger, Alan Douglas
Craig, Douglas Kenneth
Cronroy, Harvey Leonard
Crouch, Billy G
Crouse, David Austin
Cummings, John Albert
Dalrymple, Glenn Vogt
Davidson, David Edward, Jr
De Boer, Jelle
Deschner, Eleanor Elizabeth
Dethlefsen, Lyle A
Dewey, William Cornet
Djordjevic, Bozidar
Dougherty, Thomas John
Dudley, Horace Chester
DuFrain, Russell Jerome
Dugle, David L
Durand, Ralph Edward
Easterday, Otho Dunreath
Eddy, Hubert Allen
Elkind, Mortimer M
Fabrikant, Jacob I
Finian, Walter Joseph, Jr
Finkel, Miriam Posner
Fischer, James Joseph
Flint, Hollis Mitchell
Foltz, Virginia C
Forbes, Paul Donald
Friedberg, Wallace
Fritz, Thomas Edward
Fry, Richard Jeremy Michael
Gaulden, Mary Esther
Gerner, Eugene Willard
Gibbs, Samuel Julian
Gillette, Edward LeRoy
Goldman, Marvin
Grahn, Douglas
Green, Morris
Griffin, Edmond Eugene
Gutman, Paul H
Hagemann, Ronald Fred
Hahn, Eric John
Hall, Eric John
Hanson, Wayne Robert
Harrison, George H
Hart, Ronald Wilson
Hasegawa, Andrew Takeo
Heddle, John A M
Henshaw, Paul Stewart
Hodge, Frederick Allen
Hoegerman, Stanton Fred
Hogan, Gene Richard
Holtzman, Richard Beves
Howland, Joe Wiseman
Hungate, Frank Porter
Inch, William Rodger
Jacobus, David Penman
Jenkins, Vernon Kelly
Kallman, Robert Friend
Kim, Jae Ho
Kim, Lamar M
Kinnamon, Kenneth Ellis
Knill, Lamar M
Kohn, Henry Irving
Kollmann, George
Koo, Francis Keh Shing

Kornberg, Harry Alexander
Kraft, Lisbeth Martha
Kruuv, Jack
Kurohara, Samuel S
Lambremont, Edward Nelson
Lange, Christopher Stephen
Lawrence, Christopher William
Leach, William Matthew
Lebel, Jack Lucien
Lee, Young Chang
Lehnert, Shirley Margaret
Leon, Shalom A
Leone, Charles Abner
Lesher, Samuel Walter
Lett, John Terence
Lindenbaum, Arthur
Lombardi, Max H
Looney, William Boyd
Lytle, James Bert
Maier, John G
Maillie, Hugh David
Marcus, Carol Silber
Maruyama, Yosh
Maynard, Russell Hatton
McClanahan, Beatrice J
McDonnel, Gerald M
Mewissen, Dieudonne Jean
Michaelson, Solomon M
Miller, Edward Joseph
Miller, Morton W
Mitchell, Hugh Bertron
Moroson, Harold
Mulvey, Philip Francis, Jr
Nachwey, David Stuart
Nelson, Janet Sue Rasey
Nelson, Neal Stanley
Noonan, Thomas Robert
Normandin, Robert F
Norris, William Penrod
Nygaard, Oddvar Frithjof
Osborne, James William
Page, Norbert Paul
Pahl, George Leo
Painter, Robert Blair
Palmer, Ray Frederick
Patt, Harvey Milton
Pena, Hugo Gabriel
Pentel, Lajos
Phillips, Theodore Locke
Pizzarello, Donald Joseph
Powers, Edward Lawrence
Prasad, Kedar N
Proshold, Fredrick Irving
Quastel, Michael Reuben
Reichard, Sherwood Marshall
Richardson, Barry Lovell
Richmond, Chester Robert
Richmond, Robert Chaffee
Riley, Edgar Francis
Roberts, Joan Marie
Robinette, Charles Dennis
Robinson, Charles Vernon
Robinson, Gerald Arthur
Rundo, John
Rushton, Priscilla Strickland
Rust, John Howard
Sanders, Aaron Perry
Sanders, Charles Leonard, Jr
Sanders, Samuel Marshall, Jr
Schjeide, Ole Arne
Schultz, Harvey Albert
Seydel, Horst Gunter
Shani, Jashovam
Shaw, Edward Irwin
Shellabarger, Claire J
Shepherd, David Preston
Sikov, Melvin Richard
Skargard, Lloyd Donald
Skidmore, Wesley Dean
Slater, John Vernon
Smith, David Allen
Smith, Lawton Harcourt
Sobkowski, Frank J
Sondhaus, Charles Anderson
Spalding, John F
Sparrow, Rhoda Cornish
Spertzel, Richard O
Squire, Richard Douglas
Stannard, James Newell
Stearner, Sigrid Phyllis
Steele, Vernon Eugene
Stefani, Stefano
Steheny, Andrew Frank
Stevens, Walter
Still, Edwin Tanner
Stone, John Patrick
Storer, John B
Stover, Betsy Jones
Stratmeyer, Melvin Edward
Straube, Robert Leonard
Stroud, F Agnes Naranjo Schmink
Swartz, Harold M
Swingle, Karl Frederick

Talbot, John Mayo
Taves, Donald R
Thomas, Robert Glenn
Thompson, Roy Charles, Jr
Tolmach, Leonard Joseph
Townsley, Sidney Joseph
U, Raymond
Uyeki, Edwin M
Van't Hof, Jack
Varon, Myron Izak
Vetter, Richard J
Volkert, Wynn Arthur
Wachholz, Bruce William
Walters, Stanley Norman
Wampler, Stanley Norman
Watson, Joseph Alexander
Welander, Arthur Donovan
Whicker, Floyd Ward
White, David C
Whitmore, Gordon Francis
Wilson, John Dreman
Wise, Ernest George
Withers, Hubert Rodney
Wolf, Norman Sanford
Woodard, Helen Quincy
Yatvin, Milton B
Yuhas, John M

Reproductive Biology
Ainsworth, Louis
Anderson, Richard Gilpin Wood
Ansbacher, Rudi
Arthur, Alan Thorne
Barnes, Raymond D
Bernstein, Gerald Sanford
Bleier, William Joseph
Blye, Richard Perry
Boving, Bent Giede
Champlin, Arthur Kingsley
Chowdhury, Mridula
Cooper, George Wallace, Jr
Dagg, Charles Patrick
De Feo, Vincent Joseph
Diamond, Milton
Dickson, Arthur David
Dierschke, Donald Joe
Enders, Allen Coffin
Fahim, Mostafa Safwat
Fossland, Robert Gerard
Foster, Raymond Orrville
Fox, Kevin A
Fritz, Irving Bamdas
Gallo, Duane Gordon
Gates, Allen Hazen, Jr
Glasser, Stanley Richard
Gomes, Wayne Reginald
Gual, Carlos
Gustafson, Alvar Walter
Gwatkin, Ralph Buchanan Lloyd
Heath, Everett
Hintz, Marie I
Hiscoe, Helen Brush
Hoar, Richard Morgan
Hoffman, Loren Harold
Hohman, Roger Alan
Homsher, Paul John
Johnson, Erwin
Johnson, James Dean
Jones, Richard Evan
Kasprow, Barbara Ann
Kistler, Wilson Stephen, Jr
Kleinfeld, Ruth Grafman
Kulangara, Abraham Chakko
Lambert, Charles Calvin
Larkin, Lynn Haydock
Lodwick, Gwilym Savage
Mahoney, Richard Theodore
Marcus, George Jacob
Martan, Jan
McCarthy, Miles Duffield
McClurg, James Edward
Metzel, Stanley
Millette, Clarke Francis
Monson, Frederick Carlton
Morton, Bruce Eldine
Mueller, Nancy Schneider
Nakamura, Robert Masao
Oliphant, Edward Eugene
Phoenix, Charles Henry
Poirier, Gary Raymond
Prakash, Chandra
Quattropani, Steven L
Ramaley, Judith Aitken
Remeels, Gerald Gerald
Richmond, Milo Eugene
Rogers, Beverly Jane
Schantz, Ilene Sue Cottler
Scott, James Raymond
Seiman, Kelly
Shivers, Charles Alex
Spilman, Charles Hadley
Srivastava, Prakash Narain
Stencheve, Morton Albert
Tcholokian, Robert Kevork
Telfer, William Harrison
Teng, Ching Sung
Van Horn, Richard Norman
Wallace, Robin A

Wolf, Don Paul
Woodruff, Richard Ira
Yanagimachi, Ryuzo

Serology
Allen, Rovelle Harper
Anderson, Robert I
Anthony, Ronald Lewis
Boulanger, Paul
Hatgi, John Neal
Janeff, Donka Grigorova
Janeff, Jan Dimitroff
Mueller, August P
Schoenholz, Walter Kurt
Vogel, Henry
Waller, Marion Van Nostrand

Theoretical Biology
Bell, George Irving
Bergner, Per-Erik Emil
Cohn, David Lionel
Cowan, Jack David
Gatlin, Lila L
Gillespie, Colin J
Kauffman, Stuart Alan
Kerner, Edward Haskell
Koch, Arthur Louis
Lindenmayer, Aristid
Loew, Gilda M Harris
Manougian, Edward
Pattee, Howard Hunt, Jr
Perkel, Donald Howard
Rapport, David Joseph
Roti Roti, Joseph Lee
Somorjai, Rajmund Lewis
Ulanowicz, Robert Edward
Yockey, Hubert Palmer

Transplantation Biology
Ballantyne, Donald Lindsay, Jr
Beer, Alan E
Converse, John Marquis
Donawick, William Joseph
Shaffer, Charles Franklin
Steinmuller, David
Veith, Frank James

Vertebrate Biology
Blair, William Franklin
Caldwell, David Keller
Chizinsky, Walter
Coulombe, Harry N
Cunningham, Harry N, Jr
Hecht, Max Knobler
Hoppe, David Matthew
Jackson, Crawford Gardner, Jr
Layne, James Nathaniel
Ludwig, James Pinson
MacMillen, Richard Edward
Madison, Dale Martin
Mayhew, Wilbur Waldo
Natalini, John Joseph
Nickerson, Max Allen
Nordan, Harold Cecil
Scudday, James Franklin
Secoy, Diane Marie
Smith, Gary Charles
Smith, Lorraine Catherine
Snyder, David Hilton
Turner, Larry Webster
Uzzell, Thomas Marshall, Jr
Vial, James Leslie
Wake, Marvalee H
Zimmerman, Earl Graves

Veterinary Parasitology
Ah, Hyong-Sun
Andrews, John Scott
Andrews, Myron Floyd
Bergstrom, Robert Charles
Besch, Everett Dickman
Colglazier, Merle Lee
Cruthers, Larry Randall
Dewhirst, Leonard Wesley
Drudge, Junior Harold
Dunlap, Jack Sherwin
Enzie, Frank Dorr
Ewing, Sidney Alton
Ferguson, Donald Leon
Foltz, Sylvester D
Forrester, Donald Jason
Gaafar, Sayed Mohammed
Greve, John Henry
Griffiths, Henry Joseph
Hayes, Terence James
Herlich, Harry
Jordan, Helen Elaine
Knapp, Stuart Edward
Kohls, Robert E
Lindquist, William Dexter
Powers, Kendall Gardner
Rausch, Robert Lloyd
Rubin, Robert

Schlotthauer, John Carl
Shelton, George Calvin
Slater, Robert Lee
Splitter, Earl John
Szanto, Joseph
Theodorides, Vassilios John
Todd, Kenneth S, Jr
Wescott, Richard Breslich
Whitlock, John Hendrick

Wildlife Biology
Anderson, William Leno
Applegate, James Edward
Baskett, Thomas Sebree
Braun, Clait E
Callaham, Mac A
Campbell, Howard
Causey, Miles Keith
Dodds, Donald Gilbert
Erskine, Anthony J
Harder, John Dwight
Johnson, Albert Sydney, III
Klaas, Erwin Eugene
Labisky, Ronald Frank
Larson, Joseph Sbanley
Linzey, Donald Wayne
MacLulich, Duncan Alexander
Mendall, Howard Lewis
Pelton, Michael Ramsay
Raveling, Dennis Graff
Samuel, David Evan
Sowls, Lyle Kenneth
Speller, Stanley Wayne
Steinhoff, Harold William
Thomas, Jack Ward
VanDruff, Larry Wayne
Verme, Louis Joseph

BIOPHYSICS

Biophysics
Abbott, Bernard C
Abeles, Ann Lindstrom
Adelman, William Joseph, Jr
Agalides, Eugene
Aktipis, Stelios
Al-Awqati, Qais
Allen, Robert Day
Alteveer, Robert Jan George
Alvager, Torsten Karl Erik
Anderegg, John William
Anderson, Ernest Carl
Anderson, Lowell Leonard
Anderson, Nels Carl, Jr
Anderson, Thomas Foxen
Andrews, Howard Lucius
Ansevin, Allen Thornburg
Asher, Irvin Mark
Azarnia, Roobik
Bach, Sven Aage
Bahr, Gunter F
Baier, Robert Edward
Baker, Max Leslie
Baker, Richard Freligh
Baker, Robert G
Bakerman, Seymour
Bakken, George Stewart
Bale, William Freer
Bales, Barney Leroy
Baptist, Jeremy Eduard
Barany, Kate
Barnes, Edward B
Barnes, James S
Barrett, Terence William
Bartel, Allen Hawley
Bartl, Paul
Baskin, Ronald J
Bassingthwaighte, James B
Baughman, Dwight Joe
Bayley, Stanley Thomas
Bean, Charles Palmer
Bear, Richard Scott
Bearden, Alan Joyce
Beauge, Luis Alberto
Beck, James S
Becker, Milton J
Becker, Robert O
Beeman, William Waldron
Behnke, William David
Beidler, Lloyd M
Bemski, George
Benda, Stepan Vaclav
Bendet, Irwin (Jacob)
Bennun, Alfredo
Benson, Brent W
Berg, Howard Curtis
Berger, Robert Lewis
Bernhard, William Allen
Berns, Donald Sheldon
Bertsch, Walter Frank
Best, Jay Boyd
Bethune, John Lemuel
Biedebach, Mark Conrad
Bier, Milan
Bigler, Rodney Errol
Bird, Richard Putnam
Bishop, Vernon Spilman

Blackwell, John
Blanchard, Fred Ayres
Blank, Martin
Blasie, J Kent
Blattner, Frederick Russell
Baumanis, Otis Rudolf
Blum, Alvin Seymour
Blum, Harold Francis
Blum, Haywood
Blum, Jacob Joseph
Blumberg, William Emil
Bockstahler, Larry Earl
Bogart, Eliot
Bohan, Thomas Lynch
Bolton, Ellis Truesdale
Bond, Howard Edward
Borsa, Joseph
Bosmann, Harold Bruce
Bowen, John Metcalf
Bowen, Thomas Earle, Jr
Boyce, Richard P
Boyle, Mary Maurice
Brady, Allan Jordan
Brandon, Frank Bayard
Braswell, Emery Harold
Brattain, Walter Houser
Brenner, Stephen Louis
Brill, Arthur Sylvan
Brink, Frank, Jr
Brinley, Floyd John, Jr
Brinton, Charles Chester, Jr
Brody, Seymour Steven
Bronner, Felix
Brown, Arthur Charles
Brown, Arthur Morton
Brown, Harold Mack, Jr
Brown, Morden Grant
Brown, Rodney Duvall, III
Brown, Thomas Townsend
Brownell, Arnold S
Brudner, Harvey Jerome
Bruner, Leon James
Brunk, Clifford Franklin
Burke, Arthur Wade, Jr
Burke, Michael John
Burns, Victor Will
Burton, Alan Chadburn
Butler, Byron C
Butler, Charles Thomas
Butler, Keith Winston
Butler, Warren Lee
Campillo, Anthony Joseph
Candia, Oscar A
Canham, Peter Bennet
Carlson, Francis Dewey
Carlson, Roy Douglas
Carpenter, Robert Leland
Carrano, Anthony Vito
Cartwright, Thomas Edward
Casella, Alexander Joseph
Catsimpoolas, Nicholas
Celitans, Gerard John
Cellarius, Richard Andrew
Challice, Cyril Eugene
Chambers, Leslie Addison
Chance, Britton
Chandler, Louis
Chandross, Ronald Jay
Chang, Donald Choy
Chapman, John Donald
Chapman, Kent M
Chapman, Robert Earl, Jr
Chen, Yi-Der
Cheung, Herbert Chiu-Ching
Chowdhury, Tushar Kumar
Christensen, Halvor Niels
Clark, Benton C
Clark, Carl Cyrus
Clark, Eloise Elizabeth
Clarke, Alexander Mallory
Clayton, Roderick Keener
Cleary, Stephen Francis
Cleveland, Gregor George
Clinch, Norman Frederick
Cline, George Bruce
Cochran, Andrew Aaron
Cohen, Carolyn
Cohen, Robert Jay
Cohn, Mildred
Cole, Kenneth Stewart
Coleman, Joseph Emory
Coleman, Peter Stephen
Collins, Carter Compton
Colvin, John Ross
Cone, Richard Allen
Conger, Alan Douglas
Connelly, Clarence Morley
Conrad, Michael
Coohill, Thomas Patrick
Cooke, Roger
Copson, David Arthur
Corless, Joseph Michael James
Cormack, Douglas Villy
Coulter, Norman Arthur, Jr
Cowan, Jack David
Cowlishaw, John David
Cox, Andrew Chadwick
Craig, Albert Morrison

Cram, Leighton Scott
Cramer, William Anthony
Crepeau, Richard Hanes
Crowell, Jack Wesley
Crowell, John Marshall
Cummings, Donald Joseph
Curtis, Joseph C
Czerlinski, George Heinrich
Dahl, Adrian Hilman
Dallos, Peter John
Damadian, Raymond
Darden, Edgar Bascomb, Jr
David, George Berthold
Davies, David R
Davies, Phillip Wynne
Deamer, David Wilson, Jr
Dean, Phillip Nolan
De Felice, Louis John
DeLevie, Robert
De Lisi, Charles
Dennis, Warren Howard
Dennison, David Severin
DeRocco, Andrew Gabriel
DeRosier, David J
DeVault, Don Charles
Devoe, Robert
De Weer, Paul Joseph
Dickinson, Wade
Diller, Violet Marion
Dintzis, Howard Marvin
Djordjevic, Bozidar
Doane, Marshall Gordon
Dobson, Ernest L
Domanik, Richard Anthony
Dowben, Robert Morris
Downey, Harry Fred
Drees, John Allen
Dreizen, Paul
Dugas, Hermann
Dumbleton, John Herbert
Duncan, Ronald Ian
Dunham, Theodore, Jr
Dunker, Alan Keith
Dunn, Floyd
Durbin, Patricia Wallace (Mrs James T Heavey)
Durbin, Richard Paul
Dusenbery, David Brock
Easter, Stephen Sherman, (Jr)
Eaton, William Allen
Eberle, Helen I
Eberstein, Arthur
Edwards, Charles
Edwards, Harold Henry
Edwards, Merrill Arthur
Eggen, Douglas Ambrose
Ehrenstein, Gerald
Einstein, J Ralph
Eisenberg, Robert S
Eisenman, George
Eisinger, Josef
Eldredge, Donald Herbert
Elkind, Mortimer M
Engelberg, Joseph
Engelman, Donald Max
Englander, Sol Walter
Epstein, Herman Theodore
Epstein, Ludwig Ivan
Erickson, Harold Paul
Estrup, Faiza Fawaz
Evans, Evan Cyfeiliog, III
Evett, Jay Fredrick
Faber, Jan Job
Failla, Patricia McClement
Falk, Gertrude
Farley, Belmont Greenlee
Farrell, Richard Alfred
Faust, Robert Gilbert
Feher, George
Feldman, Isaac
Felsenfeld, Gary
Filmer, David Lee
Fine, Samuel
Finlayson, Birdwell
Finn, Arthur Leonard
Firth, David Richard
Fischbach, Fritz Albert
Fischbein, Irwin William
Fisher, Harold Wilbur
Fishman, Harvey Morton
Fitzhugh, Richard
Fleischman, Darrell Eugene
Floyd, Robert A
Foelsche, Trutz
Fong, Francis K
Ford, George Dudley
Ford, Norman Cornell, Jr
Forro, Frederick, Jr
Foster, Margaret C
Fotino, Mircea
Fox, Jack Lawrence
Frank, Bruce Hill
Franke, Ernst Karl
Franz, Gunter Norbert
Frauenfelder, Hans (Emil)
Fried, Jerrold
Friedman, Kenneth Joseph
Friedman, Morton Harold

Frishkopf, Lawrence Samuel
Froehlich, Jeffrey Paul
Froese, Gerd
Frost, Harold Maurice, III
Gabbay, Edmond J
Gaffey, Cornelius Thomas
Gage, Robert Stanley
Gage, Adolf Pharo
Gagliardi, L John
Galey, William Raleigh
Gardner, Reed McArthur
Garrity, Michael K
Gates, David Murray
Geacintov, Nicholas
Geduldig, Donald Stanley
Geeraets, Walter J
Gershman, Lewis C
Gerson, Donald Franklin
Ghiron, Camillo A
Giancoli, Douglas Charles
Gibbons, David Louis
Gibson, Robert John, Jr
Gilliam, James Melvin
Glaeser, Robert M
Glimcher, Melvin Jacob
Goad, Walter Benson, Jr
Goerke, Rudolph Jon
Goldman, David Eliot
Goldman, Israel David
Goldman, Lawrence
Goldstein, Byron Bernard
Goldstein, Moise Herbert, Jr
Goodall, Marcus Campbell
Goode, Melvyn Dennis
Gottlieb, Melvin Harvey
Govindjee
Graetzer, Reinhard
Graham, Dean McKinley
Gray, Irving
Gray, Joe William
Green, Keith
Greenebaum, Ben
Griffin, Robert Guy
Griffith, Owen Malcolm
Grodsky, Irvin T
Groom, Alan Clifford
Gross, Leo
Grossweiner, Leonard Irwin
Guild, Walter Rufus
Gunter, Karlene Klages
Gunter, Thomas E, Jr
Gurpide, Erlio
Gutierrez, Peter Luis
Gutstein, William H
Guttman, Rita
Haaland, John Edward
Haas, David Jean
Hademenos, James George
Hall, James Ewbank
Hall, Theodore (Alvin)
Hallett, Frederick Ross
Hamblen, David Philip
Hamilton, Thomas Charles
Hammerman, Ira Saul
Hammes, Gordon G
Hardy, William Lyle
Harmison, Charles Rice
Harper, Richard Allan
Harrington, Rodney E
Hartley, Robert William, Jr
Hartman, Richard Eugene
Hartman, Roberta Smith
Hayashi, Shuki
Hayes, Thomas L
Haymond, Herman Ralph
Haynes, Robert Hall
Hazelwood, Robert Nichols
Hazelwood, Carleton Frank
Heath, Robert Louis
Heath, Robert Thornton
Heise, John J
Helland, Jerome A
Helmstetter, Charles E
Henderson, Edward George
Herbert, Thomas James
Hersh, Robert Tweed
Herzfeld, Judith
Herzlinger, George Arthur
Heyn, Anton Nicolaas Johannes
Higgins, Joseph John
Hill, Terrell Leslie
Hill, Walter Ensign
Hille, Bertil
Himel, Chester Mora
Hinke, Joseph Anthony Michael
Hinton, Dennis Melvin
Hippensteele, James Robert
Hirsch, Henry Richard
Hirschman, Shalom Zarach
Ho, Chien
Hoff, Henry Frederick
Hoffman, Robert Alan
Hollaender, Alexander
Hollis, Donald Pierce
Holmes, Dale M
Holmes, Donald Eugene
Holzwarth, George Michael

BIOPHYSICS

Hopfield, John Joseph
Horan, Paul Karl
Horn, Leif
Horn, Lyle William
Horowicz, Paul
Horvath, William John
Horwitz, Joseph
Houk, Albert Edward Hennessee
Houk, Thomas William
Howard-Flanders, Paul
Howatson, Allan F
Hsia, Ryine Tsu-Shou
Hsieh, James Stewart
Hsu, Kwan
Huang, Huey Wen
Huang, Sylvia Lee
Huang, Wei-Tze
Huebner, Jay Stanley
Hughes, William Taylor
Hunt, John Wilfred
Hutchinson, Franklin
Hybl, Albert
Hyde, James Stewart
Incardona, Antonino L
Ingram, Forrest Duane
Inners, Lamar Daniel
Irwin, James Wesley
Isaacson, Allen
Isenberg, Irvin
Jacobson, Baruch S
Jacobson, Kenneth Allan
Jacobson, Marcus
Jacobson, Stuart Lee
Jagger, John
Jain, Mahendra Kumar
Jen-Jacobson, Linda
Jennings, William Harney, Jr
Johns, Harold E
Johns, John Michael
Johnson, C Scott
Johnson, Carroll Kenneth
Johnson, John Richard
Johnson, Michael Evart
Johnson, Ronald Gordon
Johnson, William Harding
Jones, Carol A
Jones, Janice Lorraine
Josefowicz, Jack Yitzhak
Juliano, Rudolph Lawrence
Jurist, John Michael
Kaesberg, Paul Joseph
Kan, Lou Sing
Kang, Sungzong
Kang, Yeou-Jan
Kantor, George Joseph
Kartha, Mukund K
Katzper, Meyer
Kaufman, William Carl, Jr
Keane, John Francis, Jr
Keefe, William Edward
Kehl, Theodore H
Kelly, Lola Szanto
Kempner, Ellis Stanley
Khalil, Mohamed Thanaa
Khan, Faiz Mohammad
Kidder, George Wallace, III
Kiger, John Andrew, Jr
Kikson, Rein
Kim, Carl Stephen
Kimura, Tokuji
Kisman, Kenneth Edwin
Klip, Willem
Knox, Robert Seiple
Ko, Howard Wha Kee
Koehler, James K
Koenig, Donald Frederick
Koenig, Seymour Hillel
Kolin, Alexander
Korenbrot, Juan Igal
Kornacker, Karl
Koushanpour, Esmail
Krakauer, Henry
Krebs, John S
Krimm, Samuel
Krisch, Robert Earle
Kruuv, Jack
Kubitschek, Herbert Ernest
Kuehn, Lorne Allan
Kumbar, Mahadevappa M
LaCelle, Paul (Louis)
Lang, Ronald Albert
Landowne, David
Lange, Dimitrij Adolf
Lange, Christopher Stephen
Lange, Yvonne
Langley, Kenneth Hall
Larson, Gary Eugene
Latimer, Paul Henry
Latta, Harrison
Lauffer, Max Augustus, Jr
Lavallee, Marc
Lecar, Harold
Lederman, David Mordechai
Ledley, Robert Steven
Lee, John William
Lee, Lih-Syng
Lehman, Richard Lawrence
Lehman, Robert C
Leibovic, K Nicholas
Lenhert, P Galen

Letey, John, Jr
Lett, John Terence
Levengood, William Camburn
Levich, Calman
Levine, Raphael Berg
Levinthal, Cyrus
Lewis, Aaron
Lewis, Marc Simon
Lichtman, Marshall A
Lieb, William Robert
Liebman, Paul Arno
Lindeman, Charles Benard
Lindgren, Frank Tycko
Lindley, Barry Drew
Lipetz, Leo Elijah
Lloyd, Elizabeth Luke
Loew, Gilda M Harris
Loken, Merle Kenneth
Longworth, James W
Looney, William Boyd
Lott, James Robert
Loud, Alden Vickery
Love, Warner Edwards
Lowenhaupt, Benjamin
Luftig, Ronald Bernard
Luner, Stephen Jay
Lyman, John Tompkins
Lytle, Carl David

Maas, Peter
Macey, Robert Irwin
MacHattie, Lorne Allister
Mackay, Ralph Stuart
Mackay, Michael Charles
Macnab, Robert Marshall
MacNichol, Edward Ford, Jr
Macpherson, Cullen H
Maestre, Marcos Francisco
Magnuson, James Andrew
Malich, Charles Wilson
Mandelkern, Leo
Manikoff, Jack
Manney, Thomas Richard
Manning, JaRue Stanley
Marcus, Carol Silber
Marino, Andrew Anthony
Markowitz, David
Massover, William H
Mauro, Alexander
Mauzerall, David Charles
Maxfield, Bruce Wright
Mayall, Brian Holden
McAfee, Robert Dixon
McConnell, David Graham
McConnell, Robert A
McCrady, Edward
McElhaney, Ronald Nelson
McGuire, Robert Francis
McIntyre, Thomas Woodford
McLaren, Arthur Douglas
McMichael, John Calhoun
McNatt, Eugene Melton
McNulty, Peter J
McSwain, Berah Davis
Meistrich, Marvin Lawrence
Mel, Howard Charles
Mendelson, Mortimer Lester
Menz, Leo Joseph
Messineo, Luigi
Meyn, Raymond Everett, Jr
Milanovich, Fred Paul
Mildvan, Albert S
Miller, Richard Graham
Millman, Barry MacKenzie
Milvy, Paul
Mizukami, Hiroshi
Moffat, John Keith
Monahan, Wayne Gordon
Moore, Cyril L
Moore, Dan Houston
Moore, Richard Davis
Moore, Vaughn Clayton
Morales, Manuel Frank
Moriarty, C Michael
Morken, Donald A
Mortimer, Robert Keith
Morton, Richard Alan
Moss, Alfred Jefferson, Jr
Moss, Thomas Henry
Moulton, Grace Charbonnet
Mueller, Theodore Arnold
Mullaney, Paul F
Mullins, Lorin John
Murphy, Edward Joseph
Nakamura, Robert Masao
Naughton, Michael A
Navar, Luis Gabriel
Neal, Jack Laurance
Nelson, Gary Joe
Nelson, Richard Carl
Neuman, William Frederick
Nguyen-Huu, Xuong
Nielsen, Peter Tryon
Noordergraaf, Abraham
Northrip, John Willard

Norwich, Kenneth Howard
Nossal, Ralph J
Novick, Aaron
Nussbaum, Elmer
Nutter, Robert Leland
Nyborg, Wesley LeMars
Nye, Patrick William
Oberhardt, Bruce J
Odell, Floyd Adams
Offner, Franklin Faller
Ohki, Shinpei
Ohnishi, Tsuyoshi
O'Konski, Chester Thomas
O'Leary, Dennis Patrick
Ornstein, Leonard
Ortoleva, Peter Joseph
Oster, Gerald
Owen, Charles Scott
Padilla, George M
Padlan, Eduardo Agustin
Paganelli, Charles Victor
Pagano, Richard Emil
Page, Ernest
Palatt, Paul Jay
Palaty, Vladimir
Palubinskas, Felix Stanley
Paolini, Paul Joseph, Jr
Papaefthymiou, Georgia Christou
Papahadjopoulos, Panayotis Demetrios
Pape, Leon
Parrish, Rob Gene
Parsegian, Vozken Adrian
Parsons, Donald Frederick
Parsons, Rodney Lawrence
Parthasarathy, Rengachary
Patel, Dali Jehangir
Patrick, Michael Heath
Patt, Harvey Milton
Pauler, Eugene L
Pauly, William
Pearlstein, Robert Milton
Pearlstein, Alan Stuart
Perelson, Alan Stuart
Perry, Robert Palese
Person, Stanley R
Peters, Randall Douglas
Petsko, Gregory Anthony
Peusner, Leonardo
Phibbs, Roderic H
Phillips, William Dale
Phillipson, Paul Edgar
Pickard, William Freeman
Piette, Lawrence Hector
Pinson, Ernest Alexander
Platt, John Rader
Podolsky, Richard James
Poe, Martin Turner
Pollyovee, Myron
Pooler, John Preston
Poole, Ronald John
Postow, Elliot
Powell, Michael Robert
Poznansky, Mark Joab
Pratt, Arnold Warburton
Pressman, Berton Charles
Price, Richard Walter
Prothero, John W
Puck, Theodore Thomas
Puskin, Jerome Sanford
Quay, John Ferguson
Rabinovitch, Bernard
Rabinowitz, James Robert
Rall, Wilfrid
Randall, James Edwin
Randolph, Malcolm Logan
Rautaharju, Pentti M
Rauth, Andrew Michael
Ray, Dan S
Redwood, William Raymond
Reed, George Henry
Rehm, Warren Stacy, Jr
Reichmanis, Maria
Reid, Allen Francis
Rein, Robert
Resnick, Michael Aaron
Reuben, John Philip
Reynolds, George Thomas
Ribi, Edgar
Richardson, Alfred Wendel
Richardson, Irvin Whaley
Ridgway, Ellis Branson
Riggs, Demetrios A
Riggsby, William Stuart
Rikmenspoel, Robert
Rinfret, Arthur Piers
Roach, Margot Ruth
Roark, Dennis Edward
Roberts, Richard Brooke
Robinson, Thomas Frank
Roper, Leon David
Rosen, Arthur Leonard
Rosen, Philip
Rosenberg, Alburt M
Rosenblith, Walter Alter
Rosenbloom, Joel
Rosenshein, Joseph Stanley
Rotenberg, Manuel
Rothschild, Kenneth J
Rothstein, Aser

Rougvie, Malcolm Arnold
Rowlands, Stanley
Roy, Guy
Rubinstein, Daniel
Ruby, Ronald Henry
Rupert, Claud Stanley
Rust, John Howard
Saba, George Peter, II
Sabia, Thomas Maron
Sachs, John Rogers
Sachs, Frederick
Sachs, Thomas Dudley
Salhany, Jimmy Mitchell
Salzman, Gary Clyde
Sandler, Sheldon Samuel
Sandow, Alexander
Sands, Jeffrey Alan
Sands, Richard Hamilton
Sarui, Hisashi
Sarkar, Nurul Haque
Sato, Hidemi
Sayeg, Joseph A
Scarpa, Antonio
Schaefer, Hermann Joseph
Schanne, Otto F
Scharf, Arthur Alfred
Schauf, Charles Lawrence
Scheele, Robert Blain
Scheider, Walter
Scheie, Paul Olaf
Scheltgen, Elmer
Scheuplein, Robert J
Schick, Kenneth Leonard
Schilb, Theodore Paul
Schimmel, Paul Reinhard
Schindler, Alan Michael
Schleyer, Heinz
Schmitt, Otto Herbert
Schoenborn, Benno P
Scholes, Charles Patterson
Schor, Robert
Schrank, Auline Raymond
Schwan, Herman Paul
Schwartz, Joseph A
Schwartz, Tobias Louis
Scott, David Paul
Scott, Hugh Lawrence, Jr
Scott, Jesse Friend
Sekelj, Paul
Semer, James Parker
Setlow, Jane Kellock
Setlow, Richard Burton
Seufert, Wolf D
Sha'afi, Ramadan Issa
Shah, Dinesh Ochhavlal
Shahn, Ezra
Shainoff, John Rieden
Shalaby, Ragaa Abdel Fattah
Shalek, Robert James
Shamos, Adil E
Shamos, Morris Herbert
Shanbour, Linda Livingston
Shapiro, Jacob
Sharp, David Gordon
Shear, David Ben
Sheppard, Asher R
Sherebrin, Marvin Harold
Sherman, Fred
Shichi, Hitoshi
Shipley, George Graham
Shipp, William Stanley
Shrager, Peter George
Shrauner, Barbara Abraham
Shropshire, Walter, Jr
Sibbett, Donald Joseph
Siegel, Benjamin Morton
Siegel, Edward
Silverman, David Norman
Simmovitch, Louis
Sinclair, Warren Keith
Singer, Jerome Ralph
Sinks, John Davis
Sinks, Lucius Frederick
Sinsheimer, Robert Louis
Siri, William Emil
Sittler, Orvid Dayle
Sittler, Weldon Rexer
Six, Erich Walther
Sjodin, Raymond Andrew
Skarsgard, Lloyd Donald
Skolnick, Malcolm Harris
Smith, Adolph T
Smith, Douglas Wemp
Smith, Emil L
Smith, Ian Cormack Palmer
Smith, Paul Edgar, Jr
Smith, Thomas Caldwell
Smith, William M
Snell, Fred Manget
Snipes, Wallace Clayton
Socolar, Sidney Joseph
Sokolla, Adnan
Solie, Thomas Norman
Solomon, Arthur Kaskel
Sondhaus, Charles Anderson
Song, Seh-Hoon
Sophianopoulos, Alkis John
Spangler, George Wesley

Spangler, Robert Alan
Spangler, Stanley Gordon
Spanswick, Roger Morgan
Spector, Novera Herbert
Sperelakis, Nick
Spero, Lawrence
Sperti, George Speri
Spikes, John Daniel
Stapp, John Paul
Steere, Russell Ladd
Stekiel, William John
Stephens-Newsham, Lloyd G
Stephenson, John Leslie
Stevens, Lewis Axtell
Stevenson, Dennis A
Stevenson, Heber John Richards
Stewart, Carleton C
Stewart, Peter Arthur
Stickney, John Clifford
Stillman, Irving Mayer
Stinson, Robert Henry
Stirling, Charles E
Stoeber, Werner
Stolwijk, Jan Adrianus Jozef
Strickholm, Alfred
Strickland, Erasmus Hardin
Strother, Greenville Kash
Stryer, Lubert
Studier, Frederick William
Sullivan, Walter James
Sutherland, John Clark
Swan, Algernon Gordon
Swartz, Harold M
Sweeney, William Victor
Swenberg, Charles Edward
Swez, John Adam
Sybesma, Christiaan
Szabo, Gabor
Szekely, Joseph George
Tanaka, Toyoichi
Taylor, Aubrey Elmo
Taylor, Charles Patrick Stirling
Taylor, Edwin William
Taylor, Robert E
Taylor, Stuart Robert
Taylor, William Daniel
Teaney, Dale T
Teng, Nelson N H
Terry, Thomas Milton
Tessman, Irwin
Thomas, Richard Sanborn
Thomas, Robert Glenn
Thorhaug, Anitra L
Thurber, Robert Eugene
Thurston, George Butte
Tien, Hsin Ti
Tiffany, Mary Lois
Till, James Edgar
Tobias, Cornelius Anthony
Todd, Paul Wilson
Tolles, Walter Edwin
Tong, Bok Yin
Tooney, Nancy Marion
Topham, William Sanford
Torchia, Dennis Anthony
Toribara, Taft Yutaka
Tosteson, Daniel Charles
Towe, Arnold Lester
Trautman, Rodes
Treu, Jesse Isaiah
Trkula, David
Trosper, Terry Louise
Trubatch, Sheldon L
Tsibris, John Constantine Michael
Tucker, Don
Uretz, Robert Benjamin
Vaidhyanathan, V S
Vail, William Jerald
Valeriote, Frederick Augustus
Vallee, Bert L
Van Atta, John R
Van Camp, Harlan Larue
Van Den Bosch, Frank Joseph Gerard
Viale, Richard O
Vizard, Douglas Lincoln
Vogelhut, Paul Otto
Voumakis, John Nicholas
Wagoner, Earl V, Jr
Waldren, Charles Allen
Walker, Etta Frances
Walker, Sheppard Matthew
Wall, Joseph S
Wallace, Susan Scholes
Wallace, William Edward, Jr
Wangemann, Robert Theodore
Ward, Keith Bolen, Jr
Watson, John H L
Wear, James Otto
Webb, Sydney James
Weber, Gregorio
Weeks, Charles Merritt
Weeks, Ivan Forest
Welch, Graeme P
Welling, Daniel J
West, Seymour S
Wheeler, Kenneth Theodore, Jr
White, Stephen Halley
Whittembury, Guillermo

Wickman, Herbert Hollis
Wieder, Irwin
Wiggins, James Wendell
Wilbur, David Wesley
Wiley, Don Craig
Williams, Lawrence Ernest
Williams, Robley Cook
Williams, Theodore P
Williamson, John Richard
Wilson, Jack Martin
Wilson, Michael Erich
Windhager, Erich E
Winet, Howard
Winfree, Arthur T
Wingate, Catharine L
Wise, William Curtis
Wobschall, Darold C
Wohlhieter, John Andrew
Wolbarsht, Myron Lee
Wolff, John Bruno
Wolken, Jerome Jay
Wolterink, Lester Floyd
Wong, Alan Yau Kuen
Wong, Kin-Ping
Wood, Thomas Hamil
Wood, William Otto
Woodbury, John Walter
Woodhull, Ann McNeal
Worthington, Charles Roy
Wraight, Colin Allen
Wrede, Don Edward
Wright, Allen Kent
Wright, Ann Elizabeth
Wu, Cheng-Wen
Wu, Felicia Ying-Hsiueh
Wu, Tai Te
Wunder, Charles Cooper
Wyatt, Philip Joseph
Wyckoff, Ralph Walter Graystone
Wyckoff, Robert Cushman
Wyssbrod, Herman Robert
Yamamoto, Nobuto
Yamamoto, Tomoko
Yanof, Howard Merar
Yeandle, Stephen Safford
Yeargers, Edward Klingensmith
Yen, Peter Kai Jen
Yguerabide, Juan
Yonetani, Takashi
Young, Wei
Yphantis, David Andrew
Yu, Leepo Cheng
Zablow, Leonard
Zadunaisky, Jose Atilio
Zankel, Kenneth L
Zeitz, Louis
Zelman, Allen
Zirkle, Raymond Elliott
Zucker, Robert Martin

Biodynamics
Fletcher, Edward Royce
Shelesnyak, Moses Chiam

Biophysical Chemistry
Adams, Emory Temple, Jr
Adiarte, Arthur Lardizabal
Adler, Alan David
Adler, Alice Joan
Albers, Robert Jay
Allerton, Samuel E
Amma, Elmer Louis
Applequist, Jon Barr
Armstrong, Clay M
Bahary, William S
Barton, Janice Sweeny
Baumgartner, Werner Andreas
Berkowitz, Steven Arlen
Berliner, Lawrence J
Biltonen, Rodney Lincoln
Blei, Ira
Bloomfield, Victor Alfred
Bobst, Albert M
Bock, Robert Manley
Bolton, James R
Bothner-By, Aksel Arnold
Breslow, Esther, M G
Brundage, Robert Scott
Bryant, Loren Conrad
Bush, C Allen
Butterfield, David Allan
Byrn, Stephen Robert
Campbell, Mary Kathryn
Cann, John Rusweiler
Cantor, Charles Robert
Carter, John Vernon
Cerny, Laurence Charles
Chan, Sunney Ignatius
Colen, Alan Hugh
Copeland, Edmund Sargent
Crothers, Donald M
Czeisler, Jeffrey Lance
Dahlquist, Frederick Willis
Deonier, Richard Charles
Diehn, Bodo
Donnelly, Thomas Henry
Douthart, Richard James

Eakin, Richard Timothy
Eanes, Edward David
Ehrlich, Julian
Ehrlich, Robert Stark
Eisenberg, David
Epand, Richard Mayer
Everett, Wilbur Wayne
Fabry, Mary E Riepe
Fabry, Thomas Lester
Fink, Thomas Robert
Fossel, Eric Thor
Franzen, James
Friedman, Michael E
Fuhr, Irvin
Gaffney, Betty Jean
Gemant, Andrew
Gennis, Robert Bennett
Gent, Martin Paul Neville
George, Philip
Goldsack, Douglas E
Goss, Dixie J
Gray, Donald
Griffith, O Hayes
Grigsby, Ronald Davis
Guerrero, Ariel Heriberto
Haddad, Louis Charles
Halvorson, Herbert Russell
Harpst, Jerry Adams
Harrison, Stephen Coplan
Hartzell, Charles Ross, III
Hearst, John Eugene
Hendrickson, Constance McRight
Henkens, Robert William
Hiatt, Caspar Wistar, III
Hooker, Thomas M, Jr
Horton, Aaron Wesley
Ifft, James Brown
James, Thomas Larry
Jensen, Lyle Howard
Jensen, Ronald Harry
Johnson, James Arthur
Jost, Patricia Cowan
Kahn, Leo David
Kalkwarf, Donald Riley
Kallenbach, Neville R
Karasz, Frank Erwin
Kegeles, Gerson
Keltz, Alan
Kluetz, Michael David
Knopp, James A
Koltun, Walter Lang
Kooi, Earl Robert
Kotowycz, George
Kowalsky, Arthur
Kozak, John Joseph
Kreishman, George Paul
Kropf, Allen
Krugh, Thomas Richard
Kuhlmann, Karl Frederick
Kurland, Robert John
Leslie, James
Liebe, Donald Charles
London, Robert Elliot
Macfarlane, Ronald Duncan
Madison, Vincent Stewart
Maki, August Harold
Manning, Gerald Stuart
Martin, Robert Bruce
Matthews, Charles Robert
Matthews, David Allan
McConnell, Harden Marsden
Minton, Allen Paul
Noble, Robert Warren, Jr
Nozaki, Yasuhiko
O'Brien, David F
Oldfield, Eric
Olson, Wilma King
Pease, Lila Gierasch
Pezolet, Michel
Plocke, Donald J
Polnaszek, Carl Francis
Rahn, Ronald Otto
Record, M Thomas, Jr
Roche, Rodney Sylvester
Rosenberg, Jerome Laib
Ross, Robert Talman
Rowlands, John Rhys
Rubin, John Ronald
Sass, Ronald L
Sauer, Kenneth
Scheraga, Harold Abraham
Schleyer, Walter Leo
Schmeltz, Irwin
Schmidt, Paul William
Schmidt, Paul Gardner
Schoor, Kenneth Stanley
Schoor, W Peter
Schuster, Todd Mervyn
Schwartz, Albert Truman
Seewald, David Allan
Seidah, Nabil George
Sevilla, Michael Douglas
Shaw, Barbara Ramsay
Sheetz, Michael Patrick
Simons, Elizabeth Reiman
Spivey, Howard Olin
Squire, Phil George

Starzak, Michael Edward
Stevens, Charles Le Roy
Strauss, George
Sturtevant, Julian Munson
Thomas, Charles Allen, Jr
Tollin, Gordon
Troxell, Terry Charles
Ts'o, Paul On Pong
Tsong, Tian Yow
Tu, Shu-i
Turner, Douglas Hugh
Uhlenbeck, Olke Cornelis
Umans, Robert Scott
Underkofler, Leland Alfred
Verpoorte, Jacob A
Vickery, Larry Edward
von Hippel, Peter Hans
Waggoner, Alan Stuart
Walton, Alan George
Wang, James C
Warden, Joseph Tallman
Wartell, Roger Martin
Watson, Barry
Wee, Elizabeth Liu
Weinstein, Harel
Wetmur, James Gerard
Wickstrom, Eric
Wiechelman, Karen Janice
Winkler, Marvin Howard
Worley, John David
Yang, Jen Tsi
Yapel, Anthony Francis, Jr
York, Sheldon Stafford
Yu, Nai-Teng
Zand, Robert
Zimm, Bruno Hasbrouck
Zipp, Adam Peter

Mathematical Biophysics
Harth, Erich Martin
Macy, Josiah, Jr
Matthysse, Steven William
Noble, Julian Victor
Oldfield, Daniel G
Stuehr, John Edward

Medical Biophysics
Brady, Al H
Breckenridge, John Robert
Cohen, Beverly Shapiro
Dicello, John Francis, Jr
Driscoll, Dorothy H
Greenfield, Harvey Stanley
Greenstock, Clive Lewis
Haak, Richard Arlen
Miller, James Gegan
Myers, William Graydon
Petkau, Abram
Rainbow, Andrew James
Stoll, Alice Mary
Swisher, Robert Donald
Witcofski, Richard Lou
Wyrobek, Andrew Julius
Yalow, Abraham Aaron

Molecular Biophysics
Argos, Patrick
Cohn, Gerald Edward
Deering, Reginald Atwell
Donnellan, James Edward, Jr
Duax, William Leo
Finegold, Leonard X
Fletterick, Robert John
Franklin, Richard Morris
Fung, Leslie Wo-Mei
Graham, William Rendall
Hendrickson, Wayne Arthur
Lapidus, Ivan Richard
Lattman, Eaton Edward
Maggiora, Gerald M
Norvell, John Charles
Ockman, Nathan
Oncley, John Lawrence
Quigley, Gary Joseph
Rich, Alexander
Schurr, John Michael
Seeman, Nadrian Charles
Segrest, Jere Palmer
Song, Pill-Soon
Stockton, Gerald William
Sukow, Wayne William
Szu, Shousun Chen
Thomas, David Dale
Thompson, John Darrell
Urry, Dan Wesley
Venable, John Heinz, Jr
Williams, Myra Nicol
Wyckoff, Harold Winfield
Zobel, Carl Richard

Radiation Biophysics
Achey, Phillip M
Adelstein, Stanely James
Andersen, Frank Alan
Birge, Ann Chamberlain
Bockrath, Richard Charles, Jr
Bond, Victor Potter
Braby, Leslie Alan

BIOPHYSICS

Brown, Darrell Quentin
Chapman, John Donald
Corelli, John Charles
Curtis, Stanley Bartlett
Dostal, Herbert C
Eichling, John O
Fluke, Donald John
Friesen, Benjamin S
Hoecker, Frank Edward
Johnson, James Edward
Johnson, Ronald Gene
Myers, Lawrence Stanley, Jr
Norris, Gail Royal
Okunewick, James Philip
Paterson, Malcolm Cyril
Raju, Mudundi Ramakrishna
Richards, William Robert
Rossi, Harald Herman
Roti Roti, Joseph Lee
Rowland, Robert Edmund
Schenker, Robert Alison
Wood, Robert Winfield
Yang, Chui-Hsu (Tracy)
Zimbrick, John David

BOTANY

Botany

Abbe, Ernst Cleveland
Adams, Franklin Scott
Adams, Joseph Edison
Adams, Preston
Adams, Randall Henry
Ahlgren, Isabel Fulton
Ahmadjian, Vernon
Alcorn, Gordon Dee
Aldrich, Henry Carl
Alexander, Taylor Richard
Alldridge, Norman Alfred
Ambrose, John Daniel
Anderson, Edward Frederick
Anderson, Lewis Edward
Anderson, Loran C
Anderson, Orlin
Anderson, Robert Gordon
Anderson, Roger Arthur
Anderson, Roger Clark
Anthony, Margery Stuart
Aniogninni, Joe
App, Alva Agee
Arnholt, Philip John
Arnott, Howard Joseph
Artist, Russell (Charles)
Ashworth, Ralph P

Baalman, Robert J
Babcock, Philip Arnold
Badenhuizen, Nicolaas Pieter
Bailey, Virginia Long
Bailey, Zeno Earl
Baker, Gladys Elizabeth
Baker, Robert Lewis
Baker, William Hudson
Ball, Ernest
Ballal, Srikrishna
Banerjee, Sushanta Kumar
Banks, Donald Jack
Bannan, Marvin William
Barbour, Michael G
Barclay, Harriett G
Barke, Harvey Ellis
Barnes, Burton Verne
Barrington, David Stanley
Basham, Jack Tucker
Batson, Wade Thomas
Bati, Mario Alex
Baum, Bernard R
Bausor, Sydney Charles
Beamish, Katherine I
Becker, Donald A
Belcher, Robert Orange
Bell, Clyde Ritchie
Bell, Sandra Lucille
Bellmer, Elizabeth Henry
Beneke, Everett Smith
Bennett, Ralph Edgar
Berkeley, Edmund
Biddulph, Orlin
Bigelow, Howard Elson
Billings, William Dwight
Bird, Charles Durham
Birdsey, Monroe Roberts
Bisalputra, Thana
Bischoff, Harry William
Black, Robert Cori
Blackburn, Benjamin (Coleman)
Blaser, Henry Weston
Bliss, Lawrence Carroll
Blum, Udo
Bock, Jane Haskett
Bohm, Bruce Arthur
Bohmont, Dale W
Bohning, Richard Howard
Bold, Harold Charles
Bonar, Lee
Bond, Lora
Bonner, Hazel Garrison

Bonner, Robert Dubois
Bostrack, Jack M.
Bouck, G Benjamin
Bowen, Charles Clark
Bowen, Paul Ross
Bowen, William R
Bowers, Maynard C
Bowers, Robert Charles
Boyle, William Sidney
Bozarth, Gene Allen
Bradley, Muriel Virginia
Bragg, Louis Hairston
Bragonier, Wendell Hughell
Brandhorst, Carl Theodore
Brandwein, Paul Franz
Branton, Daniel
Braun, Armin Charles
Brehm, Bertram George, Jr
Bridgers, Bernard Thomas
Brink, Royal Alexander
Bristol, Melvin Lee
Brooks, Merle Eugene
Brown, Richard Malcolm, Jr
Brown, Russell Guy
Brown, Walter Varian
Browne, Edward Tankard, Jr
Bryan, Virginia Schmitt
Bryan, Vaughn Motley
Bryant, Charles Leslie
Burbanck, Madeline Palmer
Burger, Warren Clark
Burgess, Jack D
Burton, Daniel Frederick
Burton, Verona Devine
Busby, Joe Neil
Butler, John Earl
Cain, Roy Franklin
Cain, Stanley Adair
Campbell, Robert Samuel
Caplenor, Donald
Caplin, Samuel Milton
Capon, Brian
Caponetti, James Dante
Cardillo, Frances M
Carlquist, Sherwin
Carlucci, Leeds Mario
Carothers, Zane Bland
Carter, Edward Pendleton
Carter, Jack Lee
Castenholz, Richard William
Cavaliere, Alphonse Ralph
Cellarius, Richard Andrew
Chambers, Kenton Lee
Chan, Cheung-King
Chan, Allan P
Chase, Sherret Spaulding
Cheadle, Vernon Irvin
Chen, James Che Wen
Chen, Tseh-An
Chipman, E W
Christian, James A
Christiansen, Paul Arthur
Church, George Lyle
Clark, Jasper Arnold
Clarkson, Roy Burdette
Cliburn, Joseph William
Clutter, Mary Elizabeth
Cochis, Thomas
Cody, William James
Cohen, Isadore
Coleman, Babette Brown
Collette, Alfred Thomas
Collins, Barbara Jane
Collins, Ralph Porter
Conklin, Marie Eckhardt
Conover, Robert Armine
Constantin, Milton J
Constance, Lincoln
Cook, Frankland Shaw
Cook, Philip W
Cooke, Ron Charles
Core, Earl Lemley
Cormack, Robert George Hall
Couch, William George
Couch, John Nathaniel
Cox, Donald David
Coyle, Elizabeth Eleanor
Craft, James Harvey
Crafts, Alden Springer
Creighton, Harriet Baldwin
Croasdale, Hannah Thompson
Cronquist, James
Cross, Chester Ellsworth
Crosswhite, Frank Samuel
Croxdale, Judith Gerow
Culberson, William Louis
Cumbie, Billy Glenn
Cutler, Hugh Carson
Dahl, Anthony Orville
Daily, Fay Kenoyer (Mrs William A Daily)
Dalton, Patrick Daly
Damann, Kenneth Eugene
Daniels, Gilbert S
Davidson, Darwin Ervin
Davidson, Robert A
Davis, Edward Lyon
Deal, Don Robert

Dean, Donald Stewart
Dean, Henry Lee
Decker, Robert Dean
Degener, Otto
de Laubenfels, David John
De Lisle, Albert Lorenzo
De Lisle, Donald Gordon
Delouche, James Curtis
DeMaggio, Augustus Edward
DeMott, Howard Ephraim
Denison, William Clark
Dennis, Tom Eugene
Dent, Thomas Curtis
Denton, Charles Edward
Dermen, Haig
Dever, Donald Andrew
Dieter, Stanley Gregg
Dieter, Reuben Arthur
Diehl, Norman Hudson
Dittmer, Howard James
Dobbins, David Ross
Dodd, John Durrance
Dodge, Carroll William
Dore, William George
Downs, Robert Jack
Drechsler, Charles
Drouet, Francis
Drugg, Warren Sowle
Dube, Maurice Andrew
Duke, James A
Dunlop, Douglas Wayne
Dunn, David Baxter
Dunn, Richard Hudson
Dunn, Stuart
Dyer, Hubert Jerome
Earle, Thomas Theron
Eastin, Emory Ford
Ebert, Wesley W
Ebinger, John Edwin
Edwards, Gerald Elmo
Eggler, Willis Alexander
Ehrlich, Mary Ann
Eigsti, Orie Jacob
Einhellig, Frank Arnold
Elliott, Eugene Willis
Ellison, Marion L
Embree, Robert William
Enochs, Nettie Jean
Erbe, Lawrence Wayne
Erbisch, Frederic H
Erdman, Kimball S
Erickson, Ralph O
Esau, Katherine
Esser, Robert Emmet
Estes, Edna E
Etherton, Bud
Evans, Michael Leigh
Evers, Robert August
Evert, Ray Franklin
Ewan, Joseph (Andorfer)
Ewart, Ralph Bradley
Fahselt, Dianne
Fairbrothers, David Earl
Falk, Richard H
Fearing, Olin S
Federowicz, Rose Ann
Ferchau, Hugo Alfred
Ferguson, James Roger
Feucht, James Roger
Finley, Noel Moore
Fisher, Donald B
Fisher, Jack Bernard
Fisher, T Richard
Flagg, Raymond Osbourn
Flamman, M Muriel
Flory, Walter S, Jr
Foard, Donald Edward
Fogg, John Milton, Jr
Follstad, Merle Norman
Forbes, Irvin L
Ford, Ernest Sidney
Ford, Roy Charles
Fork, Irvin J Andre
Fortin, J Andre
Franke, Robert G
Freeberg, John Arthur
Fulford, Margaret Hannah
Fuller, Melvin Stuart
Furumoto, Warren Akira
Fussell, Catharine Pugh
Gaither, Thomas Walter
Gasiorkiewicz, Eugene Constantine
Geeseman, Gordon E
Gehris, Clarence Winfred
Gelinas, Douglas Alfred
Gensel, Patricia Gabbey
Gerloff, Gerald Carl
Germann, Paul Julian
Gerrath, Joseph Fredrick
Gibbs, R Darnley
Gifford, Ernest Milton, Jr
Gilbert, Gareth E
Gilbert, Margaret Lois
Gilbert, George Willson
Gilbert, Norris W
Girouard, Ronald Maurice
Glime, Janice Mildred
Goddard, David Rockwell
Godfrey, Robert Kenneth

Goodall, David William
Goodwin, Richard Hale
Gordon, Philip Newton
Goss, James Arthur
Gould, Frank Walton
Graffius, James Herbert
Graham, Benjamin Franklin, Jr
Graham, James Carl
Graham, Shirley Ann
Grant, Michael Clarence
Grant, Verne (Edwin)
Grear, John Wesley, Jr
Green, John Irving
Green, Paul Barnett
Greenblatt, Irwin M
Greenfield, Sydney Stanley
Greenidge, Kenneth Norman Haynes
Grewe, Alfred H, Jr
Griesel, Wesley Otto
Grigsby, Buford Horace
Grillos, Steve John
Grittinger, Thomas Benjamin
Grodzinski, Bernard
Grossmann, Herbert H
Grove, Davison Greenawalt
Grove, Stanley Neal
Gunckel, James Eugene
Gunn, Charles Robert
Gupton, Oscar Wilmot
Guram, Malkiat Singh
Guttay, Andrew John Robert
Haase, Edward Francis
Hadley, Elmer Burton
Haeskaylo, Edward
Hagen, Charles William, Jr
Hagler, Thomas Benjamin
Haight, Thomas H
Hall, Gustav Wesley
Hall, Ivan Victor
Halperin, Walter
Hammond, H David
Hanks, Jess Paul
Hanzely, Laszlo
Hardberger, Florian Max
Harden, James Walker
Harding, Wallace Charles, Jr
Harper, Kimball T
Harrelson, Michael Asbury
Harrington, Dalton
Harris, Clare I
Hartman, Emily Lou
Harvey, LeRoy Hatfield
Harvey, Michael John
Haskins, Edward Frederick
Hatheway, William Howell
Haupt, Arthur Wing
Hayes, Alice Bourke
Hecht, Adolph
Heckard, Lawrence Ray
Heilman, Alan Smith
Heiner, Terry Charles
Heinig, Katherine H
Heiser, Charles Bixler, Jr
Hennen, James Douglas
Hennen, Joe Fleetwood
Henry, LeRoy Kershaw
Herman, Frederick Joseph
Herndon, Walter Roger
Herron, James Watt
Hershey, Arthur L
Hervey, Annette
Hess, Dexter Winfield
Hess, Wilford Moser Bill
Heusser, Calvin John
Hexem, Rodney Orlyn
Higgins, Daniel Joseph
Higgins, Paul Daniel
Highbotham, Noe
Hill, Eddie P
Hill, Lynn Michael
Hillier, Richard David
Hitchcock, Charles Leo
Hobbs, Clinton Howard
Hodgdon, Albion Reed
Hodgson, Richard Holmes
Hoefert, Lynn Lucretia
Hoffman, Joyce Bennett
Hoffmaster, Donald Edeburn
Hollenbeck, Irene
Holroyd, Roland
Holt, Imy Vincent
Hommersand, Max Hoyt
Hopkins, Harold Hoffman
Horner, Harry Theodore, Jr
Horton, James Heathman
Houk, Richard Duncan
Howard, Harold Henry
Howard, Richard Alden
Hu, Shiu-Ying Hsu
Huck, Morris Glen
Huckaby, John Porter
Hughes, Gilbert C
Hulbary, Robert Louis
Hunt, Gordon Ellsworth
Hunt, Kenneth Whitten
Hunter, Gordon Eugene

Hurst, Fannie Mae
Husband, David Dwight
Hutchinson, Thomas C
Hyde, Beal Baker
Irwin, Howard Samuel, Jr
Ishimoto, Tom
Jackson, Ernest Baker
Jensen, William August
Jervis, Robert Alfred
Johnson, George Thomas
Johnson, Hyrum Bennett
Johnson, John Morris
Johnson, Karen Louise
Jones, Richard Conrad
Jones, Roger
Jones, Samuel B, Jr
Jowett, David
Kaplan, Lawrence
Kapp, Ronald Ormond
Kasapligil, Baki
Kaufmann, Fred Henry
Kaul, Robert Bruce
Keefe, Thomas Leeven
Keener, Carl Samuel
Kelley, William Russell
Kennedy, Lorene Louise
Kern, Abraham K
Kingsbury, John Merriam
Kissmeyer-Nielsen, Erik
Klekowski, Edward Joseph, Jr
Klikoff, Lionel G
Knobloch, Irving William
Kochert, Gary Dean
Koeppen, Robert Carl
Koevenig, James L
Kroschewsky, Julius Richard
Kruckeberg, Arthur Rice
Kuhnen, Sybil Marie
Kullberg, Russell Gordon
LaCroix, Joseph Donald
Laetsch, Watson McMillan
LaFuze, Henry Harvey
LaHaye, Philip Arthur
Lamoureux, Charles Harrington
Lampton, Robert Koerbel
Langenauer, Haviva Dolgin
Langford, Arthur Nicol
Larsen, Arnold Lewis
Larsen, Victor Robinson, Jr
Larson, Donald Alfred
Larson, Gary Eugene
Lasseter, John Stuart
Laude, Horton Meyer
Laufersweiler, Joseph Daniel
Laurence, Maria (Maher)
Ledbetter, Myron C
Lee, Addison Earl
Leeper, George Frederick
Leisman, Gilbert Arthur
Leisner, Robert Stanley
Lemay, Yvan
LeNoir, William Cannon, Jr
Lenz, Lee Wayne
Leopold, Estella (Bergere)
Lersten, Nels R
Levering, Dale Franklin, Jr
Levetin-Avery, Estelle
Lewis, Clifford Eugene
Lewis, Frank Harlan
Lewis, Sally
Lewis, Walter Hepworth
Li, Hui-Lin
Lier, Frank George
Lieth, Helmut H F
Lieux, Meredith Hoag
Lin, Shin
Little, Elbert Luther, Jr
Little, Ruby Rice
Livingston, Luzern Gould
Livingston, Robert Blair
Lockard, J David
Locke, John Flowers
Loeffler, Robert J
Long, Terrill Jewett
Looman, Jan
Loucks, Oric Lipton
Love, Harry Schroeder, Jr
Lovett, James Satterthwaite
Lovvorn, Roy Lee
Lowry, Robert James
Loy, James Brent
Lucansky, Terry Wayne
Lyle, James Albert
Macdonald, Alastair David
Macior, Lazarus Walter
MacQuarrie, Ian Gregor
Madore, Bernadette
Maier, Charles Robert
Main, Stephen Paul
Majumder, Sanat Kumer
Maksymowich, Roman
Marengo, Norman Payson
Marks, Gayton Carl
Marr, John Winton
Marsh, Leland C
Marshall, Norton Little
Martin, Frank Winstead
Marvin, James Wallace

Mason, Charles Perry
Mason, Charles Thomas, Jr
Mason, David Lamont
Massey, Jimmy R
Mathias, Mildred Esther (Mrs Gerald L Hassler)
Mathieson, Arthur C
Maze, Jack Reiser
McAlpin, Cesaria Eugenio
McAndrews, John Henry
McChesney, James Dewey
McClaren, Milton, Jr
McCleary, James A
McClymont, John Wilbur
McCracken, Michael Dwayne
McCullough, James Matthew
McDougall, Walter Byron
McFarland, James Willis
McFeeley, James Calvin
McGregor, Ronald Leighton
McHale, John T
McIntosh, Douglas Carl
McKenna, George Finley
McKinsey, Richard Davis
McLarty, Duncan Archibald
McMullen, John Lloyd
McNeal, Dale William, Jr
McNeill, John
McVaugh, Rogers
McVeigh, Ilda
Meiselman, Newton
Menhusen, Bernadette Remus
Meredith, Farris Ray
Merry, William James
Metzner, Jerome
Meyer, Bernard Sandler
Meyer, Frederick Gustav
Miles, Philip Giltner
Miller, Gary L
Miller, Helena Agnes
Miller, Norton George
Miller, Pauline Monz
Miller, Robert Harold
Miller, Willie
Millikan, Daniel Franklin, Jr
Mills, Dallice Ivan
Mingrone, Louis V
Mish, Lawrence Bronislaw
Mitchell, William H
Mitchell, William Warren
Moir, David Ross
Montgomery, James Douglas
Moore, Harold Emery, Jr
Moore, James Marvin
Moore, Jewel Elizabeth
Mooring, John Stuart
Morgan, Delbert Thomas, Jr
Morris, Everett Franklin
Morrow, Leonard Owen
Morton, John Kenneth
Mosquin, Theodore
Moulton, James Edward
Mowbray, Thomas Bruce
Mozingo, Hugh Nelson
Mueller, Sabina Gertrude
Mullins, John Thomas
Munroe, Marian Hall
Murdy, William Henry
Murphy, Peter George
Murray, Mary Aileen
Myers, Roy Maurice
Nall, Raymond Willett
Naskali, Richard John
Nass, Hans George
Naylor, Ernst E
Naylor, James Maurice
Nedrow, Warren Wesley
Neel, James William
Neely, Florence Elizabeth
Nelson, Peter K
Neushul, Michael, Jr
Nicely, Kenneth Aubrey
Nichols, Charles
Nicholson, Nancy Lynne
Nicolson, Dan Henry
Niffenegger, Daniel Arvid
Niles, Doris Kildale
Niles, Wesley E
Nolan, James Robert
Norman, Eliane Meyer
Norris, Daniel Howard
Northen, Henry Theodore
Norton, Norman J
Obee, Donald Jennings
Oberlander, George T
O'Connell, Jesse Gordon
Ogden, Elston Gordon
Ogden, Eugene Cecil
Olive, Lindsay Shepherd
Olsen, Orvil Alva
Olson, Richard Louis
O'Neill, Thomas Brendan
Ornduff, Robert
O'Rourke, Richard Clair
Orr, Alan R
Ott, Aleta Jo Petrik
Overlease, William R

Owens, John N
Pady, Stuart McGregor
Page, Orville T
Palmatier, Elmer Arthur
Palser, Barbara Frances
Paolillo, Dominick Joseph
Parker, Kittie Fenley
Parkinson, Dennis
Parnell, Dennis Richard
Paterson, Robert Andrew
Patrick, Ruth (Mrs Charles Hodge IV)
Peterson, Curt Morris
Peterson, Robert Lawrence
Peterson, Roger Shipp
Philbrick, Ralph Nowell
Philpott, Jane
Phinney, Bernard Orrin
Phinney, Harry Kenyon
Pilcher, Benjamin Lee
Pinkava, Donald John
Plymale, Edward Lewis
Pohl, Richard Walter
Pokorny, Frank Joseph
Poole, James Plummer
Popham, Richard Allen
Porsild, Alf Erling
Porter, David
Porter, Thomas Reginald
Post, Douglas Manners
Postlethwait, Samuel Noel
Potter, Loren David
Powell, Albert Michael
Powell, Robert W, Jr
Prance, Ghillean T
Pray, Thomas Richard
Prescott, Henry Emil, Jr
Preston, Dudley A
Pringle, James Scott
Pritchard, Hayden N
Pullen, Thomas Marion
Pursell, Ronald A
Putala, Eugene Charles
Quartermann, Elsie
Quick, William Andrew
Quinn, James Amos
Rabideau, Glenn Sylvester
Radford, Albert Ernest
Radosevich, Steven Robert
Raj, Baldev
Ralston, Furman Paul, Jr
Ramseur, George Shuford
Ramus, Joseph
Rappaport, Jacques
Raspet, Mabel Wilson
Raup, Hugh Miller
Raven, Peter Hamilton
Ray, James Davis, Jr
Rayburn, William Reed
Ream, Robert Ray
Reed, Clyde F
Reed, John Frederick
Reeder, John Raymond
Reid, Hay Bruce, Jr
Reid, Milton Roy
Reines, Mervin
Reynard, George Bergin
Rice, Elroy Leon
Rice, Marion McBurney
Richardson, David H S
Richardson, F C
Rickard, William Howard, Jr
Rickett, Harold William
Ritchie, James Cunningham
Riznyk, Raymond Zenon
Roach, Archibald Wilson Kilbourne
Roberts, Edith Adelaide
Roberts, Eliot Collins
Roberts, Evan Paul
Robinson, Harold Ernest
Rodgers, Charles Leland
Rogers, Claude Marvin
Rogers, Donald Philip
Rogers, Hollis Jetton
Rollins, Reed Clark
Romans, Robert Charles
Rose, Frederick Louis, Jr
Roshal, Jay Yehudie
Rossbach, George Bowyer
Roth, Elmer Alfred
Rouffa, Albert Stanley
Rowland, Neil Wilson
Rudd, Velva Elaine
Rudolph, Emanuel David
Runk, Benjamin Franklin Dewees
St John, Harold
Sakai, William Shigeru
Salamun, Peter Joseph
Sarkissian, Igor V
Savile, Douglas Barton Osborne
Sayre, Geneva
Scheirer, Daniel Charles
Scherr, Robert Walter
Schmid, Walter Egid
Schmidt, Clifford LeRoy
Schofield, Wilfred Borden
Schramm, Jacob Richard
Schreiber, Richard William

Schrock, Gould Frederick
Schroeder, Charles Arthur
Schultes, Richard Evans
Schuster, Rudolf Mathias
Schuyler, Alfred E
Scora, Rainer W
Scott, Joseph Lee
Scott, Marvin Wade
Seibert, Russell Jacob
Seigler, David Stanley
Selsky, Melvyn Ira
Senn, Harold Archie
Sentz, James Curtis
Settle, Wilbur Jewell
Severin, Charles Hilarion
Shacklette, Hansford Threlkeld
Shapiro, Seymour
Sharitz, Rebecca Reyburn
Sharma, Govind C
Sharman, Bernard Clout
Sharp, Aaron John
Sharsmith, Carl William
Shaw, Richard Joshua
Sherman, Harry Logan
Sherman, Robert James
Shugars, Jonas P
Sifton, Harold Boyd
Silva, Paul Claude
Simpson, Donald Ray
Simpson, Jennie Laura Symons
Simpson, Marion Emma
Sinclair, Clarence Bruce
Sjolund, Richard David
Skinner, Henry Thomas
Skogstad, Kenneth
Slysh, Anton Roman
Smart, Robert Forte
Smith, Alan Reid
Smith, Albert Charles
Smith, Calvin Albert
Smith, David Kent
Smith, Douglas Roane
Smith, Eugene William
Smith, Frank Herschel
Smith, Gary Lane
Smith, Roberta Hawkins
Sohns, Ernest Reeves
Soper, James Herbert
Sorensen, Paul Davidsen
Spence, Willard Lewis
Spencer, Jack T
Sperry, Theodore Melrose
Spetzman, Lloyd Anthony
Sprague, Elizabeth F
Stalter, Richard
Stanford, Lyle Morris
Star, Aura E
Staten, Raymond Dale
Staub, Robert J
Stearns, Forest
Stebbins, George Ledyard
Steere, William Campbell
Steeves, Taylor Armstrong
Stephenson, Stephen Neil
Sterling, Clarence
Steward, Frederick Campion
Stewart, Shelton E
Stewart, Wilson Nichols
Stone, Donald Eugene
Stoudt, Harry Nathaniel
Stow, Richard W
Stroube, William Hugh
Stuckey, Ronald Lewis
Sussex, Ian Mitchell
Sutton, Dale Dinkins
Sweet, Herman Royden
Sylvester, Erhardt Paul
Szczawinski, Adam Franciszek
Taft, Clarence Egbert
Tatschl, Annehara Kathleen
Taylor, Andrew Ronald Argo
Taylor, Fred Herbert
Taylor, Marie Clark
Taylor, Raymond Leech
Taylor, Ronald
Taylor, William Randolph
Tews, Leonard L
Thaw, Richard Franklin
Thieret, John William
Thiesfeld, Virgil Arthur
Thomas, Joab Langston
Thomas, Ruth Beatrice
Thomson, Betty Flanders
Thomson, John Walter
Thomson, William Walter
Thornton, Melvin LeRoy
Thurston, Earle Laurence
Tidd, Joseph Shepard
Tiefel, Ralph Maurice
Timmermann, Dan, Jr
Tippo, Oswald
Titman, Paul Wilson
Toczek, Donald Richard
Todsen, Thomas Kamp
Tolstead, William Lawrence
Tomanek, Gerald Wayne
Tomlinson, Philip Barry
Torres, Andrew Marion

Elmstrom, Gary William
Embleton, Tom William
Emerson, Frank Henry
Enzie, Joseph Vincent
Erickson, Homer Theodore
Fazio, Steve
Ferree, David C
Feucht, James Roger
Fieldhouse, Donald John
Filinger, George Albert
Fletcher, William Ellis
Fogle, Harold Warman
Fozdar, Birendra Singh
Franklin, DeLance Flournoy
Franklin, Everett Whitney
Freeman, Fred W
Fuchigami, Leslie H
Fuck, John Edward
Furuta, Tokuji
Garren, Ralph, Jr
Garrison, Olen Branford
Gaskins, Murray Hendricks
Gaus, Arthur Edward
Geisman, Jean Richard
Gregg, Cecil Manren
Giamalva, Mike J
Gibby, David Duane
Gilbert, Franklin Andrew, Sr
Gillespie, Colin J
Girouard, Ronald Maurice
Gould, Wilbur Alphonso
Gowen, Frederick Arthur
Gregg, James Kibler, Jr
Guzman, Victor Lionel
Hackett, Wesley P
Haeseler, Carl W
Halfacre, Robert Gordon
Hall, Charles Virdus
Hammett, Harrell Lee
Hanan, Joe John
Hancock, Elizabeth Diekenberger
Hanson, Kenneth Warren
Hard, Cecil Gustav
Hardenburg, Robert Earle
Harding, James A
Harmon, Silas Albert
Harney, Patricia Marie
Harris, Richard Wilson
Hartmann, Hudson Thomas
Hatton, Thurman Timbrook, Jr
Haut, Irvin Charles
Havis, John Ralph
Hawthorne, Percy Lynnwood
Hayden, Richard Amherst
Hegwood, Donald Augustine
Hemphill, Delbert Dean
Hendershot, Charles Henry, Jr
Henderson, Warren Robert
Hensz, Richard Albert
Hepler, Paul Raymond
Hernandez, Teme P
Hertz, Leonard B
Hess, Charles
Hibbard, Aubrey D
Hilgeman, Robert Harry
Hill, Robert George, Jr
Hillyer, Irvin George
Hogan, Le Moyne
Hogue, Eugene J
Hohlt, Herman Edward
Holland, Neal Stewart
Holley, Winfred Davis
Honma, Shigemi
Hoover, Maurice Wilbur
Hopen, Herbert
Horton, Billy D
Howell, Gordon Stanley, Jr
Howell, Monticello Jefferson
Hruschka, Howard Wilbur
Hull, Jerome, Jr
Iritani, W M
Isenberg, Francis Marion Roland
Ito, Philip J
Jaworski, Casimir A
Jenkins, William Frank
Johansen, Robert H
Johnson, William Bradford
Johnstone, Francis Elliott, Jr
Joiner, Jasper Newton
Jones, Henry Albert
Jones, Lloyd George
Jones, Winston William
Kalin, Elwood Walter
Kallio, Arvo
Kamemoto, Haruyuki
Karas, James Glynn
Kattan, Ahmed A
Kawase, Makoto
Kelly, John Francis
Kemp, Gavin Arthur
Kennard, William Crawford
Kenworthy, Alvin Lawrence
Kessler, George Morton
Kiplinger, Donald Carl
Kirkpatrick, John David
Knavel, Dean Edgar
Konsler, Thomas Rhinehart
Kozel, Philip C

Kraus, James Ellsworth
Kretchman, Dale Warren
Kunkel, Robert
Kyle, Jack Hiram
Lagerstedt, Harry Bert
Lana, Edward Peter
Lane, Ronald Paton
Larsen, Fenton E
Larsen, John Elbert
Larsen, Robert Paul
Larson, Roy Axel
Lasheen, Aly M
Lauer, Florian Isidore
Lawrence, Fred Parker
Lazaruk, William
Lewis, Clarence E
Lewis, Sally
Li, Paul H
Lindstrom, Richard S
Link, Conrad Barnett
Lipe, John Arthur
Locascio, Salvador J
Lombard, Porter Bronson
Long, James Delbert
Loughton, Arthur
Mack, Harry John
Madden, George D
Madison, John Herbert, Jr
Maginnes, Edward Alexander
Malo, Simon E
Manis, Wallace Eugene
Marlowe, George Albert, Jr
Marousky, Francis John
Marshall, Ernest (Roy)
Marvel, Mason E
Mastalerz, John W
Matkin, Oris Arthur
McCollum, John Paschal
McCornack, Andrew Adams
McCown, Brent Howard
McDowell, Theodore C
McLane, Stanley Rex, Jr
Mehlquist, Gustav Arthur Leonard
Mellenthin, Walter M
Merritt, Richard Howard
Metcalf, Homer Noble
Milbocker, Daniel Clement
Miles, Neil Wayne
Miller, Knudt John
Miller, Richard Lloyd
Miller, Victor Jay
Mohr, Hubert Charles
Montelaro, James
Moore, Evon Lamar
Moore, Frank Devitt, III
Moore, James Norman
Morris, Justin Roy
Mortensen, John Alan
Moser, Bruno Carl
Mosher, Harold Elwood
Mowry, James B
Murdoch, Charles Loraine
Mustard, Margaret Jean
Nakasone, Henry Yoshiki
Nakayama, Roy Minoru
Narten, Perry Foote
Neal, Oliver Meader, Jr
Neild, Ralph E
Nelson, Donald Carl
Nelson, Paul Victor
Nelson, Stuart Harper
Nesbitt, William Belton
Newsom, Donald Wilson
Nickeson, Richard L
Nicklow, Clark W
Nightingale, Arthur Esten
Nixon, Roy Wesley
Noll, Charles Joseph
Nonnecke, Ib Libner
Norton, Joseph Daniel
Norton, Robert Alan
Nylund, Robert Einar
Oberle, George David
Oeckber, Norman Fred
Ogle, Wayne LeRoy
O'Keefe, Robert Bernard
Olson, Arthur Olaf
O'Rourke, Edmund Newton, Jr
O'Rourke, F L Steve
Overcash, Jean Parks
Paris, Clark Davis
Parsons, Jerry Montgomery
Paterson, Donald Robert
Patterson, Max E
Paul, Kamalendu Bikash
Payne, Richard N
Peacock, Neal Dow
Peck, Nathan Hiram
Pellet, Harold M
Perkins, Donald Young
Perlmutter, Frank
Perumal, Alexander
Peterson, Clinton E
Peterson, Ronald M
Pfahl, Peter Blair
Phatak, Sharad Chintaman
Phillips, Richard Lee
Pickett, William Francis

Piringer, Albert Aloysius, Jr
Pokorny, Franklin Albert
Pollard, James Edward
Pope, Daniel Townsend
Popenoe, John
Porritt, Stanley Wallace
Prashar, Paul D
Pratt, Harlan Kelley
Prend, Joseph
Preston, William Henry, Jr
Price, Hugh Criswell
Price, Philip B
Proctor, John Thomas Arthur
Proebsting, Edward Louis, Jr
Quamme, Harvey Allen
Rappaport, Lawrence
Rauch, Fred D
Raulston, James Chester
Read, Paul Eugene
Reese, Robert Lewis
Reimschussel, Ernest F
Reisch, Kenneth William
Reitz, Herman J
Reuther, Walter
Reynolds, Charles William
Richardson, Ralph William, Jr
Rickels, Jerald Wayne
Ries, Stanley K
Riggleman, James Dale
Robbins, Marion LeRon
Roberts, Alfred Nathan
Roberts, Clarence Richard
Robinson, Richard Warren
Robitaille, Henry Arthur
Rodney, David Ross
Rogers, Benjamin Lanham
Rollins, Howard A, Jr
Rom, Roy Curt
Routey, Douglas George
Rubatzky, Vincent E
Ruf, Robert Henry, Jr
Rutland, Rufus Burr
Ryan, George Frisbie
San Antonio, James Patrick
Schilletter, Julian Claude
Scholz, Earl Walter
Schubert, Oscar Edmund
Senn, Taze Leonard
Shadbolt, C Allan
Shalucha, Barbara
Shanks, James Bates
Sharpe, Ralph Harold
Sharples, George Carroll
Sheldrake, Raymond, Jr
Shen-Miller, Jane
Sherman, Wayne Bush
Sherwood, Charles H
Showalter, Robert Kenneth
Shreve, Loy William
Shutak, Vladimir Gregory
Simons, Roy Kenneth
Sims, Ernest Theodore, Jr
Singletary, Clyde C
Sites, John Wilbur
Skelton, Bobby Joe
Skinner, Henry Thomas
Smith, Carter Riley
Smith, James Elmo, Jr
Smith, Orrin Ernest
Smith, Rufus Albert, Jr
Snyder, Leon Carleton
Soule, James
Southwick, Lawrence
Sparks, Darrell
Sparks, Walter Chappel
Staby, George Lester
Stahly, Edward Arthur
Stark, Francis C, Jr
Steponkus, Peter Leo
Stevenson, Elmer Clark
Stiles, Warren Cryder
Stoltz, Leonard Paul
Stoner, Allan K
Struckmeyer, Burdean Esther
Sturrock, Thomas Tracy
Stushnoff, Cecil
Sullivan, Darrell Thornton
Swingle, Homer Dale
Taper, Charles Daniel
Tayama, Harry K
Taylor, James Clifton
Taylor, Oliver Clifton
Taylorson, Raymond Brierly
Tereshkovich, George
Thomas, Frank Bancroft
Thompson, Buford Dale
Thompson, Maxine Marie
Tibbitts, Theodore William
Tiedjens, Victor Alphons
Tinga, Jacob Hinnes
Tinga, Sik Vung
Tomkins, John Preston
Tompkins, Daniel Reuben
Torfason, Wilmer Esplin
Toyama, Thomas Kazuo
Tucker, David Patrick Hislop
Tukey, Harold Bradford, Jr
Tukey, Loren Davenport

Turnquist, Orrin Clinton
Tweedy, James Arthur
Twigg, Bernard Alvin
Unrath, Claude Richard
Vaile, Joseph Edwin
Vandemark, J S
Velez, Antonio
Vitrum, Morrill Thayer
Volz, Emil Conrad
Wander, Irvin Woodrow
Wardowski, Wilfred Francis, II
Watada, Alley E
Waters, Willie Estel
Watschke, Thomas Lee
Watson, Donald Pickett
Webster, David Henry
Welker, William V, Jr
Wells, Otho Sylvester
Welsh, Maurice Fitzwilliam
Wester, Robert Emerson
Whatley, Booker Tillman
Wheaton, Thomas Adair
Whitcomb, Carl Erwin
White, John W
Widmer, Richard Ernest
Widmoyer, Fred Bixler
Wiebe, John
Wiley, Robert Craig
Wilkins, Harold
Williams, David Douglas F
Williams, George Robertson
Williams, Max W
Wilson, Lorenzo George
Wiltbank, William Joseph
Windham, Steve Lee
Wiseman, Billy Ray
Wittwer, Sylvan Harold
Woltz, Shreve Simpson
Woodbridge, Cyril Gordon
Woodroof, Jasper Guy
Woodruff, Richard Earl
Woodward, Ralph Stanley
Worthington, John Thomas, III
Wott, John Arthur
Wright, Johnie Algie
Wurster, Richard T
Wyatt, Colen Charles
Wyman, Donald
Young, Robert Ellsworth
Young, Thomas Wilbur
Zarger, Thomas Gordon
Ziegler, Louis William
Zych, Chester Charles

Lichenology
Anderegg, Doyle Edward
Bowler, Peter Aldrich
Duewer, Elizabeth Ann
Egan, Robert Shaw
Esslinger, Theodore Lee
Hale, Mason Ellsworth, Jr
Phillips, Haskell C
Shushan, Sam
Sierk, Herbert Allen
Tavares, Isabelle Irene
Wetmore, Clifford Major

Marine Botany
Eleuterius, Lionel Numa
Foster, Michael Simmler
Fralick, Richard Allston
Gallagher, John Leslie
Johansen, Hans William
Lee, Robert K S
Liddle, Larry Brook
Menez, Eranani Guingona
Newroth, Peter Russell
Norris, James Newcome, IV
Norris, Richard Earl
Phillips, Ronald Carl
Prezelin, Barbara Berntsen
Prowse, Gerald Albert
Silberhorn, Gene Michael
Sorensen, Lazern Otto
Tiffney, Wesley Newell
Wynne, Michael

Medical Mycology
Ajello, Libero
Al-Doory, Yousef
Blank, Fritz
Bulmer, Glenn Stuart
Campbell, Charlotte Catherine
Conway, Kenneth Edward
Cozad, George Carmon
De Long, Sharon Koepcke
Erke, Keith Howard
Georg, Lucille Katherine (Mrs W L Pickard)
Gordon, Clarence Conrad
Gordon, Morris Aaron
Greer, Donald Lee
Halde, Carlyn Jean
Hall, John Walton
Hesseltine, Clifford William
Howard, Dexter Herbert
Huppert, Milton
Kaplan, William

Kaufman, Leo
Kozel, Thomas Randall
Kwon-Chung, Kyung Joo
Land, Geoffrey Allison
Larsh, Howard William
Lupan, David Martin
Mackinnon, Juan Enrique
Merz, William George
Pollack, J Dennis
Pore, Robert Scott
Rippon, John Willard
Roberts, Glenn Dale
Rogers, Alvin Lee
Romberg, Paul Frederick
Rosenthal, Stanley Arthur
Schmitt, John Arvid, Jr
Schneidau, John Donald, Jr
Shechter, Yaakov
Sinski, James Thomas
Stein, Janet Ruth
Stevens, Joseph Alfred
Stock, John Joseph
Sun, Sung Huang
Taylor, Robert Lee

Mycology
Abou-El-Seoud, Mohamed Osman
Ahearn, Donald G
Aldrich, Henry Carl
Alexopoulos, Constantine John
Allison, William Hugh
Anastasiou, Clifford J
Anderson, Mauritz Gunnar
Arnold, Mary Tryson
Backus, Edward James
Backus, Myron Port
Bandoni, Robert Joseph
Barksdale, Alma Whiffen
Barnett, Horace Leslie
Barr, Donald John Stoddart
Barra, Leh Raj
Baxter, John Wallace
Benjamin, Chester Ray
Benjamin, Richard Keith
Bereston, Eugene Sydney
Berliner, Martha D
Bianchi, Donald Ernest
Bigelow, Margaret Elizabeth Barr
Biondo, Frank X
Bland, Charles E
Brasfield, Travis Winford
Brenneman, James Alden
Brewer, Donald
Brown, Merton F
Burdsall, Harold Hugh, Jr
Campbell, Thomas Hodgen
Cantino, Edward Charles
Carroll, George C
Cavender, James C
Chick, Ernest Watson
Christenberry, George Andrew
Christensen, Martha
Clark, Jimmy Dorral
Clausz, John Clay
Clum, Floyd Myron
Collins, O'Neil Ray
Conant, Norman Francis
Cooke, John Cooper
Cooke, William Bridge
Cooper, Billy Howard
Corlett, Michael Philip
Corum, Cyril Joseph
Cowley, Gerald Taylor
Crane, Joseph Leland
Crisan, Eli Victor
Cunningham, John L
Davidson, Darwin Ervin
DeFigio, Daniel A
DeGroot, Rodney Charles
Dykstra, Michael Jack
Elizey, Joanne Tontz
Emerson, Ralph
Erdos, Gregory William
Fergus, Charles Leonard
Fuller, Melvin Stuart
Garner, Jasper Henry Barkdoll
Gauger, Wendell Lee
Gilbertson, Robert Lee
Gochenaur, Sally Elizabeth
Goldstein, Solomon
Goos, Roger Delmon
Gotelli, David M
Grappel, Sarah Fay
Gray, William Dudley
Griffin, David H
Grosklags, James Henry
Gruen, Hans Edmund
Gustafson, Ralph Alan
Haines, John Haldor
Hammill, Terrence Michael
Hanks, David V
Hanlin, Richard Thomas
Hardin, Hilliard Frances
Harding, Paul Raymond, Jr
Harrison, Kenneth Archibald
Hartmann, George Charles
Harvais, Gaetan Hugues
Haskins, Reginald Hinton

Heath, Ian Brent
Held, Abraham Albert
Hennnes, Don E
Hesten, Joe Fleetwood
Hickman, Clarence James
Hill, Eddie P
Hiratsuka, Yasuyuki
Ho, Hon Hing
Hollis, Cecil George
Huffman, Donald Marion
Hughes, Gilbert C
Hughes, Stanley John
Huneycutt, Maeburn Bruce
Hunter, Barry B
Hutchinson, James A
Hwang, Shuh-Wei
Ichida, Allan A
Illman, William Irwin
Johnson, Terry Walter, Jr
Jong, Shung-Chang
Jump, John Austin
Karve, Mohan Dattatreya
Kazama, Frederick Y
Keeping, Eleanor Silver
Keeping, Richard Paire (Dowding)
Kendrick, Bryce
Kimbrough, James W
Kirk, Paul Wheeler, Jr
Klein, Deana Tarson
Klett, Hubert Clifford
Kneebone, Leon Russell
Kobayashi, George S
Koch, William Julian
Kohlmeyer, Jan Justus
Korf, Richard Paul
Kowalski, Donald T
Kramer, Charles Lawrence
Krug, John Christian
Kuehn, Harold Herman
Kurtzman, Ralph Harold, Jr
Kurtzman, Cletus Paul
Largent, David Lee
Leathers, Chester Ray
Lee, Philip Calvin
Lentz, Paul Lewis
Levetin-Avery, Estelle
Liberta, Anthony E
Lichtwardt, Robert William
Lindhorst, Taylor Erwin
Lindsey, Julia Page
Lowe, Josiah Lincoln
Lowy, Bernard
Lubarsky, Robert
Luck-Allen, Etta Robena
Luttrell, Everett Stanley
Madhosingh, Clarence
Maiello, John Michael
Maniotis, James
Manocha, Mannohan Singh
Mayfield, John Emory
McClaren, Milton, Jr
McDonald, James Clifton
McDonough, Eugene Stowell
McGinnis, Michael Randy
McLain, Donald Davis, Jr
McNitt, Rand Edwin
Miller, James Edward
Miller, Orson K, Jr
Mims, Charles Wayne
Mislivec, Philip Brian
Mizuba, Seth Setsuo
Monoson, Herbert L
Montes, Leopoldo F
Moore, Morris
Moore, Royall Tyler
Motta, Jerome J
Nelson, Allen Charles
Nolan, Richard Arthur
Novak, Robert Otto
Olexia, Paul Dale
Orpurt, Philip Arvid
Orr, Geoffrey F
Paden, John Wilburn
Parmelee, John Aubrey
Perry, Eugene Arthur
Peterson, John Louis
Peterson, Joseph Louis
Petty, Milton Andrew, Jr
Pfister, Donald Henry
Poitras, Adrian William
Porter, John Norman
Ranzoni, Francis Verne
Raper, Kenneth Bryan
Reese, Elwyn Thomas
Reeves, Fontaine, Jr
Reinhardt, Donald Joseph
Reisert, Patricia
Reiss, Frederick
Reynolds, Don Rupert
Richard, John Lee
Ritchie, Donald Dirk
Roane, Martha Kotila
Roberts, John Maurice
Robertson, Jack Alex
Robinson, Albert Dean
Rogers, Jack David
Rogerson, Clark Thomas

Ross, Ian Kenneth
Roth, Frank J, Jr
Rothwell, Frederick Mirvan
Saltarelli, Cora G
Salvin, Samuel Bernard
Schisler, Lee Charles
Schoknecht, Jean Donze
Schwalb, Marvin N
Scott, William Wallace
Seo, John S
Seymour, Roland Lee
Shadomy, Helen Jean
Shaffer, Robert Lynn
Shanor, Leland
Shoemaker, Robert Alan
Silva-Hutner, Margarita
Simms, Emory Guy
Simms, Horace Ridgly
Singer, Rolf
Smith, Alexander Hanchett
Sorenson, William George
Sparrow, Frederick Kroeber, Jr
Spilioti, Charles Francis, Jr
Stewart, Elwin Lynn
Stuntz, Daniel Elliot
Subterkropp, Keller Francis
Sundberg, Walter James
Sussman, Alfred Sheppard
Sweet, Frank Edward
Szaniszlo, Paul Joseph
Tansey, Michael Richard
Taschdjian, Claire Louise
Tavares, Isabelle Irene
Tews, Leonard L
Therrien, Chester Dale
Thiers, Harry Delbert
Thomas, Percy LeRoy
Tiffney, Wesley Newell
Timberlake, William Edward
Trappe, James Martin
Tresner, Homer David
Tylutki, Edmund Eugene
Uecker, Francis August
Uscavage, Joseph Peter
Volz, Paul Albert
Walch, Henry Andrew, Jr
Wang, Chun-Juan Kao
Ward, John Everett, Jr
Warren, Charles O, Jr
Weeks, Robert Joe
Welden, Arthur Luna
Wells, Kenneth
Weresub, Luella Kayla
Whitney, Norman John
Whittingham, William Francis
Williams, Marion Ervin
Wilson, Charles Maye
Wilson, Kenneth Sheridan
Wilson, Ronald Wayne
Winstead, Janet
Wood, John Langille
Yu-Sun, Clare Chuan Chang
Ziller, Wolf Gunther

Olericulture
Cutcliffe, Jack Alexander
Dickerson, Chester T, Jr
Frazier, William Allen
Harrington, James Foster
Kelly, John Francis
Massey, Peyton Howard, Jr
Pollard, Leonard Heber
Shay, Junior Ralph
Smittle, Doyle Allen

Ornamental Horticulture
Brewer, James Edward
Carpenter, Edwin David
Conover, Charles Albert
Coorts, Gerald Duane
Davidson, Harold
Dirr, Michael Albert
Ealy, Robert Philip
Einert, Alfred Erwin
Evans, George Edward
Fischer, Charles Clayton
Flint, Harrison Leigh
Foret, James A
Fretz, Thomas Alvin
Gartner, John Bernard
Gibeault, Victor Andrew
Haramaki, Chiko
Harbaugh, Brent Kalen
Hasselkus, Edward R
Johnson, Charles Robert
Keen, Ray Albert
Kelley, James Durrell
Koch, Gary Martin
Lieberman, Arthur Stuart
Mahlstede, John Peter
McConnell, Dennis Brooks
McElwee, Edgar Warren
McGuire, John J
McWilliams, Edward Lacaze
Meyer, Martin Marinus, Jr
Mower, Robert G

Neel, Percy Landreth
New, Earl Hiram
Pellett, Norman Eugene
Poole, Richard Turk, Jr
Scheckel, Kurt Anthony
Shaw, Richard John
Sheehan, Thomas John
Shetler, Stanwyn Gerald
Sink, Kenneth C, Jr
Smeal, Paul Lester
Svejda, Felicitas Julia
Sydnor, Thomas Davis
Ticknor, Robert Lewis
Toop, Edgar Wesley
Van Laan, Gordon James
Waxman, Sidney
White, Donald Benjamin
Williams, Harold Hamilton

Phytochemistry
Bailey, Lowell Frederick
Bhatti, Waqar Hamid
Brown, John Kennedy
Caldwell, Carlyle Gordon
Chan, Bock G
Corse, Joseph Walters
Denford, Keith Eugene
Dorrell, Douglas Gordon
Dunkle, Larry D
Farnsworth, Norman R
Fissel, Guy Wilmer
Gottlieb, Otto Richard
Hanks, Robert William
Hassan, William Ephriam, Jr
Hixon, Ralph Malcolm
Hougen, Frithjof W
Howarth, Ronald Edward
Kelsey, Rick Guy
Khanna, Krishan L
King, Nydia Margarita
Kortschak, Hugo Peter
Mabry, Tom Joe
Martin, Susan Scott
McClure, Jerry Weldon
Miller, Joseph Edwin
Miller, Lawrence Peter
Perlin, Irwin Bernard
Romeo, John Thomas
Saunders, James Allen
Scogin, Ron Lynn
Seeley, Schuyler Draman
Segelman, Alvin Burton
Stewart, Ivan
Stewart, Robert N
Surrey, Kenneth

Paleobotany
Abbott, Maxine Langford
Andrews, Henry Nathaniel, Jr
Archangelsky, Sergio
Arnold, Chester Arthur
Ash, Sidney Roy
Axelrod, Daniel Isaac
Bartoo, Harriette Valleta (Krick)
Basson, Philip Walter
Baxter, Robert Wilson
Beals, Harold Oliver
Becker, Herman Frederick
Beyer, Arthur Frederick
Campbell, John Duncan
Cridland, Arthur A
Darrah, William Culp
Delevoryas, Theodore
Dilcher, David L
Dolph, Gary Edward
Dorf, Erling
Eggert, Donald A
Engelhardt, Donald Wayne
Felix, Charles Jeffrey
Frankenberg, Julian Myron
Fry, Wayne Lyle
Gensel, Patricia Gabbey
Gillette, Norman John
Grierson, James Douglas
Hickey, Leo Joseph
Hueber, Francis Maurice
Jarzen, David MacArthur
Kasper, Andrew E, Jr
Kosanke, Robert Max
Lowther, John Stewart
MacGinitie, Harry Dunlap
Mamay, Sergius Harry
Matten, Lawrence Charles
Nitecki, Matthew H
Penny, John Sloyan
Phillips, Tom Lee
Radforth, Norman William
Rouse, Glenn Everett
Schopf, James Morton
Scott, Richard Albert
Sidd, Benton Maurice
Spackman, William, Jr
Taylor, Thomas Norwood
Tschudy, Robert Haydn
Warter, Janet Kirchner
Willis, Jeanne Eleanor

Tang, Chung-Shih
Taylor, Elmore Hector
Tso, Tien Chioh
Tucker, Charles Leroy, Jr
Wells, John Arthur
Weybrew, Joseph Arthur

Phytogeography
Balbach, Harold Edward
Harvill, Alton McCaleb, Jr
Kuchler, August Wilhelm
Pemble, Richard Hope

Phytopathology
Abney, Thomas Scott
Adams, Peter B
Adams, Robert Evans
Agrios, George Nicholas
Aho, Paul E
Aichele, Murit Dean
Aist, James Robert
Alcorn, Stanley Marcus
Alexander, Paul Marion
Allen, Arthur (Silsby)
Allen, Ross Marvin
Allen, Thomas Cort, Jr
Allen, Wayne Robert
Allington, William B
Allison, Joseph Lewis
Allison, Patricia (Lee) (Van Burgh)
Althaus, Ralph Elwood
Altman, Jack
Alvarez, Anne Maino
Amador, Jose Manuel
Amato, Vincent Alfred
Ames, Ralph Wolfley
Andersen, Axel Langvad
Anderson, Neil Albert
Anzalone, Louis, Jr
Apple, Jay Lawrence
Apt, Walter James
Aragaki, Minoru
Areneson, Phil Alan
Arnett, James DeLos, Jr
Arnold, Gordon William
Arny, Deane Cedric
Arteman, Robert Lloyd
Asai, George Napoleon
Ashworth, Lee Jackson, Jr
Athow, Kirk Leland
Atkinson, Robert George
Atkinson, Thomas Grisedale
Averre, Charles Wilson, III
Aycock, Robert
Ayers, John E
Bagga, Harmahinder Singh
Bagnall, Richard Herbert
Baker, Kenneth Frank
Baker, R Ralph
Bald, John Grieve
Baldwin, James Gordon
Baldwin, Robert Edmund
Banttari, Ernest E
Baskin, Aaron David
Bateman, Durward F
Batson, William Edward, Jr
Baxter, John Wallace
Baxter, Luther Willis, Jr
Bean, George A
Beckman, Carl Harry
Beer, Steven Vincent
Beevers, Harry
Bega, Robert V
Bell, Alois Adrian
Bell, Carl, F
Benedict, Winfred Gerald
Bennett, Carlyle Wilson
Beraha, Louis
Berbee, John Gerard
Berger, Richard Donald
Berkenkamp, Bill Brodie
Bernier, Claude
Berry, Charles Richard
Berry, Robert Wade
Beute, Marvin Kenneth
Bever, Wayne Melville
Birchfield, Wray
Bird, Luther Smith
Bishop, Charles Franklin
Bissonnette, Howard Louis
Bitancourt, Agesilau Antonio
Black, Lowell Lynn
Blackman, Cyril Wells
Blazquez Y Servin, Carlos Humberto
Bloom, James R
Bloss, Homer Earl
Bond, William Payton
Bonde, Morris Reiner
Boone, Donald Milford

Boosalis, Michael Gus
Booth, John Austin
Boothroyd, Carl William
Borgman, Robert P
Boyer, Michael George
Boyle, John Samuel
Bozarth, Gene Allen
Bracker, Charles E, Jr
Braun, Alvin Joseph
Braverman, Samuel William
Brewer, Jesse Wayne
Bridgmon, George Harrison
Brinkerhoff, Lloyd Allen
Britton, Michael Paul
Bromfield, Kenneth Raymond
Browder, Lewis Eugene
Brown, George Eldon
Brown, William Malcolm, Jr
Browning, John Artie
Bruehl, George William
Buchenau, George William
Buddenhagen, Ivan William
Burleigh, James Reynolds
Busch, Lloyd Victor
Bushnell, William Rodgers
Bushong, Jerold Ward
Butler, Edward Eugene
Calavan, Edmond
Caldwell, Ralph Merrill
Caldwell, Roger Lee
Callahan, Kemper Leroy
Calpouzos, Lucas
Calvert, Oscar Hugh
Cameron, H Ronald
Camp, Earl D
Camp, Russell R
Campagna, Elzear (Alexandre)
Campbell, Robert Noe
Campbell, W P
Cannon, Orson Silver
Cappellini, Raymond Adolph
Carley, Harold Edwin
Carlson, Lester William
Caroselli, Nestor Edgar
Carroll, Robert Baker
Carroll, Thomas William
Carter, James Cedric
Carter, William Whitney
Castelfranco, Paul Alexander
Ceponis, Michael John
Cetas, Robert Charles
Chamberlain, Donald William
Chapman, Richard Alexander
Chi, Chien Chen
Chiarappa, Luigi
Childs, James Fielding Lewis
Chilton, St John Poindexter
Chopra, Baldeo K
Christensen, Clyde Martin
Civerolo, Edwin Louis
Clark, Raymond Loyd
Clark, Robert Vernon
Clayton, Carlyle Newton
Cobb, Fields White, Jr
Cohen, Mortimer
Cohoon, Daniel Fred
Cole, Herbert, Jr
Comstock, Jack Charles
Connor, Stephen R
Converse, Richard Hugo
Cook, Allyn Austin
Cook, Robert James
Coplin, David Louis
Corbett, M Kenneth
Corden, Malcolm Ernest
Cormack, Melville Wallace
Couch, Houston Brown
Covey, Ronald Perrin, Jr
Cowling, Ellis Brevier
Crittenden, Henry William
Crone, Lawrence John
Crossan, Donald Franklin
Curl, Elroy Arvel
Curren, Thomas
Curtis, Charles R
Daines, Robert Henry
Dale, James Lowell
Damann, Kenneth Eugene, Jr
Damsteeg, Vernon Dale
Darby, John Feaster
Darley, Ellis Fleck
Davidson, Bruce Lloyd
Davidson, Richard Shoots
Davidson, Thomas Ralph
Davis, Benjamin Harold
Davis, David
Davis, James Robert
Davis, Lily Herlinda
Davis, Robert Edward
Davis, Robert Gene
Davis, Spencer Harwood, Jr
Deahl, Kenneth Luvere
Dean, Jack Lemuel
Dean, Leslie L
Deems, Robert Eugene
Deep, Ira Washington
DeGroot, Rodney Charles
De Leon, Carlos

Delp, Charles Joseph
Demski, James Willard
Desjardins, Paul Roy
Desrosiers, Russell
DeVay, James Edson
deZeeuw, Donald John
De Zoeten, Gustaaf A
Diachun, Stephen
Dickens, Lester Emert
Dickey, Robert Shaft
Diener, Urban Lowell
Dietz, Sherl M
Dimitman, Jerome Eugene
Dochinger, Leon S
Dolan, Desmond Daniel
Douglas, Dexter Richard
Doupnik, Ben Lee, Jr
Dowler, William Minor
Drake, Charles Roy
DuCharme, Ernst Peter
Duffus, James Edward
Dukes, Philip Duskin
Duncan, Harry Ernest
Duniway, John Mason
Dunlap, Albert Atkinson
Dunleavy, John M
Duran, Ruben
Durbin, Richard Duane
Dwinell, Lew David
Echandi, Eddie
Eckert, Joseph Webster
Edmunds, Leon K
Edwards, Dale Ivan
Ehrlich, Mary Ann
Eide, Carl John
Ellett, Clayton Wayne
Ellingboe, Albert Harlan
Elliot, Arthur McAuley
Elliott, Edward Sumner
Ellis, Don Edwin
Emerson, Frank Henry
Emge, Robert George
Emmatty, Davy A
Endo, Robert Minoru
Engelhard, Arthur William
English, William Harley
Epps, William Monroe
Erwin, Donald C
Estey, Ralph Howard
Eversmeyer, Harold Edwin
Farley, James D
Feldman, Albert William
Fenwick, Harry
Filer, Theodore H, Jr
Finch, Davy C
Finley, Arthur Marion
Fisher, Kenneth D
Fitzgerald, Paul Jackson
Fleischmann, George
Flor, Harold Henry
Foley, Dean Carroll
Ford, Donald Hoskins
Ford, Richard Earl
Forsberg, Junius Leonard
Foster, Virginia
Frank, James Anthony
Frederick, Lafayette
Frederiksen, Richard Allan
Freeman, Thomas Edward
Freiberg, George William
French, Alexander Murdoch
Fridlund, Paul Russell
Frosheiser, Fred Imanuel
Froyd, James Donald
Fry, William Earl
Fulkerson, John Frederick
Fulton, Joseph Patton
Fulton, Neil Douglas
Fulton, Robert Watt
Fushtey, Stephen George
Futrell, Maurice Chilton
Gabrielson, Richard Lewis
Gallegly, Mannon Elihu
Garber, Richard Hammerle
Gardner, Wayne Scott
Garnsey, Stephen Michael
Garraway, Michael Oliver
Garren, Kenneth Howard
Gates, James Edward
Gerdemann, James Wessel
Giamalva, Mike J
Gilgut, Constantine Joseph
Gill, Denzell Leigh
Gill, Harmonhindar Singh
Gillaspie, Athey Graves, Jr
Gillespie, William Harry
Gilmer, Robert McCullough
Gilpatrick, John Daniel
Gingery, Roy Evans
Goheen, Austin Clement
Gold, Alma Herbert
Good, Harold Marquis
Goode, Monroe Jack
Gooding, Guy V, Jr
Goodman, Joseph Jacob
Goodman, Robert Merwin
Goodman, Robert Norman
Goonewardene, Hilary Felix

Gordon, Donald Theile
Gorenz, August Mark
Goth, Robert W
Gottlieb, David
Gough, Francis Jacob
Gould, Charles Jay
Graham, Joseph Harry
Graham, Shirl Orby
Grau, Craig Robert
Graves, Clinton Hannibal, Jr
Green, Gordon John
Green, Norman Edward
Green, Ralph J, Jr
Greene, George Linden
Gregory, Garold Fay
Gries, George Alexander
Griffin, Gerald D
Grimm, Gordon Ralph
Grimm, Robert Blair
Grogan, Raymond Gerald
Groth, James Vernon
Gudauskas, Robert Thomas
Gumpf, David John
Guthrie, James Warren
Haas, Jerry Henry
Hadwiger, Lee A
Hagan, William Leonard
Hagedorn, Donald James
Hager, Richard Arnold
Haglund, William Arthur
Halisky, Philip Michael
Hall, Dennis Heeley
Hall, Robert
Halloin, John McDonell
Hamilton, Richard Ian
Hampton, Michael Chisnall
Hampton, Raymond Earl
Hampton, Richard Owen
Hanchey, Penelope Jane
Hancock, Joseph Griscom, Jr
Hansen, Anton Juergen
Hansing, Earl Dahl
Hanson, Earle William
Hardison, John Robert
Hare, Woodrow Wilson
Harman, Gary Elvan
Harnish, Wayne Nelson
Harper, Frank Richard
Harrar, Jacob George
Harris, Hubert Andrew
Harrison, Arthur Leslie
Harrison, Monty DeVerl
Harry, John Boyer
Hartman, James Xavier
Hartstirn, Walter
Harvey, Alan Eric
Harvey, John Marshall
Haskett, William Courtney
Hawn, Elmer Joseph
Heagle, Allen Streeter
Healy, Michael James
Heath, Michele Christine
Hebert, Teddy T
Heggestad, Howard Edwin
Heidrick, Lee E
Helton, Audus Winzle
Henderson, Nannette Smith
Henderson, Robert Gordon
Hendrix, Floyd Fuller, Jr
Hendrix, James William
Hendrix, John Walter
Herr, Leonard Jay
Herr, Robert Roy
Heuberger, John William
Hewitt, William Boright
Hibben, Craig Rittenhouse
Hickey, Kenneth Dyer
Hiebert, Ernest
Higgins, Verna Jessie
Hilborn, Merle Tyson
Hildebrand, Donald Clair
Hildebrandt, Albert Christian
Hildreth, Robert Claire
Hilty, James Willard
Himelick, Eugene Bryson
Hine, Richard Bates
Hiratsuka, Yasuyuki
Ho, Hon Hing
Hoadley, Alfred Damon
Hobart, Oscar F, Jr
Hobbs, Clifford Dean
Hock, Winand Karl
Hodges, Charles Sasnette, Jr
Hodges, Clinton Frederick
Hoffmann, James Allen
Hoitink, Harry A J
Holcomb, Gordon Ernest
Holdeman, Quintin Lee
Hollis, John Percy, Jr
Holmes, Francis William
Holtzmann, Oliver Vincent
Hooker, Arthur Lee
Hooker, William James
Hooper, Gerald Ray
Hopkins, Donald Lee
Horn, Norman Louis, Jr
Horner, Chester Ellsworth
Horricks, Jack Stewart

BOTANY

Shigo, Alex Lloyd
Shortle, Walter Charles
Shriner, David Sylva
Shurtleff, Malcolm C. Jr
Siddiqui, Wasi Mohammad
Siegel, Malcolm Richard
Sievert, Richard Carl
Silbernagel, Matt Joseph
Sill, Webster Harrison, Jr
Silverman, William Bernard
Simons, John Norton
Simons, Marr Dixon
Simpson, William Roy
Sims, Asa C, Jr
Sinclair, James Burton
Sinclair, Wayne A
Sinden, James Whaples
Sinden, Stephen Lee
Singh, Dilbagh
Singh, Jaswant
Sinha, Ramesh Chandra
Sisler, Hugh Delane
Sitterly, Wayne R
Skilling, Darroll Dean
Skoropad, William Peter
Skotland, Calvin B
Slack, Derald Allen
Slack, Steven Allen
Sleeth, Bailey
Slykhuis, John Timothy
Smalley, Eugene Byron
Smiley, Richard Wayne
Smith, Glenn Edward
Smith, Harlan Eugene
Smith, Jeffrey Drew
Smith, Richard S, Jr
Smith, Samuel H
Smith, Thomas Earle
Smith, William Hulse
Smith, Wilson Levering, Jr
Smoot, John Jones
Snow, Gordon Franklin
Snow, Jean Anthony
Snow, Johnnie Park
Snow, Michael Dennis
Snyder, Hugh Donald
Sonoda, Ronald Masahiro
Sosebee, Ronald Eugene
Sowell, Grover, Jr
Spalding, Donald Hood
Spencer, James Alphus
Spotts, Robert Allen
Springer, John Kenneth
Sproston, Thomas, Jr
Spurr, Harvey Wesley, Jr
Stace-Smith, Richard
Stadther, Richard James
Staffeldt, Eugene Edward
Stakman, Elvin Charles
Staley, John M
Stall, Robert Eugene
Stanghellini, Michael Eugene
Staples, Richard Cromwell
Staudinger, Wilbur Leonard
Stavely, Joseph Rennie
Steadman, James Robert
Steib, Rene J
Steiner, Gary W
Stevens, Russell Bradford
Stewart, Donald Martin
Stewart, Robert Blaylock
Stienstra, Ward Curtis
Stone, William Jack Hanson
Stouffer, Martin Franklin
Stouffer, Richard Franklin
Stover, Robert Harry
Stowell, Ewell Addison
Streets, Rubert Burley
Strider, David Lewis
Strobel, Gary A
Strobel, James Walter
Sturgeon, Roy V, Jr
Stuteville, Donald Lee
Suhovecky, Albert J
Sumner, Donald Ray
Sutton, Roscoe Murray Davidson
Sutton, Turner Bond
Szkolnik, Michael
Tachibana, Hideo
Takahashi, William Noboru
Tamaoki, Taiki
Tammen, James F
Taylor, Gordon Stevens
Taylor, Jack
Te Beest, David Orien
Templeton, George Earl
Tenney, Wilton R
Teviotdale, Beth Luise
Thayer, Paul Loyd
Thomas, Herbert Rex
Thomas, John Eugene
Thomas, Walter Dill, Jr
Thompson, Hazen Spencer
Thompson, Samuel Stanley, Jr
Thornberry, Halbert Houston
Thurston, Herbert David
Thyr, Billy Dale

Tiffany, Lois Hattery
Timmer, Lavern Wayne
Tinga, Jacob Hinnes
Tinline, Robert Davies
Todd, Edwin Harkness
Tolmsoff, Walter John
Tomlinson, Harley
Torgeson, Dewayne Clinton
Townshend, John Lynden
Townsley, William W, Jr
Treshow, Michael
Troutman, Joseph Lawrence
Troxel, Allen Wendell
Tsao, Pamela Wen-Chau Wang
Tsao, Peter Hsing-Tsuen
Tuite, John F
Tweedy, Billy Gene
Unbehaun, Laraine Marie
Uyemoto, Jerry Kazumitsu
Vaartnou, Herman
Vakili, Nader Gholi
van der Zwet, Tom
Van Dyke, Cecil Gerald
Van Etten, Hans D
Van Gundy, Seymour Dean
Vanterpool, Thomas Clifford
Vargas, Joseph Martin, Jr
Varney, Reed William
Varney, Eugene Harvey
Vaughan, Edward Kemp
Veech, Joseph A
Venere, Ralph Joseph, Sr
Vest, Hyrum Grant, Jr
Viswanathan, Muri A
Volin, Raymond Bradford
Vredeveld, Nicholas Gene
Wade, Earl Kenneth
Wadley, Bryce Nephi
Wadsworth, Dallas Fremont
Wagnon, Harvey Keith
Walker, Jerry Tyler
Walkinshaw, Charles Howard, Jr
Wall, Ronald Eugene
Wallace, James Merrill
Wallen, Victor Reid
Wallin, Jack Robb
Walters, Hubert Jack
Walton, Gerald Steven
Ward, Calvin Herbert
Ward, Edmund William Beswick
Warren, Herman Lecil
Watson, Roscoe Derrick
Watterson, Jon Craig
Wave, Herbert Edwin
Weathers, Lewis Glen
Weaver, Leslie O
Weber, Darrell J
Weber, Paul Van Vranken
Webster, Robert K
Weidensaul, T Craig
Weihing, John Lawson
Weingartner, David Peter
Weinhold, Albert Raymond
Welch, Donald Ray
Wellman, Richard Harrison
Wells, Homer Douglas
Welty, Ronald Earle
Whaley, Julian Wendell
Wheeler, Harry Ernest
White, Donald Glenn
White, James Clarence
Whitehead, Marvin Delbert
Whiteside, Jack Oliver
Whitney, Elvin Dale
Whitney, Harvey Stuart
Whitney, Norman John
Wicker, Ed Franklin
Wickliff, James Leroy
Wiedman, Harold W
Wiese, Maurice Victor
Wilcoxson, Roy Dell
Wiles, Alfred Barksdale
Wilhelm, Stephen
Wilkinson, Daniel R
Wilkinson, Robert Elzworth
Williams, Albert Simpson
Williams, Edwin Bruce
Williams, Floyd James
Williams, Harold Edward
Williams, Lansing Earl
Williams, Paul Hugh
Williamson, Charles Edward
Williges, George Goudie
Willis, Carl Bertram
Willis, George Mirron
Wills, Wirt Henry
Wilson, Charles L
Wilson, Coyt Taylor
Wilson, Eugene M
Wilson, Jack Belmont
Wilson, Shirley Lane
Wilson, William Ernest
Winstead, Nash Nicks
Witcher, Wesley
Wood, Francis A
Worf, Gayle L
Worley, Joseph Francis
Wright, Norman Samuel

Wright, Theodore Richard
Wuest, Paul Joseph
Wyllie, Thomas Dean
Wynn, Willard Kendall, Jr
Wysong, David Serge
Yang, Charles (Yu-Di)
Yarwood, Cecil Edmund
Yoder, David Lee
Yoder, Donald Maurice
Yoder, Olen Curtis
Young, Harry Curtis, Jr
Young, Roy Alton
Younkin, Stuart G
Zehr, Eldon Irvin
Zentmyer, George Aubrey, (Jr)
Zettler, Francis William
Zeyen, Richard John
Zimmer, David E
Zuckerman, Bert Merton
Zummo, Natale

Plant Anatomy
Becker, Steven Allan
Berlyn, Graeme Pierce
Boke, Norman Hill
Brennan, James Robert
Byrne, John Maxwell
Calvin, Clyde Lacey
Carlson, John Bernard
Cecich, Robert Allen
Davidson, Christopher
Decker, Jane M
Derr, William Frederick
Diboll, Alfred
England, Wayne H
Eyde, Richard Husted
Freeman, Thomas Patrick
Goodman, Victor Herke
Grillos, Steve John
Harris, William M
Harrison, H Keith
Heimsch, Charles W
Hillson, Charles James
Ikenberry, Gilford John, Jr
Kuijt, Job
Kukachka, Bohumil Francis
Lott, John Norman Arthur
Mahlberg, Paul Gordon
Markle, Carrolle Anderson
McGrath, James J
Mikesell, Jan Erwin
Miller, Regis Bolden
Mogensen, Hans Lloyd
Morey, Philip Richard
Moseley, Maynard Fowle, Jr
Myers, Gerald Andy
Neubauer, Benedict Francis
Nisbet, Jerry J
Pizzolato, Thompson Demetrio
Reeve, Marian E
Reeve, Roger Mansfield
Richardson, Paul Ernest
Rost, Thomas Lowell
Saigo, Roy Hirofumi
Stern, William Louis
Susalla, Anne A
Tepper, Herbert Bernard
Towers, George Hugh Neil
Tseng, Charles C
Walker, Dan B
Wheeler, George Edward
Zimmermann, Martin Huldrych

Plant Biochemistry
Abrahamson, Lila
Albersheim, Peter
Allred, Keith Reid
Bandurski, Robert Stanley
Benedict, Chauncey
Brown, Stewart Anglin
Bryan, Ashley Monroe
Chapman, David J
Chen, Chong Maw
Cheniae, George Maurice
Christianson, Donald Duane
Clevenger, Sarah
Crean, Joseph Gaylord
Croy, Lavoy I
Daly, Joseph Michael
Dennis, David Thomas
Dollwet, Helmar Hermann Adolf
Dull, Gerald G
Dunham, Valgene Loren
Durley, Richard Charles
Ellsworth, Robert King
Finkle, Bernard Joseph
Freebairn, Hugh Taylor
French, Richard Collins
Gamborg, Oluf Lind
Hall, Timothy Couzens
Hess, John Lloyd
Humphreys, Thomas Elder
Hylin, John Walter
Ibrahim, Ragai Kamel
Jakob, Karl Michael
Johnson, Morris Alfred
Jones, John Dewi
Kahn, Joseph Stephan

Koeppe, David Edward
Kosuge, Tsune
Kuc, Joseph
Lane, Forrest Eugene
LeTourneau, Duane John
Linden, James Carl
Maclachlan, Gordon Alistair
Mangat, Bhupinder Singh
Marriage, Paul Bernard
Marsho, Thomas V
Massey, Louis Melville, Jr
Mazelis, Mendel
McCalla, Arthur Gilbert
McMahon, Vern August
Meeuse, Bastiaan Jacob Dirk
Meheriuk, Michael
Miller, Gene Walker
Norman, Arthur Geoffrey
Nozzolillo, Constance
Oliver, Jeanette Clements
Olson, Lee Charles
Peverly, John Howard
Punnett, Thomas R
Putman, Edison Walker
Racusen, David
Rebstock, Theodore Lynn
Richards, George Manning
Ross, Cleon Walter
Roy, Harry
Schrader, Lawrence Edwin
Searle, Norman Edward
Shannon, Leland Marion
Shargool, Peter Douglas
Shrimpton, Douglas Malcolm
Siegelman, Harold William
Stepka, William
Thompson, John Fanning
Thornber, James Philip
Tolbert, Robert John
Underhill, Edward Wesley
Uribe, Ernest Gilbert
Vennesland, Birgit
Wallace, James William, Jr
Wallace, Joan M
Walton, Daniel C
Wang, Dalton Ta Tung
Watkin, John Emrys
Webb, Bill D
White, Fred G
Whitehead, Eugene Irving
Williams, John Peter
Wilson, Curtis Marshall
Wilson, David George
Winget, Gary Douglas
Zacharius, Robert Marvin
Zitnak, Ambrose

Plant Breeding
Adams, Maurice Wayne
Adamson, William Charles
Albrechtsen, Rulon S
Alexander, Denton Eugene
Ali-Khan, Syed Tahir
Alston, Jimmy Albert
Anand, Satish Chandra
Anderson, Ronald Eugene
Andrews, John Edwin
Angell, Frederick Franklyn
Anstey, Thomas Herbert
Arceneaux, George
Atkins, Irvin Milburn
Axtell, John David
Aycock, Marvin Kenneth, Jr
Baenziger, H
Baker, Robert John
Baltensperger, Arden Albert
Barham, Warren Sandusky
Barker, LeRoy N
Barnes, Donald Kay
Barton, Donald Wilber
Bassett, Mark Julian
Bates, Richard Pierce
Bauman, Loyal Frederick
Beard, Benjamin H
Beard, David Franklin
Bell, Gary Milton
Berg, Clyde C
Bernard, Richard Lawson
Berry, Stanley Z
Beyer, Edgar Herman
Beyer, Elmo Monroe, Jr
Bienz, Darrel Rudolph
Bingham, Edwin Theodore
Blackmon, Clinton Ralph
Bockholt, Anton John
Bohnenblust, Kenneth E
Borchers, Edward Alan
Bouwkamp, John C
Bowen, Hollis Hulon
Brigham, Raymond Dale
Brim, Charles A
Bringhurst, Royce S
Brolmann, John Bernardus
Brooks, Stanley Nelson
Brown, Acton Richard
Brown, Charles Myers
Brown, William Lacy
Buchannon, Kenneth William

BOTANY

Buckingham, William Thomas
Buker, Robert Joe
Burrows, Vernon Douglas
Busbice, Thaddeus H
Buss, Glenn Richard
Buzzell, Richard Irving
Caffey, Horace Rouse
Campbell, Allan Barrie
Campbell, Kenneth Wilford
Canode, Chester Lang
Carbalio-Quiros, Alfredo
Carlson, Irving Theodore
Carnahan, Howard Leon
Carver, William Angus
Casady, Alfred Jackson
Chessmore, Roy A
Childers, Walter Robert
Clark, Edward Maurice
Cleveland, Richard Warren
Collins, Frederick Clinton
Collins, William Kerr
Collister, Earl Harold
Comstock, Verne Edward
Constantin, Milton J
Cooper, Richard Lee
Cope, Will Allen
Cowan, John Ritchie
Coyne, Dermot P
Crall, James Monroe
Crane, Paul Levi
Creech, Roy G
Crill, Pat
Crowder, Loy Van
Curtis, Byrd Collins
Davis, David Warren
Davis, Ralph Lanier
Davis, William Hatch
Day, Arden Dexter
Dean, Charles Edgar
Dempsey, Wesley Hugh
Dewey, Wade G
Dickson, Michael Hugh
Ditterline, Raymond Lee
Donnelly, Edward Daniel
Douglas, Alvin Gene
Downey, Richard Keith
Duclos, Leo Albert
Dudley, John Wesley
Dunn, Gerald Marvin
Duvick, Donald Nelson
Eisa, Handy Mahmoud
Elling, Laddie Joe
Ellsworth, Robert Lovell
Emery, Donald Allen
Enns, Henry
Erickson, John Robert
Eslick, Robert Freeman
Everett, Herbert Lyman
Everson, Everett Henry
Feaster, Carl Vance
Ferguson, David B
Fery, Richard Lee
Fick, Gerhardt Nelson
Findley, William Ray, Jr
Finkner, Ralph Eugene
Fitzgerald, Paul Jackson
Forbes, Ian
Foster, Albert Earl
Foster, Robert Edward, II
Francis, Charles Andrew
Frey, Kenneth John
Funk, Cyril Reed, Jr
Gardenhire, James Homer
Gasser, Heinz
Gauthier, Fernand Marcel
Geadelmann, Jon Lee
George, Donald Wayne
Gibler, John Wesley
Gibson, Pryce Byrd
Giesbrecht, John
Gilbert, James Carl
Goplen, Bernard Peter
Gorbet, Daniel Wayne
Grafius, John Edward
Granberry, Darbie Merwin
Grant, Marshall Nelson
Gray, Elmer
Greenleaf, Walter Helmuth
Greenshields, John Edward Ross
Gregory, Walton Carlyle
Gross, Delmer Ferd
Haaland, Ronald L
Hackerott, Harold Leroy
Hadley, Henry Hultman
Hagan, William Leonard
Hallauer, Arnel Roy
Hanmett, Harrell Lee
Hanson, George Peter
Hanson, Ronald Gordon
Harpstead, Dale D
Hartmann, Richard W
Harvey, Bryan Laurence
Harvey, Paul Henry
Hathcock, Bobby Ray
Haynes, Frank Lloyd, Jr
Hearn, Charles Jackson
Hecker, Richard Jacob
Heermann, Richard Martin

Heiner, Robert E
Heinz, Don J
Helgason, Sigurdur Bjorn
Helm, James Leroy
Hernandez, Teme P
Hicks, Dale R
Hikida, Hideaki Robert
Hill, Richard Ray, Jr
Hinson, Kuell
Hinze, Greg Otto
Hoff, Bert John
Hoff, John Conrad
Hogaboam, George Joseph
Holder, David Gordon
Holt, Ethan Cleddy
Hornby, Cedric Albert
Horner, Earl Stewart
Hougas, Robert Wayne
Hovin, Arne William
Hurd, Edwin Albert
Hymowitz, Theodore
Ihrke, Charles Albert
Ito, Philip J
Jaynes, Richard Andrus
Jellum, Milton Delbert
Jennings, Peter Randolph
Johannessen, George Andrew
John, Charles Alfred
Johnson, Alan Len
Johnson, Alvin A
Johnson, Glenn Richard
Johnson, Herbert Windal
Johnson, Wiley Carroll, Jr
Johnstone, Francis Elliott, Jr
Jones, Champ McMillian
Jones, Guy Langston
Jones, Jess Willard
Jones, Leonard Melvin
Josephson, Leonard Melvin
Jump, Lorin Keith
Kahler, Alex L
Kannenberg, Lyndon William
Keller, Kenneth Raymond
Kemp, Gavin Arthur
Kerr, Ernest Andrew
Khush, Gurdev S
Kiang, Yun-Tzu
Kingma, Gerbrand
Klatt, Arthur Raymond
Klinck, Harold Rutherford
Kneebone, William Robert
Knight, William Eric
Knowles, Robert Patrick
Kolar, John Joseph
Kolp, Bernard J
Konzak, Calvin Francis
Lachman, William Henry, Jr
Lambert, Howard N
Lambom, Calvin Ray
Lane, Ronald Paton
Lane, William David
Larter, Earl Nathan
Lauenroth, William Karl
Lawrence, Francis Joseph
Lawrence, Thomas
Lawson, Norman C
Layne, Richard C
Lessman, Koert J
Lewellen, Robert Thomas
Lindsey, Marvin Frederick
Livers, Ronald Wilson
Lofgren, James R
Loiselle, Roland
Lowe, Carl Clifford
Lukosevicius, Petras Povilas
Lunden, Allyn Oscar
Lyrene, Paul Magnus
Maan, Shivcharan Singh
Manning, Cleo Willard
Manwiller, Alfred
Marshall, Harold Gene
Matthews, David Livingstone
Maunder, A Bruce
Maxwell, James Donald
McCain, Francis Saxon
McCarter, States Marion
McCustion, Willis Lloyd
McDaniel, Milton Edward
McIlrath, William Oliver
McKenzie, Hugh
McNeal, Francis H
McNeill, Michael John
Mehlquist, Gustav Arthur Leonard
Melton, Billy Alexander, Jr
Merkle, Owen George
Metcalfe, David Richard
Miller, Carol Raymond
Miller, Darrell Alvin
Miller, John David
Miller, Philip Arthur
Mixon, Aubrey Clifton
Mock, James Joseph
Moore, Glenn D
Morris, John Leonard
Morris, James Joseph
Moser, Paul E
Muehlbauer, Frederick Joseph
Muramoto, Hiroshi

Murphy, Charles Franklin
Murphy, Royse Peak
Nabors, Murray Wayne
Nakayama, Roy Minoru
Nanda, Devender (Dave) Kumar
Neely, James Winston
Nelson, Lloyd Russel
Nesbitt, William Belton
Nichols, Courtland Geoffrey
Niles, George Alva
Norden, Allan James
Oldemeyer, Donald LeRoy
Oldemeyer, Robert King
Palmer, Reid G
Pardee, William Durley
Paschal, Eugene Hamer, II
Pate, James Bruce
Patterson, Fred La Vern
Peirce, Lincoln Carret
Peterson, Ronald M
Pfeifer, Robert Paul
Pickett, Robert Cooper
Pieringer, Arthur Paul
Plaisted, Robert Leroy
Pochlman, John Milton
Pollack, Bernard Leonard
Polson, David Ernest
Poneleit, Claus Gustav
Poole, Dewey Donald
Porter, Kenneth Boyd
Poster, Gerry Lynn
Potts, Howard Calvin
Povilaitis, Bronius
Prashar, Paul D
Puri, Yesh Paul
Putt, Eric Douglas
Qualset, Calvin Odell
Quesenberry, Kenneth Hays
Quick, James S
Quinones, Ferdinand Antonio
Ramming, David Wilbur
Reeves, Dale Leslie
Reinbergs, Ernests
Reitz, Louis Powers
Remley, Frank Morris
Rhodes, Ashby Marshall
Rhyne, Claude Little, Jr
Richmond, Thomas Rollin
Rigney, Jackson Ashcraft
Risius, Marvin Leroy
Roath, William Wesley
Robbins, Marion LeRon
Robertson, Larry Dee
Robinson, Harold Frank
Rogers, Owen Maurice
Rohde, Charles Raymond
Rossman, Elmer Chris
Rubis, David Daniel
Russell, Wilbert Ambrick
Ryder, Edward Jonas
Sampson, Dexter Reid
Sappenfield, William Paul
Sayers, Earl Roger
Schaefert, Robert Eugene
Schaffner, Jurgen Richard
Schonhorst, Melvin Herman
Scott, Gene E
Sechler, Dale Truman
Shannon, James Grover
Shenk, John Stoner
Sherman, Wayne Bush
Sieveking, William Earl
Simantel, Gerald M
Sisler, William Wallace
Sleper, David Allen
Smith, David Clyde
Smith, David Harrison, Jr
Smith, Glenn Sanborn
Smith, Olin Dail
Smith, Richard R
Smith, Warren Edward
Sorenson, Edgar Lavell
Southwick, Richard Arthur
Specht, James Eugene
Stafford, Roy Elmer
Stanford, Ernest Hall
Stansel, James Wilbert
Starling, James Lyne
Starling, Thomas Madison
Steele, Leon
Stefansson, Baldur Rosmund
Stith, Lee S
Strohm, Jerry Lee
Stroike, James Edward
Stucker, Robert Evan
Summers, Dennis Brian
Suzuki, Akio
Svejda, Felicitas Julia
Tai, Peter Yao-Po
Taliaferro, Charles M
Taylor, George Allan
Terrill, Thomas Robert
Thomas, James H
Thomas, Johnny Ray
Thomas, Paul Clarence
Thomas, Anson Ellis
Thompson, David J

Thompson, James Marion
Thompson, Norman Robert
Thorne, John Carl
Tigchelaar, Edward Clarence
Tomes, Dwight Travis
Topoleski, Leonard Daniel
Townley-Smith, Thomas Frederick
Trotter, Allen Richard
Trupp, Clyde Rufon
Van Elswyk, Marinus, Jr
Van Schaik, Peter Hendrik
Verhalen, Laval Mathias
Voigt, Paul Warren
Voigt, Robert Lee
Volland, Leonard Allan
Waddle, Bradford Avon
Walton, Peter Dawson
Wann, Elbert Van
Warner, Harlow Lester
Warner, John Northrup
Weaver, Gerald MacKnight
Weigle, Jack LeRoy
Weiss, Martin George
Wellhausen, Edwin John
Wells, Darrell Gibson
Wells, James Ralph
White, Thomas Galand
Widstrom, Neil Wayne
Wiggin, Henry Carvel
Wilfret, Gary Joe
Williams, Curtis
Williams, Tom Vare
Wilson, James Alexander
Wortman, Leo Sterling, Jr
Wynne, Johnny Calvin
Yermanos, Demetrios M
York, John Owen
Young, Donald Alcoe
Young, Evie Fountain, Jr
Yu, Albert Tzeng-Tyng
YungBluth, Thomas Alan
Zillinsky, Francis John
Zink, Frank W

Plant Cytology

Ahshapanek, Don Colesto
Atkinson, Lenette Rogers
Britten, Edward James
Burson, Byron Lynn
Chen, Chen Ho
Crang, Richard Earl
Dunford, Max Patterson
Fowke, Lawrence Carroll
Hickok, Leslie George
Israel, Herbert William
Kapoor, Brij M
Kasha, Kenneth John
Lorz, Albert Protus
McCollum, Gilbert Dewey, Jr
Morris, Mary Rosalind
Newcomb, Eldon Henry
Paddock, Elton Farnham
Partanen, Carl Richard
Rao, Raghavendra D
Sapra, Val T
Sarkar, Priyabrata
Sears, Ernest Robert
Smith, Benjamin Warfield
Stelter, David Albert
Ting, Yu-Chen
Tsuchiya, Takumi
Underbrink, Alan George
Van Der Woude, William Jan
Warmke, Harry Earl
Whitton, Leslie
Yu, Ming-Hung

Plant Ecology

Abrahamson, Warren Gene, II
Adams, Michael Studebaker
Arnold, Joseph Frederick
Bagley, Walter Thaine
Bard, Gily Epstein
Barton, James Don, Jr
Baskin, Jerry Mack
Beatley, Janice Carson
Bell, Katherine Lapsley
Bell, Marcus Arthur Money
Benninghoff, William Shiffer
Bernard, John Milford
Blair, Robert Marks
Bonham, Charles D
Boyce, Stephen Gaddy
Boyd, Claude Elson
Bozeman, John Russell
Branson, Farrel Allen
Bratton, Susan Power
Breeden, Johnnie Elbert
Brinson, Mark McClellan
Brotherson, Jack DeVon
Brown, Clair Alan
Brown, James Milton
Brown, Robert Thorson
Brown, Walter Howard
Buchanan, Hayle
Buck, Paul
Budowski, Gerardo

Burgess, Robert Lewis
Caldwell, Martyn Mathews
Catana, Anthony J, Jr
Catenhusen, John Alfons
Cavers, Paul Brethen
Cawley, Edward T
Chapman, Joe Alexander
Chilcote, William W
Christianson, John Dean
Clambey, Gary Kenneth
Clary, Warren Powell
Clebsch, Edward Ernst Cooper
Collins, Don Desmond
Cooper, George Raymond
Corbett, Gail Rushford
Cotter, David James
Coupland, Robert Thomas
Cribben, Larry Dean
Crockett, Jerry J
Crow, John H
Crowder, Adele A
Dale, Edward Everett, Jr
Damman, Antoni Willem Hermanus
Danks, Maureen Lee
Darrow, Robert Arthur
Daubenmire, Rexford
Dayton, Bruce R
DeSelm, Henry Rawie
Dina, Stephen James
Dodd, Jimmie Dale
Donahue, William H
Doyon, Dominique
Drew, William Brooks
Egler, Frank Edwin
Facey, Vera L
Flaccus, Edward
Fogg, George Garrett
Fonda, Richard Weston
Forman, Richard T T
Freeman, Charles Edward, Jr
Fritts, Harold Clark
Gebben, Alan Irwin
Gehris, Clarence Winfred
Glenn-Lewin, David Carl
Goder, Harold Arthur
Goetz, Harold
Good, Norma Frauendorf
Good, Ralph Edward
Gosselink, James G
Hadley, Elmer Burton
Hall, Frederick Columbus
Hanawalt, Ronald B
Hanes, Ted L
Haney, Alan William
Harcombe, Paul Albin
Harner, Richard Francis
Harthill, Marion Paul
Hartman, Richard Thomas
Heady, Harold Franklin
Heaslip, Margaret Barkley
Hermann, Richard Karl
Hesketh, J D
Hickman, James C
Hinde, Howard Parrish
Hodson, William Myron
Hoffman, George R
Houston, Walter Randolph
Hulbert, Lloyd Clair
Hull, Alvin C, Jr
Hull, James Clark
Hutcheson, Harvie Leon, Jr
Jackson, Marion T
Janke, Robert A
Jefferson, Carol Annette
Johnson, Albert W
Johnson, Michael Paul
Keammerer, Warren Roy
Keling, Ralph Walter
Kershaw, Kenneth Andrew
Kessell, Stephen Robert
Knight, Dennis Hal
Krajina, Vladimir Joseph
Kramer, Richard John
Kroh, Glenn Clinton
Kumler, Marion Lawrence
Lamp, Herbert F
Landers, Roger Q, Jr
Lang, Gerald Edward
Langenheim, Jean Harmon
Larsen, James Arthur
Launchbaugh, John L, Jr
Lavoie, Victorin
Laycock, William Anthony
Leatherman, Anna D
Levering, Dale Franklin, Jr
Lindauer, Ivo Eugene
Lindsay, Douglas Rome
Lindsey, Alton Anthony
Linn, Robert
Lipps, Emma Lewis
Lynch, Daniel Matthew
Mack, Richard Norton
Mahall, Bruce Elliott
Major, Jack
Martin, William Haywood, III
Maycock, Paul Frederick
Maysilles, James Howard
McGinnies, William Grovenor

McIntosh, Robert Patrick
McLeod, Kenneth William
McMillan, Calvin
McNaughton, Samuel J
McPherson, Donald Carman
McPherson, James King
Medve, Richard J
Mellinger, Clair
Mellinger, Michael Vance
Miller, Lee Norman
Mongelard, Joseph Cyril
Monk, Carl Douglas
Moore, John Robert
Moore, Russell Thomas
Morgan, Michael Dean
Mowbray, Thomas Bruce
Mueggler, Walter Frank
Mueller, Irene Marian
Mueller-Dombois, Dieter
Muller, Cornelius Herman
Muller, Robert Neil
Murdock, Joseph Richard
Neiland, Bonita
Newsome, Richard Duane
Niering, William Albert
Nighswonger, Paul Floyd
Nixon, Elray S
Owen, Herbert Elmer, Jr
Parsons, David Jerome
Patten, Duncan Theunissen
Patton, Ernest Gibbes
Peacock, John Talmer
Pearcy, Robert Woodwell
Pearson, Philip Richardson, Jr
Pelton, John Forrester
Penfound, William Theodore
Perino, Janice Vinyard
Pfister, Robert Dean
Phillips, Edwin Allen
Phipps, Richard L
Pittillo, Jack Daniel
Plummer, Gayther Lynn
Price, Keith Robinson
Ralston, Robert D
Raynal, Dudley Jones
Redmann, Robert Emanuel
Reed, Robert Marshall
Reid, Archibald
Rhoades, Richard Wesley
Robertson, Philip Alan
Rosenwinkel, Earl Richard
Rowe, John Stanley
Rumely, John Hamilton
Rundel, Philip Wilson
Sawyer, John Orvel, Jr
Schrock, Alta
Schultz, Arnold Max
Scott, Richard Walter
Selleck, George Wilbur
Shake, Roy Eugene
Shontz, John Paul
Shontz, Nancy Nickerson
Sigafoos, Robert Sumner
Slack, Nancy G
Smathers, Garrett Arthur
Smeins, Fred E
Smith, Alan Lyle
Smith, Alan Paul
Smith, David William
Solomon, Allen M
Spearing, Ann Marie
Stalter, Richard
Staniforth, Richard John
Stark, Nellie May
Stewart, Robert Archie, II
Swedberg, Kenneth C
Terwilliger, Charles, Jr
Thurman, Lloy Duane
Tinnin, Robert Owen
Tobiessen, Peter Laws
Tueller, Paul T
Ungar, Irwin A
VanAmburg, Gerald Leroy
Van Asdall, Willard
Van Dersal, William Richard
Viereck, Leslie A
Voigt, John Wilbur
Wagner, Richard H
Wali, Mohan Kishen
Ward, Richard Theodore
Ware, George Henry
Ware, Stewart Alexander
Waring, Richard H
Wein, Ross Wallace
Whitford, Philip Burton
Williams, George Jackson, III
Wistendahl, Warren Arthur
Witherspoon, John Pinkney, Jr
Yarranton, George Anthony
Zobel, Donald Bruce

Plant Embryology

Cass, David D
Chollet, Raymond
Christensen, Norman Leroy, Jr
Cobb, Glenn Wayne
Collins, Russell Lewis
Corson, George Edwin, Jr

Crotty, William Joseph
Dosier, Larry Waddell
Herr, John Mervin, Jr
Jordan, Edward Hill
Kuehnert, Charles Carroll
Maravolo, Nicholas Charles
Mia, Abdul Jabbar
Millington, William Frank
Prover, Corinne Bea
Raghavan, Valayamghat
Reynolds, John Dick
Smeltzer, Richard Homer
Stein, Otto Ludwig
Vanderhoef, Larry Neil
Webster, Barbara Donahue
Witmer, Carl Leslie, Jr

Plant Genetics

Acedo, Gregoria N
Allison, David C
Anderson, Robert Glenn
Arisumi, Toru
Barnett, Ronald David
Beadle, George Wells
Beckett, Jack Brown
Benepal, Parshotam S
Bliss, Fredrick Allen
Bottino, Paul James
Brawn, Robert Irwin
Brewbaker, James Lynn
Briggle, Leland Wilson
Buckner, Robert Cecil
Buss, Glenn Richard
Cameron, James Wagner
Childers, Walter Robert
Chomchalow, Narong
Chu, Yaw-En
Collins, Don Desmond
Collins, Glenn Burton
Comstock, Verne Edward
Conger, Bob Vernon
Cooper, Richard Lee
Craig, Richard
Craig, William F
Cuany, Robin Louis
Cunningham, Charles Everett
Curme, John Henry
Davis, Daniel Layten
Desborough, Sharon Lee
Devine, Thomas Edward
Doney, Devon Lyle
Fleming, Attie Anderson
Gabelman, Warren Henry
Gengenbach, Burle Gene
George, William Leo, Jr
Gorsic, Joseph
Gorz, Herman Jacob
Graham, William Doyce, Jr
Hall, Robert Lee
Hanna, Wayne William
Hanson, Warren Durward
Haskins, Francis Arthur
Haunold, Alfred
Heinz, Don J
Hernandez, Travis Paul
Holl, Frederick Brian
House, Leland Ralph
Hughes, Karen Woodbury
Janick, Jules
Jenkins, Johnie Norton
Johnson, Russell Tingey
Jones, Alfred
Jugenheimer, Robert William
Kilen, Thomas Clarence
Kleese, Roger Allen
Knott, Douglas Ronald
Kohel, Russell James
Lambert, Robert John
Latimer, Howard Leroy
Lehman, William Francis
Leng, Earl Reece
Lessman, Koert J
Liang, George H L
Lighty, Richard William
Linden, Duane B
McArthur, Eldon Durant
Myers, Oval, Jr
Nabi, Hosni Abdel
Nickell, Cecil D
Paulson, Ivan Wunder
Pavek, Joseph John
Plessers, Arthur Gerard
Poneleit, Charles Gustav
Purdy, James Lealon
Rachie, Kenneth Owen
Rajhathy, Tibor
Ramey, Harmon Hobson, Jr
Rasmusson, Donald C
Ray, Dale Allen
Robison, Earlene
Rupert, Earlene Atchison
Santamour, Frank Shalvey, Jr
Schulke, James Darrell
Shands, Henry Lee
Shannon, Michael Carlyle
Sheen, Shuh-Ji
Shumway, Lewis Kay
Sidhu, Bhag Singh

Smith, Edward Lee
Smith, Rex L
Solbrig, Otto Thomas
Squillace, Anthony Eugene
Stansel, James Wilbert
Stevens, Merwin Allen
Stuthman, Deon Dean
Taliaferro, Charles M
Theurer, Jessop Clair
Timothy, David Harry
Tipton, Kenneth Warren
Tomes, Dwight Travis
Townsend, Alden Miller
Turcotte, Edgar Lewis
Walden, David Burton
Wall, James Robert
Wallace, Donald Howard
Ward, David Justin
Washington, Willie James
Wilkes, Hilbert Garrison, Jr
Williams, John Caswell
Williams, Norman Dale
Wilson, Frank Douglas
Zoebisch, Oscar Cornelius

Plant Morphology

Abbott, Rose Marie Savelkoul
Banks, Harlan Parker
Beck, Charles Beverley
Bell, Max Ewart
Bennett, Herald Durward
Berg, Arthur R
Berg, Dwight Hillis
Bierhorst, David William
Biggs, Robert Hilton
Boole, John Allen, Jr
Burch, Charles
Canright, James Edward
Cox, Hiden Toy
deLanglade, Ronald Allan
Dickison, William Campbell
Duffy, Regina Maurice
Erspamer, Jack Laverne
Foster, Donald Bartley
Garrison, Rhoda
Giesy, Robert
Graf, Dolores Irma
Graham, Alan Keith
Gray, Lewis Richard
Greyson, Richard Irving
Hansen, Harold Westberg
Haskell, David Andrew
Henry, Robert David
Hewitson, Walter Milton
Hindman, Joseph Lee
Hostetter, Heber P, III
Hotchkiss, Arland Tillotson
Jacobs, William Paul
Jagels, Richard H
Johnson, Tillman Joseph
Kaplan, Donald Robert
Kavaljian, Leroy Gregory
Kelley, Alden Gerard
Kordan, Herbert Allen
Lommasson, Robert Curtis
Mallory, Thomas E
Martens, Jacob Louis
McGahan, Merritt Wilson
Meyer, Samuel Lewis
Miksche, Jerome Phillip
Morlang, Charles, Jr
Nickerson, Norton Hart
Posluszny, Usher
Prior, Paul Verdayne
Rahn, Joan Elma
Rappleye, Robert Du Bois
Riopel, James L
Rodin, Robert Joseph
Sattler, Rolf
Shutts, Clarence Francis
Siemer, Eugene Glen
Simone, Leo Daniel
Singh, Surendra Pratap
Smith, Bruce Barton
Spurr, Arthur Richard
Stevenson, Forrest Frederick
Swift, Lloyd Harrison
Tepfer, Sanford Samuel
Unger, James William
Vieth, Joachim
Webster, Terry R
Whittier, Dean Page
Wilson, Kenneth Allen
Wilson, Thomas Kendrick
Wyatt, Raymond L

Plant Nematology

Ayoub, Sadek M
Baldwin, James Gordon
Bell, Frank Heaton
Bergeson, Glenn Bernard
Cutler, Horace Garnett
Dickerson, Ottie J
Faulkner, Lindsey Ralph
Feldmesser, Julius
Freckman, Diana Wall
Griffin, Gerald D

BOTANY

Hamlen, Ronald Alan
Hart, Winfield Hiram
Hussey, Richard Sommers
Laughlin, Charles William
Lewis, Stephen Albert
Lownsbery, Benjamin Ferris
MacDonald, David Howard
Maggenti, Armand Richard
Malek, Richard Barry
Mercer, Edward King
Minton, Norman A
Norton, Don Carlos
Overman, Amegda Jack
Perry, Vernon G
Rhodes, Harlan Leon
Richard, David Alan
Roman, Jesse
Sandstedt, Robert Morris
Santo, Gerald Sunao
Schmidt, Donald Peter
Smart, Grover Cleveland, Jr
Smolik, James Darrell
Tarjan, Armen Charles
Thirugnanam, Muthuvelu

Plant Nutrition

Ambler, John Edward
Bennett, William Frederick
Bergman, Ernest L
Bhangoo, Mahendra Singh
Bishop, Robert Frederick
Blevins, Dale Glenn
Broyer, Theodore Clarence
Burdine, Howard William
Campbell, Joseph Dempsey
Cook, James Arthur
deMooy, Cornelis Jacobus
Embleton, Tom William
Epstein, Emanuel
Evans, Harold J
Foy, Charles Daley
Fried, Maurice
Gillingham, J T
Gupta, Umesh C
Heilman, Paul E
Jackson, William Addison
Jones, J Benton, Jr
Lee, Charles Richard
Lessman, Gary M
Lorenz, Oscar Anthony
Mason, John Leslie
McBeath, Douglas Kay
McNall, Lester R
Minotti, Peter Lee
Munns, Donald Neville
Munson, Robert Dean
Nelson, Paul Victor
Oertli, Johann Jakob
Olday, Frederick Combs
Ouellette, Gerard Joseph
Parks, Clyde Leonard
Powell, Richard Donald
Rains, Donald W
Rasmussen, Harry Paul
Romney, Evan M
Rosenau, William Allison
Schramm, Robert Johnson, Jr
Selders, Archie Arnold
Skogley, Earl O
Smith, Cyril Beverley
Tucker, Thomas Curtis
Usherwood, Noble Ransom
Varro, Stephen, Jr
Vianis, James
Voigt, Garth Kenneth
Volk, Richard James
Wallace, Arthur
Wallihan, Ellis Flower
Welch, Ross Maynard

Plant Physiology

Abdul-Baki, Aref Asad
Abdul-Ela, Mohamed Mohamed
Addicott, Fredrick Taylor
Ahlgren, George E
Akamine, Ernest Kisei
Albert, Luke Samuel
Alberte, Randall Sheldon
Albrigo, Leo Gene
Alder, Edwin Francis
Allan, John R
Allen, Paul James
Allen, Seward Ellery
Ammirato, Philip Vincent
Anderson, Robert Neils
Anderson, Irvin Charles
Anderson, James David
Ang, Jan Kee
Applegate, Howard George
Arditti, Joseph
Armstrong, Donald James
Arnison, Paul Grenville
Aronson, Jerome Melville
Asen, Sam
Ashton, Floyd Milton
Asselbergs, Edward Anton Maria

Awada, Minoru
Ayers, Alvin Dearing
Baer, Charles Henry
Baker, David Bruce
Baker, James Earl
Baker, John Bee
Ballantyne, David John
Bandel, Vernon Allan
Banting, James Daniel
Barber, John Threlfall
Barker, Allen Vaughan
Barker, William George
Barnett, James P
Barnett, Neal Mason
Barr, Richard Arthur
Barrier, George Edgar
Bartels, Paul George
Bartholomew, Duane P
Basham, Charles W
Basler, Eddie, Jr
Bass, Louis Nelson
Baum, Lawrence Stephen
Baur, Joseph Ralph
Bayer, David E
Bayer, Margret Helene Janssen
Bazzaz, Maarib Bakri
Beard, James B
Beasley, C A (Bud)
Beck, Gail Edwin
Becker, Veryl E
Becker, Wayne Marvin
Beevers, Leonard
Beinhart, Ernest George, Jr
Bell, Charles W
Benda, Gerd Thomas Alfred
Bendixen, Leo E
Bennett, Jesse Harland
Benson, Andrew Alm
Bertsch, Walter Frank
Bewley, John Derek
Beyer, Elmo Monroe, Jr
Bhattacharya, Pradeep Kumar
Bidwell, Roger Grafton Shelford
Bieber, Gene Lawrence
Biggins, John
Bing, Arthur
Bingham, Samuel Wayne
Birecka, Helena M
Bishop, Norman Ivan
Bissett, Orville R
Biswas, Prosanto K
Black, Robert Corl
Blakely, Lawrence Mace
Blaydes, David Fairchild
Blinks, Lawrence Rogers
Blumenfeld, David
Boggess, Samuel Forest
Bogorad, Lawrence
Boldue, Reginald J
Boll, William George
Bonde, Erik Kauffmann
Bonga, Jan Max
Bonner, Bruce Albert
Boole, John Allen, Jr
Borchert, Rolf
Boss, Manley Leon
Boutros, Osiris Wahba
Bowmer, Richard Glenn
Boyer, John Strickland
Brandt, Jerry Jay
Brandt, William Henry
Brecht, Patrick Ernest
Breidenbach, Rowland William
Brennan, James Alden
Brenner, Mark
Brewer, Howard Eugene
Briggs, Winslow Russell
Brown, Allan Harvey
Brown, James Wilson
Brown, John Charl
Brown, Lindsay Dietrich
Brown, Michael Mathison
Brown, Ronald Harold
Brown, Thomas Edward
Brueske, Charles H
Bryan, Herbert Harris
Buck, Paul Andrews
Budd, Thomas Wayne
Bukovac, Martin John
Buren, Lawrence Lamont
Burg, Stanley (Paul)
Burley, J William Atkinson
Burns, Joseph Charles
Burns, Robert Emmett
Burns, Russell MacBain
Burris, Joseph Stephen
Bush, Lowell Palmer
Bushnell, William Rodgers
Buttery, Brian Richard
Cailloux, Marcel Louis
Cain, John Carlton
Campbell, Carl Walter
Campbell, William Frank
Camper, Nyal Dwight
Cantliffe, Daniel James

Canvin, David T
Carns, Harry Robert
Carpenter, Bruce H
Carpenter, Will Dockery
Carroll, Arthur George
Carter, George Emmitt, Jr
Catlin, Peter Bostwick
Chancy, William R
Chang, Fa Yan
Chang, In-Kook
Chaplin, Michael H
Chasson, Robert Morton
Chen, Jane Lee
Chen, Shepley S
Chen, Tsong Meng
Cherry, Joe H
Chessin, Meyer
Chilcote, David Owen
Ching, Te May
Choe, Hyung Tae
Chrispeels, Maarten Jan
Christensen, Donald Robert
Christiansen, Meryl Naeve
Clark, Harold Eugene
Clark, Ralph B
Cleland, Charles Frederick
Cleland, Robert E
Cline, Morris George
Clum, Harold Haydn
Coggins, Charles William, Jr
Coleman, Eugene Alfred
Coleman, Robert E
Collins, Henry A
Collins, William Beck
Cooil, Bruce James
Cooke, Anson Richard
Cooper, Clee S
Cooper, William Cecil
Corbin, Frederick Thomas
Corcoran, Mary Ritzel
Cordes, William Charles
Cossins, Edwin Albert
Cotrufo, Cosimo (Gus)
Cotter, Donald James
Couch, Richard Wesley
Coulter, Murray Whitfield
Cowett, Everett R
Cowles, Joe Richard
Cox, Eugene Floyd
Craker, Lyle E
Cream, Joseph Gaylord
Cressman, Harry Keith
Crisp, Carl Eugene
Cuddy, Thomas Foster
Cumming, Bruce Gordon
Currier, Herbert Bashford
Curry, George Montgomery
Curtis, Henry L
Curtis, Otis Freeman, Jr
Cutler, Roy Walter
Cutler, Horace Garnett
Dainty, Jack
Danielson, Loran Leroy
Darrow, Robert Arthur
D'Aoust, Andre Lucien
Daugherty, Zoel W
Davies, Eric
Davies, Peter John
Davis, Daniel Layten
Davis, David G
Davis, Donald Eichard
Davis, Edwin Alden
Davis, Robert Foster, Jr
Dawson, Murray Drayton
Day, Boysie Eugene
Decker, Phares
De Hertogh, August Albert
Dekazos, Elias Demetrios
De La Fuente, Rollo K
Delmer, Deborah P
Demanche, Edna Louise
Dennis, Frank George, Jr
Dever, John E, Jr
Devlin, Robert Martin
Dickinson, David Budd
Dilley, David Ross
Dilley, Richard Alan
Dollwet, Helmat Hermann Adolf
Domanski, John Joseph, Jr
Donoho, Clive Wellington, Jr
Dostal, Herbert C
Dove, Lewis Dunbar
Duble, Richard Lee
Dugger, Willie Mack, Jr
Duke, Stephen Oscar
Dumroff, Erwin Bernard
Duncan, William Graham
Durham, Valgene Loren
Durham, James Ivey
Duysen, Murray E
Dyar, James Joseph
Dybing, Clifford Dean
Dycus, Augustus Mahon

Eaks, Irving Leslie
Ecklund, Paul Richard
Edgerton, Louis James
Edwards, Harold Herbert
Edwards, Kathryn Louise
Effer, W R
Ehlig, Carl F
Einert, Alfred Erwin
Eley, James H
Elkin, Lynne Osman
Ellis, Richard John
Elmore, Carroll Dennis
Erickson, Louis Carl
Ernest, Leland C
Eskew, David Lewis
Estermann, Eva Frances
Evans, David W
Evans, Lance Saylor
Everett, Marylee Sicklin
Ewing, Elmer Ellis
Eyster, Henry Clyde
Faris, Donald George
Farmer, Robert E, Jr
Faw, Wade Farris
Feinleib, Mary Ella (Harman)
Feltner, Kurt C
Feng, Kuo Ao
Fensom, David Strathern
Finn, James Crampton, Jr
Fiscus, Edwin Lawson
Fisher, Francis John Fulton
Fisher, John Edwin
Fletcher, John Samuel
Fletcher, Ronald Austin
Forsyth, Frank Russell
Forward, Dorothy Florence
Foutch, Harley Wayne
Fox, J Eugene
Foy, Chester Larrimore
Franklin, Albert Bernard
Fratianne, Douglas G
Freiberg, Samuel Robert
French, Charles Stacy
Frenkel, Albert W
Fritz, George John
Fry, Kenneth E
Fucik, John Edward
Funderburk, Henry Hanly, Jr
Funkhouser, Edward Allen
Gaffron, Hans
Galitz, Donald S
Gallup, Avery Housley
Galsky, Alan Gary
Galston, Arthur William
Gantt, Elisabeth
Gassman, Merrill Loren
Gassner, Edward
Geiger, Donald R
Gentile, Arthur Christopher
Gentner, Walter Andrew
George, John Ronald
Gibby, David Duane
Girton, Raymond Elwood
Goddard, David Rockwell
Goldsmith, Mary Helen Martin
Good, Norman Everett
Goodin, Joe Ray
Gordon, John C
Gorham, Paul Raymond
Gouin, Francis M
Govindjee
Gowing, Donald Proctor
Gramlich, James Vandle
Grant, Neil George
Green, Charles E, Jr
Greenblatt, Gerald A
Greene, Richard Wallace
Greene, Albert Godfrey, Jr
Greulach, Victor August
Griffith, Robert Bell
Grossenbacher, Karl Albert
Grunwald, Claus Hans
Guinn, Gene
Gussin, Arnold E S
Gusta, Lawrence V
Haard, Norman F
Haber, Alan Howard
Habermann, Helen Margaret
Hackett, Wesley P
Hagemann, Richard Harry
Hale, Maynard George
Hall, Chesley Barker
Hall, Timothy Couzens
Hall, Wayne Clark
Hallion, John McDonell
Hamilton, James Lewis
Hamilton, Robert Hillery, Jr
Hamm, Philip Curtis
Hammer, Karl Clemens
Hanan, Joe John
Hanson, John Bernard
Hare, Leonard N
Harper, James Eugene
Harris, Joseph Belknap
Harris, Robert Ernest
Harrison, Bertrand Fereday
Harrison, George Keithley

Harvey, Clark
Haun, Joseph Rhodes
Haviland, James Roger
Hawxby, Keith William
Haxo, Francis Theodore
Hayashi, Fumihiko
Heberlein, Gary T
Heck, Walter Webb
Heebner, Charles Frederick
Heeney, Harold Blair
Heichel, Gary Harold
Helgeson, John Paul
Hellmers, Henry
Henderlong, Paul Robert
Henderson, James Henry Meriwether
Hendrix, Donald Louis
Hendrix, John Edwin
Henry, Egbert Winston
Hensley, Jerry Ray
Henson, James Wesley
Herrett, Richard Allison
Hesketh, J D
Hess, Charles
Hess, Frederick Dana
Hewitt, Allan A
Hiatt, Andrew Jackson
Hill, Kenneth Lee
Hillman, William Sermolino
Hilton, Geoffrey
Hind, Geoffrey
Hiyama, Tetsuo
Hodges, Harry Franklin
Hodges, John Deavours
Hodges, Thomas Kent
Hodson, Robert Cleaves
Hoffmann, Otto Louis
Horton, Billy D
Hofstra, Gerald
Hogue, Eugene J
Holden, David Jerome
Holm, LeRoy George
Holm, Robert Eric
Holm-Hansen, Osmund
Holowinsky, Andrew Wolodymyr
Holsten, Richard David
Holt, Donald Alexander
Holton, Raymond William
Hopkins, Homer Thawley, Jr
Hopkins, William George
Hopper, Norman Wayne
Horsley, Stephen Braithwaite
Horst, Ralph Kenneth
Horton, Roger Francis
Hoshizaki, Takashi
Houghland, Geoffrey Van Clief
Housley, Thomas Lee
Howe, George Franklin
Howe, Kenneth Jesse
Howell, Golden Leon
Howell, Gordon Stanley, Jr
Howell, Robert Wayne
Hoyle, Merrill Cassius
Hruschka, Howard Wilbur
Hsiao, Theodore Ching-Teh
Huffaker, Ray C
Hull, Herbert Mitchell
Hull, Richard James
Hurtt, Woodland
Ikuma, Hiroshi
Inge, Frederick Douglass
Ingle, L. Morris
Inselberg, Edgar
Irvine, W M
Isenberg, James Estill
Isenberg, Francis Marion Roland
Isensee, Allan Robert
Islam, Nurul
Isleib, Donald Richard
Jacks, Thomas Jerome
Jackson, Earl Kenneth
Jackson, Johnny
Jackson, William Thomas
Jacobs, Mark
Jacobson, Jay Stanley
Jacobson, Louis
Jaffe, Mordecai J
Jagendorf, Andre Tridon
Janes, Byron Everett
Jeffers, Daniel L
Jeffreys, Donald Bearss
Jennings, Paul Harry
Jensen, Varon
Joham, Howard Ernest
Johanson, Lamar
Johnson, Gordon Verle
Johnson, Kenneth Duane
Johnston, Taylor Jimmie
Jonas, Herbert
Jones, Daniel David
Jones, Larry W
Jones, Leo Edward
Jones, Raymond F
Jones, Russell Lewis
Jordan, Lowell Stephen
Josephs, Melvin Jay
Joy, Kenneth Wilfred
Jung, Gerald Alvin

Jyung, Woon Heng
Kadkade, Prakash Gopal
Kamien, Ethel N
Karsten, Kenneth Stephen
Kasperbauer, Michael J
Katterman, Frank Reinald Hugh
Kaufman, Peter Bishop
Kaufmann, Merrill R
Kavanagh, Frederick
Kawase, Makoto
Keford, Noel Price
Keitt, George Wannamaker, Jr
Kendall, William Anderson
Kende, Hans Janos
Kennard, William Crawford
Kennedy, Robert Alan
Kerstetter, Rex E
Ketchie, Delmer O
Ketelapper, Hendrik Jan
Ketring, Darold L
Key, Joe Lynne
Khan, Anwar Ahmad
Khudairi, Abdul Karim
Kimmins, Warwick Charles
Kinbacher, Edward John
King, John
King, Lawrence J
Kirkham, Mary Beth
Kirkorian, Abraham D
Kirkpatrick, Hugh Charles
Kivilaan, Aleksander
Klein, Attila Otto
Klein, Richard M
Klein, William Herbert
Kleinkopf, Gale Eugene
Klepper, Elizabeth Lee (Betty)
Kliewer, Walter Mark
Klingensmith, Merle Joseph
Kochan, Walter J
Kohl, Daniel Howard
Kohl, Harry Charles, Jr
Kok, Bessel
Kollman, Gerald Eugene
Koontz, Harold Vivien
Kozel, Philip C
Kozlowski, Theodore Thomas
Kraft, Donald J
Kramer, Paul Jackson
Krauss, Beatrice Hilmer
Krauss, Robert Wallfar
Krieg, Daniel R
Krog, Norman Eiler
Ku, Han San
Kuchmak, Myron
Kulfinski, Frank Benjamin
Kurtz, Edwin Bernard, Jr
Labanauskas, Charles Kazys
Laber, Larry Jackson
LaCroix, Lucien Joseph
Lagerstedt, Harry Bert
Lakso, Alan Neil
Lambert, Roger Gayle
Lamotte, Clifford Elton
Landers, Kenneth Earl
Lane, Forrest Eugene
Lane, Harry Cleburne
Larson, Laurence Arthur
Lasheen, Aly M
Laties, George Glushanok
Lavender, Denis Peter
Lawrence, Robert Howard, Jr
Lawson, Verna Rebecca
Lee, Tsung Ting
Leese, Bernard M
Leff, Judith
Leggett, James Everett
Leinweber, Charles Lee
Leonard, Robert Thomas
Leopold, Aldo Carl
Levitt, Jacob
Lewis, Lowell N
Li, Edward Hsien-Chi
Li, Paul H
Lieberman, Morris
Liebhardt, William C
Lillehoj, Eivind B
Linck, Albert John
Lincoln, Richard G
Lipke, William G
Lippincott, Barbara Barnes
Lippincott, James Andrew
Lipton, Werner Jacob
List, Albert, Jr
Liverman, James Leslie
Lockard, Raymond G
Lockhart, James Arthur
Loe, Robert Wayne
Loercher, Lars
Loescher, Wayne Harold
Long, James Delbert
Long, Raymond Carl
Loomis, Robert Simpson
Looney, Norman E
Lopushinsky, William
Lords, James Lafayette
Loustalot, Arnaud Joseph
Lovelace, C James
Lowe, Richie Howard

Lowenhaupt, Benjamin
Lugo, Herminio Lugo
Lyman, Harvard
Lyons, James Martin
Maas, Eugene Vernon
Maass, Wolfgang Siegfried Gunther
Macdowall, Fergus D H
Mace, Kenneth Dean
Machia, Bollera Muddappa
Machlis, Leonard
MacLean, David Cameron
MacMasters, William Joseph
MacQuarrie, Ian Gregor
Madden, George D
Maier, Robert Hawthorne
Mancinelli, Alberto L
Mantai, Kenneth Edward
Manthey, John August
Marsh, Paul Bruce
Martin, Freddie Anthony
Mathes, Martin Charles
Maynard, Donald Nelson
Mayne, Berger C
McBee, George Gilbert
McBride, Landy James
McClendon, John Haddaway
McClure, George W, Jr
McCombs, Clarence Leslie
McCracken, Derek Albert
McCune, Delbert Charles
McGregor, William Henry Davis
McIlrath, Wayne Jackson
McLane, Stanley Rex, Jr
McLemore, Bobbie Frank
McLeod, Guy Collingwood
McMasters, Dennis Wayne
McNairn, Robert Blackwood
McNulty, Irving Bazil
McWhorter, Chester Gray
Mecklenburg, Roy Albert
Mederski, Henry John
Mehrlich, Ferdinand Paul
Mellor, Robert Sydney
Menser, Harry Alvin, Jr
Mertz, Dan
Meudt, Werner J
Meyer, Robert Earl
Michel, Burlyn Everett
Mielke, Eugene Albert
Miles, Charles Donald
Miles, Neil Wayne
Miller, Carlos Oakley
Miller, Charles Standish
Miller, Conrad Henry
Miller, John Henry
Miller, Knudt John
Mills, Howard Leonard
Mills, Ira Kelly
Minarik, Charles Edwin
Minshall, William Harold
Mitchell, Donald John
Mobley, Harold Morton
Moinat, Arthur David
Mongelard, Joseph Cyril
Monk, Ralph Warner
Moore, Glen
Moore, Raymond A
Moore, Thomas Carrol
Moreland, Donald Edwin
Morgan, Page Wesley
Morris, Justin Roy
Morris, Leonard Leslie
Morrison, Ralph M
Mortimer, Donald Charles
Morton, Howard LeRoy
Moss, Dale Nelson
Muir, Robert Mathew
Mulchi, Charles Lee
Muller, Walter Henry
Mullison, Wendell Roxby
Murphy, John Joseph
Murphy, Larry S
Muzik, Thomas J
Nabors, Murray Wayne
Nadler, Kenneth David
Naf, Ulrich
Nakata, Shigeru
Nance, James Francis
Naylor, Aubrey Willard
Naylor, Gerald Wayne
Nelson, Donald Carl
Nevins, Donald James
Newman, David William
Nichols, Kenneth E
Nickell, Louis G
Nielsen, Kenneth Fred
Nielsen, Peter Tryon
Nieman, Richard Hovey
Nitsos, Ronald Eugene
Nobel, Park S
Noble, Reginald Duston
Noggle, Glenn Ray
Nooden, Larry Donald
Norris, Logan Allen
Norris, Robert Francis
Norris, William Elmore, Jr

Northcraft, Richard Dunn
Oaks, B Ann
Obendorf, Ralph Louis
Ockerse, Ralph
Oelke, Ervin Albert
Oexemann, Stanley William
Ogg, Alex Grant, Jr
Ogren, William Lewis
Ohki, Kenneth
O'Kelley, Joseph Charles
Olah, Arthur Frank
O'Leary, James William
Olsen, Kenneth Laurence
O'Neal, Thomas Denny
Openshaw, Martin David
Orcutt, David Michael
Orgell, Wallace Herman
Ormrod, Douglas Padraic
Owens, Clarence Burgess
Owens, Lowell Davis
Ownby, James Donald
Ozbun, Jim L
Pack, Merrill Raymond
Paleg, Leslie G
Palevitz, Barry Allan
Pallas, James Edward, Jr
Paquin, Roger Joseph Alfred
Park, Roderic Bruce
Parker, Johnson
Parkhurst, David Frank
Pattee, Harold Edward
Patterson, Glenn Wayne
Patterson, Max E
Paul, Kamalendu Bikash
Pauli, Arland Walter
Pearce, Robert Brent
Peet, Mary Monnig
Percival, Frank William
Perley, James E
Peters, Gerald Alan
Peterson, David Maurice
Phan, Chon-Ton
Pharis, Richard Persons
Phatak, Sharad Chintaman
Phillips, Richard Lee
Pickard, Barbara Gillespie
Pickering, Ed Richard
Pickett, James M
Pierce, Wayne Stanley
Pike, Carl Stephen
Pillay, Dathathry Trichinopoly Natraj
Platt, Robert Swanton, Jr
Player, Mary Anne
Ploux, Marie Denise Madeleine
Pluenneke, Ricks Henry
Plumb, Timothy Roy, Jr
Pollard, Douglas Frederick William
Pollard, John K, Jr
Pollock, Bruce McFarland
Poole, Ronald John
Poovaiah, Bachettira Whappa
Porter, Walter Kenneth, Jr
Portz, Herbert Lester
Posner, Herbert Bernard
Powell, Loyd Earl, Jr
Powell, Robert Delafield
Pratt, Douglas Charles
Pratt, Harlan Kelley
Pratt, Lee Herbert
Pressey, Russell
Proctor, John Thomas Arthur
Purves, William Kirkwood
Quebedeaux, Bruno, Jr
Rabson, Robert
Racusen, Richard Harry
Radin, John William
Radwan, Mohamed Ahmed
Raese, John Thomas
Rappaport, Lawrence
Rasmussen, Gordon Keith
Rasmussen, Harry Paul
Rasmussen, Reinhold Albert
Rauch, Fred D
Rauser, Wilfried Ernest
Raved, Dan
Ray, Peter Martin
Rayle, David Lee
Reazin, George Harvey, Jr
Redei, Gyorgy Pal
Rediske, John Henry
Reeve, Eldrow
Rehwaldt, Charles A
Reid, Hay Bruce, Jr
Reid, Philip Dean
Rendig, Victor Vernon
Reynolds, John Horace
Rhykerd, Charles Loren
Ridley, Esther Joanne
Riekels, Jerald Wayne
Ries, Stanley K
Riggleman, James Dale
Riley, James Joseph
Rinne, Robert W
Ritenour, Gary Lee
Rizzo, Peter Jacob
Roark, Bruce (Archibald)
Robbins, William Jacob

Roberson, Ward Bryce
Roberts, Bruce R
Roberts, Clarence Richard
Roberts, David Wilfred Alan
Roberts, Donald Ray
Roberts, Lorin Watson
Roberts, Robert Michael
Robinson, Curtis
Rogers, Bruce Joseph
Rogers, James Monroe
Rogers, Robert Larry
Rogerson, Asa Benjamin
Rohrbough, Lawrence Milburn
Romani, Roger Joseph
Rosa, Nestor
Rosen, Walter George
Rosenthal, Gerald A
Ross, Cleon Walter
Rubatzky, Vincent E
Ruddat, Manfred
Ruelke, Otto Charles
Ruesink, Albert William
Runeckles, Victor Charles
Russell, George Keith
Ryder, John C, Jr
Rygg, George Leonard
Ryugo, Kay
Sacco, Paul
Sacher, Joseph Albert
Sachs, Roy M
Sadik, Sidki
Saettler, Alfred William
St John, Judith Brook
Salisbury, Frank Boyer
Samuels, George
Sandsted, Robert Morris
Sanford, Wallace Gordon
Sansing, Norman Glenn
Satter, Ruth
Saunders, James Allen
Schaedle, Michail
Scheibe, Joseph E
Schekel, Max Michael
Scherz, Joseph Francis
Schiefferstein, Robert Harold
Schiff, Jerome A
Schneider, Bernard Arnold
Schneider, Michael J
Scholz, Earl Walter
Schreiber, Marvin Mandel
Schroder, Vincent Nils
Schwartz, Martin
Schwarz, Otto John
Schweizer, Edward E
Scofield, Herbert Temple
Scott, Eion George
Scott, Peter Carlton
Scott, Tom Keck
Seibel, Prem P
Seibert, Michael
Sells, Gary Donnell
Seltmann, Heinz
Shadbolt, C Allan
Shafer, Thomas Howard
Shannon, Jack Corum
Shantz, Edgar Moore
Shaw, Warren Cleaton
Shear, Cornelius Barrett
Sheets, Thomas Jackson
Shen-Miller, Jane
Sheps, Lillian
Sheridan, Richard P
Shih, Ching-Yuan
Shimabukuro, Mary Abrahamsen
Shimabukuro, Richard Hideo
Shipp, Oliver Elmo
Shininger, Terry Lynn
Shortess, David Keen
Shrift, Alex
Shropshire, Walter, Jr
Shuel, Reginald William
Shugarman, Peter Melvin
Sij, John William
Simmovitch, David
Simpson, Graham Miller
Simpson, Marion Emma
Sims, Ernest Theodore, Jr
Singh, Bharat
Singh, Raghbir
Sirois, Jean Claude
Sisler, Edward C
Skelton, Bobby Joe
Skog, Folke
Skok, John
Slankis, Visvaldis
Sloger, Charles
Smith, Albert Ernest
Smith, Bruce Nephi
Smith, Don Wiley
Smith, Lawrence Hubert
Smith, Orrin Ernest
Smith, Richard Clark
Smith, Roberta Hawkins
Smittle, Doyle Allen
Snyder, Freeman Woodrow
Snyder, William Enoch
Sobota, Anthony E
Somers, George Fredrick, Jr

Sommer, Noel Frederick
Sorokin, Constantine Alexis
Spanswick, Roger Morgan
Sparks, Darrell
Spear, Irwin
Spearing, Ann Marie
Specht, James Eugene
Spector, Calvin
Spiers, James Monroe
Splitstoesser, Walter E
Spoerl, Edward Schnurr
Spomer, George Guy
Spomer, Louis Arthur
Staby, George Lester
Stadelmann, Eduard Joseph
Stafford, Helen Adele
Stallknecht, Gilbert Franklin
Standifer, Leonides Calmet, Jr
Stangel, Harvey J
Stanley, Ronald Alwin
Stauffer, John Frederick
Steffens, George Louis
Stehsel, Melvin Louis
Stein, Howard Jay
Steinhart, Carol Elder
Stenler, Alan James
Stephens, Henry LeRoy
Stepka, William
Steponkus, Peter Leo
Stern, Arthur Irving
Stern, Michele Suchard
Sterrett, John Paul
Stevenson, Enola L
Steward, Kerry Kalan
Stewart, Cecil R
Stewart, James McDonald
Stier, Howard Sheldon
Stiller, Mary Louise
Stocking, Clifford Ralph
Stoller, Edward Ira
Stone, Benjamin P
Stone, Edward Curry
Storey, James Benton
Stowe, Bruce Bernot
Strand, Oliver Eric
Streeter, John Gemmil
Stuckey, Irene Hawkins
Stueck, Guy Linsley
Stutte, Charles Auten
Sweet, Haven C
Sze, Heven
Taiz, Lincoln
Tanada, Takuma
Tao, Kar-Ling James
Taylor, Iain Edgar Park
Taylor, Richard Melvin
Taylorson, Raymond Brierly
Terborgh, John J
Terry, Norman
Thiesfeld, Virgil Arthur
Thimann, Kenneth Vivian
Thomas, Aubrey Stephen, Jr
Thompson, Neal Philip
Thornton, Robert Melvin
Thorpe, Trevor Alleyne
Throneberry, Glyn Ogle
Tieszen, Larry L
Ting, Irwin Peter
Tingey, David Thomas
Tinus, Richard Willard
Tjosten, John Leander
Tobin, Elaine Munsey
Tocher, Richard Dana
Todd, Glenn William
Togasaki, Robert K
Tompkins, Daniel Reuben
Torrey, John Gordon
Tregunna, E Bruce
Troyer, Robert Fulton
Troyer, James Richard
Truchelut, George Burnett
Truscott, Frederick Herbert
Tschabold, Edward Evert
Turel, Franziska Lili Margarete
Turnipseed, Glyn D
Turrell, Franklin Marion
Unrau, Abraham Martin
Upchurch, Robert Phillip
Urin, Kiyoto
Ursino, Donald Joseph
Vance, Benjamin Dwain
Vanden Born, William Henry
Vanderbeek, Leo Cornelis
Vanderhoef, Larry Neil
Van Norman, Richard Wayne
Van Overbeek, Johannes
Van Sambeek, Jerome William
Varner, Joseph Elmer
Verguin, Jacob

Vidaver, William Elliott
Vines, Herbert Max
Vitos, August John
Vogt, Albert R
Wadleigh, Cecil Herbert
Wagner, George Joseph
Waldrep, Thomas William
Wallace, Arthur
Wallner, Stephen John
Walther, Alina
Ward, Gordon Marshall
Wardell, William Lewis
Warner, Harlow Lester
Warner, Robert Lewis
Warren, Charles O, Jr
Warren, Richard Scott
Watada, Alley E
Waygood, Ernest Roy
Weaks, Thomas Elton
Weaver, Robert John
Webb, Burleigh C
Weber, John R
Weeks, Thomas F
Weete, John Donald
Wehunt, Ralph Lee
Weidner, Terry Mohr
Weinberger, Pearl
Weinstein, Leonard Harlan
Weissman, Gerard Selwyn
Welch, Ross Maynard
Welkie, George William
West, Sherlie Hill
Westwood, Melvin (Neil)
Wetherell, Donald Francis
Whatley, Booker Tilman
Wheaton, Thomas Adair
Whitaker, Ellis Hobart
White, Gordon Allan
White, William Calvin
Whitenberg, David Calvin
Whitney, John Barry, Jr
Widholm, Jack Milton
Wiebe, Herman Henry
Wiegand, Oscar Fernando
Wiggans, Samuel Claude
Wightman, Frank
Wilcox, Hugh Edward
Wiley, Lorraine
Wilkins, Harold
Wilkinson, Robert Eugene
Williamson, Ralph Edward
Williams, Gene R
Williams, Gerald Gordon
Williams, James Lovon, Jr
Williams, Max W
Williams, Miles Coburn
Williams, Robert Haworth
Williams, Stephen Edward
Willmot, Claude
Wilner, Jacob
Wilson, Alma McDonald
Wilson, Richard Hansel
Wilson, Shirley Lane
Wilson, William Curtis
Wiltbank, William Joseph
Winter, Steven Ray
Witham, Francis H
Wittenbach, Vernon Arie
Wochok, Zachary Stephen
Wohlpart, Alfred
Wolter, Karl Erich
Woodruff, Richard Earl
Woolley, Joseph Tarbet
Woolstock, Lowell Willard
Workman, Milton
Worsham, Arch Douglas
Wort, Dennis James
Wyse, Roger Earl
Yager, Robert Eugene
Yamaguchi, Masatoshi
Yang, Shang-Fa
Yaniv, Zohara
Yatsu, Lawrence Y
Yelenosky, George
Yocum, Conrad Schatte
Yoo, Bong Yul
Young, Dale W
Young, Roy E
Yu, Grace Wei-Chi Hu
Zalik, Saul
Zeevaart, Jan Adriaan Dingenis
Zeidin, Michael Hermen
Zelitch, Israel
Zimmerer, Robert P
Zimmerman, Don Charles
Zimmerman, Richard Hale
Zimmerman, Charles Edward
Zschelle, Frederick Paul, Jr

Pomology

Andersen, Emil Thorvald
Anderson, Jay LaMar
Bailey, Catherine Hayes
Blanpied, George David

Brown, Gerald Richard
Catlin, Peter Bostwick
Childs, William Henry
Claypool, Lawrence Leonard
Compton, Oliver Cecil
Crane, Julian Coburn
Cummins, James Nelson
Denby, Lyall Gordon
Dennis, Frank George, Jr
Dewey, Donald Henry
Einset, John
Ellving, Donald Carl
Faust, Miklos
Ferree, David C
Fisher, David Vince
Forshey, Chester Gene
Grierson, William
Griggs, William Holland
Haeseler, Carl W
Hartman, Fred Oscar
Hayden, Richard Amhurst
Hesse, Claron Owens
Hitz, Chester W
Kender, Walter John
Kester, Dale Emmert
Koo, Robert Chung Jen
Lakso, Alan Neil
Lamb, Robert Consay
Lilleland, Omund
Looney, Norman E
Lord, William John
Lott, Richard Vincent
MacDonald, David Howard
Martin, George C
Mielke, Eugene Albert
Miller, Sherwood Robert
Mitterling, Lloyd Alfred
Oberly, Gene Herman
Owen, Frank William
Powell, Loyd Earl, Jr
Ritter, Crum Marshall
Rom, Roy Curt
Ryugo, Kay
Savage, Earl Frederick
Schneider, George William
Smith, Carter Riley
Smock, Robert Mumford
Southwick, Franklin Wallburg
Stembridge, George Eugene
Storey, James Benton
Thompson, Arthur Howard
Titus, John S
Tukey, Ronald Bradford
Uriu, Kiyoto
Vaile, Joseph Edwin
Walker, David Rudger
Way, Roger Darlington
Westwood, Melvin (Neil)

Radiation Botany

Campbell, William Frank
Haber, Alan Howard
Moh, Carl Craig
Ursino, Donald Joseph

Systematic Botany

Alex, Jack Franklin
Altschul, Siri von Reis
Anderson, Dennis Elmo
Argus, George William
Averett, John E
Baad, Michael Francis
Bailey, Dana Kavanagh
Barclay, Arthur S
Barclay, Theodore Mitchell
Baranski, Michael Joseph
Barkley, William T
Bates, David Martin
Beaman, John Homer
Beard, Luther Stanford
Benson, Lyman David
Blasdell, Robert Ferris
Bobear, Jean B
Bradley, Ted Ray
Brashier, Clyde Kenneth
Broome, Carmen Rose
Brown, Richard McPike
Buchheim, Arno Fritz Gunther
Bunting, George Sydney, Jr
Burch, Derek George
Burk, Carl John
Cabrera, Angel Lulio
Carpenter, Irvin Watson, Jr
Carr, Gerald Dwayne
Castaner, David
Channell, Robert Bennie
Chapman, Carl Joseph
Chuang, Tsan Iang
Clausen, Robert Theodore
Clewell, Andre Franklin
Clovis, Jesse Franklin
Coffey, Janice Carlton
Cooperrider, Tom Smith
Cowan, Richard Sumner
Crawford, Daniel John
Cronquist, Arthur John
Crow, Garrett Eugene

Cruise, James E
Cutter, Lois Jotter
Davidse, Gerrit
Davidson, John Fraser
Davis, William S
DeFillipps, Robert Anthony
DeJong, Diederik Cornelis Dignus
Dempster, Lauramay Tinsley
Dent, Thomas Curtis
DeWolf, Gordon Parker, Jr
Drapalik, Donald Joseph
Dress, William John
Dudley, Theodore
Dugle, Janet Mary Rogge
Duman, Maximilian George
Durkee, LaVerne H
Dwyer, John Duncan
Ediger, Robert I
Ehrle, Elwood Bernhard
Eilers, Lawrence John
Eiser, Arthur L
Elias, Thomas S
Ellis, William Haynes
Emboden, William Allen, Jr
Esbhaugh, William Hardy
Essig, Frederick Burt
Estes, James Russell
Ezell, Wayland Lee
Faircloth, Wayne Reynolds
Fay, Marcus J
Freckmann, Robert W
Freeman, John Daniel
Freeman, Myron L
Fryxell, Paul Arnold
Fuller, Marian Jane
Furlow, John Jacob
Gastony, Gerald Joseph
Gentry, Alwyn Howard
Gillett, John Montague
Glassman, Sidney Frederick
Goldberg, Aaron
Goodman, George Jones
Gordon, Donald
Grable, Albert E
Harms, Vernon Lee
Harriman, Neil Arthur
Harrington, Harold David
Harris, Betty Wolf
Hauke, Richard Louis
Hayden, Mary Victoria
Haynes, Robert Ralph
Henrickson, James Solberg
Higgins, Larry Charles
Holmgren, Arthur Herman
Holmgren, Noel Herman
Holmgren, Patricia Kern
Holte, Karl E
Hsi, Eugene Yu-Tseng
Hutchison, Paul Clifford
Huttleston, Donald Grunert
Iltis, Hugh Hellmut
Ingram, John (William, Jr)
Isely, Duane
James, Charles William
James, Lois Elsie
Jensen, Richard Jorg
Johnson, Miles F
Johnson, Raymond Roy
Johnston, Marshall Conring
Jones, Almut Gitter
Keating, Richard Clark
Kiger, Robert William
Kirkbride, Joseph Harold, Jr
Klein, William McKinley, Jr
Koch, Rudy G
Koch, Stephen Douglas
Koelling, Alfred Cornell
Kowal, Robert Raymond
Koyama, Tetsuo
Kral, Robert
Kyhos, Donald William
Lang, Frank Alexander
Langdon, Kenneth R
Lathrop, Earl Wesley
Lawrence, George Hill Mathewson
Ledingham, George Filson
Legault, Albert
Lellinger, David Bruce
Lelong, Michel George
Lilly, Percy Lane
Lindsay, George Edmund
Lloyd, Robert Michael
Lonard, Robert (Irvin)
Longpre, Edwin Keith
Luteyn, James Leonard
Maguire, Bassett
Mahler, William Fred
Marroquin De La Fuente, Jorge Saul
Martin, William Clarence
Matuda, Eizi
McClintock, Elizabeth
Meijer, Willem
Mickel, John Thomas
Miller, Gertrude Nevada
Miller, Kim Irving
Milstead, Wayne Lavine

Mitchell, Richard Sheppard
Mohlenbrock, Robert H, Jr
Monson, Paul Herman
Moran, Reid Venable
Morley, Thomas
Murrell, James Thomas, Jr
Neher, Robert Trostle
Nevling, Lorin Ives, Jr
Ownbey, Gerald Bruce
Packard, Patricia Lois
Parks, James C
Payne, Willard William
Pfeifer, Howard William
Pickering, Jerry L
Pippen, Richard Wayne
Porter, Duncan Macnair
Preece, Sherman Joy, Jr
Ramsey, Gwynn W
Read, Robert William
Reveal, James L
Richards, Charles Davis
Richardson, Annie Louise
Roe, Keith Edward
Roland, Albert Edward
Rominger, James McDonald
Rouleau, Ernest
Sawyer, Paul Thompson
Schaeffer, Bernice Giduz
Schubert, Bernice Giduz
Schwab, Charlotte Ann
Settlemyer, Kenneth Theodore
Sieren, David Joseph
Skog, Laurence Edgar
Smith, Claude Earle, Jr
Smith, Dale Metz
Smith, Edwin Burnell
Smith, James Payne, Jr
Smith, Stanley Galen
Sohmer, Seymour H
Spellenberg, Richard (William)
Sperry, John Jerome
Spongberg, Stephen Alan
Stanford, Jack Wayne
Stern, Kingsley Rowland
Stewart, Ralph Randles
Steyermark, Julian Alfred
Stocking, Kenneth Morgan
Stolze, Robert Gardner
Stone, Margaret Hodgman
Stoutamire, Warren Petrie
Straw, Richard Myron
Stuessy, Tod Falor
Sutherland, David M
Taylor, Charles Arthur, Jr
Taylor, Raymond John
Taylor, Roy Lewis
Terrell, Edward Everett
Theobald, William L
Thomas, John Hunter
Thomas, Roy Dale
Turner, Billie Lee
Tyrl, Ronald Jay
Urbatsch, Lowell Edward
Van Faasen, Paul
Van Horn, Gene Stanley
Van Schaack, George Booth
Voss, Edward Groesbeck
Ward, Daniel Bertram
Ward, George Henry
Ware, Donna Marie Eggers
Wasshausen, Dieter Carl
Watson, James Ray, Jr
Weber, Wallace Rudolph
Webster, Grady Linder
Wedberg, Hale Levering
Weiler, John Henry, Jr
Welch, Stanley L
Wells, James Ray
Wheeler, Louis Cutter
Wilbur, Robert Lynch
Wilken, Dieter H
Williams, Kenneth Bock
Williams, Louis Otho
Windler, Donald Richard
Yates, Willard F, Jr
Zell, LaRoy W

CHEMISTRY

Chemistry
Ablard, James Elbert
Abraham, Bernard M
Abts, Mary Lavonne
Adams, Daniel Otis
Adcock, Willis Alfred
Adelson, Bernard Henry
Adelson, David E
Adinoff, Bernard
Adrian, Alan Patrick
Ager, John Winfrid, Jr
Agett, Albert Henry
Agnihotri, Ram K
Agre, Courtland LeVerne
Albert, Charles Gerald
Alberty, Robert Arnold

Albrecht, Robert H
Alexander, Allen Leander
Alexander, Guy B
Alexander, Kliem
Alfeis, Franz Juergen
Ali, Monica McCarthy
Alire, Richard Marvin
Allard, Romeo Paul
Allen, Donald Stewart
Allen, John Rybolt
Allen, Leland Cullen
Allen, Robert Edward
Alt, Leslie L
Alter, Abraham
Altman, David
Alvord, Donald C
Amberg, Carl Helmut
Ambrose, John Russell
Ambrosone, Joseph Paul
Amidon, Roger Welton
Amy, Jonathan Weekes
Anderson, Carl Martin
Anderson, Clyde Lee
Anderson, James G
Andrews, Lawrence James
Andrews, Oliver Augustus
Apt, Charles Maurice
Argabright, Perry A
Armistead, William Houston, Jr
Armstrong, Robert Thexton
Arnold, George Benjamin
Arnold, Harold Wilfred
Arnold, James Richard
Arthur, Paul, Jr
Asenjo, Conrado Federico
Asperger, Robert George
Atwood, Francis Clarke
Auerbach, Clemens
Augustine (Small), Marie
Auspos, Lawrence Arthur
Austin, Janet Evans
Avera, Fitzhugh Lee
Ayroud, Abdul-Mejid
Bachman, Gustave Bryant
Bacon, Charles Vincent
Bacon, Egbert King
Badgley, Wilfrid John
Baechler, Roy Herman
Baetz, Albert L
Bahl, Om Parkash
Bain, Ralph Lee
Bakan, Joseph A
Baker, Philip Schaffner
Baker, Richard Dean
Baldwin, Arthur Richard
Baldwin, Paul Clay
Balgley, Ely
Balis, Earl William
Balke, Claire Coddington
Ball, Donald Lee
Balnforth, Dennis
Bard, John William
Barnard, Alfred James, Jr
Barnum, Dennis W
Barr, Donald Eugene
Barrett, Fred Oliver
Bartholomew, Eleanor Rachel
Bartleson, John David
Bartlett, James Kenneth
Bartlett, Paul Doughty
Barton, James Clyde
Barusch, Maurice R
Bass, Shailer Linwood
Batchelder, Alan Coleman
Battista, Orlando Aloysius
Baulknight, Charles Wesley
Bauman, William Carrel
Baumgartner, Frederick Neil
Bavley, Abraham
Baxman, Horace Roy
Baxter, John Franklin
Bayless, Philip Leighton
Bazinet, Maurice L
Beach, George Winchester
Beach, Leland Kenneth
Beacham, Lowrie Miller, Jr
Beadle, Buell Wesley
Beattie, Robert Walter
Beaumont, Ralph Harrison, Jr

Bechtol, Lavon Dee
Beck, Lloyd Willard
Becker, Ernest I
Becker, Joseph F
Beckman, Arnold Orville
Bedenbaugh, Angela Lea Owen
Bedford, Joel S
Beech, John Alan
Beesch, Samuel C
Behr, Inga
Behrmann, Eleanor Mitts
Bender, Myron Lee
Berch, Julian
Beren, Sheldon Kuciel
Berg, John Robert
Berk, Abraham Albert
Berk, Bernard
Berkman, Michael G

Berkoff, Robert Bernard
Bernal, Ivan
Bernstein, Isidor Mayer
Berry, Clark Green
Berry, George William
Berry, Louis Milton
Betten, Cornelius, Jr
Biel, John Hans
Biester, John Louis
Bigelow, Charles C
Billica, Harry Robert
Billmeyer, Fred Wallace, Jr
Bills, John Lawrence
Bilow, Norman
Biondi, Frank Joseph
Birchenall, Charles Ernest
Birkett, Frank Elliot
Bissey, Luther Trauger
Biswas, Shib D
Bither, Tom Allen, Jr
Bixler, Harris Jacob
Bjorksten, Johan Augustus
Black, Simon
Blagg, John Creighton Lee
Blair, Grace
Blair, James Stuart
Bletzinger, John Calvin
Bliss, Laura
Blitch, Lee Wesley
Blodgett, Robert Bell
Bloor, John E
Bobbitt, James McCue
Bodamer, George Willoughby
Boggs, James Ernest
Boggs, Lawrence Allen
Bohon, Robert Lynn
Bohrer, John Junior
Bond, Bernard Batson
Bond, Howard Wissler
Bond, Robert Wallace
Bonk, James F
Bonner, Francis Truesdale
Bonner, Lyman Gaylord
Bonner, Norman Andrew
Bonner, William Andrew
Borders, Alvin Marshall
Bornong, Bernard John
Borovsky, Dov
Borovsky, Harry Herbert
Borr, Mitchell
Borrowman, S Ralph
Botros, Raouf
Bough, Wayne Arnold
Boundy, Ray Harold
Bournique, Raymond August
Bowen, Charles Verne
Bowen, David Hywel Michael
Bowerman, Ernest William
Bowman, Wilfred William
Bowne, Samuel Winter, Jr
Boyd, Charles Alexander
Boyd, John Mann
Bradley, Harris Walton
Brancone, Louis Maria
Brand, Benson Glenn
Brase, Peter Charles
Brauman, John I
Braun, Winfred Quentin
Brehm, Warren John
Bremner, John McColl
Breuer, Charles B
Brew, William Barnard
Bricker, Glenn A, Jr
Bricker, Clark Eugene
Briggs, Ben Thoburn
Bright, Gordon Stanley
Brill, Harold Clifford
Brinckerhoff, Harold Guion
Broadbent, Hyrum Smith
Brodie, Harry Joseph
Brody, Bernard B
Broege, Charles Burton
Broene, Herman Henry
Brosi, Albert Ralph
Brouns, Richard John
Brown, Eric Reeder
Brown, George Lincoln
Brown, Glenn Halstead
Brown, Harmon W, Jr
Brown, Henry
Brown, John Stewart
Brown, Leon Joseph
Brown, Phyllis R
Brown, Robert Walter
Brown, Weldon Grant
Bruce, Francis Robert
Bruening, George Emil
Bruner, Leonard Bretz, Jr
Bryan, Johannes Hadeln
Bryan, Carl Eddington
Bsharah, Lewis
Buchanan, Marion Alexander
Buckner, Ernest Jack
Bulbrook, Harry Marshall
Bump, Charles Kilbourne
Burcik, Emil Joseph
Burdick, Charles Lalor
Burdick, Everette Marshall

CHEMISTRY

Burns, George
Burt, William Enos
Butler, Keith Huestis
Butler, Thomas Arthur
Cabeen, Samuel Kirkland
Cagle, Fred William, Jr
Cahoon, Nelson Corey
Cairncross, Stanley Everett
Calandra, Alexander
Calkins, Mary Vincent
Callahan, Charles Richard
Calvert, Lauriston Derwent
Campbell, Clement, Jr
Campbell, Hallock Cowles
Cannon, Melvin Croxall
Capps, Raymond Haul
Carlson, Kenneth Theodore
Carpenter, Charles
Carpino, Louis Albert
Carr, Charles Jelleff
Carr, Donald Eaton
Carr, Duane Tucker
Carrico, James Leon
Carritt, Dayton Ernest
Carroll, Burt Haring
Carter, Albert Smith
Carter, Herbert Edmund
Carter, Irving Doyle
Cartledge, Groves Howard
Casali, Liberty
Casey, James Patrick
Caso, Marguerite Miriam
Cassaretto, Frank Philip
Cassidy, Patrick Edward
Castle, Raymond Nielson
Cauwenberg, Winfred Joseph
Chaikin, Saul William
Chamelin, Isidor Marie
Chan, Philip C
Chan, Martin
Chao, Tyng Tsair
Chap, James John
Charen, George
Charlton, David Berry
Chase, Fred Leroy
Chaubal, Madhukar Gajanan
Chen, Stephen P K
Cherayil, George Devassia
Chesnut, Clarence, Jr
Chiang, Schumann
Chinn, Clarence Edward
Chodroff, Saul
Choguill, Harold Samuel
Christensen, Roger Morris
Christian, Robert Vernon, Jr
Christman, Russell Fabrique
Chun, Edward Hing Loy
Clark, Arthur Randolph
Clark, Howard Selby
Clark, Marion Thomas
Clark, Ralph O
Clark, Raymond George
Clark, Richard Bennett
Clark, Walter
Clarke, Duane Grookett
Claus, Wilbur Scheirich
Clayton, James Oliver
Clayton, William Joseph
Cleary, Laurence Twomey
Cleek, George Kime
Cleek, Given Wood
Clegg, Lawrence Frank Levey
Clemens, Carl Frederick
Coutier, Louis
Clydesdale, Fergus Macdonald
Cobb, Thomas Berry
Coe, Kenneth Loren
Coe, Richard Hanson
Coffey, Joseph Francis
Coffman, Harold H
Cohen, Howard Joseph
Cohn, Johann Gunther Ernst
Colby, Frank Gerhardt
Cole, James Webb, Jr
Cole, John Oliver
Cole, Quintin Perry
Coll, Hans
Colton, Frank Benjamin
Colingsworth, Donald Rudolph
Compton, Charles (Daniel)
Conant, James Bryant
Condon, Francis Edward
Conlon, Daniel Rupert
Connor, Ralph (Alexander)
Conrad, Louis Johnson
Conroy, James Strickler
Cook, Alan
Cook, Lawrence Harvey
Cook, Shirl Eldon
Cook, Theodore Warren
Cooke, Lloyd Miller
Cool, Raymond Dean
Coppens, Philip
Coppoc, William Joseph
Corey, Eugene R
Corwin, James Fay
Cotton, Frank Albert

Cowles, Edward J
Cox, Edwin
Coyner, Eugene Casper
Craft, Willard Leahman, Jr
Craig, Raymond Allen
Crandall, John Lou
Craven, Charles Waller
Craven, John Kenneth
Crecelius, Robert Lee
Crentz, William Luther
Crist, Ray Henry
Cristol, Stanley Jerome
Crittenden, Alden La Rue
Croft, Wilma Janice
Crog, Richard Stanley
Crosby, Paul Faljean
Croxall, Willard (Joseph)
Croxall, James Chester
Crutcher, Harold L
Cruz, Mamerto Manahan, Jr
Cudd, Herschel Herbert
Cunningham, Frank W
Curme, George Oliver, Jr
Custer, Michael
Cutforth, Howard Glen
Cutler, Frank Allen, Jr
Cutler, Louise Marie
Dabrowiak, James Chester
Daggett, Albert Frederick
Daigle, Donald J
Dailey, Benjamin Peter
Dalton, Robert Hennah
Danuisis, Adolfas
Dance, Eldred Leroy
Danti, August Gabriel
Danzig, Meyer Hillel
Darling, Samuel Mills
Dasgupta, Sunil Priya
Dasher, Paul James
Davis, James K
Davis, Ralph Anderson
Davis, Raymond, Jr
Davis, Sherman Gilbert
Davis, Wallace, Jr
Dawe, Harold Joseph
Day, Frank, Jr
Deal, Glenn W, Jr
Dean, John Gilbert
Dean, Robert Reed
Dearing, Le Roy Matthew
de Bethune, Andre Jacques
Deck, Joseph Francis
Deischer, Claude Knauss
De La Sierra, Angell Ortiz
De Loach, Will Scott
Dell, Manuel Benjamin
deMonsabert, Winston Russel
Denckas, Milton Oliver
Denison, George Haigh
Denison, Ruth Corbet
deRosset, Armand John
Dessauer, John Hans
Detrick, Robert Sherman
Deutsch, Harold Francis
De Vries, John Edward
Dewar, Michael James Steuart
DeWitt, Bernard James
Di Carlo, Frederick Joseph
Dible, William Trotter, Jr
Dick, William Edwin, Jr
Dickison, Walter Lee
Dietemann, Allan B
Di Gangi, Frank Edward
Dille, Roger McCormick
Dills, William Leonard
Dinerstein, Robert Alvin
Dingell, James V
Dinneen, Gerald Uel
Dirkse, Thedford Preston
Di Salvo, Francis Joseph
Dittmer, Karl
Dixon, Henry Philip
Doderer, George Charles
Dodge, Durward F
Dodgen, Frank Eugene
Dolian, Frank Eugene
Dollear, Frank Gilbert
Donahue, Joseph E
Donaldson, Raymond Edwin
Dorsky, Julian
Dougherty, Patrick Henry
Dove, Ray Allen
Downs, Martin Luther
Draley, Joseph Edward
Dresdner, Richard David
Drickamer, Harry George
Drukker, Alexander Emmanuel
Dube, Harvey Albert
Duff, Robert Hodge
Duffield, Robert Brokaw
Duggan, Helen Ann
Dunlop, Andrew P
Dunlop, Edward Clarence
Dunning, Henry Armitt Brown, Jr
DuPuis, Robert Newell
Durno, William Henry
Durrill, Preston Lee

Dutton, Frederic Booth
Dutton, Herbert Jasper
Eaker, Charles Mayfield
Earle, Fontaine Richard
Eastes, John Wesley
Eby, Denise
Eckfeldt, Edgar Lawrence
Edgell, Walter Francis
Edgerton, Robert Flint
Edmonds, Sylvan Milton
Edwards, Frank C
Edwards, George
Eigen, Edward
Einich, Frederick Roland
Eiszner, James Richard
Eiger, Gerald William
Eliot, Robert S
Elkin, Eugene Mitchell
Elkins, Robert Hiatt
Ellis, Emory Leon
Elizey, Marion Lawrence, Jr
Elowe, Louis N
Engle, Robert Fry, Jr
Englert, Mary Elizabeth
English, Spofford Grady
Erickson, Wallace Alfred
Ernsdorff, Bede (Paul)
Escue, Richard Byrd, Jr
Estes, Reedus Ray
Etzler, Dorr Homer
Evans, Latimer Richard
Ewan, Maurice Albertson
Ewell, Raymond (Henry)
Failey, Crawford Fairbanks
Farihall, Arthur William
Fancher, Otis Earl
Farnholt, Larkin Hundley
Farnsworth, Marie
Faulkner, Larry Ray
Fearns, Edward Cranshaw
Fearon, Frederick William Gordon
Feldman, Isaac
Felger, Maurice Monroe
Fellmann, Robert Paul
Fenn, John Bennett
Fenstermaker, Roger William
Ferguson, Lloyd Noel
Fernelius, Willis Conard
Ferris, James Peter
Fessenden, Ralph James
Ficher, Miguel
Fiess, Harold Alvin
Field, Lamar
Filachione, Edward Mario
Filson, Malcolm Harold
Finamore, Frank Joseph
Finholt, Albert Edward
Finney, Karl Frederick
Fischbach, Henry
Fisher, Albert Madden
Fisher, Paul John
Fitch, Howard Montgomery
Fitzhugh, Andrew Fyfe
Fitzpatrick, John Thomas
Fitzwater, Robert N
Fleetwood, Charles Wesley
Fleisher, Gerard Adalbert
Flexser, Leo Aaron
Flint, Einar Philip
Fochr, Edward Gotthard
Folkers, Karl August
Follows, Alan Greaves
Forchielli, Americo Lewis
Ford, George Pratt
Ford, Milton David
Ford, Thomas Aven
Fowkes, Frederick Mayhew
Fowler, Emil Eugene
Fowler, Robert Dudley
Fowler, William Frank, Jr
Fox, Thomas G, Jr
Fraenkel, Gideon
Frahm, Elmer Edward
Frampton, Vernon Lachenous
Francel, Josef
Frankel, Richard Barry
Frazer, Jack Winfield
Freeark, Clayton Wayne
Freedman, Arthur Jacob
Freimuth, Henry Charles
French, Charles Leroy
French, Ludo Karl
Frevel, Kurt Charles
Frey, Delton Ruben
Frey, William Carl
Fried, John H
Fried, John B
Friedenstein, Hanna
Friedland, Waldo Charles
Friedman, Louis David
Frisch, Kurt Charles
Fristrom, Robert Maurice
Frost, Jackie Gene
Frost, Thomas Rogers
Fryburg, George Crumback
Fuller, John Burt
Fullhart, Lawrence, Jr

Fulmor, William
Furby, Neal Washburn
Fyrelson, Milton
Gage, Frederick Worthington
Gagewski, Fred John
Gallen, John Bryant
Gans, David Manus
Gans, Eugene Howard
Garascia, Richard Joseph
Gardner, Marjorie Hyer
Garey, Carroll Laverne
Garfield, Eugene
Garfield, Fred McKee
Garland, John Kenneth
Garner, Robert Henry
Garretson, Harold H
Garrett, Alfred Benjamin
Garst, Arthur Wilhelm
Garst, Raymond Daniel
Gavin, Dominique
Gawley, Irwin H, Jr
Gayler, Karl Herman
Geen, Henry Cory
Gehrke, Charles William
Gelberg, Alan
Gelman, Charles
George, Philip Donald
Gerecht, John Fred
Germann, Leo (Joseph Frederic Marcel)
Germann, Donald Pitt
Gero, Alexander
Gerritsen, Hendrik Jurjen
Gershon, Sol D
Gerstein, Melvin
Gerwe, Raymond Daniel
Gibbins, Betty Jane
Gibian, Thomas George
Gibson, George Herman
Gilbert, Francis Evalo
Gilbert, Richard Lapham, Jr
Gilkey, John Woodbury
Gilreath, Esmarch Senn
Gin, Jerry Ben
Gingold, Kurt
Gladding, Elinor Hartnell
Gladstone, Matthew Theodore
Glasoe, Paul Kirkwold
Gledhill, Ronald James
Glegg, Ronald Edward
Glidden, Kenneth Eugene
Gloege, George Herman
Goeldheim, Samuel Lewis
Goeldheim, Jerome
Goldin, Abraham Samuel
Goldkamp, Arthur Harvey
Goldman, Norman L
Goldthwait, Charles Francis
Goldberg, Albert Isaac
Goldblatt, Irwin Leonard
Golubovic, Aleksandar
Gomez-Ibanez, Jose Daniel
Goodloe, Paul Miller, II
Goodstein, Madeline P
Goran, Morris
Gordon, Neil E, Jr
Gordon, Samuel Morris
Gorin, Philip Albert James
Gofman, John William
Goff, Stillman R
Goering, Kenneth Justin
Grafstein, Daniel
Graham, Boynton
Granito, Charles Edward
Grant, Willard H
Gray, Kenneth Russell
Gray, Floyd Wilson
Green, Frank Orville
Green, Harry Edward
Greene, Paul E
Greider, Harold William
Grenke, Everett D
Greubel, Paul William
Greze, John Paul, Jr
Griffing, John Malcolm
Griffioen, Roger Duane
Gring, John Lukins
Groot, Cornelius
Gross, Ma'colm Edmund
Grosse, Aristid Victor
Grosz, Oliver
Gruen, Fred Martin
Grundmann, Christoph Johann
Gryting, Harold Julian
Guenther, Arthur Henry
Guenther, Frederick Oliver
Guild, Lloyd V
Gunther, Francis Alan
Gurchot, Charles
Gustafson, Donald Arvid
Gustnson, Walter Sigmund
Haake, Paul C
Haberecht, Rolf Reinhold
Habib, Emile Edward

Hackerman, Norman
Hadley, Elbert Hamilton
Haenisch, Edward Lauth
Hahn, Richard Balser
Haines, Florence Catherine
Haines, George Shuler
Haist, Grant Milford
Hakala, Neil Victor
Hale, Cecil Harrison
Hall, James Roger
Hall, Joseph Alfred
Hall, William Heinlen
Halmbacher, Paul
Halperin, Benjamin David
Halsey, John Joseph
Halstead, Bruce W
Halverson, Frederick
Halvorson, Ardell David
Halwer, Murray
Ham, Joe Strother
Hamer, Jan
Hamill, James Junior
Hamilton, Charles William
Hamlin, Kenneth Eldred, Jr
Hammett, Louis Plack
Hammond, James W
Hampel, Clifford Allen
Hanford, William Edward
Hanig, Ruth Belle Cohn
Hanks, Richard Donald
Hanna, Norman Edwin
Hannegan, John Michael
Hansch, Corwin Herman
Hansen, Harold Louis
Hansen, Waldemar Conrad
Hansen, Wilford Nels
Hanson, Milton Paul
Hanson, Robert W
Harding-Barlow, Ingeborg
Hardike, Fred Charles, Jr
Hardy, Paul Wilson
Haring, Robert Clinton
Harkins, Thomas Regis
Harpp, David Noble
Harren, Richard Edward
Harris, Warren Whitman
Harrison, Edward Merle
Harrison, Stuart Amos
Harrold, Gordon Coleson
Harter, Dana Eugene
Hartsuch, Paul Jackson
Hartwell, Jonathan Lutton
Harvey, Clarence Charles, (Jr)
Harvey, Douglas G
Haskins, Edna Ferrell
Hassell, Clinton Alton
Hatcher, John Burton
Haug, Arthur John
Hausenbuiller, Robert Lee
Hay, George William
Hayes, David Wayne
Hayes, John Robert
Haymaker, Willis Stuart
Haynes, Willis Stuart
Hazel, James Frederic
Head, Ronald Alan
Heald, Alfred Mattson
Heasley, Gene
Hecht, Sidney Michael
Hefferren, John James
Hehir, Robert M
Heiberger, Charles Adam
Hein, Richard Earl
Heines, Virginia
Heiple, Harold Rhine
Heller, Hugh Andrews
Heller, Wilfried
Hellman, Henry Martin
Hellman, Nison Norman
Helmholz, Lindsay
Helminiak, Thaddeus Edmund
Hemley, John Julian
Hendrickson, William George
Hennessy, Douglas John
Henry, Jack Leland
Henry, Ronald Andrew
Hensler, Joseph Raymond
Herdklotz, John Key
Herrick, Clifford Ernest, Jr
Herrington, Kermit (Dale)
Herron, James Dudley
Hersberger, Arthur Bucher
Hersh, Sylvan David
Hershenson, Herbert Malcolm
Herzog, Hershel Leon
Heyd, Josef William
Hibbs, Robert A
Hibbs, Roger Franklin
Hieserman, Clarence Edward
Higgins, Norton Allen
Hill, George Richard
Hill, Julian Werner
Hill, Loren Wallace
Hilliard, Roy C
Hilton, Clifford L
Hine, Jack
Hinkamp, James Benjamin

Hinkel, Robert Dale
Hirota, Noboru
Hirschfelder, Joseph Oakland
Hirtle, Donald Stephen
Hobbs, Robert Boyd
Hochstrasser, Robin
Hodge, Edward Butler
Hodgkiss, William Searles
Hodgson, Roy S
Hoegberg, Erick Ingvar
Hoekstra, Henry Raymond
Hoffmann, Gilbert Frederick
Hofrichter, Charles Henry
Hogg, Alan Mitchell
Hoglund, Paul Franklin
Holden, George Wilfrid
Holder, Charles Burt, Jr
Hollander, Max Leo
Holler, Albert Cochran
Hollister, John Hendricks
Holmes, Joseph Charles
Holmes, Owen Gordon
Holt, Wendell Levern
Holtz, David
Holtzer, Alfred Melvin
Holub, Fred F
Holzer, Walter Frank
Homeyer, August Henry
Hooper, Donald Lloyd
Horn, Christian Friedrich
Horney, Amos Grant
Hornig, Donald Frederick
Hossain, Shafi Ul
Hotten, Bruce Walter
Houghton, Augustus Sherrill
Houtz, Ray Clyde
Hovis, Louis Samuel
Howard, Edgar, Jr
Howard, Hartley Wolle
Howard, John Charles
Howsmon, John Arthur
Hrutfiord, Bjorn F
Huang, Suei-Rong
Huber, Calvin
Huffman, Hal Charles
Huggett, Clayton (McKenna)
Hughes, Edward Wesley
Hughes, Edwin R
Hughes, Hansel Leigh
Hughes, James Perry
Hume, David Newton
Hunt, Gilbert John
Hunt, Richard Henry
Hunt, Roy Edward
Hunter, James Bruce
Hunter, Norman W
Hurlbert, Bernard Stuart
Hussey, Allen Sanborn
Hydock, Joseph J
Hyne, James Bissett
Hyson, Archibald Miller
Ichniowski, Thaddeus Casimir
Idson, Bernard
Ihde, Aaron John
Ikawa, Miyoshi
Ish, Carl Jackson
Ivashkiv, Eugene
Ives, David Southwick
Iveson, Herbert Todd
Izzo, Patrick Thomas
Jacko, Michael George
Jackson, Elizabeth Burger
Jackson, George Richard
Jackson, Wendell Ford
Jackson, William Gordon
Jacober, William John
Jacobson, Donald Weldon
Jacobson, Samuel
James, Douglas Garfield Limbrey
Janota, Rudolph Benjamin
Jason, Emil Fred
Jayne, Jack Edgar
Jeanes, Jack Kenneth
Jeitschko, Wolfgang K
Jellinek, Maurice H
Jennings, Harley Young, Jr
Jensen, James Le Roy
Jensen, Otto Gerhard
Jensen, William Phelps
Jewsbury, Wilbur
Joffe, Morris H
Joffre, Stephen Paul
Johnson, Arthur Lee
Johnson, Elmer Roger
Johnson, George Chrysler
Johnson, George Dana
Johnson, Hollister, Jr
Johnson, Howard Claus Edmund
Johnson, James Donald
Johnson, James Leslie

Johnson, Kenneth Earl
Johnson, Richard D
Johnson, Robert
Johnson, Robert Morton
Johnson, Warren Charles
Johnston, Herbert Norris
Johnston, Richard S
Jones, Carl Trainer
Jones, Francis Tucker
Jones, James Homer
Jones, Jesse W
Jones, Samuel O'Brien
Jones, Thomas Oswell
Jones, William Henry, Jr
Joubin, Franc Renault
Jucaitis, Pranas Francis
Julian, Gordon Ray
Kablaoui, Mahmoud Shafiq
Kaczka, Edward Anthony
Kade, Charles Frederick, Jr
Kahn, Milton
Kaiser, Emil
Kamner, Mildred Elsie (Mrs Edward M Tolman)
Kane, Stephen Shimmon
Kano, Adeline Kyoko
Kaplan, Alex
Kaplan, Lawrence Jay
Kaplan, Phyllis Deen
Kardos, Otto
Karr, Clarence, Jr
Karrick, Neva Louise
Karsten, Kenneth Stephen
Kaska, William Charles
Katsumoto, Kiyoshi
Katz, Lenard
Kaye, Wilbur (Irving)
Keane, David Donagh
Keevil, Norman Bell
Keim, Christopher Peter
Keirns, Mary Hull
Kells, Lyman Francis
Kelly, Jeffrey John
Kempter, Charles Prentiss
Kendall, Kenneth Keese, Jr
Kennard, Kenneth Clayton
Kennedy, Frank Metler
Kent, William L
Kern, Charles James
Kern, Roland James
Keskkula, Henno
Kewish, Ralph Wallace
Khalifah, Raja Gabriel
Kharasch, Norman
Killian, Donald B
Kilpatrick, John Edgar
Kimble, Glenn Curry
Kindsvater, Howard Maxwell
King, Cecil Victor
King, James Reber
Kinsey, Victor Everett
Kirkland, Walter Dean
Kirsch, Francis William
Kirsch, Milton
Kirshenbaum, Isidor
Kitchen, Leland Joseph
Kleber, Eugene Victor
Klemme, Waldemar Arthur, Jr
Klemmer, Dorothea Elizabeth
Klopman, Gilles
Knapp, John Samuel
Knecht, Walter Ludwig
Knight, Samuel Bradley
Knopf, Daniel Peter
Knudson, George E
Koch, Edwin George
Koehn, George Willis
Koelsch, Charles Frederick
Koepfli, Joseph Blake
Kohn, Gustave K
Koken, James E
Kolb, Harry John
Konig, Otto
Koningstein, Johannes A
Konkle, George Melvin
Kostelnik, Robert J
Kraus, Frank Joseph
Kravitz, Edward
Kreider, Leonard Cale
Kreke, Cornelius W
Kremer, Chester B
Kroenke, William Joseph
Krohn, Albertine
Krouse, Howard Roy
Krummenacher, Daniel
Kruse, Jurgen M
Krynitsky, John Alexander
Kuan, Teh Soong
Kuebler, William Frank, Jr
Kuhn, Carl Sellner, Jr
Kukolich, Stephen Irvin
Kumler, Warren Donald
Kuo, Harng-Shen
Kurnick, Allen Abraham
Kuroda, Paul Kazuo
Lad, Robert Augustin
Lafornara, Joseph Philip
Lagow, Richard James

Laiderman, Donald D
Lam, Fuk Luen
Lamprey, Headlee
Land, James Edward
Landborg, Richard John
Landgrebe, Albert R
Landis, Abraham L
Landon, Donald Omar
Lanning, Francis Chowing
Lanterman, Harold H
Lappin, Gerald R
Larson, Raymond George
Larson, Thurston E
Lateef, Abdul Bari
Lau, Kenneth W
Lavin, George Israel
Lawler, Charles Wesley
Lawrence, Aubrey Wilford
Lawton, Gerald Warren
Laying, Edwin Tower
Leary, Joseph Aloysius
Leathers, Joel Monroe
Leavitt, Julian Jacob
LeBaron, Robert (Francis)
Lee, Charles Richard
Lee, Jordan Grey, Jr
Lee, Yoon Chai
Lee, Young-Jin
Lee, Yuan Tseh
Leech, William Dale
Lee-Ruff, Edward
Leffler, John Edward
Legrow, Gary Edward
Leibu, Henry J
Le Maistre, John Wesley
Lenher, Samuel
Lentz, Charles Wesley
Leo, Micah Wei-Ming
Leonard, Fred
Leonard, Margaret Ives
Leonard, Reid Hayward
Lessard, Maurice
Letkeman, Peter
LeTourneau, Robert Louis
Leutgoeb, Rosalia Aloisia
Leutritz, John, Jr
Levine, Eli Morris
Levinson, Sidney Bernard
Levy, Gabor Bela
Lew, Chel Wing
Lewis, Edward Sheldon
Lewis, George Leoutsacos
Lewis, Jasper Phelps
Lewis, Leonda Lamonte
Lewis, Warren Burton
Liang, Shou Chu
Libby, James William, Jr
Libby, Willard Frank
Liberatore, Laurence Columbus
Lieberman, Samuel Victor
Liebhafsky, Herman Alfred
Liedtke, James Dale
Light, Donald Willis
Limerick, Jack McKenzie, Sr
Lindquist, Frank Eugene
Lindquist, John Raymond
Lingafelter, Edward Clay, Jr
Lintner, Anthony Ethelbert
Lippe, Robert Lloyd
Lips, Hilaire John
Lipsicas, Max
Liska, Kenneth J
Little, Ernest Lewis, Jr
Liu, Hsing-Jang
Locke, David Creighton
Locke, David Millard
Locke, Frederic John
Lockwood, William Howard
Loehr, Thomas Michael
Lofgren, Norman Lowell
Logue, Marshall Woford
Lohuis, Delmont John
Long, James Harvey, Jr
Long, Loren Marlin
Longanbach, James Robert
Loomis, Albert Geyer
Los, Marinus
Lovell, Edwin Lister
Lowry, Charles Doak
Luetzow, Arthur Edward
Lugar, Richard Charles
Lukat, Frederic Timon
Lustgarten, Ronald Krisses
Luther, Herbert George
Luthy, Jakob Wilhelm
Lux, John Herbert
Lyman, William Ray
Lytle, Fred Edward
Maasberg, Albert Thomas
MacDonnell, Donald R
MacDowell, John Fraser
MacGregor, Ian Robertson
MacInnis, Martin Benedict
Mack, Pauline Beery
MacKay, Donald Douglas
Mackay, Johnstone Sinnott
MacKenzie, Scott, Jr
Macri, Alfred Roger

CHEMISTRY

MacVicar, Robert William
MacWilliams, Donald Gribble
MacWood, George Eugene
Mader, William John
Madson, Willard Hegland
Maender, Otto William
Magat, Eugene Edward
Maggio, Francis Xavier
Magruder, Willis Jackson
Maienthal, E June
Major, Randolph Thomas
Makin, Earle Clement, Jr
Maleeny, Robert Timothy
Malmberg, Earl Winton
Malmstadt, Howard Vincent
Maltenfort, George Gunther
Mamer, Orval Albert
Manche, Emanuel Peter
Manley, Leo William
Manning, Maurice
Marans, Nelson Samuel
Margalit, Nehemiah
Mariano, Patrick S
Marks, Tobin Jay
Maroney, William
Marsalis, Sula Johnson
Marsden, James G
Marshall, Charles Louis
Marshall, Edwin Randolph
Marshall, Harry Borden
Martell, Arthur Earl
Martens, Christopher Sargent
Martin, Arthur Francis
Martin, Frank Stephen
Martin, Irving
Martin, Jerome
Martin, Richard Blazo
Marton, Renata
Maruyama, George Masao
Massie, Samuel Proctor
Mathe, Clarence Eugene, Jr
Mathews, A L
Matsen, Frederick Albert
Matthews, Frederick White
Matthkow, Morris
Mattison, Louis Emil
Mattson, Leland Neil
Mattson, Victor Frank
Maynert, Everett William
McBay, Arthur John
McBay, Henry Cecil
McBride, Clifford Hoyt
McBride, John Alexander
McBryde, William Arthur Evelyn
McCafferty, Edward
McCarty, Charles Norman
McCarty, Lewis Vernon
McCleary, Charles David
McCloskey, James Augustus, Jr
McClung, Ronald Edwin Dawson
McCollum, John David
McConnell, Albert Lawrence
McCurry, Patrick Matthew, Jr
McDermott, Leon Anson
McDowell, Maurice James
McDuffie, Bruce
McEvoy, Francis Joseph
McEwen, Mildred Morse
McFarlane, Hugh Murray
McGandy, Edward Lewis
McGovern, John Joseph
McGraw, Leslie Daniel
McGuire, Thomas Harry
McIntyre, William Ernest, Jr
McKee, Robert Lambert
McKinnis, Charles Leslie
McKinnis, Ronald Bishop
McLain, Joseph Howard
McLaughlin, Paul John
McMillan, Graham Watson
McMullen, Eugene Joseph
McMullen, James Clinton
McMurdie, Howard Francis
McNeely, William Harold
McPhail, Andrew Tennent
McPherson, Clinton Marsud
McSweeney, Ellsworth Edward
McVey, William Henry
Mead, Darwin James
Means, Whitney Harris
Meany, John Eagleton
Medlin, William Virgil
Meehan, Edward Joseph
Meigs, Frederick Madison
Meiser, John H
Melamed, Nathan T
Melveger, Alvin Joseph
Menotti, Amel Romeo
Mercer, Walter Ashby
Merken, Melvin
Merrithew, Paul Burton
Mesrobian, Robert Benjamin
Metzger, Gershon
Metzger, Robert Melville
Metzner, Wendell Phillips
Meyer, Heinz Friedrich
Meyers, Elwood William
Meyerson, Seymour

Michael, Leslie William
Michael, Thomas Hugh Glynn
Michel, Bede Eugene
Michel, Lester Allen
Michl, Josef
Middleton, Hugh William
Mieure, James Philip
Mighton, Harold Russell
Mikes, John Andrew
Milberger, Ernest Carl
Miles, Charles Burke
Miles, Delbert Howard
Miles, James Lowell
Miller, Arthur Joel
Miller, Daniel Robert
Miller, George Tyler, Jr
Miller, Glenn Harry
Miller, Jane Alsobrook
Miller, Julian Malcolm
Miller, Richard Roy
Miller, Stanley Lloyd
Miller, Warren Widmer
Miller, William Schuyler
Milton, Kirby Mitchell
Milton, William Anthony
Minieri, P Paul
Miskel, John Albert
Mislow, Kurt Martin
Mock, Richard Armitage
Model, Frank Steven
Moe, Owen Arnold
Moehs, Peter John
Moffat, James
Mohler, Robert Allan
Mohrman, Harold W
Mohun, William Arthur
Molloy, Andrew A
Monack, Louise Charlotte
Moncrief, John William
Monson, Richard Stanley
Montagna, Amelio Emidio
Montenyohl, Victor Irl
Montieth, Richard Voorhees
Moody, Frank Baldwin
Moore, Allen Charlton
Moore, Charles Henkel
Moore, John Criswell
Moorehead, Thomas J
Morejon, Clara Baez
Morgan, Leon Owen
Morgan, Paul Winthrop
Morrell, William Egbert
Morris, Daniel Luzon
Morris, Edward C
Morris, Robert James
Morris, Robert Lyle
Morris, William Collins
Morrison, John Agnew
Mortenson, Raymond Archie
Moscony, John Joseph
Moseley, Harry Edward
Moser, Charles Edwin
Moser, James Howard
Mosier, Benjamin
Motter, Robert Franklin
Mourning, Michael Charles
Moye, Alfred Leon
Moyers, Jarvis Lee
Mukai, Cromwell Daisaku
Mulliken, Robert Sanderson
Mullins, John A
Mundy, Bradford Philip
Murphy, Cornelius Bernard
Murray, John Wolcott
Murray, William Singler
Mushak, Paul
Mutton, Donald Barrett
Myers, Raymond Reever
Mysels, Estella Katzenellenbogen
Nabedian, Kevork Vartan
Nahin, Paul Gilbert
Naibert, Zane Elvin
Naidus, Harold
Nash, Ralph Glen
Nason, Howard King
Nathan, Alan Hart
Naughton, John Joseph
Navrotsky, Alexandra
Naylor, Benjamin Franklin
Nebel, Richard Wilson
Neff, Loren Lee
Neher, Maynard Bruce
Nelson, Arthur Kendall
Nelson, Cecil Morris
Nelson, David Alan
Nelson, John Arthur
Nelson, Lawrence Barclay
Nerken, Albert
Nesbitt, Stuart Stoner
Neta, Pedatsur
Neu, Ernest Ludwig
Neubeck, Clifford Edward
Neuville, Morris Louis
Neuzil, Richard William
Newman, Bernard
Newman, Roger
Newton, Amos Sylvester

Newton, Robert Collier
Nicholson, Richard Selindh
Niederhauser, Warren Dexter
Niedrach, Leonard William
Nielsen, Norman Arnold
Niemann, Henry Ernst
Nies, Nelson Perry
Nimmo, Charles Colvin
Noe, Lewis John
Nordby, Gordon Lee
Nordmann, Joseph Behrens
Novak, John William
Novak, Arthur Francis
Nowell, Robert Louis
Nowell, John William
Nozaki, Kenzie
Nunez, Loys Joseph
Nurnberger, John Ignatius
Obermanns, Henry Ernst
O'Brien, Darrell Eugene
O'Donnell, Gordon James
O'Gorman, John Michael
O'Hare, George Alfred
Oja, Tonis
O'Konski, Chester Thomas
Olsen, Fredric Philip
Olsen, John Sylvester
Olson, Carl Marcus
Olson, Walter T
Olyuk, Paul
Omohundro, Allen Llewellyn
Oppenheim, Elliot
O'Rell, Dennis Dee
Orgel, Leslie E
Orlowski, Jan Alexander
Ostapiak, Mykola
Otis, Marshall Voigt
Ott, Donald George
Otto, John B, Jr
Owen, Walter Wycliffe
Pace, Eugene Leonard
Pace, Salvatore Joseph
Palaszek, Mary De Paul
Palmer, George David, Jr
Pande, Gyan Shanker
Papa, Domenick
Pappas, Anthony John
Parent, Paul Andrew
Park, Conrad B
Park, Joseph Dal
Parker, Edwin Davis
Parker, George Anthony
Parker, John Abel
Parker, Patrick LeGrand
Parks, Paul Franklin
Parshall, George William
Pasztor, Laszlo
Pate, Brian David
Paterson, William Gordon
Pattabhiraman, Tammanur R
Patterson, James Howard
Patton, Alva Rae
Pauling, Linus Carl
Paulsen, Grover Cleveland, Jr
Payne, Elmer Curry
Payne, Nicholas Charles
Pearce, Dennis Wiffen
Pearlson, Wilbur H
Pearson, Arthur David
Pearson, Donald Emanual
Pearson, Edward Pillsbury
Pearson, Frank Gardiner
Pearson, Ralph Gottfrid
Peck, Virgil Glenn
Peirent, Robert John
Pelletier, Gerard Eugene
Pement, Fredric William
Pensak, David Alan
Pepkowitz, Leonard Paul
Perkins, Ben Harrison
Perkins, James
Perkins, Richard W
Perrotta, James
Perry, Lloyd Holden
Peters, Charles Frederick
Peters, Edwin Francis
Peters, Franklin Traviss
Petersen, Wallace Christian
Peterson, Sigfred
Phelan, Earl Walter
Phibbs, Murray Kenneth
Phillips, David Colin
Phung, Peter Viet
Pierce, James Kenneth
Pierce, Ogden Ross
Pilato, Philip Anthony
Pimentel, George Claude
Pinkerton, Frank Henry
Piper, Douglas Edward
Plaut, Herman
Plovnick, Ross Harris
Plucknett, William Kennedy
Plump, Ralph Eugene
Podbielniak, Vincent S
Podlas, Thomas Joseph
Pokras, Harold Herbert
Polglase, William James
Pollack, Louis Rubin

Polmanteer, Keith Earl
Pomeroy, John Howard
Pomeroy, Richard Durant
Popovich, Peter
Porsche, Francis William
Porter, Galen
Porter, Charles Dale
Potempa, Sylvester Joseph
Poulter, Charles Dale
Powell, Francis X
Powell, Thomas Mabrey
Powers, Treval Clifford
Preckel, Ralph Frederick
Prescott, Keith Burns
Preston, Frank James
Preuss, Albert F
Price, Fraser Pierpont
Priest, Homer Farnum
Priest, William James
Primak, William Leo
Prindle, Aldo Martin
Pulitzer, Thomas Joseph
Punwar, Jalamsinh K
Purdue, Jack Olen
Putnam, Hamilton Wallace
Quigley, Herbert Joseph, Jr
Raber, Douglas John
Raider, Stanley Irwin
Raimondi, Donald Louis
Rainard, Leo Walter
Ralls, Jack Warner
Ramos, Lillian
Rannefeld, Clarence Edmund
Rasch, Carl Henry
Rasheed, Khalid
Rathmann, Franz Heinrich
Rauk, Arvi
Rawlings, Floyd, Jr
Raymond, Samuel
Razouk, Rashad Elias
Reas, William Harry
Redeker, Harry Erwin
Redman, Leslie Merrill
Reed, Gerald
Reed, Sherman Kennedy
Reehling, Harold Arthur
Reeves, Perry Clayton
Reichle, Walter Thomas
Reidinger, Anthony A
Reinders, Victor A
Reinisch, Ronald Fabian
Reisberg, Joseph
Rempel, Herman G
Renoll, Mary Wilhelmine
Reynolds, Charles Albert
Rhoda, Richard Noble
Rhodes, Christopher Thomas
Rhodes, Herbert Dawson
Rice, Francis Owen
Rice, Rip G
Richardson, Verlin Homer
Riches, Wesley William
Richter, George Alvin, Jr
Riddle, Edward Hollister
Rideout, Janet Litster
Riehl, Jerry A
Rietz, Edward Gustave
Rinehart, Jay Kent
Ripley, Dennis Leon
Risinger, Gerald E
Ritt, Paul Edward, Jr
Rivest, Roland
Robbins, Gordon Daniel
Roberts, Aaron Gene
Roberts, Ammarette
Roberts, Earl John
Roberts, John Edwin
Roberts, William John
Robertson, David Anthony
Robertson, James Aldred
Robertson, James Alfred
Robinette, Hillary, Jr
Robinson, Dudley Hugh
Robinson, Edwin Allin
Robinson, Wilbur Eugene
Roche, James Norman
Rochlin, Phillip
Rochow, Eugene George
Rocks, Lawrence
Roddy, William Thomas
Rodowskas, Edward Laurence
Roebuck, Alan Kitson
Roebuck, Heide
Rogers, Alan Barde
Rogers, Howard Gardner
Rogers, Howard H
Rogers, Thomas Henry, Jr
Roland, George Warren
Roldan, Luis Gonzalez
Roller, Paul S
Rollinson, Carl Linden

Ropp, Walter Shade
Rosano, Henri Louis
Rose, Arthur
Rosen, Bernard H
Rosen, Lawrence
Rosen, William M
Rosenblatt, David Hirsch
Rosenstein, Ludwig
Ross, Alberta B
Ross, John Franklin
Ross, Ronald D
Ross, Sidney David
Rostler, Fritz S
Roswell, David Frederick
Roth, Lloyd Joesph
Roth, Marie M
Roth, Shirley H
Rothen, Alexandre
Rothermel, Joseph Jackson
Rothrock, George Moore
Rothrock, Henry Shirley
Rothstein, Lewis Robert
Roussel, Philip Andrew
Rowe, Robert David
Rubin, Louis
Rubin, Thor Richard
Rudolph, Jeffrey Stewart
Ruof, Clarence Herman
Rushton, Brian Mandel
Russell, Allen Stevenson
Russell, Edwin Roberts
Russell, Robert Raymond
Russo, Michael Eugene
Ryan, Clarence Augustine, Jr
Ryan, Michael T
Ryznar, John William
Sachdev, Sham L
Sacher, Alex
Sackett, William Malcolm
Sadler, Arthur Graham
Saeman, Jerome Francis
Saeva, Franklin Donald
Safe, Stephen Harvey
St Pierre, Thomas
Sair, Louis
Saletan, Leonard Timothy
Salvin, Victor S
Sandell, Ernest Burger
Sanders, Charles Irvine
Sankey, Charles Alfred
Sardella, Dennis Joseph
Saunders, Kenneth Worden
Savedoff, Lydia Goodman
Sax, Karl Jolivette
Scallet, Barrett Lerner
Schaefer, Hugh Ferdinand
Schaefer, Vincent Joseph
Schaffel, Gerson Samuel
Schales, Otto
Scheiderbauer, Robert Albert
Scheidt, Walter Robert
Schenck, Allan
Scherer, George Allen
Schick, Jerome David
Schiessler, Robert Walter
Schilz, Carl Edward
Schlatter, Rudolph
Schmidt, Webster Raymond
Schmitt, Allen F
Schneider, Joseph
Schneider, Walter Carl
Schoen, Kurt L
Scholberg, Harold Milton
Scholes, Samuel Ray, Jr
Schramm, Robert Frederick
Schrenk, William George
Schreyer, James Martin
Schrier, Melvin Henry
Schroepfer, George John, Jr
Schroeter, Siegfried Hermann
Schuck, Robert
Schulert, Arthur Robert
Schulte, George Nicholas
Schultz, Ray Karl
Schultze, Helmuth W
Schulze, Chris Carl
Schulze, Heinz
Schurr, Garmond Gaylord
Schwartz, Donald
Schwartz, Harry
Schwartz, Leonard H
Schwendinger, George Christian
Schwendinger, Richard B
Schwing, Karl Josef
Scola, Daniel Anthony
Scott, Charles Edward
Scott, Eric James Young
Scott, John Marshall William
Scott, Peter Hamilton
Scott, Robert Crawford
Scott, William Edwin
Scribner, Leonard
Scully, Frank E, Jr
Seanor, Donald A
Sears, Karl David
Seebold, Robert Elvin
Segal, Leon
Segel, Edward

Seibert, Richard Albert
Semegen, Stephen Thomas
Semon, Waldo Lonsbury
Senkus, Murray
Seo, Stanley Toshio
Serdarevich, Bogdan
Sexsmith, Frederick Hamilton
Shackle, Dale Richard
Shakhashiri, Bassam Zekin
Shank, Lowell William
Shankland, Rodney Veeder
Shannon, Ira Lenwood
Shapiro, Isadore
Shapiro, Raymond E
Shapiro, Rubin
Shapiro, Zalman Mordecai
Sharbaugh, Amandus Harry
Sharma, Ram Karan
Sharp, Dexter Brian
Sharp, Lord Glen
Sheely, Clyde Quitman
Sheets, Donald Guy
Sheldon, John Lewis
Shen, Samuel Yi-Wen
Sheridan, Richard Collins
Sherman, Clarence Steiner
Sherman, William Reese
Shida, Mitsuzo
Shidlovsky, Igal
Shin, Kiu Hi
Shoemaker, John Daniel, Jr
Shook, Thomas Eugene
Shortridge, Robert William
Shoub, Earle Phelps
Shuger, Leroy Woodrow
Shur, E Gustave
Siddall, Thomas Henry, III
Siegel, Abraham Lazarus
Siegfried, Robert
Sievert, Carl Frank
Silleck, Clarence Frederick
Silver, Samuel Lewis
Silver, Seymour David
Simmonds, Walter Henry
Simmons, George Allen
Simmons, Glen Raymond
Simon, Eliot Morton
Simon, Wilbur
Simonsen, Stanley Harold
Simpson, William Tracy
Sims, Richard Paul Andrew
Sinclair, Edward Elliot
Singer, Solomon Elias
Singer, Stanley
Singh, Harbhajan
Singleton, Fred Gray
Skala, Hertha
Skalny, Jan Peter
Skarlos, Leonidas
Skinner, John Taylor
Skoog, Douglas Arvid
Slama, Francis J
Slansky, Cyril Method
Sleeman, Richard Alexander
Slota, Peter John, Jr
Smiley, Seymour Howard
Smith, Carolyn Jean
Smith, Charles Hooper
Smith, Charles Lea
Smith, Charles T
Smith, David Huston
Smith, Elbert George
Smith, Franklin Danford
Smith, Gerould Hammond
Smith, Ieuan Trevor
Smith, Joseph James
Smith, Manning Amison
Smith, Paul Vergon, Jr
Smith, Percy Leighton
Smith, Richard E
Smith, Ward Arden
Snow, Adolph Isaac
Snyder, Carl Edward
Snyder, Harold Herbert
Snyder, Joseph Quincy
Snyder, Milton Jack
Sobieski, James Fulton
Socolofsky, John Frederick
Socquet, Irene
Soffer, Louis M
Sollner, Karl
Solomon, David Eugene
Solomon, Miriam Grace, SC
Somers, Emmanuel
Sommers, Armiger Henry
Sonneborn, Henry, III
Sonnichsen, George Carl
Sookne, Arnold Maurice
Soper, Quentin Francis
Southwick, Philip Lee
Sowa, Frank Joseph
Spangler, Fred Walter
Spangler, John Allen
Sparberg, Esther Braun
Sparks, Morgan
Sparks, William Joseph
Spaulding, Robert Lee, Jr
Speck, Stanley Brooke

Speier, John Leo, Jr
Spiegl, Charles J
Spiegler, Kurt Samuel
Spies, Joseph Reuben
Spindel, William
Spingola, Frank
Spinks, John William Tranter
Spinner, Ernest
Spinner, Theodore
Spiro, Thomas George
Spirtes, Morris Albert
Spittler, Ernest George
Spitze, LeRoy Alvin
Spitzer, Ralph
Spoehr, Albert Frederick
Spooner, Laurence Whipple
Spradling, Stuart Leslie
Spurlock, James Josiah
Squire, Edward Noonan
Srinivasan, Sathanur Ramachandran
Stahl, William Herbert
Stambaugh, Oscar Frank
Stamm, Walter
Stanerson, Bradford Roy
Stanley, Edward Livingston
Stanley, Norman Francis
Stansbury, Harry Adams, Jr
Starbuck, Wesley Curtis
Stark, Forrest Otto
Starr, Robert I
Steckler, Robert
Steele, Arthur Burns
Steele, Clelie Truman
Steele, Richard
Steele, Sidney Russell
Steers, Edward
Steinberg, Ellis Philip
Steingiser, Samuel
Steinhardt, Ralph Gustav, Jr
Stempel, Arthur
Stephens, Roger
Sterling, John Deo
Stern, Adolph John
Stern, Eric Wolfgang
Sternberg, Heinz Walter
Stevens, Calvin Lee
Stevens, Frank Joseph
Stevenson, Chris G
Stevenson, Isaac Glenn
Stewart, Frank Edwin
Stewart, John Wray Black
Stichler, Robert Daniel
Stillings, Robert Almon
Stillwell, William Duncan
Stipanovic, Bozidar J
Stoffyn, Pierre Jules
Stoll, William Russell
Stolow, Nathan
Stonebraker, Peter Michael
Stosick, Arthur James
Stoy, William S
Strain, William Henry
Stratman, Frederick William
Stratton, Wilmer Joseph
Strauss, Simon Wolf
Strauss, Otto Peter
Strehlow, Richard Alan
Strickler, Herbert Sharpless
Strong, Walker Albert
Strouse, Charles Earl
Struve, William George
Stuart, David Marshall
Stubbs, Morris Frank
Stubbs, Ulysses Simpson, Jr
Stucker, Joseph Bernard
Stuckwisch, Clarence George
Studier, Martin Herman
Stuntz, Calvin Frederick
Sublett, Robert L
Suess, Hans Eduard
Sullivan, William Richard
Summerlin, Lee R
Summers, Donald Balch
Summers, Selby Edward
Suter, Robert Winford
Sutherland, Angus Johnston
Sutherland, Ronald George
Sutman, Frank X
Sutton, Paul Porter
Sutton, Russell Paul
Suzuki, Isamu
Sveda, Michael
Swain, Charles Gardner
Swalheim, Donald Arthur
Sweet, Thomas Richard
Swofford, Harold S, Jr
Syty, Augusta
Szinai, Stephen Slomo
Tabenkin, Benjamin
Tabet, Georges Elias
Taebel, Wilbert August
Tai, Julia Chow
Takman, Bertil Herbert
Tallman, Russell Louis
Tan, Ah-Ti Chu
Tapscott, Robert Edwin
Tarr, Betty R
Tarrant, Paul

Tarver, Harold
Tarvin, Donald
Tasker, Clinton Waldorf
Tauber, Arthur
Taunton-Rigby, Alison
Taylor, Donald Stinson
Taylor, Edward Wyllys
Taylor, Harold Nathaniel
Taylor, Howard S
Taylor, Jay Eugene
Taylor, Lloyd David
Taylor, Paul Duane
Taylor, Susan Serota
Taylor, Wendell Hertig
Tazuma, James Junkichi
Tchir, Morris Frederick
Teach, William Charles
Teague, Claude Edward, Jr
Teeters, Wilber Otis
Teichner, Robert W
Tellinghuisen, Joel Barton
Tenenbaum, Leon Edward
Teng, James
Terrant, Seldon W
Terry, Samuel Matthew
Teumac, Fred N
Thau, Marcus
Theimer, Ernst Theodore
Thelin, Jack Horstmann
Thibert, Roger Joseph
Thomas, Arthur L
Thomas, Berwyn Brainerd
Thomas, Charles Allen
Thomas, Lloyd Brewster
Thomas, Robert Joseph
Thomas, Vera
Thomas, Walter Moreland
Thompson, Albert Johnson, Jr
Thompson, Paul Woodard
Thompson, Stanley Gerald
Thornton, Charles De Wane
Thorstensen, Thomas Clayton
Thurman, Richard Gary
Timma, Donald Lee
Ting, Shih-Fan
Tinker, John Frank
Todd, David
Todd, Harold David
Toffel, George Mathias
Toguri, James M
Tomlinson, George Herbert
Tomlinson, Hazel M
Tongren, John Corbin
Toralballa, Gloria C
Tordella, John P
Toribara, Taft Yutaka
Torley, Robert Edward
Torok, Andrew, Jr
Tosoni, Anthony Louis
Towell, Edward Emerson
Townley, Robert William
Towns, Robert Lee Roy
Tranner, Frank
Traylor, Patricia Shizuko
Tremain, Henry Earl
Trent, Walter Russell
Trotz, Samuel Isaac
Trout, Paul Eugene
Trout, William Edgar, Jr
Truce, William Everett
Trueblood, Kenneth Nyitray
Truemper, Joseph Tucker
Truesdell, Alfred Hemingway
Truett, William Lawrence
Trytten, Roland Aaker
Tsao, Makepeace Uho
Tu, Chen Chuan
Tulagin, Vsevolod
Tullio, Victor
Turner, Byron
Turner, Carlton Edgar
Turner, George Robert
Turner, Vernon Lee, Jr
Tutas, Daniel Joseph
Tweed, Paul Basset
Twilley, Ian Charles
Tyler, Chaplin
Ulrich, Stephen Edgar
Unger, Israel
Urbanyi, Tibor
Urey, Harold Clayton
Urone, Paul
Urry, Wilbert Herbert
Vaala, Gordon Theodore
Vagnina, Livio L
Valentekovich, Marija Nikoletic
Van Beckum, William George
Vander Weyden, Allen Joseph
Van Dolah, Robert Wayne
Van Eaton, Robert Lee
Van Loten, Jon Clement
van Tamelen, Eugene Earl
Van Wazer, John Robert
Van Winkle, Quentin
Van Zandt, Gertrude
Varanasi, Usha Suryam
Vasilos, Thomas
Vaughan, David Evan William

CHEMISTRY

Vaughan, Wyman Ristine
Vaux, James Edward, Jr
Veber, Daniel Frank
Vedder, Willem
Veidis, Mikelis Valdis
Veleckis, Ewald
Veltman, Preston Leonard
Venable, Grant Delbert
Vennos, Mary Susannah
Via, Francis Anthony
Villani, Frank John
Vincent, George Paul
Vingiello, Frank Anthony
Voet, Andries
Vogel, Alfred Morris
Voisinet, Donald Louis
Vomhof, Daniel William
von Fischer, William
von Schriltz, Don Morris
Von Wicklen, Frederick Charles
Voss, Raymond Olson
Voter, Roger Conant
Wadinger, Robert Louis Peter
Wadman, W Hugh
Wagener, Donald David
Wagman, Donald David
Wagner, Myron L
Wagner, William Frederick
Wahl, Milton Heins
Wald, Milton M
Wasko, Peter Edmund
Waser, Jurg
Warner, John Christian
Ward, Joseph Richard
Ward, Frederick E
Wang, Chin Hsien
Walton, Charles William
Walter, Roderich
Walter, Dean Irving
Walsh, Thomas David
Walsh, Raymond Anthony
Wallace, William Eldred
Wallace, Donald Albin
Wall, Charles Ephraim
Watkins, Charles Emory
Watrous, Ralph Melvin
Watson, Dennis Ronald
Watson, James Arthur, Jr
Watson, Robert Francis
Watt, George Willard
Watters, James I
Watts, Daniel Jay
Wayne, Winston Joe
Weare, John H
Weast, Robert Calvin
Weaver, Charles R
Weaver, John Carl
Weaver, Warren Eldred
Weber, Arthur George
Weber, Joseph
Weeks, Joseph Elliott
Weeks, Owen Bayard
Wefers, Karl
Wehri, Pius Anton
Wehrmeister, Herbert Louis
Weidner, Bruce Vanscoyoc
Weiss, Herbert V
Weiss, Michael Karl
Weissler, Alfred
Weisz, Paul Burg
Welch, Steven Charles
Welcher, Richard Parke
Weldon, John William
Wells, Eugene Hadley
Wells, James Robert
Wellum, Glyn Richard
Wenaas, Paul Emil
Wendt, Theodore Mil
Wenzinger, George Robert
Werner, Rudolf
Wernimont, Grant (Theodore)
Wertwijn, George
Wertz, John Edward
West, Philip William
Westberg, Karl Rogers
Weston, Henry Morgan
Wetzel, Franklin Huff
Wheat, Franklin Huff
Wheeler, Gilbert Vernon
Wheeler, Ora Leon
Whistler, Roy Lester
Whitacre, Francis Marion
Whitcher, Wendell Jennison
White, Blanche Babette
White, Briggs Johnston
White, John Greville
White, Philip Cleaver
Whitehead, Fred
Whitfield, Robert Edward
Wickes, Glenn French
Wiedneheft, Charles John
Wiener, Robert Newman
Wiener, Karel
Wiesner, Helmut
Wiest, Emil Gabriel
Wiewiorowski, Tadeusz Karol
Wiggin, Edwin Albert

Wilber, Joe Casley, Jr
Wilcox, Archer Carl
Wilcox, Harold Edwin
Wiley, Richard Haven
Wilfong, Robert Edward
Wilkes, Charles Eugene
Wilkes, Louis Phillip
Willard, John Wesley
Williams, Cecil R
Williams, John Covington
Williams, John Frederick
Williams, Leamon Dale
Williams, Lewis David
Williams, Martin Barbour
Williamson, Stanley Morris
Wilson, Archie Spencer
Wilson, Christopher Lumley
Wilson, James Woodrow
Wilson, Kenneth MacKenzie
Wilson, Leland Leslie
Wilson, Linda S Whatley
Wilson, Martin
Wilson, Robert Curtis
Wilson, Robert G
Wilson, Thomas Putnam
Wing, Robert Edward
Winkert, John Wynia
Winkler, Carl Arthur
Wipke, Will Todd
Wirth, Henry Edgar
Wise, Edward Nelson
Wise, John Thomas
Wiseblatt, Lazare
Wissow, Lennard Jay
Wistar, Richard
Wit, Donald Reinhold
Wohleber, David Alan
Wolfsy, Leon
Wolf, Michael Joseph
Wolff, Nikolaus Emanuel
Wolford, Richard Kenneth
Wolter, Frederick John
Wolverton, Billy Charles
Wood, Darwin Lewis
Wood, Jeanie McMillin
Wood, John Stanley
Wood, Peter Douglas
Woodhouse, Edward John
Woodhouse, John Crawford
Wooding, William Minor
Woodriff, Ray Alan
Woodward, Hubert Edmund
Woody, A-Young Moon
Woody, Wayne Theodore
Wozniak, Alan Carl
Wright, Charles Cathbert
Wright, George F
Wright, William Wynn
Wu, Ming Tsung
Wu, Ting Kai
Wubbels, Gene Gerald
Wuskell, Joseph P
Wyatt, Benjamin Woodrow
Wyatt, Jeffrey Renner
Wyman, Harold Robertson
Wyman, John E
Yamada, Esther V
Yang, Nien-Chu
Yates, Shelly Gene
Yeadon, David Allou
Yeager, John Frederick
Yee, Tin Boo
Yolles, Seymour
Young, Frank Glynn
Young, Mahlon Gilbert
Young, Robert Bruce
Young, William Gould
Yourtee, John Ashby
Zabik, Matthew John
Zajcew, Mykola
Zapsalis, Charles
Zavist, Algerd Frank
Zeck, William Charles
Ziegler, Paul Fout
Ziegler, Theresa Frances
Zienty, Ferdinand B
Zimmerman, Howard Elliot
Zimmerschied, Wilford John
Zingaro, Joseph S
Zinman, Walter George
Zwarzig, Frances Ryder
Zwicker, Benjamin M G

Agricultural Chemistry

Abramitis, Walter William
Anastasiadis, Phoebus A
Ballantine, Larry Gene
Barthel, William Frederick
Beestman, George Bernard
Bingham, Robert J
Bishop, John Russell
Bjork, Carl Kenneth, Sr
Block, Michael Joseph
Brand, John S
Brewer, Arthur David
Budde, Paul Bernard
Burton, Joe Covington

Cattani, Ray August
Chan, Hak-Foon
Chin, Wei Tsung
Connon, Robert Haddon
Docks, Edward Leon
Dutra, Gerard Anthony
Erickson, David R
Fallscheer, Herman O
Fancher, Llewellyn W
Fang, Sheng Chung
Felton, Staley Lee
Fitzpatrick, Thomas Joseph
Gatterdam, Paul Esch
Greenwald, Bernard William
Halverson, Andrew Wayne
Hamaker, John Warren
Hamilton, John William
Hardesty, John Oliver
Hedrich, Loren Wesley
Hess, Earl Hollinger
Hightower, Kenneth Ralph
Ikeda, Robert Mitsuru
Inglett, George Everett
Johnson, Wayne Orrin
Jones, Merriam Arthur
Khan, Shahamat Ullah
Khayat, Ali
Kilsheimer, John Robert
Knapp, Joseph Leonce, Jr
Krass, Dennis Keith
Kuhr, Ronald John
L'Annunziata, Michael Frank
Lepage, Marius
Lichy, Charles Thorne
Loeffler, Erwin Stanley
Luckenbaugh, Raymond Wilson
Lutz, Albert William
Mabrouk, Ahmed Fahmy
Machinney, Thomas Martin
Mann, Robert Leslie
Manning, David Treadway
Maravetz, Lester L
Marks, Alfred Finlay
Martin, Edward Eugene
Masri, Merle Sid
May, Ralph Forrest
McInroy, Elmer Eastwood
Metzger, James Douglas
Mosier, Arvin Ray
Newsom, William S, Jr
Noznick, Peter Paul
Olney, Charles Edward
Owens, Daniel Kenyon
Pearce, David Archibald
Pence, James William
Peterson, Richard Grant
Polen, Percy B
Reed, John Fielding
Rigby, F Lloyd
Rockland, Louis B
Rogers, Richard Brewer
Rosenfield, Christine Ann Culp
Rupp, Eldor Gustav
Salomon, Milton
Schaefer, Charles Herbert
Schlaikjer, Carl Roger
Schoene, Dwight Lorin
Shasha, Baruch
Shaw, Ellsworth
Shinkle, Michael Paul
Skapason, Joseph Bjorn
Smit, Christian Jacobus Bester
Smith, Joseph A
Stansbury, Mack Fulton
Stein, Robert George
Stoner, Graham Alexander
Strohm, Paul F
Swift, Lyle James
Talley, Eugene Alton
Thorn, George Denis
Treichler, Ray
Treves, Gino Robert
Turner, Fred, Jr
Van Landingham, Audrey Howard
Van Runkel, Rosmarie
Walker, Howard George, Jr
Wallace, Volney
Weakley, Martin LeRoy
White, Jonathan Winborne, Jr
Wierbicki, Eugen
Witman, Eugene DeWald
Wommack, Joel Benjamin, Jr
Yaguchi, Makoto
Yih, Roy Yangming
Young, Hong Yip

Applied Chemistry

Antler, Morton
Baker, Michael Harry
Barton, Gerald Blackett
Berman, Horace Aaron
Borum, Olin H
Bresson, Clarence Richard
Brokaw, George Young
Brown, Keith Blanchard

Browning, Joe Leon
Bruenner, Rolf Sylvester
Burgess, William Howard
Carter, George Thomas
Castle, John Edwards
Clark, Walter Ernest
Connor, Daniel S
Crandlemere, Robert Wayne
Cruse, Robert Ridgely
Doss, Richard Courtland
Dunham, Kenneth Royal
Ebel, Robert Henry
Elmer, Curtis
Embree, Norris Dean
Findley, William Robert
Finn, John Martin
Frear, George Lewis
French, Kenneth William
Glendening, Norman Willard
Gregor, Harry Paul
Gwilt, John Ruff
Hart, Una Lynch
Herk, Leonard Frank, Jr
Hermann, John Alexander
Hirschmann, Robert P
Hormats, Ellis Irving
Jacobs, Albert Michael
Kaspin, Ben Louis
King, William Mattern
Kjeldgaard, Edwin Andreas
Kriege, Owen Hobbs
LaBarge, Robert Gordon
Lackey, Homer Baird
Lane, George Ashel
Linfield, Warner Max
Lloyd, William Gilbert
Mathay, William Lewis
Matson, Ted P
McIntyre, George Francis
McKenna, James Francis
Meinhard, James Edgar
Metzger, Sidney Henry, Jr
Moggio, William Aldo
Mooney, Robert Arthur
Mumbach, Norbert R
Oehmke, Richard Wallace
Orban, Edward
Paynter, John, Jr
Perilstein, Warren Louis
Pretka, John E
Rainer, Norman Barry
Raphael, Thomas
Rhodes, Donald Walter
Rosen, Milton Jacques
Sadowski, Anthony James
Schmitt, Charles Rudolph
Shaneyfelt, Duane L
Sherman, Patsy O'Connell
Skees, Hugh Benedict
Slate, Floyd Owen
Steiger, Fred Harold
Tarsey, Alexandre Rolf
Tulk, Alexander Stuart
Velturo, Anthony Francis
Waldron, Harold Francis
Williams, Robert Edward
Wintermoyer, John Paul

Atmospheric Chemistry

Bandy, Alan Ray
Berger, Jerry Eugene
Delany, Anthony Charles
de Pena, Rosa G
Duce, Robert Arthur
Farber, Robert James
Finnegan, William George
Fox, Donald Lee
Friend, James Philip
Gay, Bruce Wallace, Jr
Graedel, Thomas Eldon
Green, William Delap
Harrison, Halstead
Heaney, Robert John
Heicklen, Julian Phillip
Hochanadel, Clarence Joseph
Hudson, Frank Peter
Huntzicker, James John
Kaufman, Frederick
Lee, Richard Norman
Levy, Arthur
Levy, Hiram, II
Lipeles, Martin
Lodge, James Piatt, Jr
Mason, Allen Smith
McAfee, Kenneth Bailey, Jr
McLaren, Eugene Herbert
Mondy, Nell Irene
Newman, Leonard
Poland, Helen M
Pueschel, Rudolf Franz
Rosinski, Jan
Rowland, Frank Sherwood
Sedlacek, William Adam
Stampfer, Joseph Frederick, Jr
Stedman, Donald Hugh
Wilson, William Enoch, Jr
Yencha, Andrew Joseph

Carbohydrate Chemistry
Bachrach, Joseph
Ball, Derek Harry
BeMiller, James Noble
Bills, Alan Morris
Cifonelli, Joseph Anthony
Cottrell, Ian William
Cushing, Merchant Leroy
Doner, Landis Willard
Duke, Jodie Lee, Jr
Durette, Philippe Lionel
El Khadem, Hassan S
Friedman, Robert Bernard
Guiseley, Kenneth B
Hodge, John Edward
Isbell, Horace Smith
Jones, Duane Arnold
Kasehagen, Leo
Kovar, Jan Bernard
Larson, Roy Fred
Lineback, David R
Linker, Alfred
McGinnis, Gary David
Medcalf, Darrell Gerald
Mitchell, William Alexander
Molotsky, Hyman Max
Moser, Kenneth Bruce
Mowery, Dwight Fay, Jr
Nasir-Ud-Din, Nasir
Ness, Robert Kiracofe
Newton, John Marshall
Paschall, Eugene F
Pettit, David J
Pritchard, David Graham
Ransford, George Henry
Sandford, Paul A
Seidman, Martin
Shaw, Derek Humphrey
Short, Rolland William Phillip
Stark, John Howard
Van Cleve, John Woodbridge
Wagoner, John Allen
Walker-Nasir, Evelyne
Warren, Christopher David
Wollwage, Paul Carl
Wurzburg, Otto Bernard

Catalysis
Burwell, Robert Lemmon, Jr
Cusumano, James A
Koch, Theodore Augur

Cellulose Chemistry
Arseneau, Donald Francis
Bertram, Leon Leroy
Booth, Kenneth Gordon
Breithaupt, Lea Joseph, Jr
Brenner, Fivel Cecil
Church, John Armistead
Dobbins, Robert Joseph
Durso, Donald Francis
Gerwitz, David L
Godsay, Madhu
Gonzales, Elwood John
Hall, Frederick Keith
Hammer, Richard Benjamin
Herdle, Lloyd Emerson
MacDonald, Donald MacKenzie
Nelson, Mary Lockett (Mrs John D Guthrie)
Nelson, Robert Andrew
Portnoy, Norman Abbye
Rowland, Stanley Paul
Schenkenberg, Philip Rawson
Schwenker, Robert Frederick, Jr
Thelman, John Patrick
Turner, Rex Howell
Ward, Kyle, Jr

Cereal Chemistry
Baker, Doris
Bayfield, Edward Geoffrey
Bendelow, Victor Martin
Bushuk, Walter
Chung, Okkyung Kin
Conn, James Frederick
Cooper, Elmer James
Dahle, Leland Kenneth
D'Appolonia, Bert Luigi
Hoffman, William Howard
Hoseney, Russell Carl
Houston, David Fairchild
Kendall, John Hugh
Kite, Francis Ervin
Kosmolak, Frederick Graham
Lorenz, Klaus J
McDonald, Clarence Eugene
Meisner, Donald F
Miller, Byron Sloane
Patil, Sakharam Karsan
Pomeranz, Yeshajahu
Rankin, John Carter
Redfern, Sutton
Rohwer, Robert G
Rubenthaler, Gordon Lawrence
Sandsted, Rudolph Marion
Schanefelt, Robert Von

Shellenberger, John Alfred
Shuey, William Carpenter
Sullivan, John W
Tipples, Keith H
Waggle, Doyle H
Walker, Charles Eugene
Walsh, David Ervin
Watson, Clifford Andrew
Yamazaki, William Toshi

Chemical Embryology
Herrmann, Heinz
Mertes, David H
Wainwright, Stanley D
Weller, Edwin Matthew

Chemical Instrumentation
Abbott, Seth R
Albert, Harrison Bernard
Bloemer, William Louis
Burden, Stanley Lee, Jr
Converse, Jimmy G
Cooper, James William
Dessy, Raymond Edwin
Goodman, Philip
Hammer, Charles F
Hrubesh, Lawrence Wayne
Jackson, Darryl Dean
Johnson, Eric Robert
Johnson, Henry Wilson, Jr
Kobrin, Robert Jay
Lumney, David Clyde
Nathanson, Benjamin
Nunes, Thomas Lester
Prostak, Arnold S
Ross, Harley Harris
Vassos, Basil Harilaos
Verschingel, Roger H C
Wong, Peter Alexander
Wright, John Marlin

Chemical Kinetics
Back, Robert Arthur
Barker, John Roger
Bauer, Simon Harvey
Bogan, Denis John
Bone, Larry Iven
Bowers, Michael Thomas
Bowman, Craig Thomas
Bushey, William Raymond
Cares, William Ronald
Carmichael, Halbert Hart
Carr, Robert Wilson, Jr
Celiano, Alfred
Cole, David Le Roy
DeGraff, Benjamin Anthony
Dobson, Gerard Ramsden
Dorfman, Leon Monte
Dove, John Edward
Field, Richard Jeffrey
Gann, Richard George
Geissler, Paul Robert
Gentzler, Robert E
Gillard, Baiba Kurins
Golden, David Mark
Granoff, Barry
Grant, Edward Robert
Greiner, Norman Roy
Heatley, A Harold
Hierl, Peter Marston
Huie, Robert Elliott
Jacobson, Irven Allan, Jr
Kaufman, Frederick
Kibby, Charles Leonard
Kolb, Charles Eugene, Jr
Kreevoy, Maurice M
Laufer, Allan Henry
Legare, Richard J
LeRoy, Donald James
Lin, Sheng Hsien
Magenheimer, John Joseph
McDaniel, Robert Stewart
McGee, Thomas Howard
Mickens, Ronald Elbert
Mowery, Dwight Fay, Jr
Mutch, George William
Noyes, Richard Macy
Roscoe, John Miner
Rose, Timothy Laurence
Rowland, Frank Sherwood
Sanzone, George
Saunders, Barbara Gail Breidenbach
Schmidt, Wilfred G
Seery, Daniel J
Shortridge, Robert Glenn, Jr
Simonaitis, Romualdas
Simonelli, Anthony Peter
Slagg, Norman
Slanger, Tom George
Stein, Stephen Ellery
Swinehart, Donald Fought
Weston, Ralph E, Jr
Wiesenfeld, John Richard
Williamson, David Gadsby
Wlech, Raymond Lee
Wu, Ching-Hsong
Yergey, Alfred L, III

Chemical Metallurgy
Benner, Blair Richard
Bishop, Jay Lyman
Carter, Giles Frederick
Curtis, Ralph Wendell
El Guindy, Mahmoud Ismail
Englehart, Edwin Thomas, Jr
Felten, Edward J
Foreman, Robert Walter
Glasser, Julian
Goggin, Donald Edward
Hillner, Edward
Hollingshead, Ethan Allen
Illis, Alexander
Janzen, John
King, William Robert, Jr
Krause, Daniel, Jr
Lynd, Langtry Emmett
Macaluso, Anthony, Sr
Mears, Dana Christopher
Mears, Robert Bruce
Nafziger, Ralph Hamilton
Olympia, Pedro Lim, Jr
Paine, Robert Madison
Petersen, Quentin Richard
Pierson, Hugh Ortho
Richards, Kenneth Julian
Ricksecker, Ralph E
St Cyr, Lewis Alpha
Scharfstein, Lawrence Robert
Shafer, William McKinley
Shores, David Auth
Staudenmayer, Ralph
Tan, Francis C
Walsh, Kenneth Albert
Wynkoop, Raymond

Chemical Oceanography
Andersen, Neil Richard
Anderson, James Jay
Atkinson, Larry P
Atwood, Donald Keith
Burrell, David Colin
Calder, John Archer
Carpenter, Roy
Chau, Yiu-Kee
Codispoti, Louis Anthony
Cutshall, Norman Hollis
Fitzgerald, William Francis
Gibb, Thomas Robinson Pirie, Jr
Gordon, Louis Irwin
Green, Edward Jewett
Hochman, Harry
Kester, Dana R
Lyman, John
MacIntyre, Ferren
Mattson, James Stewart
Neihof, Rex A
Neve, Richard Anthony
Oistlund, H Gote
Owen, Robert Michael
Page, David Sanborn
Phillips, Timothy Dukes
Presley, Bobby Joe
Reeburgh, William Scott
Sayles, Frederick Livermore
Schink, David R
Schmitz, Francis John
Sharp, Jonathan Hawley
Shaw, David George
Siegel, Alvin
Stansby, Maurice Earl
Swinnerton, John W
Walton, Alan
Wiebush, Joseph Roy
Williams, Peter M
Wolfe, Douglas Arthur
Yamamoto, Sachio
Zirino, Alberto

Chemistry
Abadi, Djahangir M
Adams, Herman Ray
Adams-Mayne, Mabelle Elaine
Addanki, Somasundaram
Ahuja, Jagan N
Allen, Rovelle Harper
Altmiller, Dale Henry
Ansari, Ali
Arends, Robert Leander
Ash, Kenneth Owen
Babson, Arthur Lawrence
Baetz, Albert L
Baker, Harold Nordean
Barber, Eugene Douglas
Basinski, Daniel Henry
Beardslee, Ronald Allen
Beeler, Myrton Freeman
Bermes, Edward William, Jr
Bevill, Rardon Dixon, III
Bird, Emerson Wheat
Blaivas, Murray A
Blatt, Sylvia
Bocek, Rose Mary
Bondar, Richard Jay Laurent
Boon, Donald Arthur
Boone, Donald Joe

Briden, Roger Clarence
Brown, Harold Hubley
Brusca, Donald Richard
Buccino, Raymond, Jr
Burgess, Thomas Edward
Burnett, Robert Walter
Bush, Martin Bruce
Carroll, James Joseph
Chakrin, Alan Leonard
Chen, Shui-Chin
Chilcote, Max Eli
Clark, John Francis Bullock
Clement, Gerald Edwin
Clemson, Harry C
Cohen, Alex
Coleman, Charles Mosby
Crookshank, Herman Robert
Dalal, Fram Ruston
Danzer, Laurence Alfred
Deindoerfer, Fred H
Demetriou, James A
Deutsch, Marshall Emanuel
Di Pippo, Ascanio G
Dooley, Joseph Francis
Doumas, Basil T
Drewes, Patricia Ann
Dulkin, Sol I
Eastman, John W
Edwards, Leila
Elfbaum, Stanley Goodman
Elliott, Joseph Robert
Epstein, Emanuel
Epstein, Morton Batlan
Ertingshausen, Gerhard
Fales, Frank Weck
Feldman, William
Fernandez, Alberto Antonio
Fischer, George A
Fischer, Robert Leigh
Fleisher, Martin
Flokstra, John Hilbert
Fordice, Michael W
Foy, Robert Bastian
Francis, Dawn Elizabeth
Friedman, Max Martin
Fritsche, Herbert Ahart, Jr
Gaebler, Oliver Henry
Gambino, Salvatore Raymond
Gang, Henry
Gast, Joseph Henry
Gerlach, Howard G, Jr
Gilbert, Frederick Emerson, Jr
Gilbertson, Terry Joel
Glick, John Henry, Jr
Gorczyca, Leonard Richard
Griffiths, William C
Griswold, Robert Edward
Gruemer, Hanns-Dieter
Habig, Robert L
Haddox, Charles Hugh, Jr
Handschuh, Gerald Jay
Hanson, Thomas Lawrence
Hellerstein, Stanley
Hellwege, Herbert Elmore
Helman, Edith Zak
Henson, Carl P
Herner, Albert Erwin
Hohnadel, David Charles
Jacobs, S Lawrence
Johnson, Howard James, Jr
Juselius, Roger Elliott
Kambli, Vijaykant Bhagwan
Kaminski, Louis Alfred
Kavarnos, George James
Kay, Peter Steven
Kelley, Thomas F
Kerkay, Julius
Killingsworth, Lawrence Madison
Kream, Jacob
Laessig, Ronald Harold
La Ganga, Thomas S
Larsen, Martin Lee
Lepp, Cyrus Andrew
Levin, Robert Aaron
Levy, Robert
Levy, Samuel Wolfe
Levy, Susanna Agnes
Lo, Donald Hung-Tak
Logan, James Edward
Loveless, Loyal E
Luisada-Opper, Anita Victoria
Lusgarten, Jack Abraham
Man, Evelyn Brower
Martin, Kenneth John
Martinek, Robert George
Mason, William Burkett
McGrath, William Patrick
McGuckin, Warren Francis
McNair, Ruth Davis
Megraw, Robert Ellis
Meites, Samuel
Metlay, Max
Moffa, David Joseph
Mooney, Larry Albert
Moorehead, Wells Rufus
Moreland, Ferrin Bates
Nandedkar, Arvindkumar Narhari
Natelson, Samuel

CHEMISTRY

Nauman, Louis William
Neesby, Torben Emil
Nelson, Gilbert Harry
Nipper, Henry Carmack
Olsen, Eugene Donald
Parker, Leslie
Patton, Selma Hicks
Pearce, Richard Hugh
Penicnak, Adrian John
Perry, Billy Wayne
Pick, Robert Orville
Pliegg, Vincent Joseph
Pinto, P Vincent C
Pollard, Donald Ray
Porter, Charles Jack
Pragay, Desider Alexander
Price, Jacob Waide
Proksch, Gary J
Quinn, Joseph Freeman
Radin, Nathan
Ray, Robert Allen
Rehak, Matthew Joseph
Rice, Eugene Worthington
Richard, Douglas Warren
Rischer, Robert Louis
Rosenthal, Waldemar Arthur
Rundell, Clark Ace
Russell, James Christopher
Sage, Gloria W
Sardesai, Vishwanath M
Sasse, Edward Alexander
Savory, John
Sax, Sylvan Maurice
Schneider, Paul
Schucher, Reuben
Schwartz, Jerome Lawrence
Schwartz, Morton K
Sgoutas, Demetrios Spiros
Shepherd, Herndon Guinn, Jr
Skelley, Dean Sutherland
Smith, Steffen Wesley
Smith, Susan T
Snyder, Lloyd Robert
Sobel, Robert Edward
Spencer, Walter William
Spiegel, Herbert Eli
Spooner, George Hansford
Standefer, Jimmy Clayton
Stasiw, Roman Orest
Strandjord, Paul Edphil
Sunshine, Irving
Sutula, Chester Louis
Taylor, Kirman
Teaford, Margaret Elaine
Tietz, Norbert W
Tonks, David Bayard
Trawick, William George
Van Norman, John Donald
Walberg, Clifford Bennett
Walters, Martha I
Walwick, Earle Richard
Waner, Ann Marie
Wan, Abraham Tai-Hsin
Watts, Judith Elizabeth Culbreth
Weissman, Earl Bernard
Weissman, Norman
Wernus, Gerald R
Whisler, Kenneth Eugene
Wilson, Charles Oren
Windisch, Rita M
Wingeleth, Dale Clifford
Woodbridge, Joseph Eliot
Woronick, Charles Louis
Zabinski, Rose Marie C
Zimmerman, Gary Alan

Colloid Chemistry

Allen, Lawrence Harvey
Amick, Chester Albert
Baczewski, Alexander
Bagchi, Pranab
Bugosh, John
Cardwell, Paul H
Chamot, Walter M
Coffer, Henry Ford
Cohan, Leonard Hecht
Donoian, Haig Cadmus
Eblin, Lawrence Powell
Erickson, David Edward
Fitch, Robert McLellan
Force, Carlton Gregory
Fox, William
Friedman, Seymour K
Glazman, Yuli
Granquist, William Thomas
Greene, Bettye Washington
Greif, Mortimer
Gupta, Gian Chand
Hargitay, Bartholomew
Harrington, John Vincent

Heldman, Morris J
Hemstock, Glen Alton
Hiemenz, Paul C
Holihan, John Phillip
Holmes, James Murray
Hutchinson, Eric
Jacobsen, Robert Thomas
Katsanis, Eleftherios P
Kissa, Erik
Kolaian, Jack H
Kratohvil, Josip
Lissant, Kenneth Jordan
Lloyd, Thomas Blair
Lowden, J Alexander
Luttinger, Lionel
Luzzi, Louis A
Marra, Dorothea Catherine
Marshall, Charles Edmund
Matijevic, Egon
McAtee, James Lee, Jr
Medalia, Avrom Izak
Mendel, John Richard
Merrifield, Paul Elliott
Micale, Fortunato Joseph
Miller, Lewis F
Moore, Carl
Mukerjee, Pasupati
Mysels, Karol Joseph
O'Connor, Thomas Lee
Oppenheimer, Larry Eric
Perri, Joseph Mark
Peters, James
Povich, Michael Jean
Ray, Billy Roger
Reich, Irving
Rigterink, Merle Dale
Ross, Sydney
Sandell, Lionel Samuel
Schaeffer, William Dwight
Scheffler, George Henry
Schick, Martin J
Scott, George Vane
Shedlovsky, Leo
Sieglaff, Charles Lewis
Siriami, Aurelio Frederick
Slabaugh, Wendell Hartman
Smith, Larry
Sterman, Melvin David
Stratton, Charles Abner
Stryker, Lynden Joel
Studebaker, Merton Leland
Swank, Thomas Francis
Teicher, Harry
Tewari, Param Hans
Tsai, Tom Chung Hsiung
Tubbs, Robert Kenneth
Urbanic, Anthony Joseph
Van Dell, Robert Duane
Vanderhoff, John W
Van Olphen, Hendrik
Van Valkenburg, Jeptha Wade, Jr
Vold, Marjorie Jean
Vold, Robert Donald
Weintritt, Donald J
Zettlemoyer, Albert Charles

Corrosion

Ambrose, John Russell
Barnartt, Sidney
Caudle, Danny Dearl
Chandler, Ray James
Copson, Harry Rollason
Guttenplan, Jack David
Hausler, Rudolf H
Mertens, Frederick Paul
Pearlstein, Fred
Peterson, Charles Leslie
Rowe, Leonard C
Sinclair, James Douglas
Smith, Arthur Gerald
Thomason, William Hugh
Thompson, Charles Denison
Wilson, Karl Stuart

Cosmetic Chemistry

Anderson, Jon C
Bouchal, Alexander Wayne
Brandau, Robert Paul
Butensky, Irwin
Conrad, Lester I
Corbett, John Frank
Dean, Donald E
DeNavarre, Maison Gabriel
Faucher, Joseph A
Fioto, George Anthony
Fox, Gerald
Garcia, Mario Leopoldo
Hart, Una Lynch
Hoefelmeyer, Albert Bernard
Karg, Gerhart
Kass, Guss Sigmund
Kaufmann, Peter John
Kratohvil, Joseph Howard
Larson, Allan Bennett
Lerner, Louis Leonard
Lichtin, J Leon
Lobasso, Frank Anthony

Mackles, Leonard
Markland, William R
Marton, Oliver L
Maso, Henry Frank
Menkart, John
Misek, Bernard
Mullen, Patricia Ann
Newburger, Sylvan Henry
Oberstar, Helen Elizabeth
Parisse, Anthony John
Plechner, Sophie L
Prussak, Philip Morris
Randen, Neil Allen
Reinstein, Jerome Alan
Richardson, Earl Leroy
Schlossman, Mitchell Lloyd
Shine, Daniel Phillip
Siciliano, Arthur Anthony
Sorkin, Marshall
Steiger, Fred Harold
Tripathi, Uma Prasad
Woller, William Henry

Cytochemistry

Alfert, Max
Alperin, Richard Junius
Alpert, Morton
Berg, George G
Cohn, Norman Stanley
Darzynkiewicz, Zbigniew Dzierzykraj
DeJong, Richard Adams
Fennell, Richard Adams
Gill, James Edward
Goldfischer, Sidney L
Hand, Arthur Ralph
Lane, Nancy Jane
Levine, Rhea Joy Cottler
Long, Margaret Eleanor
Millhouse, Edward W, Jr
Moellmann, Gisela E Bielitz
Salthouse, Thomas N
Santanz, Ilene Sue Cottler
Simson, Jo Anne V
Therrien, Chester Dale
Yemma, John Joseph

Dairy Chemistry

Brunner, Jay Robert
Cardwell, Joe Thomas
Demott, Bobby Joe
Dunn, Stanley Austin
Ernstrom, Carl Anthon
Leeder, Joseph Gordon
Loewenstein, Morrison
Nickerson, Thomas Andrew
Parsons, John G
Richardson, Gary Haight
Sherbon, John Walter
Windlan, Harold Milton

Electrochemistry

Argue, Gary R
Baboian, Robert
Bandes, Herbert
Banks, William Patrick
Barnartt, Sidney
Bauer, Henry Hermann
Beacom, Seward Elmer
Behl, Wishvender K
Bharucha, Nana R
Biagetti, Richard Victor
Birke, Ronald Lewis
Bolmer, Perce W
Bonewitz, Robert Allen
Booe, James Marvin
Boonstra, Bram B
Breiter, Manfred Wolfgang
Brenner, Abner
Brooman, Eric William
Bruckenstein, Stanley
Buzzelli, Edward S
Cairns, Elton James
Camp, Eldridge Kimbel
Capriglio, Giovanni
Cels, Robert
Chambers, Lee Mason
Chao, Mou Shu
Childs, William Ves
Chin, Der-Tau
Claessens, Pierre
Compton, Kenneth Gordon
Cox, James Allen
Cunningham, Alice Jeanne
Dahms, Harald
Day, Robert James
DeLevie, Robert
Dick, James Gardner
Dietz, Paul Luther, Jr
Dignam, Michael John
Dill, Aloys John
DiMasi, Gabriel Joseph
Donovan, Sandra Steranka
Douglas, Larry Joe
Drake, Arthur Edwin
Eckert, John Andrew
Elmore, Glenn Van Ness
Enke, Christie George
Epstein, Barry D

Evans, Dennis Hyde
Evans, Richard Castleman
Farrington, Gregory Charles
Feldberg, Stephen William
Ferrell, D Thomas, Jr
Fester, Keith Edward
Foley, Robert Thomas
Fordyce, James Stuart
Foulke, Donald Gardner
Fox, Leonard P
Frankenthal, Robert Peter
Fresia, Elmo James
Garbarini, Victor C
Garner, Harry Richard
Glover, Roland Leigh
Goens, Duane N
Goffman, Martin
Goodkin, Jerome
Graves, Bruce Bannister
Grothen, Morris Paul
Gunther, Ronald George
Guttenplan, Jack David
Hall, David Alfred
Hamby, Larry Cully
Hamby, Drannan Carson
Hamer, Walter Jay
Hammerli, Martin
Hardman, Carl Charles
Harris, William Sidney
Harvey, Jay Arthur
Hatfield, Marcus Rankin
Hill, Derek Leonard
Hills, Stanley
Hine, Kenneth Ernest
Hoare, James Patrick
Holt, Matthew Leslie
Holtzen, Dwight Alan
Hubbard, Arthur T
Hull, Michael Neill
Hunger, Herbert Ferdinand
Huntzicker, Harry Noble
Hurd, Ray Merle
Ingruber, Otto Vincent
Jaffe, Sol Samson
James, Stanley D
James, William Joseph
Janes, Milton
Jennings, Charles Warren
Jordan, Wade Hampton, Jr
Kaplan, Nathan
Kegelman, Matthew Roland
Kennedy, John Harvey
Kimmerle, Frank
Kinoshita, Kimio
Kleiner, Walter Bernhard
Konrad, Dusan
Kordesch, Karl Victor
Kornfeil, Fred
Kozawa, Akiya
Kroger, Hanns H
Kronenberg, Marvin L
Kroon, James Lee
Krumbein, Simeon Joseph
Ksycki, Mary Joecile
Kugler, George Charles
Kuwana, Theodore
Kwak, Jan C T
Ladisch, Rolf Karl
Laity, Richard Warren
Laliberte, Laurent Hector
Larkin, David
Lerner, Harry
Levy, Samuel C
Lingane, Peter James
Lovecchio, Karen K
Lowenheim, Frederick Adolph
Lucas, Kenneth Ross
Ludwig, Frank Arno
Macdonald, Digby Donald
Macdonald, John James
Makowski, Mieczyslaw Paul
Malkin, Irving
Malloy, Alfred Marcus
Manley, Thomas Clinton
Mark, Harry Berst, Jr
Markwell, Dick Robert
Martinsons, Aleksandrs
Mather, William B, Jr
May, Charles Edward
McBreen, James
McKeon, Mary Gertrude
McLeod, Henry George
Meibuhr, Stuart Gene
Mentone, Pat Francis
Miles, Melvin Henry
Millard, Richard James
Miller, Barry
Mohilner, David Morris
Moore, Edward Weldon
Morehouse, Clarence Kopperl
Mucenicks, Paul Raymond
Muller, Rolf Hugo
Nakadomari, Hisamitsu
Newman, John Scott
Okinaka, Yutaka
Oldham, Keith Bentley
Oser, Willem

O'Sullivan, Thomas Denis
Owens, Boone Bailey
Oxley, James Edward
Parker, Edward Arthur
Payne, DeWitt Allen
Payton, Arthur David
Pearlstein, Fred
Peattie, Charles Gordon
Perkins, Richard Scott
Peters, Dennis Gail
Peterson, Miller Harrell
Pickett, David Franklin, Jr
Pierce, James Bruce
Plambeck, James Alan
Polowczyk, Carl John
Posey, Franz Adrian
Poslusny, Jerrold Neal
Post, Robert Elliott
Prokop, Robert A
Raleigh, Douglas Overholt
Rauh, Robert David, Jr
Recht, Howard Leonard
Reddy, Thomas Bradley
Reed, Allan Hubert
Rhodes, David R
Ricks, Herbert Elias
Rock, Peter Alfred
Roe, David Kelmer
Ruben, Samuel
Rudin, Donald Oliver
Sandifer, James Roy
Sanford, Robert Alan
Savitz, Maxine Lazarus
Schmid, Gerhard Martin
Schmitt, Anthony Paul
Schoenberg, Leonard Norman
Schultz, Franklin Alfred
Schwarz, William Merlin, Jr
Scott, William James
Sease, John William
Seegmiller, David W
Seki, Hajime
Senderoff, Seymour
Seyb, Edgar John, Jr
Shean, Gerald Michael, Jr
Shedlovsky, Leo
Shoesmith, David William
Silverman, Herbert Philip
Singer, Joseph
Sinkovic, Jelena
Slade, Arthur Laird
Slotter, Richard Arden
Smart, Wilson Harvey
Smith, Frank Roylance
Smith, Richard Neilson
Snyder, Donald Lee
Solomon, Frank I
Srinivasan, Supramaniam
Srinivasan, Vakula S
Staikos, Dimitri Nickolas
Steiner, John F
Stevens, William George
Stoner, Glenn Earl
Sturrock, Peter Earle
Swann, Sherlock, Jr
Symons, Philip Charles
Sympson, Robert F
Talaty, Erach R
Taylor, Dale Frederick
Taylor, Robert Morgan
Thacker, Raymond
Thornton, Roy Fred
Tidwell, Troy Haskell, Jr
Tobias, Charles W
Trachtenberg, Isaac
Van Rysselberghe, Pierre
Varimbi, Joseph
Wagenknecht, John Henry
Warde, Charles Glenn
Wedel, Richard Glenn
Weissbart, Joseph
Weissman, Paul Morton
Wells, Eugene Ernest, Jr
Will, Fritz Gustav
Willis, Grover C, Jr
Willson, Karl Stuart
Wilson, Claude E
Wilson, Richard Mac
Wong, Morton Min
Wright, Maurice Morgan
Yagi, Haruhiko
Yeager, Ernest Bill
Zelley, Walter Gauntt
Zetmeisl, Michael Joseph
Zuman, Petr

Environmental Chemistry
Bailey, Rodney Albert
Ballantine, David Stephen
Barthauer, Gerald Lee
Benton, Duane Marshall
Beran, Jo Allan
Berger, Selman A
Block, Arthur McBride
Brown, Robert Melbourne
Burrows, William Dickinson
Buyers, Archie Girard
Cadle, Stephen Howard

Carbone, Gabriel
Caret, Robert Laurent
Carrigan, Richard Alfred
Caruso, Sebastian Charles
Clark, Malcolm John Roy
Clark, Ronald Duane
Coleman, Curtis Burger
Cormack, James Frederick
Crosby, Donald Gibson
Day, James Meikle
Dietrich, Martin Walter
Douglass, Irwin Bruce
Dunlap, William Joe
Dzieciuch, Matthew Andrew
Eaton, James Edmonds
Edelson, David
Edwards, Raymond Richard
Erickson, Ronald E
Estes, Frances Lorraine
Feariheller, William Russell, Jr
Fisher, Gerald Lionel
Fix, Richard Conrad
Foerster, Donald Ray
Fogleman, Wavell Wainwright
Galloway, James Neville
Garrison, Arthur Wayne
Gehri, Dennis Clark
Glaze, William H
Gieiter, Melvin Earl
Gordon, Robert Julian
Gormly, James Ronald
Graefe, James Frederick
Groth, Richard Henry
Grunder, Fred Irwin
Hamersma, J Warren
Hannah, Sidney Allison
Haque, Rizwanul
Henry, William Mellinger
Hirsch, Roland Felix
Hoover, Thomas Burdett
Hopton, Frederick James
Horne, Ralph Albert
Hughes, Robert Alan
Hunt, Charles Maxwell
Hutzinger, Otto
Johar, Jogindar Singh
Johnson, Ray Leland
Jolley, Robert Louis
Keenan, Robert Gregory
Kemper, William Alexander
Kennedy, Edwin Russell
Khare, Mohan
Kilner, Scott Burgoyne
Kland, Mathilde June
Klein, David Henry
Knapp, Kenneth T
Kopfler, Frederick Charles
Kukin, Ira
Kusserow, Gerhard William
Lalancette, Jean-Marc
Lande, Sheldon Sidney
Langford, Russel Hal
Langner, Ralph Rolland
Lapp, Thomas William
Lee, Richard Fayao
Lockhart, Haines Boots, Jr
Lollar, Robert Miller
Mackinney, Herbert William
Martens, William Stephen
Mayer, Julian Richard
McAdie, Henry George
McClelland, Nina Irene
McCoubrey, James A
McKenzie, Donald Edward
McKinney, James David
Meier, Eugene Paul
Menczel, Jehuda H
Michalski, Raymond J
Micko, Michael M
Miles, Maurice Jarvis
Milham, Robert Carr
Nielsen, Julian Moyes
Oliver, Barry Gordon
O'Shea, Timothy Allan
Otto, Klaus
Patterson, Claire Cameron
Pierce, Richard Harry, Jr
Pool, Edwin Lewis
Potrafke, Earl Mark
Purcell, Thomas Charles
Pyle, James L
Ragaini, Richard Charles
Ramamoorthy, Subramaniam
Reeves, Richard Franklin
Reichel, William Louis
Rigler, Neil Edward
Rozell, Thomas Webb
Russell, Thomas Webb
Saeger, Victor William
Sarver, Emory William
Scaringelli, Frank Philip
Schermer, Eugene DeWayne
Schmauch, George Edward
Sears, Donald Richard
Seyb, Leslie Philip
Sheets, Ralph Waldo
Shendrikar, Arun D
Shuman, Mark S

Siegel, Sanford Marvin
Simons, Edward Louis
Skovronek, Herbert Samuel
Smith, Charles Francis, Jr
Smith, Ivan C
Smith, Michael James
Somers, Joseph Henry
Stepenuck, Stephen Joseph, Jr
Sterrett, Frances Susan
Stiles, David A
Stout, Virginia Falk
Tamers, Murry Allen
Taylor, Charlotte Clarke
Thompson, Fay Morgen
Tisue, George Thomas
Tompkins, Gary Alvin
Trieff, Norman Martin
Van Landingham, John W
Vind, Harold Pennington
Wahlgren, Morris A
Wallace, Robert William
Ward, Harold Roy
Watkins, Stanley Read
Weaver, Ervin Eugene
Weschler, Mary Charles
Wheeler, John Russell
Williams, David Trevor
Wilson, Lauren R
Witschonke, Charles Richard
Yang, John Yun-Wen
Young, David Matheson
Zaborowski, Leon Michael

Fluorine Chemistry
Adcock, James Luther
Borchardt, Hans J
Boudakian, Max Minas
Brace, Neal Orin
Cady, George Hamilton
Childs, William Ves
Clifford, Alan Frank
De Marco, Ronald Anthony
Denson, Donald D
DesMarteau, Darryl D
Gard, Gary Lee
Hamilton, Jefferson Merritt, Jr
Hill, Frederick Burns, Jr
Hudlicky, Milos
Irmiter, Theodore Ferer
Jackson, Harold Leonard
Kirchmeier, Robert Lynn
Milian, Alwin S, Jr
Miller, William Taylor
Naae, Douglas Gene
Noftle, Ronald Edward
Ogimachi, Naomi Neil
O'Malley, Robert Francis
Paciorek, Kazimiera J L
Passmore, Jack
Pavlath, Attila Endre
Perros, Theodore Peter
Peterson, James Oliver
Prokop, Robert A
Resnick, Paul R
Rexford, Dean R
Rosser, Robert William
Russell, Joseph Louis
Shackelford, Scott Addison
Sharts, Clay Marcus
Stump, Eugene Curtis, Jr
Tamborski, Christ
Young, Edmond Grove
Zack, Neil Richard

Food Chemistry
Acree, Terry Edward
Albert, Oscar J
Anderson, David W, Jr
Angelini, Pio
Bixby, John N
Bloom, Jack
Bolin, Harold R
Borenstein, Benjamin
Brause, Allan R
Brooker, Francis Milton
Burkwall, Morris Paton Jr
Bustead, Ronald Lorima, Jr
Campbell, Ada Marie
Carlin, Frances
Chakravarti, Diptiman
Chang, Stephen Szu Shiang
Chastain, Marian Faulkner
Chicoye, Etzer
Cowley, Milford A
Cramer, Archie Barrett
Cunningham, Hugh Meredith
Dassow, John Albert
DeMan, John Maria
Dowling, Joseph Francis
Dunn, Howard J
Emery, Donald F
Eskin, Neason Akiva Michael
Feliciotti, Enio
Fink, Kenneth Howard
Flora, Lewis Franklin
Flynn, Charles Edward
Fortney, Cecil Garfield, Jr
Fulger, Charles Von

Galetto, William George.
Germino, Felix Joseph
Hagenmaier, Robert Doller
Hammond, Earl Gullette
Hardwick, William Aubrey, Jr
Hayward, Frederick Warren
Henika, Richard Grant
Hester, Ellen Elizabeth
Heydanek, Menard George, Jr
Hillig, Fred
Hlavacek, Robert John
Hoepfinger, Lynn Morris
Holmes, Edward Lawson
Hood, Lamartine Frain
Horwitz, William
Hurst, Thomas Lighthall
Jacobson, Glen Arthur
Joiner, Robert Russell
Josephson, Ronald Victor
Joslyn, Maynard Alexander
Julien, Jean-Paul
Kean, Chester Eugene
Kimoto, Walter Iwao
Kincs, Frank Raymond
Kinsella, John Edward
Klein, Barbara P
Knapp, Frederick Whiton
Kolor, Michael Garrett
Krishnamurthy, Ramanathapur Gundachar
Lapuck, Jack Lester
Lillard, Dorris Alton
Lime, Bruce James
Lipman, Harry Jerome
Litman, Irving Isaac
Mackay, Donald Alexander-Morgan
Manley, Charles Howland
Marable, Nina Louise
Maselli, John Anthony
Matthews, Richard Finis
McGugan, Wesley Alexander
McKinney, Leonard Laurence
Mergentime, Max
Min, David Byong
Modderman, John Philip
Mohlenkamp, Marvin Joseph, Jr
Moss, Valentin G
Murphy, Elizabeth Wilcox
Myhre, David V
Nakai, Shuryo
Nawar, Wassef W
Nishida, Toshiro
Nutting, Lee
Ockerman, Herbert W
O'Connor, David Evans
O'Connor, Michael Gerald
Pareles, Stephen Ronald
Paul, Pauline Constance
Peng, Andrew Chung Yen
Petrowski, Gary E
Pomerantz, Reuben
Pons, Walter A, Jr
Powell, Eugene Loren
Pratt, Dan Edwin
Richardson, Thomas
Rosen, Joseph David
Roth, Howard
Russell, Gerald Frederick
Salwin, Harold
Samuels, Robert Bireley
Sanderson, Gary Warner
Satterlee, Lowell Duggan
Schafer, Mary Louise
Sevenants, Michael R
Shallenberger, Robert Sands
Shaver, Kenneth John
Smith, Laura Lee Weisbrodt
Smith, Lloyd Muir
Souder, Philip Walburn
Spinelli, John
Steffen, Albert Harry
Sullivan, Andrew Jackson
Swisher, Horton Edward
Tannenbaum, Steven Robert
Tiemstra, Peter J
Trelease, Richard Davis
Tressler, Donald Kiteley
Van der Kloot, Albert Peter
Van Duyne, Frances Olivia
Varsel, Charles John
Wahba, Isaac Jack
Waitman, Reuben Henry
Walter, Reginald Henry
Watking, Arthur Ernest
Wasserman, Aaron E
Weast, Clair Alexander
Weissler, Harold Edward
Westbrook, George Franklin
Whitney, Robert McLaughlin
Wilkinson, Joseph Lee
Williams, James J
Winston, James J
Woods, Alvin Edwin
Zaika, Laura Larysa

Forensic Science
Abrahamson, Dean Edwin
Arnold, Mary Tryson
Bills, Donald Duane

Bosen, Sidney Frederick
Boulet, Marcel
Brunelle, Richard Leon
Byall, Elliott Bruce
Cain, Robert Farmer
Campbell, James Alexander
Campbell, Jeptha Edward, Jr
Cantu, Antonio Arnold
Coldwell, Blake Burgess
Crandlemere, Robert Wayne
DeForest, Peter Rupert
Dwyer, James Michael
Eisner, Robert Linden
Freimuth, Henry Charles
Frank, Richard Stephen
Garner, Daniel Dee
Goertz, Grayce Edith
Gunn, John William, Jr
Hutchin, Maxine E
Jones, Peter Frank
Joseph, Alexander
Kerley, Ellis R
Kier, Lawrence Charles
Kingston, Charles Richard
Kline, Ralph Willard
Kuramoto, Simpey
Lateef, Abdul Bari
Lee, Henry C
Lipset, Frederick Roy
Lorch, Steven Kalman
Moses, Alfred James
Nauman, Louis William
Perros, Theodore Peter
Plautz, Donald Melvin
Rehling, Carl John
Roche, George William
Rudzitis, Edgars
Saferstein, Richard
Schweigert, Bernard Sylvester
Smith, Charles G
Smith, Charles Henry
Sobol, Stanley Paul
Stone, Irving Charles, Jr
Taylor, Ronald Lee
Tindall, Charles Gordon, Jr
Vonhof, Daniel William
Walton, George
Wessel, John Emmit
Whittle, Philip Rodger
Wolten, Gerard Martin
Yip, George

High Temperature Chemistry

Ackerman, Raymond J
Becker, Aaron Jay
Blackburn, Paul Edward
Buehler, Alfred
Carpenter, John Harold
Castle, Peter Myer
Corbeils, Roger
Cubiciotti, Daniel David, Jr
Dowell, Michael Brendan
Edwards, Jimmie Garvin
Epple, Robert (Paul)
Freedman, Eli (Hansell)
Gole, James Leslie
Gray, Eoin Wedderburn
Grimley, Robert Thomas
Haschke, John Maurice
Hill, Russell John
Johnson, Keith Edward
Kelly, Minton J
Kohl, Fred John
Krikorian, Oscar Harold
Krupka, Milton Clifford
Levine, Herman Saul
Lin, Sin-Shong
Lincoln, Kenneth Arnold
Lindener, Terrence Bradford
Meschi, David John
Olson, William Marvin
Papp, Cornelius Alfred
Perkins, Janet Sanford
Peterson, Dean Everett
Powers, Dana Auburn
Prentice, Jack L
Sausville, Joseph Winston
Schobert, Harold Harris
Sigai, Andrew Gary
Steiger, Roger Arthur
Sturgeon, George Dennis
Ueltz, Herbert Frank
Uy, Oscar Manuel
Valdsaar, Herbert
Wagner, Lawrence Carl
Wahlbeck, Philip Glenn
Walsh, Patrick Noel
Wang, Ke-Chin
Ward, John William
Wilks, Philip Howard
Willson, Philip James
Worrell, Wayne L

Histochemistry

Baker, John Richard
Balogh, Karoly
Becker, Norvin Howard
Brown, Stephen Clawson

Brownscheidle, Carol Mary
Connors, Natalie Ann
Cooper, John (Hanwell)
Gerold, Nicolas John
Gersh, Isidore
Glick, David
Goldberg, Benjamin
Hack, Marvin Howard
Hanker, Jacob S
Hasegawa, Junji
Higginbotham, Frances Heffin
Iyengar, V K Sundararaja
Kashiwa, Herbert Koro
Lhotka, John Francis, Jr
Longley, James Baird
Manning, John Paul
Manocha, Sohan Lall
McMillan, Paul Junior
Murphy, Krishna A S
Nakane, Paul K
Nuki, Klaus
Pinkstaff, Carlin Adam
Pritchard, Hayden N
Puchtler, Holde
Smith, Allen Anderson
Sobel, Harold John
Steeves, Harrison Ross, III

History of Chemistry

Cash, Rowley Vincent
Catsiff, Ephraim Herman
Eblin, Lawrence Powell
Haines, Florence Catherine
Hawthorne, Robert Montgomery, Jr
Hindersinn, Raymond Richard
Kim, Chung Sul
Latham, Roger Alan
Martus, Joseph Armand
Miles, Wyndham Davies
Ramsay, Ogden Bertrand
Wessing, Ritchie A

Immunochemistry

Adams, Ernest Clarence
Aladjem, Frederick
Arquembourg, Pierre Charles
Bangasser, Susan Andretta
Banovitz, Jay Bernard
Benedict, Albert Alfred
Binkley, Francis
Bishop, Claude Titus
Bishop, David C
Borek, Felix
Bradshaw, Claire Margaret
Braune, Maximilian O
Briton, Doyle
Butler, Vincent Paul, Jr
Callighan, Owen Hugh
Capone, James Joseph
Cernosek, Stanley Frank, Jr
Chambliss, Keith Wayne
Chang, Chin-Hai
Cinader, Bernhard
Clem, Lester William
Cohn, Melvin
Cote, Raymond-Henri
Cowan, Keith Morris
Creech, Hugh John
Dandliker, Walter Beach
DiCapua, Richard Anthony
Dray, Sheldon
Dubiski, Stanislaw
Egan, Marianne Louise
Engvall, Eva Susanna
Erzler, Marilynn Edith
Feigen, George Alexander
Fox, Alfred Earl
Froese, Arnold
Garvey, Justine Spring
Genco, Robert J
Gopalakrishnan, Pungampoondi Velamur
Green, Gerald
Grey, Howard M
Hakomori, Sen-Itiroh
Harowitz, Felix (Michael)
Heidelberger, Michael
Humphreys, Robert Edward
Inman, Franklin Pope, Jr
Ishizaka, Kimishige
Johnson, Brian John
Jordan, Russell Thomas
Kallestad, Steven Bix
Karol, Meryl Helene
Kreiter, Victor Peter, Jr
Kuo, Chao-Ying
Lehrer, Harris Irving
Leslie, Gerrie Allen
Levine, Lawrence
Lietze, Arthur
Liu, Chi Tan Chang
Lloyd, Kenneth Oliver
Ma, Wai-Sai
MacPherson, Catherine Frances Conway
Malkiel, Saul
Mann, Lewis Theodore, Jr
Manski, Wladyslaw J
Markowitz, Harold

Marucci, Americo Alvin
McKinney, Roger Minor
Melcher, Ulrich Karl
Michaeli, Dov
Miller, Herman T
Miller, Thomas Edward
Mudgett, Meredith
Munjal, Devidayal
Myers, Lyle Leslie
Nisonoff, Alfred
Olson, Anita Cora
Oroszlan, Stephen
Pincus, Jack Howard
Pollack, William
Polley, Margaret J
Poulik, Dave
Pressman, David
Pruzanski, Waldemar
Quinn, Richard Paul
Reben, Ralph Armand
Reisfeld, Ralph Alfred
Rule, Allyn H
Sage, Harvey J
Scalfarotto, Robert Emil
Schlamowitz, Max
Shaw, Norman Yon-Shong
Sheppard, David E
Smith, Nathan Lewis, III
Springer, Georg F
Stelos, Peter
Stemke, Gerald W
Stolfi, Robert Louis
Swanborg, Robert Harry
Sweeney, Michael Joseph
Treadway, William Jack, Jr
Tripodi, Daniel
Tyler, Jean Mary
Vannier, Wilton Emile
van Oss, Carel J
Voss, Edward William, Jr
Weetall, Howard H
Weimer, Henry Eben
Weliky, Norman
White, Gordon Justice
Wirtz, George H
Wright, George Leonard, Jr
Yoder, John Menly
Young, Janis Dillaha
Zarco, Romeo Morales

Industrial Chemistry

Abrams, Ellis
Abrams, Irving Melvin
Adams, James William
Adams, Mark F
Adler, William Benson
Aguilo, Adolfo
Alsberg, Henry
Ancona, Umberto
Anderson, Ralph F
Antonsen, Donald Hans
Anzenberger, Joseph F, Sr
Archer, Wesley Lea
Atkins, Don Carlos, Jr
Atwood, Mark Trevor
Bablec, John Stanley, Jr
Bailey, Milton
Bailey, David Ralph
Ball, Fred
Bamberger, Curt
Bancy, Ronald Howard
Barnstorff, Henry Dreses
Barth-Wehrenalp, Gerhard
Bartley, William J
Bauermeister, Herman Otto
Beard, William Quimby, Jr
Behrman, Abraham Sidney
Bell, Ian
Bennett, Harry
Benzinger, William Donald
Berkowitz, Sidney
Black, Otis Deitz
Bluestein, Claire
Blumenberg, Karl Edward
Blumenthal, Warren Barnett
Boardman, Harold
Bockstahler, Theodore Edwin
Boyle, Richard James
Brain, Devin King
Braley, Silas Alonzo, Jr
Brandner, John David
Bromberg, Milton Jay
Broun, Thorowgood Taylor, Jr
Brown, Jerome Engel
Brown, Lloyd H
Brown, Robert Getman
Bryan, Loren Aldro
Buhle, Emmett Loren
Burke, Roger E
Burney, Donald Eugene
Burrows, Walter Herbert
Busby, Hubbard Taylor, Jr
Buurman, Clarence Harold
Calbo, Leonard Joseph

Calkins, William Harold
Campbell, George Washington, Jr
Carlson, Ronald H
Carpenter, Lee Graydon
Chadwick, David Henry
Cheavens, Thomas Henry
Christena, Ray Clifford
Cipriani, Cipriano
Clarke, David Bruce
Clegg, William Josiah
Clement, Robert Alton
Cofrancesco, Anthony J
Connors, John Philip
Conrad, William Matthew
Cooper, Wilson Wayne
Copulsky, William
Cowles, Craig Schuyler
Cox, Fred Ward, Jr
Craig, Alan Daniel
Cunningham, Newlin Buchanan
Dermer, Otis Clifford
Delorme, Joachim
Deutsch, Dennis Leslie
Deviny, Edward John
Dillard, Beverly Mincey
Dixon, John Francis Clemow
Downing, Joseph Richard
Drach, John Edward
Dybalski, Jack Norbert
Dyroff, David Ray
Earle, Ralph Hervey, Jr
Edman, Dwight Douglas
Edwards, Ben E
Eiland, Ehrlich Mayo
Einne, Lester Oscar
Elden, Richard Edward
Ellendman, Merrill
Ernst, Richard Edward
Evans, Allison Bickle
Fahl, Roy Jackson, Jr
Fajans, Edgar W
Favis, Dimitrios Vasilios
Feeman, James Frederic
Feiler, William A, Jr
Ferguson, John Allen
Fisher, Elton
Fitch, Steven Joseph
Flanagan, John Vernon
Ford, Richard Westaway
Forsyth, Paul Francis
Foster, Harold Marvin
Foster, Walter Edward
Fox, Gerald
Fox, Neil Stewart
Frank, Robert Roy
Frane, Simon
Frank, Victor Samuel
Frank, William Benson
Friedman, Harold Bertrand
Fugger, Joseph
Gardner, William H
Geiger, James Woodrow
Geiger, Marion Braxton
Gilbert, Arthur Donald
Gilbert, Walter Wilson
Gillespie, Arthur Samuel, Jr
Gipson, Robert Malone
Girolami, Libero L
Glasebrook, Arthur Lawrence
Glidewell, Marvin Elmer
Gloyer, Stewart Edward
Gold, Daniel Howard
Gozdan, Walter Joseph
Gradeff, Peter S
Gregory, Arthur Stanley
Groszos, Stephen Joseph
Habib, David Peter
Halbedel, Harold Seibert
Hallada, Calvin James
Harper, Jon Jay
Harrison, Hugh Thomas
Hart, William Ardley
Hartman, Robert John
Hay, Peter Marsland
Hayer, Roy G
Hedrick, Robert Jerry
Heffron, Peter John
Heins, Richard William
Heins, Boyce Dewayne
Heneghan, Leo Francis
Henery, James Francis
Herkes, Frank Edward
Higgins, James Francis
Hoffman, Donald Oliver
Holland, Doyt K, Jr
Holland, William Frederick
Howell, John Kenneth, Jr
Huff, George Franklin
Hughes, O Richard
Humphrey, Bingham Johnson
Hunter, Byron Alexander
Idol, James Daniel, Jr
Ingwalson, Raymond Wesley
Innes, John Edwin
Ivett, Reginald William
Jabour, Ramzi Jibrail

Jaeger, Charles Wayne
Jaruzelski, John Janusz
Johnson, Fred Lowery, Jr
Johnson, Roger Alvin
Johnston, John Derland
Johnston, Katharine Gentry
Jones, Haydn
Jordan, John William
Jordon, Thomas Earl
Joseph, Solomon
Karo, Wolf
Kassner, James Edward
Kastens, Merritt Louis
Kennerly, George Warren
Kinosz, Donald Lee
Kissel, Charles Louis
Klanica, Andrew Joseph
Knapp, Malcolm Hammond
Kneisley, Joseph Wayne
Knipple, Warren Russell
Knollmueller, Karl Otto
Knott, Donald MacMillan
Komoto, Robert Gordon
Kondis, Thomas John
Kosak, John R
Kremers, Howard Earl
Kuebler, John Ralph, Jr
Kuehn, Christa Gisela
Kwiatek, Jack
LaBarge, Robert Gordon
Laity, John Lawrence
Lannert, Kent Philip
Laszlo, Tibor S
Leddy, James Jerome
Lee, Robert James
Leonard, John Alex
Lerner, Lawrence Robert
Levy, Leon Bruce
Lillwitz, Lawrence Dale
Linder, Seymour Martin
Liu, Jih-Hua
Lloyd, Thomas Blair
Lobunez, Walter
Lofton, William Milford, Jr
Logan, Ted Joe
Luberoff, Benjamin Joseph
Lundsted, Lester Gordon
Lynch, Dan K
MacIver, Donald Stuart
MacPeek, Donald Lester
Mahncke, Henry Elmore
Maizell, Robert Edward
Mallonee, James Edgar
Manning, Harold Edwin
Marchant, Cosmo
Marquis, David Maley
Martell, Michael Joseph, Jr
Martin, James Cuthbert
Massingill, John Lee, Jr
McBane, Bruce Newton
McCloskey, Allen Lyle
McCown, Joseph Dana
McDaniel, Edgar Lamar, Jr
McDuff, James Milton
McGrew, George Thomas
McGroarty, Joseph A
McGuire, Stephen Edward
McIntosh, Alexander Omar
Meadows, Geoffrey Walsh
Merner, Richard Raymond
Meyer, Garson
Milford, Alan Hackney
Millhiser, Frederick Roy
Moissides-Hines, Lydia Elizabeth
Montgomery, Stewart Robert
Moore, Earl Phillip
Moore, Ralph Bishop
Morehouse, Donald S, Jr
Moroni, Eneo C
Morris, Leo Raymond
Morse, Ronald Loyd
Mortada, Mohamed
Motz, Kaye La Marr
Mount, Lloyd Gordon
Mowry, David Thomas
Munies, Robert
Muse, Joel, Jr
Naglieri, Anthony N
Natowsky, Sheldon
Navratil, James Dale
Nazy, John Robert
Nedwick, John Joseph
Newman, Pauline
Nielsen, Robert Peter
Noack, Manfred Gerhard
Nowotny, Kurt A
Obenland, Clayton O
Pader, Morton
Palmer, John Frank, Jr
Parks, Kenneth Lee
Pecka, James Thomas
Pedroza, Gregorio Cruz
Peery, Clifford Young
Peppler, Richard Bond
Perry, Randolph, Jr
Peters, Lynn Randolph
Petke, Frederick Edward
Pettit, Paul Herschel, Jr

Planck, Ralph Waldo
Quinan, James Roger
Quinn, John Frederick
Rabold, Gary Paul
Rabourn, Warren Joseph
Railing, Wilford Edward
Rau, Eric
Reid, Kenneth Ian Gower
Renfrew, Edgar Earl
Richard, William Ralph, Jr
Richards, Joseph Dudley
Richter, Robert Freeland
Rife, Robert Seldon
Riley, Reed Farrar
Rocklin, Albert Louis
Rogers, Ralph Loucks
Rogers, Thomas Ralph
Rolston, Charles Hopkins
Rondestvedt, Christian Scriver, Jr
Rosenthal, Fritz
Rossow, Alfred George
Rowe, Jay Elwood
Rueggeberg, Walter Herman Carl
Saffer, Charles Martin, Jr
Salsbury, Jason Melvin
Savereide, Thomas J
Scardera, Michael
Schick, Margery Leone
Schlaeger, Albert Joseph
Schmidt, Eckart W
Schoenberg, Emanuel
Schroeder, Hansjuergen Alfred
Schumacher, Joseph Charles
Schwartz, Michael Muni
Schwarz, John Samuel Paul
Scott, Robert Neal
Seelbach, Charles William
Seiferle, Edwin James
Shay, Edward Griffin
Sherman, Paul Dwight, Jr
Shulman, George
Sieckhaus, John Francis
Siefker, Joseph Alphonse
Sienicki, Edward Alexander
Silvernail, Walter Lawrence
Sisco, William E
Skorcz, Joseph Anthony
Slagan, Peter Michael
Sloan, Martin Frank
Smith, William Edward
Smyrk, Charles McCahan, Jr
Solenberger, John Carl
Sowers, Edward Eugene
Steiner, Russell Irwin
Stern, Alfred
Stevens, Michael Fred
Stockburger, George Joseph
Strojny, Edwin Joseph
Strother, Corneille Osburn
Sudweeks, Walter Bentley
Summers, William Allen, Jr
Tamburin, Henry John
Tams, William P
Tanner, Alan Roger
Teague, Dwight Maxwell
Thelen, Edmund
Thompson, Quentin Elwyn
Toeniskoetter, Richard Henry
Toland, William Gridley, Jr
Toy, Arthur Dock Fon
Trimmer, Robert Whitfield
Ulrey, Stephen Scott
Unruh, Jerry Dean
Vanderkooi, William Nicholas
Vejvoda, Edward
Verburg, Robert Martin
Von, Isaiah
Vopicka, Edward
Vrieland, Gail Edwin
Walbrick, Johnny Mac
Walker, Wellington Epler
Wang, Victor Kai-Kuo
Washcheck, Paul Howard
Webb, Philip Gilbert
Wehner, Philip
Weipert, Eugene Allen
Wheat, Percy Wayne
Wheaton, Robert Miller
Whitt, Carlton Dennis
Wiles, Robert Allan
Wojtowicz, John Alfred
Wu, Ching-Yong
Wysocki, Allen John
Young, Charles Albert
Young, Herbert Lewis
Zief, Morris
Zoss, Abraham Oscar

Leather Chemistry
Bitcover, Ezra Harold
Constantin, James Michael
Lollar, Robert Miller
Merritt, Robert Edward
Panepinto, Frank William
Prentiss, William Case
Robinson, John W, Jr

Lipid Chemistry
Ansari, Ali
Archibald, Francis Magoun
Avena, Remedios M
Babayan, Vigen Khachig
Baumann, Wolfgang Josef
Bernholz, William Francis
Campbell, Kenneth Neilsen
Emken, Edward Allen
Ferguson, Karen Anne
Gardner, Harold Wayne
Haines, Thomas Henry
Kates, Morris
Kepler, Carol R
Koritala, Sanbasivaroa
Lundberg, Walter Oscar Paul
Marmer, William Nelson
McConathy, Walter James
Nevenzel, Judd Cuthbert
Pryde, Everett Hilton
Rooney, Seamus Augustine
Ross, Ronald Burns
Smith, Cecil Randolph, Jr
Sullivan, Daniel Richard
Szuhaj, Bernard F
Weber, Evelyn Joyce
Weiss, Theodore Joel
Witting, Lloyd Allen
Worthington, Robert Earl

Medicinal Chemistry
Abdel-Monem, Mahmoud Mohamed
Aboul-Enein, Hassan Yousseff
Abraham, Donald James
Abramson, Hanley N
Ackerman, James Howard
Adamski, Robert J
Advani, Shyam Bhojraj
Agrawal, Krishna Chandra
Albertson, Noel Frederick
Albrecht, William Lind
Allen, George Rodger, Jr
Anderson, LeRay J
Anderson, Wayne Keith
Archer, Sydney
Armstrong, Paul Douglas
Bajzer, William Xavier
Baker, John Keith
Balant, Charles Paul
Bambury, Ronald Edward
Bardos, Thomas Joseph
Bauer, Ludwig
Beamer, Robert Lewis
Beasley, James Gordon
Beisler, John Albert
Bell, Stanley C
Belletire, John Lewis
Bergen, John Vanderveer
Berger, Joel Gilbert
Berger, Leo
Berges, David Alan
Berkoff, Charles Edward
Berlin, Kenneth Darrell
Bernstein, Jack
Beverung, Warren Neil
Bhat, Venkatramana Kakekochi
Bigham, Eric Cleveland
Blank, Benjamin
Blankley, Clifton John
Blanton, Charles DeWitt, Jr
Bloom, Barry Malcolm
Boblitt, Robert LeRoy
Bodor, Nicolae Stefan
Bohm, Howard Allan
Boots, Marvin Robert
Borgman, Robert John
Borne, Ronald Francis
Boswell, George A, Jr
Boyd, Robert Edward, Sr
Brittelli, David Ross
Broom, Arthur Davis
Brown, Horace Dean
Brown, John Max
Brumbaugh, Richard J
Brunett, Emery W
Brunings, Karl John
Buchanan, Ronald Leslie
Burcsu, James Edward
Burger, Alfred
Cadwallader, Donald Elton
Caille, Gilles
Cain, Cornelius Kennady
Campaigne, Ernest Edwin
Campbell, Barbara Knapp
Cannon, Joseph G
Capps, David Bridgman
Carabateas, Philip M
Carr, Albert A
Carr, John B
Casey, Adria Catala
Cassady, Armand Ralph
Cassola, John Mac
Castagnoli, Neal, Jr
Chae, Kun
Chandler, Reginald Frank
Chang, Ching-Jer
Chaudhari, Bipin Bhudharlal

Cheney, Lee Cannon
Chheda, Girish B
Child, Ralph Grassing
Chinn, Leland Jew
Chorvat, Robert John
Chow, Alfred Wen-Jen
Clark, Larry P
Clark, Robert Long
Clarke, Frank Henderson
Clayton, John Mark
Clinton, Raymond Otto
Coburn, Robert A
Cocolas, George Harry
Cohen, Elliott
Cole, Jack Robert
Collins, Paul Waddell
Cook, Elton Straus
Cook, James Minton
Cory, Michael
Counsell, Raymond Ernest
Coutts, Ronald Thomson
Coward, James Kenderdine
Cragoe, Edward Jethro, Jr
Craig, Paul Norman
Cramer, Richard David, III
Crenshaw, Ronnie Ray
Cullison, David Arthur
Cunov, Carl Henry
Currie, Bruce LaMonte
Cushman, Mark
Cwalina, Gustav Edward
Czuba, Leonard J
Daniels, Troy Cook
Darling, Charles Milton
Daum, Sol Jacob
DeGrande, Gary Gaston
Degraw, Joseph Irving, Jr
Delgado, Jaime Nabor
Delia, Thomas J
Demerson, Christopher
Denton, John Joseph
DeRose, Anthony Francis
Dev, Vasu
Diamond, Julius
Dice, John Raymond
Dickinson, William Borden
Digenis, George A
Dines, Allen I
Dolfini, Joseph E
Doorenbos, Norman John
Dorman, Linneaus Cuthbert
Douglas, Bryce
Driscoll, John Stanford
Dunn, William Joseph
Duschinsky, Robert Charles
Dutta, Shib Prasad
Erhardt, Paul William
Erickson, Edward Herbert
Essery, John M
Falco, Elvira Allegra (Mrs Bass)
Farkas, Eugene
Farmer, Patrick Stewart
Fedrick, James Love
Finkelstein, Jacob
Fitzgerald, Thomas James
Fitzloff, John Frederick
Fleming, Robert Willerton
Fleysher, Maurice Henry
Fliedner, Leonard John, Jr
Foltz, Calvin Martin
Foye, William Owen
Freedman, Jules
Frey, Albert Joseph
Fried, John
Fries, David Samuel
Fryer, Rodney Ian
Fujii, Akira
Fullerton, Dwight Story
Gall, Martin
Galpin, Donald R
Garland, William Arthur
Gearien, James Edward
Gennaro, Alfonso Robert
Gilmore, William Franklin
Ginos, James Zissis
Gleason, Clarence Henry
Gobuty, Allan Harvey
Goel, Om Prakash
Gordon, Eric Michael
Gordon, Maxwell
Granchelli, Felix Edward
Grethe, Guenter
Grier, Nathaniel
Griggs, Lee Jackson
Grisar, Johann Martin
Gringauz, Alex
Grunewald, Gary Lawrence
Gschwend, Heinz W
Gupta, Shyam Kirti
Gurwara, Sweet K
Hach, Vladimir
Hager, George Philip, Jr
Hammar, Walton James
Hanna, Patrick E
Hansen, Donald Willis, Jr
Harbert, Charles A
Hardtmann, Goetz E
Harmon, John Baltzell

CHEMISTRY

Harmon, Robert E
Harrell, William Broomfield
Hartman, Kenneth Eugene
Hedrich, Loren Wesley
Heindel, Ned Duane
Heinzelman, Richard Voorhees
Henderson, Richard Elliott Lee
Henkel, James Gregory
Henry, David Weston
Heyd, William Ernst
Hirsch, Allen Frederick
Hite, Gilbert J
Hoff, Dale Richard
Hoffman, Jacob Matthew, Jr
Holcomb, George N
Holden, Kenneth George
Holland, Gerald Fagan
Hong, Chung Il
Hongberg, Irwin Leon
Hoover, Irving R
Hoover, John Russel Eugene
Horgan, Stephen William
Houlihan, William J
Howell, Charles Frederick
Huang, Chian Li
Hughes, John Lawrence
Hull, Robert Lee
Hullar, Theodore Lee
Hwang, Bruce You-Huei
Ishaq, Khalid Sulaiman
Jackson, Thomas Edwin
Jacobson, Arthur E
Jacoby, Ronald Lee
Jenkins, Herndon
Jirkovsky, Ivo
Johns, William Francis
Johnson, Howard (Laurence)
Johnson, Robert Ed
Johnson, Roland Norman
Jones, Eldon Melton
Jones, Peter Hadley
Jones, Winton D, Jr
Jorgensen, Eugene Clifford
Juby, Peter Frederick
Judd, Claude Ivan
Kagan, Fred
Kaiser, Carl
Kalm, Max John
Kanojia, Ramesh Maganlal
Kapoor, Amrit Lal
Kasparek, Stanislav Vaclav
Kaufman, Karl Lincoln
Kellert, James Clarence, Jr
Kelley, James LeRoy
Kelsey, John Edward
Kerridge, Kenneth A
Kier, Lemont Burwell
Kier, James Francis
Kingsbury, William Dennis
Klauber, Dieter Heinz
Klayman, Daniel Leslie
Klimstra, Paul D
Klingele, Harold Otto
Knaus, Edward Elmer
Knevel, Adelbert Michael
Koch, Richard Carl
Kochar, Man Mohan
Kornfeld, Edmund Carl
Korytnyk, Walter
Kramer, Stanley Phillip
Kreider, Eunice S
Kreighbaum, William Eugene
Kriesel, Douglas Clare
Krimmel, Peter
Kumkumian, Charles Simon
La Pidus, Jules Benjamin
La Rocca, Joseph Paul
Larsen, Aubrey Arnold
Lasslo, Andrew
Lattin, Danny Lee
Lawson, John Edward
Lee, Kuo-Hsiung
Lemke, Thomas Lee
Levi, Irving
Lewis, Neil Jeffrey
Loev, Bernard
Loo, Ti Li
Lu, Matthias Chi-Hwa
Luts, Heino Alfred
Magarian, Robert Armen
Magerlein, Barney John
Marshall, David Jonathan
Marshall, Winston Stanley
Martin, Alfred
Martin, Yvonne Connolly
Maryanoff, Bruce Eliot
Mason, Robert C
Mathison, Ian William
Matsumoto, Ken
May, Everette Lee
McCarthy, Walter Charles
McCarty, Frederick Joseph
McCauly, Ronald James
McFarland, James William
Mechinski, Witold
Merianos, John James
Mertes, Mathias Peter
Meschino, Joseph Albert

Meyer, Rich Bakke, Jr
Micetich, Ronald George
Mickles, James
Mikhail, Adel Ayad
Mikolasek, Dougals Gene
Milkowski, John David
Miller, Duane Douglas
Milne, George McLean, Jr
Moffett, Robert Bruce
Moissides-Hines, Lydia Elizabeth
Mokotoff, Michael
Montgomery, John Atterbury
Moon, Byong Hoon
Mooney, Paul David
Moore, Maurice Lee
Moreland, Walter Thomas, Jr
Morin, Richard Dudley
Morozowich, Walter
Moskal, Richard Edward
Murdock, Keith Chadwick
Murphey, Robert Stafford
Murphy, Charles Franklin
Murray, Wallace Jasper
Myers, Gordon Sharp
Nagasawa, Herbert Tsukasa
Nelson, Kenneth Fred
Nelson, Wendel Lane
Nematollahi, Jay
Neumann, Marguerite
Neumeyer, John L
Neustadt, Bernard Ray
Nichols, David Earl
Nieforth, Karl Allen
Ning, Robert Ye Fong
Nogrady, Thomas
Novomy, Jaroslav
Nunes, Mathews Anthony
Okamura, William H
Osman, Jack
Pandya, Mahendra Kodarial
Pappo, Raphael
Parish, Roger Cook
Parker, Roger A
Partch, Richard Earl
Peet, Norton Paul
Perchonock, Carl David
Perron, Yvon G
Petracek, Francis James
Piantadosi, Claude
Ploch, Richard Paul
Podrebarac, Eugene George
Portlock, David Edward
Portoghese, Philip S
Powers, Larry James
Prugh, John Drew
Quinn, Frank Russell
Quintana, Ronald Preston
Raabe, Austin Bauer
Rajagopalan, Parthasarathi
Ranade, Vinayak Vasudeo
Ransford, George Henry
Rasmusson, Chris Royce
Rasmusson, Gary Henry
Razdan, Raj Kumar
Redalieu, Elliot
Redl, George
Reed, Fred DeWitt, Jr
Reichenthal, Jules
Reid, Jack Richard
Remers, William Alan
Remy, David Carroll
Reynolds, Brian Edgar
Rice, Kenner Cralle
Riley, Thomas Nolan
Ringold, Howard Joseph
Robertson, Jerry Earl
Robinson, Arthur Brouhard
Roche, Edward Browning
Roderick, William Rodney
Roehrig, Gerald Ralph
Roll, David Byron
Rosati, Robert Louis
Ross, Stephen T
Rovnyak, George Charles
Ruskin, Bernice Heyman
Ruyle, William Vance
Safdy, Max Errol
Safir, Sidney Robert
Saggiomo, Andrew Joseph
Salemi, Oreste Leroy
Salvador, Romano Leonard
Santora, Norma Julian
Sastry, Bhamidipaty Venkata Rama
Schaaf, Thomas Ken
Schaefer, Howard John
Schnetter, Richard Anselm
Schnur, Rodney Caughren
Schultz, Everett Maynard
Schumacher, Ignatius
Schumann, Edward Lewis
Schwan, Thomas James
Schwartz, Charles
Schwartz, Samuel Meyer
Schwender, Charles Frederick
Scott, Francis Leslie
Scott, Kenneth Richard

Sellstedt, John H
Sestanj, Kazimir
Sharma, Ram Ashrey
Sharma, Rameshwar Kumar
Sharpe, Thomas Ray
Sheely, Yoder Fulmer
Sheehy, Richard Moats
Shelfer, Eli
Shelver, William H
Shepard, Kenneth LeRoy
Shepherd, Robert Gordon
Sherwood, Bob Edwin
Shipchandler, Mohammed Tyebji
Shiue, Chyng-Yann
Shone, Robert L
Short, Franklin Willard
Shough, Herbert Richard
Siggins, James Ernest
Simmons, Harry Dady, Jr
Simon, David Zvi
Singh, Rudra Pratap
Sinkula, Anthony Arthur
Sinsheimer, Joseph Eugene
Small, LaVerne Doryen
Smith, Charles Howard
Smith, Charles Irvel
Smith, Pierre Frank
Smith, Robert Lawrence
Smith, Terry Douglas
Snyder, Harry Raymond, Jr
Soine, Taito Olaf
Solis-Gaffar, Maria Corazon
Solo, Alan Jere
Solomons, William Ebenezer
Soloway, Albert Herman
Soloway, Harold
Stansloski, Donald Wayne
Stein, Herman H
Steinman, Martin
Sternbach, Leo H
Stiver, James Frederick
Stratford, Eugene Scott
Stubbins, James Fiske
Studt, William Lyon
Suh, John Taiyoung
Sumner, Darrell Dean
Sundelin, Kurt Gustav Ragnar
Sutton, Blaine Mote
Sweeney, Thomas Richard
Szczesniak, Raymond Albin
Tamorria, Christopher Richard
Taylor, Michael Lee
Teitel, Sidney
Telang, Vasant G
Temple, Carroll Glenn
Temple, Davis Littleton, Jr
Thio, Alan Poo-An
Thomas, Alford Mitchell
Thomas, Richard Dean
Thompson, Bobby Blackburn
Tinney, Francis John
Topliss, John G
Townsend, Leroy B
Triggle, David J
Turcotte, Joseph George
Tweit, Robert Christopher
Ulrich, Floyd Seymour
Umen, Michael Jay
Unangst, Paul Charles
Ursprung, Joseph John
Van Meter, Clarence Taylor
Varma, Ravi Kannadikovilakom
Vazakas, Aristotle John
Vida, Julius Adalbert
Vince, Robert
Voldeng, Albert Nelson
von Strandtmann, Max
Wade, James Joseph
Wade, Peter Cawthorn
Wagner, Eugene Ross
Walser, Armin
Walter, Wilbert George
Wang, Samuel S M
Ward, Frederick Edmund
Warner, Paul Longstreet, Jr
Warner, Victor Duane
Warren, James Donald
Wasson, Burton Kendall
Waters, Kenneth Lee
Watthey, Jeffrey William Herbert
Webb, William Gatewood
Webber, John Alan
Wei, Peter Hsing-Lien
Weinkam, Robert Joseph
Weisbach, Jerry Arnold
Wellings, Ian
Wells, Jack Nuk
Wheeler, William Joe
Whelton, Bartlett David
Wiley, Robert A
Willette, Robert Edmond
Williams, David Allen
Williams, Todd Robertson
Wilson, Armin Guschel
Wilson, James William
Winn, Martin
Wintter, John Ernest
Witiak, Donald T

Witkowski, Joseph Theodore
Wohl, Bernard G
Wolff, Manfred Ernst
Woltersdorf, Otto William, Jr
Worooh, Eugene Leo
Wright, George Edward
Wright, Herbert Fessenden
Wright, John Brenton
Wu, Mu Tsu
Wynn, James Elkanah
Yankee, Ernest Warren
Yevich, Joseph Paul
Yu, Terry Ta-Len
Zee-Cheng, Kwang Yuen
Zenitz, Bernard Leon

Microchemistry

Alber, Herbert Karl
Brown, Lawrence Eldon
Buckles, Marjorie Fox
Crisler, Joseph Presley
Dunkelberger, Tobias Henry
Gavrilovic, John
Huffman, Edward William Dickson
Loscalzo, Anne Grace
Ma, Tsu Sheng
Maciak, George M
Rush, Cecil Archer
Spang, Arthur William
Steyermark, Al

Natural Products Chemistry

Awad, Albert T
Bennett, Word Brown, Jr
Burreson, Burton Jay
Cane, David Earl
Chang, Ching-Jer
Clark, Charles Kittredge
Coggon, Philip
Cohen, Noal
Cole, Wayne
Cook, James Minton
Dawson, Ray Fields
Dolby, Lloyd Jay
Dominguez, Xorge Alejandro Sepulveda
Edwards, William Brundige, III
Fischer, Nikolaus Hartmut
Flath, Robert Arthur
Gager, Forrest Lee, Jr
Ganem, Bruce
Gordon, Eric Michael
Grove, Michael Dean
Hamill, Robert L
Hansen, Ralph Holm
Haskell, Theodore Herbert, Jr
Hensen, Otto Dirk
Henson, Rodger Dale
Herald, Delbert Leon, Jr
Hsu, Charles Fu-Len
Hudson, Albert Berry
Ingalsbe, David Weeden
Jennings, Walter Goodrich
Johnson, Ronald Doyle
Keeler, Richard Fairbanks
Kim, Hyeong Lak
Kingston, David George Ian
Lee, Chi-Hang
Lee, Kuo-Hsiung
MacConnell, John Griffith
Machr, Hubert
Mallams, Alan Keith
Manske, Richard Helmuth (Fred)
Marconi, Gary G
Marion, Leo (Edmond)
Michel, Gerd Wilhelm
Mikolajczak, Kenneth Lee
Miller, Thomas William
Mors, Walter B
Nair, Ramachandran Mukundalayam Sivarama
Parry, Ronald John
Porter, Lee Albert
Putter, Irving
Ranieri, Richard Leo
Renn, Donald Walter
Rickes, Edward Lawrence
Riffer, Richard
Sastry, Sindgi Dattu
Schmitz, Francis John
Shamma, Maurice
Shimizu, Yuzuru
Silverstein, Robert Milton
Singleton, Vernon Leroy
Smith, Cecil Randolph, Jr
Sodano, Charles Stanley
Soffer, Milton David
Starratt, Alvin Neil
Stevens, Kenneth Lloyd
Stipanovic, Robert Douglas
Stoessl, Albert
Stout, George Hubert
Sumner, Robert Jocelyn
Sweeny, James Gilbert
Teng, Jon Ie
Tin-Wa, Maung
Van Etten, Cecil Herman
Warthen, John David, Jr
Wenkert, Ernest

Werny, Frank
Zeiger, William Nathaniel

Neurochemistry
Albers, Robert Wayne
Anderson, Larry Ernest
Aprison, Morris Herman
Archer, Ellen Gleason
Banik, Narendra Lal
Barker, Louis Allen
Barraco, Robin Anthony
Bass, Norman Herbert
Benjamins, Joyce Ann
Benuck, Myron
Berl, Soll
Bleecker, Margit
Blume, Arthur Joel
Boulton, Alan Arthur
Brostoff, Steven Warren
Burt, Alvin Miller, III
Callahan, John William
Chao, Li-Pen
Cheng, Sze-Chuh
Churchill, Lynn
Clendenon, Nancy Ruth
Cohen, Gerald
Cohen, Stephen Robert
Costantino-Ceccarini, Elvira
Daly, John William
Datta, Ranajit Kumar
De Balbian Verster, Floris
Dettbarn, Wolf Dietrich
De Vries, George Henry
Dhopeshwarkar, Govind Atmaram
Dreiling, Charles Ernest
Dreyfus, Pierre Marc
Duffy, Thomas Edward
Dunn, Adrian John
Eiduson, Samuel
Einstein, Elizabeth Roboz
Ferris, Robert Monsour
Fidone, Salvatore Joseph
Freedman, Lewis Simon
Friedland, Joan Martha
Gardner, Sara A
Geison, Ronald Leon
Geller, Edward
Gould, Robert Michael
Greenfield, Seymour
Hartman, Boyd Kent
Held, Irene Rita
Henn, Fritz Albert
Higman, Henry Booth
Hirsch, Hilde Esther Zwirn
Ho, Ing Kang
Holt, Thomas Manning
Hornykiewicz, Oleh
Horrocks, Lloyd Allen
Hoss, Wayne Paul
Iqbal, Zafar
Jungalwala, Firoze Bamanshaw
Khan, Mozzam Ali
Kovachich, Gyula Bertalan
Kravitz, Edward Arthur
Kumar, Sudhir
Lapin, Evelyn P
Larrabee, Martin Glover
Lees, Marjorie Berman
Lim, Ramon (Khe Siong)
Lin, Sping
Ling, Alfred Soy Chou
Luine, Victoria Nall
Maker, Howard Smith
Margolis, Frank L
Margolis, Renee Kleinmann
McBride, William Joseph
McClure, William Owen
McIlwain, David Lee
Modak, Arvind T
Murthy, Mahadi Raghavandrarao V
Needleman, Saul Ben
Norton, William Thompson
O'Brien, Richard Desmond
Palmer, Eugene Charles
Pappius, Hanna M
Peterson, Rudolph Price
Prescott, David Julius
Ramsey, Robert Bruce
Rassin, David Keith
Rosenblatt, Dorrie Ellen
Rothenberg, Mortimer Abraham
Sampugna, Joseph
Samuels, Stanley
Schacht, Jochen Heinrich
Shaskan, Edward Gregory
Shuster, Louis
Siegel, Laurane Geary
Simon, Eric Jacob
Singh, Vijendra Kumar
Stahl, William Louis
Stancer, Harvey C
Suzuki, Kunihiko
Swaiman, Kenneth F
Tanaka, Ryo
Teller, David Norton
Varon, Silvio Salomone
Vrba, Rudolf
Wajda, Isabel

Wander, Joseph Day
Weber, Bruce Howard
Weinstein, Howard
Weitsen, Howard Arthur
White, Thomas David
Wilson, John Edward
Wolfgram, Frederick John
Wood, James Douglas
Yamamura, Henry Ichiro
Young, Richard Lawrence
Yu, Robert Kuan-Jen

Paper Chemistry
Abson, Derek
Anderson, Bror Ernest
Atchison, Joseph Edward
Austin, Robert Andrae
Bailey, Carl Williams, III
Barton, John Selby
Bernardin, Leo J
Bixler, A L M
Biesner, William Clark
Boehm, Robert Louis
Bradway, Keith E
Breithaupt, Lea Joseph, Jr
Brinkley, Amiel Word, Jr
Bublitz, Walter John, Jr
Carlson, Willard Emmett
Cavagna, Giancarlo Antonio
Chan, Lock Lim
Cobb, R M Karapetoff
Cordingly, Richard Henry
Cornell, Richard Henry
Czepiel, Thomas P
Davis, Gerald Titus
Doughty, Joseph Bayne
Dugal, Hardev Singh
Eames, Arnold C
Eber, Robert Joseph
Estes, Timothy King
Farewell, John P
Fleischer, Thomas B
Forman, Loren Verne
Frankle, William Ernest
Gardner, Howard Shafer
Glading, Ralph Edmond
Goldstein, Irving Solomon
Green, Robert Patrick
Greminger, George King, Jr
Gupta, Virendra Nath
Hanby, John Estes, Jr
Hay, Kenneth Dana
Hess, Cecil Lawrence
Higgins, James Joseph
Hofreiter, Bernard T
Hoge, William Henry
Howells, William Hakes
Kalisch, John Hans
Keim, Gerald Inman
Knudson, Harold William
Kohl, Harry Charles, Jr
Lang, Andrew George
Larson, Kenneth Curtis
La Vallee, William Alfred
Lawson, Louis Russell, Jr
Lea, David Chester
Leech, John G
Leopold, Bengt
Lorey, Frank William
Luce, Leonard Murray, Sr
Manning, James Harvey
McPherson, William Hakes
Miller, Fredric N
Most, David S
Nelson, George Richard
Park, Robert William
Peterson, Robert C
Procter, Alan Robert
Putnam, Stearns Tyler
Ratliff, Francis Tenney
Redd, John Coleman
Richtol, Herbert H
Robinson, James Vance
Rowe, Herbert William
Sarjeant, Peter Thomson
Scalfarotto, Robert Emil
Schoettler, James Robert
Schulz, John Hampshire
Smith, William Edmond
Sommers, Raymond A
Spiegelberg, Harry Lester
Strauss, Roger William
Su, Cheh-Jen
Taylor, David Lawrence
Tostevin, James Earle
Trotter, Patrick Casey
Verseput, Hernan Ward
Von Koeppen, Andreas
Walker, William Comstock
Wamsley, Robert Alan
Wasser, Richard Barkman
Weaver, Quentin Clifford
Whorton, Rayburn Harlen
Wise, John Thomas
Wollwage, John Carl
Zentner, Thomas Glenn

Axe, William Nelson
Bailey, Grant Carter
Ball, John Sigler
Battles, Willis Ralph
Bennett, Richard Henry
Bishop, John William
Bloch, Herman Samuel
Bombardieri, Caurino Cesar
Bosniack, David S
Brennan, James A
Brown, Lawrence Milton
Brown, Stanley Monty
Burdge, David Newman
Bush, Warren Van Ness
Caesar, Philip D
Carnell, Paul Herbert
Cavitt, Stanley Bruce
Cheavens, Thomas Henry
Christensen, Edward Richards
Condit, Paul Carr
Condo, Albert Carman, Jr
Csicsery, Sigmund Maria
Cupper, Robert Alton
Currier, Vernon Arthur
Davis, Burton H
Demkovich, Paul Andrew
Dichter, Michael
Dille, Kenneth Leroy
D'Ouville, Edmond Lawrence
Duke, Roy Burt, Jr
Edwards, Gayle Dameron
Farrauto, Robert Joseph
Feldman, Nicholas
Fenton, Donald Mason
Ferm, Robert James
Fernald, Herbert Byron
Findlay, Robert Artemas
Fischer, Paul Edgar
Fischer, Ronald Howard
Foucher, Walter David, Jr
Fox, Dale Bennett
Francis, Elliot S
Francisco, Cecil Jay, Jr
Friedman, Bernard Samuel
Frolich, Per Keyser
Gallagher, James Patrick
George, Albert El Deeb
Giannetti, Joseph Paul
Gibbons, Louis Charles
Gilbert, William Irwin
Goldstein, Marvin Sherwood
Grina, Larry Dale
Guffy, Joseph Claude
Haines, William Emerson
Harris, Samuel William
Harrison, John William
Hass, Henry Bohn
Hausler, Rudolf H
Heithaus, Joseph John, Jr
Heldman, Julius David
Hendricks, Grant Walstein
Hetzner, Stephen Aven, Jr
Hetzner, Howard Paul
Hickson, Donald Andrew
Holtmyer, Martin Dean
Holubec, Zenowie Michael
Horne, William Appler
Hosler, Peter
Houston, Robert John
Johnson, James Augustus, Jr
Johnston, Harlin Dee
Jones, Arthur Letcher
Kalinowski, Mathew Lawrence
Kang, Chia-Chen Chu
Kelso, Edward Albert
Khoobiar, Sargis
King, Laurence Frederick
Kirklin, Perry William
Knipple, Warren Russell
Kolobielski, Marjan
Koppenhoefer, Robert Mack
Korb, Ernest Lloyd
Kovach, Stephen Michael
Kray, Louis Robert
Kreglewski, Alexander
Laity, John Lawrence
Lang, William Harry
Lapporte, Seymour Jerome
Latham, Dewitt Robert
Lawson, Jimmie Brown
Leaman, Wilbur Kauffman
Lester, George Ronald
LeSuer, William Monroe
Lewis, Robert Allen
Lovett, John Robert
Lyons, Joseph F
Mallett, William Robert
Manyik, Robert Michael
Matson, Howard John
Mayer, Theodore Jack
McEvoy, James Edward
McGinnis, Edgar Lee
Meader, Arthur Lloyd, Jr
Meerbott, William Keddie
Melchiore, John J
Merrill, Howard Nicolas
Miale, Joseph Nicolas
Miller, Edward Frederick

Pathological Chemistry
Bide, Richard W
Feuer, George
Glende, Eric A, Jr
Malya, Govinda P A
Nicholson, Thomas Frederick

Pesticide Chemistry
Ajami, Alfred Michel
Alley, Earl Gifford
Anderson, Ralph F
Battershell, Robert Dean
Bellet, Eugene Marshall
Blinn, Roger C
Booth, David Layton
Bowers, Alston Gordon
Burke, Jerry Alan
Burke, Susan Schilt
Carlson, Lewis John
Caswell, Robert Little
Cruickshank, P A
Cummings, Joseph Gerard
Cupery, Willis Eli
Curless, William Toole
DiFate, Victor George
Dorman, Stephen Charles
Doyle, William Carter, Jr
Fisher, James Delbert
Fridinger, Tomas Lee
Gardiner, John Alden
Getzendaner, Milton Edmond
Glasgow, Augustus Rossell, Jr
Greenhalgh, Roy
Greichus, Yvonne A
Griffin, Thomas Scott
Gruenhagen, Richard Hamilton
Haglid, Frank Runar
Haines, Linwood Davis
Hall, Robert Turner
Han, Jerry C-Y
Helrich, Kenneth
Hill, Kenneth Richard
Hilton, James Lee
Hoffman, William Martin
Hollingworth, Robert Michael
Jenny, Neil Allan
Julin, Bruce Gustav
Kane, Gordon Philo
Katsaros, Constantine
Kesterson, Clinton Joe
Kirkpatrick, Joel Lee
Kleiman, Morton
Koestler, Robert Charles
Kolbezen, Martin (Joseph)
Kurtz, A Peter
Kurtz, David Allan
Lacoste, Rene John
LeBaron, Homer McKay
Leidy, Ross Bennett
Liddicoet, Thomas Herbert
Malhotra, Sudarshan Kumar
Marshall, William Deforrest
Miles, James William
Miller, Philip
Montgomery, Ronald Eugene
Morley, Harold Victor
Moseman, Robert Fredrick
Newsom, Herbert Charles
Norris, Logan Allen
Norton, Scotty Jim
Oloffs, Peter Christian
Pallos, Ferenc M
Portnoy, Robert Charles
Puhl, Richard James
Ragsdale, Nancy Nealy
Ramsey, Arthur Albert
Riley, Robert Charles
Saha, Jadu Gopal
Saxton, Augustus Donovan
Schattner, Robert I
Schroeder, Robert Samuel
Schultz, Donald Paul
Sears, Daniel Scott
Sheeran, Patrick Jerome
Shepard, Harold Henry
Smith, Allen Elston
Smith, Charles Aloysius
Smith, Robert Alan
Stanley, Charles William
Stolzenberg, Gary Eric
Street, Joseph Curtis
Strong, Jerry Glenn
Summers, John Clifford
Tanaka, Fred Shigeru
Thio, Alan Poo-An
Throckmorton, James Rodney
Traxler, James Theodore
Westlake, William Ellis
Whetstone, Richard Roy
Woolson, Edwin Albert
Zweig, Gunter

Petroleum Chemistry
Adams, Charles Rex
Alley, Starling Kessler, Jr
Alpert, Norman
Altgelt, Klaus H
Askevold, Robert James

75

Mills, King Louis, Jr
Mitchell, Thomas Owen
Naylor, Carter Graham
Neiswender, David Daniel
Nevitt, Thomas D
Nolan, John Thomas, Jr
Oberender, Frederick G
Obiad, Alex Golden
Pattinson, Charles Byron, Jr
Pauken, Robert John
Petersen, Joseph Claine
Peterson, Alan Herbert
Pitchford, Armin Cloyst
Pitts, Paul Miller, Jr
Plank, Charles Joseph
Pollitzer, Ernest Leo
Richter, Frederick Paul
Rigdon, Orville Wayne
Riordan, Michael Davitt
Robinson, Alfred Green
Rope, Barton Whitefield
Rosenfeld, Daniel
Rosenthal, Joel William
Rosenwald, Robert Henry
Ruchelman, Maryon Waldman (Mrs
    Roderick A Kratoville)
Ryan, Julian Gilbert
Sartor, Albin Francis, Jr
Schallenberg, Elmer Edward
Schindler, John Henry
Schlatter, Maurice Jay
Schram, Robert Wallace
Seekircher, Richard
Sheppard, Allen Anson
Silvestri, Anthony John
Smith, Brandes Henry
Smith, Robert Kinsel
Stamm, Ralph Eugene
Streiff, Anton Joseph
Sturgis, Bernard Miller
Sweeney, William Alan
Turner, William Mortimer
Udelhofen, John Henry
Van de Castle, John F
Venuto, Paul B
Warne, Thomas Martin
Weaner, George L
Weintritt, Donald J
Weisgerber, George Austin
Welch, Lester Marshall
Wetmore, David Eugene
Wheelock, Kenneth Steven
Whittemore, Irville Merrill
Wilkes, John Barker
Willette, Gordon Louis
Williams, Albert Lloyd
Wilson, John William, Jr
Woods, Warren Whitney
Zakaib, Daniel D

Pharmaceutical Chemistry

Abushanab, Elie
Ackermann, Guenter Rolf
Ahuja, Satinder
Ajami, Alfred Michel
Albert, Anthony Harold
Amsel, Lewis Paul
Amundson, Merle E
Anderson, Arthur James
Anderson, Jon C
Andrako, John
Anschel, Joachim
Antoshkiw, Thomas
Asano, Akira
Auslander, David E
Avis, Kenneth Edward
Bailey, Harold Stevens
Bailey, Leonard Charles
Balassa, Leslie Ladislaus
Ball, Edwin Lawrence
Ballard, Berton Etienne
Bandelin, Fred John
Banes, Daniel
Bapatla, Krishna M
Bariana, Dilbagh Singh
Barnstein, Charles Hansen
Barringer, William Charles
Bauguess, Carl Thomas, Jr
Becker, Clifford C
Beck, John W
Belmonte, Albert Anthony
Beltz, LeRoy Duane
Benson, Walter Roderick
Bernardy, Karel Francis
Bhat, Venkataramana Kakekochi
Bhavnani, Bhagu R
Bikin, Henry
Biles, John Alexander
Billups, Norman Frederick
Bingenheimer, Levi Edwin, Jr
Blackburn, Dale Warren
Blair, Andrew Dryden, Jr
Block, John Harvey
Blythe, Rudolph Hamma
Bodin, Jerome Irwin
Boenigk, John William

Boettner, Fred Easterday
Bope, Roger Elwood
Borke, Frank Willis
Borke, Mitchell Louis
Bray, Malcolm Davonne
Brenner, Gerald Stanley
Brenner, Ronald John
Briceaddy, Lawrence Edward
Brochmann-Hanssen, Einar
Brofazi, Fred Robert
Bryan, Hugh D
Bryan, Jack T
Bryan, Wilbur Lowell
Buehler, John David
Burgauer, Paul David
Burlage, Henry Matthew
Butensky, Irwin
Caldwell, Henry Cecil, Jr
Carter, Kenneth
Casler, David Robert
Casola, Armand Ralph
Castello, Robert Anthony
Cates, Lindley A
Chang, Charles Yu-Chun
Charnicki, Walter Francis
Chatten, Leslie George
Cheema, Zafrullah K
Chemburkar, Pramod Bhaurao
Chen, James L
Cheng, Lawrence Kar-Hiu
Chertkoff, Marvin Joseph
Childers, Ray Fleetwood
Childs, Richard Francis
Choulis, Nicolas Helias
Christman, David R
Clarke, Robert La Grone
Close, Warren James
Cohen, Edward Morton
Cohn, Robert M
Colaizzi, John Louis
Collins, Alfred Patterson
Colwell, William Tracy
Compton, Walter Ames
Conine, James William
Connors, Kenneth A
Craig, Alan Daniel
Cross, John Milton
Crowell, Wilfred J
Crouthamel, William Guy
Cureton, Glen Lee
Curtin, Lemuel Calvert
D'Adamo, Anthony
Daigle, Josephine Siragusa
Dale, Jack Kyle
Dann, Frank Warren
Darling, Charles Milton
Data, John Batiste
Davenport, Tom Forest, Jr
Dean, Donald E
Dean, Gordon Spencer
Deghenghi, Romano
Deist, Robert Paul
De Luca, Patrick Phillip
Delgado, Jaime Nabor
Dempski, Robert E
DeRose, Anthony Francis
De Silva, John Arthur F
DeWald, Eugene
DeYoung, Joyce Lewis
DiFazio, Louis T
Dill, Dale Robert
Dimmock, Jonathan Richard
Dines, Allen I
Doak, George Osmore
Doczi, John
Dodds, Alvin Franklin
Dolusio, James Thomas
Donikian, Marc Roupen
Dorn, Conrad Peter, Jr
Doyle, Thomas Daniel
Drommond, Fred George
Dugan, Gary Edwin
Dunbar, Joseph Edward
Dunker, Melvin Frederick William
Durden, John Apling, Jr
Duvall, Ronald Nash
Ebert, William Robley
Ecanow, Bernard
Eibert, John Jr
Eichman, Martin L
Eisen, Henry
Ellin, Robert Isadore
Ellis, Larry Edward
Eriksen, Stuart P
Fand, Theodore Ira
Faust, Richard Edward
Fedrick, James Love
Feldmann, Joseph Aaron
Feldmann, Edward George
Fischer, Louis
Foerzler, Ernest Carl
Fox, Chester David
Fox, Sereck Hall
Frank, Sylvan Gerald
French, Warren Neil
Fung, Ho-Leung
Gakenheimer, Walter Christian
Galinsky, Alvin M

Gardella, Libero Anthony
Garizo, John Ernest
Garrett, Edward Robert
Gauthier, George James
Geller, Milton
Gillingham, James Morris
Gimelli, Salvatore Paul
Gisvold, Ole
Glasser, Arthur Charles
Goetsch, Robert Wayne
Gottesman, Elihu
Gottman, Elihu
Goyan, Jere Edwin
Grady, Lee Timothy
Graef, Walter L
Gramling, Lea Gene
Granatek, Alphonse Peter
Grant, Ernest Walter
Green, Melvin William
Gregg, David Henry
Grim, Wayne Martin
Gross, Herbert Michael
Grossman, Edward Joseph
Grove, Donald Cooper
Haines, Bernard A, Jr
Haines, William Joseph
Hallett, Floyd Prentice
Hammer, Henry Felix
Hammer, Richard Hartman
Haney, William Garland, Jr
Haney, Glenn Herbert
Hardwidge, Edward Albert
Harris, Lewis Eldon
Harris, Loyd Ervin
Haslam, John Lee
Head, William Francis, Jr
Hennig, Arnold John
Herzog, Karl Adolph
Hester, Jackson Boling, Jr
Hewitt, Harold George
Heyd, Allen
Hiebert, John Mark
Hiestand, Everett Nelson
Higgins, Walter Mayo
Higuchi, William Iyeo
Hoffman, Allan Jordan
Hopkins, Corris Mabelle
Hoskins, Howard
Howard, Walter Hugh
Howe, Stephen Arthur
Huber, Eugene Everett
Hubsch, Harold E
Huebner, Harold Lawrence
Huitric, Charles Ferdinand
Hussain, Alain Corentin
Hutchins, Mehdi Hajiyani.
Infeld, Hastings Harold, Sr
Isaacson, Martin Howard
Isenberg, Eugene I
Ivashkiv, Allen (Charles)
Jacob, Eugene
Jacob, James Thecattil
Jaffe, Allen Leon
Jaffe, James Mark
Jahnke, Jonah
Janike, Paul Joseph
Jansen, Richard William
Jarvis, Charles I
Jarowski, Albert E
Jeffrey, James George
Jenkins, Glenn Llewelyn
Jenkins, William Wesley
Johnson, Byron Andrew
Johnson, Richard Dean
Jones, Ralph William
Joslin, Robert Scott
Kabadi, Balachandra N
Kaiser, David Gilbert
Kaistha, Krishan K
Kanig, Joseph Louis
Kapadia, Govind J
Kaplan, Allan Steven
Kaplan, Leonard Louis
Kaplan, Stanley Albert
Kapoor, Amril Lal
Karim, Aziz
Kaspar, Hans Heinrich
Katz, Irwin Alan
Katz, Martin
Kelly, Clark Andrew
Kertesz, Jean Constance
Kieffer, Robert G
King, Robert Edward
Kishimoto, Yasuo
Kline, Berry James
Klioze, Oscar
Koch, Melvin Vernon
Koda, Robert T
Kononenko, Oleg K
Korner, Milton Joseph
Koshy, Karyanil Thomas
Kowarski, Chana Rose
Krezanoski, Joseph Z
Kukla, Michael Joseph
Kuo, Norman Yu-Neng
Kushner, Samuel
Kwan, King Chiu

Lagally, Ralph Werner
Landgraf, William Charles
Lappas, Lewis Christopher
LaSala, Edward Francis
Lauer, Rudolph Frank
LaVia, Anthony F
Lawless, Gregory Benedict
Lazo-Wasem, Edgar Arthur
Leaton, John Roger
Lee, Cheuk Man
Lee, Kwan-Hua
Lehr, Hanns H
Lehr, Hanns H
Lember, August Paul
Lester, George Yohe
Letton, James Carey
Levi, Ralph Sigmund
Levin, Nathan
Leyda, James Perkins
Liao, Tsung-Kai
Lieberman, Herbert A
Lichtin, J J Leon
Lien, Eric Jung-Chi
Lintner, Carl John, Jr
Lippmann, Irwin
Lobl, Thomas Jay
Locock, Robert A
Lopez, Antonio Vincent
Lovering, Edward Gilbert
Lunsford, Carl Dalton
MacAulay, Wesley Claude
Macek, Thomas Joseph
Mackles, Leonard
Mackowiak, Elaine DeCusatis
Madison, William Leon
Magarian, Edward O
Magid, Louis
Marchisotto, Robert
Marlowe, Edward
Marshall, James R
Martin, Albert Edwin
Martin, Arnold R
Martin, Bruce Douglas
Martin, Charles Franklin
Martin, John Walter, Jr
Martinelli, Louis Carl
Massey, Beverly Wong
Masaki, Eddie H
Mathews, Shaikh Badarul
Mattocks, Hewitt William
Mattocks, Albert McLean
Mauger, John William
McClure, George Richard
McDonnell, Joseph Francis, Jr
McDowell, Wilbur Benedict
McKechan, Charles Wayne
McKinney, Myron William
McMillion, C Robert
McSheffery, John
McVean, Duncan Edward
Mehta, Nariman Bomanshaw
Merkle, F Henry
Meyer, Robert F
Meyer, Walter Edward
Mezei, Michael
Miale, Joseph Peter
Michaelis, Arthur Frederick
Miller, Larry Gene
Miller, Orville H
Miller, Theodore Charles
Misek, Bernard
Miskel, John J (Sr)
Mitchner, Hyman
Mittelstaedt, Stanley George
Mollica, Joseph Anthony
Monroe, Ezra
Montgomery, Kenneth O
Moody, Joseph E, Jr
Moran, Michael J
Morgan, Stanley J
Morrow, Duane Francis
Mourer, Kermit L
Mueller, William H
Mudson, Daniel
Murphy, Hubert William
Nash, J Frank
Nash, Robert Arnold
Needham, Thomas E, Jr
Nelson, Kenneth Gordon
Nelson, Norman Allan
Nessel, Robert J
Neumann, Helmut Carl
Nicholson, Arnold Eugene
Ninger, Fred Constant
Nobles, William Lewis
Norris, Paul Edmund
Nuessle, Noel Oliver
Nysted, Leonard Norman
Omodt, Gary Wilson
Ong, John Tjoan Ho
Osman, Jack
Owensby, Clenton Edgar
Paikoff, Myron
Parisse, Anthony John
Park, Moo Kwang
Parke, Hervey Cushman
Parker, Martin Dale
Parks, Lloyd McClain
Parrott, Eugene Lee

Patel, Appasaheb Raojibhai
Paterson, Garnet Russell
Pernarowski, Modest
Petersen, Robert Virgil
Petinato, Frank Anthony
Pflug, Gerald Ralph
Pifer, Charles William
Pikal, Michael Jon
Pinajian, John Joseph
Plakogiannis, Fotios M
Poetsch, Chester E
Poole, John William
Popli, Shankar D
Portlock, David Edward
Portmann, Glenn Arthur
Possley, Leroy Henry
Poust, Rolland Irvin
Pramoda, Maturu Krishna
Raffelson, Harold
Rapaka, Rao Sambasiva
Ratto, Peter Angelo
Reid, William Bradley
Reif, Van Dale
Rhodes, Robert Allen
Rich, Arthur Gilbert
Rifino, Carl Biaggio
Ripka, William Charles
Rippie, Edward Grant
Rising, Louis Wait
Robinson, Joseph Robert
Rogers, James Albert
Romig, Paul William
Rork, Gerald Stephen
Roscoe, Charles William
Roseman, Theodore Jonas
Rosenthal, Murray William
Rost, William Joseph
Russo, Emanuel Joseph
Ryting, Joseph Howard
Sacks, David Alan
Sadik, Farid
Saenz, Reynaldo V
Sahli, Brenda Payne
Salsbury, John Greensmith
Sam, Joseph
Saski, Witold
Schattner, Robert I
Scheindlin, Stanley
Schick, John William
Schnaare, Roger L
Scheller, George Henry
Schultz, Harry Wayne
Schwartz, Joseph Barry
Schwartz, Michael Averill
Schwartz, Thomas Werner
Sciarrone, Bartley John
Semeniuk, Fred Theodor
Senkowski, Bernard Zigmund
Shah, Ashok Chandulal
Shefter, Eli
Sheinin, Eric Benjamin
Shell, John Weldon
Sherman, Leslie
Sheth, Bhogilal
Shinkai, John H
Shroff, Arvin Pranlal
Siegal, Bernard
Siegel, Sheldon
Simonelli, Anthony Peter
Sims, Bernard
Singer, Walter
Singh, Rudra Pratap
Singiser, Robert Eugene
Sinkula, Anthony Arthur
Sinotte, Louis Paul
Sinsheimer, Joseph Eugene
Skibbe, Martin Otto
Sklar, Stanley
Smith, Stanley
Smith, Robert Victor
Somkaite, Rozalija
Sorby, Donald Lloyd
Spiegel, Allen J
Stadler, Louis Benjamin
Staiff, Donald C
Stark, John Frederick
Staum, Muni M
Stavchansky, Salomon Ayzenman
Steele, John Wiseman
Steers, Arthur Walter
Steiger, Leonard William
Stein, Gustav Albert
Stewart, James T
Stober, Henry Carl
Strike, Donald Peter
Stutz, Robert L
Su, Kenneth Shyan-Ell
Sugita, Edwin T
Sullivan, Jeremiah B
Sumerford, Wooten Taylor
Summa, A Francis
Surpuriya, Vijay B
Tan, Henry S I
Taraszka, Anthony John
Taub, Abraham
Teare, Frederick Wilson
Theimer, Edgar E
Thompkins, Leon

Thompson, Alonzo Crawford
Tice, Linwood Franklin
Tobey, Stephen Winter
Tozer, Thomas Nelson
Trager, William Frank
Tuckerman, Murray Moses
Tuthill, Harlan Lloyd
Upeslacis, Janis
Urdang, Arnold
Vacik, Dorothy Nobles
Vacik, James P
Valvani, Shri Chand
Van Meter, Clarence Taylor
Vann, Robert Lee
Vincent, Muriel C
Wagner, John Garnet
Waller, Coy Webster
Webb, Norval Ellsworth, Jr
Webber, Marion George
Weber, John Donald
Weinswig, Melvin H
Weintraub, Leonard
Weiss, Marvin
Weiss, Peter Joseph
Welles, Harry Leslie
Wheeler, Frank Carlisle
Wheeler, James Donlan
Wheeler, Larry Meade
White, Allen Ingolf
Wilken, Leon Otto, Jr
Willis, Carl Raeburn, Jr
Wilson, Charles Owens
Wolkoff, Hal Norman
Wolters, Robert John
Wood, John Henry
Wormser, Henry C
Worrell, Lee Frank
Wright, William Blythe, Jr
Yakubik, John
Yalkowsky, Samuel Hyman
Yates, Claire Hilliard
Yeh, Shu-Yuan
Yeowell, David Arthur
Zalucky, Theodore B
Zanowiak, Paul
Zderic, John Anthony
Zenker, Nicolas
Zimmer, Arthur James
Zimmerman, James Joseph
Zoglio, Michael Anthony
Zografi, George
Zuck, Donald Anton

Photochemistry

Alley, Earl Gifford
Arnold, Donald Robert
Basco, N
Binkley, Roger Wendell
Bloom, Allen
Bohning, James Joel
Burr, John Green
Calvert, Jack George
Carey, John Hugh
Chang, Catherine Teh-Lin
Charlton, James Leslie
Cohen, Abraham Bernard
Collier, Susan S
Crawford, James Worthington
Creed, David
Cunningham, Michael Paul
Dalton, James Christopher
Dauben, William Garfield
Deboer, Charles D
Dedinas, Jonas
Demas, James Nicholas
de Mayo, Paul
Dingledy, David Peter
DoMinh, Thap
Druelinger, Melvin L
Eian, Gilbert Lee
Engebrecht, Ronald Henry
Flechtner, Thomas Welch
Freed, Donald Joseph
Gano, James Edward
Golemba, Frank John
Gravel, Denis Fernand
Gruber, Gerald William
Gunning, Harry Emmet
Gupta, Amitava
Halpern, Arthur Merrill
Hancock, Kenneth George
Harbour, John Richard
Harshbarger, William Reid
Hart, Harold
Hautala, Richard Roy
Herkstroeter, William G
Hickman, James Joseph
Hill, Cliff Otis
Hirsch, Richard Henry
Houser, John J
Houston, James Grey
Irick, Gether, Jr
Jackson, Roy Joseph
Jackson, William Morgan
John, Andrew
Johnson, Donald Wayne
Kerwin, Robert Eugene
Ketley, Arthur Donald

Knight, Arthur Robert
Kushner, Arthur Simon
Kutsche, Kenneth Otto
Lamola, Angelo Anthony
Langmuir, Margaret Elizabeth Lang
Laufer, Allan Henry
Lell, Eberhard
LeRoy, Donald James
Lewis, Frederick D
Lowrey, Robert Dean
Malin, John Michael
McGee, Thomas Howard
McMillan, Garnett Ramsay
Merkel, Paul Barrett
Messer, Wayne Ronald
Millich, Frank
Murov, Steven Lee
Murphy, John Joseph
Newland, Gordon Clay
O'Connell, Edmond J, Jr
Ors, Jose Alberto
Oster, Carl Frederick
Pappas, Socrates Peter
Park, Su-Moon
Pauli, George H
Pavia, Donald Lee
Petrellis, Panayotis C
Pitts, James Ninde, Jr
Plank, Don Allen
Polowczyk, Carl John
Pond, David Martin
Raniseski, John Walter
Rauh, Robert David, Jr
Reasoner, John W
Reed, Thomas Freeman
Rennert, Joseph
Rie, John E
Robinson, George Wilse
Roscher, David Moore
Rosenberg, Ira Edward
Rumfeldt, Robert Clark
Salesin, Eugene Dennis
Satcher, Robert Lee
Scala, Alfred Anthony
Schexnayder, Mary Anne
Schuster, Gary Benjamin
Schwerzel, Robert Edward
Shaffer, Gary W
Sherman, Warren V
Shetlar, Martin David
Simonaitis, Romualdas
Simpson, George A
Stevens, Brian
Stevenson, Kenneth Lee
Stief, Louis J
Testa, Anthony Carmine
Thomas, Harold Todd
Trachtman, Mendel
Trozzolo, Anthony Marion
Tsang, Sien Moo
Valentine, Donald H, Jr
Waddell, Walter Harvey
Wagner, Michael Sidney
Walker, Michael Sidney
Waltz, William Lee
Wampler, Fred Benny
Wamser, Carl Christian
Wei, Chung-Chen
Weiss, David Steven
Weiss, Richard Gerald
Wiesenfeld, John Richard
Wijnen, Joseph M H
Winnik, Mitchell Alan
Wrighton, Mark Stephen
Young, Ainslie Thomas, Jr
Young, Robert Hayward
Yudelson, Joseph Samuel

Photographic Chemistry

Allentoff, Norman
Apellaniz, Joseph E P
Armour, Eugene Arthur
Bacon, Robert Elwin
Bard, Charleton Cordery
Barr, Charles (Robert)
Bass, Jon Dolf
Bates, James Edmund
Baum, Richard David
Baylor, Charles, Jr
Becker, Richard William
Beebe, George Warren
Bird, George Richmond
Booms, Robert Edward
Boyno, John S
Brack, Karl
Brown, James Wallace, III
Brust, David Philip
Cappel, Carl Robert
Carlson, Robert Leonard
Chang, Catherine Teh-Lin
Chapas, Richard Bernard
Charkoudian, John Charles
Childers, Robert Lee
Cohen, Jacob Isaac
Conant, Dale Holdrege
Conant, Robert Henry
Connolly, Lewis Timothy
Crane, Edward Mastin

Cunningham, Michael Paul
Dickens, John Ernest
Dickerson, Dorsey Glenn
Dieterich, David Allan
DoMinh, Thap
Downey, John Francis, Jr
Elins, Herbert Samuel
Erdmann, Duane John
Faul, William Henry
Feldman, Larry Howard
Fischer, Leewellyn C
Fix, Delbert Dale
Fleming, James Charles
Frank, William Charles
Freeman, John Paul
Goldberg, Gershon Morton
Gray, Russell Houston
Green, Milton
Guevara, Alfredo Ruben
Haase, Jan Raymond
Halfon, Marc
Hanson, Wesley Turnell, Jr
Hardham, William Morgan
Hauser, William P
Hendess, Raymond William
Hodgins, George Raymond
Holland, Andrew Brian
Holtz, Carl Frederick
Hunt, Heman Dowd
James, Thomas Howard
Jennings, Carl Anthony
Jones, Gerald Walter
Julian, Donald Benjamin
Kaplan, Mark Steven
Karlson, Richard Warren
Kasman, Sidney
Keevert, John Edward, Jr
Kliem, Peter O
Kofron, James Thomas, Jr
Koller, James Edward
Kurz, Richard Karl
Lambert, Ronald
Large, Robert F
Law, Paul Arthur
Lazaridis, Christina Nicholson
LeBlanc, Jerald Thomas
Levy, Boris
Lincoln, Lewis Lauren
Lindquist, Robert Marion
Lum, Kin K
LuValle, James Ellis
Maas, Keith Allan
Manning, Monis Joseph
McLaen, Donald Francis
Mendel, Maryann Madeliene
Merrigan, Joseph A
Millard, Frederick William
Miller, Matthew William
Milligan, Terry Wilson
Mitchell, Gary F
Moskowitz, Mark Lewis
Muenter, Annabel Adams
Neuberger, Dan
Oliver, Gene Leech
Osborn, Harland James
Patton, James Edward
Pilato, Jack Carmen
Raphael, Thomas
Rasch, Arthur Allyn
Richards, Jack Lester
Roberts, Harry Edward
Sagura, John Joseph
Sahyun, Melville Richard Valde
Schallhorn, Charles H
Sears, Timothy Stephen
Seibold, Carol Duke
Seus, Edward J
Shepp, Allan
Silverman, Robert Andrew
Simpson, William Henry
Simson, Joseph Michael
Sincius, Joseph Anthony
Smith, Gale Eugene
Stein, Samuel H
Stuart, Alfred Herbert
Sullivan, Michael Francis
Swank, Thomas Francis
Swenson, Richard Waltner
Sykes, Donald Joseph
Tan, Zoilo Cheng-Ho
Tischer, Thomas Norman
Tuite, Robert Joseph
Upson, Robert William
Van Norman, Gilden Ramon
Von Bacho, Paul Stephan, Jr
Waller, David Percival
Wayrynen, Robert Ellis
Williams, Carl James, Jr
Yamamoto, Yasushi Stephen
Yao, Jerry Shi Kuang
Youngquist, Mary Josephine
Zuckerman, Bernard
Zwick, Daan Marsh

Physiological Chemistry

Anderson, Richmond Karl
Arnow, Leslie Earle
Bates, Robert Wesley

CHEMISTRY

Baur, Fredric John, Jr
Bernstein, Sheldon
Brown, Barbara Illingsworth
Cavallito, Chester John
Chinard, Francis Pierre
Christman, Adam A
Chvapil, Milos
Clarenburg, Rudolf
Cohen, Philip Pacy
Coleman, Harold Mitchell
Crooks, George Chapman
Eyring, Edward J
Finamore, Frank Joseph
Foster, Charles David Owen
Freeland, Richard A
Freier, Esther Fay
Gans, Henry
Green, David Francis
Gruenwald, Geza
Hansen, Robert John
Harkins, Robert W
Hertelendy, Frank
Hutchin, Maxine E
Kemp, John Wilmer
Kendrick, Aaron Baker
Kinard, Fredrick William
Krahnke, Harold C
Kunin, Arthur Saul
Miller, Carl Henry, Jr
Morris, Eugene Ray
Morrow, Barry Albert
Moyer, Arden Wesley
Ogilvie, Marvin Lee
Phillips, Alvah H
Pritham, Gordon Herman
Richardson, Keith Erwin
Riley, Richard Fowble
Roberts, Joseph Linton
Rosenberg, Lawson Lawrence
Ross, Donald Joseph
Russell, James Christopher
Savchuck, William Basil
Sendroy, Julius, Jr
Shepherd, Raymond Edward
Singer, Leonard Sidney
Sisk, Lone L
Smith, Frank Ackroyd
Stetter, Joseph Robert
Tepperman, Helen Murphy
Tyler, Stanley Warren
Vars, Harry Morton
Vaughn, Clarence Benjamin
Welsch, Clifford William, Jr
Wolfersberger, Michael Gregg
Ziversmit, Donald Berthold

Plastics Chemistry
Allen, Nelson
Arends, Charles Bradford
Armbuster, David Charles
Ball, Ralph Henry
Barasch, Werner
Bender, Howard Leonard
Bennett, Richard Thomas
Benzinger, James Robert

Phytochemistry
Bailey, Lowell Frederick
Barr, Richard Arthur
Bhatti, Waqar Hamid
Brown, John Kennedy
Caldwell, Carlyle Gordon
Chan, Book G
Corse, Joseph Walters
Denford, Keith Eugene
Dorrell, Douglas Gordon
Dunkle, Larry D
Fissel, Guy Wilmer
Gottlieb, Otto Richard
Hanks, Robert William
Hassan, William Ephriam, Jr
Hixon, Ralph Malcolm
Hougen, Frithjof W
Howarth, Ronald Edward
Kelsey, Rick Guy
Khanna, Kristian L
King, Nydia Margarita
Kortschak, Hugo Peter
Kroschewsky, Julius Richard
Lau-Cam, Cesar A
Martin, Susan Scott
McClure, Jerry Weldon
Miller, Joseph Edwin
Miller, Lawrence Peter
Perlis, Irwin Bernard
Romeo, John Thomas
Scogin, Ron Lynn
Seeley, Schuyler Drannan
Stewart, Ivan
Stewart, Robert N
Surrey, Kenneth
Tang, Chung-Shih
Tso, Tien Chioh
Tucker, Charles Leroy, Jr
Tutupalli, Lohit Venkateswara
Wells, John Arthur
Weybrew, Joseph Arthur

Berta, Dominic Andrew
Bolstad, Luther
Bornstein, Leopold Frey
Brown, Stewart Cliff
Burhans, Allison Stilwell
Cabell, Pamela Whiting
Cates, Harry Louis, Jr
Chadwick, George F
Conger, Robert Perrigo
Curry, Michael Joseph
Epstein, George
Ezrin, Myer
Fick, Herbert John
Fleming, Richard Allan
Fox, Daniel Wayne
Harcsar, Francis George
Harrop, William Henry
Haxo, Henry Emile, Jr
Hornbrook, Walter John
Hunter, William Leslie
Irwin, William Edward
Klender, Gerald
Kline, Gordon Mabey
Knox, Charles Emery
Kuder, Robert Clarence
Lane, Constance A
Lewis, Ernest Eugene
Magnus, George
Marshall, Donald Irving
McTigue, Frank Henry
Mueller, Donald Scott
Nelson, John Daniel
Pyle, James Johnston
Raskin, Betty Lou
Redmond, John Peter
Reinhart, Frank Walter
Schlegel, John C
Sears, J Kern
Sechrist, Warren Doyle
Seiling, Alfred William
Smith, Eigene Arthur
Snell, John B
Tajima, Yuji Alexis
Touchette, Norman Walter
Weiner, Milton Lawrence
Wickson, Edward James
Williamson, Charles Wesley
Williger, Ervin John
Willwerth, Lawrence James
Wohnsiedler, Henry Peter
Zaremsky, Baruch

Pollution Chemistry
Auerbach, Victor
Batzar, Kenneth
Becker, Edward Samuel
Church, Ronald L
Cialdella, Cataldo
Cronkright, Walter Allyn, Jr
Deever, William Ray
Dunn, Bruce Partridge
Hager, Onslow Bonner
Howes, James E, Jr
Kawahara, Fred Katsumi
Keith, Lawrence H
Lyding, Arthur R
Moser, Frank Hans
Penrose, William Roy
Purchase, Earl Ralph
Rebagay, Teofila Velasco
Witte, Michael
Zimmt, Werner Siegfried

Polymer Chemistry
Abbott, Thomas Paul
Abere, Joseph Francis
Abernathy, Henry Herman
Abu-Isa, Ismat Ali
Aggarwal, Sundar Lal
Aharoni, Shaul Moshe
Akawie, Richard Isidore
Akeley, David Francis
Alexander, James Ernest
Aklonis, John Joseph
Alfrey, Turner, Jr
Allcock, Harry Rex
Allen, James Durwood
Allen, John Kay
Allen, Vernon R
Allen, John P
Alsberg, Henry
Altares, Timothy, Jr
Alvino, William Michael
Ambler, Michael Ray
Ambrose, Richard Joseph
Anderson, Paul George
Anderson, Arthur William
Anderson, Burton Carl
Anderson, John Norton
Anderson, Roy Scott
Andrus, Milton Henry, Jr
Ang, Tjoan-Liem
Angier, John David
Angier, Derek John

Antonucci, Frank Ralph
Apotheker, David
Arcesi, Joseph A
Arendt, Volker Dietrich
Arhart, Richard James
Arienna, Sidney
Arlt, Herbert George, Jr
Armbruster, David Charles
Armstrong, John Brian
Arnold, Charles, Jr
Arnold, Robert G
Assink, Roger Alyn
Auer, Edward Everett
Auerbach, Andrew Bernard
Auerbach, Melvin
Bach, Hartwig C
Bacskai, Robert
Baer, Eric
Bagley, George Everett
Bailey, Edward Thomas
Bair, Thomas Irvin
Baird, Richard Leroy
Baker, Leonard Morton
Baker, Richard William
Baker, William Oliver
Ball, Lawrence Ernest
Ball, Ralph Henry
Banucci, Eugene George
Barabas, Eugene S
Baranwal, Krishna Chandra
Barclay, Robert, Jr
Barie, Walter Peter, Jr
Barker, Harold Clinton
Barkhuff, Raymond Addison, Jr
Barlow, Arthur Livingston
Barney, Anthony
Barnhart, William Siddall
Barrett, Robert Earl
Barrett, Kenneth Ray
Bartovics, Albert
Basdekis, Costas H
Bata, George L
Bateman, John Hugh
Bauer, Richard G
Baughman, Glenn Laverne
Baum, Bernard
Baumann, Gert Friedrich
Baxter, Gene Francis
Baxter, James F
Beam, Charles Fitzhugh, Jr
Bean, C Thomas, Jr
Beck, Henry Nelson
Beckman, Joseph Alfred
Bedoit, William Clarence, Jr
Behrens, Ernst Wilhelm
Beindorff, Arthur Baker
Bell, Vernon Lee, Jr
Beller, Richard Joseph
Bender, Robert Joseph
Bender, Howard Sanford
Benton, Kenneth Curtis
Beradinelli, Frank Michael
Berdick, Murray
Beredjick, Nicky
Berenbaum, Morris Benjamin
Berens, Alan Robert
Berge, John Williston
Bergen, Robert Ludlum, Jr
Berger, Richard S
Bergman, Eric Arnold
Berry, Guy C
Bertozzi, Eugene R
Betso, Stephen Richard
Beyer, George Leidy
Bhatia, Sushil
Bi, Le-Khac
Bikales, Norbert M
Biletch, Norbert
Black, Carl (Ellsworth)
Black, William Bruce
Blommers, Elizabeth Ann
Bluestein, Ben Alfred
Blumstein, Alexandre
Blunt, Harry William
Bobear, William Joseph
Boettcher, F Peter
Boldebuck, Edith Maude
Bolgiano, Nicholas Charles
Bondaruk, Charles W, Jr
Bonsignore, Patrick Vincent
Bortnick, Newman Mayer
Boschan, Robert Herschel
Bouton, Thomas Chester
Bovey, Frank Alden
Bowden, Murrae John Stanley
Bower, Rafael Lee
Bower, George Myron
Boyd, Richard Hays
Brack, Karl
Brams, Stewart L
Braunstein, David Michael
Breder, Charles Vincent
Breed, Laurence Woods
Breitman, Leo
Bremner, Bart J
Brendley, William H, Jr
Breslow, David Samuel
Brett, Thomas Joseph, Jr

Brierre, Roland Theodore, Jr
Brittain, J W
Brodsky, Philip Hyman
Brown, Harold Probert
Brown, Kenneth Howard
Brown, Morton
Brown, Robert Raymond
Brown, Stewart Cliff
Browning, Horace Lawrence, Jr
Browning, William Bernard
Bruck, Stephen Desiderius
Bryant, George Macon
Bueche, Arthur Maynard
Burch, George Nelson Blair
Burke, William James
Burrell, Harry
Burton, Gilbert W
Burton, Robert Louis
Bush, Richard Wayne
Butler, John Mann
Butler, Robert Westbrook
Buzzell, John Gibson
Byrd, James Dotson
Byrd, Norman Robert
Cabanes, William Ralph, Jr
Cabasso, Israel
Cacella, Arthur Ferreira
Calderon, Nissim
Caldwell, John Richard
Calkins, William Harold
Cameron, Irvine R
Campbell, Allen James
Campbell, Gerald Allan
Campbell, Robert Wayne
Cantow, Manfred Richard
Carlson, Earl John
Carpenter, Dewey Kenneth
Carpenter, William Graham
Carraher, Charles Eugene, Jr
Carrick, Wayne Lee
Carty, Daniel T
Casella, John Francis
Cass, William Emerson
Castro, Anthony J
Cates, David Marshall
Caulfield, Daniel Francis
Causa, Alfredo G
Cavender, James Vere, Jr
Cella, Richard Joseph, Jr
Cenci, Harry Joseph
Chamberlin, Howard Allen
Chambers, Ralph Arnold
Chamot, Walter M
Chan, Look Lim
Chan, Maureen Gillen
Chang, Franklin Shih Chuan
Chang, Robert Chi-Heng
Chang, Shu-Pei
Chapman, Toby Marshall
Charbonneau, Larry Francis
Chatterjee, Pronoy Kumar
Chattha, Mohinder Singh
Chen, Catherine S H
Chen, Tsang Jan
Chen, William Kwo-Wei
Cheng, Tai Chun
Chenick, Albert George
Chipman, Gary Russell
Chiu, Jen
Christie, Peter Allan
Ciccarelli, Roger N
Citron, Joel David
Clark, Edward Shannon
Clark, Harold Arthur
Cleary, James William
Clendinning, Robert Andrew
Clough, Stuart Benjamin
Coats, Alma Winifred
Cobler, John George
Coffin, Perley Andrews
Coggeshall, A Darling
Cohen, Fredric Sumner
Cohen, Stuart Colin
Coke, Chauncey Eugene
Coleman, Lester Earl, (Jr)
Coleman, John Franklin
Coleman, Michael Murray
Collins, Edward A
Collins, Jerry Dale
Combs, Robert Leonard
Conley, Robert T
Conyne, Richard Francis
Cook, Jack E
Cooper, Jack Loring
Cooper, Terence Alfred
Coover, Harry Wesley, Jr
Cope, James Francis
Cornell, Robert Joseph
Cornet, James Oliver
Coscia, Anthony Oliver
Cosper, David Russell
Costanza, Albert James
Cotten, George Richard
Cotter, Robert James
Coury, Arthur Joseph
Craven, James Milton

Crist, Buckley, Jr
Cunningham, Robert Elwin
D'Alelio, Gaetano Francis
Daly, William Howard
Daues, Gregory W, Jr
David, Israel A
Davidson, Daniel Lee
Davis, Burns
Davis, Gerald Thomas
Davis, Gordon
Davis, Robert Bernard
Davison, John Alden
Davison, Sol
Dawes, David Haddon
Dawson, Daniel Joseph
Dawson, Robert Louis
Dawson, Thomas Larry
Deanin, Rudolph D
Dearlove, Thomas John
DeBolt, Lawrence Clifford
DeDominicis, Alex John
Deets, Gary Lee
De La Mare, Harold Elison
Delmonte, David William
Delvigs, Peter
Desai, Vinodrai Ranchhodji
Desper, Clyde Richard
De Tommaso, Gabriel Louis
DeWitt, Elmer John
Dhami, Kewal Singh
D'Ianni, James Donato
Dichter, Michael
Dickason, William Charles
Dickens, Brian
Dickie, Ray Alexander
Dickstein, Jack
Dieck, Ronald Lee
Diedrich, James Loren
Di Giacomo, Armand
Di Milo, Anthony J
Dimmig, Daniel Ashton
Dinbergs, Kornelius
Divis, Roy Richard
Dixon, George Douglass
Doak, Kenneth Worley
Dobay, Donald Gene
Dobinson, Frank
Dodge, James Stanley
Dohany, Julius Eugene
Dolenko, Allan John
Dorfman, Edwin
Dorsey, George Francis
Drake, Stevens Stewart
Drechsel, Paul David
Dreyfuss, Max Peter
Dreyfuss, Patricia
Droucas, John
Drucker, Arnold
Duddey, James E
Dudek, Thomas Joseph
Duke, June Temple
Duling, Irl Noel
Dunbar, Richard Alan
Dunbrook, Raymond Frederick
Dunlap, Lawrence Hallowell
Dunphy, James Francis
Durandetta, Donald W
Durrell, William S
Dyer, John
Dykstra, Thomas Karl
Eareckson, William Milton, III
Earing, Mason Humphry
Earl, Charles Riley
Easterbrook, Eliot Knights
Ebner, Herman George
Edelman, Leonard Edward
Edelman, Robert
Edwards, Walter Murray
Ehrhart, Wendell A
Ehrig, Raymond John
Eichinger, Bruce Edward
Elias, Hans Georg
Elliott, John Habersham
Elliott, John Raymond
Ells, Frederick Richard
Elmsie, James Stewart
Empen, Joseph A
Engel, John Hal, Jr
England, Richard Jay
Engle, Damon Lawson
Erchak, Michael, Jr
Erickson, Randall L
Ernst, John L
Essig, Henry J
Estes, Leland Lloyd
Euler, Raymond Lewis, Jr
Eustis, William Henry
Evans, Evan Franklin
Evans, Robert Morton
Evers, William C
Evers, William L
Ewart, Roswell Horr
Fadner, Thomas Alan
Fanta, George Frederick
Farago, John
Farrar, Martin Wilbur
Farrissey, William Joseph, Jr

Fawcett, Newton Creig
Fairheller, Stephen Henry
Feay, Darrell Charles
Fedors, Robert Francis
Feinberg, Stewart Carl
Feist, William Charles
Feit, Eugene David
Felicetta, Vincent Frank
Ferington, Thomas Edwin
Ferry, John Douglass
Fetscher, Charles Arthur
Fetters, Lewis
Fettes, Edward Mackay
Fick, Herbert John
Fielding-Russell, George Samuel
Firth, William Charles, Jr
Fish, John G
Fishman, Marshall Lewis
Fitch, Robert McLellan
Fitko, Chester Walter
Fitzgerald, Emerson Blanchard
Fletcher, Harry Huntington
Flexman, Edmund A, Jr
Florin, Roland Eric
Floyd, Don Edgar
Fodor, Lawrence Martin
Fogiel, Adolf W.
Fohlen, George Marcel
Foldi, Andrew Peter
Foley, Howard Kenneth
Ford, Emory A
Forester, Ralph H
Foster, Frederick Calvin
Foster, Robert Everett
Fox, Adrian Samuel
Fox, Daniel Wayne
Fox, Sidney Walter
Franta, William Alfred
Frederick, Donald Sherwood
Freeman, James Harrison
French, David Milton
French, James Edwin
Frensdorff, Hans Karl
Frey, David Allen
Frick, Neil Huntington
Frost, Lawrence William
Fuerholzer, James J
Fulmer, Glenn Elton
Funck, Dennis Light
Funt, B Lionel
Gallagher, James A
Gallivan, Robert Milo, Jr
Gander, Robert Johns
Gardlund, Zachariah Gust
Gardner, Donald Murray
Gardner, William Howlett
Garrison, William Emmett, Jr
Gates, Raymond Dee
Gaughan, Renata Rysnik
Gay, Frank P
Gaylord, Norman Grant
Gaylord, Richard J
Gebelein, Charles G
Geer, Richard P
Gehring, David Richard
Gehring, Harvey Thomas
Gelb, Leonard Louis
Gelfer, Daniel Harold
Gerhart, Howard Leon
Gerow, Clare William
Gesner, Bruce D
Ghosh, Kalyan K
Giffen, William Martin, Jr
Gilbert, Richard Dean
Gilkey, Russell
Gillen, Kenneth Todd
Gillham, John K
Giori, Claudio
Gippin, Morris
Glaser, Milton Arthur
Glasgow, David Gerald
Glenn, Furman Eugene
Glover, Leon Conrad, Jr
Gobran, Riad Hilmy
Goddu, Robert Fenno
Goebel, James Christopher
Goken, Garold Lee
Goldfinger, George
Goldman, Theodore Daniel
Goldstein, Albert
Golemba, Frank John
Golub, Morton Allan
Goodman, Alan Lawrence
Goodman, Albert
Goodman, Donald
Goodwyn, Jack Ray
Goossens, John Charles
Gordon, Gerald Arthur
Gordon, Joseph R
Gorham, William Franklin
Gosnell, Aubrey Brewer
Gosnell, Rex Beach
Gouinlock, Edward Vernon, Jr
Graham, Roger Kenneth
Grant, Warren Herbert
Grasshoff, Jurgen Michael
Graver, Richard Byrd
Gray, Don Norman

Gray, Theodore Flint, Jr
Green, Charles David
Green, Joseph
Greenley, Robert Z
Greenwald, Harold Leopold
Gross, Susan C
Grot, Walther Gustav Fredrich
Groten, Barney
Gruber, Elbert Egidius
Guggenberger, Lloyd Joseph
Guilbault, Lawrence James
Gurley, Thomas Gordon
Gutbezahl, Boris
Gutman, Charles M
Haag, Thomas Harry
Haas, Howard Clyde
Hademann, Albert Felix
Hagan, Ralph S
Hahn, Walter Leopold
Halasa, Adel Farhan
Hale, Warren Frederick
Hall, Judd Lewis
Hamb, Fredrick Lynn
Hamilton, William Lander
Hammer, Clarence Frederick, Jr
Hammer, William Frederick
Han, Charles Chih-Chao
Han, Yuri Wha-Yul
Hancock, James William
Hannon, Martin J
Hanson, Robert Burton
Hardwicke, Norman Lawson
Harmer, David Edward
Harmon, Dale George
Harmuth, Charles Moore
Harrell, Jerald Rice
Harris, Richard Lee
Harris, Robert L
Hartranft, George Robert
Hartzler, Jon David
Harwood, Harold James
Haseley, Edward Albert
Hatchard, William Reginald
Hauenstein, Jack David
Hauser, Martin
Hawkes, Stephen J
Hawkins, Walter Lincoln
Hay, Allan Stuart
Heilman, William Joseph
Heins, Conrad F
Helbert, John N
Helfand, Eugene
Helfgott, Cecil
Hellmann, Richard Jason
Henderson, John Frederick
Hendricks, James Owen
Hendrix, James Easton
Henry, Arnold William
Henry, Arthur Charles
Hepfinger, Norbert Francis
Hergenrother, William Lee
Herliczek, Siegfried H
Herndon, J Loritts, III
Hersh, Solomon Philip
Hershey, Allan Unger
Hewett, William Ainslie
Hickner, Richard Allan
Hicks, Elija Maxie, Jr
Hill, Frederick Burns, Jr
Hill, Harold Wayne, Jr
Hill, James Theo
Hill, Robert William
Hine, John Maynard
Hirsch, Stephen Simeon
Hirsty, Sylvain Max
Hirzy, John William
Ho, Chong Cheong
Hodes, William
Hodge, James Dwight
Hoehn, Harvey Herbert
Hoeschele, Guenther Kurt
Hoff, Raymond E
Hogen-esch, Theo Eltjo
Hoguet, Robert Gerard
Hoke, Donald I
Holik, Melville James
Holland, Virgil Fortune
Holmen, Reynold Emanuel
Holmes, Larry A
Holtmyer, Martin Dean
Holyoke, Caleb William, Jr
Honsberg, Wolfgang
Hoover, M Frederick
Hopfinger, Anton J
Hopkins, Allen John
Horne, Samuel Emmett, Jr
Horowitz, Carl
Horowitz, Emanuel
Hovermale, Ralph Allen
Hoyt, John Manson
Hsieh, Henry Lien
Hsu, Tsong-Han
Hu, Clyde Kuen-Hua
Huang, Samuel J
Hudson, Robert Leslie
Huemmer, Thomas Francis

Huggins, Maurice Loyal
Hulyalkar, Ramchandra K
Hundert, Murray Bernard
Hurford, Thomas Rowland
Hutchison, John Joseph
Hwa, Jesse Chia Hsi
Ikeda, Richard Masayoshi
Immediata, Tony Michael
Imperial, George Romero
Indictor, Norman
Ingram, Alvin Richard
Irvin, Howard H
Isaacs, Philip Klein
Isaacson, Henry Verschay
Isbister, Roger John
Israel, Stanley C
Jabarin, Saleh Abd El Karim
Jabloner, Harold
Jackson, Harold Leonard
Jackson, Winston Jerome, Jr
Jadrnicek, Bohumil Robert
Jahn, Alex Karl
Janssen, Arthur Gray
Jarvi, Reino A
Jarvis, Lactance Aubrey
Jendrek, John Paul, Jr
Jenkins, Sidney Hartman, Jr
Jensen, Arnold William
Jepson, Carl Henry
Jernigan, Robert Lee
Jirgensons, Arnold
Johnson, Duane Edward
Johnson, Julian Frank
Johnson, Paul Robert
Johnson, Robert William
Johnson, William Randolph, Jr
Johnston, Manley Roderick
Johnston, Norman Joseph
Johnston, Norman Wilson
Jones, Alan A
Jones, Faber Benjamin
Jones, Glenn Clark
Julinao, Peter C
Kabayama, Michiomi Abraham
Kalfayan, Sarkis Hagop
Kamath, Vasanth Rathnakar
Kamath, Yashavanth Katapady
Karasz, Frank Erwin
Karl, Curtis Lee
Karol, Frederick J
Katz, Manfred
Katz, Morton
Kaufman, Herman S
Kaufman, Martin Henry
Kaye, Howard
Keating, James T
Keck, Max Hans
Kelley, Joseph Matthew, Jr
Kelley, Philip Carlos
Kellman, Raymond
Kelly, Walter James
Kennedy, Joseph Paul
Kennedy, James Franklin
Kenyon, Allen Stewart
Keogh, Michael John
Keplinger, Orin Clawson
Kersting, Raymond James
Kesten, Yali
Kester, Dennis Earl
Kesting, Robert E
Ketley, Arthur Donald
Khan, Ausat Ali
Khanna, Ravi
Khanna, Som Nath
Killian, Frederick Luther
Kim, Ki-Soo
Kim, Yungki
King, Henry Lee
Kinstle, James Francis
Kirshenbaum, Gerald Steven
Kiss, Klara
Kissel, William John
Kitson, Robert Edward
Klapproth, William Jacob, Jr
Klein, Gerald Wayne
Kleinschuster, Jacob John
Kline, Gordon Mabey
Knight, Alan Campbell
Knobloch, Fred William
Knopf, Robert John
Knox, Jack Rowles
Kochhar, Rajindar Kumar
Koenig, Jack L
Kohan, Melvin Ira
Kohlhase, William Lawrence
Kolczynski, James Robert
Koleske, Joseph Victor
Kollar, William L
Kolyer, John M
Kopchik, Richard Michael
Kopf, Peter W
Korver, Gailerd Lee
Kraemer, John Francis
Kraiman, Eugene Alfred
Krajewski, John J
Krantz, Karl Walter
Kratzer, Reinhold

CHEMISTRY

Kresge, Edward Nathan
Kreuz, John Anthony
Kronberger, Karlheinz
Kronenthal, Richard Leonard
Kronstein, Max
Kross, Robert David
Kuan, Tiong H
Kuist, Charles Howard
Kumler, Philip L
Kun, Kenneth Allan
Kuntz, Irving
Kuo, Cheng-Yih
Kuper, Donald G
Kurath, Sheldon Frank
Kurz, James Eckhardt
Kuriner, Abraham
Kwei, Ti-Kang
Kwok, Wo Kong
Kwolek, Stephanie Louise
Kyker, Gary Stephen
Labanca, Santokh Singh
Labanca, Dominick A
Lagally, Paul
Lake, Robert D
Lakshmanan, P R
Lal, Joginder
Lando, Jerome B
Landoll, Leo Michael
Landrum, Billy Frank
Lane, Constance A
Lang, Edgar Reed
Langauer, Bruce Allen
Langley, Neal Roger
Lasky, Jack Samuel
Lavin, Edward
Lawson, Daniel David
Leavitt, Frederick Carlton
Le Blanc, John Roger
Lee, Lester Tsung-Cheng
Lee, Min-Shiu
Lee, Sung Ki
Lee, Wei-Ming
Lee, Ying Kao
Legare, Richard J
Legge, Norman Reginald
Lehr, Marvin Harold
Leininger, Robert Irvin
Leisemer, Ronald Newell
Lenz, Robert William
Leonard, Charles, Jr
Leonard, Edward Charles, Jr
Leonard, Jacques Walter
Lerner, Narcinda Reynolds
Levi, David Winterton
Levine, Ralph Manuel
Levy, Alan
Lewis, Danny Harve
Lewis, John Raymond
Lewis, Otis Griffin
Lewis, Richard Newton
Lewis, Thomas Brinley
Li, Hsueh Ming
Liberti, Frank Nunzio
Light, Rupert Edwin
Limburg, William W
Lin, Kuang-Farn
Lin, Otto Chui Chau
Lindberg, Steven Edward
Lipowitz, Jonathan
Lipowski, Stanley Arthur
Lipp, Hayden Ivan
Lipscomb, Nathan Thornton
Lisanke, Robert John, Sr
Litt, Morton Herbert
Livigni, Russell Anthony
Lo, Chien-Pen
Loan, Leonard Donald
Lockhart, Luther Bynum, Jr
Lockwood, Robert Greening
Lodon, Gary Arthur
Login, Robert Bernard
Logothetis, Anestis Leonidas
Logullo, Francis Mark
Lohr, Delmar Frederick, Jr
Long, Frank Wesley, Jr
Longworth, Ruskin
Looney, Ralph William
Lopata, Eugene Stephen
Lorensen, Lyman Edward
Lorenz, Donald H
Losekamp, Bernard Francis
Lovald, Roger Allen
Love, James Allan
Lovejoy, Elwyn Raymond
Lowery, Kirby, Jr
Lucas, Glennard Ralph
Luck, Russell M
Luckenbach, Thomas Alexander
Lukach, Carl Andrew
Lundberg, Robert Dean
Lupinski, John Henry
Luskin, Leo Samuel
Lyman, Donald Joseph
Lynch, Thomas John
Lynn, Merrill
Lyons, Peter Francis
Mackinney, Herbert William

MacLachlan, James Daniel
Macphee, Kenneth Erskine
MacWilliams, Dalton Carson
Magin, Ralph Walter
Mahiman, Bert H
Malofsky, Bernard Miles
Maloney, Daniel Edwin
Malotky, Lyle Oscar
Mandelkern, Leo
Manhart, Joseph Heritage
Mani, Inder
Manley, Rockliffe St John
Manos, Philip
Mansfield, Kevin Thomas
Manson, John Alexander
Mantell, Gerald Jerome
Marchessault, Robert Henri
Marder, Herman Lowell
Mark, James Edward
Markiewicz, Kenneth Helmut
Markowski, Henry Joseph
Markovitz, Mark
Maron, Samuel Herbert
Marshall, Richard Allen
Marshall, Thomas Ball
Martellock, Arthur Carl
Martin, Eugene Christopher
Martin, Richard Hadley, Jr
Martin, Wayne Holderness
Marvel, Carl Shipp
Mason, John Hugh
Mast, William Carlton
Matlack, Albert Shelton
Matlack, Louis Rogers
Matsuo, Keizo
Matthews, Virgil Edison
May, James Aubrey, Jr
May, Paul David
McCabe, Chester Charles
McCarthy, Neil Justin, Jr
McElroy, Wilbur Renfrew
McEwan, Ian Hugh
McFarlane, Finley Eugene
McGary, Charles Wesley, Jr
McGinnis, Vincent Daniel
McGirk, Richard Heath
McGrath, James Edward
McIntyre, Alan David
McIntyre, Donald
McKay, Jerry Bruce
McKeever, L Dennis
McNeil, Harry Daniel, Jr
McPeters, Arnold Lawrence
McPherson, James Louis
McCurdy, Orville L
Mehlhaff, Leon Curtis
Meibohm, Edgar Paul Hubert
Meier, James Archibald
Meier, John Warren
Meier, Joseph Francis
Meinhardt, Norman Anthony
Melamed, Sidney
Mendel, John Richard
Merkel, Timothy Franklin
Merken, Henry
Mermelstein, Robert
Merrill, Stewart Henry
Merz, Edmund Herman
Merz, Paul Louis
Metanomski, Wladyslaw Val
Meyer, Glen Ernest
Meyer, James Melvin
Meyer, Leo Francis
Meyer, Victor Bernard
Miksell, Sharell Lee
Milford, George Noel
Milgrom, Jack
Milkovich, Ralph
Millar, John Robert
Miller, Bernard
Miller, Ivan Keith
Miller, Kenneth Leron
Miller, Lewis F
Miller, Lewis Samuel
Miller, Robert Stephen
Miller, Walter Peter
Miller, Wilmer Glenn
Millhiser, Frederick Roy
Milich, Frank
Milone, Charles Robert
Mino, Guido
Mirabella, Francis Michael, Jr
Miranda, Thomas Joseph
Miskel, John Joseph, Jr
Mitchell, George Redmond, Jr
Moczygemba George A
Molau, Gunther Erich
Mongale, Daniel J
Monahan, Alan Richard
Moncure, Henry, Jr
Monroe, Stuart Benton
Mont, George Edward
Moore, Carl
Moore, Donald R
Moore, Eugene Roger
Moore, James Alfred
Moore, Louis Doyle, Jr

Moore, Robert Stephens
Morduchowitz, Abraham
Morosoff, Nicholas
Morris, Marion Clyde
Morton, Maurice
Mouk, Robert Watts
Moult, Roy Hepworth
Mueller, Donald Scott
Muldrow, Charles Norment, Jr
Munk, Petr
Murphey, Wilbur Alford
Murphy, John Joseph
Murphy, Walter Thomas
Murray, James Gordon
Muzycko, Thaddeus Marion
Myerholtz, Ralph W, Jr
Myers, Drewfus Young, Jr
Myers, Roger Miles
Nagel, Roger Miles
Nakagawa, T William
Nakajima, Nobuyuki
Nannelli, Piero
Naples, John Otto
Narayan, Tv Lakshmi
Narvaez, Richard
Nash, James Lewis, Jr
Naylor, Floyd Edmond
Needham, Charles D
Needles, Howard Lee
Neeley, Charles Mack
Nekutin, Vadim Constantin
Nelson, Charles Jay
Nelson, Robert Andrew
Nemphos, Speros Peter
Neuse, Eberhard Wilhelm
Nevin, Robert Stephen
Newey, Herbert Alfred
Newland, Gordon Clay
Newman, Seymour
Nielsen, Paul Livingstone
Nielsen, Lawrence Ernie
Nielsen, Stuart Dee
Niemann, Theodore Frank
Noland, James Sterling
Nolin, Joseph Arthur Benoit
Nonemaker, Larry Franklin
Nordstrom, J David
Norling, Parry McWhinnie
Norris, Forrest Harvey
Norton, Liburn Lafayette
Noshay, Allen
Notley, Norman Thomas
Novak, Ronald William
Noyes, Paul R
Nyberg, David Dolph
Nyce, Jack Leland
O'Brien, John Terence
Odian, George G
Ofstead, Eilert A
O'Leary, Kevin Joseph
O'Malley, James Joseph
O'Neill, George Joseph
O'Reilly, James Michael
O'Rell, Dennis Dee
Orphanides, Gus George
O'Shaughnessy, Marion Thomas, Jr
O'Shea, Francis Xavier
Ostfeld, Howard G
Otocka, Edward Paul
Ottaviani, Robert Augustine
Overberg, Richard Joseph
Overberger, Charles Gilbert
Owen, Donald Robertson, Jr
Oziomek, James
Pacansky, Thomas John
Paciorek, Kazimiera J L
Pacofsky, Edward Anthony
Padbury, John James
Palermo, Felice Charles
Pallen, Robert Harris
Pande, Kailash Chandra
Panzer, Hans Peter
Pappalardo, Leonard Thomas
Pappas, Nicholas
Parisek, Charles Bruce
Parish, Darrell Joe
Park, Vernon Kee
Parker, Earl Elmer
Paschke, Edward Ernest
Patra, Sushant Kumar
Patsiga, Robert A
Patterson, Donald Duke
Patterson, Gordon Derby, Jr
Patterson, William Jerry
Patton, John Thomas, Jr
Pearson, James Murray
Peascoe, Warren Joseph
Peerman, Dwight Ellsworth
Peffer, John Roscoe
Pelosi, Lorenzo Fred
Peng, Fred Ming-Sheng
Percival, Douglas Franklin
Perkel, Robert Jules
Perkins, Janet Sanford
Perry, Donald Dunham
Perry, James
Peters, Timothy Victor
Petersen, Kenneth C

Petersen, Philip Richard
Peterson, Jack Kenneth
Peterson, James Macon
Petfield, Robert Joseph
Petrich, Robert Paul
Petropoulos, Constantine Chris
Philipon, Joseph
Phillips, Paul John
Phillips, Yorke Peter
Pietrusza, Edward Walter
Pirma, Aleksander
Pirma, Irja
Pillar, Walter Oscar
Plant, William J
Platzer, Norbert
Pledger, Huey, Jr
Plessy, Boake Lucien
Plueddemann, Edwin Paul
Pollard, Robert Eugene
Pollard, Dale Flavian
Potts, James Edward
Powell, Richard James
Prapas, Aristotle George
Preston, Jack
Prevorsek, Dusan Ciril
Prober, Maurice
Prudence, Robert Thomas
Purdon, James Ralph, Jr
Putnam, Stearns Tyler
Quarles, Richard Wingfield
Quinn, Clayton Byreley
Raff, Rudolf August Victor
Raich, William Judd
Rakshys, Joseph W, Jr
Ralston, Robert Henry
Randall, Francis James
Rapoport, Lorence
Rave, Terence William
Ray-Chaudhuri, Dilip K
Raymond, Maurice A
Reardon, Joseph Edward
Rebel, William J
Reed, Thomas Freeman
Reegen, Sidney Lloyd
Reese, Cecil Everett
Reeves, Richard Allen
Reilly, Charles Bernard
Reilly, Patrick J
Reinking, Norman Herbert
Reisman, Abraham Joseph
Repka, Benjamin C, Jr
Restaino, Alfred Joseph
Reynard, Kennard Anthony
Rhein, Robert Alden
Rhum, David
Rice, David E
Riccobono, Paul Xavier
Richards, John Cadwallader
Ridger, Don Keith
Ridgway, James Stratman
Rie, John E
Riecke, Edgar Erick
Riley, Robert Lee
Rinde, James A
Ringwald, Eugene Lee
Roberts, Durward Thomas, Jr
Roberts, George P
Robins, Jack
Robinson, Donald Nellis
Robson, John Howard
Roedel, George Frederick
Rogan, John B
Rogers, Dow Albert, Jr
Roper, Robert
Rosemund, Walter Richard
Rosen, Irving
Rosen, Marvin
Rosenfeld, Jerold Charles
Ross, Stanley Elijah
Rosser, Robert William
Rossito, Conrad
Rotenberg, Don Harris
Roth, Roy William
Rowland, Stanley Paul
Rubin, Isaac D
Rudin, Alfred
Rupert, John Paul
Russell, Charles Addison
Russell, Charles Richard
Russell, James
Russell, Kenneth Edwin
Ryan, Patrick Walter
Sabia, Raffaele
Sackoff, Martin M
Sacks, William
Saegebarth, Klaus Arthur
Sahatjian, Ronald Alexander
Sahli, Brenda Payne
Sahli, Muhammad S
St Clair, Terry Lee
St Pierre, Leon Edward
Salamone, Joseph C

Sallo, Jerome Stanley
Salovey, Ronald
Saltonstall, Clarence William, Jr
Salvesen, Robert H
Sammak, Emil George
Sample, James Halverson
Samuels, Robert Joel
Samulski, Edward Thaddeus
Sandberg, Carl Lorens
Sanderson, Edwin S
Sandhu, Mohammad Akram
Sarko, Anatole
Sauerbrunn, Robert Dewey
Saunders, James Henry
Saxon, Robert
Scarpellino, Joseph
Schaefgen, John Raymond
Schatz, Robert James
Scheiber, David Hitz
Schell, William John
Scherrer, Joseph Henry
Schildknecht, Calvin Everett
Schilling, Curtis Louis, Jr
Schindler, Frederick James
Schmidt, Donald L
Schmitt, George Joseph
Schmitt, Joseph Michael
Schneider, John Arthur
Schneider, Robert L
Schober, Donald Lincoln
Schollenberger, Charles Sundy
Schrag, John L
Schreyer, Ralph Courtenay
Schroeder, Walter Arthur
Schrof, William Ernst John
Schuler, Norman William
Schultz, Herman Solomon
Schulz, Donald Norman
Schuurmans, Hendrik J L
Schwab, Frederick Charles
Schwab, Peter Austin
Schwarz, Eckhard C A
Schwenker, Robert Frederick, Jr
Scott, Franklin James
Scott, Kenneth Walter
Scruggs, Jack G
Scullin, James Philip
Seelbach, Charles William
Segal, Charles Lewis
Segall, Gordon Hart
Selman, Charles Melvin
Seltzer, Raymond
Serlin, Irving
Setterquist, Robert Alton
Seymour, Raymond Benedict
Shaneyfelt, Duane L
Shank, Charles Philip
Shannon, Frederick Dole
Sharda, Satish Chander
Sharkey, William Henry
Sharp, Louis James, IV
Sharpe, Andrew Jackson, Jr
Shaw, Montgomery Throop
Shaw, Richard Gregg
Shechter, Leon
Shelton, James Reid
Shepherd, Floyd
Sherbeck, L Adair
Sherman, Anthony Michael
Sherman, Patsy O'Connell
Shih, Hsiang
Shilling, Wilbur Leo
Shinohara, Makoto
Shivers, Joseph Clois, Jr
Short, James N
Shue, Robert Sidney
Shuford, Richard Joseph
Shulman, Sol
Sidi, Henri
Siebert, Alan Roger
Siefken, Mark William
Sieglaff, Charles Lewis
Silver, Frank Morris
Silverman, Bernard
Silverman, Joseph
Simms, John Alvin
Singh, Hakam
Singler, Robert Edward
Sircar, Anil Kumer
Skeist, Irving
Skochdopole, Richard E
Skoultchi, Martin Milton
Slade, Arthur Laird
Slade, Philip Earl, Jr
Sliemers, Francis Anthony, Jr
Slocombe, Robert Jackson
Slysh, Roman Stephan
Smid, Johannes
Smith, Anthony Gerald
Smith, Donald Arthur
Smith, Harry Andrew
Smith, James David Blackhall
Smith, Kenneth Judson, Jr
Smith, Oliver Wendell
Smith, Samuel
Smith, Shaler Gordon, Jr
Smith, Terry Edward

Smith, Thomas Woods
Smith, Thor Lowe
Smith, William Mayo, Jr
Snyder, William H
Sokol, Phillip Edward
Solomon, M Michael
Song, Won-Ryul
Sood, Satya P
Sovish, Richard Charles
Spenadel, Lawrence
Sperati, Carleton Angelo
Spitsbergen, James Clifford
Spitzer, William Carl
Squire, David R
Stackman, Robert W
Stacy, Carl J
Staengel, Vivian Thomas
Staples, Jon T
Statton, Gary Lewis
Stein, Alvin
Stein, Richard James
Stein, Richard Stephen
Steiner, Arnold Byron
Stephens, Howard L
Stephens, James Regis
Sterling, Robert Fillmore
Sterman, Melvin David
Stevens, Malcolm Peter
Stickney, Palmer Blaine
Stiehl, Roy Thomas, Jr
Stille, John Kenneth
Stone, Edward
Strand, Robert Charles
Stratton, Robert Alan
Strauss, Carl Richard
Strazdins, Edward
Strobel, Charles William
Stromberg, Robert Remson
Stryker, Harry Kane
Stueben, Kenneth Charles
Stutsman, Paul Snell
Su, Cheh-Jen
Sublett, Bobby Jones
Subramanian, Ravanasamudram
  Venkatachalam
Sullivan, Charles Irving
Sulzberg, Theodore
Sundet, Sherman Archie
Sutherland, Judith Elliott
Svoboda, Glenn Richard
Swanholm, Carl E
Sweet, Arthur Thomas, Jr
Swift, Graham
Tabb, David Leo
Tabibian, Richard
Taft, David Dakin
Taggart, William Paul
Takahashi, Akio
Takekoshi, Tohru
Takvorian, Kenneth Bedrose
Tan, Julia S
Tanikella, Murty Sundara Sitarama
Tannahill, Mary Margaret
Tanner, David
Tanquary, Albert Charles
Tarwater, Oliver Reed
Tashick, Irving
Taylor, Lynn Johnston
Temin, Samuel Cantor
Templer, David Allen
Terry, Stuart Lee
Tesoro, Giuliana C
Tess, Roy William Henry
Thames, Shelby Freland
Theil, Michael Herbert
Theis, Richard James
Thiruvengada, Seshan
Thompson, Clifford Francis
Thompson, Larry Flack
Thurmaier, Roland Joseph
Tiedeman, George Trent
Tolbert, Tommy Lyle
Tolgyesi, Eva
Tonkyn, Richard George
Tornqvist, Erik Gustav Markus
Tortorello, Anthony Joseph
Toth, William James
Toy, Madeline Shen
Traynor, Lee
Trepka, William James
Trewiler, Carl Edward
Trischler, Floyd D
Trudel, Gerald Joseph
Tsai, Tom Chung Hsiung
Tsuk, Andrew George
Tucker, Harold
Tuites, Donald Edgar
Tuites, Richard Clarence
Turner, Derek T
Turner, Robert Lawrence
Turner, William Russel
Tweedie, Adelbert Thomas
Udipi, Kishore
Uebele, Curtis Eugene
Ullman, Robert
Ulrich, Henri
Un, Howard Ho-Wei
Urbanic, Anthony Joseph

Vachon, Raymond Normand
Van de Castle, John F
Vandenberg, Edwin James
Van Den Berghe, John
Vander Hart, David Lloyd
Van der Hoff, Bernard Maria Euphemius
Vanderhoff, John W
Van Deusen, Richard L
Van Norman, Gilden Ramon
Vassallo, Donald Arthur
Vassiliades, Anthony E
Ver Strate, Gary William
Vessel, Eugene David
Vickroy, Virgil Vester, Jr
Victorius, Claus
Vigo, Tyrone Lawrence
Villars, Charles Earl
Vincent, Gerald Glenn
Vogel, Martin
Vogelfanger, Elliot Aaron
Vogelsong, Donald Clair
Vogl, Otto
Vogt, Clifford Marshall
Vogt, Herwart Curt
Volpe, Angelo Anthony
Vorchheimer, Norman
Vriesen, Calvin W
Waack, Robert
Wade, Robert Harold
Wagner, Herman Leon
Wagner, Klaus Peter
Wagner, Melvin Peter
Wagner, Richard Lloyd
Wald, Wilbur J
Walker, Charles Carey
Wallace, Thomas Patrick
Wallenberger, Frederick Theodore
Walsh, Edward John
Waltcher, Irving
Wang, Francis Wei-Yu
Wang, Jin-Liang
Ward, Frank Kernan
Warfel, David Ross
Warfield, Robert Welmore
Warner, Walter Charles
Wasserman, David
Wasserman, William Jack
Watson, William Harrison
Watterson, Arthur C, Jr
Wear, Robert Lee
Webb, Richard Lansing
Weber, Robert Emil
Wechsler, Harry C
Weddleton, Richard Francis
Weedon, Gene Clyde
Weese, Richard Henry
Weinberg, David Samuel
Weiner, Milton Lawrence
Weinstein, Arthur Howard
Weintraub, Lester
Weise, Jurgen Karl
Weiss, Douglas Eugene
Weiss, James Owen
Weiss, Jonas
Weiss, Virgil Wayne
Welch, Melvin Bruce
Wellons, Jesse Davis, III
Welsh, David Albert
Wen, Richard Yutze
Wentworth, Gary
Wentworth, Stanley Earl
Werber, Frank Xavier
Weyna, Philip Leo
Whelan, John Michael
Whelan, William Paul, Jr
White, Dwain Montgomery
White, Robert Winslow
Whitehouse, Bruce Alan
Wicker, Thomas Hamilton, Jr
Wickham, William Terry, Jr
Widmer, Zeno W, Jr
Widmer, Hans
Wiedenmann, Lynn G
Wildman, Gary Cecil
Wildnauer, Richard Harry
Wiles, David M
Wilkinson, William Kenneth
Williams, Fredrick David
Williams, Joel Lawson
Williamson, Frederick Dale
Williamson, Jerry Robert
Wilson, Donald Richard
Wilson, John Charles
Wilson, Leon William, Jr
Wilson, Lynn Harold
Wims, Andrew Montgomery
Winkler, DeLoss Emmet
Winston, Anthony
Winter, Roland Arthur Edwin
Wismer, Marco
Wissbrun, Kurt Falke
Wittbecker, Emerson Laverne
Wolf, Leslie Raymond
Wolf, Richard Eugene
Wolinski, Leon Edward
Wollner, Thomas Edward
Wolny, Friedrich Franz

Wong, Chun-Ming
Woodbrey, James Calvin
Woodward, Fred Erskine
Wooten, Willis Carl, Jr
Work, James Leroy
Work, Robert Wyllie
Wrasidlo, Wolfgang Johann
Wright, Charles Dean
Wu, Tse Cheng
Wunderlich, Bernhard
Wynne, Kenneth Joseph
Wynstra, John
Yanko, John Alexis
Yasuda, Hirotsugu
Yoon, Do Yeung
Young, Ainslie Thomas, Jr
Young, Robert Hayward
Youngman, Edward August
Yu, Arthur J
Yu, Hyuk
Yudelson, Joseph Samuel
Yuen, Po Sang
Yunick, Robert P
Zagar, Walter T
Zalar, Frank Victor
Zalewski, Edmund Joseph
Zelinski, Robert Paul
Zimmer, William Frederick, Jr
Zimmerman, Joseph
Zimmt, Werner Siegfried

Protein Chemistry
Adams-Mayne, Mabelle Elaine
Allison, William S
Anderson, Robert Lewis
Berardi, Leah Castillon
Bietz, Jerold Allen
Bing, David H
Brown, Ray Kent
Bull, Henry Bolivar
Burke, Morris
Burleigh, Bruce Daniel, Jr
Burzynski, Stanislaw Rajmund
Cardenas, Mary Janet M
Carmichael, David James
Chang, Hsing-Tze Ruan
Choong, Hsia Shaw-Lwan
Cole, Roger David
Colman, Roberta F
DeLange, Robert J
Feeney, Robert Earl
Fletcher, Paul Litton, Jr
Foltz, Thomas Roberts, Jr
Fowler, Audree Vernee
Gagen, Walter Leonard
Gallop, Paul Myron
Glover, George Irvin
Gordon, William George
Habeeb, Ahmed Fathi Sayed Ahmed
Harmison, Charles Rice
Hartman, Frederick Cooper
Hiskey, Richard Grant
Hodges, Robert Stanley
Holly, Frederick William
Hugli, Tony Edward
Inman, John Keith
Jackson, Richard Lee
Jurasek, Lubomir
Kasarda, Donald David
Kauzmann, Walter (Joseph)
Kippenstein, Gerald Lee
Kohler, Heinz
Kominz, David Richard
Levine, Philip Theodore
Levy, Daniel
Li, Lu Ku
Lowey, Susan
MacGregor, Carolyn Harvey
Maggio, Edward Thomas
Margoliash, Emanuel
Markley, John Lute
Matthews, Rowena Green
McConathy, Walter James
Mitchell, Thomas F, Sr
Moore, William Earl
Nicoli, Miriam Ziegler
Nigen, Alan Mark
Olsen, Kenneth Wayne
Pemrick, Suzanne Marie
Porzio, Michael Anthony
Prescott, David Julius
Pritchard, David Graham
Putterman, Gerald Joseph
Reitz, Henry Charles
Renthal, Robert David
Richards, Frederic Middlebrook
Romans, Robert Gordon
Schechter, Alan Neil
Schneider, Arthur Lee
Schroeder, Walter Adolph
Schultz, James Michael
Schultz, Duane Robert
Scouten, William Henry
Smith, Carol Price
Stephens, Raymond Edward
Stevenson, Frits Christiaan
Stevens, Kenneth James
Turner, Earl Wilbert

Van Kley, Harold
Vnek, John
Waterman, Michael Roberts
White, Frederick Howard, Jr
Yaguchi, Makoto

Pulp Chemistry
Abson, Derek
Allen, Gordon Ainslie
Andrews, Russell S, Jr
Atchison, Joseph Edward
Barton, John Selby
Becker, Edward Samuel
Bliesner, William Clark
Boehm, Robert Louis
Bradway, Keith E
Brinkley, Amiel Word, Jr
Bublitz, Walter John, Jr
Campbell, Robert Terry
Cordingly, Richard Henry
Danforth, Raymond Hewes
Green, Robert Patrick
Greninger, George King, Jr
Histed, John Allan
Kalisch, John Hans
Koran, Zoltan
Kubes, George Jiri
La Vallee, William Alfred
Leech, John G
Leopold, Bengt
Maranville, Lawrence Frank
Ohrn, Nils Yngve
Paulson, Jack Charles
Peterson, Robert C
Procter, Alan Robert
Redd, John Coleman
Sarkanen, Kyosti Vilho
Saxton, William Reginald
Schoettler, James Robert
Smith, William Edmond
Somsen, Roger Alan
Sun, Bernard Ching-Huey
Trotter, Patrick Casey
Tuck, Norman Gordon Maxwell
Von Koeppen, Andreas
Wollwage, Paul Carl
Zentner, Thomas Glenn

Radiation Chemistry
Ache, Hans Joachim
Allen, Augustine Oliver
Altmiller, Henry
Appleby, Alan
Askew, William Crews
Baba, Anthony John
Baker, Samuel I
Barr, Nathaniel Frank
Bennett, Gerald William
Benson, Brent W
Berkowitz, Harry Leo
Bibler, Ned Eugene
Bichsel, Hans
Bielski, Benon H J
Birstein, Seymour J
Boyd, Alan William
Buchanan, John Donald
Buck, Warren Louis
Bulusu, Suryanarayana
Burke, Gail De Planque
Burr, John Green
Burtt, Benjamin Pickering
Cathey, LeConte
Chappel, Samuel Estelle
Chen, Sow-Hsin
Christman, Edward Arthur
Chu, Keh-Chang
Chu, William Tongil
Conrad, Edward Ezra
Coulson, Larry Vernon
Coulter, Claude Alton
Davis, Thomas Pearse
Dorfman, Leon Monte
Dugle, David L
Duncer, Arthur Gustav, Jr
Dyne, Peter John
Ehrlich, Margarete
Epstein, Arnold S
Feng, Da-Fei
Freeman, Gordon Russel
Gill, Piara Singh
Gillis, Hugh Andrew
Gordon, Sheffield
Gottschall, Carl
Grant, David Graham
Guthrie, Donald Arthur
Haas, Peter Herbert
Harmer, David Edward
Hart, Edwin James
Helbert, John N
Hill, Mary Mechtilde
Holroyd, Richard Allan
Howton, David Ronald
Hunt, John Wilfred
Hurley, John Paul
Hurst, George Sam
Johnsen, Russell Harold
Jonah, Charles D
Kaplan, Michael

Kaufman, Priscilla C
Kispert, Lowell Donald
Kunz, Walter Ernest
Lee, Rupert Archibald
Lemmon, Richard Millington
Luntz, Myron
MacCallum, Crawford John
Matheson, Max Smith
McDonell, William Robert
McLaughlin, William Lowndes
Meisel, Dan
Miller, John Howard
Miller, William Robert
Miller, William Boynton, Jr
Mozumder, Asokendu
Naber, James Allen
Nagle, William Arthur
Newton, Carlos E, Jr
O'Brien, Keran
Osborn, Claiborn Lee
Osterholz, Frederick David
Owen, Terence Cunliffe
Painter, Linda Robinson
Palmer, Richard Carl
Parker, Cleofus Varren, Jr
Parrish, Clyde Franklin
Petree, Ben
Phillips, Gary Wilson
Pinson, James Wesley
Placious, Robert Charles
Potts, James Edward
Rainis, Albert Edward
Ritchie, Rufus Haynes
Ritz, Victor Henry
Rudolph, Philip S'
Russell, David Bernard
Sargent, Frederick Peter
Satcher, Myran Charles, Jr
Sauer, Myran Charles, Jr
Schillaci, Mario Edward
Schmidt, Klaus H
Schubert, Jack
Schweizer, Felix
Shah, Shirish Kalyanbhai
Sherman, Warren V
Shupe, Robert Eugene
Silverman, Joseph
Singh, Ajit
Spokas, John J
Stewart, Albert Clifton
Stinchcomb, Thomas John
Sworski, Thomas John
Torrey, Rubye Prigmore
Trachtman, Mendel
Trice, James Buckner
Trombka, Jacob Israel
Trubey, David Keith
Trudel, Gerald Joseph
Vacirca, Salvatore John
Walker, David Crosby
Waltz, William Lee
Ward, John F
Wells, Michael Byron
Woods, Robert James
Woodward, Ervin Chapman, Jr
Zimbrick, John David

Radiochemistry
Ache, Hans Joachim
Anders, Edward
Anders, Oswald Ulrich
Anghileri, Leopoldo Jose
Anselmo, Vincent C
Atrep, Moses, Jr
Balagna, John Paul
Baratta, Edmond John
Barker, Franklin Brett
Barton, George Wendell, Jr
Bingham, Carleton Dille
Blanchard, Richard Lee
Bogar, Louis Charles
Bogard, Andrew Dale
Brantley, John Calvin
Browne, Charles Idol
Bryant, Ernest Atherton
Burgus, Warren Harold
Burnett, William Thomas, Jr
Campbell, David Owen
Campbell, Donald R
Castrilon, Jose P A
Chakravarti, Dipitman
Clark, Herbert Mottram
Cohen, Donald
Cowan, George A
Currie, Lloyd Arthur
Daniels, William Richard
Deck, Charles Francis
Devoe, James Rollo
Duncan, William Perry
Ehmann, William Donald
Evans, James Bowen
Evans, John Charles, Jr
Finn, Ronald Dennet
Frank, Joan Patricia
Gleason, Geoffrey Irving
Giorgi, Angelo Louis
Goishi, Wataru
Golchert, Norbert William

Good, Mary Lowe
Gordon, Benjamin Edward
Guinn, Vincent Perry
Hardcastle, James Edward
Harris, James A
Helman, Edith Zak
Herner, Albert Erwin
Hicks, Harry Gross
Holcomb, Herman Perry
Hollbach, Natasha Coffin
Holtzman, Richard Beves
Howell, Barbara Femena
Hunt, Adolf
Irvine, John Withers, Jr
Ithakissios, Dionyssis Spiros
Jenkins, Robert Walls, Jr
Jensen, Jens Trygve
Jervis, Robert E
Jones, Owen Lloyd
Kahn, Bernd
Kanipe, Larry Gene
Kappe, David Syme
Katz, Sidney A
Kavula, Michael P, Jr
Kaye, James Herbert
Keedy, Curtis Russell
Keeley, Dean Francis
Keisch, Bernard
Kennedy, Albert Joseph
Kershner, Carl John
Kierbow, Julie Van Note Parker
King, Perry, Jr
Krey, Philip W
Lagomarsino, Raymond J
Larrabee, Graydon B
Leventhal, Leon
Liebman, Arnold Alvin
Lindner, Kenneth E
Linn, Thomas Arthur, Jr
Loranger, William Farrand
Love, Daniel Lindsley
Lyon, Walter Southern, Jr
Mahan, Kent Ira
Martell, Edward A
Matlack, George Miller
Matuszek, John Michael, Jr
McCown, John Joseph
Mehra, Mool Chand
Melgard, Rodney
Millard, Hugh Thompson, Jr
Miller, Charles Everett, Jr
Miller, George S
Montgomery, Daniel Michael
Moore, Robert Vernon
Morgan, John Walter
Neill, Warren Joseph
Nelson, Manno Fredrick
Newman, Richard Holt
Nystrom, Robert Forrest
Perlow, Mina Rea Jones
Pettijohn, Richard Robert
Potter, John Clarkson
Preswood, Rene Jesse
Rack, Edward Paul
Rainey, Robert Hamric
Reuland, Robert John
Richardson, Albert Edward
Riley, Bernard Jerome
Ritzman, Robert L
Rocco, Gregory Gabriel
Rodriguez-Fraga, Andres
Rona, Elizabeth
Rosegay, Avery
Rothman, Alan Michael
Roy, Jean-Claude
Scott, Ralph Asa, Jr
Segura, Gonzalo, Jr
Sheppard, John Clarence
Shipman, William H
Showalter, Donald Lee
Sicilio, Fred
Silker, Wyatt Burdette
Smith, Charles Francis, Jr
Stang, Louis George, Jr
Steim, Mimi M
Stehney, Andrew Frank
Stevenson, Peter Cooper
Stewart, Donald Charles
Stuart, Robert Lee
Suttle, Andrew Dillard, Jr
Tang, Yi-Noo
Tanner, James Thomas
Thomas, Gerald Andrew
Thompson, Joseph Lippard
Towne, Jack C
Tuck, Dennis George
Vandergraaf, Tjalle 'T
Van De Steeg, Garet Edward
Van Tuyl, Harold Hutchison
Varga, Louis P
Voigt, Adolf Frank
Volk, Murray Edward
Volkert, Wynn Arthur
Wang, Chih Hsing
Watters, Robert Lisle
Wechter, Margaret Ann
Welch, Michael John

Wilkniss, Peter Eberhard
Wolf, Walter
Wolfsberg, Kurt
Yaffe, Leo
Yaffe, Ruth Powers
Yanko, William Harry
Youngstrom, Richard Earl

Rubber Chemistry
Ambelang, Joseph Carlyle
Arendt, Volker Dietrich
Averill, Seward Junior
Behrens, Rudolf Adolf
Berkheimer, Henry Edward
Boozer, Charles (Eugene)
Boustany, Kamel
Boyce, Richard Joseph
Brooks, Lester Allen
Coffin, Perley Andrews
Connell, Balfour
Costanza, Albert James
Deviney, Marvin Lee, Jr
Edwards, Douglas Cameron
Erdman, John Paul
Fogiel, Adolf W.
Friedman, Emil Martin
Furry, Benjamin K
Gessler, Albert Murray
Glidden, Richard Mills
Glymph, Eakin Milton
Harcsar, Francis George
Harrop, William Henry
Hausch, Walter Richard
Haxo, Henry Emile, Jr
Healy, James C
Hicks, Arthur Earl
Hirsty, Sylvain Max
Hopkins, George Hallman
Johnston, Paul Robert
Knabeschuh, Louis Henry
Leshin, Richard
Manuel, Thomas Asbury
Mayor, Rowland Herbert
Medalia, Avrom Izak
Merz, Paul Louis
Model, Frank Steven
Morita, Eiichi
Murray, Robert Marie
Nichols, Parks Montgomery
Nyce, Jack Leland
O'Farrell, Charles Patrick
Pfeifer, Charles William
Rader, Charles Phillip
Richwine, John Robert
Saltman, William Mose
Schmiegel, Walter Werner
Schoene, Dwight Lorin
Schwartz, Norman Vincent
Shaver, Forrest Wheeler
Stevens, Harry Nelson
Stevenson, Arthur Charles
Studebaker, Merton Leland
Svetlik, Joseph Frank
Tewksbury, Charles Isaac
Tillson, Henry Charles
Trepka, William James
Trott, Gene F
Vial, Theodore Merriam
Williger, Ervin John
Wilson, Angus
Wise, Raleigh Warren
Wright, Robert L
Yanko, John Alexis
Young, Evan Johnson
Zelinski, Robert Paul
Zimmerman, Stanley Dean

Sanitary Chemistry
Denman, Wayne Leonard
Middleton, Francis Marvin
Morris, John Carell
Wright, Charles V, Jr

Soil Chemistry
Adams, Russell S, Jr
Albregts, Earl Eugene
Allison, Lowell Edward
Anderson, Warren Boyd
Babcock, Kenneth Leslie
Balam, Baxish Singh
Barber, Stanley Arthur
Barrows, Harold Lindsey
Barshad, Isaac
Bartlett, Richmond J
Bardorf, Robert Ludwig
Bennett, Allison Carr
Bingham, Frank Thomas
Blanchar, Robert W
Bohn, Hinrich Lorenz
Bower, Charles Arthur
Bowman, Bruce Tamblyn
Braids, Olin Capron
Branson, Roy
Brown, Donald A
Burns, Allan Fielding
Bushnell, Vernon Clifford

Caldwell, Alfred Craig
Caldwell, Augustus George
Campbell, Constantine Alberga
Carlson, Robert Marvin
Carstea, Dumitru Dumitru
Carter, David LaVere
Cescas, Michel Pierre
Cheng, Cheng-Yin
Cheng, Hwei-Hsien
Chesnin, Leon
Chichester, Frederick Wesley
Chien, Sen Hsiung
Clark, John S
Corey, Richard Boardman
Dantzman, Charles L
Davidson, Thomas, Jr
Dormaar, Johan Frederik
Dubrovin, Kenneth P
Dutt, Gordon Richard
Elgawhary, Salah Mohammad
Ellis, Boyd G
Ellis, Roscoe, Jr
El-Swaify, Samir Aly
Farmer, Walter Joseph
Fenster, William E
Fine, Lawrence Oliver
Fisher, Theodore Roosevelt
Fiskell, John Garth Austin
Fly, Claude Lee
Fowler, Eric Beaumont
Frink, Charles Richard
Gammon, Nathan, Jr
Gardner, Bryant Rogers
Gast, Robert Gale
Geist, Jon Michael
Geraldson, Carroll Morton
Gilliam, James Wendell
Graetz, Donald Alvin
Grunes, David Leon
Gupta, Gian Chand
Hashimoto, Isao
Hassett, John J
Heald, Walter Roland
Heddleson, Milford Raynord
Helling, Charles Siver
Hensler, Ronald Fred
Hermanson, Harvey Philip
Hill, Archie Clyde
Himes, Frank Lawrence
Ho, Clara Lin
Hoover, William L
Hortenstine, Charles C
Horton, James Henry, Jr
Hossner, Lloyd Richard
Hourigan, William R
Howe, David Orville
Hutchinson, Frederick Edward
James, Ronald Valdemar
Jenne, Everett A
Jones, Randall Jefferies
Jurinak, Jerome Joseph
Kardos, Louis Thomas
Kelley, Omer Joseph
Khasawneh, Fayez Essa
Kissel, David E
Konrad, John Grey
Kunishi, Harry Mikio
Lee, Kwang Woo
Leonard, Ralph Avery
Lewis, Glenn C
Lisk, Russell J
Lisk, Donald James
Logan, Terry James
Long, Franklin Leslie
Lowe, Lawrence E
Lowe, Philip Funk
Lutrick, Monroe Cornealous
MacKay, Donald Cyril
MacKenzie, Angus Finley
Magdoff, Frederick Robin
Marion, Giles Michael
Matthews, Burton Clare
McIntosh, Jerry Leon
McLean, Eugene Otis
McNeal, Brian Dean
Meek, Burl Dean
Menzel, Ronald George
Miller, James Roland
Miller, Raymond Jarvis
Millette, Gerard J F
Miner, Frend John
Morrill, Lawrence George
Mortland, Max Merle
Mortvedt, John Jacob
Mugwira, Luke Makore
Muir, Melvin K
Naddy, Badie Ihrahim
Nakayama, Francis Shigeru
Nash, Victor E
Naylor, Denny Ve
Nelson, Darrell Wayne
Nelson, Wesley Eugene
Newland, Leo Winburne
Nishita, Hideo
O'Connor, George Albert
Olsen, Ralph A
Onken, Arthur Blake

Oster, James Donald
Page, Albert Lee
Page, Norwood Rufus
Pearson, Robert Watt
Peck, Theodore Richard
Pendleton, John Davis
Perkins, Henry Frank
Peterson, Howard Boyd
Pionke, Harry Bernhard
Place, Gerald Alan
Pope, Alex
Porter, Lynn K
Pratt, Parker Frost
Prevatt, Robert Waldeman
Prince, Allan Bixby
Ragland, John Leonard
Reed, Lester W
Reed, Marion Guy
Reed, William Edward
Reid, Preston Harding
Reneau, Raymond B, Jr
Rennie, Donald Andrews
Rhoads, Frederick Milton
Robertson, James Alexander
Robertson, William Kitchener
Rouse, Roy Dennis
Routson, Ronald C
Rubins, Edward J
Russell, Glenn C
Ryan, James Anthony
Saiz Del Rio, Jose Francisco
Sandhu, Shingara Singh
Satchell, Donald Prentice
Scarsbrook, Clarence Edwin
Schnitzer, Morris
Scott, Albert Duncan
Sedberry, Joseph E, Jr
Seymour, Keith Goldin
Shannon, Stanton
Shelton, James Edward
Shuman, Larry Myers
Simpson, Daniel Martin Henry
Singh, Daulat
Singh, Surinder Shah
Singleton, Paul C
Smika, Darryl Eugene
Smith, Donald Henry
Smith, R L
Snyder, George Heft
Sorensen, Robert Carl
Sowden, Frederick John
Spencer, William F
Spinks, Daniel Owen
Stewart, Bobby Alton
Swoboda, Allen Ray
Tamimi, Yusuf Nimr
Thomas, Grant Worthington
Toth, Stephen John
Tullock, Robert Johns
Volk, Bob Garth
Volk, Garth William
Volk, Gaylord Monroe
Volk, Veril Van
Weber, Jerome Bernard
Webster, Gordon Ritchie
Weed, Sterling Barg
Wells, Bobby R
Westfall, Dwayne Gene
White, Ronald Paul, Sr
Whittig, Lynn D
Williams, David Emerton
Willis, Guye Henry
Wolf, Benjamin
Wright, James R

Solid State Chemistry

Adler, George
Albrecht, Andreas Christopher
Ban, Vladimir Sinisa
Bermudez, Victor Manuel
Birkeland, Stephen P
Blank, Zvi
Bloch, Aaron Nixon
Bos, William G
Brebrick, Robert Frank, Jr
Brenner, Henry Clifton
Brown, Fred
Carbajal, Bernard Gonzales, III
Chance, Ronald Richard
Chenot, Charles Frederic
Chu, Shirley Shan-Chi
Cohen, Howard Melvin
Condit, Ralph Howell
Condrate, Robert Adam
Cook, William R, Jr
Coulter, Lowell Vernon
Cox, David Ernest
Cullen, Glenn Wherry
Datta, Ranajit K
Davis, Joseph Anthony
Delbecq, Charles Jarchow
DeLuca, John Anthony
Donohue, Paul Christopher
Ekman, Carl Frederick W
Epple, Robert (Paul)
Eror, Nicholas George, Jr
Evans, Billy Joe
Exarhos, Gregory James

Ferraris, John Patrick
Ficalora, Peter
Finley, Arlington Levart
Galasso, Francis Salvatore
Gamble, Fred Ridley, Jr
Glaunsinger, William Stanley
Gleim, Paul Stanley
Gluck, Ronald Monroe
Gougoutas, Jack Zanos
Greedan, John Edward
Gruzensky, Paul M
Gutsche, Henry W
Hagemark, Kjell Ingvar
Halaby, Sami Assad
Hanak, Joseph J
Haschke, John Maurice
Hastings, Julius Mitchell
Heinz, David Murray
Herley, Patrick James
Heyding, Robert Donald
Ho, Shih Ming
Hockings, Eric Francis
Hong, Henry Yao-Pen
Hoyen, Harry Alexander, Jr
Irene, Eugene Arthur
Jacobs, Patrick W M
James, William Joseph
Janes, Donald Lucian
Janus, Alan Robert
Johnson, Rowland Edward
Johnson, Vancliff
Keester, Kenneth Lee
Keihn, Frederick George
Kellogg, Lillian Marie
Kern, Werner
Keyzer, Hendrik
Klerer, Julius
Klingsberg, Cyrus
Knop, Osvald
Koffyberg, Francois Pierre
Kokta, Milan Rastislav
Kostiner, Edward S
Kummer, Joseph T
Laudise, Robert Alfred
Lefever, Robert Allen
Leider, Herman R
LeMaster, Edwin William
Leverenz, Humboldt Walter
Libowitz, George Gotthart
Ligenza, Joseph Raymond
MacChesney, John Burnette
Masters, Burton Joseph
Mazelsky, Robert
Mazumder, Bibhuti R
McKinzie, Howard Lee
Milberg, Morton Edwin
Monchamp, Roch Robert
Mroczkowski, Stanley
Mukherjee, Tapan Kumar
Muller, Olaf
Murphy, Preston V
Naae, Douglas Gene
Narasimhan, Kalatur S V L
Nassau, Kurt
Nath, Amar
Nielsen, James Willard
Oesterreicher, Hans
O'Keeffe, Michael
Ouimet, Alfred J, Jr
Pearlman, Donald
Perlman, Morris Leonard
Perlstein, Jerome Howard
Perrino, Charles T
Perry, Paul Eberline
Phipps, Peter Beverley Powell
Pohlmann, Juergen Lothar Wolfgang
Prener, Jerome Sidney
Pringle, John Peter Scott
Rao, Vallabhajosyula U S
Revesz, Akos George
Robbins, Harry
Ropp, Richard C
Roth, Walter Lester
Rubenstein, Martin
Rummery, Terrance Edward
Sankar, Suryanarayan G
Saunders, Vernon Irving
Scheirer, Carl Latimer
Schubert, Clarence
Schwartz, Newton
Shannon, Robert Day
Shoemaker, David Powell
Simpson, William Henry
Sleight, Arthur William
Smyth, Donald Morgan
Snyder, Dexter Dean
Soled, Stuart
Stone, Bobbie Dean
Sturgeon, George Dennis
Suchow, Lawrence
Sullenger, Don Bruce
Taylor, Barry Edward
Trevoy, Donald James
Von Dreele, Robert Bruce
Wachtel, Anselm
Wakefield, Shirley Lorraine
Wallace, William Edward
Whittingham, Michael Stanley

Wilde, Richard Edward, Jr
Williams, Digby Frederick
Wilson, John Anthony
Zwicker, Walter Karl

Spectrochemistry

Clark, Leigh Bruce
Cox, Lawrence Edward
Craver, Clara Diddle (Smith)
Dorman, Douglas Earl
Dunning, Virginia Alexandria
Eikrem, Lynwood Olaf
Ellenburg, Janus Yentsch
Estep-Barnes, Patricia Anne
Finseth, Dennis Henry
Fucaloro, Anthony Frank
Gerry, Michael Charles Lewis
Hall, Ronald Henry
Hamming, Mynard C
Hutchinson, Bennett B
Jones, Richard Norman
Kniseley, Richard Newman
Manning, James Joseph
Matocha, Charles K
McAlduff, Edward J
Mettee, Howard Dawson
Moskovits, Martin
Oertel, Richard Paul
Segal, Gerald A
Shurvell, Herbert Francis
Sloane, Howard J
Spitzer, Kenneth Dale
Steinhaus, David Walter
Turse, Richard S
Wang, Maw Shiu

Steroid Chemistry

Draper, Richard William
D'Silva, Themistocles Damasceno Joaquim
Graham, Robert Earl
Herz, Josef Edward
Mathur, Rajesh Swarup
Parikh, Jekishan R
Rowland, Alex Thomas
Salmond, William Glover
Smith, Elizabeth Melva
Teng, Jon Ie
Williams, Mary Carr

Structural Chemistry

Adams, Wade J
Alexander, Leroy Elbert
Bauer, Simon Harvey
Birnbaum, George I
Boskey, Adele Ludin
Bridge, John Robert
Brock, Carolyn Pratt
Brown, Farrell Blenn
Brown, George Marshall
Brueckner, Hannes Kurt
Burow, Duane Frueh
Bushnell, Gordon William
Clardy, Jon Christel
Clemans, Stephen D
Cohen, Allen Irving
Cohen, Irwin
Coulter, Charles L
Craven, Bryan Maxwell
Dahm, Donald J
D'Antonio, Peter
Davis, Michael I
Egan, Richard Stephen
Eriks, Klaas
Fateley, William Gene
Gilardi, Richard Dean
Goldberg, Stephen Zalmund
Hencher, John Lawrence
Hoard, James Lynn
Hodgson, Derek John
Hodgson, Keith Owen
Ibers, James Arthur
Ivey, Robert Charles
Karliner, Jerrold
Kemp, Thomas Rogers
Kirchner, Richard Martin
La Mar, Gerd Neustadter
La Prade, Marie Douglas
Lawrence, Richard Manley
Lawton, Stephen Latham
Levy, Henri Arthur
Margulis, Thomas N
McCullough, James Douglas
Meyer, Edgar F
Oliver, Joel Day
Pluth, Joseph John
Radonovich, Lewis Joseph
Ross, Frederick Keith
Rudman, Reuben
Shoemaker, Clara Brink
Shoemaker, David Powell
Smith, Harold Warren
Stebbings, William Lee
Thompson, Herbert Bradford
Toomey, Joseph Edward
Traficante, Daniel Dominick
Trefonas, Louis Marco
Tulinsky, Alexander
Wahl, George Henry, Jr

Wampler, Dale Lee
Watkins, Steven F
Whitla, William Alexander
Williams, Graheme John Bramald
Williams, Jack Marvin
Wright, Christine Gerda

Sugar Chemistry
Gaddie, Robert Stanley
Hickson, John LeFever
Lim, Yong Woon
Marov, Gaspar J
McDaniel, Max Paul
Morisugu, Toshio
Pomes, Adrian Francis
Smith, Burns Ashby

Surface Chemistry
Adams, P B
Ambs, William Joseph
Bagchi, Pranab
Baier, Robert Edward
Baker, Bernard Ray
Barr, Tery Lynn
Beebe, Ralph Alonzo
Bendure, Raymond Lee
Benjamin, L
Benson, John Edward
Berkes, Jacob Stephan
Bernholz, William Francis
Bharucha, Nana R
Bhasin, Madan M
Bradley, Arthur
Brinen, Jacob Solomon
Brown, Stanley Monty
Budrys, Rimgaudas S
Cante, Charles John
Cares, William Ronald
Carter, Paul Richard
Cassel, Hans Maurice
Chakravarti, Kalidas
Chiu, Tin-Ho
Clarkson, Alfred Harris
Condon, James Benton
Cratty, Leland Earl, Jr
Csicsery, Sigmund Maria
Cuthrell, Robert Eugene
Czanderna, Alvin Warren
Dahlberg, Susan Clardy
Dasher, George Franklin, Jr
Dalla Betta, Ralph A
Dawson, Peter Thomas
Deckert, Cheryl A
Deitz, Victor Reuel
Dignam, Michael John
Drauglis, Edmund
Eberhart, James G
Eldridge, Jerome Michael
Ellis, Walton P
Ellison, Alfred Harris
Erikson, Jay Arthur
Fackler, Walter Valentine, Jr
Falkehag, S Ingemar
Feinstein, Myron Elliot
Flood, Edward Alison
Fort, Tomlinson, Jr
Frank, Sylvan Gerald
Frojmovic, Maurice Mony
Gilbreath, William Pollock
Gland, John Louis
Glazman, Yuli
Goeke, Rudolph Jon
Good, Robert James
Goodrich, Frank Chauncey
Greenlief, Charles M
Greif, Mortimer
Gruen, Dieter Martin
Haffner, Richard William
Hamilton, Willard Charlton
Harrison, George Conrad, Jr
Hen, John
Herrmann, Kenneth Walter
Herz, Arthur H
Hickson, Donald Andrew
Hillstrom, Warren W
Hindin, Saul Gerald
Hoguet, Robert Gerard
Holmes, James Murray
Houston, Robert John
Hoyen, Harry Alexander, Jr
Hsieh, Paul Yao Tong
Hudson, Alice Peterson
Huntsberger, James Robert
Imperial, George Romero
Isaac, Peter Ashley Hammond
Iyengar, Doreswamy Raghavachar
James, Laylin Knox, Jr
Jarvis, Neldon Lynn
John, Andrew
Jona, Franco Paul
Kalfoglou, George
Katsanis, Eleftherios P
Keirstead, Karl Freeman
Kennedy, George Hunt
Kerr, George Thomson
Kibby, Charles Leonard

Koiaian, Jack H
Kugler, Blanca Louise
Kumins, Charles Arthur
Lacy, Robert M
Lange, Klaus Robert
Lee, Shung-Yan Luke
Lomas, Harold
Lorenzen, Jerry Alan
Lydy, David Lee
Mackay, John Kelvin
Madey, Theodore Eugene
Magenheimer, John Joseph
Manchester, Kenneth Edward
Marra, Dorothea Catherine
Mason, Max Garrett
Matson, Ted P
May, John Walter
Mazunder, Bibhuti R
McElligott, Peter Edward
McKinzie, Howard Lee
Misra, Dwarika Nath
Morrison, Stanley Roy
Murphy, James A
Mysels, Karol Joseph
Nash, Robert Joseph
Nornes, Sherman Berdeen
Parreira, Helio Correa
Perry, Paul Eberline
Petke, Frederick David
Phillips Roger Winston
Polizzotti, Richard Samuel
Pollitzer, Ernest Leo
Prince, Leon Maximilian
Princen, Henricus Mattheus
Purchase, Mary Elizabeth
Rader, Charles Allen
Ramaker, David Ellis
Rivin, Donald
Roberts, Ronald Frederick
Rolles, Rolf
Rosen, Milton Jacques
Rosoff, Morton
Ross, Robert Anderson
Saleeb, Fouad Zaki
Sawyer, Webster Morrill, Jr
Schesser, Robert H
Schonhorn, Harold
Schrader, Malcolm Elliot
Schwartz, Anthony Max
Schwartz, Gary Paul
Schwartz, James Alan
Shuler, Robert Lee
Simmons, Gary Wayne
Singh, Hakam
Sirianni, Aurelio Frederick
Spitzer, Gene Rodell
Sparrow, Harold Edwin
Steffgen, Frederick Williams
Steinberg, Gunther
Stetter, Joseph Robert
Stowe, Robert Allen
Strathdee, Graeme Gilroy
Stryker, Lynden Joel
Tadros, Maher Ebeid
Tewari, Param Hans
Torza, Sergio
Trowbridge, James Rutherford
Tubbs, Robert Kenneth
Van Dell, Robert Duane
Vanselow, Ralf W
Van Valkenburg, Jeptha Wade, Jr
von Dohlen, Werner Claus
Voorhees, John Davidson
Voorhoeve, Rudolf Johannes Herman
Wahlbeck, Phillip Glenn
Wallace, Paul Francis
Weber, Leon
Weiher, James F
Weissman, William
Wendt, Robert Charles
Westerdahl, Carolyn Ann Lovejoy
Whalen, James William
Williams, Josephine Louise
Wilson, Peggy Mayfield Dunlap
Woodward, Fred Erskine
Yates, John Thomas, Jr
Zettlemoyer, Albert Charles
Zografi, George

Textile Chemistry
Arnold, Luther Bishop, Jr
Aspland, John Richard
Aycock, Benjamin Franklin
Baitinger, William F, Jr
Bannerman, Douglas George
Barker, Robert Henry
Bassett, Alton Herman
Bercaw, James Robert
Birkenhauer, Robert Joseph
Bixler, Dean A
Brady, Thomas E
Browne, Colin Lanfear
Calamari, Timothy A, Jr
Caroselli, Remus Francis
Chase, Vernon Lindsay
Compton, Jack

Cooke, Theodore Frederic
Cooper, Margaret Moore
Cramer, John Joseph
Davis, Richard Cecil
Dreschel, Paul David
Drelich, Arthur (Herbert)
Dupre, Edmund J
Ebert, Philip E
Elliott, John
Foght, James Loren
Franklin, William Elwood
Freeman, Richard Carl
Frishman, Daniel
Gaill, Fahmy
Goldstein, Herman Bernard
Goodson, Louie Aubrey, Jr
Greer, James Edward
Griffith, Michael Grey
Guenther, Harry Wilbert
Guion, Thomas Hyman
Hager, Glenn Frederick
Hall, Seymour Gerald
Handy, Carleton Thomas
Hershkowitz, Robert L
Horning, Roderick Henry
Horowitz, Carl
Hughes, William
Hume, Harold Frederick
Joseph, Marjory L
King, James Clarence
Koenig, Harvey Steven
Koenig, Nathan Hart
LaFleur, Kermit Stillman
Louis, Kwok Toy
Lundgren, Harold Palmer
Magee, John Robert
Machlis, Samuel
Mandel, Zoltan
Marcus, Erich
Mirhej, Michael Edward
Mizell, Louis Richard
Morbey, Graham Kenneth
Myers, Clovis D
Neal, Thomas Edward
Nuessle, Albert Christian
Olson, Arthur Russell
Ostmann, Bernard George
Otto, Wolfgang Karl Ferdinand
Panto, Joseph Salvatore
Pavlath, Attila Endre
Pemrick, Raymond Edward
Pfeiffer, Gerald Peter
Pizzarello, Roy Aloysius
Plamondon, Joseph Edward
Read, Robert E
Reeves, Wilson Alvin
Reid, John David
Reimer, Carl Clayton
Saffer, Henry Walker
Saltzman, Max
Sands, Seymour
Sargeant, Peter Barry
Scheve, Bernard Joseph
Scott, Peter John
Sello, Stephen Balthazar
Shoaf, Charles Jefferson
Smith, Betty F
Staples, Mildred Lawson
Stroud, Robert Wayne
Swanson, John Melvin
Takvorian, Kenneth Bedrose
Tucker, Charles R
Urbanik, Arthur Ronald
Vigo, Tyrone Lawrence
Wakelyn, Philip Jeffrey
Walsh, William K
Walters, Philip Marion
Wasley, William Lingel
Watson, Marshall Tredway
Wayland, Rosser Lee, Jr
Webb, Myron Quentin
Weeks, Gregory Paul
Whaley, Wilson Monroe
Wham, George Sims
Whitehouse, Bruce Alan
Williams, Ebenezer David, Jr
Williams, Michael John
Williams, Richard Anderson
Winstrom, Leon Oscar
Wittbecker, Emerson Laverne
Worsham, Walter Castine
Yeh, Kwan-Nan
Zeronian, Sarkis Haig

Theoretical Chemistry
Adams, William Henry
Alexander, Millard Henry
Allnatt, Alan Richard
Alper, Joseph Seth
Altenberger-Siczek, Aldona
Bader, Richard Frederick W
Biolsi, Louis, Jr
Birss, Fraser William
Bishop, David Michael
Bissell, Robert
Blinder, Seymour Michael
Boehnke, David Neal
Bouman, Thomas David

Bowman, Joel Mark
Brown, Nancy J
Brown, Richard Edwin
Bruns, Roy Edward
Burnell, Louis A
Carlson, Charles Merton
Carlson, Terry Scott
Caves, Thomas Courtney
Chen, Joseph Cheng Yih
Choi, Sang-Il
Choi, Yi-Der
Chung-Phillips, Alice
Clarke, Bruce Leslie
Colpa, Johannes Pieter
Conroy, Harold
Crippen, Gordon Marvin
Cukier, Robert Isaac
Curtiss, Charles Francis
Davidson, Robert Bellamy
Delos, John Bernard
Diestler, Dennis Jon
Dingle, Thomas Walter
Dunning, Thomas Harold, Jr
Dyott, Thomas Michael
Ehrenson, Stanton Jay
Epstein, Irving Robert
Ericson, Walter Carl
Ermler, Corey William
Erpenbeck, Jerome John
Evleth, Earl Mansfield
Ewig, Carl Stephen
Fink, William Henry
Fixman, Marshall
Flurry, Robert Luther, Jr
Forst, Wendell
Fraga, Serafin
Freed, Karl F
Frisch, Harry Lloyd
Garrett, Thomas Boyd
George, Thomas Frederick
Getzin, Paula Mayer
Gimarc, Benjamin M
Goddard, William Andrew, III
Golden, Sidney
Hameka, Hendrik Frederik
Harris, Robert A
Harrison, James Francis
Hegstrom, Roger Allen
Hempel, Judith Cato
Henderson, Douglas
Herrick, David Rawls
Hoffmann, Roald
Hsiung, Chi-Hua Wu
Huff, Norman Thomas
Huzinaga, Sigeru
Jaffe, Hans
Janis, F Tim
Johnson, William Randolph, Jr
Jones, Leon Lloyd
Jordan, Peter C H
Kern, Charles William
Keyes, Thomas Francis
Kim, Shoon Kyung
Kleier, Daniel Anthony
Klein, Michael Lawrence
Kolker, Harold Jerrold
Konowalow, Daniel Dimitri
Kouri, Donald Jack
Kozar, Robert Neal
Laidlaw, William George
Ladanyi, Branka Maria
Langhoff, Peter Wolfgang
Levine, Raphael David
Liebman, Joel Fredric
Livingston, Peter Moshchansky
Lohr, Lawrence Luther, Jr
Luken, William Louis
Lyon, William D
MacDonald, Carolyn Trott
Mandel, Frederic
Marron, Michael Thomas
Matcha, Robert Louis
Mazo, Robert Marc
McEwen, Kathleen Lenore
McKelvey, John Murray
McKoy, Basil Vincent
McQuarrie, Donald Allan
Mead, Chester Alden
Mehler, Ernest Louis
Miller, Kenneth John
Mortimer, Robert George
Muckenfuss, Charles
Mueller, Mary Casimira
Mueller, Charles
Nazaroff, George Vasily
Newton, Marshall Dickinson
O'Brien, Thomas Joseph
Olmsted, Richard Dale
Onsager, Lars
O'Shea, Seamus Francis
Ostlund, Neil Sinclair
Paldus, Josef
Pan, Yuh Kang
Parr, Christopher Alan
Parr, Robert Ghormley

Paul, Reginald
Pechukas, Philip
Peek, James Mack
Peterson, Carl
Pitzer, Russell Mosher
Plummer, Patricia Lynne Moore
Poirier, Jacques Charles
Politzer, Peter Andrew
Pople, John Anthony
Porter, Richard Needham
Pritchard, Huw Owen
Radel, Stanley Robert
Ree, Francis H
Rhodes, William Clifford
Ritchie, Adam Burke
Rouse, Robert Arthur
Rubin, Ephraim Leo
Ruedenberg, Klaus
Sabelli, Nora Hojvat
Sanders, William Albert
Sando, Kenneth Martin
Schaefer, Henry Frederick, III
Schmidling, David (Gilbert)
Schneider, Barry I
Schnuelle, Gary Wayne
Schwartz, Maurice Edward
Shavitt, Isaiah
Shillady, Donald Douglas
Shin, Hyung Kyu
Shipman, Lester Lynn
Shuler, Kurt Egon
Siebrand, Willem
Silver, David Martin
Sinanoglu, Oktay
Sinder, Riley Monroe
Snider, Neil Stanley
Snider, Robert Folinsbee
Spotz, Ellen Mae Lackey
Stanton, Richard Edmund
Stephens, Philip J
Stevens, Walter Joseph
Stewart, Charles Winfield
Szabo, Attila
Taylor, William Johnson
Thorson, Walter Rollier
Trindle, Carl Otis
Turner, Almon George, Jr
Tyrrell, James
Wasserman, Edel
Weber, Thomas Andrew
Wheeler, John C
White, Ronald Joseph
Whitehead, Michael Anthony
Whitman, Donald Ray
Whitten, Jerry Lynn
Williams, Denis R
Wilson, Woodrow, Jr
Wolken, George, Jr
Woodson, John Hodges
Wright, James Sherman
Wulfman, Carl E
Wyatt, Robert Eugene
Yates, Albert Carl
Zeroka, Daniel
Zielinski, Theresa Julia

Water Chemistry

Adomaitis, Vytautas Albin
Andelman, Julian Barry
Anderson, Robert Emra
Applegate, Lynn E
Bacon, Hilary Edwin
Bedford, James William
Behrman, Abraham Sidney
Block, Jacob
Bowers, Delores Maureen
Brezonik, Patrick Lee
Churchill, Ralph John
Cowen, William Frank
Daniels, John Maynard
Davies, Robert Milton
Denman, Wayne Leonard
Ditri, Frank M
Dudley, James Robert
Faust, John Philip
Fisher, Sallie Ann
Fishman, Marvin Joseph
Gardner, Wayne Stanley
Gibson, Joseph Woodward
Havlicek, Stephen
Hepler, Loren George
Hickman, Kenneth Claude Devereux
Hygh, Earl Hampton
Keighton, Walter Barker
Keilin, Bertram
Klehr, Edwin Henry
Lee, George Fred
LoSurdo, Antonio
Loy, Wayne Richard
Morgan, James John
Morton, Stephen Dana
Murphy, Thomas Joseph
Nowlin, Duane Dale
Ostroff, Anton G
Parcell, Lloyd Jamison
Petri, Lester Reinhold
Ragone, Stephen Edward
Recht, Howard Leonard

Robertson, Reed S
Salutsky, Murrell Leon
Schlessinger, Gert Gustav
Schultz, Boyd Gilbert
Singley, John Edward
Stevens, Richard Edward
Sussman, Sidney
Wootten, Michael John

Wood Chemistry

Bailey, Alan James
Bailey, Carl Williams, III
Ball, Frank Jervery
Bennett, Clifton Francis
Bolker, Henry Irving
Brink, David Liddell
Burkart, Leonard F
Chang, Hou-Min
Dorland, Rodger Malone
Eachus, Spencer William
Evans, Russell Stuart
Feist, William Charles
Goldschmid, Otto
Goldstein, Irving Solomon
Goring, David Arthur Ingham
Hart, John Henderson
Hosfeld, Ralph (Lowell)
Isenberg, Irving Harry
Keirstead, Karl Freeman
Kitazawa, George
Koch, Christian Burdick
Lai, Yuan-Zong
Landucci, Lawrence L
Larson, Kenneth Curtis
Lo, Cheng Fan
Logan, Charles Donald
Ludwig, Charles Heberle
Manville, John Fieve
McGinnes, Edgar Allan
Millett, Merrill Albert
Pearl, Irwin Albert
Rogers, Sedgwick Cookerly
Rollinson, Samuel Milton
Rowe, John Westel
Sanyer, Necmi
Sarkanen, Kyosti Vilho
Short, Paul Henry
Soltes, Edward John
Stamm, Alfred Joaquin
Stengle, William Bernard
Yao, Joe
Young, Raymond A

CHEMISTRY, ANALYTICAL

Analytical Chemistry

Abrahamson, Earl Arthur
Achey, Frederick Augustus
Ackerman, Donald Godfrey, Jr
Aczel, Thomas
Adams, Donald F
Adams, James Miller
Adams, Mark F
Adams, Martha Lovell
Adams, Ralph Norman
Adcock, Louis Henry
Adkins, John Earl, Jr
Adler, Irving Larry
Adler, Norman
Adler, Seymour Jacob
Afghan, Baderuddin Khan
Afghani, Hisham T
Ahearn, James Joseph, Jr
Alam, Mohammed Ashraful
Alberts, Gene S
Albro, Lewis Pearson
Alexander, John William
Alexander, Thomas Goodwin
Allbright, Charles Simar
Allen, Marvin Carrol
Allen, Willard Finlay
Allenstein, Richard Van
Almond, Harold Russell, Jr
Alsop, John Henry, III
Altenau, Alan Giles
Alvarez, Robert
Ambrose, Robert T
Amundson, Merle E
Anderson, Donald Hervin
Anderson, Larry Bernard
Anderson, Louis Weston
Anderson, Paul Dean
Anderson, Samuel
Andreen, Brian Herbert
Andria, George D
Anfinsen, Jon Robert
Angeloni, Francis M
Annino, Raymond
April, Robert Wayne
Aradine, Paul William
Arcand, George Myron
Archer, Vernon Shelby
Argauer, Robert John
Armentrout, David Noel
Armstrong, Alfred Ringgold
Armstrong, Robert G

Arneson, Dora Williams
Arnott, Robert A
Aronovic, Sanford Maxwell
Ashcraft, Thomas Lee
Ashley, Samuel Edward Qualtrough
Askew, William Crews
Atkins, Jaspard Harvey
Atkinson, George Francis
Attalla, Albert
Aue, Walter Alois
Auerbach, Michael Howard
Austern, Barry M
Ayres, Gilbert Haven
Babcock, Robert Frederick
Baden, David Newton
Bailey, George William
Bailey, Keith
Bair, Edward Jay
Baird, James Haythorn
Baird, Stephen Sydney
Baker, Bertsil Burgess
Baker, Doris
Baker, Harold Weldon
Baker, Harris Mitchell Jr
Baker, June Marshall
Baker, Weldon Nicholas
Balagna, John Paul
Baldwin, Jon Michael
Baldwin, Robert Charles
Baldwin, William John
Bambenek, Mark A
Bandi, William R
Banerjee, Dilip Kumar
Banerjee, Subrata
Banick, William Michael, Jr
Banta, Marion Calvin
Baratta, Edmond John
Barnes, Derek
Barnes, Edwin Ellsworth
Barnes, Lucien, Jr
Barnes, Ramon M
Barney, James Earl, II
Barrall, Edward Martin, II
Barrett, William Jordan
Barth, Howard Gordon
Barton, George Wendell, Jr
Bartschmid, Betty Rains
Bass, Virginia Carvel
Bassett, Lewis Gordon
Bastiaans, Glenn John
Bates, Roger Gordon
Bauer, Henry Hermann
Bauer, William Eugene
Baum, Harry
Baumann, Arthur Nicholas
Baumann, Elizabeth Wilson
Baumann, Frederick
Bayer, Richard Eugene
Bazzelle, William Edward
Beamish, Fred Earl
Beck, Benny Lee
Beck, John Louis
Becker, Harry Carroll
Beeman, Curt Pletcher
Behm, Roy
Beilby, Alvin Lester
Belknap, Herbert John
Bell, Rosemond Kay
Belsky, Theodore
Benson, Richard Edward
Benson, Royal H
Bentley, Kenton Earl
Bentley, Walter Roderick
Berg, Eugene Walter
Berger, Douglas G
Berger, Selman A
Bergmann, Eric Arnold
Bernard, Joseph Lionel
Bernetti, Raffaele
Berni, Ralph John
Beroza, Morton
Berry, John William
Berry, Keith O
Bertin, Eugene P
Betso, Stephen Richard
Beyerlein, Floyd Hilbert
Bhatia, Kishan
Bhatnagar, Dinech C
Bhattacharyya, Pranab K
Bickerdike, Ernest Lawrence
Bidleman, Terry Frank
Billingmeyer, Brian Arthur
Billingham, Edward J, Jr
Billo, Edward Joseph
Bingenheimer, Levi Edwin, Jr
Bingham, Carleton Dille
Binz, Carl Michael
Birch, Homer James
Birdsall, Clair Mallery
Birke, Ronald Lewis
Bishara, Rafik Hanna
Bixler, John William
Black, Arthur Herman
Black, Billy C, II
Blackburn, Thomas Roy
Blaedel, Walter John
Blanck, Harvey F, Jr

Blessel, Kenneth Wayne
Blickensderfer, Peter W
Blouin, Leonard Thomas
Bly, Donald David
Board, Robert Dennis
Boaz, Patricia Anne
Boczkowski, Ronald James
Bodre, Robert Joseph
Boggs, William Emerson
Bolleter, William Theodore
Boltz, David Ferdinand
Bond, Elizabeth Dux
Bondurant, Charles W, Jr
Booman, Glenn Lawrence
Boos, Richard Newton
Bornstein, Michael
Bosen, Sidney Frederick
Botdorf, Ruth G
Bottei, Rudolph Santo
Boulton, Mary
Bovee, Harley Howard
Bowe, Arthur Frederick
Bowers, Delores Maureen
Bowers, Raymond Harold
Bowerson, David F
Bowman, Leo Henry
Boyd, John Robert
Boyle, Walter Gordon, Jr
Bracco, Donato John
Bradley, Martin Patrick Timothy
Bradstreet, Raymond Bradford
Brady, Stephen W
Braman, Robert Steven
Brame, Edward Grant, Jr
Brandt, Manuel
Braun, Donald E
Braun, Robert Denton
Brauner, Kenneth Martin
Brauner, Phyllis Ambler
Bredeweg, Robert Allen
Bremner, Raymond Wilson
Breslin, Maureen Elizabeth
Brewer, Douglas G
Brewer, John Gilbert
Brobst, Kenneth Martin
Brofazi, Fred Robert
Bromund, Richard Hayden
Bromund, Werner Hermann
Brookman, David Joseph
Brown, Eric Richard
Brown, Ralph Andres
Brown, Theodore Llewellyn
Brownfield, Robert Bruce
Bruckenstein, Stanley
Brundin, Robert H
Brunzie, Gerald Franklin
Bryson, Theodore Cornelius
Buchanan, Edward Bracy, Jr
Buchta, Raymond Charles
Buck, Richard Pierson
Buckwold, Sidney Joshua
Budde, William L
Budke, Clifford Charles
Buell, Glen R
Burdett, Lorenzo Worth
Burg, William Robert
Burkard, Perle Nelius
Burke, Michael Francis
Burmblay, Ray Ulysses
Burnett, Bruce Burton
Burnett, John Nicholas
Burnett, Robert Walter
Burniston, George Kissam
Burroughs, James Edward
Bursey, Maurice Moyer
Burtis, Carl A, Jr
Burtner, Dale Charles
Busch, Kenneth Walter
Bush, David Graves
Bush, Karen Jean
Bush, Martin Bruce
Bushey, Albert Henry
Buswell, Robert James
Butkus, Antanas
Butler, Eliot Andrew
Butler, Lillian Ida
Buttrill, Sidney Eugene, Jr
Bydalek, Thomas Joseph
Byrne, Francis Patrick
Bystroff, Roman Ivan
Cabbiness, Dale Keith
Cadle, Stephen Howard
Cain, Carl, Jr
Caley, Earle Radcliffe
Calhoon, Stephen Wallace, Jr
Calkins, Myron Eugene
Calkins, Russel Crosby
Campbell, Bruce Henry
Campbell, Dan Norvell
Campbell, Donald Edward
Campbell, Donald R
Campbell, George Melvin
Campbell, Milton Hugh
Campbell, Robert Henry
Campion, James J
Capotosto, Augustine, Jr
Capretta, Umberto

Carley, David Wilcox
Carpenter, Robert Dean
Carpenter, T J
Carr, Edward Mark
Carr, James David
Carson, Chester Carrol
Carter, Fairie Lyn
Carter, Gerald Bate
Caskey, Albert Leroy
Caspers, Horst J
Cassidy, James Edward
Casto, Clyde Christy
Cates, Vernon E
Chadde, Frank Ernest
Chait, Edward Martin
Chakrabarti, Chuni Lal
Chambers, James Q
Chambers, Lee Mason
Chambers, William Edward
Chance, Robert L
Chandler, Carl Davis, Jr
Chandler, Dean Wesley
Chaney, Charles Lester
Chaney, Jack Che-Man
Chang, Ted Teh-Liang
Chase, Dan L
Chattoraj, Shib Charan
Chau, Alfred Shun-Yuen
Cheng, Fred Fa Wu
Cheng, Kuang Lu
Chiba, Mikio
Childers, Ray Fleetwood
Chiu, Jen
Chodos, Arthur A
Chonski, John Leo
Cincotta, Joseph John
Clarkson, Jack E
Clauss, James K
Clavan, Walter
Claiborne, Imogene B
Clapp, William Lee
Chow, Tsaihwa James
Chrepta, Stephen John
Christian, Gary Dale
Christie, Joseph Herman
Christoffersen, Donald John
Chulski, Thomas
Chong, Clyde Hok Heen
Chow, Arthur
Cholak, Jacob
Clark, Harlan Eugene
Clark, Howell R
Clark, John Francis Bullock
Clark, Patrick Joseph
Clark, Robert Paul
Clarke, George
Clarke, John Frederick Gates, Jr
Cole, Jerry Joe
Cohen, Jack
Cohen, Edward Morton
Cohen, Allen Irving
Cogswell, Howard Winwood
Coetzee, Johannes Francois
Codding, Edward George
Cockerell, Leone (Doris)
Cochran, George Thomas
Coner, Jack Payne
Conley, Jack Michael
Conkin, Robert A
Conner, Albert Z
Connors, Kenneth A
Constant, Marc Duncan
Cooke, Ronald Frank
Cooke, Samuel Leonard, Jr
Cooke, William Donald
Coolidge, Edwin Channing
Cooper, Aaron David
Cooper, Maurice D
Coppola, Elia Domenico
Cordell, Richard William
Corsini, A
Corth, Richard
Coulter, Paul (David Todd)
Cover, Richard Edward
Cowley, Thomas Gladman
Cox, James Allan
Crabtree, Eleanor Voorhees
Craig, Kenneth Alexander
Crain, Alfred V R
Cram, Stuart Proud
Creamer, Robert M
Crecely, Roger William
Crichton, David
Crocker, Jain Hay
Crosen, Robert Glenn
Cross, Charles Kenneth
Crouch, Stanley Ross
Cruickshank, Alexander Middleton
Crummett, Warren B
Crumpler, Thomas Bigelow
Cruser, Stephen Alan
Crutchfield, Charlie

Csejka, David Andrew
Cukor, Peter
Cundiff, Robert Hall
Curley, James Edward
Currah, Jack Ellwood
Curran, David James
Curran, Jerry H
Daly, Robert E
Danchik, Richard S
Daniel, Robert Eugene
Daugherty, Kenneth E
Davidson, John Edwin
Davies, Geoffrey
Davis, Abram
Davis, Donald G
Davis, Henry McRay
Davis, Joe Bill
Day, Edgar William, Jr
Day, Reuben Alexander, Jr
Day, Robert James
Dean, John Aurie
Deanhardt, Marshall Lynn
Debnath, Sadhana
DeCastro, Arthur
DeCarlo, Kenneth Harold
DeFord, Donald Dale
DeGeiso, Richard Charles
Dehne, George Clark
De Jong, Gary Joel
De Jong, Arthur F
De Schmertzing, Hannibal
De Silva, John Arthur F
Delmastro, Joseph Raymond
Delmastro, Ann Mary
Derby, James Victor
Devonshire, Leonard Norton
Dick, James Gardiner
Diebel, Robert Norman
Diehl, Harvey Clarence
Dietz, Edward Albert, Jr
Dietzman, Burton D
Dillon, Henry Kenneth
Dilts, Robert Voorhees
Dinan, Frank J
Dinen, Eugene Joseph
Dinsmore, Howard Livingstone
Dippel, William Alan
D'Itri, Frank M
Dobbelstein, Thomas Norman
Dobbins, James Talmage, Jr
Doeden, Gerald Ennen
Doerr, Robert George
Domsky, Irving Isaac
Donahoe, Pat
Douglass, Pritchard Calkins
Draga, Gerrit
Drake, Edgar Nathaniel, II
Drexler, Edward James
Drushel, Harry (Vernon)
Dubravcic, Milan Frane
Duerst, Richard William
Dullaghan, Matthew Edward
Duncan, Archibald
Dunford, Raymond A
Dunham, John Malcolm
Dunn, William Lewis
Durbin, Ronald Priestley
Durnick, Thomas Jackson
Duswalt, Allen Ainsworth, Jr
Dwyer, Robert Francis
Dyer, Frank Falkoner
Dyer, Rolla McIntyre, Jr
Easty, Dwight Buchanan
Eckert, Alfred Carl, Jr
Eckert, Joseph Nicolaus
Edsberg, Robert Leslie
Edelsen, Paul
Edwards, John C
Edwards, Kenneth Ward
Eggert, Arthur Arnold
Elder, Norman George
Eisentraut, Kent James
El Ghamry, Mohamed Tawfik
Elefsen, Bobby J
Eilenbogen, William Cromwell
Ellis, David Wertz
Eisermann, Edi
Ewing, Philip Juliber
Emerich, Donald Warren
Emerson, David Edwin
Emerson, Merle T
Emrick, Edwin Roy
England, Richard Jay
Enke, Christie George
Epp, John George
Epstein, Barry D
Erdmann, David E
Erlich, Ronald Harvey
Ernst, Robert R
Ernst, Arthur F
Espoy, Henry Marti
Estill, Wesley Boyd
Etre, Leslie Stephen
Evans, Dennis Hyde
Evans, Floyd Monte
Evenson, Merle Armin
Everett, Grover Woodrow
Evila, Ronald Frank

Evins, Charles Victor
Ewald, Fred Peterson, Jr
Ewen, Edward Francis
Ewing, Galen Wood
Eyler, Robert Wilson
Ezell, James Ben, Jr
Ezrin, Myer
Fahsel, Michael John
Fair, Frank Vernon
Fairless, Charles Michael Haskel
Fairman, William Duane
Faris, Sam Russell
Farrington, Paul Stephen
Farquhar, Mary Janet
Fassel, Velmer Arthur
Fauth, Mae Irene
Fenberg, Benjamin Allen
Feinstein, Hyman Israel
Feldman, Fredric J
Feldman, Nicholas
Fenimore, David Clarke
Ferguson, William Sidney
Fernando, Quintus
Ferrara, Louis W
Ferraro, Charles Frank
Fett, E Reinold
Fetter, Neil Ross
Fiddler, Walter
Fiehler, Harlan Edward
Fike, Winston
Fink, David Warren
Finlayson, James Bruce
Firsching, Ferdinand Henry
Fischer, Robert Blanchard
Fisher, Dale John
Fisher, John F
Fitzgerald, Jerry Mack
Fitzgerald, Maurice S
Fitzpatrick, Jimmie Doile
Flagg, John Ferard
Flanders, Clifford Auten
Flaschka, Hermenegild Arved
Fletcher, Aaron Nathaniel
Fletcher, Kenneth Steele, III
Flentge, Douglas MacArthur
Flanagan, William Vaughn
Folk, Theodore Lamson
Follweiler, Douglas MacArthur
Forcier, George Arthur
Fordyce, David Buchanan
Forman, Donald T
Forman, Earl Julian
Forrester, Frank Robert
Forsette, John Elmer
Foster, William Owen
Foster, Walter H, Jr
Fowler, Lewis
Fowler, Cecil Vernon
Francis, Richard Ernst
Frankel, Lawrence (Stephen)
Franklin, James Curry
Fram, Martin S
Franz, David Alan
Fraser, James Mattison
Freeberg, Fred E
Freed, Donald Joseph
Freeland, Max
Forrester, Frank Robert
Freeman, James Patrick
Freeman, David Haines
Fredline, Charles Eugene
Freiser, Herbert Edward
Freiser, Henry
Fresco, James Martin
Freund, Harry
Fricke, Gordon Hugh
Frierson, William Joe
Frisque, Alvin Joseph
Fritz, James Sherwood
Fritz, Klaus
Frodyma, Michael Mitchell
Frohliger, John Owen
Frye, Herschel Gordon
Fuchs, Jacob
Fujikawa, Norma Sutton
Fulda, Myron Oscar
Funkhouser, John Tower
Furse, Clare Taylor
Gagnon, John Gregory
Gahler, Arnold Robert
Gailey, Joseph A
Gainer, Frank Robert
Gajan, Raymond Joseph
Ganchoff, John Christopher
Garcia-Morin, Manuel
Gard, Leavitt Nelson
Gardner, John Alden
Gardner, Kenneth William
Garmon, Ronald Gene
Garn, Paul Donald
Garrard, Verl Grady
Gary, Julia Thomas
Geiger, William Ebling, Jr
Geisman, Raymond August, Sr
Geller, Milton
Gerard, Jesse Thomas
Gerber, James Norman
Gerhardt, Klaus Otto
Gershman, Louis Leo

Ghosh, Saroj Bandhu
Giannovario, Joseph Anthony
Gibb, Thomas Robinson Pirie, Jr
Gibbons, James Joseph
Gibson, David Michael
Gibson, Robert Harry
Gilbert, Don Dale
Gilbert, Jack Pittard
Gilbert, Theodore William, Jr
Gilbert, Thomas Rexford
Gilfrich, John Valentine
Gillespie, John Paul
Ginn, Roger Keith
Girolami, Martin E
Givand, Samuel Harold
Glanville, James Oliver
Glass, Douglas Gordon
Gleit, Chester Eugene
Glick, Charles Frey
Glickstein, Joseph
Glover, Clyde Albert
Godar, Edith Marie
Goeden, Michel G
Goetz, Charles Albert
Goetz, Rudolph W
Goode, Gerald Seymour
Goldstein, Gerald
Golton, William Charles
Gonter, Clara Ellen
Goode, Julia Pratt
Goode, Scott Roy
Goode, Benjamin Edward
Gordon, Charles Francis
Gordon, Saul
Gordus, Adon Alden
Gordon, Robert Earl
Graham, Ronald Powell
Grant, Clarence Lewis
Gray, Linsley Shepard, Jr
Green, Charles Darwin
Green, Floyd J
Greenberg, Mark Shiel
Greene, Arthur Frederick, Jr
Greenough, Ralph Clive
Greinke, Ronald Alfred
Grennan, Laurie M
Grey, Peter
Grieshammer, Lawrence Louis
Grimaldi, Frank Saverio
Griswold, Robert Edward
Grob, Robert Lee
Grogan, Michael John
Gropp, Armin Henry
Gross, Michael Lawrence
Grossman, William Elderkin Leffingwell
Groten, Barney
Groth, Joyce Lorraine
Groth, Alfred Lester
Grushka, Eli
Guerin, Michael Richard
Guerrant, Gordon Owen
Guerrero, Ariel Heriberto
Guerry, Davenport, Jr
Guertin, Donald Lucius
Guffy, Joseph Claude
Gulick, Wilson M, Jr
Gump, Barry Hemphill
Gunap, J R
Gunsberg, Ephraim
Gunter, Bobby J
Gunther, Donald Albert
Gupta, Vishnu Das
Gustin, Vaughn Kenneth
Guthrie, Frank Albert
Guthrie, Joseph D
Guymon, Ervin Park
Guyon, John Carl
Gyberg, Arlin Enoch
Haab, Walter
Haberman, Jon Phillip
Hackett, Raymond L
Haddon, William F, (Jr)
Haemi, Edward Otto
Hafford, Bradford C
Hagel, Robert B
Haggerty, William Joseph, Jr
Hakkila, Eero Arnold
Hall, Larry Cully
Hall, Randall Clark
Hall, Robert Turner
Hallett, Lawrence Trenery
Halva, Carroll J
Hammerstrom, Harold Elmore
Hanck, Kenneth William
Hankins, Bobby Eugene
Hanley, Arnold V
Hannah, Roy Barton, Jr
Hannah, Joseph Gordon
Hanneman, Walter M
Hanselman, Raymond Bush
Hansen, Robert Conrad
Haq, Mohammad Zamir-ul

Hardy, Edward Peirson, Jr
Hargis, Larry G
Harley, John Henry
Harrar, JacksonElwood
Harrell, T Gibson
Harrington, George William
Harris, Albert Zeke
Harris, Edward Lyndol
Harris, James A
Harris, Michael Joseph
Harris, Ray Edgar
Harris, Walter Edgar
Harrison, Stanley L
Harrison, Willard Wayne
Harrigan, Martin Joseph
Hartkopf, Arleigh Van
Hartley, Arnold Manchester
Hartman, Harold Beers
Hartstein, Arthur M
Harvey, Mack Creede
Hass, James Ronald
Hassler, William Woods
Hawkes, Stephen J
Hawley, Merle Dale
Haynes, William Miller
Hazdra, James Joseph
Head, William Francis, Jr
Healy, Richard Clawson
Heaton, Richard Clawson
Hedrick, Charles Edward, Jr
Hefley, Alta Jean
Heidner, Robert Hubbard
Heinrich, Kurt Francis Joseph
Heintz, Edward Allein
Helms, Boyce Dewayne
Hendricks, Russel Hyer
Henney, Robert Charles
Hercules, David Michael
Herman, Harvey Bruce
Hess, George G
Hessley, Rita K
Heveran, John Edward
Heydegger, Helmut Roland
Heyn, Arno Harry Albert
Hibbits, James Oliver
Hicks, Donald Gail
Hicks, Jackson Earl
Hieftje, Gary Martin
Higson, Harold George
Hileman, Orville Edwin, Jr
Hill, Herbert Henderson, Jr
Hill, Martha Adele
Hillis, Mary Olive
Hillsman, Overton Lindner
Hilton, Ashley Stewart
Hime, William Gene
Hines, Wallis Gartside
Hintze, Ray Everald
Hintze, Willie Lee
Hirsch, Roland Felix
Hirschfeld, Tomas Beno
Hiskey, Clarence Francis
Hitchcock, Eldon Titus
Hively, Robert Arland
Ho, Floyd Fong-Lok
Hobart, Everett W
Hobbs, John Robert
Hodgson, William Gordon
Hoff, Johan Eduard
Hoffman, Clark Samuel, Jr
Hoffman, Henry Tice, Jr
Hoffman, Israel
Hoffman, William Andrew, Jr
Hogan, David James
Holcomb, Herman Perry
Holcombe, James Andrew
Holden, John B, Jr
Holder, Paul Edward
Holland, William William
Holmes, Ivan Gregory
Holt, Ben Dance
Holtslander, William John
Homan, Joseph M
Hope, Henry Bell
Hopper, Michael James
Horan, Harold Arthur
Horner, Henry John
Horning, Evan Charles
Horton, Charles Abell
Howard, John William
Howard, Donald Eugene
Howell, James Arnold
Howick, Lester Carl
Hubbard, Arthur T
Hubbard, Willard Dwight
Huebert, Barry Joe
Huffman, Edward William Dickson, Jr
Hughes, Benjamin G
Hughes, Michael Charlea
Hughes, Raymond Hadley
Hutink, Geraldine M
Hulbert, Matthew N
Humphrey, Ray Eicken
Hunt, Albert Melvin
Hunter, George L K
Hurtt, Oscar Lee, Jr

Hurtubise, Robert John
Hurwitz, Jan Krosst
Husa, William John, Jr
Hussey, Charles Logan
Huston, Charles K
Hutchinson, Kenneth A
Hutchinson, William Marwick
Hynek, Robert James
Iddings, Frank Allen
Ihnat, Milan
Ikenberry, Luther Curtis
Illian, Carl Richard
Illmet, Ivor
Iorns, Terry Vern
Irgolic, Kurt Johann
Irving, Philip
Isaac, Robert A
Isenhour, Thomas Lee
Issaq, Haleem Jeries
Ito, Jun
Iwamoto, Reynold Toshiaki
Ja, William Yin
Jack, John James
Jackman, Donald Coe
Jackson, George Frederick, III
Jackson, Harold Woodworth
Jackson, William Andrew
Jacob, Fielden Emmitt
Jacobs, Morton Howard
Jacobs, William Donald
Jacobson, Jay Stanley
Jaffe, Marvin Richard
Jain, Naresh C
Jakob, Fredi
Jakubiec, Robert Joseph
James, Ernest P
James, Franklin Ward
James, Helen Jane
Janauer, Gilbert E
Janicki, Casimir A
Janini, George Musa
Jankovics, Lawrence Robert
Jankowski, Conrad M
Jankowski, Stanley John
Janota, Harvey Franklin
Jansing, Jo Ann
Jardine, John McNair
Jaselskis, Bruno
Javick, Richard Anthony
Jayne, Jerrold Clarence
Jefferson, Jack Howard
Jensen, Adolph Robert
Jensen, Philip Wright
Jensen, Richard Erling
Jentoft, Ralph Eugene, Jr
Jespersen, Neil David
Jezorek, John Robert
Johanson, Robert Gail
Johns, Don Herbert
Johnson, Arnold Richard, Jr
Johnson, Bruce McDougall
Johnson, Delwin Phelps
Johnson, Donald Richard
Johnson, Edward Lee
Johnson, Eric Robert
Johnson, Frank Junior
Johnson, Hilding Reynold
Johnson, Jack (Lamar)
Johnson, Ralph Alton
Johnson, Raymond Nils
Johnson, Tegner Albin
Jones, Berwyn E
Jones, Donald Eugene
Jones, Haskell Lee
Jones, Jerry Lynn
Jones, Llewellyn Claiborne, Jr
Jones, Stanley Leslie
Jones, Thomas Evan
Jordan, Joseph
Jost, Paul Douglas
Judd, Jane Harter
Juhasz, Roderick
Julin, Bruce Gustav
Jurs, Peter Christian
Juvet, Richard Spaulding, Jr
Kadin, Harold
Kadish, Karl Mitchell
Kaelble, Emmett Frank
Kagel, Ronald Oliver
Kallmann, Silve
Kane, Philip Francis
Kanzelmeyer, James Herbert
Kaplan, Ephraim Henry
Kapp, Mary Eugenia
Karasek, Francis Warren
Karchner, Jean Herschel
Karger, Barry Lloyd
Karp, Stewart
Kasler, Franz Johann
Katakkar, Garabet Haroutian
Katekaru, James
Katz, Sidney
Katz, Sidney A
Kehoe, Thomas J
Keilholtz, Gerald Watson
Keily, Hubert Joseph
Keirs, Russell John
Keller, Robert Ellis

Keller, Roy Alan
Kelley, Myron Truman
Kelly, Clark Andrew
Kelsey, David
Kender, Donald Nicholas
Kennedy, Edward Earl
Kennedy, John Harvey
Kenner, Charles Thomas
Kersey, Robert Lee, Jr
Kesner, Leo
Ketchum, Donald Frederick
Ketterer, Paul Anthony
Keyworth, Donald Arthur
Kho, Boen Tong
Kieft, Lester
Kieft, Richard Leonard
Kiehl, Samuel Jacob, Jr
Kim, Ki Hong
Kimlin, Mary Jayne
King, Donald M
King, Jimmie Ray
King, Richard Warren
King, Stanley Shih-Tung
Kipnes, Sol M
Kirby, James Ray
Kirby, Robert Emmet
Kirchmeier, Robert Lynn
Kirkland, Joseph Jack
Kirkpatrick, James W
Kirschbaum, Joel Jerome
Kiser, Donald Lee
Kissel, Thomas Robert
Klein, Hannah Ray
Kleineberg, Gerd A
Kliem, Peter O
Kline, Jerry Robert
Klingenberg, Joseph John
Kluchesky, Elmer Francis
Knecht, Laurance A
Kneip, Theodore Joseph
Kniebes, Duane Van
Knight, Homer Talcott
Koblin, Abraham
Koelsche, Charles L
Koirtyohann, Samuel Roy
Kolat, Robert S
Kolthoff, Izaak Maurits
Komyathy, Joseph Charles
Kopp, John F
Kozak, Gary S
Kramer, Raymond Arthur
Kramer, William J
Kratochvil, Byron
Krawetz, Arthur Altshuler
Krikorian, Samuel Edward, Jr
Krishen, Anoop
Krivis, Alan Frederick
Krochta, William G
Krol, George J
Kross, Robert David
Krzeminski, Stephen F
Kuempel, Donald Francis
Kuempel, John Rickey
Kugler, George Charles
Kuhn, William Frederick
Kuljian, Ernest Sam
Kusher, George Samuel
Kuwana, Theodore
Kuzel, Norbert R
Lacoste, Rene John
Lafferty, Robert Hervey, Jr
Laitinen, Herbert August
Lalancette, Roger A
Lamb, Robert Edward
Lambert, Jack Leeper
Lane, Keith Aldrich
Lane, William James
Laning, Stephen Henry
Lanterman, Elma
Lard, Edwin Webster
Large, Robert F
Larkin, Robert Hayden
Larson, Wilbur John
Larson, Wilbur S
Larson, William Day
Latshaw, David Rodney
Latterell, Joseph J
Latz, Howard W
Laughlin, Alice
Launer, Philip Jules
Law, Ronald Dee
Lawless, James George
Layloff, Thomas
Layman, Wilbur A
Leaton, John Roger
LeBlanc, Norman Francis
Lee, Alfred Tze-Hau
Lee, George H, II
Lee, James K
Lee, William Richard
Leftault, Charles Joseph, Jr
Legg, Kenneth Deardorff
Leipziger, Fredric Douglas
Leitch, Robert Edgar, Jr
Lembo, Nicholas J
Leonard, Edward H
Leonard, Guy William
Lerner, Melvin

Leslie, Wallace Dean
Lessor, Arthur Eugene, Jr
Lessor, Edith Dora
Letterman, Herbert
Leussing, Daniel, Jr
Levine, Solomon Leon
Levinson, Steven R
Levy, Arthur Louis
Levy, Paul F
Lewis, Arnold D
Lewis, Claude Irenius
Lewis, Lynn Loraine
Lewis, Phillip Albert
Leyden, Donald E
Leyon, Robert Edward
Li, Kuang-Pang
Liang, Charles Chi
Liao, Hsueh-Liang
Liao, Shu-Chung
Lichtenstein, Ivan Edgar
Liggett, Lawrence Melvin
Light, Truman S
Lin, Denis Chung Kam
Lindauer, Maurice William
Linder, Donald Ernst
Linder, Louis Jacob
Lindstrom, Frederick John
Lingane, James Joseph
Link, William B
Link, William Edward
Linn, Thomas Arthur, Jr
Little, James Noel
Liu, Chui Hsun
Ljung, Harvey Albert
Lloyd, Nelson Albert
Loach, Kenneth William
Lochmuller, Charles Howard
Locke, Harold Ogden
Loder, Edwin Robert
Lofstrom, John Gustave
Lohr, Lester Jay
Lombardo, Pasquale
Long, Calvin H
Loomis, Thomas Clement
Looyenga, Robert William
Lord, Samuel Smith, Jr
Loscalzo, Anne Grace
Lott, John Alfred
Lott, Peter F
Lotz, John Robert
Loughran, Edward Dan
Lovechio, Linda J Cline
Lovecchio, Frank Vito
Lowen, Jack
Lowen, Warren Kealoha
Lubeck, Axel John
Lubitz, Betty Baum
Lucas, Kenneth Ross
Lucchesi, Claude A
Lucke, William E
Luckmann, Frederick H
Luders, Richard Christian
Ludwig, Frederick John, Sr
Lueck, Charles Henry
Luker, William Dean
Lumpkin, Henry Earl
Luttrell, George Howard
Lyerly, Larry Alexander
Lyles, George Robert
Lysyj, Ihor
MacDonald, Alexander, Jr
MacDonald, Hubert C, Jr
MacDonald, John Chisholm
Macero, Daniel Joseph
MacFadden, Kenneth Orville
MacGee, Joseph
Machel, Albert R
Machemer, Paul Ewers
MacKellar, William John
Maddin, Charles Milford
Mahle, Nels H
Mahlman, Harvey Arthur
Mahoney, Bernard Launcelot, Jr
Mainier, Robert
Majors, Ronald E
Malaiyandi, Murugan
Maldacker, Thomas Anton
Malenfant, Arthur Lewis
Mallett, William Robert
Maloy, Joseph T
Mamantov, Gleb
Mann, Charles Kenneth
Manning, Monis Joseph
Mapes, William Henry
Margerum, Dale William
Margoshes, Marvin
Mark, Harry Berst, Jr
Markham, James J
Markunas, Peter Charles
Marple, Leland Warren
Marquardt, Roland Paul
Marsh, Frederick Leon
Marsh, Max Martin
Marsh, Richard Hayward
Marshall, Delbert Allan
Marshall, Donald D
Marshall, John Clifford

Martin, Aaron Jay
Martin, Albert Edwin
Martin, Horace F
Martin, John F
Martin, John Perry, Jr
Martin, John Robert
Martin, Kenneth John
Martin, Ronald LeKoy
Mason, John Grove
Mathisen, Maurice Earl
Mathre, Owen Bertwell
Mattes, Paul Joseph
Mattes, Frederick Henry
Mattina, Charles Frederick, Jr
Matulis, Raymond M
Maute, Robert Lewis
May, Irving
Maynard, William Rose, Jr
Maynes, Albion Donald
Mays, Robert Lewis
Mays, Rolland Lee
Mays, David Lee
McAdoo, David John
McCallum, Keith Stuart
McCarty, Billy Dean
McCauley, Gerald Brady
McClellan, Bobby Ewing
McClure, John Hibbert
McComas, Wilbur Harrison, Jr
McCormick, Patrick Gary
McCown, John Joseph
McCrone, Walter C
McCurdy, Wallace Hutchinson, Jr
McElroy, Mary Kieran
McEwen, David John
McFarren, Earl Francis
McGarry, Margaret
McGee, Charles E
McGee, William Walter
McGehee, Charles Leroy
McGill, Julian Edward
McGinness, James Donald
McGonigle, Eugene Joseph
McGuire, Francis Joseph
McGuire, John Murray
McKaveney, James P
McKay, James Brian
McKay, William Neil
McKinney, Robert Wesley
McKinney, Ted Meredith
McKinney, William Jan
McLafferty, Fred Warren
McLafferty, John J
McLean, James Dennis
McMahon, David Harold
McMasters, Donald L
McNair, Harold Monroe
McNeely, Robert Lewis
McNeil, Arthur Louis
McNutt, Ronald Clay
McSharry, William Owen
Mead, Marshall Walter
Meal, Larie L
Mechlinski, Witold
Medwick, Thomas
Mefford, David Allen
Megargle, Robert G
Megregian, Stephen
Mehlig, Joseph Parke
Mehlhaff, Leon Curtis
Meinke, William Wayne
Meisels, Gerhard George
Meites, Louis
Melgaard, Kennett Gilbert
Mell, Leroy Dayton, Jr
Melnick, Laben Morton
Meloan, Clifton E
Meloon, Daniel Thomas, Jr
Melton, James Ray
Memeger, Wesley, Jr
Mergens, William Joseph
Merritt, Charles, Jr
Merritt, Lynne Lionel, Jr
Merritt, Margaret Virginia
Merritt, Paul Eugene
Metcalfe, Lincoln Douglas
Meyer, Herbert A
Meyer, John Austin
Milano, Michael John
Milczarek, Chester J
Miles, Maurice Jarvis
Millar, John David
Miller, Charles Everett, Jr
Miller, C David
Miller, David Jacob
Miller, Francis Joseph
Miller, Herbert Crawford
Miller, James Franklin
Miller, James Monroe
Miller, Jerry K
Miller, John Walcott
Miller, Raymond Sumner
Miller, William Knight
Milliman, George Elmer
Miner, David
Mishmash, Harold Edward
Miskus, Raymond P
Mitchell, James Winfield

Mitchell, John, Jr
Mitchell, Richard Sibley
Modderman, John Philip
Mohlner, David Morris
Mollica, Joseph Anthony
Monn, Donald Edgar
Montalvo, Joseph G
Moody, John Robert
Mooney, John Bernard
Mooney, Thomas Faulkner, Jr
Moore, Carl Edward
Moore, Ralph Bishop
Morales, Raul
Morgan, Evan
Morgan, Gerald Prescott
Morris, George V
Morris, Joseph Burton
Morris, Michael D
Morris, Stephen Andrew
Morrison, Charles Freeman, Jr
Morrison, George Harold
Morrow, Norman Louis
Morrow, Roy Wayne
Mortimore, Donald Morton
Moseley, Patterson B
Mosen, Arthur Walter
Mosher, Robert Eugene
Moss, Rodney Dale
Mossotti, Victor Giovoni
Mottola, Horacio Antonio
Mountcastle, William R, Jr
Mowery, Richard Allen, Jr
Mowitz, Arnold Martin
Moyer, Hugh Anson
Moyerman, Robert Max
Mueller, Theodore Rolf
Muhs, Merrill Arthur
Muller, Donald Edward
Munger, Robert Shoop
Munson, Burnaby
Munson, James William
Muraca, Raffaele Francesco
Murie, Richard A
Murphy, Robert T
Murray, Royce Wilton
Murray, William Mozley, Jr
Musker, Warren Kenneth
Musselman, Nelson Page
Mussinan, Cynthia June
Mutha, Shantilal Chhotmal
Myers, Marcus Norville
Myers, Richard Lee
Myerson, Albert Leon
Nace, Donald Miller
Nadeau, Herbert Gerard
Nagel, Edgar Herbert
Naiman, Barnet
Neal, Thomas Edward
Nealy, Carson Louis
Neas, Robert Edwin
Nedermeyer, Peter Arthur
Neil, Thomas C
Nelson, Ivory Vance
Nesheim, Stanley
Ness, Arthur Thomas
Netterville, John T
Neumann, Fred William
Nevu, Darwin D
Newcome, Marshall Millar
Newkome, Richard Alan
Newsom, Will Roy
Newton, John Chester
Nicholson, D Allan
Nicolson, Paul Clement
Niedermayer, Alfred O
Nieman, Timothy Alan
Nikelly, John G'
Nippoldt, Bertwin W
Nix, Joe Franklin
Noel, Dale Leon
Nordquist, Paul Edgard Rudolph, Jr
Norman, Richard D
Norris, Max Valentine
Norton, Daniel Remsen
Norton, Elinor Frances
Nowak, Anthony Victor
Obinski, Russell C
Oberholtzer, James Edward
Odland, Russell Kent
Obrycki, Richard
O'Donnell, Raymond Thomas
Oglivie, James Louis
O'Haver, Thomas Calvin
Ohline, Robert Wayne
Ohnesorge, William Edward
O'Keefe, Kelly Ray
Olander, Donald Paul
Olin, Jacqueline S
Olsen, Eugene Donald
Olsen, Rodney L
Olson, Carter LeRoy
Olson, Paul B
Olver, John Walter
O'Mara, Michael Martin
O'Neal, Floyd Breland
O'Reilly, James Emil

Orna, Mary Virginia
Orzech, Chester Eugene, Jr
Osol, Arthur
Osteryoung, Robert Allen
Ott, Welland Lee
Overton, Edward Beardslee
Owens, Marvin Lee, Jr
Paabo, Maya
Pacer, Richard A
Padmanabhan, G R
Padmore, Joel M
Page, John Arthur
Palmer, Thomas Adolph
Panalaks, Thavil
Pankratz, Ronald Ernest
Pannu, Sardul S
Papariello, Gerald Joseph
Papastephanou, Constantin
Pappenhagen, James Meredith
Pardue, Harry L
Paris, Jean Philip
Park, George Bennet
Park, David J
Parker, Gordon Arthur
Parks, Albert Fielding
Parks, Archie Oliver
Parks, Ross Lombard
Parry, Edward Petterson
Parsons, James Sidney
Parsons, Michael L
Pasterczyk, William Robert
Patel, Siddharth Manial
Paterson, Arthur Renwick
Patterson, Gordon Derby, Jr
Patterson, Truett Clifton
Patton, Leo Wesley
Paul, Jeddeo
Paulsen, Paul
Paxson, John Ralph
Payne, Jimmie Sturgis, Jr
Peace, George Earl, Jr
Peard, William John
Pearse, George Anceil, Jr
Pearson, Robert Melvin
Pease, Burton Frank
Pecsok, Robert Louis
Peck, Norman Eugene
Pellizzari, Edo Domenico
Penner, Melvin H
Pennington, Lloyd Drew
Perkins, Gerald, Jr
Perkins, Robert Willard
Perone, Samuel Patrick
Perretta, Armond Thomas
Perrin, Carol Hollingsworth
Perry, John Arthur
Perry, Mary Hertzog
Peters, Dennis Gail
Peterson, Albert H
Peterson, John Ivan
Peterson, Stephen Frank
Petro, Peter Paul, Jr
Peurifoy, Paul Vastine
Pflaum, Ronald Trenda
Pfluger, Clarence Eugene
Phifer, Lyle Hamilton
Phillips, John Perrow
Phillips, Wendell Francis
Philp, Robert Herron, Jr
Pickett, Edward Ernest
Pickett, John Harold
Pickral, George Monroe, Jr
Piepmeier, Edward Harman
Pietri, Charles Edward
Pietrzyk, Donald John
Pietrzykowski, Anthony D
Pike, LeRoy
Pipenberg, Kenneth James
Piper, Carl Victor
Plambeck, James Alan
Plankey, Francis William, Jr
Plummer, Louisa Greenleaf
Pobiner, Harvey
Poddolak, Henry Andrew
Poe, Donald Patrick
Poe, Richard D
Poland, Lloyd Orville
Pons, Walter A, Jr
Pontius, Dieter J J
Pool, Edwin Lewis
Pool, Karl
Popov, Alexander Ivan
Potter, John Clarkson
Powell, William Allan
Propp, Jacob Henry
Przybylowicz, Edwin P
Pugh, Thomas L
Purdy, William Crossley
Pyle, John Tillman
Quinney, Paul Reed
Raaen, Helen Parks
Raasch, Lou Reinhart
Rabel, Fredric M
Rabenstein, Dallas Leroy
Raby, Bruce Alan
Rainey, Mary Louise
Rains, Theodore Conrad
Ramaley, Louis

Ramaswamy, H N
Ramette, Richard Wales
Ramirez-Munoz, Juan
Ramsey, Lessel Leslie
Ratti, Joginder Singh
Ray, Jesse Paul
Reber, Louis Alexander
Rechnitz, Abraham
Reed, John J R
Reed, Kenneth Paul
Reffes, Howard Allen
Rehwoldt, Robert E
Reilley, Charles Norwood
Reilly, William Leo
Rein, James Earl
Reinbold, Paul Earl
Reinmuth, William Henry
Reissmann, Thomas Lincoln
Rejto, Peter A
Rekers, Robert George
Remington, Lloyd Dean
Resnik, Frank Edward
Reuter, Wilhad
Reynolds, Samuel Allen
Rhees, Raymond Charles
Rhoads, William Denham
Rhoades, David R
Rhodes, Robert Carl
Ricca, Paul Joseph
Rice, Walter Wilburn
Richart, Herbert H
Rickard, Eugene Clark
Riddick, John Allen
Rider, Benjamin Franklin
Riggs, William McKnight
Rimes, William John
Risby, Terence Humphrey
Ritchey, John Michael
Rivera, William Henry
Rivers, Paul Michael
Rivington, Donald Erskine
Roach, Don
Roach, Paul G
Roberts, Charles Brockway
Roberts, Julian Lee, Jr
Robertson, John Harvey
Robinson, Douglas Walter
Robinson, Jack Landry
Robinson, James William
Robinson, Rex Julian
Robinson, Robert Eugene
Rochow, Theodore George
Rodell, Michael Byron
Rodgers, Robert Stanleigh
Rodgers, Sheridan Joseph
Rodriguez, Charles F
Roe, David Kelmer
Rogers, Donald Warren
Rogers, Lewis Henry
Rogers, Lockhart Burgess
Rogers, Raymond N
Rogerson, Peter Freeman
Rolf, Frederick William
Rollins, Orville Woodrow
Romano, Salvatore James
Romberger, Karl Arthur
Rommel, Marjorie Ann
Rook, Harry Lorenz
Rorabacher, David Bruce
Rose, Ira Marvin
Rose, John Logan, Jr
Rose, Stuart Alan
Rosen, Aaron A
Rosenberg, Alexander F
Rosen, Clarence James
Rosenthal, David (Walter)
Rosenthal, Donald
Rostie, Douglas McDonald
Ross, Harley Harris
Ross, Robert Anderson
Ross, Robert M
Rothbart, Herbert Lawrence
Rowan, Robert, Jr
Rowe, William A
Roy, Jean-Claude
Roy, Rabindra (Nath)
Roy, Gerald Stephen
Rozett, Richard Walter
Rubin, Leon E
Ruch, Rodney R
Rudolph, Robert Lewis
Rue, Sigurd Oscar
Rulfs, Charles Leslie
Runnels, John Hugh
Russell, Charles Addison
Russell, Douglas Stewart
Russell, Virginia Ann
Russo, William Richard
Rutgers, Jay G
Ryan, Douglas Earl
Rynasiewicz, Joseph
Sadtler, Philip
St John, Peter Alan
Salee, Eugene Merridith
Salyer, Darnell
Sams, Richard Alvin
Sandi, Emil

Sandridge, Robert Lee
Sanik, John, Jr
Santacana-Nuet, Francisco
Sargent, Roger N
Sarma, Pramod Lal
Sarneski, Joseph Edward
Sass, Samuel
Sattur, Theodore W
Sauerbrunn, Robert Dewey
Sawardeker, Jawahar Sazro
Sawyer, Donald Turner, Jr
Saxton, Augustus Donovan
Schaap, Ward Beecher
Schaefer, Robert William
Schaffer, Robert
Schalge, Alvin Laverne
Schall, Elwyn DeLaurel
Scheffer, Edward Reinhard
Schempf, John Morey
Schenker, Henry Hans
Schepers, Gerald J
Schickedantz, Paul David
Schilt, Alfred Ayars
Schleicher, John Anthony
Schmidt, Reese Boise
Schmidt, William Edward
Schneider, Frank L
Schnepfe, Marian Moeller
Schnipelsky, Paul Nicholas
Scholz, Robert George
Schroeder, Thomas Dean
Schubert, Leo
Schucker, Gerald D
Schulman, Stephen Gregory
Schultz, Franklin Alfred
Schultz, Hyman
Schulz, William
Schupp, Orion Edwin, III
Schwab, Helmut
Schwartz, Robert Donald
Schweikert, Emile Alfred
Scott, Dan Dryden
Scott, Donald Ray
Scott, Frederick Arthur
Scott, James Alan
Scott, Kenneth Richard
Scott, Lawrence William
Scott, Ralph Asa, Jr
Scott, Richard Lynn
Scott, Richard William
SCott, Robert Edward
Scroggie, Lucy E
Seabaugh, Pyrtle W
Sedlak, Michael
Segal, Hirsh Sholom
Segatto, Peter Richard
Seim, Henry Jerome
Seitz, William Rudolf
Selig, Walter
Sellers, Douglas Edwin
Senear, Allen Eugene
Senzel, Alan Joseph
Seo, Eddie Tatsu
Serfass, Frank James
Settle, Frank Alexander, Jr
Shackelford, Walter McDonald
Shadoff, Lewis Allan
Shaheen, Donald G
Shane, Norman Abraham
Shank, Ralph Chalmer
Shapras, Peter
Sharp, George Oscar
Shaw, Donald Wayne
Shaw, Elwood R
Shaw, Vernon Reed
Shearer, Charles M
Shearer, Duncan Allan
Sheer, Maxine Lana
Shepardson, John U
Sheppard, Allen Anson
Sheridan, Jane Connor
Sherma, Joseph A
Sherren, Anne Terry
Shields, Loran Donald
Shilman, Avner
Shilstone, Cecil Maxwell
Shimp, Neil Frederick
Shive, Donald Wayne
Shuba, Raymond J
Shull, Charles Morell, Jr
Shults, Wilbur Dotry, II
Siefker, Joseph Roy
Sievers, Robert Eugene
Siggia, Sidney
Sill, Claude Woodrow
Simmons, Ivor Lawrence
Simpson, Joseph Matthew
Simpson, Cohen Thomas
Simpson, Frank Martin, Jr
Simpson, Stephen Gershom
Sinclair, James Douglas
Singh, Surjit
Singleton, Bert
Siverston, John Neilos
Skelly, Norman Edward
Skogerboe, Rodney K
Skougstad, Marvin Wilmer
Slywka, Gerald William Alexander
Smith, Alan Jerrard

Smith, Albert Faris
Smith, Charles Aloysius
Smith, David Lee
Smith, David Warren
Smith, Donald E
Smith, Dwight Morrell
Smith, Earl Cooper
Smith, Edward John
Smith, Francis White
Smith, Frank Houston
Smith, Grant Warren
Smith, Isaac Litton
Smith, James Allbee
Smith, James Graham, Jr
Smith, John Elvans
Smith, Kenneth Edward
Smith, Maynard E
Smith, Michael James
Smith, Philip T
Smith, Ralph G
Smith, Robert Leonard
Smith, Robert Victor
Smullin, Charles Frederick
Snapp, Thomas Carter, Jr
Somkaite, Rozalija
Sommers, Raymond A
Southern, Thomas Martin
Spall, Walter Dale
Spell, Charles Raymond
Spence, Jack Taylor
Spiegelhalter, Roland Robert
Spielholtz, Gerald I
Spindler, Donald Charles
Spinelli, John
Spink, Charles Harlan
Sporek, Karel Frantisek
Spritzer, Michael Stephen
Srinivasan, Vakula S
Stadler, Louis Benjamin
Stalling, David Laurence
Stamm, Ralph Eugene
Stanley, Charles William
Starks, Aubrie Neal, Jr
Starnes, Paul Kiser
State, Harold M
Steele, John A
Steichen, Richard John
Steinhaus, Ralph K
Stenger, Vernon Arthur
Stephens, Bobby Gene
Stevens, Vernon Lewis
Stevens, William George
Stevens, William Harmer
Stevenson, Robert Lovell
Stewart, Reginald Bruce
Stiles, David A
Stober, Henry Carl
Stock, John Thomas
Stoffer, Robert Llewellyn
Stolzberg, Richard Jay
Stoner, Graham Alexander
Stoub, Kenneth Paul
Stratton, Cedric
Straub, William Albert
Streuli, Carl Arthur
Stricos, David Peter
Strobel, Howard Austin
Strohl, John Henry
Stromatt, Robert Weldon
Strong, Frederick Carl, III
Strong, Robert Stanley
Struck, William Anthony
Struempler, Arthur W
Strunk, Duane H
Stubblefield, Charles Bryan
Stubblefield, Robert Douglas
Sturrock, Peter Earle
Su, Yao Sin
Sudmeier, James Lee
Suffet, Irwin Henry
Suryaraman, Maruthuvakudi Gopalasastri
Sutton, John Curtis
Svec, Harry John
Swank, Howard Wigton
Swann, William B
Swanson, Donald Leroy
Swanson, Lynn Allen
Sweeney, Harold A
Sweet, David Paul
Sweet, Richard Clark
Sweetser, Philip Bliss
Swift, Ernest Haywood
Swinehart, Bruce Arden
Swingle, Robert Shelton, II
Sympson, Robert F
Szap, Peter Charles
Szutka, Anton
Tackett, James Edwin, Jr
Tackett, Stanford L
Tai, Han
Talbott, Ted Delwyn
Talley, Charles Peter
Tallman, Dennis Earl
Taraszka, Anthony John
Taylor, Albert Edward
Taylor, David Cobb
Taylor, Ella Richards
Taylor, Hugh Alan

Taylor, James Robert
Taylor, James Welch
Taylor, John Keenan
Taylor, Kirman
Taylor, Michael Lee
Taylor, Murray East
Taylor, Paul John
Taylor, Robert Morgan
Teitelbaum, Charles Leonard
Templeman, Gareth J
Thatcher, Walter Eugene
Thayer, Lewis A
Theimer, Edgar E
Thielmann, Vernon James
Thomas, Alexander Edward, III
Thomas, Carolyn Margaret
Thomas, Elizabeth Wadsworth
Thomas, Josephus, Jr
Thomas, Martha Jane Bergin
Thorne, Frederick A
Thompson, Kenneth Roy
Thompson, Richard John
Thompson, Richard Michael
Thomson, William Alexander Brown
Thumm, Byron Ashley
Tiedemann, Albert William, Jr
Tiernan, Thomas Orville
Timnick, Andrew
Tipton, George Murtha
Tishler, Frederick
Toberman, Ralph Owen
Todd, Jerry William
Tomkins, David Francis
Tompa, Albert S
Topping, Joseph John
Toren, Eric Clifford, Jr
Toren, Paul Edward
Torrey, Rubye Prigmore
Tou, James Chieh
Towle, Louis Wallace
Tremmel, Carl George
Trent, John Ellsworth
Triglia, Emil J
Tripathi, Kamala Kant
Trusell, Fred Charles
Trusk, Ambrose
Tsuji, Kiyoshi
Tuffly, Bartholomew Louis
Turi, Paul George
Turner, Anne Halligan
Turner, James Howard
Turner, William Richard
Turnquist, Truman Dale
Turse, Richard S
Tuthill, Samuel Miller
Tyler, Willard Philip
Tyson, Bruce Carroll, Jr
Uden, Peter Christopher
Ulrich, Floyd Seymour
Ulrich, William Frederick
Umbreit, Gerald Ross
Underkofler, William Leland
Upson, U Layton
Upton, Ronald P
Uznanski-Bottei, Rita Marlene
Van Atta, Robert Ernest
Vance, Hugh Gordon
Van de Poel, Josephus
Van Deren, John Medearis, Jr
Vandergraaf, Tjalle T
Vander Haar, Roy William
Van Duyne, Richard Palmer
Van Geet, Anthony Leendert
Van Hall, Clayton Edward
Van Norman, John Donald
Van Swaay, Maarten
Van't Riet, Bartholomeus
Van Vorous, Ted
Varga, Louis P
Varon, Albert
Veal, Dean Johnson
Veening, Hans
Veillon, Claude
Venkataraghavan, R
Veraguth, Arnold John
Vick, Maurice M
Vickers, Thomas J
Vickroy, David Gill
Vidaurreta, Luis E
Vilcins, Gunars
Vincent, Harold Arthur
Vincent, Robert Corbin
Vinson, Joe Allen
Vivilecchia, Richard
Volborth, Alexis
Vonderbrink, Sally Ann
Vouros, Paul
Vratny, Frederick
Wadelin, Coe William
Waehner, Kenneth Arthur
Wakeman, Irving B
Walden, George Ellis
Walker, Donald I
Walker, Joe M
Walker, Joseph
Walker, Raymond Lloyd
Wallace, Frederic Andrew
Wallace, Gerald Wayne

Walradt, John Pierce
Walsh, John Thomas
Walters, Douglas Bruce
Walters, Fred Henry
Walters, John P
Walking, Arthur Ernest
Walton, Harold Frederic
Walz, Alvin Eugene
Wang, Jin Tsai
Ward, Edward Hilson
Ward, George A
Wardi, Ahmad Hassan
Warren, Herbert Dale
Warren, Richard Joseph
Warshowsky, Benjamin
Washburn, William H
Waterbury, Glenn Raymond
Watkins, Stanley Read
Watson, Duane Craig
Watson, Jack Throck
Weaver, J Ritner
Webb, John Schurr
Weber, Charles Walton
Weber, Charles William
Weber, Dennis Joseph
Wechter, Margaret Ann
Wehry, Earl L, Jr
Weiser, Herman Joshua, Jr
Weiss, Fred Toby
Weiss, Roger Harvey
Wendel, Carlton Tyrus
Wentworth, Wayne
Werner, Thomas Clyde
West, Charles David
West, David Markham
West, Kenneth Calvin
Westneat, David French
Wetterau, Frank P
Wexler, Arthur Samuel
Weyh, John Arthur
Wharton, Harry Whitney
Whealy, Roger Dale
Wheeler, Ralph John
Whetsel, Kermit Bazil
Whidby, Jerry Frank
Whitaker, Leslie A
Whitcomb, Donald Leroy
White, James Carl
White, June Broussard
Whitehead, Thomas Hillyer
Whiteker, Roy Archie
Whiteside, Charles Hugh
Whitlock, L Ronald
Whitten, Maurice Mason
Whittle, George Patterson
Wiberley, Stephen Edward
Wickham, James Edgar, Jr
Wiebush, Joseph Roy
Wiersma, James H
Wildeman, Thomas Raymond
Wilkins, Charles Lee
Wilkinson, Joseph Ridley
Wilkinson, Ralph Russell
Will, Fritz, III
Willard, Thomas Maxwell
Willeboordse, Friso
Williams, Calvin Herndon, Jr
Williams, Frederick Wallace
Williams, Jean Paul
Williams, Max Bullock
Williams, Reed Chester
Williams, Robert Calvin
Williams, Sidney
Willis, Theodore Roosevelt
Willis, William Van
Wilson, Claude E
Wilson, George Spencer
Wilson, Larry Eugene
Wilson, Mabel Florey
Wilson, Ray Floyd
Wimer, David Carlisle
Windham, Ronnie Lynn
Winefordner, James D
Winfrey, J C
Wingender, Ronald John
Winnett, George
Winograd, Nicholas
Wiser, James Eldred
Withers, Edward Donald
Wittick, James John
Wittle, John Kenneth
Woerner, Dale Earl
Wohlfort, Sam Willis
Wolf, Thomas
Wolfe, Wayne Robert
Wolfe, Clinton Ray
Wolke, Sara Richardson
Wolkoff, Aaron Wilfred
Wolszon, John Donald
Wood, David
Woodham, Donald W
Woolley, Earl Madsen
Worthington, James Brian
Wrangell, Lewis J
Wright, Charles Hubert
Wynn, James Elkanah
Yanai, Hideyasu Steve
Yasuda, Stanley K

Yates, Ann Marie
Yates, Claire Hilliard
Yau, Wallace Wen-Chuan
Yeager, Howard Lane
Yeranian, James A
Yerick, Roger Eugene
Yoakum, Anna Margaret
Youmans, Hubert Lafay
Young, Donald Charles
Young, Harold Henry
Young, Irving Gustav
Young, Jack Phillip
Young, James Christopher F
Young, Philip Ross
Young, Sanford Tyler
Yurow, Harvey Warren
Zabin, Burton Allen
Zado, Franjo M
Zakaib, Daniel D
Zakriski, Paul Michael
Zalipsky, Jerome Jaroslaw
Zanzucchi, Peter John
Zaye, David Francis
Zellmer, David Louis
Zembrod, Anthony Raymond
Zenchelsky, Seymour Theodore
Ziegler, William Arthur
Zielinski, Walter L, Jr
Zienius, Raymond Henry
Zinner, Edward John
Zimmerman, John F
Zitomer, Fred
Zlatkis, Albert
Zobel, Henry Freeman
Zorn, Ralph Allan
Zuehlke, Carl William

Chemical Microscopy
Adler, Robert Garber
Cocks, George Gosson
Darneal, Robert Lee
Hamill, James Junior
Julian, Donald Benjamin
Krc, John, Jr
Muggli, Robert Zeno
Saylor, Charles Hamilton Proffer
Schaeffer, Harold Franklin
Walker, Donald I

Electroanalytical Chemistry
Aikens, David Andrew
Anson, Fred (Colvig)
Bard, Allen Joseph
Broman, Robert Fabel
Caton, Roy Dudley, Jr
Ellis, Samuel Benjamin
Engel, Adolph James
Fawcett, Newton Creig
Gulens, Janis
Hebert, Normand Claude
Hildebrandt, Wayne Arthur
Klatt, Leon Nicholas
Levy, Samuel C
Li, Chia-Yu
Matsuyama, George
Newton, Carolyn McCrory
Nisbet, Alex Richard
Osteryoung, Janet G
Park, George Bennet
Park, Su-Moon
Polcyn, Daniel Stephen
Shain, Irving

Mass Spectrometry
Aberth, William H
Ampulski, Robert Stanley
Andresen, Brian Dean
Barofsky, Douglas Fred
Benz, Wolfgang
Brackmann, Richard Theodore
Cameron, Angus Ewan
Caprioli, Richard Michael
Chait, Edward Martin
Christie, Warner Howard
Cone, Conrad
Costello, Catherine E
Daasch, Lester William
Desiderio, Dominic Morse, (Jr)
DoAmaral, Jefferson Ribeiro
Fenselau, Catherine Clarke
Foner, Samuel Newton
Foster, Norman George
Funke, Phillip T
Gohlke, Roland Schulz
Hassell, Clinton Alton
Hayes, John Michael
Herzog, Leonard Frederick, II
Honig, Richard Edward
Jardine, Ian
Johnson, Walter Heinrick, Jr
King, Roy Warbrick
Knudsen, Thomas Paul
Levenberg, Milton Irwin
Lewis, Jack Smith
Lin, Denis Chung Kam
Lovins, Robert E
McEwen, Charles Nehemiah
Mead, Thomas Edward

Meyerson, Seymour
Mitchum, Ronald Kem
Occolowitz, John Lewis
Podosek, Frank A
Preti, George
Rohwedder, William Kenneth
Schaffer, Robert
Schroeer, Juergen Max
Shadoff, Lewis Allan
Shew, Delbert Craig
Siegel, Melvin Walter
Simons, David Stuart
Smith, Robert Leonard
Solomon, Jerome Jay
Sparrow, Gene Rodell
Torgerson, David Franklyn
Torner, Kenneth Beamer
Wilson, Mark Ferlin
Wood, Gordon Walter
Yamdagni, Raghavendra
Zaechelein, Anthony Gabriel

Polarography
Frost, Jackie Gene
McLean, James Dennis

Inorganic Chemistry
Abbott, Edwin Hunt
Ackermann, Martin Nicholas
Adams, Arthur Curtis
Adams, Max Dwain
Adcock, James Luther
Addamiano, Arrigo
Adisesh, Setty Ravanappa
Ahmed, Ismail Yousef
Albinak, Marvin Joseph
Alexander, John J
Alexander, Martin Dale
Alfieri, Charles C
Allan, Benjamin Wilson
Allcock, Harry Rex
Allen, A D
Allen, Christopher Whitney
Allen, Joe Frank
Allenbach, Charles Robert
Allison, John Arthur Charles
Allred, Albert Louis
Almodovar, Ismael
Almond, Hy
Alton, Earl Robert
Alvarez, Vincent Edward
Anderson, Herbert Hale
Anderson, June S
Anderson, Lowell Ray
Anderson, Melvin Lee
Anderson, Samuel
Anderson, Wayne Philpott
Angelici, Robert Joe
Appelman, Evan Hugh
Aradine, Paul William
Arcand, George Myron
Archer, Ronald Dean
Argue, Gary R
Armstrong, Don Leigh
Ashby, Eugene Christopher
Ashley, Kenneth R
Asprey, Larned Brown
Atkinson, Gordon
Attig, Thomas George
Axtell, Darrell Dean
Aykan, Kamran
Baak, Tryggve
Babb, Daniel Paul
Bach, Ricardo O
Bahn, Emil Lawrence, Jr
Bailar, John Christian, Jr
Bailey, Ronald Albert
Baker, Louis Coombs Weller
Baker, Willie Arthur, Jr
Balch, Alan Lee
Baldwin, Howard Wesley
Baldwin, W George
Banks, Ephraim
Barefield, Edward Kent
Barile, Raymond Conrad
Barker, James Emory
Barkman, Erik Fredrik
Barnes, Robert Lee
Barnett, Kenneth Wayne
Barney, Duane Lowell
Barrett, Peter Fowler
Bartlett, Neil
Barton, Lawrence
Barton, Lucian Anthony
Basolo, Fred
Batey, Harry Hallsted, Jr
Bau, Robert
Baum, Parker Bryant
Baum, Stuart J
Bauman, John E, Jr
Baumann, Arthur Nicholas
Baumann, Donald Joseph
Bayer, Richard Eugene
Bayes, Alfred Lee

Beachley, Orville Theodore
Beal, James Burton, Jr
Bear, John L
Beaumont, Randolph Campbell
Beck, James Donald
Beck, Roland Arthur
Beckman, William
Beese, Ronald Eiroy
Belford, Rue Linn
Bendure, Robert J
Bennett, Larry E
Bennett, William Earl
Benson, Edmund Walter
Bereman, Robert Deane
Berg, Ernest Phillip, Jr
Berg, John Richard
Berglund, Donna Lou
Bergman, John George, Jr
Bergman, David Alvin
Bernisen, Robert Andyv
Berry, Robert Walter
Bertin, Ernest Peter
Bertolacini, Ralph James
Bertrand, Joseph Aaron
Bertsch, Charles Rudolph
Bessette, Russell Romulus
Bettinger, Donald John
Bettwy, Mary Leon
Bhatnagar, Dinesh C
Bhiwandker, Nutan C
Biagetti, Richard Victor
Bickerdike, Ernest Lawrence
Biddle, Richard Albert
Bil, Milos Sidney
Billo, Edward Joseph
Bills, Charles Wayne
Bills, James LaVar
Birden, John Harlan
Birdsall, William John
Birk, James Peter
Birnbaum, Edward Robert
Birnbaum, Ernest Rodman
Bishop, Allen David, Jr
Bissot, Thomas Charles
Blair, George Richard
Blair, Robert Paul
Blake, Daniel Melvin
Blalock, Thomas Jacks
Blass, Gerhard Alois
Blinn, Elliott L
Block, Burton Peter
Blood, Franklin Harvey
Blount, Floyd Eugene
Bobonich, Harry Michael
Boguck, Raymond Francis
Bolles, Theodore Frederick
Bond, Stephon Thomas
Boone, James Lightholder
Bos, William G
Botdorf, Ruth G
Bottei, Rudolph Santo
Bottger, William George
Bottomley, Frank
Boucher, Laurence James
Boudreaux, Edward A
Boughton, John Harland
Bouknight, Joseph Ward
Bowen, Ruth Justice
Bowkley, Herbert Louis
Boyd, Alfred Colton, Jr
Braddock-Rogers, Kenneth
Bradley, Robert Lincoln
Brady, Robert Alois
Bramlett, Christopher L
Bramlett, James Edward
Brandau, Betty Lee
Brasted, Robert Crocker
Braun, Juergen Hans
Braun, Otto Godfrey
Brauer, Phyllis Ambler
Bravo, Justo Baladjay
Breck, Donald Wesley
Brennan, Gerald L
Brenner, Douglas G
Brewer, Stephen Wiley, Jr
Breyer, Arthur Charles
Brill, Thomas Barton
Brinm, Eugene Oskar
Brinckman, Frederick Edward, Jr
Britton, William Giering
Brixner, Lothar Heinrich
Broberg, Joel Wilbur
Brothers, James Alfred
Brous, Jack
Brown, Bruce Willard
Brown, Charles Albert
Brown, David Basset
Brown, Herbert Charles
Brown, Theodore Lawrence
Brown, William Anderson
Brubaker, Carl H, Jr
Brubaker, George Randell
Bryan, Horace Alden
Bryan, Philip Steven
Buchacek, Robert Joseph
Buckley, Reginald R
Buckwold, Sidney Joshua
Bulkin, Bernard Joseph

Bull, William Earnest
Bunting, Roger Kent
Burbage, Joseph James
Burch, Wendell Dale
Burg, Anton Behme
Burgess, Thomas Edward
Burgyan, Aladar
Burkard, Pete Nelius
Burke, John A
Burmeister, John Luther
Burnett, John Lambe
Burns, Raymond Edward
Burns, Robert Tilden
Burns, Robert Price
Busch, Daryle Hadley
Bushey, William Raymond
Butler, Ian Sydney
Butler, Stephen Alan
Butler, Rudolph O
Bydalek, Thomas Joseph
Cady, George Hamilton
Cairns, John Luther
Callahan, James Clyde
Callahan, Kenneth Paul
Callis, Clayton Fowler
Campbell, Donald L
Campisi, Louis Sebastian
Cannon, John Francis
Carberry, Edward Andrew
Carfagno, Daniel Gaetano
Carlino, Richard Lewis
Carlisle, Gene Ozelle
Carlson, Gordon Andrew
Carlson, Robert Kenneth
Carlyle, David Wesley
Carrabine, John Anthony
Carter, James Clyde
Carter, Loren Sheldon
Carty, Arthur John
Castleberry, George E
Cates, Vernon E
Cavell, Ronald George
Centofanti, Louis F
Chakrabarti, Chuni Lal
Chamberlain, Phyllis Ione
Chamberland, Bertrand Leo
Chamberlin, John Macmullen
Chance, Robert L
Chandler, Alfred Bertram
Chang, James C
Charkoudian, John Charles
Chastain, Benjamin Burton
Chattoraj, Shib Charan
Chellew, Norman Raymond
Chester, Arthur Warren
Chilton, John Morgan
Chiola, Vincent
Chong, Clyde Hok Heen
Choppin, Gregory Robert
Christe, Karl Otto
Chu, Vincent Hao Kwong
Churchill, Melvyn Rowen
Claiborne, Imogene B
Clark, Howard Charles
Clark, Howell R
Clark, Ronald Jene
Clarke, Lilian A
Clase, Howard John
Clausen, Chris Anthony
Clearfield, Abraham
Cleveland, Jesse Marvin Jr
Clifford, Alan Frank
Coakley, Mary Peter
Coates, Geoffrey Edward
Cobble, James Wikle
Cochran, George Thomas
Cockerell, Leone (Doris)
Coerver, Helen Joseph
Cohen, Harvey Martin
Cohen, Irwin A
Cohen, Joseph
Cohen, Martin Allen
Cohen, Sheldon H
Cohn, Kim
Cole, Roger M
Coley, Ronald Frank
Collier, Francis Nash, Jr
Collier, Herman Edward, Jr
Collins, Alva LeRoy, Jr
Collins, Ronald William
Collman, James Paddock
Colton, Ervin
Comer, Joseph John
Condike, George Francis
Conley, Francis Raymond
Conley, Robert F
Conner, Charles Michael
Connick, Robert Elwell
Connolly, John W
Connolly, Lawrence Edward
Constant, Clinton
Cooke, Dean William
Cooke, Alden Hoopes, Jr
Cooper, John Neale
Cooper, Maurice D
Cooperstein, Raymond
Cope, Virgil W

Copeland, David Anthony
Corbett, John Dudley
Cordes, Arthur Wallace
Cordes, Eugene H
Corey, Joyce Yagla
Corfield, Peter William Reginald
Coucouvanis, Dimitri N
Cowley, Alan H
Coyle, Bernard Andrew
Coyle, Thomas Davidson
Craddock, John Harvey
Cramer, Roger Earl
Crawford, Thomas H
Crayton, Philip Hastings
Creutz, Carol Ann
Crissman, Judith Anne
Crockett, David Scott
Crook, Joseph Raymond
Crouthamel, Carl Eugene
Cruickshank, Alexander Middleton
Crum, John Kistler
Crutchfield, Charlie
Cude, Willis Augustus, Jr
Cull, Neville
Cullen, William Robert
Cullmann, Ralph E
Cummings, Sue Carol
Cummings, Jack D
Cummiskey, Charles
Cundy, Paul Franklin
Curran, Columba
Curtis, John D
Curtis, Myron David
Daane, Adrian Hill
Dalton, Richard Lee
Daly, John Matthew
Dance, Ian Gordon
Danzig, Morris Judah
Darensbourg, Donald Jude
Darensbourg, Marcetta York
Darlington, William Bruce
Das Sarma, Basudeb
Daugherty, Ned Arthur
Davenport, Derek Alfred
Davidson, John Edwin
Davies, Geoffrey
Davis, Clyde Edward
Davis, Joseph Anthony
Davis, Robert Elliott
Davison, Alan
Dawson, Gladys Quinty
Day, Marion Clyde, Jr
Dean, Walter Keith
Dean, Walter Lee
Dean, Warren Edgell
DeArmond, M Keith
Deck, Charles Francis
Decker, Jesse Smith
Deever, William Ray
De Haan, Frank P
Dehne, George Clark
Deiters, Rose Mary
Demitras, Gregory Claude
Dennis, Mary
Deshpande, Krishnanath Bhaskar
Deskin, William Arno
DesMarteau, Darryl D
Dess, Howard Melvin
Deters, John Frederick
Deutsch, Edward Allen
Devonshire, Leonard Norton
Dexter, Theodore Henry
Dickerhoof, Dean W
Dickinson, John G
Dieck, Ronald Lee
Dietz, Rudolph John
DiGiuseppe, Michael Anthony
Dillard, Clyde Ruffin
DiLorenzo, James V
Dilts, Joseph Alstyne
Dines, Martin Benjamin
Dinga, Gustav Paul
Discher, Clarence August
Dixon, Keith R
Dodson, B C
Dodson, Vance Hayden, Jr
Doedens, Robert John
Donohue, Paul Christopher
Donovan, Thomas Arnold
Douglas, Bodie Eugene
Drago, Russell Stephen
Drake, John Edward
Dreeben, Arthur B
Dreyfuss, Robert George
Drinkard, William Charles, Jr
Droll, Henry Andrew
Druding, Leonard F
DuBois, Frederick Williamson
DuBois, Thomas David
Duecker, Heyman Clarke
Duell, Paul Merwyn
Duffy, Norman Vincent, Jr
Dukes, Gary Rinehart
Dumbaugh, William Henry, Jr
Duncan, Archibald
Duncan, Leonard Clinton
Dunn, J Stanley
Durand, Edward Allen

Dzierzanowski, Frank John
Eads, Ewin Alfred
Eastland, George Warren, Jr
Eastman, Alan Dan
Eaton, Gareth Richard
Ebner, Jerry Rudolph
Eddy, Lowell Perry
Eddy, Robert Devereux
Edwards, John Oelhaf
Ehrenstorfer, Sieglinde K M
Eibeck, Richard Elmer
Eichinger, Jack Waldo, Jr
Eick, Harry Arthur
Eisenberg, Richard
Ekman, Carl Frederick W
Elder, Richard Charles
El Ghamry, Mohamed Tawfik
Elkind, Michael John
Elliott, Eugene Willis
Ellis, David Allen
Elsbernd, Helen
Elson, Robert Emanuel
Emerson, Kenneth
Endicott, John F
Enemark, John Henry
Epperson, Edward Roy
Eppler, Richard A
Erdmann, David E
Erickson, Charles Edward
Erskine, Gordon John
Espenson, James Henry
Esposito, John Nicholas
Eubanks, Isaac Dwaine
Evans, Howard Tasker, Jr
Evans, James Stuart
Evans, Roger Lynwood
Evans, William John
Everett, Ardell Gordon
Everett, Kenneth Gary
Everson, Howard E
Eyman, Darrell Paul
Faller, John William
Fanning, James Collier
Farona, Michael F
Faught, John Brian
Faut, Owen Donald
Fay, Robert Clinton
Feldman, Fredric J
Feltham, Robert Dean
Fessler, Robert Glenn
Field, Paul Eugene
Fine, Dwight Albert
Finholt, James E
Fink, Collin Ethelbert
Finlay, Gordon Roy
Finley, Arlington Levart
Fischer, Robert George, Jr
Fisher, Harold M
Flanigen, Edith Marie
Fleming, Peter B
Fleming, Suzanne M
Flynn, James Patrick
Fonseca, Anthony Gutierre
Forchheimer, Otto Louis
Ford, Peter Campbell
Forsberg, John Herbert
Forster, Denis
Fortman, John Joseph
Foss, Frederick William, Jr
Fowler, John Rayford
Frame, Harlan D
Fraser, Margaret Shirley
Frazier, Stephen Earl
Freeman, Horatio Putnam
Freeman, Wade Austin
Freidline, Charles Eugene
French, James Edwin
Frey, Frederick Wolff, Jr
Frey, John Erhart
Friedel, Arthur W
Friedman, Harold Leo
Friedman, Lawrence Boyd
Frost, Robert Edwin
Funck, Larry Lehman
Gabriel, Henry
Gagliano, Louis John
Gaines, Donald Frank
Gaiser, Romey Arthur
Galasso, Francis Salvatore
Galloway, Gordon Lynn
Ganchoff, John Christopher
Gancy, Alan Brian
Garber, Lawrence L
Gardner, William Lee
Garman, William Lee
Garrett, Barry B
Garvey, Roy George
Gaswick, Dennis C
Gatti, Anthony Roger
Geanangel, Russell Alan
Gebelt, Robert Eugene
Gebhardt, Joseph John
Gehrke, Neil Edward
Gentile, Philip
Gentke, Henry
George, James E
George, John Warren

George, T Adrian
Gerace, Paul Louis
Gerteis, Robert Louis
Ghose, Hirendra M
Gibbins, Sidney Gore
Gibbons, James Joseph
Gilbert, George Lewis
Gilbertson, Lyle Ithiel
Gilje, John
Gillespie, Ronald James
Gillman, Hyman David
Gillow, Edward William
Gimple, Glenn Edward
Ginell, William Seaman
Gingerich, Karl Andreas
Ginsberg, Alvin Paul
Ginther, Robert J
Girardot, Peter Raymond
Gititz, Melvin Hyman
Gladding, Jane B
Glaeser, Hans Hellmut
Glanville, James Oliver
Glick, Milton Don
Goddard, John Burnham
Goedken, Virgil Linus
Goehring, John Brown
Goel, Ram Gapal
Goetschel, Charles Thomas
Gogan, Niall Joseph
Gold, Marvin B
Gold, Richard Frank
Goldberg, David Elliott
Goldberg, Stephen Zalmund
Goldwhite, Harold
Gonick, Ely
Goochee, Herman Francis
Good, Mary Lowe
Gordon, Gilbert
Gordon, Joseph Grover, II
Gorman, Melville
Gortsema, Frank Peter
Gottlieb, Irvin M
Gould, Edwin Sheldon
Graham, Augustus Washington
Graham, William Arthur Grover
Grahn, Edgar Howard
Grandolfo, Marian Carmela
Grant, Louis Russell, Jr
Grantham, Leroy Francis
Gray, Harry B
Greenberg, Elliott
Greene, Arthur Frederick, Jr
Greenstadt, Melvin
Greenstone, Arthur W
Grenda, Stanley C
Grieb, Merland William
Griffiths, James Edward
Griffo, Joseph Salvatore
Grim, Samuel Oram
Grimaldi, Frank Saverio
Grimes, Russell Newell
Grimley, Eugene Burhans, III
Griswold, Ernest
Griswold, Norman Ernest
Gritmon, Timothy F
Groenweghe, Leo Carl Denis
Grubbs, Edward Howard
Gryder, John William
Guenther, William Benton
Gump, J R
Gusenius, Edwin Maurtiz
Gushee, Beatrice Eleanor
Gustison, Robert Abdon
Guth, Egbert Karl Anton
Guymon, Ervin Park
Gwinup, Paul D
Gysling, Henry J
Haas, Charles Gustavus, Jr
Haas, Terry Evans
Haas, Walter Oskar, Jr
Hadzeriga, Pablo
Haendler, Helmut Max
Hafford, Bradford C
Hafner, Harold C
Hagen, Arnulf Peder
Haight, Gilbert Pierce, Jr
Haight, Howard Lewis
Haim, Albert
Haines, Roland Arthur
Hair, Michael L
Haisty, Robert W
Hakewill, Henry, Jr
Hale, William Henry, Jr
Hall, James Louis
Hall, Richard Eugene
Hall, Stephen Kenneth
Halliday, Robert William
Halpern, Jack
Hamerski, Julian Joseph
Hamilton, Hobart Gordon, Jr
Hamilton, James Beclone
Hammaker, Geneva Sinquefield
Hammer, Averill John
Hammer, Robert Nelson
Hammerstrom, Harold Elmore
Hampton, David Clark
Hansen, Lee Duane

Hansen, Robert Conrad
Hanson, Harold Nelson
Hanson, John Elbert
Hardcastle, Kenneth Irvin
Hare, Curtis R
Hares, George Bigelow
Harrington, Marion Thomas
Harrington, Roy Victor
Harris, Arlo Dean
Harris, Robert Hutchison
Hart, Haskell Vincent
Hart, William James
Hartman, John Stephen
Hartwell, George E, Jr
Hastie, John William
Hatch, Conrad V
Hatfield, William E
Haworth, Daniel Thomas
Hawthorne, Marion Frederick
Hayles, William Joseph
Haymore, Barry Lant
Hazen, Wayne Colby
Hazlehurst, David Anthony
Healy, Robert Michael
Heath, Roy Elmer
Heino, Walden Leo
Heitsch, Charles Weyand
Hellwege, Herbert Elmore
Hempel, Judith Cato
Hendricker, David George
Hentz, Forrest Clyde, Jr
Hepworth, John Leonard
Herlinger, Albert William
Hertenstein, Harold (Nelson)
Hertzberg, Elliot Paul
Hess, Richard William
Hess, Wendell Wayne
Heyman, Laurel Elaine
Hicks, Donald Gail
Higgins, Dorothy
Hill, Mary Ann Gertrude
Hill, Mary Mechtilde
Hill, Roy Dean
Hiller, Frederick W
Hintze, Ray Everald
Hirayama, Chikara
Hobrock, Don Leroy
Hodgson, Derek John
Hoffman, Lewis Charles
Hofman, Emil Thomas
Hogan, David James
Hohman, William H
Hohnstedt, Leo Frank
Holland, Hans J
Holmes, Robert Richard
Holoway, Michael O
Holt, Smith Lewis, Jr
Holtzclaw, Henry Fuller, Jr
Holzmann, Richard Thomas
Homan, Joseph M
Homman, Guy Burger
Honaker, Carl Boggess
Hood, Robert Luther
Hooper, Robert John
Hoover, William L
Hopkins, Paul Donald
Horner, Sally Melvin
Horner, William Wesley
Horner, Howard Chester
Horrigan, Philip Archibald
Horrocks, William DeWitt, Jr
Horwitz, Earl Philip
Hoster, Donald Paul
Hough, William Vernon
Houk, Clifford C
Houk, Larry Wayne
Howard, Charles
Howatson, John
Howell, John Emory
Huchital, Daniel H
Huckle, William George
Hudson, Robert McKim
Hufstedler, Robert Sloan
Hughes, Benjamin G
Hughes, William Bond
Huheey, James Edward
Humiec, Frank S, Jr
Hummers, William Strong, Jr
Hunt, Dominic Joseph
Hunt, Harold Russell, Jr
Hunt, John Baker
Hunt, John Philip
Hunt, Richard Lee
Hurley, Carl Robert
Hurley, Forrest Reyburn
Hurst, Peggy Morison
Huston, John Lewis
Hutto, Francis Baird, Jr
Hutton, Wilbert, Jr
Hyatt, David Ernest
Hyde, Kenneth E
Iloff, Phillip Murray, Jr
Interrante, Leonard V
Isied, Stephan Saleh
Izatt, Reed McNeil
Jabalpurwala, Kaizer E
Jache, Albert William
Jackman, Donald Coe

Jackson, Carey Birdsong
Jackson, Julius
Jackson, Margaret E
Jackson, William Morrison
Jacobs, Sanford
Jaecker, John Alvin
Jaeger, Ralph R
Jaffe, Philip Monlane
James, Brian Robert
Janzen, Alexander Frank
Jaquith, Richard Herbert
Jayne, Jerrold Clarence
Jekel, Eugene Carl
Jenkins, William A
Johannesen, Rolf Bradford
Johannsen, Herman Andrew
Joesten, Melvin D
Job, Robert Charles
Jezorek, John Robert
Jeremias, Charles George
Jenkins, Wilmer Atkinson, II
John, Lucille
Johnson, Dale A
Johnson, Frederic Allan
Johnson, K Jeffrey
Johnson, Malcolm Pratt
Johnson, Mary Frances
Johnson, Ronald Carl
Johnson, Tegner Albin
Johnston, Charles Paul
Johnston, William Jacob
Johnston, William Dwight
Jolly, William Lee
Jonassen, Hans Boegh
Jones, Mark Martin
Jones, Richard Evan, Jr
Jones, Thomas Evan
Jones, Wilber Clark
Jonte, John Haworth
Jordan, Robert Beatty
Joy, George Cecil, III
Joyner, Ralph Delmer
Jungbauer, Mary Ann
Kaczmarczyk, Alexander
Kallen, Thomas William
Kalnajs, Janis Arvids
Kanamueller, Joseph M
Kane-Maguire, Noel Andrew Patrick
Karayannis, Nicholas M
Karipides, Anastas
Karraker, Robert Harreld
Kask, Uno
Kauffman, George Bernard
Kaufman, Jay Victor Richard
Kay, Eric
Kestigian, Michael
Kidd, Robert Garth
Keeley, Dean Francis
Keenan, Charles William
Keenan, Thomas K
Kieft, Richard Leonard
Kifer, Edward W
King, Edward Frazier
King, Edward Louis
King, Reatha Clark
King, Robert Bruce
King, Walter Bernard
Kelmers, Andrew Donald
Kelly, Henry Curtis
Keller, Raymond Nevoy
Keller, Philip Charles
Keiter, Richard Lee
Kennelly, Mary Marina
Kenney, Malcolm Edward
Keppel, Charles Robert
Kern, Werner
Kirchner, Richard Martin
Kirsch, Warren Bernard
Kirschner, Stanley
Kiser, Robert Wayne
Kistenmacher, Thomas John
Kistner, Clifford Richard
Kivilghn, Herbert Daniel, Jr
Klabunde, Kenneth John
Klassen, David Morris
Klein, Donald Lee
Klein, Morton Joseph
Klein, Richard M
Kleinberg, Jacob
Kline, Robert Joseph
Kluiber, Rudolph W
Kluksdahl, Harris Eudell
Knight, James Aldon
Knockemus, Ward Wilbur
Knop, Charles Philip
Knop, Osvald
Knoth, Walter Henry, Jr
Knox, Kerro
Kodama, Goji
Koehler, William Henry
Koesche, Charles L
Koerner, Frederick William
Kogut, Leonard S
Kokalis, Soter George
Kokenge, Bernard Russell
Koknat, Friedrich Wilhelm
Komyathy, Joseph Charles

Korst, William Lawrence
Kotz, John Carl
Koubek, Edward
Kowalski, Stephen Wesley
Kraihanzel, Charles S
Krannich, Larry Kent
Krause, Adrienne Wickenden
Krause, Horatio Henry
Krause, Ronald Alfred
Kreidler, Eric Russell
Kriner, William Arthur
Krivak, Thomas Gerald
Kroger, James Harry
Krueger, Hanns H
Krueger, Paul Carlton
Kruse, Howard Wendell
Kruse, Walter
Kubas, Gregory Joseph
Kubota, Mitsuru
Kuempel, John Rickey
Kuhl, Gunter Hinrich
Kupel, Richard E
Kury, John William
Kust, Roger Nayland
Kustin, Kenneth
Kutal, Charles Ronald
Kuznesof, Paul Martin
Kwitowski, Paul Thomas
Kyker, Gary Stephen
Laferriere, Arthur L
Lafferty, Robert Hervey, Jr
Lagowski, Joseph John
Lalancette, Jean-Marc
Lamba, Ram Sarup
Lambert, Jack Leeper
Landesman, Herbert
Landis, Vincent J
Lange, Gerhard Paul
Langer, Horst Gunter
Lanoux, Sigred Boyd
La Prade, Marie Douglas
Larkins, Thomas Hassell, Jr
Larson, Edwin Merritt
Larson, Wilbur S
Latham, Ross, Jr
Latham, Wilbur R
Laubengayer, Albert Washington
Lauzau, Wilbur R
Layde, Durward Charles
Leaders, William M
Lefield, Robert Francis
Lever, Alfred B P
Levitt, Leonard Sidney
Levy, Ezra
Lewis, B Kenneth
Lewis, Leroy Crawford
Le Blanc, Robert Bruce
Le Duc, J Adrien Maher
Lee, Kah-Hock
Lee, Ralph Hewitt
Lee, Yat-Shir
Leffler, Amos J
Legal, Casimer Claudius, Jr
Lehman, Dennis Dale
Lehman, Duane Stanley
Lichtenstein, Ivan Edgar
Linck, Robert George
Ling, Harry Wilson
Linton, Robert Walter
Lipp, Steven Alan
Lippard, Stephen J
Lipsett, Solomon George
Lister, Maurice Wolfenden
Little, Robert Greenwood
Liu, Chong Tan
Liu, Chui Fan
Liu, Chui Hsun
Lo, George Albert
Lockhart, William Lafayette
Loeliger, David A
Lofquist, Marvin John
Lonadier, Frank Dalton
Long, George Gilbert
Long, Kenneth Maynard
Longhi, Raymond
Longo, John M
Love, Calvin Miles
Lucid, Michael Francis
Ludwick, Larry Martin
Luehrs, Dean C
Lussier, Roger Jean
Lustig, Max
Lutz, Charles William
Lynde, Richard Arthur
Lyon, Donald Wilkinson
MacDiarmid, Alan Graham
MacDonald, David J
Machel, Albert R
MacInnes, David Fenton, Jr
MacKenzie, Donald Hershey
Mackey, John Linn
Mackiw, Vladimir Nicolaus
Macklin, John Welton
Madan, Stanley Krishen
Madger, Jules
Magee, Charles Brian
Magee, John Storey, Jr
Magin, George Benedict, Jr

Magnell, Kenneth Robert
Magnuson, Vincent Richard
Magnuson, Winifred Lane
Mague, Joel Tabor
Maguire, Keith Dean
Mahler, Walter
Mahlman, Harvey Arthur
Maier, Thomas O
Maier, Mary Louise
Maik, Jim Gorden
Malik, John Michael
Malkin, Irving
Malone, Leo Jackson, Jr
Mamantov, Gleb
Mango, Frank Donald
Marganin, Dale William
Maricondi, Chris
Marianelli, Robert Silvio
Marking, Ralph H
Marley, James Aloysius
Marshall, Donald D
Marshall, Robert Herman
Martin, Dean Frederic
Martin, Donald Ray
Martin, George Lloyd
Martin, John Elmslie
Maruca, Robert Eugene
Maselli, James Michael
Mason, Caroline Faith Vibert
Mason, W Roy, III
Mast, Roy Clark
Matheson, Arthur Ralph
Mathew, Mathai
Mathiason, Dennis R
Matkovich, Vlado Ivan
Matsuguma, Harold Joseph
Mattair, Robert
Mattern, Kenneth Lawrence
Matuszko, Anthony Joseph
Maya, Leon
Mayer, Walter
Mayfield, Harold Gordon
Mayper, Stuart Allan
McAllister, Warren Alexander
McBride, William Robert
McCarley, Robert Eugene
McCarroll, William Henry
McClelland, Alan Lindsey
McColough, Fred, Jr
McCormick, Bailie Jack
McCullough, James Douglas
McCullough, John Franklin
McDonald, John E
McDugle, Woodrow Gordon, Jr
McElroy, Mary Kieran
McFarlin, Richard Francis
McGavock, William Crews
McGinnis, Charles Joseph
McGrath, John F
McLean, John A, Jr
McMahan, William H
McMordie, Warren C, Jr
McNaney, John A
McNutt, Ronald Clay
Mead, Edward Jairus
Mecay, William Lloyd
Mee, Jack Everett
Meek, Devon Walter
Meek, Violet Imhof (Mrs Devoon W)
Mehal, Edward Walter
Mehra, Mool Chand
Melaven, Arthur David
Mellon, Edward Knox, Jr
Meloni, Edward George
Melson, Gordon Anthony
Menashi, Jameel
Mentone, Pat Francis
Mericola, Francis Carl
Meshri, Dayaldas Tanumal
Messenger, Joseph Umiah
Metz, Florence Irene
Meyer, Herbert A
Meyer, Thomas J
Meyers, M Douglas
Mezey, Eugene Julius
Michael, Norman
Michmayr, Manfred
Middaugh, Richard Lowe
Mikulski, Chester Mark
Milburn, Ronald McRae
Miczarek, Chester J
Miles, Francis Turquand
Millard, George Buente
Miller, Arthur
Miller, Gordon Smith
Miller, Jack Martin
Miller, Joel Steven
Miller, Norman E
Miller, Paul Thomas
Miller, Phillip Allen
Mills, Jerry Lee
Milne, John B
Miner, Frend John
Miniatas, Birute Ona
Minne, Ronn N
Mitra, Grihapati

Moberly, Lawrence Ervin
Moczygemba, George A
Mode, Vincent Alan
Moeller, Carl William, Jr
Moeller, Therald
Molnar, Stephen P
Moody, David Coit, III
Moore, Edward Lee
Moore, John Ward
Moore, Lawrence Edward
Moore, Michael Cabot
Moore, Thomas Edwin
Moran, Edward Francis, Jr
Morehouse, Sheila McEnness
Morgan, George L
Morgan, Wyman
Morrey, John Ralph
Morris, Albert Gregory
Morris, Joseph Grant
Morrison, William Alfred
Morrison, William Harvey, Jr
Morrow, Scott
Morse, Karen W
Mosbo, John Alvin
Moss, Frank Anthony James
Moss, Herbert Irwin
Movius, William Gust
Moyer, James Ward
Moyer, John Raymond
Moyer, Ralph Owen, Jr
Muetterties, Earl Leonard
Mulay, Laxman Nilakantha
Muller, Olaf
Murbach, Earl Wesley
Murch, Robert Matthews
Murmann, R Kent
Murphy, Clarence John
Murray, Edward Conley
Musgrave, Ted Russell
Myers, Clifford Earl
Myers, William Howard
Naeser, Charles Rudolph
Nagel, Terry Marvin
Nakon, Robert Steven
Nancollas, George H
Nathan, Lawrence Charles
Neal, John Alexander
Nebergall, William Harrison
Nebgen, John William
Nechamkin, Howard
Neckers, James Warwick
Nehls, James Warwick
Neithamer, Richard Walter
Nelson, Gregory Victor
Nelson, John Henry
Nelson, Wilfred H
Neptune, John Addison
Nesbitt, Lyle Edwin
Netherton, Lowell Edwin
Neumann, Henry Matthew
Neuvar, Erwin W
Newkirk, Herbert William
Newlands, Michael John
Newton, Thomas William
Newton, William Edward
Ng, George
Ng, George
Nicholson, Douglas Gillison
Nicholson, Geoffrey Charles
Niedenzu, Kurt
Nikodem, Robert Bruce
Niu, Joseph H Y
Noack, Manfred Gerhard
Noftle, Ronald Edward
Noller, David Conrad
Nordmeyer, Francis R
Nordquist, Paul Edgard Rudolph, Jr
Norman, Andrea Hausman
Norman, Arlan Dale
Norris, A R
Nowotny, Kurt A
Nyman, Carl John
Obear, Frederick W
Oberle, Thomas M
O'Brien, James Francis
O'Brien, Thomas Doran
Odom, Luke Ralph
Odone, Jerome David
Oei, Djong-Gie
Oestreich, Charles Henry
Ogle, Pearl Rexford, Jr
Oliver, John Preston
Olander, James Alton
Olson, Maynard Victor
Onak, Thomas Philip
Onstott, Edward Irvin
Onyszchuk, Mario
Opitz, Herman Ernest
Ortego, James Dale
Ostap, Stephen
Ostdick, Thomas
Osthoff, Robert Charles
Otley, Kurt O
Overton, Edward Beardslee
Owen, James Emmet
Owen, John Harding
Paine, Robert Treat, Jr

Palilla, Frank C
Palmer, Bryan D
Palmer, Jay
Palmer, Richard Alan
Park, Kyu Chang
Parker, William Evans
Parry, Robert Walter
Parsons, James Bayard
Parsons, Theran Duane
Partenheimer, Walter
Partridge, Jerry Alvin
Passmore, Jack
Patmore, Edwin Lee
Patterson, Howard Hugh
Patterson, Truett Clifton
Patton, Armine Deane
Paul, Armine Deane
Pavkovic, Stephen F
Paxson, John Ralph
Payne, Dwight Arthur
Peach, Michael Edwin
Peard, William John
Pearson, Robert Stanley
Peavler, Robert Jean
Pence, Harry Edmond
Pennington, David Eugene
Penneman, Robert Allen
Peppard, Donald Francis
Perlow, Mina Rea Jones
Perry, William Daniel
Perumareddi, Jayarama Reddi
Peters, Till Justus Nathan
Petersen, John David
Peterson, Louis K
Phillips, Bert
Phillips, John R
Pickral, George Monroe, Jr
Pierpont, Cortlandt Godwin
Pignolet, Louis H
Pilger, Richard Christian, Jr
Pinnavaia, Thomas J
Pinnell, Robert Peyton
Pitha, John Joseph
Pizzolato, Philip Joseph
Place, Robert Daniel
Pleasants, Elsie W
Plymale, Donald Lee
Poe, Anthony John
Poetz, Robert George
Pohlmann, Hans Peter
Poland, Lloyd Orville
Poncha, Rustom Pestonji
Pond, Judson Samuel
Pope, Michael Thor
Popov, Alexander Ivan
Popp, Carl John
Popp, Gerhard
Porter, Vernon Ray
Porterfield, William Wendell
Poskozim, Paul Stanley
Post, Elroy Wayne
Postmus, Clarence, Jr
Potenza, Joseph Anthony
Potter, Norman D
Potts, John Calvin
Powell, Robert Allen
Powell, Howard B
Powell, Jack Edward
Preer, James Randolph
Pribble, Mary Jo
Price, Harry James
Pundsack, Frederick Leigh
Purcell, Keith Frederick
Purdie, Neil
Pylewski, Louis Lawrence
Quagliano, James Vincent
Quail, John Wilson
Quane, Denis Joseph
Quass, La Verne Carl
Quill, Laurence Larkin
Quinlan, Kenneth Paul
Quinn, Lawrence Paul
Radimer, Kenneth John
Radzikowski, M St Anthony
Ragsdale, Ronald O
Raisen, Elliott
Rakestraw, Lawrence Frederick
Ramaswamy, H N
Ramsey, William James
Randall, John J, Jr
Randall, Rogers Ellis, Sr
Rasmussen, Russell Lee
Ratcliffe, Charles Thomas
Raymond, Kenneth Norman
Rayner-Canham, Geoffrey William
Reade, Richard Francis
Reagan, William Joseph
Reed, A Thomas
Reed, Christopher Alan
Reeder, Ray Robert
Reiff, William Michael
Reinbold, Paul Earl
Reinhardt, Richard Alan
Reis, Arthur Henry, Jr
Reisman, Arnold
Rempel, Garry Llewellyn
Resnik, Robert Kenneth
Reynard, John William

Reynard, Kennard Anthony
Reynolds, Warren Lind
Rhees, Raymond Charles
Rhein, Robert Alden
Rice, William Edward
Rich, Ronald Lee
Richardson, Mary Frances
Richason, George R, Jr
Richter, Edward Eugene
Richter, G Paul
Richter, Harold Gene
Rila, Charles Clinton
Riley, John Thomas
Ring, Morey Abraham
Risen, William Maurice, Jr
Ritcey, Gordon M
Ritchey, John Michael
Ritter, David Moore
Ritter, Ronald Dale
Rivela, Louis John
Robbins, Murray
Roberts, Charles Brockway
Robertson, James Allen
Robertson, Wilbert Joseph, Jr
Robinson, Edward Arthur
Robinson, Martin Alvin
Robinson, William Robert
Rodesiler, Paul Frederick
Rodgers, Glen Ernest
Rogers, Donald Richard
Rohrback, Gilson Henry
Rohrer, Charles Stephen
Rollins, Orville Woodrow
Rollmann, Louis Deane
Root, Charles Brian
Rorabacher, David Bruce
Rosenberg, Richard Martin
Rosser, Edward Barry
Rothermel, Daphne Land
Roubal, Ronald Keith
Royer, Donald Jack
Rubenstein, Martin
Rubino, Andrew M
Rudd, DeForest Porter
Rudolph, Ralph W
Rudzitis, Edgars
Ruff, John K
Rulfs, Charles Leslie
Rumfeldt, Robert Clark
Rumpel, Max Leonard
Rund, John Valentine
Rupp, John Jay
Russ, Charles Roger
Russell, John Blair
Russell, Virginia Ann
Rustad, Douglas Scott
Ryan, Douglas Earl
Ryan, Jack Lewis
Ryan, James Arthur
Rycheck, Mark Rule
Ryschkewitsch, George Eugene
Saalfeld, Fred Eric
Sacks, Lawrence J
Sadana, Yoginder Nath
Safford, Edward LaPorte
Sager, Ray Stuart
Saillant, Roger Barry
Salmon, James F
Salot, Stuart Edwin
Sams, Lewis Calhoun, Jr
Sanders, Robert N
Sanderson, Robert Thomas
Sanger, Alan Rodney
Sangster, Raymond Charles
Saraceno, Anthony Joseph
Sauer, Dennis Theodore
Savage, Robert O, Jr
Sawin, Stephen Sanford
Scaife, Charles Walter John
Schaeffer, Riley
Schafer, Lothar
Schaff, Mary Ellen
Schelar, Virginia Mae
Schenkenberg, Philip Rawson
Schiefelbein, Benedict
Schlemper, Elmer Otto
Schmeckenbecher, Arnold F
Schmitt, Anthony Paul
Schmulbach, Charles David
Schnable, George Luther
Schnepfe, Marian Moeller
Schoonover, Irl Corley
Schram, Eugene P
Schrauzer, Gerhard N
Schreiner, Anton Franz
Schreiner, Felix
Schubert, Leo
Schuele, William John
Schug, Kenneth
Schugar, Harvey
Schultz, Arthur Ray
Schultz, Donald Raymond
Schultz, Linda Dalquest
Schulz, Wallace Wendell
Schumn, Robert Allen
Schwan, Theodore Carl
Schweitzer, George Keene
Schweizer, Albert Edward

Scott, Arthur Ferdinand
Scott, Bruce Albert
Scott, Earle Stanley
Sears, Curtis Thornton, Jr
Sears, Mildred Bradley
Seaton, Jacob Alif
Selbin, Joel
Senior, John Brian
Senoff, Caesar V
Seufert, Ludwig E
Seufzer, Paul Richard
Shannon, Robert Day
Sharkey, John Bernard
Sharp, Kenneth George
Shaver, Richard Cornell
Shaw, C Frank, III
Shaw, Wilfrid Garside
Sheard, John Leo
Sheppard, John Clarence
Sherer, Sankey
Sheridan, Peter Sterling
Sherockman, Andrew Antolcik
Sherwood, Calder Smith, III
Shin, Yong Ae Im
Shineman, Richard Shubert
Shinn, Dennis Burton
Shoemaker, Carlyle Edward
Shore Sheldon Gerald
Shozda, Raymond John
Shriver, Duward F
Sicilio, Fred
Siebring, Barteld Richard
Siefert, August Carl
Siefker, Joseph Roy
Siegel, Bernard
Sievers, Robert Eugene
Silber, Herbert Bruce
Silverman, Morris Bernard
Simplicio, Jon
Singleton, David Michael
Sink, Donald Woodfin
Sistrunk, Thomas Olloise
Skelcey, James Stanley
Slater, James Louis
Slepetys, Richard Algimantas
Slinkard, William Earl
Slotter, Richard Arden
Smalley, Edmund Walter
Smart, James Conrad
Smith, Arthur Leo
Smith, Herbert L
Smith, Joseph Harold
Smith, Robert Kingston
Smith, Rodger Chapman
Smith, Wayne Lee
Smithson, George Raymond, Jr
Snavely, Fred Allen
Sobon, Leon Edward
Soderberg, Roger Hamilton
Solomon, Irvine Jerome
Sopp, Samuel William
Sorensen, David T
Southern, Thomas Martin
Sowards, Donald Maurice
Spanier, Edward J
Spees, Steven Tremble, Jr
Spence, Jack Taylor
Spencer, Jesse G
Spielman, John Russel
Spielvogel, Bernard Franklin
Spike, Clark Ghael
Splitgerber, George H
Sprague, Robert W
Springer, Charles S, Jr
Stange, Hugo
Staniforth, Robert Arthur
Stanitski, Conrad Leon
Stanko, Joseph Anthony
State, Harold M
Stear, Adrian N
Stearns, Robert Inman
Steele, Jack
Stehly, David Norvin
Stein, Lawrence
Steinbrecher, Lester
Steindler, Martin Joseph
Steinman, Robert
Stephen, Keith H
Steunenberg, Robert Keppel
Stevens, Laurence Guy
Steward, Omar Waddington
Stingl, Hans Alfred
Stoker, Howard Stephen
Stone, Bobbie Dean
Stonehouse, Albert James
Storhoff, Bruce Norman
Stoufer, Robert Carl
Strahs, Gerald
Strange, Ronald Stephen
Strathdee, Graeme Gilroy
Stratton, Carol
Straub, Darel K
Strecker, Harold Arthur
Strommen, Dennis Patrick
Stuckey, John Edmund
Stucky, Galen Dean
Suchow, Lawrence
Sujishi, Sei

Sullivan, Edward Augustine
Sullivan, Mary Louise
Sutton, Derek
Swanson, Basil Ian
Swartz, Marjorie Louise
Sweeny, Daniel Michael
Sweet, Roger George
Swinehart, Carl Francis
Swinehart, James Herbert
Tanaka, John
Tannenbaum, Stanley
Tappmeyer, Wilbur Paul
Tarr, Donald Arthur
Taube, Henry
Taylor, Donald Francis
Taylor, Moddie Daniel
Taylor, Robert Craig
Tecotzky, Melvin
Teggins, John E
Templeton, John Charles
Tennant, Charles Beard
Terada, Kazuji
Ter Haar, Gary L
Tharp, A G
Thayer, John Stearns
Theriot, Leroy James
Thibeault, Jack Claude
Thoma, Roy E
Thomas, Forrest Dean, II
Thomas, Frank Harry
Thomas, William Benjamin
Thompson, David Wallace
Thompson, Gary Haughton
Thompson, James Charlton
Thompson, Lancelot Churchill Adalbert
Thompson, Larry Clark
Thompson, Major Curt
Thompson, Martin Leroy
Thompson, Phillip Gerhard
Thompson, Richard Claude
Thurner, Joseph John
Tillay, Eldrid Wayne
Titus, Donald Dean
Todd, Lee John
Tolman, Chadwick Alma
Tomic, Ernst Alois
Tomlinson, Richard Howden
Toney, Joe David
Toogood, Gerald Edward
Torop, William
Torp, Bruce Alan
Tracy, Joseph Walter
Treffner, Walter Sebastian
Treichel, Paul Morgan, Jr
Treptow, Richard S
Trevorrow, Laverne Everett
Trigg, William Walker
Trimble, Russell Fay
Tripathi, Uma Prasad
Truex, Timothy Jay
Tschudi, Wilbur James
Tsigdinos, George Andrew
Tuck, Dennis George
Turner, Almon George, Jr
Tuve, Richard Larsen
Tyree, Sheppard Young, Jr
Tyson, George Noblit, Jr
Ucko, David A
Ulrich, William Frederick
Upchurch, Donald Gene
Urbach, Frederick Lewis
Urdy, Charles Eugene
Urry, Grant Wayne
Utke, Allen R
Vallee, Richard Earl
Van Alten, Lloyd
Van Artsdalen, Ervin Robert
Vance, Robert Floyd
Van Doorne, William
Van Dyke, Charles H
Van Meter, Wayne Paul
Van Osdall, Thomas Clark
Vaska, Lauri
Vaughn, Joe Warren
Venezky, David Lester
Vernon, Gregory Allen
Villa, Juan Francisco
Vinal, Richard S
Vinton, William Howells
Violante, Michael Robert
Vogel, Glenn Charles
Vogt, Lester Herbert, Jr
Von Winbush, Samuel
Wade, Robert Charles
Wadley, Margil Warren
Waggoner, William Cole
Waggoner, William Horace
Wagner, Ross Irving
Wainer, Eugene
Waldo, Willis Henry
Walker, Ian Munro
Wallace, William J
Waller, Mary Concetta
Walmsley, Frank
Walmsley, Judith Abrams
Walsh, Edward John
Walter, Joseph L

CHEMISTRY, INORGANIC

Walter, Paul Hermann Lawrence
Wampler, Dale Lee
Wamser, Christian Albert
Wang, Jin Tsai
Ward, Edward Hilson
Ward, Laird Gordon Lindsay
Wardeska, Jeffrey Gwynn
Warf, James Curen
Warren, Clarence Gerald
Warren, Herbert Dale
Wartik, Thomas
Wasfi, Sadiq Hassan
Wasson, John R
Watkins, Kay Orville
Watt, William Joseph
Watters, Kenneth Lynn
Waterson, Kenneth Lynn
Wayland, Bradford B
Weaver, John Arthur
Weber, Charles Walton
Weber, James Harold
Webster, Clyde Leroy, Jr
Webrecht, Walter Eugene
Weick, Charles Frederick
Weidig, Charles Ferdinand
Weil, Thomas Andre
Weiser, David W
Weiss, Gerald S
Weiss, Harold Gilbert
Weissberger, Edward
Weisz, Robert Stephen
Welch, Garth Larry
Welch, Melvin Bruce
Weller, Paul Franklin
Wendlandt, Wesley W
Wentworth, Rupert A D
Werneke, Michael Francis
Wertz, David Lee
West, Douglas Xavier
Westland, Alan Duane
Westmore, John Brian
Wexell, Dale Richard
Wheatland, David Alan
Wheelock, Kenneth Steven
White, Alan Jonathon
White, David Gover
White, June Broussard
Whiteway, Stirling Giddings
Whitaker, Mack Page
Whitten, Kenneth Wayne
Nicholas, Mark L
Wickham, Donald G
Wickham, James Edgar, Jr
Wiedemeier, Herbert
Wiersema, Richard Joseph
Wiggins, James William
Wiles, Donald Roy
Wilhelm, Dale Leroy
Wik, William David
Wilkes, Glenn Richard
Wilkie, Charles Arthur
Wilkins, Ralph G
Willard, Thomas Maxwell
Williams, Colin James
Williams, Donald Howard
Williams, Jack Marvin
Williams, Loring Rider
Williams, Oren Francis
Williamson, Frank Shaver, Jr
Willis, Christopher John
Willis, William Van
Wilson, Byron J
Wilson, Dean George
Wilson, John Shirley
Wilson, Lauren R
Wilson, Lauren Edward
Wingeleth, Dale Clifford
Winkley, Donald Charles
Wiseman, George Edward
Witmer, William Byron
Wittle, John Kenneth
Wold, Aaron
Woltermann, Gerald M
Wood, James Lee
Woods, Joe Darst
Woods, Mary
Woodward, Glenn Jones
Workman, Marcus Orrin
Wrathall, Jay W
Wrighton, Mark Stephen
Wyma, Richard J
Wymore, Charles Elmer
Wynne, Kenneth Joseph
Yalman, Richard George
Yamauchi, Masanobu
Yang, Julie Chi-Sun
Yingst, Ralph Earl
Yocom, Perry Niel
Yoder, Claude H
Yoke, John Thomas
Yost, Don M
Young, Archie Richard, II
Young, Donald C
Zabin, Burton Allen
Zack, Neil Richard
Zatko, David A
Zeldin, Martel

Zeleznik, Pauline
Ziegler, Robert G
Zielen, Albin John
Zimmerman, John Gordon
Zingaro, Ralph Anthony
Ziolo, Ronald F
Zompa, Leverett Joseph
Zuckerman, Jerold J
Zundahl, Steven Stanford

Physical Inorganic Chemistry

Adin, Anthony
Anbar, Michael
Anderson, Charles Thomas
Armor, John N
Austin, Alfred Ells
Baca, Glenn
Bacon, Frank Rider
Baes, Charles Frederick, Jr
Bailin, Lionel J
Bains, Malkiat Singh
Bamberger, Carlos Enrique Leopoldo
Bancroft, George Michael
Barber, Eugene John
Barney, Gary Scott
Beckering, Willis
Begun, George Murray
Belitskus, David
Bernays, Peter Michael
Birdwhistell, Ralph Kenton
Blank, Charles Anthony
Blau, Henry Hess
Boardman, William Walter, Jr
Bonnell, David William
Bourgon, Marcel
Brand, John Robert
Brault, Albert Thomas
Bredig, Max Albert
Brown, Patrick Michael
Brunschwig, Bruce Samuel
Bulloff, Jack John
Burger, Leland Leonard
Byler, David Michael
Campbell, Larry Edwin
Celiano, Alfred
Chaffee, Eleanor
Chakrabarty, Manoj R
Chock, Ernest Phayman
Clayton, John Charles (Hastings)
Clough, Francis Bowman
Clough, Philip James
Colburn, Charles Buford
Coleman, William Fletcher
Court, Anita
Cubicciotti, Daniel David, Jr
Dahl, Lawrence Frederick
Darnell, Alfred Jerome
Debye, Nordulf Wiking Gerud
DeLorenzo, Ronald Anthony
Downey, Joseph Robert, Jr
Dutta-Ahmed, Akhtar
Earley, Joseph Emmett
Eachus, Raymond Stanley
El-Awady, Abbas Abbas
Erbacher, John Kornel
Falk, Charles David
Fehner, Thomas Patrick
Ferrante, Michael John
Fischer, Albert Karl
Fisher, James Russell
Fleischauer, Paul Dell
Fogel, Norman
Fox, Robert Kriegbaum
Fox, William B
Franco, Nicholas Benjamin
Fryxell, Robert Edward
Garrett, Michael Benjamin
Green, Agnes Ann
Greenhouse, Harold Mitchell
Greifer, Aaron Philip
Grunthaner, Frank John
Guertin, Jacques F
Haden, Walter Linwood, Jr
Hahn, Harold Thomas
Hammer, Robert Russell
Harris, Gordon McLeod
Harford, Winslow Hopper
Hicks, Kenneth Ward
Hoffman, Alan Bruce
Hoffman, Charles John
Hoffman, Morton Z
Hoggard, Patrick Earle
Homeier, Edwin H, Jr
House, James Evan, Jr
Hoppenjans, Donald William
Howell, Peter Adam
Howell, Harold Eugene
Hubbard, Richard Alexander, II
Hudson, Alice Brandon
Hutchison, James Robert
Iannuzzi, Melanie Mary
Israeli, Julius Yigal
Jackowitz, John Franklin
James, Dean B
Johnson, David Alfred
Kafalas, James A
Kaplan, Roy Irving
Katzin, Leonard Isaac
Keder, Wilbert Eugene

Keeler, Robert Adolph
Kelley, Kenneth K
Kendig, Martin William
King, James P
King, Thomas Morgan
Kirshenbaum, Abraham David
Klein, Philipp Hillel
Klingen, Theodore James
Kokoszka, Gerald Francis
Kowalak, Albert Douglas
Koziowski, Theodore R
Kuczkowski, Robert Louis
Langford, Cooper Harold, III
Laraeh, Simon
Lawless, Edward William
Lazarus, Marc Samuel
Lefkowitz, Stanley A
Lintvedt, Richard Lowell
Lock, Colin James Lyne
Long, Gary John
Lyon, Luther Lawrence, Jr
Macalady, Donald Lee
Magnuson, Lawrence Bersell
Manasevit, Harold Murray
Margrave, John Lee
Martin, Don Stanley, Jr
May, Walter Ruch
McCarthy, Paul James
McDonald, Hector O
McGill, Robert Mayo
McNamara, John Edward
Mercer, Edward Everett
Meriden, Charles Waymond
Meyer, Carl Beat
Milligan, Winfred Oliver
Moore, Robert Earl
Morris, Kelso Bronson
Morrow, Jack I
Moss, Lester Robert
Moss, Robert Henry
Myers, Ralph Thomas
Neely, James
Norris, Thomas Hughes
O'Neill, Richard Thomas
Orebaugh, Errol Glenn
Osterheld, Robert Keith
Owens, Clifford
Pagenkopf, Gordon K
Panzer, Richard Earl
Pasternack, Robert Francis
Petersen, Harold, Jr
Petersen, Jeffrey Lee
Peterson, Joseph Richard
Petrocelli, Americo W
Phelan, Nelson Flagge
Piehl, Donald Herbert
Pinch, Harry Louis
Porter, Richard Francis
Post, Roy G
Potts, Melvin Lester
Quinn, Rod King
Radtke, Douglas Dean
Rasmussen, Paul
Readnour, Jerry Michael
Reed, William Robert
Reising, Richard F
Rhyne, Thomas Crowell
Ringwald, Owen Edward
Roberts, Elliott John
Roscoe, John Stanley, Jr
Rosenthal, Michael R
Rowley, David Alton
Schildcrout, Steven Michael
Schmidt, Donald Dean
Schnizlein, John Glenn
Sen, Buddhadev
Sherry, Howard S
Shoup, Robert D
Sienko, Michell Joseph
Silverman, Meyer David
Sinclair, William Robert
Skinner, James F
Sloan, Gilbert Jacob
Smardzewski, Richard Roman
Smart, James Blair
Smith, Jean Gillen
Smith, Jerry Joseph
Solomon, Edward I
Sprowles, Jolyon Charles
Stafford, Fred E
Stagg, William Ray
Staples, Bert Roland
Stern, Kurt Heinz
Sullivan, William Francis
Sutin, Norman
Swift, Harold Eugene
Tamres, Milton
Teague, Marion Warfield
Thomas, David Gilbert
Thomas, John Paul
Thompson, Joseph Kyle
Thompson, Mary E
Tong, James Ying-Peh
Tucci, Edmond Raymond
Uchtman, Vernon Albert
Vander Wall, Eugene
Vandersport, Thomas Henry
Vigee, Gerald S

Voorhoeve, Rudolf Johannes Herman
Wason, Satish Kumar
Waugh, John Lodovick Thomson
Webb, Thomas Joseph, Jr
Weeks, Thomas Norman
Welch, Cletus Norman
Wells, James Edward
Weschler, Charles John
Westman, Albert Ernest Roberts
Whitfill, Jesse Edmund
Wicker, Donald Lee
Wicker, Robert Kirk
Williams, Rickey Jay
Willson, Donald Bruce
Woods, Clifton, III
Worrell, Jay H
Wright, Kenneth James
Yeats, LeRoy Brough, Jr
Yellin, Wilbur
Yingst, Harvey Austin
Yunker, Wayne Harry
Zimmerman, Donald Nathan
Zipp, Arden Peter

CHEMISTRY, ORGANIC

Organic Chemistry

Aaron, Herbert Samuel
Aaronoff, Burton Robert
Abashian, Steven
Abbott, John Richards
Abegg, Victor Paul
Abell, Jared
Abell, Paul Irving
Abend, Phillip Gary
Aberhart, Donald John
Abramovitch, Rudolph Abraham Haim
Abrash, Henry J
Abu-Hajj, Jusuf J
Acerbo, Samuel Nicholas
Ackermann, Hervey Winfield, Jr
Ackermann, Joseph Francis
Ackerman, Guenter Rolf
Ackerman, Robert George
Adamcik, Joe Alfred
Adams, Alden Ross
Adams, Elizabeth
Adams, John Howard
Adams, Kenneth Allen Harry
Adams, Kenneth Howard
Adams, Leon Milton
Adams, Phillip
Addinall, Carl Rupert
Addison, Leslie Mandeville
Adduci, Jerry M
Adickes, H Wayne
Adler, Irving Larry
Adolph, Horst Guenter
Afonso, Adriano
Aft, Harvey
Aftergut, Siegfried
Agnello, Eugene Joseph
Agosta, William Carleton
Aguiar, Adam Martin
Ainbinder, Zarah
Ainsworth, Cameron
Akawie, Richard Isidore
Alaimo, Robert J
Alapovic, Petar
Albach, Roger Fred
Albers-Schonberg, Georg
Alberts, Arnold A
Albright, James Andrew
Albright, Jay Donald
Albright, Robert Lee
Aldous, Duane Leo
Aldrich, Paul E
Aldridge, Mary Hennen
Alexander, Benjamin H
Alexander, Edward Cleve
Alexander, Ernest John
Alferi, Charles C
Alford, Harvey Edwin
Allan, George Graham
Allen, Charles Francis Hitchcock
Allen, Charles Freeman
Allen, Duff Shederic, II
Allen, Lewis Edwin
Allen, Paul, Jr
Allinger, Norman Louis
Allison, Jerry Robert
Allison, John Arthur Charles
Allison, John P
Allred, Evan Leigh
Alm, Robert M
Almond, Harold Russell, Jr

Alper, Howard
Alt, Gerald Horst
Althuis, Thomas Henry
Altland, Henry Wolf
Altman, Lawrence Jay
Altscher, Siegfried
Altwicker, Elmar Robert
Amai, Robert Lin Sung
Ambelang, Joseph Carlyle
Ambrus, Laszlo
Ammon, Herman L
Amstutz, Edward Delbert
Amundsen, Lawrence Hardin
Andersen, Kenneth K
Andersen, Niels Hjorth
Anderson, Albert Edward
Anderson, Amos Robert
Anderson, Arthur Bernhardt
Anderson, Arthur G, Jr
Anderson, Curtis Benjamin
Anderson, Floyd Edmond
Anderson, George Washington
Anderson, Gloria Long
Anderson, Hugh Verity
Anderson, LeRay J
Anderson, Paul LeRoy
Anderson, Richard C
Anderson, Robert Christian
Anderson, Robert Griffin
Anderson, Roy Scott
Anderson, Wayne Keith
Andresen, Brian Dean
Andrews, Eugene Raymond
Anet, Frank Adrien Louis
Angel, Henry Seymour
Angeline, John Frederick
Angelo, Rudolph J
Angerer, John David
Angier, Robert Bruce
Ariyan, Zaven S
Arkell, Alfred
Anschel, Morris
Anselme, Jean-Pierre L M
Anspon, Harry Davis
Ansul, Gerald R
Anthes, John Allen
Antia, Naval Jamshedji
Antonius (Kennelly), Mary
Applequist, Douglas Einar
Applewhite, Thomas Hood
Apsimon, John W
Arcesi, Joseph A
Archer, Robert Allen
Arganbright, Robert Philip
Argoudelis, Chris J
Ariemma, Sidney
Arimoto, Fred Shunji
Arit, Herbert George, Jr
Armbruster, Charles William
Armour, Eugene Arthur
Armstrong, Paul Douglas
Armstrong, Robert Krick
Armstrong, William Lawrence
Arnold, Allen Parker
Arnold, Donald Robert
Arntzen, Clyde Edward
Arrington, Jack Phillip
Artman, Neil Ross
Asato, Goro
Ash, Arthur Burr
Ashburn, Gilbert
Ashby, Bruce Allan
Ashby, Eugene Christopher
Ashe, Arthur James, III
Ashford, Theodore Askounes
Ashford, Walter Rutledge
Ashley, Warren Cotton
Ashton, Joseph Benjamin
Aspelin, Gary Bertil
Aspinall, Gerald Oliver
Aspinall, Samuel Rusmisell
Assony, Steven James
Astle, Melvin Jensen
Atassi, Zouhair
Atchley, Ralph Warren
Athey, Robert Jackson
Atkins, Kenneth Earl
Atkins, Robert Charles
Atkins, Ronald Leroy
Atkinson, Edward Redmond
Atkinson, Joseph George
Attaway, John Allen
Atwater, Norman Willis
Atwell, William Henry
Atwood, Jerry Lee
Au, Andrew Taichiu
Aue, Donald Henry
Aue, Walter Alois
Auerbach, Irving
Auerbach, Melvin
Aufdermarsh, Carl Albert, Jr
Augl, Joseph Michael
Augustine, Robert Leo
Ault, Addison
Austin, Paul Rolland
Austin, Thomas Howard
Autenrieth, John Stork
Autrey, Robert Luis

Auyang, King
Aviram, Ari
Avonda, Frank Peter
Axelrad, George
Axen, Udo Friedrich
Axenrod, Theodore
Ayer, William Alfred
Ayers, Orval Edwin
Baarda, David Gene
Babad, Harry
Babb, Robert Massey
Babcock, John Claude
Babson, Robert Daniel
Bach, Frederick Louis
Bach, Shirley
Bacha, John D
Bachelor, Frank William
Bachman, Gerald Lee
Bachman, Paul Lauren
Baclawski, Leona Marie
Bacskai, Robert
Badding, Victor George
Bader, Alfred Robert
Bader, Henry
Badertscher, Darwin Earl
Badger, Rodney Allan
Baechler, Raymond Dallas
Baer, Erich
Baer, Hans Helmut
Baer, Massimo
Bagby, Marvin Orville
Bagli, Jenanbux Framroz
Bahner, Carl Tabb
Baiamonte, Vernon D
Bailes, Richard Hazel
Bailey, David Tiffany
Bailey, Dennis Mahlon
Bailey, Donald Leroy
Bailey, Edward Thomas
Bailey, Philip Sigmon
Bailey, Thomas Daniel
Bailey, William Francis
Bailey, William John
Baird, Merton Denison
Baisted, Derek John
Baizer, Manuel M
Baizer, William Xavier
Bak, David Arthur
Baker, Don Robert
Baker, Earl Wayne
Baker, Frank Weir
Baker, Joseph Willard
Baker, Leonard Morton
Baker, Robert Henry
Bakker, Gerald Robert
Baldoni, Andrew Ateleo
Baldwin, Jack Edward
Baldwin, Maynard Martin
Baldwin, Roger Allan
Baldwin, Willis Harford
Baldwin, Winfield Morgan, Jr
Ball, Frances Louise
Ball, Frank Jervery
Ball, Fred
Ball, Lawrence Ernest
Ball, William Lee
Ballina, Rudolph August
Ballinger, Peter Richard
Balmer, Clifford Earl
Balmer, Louis Whiteside
Balsley, Richard Benjamin
Baltazzi, Evan S
Balthazor, Terrell Mack
Balthis, Joseph Hendrickson, Jr
Banigan, Thomas Franklin, Jr
Banitt, Elden Harris
Bankert, Ralph Allen
Banks, Clarence Kenneth
Banks, Harold Douglas
Banks, Richard C
Bann, Robert (Francis)
Bannerman, Robert Alexander Brock
Bannister, Brian
Bar, Hans-Peter
Barabas, Eugene S
Barclay, Lawrence Ross Coates
Bardolph, Marinus Peter
Bare, Paul Orville
Bariana, Dilbagh Singh
Barker, George Ernest
Barker, Harold Clinton
Barker, Marvin Windel
Barker, Richard Gordon
Barkey, Kenneth Thomas
Barkhurst, Rodney Charles
Barkley, Lloyd Blair
Barks, Paul Allan
Barnes, Charlie James
Barnes, Carl Edmund
Barnes, David Kennedy
Barnes, Garrett Henry, Jr
Barnes, Robert Keith
Barnett, Roderick Arthur
Barnett, Charles Jackson
Barnett, Kenneth Wayne
Barnum, Emmett Raymond
Baron, Frank A

Barone, John A
Barr, John Tilman
Barreras, Raymond Joseph
Barrett, Edward Joseph
Barrett, Mary Olivia
Barron, Eugene Roy
Barrueto, Richard Benigno
Barry, Arthur John
Barry, Guy Thomas
Barry, Roger Donald
Bartels, George William, Jr
Bartels-Kieth, James Richard
Barton, Kenneth Ray
Barton, Lucian Anthony
Barton, Thomas J
Bartron, Lester Ray
Bartsch, Richard Allen
Bashour, Joseph Tamir
Basinski, John Edward
Bass, Robert Gerald
Bassett, Joseph Yarnall, Jr
Bassler, Gerald Clayton
Batcho, Andrew David
Bateman, John Hugh
Bates, Robert Brown
Batorewicz, Wadim
Bauchwitz, Peter Siegbert
Bauer, Albert Webb
Bauer, Frederick William
Bauer, Ludwig
Bauer, Richard G
Bauer, Ronald Sherman
Bauer, Stewart Thomas
Bauer, William, Jr
Bauld, Nathan Louis
Baum, Arthur Aloysius
Baum, Bruton Murry
Baum, John William
Bauman, Bernard D
Baumann, Robert Andrew
Baumgarten, Jacob Bruce
Baumgarten, Ronald J
Baumgartner, George Julius
Baur, Fredric John, Jr
Bausher, Larry Paul
Baxter, Warren Nesmith
Bayer, Horst Otto
Baylor, Charles, Jr
Baylouny, Raymond Anthony
Bays, James Philip
Beachem, Michael Thomas
Beak, Peter
Beal, Philip Franklin, III
Bealor, Mark Dabney
Beam, Charles Fitzhugh, Jr
Bean, Gerritt Post
Bearce, Winfield Hutchinson
Bearse, Arthur Everett
Beasley, James Gordon
Beattie, Thomas Robert
Beavers, Dorothy (Anne) Johnson
Beavers, Ellington McHenry
Beavers, Leo Earice
Bebb, Robert Lloyd
Bechara, Ibrahim
Bechtle, Gerald Francis
Beck, Curt Werner
Beck, Harris Graybill
Beck, Karl Maurice
Beck, Mae Lucille
Beck, Paul Edward
Becker, Edward Brooks
Becker, Robert Hugh
Beckerbauer, Richard
Beckwith, Newell Pierce
Bedenbaugh, John Holcombe
Bederka, John Paul, Jr
Beede, Charles Herbert
Beel, John Addis
Beelik, Andrew
Beereboom, John Joseph
Behr, Lyell Christian
Behrmann, Eleanor Mitts
Beichl, George John
Beiler, Theodore Wiseman
Beinfest, Sidney
Beispiel, Myron
Beitchman, Burton David
Beitzel, Richard Earl
Belew, John Seymour
Bell, Harold Morton
Bell, John Barr, Jr
Bell, Malcolm Rice
Bell, Russell A
Bell, Stanley C
Belleau, Bernard Roland
Bellina, Russell Frank
Bellino, Vito Victor
Bellis, Ernest Anthony
Belloli, Robert Charles
Bendall, Victor Ivor
Bender, Howard Sanford
Bender, Reinhold
Benedict, Joseph T
Benington, Frederick
Benitez, Allen

Benjaminov, Robert Benjamin S
Benkeser, Robert Anthony
Benn, Walter R
Bennett, James Gordy, Jr
Bennett, Ovell Francis
Bennett, Richard Bond
Bennett, Robert Putnam
Benning, Calvin James
Benson, Barrett Wendell
Benson, Frederic Rupert
Benson, Harriet
Benson, Robert Franklin
Bent, Richard Lincoln
Bentley, Michael David
Benton, Francis Lee
Bentrude, Wesley George
Benz, Ralph Wagner
Benz, George William
Berchtold, Glenn Allen
Berdahl, James Maynard
Beredjick, Nicky
Berenbaum, Morris Benjamin
Berg, Jeffrey Howard
Berger, Arthur
Berger, Daniel Richard
Berger, S Edmund
Bergen, David Alan
Bergman, Elliot
Bergman, Robert George
Bergmark, William R
Bergomi, Angelo
Bergstrom, Clarence George
Beringer, Frederick Marshall
Berkelhammer, Gerald
Berkoff, Charles Edward
Berkowitz, Lewis Maurice
Berlin, Kenneth Darrell
Berliner, Ernst
Berliner, Frances (Bondhus)
Berman, Elliot
Berman, Lawrence Uretz
Berneis, Hans Ludwig
Bernetti, Raffaele
Bernhard, Richard Allan
Bernstein, Seymour
Berry, David A
Berry, James Wesley
Berry, Keith O
Berry, Roy Alfred, Jr
Berry, William Lee
Berson, Jerome Abraham
Bertani, Laura Marie
Bertoniere, Noelie Rita
Besozzi, Alfio Joseph
Besso, Michael M
Beug, Michael William
Beyler, Roger Eldon
Bharucha, Keki Rustomji
Bhatia, Kishan
Bhatnagar, Ajay Sahai
Bicking, John Beeh
Bidlack, Verne Claude, Jr
Bieber, Theodore Immanuel
Biehl, Edward Robert
Bieman, Klaus
Bieron, Joseph F
Biersmith, Edward L
Bigelow, Melvin Jerome
Biggerstaff, Warren Richard
Bikales, Norbert M
Bil, Milos Sidney
Billig, Franklin A
Billman, John Henry
Bills, Alan Morris
Bills, Charles Wayne
Billups, W Edward
Bimber, Russell Morrow
Bindra, Jasjit Singh
Bingham, Richard Charles
Binkley, Roger Wendell
Binnell, James Monroe
Bird, Charles Norman
Birnbaum, Hermann
Bischoff, Fritz Emil
Bishop, Charles Anthony
Bishop, Jay Lyman
Bishop, Muriel Boyd
Bissell, Eugene Richard
Bissell, Robert
Bissing, Donald Eugene
Bjornson, August Sven
Black, Donald K
Black, Howard Charles
Blackburn, Edward Victor
Blackwood, Samuel William
Blackwood, Robert Keith
Blades, Charles Ernest
Blaha, Eli William
Blair, Charles Melvin, Jr
Blair, Eicyl Howell
Blair, Robert Paul
Blake, Jules
Blake, James J
Blank, Benjamin
Blankenstein, William E
Blankespoor, Ronald Lee

Blanton, Charles DeWitt, Jr
Blatchford, John Kerslake
Blatt, Albert Harold
Blay, Jorge Albert
Bleasdale, James Lewis
Blecker, Harry Herman
Bledsoe, Harry Herman
Blewett, Charles William
Bliss, Arthur Dean
Blickenstaff, Robert Theron
Blizzard, Richard Reese, Sr
Block, Eric
Block, Michael Joseph
Block, Paul, Jr
Blomberg, Richard Nelson
Blomquist, Alfred Theodore
Blomquist, Richard Frederick
Blood, Charles Allen, Jr
Bloom, Allen
Bloom, Barry Malcolm
Bloomer, James L
Bloomfield, Jordan Jay
Blouin, Florine Alice
Bluestein, Allen Channing
Bluestein, Bernard Richard
Bluestone, Henry
Bluhm, Aaron Leo
Bluhm, Harold Frederick
Blum, Stanley Walter
Boardway, Nancy Louise
Boatman, Sandra
Bobbitt, Jeffrey L
Bobe, Ernest Christoph
Bobko, Edward
Bocksch, Robert Donald
Bodanszky, Agnes Adrienne
Boden, Herbert
Boekelheide, Virgil Carl
Boehme, Werner Richard
Boerwinkle, Fred P
Bogard, Terry L
Bogart, William Hawkins, Jr
Bogdanowicz, Mitchell Joseph
Boger, Eliahu
Boggs, Norman Towar, III
Bohen, Joseph Michael
Bohner, James Calvin
Boikess, Robert S
Bolhofer, William Alfred
Bolinger, Frederick William
Bolon, Donald A
Boots, Sharon G
Booth, Gary Edwin
Booth, Robert Edwin
Booth, William Thomas, Jr
Boothe, James Howard
Bolt, Robert O'Connor
Bolton, B A
Bonaventura, Maria Migliorini
Bond, Frederick Thomas
Bond, William Bradford
Bonner, William Hallam, Jr
Bonvicino, Guido Eros
Boone, James Ronald
Boorman, Philip Michael
Borchert, Peter Jochen
Borden, George Wayne
Borden, Weston Thatcher
Bordenca, Carl
Borders, Donald B
Bordner, Jon D B
Bordwell, Frederick George
Borkovec, Alexej B
Borkowski, Walter Leonard
Borne, Ronald Francis
Borowitz, Grace Burchman
Borror, Alan L
Bosch, Warren Luther
Boschan, Robert Herschel
Bose, Ajay Kumar
Boshart, Gregory Lew
Bosin, Talmage R
Bosniack, David S
Bossert, Roy Garner
Bost, Howard William
Boswell, Charles Leland
Boswell, Donald Eugene
Botimer, Laurence Wallace
Botteron, Donald George
Bottini, Albert Thomas
Bottorff, Edmond Milton
Bouboulis, Constantine Joseph
Boucher, Raymond Edward
Boucher, Roger
Bouis, Paul Andre
Bourns, Arthur Newcombe
Bowen, Douglas Malcomson
Bowers, George Henry, III
Bowman, Lewis Wilmer
Bowman, Max I
Bowman, Newell Stedman
Bowman, Robert Samuel
Bown, Delos Edward
Boxer, Robert Jacob
Boyack, Gerald (Arthur)
Boyd, Robert Neilson
Boyd, Samuel Neil, Jr

Boye, Frederick C
Boyer, Alvin C
Boyer, Joseph Henry
Boyer, William Montgomery
Boykin, David Withers, Jr
Bozak, Richard Edward
Brabander, Herbert Joseph
Brace, Neal Orin
Brackenridge, David Ross
Brader, Walter Howe, Jr
Bradlow, Herbert Leon
Bradshaw, Jerald Sherwin
Bradsher, Charles Kilgo
Bradstreet, Raymond Bradford
Brady, Leonard Everett
Brady, Robert Frederick, Jr
Brady, Stephen Francis
Bragg, Philip Dell
Bragole, Robert Anthony
Braid, Milton
Braidwood, Clinton Alexander
Brake, Jon Michael
Braley, James Alexander
Brandenberger, Stanley George
Brandman, Harold A
Brannen, Cecil Gray
Brannen, William Thomas, Jr
Brannon, Donald Ray
Brasel, Robert George
Braun, Loren L
Braunstein, David Michael
Breazeale, Robert David
Breitbeil, Fred W, III
Breiter, Jerome John
Bremer, Keith George
Bremner, Bart J
Brennan, Michael Edward
Brenner, Gerald Stanley
Brenner, Charles Ray
Breslow, David Samuel
Breslow, Ronald
Brewster, James Henry
Brickman, Leo
Bridger, Robert Frederick
Brigeo, William Alphonsus
Brieaddy, Lawrence Edward
Brieger, Gottfried
Briggs, Paul Clayton, Jr
Briggs, Richard M
Briggs, William Scott
Briles, George Herbert
Brill, Earl
Brill, William Franklin
Brinker, Keith Clark
Bristol, Douglas Walter
Brittelli, David Ross
Broaddus, Charles D
Brockington, James Wallace
Brod, John Sydney
Brodasky, Thomas Francis
Brodhag, Alex Edgar, Jr
Brodmann, John Milton
Brodway, Nicolas
Brokke, Mervin Edward
Bromund, Werner Hermann
Brook, Adrian Gibbs
Brooker, Robert Munro
Brookhart, Maurice S
Brooks, Robert Alan
Brotherton, Robert John
Brower, Kay Robert
Brown, Alfred Edward
Brown, Bernard Beau
Brown, Bruce Willard
Brown, Donald John
Brown, Ellis Vincent
Brown, Frances Campbell
Brown, George Earl
Brown, Gordon Manley
Brown, Henry Clay, III
Brown, Herbert Charles
Brown, John Angus
Brown, Peter
Brown, Richard Emery
Brown, Robert Raymond
Brown, Ronald Frederick
Brown, Stuart Houston
Brown, William Henry
Brownell, George L
Brownfield, Robert Bruce
Browning, Daniel Dwight
Brownlee, Paula Pimlott
Brownstein, Sydney Kenneth
Bruce, John MacMillan, Jr
Bruesch, John F
Brundage, Donald Keith
Brunelle, Thomas E
Bruning, Karl John
Bruno, Arthur John
Bruson, Herman Alexander
Brust, Harry Francis
Brutcher, Frederick Vincent, Jr
Bryant, Rhys
Bublitz, Donald Edward
Buchanan, Gerald Wallace
Buchanan, James Balfour
Buchanan, James Wesley
Buchi, George

Buckler, Sheldon A
Buckwalter, Howard McWilliams
Bucolo, Giovanni
Bucovaz, Edsel Tony
Buehler, John A
Buell, George Christopher
Buell, Glen R
Buess, Charles Merlyn
Bukovnik, Peter Anthony
Bulbenko, George Fedir
Bullock, Austin Larnel
Bullock, Eric
Bullough, Vaughn Lynn
Bulusu, Suryanarayana
Bumgardner, Carl Lee
Bumpus, Francis Merlin
Bunce, Nigel James
Bunce, Stanley Chalmers
Buncel, Erwin
Bundy, Gordon Leonard
Bunnett, Joseph Frederick
Bunting, Albert L
Bunting, John William
Bunton, Clifford A
Buntrock, Robert Edward
Buonocore, Michael
Buras, Edmund Maurice
Burch, Joseph Eugene
Burckhart, Robert H
Burg, Marion
Burgauer, Paul David
Burge, Robert Ernest, Jr
Burgert, Bill E
Burgess, Edward Meredith
Burgmaier, George John
Burgoyne, Edward Eynon
Burgstahler, Albert William
Burke, Hanna Suss
Burke, Morris
Burke, Richard David
Burkett, Howard (Benton)
Burkhard, Charles (Austin)
Burks, Robert Elbert, Jr
Burlant, William Jack
Burleson, James C
Burmel, Robert H
Burness, Donald Mac Arthur
Burnette, Marvin Clifton
Burns, Hugh Donald
Burnside, Charles H
Burrell, Harold Paul Charles
Burrous, Mervyn Lee
Burson, Sherman Leroy, Jr
Burt, Gerald Dennis
Burti, Umbay H
Burton, Donald Joseph
Burton, Gilbert W
Burton, Willard White
Burton, Llewellyn Wilson
Busby, Hubbard Taylor, Jr
Buser, Kenneth Rene
Bush, David Clair
Bush, Walter Monroe
Busse, Robert Franklyn
Bussert, Jack Francis
Butler, Donald Eugene
Butler, George Bergen
Butler, John Mann
Butler, Robert Westbrook
Button, Allan Clifford
Buyske, Donald Albert
Byers, Donald James
Byrd, David Shelton
Byrnes, Eugene William
Cabanes, William Ralph, Jr
Cadotte, John Edward
Caflisch, Edward George
Cahn, Arno
Cahnmann, Hans Julius
Cahoy, Roger Paul
Caine, Charles Eugene
Caine, Drury Sullivan, III
Cairns, Robert Edward
Cairns, Theodore L
Calaway, Paul Kenneth
Caldwell, Henry Cecil, Jr
Calfee, John Elton
Canfield, James Howard
Callen, George Milton
Callen, Joseph Edward
Calvin, Melvin
Cameron, Margaret Davis
Camp, David Bennett
Campbell, Alfred
Campbell, Allen James
Campbell, Bruce (Nelson), Jr
Campbell, Clyde Del
Campbell, Hugh John
Campbell, Paul Gilbert

Cappel, Carl Robert
Capps, Julius Daniel
Capstack, Ernest
Carabateas, Philip M
Carboni, Rudolph A
Cardenas, Carlos Guillermo
Caress, Edward Alan
Caret, Robert Laurent
Cargill, Francis Arthur
Cargill, Robert Lee, Jr
Carlile, Clayton George
Carlin, Robert Burnell
Carlin, Charles Herrick
Carlo, Bruce Arne
Carls, Dana Peter
Carlsen, Emil Herbert
Carlson, Gustaf Harry
Carlson, Mildred V
Carlson, Norman Arthur
Carlson, Robert Gideon
Carlson, Robert M
Carlson, Steven Allen
Carmack, Marvin
Carnahan, Robert Edward
Carnes, Joseph John
Carney, Albert Stricker
Carney, James Joseph
Carney, Richard William James
Carpenter, Barry Keith
Carpenter, James William
Carpenter, Paul Gershom
Carpenter, Sammy
Carpenter, T J
Carr, Albert A
Carr, Russell L K
Carrick, Wayne Lee
Carrick, Frederick E
Carroll, F Ivy
Carroll, Robert Baker
Carson, Frederick Wallace
Carter, Kenneth Nolon
Carter, Mary Eddie
Cartledge, Frank
Carty, Daniel T
Casagrande, Daniel Joseph
Casanova, Joseph
Caserio, Marjorie C
Casey, Adria Catala
Casey, Charles P
Casey, John Edward, Jr
Cason, James, Jr
Casparian, Sarkis Manoug
Caspi, Eliahu
Cassady, John Mac
Cassidy, Harold Gomes
Ceprini, Mario Q
Cerefice, Steven A
Cessna, John Curtis
Cevasco, Albert Anthony
Chafetz, Harry
Chaffin, Tommy L
Chamberlain, David Leroy, Jr
Chamberlain, Malcolm
Chamberlin, Earl Martin
Chamberlin, James Wesley
Chambers, James Richard
Chambers, Robert Rood
Chambers, Vaughan Crandall, (Jr)
Champion, William (Clare)
Chan, Tak-Hang
Chandler, Roger Eugene
Chandross, Edwin A
Chaney, David Webb
Chapman, Douglas Wilfred
Chapman, Orville Lamar
Chapman, Richard David
Chaplow, Cecil Clendis, Jr
Charlton, James Leslie
Chassy, Bruce Matthew
Chauffe, Leroy
Chaykovsky, Michael
Cheema, Zaffrullah K
Cheer, Clair James

Chemerda, John Martin
Chen, Ming Chih
Cheng, Chia-Chung
Cheng, Fred Fa Wu
Cheng, Shu-Sing
Cherkofsky, Saul Carl
Chiasson, Bertrand Arnold
Chibnik, Sheldon
Chickos, James S
Chiddix, Max Eugene
Chien, Ping-Lu
Chignell, Colin Francis
Childress, Scott Julius
Chilton, William Scott
Chipman, Wilmon B
Chitharanjan, Dakshinamurthy
Chitwood, Henry Cady
Chitwood, James Leroy
Chmurny, Alan Bruce
Cho, Arthur Kenji
Choi, Yong Chun
Choong, Hsia Shaw-Lwan
Chopra, Naiter Mohan
Chow, Alfred Wen-Jen
Chow, Sui-Wu
Chow, Yuan Lang
Christensen, Larry Wayne
Christian, Curtis Gilbert
Christian, Samuel Terry
Christiansen, Robert George
Christman, David R
Chu, Edith Ju-Hwa
Chubb, Francis Learmonth
Chudd, Cletus Charles
Church, Robert Fitz (Randolph)
Churchill, Constance Louise
Chute, Walter John
Cialdella, Cataldo
Ciskowski, Joseph M
Cislak, Francis Edward
Citron, Joel David
Ciula, Richard Paul
Claff, Chester Eliot, Jr
Clagett, Donald Carl
Clapp, Leallyn Burr
Clapper, Thomas Wayne
Clark, Allen Keith
Clark, Carroll Thomas
Clark, Charles Austin
Clark, David Ellsworth
Clark, Frank S
Clark, Larry P
Clark, Raymond Donald
Clark, Robert A
Clark, Ronald David
Clark, Ronald Duane
Clark, Samuel Friend
Clark, Sidney Gilbert
Clark, Stephen Darrough
Clarke, Frank Henderson
Clarke, Wilbur Bancroft
Class, Jay Bernard
Claypool, Don Pearson
Clayton, Anthony Broxholme
Clayton, David Walton
Cleary, James William
Cleland, George Horace
Clemans, George Burtis
Clemens, David Henry
Clemens, Lawrence Martin
Clement, William H
Clendinning, Robert Andrew
Cleveland, Elonza Alexander, Jr
Cleveland, James Perry
Cleveland, Thomas Hilburn
Clevenger, Richard Lee
Cline, Edward Terry
Cline, Warren Kent
Closs, Gerhard Ludwig
Closson, William Deane
Clough, Stuart Chandler
Cluff, Edward Fuller
Coan, Stephen B
Coates, Robert Mercer
Cobb, Emerson Gillmore
Cobb, Raymond Lynn
Coburn, Everett Robert
Coburn, Michael Doyle
Cochran, John Charles
Cochrane, Chappelle Cecil
Coe, Beresford
Coffey, Michael Dewayne
Cogdell, Thomas James
Cogliano, Joseph Albert
Cogswell, George Wallace
Cohen, Abraham Bernard
Cohen, Alex
Cohen, Hyman L
Cohen, Irwin
Cohen, Merrill
Cohen, Murray Samuel
Cohen, Saul G
Cohen, Saul Mark
Cohen, Sidney
Cohen, Theodore
Coke, James Logan
Cole, Thomas Winston, Jr

Cole, Walter Earl
Coleman, Denis
Coleman, John Franklin
Coleman, Lester Earl, (Jr)
Coleman, Richard J
Coleman, William Earl
Collette, John Wilfred
Collins, Frances Wilmoth
Collins, Jerry Dale
Collins, John W
Collins, Joseph Charles, Jr
Colter, Allan Kennedy
Comen, Alan Lee
Comer, William Timmey
Commerford, John D
Compere, Edward L, Jr
Compton, Ell Dee
Compton, William David
Conca, Romeo John
Conciatori, Anthony Bernard
Condit, Paul Brainard
Condray, Ben Rogers
Cone, Conrad
Conigliaro, Peter James
Conkin, Robert A
Conley, Robert T
Connor, David Thomas
Conover, Lloyd Hillyard
Conrad, Walter Edmund
Conte, John Salvatore
Controulis, John
Convery, Robert James
Conway, Walter Donald
Cook, Addison Gilbert
Cook, Clarence Edgar
Cook, Donald Jack
Cook, Gordon Smith
Cook, Kenneth Emery
Cook, Lawrence C
Cook, Newell Choice
Cook, Paul Laverne
Cook, Richard Sherrard
Cook, Wendell Sherwood
Cook, William Boyd
Cooke, Manning Patrick, Jr
Cooley, James Hollis
Coolidge, Edwin Channing
Coombs, Renate Bangert
Coombs, Robert Victor
Cooney, Robert Clair
Cooper, David John
Cooper, Douglas Elhoff
Cooper, Emerson Amenhotep
Cooper, Glenn Dale
Cooper, Paul David
Cooper, Robin D G
Copelin, Harry B
Copenhaver, John William
Copes, Joseph Paul
Coppock, William Homer
Coran, Aubert Y
Coraor, George Robert
Corbett, John Frank
Corbin, James Lee
Cordon, Martin
Cords, Donald Philip
Corey, Albert Eugene
Corey, Elias James
Corkern, Walter Harold
Corrigan, John Raymond
Corwin, Alsoph Henry
Cosby, John Norman
Cosmatos, Alexandros
Cosper, David Russell
Cote, Philip Norman
Cotruvo, Joseph Alfred
Cotton, Wyatt Daniel
Couch, Margaret Wheland
Coulson, Dale Robert
Counts, Wayne Boyd
Coury, Arthur Joseph
Coutts, Ronald Thomson
Couvillion, John Lee
Covey, Rupert Alden
Cowan, Dwaine O
Cowan, John C
Cox, David Buchtel
Cox, Eugene Floyd
Cox, James Carl, Jr
Cox, James Reed, Jr
Cox, Lionel Audley
Cox, Richard Harvey
Cox, William Lester
Crabtree, Eleanor Voorhees
Cragg, Hoyt J
Cragoe, Edward Jethro, Jr
Craig, John Cymerman
Craig, John Horace
Craig, Louis Elwood
Cram, Donald James
Cramer, Francis Barnard
Crandall, Elbert Williams
Crandall, Jack Kenneth
Crane, Grant
Crano, John Carl
Crary, James Walter
Craven, Robert Lee
Crawford, James Dalton

Crawford, James Worthington
Crawford, Jean Veghte
Crawford, Robert James
Crawford, Thomas Charles
Crean, Patrick J
Creger, Paul LeRoy
Creighton, Robert Hervey Jermain
Creighton, Stephen Mark
Cremer, Sheldon E
Cresswell, Arthur
Crews, Philip O
Crimmins, Timothy Francis
Cripps, Harry Norman
Crivello, James V
Croce, Louis J
Croft, Thomas Stone
Cromwell, Norman Henry
Cronin, Timothy H
Cronyn, Marshall William
Crooks, Harry Means, Jr
Crosby, Alan Hubert
Cross, Alexander Dennis
Cross, David Ralston
Crounse, Nathan Norman
Crouse, Dale McClish
Crovetti, Aldo Joseph
Crow, Edwin Lee
Crowe, Bernard Francis
Crowshaw, Keith
Crum, James Davidson
Crump, John William
Cryberg, Richard Lee.
Csallany, Agnes Saari
Capilla, Joseph
Csendes, Ernest
Cubberley, Adrian H
Culbertson, Chicita Frances
Culbertson, Billy Muriel
Cullen, William Charles
Cummings, Thomas Fulton
Cummins, Earl Wesley
Cunico, Robert Frederick
Cunningham, Earlene Brown
Cunningham, Howard Charles
Cunov, Carl Henry
Cupas, Chris Angelo
Cupper, Robert Alton
Curphey, Thomas John
Curran, William Vincent
Currell, Douglas Leo
Currie, Bruce LaMonte
Curry, Howard Millard
Curry, Thomas Harvey
Curtin, David Yarrow
Cuscurida, Michael
Cushley, Robert John
Cushman, Mark
Cutler, Royal Anzly, Jr
Cutshall, Theodore Wayne
Cywinski, Norbert Francis
Dacons, Joseph Carl
Da Costa, William A
Daemiker, Hans Ulrich
Daessle, Claude
Dahill, Robert T, Jr
Dahlgard, Muriel Genevieve
Dailey, Joseph Patrick
Dain, Jeremy George
Dale, John Irvin, III
Dale, Wesley John
Dalman, Gary
Dalton, David Robert
Dalton, Jack L
Dalton, James Christopher
Dalton, Philip Benjamin
Daly, John Joseph, Jr
Daly, William Howard
D'Amico, John J
Damle, Suresh B
Danehy, James Philip
Danen, Wayne C
Daniel, Daniel S
Danieley, Earl
Daniels, Ralph
Daniels, Wiley Edgar
Danishefsky, Isidore
Dann, John Ralph
Dannley, Ralph Lawrence
Danzig, Morris Judah
Dappen, Glen Marshall
Darby, Robert Albert
Dardoufas, Kimon C
Darlage, Larry James
Darlak, Robert
Darling, Stephen Deziel
Darling, Stephen Foster
Daub, Guido Herman
Daues, Gregory W, Jr
Daul, George Cecil
Daum, Sol Jacob
Davenport, Tom Forest, Jr
Daves, Glenn Doyle, Jr
Davey, William Raymond
Davies, Richard Edgar
Davis, Burns
Davis, Dennis Duval
Davis, Edward Melvin
Davis, Edwin Griffith

Davis, Franklin A
Davis, George Thomas
Davis, Harry Willard
Davis, Horace Raymond
Davis, Martin Arnold
Davis Pauls
Davis, Robert Bernard
Davis, Selby Brinker
Dawes, John Leslie
Dawson, Charles Reginald
Dawson, John William
Day, Bruce Frederick
Day, Jack Calvin
Dayan, Jason Edward
DeAcetis, William
Dean, Walter Lee
Deans, Sidney Alfred Vindin
DeBardeleben, John F
Debono, Manuel
DeBow, Lee Richard
DeBruin, Kenneth Edward
DeBrunner, Marjorie R
DeBrunner, Ralph Edward
DeCamp, Mark Rutledge
Dechary, Joseph Martin
Deger, Thomas Edward
Deghenghi, Romano
DeGrande, Gary Gaston
Dehm, Henry Christopher
Dehn, Joseph William, Jr
Deinet, Adolph Joseph
DeJongh, Don C
DeKorte, John Martin
DeLaitsch, Dale M
De La Mare, Harold Elison
DeLaMater, George (Bearse)
de Langre, John Paul
Del Bel, Elsio
Delcamp, Robert Mitchell
Delia, Thomas J
Delmonte, David William
Delton, Mary Helen
De Luca, Donald Carl
DeMarinis, Robert Michael
De Matte, Michael L
de Mauriac, Richard Arthur
de Mayo, Paul
DeMedics, E M J A
De Member, John Raymond
Demuth, John Robert
Denham, Joseph Milton
Denison, M Carl
Denk, Ronald H
Denkewalter, Robert George
Denney, Donald Berend
Denney, George Hutcheson
Denoon, Clarence England, Jr
Denson, Donald D
DePree, David Otte
De Puy, Charles Herbert
Derieg, Michael E
Dermody, William Joseph
De Selms, Roy Charles
Deslongchamps, Pierre
Dessauer, Rolf
de Stevens, George
DeTar, DeLos Fletcher
Detert, Francis Lawrence
De Tommaso, Gabriel Louis
Detty, Wendell Eugene
Detweiler, William Kenneth
Deutsch, Daniel Harold
Deutschman, Archie John, Jr
DeVenuto, Frank
DeWald, Horace Albert
De Wall, Gordon
Dewey, Fred McAlpin
De Witt, Hobson Dewey
Dewolfe, Robert Hill
DeYoung, Jacob J
Deyrup, James Alden
Dhami, Kewal Singh
Dial, William Richard
Diamond, Martin J
Diassi, Patrick Andrew
Diaz, Arthur Fred
DiBella, Eugene Peter
Dickason, Alan Frederick
Dickason, William Charles
Dickerman, Stuart Carlton
Dickerman, Charlesworth Lee
Dickerson, James Perry
Dickie, John Peter
Dickinson, Clifford Lee, Jr
Dicksten, Jack
Dien, Chi-Kang
Dietrich, Joseph Jacob
DiFate, Victor George
Dighe, Shrikant Vishwanath
DiGiovanna, Charles V
Digman, Robert V
Dilgen, St Francis
Dill, Charles William
Dill, Dale Robert
Dillard, John Gammons
Dille, Kenneth Leroy
Dilling, Wendell Lee

Dills, Charles E
Dimitroff, Edward
Dimond, Harold Lloyd
Dinan, Frank J
Dineen, Eugene Joseph
Dinner, Alan
Dinwiddie, Joseph Gray, Jr
Di Pippo, Ascanio G
Dishart, Kenneth Thomas
Dittmer, Donald Charles
Diuguid, Lincoln Isaiah
Diveley, William Russell
Dix, James Seward
Dixon, Joseph Ardiff
Djerassi, Carl
DoAmaral, Jefferson Ribeiro
Doane, William M
Dobinson, Frank
Dobson, Raymond Monroe
Dodson, Raymond Monroe
Doering, William Von Eggers
Doerr, Marvin LeRoy
Doerr, Robert George
Doerschuk, Albert Peter
Dolbier, William Read, Jr
Dolby, Lloyd Jay
Dolfini, Joseph E
Dominianni, Samuel James
Domning, Beryl W
Donaruma, Lorraine Guy
Doncan, Alvan
Doolittle, Dortha Bailey
Doornbos, Earl
Doorenbos, Harold E
Doran, Thomas J, Jr
D'Orazio, Vincent T
Dorfman, Edwin
Dorion, George Henry
Dority, Guy Hiram
Dorko, Ernest A
Dorrence, Samuel Michael
Doshan, Harold David
Doss, Nagib A
Dougherty, Charles Michael
Doughty, Mark
Douglas, James Edward
Douglass, John Richmond
Doukas, Harry Michael
Doumaux, Arthur Roy, Jr
Dowbenko, Rostyslaw
Dowd, Paul
Downing, Ralph Charlson
Downs, James Joseph
Doyle, Richard Robert
Doyle, Thomas Daniel
Dragun, Henry L
Draper, John Daniel
Dreby, Edwin Christian, III
Dreisbach, Paul Franklin
Dresden, Carlton F
Dressler, Hans
Dressler, Robert Louis
Drew, Howard Felshaw
Dreyer, David
Driesch, Albert John
Driscoll, Gary Lee
Drisko, Richard Warren
Droucas, John
Drucker, Arnold
Druelinger, Melvin L
Drumm, Manuel Felix
Drummond, Paul Edward
Drysdale, John Jay
Dub, Michael
Dubois, Ronald Joseph
Duddey, James E
Dudley, James E
Duffey, Donald Creagh
Duffy, Kenneth Harrison
Dugan, LeRoy, Jr
Dugas, Hermann
Duggins, Harry A
Duggins, William Edgar
Duke, June Temple
Dullaghan, Matthew Edward
Dunathan, Harmon Craig
Dunkel, Morris
Dunlap, William Joe
Dunn, Danny Leroy
Dunn, George Lawrence
Dunn, Howard Eugene
Dunning, John Walcott
Dunworth, William Paul
Dupont, Paul Emile
Durand, Marc L
Durante, Anthony Joseph
Durden, John Apling, Jr
Durette, Philippe Lionel
Durham, Lois Jean
Durrell, William S
Durrell, Albert Matthew, Jr
Dursch, Friedrich
Durst, Tony
Dutra, Gerard Anthony
Dutra, Raniro Carvalho
Dutta, Shib Prasad
Dutton, Guy Gordon Studdy
Duval, Robert C
Duvall, Harry Marean

Duvall, Jacque L
Dvornik, Dushan Michael
Dyar, Robert Matthew
Dybvig, Douglas Howard
Dyck, Arnold Wolff Jan
Dye, William Thomson, Jr
Dye, Elizabeth
Dykstra, Stanley John
Dykstra, Thomas Karl
Dymicky, Michael
Dyson, Ian Fraser
Eadon, George Albert
Easterbrook, Eliot Knights
Eastes, Frank Elisha
Eastham, Arthur Middleton
Eastham, Jerome Fields
Eastman, Richard Hallenbeck
Easton, Nelson Roy
Eaton, Philip Eugene
Eberhard, Anatol
Eberhardt, Manfred Karl
Eberle, Marcel Karl
Ebetino, Frank Frederick
Ebner, Herman George
Eby, Charles J
Eby, Harold Hildenbrandt
Eby, John Martin
Eby, Lawrence Thornton
Eck, David Lowell
Ecke, George Graff
Eckroth, David Raymond
Economy, James
Edamura, Fred Y
Edelman, Robert
Edgerton, Richard Oliver
Egerton, William John
Eich, John Joseph
Eisch, John Joseph
Edmonds, John Harold
Edmondson, Morris Stephen
Edmison, Marvin Tipton
Edwards, Joseph D, Jr
Edwards, Oliver Edward
Edwards, Warrick Rigeley, Jr
Edwards, William Brundige, III
Egberg, David Curtis
Ege, Seyhan Nurettin
Ehrenfeld, Robert Louis
Ehrlich, Felix Frederick
Eian, Gilbert Lee
Eichel, Herman Joseph
Eicher, John Harold
Eisch, John William
Elam, Edward Underwood
Elbling, Irving Nelson
Elder, John William
Eisch, John Joseph
Eiseman, Fred S
Eisenbraun, Allan Alfred
Eisenbraun, Edmund Julius
Eisenhardt, William Anthony, Jr
Eisenhauer, Hugh Ross
Elakovich, Stella Daisy
El Khadem, Hassan S
Elkins, John Rush
Elefson, Ralph Donald
Elderfield, Robert Cooley
Eldred, Nelson Richards
Eleuterio, Herbert Sousa
Elia, Raymond J
Eiel, Ernest Ludwig
Eliel, Ernest Ludwig
Ellestad, George A
Ellingson, Rudolph Conrad
Ellingboe, Ellsworth Knowlton
Elliott, Irvin Wesley
Elliott, Robert Daryl
Elliott, William H
Ellis, Alan F
Ellis, Jerry William
Ellis, Larry Edward
Ellis, Leonard Culberth
Elizey, Samuel Edward, Jr
Elmer, Otto Charles
Elofson, Richard Macleod
Elslager, Edward Faith
Elson, William O
Elwood, James Kenneth
Embree, Harland Dumond
Emerman, Sidney
Emerson, David Winthrop
Emert, Jack Isaac
Emling, Bertin Leo
Emmick, Robert D
Emmick, Thomas Lynn
Emmons, William David
Emrich, William Oscar
Emrich, Donald Day
Endres, Hans Henry
Enders, Leland Sander
Engel, Charles Robert
Engel, John Francis
Engel, John Hal, Jr
Engelhardt, Edward Louis
Engelhardt, Vaughn Arthur
Engle, Robert Rufus
Engler, Reto Arnold
English, Jackson Pollard
Englund, Charles R
Ennor, Kenneth Stafford
Enos, Herman Isaac, Jr
Ensor, Elwood Henderson
Epstein, Joseph

Epstein, Martin Eden
Erby, William Arthur
Erdman, Timothy Robert
Erickson, Bruce Wayne
Erickson, John Gerhard
Erickson, Karen Louise
Erickson, Raymond Leroy
Erickson, Wayne Francis
Erikson, J Alden
Erkkila, Armas Victor
Erman, William F
Ernsberger, Maurice Leon
Ertel, Henry Robinson
Esack, Ashmeed
Esayian, Manuel
Eschinasi, Emile Haviv
Eskelson, Cleamond D
Esmay, Donald Levern
Essery, John M
Esse, Robert Carlyle
Esslinger, William Glenn
Estes, John H
Estrin, Norman Frederick
Etter, Robert Miller
Evanega, George R
Evans, Charles P
Evans, David Albert
Evans, Francis Eugene
Evans, Franklin James, Jr
Evans, Gordon Goodwin
Evans, Robert John
Evans, Taylor Herbert
Evensen, Thomas James
Everett, John Edward
Everly, Charles Ray
Evers, William John
Evert, Henry Earl
Eveslage, Sylvester Lee
Evleth, Earl Mansfield
Ewart, Hugh Wallace, Jr
Faessinger, Robert William
Fagan, Paul V
Faggan, Joseph Edward
Fahey, Charlotte Wieghard
Fahey, Dennis Martin
Fahey, Robert C
Fahrenholtz, Kenneth Earl
Falk, John Carl
Falkehag, S Ingemar
Fallis, Alexander Graham
Fan, Joyce Wang
Fancher, Llewellyn W
Fand, Theodore Ira
Fang, Fabian Tien-Hwa
Fanta, Paul Edward
Farber, Hugh Arthur
Farber, Sergio Julio
Farkas, Eugene
Farmer, Larry Bert
Farnsworth, Carl Leon
Farnum, Bruce Wayne
Farnum, Donald G
Farquhar, Gale Burton
Farrar, Grover Louis
Farrar, Martin Wilbur
Farrar, Ralph Coleman
Farrell, James Kenneth
Fassnacht, John Hartwell
Fath, Joseph
Fatiadi, Alexander Johann
Fatora, Frank Charles, Jr
Faul, William Henry
Faulk, Dennis Derwin
Faulkner, D John
Favre, Henri Albert
Fawcett, Mark Stanley
Fayter, Richard George, Jr
Fearing, Ralph Burton
Feazel, Charles Elmo, Jr
Fechter, Robert Bernard
Fedor, Leo Richard
Fehnel, Edward Adam
Feig, Gerald
Feigl, Dorothy Marie
Felton, Stephen M
Fein, Marvin Michael
Feinstein, Allen Irwin
Fenglio, Richard Andrew
Fenselau, Allan Herman
Fenton, Stuart William
Ferguson, James Williams
Ferguson, Philip Rex
Fernandez, Jose Martin
Fernandez, Jack Eugene
Ferraris, John Patrick
Ferraro, John J
Ferrell, William James
Ferretti, Aldo
Ferrigno, Thomas Howard

Ferris, James Peter
Ferstandig, Louis Lloyd
Feuer, Henry
Field, George Francis
Fields, Clark Leroy
Fields, Ellis Kirby
Fields, Melvin
Fields, Thomas Lynn
Fieser, Louis Frederick
Figdor, Sanford Kermit
Figueras, John
Figueras, Patricia Ann McVeigh
Fike, Harold Lester
Filandro, Anthony Salvatore
Filbey, Allen Howard
Filipescu, Nicolae
Filler, Robert
Finch, Gaylord Kirkwood
Finch, Neville
Findlay, John A
Fine, Leonard W
Finegold, Harold
Finelli, Anthony Francis
Finkbeiner, Herman Lawrence
Finkelstein, Jacob
Finkelstein, Manuel
Finley, Joseph Burton
Finnegan, Richard Allen
Firestone, Raymond A
Fischback, Bryant C
Fischbeck, Robert Louis, Jr
Flachskam, Robert Louis, Jr
Fischman, Morris
Fish, Richard Wayne
Fishbein, Lawrence
Fisher, Thomas Welch
Fisher, Thomas Henry
Fisher, Norman Gail
Fisher, John Gatewood
Fisher, Gene Jordan
Fisher, Earl Eugene
Fisher, Charles Harold
Fishel, Derry Lee
Fishel, John B
Fitch, Calvin L
Fitchett, Glimer Trover
Fitzgerald, Conelius Gilbert
Fitzpatrick, Jimmie Doile
Fletcher, Thomas Lloyd
Flitter, David
Flood, Thomas Charles
Flores-Gallardo, Hector
Florey, Klaus
Floyd, Joseph Calvin
Flynn, John Joseph, Jr
Flynn, Kenneth G
Fodor, Gabor
Fodor, George Emeric
Foglia, Thomas Anthony
Fohlen, George Marcel
Foldi, Andrew Peter
Foldvary, Elmer
Foley, Henry Grant
Foltz, Calvin Martin
Foltz, Rodger L
Fonken, Gerhard Joseph
Fono, Andrew
Foote, Carlton Dan
Foote, Christopher S
Foote, Gordon Lee
Forbes, Malcolm Holloway
Ford, John Albert, Jr
Ford, Warren Thomas
Forker, Robert Fencil
Forkey, David Medrick
Fornefeld, Eugene Joseph
Forrest, Thomas Peter
Forrester, Sherri Rhoda
Forsblad, Ingemar Bjorn
Forshey, William Osmond, Jr
Foscante, Raymond Eugene
Foster, Warren Schumann
Foster, Charles Howard
Foster, Gerald Lawrence
Foster, Robbie T
Foulke, Donald Gardner
Fouron, Yves
Fowler, Frank Wilson
Fowler, Joanna S
Fox, Adrian Samuel
Fox, Charles Junius
Fox, Bernard Lawrence
Fox, Donald E
Fozzard, George Broward
Francis, John Elsworth
Francis, William Connett
Franck, Richard W
Frank, Charles Edward
Frank, Forrest Jay
Frank, Jean Ann
Frankel, William Charles
Frankel, Edwin N
Frankenberg, Peter Edgar
Frankenfeld, John William
Frankland, Albert Ernest
Franklin, Richard Crawford
Franklin, Robert Louis

Franko-Filipasic, Borivoj Richard Simon
Franks, John Anthony, Jr
Franzus, Boris
Fraser-Reid, Bertram Oliver
Frauenglass, Elliott
Frazza, Everett Joseph
Frederick, Marvin Ray
Fredericksen, James Monroe
Freed, Elisabeth Hertz
Freed, Meier Ezra
Freedman, Harold Hersh
Freedman, Robert Wagner
Freeman, James Harrison
Freeman, Jeremiah Patrick
Freeman, Kenneth Alfrey
Freeman, Peter Kent
Freeman, Robert Clarence
Freeman, Stanley Knoel
Freidinger, Roger Merlin
French, Berlin Carson
French, James C
Freter, Kurt Rudolf
Frey, Albert Joseph
Frey, David Allen
Frey, Sheldon Ellsworth
Frey, Thomas G
Freyermuth, Harlan Benjamin
Fridinger, Tomas Lee
Fried, John
Fried, Josef
Friedlander, Henry Z
Friedlander, William Sheffield
Friedman, Harris Leonard
Friedman, Lester
Friedman, Mendel
Friedman, Paul
Friedman, Seymour K
Friedman, Sidney
Friedrich, Edwin Carl
Friedrich, Louis Elbert
Friess, Seymour Louis
Frilette, Vincent Joseph
Fritz, Henry Edward
Froemsdorf, Donald Hope
Froning, Joseph Fendall
Frost, Kenneth Almeron, Jr
Frostick, Frederick Charles, Jr
Fry, Albert Joseph
Fry, Arthur James
Fry, James Leslie
Fry, John Sedgwick
Frye, Cecil Leonard
Fu, Wallace Yamtak
Fuchs, Julius Jakob
Fuchs, Richard
Fuerholzer, James J
Fuhlhage, Donald Wayne
Fuhrman, Albert William
Fukuto, Tetsuo Roy
Fuller, Glenn
Funke, Phillip T
Furey, Robert Lawrence
Furrow, Clarence Lee
Fusco, Gabriel Carmine
Fuson, Reynold Clayton
Gabrielsen, Bjarne
Gadek, Frank Joseph
Gaertner, Van Russell
Gaffney, Barbara Jean
Gage, Clarke Lyman
Gagen, James Edwin
Gager, Forrest Lee, Jr
Gainer, Gordon Clements
Gaines, Jack Raymond
Gaj, Bernard Joseph
Gajan, Raymond Joseph
Gajewski, Joseph J
Gal, George
Galbraith, Harry Wilson
Galkowski, Theodore Thaddeus
Gall, Martin
Gall, Walter George
Gallagher, George Arthur
Gallopo, Andrew Robert
Galloway, Ethan Charles
Galton, Suzanne A
Gamble, Dean Franklin
Gano, Robert Daniel
Ganti, Venkat Rao
Garbacik, Theodore John
Garber, John Douglas
Garbrecht, William Lee
Garcia, Edward Ernest
Gardiner, John Brooke
Gardlund, Zachariah Gust
Gardner, Frederick Albert
Gardner, Joseph Arthur Frederick
Gardner, Pete D
Garin, David L
Garland, Charles E
Garland, Robert Bruce
Garmaise, David Lyon
Garman, John Andrew
Garneau, Francois Xavier
Garner, Albert Y
Garrett, James M

Garrison, William Emmett, Jr
Garsky, Victor Michael
Garst, John Fredric
Gartner, Edward A
Garty, Kenneth Thomas
Garven, Floyd Charles
Garwood, William Everett
Gash, Kenneth Blaine
Gash, Virgil Walter
Gasser, William
Gassman, Paul G
Gast, Lyle Everett
Gaston, Lyle Kenneth
Gates, John Warburton, Jr
Gates, Marshall DeMotte, Jr
Gaudry, Roger
Gaul, Richard Joseph
Gavlin, Gilbert
Gear, James Richard
Gearien, James Edward
Gebauer, Peter Anthony
Gebelein, Charles G
Geels, Edwin James
Geer, Richard P
Geering, Emil John
Geiger, Robert Warren
Geipel, Lothar Ernst
Genet, Rene P H
Gensler, Walter Joseph
Gentry, Willard Max, Jr
Genzer, Jerome Daniel
George, Daniel Eugene
George, Paul John
Georgian, Vlasios
Gerber, Nancy Nichols
Gerber, Samuel Michael
Gerhardt, George William
Gerhardt, Klaus Otto
Germany, Archie Herman
Gerns, Fred Rudolph
Gerry, Henry
Gershbein, Leon Lee
Gerteisen, Thomas Jacob
Gervay, Joseph Edmund
Gerwe, Roderick Daniel
Gever, Gabriel
Gewanter, Herman Louis
Ghirardelli, Robert George
Giacobbe, Thomas Joseph
Giannotti, Ralph Alfred
Giants, Thomas William
Gianturco, Maurizio
Giarrusso, Frederick Frank
Gibbons, Loren Kenneth
Gibbs, Hugh Harper
Gibbs, James Albert
Gibson, David Michael
Gibson, Dorothy Hinds
Gibson, Gerald W
Gibson, James Donald
Gibson, Thomas William
Gifford, David Stevens
Gilde, Hans-Georg
Giles, Jesse Albion, III
Gilham, Peter Thomas
Gillespie, Jesse Samuel, Jr
Gillespie, John Paul
Gillespie, Robert Howard
Gilliom, George Alexander
Gilliom, Richard D
Gillis, Bernard Thomas
Gilman, Henry
Gilman, Norman Washburn
Gilmont, Robert Edward
Gilmont, Ernest Rich
Gilow, Helmuth Martin
Giner-Sorolla, Alfreso
Ginger, Leonard George
Gingras, Bernard Arthur
Ginos, James Zissis
Ginsburg, David
Girard, James Emery
Gisin, Balthasar Friedrich
Gist, Lewis Alexander, Jr
Giuffrida, Robert Eugene
Givens, Richard Spencer
Giza, Chester Anthony
Gladstone, Harold Maurice
Gladysz, John A
Glamkowski, Edward Joseph
Glaros, George Raymond
Glaser, Robert
Glass, Dudley Brewer
Glass, Richard Steven
Glassick, Charles Etzweiler
Glazier, Robert Henry
Gleason, Edward Hinsdale, Jr
Gleason, Robert Willard
Gleicher, Gerald Jay
Gleim, Clyde Edgar
Gletsos, Constantine
Glick, Francis James
Glickman, Samuel Arthur
Gloth, Richard Edward
Glover, Allen Donald
Glover, George Irvin
Glover, Leon Conrad, Jr
Gluntz, Martin Lucius

Gobran, Ramsis
Godar, Edith Marie
Godefroi, Erik Fred
Godfrey, John Carl
Godt, Henry Charles, Jr
Goe, Gerald Lee
Goebel, Charles Gale
Goeckner, Norbert Anthony
Goeke, George Leonard
Goel, Om Prakash
Goering, Harlan Lowell
Goetschel, Charles Thomas
Goetz, Rudolph W
Goheen, David Wade
Goheen, Gilbert Earl
Gokel, George William
Goken, Garold Lee
Gold, Allen Morton
Gold, Elijah Herman
Goldberg, Eugene P
Goldberg, Joseph Louis
Goldberg, Stanley Irwin
Goldblatt, Leo Arthur
Goldish, Dorothy May (Bowman)
Goldmacher, Joel E
Goldsby, Arthur Raymond
Goldschmidt, Alfred
Goldschmidt, Eric Nathan
Goldstein, Ernst Moritz
Goldstein, Melvin Joseph
Goldstein, Theodore Philip
Goller, Edwin John
Gomez, Ildefonso Luis
Gonzalez De Alvarez, Genoveva
Good, Pearl
Goode, William Edward
Gooding, Chester Martin (Briggs)
Goodman, Alan Lawrence
Goodman, Leon
Goodman, Murray
Goodrich, Judson Earl
Goodson, Leslie Alan
Goodson, Louis Hoffman
Goodwin, John Thomas, Jr
Gorbunoff, Marina J
Gorden, Berner J
Gordon, Annette Waters
Gordon, Arnold J
Gordon, Chester Duncan
Gordon, Irving
Gordon, John Edward
Gordon, Joseph R
Gordon, Kenneth Milton
Gordon, Myra
Gore, William Earl
Goren, Mayer Bear
Goretta, Louis Alexander
Gorham, William Franklin
Gorman, Susan B
Gormley, William Thomas
Gortler, Leon Bernard
Gorton, Bert SoRelle
Gosling, John William
Gosnell, Rex Beach
Gosselink, Eugene Paul
Gottesman, Roy Tully
Gotz, Manfred
Gough, Robert George
Gould, David Huntington
Gould, Edwin Sheldon
Gould, Jack Richard
Gould, Kenneth Alan
Gourse, Jerome Allen
Graber, Robert Philip
Grabiel, Charles Edward
Graef, Walter L
Graham, Aloysius
Graham, Bruce Allan
Graham, David E
Graham, Eric Stanley
Graham, Harold Nathaniel
Graham, Jack Raymond
Graham, Joseph H
Graham, Paul Roger
Graham, Robert Leslie
Gramera, Robert Eugene
Grams, Gary Wallace
Grand, Paul Sheldon
Grand, Donald R
Grant, Edwin Allen, Jr
Grant, Frederick Warren, Jr
Grant, Peter Malcolm
Grasshoff, Jurgen Michael
Gratz, Roy Fred
Gray, Allan P
Gray, Frederick William
Grdinic, Marcel Rudolph
Greco, Claude Vincent
Green, Brian
Green, Floyd J
Green, John Wilson
Green, Michael John
Greenbaum, Sheldon Boris
Greenberg, Herman Samuel
Greene, Charles Edwin
Greene, Charles Richard
Greene, Frederick Davis, II
Greene, Hoke Smith

Greene, Janice L
Greenfield, Harold
Greenfield, Stanley A
Greenspan, Frank Philip
Greenwald, Richard B
Greenwood, Fred Laurel
Greer, Albert H
Gregg, David Henry
Gregg, Donald Crowther
Gregg, Earl Charles, Jr
Greiner, Richard William
Greizerstein, Walter
Grethe, Guenter
Grettie, Donald Pomeroy
Greyson, William Lawrence
Gribble, Gordon W
Griffin, Claibourne Eugene, Jr
Griffin, Dale Miller, Jr
Griffin, Gary Walter
Griffin, George Robert
Griffin, Rodger W, Jr
Griffin, William Dallas
Griffith, Martin G
Grillot, Gerald Francis
Grimm, Charles Henry
Grina, Larry Dale
Grindstaff, Wyman Keith
Grisar, Johann Martin
Griswold, Paul Hulet, Jr
Gritter, Roy John
Grivsky, Eugene Michael
Groeger, Theodore Oskar
Grose, Herschel Gene
Gross, Benjamin Harrison
Gross, Erhard
Gross, Michael Lawrence
Gross, Paul Hans
Gross, Peter Fredrick
Grosser, Frederick
Grossert, James Stuart
Grostic, Marvin Ford
Grothaus, Clarence (Edward)
Grotta, Henry Monroe
Grovenstein, Erling, Jr
Groves, John Taylor, III
Grubbs, Robert Howard
Gruber, Wilhelm F
Grummitt, Oliver Joseph
Gruskin, Bernard
Grutzner, John Brandon
Gschwend, Heinz W
Guehler, Paul Frederick
Guerrant, William Barnett, Jr
Gugig, William
Guidry, Carlton Levon
Guiducci, Mariano A
Guile, Ralph Lawrence
Gum, Wilson Franklin, Jr
Gumprecht, William Henry
Gut, Marcel
Gunberg, Paul F
Gunther, Wolfgang Hans Heinrich
Gupta, Shyam Kirti
Gurien, Harvey
Gurst, Jerome E
Guss, Cyrus Omar
Gust, John Devens, Jr
Gustafson, Carl Gustaf, Jr
Gustafson, David Harold
Guterbel, Louis Charles
Guthrie, David Burrell
Guthrie, Donald Arthur
Guthrie, James Leverette
Guthrie, Roger Thackston
Gutsche, Carl David
Guttmann, Andrew Titus
Gwynn, Bernard Henry
Haberfield, Paul
Hach, Vladimir
Haddon, Virginia Rae
Haensel, Vladimir
Haffley, Philip Gene
Hager, Robert B
Hagopian, Miasnig
Hahn, Roger C
Haines, Paul Gordon
Hajos, Zoltan George
Hakewill, Henry, Jr
Haley, Robert Currie
Hall, David Warren
Hall, George E
Hall, Henry Kingston, Jr
Hall, J Herbert
Hall, John B
Hall, Kimball Parker
Hall, Luther Axtell Richard
Hall, Philip Layton
Hall, Richard Coumains
Hall, Richard Harold
Hall, Stan Stanley
Hall, Thomas Kenneth
Ham, George Edward
Hamel, Coleman Rodney
Hamel, Edward E
Hamer, Martin
Hamilton, Chester Eugene
Hamilton, James Chipman
Hamilton, Lewis R

Hamilton, William Lander
Hamlet, Zacharias
Hamm, Donald Ivan
Hamm, Kenneth Lee
Hamm, Thomas Edward
Hammaker, Ellwood Meacham
Hammaker, William Curl
Hammer, Charles Rankin
Hammer, Richard Benjamin
Hammer, Richard Hartman
Hammond, George Simms
Hammond, James Alexander, Jr
Hammond, Willis Burdette
Hampton, Alexander
Hampton, Burt Laurent
Hampton, David Clark
Han, Yuri Wha-Yul
Hance, Paul D
Hancock, Charles Kinney
Hancock, John Edward Herbert
Hancock, Kenneth George
Handrick, George Richard
Hanley, James Richard, Jr
Hanley, John Joseph
Hanlon, Harry Thomas
Hanlon, David Paul
Hannah, John
Hanneman, Walter W
Hansen, Donald Willis, Jr
Hansen, Holger Victor
Hansen, John Frederick
Hanson, Harry Thomas
Hanson, Marvin Wayne
Hansrote, Charles Johnson, Jr
Hanzel, Robert Stephen
Haq, Mohammad Zamir-ul
Harbert, Charles A
Harbort, Kenn E
Harding, Maurice James Charles
Hardman, Goetz E
Hardwicke, James Ernest, Jr
Hardy, Edgar Erwin
Hardy, Elizabeth MacGregor
Hardy, William Baptist
Hare, Peter Edgar
Harfenist, Morton
Hargis, J Howard
Hargraves, Chester Arthur II
Hayes, Clarence Frank
Harkin, John McLay
Harlan, Horace David
Harmon, Kenneth Millard
Harmon, Robert E
Harmuth, Charles Moore
Harness, Grant Hopkins
Harnish, Donald Philip
Harowitz, Charles Lichtenberg
Harper, Edwin T
Harrell, Bryant (Eugene, Jr)
Harriman, Benjamin Ramage
Harris, Edwin Randall
Harris, Francis Laurie, Jr
Harris, George Christie
Harris, Guy H
Harris, Henry Earl
Harris, James Joseph
Harris, John Ferguson, Jr
Harris, Thomas Munson
Harrison, Alexander George
Harrison, Ernest Augustus, Jr
Harrison, James Beckman
Harrison, William Ashley
Hart, Harold
Hart, Herbert Dorian
Hart, Phillip A
Hart, William Forris
Hartline, Richard
Hartman, John Alan
Hartman, Kenneth Eugene
Hartman, Paul Francis
Hartzel, Lawrence Woodring
Hartzfeld, Howard Alexander
Hartzler, Harris Dale
Harvey, John, Jr
Harvey, Robert Joseph
Harvey, Ronald Gilbert
Harvey, George Ranson
Hasbrouck, Richard Berend
Hasek, William Robert
Hasek, Robert Hall
Haske, Bernard Joseph
Hassler, William Robert
Hassner, Alfred
Hasty, Noel Marion, Jr
Hatch, Lewis Frederic
Hatch, Melvin (Jay)
Haubein, Albert Howard
Haupschein, Murray
Hause, Norman Laurance
Hauser, Jack W
Hauser, Paul Matthew
Hautala, Richard Roy
Havel, James Joseph
Havlicek, Stephen
Havran, Robert Thomas
Hawbecker, Byron L

Hawkins, Richard Thomas
Hawks, George H, III
Hawley, Lewis Burton, Jr
Hawley, Thomas G, Jr
Hawthorne, Robert Montgomery, Jr
Hay, Allan Stuart
Hayasu, Ryoichi
Hayek, Mason
Hayes, Francis Newton
Hayes, Kenyon (Joseph)
Hayes, Robert Arthur
Haynes, Leroy Wilbur
Hays, Donald R
Hays, John Thomas
Hazen, George Gustave
Heacock, James Flaviel
Heacock, Ronald A
Hearn, Robert Arthur
Hearst, Peter Jacob
Heaton, Charles Daniel
Heberling, Jack Waugh, Jr
Hechenbleikner, Ingenuin Albin
Hecht, Stephen Samuel
Heckman, Robert Arthur
Hedenburg, John Frederick
Hedrick, Glen Willard
Hedrick, Ross Melvin
Heeren, James Kenneth
Hegedus, Louis Stevenson
Heggie, Robert
Heggie, Robert Murray
Heiba, El-Ahmadi Ibrahim
Heiberger, Philip
Heimer, Rudolph Louis
Heimer, Norman Eugene
Heindel, Ned Duane
Heine, Harold Warren
Heiner, Mary Grace
Heininger, Samuel Allen
Heinle, Preston Joseph
Heisey, Lowell Vernon
Heitmeier, Donald Elmer
Heldt, Walter Z
Helgeson, Jorge
Helmick, Larry Scott
Helmkamp, George Kenneth
Helms, John F
Henderson, David Rippey
Henderson, Robert Burr
Henderson, Thomas Ranney
Henderson, Ulysses Virgil, Jr
Hendon, Joseph D
Hendon, William Arthur, Jr
Hendrickson, James Easton
Hendrix, James Easton
Hendrickson, Yngve Gust
Henery-Logan, Kenneth Robert
Henion, Richard S
Henkel, James Gregory
Hennecke, Henry Fred
Hennion, George Felix
Henrici, Henry Emil
Henrick, Clive Arthur
Henry, Joseph Peter
Henson, Paul D
Henzel, Richard Paul
Hepfinger, Norbert Francis
Herber, Roland Eugene
Herber, John Frederick
Herbrandson, Harry Fred
Herbst, Robert Max
Herbstman, Sheldon
Hergert, Herbert L
Herman, Daniel Francis
Hermann, Joseph Leonard
Herold, Robert Johnston
Herr, Ross Robert
Herrick, Elbert Charles
Herrick, Franklin Willard
Herrick, Guy Scott
Herriott, Arthur W
Hershberg, Emmanuel Benjamin
Hershey, John William Baker
Hertler, Walter Raymond
Hertzler, Donald Vincent
Herz, Werner
Hess, Bernard Andes, Jr
Hess, Hans-Jurgen Ernst
Hess, Patrick Henry
Hess, Ronald Eugene
Hess, William Wilson
Hessel, Donald Wesley
Hester, Jackson Boling, Jr
Hetzel, Donald Stanford
Heuberger, Oscar
Heumann, Karl Fredrich
Heuser, Leon John
Hewett, James Veith
Hewson, William Bell
Heyd, Charles E
Heyd, William Ernst
Heying, Theodore Louis
Heyman, Karl
Hickman, Howard Minor
Hicks, Arthur M

Hider, Shibley A
High, LeRoy Bertolet
Highet, Robert John
Higuchi, Takeru
Hileman, Robert E
Hilfiker, Franklin Roberts
Hill, Ada Sinz
Hill, Arthur Joseph, Jr
Hill, Carl McClellan
Hill, Elgin Alexander
Hill, Henry Aaron
Hill, James Aubrey
Hill, James Wagy
Hill, John Hamon Massey
Hill, John William
Hill, Marion Elzie
Hill, Richard Keith
Hill, Trevor Bruce
Hillstrom, Warren W
Hilton, H Wayne
Hines, Paul Steward
Hines, Wallis Gartside
Hinkes, Thomas Michael
Hinkamp, Paul Eugene, (II)
Hinman, Charles Wiley
Hinshaw, Jerald Clyde
Hirsch, Christophe Henri Werner
Hirsch, Arthur
Hirsch, Jerry Allan
Hirsch, Richard Henry
Hirsch, Stephen Simeon
Hirschberg, Albert I
Hirschy, Harlan W
Hisey, Robert Warren
Hix, Dawes Nyukieu
Hix, Homer Bennett
Hlavka, Joseph John
Hobbs, Charles Clifton, Jr
Hobbs, M Floyd
Hoblit, Louis Douglas
Hoch, Paul Edwin
Hochwalt, Carroll Alonzo
Hodes, William
Hodnett, Ernest Matelle
Hoeger, Erhard Fritz
Hoehle, Milton Louis
Hoeksema, Herman, Jr
Hoff, Wilford J, Jr
Hofferth, Burt Frederick
Hoetzel, Charles Bernard
Hoffman, Dorothea Heyl
Hoffman, Henry Allen, Jr
Hoffman, Norman Edwin
Hoffman, Robert Vernon
Hoffman, Roger Allen
Hoffman, Theodore P
Hoffmann, Warren E
Hoffmann, Arthur Kentaro
Hoffmann, Dietrich
Hoffmann, Sandor Alexander
Hoffmeister, Elaine Helen
Hofreiter, Bernard T
Hofsommer, John C
Hogan, John Paul
Hogan, Philip
Hogsed, Milton Jones
Hoh, George Lok Kwong
Hoiness, Connie Marquez
Hoiness, David Eldon
Hokama, Takeo
Hoke, Donald I
Holcomb, Walter Floyd
Holden, Winfried Thomas
Holland, Robert Joseph
Holstein, Ulrich
Holm, Myron Jones
Holm, Reynold Emanuel
Holmes, David Eldon
Holmes, Donald A
Holmes, Jerry Dell
Holmes, R H Lavergne
Holmes, Richard Remsen
Holmquist, Howard Emil
Holoway, Clayton Frank
Holst, Edward Harland
Holt, Robert Louis
Holt, John Robert
Holtkamp, Freddy Henry
Holty, David Webster
Holum, John Robert
Holy, Norman Lee
Holysz, Roman Paul
Homberg, Otto Albert
Homberg, George
Honig, Milton Leslie
Honsberg, Wolfgang
Hood, Horace Edward
Hook, Edwin Oscar
Hoops, Stephen C
Hoover, Fred Wayne
Hoover, John
Hooz, John
Hopkins, Clarence Yardley
Hopkins, Thomas Robert
Hoppens, Harold Albert
Hopper, Paul Frederick
Horeczy, Joseph Thomas
Hornbach, Joseph Michael
Hornbaker, Edwin Dale

Hornberger, Carl Stanley, Jr
Horner, Samuel Emmett, Jr
Hornemann, Ulfert
Horning, James William, Jr
Horning, William Clarke
Horowitz, Robert Miller
Horrocks, Robert H
Hort, Eugene Victor
Hortmann, Alfred Guenther
Horton, Derek
Horton, Joseph William
Horton, Norman Hagood
Horton, Robert Louis
Horwitz, Jerome Philip
Hosansky, Norman Leon
Hosler, Charles Frederick, Jr
Hosler, John Frederick
Hoster, Donald Paul
Hoteling, Eric Bell
Houff, Alva Leroy
Houghlan, William J
Houk, Kendall Newcomb
House, Herbert Otis
House, William T
Houston, James Grey
Houtman, Thomas, Jr
Howald, Jeremiah Mark
Howard, Kenneth Leon
Howard, William Lowry
Howe, Robert Kenneth
Howe, William Jeffrey
Howell, David Moore
Howell, Edward Tillson
Howell, Walter Colston
Howell, William Salem
Hower, Frdolin Alfonse
Hoyer, Vinton Asbury, Jr
Hoyt, Earle B, Jr
Hoyt, Robert Mikell
Hsi, Richard S P
Hsu, Roger Yun Kung
Huang, Charles T L
Huang, Der-Shing
Hubbard, Samuel J
Hubbard, Forest Craven
Hubbard, James K
Huber, Joel E
Huber, Melvin Lefever
Huber, Wilson Frederick
Hudak, Norman John
Hudgin, Donald Edward
Hudlicky, Milos
Hudnall, Phillip Montgomery
Hudson, Frederick Mitchell
Hudson, Glenn Vincent
Huebner, Charles Vincent
Huestis, Laurence Dean
Huffman, Allan Murray
Huffman, Clarence W
Huffman, George Wallen
Huffman, John William
Huffman, John William, Jr
Huffman, Kenneth Robert
Huffman, Robert Wesly
Hughes, David William
Hughes, John Lawrence
Hughes, Mark
Hughes, Robert David
Hull, Carl Max
Hull, Clarence Joseph
Huller, Prentice Roy
Hull, Theodore Lee
Hultquist, Martin Everett
Humber, Leslie George
Hummel, Donald George
Hummel, James Knight
Hundert, Murray Bernard
Hung, William Mo-Wei
Hunger, Gunther K
Hunsberger, Isaac Moyer
Hunsucker, Jerry H
Hunt, Charles Kellogg
Hunt, Donald F
Hunt, John Meacham
Hunt, Russell Aubrey, Jr
Hunt, William Cecil
Hunter, George L K
Hunter, Frank Ray
Hunter, Wood E
Huntsman, William Duane
Hurd, Richard Nelson
Hurdis, Everett Cushing
Hurley, Rupert B
Hurwitz, Melvin David
Husband, Robert Murray
Huskins, Chester Walker
Hussey, Edward Walter
Hussung, Karl Frederick
Hutchcroft, Alan Charles
Hutchinson, Robert Owen, Sr
Hutchinson, James Herbert, Jr
Hutchison, John Joseph
Hutchison, Robert B
Huttenlocher, Dietrich F
Hutton, Thomas Watkins
Huyser, Earl Stanley
Hwang, Bruce You-Huei
Hyatt, Asher Angel
Hyde, Robert Wallace
Hylander, David Peter

Hylton, Thomas Anthony
Hymans, William E
Hyndman, Harry Lester
Hynes, John Barry
Iannicelli, Joseph
Idelson, Martin
Ieyoub, Kalil Phillip
Iffland, Don Charles
Ihndris, Raymond Will
Ihrman, Kryn George
Iloff, Phillip Murray, Jr
Imhoff, Michael Andrew
Immediata, Tony Michael
Impastato, Fred John
Indelicato, Joseph Michael
Indictor, Norman
Ingham, Robert Kelly
Inglessis, Criton George S
Ingraham, Lloyd Lewis
Ingram, Sammy Walker, Jr
Inman, Charles Gordon
Insalaco, Michael Anthony
Inscoe, May Nilson
Inskip, Harold Kirkwood
Iob, Vivian
Ireland, Robert Ellsworth
Isaacson, Eugene I
Isbell, Arthur Furman
Isbell, Raymond Eugene
Isenberg, Norbert
Isensee, Robert William
Izydore, Robert Andrew
Jabloner, Harold
Jablonski, Werner Louis
Jackisch, Philip Frederick
Jackman, Lloyd Miles
Jackson, Peter H
Jackson, Thomas Gerald
Jacob, Theodore August
Jacobs, Harvey
Jacobs, Thomas Lloyd
Jacobson, Bernard Howard
Jacobus, Otha John
Jacoby, Lawrence John
Jacoby, Thomas Franklin
Jaeger, David Allen
Jaeger, Herbert Karl
Jaffe, Edward E
Jaffe, Fred
Jaffe, Marvin Richard
Jahn, Alex Karl
Jahn, Edwin Cornelius
James, David Eugene
James, Floyd Lamb
James, John Cary
James, Philip Nickerson
Jamieson, Norman Clark
Jamison, Joel Dexter
Jampolsky, Lester Mischa
Jankowski, Christopher K
Janssen, Jerry Frederick
Janzen, Edward George
Jaouni, Taysir M
Jaquiss, Donald B G
Jarvis, Bruce B
Jasinski, Jerry Peter
Jayaraman, H
Jean, George Noel
Jeffs, Peter W
Jelinek, Charles Frank
Jen, Yun
Jendrek, John Paul, Jr
Jenkins, Alfred Martin
Jenkins, Herndon
Jenkins, James William
Jenkins, Lloyd Theodore
Jenkins, Philip Winder
Jennings, Bojan Hamlin
Jennings, Carl Anthony
Jensen, Arnold William
Jeremias, Charles George
Jerina, Donald M
Jernow, Jane L
Jerome, Joseph Benedict
Jeskey, Harold Alfred
Jewell, Richard A
Jewett, John Gibson
Jezl, James Louis
Jirkovsky, Ivo
Jobin, Ralph Alfred
Johanson, Alva Joseph
Johns, William Francis
Johnson, Alexander Lawrence
Johnson, Alyn William
Johnson, Arnold Nathaniel
Johnson, Calvin Keith
Johnson, Carl Arnold
Johnson, Carl Edwin
Johnson, Carl Erick
Johnson, Carl Randolph
Johnson, David Aaron
Johnson, Donald Curtis
Johnson, Donald Wayne
Johnson, Fatima Nunes
Johnson, Francis
Johnson, George Frederick
Johnson, Harry William, Jr
Johnson, James Elver

Johnson, John Arnold
Johnson, John Hal
Johnson, John Harpster
Johnson, John Raven
Johnson, John Webster, Jr
Johnson, Kenneth
Johnson, Kenneth Eugene
Johnson, LaVell R
Johnson, Leroy Dennis
Johnson, Michael Ross
Johnson, Oscar Hugo
Johnson, Rayner Selby
Johnson, Richard Lawrence
Johnson, Robert Gudwin
Johnson, Robert Hall
Johnson, Robert Reiner
Johnson, Robert William, Jr
Johnson, Roland Norman
Johnson, Sharon Leijoy
Johnson, Thomas Lynn
Johnson, Wallace Delmar
Johnson, William H
Johnson, William K
Johnson, William Summer
Johnston, Frederick Lewis
Johnston, Gordon Robert
Johnston, Hugh William
Johnston, James Douglas
Johnston, Jean Vance
Jonas, John Joseph
Jones, Alfred Russell
Jones, Daniel Gethen
Jones, David A, Jr
Jones, Edward Stephen
Jones, Elmer Everett
Jones, Evan Thomas
Jones, Frederick Mason, III
Jones, George Francis
Jones, Giffin Denison
Jones, Glenn Clark
Jones, Gordon Henry
Jones, Harold Lester
Jones, Howard
Jones, James Holden
Jones, Jean Elmore
Jones, Jennings Hinch
Jones, John Bryan
Jones, John Kenyon Netherton
Jones, Lee Bennett
Jones, Maitland, Jr
Jones, Maurice Harry
Jones, Merriam Arthur
Jones, Paul Raymond
Jones, Peter Hadley
Jones, Roger Alan
Jones, Ronald Goldin
Jones, Stephem Thomas
Jones, William Howry
Jones, William Jonas, Jr
Jones, William Maurice
Jones, William Norton, Jr
Jordan, David M
Jordan, Manuel Albert
Jorgensen, Eugene Clifford
Joshua, Henry
Joullie, Madeleine M
Joyce, Maurice Michael
Joyce, Richard Evans
Judge, Joseph Malachi
Judson, Horace Augustus
Jules, Leonard Herbert
Jung, James Moser
Jung, John Andrew, Jr
Jungermann, Eric
Jungman, Richard A
Jurewicz, Anthony Theodore
Just, George
Juster, Norman Joel
Kaback, Stuart Mark
Kabara, Jon Joseph
Kabbe, Frederick Carl
Kadaba, Pankaja Kooveli
Kadesch, Richard Gilmore
Kaeding, Warren William
Kagan, Benjamin
Kagan, Fred
Kagan, Jacques
Kagan, Herbert Paul Nash
Kahn, Donald Jay
Kahlenberg, Eilhard Nash
Kaiser, Edward William
Kaiser, Edwin Michael
Kaizerman, Samuel
Kalenda, Norman Wayne
Kallianos, Andrew George
Kalota, Dennis Jerome
Kaltenbronn, James S
Kamienski, Conrad William
Kaminski, James Joseph
Kaminski, Joan Mary
Kampmeier, Jack A
Kane, Peter Tai Yuen
Kane, Bernard James
Kane, Howard L
Kane, James Joseph
Kane, William Paul

Kang, Jung Wong
Kaniecki, Thaddeus John
Kapff, Sixt Frederick
Kaplan, Fred
Kaplan, Julius Frank
Kaplan, Leonard
Kaplan, Melvin
Kaplan, Ralph Benjamin
Kaplan, William
Karabatsos, Gerasimos J
Karabinos, Joseph Vincent
Karady, Sanor
Karl, Curtis Lee
Karll, Robert E
Karmas, George
Karnatz, Frank Albert
Karten, Marvin J
Kaslow, Christian Edward
Kasparek, Stanislav Vaclav
Kassal, Robert James
Kasubick, Robert Valentine
Katchman, Arthur
Katner, Allen Samuel
Katz, Leon
Katz, Morton
Katz, Thomas Joseph
Kauder, Otto Samuel
Kauer, James Charles
Kauffman, Harry Frey
Kauffman, Joel Mervin
Kaufman, Daniel
Kaufman, Harold Alexander
Kaufman, Kurt Dunn
Kaufman, Priscilla C
Kaufmann, Esteban
Kavanagh, Kevin Enda
Kay, Edward Leo
Kay, Irving Allan
Kaye, Samuel
Kealy, Thomas Joseph
Keana, John F W
Kearley, Francis Joseph, Jr
Keating, James T
Keaveney, William Patrick
Keefer, Larry Kay
Keeffe, James Richard
Keene, Frederick J
Kehoe, Lawrence Joseph
Kehr, Clifton Leroy
Keiser, Jeffrey E
Keith, Dennis Dalton
Keller, James Lloyd
Keller, William John
Kelley, Alec Ervin
Kelley, Maurice Joseph
Kellgren, John
Kellman, Raymond
Kellogg, Craig Kent
Kellogg, Richard Morrison
Kelly, Floyd W, Jr
Kelly, Robert Charles
Kelly, Robert James
Kelly, Ronald Burger
Kelly, Thomas Edward
Kelly, Thomas Ross
Kelly, Walter James
Kemp, Daniel Schaeffer
Kende, Andrew S
Kendrick, Lawrence W, Jr
Kenley, Richard Alan
Kennedy, Flynt
Kennedy, John Elmo, Jr
Kennedy, Richard J
Kennedy, Robert Michael
Kennedy, Robert Wilson
Kennelly, Mary Marina
Kennelly, Edward Joseph
Kenny, David Herman
Kent, Frank William
Kent, Robert Eugene
Keogh, Michael John
Keough, Allen Henry
Keown, Robert William
Kepner, Richard Edwin
Kerber, Robert Charles
Kercheval, James William
Kerr, Ralph Oliver
Kerr, Richard John
Kerr, Richard A
Kerridge, Kenneth A
Kertesz, Dennis Jay
Kessel, William George
Kesslin, George
Kestner, Melvin Michael
Ketcham, Roger
Ketterer, Charles Clifford
Khana, Ausat Ali
Khorana, Har Gobind
Kice, John Lord
Kiefer, Edgar Francis
Kiehlmann, Eberhard
Kienle, Robert Nelson
Kierstead, Richard Wightman
Kilday, Warren D
Kilsheimer, Sidney Arthur
Kim, Chung Sul
Kim, Leo
Kim, Yungki

King, Ellis Gray
King, Henry Lee
King, James Frederick
King, John Mathews
King, Lafayette Carroll
King, Michael M
King, Te Piao
Kingsbury, Charles Alvin
Kingsbury, William Dennis
Kinnel, Robin Bryan
Kinstle, Thomas Herbert
Kintner, Robert Roy
Kirby, Ben Harrison
Kirch, Lawrence S
Kircher, Henry Winfried
Kircher, Frederick Karl
Kirchner, Justus George
Kirdani, Rashad Y
Kirk, James Curtis
Kirk, Philip Moore
Kirkpatrick, Joel Lee
Kirst, Herbert Andrew
Kise, Mearl Alton
Kiskis, Ronald Clements
Kissman, Henry Marcel
Kittila, Richard Sulo
Kittle, Paul Alvin
Klaas, Nicholas Paul
Klabunde, Kenneth John
Klacsmann, John Anthony
Klager, Karl
Klanderman, Bruce Holmes
Klass, Donald Leroy
Klayman, Daniel Leslie
Kleibacker, Wilson McAlarney
Kleiman, Morton
Klein, Bernard
Klein, David Xavier
Klein, Francis Michael
Klein, Howard Clarason
Klein, Richard M
Klein, William Arthur
Kleinman, Roberta Wilma
Kleinschmidt, Albert Willoughby
Kleinschmidt, Roger Frederick
Kleinspehn, George Gehret
Klemchuk, Peter Paul
Klemm, LeRoy Henry
Kliegman, Jonathan Morris
Klijanowicz, James Edward
Klimisch, Richard L
Kline, Richard Henry
Klinedinst, Paul Edward, Jr
Klingele, Harold Otto
Klingensmith, George Bruce
Klingsberg, Erwin
Klink, Joel Richard
Klinke, David J
Kloetzel, Milton Carl
Klopotek, David L
Klopping, Hein Louis
Klose, Thomas Richard
Kluger, Ronald H
Klundt, Irwin Lee
Kmiecik, James Edward
Knaggs, Edward Andrew
Knapman, Fred William
Knapp, Gordon Grayson
Knapp, Robert Lester
Knauer, Bruce Richard
Knechel, William Franklin
Knee, Terence Edward Creasey
Knell, Martin
Knight, David Bates
Knight, James Albert, Jr
Knipmeyer, Hubert Elmer
Knobl, George Martin, Jr
Knobloch, Fred William
Knobloch, James Otis
Knock, Frances Engelmann
Knopf, Robert John
Knowles, Cecil Martin
Knowles, M B
Knowles, Richard N
Knox, Walter Robert
Knox, William Tyndall
Knutson, Clarence Arthur, Jr
Kober, Ehrenfried H
Koch, Kay Frances
Koch, Stanley D
Koch, Tad H
Koch, Walter Theodore
Koe, B Kenneth
Koenig, Fred R
Koenig, Paul Edward
Koenig, Thomas W
Kofron, William G
Kogon, Irving Charles
Kohlbrenner, Philip John
Kohlhase, William Lawrence
Kolb, Doris Kasey (Mrs K E Kolb)
Kolb, Kenneth Emil
Kolka, Alfred Jerome
Kolyer, John M
Komarmy, Julius Michael
Konecky, Milton Stuart
Konigsbacher, Kurt S

CHEMISTRY, ORGANIC

Konigsberg, Moses
Kononenko, Oleg K
Konzelman, Leroy Michael
Koob, Robert Philip
Koons, Charles Bruce
Kopecky, Karl Rudolph
Koppel, Gary Allen
Korach, Malcolm
Kornblum, Nathan
Kornet, Milton Joseph
Korst, James Joseph
Korte, William David
Kosak, Alvin Ira
Kosolapoff, Gennady Michael
Kossoy, Aaron David
Koster, Robert Allen
Kotch, Alex
Kotick, Michael Paul
Kotnik, Louis John
Kovacic, Eugene George
Kovacic, Joseph Edward
Kovacic, Peter
Kovacs (Nagy), Hanna
Kowalkany, George Norman
Kowanko, Nicholas
Koziowski, Robert H
Krackov, Mark Harry
Kraemer, John Francis
Kraft, Elise
Krahler, Stanley Earl
Krakower, gerald W
Kramer, John Karl Gerhard
Kramer, William J
Kranti, Michael Joseph Anthony
Krantz, Allen
Krantz, Karl Walter
Krantz, Reinhold John
Kranzfelder, Arthur Leonard
Krapcho, Andrew Paul
Kraus, Kenneth Wayne
Krause, Josef Gerald
Kraychy, Stephen
Krbechek, Leroy O
Kreevoy, Maurice M
Kreh, Donald Willard
Kreibich, Roland
Kreider, Henry Royer
Krespan, Carl George
Kress, Bernard Hiram
Kress, Thomas Joseph
Kresse, Jerome Thomas
Kreuzer, James Leon
Kreysa, Frank Joseph
Kriens, Richard Duane
Krimen, Lewis Irvin
Krishnamurthy, Sundaram
Kritchevsky, David
Krogh, Lester Christensen
Kroll, Harry
Kronberger, Karlheinz
Kropp, James Edward
Kropp, Paul Joseph
Kruh, Daniel
Krubiner, Alan Martin
Krubsack, Arnold J
Krueger, James Elwood
Krueger, John W
Krueger, Paul A
Krueger, Robert A
Krueger, Spencer M
Krueger, William E
Krug, Robert Charles
Krumel, Edward William
Kruse, Carl William
Krutak, James John, Sr
Krysiak, Henry R
Kubler, Donald Gene
Kucera, Clare H
Kucera, Thomas J
Kuceski, Vincent Paul
Kuck, Julius Anson
Kuder, James Edgar
Kudzin, Stanley Francis
Kuehl, Frederick Albert, Jr
Kuehne, Martin Eric
Kuhlmann, George Edward
Kuhn, Hans Heinrich
Kuivila, Henry Gabriel
Kulka, Marshall
Kulp, Stuart S
Kumar, Surinder
Kumler, Philip L
Kumli, Karl F
Kundu, Nakuleswar
Kuo, Chan-Hwa
Kuo, Norman Yu-Neng
Kupchan, S Morris
Kupel, Richard E
Kupstas, Edward Eugene
Kurath, Paul
Kurchacova, Elva S
Kurihara, Norman Hiromu
Kuritzkes, Alexander Mark
Kurkov, Victor Peter
Kurtz, Abraham V
Kuryla, William C

Kurz, Michael E
Kushner, Arthur Simon
Kutik, Leon
Kutney, James Peter
Kwalwasser, William David
Kwart, Harold
Kwartler, Charles Edward
Kwok, Wo Kong
Kyba, Evan Peter
Kyung, Jai Ho
Labana, Santokh Singh
Labanca, Dominick A
Labows, John Norbert, Jr
LaCount, Robert Bruce
Lada, Arnold
Laemmle, Joseph
Laferriere, Arthur L
Lagally, Ralph Werner
Lagowski, Jeanne Mund
Lakritz, Julian
Lalonde, Robert Thomas
Lamb, Robert Charles
Lamb, Robert W
Lamb, Sandra Ina
Lamba, Ram Sarup
Lambert, Frank Lewis
Lambert, James LeBeau
Lambert, Joseph B
Lambert, Rogers Franklin
Lampman, Gary Marshall
Land, Anthony Hamilton
Landau, Edward Frederick
Lande, Saul
Lander, Harold Paul
Landesberg, Joseph Marvin
Landgrebe, John A
Landis, Phillip Sherwood
Landolt, Robert George
Landucci, Lawrence L
Lane, Charles A
Lane, Philip Charles
Lang, Stanley Albert, Jr
Lange, Gordon Lloyd
Langeland, William Enberg
Langford, Robert Bruce
Langford, Arthur Walter, Jr
Langkammerer, Carl Martin
Langley, Robert Charles
Langsjoen, Arne Nels
Langston, James Horace
Langworthy, William Clayton
Lann, Joseph Sidney
Lanson, Herman Jay
Lapporte, Seymour Jerome
Largman, Theodore
Larkin, John Michael
Larock, Richard Craig
Larrabee, Clifford Everett
Larrabee, Richard Brian
Larrison, Millard Samuel
Larsen, Eric Russell
Larsen, John W
Larsen, Gerald Willis
Larson, Gustav Olof
Larson, Harold Olaf
Larson, Leslie L
Larson, Lester Mikkel
Larson, Thomas E
LaSala, Edward Francis
Lasthuysen, Willem
Laszlo, Pierre
Latimer, Steve B
Latourette, Harold Kenneth
Lau, Philip T S
Laubach, Gerald David
Laubscher, Aner Nearhood
Laufer, Daniel A
Laughton, Paul Herbert
Laughton, Paul MacDonell
Laursen, Paul Herbert
Lautenschlager, Friedrich Karl
Lavagnino, Edward Ralph
Lavanish, Jerome Michael
Laver, Murray Lane
Laverty, John Joseph
Lavigne, Andre Andre
Lavine, Theodore Frederick
Lawhead, James Stout
Lawrence, Franklin Isaac Latimer
Lawson, Daniel David
Lawson, David Francis
Lawson, Julian Keith, Jr
Layer, Robert Wesley
Layton, Roger
Lazardis, Christina Nicholson
Lazarus, Allan Kenneth
Leach, James Moore
Leaf, Clyde William
Leahy, Sidney Marcus
Leak, John Clay, Jr
Leake, Preston Hildebrand
Leake, William Walter
Leal, Joseph Rogers
Leary, Ralph John
LeBel, Norman Albert
LeBlanc, Jerald Thomas
LeBleu, Ronald Eugene
LeClaire, Claire Dean

Ledbetter, Harvey Don
Ledlie, David B
Lednicer, Daniel
Lee, Alfred Tze-Hau
Lee, Cheuk Man
Lee, Choi Chuck
Lee, Dean Ralph
Lee, Donald Garry
Lee, Lester Tsung-Cheng
Lee, Lieng-Huang
Lee, Richard Jui-Fu
Lee, Shui Lung
Lee, Sung Ki
Lee, Tzoong-Chyh
Lee, William Wei
Lee, Young-Jin
Lee, Yu-Sun
Leeds, Morton W
Leekley, Robert Michell
Leeper, Robert Walz
Leete, Edward
Lefar, Morton Saul
Lefebvre, Yvon
Leffingwell, John C
Leffler, Martin Templeton
LeGoff, Eugene
Lehman, Joe Junior
Lehne, Richard Karl
Lehr, Roland E
Lehrman, Leo
Lehman, Lawrence Fred
Leiby, Robert William
Leiff, Morris
Leimgruber, Willy
Leis, Donald George
Leiserson, Lee
Leitch, Leonard Christie
Lemal, David M
LeMay, Harold E, Jr
Lemieux, Raymond Urgel
Lemke, Thomas Franklin
Lemper, Anthony Louis
Lengyel, Istvan
Lenhert, Anne Gerhardt
le Noble, William Jacobus
Lenox, Ronald Sheafter
Lenz, George Richard
Leonard, John Edward
Leopold, Robert Summers
Lepse, Paul Arnold
Le Quesne, Philip William
Leshin, Richard
Lessard, Jean
Lester, Charles Turner
Leston, Gerd
Letsinger, Robert Lewis
Leubner, Gerhard Walter
Levand, Oscar
Lever, Cyril, Jr
Levesque, Charles Louis
Levin, Robert Harold
Levin, Ronald Harold
Levine, Aaron William
Levine, Robert
Levine, Samuel Gale
Levine, Stephen Alan
Levinson, Alfred Stanley
Levis, William Walter, Jr
Levitt, George
Le Von, Ernest Franklin
Levy, Edward Robert
Levy, Jack Benjamin
Levy, Joseph
Levy, Joseph Benjamin
Levy, Louis A
Levy, Morton Frank
Lew, Baak Wai
Lewars, Errol George
Lewenz, George F
Lewin, Seymour Z
Lewis, Cameron David
Lewis, Charles Edward
Lewis, Daniel William
Lewis, Dennis Allen
Lewis, Dennis Osborne
Lewis, Elmer James
Lewis, George Edwin
Lewis, Morton
Lewis, Neil Jeffrey
Lewis, Robert Glenn
Lewis, Sheldon Noah
Lewis, Silas Davis
Leyland, Harry Mours
Leznoff, Clifford Clark
Li, Tao Ping
Liao, Hsiang Peng
Liao, Tsung-Kai
Liauw, Koei-Liang
Libbey, William Jerry
Libby, William Harris
Liberles, Arno
Licari, James John
Lichtenwalter, Glen
Lichter, Robert (Louis)
Liddell, Robert William, Jr
Lieberman, Hillel

Liebman, Arnold Alvin
Lien, Arthur Philip
Liepins, Raimond
Lies, Thomas Andrew
Liggett, Lawrence Melvin
Light, Kenneth Karl
Light, Robley Jasper
Lightner, David A
Lilienfeld, Irving
Lillya, Clifford Peter
Lilyquist, Marvin Russell
Lin, Kang
Lindbeck, Wendell Arthur
Lindberg, James George
Lindberg, Steven Edward
Linden, Robert Marion
Lindquist, Robert Nels
Lindsay, Jacque K
Lindsay, Kenneth Lawson
Lindsey, Rico Vernon, Jr
Lindsey, William B
Link, William B
Linn, Bruce Oscar
Linn, Carl Barnes
Linn, William Joseph
Linsay, Ernest Charles
Linsk, Jack
Linstromberg, Walter William
Linton, Howard Richard
Lipinski, Christopher Andrew
Lipka, Benjamin
Lipkin, David
Lippert, Arnold Leroy
Lipscomb, Robert DeWald
Lipsitz, Paul
Lipton, Samuel Harry
Lira, Emil Patrick
Lisk, George F
Little, Edwin Demetrius
Little, John Clayton
Little, John Stanley
Little, Randel Quincy, Jr
Little, Raymond Daniel
Little, William Frederick
Liu, Robert Shing-Hei
Liu, Sophia Yan
Livengood, Samuel Miller
Lloyd, William Gilbert
Lloyd, Winston Dale
Lo, Elizabeth Shen
LoCicero, Joseph Castelli
Lockwood, Karl Lee
Loeb, Melvin Lester
Loeffler, Larry James
Loening, Kurt L
Loeppert, Richard Henry
Loeppky, Richard N
Loev, Bernard
Loftfelman, Frank Fred
Loftfield, Robert Berner
Login, Robert Bernard
Lohmann, Karl H
Lohner, Donald J
Lohr, Lester Jay
Loire, Norman Paul
Lok, Roger
Lokensgard, Jerrold Paul
Lomax, Eddie
Lombardino, Joseph George
Lombardo, Anthony
Loncrini, Donald Francis
Long, Frank Wesley, Jr
Long, Louis, Jr
Long, Norman Oliver
Longenecker, William Hilton
Longhi, Raymond
Longone, Daniel Thomas
Longroy, Allan Leroy
Looker, James Howard
Looker, Jerome J
Lorenz, Carl Edward
Lorenz, Donald H
Lorenz, John Clark
Lorenz, Roman R
Losee, Michael Leonard
Losekamp, Bernard Francis
Lothrop, Warren Craig
Loucas, Spiro P
Loudfoot, James Herbert
Loudon, Gordon Marcus
Lourie, Alan David
Loughran, Gerard Andrew, Sr
Love, George M
Love, Jim
Lovett, Eva G
Low, Hans
Lowe, James N
Lowe, Orville G
Lowe, Warren
Lowrance, William Wilson, Jr
Lowric, Harman Smith
Lowry, Betty Jean Ragle
Lowry, Thomas Hastings
Lowstuter, William Robert
Lu, Mary Kwang-Ruey Chao
Lu, Peter Herman
Lubitz, Betty Baum

Lucier, John J
Luck, Russell M
Ludwick, Adriane Gurak
Ludwig, Bernard John
Ludwig, Frederick John, Sr
Ludwig, Jerome Howard
Ludwig, Richard Eli
Lufkin, James E
Luibrand, Richard Thomas
Lukach, Carl Andrew
Lukas, George
Lukes, Robert Michael
Lukin, Marvin
Lundeen, Allan Jay
Lurie, Arnold Paul
Luskin, Leo Samuel
Lusskin, Robert Miller
Lutz, Wilson Boyd
Lwowski, Walter Wilhelm Gustav
Lyle, Gloria Gilbert
Lyle, Robert Edward
Lyman, William Chester, Jr
Lynch, Darrel Luvene
Lynch, Don Murl
Lynn, John Wendell
Lyon, Cameron Kirby
Lyons, James Edward
Lyznicki, Edward Peter, Jr
Ma, Tsu Sheng
Mabry, Tom Joe
MacAvoy, Thomas Coleman
MacDonald, Alan Angus
MacDonald, Robert Neal
MacDonald, Stewart Ferguson
MacDowell, Denis W H
Maddox, V Harold, Jr
Madhav, R
Madoff, Milton
Machiele, Delwyn Earl
Machleder, Warren Harvey
MacKay, Francis Patrick
MacKay, Kenneth Donald
MacKellar, Donald Gordon
Mackenzie, Neil Mitchill
MacLachlan, James Daniel
MacLean, David Bailey
Macmorine, Hilda Mildred Grace
McMullen, Clinton William
Macnair, Richard Nelson
Macphee, Kenneth Erskine
Magid, Ronald
Magnani, Arthur
Magnien, Ernest
Magnuson, Eugene Robert
Maerov, Sidney Benjamin
Magarian, Robert Armen
Magee, Richard Joseph
Magrane, John Kearns, Jr
Mahajan, Kishan Paul
Mahan, John Elmer
Mahesh, Virendra B
Mahler, Walter
Maienthal, Millard
Mainen, Eugene Louis
Maizel, Benjamin Leo
Majer, Jaroslav
Majewski, Robert Francis
Majewski, Theodore E
Malaiyandi, Murugan
Malan, Rodwick Lapur
Malchick, Sherwin Paul
Malhotra, Sudarshan Kumar
Malkemus, John David
Mallory, Clelia Wood
Mallory, Frank Bryant
Malloy, Thomas Patrick
Malone, William Maxton
Malter, Margaret Quinn
Malzahn, Ray Andrew
Manatt, Stanley L
Mancera, Octavio
Manchester, Donald Fraser
Mandell, Leon
Mandlik, Jayant V
Mange, Franklin Edwin
Manger, Martin C
Mangham, Jesse Roger
Mango, Frank Donald
Manhas, Maghar Singh
Mani, Rama I
Manion, Jerald Monroe
Maniscalco, Ignatius Anthony
Manly, Donald G
Mann, David Jacob
Manning, Joseph Victor
Mansfield, Joseph Victor
Marantz, Laurence Boyd
Marascia, Frank Joseph
Marburg, Stephen
Marcelli, Joseph F
March, Louis Charbonnier
Marco, Gino Joseph
Marcy, Willard
Marfey, Sviatopolk Peter

Margerison, Richard Bennett
Maricich, Tom John
Maricq, John
Mariella, Raymond Peel
Marino, Joseph Paul
Mark, Robert Vincent
Mark, Victor
Markees, Diether Gaudenz
Markey, Sanford Philip
Markgraf, John Hodge
Markiw, Roman Teodor
Markley, Lowell Dean
Markovitz, Mark
Marks, Gerald Samuel
Marmer, William Nelson
Marmor, Solomon
Marquardt, Dawn Nilan
Marquardt, Roland Paul
Marquis, Edward Thomas
Marr, Eleanor B
Marsh, Nat Huyler
Marshall, Charles Wheeler
Marshall, Fred Taylor
Marshall, Frederick J
Marshall, James Arthur
Marshall, James Lawrence
Marshall, Maud Alice
Martellock, Arthur Carl
Martin, Charles Edward
Martin, James Cullen
Martin, James D
Martin, Robert Allan
Martin, Robert O
Martin, Tellis Alexander
Martin, William Harry
Martin, William Royall, Jr
Martinez, Alberto Magin
Martinez Nadal, Noemi G
Marullo, Nicasio Philip
Marvel, John Thomas
Marvell, Elliot Nelson
Marx, John Norbert
Marx, Michael
Masamune, Satoru
Masciantonio, Philip (X)
Mascioli, Rocco Lawrence
Maslow, Philip Herman
Mason, Perry Shipley, Jr
Massiah, Thomas Frederick
Masters, John Edward
Masuelli, Frank John
Mathers, Alexander Pickens
Mathews, David A
Mathews, Frederick John
Mathews, Walter Kelly
Mathison, Ian William
Matray, Otto Jack
Matson, Howard John
Matsuda, Ken
Matsumoto, Ken
Matt, Joseph
Matta, Michael Stanley
Mattano, Leonard August
Matthews, Clifford Norman
Matthews, Demetreos Nestor
Matthews, Gary Joseph
Mattison, Phillip LeRoy
Mattox, Vernon Ross
Mattson, Guy C
Mattson, Raymond Harding
Matuszko, Anthony Joseph
Matzner, Edwin Arthur
Matzner, Markus
Maul, James Joseph
Maurer, John Edward
Maury, Lucien Garnett
Mavity, Julian Maris
May, Ernest Max
May, Walter
Maycock, Jerry Ray
Mayer, Raymond Parm
Mayfield, Darwin Lyell
Maynard, Carl Wesley, Jr
Maynard, John Thomas
Mayo, Dana Walker
Mazur, Octavio
Mazur, Stephen
Mazzeno, Laurence William
Mazzocchi, Paul Henry
McAleer, William Joseph
McAlpine, James Bruce
McArthur, Colin Richard
McArthur, Richard Edward
McBride, Edward Francis
McBride, Joseph James, Jr
McBurney, Lane Fordyce
McCain, George Howard
McCain, James Herndon
McCaleb, Kirtland Edward
McCall, Marvin Anthony
McCane, Donald Irwin
McCarthy, James Francis
McCarthy, James Ray, Jr
McCarty, John Randolph
McCarty, Frederick Joseph
McCarty, John Edward
McCasland, Gifford Ewing
McCaully, Ronald James

McClelland, Charles Paul
McClenachan, Ellsworth
McClenahan, William St Clair
McClenon, John R
McCloskey, Chester Martin
McClure, George Richard
McClure, Judson P
McCollum, Anthony Wayne
McConnell, James Francis
McConnell, Virginia Fenner
McCormack, John Joseph, Jr
McCormack, William Brewster
McCoy, David Ross
McCoy, George
McCoy, Joseph Wesley
McCoy, Layton Leslie
McCoy, V Eugene, Jr
McCreight, Robert Willis
McCullough, James Douglas, Jr
McCullough, John James
McCullough, Thomas F
McDermott, John Patrick
McDonald, David William
McDonald, Richard Norman
McDonough, Everett Goodrich
McDowell, Wilbur Benedict
McElheny, George Clark
McEwen, William Edwin
McFarland, John William
McGahren, William James
McGeer, Edith Graef
McGinn, Clifford
McGonigal, William E
McGovern, Terrence Phillip
McGrath, Michael Glennon
McGrath, Thomas Frederick
McGraw, Gerald Wayne
McGraw, Hugo Richard
McGreer, Donald Edward
McGregor, Donald Neil
McGregor, Stanley Dane
McGuinness, James Anthony
McGuire, Francis Joseph
McGurk, Donald J
McHugh, Kenneth Laurence
McIntyre, Thomas William
McKague, Allan Bruce
McKay, Donald Edward
McKay, Jerry Bruce
McKee, James Robert
McKeever, Charles H
McKelvie, Neil
McKeon, James Edward
McKinney, Roger Minor
McKinnon, David M
McKirahan, Richard Duncan
McLaen, Donald Francis
McLaughlin, Joseph, Jr
McLaughlin, Robert Lawrence
McLean, Stewart
McLellan, Crawford Reid
McManus, James Michael
McManus, Samuel Plyler
McMaster, Paul D
McMichael, Kirk Dugald
McMillian, Frank Lebarron
McMillion, C Robert
McMullen, Charles Henry
McMullen, Warren Anthony
McNairy, Sidney A, Jr
McNally, James Green, Jr
McNamara, James Henry
McNeil, Harry Daniel, Jr
McNelis, Edward Joseph
McNiven, Neal Lindsay
McPherson, Charles Allen
McPherson, James Beverley, Jr
McRowe, Arthur Watkins
McSweeney, Jean
McCurdy, Orville L
McWhorter, Earl James
Mead, John Marcus
Mead, Thomas Edward
Mears, Thomas Wood
Mechanic, Gerald
Mecklenborg, Kenneth Thomas
Medeiros, Robert Whippen
Mednick, Morton L
Meeks, John Sawyers
Meeks, Benjamin Spencer, Jr
Megson, Frederic Houghton
Mehmedbasich, Enver
Mehta, Avinash C
Meienhofer, Johannes Arnold
Meierhoefer, Alan William
Meikle, Richard William
Meinert, Walter Theodore
Meinke, Wilmon William
Meinschein, Warren G
Meinwald, Jerrold
Meinwald, Yvonne Chu
Meisel, Seymour Lionel
Meisenheimer, John Long
Meislich, Herbert
Meister, Peter Dietrich
Mekler, Arlen B
Melby, Lester Russell
Melchior, Norten Cass

Melera, Attilio
Meloy, Carl Ridge
Melton, Thomas Mason
Meltzer, Robert Israel
Melville, George S, Jr
Melville, Marjorie Harris
Melvin, Horace Willis
Memeger, Wesley, Jr
Menapace, Lawrence William
Mench, John Warren
Mendel, Arthur
Mendelson, Wilford Lee
Menger, Fred M
Menz, William Wolfgang
Mercer, Gerald Dean
Meresz, Otto
Merijanian, Aris
Merriwether, Lewis Smith
Merritt, Richard Foster
Merrow, Raymond Theodore
Mertel, Holly Edgar
Merten, Helmut Ludwig
Mertens, Edward William
Meschino, Joseph Albert
Messer, Wayne Ronald
Messing, Sheldon Harold
Metz, Fred Lewis
Metzger, James Douglas
Meuly, Walter C
Meyer, Delbert Henry
Meyer, Ferdinand Clark
Meyer, Gregory Carl
Meyer, Rich Bakke, Jr
Meyer, Walter Leslie
Meyer, William Paul
Meyers, Albert Irving
Meyers, Cal Yale
Meyers, Martin Bernard
Meyers, Robert Allen
Miale, Joseph Peter
Micetich, Ronald George
Mich, Thomas Frederick
Michael, James Richard
Michaelis, Carl I
Michel, Rudolph Henry
Micheli, Robert Angelo
Michell, John Humfrey
Michels, Julian Getz
Michelson, Malvin J
Middleton, William Joseph
Midland, Michael Mark
Miedema, Eddy
Miesel, John Louis
Mighton, Charles Joseph
Mihailovski, Alexander
Mihajlov, Vsevolod S
Mihina, Joseph Stephen
Mijal, Chester Francis
Mikulec, Richard Andrew
Milakofsky, Louis
Miles, George Benjamin
Miles, Marion Lawrence
Milewich, Leon
Milionis, Jerry Peter
Mill, Theodore
Miller, Bernard
Miller, Charles Alexis
Miller, Charles Edward
Miller, Edsel Leo
Miller, Edward George
Miller, Emery B
Miller, Francis Marion
Miller, Frank
Miller, Franklyn David
Miller, Glen Russel
Miller, Harry Brown
Miller, Howard Anthony
Miller, James L
Miller, James Richard
Miller, James Robert
Miller, Jn Joseph
Miller, Larry Lee
Miller, Leonard Edward
Miller, Leroy Jesse
Miller, Nathan C
Miller, Philip
Miller, Robert Clay
Miller, Robert Dennis
Miller, Robert Witherspoon
Miller, Roy G, Jr
Miller, Russel Bryan
Miller, Stanley Johnson
Miller, Thomas Gore
Miller, Wilbur Hobart
Miller, William Martin
Miller, William Riedel
Miller, William Taylor
Miller, William Theodore
Milligan, Barton
Milligan, Terry Wilson
Millington, James E
Mills, Frank D
Mills, Gordon Frederick
Mills, Jack F
Milne, Henry Bayard
Minard, Robert David
Minckler, Leon Sherwood, Jr
Miner, Carl Shelley, Jr

Miner, Robert Scott, Jr
Minesinger, Richard Rockwell
Minn, James
Minor, John Threecvelous
Minyard, James Patrick
Miron, Simon
Mirza, John
Miskel, John J, (Sr)
Miskel, John Joseph, Jr
Miskus, Raymond P
Misner, Robert E
Mitch, Frank Allan
Mitchell, Roy Ernest
Mitchell, Thomas Owen
Miwa, Thomas Kanji
Mizzoni, Renat Herbert
Moates, Robert Franklin
Mochel, Gene Vernon
Mochel, Walter Edwin
Modic, Frank Joseph
Modic, William L
Moe, George
Moersch, George William
Moffett, John Gilbert
Moffett, Eugene Wilkin
Moffett, Robert Bruce
Moffett, Samuel McKee
Mohacsi, Erno
Mohan, Arthur G
Mohrig, Jerry R
Moir, Robert Young
Mokotoff, Michael
Moll, Harold Westbrook
Molnar, Nicholas M
Molnar, Stephen P
Molyneux, Russell John
Monagle, John Joseph, Jr
Monahan, Audrey Small
Monroe, Harold W
Monroe, Bruce Malcolm
Monroe, Pearle Arvel
Monsimer, Harold Gene
Montana, Andrew Frederick
Montgomery, John Atterbury
Montgomery, Lawrence Kernan
Montgomery, Ronald Eugene
Monti, Stephen Arion
Moon, Sung
Mooney, David Samuel
Mooney, Richard Warren
Moore, Alexander Mazyck
Moore, Donald R
Moore, Gordon George
Moore, Harold W
Moore, James Alexander
Moore, James Alfred
Moore, Leonard Oro
Moore, Leslie David
Moore, Maryalice Conley
Moore, Perry Alldredge
Moore, Ralph Gower Davies
Moore, Richard Anthony
Moore, Richard E
Moore, Richard Newton
Moore, Theron Langford
Moore, James Robert
Moore, William Robert
Moores, Mead Stephen
Moores, Gilbert Ellsworth
Moppett, Charles Edward
Moran, Juliette May
Moran, William Joseph
Moran, Simon Joseph
Morand, Peter
Morath, Richard Joseph
Morello, Edwin Francis
Morgan, Marcus S
Morgan, Robert Lee
Mori, Peter Taketoshi
Mori, Raymond I
Moriarty, Robert M
Morin, Robert Bennett
Morita, Hirokazu
Moritz, Terence Clark
Morill, Cletus Eugene
Morris, David Julian
Morris, Gene Franklin
Morris, Rupert Clarke
Morris, Quentin L
Morrison, Glenn C
Morrison, Harry
Morrison, James Daniel
Morrison, Robert W, Jr
Morrison, Wiley Herbert, III
Moritz, Fred Leonard
Morrow, Homer Nicholas, Jr
Mortimer, Charles Edgar
Morton, John West, Jr
Morton, Thomas Hellman
Mosby, William Lindsay
Moschopedis, Speros E
Moses, Ronald Elliot
Mosher, Harry Stone
Moss, Robert Allen
Moss, Rodney Dale
Mottus, Edward Hugo
Moulton, Wilbur Norton
Mount, Ramon Albert
Mowat, John Halley

Mower, Howard Frederick
Moyer, Calvin Lyle
Moyer, Joseph Donald
Moyer, Melvin Isaac
Moyer, Patricia Helen
Moyerman, Robert Max
Moyle, Clarence Llewellyn
Mrozik, Helmut
Muchowski, Joseph Martin
Mudrak, Anton
Mueller, George Peter
Mueller, Max Best
Muffley, Harry Chilton
Muhs, Merrill Arthur
Mukherjee, Tapan Kumar
Muller, Thomas C
Mulineaux, Richard Denison
Mulvaney, James Edward
Mulvey, Dennis Michael
Munavalli, Somashekhar
Munch, John Howard
Munchausen, Linda Lou
Mundell, Percy Meldrum
Munn, George Edward
Munsell, Monroe Wallwork
Munson, H Randall, Jr
Murai, Kotaro
Murai, Ronald Lee
Murchison, John Taynton
Murdoch, Arthur
Murdoch, Joseph Richard
Murphy, Clarence John
Murphy, Daniel Barker
Murphy, James Gilbert
Murphy, Robert Carl
Murphy, Walter Thomas
Murphy, Zatis Luain
Murray, John Joseph
Murray, Joseph
Murray, Leo Thomas
Murray, Thomas J
Murray, Thomas Pinkney
Murty, Dasika Radha Krishna
Musser, David Musselman
Musser, Michael Tuttle
Muth, Chester William
Mutsch, Edward L
Myerly, Richard Crebs
Myers, Dale Kamerer
Myers, Donald Royal
Myers, John Albert
Myhre, Philip C
Myles, William John
Nace, Harold Russ
Nachtigall, Guenter Willi
Nader, Allan E
Naegele, Edward Wister, Jr
Naff, Marion Benton
Nafissi-Varchei, Mohammad Mehdi
Nagel, Donald Lewis
Nagel, Fritz John
Nager, Urs Felix
Nagler, Robert Carlton
Nagvary, Joseph
Nair, Vasu
Nakagawa, T William
Nakanishi, Koji
Napier, Roger Paul
Naples, Felix John
Naples, John Otto
Naqvi, Saiyid Mahmoodul Hasan
Naragon, Ernest Ashley
Narang, Saran A
Narayan, Tv Lakshmi
Naro, Paul Anthony
Narske, Richard Martin
Nash, Edmund Garrett
Nave, Paul Michael
Naves, Renee G
Naylor, Marcus A, Jr
Naylor, Robert S
Neale, Robert S
Nealey, Richard John
Nebenzahl, Linda Levine
Nebenzahl, Lakshmanan
Neeman, Moshe
Negishi, Ei-Ichi
Neher, Clarence M
Neidig, Howard Anthony
Neil, Gary Lawrence
Neil, Thomas C
Neill, Alexander Bold
Neklutin, Vadim Constantin
Nelan, Donald Royce
Nelb, Robert Gilman
Nelsen, Stephen Flanders
Nelsen, Thomas Robert
Nelson, Bernard Andrew
Nelson, Aaron Louis
Nelson, Don B
Nelson, Edward R
Nelson, George Leonard
Nelson, Jerry Allen
Nelson, John Archibald
Nelson, Nils Keith
Nelson, Reginald David

Nelson, Roger Peter
Nelson, Wayne Franklin
Nemec, Josef
Nemec, Joseph William
Nersassian, Arthur
Nesty, Glenn Albert
Nethcut, Philip Edwin
Netleton, Donald Edward, Jr
Neufeld, Cornelius Herman Harry
Neumann, Fred William
Neumann, Marguerite
Neumiller, Harry Jacob, Jr
Neven, Maurice C
Nevill, William Albert
Nevill, Robert Stephen
Nevin, Thomas D
Newbold, William Edward
Newbury, Jack
Newcombe, George R
Newell, Marjorie Pauline
Newkome, George R
Newman, Melvin Spencer
Newman, Norman
Newman, Stanley Ray
Newmark, Harold Leon
Newsom, Herbert Charles
Newson, Raymond A
Newton, Thomas Allen
Nicholas, Paul Peter
Nicholls, Robert Van Vliet
Nichols, Joseph
Nichols, Samuel Harding, Jr
Nicholson, Isadore
Nicholson, John Angus
Nickerson, Mortimer Henderson
Nickon, Alex
Nicolaides, Ernest D
Nicolaides, Nicholas
Nieh, Marjorie T
Nieh, Arnold Thor
Nielsen, Donald R
Nielsen, Lawrence Arthur
Nielsen, Stuart Dee
Nienhouse, Everett J
Nightingale, Dorothy Virginia
Ning, Robert Ye Fong
Nisbet, Michael Alan
Nitz, Otto William Julius
Niu, Joseph H Y
Nix, Sydney Johnston, Jr
Nobis, John Francis
Nodiff, Edward Albert
Nordby, Harold Edwin
Nordin, Ivan Conrad
Nordlander, John Eric
Nordstrom, J David
Norell, John Reynolds
Noe, Eric Arden
Noe, James L
Noland, Wayland Evan
Noll, Clarence Irwin
Nonemaker, Larry Franklin
Norcross, Bruce Edward
Nordby...
Norman, George Russel
Norman, Oscar Loris
Norris, Terry Orban
Norris, William Philip
Northcott, Jean
Northington, Dewey Jackson, Jr
Norton, Charles J
Norton, Richard Vail
Norton, Ted Raymond
Noshay, Allen
Novak, Ernest Richard
Novak, Ronald William
Novello, Frederick Charles
Nowak, Robert Michael
Noyce, Donald Sterling
Noyes, Paul R
Nugent, Maurice Joseph, Jr
Nunn, William Ralph
Nunmy, Leslie Grey, Jr
Nussbaum, Alexander Leopold
Nussbaum, Siegfried
Nyi, Kayson
Nyilas, Emery
Nyquist, Harlan LeRoy
Nystrom, Robert Forrest
Oakes, Billy Dean
Obbink, Russell C
Oberender, Frederick G
Obermayer, Arthur S
Oberster, Arthur Eugene
O'Brien, Daniel H
O'Brien, Michael Harvey
Obrycki, Richard
Ochrymowycz, Leo Arthur
Ocken, Paul Robert
O'Connell, Edmond J, Jr
O'Connor, James J
O'Connor, Joseph Michael
O'Dell, Durward George
Odell, Norman Raymond
Odioso, Raymond C
Odom, Homer Clyde, Jr
O'Donohue, Cynthia H
Oehlschlaeger, Herman Fred

Offenhauer, Robert Dwight
Offner, Peter
Oftedahl, Marvin Loren
Ogilvie, James William, Jr
Oh, Chan Soo
O'Hara, Elizabeth Mary
Oishi, Masayoshi
Ojakaar, Leo
Olah, George Andrew
Olah, Judith Agnes
Olberg, Ralph Charles
Olechowski, Jerome Robert
Oleesen, John Allen
O'Leary, Marion Hugh
Oliver, Gene Paul
Oliver, Kenneth Leo
Olin, Arthur David
Olin, Jacqueline S
Olinger, Janet
Olofson, Roy Arne
Olsen, Carl John
Olsen, Richard Kenneth
Olsen, Ronald G
Olson, Danford Harold
Olson, Edwin S
Olson, Melvin Martin
O'Mara, Michael Martin
O'Meara, Desmond
Onak, Thomas Philip
Ondetti, Miguel Angel
O'Neal, Grady Malcolm
O'Neal, Hubert Ronald
Onopchenko, Anatoli T
Opic, Joseph Wendell
Opie, Thomas Ranson
Oppelt, John Christian
Orchin, Milton
Orlando, Charles M
Orloff, Harold David
Oroshnik, William
Orphanos, Demetrius George
Orr, James Cameron
Orr, Robert S
Orr, Wilson Lee
Orth, George Otto, Jr
Osawa, Eiji
Osawa, Yoshio
Osborn, Claiborn Lee
Osborne, David Wendell
Osman, Elizabeth Mary
Ospenson, Joseph Nils
Ostercamp, Daryl Lee
Osterholtz, Frederick David
Osuch, Carl
Otey, Felix Harold
Ottenbrite, Raphael Martin
Otto, Robert Crittenden
Otto, Ferdinand Philip
Ouderkirk, John Thomas
Ouellette, Robert J
Owen, John Thomas
Owen, John Reindel
Owen, Louis John
Owen, Terence Cunliffe
Owens, Frederick Hammann
Oyama, Vance I
Oziomek, James
Ozog, Francis Joseph
Pace, Henry Alexander
Pace, William Theodore
Pace Asciak, Cecil
Pachter, Irwin Jacob
Pacofsky, Edward Anthony
Padgett, Algie Ross
Padra, Frank George
Padwa, Albert
Pagano, Alfred Horton
Page, Harold Alfred
Pagni, Richard
Paine, John Raymond
Painter, Edgar Page
Paisley, David M
Pal, Bimal Chandra
Palchak, Robert Joseph Francis
Palethorpe, George
Pallos, Ferenc M
Palmer, Fred Shank
Palmer, Glenn Earl
Palopoli, Frank Patrick
Pandell, Alexander Jerry
Panetta, Charles Anthony
Panici, Ronald J
Panson, Gilbert Stephen
Panzer, Jerome
Panzera, Pete
Papa, Anthony Joseph
Papadopoulos, Eleftherios Paul
Papaioannaou, Christos George
Papanikolaou, Nicholas E
Pappalardo, Leonard Thomas
Pappas, Betty Colleen
Pappas, James John
Pappas, Nicholas
Pappas, Raphael
Paquette, Leo Armand

Parcell, Robert Ford
Parent, Richard Alfred
Parham, William Eugene
Parish, Roger Cook
Park, Chung Ho
Parker, Richard Ghrist
Parker, William Lawrence
Parkinson, Gilbert Gordon, Jr
Parks, Terry Everett
Parmerter, Stanley Marshall
Parsons, Robert W, Jr
Parsons, Timothy F
Partch, Richard Earl
Partridge, John Jerome, Jr
Pasto, Daniel Jerome
Pataki, John
Patchett, Arthur Allan
Patel, Kalyanji U
Patinkin, Seymour Harold
Patmore, Edwin Lee
Patrick, George Robert
Patrick, James Burns
Patrick, Timothy Benson
Patrick, Tracy Minard, Jr
Patsiga, Robert A
Patterson, Dennis Ray
Patterson, George Harold
Patterson, John Arthur
Patterson, John Miles
Patterson, Ralph Francis
Patterson, Ronald Brinton
Patton, James Winton
Patton, Leo Wesley
Patton, Tad LeMarre
Paudler, William W
Paukstelis, Joseph V
Paul, Albert P
Paul, Edward Gray
Paul, Rolf
Paulshock, Marvin
Paulson, Donald Robert
Paulson, Mark Clements
Paustian, John Earle
Pavia, Donald Lee
Paviak, Stanley C
Pavlic, Albert Alan
Pavlos, John
Pawlowski, Norman E
Payne, George Bernson
Peake, Clinton J
Peakes, Lawson Vernon
Pearson, Myrna Schmidt
Pearson, Robert Edward
Pearson, Tillmon Henry
Pearson, Wesley A
Peascoe, Warren Joseph
Peck, Richard Merle
Peet, Norman Eugene
Pegolotti, James Alfred
Pelczar, Francis A
Pellegrini, Frank C
Pellegrini, John P, Jr
Pelletier, S William
Pelley, Ralph L
Pellon, Joseph
Pelosi, Stanford Salvatore, Jr
Penner, Siegfried Edmund
Pennington, Frank Cook
Pepoy, Louis John
Pepper, James Morley
Pepperman, Armand Bennett, Jr
Pera, John Dominic
Perchonock, Carl David
Percival, William Colony
Perkel, Robert Jules
Perkins, Edward George
Perlin, Arthur Saul
Perlman, Kato (Katherine) Lenard
Perlmutter, Howard D
Pernert, John Carl
Perozzi, Edmund Frank
Perrin, Charles Lee
Perron, Yvon G
Perry, Donald Dunham
Perry, Edward Mahlon
Perry, Mary Hertzog
Perry, Robert Hood, Jr
Perry, Rufus Patterson
Personeus, Gordon Rowland
Perun, Thomas John
Peters, Howard McDowell
Peters, John Thomas
Petersen, Ingo Hans
Petersen, Joseph Claine
Peterson, Robert J
Peterson, Donald J
Peterson, James Oliver
Peterson, Janet Brooks
Peterson, Melbert Eugene
Peterson, Paul E
Peterson, William Hampton
Peterson, William Roger
Petit, Michael Geoffrey
Petrarca, Anthony Edward
Petrellis, Panayotis C
Petterson, Robert Carlyle
Pettit, George Robert

Pettit, Rowland
Pews, Richard Garth
Pfeiffer, Francis Richard
Pheasant, Richard
Phillips, Judson Christopher
Phillips, Arthur Page
Phillips, Barry Allen
Phillips, Benjamin
Phillips, Bruce Edwin
Phillips, Donald David
Phillips, Joseph
Phillips, Lee Vern
Phillips, Marshall
Phillips, Richard Fifield
Phillips, Robert Edward
Pian, Charles Hsueh Chien
Piatak, David Michael
Piccolini, Richard John
Piche, Lucien
Pickard, Porter Louis, Jr
Pickart, Don Edward
Piehl, Frank John
Piel, Elmar Viking
Pier, Harold William
Pierce, Arleen Cecilia
Pierson, William Grant
Pietrusza, Edward Walter
Pietsch, Gerhard Josef
Pike, Ronald Marston
Pike, Roscoe Adams
Pincock, Richard Earl
Pincus, Irving
Pinder, Albert Reginald
Pine, Lloyd A
Pine, Stanley H
Pines, Herman
Pines, Seemon H
Pinkney, Paul Swithin
Pinkus, Jack Leon
Pino, Lewis Nicholas
Pinto, Frank G
Piper, James Robert
Piper, James Underhill
Pirkle, William H
Piroue, Robert Paul
Pittman, Charles U, Jr
Pivonka, William
Pizzini, Louis Celeste
Plambeck, Louis, Jr
Plank, Don Allen
Plant, Howard Leon
Plapinger, Robert Edwin
Platau, Gerard Oscar
Platt, Alan Edward
Pleasants, Elsie W
Pledger, Huey, Jr
Pletcher, Wayne Albert
Plimmer, Jack Reynolds
Plonsker, Larry
Plostnieks, Janis
Plue, Arnold Frederick
Plummer, Benjamin Frank
Plunkett, Roy Joseph
Pluscec, Josip
Podrebarac, Eugene George
Poe, James Edgar
Poe, Richard D
Poel, Russell J
Pohland, Albert
Pohland, Hermann W
Polevy, John Henry
Politzer, Ieva Ruks
Pollack, Maxwell Aaron
Pollak, Kurt
Pomerantz, Irwin Herman
Pomerantz, Martin
Pomonis, James George
Pond, David Martin
Ponder, Billy Wayne
Ponticello, Gerald S
Ponticello, Ignazio Salvatore
Pontius, Dieter J J
Poon, Bing Toy
Poos, George Ireland
Popoff, Ivan Christoff
Popp, Frank Donald
Popper, Thomas Leslie
Port, William Solomon
Porter, Herschel Donovan
Porter, Ned Allen
Portnoy, Norman Abbye
Poshkus, Algirdas C
Posner, Gary Herbert
Postelnek, William
Postl, Anton
Posvic, Harvey Walter
Potter, Howard A
Potter, Neil H
Potts, Kevin T
Poulos, Nicholas A
Powell, Burwell Frederick
Powell, Leo S
Powell, Warren Howard
Powers, Daniel D
Powers, Edward James
Powers, Jack W
Powers, John Clancey, Jr

Powers, John Michael
Poziomek, Edward John
Prager, Julianne Heller
Prahl, Helmut Ferdinand
Prasad, Raj Nandan
Pratt, Ernest Fay
Pratt, Richard J
Pratt, Yolanda Tota
Precopio, Frank Mario
Preiss, Donald Merle
Press, Jeffery Bruce
Preston, Jack
Preston, Robert Kreig
Preti, George
Price, Charles Coale
Price, Edward Hector
Price, Harold Anthony
Price, Howard Charles
Price, Leonard
Price, Martin Burton
Prichard, William W
Priesing, Charles Paul
Pritchett, Ervin Garrison
Privett, James E
Prober, Maurice
Prochaska, Robert Joseph
Proffitt, Thomas Jefferson, Jr
Prokipcak, Joseph Michael
Proops, William Robert
Prosser, Robert M
Prosser, Thomas John
Prout, Franklin Sinclair
Pruckmayr, Gerfried
Pryor, William Austin
Pucknat, John Godfrey
Puerckhauer, Gerhard Wilhelm Richard
Pugliese, Michael
Pummer, Walter John
Punderson, John Oliver
Purrington, Suzanne T
Puterbaugh, Walter Henry
Putnam, Robert Conrad
Putnam, Robert Ervin
Putzig, Donald Edward
Pyle, James L
Pyler, Richard Ernst
Pynadath, Thomas I
Pyron, Raymond Scott
Quin, Louis Dubose
Quinn, Edwin John
Quinn, James Gerard
Quisenberry, Richard Keith
Quo, Sih-Gwan
Raabe, Austin Bauer
Raaen, Vernon F
Raasch, Maynard Stanley
Raban, Morton
Rabideau, Peter W
Rabiger, Dorothy June
Rabjohn, Norman
Rademacher, Leo Edward
Rader, Charles Phillip
Radford, Herschel Donald
Radford, Terence
Radlick, Philip Chris
Rae, Louis Otto
Raether, Louis Otto
Raffauf, Robert Francis
Rafuse, Mary Jane Lounsbury
Raha, Chitta Ranjan
Rainey, William Thomas, Jr
Raizman, Paula
Rajagopalan, Parthasarathi
Rakhit, Sumanas
Rakoff, Henry
Raleigh, James Arthur
Raley, Charles Francis, Jr
Ramachandran, Subramania
Ramey, Bobbie Joe
Ramey, Chester Eugene
Ramirez, Fausto
Ramler, Edward Otto
Ramp, Floyd Lester
Ramsay, Ogden Bertrand
Ramsden, Hugh Edwin
Ranck, Ralph Oliver
Rand, Leon
Randen, Neil Allen
Ranieri, Richard Leo
Ranney, Maurice William
Ransley, Derek Leonard
Ranson, William Wade
Rao, Pemmaraju Narasimha
Rao, Yedavalli Shyamsunder
Rapaport, Eliezer
Rapoport, Henry
Rapp, Robert Dietrich
Rapp, Waldean G
Rastetter, William Harry
Ratchford, William Paul
Rathenberg, Michael William
Rattenbury, Kenneth Harrison
Ratts, Kenneth Wayne
Rauch, Emil Bruno
Rauch, Stewart Emmart, Jr
Rauhut, Michael McKay
Raulins, Nancy Rebecca
Raunio, Elmer Kauno

Rausch, David John
Rausch, Douglas Alfred
Rausch, Gerald
Rausch, Marvin D
Raut, Kamalakar Bakrishna
Ravel, Terence William
Ravely, Melville Fuller
Ravindran, Nair Narayanan
Rawalay, Surjan Singh
Rawlinson, David John
Ray, William Jackson, Jr
Ray-Chaudhuri, Dilip K
Raymond, Maurice A
Raynolds, Stuart
Razniak, Stephen L
Razdan, Raj Kumar
Read, Donald Earle
Ream, Bernard Claude
Reap, James John
Reardon, Joseph Edward
Rebek, Julius, Jr
Rebenfeld, Ludwig
Rebstock, Mildred Catherine
Reckhow, Warren Addison
Recsei, Andrew A
Redlich, Dorothy Von
Redmore, Derek
Ree, Buren Russel
Reeder, Charles Edgar
Rees, Alun Hywel
Rees, Thomas Charles
Rees, William Wendell
Reese, Floyd Ernest
Reese, Millard Griffin, Jr
Reeve, Edward Wilkins
Reeves, Richard Edwin
Reeves, W Preston
Regan, Thomas Hartin
Reggel, Leslie
Rehberg, Chessie Elmer
Rehfuss, Mary
Reiber, Harold George
Reich, Charles
Reich, Donald Arthur
Reich, Hans
Reich, Hans Jurgen
Reich, Ieva Lazdins
Reichard, Douglas Warren
Reid, Donald Eugene
Reid, Evans Burton
Reid, Jack Richard
Reid, James Cutler
Reid, Sidney George
Reid, Stanley Lyle
Reid, Thomas S
Reider, Malcolm John
Reidies, Arno H
Reif, Donald John
Reiff, Harry Elmer
Reifschneider, Walter
Reilly, Charles Bernard
Reilly, Edward Leo
Reilly, William Leo
Reily, William Singer
Reimann, Hans
Reinecke, Charles Everett
Reinecke, Manfred Gordon
Reinhardt, Robert Milton
Reinheimer, John David
Reintjes, Marten
Reist, Elmer Joseph
Reitz, John Marsteller
Reitz, Robert Rex
Relyea, Douglas Irving
Remar, Joseph Francis
Remers, William Alan
Remes, Nathaniel L
Remington, William Roscoe
Renfrow, William Burns, Jr
Rennhard, Hans Heinrich
Rentmeester, Kenneth R
Replogle, Lanny Lee
Rerick, Mark Newton
Resconich, Samuel
Ressler, Charlotte
Reyes, Zoila
Reynolds, Brian Edgar
Reynolds, Francis Joseph
Reynolds, George Arthur
Reynolds, Rosalie Dean (Sibert)
Reynolds-Warnhoff, Patricia
Rhinesmith, Herbert Silas
Rhoads, Sara Jane
Rhodes, Robert Carl
Rhum, David
Riccobono, Paul Xavier
Rice, David E
Rice, Frederick Anders Hudson
Rich, Richard Douglas
Richards, George Martin
Richards, Jack Lester
Richards, Marvin Sherrill
Richardson, Alfred, Jr
Richardson, Arlan Gilbert
Richardson, Graham McGavock
Richardson, Kathleen Schueller

Richardson, Paul Noel
Richardson, Wallace Lloyd
Richardson, William Harry
Richer, Jean-Claude
Richmond, Charles William
Richter, Edward Eugene
Richter, George Holmes
Richter, Reinhard Hans
Richter, Sidney Bernard
Richwine, John Robert
Rickborn, Bruce Frederick
Ricker, Donald Oscar
Ridgway, Robert Worrell
Rieger, Anne Lloyd
Rieger, William Holley
Riehl, Mary Agatha
Rieke, Reuben Dennis
Riemann, James Michael
Rife, William C
Rigdon, Orville Wayne
Rigney, James Arthur
Rike, Zeb W, III
Rila, Charles Clinton
Riley, Robert Lee
Rim, Yong Sung
Rimmer, Robert W
Rinderknecht, Heinrich
Rinehart, Kenneth Lloyd, Jr
Rinehart, Robert Eugene
Ripka, William Charles
Rislove, David Joel
Ritcey, R Richard
Ritchie, Gordon M
Ritchie, Calvin Donald
Rivers, Paul Michael
Rivers, Prince
Rizzi, George Peter
Roach, J Robert
Roach, Paul G
Robb, Ernest Willard
Robbins, Paul Edward
Robbins, Robert
Robbins, Thomas Ennis, Jr
Roberson, Elbert B, Jr
Roberts, Bryan Wilson
Roberts, Carleton W
Roberts, Floyd Edward, Jr
Roberts, Francis Donald
Roberts, Harry Edward
Roberts, John D
Roberts, Richard W
Roberts, Royston Murphy
Roberts, Thomas David
Robertson, Dale Norman
Robertson, Donald Edwin
Robertson, Jerry Earl
Robertson, Ross Elmore
Robin, Burton Howard
Robin, Michael
Robins, Janis
Robins, Morris Joseph
Robins, Richard Dean
Robins, Roland Kenith
Robinson, Cecil Howard
Robinson, Charles Albert
Robinson, Charles Nelson
Robinson, John Howard
Robinson, John W, Jr
Robinson, Kenneth Robert
Robinson, Robert Earl
Roder, Donald Mason
Roderick, William Rodney
Rodewald, Lynn B
Rodewald, Paul Gerhard, Jr
Rodgman, Alan
Rodig, Oscar Rudolf
Roe, Arthur
Roehrig, Gerald Ralph
Roelofs, Wendell Lee
Rogan, John B
Rogers, Edward Franklin
Rogers, Stearns Walter
Rogers, Tommie Gene
Rogerson, Thomas Dean
Rogic, Milorad Mihailo
Roha, Max Eugene
Rohde, Kenneth Lincoln
Roitman, James Nathaniel
Rolih, Robert J
Roller, Peter Paul
Romain, Charles B
Romo, Jesus
Ronald, Bruce Pender
Ronald, Robert Charles
Roobol, Norman R
Root, Charles Arthur
Ropp, Gus Anderson
Rosamond, James Donald
Rose, Charles Buckley
Rose, Ira Marvin
Rose, James Stephenson
Rose, Norman Carl

Rosegay, Avery
Rosen, Aaron A
Rosen, Marvin
Rosen, Perry
Rosen, William Edward
Rosenberg, Hans Reinhard
Rosenberg, Joseph
Rosenblum, Myron
Rosenbrook, William, Jr
Rosenburg, Dale Weaver
Rosene, Clarence James
Rosene, Robert Bernard
Rosenfeld, Jerold Charles
Rosenkranz, George
Rosenman, Edward P
Rosenquist, Edward P, Jr
Rosenstock, Paul Daniel
Rosenthal, Alex
Rosenthal, Arnold Joseph
Rosi, David
Rosi, Rudolph
Rosowsky, Andre
Ross, Alexander
Ross, Camilla Brems
Ross, Daniel Louis
Ross, Donald Lewis
Ross, Joseph Hansbro
Ross, Robert Edward
Ross, Robert M
Ross, Ronald Burns
Rossi, Louis J
Rossiter, Bryant William
Rossito, Conrad
Roth, Barbara
Roth, Jerome A
Roth, Marie M
Roth, Philip B
Roth, Robert George
Roth, Ronald John
Rothman, Edward Samuel
Rothrock, Thomas Stephenson
Roura, Miguel Jacinto
Rouse, Robert S
Rousseau, Viateur
Rovnyak, George Charles
Rowe, James Linoln
Rowland, Charles Sherman
Rowland, Ralph L
Rowton, Richard Lee
Roy, Dibyendu Nath
Roy, Marie L
Royals, Edwin Earl
Rubenstein, Kenneth E
Rubin, Alan Barry
Rubin, Isaac D
Rubin, Mordecai B
Rubinstein, Nathan
Rubinstein, Harry
Rubottom, George M
Ruby, Philip Randolph
Ruddy, Arlo Wayne
Rudel, Harry William
Rudesill, James Turner
Rudner, Bernard
Rudy, Thomas Philip
Rueman, Sven Helmuth
Rugen, Donald Frederick
Runquist, Olaf A
Ruoff, William (David)
Rush, James E
Rush, Kent Rodney
Ruskin, Bernice Heyman
Russell, Charlotte Sananes
Russell, Glen Allan
Russell, Henry Franklin
Russell, John George
Russell, Peter Byron
Russey, William Edward
Russo, Thomas Joseph
Rust, Frederick Farlow
Rutenberg, Morton Wolf
Rutherford, Henry Ames
Rutherford, Kenneth Gerald
Ruthruff, Robert Freeborn
Rutkin, Philip
Rutkowski, Alfred John
Rutledge, Thomas Franklin
Rutter, Jerry L
Ruyle, William Vance
Ryan, Anne Webster
Ryan, John William
Ryan, Joseph Dennis
Ryan, Richard Patrick
Ryder, Bernard Leroy
Ryder, Elliott Elkington, Jr
Ryerson, George Douglas
Saari, Walfred Spencer
Sabacky, M Jerome
Sacco, Louis Joseph, Jr
Saegebarth, Klaus Arthur
Saffy, Max Errol
Sager, William Frederick
Sahli, Muhammad S
St John, Wayne Lloyd
Salamon, Ivan Istvan
Salamone, Joseph C
Salce, Ludwig
Salemi, Oreste Leroy

Salisbury, Lynn
Salivar, Charles Joseph
Sallay, Stephen
Salminen, Ilmari Fritiof
Salomone, Ramon Angelo
Saltiel, Jack
Saltonstall, Clarence William, Jr
Saltzman, Martin D
Salvesen, Robert H
Sanchez, Robert A
Sanderfer, Paul Otis
Sanderson, Edwin S
Sandri, Joseph Mario
Sands, Richard Dayton
San Filippo, Joseph, Jr
Sanford, Robert Alois
Santaniello, Anthony Frank
Santelli, Thomas Robert
Santi, James Owen
Santi, Daniel V
Santilli, Arthur A
Sapino, Chester, Jr
Sarantakis, Dimitrios
Sarbach, Donald Victor
Sareff, Lewis Hastings
Sarges, Reinhard
Saroff, Harry Arthur
Sartoris, Nelson Edward
Sasin, Richard
Satkowski, William Briscoe
Sauer, John Carl
Sauer, Carl Kilbourne
Sauers, Richard Frank
Sauers, Ronald Raymond
Saul, George Archer
Saunders, Charles Richard
Saunders, James Henry
Saunders, Martin
Sause, Henry William
Sauter, Frederick Joseph
Savage, Dennis Jeffrey
Savitz, Maxine Lazarus
Sawdey, George Washington
Saxe, Bernhard David
Sayigh, Adnan Abdul Rida
Scaiola, Luciano Carol
Scalera, Mario
Scalzi, Francis Vincent
Scamehorn, Richard Guy
Scanio, Charles John Vincent
Scanlan, Mary Ellen
Scarpellino, Joseph
Schaaf, Kurt Herbert
Schaaf, Robert Lester
Schaafsma, Bernard Richard
Schaap, Arthur Paul
Schaap, Luke Anthony
Schaap, Ward Beecher
Schach, Von Witenau, Manfred
Schack, Carl J
Schadt, Frank Leonard, III
Schaefer, Arthur Edward
Schaefer, Frederic Charles
Schaeffer, James Robert
Schaefgen, Robert Eben
Schaleger, Larry L
Schappell, Frederick George
Scharver, Jeffrey Douglas
Schauble, J Herman
Schearer, William Richard
Schechter, Milton Seymour
Scheidt, Francis Matthew
Scheiner, Peter
Scheirer, Kirby Vaughn, Jr
Scherer, Robert Allan
Scheuer, Paul Josef
Schenyder, Mary Anne
Schickedantz, Paul David
Schiff, Sidney
Schiffmann, Elliot
Schilling, Curtis Louis, Jr
Schilling, Curtis Louis
Schimelpfenig, Clarence William
Schink, Chester Albert
Schipper, Edgar
Schirch, Laverne Gene
Schirmer, Joseph P, Jr
Schisla, Robert M
Schlameus, Herman Wade
Schleigh, William Robert
Schlenk, Herbert
Schlessinger, Richard H
Schlossman, Irwin S
Schmalz, Alfred Chandler
Schmelli, Francis Lawrence
Schmerling, Louis
Schmid, George Henry
Schmidt, Paul J
Schmit, Gaston L
Schmitt, William Joseph
Schmitz, John Vincent
Schmukler, Seymour
Schnabel, Wilhelm J
Schnack, Larry G

Schneider, Charles Aloysius
Schneider, Henry Joseph
Schneider, John Arthur
Schneider, Ronald Alan
Schneider, William Paul
Schnettler, Stewart Wright
Schnizer, Richard Anselm
Schnoes, Arthur Wallace
Schnoes, Heinrich Konstantin
Schock, Richard Unger, Jr
Schoellmann, Guenther
Schoenholz, Daniel
Schoenthaler, Arnold Charles
Scholl, Allen M
Scholl, Philip Cornelius
Schollenberger, Charles Sundy
Scholnick, Frank
Schramm, Charles H
Schreck, James Otto
Schreiber, Eric Christian
Schreiber, William Lewis
Schriesheim, Alan
Schroeder, Herbert August
Schroeder, Herman Elbert
Schroeder, Juel Pierre
Schroeder, Leland Roy
Schroeder, Walter Arthur
Schroeder, William, Jr
Schroll, Gene E
Schrotenboer, Gordon Harvey
Schubert, Walter John
Schubert, Wolfgang Manfred
Schubert, Bruno Otto Gottfried
Schuerch, Conrad
Schuetz, Robert David
Schuetze, Clarke E
Schuler, Elmer Christian
Schuler, Robert Frederick
Schulley, John Damian
Schultz, Arthur George
Schultz, Frederick John
Schultz, Harry Pershing
Schultz, John E
Schultz, Robert George
Schultz, Thomas Henry
Schultz, Donald Norman
Schultz, Johann Christoph Friedrich
Schumacher, Ignatius
Schumacher, Joseph Nicholas
Schumacher, Roy Joseph
Schumann, Edward Lewis
Schuster, David Israel
Schwab, Arthur William
Schwab, Theodore Carl
Schwartz, Joseph Robert
Schwartz, Martin Alan
Schwartz, Hans Jakob
Schwartz, Maurice Jacob
Schwarzer, Carl G
Schweizer, Carl Earle
Schweizer, Edward Ernest
Schwoegler, Edward John
Sciarini, Louis John
Scott, Alastair Ian
Scott, Andrew Edington
Scott, Donald Albert
Scott, Francis Leslie
Scott, Franklin James
Scott, George Prescott
Scott, George William
Scott, Peter Michael
Scott, Robert Blackburn, Jr
Scott, Samuel LeRoy
Scozzie, James Anthony
Scribner, John David
Seagers, William James
Searle, Norman Edward
Searle, Roger
Searles, Arthur Langley
Searles, Scott, Jr
Sedlak, John Andrew
Sedlak, Michael
Sedor, Edward Andrew
Seeley, Millard Garfield
Seffl, Raymond James
Segall, Stanley
Seger, Francis Michael
Seib, Paul A
Seiber, James N
Seifert, Wolfgang K
Seigler, David Stanley
Sellas, James Thomas
Seligman, Robert Bernard
Sellers, John William
Selover, James Carroll
Selter, Gerald A
Seltzer, Raymond
Seltzer, Stanley
Seiwitz, Charles Myron
Semenuk, Nick Sarden
Semmelhack, Martin F
Sengupta, Sisir K
Senn, Oliver Frederic
Sepkoski, Joseph John
Serve, Munson Paul

Servis, Kenneth L
Sestanj, Kazimir
Setliff, Frank Lamar
Settine, Robert Louis
Seubold, Frank Henry, Jr
Seufert, Ludwig E
Seus, Edward J
Seven, Raymond Peter
Severson, Roland George
Sexsmith, David Randal
Seymour, Keith Morton
Shabica, Anthony Charles, Jr
Shackelford, James Marshall
Shackelford, Scott Addison
Shafer, Comer Drake
Shafer, Paul Richard
Shaffer, Gary W
Shafizadeh, Fred
Shalit, Harold
Shand, Edwin William
Shanks, John Amos
Shannon, Frederick Dole
Shapiro, Bernard Lyon
Shapiro, Hymin
Shapiro, Leonard
Shapiro, Robert
Shapiro, Robert Howard
Shapiro, Sydney Harold
Sharefkin, David Michael
Sharefkin, Jacob George
Sharkey, William Henry
Sharma, Gurdial Mal
Sharma, Ram Krishan
Sharp, Alvin George
Sharp, Richard Lee
Sharpe, Andrew Jackson, Jr
Sharpless, Karl Barry
Sharts, Clay Marcus
Shaw, James Edward
Shaw, John Thomas
Shaw, Philip Eugene
Shealy, Otis Lester
Shealy, Yoder Fulmer
Shearer, Charles M
Shearer, Newton Henry, Jr
Shechter, Harold
Shechter, Leon
Sheehan, Desmond
Sheehan, John Clark
Sheehan, John Timothy
Sheehan, William C
Sheeley, Richard Moats
Sheeran, Stanley Robert
Shepard, Edwin Reed
Shepard, Robert Andrews
Shepherd, James Willis
Shepherd, Robert Gordon
Sheppard, Chester Stephen
Sheppard, William Arthur
Sheps, Louis Jack
Sherer, James Pressly
Sherman, Albert Herman
Sherman, Edward
Sherr, Allan Ellis
Shevlin, Philip Bernard
Shields, Joan Esther
Shiley, Richard H
Shilling, Wilbur Leo
Shillington, James Keith
Shine, Henry Joseph
Shine, Robert John
Shine, Timothy D
Shine, William Morton
Shiner, Edward Arnold
Shipchandler, Mohammed Tyebji
Shirley, David Allen
Shirley, Robert Louis
Shiue, Chyng-Yann
Shoemaker, Clarence Jay
Shoemaker, Gradus Lawrence
Shone, Robert L
Shore, Fred L
Short, William Arthur
Shotton, James Arthur
Shotts, Adolph Calveran
Shotwell, Odette Louise
Showell, John Sheldon
Shozda, Jean'ne Marie
Shreeve, Jean'ne Marie
Shriner, Ralph Lloyd
Shryock, Gerald Duane
Shulman, Sol
Shumate, Kenneth McClellan
Shupe, Russell Dwayne

Sickels, Jackson Pyburn
Siconolfi, Carmine Anthony
Siddiqui, Iqbal Rafat
Sidi, Henri
Sidler, Jack D
Siedschlag, Karl Glenn, Jr
Sieg, Albert Louis
Siegart, William Raymond
Siegel, Herbert
Siegel, Maurice L
Siegel, Samuel
Siek, Theodore John
Sifford, Dewey H
Sigel, Carl William
Signorino, Charles Anthony
Silberman, Robert G
Silbert, Leonard Stanton
Sill, Arthur DeWitt
Silva, Ricardo
Silveira, Augustine, Jr
Silver, Frank Morris
Silverman, Bernard
Silversmith, Ernest Frank
Simmons, Howard Ensign, Jr
Simmons, Jean Elizabeth Margaret
Simmons, Thomas Carl
Simon, Myron Sydney
Simonoff, Robert
Simonson, David Michael
Simonson, Donald Raymond
Simpson, Billy Doyle
Simpson, John Ernest
Sims, James Joseph
Sims, Rex J
Sims, Victor A
Sinclair, Henry Beall
Sinclair, Richard Glenn, II
Singer, Lawrence Alan
Singh, Prithipal
Singh, Udai Pratap
Singleton, Bert
Singleton, David Michael
Singleton, Tommy Clark
Sinotte, Louis Paul
Sircar, Jagadish Chandra
Sisenwine, Samuel Fred
Sisk, Lone L
Sisti, Anthony Joseph
Sitrin, Robert David
Sjolander, John Rogers
Skell, Philip S
Skinner, Charles Gordon
Skinner, Wilfred Aubrey, Jr
Skolnik, Herman
Skoog, Ivan Hooglund
Skoultchi, Martin Milton
Skovronek, Herbert Samuel
Skrypa, Michael John
Slade, Landry Thomas
Slagel, Robert Clayton
Slater, Carl David
Slater, George P
Slates, Harry Lovell
Sleezer, Paul David
Slessor, Keith Norman
Sletzinger, Meyer
Slezak, Frank Bier
Slifkin, Sam Charles
Slocum, Donald Warren
Sluder, John Cochran
Slusarchyk, William Allen
Slutsky, Joel
Smalheer, Calvin Van Laaten
Smalley, Arnold Winfred
Smart, Charles William
Smart, G N Russell
Smart, William Donald
Smat, Robert Joseph
Smedley, William Michael
Smiley, Robert Arthur
Smith, Allan Edward
Smith, Benjamin Harper, Jr
Smith, Charles Howard
Smith, Claibourne Davis
Smith, Craig La Salle
Smith, Curtis Page
Smith, David Hibbard
Smith, Delmont K
Smith, Douglas Stewart
Smith, Earl Westley
Smith, Edgar Dumont
Smith, Francis Xavier
Smith, Frederick Albert
Smith, George Henry
Smith, Grant Gill
Smith, Grant Warren, II
Smith, Harold Ladd, Jr
Smith, Harry Andrew
Smith, Howard E
Smith, Howard Leroy
Smith, James Doyle
Smith, James Graham
Smith, James Luther
Smith, James Miller, Jr
Smith, Jerry Howard
Smith, Joshua Daniel
Smith, Leland Leroy
Smith, Leonard
Smith, Lewis Oliver, Jr

Smith, Lois C
Smith, Lowell R
Smith, Norman H
Smith, Perrin Gary
Smith, Peter Alan Somervail
Smith, Richard Frederick
Smith, Robert Bruce
Smith, Robert Johnson
Smith, Robert Verne
Smith, Robert Warren
Smith, Stanford Lee
Smith, Stanley Glen
Smith, Victor Herbert
Smith, Walter Thomas, Jr
Smith, Wesley Earl
Smith, William Burton
Smolinsky, Gerald
Smook, Malcolm Andrew
Smutny, Edgar Josef
Snapp, Thomas Carter, Jr
Sneen, Richard Allen
Snell, Robert L
Snieckus, Victor A
Snow, John Elbridge
Snow, John Thomas
Snyder, Carl Henry
Snyder, Harold Ray
Snyder, Harry Raymond, Jr
Snyder, Robert Harvey
Snyder, William H
Sobolev, Igor
Soday, Frank John
Soeder, Robert W
Sokol, Herman
Sokol, Phillip Edward
Solar, Samuel Louis
Soldati, Gianluigi
Soll, Dieter Gerhard
Sollman, Paul Benjamin
Sollott, Gilbert Paul
Solmssen, Ulrich Volckmar
Solo, Alan Jere
Solodar, Arthur John
Solodar, Warren E
Solomon, Irvine Jerome
Solomon, M Michael
Solomon, Malcolm David
Solomon, Thomas William Graham
Solomons, William Ebenezer
Soloway, Harold
Soloway, Samuel Barney
Soloway, Saul
Sommer, Leo Harry
Sommers, Jay Richard
Sonnenberg, Joseph
Sonnenfeld, Richard John
Sowa, John Robert
Sonnet, Philip E
Sonnichsen, Harold Marvin
Sonntag, Norman Oscar Victor
Sonntag, Roy Windham
Sorell, Henry P
Sorensen, David Perry
Sorkin, Howard
Sorter, Peter F
Sosis, Paul
Sosnovsky, George
Soto, Aida R
Soukup, Victor Gerald
Soulen, Robert Lewis
Southwick, Everett West
Sovish, Richard Charles
Sowa, Walter
Spadafino, Leonard Peter
Spaght, Monroe Edward
Spande, Thomas Frederick
Spangler, Martin Ord Lee
Spanninger, Philip Andrew
Spannuth, Hiram Troutman
Spano, Francis A
Sparacino, Charles Morgan
Sparks, Allen Kay
Spatz, Sydney Martin
Spayd, Richard W
Spears, Alexander White, III
Speck, Rhoads McClellan
Speers, Louise (Mrs Henry Croix)
Speight, James G
Spence, Gavin Gary
Spence, John A
Spencer, Claude Franklin
Spencer, Elvins Yuill
Spencer, John Lawrence
Spencer, Ralph Donald
Spencer, Thomas A
Spenger, Robert E
Sperber, Nathan
Sperley, Richard Jon
Spessard, Dwight Rinehart
Speziale, Angelo John
Spialter, Leonard
Spilker, Clarence William
Spindt, Roderick Sidney
Spitzer, Jeffrey Chandler
Spitzer, Penn Fulton, Jr
Splies, Robert Glenn
Spliethoff, William Ludwig
Spooner, Alfred Brent

Sporek, Karel Frantisek
Sporzynski, Adam Przemyslaw
Spraggins, Robert Lee
Sprague, Robert Hicks
Sprague, Cornelius Austin
Spriggs, Alfred Samuel
Springer, Charles Havice
Sprung, Joseph Asher
Spurlock, Langley Augustine
Squibb, Samuel Dexter
Sroog, Cyrus Efrem
Staab, Frank William
Stacey, Francis Wilfred
Stackman, Robert W
Stacy, Gardner Wesley
Staff, Charles Hubert
Stahl, Roland, Edgar
Stahly, Eldon Everett
Staiger, Roger Powell
Staker, Donald David
Stakis, Andris A
Staley, Stuart Warner
Stalick, Wayne Myron
Stanin, Theodore E
Stalling, David Laurence
Stallings, James Cameron
Stammer, Charles Hugh
Stamper, Martha C
Stanaback, Robert John
Standish, Norman Weston
Stanfield, James Aarmond
Stanfield, Manie K
Stanford, John W
Stang, Peter John
Stange, Hugo
Stanley, Lester Nelson
Stanley, William Lyons
Stanonis, David Joseph
Stansfield, Roger Ellis
Stanton, Garth Michael
Stanton, William Alexander
Staples, Jon T
Stark, James Cornelius
Starke, Albert Carl, Jr
Starkovsky, Nicolas Alexis
Starks, Fred W
Starnes, Paul Kiser
Starr, Leon
Stecher, Emma Dietz
Steck, Edgar Alfred
Steckler, Bernard Michael
Stedman, Robert John
Steele, John A
Steelink, Cornelius
Stefancsik, Ernest Anton
Stefani, Andrew Peter
Stehouwer, David Mark
Stehsel, Melvin Louis
Stein, Charles W C
Stein, Harvey Philip
Stein, Reinhardt P
Steinbach, Leonard
Steinberg, David H
Steinberg, Eliot
Steinberg, Howard
Steiner, Werner Douglas
Steinhardt, Charles Kendall
Steinle, Edmund Charles, Jr
Steinman, Harry Gordon
Steinman, Martin
Steinman, Robert
Steinmetz, Walter Edmund
Steller, Kenneth Eugene
Stemniski, John Roman
Stemniski, Michael Andrew
Stenberg, Virgil Irvin
Stenseth, Raymond Eugene
Stepan, Alfred Henry
Stephens, Dale Nelson
Stephens, John Arnold
Stephens, Lawrence James
Stephens, William D
Stephenson, Danny Lon
Stephenson, Samuel Edward, Jr
Sterken, Gordon Jay
Stermitz, Frank
Stern, Max Herman
Stern, Robert Louis
Sternbach, Daniel David
Stevens, Henry Conrad
Stevens, Robert Velman
Stevens, Travis Edward
Stevenson, Charles Edward
Stevenson, David
Stevenson, Eugene Hamilton
Stevenson, Robert William
Stewart, Alva Theodore, Jr
Stewart, D K R
Stewart, John Mathews
Stewart, Roberta A
Stewart, William Thomas
Sticker, Robert Earl
Sticker, William Carl
Stiles, Martin
Still, Gerald G
Still, Ian William James
Stille, John Kenneth
Stillwell, Richard Newhall

Stimson, Miriam Michael
Stine, William R
Stinson, Stephen Charles
Stirewalt, Edward Neale
Stobaugh, Robert Earl
Stock, Leon M
Stockel, Richard F
Stocker, Fred Butler
Stocker, Jack Hubert
Stockton, Mary Rose
Stoffer, James Osber
Stoffer, Gerhard
Stokes, William Moore
Stolberg, Marvin Arnold
Stolow, Robert David
Stone, Edward
Stone, Herbert
Stone, George Green
Stoner, Marshall Robert
Storfer, Stanley J
Stothers, John Bailie
Stout, Charles Allison
Stout, Edward Irvin
Stout, Mason Gardner
Stoutamire, Donald Wesley
Stowell, John Charles
Stradling, Samuel Stuart
Straley, James Madison
Stratton, Charlotte Dianne
Straus, Alan Edward
Strause, Sterling Franklin
Straw, Harry Arthur
Strem, Michael Edward
Struve, William Scott
Strycker, Stanley Julian
Stribley, Rexford Carl
Strickler, Paul Donovan
Stricklin, Buck
Strier, Murray Paul
Strigh, Paul Leonard
Strobach, Donald Roy
Stromberg, Verner L, Jr
Struck, Robert Frederick
Strunk, Richard John
Sturzenegger, August
Stutz, Robert L
Stux, Paul
Su, George Chung-Chi
Su, Helen Chien-Fan
Su, Tah-Mun
Sublett, Bobby Jones
Subramanian, Pallatheri Manackal
Sugathan, Kenneth Kochappan
Sugihara, James Masanobu
Sugimoto, Roy
Suh, John Taiyoung
Sullivan, Daniel Richard
Sullivan, Jeremiah B
Sullivan, John M
Sullivan, Thomas Frederick
Sumerford, Wooten Taylor
Summers, James Thomas
Summers, Lawrence
Sumner, Thomas
Sunrell, Gene
Sund, Eldon H
Sundberg, Richard J
Sundberg, Robert Lee
Sundeen, Joseph Edward
Sundelin, Kurt Gustav Ragnar
Sundelin, Charles Eugene
Sunderwirth, Stanley George
Sundholm, Norman Karl
Supple, Jerome Henry
Surbey, Donald Lee
Surmatis, Joseph D
Surrey, Alexander Robert
Susi, Peter Vincent
Suter, Stuart Ross
Sutherland, George Leslie
Sutton, John Curtis
Suydam, Frederick Henry
Suzuki, Shigeto
Swackhamer, Farris Saphar
Swain, Ansel Parrish
Swakon, Edward Antone
Swamer, Frederic Wurl
Swaningen, Roy Archibald, Jr
Swartzentruber, Paul Edwin
Sweden, Frank
Sweeley, Charles Crawford
Sweeny, Richard F
Sweeny, Arthur, Jr
Sweet, Frederick
Sweet, Ronald Lancelot
Sweeting, Linda Marie
Sweeting, Orville John
Sweetman, Brian Jack
Swenson, Jack Spencer
Swerlick, Isadore
Swern, Daniel

Swicklik, Leonard Joseph
Swift, Abbot Montague
Swindell, Robert Thomas
Swinehart, James Stephen
Swisher, Joseph Vincent
Swiss, Jack
Szabo, Arthur Gustav
Szabo, C Karoly
Szarek, Walter Anthony
Szmant, Herman Harry
Taber, David
Tabor, Theodore Emmett
Taft, David Dakin
Taft, Robert Wheaton, Jr
Takekoshi, Tohru
Takemura, Kaz Horace
Talaty, Erach R
Talbot, Preston Tidball
Talbot, Richard Lloyd
Tallent, William Hugh
Tallman, Ralph Colton
Tamborski, Christ
Tan, Henry Harry
Tanabe, Masato
Tanner, Dennis David
Tapp, William Jouette
Tarbell, Dean Stanley
Tarczy, E K
Tarney, Robert Edward
Tashick, Irving
Tate, Bryce Eugene
Tate, David Paul
Tate, Fred Alonzo
Tatum, William Earl
Taub, David
Taufen, Harvey James
Taurins, Alfred
Tauss, Kurt H
Tavares, Donald Francis
Taves, Milton Arthur
Taylor, Charles William
Taylor, Kenneth Grant
Taylor, Lynn Johnston
Taylor, Robert Burns, Jr
Taylor, Stephen Keith
Taylor, Edward Alan
Taylor, Edward Curtis
Taylor, Gary N
Taylor, James Lester
Taylor, John H
Taylor, Keith Mar
Teach, Eugene Gordon
Teague, Harold Junior
Teague, Peyton Clark
Teasdale, William Brooks
Tedeschi, Robert James
Teegarden, David Morrison
Teeter, Howard Maple
Teeter, Richard Malcolm
Tekel, Ralph
Teller, Daniel Myron
Temple, Carroll Glenn
Temple, Robert Dwight
Temple, Stanley
Tencza, Thomas Michael
Tennyson, Richard Harvey
Teranishi, Roy
Ternay, Andrew Louis, Jr
Terrell, Ross Clark
Terry, Daniel Hetfield
Terry, Ona Joy
Terry, Paul H
Terss, Robert H
Tertzakian, Gerard
Tessler, Martin Melvyn
Tetenbaum, Marvin Theodore
Thalacker, Victor Paul
Thames, Warren Alan
Thames, Shelby Freland
Thamm, Richard C, Jr
Theile, Fred Charles
Theilheimer, William
Theine, Alice
Theisen, Cynthia Theres
Thekkekandam, Joseph Thomas
Thelen, Charles Walter
Theobald, Clement Walter
Theuer, William John
Thibault, Thomas Delor
Thielen, Lawrence Eugene
Thies, Richard William
Thiessen, William Ernest
Thirtle, John Robson
Thomas, Charles J
Thomas, Alexander Edward, III
Thomas, Ann P
Thomas, George Richard
Thomas, Joseph Calvin
Thomas, Lewis Edward
Thomas, McCalip Joseph
Thomas, Paul David
Thomas, Robert Malcolm
Thomas, Telfer Lawson
Thomas, Walter William
Thompson, Alonzo Crawford
Thompson, Bobby Blackburn
Thompson, Crayton Beville

Thompson, Gerald Lee
Thompson, Grant
Thompson, Harold G
Thompson, Henry Theron
Thompson, Howard E
Thompson, Hugh Walter
Thompson, James Edwin
Thompson, Malcolm J
Thompson, Norman Strom
Thompson, Wynelle Doggett
Thornton, Edward Ralph
Thornton, Roger Lea
Throckmorton, Peter E
Thurber, William Samuels
Thurmaier, Roland Joseph
Thweatt, John G
Thyagarajan, B S
Tibbetts, Merrick Sawyer
Tidwell, Thomas Tinsley
Tiedtke, Harlan E
Tieman, Charles Henry, Jr
Tien, Rex Yuan
Tiers, George Van Dyke
Tietz, Emery G
Tilbert, Tommy Lyle
Tilles, Harry
Tillman, Richard Milton
Timell, Tore Erik
Timony, Peter Edward
Tinsley, Samuel Weaver
Tishler, Max
Titus, Elwood Owen
Titus, Richard Lee
Tkachuk, Russell
Tobkes, Martin
Tobolsky, Laszlo Gyula
Tokoli, Emery G
Tolbert, Tommy Lyle
Tolgyesi, Eva
Tolgyesi, William Steven
Tokmith, Henry
Tom, Theodore Benton
Tomasz, Maria
Tomasi, Gordon Ernest
Tomboulian, Paul
Tomcufcik, Andrew Stephen
Tomer, Kenneth Beamer
Tomesko, Andrew M
Tomomatsu, Hideo
Tong, Yulan Chang
Tonkyn, Richard George
Tomlis, John A
Toohill, Richard B
Toren, George Anthony
Totton, Ezra Lester
Touchstone, Joseph Cary
Towle, Philip Hamilton
Toy, Madeline Shen
Trachtenberg, Edward Norman
Tracy, David J
Trahanovsky, Walter Samuel
Traise, Thomas
Tramondozzi, John Edmund
Traumann, Klaus Friedrich
Traylor, Teddy G
Traynelis, Vincent John
Traynham, James Gibson
Traynor, Lee
Trepka, Robert Dale
Tretter, James Ray
Trevillyan, Alvin Earl
Trifan, Daniel Siegfried
Trimitsis, George B
Trinler, William A
Trischler, Floyd D
Trisler, John Charles
Trix, Phelps
Trofimenko, Swiatoslaw
Troll, Walter
Trost, Barry M
Trost, Henry Biggs
Trotter, Claude Henry
Trowbridge, Dale Brian
Trozzolo, Anthony Marion
Trucker, Donald Edward
Truitt, B Price
Trumbull, Elmer Roy, Jr
Tryon, Sager
Tsai, Lin
Tscharner, Christopher J
Tse, Rose (Lou)
Tseng, Chien Kuei
Tsong, Yun Yen
Tsou, Kwan Chung
Tucker, Irwin William
Tucker, William Preston
Tucker, Willie George
Tufariello, Joseph James
Tuites, Richard Clarence
Tuleen, David L
Tull, Roger James
Tulloch, Alexander Patrick
Tulloch, Charles William
Turbak, Albin Frank
Turck, Joseph Abraham Valentine, Jr
Turcotte, Joseph George
Turer, Jack
Turk, Amos

Turkel, Rickey Martin
Turnblom, Ernest Wayne
Turnbull, Lennox Birckhead
Turner, Andrew B
Turner, Fred Allen
Turner, Robert James
Turnquest, Byron W
Turro, Nicholas John
Turwiler, Frank Bryan
Tweedie, Virgil Lee
Twelves, Robert Ralph
Tyczkowski, Edward Albert
Tyner, David Anson
Uebel, Jacob John
Uehling, Edwin Fisher
Ullman, Edwin Fisher
Ullyot, Glenn Edgar
Ulshafer, Paul R
Umen, Michael Jay
Undeusch, William Charles
Ungnade, James M
Untch, Karl George
Updegraff, Ivor Heberling
Upham, Roy Herbert
Upson, Robert William
Urbach, Arthur Ronald
Urbanski, Arthur Ronald
Urbas, Branko
Ursino, Joseph Anthony
Ursprung, Joseph John
Vachon, Raymond Normand
Vacik, Dorothy Nobles
Vail, Sidney Lee
Vail, Thomas Michael
Valega, Zdenek
Valentine, Frank Rossiter
Valiaveedan, George Devasia
van Aller, Robert Thomas
Van Dam, M
Vandegaer, Jan Edmond
Van Denheuvel, Franz Anne
Van de Poel, Josephus
Vandepitte, John
Van der Burg, Sjirk
Vanderhoek, Jack Yehudi
Vander Stouw, Gerald Gordon
Vander Valk, Paul David
van der Veen, James Morris
Vanderwerff, Calvin Anthony
Vanderwerff, William D
van Duuren, Benjamin Louis
Van Dyke, John William, Jr
Vanelli, Ronald Edward
van Fossen, Paul
Van Gulick, Norman Martin
Van Handel, Emile
Van Heyningen, Earle Marvin
Van Horn, Ruth Warner
Van Landuyt, Dennis Clarke
Van Meter, James P
Van Orden, Harris O
Van Order, Robert Bruce
Van Rheenen, Verlan H
Van Strien, Richard Edward
Van Verth, James Edward
Varkey, Thanakamma Eapen
Varma, Rajendra
Vasiliauskas, Edmund
Vastine, Frederick Davidson
Vaughn, Thomas Hunt
Veazey, Thomas Mabry
Vedejs, Edwin
Veldhuis, Benjamin
Venerable, James Thomas
Venier, Clifford George
Venkatachalam, Taracad Krishnan
Verbanac, Frank
Verbanc, Anthony James
Verbit, Lawrence
Verbrugge, Calvin James
Vercellotti, John R
Verell, Ruth Ann
Vernon, John Ashbridge
Verter, Herbert Sigmund
Vesting, Martha Meredith
Vick, Gerald Kieth
Vignale, Michael Joseph
Vignes, Robert Paul
Vincent, Donald Leslie
Vineyard, Billy Dale
Vinton, William Howells
Viola, Alfred
Viohl, Paul
Vitcha, James F
Vitullo, Victor Patrick
Voedisch, Robert W
Vogel, George
Vogel, Martin
Vogel, Paul William
Vogelfanger, Elliot Aaron
Vogt, Clifford Marshall
Vogt, Charles Frederick
Volk, Murray Edward
Volker, Eugene Jeno
Vollmar, Arnulf R
Volpe, Angelo Anthony
Volpp, Gert Paul Justus

Vona, Joseph Albert
Von Ostwalden, Peter Weber
Von Riesen, Daniel Dean
Von Rosenberg, Joseph Leslie, Jr
Von Rudloff, Ernst Max
Von Schuching, Susanne
Von Stryk, Frederick George
Vosburgh, William George
Vosti, Donald Curtis
Vouros, Paul
Vozza, John F
Vratsanos, Spyros M
Vullo, William Joseph
Vygantas, Auste Marija
Waddle, Howard Meffert
Wade, Clarence W R
Wade, Peter Cawthorn
Wade, Robert Harold
Wagenknecht, John Henry
Waggoner, Terry Bill
Wagner, Arthur Franklin
Wagner, Charles Roe
Wagner, Frank A, Jr
Wagner, Frank S, Jr
Wagner, Gerald Roy
Wagner, Hans
Wagner, John Garnet
Wagner, Melvin Peter
Wagner, Peter J
Wagner, Robert Edwin
Wagner, Romeo Barrick
Wagner, William Sherwood
Waite, Moseley
Walba, Harold
Walborsky, Harry M
Walecka, Jerrold Alberts
Walia, Jasjit Singh
Walker, Francis H
Walker, John J
Walker, Joseph
Walker, Russell Wagner
Walker, Ruth Angelina
Walker, Wellington Epler
Wall, Robert Gene
Wallace, Edwin Garfield
Wallace, Robert Allan
Waller, David Percival
Waller, Francis Joseph
Walles, Wilhelm Egbert
Walsh, Edward Joseph
Walsh, Edward Joseph, Jr
Walsh, Edward Nelson
Walsh, John Paul
Walsh, William Louis
Walter, Charles Robert, Jr
Walter, Gerald Joseph
Walter, Henry Alexander
Walter, Henry Clement
Walter, Thomas James
Walters, Lee Rudyard
Walton, Edward
Walton, Henry Miller
Walton, Theodore Ross
Walton, Warren Lewis
Wamser, Carl Christian
Wang, Chi-Hua
Wang, Chun Shan
Wang, Nancy Yang
Wang, Shih Yi
Ward, Frank Kernan
Ward, John Edward
Ward, Richard Bernard
Wardner, Carl Arthur
Warfel, David Ross
Warfield, Albert Harry
Warfield, Peter Foster
Waring, Derek Morris Holt
Warkentin, John
Warne, Thomas Martin
Warner, Charles D
Warner, John Scott
Warner, Philip Mark
Warnhoff, Edgar William
Waroblak, Michael Theodore
Warren, Harold Hubbard
Warren, Mitchum Ellison, Jr
Wartman, William Benjamin, Jr
Wasacz, John Peter
Washburn, Lee Cross
Washburne, Stephen Shepard
Wasserman, Edel
Wasserman, Harry H
Wasson, Richard Lee
Wat, Edward Koon Wah
Watanabe, Kyoichi A
Waters, James Augustus
Waters, William Lincoln
Watkins, Spencer Hunt
Watson, Harold John
Watson, Richard Noble
Watson, Robert C
Watt, William Russell
Watterson, Arthur C, Jr
Wattchy, Jeffrey William Herbert
Watts, Exum DeVer

Waugh, Richard Campbell
Wawzonek, Stanley
Wayman, Morris
Wearn, Richard Benjamin
Weatherbee, Carl
Weaver, Jeremiah William
Weaver, Leo James
Weaver, William Michael
Webb, Irving D
Webb, James L A
Webb, Robert Lee
Webb, Thomas Howard
Webb, William Paul
Webber, Gayle Milton
Webber, Thomas Gray
Weber, Karl Hansel
Weberg, Berton Charles
Webers, Vincent Joseph
Webster, Eleanor Rudd
Webster, James Albert
Webster, Owen Wright
Wechter, William Julius
Weck, Friedrich Josef
Wedegaertner, Donald K
Weesner, William Eldred
Weetman, David G
Wehman, Anthony Theodore
Wehr, Henry William, Jr
Wei, Chung-Chen
Weidlein, Edward Ray, Jr
Weier, Richard Mathias
Weil, Edward David
Weiland, Henry Joseph
Weiler, Ernest Dieter
Weiler, Lawrence Stanley
Weill, Carol Edwin
Weilmuenster, Earl Adam
Weinberg, Norman L
Weinberger, Harold
Weinberger, Lester
Weiner, Henry
Weininger, Stephen Joel
Weinreb, Steven Martin
Weinshenker, Ned Martin
Weinstein, Boris
Weinstock, Joseph
Weinstock, Leonard M
Weintraub, Leonard
Weintraub, Philip Marvin
Weisbach, Jerry Arnold
Weisblat, David Irwin
Weisenborn, Frank L
Weisgerber, Cyrus Aaron
Weisgraber, Karl Heinrich
Weisler, Leonard
Weiss, Benjamin
Weiss, David Steven
Weiss, Irma Tuck
Weiss, James Allyn
Weiss, James Owen
Weiss, Martin Joseph
Weiss, Philip
Weiss, Ulrich
Welch, Clark Moore
Welch, Dean Earl
Welch, Frank Joseph
Welch, Thomas Harris
Welch, Willard McKowan, Jr
Welch, Zara D
Welky, Norman
Welldon, Paul Burke
Weller, Lowell Ernest
Wellings, Ian
Wellman, William Edward
Wells, Darthon Vernon
Wells, Eugene Ernest, Jr
Wells, Franklin Burnham
Wells, Paula Parker
Welsh, David Albert
Welstead, William John, Jr
Weltman, Clarence A
Wempe, Lawrence Kyran
Wendland, Ray Theodore
Wendler, Norman Lord
Wentland, Mark Philip
Wentland, Stephen Henry
Wentworth, Gary
Werbel, Leslie Morton
Werber, Frank Xavier
Werner, Ervin Robert, Jr
Werner, Lincoln Harvey
Werner, Raymond Edmund
Werth, Richard George
Weschler, Joseph Robert
Wescott, Lyle DuMond, Jr
West, Charles P
West, Richard Lowell
West, William Alvin
Westfahl, Jerome Clarence
Westheimer, Frank Henry
Westlake, Harry Edward, Jr
Westley, John William
Westman, Thomas Louis
Weston, Arthur Walter
Westover, James Donald
Wetmore, David Eugene
Wetmore, Stanley Irwin, Jr

Weyna, Philip Leo
Whaley, Howard Arnold
Whaley, Thomas Williams
Wharton, Peter Stanley
Wheatley, William Bacon
Wheeler, Desmond Michael Sherlock
Wheeler, Edward Norwood
Wheeler, James William, Jr
Wheeler, Thomas Neil
Wheeler, William Raleigh
Whelan, Barbara Jean King
Whitaker, Robert Dallas
Whitcomb, Gordon Putnam
White, David Raymond
White, Dwain Montgomery
White, Emil Henry
White, Halbert Constantine
White, Harold Keith
White, Harold McCoy
White, Harry Joseph
White, Howard Sorrel
White, James David
White, Jerry Eugene
White, Julius
White, Robert Winslow
White, William Harold
Whitehurst, Darrell Duayne
Whiteley, Thomas Edward
Whitesides, George McClelland
Whitman, Gerald Messner
Whitney, Ambrose Grunhagen
Whitney, George Stephen
Whitney, Joel Gayton
Whitney, Robert Byron
Whitney, Thomas Allen
Whittemore, William Bernard
Whittemore, Charles Alan
Whittingham, David James
Whittle, John Antony
Whittle, Philip Rodger
Whyte, Donald Edward
Wiberg, Kenneth Berle
Wich, Grosvenor Searles
Wick, Emily Lippincott
Wick, Lawrence Bernard
Wicker, Thomas Hamilton, Jr
Wickrema Sinha, Asoka J
Wideman, Lawson Gibson
Wiedenmann, Lynn G
Wiegand, Gayl
Wiegert, Philip E
Wierenga, Wendell
Wierengo, Cyril John, Jr
Wiesler, Donald Paul
Wigfield, Donald Compston
Wightman, Robert Harlan
Wilbur, James Myers, Jr
Wilcox, Charles Frederick, Jr
Wilder, Pelham, Jr
Wildi, Bernard Sylvester
Wildman, William Cooper
Wilds, Alfred Lawrence
Wilen, Samuel Henry
Wiley, Douglas Walker
Wiley, Michael David
Wiley, Paul Fears
Wiley, William Lee
Wilgus, Donovan Ray
Wilip, Elmar Konstantin
Wilkin, Louis Alden
Wilkins, Raymond Leslie
Wilkinson, Christopher Foster
Wilkinson, Raymond George
Wilkinson, William Kenneth
Willard, Joe Raymond
Willard, John Jay
Willard, Paul Edwin
Willcott, Mark Robert, III
Willette, Robert Edmond
Williams, Albert Lloyd
Williams, Allan, Rawson
Williams, Byron Lee, Jr
Williams, Francis Trueman
Williams, Harry Douglas
Williams, Jack L R
Williams, James Horace
Williams, Robert Hackney
Williams, Roy Lee
Williams, William Michael
Williams, William Wilson
Williamson, Hugh A
Williamson, Jerry Robert
Williamson, Kenneth Lee
Williamson, Thurmond A
Willis, Victor Max
Willits, Charles Haines
Willner, David
Willson, Clyde D
Wilson, Armin Guschel
Wilson, Burton David
Wilson, Donald Richard
Wilson, Elwood Justin, Jr
Wilson, Evelyn H
Wilson, Glenn Rhodes
Wilson, Gordon, Jr
Wilson, Harold Frederick
Wilson, James D
Wilson, James Dennis

Wilson, Joseph William
Wilt, James William
Wiman, Robert Edgar
Wimer, David Carlisle
Winch, Bradley Louis
Windholz, Thomas Bela
Winestock, Claire Hummel
Winfrey, J C
Wingrove, Alan Smith
Winicov, Edith
Winicov, Herbert
Winkel, Cleve R
Winkler, Robert Randolph
Winn, A Vernon
Winner, Bernard Mark
Winnik, Mitchell Alan
Winslow, Alfred Edwards
Winslow, Field Howard
Winstead, Meldrum Barnett
Winter, Roland Arthur Edwin
Winter, Rudolph Ernst Karl
Winters, Lawrence Joseph
Winthrop, Stanley Oscar
Wintner, Claude Edward
Wirth, John Christian
Wirth, Joseph Glenn
Wise, Gene
Wise, Hugh Edward, Jr
Wise, Lawrence David
Wise, Paul Henry
Wise, Richard Melvin
Wiseman, George Edward
Wiseman, John R
Wiseman, Park Allen
Wisner, Jackson Ward, Jr
Wisner, Allan
Wither, Ross Plummer
Witherell, Donald Ray
Witiak, Donald T
Witkop, Bernhard
Witkoski, Francis Clement
Witkowski, Joseph Theodore
Witman, Robert Charles
Witschard, Gilbert
Witt, Enrique Roberto
Witt, John, Jr
Witte, Michael
Wittekind, Raymond Richard
Witterholt, Vincent Gerard
Witten, Benjamin
Wittman, William F
Witzel, Frank
Wlech, Raymond Lee
Wohl, Bernard G
Wohlers, Herbert C
Wojcicki, Andrew
Wolf, Alfred Peter
Wolf, Donald Edwin
Wolf, Frank James
Wolf, Leslie Raymond
Wolf, Richard Eugene
Wolf, Walter Alan
Wolfe, James F
Wolfe, James Richard, Jr
Wolfe, John Kavanaugh
Wolfe, Roger Thomas
Wolff, Ivan A
Wolff, Steven
Wolff, William Francis
Wolfhagen, James Langdon
Wolinsky, Joseph
Wollensak, John Charles
Wollner, Thomas Edward
Wolsey, Wayne C
Womer, Walter Dale
Wong, Chiu Ming
Wong, Shi-Yin
Woo, Gar Lok
Woo, James T K
Wood, Burrell Lusha, Jr
Wood, Gordon Walter
Wood, Harry Burgess, Jr
Wood, John Edward, III
Wood, John Martin
Wood, Louis L
Wood, Randall Dudley
Woodbridge, Joseph Eliot
Woodborn, Henry Milton
Woodburn, Charles William
Woods, Robert Claude
Woods, Thomas Stephen
Woodward, David Willcox
Woodward, Robert Burns
Woodward, Curtis Wilmer
Woodyard, James Douglas
Woodyard, William T
Woolford, Robert Graham
Woolsey, Neil Franklin
Woosley, Royce Stanley
Work, Stewart D
Workman, Wesley Ray
Worman, James John
Worrall, Winfield Scott
Wotiz, Herbert Henry
Wrenn, Henry K
Wrenn, Samuel Nathaniel

Wriede, Peter Artur
Wright, Charles Joseph
Wright, Everett James
Wright, Howard Edwards, Jr
Wright, James Roscoe
Wright, Joe Carrol
Wright, John Collins
Wright, Oscar Lewis
Wright, Robert L
Wrigley, Arthur Nelson
Wristers, Harry (Jan)
Wu, Chisung
Wu, Mu Tsu
Wu, Tse Cheng
Wu, Yao Hua
Wuchter, Richard B
Wulfers, Thomas Frederick
Wunz, Paul Richard, Jr
Wurth, Michael John
Wyman, James Calvin
Wyman, George Martin
Wynn, Charles Martin, Sr
Wynne, Benjamin Delaney
Wystrach, Vernon Paul
Yaffe, Roberta
Yagi, Haruhiko
Yale, Harry Louis
Yamada, Yoshikazu
Yamagishi, Frederick George
Yamamoto, Yasushi Stephen
Yamamoto, Dominic Tsung-Che
Yankee, Ernest Warren
Yanko, William Harry
Yankeelov, John Allen, Jr
Yarbrough, Arthur C, Jr
Yarian, Dean Robert
Yates, Peter
Yeager, Sandra Ann
Yeakey, Ernest Leon
Yeats, Ronald Bradshaw
Yee, Tucker Tew
Yeh, Kuo-Chen
Yip, Roderick Wing
Yohe, Gail Robert
Yoho, Clayton W
York, Owen Jr
Yorton, Joan Bannister
Yost, John Franklin
Yost, Robert Stanley
Yost, William Lassiter
Young, David Caldwell
Young, DeWalt Secrist
Young, Sanford Tyler
Young, Thomas Edwin
Youngblood, Bettye Sue
Youngdale, Gilbert Arthur
Youngman, Edward August
Youngs, Vernon Leroy
Youngstrom, Richard Earl
Yourtee, Lawrence Karn
Yu, Chia-Nien
Yu, Shiu Yeh
Yu, Terry Ta-Jen
Yuen, Po Sang
Yuska, Henry
Zabriskie, John Lansing, Jr
Zacharias, David Edward
Zaczek, Norbert Marion
Zajac, Walter William, Jr
Zajacek, John George
Zalar, Frank Victor
Zalay, Ethel Suzanne
Zaikow, Leon Harry
Zambito, Arthur Joseph
Zarrella, William Michael
Zauge, Harold Elmer
Zavarin, Eugene
Zawadzki, Joseph Francis
Zaweski, Edward F
Zee-Cheng, Kwang Yuen
Zeftel, Leo
Zehner, Lee Randall
Zehrung, Winfield Scott, III
Zeiss, Harold Hicks
Zeil, Howard Charles
Ziegler, Carl Naeher
Zemlicka, Jiri
Zentmyer, David Taylor
Zervekh, Charles Ezra, Jr
Zey, Robert L
Zicarbarth, Timothy Dean
Zhivadinovich, Milka Radoicich
Zieger, Herman Ernst
Ziegler, Frederick Edward
Ziegler, Peter
Zienty, Mitchell Frank
Ziering, Albert
Ziffer, Herman
Zilch, Karl T
Zimmer, Hans
Zimmerman, Barry
Zinkel, Duane Forst
Zinnes, Harold
Zisson, James

Zlatkis, Albert
Zletz, Alex
Zollinger, Joseph LaMar
Zoltewicz, John A
Zon, Gerald
Zook, Harry David
Zuckerman, Samuel
Zuech, Ernest A
Zukas, Danute
Zuzack, John W
Zvejnieks, Andrejs
Zwecker, William R
Zweifel, George

Industrial Organic Chemistry

Anzenberger, Joseph F, Sr
Archer, Wesley Lea
Atwood, Mark Trevor
Ball, David Ralph
Bamberger, Curt
Barclay, Roger E
Barnstorff, Henry Dreses
Bauermeister, Herman Otto
Bell, Ian
Berkowitz, Sidney
Bluestein, Claire
Bockstahler, Theodore Edwin
Boyle, Richard James
Brain, Devin King
Bromberg, Milton Jay
Brown, Lloyd H
Buhle, Emmett Loren
Burke, Roger E
Burney, Donald Eugene
Buurman, Clarence Harold
Calbo, Leonard Joseph
Chadwick, David Henry
Chang, Shu-Pei
Clarke, David Bruce
Clegg, William Josiah
Clement, Robert Alton
Cofrancesco, Anthony J
Cowles, Craig Schuyler
Derfer, John Mentzer
Dermer, Otis Clifford
Deutsch, Dennis Leslie
Deviny, Edward John
Drach, John Edward
Dybalski, Jack Norbert
Edmans, Dwight Douglas
Edwards, Ben E
Feeman, James Frederic
Fischer, Robert George, Jr
Flanagan, John Vernon
Foster, Walter Edward
Frame, Robert Roy
Frank, Victor Samuel
Gipson, Robert Malone
Gloyer, Stewart Edward
Goetz, Richard W
Gradeff, Peter S
Gregory, Arthur Stanley
Habib, David Peter
Hall, Charles Mack
Harper, Jon Jay
Hartman, Robert John
Harwell, Kenneth Elzer
Hedrick, Robert Jerry
Heffron, Peter John
Hein, Richard William
Herkes, Frank Edward
Humphrey, Bingham Johnson
Hunter, Byron Alexander
Ingwalson, Raymond Wesley
Innes, John Edwin
Jaeger, Charles Wayne
Jaruzelski, John Janusz
Johnson, Fred Lowery, Jr
Johnson, Roger Alvin
Johnston, John Derland
Johnston, Katharine Gentry
Jordon, Thomas Earl
Karo, Wolf
Kassner, James Edward
Kaye, Howard
Kissel, Charles Louis
Knapp, Malcolm Hammond
Knollmueller, Karl Otto
Knott, Donald MacMillan
Komoto, Robert Gordon
Kosak, John R
Kwiatek, Jack
Lamert, Kent Philip
Lee, Robert James
Lerner, Lawrence Robert
Levy, Leon Bruce
Lillwitz, Lawrence Dale
Linder, Seymour Martin
Linfield, Warner Max
Liu, Jih-Hua
Lofton, William Milford, Jr
Lundsted, Lester Gordon
MacPeek, Donald Lester
Manning, Harold Edwin
Marquis, David Maley
Martin, James Cuthbert
Massingill, John Lee, Jr

McCown, Joseph Dana
McDaniel, Edgar Lamar, Jr
McDuff, James Milton
Merner, Richard Raymond
Moroni, Eneo C
Morris, Leo Raymond
Morse, Ronald Loyd
Motz, Kaye La Marr
Mount, Lloyd Gordon
Muse, Joel, Jr
Naglieri, Anthony N
Natowsky, Sheldon
Nazy, John Robert
Nedwick, John Joseph
Palmer, John Frank, Jr
Pedroza, Gregorio Cruz
Peery, Clifford Young
Peters, Lynn Randolph
Pryde, Everett Hilton
Rabourn, Warren Joseph
Railing, Wilford Edward
Renfrew, Edgar Earl
Richard, William Ralph, Jr
Richter, Robert Freeland
Rife, Robert Seldon
Rogers, Ralph Loucks
Rolston, Charles Hopkins
Rondestvedt, Christian Scriver, Jr
Rowe, Jay Elwood
Salsbury, Jason Melvin
Saveride, Russell Irwin
Scardera, Michael
Schwarz, John Samuel Paul
Sherman, Paul Dwight, Jr
Shulman, George
Sisco, Joseph Anthony
Skorcz, Joseph Michael
Slagan, Peter Michael
Sloan, Martin Frank
Smith, William Edward
Smyrk, Charles McCahan, Jr
Sowers, Edward Eugene
Steiner, Russell Irwin
Stern, Alfred
Stockburger, George Joseph
Summers, William Allen, Jr
Tamborin, Henry John
Tams, Alan Roger
Tanner, Alan Roger
Thompson, Quentin Elwyn
Trimmer, Robert Whitfield
Ulrey, Stephen Scott
Unruh, Jerry Dean
Victorius, Claus
Von, Isaiah
Vopicka, Edward
Walbrick, Johnny Mac
Washcheck, Paul Howard
Webb, Philip Gilbert
Weiner, Philip
Weipert, Eugene Allen
Wheat, Percy Wayne
Wiles, Robert Allan
Wu, Ching-Yong
Wysocki, Allen John
Young, Charles Albert
Young, Herbert Lewis

Organic Polymer Chemistry

Alexander, James Ernest
Allen, James Durwood
Anderson, Burton Carl
Andrus, Milton Henry, Jr
Antonucci, Frank Ralph
Arhat, Richard James
Armitage, John Brian
Banucci, Eugene George
Barie, Walter Peter, Jr
Barnhart, William Siddall
Baughman, Glenn Laverne
Baumann, Gert Friedrich
Berger, Richard S
Bletch, Harry
Black, William Bruce
Blommers, Elizabeth Ann
Bolgiano, Nicholas Charles
Bower, George Myron
Breder, Charles Vincent
Breed, Laurence Woods
Brown, Harold Probert
Brown, Kenneth Howard
Byrd, James Dotson
Byrd, Norman Robert
Campbell, Gerald Allan
Cass, William Emerson
Castro, Anthony J
Cavender, James Vere, Jr
Charbonneau, Larry Francis
Christie, Peter Allan
Cobler, John George
Cohen, Stuart Colin
Cook, Jack E
Corner, James Oliver
David, Israel A
Dawson, Daniel Joseph
DeVigs, Peter
Desai, Vinodrai Ranchhodji
DeWitt, Elmer John

Dinnig, Daniel Ashton
Dinbergs, Kornelius
Dolenko, Allan John
Durandetta, Donald W
Eareckson, William Milton, III
Earing, Mason Humphry
Elliott, John Raymond
Elmslie, James Stewart
Empen, Joseph A
Ernst, John L
Essig, Henry J
Estes, Leland Lloyd
Etter, Raymond Lewis, Jr
Fanta, George Frederick
Fantassey, William Joseph, Jr
Feairheller, Stephen Henry
Feinberg, Stewart Carl
Feit, Eugene David
Felicetta, Vincent Frank
Fetscher, Charles Arthur
Firth, William Charles, Jr
Fitko, Chester Walter
Fletcher, Harry Huntington
Floyd, Don Edgar
Ford, Emory A
Fox, Sidney Walter
Gehrig, Neil Edward
Gehring, Harvey Thomas
Geib, Leonard Louis
Gelfer, Daniel Harold
Gerow, Clare William
Gesner, Bruce D
Gilbert, Eugene Charles
Glaser, Milton Arthur
Gleim, Furman Eugene
Goldman, Theodore Daniel
Goossens, John Charles
Graver, Richard Byrd
Gray, Don Norman
Green, Charles David
Grot, Walther Gustav Fredrich
Gurley, Thomas Gordon
Hamb, Fredrick Lynn
Hancock, James William
Harris, Richard Lee
Harris, Robert L
Hartranft, George Robert
Hartzler, Jon David
Heins, Conrad F
Hickner, Richard Allan
Hicks, Elija Maxie, Jr
Hindersinn, Raymond Richard
Hine, James Maynard
Hirzy, John William
Hoehn, Harvey Herbert
Hoeschele, Guenther Kurt
Hogen-esch, Thieo Eltjo
Holik, Melville James
Holyoke, Caleb William, Jr
Hoover, M Frederick
Hsu, Tsong-Han
Hu, Clyde Kuen-Hua
Hudson, Robert Leslie
Idol, James Daniel, Jr
Ingram, Alvin Richard
Isaacson, Henry Verschay
Jarvis, Lactance Aubrey
Jenkins, Sidney Hartman, Jr
Johnson, Duane Edward
Johnson, Manley Roderick
Johnston, Norman Joseph
Kester, Dennis Earl
Kleinschuster, Jacob John
Kohan, Melvin Ira
Kolczynski, James Robert
Kopchik, Richard Michael
Kreuz, John Anthony
Kutner, Abraham
Lakshmanan, P R
Le Blanc, John Roger
Limburg, William W
Lipowski, Stanley Arthur
Lisanke, Robert John, Sr
Lo, Chien-Pen
Lockhart, Luther Bynum, Jr
Lockwood, Robert Greening
Logullo, Francis Mark
Lorensen, Lyman Edward
Lovald, Roger Allen
Lucas, Glennard Ralph
Maiofisky, Bernard Miles
Manhart, Joseph Heritage
Manos, Philip
Mansfield, Kevin Thomas
Mantell, Gerald Jerome
Marder, Herman Lowell
Markowski, Henry Joseph
Martin, Eugene Christopher
Marvel, Carl Shipp
Matlack, Albert Shelton
Matthews, Virgil Edison
McCarthy, Neil Justin, Jr
McGirk, Richard Heath

McGrath, James Edward
McKelvey, John Murray
Meinhardt, Norman Anthony
Merkel, Timothy Franklin
Merrill, Stewart Henry
Meyer, Leo Francis
Meyer, Victor Bernard
Millar, John Robert
Miller, Lewis Samuel
Miller, Walter Peter
Milone, Charles Robert
Miranda, Thomas Joseph
Mitchell, George Redmond, Jr
Monagle, Daniel J
Moncure, Henry, Jr
Morduchowitz, Abraham
Moult, Roy Hepworth
Murray, James Gordon
Myers, Drewfus Young, Jr
Newey, Herbert Alfred
Noland, James Sterling
Noren, Gerry Karl
Norton, Lilburn Lafayette
O'Brien, John Terence
Orphanides, Gus George
O'Shea, Francis Xavier
Ostfield, Howard G
Ottaviani, Robert Augustine
Padbury, John James
Palermo, Felice Charles
Panek, Edward John
Panzer, Hans Peter
Parisek, Charles Bruce
Parish, Darrell Joe
Parker, Earl Elmer
Patterson, William Jerry
Patton, John Thomas, Jr
Pelosi, Lorenzo Fred
Percival, Douglas Franklin
Petersen, Kenneth C
Petersen, Philip Richard
Petfield, Robert Joseph
Petropoulos, Constantine Chris
Plant, William J
Prapas, Aristotle George
Prudence, Robert Thomas
Quinn, Clayton Byerley
Rapoport, Lorence
Reinking, Norman Herbert
Rider, Don Keith
Riecke, Edgar Erick
Ringwald, Eugene Lee
Roberts, Durward Thomas, Jr
Ross, Stanley Elijah
Russell, Charles Irving
Ryan, Patrick Walter
Sackoff, Martin M
Sallo, Jerome Stanley
Sample, James Halverson
Sandhu, Mohammad Akram
Sandridge, Robert Lee
Schick, Margery Leone
Schmidt, Donald L
Schober, Donald Lincoln
Schreyer, Ralph Courtenay
Schrof, William Ernst John
Scullin, James Philip
Segall, Gordon Hart
Selman, Charles Melvin
Smith, Oliver Wendell
Smith, Thomas Woods
Smith, William Mayo, Jr
Spitzer, William Carl
Steiner, Arnold Byron
Stephens, James Regis
Sterling, Robert Fillmore
Stevens, Malcolm Peter
Strand, Robert Charles
Stutsman, Paul Snell
Sullivan, Charles Irving
Sulzberg, Theodore
Swanholm, Carl E
Swift, Graham
Tesoro, Giuliana C
Theis, Richard James
Thompson, Clifford Francis
Turner, Robert Lawrence
Ulrich, Henri
Un, Howard Ho-Wei
Van Den Berghe, John
Vessel, Eugene David
Villars, Calvin W
Vriesen, Calvin W
Wagner, Klaus Peter
Walker, Charles Carey
Wallenberger, Frederick Theodore
Wasserman, William Jack
Wear, Robert Lee
Weese, Richard Henry
Weinstein, Arthur Howard
Weintraub, Lester
Weiss, Douglas Eugene
Wen, Richard Yutze
Wentworth, Stanley Earl

Whelan, William Paul, Jr
Whitt, Carlton Dennis
Wilson, John Charles
Wolinski, Leon Edward
Wolny, Friedrich Franz
Wong, Chun-Ming
Wynstra, John
Yunick, Robert P
Zalewski, Edmund Joseph
Zimmer, William Frederick, Jr

Organometallic Chemistry
Alexander, John J
Alich, Agnes Amelia
Alper, Howard
Armbrecht, Frank Maurice, Jr
Attig, Thomas George
Barefield, Edward Kent
Belloli, Robert Charles
Bergman, Robert George
Blake, Daniel Melvin
Bluestein, Ben Alfred
Brewer, Stuart Dexter
Brubaker, Carl H, Jr
Buchanan, David Hamilton
Calabretta, Peter Joseph
Calderon, Nissim
Cannon, John Burns
Carberry, Edward Andrew
Carlson, Bruce Arne
Carpenter, Barry Keith
Carty, Arthur John
Cerefice, Steven A
Cheng, Tai Chun
Coates, Geoffrey Edward
Collins, Alva LeRoy, Jr
Conder, Harold Lee
Coover, Harry Wesley, Jr
Cope, James Francis
Copeland, Richard Franklin
Coulson, Dale Robert
Coyle, Thomas Davidson
Cramer, Richard (David)
Curtis, Myron David
Darensbourg, Marcetta York
Darensbourg, Donald Jude
Davis, Dennis Duval
Dawes, John Leslie
Desio, Peter John
DeYoung, Edwin Lawson
Dines, Martin Benjamin
Dobson, Gerard Ramsden
Doyle, John Robert
Dub, Michael
Dunks, Gary Burr
Dunny, Stanley
Efraty, Avi
Erskine, Gordon John
Fahey, Darryl Richard
Fain, John William
Faller, John William
Felton, John James
Fitch, John William, III
Flood, Thomas Charles
Freedman, Leon David
Gaidis, James Michael
Gardner, Sylvia Alice
Geoffroy, Gregory Lynn
George, T Adrian
Gibson, Dorothy Hinds
Giering, Warren Percival
Gilman, Henry
Gitlitz, Melvin Hyman
Gladysz, John A
Gloth, Richard Edward
Gocke, George Leonard
Griffith, Elizabeth Ann Hall
Grim, Samuel Oram
Grimes, Russell Newell
Guggenberger, Lloyd Joseph
Gysling, Henry J
Halm, James Maurice
Halpern, Donald F
Halpern, Jack
Hartgerink, Ronald Lee
Hartwell, George E, Jr
Haymore, Barry Lant
Hegedus, Louis Stevenson
Heiling, John Frederic
Helquist, Paul M
Henry, Patrick M
Herskovitz, Thomas
Hess, George G
Hopper, Steven Phillip
Husk, George Ronald
Hyatt, David Ernest
Irgolic, Kurt Johann
Jack, Thomas Richard
Janzen, Alexander Frank
Jennings, Paul W
Jones, Paul Ronald
Juenge, Eric Carl
Kaesz, Herbert David
Kim, Leo
Klemann, Lawrence Paul
Knoth, Walter Henry, Jr
Kochi, Jay Kazuo
Kotz, John Carl

Kupchik, Eugene John
Laemmle, Joseph Thomas
Lehman, Dennis Dale
Levy, Alan B
Lipowitz, Jonathan
Lukehart, Charles Martin
Lyons, James Edward
MacLaury, Michael Risley
Mague, Joel Tabor
Manzer, Leo Ernest
Martinez, Nilda
Matteson, Donald Stephen
May, James Aubrey, Jr
McClellan, William Robert
McClure, James Douglas
Menke, Andrew Giedrius
Meyer, Carol Diane
Midland, Michael Mark
Moedritzer, Kurt
Morris, Donald Eugene
Moser, William Ray
Mrowca, Joseph J
Muntz, Ronald Lee
Murch, Robert Matthews
Nelson, John Henry
Neuse, Eberhard Wilhelm
Nicholson, D Allan
Nile, Terence Anthony
Opie, Thomas Ranson
Pande, Kailash Chandra
Panek, Edward John
Pannell, Keith Howard
Paulik, Frank Edward
Pitt, Colin Geoffrey
Poist, John Edward
Post, Elroy Wayne
Prince, Martin Irwin
Prokai, Bela
Quirk, Roderic Paul
Rakita, Philip Erwin
Rausch, Marvin D
Reed, Joseph
Reger, Daniel Lewis
Reifenberg, Gerald H
Rheingold, Arnold L
Richey, Herman Glenn, Jr
Rosenberg, Sanders David
Sahatjian, Ronald Alexander
Salinger, Rudolf Michael
Sams, John Robert, Jr
San Filippo, Joseph, Jr
Sanger, Alan Rodney
Schaumberg, Gene David
Schmidling, David (Gilbert)
Scholer, Frederick Richard
Schrock, Richard Royce
Schwartz, Jeffrey
Sears, Curtis Thornton, Jr
Setterquist, Robert Alton
Seyferth, Dietmar
Sheats, John Eugene
Shubkin, Ronald Lee
Shue, Robert Sidney
Sick, Lowell Victor
Singh, Gurdial
Slocum, Donald Warren
Smith, Homer Alvin, Jr
Steward, Omar Waddington
Tennent, Howard Gordon
Thayer, John Stearns
Thompson, David Wallace
Tobias, Russell Stuart
Tomaja, David Louis
Torkelson, Arnold
Tsutsui, Minoru
Valentine, Donald H, Jr
Van Dyke, Charles H
Vastine, Frederick Davidson
Verkade, John George
Waters, William Lincoln
Wegner, Patrick Andrew
Wehman, Anthony Theodore
Weissberger, Edward
Weissman, Paul Morton
Wendel, Samuel Reece
Wender, Irving
West, Robert
White, Alan Jonathon
Wiesboeck, Robert A
Wojcicki, Andrew
Zaremsky, Baruch
Zuckerman, Jerold J

Physical Organic Chemistry
Adam, Waldemar
Adams, Otis William
Addy, John Keith
Adler, George
Albrecht, Frederick Xavier
Alekman, Stanley L
Allara, David Lawrence
Allen, Gary William
Altschul, Rolf
Altston, Peter Van
Andose, Joseph David
Andrist, Anson Harry
Andrulis, Peter Joseph, Jr
Archie, William C, Jr

Arnett, Edward McCollin
Arnold, David Brown
Arnold, Richard Thomas
Arnowich, Beatrice
Augood, Derek Raymond
Ayers, Paul Wayne
Badin, Elmer John
Bailey, Roy Horton, Jr
Bair, Thomas Irvin
Bakule, Ronald David
Baldwin, John E
Ballentine, Alva Ray
Bank, Shelton
Bannister, William Warren
Barnett, Ronald E
Baumgarten, Reuben Lawrence
Bearden, William Harlie
Beishline, Robert Raymond
Bell, Charles E, Jr
Bender, Daniel Frank
Benfey, Otto Theodor
Beres, John Joseph
Bernasconi, Claude Francois
Berndt, Donald Carl
Bernstein, Stanley Carl
Bertin, Henry John, Jr
Blackham, Angus Udell
Bly, Robert Stewart
Bongiorni, Domenic Frank
Booman, Keith Albert
Borchardt, John Keith
Bostick, Edgar E
Brady, William Thomas
Brauman, Sharon Kruse
Brown, John Francis, Jr
Brown, Keith Charles
Brown, Robert Stanley
Bruck, Peter
Bryan, Mary Leo
Buchholz, Allan C
Buckles, Lawrence Calvin
Buckles, Robert Edwin
Bunburg, David Leslie
Burlinson, Nicholas Edward
Burnett, Leo Seth
Bushwhaler, Charles Hackett
Cambray, Joseph
Cammarata, Arthur
Cammarata, Peter S
Campbell, Hugh John
Campbell, Thomas Cooper
Caputo, Joseph Anthony
Carle, Kenneth Roberts
Carter, Robert Everett
Caserio, Frederick F, Jr
Cawley, John Joseph
Cengel, John Anthony
Chandler, William David
Chang, Kuang-Chou
Charton, Marvin
Chau, Michael Ming-Kee
Chaudhary, Sohan Singh
Childs, Ronald Frank
Clark, Louis Watts
Clovis, James S
Coburn, William Carl, Jr
Cohen, Gordon Mark
Cohen, Lester Allan
Cohen, Louis Arthur
Colebrook, Lawrence David
Collins, Clair Joseph
Connor, James Edward, Jr
Cook, Richard James
Cooperman, Barry S
Craig, Arnold Charles
Cramer, Richard David, III
Creagh, Linda Truitt
Creamer, Robert M
Crist, DeLanson Ross
Crowell, Thomas Irving
Crumrine, David Shafer
Curtice, Jay Stephen
Dalrymple, David Lawrence
Dannenberg, Joseph
Daughenbaugh, Randall Jay
Davis, Brian Clifton
Dean, David Lee
De Camp, Wilson Hamilton
Decora, Andrew Wayne
Dennis, Edward A
Deno, Norman C
Dias, Jerry Ray
DiGiorgio, Joseph Brun
Dixon, Marvin Porter
Doubleday, Charles E, Jr
Dougherty, Ralph C
Doyle, Michael P
Duncan, Charles Donald
Duncan, James Alan
Dunn, Gerald Emery
Dunn, John Robert
Dunne, Thomas Gregory
Duvall, Scott G
Dwyer, Sean G
Eaton, David Fielder
Edward, John Thomas
Ehrenson, Stanton Jay
Eikenberry, Jon Nathan

Elliot, J Lell
Engel, Leslie
Engel, Paul Sanford
Engler, Edward Martin
Eping, Gary Arnold
Espy, Herbert Hastings
Esteve, Ramon M, Jr
Ewing, Sheila Pauline
Factor, Arnold
Fainberg, Arnold Harold
Fall, Harry H
Fearn, James Ernest
Feiler, Robert Livingston
Fendler, Eleanor Johnson
Fenoglio, David John
Ferren, Richard Anthony
Finley, Kay Thomas
Firkins, John Lionel
Fitzgerald, Patrick Henry
Fliszar, Sandor
Ford, Richard Alan
Forster, Eric Otto
Fort, Raymond Cornelius, Jr
Foss, Robert Paul
Francis, Peter Schuyler
Franz, James Alan
Fraser, Robert Rowntree
Freeman, Fillmore
Friedlander, Herbert Norman
Fuhrmann, Robert
Gale, Laird Housel
Gandour, Richard David
Gaspar, Peter Paul
Gayle, John Ben
Gettler, Joseph Daniel
Giam, Choo-Seng
Gibian, Morton J
Gibson, Harry William
Giddings, William Paul
Gilbert, John Carl
Glaspie, Peyton Scott
Golinkin, Herbert Sheldon
Goon, David James Wong
Gordon, Marshall
Gorski, Robert Alexander
Gosser, Lawrence Wayne
Graminski, Edmond Leonard
Granger, Maurice Roy
Graybill, Bruce Myron
Haggard, Richard Allan
Griffin, Anselm Clyde, III
Griffiths, David Warren
Grubbs, Edward
Grunwald, Ernest Max
Gupta, Amtiava
Gurudata, Neville
Guthrie, Robert D
Hakka, Leo Ernest
Hall, Kenneth Lynn
Hammer, Gary D
Hammons, James Hutchinson
Hanstein, Walter Georg
Harding, Charles Enoch
Hardman, Bruce Bertolette
Harris, Leland
Harrison, Arnold Myron
Hart, Donald John
Haupt, Frederic Curt
Hayward, Lloyd Douglas
Haywood-Farmer, John
Hazlett, Robert Neil
Heckert, David Clinton
Hendry, Dale Glenn
Henrichs, Paul Mark
Herndon, William Cecil
Herz, Jack L
Herz, Matthew Lawrence
Hiatt, Richard Rowls
Hodgdon, Russell Bates, Jr
Hoeg, Donald Francis
Hoffman, Michael K
Holtz, Hans Dietrich
Horan, Francis E
Howe, King Lau
Howe, Norman Elton, Jr
Howell, Thomas James
Idoux, John Paul
Illingworth, George Ernest
Ingold, Keith Usherwood
Irwin, Philip George
Isaks, Martin
Israel, Stanley C
Jaffe, Annette Bronkesh
Jensen, James Leslie
Jesaitis, Raymond G
Jex, Victor Bird
Ji, Sungchul
Jochsberger, Theodore
Johnson, John Enoch
Johnson, Richard Stebbins
Jones, Richard Hamilton
Jones, Guilford, II
Jurch, George Richard, Jr
Kaiser, Emil Thomas
Kakis, Frederic Jacob
Kamego, Albert Amil
Kamlet, Mortimer Jacob
Kapecki, Jon Alfred

Kaplan, Lloyd Allan
Kaplan, Martin L
Kauffman, Glenn Monroe
Kemp, Kenneth Courtney
Kevill, Dennis Neil
Kim, John Poong-Kil
Kinnman, Judith Pollock
Klopfenstein, Charles E
Knudsen, George Andrew, Jr
Koch, Heinz Frank
Konizer, George Burr
Kouba, Rudolph Frank
Kovar, Jan Bernard
Krabacher, Bernard
Kramer, Brian Dale
Kramer, George Mortimer
Kresge, Alexander Jerry
Krieger, Jeanne Kann
Kriz, George Stanley, Jr
Kryger, Roy George
Kulling, Rudolph K
Kumamoto, Junji
Kurland, Jonathan Joshua
Kurz, Joseph Louis
Kurz, Judith Leah
Ladenheim, Harry
Langford, Paul Brooks
Langsdorf, William Philip
Latimer, Donald Andrew
Latta, Bruce McKee
Lauderback, Sanford Keith
Laughlin, Robert Gene
Lawler, Ronald George
Lee, Do-Jae
Lee, Warren G
Leffek, Kenneth Thomas
Leftin, Harry Paul
Leonard, John Joseph
Lepley, Arthur Ray
Levitt, Leonard Sidney
Levy, George Charles
Lewin, Anita Hana
Lewin, James William
Li, Wu-Shyong
Lichtin, Norman Nahum
Liggero, Samuel Henry
Livant, Peter David
Loew, Leslie Max
Lorand, John Peter
Lorenzen, Keith Eden
Losin, Edward Thomas
Lowe, James Urban, Jr
Lown, James William
Lowry, Charles Boyce
Lowry, Nancy
Lura, Richard Dean
Lutz, Raymond Paul
Lynch, Brian Maurice
Mabey, William Ray
Mach, Martin Henry
Magee, Philip Stewart
Magid, Linda Jenny
Marchand, Alan Philip
Mark, Harold Wayne
Marshall, Henry Peter
Marsi, Kenneth Larue
Martin, William Butler, Jr
Marton, Joseph
Maskornick, Michael J
Mateer, Richard Austin
Matuszak, Charles A
Mayo, Frank Rea
McBride, James Michael
McConaghy, John Stead, Jr
McKelvey, Donald Richard
McKelvey, Ronald Deane
McLeod, Richard Kenneth
Mequerian, Garbis H
Mendenhall, George David
Merrifield, D Bruce
Michejda, Christopher Jan
Miller, David Lee
Miller, Sydney Israel
Minch, Michael Joseph
Moore, Cecilia Louise
Morgan, Charles Robert
Moses, Francis Guy
Mosher, Melvyn Wayne
Moss, Ernest Kent
Moye, Anthony Joseph
Muck, Darrel Lee
Murr, Brown L, Jr
Murray, Robert Wallace
Nelson, Kay Leroi
Neuman, Robert C, Jr
Neumann, Calvin Lee
Newitt, Edward James
Newton, Robert Andrew
Nigh, Wesley Gray
Ogliaruso, Michael Anthony
Orvik, Jon Anthony
Osbahr, Albert J, Jr
Otken, Charles Clay
Pacifici, James Grady
Palaitis, Waldemar
Panar, Manuel
Parkany, Cyril
Parker, Winfred Evans
Peck, David W
Perdue, Edward Michael

Perettie, Donald Joseph
Pfeiffer, Joseph George
Pinkus, Albin George
Pinnick, Herbert Robert, Jr
Pocker, Yeshayau
Pollack, Ralph Martin
Porter, John J
Posner, Tamar Beatrice
Powell, Justin Christopher
Puar, Mohindar S
Quimby, Daniel Joseph
Raciszewski, Zbigniew
Raley, John Howard
Ramsey, Brian Gaines
Raniseski, John Walter
Read, David Hadley
Read, Kenneth Joseph
Reuwer, Joseph Francis, Jr
Rhodes, Yorke Edward
Ricci, Robert William
Richards, Charles Norman
Rieger, Martin Max
Robbins, Harry
Roberts, Donald Duane
Robertson, Jerold C
Rocek, Jan
Rodemeyer, Stephen A
Rodgers, James Edward
Roebuck, Albert Henry
Roos, Leo
Roscher, Nina Matheny
Rosenberg, Ira Edward
Rosenfeld, Stuart Michael
Ross, David Samuel
Ross, Lawrence James
Rowe, Paul E
Rowell, Charles Frederick
Rudkin, George Osborne
Ruhnke, Edward Vincent
Sample, Thomas Earl, Jr
Sand, Ralph E
Sande, Vernon Ralph
Sarasohn, Ilya M
Sarge, Theodore William
Saunders, William Hundley, Jr
Schiavelli, Melvyn David
Schleyer, Paul Von Rague
Schmidt, Francis John
Schmidt-Collerus, Josef Johannes
Schowen, Richard Lyle
Schreiber, Kurt Clark
Schueler, Paul Edgar
Schuster, Ingeborg I M
Schwerzel, Robert Edward
Sebastian, John Francis
Selker, Milton Leonard
Senkler, George Henry, Jr
Shafer, Jules Alan
Shearer, Greg Otis
Sheats, John Eugene
Shiner, Vernon Jack, Jr
Simpson, David Alexander
Singh, Ajaib
Skan, Evald Laurids
Smart, Bruce Edmund
Smith, Eileen Patricia
Smith, Gerard Vinton
Smith, Homer Alvin, Jr
Smith, Louis Charles
Smith, Peter James
Smith, Vincent Francis, Jr
Smith, Wendell Franklyn, Jr
Smith, William Novis, Jr
Sojka, Stanley Anthony
Sorensen, Theodore Strang
Sottery, Theodore Walter
Sovocool, George Wayne
Spangler, Charles William
Spertt, Arnold
Stein, Allan Rudolph
Stewart, Ross
Stone, Joe Thomas
Streitwieser, Andrew, Jr
Strom, Edwin Thomas
Stuber, Fred A
Stump, Billy Lee
Sturmer, David Michael
Sullivan, Lloyd John
Swenton, John Stephen
Taagepera, Mare
Takeshita, Tsuneichi
Taller, Robert Arthur
Thompson, Evan M
Thompson, Robert Gene
Thomson, Tom Radford
Thornton, Elizabeth K
Thorpe, Martha Campbell
Tolgyesi, Guy J
Tomalia, Donald Andrew
Towns, Donald Lionel
Tremelling, Michael, Jr
Vander Burgh, Leonard F
van Dijk, Christiaan Pieter
Van Lanen, Robert Jerome
Van Sickle, Dale Elbert
Varga, Charles E
Vill, John Joseph

Vogel, Philip Christian
Voorhees, Kent Jay
Wagner, Robert Marvin
Walling, Cheves
Walsh, James Aloysius
Walter, Robert Irving
Walters, Edward Albert
Warren, Craig Bishop
Warrick, Percy, Jr
Weiner, Steven Allan
Weisman, Kenneth C
Weiss, David Halbert
Wewerka, Eugene Michael
Weisfeld, Lewis Bernard
Weiss, Richard Gerald
Westaway, Kenneth C
White, Joe Wade
White, William North
Whitten, David G
Williams, Joel Mann, Jr
Willis, Roland George
Williamson, Martin John
Wolf, Philip Frank
Wolfarth, Eugene F
Woods, William George
Wright, Robert W
Yager, Billy Joe
Yates, Keith
York, John Lyndal
Zanger, Murray
Zavitsas, Daniel Maximilian
Zavitsas, Andreas Athanasios
Zweig, Arnold

Synthetic Organic Chemistry

Anderson, Hugh John
Babler, James Harold
Bailey, Keith
Bair, Kenneth Walter
Barr, Thomas M
Baumgarten, Henry Ernest
Beck, Keith Russell
Bender, Paul Elliot
Berrier, John Vincent
Bersohn, Malcolm
Bertelson, Robert Calvin
Black, Martin Luther
Borchardt, John Keith
Bornstein, Joseph
Bornstein, Newman Mayer
Bortnick, Robert Mathews
Bowman, Robert Mathews
Brady, Thomas E
Burke, Susan Schlit
Burow, Kenneth Wayne, Jr
Carlson, Glenn Richard
Chang, Charles Hung
Chang, Chin Hsin
Chong, Berni Patricia
Chong, Joshua Anthony
Christensen, Bert Einar
Cohen, Noal
Colwell, William Tracy
Connor, Daniel S
Cory, Robert Mackenzie
Crumrine, Ann Louise
Cummins, Richard Williamson
Davis, Marvin Lester
Decker, Quintin William
Deem, Mary Lease
Dickinson, William Borden
D'Silva, Themistocles Damasceno Joaquim
Duncan, William Perry
Eckler, Paul Eugene
Engel, James Francis
Engel, Robert Ralph
Evans, Ben Edward
Fenyes, Joseph Gabriel Egon
Fieldhouse, John W
Fischer, Nickolaus Hartmut
Fisher, Richard Paul
Fitzpatrick, Joseph Michael
Flaugh, Michael Edward
Foltz, George Edward
Friedrich, John Philip
Gallivan, Robert Milo, Jr
Ganem, Bruce
Gavin, David Francis
Gilbert, Allan Henry
Gilbert, Eugene Charles
Gisin, Balthasar Friedrich
Giza, Chester Anthony
Goldsmith, David Jonathan
Goldstein, Albert
Goodman, Henry Gaines
Grant, Barbara Dianne
Greene, Joseph Lee, Jr
Grieco, Paul Anthony
Griscom, Richard William
Gruenbaum, William Tod
Gruett, Monte Deane
Gwynn, Donald Eugene
Hagan, Charles Patrick
Hagan, Gary Ralph
Harper, Richard Waltz
Haynes, George Rufus
Heathcock, Clayton Howell
Heather, James Brian
Helquist, Paul M

Hendrickson, James Briggs
Herron, David Kent
Hess, Lawrence George
Hirwe, Ashalata Shyamsunder
Hiskey, Richard Grant
Hochstetler, Alan Ray
Hoffman, Jacob Matthew, Jr
Horton, Walter James
Hudriik, Anne Marie
Hudriik, Paul Frederick
Hyatt, John Anthony
Jacobi, Peter Alan
Jacobi, Martin John
Jacobson, Richard Martin
Jelinek, Arthur Gilbert
Johnson, Richard Carl
Jung, Michael Ernest
Katzenellenbogen, John Albert
Kidwell, Roger Lynn
Kiely, Donald Edward
Klutchko, Sylvester
Koster, William Henry
Krajewski, John J
Krass, Dennis Keith
Kretchmer, Richard Allan
Kulier, Charles Peter
Kurtz, Richard Robert
Lawton, Richard G
Lesher, George Yohe
Lundberg, Charles Andrew, Jr
Lyons, Harold Dwight
MacLeay, Ronald E
Marcus, Erich
Martin, Stephen Frederick
Maryanoff, Bruce Eliot
Mason, James Willard
Mattor, John Alan
Maulding, Donald Roy
McEntee, Thomas Edward
McIntosh, John McLennan
McMurry, John Edward
Meen, Ronald Hugh
Mendel, Maryann Madeliene
Merrill, Ronald Eugene
Meteyer, Thomas Edward
Mirviss, Stanley Burton
Mooberry, Jared Ben
Moon, Neil Sennett
Mouk, Robert Watts
Murdock, Keith Chadwick
Nash, William Donald
Okamura, William H
Olsen, Robert Thorvald
Osborn, Harland James
Osborne, Charles Edward
Ott, Donald George
Owsley, Dennis Clark
Parker, Kathlyn Ann
Pelavin, Lawrence
Perry, Clark William
Petrillo, Edward William
Portnoy, Robert Charles
Poslusny, Jerrold Neal
Raghu, Sivaraman
Reasenberg, Julian Robert
Reitman, Larry N
Rice, Kenner Cralle
Robey, Roger Lewis
Ronzio, Anthony Rose
Sagar, William Clayton
Salmond, William Glover
Schell, Fred Martin
Schleppnik, Alfred Adolf
Schlicht, Raymond Charles
Schnur, Rodney Caughren
Schoenewaldt, Erwin Frederick
Schultz, William Clinton
Schulze, William Eugene
Scott, Lawrence Tressler
Sherwood, Bob Edwin
Sipos, Frank
Sircar, Ila
Smith, Elizabeth Melva
Soffer, Milton David
Spessard, Gary Oliver
Stevens, Sandra
Still, W Clark, Jr
Stork, Gilbert Jesse
Struble, Dean L
Surapaneni, Chalapathi Rao
Tarwater, Oliver Reed
Toomey, Joseph Edward
Tortorello, Anthony Joseph
Traxler, James Theodore
Turek, William Norbert
Vinick, Fredric James
Walker, Jerry Arnold
Warren, James Donald
White, Ralph Lawrence, Jr
Whitesell, James Keller
Wight, Hewitt Glenn
Williams, John Roderick
Wilson, Stephen Ross
Wood, James Kenneth
Wright, George Carlin
Wright, Ian Glaisby
Wulfman, David Swinton

Yiannios, Christ Nicholas
Ziegler, John Benjamin

CHEMISTRY, PHYSICAL

Physical Chemistry

Abbott, Andrew Doyle
Abel, Alan Wilson
Abel, William T
Abelson, Philip Hauge
Abrahamson, Edwin William
Abrams, Albert
Abrams, Lloyd
Abrams, Marvin Colin
Abu-Isa, Ismat Ali
Acrivos, Juana Luisa Vivo
Adam, Frank Cuthbert
Adamczak, Robert L
Adams, George Baker, Jr
Adams, John Quincy
Adams, Richard Melverne
Adamson, Arthur Wilson
Addamiano, Arrigo
Adelman, Albert H
Adelstein, Peter Z
Adisesh, Setty Ravanappa
Adler, Norman
Adrian, Frank John
Ahlers, Guenter
Ahmad, Iqbal
Ahrens, Rolland William
Airee, Shakti Kumar
Akins, Daniel L
Aklonis, John Joseph
Alam, Mohammed Ashraful
Albers, Edwin Wolf
Albrecht, Steven Harold
Albrecht, William Lloyd
Albrecht, William Melvin
Alexander, Robert Benjamin
Alfieri, Gaetano T
Alger, Terry Dean
Ali, Mahamed Asgar
Allan, Barry David
Allan, Benjamin Wilson
Allen, Anneke S
Allen, John W
Allison, Jean Batchelor
Allnatt, Alan Richard
Altares, Timothy, Jr
Alter, Harvey
Alter, Henry Ward
Alter, John Emanuel
Altgelt, Klaus H
Altman, Robert Leon
Altmiller, Henry
Alvarez, Robert
Alyea, Hubert Newcombe
Amata, Charles David
Amborski, Leonard Edward
Ambrosiani, Vincent F
Amell, Alexander Renton
Ames, Donald Paul
Ames, Lynford Leonhart
Amey, Ralph Leonard
Amick, Chester Albert
Amis, Edward Stephen
Ammondson, Clayton John
Amster, Adolph Bernard
Ander, Paul
Andersen, Donald Edward
Andersen, Hans Christian
Andersen, Terrell Neils
Andersen, Wilford Hoyt
Anderson, Adolph (Gustof)
Anderson, Frank Hamel
Anderson, Frank Wallace
Anderson, George Robert
Anderson, Harold J
Anderson, Herbert Rudolph, Jr
Anderson, James E
Anderson, Jay Martin
Anderson, Keith Phillips
Anderson, Richard Louis
Anderson, Robbin Colyer
Anderson, Robert Bernard
Anderson, Robert Hunt
Anderson, Roger E
Anderson, Roger W
Andrews, Arthur Clinton
Andrews, John Timothy Sawford
Andrews, Rodney Denlinger, Jr
Andrews, William Lester Self
Anex, Basil Gideon
Angell, C A

Angell, Charles Leslie
Anhorn, Victor John
Annino, Raymond
Anselmo, Vincent C
April, Robert Wayne
Arabian, Karekin Gaspar
Aranow, Ruth Lee Horwitz
Arendt, Ronald H
Arents, John (Stephen)
Argersinger, William John, Jr
Arledter, Hanns Ferdinand
Armanini, Louis Anthony
Armendarez, Peter X
Armington, Alton
Armstrong, Andrew Thurman
Armstrong, David Anthony
Armstrong, George Thomson
Armstrong, William David
Arnell, John Carstairs
Aronson, Seymour
Arquette, Gordon James
Arrington, Charles Hammond, Jr
Arthur, Jett Clinton
Asakawa, George
Ascah, Ralph Gordon
Ashby, Carl Toliver
Asprey, Larned Brown
Asunmaa, Saara K
Atack, Douglas
Aten, Carl Faust, Jr
Atkins, Jaspard Harvey
Atkinson, Gordon
Atkinson, Russell H
Atoji, Masao
Attalla, Albert
Atwood, Gilbert Richard
Atwood, Jerry Lee
Atwood, Kenton
Auborn, James John
Auerbach, Irving
Auld, David Stuart
Auld, Edward George
Avery, William Hinckley
Avgeropoulos, George N
Axworthy, Arthur Edward, Jr
Ayers, Caroline LeRoy
Aziz, Philip Michael
Baak, Tryggve
Bach, Ricardo O
Bacher, Frederick Addison
Bachman, Kenneth Charles
Bachmann, John Henry
Back, Margaret Helen
Backus, John King
Baetzold, Roger C
Baglio, Joseph Anthony
Bahe, Lowell W
Baidins, Andrejs
Bailes, Richard Hazel
Bailey, Charles Edward
Bailey, F Wallace
Bailey, Frederick Eugene, Jr
Bailey, Samuel David
Bair, Edward Jay
Baird, Donald Heston
Baird, James Clyde
Baird, Norman Colin
Bak, David Arthur
Baker, Floyd B
Baker, Harold Theodore
Baker, John Keith
Baker, Louis, Jr
Baker, Richard William
Baker, William Oliver
Bald, Kenneth Charles
Baldwin, Bernard Arthur
Baltz, Alfred
Bangerter, Benedict W
Bansal, Krishan Murari
Banter, John C
Baranwal, Krishna Chandra
Barany, Kate
Barany, Ronald
Barber, Patrick George
Barber, William Austin
Bard, Allen Joseph
Bard, Richard James
Bardwell, John Alexander Eddie
Barfield, Michael
Barger, John Walter
Bargon, Joachim
Barieau, Raymond Conrad
Barile, Raymond (Eugene)
Barlow, Irmela Christiane
Barnard, William Sprague
Barnes, Charlie James
Barnes, Donald George
Barnett, Herald Alva
Barnhart, David M
Barr, James K
Barr, John Baldwin
Barradas, Remigio Germano
Barrante, James Richard
Barrett, Jerry Wayne
Barrett, Robert Earl
Barrett, Wayne Thomas

Barrow, Gordon M
Barry, Arthur John
Bartell, Lawrence Sims
Barth, Max
Bartholomew, Roger Frank
Bartok, William
Barton, Randolph, Jr
Barton, Richard J
Bascom, Willard D
Basila, Michael Robert
Bass, Jonathan Langer
Basseches, Harold
Bassett, David R
Batchelor, William Henry
Bates, Francis Leslie
Bates, Richard Doane, Jr
Bates, William Wannamaker, Jr
Batha, Howard Dean
Batt, Russell Howard
Batten, Charles Francis
Battino, Rubin
Battiste, Merle Andrew
Bauchwitz, Peter Siegbert
Bauer, Eldon Eugene
Bauer, Walter Hermann
Baum, Parker Bryant
Bauman, Richard Gilbert
Baumbach, Donald Otto
Baumgartner, Werner Andreas
Baur, Mario Elliott
Baurer, Theodore
Baxter, John Edwards
Bayes, Kyle D
Beach, John Youngs
Beachell, Harold Charles
Beamer, William Howard
Beamesderfer, John William
Beare, Steven Douglas
Bearman, Richard John
Beattie, Willard Horatio
Beatty, James Wayne, Jr
Beauchamp, Jesse Lee
Beaudet, Robert A
Beaver, Earl Richard
Becher, Paul
Bechtold, Max Fredrick
Beck, Paul W
Beck, Roland Arthur
Becker, Aaron Jay
Becker, Edwin Demuth
Becker, Edwin Norbert
Beckman, William
Beer, Sylvan Zavi
Beese, Ronald Elroy
Beetch, Ellsworth Benjamin
Begala, Arthur James
Behl, Wishvender K
Beichl, George John
Beineke, Thomas Andrew
Beischer, Dietrich Eberhard
Beistel, Donald W
Belford, Rue Linn
Bell, Jerry Alan
Bell, Jimmy Todd
Bell, Thomas Norman
Bellin, Judith Schryver
Belner, Robert Joseph
Ben, Manuel
Benda, Stepan Vaclav
Bender, Max
Bender, Paul J
Bendure, Raymond Lee
Benerito, Ruth Rogan
Benson, George Campbell
Benson, Herbert Linne, Jr
Benson, Loren Allen
Benson, Sidney William
Benston, Margaret Lowe
Bent, Henry Albert
Bentz, Ralph Wagner
Ben-Zvi, Ephraim
Beresniewicz, Aleksander
Berg, John Joseph
Berg, John Richard
Bergh, Arpad A
Bergna, Horacio Enrique
Berkey, Edgar
Berkowitz, Joan B
Berkowitz, Joseph
Bernard, Walter Joseph
Berndt, Alan Fredric
Berne, Bruce J
Bernecker, Richard Rudolph
Berneis, Hans Ludwig
Bernheim, Robert A
Berns, Donald Sheldon
Bernstein, Benjamin Tobias
Bernstein, Harold Joseph
Bernstein, Richard Barry
Berry, Michael James
Berry, Myron Garland
Berry, Richard Stephen
Bertie, John E
Bertin, Ernest Peter
Bertrand, Gary Lane
Bertrand, Rene Robert

Besserman, Marion
Bestul, Alden Beecher
Bethune, John Lemuel
Bett, John Alexander Stuart
Bettelheim, Frederick A
Bettman, Max
Bettis, Robert Holladay
Beuhler, Robert James, Jr
Bevacqua, Joseph Perry
Beveridge, David L
Bezman, Richard David
Bhasin, Madan M
Bhattacharyya, Pranab K
Biagas, Wilfred Michael
Bidinosti, Dino Ronald
Biefer, Gregory James
Biehl, Edward Robert
Biehl, Edward Robert
Bien, George Sung-Nien
Bien, Paul Beh Nien
Bierlein, Theo Karl
Biermann, Wendell J
Bigelow, Wilbur Charles
Bigeleisen, Jacob
Bikerman, Jacob Joseph
Binford, Jesse Stone, Jr
Biordi, Joan Concetta
Birdsall, Henry Alfred
Birkett, James Davis
Birky, Merritt Merle
Bishop, Allen David, Jr
Bishop, Charles Anthony
Bissett, Charles Lynn
Bisson, Thomas Charles
Bivens, Richard Lowell
Black, Donald Lee
Black, Graham
Black, Harvey F, Jr
Black, James Francis
Black, Rodney Elmer
Blackburn, Thomas Roy
Blackman, Carl F, Jr
Blades, Arthur Taylor
Blair, Charles Melvin, Jr
Blair, James Edward
Blaker, Robert J
Blaker, Robert Hockman
Blakeslee, A Eugene
Blanc, Joseph
Blanck, Harvey F, Jr
Blander, Milton
Blanke, Bertram Charles
Blankenship, Floyd Allen
Blankenship, Forrest (Farley)
Blaustein, Bernard Daniel
Blincoe, Clifton (Robert)
Bloch, Ronald Edward
Block, Stanley
Bloemer, William Louis
Blomgren, George Earl
Blount, Floyd Eugene
Blum, Samuel Emil
Blumberg, Avrom Aaron
Blurton, Keith F
Blyholder, George Donald
Blytas, George Constantin
Boaz, Patricia Anne
Bobbitt, Jeffrey L
Bobka, Rudolph J
Bock, Ernst
Bochkoff, Frank James
Bockris, John O'Mara
Boddy, Philip J
Bode, James Daniel
Bodi, Lewis Joseph
Bodily, David Martin
Bodson, Herman
Boehnke, David Neal
Boggio, Joseph E
Boggs, William Emerson
Bohmfalk, Erwin Frederick, Jr
Bohn, Conrad Rathmann
Bohn, Robert K
Boldridge, William Franklin
Bolles, Theodore Frederick
Bolton, Anthony Peter
Bolton, Arthur Chalmer
Bond, Donald C
Bond, William Bradford
Bond, Walter D
Bone, Larry Irvin
Bonner, Billy Edward
Bonner, Oscar Davis
Booth, Max Howard
Bopp, Thomas Theodore
Borchers, Curtis Edward
Borden, Kenneth Duane
Borg, Richard John
Borkowski, Raymond P
Bornmann, John Arthur
Bosley, David Emerson
Boss, Bruce David
Boston, Charles Ray
Both, Eberhard
Botger, Gary Lee
Boudart, Michel
Bowen, Lawrence Hoffman
Bowen, Ruth Justice

Bowen, John Edwin
Bowers, Peter George
Bowers, Spotswood D, Jr
Bowerson, David F
Bowman, Allen Lee
Bowman, Ray Douglas
Bowman, Robert Samuel
Boyd, George Edward
Boyd, Richard Hays
Boyd, Robert Henry
Boyko, Edward Raymond
Boyles, James Glenn
Brabson, George Dana, Jr
Bracco, Donato John
Bracken, Ronald Clay
Brady, Edward Lewis
Brady, George W
Brady, Ruth Mary
Bragin, Joseph
Bramwell, Fitzgerald Burton
Brand, John C
Brand, James Lewis
Brandt, Luther Warren
Brandt, Werner Wilfried
Brands, John Frederick
Brandvold, Donald Keith
Brantley, Lee Reed
Brash, John Law
Brashier, Gary Kermit
Braun, Juergen Hans
Braun, Robert Leore
Braunstein, Jerry
Bray, Norman Francis
Breazeale, William Horace
Brecher, Charles
Breck, Wallace Graham
Bree, Alan V
Breisacher, Peter
Breiter, Manfred Wolfgang
Breitman, Leo
Bremer, Robert F
Brendley, William H, Jr
Brennan, Harry Michael
Brinton, Robert K
Brion, Christopher Edward
Brisey, Ruben Marion
Bristow, William Warren
Britt, A D
Britton, William Giering
Broach, Wilson J
Brockway, Lawrence Olin
Brodale, Gary Edward
Brodasky, Thomas Francis
Brodd, Ralph James
Broge, Robert Walter
Broido, Abraham
Brokaw, Richard Spohn
Brom, Joseph March, Jr
Bromberg, J Philip
Bromels, Edward
Bronfin, Barry Robert
Broodo, Archie
Brooks, Alfred Austin, Jr
Brooks, Clyde S
Brooks, Philip Russell
Brooks, Wendell V F
Brous, Jack
Brown, Charles Julian, Jr
Brown, Charles Thomas
Brown, Christopher W
Brown, Duane
Brown, Henry Trueheart
Brown, James Douglas
Brown, John Angus
Brown, John Boyer
Brown, Lloyd Leonard
Brown, Lowell Severt
Brown, Norman Louis
Brown, Oliver Leonard Inman
Brown, Paul Edmund
Brown, Richard Julian Challis
Brown, Robert Eugene
Browning, Horace Lawrence, Jr
Brownscombe, Eugene Russell
Broyde, Baret
Brugger, John Edward
Brumberger, Harry
Brummer, S Barry
Brusic, James Powers
Bryant, Herman Grey, Jr
Bryce, Hugh Glendinning
Buchanan, James Wesley
Buchler, Alfred
Buck, Richard Pierson
Buck, Warren Howard
Buckingham, John Herbert
Buckley, Bernard Patrick

Buckley, Reginald R
Bueche, Arthur Maynard
Buell, Wayne H
Buenker, Robert J
Buff, Frank Paul
Bugg, Charles Edward
Bugosh, John
Bujake, Jahn Edward, Jr
Bulas, Romuald
Bulgrin, Vernon Carl
Bulkin, Bernard Joseph
Bull, Henry Bolivar
Bullock, Jonathan S, IV
Bundschuh, James Edward
Bunger, William Boone
Bunker, Don Louis
Bupp, Lamar Paul
Burchill, Charles Eugene
Burford, Mortimer Gilbert
Burgess, Kenneth Alexander
Burkhart, Richard Delmar
Burlew, John Swain
Burley, Gordon
Burley, David Richard
Burlingame, Alma L
Burnham, John
Burns, John Howard
Burns, Richard Price
Burow, Duane Frueh
Burr, Horace Kelsey
Burrell, Elliot Joseph, Jr
Burris, Robert Tilden
Bursh, Talmage Poutau
Burton, James Samuel
Burton, Louis Lasseter
Burton, Milton
Burtt, Benjamin Pickering
Burwasser, Herman
Busey, Richard Hoover
Bush, Stewart Fowler
Bushey, Albert Henry
Bushey, Gordon Lake
Busing, William Richard
Butcher, Samuel Shipp
Butera, Richard Anthony
Butler, Edward Byron
Butler, James Newton
Butzer, Karl Wilhelm
Byrd, David Shelton
Byrd, Willis Edward
Byrn, Ernest Edward
Cabana, Aldee
Cadbury, William Edward, Jr
Cadenhead, David Allan
Cadle, Richard Dunbar
Cadogan, Kevin Denis
Cady, Howard Hamilton
Caesar, Cameron Hull
Cahill, Charles L
Cairns, Elton James
Cairns, Robert William
Calcote, Hartwell Forrest
Calhoun, Gordon Maxwell
Callahan, John Joseph
Callanan, Margaret Joan
Callender, Wade Lee
Calloway, E Dean
Calvert, Jack George
Calvo, Crispin
Campbell, Alan Newton
Campbell, Francis James
Campbell, George Melvin
Campbell, Jack Allen
Campbell, James Arthur
Campbell, John Hyde
Campbell, William Joseph
Canham, Richard Gordon
Cannon, Peter
Cannon, Richard Francis
Cante, Charles John
Cantow, Manfred Josef Richard
Cantrell, Joseph Sires
Cappas, C
Caputi, Roger William
Capwell, Robert J
Carfagno, Daniel Gaetano
Carlson, Douglas W
Carlson, Gary Alden
Carlson, Gerald Leroy
Carlson, Keith Douglas
Carnall, William Thomas
Caron, Aimery Pierre
Carpenter, Dewey Kenneth
Carpenter, Gene Blakely
Carr, Clide Isom
Carraher, Charles Eugene, Jr
Carreira, Lionel Andrade
Carrington, Tucker
Carroll, Benjamin L
Carroll, Frederick E
Carroll, Harvey Franklin
Carroll, Murray Norman
Carroll, Robert William
Carroll, Willis Lee
Carter, Loren Sheldon
Carter, Melvin K

Carter, Orvin Lee
Carter, Owen, Jr
Carter, Peter G
Casassa, Edward Francis
Casper, John Matthew
Caspers, Horst J
Cassen, Thomas Joseph
Castellan, Gilbert William
Castro, George
Catcott, Earle David
Cater, Duane Stewart
Catlett, Ephraim Herman
Catsiff, Danny Dearl
Caughlan, Charles Norris
Cave, William Thompson
Cella, Richard Joseph, Jr
Cells, Robert
Cerankowski, Leon Dennis
Chackerian, Charles, Jr
Chakravarti, Kalidas
Chamberlain, David Leroy, Jr
Chamberlin, John Macmullen
Chandler, Raymond Kai-Chow
Chandler, Dean Wesley
Chandross, Edwin A
Chandross, Edwin A
Chang, Charles C
Chang, Ching-Jen
Chang, James C
Chang, Joseph Yung
Chang, Raymond
Chang, Shu-Sing
Chang, Ted Teh-Liang
Chang, Yeong-Jen Peter
Chao, Jing
Chapin, Douglas Scott
Chapman, Thomas Shelby
Charles, Donald Foster
Chasanov, Martin Gerson
Chase, Grafton D
Chatterjee, Pronoy Kumar
Chiang, Joseph Fei
Chiang, Yuen-Sheng
Chien, James C W
Child, William Clark, Jr
Chiotti, Premo
Chittick, Donald Ernest
Chiu, Tai-Woo
Chiu, Ying-Nan
Cho, Kon Ho
Chong, Delano Pun
Chou, David Yuan Pin
Chow, Che Chung
Chow, Fu Ho Chen
Chow, Lawrence Chung-Lung
Christ, Charles Louis
Christie, Karl Otto
Christensen, Carl Joseph
Christensen, Sherril Duane
Christoffersen, Ralph Earl
Christofferson, Glen Davis
Chrysochoos, John
Chu, Benjamin Peng-Nien
Chu, Victor Fu Hua
Chu, Vincent Hao Kwong
Chupka, William Andrew
Church, John Armistead
Church, Shepard Earll, (Jr)
Cichowski, Robert Stanley
Ciric, Julius
Clardy, LeRoy
Clark, Harlan Eugene
Clark, Milton B
Clark, Patricia Ann
Clark, Paul Enoch
Clark, Robert Paul
Clark, Ronald Keith
Clark, Salem Thomas
Clark, William Dempsey
Clarke, George A
Clarke, Richard Henry
Clarke, Richard Penfield
Clarkson, Robert Breck
Clauss, James K
Claussen, Walter Frederick
Clay, John Paul
Cleaves, Duncan Worster
Cleland, Robert Lindbergh
Clever, Henry Lawrence
Clifford, Howard James
Clifton, David Geyer
Clingman, William Herbert, Jr
Cloney, Robert Dennis
Coakley, Mary Peter
Coates, Arthur Donwell
Cobb, Carolus M
Cobble, James Wikle

Cochran, Charles Norman
Cochrane, Hector
Cocivera, Michael
Coffey, Charles Eugene
Cogan, Harold Louis
Cohen, Edward David
Cohen, George Lester
Cohen, Gerson H
Cohen, Morris
Cohen, Norman
Coker, Earl Howard, Jr
Cole, David F
Cole, David Le Roy
Cole, Richard
Coleman, Charles Franklin
Coleman, James Edward
Coles, James Stacy
Coley, Ronald Frank
Colgate, Samuel Oran
Colichman, Eugene Louis
Collins, Edward A
Collins, Frank Charles
Collins, Timothy Leo, Jr
Colmenares, Carlos Adolfo
Colodny, Paul Charles
Colvin, Clair Ivan
Combs, Robert Leonard
Compere, Edgar Lattimore
Compton, Leslie Ellwyn
Conan, Robert James, Jr
Conary, Robert Ekvall
Condon, James Benton
Conley, Harry Lee, Jr
Conn, Paul Joseph
Conner, Jack Michael
Conway, Brian Evans
Conway, Dwight Colbur
Cook, Edward Hoopes, Jr
Cook, Evin Lee
Cook, Gerhard Albert
Cook, Glenn Melvin
Cooke, Derry Douglas
Cooke, Samuel Leonard, Jr
Cooley, Stone Deavours
Cooney, John Leo
Cooper, Gerald Rice
Cooper, John Neale
Cooper, Walter
Coots, Alonzo Freeman
Copeland, James Lewis
Copeland, Richard Franklin
Copes, Joseph Paul
Coplan, Michael Alan
Coppel, Claude Peter
Coppola, Patrick Paul
Cordes, Herman Fredrick
Coriell, Sam Ray
Cornell, Paul Hampton
Corning, Mary Elizabeth
Corrin, Myron Lee
Corsaro, Robert Dominic
Corth, Richard
Cotten, George Richard
Cotton, John Edward
Coulter, Lowell Vernon
Courchene, William Leon
Courtney, Welby Gillette
Coutts, John Wallace
Cover, Herbert Lee
Coward, Nathan A
Cowgill, James Joseph
Cox, Anna Lucile
Cox, James Carl, Jr
Cox, Lionel Audley
Cox, William Lester
Cozzens, Robert F
Craig, James Porter, Jr
Craig, Norman Castleman
Craig, Raymond S
Cramer, William Herbert
Cratin, Paul David
Cratty, Leland Earl, Jr
Crawford, Bryce (Low), Jr
Crawford, Clayton McCants
Crawford, Oakley H
Creighton, Stephen Mark
Crespi, Henry Lewis
Crider, Fretwell Goer
Criss, Cecil M
Croce, Ian Murray
Croll, Lau Murray
Crompton, Charles Edward
Crosby, Glenn Arthur
Crosley, David Risdon
Cross, Jon Byron
Crowder, Gene Autry
Crowe, George A
Cruser, Stephen Alan
Crutchfield, Marvin Mack
Csejka, David Andrew
Culbertson, Billy Muriel
Cummin, Alfred Samuel
Cummings, Thomas Fulton
Cunniff, Patricia A
Cunningham, Clarence Marion
Cunningham, Geoffrey Everett

Cunningham, George Lewis, Jr
Cunningham, Paul Thomas
Curl, Juan Daniel
Curl, Robert Floyd, Jr
Curne, Henry Garrett
Curnutt, Jerry Lee
Current, Jerry H
Currie, Lloyd Arthur
Curtis, Earl Clifton, Jr
Custard, Herman Cecil
Cuthbertson, George Raymond
Cutler, Janice Ann
Cutnell, John Daniel
Cutter, Paul Ramey
Cvetanovic, Ratimir J
Dacey, John Robert
Dahl, Alton
Dahler, John S
Dahlgren, George
Dahlstrom, Bertil Philip, Jr
Dahm, Donald B
Dahms, Harald
Dakin, Thomas Wendell
Daley, Henry Owen, Jr
D'Amato, Richard John
Dames, Charlotte A
Damour, Paul Lawrence
Danforth, Joseph Davis
Daniel, Daniel S
Daniel, Robert Eugene
Daniels, Malcolm
Daniels, William Ward
Dannhauser, Walter
Danzer, Laurence Alfred
Daoust, Hubert
Darken, Lawrence Stamper
Darrow, Frank William
Darwent, Basil de Baskerville
Dasher, George Franklin, Jr
Davenport, John Eaton
David, Carl Wolfgang
Davidson, Donald West
Davidson, Ernest Roy
Davidson, William John
Davies, David Huw
Davies, Donald Harry
Davis, Bruce W
Davis, Burton H
Davis, Henry Mauzee
Davis, Henry McRay
Davis, Howard Ted
Davis, Hubert Greenidge
Davis, Jefferson Clark, Jr
Davis, Merton Louis
Davis, Philip Seals
Davis Raymond E
Davis, Stanley Gannaway
Davis, Thomas Wilders
Davis, William Donald
Davison, Peter Fitzgerald
Davison, Robert Wilder
Davisson, Edwin Orlando
Dawes, David Haddon
Dawson, Thomas Larry
Day, Herman O'Neal, Jr
Day, Jesse Harold
Day, Marion Clyde, Jr
Dea, Phoebe Kin-Kin
Deal, Bruce Elmer
Deal, Carl Hosea, Jr
Deal, Ralph Macgill
Dean, Anthony Marion
Dean, Warren Edgell
Dearman, Henry Hursell
DeArmond, M Keith
Debnath, Sadhana
DeBoer, Frank Edward
Decius, John Courtney
Deck, Joseph Charles
Deckers, Jacques (Marie)
DeDecker, Hendrik Kamiel Johannes
Dedinas, Jonas
Dee, Diana
Degginger, Edward R
De Haan, Frank P
de Heer, Joseph
Dehl, Ronald
DeHollander, William Roger
De Korte, Aart
Delaney, Charles MacGregor
DeLap, James Harve
Del Duca, Betty Spahr
De Lorenzo, Eugene Joseph
Dempsey, John Nicholas
Den Besten, Ivan Eugene
Denison, Jack Thomas
Denison, M Carl
Denney, Donald John
Dennis, Kent Seddens
DePhillips, Henry Alfred, Jr
Derrick, Mildred Elizabeth
Desai, Raman Lalbhai
De Sando, Richard John
Deshpande, Krishnanath Bhaskar
Desiderato, Robert, Jr
DeSieno, Robert P

Desnoyers, Jacques Edouard
Dettre, Robert Harold
Deutch, John Mark
Deutsch, John Ludwig
DeVault, Don Charles
Dever, David Francis
Deviney, Marvin Lee, Jr
Devlin, Joseph Paul
DeVoe, Howard Josselyn
de Vries, Adriaan
De Vries, Dale Byron
Dewald, Robert Reinhold
Dewhurst, Harold Ainslie
Deyrup, Alden Johnson
Diamond, Jacob Joseph
Dickerson, Richard Earl
Dickie, Ray Alexander
Dickinson, Alan Charles
Dickinson, Leonard Charles
Dickson, Arthur Donald
Didwania, Hanuman Prasad
Dierenfeldt, Karl Emil
Dierolf, Jack
Diesen, Ronald W
Dietrick, Harry Joseph
Dietz, Paul Luther, Jr
Dillard, John Gammons
Dillemuth, Frederick Joseph
Dillon, Henry Kenneth
Dills, Charles E
DiLorenzo, James V
Dimitroff, Edward
Dinegar, Robert Hudson
Dingeldy, David Peter
Disch, Raymond L
Discher, Clarence August
Dismukes, Edward Brock
Dismukes, Gerard Charles
Ditman, John Gordon
Dixon, William Brightman
Dobbs, Frank W
Dobbs, Gregory Melville
Dobrott, Robert D
Dobry, Alan (Mora)
Dobyns, Leona Danette
Dodd, Charles Gardner
Dodge, Richard Patrick
Dodson, Charles Leon, Jr
Dodson, Richard Wolford
Doedens, Robert John
Doepker, Richard DuMont
Doigan, Paul
Dolch, William Lee
Dole, Malcolm
Donath, Ernest E
Dondes, Seymour
Donoian, Haig Cadmus
Donovan, John W
Donovan, Sandra Steranka
Doody, John Stuart
Dooling, John Stuart
Dorain, Paul Brendel
Doremus, Robert Heward
Dorko, Ernest A
Dormant, Leon M
Dorough, Gus Downs, Jr
Dorris, Kenneth Lee
Douglas, David Lewis
Douglas, John Edward
Douglas, Larry Joe
Douville, Philip Raoul
Downey, Bernard Joseph
Downing, George V, Jr
Downs, James Joseph
Dows, David Alan
Dragun, Henry L
Drake, Edgar Nathaniel, II
Draper, Arthur Lincoln
Dreby, Edwin Christian, III
Dreikorn, Russell E
Drenan, James Warner
Drexhage, Karl Heinz
Drinkard, Russell Drew
Duchamp, David James
Duck, William N, Jr
Duedall, Iver Warren
Duerst, Richard William
Duffey, Donald Creagh
Duisman, Jack Arnold
Duke, Frederick Robert
Dulmage, William James
Dulz, Gunther
Dumbaugh, William Henry, Jr
Dumke, Walter Henry
Dumke, Warren Lloyd
Dunbar, Phyllis Marguerite
Dunbar, Robert Copeland
Duncan, Budd Lee
Dunicz, Boleslaw Ludwik
Dunkelberger, Tobias Henry
Dunkle, Michael Patrick
Dunlap, Robert D
Dunn, Thomas M
Dunning, Robert Lewis
Durham, Elford Sturtevant
Durham, George Stone
Durig, James Robert
Dusenbury, Joseph Hooker

Duty, Robert C
Duvall, Jacque L
Duwell, Ernest John
Dux, James Philip
Dwiggins, Claudius William, Jr
Dwyer, James Michael
Dwyer, Robert Francis
Dybowski, Cecil Ray
Dybwad, Jens Peter
Dyck, Rudolph Henry
Dye, James Louis
Dyer, Lawrence D
Dyke, Thomas Robert
Dymerski, Paul Peter
Dymicky, Michael
Dzidic, Ismet
Dzombak, William Charles
Eager, Richard Livingston
Earl, Charles Riley
Easley, Warren C
Eastland, George Warren, Jr
Eastman, Michael Paul
Eastwood, DeLyle
Eaton, Donald Rex
Eatough, Norman L
Ebbing, Darrell Delmar
Ebdon, David William
Eberhardt, William Henry
Echols, Joseph Todd, Jr
Eckert, George Frank
Eckhardt, Craig Jon
Eckstein, Bernard Hans
Edmiston, Clyde
Edwards, Gerald Alonzo
Edwards, Harry Wallace
Edwards, Kenneth Ward
Edwards, Oscar Wendell
Egan, James John
Eggers, David Frank, Jr
Ehlert, Thomas Clarence
Ehrenstorfer, Sieglinde K M
Ehrlich, Paul
Eichelberger, Robert Leslie
Eick, Harry Arthur
Eidinoff, Maxwell Leigh
Eisenberg, Adi
Eissler, Robert L
Ekler, Kurt
Ekstrom, Lincoln
El-Bayoumi, Mohamed Ashraf
Elder, Fred A
Eliason, Morton A
Elleman, Thomas Smith
Elliott, Ralph Benjamin
Ellis, David Allen
Ellis, Richard Bassett
Ells, Victor Raymond
El-Sayed, Mostafa Amr
Elsdon, William Lloyd
El-Shimi, Ahmed Fayez
Elson, Jesse
Emerson, Kenneth
Emerson, Merle T
Emery, Edward Mortimer
Emmett, Paul Hugh
Endres, Leland Sander
Endres, Paul Frank
Engebretson, Gordon Roy
Engelke, John Leland
Engleman, Rolf, Jr
English, Albert Charles
Epel, Joseph Norman
Epstein, Joseph
Epstein, Lawrence Melvin
Epstein, Leo Francis
Erb, Robert Allan
Erickson, David Edward
Erickson, John M
Erickson, Luther E
Erickson, Randall L
Ericson, Corey William
Erkkila, Armas Victor
Ernsberger, Fred Martin
Everard, Martin Edward
Everson, Howard E
Ewing, Galen Wood
Ewing, George Edward
Ewing, Gordon J
Ewing, Richard Everett
Exarhos, Gregory James
Eyler, John Robert
Eyring, Henry
Eyring, LeRoy
Eyster, Eugene Henderson
Fackenthal, Edward
Fagley, Thomas Fisher

CHEMISTRY, PHYSICAL

Fairless, Billy J
Fajer, Jack
Falconer, Warren Edgar
Falk, Michael
Falkiewicz, Michael Joseph
Fallgatter, Michael
Farber, Milton
Farewell, John P
Faris, Sam Russell
Farmer, James Bernard
Farone, William Anthony
Farrar, Charles Frederick
Farrar, David Turner
Farrar, John
Farrar, Robert Lynn, Jr
Farrar, Thomas C
Farrington, Gregory Charles
Fass, Richard A
Fassel, Velmer Arthur
Faubion, Billy Don
Favis, Dimitrios Vasilios
Fehlner, Francis Paul
Feighan, Maria Josita
Felder, William
Feldman, Milton H
Feldstein, Nathan
Feltham, Robert Dean
Felty, Evan J
Fenrick, Harold William
Ferguson, Raymond Craig
Fessenden, Richard Warren
Fessler, Robert Glenn
Fetsko, Jacqueline Marie
Fetters, Lewis
Feuer, Irving
Ficalora, Peter
Field, Frank Henry
Field, Paul Eugene
Fifer, Robert Alan
Filbert, James E
Filipescu, Nicolae
Filseth, Stephen V
Filson, Don P
Finch, Jack Norman
Finegold, Harold
Fineman, Morton A
Finholt, James E
Fink, Richard David
Finkbeiner, Herman Lawrence
Finke, Herman Louis
Firestone, Richard Francis
Fischer, Arthur H
Fischer, David John
Fischer, George Morrison
Fisher, Thomas Henry
Fitts, Donald Dennis
Flanagan, Ted Benjamin
Flank, William H
Flannery, John B, Jr
Flato, Jud B
Flautt, Thomas Joseph, Jr
Fleck, James Burton
Fleischer, Joseph
Fleming, Sydney Winn
Fletcher, Peter C
Fletcher, William Henry
Flierl, Donald William
Flom, Donald Gordon
Flory, Paul John
Flowers, Ralph Grant
Flynn, George Patrick
Flynn, James Patrick
Flynn, Joseph Henry
Fontijn, Arthur
Foos, Raymond Anthony
Foote, John K
Forcheimer, Otto Louis
Fordham, James Lynn
Fordyce, James Stuart
Foreman, Dennis Walden, Jr
Forest, Edward
Forest, Harvey
Forgacs, Otto Lionel
Forst, Wendell
Forster, Leslie Stewart
Fortnum, Donald Holly
Forziati, Alphonse Frank
Foster, Alfred Field
Foster, Melvin S
Foster, Norman Francis
Foster, Norman George
Foster, Perry Alanson, Jr
Foster, Walter H, Jr
Foster, Wilfred John Daniel
Foster, William Roderick
Fowler, Lewis
Fox, William
Foy, Walter Lawrence
Frame, Harlan D
Francis, Stanley Arthur
Frank, Henry Sorg
Frankel, Lawrence (Stephen)
Franklin, Thomas Chester
Franzen, Hugo Friedrich
Franzosa, Edward Sykes

Franzus, Boris
Frasco, David Lee
Fraser, Donald Boyd
Fraser, James Mattison
Frashier, Loyd Dola
Fratiello, Anthony
Frazee, Jerry D
Freasier, Ben Forest
Frech, Roger
Frederick, John Edgar
Frederick, Kenneth Jacob
Fredericks, William John
Freed, Karl F
Freed, David Haines
Freeman, Gordon Russel
Freeman, John Jerome
Freeman, Mark Phillips
Freeman, Raymond
Freeman, Robert David
Freiling, Edward Clawson
Freiser, Stephan T
French, William George
Fremdorff, Hans Karl
Frey, Harold Joseph
Frick, Neil Huntington
Fried, Vojtech
Friedman, Harold Bertrand
Friedman, Harold Leo
Friedman, Robert Harold
Friedrich, Bruce H
Friel, Patrick Joseph
Fries, James Andrew
Fries, Ralph Jay
Friess, Seymour Louis
Frisque, Alvin Joseph
Fritsch, James John
Fritsche, Herbert William
Fritz, James John
Frohnsdorff, Geoffrey James Carl
Froix, Michael Francis
Frosch, Robert Peter
Frost, Arthur Atwater
Frost, David Cregreen
Frystak, Ronald Wayne
Fucaloro, Anthony Frank
Fugassi, James Paul
Fuget, Charles Robert
Fuller, Edward C
Fuller, Everett J
Fuller, Martin Emil, II
Fuller, Milton E
Fuoss, Raymond Matthew
Furrow, Stanley Donald
Furtell, Jean H
Fuzek, John Frank
Fye, Paul McDonald
Gabelnick, Stephen David
Gable, Ralph William
Gabor, Thomas
Gabriel, Henry
Gager, Helen McClure
Gaines, George Loweree, Jr
Galiano, Robert Joseph
Gall, James William
Gall, John Frederick
Gallegos, Emilio Juan
Galligan, James Bernard
Gallivan, James Bernard
Galperin, Irving
Gancy, Alan Brian
Gangwer, Thomas E
Gann, Richard George
Gannon, David John
Gantzel, Peter Kellogg
Garber, Calvin Samuel
Garcia, Mario Leopoldo
Garcia-Morin, Manuel
Gardner, Kenneth William
Gardner, William Cecil, Jr
Gardner, David Milton
Gardner, Philip John
Gardner, Ralph Alexander
Gardner, William Lee
Gardner, William Reavis
Garfield, Lawrence James
Garfinkel, Harmon M
Garik, Valdemar L
Garland, Carl Wesley
Garn, Paul Donald
Garrard, Verl Grady
Garrett, Barry B
Garrett, Bowman Staples
Garrett, Robert Roth
Garrison, Warren Manford
Garst, John Fredric
Garvin, David
Gatz, Carole R
Gatz, David Lawrence
Gaudioso, Stephen Lawrence
Gaughan, Renata Rysnik
Gavin, Robert M, Jr
Gawer, Albert Henry
Gay, David Lawrence
Gayles, Joseph Nathan, Jr
Geddes, Amos Leslie
Gee, Allen
Gehatia, Matatiahu T
Gemant, Andrew
Gendron, Pierre Raoul
Genge, Colin Arthur

Gentner, Robert F
Gentry, William Ronald
George, James Henry Bryn
George, Raymond Arthur
Gerbacia, William Edward
Gerfen, Charles Otto
Gerhold, George A
Gershinowitz, Harold
Gerstein, Bernard Clemence
Gesser, Hyman Davidson
Getzen, Forrest William
Ghandehari, Mohammad Hossein
Ghose, Hirendra M
Ghosh, Kalyan K
Ghosh, Saroj Bandhu
Giauque, William Francis
Gibbard, H Frank, Jr
Gibbons, David Louis
Gibbs, Julian Howard
Gibbs, Robert John
Gibson, William Eugene
Gibson, Ralph Edward
Giddings, John Calvin
Giddings, Lorrain Eugene, Jr
Gierke, Timothy Dee
Giguere, Paul Antoine
Gilbert, John Barry
Gilbert, Nathan
Gilby, Anthony Christopher
Gildseth, Wayne
Giles, Waldron
Gilkerson, William Richard
Gill, Piara Singh
Gill, Robert Anthony
Gill, Robert F, Jr
Gill, Stanley Jensen
Gillen, Raymond Daniel
Gilles, Paul Wilson
Gillespie, Ronald James
Gillow, Edward William
Gilman, Paul Brewster, Jr
Gilmore, Forrest Cubley
Gilmour, Hugh Stewart Allen
Ginell, William Seaman
Gingerich, Karl Andreas
Ginn, Martin E
Gislason, Eric Arni
Gittler, Franz Ludwig
Giulianelli, James Louis
Givens, William Geary
Gladden, James Kelly
Gladow, Elroy Merle
Glarum, Sivert Herb
Glarum, Sivert N
Glasel, Jay Arthur
Glass, Graham Percy
Glassbrook, Clarence I
Glasstone, Samuel
Glaunsinger, William Stanley
Gleaves, John Thompson
Gliew, David Neville
Glick, Milton Don
Glick, Richard Edwin
Gliogovsky, Robert L
Gluyas, Richard Edwin
Goates, James Rex
Godbey, William Givens
Godycki, Ludwig Edward
Goens, Duane N
Gold, Lewis Peter
Goldberg, Ira Barry
Goldberg, Paul
Goldberg, Robert Nathan
Golden, Sidney
Goldenberg, Neal
Goldfarb, Theodore D
Goldish, Elihu
Goldman, Stephen Allen
Goldring, Lionel Solomon
Goldsmith, Harry L
Goldsmith, Henry Arnold
Goldstein, Jacob Herman
Goldstein, Martin
Golob, Fred
Golob, Morton Allan
Goll, Robert John
Golike, Ralph Crosby
Gomer, Robert
Gonet, Frank
Gonzalez, Richard D
Goodfriend, Paul Louis
Goodings, John Martin
Goodkin, Jerome
Goodman, Jerome
Goodman, Philip
Goodman, Seymour
Gordon, Alvin S
Gordon, Barry Maxwell
Gordon, George Selbie
Gordon, John Edward
Gordon, Robert Dixon
Gordon, Robert Jay
Gordon, William Cummins
Gordon, William Edwin
Gore, Robert Cummins
Gore, Wilbert Lee
Gorin, Everett

Gornick, Fred
Gouinlock, Edward Vernon, Jr
Goyan, Frank Mayer
Grabenstetter, Robert John
Grady, Harold Roy
Graf, Peter Emil
Graham, John David
Graham, Laurine LaPlanche
Graham, Raymond
Graham, Robert Leslie
Granquist, William Thomas
Grant, David Morris
Grant, Douglas Hope
Grant, Richard J
Grant, Robert Charles S
Grant, Warren Herbert
Grantham, Leroy Francis
Grasselli, Robert Karl
Gray, John Augustus, III
Gray, June Pfister
Gray, Thomas James
Graybeal, Jack Daniel
Graydon, William Frederick
Greager, Oswald Herman
Greeley, Richard Stiles
Green, David William
Green, John William
Green, Michael Enoch
Green, Richard Lee
Greene, Charles Richard
Greene, Edward Forbes
Greene, Hoke Smith
Greene, Kenneth Titsworth
Greenshields, John Bryce
Greenspan, Joseph
Greenstein, Leon M
Greenberg, Daniel
Greenberg, Elliott
Greenberg, Sidney Abraham
Greer, Sandra Charlene
Gregory, M Duane
Gregory, Norman Wayne
Greig, Joseph Wilson
Greiner, Richard William
Grey, James Tracy, Jr
Grieshammer, Lawrence Louis
Griffel, Maurice
Griffin, John Leander
Griffin, Robert Guy
Griffith, Edward Jackson
Griffith, Elizabeth Ann Hall
Griffith, Martin G
Griffo, Joseph Salvatore
Grindstaff, Teddy Hodge
Grogan, Michael John
Groh, Harold John
Gronholz, LeRoy Frederick
Gropp, Armin Henry
Gross, Frederick Bruce
Gross, Paul Hans
Gross, Paul Magnus
Gross, Paul Magnus, Jr
Grosser, Arthur Edward
Grossman, Jack Joseph
Grotz, Leonard Charles
Groves, Ewart Lester
Groves, Warren Olley
Growcock, Frederick Bruce
Grubb, Willard Thomas
Grubb, Thomas Albin
Grundmeier, Ernest Winston
Gruntfest, Irving James
Grushkin, Bernard
Gruszka, Eli
Gruss, Leonard Louis
Gryder, John William
Guenther, William Benton
Guillet, James Edwin
Guillory, Jack Paul
Guillory, William Arnold
Gulbransen, Earl Alfred
Gulrich, Leslie William, Jr
Gunn, George Bradford
Gunn, Stuart Richard
Gunnell, Thomas Jefferson
Gunning, Harry Emmet
Gunmke, Raymond
Gusman, Samuel
Gustavson, Marvin Ronald
Gutmann, David
Gutowsky, Herbert Sander
Gutschow, Nathan Robert
Guth, Egbert Karl Anton
Gupta, Vijay Kumar
Gurry, Robert Wilton
Gwinn, William Dulaney
Gwinup, Paul D
Gyberg, Arlin Enoch
Haag, Robert Marlay
Haaland, David Michael
Haas, John William, Jr
Haas, Trice Walter
Habermann, Clarence E
Hach, Edwin E, Jr
Hadermann, Albert Felix

Hadley, Steven George
Hafner, Harold C
Hagan, Lucy Gay
Hagen, Richard Martin
Hager, Stanley Lee
Hagstrom, Stanley Alan
Hahne, Rolf Mathieu August
Hair, Michael L
Halaby, Sami Assad
Halberstadt, Marcel Leon
Halford, Ralph Stanley
Hall, Edward Duncan
Hall, Elton Harold
Hall, George Arthur, (Jr)
Hall, Henry Kingston, Jr
Hall, James Lester
Hall, John Henry, Jr
Hall, Lowell Headley, II
Hall, Lyle Clarence
Hall, Wade Eckes
Hall, William Earl
Haller, Gary Lee
Haller, Ivan
Haller, Wolfgang Karl
Halpern, Arthur Merrill
Halsey, George Dawson, Jr
Haltner, Arthur John
Hamer, Walter Jay
Hamill, William Henry
Hamilton, Janet V
Hamilton, Walter S
Hamilton, Willard Charlson
Hamm, Randall Earl
Hammaker, Robert Michael
Hammel, Edward Frederic
Hamori, Eugene
Hampson, Robert F, Jr
Hance, Robert Lee
Hand, Clifford Warren
Handy, Lyman Lee
Hanley, Howard James Mason
Hanlon, Thomas Lee
Hanna, Melvin Wesley
Hannay, Norman Bruce
Hannum, Steven Earl
Hanrahan, Edward S
Hanrahan, Robert Joseph
Hansen, Charles M
Hansen, Donald Joseph
Hansen, Gerald Delbert, Jr
Hansen, Paul Vincent, Jr
Hansen, Robert Douglas
Hansen, Robert Suttle
Hansford, Rowland Curtis
Hanson, Allen Louis
Hanson, Harold Nelson
Hanson, Mervin Paul
Hansrote, Charles Johnson, Jr
Hanst, Philip Lincoln
Hard, Thomas Michael
Hardcastle, Kenneth Irvin
Hardgrove, George Lind, Jr
Hare, Curtis R
Hargis, I Glen
Hargitay, Bartholomew
Harju, Philip Herman
Harkins, Carl Girvin
Harlan, Horace David
Harmony, Martin D
Harney, Brian Michael
Harnsberger, Hugh Francis
Harrington, Rodney E
Harris, Albert Zeke
Harris, Charles Bonner
Harris, David R
Harris, Harold H
Harris, Jesse Ray
Harris, Murray T
Harris, Preston Mayne
Harris, Robert Laurence
Harris, William Charles
Harrison, Alexander George
Harrison, Anna Jane
Harrison, Jonas P
Harrison, Lionel George
Harrison, Thomas Southworth
Harriss, Donald K
Harrod, John Frank
Hart, Haskell Vincent
Hart, Maurice I, Jr
Hart, Robert George
Hart, William James
Harteck, Paul
Hartman, Karl August, Jr
Hartman, Kenneth Owen
Hartstein, Arthur M
Hartung, Homer Arthur
Hartzog, James Victor
Harvey, Albert Bigelow
Harvey, Ross Buschlen
Harwood, Walter William
Harwood, William H
Haseley, Edward Albert
Hashmall, Joseph Alan
Haskell, Vernon Charles
Hassell, Robert John
Hatala, Conrad V
Hatch, Conrad V

Hatch, Richard C
Hatfield, John Dempsey
Hatzenbuhler, Douglas Albert
Haubach, Walter Jennings, Jr
Hauenstein, Jack David
Haugen, Gilbert R
Haugh, Eugene (Frederick)
Hauk, Peter
Hauser, Edward Russell
Hawkes, Arthur Stanley
Hawkins, Peter Jack
Hawton, Larry David
Hay, Alden Wendell
Hayes, Kenneth Edward
Hayes, Robert Green
Hayles, William Joseph
Hazeltine, James Ezra, Jr
Hazlegrove, Leven Savage
Hazlehurst, David Anthony
Heald, Emerson Francis
Healey, Frank Henry
Healy, Robert Michael
Heberger, John M
Hebert, Alvin Joseph
Hecht, Harry George
Heckel, Edgar
Heckelsberg, Louis Fred
Heckler, George Earl
Hedberg, Kenneth Wayne
Hedges, Richard Marion
Hedman, Fritz Algot
Heeschen, Jerry Parker
Heiberger, Philip
Heilweil, Israel Joel
Heinemann, Heinz
Heininger, Clarence George, Jr
Held, Kalman M
Held, Robert Paul
Heldt, Walter Z
Helfferich, Friedrich Georg
Helfgott, Cecil
Helgeson, Harold Charles
Heller, Adam
Heller, Carl A
Heller, Hanan Chonon
Helman, William Phillip
Helwig, Lawrence E
Helz, Armin Werner
Hembree, George Hunt
Hemmes, Paul Richard, Jr
Henchman, Michael J
Henderson, Giles Lee
Henderson, John Frederick
Hendricks, Robert William
Hendricks, Sterling Brown
Henis, Jay Myls Stuart
Henkin, Hyman
Henley, Melvin Brent, Jr
Hennelly, Edward Joseph
Henry, Bryan Roger
Hentz, Forrest Clyde, Jr
Hepler, Loren George
Herczog, Andrew
Herd, Allen K, III
Herdklotz, Richard James
Herglotz, Heribert Karl Josef
Heric, Eugene Leroy
Herm, Ronald Richard
Herman, Jan Aleksander
Hermans, Jan Joseph
Hermans, Richard Anthony
Herr, Frank Leaman, Jr
Herrick, Claude Cummings
Herrmann, Kenneth Walter
Herron, John Thomas
Hersh, Herbert N
Herskovits, Theodore Tibor
Hertel, George Robert
Hertl, William
Hertzberg, Martin
Herzfeld, Simon Herman
Hess, John Monroe Converse
Hess, LaVerne Derryl
Heston, William May, Jr
Heuring, Vincent Paul
Hexter, Robert Maurice
Heydegger, Helmut Roland
Hickman, James Blake
Hicknott, Thomas Ward
Hickok, Robert Lee
Hicks, John Frederick Gross, (Jr)
Hicks, William Thomas
Hicks-Bruun, Mildred M
Hidy, George Martel
Hiebert, Gordon Lee
Hiemenz, Paul C
Higuchi, Takeru
Hildebrand, Joel Henry
Hill, Cliff Otis
Hill, Derek Leonard
Hill, Eric Stanley
Hill, Mary Ann Gertrude
Hill, Russell John
Hiller, Dale Murray
Hiller, Frederick W
Hilton, Ray
Hiltz, Arnold Aubrey
Hinckley, Conrad Cutler

Hinds, Nancy Webb
Hines, William Grant
Hinkebein, John Arnold
Hinks, David George
Hinkson, Thomas Clifford
Hinton, James Faulk
Hiraoka, Hiroyuki
Hirayama, Chikara
Hirko, Ronald John
Hirsch, Ernest
Hirst, Robert Charles
Hisatsune, Isamu Clarence
Ho, James Chien Ming
Hoback, John Holland
Hobbs, Anson Parker
Hobbs, John Robert
Hobbs, M Floyd
Hobbs, Marcus Edwin
Hobgood, Richard Troy, Jr
Hobrock, Don Leroy
Hobson, James Harvey
Hobson, Melvin Clay, Jr
Hoch, Hans
Hochanadel, Clarence Joseph
Hochberg, Melvin
Hodgen, Gary Dean
Hodgins, John Willard
Hodgson, Gordon Wesley
Hockje, Howard Hail
Hockstra, John Junior
Hoerr, Charles William
Hoeve, Cornelis Abraham Jacob
Hofer, Lawrence John Edward
Hoffman, Brian Mark
Hoffman, Everett John
Hogge, Ernest
Hogle, Donald Hugh
Holburn, Ruth Robertson
Holcomb, David Nelson
Holden, Geoggrey
Holden, Harold William
Holden, James Richard
Hollahan, John Ronald
Hollenberg, J Leland
Holleran, Eugene Martin
Hollibaugh, William Calvert
Hollies, Norman Robert Stanley
Hollinger, Henry Boughton
Hollingsworth, Charles Alvin
Hollingsworth, Ernest Howard
Holloman, Miles Edward
Holloway, Homer Lawton
Holman, James Lawson
Holmes, Howard Frank
Holmes, John Leonard
Holmes, Robert Edward
Holt, Eugene Lawrence
Holt, William Carl, Jr
Holtslander, William John
Holtzberg, Frederick
Holzmann, Richard Thomas
Holzwarth, George Michael
Hood, George Clement
Hood, Harrison Porter
Hoogsteen, Karst
Hooley, Joseph Gilbert
Hopkins, Elbert Erskine
Hopkins, Esther Arvilla Harrison
Hopkins, Harry P, Jr
Hopkins, Horace H, Jr
Hopkins, Theodore Emo
Hornack, Frederick Mathew
Horne, Frederick Herbert
Hornig, Howard Chester
Hornig, James Frederick
Horning, William Clarke
Hornung, Erwin William
Horowitz, Paul Martin
Horsma, David August
Horton, Robert Louis
Horton, William Sheldon
Horvath, Csaba Gyula
Hoskins, Leo Claron
Hostettler, John Davison
House, Edward Holcombe
Houseman, Barton L
Houser, Thomas J
Hovey, Richard John
Howald, Reed Anderson
Howard, Barbara Yoder
Howe, John A
Howell, Barbara Fennema
Howell, Daniel Bunce
Howery, Darryl Gilmer
Hoyer, Horst Walter
Hoyt, Justus
Hsia, Yu-Ping
Hsu, Edward Ching-Sheng
Htoo, Maung Shwe
Huang, Thomas Tao Shing
Hubbard, Colin D
Hubble, Billy Ray
Huckaby, Dale Alan
Huddle, Benjamin Paul, Jr
Huddleston, George Richmond, Jr
Hudgins, Charles Milton, Jr
Hudson, Robert McKim

Huebert, Barry Joe
Huffman, Ernest Otto
Huffman, Robert Eugene
Huggins, Charles Marion
Huggins, Maurice Loyal
Hughes, Robert Edward
Hull, Harry H
Humphrey, George Louis
Hung, John Hui-Hsiung
Hunston, Donald Lee
Hunt, Ann Hampton
Hunt, Graham R
Hunt, John Philip
Hunt, Paul Payson
Huntzicker, Harry Noble
Hurley, Rupert B
Huston, John Lewis
Hutcheon, Alan Thomson
Hutchison, Clyde Allen, Jr
Hutton, Harold M
Hyndman, John Robert
I, Ting-Po
Iden, Charles R
Ignatowski, Albert J
Ihrig, Judson La Moure
Ikels, Kenneth G
Ilardi, Joseph Michael
Illinger, Karl Heinz
Ilmet, Ivor
Ilten, David Frederick
Ingalls, Ronald Boyd
Ingersoll, Henry Gilbert
Ingle, George William
Inglefield, Paul T
Ingraham, Lloyd Lewis
Inman, Ross
Innes, William Beveridge
Inskeep, Richard Guy
Insley, Earl Glendon
Ionescu, Lavinel G
Irani, N F
Irani, Riyad Rida
Ireland, Carol Beard
Iribarne, Julio Victor
Irish, Donald Edward
Irvine, James Bosworth
Isbrandt, Lester Reinhardt
Ishida, Takanobu
Iyengar, Doreswamy Raghavachar
Jabalpurwala, Kaizer E
Jach, Joseph
Jacknow, Joel
Jackobs, John Joseph
Jackson, Earl Graves
Jackson, George Frederick, III
Jackson, Margaret E
Jackson, William Morrison
Jacob, Fielden Emmitt
Jacobs, Donald
Jacobs, Gerald Daniel
Jacobs, Harvey
Jacobs, Patrick W M
Jacobson, Ada Leah Hyne
Jacobson, Harold
Jacobson, Irven Allan, Jr
Jacobson, Robert Andrew
Jacox, Marilyn Esther
Jaffe, Hans
Jaffe, Michael
Jaffe, Sigmund
Jakobsen, Robert John
James, Herbert I
James, MarLynn Rees
Jameson, A Keith
Jamieson, William David
Janauer, Gilbert E
Janini, George Musa
Jankovics, Lawrence Robert
Jansing, Jo Ann
Janson, Thomas Ralph
Janz, George John
Janzen, Jay
Janzen, Wayne Roger
Jarnagin, Richard Calvin
Jehle, Leon Paul
Jellinek, Hans Helmut Gunter
Jenkins, William A
Jenkins, Glenn Herbert
Jenkins, Wilmer Atkinson, II
Jentoft, Ralph Eugene, Jr
Jernigan, Robert Lee
Joffe, Joseph
Johansen, Robert Torolf
Johnson, Carl Emil, Jr
Johnson, Charles Sidney, Jr
Johnson, Clarence Albert
Johnson, Dale A
Johnson, David Russell
Johnson, Edwin Wallace
Johnson, Frederic Allan
Johnson, Grover Leon
Johnson, Harlan Bruce
Johnson, Irving
Johnson, James Edwin
Johnson, James Steven, Jr
Johnson, Kenneth Eugene
Johnson, Lewis Benjamin, Jr
Johnson, Marvin Francis Linton

117

Johnson, Philip M
Johnson, Ralph E
Johnson, Ray Leland
Johnson, Richard Allen
Johnson, Rulon Edward, Jr
Johnson, Warren Frederick
Johnson, William Thomas Mitchell
Johnston, Christian William
Johnston, David Owen
Johnston, Francis J
Johnston, Harold Sledge
Johnston, Helen
Johnston, Milton Dwynell, Jr
Johnston, Stewart Archibald
Jolley, John Eric
Jonas, Leonard Abraham
Jonas, Alan Richard
Jones, Dane Robert
Jones, Daniel Elven
Jones, Francis Thomas
Jones, George Lett, Jr
Jones, Leon Lloyd
Jones, Lester Tyler, Jr
Jones, Llewellyn Hosford
Jones, Marvin Thomas
Jones, Maurice Harry
Jones, Morton Edward
Jones, Owen Lloyd
Jones, Peter Frank
Jones, Richard Evan, Jr
Jones, Samuel Simpson
Jones, Thomas Hubbard
Jones, Wesley Morris
Jones, William Ernest
Jongenburger, Huibert S
Jonnard, Raymond
Jordan, John Emnett
Jordan, Kenneth Gary
Jordan, Truman H
Jordan, Wade Hampton, Jr
Jorgensen, Helmuth Erik Milo
Joshi, Bhairav Datt
Jost, Ernest
Joyce, Blaine R
Joyce, Ronald Stone
Judkins, Roddie Reagan
Judson, Charles Morrill
Julian, Maureen M
Julien, Hiram Paul
Julien, Larry Martin
Jumper, Charles Frederick
Jung, Hilda Zifle
Jura, George
Kaelble, David Hardie
Kaiser, Edward William, Jr
Kallo, Robert Max
Kalnajs, Janis Arvids
Kamath, Yashavanth Katapady
Kambour, Roger Peabody
Kamen, Martin David
Kammeyer, Carl William
Kan, Lou Sing
Kana'an, Adil Sadeq
Kandel, Richard Joshua
Kanipe, Larry Gene
Kanter, Manuel Allen
Kantro, David Leon
Kao, Oranda Hai-Wen
Kaplan, Leon H
Kaplan, David Gilbert
Kappe, David Syne
Kapral, Raymond Edward
Karg, Gerhart
Karickhoff, Samuel Woodford
Karipides, Anastas
Karl, David Joseph
Karplus, Martin
Karpovich, John
Kasai, Paul Haruo
Kasner, Fred E
Kasper, John Simon
Katan, Theodore
Katz, John Edward
Katz, Joseph J
Katz, Lewis
Katz, Sidney
Kaufman, Ernest D
Kaufman, Myron Jay
Kaufman, Richard Gilbert
Kauzmann, Samuel
Kauzmann, Walter (Joseph)
Kay, Eric
Kay, Jack Garvin
Kay, Robert Leo
Kazanjian, Armen Roupen
Kearns, David R
Keaton, Clark M
Kebarle, Paul
Keedy, Curtis Russell
Keeler, Raymond Marsh
Keeling, Charles David
Keely, William Martin
Keenan, Arthur George
Keidel, Frederick Andrew
Keil, Robert Gerald
Keilholtz, Gerald Watson
Keith, Charles Herbert

Keizer, Clifford Richard
Keizer, Joel Edward
Kelemen, Dennis George
Keller, Douglas Vern, Jr
Keller, James Lloyd
Keller, Joseph Herbert
Keller, Oswald Lewin
Keller, Philip Joseph
Keller, Richard Alan
Keller, Wayne Hicks
Kellerman, Martin
Kellner, Jordan David
Kellner, Stephan Maria Eduard
Kells, Milton Carlisle
Kelly, Edward Joseph
Kelly, Francis John
Kelly, John, Jr
Kelsh, Dennis J
Kemp, Jacob David
Kemp, Marvin K
Kendall, James Tyldesley
Keneshea, Francis Joseph
Kennard, Divaker B
Kennedy, William Dempsey
Kennedy, Donald J
Kenney, John T
Kennicott, Philip Ray
Kenson, Robert Earl
Kent, James Woodward
Kereszies-Nagy, Steven
Kerker, Milton
Kerley, Gerald Irwin
Kern, Ralph Donald, Jr
Kern, Raymond Adrian
Kerr, Edwin Robert
Kerr, Eugene Charles
Kershner, Carl John
Kerwin, Julie Van Note Parker
Kiernan, William John
Kiefer, Edward W
Killingbeck, Stanley
Keyes, Richard Taylor
Keyzer, Hendrik
Khare, Bishun Narain
Khare, Mohan
Kidd, Richard Wayne
Kimlin, Mary Jayne
Kimmel, Elias
Kimmel, Howard S
Kimmel, Paul Issac
King, Arthur Bruce
King, Delbert Leo
King, James, Jr
King, Lowell Alvin
King, Peter Foster
King, Reatha Clark
King, Richard Warren
King, Stanley Shih-Tung
King, William Travers
Kinney, Gilbert Ford
Kinsey, James Lloyd
Kinsey, Philip A
Kinsinger, Jack Burl
Kipp, Egbert Mason
Kirby, George Francis
Kircher, John Frederick
Kirchhoff, William Hayes
Kirk, Alexander David
Kirk, David Clark, Jr
Kirkien-Rzeszotarski, Alicia M
Kirkpatrick, William John
Kirn, James Frederick
Kirsch, Joseph Lawrence, Jr
Kirtman, Bernard
Kiss, Klara
Kistakowsky, George Bogdan
Kittelberger, John Stephen
Kittila, Allan Benona
Kittle, Paul Alvin
Kittsley, Scott Loren
Kiviat, Fred E
Kivligin, Herbert Daniel, Jr
Klainer, Stanley M
Klassen, Norman Victor
Klein, Daniel Anthony
Klein, Elias
Klein, Gerhart Paul
Klein, Nathan
Klein, Ralph
Kleinsteuber, Tilmann Christoph Werner
Klemperer, William
Klempner, Daniel
Kleppa, Ole Jakob
Klier, Julius
Klier, Kamil
Kliger, David Saul
Kliman, Harvey Louis
Klinck, Ross Edward
Kline, Charles Howard

Klingensmith, George Bruce
Klinkhammer, Michael Dennis
Klug, Dennis Dwayne
Klug, Harold Philip
Klute, Charles Henry
Knecht, Laurance A
Knecht, Terence Edward Creasey
Knight, Jere Donald
Knight, Lon Bishop, Jr
Knobler, Charles Martin
Knox, Andrew Gibson
Knudsen, Dennis Ralph
Knudsen, Oran Milton
Ko, Hon-Chung
Ko, Theodore Augur
Koch, William George
Koenig, Charles Louis
Koenig, Inge Rabes
Koenig, Jack L
Koerner, William Elmer
Koetzle, Thomas F
Kohl, Fred John
Kohler, Bryan Earl
Kohn, Erwin
Kolodny, Nancy Harrison
Komarmy, Julius Michael
Komp, Richard Joseph
Konde, Anthony Joseph
Koob, Robert Duane
Koons, Lawrence Franklin
Koopmann, Henry Ferdinand
Kooser, Robert Galen
Kopelman, Raoul
Kopf, Peter W
Kornfeil, Fred
Kornick, George Jiri
Kornblum, Saul S
Korst, William Lawrence
Koshba, Walter Louis
Koski, Walter S
Kossiakoff, Alexander
Koster, David F
Kotin, Leonard
Kottlar, Abraham Morris
Kottle, Sherman
Kowert, Bruce Arthur
Kozak, Gary S
Kraemer, Louise Margaret
Krakow, Burton
Krasnow, Marvin Ellman
Kraus, Douglas Lawrence
Kraus, Gerard
Krause, Paul Frederick
Krause, Sonja
Krawetz, Arthur Altshuler
Kreider, Henry Royer
Kreidl, Ekkehard Ludwig
Kreilick, Robert W
Kreis, Ronald W
Kremheller, Alfred
Kresch, Alan J
Krieger, Firmin Joseph
Krieger, Irvin Mitchell
Krier, Carol Alnoth
Krigbaum, William Richard
Krol, George J
Kromann, Paul Roger
Kronenberg, Marvin L
Kronman, Martin Jesse
Kropp, John Leo
Krouskop, Ned Carter
Krueger, Paul Carlton
Krueger, Peter J
Krueger, Robert Harold
Kruh, Robert Frank
Kruse, Ferdinand Hobert
Kruse, Walter
Krusic, Paul Joseph
Krutak, James John, Sr
Kruus, Peeter
Krzeminski, Stephen F
Kubose, Don Akeru
Kubu, Edward Thomas
Kuecker, John Frank
Kuffner, Roy Joseph
Kuhlmann, George Edward
Kukolich, Stephen George
Kulevsky, Norman
Kumar, Mahadevappa M
Kung, Harold Hing Chuen
Kunin, Robert
Kunin, Irwin Douglas, Jr
Kuntz, Robert Roy
Kunz, Hans Joseph
Kunzler, John Eugene
Kuppers, James Richard
Kurfman, Virgil Benson
Kury, James Eckhard
Kury, John William
Kushner, Lawrence Maurice
Kushner, Richard Allan
Kuska, Henry (Anton)
Kust, Roger Nayland
Kustin, Kenneth
Kuzmak, Joseph Milton
Kwak, Jan C T

Kwart, Harold
Kwei, Ti-Kang
Kwiram, Alvin L
Kybett, Brian David
Kyger, Jack Adolphus
Laane, Jaan
Labuza, Theodore Peter
Lacher, John Robert
Lachman, Alfred
Ladd, John Herbert
Ladner, Jane Ellen Crawford
Ladner, Sidney Jules
Lafferty, Walter J
Lai, David Ying Fat
Laidler, Keith James
Laity, Richard Warren
Lakshminarayanaiah, Nallanna
Lam, Vinh-Te
Lamb, Frank Wyman
Lamb, Robert W
Lambert, Maurice C
Lambert, Sheldon Marvin
Lambrecht, Richard Merle
Lampe, Frederick Walter
Lancaster, John Edgar
Landel, William Everett
Landel, Robert Franklin
Landskroener, Peter Armstrong
Landsman, Douglas Anderson
Lane, George Ashel
Lane, Conrad Marvin
Lang, Frank Theodore
Lang, Robert Eugene
Lang, Robert Phillip
Langer, Klaus Robert
Langer, Sidney
Langerman, Neal Richard
Langlois, Gordon Ellerby
Langmuir, Margaret Elizabeth Lang
Lanning, William Clarence
Lanzerotti, Mary Yvonne DeWolf
Lanzl, George Frank
LaPietra, Richard Andrew
LaPosa, Joseph David
La Rochelle, John Hart
Larry, John Robert
Larsen, David W
Larsen, Elmer Conrad
Larsen, John W
Larsen, Robert Peter
Larson, Clarence Edward
Larson, Denis Wayne
Larson, Sager Daryl
Larson, Thomas E
Larson, William Day
Lasker, Sigmund E
Latscher, Carl Ernest
Lauer, George
Lauer, James Lothar
Laurie, Victor William
Lautenberger, William J
Lauterbach, George Ervin
Lauterbach, Paul Christian
Lauver, Richard William
Lauzau, Wilbur R
Lavery, Bernard James
LaVilla, Robert E
Lawani, Samuel Adetunji
Layman, Wilbur A
Layne, Ennis C
Layton, Lionel H
Leader, Gordon Robert
Lebowitz, Jacob
LeBlanc, Oliver Harris, Jr
LeBlanc, Ronald Guy
LeBel, Ronald Guy
le Duc, J Adrien Maher
Lee, Chuan-Pu
Lee, Donald Garry
Lee, Edward Kyung Chai
Lee, Emerson Howard
Lee, Frederick Strube
Lee, Garth Lorraine
Lee, George H, II
Lee, James K
Lee, Min-Shiu
Lee, Pang-Kai
Lee, Raymond Curtis
Lee, Robert E, Jr
Lee, Wei-Ming
Leed, Russell Ernest
Lefler, Amos J
Leffler, Esther Barbara
Legg, John Wallis
LeGrand, Donald George
Lebowitz, Leonard
Leidheiser, Henry, Jr
Leifer, Leslie
Leiga, Algird George
Leineweber, James Peter
Leininger, Paul Miller
Leiserson, Lee
Leister, Harry M
Leitz, Fred John, Jr
Leland, Frances Elbridge
Lemmerman, Karl Edward
Leniart, Daniel Stanley
Leon, Burke
Leonard, Jacques Walter

Johnson, Myrtle F
Leonard, John Edward
Leonard, William J. Jr
Leone, James A
Lepie, Albert Helmut
Lester, George Ronald
Lester, Joseph Eugene
Lester, William Alexander, Jr
Le Surf, Joseph Eric
Leubner, Ingo Herwig
Levey, Gerrit
Levin, Irvin
Levin, Isador
Levine, Charles (Arthur)
Levine, Herman Saul
Levine, Howard Bernard
Levine, Ira Noel
Levine, Oscar
Levine, Samuel
Levine, Samuel W
Levine, Sumner Norton
Levkov, Jerome Stephen
Levy, Arthur Louis
Levy, Ezra
Levy, Joseph Benjamin
Lewin, Seymour Z
Lewis, Armand Francis
Lewis, Bernard
Lewis, Charles William
Lewis, David Kenneth
Lewis, Donald Richard
Lewis, James Edward
Lewis, Leroy Crawford
Lewis, Milton
Lewis, Paul Herbert
Lewis, Richard Thomas
Lewis, Sheldon Noah
Li, Chi-Tang
Li, Norman Chung
Liang, Charles Chi
Liao, Shu-Chung
Libackyj, Anfir
Liberti, Frank Nunzio
Lichtman, Irwin A
Lieberman, Morton Leonard
Liebeskind, Herbert
Liebman, Samuel
Liengme, Bernard V F
Lietzke, Milton Henry
Likes, Carl James
Lin, Edward C
Lin, Jeong-Long
Lin, Stephen Fang-Maw
Lind, Charles Douglas
Lind, Maurice David
Lindauer, Maurice William
Linde, Peter Franz
Lindenbaum, Siegfried
Lindenberg, Katja Lakatos
Lindenmeyer, Paul Henry
Lindfors, Karl Russell
Lindgren, Robert M
Lindholm, Robert D
Lindquist, Robert Henry
Lindsay, James Gordon
Lindsay, William Tenney, Jr
Line, Lloyd Ernest, Jr
Linford, John Herbert
Link, Gordon Littlepage
Linke, William Finan
Linner, Edward Robert
Linschitz, Henry
Linton, Everett Percival
Linzer, Melvin
Lippmann, David Zangwill
Lipschutz, Michael Elazar
Lipscomb, William Nunn, Jr
Lipsett, Solomon George
Lipsig, Joseph
Lipsky, Sanford
Lipsky, Seymour Richard
Liu, Fred Wei Jui
Liu, Michael T H
Livingston, Daniel Isadore
Livingston, Robert Louis
Loening, Kurt L
Loepfky, Richard N
Lokken, Stanley Jerome
Lonadier, Frank Dalton
Londergan, Martin Christoper
Long, Arthur Owen
Long, Franklin A
Long, Michael Edgar
Long, Robert William
Longfield, James Edgar
Longmire, Martin Shelling
Longo, Frederick R
Longworth, Lewis Gibson

Lonsdale, Harold Kenneth
Looney, Ralph William
Loos, Karl Rudolf
Loprest, Frank James
Lord, Richard Collins, Jr
Lorenz, Max Rudolph
Lorenz, Philip Boalt
Lorimer, John William
Lory, Earl Christian
LoSurdo, Antonio
Lott, Peter F
Lougher, Edwin Henry
Lovejoy, Roland William
Low, Manfred Josef Dominik
Lowell, Seymour
Lowenstein, Michael Zimmer
Lower, Stephen K
Lowry, Eric G
Lowry, George Gordon
Lozier, Gerald Scott
Lublin, Paul
Luborsky, Fred Everett
Luborsky, Samuel William
Lucchesi, Claude A
Lucchesi, Peter J
Luckenbach, Thomas Alexander
Luckey, George William
Ludlum, Kenneth Hills
Ludwig, Frank Arno
Ludwig, Oliver George
Luebbe, Ray Henry, Jr
Luft, Ludwig
Lui, Yiu-Kwan
Lumb, Ralph F
Lumry, Rufus Worth
Lund, John Turner
Lundell, O Robert
Luner, Charles
Luner, Philip
Lunsford, Jack Horner
Luoma, John Robert Vincent
Lupton, John Madison
Lustig, Stanley
Luttinger, Lionel
Lutton, Edwin Scott
Lutz, Charles William
Lutz, George John
LuValle, James Ellis
Lykos, Peter George
Lyon, David N
Lyon, Richard Kenneth
Lyon, William Graham
Lyons, Margaret S
Lyons, Peter Francis
Lyons, Philip Augustine
Maass, George Joseph
Maatman, Russell Wayne
Mabis, Alton John
Macaluso, Pat
MacColl, Robert Joseph
MacDonald, David J
MacDonald, Digby Donald
MacDonald, Hubert C, Jr
MacDougall, Duncan Peck
Macek, Andrej
MacFarlane, Robert, Jr
Maciel, Gary Emmet
Mack, Lawrence Lloyd
MacKay, Colin Francis
MacKay, Raymond Arthur
MacKenzie, Donald Robertson
Mackey, John Linn
Mackiw, Vladimir Nicolaus
Mac Knight, William John
MacLaren, Richard Oliver
MacLauchlan, Donald Wells
Maclay, William Nevin
Maclean, Donald Isadore
Macur, George J
MacWilliam, Edgar Alexander
Magder, Jules
Magee, Charles Brian
Magee, Ellington McFall
Magee, John Lafayette
Magee, John Storey, Jr
Magofin, James Edward
Maguire, Mildred May
Mahan, Bruce Herbert
Mahan, Kent Ira
Maher, Philip Kenerick
Mahoney, Bernard Launcelot, Jr
Mains, Gilbert Joseph
Mair, Robert Dixon
Maisch, William George
Maki, Arthur George, Jr
Makowski, Mieczyslaw Paul
Malik, Jim Gorden
Malin, Murray Edward
Malinauskas, Anthony Peter
Malinowski, Edmund R
Malloy, Alfred Marcus
Malloy, Herbert Dean
Malmberg, Marjorie Schooley
Malone, Creighton Paul
Mandelcorn, Lyon
Manes, Milton
Mangelson, Nolan Farrin
Manger, Charles Walter

Manning, James Joseph
Manning, Robert Joseph
Mannis, Fred
Manson, John Alexander
Mantei, Kenneth Alan
Maple, Telford Grant
March, Raymond Evans
Marchessault, Robert Henri
Marchetti, Alfred Paul
Marchi, Raymond Paul
Marcinkowsky, Arthur Ernest
Marcotte, Ronald Edward
Marcus, Robert Boris
Marcus, Rudolph Arthur
Marcus, Rudolph Julius
Mariani, Henry A
Marion, Alexander Peter
Mark, Earl Larry
Mark, Herman Francis
Marker, Leon
Markham, M Clare
Markle, H Chester, Jr
Markovitz, Hershel
Markowitz, Joseph Morris
Maron, Samuel Herbert
Marquart, John R
Marr, Harold Everett, III
Marron, Michael Thomas
Marsh, Frederick Leon
Marsh, Glenn Anthony
Marsh, Max Martin
Marsh, Richard Edward
Marshall, Maryan Lorraine
Marshall, Philip Richard
Marshall, Walter Lincoln
Marshall, William Joseph
Marshall, William Leitch
Marston, Alfred Lawrence
Martin, Alfred
Martin, Edward Shaffer
Martin, Francis W
Martin, Frederick Johnson
Martin, Glenn Ellis
Martin, John Perry, Jr
Martin, John Scott
Martin, Richard Hugo
Martin, Richard McKelvy
Martin, Stanley Buel
Martin, Thomas Waring
Martin, Willard John
Martin, William Gerald
Martinez-Pico, Jose Luis
Martini, Catherine Marie
Martire, Daniel Edward
Marvin, Henry Howard, Jr
Marvin, Robert Sidney
Marx, Paul Christian
Marzzacco, Charles Joseph
Masciantonio, Philip (X)
Masi, Joseph Francis
Maskal, John
Mason, Charles Morgan
Mason, Donald Frank
Mason, Harold Frederick
Mason, Stanley George
Massa, Dennis Jon
Masson, Charles Robb
Mast, Roy Clark
Masterson, William Lewis
Mastrangelo, Sebastian Vito Rocco
Matesich, Mary Andrew
Matheson, Harry
Mathews, Collis Weldon
Matijevic, Egon
Matlack, Louis Rogers
Matovich, Edwin
Mattax, Calvin Coolidge
Matz, William Howard
Maulding, Hawkins Valliant, Jr
Maury, Lucien Garnett
Mausteller, John Wilson
Mavrodineanu, Radu
Mayberry, Paul Calvin
Mayer, Stanley Wallace
Mayer, William John
Mayo, Ralph Elliott
McAdoo, David John
McAllister, Stuart Allan
McAllister, William Albert
McBrady, John J
McBride, David Warren
McCall, David Warren
McCallum, Kenneth James
McCarroll, Bruce
McCartney, John Richard
McCarty, Clark William
McCarty, Maclyn, Jr
McCauley, James A
McCleary, Harold Russell
McClellan, Aubrey Lester
McCloskey, Kenneth Emory
McClure, David Warren
McClure, Donald Stuart
McColloch, Robert James
McCoy, Charles Ralph
McCoy, Joseph Hamilton
McCoy, Raymond Duncan
McCoy, William Harrison

McCullough, James Douglas, Jr
McCullough, John Price
McCullough, Roy Lynn
McCune, Homer Wallace
McCurdy, Keith G
McDaniel, Carl Vance
McDevit, William Ferris
McDonald, Charles Cameron
McDonald, Hugh Joseph
McDonald, Jimmie Reed
McDonald, John E
McDonald, Joseph Kyle
McDonald, Ray Locke
McDonald, Robert Skillings
McDowell, Charles Alexander
McDowell, Hershel
McDowell, John Robert
McDugle, Woodrow Gordon, Jr
McEicheran, Donald Elmo
McEwan, William Shelley
McGarvey, Bruce Ritchie
McGeer, James Peter
McGinnis, William Joseph
McGlynn, Sean Patrick
McGonigal, Paul J
McGrath, John F
McGraw, Gary Earl
McGuinness, Michael Joseph, Jr
McHale, Edward Thomas
McHugh, James Anthony, Jr
McIntosh, Robert Lloyd
McIntyre, James Douglass Edmonson
McKay, William Neil
McKee, Douglas William
McKeever, L Dennis
McKenney, Donald Joseph
McKenzie, Donald Edward
McKeown, James John
McKinney, Charles Dana, Jr
McKinney, David Scroggs
McKinney, Paul Caylor
McLeod, Kenneth Neil
McLeod, Lloyd Alexander
McMahon, Howard Oldford
McMahon, Paul E
McMenamy, Rapier Hayden
McMillan, Clara Albertina
McMillan, Juan Alejandro
McNesby, James Robert
McQuigg, Robert Duncan
McRae, Eion Grant
McRae, Wayne Alan
Meadors, Victor Gerald
Meadows, James Wallace, Jr
Meads, Philip F
Meagher, James Francis
Meakin, Paul
Meal, Harlan C
Meal, Larie L
Meeker, Thrygve Richard
Meeks, Frank Robert
Meelheim, Richard Young
Mehta, Mahesh J
Meisel, Dan
Meisels, Gerhard George
Meites, Louis
Mellor, John
Melrose, James C
Melton, Lynn Ayres
Menashi, Jameel
Mencik, Zdenek
Menefee, Emory
Mengenhauser, James Vernon
Mennitt, Philip Gary
Menzinger, Michael
Mercier, Philip Laurent
Meriwether, Lewis Smith
Merrill, Jerald Carl
Merrill, John Richard
Merrin, Seymour
Merten, Ulrich
Mertens, Frederick Paul
Meselson, Matthew Stanley
Meshri, Dayaldas Tanumal
Messenger, Joseph Umlah
Messer, Charles Edward
Messerly, George Henry
Metcalfe, Joseph Edward, III
Method, Peter Francis
Metz, Clyde
Metz, Donald J
Metz, Florence Irene
Metzger, William Henry, Jr
Meyer, Albert William
Meyer, Edmond Gerald
Meyer, Edwin F
Meyer, Eugene Frank, Jr
Meyer, Leon Herbert
Meyer, Norman James
Meyer, Richard Thomas
Meyer, Vincent D
Meyers, Earl Lawrence
Meyers, Gene Howard
Miceli, Fortunato Joseph
Miceli, Angelo Sylvestro
Michael, Joe Victor
Michaelis, Arthur Frederick
Michaels, Adlai Eldon

Mickelsen, John Raymond
Mihajlov, Vsevolod S
Mijal, Chester Francis
Mikula, James J
Mikus, Felix F
Millard, Ben
Miller, Melvin Henry
Miller, Alexander Andrew
Miller, Arild Justesen
Miller, Arnold
Miller, Arthur
Miller, Carl Stinson
Miller, Donald Milroy
Miller, Donald Gabriel
Miller, Eugene D
Miller, Floyd Laverne
Miller, Foil Allan
Miller, Franklyn David
Miller, Frederick Arnold
Miller, George Alford
Miller, George E
Miller, Gerald Ray
Miller, Gordon Smith
Miller, Harold A
Miller, James Richard
Miller, John Clark
Miller, John George
Miller, John Robert
Miller, Kenneth Jay
Miller, Melvin P
Miller, Meredith
Miller, Richard Edward
Miller, Richard J
Miller, Robert L
Miller, Roy Richard
Miller, Steven Ralph
Miller, Theodore Lee
Miller, Thomas Anderson
Miller, Clifford E
Miller, Paul Chambers
Milner, Roy Ernest
Mital, Kashmiri Lal
Milz, Wendell Collins
Miner, Bryant Albert
Miniatas, Birute Ona
Miodozenice, Arthur Roman
Moberly, Lawrence Ervin
Mochel, Walter Edwin
Mochel, Virgil Dale
Modic, Frank Joseph
Moe, George
Moffat, John Blain
Monahan, Alan Richard
Mohacsi, Erno
Moniz, William Bettencourt
Monse, Ernst Ulrich
Montgomery, Francis Bertram
Montgomery, George Edmund
Montgomery, Andrew Harrison
Montgomery, Lawrence Kernan
Montgomery, Peter Williams
Monjar, Monty Jack
Moody, Leroy Stephen
Moolenaar, Robert John
Moon, Tag Young
Mooney, Richard Warren
Moore, Eunice Martha
Moore, Francis Bertram
Moore, George Edmund
Moore, George Edward
Moore, Gordon Earle
Moore, John Hays, Jr
Moore, John Ward
Moore, John Williamson
Moore, Perry Alldredge
Moore, Robert Lee
Moore, Walter John
Moore, William Marshall
Moorti, Varahur R Guru
Moos, Anthony Manuel
Mopsik, Frederick Israel
Moran, Thomas Francis
Morawetz, Herbert
Morehead, Frederick H
Moreland, Charles Glen
Morgan, Charles Ebenezer
Morgan, James Glen
Morokuma, Keiji
Morosin, Bruno
Morosoff, Nicholas

Morrey, John Rolph
Morris, George V
Morris, Harris Lee
Morris, Humbert
Morrison, James Alexander
Morrow, Bruce William
Morrow, John Charles, III
Morse, Norman Louis
Morse, Erwin Emerson
Mortensen, Earl Miller
Mortimer, Robert George
Morton, Stephen Dana
Moscowitz, Albert
Moseley, William David, Jr
Moser, Herbert Charles
Mosesman, Max Abe
Mosley, Jules Warren
Moskowitz, John Ross
Moss, Frank Anthony James
Mosser, John Snavely
Mossotti, Victor Giovoni
Most, Joseph Morris
Motsavage, Vincent Andrew
Mottley, Carolyn
Moule, David
Mountcastle, William R, Jr
Moynihan, Cornelius Timothy
Moynihan, Robert Edward
Mucci, Joseph Francis
Muchow, Gordon Mark
Mueller, Charles Richard
Mueller, Manfred Ernst
Mueller, Mary Casimira
Mueller, Walter A
Muenow, David W
Muenter, Annabel Adams
Muenter, John Stuart
Muir, Donald Ridley
Mukerjee, Pasupati
Mulas, Pablo Marcelo
Mulay, Laxman Nilakantha
Mulford, Robert Neal Ramsay
Mullen, Robert Terrence
Mullen, Norbert
Mullhaupt, Joseph Timothy
Mullin, Charles R
Mullins, Lawrence J, Jr
Munch, Ralph Howard
Mundy, Belvey Washington
Munson, Burnaby
Munson, Ronald Alfred
Murad, Edmond
Murchison, Pamela W
Murdock, Gordon Alfred
Murphy, George Washington
Murphy, James A
Murphy, James Francis
Murphy, James Wallace
Murray, Francis E
Murray, Linwood Asa, Jr
Murmann, Richard P
Muschlitz, Earle Eugene, Jr
Music, Jack Farris
Much, George William
Myers, Albert Leroy
Myers, Benjamin Franklin, Jr
Myers, Claude Grenville
Myers, Clifford Earl
Myers, Gardiner Hubbard
Myers, George E
Myers, Richard Showse
Myers, Rollie John, Jr
Nachod, Frederick Constantine
Nachtrieb, Norman Harry
Naddy, Badie Ihrahim
Nadeau, Herbert Gerard
Nadler, Melvin Philip
Nafie, Laurence Allen
Nagel, Terry Marvin
Nakamoto, Kazuo
Nakanilas, George H
Nancollas, George H
Nancey, Thomas Ray
Naragon, Ernest Ashley
Naro, Paul Anthony
Nash, Charles Presley
Nash, Leonard Kollender
Nath, Amar
Nath, Murray L
Nathans, Marcel Willem
Nauman, Robert Vincent
Nauman, Robert Alexander
Nazarian, Girair Mihran
Negepen, Ernest John
Neddenriep, Richard Joe
Neece, George A
Neely, Charles Mack
Neely, Brock Wesley
Neely, Stanley Carrell
Neely, William Charles
Neff, Laurence D
Neff, Vernon Duane
Neidig, Howard Anthony
Neill, Donald E
Neill, Warren Joseph
Neilson, George Francis, Jr

Nelles, Maurice
Nelson, David Lynn
Nelson, Richard David
Nelson, Robert Norton
Nelson, Roger Edwin
Nelson, Wayne Franklin
Nemeth, Ronald Louis
Nemethy, George
Neptune, William Everett
Nesbit, Lyle Edwin
Nesler, F H Max
Netherton, Lowell Edwin
Neti, Radhakrishna Murty
Netzel, Daniel Anthony
Neu, John Ternay
Neubert, Theodore John
Neuffer, John E
Neugebauer, Constantine Aloysius
Neven, Maurice C
Newby, William Armon, Jr
Newby, Frank Edward
Newkirk, Herbert William
Newman, David S
Newschwander, Wilfrid Williams
Newton, James Henry
Newton, Thomas William
Nibler, Joseph William
Nicely, Vincent Alvin
Nichol, James Charles
Nichols, Ambrose Reuben, Jr
Nichols, George Morrill
Nichols, James Randall
Nicholson, Daniel Elbert
Nicholson, Margie May
Nickerson, John David
Nieman, George Frederick
Nordman, Christer Eric
Norris, A R
Northcott, Jean
Norton, Francis James
Norton, Peter Robert
Notley, Norman Thomas
Novotny, Donald Bob
Noyes, Richard Macy
Noyes, William Albert, Jr
Nozik, Arthur Jack
Nuttall, Ralph Leslie
Nutter, James Douglas
Nylund, Robert E
Obrenski, Robert John
O'Brien, James Francis
O'Brien, Robert Neville
O'Brien, Thomas Joseph
O'Connell, James Anthony
Odiorne, Truman J
O'Donnell, Raymond Thomas
Oei, Djong-Gie
Oeike, William C
Oene, Henk Van
Oetting, Franklin Lee
Often, Henry William
Offenhartz, Peter O'Donnell
Ogle, Pearl Rexford, Jr
Ogren, Paul Joseph
Ogryzlo, Elmer Alexander
O'Hare, Patrick
Ohlberg, Stanley Miles
Ohta, Masao
Okabe, Hideo
Olderman, Gerald M
Oldham, Keith Bentley
Olechowski, Jerome Robert
Oliver, James Russell
Olmsted, John, III
Olsen, Douglas Alfred
Olson, David Harold
Olson, William Bruce
O'Mara, James Herbert
O'Neal, Harry E
O'Neal, Harry Roger
O'Neil, James R
Onsott, Edward Irvin
Onwood, David P
Ordway, Fred (Delancy)
Orem, Michael William
Oriani, Richard Anthony
Oriel, Patrick John
Orieman, Edwin Franklin
Orlick, Charles Alex
Orna, Mary Virginia
Orttung, William Herbert
Orwoll, Richard David
Orwoll, Robert Arvid
Ory, Horace Anthony
Orzechowski, Adam
Osborg, Hans

Osborne, Darrell Wayne
Osborne, J Scott, Jr
Oser, Willem
O'Shaughnessy, Marion Thomas, Jr
Osipow, Lloyd Irving
Oster, Gerald
Osterhoudt, Hans Walter
O'Sullivan, Thomas Denis
Ott, J Bevan
Otto, Wolfgang Karl Ferdinand
Otvos, John William
Ouderkirk, John Thomas
Ouellet, Ludovic
Ouellet, Derick William
Overend, John
Overman, Joseph DeWitt
Owen, James Robert
Owen, John Harding
Owens, Charles Wesley
Owens, Kenneth Eugene
Ozarow, Vernon
Ozog, Francis Joseph
Padgett, Alice Adams
Padmanabhan, G R
Padnos, Norman
Padrta, Frank George
Pak, Charles Y
Palik, Emil Samuel
Pall, David B
Pallen, Robert Harris
Palmer, Alan Blakeslee
Palmer, Howard Benedict
Palmer, Jay
Palmer, Kenneth James
Panson, Gilbert Stephen
Papadopoulos, Michael N
Papazian, Harold Aram
Papazian, Louis Arthur
Papee, Henry Michael
Papineau-Couture, Gilles
Parchen, Frank Raymond, Jr
Pariser, Rudolph
Parisi, George I
Park, Kisoon
Parker, David J
Parker, John Hilliard
Parker, Richard C
Parker, Vivian
Parkinson, William Walker, Jr
Parkins, John Alexander
Parks, Eric K
Parks, Norris Jim
Parlee, Norman Allen Devine
Parmenter, Charles Stedman
Parrill, Irwin Homer
Parrish, Clyde Franklin
Parrish, Robert G
Pasfield, William Horton
Passchier, Arie Anton
Passis, Angelo Vlasios
Pattengill, Merle Dean
Patterson, Andrew, Jr
Patterson, Howard Hugh
Patterson, William Alexander
Patton, Elizabeth VanDyke
Patton, Hugh Wilson
Paul, Iain C
Pauley, James C
Paulsen, Duane E
Paulson, John Frederick
Pawlowski, Anthony T
Payne, Donald Hughel
Payton, Patrick Herbert
Pearlman, Harry
Pearson, Earl Freeman
Pearson, James Murray
Pearson, Robert Melvin
Peatman, William Burling
Pecora, Robert
Pedersen, Lee G
Peek, Harry Milton
Peller, Leonard
Perez-Albuerne, Evelio A
Peri, John Bayard
Perkins, Alfred James
Perkins, Gerald, Jr
Perkins, Harolyn King
Perkins, Walter George
Perloff, Alvin
Perry, Edmond S
Perry, Ernest John
Perry, Reeves Baldwin
Person, Willis Bagley
Pertel, Richard
Perumareddi, Jayarama Reddi
Perzak, Frank John
Pessen, Helmut
Petersen, Donald H
Petersen, Donald Ralph
Petersen, Raymond Carl
Peterson, Arthur F
Peterson, Axel Harding
Peterson, Charles Leslie
Peterson, Donald Bruce
Peterson, Donald Lee

Peterson, Jack Kenneth
Peterson, Lowell E
Peterson, Norman Cornelius
Peterson, Selmer Wilfred
Petit, Michael Geoffrey
Petrakis, Leonidas
Petrella, Ronald Vincent
Petricciani, John C
Petro, Anthony James
Petrucci, Ralph Herbert
Petrucci, Sergio
Phifer, Harold Edwin
Phillips, Brian Ross
Phillips, David Berry
Phillips, George Wygant, Jr
Phillips, Norman Edgar
Phillips, Norman William Frederick
Phillips, Travis J
Piccolini, Richard John
Pickard, Hugh Brown
Pickett, Herbert McWilliams
Pickett, Lucy Weston
Pierce, Louis
Pierce, Marion Armbruster
Pierce, Percy Everett
Pierce, Robert Charles
Pierce, Robert Henry Horace, Jr
Piermarini, Gasper J
Pierotti, Robert Amadeo
Piersma, Bernard J
Pierson, William Grant
Pierson, William Roy
Pieski, Edwin Thomas
Pignocco, Arthur John
Pikal, Michael Jon
Pilar, Frank Louis
Pilger, Richard Christian, Jr
Pinkham, Chester Allen, III
Pinkston, John Turner
Pinsler, Heinz Willi
Pinson, James Wesley
Piper, Roger D
Pirkle, Willis Nathaniel
Pish, George
Pitt, Donald Alfred
Pitts, James Ninde, Jr
Pitzer, Kenneth Sanborn
Place, Robert Daniel
Platford, Robert Frederick
Platko, Frank Edward
Plaush, Albert Charles
Plazek, Donald John
Plessy, Boake Lucien
Plodinec, Matthew John
Plumb, Robert Charles
Plummer, William Allan
Po, Henry Ng
Poole, Donald Ray
Poole, John Anthony
Pobereskin, Meyer
Poppe, Wassily
Pobiner, Harvey
Pochan, John Michael
Pocius, Alphonsus Vytautas
Podgurski, Harry Howard
Poe, Martin Turner
Pohlmann, Hans Peter
Poirier, Jacques Charles
Polanyi, John Charles
Polestak, Walter John S
Pollack, Gordon (Paul)
Pollnow, Gilbert Frederick
Pollock, Bernard David
Poncha, Rustom Pestonji
Poranski, Chester F, Jr
Porter, Gerald Bassett
Porter, John T
Porter, Paul Edward
Porter, Raymond P
Porter, Richard A
Porter, Richard Needham
Porter, Roger Stephen
Posey, Franz Adrian
Posner, Aaron Sidney
Potenza, Joseph Anthony
Potter, Norman D
Potter, Ralph Miles
Potts, John Calvin
Poulson, Richard Edwin
Powell, Arnet L
Powell, David Lee
Powell, George Louis
Powell, Jack Edward
Powell, Manly Joy
Powell, Maurice Green
Powell, Ralph Robert
Powell, Richard Edward
Powers, Robert William
Poynter, James William
Prager, Stephen
Praissman, Melvin
Pratt, David W
Pratt, Thomas Herring, Jr
Prentice, Jack L
Prenzlow, Carl Frederick
Presley, Cecil Travis
Price, Donna
Price, Elton

Price, Robert Harper
Price, Stanley James Whitworth
Price, Whitfield
Priepke, Rudolf Julius
Princen, Henricus Mattheus
Princen, Lambertus Henricus
Pringle, Wallace C, Jr
Pritchard, Glyn O
Pritchard, Huw Owen
Prock, Alfred
Proskauer, Eric S
Provder, Theodore
Provost, Ronald Harold
Prud Homme, Jacques
Prusaczyk, Joseph Edward
Pryor, Joseph Ehrman
Pugmire, Ronald J
Purdie, Neil
Purdon, William Andrew Bowie
Pye, Earl Louis
Pyper, James William
Pysh, Eugene Stephen
Pytkowicz, Ricardo Marcos
Pyun, Chong Wha
Quinlan, John Edward
Quist, Arvin Sigvard
Rabalais, John Wayne
Rabideau, Sherman Webber
Rabinovitch, Benton Seymour
Rabitz, Herschel Albert
Rack, Edward Paul
Radspinner, John Asa
Raff, Lionel M
Ragle, John Linn
Raich, Henry
Raines, Thaddeus Joseph
Raisen, Elliott
Raizen, Senta Amon
Rakowsky, Frederick William
Ramette, Richard Wales
Ramsley, Alvin Olsen
Ranck, John Philip
Rand, Myron Joel
Rand, Salvatore John
Randall, Duncan
Randall, James Carlton, Jr
Randall, John J, Jr
Randall, William Carl
Rangra, Avinash K
Ranney, Maurice William
Ransil, Bernard Jerome
Rapean, John C
Rapp, Donald
Raridon, Richard Jay
Rasaiah, Jayendran C
Rashkin, Jay Arthur
Ratchford, Robert James
Ratchford, William Paul
Rathe, Pierre
Rathnamma, Dasara V
Ratner, Mark A
Rauscher, Grant K
Raw, Cecil John Gough
Rawls, Henry Ralph
Ray, Wendell Augustus
Raymonda, John Warren
Read, John Frederick.
Readdy, Arthur F, Jr
Reade, Richard Francis
Reardon, Joseph Daniel
Reber, Louis Alexander
Rebick, Charles
Reck, Gene Paul
Recktenwald, Gerald William
Redden, Patricia Ann
Redding, Rogers Walker
Reddy, Gade Subbarami
Redington, Richard Lee
Redlich, Otto
Ree, Alexius Taikyue
Reed, Allan Hubert
Reed, John Francis
Reeder, Charles Edgar
Reeder, Ray Robert
Reeves, Leonard Wallace
Reffner, John A
Rehner, John, Jr
Reichardt, Charles Henry
Reichman, Sandor
Reid, Charles Edward
Reid, James Alexander
Reid, William John
Reinsborough, Vincent Conrad
Reisman, Arnold
Reiss, Howard
Rembold, Eugene Albert
Renfrew, Malcolm MacKenzie
Renich, Paul William
Renner, Terrence Alan
Resing, Henry Anton
Restaino, Alfred Joseph
Reuland, Donald John
Reynolds, Allan Eastman
Reynolds, Garth Fredric
Reynolds, Jefferson Wayne
Reynolds, John Hughes, IV
Reynolds, Melvin Ferguson
Reynolds, William Walter

Rhee, Jay Jea-yong
Rhee, Kee Hyun
Ricci, John Ettore
Rice, Bernard
Rice, Dale Wilson
Rice, Oscar Knefler
Rice, Stuart Alan
Rice, William Edward
Richard, Alfred Joseph
Richards, John Watson
Richards, Lorenzo Willard
Richards, R Ronald
Richardson, Albert Edward
Richardson, Frederick S
Richardson, Jeffery Howard
Richardson, Peter Charles
Richardson, Ronald John
Richey, Willis Dale
Richlin, Jack
Richter, Helen Wilkinson
Ricketts, John Adrian
Rider, Paul Edward
Ridge, Douglas Poll
Rieger, Philip Henri
Rieke, James Kirk
Ries, Herman Elkan, Jr
Riesz, Peter
Riggs, William McKnight
Riley, Bernard Jerome
Riley, Clyde
Riley, Stephen James
Rinse, Jacobus
Rippere, Ralph Elliott
Risen, William Maurice, Jr
Ristow, Bruce W
Ritchey, William Michael
Riter, John Randolph, Jr
Ritter, David Moore
Ritter, Hartien Sharp
Ritter, Robert L
Ritzman, Robert L
Roach, Donald Vincent
Roake, William Earl
Robbins, Omer Ellsworth, Jr
Robbins, Robert Crowell
Roberti, Dominic M
Roberts, Edwin Kirk
Roberts, Grady Leon
Roberts, Reginald Francis
Roberts, Richard William
Robertson, Alexander Allen
Robertson, Bobby Ken
Robertson, Nat Clifton
Robertson, Richard Earl
Robertson, Robert Frank Struan
Robertson, Roderick Francis
Robertson, Ross Elmore
Robins, Jack
Robins, Janis
Robinson, Dean Wentworth
Robinson, Edward Arthur
Robinson, Glen Moore, III
Robinson, Joseph Dewey
Robinson, Press L
Robinson, Robert Reid
Roboz, John
Robson, Harry Edwin
Rock, Elizabeth Jane
Rocklin, Albert Louis
Rodgers, Alan Shortridge
Roe, Ryong-Joon
Roebber, John Leonard
Rogers, Charles Edwin
Rogers, Donald Warren
Rogers, Horace Elton
Rogers, Jesse Wallace
Rogers, Max Tofield
Rohrbach, Gilson Henry
Rohrer, Charles Stephen
Rollino, John A
Romans, James Bond
Romary, John Kirk
Romberger, Karl Arthur
Rome, Martin
Ronayne, Michael Richard
Roof, Jack Glyndon
Root, Charles Brian
Root, John Walter
Roper, Gerald C
Roquite, Bimal C
Rose, Donald Glenn
Rose, Philip I
Rose, Wayne Burl
Rosenbaum, Eugene Joseph
Rosenberg, Allan (Herbert)
Rosenblatt, Gerd Matthew
Rosenblum, Charles
Rosenfield, Joan Samour
Rosenquist, Edward P
Rosenthal, Arnold Joseph
Rosenthal, Donald
Rosenwasser, Hyman B
Rosevear, Francis Burt
Roskos, Roland R
Rosman, Howard
Rosner, Anthony Leopold
Rosoff, Morton
Ross, John

Rossington, David Ralph
Rossini, Frederick Dominic
Rossmassler, Stephen Atwater
Rotariu, George Julian
Roth, James Frank
Roth, Walter
Rothbart, Herbert Lawrence
Rothman, Alan Bernard
Rothman, Sam
Rouse, Prince Earl, Jr
Rousseau, Denis Lawrence
Rowe, Carleton Norwood
Rowell, Robert Lee
Rowen, John William
Rowland, George Peabody, Jr
Rowland, Richard Lloyd
Roy, Rabindra (Nath)
Rozelle, Lee Theodore
Rozelle, Ralph B
Rubin, Bernard
Rubin, Herbert
Rubin, Robert Joshua
Rubino, Andrew M
Ruby, Stanley
Ruch, Richard Julius
Rudolph, Philip S
Ruhle, George Cornelius
Ruland, Norman Lee
Rulon, Richard M
Rummel, Robert Edwin
Rummens, F H A
Rump, Ellis Samuel, Jr
Ruoff, Arthur Louis
Rupert, Joseph Paul
Ruppel, Thomas Conrad
Rush, John Joseph
Rush, Richard Marion
Russell, Charles Daniel, Jr
Russell, David Bernard
Russell, James
Russell, Joel W
Russell, Kenneth Homer
Russell, Morley Egerton
Russin, Nicholas Charles
Rustad, Douglas Scott
Rutenberg, Aaron Charles
Ruth, John Moore
Rutkowski, Beverly Jean
Rutledge, Gene Preston
Rutledge, Robert L
Ryan, John Peter
Ryason, Porter Raymond
Ryce, Stephen A
Rytting, Joseph Howard
Rzad, Stefan Jacek
Saad, Hosny Younes A
Saalfeld, Fred Eric
Saba, William George
Sabatini, Joseph Francis
Sacher, Edward
Sadler, Monroe Scharff
Sadowski, Chester M
Saffer, Alfred
Saffer, Henry Walker
Safron, Sanford Alan
Sagal, Matthew Warren
Sage, Gloria W
Sager, Ray Stuart
St John, Daniel Shelton
St Louis, Robert Vincent
Sakano, Theodore K
Sakhnovsky, Alexander Alexandrovitch
Saldick, Jerome
Saleeb, Fouad Zaki
Sallavanti, Robert Armando
Salmon, James F
Salmon, Oliver Norton
Salomon, Mark
Salotto, Robert Ephriam
Saloutos, Anthony W
Salovey, Ronald
Saltsburg, Howard Mortimer
Salzberg, Hugh William
Salzman, William Ronald
Samuels, Robert Joel
Samulski, Edward Thaddeus
Samworth, Eleanor A
Sanborn, Russell Hobart
Sancier, Kenneth Martin
Sandell, Dewey Jay
Sanders, Robert N
Sanderson, Benjamin S
Sandler, Yehuda Ludwig
Sands, Donald Edgar
Sands, George Dewey
Sandus, Oscar
Sanford, Robert Alan
Santaniello, Anthony Frank
Saperstein, David Dorn
Sappenfield, Dale S
Sardisco, John Baptist
Sargent, Lowrie Barnett, Jr
Sargent, Roger N
Sarkar, Bibudhendra
Sarkar, Nitis
Sarmousakis, James Nicholas
Sarner, Stanley Frederick
Sasmor, Daniel Joseph

Satkiewicz, Frank George
Satterthwaite, Cameron B
Sattizahn, James Edward, Jr
Saunders, Frank Linwood
Saunders, Peter Reginald
Saupe, Alfred (Otto)
Saur, Roger Leo
Savitsky, George Boris
Savitzky, Abraham
Savoie, Rodrigue
Saylor, Charles Hamilton Proffer
Sayre, Edward Vale
Schaad, Lawrence Joseph
Schadel, Jacob Franklin
Schaefer, Theodore Peter
Schaeffer, Harold Franklin
Schafer, Lothar
Schaffer, Samuel Robert, Jr
Scharpen, LeRoy Henry
Schatz, Paul Nanon
Scheer, Milton David
Scheffler, George Henry
Scheirer, Carl Latimer
Schellman, John Anthony
Schelly, Zoltan Andrew
Scherz, Thnothy William
Scherer, James R
Schettler, Paul Davis, Jr
Scheuplein, Robert J
Schiesser, Robert H
Schiff, Harold Irvin
Schiffman, Louis F
Schimitschek, Erhard Josef
Schissler, Donald Owen
Schlag, Edward William
Schlegel, James M
Schmelzie, Ambrose Francis
Schmidt, Arthur
Schmidt, Hartland H
Schmidt, Raymond LeRoy
Schmitt, John Aloysius
Schmitt, Joseph Lawrence, Jr
Schmitz, Roman A
Schmude, Keith E
Schneider, Allan Stanford
Schneider, David Martin
Schneider, Friedemann W
Schneider, Henry
Schneider, Maxine Dorothy
Schneider, William George
Schnepp, Otto
Schochet, Melvin Leo
Scholz, John Joseph, Jr
Schomaker, Verner
Schonhorn, Harold
Schoonmaker, Richard Clinton
Schott, Garry Lee
Schott, Hans
Schottmiller, John Charles
Schrader, David Martin
Schrader, Malcolm Elliot
Schrage, Samuel
Schram, Alfred Francis
Schreiber, Henry Peter
Schreiner, Felix
Schreiner, Heinz Rupert
Schremp, Frederic William
Schreurs, Jan Willem Herman
Schrier, Eugene Edwin
Schrodt, Ariel Gilbert
Schroeder, LeRoy William
Schroeder, Rudolph Alrud
Schufle, Joseph Albert
Schug, John Charles
Schuh, Merlyn Duane
Schuldiner, Sigmund
Schuler, Robert Hugo
Schultz, John Wilfred
Schultz, Karlo Francis
Schurr, John Michael
Schwab, Helmut
Schwartz, Charles Marvin
Schwartz, Geraldine Cogin
Schwartz, Joseph Robert
Schwartz, Lowell Melvin
Schwartz, Robert Nelson
Schwartz, Stephen Eugene
Schwartz, Harold A
Schwenderman, Richard Henry
Schwartz, William Merlin, Jr
Scott, Allen Brewster
Scott, Bruce Albert
Scott, Donald Ray
Scott, Donald William
Scott, Robert Lane
Scott, Roy Albert, III
Scott, Troy Alexander, Jr
Scoville, Herbert, Jr
Scribner, Bourdon Francis
Seager, Spencer Lawrence
Searle, Norma Zizmer
Sears, Alan Roy
Sears, Paul Gregory
Sears, Timothy Stephen
Secco, Etalo Anthony
Secrest, Donald H
Seddon, William Arthur
Sedlet, Jacob

Seeley, Robert D
Seely, Gilbert Randall
Seevers, Robert Edward
Seff, Karl
Segal, Bernice G
Segato, Peter Richard
Seifert, Ralph Louis
Seiger, Harvey Norman
Selby, Theodore W
Selover, Theodore Briton, Jr
Seltzer, Stanley
Semeluk, George Peter
Senkus, Raymond
Senozan, Nail Mehmet
Servis, Kenneth L
Servos, Reva R
Seshadri, Kaikunte S
Sethi, Dhanwant S
Setser, Donald W
Sevilla, Michael Douglas
Shaffer, Lloyd Hamilton
Shah, Shirish Kalyanbhai
Shahin, Michael Mahmood
Shahin, Albert Leopold
Shallcross, Frank Van (Loon)
Shanefield, Daniel J
Sharp, James H
Sharp, John Arthur
Sharp, Robert Richard
Sharpe, Louis Haughton
Sharpless, Norman Edward
Sheetz, David P
Sheetz, Maxine Lana
Sheffer, Harry
Sheffield, Oliver Epravel
Shearer, Edmund Cook
Shearer, Steven Alan
Shaya, Steven Alan
Shaw, Don W
Shatz, Malcolm
Sheasley, William David
Sheats, George Frederic
Shedlovsky, Theodore
Shellenbarger, Robert Martin
Shechan, William Francis
Shen, Lawrence Y L
Shen, Mitchel C
Sheng, Shan-Jen
Shepard, Joseph William
Sheppard, Ervin
Sherburne, Russell Knight
Sherer, Stanley
Sherman, Robert Howard
Sherry, Peter Burum
Sherwood, A Gilbert
Shewmaker, James Edward
Shiao, Daniel Da-Fong
Shigeishi, Ronald A
Shih, Kwang Kuo
Shilman, Avner
Shin, Hyung Kyu
Shin, Benjamin Kin Chong
Shipsey, Edward Joseph
Shirk, James Siler
Shoup, Charles Samuel, Jr
Shreve, George Wilcox
Shuler, Woodfin Epps
Shoat, Mary La Salle
Shock, D'Arcy Adriance
Shombert, Donald James
Shoor, Arthur Joseph
Shoolery, James Nelson
Shull, Charles Morell, Jr
Shull, Don Louis
Shultz, Allan R
Shurz, Kalman
Sibbett, Donald Joseph
Sicotte, Yvon
Siebert, Alan Roger
Sieck, L Wayne
Siegel, Bernard
Siegel, Georges Giovanni
Siegel, Marshall Mayer
Signorino, Charles Anthony
Silbert, Leonard Stanton
Silbey, Robert James
Silker, Wyatt Burdette
Silver, Herbert Graham
Silverman, Jacob
Silverton, James Vincent
Silverman, Sam M
Simard, Gerald Lionel
Sime, Rodney J
Sime, Ruth Lewin
Sinha, Robert
Simmons, Eugene Lynn
Simmons, Gary Wayne
Simmons, Ivor Lawrence
Simmons, Joe Denton
Simon, Dorothy Martin
Simon, Frederick Tyler
Simon, Verne A
Simons, Harold Lee
Simonsen, David Raymond
Simpson, Clifford Carlton, Jr
Simpson, Frank Martin, Jr
Simpson, Paul Gravis

Simpson, Warren Candler
Sims, Bernard
Sims, Leslie Berl
Singer, Joseph Marcus
Singh, Surjit
Singleton, Jack Howard
Sink, Woodford Grady
Sipe, Herbert James, Jr
Siska, Peter Emil
Sisko, Arthur William
Sitney, Lawrence Raymond
Skarulis, John Anthony
Skelly, Norman Edward
Skewis, John David
Skinner, Gordon Bannayne
Skochdopole, Richard E
Skrabek, Emanuel Andrew
Slagg, Norman
Slattery, Charles Wilbur
Slepetys, Richard Algimantas
Slichter, William Pence
Sloane, Thompson Milton
Slomp, George
Slotnick, Herbert
Slowinski, Emil J, Jr
Slutsky, Leon Judah
Sly, William Glenn
Smellie, Robert Henderson, Jr
Smid, Johannes
Smiley, Harry M
Smith, Alan Wayne
Smith, Albert Lee
Smith, Darwin Waldron
Smith, Allan Laslett
Smith, Dwight Morrell
Smith, Francis White
Smith, Frank Roylance
Smith, Frederick Albert
Smith, George Byron
Smith, George David
Smith, Donald Foss
Smith, Donald E
Smith, David Lee
Smith, Donald Reed
Smith, Dudley Cozby
Smith, Gordon Meade
Smith, George Pedro
Smith, Grant Warren
Smith, Harlan Millard
Smith, Hilton Albert
Smith, Homer Pine
Smith, Joseph Harold
Smith, Kenneth Judson, Jr
Smith, Kenneth McGregor
Smith, Leslie E
Smith, Martin Bristow
Smith, Michael
Smith, Norman Obed
Smith, Paul Kent
Smith, Peter
Smith, Philip T
Smith, Richard Neilson
Smith, Richard Pearson
Smith, Ronald Nelson
Smith, Ronald W
Smith, Sidney Ruven
Smith, Thomas David
Smith, Trudy Enzer
Smith, Walter MacFarlane
Smith, Warren Harvey
Smith, Wayne Lee
Smith, Wendell Vandervort
Smith, William Thomas, Jr
Smoot, Charles Richard
Smyth, Charles Phelps
Snavely, Earl Samuel, Jr
Snead, Claybourne C
Snelgrove, James Arthur
Snipp, Robert Leo
Snow, Milton Leonard
Snow, Richard L
Snyder, Dexter Dean
Snyder, Paul Edwin
Soffer, Irving Herbert
Sokoloski, Theodore Daniel
Solc, Karel
Soldano, Benny A
Solie, Thomas Norman
Solomon, Jack
Solomon, Jerome Jay
Solomon, Joseph Alvin
Solomons, Cyril
Soltzberg, Leonard Jay
Sommerfeldt, Theron G
Sonorjai, Gabor Arpad
Sonnessa, Anthony J
Sood, Satya P
Sood, Zoltan Geza
Sopp, Samuel William
Soth, Glenn Carroll
Soulen, John Richard
Soulen, Frederick William
Sovers, Ojars Juris
Sowards, Donald Maurice
Spaght, Monroe Edward
Spahr, Alexander White, III
Spears, Frank Harold
Spedding, Frank Harold

Speiser, Rudolph
Spell, Aldenlee
Spenadel, Lawrence
Spencer, Harold Garth
Spencer, Harry Edwin
Spencer, Hugh Miller
Spencer, James Nelson
Spencer, Jesse G
Spencer, John Brockett
Spencer, Robert Shirley
Sperling, Leslie Howard
Spicer, Leonard Dale
Spicer, William Monroe
Spinar, Leo Harold
Spink, Charles Harlan
Spinner, James Clifford
Spitsbergen, James Clifford
Spiwak, Lazarus
Sporer, Alfred Herbert
Sprague, Estel Dean
Spratley, Richard Denis
Sprokel, Gerard J
Spurgeon, William Marion
Spurlin, Harold Morton
Spurr, Orson Kirk, Jr
Squire, David R
Squires, Arthur Morton
Staats, Percy Anderson
Stacy, Ronald Paul
Stafford, Charles Henry
Staats, Robert Ardagh
Staley, Ralph Horton
Stamm, Alfred Joaquin
Stamm, Robert Franz
Stammelman, Mortimer Jacob
Stanbrey, John Joseph
Stanley, William Gordon
Starkweather, Howard Warner, Jr
Starr, Duane Frank
Staskiewicz, Bernard Alexander
Stauffer, Charles Henry
Stauffer, Robert Eliot
Stearns, Edwin Ira
Stearns, Richard S
Steel, Colin
Steele, Jack
Steele, Warren Cavanaugh
Steele, William A
Stefani, Andrew Peter
Stein, Ronald Paul
Stein, Sidney J
Stein, Stephen Ellery
Steinberg, Gunther
Steinberg, Martin
Steiner, Bruce
Steinfeld, Jeffrey Irwin
Steinfink, Hugo
Steinhardt, Jacinto
Steinmetz, Wayne Edward
Steinrauf, Larry King
Stejskal, Edward Otto
Stengle, Thomas Richard
Stephens, Edgar Ray
Stephenson, Clark Conkling
Sterman, Samuel
Stetson, Alvin Rae
Stevens, John Gebret
Stewart, George Hudson
Stewart, Gerald Walter
Stewart, James Allen
Stewart, Robert Daniel
Stewart, Robert F
Stidham, Howard Donathan
Stigliani, William Michael
Stimler, Suzanne Stokes
Stivala, Salvatore Silvio
Stockman, David Lyle
Stockmayer, Walter Hugo
Stoeber, Werner
Stokes, Charles Sommers
Stolten, Hans Joseph
Stone, John Ernest
Stone, Robert Marion
Stonehouse, Albert James
Stoner, Allan Wilbur
Stoner, Elaine Carol Blatt
Stookey, Stanley Donald
Stoughton, Raymond Woodford
Stout, John Willard
Strauss, Herbert L
Strauss, Mary Jo
Strauss, Ulrich Paul
Strazdins, Edward
Streetman, John Robert
Strehlow, Roger Albert
Strehlow, Wolfgang Hans
Streib, William E
Streng, Alex G
Streng, William Harold
Stricker, Stewart Jeffery
Strickland, William
Strier, Murray Paul
Stricter, Grederick John
Strobel, Howard Austin
Strong, Judith Ann
Strong, Laurence Edward

Strong, Robert Lyman
Stross, Fred Helmut
Stubberman, Robert Frank
Stubbings, Robert Lamb
Stubblefield, Cedric Taylor
Stuckey, John Edmund
Stull, Daniel Richard
Stultz, Robert Lee, Jr
Swalley, William Calvin
Su, Lao-Sou
Sudbury, John Dean
Suggitt, Robert Murray
Sukava, Armas John
Sullivan, James Thomas, Jr
Sullivan, John Henry
Sullivan, Miles Vincent
Sullivan, Richard Christopher
Sumner, George Gardner
Sun, Siao Fang
Sundberg, Michael William
Supinskas, Raymond Joseph
Suryaraman, Maruthuvakudi Gopalasastri
Sutter, John Ritter
Sutula, Chester Louis
Svec, Harry John
Swain, Howard Aldred, Jr
Swanson, Donald Leroy
Swanson, Jack Lee
Swanson, James A.
Swanson, John William
Swanson, Lynwood Walter
Swanson, Sigurd Arthur
Sward, Edward Lawrence, Jr
Swarts, Elwyn Lowell
Sweeney, Michael Anthony
Sweeney, Keith Holcomb
Sweet, Richard Clark
Sweet, Roger George
Sweeton, Frederick Humphrey
Swenson, Charles Allyn
Swift, Robinson Marden
Swift, Terrence James
Swiger, Elizabeth Davis
Swindells, Frank Evans
Swingley, Charles Stephen
Swoboda, Thomas James
Sykes, Brian Douglas
Synek, Miroslav (Mike)
Szasz, Stephen E
Szekely, Andrew Geza
Szerenyi, Peter
Szwarc, Michael
Szymanski, Herman A
Taber, Joseph John
Taft, Robert Wheaton, Jr
Takano, Masaharu
Talley, Claude Parks
Tamers, Murry Allen
Tamsky, Morgan Jerome
Tan, Julia S
Tang, Kwong-Tin
Tang, Yi-Noo
Tanikella, Murty Sundara Sitarama
Tannenbaum, Harvey
Tanner, John Eyer, Jr
Tao, Shu-Jen
Taraszka, Mildred J
Tarkow, Harold
Tarpley, Anderson Ray, Jr
Tatum, James Patrick
Tauer, Kenneth J
Taylor, Alfred Henry, Jr
Taylor, Edward Donald
Taylor, Edward Godfrey
Taylor, Ellison Hall
Taylor, George Russell
Taylor, John Keenan
Taylor, Kathleen C
Taylor, Morris D
Taylor, Peter Anthony
Taylor, Robert Cooper
Taylor, William Johnson
Teal, Gordon Kidd
Teeters, Wilbur Oldroyd
Templeman, Gareth J
Templeton, Charles Clark
Templeton, David Henry
Tensmeyer, Lowell George
Testa, Anthony Carmine
Tetenbaum, Marvin
Tevebaugh, Arthur David
Thacker, Raymond
Thiruvengada, Seshan
Thode, Henry George
Thoma, Roy E
Thomas, Estes Centennial, III
Thomas, Henry Carrison
Thomas, Howard Major
Thomas, John Kerry
Thomas, John Richard
Thomas, Lazarus Daniel
Thomas, Martha Jane Bergin
Thomas, Robert
Thomas, Timothy Farragut
Thomas, Tudor Lloyd

Thomas, William G
Thomason, William Hugh
Thompson, Arthur Carsten
Thompson, Clifton C
Thompson, Donald Leo
Thompson, Douglas Stuart
Thompson, Edward Valentine
Thompson, Gary Haughton
Thompson, James Oliver
Thompson, Joseph Lippard
Thompson, Kenneth Roy
Thompson, Peter Trueman
Thompson, Ralph J
Thompson, Robert John, Jr
Thompson, Warren Elwin
Thorne, James Meyers
Throckmorton, Morford Church
Tice, Russell L
Tichenor, Robert Lauren
Tickner, Alfred William
Ticknor, Leland Bruce
Tiers, George Van Dyke
Tietjen, James Joseph
Tilley, George Lewis
Timmons, Richard B
Tincher, Wayne Coleman
Tingey, Garth Leroy
Tinker, David Owen
Tinoco, Ignacio, Jr
Tinti, Dino S
Tipton, Ann Baugh
Tirman, Alvin
Tobias, Irwin
Tobiason, Frederick Lee
Tobin, Marvin Charles
Toby, Sidney
Todd, Samuel Spaulding
Tokuhiro, Tadashi
Tolbert, Thomas Warren
Tollefson, Eric Lars
Tolles, William Marshall
Tolman, Chadwick Alma
Tomezsko, Edward Stephen John
Tomic, Ernst Alois
Tomimatsu, Toshio
Tomkiewicz, Micha
Tomlinson, Michael
Tomlinson, Richard Howden
Tompa, Albert S
Toney, Frank Morgan
Tong, Lee Karl Jan
Toome, Voldemar
Topol, Leo Eli
Topp, Allan Crickington
Toro-Goyco, Efrain
Torre, Frank John
Tou, James Chieh
Trambarulo, Ralph
Transue, Laurence Frederick
Trapp, Charles Anthony
Trawick, William George
Treadway, Robert Holland
Treffner, Walter Sebastian
Tregillus, Leonard Warren
Treitler, Theodore Leo
Tremaine, Peter Richard
Treumann, William Borgen
Trevelyan, Benjamin John
Trevorrow, Laverne Everett
Trick, Gordon Staples
Trifan, Daniel Siegfried
Trifunac, Alexander Dimitrije
Trivich, Dan
Troisi, Raphael Angelo
Trotter, James
Trotter, Philip James
Truhlar, Donald Gene
Trumbore, Conrad Noble
Trumbore, Forrest Allen
Turer, Jack
Turkevich, John
Turnbull, Bruce Felton
Turnbull, David
Turner, Derek T
Turner, Edwin Morris
Turner, Noel Hinton
Turner, Ralph Waldo
Turner, Robert Chapman
Turrell, George Charles
Turrell, Sylvia Jones
Tuttle, Thomas R, Jr
Tutwiler, Frank Bryan
Tuve, Richard Larsen
Tykodi, Ralph John
Ucci, Pompelio Angelo
Ullman, Robert

Ulmer, David Clyde
Ulmer, Richard Clyde
Ultee, Casper Jan
Underwood, Donald Lee
Unger, Lloyd George
Unland, Mark Leroy
Urbain, Walter Mathias
Uy, Oscar Manuel
Vail, Charles Brooks
Vaisnys, Juozas Rimvydas
Vala, Martin Thorvald, Jr
Valencich, Trina J
Vallee, Richard Earl
Valletta, Robert M
Van Artsdalen, Ervin Robert
Vanas, Don Woodruff
VanCleave, Allan Bishop
Vandegaer, Jan Edmond
Vandenbelt, John Melvin
Vander Hart, David Lloyd
van der Helm, Dick
Vanderkooi, Nicholas, Jr
Vanderryn, Jack
Vanderslice, Joseph Thomas
Vanderslice, Thomas Aquinas
Vanderzee, Cecil Edward
Van De Steeg, Garet Edward
Van Dyk, John William
Van Geet, Anthony Leendert
Van Hecke, Gerald Raymond
Van Hise, James R
Van Holde, Kensal Edward
Van Hook, Andrew
Van Hook, James Paul
Van Hook, William Alexander
Van Lente, Kenneth Anthony
Van Lier, Jan Antonius
Vanpee, Marcel
Vanselow, Clarence Hugo
Vanselow, Ralf W
Van Zeggeren, Frederik
Varadi, Peter Ferencz
Varin, Roger Robert
Varnerin, Robert E
Vasile, Michael Joseph
Vassiliades, Anthony E
Vassiliou, Eustathios
Vastola, Francis J
Vaudo, Anthony Frank
Vaughan, John Dixon
Vaughan, Philip Alfred
Vaughan, Worth E
Vaughn, Joe Warren
Veis, Arthur
Verdier, Peter Howard
Verhoek, Frank Henry
Vernardakis, Theodore Galaction
Vernon, Lonnie William
Verrall, Ronald Ernest
Vickroy, Virgil Vester, Jr
Vidulich, George A
Viera, Ernest Charles
Vier, Dwayne Trowbridge
Viers, Jimmy Wayne
Vijayendran, Bheema R
Vikis, Andreas Charalambous
Vilbrandt, Charles Frank
Vincent, James Sidney
Vinciguerra, Michael Joseph
Vincow, Gershon
Viola, John Thomas
Violante, Michael Robert
Virmani, Yash Paul
Voeks, John Forrest
Vogel, Richard Clark
Vogelsong, Donald Clair
Vogt, Thomas Clarence, Jr
Voigt, Eva-Maria
Volman, David H
Voltz, Sterling Ernest
Von Bodungen, George Anthony
Vonnegut, Bernard
Von Weyssenhoff, Hanns
Von Winbush, Samuel
Vorres, Karl S
Vos, Kenneth Dean
Vratny, Frederick
Vronen, Benjamin H
Waack, Richard
Waage, Edward Vern
Wade, Charles Gordon
Wade, William H
Waech, Theodore G
Wagner, Eugene Stephen
Wagner, Herman Block
Wagner, James Bruce, Jr
Wagner, Paul
Wahrhaftig, Austin Levy
Wait, Samuel Charles, Jr
Watkins, George Raymond
Wakefield, Gene F
Wakeham, Helmut
Walden, George Ellis
Walker, Bennie Frank
Walker, Bernard Forestier
Walker, Jack
Walker, Robert Winn

Walker, Ronald Elliot
Walker, Russell Wagner
Walker, William Comstock
Walkup, John Harper
Wall, Frederick Theodore
Wall, Robert Allen
Wallace, Frederic Andrew
Wallace, Gerald Wayne
Wallace, Richard Maitha
Wallace, Terry Charles
Wallace, Thomas Patrick
Wallace, William James Lord
Walsh, Joseph Matthew
Walsh, Robert Michael
Walters, John Philip
Walton, George
Walz, Alvin Eugene
Wampler, Fred Benny
Wan, Jeffrey Kwok-Sing
Wang, Chih Chun
Wang, Victor Kai-Kuo
Ward, Anthony Thomas
Ward, Charlotte Reed
Ward, Curtis Howard
Ward, James Andrew
Ward, John William
Ward, Thomas Marsh
Ward, Truman L
Ware, William Romaine
Waring, Charles Emmett
Waring, Worden
Warner, Theodore Baker
Warrick, Earl Leathen
Warrington, Terrell L
Washington, Elmer L
Wasser, Richard Barkman
Wasserman, Moe Stanley
Wasson, John R
Watermeier, Leland A
Watkins, Kay Orville
Watkins, Kenneth Walter
Watson, Marshall Tredway
Watson, William Harold, Jr
Watts, Harry
Waugh, John Stewart
Weakliem, Herbert Alfred, Jr
Wear, James Otto
Weaver, Edwin Snell
Weaver, Henry D, Jr
Weaver, John Richard
Webb, Allen Nystrom
Webb, Charles Alan
Webber, Stephen Edward
Webber, Dennis Joseph
Weber, Leon
Weck, Friedrich Josef
Wedemeyer, Robert E
Weed, Homer Clyde
Wei, Pax Samuel Pin
Weigand, Oscar Emil, Jr
Weigl, John Wolfgang
Weiland, Henry Joseph
Weiner, Eugene Robert
Weiner, Robert Samuel
Weinstein, Allan
Weinstock, Bernard
Weir, William David
Weismann, Theodore James
Weiss, Harold Gilbert
Weiss, Jerome
Weiss, Karl H
Weiss, Morris J
Weissman, Samuel Isaac
Weitz, Eric
Wellman, Russel Elmer
Wells, Adoniram Judson
Wells, Edward Joseph
Weltman, Clarence A
Weltner, William, Jr
Wen, Wen-Yang
Wendricks, Roland N
Wendt, Richard P
Wenger, Franz
Wentink, Tunis, Jr
Wentorf, Robert H, Jr
Wentworth, Wayne
Werkema, George Jan
Werkema, Marilyn S
Wertz, David Lee
Wertz, Dennis William
Wessling, Ritchie A
West, Rose Gayle
Westenbarger, Gene Arlan
Westerdahl, Raymond P
Westmore, John Brian
Westrum, Edgar Francis, Jr
Wettack, F Sheldon
Whalley, Edward
Whan, Ruth Elaine
Wharton, James Henry
Wharton, Lennard
Wharton, Walter Washington
Whatley, Alfred T
Whatley, Thomas Alvah
Whealy, Roger Dale
Wheeler, Charles Mervyn, Jr
Whidby, Jerry Frank
Whidden, Helen Louise

Pietri, Charles Edward
Pillay, K K Sivasankara
Plasil, Franz
Poggenburg, John Kenneth, Jr
Porile, Norbert Thomas
Poskanzer, Arthur M
Powers, James Allen
Preiss, Ivor Louis
Prohaska, Charles Anton
Prussin, Stanley Gerald
Rasmussen, John Oscar, Jr
Rayudu, Garimella V S
Reavis, James Gene
Reed, George W, Jr
Reed, Mary Frances
Reeder, Paul Lorenz
Reedy, Robert Challenger
Remsberg, Louis Philip, Jr
Rengan, Krishnaswamy
Rightmire, Robert
Roesmer, Josef
Rubinson, William
Ruddy, Francis Henry
Ruiz, Carl P
Runnalls, Nelva Earline Gross
Russell, Irving James
Ryan, Victor Albert
Sabu, Dwarka Das
Santry, D C
Sarantites, Demetrios George
Schell, William R
Schuman, Robert Paul
Seaborg, Glenn Theodore
Sheline, Raymond Kay
Sher, Alvin Harvey
Shudde, Rex Hawkins
Silva, Robert Joseph
Smith, Francis Marion
Sodd, Vincent J
Stephens, Frank Samuel
Stewart, Robert Francis
Stone, John Austin
Stoughton, Raymond Woodford
Sugarman, Nathan
Sugihara, Thomas Tamotsu
Sullivan, Thomas Frederick
Sun, Kuan-Han
Swanson, David G, Jr
Tewes, Howard Allan
Thomas, Thomas Darrah
Thompson, Ronald Hobart
Townley, Charles William
Troutner, David Elliott
Turkevich, Anthony
Turner, Stanley Eugene
Uhl, Dale Lynden
Unik, John Peter
Vandenbosch, Robert
Van Hise, James R
Viola, Victor E, Jr
Wahl, Arthur Charles
Wahl, Werner Henry
Wai, Chien Moo
Walters, William Ben
Ward, Thomas Edmund
Watson, Rand Lewis
Wikjord, Alfred George
Wild, John Frederick
Wiley, John Robert
Wilhelmy, Jerry Barnard
Williams, David Cary
Williams, Evan Thomas
Williams, Robert Allen
Wing, James
Wischow, Russell P
Wogman, Ned Allen
Wolfsberg, Kurt
Wolke, Robert Leslie
Yates, Steven Winfield
Yule, Herbert Phillip
Zoller, William H

Physical Inorganic Chemistry
Adin, Anthony
Anbar, Michael
Anderson, Charles Thomas
Armor, John N
Austin, Alfred Ells
Bacon, Glenn
Bacon, Frank Rider
Baes, Charles Frederick, Jr
Bailin, Lionel J
Bains, Malkiat Singh
Bamberger, Carlos Enrique Leopoldo
Bancroft, George Michael
Barber, Eugene John
Barney, Gary Scott
Beckering, Willis
Begun, George Murray
Beliskus, David
Bernays, Peter Michael
Birdwhistell, Ralph Kenton
Blank, Charles Anthony
Blau, Henry Hess
Boardman, William Walter, Jr
Bonnell, David William
Bourgon, Marcel
Brand, John Robert

Brault, Albert Thomas
Bredig, Max Albert
Brown, Patrick Michael
Brunschwig, Bruce Samuel
Bulloff, Jack John
Burger, Leland Leonard
Byler, David Michael
Campbell, Larry Edwin
Chaffee, Eleanor
Chakrabarty, Manoj R
Chock, Ernest Phaynan
Clayton, John Charles (Hastings)
Clough, Francis Bowman
Clough, Philip James
Colburn, Charles Buford
Coleman, William Fletcher
Dahl, Lawrence Frederick
Darnell, Alfred Jerome
Debye, Nordulf Wiking Gerud
DeLorenzo, Ronald Anthony
Downey, Joseph Robert, Jr
Dutta-Ahmed, Akhtar
Eachus, Raymond Stanley
Earley, Joseph Emmett
El-Awady, Abbas Abbas
Erbacher, John Kornel
Falk, Charles David
Fehlner, Thomas Patrick
Ferrante, Michael John
Fischer, Albert Karl
Fleischauer, Paul Dell
Fogel, Norman
Fox, Robert Kriegbaum
Fox, William B
Franco, Nicholas Benjamin
Fryxell, Robert Edward
Garrett, Michael Benjamin
Green, Agnes Ann
Greenhouse, Harold Mitchell
Greifer, Aaron Philip
Grunthaner, Frank John
Guertin, Jacques P
Haden, Walter Linwood, Jr
Hahn, Harold Thomas
Hammer, Robert Russell
Harris, Gordon McLeod
Hartford, Winslow Hopper
Hicks, Kenneth Ward
Hoffman, Alan Bruce
Hoffman, Charles John
Hoffman, Morton Z
Hoggard, Patrick Earle
Homeier, Edwin H, Jr
Hoppenjans, Donald William
House, James Evan, Jr
Howell, Peter Adam
Hubbard, Richard Alexander, II
Hudson, Alice Brandon
Hutchison, James Robert
Iannuzzi, Melanie Mary
Israeli, Julius Yigal
Jackovitz, John Franklin
James, Dean B
Johnson, David Alfred
Kafalas, James A
Kaplan, Roy Irving
Katzin, Leonard Isaac
Keder, Wilbert Eugene
Keeler, Robert Adolph
Kelley, Kenneth K
Kendig, Martin William
King, James P
King, Thomas Morgan
Kirshenbaum, Abraham David
Klein, Philipp Hillel
Klingen, Theodore James
Kokoska, Gerald Francis
Kowalak, Albert Douglas
Kozlowski, Theodore R
Kuczkowski, Robert Louis
Langford, Cooper Harold, III
Larach, Simon
Lazarus, Marc Samuel
Lefkowitz, Stanley A
Lintvedt, Richard Lowell
Lock, Colin James Lyne
Long, Gary John
Lyon, Luther Lawrence, Jr
Macalady, Donald Lee
Magnusson, Lawrence Bersell
Manasevit, Harold Murray
Margrave, John Lee
Martin, Don Stanley, Jr
May, Walter Ruch
McCarthy, Paul James
McDonald, Hector O
McGill, Robert Mayo
McNamara, John Edward
Mercer, Edward Everett
Merideth, Charles Waymond
Meyer, Carl Beat
Milligan, Winfred Oliver
Moore, Robert Earl
Morris, Kelso Bronson
Morrow, Jack I
Morss, Lester Robert
Moss, Robert Henry
Myers, Ralph Thomas

Neely, James W
Norris, Thomas Hughes
O'Neill, Richard Thomas
Orebaugh, Errol Glen
Osterheld, Robert Keith
Owens, Clifford
Pagenkopf, Gordon K
Panzer, Richard Earl
Pasternack, Robert Francis
Petersen, Harold, Jr
Petersen, Jeffrey Lee
Peterson, Joseph Richard
Petrocelli, Americo W
Phelan, James Richard
Piehl, Donald Herbert
Pinch, Harry Louis
Porter, Richard Francis
Post, Roy G
Potts, Melvin Lester
Quinn, Rod King
Radtke, Douglas Dean
Rasmussen, Paul
Readnour, Jerry Michael
Reed, William Robert
Reising, Richard F
Rhyne, Thomas Crowell
Ringwald, Owen Edward
Roberts, Elliott John
Roscoe, John Stanley, Jr
Rosenthal, Michael R
Rowley, David Alton
Schildcrout, Steven Michael
Schmidt, Donald Dean
Schnizlein, John Glenn
Sen, Buddhadev
Sherry, Howard S
Shoup, Robert D
Sienko, Michell Joseph
Silverman, Meyer David
Sinclair, William Robert
Skinner, James F
Sloan, Gilbert Jacob
Smardzewski, Richard Roman
Smart, James Blair
Smith, Jean Gillen
Smith, Jerry Joseph
Solomon, Edward I
Sprowles, Jolyon Charles
Stafford, Fred E
Stagg, William Ray
Staples, Bert Roland
Stern, Kurt Heinz
Sullivan, William Francis
Sutin, Norman
Swift, Harold Eugene
Tamres, Milton
Teague, Marion Warfield
Thomas, David Gilbert
Thomas, John Paul
Thompson, Joseph Kyle
Thompson, Mary E
Tong, James Ying-Peh
Tucci, Edmond Raymond
Uchtman, Vernon Albert
Vanderspurt, Thomas Henry
Vander Wall, Eugene
Vigee, Gerald S
Wason, Satish Kumar
Waugh, John Lodovick Thomson
Webb, Alan Wendell
Weeks, Thomas Joseph, Jr
Welch, Cletus Norman
Wells, James Edward
Weschler, Charles John
Westman, Albert Ernest Roberts
White, Jesse Edmund
Whitfill, Donald Lee
Wicker, Robert Kirk
Williams, Rickey Jay
Willison, Donald Bruce
Wittenberg, Layton Junior
Woods, Clifton, III
Worrell, Jay H
Wright, Kenneth James
Yeatts, LeRoy Brough, Jr
Yingst, Harvey Austin
Yunker, Wayne Harry
Zimmerman, Donald Nathan
Zipp, Arden Peter

Physical Organic Chemistry
Abramovici, Miron
Adam, Waldemar
Adams, Otis William
Addy, John Keith
Albrecht, Frederick Xavier
Alekman, Stanley L
Allara, David Lawrence
Allen, Gary William
Alston, Peter Van
Altschul, Rolf
Andose, Joseph David
Andrist, Anson Harry
Andrulis, Peter Joseph, Jr
Archie, William C, Jr
Arnett, Edward McCollin
Arnold, David Brown
Arnold, Richard Thomas

Arnowich, Beatrice
Augood, Derek Raymond
Ayers, Paul Wayne
Badin, Elmer John
Bailey, Roy Horton, Jr
Bakule, Ronald David
Baldwin, John E
Ballentine, Alva Ray
Bank, Shelton
Bannister, William Warren
Barnett, Ronald E
Baumgarten, Reuben Lawrence
Bearden, William Harlie
Bedoit, William Clarence, Jr
Beishline, Robert Raymond
Bell, Charles E, Jr
Bender, Daniel Frank
Benfey, Otto Theodor
Beres, John Joseph
Bernasconi, Claude Francois
Berndt, Donald Carl
Bernstein, Stanley Carl
Bertin, Henry John, Jr
Blackham, Angus Udell
Bly, Robert Stewart
Bongiorni, Domenic Frank
Booman, Keith Albert
Bostick, Edgar E
Brady, William Thomas
Brauman, Sharon Kruse
Brown, John Francis, Jr
Brown, Keith Charles
Brown, Robert Stanley
Bruck, Peter
Bryan, Mary Leo
Buchholz, Allan C
Buckles, Lawrence Calvin
Buckles, Robert Edwin
Bunbury, David Leslie
Burfinson, Nicholas Edward
Burnett, Leo Seth
Bushweller, Charles Hackett
Cambray, Joseph
Cammarata, Arthur
Campbell, Thomas Cooper
Caputo, Joseph Anthony
Carle, Kenneth Roberts
Carter, Robert Everett
Caserio, Frederick F, Jr
Cawley, John Joseph
Cengel, John Anthony
Chandler, William David
Chang, Kuang-Chou
Charton, Marvin
Chau, Michael Ming-Kee
Chaudhary, Sohan Singh
Chickos, James S
Childs, Ronald Frank
Clark, Louis Watts
Clovis, James S
Coburn, Robert A
Coburn, William Carl, Jr
Cohen, Gordon Mark
Cohen, Lester Allan
Cohen, Louis Arthur
Colebrook, Lawrence David
Collins, Clair Joseph
Connor, James Edward, Jr
Cook, Richard James
Cooperman, Barry S
Craig, Arnold Charles
Creagh, Linda Truitt
Crist, DeLanson Ross
Crowell, Thomas Irving
Crumrine, David Shafer
Curtice, Jay Stephen
Dalrymple, David Lawrence
Dannenberg, Joseph
Daughenbaugh, Randall Jay
Davis, Brian Clifton
Dean, David Lee
De Camp, Wilson Hamilton
Decora, Andrew Wayne
Deno, Norman C
Dias, Jerry Ray
DiGiorgio, Joseph Brun
Dixon, Marvin Porter
Doubleday, Charles E, Jr
Dougherty, Ralph C
Doyle, Michael P
Duncan, Charles Donald
Duncan, James Alan
Dunn, Gerald Emery
Dunn, John Robert
Dunne, Thomas Gregory
Duvall, John Joseph
Dwyer, Sean G
Eaton, David Fielder
Edward, John Thomas
Eikenberry, Jon Nathan
Elliott, J Lell
Eng, Leslie
Engel, Paul Sanford
Engler, Edward Martin
Espy, Herbert Hastings
Esteve, Ramon M, Jr
Ewing, Sheila Pauline

CHEMISTRY, PHYSICAL

Factor, Arnold
Fainberg, Arnold Harold
Fall, Harry H
Fearn, James Ernest
Feller, Robert Livingston
Fendler, Eleanor Johnson
Fenoglio, David John
Ferren, Richard Anthony
Finley, Kay Thomas
Firkins, John Lionel
Fitzgerald, Patrick Henry
Fliszar, Sandor
Ford, Richard Alan
Forster, Eric Otto
Fort, Raymond Cornelius, Jr
Foss, Robert Paul
Francis, Peter Schuyler
Franz, James Alan
Fraser, Robert Rowntree
Freeman, Fillmore
Friedlander, Herbert Norman
Fuhrmann, Robert
Gale, Laird Housel
Gandour, Richard David
Gaspar, Peter Paul
Gayle, John Ben
Gettler, Joseph Daniel
Giam, Choo-Seng
Gibian, Morton J
Gibson, Harry William
Giddings, William Paul
Gilbert, John Carl
Giaspie, Peyton Scott
Golinkin, Herbert Sheldon
Goon, David James Wong
Gordon, Marshall
Gorski, Robert Alexander
Gosser, Lawrence Wayne
Graminski, Edmond Leonard
Granger, Maurice Roy
Graybill, Bruce Myron
Griffin, Anselm Clyde, III
Griffiths, David Warren
Grubbs, Edward
Grunwald, Ernest Max
Gurudata, Neville
Guthrie, Robert D
Guttmann, Andrew Titus
Haggard, Richard Allan
Hakka, Leo Ernest
Hall, Kenneth Lynn
Halm, James Maurice
Hammer, Gary G
Hammons, James Hutchinson
Harding, Charles Enoch
Hardman, Bruce Bertolette
Harris, Leland
Harrison, Arnold Myron
Hart, Donald John
Haupt, Frederic Curt
Hayward, Lloyd Douglas
Haywood-Farmer, John
Hazlett, Robert Neil
Hecker, David Clinton
Heiba, El-Ahmadi Ibrahim
Hendry, Dale Glenn
Henrichs, Paul Mark
Herndon, William Cecil
Herz, Jack L
Herz, Matthew Lawrence
Hiatt, Richard Rowis
Hodgdon, Russell Bates, Jr
Hoeg, Donald Francis
Hoffman, Michael K
Holtz, Hans Dietrich
Horan, Francis E
Howe, King Lau
Howe, Norman Elton, Jr
Howell, Thomas James
Illingworth, George Ernest
Ingold, Keith Usherwood
Irwin, Philip George
Isaks, Martin
Jaffe, Annette Bronkesh
Janzen, Edward George
Jensen, James Leslie
Jesaitis, Raymond G
Jex, Victor Bird
Ji, Sungchul
Jochsberger, Theodore
Johnson, John Enoch
Johnson, Richard Stebbins
Jones, Guilford, II
Jones, Richard Hamilton
Jurch, George Richard, Jr
Kaiser, Emil Thomas
Kakis, Frederic Jacob
Kamego, Albert Amil
Kamlet, Mortimer Jacob
Kapecki, Jon Alfred
Kaplan, Martin L
Kaufmann, Glenn Monroe
Kemp, Kenneth Courtney
Kevill, Dennis Neil
Kim, John Poong-Kil
Kinman, Judith Pollock
Klopfenstein, Charles E
Knudsen, George Andrew, Jr

Knudsen, Thomas Paul
Koch, Heinz Frank
Kochi, Jay Kazuo
Konizer, George Burr
Krabacher, Bernard
Kramer, Brian Dale
Kramer, George Mortimer
Krege, Alexander Jerry
Krieger, Jeanne Kann
Kritz, George Stanley, Jr
Kryger, Roy George
Kulling, Rudolph K
Kumamoto, Junji
Kurland, Jonathan Joshua
Kurz, Joseph Louis
Ladenheim, Harry
Langsdorf, William Philip
Latimer, Donald Andrew
Latta, Bruce McKee
Lauderback, Sanford Keith
Laughlin, Robert Gene
Lawler, Ronald George
Lee, Warren G
Lee, Do-Jae
Leffek, Kenneth Thomas
Leonard, John Joseph
Lepley, Arthur Ray
Levy, George Charles
Lewin, Anita Hana
Li, Wu-Shyong
Lichtin, Norman Nahum
Liggero, Samuel Henry
Livant, Peter David
Lorand, John Peter
Lorentzen, Keith Eden
Losin, Edward Thomas
Lowe, James Urban, Jr
Lowry, James William
Lowry, Charles Boyce
Lowry, Nancy
Lura, Richard Dean
Lutz, Raymond Paul
Lynch, Brian Maurice
Mabey, William Ray
Mach, Martin Henry
Magee, Philip Stewart
Magid, Linda Jenny
Marchand, Alan Philip
Mark, Harold Wayne
Marsh, Henry Peter
Marsi, Kenneth Larue
Martin, William Butler, Jr
Martin, Joseph
Maskornick, Michael J
Mateer, Richard Austin
Matuszak, Charles A
Mayo, Frank Rea
McBride, James Michael
McConaghy, John Stead, Jr
McKelvey, Donald Richard
McKelvey, Ronald Deane
McLeod, Richard Kenneth
Meguerian, Garbis H
Mendenhall, George David
Merrifield, D Bruce
Michejda, Christopher Jan
Miller, David Lee
Miller, Sydney Israel
Minch, Michael Joseph
Moore, Cecilia Louise
Morgan, Charles Robert
Moses, Francis Guy
Mosher, Melvyn Wayne
Moss, Ernest Kent
Muck, Darrel Lee
Murr, Brown L, Jr
Murray, Robert Wallace
Nebenzahl, Linda Levine
Nelson, Kay Leroi
Neuman, Robert C, Jr
Neuman, Calvin Lee
Newitt, Edward James
Newton, Robert Andrew
Nigh, Wesley Gray
Ogliaruso, Michael Anthony
Orvik, Jon Anthony
Otken, Charles Clay
Pacifici, James Grady
Palaitis, Waldemar
Panar, Manuel
Parkanyi, Cyril
Parker, Winfred Evans
Peck, David W
Perdue, Edward Michael
Perettie, Donald Joseph
Pfeiffer, Joseph George
Pinkus, Albin George
Pinnick, Herbert Robert, Jr
Platt, Alan Edward
Pocker, Yeshayau
Pocker, John J
Porter, Tamar Beatrice
Posner, John J
Powell, Justin Christopher
Puar, Mohindar S
Quimby, Daniel Joseph
Raciszewski, Zbigniew

Raley, John Howard
Ramsey, Brian Gaines
Read, David Hadley
Reed, Kenneth Joseph
Reuwer, Joseph Francis, Jr
Rhodes, Yorke Edward
Ricci, Robert William
Richards, Charles Norman
Rieger, Martin Max
Roberts, Donald Duane
Robertson, Jerold C
Rocek, Jan
Rodemeyer, Stephen A
Rodgers, James Edward
Roebuck, Albert Henry
Roos, Leo
Roscher, Nina Matheny
Rosenfeld, Stuart Michael
Ross, David Samuel
Ross, Lawrence James
Rowe, Paul E
Rowell, Charles Frederick
Rudkin, George Osborne
Ruhnke, Edward Vincent
Sample, Thomas Earl, Jr
Sand, Ralph E
Sandel, Vernon Ralph
Sarasohn, Ilya M
Sarge, Theodore William
Saunders, William Hundley, Jr
Schiavelli, Melvyn David
Schleyer, Paul Von Rague
Schmidt, Francis Henry
Schmidt-Collerus, Josef Johannes
Schowen, Richard Lyle
Schreiber, Kurt Clark
Schueler, Paul Edgar
Schuster, Ingebog I M
Sebastian, John Francis
Selker, Milton Leonard
Senkler, George Henry, Jr
Shackleit, Comer Drake
Shafer, Jules Alan
Shearer, Greg Otis
Shiner, Vernon Jack, Jr
Simpson, David Alexander
Singh, Ajaib
Skau, Evald Laurids
Smart, Bruce Edmund
Smith, Eileen Patricia
Smith, Gerard Vinton
Smith, Louis Charles
Smith, Peter James
Smith, Vincent Francis, Jr
Smith, Wendell Franklyn, Jr
Smith, William Novis, Jr
Sojka, Stanley Anthony
Sorensen, Theodore Strang
Sottery, Theodore Walter
Sovocool, George Wayne
Spangler, Charles William
Speert, Arnold
Stein, Allan Rudolph
Stein, Peter James
Stewart, Ross
Stone, Joe Thomas
Streitwieser, Andrew, Jr
Strom, Edwin Thomas
Stuber, Fred A
Stump, Billy Lee
Sturmer, David Michael
Swenton, John Stephen
Taagepera, Mare
Takeshita, Tsuneichi
Taller, Robert Arthur
Thompson, Evan M
Thompson, Robert Gene
Thornton, Tom Radford
Thornton, Elizabeth K
Thorpe, Martha Campbell
Tomalia, Donald Andrew
Tourigny, Guy J
Towns, Donald Lionel
Tremelling, Michael, Jr
Vander Burgh, Leonard F
van Dijk, Christiaan Pieter
Van Sickle, Dale Elbert
Varga, Charles E
Vill, John Joseph
Vogel, Philip Christian
Voorhees, Kent Jay
Wagner, Peter J
Wagner, Robert Marvin
Walling, Cheves
Walsh, James Aloysius
Walters, Robert Irving
Walters, Edward Albert
Warren, Craig Bishop
Warrick, Percy, Jr
Weiner, Steven Allan
Weisfeld, Lewis Bernard
Westaway, Kenneth C
Wewerka, Eugene Michael
White, David Halbert
White, Joe Wade
White, William North
Williamson, Martin John
Willis, Roland George

Wolf, Philip Frank
Wolfarth, Eugene F
Woods, William George
Wright, Robert W
Wulfman, David Swinton
Yager, Billy Joe
Yates, Keith
Yee, Tucker Tew
Zanger, Murray
Zavitsas, Daniel Maxmillian
Ziebarth, Andreas Athanasios
Zweig, Arnold

Quantum Chemistry

Albrecht, Andreas Christopher
Ali, Mahamed Asgar
Avery, John Scales
Avgeropoulos, George N
Bagus, Paul Saul
Bardo, Richard Dale
Bardsley, James Norman
Bartlett, Rodney Joseph
Beck, Donald Richardson
Bock, Charles Walter
Cantu, Antonio Arnold
Chang, Shih-Yung
Chang, Tai Yup
Chipman, Daniel Myron
Chiu, Luc-Yung Chow
Chung-Phillips, Alice
Companion, Audrey (Lee)
Cusachs, Louis Chopin
Dash, Harriman Harvey
Del Bene, Janet Elaine
Dwyer, Rowland William, Jr
Ellison, Frank Oscar
Ellzey, Marion Lawrence, Jr
Gordon, Mark Stephen
Graves, John Lowell
Grein, Friedrich
Harriman, John E
Harrison, James Francis
Hashmall, Joseph Alan
Heaton, Maria Malachowski
Hollister, Charlotte Ann
Howell, James MacGregor
Hunter, Geoffrey
Johnson, Keith Huber
Julienne, Paul Sebastian
Kaufman, Joyce J
Kenney-Wallace, Geraldine Anne
Kim, Hyunyong
Klein, Douglas J
Kollman, Peter Andrew
Lin, Che-Shung
Lowe, John Philip
Malli, Gulzari Lal
McIver, James W, Jr
Mies, Frederick Henry
Moffat, John Blain
Morokuma, Keiji
Orloff, Malcolm Kenneth
Raffenetti, Richard Charles
Ramaker, David Ellis
Rein, Robert
Richardson, James Wyman
Rothenberg, Stephen
Schnaare, Roger L
Shull, Harrison
Sichel, John Martin
Silverstone, Harris Julian
Simons, Gary
Switkes, Eugene
Walnut, Thomas Henry, Jr
Whitehead, Michael Anthony
Zemke, Warren T
Zung, Joseph T

COMMUNICATIONS SCIENCE

Communications Science

Altman, Philip Lawrence
Armitage, John Denton, Jr
Beranek, Leo Leroy
Black, John Wilson
Borland, John Raymond
Boyd, John Paul
Brown, John Lawrence, Jr
Buckingham, William Thomas
Byrnes, Francis Clair
Caldwell, Melba Carstarphen
Cohen, Leonard George
Colton, Raymond H
Cooper, Franklin Seaney
Cornett, Richard Orin
DeBakey, Selma
Farmer, Malcolm French
Ferguson, Mary Hobson
Gary, Robert
Gilbert, Myron B
Golden, Robert K
Grover, Paul L, Jr
Hagelbarger, David William
Hammond, William Marion
Hicks, Patricia Fain

Houde, Robert A
Hubbard, William Marshall
Kadota, T Theodore
Ling, Daniel
Minkoff, John
Moran, Thomas Patrick
Natt, Michael Philip
Olive, Joseph P
Ostwald, Peter Frederic
Provencio, Jesus Roberto
Ray, Howard Eugene
Reiter, Elmar Rudolf
Roland, Charles Gordon
Rothman, Howard Barry
Rubin, Herbert
Silverman, Sol Richard
Spiegel, Allen David
Tamorria, Christopher Richard
Vervoort, Gerardus
Watanabe, Akira
Welbeck, Paa-Bekoe Henry
Wellar, Barry Sheldon
White, George Matthews

COMPUTER & INFORMATION SCIENCES

Computer Science
Abbott, Robert Classie
Abdali, Syed Kamal
Ackerman, Eugene
Adams, J Mack
Agrawala, Ashok Kumar
Agresti, William W
Aiken, Jack David
Alanen, Jack David
Alter, Ronald
Alton, Donald Alvin
Anderson, Roger E
Anschel, Morris
Anshel, Michael
Arbib, Michael A
Ashenhurst, Robert Lovett
Ashman, Michael Nathan
Asprey, Winifred Alice
Atchison, William Franklin
Atwood, John William
Aufenkamp, Don
Bajscy, Ruzena K
Baker, Brenda Sue
Baker, James Addison
Baker, Robert Henry, Jr
Baker, Theodore Paul
Baldwin, Lynne Juedeman
Barnard, Anthony C L
Barnes, Bruce Herbert
Barr, Roger Coke
Bartels, Peter H
Bartels, Richard Harold
Bass, Leonard Joel
Bateman, Barry Lynn
Batson, Alan Percy
Bayer, Douglas Leslie
Beckman, Frank Samuel
Bednar, Jonnie Bee
Behforooz, Ali
Bell, James Richard
Bell, Stoughton
Bennett, James Hallam
Berge, Truman Kent
Berger, Jay Manton
Berger, William J
Bergquist, James William
Berk, Toby Steven
Berning, Jean Ackerman
Bernstein, Herbert Jacob
Bersohn, Malcolm
Bertaut, Edgard Francis
Beyer, Wendell T
Bhandarkar, Dileep Pandurang
Biermann, Alan Wales
Blackwell, Paul K, II
Blass, William Errol
Blattner, Meera McCuaig
Blevins, Maurice Everett
Board, Robert Dennis
Bobrow, Daniel G
Bowlden, Henry James
Bowman, Carlos Morales
Boyle, James Martin
Brackett, John Washburn
Brady, Allen H
Brainerd, Walter Scott
Breedlove, James Robby, Jr
Britt, Patricia Marie
Brown, Harold David
Brown, Richard Maurice
Brown, William Stanley
Browne, James Clayton
Brundage, Robert Earl
Buchanan, Bruce G
Bui, Tien Dai
Burger, Robert Thornton
Burkhardt, Walter H
Butler, Daniel Knowles
Butler, Harold S

Butterfield, Veloy Hansen, Jr
Callender, E David
Campaigne, Howard Herbert
Campbell, Graham Hays
Cantor, David Geoffrey
Carey, Bernard Joseph
Carlson, Gary
Carlstead, Edward Meredith
Carney, Edward J
Carol, Bernard
Carter, Elmer Buzby
Carter, Harvey Pate
Caviness, Bobby Forrester
Cercone, Nicholas Joseph
Chambers, John McKinley
Chen, I-Ngo
Chen, Robert Chia-Hua
Chen, Tien Chi
Chester, Daniel Leon
Chin, Yeh-Hao
Chow, Wen Mou
Christensen, Gerhardt C
Clementson, Gerhardt C
Close, Richard Thomas
Coling, Forrest L
Collins, George Edwin
Comba, Paul Gustavo
Conrow, Kenneth
Constable, Robert L
Conway, Lynn Ann
Coriell, Kathleen Patricia
Costello, Donald F
Coulter, Neal Stanley
Cousineau, Leo
Crenshaw, Jack Westcott
Cryer, Colin Walker
Cullen, Dermott Edward
Cusachs, Louis Chopin
Cutlip, William Frederick
Dalrymple, Stephen Harris
Daniel, James Wilson
Daniel, Leonard Rupert
Davis, Chester L
Davis, Henry Werner
Davis, Ruth Margaret
Daykin, Philip Norman
Dean, Charles Edwin
De Lillo, Nicholas Joseph
De Maine, Paul Alexander Desmond
DeMillo, Richard A
Denenberg, Charlotte Goryn
Dennis, Martha Greenberg
Dershem, Herbert L
Desautels, Edouard Joseph
Deutsch, Murray Lewis
Dolch, John Parker
Dolin, Stanley A
Drew, Dan Dale
Drew, Daniel L
Drossman, Melvyn Miles
Dryden, Warren Arnold
Dudziak, Walter Francis
Dunaway, Donna Kastle
Dwyer, Thomas A
Easton, William Bigelow
Eberlein, Patricia James
Ecklund, Earl Frank, Jr
Eggert, Arthur Arnold
Ehle, Byron Leonard
Eicher, David G
Eigot, Calvin C
Elkind, Jerome I
Ellis, Glen Edward
Engel, Frank August, Jr
Epstein, George
Epstein, Samuel David
Esack, Ashmeed
Evans, Thomas George
Fabrey, James Douglas
Falconer, David G
Falconer, David Ross
Federighi, Francis D
Fernbach, Sidney
Ferrante, Jeanne
Feyock, Stefan
Findler, Nicholas Victor
Fischer, Patrick Carl
Fisher, Donald D
Fleck, Arthur C
Flegal, Robert Melvin
Fleming, James Joseph
Fletcher, Charles Howard
Fletcher, John George
Floyd, Robert W
Floyd, William Beckwith
Forker, Robert Fencil
Fosberg, Mary Dee Harris
Fox, Phyllis
Frank, Thomas Stolley
Freeman, Martin
Frieder, Gideon
Friedman, Jack P
Friedman, Joyce Barbara
Fulytn, Robert Victor
Fushimi, Fred Chikashi
Futrelle, Robert Peel
Gabriel, John R
Gajendan, Nandigam

Gallaher, Lawrence Joseph
Galler, Bernard Aaron
Gallie, Thomas Muir
Garey, Michael Randolph
Garwick, Jan Vaumund
Gary, John Mitchell
Gatewood, Lael Cranmer
Gear, Charles William
Gelernter, Herbert Leo
Gentleman, William Morven
George, Ronald Edison
Gersting, John Marshall, Jr
Geschke, Charles Matthew
Ghosh, Sakti P
Gibbs, Norman Edgar
Gladney, Henry M
Glaser, Edmund M
Glaze, Richard Michael
Gleissner, Gene Heiden
Goddard, Alton R
Goering, Orville
Goguen, Joseph A, Jr
Goldberg, Conrad Stewart
Golden, Robert K
Goldstein, Paul
Golin, Stuart
Gonzalez-Arce, Teofilo Francisco
Goodman, Richard Henry
Goodman, Seymour
Gorn, Saul
Gorsline, George William
Gotterer, Malcolm Harold
Gould, William Allen
Graham, James W
Graham, Roger Neill
Graham, Susan Lois
Grau, Albert A
Green, Claude Cordell
Gregory, James McKanna
Greibach, Sheila Adele
Gries, David
Groenweghe, Leo Carl Denis
Gross, Michael Ralph
Groves, William Ernest
Gustavson, Fred Gehrung
Haag, James Norman
Habermann, Arie Nicolaas
Haddix, George Franklin
Halstead, Maurice Howard
Halton, John Henry
Hamblen, John Wesley
Hamlin, Griffith Askew, Jr
Hammer, Carl
Hammer, Preston Clarence
Hamming, Richard W
Hammond, William Edward
Hammond, William Marion
Hampel, Viktor Erwin
Hansen, William Anthony
Hardy, John W, Jr
Harrison, Malcolm Charles
Harrison, Michael A
Harrop, Ronald
Hart, John Francis
Hartmanis, Juris
Hausner, Arthur
Haverty, John Patrick
Heaps, Harold Stanley
Heller, Jack
Heller, Stephen Richard
Hemmerle, William J
Henderson, Donald Lee
Henderson, Madeline M Berry
Henschen, Lawrence Joseph
Hermans, Hans J
Herriot, John George
Hightower, James K
Hildebrandt, Theodore Ware
Hodes, Louis
Hoffman, Alexander A J
Holdeman, Jonas Tillman, Jr
Holland, John Henry
Holmes, William Farrar
Holoien, Martin Olaf
Holst, William Frederick
Horowitz, Ellis
Howe, William Jeffrey
Howell, Jo Ann Shaw
Hsu, John Y
Hu, Te Chiang
Hull, Thomas Edward
Hurst, Rex LeRoy
Hurwitz, Alexander
Hutcheson, Paul Henry
Hutchinson, George Keating
Huthnance, Edward Dennis, Jr
Ingram, Glenn R
Israel, Jay Elliot
Itoga, Stephen Yokio
Jacobs, Eugene Howard
Jacobs, Walter William
Jamison, Steven Lyle
Jarvis, John Frederick
Jefferson, David Kenoss
Jennemann, Vincent Francis
Jensen, Clayton Everett
Johnson, David Stifler
Johnson, Stephen Curtis

Johnson, Wallace E
Johnson, Whitney Larsen
Jones, Anita Katherine
Jones, Daniel Elven
Jones, Louise Hinrichsen
Jones, Ronald Dale
Juliussen, J Egil
Kahan, William M
Kaiser, Christopher B
Kallander, John William
Kanal, Laveen Nanik
Kaplan, Ronald M
Katz, Louis
Katz, Sidney Marco
Katzper, Meyer
Kaufman, Linda Carol
Keddy, James Richard
Keenan, Thomas Aquinas
Kehl, Theodore H
Kehl, William Brunner
Keller, Mary Kenneth
Keller, Roy Fred
Kendall, Burton Nathaniel
Kennedy, James M
Kerr, Douglas S
Killion, Lawrence Eugene
Kimball, Ralph B
Kirkwood, Charles Edward, Jr
Knuth, Donald Ervin
Knutson, Charles Dwaine
Koh, Young O
Kolsky, Harwood George
Konstam, Aaron Harry
Korfhage, Robert R
Korsh, James F
Kortzeborn, Robert Neal
Krueger, Eugene Rex
Kuo, Shan Sun
Kurtz, Thomas Eugene
Kusak, Lloyd James
La Bonte, Anton Edward
LaMonica, Carl J
Lampson, Butler Wright
Landauer, Rolf William
Laurance, Neal L
Lawson, Mildred Wiker
Ledin, George, Jr
Lee, Ralph Edward
Letcher, John Henry
Levenberg, Milton Irwin
Levow, Roy Bruce
Levy, Allan Henry
Levy, Leon Sholom
Lichter, James Joseph
Liedtke, Claus-Eberhard
Lien, Yeong-Chung Edmund
Lin, Benjamin Ming-Ren
Lin, Shen
Lindberg, Donald Allan Bror
Linn, John Charles
Linz, Peter
Littlewood, Roland Kay
Liu, Chung Laung
Loveland, Donald William
Lovell, Stuart Estes
Lowther, John Lincoln
Luckham, David Comptom
Luehrmann, Arthur Willett, Jr
Lutz, Robert William
Lycklama, Heinz
Lynch, William C
Ma, Cynthia Sanman
Macaluso, Pat
Mackey, Karen Ethel
MacLaren, Malcolm Donald
Madden, Stephen James, Jr
Magee, Michael Jack
Maisel, Herbert
Maissel, Leon I
Malcolm, Michael Alexander
Mamelak, Joseph Simon
Mamrak, Sandra Ann
Mancusi, Michael D
Marsaglia, George
Mathis, Robert Fletcher
Mattison, Roland Lees
McAllister, Marialuisa N
McCarthy, John
McCasland, Gifford Ewing
McCoy, John Harold
McCreight, Edward M
McGloin, Paul Arthur
McGowan, Clement Leo, III
McIntyre, Thomas Woodford
McNaughton, Robert
McQuarrie, Donald G
Meads, Philip Francis, Jr
Meagher, Ralph Ernest
Meissner, Loren Phillip
Mello, James Francis
Meriwether, John R
Meteer, James William
Meyer, Harvey John
Meyers, Gene Howard
Michalski, Ryszard Stanislaw
Miller, James Milton
Miller, William Frederick
Mishelevich, David Jacob

128

Zacharias, David Edward

Howse, Harold Darrow
Hsie, Abraham Wuhsiung
Hubbard, Ann Louise
Hull, Robert Normond
Humphreys, Susie Hunt
Humphries, Asa Alan, Jr
Hunter, Jerry Don
Hurley, Maureen
Hymer, Wesley C
Ignarro, Louis Joseph
Jacobson, Ann Beatrice
Jakob, Karl Michael
Jamieson, James Douglas
Jasper, Donald Kohen
Jenkins, Robert Allan
Jensen, Cynthia G
Jensen, Thomas E
Jeon, Kwang Wu
Johnson, Byron F
Johnson, James Dean
Johnson, Ross Glenn
Johnson, Thomas Raymond
Jones, Gary Edward
Jost, Jean-Pierre
Juergensmeyer, Elizabeth B
Juliano, Rudolph Lawrence
Kalt, Marvin Robert
Kane, Robert Edward
Kaneshiro, Edna Sayomi
Kaplan, Jacob Gordin
Kaplan, Joel Howard
Kapp, Leon Neal
Kasten, Frederick H
Kaye, Jerome Sidney
Keefe, John Richard
Kennedy, James Cecil
Keogh, Richard Neil
Kessler, Dietrich
Kirschner, Marc Wallace
Kischer, Clayton Ward
Kleinfeld, Ruth Grafman
Kleinsmith, Lewis Joel
Klem, Edward Benson
Klevecz, Robert Raymond
Kloetzel, John Arthur
Kluss, Byron Curtis
Knight, Robert Hallowell
Knowles, Barbara B
Kollmorgen, G Mark
Kollros, Jerry John
Korn, Edward David
Kornberg, Thomas B
Koroly, Mary Jo
Koros, Aurelia M Carissimo
Kort, Margaret Alexander
Kraemer, Paul Michael
Krause, Margarida Oliveira
Kreibich, Gert
Kulfinski, Frank Benjamin
Kumar, Ajit
Kuntz, Eloise
Lafontaine, Jean-Gabriel
Laird, Charles David
Lala, Peeyush Kanti
Lampidis, Theodore James
LaVail, Matthew Maurice
Lawford, George Ross
Leak, Lee Virn
Ledbetter, Myron C
Leduc, Elizabeth
Lee, Young Chang
Leffert, Hyam Lerner
Leffingwell, Thomas Pegg, Jr
Leiter, Edward Henry
Leith, John Douglas, Jr
Lemanski, Larry Fredrick
Leppard, Gary Grant
Lerner, Michael Paul
Lesher, Samuel Walter
Levine, Laurence
Levinson, Warren E
Levy, Michael R
Li, Edward Hsien-Chi
Lieberman, Irving
Lillich, Thomas Tyler
Linck, Richard Wayne
Lipkin, George
Lizardi, Paul Modesto
Lockwood, Arthur H
Loizzi, Robert Francis
Long, Cedric William
Longenecker, Bryan Michael
Longo, Frank Joseph
Lorch, Joan
Loud, Alden Vickery
Lu, Benjamin Chi-Ko
Lubin, Martin
Lumsden, Richard
Lung, Ben
Lydon, Carol Guze Konrad
Lyons, Richard Bernard
MacDonald, Eve Lapeyrouse
Mack, James Patrick
Malamed, Sasha
Malamud, Daniel F
Malhotra, Sudarshan Kumar
Manasek, Francis John
Maniatis, George Marinos

Mankovitz, Ralph
Margulis, Lynn
Martin, Billy Joe
Martin, Terence Edwin
Maser, Morton D
Maslow, David E
Massey, Linda Kathleen Locke
Massover, William H
Matheson, Alastair Taylor
Matioli, Gastone
Matsudo, Hitoshi
Mayall, Brian Holden
Mayberry, Lillian Faye
Mayhew, Eric George
McAtee, Lloyd Thomas
McCalla, Dennis Robert
McClure, Michael Edward
McGraw, John Leon, Jr
McIntosh, John Richard
McKinney, Ralph Vincent, Jr
Meetz, Gerald David
Meistrich, Marvin Lawrence
Melera, Peter William
Melnykovych, George
Messmer, Trudy Ottilia
Meszler, Richard M
Metevia, Louis Anthony
Meyer, Diane Hutchins
Meyer, Ralph Roger
Meyers, Kenneth Purcell
Michaels, John Edward
Miller, Kenneth Raymond
Miller, Oscar L, Jr
Millis, Albert Jason Taylor
Mironescu, Stefan Gheorghe Dan
Misch, Donald William
Mitchell, Ann Denman
Moehring, Joan Marquart
Moellmann, Gisela E Bielitz
Moens, Peter B
Montes De Oca, Hector
Monty, Kenneth James
Moore, Bobby Graham
Morgan, Juliet
Moribor, Louis G
Morris, Gerald Patrick
Moses, Montrose James
Mukherjee, Anil B
Mundell, Robert David
Munger, Bryce Leon
Murison, Gerald Leonard
Murphy, Donald G
Nasatir, Maimon
Nations, Claude
Neaves, William Barlow
Nelson, Phillip Gillard
Newstead, James Duncan MacInnes
Nichols, Barbara Ann
Nicklas, Robert Bruce
Nigam, Vijai Nandan
Noe, Bryan Dale
Noonan, Kenneth Daniel
Norin, Allen Joseph
Novales, Ronald Richards
Nyquist, Sally Elizabeth
Oaks, John Adams
O'Brien, John Aloysius
O'Brien, Richard Lee
O'Connell, Kevin Marshall
Odell, Theodore Tellefsen, Jr
Oldfield, Daniel G
Olmsted, Joanna Belle
Olson, Rodney Andreen
Orlic, Donald
Ornstein, Leonard
Orr, Alan R
Orsi, Ernest Vinicio
Outka, Darryll E
Owczarzak, Alfred
Padawer, Jacques
Padykula, Helen Ann
Page, Clayton R, III
Palade, George E
Palevitz, Barry Allan
Panessa, Barbara Jean
Pardue, Mary Lou
Parsons, John Arthur
Partin, John Calvin
Pawlowski, Philip John
Peachey, Lee DeBorde
Pekarthy, James Maurice
Perlman, Philip Stewart
Pfeiffer, Steven Eugene
Phillips, David Mann
Phillips, Stephanie Gordon
Philpott, Charles William
Pickett, Patricia Booth
Pickett-Heaps, Jeremy David
Pitlick, Frances Ann
Plaut, Walter (Sigmund)
Pokorny, Kathryn Stein
Polet, Herman
Pollack, Robert Elliot
Pollard, Thomas Dean
Pollock, Edward G
Poole, Brian Howard
Porter, Keith Roberts
Pourcho, Roberta Grace

Preisler, Harvey D
Prensky, Wolf
Press, Newtol
Price, Paul Jay
Price, Richard Walter
Priest, Robert Eugene
Puszkin, Saul
Rabinovitch, Michel Pinkus
Rae, Peter Murdoch McPhail
Rafferty, Nancy S
Ramsey, James Carroll
Rao, Potu Narasimha
Rasmussen, Howard
Rebhun, Lionel Israel
Reedy, Michael K
Reilly, Richard W
Revel, Jean Paul
Richter, Goetz Wilfried
Riley, Edward Eddy, Jr
Ripley, Robert Clarence
Ris, Hans
Risby, Edward Louis
Ritter, Hope Thomas Martin, Jr
Rizki, Tahir Mirza
Rizzo, Peter Jacob
Robbins, Jay Howard
Roberts, John Fredrick
Robinson, David Mason
Robison, Wilbur Gerald, Jr
Roddy, Martin Thomas
Rodewald, Richard David
Rogers, Steffen Harold
Rogers, Thomas D
Romanovicz, Dwight Keith
Ronzio, Robert A
Rose, Patricia McGovern
Rosen, Steven David
Rosenberg, Murray David
Ross, Ian Kenneth
Ross, Michael H
Ross, Russell
Rotermund, Albert J, Jr
Roth, Thomas Frederic
Rottmann, Warren Leonard
Ruben, Robert Joel
Rubin, Arnold David
Rubin, Harry
Rubin, Walter
Ruddle, Francis Hugh
Runyan, William Scottie
Rutherford, Charles
Ryan, Jon Michael
Ryan, Una Scully
Ryser, Hugues Jean-Paul
Sabatini, David Domingo
Sadava, David Eric
Salmon, Edward Dickinson
Sapp, Walter J
Sarin, Prem S
Sauer, Leonard A
Savage, Howard Edson
Savage, Robert E
Scharff, Matthew Daniel
Schechter, Joel Ernest
Schjeide, Ole Arne
Schleicher, Joseph Bernard
Schmitter, Arline Catherine
Schmitter, Ruth Elizabeth
Schraer, Harald
Schraer, Rosemary
Schreiber, Richard William
Schuel, Herbert
Schuetz, Allen W
Schuit, Kenneth Edward
Schulman, Herbert Michael
Schuster, George Sheah
Schwartz, Arthur Gerald
Schweber, Miriam Schurin
Schwelitz, Faye Dorothy
Segrest, Jere Palmer
Seiden, David
Selman, Kelly
Sercarz, Eli
Sergeant, Tom Paul
Shafiq, Saiyid Ahmad
Sharkey, Margaret Mary
Sharp, William R
Sheffield, Joel Benson
Shelton, Emma
Sheppard, John Richard
Sheridan, Judson Dean
Sheridan, William Francis
Shin, Seung-il
Shires, Thomas Kay
Shivers, Richard Ray
Shnitka, Theodor Khyam
Shodell, Michael J
Sicko-Goad, Linda May
Siegel, Elsie P
Silverman, David J
Simard, Rene
Simpson, Larry P
Simson, Jo Anne V
Sinclair, John Henry
Sinclair, Ronald
Singer, Irwin I
Singer, Robert Mark
Sirbasku, David Andrew

Sisken, Jesse Ernest
Skoultchi, Arthur
Slautterback, David Buell
Smith, David Spencer
Smith, Roy George
Smith-Sonneborn, Joan
Snyder, Judith Armstrong
Socher, Susan Helen
Soeiro, Ruy
Soifer, David
Spear, Patricia Gail
Speer, Henry Lee
Spolsky, Christina Maria
Spooner, Brian Sandford
Staehelin, Lucas Andrew
Stahl, Philip Damien
Stanley, Evan Richard
Stanley, Hugh P
Stein, Gary S
Stein, Gretchen Herpel
Steinberger, Anna S
Stephens, Raymond Edward
Stern, Herbert
Stevens, Ann Rebecca (Larkin)
Stevens, Dean Finley
Steward, Frederick Campion
Stich, Hans F
Stone, Gordon Emory
Strauss, Bernard S
Stromer, Marvin Henry
Stubblefield, Travis Elton
Studzinski, George P
Sun, Alexander Shihkaung
Sun, Nai Chau
Sutherland, Robert Melvin
Sutton, William Wallace
Swanson, Harold Dueker
Swinton, David Charles
Szabo, Arlene Slogoff
Szalay, Jeanne
Szego, Clara Marian
Szollosi, Daniel Gabriel
Talbot, Prudence
Telser, Alvin Gilbert
Teng, Christina Wei-Tien Tu
Teng, Nelson N H
Thomas, Mary Beth
Thomas, Robert Joseph
Thompson, Edward Ivins Bradbridge
Thompson, John Eveleigh
Thompson, Lawrence Hadley
Thurston, Earle Laurence
Till, James Edgar
Tobey, Robert Allen
Togasaki, Robert K
Tolman, Leonard Joseph
Tolnai, Susan
Tomasz, Alexander
Topp, William Carl
Travis, Dorothy Frances
Trelease, Richard Norman
Trump, Benjamin Franklin
Unakar, Nalin J
Urban, Paul
Vail, William Jerald
Van Frank, Richard Mark
Vankin, George Lawrence
Van't Hof, Jack
Venketeswaran, S
Viceps-Madore, Dace I
Vigil, Eugene Leon
Vincent, Monroe Mortimer
Wachsberger, Phyllis Rachelle
Wachtel, Allen W
Wagenaar, Emile B
Wagner, Roger Curtis
Walker, Glenn Kenneth
Walker, Ian Gardner
Walters, David Royal
Wang, Richard J
Warner, Jonathan Robert
Warren, Robert Holmes
Waters, Michael Dee
Watters, Christopher Deffner
Waymouth, Charity
Webber, Mukta Mala
Weber, Michael Joseph
Weigle, William O
Weinreb, Eva Lurie
Weinsieder, Alan
Weinstock, Alfred
Weisenberg, Richard Charles
Weiss, Leonard
Weissmann, Gerald
Weistrop, Jessie Syd
Welch, Robert McClam
Wergin, William Peter
Weston, James A
Widnell, Christopher Courtenay
Wiener, Joseph
Wikswo, Muriel Anastasia
Wille, John Jacob, Jr
Williams, Daniel Charles
Williams, Robert K
Wilson, Barry William
Wilson, Frank Joseph
Wilson, Michael John
Wilson, Onslow Harus

Wise, Benjamin Nathan
Wise, Gary E
Witman, George Bodo, III
Wolff, Stephen Landis
Wolff, David A
Wolsky, Maria de Issekutz
Wolstenholme, David Robert
Wood, William Otto
Wood, Eunice Marjorie
Woodcock, Christopher Leonard Frank
Woods, Philip Sargent
Worgul, Basil Vladimir
Wourms, John Barton
Wrathall, Jean Rew
Wray, Granville Wayne
Wright, Richard Donald
Wu, Roy Shih-Shyong
Wu, James Shung-Jun
Yager, James Donald, Jr
Yang, Shung-Jun
Yoo, Bong Yul
Young, Charity Louise
Youssef, Nabil Naguib
Yu, Grace Wei-Chi Hu
Zalik, Sara E
Zeldin, Michael Hermen
Zickle, Horst Ronald
Zilz, Melvin Leonard
Zimmerman, Arthur Maurice
Zimmerman, William Frederick
Zucker-Franklin, Dorothea
Zull, James E

Cell Physiology
Albach, Richard Allen
Allison, Betzabe M
Alscher, Ruth Paula
Anderson, Neil Owen
Andrus, William DeWitt, Jr
Azarnia, Roobik
Barr, Charles E
Barry, Sue-ning C
Bartell, Cleimer Kay
Barton, Jay, II
Bassett, Charles Andrew Loockerman
Baum, Lawrence Stephen
Bell, Rondal E
Benolken, Robert Marshall
Berryhill, Virginia Farmer
Bianchi, Carmine Paul
Breil, Sandra J
Bronk, John Ramsey
Burns, Victor Will
Cahoon, Mary Odile
Calleau, Relda
Cantor, Marvin H
Cardullo, Maria Ann
Cascarano, Joseph
Chambers, Edward Lucas
Chang, Donald Choy
Cline, Sylvia Good
Cobb, Howell Dee, Jr
Cohen, Natalie Shulman
Conte, Frank Philip
Cook, James Richard
Cook, John Samuel
Cooperstein, Sherwin Jerome
Costello, Leslie Carl
Couillard, Pierre
Coulson, Patricia Bunker
Cox, Dudley
Cushman, Samuel Wright
D'Aoust, Brian Gilbert
Davidson, Samuel James
DeCaro, Thomas F
Dembitzer, Herbert
Donohoo, John T
Downey, Ronald J
Dyson, Robert Duane
Dziak, Rose Mary
Entman, Mark Lawrence
Faust, Robert Gilbert
Fisher, William David
Folinas, Helen
Frank, Martin
Freed, James Melvin
Fritz, Michael E
Ganguly, Pankaj
Glinos, Andre Dimitri
Goldring, Irene P
Goldstein, Stuart Frederick
Goodfellow, Robert David
Griffith, Donal Louis
Guthwin, Hyman
Hackett, Nora Reed
Hall, Iris Beryl Haddon
Harding, Clifford Vincent, Jr
Hay, Robert J
Hilden, Shirley Ann
Honda, Shigeru Irwin
Horowitz, Samuel Boris
Hunter, Jerry Don
Jones, C Robert
Jones, Janice Lorraine
Kalant, Harold
Kassel, Robert Lawrence
Kelly, Robert P
Kempner, Walter

King, Elizabeth Norfleet
King, Gladys Smith
King, Mary Margaret
Klein, Richard Lester
Kleinzeller, Arnost
Kodama, Robert Makoto
Kundje, Fredericka Dodyk
Lang, Dennis Robert
Lanigan, M Regina
La Noue, Kathryn F
Lawrence, Addison Lee
Leach, William Matthew
Le Fevre, Paul Green
Levinson, Charles
Lin, James C H
Lindemann, Charles Benard
Malanga, Carl Joseph
Mawe, Richard C
Mazur, Peter
McCashland, Benjamin William
McManus, Thomas (Joseph)
Melnick, Ronald L
Min, Hong Shik
Mitchell, John Laurin Amos
Moner, John George
Monette, Francis C
Moos, Carl
Murphy, Marjory Beth
Neufeld, Gaylen Jay
Nielsen, Peter James
Noonan, Sharon Mariella
Norman, William Harvey
Ohr, Eleonore A
Olive, John H
Oliver, Janet Mary
Pace, Donald M
Padilla, George M
Paine, Philip Lowell
Paschall, Homer Donald
Perry, Kenneth W
Peterson, Gary Lee
Picciolo, Grace Lee
Pisano, Joseph Carmen
Poisner, Alan Mark
Pribor, Donald B
Price, Steven
Pruett, Patricia Onderdonk
Pryor, Marilyn Ann Zirk
Quarles, Thomas Stephen
Quinton, Paul Marquis
Ragsdale, Nancy Nealy
Reddan, John R
Rinaldi, Robert Arthur
Rodenberg, Sidney Dan
Rohrs, Harold Clark
Rose, Birgit Loewenstein
Rotermund, Albert J, Jr
Rothstein, Aser
Rothstein, Howard
Rourke, Arthur W
Saks, Norman Martin
Sanders, Aaron Perry
Sanford, William Corbin
Sanui, Hisashi
Schaefer, Gideon W
Schirmer, Horst K A
Searcy, Dennis Grant
Seefried, Adolf Von
Severson, Arlen Raynold
Shank, Brenda Mae Buchhold
Sheridan, Judson Dean
Sloboda, Adolph Edward
Smith, Donald Eugene
Smith, Willard Newell
Socolar, Sidney Joseph
Somjyo, Avril Virginia
Spoerl, Edward Schnurr
Stadelmann, Eduard Joseph
Stevens, William Clark
Stonier, Tom Ted
Storey, Bayard Thayer
Strauss, Phyllis R
Swenson, Paul Arthur
Sylvia, Avis Latham
Terry, Robert Lee
Ting-Beall, Hie Ping
Tomlinson, Gus
Tonna, Edgar Anthony
Townes, Mary McLean
Trubowitz, Sidney
Turley, Hugh Patrick
Ulrich, Frank
Van Rossum, George Donald Victor
Waithe, William Irwin
Wallen, Donald George
Wentraub, Robert Louis
Wells, Marion Robert
Whitfield, James F
Wiercinski, Floyd Joseph
Williams, William Thomas
Wilson, David Everett
Winicur, Sandra
Wolfson, Nancy Dolly
Yu, Byung Pal
Zuzolo, Ralph C

Cytochemistry
Alfert, Max
Alpern, Richard Junius
Bauer, Gustav Morton
Bogitsh, Burton Jerome
Cordes, William Charles
DeJong, Donald Warren
Fennell, Richard Adams
Gill, James Edward
Lane, Nancy Jane
Long, Margaret Eleanor
Martin, Billy Joe
Olkowski, Zbigniew L
Patterson, Elizabeth Knight
Rasch, Ellen M
Shannon, Wilburn Allen, Jr
Yemma, John Joseph

Cytogenetics
Aalders, Lewis Eldon
Altenburg, Lewis Conrad
Avers, Charlotte Jo
Bashaw, Elexis Cook
Bell, Sandra Lucille
Bender, Michael A
Boutros, Susan Noblit
Brandom, William Franklin
Braungart, Dale Carl
Bregman, Allyn Aaron
Briton, Donald MacPhail
Brooks, Antone L
Brown, Meta (Louise) Suche
Brown, Spencer Wharton
Buck, Raymond Wilbur, Jr
Burdick, Allan Bernard
Byrd, J Rogers
Carlson, Wayne R
Carr, David Harvey
Carrano, Anthony Vito
Chen, Andrew Tai-Leng
Chen, Tchaw-Ren
Chrisman, Charles Larry
Cohen, Marmon Moses
Cooper, Kenneth Willard
Craddock, Elysse Margaret
Davis, Herbert L, Jr
Dewey, Douglas R
Dollinger, Elwood Johnson
Duck, Bobby Neal
Dvorak, Jan
Eisen, James David
Endrizzi, John Edwin
Erickson, John (Elmer)
Evans, Laurie Edward
Freire-Maia, Dertia Villalba
French, Wilbur Lile
Fu, Wei-ning
Garber, Edward David
Garner, James Gregory
Gersh, Eileen Sutton
Glover, David Val
Goh, Kong-Oo
Grant, William Frederick
Grell, Mary
Gwynn, Edgar Percival
Hall, Marion Trufant
Heineman, Richard Leslie
Hittelman, Walter Nathan
Ho, Keh Ming
Hoff, Victor John
Huang, Chester Chen-Chiu
Hull, John Winter
Humphrey, Donald Glen
Jackson, Raymond Carl
Jalal, Syed M
Jenkins, Burton Charles
Johnson, George Robert
Joneja, Madan Gopal
Kadanka, Zdenek Karel
Kaltsikes, Pantouses John
Kamra, Om Perkash
Kerber, Erich Rudolph
Kidd, Harold J
Kikudome, Gary Yoshinori
Kimber, Gordon
Kohl, John C, Jr
Kreutzer, Richard D
Krim, Mathilde
Kuspira, J
Lam, Shue-Lock
Larson, Ruby Ila
Lesins, Karlis A
Longwell, Arlene Crosby (Mazzone)
Love, Robert Merton
Ma, Nancy Shui Fong
Ma, Te Hsiu
MacDonald, Malcolm Duncan
Macintyre, Malcolm Neil
Maguire, Marjorie Paquette
McFee, Alfred Frank
McFeely, Richard Aubrey
Mehyk, John H
Menzel, Margaret Young
Merz, Timothy
Michel, Kenneth Earl

Mickey, George Henry
Miller, Dorothy Anne Smith
Mohandas, Thuluvancheri
Moore, Charleen Morizot
Moorhead, Paul Sidney
Morrison, Jack William
Mottinger, John P
Mukherjee, Asit B
Murray, Beatrice E
Nasjleti, Carlos Eduardo
Newman, Lester Joseph
Nichols, Warren Wesley
Nur, Uzi
Ohnuki, Yasushi
Okada, Tadashi A
Palmer, Catherine Gardella
Pathak, Sen
Patterson, Rosalyn Mitchell
Pendse, Prapasinha C
Petersen, Kyle W
Petersen-Adkisson, Karen
Phillips, Lyle Llewellyn
Phillips, Ronald Lewis
Prasad, Naresh
Preston, Robert Julian
Rai, Karamjit Singh
Raghunathan, Lalitha
Ramon, Serafin
Rao, Potu Narasimha
Rick, Charles Madeira, Jr
Riley, Herbert Parkes
Rodman, Toby C
Rotar, Peter P
Rothwell, Norman Vincent
Rowley, Janet Davison
Sagawa, Yoneo
St. Amand, Wilbrod
Sarvella, Patricia Ann
Schank, Stanley Cox
Schertz, Keith Francis
Serota, Cornelia Ann Roach
Shaver, Evelyn Louise
Shepard, June Smith
Singh, Dharmdeo Narayan
Soudek, Dusan Edward
Srivastava, Probodh K
Stevenson, Harlan Quinn
Stringam, Gary Rice
Swanson, Carl Pontius
Tai, William
Tantravahi, Ramana V
Tjio, Joe Hin
Triantaphyllou, Anastasios Christos
Vig, Baldev K
Virkki, Niilo
Walters, James Lee
Walters, Marta Sherman
Warren, Richard Joseph
Weber, David Frederick
Weinstein, David
Wellwood, Arnold Augustus
Welter, Dave Allen
Williams, John Watkins, III
Wilson, Arthur Charles
Wise, Dwayne Allison
Wolff, Sheldon
Wurster-Hill, Doris Hadley
Wyandt, Herman Edwin, Jr
Ying, Kuang Lin

Cytopathology
Bibbo, Marluce
Eastwood, Abraham Bagot
Gorthy, Willis Charles
Greenberg, Stanley Donald
Hall, Raymond G, Jr
Kalnins, Zelma A Grinfelds
Liss, Robert H
Roszel, Jeffie Fisher
Schreiber, Hans
Sheehan, John Francis

Cytotaxonomy
Dietz, Robert Austin
Love, Askell
Love, Doris
Moore, Raymond John

Neurocytology
Adinolff, Anthony M
Bennett, Kimberly D
Blanks, Janet Marie (Clarenbach)
Brawer, James Robin
Hattori, Toshiaki
Kaiserman-Abramof, Ita Rebeca
Phelps, Creighton Halstead
Pysh, Joseph John

Plant Cytology
Ahshapanek, Don Colesto
Atkinson, Lenette Rogers
Britten, Edward James
Burson, Byron Lynn
Campbell, Kenneth Wilford
Chen, Chen Ho
Crang, Richard Earl

Dunford, Max Patterson
Fowke, Lawrence Carroll
Hickok, Leslie George
Israel, Herbert William
Kapoor, Brij M
Kasha, Kenneth John
Lorz, Albert Protus
McCollum, Gilbert Dewey, Jr
Morris, Mary Rosalind
Newcomb, Eldon Henry
Paddock, Elton Farnham
Partanen, Carl Richard
Rao, Raghavendra D
Sapra, Val T
Sarkar, Priyabrata
Sears, Ernest Robert
Smith, Benjamin Warfield
Stetler, David Albert
Ting, Yu-Chen
Tsuchiya, Takumi
Underbrink, Alan George
Van Der Woude, William Jan
Warmke, Harry Earl
Whitton, Leslie
Yu, Ming-Hung

DAIRY SCIENCE

Dairy Science

Allaire, Francis Raymond
Arbuckle, Wendell Sherwood
Autrey, Kenneth Maxwell
Barnhart, John Love
Bassette, Richard
Beam, John E
Blosser, Timothy Hobert
Boyd, Earl Neal
Bradley, Robert Lester, Jr
Brown, Murray Allison
Buckalew, John McKinney
Cannon, Robert Young
Cobble, James William
Conlin, Bernard Joseph
Cosgrove, Clifford James
Davis, Herbert Clarence
Egermeier, Edward R
Erb, John Hoffman
Eigen, William M
Flake, John C
Flipse, Robert Joseph
Foreman, Charles Frederick
Fosgate, Olin Tracy
Freeman, Theodore Russell
Frye, Jennings Bryan, Jr
Fryman, Leo Ray
Gilmore, Herbert Clarence
Grill, Herman, Jr
Gwazdauskas, Francis Charles
Harper, Willis James
Harris, Barney, Jr
Harris, Ralph Rogers
Harshbarger, Kenneth E
Hartman, Grant Henry
Head, H Herbert
Heald, Charles William
Helm, Raymond E
Henningson, Robert Walter
Herreid, Ernest Oliver
Hetrick, John Henry
Hoffman, William F
Holter, James Burgess
Huber, John Talmage
Hunsaker, Lloyd R
Irvine, Donald McLean
Jolly, Ramesh C
Jones, Gerald Murray
Keeney, Mark
Kesler, Earl Marshall
King, Willis Alonzo
Kleyn, Dick Henry
Knodt, Cloy Bernard
Kunkel, Reinold Walter
Kurtz, George Wilbur
Lamb, Robert Cardon
Lassiter, Charles Albert
Lazar, James Tarlton, Jr
Leighton, Rudolph Elmo
Lundquist, Norman Stanley
Lusk, John William
Manus, Louis John
Marshall, James Tilden, Jr
Martz, Fredric A
McCarthy, Robert David
McGilliard, Lon Dee
McGilliard, Michael Lon
Moore, Clarence L
Morrill, James Lawrence, Jr
Moss, Buelon Rexford
Muck, George A
Murthy, Gopala Krishna
Natzke, Roger Paul
Nelson, Harvard G

Nilson, Kay Milligan
Norton, Charles Lawrence
O'Dell, Wayne Talmage
Okarma, Theodore Joseph
Olbrich, Steven Emil
Olds, Durward
Oliver-Padilla, Fernando Luis
Olson, Howard H
Ormiston, Emmett Ezekiel
Pearson, Ronald Earl
Porter, Arthur R
Price, Walter Van Vranken
Radloff, Harold David
Ralston, Noel Printiss
Ramsey, Dero Saunders
Reaves, Paul Marvin
Richardson, Don Orland
Rippen, Alvin Leonard
Ross, Richard Henry
Salisbury, Glenn Wade
Schultz, Loris Henry
Sendelbach, Anton G
Shellenberger, Paul Robert
Slatter, Walter LeClare
Smith, Arnold Chauncey
Smith, James W
Smith, Kenneth Larry
Spahr, Sidney Louis
Starkey, Eugene Edward
Staubus, John Reginald
Steele, Robert L
Stewart, Gordon Arnold
Stott, Gerald H
Stull, John Warren
Swanson, Eric Wallace
Thomas, John William
Tobias, Joseph
Trautman, Jack Carl
Trimberger, George William
Tuckey, Stewart Lawrence
Van Horn, Harold H, Jr
Vinson, William Ellis
Wallace, Willie Robert
Ward, George Merrill
Wells, Phillip Richard
Wilcox, Charles Julian
Wilson, Edward Matthew

Dairy Bacteriology

Bennett, Frederick William
Deane, Darrell Dwight
Gibson, Douglas (Lorne)
Lawton, Wallace Clayton
Luedecke, Lloyd O
Morgan, Max Eugene
Olson, Harold Cecil
Overcast, Woodrow Webb
Schultze, Walter Donald
Wagenaar, Raphael Omer
Ward, Paul J
Wilkowske, Howard Hugo

Dairy Chemistry

Cardwell, Joe Thomas
Demott, Bobby Joe
Ernstrom, Carl Anthon
Leeder, Joseph Gordon
Loewenstein, Morrison
Nickerson, Thomas Andrew
Parsons, John G
Potter, Frank Elwood
Richardson, Gary Haight
Sherbon, John Walter
Windlan, Harold Milton

Dairy Husbandry

Adams, Richard Sanford
Anderson, Ralph Robert
Beck, Glenn Hans
Brundage, Arthur Lain
Cairns, Gordon Mann
Campbell, John Roy
Cason, James Lee
Clifton, Carl Moore
Dickey, Howard Chester
Dracy, Arthur E
Edgerly, Charles George Morgan
Fisher, George Robert
Gaunya, William Stephen
Goings, Richard Lewis
Heizer, Edwin Elbert
Herman, Harry August
Hesseltine, Wilbur R
Hodgson, Ralph Edward
Horton, Otis Howard
Howard, W Terry
Hurst, Victor
Hyatt, George, Jr
Kennedy, Wadaran Latamore
Ludwick, Thomas Murrell
Marshall, Sidney Paul
Merilan, Charles Preston
Mitchell, Ralph Gerald
Muller, Lawrence Dean
Niedermeier, Robert Paul
Okarma, Theodore Joseph
Padgitt, Dennis Darrell
Parsons, George E

Plowman, Ronald Dean
Porterfield, Ira Deward
Richards, Clyde Rich
Rollins, Gilbert Horace
Schabinger, John Robert
Schmidt, Glen Henry
Sigrist, Jacob C
Stallcup, Odie Talmadge
Thomas, Roy Orlando
Turk, Kenneth Leroy
Watrous, George H, Jr
Wilcox, Clifford LaVar
Young, Charles Wesley

Dairy Industry

Atherton, Henry Vernon
Boyd, James C
Coulter, Samuel Todd
Emmons, Douglas Byron
Gould, Ira A, (Jr)
Kielsmeier, Elwood William
Reinbold, George W
Seals, Rupert Grant
Stahlman, Clarence L
Webb, Byron Horton
Wilner, Jerome
Winder, William Charles

Dairy Microbiology

Martin, James Harold
Mikolajcik, Emil Michael
Nath, K Rajinder
Smith, Kenneth Leroy
White, Charles Henry
Willits, Richard Ellis

Dairy Nutrition

Baham, Arnold
Britt, Danny Gilbert
Bush, Linville John
Chance, Charles Marion
Goings, Richard Lewis
Hawkins, George Elliott
Hemken, Roger Wayne
Hillman, Donald
Larsen, Howard James
McCullough, Marshall Edward
Owen, Foster Gamble
Polan, Carl E
Rakes, Allen Huff
Ramage, Carroll Herbert
Schingoethe, David John
Sudweeks, Earl Max

DENTISTRY

Dentistry

Albright, John T
Alling, Charles Calvin III
Ambrose, Ernest R
Andrews, James Tucker
Angelopoulos, Angelos Panayotis
Appleby, Ralph Carson
Applebaum, Edmund
Askew, Harold Cochran
Ast, David Bernard
Barber, Thomas King
Baum, Lloyd
Bawden, James Wyatt
Bennett, Carroll G
Bennett, Ian Cecil
Benoit, Peter Wells
Bensinger, David August
Besic, Frank Charles
Bhussry, Baldev Raj
Bickley, Harmon C
Biddington, William Robert
Bixler, David
Bjorndal, Arne Magne
Bogan, Robert L
Bolden, Theodore Edward
Bolender, Charles L
Boucher, Louis Jack
Bowen, Rafael Lee
Bradley, Richard E
Branstad, William
Brewer, Allen A
Brightman, Vernon
Brown, William Ernest
Bruckner, Robert Joseph
Brudevold, Finn
Buck, Douglas L
Buhler, John Embich
Burch, William Paul
Burnett, George Wesley
Burrill, Dan Y
Byrd, David Lamar
Canby, Henry Fawcett
Castaldi, Cosmo Raymond
Cedar, Warren Richard
Chan, Kai Chiu
Charbeneau, Gerald T
Chasens, Abram I
Chinea, Jose Juan
Clark, James William
Clough, Oliver Wendell

Cobb, Charles Madison
Cohen, David Walter
Collings, Charles Kenneth
Coughlin, John W
Coy, Richard Eugene
Crandell, Clifton E
Cunningham, William John
Custer, Frederic
Cutitta, Joseph Anthony
Davis, Carl O
DeRisi, Mary Christine
Dew, William Calland
Dirksen, Thomas Reed
Dogon, Leon I
Douglas, Bruce L
Dowson, John
Dunning, James Morse
Eames, Wilmer B
Elwood, William K
Elzay, Richard Paul
Engel, Milton Baer
Englander, Harold Robert
English, James Andrew
Ennever, John Joseph
Farber, Paul Alan
Faust, Homer Edward
Feagin, Frederick F
Feller, Ralph Paul
Ferencz, Nicholas
Ferrigno, Peter D
Fingar, Walter Wiggs
Finn, Sidney Bernard
Fischer, Theodore E
Fore, Harry Waugh, Jr
Fox, Lewis
Frankel, John Martin
Frechette, Arthur Roy
Garrington, George Everett
George, William Arthur
Gerhard, Rinert J
Gibilisco, Joseph
Gier, Ronald E
Goaz, Paul William
Goepp, Robert August
Going, Robert Ernest
Goldhaber, Paul
Gottsegen, Robert
Gowgiel, Joseph Michael
Graham, William Lee
Grainger, Robert Moore
Grandel, Eugene Robert
Griffiths, Norman Henry Campbell
Grigsby, William Redman
Grodums, Emma Irene
Gron, Poul
Grossman, Louis Irwin
Gupta, Om Prakash
Hammons, Paul Edward
Harris, Melvyn H
Hartsook, Joseph Thurman
Harvold, Egil
Hayden, Jess, Jr
Hayes, Richard Lloyd
Hazen, Stanley P
Hein, John William
Hendershot, Leland Clifford
Hennon, David Kent
Henry, Clay Allen
Henry, Joseph L
Herzberg, Fred
Hine, Maynard Kiplinger
Hoffman, Robert
Huffman, Richard William
Hyman, Milton
Ingle, John Ide
Israel, Harry, III
Jennings, Richard Eugene
Johannessen, Leif Bertram
Johnson, Clinton Charles
Johnson, Robert H
Johnson, Wallace W
Kaellis, Eugene
Kakehashi, Samuel
Kaufman, Edward Godfrey
Keller, Stanley E
Kemmet, Wilfred J
Kendrick, Francis Joseph
Khanna, Shadi Lall
Kinersly, Thorn
Kinzer, Robert Leroy
Klavan, Bennett
Kleinman, Daniel J
Knighton, Holmes Tutt
Koulourides, Theodore I
Krajicek, Dayton Dunbar
Kramer, Gerald M
Kramer, William S
Krol, Arthur J
Kronman, Joseph Henry
Krutchkoff, David James
Kutscher, Austin Harrison
Laffitte, Herbert Bonell
Lambert, Joseph Parker
Langland, Olaf Elmer
Lantz, Harold J
Largent, Max Dale
Law, David Barclay
Leraas, Harold J

Leung, So Wah
Lillie, John Howard
Lindahl, Roy Lawrence
Listgarten, Max
Lobene, Ralph Rufino
Luca, John J
Lund, Melvin Robert
Lussier, Jean Paul
MacKenzie, Richard Stanley
MacRae, Patrick Daniel
Madden, Richard M
Mann, Wallace Vernon, Jr
Manson-Hing, Lincoln Roy
Marble, Howard Bennett, Jr
Martens, Leslie Vernon
Massler, Maury
Maurice, Charles George
Mazur, Boleslaw
McCarthy, Frank Martin
McDonald, Ralph Earl
McElroy, Donald L
McKnight, James Pope
McLean, James Douglas
McPhail, Charles Herbert
Meckel, Alfred Hans
Medina, Jose Enrique
Meyer, Maurice Wesley
Mezl, Zdenek
Millard, Herbert Dean
Miller, Clifford H
Mink, John R
Mitchell, David Farrar
Mohammed, Clive Inram
Moller, Palmi
Moore, Alton Wallace
Morris, Alvin Leonard
Morris, Melvin Lewis
Morrison, Richard Donald
Morrison, Kenneth N
Moulton, George Herbert
Myall, Robert William T
Naidorf, Irving Joseph
Neilson, John Warrington
Nikiforuk, Gordon
Niswander, Jerry David
Norman, Richard David
Nuckles, Douglas Boyd
Oaks, J Howard
Ogden, Ingram Wesley
Ohlenbusch, Robert Eugene
Oliet, Seymour
Olsen, Edmund Severn, Jr
Olson, Donald Lee
Olson, John Victor
Ortman, Harold R
Overberger, James Edwin
Paffenbarger, George Corbly
Parfitt, Gilbert J
Parker, Leroy A, Jr
Patterson, Samuel S
Pavone, Ben W
Pearlman, Sholom
Perman, Dorothy
Person, Philip
Peterson, Shailer Alvarey
Peyton, Floyd Avery
Ping, Ronald Stanley
Popovich, Frank
Porter, Chastain Kendall
Powell, Richard Anthony
Quartararo, Ignatius Nicholas
Ramfjord, Sigurd
Rapuano, Joseph A
Rayson, Jack Henry
Reed, Homer Vernon
Reeves, Robert Lloyd
Reynolds, Nancy Miller
Riedel, Richard Anthony
Ringsdorf, Warren Marshall, Jr
Rinne, Vernon Wilmer
Rivetti, Henry Conrad
Rizzo, Anthony Augustine
Robinson, John E, Jr
Robinson, Leonard H
Roistacher, Seymour Lester
Rosenstein, Solomon Nathan
Rossi, Edward P
Roth, Genevieve D
Rovelstad, Gordon Henry
Roydhouse, Richard Heim
Ryge, Gunnar
Sabiston, Charles Barker, Jr
Sanfilippo, Francis Anthony
Sassouni, Viken
Savara, Bhim Sen
Saxe, Stanley Richard
Scapino, Robert Peter
Scheman, Paul
Schilder, Herbert
Schmitt, Kenneth Frederick
Schoen, William P
Schuchard, Alfred
Schuessler, Carlos Francis
Schweiger, James W
Segal, Alan H

Seiger, William George
Selzer, Samuel
Shankle, Robert Jack
Shapiro, Stewart
Sharry, John Joseph
Shiere, Frederic Roland
Silha, Robert Emmett
Silva, George Douglas
Singh, Inder Jit
Sinkford, Jeanne C
Slavin, Ovid
Smith, Charles T
Smith, Roy Martin
Smudski, James W
Sognnaes, Reidar Fauske
Soper, Edward Henry
Sorenson, Fred M
Spedding, Robert H
Speed, Edwin Maurice
Sperber, Geoffrey Hilliard
Staple, Peter Hugh
Stark, Marvin Michael
Starkey, Paul Edward
Steffek, Anthony J
Steinman, Henry Robert
Steinman, Ralph R
Stewart, Arthur Van
Stewart, Jack Lauren
Stibbs, Gerald Denike
Stiff, Robert H
Stookey, George K
Stout, Walter Clay
Strain, Jerome Chamberlain
Striffler, David Frank
Sullivan, Russell William
Sunnich, Russell William
Swartz, Walter H
Swenson, Henry Maurice
Swerdlow, Herbert
Swords, Ruth Riley
Sykora, Oscar P
Tanzer, Jason Michael
Ten Cate, Arnold Richard
Terkla, Louis Gabriel
Teuscher, George William
Thayer, Keith Evans
Thomas, Norman Randall
Tingey, Ward M
Tobin, Daniel F
Tocchini, John Joseph
Tomich, Charles Edward
Trott, John Richard
Tuckson, Coleman Reed, Jr
Turesky, Samuel Saul
Urbanek, Vincent Edward
Van Hassel, Henry John
Vasileff, Vasil
Vincent, Gordon Ross
Vinton, Paul Wesley
Volker, Joseph Francis
Waggener, Donald Todd
Waldren, Alfred Carson, Jr
Warren, Donald W
Webber, Karl Keen
Weber, Vinson M
Webster, William Philip
Weiss, Marvin B
Wesley, Robert Cook
Wessels, Kenneth Edwin
Westmoreland, Winfred William
Winkler, Sheldon
Wisotzky, Joel
Witkin, George Joseph
Wittemann, Joseph Klaus
Woelfel, Julian Bradford
Wolcott, Robert B
Wong, Walter Mun-Fay
Woodward, James D
Woolley, LeGrand H
Worman, Stephen
Wright, Wellesley Horton
Wuehrmann, Arthur H
Wycoff, Samuel John
Wykhuis, Walter Arnold
Yen, Peter Kai Jen
Yoder, John L
Young, John Kiger
Young, Wesley O
Yurkstas, A Albert
Zamikoff, Irving Ira
Zander, Helmut A
Zegarelli, Edward Victor
Zinner, Doran David
Zumbrunnen, Charles Edward

Dental Anthropology
Dahlberg, Albert A
Greene, David Lee
Mayhall, John Tarkington
Turner, Christy Gentry, II

Dental Epidemiology
Burt, Brian Aubrey
Donnelly, Charles Joseph
Duany, Luis F, Jr
Lopez, Hady
Meskin, Lawrence Henry
Russell, Albert Lee

Dental Hygiene
Malvitz, Dolores Marie
Ragland, Ruth Hines
Swords, Ruth Riley

Dental Materials
Anthony, David Henry
Cassel, James Martin
Cheung, Peter Pak Lun
Dickson, George
Forbes, James Franklin
Greenberg, Orrin
Hudson, Donald Charles
Huget, Eugene F
Johnson, Leonard N
Korostoff, Edward
Langeland, Kaare
Mabie, Curtis Parsons
Mohammed, M Hamdi A
Moon, Peter Clayton
O'Brien, William Joseph
Phillips, Ralph W
Powers, John Michael
Rodriguez, Mario Santos
Rootare, Hillar Muldar
Smith, Dennis Clifford
Sandrik, James Leslie
Tateosian, Louis Hagop
Taylor, Duane Francis
Tillitson, Edward Walter

Dental Pathology
Kreshover, Seymour J
Wheatcroft, Merrill Gordon
White, Dean Kincaid

Dental Radiology
Buchholz, Robert E
Gibbs, Samuel Julian
Mourshed, Farouk Ali
Richards, Albert Gustav

Dental Research
Alam, Syed Qamar
Bahn, Arthur Nathaniel
Berndt, Alan Fredric
Brauer, Gerhard Max
Burdi, Alphonse R
Calhoun, Noah Robert
Carlson, David Sten
Catalanotto, Frank Alfred
Chow, Laurence Chung-Lung
Cooley, William Edward
Craig, Robert George
DePaola, Dominick Philip
Farah, Jean William
Gartner, Leslie Paul
Gibbs, Charles Howard
Gilmour, Marion Nyholm H
Goggins, John Francis
Gold, William
Goldberg, Louis J
Goodson, Jo Max
Hobart, Donald James
Hoffman, Jerry Irwin
Holmstedt, Jan Olle Valter
Hylander, William Leroy
Irving, James Tutin
Johansen, Erling
Jurecic, Anton
Kinersly, Thom
Kusy, Robert Peter
Larson, Rachel Harris (Mrs John Watson Henry)
Levenson, Gordon Edward
Loebenstein, William Vaille
Lyon, Harvey William
Manly, Marian LeFevre
McConnell, Duncan
Messer, Harold Henry
Moffett, Benjamin Charles, Jr
Moss-Salentijn, Letty
Nylen, Marie Ussing
Osborne, John William
Patel, Pratull Raojibhai
Pianotti, Roland Salvatore
Pitts, Robert Gary
Potter, Rosario H Yap
Rawls, Henry Ralph
Reussner, George Henry
Ritchey, Thomas William
Rupp, Nelson Woodward
Schachtele, Charles Francis
Seibel, Werner
Tesk, John A
Williams, David Lloyd
Wisotzky, Joel
White, Edward

Endodontics
Auerbach, Morris Baline
Clem, William Henry
Gurney, Benjamin Franklin
Guttuso, James
James, Garth A
LoMonaco, Carmine Joseph
Shovlin, Francis Edward
Siskin, Milton
West, Robert Cooper

Periodontics
Berdon, John Kenneth
Bishara, Samir Edward
Catalanotto, Frank Alfred
Cinotti, William Ralph
Collord, James
Corcoran, John W
Deasy, Michael Joseph
Fultz, R Paul
Hennon, David Kent
Hill, Clement Joseph
Morawa, Arnold Peter
Nowak, Arthur John
Oldenburg, Theodore Richard
Parkins, Frederick Milton
Poole, Andrew E
Rapp, Robert
Starkey, Paul Edward
Stewart, Ray Edward
Sullivan, Robert Emmett
Taylor, Paul Peak
Till, Michael John
Tracy, William E
Venham, Larry Lee
Wei, Stephen Hon Yin
Wells, Jack E

History of Dentistry
Orland, Frank J

Orthodontics
Ackerman, James L
Aduss, Howard
Behrents, Rolf Gordon
Biggerstaff, Robert Huggins
Buchin, Irving D
Burstone, Charles Justin
Christiansen, Richard Louis
Davidson, William Martin
Deuschle, Frederick Marion
Di Paolo, Rocco John
Di Salvo, Nicholas Armand
Dougherty, Harry L
Furstman, Lawrence L
Gainsforth, Burdette Livingston
Garner, LaForrest D
Gianelly, Anthony Alfred
Graber, Lee Winn
Graber, Touro Mor
Green, Larry J
Grewe, John Mitchell
Grossman, Richard C
Harris, James Edward
Haryett, Rowland D
Higley, Lester Bodine
Hoffman, Robert
Horowitz, Sidney Lester
Hunter, William Stuart
Isaacson, Robert John
Jacobs, Richard M
Lear, Clement S C
Marshall, Kenneth Chenery
Mathews, J Rodney
McNamara, James Alyn, Jr
Merow, William Wayne
Moorrees, Coenraad Frans August
Moyers, Robert Edison
Nanda, Ravindra
Nelson, Robert Mellinger
Nikolai, Robert Joseph
Olin, William (Harold)
Profit, William R
Pruzansky, Samuel
Quinn, Galen Warren
Raeder, Arthur O
Renfroe, Earl Wiley
Richardson, Elisha Roscoe
Rosenstein, Sheldon William
Sassouni, Viken
Smiley, Gary Ray
Speidel, Thomas Michael
Staley, Robert Newton
Stemm, Robert Marvin
Storey, Arthur Thomas
Via, William Fredrick, Jr
Weber, Faustin N
Weinstein, Sam
Williams, Benjamin Hayden
Woodside, Donald G
Zingeser, Maurice Roy
Zwemer, Thomas J

Pedodontics

Periodontics
Alfano, Michael Charles
Allen, Don Lee
Ash, Major McKinley, Jr
Baer, Paul Nathan
Bensinger, David August
Blanton, Patricia Louise
Bradley, Richard E
Burman, Louis Robert
Chasens, Abram I
Ciancio, Sebastian Gene
Cobb, Charles Madison
Conroy, Charles William

Spillett, James Juan
Stewart, Robin Kenny
Tanner, James Taylor
Terman, Charles Richard
Tomich, Prosper Quentin
Verner, Jared
Weller, Milton Webster
Wiens, John Anthony
Williams, Eliot Churchill

Aquatic Ecology

Alexander, Vera
Anderson, Norman Herbert
Anderson, Richard Orr
Arnold, Dean Edward
Bardach, John E.
Barko, John William
Barnes, James Ray
Barnes, Stephen Noble
Bellis, Vincent J., Jr
Blinn, Dean Ward
Bouck, Gerald R
Bowen, Stephen H
Bozniak, Eugene George
Buchanan, Ronald James
Burton, Thomas Maxie
Cherry, Donald Stephen
Chisholm, Sallie Watson
Clark, William Jesse
Cole, Richard Allen
Collins, Gary Brent
Confer, John L.
Cooke, George Dennis
Dale, Hugh Monro
Davies, Robert Milton
Davis, Graham Johnson
Dean, John Mark
Del Fosse, Ernest Sheridan
Dillard, Gary Eugene
Durham, Leonard
Eckblad, James Wilbur
Erman, Don Coutre
Fraleigh, Peter Charles
Gale, William F.
Goldstein, Robert Martin
Gray, Robert H
Haase, Bruce Lee
Haertel, Lois Steben
Haines, Terry Alan
Hansmann, Eugene William
Hartland-Rowe, Richard C B
Helm, William Thomas
Hornuff, Lothar Edward, Jr
Howell, Henry Howze
Hummon, William Dale
Hurst, Elaine H
Kelly, Mahlon George, Jr
Knight, Allen Warner
Lane, Edwin David
Lanza, Guy Robert
LaRow, Edward J
Lauer, Gerald J
Likens, Gene Elden
Linton, Kenneth Jack
Long, Charles James
Low, Edward B
Main, Stephen Paul
Manny, Bruce Andrew
Mayer, Foster Lee, Jr
McIntire, Charles David
Mercando, Neil Aldo
Minshall, Gerry Wayne
Mozley, Samuel Clifford
Murvosh, Chad M
Naiman, Robert Joseph
Nebeker, Alan V
Neuhold, John Mathew
Normandeau, Donald Arthur
O'Brien, William John
Otto, Robert George
Prins, Rudolph
Prophet, Carl Wright
Quertermus, Carl John, Jr
Rachlin, Joseph Wolfe
Reid, George Kell
Rennie, Thomas Howard
Richman, Sumner
Ringler, Neil Harrison
Robertson, Andrew
Rosenberg, David Michael
Sanger, Jon Edward
Schaffner, William Robert
Seaman, Elwood Armstrong
Shan, Robert Kuo-Cheng
Shuster, Carl Nathaniel, Jr
Sissom, Stanley Lewis
Spencer, Larry T
Sprules, William Gary
Stanton, George Edwin
Stasko, Aivars B
Stewart, Kenneth Wilson
Stober, Quentin Jerome
Storr, John Frederick
Thornton, Kent W
Triska, Frank John
Vanderploeg, Henry Alfred
Vannote, Robin L
Walker, William Howard
Waters, Thomas Frank
White, David Arnold
Whiteside, Melbourne C
Wiltshire, Charles Thomas
Wiser, Cyrus Wymer
Woodall, William Robert, Jr
Yongue, William Henry

Conservation

Clark, Wilson Farnsworth
Clough, Garrett Conde
Coolidge, Harold Jefferson
Davis, William B
Disinger, John Franklin
Ehrenfeld, David W
Gerard, Cleveland Joseph
Gottschalk, John Simison
Hamilton, Lawrence Stanley
Hoffmaster, Donald Edeburn
Hubbard, John Patrick
Hunt, Kenneth Whitten
Joranson, Philip Nathaniel
Kale, Herbert William II
Nickum, John Gerald
Niemi, Alfred Otto
Packard, Fred Mallery
Pierson, David W
Rateaver, Bargyla
Salter, Paul Sanford
Sauer, Pauline Louise
Schofield, Edmund Acton, Jr
Snyder, Harold
Spivak, Monroe Leon
Stapp, William B
Taylor, Merrel Arthur
Yambert, Paul Abt
Zwerman, Paul Joseph

Desert Ecology

Crosswhite, Carol D
Muna, Martin Hammond
Sheps, Lillian

Environmental Chemistry

Afghan, Baderuddin Khan
Alter, Harvey
Ambrose, Robert T
Andelman, Julian Barry
Baarda, David Gene
Bailey, Rodney Albert
Ballantine, David Stephen
Barthauer, Gerald Lee
Baumgarten, Ronald J
Beran, Jo Allan
Beug, Michael William
Bidleman, Terry Frank
Block, Arthur McBride
Bromels, Duane Marshall
Brown, George Willard, Jr
Brown, Robert Melbourne
Burrows, William Dickinson
Buyers, Archie Girard
Carbone, Gabriel
Carrigan, Richard Alfred
Carter, Melvin K
Caruso, Sebastian Charles
Churchill, Ralph John
Clark, Malcolm John Roy
Coleman, Curtis Burger
Cormack, James Frederick
Crosby, Donald Gibson
Day, James Meikle
Dietrich, Martin Walter
Doig, Marion Tilton, III
Douglas, Irwin Bruce
Dunkle, Michael Patrick
Dziecuch, Matthew Andrew
Eaton, James Edmonds
Edelson, David
Edwards, Raymond Richard
Erickson, Ronald E
Estes, Frances Lorraine
Fearheller, William Russell, Jr
Fisher, Gerald Lionel
Foerster, Donald Ray
Fogleman, Wavell Wainwright
Fries, James Andrew
Frystak, Ronald Wayne
Galloway, James Neville
Garrison, Arthur Wayne
Gehri, Dennis Clark
Glaze, William H
Goffman, Martin
Gordon, Robert Julian
Gormly, James Ronald
Graefe, Allen Frederick
Groth, Richard Henry
Grunder, Fred Irwin
Hall, Stephen Kenneth
Hamersma, J Warren
Hannah, Sidney Allison
Haque, Rizwanul
Hecht, Stephen Samuel
Henry, William Melinger
Hoover, Thomas Burdett
Hopton, Frederick James
Horne, Ralph Albert
Horton, Joseph William
Hughes, Robert Alan
Hunt, Charles Maxwell
Hutzinger, Otto
Johar, Joginder Singh
Jolley, Robert Louis
Keenan, Robert Gregory
Kemper, William Alexander
Kennedy, Edwin Russell
Kilner, Scott Burgoyne
Kland, Mathilde June
Klein, David
Knapp, Kenneth T
Kneip, Theodore Joseph
Kopfler, Frederick Charles
Kukin, Ira
Kusserow, Gerhard William
Lande, Sheldon Sidney
Langner, Ralph Rolland
Langworthy, William Clayton
Lapp, Thomas William
Lee, Richard Fayao
Lockhart, Haines Boots, Jr
Marano, Frederick Francis
Martens, William Stephen
McAdie, Henry George
McClelland, Nina Irene
McCoubrey, James A
McKinney, James David
Meier, Eugene Paul
Menczel, Jehuda H
Meresz, Otto
Michalski, Raymond J
Micko, Michael M
Milham, Robert Carr
Morley, Harold Victor
Nielsen, Julian Moyes
Oliver, Barry Gordon
O'Shea, Timothy Allan
Otto, Klaus
Parry, Edward Peterson
Pierce, Richard Harry, Jr
Potrafke, Earl Mark
Purcell, Thomas Charles
Ragaini, Richard Charles
Ramamoorthy, Subramaniam
Reeves, Richard Franklin
Reichel, William Louis
Rigler, Neil Edward
Russell, Thomas Webb
Saeger, Victor William
Scaringelli, Frank Philip
Schermer, Eugene DeWayne
Schmauch, George Edward
Sears, Donald Richard
Seyb, Leslie Philip
Sheets, Ralph Waldo
Shendrikar, Arun D
Shuman, Mark S
Simons, Edward Louis
Smith, Ivan C
Somers, Joseph Henry
Stepenuck, Stephen Joseph, Jr
Sterrett, Frances Susan
Stoker, Howard Stephen
Stout, Virginia Falk
Suffet, Irwin Henry
Swisher, Robert Donald
Taylor, Charlotte Clarke
Tisue, George Thomas
Trieff, Norman Martin
Van Landingham, John W
Vind, Harold Pennington
Wahlgren, Morris A
Ward, Harold Roy
Weaver, Ervin Eugene
Weiss, Fred Toby
Weschler, Mary Charles
Wilding, Raymond Earl
Williams, David Trevor
Williams, Ronald Lloyde
Winschonke, Charles Richard
Wolkoff, Aaron Wilfred
Yang, John Yun-Wen
Young, David Matheson
Zaborowski, Leon Michael

Environmental Geology

Anderson, Howard T
Arnold, Ralph Gunther
Baker, Victor Richard
Baker-Blocker, Anita Linda
Bartlett, Grant Aulden
Beaulieu, John David
Buchwald, Caryl Edward
Bush, Alfred Lerner
Bushnell, Kent O
Campbell, Ian
Clark, Howard Charles, Jr
Cleaves, Emery Taylor
Colton, Roger Burnham
Costa, John Emil
Crittenden, Max Dermont, Jr
Deal, Dwight Edward
Dochring, Donald O
Eastler, Thomas Edward
El-Ashry, Mohamed T
Emrich, Grover Harry
Fakundiny, Robert Harry
Fulmer, Charles V
Fyles, John Gladstone
Geyer, Alan Raymond
Ghuman, Gian Singh
Goebel, Edwin DeWayne
Grant, Douglas Roderick
Groff, Donald William
Harrison, Wyman
Hill, Patrick Arthur
Hollenbaugh, Kenneth Malcolm
Howd, Frank Hawver
Janke, Norman C
Johnson, Ross Byron
Knutson, Carroll Field
Kunkle, George Robert
Lessing, Peter
Lundgren, Lawrence William, Jr
Martinez, Joseph Didier
McClellan, William Alan
Miranda, Henry A, Jr
Mirsky, Arthur
Montagne, John M
Nriagu, Jerome Okonkwo
Paulson, Richard Rudolph
Pestrong, Raymond
Pipkin, Bernard Wallace
Rae, George Ramsay
Ruedisili, Lon Chester
Sage, Orrin Grant, Jr
Schleicher, David Lawrence
Temple, Kenneth Loren
Upchurch, Sam Bayliss
Wait, James Richard
Wentworth, Carl M, Jr
Williams, Julian
Winkler, Erhard Mario
Workman, William Edward

Environmental Health

Anderson, David Martin
Barabas, Silvio
Barton, Charles Julian, Sr
Beard, Rodney Rau
Beck, William J
Blackwell, Floyd Oris
Bond, Richard Guy
Brenniman, Gary Russell
Brown, Harold Victor
Campbell, Kirby I
Carnow, Bertram Warren
Chan, Leland
Cheever, Charles Lyle
Cody, Terence Edward
Cole, Jerome F
Coluci, Anthony Vito
Conar, Cyril Lewis
Dahncke, Barton Eugene
Deininger, Rolf A
de Serres, Frederick Joseph
Dreisbach, Robert Hastings
Ehrlich, Richard
Eisenbud, Merril
Elia, Victor John
Emmett, Edward Anthony
Falk, Hans Ludwig
Feasley, Charles Frederick
Ferris, Benjamin Greeley, Jr
Flowers, Earl Shederick
Freudenthal, Peter
Frigerio, Norman Alfred
Garner, Reuben John
Gori, Gio Batta
Gray, Robert Howard
Hake, Carl (Louis)
Hemphill, Delbert Dean
Hickman, John Roy
Hilbert, Morton S
Horstman, Sanford W
Humphrey, Harold Edward Burton, Jr
Iglar, Albert Francis, Jr
Jacobson, Alvin Raymond
Johnson, Jerry Michael
Judd, Stanley H
Kitzke, Eugene David
LeMunyan, Cobert Duane
Levin, Gilbert Victor
Lippmann, Morton
Long, Keith Royce
Lovett, Joseph
Lower, William Russell
Lucis, Ojars Janis
Mallison, George Franklin
Malone, Winfred Francis
Mattheis, Eula Bingham
McNerney, James Murtha
Meyer, Alvin F., Jr
Morgan, Monroe Talton, Sr
Morton, John Dudley
Nicholson, William Jamieson
Niemeier, Richard William
Parker, Robert Davis Rickard
Patel, Dhun Burjor

Paulus, Harold John
Peters, Howard August
Pier, Stanley Morton
Pierce, James Otto, II
Pogrund, Edward Seymour
Radford, Edward Parish, Jr
Roslinski, Lawrence Michael
Saltzman, Bernard Edwin
Sax, Newton Irving
Schaub, Stephen Alexander
Schiager, Keith Jerome
Severs, Richard Keith
Shiffman, Morris A
Singh, Harpal P
Spector, Bertram
Strehlow, Clifford David
Swain, Wayland Roger
Tiller, Richard Edward
Van Pelt, Wesley Richard
Vesley, Donald
Wagner, Vaughn Edwin
Wayne, Lowell Grant
Wilkening, George Martin
Wilson, Robert Hallowell
Wolf, Harold William
Wolff, Arthur Harold
Wrenn, McDonald Edward
Young, Keith Preston
Yu, Ming-Ho

Environmental Management

Adamek, Eduard Georg
Adrounie, V Harry
Allen, Gordon Ainslie
Arnold, Joseph Frederick
Atkisson, Arthur Albert
Baumann, Duane Dennis
Becker, Edward Brooks
Behrend, Donald Fraser
Bendix, Selina (Weinbaum)
Bennett, James Marvin
Biswas, Asit Kumar
Bowman, Mary Lynne
Bowman, Wallace Deal
Brathovde, James Robert
Bregman, Jacob Israel
Brewer, William Augustus
Brower, Frank M
Buscemi, Philip Augustus
Campanella, Paul Joseph, II
Chessin, Meyer
Clapham, Wentworth B, Jr
Cole, Richard
Compton, Eli Dee
Copeland, Otis Lee
Cox, Doak Carey
Craig, Roy Phillip
Deju, Raul A
DeSanto, Robert Spilka
Efford, Ian Ecott
Eigner, Joseph
French, William Edwin
Frey, John H
Gilbert, Robert Arthur
Gold, Richard Frank
Goodland, Robert James
Graham, Paul Roger
Greenfield, Stanley Marshall
Hahne, Rolf Mathieu August
Hall, Ross Hume
Hennigan, Robert Dwyer
Holt, Vernon Emerson
Horowitz, Joel Lawrence
Jenkins, Robert Ellsworth, Jr
Jorgensen, Erik
Kalt, Melvyn Barry
Kasperson, Roger Eugene
Kazmaier, Harold Eugene
Lane, Robert K
Lansinger, John Marcus
Lasday, Albert Henry
Lau, Norman Eugene
Lee, Kai Nien
Leeson, Bruce Frank
MacArthur, Donald M
Maier, Robert Hawthorne
Mayer, Julian Richard
McCormick, Jack Sovern
McKirahan, Richard Duncan
McLellan, Alden, IV
Medz, Robert B
Michaud, Howard H
Morgan, Millett Granger
Morhardt, Sylvia Staehle
Mosser, John Snavely
Mount, Wayne Delano
Narten, Perry Foote
Parsons, David Jerome
Picker, Robert
Robertson, William, IV
Rockwell, Julius, Jr
Sarakwash, Michael
Schaeffer, David Joseph
Scoville, John J
Seitz, Wesley Donald
Shapiro, Jewel Templeton
Stanford, Geoffrey Brian
Stewart, Donald Borden

Tester, Cecil Fred
Thompson, James Arthur
Thompson, Milton Avery
Vetrano, James Bond
Ware, Stanton James
Warkentin, Benno Peter
Watterston, Kenneth Gordon
Wayman, Cooper H
Weyl, Peter K
Wilcox, Howard Albert
Wilkinson, Ralph Russell

Environmental Medicine

Albert, Roy Ernest
Bromberger-Barnea, Baruch (Berthold)
Dinman, Bertram David
Eisenberg, Rita B
Ferin, Juraj
Frazier, John Melvin
Giel, Bohdan Gielecinski
Goldstein, David
Gunnison, Albert Farrington
Hamilton, Robert William, Jr
Heimann, Harry
Holdren, John Paul
Hubbard, Roger W
Jones, LeeRoy George
Nelson, Norton
Orris, Leo
Proctor, Donald Frederick
Sagan, Leonard A
Selikoff, Irving John
Swift, David Leslie
Wilson, John Thomas, Jr
Zaki, Mahfouz H

Environmental Physics

Barnett, Albert Gerald
Breed, Benny Ray
Brown, George Raymond
Chernosky, Edwin Jasper
Chopra, Kuldip P
Constant, Frank Woodbridge
Crawford, George Wolf
de Latour, Christopher
English, Bruce Vaughan
Fricke, Werner
Garrell, Martin Henry
Herndon, Roy C
Hyder, Charles Latif
John, Walter
Kester, William Lee
Ku, Peh Sun
Loos, Hendricus G
Martin, John H
Matta, Joseph Edward
Milford, Sidney Nevil
Nelson, John William
Olson, Willard Paul
Rae, Stephen
Rosen, Leonard Craig
Sautter, Chester A
Seaman, Gregory G
Soberman, Robert K
Swissler, Thomas James
Sydor, Michael
Warren, Mashuri Laird
Watson, Robert Dale

Environmental Physiology

Aleksiuk, Michael
Alliston, Charles Walter
Besch, Emerson Louis
Blume, Frederick Duane
Buikema, Arthur L, Jr
Burlington, Roy Frederick
Carlson, Gerald Eugene
Chute, Robert Maurice
Coburn, Corbett Benjamin, Jr
Curtis, Stanley Evan
Dieter, Michael Phillip
Duman, John Girard
Ekberg, Donald Roy
Feldmeth, Carl Robert
Ferguson, James Homer
Gatten, Robert Edward, Jr
Gelderloos, Orin Glenn
Hadley, Neil F
Hall, Arthur Lee
Hayward, John S
Henshaw, Robert Eugene
Heroux, Olivier Joseph Paul
Hertig, Bruce Allerton
Hixson, Floyd Marcus
Holmes, Kenneth Robert
Hoptman, Julian
Houston, Arthur Hillier
Hurst, Robert Nelson
Johnson, Donald William
Joyce, Elaine C Elder
Kelly, Herbert Barrett, Jr
Kilgore, Delbert Lyle, Jr
Kreider, Marlin Books
Krizek, Donald Thomas
Leon, Henry A
Lustick, Sheldon Irving
Mackay, William Charles
Miller, Lyster Keith

Minnich, John Edwin
Morrison, Peter Reed
Morse, John Thomas
Moss, Sanford Alexander, III
Mullins, Jeanette Somerville
Nagy, Kenneth Alex
Newsom, Bernard Dean
Paim, Uno
Pandolf, Kent Barry
Panuska, Joseph Allan
Parsons, Lawrence Reed
Payne, Jeremiah Frederick
Percy, Jonathan Arthur
Pereira, Martin Rodrigues
Phillips, Richard Dean
Platner, Wesley Stanley
Pough, Frederick Harvey
Puglia, Charles David
Ramsey, James Marvin
Reid, Donald House
Richards, Oscar White
Ridgway, Sam H
Robinson, Sumner M
Segal, Earl
Sellers, Cletus Miller, Jr
Shields, Jimmie Lee
Smiles, Kenneth Albert
Stones, Robert C
Sudia, Theodore William
Taylor, Charles Richard
Tibbitts, Theodore William
Vernberg, Winona B
Weathers, Wesley Wayne
Weinstein, Leonard Harlan
Whitten, Bertwell Kneeland
Wood, Jack Sheehan
Woodring, Jay Porter
Yousef, Mohamed Khalil

Forest Ecology

Arno, Stephen Francis
Art, Henry Warren
Bakuzis, Egolfs Voldemars
Barney, Charles Wesley
Berglund, John Verne
Boyer, William Davis
Brown, Bruce Antone
Brown, James Harold
Brown, James Henry, Jr
Brush, Grace Somers
Burger, Dionys
Callaham, Robert Zina
Carvell, Kenneth Llewellyn
Challinor, David
Cobbe, Thomas James
Coffman, Michael S
Crankshaw, William Bliss
Crow, Alonzo Bigler
Day, Gordon Malcolm
DeBrunner, Louis Earl
Edgren, James W
Eriksen, Clyde Hedman
Ferrill, Mitchell
Garrison, George Alfred
Grandtner, Miroslav Marian
Grigal, David Francis
Griffin, Ralph Hawkins
Harris, Robert Wilson
Hebb, Edwin Atkins
Heinselman, Miron L
Hodgkins, Earl Joseph
Holt, Harvey Allen
Hutnik, Russell James
Jayne, Benjamin A
Johnson, Frederic Duane
Jurdant, Michel Louis
Kaufman, Clemens Marcus
Ketchledge, Edwin Herbert
Kilgore, Bruce Moody
Kimmins, James Peter
Long, Alan Jack
Lorio, Peter Leonce Jr
Marquis, David Alan
Meeuwig, Richard O'Bannon
Miller, Neil Austin
Moehring, David Marion
Newton, Michael
Plass, William T
Reid, Charles Phillip Patrick
Romancier, Robert Marshall
Schneider, Gerhardt
Shipman, Robert Dean
Silker, Theodore Henry
Smith, David Martyn
Smith, Ernest Chalmers
Stout, Benjamin Boreman
Strand, Robert Fenton
Strang, Robert M
Switzer, George Lester
Tanaka, Yasuomi
Tappeiner, John Cummings, II
Teate, James Lamar
Thornburgh, Dale A
Tierson, William Cornelius
Van Wagtendonk, Jan Willem
Weaver, George Thomas
Winget, Carl Henry

Winjum, Jack Keith
Woods, Frank Wilson

Fresh Water Ecology

Bradshaw, Aubrey Swift
Chesnut, Thomas Lloyd
Cushing, Colbert Ellis, Jr
Danforth, William (Frank)
DePoe, Charles Edward
Diggins, Maureen Rita
Fitzgerald, George Patrick
Harman, Willard Nelson
Harper, Pierre Paul
Hassler, Thomas J
Herbst, Richard Peter
Hiltunen, Jarl Kalervo
Hynes, Hugh Bernard Noel
Johnson, Howard Ernest
Lang, Kenneth Lyle
Lotspeich, Frederick Benjamin
Lush, Donald Lawrence
Mackay, Rosemary Joan
Miller, Kenneth Melvin
Montroy, Leo Dennis
Tennessen, Kenneth Joseph
Wallace, Ronald Richard
Williams, Louis Gressett
Wissing, Thomas Edward

Human Ecology

Bajema, Carl J
Dunsing, Marilyn Magdalene
Esser, Aristide Henri
Haase, Bruce Lee
Hickey, Richard James
Holway, James Gary
Lee, James A
Mecom, John Oden
Miller, George Tyler, Jr
Mosher, John Ivan
Pierson, Dolores Lehmann
Purchase, Mary Elizabeth
Richerson, Peter James
Roze, Janis Arnold
Sargent, Frederick, II
Shurley, Jay Talmadge
Teal, John Jerome, Jr
Vayda, Andrew Peter
Williams, Nelson Noel
Williamson, Penelope Rose

Insect Ecology

Allen, Douglas Charles
Brust, Reinhart A
Burleigh, Joseph Gaynor
Carolin, Valentine Mott, Jr
Cothran, Warren Roderic
Coulson, Robert N
Davis, Robert
De Bach, Paul (Hevener)
Evans, William George
Frank, John Howard
Frankie, Gordon William
Frazer, Bryan Douglas
Harcourt, Douglas George
Haverty, Michael Irving
Hays, Kirby Lee
Hendrick, Rodney Douglas
Hoy, Marjorie Ann
Huber, Roger Thomas
Huffaker, Carl Barton
Kay, Carol Ann
Kieckhefer, Robert William
Knight, Clifford Burnham
Krogstad, Blanchard Orlando
Lee, Siu-Lam
LeRoux, Edgar Joseph
MacLean, David Belmont
Mason, Charles Eugene
Pass, Bobby Clifton
Rust, Richard W
Ryan, Roger Baker
Schultz, John Charles
Sears, Markham Karli
Semtner, Paul Joseph
Sheldon, Joseph Kenneth
Sinha, Ranendra, Nath
Smith, Grahame J C
Streams, Frederick Arthur
Tarpley, Wallace Armell
Ueckert, Darrell Neal
Uhler, Lowell Dohner
Wellington, William George

Invertebrate Ecology

Berry, James William
Crawford, Clifford Smeed
Driscoll, Egbert Gotzian
Dundee, Dolores Saunders
Gable, Michael
Halgren, Lee A
Keith, Donald Edwards
Kraft, Kenneth J
McMahon, Robert Francis, III
Radwin, George E
Rutherford, James Charles
Ryals, George Lynwood, Jr
Stanton, George Edwin

Limnology

Bachmann, Roger Werner
Bahr, Thomas Gordon
Baker, Clinton Lyle
Barrett, Paul Howard
Baumgardner, Ray K
Bayne, David Roberge
Bean, Daniel Joseph
Beatty, Kenneth Wilson
Beeton, Alfred Merle
Benson, Norman G
Blanton, Jackson Orin
Borst, Daryl C
Bothwell, Max Lewis
Boyd, Claude Elson
Bresnick, Gerald Irwin
Brezonik, Patrick Lee
Britt, N Wilson
Brunskill, Gregg J
Buscemi, Philip Augustus
Cairns, John, Jr
Campbell, Robert Seymour
Clark, William Jesse
Clifford, Hugh Fleming
Coffman, William Page
Cole, Gerald Ainsworth
Coler, Robert A
Cowell, Bruce Craig
Cummins, Kenneth William
Davis, Charles (Carroll)
Dendy, John Stiles
De Witt, John William, Jr
Donaldson, John Russell
Dorris, Troy Clyde
Eberly, William Robert
Edmondson, W Thomas
Edmondson, Yvette Hardman
Emery, Richard Meyer
Everett, Lorne Gordon
Findley, Diane Ingram
Fisher, Stuart Gordon
Fremling, Calvin R
Frey, David Grover
Fuhs, George Wolfgang
Funk, William Henry
Galler, Sidney Roland
Gallup, Donald Noel
Giesy, John Paul, Jr
Glooschenko, Walter Arthur
Goldman, Charles Remington
Gorham, Eville
Goulden, Clyde Edward
Halsey, Thomas Gordon
Haney, James Filmore
Hannan, Herbert Herrick
Harmsworth, Rodney V
Harp, George Lemauil
Harvey, Harold H
Haynes, Robert C
Hickman, Michael
Hille, Kenneth R
Hobbs, Horton Holcombe, III
Hoffman, Dale A
Hofstetter, Ronald Harold
Hohn, Matthew Henry
Hough, Richard Anton
Howard, Harold Henry
Huntsinger, Karolyn Regina
Hurlbert, Stuart Hartley
Jackson, Daniel Francis
John, Kenneth Rydal
Johnson, Lionel
Jones, John Richard
Kilham, Peter
King, Darrell Lee
Klemm, Donald J
Knudson, Vernie Anton
Koob, Derry Delos
Kunny, Bartholomew Kenneth
Langford, Raymond Robert
Lauer, Gerald J
Lauff, George Howard
Lavelle, James W
Legge, Thomas Nelson
Likens, Gene Elden
Lind, Owen Thomas
Linn, DeVon Wayne
Longley, Glenn Jr
Loomis, Robert Henry
Lush, Donald Lawrence
Main, Robert Andrew
Malueg, Kenneth Wilbur
Makarewicz, Joseph Chester
Manny, Bruce Andrew
Marshall, Jack Stanton
Marzolf, George Richard
Mason, David Thomas
Mathis, Billy John
McCombie, Alen Milne
McConville, David Raymond
McCullough, Jack Dennis
McNabb, Clarence Duncan, Jr
McNaught, Donald Curtis
Milligan, Troy Bea
Moll, Russell Addison
Mortimer, Clifford Hiley
Moshiri, Gerald Alexander
Nall, Raymond Willett
Namminga, Harold Eugene
Neel, Joe Kendall, Sr
O'Brien, William John
Olsen, Sigurd
Ostrander, Darl Reed
Page, Thomas Lee
Parker, Michael
Parker, Richard Alan
Parsons, John David
Patalas, Kazimierz
Peterson, Richard Randolph
Peterson, Spencer Alan
Potash, Milton
Powers, Charles F
Ratzlaff, Willis
Reed, Edward Brandt
Rich, Peter Hamilton
Richardson, Jonathan L
Richerson, Peter James
Rosing, Lorraine Morin
Ruschmeyer, Orlando R
Sadler, William Otho
Saether, Ole Anton
Savitz, Jan
Scheske, Claire L
Schindler, David William
Schmitz, William Robert
Scott, Donald Charles
Seilheimer, Jack Arthur
Shapiro, Joseph
Silvey, J K Gwynn
Simmons, George Matthew, Jr
Slack, Keith Vollmer
Smith, Charles Edward, Jr
Sobacki, Leonard Paul
Solero, Raymond Arthur
Solman, Victor Edward Frick
Spacie, Anne
Stahl, John Benton
Stanford, Jack Arthur
Stern, Daniel Henry
Stewart, Kenton M
Stockner, John G
Stoermer, Eugene F
Stull, Elisabeth Ann
Sweeney, Robert Anderson
Tarapchak, Stephen J
Trama, Francesco Biagio
Tubb, Richard Arnold
Tuinstra, Kenneth Eugene
Vollenweider, Richard A
Waller, William T
Ward, Fredrick George
Wetzel, Robert George
Wilhm, Jerry L
Wise, Charles Davidson
Wohler, James Richard, II
Wood, Kenneth George
Wright, John Clifford
Wright, Thomas Dodson

Marine Ecology

Adelson, Lionel Morton
Ainley, David George
Aldrich, David Virgil
Bakus, Gerald Joseph
Bane, Gilbert Winfield
Banus, Mario Douglas
Baylor, Edward Randall
Boynton, Walter Raymond
Bright, Donald Bolton
Brunel, Pierre
Brusca, Gary J
Calabrese, Anthony
Carr, William Edward Statter
Cervonka, Robert Henry
Chestnut, Alphonse F
Cirino, Elizabeth Fahey
Croker, Robert Arthur
Cronin, Lewis Eugene
Dahl, Arthur Lyon
Dauer, Daniel Martin
Davis, Luckett Vanderford
Diaz, Robert James
Dragovich, Alexander
Druehl, Louis D
Dupuy, John L
Eberhardt, Robert Louis
Engstrom, Norman Ardell
Farris, Richard Austin
Flittner, Glenn Arden
Ford, Richard Fiske
Foreman, Ronald Eugene
Funicelli, Nicholas Anthony
George, Carl Joseph Winder
Glynn, Peter W
Goering, John James
Goshall, Daniel Warren
Haines, Evelyn Brown
Hendler, Gordon Lee
Hidu, Herbert
Hollister, Charles Davis
Houston, Roy Seamands
Hurley, Ann Catherine
Hyer, Paul Vincent
Johannes, Robert Earl
Johnson, Samuel Edgar, II
Jones, Gilbert Fred
Kelso, Donald Preston
Kolipinski, Milton Charles
Lacroix, Guy
Lee, Welton Lincoln
Levandowsky, Michael
Lowe, Jack Ira
Marsh, James Alexander, Jr
Maurer, Donald Leo
Maynard, Nancy Gray
McAlister, William Bruce
McCormick, Jon Michael
Miller, Robert Joseph
Moore, Johnes Kittelle
Morris, Byron Frederick
Newkirk, Gary Francis
Nybakken, James W
Ogden, John Conrad
Osman, Richard William
Powles, Percival Mount
Rice, Theodore Roosevelt
Rose, Curt D
Schreiber, Ralph Walter
Scott, Kenneth John, Jr
Setzler, Eileen Marie
Singletary, Robert Lombard
Smith, David Francis
Stancyk, Stephen Edward
Stickney, Alden Parkhurst
Summers, William Clarke
Sutherland, John Patrick
Taylor, Peter Berkley
Terry, Orville Whitfield
Thomson, Donald A
Thum, Alan Bradley
Tolderlund, Douglas Stanley
Trott, Lamar Brice
Vadas, Robert Louis
Virnstein, Robert W
Warme, John Edward
Wenner, Robert Ronald
Whitney, David Earle
Wigley, Roland L
Williams, Richard Birge
Willingham, Charles Allen
Wohlschlag, Donald Eugene
Wood, Carl Eugene
Woodmansee, Robert Asbury
Worrest, Robert Charles

Microbial Ecology

Adams, John Collins
Azam, Farooq
Bartha, Richard
Belly, Robert T
Bibel, David Jan
Bounds, Harold C
Brock, Katherine Middleton
Bryant, Marvin Pierce
Burnison, Bryan Kent
Carlucci, Angelo Francis
Cassin, Joseph M
Cullimore, Denis Roy
Fletcher, Donald Warren
Gillespie, Paul Albert
Gordon, Ronald Claire
Grula, Mary Mucelroy
Guthrie, Rufus Kent
Hoffman, Harrison Adolph
Jay, James Monroe
Kanakanati, Antony
Kodischek, Leah K
Lange, Willy
Pfaender, Frederic Karl
Post, Frederick Just
Praner, David
Pritchard, Parmely Herbert
Robinson, John Bertram
States, Jack Sterling
Stotzky, Guenther
Tiedje, James Michael
Triska, Frank John
Tyler, Max Ezra
Vishniac, Helen Simpson
Weiler, William Alexander
Whitt, Dixie Dailey
Wolin, Meyer Jerome

Medical Ecology

Cech, Irina
Haddon, William, Jr
Krueger, Albert Paul
Strain, Boyd Ray

Paleoecology

Alexander, Richard Raymond
Anderson, Edwin J
Baer, James L
Buzas, Martin A
Cameron, Barry Winston
Camp, Mark Jeffrey
Clark, George Richmond, II
Crouch, Robert Wheeler
Curran, Harold Allen
Dodd, James Robert
Ecdale, Allan Anton
Forester, Richard Monroe
Gray, Jane
Howard, James F
Kissling, Don Lester
Kontrovitz, Mervin
Logan, Alan
Nations, Jack Dale
Parks, James Marshall, Jr
Rhoads, Donald Cave
Richards, Pierre Joseph Herve
Richards, Richard Peter
Sarjeant, William Antony Swithin
Shaak, Graig Dennis
Shourd, Melvin Lee
Stevens, Calvin H
Tuthill, Samuel James
Vincent, Jerry William
Walker, Kenneth Russell
Wright, Herbert Edgar, Jr
Wright, Robert Paul

Physiological Ecology

Amen, Ralph DuWayne
Amundsen, Clifford C
Anderson, Gary
Apley, Martyn Linn
Ashby, William Clark
Bakken, George Stewart
Barnett, Leland Bruce
Behan, Mark Joseph
Bettinger, Thomas Lee
Bennett, James Peter
Bergman, Harold Lee
Bjorkman, Olle
Brower, James E
Bunce, James Arthur
Burky, Albert John
Calder, William Alexander, III
Caldwell, Richard Stanley
Cech, Joseph Jerome, Jr
Claussen, Dennis Lee
Coles, Richard Warren
Congleton, James Lee
Coyne, Patrick Ivan
Cupp, Paul Vernon, Jr
DeWitt, Calvin Boyd
Dimock, Ronald Vilroy, Jr
Einhellig, Frank Arnold
Elias, Robert William
Eriksen, Clyde Hedman
Feng, Sung Yen
Gessaman, James A
Graham, Terry Edward
Grubbs, David Edward
Hackbarth, Winston (Philip)
Hamby, Robert Jay
Hansen, David Henry
Hawke, Scott Dransfield
Hays, Robert Leroy
Helms, Carl Wilbert
Herreid, Clyde F, II
Hicks, David L
Higgins, Paul Daniel
Hinds, David Stewart
Hunter, Robert Douglas
Hutchison, Victor Hobbs
Kaufmann, Merrill R
Kay, Fenton Ray
King, James Roger
Leffler, Charles William
MacMillen, Richard Edward
McCown, Brent Howard
McDonough, Walter Thomas
Moorcock, Robert Edward
Moshier, James Arthur
Mullen, Robert Keech
Norton, David William
Oldfield, Thomas Edward
Packard, Gary Claire
Paterson, David Thomas
Polohowich, John Jacob
Ritchie, Gary Alan
Rothwell, Frederick Mirvan
Sadleir, Richard Michael Francis Stuart
Saks, Norman Marrin
Schneider, David Edwin
Schneider, Mark Joseph
Secor, Jack Behrent
Sevilla-Gardiner, Josefina Zialcita
Sharp, Gary Duane
Stainken, Dennis M
Studier, Eugene H
Szarek, Stanley Richard
Trost, Charles Henry
Underwood, Lawrence Statton
West, George Curtiss
Willemsen, Roger Wayne

Wunder, Bruce Arnold
Zimmerman, Craig Arthur

Plant Ecology
Abrahamson, Warren Gene, II
Adams, Michael Studebaker
Bard, Gily Epstein
Barton, James Don, Jr
Baskin, Jerry Mack
Beatley, Janice Carson
Becker, Donald A
Bell, Katherine Lapsley
Bell, Marcus Arthur Money
Benninghoff, William Shiffer
Bernard, John Milford
Blair, Robert Marks
Bleak, Alvin Thomas
Bonham, Charles D
Boyce, Stephen Gaddy
Bozeman, John Russell
Branson, Farrel Allen
Bratton, Susan Power
Breeden, Johnnie Elbert
Brinson, Mark McClellan
Brotherson, Jack DeVon
Brown, James Milton
Brown, Robert Thorson
Brown, Walter Howard
Buchanan, Hayle
Buck, Paul
Burgess, Robert Lewis
Caldwell, Martyn Mathews
Catana, Anthony J, Jr
Catenhusen, John Alfons
Cavers, Paul Brethen
Cawley, Edward T
Chapman, Joe Alexander
Chilcote, William W
Christensen, Norman Leroy, Jr
Christiansen, Paul Arthur
Christianson, John Dean
Churchill, Algernon Coolidge
Clambey, Gary Kenneth
Clary, Warren Powell
Clebsch, Edward Ernst Cooper
Cooper, George Raymond
Cotter, David James
Coupland, Robert Thomas
Crockett, Jerry J
Crow, John H
Crowder, Adele A
Dale, Edward Everett, Jr
Damman, Antoni Willem Hermanus
Danks, Maureen Lee
Daubenmire, Rexford
Dayton, Bruce R
DeSelm, Henry Rawie
Dina, Stephen James
Dodd, Jimmie Dale
Donahue, William H
Doyon, Dominique
Drew, William Brooks
Egler, Frank Edwin
Facey, Vera L
Flaccus, Edward
Fogg, George Garrett
Fonda, Richard Weston
Forman, Richard T T
Freeman, Charles Edward, Jr
Fritts, Harold Clark
Gebben, Alan Irwin
Glenn-Lewin, David Carl
Goder, Harold Arthur
Goetz, Harold
Good, Norma Frauendorf
Good, Ralph Edward
Gosselink, James G
Graham, Benjamin Franklin, Jr
Haase, Edward Francis
Hall, Frederick Columbus
Hamilton, Ernest Scovell
Hanawalt, Ronald B
Hanes, Ted L
Haney, Alan William
Harcombe, Paul Albin
Harner, Richard Francis
Harthill, Marion Paul
Hartman, Richard Thomas
Hazell, Don Bliss
Heady, Harold Franklin
Heaslip, Margaret Barkley
Hermann, Richard Karl
Hinde, Howard Parrish
Hodson, William Myron
Hoffman, Gerald
Hofstra, Gerald R
Houston, Walter Randolph
Hulbert, Lloyd Clair
Hull, Alvin C, Jr
Hull, James Clark
Hutcheson, Harvie Leon, Jr
Jackson, Marion T
Janke, Robert A
Jefferson, Carol Annette
Jenkins, Robert Walls, Jr
Johnson, Albert W
Johnson, Karen Louise
Johnson, Michael Paul

Jones, John Robert
Keammerer, Warren Roy
Kelting, Ralph Walter
Kershaw, Kenneth Andrew
Kessell, Stephen Robert
Knight, Dennis Hal
Krajina, Vladimir Joseph
Kramer, Richard John
Kroh, Glenn Clinton
Kumler, Marion Lawrence
Lamp, Herbert F
Landers, Roger Q, Jr
Lang, Gerald Edward
Langenheim, Jean Harmon
Larsen, James Arthur
Lauenroth, William Karl
Launchbaugh, John L, Jr
Lavoie, Victorin
Laycock, William Anthony
Leatherman, Anna D
Lindauer, Ivo Eugene
Lindsay, Douglas Rome
Lindsey, Alton Anthony
Linn, Robert
Lipps, Emma Lewis
Lynch, Daniel Matthew
Mack, Richard Norton
Mahall, Bruce Elliott
Major, Jack
Martin, William Haywood, III
Maycock, Paul Frederick
McAndrews, John Henry
McGinnies, William Grovenor
McGinnies, William Joseph
McIntosh, Robert Patrick
McLeod, Kenneth William
McMillan, Calvin
McNaughton, Samuel J
McPherson, Donald Carman
McPherson, James King
Medve, Richard J
Mellinger, Clair
Mellinger, Michael Vance
Miller, Lee Norman
Monk, Carl Douglas
Moore, John Robert
Moore, Russell Thomas
Morgan, Michael Dean
Muegler, Walter Frank
Mueller, Irene Marian
Mueller-Dombois, Dieter
Muller, Cornelius Herman
Muller, Robert Neil
Murdock, Joseph Richard
Neiland, Bonita
Newsome, Richard Duane
Niering, William Albert
Nighswonger, Paul Floyd
Nixon, Elray S
Owen, Herbert Elmer, Jr
Patten, Duncan Theunissen
Patton, Ernest Gibbes
Peacock, John Talmer
Pearcy, Robert Woodwell
Pearson, Philip Richardson, Jr
Pelton, John Forrester
Penfound, William Theodore
Perino, Janice Vinyard
Phillips, Edwin Allen
Phipps, Richard L
Pieper, Rex Delane
Pittillo, Jack Daniel
Plummer, Gayther Lynn
Price, Keith Robinson
Ralston, Robert D
Raynal, Dudley Jones
Redmann, Robert Emanuel
Reed, Robert Marshall
Reid, Archibald
Rhoades, Richard Wesley
Robertson, Philip Alan
Rosenwinkel, Earl Richard
Rowe, John Stanley
Rumely, John Hamilton
Rundel, Philip Wilson
Schrock, Alta
Schultz, Arnold Max
Scott, Richard Walter
Selleck, George Wilbur
Shake, Roy Eugene
Shontz, John Paul
Shontz, Nancy Nickerson
Sigafoos, Robert Sumner
Slack, Nancy G
Smathers, Garrett Arthur
Smeins, Fred E
Smith, Alan Lyle
Smith, Alan Paul
Smith, David William
Solomon, Allen M
Sosebee, Ronald Eugene
Staniforth, Richard John
Stark, Nellie May
Swedberg, Kenneth C
Terwilliger, Charles, Jr
Thurman, Lloy Duane
Timin, Robert Owen
Tobiessen, Peter Laws

Tueller, Paul T
Ungar, Irwin A
VanAmburg, Gerald Leroy
Van Asdall, Willard
Van Dersal, William Richard
Viereck, Leslie A
Voigt, John Wilbur
Volland, Leonard Allan
Wagner, Richard H
Wali, Mohan Kishen
Walther, Alina
Ward, Richard Theodore
Ware, George Henry
Ware, Stewart Alexander
Waring, Richard H
Warner, James Howard
Wein, Ross Wallace
Whitcomb, Carl Erwin
Whitford, Philip Burton
Williams, George Jackson, III
Williams, John Simeon
Wistendahl, Warren Arthur
Yarranton, George Anthony
Youngner, Victor Bernarr
Zobel, Donald Bruce

Pollution Biology
Bender, Michael E
Benedict, Harris Miller
Brungs, William Aloysius, Jr
Clark, Donald Ray, Jr
Feder, William Adolph
Hancock, Elizabeth Diekenberger
Jackson, Herbert William
Kudenov, Jerry David
LeGore, Richard Stephen
Leone, Ida Alba
Martin, John Holland
Martin, Michael
Merkley, Wayne Bingham
Reese, Weldon Harold
Rice, Stanley Donald
Stendell, Rey Carl
Stickel, William Henson
Strand, John A, III
Tarzwell, Clarence Matthew

Pollution Chemistry
Auerbach, Victor
Bacon, Hilary Edwin
Batzar, Kenneth
Bendure, Robert J
Cronkright, Walter Allyn, Jr
Fisher, Sallie Ann
Gahler, Arnold Robert
Hager, Onslow Bonner
Howes, James E, Jr
Hygh, Earl Hampton
Kawahara, Fred Katsumi
Keith, Lawrence H
Lyding, Arthur R
Moser, Frank Hans
Patel, Siddharth Manilal
Penrose, William Roy
Purchase, Earl Ralph
Rebagay, Teofila Velasco

Population Ecology
Chitty, Dennis Hubert
Collins, Nicholas Clark
Confer, John L
Corbin, Kendall Wallace
Downhower, Jerry F
Edwards, Dallas Craig
Fretwell, Steve D
Fujii, Koichi
Gibbons, J Whitfield
Goldstein, Melvin C
Greenlaw, Jon Stanley
Hansen, David Henry
Healey, Michael Charles
Healy, William Ryder
Horn, Henry Stainken
Jaeger, Robert Gordon
Jensen, Paul
Jones, Kirkland Lee
Koplin, James Ray
Mathisen, Ole Alfred
McFadden, James Thompson
Paris, Oscar Hall
Park, Thomas
Quick, Horace Floyd
Reid, William Harper
Rosenzweig, Michael Leo
Schwalbe, Paul Wayman
Skillman, Robert Allen
Slade, Norman Andrew
Smith, Dwight Glenn
Stinner, Ronald Edwin
Taylor, Robert Joe
Turnock, William James
Tuttle, Merlin Devere
Weise, Charles Martin
Wise, David Haynes
Yamada, Sylvia Behrens

Radiation Ecology
Anthony, Margery Stuart

Brower, John Harold
Crossley, DeRyee Ashton, Jr
Emery, Richard Meyer
Erdman, Howard E
Hanson, Wayne Carlyle
Levy, Charles Kingsley
Montgomery, Daniel Michael
Ophel, Ivan Lindsay
Pendleton, Robert Cecil
Rhoads, William Anderson
Seymour, Allyn H
Templeton, William Lees
Watson, Donald Gordon
Witherspoon, John Pinkney, Jr

Range Ecology
Blaisdell, James Pershing
Dittberner, Phillip Lynn
Dodd, Jerrold Lowell
Duvall, Vinson Lamar
Dyksterhuis, Edsko Jerry
Eddleman, Lee E
Haas, Robert Henry
Halls, Lowell Keith
Hedrick, Donald Ward
Johnston, Alex
Keller, Wesley
Kinsinger, Floyd Elton
Klemmedson, James Otto
Martin, Samuel Clark
May, Morton
McLean, Alastair
Nelson, Jack Raymond
Pettit, Russell Dean
Pyott, William Tucker
Ries, Ronald Edward
Springfield, Harry Wayne
Sykes, Dwane Jay
Thomas, Gerald Waylett
Ueckert, Darrell Neal
Wasser, Clinton Howard
Wilson, David George
Woodmansee, Robert George

Resource Management
Adams, David A
Adelson, Harold Ely
Ahearne, John Francis
Akins, Glenn John
Alexander, Earl Glynn
Alexander, Madeline J
Anderson, James Arthur
Antonini, Gustavo Arthur
Beatty, Marvin Theodore
Berry, Leonard
Berryman, Jack Holmes
Blaser, Robert U
Blount, Stanley Freeman
Buerle, David E
Cain, John Manford
Copeland, Otis Lee
Dils, Robert Earl
Dingman, Stanley Lawrence
Doede, John Henry
Flaschen, Stewart Samuel
Gibbons, John Howard
Glick, Phillip Ray
Gould, Ernest Morton, Jr
Gregersen, Hans Miller
Griffiths, George Motley
Hanson, Joe A
Johnson, Carl Sand
Johnson, Frederick Carroll
Khan, Mohammad Asad
Knorr, Philip Noel
Krejsa, Richard Joseph
Kreysa, Frank Joseph
Lapping, Mark Barry
Lehmann, Eiroy Paul
Levin, Michael Howard
Lien, Arthur Philip
Malcolm, Janet May
Malina, Marshall Albert
McLellan, Alden, IV
Murnane, Thomas George
Nash, Colin Edward
Patterson, Richard L
Pease, James Robert
Peterson, John William
Ramsay, William Charles
Raymond, Frank Leroy
Ross, William Michael
Royce, William Francis
Rue, Edward Evans
Schiff, Daniel
Shrier, Adam Louis
Silvert, William Lawrence
Smedes, Harry Wynn
Steele, Timothy Doak
Stout, Paul Richard
Suarez, Thomas H
Szekely, Ivan J
Taylor, Robert Franklin
Verkhovsky, Boris Samuel
Vertrees, Robert Layman
Watson, Jeffrey
Whipple, Francis Oliver
Wilkins, Bruce Tabor

Gillingham, Robert J
Gnau, Donald Vaughn
Green, Edward H
Green, Robert Eugene
Gregory, Brian Charles
Gunter, Roy Chalmers, Jr
Hamman, Donald Jay
Hanson, Per Roland
Helmer, John
Hernqvist, Karl Gerhard
Hobson, John Peter
Hoffman, James Tracy
Horak, Jerry Robert
Howell, John Foss
Huebner, Russell Henry
Hult, John Luther
Hupert, Julius Jan Marian
Hyams, Henry C
Isaacson, Michael Saul
Jensen, Arthur Seigfried
Johnson, Robert Glenn
Joyner, Weyland Thomas, Jr
Kadaba, Prasad Krishna
Kanter, Helmut
Kim, Dae Mann
Kinsel, Tracy Stewart
Kogelnik, H W
Kompfner, Rudolf
Korneff, Theodore
Kundu, Mukul Ranjan
Kuper, J B Horner
Kupferberg, Kenneth Maurice
Ladd, William Alexander
Laden, Hyman Nathaniel
Lally, Vincent Edward
Laponsky, Alfred Baer
Levenson, Marc David
Lockwood, Grant John
Lonky, Martin Leonard
Mackay, Ralph Stuart
Manson, Donald Joseph
Marshall, J Howard, III
Martineau, Robert Jean
Mayo, Santos
McClintock, Michael
McClure, Benjamin Thompson
McIntosh, Bruce Andrew
McNutt, John Dewight
McRae, Lorin Post
Medicus, Gustav Konrad
Middleton, Arthur Everts
Mihran, Theodore Gregory
Miller, John Henry, III
Miller, William Robert, Jr
Mitchell, George Weston
Mueller, Erwin W
Mulson, Joseph F
Navon, David H
Nelson, Vernon Ronald
Newman, Theodore Joseph
Okaya, Akira
Oman, Robert Milton
Ortel, William Charles Gormley
Palocz, Istvan
Parker, William Edward
Pashler, Peter Edward
Pearlman, Michael R
Rice, Philip Joseph, Jr
Richard, Patrick
Rigrod, William Walter
Robbins, Gordon Daniel
Robinson, Aaron Zed, Jr
Roellig, Leonard Oscar
Rogers, Howard H
Rogers, Thomas F
Rosen, Paul
Rowe, Irving
Russell, James Torrance
Rutledge, James Luther
Sadowski, Henry
Savet, Paul H
Schneider, Sol
Schoenfeld, Robert Louis
Schuster, Nick August
Schwartz, Helmut Julius
Schwetman, Herbert Dewitt
Seto, Yeb Jo
Sewell, Frank Anderson, Jr
She, Chiao-Yao
Shelton, Wilford Neil
Shive, John Northrup
Shkarofsky, Issie Peter
Shmoys, Jerry
Short, Oliver Alton
Smith, George Foster
Snavely, Benjamin Breneman
Spielman, Harold S
Starr, Merle Arthur
Stehle, Philip McLellan
Stein, William Earl
Stewart, Thomas William Wallace
Tamir, Theodor
Taylor, Donald Rudolph, Jr
Vanier, Jacques
Vlcek, Donald Henry
Wagener, Johann Christian Siegfried

Wagner, William Gerard
Walter, William Trump
Ward, Samuel Abner
Warters, William Dennis
Watkins, Robert Arnold
Waugh, John Blake-Steele
Wiens, Jacob Henry
Wilson, Robert Gray
Witts, James Reed
Wittenberg, Albert M
Wolff, Nikolaus Emanuel
Woods, Roy Alexander
Wright, Kenneth Arthur
Yeh, Yin
Zitter, Robert Nathan

Electrooptics
Almeida, Silverio Pedro
Anthony, Romuald
Ashby, Val Jean
Astheimer, Robert W
Barrekette, Euval S
Bartolini, Robert Alfred
Beall, Horace Ansley
Brienza, Michael Joseph
Buckman, Alvin Bruce
Chow, Christopher N
Christensen, Niels Gunnar
Coleman, Howard S
Cooper, Howard Gordon
Dougherty, Joseph Patrick
Dressel, Herman Otto
Free, John Marshall
Geller, Myer
Gurski, Thomas Richard
Han, Ki Sup
Hansen, J Richard
Harvey, George Lloyd
Henning, Harley Barry
Honey, Richard Churchill
Hummer, Robert Franklin
Hutter, Edwin Christian
Jablonski, Frank Edward
Jaffe, Hans
Kahn, Frederic Jay
Kinzly, Robert Edward
Knight, Gordon Raymond
Kornstein, Edward
Kozma, Adam
Kurtz, Stewart K
Maydan, Dan
McMahon, Thomas Joseph
Medved, David Bernard
Nelson, Kyler Fischer
O'Brien, Brian
Penz, P Andrew
Richman, Isaac
Ruderman, Irving Warren
Sauermann, Gerhard Otto
Shackelford, Robert G
Sheridan, Nicholas Keith
Silverstein, Elliot Morton
Sneed, Richard J
Soref, Richard Allan
Spaulding, Richard Alan
Starkweather, Gary Keith
Stegelmann, Erich J
Titterton, Paul James
Traub, Alan Cutler
Vahey, David William
Vance, Dennis William
Volz, William Beckham
Whitney, Colin Gordon
Williamson, Arthur Elridge, Jr
Wunderman, Irwin
Zarem, Abe Mordecai

EMBRYOLOGY

Embryology
Alexander, Lloyd Ephraim
Algard, Franklin Thomas
Allan, Frank Duane
Armstrong, Rosa Mae
Arnold, John Miller
Avery, Gordon B
Avery, James Knuckey
Bagnara, Joseph Thomas
Barber, Mary Lee
Barry, Alexander
Bates, M Noble
Beitch, Irwin
Black, John Larry
Black, Robert Earl Lee
Block, Matthew Harold
Bodemer, Charles William
Bongard, Steven J
Borle, Andre Bernard
Bransome, Edwin D, Jr
Brent, Robert Leonard
Briggs, Robert William
Brinster, Ralph Lawrence
Bueker, Elmer Daniel
Burns, Robert Kyle
Butler, Harry
Cairns, John Mackay

Caston, J Douglas
Cavey, Michael John
Chalkley, Donald Thomas
Chan, An Soo
Church, Gilbert
Clark, Hugh
Clawson, Robert Charles
Clement, Anthony Calhoun
Cohen, Alan Mathew
Collier, Jack Reed
Coulombre, Alfred Joseph
Counce, Sheila Jean
Cowden, Ronald Reed
Creager, Joan Guynn
Dalton, Howard Clark
Davies, Jack
Deck, James David
Decker, John D
DeHaan, Robert Lawrence
Derrick, Grace Ethel
Dossel, William Edward
Dunnebacke-Dixon, Thelma Hudson
Duwe, Arthur Edward
Ebert, James David
Edidin, Michael Aaron
Edwards, Nancy Claire
Erickson, Alan Eric
Eschenberg, Kathryn (Marcella)
Evans, Hiram John
Fabian, Michael William
Fedoroff, Sergey
Ferm, Vergil Harkness
Fernald, Robert Leslie
Fish, William Arthur
Fisher, Don Lowell
Foote, Florence Martindale
Ford, Peter
Francoeur, Robert Thomas
Friedman, Harvey Paul
Froelich, Jean Small
Frost, David
Fuller, Eugene George
Gier, Herschel Thomas
Glade, Richard William
Glucksohn-Waelsch, Salome
Goldsberry, Steve
Grabowski, Casimer Thaddeus
Graves, Artis P
Grodner, Mary Laslie
Harris, Albert Kenneth, Jr
Harris, Thomas Mason
Harrison, John Robert
Harvey, Elmer Bostwick
Hayes, Raymond L, Jr
Healy, Eugene A
Hendrickx, Andrew George
Henley, Catherine
Herschler, Michael Saul
Hertig, Arthur Tremain
Hilliard, Stephen Dale
Hinsch, Gertrude Wilma
Hirt, Bethold Joseph
Hisaoka, Kenichi Kenneth
Holtfreter, Johannes Friedrich Karl
Holtzer, Howard
Hopper, Arthur Frederick
Houston, Marshall Lee
Humphries, Asa Alan, Jr
Hunter, Alice S (Baker)
Hunter, Ruth Macmillan
Hwang, Ung Kee
Idzikowsky, Henry Joseph
Ivey, William Dixon
Johnson, Leland Gilbert
Kaighn, Morris Edward
Kaye, Nancy Weber
King, Robbins Sydney
King, Thomas Joseph
Kischer, Clayton Ward
Kitchin, Irwin Clark
Klapper, Clarence Edward
Kleiss, Ekkehard
Kochhar, Devendra M
Kollar, Edward James
Kollros, Jerry John
Krause, James Barber
Krehbiel, Eugene B
Laham, Quentin Nadime
Lawrence, Irvin E, Jr
Lehman, Harvey Eugene
Lehman, Lillian Margot Youngs
Leibo, Stanley Paul
Leung, Christopher Chung-Kit
Long, Sally Yates
Lowe, Janet Marie
Luckett, Winter Patrick
Lumb, Ethel Sue
Lund, Douglas E
Martin, Edward Williford
Maslow, David E
Mathews, Willis Woodrow
Matulionis, Daniel H
Mayo, Marie Joiner
McCallion, David John
McCrady, Edward
McCurdy, Harriet Mace
McKenzie, John Ward
Menees, James H

Mezger-Freed, Liselotte
Mizejewski, Gerald Jude
Moore, Betty Clark
Narbaitz, Roberto
Naughten, John Charles
Neff, William Medina
Nelsen, Olin E
Orsini, Margaret Ward (Giordano)
Owen, Alice Koning
Paff, George Hugo
Pai, Anna Chao
Pollack, Emanuel Davis
Popp, Raymond Arthur
Rafferty, Keen Alexander, Jr
Ramm, Gordon Morley
Rappaport, Raymond, Jr
Reynolds, Wynetka Ann King
Richter, Kenneth Murrel
Rieck, Alvin Frank
Rollason, Grace Saunders
Rolle, Gloria Katharine
Rose, Sylvan Meryl
Rounds, Donald Edwin
Rowell, Lyman Smith
Rudnick, Dorothea
Runner, Meredith Noftzger
Ruth, Royal Francis
Saunders, John Warren, Jr
Schlesinger, Allen Brian
Schmidt, Anthony John
Schuetz, Allen W
Searls, Robert L
Seegmiller, Robert Earl
Segal, Sheldon Jerome
Severn, Charles B
Shaver, John Rodney
Shelden, Robert Merten
Shell, Lester Crane
Smail, James Richard
Smith, Stephen D
Smithberg, Morris
Snook, Theodore
Snyder, George Edward
Sorokin, Sergei Pitirimovitch
Spence, Alexander Perkins
Spiegelman, Martha
Spiroff, Boris E N
Staugaard, Burton Christian
Steinberg, Malcolm Saul
Stephens, Lee Bishop, Jr
Swartz, Gordon Elmer
Szebenyi, Emil
Szollosi, Daniel Gabriel
Taber, Elsie
Thompson, Michael Bruce
Tibbitts, Forrest Donald
Tobin, Charles Emil
Todd, Robert Emerson
Trelstad, Robert Laurence
Tucker, Marie
Van Alten, Pierson Jay
Vaupel, Martin Robert
Waterman, Allyn Jay
Watterson, Ray Leighton
Weis, Peddrick
Weisz, Paul B
Weston, John Colby
White, Lewis L
Whittaker, John Richard
Williams, Mary Louise Monica Fritts
Wilson, Doris Burda
Wilson, James Graves
Wilson, Wilfred J
Yow, Francis Wagoner
Zimmerman, Selma Blau

Chemical Embryology
Berryhill, Virginia Farmer
Herrmann, Heinz
Mertes, David H
Wainwright, Stanley D
Weller, Edwin Matthew

Experimental Embryology
Anderton, Laura Gaddes
Beaudoin, Allan Roger
Blandau, Richard Julius
Criley, Bruce
Duffey, Lowell Myers
Eichler, Victor B
Finnegan, Cyril Vincent
Fowler, Ira
Goff, Richard Allen
Hand, George Samuel, Jr
Hopkins, Betty Jo Henderson
Jaffee, Oscar Charles
Jensh, Ronald Paul
Jones, Roy Winfield
Landauer, Walter
Mallette, John M
McLaughlin, Ellen Winnie
Mitchell, John Taylor
Noden, Drew M
Pogany, Gilbert Claude
Rohner, Mary Christopher
Seale, Raymond Ulric
Smith, Ellen Marian Evans
Spears, James Richard

Tucker, Gail Susan
White, Elizabeth Lloyd

Invertebrate Embryology
Fell, Paul Erven
Potswald, Herbert Eugene

Neuroembryology
Baird, John Jeffers
Bekoff, Anne Laurens
Das, Gopal Dwarka
Grant, Philip
Heaton, Marieta Barrow
Kosmahl, Henry G
Lyser, Katherine May (Mrs E Shouby)
Meyer, Ronald Leo
Noden, Drew M
Sohal, Gurkirpal Singh
Woods, Geraldine Pittman

Plant Embryology
Cass, David D
Cobb, Glenn Wayne
Corson, George Edwin, Jr
Crotty, William Joseph
Cummins, James Nelson
Dosier, Larry Waddell
Galinat, Walton Clarence
Herr, John Mervin, Jr
Hickman, James C
Jones, Leo Edward
Jordan, Edward Hill
Kuchner, Charles Carroll
Maravolo, Nicholas Charles
Mia, Abdul Jabbar
Millington, William Frank
Prover, Corinne Bea
Raghavan, Valayanghat
Reynolds, John Dick
Smeltzer, Richard Homer
Stein, Otto Ludwig
Webster, Barbara Donahue
Withner, Carl Leslie, Jr

Vertebrate Embryology
Ballard, William Whitney
Boylan, Elizabeth Shippee
Fowler, James A
Goudsmit, Esther Marianne
Heim, Werner George
Johnston, Perry Max
Keys, Charles Everel
Schreiweis, Donald Otto
Walton, Barbara Ann

ENDOCRINOLOGY

Endocrinology
Adams, Walter Church
Ahluwalia, Balwant Singh
Akbar, Abullatah Maksood
Akins, Ervin Loraine
Albert, Alexander
Allen, Willard M
Alszuler, Norman
Amoss, Max St Clair
Anderson, David Gene
Anderson, Nels Carl, Jr
Arguelles, Amilcar Emilio
Arimura, Akira
Assaykeen, Tatiana Anna
Astwood, Edwin Bennett
Avioli, Louis
Bacchus, Habeeb
Bellabarba, Diego
Bennett, John Philip
Berg, Olga Aronowitz
Berger, Sheldon
Bern, Howard Alan
Bethune, John Edmund
Betz, George
Betz, Thomas William
Beyler, Arthur Lewis
Bhalla, Vinod Kumar
Biglieri, Edward George
Binkley, Sue Ann
Bird, Charles Edward
Birnbaumer, Lutz
Black, John B
Blackard, William Griffith
Blair, A James, Jr
Bo, Walter John
Bogdanove, Emanuel Mendel
Bolt, Douglas John
Borth, Rudi
Boshell, Buris Raye

Bottoms, Gerald Doyle
Boyar, Robert Martin
Bradbury, James Thomas
Braselton, Webb Emnett, Jr
Brecher, Peter I
Breitenbach, Robert Peter
Breneman, David Michael
Brinck-Johnson, Truls
Brinn, Jack Elliott, Jr
Bronsky, David
Brooks, Samuel Carroll
Brown, Barry Lee
Brown, Patricia Stocking
Browne, John Symonds Lyon
Bruchovsky, Nicholas
Bukovsky, William
Bullock, John
Bullock, Leslie Patricia
Burns, John Mitchell
Burns, Thomas Wade
Burrell, Craig Donald
Byrnes, William Winfield
Callantine, Merritt Reece
Campbell, Bonnalie Oetting
Canary, John Joseph
Canfield, Robert E
Carbonneau, Roch
Carithers, John Reynolds
Carter, Anne Cohen
Carter, Jeanine Rutherford
Cassidy, Carl Eugene
Castro, Alberto
Cavalieri, Ralph R
Challoner, David Reynolds
Chambers, John William
Chang, Chin-Chuan
Chang, Nada
Channing, Cornelia Post
Charr, Jerome James
Charters, Elaine Mary
Chatterton, Robert Treat, Jr
Chavin, Walter
Chopra, Inder Jit
Choudary, Jasti Bhaskararao
Chretien, Michel
Christian, John Jermyn
Chu, Luke Lo-Hwa
Clark, James Henry
Claus-Walker, Jacqueline Lucy
Clayton, Raymond Brazenor
Clayton-Hopkins, Judith Ann
Clitheroe, H John
Cochrane, Robert Lowe
Coffey, James Cecil, Jr
Cohen, Margo Nita Panush
Cohen, Rochelle Sandra
Cohen, Saul Louis
Colby, Howard David
Collins, Elliott Joel
Collins, William Edgar
Compher, Marvin Keen, Jr
Convey, Edward Michael
Cook, Harry
Coppola, John Anthony
Cornell, James S
Costello, Leslie Carl
Costoff, Allen
Coy, David Howard
Coyne, Mary Frances D
Cramer, Eva Brown
Crawford, John Douglas
Creange, John Ellyson
Creek, Robert Omer
Crisp, Thomas Mitchell, Jr
Criss, Wayne Eldon
Crutchfield, Floy Love
Cunningham, Russell D
Currie, Gustavus Noel, II
Cushman, Paul, Jr
Dale, Edwin
D'Angelo, Savino Albert
Davidson, Mayer B
Davis, Paul Joseph
Davis, Robert Harry
Davis, Steven Lewis
Davis, William James
Dawson, J W
De Feo, Vincent Joseph
De Groot, Jack
DeGroot, Leslie Jacob
Deller, John Joseph, Jr
Demers, Laurence Maurice
Derby, Albert
DeSombre, Eugene Robert
Di George, Angelo Mario
Doe, Richard P
Doering, Charles Henry
Dominguez, Oscar V
Doyle, Lee Lee
Drucker, William D
Duckworth, William Clifford
Duncan, Gordon W
Dunn, Arnold Samuel
Dunn, John Thornton
Dunn, Thomas Guy
Dustman, John Henry
Dziuk, Philip J
Edgren, Richard Arthur

Eisenstein, Albert Bernard
Elders, Minnie Joycelyn
Eletheriou, Basil E
Eliel, Leonard Paul
Enesco, Mircea Aaron
England, Barry Grant
Ensinck, John William
Erpino, Michael James
Ertel, Norman H
Ertel, Robert James
Estep, Herschel Leonard
Evans, John Stone
Eversole, Wilburn John
Ewing, Larry Larue
Ezrin, Calvin
Faber, Lee Edward
Faiman, Charles
Fajer, Abram Benojan
Faludi, Georgina
Fand, Sally Bogolub
Fanslow, Don J
Farmer, Susan Walker
Farmer, T Albert, Jr
Fawcett, Colin Peter
Felig, Philip
Fellows, Robert Ellis, Jr
Fevold, Harry Richard
Ficher, Miguel
Fiske, Virginia Mayo
Fletcher, William Henry
Florsheim, Warner Hans
Flourer, George R
Foreman, Dathl Lois
Forsham, Peter Hugh
Foster, Giraud Vernam
Fougeron, Myron George
Freinkel, Norbert
Freund, Gerhard
Frieden, Edward Hirsch
Frieders, Robert B Jr
Fritz, George Richard, Jr
Frohman, Lawrence Asher
Furth, Eugene David
Gaia, Richard R
Gallo, Robert Vincent
Galton, Valerie Anne
Garren, Henry Wilburn
Gass, George Hiram
Gaunt, Robert
Genest, Jacques
Genuth, Saul M
Geschwind, Irving I
Gibson, Kenneth David
Gilgore, Sheldon G
Gill, Gordon Nelson
Gittes, Ruben Foster
Glasser, Stanley Richard
Glick, David M
Gold, Allen
Gold, Jay Joseph
Goldfarb, Alvin F
Goldman, Max
Goldsmith, Ralph Samuel
Goldstein, Maurice Sabin
Goldzieher, Joseph William
Gona, Amos G
Goodman, David Barry Poliakoff
Goodman, Henry Maurice
Goodman, Charles Joseph
Gorell, Thomas Andrew
Gorman, Colum A
Gorski, Jack
Graff, Morris Morse
Grauer, Robert Coleman
Graves, William Earl
Green, Orville
Greenblatt, Robert Benjamin
Greenman, David Lewis
Greenwood, Frederick C
Greer, Monte Arnold
Gregerman, Robert Isaac
Griffith, David R
Grodsky, Gerold Morton
Gross, Jack
Grosvenor, Clark Edward
Grunt, Jerome Alvin
Gual, Carlos
Gumbreck, Laurence Gable
Habener, Joel Francis
Hadd, Harry Earle
Haddock, Lillian
Hafs, Harold David
Hagadon, Irvine R
Hahn, DoWon
Hahn, Harold Frank
Halkerson, Ian D K
Hall, Peter Francis
Hall, Sam Rutherford
Halmi, Nicholas Stephen
Hansen, Robert John
Harding, Boyd W
Harding, Homer Robert
Harper, Michael John Kennedy
Hauntz, Edgar Alfred
Hay, Eleanor (Louise) Clarke
Hayashida, Tetsuo
Hazelwood, Robert Leonard
Heine, Melvin Wayne

Heller, Carl George
Hendrich, Chester Eugene
Henricks, Donald Maurice
Henry, Walter Lester, Jr
Henzl, Milan Rastislav
Hersberger, Lee George
Hershman, Jerome Marshall
Hess, Melvin
Hirsch, Michael Allen
Hisaw, Frederick Lee, Jr
Hobkirk, Ronald
Hoch, Frederic Louis
Hoffmann, Frederick Gustave
Hoffmann, Daniel Louis
Holland, Emil Otto
Holland, James Philip
Hollman, Gerald Hall
Holmes, William Neil
Holmes, Dorsey Emil
Holtkamp, Dorsey Neil
Horrobin, David Frederick
Horwitz, David Larry
Houlihan, Rodney T
Hubbard, William Ralph
Humes, John Leroy
Hummel, Brian Christopher W
Hymer, Wesley C
Idzkowsky, Henry Joseph
Inskeep, Emmett Keith
Irwin, Glenn Ward, Jr
Jaanus, Siret Desiree
Jackson, Gary Loucks
Jacobs, Barbara B
Jaffe, Robert B
Jasper, Robert Lawrence
Jeffrey, John J
Jellinck, Peter Harry
Jirku, Helmut
Johnson, Donald Charles
Johnston, Cyrus Conrad, Jr
Jones, John Evan
Joshi, Madhusudan Shankarrao
Kahn, Raymond Henry
Kalant, Norman
Kaplan, Selna L
Katz, Fred H
Kaye, Nancy Weber
Keele, Doman Kent
Keil, Lanny Charles
Kendall, John Walker, Jr
Kenny, Alexander Donovan
Kerlan, Joel Thomas
Khairallah, Edward A
King, Cecil Thomas G
Kinson, Gordon A
Kipnis, David Morris
Kirschner, Marvin Abraham
Kitabchi, Abbas E
Klaiber, Edward L
Knobil, Ernst
Kochakian, Charles Daniel
Koerner, Diona Heather
Kogut, Maurice D
Korenman, Stanley G
Kornel, Ludwig
Kostyo, Jack Lawrence
Kowalewski, Konstanty Piotr
Kowarski, A Avinooam
Kraicer, Peretz Freeman
Krainitz, Leon
Krebiel, Eugene B
Kruger, Fred Albert
Kumar, Manjula Satyendra
Kupperman, Herbert Spencer
La Ganga, Thomas S
Ladinsky, Judith L
Ladislaw, John Coleman
Lakshman, Akarasi Bhojaraj
Lambert, Helen Haynes
Landau, Richard Louis
Lauderdale, James W, Jr
Lavenda, Nathan
Leach, Carolyn Sue
Leathem, James Howard
Leavitt, Wendell W
Lederis, Karl
Lee, James B
Lee, Kai-Lin
Leeper, Robert Dwight
Lehoux, Jean-Guy
Leibach, Fredrick Hartmut
Leland, Thomas Mikell
Lennon, Harry Degener
Lenz, Paul Heins
Leonora, John
Lerner, Leonard Joseph
Leung, Benjamin Shuet-Kin
Levenstein, Irving
Levey, Gerald Saul
Levey, Harold Abram
Levine, Jon Howard
Levine, Rachmiel
Levitz, Mortimer
Lewis, Urban James
Liao, Shutsung
Liddle, Grant Winder

Liechty, Richard D
Lindall, Arnold Walfred
Linfoot, John Ardis
Linkie, Daniel Michael
Lippman, Marc Estes
Lisk, Robert Douglas
Liss, Robert H
Lloyd, Charles Wait
Lloyd, James Armon
Lobl, Richard Tolstoi
Lobo, Luiz Carlos Galvao
Loewenstein, Joseph Edward
London, William Thomas
Longley, William Joseph
Lorscheider, Fritz Louis
Lucis, Ojars Janis
Lufkin, Edward Gwynne
Lutwak, Leo
Macchi, I Alden
MacGillivray, Margaret Hilda
MacLeod, Robert Meredith
Madison, Leonard Lincoln
Mahajan, Damodar K
Mahajan, Satish Chander
Mahesh, Virendra B
Mahler, Richard Joseph
Mallette, John M
Maloof, Farahe
Mansour, A Maher
Maraspin, Lyno Evelino
Margolin, Solomon
Marks, Leon Joseph
Martin, Constance Rigler
Martin, Loren Gene
Marvin, Horace Newell
Marx, Stephen John
Masson, Georges Marie Charles
Mayberry, William Eugene
Mayer, George Pat
Mayo, Marie Joiner
Mazer, Ronald Steven
McArthur, Janet W
McConahey, William McConnell, Jr
McCormick, George M, II
McCracken, John Aitken
McDonald, John Kennely
McEvoy, Donald
McFarland, Kay Flowers
McGarry, Eleanor E
McGregor, Robert Finley
McKenzie, John Maxwell
McKinney, Gordon R
McShan, William Hartford
Mead, Rodney A
Meador, Clifton Kirkpatrick
Means, Anthony R
Medici, Paul T
Medlen, Ammon Brown
Meek, Joseph Chester, Jr
Metz, Ralph A, Jr
Meyers, Kenneth Purcell
Midgley, A Rees, Jr
Migeon, Claude Jean
Miller, Myron
Mischler, Richard Avery
Mittier, James Carlton
Moldawer, Marc
Molinari, Pietro Filippo
Molnar, George D
Money, William Lang
Moon, Richard C
Moore, Frank Ludwig
Moorhouse, John A
Morato, Tomas
Mori, Kanaka Fred
Morris, David Julian
Mosier, H David, Jr
Moskowski, Erica F
Moulton, Bruce Carl
Mounib, M Said
Mowles, Thomas Francis
Mueller, Joseph Robert
Mullinix, Kathleen Patricia
Munck, Allan Ulf
Murphy, Beverley (Elaine) Pearson
Nadler, Norman Jacob
Nandi, Satyabrata
Nayak, Ramnath V
Nayfeh, Shihadeh Nasri
Neill, Jimmy Dyke
Nellor, John Ernest
Nelson, Marita Lee
Nelson, Norman Crooks
Neri, Rudolph Orazio
Ney, Robert Leo
Nickerson, Peter Ayers
Nicolette, John Anthony
Nicoll, Charles S
Nicolls, Ken E
Nikitovitch-Winer, Miroslava B
Norris, James Scott
Northrop, Gretajo
Notation, Albert David
Notides, Angelo C
Nowaczynski, Wojciech
Nugent, Charles Arter, Jr

Nussbaum, Noel Sidney
Nusynowitz, Martin Lawrence
Nuting, Ehard Forrest
O'Connell, Kevin Marshall
O'Connor, John Francis
Oddis, Leroy
Odell, William Douglas
Okey, Allan Bernhardt
Oldham, Susan Banks
Olsen, Gjerding
O'Malley, Bert W
Ontjes, David A
Opel, Howard
Oppenheimer, Jack Hans
Orgebin-Crist, Marie-Claire
Osawa, Yoshio
Ottis, Kenneth
Pak, Charles Y
Pancoe, William Louis, Jr
Pang, Peter Kai To
Parker, Donal C
Parker, Mary Langston
Parkhie, Mukund Raghunathrao
Parks, John S
Pasley, James Neville
Pastan, Ira Harry
Pauk, George Lyon
Paulsen, Charles Alvin
Paulsen, Elsa Proehl
Paulson, Stuart R
Payne, Anita H
Peake, Robert Lee
Pearson, John Richard
Peaslee, Margaret H
Peck, William Arno
Peckham, William Dierolf
Pederson, Vernon Clayton
Pelletier, Georges H
Penhos, Juan Carlos
Pento, Joseph Thomas
Perryman, Elizabeth Kay
Persky, Harold
Peterson, Ralph Edward
Peterson, Roy Phillip
Petit, Donald William
Philibert, Robert Lawrence
Piacsek, Bela Emery
Pickford, Grace Evelyn
Pietras, Richard Joseph
Piliero, Sam Joseph
Pi-Sunyer, F Xavier
Pittman, James Allen, Jr
Plotka, Edward Dennis
Pointer, Richard Hamilton
Ponthier, Roy Leonce, Jr
Porter, Johnny Ray
Post, Theodore B
Potts, Gordon Oliver
Potts, John Thomas
Powell, Richard Cinclair
Premachandra, Bhartur N
Preslock, James Peter
Primack, Marshall Philip
Pritchett, John Franklyn
Prokesch, Jeanne Chase
Puett, John David
Raacke, Ilse Dorothea
Raisz, Lawrence Gideon
Ramachandran, Janakiraman
Ramaley, Judith Aitken
Ramey, Estelle R
Rathnam, Premila
Ratner, Albert
Raud, Heinz Randar
Read, Charles H
Read, Virginia Hall
Reddy, William John
Reel, Jerry Royce
Reeves, Jerry John
Reichert, Leo E, Jr
Reit, Barry
Relkin, Richard
Rennels, Edward Gerald
Renzi, Alfred Arthur
Revesz, Clara Rona
Rich, Clayton
Richardson, George S
Riegle, Gail Daniel
Riekstniece, Emilija Katrina
Rifkind, Arleen B
Rillema, James Alan
Rinard, Gilbert Allen
Rivera, Evelyn Margaret
Robbins, Jacob
Roberts, Andre
Roberts, John Stephen
Robertson, Kenneth David
Robertson, Douglas Reed
Robinson, Gerald Garland
Robinson, Jerry Allen
Roche, Marcel
Rockwood, Joseph Guy
Rockwood, William Philip
Rogers, Philip Virgilius
Rose, David Peter
Rose, Edward
Rosen, Jeffrey Mark

Rosen, Ora Mendelsohn
Rosen, Saul W
Rosenbloom, Arlan Lee
Rosenthal, Judith Wolder
Rosoff, Betty
Roth, Jesse
Rothschild, Edmund Otto
Rovner, David Richard
Roy, Arun K
Royce, Paul Chadwick
Rubenstein, Arthur Harold
Rubinstein, Lydia
Ruh, Mary Frances
Ruhmann-Wennhold, Ann Gertrude
Ryan, Kenneth John
Ryan, Robert J
Saiduddin, Syed
Samaan, Naguib A
Sandor, Thomas
Santamarina, Enrique
Saroff, Jack
Saunders, Francis Joseph
Savard, Francis Gerald Kenneth
Savery, Harry P
Saxena, Brij B
Sayers, George
Schally, Andrew Victor
Schimmer, Bernard Paul
Schindler, William Joseph
Schrader, William Thurber
Schultz, Alvin Leroy
Schwartz, Harold Leon
Scott, Joseph Lybrand
Scow, Robert Oliver
Scurry, Murphy Townsend
Segal, Sheldon Jerome
Segaloff, Albert
Sehe, Charles Theodore
Sellner, Ronald George
Senior, Boris
Shade, Robert Eugene
Shaw, Ralph Arthur
Shellabarger, Claire J
Sherman, Lawrence
Sherwood, Louis Maier
Shipley, Meyer, Elva G
Shirley, Barbara Anne
Shome, Basudev
Shum, Wan-Kyng Liu
Sidky, Younan Abdel Malik
Siegel, Edward T
Siegel, Elsie P
Silva, Omega Logan
Simard, Sylvain J
Simkin, Benjamin
Simmons, David J
Simopoulos, Artemis Panageotis
Simpson, Everett Coy
Singh, Sant Parkash
Sinha, Dilip
Sinha, Yagya Nand
Skahen, Julia Goodsell
Skelley, Dean Sutherland
Smith, Carl Walter, Jr
Smith, Donald Eugene
Smith, Olive Watkins
Smith, Richmond Watson, Jr
Smith-Gill, Sandra Joyce
Soderwall, Arnold Larson
Soloff, Melvyn Stanley
Solomon, David Harris
Sonenberg, Martin
Southren, A Louis
Spaulding, Stephen Waasa
Spaziani, Eugene
Squires, Bruce Paul
Srivastava, Laxmi Shanker
Stabenfeldt, George H
Stabler, Timothy Allen
Stafford, Robert Oppen
Stagg, Ronald M
Stancel, George Michael
Stanley, Allan John
Steele, John Earle
Steiner, Donald Frederick
Steinetz, Bernard George, Jr
Stevenson, Joseph Ross
Stewart, Jennifer Keys
Stith, Rex David
Stolc, Viktor
Stolpe, Stanley George
Stolzenberg, Sidney Joseph
Stone, Daniel Boxall
Stone, John Patrick
Stucki, Jacob Calvin
Sussman, Karl Edgar
Swanson, Robert James
Swanson-Eartly, Heidi H
Swislocki, Norbert Ira
Talbert, George Brayton
Tavolga, Margaret Cordsen
Teague, Robert Sterling
Tenenhouse, Alan M
Thau, Rosemarie B Zischka
Thomas, Patricia Zeis
Thomas, William Clark, Jr
Thompson, Phebe Kirsten
Toews, Cornelius J

Toverud, Svein Utheim
Trachewsky, Daniel
Traurig, Harold Henry
Tsang, Charles Pak Wai
Tullner, William W
Tutwiler, Gene Floyd
Tyler, Jean Mary
Tzagournis, Manuel
Ulvedal, Frode
Underwood, Louis Edwin
Ungar, Frank
Urquhart, John, III
Valenta, Lubomir Jan-Vaclav
Vallowe, Henry Howard
Vandenbergh, John Garry
Vanderlaan, Willard Parker
Van Wyk, Judson John
Varandani, Partab T
Veneziale, Carlo Marcello
Vernikos-Danellis, Joan
Villee, Dorothy
Voigt, Walter
Volk, Thomas Lewis
Volpe, Robert
Volpert, Eugene M
Voorhess, Mary Louise
Vranic, Mladen
Wagner, Thomas Edwards
Walker, Richard Francis
Wallace-Haagens, Mary Jean
Walsh, Scott Wesley
Wan, Abraham Tai-Hsin
Ward, Walter Frederick
Warner, Eldon Dezelle
Warner, Marlene Ryan
Warren, James C
Watnick, Arthur Saul
Weinstein, Ira
Weintraub, Bruce Dale
Weitzman, Mary C
Welsh, George W, III
Weltman, A Stanley
Werthessen, Nicholas Theodore
Weymouth, Richard J
Wiebe, John Peter
Wigand, Jeffrey Stephen
Wilcox, Ronald Bruce
Wilhelmi, Alfred Ellis
Williams, Gerald Albert
wilson, Everett D
Wilson, Freddie Elton
Wilson, Lowell D
Wilson, Michael John
Winter, Jeremy Stephen Drummond
Wiswell, John Gordon
Wolf, Richard Clarence
Wolff, Jan
Wong, Harry Yuen Chee
Wood, Francis C, Jr
Woods, James E
Wright, Peter Hedley
Wurtman, Richard Jay
Yasumura, Seiichi
Yip, Cecil Cheung-Ching
Yochim, Jerome M
Yodaiken, Ralph Emile
Yoder, John Menly
Yoshinaga, Koji
Zuccarello, William A

Comparative Endocrinology
Avila, Vernon Lee
Barnawell, Earl B
Baskin, Denis George
Callard, Gloria Vincz
Chiasson, Robert Breton
Clark, Nancy Barnes
de Vlaming, Victor Lynn
Donaldson, Edward Mossop
Epple, August Wilhelm
Fenwick, James Clarke
Hadley, Mac Eugene
Higgs, David Archibald
Hoffman, Roger Alan
Holloway, John Robert
Holtzman, Seymour
Hunt, Ernest Lowell
Jegla, Thomas Cyril
Jones, Richard Evan
Kelsey, Ruben Clifford
Klicka, John Kenneth
Lucis, Ruta
Maher, Michael John
McWhinnie, Dolores J
Norris, David Otto
Novales, Ronald Richards
Patent, Gregory Joseph
Platt, James Earl
Rao, Krothapalli Ranga
Schreibman, Martin Paul
Tassava, Roy A
Tullis, Richard Eugene
Weisbart, Melvin

Neuroendocrinology
Alvarez-Buylla, Ramon
Ben-Jonathan, Nira
Borer, Katarina Tomljenovic

Burgus, Roger Cecil
Campbell, Gary Thomas
Chakraborty, Prabir Kumar
Clemens, James Allen
Corbin, Alan
Cramer, Oneida Morningstar
Curry, John Joseph, III
Dellmann, Horst-Dieter
Dunn, Jon D
Fasano, Anthony Vincent
Fowler, Dona Jane
Gorski, Roger Anthony
Grosser, Bernard Irving
Grota, Lee J
Haller, Edwin Wolfgang
Ifft, John Dempster
Innes, David Lyn
Jacobs, John Joseph
Kaira, Satya Paul
Kizer, John Stephen
Knigge, Karl Max
Kozlowski, Gerald P
Krey, Lewis Charles
Lamperti, Albert A
Litteria, Marilyn
Lynch, Harry James
Malven, Paul Vernon
Mason, John Wayne
McCann, Samuel McDonald
Milmore, John Edward
Mitchell, J Andrew
Moberg, Gary Philip
Nishioka, Richard Seiji
Norman, Reid Lynn
Ondo, Jerome G
Passo, Stanley Samuel
Peter, Richard Ector
Pohorecky, Larissa Alexandra
Quadagno, David Michael
Quadri, Syed Kaleemullah
Quinn, David Lee
Reiter, Russel Joseph
Rhees, Reuben Ward
Santisteban, George Anthony
Sawyer, Charles Henry
Saxena, Anjali
Smith, Erla Ring
Sokol, Hilda Weyl
Sorrentino, Sandy, Jr
Spies, Harold Glen
Stouter, Vincent Paul
Stumpf, Walter Erich
Timiras, Paola Silvestri
Turgeon, Judith Lee
Tyrey, Lee
Ulrich, Renee Sandra
Voogt, James Leonard
Weiner, Richard Ira
Weisz, Judith
Wheaton, Jonathan Edward
Wilbur, Donald Lee
Winter, Robert John
Zucker, Irving

Pediatric Endocrinology
Archibald, Reginald MacGregor
Arnold, Mary B
Brasel, Jo Anne
Grossman, Milton S
Hollowell, Joseph Gurney, Jr
Kohler, Elaine Eloise Humphreys
MacMillan, Duncan Robert
Martin, James Frederic
Root, Allen William
Rosenfield, Robert Lee
Ruvalcaba, Rogelio H A
Weldon, Virginia V

Reproductive Endocrinology
Adkins, Elizabeth Kocher
Ahmad, Nazir
Andersen, Richard Nicolaj
Anderson, Mary Loucile
Bhatnagar, Ajay Sahai
Billiar, Reinhart Billie
Bingel, Audrey Susanna
Brinkley, Howard J
Brooks, Jerry R
Burden, Hubert White
Chalamalasetty, Venkateswara Rao
Chatoraj, Sati Charan
Chowdhury, Ajit Kumar
Chung, Kyung Won
Cole, Harold Harrison
Coulson, Patricia Bunker
Dermody, William Christian
Dubin, Norman H
Ellis, LeGrande Clark
Foster, Douglas Layne
France, Evelyn S (Kalagher)
Freeman, Marc Edward
Givens, James Robert
Gwazdauskas, Francis Charles
Hargrove, James Lee
Hopkins, Thomas Franklin
Jackanicz, Theodore Michael
Jacobson, Herbert (Irving)
Johnston, John O'Neal
Jones, Sanford L
Kalland, Gene Arnold
Karsch, Fred Joseph
Katzenellenbogen, Benita Schulman
Kaye, Alvin Maurice
Keyes, Paul Landis
Kimball, Frances Adrienne
Kimmel, Gary Lewis
Kling, Orzo Ray
Kulkarni, Bidy D
Kulshestwar, Anant Pandurang
Leathem, James Hain
Ledwitz-Rigby, Florence Ina
Lee, Chung
Lipner, Harry Joel
Louis, Thomas Michael
Mathur, Rajesh Swarup
Maurer, Ralph Rudolf
Mills, Thomas Marshall
Niswender, Gordon Dean
Ogle, Thomas Frank
Pitkow, Howard Spencer
Pollard, Jeffrey William
Rayford, Phillip Leon
Reyes, Francisco Ismael
Robertson, Hamish Alexander
Rothchild, Irving
Spears, James Richard
Stetson, Milton H
Stewart, Sheila Frances
Stormshak, Fredrick
Swanson, Lloyd Vernon
Taylor, Rhoda E
Thatcher, William Watters
Thompson, Frederick Nimrod, Jr
Wagner, William Charles
Weiss, Gerson
Wilks, John William
Wright, Paul Albert
Yen, Samuel Show-Chih
Zeiler, Frank Jacob
Zimbelman, Robert George

ENTOMOLOGY

Entomology
Abercrombie, Jay
Adams, Curtis H
Adams, Phillip A
Adkins, Randall Henry
Adkins, Theodore Roosevelt, Jr
Adler, Victor Eugene
Aderz, Warren Clifford
Akre, Roger David
Alee, Marshall Craig
Allen, George E
Allen, Harry Willis
Allen, Richard K
Aller, Harold Ernest
Allison, William Earl
Allred, Dorald Mervin
Anderson, John Fredric
Anderson, Lauren Davis
Anderson, Norman Herbert
Anderson, Robert Curtis
Anderson, Russell D
Andrawes, Nathan R
Andres, Lloyd A
Andrews, Gordon Louis
Angalet, George William
Anthon, Edward W
Apple, James Wilbur
Appleby, James E
Applegate, Richard Lee
Arbogast, Richard Terrance
Arnaud, Paul Henri, Jr
Arnett, Ross Harold, Jr
Arnold, John Walter
Arthur, B Wayne
Ashdown, Donald
Atwood, Carl Edmund
Atwood, Mark Wyllie
Atyeo, Warren Thomas
Bacon, John Alvin
Bacon, Oscar Gray
Bailey, Donald Leroy
Bailey, Jack Clinton
Bailey, Leo L
Bailey, Stanley Fuller
Baird, Craig Riska
Baker, Edward William
Baker, Griffin Jonathan
Ballard, Ralph Campbell
Balsbaugh, Edward Ulmont, Jr
Bancroft, Harold Ramsey
Banks, William Alden
Baranowski, Richard Matthew
Barker, Philip Shaw
Barker, Roy Jean
Barnes, Ralph Craig
Barnett, Douglas Eldon
Barnhart, Clyde Sterling, Sr
Baron, Ronald L
Barr, William Frederick
Barron, John Robert
Barry, Billy Dean
Barry, Cornelius
Bartholomai, C W
Bartlett, Blair Ralph
Bass, Max Herman
Bath, James Edmond
Batra, Suzanne Wellington Tubby
Batzer, Harold Otto
Bay, Darrell Edward
Beal, Richard Sidney, Jr
Beard, Raimon Lewis
Beardsley, John Wyman, Jr
Beatty, Alice Ferguson
Bechtel, Robert Christy
Becker, Edward Coulton
Becker, Irwin Bernard
Becnel, Irwin Joseph
Beer, Robert Edward
Beinert, Helmut
Belkin, John Nicholas
Bellinger, Peter F
Bender, Harvey Alan
Benedict, John Howard, Jr
Bennett, Gordon Fraser
Benson, Robert Leland
Berner, Lewis
Berry, Ralph Eugene
Berryman, Alan Andrew
Berthold, Robert, Jr
Bess, Henry Alver
Bick, George Herman
Bickley, William Elbert
Bidlingmayer, William Lester
Birch, Martin Christopher
Bishop, Jack Lynn
Bishop, Guy William
Black, Emmett Russell, Jr
Blair, Billie D
Blake, Doris Holmes
Blake, George Henry, Jr
Blakeley, Phillip Earl
Blakeslee, Theodore Edwin
Blanc, Francis Louis
Bland, Roger Gladwin
Blickle, Robert Louis
Bliss, Myron, Jr
Blum, Murray Sheldon
Bobb, Marvin Lester
Boddy, Dennis Warren
Bode, William Morris
Boesel, Marion Waterman
Bohart, Richard Mitchell
Booth, Gary Melvon
Borchers, Harold Allison
Boren, Roger Boatner
Borror, Donald Joyce
Boston, James D
Boudreaux, Henry Bruce
Boush, George Mallory
Bowen, Myles Foster
Bowling, Clarence C
Bowman, James Sheppard
Braasch, Norman L
Bradley, Julius Roscoe, Jr
Bram, Ralph A
Branson, Terry Fred
Bratt, Albertus Dirk
Bray, Dale Frank
Brazzel, James Roland
Breeland, Samuel Glover
Breland, Osmond Philip
Brett, Charles H
Brewer, Franklin Douglas
Brewer, Jesse Wayne
Brezner, Jerome
Britt, N Wilson
Broersma, Delmar B
Bronskill, Joan Frances
Brook, Ted Stephens
Brookes, Victor Jack
Brookes, Derl
Brooks, Marion Alice
Brooks, Robert Franklin
Brower, John Harold
Brown, Harley Procter
Brown, Carl Dee
Brown, Leland Ralph
Brust, Reinhart A
Brusven, Merlyn Ardel
Bryan, Douglas Everett
Bucher, Gordon Edwards
Buffington, John Douglas
Bull, Don Lee
Bullock, Howard R
Bullock, Robert Crossley
Burbutis, Paul Philip
Burkhardt, Christian Carl
Burkholder, Wendell Eugene
Burks, Barnard De Witt
Burrage, Robert Columbus
Burrage, R H
Burts, Everett C
Bushland, Raymond Cecil
Butcher, James Walter
Butler, George Daniel, Jr
Butler, Linda
Butts, William Lester
Buxton, Jay A
Byers, George William
Byers, John Robert
Byers, Robert Allan
Byrne, Hugh Desmond
Calkins, Carrol Otto
Callahan, Philip Serna
Callenbach, John Anton
Caltagirone, Leopoldo Enrique
Campau, Edward Junior
Campbell, Alan
Campbell, Frank Leslie
Campbell, Ian Maclean
Campbell, John Bryan
Campbell, William Robert
Campbell, William Vernon
Cannon, William Nelson, Jr
Cantelo, William Wesley
Cantrall, Irving James
Carde, Ring Richard Tomlinson
Carlson, Oscar Verdell
Carlson, Robert Bruce
Carlson, Stanley David
Carman, Glenn Elwin
Carnahan, James Elliot
Caron, Dewey Maurice
Carpenter, Frank Morton
Carruth, Laurence Adams
Carter, Robert Duncan
Cartier, Jean Jacques
Cartwright, Oscar Ling
Cate, James Richard, Jr
Chalfant, Richard Bruce
Chamberlin, Derrell Lynn
Chambers, Leland
Chandler, Satish Raman
Chang, Franklin
Chang, Shen Chin
Chant, Donald A
Chapman, Harold Clyde
Chapman, Paul Jones
Chapman, R Keith
Chemsak, John A
Cheng, Tien-Hsi
Cherry, Edward Taylor
Chesnut, Thomas Lloyd
Chiang, Huai C
Childs, Dana Pitt
Chiykowski, Lloyd Nicholas
Church, Norman Stanley
Cibula, Adam Burt
Clarke, John Frederick Gates
Clay, Mary Ellen
Clench, Harry Kendon
Cleveland, Merrill L
Clouter, Elmer Joseph
Clower, Dan Frederic
Cmejla, Howard Edward
Coffey, Marvin Dale
Cogburn, Ray
Cole, Arthur Charles, Jr
Cole, Walter Eckle
Collins, Richard Cornelius
Collins, William John
Colwell, Robert Knight
Combs, Robert L, Jr
Compton, Charles Chalmer
Cone, Wyatt Wayne
Connor, Lawrence John
Cook, David Russell
Cook, Edwin Francis
Cooke, Herman Glenn
Coon, Beckford Feddersen
Cooper, Murray Irving
Copeland, Thompson Preston
Coppel, Harry Charles
Corbet, Philip Steven
Corey, Robert Arden
Corkins, Jack Philips
Couch, Terry Lee
Coulson, Jack Richard
Counselman, C J
Couser, Raymond Dowell
Covell, Charles Van Orden, Jr
Cox, H C
Cox, D D
Craig, George Brownlee, Jr
Craig, Wilfred Stuart
Cram, William Thomas
Creighton, Charlie Scattergood
Creighton, John Thomas
Cress, Donald Chauncey
Cromroy, Harvey Leonard
Cross, William Henley
Crosswhite, Carol D
Crowell, Hambin Howes
Crowell, Robert Merrill
Cruz, Carlos
Cupp, Eddie Wayne
Curtin, Thomas J
Curtis, Charles Elliott

Daggy, Richard Henry
Dahm, Paul Adolph
Dakin, Matt Eitel
Daly, Howell Vann
Dame, David Allan
Daniels, Norris Eugene
Darby, Rollo E
Darlington, Philip Jackson, Jr
Dasch, Clement Eugene
Daugherty, David M
Davenport, Demorest
Davidson, John Angus
Davidson, Ralph Howard
Davies, Douglas Mackenzie
Davis, Donald Ray
Davis, Donald Walter
Davis, Harry Glenwood
Davis, Norman Thomas
Deal, Andrew Stuart
Dean, Herbert A
De Bach, Paul (Hevener)
DeCoursey, Russell Myles
DeFoliart, Gene Ray
Dekle, George Wallace
Delfin, Eliseo Dais
Del Fosse, Ernest Sheridan
DeLoach, Culver Jackson, Jr
Dennis, Clifford John
Dennis, Norman McLeod
Deonier, D L
Dewey, James Edwin
Dickason, Elvis Arnie
Dicke, Ferdinand Frederick
Dietz, Alfred
Dimond, John Barnet
Dixon, Carl Franklin
Dixon, John Charles
Dixon, Stuart Edward
Doane, Charles Chesley
Dobrovsky, Todor Manoloff
Dogger, James Russell
Dolphin, Robert Earl
Donley, David Edward
Donohoe, Heber Clark
Doutt, Richard Leroy
Dow, Richard Phelps
Downes, John Antony
Downey, John Charles
Drea, John James, Jr
Duda, Edward John
Dunbar, Dennis Monroe
Durant, John Alexander, III
Dyer, Judith Gretchen
Dysart, Richard James
Earle, Norman Williston
Ebeling, Walter
Eddy, Thomas A
Edmunds, George Francis, Jr
Edmunds, Lafe Rees
Edwards, J Gordon
Edwards, John S
Eickwort, George Campbell
Eiben, Galen J
Eighme, Lloyd Elwyn
Eikenbary, Raymond Darrell
Ellington, Joe J
Elsey, Kent D
Elzinga, Richard John
Emsley, Michael Gordon
Enns, Wilbur Ronald
Eschle, James Lee
Eskafi, Fred M
Estep, Charles Blackburn
Evans, Howard Ensign
Fagan, Ernest Brad
Fairchild, Graham Bell
Fairchild, Mahlon Lowell
Falcon, Louis A
Falter, John Max
Farr, Thomas Howard
Fast, Paul Gerhardt
Feeny, Paul Patrick
Ferguson, George Ray
Ferguson, William E
Finlayson, Douglas Gordon
Fiori, Bart J
Fischang, William John
Fischer, Roland Lee
Fisher, Ellsworth Henry
Fisher, Theodore William
Fitzpatrick, George
Fleming, Richard Cornwell
Fleischer, Charles Anthony
Fletcher, ed Walker
Fletcher, Lowell W
Flint, Hollis Mitchell
Flowers, Ralph Wills
Floyd, Ernest Hazel
Fluke, Sam Spruill
Fluno, John Arthur
Flynn, Arthur David
Folts, Dwight David
Foote, Benjamin Archer
Foote, Richard Herbert
Foott, William Henry
Forbes, Albert Ronald

Forbes, James
Force, Don Clement
Forgash, Andrew John
Forsythe, Howard Yost, Jr
Foster, James Russell
Foster, John Edward
Fraenkel, Gottfried Samuel
Franclemont, John George
Franklin, Rudolph Thomas
Frazier, James Lewis
Frazier, Norman Walter
Frederickson, Richard William
Freele, Hugh W
Freeman, Jeffrey Van Duyne
Freitag, Julius Herman
Fremling, Calvin R
French, Allen Lee
French, Ellery Walter
French, Frank Elwood, Jr
Friauf, James Joseph
Frick, Kenneth Eugene
Frisbie, Raymond Edward
Frishman, Austin Michael
Fritz, Roy Fredolin
Froeschner, Richard Charles
Frohne, William Carrington
Furman, Deane Philip
Fusco, Robert Angelo
Fye, Robert Eaton
Galindo, Pedro
Gallun, Robert Louis
Garay, Gustav John
Garmus, Ralph David
Garner, William Vaughn
Garth, John Shrader
Gary, Norman Erwin
Gentry, Joseph Wesley
George, Boyd Winston
George, Charles Redgenal
George, John Allen
George, John Edwin
Georghiou, George Paul
Gerber, Charles Edwin
Gerberg, Eugene Jordan
Gerhardt, Paul Donald
Gerhard, Reid Richard
Getzin, Louis William
Gibson, William Wallace
Gilbert, Edward E
Gillaspy, James Edward
Gilliland, Floyd Ray, Jr
Gilmore, James Eugene
Gilstrap, Franklin Ephriam
Gittins, Arthur Richard
Gladney, William Jess
Glass, Edward Hadley
Glen, Robert
Glover, Sandra Jean
Gloyd, Leonora Katherine
Goddin, Avery Howe
Goeden, Richard Dean
Gooding, Ronald Harry
Gojmerac, Walter Louis
Goodrich, Michael Alan
Goodwin, James Thomas
Goonewardene, Hilary Felix
Gordh, Gordon
Gould, George Edwin
Goulding, Robert Lee, Jr
Graham, Charles Lee
Graham, Harry Morgan
Graham, Kenneth
Graham, Lewis Texada
Graves, Jerry Brook
Graves, Robert Charles
Gray, Henry Emil
Grayson, James McDonald
Green, Henry Burwell
Greenberg, Bernard
Greene, Gerald L
Griffith, Melvin Eugene
Grigarick, Albert Anthony, Jr
Grimm, James K
Grissell, Edward Eric Fowler
Gupta, Ayodhya P
Gurney, Ashley Buell
Gustafson, Joel Frank
Guthrie, Frank Edwin
Guthrie, Wilbur Dean
Guyer, Gordon Earl
Gyrisco, George Gordon
Habeck, Dale Herbert
Haberman, Warren Otto
Hackwell, Glenn Alfred
Hagen, Kenneth Sverre
Hagmann, Lyle Everest
Haines, Kenneth A
Haines, Robert Gordon
Hair, Jakie Alexander
Halfhill, John Eric
Halgren, Lee A
Hall, Franklin Robert
Hamilton, Eugene W
Hamilton, Robert W
Hamlen, Ronald Alan
Hammond, Abner M, Jr
Hamrum, Charles Lowell
Hanec, William

Hanna, Ralph Lynn
Hansen, Harry Louis
Hansens, Elton J
Hanson, John Francis
Haramoto, Frank H
Hardee, Dicky Dan
Harden, Philip Howard
Harding, Wallace Charles, Jr
Hardy, D Elmo
Harein, Phillip Keith
Harris, Charles Ronald
Harris, Ernest James
Harris, Frank Aubrey
Harris, Jesse Max
Harris, Marvin Kirk
Harris, Robert Lee
Harrison, James Ostelle
Harrison, Robert J
Harriss, Thomas T
Hart, Elwood Roy
Hart, Richard Allen
Hartberg, Warren Keith
Harvey, Thomas Larkin
Harwood, Robert Frederick
Hasbrouck, Frank Flinn
Hastings, Ellsworth (Bernard)
Hatch, Melville Harrison
Hatchett, Jimmy Howell
Haynes, Harry Leonard
Hays, Donald Brooks
Hays, Sidney Brooks
Head, Robert Berturm
Heimpel, Arthur MacLeod
Helgesen, Robert Gordon
Heming, Bruce Sword
Hendricks, Donavan Edward
Hendrickson, Robert Mark, Jr
Henneberry, Thomas James
Henry, Charles Stuart
Hensley, Sess D
Henson, Joseph Lawrence
Henson, Walter Robert
Hepner, Leon Wilburne
Herring, Jon Lamar
Hetrick, Lawrence Andrew
Hewitt, George Berlyn
Hibbs, Edwin Thompson
Hickey, William August
Hildebrand, John Grant, III
Hill, Clarence Howell
Hill, Kenneth Lee
Hill, Roscoe Earle
Hill, Stuart Baxter
Hilliard, John Roy, Jr
Hilsenhoff, William Leroy
Himel, Chester Mora
Hinckley, Alden Dexter
Hink, Walter Fredric
Hintz, Howard William
Hixson, Ephriam
Hoage, Terrell Rudolph
Hodson, Alexander Carlton
Hoffman, Jarett David
Hoffmann, Julius R
Hoffmann, Clarence Howard
Hofmaster, Richard Namon
Hogue, Charles Leonard
Holdsworth, Robert Powell, Jr
Holland, George Pearson
Holloway, Rodney Leon
Hoopingarner, Roger A
Hopkins, Theodore Louis
Horber, Ernst Konrad
Horn, David Jacobs
Horner, Norman V
Hornuff, Lothar Edward, Jr
Horsfall, William Robert
Houk, Wallace Eugene
Hovey, Charles Louis
Howden, Henry Fuller
Howe, Robert George
Howe, Wayne Lamoyne
Howell, Dariel Elza
Hower, Arthur Aaron, Jr
Howitt, Angus Joseph
Hoy, Stanley Charles
Hoyt, Stanley Charles
Hubbell, Theodore Huntington
Huber, Ivan
Huffman, Edward Wight
Huggans, James Lee
Hull, George, Jr
Humphreys, Jan Gordon
Hunter, Preston Eugene
Hurd, Paul David, Jr
Husband, Robert W
Hyland, Kerwin Ellsworth, Jr
Iglinsky, William
Ignoffo, Carlo Michael
Ihde, Keith Desmond
Jacklin, Stanley William
Jackson, Robert Dewey
Janes, Melvin Joseph
Janes, Ray Low
Jantz, Orlo Kenneth
Jaycox, Elbert Ralph
Jefferson, Roland Newton
Jeppson, Lee Ralph

Joachim, Frank G
Johansen, Carl August
Johansson, Mildred P
Johnson, Albert Wayne
Johnson, Donald Ross
Johnson, Warren Thurston
Jones, Philip Arthur
Jones, Walter Larue
Jordan, Cedric Roy
Jorgensen, Clive D
Jubb, Gerald Lombard, Jr
Judd, William Wallace
Kadoum, Ahmed Mohamed
Kaloostian, George H
Kamran, Mervyn Arthur
Kantack, Benjamin H
Kaplanis, John Nicholas
Kauffeld, Norbert M
Keaster, Armon Joseph
Keith, David Lee
Kelsey, Lewis Preston
Kelsheimer, Eugene Gillespie
Kennedy, Maldon Keith
Kerr, Stratton H
Kerr, Theodore William, Jr
Kethley, John Bryan
Kevan, Douglas Keith McEwan
Kido, George Seiji
Kincade, Robert Tyrus
Kindler, Sharon Dean
King, Charles C
King, Donald Roy
King, Edwin Wallace
King, Herman (Lee)
Kingsolver, John Mark
Kinzer, H Grant
Kistner, David Harold
Klassen, Waldemar
Klaus, Ewald Fred, Jr
Klein, Michael Gardner
Klostermeyer, Edward Charles
Klun, Jerome Anthony
Knapp, Fred William
Knight, Harry Hazelton
Knight, Kenneth Lee
Knipling, Edward Fred
Knoke, John Keith
Knowles, Charles Otis
Knowlton, Carroll Babbidge, Jr
Knowlton, George Franklin
Knutson, Herbert Claus
Knutson, Lloyd Vernon
Koerber, Thomas William
Kok, Loke-Tuck
Kosztarab, Michael
Kouskolekas, Costas Alexander
Koval, Charles Francis
Kraft, Gerald F
Krantz, Gerald William
Kreasky, Joseph Bernard
Krekeler, Carl Herman
Krestensen, Elroy R
Kring, James Burton
Krombein, Karl Von Vorse
Kuitert, Louis Cornelius
Kulman, Herbert Marvin
Kurczewski, Frank E
Kuyper, C Keith
Kvenberg, John Eide
Kwolek, William F
LaBerge, Wallace E
Ladd, Thyril Leone, Jr
Laing, John E
Lambdin, Paris Lee
Lancaster, Jessie Leonard, Jr
Lange, William Harry, Jr
Langford, George Shealy
Lanham, Urless Norton
Lashomb, James Harold
Laster, Marion Logan
Lattin, John D
Lauck, David R
Laveglia, James Gary
Laveglia, Robert James
Lawrence, Fred Parker
Lawrence, John Francis
Lawrence, Vinnedge Moore
Lawson, Fred Avery
Lea, Arden Otterbein
LeCato, George Leonard, III
Leeper, John Robert
Lefkovitch, Leonard Philip
Leggett, Joseph Edwin
Legner, E Fred
Lehmkuhl, Dennis Merle
Leigh, Thomas Francis
Lener, Walter
Leonard, David E
Leopold, Roger Allen
Levine, Marshall Robert
Levine, Harvey Robert
Lewis, Robert Earl
Lewis, Standley Eugene
Lewis, Wallace Joe
Libby, John Lester
Lichtenstein, Emanuel Paul
Lienk, Siegfried Eric

ENTOMOLOGY

Lilly, John Henry
Limpel, Lawrence Eugene
Lin, Cheng Shan
Linam, Jay H
Lincoln, Charles Gatewood
Lindgren, David Leonard
Lindquist, Donald Arthur
Lindquist, Richard Kenneth
Linley, John Roger
Linsley, Earle Gorton
Liscombe, Ernest A R
Little, Harold Franklin
Lloyd, Edwin Phillips
Lloyd, John Edward
Lofgren, Clifford Swanson
Long, William Henry
Louloudes, Spiro James
Lovell, James Byron
Lowe, James Harry, Jr
Lowe, Ronald Edsel
Luckmann, William Henry
Ludvik, George Franklin
Ludwig, Carl Edward
Lugthart, Garri John, Jr
Lum, Patrick Tung Moon
Lunn, Horace Odin
Lunn, Steele Ray
MacArthur, Kenneth William
MacCarthy, Hubert Reagh
MacCollom, George Butterick
MacCreary, Donald
MacGillivray, M Ellen
Mackauer, Manfred
MacLean, Bonnie Kuseske
MacLellan, Charles Roger
MacLeod, Ellis Gilmore
MacLeod, Guy Franklin
MacPhee, Albert William
Madsen, Harold F
Magnarelli, Louis Anthony
Maksymiuk, Bohdan
Malcolm, David Robert
Mallis, Arnold
Mampe, Charles Douglass
Mangat, Baldev Singh
Mangitz, George Rudolph
Manzelli, Manlio Arthur
Maramorosch, Karl
March, Ralph Burton
Marks, Louis Sheppard
Marsh, Paul Malcolm
Marshall, James Dale
Marston, Norman Lee
Martin, Dial Franklin
Martin, Ethelbert Cowley
Martin, Lloyd Milo
Martorell, Luis Felipe
Marucci, Philip Edward
Mason, Richard Randolph
Masteller, Edwin C
Mathewson, John Angell
Mauston, Glenn Warren
Maxwell, Fowden Gene
Maxwell, Kenneth Eugene
Mayer, Marion Sidney
Mayo, Z B
McAlpine, James Francis
McClanahan, Robert Joseph
McClure, Mark Stephen
McCoy, Clayton William
McDaniel, Burruss, Jr
McDonald, Ian Cameron Crawford
McEwen, Freeman Lester
McFadden, Max Wulfsohn
McGaughey, William Horton
McGill, John Joseph
McHaffey, David George
McIver, Susan Bertha
McKelvey, John Jay, Jr
McKnight, Melvin Edward
McLaughlin, Roy Earl
McLean, Donald Lewis
McLeod, John Malcolm
McMillan, William Stirling
McMullian, Wallard
McMullian, James A
McMurtry, James A
McPherson, John Edwin
McWilliams, Kenneth Leroy
Mead, Frank Waldreth
Meadows, Charles Milton
Medler, John Thomas
Menke, Arnold Stephen Ernst
Menn, Julius Joel
Menon, Maya Devi
Meola, Shirlee May
Merkl, Marvin Eugene
Merritt, Richard William
Messenger, Powers Slater
Messersmith, Donald Howard
Metcalf, Robert Lee
Michener, Charles Duncan
Middlekauff, Woodrow Wilson
Miller, Charles Douglas F
Miller, Douglass Ross
Miller, Howard Charles

Miller, Richard Lloyd
Miller, Thomas Albert
Mills, Robert Barney
Milne, David Hall
Milner, Floyd Duane
Minnick, Danny Richard
Mistric, Walter J, Jr
Mitchell, Earl Bruce
Mitchell, Everett Royal
Mitchell, Henry Cooper
Mitchell, Joseph Christopher
Mitchell, Wallace Clark
Mockford, Edward Lee
Moffett, Joseph Orr
Monroe, Ronald Eugene
Moody, Julius Reynard
Moore, Glenn D
Moore, Harry Ballard, Jr
Moore, Leon
Moore, Raymond F, Jr
Moore, Stevenson, III
Moore, Thomas Edwin
Morris, Glenn Karl
Morrison, Eston Odell
Morrison, Frank Orville
Morse, Roger Alfred
Moss, W Wayne
Muka, Arthur Allen
Mulkern, Gregory Benedict
Mulla, Mir S
Munma, Ralph O
Munroe, Eugene Gordon
Murdoch, Wallace Pierce
Murphey, Milledge
Musgrave, Carol Ann
Music, Gerald Joe
Naegele, John Adam
Namba, Ryoji
Nash, Robley Wilson
Nault, Lowell Raymond
Neal, John William, Jr
Nebeker, Alan V
Neel, William Wallace
Neilson, Charles Henry
Nelson, Eric V
Nelson, John Marvin, Jr
Nelson, Richard Douglas
Nelson, Vernon A
Nelson, William Arnold
Nesbitt, Herbert Hugh John
Nettles, William Carl, Jr
Neunzig, Herbert Henry
Newsom, Leo Dale
Nichols, Philip Ray
Nicholson, Harry Page
Nickel, John L
Nielsen, David Gary
Nielsen, Lewis Thomas
Nielsen, Mervin William
Niemczyk, Harry D
Nigg, Herbert Nicholas
Noetzel, David Martin
Nuting, William Leroy
Nye, William Preston
Oatman, Earl R
Ode, Philip E
O'Keeffe, Lawrence Eugene
Olive, Aulsey Thomas
Oliver, Abe D, Jr
Oman, Paul Wilson
O'Neil, Louis C
Onsager, Jerome Andrew
Oonnithan, Easwaran Sukumaran
Ordway, Ellen
Ortman, Eldon Emil
Osgood, Charles Edgar
Osmun, John Vincent
Ota, Asher Kenhachiro
Owen, Bernard Lawton
Owen, William Bert
Owsley, William Burr
Ozburn, George W
Palm, Charles Edmund
Paradis, Rodolphe Omer
Parencia, Charles R
Parrott, William Lamar
Paschke, John Donald
Patrick, Charles Russell
Pauly, Ludwig K
Payne, Jerry Allen
Peacock, John William
Pechuman, Laverne Leroy
Peck, Stewart Blaine
Pedigo, Larry Preston
Pepper, James Hubert
Persing, Charles Oscar
Peters, Don Clayton
Peters, Harold Truman
Peters, Leroy Lynn
Peters, Thomas Michael
Peters, William Lee
Peterson, Allan George
Peterson, James J
Peterson, Bobbie Vern (Robert)
Petty, Howard B
Pfadt, Robert E
Pfrimmer, Theodore Roscoe

Philip, Cornelius Becker
Phillips, Jacob Robinson
Phillips, John Henry Howard
Phillips, William George
Phillis, William Avery, III
Pielou, Douglas Patrick
Pienkowski, Robert Louis
Pieper, Gustav Rene
Pilon, Jean-Guy
Pimentel, David
Pinto, John Darwin
Pipa, Rudolph Louis
Pitman, Gary Boyd
Pitre, Henry Nolle, Jr
Pletsch, Donald James
Poe, Sidney LaMarr
Polhemus, John Thomas
Post, Richard Lewis
Poston, Freddie Lee, Jr
Powell, Donnie Melvin
Powell, Jerry Alan
Pratt, Harry Davis
Pratt, John Jacob, Jr
Presser, Bruce Douglas
Price, Peter Wilfrid
Price, Richard Graydon
Price, Roger DeForrest
Pritchard, Gordon
Proshold, Fredrick Irving
Pruess, Kenneth Paul
Purcell, Alexander Holmes, III
Putler, Benjamin
Pyenson, Louis L
Quisenberry, Benson F
Rabb, Robert Lamar
Race, Stuart Rice
Radcliffe, Edward B
Radford, Robert Earl
Redinger, Leonard Maurice
Redmond, Billy Lee
Reed, John K
Reed, Walter T
Rees, Bryant Eugene
Rees, Don Merrill
Reeves, Roger Marcel
Rehn, John William Holman
Reichart, Charles Valerian
Reid, Milton Roy
Reinecke, John Philip
Reinert, James Arnold
Retenmeyer, Carl William
Reynolds, Harold Truman
Richardson, Henry Howe
Richerson, Jim Vernon
Ridgway, Richard L
Riegel, Garland Tavner
Riehl, Louis Adam
Riemann, John G
Riley, Robert C
Ring, Richard Alexander
Rings, Roy Wilson
Ritcher, Paul Osborn
Roach, William Kenney
Roback, Selwyn
Robbins, William E
Roberts, Howard Radclyffe
Roberts, John Harvey
Robertson, Clyde Henry
Robertson, Robert L
Robinson, Arthur Grant
Robinson, Harold Ernest
Robinson, James Vance
Robinson, William H
Rodriguez, Juan Guadalupe
Roehmild, George M
Rogers, Charlie Eilic
Rolston, Lawrence H
Romoser, William Sherburne
Rosenberger, Charles Rupley
Ross, Anthony
Ross, Edward Shearman
Ross, Herbert Holdsworth
Ross, Mary Harvey
Roussel, John S
Rowley, Wayne A
Ruppel, Robert Frank
Russell, Charles Clayton
Russell, Louise May
Russell, Mercer P
Rutschky, Charles William
Sabrosky, Curtis Williams
Saether, Ole Anton
Sailer, Reece Ivan
Saini, Rajinder S
Salt, Reginald Wilson
Sampson, William Wilson
Sanders, Darryl Paul
Sanderson, Milton William

Sartor, Clyde Flake
Sauer, John Robert
Sauer, Richard John
Saunders, Joseph Lloyd
Savos, Milton George
Schaefer, Carl W, II
Schaefer, Charles Herbert
Schaefers, George Albert
Schaffner, Joseph Clarence
Schalk, James Maximillian
Scharff, Donald Kenneth
Scheel, Carl Alfred
Scheibner, Rudolph A
Schlinger, Evert Irving
Schmidt, Claude Henri
Schneider, Donald Lloyd
Schuder, John Theodore
Schuh, Joe Anthony
Schuster, David J
Schwartz, Paul Henry
Scott, Harold George
Scott, Harry Eldon
Scott, James Allan
Scudder, Geoffrey George Edgar
Seawright, Jack Arlyn
Seay, Thomas Nash
Sedman, Yale S
Selander, Richard Brent
Semel, Maurice
Sevacherian, Vahram
Sferra, Pasquale Richard
Shaddy, James Henry
Shambaugh, George Franklin
Shanks, Carl Harmon, Jr
Shapiro, Martin
Sharma, Madan Lal
Sharp, Silas
Shaw, Frank Robert
Shaw, John Gilbert
Shenefelt, Roy David
Shepard, Buford Merle
Shepard, Harold Henry
Sherman, Martin
Shinkle, Michael Paul
Shinn, Alvin Fleetwood
Shipp, Oliver Elmo
Shorey, Harry Haslam
Shorter, Daniel Albert
Shubeck, Paul Peter
Silberglied, Robert Elliot
Silhacek, Donald Le Roy
Simanton, William Aldrich
Simkover, Harold George
Simons, John Norton
Simpson, Geddes Wilson
Simpson, Karl William
Simpson, Robert Gene
Skelsey, James Jeremiah
Skelton, Thomas Eugene
Slater, James Alexander
Sleeper, David Allanbrook
Sleeper, Elbert Launce
Sloan, Miner Joe
Slosser, Jeffrey Eric
Sluss, Robert Reginald
Smetana, Ales
Smilowitz, Zane
Smith, Clyde F
Smith, David Rollins
Smith, Dean Seyward
Smith, Floyd Franklin
Smith, Howard Weedon
Smith, James Willie, Jr
Smith, John Cole
Smith, Marion Estelle
Smith, Ray Fred
Smith, Stanford Dennis
Smith, Virgil Kirkland, Jr
Smittle, Burrell Joe
Smythe, Richard Vincent
Soderstrom, Edwin Loren
Sollers-Riedel, Helen
Sommerman, Kathryn Martha
Sorensen, Kenneth Alan
Spangler, Hayward Gosse
Spangler, Paul Junior
Sparks, Alton Neal
Spear, Philip James
Specht, Harold Balfour
Speers, Charles Frederick
Sperka, Christina Kanschat
Spink, William T
Spooner, John D
Sprenkel, Richard Keiser
Spyhalski, Edward James
Staal, Gerardus Benardus
Standifer, Lonnie Nathaniel
Stannard, Lewis Judson, Jr
Starks, Kenneth James
Starnes, Ordway
Stark, Harold Emil
Staples, Robert
Steffan, Wallace Allan
Stehr, Frederick William
Steiner, Loren Franklin
Steinhauer, Allen Laurence
Steitzer, Lorin Roy
Stephen, William Procuronoff

Sterling, Winfield Lincoln
Stern, Vernon Mark
Sternburg, James Gordon
Stevens, Thomas McConnell
Stewart, Kenneth Wilson
Stockhammer, Karl Adolf
Stockton, William Denis
Stoetzel, Manya Brooke
Stoffolano, John George, Jr
Stoltz, Robert Lewis
Stoltzfus, William Bryan
Stone, Jay D
Stoner, Adair
Stoner, Warren Norton
Storch, Richard Harry
Streu, Herbert Thomas
Strong, Rudolph Greer
Sudia, William Daniel
Sullivan, Daniel Joseph
Summers, Francis Marion
Summers, Thomas Eugene
Sutter, Gerald Rodney
Swailes, George Edward
Swenson, Knud George
Swift, Fred Calvin
Swisher, Ely Martin
Sylvester, Edward Sanford
Tafuri, John Francis
Tamaki, George
Tanner, Gary Dale
Taschenberg, Emil Frederick
Tassoni, Joseph Paul
Tauber, Maurice Jesse
Taylor, Philip Seyfang
Terranova, Andrew Charles
Tetrault, Robert Close
Thatcher, Theodore Ossip
Thirugnanam, Muthuvelu
Thomas, Gustave Daniel
Thomas, Hollis Allen
Thompson, Clarence Garrison
Thompson, Frederic Christian
Thompson, Hugh Erwin
Thompson, Patrick Haley
Thompson, Robert Kruger
Thompson, Victor Carl
Thorsteinson, Asgeir Jonas
Thurman, Duane Edward
Thurston, Richard
Tietjen, William Leighton
Tilden, James Wilson
Toba, Hachiro Harold
Todd, James Wyatt
Townes, Henry Keith, Jr
Townes, Marjorie Chapman
Townsend, Howard Garfield, Jr
Townsend, Kenneth
Trammel, Kenneth
Travis, Asher Eugene
Treat, Asher Eugene
Triplehorn, Charles A
Truxal, Fred Stone
Tsao, Ching Hsi
Tunis, William David
Turner, Ernest Craig, Jr
Turner, William Jr
Turnipseed, Samuel Guy
Tuttle, Donald Monroe
Ulagaraj, Muniyandy Seydunganallur
Unzicker, John Duane
Valcarce, Arland Casiano
Valentine, Barry Dean
Van Cleave, Horace William
Vandehey, Robert C
van den Bosch, Robert
Van Geluwe, John David
Vickers, David Hyle
Vickery, Vernon Randolph
Vinson, S Bradleigh
Visscher, Saralee Neumann
Waddill, Van Hulen
Wade, William Howard
Wagner, John Alexander
Waites, Robert Ellsworth
Waldbauer, Gilbert Peter
Waldron, Acie Chandler
Walgenbach, David D
Walker, David Whitman
Walker, John Robert
Walker, Thomas Jefferson
Wall, William James
Wallace, George Egbert
Wallace, James Bruce
Wallner, William E
Walstrom, Robert John
Walton, Cyprian James
Ward, Charles Richard
Ward, Gertrude Luckhardt
Ware, George Whitaker, Jr
Warren, Lloyd Oliver
Washino, Robert K
Waters, Norman Dale
Watkins, Julian F, II
Watson, David Livingston
Watson, Robert Lee
Watson, Theo Franklin
Watson, Wynnfield Young
Watters, Frederick Lewis

Watts, John Gordon
Wave, Herbert Edwin
Weathersby, Augustus Burns
Weaver, Andrew Albert
Webb, Donald Wayne
Webb, Morgan Chofield, III
Weber, Neal Albert
Weber, Richard Gerald
Webster, James Allan
Weekman, Gerald Thomas
Weems, Howard Vincent, Jr
Weidhaas, John August, Jr
Wellso, Stanley Gordon
Werner, Floyd Gerald
Werner, Richard Allen
Weseloh, Ronald Mack
Wessel, Richard Deaton
Wheeler, Alfred George, Jr
Wheeler, George Carlos
Wheeler, Jeanette Norris
Whipp, Arthur Andrew
Whitcomb, Willard Hall
White, Albert Cornelius
White, Leland Darrell
Whitehead, Armand T
Whitesell, James Judd
Whitney, Wendell Keith
Whittemore, Frederick Winsor
Wick, James Roy
Wiggins, Glenn Blakely
Wilde, Gerald Eldon
Wilkinson, Christopher Foster
Wilkinson, Paul R
Wilkinson, Robert Cleveland, Jr
Willard, John Royal
Willey, Ruth Lippitt
Williams, Michael Ledell
Williams, Robert K
Willis, Edwin Roy
Willis, Harold Lester
Wilson, Billy Ray
Wilson, Clifton Artie
Wilson, Louis Frederick
Wilson, Mark Curtis
Wilson, Richard Lee
Wing, Merle Wesley
Wingo, Curtis W
Wirth, Willis Wagner
Wiseman, Billy Ray
Witkowski, John Frederick
Wolfenbarger, Dan A
Wolfenbarger, Daniel Otis
Wolff, Theodore Albert
Womack, Herbert
Wong, Horne Richard
Wood, Francis Eugene
Wood, Stephen Lane
Woodbury, Elton Norris
Woodruff, Robert Eugene
Wooldridge, David Paul
Wright, Charles Gerald
Wright, Kenneth Harold
Wright, Wayne Gordon
Wylie, William Dickey
Yadav, Raghunath P
Yamamoto, Robert Takaichi
Yearian, William C
Yendol, William G
Yoder, Wayne Alva
Yoho, Timothy Price
Yonke, Thomas Richard
Yoshimoto, Carl Masaru
Young, James Christopher F
Young, Jerry H
Young, Seth Yarbrough, III
Yun, Young Mok
Zavortink, Thomas James
Zimmerman, Elwood Curtin
Zimmerman, James Roscoe
Zukel, John William
Zwick, Robert Ward

Obenchain, Frederick DeCroes
Oldfield, George Newton
Oliver, James Henry, Jr
Peterson, Paul Constant
Radovsky, Frank Jay
Rohde, Charles John, Jr
Scarborough, Charles Spurgeon
Singer, George
Walker, Neil Allan
Wilson, Nixon Albert

Arachnology

Archer, Allan Frost
Drew, William Arthur
Drucker, Philip
Duffield, Lathel Flay
Eberhard, William Granville
Horner, Norman V
Levi, Herbert Walter
Muchmore, William Breuleux
Muma, Martin Hammond
Nelson, Sigurd Oscar, Jr
Peck, William B
Reiskind, Jonathan
Roth, Vincent Daniel
Rovner, Jerome Sylvan
Stahnke, Herbert Ludwig
Thompson, Raymond Harris
Unzicker, John Duane
Williams, Stephen

Economic Entomology

Adkisson, Perry Lee
Allen, William Westhead
Anderson, Norman Lewis, Jr
Arnett, William Harold
Asquith, Dean
Bariola, Louis Anthony
Barke, Harvey Ellis
Barnes, Douglas
Barnes, Martin McRae
Barges, Rex J
Barton, Harvey Eugene
Baumhover, Alfred Henry
Beirne, Bryan Patrick
Beland, Gary LaVern
Bottger, Gilbert Ted
Boyer, William Paul
Brunner, Jay F
Burdit, Arthur Kendall, Jr
Carlson, Elmer Carl
Carpenter, Gene Paul
Christensen, Christian Martin
Cohick, A Doyle, Jr
Dahms, Reynold George
Darst, Philip High
Davich, Theodore Bert
Davis, Alexander Cochran
Day, William H
DeBoo, Robert Ford
Eden, William Gibbs
Ellis, Clifford Roy
Ewart, William Howard
Fronk, William Don
Gentile, Adrian George
Goleman, Denzil Lyle
Granett, Philip
Gregory, Wesley Wright, Jr
Harding, James Alfred
Harper, Alex Maitland
Harris, Emmett Dewitt, Jr
Harrison, Floyd Perry
Haws, Byron Austin
Highland, Henry Arthur
Hoffman, Robert A
Holmes, Neil Delvin
Hopkins, Lemac
Huddleston, Ellis Wright
Kamm, James A
Kennedy, George Grady
Kirk, Vernon Miles
Koehler, Carlton Smith
LaPlante, Albert Aurel, Jr
Larson, Noal P
Laudani, Hamilton
Lentz, Gary Lynn
Lopez, Genaro
Lyon, William Francis
Matteson, John Warren
Meyer, Ronald Harmon
Moore, Joseph B
Nettles, William Carl
Norton, Robert James
Ortega, Alejandro
Owens, John Charles
Partida, Gregory John, Jr
Pass, Bobby Clifton
Pausch, Robert Dale
Rawlins, William Arthur
Richards, Russell Fayette
Roberts, James Ernest, Sr
Robinson, John Frank
Roney, James Neville
Ryder, John C, Jr
Sanford, James Walker
Schuster, Michael Frank
Sechriest, Ralph Earl
Semtner, Paul Joseph

Acarology

Allen, William Westhead
Anastos, George
Bohnsack, Kurt K
Cross, Hansell Flynn
Denmark, Harold Anderson
Drummond, Roger Otto
Farrier, Maurice Hugh
Fashing, Norman James
Gladney, William Jess
Gless, Elmer E
Hall, Clarence Coney, Jr
Holman, Leta Jane
Homsher, Paul John
Hunter, Preston Eugene
Jalil, Mazhar
Lindquist, Evert E
Loomis, Edmond Charles
Loomis, Richard Biggar
Moffitt, Harold Roger
Moser, John C
Nevin, Floyd Reese
Newell, Irwin Mayer
Newkirk, Richard Albert Michael
Nickel, Phillip Arnold
Nutting, William Brown

Smith, Edward Holman
Smith, Omar Ewing, Jr
Snapp, Oliver Irvin, Sr
Snetsinger, Robert J
Staetz, Charles Alan
Stockdale, Harold James
Summers, Charles Geddes
Tashiro, Haruo
Tozloski, Albert Henry
Treece, Robert Eugene
Turpin, Frank Thomas
Walz, Arthur Joseph
Washburn, Richard Hancorne
Workman, Ralph Burns
Young, William Robert

Forest Entomology

Amman, Gene Doyle
Anderson, Roger Fabian
Barras, Stanley J
Benjamin, Daniel Marshall
Brown, Norman Rae
Cameron, Edward Alan
Connola, Donald Pascal
Dahlsten, Donald L
Daterman, Gary Edward
DeBoo, Robert Ford
De Mars, Clarence John, Jr
Drooz, Arnold T
Esenther, Glenn R
Farrier, Maurice Hugh
Fatzinger, Carl Warren
Fellin, David Gene
Fettes, James Joseph
Finnegan, Raymond Joseph
Fox, Richard Charles
Giese, Ronald Lawrence
Hain, Fred Paul
Harman, Dan M
Haynes, Dean L
Heikkenen, Herman John
Jennings, Daniel Thomas
Kearby, William H
Knight, Fred Barrows
Lanier, Gerald Norman
Lyon, Robert Lyndon
Macdonald, Duncan Ross
Merkel, Edward Paul
Miller, William Eldon
Mitchell, Russel Gene
Mott, David Gordon
Muldrew, James Archibald
Nebeker, Thomas Evan
Nichols, James Otis
Nord, John C
Olson, Robert Edward
Rennels, Robert Gossett
Rudinsky, Julius Alexander
Safranyik, Laszlo
Schenk, John Albright
Schmiege, Donald Charles
Simeone, John Babtista
Simmons, Gary Adair
Sippell, William Lloyd
Sloan, Norman F
Smith, Richard Harrison
Solomon, James Doyle
Stark, Ronald William
Stevens, Robert E
Thatcher, Robert Clifford
Varty, Isaac William
Waters, William E
Williams, Carroll Burns, Jr
Witter, John Allen
Wood, David Lee
Yates, Harry Orbell, III

Insect Ecology

Allen, Douglas Charles
Burleigh, Joseph Gaynor
Carolin, Valentine Mott, Jr
Cothran, Warren Roderic
Coulson, Robert N
Davis, Robert
Evans, William George
Frank, John Howard
Frankie, Gordon William
Frazer, Bryan Douglas
Harcourt, Douglas George
Havery, Michael Irving
Hays, Kirby Lee
Hendrick, Rodney Douglas
Hoy, Marjorie Ann
Huffaker, Carl Barton
Kay, Carol Ann
Kieckhefer, Robert William
Knight, Clifford Burnham
Krogstad, Blanchard Orlando
Lee, Siu-Lam
LeRoux, Edgar Joseph
MacLean, David Belmont
Mason, Charles Eugene
Rust, Richard W
Ryan, Roger Baker
Schultz, John Charles
Sears, Markham Karli

Sheldon, Joseph Kenneth
Sinha, Ranendra, Nath
Smith, Grahame J C
Streams, Frederick Arthur
Tarpley, Wallace Armell
Uhler, Lowell Dohner
Wellington, William George

Insect Morphology
Craig, Douglas Abercrombie M
Dicke, Robert Jerome
Drecktrah, Harold Gene
Helms, Thomas Joseph
Hermann, Henry Remley, Jr
Parsons, Margaret Cranston
Stay, Barbara
Woolever, Patricia S
Zacharuk, R Y

Insect Pathology
Adams, Jean Ruth
Angus, Thomas Anderson
Bell, Marion Randolph
Briggs, John Dorian
Brooks, Wayne Maurice
Brown, Anthony William Aldridge
Cantwell, George E
Castillo, Jessica Maguila
Clark, Truman Benton
Cunningham, John Castel
Dulmage, Howard Taylor
Friedman, Stanley
Goodwin, Ronald Hayse
Hall, Irvin Monroe
Harper, James Douglas
Holbrook, Frederick R
Hughes, Kenneth Marion
Jaques, Robert Paul
Kaplan, Martin L
Kaya, Harry Kazuyoshi
Kramer, John Paul
Kurtti, Timothy John
Maddox, Joseph Vernard
Nordin, Gerald LeRoy
Ramoska, William Allen
Reed, David Kent
Rinderer, Thomas Earl
Roberts, Donald Wilson
Shimanuki, Hachiro
Splittstoesser, Clara Quinnell
Stairs, Gordon R
Tamashiro, Minoru
Tanada, Yoshinori
Vaughn, James L
Wittig, Gertraude Christa
Zimmack, Harold Lincoln

Insect Physiology
Adams, Jean Burnham
Berry, Spencer Julian
Bhaskaran, Govindan
Butz, Andrew
Chevone, Boris Ivan
Clark, Edgar William
Cochran, Donald Gordon
Collins, Janet Valerie
Cook, Benjamin Jacob
Crowder, Larry A
Dahlman, Douglas Lee
Dasch, Gregory Alan
Davis, Gordon Richard Fuerst
Denlinger, David Landis
Dethier, Vincent Gaston
Downer, Roger George Hamill
Eaton, John LeRoy
Ewen, Alwyn Bradley
Fain, Margery Jones
Feir, Dorothy Jean
Fisk, Frank Wilbur
Flemings, Milton Baker
Florentine, Gerald Joseph
Friend, William George
Happ, George Movius
Holman, Grant Mark
House, Howard Leslie
Hsiao, Ting Huan
Jones, Jack Colvard
Jones, Richard Lamar
Judson, Charles LeRoy
Judy, Kenneth James
Kadner, Carl George
Kamal, Adel S
Keeley, Larry Lee
Khan, Mohammed Abdul Quddus
Koeppe, John K
Krysan, James Louis
Larsen, Joseph Reuben
Leahy, Mary Gerald
Liles, James Neil
Linley, John Roger
Locke, Michael
Loschiavo, Samuel Ralph

McFarlane, John Elwood
Mittler, Thomas E
Mullins, Donald Eugene
Muchmor, John A
Nappi, Anthony Joseph
Nation, James Lamar
Nayar, Jai Krishen
Nijhout, H Frederick
Nijhout, Mary McAllister
Noblet, Raymond
Pappas, Larry George
Raisbeck, Barbara
Rao, Balakrishna Raghavendra
Retnakaran, Arthur
Richards, Albert Glenn
Riddiford, Lynn Moorhead (Mrs James W Truman)
Riegert, Paul William
Robertson, Jacqueline Lee
Roth, Louis Marcus
Rousell, Gerald
Salkfeld, E Helen
Schroeder, Mark Edwin
Seligman, Isaac Morris
Smyth, Thomas, Jr
Svoboda, James Arvid
Truman, James William
Weaver, Nevin
Wilkens, Jewel L
Wimer, Larry Thomas
Woodruff, Laurence Clark
Wyatt, Gerard Robert

Insect Toxicology
Armstrong, John Alexander
Barker, Philip Shaw
Berger, Robert S
Brady, Ullman Eugene, Jr
Cutkomp, Laurence Kremer
Dauterman, Walter Carl
Dorough, Hendley Wyman
Goodwin, William Jennings
Hastings, Felton Leo
Johnson, Richard Emanuel
Mehendale, Harihara Mahadeva
Menzer, Robert Everett
Nakatsugawa, Tsutomu
Nigam, Prakash Chandra
Page, Louise
Perry, Albert Solomon
Quraishi, Mohammed Sayeed
Roberts, Richard Bruce
Scoggin, John Kyle
Soboczenski, Edward John
Sun, Yun Pei
Tate, Laurence Gray
Turnquist, Richard Lee
Weiden, Mathias Herman Joseph
White, Jane Vicknair
Wittman, James Smythe, III
Zaharis, John Louis

Medical Entomology
Aitken, Thomas Henry Gardiner
Amin, Omar M
Andersen, Dean Martin
Anderson, John Richard
Axtell, Richard Charles
Baerg, David Carl
Barnes, William Wayne
Barnett, Herbert Chester
Bay, Ernest C
Beadle, Leslie Dewey
Benach, Jorge L
Berry, Richard Lee
Blanton, Franklin Sylvester
Bonnet, David Dudley
Boreham, Melvin Murray
Burden, George Stanley
Butler, Joseph Miles
Chamberlain, Roy William
Christensen, Howard Anthony
Clements, Burie Webster
Darsie, Richard Floyd, Jr
Dowell, Frank Herbert
Eads, Richard Bailey
Edman, John David
Eldridge, Bruce Frederick
Elbel, Robert E
Emerson, Kary Cadmus
Evans, Burton Robert
Ezell, William Bruce, Jr
Fairchild, Homer Eaton
Favorite, Frank G, Jr
Fleming, Glenn Allen
Fontaine, Russell Edgar
Fowler, Harland Wade, Jr
Fox, Irving
Gingrich, Richard Earl
Gorham, John Richard
Gouck, Harry (Kydd)
Gould, Douglas Jay
Graham, Owen Hugh
Gratz, Norman G
Grimstad, Paul Robert
Grothaus, Roger Harry
Gwadz, Robert Walter

Haines, Thomas Walton
Hall, Donald William
Havertz, David S
Hayes, George Roy, Jr
Hayes, Richard Oliver
Herrin, Charles Selby
Hill, Alfred, Jr
Hilton, Donald Frederick James
Hitchcock, James Carroll, Jr
Hobbs, Jesse H
Holway, Richard Thomas
Hopla, Cluff Earl
Hu, Stephen Moi Kee
Hull, William Ballou
Hurlbut, Herbert Sumner
Jambach, Hugo Andrew, Jr
Johnson, William E, Jr
Joseph, Stanley Robert
Kappus, Karl Daniel
Keegan, Hugh Lawrence
Keirans, James Edward
Khalaf, Kamel T
Kliewer, Marc Jeffrey
Klowden, Marc Jeffrey
La Brecque, Germain C
Lacaillade, Charles William
Laird, Marshall
Lane, Robert Sidney
Lee, Vernon Harold
Lewis, Franklin Beach
Lewis, Leyburn F
Lungstrom, Leon
McClelland, George Anderson Hugh
McDaniel, Ivan Noel
McLintock, John James Reid
Miller, Albert
Miura, Takeshi
Morgan, Neal O
Mount, Gary A
Mullen, Gary Richard
Murray, William Donald
Nelson, Bernard Clinton
Nelson, Robert LeRoy
Newhouse, Verne Frederic
Newson, Harold Don
Nowell, Wesley Raymond
O'Connor, Charles Timothy
Olson, Jimmy Karl
Ouzts, Johnny Drew
Parrish, Dale Wayne
Patterson, Richard Sheldon
Patterson, William Junior
Peck, John Hubert
Pitts, Charles W
Provost, Maurice Wilfred
Radovsky, Frank Jay
Rathbun, Carlisle Baxter, Jr
Reed, Horace Beecher, Jr
Rice, Paul LaVerne
Rigby, Paul T
Roberts, Richard Harris
Rogers, Andrew Jackson
Rozeboom, Lloyd Eugene
Ryckman, Raymond Edward
Scanlon, John Earl
Siverly, Russell Emmett
Smith, Carroll N
Smith, Omar Ewing, Jr
Snoddy, Edward L
Spielman, Andrew
Steelman, Carrol Dayton
Taylor, Robert Tieche
Terwedow, Henry Albert, Jr
Thurber, George A
Thurman-Swartzwelder, Ernestine H
Tipton, Vernon John
Tonn, Robert J
Traub, Robert
Travis, Bernard Valentine
Trpis, Milan
Vanderberg, Jerome Philip
Wallis, Robert Charles
Ward, Ronald Anthony
Weidhaas, Donald E
Williams, Roger Wright
Wood, Sherwin Francis
Wright, James Elbert
Wright, Russell Emery
Zimmerman, John Harvey

Systematic Entomology
Abbasi, Qamar Ali
Allen, Robert Thomas
Anderson, Donald Morgan
Arnell, J Hal
Ashlock, Peter Dunning
Baumann, Richard William
Benoit, Paul
Blocker, Henry Derrick
Brown, Calvin Hugh
Cazier, Mont Adelbert
Christian, Paul Jackson
Connell, Walter Anthony
Cross, Earle Albright, Jr
Curry, La Verne Leon
Dalgleish, Robert Campbell

Duckworth, Walter Donald
Ferguson, Douglas Campbell
Flint, Oliver Simeon, Jr
Freytag, Paul Harold
Gagne, Raymond J
Gill, Gordon Drew
Gotwald, William Harrison, Jr
Gregg, Robert Edmond
Haag, William George
Hambleton, Edson Jorge
Hobbs, Kenneth Rollin
Hodges, Ronald William
James, Maurice Theodore
Johnson, Clarence Daniel
Kavanagh, David Henry
Kim, Ke Chung
Kramer, James Philip
La Rivers, Ira John
Mason, William Richardson Miles
Miller, David Clair
Peterson, Lance George
Rentz, David Charles
Rozen, Jerome George, Jr
Schuh, Randall Tobias
Shaffer, Jay Charles
Smith, Eric Howard
Stage, Gerald Irving
Thorp, Robbin Walker
Valley, Karl Roy
Wasbauer, Marius Sheridan
White, Richard Earl
Young, David Allan

Veterinary Entomology
Butler, Jerry Frank
Crystal, Maxwell Melvin
Dobson, Richard Cecil
Rogoff, William Milton

FISH & WILDLIFE SCIENCES

Fish Biology
Bardach, John E
Batts, Billy Stuart
Baxter, George T
Beardsley, Grant Lindley, Jr
Bevan, Donald Edward
Bond, Lyndon Herrick
Borgeson, David P
Brewer, Gary David
Broadhead, Gordon Clifford
Brown, Bradford E
Bulow, Frank Joseph
Burgner, Robert Louis
Cahn, Phyllis Holstein
Caillouet, Charles W
Campbell, Charles J
Cleaver, Frederick Charles
Cole, Charles Franklyn
Collins, Richard Arlen
Cooper, Edwin Lavern
Cooper, Gerald Paul
Cope, Oliver Brewem
Davis, William Spencer
Dizon, Andrew Russell
Donaldson, John Russell
Dovel, William Lawrence
Dovel, Virgil Eugene
Duncan, Thomas O
Ebert, Earl Ernest
Eckroat, Larry Raymond
Eicher, George J
Eipper, Alfred Ward
Elson, Paul Frederick
Flickinger, Stephen Albert
Fore, Paul Lewis
Gerking, Shelby Delos
Greene, George Nystrom
Griswold, Bernard Lee
Grossiein, Marvin Darrel
Grover, John Harris
Gunning, Gerald Eugene
Hagen, Harold Kolstoe
Hales, Donald Caleb
Hall, James Dane
Hamilton, James Arthur Roy
Hatch, Richard Wallace
Hauser, William Joseph
Haven, Dexter Stearns
Heard, William R
Henry, Kenneth Albin
Hergenrader, Gary Lee
Heyerdahl, Eugene Gerhardt
Hodder, Vincent MacKay
Hokanson, Kenneth Eric Fabian
Horton, Howard Franklin
Hunt, Robert L
Huntsman, Gene Raymond
Idyll, Clarence Purvis
Iversen, Edwin Severin
Jenkins, Robert M
Johnson, James Edward
Jones, Bernard R
Juhl, Rolf
June, Fred C
Kelcher, James J

Trainer, Daniel Olney

Kennedy, William Alexander
Kilpatrick, Earl Buddy
Kimsey, Jerry Bruce
Kinney, Edward Coyle, Jr
Kjelson, Martin Anton, III
Klaassen, Harold Eugene
Klawe, Witold L
Koo, Ted Swei-Yen
Kuehn, Jerome H
Larimore, Richard Weldon
Latta, William Carl
Lennon, Robert Earl
Lennon, Herbert Lee
Leon, Kenneth Allen
Lewis, Robert Minturn
Marfin, William Robert
Mathur, Dilip
Maxfield, Galen Harry
May, Arthur William
McCauley, Robert William
McLain, Albertson Lamson
Mechan, William Robert
Merrell, Theodore Reed, Jr
Merriner, John Vennor
Momot, Walter Thomas
Moss, Donovan Deaft
Mount, Donald I
Murphy, Garth Ivor
Nakatani, Roy E
Neill, William Harold
Orcutt, Harold George
Pacheco, Anthony Louis
Patriarche, Mercer Harding
Power, Geoffrey
Purkett, Charles A, Jr
Rajagopal, P K
Raleigh, Robert Franklin
Reed, James Brooker
Reed, Roger J
Reynolds, James Blair
Ricker, William Edwin
Ridenhour, Richard Lewis
Roessler, Martin A
Rogers, Donald Eugene
Rothschild, Brian James
Saila, Saul Bernhard
Scalet, Charles George
Scattergood, Leslie Wayne
Schmitz, William Robert
Schmulbach, James C
Schoettger, Richard A
Scott, Donald Charles
Southward, Glen Morris
Stalnaker, Chair B
Stauffer, Thomas Miel
Shaklee, James Brooker
Shell, Eddie Wayne
Shomura, James Harold
Stevenson, James Harold
Strasburg, Donald Wishart
Struhsaker, Jeannette Adair Whipple
Summerfelt, Robert C
Tack, Peter Isaac
Tait, James Simpson
Talbot, Gerald Byron
Thompson, Richard Baxter
Toetz, Dale W
Verhoeven, Leon A
Wagner, Harry Henry
Walburg, Charles Herman
Walford, Lionel Albert
Walker, Charles R
Warren, Charles Edward
West, Jerry Lee
Whitney, Richard Ralph
Wisby, Warren Jensen
Wise, John P

Fisheries

Abramson, Norman Jay
Allee, Brian James
Allen, George Herbert
Applegate, Vernon Calvert
Barclay, Lee Armstead, (Jr)
Barrett, Izadore
Bayliff, William Henry
Becker, Clarence Dale
Behmer, David J
Brett, John Roland
Buck, David Homer
Bulkley, Ross Vivian
Burrell, Victor Gregory, Jr
Butler, Robert Lee
Campbell, DeWayne E
Carlander, Kenneth Dixon
Carr, John Frank
Chadwick, Harold King
Chance, Charles Jackson
Clark, John Harlan
Clark, Minor E

Cobb, Bryant Franklin, III
Colby, Peter J
Congleton, James Lee
Copes, Frederick Albert
Dahlberg, Michael Lee
Dammann, Arthur Erle
De Witt, John William, Jr
Donaldson, Edward Mossop
Duncan, Bryan Lee
Erman, Don Coutre
Farris, David Allen
Fetterolf, Carlos De La Mesa, Jr
Fingerman, Sue Whitsell
Firestone, Alexander
Fox, William Walter, Jr
Fredin, Reynold A
Fryer, John Louis
Fukuhara, Francis M
Funicelli, Nicholas Anthony
Gard, Richard
Gregory, Richard Wallace
Hanser, Donald Frary
Hare, Gerard Murdock
Harvey, Harold H
Hassler, Thomas J
Hastings, Waldon Houston
Hendricks, Lawrence Joseph
Hester, Frank J
Higman, James B
Hoagman, Walter John
Horak, Donald L
Horton, Donald Bion
Houde, Edward Donald
Houston, Alan Stewart
Hunn, Joseph Bruce
Johnson, James Howard
Johnson, Lionel
Jones, Albert Cleveland
Kask, John Laurence
Katz, Max
Kelley, John Richard, Jr
Kessler, Dan
Kilambi, Varad Raj
King, Willis
Kukuhn, Joseph Henry
Lagler, Karl Frank
Lane, Edwin David
Larkin, Peter Anthony
Laurs, Robert Michael
Lawler, George Herbert
Lawrence, John Medlock
Lawrence, William Mason
Leduc, Gerard
LeMier, Emanuel H
Lewis, William Madison
Lievense, Stanley James
Linton, Thomas LaRue
Loosanoff, Victor L
Lovell, Richard Thomas
Low, Loh-Lee
Maltzeff, Eugene M
McCann, James Alwyn
McCombie, Alen Milne
McConville, David Raymond
McNeil, William J
Mearns, Alan John
Messieh, Shoukry Naseef
Meyer, Fred Paul
Mikenberg, Giora
Moorman, Robert Bruce
Neal, Richard Allan
Nelson, Philip R
Nicholson, William Robert
O'Connor, Joel Sturgens
Patalas, Kazimierz
Pauley, Gilbert Buckhannan
Pearson, William Dean
Pella, Jerome Jacob
Pereyra, Walter T
Perry, Lorin Edward
Phelps, Ronald P
Platts, William Sidney
Plumb, John Alfred
Pritchard, G Ian
Pruter, Alonzo Theodore
Quarles, Carroll Adair, Jr
Rathjen, Warren Francis
Regier, Henry Abraham
Regier, Lloyd Wesley
Rockwell, Julius, Jr
Roelofs, Terry Dean
Rosebery, Dean Arlo
Rounsefell, George Armytage
Royce, William Francis
Rucker, Robert Raymond
Sakagawa, Gary Toshio
Salo, Ernest Olavi
Schneider, Philip William, Jr
Schoning, Robert Whitney
Seguin, Louis-Roch
Shelton, William Lee
Sidwell, Virginia DeCecco
Sigler, William Franklin
Smith, Lloyd Lyman
Smith, Roderick MacDowell
Smoker, William Alexander
Stasko, Alvars B
Stauffer, Gary Dean

Steinberg, Maynard Albert
Sumner, Richard Lawrence
Sykes, James Enoch
Tarr, Hugh Lewis Aubrey
Taube, Clarence Martin
Tibbles, John James
Trefethen, Parker S
Tubb, Richard Arnold
Tyler, Albert Vincent
Uthe, John Frederick
Van Cleve, Richard
Warner, Kendall
Waters, Thomas Frank
Wedermeyer, Gary Alvin
Welander, Arthur Donovan
Whiteside, Bobby Gene
Williams, Francis
Williams, Wells Eldon
Woelke, Charles Edward
Wydoski, Richard Stanley

Fisheries Management

Baxter, John Lewis
Byrd, Isaac Burlin
Clothier, William Delbert
Davies, William Donald
Day, Lewis Rodman
Durham, Leonard
Edson, Quentin A
Grizzell, Roy Ames, Jr
Halsey, Thomas Gordon
Hanks, Robert William
Harville, John Patrick
Herrington, William Charles
Hughes, Eldon Parker
Jeffers, Keith Bartlett
Jeffrey, Norris Boddie
Jester, Douglas Brewer
Lackey, Robert T
Lovshin, Leonard Louis, Jr
McHugh, John Laurence
Mercer, Malcolm Clarence
Messersmith, James David
Myhre, Richard John
O'Gara, Bartholomew Willis
Pagan-Font, Francisco Alfredo
Powers, Joseph Edward
Roos, John Francis
Russell, Thomas Randall
Schumacher, Robert E
Skillman, Robert Allen
Stevenson, William Henry
Vanicek, C David
Wilder, Donald George
Wilson, James Larry

Fish Pathology

Amend, Donald Ford
Bullock, Graham Lambert
Gibson, George G
Landolt, Marsha LaMerle
Lewis, Donald Howard
Margolis, Leo
Post, George
Rogers, Wilmer Alexander
Walker, Roland
Wobeser, Gary Arthur

Wildlife Biology

Anderson, William Leno
Applegate, James Edward
Baskett, Thomas Sebree
Braun, Clat E
Callaham, Mac A
Campbell, Howard
Causey, Miles Keith
Dodds, Donald Gilbert
Erskine, Anthony J
Harder, John Dwight
Johnson, Albert Sydney, III
Johnson, Eric Van
Klaas, Erwin Eugene
Krull, John Norman
Labisky, Ronald Frank
Larson, Joseph Sbanley
Linzey, Donald Wayne
MacLulich, Duncan Alexander
Mendall, Howard Lewis
Pelton, Michael Ramsay
Raveling, Dennis Graff
Samuel, David Evan
Short, Henry Laughton
Skovlin, Jon Matthew
Sowls, Lyle Kenneth
Speller, Stanley Wayne
Steinhoff, Harold William
Thomas, Jack Ward
VanDruff, Larry Wayne
Verme, Louis Joseph

Wildlife Diseases

Cowan, Archibald B
Dieterich, Robert Arthur
Hayes, Frank Alfred
Lange, Robert Echlin, Jr
McCoy, John Neal
Stuht, John
Thorpe, Bert Duane

Wildlife Ecology

Ables, Ernest D
Anderson, Daniel William
Anderson, Raymond Kenneth
Andrews, Charles Lawrence
Anthony, Robert Gene
Bailey, James Allen
Banko, Winston Edgar
Baumgartner, Frederick Milton
Bolen, Eric George
Buckner, Charles Henry
Burger, George Vanderkarr
Case, Ronald Mark
Cauley, Darrell Lee
Chesemore, David Lee
Cornwell, George William
Coulter, Malcolm Wilford
Crawford, Hewlette Spencer, Jr
Crawford, Richard Dwight
Davis, Charles A
Ditberner, Philip Lynn
Doerr, Philip David
Elder, William Hanna
Evenden, Fred George
Flinders, Jerran Trueman
Flyger, Vagn Folkmann
Fredrickson, Leigh H
Gipson, Philip Sharon
Glading, Ben
Golet, Francis Charles
Hamerstrom, Frederick Nathan, Jr
Hamilton, Robert Bruce
Harlow, Richard Fessenden
Hewston, John G
Hickey, Joseph James
Hine, Ruth Louise
Holsworth, William Norton
Hungerford, Charles Roger
Hungerford, Kenneth Eugene
Keith, James Oliver
Kennamer, James Earl
Kirkpatrick, Charles Milton
Kirkpatrick, Ralph Donald
Kutilek, Michael Joseph
Lewis, James Chester
Linder, Raymond
Longrie, Dean Paul
Low, Jessop Budge
Lyon, Leonard Jack
Maguder, Theodore Leo, Jr
Manuwal, David Allen
Marchinton, Robert Larry
Marriage, Lowell Dean
Mech, Lucyan David
Montague, Fredrick Howard, Jr
Moore, Jerry Arnold
Owen, Ray Bucklin, Jr
Peek, James Merrell
Peterson, Rolf Olin
Pimlott, Douglas Humphreys
Porter, Richard Dee
Richens, Voit B
Robinson, William Laughlin
Rogers, John Gilbert, Jr
Rongstad, Orrin James
Rosene, Walter, Jr
Ruff, Robert LaVerne
Samson, Fred Burton
Schmidt, John Lancaster
Shaw, James Harlan
Shaw, William Wesley
Smith, Norman Sherrill
Stewart, Robert Earl
Streeter, Robert Glen
Swanson, Gustav Adolph
Taber, Richard Douglas
Thompson, Daniel Quale
Thompson, Marvin Pete, Jr
Weeks, Harmon Patrick, Jr
Wight, Howard Morgan
Wing, Larry Dean
Wood, Gene Wayne
Wright, Bruce Stanley
Zwickel, Fred Charles

Wildlife Management

Arner, Dale H
Balser, Donald S
Barske, Philip
Bartonek, James Cloyd
Behrend, Donald Fraser
Berryman, Jack Holmes
Blouch, Ralph Irving
Chabreck, Robert Henry
Christensen, Glen C
Coulter, Llewellyn Legrande
Cowan, Archibald B
Dammann, Arthur Erle
Davis, Frederic Whitlock
Davis, William B
Dean, Frederick Chamberlain
Eddy, Thomas A
Ellarson, Robert Scott
Gallizioli, Steve
Giles, Robert H, Jr
Glasgow, Leslie Lloyd

Glazener, William Caleb
Glover, Fred Arthur
Good, Ernest Eugene
Gould, Walter Phillip
Greeley, Frederick
Gysel, Leslie William
Gullion, Gordon W
Hanson, Wayne Carlyle
Harris, John Tom
Hein, Dale Arthur
Huey, William S
Jenkins, David H
Jenkins, James Hobart
Jordan, Peter Albion
Kadlec, John A
Keith, Lloyd Burrows
Klimstra, Willard David
Kozicky, Edward Louis
Lawrence, William Hobart
Longhurst, William Murray
Mackie, Richard John
Marshall, William Hampton
McCabe, Robert Albert
McClure, Howe Elliott
McCullough, Dale Richard
Michael, Edwin Daryl
Monson, Gale (Wendell)
Morrison, John Albert
Mosby, Henry Sackett
Petrides, George Athan
Reynolds, John Keith
Rudersdorf, Ward J
Ryder, Ronald Arch
Severinghaus, Charles William
Smith, Robert Leo
Tanner, Ward Dean, Jr
Wade, Dale A
Webb, James Woodrow
Wood, Roy Kellum
Yocom, Charles Frederick

Wildlife Pathology
Cheatum, Evelyn Leonard

Wildlife Research
Blankenship, Lytle Houston
Bookhout, Theodore Arnold
Dahlgren, Robert Bernard
Downing, Robert Lee
Drawe, D Lynn
Erickson, Ray Charles
Gilmer, David Seeley
Howard, Volney Ward, Jr
Knowlton, Frederick Frank
Krapu, Gary Lee
Marks, Stuart A
Mautz, William Ward
McGinnes, Burd Sheldon
Meslow, E Charles
Ohlendorf, Harry Max
Patton, David Roger
Scanlon, Patrick Francis
Schemnitz, Sanford David
Scott, Thomas George
Springer, Paul Frederick
Sickel, William Henson
Stone, Charles Porter
Varner, Larry Weldon
Ward, Angus Lorin

FOOD SCIENCES

Food Science
Aberle, Elton D
Acker, Geraldine Enod
Acosta, Phyllis Brown
Adams, Angus Macaulay
Addis, Paul Bradley
Adler, Frederick E W
Ahmed, Esam Mahmoud
Akerboom, Jack
Alexander, Earl Glynn
Alton, Alvin John
Amen, Ronald Joseph
Amundson, Clyde Howard
Anderson, Edward Everett
Anderson, Ray Harold
Andres, Cal L
Ang, Catharina Yung-Kang Wang
Anglemier, Allen Francis
Appledorf, Howard
Armbruster, Gertrude D
Arnold, Roy Gary
Ashworth, Ural Stephen
Auerbach, Earl
Augustin, Jorg A L
Avera, Fitzhugh Lee
Ayres, John Clifton
Ayuso, Katharine
Babayan, Vigen Khachig
Babbitt, Jerry
Badenhop, Arthur Fredrick
Baker, Robert Carl
Baldwin, Heber Ross
Barber, Franklin Weston

Bard, John C
Barkate, John Albert
Bass, Mary Anna
Bates, Charles
Bauman, Howard Eugene
Beach, Betty Laura
Beam, John E
Bechtle, Robert M
Bedford, Clifford Levi
Bednarcyk, Norman Earle
Bedrosian, Karakian
Beelman, Robert B
Behnke, James Ralph
Beisel, Clifford Gordon
Bennion, Marion
Berry, Joe Gene
Beyer, William W
Billerbeck, Fred William, Jr
Binkerd, Evan Francis
Birdsall, John J
Blackwood, Chesley M
Bluhm, Leslie
Borgstrom, Georg Arne
Bourland, Charles Thomas
Bourne, Malcolm Cornelius
Bowers, Jane Ann (Raymond)
Bowman, Ferne
Boyd, Glenn D
Bradley, Robert Lester, Jr
Branen, Alfred Larry
Breene, William Michael
Brekke, Orville Neil
Briskey, Ernest J
Brockmann, Maxwell Curtis
Brody, Aaron Leo
Brogle, Richard Charles
Brown, Delos D
Brown, William E
Brunner, Jay Robert
Buchanan, Robert Lester, Jr
Buck, Paul Andrews
Buck, Robert Edward
Bujake, Jahn Edward, Jr
Burgess, Hovey Mann
Burnette, Mahlon Admire, III
Burns, Edward Eugene
Burr, Horace Kelsey
Buss, Charles Delevan
Busted, Ronald Lorima, Jr
Caldwell, Elwood F
Campbell, Colin Arthur
Campbell, Michael Floyd
Carbonell, Robert Joseph
Carpenter, Zerle Leon
Caul, Jean Frances
Cecil, Sam Reber
Chambers, James Vernon
Chan, Harvey Thomas, Jr
Chandan, Ramesh Chandra
Chapman, Ross Alexander
Chichester, Clinton Oscar
Chism, Grady William, III
Christianson, George
Chung, Jiwhey
Chung, Ronald Aloysius
Clark, Allen Varden
Clark, Walter Leighton, III
Clydesdale, Fergus Macdonald
Cobb, Bryant Franklin, III
Cole, Morton S
Coleman, Richard J
Collins, Jimmie Lee
Collins, William F
Colmey, John C
Constantinides, Spiros Minas
Cooter, Frederick William
Cooper, Owen
Corliss, Glenn Arthur
Cornell, Alan
Cosgrove, Clifford James
Cotton, Robert Henry
Craig, Theodore Warren
Crawford, David Lee
Crawford, Thomas Michael
Creinin, Howard Lee
Crevasse, Gary A
Cronk, Ted Clifford
Cross, Hiram Russell
Cunningham, Franklin E
Curwen, David
Daravingas, George Vasilios
Darragh, Richard T
Daun, Henryk
Dave, Bhaichandra A
David, Jean
Davies, Donald Leslie
Davis, Paul A
Dawson, Lawrence E
Dean, Robert Waters
Deceles, George Arthur, Jr
de Figueiredo, Mario P
Detazos, Elias Demetrios
Dennison, Raymond Alexander
Dibble, Marjorie Veit
Dienst, Carl Sedgwick
Dill, Charles William

Dimick, Paul Slayton
Downes, Theron Winship
Downing, Donald Leonard
Drake, Stephen Ralph
Dubravcic, Milan Frane
Dunkley, Walter Lewis
Dunn, Cecil Gordon
Dunshee, Bryant R
Durst, Jack Rowland
Duxbury, Dean David
Dymsza, Henry A
Earl, Allan Edwin
Ebert, Andrew Gabriel
Einset, Eystein
Eitenmiller, Ronald Ray
Ellerton, Melvin Elroy
Ellinger, Rudolph H
Endres, Joseph George
Epley, Richard Jess
Erdman, John Wilson, Jr
Eshenour, Terry Ray
Esselen, William B
Fagerson, Irving Seymour
Farkas, Daniel Frederick
Felters, David Anthony
Fennema, Owen Richard
Ferrier, Leslie Kenneth
Fields, Marion Lee
Finley, John Westcott
Fischbach, Eugene
Fleming, Henry Pridgen
Flick, George Joseph
Flora, Lewis Franklin
Ford, Robert Sedgwick
Fram, Harvey
Franceschini, Remo
Francis, Frederick John
Franzosa, Edward Sykes
Freeman, Lawrence Reed
Freund, Peter Richard
Fritzsche, Herbert William
Frodey, Ray Charles
Froning, Glenn Wesley
Fulde, Roland Charles
Fuleki, Tibor
Fung, Daniel Yee Chak
Gagnon, Marcel
Gallander, James Francis
Garner, Richard Gordon
Giggard, Earl David
Gilbert, Seymour George
Gillette, Tedford A
Gizis, Evangelos John
Glenister, Paul Robson
Gnaedinger, Richard H
Goldblith, Samuel Abraham
Goodenough, Eugene Ross
Goodman, Louis P
Goodwin, Tommy Lee
Gordon, Joan
Goresline, Harry Edward
Gould, Max Randall
Gould, Wilbur Alphonso
Grab, Eugene Granville, Jr
Graham, Dee McDonald
Graham, Donald C W
Greene, Barbara E
Gregory, Max Edwin
Grodner, Robert Maynard
Gross, Charles Ezra
Guy, Eugene James
Hadziyev, Dimitri
Hagberg, Elroy Carl
Hagen, Richard Eugene
Hahn, Richard Ray
Hale, Kirk Kermit, Jr
Hall, Kenneth Noble
Hall, Richard Leland
Hamm, Douglas
Hamre, Melvin L
Hang, Yong Deng
Hankinson, Denzel J
Hanley, Joseph Wall
Hansen, Poul M T
Hard, Margaret McGregor
Harmon, Laurence George
Harris, Natholyn Dalton
Harrison, Dorothy Lucile
Hart, William James, Jr
Hartung, Theodore Eugene
Hartzell, Thomas H
Hasegawa, Shin
Hastings, Carl Wayne
Hayakawa, Kan-Ichi
Hayes, Kirby Maxwell
Heath, James Lee
Hedrick, Theodore Isaac
Heidelbaugh, Norman Dale
Henry, Wayne E
Highlands, Matthew Edward
Hildebrand, Frank Childs
Hills, Claude Hibbard
Ho, Genevieve Po-Ai
Hoersch, Theodore Matthew
Hoff, Johan Edward
Holahan, John L, Sr
Holland, Robert Francis

Holmes, David G
Holzinger, Thomas Walter
Hoover, William Jay
Hoover, Samuel Randolph
Hopper, Paul Frederick
Hoskins, Frederick Hall
Huber, Clayton Shirl
Hugunin, Alan Godfrey
Hultin, Herbert Oscar
Hunnell, John Wesley
Hunt, Fern Ensminger
Hurley, William Charles
Husaini, Saeed A
Ingalsbe, David Weeden
Ingle, James Davis
Irwin, William Elliot
Ishler, Norman Hamilton
Ivey, Francis James
Jackson, Harold
Jackson, John Mathews
Jackson, Richard Field
Jackson, Robert Howard
James, William Holden
Jaye, Murray Joseph
Jaynes, Hugh Oliver
Jaynes, John Alva
Jen, Joseph Jwu-Shan
Jezeski, James John
Jimenez, Miguel Angel
Jobin, Ralph Alfred
Joffe, Frederick M
Johnson, Bobby Ray
Johnson, Dale Waldo
Johnson, Karl Robert
Johnson, Littleton Wales
Johnson, Ogden Carl
Johnson, Melvin Roscoe
Johnston, William Kirby
Josephson, Edward Samuel
Judge, Max David
Kaffezakis, John George
Kainski, Mercedes H
Kapsalis, John George
Karel, Marcus
Karmas, Endel
Kattan, Ahmed A
Katz, Ira
Katz, Morris Howard
Kaufmann, Henry Hans
Keenan, Thomas William
Keeney, Philip G
Kendall, John Hugh
Kendall, Albert Raymond
Keyser, William Lacy
Khan, Mahmood Ahmed
Khan, Paul
Kifer, Paul Edgar
Kilbuck, John Henry
King, Frederick Jessop
King, Raymond Leroy
Kissmeyer-Nielsen, Erik
Kitaka, Robert Shinnosuke
Kitson, John Aidan
Koehler, Philip Edward
Kohn, Gaston G
Kornetsky, Frank Vincent
Kosikowski, Frank Vincent
Koula, Anthony W
Kraft, Allen Abraham
Kramer, Amihud
Kristoffersen, Thorvald
Kroenberg, Bernd
Kroger, Manfred
Krum, Jack Kern
Kuesel, Donald Charles
Kumar, Surinder
Kummerow, Fred August
Labbee, Marcel D
LaBelle, Robert Lawrence
LaBree, Theodore Robert
Lachance, Paul Albert
Lampi, Rauno Andrew
Lampech, Earl Diwuin
Langier, James Edward
Lauro, Gabriel Joseph
Lebermann, Kenneth Wayne
Lechowich, Richard V
Lee, Chang Yong
Lee, Wei Hwa
Lee, William Glen
Leichter, Joseph
Lessard, Maurice
Leverton, Ruth Mandeville
Levin, Robert E
Lewis, Burnadine Langston
Libbey, Leonard Morton
Lindsay, Robert Clarence
Liska, Bernard Joseph
Litchfield, John Hyland
Little, Angela C
Livingston, G E
Lockhart, Ernest Earl
Loeffler, Harold Julius

Longree, Karla
Lopez, Anthony
Lougheed, Thomas Crossley
Luckmann, Frederick H
Luh, Bor Shiun
Luizzo, Joseph Anthony
Lukes, Thomas Mark
Lund, Daryl B
Lundeen, Glen Alfred
Lusas, Edmund W
Lynch, Gerald John
MacGregor, Dugal
MacNeil, Joseph H
Maga, Joseph Andrew
Malaspina, Alex
Mangel, Margaret
March, Richard Pell
Marion, James Edsel
Marion, William W
Markakis, Pericles
Marsh, Alice Garrett
Marsh, Richard Riley
Marshall, James Tilden, Jr
Marshall, Robert T
Mason, Michael E
Mast, Morris Glen
Matches, Jack Ronald
Matthews, Ruth Hastings
Mattil, Karl Frederick
Matz, Samuel Adam
Mavis, James Osbert
Maxcy, Ruthford Burt
May, Kenneth Nathaniel
Mayer, Warren Clifford
McCloskey, Kenneth Emory
McConnell, John Earl Willard
McDivitt, Maxine Estelle
McGill, Lois Sather
Meade, Reginald Eson
Meggison, David Laurence
Mehrlich, Ferdinand Paul
Melachouris, Nicholas
Melcer, Irving
Mellor, David Bridgwood
Mendelhall, Von Thatcher
Meyer, Richard Irwin
Mickelberry, William Charles
Midura, Thaddeus
Miller, Gary A
Miller, Kenneth Melvin
Miller, Ralph Albert
Mills, William Carlos
Milner, Max
Milner, Reid Thompson
Mitchell, Donald Gilman
Mitchell, William Alexander
Moeller, Theodore William
Mohr, Willard Phillip
Montgomery, Morris William
Moore, Edwin Lewis
Moreau, Jean Raymond
Moriarty, John Henry
Morris, Howard Arthur
Morse, Roy Earl
Moser, Roy Edgar
Mosby, Raymond Joseph
Motawi, Kamal El-Din Hussein
Moyer, James Charles
Mullen, Joseph David
Mullins, Auttis Marr
Mulvaney, Thomas Richard
Muneta, Paul
Mykleby, Ray W
Nagel, Charles William
Nakayama, Tommy
Nanz, Robert Augustus Rollins
Narayan, Krishamurthi Ananth
Nash, Nat H
Nelson, Alvin I
Nelson, John Howard
Nelson, John Howard
Nelson, Woodrow Ensign
Newton, Stephen Bruington
Ng, Henry
Nicholas, Richard Carpenter
Nielsen, Verner Henry
Niven, Charles Franklin, Jr
Noble, Ann Curtis
Ogilvy, Winston Stowell
Olcott, Harold Saft
Olson, Norman Fredrick
O'Mahony, John Patrick
O'Meara, John Pierce
Opie, Joseph Wendell
Packard, Vernal Sidney, Jr
Palumbo, Samuel Anthony
Pan, Bonnie Sun
Pariser, Ernest Reinhard
Parmelee, Carlton Edwin
Passey, Chand Arjun
Patel, Savinay S
Pavey, Robert Louis
Pearson, Albert Marchant
Pedraja, Rafael Rodobaldo
Perrozzi, Joseph Richard
Perry, Margaret Nutt
Pflug, Irving John
Phaff, Herman Jan

Philip, Thomas
Philips, Jean Allen
Pierson, Merle Dean
Pigott, George M
Pilkington, Dwain H
Plantz, Philip Edward
Ponte, Joseph G, Jr
Potter, Frank Elwood
Potter, Norman N
Powell, William Edward, III
Powers, John Joseph
Powrie, William Duncan
Prater, Arthur Nickolaus
Prescott, Henry Emil, Jr
Price, James F
Price, Ralph Lorin
Prothro, Johnnie W
Prudhomme, Edward Louis
Prudhomme, Ronald Edward
Purvis, George Allen
Puski, Gabor
Putnam, Hamilton Wallace
Pyne, Alvan Wesley
Quinn, John R
Rackis, Joseph John
Radanovics, Charles
Rakosky, Joseph, Jr
Ramsey, Lessel Leslie
Ramstad, Paul Ellertson
Ranadive, Arvind S
Ranalli, Anthony William
Rand, Arthur Gorham, Jr
Randolph, Henry England
Rao, Akkinapally V
Rao, Ramachandra M R
Rasekh, Jamshid G
Reddy, Sunki, Gopal
Regenstein, Joe Mac
Regier, Lloyd Wesley
Reineccius, Gary (Aubrey)
Rhee, Khee Choon
Richberg, Carl George
Riel, Rene Rosaire
Riha, William E, Jr
Rikert, John A
Roberts, William Milner
Robinson, Radcliffe Franklin
Robinson, Willard Bancroft
Rolfes, Thomas J
Rooney, Lloyd William
Ross, Robert Edgar
Rubin, Leon Julius
Ryan, Dale Scott
Rymal, Kenneth Stuart
Salant, Abner
Salunkhe, Datta K
Sapers, Gerald M
Saunders, Robert Montgomery
Sauter, Erwin Andrew
Sawyer, Frederick Miles
Scanlan, Richard Anthony
Schaller, Daryl Richard
Schanefelt, Robert Von
Schiffmann, Robert F
Schlimme, Donald Vincent
Schmiege, Clement Carl
Schuldt, Erich Henry
Schuler, George Albert
Schultz, Harold William
Schuytema, Carl G
Schwall, Donald V
Scott, Don
Scott, Lawrence William
Secrist, John Leonard
Seehafer, Marlyn E
Segall, Stanley
Servadio, Gildo Joseph
Setser, Carole Sue
Shahani, Khem Motumal
Shannon, Edward Leo
Sherck, Charles Keith
Sherman, Joseph E
Shewfelt, Albert Lorne
Shipe, William Franklin
Showalter, Robert Kenneth
Siedler, Arthur James
Silberstein, Otmar Otto
Silverman, Gerald
Simpson, Kenneth L
Singh, Bharat
Sinnhuber, Russell Otto
Sinskey, Anthony J
Sistrunk, William Allen
Skelton, Marilyn Mae
Slabyj, Bohdan M
Sleeth, Rhule Bailey
Smit, Christian Jacobus Bester
Smith, Malcolm Crawford, Jr
Smouse, Thomas Hadley
Snyder, Harry E
Solberg, Myron
Somogyi, Laszlo P
Son, Chung Hyun
Spencer, John Valentine
Stadelman, William Jacob
Staff, Charles Hubert
Stanley, David Warwick
Stapf, Robert Joseph

Steinberg, Marvin Phillip
Steinberg, Maynard Albert
Sternberg, Moshe
Stewart, George Franklin
Stine, Charles Maxwell
Stoll, William Francis
Stone, Charles Dean
Stone, Warren Kenneth
Strock, Herman
Strong, Dorothy Hussemann
Stumbo, Charles Raymond
Sulzbacher, William Louis
Supran, Michael Kenneth
Swanson, Arthur Martin
Swanson, Barry Grant
Swope, Fred C
Tarver, Fred Russell, Jr
Tatini, Sita Ramayya
Taylor, Betty
Taylor, Steve L
Thomas, Alan
Thomas, Elmer Lawrence
Thomas, Frank Bancroft
Thomas, William Robb
Tillotson, James E
Tinklin, Gwendolyn L
Tischer, Robert George
Titus, Dudley Seymour
Tobias, Joseph
Touba, Ali R
Tuma, Harold J
Tung, Marvin Arthur
Turkki, Pirkko Reetta
Turney, Lawrence Joseph
Twigg, Bernard Alvin
Tybor, Philip Thomas
Unklesbay, Nan F
Urbain, Walter Mathias
Van Den Berg, L
Van Hulle, Glenn Joseph
Vaughn, Moses William
Vedamuthu, Ebenezer Rajkumar
Vetter, James Louis
Vickers, Zata Marie
Vondell, Richard M
Voss, Gordon D
Vosti, Donald Curtis
Wabeck, Charles J
Wallander, Jerome F
Wallenfeldt, Evert
Walradt, John Pierce
Walter, William Mood, Jr
Ward, Arlin Bruce
Warnecke, Melvin Oscar
Weber, Frank E
Webster, John Robert
Weckel, Kenneth Granville
Wei, Lun-Shin
Weiss, Ronald
Welch, Rodney Channing
Wells, Frank Edward
Wells, Phillip Richard
Wesley, Roy Lewis
Westcott, Donald Elvin
Whipple, Royson Newton
Whitehair, Leo A
Wilcox, Joseph Clifford
Wiley, Robert Craig
Wilkinson, Raleigh James
Williams, Leamon Dale
Williams, Marion Porter
Wilson, John M
Winder, William Charles
Wishnetsky, Theodore
Wistreich, Hugo Eryk
Wodicka, Virgil Orville
Wolin, Alan George
Wood, Darrell Fenwick
Woodburn, Margy Jeanette
Woodcock, Charles Martin
Worthington, Robert Earl
Wrolstad, Ronald Earl
Wu, Ming Tsung
Yamins, Jacob Louis
Yang, Hoya Y
Yeatman, John Newton
Younathan, Margaret Tims
Young, Clyde Thomas
Young, Louis Lee
Zaehringer, Mary Veronica
Zall, Robert Rouben
Zallen, Eugenia Malone
Zapsalis, Charles
Zottola, Edmund Anthony
Zoumas, Barry Lee
Zubeckis, Edgar

Cereal Chemistry

Abbott, Donald Clayton
Bayfield, Edward Geoffrey
Bendelow, Victor Martin
Bushuk, Walter
Chung, Okkyung Kin
Conn, James Frederick
Cooper, Elmer James
Dahle, Leland Kenneth
D'Appolonia, Bert Luigi
Hoffman, William Howard

Hoseney, Russell Carl
Houston, David Fairchild
Kite, Francis Ervin
Kosmolak, Frederick Graham
Lorenz, Klaus J
McDonald, Clarence Eugene
McGuire, Charles Francis
Meisner, Donald F
Miller, Byron Sloane
Patil, Sakharam Karsan
Rankin, John Carter
Redfern, Sutton
Rohwer, Robert G
Rubenthaler, Gordon Lawrence
Sandsted, Rudolph Marion
Seitz, Larry Max
Shellenberger, John Alfred
Shuey, William Carpenter
Sullivan, John W
Tipples, Keith H
Waggle, Doyle H
Walker, Charles Eugene
Walsh, David Ervin
Watson, Clifford Andrew
Yamazaki, William Toshi

Food Bacteriology

Appleman, Milo Don
Elliot, Robert Paul
Fishbein, Morris
Lipman, Harry Jerome
Rice, Andrew Cyrus
Wilson, Charles Marshall

Food Biochemistry

Cherry, John Paul
Clements, Robert Lawrence
Deatherage, Fred E
Goepfert, John McDonnell
Honig, David Herman
Hynes, John Thomas
Moats, William Alden
Palmer, James Kenneth
Patton, Stuart
Youngquist, Rudolph William

Food Chemistry

Acree, Terry Edward
Anderson, David W, Jr
Angelini, Pio
Bixby, John N
Bloom, Jack
Bolin, Harold R
Borenstein, Benjamin
Brause, Allan R
Brooker, Francis Milton
Campbell, Ada Marie
Carlin, Frances
Chang, Stephen Szu Shiang
Chastain, Marian Faulkner
Chicoye, Etzer
Cowley, Milford A
Cramer, Archie Barrett
Cunningham, Hugh Meredith
Dassow, John Albert
DeMan, John Maria
Dowling, Joseph Francis
Dunn, Howard J
Dutra, Ramiro Carvalho
Emery, Donald F
Eskin, Neason Akiva Michael
Feliciotti, Enio
Fink, Kenneth Howard
Flynn, Charles Edward
Fortney, Cecil Garfield, Jr
Fulger, Charles Von
Galetto, William George
Germino, Felix Joseph
Gluntz, Martin Lucius
Hagenmaier, Robert Doller
Hammond, Earl Gullette
Hayward, Frederick Warren
Hefley, Alta Jean
Henika, Richard Grant
Hester, Ellen Elizabeth
Heydanek, Menard George, Jr
Hillig, Fred
Hlavacek, Robert John
Hoepfinger, Lynn Morris
Holmes, Edward Lawson
Horwitz, William
Hurst, Thomas Walthall
Hyde, Robert Wallace
Jackson, Earl Roger
Jacobson, Glen Arthur
Johnson, John Hal
Joiner, Robert Russell
Josephson, Ronald Victor
Joslyn, Maynard Alexander
Julien, Jean-Paul
Kean, Chester Eugene
Kimoto, Walter Iwao
Kinsella, Frank Raymond
Kinsella, John Edward
Klein, Barbara P
Knapp, Frederick Whiton
Kolor, Michael Garrett

FOOD SCIENCES

Krishnamurthy, Ramanathapur Gundachar
Lapuck, Jack Lester
Lee, Yuen San
Lillard, Doris Alton
Lime, Bruce James
Lindsay, Robert Clarence
Litman, Irving Isaac
Mackay, Donald Alexander-Morgan
Manley, Charles Howland
Marable, Nina Louise
Maselli, John Anthony
Matthews, Richard Finis
McGugan, Wesley Alexander
McKinney, Leonard Laurence
Mergentine, Max
Mina, David Byong
Mohlenkamp, Marvin Joseph, Jr
Morr, Charles Vernon
Moss, Valentin G
Myhre, David V
Nakai, Shuryo
Nawar, Wassef W
Nishida, Toshiro
Nutting, Lee
Ockerman, Herbert W
O'Connor, David Evans
O'Connor, Michael Gerald
Pareles, Stephen Ronald
Paul, Pauline Constance
Peng, Andrew Chung Yen
Petrowski, Gary E
Pomerantz, Reuben
Powell, Eugene Loren
Pratt, Dan Edwin
Richardson, Thomas
Roach, J Robert
Rosen, Joseph David
Roth, Howard
Russell, Gerald Frederick
Salwin, Harold
Samuels, Robert Bireley
Sanderson, Gary Warner
Satterlee, Lowell Duggan
Schafer, Mary Louise
Settoon, Patrick Delano
Sevenants, Michael R
Shallenberger, Robert Sands
Shaver, Kenneth John
Smith, Laura Lee Weisbrodt
Smith, Lloyd Muir
Souder, Philip Walburn
Steffen, Albert Harry
Swisher, Horton Edward
Tannenbaum, Steven Robert
Tiemstra, Peter J
Trelease, Richard Davis
Tressler, Donald Kiteley
Van der Kloot, Albert Peter
Van Duyne, Frances Olivia
Varsel, Charles John
Wahba, Isaac Jack
Waitman, Reuben Homer
Walter, Reginald Henry
Wasserman, Aaron E
Weast, Clair Alexander
Weissler, Harold Edward
Westbrook, George Franklin
Whitney, Robert McLaughlin
Williams, Joseph Lee
Winston, James J
Woods, Alvin Edwin
Zaika, Laura Larysa

Food Microbiology

Alford, John Abright
Ashton, David Hugh
Avens, John Stewart
Babel, Frederick John
Banwart, George J
Beuchat, Larry Ray
Bothast, Rodney Jacob
Brown, William Lewis
Bryan, Frank Leon
Bullerman, Lloyd Bernard
Busta, Francis Frederick
Cameron, Donald Eugene
Clark, David Sedgefield
Collins, Edwin Bruce
Dahiya, Raghunath S
Davidson, Charles Mackenzie
Denny, Cleve B
Eklund, Melvin Wesley
El-Bisi, Hamed Mohamed
Elinger, Rudolph H
Elliott, James Angus
Fox, Kenneth Ian
Frank, Hilmer Aaron
Gabis, Damien Anthony
Gilliland, Stanley Eugene
Hartman, Paul Arthur
Hauschild, Andreas H W
Heimsch, Richard Charles
Hickernell, Gary L
Huskey, Glen E
Idziak, Edmund Stefan
King, Alfred Douglas, Jr
Koburger, John Alfred
LaGrange, William Somers

Langlois, Bruce Edward
Larkin, Edward P
Lechowich, Richard V
Ledford, Richard Allison
Maisch, Weldon Frederick
Marth, Elmer Herman
McAnelly, John Kitchel
Mercun, Arthur J
Morgan, Bruce Henry
Morita, Toshiko N
Mundt, John Orvin
Naumann, Hugh Donald
Nelson, Frank Eugene
Nickelson, Ranzell, II
O'Leary, Virginia Sawyer
Park, Chong Eel
Patel, Girishchandra Babubhai
Perkins, William Edward
Powers, Edmund Maurice
Rayman, Mohamad Khalil
Rowley, Durwood B
Segner, Wayne Philip
Smittle, Richard Baird
Spira, William Martin
Stevenson, Kenneth Eugene
Stine, James Bryan
Tompkin, Robert Bruce
Vanderzant, Carl
Vaughn, Reese Haskell
Walker, Homer Wayne
Wehrle, Louis, Jr
Westhoff, Dennis Charles
White, James Carrick
Witter, Lloyd David
Yates, Alfred Randolph

Meat Science

Aberle, Elton D
Adams, Charles Henry
Allen, Deloran Matthew
Alsmeyer, Richard Harvey
Bray, Robert Woodbury
Buck, Ernest Mauro
Cahill, Vern Richard
Costello, William James
Dalrymple, Ronald Howell
Dutson, Thayne R
Epley, Richard Jess
Field, Ray A
Forrest, John Charles
Gillette, Tedford A
Grant, Darroll Lee
Guenther, John James
Huffman, Dale L
Jeremiah, Lester Earl
Kauffman, Robert Giller
Kennick, Walter Herbert
King, General Tye
Kropf, Donald Harris
Lewis, Paul Kermith, Jr
Link, Bernard Alvin
Marsh, Benjamin Bruce
Merkel, Robert Anthony
Mullins, Anttis Mearl
Murphy, Robert Emmett
Neer, Keith Lowell
Orts, Frank Ausut
Parrett, Ned Albert
Pierce, John Cleve, Jr
Plimpton, Rodney F, Jr
Rea, Ronald Howard
Reagan, James Oliver
Schmidt, Glenn Roy
Sebranek, Joseph George
Smith, Gary Chester
Suess, Gene Guy
Tittiger, Franz
Tom, Baldwin Heng
Tuma, Harold J
Usborne, William Ronald
Ziegler, John Henry, Jr

Sugar Chemistry

Gaddie, Robert Stanley
Hickson, John LeFever
Lim, Yong Woon
Marov, Gaspar J
McDaniel, Max Paul
Morisugu, Toshio
Pomes, Adrian Francis
Smith, Burns Ashby

FORESTRY

Forestry

Abbott, Herschel George
Adair, Kent Thomas
Ahlgren, Clifford Elmer
Aldon, Earl F
Allen, Robert Max
Allen, Alvin Arthur
Andrus, Charles Frederick
Arnold, R Keith
Bagley, Walter Thaine
Baldwin, Henry Ives
Barnes, Burton Verne
Barrett, James Passmore
Barrett, John William
Basham, Jack Tucker
Baskerville, Gordon Lawson
Bay, Roger Rudolph
Bayly, George Henry Uniacke
Berry, Richard Wallace
Bissett, Orville R
Biswell, Harold Hubert
Blake, George Marston
Boggess, William Clark
Boldt, Charles Eugene
Bolle, Arnold William
Bourchier, Robert James
Boyd, Charles Curtis
Bramble, William Clark
Brandt, Robert William
Briegleb, Philip Anthes
Brundage, Roy Charles
Bruns, Paul Eric
Bryant, Robert L
Budowski, Gerardo
Burns, Paul Yoder
Byrnes, William Richard
Cable, Dwight Raymond
Carow, John
Carter, Mason Carlton
Castagnozzi, Daniel M
Chamberlin, Henry Howard
Chaney, William R
Chapline, William Ridgely, (Jr)
Cheston, Charles Edward
Choong, Elvin T
Clark, Floyd Bryan
Coleman, Donald George
Cool, Bingham Mercur
Cooper, Robert Warren
Core, Harold Addison
Crowther, C Richard
Cunningham, Gordon Rowe
Davenport, O Malcolm
DeBell, Dean Shaffer
DeByle, Norbert V
Devall, Wilbur Bostwick
Dilworth, John Richard
Douglass, James Edward
Duffield, John Warren
Duncan, Donald Pendleton
Echols, Robert M
Ehrenreich, John Helmuth
Eikenbary, Raymond Darrell
Ek, Alan R
Engelhard, Robert J
Eschner, Arthur Richard
Fahnestock, George Reeder
Fairfax, Sally Kirk
Farmer, Robert E, Jr
Farrar, John Laird
Ferrell, William Kreiter
Fleischer, Herbert Oswald
Foil, Robert Rodney
Franklin, Jerry Forest
Fredriksen, Richard L
Freeland, Forrest Dean, Jr
Freeman, Jeffrey Van Duyne
Freeman, Fred W
Furnival, George Mason
Gaertner, Erika Eva
Garratt, George Alfred
Gessel, Stanley Paul
Gibbs, Carter B
Gilmore, Alvan Ray
Goggans, James F
Gould, Walter Philip
Grah, Rudolf Ferdinand
Greene, James T
Grizzell, Roy Ames, Jr
Grober, Samuel
Gysel, Leslie William
Haddock, Philip George
Hafley, William LeRoy
Hallgren, Alvin Roland
Hamilton, Lawrence Stanley
Hanne, Christopher John
Hansen, Henry Leo
Hargreaves, Leon Abraham, Jr
Harlow, William Morehouse
Hart, Clarence Arthur
Hart, George Emerson, Jr
Hartesveldt, Richard J
Hebner, Charles Frederick
Heiner, Terry Charles
Helms, John Andrew
Herrick, Allyn Marsh
Herrick, Lawrence Andrew
Holcomb, Carl James
Holland, Israel Irving
Hook, Donald
Hopkins, Frederick Sherman, Jr
Hosner, John Frank
Hough, Walter Andrew
Howe, John Prentice
Huffman, Jacob Brainard
Ifju, Geza
Irving, Frank Dunham
James, Lee Morton
Jemison, George Meredith
Jensen, Keith Frank
Johnson, Edward A
Jones, John Robert
Jorgensen, Erik
Kayll, Albert James
Keister, Thomas Dwight
Kelley, Allen Frederick, Jr
Ker, John William
Klein, Edward Lawrence
Knoerr, Kenneth Richard
Koch, Christian Burdick
Koelling, Melvin R
Kraemer, J Hugo
Kramer, Paul R
Ku, Timothy Tao
Kurmis, Ernest A
Kurmes, Vilis
Lamb, Frank Bruce
Lammi, Joe Oscar
Langevin, Raymond J Francois
Lapping, Mark Barry
Larsen, Harry Stites
Lassen, Laurence E
Lawrence, William Hobart
Leach, Charles Willard
Lee, Chen Hui
Leopold, Aldo Starker
Lester, Donald Thomas
Lewis, Gordon Depew
Libby, William John, (Jr)
Lindmark, Ronald Dorance
Little, Silas, Jr
Livingston, Knox W
Loewenstein, Howard
Loomis, Robert Morgan
Lorenz, Ralph William
Lotan, James E
Lowry, Gerald Lafayette
Lutz, Harold John
MacConnell, William Preston
Macleod, John Campbell
Maksymiuk, Bohdan
Maloney, James Eugene
Manthy, Robert Sigmund
Maple, William Robert
Martin, Clifton Boyd
Martin, Robert Edward, Jr
Martinelli, Mario, Jr
Mason, Richard Randolph
May, Jack Truett
McArdle, Richard Edwin
McComb, Andrew Logan
McCormack, Maxwell Leland, Jr
McDermid, Robert Wesson
McDermott, Robert Emmet
McGregor, William Henry Davis
McKillop, William L M
Megraw, Robert Arthur
Melton, Rex Eugene
Mercer, Clair
Merritt, Clair
Meyer, Merle P
Mignery, Arnold Louis
Michell, Arthur Stephen
Miles, William Raymond
Miller, Leon Sherwood
Minor, Charles Oscar
Moak, James Emanuel
Mogren, Edwin Walfred
Morison, Ian George
Mosby, Henry Sackett
Muench, John, Jr
Murphy, James L
Myers, Charles Christopher
Myers, Wayne Lawrence
Nelson, Robert Eldon
Nelson, Thomas Charles
Nutting, Albert Deane
Ohman, John Hamilton
O'Keefe, Robert Bernard
Oliver, Abe D, Jr
Ostrom, Carl Eric
Packer, Paul Earl
Parker, George Ralph
Pentoney, Richard Ellis
Philpot, Charles Walter
Post, Boyd Wallace
Preston, Richard Joseph, Jr
Pringle, Stanley Leroy
Ramke, Thomas Franklin
Redmond, Douglas Rollen
Rhodes, Arnold Densmore
Richards, Dean Boyd
Robinson, Dan D
Rohde, Wayne G
Roller, Kalman Joseph
Rothwell, Richard Lee
Row, Clark
Roy, Douglass Fielding
Rutherford, William, Jr
Sachs, Irving Benjamin
Safford, Lawrence Oliver
Satterlund, Donald Robert
Schaefer, Walter Howard
Schallau, Con H
Schubert, Thomas Herman
Schultz, John David

Seale, Robert Henry
Settergren, Carl David
Shade, Elwood B
Shain, William Arthur
Shanklin, John Ferguson
Sharp, John Buckner
Sharpe, Grant William
Shearer, Raymond Charles
Shreve, Loy William
Siegel, William Carl
Silverborb, Savel Benhard
Skok, Richard Arnold
Sluder, Earl Ray
Smith, David William
Smith, John Harry Gilbert
Smith, Richard Chandler
Smith, Wayne H
Smoker, William Alexander
Snoke, Lloyd Randolph
Sopper, William Edward
Sowder, Arthur Merrill
Spurr, Stephen Hopkins
Stairs, Gerald Ray
Steensen, Donald H J
Stein, William Ivo
Steinbeck, Klaus
Steinbrenner, Eugene Clarence
Stiteler, William Merle III
Stoehr, Henry Arthur
Stoltenberg, Carl H
Stonecypher, Roy W
Storey, Theodore George
Strohmann, Rudolph Otto
Sturgeon, Edward Earl
Suddarth, Stanley Kendrick
Sullivan, Alfred Dewitt
Sullivan, John Dennis
Sutherland, Charles F
Swan, Henry Stewart Drummond
Swinford, Kenneth Roberts
Thirgood, Jack Vincent
Thomas, David Phillip
Thompson, Emmett Frank
Thompson, George Willis
Tocher, Stewart Ross
Trefethen, Parker S
Vaartaja, Olli
Van Deusen, James Lowell
Van Lear, David Hyde
Van Slyke, Arthur Lawton
Van Wagner, Charles Edward
Vaux, Henry James
Vimmerstedt, John P
Wadsworth, Frank H
Wagle, Robert Fay
Wallace, Oliver P, Sr
Walters, Charles Sebastian
Ward, Wilber W
Weetman, Gordon Frederick
West, Richard Fussell
West, William Irvin
Westveld, Ruthford Henry
Wheeler, Philip Ridgly
White, Donald Perry
White, Edwin Henry
White, Zebulon Waters
Whitmore, Roy Alvin, Jr
Whittaker, James Curtiss
Wiant, Harry Vernon, Jr
Williams, David Edward
Wilm, Harold Gridley
Wilson, Brayton F
Winch, Fred Everett, Jr
Winer, Herbert Isaac
Wohletz, Ernest W
Wygant, Noel Darwin
Wylie, Aubrey Evans
Young, Harold Edle
Youngs, Robert Leland
Zarger, Thomas Gordon
Zasada, Zigmond Anthony
Zinke, Paul Joseph

Dendrology
Benseler, Rolf Wilhelm
Blackburn, Benjamin (Coleman)
Cockrell, Robert Alexander
Little, Elbert Luther, Jr
Worrall, John Gatland

Forest Biometry
Ashley, Marshall Douglas
Bickford, Charles Allen
Bower, David Roy Eugene
Burkhart, Harold Eugene
Dell, Tommy Ray
Kovner, Jacob L
Kozak, Antal
Martin, Margaret Pearl
Moser, John William, Jr
Norick, Nancy Xavier
Pienaar, Leon Visser
Van Slyke, Arthur Lawton

Forest Ecology
Arno, Stephen Francis
Art, Henry Warren
Baker, John Richard

Bakuzis, Egolfs Voldemars
Barney, Charles Wesley
Boyer, William Davis
Brown, Bruce Antone
Brown, James Harold
Brown, James Henry, Jr
Brush, Grace Somers
Burger, Dionys
Callaham, Robert Zina
Carvell, Kenneth Llewellyn
Challinor, David
Cobbe, Thomas James
Coffman, Michael S
Crankshaw, William Bliss
Crow, Alonzo Bigler
DeBrunner, Louis Earl
Edgren, James W
Ferrill, Mitchell
Garrison, George Alfred
Grandtner, Miroslav Marian
Harris, Robert Wilson
Hebb, Edwin Atkins
Heinselman, Miron L
Hodgkins, Earl Joseph
Holt, Harvey Allen
Hutnik, Russell James
Jurdant, Michel Louis
Ketchledge, Edwin Herbert
Kilgore, Bruce Moody
Kimmins, James Peter
Long, Alan Jack
Marquis, David Alan
McClenahen, James Richard
Meeuwig, Richard O'Bannon
Messenger, Aubrey Steven
Miller, Neil Austin
Newton, Michael
Plass, William T
Reid, Charles Phillip Patrick
Romancier, Robert Marshall
Schneider, Gerhardt
Shipman, Robert Dean
Silker, Theodore Henry
Smith, Ernest Chalmers
Stout, Benjamin Boreman
Strand, Robert Fenton
Strang, Robert M
Switzer, George Lester
Tanaka, Yasuomi
Tappeiner, John Cummings, II
Teate, James Lamar
Thornburgh, Dale A
Tierson, William Cornelius
Van Wagtendonk, Jan Willem
Weaver, George Thomas
Winget, Carl Henry
Winjum, Jack Keith
Woodmansee, Robert George

Forest Economics
Adams, Thomas C
Anderson, Walter Clinton
Baker, Harold Lawrence
Beuter, John H
Bond, Robert Sumner
Bruce, Richard W
Buongiorno, Joseph
Chappelle, Daniel Eugene
Davis, Lawrence S
Dyson, Peter John
Gregersen, Hans Miller
Knorr, Philip Noel
Knudson, Douglas Marvin
Kurtz, William Boyce
Marty, Robert Joseph
McGuire, John R
Partain, Gerald Lavern
Petriceks, Janis
Reimer, Don R
Schreuder, Gerard Fritz
Sullivan, Edward T
Teeguarden, Dennis Earl
Tombaugh, Larry William
Waggener, Thomas Runyan
Walker, Nathaniel
Wambach, Robert F
Wells, Garland Ray
Whaley, Ross Samuel
White, David Evans
Worrell, Albert Cadwallader
Yoho, James Gibson
Zinn, Gary William
Zivnuska, John Arthur

Forest Entomology
Amman, Gene Doyle
Anderson, Roger Fabian
Barras, Stanley J
Benjamin, Daniel Marshall
Brown, Norman Rae
Cameron, Edward Alan
Connola, Donald Pascal
Dahlsten, Donald L
Daterman, Gary Edward
De Mars, Clarence John, Jr
Drooz, Arnold T
Esenther, Glenn R
Fatzinger, Carl Warren

Fellin, David Gene
Fettes, James Joseph
Finnegan, Raymond Joseph
Fox, Richard Charles
Giese, Ronald Lawrence
Hain, Fred Paul
Harman, Dan M
Haynes, Dean L
Heikkenen, Herman John
Jennings, Daniel Thomas
Kearby, William H
Knight, Fred Barrows
Lanier, Gerald Norman
Lyon, Robert Lyndon
Macdonald, Duncan Ross
Merkel, Edward Paul
Miller, William Eldon
Mitchell, Russel Gene
Muldrew, James Archibald
Nebeker, Thomas Evan
Nichols, James Otis
Nord, John C
Olson, Robert Edward
Rennels, Robert Gossett
Rudinsky, Julius Alexander
Safranyik, Laszlo
Schenk, John Albright
Schmiege, Donald Charles
Simeone, John Babtista
Simmons, Gary Adair
Sippell, William Lloyd
Sloan, Norman F
Smith, Richard Harrison
Smith, Russell K
Solomon, James Doyle
Stark, Ronald William
Stevens, Robert E
Thatcher, Robert Clifford
Varty, Isaac William
Waters, William E
Williams, Carroll Burns, Jr
Witter, John Allen
Wood, David Lee
Woods, Frank Wilson
Yates, Harry Orbell, III

Forest Genetics
Barber, John Clark
Barnett, Paul Edward
Beineke, Walter Frank
Blair, Roger L
Canavera, David Stephen
Cech, Franklin Charles
Clausen, Knud Erik
Critchfield, William Burke
Daniels, Jess Donald
Dawson, David H
Dinus, Ronald John
Einspahr, Dean William
Fechner, Gilbert Henry
Franklin, Edward Carlyle
Funk, David Truman
Garrett, Peter Wayne
Gerhold, Henry Dietrich
Gladstone, William Turnbull
Goddard, Ray Everett
Guries, Raymond Paul
Hall, Richard Brian
Hanover, James W
Illingworth, Keith
Jokela, Jalmer John
Karnosky, David Frank
Kellison, Robert Clay
Khalil, Muhammad Ahsan Khan
Kraus, John Franklyn
Kriebel, Howard Burtt
Krugman, Stanley Liebert
La Farge, Timothy
Lanner, Ronald Martin
Long, Ernest M
Manley, Stephen Alexander
McElwee, Robert L
Mergen, Francois
Morgenstern, Erwin Kristian
Namkoong, Gene
Nienstaedt, Hans
Orr-Ewing, Alan Lindsay
Perry, Thomas Oliver
Posey, Clayton Eugene
Rink, George
Ruby, John L
Schoenike, Roland Ernest
Schreiner, Ernst Jefferson
Selders, Archie Arnold
Steiner, Kim Carlyle
Stettler, Reinhard Friederich
Taft, Kingsley Arter, Jr
Thielges, Bart A
Thor, Eyvind
Valentine, Fredrick Arthur
Van Buijtenen, Johannes Petrus
Wang, Chi-Wu
Wilkinson, Ronald Craig
Woessner, Ronald Arthur
Wright, Jonathan William
Yeatman, Christopher William
Zobel, Bruce John

Forest Hydrology
Belt, George Harley, Jr
Bethlahmy, Nedavia
Brown, George Wallace
Brown, Harry Esmond
Chang, Mingteh
Coltharp, George B
Golding, Douglas Lawrence
Harper, Warren Charles
Hibbert, Alden R
Mace, Arnett C, Jr
Mader, Donald Lewis
Maki, Tenho Ewald
Merriam, Robert Arnold
O'Hayre, Arthur Paul
Patric, James Holton
Singh, Teja
Slaughter, Charles Wesley
Striffler, William D
Swanson, Robert Harold
Ursic, Stanley John
Wheeler, Richard Hunting
Willington, Robert Peter
Zwolinski, Malcolm John

Forest Management
Adams, Darius Mainard
Atkinson, William Allen
Baker, Robert Donald
Bare, Barry Bruce
Bell, John Frederick
Bender, Walter Louis
Bourdo, Eric A, Jr
Brender, Ernst Victor
Bryant, Ralph Clement
Countryman, David Wayne
Daniel, Theodore William
Davis, Kenneth Pickett
Deitschman, Glenn Howard
Drew, T John
Duerr, William Allen
Gara, Robert
Godman, Richard M
Gray, John Lewis
Gruschow, George F
Haines, Harry Caum
Harper, Verne Lester
Heninger, Ronald Lee
Horn, Allen Frederick, Jr
Johnson, Norman Elden
Kissick, Norman Lennox
Lane, Richard Dale
Lange, Robert Walter
Larson, Charles Conrad
Larson, Frederic Roger
Lewis, David Kent
McElwee, Robert L
Merriam, Lawrence Campbell, Jr
Miller, Roswell Kenfield
Mosher, Milton Monroe
Myers, Clifford Albert, Jr
Newnham, Robert Montague
Page, Alan Cameron
Patterson, Archie Edgar
Pierce, William R
Plumb, Timothy Roy, Jr
Roach, Benjamin Arthur
Ronco, Frank, Jr
Rudolph, Victor J
Shirley, Frank Connard
Shirley, Hardy Lomax
Steele, Robert Wilbur
Stenzel, George
Walker, Nathaniel
Wilson, Carl C

Forest Mensuration
Aldrich, Robert Clement
Beers, Thomas Wesley
Bell, John Frederick
Bruce, David
Colwell, Robert Neil
Curtis, Robert Orin
Grosenbaugh, Lewis Randolph
Herman, Francis Robert
Johnson, Evert William
Leary, Rolfe Albert
Mitchell, Kenneth John
Myers, Clifford Albert, Jr
Newnham, Robert Montague
Paine, David Philip

Forest Meteorology
Federer, C Anthony
Fosberg, Michael Allen
Herrington, Lee Pierce
Reifsnyder, William Edward
Shea, Keith Raymond
Whitney, Roy Davidson

Forest Pathology
Amburgey, Terry L
Berry, Frederick Hamer
Blakeslee, George M
Boyce, John Shaw, Jr
Brown, Merton F
Campana, Richard John

Campbell, William Andrew
Canfield, Elmer Russell
Crosby, Emory Spear
Davidson, Alexander Grant
Davidson, Ross Wallace
Davis, Terry Chaffin
Driver, Charles Henry
Eslyn, Wallace Eugene
Etheridge, David Elliott
French, David Weston
Henry, Berch Waldo
Hawksworth, Frank Goode
Hepting, George Henry
Hesterberg, Gene Arthur
Highley, Terry L
Houston, David Royce
Howe, Virgil K
Hunt, Richard Stanley
Kimmey, James William
Kondo, Edward Shin-Chi
Krebill, Richard G
Lachance, Denis
Leaphart, Charles Donald
Lightle, Paul Charles
McCracken, Francis Irvin
McGrath, William Thomas
McKenzie, Malcolm Arthur
McNabb, Harold Sanderson, Jr
Merrill, William
Morton, Harrison Leon
Nordin, Vidar John
Paine, Lee Alfred
Patton, Robert Franklin
Phelps, William Robert
Rice, Peter (Franklin)
Ross, Eldon Wayne
Roth, Lewis Franklin
Rowan, Samuel James
Setliff, Edson Carmack
Shaw, Charles Gardner, III
Smith, Richard Barrie
Stambaugh, William James
Tainter, Franklin Hugh
Toole, Eben Richard
Towers, Barry
Trappe, James Martin
Wilcox, Webster Wayne
Zabel, Robert Alger
Zak, Bratislav
Ziller, Wolf Gunther

Forest Physiology
Barnes, Robert Lloyd
Bilan, M Victor
Dickmann, Donald Irvin
Elam, William Warren
Fitzgerald, Charles H
Gatherum, Gordon Elwood
Hellmers, Henry
Kozlowski, Theodore Thomas
Larson, Merlyn Milfred
Larson, Philip Rodney
Owston, Peyton Wood
Phares, Robert Eugene
Rietveld, Willis James
Sommer, Harry Edward
Wang, Ben Shih-pin
Whitmore, Frank William
Winton, Lawson Lowell
Zaerr, Joe Benjamin

Forest Products
Adams, Daniel Otis
Alexander, Stuart David
Anderson, Arthur Bernhardt
Baechler, Roy Herman
Barefoot, Aldos Cortez, Jr
Beall, Francis Carroll
Beals, Harold Oliver
Behr, Eldon August
Bensend, Dwight Winfred
Bethel, James Samuel
Biblis, Evangelos J
Blankenhorn, Paul Richard
Bodig, Jozsef
Brown, John Henry
Bryant, Ben S
Carter, Roy Merwin
Choong, Elvin T
Clark, James d'Argaville
Cockrell, Robert Alexander
Comstock, Gilbert Leroy
Cooper, Glenn Adair, Jr
Cote, Wilfred Arthur, Jr
Davidson, Robert W
De Zeeuw, Carl Henri
Dickinson, Fred Eugene
Dobbins, Thomas Edward
Eckelman, Carl A
Ellis, Everett Lincoln
Ellwood, Eric Louis
Erickson, Harvey D
Erickson, Robert W
Ethington, Robert Loren
Fogg, Peter John
Franz, Norman Charles
Frost, Thomas Rogers
Garland, Hereford

Forest Soils
Baker, James Bert
Blackmon, Bobby Glenn
Campbell, Ralph Edmund
Carmean, Willard Handy
Cochran, Patrick Holmes
Cole, Dale Warren
Cox, Gene Spracher
Doolittle, Warren Truman
Dyrness, Christen Theodore
Fletcher, Peter Whitcomb
Hatchell, Glyndon Elbert
Hauxwell, Donald Lawrence
Heilman, Paul E
Henderson, Gray Stirling
Ike, Albert Francis
Kennedy, Harvey Ellis, Jr
Klock, Glen Orval
Krause, Helmut
Lane, Carl Leaton

Gaslick, Harold Bailey
Gerljansen, Roland O
Goodyear, William Frederick, Jr
Gorbatsevich, Serge N
Goulet, Marcel
Guiher, John Kenneth
Guthrie, Franklin Kirney
Hamilton, John Robert
Hann, Robert A
Haygreen, John G
Hill, John Ledyard
Hunt, Michael O'Leary
Isebrands, Judson G
Jayne, Benjamin A
Jenks, Theodore Eugene
Kalnins, Otto Julius
Kaufert, Frank Henry
Kenaga, Duane Leroy
Kennedy, Robert William
King, Woodrow Wilson
Klein, Edward Lawrence
Koch, Peter
Koran, Zoltan
Krahmer, Robert Lee
Krier, John Peter
Kubler, Hans Jakob
Kutscha, Norman Paul
Langwig, John Edward
Lehmann, William Fredrick
Leney, Lawrence
Levi, Michael Phillip
Lyon, Duane Edgar
Manwiller, Floyd George
Marra, Alan A
Mathur, Vishwa Nath Prasad
McAlister, Robert Hardy
McKean, Herbert Baldwin
McKimmy, Milford D
McMillin, Charles W
Moslemi, Ali A
Muench, John, Jr
Murphey, Wayne K
Nearn, William Thomas
Nelson, George Richard
Noskowiak, Arthur Fredrick
Pillar, Walter Oscar
Preston, Stephen Boylan
Quirk, John Thomas
Raphael, Harold James
Resch, Helmuth
Rice, James Thomas
Saeman, Jerome Francis
Sastry, Cherla Bhaskara Rama
Scheffer, Theodore Comstock
Schniewind, Arno Peter
Sebastian, Leslie Paul
Senft, John Franklin
Short, Paul Henry
Shottafer, James Edward
Shuler, Craig Edward
Simmons, Frederick Charles
Skaar, Christen
Skolmen, Roger Godfrey
Suliker, Alan
Smith, Walton Ramsey
Stark, Eric Walter
Suchsland, Otto
Sun, Bernard Ching-Huey
Tang, Ruen Chiu
Taras, Michael Andrew
Taylor, Fred William
Thomas, Richard Joseph
Thompson, Warren Slater
Troxell, Harry Emerson, Jr
Urling, Gerard Phelps
Van Vliet, Antone Cornelis
Vick, Charles Booker
Wahlgren, Harold Emil
Wangaard, Frederick Field
Wellons, Jesse Davis, III
Wellwood, Robert William
West, Richard Fussell
Whipple, Francis Oliver
Yavorsky, John Michael
Zerbe, John Irwin

Silviculture
Armson, Kenneth Avery
Baker, James Bert
Berglund, John Verne
Blum, Barton Morrill
Burns, Russell MacBain
Collins, Paul Everett
Dingle, Richard William
Doolittle, Warren Truman
Gordon, John C
Griffin, Ralph Hawkins
Gruschow, George F
Hatchell, Glyndon Elbert
Johnson, John William
Johnson, Norman Elden
Kaufman, Clemens Marcus
Knudson, Douglas Marvin
Merrifield, Robert G
Powell, Graham Reginald
Roberts, Edward Guernsey
Rudolf, Paul Otto
Schlesinger, Richard Cary
Schmidt, Wyman Carl
Scott, David Robert Main
Smith, David Martyn
Stransky, John Janos
Tackle, David
Tryon, Henry Haven
Van Haverbeke, David F
Walker, Laurence Colton
Winch, Fred Everett, Jr
Woodman, James Nelson

Leaf, Albert Lazarus
Linnartz, Norwin Eugene
Lorio, Peter Leonce Jr
Mader, Donald Lewis
Mahendrappa, Mukkatira Kariappa
Maki, Tenho Ewald
Marion, Giles Michael
McFee, William Warren
Moehring, David Marion
Moore, Duane Grey
Nimlos, Thomas John
Pritchet, William Lawrence
Ralston, Charles William
Riekerk, Hans
Scott, William
Smalley, Glendon William
Stone, Earl Lewis, Jr
Waterson, Kenneth Gordon
Webster, Stephen Russell
Wooldridge, David Dilley
Youngberg, Chester Theodore

GENETICS

Genetics
Abplanalp, Hans
Abrahamson, Seymour
Adams, Jack Donald
Adams, James Norman
Adams, Jane N
Adelberg, Edward Allen
Adler, Julius
Albrechtsen, Rulon S
Alexander, Mary Louise
Ali-Khan, Syed Tahir
Allard, Robert Wayne
Allen, Sally Lyman
Anderson, Ronald Eugene
Anderson, W French
Andrews, John Edwin
Angell, Frederick Franklyn
Annan, Murvel Eugene
Anstey, Thomas Herbert
Armstrong, Robert John
Ashton, Geoffrey C
Atherly, Alan G
Atwood, Sanford Soverhill
Austin, Mary Lellah
Axelrod, David E
Axtell, John David
Ayala, Francisco Jose
Baenziger, H
Baharach, Martin Max
Baer, Adela Dee
Bailey, Zeno Earl
Baker, Percy Hayes
Baker, Raymond Milton
Baker, Richard H
Baker, Robert Frank
Baker, William Kaufman
Balbinder, Elias
Baldy, Marian Wendorf
Bamrick, John Francis
Band, Henrietta Trent
Barbour, Stephen D
Barnes, Donald Kay
Barnett, Audrey
Barnhart, Benjamin J
Barr, Harry L
Barry, Edward Gail
Bartlett, Alan C
Baumgardner, Kandy Diane
Baumiller, Robert Cahill

Baylis, John Robert, Jr
Beamish, Katherine I
Beardsley, Robert Eugene
Becker, Walter Alvin
Behne, Ronald John
Bell, Audra Earl
Belser, William Luther, Jr
Bemis, William Putnam
Bender, Harvey Alan
Bennett, Dorothea
Berg, Claire M
Berger, Edward Michael
Bergh, Berthold Orphie
Bergmann, Fred Heinz
Berlyn, Mary Berry
Bernstein, Harris
Bernstein, Aleck
Bernstein, Julian L
Bhalla, Satish Chander
Biggers, Charles James
Bigley, Robert Harry
Birky, Carl William, Jr
Bishop, Jack Belmont
Bixler, David
Bleyman, Lea Kanner
Bloom, Arthur David
Blumm, Cecil Thomas
Boche, Robert DeVore
Boguslawski, George
Boyes, John Wallace
Boy- Boynton, Thomas P
Borgaonkar, Digamber Shankarrao
Bostian, Carey Hoyt
Bott, Kenneth F
Boulware, Ralph Frederick
Bowen, Hollis Hulon
Bowen, Howard S
Bowman, Barbara Hyde
Bowman, Sarane Thompson
Bowne, Samuel Winter, Jr
Boyer, Samuel Talton
Brady, Sterling Gaylen
Bradford, Gordon Eric
Braver, Gerald
Brierley, Jean
Briggs, Robert Wilbur
Brink, Royal Alexander
Brockman, Herman E
Broderick, Lynne Sechrist
Brody, Stuart
Brosseau, George Emile, Jr
Brown, Douglas Fletcher
Brown, Vernon
Brown, Howard S
Brown, Russell Vedder
Brown, William Lacy
Brown, William Paul
Bruck, David Lewis
Bruneau, Leslie Herbert
Bryan, Clifford Randall
Bryson, Vernon
Buchanan, Kenneth William
Bukhari, Ahmad Iqbal
Bukovsan, Laura A
Bullen, Miles Rex
Burdick, Allan Bernard
Burfening, Peter J
Buri, Peter Frederick
Burk, Lawrence G
Burns, George W
Burns, Martin Joe
Busbice, Thaddeus H
Busch, Robert Henry
Butler, Leonard
Butterworth, Francis M
Butzel, Henry M, Jr
Byrd, Wilbert Preston
Bywaters, James Humphreys
Caldecott, Richard S
Campbell, David Paul
Campbell, Ian Maclean
Carlson, Elof Axel
Carlson, James H
Carlson, Bruce Charles
Carmon, James Lavern
Carnahan, Howard Leon
Carpenter, Adelaide Trowbridge Clark
Carson, J David David
Carte, Ira F
Carter, Robert Clifton
Cashel, Michael
Center, Elizabeth M
Chaganti, Raju Sreerama Kamalasana
Chai, Chen Kang
Champe, Sewell Preston
Champion, Arthur Kingsley
Chang, Tien-ding
Chaplin, James Ferris
Chapman, Verne M
Chase, Sherret Spaulding
Chi, Kuo-Ruey
Chiang, Morgan S
Chiasson, Leo Patrick
Chinnici, Joseph (Frank) Peter
Chisler, John Adam
Chovnick, Arthur

Chown, H Bruce
Christian, James A
Christian, Ross Edgar
Chu, Ernest Hsiao-Ying
Clancy, Clarence William
Clark, Alvin John
Clark, Arnold M
Clark, Flora Mae
Clark, Roger William
Clayberg, Carl Dudley
Clayton, Frances Elizabeth
Clement, William Madison, Jr
Clise, Ronald Leo
Close, Perry
Clowes, Royston Courtenay
Coe, Edward Harold, Jr
Coe, Gerald Edwin
Cohen, Larry William
Cole, Thomas A
Collier, Jesse Wilton
Collins, William Kerr
Compton, William A
Comstock, Ralph Ernest
Conklin, Marie Eckhardt
Cook, Robert Carter
Cook, Robert Edward
Cooke, Fred
Cooper, Harold Eugene
Cooper, Jane Elizabeth
Cooper, Stephen
Cope, Will Allen
Corcoran, Mary Ritzel
Corcos, Alain Francois
Corwin, Harry O
Cotter, Mary Virginia
Cotter, William Bryan, Jr
Coulter, Murray Whitfield
Counce, Sheila Jean
Cox, Edward Charles
Craig, James Verne
Creech, Roy G
Creel, Gordon C
Cribbs, Richard Madison
Crittenden, Lyman Butler
Crow, James Franklin
Crumpacker, David Wilson
Cullen, Mary Urban
Cummins, Joseph E
Dachtler, Sally Louise
Dahmen, Jerome J
Dalton, Howard Clark
Dalton, Lonnie Gene
Daly, Kevin Richard
Darlington, Gretchen Ann Jolly
Datta, Surinder P
Daugherty, Patricia A
Davidson, Richard Laurence
Davidson, Ronald G
Davis, David Gale
Davis, Elmo Warren
Davis, Ralph Lanier
Davis, Rowland Hallowell
Davis, William Hatch
Dawson, Peter Sanford
Dawson, Wallace Douglas, Jr
Day, Peter Rodney
Dean, Charles Edgar
Dearden, Douglas Morey
DeBusk, Aron Gib
Delisle, Albert Lorenzo
DeMarinis, Frank
Dempsey, Wesley Hugh
Denell, Robin Ernest
Dennett, Robert Kingsley
Dessureaux, Lionel
Dev, Vaithilingam Gangathara
Dewey, Jan M J
Dewey, Wade G
Dickerman, Richard Curtis
Dickinson, Frank N
DiPaolo, Joseph Amadeo
Dippell, Ruth Virginia
Dodson, Edward O
Doe, Frank Joseph
Doerder, F Paul
Doolittle, Donald Preston
Dorfman, Ben-Zion
Dorn, Gordon L
Doudney, Charles Owen
Dove, William Franklin
Doyle, Roger Whitney
Drake, John W
Dronamraju, Krishna Rao
Druger, Marvin
Dubes, George Richard
Duckworth, Donna Hardy
Dudgeon, Edna
Dudley, John Wesley
Dunn, Gary Raymond
Dunn, Gerald Marvin
Durham, Ralph Marion
Dutta, Sisir Kamal
Duvick, Donald Nelson
Dyck, Peter Leonard
Eberhart, Steve A
Ebert, Wesley W
Eckroat, Larry Raymond
Egell, Marshall Hall

Edington, Charles W
Edward, Cosmas
Edwards, Lewis Hiram
Edwardson, John Richard
Eicher, Eva Mae
Eisen, Eugene J
Ellgaard, Erik G
Emara, Yehia Abdelaziz Saleh
Emerson, Sterling (Howard)
Emmel, Thomas C
Enfield, Franklin D
Engh, Helmer A, Jr
Englert, Du Wayne Cleveland
Englesberg, Ellis
English, Darrel Starr
Enns, Henry
Ensign, Stewart Ellery
Epler, James L
Erdman, Howard E
Erk, Frank Chris
Erlick, Barry J
Esposito, Michael Salvatore
Esposito, Rochelle E
Evans, Raeford G
Everett, Herbert Lyman
Everson, Everett Henry
Faberge, Alexander Cyril
Fahmy, Mohamed Hamed
Falls, Harold Francis
Faris, Donald George
Farnsworth, Marjorie Whyte
Fattig, W Donald
Fejer, Stephen Oscar
Feldman, Jerry F
Fery, Richard Lee
Fick, Gerhardt Nelson
Ficsor, Gyula
Finger, Irving
Fink, Gerald Ralph
Fischer, Glenn Albert
Fisher, Kathleen Mary
Florian, Svatopluk Fred
Flory, Walter S, Jr
Fogel, Seymour
Foltz, Virginia C
Forbes, Oliver Clifford
Forster, Jean Lois
Forsthoefel, Paulinus Frederick
Fowler, Gregory L
Fox, Allen Sander
Fox, Maurice Stanford
Fox, Richard Romaine
Fradkin, Cheng-Mei Wang
Francis, Bettina Magnus
Friedman, Lawrence David
Frischer, Henri
Fristrom, James W
Fuerst, Robert
Fujimoto, Atsuko Ono
Fuller, Marian Jane
Fuscaldo, Anthony Alfred
Fuscaldo, Kathryn Elizabeth
Futch, David Gardner
Gaines, James Abner
Galinat, Walton Clarence
Galinsky, Irving
Galletta, Gene John
Gally, Joseph Anthony
Galsworthy, Peter Robert
Ganesan, Adayapalam T
Ganschow, Roger Elmer
Gardenhire, James Homer
Gardner, Arthur Wendel
Gardner, Eldon John
Gartler, Stanley Michael
Gasser, David Lloyd
Gates, Allen Hazen, Jr
Geer, Billy W
Geeseman, Gordon E
Generoso, Walderico Malinawan
Gerdes, Raymond A
Gerdy, James Robert
Gerstel, Dan Ulrich
Gethmann, Richard Charles
Gfeller, Barbara
Giesbrecht, John
Gilden, Raymond Victor
Giles, Norman Henry
Gilham, Nicholas Wright
Gilman, Lauren Cundiff
Gilmore, Earl C
Ginsburg, Benson Earl
Girvin, Eb Carl
Glass, Arthur Warren
Glass, Hiram Bentley
Glick, Bruce
Glucksohn-Waelsch, Salome
Godley, Willie Cecil
Goetinck, Paul Firmin
Goggans, James F
Gold, John Rush
Goldman, Ben Roy
Goldman, Stephen L
Gonsalves, Neil Ignatius
Goodenough, Ursula Wiltshire
Gooder, Harry
Goodal, Sol Howard

Goodlin, Robert Clair
Goodman, Harold Orbeck
Goodman, Major M
Goodwin, Kenneth
Goplen, Bernard Peter
Gordon, Manuel Joe
Gottlieb, Frederick Jay
Gould, Adair Brasted
Gowans, Charles Shields
Grafius, John Edward
Grahn, Douglas
Grant, Bruce S
Grant, Verne (Edwin)
Gratzner, Howard G
Gravett, Howard L
Green, Donald MacDonald
Green, Earl Leroy
Green, Margaret Creighton
Green, Melvin Martin
Greenblatt, Irwin M
Gregg, Thomas G
Grell, Ellsworth Herman
Grell, Rhoda Frank
Griesbach, Robert Anthony
Griffing, J Bruce
Griffiths, Anthony J F
Groner, Miriam Georgia
Grosch, Daniel Swartwood
Gross, Samson Richard
Grunder, Allan Angus
Guest, William C
Gupton, Creighton Lee
Gyles, Nicholas Roy
Haas, Felix Levere
Hackerott, Harold Leroy
Hackett, Adeline J
Hadley, Henry Hultman
Haggard, Bruce Wayne
Hagy, George Washington
Halaban, Ruth
Haley, Leslie Ernest
Hall, Benedict Mark
Hammett, James Roy
Hammons, Ray Otto
Hampton, Suzanne Harvey
Handler, Shirley Wolz
Hanks, George D
Hansen, Carl Tams
Harding, James A
Harlan, Jack Rodney
Harm, Walter
Harney, Patricia Marie
Harpstead, Dale D
Harrington, Francis Eugene
Harris, James Edward
Harris, Robert Ernest
Harris, Robert Martin
Hart, Gary Elwood
Hartberg, Warren Keith
Hartwell, Leland Harrison
Harvey, Bryan Laurence
Haskins, Caryl Parker
Hatch, Frederick Tasker
Hawthorne, Donald Clair
Hayes, Donald H
Healy, Eugene A
Hecht, Frederick
Hedgecock, Dennis
Hedman, Stephen Clifford
Hedrick, Philip William
Heiner, Robert E
Helinski, Donald Raymond
Helling, Robert Bruce
Helm, James Leroy
Herforth, Robert S
Herskowitz, Irwin Herman
Hertel, Elmer William
Herzenberg, Leonard Arthur
Hess, John Berger
Heston, Leonard L
Heston, Walter Enoch
Hexter, William Michael
Hickey, William August
Higgs, Roger L
Hildemann, William Henry
Hildreth, Philip Elwin
Hill, Charles Whitacre
Hill, Helene Zimmermann
Hill, Jack F
Hill, Richard Norman
Hill, Richard Ray, Jr
Hillers, Joe Karl
Hinson, Kuell
Hinton, Claude Willey
Hoage, Terrell Rudolph
Hoch, James Alfred
Hochman, Benjamin
Hoefnagel, Dick
Hoffman, Harold A
Hogg, Robert W
Holland, Lewis
Hollander, Willard Fisher
Holm, David George
Honeyman, Merton Seymour
Hood, Cornelius Henry
Hooker, Arthur Lee
Hoornbeek, Frank Kent
Hopper, Anita Klein

Horner, Theodore Wright
Horowitz, Norman Harold
Hotta, Yasuo
Hougas, Robert Wayne
House, Verl Lee
Hovin, Arne William
Howe, Henry Branch, Jr
Howes, Cecil Edgar
Hsie, Abraham Wuhsiung
Hsu, Tao-chiuh
Huang, Pien-Chien
Hubby, John L
Hufham, James Birk
Hughes, Morris Burdette
Hujing, Frans
Hull, Inez Mary
Humes, Paul Edwin
Hurst, Donald D
Huskey, Robert John
Hutchison, Clyde Allen, III
Hyer, Angus Hillyard
Hymowitz, Theodore
Ihrke, Charles Albert
Ihssen, Peter Edward
Infanger, Ann Martin
Irr, Joseph David
Irwin, Malcolm Robert
Ives, Philip Truman
Jackson, John Fenwick
Jackson, Sharon Wesley
Jagiello, Georgiana Mary
Jainchill, Jerome
James, Allen Pinsent
Jarvis, Floyd Eldridge, Jr
Jeffery, Duane Eldro
Jenkins, John Bruner
Jessen, Carl Roger
Jimenez-Marin, Daniel
Johnsen, Roger Craig
Johnson, August S
Johnson, Bertil Lennart
Johnson, F Clifford
Johnson, Franklin M
Johnson, Freeman Keith
Johnson, Herbert Windal
Johnson, Lewis Warren
Johnson, Melvin Walter, Jr
Johnson, William Wayne
Johnston, John Spencer
Jones, Gary Edward
Jones, Theodore Charles
Jones, William F
Joppa, Leonard Robert
Josephson, Aaron Mortimer
Judd, Burke Haycock
Juergensmeyer, Elizabeth B
Jump, Lorin Keith
Kabat, David
Kafer, Etta (Mrs E R Boothroyd)
Kahler, Alex L
Kaiser, Armin Dale
Kallio, Arvo
Kallman, Klaus D
Kalter, Harold
Kan, James Hung-Kei
Kayhart, Marion
Kelleher, Raymond Joseph, Jr
Kelley, William Nimmons
Kelton, Diane Elizabeth
Keppler, William J
Kermicle, Jerry Lee
Kernaghan, Roy Peter
Kerr, Warwick Estevam
Kerschner, Jean
Kessler, Seymour
Kieffer, Nat
Kiger, John Andrew, Jr
King, Jack Lester
King, James Clement
King, Mary-Claire
King, Robert Charles
Kinloch, Bohun Baker, Jr
Kinsey, John Aaron, Jr
Kirsjansson, F K
Kittilson, Harold Lee
Kitzmiller, James Blaine
Klassen, Waldemar
Kleinhofs, Andris
Klekowski, Edward Joseph, Jr
Klinger, Harold P
Klotz, John William
Knowles, Barbara B
Knudson, Alfred George, Jr
Koerting, Lola Elisabeth
Koger, Marvin
Kontras, Stella B
Konzak, Calvin Francis
Koo, Francis Keh Shing
Kowles, Richard Vincent
Kraay, Gerrit Jacob
Kramer, Nicholas William
Kreizinger, Jean Dolloff

Sparrow, Arnold Hicks
Specht, Lawrence W
Spector, Calvin
Spelsberg, Thomas Coonan
Spiegelman, Solomon
Spieler, Richard Arno
Spiess, Eliot Bruce
Spofford, Janice Brogue
Sprague, Lucian Matthew
Squire, Richard Douglas
Srb, Adrian Morris
Srivastava, Satish Kumar
Stadler, David Ross
Stahl, Franklin William
Stairs, Gerald Ray
Stalker, Harrison Dailey
Stansfield, William D
Stauffer, James
Steffensen, Dale Marriott
Steiner, Erich E
Stich, Hans F
Stimpfling, Jack Herman
Stinson, Harry Theodore, Jr
Stock, David Allen
Stone, David
Stone, Donald Eugene
Stone, Howard Anderson
Stone, William Harold
Stonecypher, Roy W
Stormont, Clyde J
Strelein, Jacob Wayne
Streisinger, George
Strickberger, Monroe Wolf
Strickland, Walter Nicholas
Strickler, Dwight Johnston
Strohm, Jerry Lee
Stuber, Charles William
Stutz, Howard Coombs
Styles, Ernest Derek
Styles, Salma Mahmoud
Sueoka, Noboru
Sulerud, Ralph L
Sullivan, Robert Little
Sun, Nai Chau
Susman, Millard
Suzuki, David Takayoshi
Takats, Stephen Tibor
Tallan, Irwin
Tamarin, Robert Harvey
Tan, James Chien-Hua
Tan, Yin Hwee
Tartof, Kenneth D
Taschdjian, Edgar
Taub, Stephen Robert
Taylor, Percy LeRoy
Taylor, Boyd Eugene
Taylor, George Allan
Taylor, Kenneth Monroe
Taylor, Milton William
Taylor, Ronald
Teas, Howard Jones
Tegenkamp, Thomas Richard
Temin, Rayla Greenberg
Templeton, Joe Wayne
Terrill, Mary Judith (Robinson)
Tevethia, Mary Judith (Robinson)
Thelen, Thomas Harvey
Thomas, Charles Hill
Thomas, James H
Thomas, Percy LeRoy
Thomas, Walter Ivan
Thompson, David J
Thompson, James Neal, Jr
Thompson, James Scott
Thompson, Jerry Nelson
Thompson, Margaret A Wilson
Thompson, Maxine Marie
Thompson, Norman Robert
Thompson, Peter Ervin
Thompson, Steven Risley
Thorne, John Carl
Threlkeld, Stephen Francis H
Thwaites, William Mueller
Tigchelaar, Edward Clarence
Tips, Robert Leonard
Todd, Neil Bowman
Tokunaga, Chiyoko
Tomes, Mark Louis
Tonzetich, John
Townes, Philip Leonard
Townsend, Charley E
Townsend, Joel Ives
Tracey, Martin Louis, Jr
Trasler, Daphne Gay
Travis, Dennis Michael
Treanor, Katherine P
Treece, Jack Milan
Trimble, Robert Bogue
Tritz, Gerald Joseph
Trommershausen-Smith, Ann L
Trosko, James Edward
Trout, William Edgar, III
Tsai, Chia-Yin
Tullis, James Earl
Tuveson, Robert Williams
U, Raymond
Uchida, Irene Ayako
Ulrich, Valentin
Upadhyaya, Rajarama Belle

Uphoff, Delta Emma
Van Elswyk, Marinus, Jr
Van Vleck, Lloyd Dale
Vella, Francis
Verghese, Margrith Wehrli
Verley, Frank A
Vesely, John Anthony
Vigfusson, Norman V
Vlahakis, George
Vodkin, Michael Harold
Voelker, Robert Allen
Von Borstel, Robert Carsten
Vyas, Girish Narmadashankar
Waelsch, Salome Gluecksohn
Wagenaar, Emile B
Wagner, Robert Philip
Wagoner, Dale E
Wainwright, Lillian K (Schneider)
Waite, Albert B
Wallace, Bruce
Wallace, Douglas Cecil
Walter, Mable Ruth
Wang, Richard J
Wann, Elbert Van
Ward, Calvin Lucian
Ward, Oscar Gardien
Warren, Don Cameron
Warters, Mary
Waters, Nelson Fenn
Watson, Jack Ellsworth
Watson, Margaret Liebe
Weaver, Ellen Cleminshaw
Weaver, James Bode, Jr
Webb, Robert Bradley
Webber, Brooke Bland
Weeks, Leo
Wegmann, Thomas George
Weibust, Robert Smith
Weijer, Jan
Weinstein, Alexander
Weir, John Arnold
Welch, James Edward
Welch, Quintin B
Welch, Robert McClam
Wells, Stewart Alderson
Welshons, William John
Wesenberg, Darrell
Wessling, Wolfgang Heinrich
West, David Armstrong
Wheeler, Donald Alsop
Wheeler, Marshall Ralph
Wheelis, Mark Lewis
Whitaker, Thomas Wallace
White, Donald Benjamin
White, John Marvin
Whited, Dean Allen
Whiting, Anna Rachel
Wiberg, John Samuel
Widholm, Jack Milton
Widstrom, Neil Wayne
Wiggin, Henry Carvel
Wilcox, Frank H
Wilcox, James Raymond
Wilder, William Baylor
Wiley, William Henry
Wilfret, Gary Joe
Williams, Charles Melville
Williams, Larry Gale
Williamson, David Lee
Williamson, John Hybert
Wills, Christopher J
Wilson, James Franklin
Wilson, John Howard
Wilson, Katherine Schmitkons
Wilson, Vernon Eldridge
Wolf, Thomas Michael
Wolff, George Louis
Womack, Frances C
Womack, James E
Wong, Patrick Tin-Choi
Wood, Henderson Kingsberry
Wood, William Barry, III
Woodruff, Ronny Clifford
Woodward, Ray R
Woodward, Val Waddoups
Woolf, Charles Martin
Woolley, George Walter
Workman, Peter L
Wrathall, Jean Rew
Wright, Clarence Paul
Wright, David Anthony
Wright, James Everett, Jr
Wright, Sewall
Wu, Ching Kuei
Yanders, Armon Frederick
Yao, Kenneth Tsoong-Sieu
Yarbrough, Karen Marguerite
Yehle, Clifford Omer
Yen, Terence Tsin Tsu
Yermanos, Demetrios M
Yoon, Chai Hyun
Yoshida, Akira
Young, Bobby Gene
Young, Donald Alcoe
Young, Michael Warren
Young, William Johnson, II
Yu, Pao-Lo
Yungbluth, Thomas Alan

Yunis, Jorge J
Yu-Sun, Clare Chuan Chang
Zarcaro, Robert Michael
Zartman, David Lester
Zelle, Max Romaine
Zimm, Georgianna Grevatt
Zimmering, Stanley

Animal Genetics

Ash, William James
Bennett, Cecil Jackson
Berg, Roy Torgny
Blackwell, Robert Leighton
Bohren, Bernard Benjamin
Brinks, James S
Busch, Robert Edward
Byrne, Barbara Jean McManamy
Byrne, Bruce Campbell
Chambers, Doyle
Chase, Herman Burleigh
Cole, Randall Knight
Collins, Walter Marshall
Crockett, Joe Richard
Crowl, Robert Harold
Dillard, Emmett Urcey
Eastwood, Basil R
Eldridge, Franklin Elmer
Emsley, Alan Burns
Fausch, Homer David
Fechheimer, Nathan S
Fox, James David
Fox, Thomas Walton
Franke, Donald Edward
Fredeen, Howard T
Garnett, Ian
Garwood, Vernon Abington
Gaunt, Stanley Newkirk
Harris, Dewey Lynn
Havenstein, Gerald B
Hickman, Charles Garner
Hrubant, Henry Everett
Jerome, Frederick Nelson
King, Steven Clarence
Legates, James Edward
Ludwig, Isadore
Lush, Jay Laurence
Marcum, James Benton
Marks, Henry L
Marlowe, Thomas Johnson
Martin, Truman Glen
McDaniel, Benjamin Thomas
McGilliard, Michael Lon
McGuire, John Albert
Morgan, Walter Clifford
Nicholson, Hugh Hampson
Patterson, Troy B
Quevedo, Walter Cole, Jr
Rasmusen, Benjamin Arthur
Roubicek, Carl Ben
Shelby, Charles Edwin
Shirley, Herschel Vincent, Jr
Shreffler, Carol Kauffman
Somes, Ralph Gilmore, Jr
Swiger, Louis Andre
Thompson, Carl Eugene
Towner, Richard Henry
Vogt, Dale William
Walton, Robert Eugene
Wheat, John David

Bacterial Genetics

Andersen, William Ralph
Brownell, George H
Duggan, Dennis E
Elliott, Rosemary Eskridge
Ihler, Karin Ippen
Roth, John R
Sunshine, Melvin Gilbert

Behavioral Genetics

Anderson, Victor Elving
Beaudoin, Jacques
Benzer, Seymour
Beyer, Edgar Herman
Caspari, Ernst Wolfgang
DeFries, John Clarence
Dixon, Linda Kay
Ebert, Patricia Dorothy
Grossfield, Joseph
Hirsch, Jerry
Kidd, Kenneth Kay
Lynch, Carol Becker
Pruzan, Anita M
Reed, Thomas Edward, Jr
Ringo, John Moyer
Schlesinger, Kurt
Simmel, Edward Clemens
Vandenberg, Steven Gerritjan
Wimer, Richard E

Biochemical Genetics

Ambrose, John Augustine
Ames, Bruce Nathan
Baglioni, Corrado
Barrett, Dennis
Benson, Charles Everett
Benzinger, Rolf Hans
Bode, Vernon Cecil

Bonner, James (Fredrick)
Brooks, Margaret Hoover
Brot, Frederick Elliot
Brown, Loretta Ann Port
Calvert, Allen Fisher
Chalmers, John Harvey, Jr
Champney, William Scott
Chasin, Lawrence Allen
Chen, Shi-Han
Colby, Clarence
Cortner, Jean A
Dalby, Arthur
Daniel, William L
Del Vecchio, Vito Gerard
Eberhart, Bruce Maclean
Farmer, James Lee
Ferrell, Robert Edward
Folk, William Robert
Fraser, Murray Judson
Friedman, Selwyn Marvin
Fuchs, Morton S
Gall, Graham A E
Goff, Christopher Godfrey
Grimwood, Brian Gene
Hall, Benjamin Downs
Henke, Randolph Ray
Holmes, Edward Warren
Hudock, George Anthony
Kaczmarczyk, Walter J
Kadner, Robert Joseph
Kambysellis, Michael Panagiotis
Koch, Elizabeth Anne
LaBrie, David Andre
Landy, Arthur H
Lightfoot, Donald Richard
Ling, Hubert
Liu, Houng-Zung
Lucas, Myron Cran
MacKay, Vivian Louise
Malcolm, Alexander Russell
Maroni, Gustavo Primo
Matalon, Reuben
Matsushita, Tatsuo
Meade, Linda Celida
Mitchell, Ann Denman
Nash, David
Paigen, Kenneth
Patterson, David
Sargent, Malcolm Lee
Sass-Kortsak, Andrew
Schneider, Julius Edward
Simonds, Josephine Abigail
Sorger, George Joseph
Summers, Wilma Poos
Tashian, Richard Earl
Tourian, Ara Yervant
Utter, Fred Madison
Wappner, Rebecca Sue
Wenger, David Arthur
Weyter, Frederick William
Wilcox, Gary Lynn

Cytogenetics

Aalders, Lewis Eldon
Alexander, Denton Eugene
Altenburg, Lewis Conrad
Anderton, Laura Gaddes
Avers, Charlotte Jo
Bashaw, Elexis Cook
Bender, Michael A
Bloom, Stephen Earl
Boutros, Susan Noblit
Braungart, Dale Carl
Breg, William Roy
Bregman, Allyn Aaron
Britton, Donald MacPhail
Brooks, Antone L
Brown, Meta (Louise) Suche
Brown, Spencer Wharton
Buck, Raymond Wilbur, Jr
Bullen, Miles Rex
Byrd, J Rogers
Carlson, Wayne R
Chen, Andrew Tat-Leng
Chrisman, Charles Larry
Cohen, Maimon Moses
Cooper, Kenneth Willard
Da Cunha, Antonio Brito
Davis, Herbert L, Jr
Deaven, Larry Lee
Dewey, Vaithilingam Gangathara
Dollinger, Elwood Johnson
Duck, Bobby Neal
Dvorak, Jan
Endrizzi, John Edwin
Erickson, John (Elmer)
Evans, Laurie Edward
Farber, Phillip Andrew
French, Wilbur Lile
Fu, Wei-ning
Garber, Edward David
Garner, James Gregory
Gersh, Eileen Sutton
Glover, David Val
Grant, William Frederick
Grell, Mary
Grun, Paul

Gwynn, Edgar Percival
Hayes, Thomas G
Heinemann, Richard Leslie
Helgason, Sigurdur Bjorn
Herschler, Michael Saul
Hite, Mark
Hoff, Victor John
Hull, John Winter
Humphrey, Donald Glen
Hungerford, David A
Jackson, Raymond Carl
Jalal, Syed M
Jenkins, Burton Charles
Johnson, George Robert
Joneja, Madan Gopal
Kadanka, Zdenek Karel
Kaitsikes, Pantouses John
Kerber, Erich Rudolph
Kidd, Harold J
Kikudome, Gary Yoshinori
Kimber, Gordon
Kohl, John C, Jr
Kreutzer, Richard D
Krim, Mathilde
Kuspira, J
Lam, Shue-Lock
Larson, Ruby Ila
Lesins, Karlis A
Littlefield, Gayle
Love, Robert Merton
Ma, Nancy Shui Fong
Ma, Te Hsiu
MacDonald, Malcolm Duncan
Macintyre, Malcolm Neil
Maguire, Marjorie Paquette
McFee, Alfred Frank
McQuade, Henry Alonzo
Melnyk, John H
Menzel, Margaret Young
Merz, Timothy
Michel, Kenneth Earl
Mickey, George Henry
Miller, Morton W
Moens, Peter B
Moore, Charleen Morizot
Moorhead, Paul Sidney
Morishima, Akira
Morrison, Jack William
Mottinger, John P
Mukherjee, Asit B
Murray, Beatrice E
Nasjleti, Carlos Eduardo
Newman, Lester Joseph
Nichols, Warren Wesley
Nur, Uzi
Ohno, Susumu
Ohnuki, Yasushi
Okada, Tadashi A
Oliver, James Henry, Jr
Palmer, Catherine Gardella
Pathak, Sen
Patterson, Rosalyn Mitchell
Pendse, Pratapsinha C
Pesetsky, Irwin
Petersen, Kyle W
Petersen-Adkisson, Karen
Phillips, Lyle Llewellyn
Phillips, Ronald Lewis
Powers, Harold O
Preston, Robert Julian
Rai, Karamjit Singh
Ramon, Serafin
Rick, Charles Madeira, Jr
Riley, Herbert Parkes
Rodman, Toby C
Rothwell, Norman Vincent
Rowley, Janet Davison
Sagawa, Yoneo
St-Amand, Wilbrod
Sarvella, Patricia Ann
Schaeffer, Jurgen Richard
Schank, Stanley Cox
Schertz, Keith Francis
Schroeter, Gilbert Loren
Sergovich, Frederick Raymond
Shaver, Evelyn Louise
Soudek, Dusan Edward
Srivastava, Probodh K
Stevenson, Harlan Quinn
Stringam, Gary Rice
Swanson, Carl Pontius
Tai, William
Tjio, Joe Hin
Triantaphyllou, Anastasios Christos
Vig, Baldev K
Virkki, Nilo
Walters, James Lee
Walters, Marta Sherman
Weber, David Frederick
Weinstein, David
Wellwood, Arnold Augustus
Williams, John Watkins, III
Wilton, Arthur Charles
Wise, Dwayne Allison
Wolman, Sandra R
Wolff, Sheldon
Wurster-Hill, Doris Hadley
Wyandt, Herman Edwin, Jr

Ying, Kuang Lin

Developmental Genetics
Abbott, Ursula K
Ahearn, Jayne Newton
Arking, Robert
Artzt, Karen
Bass, Sukh D
Beck, Sidney L
Benjaminson, Morris Aaron
Bewley, Glenn Carl
Brumbaugh, John (Albert)
Caspari, Ernst Wolfgang
Cummings, Michael R
Dickinson, William Joseph
Doane, Winifred Walsh
Donady, John James
Elmer, William Arthur
Engstrom, Lee Edward
Erway, Lawrence Clifton, Jr
Falk, Darrel Ross
Frankel, Joseph
Friedman, Thomas Baer
Froiland, Thomas Gordon
Galbraith, Donald Barrett
Grimes, Gary Wayne
Graham, James Douglas
Hanly, Edward William
Hartl, Daniel L
Hartung, Ernest William
Hearney, Elaine Frances
Hodgetts, Ross Birnie
Holstein, Thomas James
Hughes, Norman
Imberski, Richard Bernard
Johnson, Thomas Eugene
Kalicki, Henrietta
Kessin, Richard Harry
King, Jonathan Alan
Loy, James Brent
Lucchesi, John Charles
Lyerla, Timothy Arden
Markert, Clement Lawrence
Mericle, Rae Phelps
Mizell, Merle
Moore, Jay Winston
Mullen, Richard Joseph
Newton, (William) Austin
Pedersen, Roger Arnold
Peeples, Earle Edward
Rizki, Tahir Mirza
Schneiderman, Howard Allen
Schwinck, Ilse
Shearn, Allen David
Sheraid, Allen Franklin
Smith, Patricia Anne
Smith-Gill, Sandra Joyce
Sterling, Anne
Sullivan, David Thomas
Surver, William Merle, Jr
Tompkins, Robert
Whitt, Gregory Sidney
Wolfe, Herbert Glenn
Wright, Theodore Robert Fairbank
Zalokar, Marko

Forest Genetics
Barber, John Clark
Barnett, Paul Edward
Beineke, Walter Frank
Blair, Roger L
Canavera, David Stephen
Cech, Franklin Charles
Clausen, Knud Erik
Critchfield, William Burke
Daniels, Jess Donald
Dawson, David H
Dinus, Ronald John
Drew, T John
Einspahr, Dean William
Fechner, Gilbert Henry
Franklin, Edward Carlyle
Funk, David Truman
Garrett, Peter Wayne
Gerhold, Henry Dietrich
Gladstone, William Turnbull
Goddard, Ray Everett
Guries, Raymond Paul
Hall, Richard Brian
Hanover, James W
Illingworth, Keith
Jokela, Jalmer John
Kellison, Robert Clay
Khalil, Muhammad Ahsan Khan
Kraus, John Franklyn
Kriebel, Howard Burtt
Krugman, Stanley Liebert
La Farge, Timothy
Lanner, Ronald Martin
Long, Ernest M
Manley, Stephen Alexander
Mergen, Francois
Morgenstern, Erwin Kristian
Namkoong, Gene
Nienstaedt, Hans
Orr-Ewing, Alan Lindsay
Perry, Thomas Oliver
Posey, Clayton Eugene

Rink, George
Ruby, John L
Schoenike, Roland Ernest
Schreiner, Ernst Jefferson
Steiner, Kim Carlyle
Stettler, Reinhard Friederich
Szikiai, Oscar
Taft, Kingsley Arter, Jr
Thielges, Bart A
Thor, Eyvind
Valentine, Fredrick Arthur
Van Buijtenen, Johannes Petrus
Wang, Chi-Wu
Wilkinson, Ronald Craig
Woessner, Ronald Arthur
Wright, Jonathan William
Yeatman, Christopher William
Zobel, Bruce John

Human Genetics
Allen, Gordon
Amato, R Stephen S
Anderson, Victor Elving
Barnett, Donald Ray
Battle, Helen Irene
Bell, Alexander Graham
Bias, Wilma B
Bowen, Peter
Bowman, James E
Caskey, Charles Thomas
Chakraborty, Ranajit
Chan, Teh-Sheng
Chase, Gary Andrew
Chung, Chin Sik
Cohen, Bernice Hirschhorn
Collette, Alfred Thomas
Condell, Yvonne C
Cox, Diane Wilson
Daniel, William J
Donald, Lynda Joan
Dukepoo, Frank Charles
Eisen, James David
Eisenstadt, Jerome Melvin
Falek, Arthur
Freire-Maia, Ademar
Freire-Maia, Dertia Villalba
Friedman, Thomas Baer
German, James Lafayette, III
Gershowitz, Henry
Gilbert, Fred
Grundbacher, Frederick John
Hackel, Emanuel
Hamerton, John Laurence
Hawkins, Morris, Jr
Headings, Verle Emery
Herndon, Claude Nash
Higgins, James Victor
Hirschhorn, Kurt
Howard-Peebles, Patricia Nell
Hughes, Byron Orville
Hutton, Elaine Myrtle
Jarvik, Lissy F
Juberg, Richard Caldwell
Kaplan, Arnold Raymond
Kimberling, William J
Klebe, Robert John
Kloepfer, Henry Warner
Kucherlapati, Raju Suryanarayana
Kurczynski, Thaddeus Walter
Long, Walter K
MacCluer, Jean Walters
MacDiarmid, William Donald
Mayes, Jary S
McAlpine, Phyllis Jean
McCormack, Michael Kevin
McKenzie, Wendell Herbert
Milham, Samuel, Jr
Miller, Orlando Jack
Mohandas, Thuluvancheri
Muir, William A
Muscho, Henry M
Myrianthopoulos, Ninos
Nadler, Henry Louis
Nance, Walter Elmore
Neufeld, Elizabeth Fondal
Newcombe, Howard Borden
Niswander, Jerry David
Osborne, Richard Hazelet
Palm, John Daniel
Pelzer, Charles Francis
Poole, Andrew E
Powers, Harold O
Prince, John Edward
Rattazzi, Mario Cristiano
Reed, Elizabeth Wagner
Reed, Thomas Edward, Jr
Regan, James Dale
Ricciuti, Florence Christine
Rowley, Peter Templeton
Rucknagel, Donald Louis
Salzano, Francisco Mauro
Schacht, Lee Eastman
Schneider, Edward Lewis
Schull, William Jackson
Shapiro, Larry Jay
Shaw, Margery Wayne
Simpson, Nancy E
Sing, Charles F

Singh, Dharmdeo Narayan
Sinnock, Pomeroy
Smouse, Peter Edgar
Solish, George Irving
Sparkes, Robert Stanley
Spence, Mary Anne
Stencheven, Morton Albert
Strong, Louise Connally
Sutton, Harry Eldon
Tantravahi, Ramana V
Taylor, Harold Allison, Jr
Tedesco, Thomas Albert
Turner, J Howard
Warburton, Dorothy
Warren, Richard Joseph
Watson, Jack Ellsworth
Weaver, David Dawson
Welch, J Philip
Whissell-Buechy, Dorothy Y E
Willey, Ann Morris
Witkop, Carl Jacob, Jr

Immunogenetics
Aminoff, David
Amos, Dennis Bernard
Bacon, Larry Dean
Bailey, Donald Wayne
Bell, Clara G
Boraker, David Kenneth
Breyer, Edward Joseph
Briles, Connally Oran
Briles, Worthie Elwood
Cherry, Marianna
Cohen, Carl
David, Chelladurai S
Dorf, Martin Edward
Fanguy, Roy Charles
Foster, Morris
Fraser, Blair Allen
Frelinger, Jeffrey
Giblett, Eloise Rosalie
Haughton, Geoffrey
Hines, Harold C
Hood, Leroy E
Hsu, Susan Hu
Johnson, Armead H
Kindt, Thomas James
Marsh, David George
Mittal, Kamal Kant
Mobraaten, Larry Edward
Moor-Jankowski, Jan K
Passmore, Howard Clinton
Sanders, Bobby Gene
Schacter, Bernice Zeldin
Schanfield, Moses Samuel
Schierman, Louis W
Schultz, Jane Schwartz
Session, John Joe
Solomon, Joel Martin
Verghese, Margrith Wehrli
Wang, An-Chuan
Ward, Frances Ellen

Medical Genetics
Andrews, Lucy Gordon
Bannerman, Robin Mowat
Bartalos, Mihaly
Bernstein, Seldon Edwin
Boyer, Samuel H, IV
Burgeson, Robert Eugene
Caldwell, Jerry
Christian, Joe Clark
Comings, David Edward
Cox, Rody Powell
Dallaire, Louis
Danner, Dean Jay
Donahue, Roger Purtee
Edwards, John Anthony
Eldridge, Roswell
Elsas, Louis Jacob, II
Epstein, Charles Joseph
Erlenmeyer-Kimling, L
Finley, Sara Crews
Finley, Wayne House
Fischman, Harlow Kenneth
Fraser, Frank Clarke
Friedman, J Robert David
Garza-Chapa, Raul
Gordon, Hymie
Guthrie, Robert
Hillman, Richard Ephraim
Hodes, Marion Edward
Hollister, David William
Holtzman, Neil Anton
Hsia, Yujen Edward
Kaufmann, Berwind Norman
Koler, Robert Donald
Lozzio, Carmen Bertucci
Lynch, Henry T
Mellman, William Jules
Metrakos, Julius Demetrius
Migeon, Barbara Ruben
Mitchell, Arno Gunther
Motulsky, Arno Gunther
Murray, Robert Fulton, Jr
Nightingale, Elena Ottolenghi
Omenn, Gilbert Stanley

Opitz, John Marius
Patterson, Donald Floyd
Pierce, Edward Ronald
Prasad, Rupi
Prescott, Gerald H
Rainer, John David
Reed, Terry Eugene
Rimoin, David (Lawrence)
Sanger, Warren Glenn
Schaible, Robert Hilton
Shokeir, Mohamed Hassan Kamel
Sly, William S
Soltan, Hubert Constantine
Stamatoyannopoulos, George
Steinberg, Arthur Gerald
Stewart, Ray Edward
Summit, Robert L
Thurmon, Theodore Francis
Tishler, Peter Verveer
Vigfusson, Norman V
Williams, Hibbard E
Yamauchi, Toshio
Young, William Irving

Microbial Genetics
Al-Aidroos, Karen Messing
Allen, Marcia Katzman
Allen, Wendall E
Atkins, Charles Gilmore
Bachmann, Barbara Joyce
Barratt, Raymond William
Bear, Carl Adams
Beck, Doris Jean
Bernheimer, Harriet P
Bhattacharjee, Jnanendra K
Brescia, Vincent Thomas
Bullas, Leonard Raymond
Campbell, Allan McCulloch
Case, Mary Elizabeth
Chakrabarty, Ananda Mohan
Cole, Michael Allen
Copeland, James Clinton
Crawford, Irving Pope
Cross, Ronald Allan
Curtiss, Roy, III
DeGiovanni-Donnelly, Rosalie F
Diehl, William Paul
Ely, Berten E, III
Falkinham, Joseph Oliver, III
Fantini, Amedeo Alexander
Felkner, Ira Cecil
Folsome, Clair Edwin
Garro, Anthony Joseph
Glatzer, Louis
Goldberg, Ivan D
Goldschmidt, Raul Max
Gorini, Luigi Costantino
Gough, Michael
Gutz, Herbert
Harriman, Philip Darling
Iha, Thomas H
Johnson, Ben Francis
Johnson, Edward Miles, Jr
Jollick, Joseph Darryl
Jones, Lily Ann
Kelly, Beatrice L
Kowalski, John Bernard
Lacy, Ann Matthews
Lancaster, John
Landman, Otto Ernest
Lasure, Linda Lee
Levin, Barbara Chernov
Levinthal, Mark
Liss, Alan
Lo, Theodore Ching-Yang
Marcus, Leon
Marinus, Martin Gerard
McDougall, Kenneth J
Mehta, Bipinchandra Mohanlal
Michalka, Jack
Middleton, Richard B
Miller, Lynn
Miller, Robert Verne
Morrow, Terry Oran
Nakamura, Kazuo
Nasser, DeLill
Novick, Richard P
Ogg, James Elvis
Parker, John Hamilton
Pattee, Peter A
Pene, Jacques Jean
Prasad, Ishwari
Puhalla, John Edward
Rae, Margaret Engel
Raizen, Carol Eileen
Reilly, Bernard Edward
Rotheim, Minna B
Sanderson, Kenneth Edwin
Schoenhard, Delbert E
Schwinghamer, Erwin A
Scott, June Rothman
Siegel, Eli Charles
Sinha, Raj P
Smith, Hamilton Othanel
Stewart, Charles Ranous
Stine, Gerald James

Stocker, Bruce Arnold Dunbar
Takahashi, Francois Iwao
Thorne, Curtis Blaine
Tien, Weichen
Unowsky, Joel
Voll, Mary Jane
Wassermann, Felix Emil
Wax, David (Lawrence)
Widmayer, Dorothea Jane
Yourno, Joseph Dominic
Zahler, Stanley Arnold

Molecular Genetics
Aaron, Charles Sidney
Abelson, John Norman
Ahmed, Asad
Angerer, Lynne Musgrave
Ausubel, Frederick Michael
Avadhani, Narayan G
Berger, Hillard
Bernstein, Kenneth
Bertani, Giuseppe
Bleyman, Michael Alan
Boice, Lu Belle
Botstein, David
Broker, Thomas Richard
Campbell, Douglas Arthur
Carl, Philip Louis
Crasemann, Jean M
Davern, Cedric I
Davis, Ronald Wayne
Dean, Donald Harry
DeKloet, Siwo R
Edelman, Marvin
Edgar, Robert Stuart
Elgin, Sarah Carlisle Roberts
Falke, Ernest Victor
Fisher, Kenneth Walter
Ganesan, Ann K
Godson, Godfrey Nigel
Goodman, Fred
Greer, Sheldon
Grodzicker, Terri Irene
Gurney, Elizabeth Tucker Guice
Haber, James Edward
Hall, Dwight Hubert
Hardman, John Kemper
Herskowitz, Ira
Hogness, David Swenson
Howe, Martha Morgan
Hunt, John A
Jacobson, James William
Jargiello, Patricia
Jerkofsky, Maryann
Jervis, Herbert Hunter
Klock, Peter Allan,
Kushner, Sidney Ralph
Landry, Edward F
Laycock, David Gerald
Lomax, Margaret Irene
Maas, Werner Karl
Meagher, Richard Brian
Moody, Eric Edward Marshall
Parkinson, John Stansfield
Pearson, Mark Landell
Pollard, Jeffrey William
Prakash, Louise
Reznikoff, William Stanton
Riley, Monica
Rose, Raymond Wesley, Jr
Rosenberg, Marvin J
Rothstein, Rodney Joel
Rupp, W Dean
Russell, Peter James
Schaeffer, Gideon W
Scheltgen, Elmer
Schwartz, Alice Griffin
Sega, Gary Andrew
Shapiro, James Alan
Siegel, Albert
Slavik, Nelson Sigman
Smith, Oliver Hugh
Stiles, John I, (Jr)
Strauss, Ellen Glowacki
Watson, Maxine Amanda
Webster, Robert Edward
Weil, Jon David
Zamenhof, Patrice Joy
Zinder, Norton David
Zipser, David
Zuccarelli, Anthony Joseph

Plant Genetics
Acedo, Gregoria N
Allison, David C
Anderson, Robert Glenn
Arisumi, Toru
Aycock, Marvin Kenneth, Jr
Beadle, George Wells
Beckett, Jack Brown
Benepal, Parshotam S
Berg, Clyde C
Bliss, Fredrick Allen
Bottino, Paul James
Brawn, Robert Irwin
Brewbaker, James Lynn
Briggle, Leland Wilson
Buckner, Robert Cecil

Buzzell, Richard Irving
Cameron, James Wagner
Chu, Yaw-En
Collins, Frederick Clinton
Collins, Glenn Burton
Conger, Bob Vernon
Craig, Richard
Cuany, Robin Louis
Cunningham, Charles Everett
Curme, John Henry
Curtis, Byrd Collins
Desborough, Sharon Lee
Devine, Thomas Edward
Doney, Devon Lyle
Fleming, Attie Anderson
Forsberg, Robert Arnold
Gabelman, Warren Henry
Gengenbach, Burle Gene
George, William Leo, Jr
Gorsic, Joseph
Gorz, Herman Jacob
Graham, William Doyce, Jr
Green, Charles E, Jr
Hall, Robert Lee
Hanna, Wayne William
Hanson, Warren Durward
Haskins, Francis Arthur
Haunold, Alfred
Hernandez, Travis Paul
Hikida, Hideaki Robert
Holl, Frederick Brian
House, Leland Ralph
Hughes, Karen Woodbury
Janick, Jules
Jenkins, Johnie Norton
Johnson, Russell Tingey
Jones, Alfred
Jugenheimer, Robert William
Kahlon, Prem Singh
Kilen, Thomas Clarence
Kleese, Roger Allen
Knott, Douglas Ronald
Kohel, Russell James
Lambert, Howard John
Latimer, Howard Leroy
Lehman, William Francis
Leng, Earl Reece
Liang, George H L
Lighty, Richard William
Linden, Duane B
Lunden, Allyn Oscar
Marshall, Harold Gene
McArthur, Eldon Durant
McKenzie, Hugh
McNeal, Francis H
Miller, John David
Myers, Oval, Jr
Nabi, Hosni Abdel
Nickell, Cecil D
Norden, Allan James
Oberle, George David
Paulson, Ivan Wunder
Pavek, Joseph John
Plessers, Arthur Gerard
Polson, David Ernest
Purdy, James Lealon
Rachie, Kenneth Owen
Rajhathy, Tibor
Ramey, Harmon Hobson, Jr
Rasmusson, Donald C
Ray, Dale Allen
Robison, Laren R
Rupert, Earlene Atchison
Santamour, Frank Shalvey, Jr
Schulke, James Darrell
Shands, Henry Lee.
Shannon, Michael Carlyle
Sheen, Shuh-Ji
Shumway, Lewis Kay
Sidhu, Bhag Singh
Smith, Edward Lee
Smith, Rex L
Smith, Richard R.
Solbrig, Otto Thomas
Squillace, Anthony Eugene
Stevens, Merwin Allen
Stroike, James Edward
Stuthman, Deon Dean
Theurer, Jessop Clair
Thompson, Anson Ellis
Thompson, James Marion
Timothy, David Harry
Tipton, Kenneth Warren
Townsend, Alden Miller
Turcotte, Edgar Lewis
Verhalen, Laval Mathias
Walden, David Burton
Wall, James Robert
Wallace, Donald Howard
Ward, David Justin
Washington, Willie James
Welsh, James Ralph
Wilkes, Hilbert Garrison, Jr
Williams, John Caswell
Williams, Norman Dale
Wilson, Frank Douglas
Wynne, Johnny Calvin
York, John Owen

Zoebisch, Oscar Cornelius

Population Genetics
Adams, Julian Philip
Allendorf, Frederick William
Anderson, Wyatt W
Burnside, Edward Blair
Cameron, David Glen
Carmody, George R
Chakraborty, Ranajit
Chapman, Stephen R
Cockerham, Columbus Clark
Crenshaw, John Walden, Jr
Cundiff, Larry Verl
Da Cunha, Antonio Brito
Dewees, Andre Aaron
Ehrman, Lee
Ewens, Warren John
Falk, Catherine T
Farish, Donald James
Felsenstein, Joseph
Frahm, Richard R
Goodwill, Robert
Grassle, Judith Payne
Griswold, Kenneth Edwin, Jr
Grossman, Michael
Hartl, Daniel L
Highton, Richard
Huether, Carl Albert, Jr
Jensen, John Neil
Kashyap, Tapeshwar S
Katz, Alan Jeffrey
Kiang, Yun-Tzu
Kidd, Kenneth Kay
Kinney, Terry B, Jr
Kirby, Karen Marie
Koehn, Richard Karl
Krause, Eliot
Leigh, Egbert G, Jr
Lester, Larry James
Li, Ching Chun
Li, Wen-Hsiung
MacCluer, Jean Walters
Merritt, Edison S
Merritt, Robert Buell
Mitton, Jeffry Bond
Mode, Charles J
Morton, Newton Ennis
Moyer, Samuel Edward
Nagylaki, Thomas Andrew
Nei, Masatoshi
Parker, Robert J
Richmond, Rollin Charles
Schalles, Robert R
Scheinberg, Eliyahu
Sharp, Gary Duane
Siers, David Gard
Spielman, Richard Saul
Spieth, Philip Theodore
Taylor, Charles Ellett
Vinson, William Ellis
Warburton, Frederick E
Weisbrot, David R
Williams, Bobby Joe
Wilson, Stanley Porter
Wunder, William W
Yardley, Darrell Gene
Young, Sydney Sze Yih
Zouros, Eleftherios

Poultry Genetics
Berg, Robert W
Bernier, Paul Emile
Buckland, Roger Basil
Buss, Edward George
Crawford, Roy Douglas
Kondra, Peter Alexander
Law, George Robert John
Moore, Claude Henry
Proudfoot, F G
Shoffner, Robert Nurman
Washburn, Kenneth W

Quantitative Genetics
Baker, Robert John
Ball, Mary Uhrich
Boylan, William J
Calhoon, Robert Ellsworth
Carballo-Quiros, Alfredo
Cockerham, Columbus Clark
Cress, Charles Edwin
Gardner, Charles Olda
Goodwill, Robert
Grossman, Michael
Hargrove, George Lynn
Hegmann, Joseph Paul
Katz, Alan Jeffrey
Kolakowski, Donald Louis
Leamy, Larry Jackson
Matzinger, Dale Frederick
Nyquist, Wyman Ellsworth
Robinson, Harold Frank
Sargent, Frank Dorrance
Young, Sydney Sze Yih

Radiation Genetics
Boyd, James Brown
Cooper, Philip Harlan

Cyr, Wilbur Howard
Dillon, Richard Thomas
Kamra, Om Perkash
Newcombe, Howard Borden
Yost, Henry Thomas, Jr

GEOCHEMISTRY

Geochemistry

Adams, John Allan Stewart
Allen, Clifford Marsden
Allen, Gary Curtiss
Alt, David D
Ames, Roger Lyman
Anderson, Gregor Munro
Anderson, Thomas Frank
Angino, Ernest Edward
Arth, Joseph George
Attrep, Moses, Jr
Ault, Wayne Urban
Baadsgaard, Halfdan
Bada, Jeffrey L
Baird, Alexander Kennedy
Banford, Robert Wendell
Banks, Philip Oren
Barnard, Walther M
Barnes, Hubert Lloyd
Barnes, Ivan
Barsdate, Robert John
Barsky, Constance Kay
Bass, Manuel N
Beall, George Halsey
Beane, Richard Edward
Bell, Peter M
Bennington, Kenneth Oliver
Bickford, Marion Eugene, Jr
Bischoff, James Louden
Bisque, Ramon Edward
Bitz, Miriam L
Blumer, Max
Bodine, Marc Williams, Jr
Boettcher, Arthur Lee
Bogard, Donald Dale
Bohlmann, Edward Gustav
Bolter, Ernst A
Bostrom, Kurt G V
Botino, Michael Louis
Bowser, Carl
Boyle, Robert William
Boynton, William Vandegrift
Bricker, Owen P, III
Broecker, Wallace
Brooke, John Percival
Brookins, Douglas Gridley
Brown, Alexander Cyril
Brown, Harrison Scott
Brown, Seward Ralph
Brueckner, Hannes Kurt
Brunskill, Gregg J
Buckley, Dale Eliot
Burkart, Burke
Burnett, Donald Stacy
Burnham, Clifford Wayne
Burns, Roger George
Buseck, Peter R
Cabri, Louis J
Cannon, Helen Leighton
Carl, James Dudley
Carman, John Homer
Carmichael, Ian Stuart
Carpenter, Alden B
Carpenter, John Richard
Carr, Michael H
Cassidy, William Arthur
Castano, John Roman
Catanzaro, Edward John
Chaffee, Maurice Ahlborn
Chao, Tsun Tien
Charles, Robert Wilson
Chodos, Arthur A
Chou, Chen-Lin
Chow, Tsaihwa James
Church, Stanley Eugene
Clanton, Uel S, Jr
Clark, Allen LeRoy
Clark, Benton C
Clark, Lloyd Allen
Clark, Roy Slayton, Jr
Clayton, Robert Norman
Clemency, Charles V
Cloke, Paul LeRoy
Coe, Robert Stephen
Cohen, Alvin Jerome
Cohen, Lewis H
Collins, Arlee Gene
Condie, Kent Carl
Conomos, Tasso John
Copeln, Edward Casimere
Corbett, Robert G
Cowgill, Ursula Moser
Craig, Harmon
Craig, James Roland
Crawford, William Arthur

Crisp, Edward Lee
Crocket, James Harvie
Cullers, Robert Lee
Czamanske, Gerald Kent
Dachille, Frank
Damon, Paul Edward
Dasch, Ernest Julius
Davies, Tudor T
Davis, Nicholas Falconer
Dean, Walter E, Jr
Deines, Peter
Delavault, Robert Edmund
Dennen, William Henry
Dickinson, Stanley Key, Jr
Dolloff, Norman
Drake, Michael Julian
Earley, James William
Eastwood, Raymond L
Eckelmann, Walter R
Edgar, Alan D
Ehmann, William Donald
El Wardani, Sayed Aly
Engel, Albert Edward John
Epstein, Samuel
Ernst, Wallace Gary
Eugster, Hans Peter
Ewing, Rodney Charles
Faure, Gunter
Feely, Herbert William
Feiss, Paul Geoffrey
Ferrell, Ray Edward, Jr
Field, Cyrus West
Finlayson, James Bruce
Fisher, Frederick Stephen
Fisher, James Russell
Fleischer, Michael
Fournier, Robert Orville
Fredrikson, Kurt A
French, Bevan Meredith
Frey, Frederick August
Friedman, Irving
Fritz, Peter
Frye, Keith
Fudali, Robert F
Furlong, Robert B
Gaines, Alan McCulloch
Ganguly, Jibamitra
Garlick, George Donald
Garrels, Robert Minard
Gaudette, Henri Eugene
Gentry, Robert Vance
George, Albert El Deeb
Gibbs, Ronald John
Giletti, Bruno John
Glover, Everett D
Goldsmith, Julian Royce
Goldstein, Theodore Philip
Goles, Gordon George
Govett, Gerald James
Graf, Donald Lee
Gresens, Randall Lee
Grossman, Lawrence
Grubbs, Donald Keeble
Gunn, Bernard M
Haas, Herbert
Halbig, Joseph Benjamin
Hall, Henry Thompson
Halva, Carroll J
Hanna, William Jefferson
Hanshaw, Bruce Busser
Hanson, Hiram Stanley
Hart, Stanley Robert
Haskin, Larry A
Hatcher, Robert Dean, Jr
Hawkins, Daniel Ballou
Heald, Emerson Francis
Helz, George Rudolph
Hem, John David
Henry, Christopher Duval
Hermes, O Don
Herr, Frank Leaman, Jr
Hess, Paul C
Heyl, Allen Van, Jr
Hill, Walter Edward, Jr
Hiltrop, Carl L
Hinners, Noel W
Hitchon, Brian
Hofmann, Albrecht Werner
Holdaway, Michael Jon
Holland, Heinrich Dieter
Holley, Charles Elmer, Jr
Horn, Myron Kay
Horvitz, Leo
Hostetler, Paul B
Howard, James Hatten, III
Hower, John, Jr
Huang, Wen Hsing
Huebner, John Stephen
Hurst, Richard William
Ingerson, Fred Earl
Isphording, Wayne Carter
James, Richard Stephen
Johns, William Davis
Johns, Willis Merle
Jonte, John Haworth
Jones, Lois Marilyn
Kamili, Diana Chapman
Kaplan, Issac R

Kay, Robert Woodbury
Keighin, Charles William
Keith, Mackenzie Lawrence
Kennedy, Vance Clifford
Kerrick, Derrill M
Kharaka, Yousif Khoshu
Kilham, Peter
Kiline, Attila Ishak
King, Elbert Aubrey, Jr
Kinsman, David John James
Klusman, Ronald William
Koster van Groos, August Ferdinand
Kothny, Evaldo Luis
Koucky, Frank Louis, Jr
Kramer, James Richard
Krauskopf, Konrad Bates
Kretz, Ralph
Krogh, Thomas Edvard
Kroopnick, Peter Michael
Ku, Teh-Lung
Kujawa, Frank B
Kulaerud, Gunnar
Kulp, John Laurence
Lafon, Guy Michel
Lal, Devendra
Lambert, Richard St John
Land, Lynton S
Lange, Ian M
Langmuir, Donald
Lanphere, Marvin Adler
Larimer, John Milton
Laughlin, Alexander William
Leahy, Richard Gordon
Learned, Robert Eugene
Lemish, John
Lewis, John Simpson
Linkletter, George Onderdonk
Long, Austin
Long, Leon Eugene
Louden, L Richard
Luce, Robert William
Ludington, Stephen Dean
Luedemann, Lois W
Lukert, Michael T
Luth, William Clair
Macdougall, John Douglas
MacGregor, Ian Duncan
MacLean, Wallace H
Magin, George Benedict, Jr
Majmundar, Hasmukhrai Hiralal
Manghnani, Murli Hukumal
Manheim, Frank T
Mantei, Erwin Joseph
Mao, Ho-Kwang
Marland, Gregg (Hinton)
Marranzino, Albert Pasquale
Martell, Edward A
Mason, Brian Harold
Maxwell, John Alfred
May, Irving
McDowell, Fred Wallace
McGannon, Donald E, Jr
McIver, Richard Donald
McKague, Herbert Lawrence
McNutt, Robert Harold
Megrue, George Henry
Melson, William Gerald
Merrin, Seymour
Meyer, Henry Oostenwald Albertijn
Michalowski, Joseph Thomas
Mielke, James Edward
Millard, Hugh Thompson, Jr
Miller, Donald Spencer
Miller, William Robert
Mitterer, Richard Max
Modreski, Peter John
Moiseyev, Alexis N
Moll, William Francis, Jr
Monaghan, Patrick Henry
Moore, Carlton Bryant
Moore, Duane Milton
Moore, Raymond Kenworthy
Moore, Thomas Francis
Morgan, John Walter
Morris, David Albert
Morrison, Garrett Louis
Morse, John Wilbur
Mose, Douglas George
Mowat, Thomas C
Moxham, Robert Lynn
Muan, Arnulf
Mueller, Paul Allen
Muenow, David W
Muffler, Leroy John Patrick
Munoz, James Loomis
Murphy, Mary Teresa Joseph
Murray, Marc Michael
Murthy, Varanasi Rama
Nash, John Thomas
Naylor, Richard Stevens
Nix, Joe Franklin
Noakes, John Edward
Nordlie, Bert Edward
Northrop, David A
Northrop, Clyde John Marshall, Jr
Norton, Denis Locklin
Nussmann, David George

Ohmoto, Hiroshi
Olson, Edwin Andrew
O'Neil, James R
Pajari, George Edward
Parker, Patrick LeGrand
Parry, William Thomas
Patterson, Claire Cameron
Patton, James Winton
Pearce, Thomas Hulme
Pearson, Frederick Joseph
Perhac, Ralph Matthew
Perry, Eugene Carleton, Jr
Phair, George
Philpotts, John Aldwyn
Pollard, Charles Oscar, Jr
Pollard, Lin Davis
Polzer, Wilfred L
Post, Edwin Van Horn
Potts, Mark John
Powell, James Lawrence
Presnall, Dean C
Puffer, John H
Pushkar, Paul
Putman, George Wendell
Ragland, Paul C
Ragone, Stephen Edward
Raup, Omer Beaver
Reitan, Paul Hartman
Reitsema, Robert Harold
Renton, John Johnston
Reynolds, Peter Herbert
Reynolds, Robert Coltart, Jr
Roberson, Charles Elmer
Roe, Glenn Dana
Roedder, Edwin Woods
Roeder, Peter Ludwig
Rogers, John James William
Rogers, Marion Alan
Romberger, Samuel B
Roseboom, Eugene Holloway, Jr
Ross, Malcolm
Rouse, George Elverton
Rowe, Marvin W
Runnells, Donald DeMar
Rutstein, Martin S
Rye, Robert O
Sackett, William Malcolm
Sato, Motoaki
Savin, Samuel Marvin
Sayles, Frederick Livermore
Schutz, Donald Frank
Schwarcz, Henry Philip
Schwarzer, Theresa Flynn
Scotford, David Matteson
Scott, Steven Donald
Severson, Ronald Charles
Shade, John William
Shanks, Wayne C, III
Shaw, Denis Martin
Sherwood, William Cullen
Siever, Raymond
Silberman, Miles Louis
Silver, Leon Theodore
Silverman, Melvin Philip
Skinner, Brian John
Slawson, William Francis
Slentz, Loren William
Smalley, Robert Gordon
Smith, Frederick Gordon
Smith, James Beaver
Smith, Richard Elbridge
Smith, Terence E
Snetsinger, Kenneth George
Sokoloff, Vladimir P
Sommer, Michael Anthony, II
Sommer, Sheldon A
Sommerfeld, Richard Arthur
Sood, Manmohan K
Speidel, David H
Steele, Kenneth F
Stensrud, Howard Lewis
Stevens, Nelson Pierce
Stonehouse, Harold Bertram
Stormer, John Charles, Jr
Stow, Stephen Harrington
Stueber, Alan Michael
Suhr, Norman Henry
Sutherland, Jeffrey C
Swanson, Samuel Edward
Szabo, Barney Julius
Takahashi, Taro
Tanner, Allan Bain
Tappe, John
Tatsumoto, Mitsunobu
Taylor, G Jeffrey
Taylor, Hugh P, Jr
Taylor, Lawrence August
Thomas, Herman Hoit
Thompson, Geoffrey
Thompson, James Burleigh, Jr
Thompson, Mary Eleanor
Thorstenson, Donald Carl
Thrailkill, John Vernon
Tilton, George Robert
Titley, Spencer Rowe
Towell, David Garrett
Tupper, William Macgregor

Turek, Andrew
Turekian, Karl Karekin
Ulmer, Gene Carleton
van de Kamp, Peter Cornelis
Vaughan, David Evan William
Volborth, Alexis
Volchok, Herbert Lee
Walter, Louis S
Walters, Lester James, Jr
Wampler, Jesse Marion
Warburton, David Lewis
Warshaw, Charlotte Marsh
Weaver, Charles Edward
Weber, Jon
Webster, Clyde Leroy, Jr
Weidner, Jerry R
Weiler, Roland R
Weill, Daniel Francis
Wenner, David Bruce
Wershaw, Robert Lawrence
West, Walter Scott
White, William Blaine
Whitney, Philip Roy
Whitmore, Donald Osgood
Wickman, Frans Erik
Wiband, John Truax
Wildeman, Thomas Raymond
Williams, Richard John
Willis, Eric Herbert
Wiltse, Milton Adair, Jr
Wones, David R
Wood, John Armstead, Jr
Wood, Warren Wilbur
Wylie, Peter John
Young, Edward Joseph
Young, William Allen
Zeller, Edward Jacob

Marine Geochemistry
Berner, Robert A
Biscaye, Pierre Eginton
Burnett, William Craig
Chambers, Richard Lee
Dayal, Ramesh
Deuser, Werner Georg
Farrington, John William
Goldberg, Edward D
Huggett, Robert James
Lerman, Abraham
Loring, Douglas Howard
Man, Eugene Herbert
Mangelsdorf, Paul Christoph, Jr
Moore, Willard S
Ristvet, Byron Leo
Schultz, David Michael
Scott, Martha Richter
Whelan, Thomas, III
Wolgemuth, Kenneth Mark
Wood, Elwyn Devere

Organic Geochemistry
Adelman, Robert Leonard
Baker, Donald Roy
Barker, Colin G
Breger, Irving A
Brown, Frederick S
Cooper, James Erwin
Erdman, John Gordon
Felbeck, George Theodore, Jr
Ferris, Bernard Joe
Fields, Donald Lee
Given, Peter Hervey
Harrison, William Earl
Hayes, John Michael
Ho, Thomas Tong-Yun
Hood, Archie
Jackson, Togwell Alexander
Jacobs, Richard Joe
James, Alan Thomas
James, Daniel Shaw
Kvenvolden, Keith Arthur
Lucas, George Bond
Meyers, Philip Alan
Mueller, George
Nagy, Bartholomew Stephen
Palacas, James George
Philp, Richard Paul
Reuter, Johannes Helmut
Schorno, Karl Stanley
Schrayer, Grover L Jr
Schwartz, Alan William
Silverman, Sol Robert
Thompson, Robert Richard
Waples, Douglas Wendle
Williams, Jack A

GEOGRAPHY

Geography
Abler, Ronald Francis
Acker, William James
Adams, Russell Blair
Adams, William Peter
Ahsan, S Reza
Albaum, Melvin
Aldrich, Frank Thatcher

Alexander, Charles Stevenson
Alexander, Lewis McElwain
Allen, Robert Houston
Allen, James Paul
Anderson, Thomas Dale
Anstey, Robert L
Anthrop, Donald F
Armstrong, R Warwick
Arnold, Brigham Alicen
Aschmann, H Homer
Ashbaugh, James G
Ayala, Reynaldo
Bacon, Phillip
Bajza, Charles C
Baker, Simon
Baker, William Bryan
Ball, John Miller
Barton, Byron Kurtz
Basile, David Giovanni
Batchelder, Robert Bruce
Bauer, Francis Harry
Baumann, Duane Dennis
Baxevanis, John James
Beaty, Chester Broomell
Beauregard, Ludger
Bechtol, Bruce Emerson
Bederman, Sanford Harold
Beishlag, George Albert
Belthuis, Lyda Carol
Benner, Velma
Bennett, Don C
Bennett, Iven
Bergen, John Victor
Berger, Rainer
Bergmann, John Francis
Best, Alan C G
Betz, Gabriel Pohl
Beyer, Jacquelyn L
Bingham, Edgar
Bird, John Brian
Bishop, Richard Lawrence
Bjorklund, Elaine M
Blair, Alexander Marshall
Blick, James Donald
Blouet, Brian Walter
Blount, Stanley Freeman
Boas, Charles William
Boley, Robert Eugene
Bone, Robert M
Booth, Alfred Whaley
Borchert, John Robert
Bordne, Erich Fred
Boswell, Thomas D
Bounds, John Howard
Bowden, Leonard Walter
Bowen, Marshall Everett
Bowen, Neal Monroe
Bowman, Robert Goldthwait
Bradley, Virginia
Brand, Donald Dilworth
Brand, Richard Robert
Brehob, Kenneth Raymond
Brinkman, Leonard W, Jr
Brooke, Clarke Harding
Brooks, Nathan Cyrus
Brown, Ralph Clarence
Brown, Roger James Evan
Brown, William Jeffrey
Brownell, Joseph William
Brumm, Stanley David
Brunnschweiler, Dieter Heinz
Bryant, Nevin Arthur
Buchanan, W C
Burch, Erwin Julius
Burghardt, Andrew Frank
Burke, Terence
Burns, Bert E
Burrill, Meredith Frederic
Bush, Everett Homer
Bush, Oakleigh Ross
Bushong, Allen David
Buss, Walter Richard
Byron, William Glenn
Cailleux, Andre Paul
Campbell, Robert Dale
Cargo, Douglas Bruce
Carmin, Robert Leighton
Casetti, Emilio
Castner, Henry Walker
Chambers, Jack Virgil
Chang, Jen Hu
Chang, Kuei-Sheng
Chang, Tsuen Kung
Chapman, Albert Simonds
Chapman, John Doneric
Chardon, Roland E
Charlier, Roger Henri
Chase, Harrison Vernon
Chesnutwood, Charles Mark
Clarkson, James David
Clay, James William
Cohen, Saul Bernard
Cole, Chester F
Collier, James Eli
Conkling, Edgar Clark
Connors, Theodore Thomas

Conoyer, John Weedon
Cook, Earl Ferguson
Cooper, Paul David
Corbet, John Harry
Corry, Martha Lucille
Cossaboom, Robert T
Cotton, James V
Coulson, Michael Robert Cummins
Courtney, Dale Elliott
Cox, Kevin Robert
Craig, Alan Knowlton
Crary, Douglas Dunham
Crisler, Robert Morris
Critchfield, Howard John
Cross, Clark Irwin
Crowley, John Max
Cunningham, Frank Firman
Cunningham, William Glenn
Curry, Leslie
Curti, Gabriel Philip
Cutler, Irving Herbert
Cutler, Richard Oscar
Cushall, Alden Denzel
Dacey, Michael Francis
Dahms, Fredric Arthur
Dambaugh, Luella N
Dando, William Arthur
Dart, John Olney
Davis, James Leslie
Day, Richard Lorey
Dean, William George
Deasy, George F
De Blij, Harm Jan
Demko, Donald
Demko, George Joseph
Detwyler, Thomas Robert
Diem, Aubrey
Dienes, Leslie Dennis
Dierickx, Charles Wallace
Dillon, Lowell Ivan
Dionne, Jean-Claude
Dixon, James William, Jr
Dodd, Arthur V
Dohrs, Fred E
Dojcsak, Gyozo Victor
Donley, Michael William
Dordick, Isadore L
Downs, Roger Michael
Dozier, Craig Lanier
Drewes, Wolfram Ulrich
Drummond, Robert Norman
Dunbar, Gary Seamans
Dunbar, Isobel Moira
Durrenberger, Robert W
Earick, Arthur David
Edie, Milton James
Edwards, Clinton R
Eichenlaub, Val L
Eidt, Robert C
Eiselen, Elizabeth
Emery, Byron Elwyn
Emory, Samuel Thomas
English, Van Harvey
Enman, John A
Epperson, Terry Elmer, Jr
Erb, David Kinsey
Ericksen, Sheldon Danielsen
Ervin, Roger Edward
Evans, Brian M
Eydal, Astvaldur
Eyre, John D
Farley, Albert Leonard
Fellmann, Jerome Donald
Fernstrom, John Richard
Field, Neil Collard
Finley, Robert William
Finney, Mildred Irene
Firman, David
Fite, Robert Carl
Fitzgerald, Denis Patrick
Fitzsimmons, James G
Ford, Derek Clifford
Fraser, John Keith
Frazer, William James
Frederic, Paul Burgess
Frenkel, Robert Edgar
Fullerton, Ralph O
Fuson, Robert Henderson
Gallagher, James Weldon
Gandre, Donald Alfred
Garland, John Henry
Garrigues, Woodford McDowell
Gassaway, Alexander Ramsey
Gault, Thomas Gower
Gehrke, Willis Timothy
Gersmehl, Philip J
Ghobrial, Girgis Bakhoum
Gladfelter, Bruce G
Glassner, Martin Ira
Goldberg, Jerald Melvin
Gomez-Ibanez, Daniel Alexander
Goodman, James Marion
Gore, Dorothy J
Gosling, Lee Anthony Peter
Gould, Peter Robin
Gould, Robert Barris
Greco, Peter V
Greenwood, Ned H

Greer, Deon Carr
Grenier, Fernand
Gressitt, Judson Linsley
Griffin, Donald William
Grosvenor, Melville Bell
Grove, Arthur M
Gulick, Luther Halsey, Jr
Gwynne-Thomas, Eric Hubert
Haase, Ynez Durnford
Hacker, Walter Rudolf
Hall, John Whiting
Hamelin, Louis-Edmond
Hamming, Edward
Hansen, Carl Lough
Hanson, Earl Parker
Hanson, Perry Oliver
Hanson, Susan Easton
Hardy, Ernest Edward
Hare, Frederick Kenneth
Harman, Jay Reginald
Harris, Chauncy Dennison
Harris, Richard Colebrook
Harris, Stuart Arthur
Harris, William N
Hartshorn, Truman Asa
Hartshorne, Richard
Haupert, John Selby
Havill, Thomas Lampert
Hawley, Dorothea Burton
Haynes, Kingsley Edwin
Hecock, Richard Douglas
Hegen, Edmund Eduard
Heimonen, Henry Samuel
Heintzelman, Oliver Harry
Helburn, Nicholas
Heller, Charles Frederick, Jr
Hellman, Allen David
Henry, James Thurman
Hereford, Joseph Pierce
Herman, Theodore
Hess, Charles F
Hewes, Leslie
Higbee, Edward
Hill, Alton David, Jr
Hilliard, Sam Bowers
Hiltner, John, Jr
Hoag, Leverett Paddock
Hodgkins, Jordan Atwood
Hoffman, Lawrence Arnes
Hole, Francis Doan
Holtgrieve, Donald Gordon
Holzner, Lutz Ernest
Honkala, Rudolf A
Hooson, David John Mahler
Hoover, W Farrin
Hordon, Robert Marshall
Horvath, Ronald J
Howe, George Marvel
Hoy, Harry Eugene
Hoyt, Joseph Bixby
Hu, Charles Y
Huddlestun, Dwight Leroy
Hughes, John Derek
Huke, Robert Edward
Hunker, Henry L
Hunter, James Murry
Innes, Frank Cecil
Innis, Donald Quayle
Jackson, Charles Ian
Jackson, W A Douglas
Jacobson, Daniel
James, Preston Everett
Jenks, George Frederick
Jett, Stephen Clinton
Johnson, Donald Lee
Johnson, Gary Edwin
Johnson, Ladd Lind
Johnson, Paul Timothy
Johnson, Robert Bethune
Johnson, Warren Arthur
Jones, Reece Alexander
Jones, Stephen Barr
Jordan, Terry G
Jost, Tadeusz Piotr
Jumper, Sidney Roberts
Just, Mary Agnes
Kakiuchi, Hiroaki George
Kallay, Ferencz Paul
Kane, Henry Edward
Karabenick, Edward
Kasperson, Roger Eugene
Kazeck, Melvin E
Kearns, Kevin Corrigan
Kemler, John Hughes
Kennelly, Robert Andrew
Kennison, Lawrence Sanford
Kenyon, James Byron
Kersten, Earl William
Kesel, Richard Herman
Kiefer, Wayne Eugene
Kilheffer, Esther May
Kimber, Clarissa Therese
King, James Wilhelmsen
Kish, George
Kline, Hibberd Van Buren, Jr
Klove, Robert Charles
Knipmeyer, William B
Koch, Walter Ferdinand

GEOGRAPHY

Kohn, Clyde Frederick
Korsok, Albert Joseph
Kostbade, J Trenton
Kovacik, Charles Frank
Kramer, Fritz Louis
Kreske, Richard Daniel
Kress, Warren Donald
Krueger, Ralph R
Laidig, Kermit McClellan
LaMont, Robert Ellis
Lantis, David W
Latham, James Parker
Laycock, Arleigh Howard
Layton, Robert L
Leahy, Edward Prior
LeBlanc, Robert George
Leigh, Roger
Leinbach, Thomas Raymond
Lemaire, Minnie Ethel
Lewis, George Knowlton
Lewis, Lawrence A
Lier, John
Lindberg, James Beckwith
Lindgren, David Treadwell
Lineback, Neal Gambill
Lloyd, Trevor
Lloyd, Howell Clevenger
Lockmann, Ronald Frederick
Logan, Richard Fink
Lohr de Irizarry, Mildred Tucker
Long, Harriet Ruth
Loring, Robert David
Lovingood, Paul Evans, Jr
Loy, William George
Lycan, D Richard
Lydolph, Paul E
Ma, Laurence Jun-Chao
MacFadden, Clifford Herbert
Macfarlane, Sidney Kerr
MacGraw, Frank Moss
Macinko, George
MacPhail, Donald Dougald
Malmstrom, Vincent Herschel
Manners, Ian Robert
Mannion, John Joseph
Marble, Duane F
March, Andrew Lee
Markham, Charles G
Marple, Robert P
Marshall, John U
Mason, Dorothy Stafford
Matley, Ian Murray
Maxfield, Ollie Orland
McArthur, Neil M
McBryde, Felix Webster
McCune, Shannon
McDowell, Horace Greeley
McGaugh, Maurice Edron
McIntire, William Grant
McIntosh, Charles Barron
McKnight, Tom
McNeal, Roy Wilson
McNulty, Michael Leigh
Meade, Melinda Sue
Meeks, Harold Austin
Melezin, Abraham
Mensoian, Michael George, Jr
Merrens, Harry Roy
Merriam, Willis Bungay
Merrill, Gordon Clark
Mettner, William Joseph
Metz, William Clinton
Meyer, Alfred Herman Ludwig
Meyer, Morton A
Michel, Aloys Arthur
Micklewright, Malcolm A
Mihaly, Louis James
Miles, Edward Jervis
Milfred, Clarence James
Miller, E Willard
Miller, Elbert Ernest
Miller, Vincent Paul, Jr
Mitchell, Lloyd Vernon
Mitchell, Robert Davis
Momsen, Janet Henshall
Monahan, Robert Leonard
Moore, Conrad Taylor
Morrill, Richard Leland
Morris, John Wesley
Morris, Rita, M L
Morrissette, Hugues
Mote, Victor Lee
Moulton, Benjamin
Murphy, Don Robison
Murray, Malcolm Arthur
Myers, Merle Wentworth
Nasse, George Nicholas
Natoli, Salvatore John
Nelson, Charles Edward
Nelson, Howard Joseph
Nelson, Ronald Eugene

Newman, James L
Nicholson, Norman Leon
Niddrie, David Lawrence
Noble, Allen George
Nostrand, Richard Lee
Nunley, Richard E
Nutt, David Clark
Nystuen, John David
Oakes, Lester Charles
Odell, Clarence Burt
Ogilvie, Bruce Crossan
Olson, Ralph Eugene
Olson, Walter
Orr, Douglas Milton, Jr
Orth, Donald Joseph
Osborne, Brian S
Owen, Edgar Wesley
Paige, Richard Joseph
Pannell, Clifton Wyndham
Parmenter, Guy Norris
Parry, John Trevor
Patten, George Philip
Patton, Clyde Perry
Patton, Donald John
Pattison, William David
Pederson, Leland Roger
Peet, Richard
Peltier, Louis Cook
Peltzer, Karl Josef
Pitts, Forrest Ralph
Pennington, Campbell White
Perret, Maurice Edmond
Perry, Robert F, Jr
Petrov, Victor P
Peucker, Thomas K
Phelps, Jewell A
Philbrick, Allen Kellogg
Phillips, Yvonne
Picker, Robert
Poulsen, Thomas Martin
Pounds, Norman John Greville
Prentice, Virginia Lee
Preston, Richard Ellis
Price, Dallas Adolph
Price, Edward Thomas
Proudfoot, Vincent Bruce
Pruitt, Evelyn Lord
Prunty, Merle Charles
Pruser, Etha Marie
Pryde, Philip Rust
Psuty, Norbert Phillip
Putnam, Donald Fulton
Radell, David
Rae, George Ramsay
Rahman, Mushtaq-Ur
Randall, Richard Rainier
Randall, John Reed
Raphael, C Nicholas
Rappenecker, Caspar
Raup, Hallock Floyd
Rayner, John Norman
Reilly, Timothy Frank
Reinan, Robert Ellis
Reinemann, Martin
Reith, Gertrude McKean
Ressler, John Quenton
Reynolds, David Reid
Rheumer, George Alfred
Rizza, Paul Frederick
Richards, John Howard
Richardson, Robert William
Richason, Benjamin Franklin, Jr
Ridd, Merrill Kay
Riddell, John Barry
Ridge, Frank Gerald
Riess, Richard O
Rizza, Paul Frederick
Roberts, Michael Charles
Robinson, John Lewis
Robinson, Malcom Emerson
Robinson, William Condit
Rodgers, Allan J
Romsa, Gerald Harry
Ronne, Finn
Rooney, John Francis
Rosevall, Lynn Albert
Ross, William Gillies
Rothwell, Stuart Clark
Rugg, Dean Sprague
Ruggles, Richard Irwin
Rumage, Kennard Walter
Rumney, George Richard
Russell, Joseph Albert
Russwurm, Lorne H
Ryan, Carolyn J
Ryan, Kenneth Bruce
Sabbagh, Michael E
Salisbury, Howard Graves
Salisbury, Neil Elliot
Sangree, Anne Coates
Sas, Anthony
Satterthwaite, Ridgway
Schadegg, Francis John
Schilz, Gordon B
Scholer, Jerry Parker
Schultz, Gwendolyn Monett
Schwartzberg, Joseph Emanuel
Schwendenwein, Gerd (Anton)

Scripter, Morton W
Seawall, Frank
Seitz, Kerlin (McCullough)
Shaffer, Ralph Gunter
Sharp, Virginia Leah
Shaudys, Vincent Kirkbride
Shear, James Algan
Sherman, John Clinton
Shin, Suk-Han
Shirey, George S
Shlemon, Roy J
Shrestha, Mohan Narayan
Silvernail, Richard George
Simkins, Paul Dean
Simmons, James William
Simpson, Robert Bonebrake
Sinclair, Robert
Slick, Max Harrell
Smith, David August
Smith, Samuel Ivan
Smith, Albert W
Snead, Rodman E
Sobol, John Andrew
Sommers, Lawrence Melvin
Sopher, David E
Spayne, Robert William
Spelt, Jacob
Stanley, Emilio Joseph
Steiner, Dieter
Steinhauser, Fredric R
Sterling, Henry Somers
Stevens, Richard Edward
Stoddard, Robert Hugh
Stohr, Walter B
Stokes, George Alwin
Stone, Kirk Haskin
Stover, Stephen Leech
Strack, Charles Miller
Stuart, Merrill M
Sublett, Michael Dean
Summers, William Francis
Taaffe, Edward James
Talbot, Ashley Frederick
Tan, Kok-Chiang
Taylor, David Ruxton Fraser
Taylor, Gordon
Taylor, James Addison
Taylor, Robert Martin
Terjung, Werner Heinrich
Terrell, Roy Paul
Thomas, Benjamin Earl
Thomas, Frank Henry
Thomas, Robert Nelson
Thompson, Kenneth
Thoren, Conrad Joseph
Tiedemann, Clifford E
Tilmann, Jean Paul
Timofeeff, Nicolay P
Tirtha, Ranjit
Tobler, Waldo Rudolph
Toni, Youssef Tanious
Toy, Terrence Joseph
Trotter, John Ellis
Trussell, Margaret Edith
Tuller, Stanton Ernest
Turner, Billie Lee, II
Tyman, John Langton
Ullman, Edward Louis
Urbscheit, Peter
Vance, James Einon, Jr
Vanderhill, Burke Gordon
Vandermeer, Canute
Vann, John Herman
Van Riper, Joseph Edwards
Varney, Charles Broadwell
Vermeer, Donald E
Viletto, John, Jr
Villmow, Jack N
Vogel, Philip E
Vouras, Paul Peter
Wake, William Henry
Walker, Harley Jesse
Walmsley, Mildred Marie
Warkentin, John Henry
Warn, George Frederick
Warntz, William
Washburn, Henry Bradford, Jr
Webb, John W
Webb, Kempton Evans
Wegend, Guido Gustav
Welch, Roy Allen
Werner, Christian
Wesche, Rolf Juergen
Wester, Lyndon Leonard
Whebell, Charles Frederick John
Wheeler, Jesse Harrison, Jr
Whitaker, Joe Russell
White, Kenneth L, Sr
White, Russell Alan
Wibling, Robert Kenton
Wibanks, Thomas John
Wilkie, Gene C
Wilkie, Richard W
Williams, Anthony Vearncombe
Williams, Charles Edwin
Wills, Bernt Lloyd

Wilson, Andrew Wilkins
Wilson, Eugene Murphey
Wilson, James Newton
Wilson, Michael Robert
Winnie, William M, Jr
Winsberg, Morton Daniel
Winters, Harold Abraham
Winters, David Clinton
Witmer, Richard Everett
Withuhn, Burton Orrin
Witzig, Frederick Theodore
Wolfe, Roy Israel
Wonders, William Clare
Wong, Shue Tuck
Wood, Harold Arthur
Wood, John David
Wood, Walter Abbott
Woodruff, James F
Wright, Marion Irene
Yoder, Julian Clifton
Young, Bruce C
Youngmann, Carl Ernst
Zadrozny, Mitchell G
Zakrzewska, Barbara (Mrs M Borowiecki)
Zobler, Leonard

Biogeography

Andrie, Robert Francis
Bennett, Charles Franklin
Burrill, Robert Meredith
Cornell, Howard Vernon
Fonaroff, Leonard Schuyler
Foster, Robert H
Fraser, Donald Alexander
Gaines, John Franklin
Gordon, Burton LeRoy
Holzatel, Christina Marie
Johannessen, Carl L
McMillan, R Bruce
Parsons, James Jerome
Powell, John Martin
Rees, John David
Rounds, Richard Clifford
Sauer, Jonathon Deninger
Schiel, Joseph Bernard, Jr
Schmid, James Addison
Stillwell, Harold Daniel
Street, John Malcolm
Vermeij, Geerat Jacobus
Walter, Hartmut
Wilhelm, Eugene J, Jr

Agricultural Geography

Blaut, James Morris
Chakravarti, Aninda Kumar
Hills, Theodore Lewis
Kollmorgen, Walter Martin
Olmstead, Clarence Walter
Reeds, Lloyd George
Tideman, Philip Lundsten

Cartography

Archer, Alford
Bard, Robert Charles
Bouchard, Louis-Marie
Chovitz, Bernard H
Clay, James William
Coulson, Michael Robert Cummins
Cramer, Robert Eli
Crawford, Paul Vincent
Cuff, David J
Dahlberg, Richard Ernest
Davies, Merton Edward
Draeger, William Charles
Drummond, Robert Norman
Duffett, Walter N
Espenshade, Edward Bowman, Jr
Gillis, James E, Jr
Head, Clifford Grant
Hellman, Allen David
Higgs, Gary Kent
Ilick, John Rowland
Johnson, Evert William
Karinen, Arthur Eli
Klove, Robert Charles
Knox, Arthur Stewart
Lee, James Stewart
Long, Robert Grant
Loy, William George
Mackay, John Ross
Maune, David Francis
Mettner, William Joseph
Odell, Clarence Burt
Ogilvie, Bruce Crossan
Phillips, Brian Antony Morley
Quinn, Alfred Otto
Thoren, Conrad Joseph
Tobler, Waldo Rudolph
Voglesong, William Frederick
Welch, Roy Allen
Woodward, David Alfred
Youngmann, Carl Ernst

Cultural Geography

Adams, John Edward
Aiken, Charles S
Allen, John Logan

Anderson, Jeremy
Augelli, John Pat
Barton, M Xaveria
Bayr, Klaus J
Brooks, Hugh Campbell
Bruman, Henry John
Carney, George Olney
Carter, George Francis
Comeaux, Malcolm Louis
Coppock, Henry Aaron
Denevan, William Maxfield
Deshler, Walter W
Doran, Edwin, Jr
Eder, Herbert Michael
English, Paul Ward
Ferdon, Edwin Nelson, Jr
Field, Chris
Field, Thomas Parry
Fuller, Gary Albert
Gade, Daniel Wayne
Glacken, Clarence James
Good, Charles Munden
Guzman, Louis Enrique
Hafner, James Allan
Hart, John Fraser
Heidenreich, Conrad Edmund
Herold, Laurance Carter
Hsieh, Chiao-Min
Icenogle, David William
Isaac, Erich
Jackson, Richard H
Karan, Pradyumna P
Kariel, Herbert George
Kaups, Matti
Knapp, Ronald Gary
Kniffen, Fred Bowerman
Kolars, John F
Koroscil, Paul Michael
Kyser, Forrest DeWayne
Landing, James Edward
Larson, Albert Jo
Lee, David Raymond
Lewis, Peirce Fee
Mather, Eugene Cotton
Mayfield, Robert Charles
McIntire, Elliot Gregor
Meinig, Donald William
Meyer, Douglas Kermit
Mikesell, Marvin Wray
Miller, Edith Joan Wilson
Minnick, Robert Fletcher
Moline, Norman Theodore
Myers, Sarah Kerr
Nissly, Charles Martin
Noble, William Allister
Pearson, Ross Norton
Perkins, Lee E
Peterson, Edwin Loose
Raitz, Karl Bennett
Rauch, Dolores
Reed, Robert Ronald
Rehder, John Burkhardt
Richardson, Bonham Churchill
Richtik, James Morton
Salter, Christopher Lord
Sawatzky, Harry Leonard
Scantling, Frederick Holland
Schoen, Meera
Scott, James William
Sheck, Ronald Calvin
Shimkin, Demitri Boris
Sonnenfeld, Joseph
Spencer, Joseph Earle
Stewart, Norman Reginald
Strietelmeier, John Henry
Thomas, William LeRoy
Totten, Don Edward
Trindell, Roger Thomas
Tuan, Yi-Fu
Tweedie, Stephen William
Vander Meer, Paul
Vander Velde, Edward Jay
Wacker, Peter Oscar
Waddell, Eric Wilson
Wagner, Philip Laurence
Watson, James Wreford
Webb, George Willis
Winslow, John Hathaway
Wishart, David John
Wolf, Laurence Grambow
Yahr, Charles Corbin
Zelinsky, Wilbur

Economic Geography

Ackerman, William Vaughn
Aiken, Charles S
Allard, Giles Oliver
Ante, Robert
Avcin, Matthew John, Jr
Ballert, Albert George
Barber, Melvin Clyde, III
Bean, John Lewis
Beyers, William Bjorn
Bhatia, Shyam Sunder
Blackbourn, Anthony
Bloomfield, Gerald Taylor
Boucher, Bertrand Phillip
Brooks, James Eugene
Brown, Robert Charles
Butler, Joseph Herbert
Classen, Harold Arthur
Collier, Gerald Loyd
Davis, John Tait
Dyer, Donald Ray
Eiselen, Elizabeth
Ekblaw, Sidney Everette
Eliot Hurst, Michael Eliot
Engass, Peter Maurice
Found, William Charles
Fuchs, Roland John
Gibson, Lay James
Griffin, Paul F
Hafner, James Allan
Hall, Robert Burnett, Jr
Hance, William Adams
Harper, Robert Alexander
Harrison, Peter
Henkel, Ray
Higgs, Gary Kent
Hoffman, George W
Horst, Oscar Heinz
Hoy, Don Roger
Huddlestun, Dwight Leroy
Hung, Frederick Fu
Inaba, Masaharu George
Ironside, Robert Geoffrey
Irving, Robert McCardle
Jarochowski, Maria Anna
Jensen, Mead Leroy
Jensen, Robert Granville
Johnson, Eric Shepherd
Jones, Donald Wade
Jumper, Sidney Roberts
Karan, Pradyumna P
Kircher, Harry Bertram
Krumme, Gunter
Lall, Amrit
Lankford, Philip Marlin
Lentnek, Barry
Lewis, Donald W
Lewis, James Eldon
Lonsdale, Richard Ellis
Lounsbury, John Frederick
MacDougall, Edward Bruce
Mandell, Paul Irving
Martinson, Tom L
Mattingly, Paul Fredrick
McCarty, Harold Hull
McConnell, James E
McElhoe, Forrest Lester, Jr
McGregor, John Robert
McIntyre, Wallace Edward
McMillion, Ovid Miller
McNee, Robert Bruce
Melamid, Alexander
Meuter, Ralph F
Mings, Robert Charles
Minkel, Clarence Wilbert
Mitchell, Lisle Series
Moore, Charles Wayne
Newling, Bruce Edgar
Nystrom, J Warren
Olsson, Gunnar
Oppel, Edwin Irving
Osborn, David Gordon
Otte, Herman Frederick
Papageorgiou, George John
Pelletier, Raymond Marcel
Powell, Grace L
Powell, Lanny C
Rau, Herbert Lawrence, Jr
Ray, David Michael
Roder, Wolf
Roepke, Howard George
Rushton, Gerard
Schultz, Ronald Richard
Sebor, Milos Marie
Semple, Robert Keith
Sen, Lalita
Sewell, W R Derrick
Shaw, Earl Bennett
Siddiqi, Akhtar Husain
Smith, Bruce Wayne
Smith, Helen Leonore
Smith, Thomas Russell
Stafford, Howard A
Starkey, Otis Paul
Steed, Guy Percy F
Stevens, George Putnam
Stutz, Frederick Paul
Taaffe, Robert Norman
Tata, Robert Samuel
Thoman, Richard Samuel
Thomas, Morgan D
Thompson, Gary Lynn
Thouez, Jean-Pierre Mary
Tuthill, Richard Lovejoy
Underwood, Robert Marshall
Vinge, Clarence L
Visher, Halene Hatcher
Wagstaff, H Reid
Wallace, William Huston
Webb, George Willis
Wheeler, James Orton
Wohlenberg, Ernest Harold
Yeates, Maurice
Yuill, Robert Stanley

Historical Geography

Adams, Robert McCormick
Allen, John Logan
Barnum, Horace Gardiner
Boyce, Ronald Reed
Carney, George Olney
Clarke, John
Code, William Robert
DeVorsey, Louis, Jr
Engass, Peter Maurice
Fatzinger, Dale Roger
Goheen, Peter George
Hamburg, James F
Head, Clifford Grant
Heidenreich, Conrad Edmund
Innes, Frank Cecil
Johnson, Hildegard Binder
Koroscil, Paul Michael
Landing, James Edward
Larson, Albert Jo
Lewthwaite, Gordon Rowland
Matthews, James Swinton
McCarthy, Albert Joseph Patrick
McManis, Douglas R
Moore, Conrad Taylor
Scott, James William
Stanislawski, Dan
Thompson, John
Tracie, Carl Joseph
Trindell, Roger Thomas
Vinge, Clarence L
Wacker, Peter Oscar
Ward, David
Westfall, John Edward
Wheeler, David Laurie
Wheeler, Jesse Harrison, Jr
Winslow, John Hathaway

Marine Geography

Bryan, Wilfred Bottrill
Ray, John Bernard

Physical Geography

Adams, Bruce Edward
Ahnert, Frank
Akin, Wallace Elmus
Arnfield, Anthony John
Aufdemberge, Theodore Paul
Balbach, Harold Edward
Basile, Robert Manlius
Berlin, Graydon Lennis
Berry, Leonard
Borowiecki, Barbara Zakrzewska
Bunting, Brian Talbot
Carroll, June Starr
Centorino, James Joseph
Crawford, Paul Vincent
Cross, Ralph Donald
Cuff, David J
De Percin, Fernand
Drummond, Robert Roland
Duffett, Walter N
Ebert, Charles H V
Ekblaw, Sidney Everette
Erhart, Rainer R
Espenshade, Edward Bowman, Jr
Falconer, Allan
Gabler, Robert Earl
Gagliano, Sherwood Moneer
Grey, Alan Hopwood
Griffiths, Thomas Melvin
Gross, Herbert Henry
Hammer, Richard M
Hammond, Edwin Hughes
Hastings, Andrew Dewey, Jr
Haugen, Richard Kenneth
Hidore, John J
Higgins, Charles Graham
Holder, Virgil Harold
Hook, John Clinton
Houghton, John G
Ives, John David
Jackman, Albert Havens
Johnson, John Peter
Kaatz, Martin Richard
Kellman, Martin C
Kesik, Andrzej B
Knox, James Clarence
Kyser, Forrest DeWayne
Lahey, James Frederick
Lane, Charles Franklin
Langdon, George L J
Lehrer, Paul Lindner
Lemons, Hoyt
Lewis, Anthony James
Linehan, Urban Joseph
Lougeay, Ray Leonard
Lounsbury, John Frederick
Mahaney, William C
Maier, Emanuel
Marcus, Melvin Gerald
Marcus, Robert Brown
Mausel, Paul Warner
McArthur, David Samuel
McBoyle, Geoffrey Reid
McCoy, Roger Michael
McElhoe, Forrest Lester, Jr
McIntyre, Michael Perry
McKinney, William Mark
McPherson, Harold James
Miller, Jesse William, Jr
Muller, Robert Albert
Murphy, Richard Ernest
Nunnally, Nelson Rudolph
Oliver, John Edward
Outcalt, Samuel Irvine
Patterson, James Edward
Pease, Robert Wright
Perejda, Andrew Daniel
Phillips, Brian Antony Morley
Pike, Richard Joseph, Jr
Place, John Louis
Randall, Duncan Peter
Reider, Richard Gary
Rudd, Robert Dean
Sakalowsky, Peter Paul, Jr
Sanderson, Marie Elizabeth
Saucier, Roger Thomas
Schmudde, Theodore Henry
Schroeder, Carlton Raymond
Schwenderman, Joseph Raymond, Jr
Schwenderwein, Gerd (Anton)
Sharpe, David McCurry
Sparrow, Christopher John
Stallings, Emmett Francis
Stauffer, Truman Parker, Sr
Steiner, Rodney
Stevens, Dale John
Stillwell, Harold Daniel
Taylor, James Woodall
Webb, Robert MacHardy
Welsted, John Edward
Wilkinson, Thomas Preston
Williams, Aaron, Jr

Phytogeography

Harvill, Alton McCaleb, Jr
Kuchler, August Wilhelm

Population Geography

Archer, Alford
Bennett, David Gordon
Brandt, Donald Paul
Brush, John Edwin
Bruyere, Donald Eugene
Dyer, Donald Ray
Fuller, Gary Albert
Gonzalez, Alfonso
Karinen, Arthur Eli
McDonald, James Robert
McTaggart, William Donald
Meade, James Montgomery
Moore, Eric G
Pirie, Peter Nigel Douglas
Quick, Horace Floyd
Rengert, George Frederick
Rowland, Richard Hugh
Schnell, George Adam
Schofer, Jerry Parker
Smith, Richard Vergon
Wang, I-Shou
Zelinsky, Wilbur

Resource Geography

Anderson, James Richard
Antonini, Gustavo Arthur
Brown, Robert Charles
Brueckheimer, William Rogers
Calef, Wesley Carr
Chubb, Michael
Earney, Fillmore Christy Fidelis
Francis, Karl Earvil
Gates, Gary Rickey
Gildea, Ray Yeakle
Guernsey, James Lee
Guest, Buddy Ross
Hammond, Kenneth Allen
Hegen, Edmund Eduard
Highsmith, Richard Morgan, Jr
Hollingshead, Anne Huston
Jensen, John Granville
Kates, Robert William
Keen, Elmer A
Kennamer, Lorrin Garfield
Kromm, David Elwyn
Lee, James Stewart
Limbird, Arthur George
Luten, Daniel B, Jr
Lynch, Donald Francis
McPherson, Harold James
Meade, James Montgomery
Micklin, Philip Patrick
Moline, Norman Theodore
Nelson, James Gordon
Norsworthy, Stanley Frank
Nowak, Wieslaw Stanislaw Wladyslaw
O'riordan, Timothy
Raup, Henry Armstrong
Ray, John Robert
Roepke, Howard George
Rowntree, Rowan Allen
Saarinen, Thomas Frederick

Sawatzky, Harry Leonard
Schmudde, Theodore Henry
Schroeder, Carlton Raymond
Schwendeman, Joseph Raymond, Jr
Searcy, Nelson Donald
Stansfield, Charles Arthur
Tideman, Philip Lundsten
Visher, Halene Hatcher
Wallace, Wayne Alexander
White, Gilbert Fowler

Transportation Geography
Bierman, Don Edward
Black, William Richard
Dickason, David Gordon
Eliot Hurst, Michael Eliot
Heiges, Harvey Eric
Kansky, Karel Joseph
Lancaster, Jane
Lewis, James Eldon
Mohamed, Harold
Scott, Thomas William
Sen, Lalita
Siddall, William Richard
Starr, John Thornton, Jr
Wheeler, James Orton

Urban Geography
Adams, John Stephen
Anton, John Ralph
Barber, Melvin Clyde, III
Bean, John Lewis
Berry, Brian Joe Lobley
Birchard, Ralph Edwin
Bouchard, Louis-Marie
Boyce, Ronald Reed
Brandt, Donald Paul
Brown, Lawrence Alan
Brush, John Edwin
Butimer, Anne
Carey, George Warren
Casper, Berenice Margaret
Chang, Sen-Dou
Clark, William Arthur Valentine
Code, William Robert
Crosby, John Albert
Darden, Joe Turner
Davies, Christopher Shane
Denis, Paul Yves
Dickinson, Robert Eric
Drummond, Robert Roland
Dueker, Kenneth John
Dutt, Ashok Kumar
Erickson, Robert James
Epstein, Bart Jacob
Evenden, Leonard Jesse
Fusch, Richard Dennis
Gaede, Herbert Lawrence
George, John L
Getis, Arthur
Gibson, Lay James
Goheen, Peter George
Golledge, Reginald George
Guernsey, James Lee
Hanten, Edward W
Harper, Robert Alexander
Hartshorn, Truman Asa
Herold, Laurance Carter
Hirt, Howard Franklin
Hoffman, Wayne Larry
Horton, Frank Elba
Hughes, James Charles
Hume, Valerie Elizabeth
Jackson, John Nicholas
Jacobs, John Francis
Johnson, Lane Joseph
Jones, James Patrick
Kansky, Karel Joseph
King, Leslie John
Kiang, Ying Cheng
Kirchherr, Eugene Carl
Kornhauser, David Henry
Kubiak, Timothy James
Kulkarni, Gopal Shrinivas
Lai, David Chuen Yan
Lanegran, David Andrew
Lankford, Philip Martin
Lemon, James Thomas
Ludwig, Armin K
Marr, Paul Donald
Mattingly, Paul Fredrick
Mayer, Harold M
McNee, Robert Bruce
Melvin, Ernest Eugene
Meuter, Ralph F
Mitchell, Lisle Series
Mohamed, Harold
Mookherjee, Debnath
Moore, Eric G
Muraco, William Anthony
Murphy, Raymond Edward
Nash, Peter Hugh, Sr
Newling, Bruce Edgar
Northam, Ray Mervyn
Odland, John
Palm, Risa Ileen
Papageorgiou, George John

Perle, Eugene Daniel
Porteous, John Douglas
Pyle, Gerald Fredric
Radell, David
Ricour-Singh, Francoise
Ritter, Fredric Arnold
Roznowski, Donald Martin
Rushton, Gerard
Salter, Paul Sanford
Sargent, Charles S, Jr
Schultz, Ronald Richard
Schwind, Paul Jackson
Smith, Bruce Wayne
Smith, Peter John
Smith, Richard Vergon
Stadel, Christoph
Starr, John Thornton, Jr
Sternberg, Rolf
Stutz, Frederick Paul
Swain, Harry Sheldon
Swartz, Robert David
Thoman, Richard Samuel
Thouez, Jean-Pierre Mary
Trott, Charles Eugene
Villeneuve, Paul Yvon
Vuicich, George
Wang, I-Shou
Ward, David
Wellar, Barry Sheldon
Wolf, Laurence Granbow
Wright, Robert Crumn
Yeates, Maurice
Yuill, Robert Stanley
Zieber, George Henry

Zoogeography
Genoways, Hugh Howard
Macpherson, Andrew Hall
Simoons, Frederick John
Smetana, Ales
Wilson, John William, III

GEOGRAPHY, REGIONAL

Geography of Africa
Bayr, Klaus J
Birchard, Ralph Edwin
Deshler, Walter W
Gaede, Herbert Lawrence
Good, Charles Munder
Hance, William Adams
Holz, Robert Kenneth
Houk, Richard J
Kirchherr, Eugene Carl
Lee, David Raymond
Simko, Robert Alexander
Tuthill, Richard Lovejoy

Geography of Asia
Ante, Robert
Beatty, George Franklin
Bhatia, Shyam Sunder
Classen, Harold Arthur
Dickason, David Gordon
Dutt, Ashok Kumar
Doerr, Arthur Harry
Frank, Ralph William
Ginsburg, Norton Sydney
Hall, Robert Burnett, Jr
Hirt, Howard Franklin
Hsieh, Chiao-Min
Hung, Frederick Fu
Illick, John Rowland
Keen, Elmer A
Kolars, John F
Kulkarni, Gopal Shrinivas
Mayfield, Robert Charles
McColl, Robert William
McTaggart, William Donald
Mookherjee, Debnath
Reed, Robert Ronald
Schoen, Meera
Smith, Helen Leonore
Spencer, Joseph Earle
Sivasrava, Harishankar Prasad
Thomas, William LeRoy
Wernstedt, Frederick Lage
Withington, William Adriance
Yahr, Charles Corbin

Geography of Australia
Field, Thomas Parry

Geography of Canada
Irving, Robert McCardle
Ray, David Michael
Raymond, Charles Wyatt
Tracie, Carl Joseph

Geography of China
Chang, Sen-Dou
Kiang, Ying Cheng
Knapp, Ronald Gary
Lai, David Chuen Yan

Salter, Christopher Lord
Wu, Cheng-Tsu

Geography of the Caribbean
Adams, John Edward
Hecht, Melvin Edwin
Richardson, Bonham Churchill

Geography of Europe
Adams, Bruce Edward
Dickinson, Robert Eric
Heiser, Willard Wayne
Houk, Valerie Elizabeth
Hune, Valerie Elizabeth
Koller, Ernst Frank
Kostanick, Huey Louis
Matthews, James Swinton
McCarthy, Albert Joseph Patrick
McDonald, James Robert
Mingh, Julian Vincent
Schulz, Peter
Scott, Thomas William
Tatham, George

Geography of Great Britain
Strietelmeier, John Henry

Geography of India
Goodman, Robert Joseph

Geography of Japan
Kornhauser, David Henry

Geography of Latin America
Ackerman, William Vaughn
Augelli, John Pat
Bruman, Henry John
Clark, William Thomas, Jr
Denevan, William Maxfield
Denis, Paul Yves
Dillman, Charles Daniel
Eder, Herbert Michael
Fatzinger, Dale Roger
Frost, Melvin Jesse
Gonzalez, Alfonso
Horst, Oscar Heinz
Hoy, Don Roger
Hughes, James Charles
Johnson, Eric Shepherd
Langdon, George L J
Lentnek, Barry
Long, Robert Grant
Loring, Robert David
Ludwig, Armin K
Martinson, Tom L
Minkel, Clarence Wilbert
Myers, Sarah Kerr
Nissly, Charles Martin
Parsons, James Jerome
Pearson, Ross Norton
Read, Mary Jo
Rengert, George Frederick
Sheck, Ronald Calvin
Stanley, Raymond Wallace
Sternberg, Hilgard O'Reilly
Sternberg, Rolf
Tata, Robert Joseph
Thompson, John
Turk, Jessie Rose
West, Robert Cooper

Geography of the Mediterranean
Wheeler, David Laurie

Geography of the Middle East
Boucher, Bertrand Phillip
English, Paul Ward
Foster, Fred William
Hallgren, Alvin Roland
Paige, Richard Joseph

Geography of North America
Ballert, Albert George
Cramer, Robert Eli
Dow, Maynard Weston
Gregor, Howard Frank
Hart, John Fraser
McMillion, Ovid Miller
Zobel, Herbert Lawrence

Geography of Northern Lands
Jackman, Albert Havens
Kaups, Matti
Lynch, Donald Francis
Wonders, William Clare

Geography of the Pacific
Doran, Edwin, Jr
Farrell, Bryan Henry
Lewthwaite, Gordon Rowland
McIntyre, Michael Perry

Geography of Polar Regions
Anderson, Lawrence Conrad
Bertrand, Kenneth John

Geography of the Soviet Union
Bruyere, Donald Eugene

Chamberlin, Thomas Wilson, Sr
Fuchs, Roland John
Jensen, Robert Granville
Knadler, George Arthur
Kostanick, Huey Louis
Lonsdale, Richard Ellis
Mieczkowski, Zbigniew Ted
Pereyda, Andrew Daniel
Rowland, Richard Hugh
Thompson, Gary Lynn
Underwood, Robert Marshall
Wallace, Wayne Alexander

Geography of the United States
Ahnert, Frank
Carroll, June Starr
Hamburg, James F
McCarty, Harold Hull
Patterson, James Edward
Raitz, Karl Bennett
Roznowski, Donald Martin
Searcy, Otis Paul
Wallace, William Huston

GEOLOGY
Abdel-Gawad, Monem
Adams, John Kendal
Adams, Robert W
Adams, William L
Adler, Hans Henry
Agatston, Robert Stephen
Agenbroad, Larry Delmar
Agnew, Allen Francis
Agron, Sam Lazrus
Aitken, Janet Mora
Albee, Arden Leroy
Albee, Howard Franklin
Alberding, Herbert
Albers, John P
Albritton, Claude Carroll, Jr
Alfors, John Theodore
Allard, Giles Oliver
Allen, Arthur T, Jr
Allen, Clarence Roderic
Allen, Clifford Marsden
Allen, Jack C, Jr
Allen, John Eliot
Allen, Rhesa McCoy, Jr
Allingham, John Wing
Allison, Richard C
Alt, David D
Amsbury, David Leonard
Andersen, Harold Veral
Anderson, Charles Alfred
Anderson, Duwayne Marlo
Anderson, Francis David
Anderson, John Jerome
Anderson, Norman Roderick
Anderson, Richard Charles
Anderson, Richard Jasper
Anderson, Roger Yates
Anderson, Wayne I
Andrews, George William
Anikouchine, William A
Anthony, John Williams
Appleman, Daniel Everett
Appleyard, Edward Clair
Applin, Paul Livingston
Arbenz, Johann Kaspar
Arce, Jose Edgar
Archbold, Norbert L
Arkle, Thomas, Jr
Armstrong, Frank Clarkson
Armstrong, Herbert Stoker
Arnal, Robert Emile
Arnott, Ronald James
Arnow, Theodore
Arper, William Burnside
Ash, Sidney Roy
Atherton, Elwood
Austin, Carl Fulton
Austin, George Stephen
Austin, Roger Seth
Avent, Jon C
Badgley, Peter Coles
Bailey, Edgar Herbert
Bailey, Roy Alden
Bailey, Thomas Laval
Bain, Roger J
Baird, David McCurdy
Baker, Arthur Alan
Baker, Donald Roy
Baker, Glenn Jackson
Baldwin, Arthur Dwight, Jr
Baldwin, Brewster
Baldwin, Ewart Merlin
Ball, Clayton Garrett
Ball, Mahlon M
Ball, Stanton Mock
Ballmann, Donald Lawrence

DuMontelle, Paul Bertrand
Dunham, Robert Jacob
Dunn, James Robert
Durham, Clarence Olson, Jr
Durrell, Cordell
Dutcher, Russell Richardson
Dutro, John Thomas, Jr
Eakins, Peter Russell
Earll, Fred Nelson
Easterbrook, Don J
Eberlein, George Donald
Echols, Dorothy Jung
Eckel, Edwin Butt
Eckelmann, Frank Donald
Edmund, Rudolph William
Ehlig, Perry Lawrence
Ehrlich, Robert
Ehrreich, Albert LeRoy
Eicher, Don Lauren
Eifler, Gus Kearney, Jr
Eisenstat, Phillip
Elam, Jack Gordon
Elberty, William Turner, Jr
Elder, Wilfred Allan
Elders, Eugene Rudolph
Ellinwood, Howard Lyman
Eliot, David Hawksley
Eliott, Jane Elizabeth Inch
Ellison, Ross Courtland
Ellison, Robert L
Ellison, Samuel Porter, Jr
Emerson, Donald Orville
Emerson, John Wilford
Emery, Philip Anthony
Emmons, Richard Conrad
Emslie, Ronald Frank
Engel, Albert Edward John
Enos, Paul (Portenier)
Enzmann, Robert D
Epis, Rudy Charles
Epstein, Jack Burton
Erb, David Kinsey
Erdman, Oscar Alvin
Erickson, Ralph Leroy
Ern, Ernest Henry
Erskine, Christopher Forbes
Erwin, Robert Bruce
Espenshade, Gilbert Howry
Esthinton, Raymond Lindsay
Eveland, Harmon Edwin
Evensen, James Millard
Everett, Ardell Gordon
Evitt, William Robert
Faas, Richard William
Fagan, John J
Fahnestock, Robert Kendall
Fairbridge, Rhodes Whitmore
Fairley, William Merle
Faizi, Salih
Fan, Paul Hsiu-Tsu
Fan, Pow-Foong
Farmer, George Thomas, Jr
Farnham, Paul Rex
Farnsworth, Roy Lothrop
Farr, John B
Fary, Raymond W, Jr
Faust, George Tobias
Fawcett, James Jeffrey
Feininger, Tomas
Felsher, Murray
Felts, Wayne Moore
Fenner, Peter
Ferguson, Laing
Ferm, John Charles
Fernald, Arthur Thomas
Ferrians, Oscar John, Jr
Feth, John Henry
Fett, John D
Fidlar, Marion Moore
Fiero, George William, Jr
Filippone, Walter R
Finch, Warren Irvin
Finney, Joseph J
Fischer, Alfred George
Fisher, Frederick Stephen
Fisher, George Wescott
Fisher, Irving Sanborn
Fisher, Richard Virgil
Fisher, William Lawrence
Fiske, Richard Sewell
Fitzsimmons, John Paul
Flawn, Peter Tyrrell
Flemal, Ronald Charles
Fleming, Robert William
Fletcher, Frank William
Flint, Norman Keith
Flores, Romeo M
Flower, Rousseau Hayner
Folinsbee, Robert Edward
Foose, Richard Martin
Forbes, Robert Briedwell
Ford, Arthur B
Ford, Derek Clifford
Ford, John Philip
Forrester, James Donald

Fortier, Yves Oscar
Foster, Helen Laura
Foster, Norman Holland
Foster, Robert L
Fox, Frederick Glenn
Fox, James Ellison
Fox, Stephen Knowlton, Jr
Fox, William Templeton
Frank, Glenn William
Franklin, George Joseph
Franks, Paul C
Frarey, Murray James
Fraunfelter, George H
Frautschy, Jeffery Dean
Frebold, Hans (Wilhelm Ludwig August Herman)
Frederickson, Arman Frederick
Frederickson, Edward Arthur
Freeman, Jacob
Freeman, Thomas J
Friedman, Jules Daniel
Friedman, Samuel Arthur
Frimpter, Michael Howard
Fritz, Peter
Froelich, Albert Joseph
Frondel, Judith W
Frost, Stanley H
Frye, John Chapman
Fryklund, Verne Charles, Jr
Fryxell, Fritiof Melvin
Fullagar, Paul David
Fuller, John William
Fuller, Richard Eugene
Fulton, Robert John
Funkhouser, John William
Furlong, Ira E
Furnish, William Madison, Jr
Furnival, George Mitchell
Fyles, James Thomas
Gaal, Robert
Gabelman, John Warren
Gabrielse, Hubert
Gait, Robert I
Ganguly, Jibamitra
Garbarini, George S
Gardner, Murray Curtis
Garnier, Benjamin John
Garrison, Louis Eldred
Garrison, Robert Edward
Gastil, R Gordon
Gates, George Oscar
Gates, Olcott
Gates, Robert Maynard
Gaudette, Henri Eugene
Gawarecki, Stephen Jerome
Gazin, Charles Lewis
Gealy, Elizabeth Lee
Gealy, John Robert
Gealy, William James
Gee, David Easton
Genes, Andrew Nicholas
Geraghty, James Joseph
Gerrard, Thomas Aquinas
Ghaly, Tharwat Shahata
Ghent, Edward Dale
Gheith, Mohamed A
Gibb, Richard A
Gilderslewe, Benjamin
Gill, James Edward
Gillett, Lawrence B
Gilliland, William Nathan
Gillis, James E, Jr
Gilluly, James
Gilman, Richard Atwood
Gimbrede, Louis de Agramonte
Gin, Thon Too
Gipson, Mack, Jr
Gittins, John
Glaeser, John Douglas
Gloeksky, Yvon Raoul
Glock, Waldo Sumner
Gluskoter, Harold Jay
Goetz, Alexander Franklin Hermann
Gold, David Percy
Goldberg, Jerald Melvin
Goldich, Samuel Stephen
Goldsmith, Julian Royce
Goldsmith, Richard
Goldthwait, Richard Parker
Golightly, John Paul
Gonzales, Serge
Goode, Harry Donald
Goodell, Horace Grant
Gooding, Ansel Miller
Goodspeed, Robert Marshall
Goodwin, Bruce K
Goranson, Leonard D
Gore, Dorothy J
Gould, Howard Ross
Gould, Laurence McKinley
Gower, Howard Dale
Graham, Charles Edward
Grant, Stanley Cameron
Grant, Willard H
Grantz, Arthur

Gravenor, Conrad Percival
Graves, Howard Bradley, Jr
Graves, Roy William, Jr
Gray, Clifton Herschel, Jr
Gray, Henry Hamilton
Gray, Ralph J
Greeley, Ronald
Green, Francis Earl
Green, John Chandler
Greenberg, Seymour Samuel
Greenblatt, Jayson Herschel
Greene, Robert Carl
Greenman, Norman Nathan
Greenwood, Hugh J
Greenwood, William R
Gregor, Clunie Bryan
Greiner, Hugo R
Grener, Gordon Conrad
Grenier, Paul Emile
Grew, Priscilla Croswell Perkins
Grice, Reginald Hugh
Gries, John Paul
Griffin, Villard Stuart, Jr
Griggs, Allan Bingham
Gromme, Charles Sherman
Groff, Sidney Lavern
Groot, Johan Jacob
Grose, Lucius Trowbridge
Gross, David Lee
Grossman, William Lewis
Grow, George Copernicus, Jr
Grubbs, Donald Keeble
Gryc, George
Guber, Albert Lee
Gucwa, Paul Ramon
Guidotti, Charles V
Gulbrandsen, Robert Allen
Guldenzopf, Emil Charles
Gustavson, Thomas Carl
Gutschick, Raymond Charles
Gutstadt, Allan Morton
Guyton, James W
Guven, Necip
Haas, Otto Henry
Hack, John Tilton
Hackett, Orwoll Milton
Hacquebard, Peter Albertus
Haddock, Gerald Hugh
Haefner, Richard Charles
Hagan, Wallace Woodrow
Hagner, Arthur Feodor
Hale, William Ernest
Hale, Clarence Albert, Jr
Hall, John Frederick
Hall, Leo M
Hall, Wayne Everett
Hall, William Bartlett
Hallet, Bernard
Halsey, Jonathan Horace
Hambleton, William Weldon
Hamblin, William Kenneth
Hamburger, Richard
Hamill, Louis
Hamilton, Charles Leroy
Hamilton, Daniel Kirk
Hamilton, Douglas Holmes
Hamilton, Warren (Bell)
Hamlin, William Henry
Hammond, Paul Ellsworth
Haney, Donald C
Hanscom, Roger H
Hansen, Alan Ray
Hansen, George Henry
Hansen, Robert Herbert
Hansen, Hiram Stanley
Happ, Stafford Coleman
Harbour, Bob Adrian
Hardage, James Otto
Harder, James Otto
Hardin, George C, Jr
Hargraves, Robert Bero
Harker, Robert Ian
Harker, Peter
Harrington, John Conrad
Harrington, John Wilbur
Harris, Ann Graetsch
Harris, C Earl, Jr
Harris, David Vernon
Harris, Rae Lawrence, Jr
Harris, Stanley Edwards, Jr
Harris, Jack Lamar
Harrison, James Merritt
Harshbarger, John W
Hart, Lyman Herbert
Hart, Richard Royce
Hart, Stanley Robert
Harvey, John Frank
Harvey, James Austin
Harwood, David Smith
Hase, Donald Henry
Haselton, George Montgomery
Hatfield, Craig
Hatheway, Richard Brackett
Hattin, Donald Edward
Havenor, Kay Charles
Hawkins, James Wilbur, Jr
Hawley, John William
Haworth, Alfred John

Hayes, John Bernard
Hayes, John Robert
Hayes, Miles O
Haynes, Caleb Vance, Jr
Haynes, Donald Duane
Hays, James Henry
Hayward, Oliver Thomas
Heard, Bernard Carter, Jr
Heath, Ralph Carr
Heck, Edward Timmel
Heckel, Philip Henry
Hedberg, Hollis Dow
Hedlund, Richard Warren
Hedlund, David C
Heede, Burchard Heinrich
Heezen, Bruce Charles
Heinrich, Richard Allen
Heinonen, Henry Samuel
Heindl, Leopold Alexander
Heinrich, Eberhardt William
Helgeson, Harold Charles
Helsley, Charles Everett
Hemley, James A
Hemley, John Julian
Helwig, James A
Henderson, Arnold Richard
Henderson, Gerald Gordon Lewis
Henderson, Gerald Vernon
Hendricks, Thomas Andrews
Hendricks, Herbert Edward
Hendrix, Thomas Eugene
Hendry, Charles Walter, Jr
Hendry, Hugh Edward
Henricksen, Thomas Alva
Henry, Vernon James, Jr
Herdendorf, Charles Edward, III
Hereford, Joseph Pierce
Heron, Duncan
Heroy, William Bayard, Jr
Herrick, Stephen (Marion)
Herrmann, Leo Anthony
Hershey, Howard Garland
Herz, Norman
Hess, David Filbert
Hewett, Charles Hayden
Hewitt, Donald Francis
Hewitt, Philip Cooper
Heyl, Allen Van, Jr
Heyl, George Richard
Heylmun, Edgar B
Hibbard, Malcolm Jackman
Hickox, John Ekstrom
Hietanen-Makela, Anna (Marta)
Higgs, Donald Val
Higgins, Gary Hoyt
Higgins, James Woodrow
Hills, Brenda Baer
Hills, John Moore
Hills, Leonard Vincent
Hinchey, Norman Shreve
Hinrichs, Noel W
Hinners, Frederick Woods
Hinze, Lehi Ferdinand
Hirschfeld, Sue Ellen
Hoagland, Alan D
Hobbs, Charles Roderick Bruce, Jr
Hobbs, Samuel Warren
Hobson, John Peter, Jr
Hobson, Richard David
Hodder, Robert William
Hoffer, Jerry M
Hoffman, James Irvie
Hoffman, Paul Fajvel
Hogberg, Rudolph Karl
Hogan, Roger D
Hoke, John Humphreys
Hole, Gilbert Lee
Hollister, Lincoln Steffens
Holmes, Clifford Newton
Holmes, Ralph Jerome
Holt, Charles Lee Roy, Jr
Holton, Adolphus
Holwerda, James G
Holyk, Walter
Holzer, Thomas Lequear
Hood, William Calvin
Hooke, Robert Lebaron
Hooker, Marjorie
Hoots, Harold William
Hoover, Linn
Hoover, W Farrin
Hopkins, David Moody
Hopkins, M E
Hopkins, William Stephen, Jr
Hoppin, Richard H
Hopson, Clifford Andree
Horn, Myron Kay
Horowitz, Alan Stanley
Horstman, Arden William

Horton, Robert Carlton
Horz, Friedrich
Hoskins, Donald Martin
Hosmer, Henry Liggett
Hostetler, Paul B
Hotz, Preston Enslow
Hough, Jack Luin
Hovey, Richard Dean
Howard, Arthur David
Howard, James Dolan
Howard, James Michael
Howard, John Hall
Howard, Keith Arthur
Howe, Wallace Brady
Hriskevich, Michael Edward
Hsu, Kenneth Jinghwa
Hubbert, Marion King
Huber, Norman King
Hubert, John Frederick
Huddle, John Warfield
Huff, Warren D
Huffington, Roy Michael
Huffman, George Garrett
Hughes, Jack Thomas
Hughes, Paul Warren
Hughes, Richard David
Hughes, Travis Hubert
Huh, Oscar Karl
Hulsey, Jess Dale
Hume, James David
Humphrey, William Elliott
Hunt, Charles Butler
Hunt, Charles Warren, III
Hunt, Lee McCaa
Hunter, Hugh Edwards
Hunter, Ralph Eugene
Hurley, Neal Lilburn
Hurst, Richard William
Hurst, Vernon James
Hussey, Arthur M, II
Husted, John E
Hutchinson, Richard William
Hutchinson, Robert Maskiell
Hutchison, David M
Hutchison, William Watt
Ingerson, Fred Earl
Ingham, Albert Irwin
Ingle, James Chesney, Jr
Ingram, Roy Lee
Inkster, John Wibirt
Inman, Douglas Lamar
Inners, Jon David
Irish, Ernest James Wingett
Irvine, T Neil
Isachsen, Yngvar William
Jablonski, Eugene
Jackson, Everett Dale
Jackson, Kern Chandler
Jackson, Philip Larkin
Jackson, Togwell Alexander
Jacobs, Alan Martin
Jacobsen, Lynn C
Jaffe, Howard William
Jahns, Richard Henry
James, Harold Lloyd
James, Jack Alexander
James, Laurence Beresford
Jeffords, Russell Macgregor
Jenkins, Stuart Edward
Jenness, Stuart Edward
Jennings, Albert Ray
Joensuu, Oiva I
Johnsen, John Herbert
Johnson, Frank Walker
Johnson, Frederick Arthur, Jr
Johnson, Gary Dean
Johnson, Hamilton McKee
Johnson, Kenneth Sutherland
Johnson, Robert Britten
Johnson, Robert W, Jr
Johnson, Vard Hayes
Jolliffe, Alfred Walton
Jones, Blair Francis
Jones, Daniel John
Jones, David Lawrence
Jones, David Lloyd
Jones, Douglas Epps
Jones, Eugene Laverne
Jones, Gareth Hubert Stanley
Jones, Lois Marilyn
Jones, Paul Hastings
Jones, Robert William
Jordan, James N
Jordan, Robert R
Jordan, William Malcolm
Judson, Sheldon, (Jr)
Jull, Mary Agnes
Just, Mary Agnes
Justus, Philip Stanley
Kahle, Charles F
Kahn, James Steven
Kaikow, Julius
Kaiser, Russell Florentine
Kane, Henry Edward
Kanizay, Stephen Peter
Karig, Daniel Edmund
Karklins, Olgerts Longins
Karlstrom, Thor Nels Vincent

Karner, Frank Richard
Karpinski, Robert Whitcomb
Karrow, Paul Frederick
Kaska, Harold Victor
Kassander, Arno Richard, Jr
Kauffman, Marvin Earl
Kaufman, Robert Frank
Kays, M Allan
Keefer, William Richard
Keller, Kenneth Frank
Keller, Walter David
Kemp, Anthony Lionel
Kendall, Kenneth Keese, Jr
Kent, Deane Frederick
Kent, Douglas Charles
Kent, Lois Schoonover
Kerr, James William
Kieffer, Susan Werner
Kiersch, George Alfred
Kindle, Cecil Haldane
Kindle, Edward Darwin
King, Arthur Francis
King, John Stuart
King, Philip Burke
King, Robert Evans
King, William Roy, Jr
Kinney, Douglas Merrill
Kinsman, David John James
Kirkemo, Harold
Kirkpatrick, Robert James
Kistler, Ronald Wayne
Kittleman, Laurence Roy, Jr
Klassen, Rudolph Waldemar
Kleen, Harold J
Kleinpell, Robert Minssen
Kleist, John Raymond
Klement, Karl Walter
Klemic, Harry
Klepper, Montis Ruhl
Klepper, Harry John
Klovan, John Edward
Knight, Fred G
Knowles, David M
Knox, Burnal Ray
Kocurko, Michael John
Koenig, Karl Joseph
Kolb, Charles Rudolph
Kopp, Otto Charles
Kornicker, Louis Sampson
Kovach, Jack
Kozak, Samuel J
Kraft, John Christian
Kramer, James Richard
Kratchman, Jack
Kremp, Gerhard Otto Wilhelm
Krieger, Medora Hooper
Krumbein, William Christian
Krumenacher, Daniel
Krushensky, Richard D
Krutak, Paul Russell
Kuenzi, W David
Kuhn, Truman Howard
Kuntz, Mel Anton
Kupsch, Walter Oscar
LaBerge, Gene L
Ladd, Harry Stephen
LaFountain, Lester James, Jr
Laird, Wilson Morrow
Lajoie, Jean
Lajtai, Emery Zoltan
Lamar, Donald Lee
LaMarche, Valmore Charles, Jr
LaMoreaux, Philip Elmer
Lance, John Franklin
Land, Lynton S
Landes, Kenneth Knight
Landon, Robert E
Lange, Ian M
Langway, Chester Charles, Jr
Lanphere, Marvin Adler
Laporte, Leo Frederic
La Prade, Kerby Eugene
Larimore, Ann Evans
LaRocque, Joseph Alfred Aurele
Larson, Edward Richard
Larson, Edwin E
Laswell, Troy James
Lattman, Laurence Harold
Laudon, Laurence Robert
Laudon, Thomas S
Laughon, Robert Bush
Laurent, Roger
Lautenschlager, Herman Kenneth
Laval, William Norris
Lawrence, David Reed
Lawrence, Robert D
Lawson, Ralph Willard
Leavy, Thomas A
LeBlanc, Arthur Edgar
LeBlanc, Rufus Joseph
Lee, Hulbert Austin
Lee, James William
Lee, Keenan
Leech, Geoffrey Bosdin
Lefebvre, Richard Harold
Lehman, David Hershey
Leighton, Freeman Beach
Leith, Carlton James

LeMasurier, Wesley Ernest
Lemmon, Dwight Moulton
Lemon, Roy Richard Henry
Lenz, Alfred C
Leonard, Benjamin Franklin
Lepp, Henry
Lerbekmo, John Franklin
Lesure, Frank Gardner
Leutze, Willard Parker
Levandowski, Donald William
Levin, Harold Leonard
Levings, William Stephen
Levinson, Stuart Alan
Lewis, Arthur Edward
Lewis, John Hubbard
Lian, Harold Maynard
Libby, Willard Gurnea
Lidiak, Edward George
Liebe, Richard Milton
Lindholm, Roy Charles
Lindsey, David Allen
Lineback, Jerry Alvin
Link, Peter K
Linn, Kurt O
Lintz, Joseph, Jr
Liou, Juhn G
Lipman, Peter Waldman
Lipps, Jere Henry
Little, Robert Lewis
Lloyd, Joel Joseph
Lockwood, John Paul
Loetterle, Gerald John
Lohman, Kenneth Elmo
Lohman, Stanley William
Lohrengel, Carl Frederick, II
Long, Clarence Sumner, Jr
Longley, William Warren
Loomis, Alden Albert
Loring, Arthur Paul
Lougheed, Milford Seymour
Lounsbury, Richard William
Love, John David
Lovingood, Paul Evans, Jr
Lowell, Gary Richard
Lowry, Jean
Lowry, Wallace Dean
Lozo, Frank Edgar
Lucas, James Robert
Lucke, John Becker
Ludlum, John Charles
Ludman, Allan
Lueninghoener, Gilbert Carl
Luft, Stanley Jeremie
Lugn, Alvin Leonard
Lukert, Michael T
Lumbers, Sydney Blake
Lumsden, David Norman
Lumsden, William Watt, Jr
Lund, Ernest Howard
Lundin, Robert Folke
Lustig, Lawrence Kenneth
Luth, William Clair
Luther, Edward Turner
Lynts, George Willard
Lyon, Ronald James Pearson
Lyons, Erwin Kim
Lyons, John Bartholomew
Maccini, John Andrew
Macdonald, Gordon Andrew
MacDonald, Harold John Carleton
MacDowell, John Fraser
MacFadyen, John Archibald, Jr
MacKenzie, David Brindley
MacLachlan, James Crawford
MacLean, Wallace H
MacMahan, Horace Arthur, Jr
MacNeill, Rupert Heath
Macquown, Roger Webb
MacQuown, William Charles, Jr
MacRae, Neil D
Macri, Alfred Roger
MacTavish, John N
Maddocks, Rosalie Frances
Madison, James Ambrose
Madsen, James Henry, Jr
Maher, John Charles
Maher, Stuart Wilder
Majmundar, Hasmukhrai Hiralal
Malahoff, Alexander
Malde, Harold Edwin
Mallory, Bob Franklin
Mallory, Virgil Standish
Mallory, William Wyman
Mamet, Bernard Leon
Mandra, York T
Manger, George Edward
Mangus, Marvin D
Mankin, Charles John
Mann, Christian John
Mann, John Allen
Manning, John Francis, Jr
Manspeizer, Warren
Marcantel, Emily Laws
Marine, Ira Wendell
Marks, Jay Glenn
Marsden, Ralph Walter

Marsh, Bruce David
Martignole, Jacques
Martin, George Carlyle
Martin, Richard Harold
Martin, Robert Francois Churchill
Martin, Roger Charles
Martin, Wayne Dudley
Masters, Charles Day
Mather, Katharine Kniskern
Mathews, Geoffrey William
Mathews, William Henry
Matsch, Charles Leo
Matthews, Charles Sedwick
Matthews, William Henry, III
Mattox, Richard Benjamin
Mattson, Peter Humphrey
Matuszak, David Robert
Mauger, Richard L
Maughan, Edwin Kelly
Maurer, Robert Eugene
Maxwell, Charles Henry
Maxwell, James Christie
Maxwell, John Crawford
Mayer, Victor James
Maynard, Robert G
Mayo, Evans Blakemore
McAllister, Arnold Lloyd
McAllister, James Franklin
McAndrews, Harry
McAnulty, William Noel
McBee, William, Jr
McBirney, Alexander Robert
McCabe, Louis Cordell
McCallum, Malcolm E
McCammon, Richard B
McCampbell, John Caldwell
McClellan, Guerry Hamrick
McCollough, Edward Heron
McConnell, Robert Kendall
McCormick, James E
McCrone, Alistair William
McCulloch, David Sears
McDaniel, Willard Rich
McDowell, Robert Carter
McGannon, Donald E, Jr
McGeary, David F R
McGehee, Richard Vernon
McGetchin, Thomas R
McGill, George Emmett
McGill, John Thomas
McGowen, Joseph Hobbs
McGregor, Duncan J
McGrew, Paul Orman
McGugan, Alan
McGuire, Odell
McIntyre, David H
McIntyre, Donald B
McIntyre, John Bowie
McIver, Norman L
McKague, Herbert Lawrence
McKee, Edwin Dinwiddie
McKee, Edwin H
McKelvey, Vincent Ellis
McKenzie, Garry Donald
McKillop, J H
McKim, Harlan L
McLaren, Digby Johns
McLaughlin, Kenneth Phelps
McLaughlin, Thad Gerald
McManus, Dean Alvis
McMillan, Neil John
McMurchy, Robert Connell
McNair, Andrew Hamilton
McNaughton, Duncan Anderson
McNitt, James R
McWhirter, Nolan
Meade, Grayson Eichelberger
Meade, Robert Heber, Jr
Meissner, Charles Roebling, Jr
Mellen, Frederic Francis
Mellon, George Barry
Melvin, John Harper
Menard, Henry William, Jr
Mencher, Ely
Merifield, Paul M
Merriam, Daniel Francis
Merriam, Richard Holmes
Merrill, Glen Kenton
Merrill, William Meredith
Meyers, Theodore Ralph
Meyers, James Harlan
Michener, Charles Edward
Mickelson, John Chester
Middleton, Gerard Viner
Miesch, Alfred Thomas
Milici, Robert Calvin
Miller, Daniel Newton, Jr
Miller, Fred Key
Miller, Halsey Wilkinson, Jr
Miller, Mary H
Miller, Maynard Malcolm

GEOLOGY

Winkler, Virgil Dean
Winslow, John Durfee
Winston, Donald
Winters, Harold Abraham
Winters, Stephen Samuel
Wise, Donald U
Wise, Sherwood Willing, Jr
Wobus, Reinhard Arthur
Wohlford, Duane Dennis
Wolf, Karl Heinz
Wolman, Markley Gordon
Wolfe, Caleb Wroe
Wolfe, Edward W
Wolfe, Peter Edward
Wolff, Manfred Paul
Wolff, Roger Glen
Wood, Leonard Alton
Wood, Spencer Hoffman
Wood, William Hulbert
Woodard, Henry Herman, Jr
Woodling, Wendell Phillips
Woods, Raymond Douglas
Worsfold, Richard John
Worts, George Frank, Jr
Woyski, Margaret Skillman
Wray, John Lee
Wright, Jerome J
Wright, Lauren Albert
Wright, Robert James
Wright, Thomas L
Wrucke, Chester Theodore, Jr
Wyder, John Ernest
Wyllie, Peter John
Wynne-Edwards, Hugh Robert
Yasso, Warren E
Yatsu, Eiju
Yeats, Robert Sheppard
Yeend, Warren Ernest
Zeigler, John Milton
Young, John Cannon
Young, Leonard M
Young, Richard A
Young, Robert Spencer
Youngquist, Walter
Zadnik, Valentine Edward
Zakrzewska, Barbara (Mrs M Borowiecki)
Zen, E-An
Zenger, Donald Henry
Zimmer, Louis George
Ziony, Joseph Israel
Zygmunt, Warren W

Biospeleology
Foldes, Francis Ferenc
Mitchell, Robert W

Dendrochronology
Bannister, Bryant
Dean, Jeffrey Stewart
Ferguson, Charles Wesley
Michael, Henry N
Robinson, William James

Economic Geology
Adams, John Wagstaff
Aho, Aaro E
Alvord, Donald C
Anderson, James Arthur
Ashley, Roger Parkman
Averitt, Paul
Bailey, Robert Vernon
Bamford, Robert Wendell
Banfield, Armine Frederick
Bassett, Allen Mordorf
Bates, Robert Latimer
Baxter, James Watson
Beane, Richard Edward
Belt, Charles Banks, Jr
Bergendahl, Maximilian Hilmar
Blais, Roger A
Boardman, Robert Leland
Bonnichsen, Bill
Bowyer, Ben
Boyle, Robert William
Brady, Lawrence Lee
Braun, Lewis Timothy
Brobst, Donald Albert
Broderick, Alan Thomas
Broughton, William Albert
Brown, Alexander Cyril
Brown, Edwin Augustus
Brown, Henry Seawell
Brown, Severn Parker
Bruce, Wayne Royal
Brummer, Johannes J
Buie, Bennett Frank
Buseck, Peter R
Bush, Alfred Lerner
Butler, Arthur Pierce, Jr
Callaghan, Eugene
Cameron, Eugene Nathan
Carlisle, Donald
Carlson, Hugh Douglas
Chaffee, Maurice Ahlborn
Cheney, Eric Swenson
Chico, Raymundo Jose
Clark, Allen LeRoy

Clark, Kenneth Frederick
Clark, Lloyd Allen
Clemons, Russell Edward
Cleveland, John H
Cofer, Harland E, Jr
Corbett, Robert G
Cornwall, Henry Rowland
Cox, Dennis Purver
Craig, Lawrence Carey
Danser, James Weart
Davidson, David Francis
Davies, James Frederick
Davies, George Herbert
Davis, James Howell
Davis, Robert Irving
Doelling, Helmut Hans
Dole, Hollis Mathews
Donovan, John Francis
Dorheim, Fredrick Houge
Douglas, Hugh
Dreyer, Robert Marx
Dudar, John S
Durkee, Edward Fleming
Einaudi, Marco Tullio
Elston, Wolfgang Eugene
Ericksen, George Edward
Evans, James R
Everhart, Donald Lough
Farquhar, Oswald Cornell
Feiss, Paul Geoffrey
Ferguson, Herman White
Field, Cyrus West
Fischer, Frederick Thomas
Fischer, Richard Philip
Flint, Arthur Emerson
Flint, Delos Edward
Franklin, James McWillie
Frezon, Sherwood Earl
Fulton, Robert Burwell, III
Gaines, Richard Venable
Groeneveld Meijer, Willem Otto Jan
Grogan, Robert Mann
Grotewold, Andreas
Guild, Philip White
Gundersen, James Novotny
Gutierrez, Luis Garcia
Hagni, Richard D
Hanley, John Bernard
Harbaugh, John Warvelle
Harshman, Elbert Nelson
Hawkes, Herbert Edwin, Jr
Hawkins, William Max
Hay, Robert E
Haynes, Simon John
Helton, Walter Lee
Henrickson, Eiler Leonard
Henshaw, Paul Carrington
Heyl, George Richard
Hill, Patrick Arthur
Hilpert, Lowell Sinclair
Houston, Robert S
Howd, Frank Hawver
Hunt, Graham Hugh
Jacob, Arthur Frank
Jacob, Leonard, Jr
Jenks, William Furness
Jicha, Henry Louis, Jr
Jindrich, Vladimir
Jinks, Douglas David
Johns, Willis Merle
Johnson, Henry Stanley, Jr
Jolly, Janice Laurene Willard
Jones, Michael Baxter
Jones, Russell Heber Blade
Kadey, Frederic L, Jr
Kalliokoski, Jorma Osmo Kalervo
Kamilli, Robert Joseph
Keighin, Charles William
Kelley, Vincent Cooper
Kelly, William Crowley
Kesler, Stephen Edward
Khawaja, Ikram Ullah
Kiilsgaard, Thor H
King, Ralph Hughes
Kisvarsanyi, Geza
Koch, George Schneider, Jr
Konig, Ronald H
Kotlowski, Frank Edward
Kruger, Fredrick Christian
Kunasz, Ihor Andrew
Lang, Arthur (Hamilton)
Langford, Fred F
Larson, Lawrence T
Laughlin, Alexander William
Learned, Robert Eugene
Lee, Charles Albert
Lemish, John
Lewis, Richard Wheatley, Jr
Little, Howard Wallace
Loring, William Bacheller
Lovering, Thomas Seward
Lowell, Wayne Russell

Main, Frederic Hall
Mancuso, Joseph J
Mather, William Bardwell
McCartney, William Douglas
McDivitt, James Frederick
Meyer, Richard Fastabend
Michell, Wilson Doe
Miller, Franklin Stuart
Miller, Ralph Leroy
Milligan, George Clinton
Mills, Joseph William
Moore, John Carman Gailey
Muessig, Siegfried Joseph
Murray, Haydn Herbert
Newcomb, Edward Lindsay
Nielsen, Richard Leroy
Nininger, Robert D
Ninniger, Dennis Lon
Ohle, Ernest Linwood
Olson, Richard Hubbell
Park, Won Choon
Parton, Priscilla Candace
Payne, Anthony Luke
Petersen, Ulrich
Pirkle, Earl Conly
Post, Edwin Van Horn
Prager, Gerald David
Pride, Douglas Elbridge
Prinz, William Charles S
Proctor, Paul Dean
Purdom, William Berlin
Pye, Edgar George
Radtke, Arthur Sears
Reeves, Robert Grier (Lefevre)
Ridge, John Drew
Riley, Charles Marshall
Robertson, Forbes
Robertson, James Magruder
Romberger, Samuel B
Rooney, Lawrence Frederick
Ross, Howard Persing
Royce, Josiah
Sahinen, Uuno Mathias
Sato, Motoaki
Scheid, Vernon Edward
Schmidt, Richard Arthur
Scott, Steven Donald
Shannon, Spencer Sweet, Jr
Sims, Samuel John
Sinclair, Alastair James
Singewald, Quentin Dreyer
Skinner, Brian John
Smith, Arthur R
Smith, John M
Socolow, Arthur A
Staatz, Mortimer Hay
Stephens, Maynard Moody
Stoiber, Richard Edwin
Sumartojo, Jojok
Sutherland-Brown, Atholl
Swinney, Chauncey Melvin
Teague, Kefton Harding
Theodore, Ted George
Thomas, Blakemore Ewing
Thompson, Tommy Burt
Thompson, Kenneth Clair
Thorpe, Ralph Irving
Tolbert, Gene Edward
Tooker, Edwin Wilson
Turner, Mortimer Darling
Tweto, Ogden
Van Alstine, Ralph Erskine
Venuto, Carmine Joseph
Walker, Eugene Hoffman
Wallace, Roberts Manning
Weitz, John Hills
West, Walter Scott
Whelan, James Arthur
White, Walter Stanley
Wier, Charles Eugene
Wilmarth, Verl Richard
Wilson, Harry David Bruce
Wilson, Robert Lee
Witkind, Irving Jerome
Worl, Ronald Grant

Environmental Geology
Anderson, Howard T
Arnold, Ralph Gunther
Baker-Blocker, Anita Linda
Bartlett, Grant Auden
Bickel, Edwin David
Buchwald, Caryl Edward
Bushnell, Kent O
Campbell, Ian
Caruccio, Frank Thomas
Cox, Doak Carey
Crittenden, Max Dermont, Jr
Deal, Dwight Edward
Dudar, John S
Easter, Thomas Edward
Emrich, Grover Harry
Fakundiny, Robert Harry
Fellows, Larry Dean
Fulmer, Charles V
Fyles, John Gladstone
Geyer, Alan Raymond

Ghuman, Gian Singh
Goebel, Edwin DeWayne
Grant, Douglas Roderick
Grear, Philip French-Carson
Groff, Donald William
Halbig, Joseph Benjamin
Harrison, Wyman
Hollenbaugh, Kenneth Malcolm
Janke, Norman C
Johnson, Ross Byron
Knutson, Carroll Field
Kunkle, George Robert
Lessing, Peter
Lundgren, Lawrence William, Jr
Martinez, Joseph Didier
McClellan, William Alan
Miranda, Henry A, Jr
Mirsky, Arthur
Nichols, Donald Ray
Nriagu, Jerome Okonkwo
Palmquist, Robert Clarence
Pessl, Fred, Jr
Pestrong, Raymond
Pipkin, Bernard Wallace
Remson, Irwin
Sage, Orrin Grant, Jr
Saucier, Roger Thomas
Schleicher, David Lawrence
Stow, Stephen Harrington
Temple, Kenneth Loren
Upchurch, Sam Bayliss
Wait, James Richard
Weiss, Dennis
Wenworth, Carl M, Jr
Williams, Julian
Winkler, Erhard Mario
Workman, William Edward

Geochronology
Alexander, Emmit Calvin, Jr
Armstrong, Richard Lee
Bada, Jeffrey L
Bikerman, Michael
Boellstorff, John David
Dallmeyer, R David
Damon, Paul Edward
Farquhar, Ronald McCunn
Faure, Gunter
Fisher, David E
Haas, Herbert
Halpern, Martin
Hartung, Jack Burdair
Henry, Christopher Duval
Hills, Francis Allan
Holloway, John Reque
Hurley, Patrick Mason
Johnson, Noye Monroe
Ku, Teh-Lung
Mattinson, James Meikle
Mose, Douglas George
Nunes, Paul Donald

Exploration Geology
Arden, Daniel Douglas
Bromberger, Samuel H
Brummer, Johannes J
Cameron, Robert Alan
Campbell, Michael David
Dahl, Harry Martin
Davis, Nicholas Falconer
Duschatko, Robert William
Erickson, John William
Evans, James Eric Lloyd
Eyer, Jerome Arlan
Ferris, Clinton S, Jr
Goranson, Edwin Alexander
Grutt, Eugene Wadsworth, Jr
Haeberle, Frederick Roland
Halferdahl, Laurence Bowes
Hiestand, Thomas Cleon
Jindrich, Vladimir
Jinks, Douglas David
Jizba, Zdenek Vaclav
Jodry, Richard L
Johnson, Henry Stanley, Jr
Kadey, Frederic L, Jr
Kelley, Dana Robineau
Kohls, Donald W
Lawrence, Edmond Francis
Leighton, Morris Wellman
Marr, John Douglas
Merrill, Robert Kimball
Morris, David Albert
Murany, Ernest Elmer
Nielson, Dianne Ruth Gerber
Quirke, Terence Thomas, Jr
Rall, Raymond Wallace
Rex, Robert Walter
Rogers, James Kenneth
Rue, Edward Evans
Saunders, Donald Frederick
Schneider, William T
Slodowski, Thomas R
Smith, William Lee
Taranik, James Vladimir
Wade, Franklin Alton
Watson, Ralph A
Wilson, L Kenneth

Reynolds, Peter Herbert
Smiley, Terah Leroy
Stern, Thomas Whital
Sutter, John Frederick
Tatsumoto, Mitsunobu
Turner, Donald Lloyd
Van Couvering, John Anthony
Wampler, Jesse Marion
Zartman, Robert Eugene

Geological Oceanography
Anderson, Franz Elmer
Anderson, John B
Andrews, James Einar
Andrews, Robert Sanborn
Bailey, James Stuart
Barnes, Burton B
Behrens, Earl William
Berry, Richard Warren
Biggs, R B
Bonatti, Enrico
Bouma, Arnold Heiko
Bryant, William Richards
Buffington, Edwin Conger
Carlson, Paul Roland
Christofferson, Eric
Coch, Nicholas Kyros
Creager, Joe Scott
Curray, Joseph Ross
Difford, Winthrop Cecil
Dill, Robert Floyd
Dillon, William Patrick
Donohue, John J
Doyle, Larry James
Duane, David Bierlein
Edgar, Norman Terence
Emery, Kenneth Orris
Field, Michael Ehrenhart
Fink, Loyd Kenneth, Jr
Fischer, Robert Lloyd
Fleischer, Peter
Folger, David W
Galehouse, Jon Scott
Gardner, James Vincent
Glass, Billy Price
Goldsmith, Victor
Gorsline, Donn Sherrin
Grant, Alan Carson
Griffin, George Melvin, Jr
Griggs, Gary B
Grinnell, Robert S, Jr
Hamilton, Edwin Lee
Harding, James Lombard
Hay, William Winn
Heath, G Ross
Hesse, Reinhard
Holmes, Charles Ward
Hulsemann, Jobst
Hurley, Robert Joseph
Hyne, Norman John
Inderbitzen, Anton Louis, Jr
Johnson, George Leonard
Jones, James I
Keller, George H
King, Lewis H
Krause, Dale Curtiss
Kroenke, Loren William
Kulm, Laverne Duane
Lavelle, John William
Lill, Gordon Grigsby
Lins, Thomas Wesley
Loughridge, Michael Samuel
Ludwick, John Calvin, Jr
Marlowe, James Irvin
Matthews, Jerry Lee
McIntyre, Andrew
Meisburger, Edward Paul
Meyerson, Arthur Lee
Molnia, Bruce Franklin
Moore, David Gillis
Moore, George Thomas
Moore, Theodore Carlton, Jr
Nayudu, Y Rammohanroy
Nichols, Maynard M
Normark, William Raymond
Oetking, Philip
Oostdam, Bernard Lodewijk
Palmer, Harold Dean
Pelletier, Bernard Roderick
Perkins, Ronald Dee
Peterson, Melvin Norman Adolph
Pilkey, Orrin H
Pyle, Thomas Edward
Rea, David Kenerson
Rezak, Richard
Rodolfo, Kelvin S
Rona, Peter Arnold
Ross, David A
Schmalz, Robert Fowler
Schmidt, Volkmar
Schubel, Jerry Robert
Schwartz, Maurice Leo
Sen Gupta, Barun Kumar
Sharma, Ghanshyam D
Shideler, Gerald Lee
Shykind, Edwin B
Sly, Peter G
Smith, Riley Seymour, Jr

Southam, John Ralph
Spencer, Derek W
Stanley, Daniel Jean
Sternberg, Richard Walter
Stevens, Richard S
Taber, Robert William
Taft, William H
Thayer, Paul Arthur
Thiede, Jorn
Vallier, Tracy L
Van Andel, Tjeerd Hendrik
Walton, William Ralph
Wang, Frank Feng Hui
Wanless, Harold Rogers
Wilson, Philo Calhoun
Wimberley, Stanley
Wright, Frederick Fenning

Geomorphology
Alexander, Charles Stevenson
Allen, James Richard
Anderson, Norman Roderick
Andrews, John Thomas
Atwood, Wallace Walter, Jr
Baker, Victor Richard
Bates, Robert Elery
Bonnett, Richard Brian
Brice, James Coble
Bull, William Benham
Caine, T Nelson
Campbell, Catherine Chase
Cannon, Philip Jan
Carson, Robert James, III
Chapman, Donald Harding
Chapman, Lyman John
Clausen, Eric Neil
Coleman, James Malcolm
Cooley, Richard Lewis
Costa, John Emil
Crowl, George Henry
Curry, Robert Rodney
Doehring, Donald O
Dolliver, Claire Vincent
Donley, Michael William
Dort, Wakefield, Jr
Dunne, Thomas
Dury, George H
Eschman, Donald Frazier
Fleisher, Penrod Jay
Flint, Jean-Jacques
Foster, Harold Douglas
Garner, Hessie Filmore
Godfrey, Andrew Elliott
Greenwood, Brian
Hadley, Richard Frederick
Hansen, Carl Lough
Harrison, Samuel S
Harrison, Wilks Douglas
Haselton, George Montgomery
Hattersley-Smith, Geoffrey Francis
Hawley, John William
Hill, Mary Rae
Hughes, John Derek
Hussey, Keith Morgan
Johnson, Kenneth George
Johnson, Peter Graham
Jopling, Alan Victor
Keller, Edward Anthony
Kiver, Eugene P
Knox, James Clarence
Koons, Donaldson
Lackey, Laurence
LaFleur, Robert George
Lambert, Paul Wayne
Larsen, Frederick Duane
Lasalle, Pierre
Lasca, Norman P, Jr
LaValle, Placido Dominick
Leopold, Luna Bergere
Lovejoy, Earl Mark Paul
Mackay, John Ross
Mahard, Richard Harold
Mayewski, Paul Andrew
McCulloch, David Sears
Mears, Brainerd, Jr
Melhorn, Witon Newton
Miller, Jesse William, Jr
Miller, Victor Charles
Milling, Marcus Eugene
Morisawa, Marie
Myers, Arthur John
Nave, Floyd Roger
Novak, Irwin Daniel
Orvos, Ervin George
Owens, Ervin George
Palmquist, Robert Clarence
Pewe, Troy Lewis
Pierce, Kenneth Lee
Rahn, Perry H
Raphael, C Nicholas
Reams, Max Warren
Reshkin, Mark
Rhodes, Dallas D
Roberts, Michael Charles
Rosalsky, Maurice B
Ross, Alex R
Roth, Eldon Sherwood
Schmaltz, Lloyd John

Schumm, Stanley Alfred
Sharp, Robert Phillip
Shimer, John Asa
Shlemon, Roy J
Shreve, Ronald Lee
Shroder, John Ford, Jr
Stephenson, Richard Allen
Stewart, David Perry
Stone, Richard O'Neill
St-Onge, Denis Alderic
Taylor, Lawrence Dow
Toy, Terrence Joseph
Trenhaile, Alan Stuart
Tuttle, Sherwood Dodge
Van Burkalow, Anastasia
Viletto, John, Jr
Wallace, Ronald Gary
Washburn, Albert Lincoln
Welder, Frank A

Glaciology
Adams, William Alfred
Adams, William Peter
Armstrong, John Edward
Bader, Henri
Benson, Carl Sidney
Bentley, Charles Raymond
Blake, Weston, Jr
Bleuer, Ned Kermit
Chapman, Donald Harding
Chapman, William Frank
Clarke, Garry K C
Craig, Bruce Gordon
Crowl, George Henry
Evenson, Edward B
Fleisher, Penrod Jay
Forsyth, Jane Louise
Giovinetto, Mario Bartolome
Gold, Lorne W
Hallet, Bernard
Hamilton, Thomas Dudley
Hartshorn, Joseph Harold
Hattersley-Smith, Geoffrey Francis
Hodge, Steven McNiven
Kamb, Walter Barclay
Knoll, Kenneth Mark
Langleben, Manuel Phillip
Matsch, Charles Leo
Mayewski, Paul Andrew
Meier, Mark Frederick
Newman, William Alexander
Nobles, Laurence Hewit
Osterkamp, Thomas Eugene
Pessl, Fred, Jr
Post, Austin
Reger, Richard David
Reid, John Reynolds, Jr
Schmaltz, Lloyd John
Schneider, Allan Frank
Shreve, Ronald Lee
Stalker, Archibald MacSween
Taylor, Lawrence Dow
Untersteiner, Norbert
Weeks, Wilford Frank

Historical Geology
Boyd, Donald Wilkin

Meteoritics
Black, David Charles
Carver, Eugene Arthur
Chou, Chen-Lin
DeGasparis, Aurelio Alfonso A
Dence, Michael Robert
Fisher, David E
Halliday, Ian
Hofmann, Albrecht Werner
Houston, Walter Scott
Huss, Glenn I
Keil, Klaus
Lewis, John Simpson
Lewis, Roy Stephen
Macdougall, John Douglas
Marvin, Ursula Bailey
Moore, Carlton Bryant
Neste, Sherman Lester
Podosek, Frank A
Van Schmus, William Randall
Wilkening, Laurel Lynn
Zimmerman, Peter David

Mining Geology
Bray, Joseph Moyer
Chace, Frederic Mason
Chico, Raymundo Jose
Coats, Robert Roy
Cooke, Hermon Richard, Jr
Davis, James Howell
Davis, Robert Irving
Greenwood, Robert
Hastings, Earl L
Nowlan, James Parker
Parker, Pierre E
Rapaport, Irving
Rove, Olaf Norberg
Savanick, George Adrian
Tonking, William Harry
Volbrecht, Stanley Gordon

Wayland, Russell Gibson
Wilson, Robert Lee

Petrography
Barnett, Charles William Henry
Berry, William Francis
Councill, Richard J
Enlows, Harold Eugene
Erickson, Edwin Sylvester, Jr
Friedlaender, Carlo Gotthelf Immanuel
Gray, Ralph J
Horz, Friedrich
Hounslow, Arthur William
Scholle, Peter Allen

Petroleum Geology
Alpert, Norman
Alpha, Andrew Gray
Andrichuk, John Michael
Arabian, Karekin Gaspar
Barrow, Thomas D
Bieberman, Robert Arthur
Blair, Robert William
Brady, Robert Townsend
Burroughs, Richard Lee
Cardwell, Dudley H
Chamberlin, Thomas Leland
Chawner, William Donald
Chuber, Stewart
Cohee, George Vincent
Connors, Theodore Thomas
Conselman, Frank Buckley
Cordell, Robert James
Cree, Allan
Crews, Lowell Thomas
Crisp, Edward Lee
Dahl, Harry Martin
Dickey, Parke Atherton
Donohue, David Arthur Timothy
Doumani, George Alexander
Durkee, Edward Fleming
Edwards, John D
Erdman, John Gordon
Fanshawe, John Richardson, II
Fay, Philip S
Ferris, Bernard Joe
Fisher, James Harold
Fisher, Stanley Parkins
Forgotson, James Morris, Jr
Fox, John R
Frantz, Wendelin R
Frenzel, Hugh N
Frey, Maurice G
Furnival, George Mitchell
Goldstein, August, Jr
Goldthwaite, Duncan
Gorton, Kenneth Arnold
Green, Thom Henning
Haas, Merrill Wilber
Haglund, David Seymour
Halbouty, Michael Thomas
Hall, Hubert Handel
Haller, Charles Regis
Haun, John Daniel
Hay-Roc, Hugh
Hirsch, John Michele
Holden, Frederick Thompson
Hoskins, Cortez William
Huey, Arthur S
Hull, Joseph Poyer Deyo, Jr
Hunt, Robert Elton
Johnson, Carlton Robert
Jones, Theodore Sidney
Keenmon, Kendall Andrews
Kidwell, Albert Laws
Klemme, Hugh Douglas
Knoll, Kenneth Mark
Konkel, Philip M
Lane, Donald Wilson
Lehmann, Elroy Paul
Longacre, Susan Ann Burton
Lowell, James Diller
Luttrell, Eric Martin
MacDougall, Robert Douglas
MacKnight, Franklin Collester
Mason, John Frederick
Mattis, Allen Francis
McConnell, Andrew Pollock, Jr
McCrossan, Robert George
McCubbin, Donald Gene
McCulloh, Thane H
McFarlan, Edward, Jr
McGookey, Donald Paul
Miller, Betty M (Tinklepaugh)
Moore, Wayne Elden
Mosher, Loren Cameron
Mundt, Philip A
Myers, Earl Eugene
Nine, Ogden Wells, Jr
Oglesby, Gayle Arden
Olson, Walter Sigfrid
Oros, Margaret Olava (Erickson)
Oxley, Philip
Palmer, Richard Bradbury
Paulken, Robert John
Paulson, Oscar Lawrence
Payne, Myron William
Record, Walter Ross

GEOLOGY

Redmond, John Lynn
Reed, Eugene Clifton
Reid, Kenneth Ian Gower
Reso, Anthony
Ries, Edward Richard
Greig, Joseph Wilson
Robinson, William John
Roehl, Perry Owen
Rogers, Marion Alan
Rose, Peter R
Rose, William Dake
Rosen, Norman Charles
Rothwell, William Thomas, Jr
Rue, Sigurd Oscar
Ryder, Robert Thomas
Sarmiento, Roberto
Sirrine, George Keith
Sneider, Robert Morton
Stephens, Raymond Weathers, Jr
Strickland, John Willis
Swanson, Donald Charles
Tanner, Joseph Jarratt
Thompson, Samuel, III
Townley, John Lewis, III
Tucker, Paul William
Turner, Daniel Stoughton
Wheeler, Robert Reid
Woodward, Harry W
Young, John Albion, Jr
Young, William Allen

Petrology

Aho, Aaro E
Allen, Gary Curtiss
Alper, Allen Myron
Amos, Dewey Harold
Anderson, Alfred Titus, Jr
Arculus, Richard J
Asquith, George Benjamin
Baars, Donald Lee
Baird, Alexander Kennedy
Barnes, Charles Winfred
Bateman, Paul Charles
Bickford, Marion Eugene, Jr
Blackburn, William Howard
Blatt, Harvey
Bloomer, Robert Oliver
Boettcher, Arthur Lee
Boles, James Richard
Bonnichsen, Bill
Boone, Gary M
Bothner, Wallace Arthur
Briggs, Louis Isaac, Jr
Brooks, Elwood Ralph
Bryan, Wilfred Bottrill
Burt, Donald McLain
Butler, James Robert
Campbell, John Arthur
Carlton, Richard Walter
Carman, John Homer
Carmichael, Ian Stuart
Carver, Robert E
Chayes, Felix
Chen, Chih Shan
Chown, Edward Holton
Clark, John
Cohen, Arthur David
Colberson, Kenneth David
Condie, Kent Carl
Connally, George Gordon
Cornwall, Henry Rowland
Crawford, Maria Luisa Buse
Crosby, Percy
Darby, Dennis Arnold
Davies, James Frederick
Dawson, James Clifford
Dengo, Gabriel
Deininger, Robert Wade
Dick, Henry Jonathan Biddle
Dickey, John Sloan, Jr
Dickinson, Kendell A
Dietrich, Richard Vincent
Dodd, Robert Taylor
Duesing, Constantin Michael
Duffield, Wendell Arthur
Eastwood, Raymond L
Edgar, Alan D
Ehlig, Perry Lawrence
Elliot, David Hawksley
Emerson, Donald Orville
Emslie, Ronald Frank
Ern, Ernest Henry
Ernst, Wallace Gary
Essene, Eric J
Evans, Bernard William
Fairbairn, Harold Williams
Folk, Robert Louis
Ford, Arthur B
Friedman, Gerald Manfred
Frye, Charles Isaac
Fudali, Robert F
Garlick, George Donald
Gavasci, Anna Teresa
Ghaly, Tharwat Shahata
Gilbert, Murray Charles
Glassley, William Edward
Godchaux, Martha Miller
Gold, David Percy

Granata, Walter Harold, Jr
Grant, James Alexander
Grant, Sheldon Kerry
Greenwood, Hugh J
Gresens, Randall Lee
Grew, Edward Sturgis
Griffiths, John Cedric
Guidotti, Charles V
Hagni, Richard D
Hanson, Gilbert N
Harvey, Richard David
Harwood, David Smith
Hathway, Richard Brackett
Hay, Richard Le Roy
Hays, James Fred
Heald, Milton Tidd
Hermes, O Don
Hess, David Filbert
Hess, Paul C
Hewins, Roger Herbert
Higgins, Ralph Edward
Himmelberg, Glen Ray
Hodge, Dennis
Hoge, Harry Porter
Holdaway, Michael Jon
Hollister, Lincoln Steffens
Holloway, John Requa
Houston, Robert S
Howland, Arthur Lloyd
Hsu, Liang-Chi
Huebner, John Stephen
Hughes, Charles James
Hyndman, Donald William
Jacob, Arthur Frank
James, Richard Stephen
Jennings, Ted Vernon
Johnson, Clayton Henry, Jr
Jolly, Janice Laurene Willard
Jolly, Wayne Travis
Kamilli, Diana Chapman
Kay, Suzanne Mahlburg
Kays, M Allan
Keil, Klaus
Keith, Brian Duncan
Kepper, John C
Kerrick, Derrill M
Kirkpatrick, Robert James
Klein, Cornelis
Koepnick, Richard Borland
Kohland, William Francis
Kretz, Ralph
Kudo, Albert Masakiyo
Kuellmer, Frederick John
Kuntz, Mel Anton
Lambert, Richard St John
Laurin, Andre Frederic
Lee, Kwang-Yuan
Liese, Homer C
Lipin, Bruce Reed
Loney, Robert Ahlberg
Longshore, John David
Loomis, Timothy Patrick
Lowe, Kurt
Ludman, Allan
Lumbers, Sydney Blake
Lyons, Paul Christopher
MacGregor, Ian Duncan
Malcuit, Robert Joseph
Martin, Charles Wellington, Jr
Martin, DeWayne
Mathews, Geoffrey William
Mattinson, James Meikle
McBirney, Alexander Robert
McBride, Earle Francis
McCormick, George R
McDowell, John Parmelee
McNutt, Robert Harold
McTaggart, Kenneth C
Medaris, L Gordon, Jr
Melson, William Gerald
Meyer, Joachim Dietrich
Miyashiro, Akiho
Moore, John Marshall, Jr
Moores, Eldridge Morton
Morrison, Donald Allen
Morse, Stearns Anthony
Mossop, Grant Dilworth
Moussa, Mounir Tawfik
Murray, Marc Michael
Murray, Raymond Carl
Mutis-Duplat, Emilio
Nagle, Frederick, Jr
Namy, Jerome Nicholas
Nash, William Purcell
Neal, William Joseph
Newton, Robert Chaffer
Nichols, Kathryn Marion
Odom, Ira Edgar
Page, Norman J
Pajari, George Edward
Payton, Charles Ellis
Pearce, Thomas Hulme
Perkins, Ronald Dee
Phair, George
Philpots, Anthony Robert

Picard, M Dane
Pilkington, Harold Dean
Piotrowski, Joseph Martin
Pittman, Edward N
Plummer, Charles Carlton
Powell, Benjamin Neff
Presnall, Dean C
Prinz, Martin
Puffer, John H
Quinn, Alonzo Wallace
Ragland, Paul C
Raymond, Loren Arthur
Reams, Max Warren
Reitan, Paul Hartman
Reynolds, Robert Coltart, Jr
Riggs, Karl A
Robinson, Peter
Roeder, Peter Ludwig
Roepke, Harlan Hugh
Rubel, Daniel Nicholas
Rutherford, Malcolm John
Schenk, Paul Edward
Schwarcz, Henry Philip
Scofford, David Matteson
Seyfert, Carl K Jr
Sheridan, Michael Francis
Sibley, Duncan Fawcett
Siever, Raymond
Sigurdsson, Haraldur
Silver, Leon Theodore
Simkin, Thomas Edward
Simmons, William Bruce, Jr
Simpson, Dale R
Skinner, William Robert
Smith, Eugene Irwin
Smith, Terence E
Smithson, Scott Busby
Smoke, Arthur Wilmot
Sood, Manmohan K
Speidel, David H
Stanonis, Francis Leo
Stephens, Maynard Moody
Stevenson, Ralph Girard, Jr
Stormer, John Charles, Jr
Strahl, Erwin Otto
Sumartojo, Jojok
Swanson, Samuel Edward
Swinney, Chauncey Melvin
Taubeneck, William Harris
Taylor, Edward Morgan
Taylor, G Jeffrey
Taylor, Hugh P Jr
Temyson, Marilyn Elizabeth
Tewhey, John David
Thompson, Alan Bruce
Thompson, James Burleigh, Jr
Thornton, Tommy Burt
Thornton, Charles Perkins
Treves, Samuel Blain
Turner, Francis John
Valentine, Wilbur Goodrich
Wadsworth, William Bingham
Wahlstrom, Ernest Eugene
Waiawender, Michael John
Walker, David
Wallace, Chester Alan
Ward, William Cruse
Weiblen, Paul Willard
Weill, Daniel Francis
White, Stanton M
White, William Arthur
Whitney, Philip Roy
Wiebe, Robert A
Wilband, John Truax
Wilbanks, John Randall
Wilshire, Howard Gordon
Wise, William Stewart
Wohlford, Duane Dennis
Wones, David R
Woo, Ching Chang
Wright, Thomas L
Yoder, Hatten Schuyler
Young, Davis Alan
Young, Steven Wilford
Zen, E-An

Photogeology

Collins, Lorence Gene
Lowman, Paul Daniel, Jr
Taranik, James Vladimir

Physical Geology

Bard, Robert Charles
Barlow, James A, Jr
Bowen, Richard Lee
Calver, James Lewis
Eaton, Gordon Pryor
Gilbert, Charles Mervin
Graham, Jack Bennett
Ivey, Marvin
Montagne, John M
Murphy, Michael Joseph
Olive, Wilds Williamson
Osterwald, Frank William
Rice, William Abbott

Ross, Alex R
Russell, Martin
Snyder, John L
Wingard, Paul Sidney
Woodruff, Charles Marsh, Jr

Quaternary Geology

Armstrong, John Edward
Baker, Richard Graves
Beck, Henry V
Borns, Harold William, Jr
David, Peter P
Dort, Wakefield, Jr
Dreimanis, Aleksis
Elson, John Albert
Farrand, William Richard
Fullerton, David Stanley
Gorman, William Alan
Hallberg, George Robert
Hamilton, Thomas Dudley
Johnson, William Hilton
Lackey, Laurence
Lambert, Paul Wayne
Lasca, Norman P, Jr
Linkletter, George Onderdonk
Maher, Louis James, Jr
Melhorn, Wilton Newton
Miller, Clifford Daniel
Newman, Walter S
Pierce, Kenneth Lee
Porter, Stephen Cummings
Reger, Richard David
Richmond, Gerald Martin
Rutter, Nathaniel Westlund
Semken, Holmes Alford, Jr
Sims, John David
Smith, George Irving
Steen-McIntyre, Virginia Carol
Ten Brink, Norman Wayne
Washburn, Albert Lincoln
Weber, William Mark
Williams, Richard Sugden, Jr
Zumberge, James Herbert

Regional Geology

Blank, Horace Richard, Jr
Bond, John Gilbert
Bryant, Bruce Hazelton
Dellwig, Louis Field
Eade, Kenneth Edgar
Glover, Lynn, III
Harrison, Jack Edward
McKee, Bates
Muller, Jan Engelbert
Norman, Carl Edgar
Pincus, Howard Jonah
Pratt, Howard Riley
Riger, James Aloysius
Spetzler, Hartmut August Werner

Rock Mechanics

Brown, Jim McCaslin
Coursen, David Linn
Dunn, David Evan
Friedman, Melvin
Heard, Hugh Corey
Judd, William Robert
Poole, William Hope

Sedimentology

Abbott, Patrick Leon
Ahlbrandt, Thomas Stuart
Andrichuk, John Michael
Basan, Paul Bradley
Bebout, Don Gray
Belt, Edward Scudder
Boosman, Jaap Wim
Bowman, James Floyd, II
Briggs, Garrett
Caty, Jean Louis
Charlesworth, Lloyd James, Jr
Coch, Nicholas Kyros
Cotter, Edward
Council, Richard J
Cross, Timothy Aureal
Dawson, James Clifford
Dean, Walter E, Jr
Dott, Robert Henry, Jr
Droste, John Brown
Ehrlich, Robert
Ethridge, Frank Guide
Eugster, Hans Peter
Fields, Robert William
Flores, Romeo M
Frakes, Lawrence Austin
Fraser, Gordon Simon
Friedman, Gerald Manfred
Fruth, Lester Sylvester, Jr
Gauri, Kharaiti Lal
Ginsburg, Robert Nathan
Greenwood, Brian
Halley, Robert Bruce
Hand, Bryce Moyer
Hanson, Henry W. A, III
Harper, John David
Henniger, Bernard Robert
Hesse, Reinhard

Hoskin, Charles M
Jackson, Roscoe George, II
Jones, James Ogden
Kanes, William H
Kehler, Philip Leroy
Klein, George deVries
Kocurko, Michael John
Krinsley, David
Lawson, David Edward
Lee, Kwang-Yuan
Lowe, Donald Ray
Lowright, Richard Henry
Mackenzie, Frederick Theodore
Manos, Constantine T
Martini, Ireneo Peter
Matthews, Robley Knight
McCubbin, Donald Gene
McFarlan, Edward, Jr
McKee, James W
McMullen, Robert Michael
Milling, Marcus Eugene
Moiola, Richard James
Morton, Robert Alex
Mossop, Grant Dilworth
Naidu, Angi Satyanarayan
Nelson, Bruce Warren
Oertel, George Frederick
Ore, Henry Thomas'
Orvos, Ervin George
Owens, Edward Henry
Payne, Myron William
Pirie, Robert Gordon
Pittman, Edward D
Rosen, Norman Charles
Rosenfeld, Melvin Arthur
Rukavina, Norman Andrew
Russell, Richard Dana
Savage, E Lynn
Schneidermann, Nahum
Scholle, Peter Allen
Shevenell, Frederic Lyon
Shevenell, Thomas Cortland
Shideler, Gerald Lee
Shrock, Robert Rakes
Sims, John David
Slatt, Roger Malcolm
Smith, Norman Dwight
Swanson, Donald Charles
Swift, Donald J P
Thompson, Allan M
Wanless, Harold Rogers
Watson, Richard Lee
Weaver, Charles Edward
Whisonant, Robert Clyde
Wilkinson, Bruce H
Wolfbauer, Claudia Ann
Wolff, Manfred Paul
Woodrow, Donald L
Yole, Raymond William
Young, Grant McAdam

Soil Classification

Frazee, Charles Joseph
Hajek, Benjamin F
Peterson, Frederick Forney
Pomerening, James Albert
Soto, Gerardo H
Yahner, Joseph Edward

Soil Genesis

Allen, Bonnie L
Cunningham, Robert Lester
Daniels, Raymond Bryant
Fenton, Thomas E
Foss, John E
Franzmeier, Donald Paul
Gile, Leland Henry
Hutchings, Theron Bird
McDole, Robert E
McKeague, Justin Alexander
Parsons, Roger Bruce

Speleology

Christiansen, Kenneth Allen
Nicholas, Gerardus
Sullivan, Nicholas

Stratigraphy

Avcin, Matthew John, Jr
Baars, Donald Lee
Baillie, Andrew Dollar
Bassett, Henry Gordon
Beaulieu, John David
Block, Douglas Alfred
Boellstorff, John David
Bond, John Gilbert
Boone, Peter Augustine
Branson, Carl Colton
Braun, Willi Karl
Braunstein, Jules
Bremen, Daniel Joseph
Brent, William B
Briggs, Darinka Zigic
Catacosinos, Paul Anthony
Chen, Chih Shan
Chenoweth, Philip Andrew
Clark, Thomas Henry
Coates, Anthony George

Colburn, Ivan Paul
Copeland, Charles Wesley, Jr
Craig, Lawrence Carey
Cramer, Howard Ross
Donahue, Jack David
Doumani, George Alexander
Fellows, Larry Dean
Ford, John Philip
Fournier, George Richard
Franklin, James McWillie
Frantz, Wendelin R
Frye, Charles Isaac
Gentile, Richard J
Gerhard, Lee C
Goodwin, Peter Warren
Granata, Walter Harold, Jr
Gutstadt, Allan Morton
Hale, Lyle A
Hanson, Alvin Maddison
Harker, Peter
Harper, John David
Harrington, Jonathan W
Hart, Richard Royce
Hawley, David
Head, James William, III
Helfrich, Charles Thomas
Heller, Robert Leo
Helton, Walter Lee
Henniger, Bernard Robert
Hinman, Eugene Edward
Hooks, William Gary
Horne, Gregory Stuart
Howard, James F
Howe, Herbert James
Huff, William J
Jones, James Ogden
Kanes, William H
Kehler, Philip Leroy
Keller, Allen S
Kent, Harry Christison
Klepser, Harry John
Koch, Donald Leroy
Kolata, Dennis Robert
Kopf, Rudolph William
Kraft, John Christian
Kurtz, Vincent E
Lamb, George Marion
Lewis, George Edward
Liberty, Bruce Arthur
Long, William Ellis
McWilliams, Robert Gene
Metz, Robert
Mintz, Leigh Wayne
Mountjoy, Eric W
Moussa, Mounir, Tawfik
Murphy, Michael A
Murray, Frederick Nelson
Namy, Jerome Nicholas
Nelson, Samuel James
Nichols, Kathryn Marion
Nodine-Zeller, Doris Eulaia
Nordeng, Stephan C
Olsen, Rex E
Osborne, Robert Howard
Owen, Donald Edward
Palmer, Katherine Van Winkle
Pipiringos, George Nicholas
Pitrat, Charles William
Puri, Harbans Singh
Rall, Elizabeth Pretzer
Reso, Anthony
Rickard, Lawrence Vroman
Roberts, Thomas Glasdir
Rose, Peter R
Ryder, Robert Thomas
Savage, Norman Michael
Schafroth, Don W
Schenk, Paul Edward
Silberling, Norman John
Sloan, Robert Evan
Smosna, Richard Allan
Spencer, Randall Scott
Spink, Walter John
Spreng, Alfred Carl
Stearn, Colin William
Steele, Grant
Stokes, William Lee
Sturgeon, Myron Thomas
Suhm, Raymond Walter
Tesmer, Irving Howard
Thompson, Samuel, III
Trexler, John Peter
Van Couvering, John Anthony
Vokes, Harold Ernest
Vosburg, David Lee
Walker, Kenneth Russell
Wallace, Chester Alan
Wang, Kia K
Washburn, Robert Henry
Weart, Richard Claude
Welby, Charles William
Welder, Frank A
Wheeler, Harry Eugene
Whisonant, Robert Clyde
Wilde, Garner Lee
Wilson, Philo Calhoun
Wolleben, James A
Woodrow, Donald L

Williams, Lyman O
Woodland, Bertram George
Woodward, Lee Albert
Wright, William Herbert, III
Younce, Gordon Baldwin

Volcanology

Alvarez, Walter
Anderson, Alfred Titus, Jr
Bullard, Fred Mason
Coats, Robert Roy
Elston, Wolfgang Eugene
Jolly, Wayne Travis
Krushensky, Richard D
Kudo, Albert Masakiyo
McKay, David Stewart
Miller, Clifford Daniel
Rankin, Douglas Whiting
Rose, William Ingersoll, Jr
Rubel, Daniel Nicholas
Sheridan, Michael Francis
Sigurdsson, Haraldur
Smith, Alan Lewis
Smith, Robert Leland
Stoiber, Richard Edwin
Treves, Samuel Blain
Ward, Peter Langdon

Structural Geology

Ave Lallemant, Hans Gerhard
Babcock, Elkanah Andrew
Baer, Alec Jean
Baker, David Warren
Barnes, Charles Winfred
Bateman, Paul Charles
Brent, William B
Brocoum, Stephen John
Brown, Severn Parker
Burford, Arthur Edgar
Burger, Henry Robert, III
Bursnall, John Treharne
Cardwell, Dudley H
Cebull, Stanley Edward
Chapple, William Massee
Chown, Edward Holton
Christie, John McDougall
Colburn, Ivan Paul
Craddock, (John) Campbell
Crosby, Gary Wayne
Cserna, Eugene George
Dalziel, Ian William Drummond
Davidson, Donald Miner, Jr
Davis, George Herbert
Dengo, Gabriel
Dennis, John Gordon
Dewey, John Frederick
Dillon, John Thomas
Dover, James H
Duffield, Wendell Arthur
Dunn, David Evan
Edwards, Jonathan, Jr
Elliott, David
Faill, Rodger Tanner
Farrington, William Benford
Frey, Maurice G
Friedman, Melvin
Goddard, Edwin Newell
Gries, John Charles
Hale, Lyle A
Haller, John
Hamil, Martha M
Hardy, Clyde Thomas
Hatch, Norman Lowrie, Jr
Hatcher, Robert Dean, Jr
Hawley, David
Jennings, Ted Vernon
Keller, Allen S
Kennedy, Michael John
Kopf, Rudolph William
Kukolich, Stephen George
Kupfer, Donald Harry
Laurin, Andre Frederic
Logan, John Merle
Loney, Robert Ahlberg
Long, William Ellis
Lovejoy, Earl Mark Paul
Lowell, James Diller
MacDonald, William David
McCullough, Edgar Joseph, Jr
Milligan, George Clinton
Moores, Eldridge Morton
Murray, Frederick Nelson
Nelson, Richard B
Nickelsen, Richard Peter
Norman, Carl Edgar
Oertel, Gerhard Friedrich
Oriel, Steven S
Parker, Ronald Bruce
Pierce, William G
Pincus, Howard Jonah
Pratt, Howard Riley
Price, Raymond Alex
Proctor, Paul Dean
Prucha, John James
Ragan, Donal Mackenzie
Rance, Hugh
Raymond, Loren Arthur
Richard, Benjamin H
Riva, John
Roper, Paul James
Rousell, Don Herbert
Sales, John Keith
Schafroth, Don W
Schwerdtner, Walfried Martin
Seeger, Charles Ronald
Seyfert, Carl K, Jr
Skinner, William Robert
Spink, Walter John
Starkey, John
Stauffer, Mel R
Stevens, George Richard
Stoever, Edward Carl, Jr
Tobisch, Othmar Tardin
Tullis, Julia Ann
Tullis, Terry Edson
Underwood, James Ross, Jr
Walker, Roger Geoffrey
Washburn, Robert Henry
Wenk, Hans-Rudolf
Wilbanks, John Randall

Yole, Raymond William
Young, Frederick Griffin
Young, Grant McAdam
Ziegler, Alfred M

GEOPHYSICS

Geophysics

Ab Iorwerth, Hefin
Acheson, Cyrus Harold
Adams, William Mansfield
Adkins, John Nathaniel
Ahman, Moid Uddin
Ahrens, Thomas J
Aldrich, Lyman Thomas
Aldridge, Keith Douglas
Algermissen, Sylvester Theodore
Alksne, Alberta Yearian
Alldredge, Leroy Romney
Allenby, Richard John, Jr
Allingham, John Wing
Alsop, Leonard E
Anderson, Don Lynn
Anderson, Orson Lamar
Angona, Frank Anthony
Arce, Jose Edgar
Assur, Andrew
Backus, George Edward
Backus, Milo M
Bailey, Dana Kavanagh
Bainbridge, Arnold Ernest
Baker, William Laird
Balch, Alfred Hudson
Ball, Mahlon M
Balsley, James Robinson, Jr
Balton, Isidore Alfred
Banaugh, Robert Peter
Banerjee, Subir Kumar
Bardeen, Thomas
Barnes, David Fitz
Barringer, Anthony R
Bassett, William Akers
Bayhi, Joseph Franklin
Bean, Robert Jay
Beck, Alan Edward
Beck, Myrl Emil, Jr
Becker, Adrian Anthony
Bell, Maurice Evan
Bell, Peter M
Bentley, Charles Raymond
Benton, Edward Rowell
Berg, Eduard
Berg, Joseph Wilbur, Jr
Berkhout, Aart W J
Biehler, Shawn
Birch, Albert Francis
Blakely, Robert Fraser
Blanchard, Jonathan Ewart
Bledsoe, John D
Blum, Victor Joseph
Bodvarsson, Gunnar
Bollinger, Gilbert A
Bonini, William Emory
Booker, John Ratcliffe
Bostrom, Robert Christian
Boucher, Gary Wynn
Bradner, Hugh
Brant, Arthur Albert
Breiner, Sheldon
Bromery, Randolph Wilson
Brooke, John Percival
Brown, Maurice Vertner
Brown, Robert Henry
Brown, Robert Lawrence
Brune, James N
Brynjolfsson, Ari
Bull, Colin Bruce Bradley
Bullard, Edward Crisp
Bullard, Edwin Roscoe, Jr
Bunce, Elizabeth Thompson
Burke, Kenneth B S
Burrows, John Ronald

GEOPHYSICS

Byerlee, James Douglas
Byerly, Perry Edward
Cabaniss, Gerry Henderson
Cain, Joseph Carter
Cantwell, Thomas
Carmichael, Robert Stewart
Carr, Jerome Brian
Carter, Neville Louis
Case, James Edward
Cathey, Everett Henry
Cathles, Lawrence MacLagan, III
Cheney, Monroe G
Chiburis, Edward Frank
Chinnery, Michael Alistair
Chisholm, Roderick G
Chowdhury, Dipak Kumar
Christensen, Nikolas Ivan
Christian, Wayne Gillespie
Chubb, Talbot Albert
Clark, Howard Charles, Jr
Clark, Sydney P, Jr
Clay, Clarence Samuel
Clayton, Neal
Clement, William Glenn
Clough, John Wendell
Coe, Robert Stephen
Collett, Leonard Stanier
Cook, John Call
Cook, Kenneth Lorimer
Cook, Richard Kaufman
Corbato, Charles Edward
Costain, John Kendall
Costanino, Marc Shaw
Coster, Hendrik Paulus
Couch, Richard W
Cox, Allan Verne
Crary, Albert Paddock
Crosby, Gary Wayne
Crossley, David John
Crosson, Robert Scott
Crowe, Christopher
Czapski, Ulrich Hans
Dahlen, Francis Anthony, Jr
Dandekar, Balkrishna S
Danes, Zdenko Frankenberger
Darby, Edsel Kenneth
Dash, Bibhu Prasad
Davis, Briant LeRoy
Davis, Thomas Neil
Davis, Willard E
DeBoer, Jelle
De Bremaecker, Jean-Claude
Decker, Robert Wayne
DeGasparis, Aurelio Alfonso A
Dehlinger, Peter
DeNoyer, John M
Deutsch, Ernst Robert
Dewart, Gilbert
Diment, William Horace
Dix, Charles Hewitt
Dobecki, Thomas Lee
Dobrin, Milton Burnett
Doell, Richard Rayman
Doig, Ronald
Donoho, Horrie Van Waldo
Dorman, Henry James
Dorman, Leroy Myron
Dosso, Harry William
Doty, William Earl Neal
Douthit, Thomas D Nathan
Dowling, Forrest Leroy
Dowling, John J
Drake, Charles Lum
Driver, Edgar Steward
Du Bois, Robert Lee
Dunlap, Henry Francis
Eaton, Jerome F
Eckhardt, Donald Henry
Ellis, Robert Malcolm
Elsasser, Walter M
Engdahl, Eric Robert
England, Anthony W
Everenden, Jack Foord
Ewing, John I
Fahlquist, Davis A
Farnham, Paul Rex
Farr, John B
Faul, Henry
Feagin, Frank J
Fergusson, Gordon John
Ferris, Craig
Fertl, Walter Hans
Fett, John D
Fillippone, Walter R
Fink, Don Roger
Fitzpatrick, Michael Morson
Fitzpatrick, Robert Charles
Fleischer, Robert Louis
Flinn, Edward Ambrose
Fogelson, David Eugene
Forester, Robert Donald
Foster, Manus R
Fougere, Paul Francis
Fraser-Smith, Antony Charles
Freden, Stanley Charles
Fredericks, Robert W
Frey, John H
Frischknecht, Frank C

Frisillo, Albert Lawrence
Fuller, Michael D
Gage, Kenneth Seaver
Galbraith, James Nelson, Jr
Gangi, Anthony Frank
Gans, Roger Frederick
Garland, George David
Gee, Susan McMillin
Geldart, Lloyd
Gentry, Robert Vance
Gerdel, Robert Wallace
Gerlach, Alan Meyer
Gibb, Richard A
Gibbs, James Freeman
Gimlett, James I
Ginsburg, Merrill Stuart
Goetz, Alexander Franklin Hermann
Goetze, Christopher
Goforth, Thomas Tucker
Gordon, Robert Boyd
Gough, Denis Ian
Gouin, Pierre Laurier
Green, Harry Western, II
Greene, Gordon William
Greenewalt, David
Greenfield, Roy Jay
Gregory, Carter H
Grine, Donald Reaville
Griscom, Andrew
Grobecker, Alan J
Gross, Gerardo Wolfgang
Grossing, Bernardo Fruedenburg
Gupta, Indra Narayan
Hales, Anton Linder
Hall, David Joseph
Haubrich, Richard August
Hauge, Paul Stephen
Hawkins, James Edward
Hays, James Fred
Hays, John Robert
Heacock, Richard Ralph
Healy, John H
Heinrich, Ross Raymond
Heinrichs, Donald Frederick
Heirtzler, James Ransom
Helsley, Charles Everett
Henderson, Roland George
Heppner, James P
Hermance, John Francis
Heroy, William Bayard, Jr
Herron, Eugene Thornton, Jr
Herron, Thomas J
Herzog, Leonard Frederick, II
Hessler, Victor Peter
Hicks, Grady Thomas
Higgs, Robert Hughes
Hilchie, Douglas Walter
Hill, David Paul
Hill, Donald Gardner
Hinze, William James
Hirshman, Julius
Hodge, Dennis
Hoekstra, Pieter
Hoffer, Abraham
Hoke, John Humphreys
Holmes, Charles Robert
Hoover, Donald Brunton
Hopkins, George H, Jr
Hopkins, John Raymond
Horai, Ki-iti
Horita, Robert Eiji
Hoskins, Hartley
Hovey, Richard Dean
Howell, Benjamin F, Jr
Hsui, Albert Tong-Kwan
Hubbert, Marion King
Huebsch, Ian O
Hurley, Neal Lilburn
Hyndman, Roy D
Innes, Morris James Sage
Ives, Ronald Lorenz
Iyer, Hariharaiyer Mahadeva
Jackson, David Diether
Jackson, Philip Larkin
Jacobs, Stanley J
James, David Evan
Jamieson, John Calhoun
Jessop, Alan Michael
Johns, Roman Karol Chrzaszcz
Johnson, Hamilton McKee
Johnson, James Franklin
Johnson, John Charles
Johnson, Leonard Evans
Johnson, Robert W, Jr

Johnson, William W
Joiner, Harry Francis, II
Jones, Hal Joseph
Jones, Stanley Bennett
Jordan, Thomas Hillman
Joyner, William B
Kaarsberg, Ernest Andersen
Kahle, Anne Bettine
Kanasewich, Ernest Raymond
Kane, Martin Francis
Kaplan, Joseph
Kaplan, Lewis David
Katz, Samuel
Kaufman, Sidney
Kaula, William Mason
Keen, Michael J
Kehle, Ralph Otmar
Kellerstrass, Ernst Junior
Kennedy, George Clayton
Kennedy, James Francis
Kern, John W
Kibler, Kenneth G
Kim, Daniel Y
King, Chi-Yu
King, Elizabeth Raymond
Kirkham, Don
Kissinger, Carl
Kleinkopf, Merlin Dean
Knaflich, Helen B
Knopoff, Leon
Kolstad, George Andrew
Kothe, Kenneth Ralph
Krivoy, Harold Lloyd
Kuiper, Logan Keith
Kunze, Adolf Wilhelm Gerhard
Kuo, John Tsung (Fen)
Lachenbruch, Arthur Herold
La Fehr, Thomas Robert
Lamar, Donald Lee
La Mori, Phillip Noel
Lander, James French
Landrum, Ralph Avery, Jr
Langel, Robert Allan
Langseth, Marcus G, Jr
Lansinger, John Marcus
Lanzerotti, Louis John
Larson, Edwin E
Larson, Jerome Valjean
Laster, Stanley Jerral
Latham, Gary V
Lavin, Peter Masland
Lawrence, Philip Linwood
Leblanc, Gabriel
Lee, Jean T
Lee, Tien-Chang
Lee, William Hung Kan
Leith, Thomas Henry
Leonard, Robert Stuart
Lettau, Heinz Helmut
Lewis, Trevor John
Liebermann, Robert C
Lincoln, Jeannette Virginia
Linde, Alan Trevor
Lippitt, Louis
Lister, Clive R B
Lokken, John Erwin
Long, Leland Timothy
Longacre, William Atlas
Low, Robert James
Lowell, Robert Paul
MacDonald, Gordon James Fraser
MacKenzie, Glenn S
Madden, Theodore Richard
Magoran, Thomas R
Maier, Eugene Jacob Rudolph
Malahoff, Alexander
Maloof, Giles Wilson
Manchee, Eric Best
Manghnani, Murli Hukumal
Mansinha, Lalatendu
Martin, James Milton
Martner, Samuel (Theodore)
Mason, Ronald George
Mateker, Emil Joseph, Jr
Mather, Keith Benson
Matsushita, Sadami
Matsumoto, Tosimatu
Maxwell, James Christie
Mayne, William Harry
McConnell, Robert Kendall
McCormack, Harold Robert
McEuen, Robert Blair
McEvilly, Thomas V
McGinnis, Lyle David
McMurry, Earl William
McPherron, Robert Lloyd
Mead, Gilbert Dunbar
Mead, Judson
Mercado, Edward J
Mereu, Robert Frank
Merrill, Ronald Thomas
Merritt, Melvin Leroy
Meyer, Richard Ernst
Meyers, Herbert
Miatech, Gerald James
Michael, William Herbert, Jr
Miles, John Wilder
Miller, Edward Titus

Miller, Max K
Milne, Allen Ritchie
Minear, John W
Mitchell, Brian James
Mogro-Campero, Antonio
Mooney, Harold Morton
Morgan, William Jason
Morris, Gerald Brooks
Morrison, Huntly Frank
Mossman, Reuel Wallace
Mudie, John David
Mukherji, Kalyan Kumar
Musgrave, Albert Wayne
Mut, Stuart Creighton
Nabighian, Misac N
Nafe, John Elliott
Neidell, Norman Samson
Nestvold, Elwood Olaf
Niblett, Edward Ronald
Noltimier, Hallan Costello
Norden, John Alexander
Northrop, John
Nosal, Eugene Adam
Nur, Amos M
Nuttli, Otto William
Obradovich, John Dinko
Odishaw, Hugh
O'Keefe, John Aloysius
Oliver, Jack Ertle
Olsen, Kenneth Harold
Opdyke, Neil
Orange, Arnold
Orlanski, Isidoro
Orlin, Hyman
Page, Robert Alan, Jr
Pakiser, Louis Charles, Jr
Palmer, H Currier
Palmer, James E
Pan, Poh-Hsi
Parks, George Kung
Pavey, George Madison, Jr
Pawlowicz, Edmund F
Peeples, Wayne Jacobson
Pekeris, Chaim Leib
Peltier, William Richard
Peoples, James A, Jr
Perkins, William Enfield
Perret, William Riker
Peter, George
Petersen, Jack Warren
Petersen, Carl Frank
Peterson, Donald Neil
Petty, Scott, Jr
Pfluke, John H
Phillips, Joseph D
Phinney, Robert A
Pickett, George R
Pilant, Walter L
Pollack, Henry Nathan
Porter, Lawrence Delpino
Posmentier, Eric S
Press, Frank
Raitt, Russell Watson
Raleigh, Cecil Baring
Ramirez, Jesus Emilio
Rankin, David
Rasmussen, William Otto
Ravindra, Ravi
Raymond, Charles Forest
Rechtien, Richard Douglas
Redmond, John Charles
Reeves, Robert Grier (Lefevre)
Regan, Robert David
Reiter, Marshall Allan
Revetta, Frank Alexander
Rice, Robert Bruce
Richard, Benjamin H
Riecker, Robert E
Robinson, Allan Richard
Robinson, Edwin S
Robinson, Enders Anthony
Robinson, James H
Robinson, William John
Rochester, Michael Grant
Roddy, David John
Romig, Philip Richardson
Ross, John Creighton
Ross, David I
Ross, Howard Persing
Roy, Robert Francis
Rudman, Albert Julius
Runge, Richard John
Rupert, Gerald Bruce
Russell, Christopher Thomas
Russell, Richard Doncaster
Ryan, Jack A
Sacks, I Selwyn
St Amand, Pierre
Salisbury, Matthew Harold
Sanford, Allan Robert
Sangree, John Brewster, Jr
Sass, John Harvey
Sauck, William August
Saull, Vincent Alexander
Savage, James Crampton
Savit, Carl Hertz
Sax, Robert Louis
Scharon, Harry Leroy

Scheidegger, Adrian Eugen
Schilling, Gerhard Friedrich
Schmidt, Victor Adolf
Schneider, William Aeppli
Schock, Robert Norman
Schoenberger, Michael
Scholz, Christopher Henry
Schreiber, Edward
Schubert, Gerald
Schwerdtfeger, Werner
Sclater, John George
Scott, James Henry
Secor, Donald Terry, Jr
Seeger, Charles Ronald
Seguin, Maurice Krisholm
Sendlein, Lyle V A
Seriff, Aaron Jay
Shapiro, James Norman
Shapley, Alan Horace
Sherwood, John William Charles
Shillibeer, Harry Albert
Shive, Peter Northrop
Sigal, Richard Frederick
Sill, William Robert
Silverman, Sam M
Simmons, Gene
Simon, Ruth B
Singer, Siegfried Fred
Singh, Harinder
Skiles, Durward DeWitt
Slack, Howard Addison
Slemmons, David Burton
Slichter, Louis Byrne
Small, John, Jr
Smith, Bruce Dyfrig
Smith, Leslie Garrett
Smith, Newbern
Smith, Robert Baer
Smith, Stewart W
Smith, Thomas Jefferson
Smithson, Scott Busby
Smylie, Douglas Edwin
Snyder, Howard Arthur
Solomon, Sean Carl
Souder, Wallace William
Spall, Henry Roger
Speiser, Theodore Wesley
Spencer, Joseph Walter
Spencer, Terry Warren
Spilhaus, Athelstan Frederick, Jr
Srivastava, Surat Prasad
Stanley, Glenn M
Stauder, William
Steele, William Kenneth
Steeples, Donald Wallace
Steinhart, John Shannon
Stephenson, Lee Palmer
Sterrett, Kay Fife
Stocker, Richard Louis
Stone, David B
Straley, H W, III
Strangway, David W
Stratton, Everett Franklin
Strick, Ellis
Sturgul, John Roman
Sumner, John Randolph
Sumner, Roger D
Sun, James Ming-Shan
Suppe, John Edward
Sutton, George Harry
Svendsen, Kendall Lorraine
Sykes, Lynn Ray
Symons, David Thorburn Arthur
Takahashi, Taro
Talwani, Manik
Tanner, Allan Bain
Tanner, James Gordon
Telford, William Murray
Teng, Ta-Liang
Thapar, Mangat Rai
Thiruvathukal, John Varkey
Thomas, Michael Durst
Thome, George Durst
Thompson, George Albert
Thompson, Loren Edward, Jr
Thompson, William Baldwin
Thomson, Ker Clive
Tilley, Aubra Everett
Tixier, Maurice Pierre
Todd, Terrence Patrick
Toksoz, Mehmet Nafi
Tolstoy, Ivan
Toth, Jozsef
Treitel, Sven
Tripp, Russell Maurice
Tryggvason, Eysteinn
Tsao, Peter
Tullis, Terry Edson
Tuman, Vladimir Shlimon
Turcotte, Donald Lawson
Uffen, Robert James
Ulrych, Tadeusz Jan
Unger, John Duey
Vacquier, Victor
Vajk, Raoul
Vanicek, Petr
van der Voo, Rob
Van Nostrand, Robert Gaige

Verhoogen, John
Verosub, Kenneth Lee
Viksne, Andy
Vincenz, Stanislaw Aleksander
Vitousek, Martin J
Vozoff, Keeva
Wadsworth, Donald van Zelm
Waff, Harve S
Wahl, William George
Walker, James Callan Gray
Wall, Robert Ecki
Walsh, Joseph Broughton
Walter, Edward Joseph
Wang, Herbert Fan
Ward, Stanley Harry
Warren, David Henry
Wasserburg, Gerald Joseph
Watkins, Joel Smith, Jr
Watkins, Norman David
Watson, Kenneth
Watson, Robert Joseph
Weart, Wendell D
Weaver, John Scott
Weaver, John Trevor
Weber, Jean Robert
West, Gordon Fox
Wetherill, George West
White, James Edward
Whitham, Kenneth
Wideman, Charles James
Widess, Moses B
Wiggins, Ralph Ambrose
Williams, David Lee
Williams, Owen Wingate
Williams, Philip Sidney
Willis, David Edwin
Willis, Eric Herbert
Wilson, Charles R
Wilson, James Tinley
Wilson, John Tuzo
Wong, How-Kin
Wood, Fergus James
Woods, John Price
Woollard, George Prior
Worth, James Judson Blackley
Worzel, John Lamar
Wright, James Arthur
Wuenschel, Paul Clarence
Wyble, D O
Wyckoff, Peter Hines
Wyder, John Ernest
Wyllie, Malcolm Robert Jesse
York, Derek H
Young, Richard Evans

Aeronomy
Akasofu, Syun-Ichi
Belon, Albert Edward
Biondi, Manfred Anthony
Bowhill, Sidney Allan
Carlson, Robert Warner
Chamberlain, Joseph Wyan
Chameides, William Lloyd
Chen, Abel Jer-Jiunn
Cicerone, Ralph John
Clark, Kenneth Courtright
Dandekar, Balkrishna S
Deepak, Adarsh
Degen, Vladimir
Donn, William L
Ferguson, Eldon Earl
Fogle, Benson Tarrant
Fremouw, Edward Joseph
Gattinger, Richard Larry
Gregory, John B
Heicklen, Julian Phillip
Hennessey, James J
Hines, Colin Oswald
Hoch, Richmond Joel
Hudson, Robert Douglas
Hunsucker, Robert Dudley
Jones, Arthur Vallance
Leonard, Robert Stuart
Lowe, Robert Peter
Marmo, Frederick Francis
Mauersberger, Konrad
McKinley, Donald William Robert
McNamara, Allen Garnet
Meriwether, John Williams, Jr
Meier, Robert R
Minzner, Raymond Arthur
Nagy, Andrew F
Narcisi, Rocco S
Olivero, John Joseph, Jr
Olson, Raymond Verlin
Paghis, Irvine
Peterson, Vern Leroy
Popoff, Ilia George
Porter, Hayden Samuel, Jr
Rees, Manfred Hugh
Romick, Gerald J
Sechrist, Chalmers Franklin, Jr
Shepherd, Gordon Greeley
Sivjee, Gulamabas Gulamhusen
Slowey, Jack William
Smith, Robert Earl
Swider, William, Jr

Turco, Richard Peter
Van Tassel, Roger A
Van Zandt, Thomas Edward
Vaughan, William Walton
Watanabe, Tomiya
Watson, Michael Douglas
Whitten, Robert Craig, Jr

Atmospheric Physics
Allee, Paul Andrew
Amme, Robert Clyde
Anger, Clifford D
Atreya, Sushil Kumar
Ballard, Harold Noble
Banks, Peter Morgan
Barasch, Guy Errol
Bauer, Ernest
Berry, Edwin X
Bhartendu
Birchfield, Gene Edward
Blais, Roger Nathaniel
Blau, Henry Hess, Jr
Borst, Walter Ludwig
Borucki, William Joseph
Brewer, Alan West
Brown, Robert Alan
Buck, Richard F
Bushnell, Robert Hempstead
Caffee, Robert Francis
Calvert, Wynne
Campbell, Wallace Hall
Carlon, Hugh Robert
Chahine, Moustafa Toufic
Chappell, Charles Franklin
Chiu, Chin-Shan
Chiu, Lue-Yung Chow
Chiu, Yam-Tsi
Christie, Alistair D
Chylek, Petr
Collis, Ronald Thomas
Cooney, John Anthony
Coulson, Kinsell Leroy
Cox, Stephen Kent
Czepyha, Chester George Reinhold
Das, Phanindramohan
Deirmendjian, Diran
Derr, Vernon Ellsworth
Dews, Edmund
Dolezalek, Hans
Donaldson, Robert Evans
Douglas, Richard Herbert
Drayson, Sydney Roland
East, Conrad
Eaton, Larry Rodney
Engelman, Arthur
Ezemenari, Fidel Rex Chukwuemeka
Fejer, Jules A
Feldman, Paul Donald
Fenstermacher, Robert Lane
Fitz, Harold Carlton, Jr
Fitzgerald, Donald Ray
Fitzgerald, James W
Fraser, Alistair Bisson
Freier, George David
Fritz, Richard Blair
Fukuta, Norihiko
Fullerton, Charles Michael
Geehern, Margaret Kennedy
Gibbons, Mathew Gerald
Glover, Kenneth Merle
Gokhale, Narayan Ramchandra
Goldan, Paul David
Goldberg, Richard Aran
Hall, Freeman Franklin, (Jr)
Hall, Richard Travis
Hamlin, Daniel Allen
Hanel, Rudolf A
Hardy, Kenneth Reginald
Harper, Robert M
Harris, Kent Karren
Heer, Raymond Robert, Jr
Hernandez, Gonzalo J
Hess, Wilmot Norton
Hicks, Bruce Boundy
Hobbs, Peter Victor
Hoffman, John Harold
Hofmann, David John
Honsaker, John Leonard
Hooke, William Hines
Hord, Charles W
Israel, Gerhard Wilhelm
Jastrow, Robert
Jurica, Gerald Michael
Justus, Carl Gerald
Katsaros, Kristina
Kim, Jai Soo
Kleis, William Delong
Konigsberg, Alvin Stuart
Kraakevik, James Henry
Kyle, Herbert Lee
List, Roland
Little, Charles Gordon
Lorenz, Philip Jack
MacCracken, Michael Calvin
Mandics, Peter Alexander
Markson, Ralph Joseph
Martin, John David

Maybank, John
McClatchey, Robert Alan
McCormick, Michael Patrick
McIntyre, Donald Patrick
McPeters, Richard Douglas
Megaw, William James
Mejia, Gaston Rene
Miller, August
Moe, Mildred Minasian
Mohnen, Volker A
Moore, Charles Bachman, Jr
Murcray, David Guy
Nagy, Andrew F
Neff, William David
Old, Thomas Eugene
Orgill, Montie M
Orville, Richard Edmonds
Pallmann, Albert J
Peek, Harry Milton
Penndorf, Rudolf
Peterson, Kendall Robert
Peterson, Lennart Rudolph
Philbrick, Charles Russell
Pike, Julian M
Plooster, Myron Nieveen
Pope, Joseph Horace
Potter, John Fred
Prag, Arthur Barry
Procunier, Richard Werner
Pueschel, Rudolf Franz
Randall, Charles McWilliams
Rao, Ganti Lakshminarayana
Reck, Ruth Annette
Remsberg, Ellis Edward
Reynolds, George William, Jr
Richmond, Arthur Dean
Robinson, George David
Roper, Robert George
Rudd, Millard Eugene
Ruhnke, Lothar Hasso
Ruskin, Robert Edward
Rust, Walter David
Ruttenberg, Stanley
Schwiesow, Ronald Lee
Scott, William Taussig
Semonin, Richard Gerard
Shah, Govindlal M
Shardanand
Shaw, Glenn Edmond
Shemansky, Donald Eugene
Shettle, Eric Payson
Shimazaki, Tatsuo
Shlanta, Alexis
Silverman, Bernard Allen
Sinclair, Peter C
Smith, Paul Letton, Jr
Stakutis, Vincent John
Stephens, James Briscoe
Stephens, Timothy Lee
Stern, Sidney Charles
Stewart, Richard Willis
Straus, Joe Melvin
Swenson, Gary Russell
Taffe, William John
Tarpley, Jerald Dan
Telford, James Wardrop
Thomas, Gary E
Tisone, Gary C
Todd, Edward Payson
Turner, Robert Elwood
Vali, Gabor
Van Zandt, Thomas Edward
Veirs, Val Rhodes
Vickers, William W
Volz, Frederic Ernst
Walton, John Joseph
Ward, Gray (Ganesh)
Weickmann, Helmut K
Weinberg, Jerry L
Weinreb, Michael Philip
Welch, Ronald Maurice
Whitney, Cynthia Kolb
Wilkening, Marvin Hubert
Wukelic, George Edward
Yarger, Douglas Neal
Young, Stephen James
Ziauddin, Syed

Exploration Geophysics
Anderson, Christian Donald
Applegate, James Keith
Barnes, Burton B
Clee, Thomas Edward
Cook, Ernest Ewart
Cronk, Caspar
Desai, Kantilal Panachand
Eisner, Elmer
Elliott, Sheldon Ellwood
Eyer, Jerome Arlan
Foulks, Sidney Marshall
French, William Stanley
Hilterman, Fred John
Holmer, Ralph Carrol
Hsu, I-Chi
Jennemann, Vincent Francis
Johnston, David
Larner, Kenneth Lee
Levin, Franklyn Kussel

177

Liu, Luke Lokia
Long, James Alvin
Marr, John Douglas
Meidav, Tsvi
O'Donnell, Thomas John
Reed, Dale Hardy
Rydelek, Paul Anthony
Ryu, Jisoo Vinsky
Sheriff, Robert Edward
Summer, John Stewart
Tanis, James Iran
Trembly, Lynn Dale
Waters, Kenneth Harold
West, Robert Elmer
Westphal, Warren Henry
Whittlesey, John Rb
Wilson, John Human
Wolfe, Paul Jay
Wu, Changsheng
Yarger, Harold Lee
Yungul, Sulhi Hasan

Forest Hydrology
Bethlahmy, Nedavia
Brown, George Wallace
Chang, Harry Esmond
Chang, Mingteh
Coltharp, George B
Golding, Douglas Lawrence
Harper, Warren Charles
Hibbert, Alden R
Mace, Arnett C, Jr
Merriam, Robert Arnold
O'Hayre, Arthur Paul
Patric, James Holton
Singh, Teja
Slaughter, Charles Wesley
Swanson, Robert Harold
Swift, Lloyd Wesley, Jr
Wheeler, Richard Hunting
Willington, Robert Peter
Wooldridge, David Dilley
Zwolinski, Malcolm John

Geodesy
Baker, Leonard Samuel
Carter, William Eugene
Chovitz, Bernard H
Clark, Thomas Arvid
De Jong, Sybren Hendrik
Ewing, Clair Eugene
Fischer, Irene Kaminka
Fliegel, Henry Frederick
Gaposchkin, Edward Michael
Hamilton, Angus Cameron
Hunt, Mahlon Seymour
Khan, Mohammad Asad
King, Robert Wilson, Jr
Kivioja, Lassi A
Laurila, Simo Heikki
Madden, Stephen James, Jr
Mourad, A George
Mueller, Ivan I
Ortin, Hyman
Rapp, Richard Henry
Smith, Paul Albert
Stearn, Joseph Leonard
Taylor, Eugene Alfred
Ugincius, Peter
Uotila, Urho Antti
Vanicek, Petr
Williams, Owen Wingate

Hydrogeology
Ahman, Moid Uddin
Aley, Thomas John
Anderson, Mary Pikul
Bean, Robert Taylor
Beck, Henry V
Bernes, Boris John
Bianchi, William C
Campbell, Michael David
Carnahan, Chalon Lucius
Cartwright, Keros
Caruccio, Frank Thomas
Cohen, Philip
Cooley, Richard Lewis
Cushman, Robert Vittum
Davis, Stanley Nelson
Drescher, William James
Dunn, Darrel Eugene
Emery, Philip Anthony
Farvolden, Robert Norman
Fenske, Paul Roderick
Fetter, Charles Willard, Jr
Fischer, Frederick Thomas
Flint, Jean-Jacques
Foley, Frank Clingan
Gates, Joseph Spencer
Getzen, Rufus Thomas
Gordon, Ellis Davis
Harsh, John F
Heigold, Paul C
Helm, Donald Cairney
Hughes, George M
Johnson, Carlton Robert

Johnson, Noye Monroe
Kaufmann, Robert Frank
Kempton, John Paul
Kharaka, Yousif Khoshu
Koch, Donald Leroy
Kuiper, Logan Keith
Lehr, Jay H
Lissey, Allan
Manera, Paul Allen
McComas, Murray Ratcliffe
Mercer, James Wayne
Meyboom, Peter
Miller, David W
Montgomery, Errol Lee
Moore, John Ezra
Neuman, Shlomo Peter
Nork, William Edward
Nyman, Dale James
Palmer, Arthur N
Papadopulos, Stavros Stefanu
Parizek, Richard Rudolph
Pearson, Frederick Joseph
Pederson, Darryll Thoralf
Pfannkuch, Hans Olaf
Pinder, George Francis
Poland, Joseph Fairfield
Reed, Eugene Clifton
Rosenshein, Joseph Samuel
Ruedisili, Lon Chester
Rundells, Donald DeMar
Saleem, Zubair A
Schmidt, Ronald Grover
Schneider, Robert
Scott, John Malcolm, Jr
Sharp, John Van Alstyne
Smith, Ralph Emerson
Stringfield, Victor Timothy
Trescott, Peter Chapin
Trexler, Bryson Douglas, Jr
Waller, Roger Milton
Warner, Don Lee
Warren, James Clark
Welby, Charles William
Wilson, William Edward, III
Wright, Jerome J

Hydrology
Anderson, Henry Walter
Babcock, Horace Maxson
Back, William
Barksdale, Henry Compton
Bredehoeft, John Dallas
Brown, Ira Charles
Brown, Stuart Greene
Brutsaert, Wilfried
Buckingham, Forrest Morgan
Buddemeier, Robert Worth
Caine, T Nelson
Case, Clinton Meredith
Clark, Robert H
Clarke, Frank Eldridge
Cluff, Carwin Brent
Cornish, John Henry
Csallany, Sandor Csergo
Curtis, Bruce Franklin
Daugherty, Franklin W
Davis, George H
DeCook, Kenneth James
Deju, Raul A
Dickinson, William Trevor
Dingman, Stanley Lawrence
Dodson, Chester Lee
Doyle, Frank Lawrence
Dragoun, Frank J
Dreeszen, Vincent Harold
Dunne, Thomas
Eagleson, Peter Sturges
Eschner, Arthur Richard
Evans, Daniel Donald
Everett, Lorne Gordon
Fetter, Charles Willard, Jr
Foster, Harold Douglas
Foxworthy, Bruce L
Freeland, Forrest Dean, Jr
Freeze, Roy Allan
Frimpter, Michael Howard
Gordon, Ellis Davis
Hall, Francis Ramey
Hammer, Richard M
Harr, Robert Dennis
Harshbarger, John W
Hart, George Emerson, Jr
Haushild, William Leland
Heindl, Leopold Alexander
Hendricks, Ernest LeRoy
Hewlett, John David
Hornberger, George Milton
Hubbard, John Edward
Hudlow, Michael Dale
Jarvis, Richard Stanley
Jones, Everett Bruce
Jones, James Richard
Kakela, Peter John
Kennedy, Vance Clifford
Koopman, Francis Christian
Krimgold, Dov B

Kunkel, Fred
LaMoreaux, Philip Elmer
Lane, Leonard James
Lang, Solomon Max
Langbein, Walter
Langford, Russel Hal
Laycock, Arleigh Howard
Lee, Richard
Lehr, Jay H
LeGrand, Harry E
Lennox, Donald Haughton
LeRoux, Edmund Frank
Limpert, Frederick Arthur
Marine, Ira Wendell
Maxey, George Burke
McGuinness, James L
McMillion, Leslie Glen
Miflin, Martin David
Mink, John F
Moiz, Fred John
Moody, David Wright
Moore, John Ezra
Morgan, Charles O
Murray, Charles Richard
Nace, Raymond Lee
Neuman, Shlomo Peter
Osborn, Herbert B
Pashley, Emil Frederick, Jr
Peck, Eugene Lincoln
Peckham, Alan Embree
Peterson, Frank Lynn
Petri, Lester Reinhold
Pettyjohn, Wayne A
Phoenix, David A
Pickering, Ranard Jackson
Plebuch, Raymond Otto
Pluhowski, Edward John
Powell, John Carleton, Jr
Prescott, Glenn Carleton, Jr
Qashu, Hasan Khalil
Rahn, Perry H
Reisenauer, Andrew E
Remson, Irwin
Resnick, Sol Donald
Richards, Marshall Monroe
Riley, James Joseph
Rosenshein, Joseph Samuel
Salomonson, Vincent Victor
Schneider, Robert
Signor, Donald C
Simpson, Eugene Sidney
Smoot, George Fitzgerald
Snider, Robert Gordon
Snyder, Charles Theodore
Snyder, Willard Monroe
Sommers, David A
Spieker, Andrew Maute
Steele, Timothy Doak
Stephenson, David A
Sutherland, Jeffrey C
Swenson, Frank Albert
Swenson, Herbert Alfred
Tabler, Ronald Dwight
Tarble, Richard Douglas
Tromble, John M
Turk, Leland Jan
Tyagi, Avdhesh Kumar
Visher, Frank N
Weeks, Wilford Frank
Weist, William Godfrey, Jr
Wershaw, Robert Lawrence
Westfall, Arthur Oscar
White, Walter Finch
Wiesnet, Donald Richard
Williams, Roy Edward
Wolff, Roger Glen
Wood, Warren Wilbur
Zand, Siavosh Marc
Zetzel, Eugene Paul
Zobler, Leonard

Ionospheric Physics
Aarons, Jules
Benson, Robert Frederick
Brace, Larry Harold
Davies, Kenneth
Farley, Donald T Jr
Francis, Samuel Hopkins
Gerson, Nathaniel Charles
Hayden, Leonard Octavius
Hedlund, Donald A
Herman, John R
Kerr, Donald M, Jr
Kilb, Ralph Wolfgang
Koib, Edward John
LaBahn, Raymond Willis
McDiarmid, Donald Ralph
Millman, George Harold
Ostrow, Sidney Maurice
Quinn, Thomas Patrick
Rufenach, Clifford L
Rush, Charles Merle
Sagalyn, Rita C
St Maurice, Jean-Pierre
Sechrist, Chalmers Franklin, Jr
Sharp, Richard Dana
Tanenbaum, Basil Samuel
Toman, Kurt

Wallis, Donald Douglas James H
Wand, Ronald Herbert

Marine Geophysics
Anderson, George Boine
Atwater, Tanya Maria
Behrendt, John Charles
Bennett, Lee Cotton, Jr
Frosch, Robert Alan
Geddes, Wilbur Hale
Grim, Paul J
Hammond, Stephen Randolph
Haworth, Richard Thomas
Hussong, Donald MacGregor
Johnson, Rockne Hart
Keen, Charlotte Elizabeth
Knott, Sydney T
Kuperman, William Aaron
Loncarevic, Bosko (D)
Ludwig, William Jackson
Luyendyk, Bruce Peter
Molnar, Peter Hale
Odegard, Mark Erie
Ostenso, Ned Allen
Riggin, John Dewitt
Shor, George G, Jr
Taylor, Patrick Timothy
Vogt, Peter Richard
Von Herzen, Richard P
Wold, Richard John

Paleomagnetism
Banerjee, Subir Kumar
Brecher, Aviva
Butler, Robert Franklin
Elston, Donald (Parker)
Irving, Edward
McMahon, Beverly Edith
Murthy, Gummuluru Satyanarayana
Robertson, William Archer
Schmidt, Victor Adolf

Seismology
Aki, Keiiti
Allen, Clarence Roderic
Anderson, Don Lynn
Berry, Michael John
Blandford, Robert Roy
Bollinger, Gilbert A
Bolt, Bruce A
Boucher, Gary Wynn
Brune, James N
Bufe, Charles Glenn
Byerly, Perry
Chinnery, Michael Alistair
Clarke, Garry K C
Clayton, Neal
Cloud, William K
Clowes, Ronald Martin
Couch, Richard W
Crosson, Robert Scott
Dahm, Cornelius George
Dainty, Anton Michael
Dalins, Ilmars
Dziewonski, Adam Marian
Eaton, Jerry Paul
Evernden, Jack Foord
Frantti, Gordon Earl
Furumoto, Augustine S
Gershanik, Simon
Gilbert, James Freeman
Gupta, Indra Narayan
Harkrider, David Garrison
Hart, Pembroke J
Hermann, Robert Bernard
Hodgson, John Humphrey
Isacks, Bryan L
Iyer, Hariharaiyer Mahadeva
Keller, George Randy, Jr
Knott, Sydney T
Lahr, John Clark
Lander, James French
Leblanc, Gabriel
Lomnitz, Cinna
Long, Leland Timothy
Madariaga, Raul Ivan
McEvilly, Thomas V
McGarr, Arthur
Mei, Alexis Italo
Mercado, Edward J
Mereu, Robert Frank
Molnar, Peter Hale
Nason, Robert Dohrmann
Northwood, Thomas David
Odegard, Mark Erie
Pfluke, John H
Richter, Charles Francis
Romig, Phillip Richardson
Romney, Carl Fredrick
Ryall, Alan S, Jr
Shurbet, Deskin Hunt, Jr
Slichter, Louis Byrne
Snoke, J Arthur
Stauder, William
Sutton, George Harry
Thapar, Mangat Rai
Thatcher, Wayne Raymond
Thomson, Ker Clive

Tocher, Don
Turpening, Roger Munson
Unger, John Duey
Walker, Daniel Alvin
Ward, Peter Langdon
Warren, David Henry
Watkins, Joel Smith, Jr
Weichert, Dieter Horst
Wells, Frederick Joseph
Wong, David C
Wyss, Max

Soil Physics
Adams, John Edgar
Bertrand, Anson Rabb
Blake, George Rowland
Bourget, Sylvio-J
Box, James Ellis, Jr
Brandt, Gerald H
Bruce, Robert Russell
Cannell, Glen H
Cassel, D Keith
Corey, John Charles
Danielson, Robert Eldon
Davidson, James Melvin
Day, Paul Russell
Dirksen, Christiaan
Eagleman, Joe R
Ekern, Paul Chester
Elrick, David Emerson
Enfield, Carl George
Epstein, Eliot
Ferguson, Albert Hayden
Flocker, William Jack
Forsythe, Warren M
Fritton, Daniel Dale
Gardner, Walter Hale
Gifford, Richard Oliver
Green, Richard E
Grover, Ben Leo
Hanks, Ronald John
Hoadley, Robert Bruce
Hornsby, Arthur Grady
Horton, Maurice Lee
Jensen, Creighton Randall
Kimball, Bruce Arnold
Kohl, Robert A
Kunze, Raymond J
Lugo-Lopez, Miguel Angel
Mansell, Robert Shirley
Mazurak, Andrew Peter
McCauley, Garry Nathan
Meredith, Harvey L
Miller, Robert Demorest
Miyamoto, Seiichi
Molz, Fred John
Nielsen, Donald R
Olson, Tamlin Curtis
Page, John Boyd
Patten, Gaylord Penrod
Peters, Doyle Buren
Powers, William L
Raney, William Andrew
Reicosky, Donald Charles
Ritchie, Joe T
Rogowski, Andrew S
Rolston, Dennis Eugene
Rubin, Jacob
Runkles, Jack Ralph
Shawcroft, Roy Wayne
Shaykewich, Carl Francis
Starr, James LeRoy
Stevenson, David Stuart
Stewart, Gordon LeRoy
Stolzy, Lewis Hal
Stone, Loyd Floyd
Stone, Loyd Raymond
Swartzendruber, Dale
Tackett, Jesse Lee
Taylor, George Stanley
Thiel, Thomas J
Tsuji, Gordon Yukio
Van Doren, David Miller, Jr
Warrick, Arthur W
Wendt, Charles William
Whisler, Frank Duane
Whiteley, Eli Lamar
Whitt, Darnell Moses
Wiegand, Craig Loren
Wierenga, Peter J
Willey, Cliff Rufus
Willis, Wayne Owen
Wilson, Clyde Livingston
Wilson, Lorne Graham

Watershed Management
Anderson, Henry Walter
Aubertin, Gerald Martin
Black, Peter Elliott
Blackburn, Wilbert Howard
DeByle, Norbert V
DeWalle, David Russell
England, Charles Bennett
Frere, Maurice Herbert
Gay, Lloyd Wesley
Heede, Burchard Heinrich
Johnson, Edward A
Kitchen, Joseph Henry

McComb, Andrew Logan
Meiman, James R
Rango, Albert
Rothwell, Richard Lee
Schumm, Stanley Alfred
Settergren, Carl David
Smith, James LeRoy
Tabler, Ronald Dwight
Teller, Leo
Thames, John Long
Ursic, Stanley John

HEALTH SCIENCES

Health Science
Appelgren, Walter Phon
Baumann, Thiema Marie Wolf
Bland, Hester Beth
Blockstein, William Leonard
Bowery, Thomas Glenn
Brink, Norman George
Cornesky, Robert Andrew
Doran, Peter Cobb
Dornbush, Rhea L
Doscher, Nathan
Ducker, Thomas Barbee
Efron, Herman Yale
Emlet, Harry Elsworth Jr
Eppright, Margaret Anne
Epstein, Lois Barth
Gaik, Geraldine Catherine
Glick, J Leslie
Gortatowski, Melvin Jerome
Harris, Raymond
Henzlik, Raymond Eugene
Hine, Gerald John
Jacobs, Donald
Levine, Jules Ivan
Levison, Matthew Edmund
Liedtke, Claus-Eberhard
Lindberg, David Seaman, Sr
Lindsey, Dortha Ruth
Liskey, Nathan Eugene
LoGerfo, John J
Love, Robert Lyman
Lucas, Stephen Bernard
MacArthur, Donald M
Malinowski, Henry John
Marshall, James R
Novak, Alfred
Pierce, Edward Ronald
Pigman, Ward
Rahman, Yueh Erh
Sebrell, William Henry
Silverstein, Martin Elliot
Smith, Vivian Sweibel
Suskind, Raymond Robert
Synovitz, Robert J
Winkler, Charles Herman, Jr
Wood, Thomas Ross
Yang, Shiang-Ping
Yellin, Herbert

Community Health
Abramson, Fredric David
Bowns, Beverly David
Brown, Edwin Wilson, Jr
Cross, Richard James
Drosness, Daniel Leed
Falk, Leslie Alan
Frank, Andrew Julian
Grubb, Alan S
Haug, Norman L
Hughes, Edward Francis Xavier
Kataja, Eva I
Lewis, Irving James
Lichter, Edward A
Martin, Samuel Preston, III
McAtee, Patricia Rooney
McNamara, Michael Joseph
Mindlin, Rowland L
Olsen, Donna Mae
Roberts, Dean Winn
Salber, Eva Juliet
Schwartz, Jerome Leroy
Sidel, Victor William
Tarail, Robert
Verby, John E
Vermeersch, Joyce Ann
Werdegar, David
Yamanaka, William Kiyoshi
Zimmer, James Griffith

Environmental Health
Anderson, David Martin
Barton, Charles Julian, Sr
Beck, Milan F
Blackwell, Floyd Oris
Bond, Richard Guy
Breniman, Gary Russell
Brown, Harold Victor
Cabelli, Victor Jack
Campbell, Kirby I
Carnow, Bertram Warren
Chan, Leland
Cody, Terence Edward

Cole, Jerome F
Colucci, Anthony Vito
Comar, Cyril Lewis
Dahneke, Barton Eugene
Deininger, Rolf A
de Serres, Frederick Joseph
Dreisbach, Robert Hastings
Ehrlich, Richard
Eisenbud, Merril
Elia, Victor John
Emmett, Edward Anthony
Falk, Hans Ludwig
Feasley, Charles Frederick
Ferris, Benjamin Greeley, Jr
Finelli, Vincent Nicola
Flowers, Earl Shederick
Freudenthal, Peter
Garner, Reuben John
Gray, Robert Howard
Hake, Carl (Louis)
Hall, William A
Harley, John Henry
Harris, Elliott Stanley
Hass, James Ronald
Hickman, John Roy
Hilbert, Morton S
Horstman, Sanford W
Humphrey, Harold Edward Burton, Jr
Iglar, Albert Francis, Jr
Jacobson, Alvin Raymond
Johnson, Jerry Michael
Judd, Stanley H
Kitzke, Eugene David
Leach, Leonard Joseph
LeMunyan, Cobert Duane
Levin, Gilbert Victor
Lippmann, Morton
Long, Keith Royce
Lovett, Joseph
Makielski, Sally Kimball
Mallison, George Franklin
Malone, Winfred Francis
Mattheis, Eula Bingham
Meyer, Alvin F, Jr
Michael, Paul Lee
Morgan, Monroe Talton, Sr
Morton, John Dudley
Niemeier, Richard William
Parker, Robert Davis Rickard
Patel, Dhun Burjor
Paulus, Harold John
Peters, Howard August
Pier, Stanley Morton
Pierce, James Otto, II
Pogrund, Robert Seymour
Saltzman, Bernard Edwin
Schiager, Keith Jerome
Severs, Richard Keith
Singh, Harpal P
Smith, Carroll Ward
Sobsey, Mark David
Spector, Bertram
Spence, John A
Strehlow, Clifford David
Tiller, Richard Edward
Vesley, Donald
Wedlick, Harold Lee
Wilkening, George Martin
Wilson, Robert Hallowell
Wolf, Harold William
Wolff, Arthur Harold
Wrenn, McDonald Edward
Yu, Ming-Ho

Epidemiology
Abramson, Fredric David
Albrecht, Robert Michael
Alexander, Edward Russell
Allen, Alfred Marston
Ames, Wendell Russell
Anderson, Donald Oliver
Archer, Victor Eugene
Arthur, Ransom James
Asal, Nabih Rafia
Askew, Cornelius, Jr
Azar, Joseph E
Babbott, Frank Lusk, Jr
Babione, Robert William
Baer, George M
Baker, Susan Pardee
Banta, James E
Benarde, Melvin Albert
Bennett, Peter Howard
Blackburn, Henry Webster, Jr
Brody, Jacob A
Buck, Alfred A
Burch, Thomas Adams
Bures, Milan F
Cassel, John Charles
Chin, Tom Doon Yuen
Choi, Nung Won
Cobb, Sidney
Cohen, Bernice Hirschhorn
Comstock, George Wills
Crook, James Richard
Curnen, Mary G McCrea
D'Atri, David Albert
Davenport, Fred M

Davies, Jack Neville Phillips
Day, Robert Winsor
Dean, Andrew Griswold
Dick, Elliot C
Doege, Theodore Charles
Donnell, Henry Denny, Jr
Dunn, Frederick Lester
Ehrlich, S Paul, Jr
Eklund, Carl Milton
Eller, Charles Howe
Emanuel, Irvin
Evans, Alfred Spring
Fagan, Raymond
Fischer, Diana Bradbury
Fisher, James Wallwin
Fooge, William Herbert
Fox, John Perrigo
Friedman, Lawrence Abraham
Gale, James Lyman
Gangarosa, Eugene J
Garfinkel, Lawrence
Geiger, H Jack
Gelfand, Henry Morris
Glass, Robert Loring
Goldfield, Martin
Goldstein, Inge F
Goodwin, Melvin Harris, Jr
Gordis, Leon
Gottlieb, Marise Suss
Gruenberg, Ernest Matsner
Guirgis, Hoda A
Gullen, Warren Hartley
Hamblet, Frederick Edwards
Hammon, William McDowell
Heath, Clark Wright, Jr
Heeren, Ralph Heinrich
Henderson, Maureen McGrath
Herman, Bertram
Higgins, Ian T
Higgins, Millicent Williams Payne
Hirohata, Tomio
Hollinger, Frederick Blaine
Hollister, Arthur Clair, Jr
Hook, Ernest Benjamin
Horstmann, Dorothy Millicent
Horton, Robert John Munro
Houser, Harold Byron
Hrubec, Zdenek
Hutchison, George B
Ibrahim, Michel A
Ingalls, Theodore Hunt
Ingraham, Hollis Steadman
Jamison, Homer Claude
Jekel, James Franklin
Johnson, Benjamin C
Joly, Daniel Jose
Kahn, Harold A
Kaiser, Robert L
Kapikian, Albert Zaven
Kaufmann, Arnold Francis
Keil, Julian Eugene
Keller, Martin David
Kelsey, Jennifer Louise
Kissinger, David George
Kokernot, Robert Hutson
Korns, Robert Fulton
Kraus, Arthur Samuel
Kraus, Jess F
Labarthe, Darwin Raymond
LaMotte, Louis Cossitt, Jr
Landau, Emanuel
Langmuir, Alexander Duncan
Langner, Thomas S
Last, John Murray
Lebowitz, Michael David
Lee, John Alexander Hugh
Leiby, George Martin
LeMaistre, Charles Aubrey
Lemon, Frank Raymond
Lennette, Edwin Herman
Lilienfeld, Abraham Morris
Lundin, Frank E, Jr
Luoto, Lauri
Maassab, Hunein Fadlo
Macdonald, Eleanor Josephine
Mackenzie, Cortlandt John Gordon
MacMahon, Brian
Magoffin, Robert Louis
Martin, Russell James
Masi, Alfonse Thomas
Mason, John
Matanoski, Genevieve M
Mausner, Judith S
McCollum, Robert Wayne
McCorkle, Lois Pake
McCroan, John Edgar, Jr
McCusker, Jane
McDonald, Alison Dunstan
McDonald, John Corbett
McQueen, James Lee
Millar, James Donald
Miller, Anthony Bernard
Milmore, Robert Warwick
Monk, Mary Alice
Monto, Arnold Simon
Moore, Roscoe Michael, Jr
Morton, William Edwards

Mosley, James W
Mosley, Wiley Henry
Mulvihill, John Joseph
Munan, Louis
Neff, Beverly Jean
Nelson, Kenneth Wyatt
Newell, Vaun Archie
Niederman, James Corson
Oalmann, Margaret Claire
Oberman, Albert
O'Brien, William M
Omohundro, Richard E
Omran, Abdel Rahim
Oseasohn, Robert
Ostfeld, Adrian Michael
Paffenbarger, Ralph Seal, Jr
Palmer, Alan
Pasternack, Bernard Samuel
Payne, Fred J
Peacock, Peter N B
Peters, James Alexander
Peterson, Donald Richard
Philip, Robert Neil
Portnoy, Bernard
Purshottam, Natesaer
Ravenholt, Reinert Thorolf
Reed, Dwayne (Milton)
Reeves, William Carlisle
Reich, George Arthur
Rose, Gordon Wilson
Rosen, Leon
Rotkin, Isadore David
Rushforth, Norman B
Rycheck, Russell Rule
Sackett, David Lawrence
Saiger, George Lewis
Sartwell, Philip Earl
Scholtens, Robert George
Schultz, Myron Gilbert
Schuman, Leonard Michael
Schuman, Stanley Harold
Schwabe, Calvin Walter
Schwartz, John T
Schweitzer, Morton David
Selevan, Sherry Gail
Shapiro, Sam
Sheehe, Paul Robert
Shekelle, Richard Barten
Silberg, Stanley Louis
Silverman, Charlotte
Slusky, Herbert L
Sohler, Katherine Berridge
Speizer, Frank Erwin
Spiers, Philip Sackville
Stallones, Reuel Arthur
Stark, Charles R
Steele, Robert
Stein, Zena
Stern, Elizabeth
Susser, Mervyn W
Taylor, Carl Ernest
Terris, Milton
Tesh, Robert Bradfield
Thomas, Robert Joseph
Thompson, Gordon William
Thorne, Melvyn Charles
Tokuhata, George K
Top, Franklin Henry, Sr
Vaughn, John B
Vianna, Nicholas Joseph
Villarejos, Victor Moises
Vobecky, Josef
Voors, Antonie Wouter
Vought, Robert Louis
Wallace, Gordon Dean
Walter, William Arnold, Jr
Watson, Robert Lee
Watt, James
White, Kerr Lachlan
Winkelstein, Warren, Jr
Winsberg, Gwynne Roeseler
Winter, Phillip E
Work, Telford Hindley
Worth, Robert McAlpine
Wright, Harry Tucker, Jr
Wynder, Ernst Ludwig
Wyon, John Benjamin
Zimmerman, Robert A

Exercise Physiology
Allen, Paul D
Benison, Betty Bryant
Bonner, Hugh Warren
Cafarelli, Enzo Donald
Dukes-Dobos, Francis N
Fox, Edward L
Glaser, Roger Michael
Golnick, Philip D
Greenleaf, John Edward
Harrison, Aix B
Landry, Fernand
Maksud, Michael George
Nagle, Francis J
Oscai, Lawrence B
Pargman, David
Pollock, Michael L
Robertson, Robert James

Ruhling, Robert Otto
Soule, Roger Gilbert
Spence, Dale William
Vogel, James Alan
Wright, James Edward
Zauner, Christian Walter

Health Physics
Augustine, Robertson J
Auxier, John Alden
Baltzo, Ralph M
Barber, Donald E
Barry, P J S
Bogar, Louis Charles
Bradley, Francis J
Brodsky, Allen
Buchanan, John Donald
Caruthers, Leo Thomas, Jr
Cember, Herman
Chaney, Edward L
Clouter, Roger Joseph
Cowan, Frederick Pierce
Cowser, Kenneth Emery
Drew, Robert Taylor
Dunning, Gordon Merrill
Finston, Roland A
Friedman, Helen Lowenthal
Furr, Aaron Keith
Furtado, Victor Cunha
George, Robert Eugene
Gesell, Thomas Frederick
Gibson, Thomas Alvin, Jr
Gollon, Peter J
Harvey, John Wilcox
Haughey, Francis James
Herrington, James Roland
Heslep, John McKay
Holeman, George Robert
Holford, Richard Moore
Hsieh, Jen-Shu
Hubbell, Harry Hopkins, Jr
Hull, Andrew P
Johnson, John Richard
Kantrowitz, Arthur (Robert)
Kaplan, Arthur Lewis
Kaufman, George Emmett Clarence
Kennete, Albert Patrick
Klement, Alfred William, Jr
Larson, Harold Vincent
Lawrence, James Neville Peed
Lenhard, Joseph Andrew
Lott, Sam Houston, Jr
Luetzelschwab, John William
Maillie, Hugh David
Manly, Philip James
Mann, Bruce Jameson
Martin, Thomas George, III
McCaslin, John Weaver
Meinhold, Charles Boyd
Mercer, Thomas T
Milne, Walter Leroy
Moore, Craig Damon
Morgan, Karl Ziegler
Neff, Richard D
Nelson, Walter Ralph
Oliver, George Joseph
Osborne, Richard Vincent
Ostrom, Thomas Ross
Parker, Herbert Myers
Pelletier, Charles A
Poston, John Ware
Putzier, Edward Anthony
Reing, William Charles
Roberts, Carlyle Jones
Roessler, Charles Ervin
Ross, John Edward
Sanders, Samuel Marshall, Jr
Sashin, Donald
Shleien, Bernard
Siebentritt, Carl R, Jr
Sliney, David H
Snyder, Walter Stephen
Soldat, Joseph Kenneth
Sprafka, Robert J
Tanner, Raymond Lewis
Teresi, Joseph Dominic
Thomas, Ralph Harold
Thompson, Donald Leroy
Van Pelt, Wesley Richard
Vetter, Richard J
Watters, Robert Lisle
Wedlick, Harold Lee
Werkema, George Jan
Wilson, John William
Wynveen, Robert Allen
Yaniv, Shlomo Stefan
Yoder, Robert E
Ziemer, Paul L

Hygiene
Anderson, Carl Leonard
Frappier, Armand
Gershenfeld, Louis
Grant, John Gray
MacDonald, Kenneth
Norman, Edward Cobb

Industrial Health
Mallery, Otto Tod
McFee, Donald Ray
Xintaras, Charles

Industrial Hygiene
Ayres, Paul E
Baier, Edward John
Berry, Clyde Marvin
Bradley, William Robinson
Brown, Robert Don
Burg, William Robert
Byers, Dohrman Harold
Campbell, Evan Edgar
Caplan, Paul E
Cheever, Charles Lyle
Cook, Warren Ayer
Demen, John McCay
Eitinger, Harry Joseph
Farrah, George Henry
Ferber, Kelvin Halket
Fredrick, William George
Giever, Paul Mathew
Johnson, Kenneth Delford
Kenney, Gary Dale
Krebs, William H
Long, James Earl
Lundquist, Marjorie Ann
MacFarland, Harold Noble
Mercer, Thomas T
Milburn, Burton Burnett
Morgan, James Frederick
Nelson, Kenneth William
Powell, Charles Herbert
Rappaport, Stephen Morris
Ratney, Ronald Steven
Reichenbach, George Suter, Jr
Reist, Parker Cramer
Ross, Donald Morris
Ross, John Edward
Rowe, Verald Keith
Sansone, Eric Brandfon
Schreibers, William J
Schulte, Harry Frank
Smith, Ralph G
Steffey, Oran Dean
Stephens, John Firth
Stofer, Robert Llewellyn
Thompson, Vinton Newbold
Van Atta, Floyd August
Van Ert, Mark Dewayne
Wheeler, Elmer Perley
Wood, Gerry Odell
Yoder, Robert E

Nutrition
Aberle, Sophie Bledsoe
Abernathy, Richard Paul
Acker, Geraldine Enod
Ackerman, Clemens John
Adibi, Siamak A
Afergood, Lilla
Ahrens, Richard August
Aines, Philip Deane
Alexander, Herman Davis
Alfano, Michael Charles
Alford, Betty Bohon
Allen, Lindsay Helen
Allred, John B
Altschul, Aaron Mayer
Amen, Ronald Joseph
Anderson, Helen Lester
Anderson, Ingrid
Anderson, John Joseph Baxter
Anderson, Joseph Tomlinson
Anderson, Robert L
Anderson, Thomas Alexander
Anderson, Zoe Estelle
Angel, Joseph Francis
Anthony, Luean Evangeline
Apgar, Barbara Jean
Appledorf, Howard
Arata, Dorothy
Armich, Lotte
Austic, Richard Edward
Autrey, Kenneth Maxwell
Babcock, MacLean Jack
Baile, Clifton Augustus, III
Baird, Derwood McVey
Baker, Herman
Baldini, James Thomas
Baldwin, Ransom Leland, Jr
Banks, William Louis
Barbour, Helen F
Barnes, Richard Henry
Barnes, Lewis Abraham
Bass, Mary Anna
Bates, Margaret Westbrook
Bauernfiend, Jacob Christopher
Bavetta, Lucien Andrew
Bayley, Henry Shaw
Beach, Betty Laura
Beal, Virginia Asta
Beaton, George Hector
Beaton, John Rogerson
Beauchene, Roy E
Beigler, Myron Arnold

Bell, Roma Raines
Benevenga, Norlin Jay
Bennion, Marion Jay
Bensadoun, Andre
Berryman, George Hugh
Bhagavan, Hemmige
Bieri, John Gunther
Bierenbaum, Marvin L
Bird, Francis Howe
Blackwood, Unabelle Boggs
Blamberg, Donald Lee
Block, Walter David
Bodwell, Clarence Eugene
Boehne, John William
Bollinger, James Norman
Bond, Jenny Taylor
Bowering, Jean
Boykin, Lorraine Stith
Bragg, Darrell
Branion, Hugh Douglas
Brewer, Wilma Denell
Briggs, George McSpadden
Brink, Marion Francis
Brinkman, Gail Lynn
Brisson, Germain J
Broquist, Harry Pearson
Browe, John Harold
Brown, Helen Bennett
Brown, Myrtle Laurestine
Brown, Richard Edward
Brown, Robert Glenn
Brush, Miriam Kelly
Bucy, LaVerne
Bunce, George Edwin
Burgess, Hovey Mann
Burnette, Mahlon Admire, III
Burton, Benjamin Theodore
Calhoun, William Kenneth
Calloway, Doris Howes
Campbell, James Alexander
Campbell, Thomas Colin
Canham, John Edward
Canolty, Nancy Lemmon
Cardon, Bartley Pratt
Carroll, Catherine
Carroll, Harold Wilson
Carruth, Betty Ruth
Carruth, Kayla Bernard
Cason, James Lee
Castanera, Esther Goossen
Caster, William Oviatt
Cederquist, Dena Caroline
Chang, Irene Ching Lai
Chang, John Wan-Yuin
Chaplin, Michael H
Cheeke, Peter Robert
Chen, Linda Li-Yueh Huang
Childs, George Richard
Christiansen, Marjorie Miner
Clark, Helen Edith
Coccodrilli, Gus D, Jr
Cohen, Saul Israel
Combs, Gerald Fuson, Jr
Conrad, Harry Russell
Conrad, Herbert M
Consolazio, Carlo Frank
Contento, Isobel Corneli
Cooley, Maxwell Louis
Coolidge, Ardath Anders
Coon, Craig Nelson
Cords, Richard Henry, Jr
Costain, Robert Anthony
Cotton, Robert Henry
Couch, James Russell
Cousins, Robert John
Cowan, James W
Cox, Allan Clayton
Crampton, Earle Wilcox
Cravens, William Windsor
Crosby, Lon Owen
Cullen, Marion Permilla
Cunningham, Glenn N
Dakshinamurti, Krishnamurti
Dam, Richard
Darby, William Jefferson
Davis, Carl Lee
Davis, Elizabeth Young
Davis, Larry E
Davis, Peyton Nelson
Davison, Kenneth Lewis
Dawson, Earl B
Day, Mary-Lou
Daza, Carlos Hernan
Debes, Sue Ann
De Nisco, Stanley Gabriel
DePaola, Dominick Philip
Desai, Indrajit Dayalji
Devine, Marjorie M
Devoe, Charles W
Dibble, Marjorie Veit
Dinning, James Smith
Dirige, Ofelia Villa
Doberenz, Alexander R
Doctor, Bhupendra Pannalal
Donald, Elizabeth Ann
Doyle, Margaret Davis
Draper, Harold Hugh

Dreizen, Samuel
Driskell, Judy Anne
Dumm, Mary Elizabeth
Dunkley, Colleen Rose
Dupont, Jacqueline (Louise)
Dutra Oliveira, Jose Eduardo
Ebbs, Jane Cotton
Eckstein, Eleanor Foley
Edwards, Cecile Hoover
Elson, Charles
Emerson, Gladys Anderson
Ershoff, Benjamin H
Evans, Edwin Victor
Evans, Gary William
Evans, Joseph Liston
Evans, Robert John
Farmer, Florence Amelia
Feigenbaum, Abraham Samuel
Feniak, Elizabeth
Fick, Bessie Davey
Filer, Lloyd Jackson, Jr
Fillios, Louis Charles
Fincke, Margaret Louise
Finkelstein, Beatrice
Fisher, Hans
Flynn, Margaret A
Fomon, Samuel Joseph
Forbes, Richard Mather
Ford, Clinita Arnsby
Foss, Donald C
Fox, Hazel Metz
Frakes, Elizabeth (McCune)
Frank, Oscar
Frank, Robert Loeffler
Franklin, Ruth Ann
Fritz, Herbert Ira
Frobish, Lowell T
Fry, Peggy Crooke
Fryer, Elsie Beth
Fuqua, Mary Elizabeth
Futrell, Mary Feltner
Garcia, Pilar A
Garlich, Jimmy Dale
Gaut, Zane Noel
Geer, Billy W
Goldbloom, David Ellis
Gordon, Joan
Gormican, Annette
Gortner, Willis Alway
Goyco Daubon, Jose A
Graham, Donald C W
Graham, George G
Grainger, Robert Ball
Gram, Mary Rose
Green, Mary Eloise
Greene, Barbara E
Greene, Daryle E
Greenwood, Mary Rita Cooke
Gunning, Barbara E
Guthrie, Helen A
Guzman Foresti, Miguel Angel
Hackett, Patricia Lou
Hackler, Lonnie Ross
Hall, Kenneth Noble
Halver, John Emil
Hankin, Jean H
Hansard, Samuel Leroy
Hard, Margaret McGregor
Harkins, Robert W
Harland, Barbara Ferguson
Harper, Laura Jane
Harrill, Inez Kemble
Hartman, David Robert
Hartoft, Walter Stanley
Hathcock, John Nathan
Hawkins, Winthrop Wesley
Hawthorne, Betty Eileen
Hayes, Kenneth Cronise
Hegarty, Patrick Vincent
Hegsted, David Mark
Heinicke, Herbert Raymond
Helm, Raymond E
Herbert, Jack Durnin
Herndon, John Francis
Hertzler, Ann Atherton
High, Edward Garfield
Hilker, Doris M
Hill, Charles Horace, Jr
Hill, Fredric William
Hill, Roberta Bieler
Hilston, Neal William
Ho, Genevieve Po-Ai
Hoar, Donald Wayne
Holman, Ralph Theodore
Hoover, Loretta White
Hopkins, Leon Lorraine, Jr
Hopper, John Henry
Horrington, Emily Mae
Horwitt, Max Kenneth
Hoskins, Frederick Hall
House, William Burtner
Hsu, Jeng Mein
Hubbard, Daniel Willis
Huenemann, Ruth L
Hulan, Howard Winston
Hundley, James Manson
Hunscher, Helen Alvina (Mrs H P Wilkinson)

Hunt, Sara McClanahan
Hurley, Lucille Shapson (Mrs Kenneth Thompson)
Hursh, Laurence M
Hurt, H David
Hutjens, Michael Francis
Ingram, William Prentiss, Jr
Jackson, Thad Marshall
Jacob, Robert Allen
Jacobson, Don Richard
Jacobson, Marion
Jaffe, Werner G
Jansen, Gustav Richard
Jelliffe, Derrick Brian
Jenkins, Kenneth James William
Jerome, Norge Winifred
Johnson, B Connor
Johnson, Dewey, Jr
Johnson, Doris
Johnson, Elizabeth Cox
Jomain-Baum, Mireille
Jordan, Constance (Louise) Brine
Jukes, Thomas Hughes
Kahn, Samuel George
Kainski, Mercedes H
Kallfelz, Francis A
Kaplan, Murray Lee
Keane, Kenneth William
Kellogg, David Wayne
Kelsay, June Lavelle
Kenney, Mary Alice
Keys, Ancel (Benjamin)
Khan, Mahmood Ahmed
Kies, Constance
Kilgore, Lois Taylor
King, Dorothy Wei (Cheng)
King, Janet Carlson
King, Kendall Willard
Kirksey, Avanelle
Kittrell, Flemmie P
Klevay, Leslie Michael
Knox, Kirvin L
Kokatnur, Mohan Gundo
Konishi, Frank
Konlande, James Edward
Kopito, Louis Eliezer
Koyal, Sankar Nath
Kratzer, Frank Howard
Kraynak, Matthew Edward
Krehl, Willard Arthur
Kreitzman, Stephen Neil
Krishna, Gopal X
Kroenberg, Bernd
Kronfeld, David Schultz
Krumdieck, Carlos L
Kubena, Leon Franklin
Kunkel, Harriott Orren
Kurnick, Allen Abraham
Kurzynske, Janet Stickley
Lachance, Paul Albert
Lamb, Mina Marie Wolf
Landes, Doelas Randy
Langford, Florence
Larkin, Frances Ann
Latham, Michael Charles
Lee, Chung
Lee, Melvin
Leevy, Carroll M
Lei, Kai Yui
Leichter, Joseph
Leitzmann, Claus
Leklem, James Erling
Leung, Philip Min Bun
Leung, Woot-Tsuen Wu
Leveille, Gilbert Antonio
Leverton, Ruth Mandeville
Levy, Robert Sigmund
Lewis, Burnadine Langston
Lewis, Harvye Fleming
Lewis, Jane Sanford
Lieber, Charles Saul
Liener, Irvin Ernest
Light, Amos Ellis
Linkswiler, Hellen
Lis, Elaine Walker
Little, Perry L
Lloyd, Lewis Ewan
Lockhart, Haines Boots
Longenecker, John Bender
Lorenzen, Evelyn June
Luckey, Thomas Donnell
Luecke, Richard William
Luick, Jack Roger
Lushbough, Channing Harden
Lutwak, Leo
Lyman, Richard Lee
Lynch, Gerald John
Mafarachisi, Boaz Amon
Magee, Aden Combs, III
Mahaffey, Kathryn Rose
Mangel, Margaret
Mangelson, Farrin Leon
Mann, George Vernon
Margen, Sheldon
Marion, James Edsel
Marlatt, Abby Lindsey
Marlett, Judith Ann

Marsh, Alice Garrett
Martin, Jack E
Martin, Roy Joseph, Jr
Martin, William Gilbert
Mason, Karl Ernest
Mason, Marion
Mata, Leonardo J
Mathias, Melvin Merle
Matthews, Ruth Hastings
Matz, John J, Jr
Mayer, Jean
McClure, Frank James
McDivitt, Maxine Estelle
McDonald, Bruce Eugene
McDonald, James Lee, Jr
McDonnell, Joseph Francis, Jr
McDowell, Marion Edward
McGinnis, James
McKigney, John Ignatius
McKnight, William Fraulene
McPherson, Clara
McWilliams, Margaret Ann
Meade, Robert J
Meadows, Jerriane Kujie Stafford
Mehring, Jeffrey Scott
Melliere, Alvin L
Menzel, Daniel B
Mertz, Walter
Meyer, James Henry
Meyer, Walter H
Mgbodile, Marcel Ume
Michaud, Laurent
Miller, Agnes Chambless
Miller, Donald Fletcher
Miller, Gary A
Miller, Lorraine Theresa
Miller, Oscar Neal
Miller, Sanford Arthur
Milligan, Larry Patrick
Milner, John Austin
Miner, James Joshua
Minnich, Virginia
Mitchell, Madeleine Enid
Moore, Aimee N
Moore, Jerry Lamar
Morgan, Barbara Louise
Morley, Nina Hope
Morris, Eugene Ray
Morris, Harold Paul
Morris, Rosemary Shull
Morrison, Alexander Baillie
Morse, Ellen Hastings
Morse, Lura Myra
Motzok, Ilary
Moxon, Alvin Lloyd
Mraz, Frank Rudolph
Munro, Hamish N
Murai, Mary Miyeko
Murphy, Elizabeth Wilcox
Myron, Duane R
Nagy, Julius G
Nance, John Arthur
Navia, Juan Marcelo
Neff, Carroll Forsyth
Nelson, Eldon Carl
Nelson, Ralph A
Nesheim, Malden Charles
Nesheim, Robert Olaf
Neville, Janice Nelson
Newberne, Paul M
Newell, Kathleen
Nichoalds, George Edward
Nichols, Buford Lee, Jr
Nielsen, Forrest Harold
Nimni, Marcel Efraim
Nizel, Abraham Edward
Nordsiek, Frederic William
Norvell, Michael Jimmy
Oace, Susan M
Obert, Jessie C
O'Dell, Boyd Lee
Odland, Lura Mae
Ohlson, Margaret Alexander
Oien, Helen Grossbeck
Ory, Robert Louis
Osborn, Margaret Olive
Oser, Bernard Levussove
Ostenso, Grace Laudon
Ostwald, Rosemarie
Otterby, Donald Eugene
Ousterhout, Lawrence Elwyn
Owen, George Murdock
Pagenkopf, Andrea L
Palafox, Anastacio Laida
Palmer, Grant H
Parks, Paul Franklin
Parrish, Donald Baker
Patrick, Homer
Payne, Irene R
Pearson, Paul Brown
Peeler, Herbert Tremble
Peng, Yeh-Shan
Peterson, Daniel Walter
Pettit, L Archer
Pfander, William Harvey
Phansalkar, Sadashiv Vinayak
Phelps, Richard A
Phillips, Jean Allen

Picciano, Mary Frances Ann
Pike, Ruth Lillian
Pipkin, George Erwin
Pi-Sunyer, F Xavier
Platzer, Edward George
Plumlee, Millard P, Jr
Poling, Clyde Edward
Pories, Walter J
Prater, Arthur Nickolaus
Prather, Mary Elizabeth Sturkie
Price, Jack D
Pringle, Dorothy Jutton
Proctor, Jerry Franklin
Prothro, Johnnie W
Purdom, Martha Elda
Purvis, George Allen
Pye, Orrea Florence
Rambaut, Paul Christopher
Rasmussen, Oscar Gustav
Rathmann, Dorothy Marie
Read, Merrill Stafford
Reber, Elwood Frank
Reddy, Bandaru Sivarama
Register, Ulma Doyle
Reid, Bobby Leroy
Renner, Ruth
Reussner, George Henry
Reynolds, Marjorie Lavers
Rice, Eldon Emerson
Richardson, Martha
Riggs, Thomas Rowland
Rinehart, Keith Edward
Ringsdorf, Warren Marshall, Jr
Ritchey, Sanford Jewell
Rivers, Jerry Margaret
Robbins, Ralph Compton
Robertson, Elizabeth Chant
Robinson, Corinne (Hogden)
Robson, John Robert Keith
Roderuck, Charlotte Elizabeth
Roe, Daphne A
Rogers, Quinton Ray
Rogers, William Edward, Jr
Romsos, Dale Richard
Rosen, Fred
Rosenberg, Hans Reinhard
Rosenberg, Irwin Harold
Rosenfield, Daniel
Rousseau, Joseph Edward, Jr
Runyan, Thora J
Runyan, William Scottie
Rusoff, Irving Isadore
Rusoff, Louis Leon
Sabo, Dennis John
Sabry, Zakaria I
Sanchez, Albert
Sandstead, Harold Hilton
Sansone-Bazzano, Gail
Sargent, Frederick, II
Sassoon, Humphrey Frederick
Savage, Jane Ramsdell
Sawits, Marie Schlam
Sawyer, Frederick Miles
Schaefer, Arnold Edward
Schelling, Gerald Thomas
Schemmel, Rachel A
Schenk, Roy Urban
Schneider, Donald Louis
Schneider, Howard Albert
Schofield, Frances Armistead
Schulz, Arthur R
Schwartz, Ruth
Scott, Milton Leonard
Scrimshaw, Nevin Stewart
Seelig, Mildred Sylvia
Seifter, Eli
Serafin, John Augustus
Sexton, Edwin Leon
Shank, Fred Ross
Shapiro, Ralph
Shaw, Emil Gilbert
Sheffy, Ben Edward
Shils, Maurice Edward
Sibbald, Ian Ramsay
Sidwell, Virginia DeCecco
Siedler, Arthur James
Skala, James Herbert
Slack, Samuel Thomas
Slyter, Leonard L
Smith, Edwin Barkley, Jr
Smith, Irvin Darrow
Smith, Jack Louis
Smith, Janice Minerva
Smith, Mary Ann Harvey
Smith, Robert Ewing
Smith, Sam Corry
Soldo, Anthony Thomas
Sosulski, Frank Walter
Souders, Helen Jeanette
Spannuth, Hiram Troutman
Speckmann, Elwood W
Speer, Vaughn C
Speirs, Mary
Spirakis, Charles N
Standal, Bluebell R S'Yiem
Stanley, Barry Chapin
Starcher, Barry Chapin
Stare, Fredrick John

Yankauer, Alfred
Yerby, Alonzo Smythe
Yogore, Mariano G, Jr
Zusman, Jack

Public Health Administration
Akers, Robert Preston
Allen, Ernest Mason
Ast, David Bernard
Beckjord, Philip Rains
Bernstein, Leon
Boss, Willis Robert
Bowen, Cornelius Monroe
Casselman, Warren Gottlieb Bruce
Comer, James Pierpont
Counts, Jon Milton
Crouch, Madge Louise
Crout, John Richard
Dalmat, Herbert Theodore
Dever, G E Alan
Donaldson, Alan Weston
Eagles, Eldon Lewis
Gerring, Irving
Harlin, Vivian Krause
Harris, Maureen Isabelle
Harris, Saul Joseph
Hatchett, Stephen Pinckney
Hausler, William John, Jr
Hayes, Guy Scull
Henley, William Ballentine
Heustis, Albert Edward
Howard, Lee Milton
Jay, George Edgar, Jr
Jolley, Homer Richard
Lambert, Paul Dudley
Lane, John Edward
Long, Earl Ellsworth
Lyons, Joseph Paul
Mazade, Noel Andre
Moore, Robert Conley
Pahl, Herbert Bowen
Price, David Edgar
Ranta, Lawrence Edward
Seubold, Frank Henry, Jr
Sibley, Hiram
Sox, Ellis Dean
Steinwachs, Donald Michael
Wenzel, Richard Louis
Wertheimer, Albert I
Westman, Ragnar Theophile
Williamson, Francis Sidney Lainer
Worden, John Lorimer, Jr
Zerzavy, Frederick M

Public Health Education
Green, Lawrence Winter
Hinrichs, Marie Agnes
Kent, Rosemary (Christine) May
Lonergan, Lester Harold
Tavano, Donald C
Welbeck, Paa-Bekoe Henry
Yoho, Robert Oscar

Radiological Health
Bair, William J
Brantley, John Calvin
Bushong, Stewart Carlyle
Cheung, Augustine Y
Ellett, William H
Fairman, William Duane
Foreman, Harry
Guilmette, Raymond Alfred
Jenkins, Vernon Kelly
Johnson, Ronald Gene
Lea, William Lief
Lee, Arthur Clair
Love, Robert Alexander
Matuszek, John Michael, Jr
McGee, Charles E
Meredith, Orsell Montgomery
Milne, Walter Leroy
Mooney, Richard T
Paras, Peter
Parrott, Marshall Ward
Richardson, Allan Charles Barbour
Shapiro, Jacob
Skrable, Kenneth William
Wegst, Walter F, Jr
Whipple, George Hoyt, Jr
Willhoit, Donald Gillmor
Zelac, Ronald Edward
Zirkes, Al

Tropical Public Health
Cash, Richard Alan
Dysinger, Paul William
Jobin, William Roger
Latham, Michael Charles
Legters, Llewellyn J
Robinson, Henry William

Veterinary Public Health
Beran, George Wesley
Blenden, Donald
Diesch, Stanley L
Dorn, Charles Richard
Dreesen, David W
Fluharty, Dean Milton

Fowler, James Lee
Hubbard, Harold Benson
Jones, Russell K
Kemp, Graham Elmore
Konnerup, Nels Millard
Moon, Jay Ryong
Neurauter, Lloyd Joseph
Pope, Robert Eugene
Schantz, Peter Mullineaux
Schnurrenberger, Paul Robert
Tierkel, Ernest Shalom

HISTOLOGY

Histology
Ahmad, Nazir
Anderson, John Walberg
Armstrong, Rosa Mae
Axelrad, Arthur Aaron
Bates, M Noble
Beckert, William Henry
Belanger, Leonard Francis
Beresford, William Anthony
Berk, Harold
Bertalanffy, Felix D
Bevelander, Gerrit
Blaha, Gordon C
Bourne, Earl Whitfield
Buck, Robert Crawford
Buno, Washington Hector
Burton, Alexis Lucien
Calhoun, Mary Lois
Carpenter, Russell Le Grand
Carriere, Rita Margaret
Caso, Louis Victor
Chan, An Soo
Chang, William Wei-Lien
Chapman, Arthur Owen
Chinea, Jose Juan
Clawson, Robert Charles
Cominsky, Catherine
Connors, Natalie Ann
Crummy, Pressley Lee
Cutts, James Henry
De Bruyn, Peter Paul Henry
Deck, James David
Dellmann, Horst-Dieter
Dockum, Norman Leslie
Dossel, William Edward
Dougherty, William J
Dropp, John Jerome
Eagles, Jan
Emmel, Victor Meyer
Erlandsen, Stanley L
Erpino, Michael James
Fahrenbach, Wolf Henrich
Farbman, Albert Irving
Fawcett, Don Wayne
Fedoroff, Sergey
Finn, James Bernard
Ford, Peter
Fuller, Eugene George
Gartner, Leslie Paul
Geissinger, Hans Dieter
Going, Robert Ernest
Goldstein, Abraham M B
Grimm, Arthur F
Grodums, Emma Irene
Grosso, Leonard
Gustafson, Alvar Walter
Habowsky, Joseph Edmund Johannes
Ham, Arthur Worth
Hardy Falilding, Margaret Hurlstone
Harris, Thomas Mason
Hartman, Mary Ellen
Hayashida, Kaye
Hayes, Thomas G
Hearney, Elaine Frances
Heath, Gordon Wayne
Hibbs, Richard Guythal
Hirschman, Albert
Hoffman, Loren Harold
Holloway, Clarke L
Horst, G Roy
Houston, Elsie Washburn
Houston, Marshall Lee
Humphrey, Rufus R
Hung, Kuen-Shan
Isler, Henri Gustave
Jande, Sohan Singh
Jensh, Ronald Paul
Johannessen, Leif Bertram
Johnson, Dorothy Dole
Johnson, Oscar Walter
Julyan, Frederick John
Kahn, Raymond Henry
Kallenbach, Ernst Adolf Theodor
Kaltenbach, Jane Couffer
Kapur, Shakti Prakash
Keis, Adelbert Ferdinand Richard
Kelsall, Margaret Aston
King, Gladys Smith
King, John Edward
Kopriwa, Beatrix Markus
Krause, William John
Laham, Quentin Nadime

Lamberg, Stanley Lawrence
LaVelle, Faith Wilson
Lawrence, Irvin E, Jr
Lewis, Carmie Perrota
Loevy, Hannelore Taschini
Luft, John Herman
Lung, Ben
MacCallum, Donald Kenneth
Marshall, Frederick James
McKenzie, John Ward
McMillan, Donald Burley
Medlen, Ammon Brown
Melcher, Antony Henry
Menees, James H
Menton, David Norman
Mitchell, Ormond Glenn
Mohn, Melvin P
Momberg, Harold Leslie
Monson, Frederick Carlton
Morgans, Leland Foster
Murphy, Henry D
Murray, Raymond Gorbold
Newstead, James Duncan MacInnes
Nixon, Charles William
Nunnemacher, Rudolph Fink
O'Morchoe, Patricia Jean
Paff, George Hugo
Palade, George E
Peach, Roy
Penney, David P
Pereira, Gerard P
Pinero, Gerald Joseph
Pinkstaff, Carlin Adam
Pollock, Robert J, Jr
Putnam, Jerry L
Quinton-Cox, Robert
Rapp, Robert
Rhodes, Rondell H
Richter, Kenneth Murrel
Robinson, Hamilton Burrows Greaves
Robinson, Roy Garland, Jr
Rollason, Herbert Duncan
Rolle, Gloria Katharine
Roth, Owen
Safanie, Alvin H
St Pierre, Ronald Leslie
Salt, Walter Raymond
Salthouse, Thomas N
Schatz, Leo
Shea, John Raymond Michael, Jr
Sisca, Rodger Franklin
Skjonsby, Harold Samuel
Smithson, Janet Eleanor
Snook, Theodore
Sorokin, Sergei Pitirimovitch
Strachan, Donald Stewart
Strautz, Robert Lee
Suciu-Foca, Nicole M
Swanson, Ernest Allen, Jr
Tamarin, Arnold
Terracio, Louis
Tillman, Larry Jaubert
Tillson, Albert Holmes
Towns, Clarence, Jr
Trombetta, Louis David
Van Breemen, Verne Leroy
Van Sickle, David C
Warshawsky, Hershey
Waterhouse, John P
Weary, Marlys E
Webber, William A
Webster, Richard Curtis
Wei, Stephen Hon Yin
Weisman, Harvey
Weitzman, Mary C
Werner, Henry James
West, William T
Weston, John Colby
Wood, Pauline J
Yaeger, James Amos
Zaki, Abd El-Moneim Emam
Zeit, Walter

Histochemistry
Belanger, Leonard Francis
Bolduc, Reginald J
Breil, Sandra J
Brown, Stephen Clawson
Brownscheidle, Carol Mary
Evans, Lance Saylor
Gerold, Nicolas John
Gersh, Isidore
Glick, David
Goldberg, Benjamin
Hack, Marvin Howard
Hanker, Jacob S
Iyengar, V K Sundararaja
Leopold, Roger Allen
Lhotka, John Francis, Jr
Lillie, Ralph Dougall
Longley, James Baird
Manning, John Paul
Manocha, Sohan Lall
Murthy, Krishna A S
Nakane, Paul K
Nuki, Klaus
Puchtler, Holde
Shellhamer, Robert Howard

Steeves, Harrison Ross, III
Woods, James E

Histopathology
Artizzu, Maria
Daoust, Roger
Harris, Norman Oliver
Jewell, Frederick Forbes
Ouellette, Guillemond Benoit
Smith, James Knox
Yevich, Paul Peter

Histophysiology
De Robertis, Eduardo Diego P
Desmond, Alton Harold
Grob, Howard Shea
Martin, Kathryn Helen
Quay, Wilbur Brooks
Riley, Edward Eddy, Jr
Wimsatt, William Abell

IMMUNOLOGY

Immunology
Abdou, Nabih I
Abrahams, Irving
Abramoff, Peter
Ackerman, Neil Richard
Adamkiewicz, Vincent Witold
Adams, Dolph Oliver
Adler, Frank Leo
Affronti, Lewis Francis
Ainis, Herman
Al-Askari, Salah
Alexander, James Wesley
Alexander, Nancy J
Allen, Peter Zachary
Alms, Thomas H
Altman, Leonard Charles
Ambrose, Charles T
Amkraut, Alfred A
Anderson, Burton
Anderson, Byron
Anderson, Robert Simpers
Anthony, Ronald Lewis
Appella, Ettore
Archer, Stanley J
Argall, Clifford Irving
Arnason, Barry Gilbert Wyatt
Arquembourg, Pierre Charles
Arquilla, Edward R
Asofsky, Richard Marcy
Atchison, Robert Wayne
Austen, K Frank
Azar, Miguel M
Bach, Michael Klaus
Baer, Harold
Baer, Rudolf L
Bailey, Garland Howard
Bajpai, Praphulla K
Baker, Phillip John
Bankert, Richard Burton
Baram, Peter
Bardana, Emil John, Jr
Barker, Clyde Frederick
Barnes, Glover William
Barrett, James Thomas
Barriga, Omar Oscar
Bartfeld, Harry
Barth, Rolf Frederick
Bass, Joseph Alonzo
Batista, Jack Richard
Bauer, Dietrich Charles
Baum, John
Bauman, Norman
Baumgarten, Alexander
Bawden, Monte Paul
Bean, Michael Arthur
Beaulieu, J A E M
Becker, Elmer Lewis
Bell, Clara G
Benacerraf, Baruj
Benjamin, Eliezer
Bennett, Joe Claude
Berczi, Istvan
Bergman, Robert Kaye
Berney, Steven
Bernstein, Stanley H
Bibb, William Robert
Bice, David Earl
Bigley, Nancy Jane
Billingham, Rupert Everett
Bing, David H
Blaese, Robert Michael
Blair, Phyllis Beebe
Blaker, Robert Gordon
Blitz, Ruth R
Biznakov, Emile George
Bloch, Jerome Bernard
Bloom, Barry R
Bloom, Eda Terri
Bogden, Arthur Eugene
Boley, Robert B
Bolognesi, Dani Paul
Boney, William Arthur, Jr
Boraker, David Kenneth

Boros, Dov Lewis
Borsos, Tibor
Bowen, James Milton
Bowen, William H
Boyd, William Clouser
Brackett, Robert Giles
Bradshaw, Claire Margaret
Brandiss, Michael W
Brennan, Patricia Conlon
Brenner, Lorry Jack
Brockman, John A, Jr
Broder, Irvin
Brown, George L
Brown, John Clifford
Brown, Russell Vedder
Burdash, Nicholas Michael
Burkholder, Peter M
Burnham, Thomas K
Burrell, Robert Guthrie
Burton, Robert McMahon
Butchko, Gregory Michael
Butler, John Edward
Byfield, Patricia E
Byrd, William Joseph
Cabrera, Edelberto Jose
Cain, William Aaron
Camp, Frank Rudolph, Jr
Campbell, Priscilla Ann
Campbell, Samuel Gordon
Cannon, Donald Charles
Cantor, Ena D
Capra, J Donald
Carpenter, Philip Lewis
Carr, Ronald Irving
Carski, Theodore Robert
Casey, Helen Liles
Caso, Louis Victor
Castro, Anthony Edward
Cathou, Renata Egone
Cerini, Costantino Peter
Chaffee, Elmer Fenn
Chang, Ik-Chin
Chaparas, Sotiros D
Chaperon, Edward Alfred
Chase, Randolph Montieth, Jr
Chen, James Pai-Fun
Childress, Evelyn Tutt
Christian, Charles L
Christian, Walter
Chu, Yang-Ming
Chused, Thomas Morton
Claman, Henry Neumann
Clark, Junius Manson
Clem, Lester William
Coe, John Emmons
Cohen, Elias
Cohen, Flossie
Cohen, Martin William
Cohen, Nicholas
Cohen, Pinya
Cohen, Sheldon Gilbert
Cohen, Stanley
Cole, Gerald Alan
Coleman, Robert Marshall
Colley, Daniel George
Collins, Amy L Tsui
Colten, Harvey Radin
Conant, Robert M
Condoulis, William V
Cone, Robert Edward
Connolly, John Joseph
Considine, Richard George
Contento, Isobel Corneil
Cook, Eula Belle Maley
Cook, Albert Hewett
Coons, Albert Hewett
Cooper, Edwin Lowell
Cooper, Max Dale
Cooperband, Sidney R
Coyne, Veronica E
Crandall, Richard B
Cravitz, Leo
Cremer, Natalie E
Criswell, Bennie Sue
Crowle, Alfred John
Cruse, Julius Major, Jr
Curtis, Gary Lynn
Cusumano, Charles Louis
Daguillard, Fritz
D'Alesandro, Philip Anthony
Dalmasso, Agustin Pascual
Damian, Raymond T
Danneberg, Arthur Milton, Jr
Dardas, Terry Jay
Datta, Surinder P
Daul, Carolyn Virginia Beach
Davenport, Calvin Armstrong
David, John R
Davie, Joseph Myrten
Davis, William C
Dawson, Jeffrey Robert
Day, Eugene Davis
DeBoer, Carl John
De Fazio, Sally Ruth
De Goes, Paulo
Del Villano, Bert Charles
Dent, Peter Boris
De Repentigny, Jacques
Desowitz, Robert

Despommier, Dickson
de Weber, Leverett L
Devlin, Richard Gerald, Jr
DeWitt, Charles Wayne, Jr
DiLeone, Gilbert Robert
Di Sabato, Giovanni
Dixon, Frank James
Domingue, Gerald James
Donaldson, David Miller
Donaldson, Paul
Dorrington, Keith John
Draper, Laurence Rene
Dreesman, Gordon Ronald
Dreyer, William J
Dupuy, Harstry Joseph
Duquesnoy, Rene J
Dutton, Richard W
Duwe, Arthur Edward
Dyer, John Kaye
Eaton, Monroe Davis
Edberg, Stephen Charles
Edgington, Thomas S
Edidin, Michael Aaron
Edsall, Geoffrey
Edwards, Joshua Leroy
Edwards, Robert Bryce
Eidinger, David
Eisen, Herman Nathaniel
Eisenstein, Toby K
Elgert, Klaus Dieter
Eng, Chee Ping
English, Leonard Stanley
Epps, Anna Cherrie
Esber, Henry Jemil
Escobar, Mario R
Esposito, Vito Michael
Evans, Richard Todd
Everhart, Donald Lee
Eyre, Peter
Fahey, John Leslie
Fahlberg, Willson Joel
Falk, Lawrence A, Jr
Falkler, William Alexander, Jr
Falletta, John Matthew
Fanale, Louisa P
Fanger, Michael Walter
Faust, Charles Harry, Jr
Feeley, John Cornelius
Feingold, Ben F
Feit, Carl
Fenton, John William, II
Fife, Earl Hanson, Jr
Fink, Jordan Norman
Finke, James Harold
Fireman, Philip
Fishman, Marvin
Flanagan, Thomas Donald
Fletcher, Mary Ann
Flick, John A
Fogel, Bernard J
Folds, James Donald
Fong, Jack Sun-Chik
Fontana, Vincent J
Forsdyke, Donald Roy
Foster, John Wallace
Frank, Michael M
Freedman, Henry Hillel
Freedman, Murray H
Freedman, Samuel Orkin
Frey, James R
Frick, Oscar L
Friedman, Eli A
Friedman, Harvey Paul
Friedman, Herman
Friou, George Jacob
Fritz, Katherine Elizabeth
Fudenberg, H Hugh
Fugmann, Ruth Adele
Fulginiti, Vincent Anthony
Fuller, Laphalle
Fung, Henry C, Jr
Fuson, Roger Baker
Gabrielsen, Ann Emily
Gardner, Edward, Jr
Gasser, David Lloyd
Gaumer, Herman Richard
Getzer, Justus
Gengozian, Nazareth
Gershon, Richard K
Ghosh, Chitta Ranjan
Giaver, Ivar
Gigli, Irma
Gilden, Raymond Victor
Gill, Thomas James, III
Girard, Kenneth Francis
Gleich, Gerald J
Godfrey, Henry Philip
Goldberg, Leonard Stephen
Golden, Carole Ann
Goldust, Marvin Bertram
Goldman, Armond Samuel
Goldstein, Gideon
Golub, Edward S
Golub, Sidney Harris
Gooding, Linda R
Goodman, Howard Charles
Goodman, Joel Warren
Goodman, Morris

Gordon, Julius
Gotoff, Samuel P
Gottlieb, A Arthur
Graber, Charles David
Graf, Liselotte
Grappel, Sarah Fay
Gray, Gary D
Greenberg, Arnold Harvey
Greenberg, Leonard Jason
Greenberg, Louis
Greene, Elias Louis
Greene, Joyce Marie
Greene, Nathan Doyle
Greeson, Phillip Doyle
Griffin, Diane Edmund
Grossberg, Allan Louis
Groves, David Lynn
Gruenwald, Ruben
Guenther, Donna Marie
Guerault, Armand
Gupta, Sudhir
Gusdon, John Paul
Guttmann, Ronald D
Guy, Leona Ruth
Gyulai, Eugene Jeno
Habeeb, Ahmed Fathi Sayed Ahmed
Haber, Edgar
Habich, John Harold
Hadden, John Winthrop
Hall, Charles Thomas
Han, Tin
Hanes, Deanne Meredith
Hanks, John Harold
Hanly, W Carey
Hanna, Edgar Ethelbert, Jr
Hardegree, Mary Carolyn
Harris, Jules Eli
Harris, Tzvee N
Harrison, Bettina Hall
Harrison, David Ellsworth
Hartler, Brack Gillium, Jr
Hawkins, David Geoffrey
Hayes, Sheldon P
Hehre, Edward James
Heidrick, Margaret Louise
Heim, Lyle Raymond
Heise, Eugene Royce
Hellstrom, Ingegerd Elisabet
Hellstrom, Karl Erik Lennart
Henley, Walter L
Henney, Christopher Scot
Henson, Claudia
Henson, Peter Michell
Heppner, Gloria Hill
Herr, Harry Wallace
Herscowitz, Herbert Bernard
Hersh, Evan Manuel
Herzenberg, Leonard Arthur
Hess, Evelyn V
Heuer, Ann Elizabeth
Hewetson, John Francis
Hicks, Edward James
Hicks, Harold Richard
Hildemann, William Henry
Hillyer, George Vanzandt
Hinton, Dennis Melvin
Hiramoto, Raymond Natsuo
Hirata, Arthur Atsunobu
Hirsch, James Gerald
Hirsch, Martin Stanley
Hoff, Gail Richards
Hoff, Richard Lee
Hoffmann, Donald Richard
Hoffmann, Edward Marker
Hoffmann, Louis Gerhard
Hogg, Richard A
Hokama, Yoshitsugi
Holman, Halsted Reid
Holme, George
Holper, Jacob Charles
Holzman, Robert Stephen
Hong, Richard
Horning, Maria O'C
Horton, John Edward
Horton, Richard E
Houck, John Candee
Howard, Ronald M
Howe, Michael Luray
Howell, Stephen Barnard
Hsu, Hsi Fan
Hsu, Shu Ying Li
Hubbard, William T
Hubscher, Thomas
Hudson, Bruce William
Hunt, William B, Jr
Hunter, Robert L
Hyde, Richard Moorehead
Ichiki, Albert Tatsuo
Ing, Wai Kwok
Ingraham, Joseph Sterling
Isa, Abdallah Mohammad
Ishizaka, Kimishige
Jackson, Anne Louise
Jackson, Thad Marshall
Jacobs, Alma Alice
Jacobs, Barbara B
Jacobs, Diane Margaret

Jacoby, Robert Ottinger
Janicki, Bernard William
Janssen, Robert (James) J
Jaroslow, Bernard Norman
Jasin, Hugo E
Jennings, Billy Ray
Jensen, Richard Harvey
Jeska, Edward Lawrence
Joel, Darrel Dean
Johnson, Alice Ruffin
Johnson, Arthur Gilbert
Johnson, Howard M
Johnson, Rother Rodenious
Johnston, Theodore Reynold
Johnston, Marilyn Frances Meyers
Jordon, Robert Earl
Josephson, Alan S
Jurand, Jerry George
Justus, David Eldon
Jutila, John W
Kagan, Irving George
Kahan, Barry D
Kaliss, Nathan
Kaplan, Carolyn Marie
Kantor, Fred Stuart
Kaplan, Alan Marc
Kaplan, Joel Howard
Kaplan, Melvin Thomas
Kaplan, Morris Aaron
Karl, Seung Chul
Karp, Richard Dale
Karush, Fred
Kassel, Robert Lawrence
Kelleher, James Joseph
Kelly, Michael Thomas
Kemp, Walter Michael
Kennedy, James Cecil
Kenney, Michael Thomas
Kenyon, Alan J
Kessel, Rosslyn William Ian
Kettman, John Rutherford, Jr
Kenyon, Richard H
Kim, Byung Suk
Kim, Charles Wesley
Kim, Yoon Berm
Kim, Young Tai
Kind, Phyllis Dawn
Kirkpatrick, Charles Harvey
Kiszkiss, David F
Kite, Joseph Hiram, Jr
Klein, Paul Alvin
Kligman, Lorraine H L
Knight, Katherine Lathrop
Kniker, William Theodore
Kochan, Ivan
Kohler, Heinz
Kollmorgen, G Mark
Kong, Yi-Chi Mei
Koppenhefer, Thomas Lynn
Korngold, Leonhard
Koros, Aurelia M Carissimo
Koshland, Marian Elliott
Krakow, Gladys
Kramer, Norman Clifford
Krause, Richard Michael
Krueger, Robert George
Kubo, Ralph Teruo
Kuhns, William Joseph
Kulkarni, Anant Sadashiv
Kumar, Manjula Satyendra
Kwapinski, J B George
Kyriazis, Andreas P
Lambert, Reginald Max
Lamm, Michael Emanuel
Lamon, Eddie William
Lamoureux, Gilles
Lang, Bruce Z
Lang, Raymond W
Lange, Charles Ford, Jr
Larsen, Bodil Astrid
Larson, Kenneth Allen
LaSalle, Marjorie
Laux, David Charles
Lawrence, David A
Lazda, Velta Abuls
Le Beau, Leon Joseph
Leddy, John Plunkett
Lee, Chang Ling
LeFor, William Mathew
Legler, Donald Wayne
Lennon, Vanda Alice
Leon, Myron A
Leone, Charles Abner
Lepow, Irwin Howard
Lerner, Stephen Paul
Lerner, Richard Alan
Leskowitz, Sidney
Leslie, Gerrie Allen
Leu, Richard William
Leung, Christopher Chung-Kit
Levine, Leo
Levine, Philip
Levitt, Neil Hilliard
Levy, David Alfred
Levy, Nelson Louis
Lewis, Robert Miller
Leznoff, Arthur

Warner, George S
Watson, Edna Sue
Watson, Ronald Ross
Waxdal, Myron John
Webb, Phyllis Marie
Webster, Robert G
Wegmann, Thomas George
Weidanz, William P
Weigle, William O
Weiler, Eberhardt
Weiler, Ivan-Jeanne Mayfield
Weiner, Lawrence Myron
Weiner, Russell Shirley
Weislow, Owen Stuart
Weiss, Charles
Weiss, David Walter
Weksler, Marc Edward
Welter, C Joseph
Weltman, Joel Kenneth
Weseli, Donald Fenton
West, Clark Darwin
Wheeler, Harry Ogden
White, Edward
Whitehouse, Frank, Jr
Whiteside, Roberta Emerson
Whiteside, Theresa L
Whitten, Harrell David
Widder, James Stone
Widmark, Rudolph M
Wiggin, Norman Jack Bridgman
Wiley, Bill Beauford
Wilhelm, Rudolf Ernst
Williams, Jeffrey F
Williams, Ralph C, Jr
Willoughby, William Franklin
Wilson, Darcy Benoit
Wilson, Robert James
Windhorst, Dorothy Baker
Winfield, John Buckner
Winn, Henry Joseph
Winters, Wendel Delos
Wishart, Franklyn Ogilvie
Wittig, Heinz Joseph
Woehler, Michael Edward
Wofsy, Leon
Wolberg, Gerald
Wolcott, Robert Michael
Wolf, Sheldon Malcolm
Wong, Donald Tai On
Woods, Alexander Hamilton
Wright, George Green
Wust, Carl John
Yanari, Sam Satomi
Yang, Ovid Y H
Yen-Watson, Belinda R S
Yokoyama, Mitsuo
Yoo, Tai-June
Zarkower, Arian
Zatz, Marion M
Zeleznick, Lowell D
Zeschke, Richard Herman
Zeya, Hasan Ismail
Ziccardi, Robert John
Zimmerman, Daniel Hill
Zimmerman, Sarah E
Zmijewski, Chester Michael
Zolla-Pazner, Susan Beth
Zurawski, Vincent Richard, Jr
Zvaifler, Nathan J

Immunobiology

Aden, David Paul
Agnew, Robert Morson
Arala-Chaves, Mario Passalaqua
Badger, Alison Mary
Baxter, Joseph George
Baxter, William D
Bellone, Clifford John
Blakeslee, Dennis Lauren
Blazkovec, Andrew A
Borysenko, Myrin
Byers, Vera Steinberger
Carter, Bettina Bush (Mrs Daniel F Jackson)
Carter, Brian Geoffrey
Cebra, John Joseph
Chiscon, Martha Oakley
Claflin, Alice J
Click, Robert Edward
Craig, Susan Walker
Cudkowicz, Gustavo
Day, Noorbibi Kassam
Feldbush, Thomas Lee
Field, Arthur Kirk
Fraker, Pamela Jean
Franzl, Robert E
Freeman, Max James
Gale, Robert Peter
Hanna, Michael G, Jr
Harris, Nick Steven
Hirschhorn, Rochelle
Houston, William Eddie
Hsu, Konrad Chang
Jensen, Joerg
Johnston, Muriel Evelyn
Kent, Naim Hasan
Killion, Jerald Jay

Kornfeld, Lottie
Kuhn, Raymond Eugene
Loan, Raymond Wallace
Lopez, Carlos
Loughman, Barbara Ellen Evers
Lukasewycz, Omelan Alexander
Mandy, William John
Manson, Lionel Arnold
Maurer, Bruce Anthony
McBride, Raymond Andrew
McConnachie, Peter Ross
McElree, Helen
Miller, Gary William
Miller, Ronald Kent
Mitchell, Kenneth Frank
Nash, Donald Robert
Nathenson, Stanley G
Pascal, Theresa A
Patterson, Ronald James
Pearson, David Francis
Ranney, David Francis
Rice, Thomas Kenneth
Rich, Robert Regier
Sabatini, Edris Rinaldo
Sabet, Tawfik Younis
Segre, Mariangela Bertani
Sellin, Helen Gill
Simnoff, Paul
Smart, Keith Lorenzo
Speirs, Robert Sisson
Stringfellow, Dale Alan
Szakal, Andras Kalman
Terres, Geronimo
Thomas, Paul Milton
Thorn, Richard Mark
Toben, Howard Ray
Vann, Douglas Carroll
Vincent, Monroe Mortimer
Willis, Judith Ione
Winter, Alexander J
Wojnar, Robert John
Zighelboim, Jacob
Zwilling, Bruce Stephen

Immunochemistry

Aladjem, Frederick
Allen, Peter Zachary
Amerault, Thomas Eugene
Amkraut, Alfred A
Bangasser, Susan Andretta
Banovitz, Jay Bernard
Benedict, Albert Alfred
Bishop, Claude Titus
Bishop, David C
Borek, Felix
Brundage, William Gregory
Callaghan, Owen Hugh
Capone, James Joseph
Cawley, Leo Patrick
Cernosek, Stanley Frank, Jr
Chambliss, Keith Wayne
Cheng, Frank Hsieh Fu
Cinader, Bernhard
Cote, Melvin
Cowan, Raymond-Henri
Creech, Hugh Morris
Dandliker, Walter Beach
Davis, Gary Samuel
Davis, Raymond Vincent
DiCapua, Richard Anthony
Dubiski, Stanislaw
Egan, Marianne Louise
Elliott, Willard Buford
Engvall, Eva Susanna
Esser, Alfred F
Etzler, Marilyn Edith
Fleisher, Martin
Fox, Alfred Earl
Fredricks, Walter William
Froese, Arnold
Garvey, Justine Spring
Genco, Robert J
Gopalakrishnan, Pungampoondi Velamur
Gough, Patricia Marie
Green, Gerald
Grey, Howard M
Gupta, Krishana Chandara
Hatheway, Charles Louis
Heidelberger, Michael
Humphreys, Robert Edward
Inman, Franklin Pope, Jr
Johnson, Brian John
Jordan, Russell Thomas
Karol, Meryl Helene
Kochwa, Shaul
Kreiter, Victor Peter, Jr
Kuo, Chao-Ying
Levine, Lawrence
Lietze, Arthur
Liu, Chi Tan Chang
Ma, Wai-Sai
MacPherson, Catherine Frances Conway
Maizel, Saul
Marucci, Americo Alvin
Mayers, George Louis
Menninger, Florian Francis, Jr

Immunogenetics

Amos, Dennis Bernard
Bacon, Larry Dean
Bailey, Donald Wayne
Breyere, Edward Joseph
Briles, Connally Oran
Briles, Worthie Elwood
Caldwell, Jerry
Cherry, Marianna
Cohen, Carl
David, Chelladurai S
Dorf, Martin Edward
Fanguy, Roy Charles
Foster, Morris
Fraser, Blair Allen
Frelinger, Jeffrey
Haughton, Geoffrey
Hood, Leroy E
Hsu, Susan Hu
Johnson, Armead H
Klein, Jan
Marsh, David George
Mittal, Kamal Kant
Mobraaten, Larry Edward
Moor-Jankowski, Jan K
Passmore, Howard Clinton
Sanders, Bobby Gene
Schacter, Bernice Zeldin
Schanfield, Moses Samuel
Scherman, Louis W
Schultz, Jane Schwartz
Session, John Joe
Solomon, Joel Martin
Wang, An-Chuan
Ward, Frances Ellen

Immunohematology

Allen, Fred Harold, Jr
Alsever, John Bellows
Baker, William J
Costea, Nicolas V
Crookston, Marie Cutbush
Graham, Henry Alexander, Jr
Issitt, Peter David
LaSalle, Marjorie
Leif, Robert Cary
Marsters, Roger Westcott
Masouredis, Seraeim Panogiotis
McMillan, Robert
Murphy, Martin Joseph, Jr
Myhre, Byron Arnold
Oishi, Noboru
Pirofsky, Bernard
Soules, David Edward
Wiener, Alexander S
Yokoyama, Mitsuo

Immunopathology

Bennett, William Ernest
Brassard, Andre
Chase, William Henry
Duncan, John Robert
Dvorak, Ann Marine-Tompkins
Fierer, Joshua A
Fleetwood, Mildred Kaiser
Halonen, Marilyn Jean
Hammond, Mary Elizabeth

Mestecky, Jiri
Michaeli, Dov
Miller, Thomas Edward
Moskowitz, Merwin
Mudgett, Meredith
Munjal, Devidayal
Nisonoff, Alfred
Nowotny, Alois Henry
Oroszlan, Stephen
Polley, Margaret J
Poretz, Ronald David
Pressman, David
Quinn, Richard Paul
Rebers, Paul Armand
Reisfeld, Ralph Alfred
Renn, Donald Walter
Rockey, John Henry
Rule, Allyn H
Sage, Harvey J
Small, Parker Adams, Jr
Smyth, Robert Daniel
Springer, Georg F
Stelos, Peter
Stemke, Gerald W
Stolfi, Robert Louis
Stone, Stanley S
Tengerdy, Robert Paul
Treadway, William Jack, Jr
Tripodi, Daniel
Vamier, Wilton Emile
van Oss, Carel J
Voss, Edward William, Jr
Weetall, Howard H
Weimer, Henry Eben
White, Gordon Justice
Wirtz, George H
Wright, George Leonard, Jr
Zarco, Romeo Morales

Transplantation Immunology

Bealmear, Patricia Maria
Codd, John Edward
Dosseter, John Beamish
Elkins, William L
Etheredge, Edward Ezekiel
Hackel, Emanuel
Norin, Allen Joseph
Rapaport, Felix Theodosius
Ritzmann, Stephan E
Rosenberg, Jerry C
Truitt, Robert Lindell

Veterinary Immunology

Belden, Everet Lee
Buening, Gerald Mathew
Cunliffe, Harry R
Fraser, C E Ovid
Ingram, Donald George
Morgan, Donald O'Quinn
Ruckerbauer, Gerda Margareta

Haskill, John Stephen
Jones, David Stevens
Koffler, David
Locke, Robert F
Majde, Jeannine Adkins
Marx, James John, Jr
Monjan, Andrew Arthur
Pan, In-Chang
Perper, Robert J
Pinckard, Robert Neal
Smith, Henry John
Valenzuela, Rafael
White, Thomas George
Yoshida, Takeshi

INSTRUMENTATION

Instrumentation

Acheson, Donald Theodore
Alles, Harold Gene
Anger, Hal Oscar
Atkinson, George Francis
Barrett, Harrison Hooker
Baum, John Joseph
Becker, Randolph Armin
Berger, Robert Lewis
Bernard, Joseph Lionel
Blatt, Joel Herman
Bowman, Robert Lewis
Braman, Robert Steven
Callahan, Mary Vincent
Clemens, Anton Hubert
Conlon, Daniel Rupert
Cropper, Walter V
Diem, Hugh Egbert
Earley, Laurence E
Ehrenthal, Harry Charles
Elings, Virgil Bruce
Falk, Edward D
Fisher, Dale John
Gheorghiu, Paul
Greenspan, Joseph
Greenwood, Ivan Anderson
Haas, David Jean
Helmick, Herbert Hanna
Hill, David Lawrence
Hopkins, Robert Charles
John, Joseph
Kaminski, Louis Alfred
Kaye, Wilbur (Irving)
Keepin, George Robert, Jr
Klainer, Stanley M
Knapp, Stephen Laurence
Kurz, Richard J
Kuzel, Norbert R
Leone, James A
Levy, Paul F
McAllister, Robert Wallace
McCandless, Robert William
McCoy, Raymond Duncan
McCrea, Peter Frederick
McCuckin, Warren Francis
Monforee, Gervaise Edwin
Mueller, Theodore Rolf
Paolini, Francis Rudolph
Parkes, Alan Schofield
Peacock, Roy Norman
Pearlman, Michael R
Pinney, Jack Erwin
Raible, Raymond W
Rayside, John Stuart
Read, Philip Lloyd
Rikmenspoel, Robert
Rollwitz, William Lloyd
Rotariu, George Julian
Rowland, Fred W
Rubin, Lawrence G
Sawers, James Richard, Jr
Schanne, Otto F
Sharkey, John Bernard
Shawhan, Elbert Nell
Silverman, Howard L
Smith, Malcolm (Kinmonth)
Spergel, Philip

Sullivan, Harris Martin
Sutter, David Franklin
Taylor, Bruce Cahill
Thompson, Robert Deane
Thompson, Ronald Halsey
Tingle, William Herbert
Tryon, Max
Vogel, Alfred Morris
Vroom, Alan Heard
Weber, Richard Rand
Wehrli, Robert L
Whatley, Thomas Alvah
White, Eugene Wilbert
Williamson, Donald Elwin
Woodruff, Joseph Franklin
Wotherspoon, Neil
Zimmerer, Robert W

Chemical Instrumentation
Abbott, Seth R
Albert, Harrison Bernard
Burden, Stanley Lee, Jr
Conigliaro, Peter James
Converse, Jimmy G
Cooper, James William
Dessy, Raymond Edwin
Giannovario, Joseph Anthony
Hrubesh, Lawrence Wayne
Jackson, Darryl Dean
Jankowski, Conrad M.
Johnson, Henry Wilson, Jr
Klatt, Leon Nicholas
Kobrin, Robert Jay
Lunney, David Clyde
Nunes, Thomas Lester
Prostak, Arnold S
Vassos, Basil Harilaos
Verschingel, Roger H C
Wong, Peter Alexander
Wright, John Martin

Electrocardiology
Dower, Gordon Ewbank
Fischmann, Eugene J

Electromyography
Ebel, Alfred
Gray, Edwin R
Melvin, John Lewis

Electroencephalography
Bankier, Robert G
Blumenthal, Irving Jack
Henry, Charles Eric
Hughes, John Russell
Kooi, Kenneth Ashley
Obrist, Walter Dorn

Electron Microscopy
Abram, Dinah
Anderson, Frank Wallace
Baba, Nobuhisa
Ball, Frances Louise
Barlow, Irmela Christiane
Bills, Robert F
Black, Donald Lee
Blystone, Robert Vernon
Boatman, Edwin S
Brinn, Jack Elliott, Jr
Brown, Robert Francis
Buschmann, Robert John
Callas, Gerald
Campbell, William (Aloysius)
Chamberlain, Nugent Francis
Chambers, John Edward
Christie, John McDougall
Cohen, Arthur Leroy
Comer, Joseph John
Constantinides, Paris
Coons, Lewis Bennion
Cote, Wilfred Arthur, Jr
Daniel, Jack Leland
Darlington, Robert Wells
Davis, Ronald Wayne
DeLamater, Edward Doane
Dessouky, Dessouky Ahmad
Dickey, Joseph Freedman
DiDio, Liberato John Alphonse
Doane, Frances Whitman
Dodson, Ronald Franklin
Duff, Robert Hodge
Dunn, Robert Fowler
Eastwood, Abraham Bagot
Egar, Margaret Wells
Engster, Maryann Sandra
Erickson, Harold Paul
Erickson, John Otto
Essner, Edward Stanley
Fahimi, Hossein Dariush
Federman, Micheline
Fischlschweiger, Werner
Fraser, David Allison
Gardner, John Owen
Geiss, Roy Howard
Goodman, Joseph Robert
Goodman, Tine
Gordon, Albert Raye
Gordon, Gloria

Haar, Jack Luther
Hackenbrock, Charles Robert
Harris, Patricia J
Hart, Raymond Kenneth
Hasegawa, Ichiro
Hashimoto, Ken
Hay, Ian Leslie
Heckman, Francis Austin
Henry, Timothy James
Herbener, George Henry
Herman, Lawrence
Holmgren, Paul
Hormats, Ellis Irving
Hostetler, Jeptha Ray
Hugon, Jean S
Isaacson, Michael Saul
Ito, Susumu
Johnson, Charles F
Kallenbach, Ernst Adolf Theodor
Kampe, Dennis James
Krupa, Paul L
Lafontaine, Jean-Gabriel
Lawson, Kenneth Dare
Leak, Lee Virn
Lee, Chin-Chiu
Le Fever, Hermon Michael
Leifer, Calvin
Lim, David J
Lynn, Joseph Alden
Maldonado, Jorge Eusebio
McGee-Russell, Samuel M
Melton, Carl Wesley
Metevia, Louis Anthony
Milburn, Nancy Stafford
Miller, Carl Henry, Jr
Millhouse, Edward W, Jr
Moses, Montrose James
Nalbandian, John
Nazerian, Keyvan
Neale, Elaine Anne
Niewenhuis, Robert James
Noonan, Sharon Mariella
Nopanitaya, Waykin
Normandin, Diane Kilbourne
Nylen, Marie Ussing
O'Brien, Elinor Murray
Olins, Ada Levy
Oshiro, Lyndon Satoru
Oster, Carl Frederich
Pangborn, Jack
Pease, Daniel Chapin
Penney, David P
Ranganathan, Vajapeyam S
Reimann, Bernhard Erwin Ferdinand
Reissig, Magdalena
Retsky, Michael Walter
Richardson, Mary Elizabeth
Richter, Ward Robert
Roth, Lawrence Max
Roth, Sanford Irwin
Rouze, Stanley Rupie
Sakai, William Shigeru
Schechter, Joel Ernest
Schidlovsky, George
Schiefelbein, Benedict
Schnuda, Nasr Danial
Schreiber, Thomas Paul
Schuster, Frederick Lee
Seibel, Hugo Rudolf
Shelton, Emma
Sheridan, Michael N
Silver, Burton Barney
Simon, Gerard Theodor
Stachelin, Lucas Andrew
Stanley, Hugh P
Stenback, Wayne Albert
Stenger, Richard J
Sternlieb, Irmin
Tandler, Bernard
Todd, Gordon Livingston
Tomlinson, Gus
Towe, Kenneth McCarn
Van Horn, Diane Lillian
Watabe, Norimitsu
Weinstein, Ronald S
Whitehouse, Ronald Leslie S
Williams, Daniel Charles
Wistreich, George A
Wolfe, Stephen Landis
Wood, Eunice Marjorie
Wynne-Roberts, Caroline Rosales
Zajac, Barbara Ann
Zamboni, Luciano
Zimmerman, Stanley Dean

Electrophotography
Andrus, Paul Grier
Dehm, Richard Lavern
Elter, John Frederick
Felty, Evan J
Gerace, Paul Louis
Groner, Carl Fred
Kosel, George Eugene
Liebman, Alan Joel
Rubin, Bruce Joel
Savit, Joseph
Van Der Voorn, Peter C
Weinberger, Lester

Microscopy
Barish, Leo
Crossman, Germain Charles
Diehl, Stanley Gregg
Ellis, Sidney Glenn
Geissinger, Hans Dieter
Gray, Peter
Haberfeld, Joseph Lennard
Heckman, Francis Austin
Hock, Charles William
Johnson, Robert Ivar
Kaeppner, Werner Martin
Kinsinger, William George
Ladd, William Alexander
Lawson, Katheryn Emanuel
Leney, Lawrence
Morrow, Scott
Orzechowski, Adam
Parker, Bernard
Thomas, Virginia Lee
Trotter, Nancy Louisa
Tucker, Paul Arthur
Van Cott, Harrison Corbin
Wilson, John William, Jr

Photography
Behrens, Herbert Ernest
Burgmaier, George John
Demers, Pierre (A E)
Eyer, James Arthur
Farran, Charles Frederick
Gilmour, Hugh Stewart Allen
Gosling, John William
Henisch, Heinz Kurt
Holland, Russell Sedgwick
Kaprelian, Edward Karnig
Kriss, Michael Allen
Land, Edwin Herbert
Levinson, Steven R
Martinez, Alberto Magin
Millikan, Allan G
Motter, Robert Franklin
Olsen, Ronald G
Pearson, George John
Rickmers, Albert D
Ross, Edward Shearman
Skillman, David Corwin
Starr, John Edward
Todd, Hollis N
Van Dam, M
Weber, Julius
Wurth, Michael John

Photometry
Cunningham, Alice Jeanne
Grum, Franc
Ramirez-Munoz, Juan
Rennilson, Justin J
Sanders, Corey Leroy
Tobey, Frank Lindley, Jr
Voglesong, William Frederick

Roentgenology
Etter, Lewis Elmer
Fixott, Henry Cline
Lalli, Anthony
Margulies, Milton
Tuddenham, William J
Wald, Samuel Stanley

Ultrasound
Barthel, Romard
Belford, John F
Eggleton, Reginald Charles
Fishman, Sherman Sampson
Hildebrand, Bernard Percy
Linzer, Melvin
Miller, James Gegan
Ryan, Robert Pat

INTERDISCIPLINARY SCIENCES

Academic Administration
Anderson, Adolph (Gustof)
Atkinson, Gene
Bader, Kenneth L
Baker, Louis Coombs Weller
Benedict, Joseph T
Biester, John Louis
Blackwell, Lawrence A
Blockstein, William Leonard
Bohmont, Dale W
Bond, Richard Randolph
Bradford, Spencer Graves
Brase, Peter Charles
Bridgers, William Frank
Briggs, William Egbert
Brownlee, Paula Pimlott
Bryan, Virginia Schmitt
Byrnes, Francis Clair
Camerino, Pat William
Christenberry, George Andrew
Christensen, David Emun

Clark, Ronald David
Conklin, James Byron, Jr
Connors, Philip Irving
Cornell, Samuel Douglas
Cottam, Grant
Crumb, Glenn Howard
Davis, Richard Elden
Davis, William Potter, Jr
Day, William W
De Carlo, Charles R
Deckert, Gordon Harmon
Delaney, Patrick Francis, Jr
DeProspo, Nicholas Dominick
Dietz, Frank Tobias
Dils, Robert Earl
Doerr, Arthur Harry
Dooling, John Stuart
Dutton, Frederic Booth
Edmison, Marvin Tipton
Ellis, Robert Homer
Fischer, Robert Blanchard
Fisher, Leon Harold
Flawn, Peter Tyrrell
Friedman, Edward Alan
Fry, Jack L
Gabriel, Mordecai Lionel
Giese, David Lyle
Gillespie, Robert Gordon
Gist, George Reinecker
Gluckstern, Robert Leonard
Gogan, Niall Joseph
Goldman, James Allan
Gould, Harold A
Greenstein, Julius S
Griffin, William Dallas
Grobman, Arnold Brams
Grobstein, Clifford
Groszos, Stephen Joseph
Hagen, Charles William, Jr
Hanson, Harold Palmer
Harris, Chauncy Dennison
Harris, Henson
Harris, Lawrence
Hausman, William
Heldman, Morris J
Heller, David Harold
Hermann, Theodore S
Hershenson, Benjamin R
Hibbs, Leon
Hobbs, Kenneth Burkett
Hogness, John Rusten
Holmes, Charles Henry
Hornig, Lilli Schwenk
Horton, Frank Elba
Houle, Joseph E
Hyde, Walter Lewis
Ike, Albert Francis
Ivany, J W George
Jacobsen, Neil Soren
Jamrich, John Xavier
Johnsen, Russell Harold
Johnston, Robert Ward
Kano, Adeline Kyoko
Kelley, John Francis, Jr
Klivington, Kenneth Albert
Kogel, Marcus David
Krill, Karl Emil
Kugel, Robert Benjamin
Larson, Russell Edward
Larson, Vernon C
Lee, Ralph Hewitt
Lembeck, William Jacobs
Lewis, Willard Deming
Lindberg, David Seaman, Sr
Linnell, Robert Hartley
Lynton, Ernest Albert
Malzahn, Ray Andrew
Manley, Lillian C
Mapother, Dillon Edward
Massey, Peyton Howard, Jr
McBee, James Leonard, Jr
McIntire, Sumner Harmon
McMillan, Harlan L
Merritt, Richard Howard
Meyer, Edmond Gerald
Miller, Robert L
Miller, William Boynton, Jr
Miller, William Frederick
Monnier, Dwight Chapin
Moye, Anthony Joseph
Murray, Donald Shipley
Myers, Richard F
Nanzetta, Philip Newcomb
Nash, Peter Hugh, Sr
Netzel, Richard G
Nicholson, Norman Leon
Noble, Robert Vernon
Noland, George Bryan
Officer, James Eoff
Okashimo, Katsumi
Ostdick, Thomas
Paradise, Michael Emmanuel
Pennington, Lloyd Drew
Perry, Margaret Nutt
Petrucci, Ralph Herbert
Price, Glenn Albert
Robinson, Leonard H
Rosenfeld, Melvin Arthur

Rothenbury, Raymond Albert
Scheel, Nivard
Schofield, James Roy
Schwartz, Arthur Harold
Schwartz, Douglas Wright
Shacklett, Robert Lee
Shields, Loran Donald
Shull, Harrison
Sibley, Willis Elbridge
Sinclair, D G
Smith, Robert Bruce
Spriggs, Richard Moore
Stebbins, Dean Waldo
Sugihara, James Masanobu
Swoyer, Vincent Harry
Talley, Lawrence Horace
Thomas, Charles Carlisle, Jr
Toft, Robert Jens
Tucker, Woodson Coleman, Jr
Verbrugge, Frank
Walker, Augustus Chapman
Watts, William Wilbur
Weaver, Albert Bruce
Weber, Lavern J
Webster, Thomas G
Welch, Garth Larry
Weller, Richard Irwin
Wick, Emily Lippincott
Williams, Denis R
Wynn, Charles Martin, Sr
Zivnuska, John Arthur
Zook, Harry David

Futuristics
Craver, John Kenneth
Freeman, Jere Evans
Gilbreath, William Pollock
Logan, Robert Kalman
McGinn, Edward James
Mendell, Jay Stanley
Sangster, Raymond Charles
Schiessler, Robert Walter

History of Science
Abegg, Victor Paul
Abrams, John Werner
Allen, Garland Edward, III
Beidleman, Richard Gooch
Berendzen, Richard Earl
Blanpied, William Antoine
Brush, Stephen George
Burstyn, Harold Lewis
Cohen, I Bernard
Daub, Edward E
Drennan, Ollin Junior
Dudley, Underwood
Fisher, Gordon McCrea
Fullmer, June Zimmerman
Garmon, Lucille Burnett
Gershenowitz, Harry
Giancoli, Douglas Charles
Gienapp, John Charles
Gorman, Melville
Gosselin, Edward Alberic
Grabiner, Judith Victor
Gruber, Jacob William
Hawkins, Thomas William, Jr
Hazen, Robert Miller
Hiebert, Erwin Nick
Holmes, Ivan Gregory
Ihde, Aaron John
Jones, Daniel Patrick
Kaufman, George Bernard
Kennedy, Fred Henry
Kopperl, Sheldon Jerome
Kuslan, Louis Isaac
Layton, Edwin Thomas, Jr
Leake, Chauncey D
Lerner, Lawrence S
Loud, Oliver Schule
Lutzker, Edythe
Maurer, Edward Robert
Mayr, Ernst
McKnight, John Lacy
Mercer, Sherwood Rock
Miller, Arthur I
Mindel, Joseph
Mortimer, Charles Edgar
Oswalt, Dallas Leon
Parascandola, John Louis
Phillips, Esther Rodlitz
Raman, Varadaraja Venkata
Roller, Duane Henry DuBose
Rosen, Sidney
Sparberg, Esther Braun
Spencer, James Alan
Schrage, Samuel
Schufle, Joseph Albert
Shapere, Dudley
Shapiro, Alan Elihu
Siegfried, Robert
Stahlman, William Duane
Stauffer, Robert Clinton
Steckler, Bernard Michael
Suppe, Frederick (Roy)
Theisen, Wilfred Robert

Van Ryzin, Marina
Waggoner, Margaret Ann
Webster, Eleanor Rudd
Weinstein, Alexander
Weymouth, Patricia Perkins
Whitten, Maurice Mason
Wise, Matthew Norton
Zei, Dino

Institutional Research
Gehle, Marvin Harlan
Kissler, Gerald Ray
Linnell, Robert Hartley
Montgomery, David Carey
Phillips, Don Irwin
Ross, David Paul
Sheehan, Bernard Stephen
Watts, William Wilbur

Natural History
Bryan, Edwin Horace, Jr
Cromartie, William James, Jr
Douglass, John Richmond
Griffo, James Vincent, Jr
Gunnerson, James Howard
Lewis, Charles Bernard, Jr
McKinley, Daniel Lawson
Merzbacher, Claude F
Sauer, Pauline Louise
Stewart, Glenn Raymond
Thaw, Richard Franklin
Waldbauer, Eugene Charles
Wirtz, John Harold

Philosophy of Science
Buzzelli, Donald Edward
Cohen, Robert Sonne
Fezer, Karl Dietrich
Gershenowitz, Harry
Glover, Elsa Margaret
Godel, Kurt
Goldman, Jack Leslie
Goldman, Jack Allan
Kaiser, Christopher B
Klotz, John William
Katsoff, Louis Osgood
Leith, Thomas Henry
Loud, Oliver Schule
Lucey, Carol Ann
O'Rourke, Richard Clair
Pappademos, John Nicholas
Quay, Paul Michael
Rescher, Nicholas
Ruffner, James Alan
Schlegel, Richard
Schweitzer, George Keene
Scott, George Prescott
Shapere, Dudley
Shimony, Abner
Smith, Gerrit Joseph
Suppe, Frederick (Roy)
Williams, Mary Bearden
Zimmerman, Edward John
Zusy, Dennis

Research Administration
Abrams, Edward
Aepli, Otto Theodore
Albers, Walter Anthony, Jr
Allen, LeRoy Richard
Andersen, Donald Edward
Andrew, Kenneth J
Andrew, Barbara Jean
Arnow, Leslie Earle
Aspinall, Samuel Rusmisell
Ausman, Robert K
Babbee, John Stanley, Jr
Bader, Michel
Baer, Donald Robert
Bailar, John Christian, III
Baker, Carl Gwin
Baker, Charles Albert
Balant, Charles Paul
Bates, Howard E
Baril, Albert, Jr
Barnes, Derek
Barr, James K
Bates, Philip Knight
Bauer, Frederick William
Beckhorn, Edward John
Beigler, Myron Arnold
Bell, James Richard
Belohlav, Leo Rudolf
Ben Daniel, David J
Benedict, Donald Lee
Berryman, George Hugh
Blair, Allen G
Blake, Francis Gilman
Blaser, Robert U
Blaunstein, Robert P
Bogaty, Herman
Borum, Olin H
Bowman, Wallace Deal
Boyd, George Addison
Bresler, Jack Barry
Brody, Gerald
Broida, Theodore Ray
Bronfin, Barry Robert

Brotherton, Robert John
Brown, John Rowland, Jr
Bullock, Paul David
Burns, Allan Fielding
Burns, Denver Peeper
Burrows, Leslie Raymond
Butler, John Parkman
Callen, Joseph Edward
Camougis, George
Carr, Clide Isom
Casady, Robert Barnes
Casady, George
Castro, George
Charlton, Allan P
Charlton, Gordon Randolph
Chase, Fred Leroy
Clark, Floyd Bryan
Clark, Joe Haller
Clyner, Edward Nielsen
Cohen, Arthur Leroy
Condell, William John, Jr
Condiffe, Peter George
Coogan, John Michael
Cook, Elbert Gary, Jr
Cooper, Howard Gordon
Cooper, Raymond David
Copenhaver, John William
Cox, Fred Ward, Jr
Craig, John R
Crane, Langdon Teachout, Jr
Cravens, William Windsor
Croach, Jesse William, Jr
Crog, Richard Stanley
Cropper, Walter V
Croxton, Frank Cushaw
Crump, Stuart Faulkner
Cubberley, Adrian H
Cundiff, Robert Hall
Curtin, Thomas J
Cutler, Frank Allen, Jr
Dallon, Dale Sherman
Daniels, William Ward
Davis, James K
Davis, Joseph Berry
Davis, Selby Brinker
Davis, Wilford Lavern
De La Burde, Roger Z
deMonsabert, Winston Russel
De Sesa, Michael Anthony
Dickas, Albert Binkley
Domanski, Thaddeus John
Doschek, Wardella Wolford
Douglas, Charles Herbert
Duga, Jules Joseph
Eccleshall, Donald
Edman, James Richard
Efron, Herman Yale
Ellenbogen, William Cromwell
Ellis, Robert Homer
Englert, Joseph
English, Spofford Grady
Fawcett, Sherwood Luther
Feldman, William
Ferguson, Joseph Gantt
Fiege, Herbert Reynold, Jr
Finklea, John F
Finn, John Martin
Fischer, Eugene Charles
Fisher, Earl Eugene
Flagg, John Ferard
Fleming, Lawrence Thomas
Fonken, Gunther Siegfried
Fordham, James Lynn
Foster, Leigh Curtis
Foster, Robert Everett
Fox, John R
Fricken, Raymond Lee
Friedman, Orrie Max
Fulton, James McCullough
Furman, Robert Howard
Fusfeld, Herbert Irving
Gagen, Walter Leonard
Ganz, Aaron
Garber, David Harrison
Garman, John Andrew
Gatti, Anthony Roger
Gever, Gabriel
Glsten, Jacob Burrill
Gladney, Henry M
Golding, Lionel Solomon
Goldsmith, Eli David
Goldwater, William Henry
Govier, William Charles
Greene, Michael P
Gudmundsen, Richard Austin
Gun, Wilson Franklin, Jr
Gundersen, Larry Edward
Gustavson, Marvin Ronald
Guthe, Otto Emmor
Guthrie, Franklin Kirmey
Haas, Ward John
Hamlin, Kenneth Eldred, Jr
Hammond, Kenneth Allen
Hanks, Robert William
Hanna, Norman Edwin
Hardwick, William Aubrey, Jr
Harper, Verne Lester
Harris, Barton A

Harris, Elliott Stanley
Hartop, William Lionel, Jr
Haun, Robert Dee, Jr
Haussmann, R K Dieter
Havemeyer, Ruth Naomi
Hayes, Joseph Edward, Jr
Hayes, Willis Stuart
Heitkamp, Norman Denis
Henderson, Robert Wesley
Herberon, Stephen Aven, Jr
Herman, Samuel Sidney
Herrman, Marion Frank
Heying, Theodore Louis
Hietbrink, Earl Henry
Hill, Richard William
Hillier, James
Hinmelfarb, Philip
Hirschy, Harlan W
Ho, Cho-Yen
Hobbs, Kenneth Burkett
Hockenga, Mark T
Hoerger, Fred Donald
Hoffman, Jerry Irwin
Hogan, Edward Merrick
Holland, Melvin Gerald
Holtz, David
Howell, Donald Lavern
Howie, Mary Gertrude
Howson, John Arthur
Hubbard, Harold Mead
Huggins, Charles Marion
Hume, William
Huston, Keith Arthur
Igoe, John Waite
Ingle, George William
Jengem, Richard Louis
Jezl, James Louis
Johnson, Irving Stanley
Johnson, Jack Donald
Johnson, Kenneth
Johnson, Ogden Carl
Jolley, Homer Richard
Jones, Jerry Lynn
Josephson, Edward Samuel
Joyner, Powell Austin
Kammeraad, Adrian
Kaplan, Nathan
Kaufman, Ann Anderson
Keller, Eldon Lewis
Keller, Oswald Lewin
Kemp, Gordon Arthur
Kennedy, James Vern
Kerr, Donald M, Jr
Kincaid, David Reed
King, John Albert
Kinney, Terry B, Jr
Kinnie, Irvin Gray
Kniebes, Duane Van
Kolb, Edward John
Kolmpeyer, Hubert Elmer
Kranz, Alan Zelig
Krieger, Carl Henry
Krier, Carol Alnoth
Krog, Norman Eiler
Kuhns, Ellen Swomley
Lahr, Roy (Jeremy)
Lazen, Alvin Gordon
Leaning, William Henry Dickens
LeBlanc, Norman Francis
Levy, Mortimer
Lindsay, Dale Richard
Loebner, Egon Erziel
Lohring, Ronald Keith
Lougher, Edwin Henry
Lueck, Roger Hawks
Lundegard, Robert James
MacKinney, Arland Lee
Maclay, William Nevin
Malkin, Martin F
Marcus, Sheldon H
Marlin, Robert Lewis
Marsden, Halsey M
Marshall, David Jonathan
Maune, David Francis
Maure, Robert Joseph
Mayeux, Jerry Vincent
McClelland, Alan Lindsey
McFadden, Max Wulfsohn
McKeehan, Charles Wayne
McKenzie, Walter Lawrence
McLaughlin, Gerald Wayne
McLaughlin, Robert Lawrence
McLeod, Lloyd Alexander
Meers, Joseph Tinsley
Megrue, George Henry
Meisel, Seymour Lionel
Menkart, John
Merritt, Doris Honig
Metzger, Gershon
Meyboom, Peter
Miller, A Eugene
Miller, David Robert
Miller, Daniel Robert
Miller, Irwin
Miller, Theodore Charles
Moore, Allen Charlton

Moore, Leonard Oro
Moores, Eugene Albert
Moseman, Albert Henry
Moulton, James Frank, Jr
Muir, Donald Ridley
Munies, Robert
Murino, Vincent S
Murphey, Wilbur Alford
Narin, Francis
Nash, Peter Howard
Neubert, Ralph Lewis
Nickerson, Mortimer Henderson
Nolan, John Thomas, Jr
Nordberg, William
O'Conor, Gregory Thomas
Odell, Floyd Adams
Oehmke, Richard Wallace
Oesterling, James Frederick
Oxman, John Hamilton
Oxman, Michael Allan
Passman, Sidney
Payne, Holland I
Perry, Robert Hood, Jr
Peterson, Merlin Henry
Pheasant, Richard
Pickard, Porter Louis, Jr
Plough, Irvin Chaffee
Plowman, Ronald Dean
Powers, Robert Allen
Poziomek, Edward John
Prescott, Paul Ithel
Pry, Robert Henry
Pullan, George Thomas
Rabin, Robert
Raizen, Senta Amon
Rall, Waldo
Ram, J Sri
Ramsey, Jerry Warren
Randolph, Philip L
Recktenwald, Gerald William
Reich, Charles
Richardson, Billy
Riegel, Kurt Wetherhold
Riley, N Allen
Rogers, William Irvine
Rosenberg, Richard Martin
Ross, Sidney
Rothman, Sam
Rubin, Bernard
Rutledge, Gene Preston
Sachs, Robert Green
Sage, Nathaniel McLean, Jr
Saggiomo, Andrew Joseph
Sands, Daniel Edward
Sannella, Joseph L
San Soucie, Robert Louis
Sarett, Lewis Hastings
Sarles, Lynn Redmon
Schalit, Lewis Martin
Schallenberg, Elmer Edward
Scheinok, Perry Aaron
Schiaffino, Silvio Stephen
Schmehl, Francis Lawrence
Schmidt, Jack Russell
Schmiedeshoff, Frederick William
Schultz, Frederick John
Schurman, Jack Vair
Schwartz, Sidney A
Schweiker, George Christian
Scott, William Caswell, Jr
Secrest, Everett Leigh
Seufzer, Paul Richard
Shapiro, Sam
Sherman, Edward
Shoemaker, Clarence Jay
Short, James N
Shriver, Ellsworth Harold
Sidwell, Robert William
Sieling, Dale Harold
Simmonds, Richard Carroll
Singiser, Robert Eugene
Singurwalla, Nozer Drabsha
Smith, Bernard
Smith, Douglas Stewart
Smith, Joseph James
Snaper, Alvin Allyn
Snider, Robert Gordon
Sobol, Stanley Paul
Solomon, David Eugene
Spokes, Gilbert Neil
Steinberg, Eliot
Steindler, Martin Joseph
Stelzer, Lorin Roy
Sternfeld, Leon
Stevens, Raymond Sawtell
Stockton, John Richard
Stotz, Robert William
Strehlow, Wolfgang Hans
Stuart, David Edward
Stucki, Jacob Calvin
Sugarman, Meyer Louis, Jr
Sunderman, Duane Neuman
Sung, Chien-Bor
Swartout, John Arthur
Swindale, Leslie D
Talbot, Timothy Ralph, Jr
Tate, Parr Allen

Taylor, Howard Lawrence
Taylor, James Earl
Taylor, Robert Franklin
Tessieri, John Edward
Thorn, John Paul
Throw, Francis Edward
Tickner, Alfred William
Tower, Donald Bayley
Trice, William Henry
Tuesday, Charles Sheffield
Tuite, Robert Joseph
Turbyfill, Charles Lewis
Valerio, David Allen
Vanderryn, Jack
Van Lopik, Jack Richard
van Raalte, John A
Veigel, Jon Michael
Vincent, Gerald Glenn
Vlcek, Donald Henry
Wade, Robert Simson
Wakefield, Gene F
Walker, Augustus Chapman
Walker, John Robert
Warner, Jacob Larue
Warren, John Lucius
Weakley, Martin LeRoy
Weinberg, Myron Simon
Whedon, George Donald
Wheeler, Edward Stubbs
Whiter, Paul Francis
Whitman, Erwin N
Wile, Howard P
Wilson, David George
Wilson, Linda S Whatley
Wineland, William Clemard
Winter, David Leon
Wise, Edward Nelson
Wise, Raleigh Warren
Wohlers, Herbert C
Wolfe, Roger Thomas
Wolff, Ivan A
Wood, George William
Woodard, Ralph Emerson
Wymer, Raymond George

Science Administration

Ambler, Ernest
Arpke, Charles Kenneth
Arrington, Wendell S
Aull, Luther Bachman, III
Bannister, Bryant
Bartl, Paul
Bartley, William Call
Bautz, Laura Patricia
Beckler, David (Zander)
Berg, Winfred Emil
Bierly, Eugene Wendell
Blatt, Sylvia
Blatt, Lewis Martin
Blood, Benjamin Donald
Bray, David Frederick
Brent, J Allen
Buccino, Alphonse
Burley, Gordon
Carpenter, Richard A
Carroll, James Barr
Chenery, Peter Jaspersen
Clement, Duncan
Cooper, Martin Jacob
Cornell, Samuel Douglas
Croom, Herman Lee
Crum, John Kistler
Davidson, Gilbert
Dees, Bowen Causey
Delappe, Irving Pierce
Demetriades, Sterge Theodore
Dermody, William Joseph
Diamond, Jacob Joseph
Earle, Ralph Hervey, Jr
Estrin, Norman Frederick
Etzel, Howard Wesley
Feinstone, Wolffe Harry
Flagg, Raymond Osbourn
Fleischer, Robert
Fox, Thomas G, Jr
Frank, Richard Stephen
Freden, Stanley Charles
Gale, George Osborne
Gary, Robert
Gilbert, Myron B
Gist, Lewis Alexander, Jr
Goldman, David Tobias
Golub, Abraham
Grant, Stanley Cameron
Grisamore, Nelson Thomas
Gunter, John William, Jr
Gutter, Frederick Jay
Gwilt, John Ruff
Haaland, John Baltzell
Harmon, John Baltzell
Held, Victor Maxwell
Hellmann, Max
Heydrick, Fred Painter
Hicks, Donald A
Hosley, Robert James
Hubregs, William Henry
Hyatt, Asher Angel
Johnson, Alvin A

Knappenberger, Paul Henry, Jr
Kovach, Eugene George
Kyker, Granvil Charles
Laufer, Arthur Russell
Lawson, Louis Russell, Jr
Leach, Berton Joe
Leavitt, Milo David, Jr
Lepovetsky, Barney Charles
Lovell, James F
Mallinson, George Greisen
Manchee, Eric Best
Martin, Richard Blazo
Martino, Joseph Paul
Mathis, John Buell
Maycock, Paul Dean
McCoubrey, Arthur Orland
Michael, Thomas Hugh Glynn
Miller, Bennett
Morris, Rosemary Shull
Moss, Melvin Lane
Moss, Woodrow Glen
Murray, William Sparrow
Oakley, David Charles
Patterson, Ralph Francis
Raheja, Manu Chatrumal
Ries, Richard Ralph
Roscoe, Henry George
Rossmassler, Stephen Atwater
Scharer, Sherwood Bruce
Schoen, Richard Isaac
Sheldon, Donald Russell
Silverman, Jacob
Simpson, John Arol
Speer, Donald Arthur
Stephenson, Norman Robert
Stevenson, Heber John Richards
Stevenson, Robert Edwin
Stewart, Alva Theodore, Jr
Tape, Gerald Frederick
Tepper, Morris
Teumac, Fred N
Tiedemann, Albert William, Jr
Todsen, Thomas Kamp
Tomas, Francisco
Torio, Joyce Clarke
Upholt, William Martin
Wahl, Milton Heins
Walden, Clyde Harrison
Walter, Fred John
Watkins, Allen Harrison
Weiss, Marvin
Weissberg, Alfred
White, Howard Julian, Jr
Wingfield, Edward Christian
Winn, Edward Barriere

Science Education

Adams, David Lawrence
Adams, Sam
Alfke, Dorothy
Allen, Joe Frank
Anderson, June S
Anderson, Lee Roy
Andrews, Oliver Augustus
Atkinson, Gene
Autrey, Robert Luis
Ayers, Jerry Bart
Babcock, William James Verner
Bailey, Donald Etheridge
Bailey, John H
Bailey, Maurice Eugene
Baker, David Thomas
Baker, Robert Charles
Baldauf, Richard John
Bard, Eugene Dwight
Bass, J Carl
Beard, Jean
Bennett, John Francis
Bennett, Lloyd M
Bergen, Catherine Mary
Berkland, Terrill Raymond
Berman, Arthur Irwin
Billig, Franklin A
Blanchet, Waldo W E
Blatt, Jeremiah Lion
Blinn, Walter Craig
Boercker, Fred D
Boosman, Jaap Wim
Borom, John Lee
Brandou, Julian Robert
Breedlove, Charles B
Brown, Willard Andrew
Bruno, Merle Sanford
Buchanan, George Dale
Bullington, Robert Adrian
Burger, Ambrose William
Burkman, Ernest
Butts, David
Carr, George Leroy
Cathey, Everett Henry
Cignetti, Jess A
Citron, Irvin Meyer
Clague, William Donald
Cobb, Walter V
Come, Thomas V
Cox, Louis Thomas, Jr
Craik, Eva Lee
Crosen, Robert Glenn

Cummins, Ernie Lee
Custer, Robert Louis
Deal, Don Robert
Demanche, Edna Louise
Dils, Robert James
Dodson, B C
Donaldson, Robert Rymal
Dough, Robert Lyle
Driskill, William David
Dubins, Mortimer Ira
Dubois, Donald Ward
Dudley, Frank Mayo
Dunning, Gordon Merrill
Edwards, Thomas F
Eiss, Albert Frank
Elkin, Lynne Osman
Elliott, James McFarland
Estin, Robert William
Faber, Shepard Mazor
Falls, William Randolph, Sr
Farmer, Walter Ashford
Feher, Elsa
Fenner, Peter
Fezer, Karl Dietrich
Fisher, Kathleen Mary
Fitzsimmons, James G
Foecke, Harold Anthony
Foster, Norman Francis
Fowler, Horatio Seymour
Fraser, Thomas Petigru
Fredenburg, Robert Love
Fribourgh, James H
Friedrich, Benjamin C
Gardner, Marjorie Hyer
Garmon, Lucille Burnett
Gates, Henry Stillman
Gee, Charles William
Geer, Ira W
Gering, Robert Lee
Gibbins, Betty Jane
Gillespie, Walter Lee
Glidden, Harley Fremont
Goldhor, Susan
Gonzalez, Paula
Goodstein, Madeline P
Gosselin, Edward Alberic
Green, Ben Arthur, Jr
Greenberg, Herman Samuel
Greenstadt, Melvin
Gross, Phyllis P
Haakonsen, Harry Olav
Haber-Schaim, Uri
Hademenos, James George
Hafer, Paul Egan
Haight, Gilbert Pierce, Jr
Han, Seong S
Hanson, Robert W
Harke, Douglas J
Henkel, Elmer Thomas
Henshaw, Clement Long
Herron, James Dudley
Hinerman, Charles Ovalee
Holmes, Neal Jay
Homman, Guy Burger
Humberd, Jesse David
Hurd, Paul DeHart
Hutton, Wilbert, Jr
Immergut, Edmund Heinz
Irby, Bobby Newell
Ivany, J W George
Jacobs, Joseph H
Jacobson, Willard James
Jeanmaire, Robert L
Johnson, Lloyd Kenneth
Johnson, Philip Gustav
Johnston, James Baker
Jones, Franklin M
Jones, Haskell Lee
Jones, Katherine Maurice
Jones, William F
Kamrin, Michael Arnold
Kapiloff, Paul Louis
Kaplan, Eugene Herbert
Katzin, Joel Carl
Kaufman, Herman S
Kelly, Richard Delmer
Kirksey, Howard Graden, Jr
Klinge, Paul E
Koch, Rudy G
Koppa, F Russell
Kowalski, Stephen Wesley
Lambert, Frank (Edwin), Jr
Landborg, Richard John
Laughlin, Ethelreda R
Lee, Addison Earl
Lee, Donald E
Lisonbee, Lorenzo Kenneth
Little, Robert Narvaez, Jr
Livermore, Arthur Hamilton
Lockard, J David
Lockwood, Linda Gail
Lofgren, Ruth
Lucas, Stephen Bernard
Maben, Jerrold William
MacCracken, Elliott b
Mackay, Lottie Elizabeth Bohm
Magruder, Willis Jackson
Mallinson, George Greisen

Mandell, Alan
Marek, Jerry William
Matheis, Floyd E
Maurer, Edward Robert
May, Kenneth Ownsworth
Mayer, Victor James
Mayfield, Melburn Ross
McBurney, Wendell Faris
McCombs, Freda Siler
McDermott, Lillian Christie
McElhinney, Margaret M (Cocklin)
McKae, Robert James
Mellon, Edward Knox, Jr
Menefee, Robert William
Merken, Melvin
Mertens, Thomas Robert
Meyer, Goldye W
Milgrom, Harry
Miller, Franklin, Jr
Mindel, Joseph
Montean, John J
Moon, Thomas Charles
Moorehead, William Douglas
Morgan, Ashley Grantham, Jr
Morrell, William Egbert
Munch, Theodore
Munger, Edward Arnold
Muto, Peter
Naibert, Zane Elvin
Navarra, John Gabriel
Neal, Louise Adelaide
Neie, Van Elroy
Neumann, Herschel
Newhouse, W Jan
Nicolas, Jesus
Nisbet, Jerry J
Norman, Billy Ray
Norman, William Harvey
Odell, Lois Dorothea
O'Hearn, George Thomas
Oliver, Lawrence Hampton
Olson, John Bennet
Olstad, Roger Gale
Oshima, Eugene Akio
Owens, John Harold
Pafford, William N
Page, Allen D
Paldy, Lester George
Parke, Edward Charles, Jr
Parks, Terry Everett
Paske, William Charles
Passer, Moses
Pearson, Robert Edward
Pella, Milton Orville
Phillips, Don Irwin
Pierson, David W
Poel, Robert Herman
Preger, Fred Titus
Rachinsky, Michael Richard
Ramos, Lillian
Rayner-Canham, Geoffrey William
Renner, John Wilson
Reuss, Ronald Merl
Riban, David Michael
Rice, Robert Arnot
Richardson, Rayman Paul
Richardson, Verlin Homer
Riggsby, Ernest Duward
Robinson, Kent
Romer, Alfred
Romey, William Dowden
Roth, Robert Earl
Rowe, Mary Budd
Russell, Harvey R
Rutherford, Floyd James
Sachteben, Clyde Clinton
St Amand, Joseph
Salamon, Richard Joseph
Schenberg, Samuel
Schillinger, Edwin Joseph
Schmucker, Joseph S
Schneiderwent, Myron Otto
Schonbeck, Niels Daniel
Schultz, Ida Beth
Schultze, Lothar Walter
Schwartz, Brian B
Schwartz, Judah Leon
Schwartz, Maurice Leo
Schwenck, Julius Rae
Scott, Dan Dryden
Seeburger, George Harold
Seiser, Will Lindsey
Semon, John H
Shannon, Jerry A, Jr
Shaver, Paul Merl
Sheppard, Moses Maurice
Shrader, John Stanley
Shrigley, Robert Leroy
Shrum, John W
Shumate, Kenneth McClellan
Siemankowski, Francis Theodore
Silber, Robert L
Sipe, Harry Craig
Siesnick, Irwin Leonard
Smith, Alden Ernest
Smith, Richard Avery
Smith, Roy E
Snigen, Donald Albert

Snyder, Ernest Elwood, Jr
Spielman, Harold S
Sprain, Wilbur
Stekel, Frank D
Stephens, Lawrence James
Stewart, Herbert
Stoever, Edward Carl, Jr
Stokes, Jimmy Cleveland
Stolberg, Robert
Stratton, Charles Abner
Stronck, David Richard
Summerlin, Lee R
Sund, Robert B
Talley, Lawrence Horace
Tannebaum, Harold E
Tanzer, Charles
Taylor, Morris D
Thomas, Joseph Calvin
Tillery, Bill W
Torop, William
Towe, George Coffin
Trout, Verdine Eliza
Trowbridge, Leslie Walter
Turner, George Cleveland
Twiest, Gilbert Lee
Tyndall, Jesse Parker
Uhlhorn, Kenneth W
Urban, John
Utgard, Russell Oliver
Venerable, Grant Delbert
Vessel, Matthew F
Vitrogan, David
Voelker, Alan Morris
Wald, Francine Joy
Wallace, Robert William
Walton, William Upton
Washton, Nathan Seymour
Watson, Fletcher Guard
Watson, Robert Francis
Werking, Robert Junior
West, Felicia Emminger
Westmeyer, Paul
White, Alvin Murray
Whitney, Robert C
Wilber, Joe Casley, Jr
Wilcox, Floyd Lewis
Wilcox, Lee Roy
Williamson, Hugh A
Williamson, Stanley Ellsworth
Winter, Stephen Samuel
Winter, Gerald Lee
Wood, Lowell Thomas
Woodard, Robert Louis
Yager, Robert Eugene
Yale, Francis Gaymon
Yap, Fung Yen
Yarbrough, Arthur C, Jr
Yoesting, Clarence C
Yohe, Daniel Charles
Young, David Paris
Zacharias, Jerrold Reinach
Zimmer, George P
Zingaro, Joseph S

Science Policy

Auerbach, Lewis Edward
Bartley, William Call
Blanpied, William Antoine
Branscomb, Lewis McAdory
Buzzelli, Donald Edward
Carroll, James Barr
Chalkley, Donald Thomas
Charlebois, Clarence Thomas
Chartock, Michael Andrew
Coogan, John Michael
Craig, Paul Palmer
De Simone, Daniel Van
Dickens, Charles Henderson
Drew, Russell Cooper
Duga, Jules Joseph
Elliott, David Duncan
Falk, Charles Eugene
Flajser, Steven Henry
Forman, Sydney Alexander
Fry, John Craig
Goldfarb, Theodore D
Greenlee, Lorance Lisle
Gusman, Samuel
Hartman, Lawton Mervale, III
Jackson, Ray Weldon
Kates, Josef
Katz, Leon
Kelly, Francis John
Kelly, Henry Charles
Laster, Leonard
Lawless, Edward William
Lee, Kai Nien
Littlewood, William Herbert
Long, Franklin A
Lowrance, William Wilson, Jr
McGuire, David Kelly
McRae, Vincent Vernon
McTaggart-Cowan, Patrick Duncan
Mencher, Alan George
Moravcsik, Michael Julius
Morgan, Millett Granger
Morrissey, Arthur Charles

Murphy, Brian Logan
Neureiter, Norman Paul
Otway, Harry John
Paldy, Lester George
Reiss, James Joseph, Jr
Robertson, William, IV
Rosen, Stephen
Salkovitz, Edward Isaac
Shapero, Donald Campbell
Steinhart, John Shannon
Stonier, Tom Ted
Thomas, Charles S
Weiss, Charles, Jr
Werdel, Judith Ann
Wilson, Andrew Hastie
Wolfle, Dael (Lee)
Wright, Robert Raymond
York, Herbert Frank
Zwolenik, James Joseph

Science Writing

Asimov, Isaac
Auerbach, Lewis Edward
Baker, Jeffrey John Wheeler
Bengelsdorf, Irving Swem
Berg, Dana B
Bernstein, Elaine Katz
Blakeslee, Alton Lauren
Brown, John Max
Braunstein, Helen Mentcher
Campbell, James Fulton
Cassidy, Harold Gomes
Craig, Roy Philip
Cromer, Alan H
Dietz, David (Henry)
Durand, Edward Allen
Elder, Joseph Denison
Ewart, Ralph Bradley
Fraser, Thomas Petigru
Garone, John Edward
Greenstone, Arthur W
Hall, Luther Axtell Richard
Hanson, Joe A
Harris, Miles Fitzgerald
Hart, Charles Willard, Jr
Hart, Dabney Gardner
Kravitz, Edward
Leerburger, Benedict Alan, Jr
Locke, David Millard
Loeffler, Harold Julius
Maben, Jerrold William
Mackay, Lottie Elizabeth Bohm
Mann, Kingsley M
Martin, Eric Wentworth
Metzger, H Peter
Milgrom, Harry
Oehser, Paul Henry
Riggsby, Ernest Duward
Rogers, Bruce Joseph
Rose, Elizabeth Gates
Rosen, Stephen
Schifferes, Justus J
Schwenck, Julius Rae
Sullivan, Walter Seager
Tacker, Martha McClelland
Vietmeyer, Noel Duncan
Wilson, Jerry Dick

Scientific Bibliography

Blau, Edmund Justin
Chen, Ching-Chih
Locanthi, Dorothy Davis
Newkirk, Richard Albert Michael
Paylore, Patricia Paquita
Pursglove, Laurence Albert
Shechan, John Timothy

MATERIALS SCIENCE

Materials Science

Agresti, David George
Andrade, Joseph D
Bacon, Roger
Band, Hans Edward
Bennett, Lawrence Herman
Biver, Carl John, Jr
Borsenberger, Paul Michael
Brady, Brian T
Bube, Richard Howard
Carruthers, John Robert
Cezarilyan, Ared
Chen, Tu
Chynoweth, Alan Gerald
Conn, Paul Kohler
Corak, William Sydney
Crawford, Roy Kent
Daniel, Jack Leland
Davidson, Theodore
Diem, Hugh Egbert
Dodd, Charles Gardner
Epstein, George
Eror, Nicholas George, Jr
Feather, David Hoover
Folweiler, Robert Cooper
Frohsdorff, Geoffrey James Carl
Gentile, Anthony L

Harkins, Carl Girvin
Hart, Edward Walter
Henry, Leonard Francis, III
Hickok, Robert Lee
Hubbell, Wayne Charles
Hynes, Thomas Vincent
Isserow, Saul
Jesser, William Augustus
Johnson, Robert Alan
Kalnin, Ilmar L
Keezer, Richard Clark
Kemper, Charles Prentiss
Khattak, Chandra Prakash
Kimmel, Robert Michael
Koepke, Barry George
Lagally, Max Gunter
Lauer, Robert B
Lawless, Kenneth Robert
Lefever, Robert Allen
Lindenmeyer, Paul Henry
Luedemann, Lois W
Lugassy, Armand Amram
MacVicar, Margaret Love Agnes
Mark, Peter Herman
Mattox, Donald Moss
Mattraw, Harold Claude
McCandless, Robert William
McCauley, James Weymann
McConnell, Duncan
Meyer, Robert Bruce
Miller, Gerald R
Mlavsky, Abraham Isaac
Moser, John Benedikt
Mosley, Wilbur Clanton, Jr
Mueller, Herbert J
Myers, Mark B
Newman, Sanford Bernhart
O'Connell, John Joseph, Sr
O'Leary, Robert Kent
Page, Derek Howard
Parrish, William
Perloff, David Steven
Petersen, Donald H
Petersen, Pierre Marc
Postchek, William
Powers, Joseph
Pry, Robert Henry
Roobol, Norman R
Rosauer, Elmer Augustine
Routbort, Jules Lazar
Rowland, Sattley Clark
Roy, Rustum
Ruby, Stanley
Sara, Raymond Vincent
Sasnor, Daniel Joseph
Schroder, Klaus
Schwerer, Frederick Carl
Scola, Daniel Anthony
Shapiro, Isadore
Shobert, Erle Irwin, II
Shottafer, James Edward
Skalny, Jan Peter
Sosin, Abraham
Steinitz, Michael Otto
Stevenson, David Austin
Stringfellow, Gerald B
Sullivan, Harris Martin
Swartz, John Croucher
Taylor, William Edwin
Thelen, Edmund
Thomas, Joseph Francis, Jr
Thompson, Ronald Halsey
Thrower, Peter Albert
Tisone, Thomas C
Tryon, Max
Tuler, Floyd Robert
Vedam, Kuppuswamy
Vernardakis, Theodore Galaction
Vroom, Alan Heard
Wang, Chih Chun
Wawner, Franklin Edward, Jr
Wells, Ralph Gordon
White, Jack Lee
White, William Blaine
Wiener, Sidney
Wilsdorf, Doris Kuhlmann
Wohlfort, Sam Willis
Woodruff, Joseph Franklin
Young, Franklin Alden, Jr

Ceramics

Alexander, Alexandre Emil

Biomaterials

Brown, Stanley Alfred
Cowsar, Donald Roy
Kim, Sung Wan
Krouskop, Thomas Alan
Lautenschlager, Eugene Paul
Lyman, Donald Joseph
Marshall, Grayson William, Jr
Rechtien, James Joseph
Refojo, Miguel Fernandez
Rijke, Arie Marie
Seaman, Geoffrey Vincent F
Walker, Thomas Carl
Wineman, Robert Judson

Armistead, William Houston, Jr
Beall, George Halsey
Bhiwandker, Nutan C
Boyd, David Charles
Britton, Marvin Gale
Burn, Ian
Byers, Stanley A
Campbell, Donald Edward
Cichowski, Robert Stanley
Dutta, Sunil
Eppler, Richard A
Ervin, Guy, Jr
Fletcher, Peter C
Goodwin, Charles Arthur
Greskovich, Charles David
Gupta, Tapan Kumar
Henry, Edward Carleton
Hoagland, Mahlon Bush
Hurst, Thomas Leonard
Keller, Walter David
Klemm, Waldemar Arthur, Jr
Klingsberg, Cyrus
Kreidler, Eric Russell
Land, Peter L
Lee, Haynes A
Lindsay, James Gordon
McGowan, H Christopher
Medrud, Ronald Curtis
Mikami, Harry M
Moss, Herbert Irwin
Myers, Mark B
Newkirk, Terry Franklin
Nicholson, Geoffrey Charles
Noone, Michael John
Palm, John Andrew
Patchett, Joseph Edmund
Prindle, William Roscoe
Prochazka, Svante
Radford, Kenneth Charles
Rockett, Thomas John
Rulon, Richard M
Sadler, Arthur Graham
Schreiber, Edward
Stull, John Leete
Tiernan, Robert Joseph
Vasilos, Thomas
Wang, Ke-Chin
Willson, Philip James
Yee, Tin Boo
Yoldas, Bulent Erturk

Dental Materials

Anthony, David Henry
Cassel, James Martin
Cheung, Peter Pak Lun
Dickson, George
Forbes, James Franklin
Greenberg, Orrin
Hudson, Donald Charles
Huget, Eugene F
Johnson, Leonard N
Korostoff, Edward
Langeland, Kaare
Mabie, Curtis Parsons
Mohammed, M Hamdi A
Moon, Peter Clayton
Moser, John Benedikt
Moskowitz, Harvey D
O'Brien, William Joseph
Phillips, Ralph W
Powers, John Michael
Rootare, Hillar Muidar
Sandrik, James Leslie
Schoen, William P
Smith, Dennis Clifford
Tateosian, Louis Hagop
Taylor, Duane Francis
Teteruck, Walter R
Tillitson, Edward Walter
Vincent, Gordon Ross

Fuel Science

Abel, William T
Affens, Wilbur Allen
Armington, Alton
Bachman, Kenneth Charles
Baldwin, Roger Allan
Bauer, Hans Fred
Belles, Frank Edward
Boodman, Norman S
Brown, Frederick Ronald
Coates, Arthur Donwell
Conroy, James Strickler
Corbeels, Roger
Cutforth, Howard Glen
Davis, Marshall Earl
Dolbear, Geoffrey Emerson
Droege, John Walter
Eccleston, Barton Henry
Frost, Kenneth Almeron, Jr
Geller, Irwin
Giannetti, Joseph Paul
Given, Kenneth Hervey
Gould, Kenneth Alan
Granoff, Barry
Growcock, Frederick Bruce
Gruner, Joseph

Harney, Brian Michael
Hill, George Richard
Johnson, Mary Lynn Miller
Kang, Chia-Chen Chu
Koehl, William John, Jr
Kolobielski, Marjan
Lang, William Harry
Leonard, Joseph Thomas
Lovell, Harold Lemuel
McQuaid, Richard William
Metcalfe, Joseph Edward, III
Miller, James Frederick
Mills, George Alexander
Neavel, Richard Charles
Palmer, Howard Benedict
Robinson, Sanders David
Rosenthal, Joel William
Schmidt, Eckart W
Schowalter, Kenneth Arthur
Seery, Daniel J
Sheppard, William James
Steffgen, Frederick Williams
Tipton, Ann Baugh
Vernon, Lonnie William
Wender, Irving
Williams, Joel Mann, Jr

MATHEMATICS

Mathematics

Aaboe, Asger (Hartvig)
Aagard, Roger L
Abel, William Robert
Abeles, Francine
Abernathy, Robert O
Aberth, Oliver George
Abhyankar, Shreeram
Abikoff, William
Ablow, Clarence Maurice
Abraham, Ralph Herman
Abungu, Cornelio Oyola
Adams, Helen Elizabeth
Adams, Robert D
Adenstedt, Rolf Karl
Adkins, Julia Elizabeth
Adler, Alfred
Adler, Roy Lee
Adney, Joseph Elliott, Jr
Adorno, David Samuel
Aeppli, Alfred
Agins, Barnett Robert
Agnew, Jeanne Le Caine
Agoston, Max Karl
Ahlbrandt, Calvin Dale
Ahlfors, Lars Valerian
Ahmad, Shair
Ahuja, Jagdish C
Ainsworth, Oscar Richard
Aissen, Michael Israel
Aizley, Paul
Akers, Sheldon Buckingham, Jr
Alaka, Mikhail A
Alaoglu, Leonidas
Alavi, Yousef
Albaugh, A Henry
Albert, Eugene
Albrecht, Norman Edward
Albright, James Curtice
Alder, Henry Ludwig
Alexander, Forrest Doyle
Alexander, Howard Wright
Alexander, James Crew
Alexander, Madeline J
Alexanderson, Gerald Lee
Aleksne, Alberta Yearian
Allan, David Wayne
Allard, Nona Mary
Allen, Arnold Oral
Allen, Edward Switzer
Allen, Frank B
Allen, Harry Prince
Allen, John Ed
Allen, Stephen Ives
Allgower, Eugene L
Alling, Norman Larrabee
Almgren, Frederick Justin, Jr
Alo, Richard Anthony
Alperin, Jonathan L
Aismiller, Rufard G, Jr
Alt, Franz Leopold
Alter, Ronald
Alvarez, Laurence Richards
Amemiya, Frances (Louise) Campbell
Ames, Dennis Burley
Amir-Moez, Ali E
Amos, Donald E
Anastasio, Salvatore
Andalafte, Edward Ziegler
Anderle, Richard
Anderson, A Keith
Anderson, Allan George
Anderson, David H
Anderson, Donald Werner
Anderson, Edmund Hughes
Anderson, Frank Wylie

Anderson, George Albert
Anderson, Glen Douglas
Anderson, R F V
Anderson, Rodney Ebon
Anderson, Rowland C
Anderson, Virgil Lee
Andrea, Stephen Alfred
Andree, Richard Vernon
Andrews, Alice E (Mrs Theodore B Hunt)
Andrews, George Eyre
Andrews, George Harold
Andrews, John Jacob
Andrushkiw, Joseph Wasyl
Angell, Thomas Strong
Angotti, Rodney
Animalu, Alexander Obiefoka Enukora
Anshel, Michael
Anton, Howard
Apostol, Tom M
Appel, Kenneth I
Appling, William David Love
Archibald, Ralph George
Arens, Richard Friederich
Arenson, Donald Lewis
Argabright, Loren N
Arkowitz, Martin Arthur
Armacost, David Lee
Armacost, William L
Armijo, Larry
Armstrong, James Walter, Jr
Armstrong, Kenneth William
Arnoff, E Leonard
Arnold, Hubert Andrew
Arnold, Jimmy Thomas
Arnold, Leslie K
Arnold, Robert Fairbanks
Aronson, Donald Gary
Aronszajn, Nachman
Arsove, Maynard Goodwin
Arteaga, Lucio
Arveson, William Barnes
Arzt, Sholom
Asadulla, Syed
Ash, Robert B
Ashton, Joseph Benjamin
Ashy, Peter Jawad
Asprey, Winifred Alice
Assadourian, Fred
Astrachan, Max
Astromoff, Andrew
Atalla, Robert E
Atchison, William Franklin
Atkins, Ferrel
Atkins, Henry Pearce
Atkinson, Kendall E
Attebery, Billy Joe
Auer, Jan Willem
Aufenkamp, Don
Aull, Charles Edward
Ault, John Willard
Auslander, Bernice Liberman
Auslander, Joseph
Auslander, Louis
Austin, Charles Ward
Austin, Donald Guy
Austin, Thomas LeRoy, Jr
Authement, Ray Paul
Avann, Sherwin Parker
Ayoub, Christine Williams
Ayoub, Raymond G Dimitri
Ayres, William Leake
Azpeitia, Alfonso Gil
Bachelis, Gregory Frank
Bachman, George
Bacon, Harold Maile
Bade, William George
Badger, Blanche Crisp
Baer, Robert M
Baeumler, Howard William
Baggett, Larry W
Bagley, Robert Waller
Bailey, Donald Forest
Bailey, Duane W
Bailey, Herbert R
Bailey, Paul Bernard
Bailey, William T
Baily, Walter Lewis, Jr
Baird, Leemon Claude
Baker, Edward George
Baker, Kirby Alan
Balbes, Raymond
Baldwin, Joe G, Jr
Baldwin, Joseph
Balint, Francis Joseph
Balka, Don Stephen
Ball, Billy Joe
Ball, Ralph Wayne
Ball, Richard William
Ballantine, C S
Ballou, Donald Henry
Balogh, Charles B
Bandes, Dean
Bank, Steven Barry
Barback, Joseph
Barbeau, Edward Joseph, Jr
Barber, Sherburne Frederick

Barcus, William Dickson, Jr
Bardwell, George
Baric, Lee Wilmer
Barksdale, James Bryan, Jr
Barlaz, Joshua
Barnes, Bruce Herbert
Barnes, Earl Russell
Barnes, John Landes
Barnes, Wallace Edward
Barnes, Wilfred E
Barr, Alvin Francis
Barr, Michael
Barrar, Richard Blaine
Barrett, Lida Kittrell
Barrett, Louis Carl
Barry, John Young
Bartels, Robert Christian Frank
Bartle, Robert Gardner
Bartley, Edward Francis
Bartlow, Thomas L
Bartoo, James Breese
Basavappa, Parannara
Baskervill, Margaret (Malone)
Basoco, Miguel Antonio
Bass, Hyman
Bateman, Felice Davidson
Bateman, Paul Trevier
Bates, Grace Elizabeth
Batho, Edward Hubert
Batten, George Washington, Jr
Baum, John Daniel
Baum, Paul Frank
Baumgart, John Keppler
Baumslag, Gilbert
Bauserman, Thomas
Baussus-Von Luetzow, Hans Gerhard
Baxter, Glen Earl
Baxter, Willard Ellis
Bazer, Jack
Beach, James Wilson
Bean, Ralph J
Bean, William Clifton
Beane, Donald Gene
Bear, Herbert S, Jr
Beaty, Marjorie Heckel
Beauchamp, John J
Beaumont, Ross Allen
Bechtel, Robert D
Beck, Anatole
Beck, Jonathon Mock
Beck, William Nelson
Becker, Gerald Anthony
Becker, Jerry Page
Beckman, Frank Samuel
Bedgood, Dale Ray
Bedient, Jack DeWitt
Bedient, Phillip E
Bednar, Jonnie Bee
Bednarek, Alexander R
Beechler, Barbara Jean
Beekman, John Alfred
Beelis, Harold E
Bellman, Richard Ernest
Belluce, Lawrence P
Beltrami, Edward J
Bender, Edward Maurice
Bender, Edward Anton
Bender, Phillip R
Behara, Minaketan
Bein, Donald
Benes, Vaclav Edvard
Beinke, Lowell Wayne
Beiter, Marion
Bell, Charles Bernard, Jr
Bell, Curtis Porter
Bell, Harold
Bell, Howard E
Bell, John Clarence
Bell, Stoughton
Bellis, Harold E
Benington, Frederick
Bennett, James Hallam
Bennett, Lonnie Truman
Bennett, Mary Katherine
Benson, Dean Clifton
Benson, Donald Charles
Benson, William Howard
Bentley, Herschel Lamar
Benzinger, Harold Edward, Jr
Beougher, Elton Earl
Berg, Ira David
Berg, Paul Walter
Bergdall, Irene Floy
Berger, Melvyn Stuart
Berger, William J
Bergman, Stefan
Berk, Kenneth N
Berkovitz, Leonard David
Berkovitz, Jerome
Berkson, Earl Robert
Berman, Gerald
Berman, Joel David
Bernardi, Salvatore Dante
Bernau, Simon J

Berndt, Bruce Carl
Bernhardt, Robert L, III
Bernhart, Arthur
Berning, Peter H
Bernkopf, Michael
Bernstein, Allen Richard
Bernstein, Barbara Elaine
Bernstein, Dorothy Lewis
Bernstein, Irwin S
Berri, Manuel Philip
Berry, Andrew Campbell
Bers, Lipman
Berstein, Israel
Bertholf, Dennis E
Bethel, Edward Lee
Betz, Ebon Elbert
Beyer, William A
Beyer, William Hyman
Bhatia, Nam Parshad
Bhattacharya, Ramendra Kumar
Bichteler, Klaus Richard
Bick, Theodore A
Bickel, Robert John
Bickerstaff, Thomas Alton
Bidwell, Leonard Nathan
Biesele, Ferdinand Charles
Biesterfeldt, Herman John, Jr
Biles, Charles Morgan
Billera, Louis Joseph
Billingsley, Patrick Paul
Billotti, Joseph Eugene
Bing, Kurt
Bing, R H
Bird, Marion Taylor
Biriuk, George
Birkhoff, Garrett
Birman, Joan Sylvia
Birnbaum, Sidney
Birnbaum, Zygmunt William
Birtel, Frank T
Bishop, Errett A
Bissinger, Barnard Hinkle
Bittinger, Marvin Lowell
Bitzer, Carl Wilfrid
Blackadar, Bruce Evan
Blackburn, Thomas Henry
Blackett, Donald Watson
Blackman, Shirley Allart
Blackman, Jerome
Blackwell, David (Harold)
Blackwell, Paul K, II
Blair, Robert Louie
Blake, Louis Harvey
Blake, Robert George
Blakers, Albert Laurence
Blakley, George Robert
Blanch, Gertrude
Blanche, Ernest Evred
Blau, Julian Herman
Bledsoe, Woodrow Wilson
Blecher, Michael Nathaniel
Blizzard, Richard Reese, Sr
Block, Richard Earl
Blomquist, Mary M Osborn
Blum, Edward K
Blum, Julius Rubin
Blum, Manuel
Blum, Richard
Blumberg, John Otto
Blumberg, Victor Michael
Blumenthal, Leonard Mascot
Blumenthal, Robert McCallum
Blyth, Mary Isobel
Boardman, John Michael
Boas, Ralph Philip, Jr
Bobisud, Larry Eugene
Bobo, Edwin Ray
Bobonis, Augusto
Bochner, Salomon
Boes, Ardel J
Bogart, Kenneth Paul
Bogdan, Victor Michael
Bohn, Sherman Elwood
Bohnenblust, Frederic
Bohrer, Robert Edward
Bolander, Richard
Bolger, Edward M
Bolker, Ethan D
Bompart, Billy Earl
Bonar, Daniel Donald
Bond, Albert F
Boone, James Robert
Boothby, William Munger
Bordelon, Derrill Joseph
Borel, Armand
Borges, Carlos Rego
Borofsky, Samuel
Borosh, Itshak
Borrego, Joseph Thomas
Borwein, David
Bosch, Warren Luther
Boswell, Rupert Dean, Jr
Bott, Raoul H
Botts, Truman Arthur
Bouldin, Richard Hindman
Boullion, Thomas L
Bourgin, David Gordon
Bourne, Samuel G
Bouwsma, Ward D

Bower, Oliver Kenneth
Bowie, Harold E
Bowling, Floyd E
Bowman, Eugene W
Bowman, Robert Hunt
Bownds, John Marvin
Boyce, Donald Joe
Boyce, Stephen Scott
Boyce, William Martin
Boyer, Carl Benjamin
Boyer, Delmar Lee
Boyer, Lee Emerson
Boylan, Edward
Braden, Charles McMurray
Bradford, James C
Bradley, Allen H
Bradley, John Spurgeon
Brady, Wray Grayson
Brady, Louis Richard
Bragg, Louis Richard
Brahana, Thomas Roy
Brainerd, Barron
Brainerd, Walter Scott
Bram, Joseph
Bram, Leila Dragonette
Branges, Louis De
Brannen, Joseph P
Branovan, Leo
Brauer, Fred
Brauer, George Ulrich
Brauer, Richard Dagobert
Braunfeld, Peter George
Braunschweiger, Christian Carl
Brechner, Beverly Lorraine
Bredon, Glen E
Breeman, Donald George
Bressler, David Wilson
Breusch, Robert Hermann
Bridgman, George Henry
Briggs, Charles Francis
Brigham, Nelson Allen
Briney, Robert Edward
Britton, Jack Rolf
Brixey, John Clark
Brockman, Harold W
Brooks, Foster (Lindsey)
Brooks, James Keith
Brooks, Robert M
Brooks, Sam Raymond
Brookshear, James Glenn
Brothers, John Edwin
Brown, Arlen
Brown, Austin Robert, Jr
Brown, Charles A
Brown, Claude Harold
Brown, David T
Brown, Dennison Robert
Brown, Edward Allan
Brown, Edward Martin
Brown, Ferdinand Louis
Brown, Francis Robert
Brown, George William
Brown, John A
Brown, John Edwin
Brown, John Wesley
Brown, Joseph Ross
Brown, Kenneth E
Brown, Morton
Brown, Richard Harland
Brown, Richard Kettel
Brown, Robert Bruce
Brown, Robert C
Brown, Robert Dillon
Brown, Robert Wallace
Brown, Thomas Andrew
Brown, Thomas Craig
Brownell, Frank Herbert, III
Brualdi, Richard Anthony
Bruck, Richard Hubert
Bruckner, Andrew M
Brunfiel, Gloria Florette
Brunfiel, Charles
Brunk, Hugh Daniel
Bruyr, Donald Lee
Bryan, Charles A
Bryan, Robert Neff
Bryant, Billy Finney
Bryant, John Logan
Buchanan, David Alvin
Buchsbaum, David Alvin
Buck, Charles (Carpenter)
Buchi, Julius Richard
Buckholtz, James Richard
Buckland, Golden Thaddeus
Buckeye, Donald Andrew
Buckley, Joseph Thaddeus
Buell, Carleton Eugene
Buell, Elliott Lyndon
Buesking, Clarence W
Bullen, Peter Southcott
Bullis, George LeRoy
Bullock, Robert M, III
Bullock, Roberts Cozart
Bumcrot, Robert J
Bunge, Marta C
Bunn, Lucas N H
Buntinas, Martin George
Burch, Benjamin Clay
Burcham, Paul Baker
Burchard, Hermann Georg
Burckel, Robert Bruce
Burdette, Albert Clark

Burdge, Donald Eugene
Bures, Donald John (Charles)
Burford, Thomas Maynard
Burgess, Cecil Edmund
Burgstahler, Sylvan
Burke, Leonarda
Burkhart, Sarah Maybelle
Burkholder, Donald Lyman
Burling, James P
Burr, Irving Wingate
Burrill, Claude Wesley
Burton, John Henry William
Burton, Leonard Pattillo
Burton, Theodore Allen
Busby, Robert Clark
Buschman, Herbert
Busenberg, Stavros Nicholas
Buser, Mary Paul
Bush, George Clark
Bush, Kenneth Arthur
Bush, Laurens Earle
Bushaw, Donald (Wayne)
Butcher, John Harvey
Butler, Lewis Clark
Butler, Margaret K
Butler, Ronald G
Butler, Terence
Butler, Walter Cassius
Button, Alton Thomas
Butterworth, Allen Virgil
Byers, Hubert S
Byers, Gordon Cleaves
Bynum, William Lee
Byrd, Richard Dowell
Byrne, John Richard
Bzoch, Ronald Charles
Cain, George Lee, Jr
Cairns, Stewart Scott
Cairns, Thomas W
Calabi, Eugenio
Calabrese, Philip G
Caldwell, William V
Calvert, Ralph Lowell
Cameron, Edward Alexander
Cameron, Robert Horton
Campaigne, Howard Herbert
Campbell, Howard Ernest
Campbell, James Dow
Campbell, Louis Lorne
Campbell, William H
Canavan, Robert I 1
Canfield, Earle Lloyd
Cannon, Edward Whitney
Cannon, John Rozier
Cannon, Lawrence Orson
Cannon, Raymond Joseph, Jr
Cantor, David Geoffrey
Cantrell, Grady Leon
Cantwell, John Christopher
Capobianco, Michael F
Cappas, C
Carasso, Alfred Samuel
Carlitz, Leonard
Carlson, David Hilding
Carlson, John W
Carlson, Kermit Howard
Carlson, Philip R
Carpenter, Dorothy Irene
Carr, John Weber, III
Carroll, Robert Leon
Carroll, Robert Wayne
Carruth, Philip Wilkinson
Carson, Albert B
Carson, George Stephen
Carson, George Walter
Carson, Robert Cleland
Carter, David Southard
Carter, Frederick J
Carter, Harry Nelson
Carter, William Caswell
Cassidy, Samuel H
Castagna, Frank
Castellani, Maria
Cater, Frank Sydney
Catin, Donald E
Cato, Benjamin Ralph, Jr
Catrambone, Joseph Anthony, Sr
Cavanagh, Timothy D
Cecil, David Rolf
Celauro, Francis L
Celeste, Vincent
Chacon, Rafael Van Severen
Chaiken, Jan Michael
Chakerian, Gulbank Donald
Chalkley, Roger
Chamberlain, Erling William
Chamberlin, Richard Eliot
Chambers, Barbara Mae Fromm
Chambers, Charles McKay, Jr
Chandler, Richard Edward
Chandra, Jagdish
Chaney, George L
Chang, Bomshik
Chang, Shao-chien
Chao, Chong-Yun

Chao, Jia-Arng
Charlap, Leonard Stanton
Charlton, Harvey Johnson
Charnes, Abraham
Chartrand, Gary
Chatland, Harold
Cheema, Mohindar Singh
Chen, Orin Nathaniel
Chen, Kuo-Tsai
Chen, Yu Why
Cheney, Elliott Ward, (Jr)
Cherack, Paul Francis
Cherlin, George (Yale)
Chess, Karin V T
Chester, Clive Ronald
Chicks, Charles Hampton
Childs, Lindsay Nathan
Chilton, Bruce L
Chilton, Phyllis Zweig
Chinn, Richard Leigh
Chitwood, Howard
Chivukula, Ramamohana Rao
Chopra, Dharam-Vir
Chover, Joshua
Chow, Bryant
Chow, Tseng Yeh
Chow, Yuan Shih
Chow, Yutze
Chrestenson, Hubert Edwin
Christensen, Ermine America
Christian, Robert Roland
Christman, Marvin Henry
Christopher, John
Chrisy, John Harlan
Chu, Sherwood Cheng-Wu
Chuckrow, Vicki G
Chui, Charles Kam-Tai
Chung, Fan Rong King
Chung, Kai Lai
Church, Alonzo
Church, Charles Alexander, Jr
Churchill, Edmund
Civin, Paul
Clanton, Donald Henry
Clark, Allan H
Clark, Charles Lester
Clark, Crosman Jay
Clark, Frank Eugene
Clark, William Edwin
Clarke, Allen Bruce
Clarke, Robert Alma
Clarkson, Donald R
Claxton, William Eugene
Clay, James Ray
Cleaver, Charles E
Cleaver, Frank L
Clemens, Charles Herbert
Clemens, Stanley Ray
Clement, Paul Arnold
Clements, George Francis
Clemmons, John B
Clifford, Alfred Hoblitzelle
Clifton, Yeaton Hopley
Clinnick, Mansfield
Clough, Robert Ragan
Clutterham, David Robert
Coddington, Earl Alexander
Coffman, Charles Vernon
Cohen, David Warren
Cohen, Haskell
Cohen, Herman Jacob
Cohen, Leon Warren
Cohen, Paul Joseph
Cohn, Harvey
Cohn, Leslie
Cole, Nancy
Cole, Randal Hudie
Coleman, Donald Brooks
Coleman, Ralph H
Coles, William Jeffrey
Collins, George Edwin
Collins, Heron Sherwood
Colquitt, Landon Augustus
Colton, David L
Colvin, Burton Houston
Combellack, Wilfred James
Comfort, William Wistar
Comstock, Dale Robert
Conlan, James
Conley, Charles Cameron
Connelly, Damian
Conner, Howard Emmett
Conner, Pierre Euclide, Jr
Connor, Frank Field
Conrad, Paul
Constable, Robert L
Consul, Prem Chandra
Cook, Clarence Harlan
Cook, Frederick Lee
Cook, Thurlow Adrean
Cooke, Henry Charles
Cooke, Roger Lee
Cooley, Hollis Ray
Cooley, William Peyton, Jr
Cooley, Robert Lee
Coon, Geraldine Alma
Coon, Lewis Hulbert

Coonce, Harry B
Cooney, Miriam Patrick
Cooperman, Philip
Copeland, Arthur Herbert, Jr
Corley, Glyn Jackson
Cornette, James L
Corson, Harry Herbert
Cottle, Richard W
Courtney, Cameron B
Cowling, Vincent Frederick
Cox, Henry Miot
Cox, Raymond H
Coxeter, Harold Scott MacDonald
Crabtree, Douglas Everett
Crabtree, James Bruce
Craft, George Arthur
Craig, Cecil Calvert
Crampton, Theodore Henry Miller
Crane, Roger L
Cranford, Robert Henry
Crapo, Henry Howland
Craswell, Keith J
Craw, Alexander R
Crawford, William Stanley Hayes
Creese, Thomas Morton
Crim, Sterling Cromwell
Croom, Frederick Hailey
Crouch, Ralph Boyett
Crow, Edwin Louis
Crowe, Donald Warren
Crowell, Julian
Crowell, Richard Henry
Crowley, Thomas Henry
Crowson, Henry L
Crumley, Richard D
Crump, Kenny Sherman
Csorgo, Miklos
Culbertson, George Edward
Cullen, Charles G
Cullen, Helen Frances
Cullen, Theodore John
Culpepper, Benjamin Hays
Cummings, Gideon Alston
Cummings, Kenneth Burdette
Cunkle, Charles Henry
Cunnea, William M
Cunningham, Allen Byron
Cunningham, Frederic, Jr
Cunningham, Robert Stephen
Curd, Rudy Leroy
Curjel, Caspar Robert
Curran, Daniel R
Currier, Albert (Eldred)
Curtis, Charles Whittlesey
Curtis, Herbert John
Curtis, Howard Benton, Jr
Curtis, Morton Landers
Curtis, Philip Chadsey, Jr
Curtz, Thaddeus Bankson
Cuthill, Elizabeth
Cutler, Doyle O
Cutler, Leonard Samuel
Cutlip, William Frederick
Czerniakiewicz, Anastasia Juana
Dahiya, Rajbir Singh
Daly, Frederick Thomas
D'Ambrosio, Ubiratan
Danese, Arthur E
Daniel, Victor Wayne
Dank, Milton
Danskin, John Moffatt, Jr
Darling, Eugene Merrill, Jr
Darsow, William Frank
Darst, Richard B
Darwin, James T, Jr
Dauer, Jerald Paul
David, Florence N
Davies, Robert
Davies, Chandler
Davis, Edward Allan
Davis, Edward Dewey
Davis, Harry Floyd
Davis, Herbert Thaddeus, III
Davis, Jeffrey Robert
Davis, Kenneth Joseph
Davis, Larry Wallace
Davis, Morton David
Davis, Myrtis
Davis, Paul Lawrence
Davis, Philip
Davis, Robert Benjamin
Davis, Robert Dabney
Davis, Robert Lloyd
Davis, Thomas Austin
Davis, Watson M
Davison, Thomas Matthew Kerr
Davison, Walter Francis
Davitt, Richard Michael
Dawson, Donald Andrew
Day, James Thomas
Day, Mahlon Marsh
Daykin, Philip Norman
Deal, Ervin R
Dean, David W
Dean, Richard Albert
Dean, Robert Yost

Deaton, Edmund Ike
de Boor, Carl (Wilhelm) Reinhold
De Carlo, Charles R
De Cicco, Henry
De Cicco, John
Deddens, James Albert
Deeds, Joseph Bird
De Gray, Ronald Willoughby
Dekker, David Bliss
Dekker, Jacob Christoph Edmond
De Korvin, Andre
DeLand, Edward Charles
Delaney, Frank Michael
DeLeeuw, Karel
Deliyannis, Platon Constantine
DeLury, Daniel Bertrand
DeMarr, Ralph Elgin
DeMeyer, Frank R
DeMillo, Richard A
Deming, Robert W
Demos, Miltiades Stavros
Denbow, Carl (Herbert)
Dennemeyer, Rene Felix
Dennis, John Emory, Jr
Denzel, George Eugene
DePalma, James John
de Pillis, John
DePree, John Deryck
DePrima, Charles Raymond
Derrick, William Richard
De Sapio, Rodolfo Vittorio
Deskins, Wilbur Eugene
DeSua, Frank Crispin
Devinatz, Allen
Devries, David J
Dinkines, Flora
De Witt, Warren Peyton
Diamond, Ainsley Herbert
Diamond, Harold George
Dickinson, Alice B
Dickinson, David (James)
Dickson, Spencer E
Diekhans, Herbert Henry
Diem, John Edwin
Diestel, Joseph
Dilworth, Robert Palmer
Dimsdale, Bernard
Dinneen, Gerald Paul
Divinsky, Nathan Joseph
Divis, Bohuslav B
Dixon, Lyle Junior
Dobyns, Roy A
Dodson, Norman Elmer
Dolan, Winthrop Wiggin
Dollard, John D
Donaldson, James A
Donoghue, William F, Jr
Donsker, Monroe David
Doob, Joseph Leo
Dor, Leonard Eliezer
Dorgan, William E
Dorn, William S
Dorroh, James Robert
Doss, Raouf
Douglas, Robert James
Douglas, Ronald George
Douglass, Raymond Donald
Douglass, Roger Thackery
Douglis, Avron
Dowds, Richard E
Dowling, Marie Augustine
Downing, John Scott
Downing, Reginald Horton
Doyle, Thomas Carlson
Doyle, Worthie Lefler, Jr
Draeger, Sidney S
Drake, David Allyn
Drazin, Michael Peter
Drennan, James Elliott
Dreskin, Sanford A
Dressler, Francis George
Dressler, Raymond Donald
Drew, Dan Dale
Drew, John H
Driscoll, John G
Drobnies, Saul Isaac
Drobot, Stefan
Drozda, William
Drucker, Bertram Morris
Drufenbrock, Diane
Dubay, George Henry
Dubins, Lester Eli
Dubinsky, Edward Leonard
Dubisch, Roy
Dubois, Donald Ward
Dubuc, Serge
Duda, Edwin
Dudley, Underwood
Duff, George Francis Denton
Duffin, Richard James
Dugundji, James
Duke, John Walter
Duke, Richard Alter
Dull, Martin Honer
Duncan, Cecil Eugene
Duncan, David Gordon

Duncan, Donald Lee
Duncan, Richard Dale
Dunham, Charles Burton
Dunkl, Charles Francis
Dunn, Donald William
Dunn, Samuel L
Dunton, Marguerite Elizabeth
Dupree, Daniel Edward
Duquette, Alfred L
Duran, Benjamin S
Duren, Peter Larkin
Duren, William Larkin, Jr
Duret, Maurice Francis
Durfee, William Hetherington
During, Frederick Charles
Dutka, Jacques
Dwight, Leslie Alfred
Dwyer, Thomas Aloysius Walsh, III
Dwyer, Wendell Arthur
Dye, Henry Abel
Dyer, Eldon
Dyer, James Arthur
Dyer, Joan L
Eagon, John Alonzo
Eakin, Richard R
Earle, Clifford John, Jr
Easterday, Kenneth E
Eastham, James Norman
Easton, Robert Walter
Eaton, William Thomas
Eaves, David Magill
Eberhart, Paul
Eberlein, William Frederick
Ebin, David G
Ecker, Edward Stuart, Sr
Ecker, Edwin D
Edelson, Allan L
Edelson, Sidney
Edison, Larry Alvin
Edmundson, Don Elton
Edmundson, Harold Parkins
Edwards, Charles Henry, Jr
Eells, James, Jr
Effros, Edward George
Egan, Francis P
Egan, Howard L
Egar, Joseph Michael
Eggan, Lawrence Carl
Ehrenpreis, Leon
Ehrlich, Gertrude
Eilenberg, Samuel
Eisele, Carolyn
Eisen, Martin
Eisenberg, Sheldon Merven
Eisenstadt, Bertram Joseph
Eklof, Paul Christian
Elandt-Johnson, Regina C
Elder, Alexander Stowell
Eldridge, Klaus Emil
Elgot, Calvin C
Elich, Joe
Elkins, Thomas Anthony
Elliott, Joanne
Elliott, William Whitefield
Ellis, Homer Godsey
Ellis, Richard Steven
Ellis, Robert L
Ellis, Wade
Elsey, Margaret Grace
Embree, Earl Owen
England, James Walton
Enig, Julius William
Enneking, Marjorie
Ennulat, Reinhard D
Enochs, Edgar Earle
Epstein, Bernard
Epstein, George
Epstein, Marvin Phelps
Ericksen, Wilhelm Skjetstad
Erkiletian, Dickran Hagop, Jr
Ernest, John Arthur
Ernsdorff, Louis Edward
Eigen, Garret Jay
Evans, Arwel
Evans, Edward William
Evans, Jacqueline P
Evans, Noel Dee
Evans, Thomas George
Evans, Trevor
Evans, William Buell
Evens, Leonard
Everett, Cornelius Joseph
Eves, Howard (Whitley)
Ewing, George McNaught
Faber, Richard Leon
Fadell, Albert George
Fadell, Edward Richard
Fairchild, William Warren
Faith, Carl Clifton
Fan, Ky
Fang, Joong
Fang, So-Fei
Farkas, Hershel M
Farley, Reuben William
Farnell, Albert Bennett
Fary, Istvan
Fasenmyer, Mary Celine

Fass, Arnold Lionel
Faucett, William Munroe
Faught, Donald Thomas
Faust, Claude Marie
Feeman, George Franklin
Feit, Sidnie Marilyn
Feit, Walter
Felder, Virginia Isabelle
Feldman, Chester
Feldman, Hyman Morris
Feldman, Jacob
Feldman, Leonard
Feliciano-Dodonoff, Manuel
Fennell, Robert E
Ferguson, David John
Ferguson, Harry
Ferguson, Helaman Rolfe Pratt
Ferguson, Le Baron O
Ferguson, William Allen
Ferling, John Albrecht
Ferrar, Joseph C
Fialkow, Aaron David
Ficken, Frederick Arthur
Fields, Jerry L
Fife, Paul Chase
Filano, Albert Eugene
Finch, John Vernor
Finco, Arthur A
Fine, Nathan Jacob
Fink, Arlington M
Finkbeiner, Daniel Talbot, II
Finkelstein, Mark
Firey, William James
Fisch, Forest Norland
Fischer, Irwin
Fischer, John Eugene
Fishback, William Thompson
Fisher, Donald D
Fisher, Gordon McCrea
Fisher, Newman
Fisher, Robert Charles
Fisher, Stephen D
Fishman, Robert Sumner
Fisk, Donald
Fitzpatrick, Ben, Jr
Flanagan, Carroll Edward
Flanagan, Joseph Edward
Flanagan, Robert Joseph
Flatt, Horace Perry
Flatto, Leopold
Fleisher, Harold
Fleming, Richard Joseph
Fleming, Walter
Fleming, Wendell Helms
Fletcher, William Thomas
Flood, Merrill Meeks
Flusser, Peter R
Foisy, Hector B
Foland, Neal Eugene
Folkert, Jay Ernest
Folland, Gerald Budge
Folley, Karl Wilmot
Foos, Barbara Ann
Forbes, Jack Edwin
Ford, David A
Ford, Lester Randolph, Jr
Ford, Patrick Lang
Forelli, Frank John
Foreman, Calvin
Forman, William
Forray, Marvin Julian
Fors, Elton W
Fossum, Robert
Fossum, Robert Merle
Foster, Alfred Leon
Foster, Lorraine L
Foster, Mary Lee
Foulis, David James
Foulser, David A
Fowler, Kenneth Arthur
Fox, Augustus Henry
Fox, Bennett L
Fox, David William
Fox, Phyllis
Fox, William Cassidy
Frame, James Sutherland
Francis, Eugene A
Francis, George Konrad
Francis, Richard L
Frandsen, Henry
Frank, Evelyn
Frank, Thomas Stolley
Frankel, Theodore Thomas
Franz, Edgar Arthur
Fraser, Grant Adam
Fraser, Robert B
Fray, Robert Dutton
Frazer, Lowell Keith
Frazier, Robert Carl
Fredricks, David Alan
Freed, Aubyn
Freedman, Allen Roy
Freedman, Marvin I
Freeman, John Clinton, Jr
Freeman, Robert
Freier, Jerome Bernard
Freilich, Gerald
Freimer, Marshall Leonard

MATHEMATICS

Fremont, Herbert Irwin
Freyd, Peter John
Freyre, Raoul Manuel
Frick, Charles Harold
Fridy, John Albert
Friedman, Avner
Friedman, Charles Nathaniel
Friedman, Henry David
Friedrichs, Kurt Otto
Frink, Aline
Frisinger, H Howard, II
Fritsche, Richard T
Fritz, Carol S
Froemke, Jon
Fronabarger, Carl Valentine
Fry, Cleota Gage
Fry, Thornton Carl
Fryling, Robert Howard
Fryxell, Ronald C
Fu, Lorraine Shao-Yen
Fuchs, Laszlo
Fuhrken, Gebhard
Fulkerson, Delbert Ray
Fulks, Watson
Fuller, Francis Brock
Fuller, Kent Ralph
Fuller, Leonard Eugene
Fuller, Roy Joseph
Fuller, William Richard
Fulton, Dawson Gerald
Fulton, John David
Funkenbusch, Walter William
Furman, Walter L
Fusaro, Bernard A
Gans, David
Garabedian, Paul Roesel
Garcia, Mariano
Gardner, Gerald Henry Fraser
Garey, Michael Randolph
Garfunkel, Irving Minturn
Garg, Krishna Murari
Garland, Howard
Garner, Jackie Bass
Garner, Lynn E
Garner, Meridon Vestal
Garnet, Hyman
Garrett, James Richard
Garrett, Robert Lee
Garrick, Isadore Edward
Garriga-Rodriguez, Francisco
Garrow, Robert Joseph
Garsia, Adriano Mario
Gary, John Mitchell
Gaskill, Irving E
Gass, Clinton Burke
Gass, Frederick Stuart
Gass, Saul Irving
Gately, Wilson York
Gates, Leslie Dean, Jr
Gatewood, Buford Echols
Gauvin, Hervey Paul
Gaver, Donald Paul
Gee, Susan McMillin
Geeslin, Roger Harold
Gehring, Frederick William
Geisler, Murray Aaron
Geissinger, Ladnor Dale
Gelbart, Abe
Gelbaum, Bernard Russell
Genevese, Frank
Gentleman, William Morven
Gentry, Ivey Clenton
Gentry, Karl Ray
Georgakis, Constantine
George, Dick Leon
George, John Harold
George, Melvin Douglas
Geraghty, Michael A
Gerlach, Eberhard
Gerneth, Dal Charles
Gershenson, Hillel Halkin
Gerst, Irving
Gerstein, Larry J
Gerstenhaber, Murray
Geshner, Robert Andrew
Getoor, Ronald Kay
Gewirtz, Allan
Gewirtzman, Leonard
Ghent, Kenneth Smith
Gibbons, M Seraphim
Gibson, Peter Murray
Giese, John H
Giever, John Bertram
Gilbert, Edgar Nelson

Gilbert, Paul Wilner
Gilbert, Robert Pertsch
Gil de Lamadrid, Jesus
Gilfeather, Frank L
Gill, Gurcharan S
Gill, Basil Early
Gillam, James Horace
Gillette, Dean
Gilliland, Dennis Crippen
Gillman, David
Gillman, Leonard
Gilman, Albert F, III
Gilman, John Frances
Gilmore, Earl Howard
Gilmore, Maurice Eugene
Ginivan, Francis Joseph
Ginnings, Gerald Keith
Ginsburg, Michael
Girard, Dennis Michael
Gitteman, Arthur P
Givens, Wallace, (Jr)
Gladman, Charles Herman
Glassner, Moses
Glauberman, George Isaac
Gleason, Andrew Mattei
Gleissner, Gene Heiden
Gleyzal, Andre
Glickfeld, Barnett W
Glicksberg, Irving Leonard
Glimm, James Gilbert
Glynn, William Allen
Gobirsch, Richard Paul
Goddard, Alton R
Godino, Charles F
Goebel, Jack Bruce
Goedecker, Peter V
Goetz, Abraham
Goheen, Harry Earl
Goldberg, Estelle Maxine
Goldberg, Merrill B
Goldberg, Richard
Goldberg, Richard Robinson
Goldberg, Samuel
Goldberg, Samuel I
Goldberg, Seymour
Goldman, Alan Joseph
Goldman, Charles C
Goldman, Malcolm
Goldman, Oscar
Goldsmith, Donald Leon
Goldstein, Max
Goldstein, Myron
Golomar, Dorothy May
Golomb, Michael
Golomb, Solomon Wolf
Golub, Gene H
Goman, Edward Gordon
Gonzalez-Fernandez, Jose Maria
Good, Irving John
Goodman, Adolph Winkler
Goodrich, Robert Kent
Goodrick, Richard Edward
Goodspeed, Frederick Maynard (Cogswell)
Goodwin, Bruce Edward
Goodwin, Robert Joseph
Gordon, Samuel Robert
Gordon, William Bernard
Gorman, John Richard
Goslin, Roy Nelson
Goss, Robert Nichols
Gottlieb, Daniel Henry
Gottschalk, Walter Helbig
Gough, Lillian
Gould, Henry Wadsworth
Gould, Sydney Henry
Gould, William Allen
Gould, William E
Govindarajulu, Zakkula
Gowdy, Spencer O
Graber, Leland D
Grabiner, Sandy
Grad, Arthur
Gragg, William Bryant, Jr
Graham, Colin C
Graham, James W
Graham, John Elwood
Graham, Malcolm
Graham, Ray Logan
Graham, Ronald Lewis
Gramick, Jeannine
Grams, Anne P
Grandchamp, Yvon
Grandy, Charles Creed
Gratzer, George
Grau, Albert A
Graue, Louis Charles
Graver, Jack Edward
Graves, Glenn William
Graves, Robert Lawrence
Gray, Allan, Jr
Gray, Brayton
Gray, John Walker
Gray, Mary Wheat
Graybeal, Walter Thomas
Grechie, Richard Joseph
Green, Daniel Thomas
Green, John Willie
Green, Leon William

Green, Robert Eugene
Green, Simon
Greenberg, Marvin Jay
Greene, Curtis
Greene, Joseph S
Greenleaf, Newcomb
Greenleaf, Frederick
Greenspan, Bernard
Greenspan, Donald
Greenstein, David Snellenburg
Greenwood, Joseph Albert
Greenwood, Robert Ewing
Greer, Earl Vincent
Greer, Edison
Greever, John
Greiner, Peter Charles
Greinke, Everett D
Gretsky, Neil E
Greville, Thomas Nall Eden
Griego, Richard Jerome
Griesmer, James Hugo
Griffin, Ernest Lyle
Griffith, (Jr)
Griffith, H C
Griffith, Philip A
Grimm, Carl Albert
Groemer, Helmut (Johann)
Gronski, Jan Maksymilian
Gross, Fletcher
Gross, Fred
Gross, Jonathan Light
Gross, Leonard
Grossman, George
Grossman, Mildred Lucile
Grossman, Nathaniel
Grosswald, Emil
Grove, Larry Charles
Gruber, Gary Richard
Gruber, H Thomas
Gudder, Stanley Philip
Guderley, Karl Gottfried
Guenther, Paul Ernest
Guggenbuhl, Laura
Guggenheimer, Heinrich Walter
Guilfoyle, Richard Howard
Guinand, Andrew Paul
Guinn, Theodore
Gulick, Sidney L, III
Gundersen, Roy Melvin
Gunderson, Norman Gustav
Gunji, Hiroshi
Gunning, Robert Clifford
Gupta, Shanti Swarup
Gurk, Herbert Morton
Gurland, John
Guseman, Lawrence Frank, Jr
Gustafson, Karl Edwin
Gustin, William
Gutzman, Wayne Wallace
Guy, Richard K
Guy, William Thomas, Jr
Haake, James Walter
Haas, Felix
Haase, Kurt Harald
Haber, Robert Morton
Hachigian, Jack
Haddix, George Franklin
Hadlock, Charles Robert
Hadlock, Edwin Harold
Haenisch, Siegfried
Haeuslein, Guenter Karl
Hafstrom, John Edward
Hagan, Melvin Roy
Hageman, Louis A
Haggard, J D
Haggard, Paul Wintzel
Hagis, Peter, Jr
Hahn, Hwa Suk
Hahn, Liang-Shin
Hahn, Samuel Wilfred
Haigh, William E
Halpern, Raoul
Haimo, Franklin
Haimo, David Clark
Hajek, Otomar
Hajian, Arshag B
Haken, Wolfgang
Halberg, Charles John August, Jr
Hale, Jack Kenneth
Hale, Alfred Washington
Hales, Raleigh Stanton, Jr
Halkin, Hubert
Hall, Carl Eldridge
Hall, James Emerson
Hallam, Thomas Guy
Hallerberg, Arthur Edward
Halperin, Israel
Halton, John Henry
Hamilton, Elbert W
Hamilton, Hugh James
Hamilton, William Wingo
Hammer, Preston Clarence
Hammett, Michael E
Hammond, Robert Grenfell
Hancock, Vernon Ray
Hanes, Harold
Hanhauser, Martin A
Hanna, James Ray
Hanna, Martin Slater
Hanneken, Clemens
Hannon, Herbert Harold

Hansen, Donald Joseph
Hansen, Rodney Thor
Hansman, Margaret Mary
Hanson, James Edward
Hanson, David Lee
Hanson, Robert Harold
Harary, Frank
Hardell, William John
Hardgrove, Clarence Ethel
Hardman, Dennis Hunter
Hardy, Flournoy Lane
Hardy, John W, Jr
Hare, William Ray, Jr
Harley, Peter W, III
Harper, Lawrence Raymond, Jr
Harper, Lawrence Hueston
Harrington, Walter Joel
Harris, Bruno
Harris, Henson
Harris, J Douglas
Harris, Reece Thomas
Hartley, Miles C
Harris, Theodore Edward
Harrison, David Kent
Harrison, Malcolm Charles
Harrison, Ross Arthur
Harrold, Ronald Baxter
Harrold, Scott Boynton
Harrold, Orville Grover, Jr
Hart, Alice Goodson
Harshbarger, Frances
Hart, Garry Dewaine
Hart, Charles, Jr
Hartfiel, Darald Joe
Hattemer, Jimmie Ray
Hatke, Mary Agnes
Hartley, Dallas T
Hartman, Philip
Hartmanis, Juris
Hartmann, Frederick W
Hartnett, William Edward
Haruki, Hiroshi
Harvey, Albert Raymond
Harvey, Charles Arthur
Harvey, F Reese
Harvey, John Grover
Hashisaki, Joseph
Haskell, Charles Thomson
Hastings, Stuart Pendleton
Hausdoerffer, William H
Hausner, Jutta
Hausner, Alvin Stanley
Hawkins, Thomas William, Jr
Hawley, Newton Seymour, Jr
Hawthorne, Frank Sylvester
Hay, George Edward
Hayden, Seymour
Hayes, Murray Lawrence
Hayes, Robert Mayo
Hayes, Charles Amos, Jr
Hayes, Dallas T
Hayes, Thomas Lee
Head, Thomas James
Headley, Velmer Bentley
Healy, Paul William
Hearne, Horace Clark, Jr
Hersey, Bryan Vandiver
Heath, David Clay
Heath, Robert Winship
Heatherly, Henry Edward
Heckscher, Stevens
Hedayat, Abdossamad
Hedberg, Marguerite Zeigel
Hedlund, Gustav Arnold
Hedlund, James Howard
Hedstrom, Gerald Walter
Hedstrom, John Richard
Hee, Christopher Edward
Heed, Joseph James, Jr
Heerema, Nickolas
Heidel, John Willard
Heikkinen, Donald D
Heil, Wolfgang Heinrich
Heinig, Hans Paul
Heins, Albert Edward
Heins, Maurice Haskell
Helgason, Sigurdur
Helleseth, Joseph Patrick
Heller, Alex
Heller, Robert
Hellman, Morton J
Helmbold, Robert Lawson
Helmer, Olaf
Helms, Lester LaVerne
Helton, Burrell W
Helton, John W
Helton, Floyd Franklin
Hemmingsen, Erik
Henderson, David Wilson
Hendricks, Walter James
Hendrickson, Morris S
Henkelman, James Henry
Henney, Dagmar Renate
Henney, James Perry
Henrici, Peter K
Henriksen, Melvin

Henry, Boyd (Herbert)
Hensel, Gustav
Herbst, Robert Taylor
Herda, Hans-Heinrich Wolfgang
Herman, Eugene Alexander
Hermes, Henry Gustav
Herpel, Coleman
Herr, David Guy
Herschorn, Michael
Hershenov, Joseph
Hershner, Ivan Raymond, Jr
Herstein, Israel Nathan
Hertz, Douglas Nelson
Hertzig, David
Herwitz, Paul Stanley
Herz, Carl Samuel
Herzog, Fritz
Herzog, John Orlando
Hesch, Elizabeth Beaman
Hestenes, Magnus Rudolph
Heuer, Charles Vernon
Heuer, Gerald Arthur
Heuvers, Konrad John
Heverly, John Ross
Hewgill, Denton Elwood
Hewitt, Edwin
Heynick, Louis Norman
Hibbs, Leon
Hicks, Noel J
Hicks, Troy L
Higdem, Roger Leon
Higgins, John Clayborn
Higgins, Theodore Parker
Higgins, William Brouillard
Hight, Donald Wayne
Higman, Donald Gordon
Hilbert, Stephen Russell
Hildebrand, Francis Begnaud
Hildebrand, James Leslie
Hildebrandt, Emanuel Henry Carl
Hildebrandt, John Joseph
Hildebrandt, Theodore Ware
Hilding, Stephen R
Hill, David Byrne
Hill, Edward T
Hill, Paul Daniel
Hill, Robert Joe
Hillam, Kenneth L
Hillman, Abraham P
Hilton, Peter John
Hind, Alfred Thomas, Jr
Hinkkanen, Donald William
Hinrichs, Lowell A
Hinrichsen, John James Luett
Hinton, Don Barker
Hirsch, Morris William
Hirsch, Peter M
Hirsch, Warren Maurice
Hirschfelder, John Joseph
Hirschman, Isidore Isaac, Jr
Hirshon, Ronald
Ho, Chung-Wu
Ho, Grace Ping-Poo
Ho, Shung Pun
Hobbs, Billy Frankel
Hobby, Charles R
Hochberg, Murray
Hocking, John Gilbert
Hocking, Ronald Raymond
Hodel, Richard Earl
Hodge, James Edgar
Hodges, Billy Gene
Hodges, John Herbert
Hodgson, Jonathan Peter Edward
Hoel, Paul Gerhard
Hoffer, Alan R
Hoffman, Alan Jerome
Hoffman, Frederick
Hoffman, Kenneth Myron
Hoffman, Ruth I
Hoffmann, Karl Heinrich
Hogan, Guy T
Hogan, John Wesley
Hoggatt, Verner Emil, Jr
Holder, Leonard Irvin
Hole, John J
Holland, Samuel S, Jr
Holland, Wilbur Charles, Jr
Holley, Richard Andrew
Hollingsworth, Jack W
Hollingsworth, John Gressett
Holmes, Calvin Virgil
Holt, Everett William
Holt, John Melvin
Holter, Marvin Rosenkrantz
Holub, James Robert
Holzinger, Joseph Rose
Homann, Frederick Anthony
Homer, Roger Harry
Hoo, Cheong Seng
Hood, Rodney Taber
Hooke, Robert
Hopper, Grace Murray
Horn, Roger Alan
Hornback, Joseph Hope
Horne, James Grady, Jr
Horner, James M
Horton, Robert Eugene

Horton, Thomas Roscoe
Horwitz, Harold M
Houh, Chorng Shi
Houle, Joseph E
Hourston, Rognvald Charles Nicol
Householder, Alston Scott
Householder, James Earl
Howard, Aughtum Smith
Howard, Henry Cobourn
Howe, Roger
Howell, James Levert
Howland, James Secord
Howland, Richard A
Hoyle, Hughes Bayne, Jr
Hsieh, Po-Fang (Philip)
Hsiung, Chuan Chih
Hsu, Nai-Chao
Hsu, Yu Kao
Hsu, Yu-Sheng
Hu, Sze-Tsen
Huckleberry, Alan Trinler
Hudson, Anne Lester
Hudson, Frank M
Hudson, George Elbert
Hudson, Sigmund Nyrop
Huffman, Louie Clarence
Hufford, George (Allen)
Huffstutler, Ronald
Huggins, Frank Norris
Hughart, Stanley Parlett
Hughes, David Knox
Hughes, Howard Kibble
Humberd, Jesse David
Hummel, James Alexander
Humphreys, Mabel Gweneth
Hundhausen, Joan Rohrer
Huneke, Harold Vernon
Huneke, John Philip
Hungerford, Thomas W
Hunsaker, Neville Carter
Hunt, Beryl Eleanor
Hunter, Larry Clifton
Hunter, Robert P
Hunzeker, Hubert LaVon
Hurd, Albert Emerson
Hurd, Cuthbert Corwin
Hurt, Norman Edward
Hurwitz, Solomon
Husain, Delawar
Husain, Syed Alamdar
Huston, Antoinette Killen
Hutchings, William Lawrence
Hutchinson, Lee Pressley
Hutchinson, Jeanne S
Hutzenlaub, John F
Hyde, Kendell Heman
Hyers, Donald Holmes
Iglarsh, Harvey Jerome
Ignatia (Frye), Mary
Igusa, Jun-Ichi
Ikenberry, Jesse Emmert
Iltis, Donald Richard
Immel, Eric Robert
Ingraham, Mark Hoyt
Ingram, William Thomas
Innis, George Seth
Insel, Arnold J
Ionescu Tulcea, Alexandra
Irwin, John McCormick
Irwin, Richard Eugene
Isaac, Richard Eugene
Isaacs, Irving Martin
Isaacson, Eugene
Isbell, John Rolfe
Isenecker, Lawrence Elmer
Isles, David Frederick
Ito, Kiyoshi
Ito, Noboru
Itzkowitz, Gerald Lee
Iwasawa, Kenkichi
Izzo, Joseph Anthony, Jr
Jackson, Lloyd K
Jackson, Prince A, Jr
Jackson, Robert Bruce, Jr
Jacob, Henry George, Jr
Jacobowitz, Ronald
Jacobs, Marc Quillen
Jacobson, Bernard
Jacobson, Florence Dorfman
Jacobson, Nathan
Jacoby, Alexander Robb
Jaffa, Robert E
Jagerman, David Lewis
James, Donald Gordon
James, Ralph L
Jamieson, Derek Maitland
Jamison, Herman Free
Jamison, King W, Jr
Jamison, Robert Edward
Jamison, Steven Lyle
Jamison, William H
Janos, Ludvik
Jans, James Patrick
Jansen, Bernard Joseph
Janusz, Gerald Joseph
Jarnagin, Milton Preston, Jr

Jarvinen, Richard Dalvin
Jasper, Samuel Jacob
Jaswal, Sitaram Singh
Javaher, James N
Jeffery, Ralph Lent
Jenkins, Howard Bryner
Jenkins, James Allister
Jenkins, Joe Wiley
Jenkins, Ted R
Jenkins, Terry Lloyd
Jenks, Richard D
Jenner, William Elliott
Jennings, Stephen Arthur
Jensen, Bruce A
Jerison, Meyer
Jodeit, Max A, Jr
Joffe, Anatole
Johannes, Karl A
John, Fritz
Johnson, Augustus Clark
Johnson, Carlos Sigfrid, Jr
Johnson, Charles Andrew
Johnson, David Stifler
Johnson, Diane Mary
Johnson, Donald Elwood
Johnson, Donald Glen
Johnson, Donovan Albert
Johnson, Dudley Paul
Johnson, Elgy Sibley
Johnson, Frankford Milam
Johnson, Gordon Gustav
Johnson, Harold Hunt
Johnson, Howard Arthur, Sr
Johnson, James Stephen
Johnson, Johnny Albert
Johnson, Joseph Lemuel, Jr
Johnson, Larry K
Johnson, Lee Murphy
Johnson, Lee W
Johnson, Millard Wallace, Jr
Johnson, Paul Bennett
Johnson, Phillip Eugene
Johnson, Ralph Bernard
Johnson, Ralph T, Jr
Johnson, Raymond Lewis
Johnson, Richard Edward
Johnson, Robert Leroy
Johnson, Robert P
Johnson, Robert S
Johnson, Robert Wells
Johnson, Roger D, Jr
Johnson, Roy Andrew
Johnson, Selmer Martin
Johnson, Wendell Gilbert
Johnson, William Lee
Johnston, Andrea
Johnston, Ernest Raymond
Johnston, Richard Milton
Johnston, Robert Howard
Joiner, Harry Francis, II
Jones, Benjamin Franklin, Jr
Jones, Burton Wadsworth
Jones, Eleanor Green
Jones, Grover Stephen
Jones, J P
Jones, John, Jr
Jones, John Paul
Jones, Major Boyd
Jones, Phillip Sanford
Jones, R E Douglas
Jones, Roy Carl, Jr
Jones, William B
Jonsson, Bjarni
Jonsson, Wilbur Jacob
Jordan, George Samuel
Jordan, James Henry
Jordan, Steven Lee
Juberg, Richard Kent
Juola, Robert C
Kac, Mark
Kadish, Abraham
Kadison, Richard Vincent
Kadota, T Theodore
Kagan, Joel (David)
Kahan, William M
Kahn, Daniel Stephen
Kahn, Donald W
Kahn, Louis B
Kahn, Peter Jack
Kainen, Paul Chester
Kakutani, Shizuo
Kales, Morris L
Kalin, Robert
Kalinowski, Walbert C
Kalisch, Gerhard Karl
Kallaher, Michael Joseph
Kaller, Cecil Louis
Kallianpur, Gopinath
Kalman, Ralph Arthur
Kalman, Rudolf Emil
Kalmanson, Kenneth
Kalme, John S
Kaltenborn, Howard Scholl
Kammerer, William John
Kamowitz, Herbert M
Kanarik, Rosella
Kane, Robert B

Kannappan, Palaniappan L
Kanter, Louis Harold
Kantorovitz, Shmuel
Kaplan, Edward Lynn
Kaplan, James
Kaplan, Samuel
Kaplan, Wilfred
Kaplansky, Irving
Kapoor, S F
Karel, Martin Lewis
Karnes, Houston Thurman
Karst, Otto Jacob
Karush, William
Kas, Arnold
Kascic, Michael Joseph, Jr
Kasriel, Robert H
Kato, Tosio
Katsoff, Louis Osgood
Katz, Irving
Katz, Israel Norman
Kauffman, Ellwood
Kaufman, Hyman
Kaufmann, Alvern Walter
Kazarinoff, Nicholas D
Kearns, Donald Allen
Keedy, Mervin Laverne
Keen, Linda
Keenan, Edward Milton
Keesee, John William
Keisler, James Edwin
Kelemen, Charles F
Kelisky, Richard Paul
Keller, Joseph Bishop
Keller, Marion Wiles
Keller, Mary Kenneth
Keller, Roy Fred
Kelley, Allen Frederick, Jr
Kelley, John Ernest
Kelley, John Le Roy
Kellogg, Charles Nathaniel
Kelly, Edgar Preston, Jr
Kelly, John Beckwith
Kelly, Leroy Milton
Kelly, Paul J
Kelman, Robert Bernard
Kemeny, John George
Kemp, Louis Franklin, Jr
Kemp, Robert Richard Dingle
Kemperman, Johannes Henricus Bernardus
Kenelly, John Willis, Jr
Kennedy, Edward Stewart
Kenner, Morton Roy
Kennison, John Frederick
Kent, Clement F
Kenyon, Hewitt
Keogh, Frank Richard
Kerce, Robert H
Kerr, Carl E
Kerr, Donald R
Kershner, Richard Brandon
Kertz, George J
Kerzman, Norberto Luis
Kesarwani, Roop Narain
Keyes, James Frazier
Keynes, Harvey Bayard
Kezlan, Thomas Phillip
Khabbaz, Samir Anton
Khalil, Mohamed Thanaa
Khurana, Surjit Singh
Kibbey, Donald Eugene
Kieval, Harry Sears
Killam, Eleanor
Kim, K-ang
Kim, Soon-Kyu
Kim, Woo Jong
Kimeldorf, George S
Kimme, Ernest Godfrey
Kimura, Naoki
Kincaid, Wilfred Macdonald
Kincannon, John Alvin
Kinderlehrer, David (Samuel)
King, Amy P
King, Calvin Elijah
King, Jerry Porter
King, Larry Michael
King, R Maurice, Jr
King, Robert William
Kingston, John Maurice
Kinney, David Webster
Kinsey, David Webster
Kipps, Thomas Charles
Kirby, Robion C
Kirch, Murray R
Kirk, David Blackburn
Kirk, Joe Eckley, Jr
Kirk, William Arthur
Kiser, Lola Frances
Kist, Joseph Edmund
Kister, James Milton
Klaasen, Gene Allen
Klamkin, Murray S
Klatt, Gary Brandt
Klee, Victor La Rue, Jr
Kleiner, Alexander F, Jr
Kleinfeld, Margaret Humm
Kleitman, Daniel J
Klemola, Tapio
Klimczak, Walter John

MATHEMATICS

Kimko, Eugene M
Klinger, William Russell
Klopfenstein, Kenneth F
Klotz, George Arthur
Knapp, Anthony William
Kneale, David Isaac
Knee, Samuel George
Knight, Frank B
Knight, Lyman Coleman
Knopp, John Charles
Knopp, Paul Joseph
Knutson, Donald Ervin
Kobayashi, Shoshichi
Koch, Charles Frederick
Koch, Robert Jacob
Kocher, Frank T
Kocher, Simon Bernard
Kochman, Stanley Oscar
Kodres, Uno Robert
Koehler, Truman Lester
Koh, Kwangil
Kohlbecker, Eugene Edmund
Kohls, Carl William
Kolchin, Ellis Robert
Kolchin, Joseph John
Kolman, Bernard
Kolodner, Ignace Izaak
Komm, Horace
Konguetsof, Leonidas
Konhauser, Joseph Daniel Edward
Konheim, Alan G
Koo, Delia Wei
Koopman, Elaine
Koppelman, Bernard Osgood
Koranyi, Adam
Korevaar, Jacob
Korfhage, Robert R
Korgen, Reinhard Lunde
Kork, John O
Kosier, Frank J
Kosinski, Antoni A
Kostant, Bertram
Kotiah, Thoddi Chandrasekara
Kotin, Leon
Kottarski, Ignacy Icchak
Kottman, Clifford Alfons
Kovach, Ladis Daniel
Koval, Daniel
Kowalik, Virgil C
Kra, Irwin
Krabill, David Milton
Krakowski, Fred
Krakowski, Martin
Krall, Allan M
Krall, Harry Levern
Krauss, Peter H
Krebiel, Jesse D
Kreider, Orlando Clark
Kreiling, Daryl
Kreimer, Herbert Frederick, Jr
Kreith, Kurt
Krieger, Henry Alan
Kriegsman, Helen
Krishnan, Viakalathur Sankrithi
Krom, Melven R
Kronk, Hudson V
Krueger, Eugene Rex
Kruse, Arthur Herman
Kruskal, Joseph Bernard
Kuang, Huan Pao
Kueker, David William
Kuenzi, Norbert James
Kuhn, Harold William
Kuipers, Jack
Kuipers, Lauwerens
Kujala, Robert Oakes
Kuller, Robert G
Kullman, David Elmer
Kuntz, Richard A
Kunze, Ray A
Kurss, Herbert
Kurth, Rudolf
Kurtz, Thomas Gordon
Kusak, Lloyd James
Kuzawa, Mary Grace
Kwon, Young Koan
Kwun, Kyung Whan
Labarre, Anthony E, Jr
L'Abbe, Maurice
Labute, John Paul
Lacey, Howard Elton
Lacroix, Norbert Hector Joseph
Laden, Hyman Nathaniel
Ladendorf, Agnes J
Laderman, Jack
Laframboise, Marc Alexander
Lagnese, John Edward
Lai, Tze Leung

Lakein, Richard Bruce
Lakshmikantham, Vangipuram
Lal, Mohan
Lam, Ping-Fun
Lambek, Joachim
Lambert, Robert J
Lambert, William M, Jr
Lamperti, John Williams
Lancaster, George Maurice
Lancaster, Otis Ewing
Landau, Henry Jacob
Landers, Aubrey Wilfred
Landers, Mary Kenny
Landesman, Edward Milton
Landin, Joseph
Landrum, Bobby L
Landweber, Peter Steven
Lane, Bennie Ray
Lane, Leo Jerome
Lange, Lester Henry
Langebartel, Ray Gartner
Langenhop, Carl Eric
Langevin, Robert Arthur
Langford, Eric Siddon
Langlands, Robert P
Langworthy, Harold Frederick
Lanski, Charles Philip
Lapidus, Arnold
Lapidus, Leo
Laplaza, Miguel Luis
Larguier, Everett Henry
Larkin, James Richard
Larmore, Lawrence Louis
Larney, Violet Hachmeister
Larsen, Charles McLoud
Larson, Richard Gustavus
Larson, Roland Edwin
Larson, Robert Dustin
Lavalle, Lorraine Doris
Lavender, DeWitt Earl
Laverell, William David
Lavroff, Viacheslav V
Lawrence, James Franklin
Laws, Leonard Stewart
Lawson, Herbert Blaine, Jr
Lawson, Jimmie Don
Lawson, John Douglas
Lawson, Mildred Wiker
Lawson, John Allen
Lawver, Donald Allen
Lawwill, Stanley Joseph
Lax, Anneli
Lax, Peter David
Lay, David Clark
Lay, Steven R
Leach, Ernest Bronson
Leader, Solomon
Leadley, John David
Leake, Lowell, Jr
LeBel, Jean Eugene
Lebow, Arnold
Ledley, Robert Steven
Lee, Chung N
Lee, Hua-Tsun
Lee, Talmage Hoyle
Leela, Srinivasa (G)
Leese, Eric Leslie
Leetch, James Frederick
Leening, David John
LeFkowitz, Ruth Samson
Lefton, Phyllis
Lehman, Alfred Baker
Lehman, Russell Sherman
Lehmer, Derrick Henry
Leibler, Richard Arthur
Leighton, Walter (Woods)
Leininger, Charles W
Leisenring, Kenneth Baylis
Leitzel, James Robert C
Leitzel, Joan Phillips
Leser, Tadeusz
Lesley, Frank David
Lesniak, Linda Marie
LeVan, Marijo O'Connor
Levenson, Morris E
LeVeque, William Judson
Levi, Howard
Levin, Gerson
Levin, Simon Asher

Levine, Harold Irving
Levine, Howard Allen
Levine, Jack
Levine, Jeffrey
Levine, Leo Meyer
Levinger, Bernard Werner
Levinson, Michael Leonard
Levit, Norman Jay
Levitt, Hilbert
Levow, Roy Bruce
Levy, Harry
Lewis, Charles Joseph
Lewis, Daniel Ralph
Lewis, Donald John
Lewis, James Vernon
Lewis, Jesse C
Lewis, Paul Edwin
Lewis, Paul Weldon
Lewis, William James
Lewy, Hans
Leysieffer, Frederick Walter
Li, Ming Chiang
Li, Tien-Yien
Liang, Joseph Jen-Yin
Libera, Richard Joseph
Lick, Don R
Lieberman, Burton Barnet
Liebmann, Paul W
Liggett, Thomas Milton
Light, John Henry
Ligh, Steve
Lightner, James Edward
Lightstone, Albert Harold
Lin, Bor-Luh
Lin, Pi-erh
Lin, Shen
Lin, Sue Chin
Lin, Tung-Po
Lindahl, Clarence Homer
Lindgren, Bernard William
Lindgren, William Frederick
Lindsay, Charles McCown
Lindstrom, Andrew O, Jr
Line, John Paul
Ling, Donald Percy
Lininger, Lloyd Lesley
Linis, Viktors
Linnstaedter, Jerry Leroy
Linton, Fred E J
Lipman, Joseph
Lipsich, H David
Lisman, Henry
Lissner, David
Lister, Frederick Monie
Little, Charles Edward
Liverman, Solomon Elieser
Liverman, Thomas Phillip George
Livesay, George Roger
Livingood, John N B
Lloyd, Joel Joseph
Lloyd, Justin Thomas
Lloyd, Stuart Phinney
Lobb, Barry Lee
Locke, John Franklin
Locke, Philip M
Locke, Stanley
Locker, John L
Lockhart, Brooks Javins
Loeb, Peter Albert
Loewen, Kenneth Leroy
Loewe, Michel
Loflin, Zacharaih Lowe
Lofquist, George W
Lohwater, A J
Loman, Laverne
Lomonaco, Samuel James, Jr
Long, Andrew Fleming, Jr
Long, Calvin Thomas
Long, Richard Gene
Loomis, Harold George
Loomis, Lynn H
Lorch, Edgar Raymond
Lorch, Lee (Alexander)
Lotkin, Mark Max
Lott, Fred Wilbur, Jr
Loud, Warren Simms
Loustaunau, Joaquin
Lovaglia, Anthony Richard
Love, Theodore Arceola
Loveland, Donald William
Loveless, David
Lovingood, Judson Allison
Low, Marc E
Lowengrub, Morton
Lowenthal, Franklin
Lowman, Bertha Pauline
Lu, Mary Kwang-Ruey Chao
Lubell, David
Lubin, Jonathan Darby
Lucas, Thomas Ramsey
Luchins, Edith Hirsch
Luckey, Robert Ruel Raphael
Ludden, Gerald D
Ludwig, Donald A

Ludwig, Garry (Gerhard Adolf)
Ludwig, Hubert Joseph
Lukacs, Eugene
Lukas, Joan Donaldson
Luke, Stanley D
Lumer, Gunter
Lundell, Albert Thomas
Lurye, Jerome Robert
Luther, Chester Francis
Luther, Herbert Adesla
Luttmann, Frederick William
Lutz, Donald Alexander
Lyjak, Robert Fred
Lynch, William C
Lyndon, Roger Conant
Lynn, Melvyn Stuart
Lytle, Raymond Alfred
MacCamy, Richard C
MacCracken, Elliott b
MacDonald, John Lauchlin
MacDonell, John Joseph
MacDowell, Robert W
Macey, Wade Thomas
MacGregor, Thomas Harold
Mach, George Robert
Machtinger, Lawrence Arnold
Machusko, Andrew Joseph, Jr
Macki, Jack W
MacKenzie, Robert Earl
Mackey, George Whitelaw
Maclay, Charles Wylie
MacLane, Saunders
MacNeill, Ian B
MacPhail, Moray St John
MacRae, Robert E
MacWherter, John Baird
MacWilliams, Florence Jessie
Maddox, Billy Hoyte
Madden, Bernard L
Maffett, Andrew L
Magee, Gordon Richey
Magee, John Francis
Magid, Andy Roy
Magill, Kenneth Derwood, Jr
Maghveras, Spyros Simos
Magnus, Arne
Mahavier, William S
Mahowald, Mark Edward
Maier, Eugene Alfred
Mairhuber, John Carl
Maker, Philip T
Makowski, Gary George
Maksoudian, Y'Leon
Malbon, Wendell Endicott
Maletsky, Evan M
Maloof, Giles Wilson
Maltese, George J
Mancuso, Vincent J
Mandelker, Mark
Manduley, Ilma Morell
Maneri, Carl C
Manheim, Jerome Henry
Manis, Merle E
Manjarrez, Victor M
Manley, Leo William
Mann, Henry Berthold
Manoharan, A Chelvanayakam
Manougian, Manoug N
Mansfield, Larry Everett
Manz, Bruno Julius
Marchand, Margaret O
Marcou, Rene Joseph
Marcus, Michael Barry
Marcus, Philip Selmar
Mardellis, Anthony
Marden, Morris
Margaris, Angelo
Margules, Gabriel
Margulies, William George
Mark, J Carson
Markham, Elizabeth Mary
Markle, Gerald E
Markley, William A, Jr
Markus, Lawrence
Marley, Gerald C
Marlow, William Henry
Marr, John Maurice
Marquardt, Donald Wesley
Marsaglia, George
Marsden, Jerrold Eldon
Marshall, Albert Waldron
Marth-Snader, Ella Carolyn
Martin, Abram Venable
Martin, Bernard Loyal
Martin, Charles K
Martin, Frank Burke
Martin, Kenneth Edward
Martin, LeRoy Brown, Jr
Martin, Norman Marshall
Martin, William Ted
Martin, William Frizzel Grafton
Marx, Morris Leon
Masani, Pesi Rustom
Masatis, Ceslovas
Maserick, Peter H

Maskit, Bernard
Mason, Jesse David
Mason, Lysle C
Massey, William S
Massey, Winston Louis
Masters, Christopher Fanstone
Mathews, Harry T
Mathews, Jerold Chase
Mathis, Robert Fletcher
Mathsen, Ronald M
Matlis, Eben
Matthies, Karl Heinrich
Mattingly, Glen E
Mauldin, Richard Daniel
Maxfield, Margaret Waugh
Maxson, Carlton J
Maxwell, Charles Neville
Maxwell, Glenn
May, Kenneth Ownsworth
Maybee, John Stanley
Mayer, Joerg Werner Peter
Maynard, Hugh Bardeen
Mayor, John Roberts
McAdam, Terry Donald
McAlister, Donald Beaton
McAllister, Byron Leon
McAllister, Cyrus Ray
McAllister, Gregory Thomas, Jr
McAlpin, Marialuisa N
McAlpin, John Harris
McArthur, Charles Wilson
McArthur, William George
McAuley, Louis Floyd
McBride, Woodrow H
McBrien, Vincent Owen
McCabe, John Patrick
McCabe, Robert Lyden
McCallion, William James
McCammon, Mary
McCandless, Byron Howard
McCann, Roger C
McCoy, Ronald Eugene
McCoy, Thomas LaRue
McCready, Thomas Arthur
McCrossen, Garner
McCue, Edmund Bradley
McCully, Joseph C
McDaniel, Wilbur Charles
McDonald, Bernard Robert
McDonald, Janet
McDonald, William John
McDougle, Paul E
McDowell, Robert Hull
McEwen, William Robert
McFadden, Leonard
McFadden, Robert
McGaughey, Albert Wayne
McGavock, William Gillespie
McGehee, Oscar Carruth
McGehee, Richard Paul
McGill, Suzanne
McGlasson, Alvin Garnett
McGloin, Paul Arthur
McGuigan, Robert Alister, Jr
McIntosh, William David
McKay, Harley Ellsworth
McKean, Hartley Ellsworth
McKellips, Terral Lane
McKelvey, Robert William
McKenna, James
McKenzie, Harvey
McKenzie, Ralph Nelson
McKibben, John Joseph
McKiernan, Michel Amedee
McKillop, Lucille Mary
McKinley, James Ernest
McKinney, Max Terral
McKinney, Richard Leroy
McKnight, James Dawson, Jr
McLachlan, Jack Enloe
McLaughlin, Eugene Kay
McLaughlin, James Joseph
McLaughlin, Thomas G
McLeod, James Ernest
McLeod, Edward Blake
McLinden, Lynn
McMillan, Brockway
McMillan, Louis Kelly, Jr
McMillin, Kenneth M
McMinn, Trevor James
McMorris, Fred Raymond
McNaughton, Robert
McNeill, Robert Bradley
McRae, Daniel George
McShane, Edward James
Meacham, Robert Colegrove
Mears, Florence Marie
Mech, William Paul
Mecklosky, Morton

Medlin, Gene Woodard
Medsger, Gerald William
Meek, James Latham
Megibben, Charles Kimbrough
Mehr, Cyrus B
Meisters, Gary Hosler
Melnick, Edward Lawrence
Melter, Robert Alan
Mendel, Clifford William
Mendelsohn, Nathan Saul
Mendelson, Bert
Menon, Manavazhi Vijaya Krishna
Menzin, Margaret Schoenberg
Mericle, R Bruce
Merkes, Edward Peter
Merrill, David McCray
Merrill, Samuel, III
Meserve, Bruce Elwyn
Mesirov, Jill Portner
Meux, John Wesley
Meyer, Burnett Chandler
Meyer, Charles Frederick
Meyer, Henry Irving
Meyer, Herman
Meyer, Jean-Pierre
Meyer, Richard Ernst
Meyer, Walter Joseph
Meyer, Werner Franz
Meyer, William Herman Lewis
Meyers, Leroy Frederick
Meyers, Norman George
Mezzino, Michael Joseph, Jr
Michael, William Alexander
Michalowicz, Joseph Victor
Mickle, Earl John
Micklich, John R
Middlemiss, Ross Raymond
Miech, Ronald Joseph
Miehle, William
Mielke, Paul Theodore
Mielke, Paul W, Jr
Mientka, Walter Eugene
Miles, Ernest Percy, Jr
Miles, Frank Belsley
Miles, Henry Jarvis
Miles, Joseph Belsley
Milgram, Richard James
Milkman, Joseph
Miller, A Eugene
Miller, Bruce Linn
Miller, Charles Frederick, III
Miller, Donald Smith
Miller, Donald Wright
Miller, Gerson Harry
Miller, Gordon Lee
Miller, John Grier
Miller, Kenneth Sielke
Miller, Max K
Miller, Richard Albert
Miller, Richard Keith
Miller, Richard Roy
Miller, William Brunner
Millett, Kenneth Cary
Milligan, Merle Wallace
Mills, William Harold
Milner, Eric Charles
Milnor, John Willard
Milnor, Tilla Savanuck Klotz
Minassian, Donald Paul
Minker, Jack
Minsky, Marvin Lee
Minty, George James, Jr
Miracle, Chester Lee
Miranker, Willard Lee
Mishoe, Luna I
Mislove, Michael William
Mitchell, Barry Miller
Mitchell, Benjamin Evans
Mitchell, Josephine Margaret
Mitchell, Merle
Mitchell, Roger W
Mitchell, Theodore
Mitchell, Wilbur Leonard
Mitchem, John Alan
Mittleman, Don
Mobley, Jean Bellingrath
Mock, Gordon Duane
Mode, Charles J
Moenter, Richard Luther
Mohanty, Sri Gopal
Mohat, John Theodore
Moise, Edwin Evariste
Moler, Cleve B
Molloy, Raymond William
Molloy, Marilyn
Mond, Bertram
Monk, James Donald
Montague, Harriet Frances
Montague, Patricia Tucker
Montague, Stephen
Montgomery, John C.
Montgomery, Mabel D
Montgomery, Richard Glee
Montroll, Elliott Waters
Montzingo, Lloyd J, Jr
Mooers, Calvin Northrup

Moon, John Wesley
Moon, Wilchor David
Moore, Berrien, III
Moore, Bill C
Moore, Calvin C
Moore, Charles Godat
Moore, Hal G
Moore, John Coleman
Moore, John Douglas
Moore, John Douglas
Moore, John Thomas
Moore, Marion E
Moore, Marvin G
Moore, Nadine Hanson
Moore, Ramon Edgar
Moore, Richard Allan
Moore, Robert Alonzo
Moore, Robert H
Moore, Theral Orvis
Moore, Warren Keith
Moran, Charles William
Mordeson, John N
Mordue, Dale Lewis
Morelock, James Crutchfield
Moreno, Carlos Julio
Morgan, Kathryn A
Morgan, Raymond Victor, Jr
Morgenthaler, George William
Moritz, Roger Homer
Morrel, Bernard Baldwin
Morrell, Joseph Salvador
Morrey, Charles Bradfield, Jr
Morrill, John Elliott
Morris, George William
Morris, Halcyon Ellen McNeil
Morris, Herbert Allen
Morris, Peter Craig
Morris, William Lewis
Morrison, Donald Ross
Morrison, John Joseph
Morrow, David Clarence
Morrow, James Allen, Jr
Morse, Anthony Perry
Morse, Marston
Moschovakis, Joan Rand
Moschovakis, Yiannis N
Moser, Jurgen (Kurt)
Moser, Louise Elizabeth
Moser, William O J
Mosesson, Zehman I
Mosher, Robert E
Moskowitz, Martin A
Mostert, Paul Stallings
Mostow, George Daniel
Mott, Thomas
Motteler, Zane Clinton
Moulis, Edward Jean, Jr
Mount, Bertha Lauritzen
Mount, Kenneth R
Moyer, John Clarence
Muecke, Herbert Oscar
Muench, Donald Leo
Muggli, Joanne
Muir, Donald Earl
Mukherjea, Arunava
Mulcrone, Thomas Francis
Mulhern, Thomas Patrick
Mullen, Kenneth
Muller, Elsie
Muller, Eric Rene
Mulligan, James Edward, Jr
Mullin, Ronald Cleveland
Mullins, Edgar Raymond, Jr
Munkres, James Raymond
Munroe, Marshall Evans
Murasugi, Kunio
Murdoch, David Carruthers
Mureika, Roman A
Murphy, Grattan Patrick
Murphy, James Lee
Murray, Christopher Brock
Murray, Francis Joseph
Murray, George Graham, Jr
Muses, Charles Arthur
Mycielski, Jan
Myers, Donald Earl
Myers, Franklin Guy
Myers, William Howard
Myhill, John
Myint-U, Tyn
Myrick, Alvin Grant
Nachbin, Leopoldo
Nafoosi, A Aziz
Nahikian, Howard Movess
Najar, Rudolph Michael
Nanda, Jagdish L
Nanzetta, Philip Newcomb
Narayana, Tadepalli Venkata
Nashed, Mohammed Zuhair Zaki
Nathanson, Melvyn Bernard
Nathanson, Weston Irwin
Nau, Richard William
Naymik, Daniel Allan
Nee, M Coleman
Neff, John David
Neff, Mary Muskoff

Neggers, Joseph
Negrepontis, Stylianos
Nehari, Zeev
Nelsen, Roger Bain
Nelson, A Carl, Jr
Nelson, Bill C
Nelson, Harry Ernest
Nelson, James Donald
Nelson, Joseph Edward
Nelson, Larry Dean
Nelson, Lloyd Steadman
Nelson, Oscar Tivis, Jr
Nelson, Raymond John
Nelson, Robert John
Nelson, Theodora S
Nemeth, Abraham
Neri, Umberto
Nering, Evar Dare
Nerode, Anil
Nesbeda, Paul
Ness, Linda Ann
Nestell, Merlynd Keith
Neuberger, John William
Neugebauer, Christoph Johannes
Neuhouser, David Lee
Neumann, Richard K
Neustadter, Siegfried Friedrich
Neuwirth, Jerome H
Neuzil, John Paul
Newberger, Edward
Newberger, Stuart Marshall
Newbery, A Chris
Newhouse, Albert
Newman, James Martin
Newman, Morris
Newman, Rogers J
Ney, Peter E
Niccolai, Nilo Anthony
Nichols, Eugene Douglas
Nichols, John C
Nichols, Joseph Caldwell
Nicholson, Eugene Haines
Nicholson, Victor Alvin
Nickerson, Helen Kelsall
Nico, William Raymond
Nicol, Charles Albert
Nicolaenko, Basil
Nielsen, G Howard
Nielsen, Kaj Leo
Nielson, George Marius
Nielson, Howard Curtis
Niemann, Ralph Henry
Nigam, Lakshmi Narayan
Nijenhuis, Albert
Nikolai, Paul John
Nilson, Edwin Norman
Nirenberg, Louis
Niven, Ivan (Morton)
Nohel, John Adolph
Noll, Walter
Nolstad, Arnold Ragnvald
Noonan, James Waring
Nooney, Grove C
Nordhaus, Edward Alfred
Norman, Edward
Norman, Robert Daniel
Norman, Robert Zane
Norris, Eugene Michael
Norris, Fletcher R
Northam, Edward Stafford
Northcutt, Robert Allan
Norton, Donald Alan
Norton, Karl Kenneth
Norvig, Torsten
Novosad, Robert S
Nower, Leon
Nuber, John Arthur
Nunke, Ronald John
Nussbaum, Adolf Edward
Nymann, DeWayne Stanley
Nymann, James Eugene
Oberbeck, Thomas Edmond
Oberg, Edwin Nathaniel
Oberhettinger, Fritz
O'Brien, Katharine Elizabeth
O'Brien, Redmond R
O'Connor, Robert Eric
Oddson, John Keith
Odle, John William
Oehmke, Robert H
Oestreicher, Hans Laurenz
Ogawa, Hajimu
Ogden, Robert David
Ogden, William Frederick
Ogilvy, C Stanley
Oh, Yoon Yong
Ohi, Donald Gordon
Ohm, Jack Elton
Ohmer, Paul Adolph
Ohmer, Merlin Maurice
Okashimo, Katsumi
Olds, Carl Douglas
Oler, Norman
Olinick, Michael
Oliphant, Malcolm William
Oliver, Gloria
Oliver, Carl Edward
Oliver, Henry William
Ollmann, Loyal Taylor

MATHEMATICS

Olmer, Jane Chasnoff
Olmsted, John Meigs Hubbell
Olshen, Richard Allen
Olson, Frank R
Olum, Paul
O'Malley, Mary Therese
O'Meara, O Timothy
O'Neil, Patrick Eugene
O'Neil, Peter Vincent
O'Neill, John Dacey
Oppelt, John Andrew
Ord, John Allyn
Ordman, Edward Thorne
Orey, Steven
Orland, George H
Orlik, Peter Paul
Orr, Richard Clayton
Ortiz-Suarez, Augusto Hermino
Orton, William R, Jr
Orzech, Morris
Osborn, Howard Ashley
Osborn, J Marshall, Jr
Osborn, John Edward
Osborn, Roger (Cook)
Osgood, Charles Freeman
Osher, Stanley Joel
Ossesia, Michel Germain
Ostberg, Donald Ross
Ostrand, Phillip Arthur
Ostrom, Theodore Gleason
Otter, Richard Robert
Ottinger, Carol Blanche
Otto, Albert Dean
Otoson, Harold
Oursler, Clelie Curtis
Outcalt, David L
Owen, Guillermo
Owens, Alvin Jewel
Owens, Glynn
Owens, Robert Hunter
Oxley, Theron D, Jr
Oxtoby, John Corning
Ozimkoski, Raymond Edward
Pace, Wesley Emory
Paciorek, Joseph Walter
Padberg, Harriet A
Padgett, William Jowayne
Page, Nelson Franklin
Page, Rector Lee
Page, Robert Leroy
Pager, David
Paine, Alan Henry
Paine, Dwight Milton
Painter, Page Robbins
Painter, Richard J
Palais, Richard Sheldon
Palas, Frank Joseph
Pall, Gordon
Palmer, Edgar M
Palmer, Theodore Paine
Palmore, Julian Ivanhoe, III
Panda, Rekha
Pandey, Jagdish Narayan
Papakyriakopoulos, Christos Dimitriou
Papanicolaou, George Constantine
Papert, Seymour A
Papp, Francis Joseph, IV
Papp, Zoltan
Pardee, Otway O'Meara
Park, Chull
Park, Won Joon
Parker, Ernest Tilden
Parker, Rodger D
Parker, Sidney Thomas
Parker, Willard Albert
Parks, Daniel K
Parlett, Beresford
Parnes, Milton N
Parr, Phyllis Graham
Parrott, Stephen Kinsley
Parsons, Louise Alayne
Parsons, Torrence Douglas
Partington, Carl Ralph
Passow, Eli (Aaron)
Patil, Ganapati P
Patil, Surgounda A
Patt, Charles Richard
Paulie, M Catherine Therese
Payne, Mary Hewlett
Payne, Stanley E
Peacy, Carl Mark, Jr
Pearl, Martin Herbert
Pearson, Bennie Jake
Pearson, Hans Lennart
Pease, Marshall Carleton, III
Peck, Lyman Colt
Pedersen, Franklin D
Pedersen, Roger Noel
Pedoe, Daniel
Peeples, William Dewey, Jr
Peglar, George W
Peisakoff, Melvin Philip
Pelliciaro, Edward Joseph
Peluso, Ada
Pennington, Arthur Wesley
Pepper, Paul Milton

Pereira, Carlos Martin
Perel, William Morris
Perkins, Peter
Perlin, Irwin Earl
Perlis, Alan Jay
Perry, Charles Rufus, Jr
Perry, E L, Jr
Perry, John Murray
Perry, Norman Conrad
Perry, William James
Persyn, Gilbert A
Pervin, William Joseph
Peters, Iland Dee
Peters, Joseph William
Petersen, Stefan
Petersen, Bent Edvard
Peterson, Bruce Bigelow
Peterson, Donald Palmer
Peterson, Elmor Lee
Peterson, Franklin Paul
Peterson, Harold Leroy
Peterson, John Alvin
Peterson, John M
Petryshyn, Walter Volodymyr
Pettis, Billy James
Pettofrezzo, Anthony J
Peyser, Gideon
Pfaff, Donald Chesley
Pfaffenberger, William Elmer
Pfaltzgraff, John Andrew
Pfeffer, Washek F
Pfeifer, George William
Pfeifer, Paul Edwin
Pflugfelder, Hala
Phelps, Dean G
Phelps, Jack
Phelps, Robert Ralph
Philipp, Walter V
Philippon, Lloyd L
Phillips, David Lowell
Phillips, Esther Rodlitz
Phillips, George Herbert, Jr
Phillips, Richard E, Jr
Phillips, Robert Bass, Jr
Phillips, Robert Gibson
Pickands, James, III
Pickens, Charles Glenn
Piech, Margaret Ann
Pierce, Keith Robert
Pignani, Tullio Joseph
Pilgrim, Donald
Pillai, Raman Narayana
Pimbley, George Herbert, Jr
Pinsky, Mark A
Pinter, Charles Claude
Pinzka, Charles Frederick
Piper, William Stephen
Pippert, Raymond Elmer
Pirenian, George
Pirenian, Zareh Meguerditch
Pitcher, Arthur Everett
Pitcher, Tom Stephen
Pitt, Loren Dallas
Pittenger, Arthur O
Pittman, Chaty Roger
Pixley, Alden F
Pixley, Emily Chandler
Piziak, Robert
Plaat, Otto
Plank, Donald Leroy
Pleasant, James Carroll
Plemmons, Robert James
Pless, Vera Stepen
Plotkin, Jacob Manuel
Plummer, Michael David
Pogorzelski, Henry Andrew
Pohl, William Francis
Poiani, Eileen Louise
Poliferno, Mario Joseph
Polimeni, Albert D
Polivka, Raymond Peter
Polking, John C
Pollack, Richard M
Pollak, Henry Otto
Pollard, Harry
Pollatsek, Harriet Suzanne
Pollin, Jack Murph
Polonsky, Ivan Paul
Polya, George
Pomerance, Carl
Pomeranz, Janet Bellcourt
Ponnapalli, Ramachandramurty
Ponomarev, Paul
Poole, Albert Roberts
Poole, George Douglas
Poole, John Terry
Poorman, Alan Gene
Pope, Noel Kynaston
Pope, Paul Terrell
Popejoy, William Dean
Pop-Stojanovic, Zoran Rista
Porsching, Thomas August
Port, Sidney C
Porta, Horacio A
Porter, A Duane
Porter, Donald Henry
Porter, Gerald Joseph
Porter, James Franklin
Porter, John Robert

Portmann, Walter Oddo
Posey, Eldon Eugene
Posner, Edward Charles
Poss, Richard Leon
Postley, John Appel
Postman, Robert Derek
Potoczny, Henry Basil
Potter, Thomas Franklin
Potts, Donald Harry
Pour-El, Marian Boykan
Powell, Robert Ellis
Pownall, Malcolm Wilmor
Prabhu, Narahari Umanath
Pressman, Irwin Samuel
Preston, Gerald Cowles
Price, Charles Morton
Price, James Ferris
Price, Kenneth Hugh
Priddy, Stewart Beauregard
Prielipp, Robert Walter
Prikry, Karel Libor
Prim, Robert Clay
Proctor, Clarke Wayne
Proctor, Thomas Gilmer
Prosser, Reese Trego
Prouse, Ervin Joseph
Pruitt, Murray Harold
Pruitt, Ralph L
Puckette, Stephen Elliott
Purdue, Peter
Purvis, Colbert Thaxton
Putnam, Alfred Lunt
Putnam, Calvin Richard
Pyle, Henry Randolph
Pyle, Leonard Duane
Quade, Edward Schaumberg
Querry, John William
Quigley, Frank Douglas
Quinn, John Phillip
Raab, Joseph A
Rabenstein, Albert Louis
Rabina, Manuel Jose
Rabinowitz, Philip
Rabson, Gustave
Radcliff, Henry Herbert, Jr
Radford, David Eugene
Radnitz, Alan
Raghavan, Thirukkannamangai E S
Raisbeck, Gordon
Rajnak, Stanley L
Rakestraw, Roy Martin
Ralley, Thomas G
Ralston, James Vickroy, Jr
Ramaley, James Francis
Ramanujan, Melapalayam Srinivasan
Ramirez, Donald Edward
Ramras, Mark Bernard
Ramsay, Arlan (Bruce)
Ramsay, O C, Jr
Raney, George Neal
Rao, Kannegami Nageswara
Rao, Koneru Venkata Rajeswara
Rao, Malempati Madhusudana
Rao, Poduri S R S
Raphael, Bertram
Ratliff, Louis Jackson, Jr
Ratner, Lawrence Theodore
Rattray, Basil Andrew
Rauch, Harry Ernest
Ray, David Scott
Raymon, Louis
Read, Ronald Cedric
Reade, Maxwell Ossian
Rearick, David F
Reaves, Harry Lee
Reay, John R
Rebman, Kenneth Ralph
Rechard, Ottis William
Rector, Robert Wayman
Reddy, Aluru Raghuram
Redhefter, Raymond Moos
Reed, Coke S
Reed, Dennis Keith
Reed, Ellen Elizabeth
Reeder, John Hamilton
Rees, Charles Sparks
Rees, Mina Spiegel
Rees, Paul Klein
Reeves, Roy Franklin
Regan, Francis
Reich, Daniel
Reich, Edgar
Reich, Kenneth Brooks
Reid, James Dolan
Reid, Lois Jean
Reiner, Irma Moses
Reiner, Irving
Reingold, Haim
Reinhardt, William Nelson
Reinhart, Bruce Lloyd
Reisel, Robert Benedict

Reister, David Bryan
Reiter, Harold Braun
Remage, Russell, Jr
Rempfer, Robert Weir
Reschovsky, Helene J
Resnick, Sidney I
Resnikoff, Howard L
Restrepo, Rodrigo Alvaro
Retherford, James Ronald
Reudink, Douglas O
Reyes, George Everett
Reynolds, Thomas Lee
Reyblatt, Zinovy V
Rhee, Haewun
Rhoades, Billy Eugene
Rhodes, John Lewis
Ribenboim, Paulo
Ribeiro, Hugo B
Ribler, Henry James
Rice, Barbara Slyder
Rice, Earl Clifton
Rice, Jimmy Marshall
Rice, Norman Molesworth
Rice, Orville Millard
Rice, Robert Bruce
Rich, Michael
Rich, Robert Peter
Richards, Jonathan Ian
Richards, Paul Bland
Richardson, Henry Russell
Richter, Hans E
Richter, Wayne H
Rickart, Charles Earl
Rickey, Neil William
Rickey, Frank Atkinson
Riddell, James
Riddell, Ronald Cameron
Rideout, Donald Eric
Ridge, William Clayton
Rieffel, Marc A
Riedl, John Orth, Jr
Rieke, Carol Anger
Riemenschneider, Sherman Delbert
Riess, Ronald Dean
Rigby, Fred Durnford
Rigby, John H
Riggle, Timothy A
Riggs, Charles Lathan
Rim, Dock Sang
Rinaldi, Leonard Daniel
Rinehart, Robert Fross
Riner, John William
Ring-Carroll, Rose
Ringeisen, Richard Delose
Ringenberg, Lawrence Albert
Rio, Sheldon T
Riordan, William J
Rishel, Raymond Warren
Rissanen, Jorma Johannes
Ritcey, Leland Frederick Samuel
Ritchie, Robert Wells
Ritter, Eugene Kerfoot
Ritter, Irving Frederick
Rivlin, Theodore J
Robbin, Joel W
Roberts, A Wayne
Roberts, George Gilbert
Roberts, Joel Laurence
Roberts, John Henderson
Roberts, Joseph Buffington
Roberts, Leslie Gordon
Roberts, Robert Abram
Robertson, Jack M
Robertson, James Byron
Robertson, Malcolm Slingsby
Robertson, Thomas N
Robinson, Charles Dee
Robinson, Derek Scott
Robinson, Donald Wilford
Robinson, Gilbert De Beauregard
Robinson, Ivor
Robinson, Louis
Robinson, Raphael Mitchel
Robinson, Stewart Marshall
Robinson, Thomas John
Robinson, Wilbur Judson
Robold, Alice Ilene
Robold, C Carl
Rochberg, Richard Howard
Rockafellar, Ralph Tyrrell
Rocke, David M
Rockhill, Theron D
Rockoff, Maxine Lieberman
Rod, David Lawrence
Rodin, Burton
Rodriguez, C Carl
Rodriguez, Anibal
Rodriguez, Argelia Velez
Rodriguez, Dennis Milton
Rodriguez, Haydee C
Rogak, Earl
Rogers, Hartley, Jr
Rogers, Jack Cree
Rogers, James Ted, Jr
Rognlie, Dale Murray
Rohrl, Helmut
Rolf, Howard Leroy

Rolwing, Raymond H
Rooney, Paul George
Roosenraad, Cris Thomas
Root, David Harley
Rorres, Chris
Rose, Donald Clayton
Rose, Gene Fuerst
Rose, Israel Harold
Rose, Milton Edward
Rose, Nicholas John
Roselle, David Paul
Roseman, Joseph Jacob
Rosen, David
Rosen, Michael Ira
Rosen, Ronald Haiam
Rosen, Saul
Rosen, William G
Rosenbaum, Robert Abraham
Rosenberg, Alex
Rosenberg, Herman
Rosenberg, Ivo George
Rosenberg, Paul Charles
Rosenbloom, Martin Jacob
Rosenblum, Marvin
Rosencrans, Steven I
Rosenfeld, Melvin
Rosenfeld, Norman-Samuel
Rosenlicht, Maxwell
Rosenstein, George Morris, Jr
Rosenthal, John William
Rosenthal, Peter (Michael)
Rosenthall, Edward
Rose (Rauen), Mary
Rosier, Ronald Crosby
Rosman, Bernard Harvey
Rothaus, Oscar Seymour
Rothenberg, Melvin G
Rothenberg, Ronald Isaac
Rothlisberger, Hazel Marie
Rothschild, Bruce Lee
Rotman, Joseph Jonah
Roulier, John Arthur
Rouse, Herbert Ronald
Rovnyak, James L
Rowe, Paul Preston
Rowland, John H
Roxin, Emilio O
Royall, Norman Norris, Jr
Royster, Wimberly Calvin
Royston, Robert Winter
Rozsnyai, Balazs
Rubel, Lee Albert
Rubin, Herman
Rubin, Jean E
Ruckle, William Henry
Rudd, David
Rudin, Mary Ellen
Rudin, Walter
Rudolph, Ray Ronald
Rudolph, William Brown
Rue, James Sandvik
Rueve, Charles Richard
Ruggles, Ivan Dale
Rund, Hanno
Rush, David Eugene
Russell, Charles Bradley
Russell, David L
Russell, Dennis C
Russell, Jack Unger
Rust, Charles Harry
Rutland, Leon W
Rutledge, Dorothy Stallworth
Rutledge, Joseph Dela
Rutledge, Edgar B
Rutter, Edgar A, Jr
Ryan, Donald Edwin
Ryan, Frank Beall
Ryan, Peter Michael
Ryan, Robert Dean
Rygg, Paul Theodore
Rymer, Harry
Ryser, Herbert John
Saade, John Marshall
Saalfrank, Charles W
Saari, Donald Fene
Saaty, Thomas L
Sabel, Clara Ann
Sabharwal, Chaman Lal
Sabidussi, Gert Otto
Sachs, David
Sachs, Jerome Michael
Sacks, Gerald Enoch
Sacks, Jerome
Sacksteder, Richard Carl
Sagan, Hans

Sagan, Leon Francis
Sah, Chih-Han
Saini, Girdhari Lal
St Mary, Donald Frank
Sakai, Shoichiro
Sakhare, Vishwa M
Sallee, G Thomas
Saltz, Daniel
Saltzer, Charles
Samelson, Hans
Samn, Sherwood
Sampson, Joseph Harold
Sams, Burnett Henry, III
Samuelson, Donald James
Sanchez, David A
Sanders, Oliver Paul
Sanders, William Mack
Sanderson, Judson
Sansing, Raymond Clayton
Sansone, Fred J
San Soucie, Robert Louis
Santos, Eugene (Sy)
Sarason, Donald Erik
Sarason, Leonard
Sard, Arthur
Sario, Leo Reino
Saslaw, Samuel
Sather, Duane Paul
Sato, Daihachiro
Saunders, Frank Wendell
Saunders, Roy Bly
Sauve, James Willard
Savage, Nevin William
Savitch, Walter John
Sawyer, Jane Orrock
Sawyer, Stanley Arthur
Sawyer, W Warwick
Sayeki, Hidemitsu
Saylor, Richard Lynn
Scalora, Frank Salvatore
Scandura, Joseph M
Scarborough, Charles T, Jr
Schaefer, Donald John
Schaefer, Paul Theodore
Schaeffer, David George
Schaer, Jonathan
Schaef, Henry Maximilian
Schafer, James A
Schafer, Richard Donald
Schaffer, Juan Jorge
Schanuel, Stephen Hoel
Scharn, Herman Otto Friedrich
Schatten, Robert
Schatz, Joseph Arthur
Schaumberger, Norman
Scheerer, Anne Elizabeth
Scheid, Francis
Scheinok, Perry Aaron
Schell, Emil Daniel
Schellenberg, Paul Jacob
Schild, Albert
Schilling, John Joseph
Schindler, Guenter Martin
Schindler, Susan
Schirmer, Helga H
Schlesinger, Stewart Irwin
Schleusner, John William
Schlissel, Arthur
Schlomiuk, Dana
Schlomiuk, Norbert
Schlosser, Jon A
Schmid, Wilfried
Schmidt, Harvey John, Jr
Schmidt, Jurgen Volkmar
Schmidt, Richard Nicholas
Schmidt, Wolfgang M
Schmitt, Klaus
Schmittroth, Louis Anthony
Schnare, Paul Stewart
Schneckenburger, Edith Ruth
Schneeberger, Charles Michael
Schneider, Harold O
Schneider, Ross Nelson
Schober, Glenn E
Schochet, Claude Lewis
Schoen, Kenneth
Schoenberg, Isaac Jacob
Scholz, Dan Robert
Schrader, Dorothy Virginia
Schrader, Keith William
Schraut, Kenneth Charles
Schreiner, Erik Andrew
Schriber, Thomas J
Schrot, Mary Dolores
Schubert, Cedric F
Schubert, Jewell Emma
Schubert, Blanche Beatrice
Schultz, James Edward
Schultz, Reinhard Edward
Schumaker, John Abraham
Schumaker, Larry L
Schumaker, Robert Louis
Schupp, Paul Eugene
Schurle, Arlo Willard
Schurrer, Augusta
Schuster, Seymour

Schuur, Jerry D
Schuurmann, Frederick James
Schwartz, Abraham
Schwartz, Alan Lee
Schwartz, Herman Meyer
Schwartz, Jacob Theodore
Schwartzman, Sol
Schweizer, Berthold
Schweppe, Earl Justin
Schwerdtfeger, Hans Wilhelm Eduard
Scorsone, Francesco G
Scott, Edward Joseph
Scott, Leland Latham
Scott, Walter Tandy
Scoville, Richard Arthur
Scrimger, Edward Brantly, Jr
Scroggs, James Edward
Sealander, Carl Elof
Searcy, Charles Jackson
Seber, Robert Charles
Sebesta, Charles Frederick
Seckler, Bernard David
Secrest, Bruce Gill
Seebach, J Arthur, Jr
Seebeck, Charles Louis, Jr
Seekins, Charles William
Seely, Justus Frandsen
Seever, Galen Lathrop
Segal, Arthur Cherny
Segal, Irving Ezra
Segal, Sanford Leonard
Segers, Richard George
Seibert, Peter
Seidel, Wladimir
Seidenberg, Abraham
Seifert, Ralph Louis, Jr
Seiken, Arnold
Selberg, Atle
Selden, Dudley Byrd
Selfridge, Ralph Gordon
Seligman, George Benham
Sell, George Roger
Sellers, Peter Hoadley
Sells, Jean Thurber
Semon, Warren Lloyd
Senechal, Lester John
Senechalle, David Albert
Senge, George H
Senseig, Chester
Senter, Harvey
Sentilles, F Dennis, Jr
Seppi, Edward Joseph
Serrin, James B
Seshu, Lilly Hannah
Sethares, George C
Settles, Ronald Dean
Seybold, Mary Anice
Shabanowitz, Harry
Shader, Leslie Elwin
Shader, Melvin A
Shaffer, Dorothy Browne
Shaffer, Douglas Howerth
Shaftman, David Harry
Shahin, Jamal Khalil
Shale, David
Shalen, Peter Brock
Shanahan, Patrick
Shank, Herbert S
Shanks, Daniel
Shanks, Eugene Baylis
Shanks, Merrill Edward
Shapiro, Edwin Seymour
Shapiro, George
Shapiro, Harvey Lee
Shapiro, Jack Sol
Shapiro, Jesse Marshall
Shapiro, Leonard David
Shapiro, Marvin Benjamin
Shapiro, Norman Zalman
Shapley, Lloyd Stowell
Sharma, Mahesh Chandra
Sharp, Edward A
Sharp, Henry, Jr
Sharpe, Michael John
Sharples, Alan
Shatoff, Larry David
Shatz, Stephen S
Shauck, Maxwell Eustace, Jr
Shaw, Joe William, Jr
Shawhan, Gerald L
Shea, Daniel Francis
Sheffield, Roy Dexter
Shell, Donald Lewis
Shelton, Eugene Paul
Shelton, Ronald M
Shepherd, William Lloyd
Shepp, Lawrence Alan
Sheridan, Laurence Ward
Sherk, Frank Arthur
Sherman, Gary Joseph
Sherman, Malcolm J
Sherman, Seymour
Sherman, Thomas Lawrence
Sherman, Thomas Oakley
Sherrod, John
Sherry, Edwin J
Shershin, Anthony Connors

Shields, Allen Lowell
Shiffman, Bernard
Shiffman, Max
Shiflett, Lilburn Thomas
Shiflett, Ray Calvin
Shilepsky, Arnold Charles
Shimamoto, Yoshio
Shimura, Goro
Shinbrot, Marvin
Shive, Robert Allen, Jr
Shively, Ralph Leland
Shklov, Nathan
Shockley, James Edgar
Shoemaker, Richard W
Shoenfield, Joseph Robert
Sholander, Marlow
Shonkwiler, Ronald Wesley
Shore, Samuel David
Shores, Thomas Stephen
Short, Donald Ray, Jr
Shreve, David Carr
Shreve, Darrell Rhea
Shriner, Walter Owen
Shryock, A Jerry
Shub, Michael I
Shubert, Bruno Otto
Shuck, John Winfield
Shulman, Harold
Shults, Mayo Glenwood
Sibert, Elbert Ernest
Sibuya, Yasutaka
Sichler, Jiri Jan
Sidman, Robert David
Siegel, Martha J
Sigler, Laurence Edward
Sigmon, Kermit Neal
Silber, Robert
Silberger, Allan Joseph
Silberger, Donald Morison
Silva, Joseph A
Silver, Murray
Silverman, Edward
Silverman, Robert
Simmons, David Rae
Simmons, George Finlay
Simmons, Harold Franklyn
Simon, Arthur Bernard
Simon, Carl Paul
Simons, Stephen
Simons, William Haddock
Simons, William Harris
Simpson, James Edward
Sims, Benjamin Turner
Sims, Stillman Austin
Sinclair, Annette
Sinden, Frank William
Singer, James
Singh, Sankatha Prasad
Singh, Shri Krishna
Singleton, Robert Richmond
Singley, Donald Heath
Sinha, Indranand
Sinke, Carl
Sinkhorn, Abraham
Sinkov, Abraham
Sion, Maurice
Siry, Joseph William
Skaff, Michael Samuel
Skarda, Ralph V, Jr
Skarlos, Leonidas
Skeath, J Edward
Skinner, Richard Emery
Sklar, Abraham
Slater, Peter John
Slaughter, Frank Gill, Jr
Sledd, Marvin Banks
Sledd, William T
Slepian, Paul
Slesnick, William Ellis
Sloan, Albert Russell
Sloan, Robert W
Sloane, Neil James Alexander
Sloat, Floyd Brooksher
Sloss, Frank Brooke
Slotterbeck, Oberta Ann
Sloyan, Mary Stephanie
Smale, Stephen
Small, Donald Bridgham
Small, William Andrew
Smiley, Malcolm Finlay
Smith, Armand Verne, Jr
Smith, Charles Bassel
Smith, Charles R
Smith, David Alexander
Smith, Donald Luke
Smith, Donald Ray
Smith, Frank A
Smith, Frank Engelbert
Smith, Franklin Chapin, Sr
Smith, Gaston
Smith, Harvey Alvin
Smith, James Clarence, Jr
Smith, James L
Smith, James Thomas
Smith, Jesse Leo
Smith, Joe K
Smith, John Howard
Smith, John Wolfgang

Smith, Kennan Taylor
Smith, Kirby Campbell
Smith, Larry
Smith, Lehi Tingen
Smith, Luther W
Smith, Marianne (Ruth) Freundlich
Smith, Marion Bush, Jr
Smith, Martha Kathleen
Smith, Norman Hankele
Smith, Oliver King
Smith, Paul Althaus
Smith, Peter David
Smith, Robert Seton
Smith, Robert Paul
Smith, Shelby Dean
Smith, Thomas Jefferson
Smith, Wayne Earl
Smith, William Allen
Smith, William K
Smith, William R
Smith, William Walker
Smoke, William Henry
Smoller, Joel A
Snader, Daniel Webster
Snell, James Laurie
Snell, Robert Isaac
Snider, Arthur David
Snover, James Edward
Snow, Donald Ray
Snow, Douglas Oscar
Snow, Wolfe
Snyder, Andrew Kagey
Snyder, Walter Stephen
Soare, Robert I
Sobczyk, Andrew
Sobey, Arthur Edward, Jr
Sohmer, Bernard
Sokolowski, Henry Alfred
Sokolowsky, Daniel
Solomon, Jimmy Lloyd
Solomon, Louis
Solovay, Robert M
Sonn, Jack
Sonneborn, Lee Meyers
Sonner, Johann
Sons, Linda Ruth
Sopka, John J, Jr
Sorgenfrey, Robert Henry
Southard, Thomas Hollister
Sowell, Katye Marie Oliver
Spahr, Carlos G, II
Spanier, Edwin Henry
Spanier, Jerome
Sparks, Arthur Godwin
Specht, Edward John
Specht, Robert Dickerson
Specht, William Harold
Speece, Herbert E
Spelmann, John W
Spencer, Armond E
Spencer, Domina Eberle (Mrs Parry Moon)
Spencer, Donald Clayton
Spencer, Guilford Lawson, II
Spicer, Donald Z
Spicer, John A
Spiegel, Eugene
Spielberg, Stephen E
Spira, Robert Samuel
Spitzbart, Abraham
Spitzer, Frank L
Spohn, William Gideon, Jr
Spragens, William Henry, Jr
Sprague, Richard Howard
Spraker, Harold Stephen
Sprecher, David A
Spring, Ray Frederick
Springer, Charles Eugene
Springer, George
Sprinkle, H D
Squier, Donald Platte
Sra, Kewal Singh
Srinivasacharyulu, Kilambi
Srinivasan, Bhama
Srinivasan, Hari Mohan
Srivastava, Jagdish Narain
Stackelberg, Olaf Patrick
Stahl, Saul
Staknis, Victor Richard
Staley, David H
Staley, Robert Delmer
Stallmann, Friedemann Wilhelm
Stamey, William Lee
Standish, Charles Junior
Stanek, Jean Chan
Stanek, Peter
Stanley, Nathaniel Richard
Stanley, Robert Lauren
Stannard, William A
Stanojevic, Caslav V
Stanton, Nancy Kahn
Starcher, George William
Stark, Jeremiah Milton
Starke, Emory Potter
Starr, David Wright
Starr, Norton
Starret, Austin Leroy
Stasheff, James Dillon

Stauder, M Francis Borgia
Stavroudis, Orestes Nicholas
Stearns, Richard Edwin
Steele, Mary Philip
Steele, William F
Steen, Frederick Henry
Steen, Lynn Arthur
Stein, Elias M
Stein, Frederick Max
Stein, James DeWitt, Jr
Stein, Junior
Stein, Marjorie Leiter
Stein, Marvin L
Stein, Michael Roger
Stein, Norman
Stein, Sherman Kopald
Steinberg, Herbert Aaron
Steinberg, Robert
Steinberg, Stuart Alvin
Steiner, Anne Kercheval
Steiner, Eugene Francis
Steiner, Gilbert
Steinlage, Ralph Cletus
Stell, George Roger
Stelson, Hugh Eugene
Stemple, Joel G
Stenger, William
Stephany, Edward O
Stephens, Clarence Francis
Stephens, Harold W
Stephens, Stanley LaVerne
Stephenson, Robert Moffatt, Jr
Sterling, Daniel J
Sterling, Nicholas J
Stern, Samuel T
Sternstein, Martin
Sterrett, Andrew
Sterrett, John Kenneth
Stevens, Robert R
Stevenson, Everett E
Stevenson, John Crabtree
Steward, James Gordon
Steward, Robert F
Stewart, Bonnie Madison
Stewart, Donald George
Stewart, Elmo Joseph
Stewart, Frank Moore
Stewart, James Collier
Stewart, Joseph Kyle
Stewart, Robert Clarence
Stewart, Ruth Carol
Stiel, Edsel Ford
Stiles, Wilbur J
Stillman, Richard Ernest
Stilwell, Kenneth James
Stipanowich, Joseph J
Stockmeyer, Paul Kelly
Stocks, Douglas Roscoe, Jr
Stockton, Doris S
Stoddard, Dan Warren
Stoddard, James H
Stoker, James Johnston
Stokes, Joseph Franklin
Stokes, Russell Aubrey
Stokes, William Glenn
Stolarsky, Kenneth B
Stoll, Robert Roth
Stoll, Wilhelm
Stone, Alexander Glatstein
Stone, Alexander Paul
Stone, Charles Joel
Stone, Dorothy Maharam
Stone, Erika Mares
Stone, Lawrence David
Stone, Marshall Harvey
Stone, Michael Gates
Stone, Samuel Arthur
Stone, William Matthewson
Stoneham, Richard George
Stopher, Emmet Carson
Storer, Thomas
Stortz, Clarence A
Storvick, David A
Stout, Edgar Lee
Stout, William F
Strait, Peggy
Straley, Tina
Stralka, Albert R
Strange, William Gilbert
Strang, William Ernest
Strasser, Elvira Rapaport
Stratopoulos, George
Strauch, Ralph Eugene
Straus, Ernst Gabor
Strauser, Wilbur Alexander
Strauss, Aaron Solomon
Strauss, Frederick Bodo
Strauss, Monty Joseph
Strauss, Walter A
Strebe, David Dietrich
Strecker, George Edison
Strecok, Anthony J
Strehler, Allen Frederick
Strichartz, Robert Stephen
Stright, I Leonard

Strimling, Walter Eugene
Strodt, Walter Charles
Strohl, George Ralph, Jr
Stromberg, Karl Robert
Strother, Wayman L
Struble, Raimond Aldrich
Struik, Ruth Rebekka
Stubbe, John Sunapee
Stueben, Edwin Frank
Stuelpnagel, John Clay
Sturley, Eric Avern
Sudbarao, Mathukumalli Venkata
Sudderth, William David
Sullivan, Donald
Sullivan, Hugh D
Sullivan, Joseph Arthur
Sullivan, Michael Joseph, Jr
Sulski, Leonard C
Summer, Donovan Bradshaw
Summers, William Hunley
Summerville, Richard Marion
Suprunowicz, Konrad
Sussman, Irving
Sussman, Charles Samuel
Sutton, Louise Nixon
Suzuki, George
Suzuki, Noboru
Svoboda, Rudy George
Swaminathan, Srinivasa
Swan, Richard Gordon
Swank, Rolland Laverne
Swann, Howard Story Gray
Swanson, Charles Andrew
Swanson, Clarence A E
Swanson, Leonard William
Swart, William John
Swartz, William Lee
Sweedler, Moss Eisenberg
Sweeney, William John
Swerling, Peter
Swick, Kenneth Mark
Swenson, Donald Adolph
Swift, George Herbert, Jr
Swift, Jonathan Dean
Swift, William Clement
Syski, Ryszard
Szeptycki, Pawel
Szeto, George
Szusz, Peter
Tabatabaian, Ali Mohammad
Taibleson, Mitchell H
Takahashi, Shuichi
Takesaki, Masamichi
Talbot, Walter Richard
Tamari, Dov
Tan, Kok-Keong
Tang, Victor Kuang-Tao
Tanimoto, Taffee Tadashi
Tanis, Elliot Alan
Tarski, Alfred
Taub, Abraham Haskel
Taucci, Enes Barbara
Taussky, Olga (Mrs John Todd)
Taylor, Angus Ellis
Taylor, Donald Curtis
Taylor, Floyd Heckman
Taylor, Francis B
Taylor, Howard Edward
Taylor, Jerry Duncan
Taylor, John Christopher
Taylor, John Joseph
Taylor, Michael Dee
Taylor, Robert Lee
Taylor, Walter Fuller
Teichroew, Daniel
Tennison, Robert L
Teresine (Lewis), Mary
Terry, Raymond Douglas
Terzuoli, Andrew Joseph
Textor, Robin Edward
Thalgott, Fred William
Thatcher, James W
Theile, Fred Charles
Theilheimer, Feodor
Thickstun, William Russell, Jr
Thierrin, Gabriel
Thieme, Melvin T
Thomas, Charles Gomer
Thomas, Clayton James
Thomas, Edward Sandusky, Jr
Thomas, George Brinton, Jr
Thomas, Harold Lee
Thomas, James William
Thomas, John Pelham
Thomas, Joseph Charles
Thomas, Robert Spencer
Thomas, Tracy Yerkes
Thomas a Kempis, Mary
Thomeier, Siegfried
Thompson, Joel Douglas
Thompson, Anthony C
Thompson, Howard E
Thompson, Maynard
Thompson, James Robert
Thompson, Richard Bruce
Thompson, Robert Charles

Thompson, Robert Harry
Thompson, Robert James
Thornton, Melvin Chandler
Thorp, Edward O
Thrall, Robert McDowell
Thron, Wolfgang Joseph
Thurman, George Raymond
Thyer, Norman Harold
Tidd, Robert Frederick
Tierney, John A
Tikson, Michael
Tilley, John Leonard
Tillman, Stephen Joel
Tillotson, Donald Bearse
Tilson, Bret Ransom
Timourian, James Gregory
Tindell, Ralph S
Ting, Tsuan Wu
Tingley, Arnold Jackson
Tintner, Gerhard
Tirman, Alvin
Titus, Charles Joseph
Toews, Kornelius Gerhard
Tolle, Jon Wright
Tolleson, Jeffrey L
Tolsted, Elmer Beaumont
Tomlinson, Michael Bangs
Tondeur, Philippe
Tondra, Richard John
Toney, Fred, Jr
Tong, Mary Powderly
Tong, Yung Liang
Tonne, Philip Charles
Topp, William Robert
Tornheim, Leonard
Toralballa, Leopoldo Vasquez
Torrance, Esther McCormick
Toskey, Burnett Roland
Toubassi, Elias Hanna
Townsend, Ralph N
Trahan, Donald Herbert
Trampus, Anthony
Trauth, Charles Arthur, Jr
Trench, William Frederick
Treybig, Leon Bruce
Trimble, Harold Callander
Trine, Franklin Dawson
Tripp, Ralph Harry
Tropp, Henry S
Tross, Carl Henry
Trott, S M
Trotter, Hale Freeman
Troy, Daniel Joseph
Troyer, Robert James
Truax, Robert Lloyd
Truesdell, Clifford Ambrose, III
Trutt, David
Trytten, George Norman
Tserpes, Nicolas A
Tucciarone, John Peter
Tuchinsky, Philip Martin
Tuckerman, Bryant
Tull, Jack Phillip
Tuller, Annita
Tulloch, Lynn Hardyn
Tully, Edward Joseph, Jr
Tung, John Shih-Hsiung
Tupac, James Daniel
Turgeon, Jean
Turner, Alice Willard
Turner, Lincoln Hulley
Turner, Nura Dorothea
Turner, Ralph Waldo
Turner, Robert E L
Turner, Veras D
Turrittin, Hugh Lonsdale
Tuttle, William Thomas
Tuzar, Jaroslav
Tymchatyn, Edward Dmytro
Ucci, Pompeiio Angelo
Uehara, Hiroshi
Uhl, John Jerry, Jr
Uhlenbeck, Karen K
Ulam, Stanislaw Marcin
Ullman, Arthur William James
Ullman, Joseph Leonard
Ullom, Stephen Virgil
Ullrich, David Frederick
Ulmer, Gilbert
Underwood, Douglas Haines
Ungar, Gerald S
Ursell, John Henry
Uschold, Richard L
Vaillancourt, Remi Etienne
Valentine, Frederick Albert
Valentine, Joseph Earl
Van Alstyne, John Pruyn
Vanaman, Sherman Benton
Vance, Elbridge Putnam
Vance, Joseph Francis
Vancko, Robert Michael

Vanderlin, Carl Joseph, Jr
Van Der Linde, Reinhoud H
van der Vaart, Hubertus Robert
Van de Wetering, Richard Lee
Van Eeden, Constance
Van Enkevort, Ronald Lee
Van Horn, David Downing
Van Ryzin, Martina
Van Schaack, George Booth
Vanstone, J R
Vanstone, Scott Alexander
Van Straten, Mary Petronia
Van Tuyl, Andrew Heuer
Van Veldhuizen, Philip Androcles
Van Vleck, Fred Scott
Van Wyk, Rodney
Varadarajan, Kalathoor
Varberg, Dale Elthon
Varga, Richard S
Varineau, Verne John
Varma, Arun Kumar
Varnhorn, Mary Catherine
Vasquez, Alphonse Thomas
Vaughan, Herbert Edward
Vaughan, Jerry Eugene
Vaughan, Loy Ottis, Jr
Vaught, Nick Hampton
Vaught, Robert L
Veatch, Ralph Wilson
Veech, William Austin
Vegh, Emanuel
Velesz, Dunstan George
Velez, William Yslas
Venit, Stewart Mark
Verdina, Joseph
Verhey, Roger Frank
Verner, James Hamilton
Vichich, Thomas E
Vick, James Whitfield
Villemure, M Paul James
Vinograde, Bernard
Vinson, Richard G
Vinsonhaler, Charles I
Vitale, Richard Albert
Vobach, Arnold R
Vogeli, Bruce R
Voichick, Michael
Vonbun, Friedrich Otto
von Holdt, Richard Elton
Votaw, Charles Isaac
Vroom, Kenneth Edwin
Vuckovic, Vladeta
Wachman, Murray
Waddell, Mathews Cary
Wade, Luther Irwin
Wade, Thomas Leonard, Jr
Wade, William Raymond, II
Wadsworth, George Proctor
Waggoner, Wilbur J
Wagner, Carl George
Wagner, Daniel Hobson
Wagner, Eric G
Wagner, Harry Mahlon
Wagner, Raphael Darrel
Wagner, Richard Carl
Wagner, Robert Wanner
Wagoner, Samuel Standfield, Jr
Wahab, James Hatton
Wahl, Jonathan Michael
Wahlstrom, Lawrence F
Waite, Alan C
Waits, Bert Kerr
Wakerling, Raymond Kornelious
Waksberg, Armand L
Walczak, Hubert R
Walden, Ralph Eldon
Walden, William Earl
Walder, Orlin E
Waldinger, Hermann V
Wales, David Bertram
Wall, Donald Dines
Wall, James Graham
Wallace, Alexander Doniphan
Wallace, Andrew Hugh
Wallace, Kyle David
Wallach, Sylvan
Wallen, Clarence Joseph
Walling, Derald Dee
Wallis, Anthony
Walls, Robert Clarence
Walpole, Ronald Edgar
Walston, Dale Edouard
Waltcher, Azelle Brown
Walter, Everett L
Walters, Eleanor Boyd
Waltman, William Lee

Walton, Lewis F
Walum, Herbert
Wampler, Joe Forrest
Wang, Hao
Wang, Hsien Chung
Wang, Ju-Kwei
Wang, Ya-Yen Lee
Wantland, Evelyn Kendrick
Ward, Frederick Roger
Ward, Harold Nathaniel
Ward, James Audley
Ward, James Edward, III
Ward, Lewis Edes, Jr
Wardlaw, William Patterson
Wardwell, James Fletcher
Ware, James Gareth
Warga, Jack
Warlick, Charles Henry
Warne, Ronson Joseph
Warner, Charles Robert
Warner, Frank Wilson, III
Warner, John Ward, Jr
Warnock, Walter George
Warren, Peter
Warren, Richard Hawks
Warschawski, Stefan Emanuel
Warten, Ralph Martin
Wasow, Wolfgang Richard
Wasserman, Arthur Gabriel
Waterhouse, William Charles
Waterman, Daniel
Waterman, Michael S
Watkins, Mark E
Watkins, William
Watson, Martha F
Wattenberg, Franklin Arvey
Watts, Charles Edward
Weaver, Milo Wesley
Weaver, Warren
Webb, William Albert
Webber, Carroll A, Jr
Weber, Waldemar Carl
Webster, Merritt Samuel
Webster, Porter Grigsby
Wechsler, Martin T
Wedel, Arnold Marion
Wegner, Kenneth Warren
Weibel, Armella
Weichsel, Paul M
Weidlich, John Edward, Jr
Weidman, Donald Robert
Weigle, Robert Edward
Weihe, Joseph William
Weil, Andre
Weil, Clifford Edward
Weinacht, Richard Jay
Weinbaum, Carl Mattin
Weinberg, Elliot Carl
Weinberger, Peter Jay
Weinless, Michael Howard
Weinstein, Alexander
Weinstein, Joseph M
Weinstein, Stanley Edwin
Weinstock, Barnet Mordecai
Weintraub, Sol
Weinzweig, Avrum Israel
Weiser, Daniel
Weisner, Louis
Weiss, Guido Leopold
Weiss, Max Leslie
Weiss, Michael David
Weiss, Robert John
Weiss, Sol
Weissberg, Alfred
Welch, Lloyd Richard
Welland, Grant Vincent
Welland, Robert Roy
Weller, Glenn Peter
Wells, Charles Prentiss
Wells, Edward Henry
Wells, James Howard
Wells, Jane Frances
Wells, Raymond O'Neil, Jr
Welmers, Everett Thomas
Welna, Cecilia
Wend, David Van Vranken
Wendel, James Gutwillig
Wendroff, Burton
Wendt, Arnold
Wenger, Ronald Harold
Wenjen, Chien
Wenska, Tom Marion
Wente, Henry Christian
Wernick, Robert J
Wernick, William
Werth, John St Clair, Jr
Wesler, Oscar
Wesson, James Robert
West, James Edward
Wester, Elbert Truman
Western, Donald Ward
Westlake, Wilfred James
Westwick, Roy
Wetsman, Allen William
Wetzel, John Edwin
Wexler, Charles
Weyl, F Joachim

Whaples, George William
Wheeden, Richard Lee
Wheeler, Charles Horatio, III
Wheeler, Robert Lee
Whipkey, Kenneth Lee
Whitaker, Mack Lee
White, Albert George, Jr
White, Alvin Murray
White, Arthur Thomas II
White, Christopher Clarke
White, John Thomas
White, Myron Edward
White, Paul A
White, Warren Humphreys
Whitehead, George William
Whiteman, Albert Leon
Whitesitt, John Eldon
Whitfield, John Howard Mervyn
Whitley, William Thurmon
Whitman, Andrew Peter
Whitman, Philip Martin
Whitman, Walter William
Whitmore, William Francis
Whitney, Hassler
Whittaker, James Victor
Whitlesey, Emmet Finlay
Whyburn, Lucille Enid
Wicht, Marion Cammack
Wicker, Berthold Robert
Wickes, Harry E
Widder, David Vernon
Widom, Harold
Wiginton, Carroll Lamar
Wilansky, Albert
Wilcox, Calvin Hayden
Wilcox, Howard Joseph
Wild, Wayne Grant
Wilde, Edwin Frederick, Jr
Wilder, Donald Richard
Wilder, Raymond Louis
Wilf, Herbert
Wilke, Frederick Walter
Wilkerson, Robert C
Wilkins, J Ernest, Jr
Wilkinson, Arthur
Wilkinson, Jack Dale
Wilks, Charles Edward
Willard, Stephen
Willcox, Alfred Burton
Willerding, Margaret Frances
Willett, Douglas W
Willett, Richard Michael
Williams, Bennie B
Williams, Charles Wiley
Williams, Earl R
Williams, Eddie Robert
Williams, George Kenneth
Williams, Hugh Cowie
Williams, James Garner
Williams, Joseph John
Williams, Kenneth Stuart
Williams, Leland Hendry
Williams, Lynn Roy
Williams, Mary Elizabeth
Williams, Richard Kelso
Williams, Robert Elvin
Williams, Scott Warner
Williams, William Howard
Williams, William Orville
Williamson, Richard Edmund
Williamson, Stanley Gill
Williamson, Susan
Willoughby, Ralph Arthur
Wilson, Carroll Kipper
Wilson, David E
Wilson, Eric LeRoy
Wilson, F Wesley Jr
Wilson, Howard Le Roy
Wilson, Howell Kenneth
Wilson, John Phillips
Wilson, Paul Robert
Wilson, Richard Michael
Wilson, Robert Lee
Wilson, Robert Lee, Jr
Wilson, Walter Lucien, Jr
Windham, Michael Parks
Windholz, Walter M
Windolph, Joseph R
Winger, Milton Eugene
Winograd, Shmuel
Winslow, Richard Edward
Winter, David John
Winton, Charles Newton
Winton, Lowell Sheridan
Winzenread, Marvin Russell
Wirszup, Izaak
Wiser, Horace Clare
Witz, Klaus Gerhard
Wixson, Eldwin A, Jr
Wogen, Warren Ronald
Wohl, Philip R
Wohlgelernter, Devora Kasachkoff
Wolf, Charles Trostle
Wolf, Frank Louis
Wolf, Frantisek
Wolf, Ira Kenneth
Wolf, Joseph Albert

Wolfe, Carvel Stewart
Wolfe, Dorothy Wexler
Wolfe, Jack
Wolfe, James H
Wolfe, Stephen James
Wolfowitz, Jacob
Wolfson, Kenneth Graham
Wolinsky, Albert
Wolk, Elliott Samuel
Woll, John William, Jr
Wollan, Gerhard Norval
Wolontis, Vidar Michael
Wolsson, Kenneth
Wolstenholme, William Ernest
Womble, Eugene Wilson
Wonenburger, Maria Josefa
Wong, James Chin-Sze
Wong, Pui Kei
Wong, Tin Kin
Wong, Yuen-Fat
Woo, Norman Tzu Teh
Wood, Bruce
Wood, Craig Adams
Wood, James Thornton
Wood, William Hulbert
Woodby, Lauren G
Woodriff, Roger L
Woods, Cecil Lamborn
Woods, Dale
Woods, Jimmie Dale
Woodside, William
Woodworth, Wayne Leon
Wooldridge, Elizabeth Taylor
Woolf, William Blauvelt
Worrell, John Mays, Jr
Wouk, Arthur
Wray, Joe Willie
Wrede, Robert C, Jr
Wren, Frank Lynwood
Wright, David Lee
Wright, Elisabeth Muriel Jane
Wright, Farroll Tim
Wright, Fred Boyer
Wright, Fred Marion
Wright, Harvel Amos
Wright, Martin
Wu, Hung-Hsi
Wulbert, Daniel Eliot
Wulff, John Leland
Wunderlich, Marvin C
Wyler, Oswald
Wyman, Bostwick Frampton
Wyman, Max
Wyse, Frank Oliver
Wyss, Walter
Wyszecki, Gunter
Yackel, James W
Yagi, Fumio
Yale, Irl Keith
Yale, Paul B
Yang, Chao-Hui
Yang, Chung-Tao
Yantis, Richard P
Yarnelle, John E
Yaspan, Arthur
Yates, Richard Lee
Yau, Shing-Tung
Yeardley, Nelson Paul
Yearout, Paul Harmon, Jr
Yeh, James Jui-Tin
Yff, Peter
Yntema, Mary Katherine
Yohe, James Michael
Yonda, Alfred William
Yonda, David Monaghan, Jr
Young, Frank Hood
Young, Gail Sellers, Jr
Young, James Howard
Young, John William
Young, Paul McClure
Young, Ralph Waldo
Younger, Daniel H
Younglove, James Newton
Yozwiak, Bernard James
Zahler, Raphael
Zaidman, Samuel
Zame, Alan
Zame, William Robin
Zander, Vernon Emil
Zaphyr, Peter Anthony
Zaring, Wilson Miles
Zariski, Oscar
Zassenhaus, Hans J
Zechmann, Albert W
Zedek, Mishael
Zeigler, Royal Keith
Zeitlin, Joel Loeb
Zeller, Mary Claudia
Zelmanowitz, Julius Martin
Zeoli, Harold Wilson
Zerla, Fredric James
Zettl, Larry Joseph
Zettl, Anton J
Ziebur, Allen Douglas
Ziemer, William P
Zierler, Neal
Zilber, Joseph Abraham
Zimmerberg, Hyman Joseph

Zimmerman, Henry B
Zimmerman, Lester J
Zindler, Richard Eugene
Zink, Robert Edwin
Zirakzadeh, Aboulghassem
Zitarelli, David Earl
Zlot, William Leonard
Zuckerberg, Hyam L
Zukowski, Lucille Pinette
Zumbrun, John Robert
Zund, Joseph David
Zusman, Fred Selwyn
Zvengrowski, Peter Daniel
Zwick, Earl J
Zwier, Paul J
Zygmund, Antoni

Actuarial Mathematics
Garfin, Louis
Hubbart, Wayland Michael
Jones, Donald Akers
Knowler, Lloyd A
Kormes, Mark
Nesbitt, Cecil James
Torrance, Ellen McCormick
Zubay, Eli Alan

Algebra
Adelberg, Arnold M
Akasaki, Takeo
Alin, John Suemper
Altman, Allen Burchard
Anderson, Michael Peter
Arendt, Billy Dean
Armendariz, Efraim Pacillas
Artzy, Rafael
Assmus, Edward Ferdinand, Jr
Ault, Janet E (Mills)
Avelsgaard, Roger A
Azumaya, Goro
Beauregard, Raymond A
Beck, Robert Edward
Bell, James Henry
Benard, Mark
Bercov, Ronald David
Berman, Elizabeth Alexandra
Bevis, Jean Harwell
Bickel, Thomas Fulcher
Blum, Peter
Blumenthal, Robert George
Boone, William Werner
Boorman, Evelyn Hutterer
Borm, Alfred Ervin
Brackin, Eddy Joe
Brande, Edward Woodrow
Brawley, Joel Vincent, Jr
Brewer, James W
Brizolis, Demetrios
Brooks, James O
Buccino, Alphonse
Byrd, Kenneth Alfred
Cable, Charles Allen
Cannonito, Frank Benjamin
Carlson, Jon Frederick
Cassel, David Wayne
Celik, Hasan Ali
Chalkley, Roger
Clark, William Glenn
Cohen, Eckford
Cohn, Richard Moses
Comer, Stephen Daniel
Coppage, William Eugene
Coughlin, Raymond Francis
Countryman, James Joseph
Cox, Milton D
Crittenden, Rebecca Slover
Cude, Joe E
Cummings, Larry Jean
Cutler, Janice Zemanek
D'Alarcao, Hugo T
Dalla, Ronald Harold
Daly, John Francis
Darden, Geraldine C
Dauns, John
Davis, Clyde William, Jr
Davis, Elwyn H
Di Franco, Roland B
Dimitroff, George Ernest
Dobbs, David Earl
Dribin, Daniel Maccabaeus
Durbin, John Riley
Dyer-Bennet, John
Eaton, James Edmonds
Eaves, James Clifton
Eisenbud, David
Eke, Boniface Ihemeotuonye
Emerson, Marion Preston
Falconer, Etta Zuber
Faudree, Ralph Jasper, Jr
Fauntleroy, Amassa Courtney
Federighi, Enrico Thomas
Fenrick, Maureen Helen
Fincher, Bobby Lee
Fisher, James Louis
Floyd, Denis Ragan
Fogarty, John Charde

Formanek, Edward William
Fossum, Timothy V
Franke, Charles H
Friesen, Donald Kent
Fuelberth, John Douglas
Fulp, Ronald Owen
Gaal, Ilse Lisl Novak
Gasti, George Clifford
Geller, Susan Carol
Gersten, Stephen M
Gerth, Frank Emmett, III
Gibson, Robert Wilder
Gilbert, Jimmie Dale
Gilbert, Robert William, Jr
Goldhaber, Jacob Kopel
Good, Richard Albert
Grabois, Neil
Graves, William Howard
Gray, James F
Green, Edward Lewis
Green, Sherry Merrill
Greim, Barbara Ann
Griess, Robert Louis, Jr
Grigori, Artur
Grossman, Edna K
Gustafson, William Howard
Guterman, Martin Mayr
Hall, Marshall, Jr
Hampton, Charles Robert
Harris, Morton E
Harrison, Nancy Evelyn
Hetzer, Garry A
Hildebrant, John A
Hill, Victor Ernst
Holyoke, Thomas Campbell
Horn, Alfred
Horn, Rajinder Bir
Hostinsky, Lois Aileen
Hou, Roger Hsiang-Dah
Howland, Richard A
Hsia, John S
Huckaba, James Albert
Hughes, Daniel Richard
Hurley, James Frederick
Hutchinson, George Allen
Idowu, Elayne Arrington
Jaffee, Harris Alexander
Janowitz, Melvin Fiva
Jenkins, Emerson D
Johnson, Eugene Carlyle
Johnson, Charles Royal
Johnson, Eugene W
Kambayashi, Tatsuji
Kapp, Kenneth M
Kass, Seymour
Kearns, Thomas J
Keigher, William Francis
Kim, Jin Bai
Kinloch, John
Kleiman, Howard
Kleinfeld, Erwin
Knighten, Robert Lee
Koehler, Anne Bramble
Kokoris, Louis A
Kovacic, Jerald J
Krause, Guenter
Kronstein, Karl Martin
Kruse, Robert Leroy
Kuczkowski, Joseph Edward
Kundert, Esayas G
Laible, Jon Morse
Lam, Tsit-Yuen
Lamont, Patrick John Coll
Lando, Barbara Ann
Lange, Gail Laura
Larsen, Leland Malvern
Larsen, Max Dean
Latch, Dana May
Lawvere, Francis William
Lea, James Wesley, Jr
Levy, Lawrence
Liebert, Wolfgang
Losey, Gerald Otis
Luedeman, John Keith
Luh, Jiang
Lundgren, J Richard
Mader, Adolf
Magarian, Elizabeth Ann
Malone, Joseph James
Marcus, Marvin
Markham, Thomas Lowell
Maxfield, John Edward
Mazeres, Reginald Merle
McCarthy, Donald John
McCleary, Stephen Hill
McCown, Malcolm G
McKee, Ruth Stauffer
McLean, Robert T
McNeil, Phillip Eugene
McQuarrie, Bruce Cale
Mena, Roberto Abraham
Merris, Russell Lloyd
Newborn, Ancel Clyde
Miller, David Clyde
Miller, Don Dalzell
Minc, Henryk
Montgomery, Susan
Mott, Joe Leonard
Moyls, Benjamin Nelson

Myers, Carol Bruce
Nail, Billy Ray
Nelson, Evelyn Merle
Newman, Kenneth Wilfred
Nobusawa, Nobuo
O'Callaghan, Robin Kuebler
O'Malley, Matthew Joseph
O'Nan, Michael Ernest
Orzech, Grace Geist
Orzech, Joseph Harold
Osofsky, Barbara Langer
Paley, Hiram
Park, Hubert Vern
Parr, James Theodore
Parry, Charles J
Passman, Donald Steven
Patenaude, Robert Alan
Peake, Edmund James, Jr
Peinado, Rolando E
Perlis, Sam
Petrich, Mario
Petro, John William
Phan, Kok-Wee
Phillips, Veril LeRoy
Picgari, George
Pierce, Thomas Scott
Pirnot, Thomas Leonard
Pohl, Victoria Mary
Poland, John C
Pollak, Barth
Price, David Thomas
Propes, Ernest (A)
Racine, Michel Louis
Ralston, Elizabeth Wall
Real, Cathleen Clare
Reynolds, William Francis
Ribes, Luis
Riehm, Carl Richard
Riles, James Byrum
Robinson, Daniel Alfred
Rodabaugh, David Joseph
Rossa, Robert Frank
Rudvalis, Arunas
Satyanarayana, Motupalli
Schafer, Alice Turner
Schenkman, Eugene Victor
Schiessinger, Michael
Schue, John R
Scott, Leonard Lewy, Jr
Scott, William Raymond
Sehgal, Surinder K
Sharp, Thomas Joseph
Shock, Robert Charles
Shult, Ernest E
Sit, William Yu
Slaby, Harold Theodore
Smith, Harry Francis
Smith, James Reaves
Smith, Velma Merrilene
Spitznagel, Edward Lawrence, Jr
Stander, Joseph W
Stanley, Richard Peter
Stark, Betty Salzberg
Stone, David Ross
Stroud, Junius Brutus
Stuth, Charles James
Suh, Tae-Il
Sun, Hugo Sui-Hwan
Svanes, Torgny
Taft, Earl J
Tang, Alfred Sho-Yu
Tarwater, Jan Dalton
Teague, Tommy Kay
Teller, John Roger
Teply, Mark Lawrence
Thaler, Alvin Isaac
Thedford, William Andrew
Thurston, Hugh Ansfrid
Thwing, Henry Warren
Tierney, Myles
Timmer, Kathleen Mae
Tomber, Marvin L
Tsai, Chester E
Vaughan, Theresa Phillips
Ventriglia, Anthony E
Verma, Sadanand
Vick, George R
Walter, John Harris
Ware, Roger Perry
Warfield, Robert Breckinridge, Jr
Warner, Seth L
Weiner, Louis Max
Weiss, Edwin
Wells, Charles Frederick
Wells, Jacqueline Gaye
Weston, Kenneth W
Wheaton, Burdette Carl
Wiegand, Sylvia Margaret
Wiegmann, Norman Arthur
Wilcox, Lee Roy
Wilson, Jack Charles
Wilson, Robert Lee
Wisner, Robert Joel
Wojcik, Anthony Stephen
Wong, Warren James
Wright, Charles R B
Yaqub, Adil Mohamed
Yellen, Jay

Yohe, Cleon Russell
Zelinsky, Daniel
Zelmanowitz, Julius Martin
Zemmer, Joseph Lawrence, Jr

Analytical Mathematics
Benedetto, John
Driscoll, Richard James
Eaves, Edgar Dewey
Friberg, Martin Samuel
Gulliver, Robert David, II
Laha, Radha Govinda
Osborn, James Maxwell
Peressini, Anthony L
Shah, Ghulam M
Solomon, Donald W
Stanaitis, Otonas Edmundas
Stanton, Charles Madison
Stellmacher, Karl L
Su, Ruth Wolf
Wu, Jang-Mei Gloria

Applied Mathematics
Abu-Shumays, Ibrahim Khalil
Acheson, Willard Phillips
Adam, Julian
Ader, Olin Blair
Adomian, George
Aggarwala, Bhagwan D
Agresta, Joseph
Agrawal, Jagdish Chandra
Ahlberg, John Harold
Ahluwalia, Daljit Singh
Albert, Arthur Edward
Allen, Richard Crenshaw, Jr
Alverson, Roy Carl
Amazigo, John C
Ames, William F
Anderson, Jan Frederick
Anderson, Donald Gordon Marcus
Anderson, Howard Benjamin
Anderson, Willard Eugene
Andrew, Merle M
Andrus, Stuart S
Aronofsky, Julius S
Asch, Michael Edward
Ashenhurst, Robert Lovett
Asner, Bernard A, Jr
Auchmuty, James Francis Giles
Averill, Frank Wallace
Babuska, Ivo Milan
Baildon, John David
Bailey, James L
Baker, Charles Ray
Bakhshi, Vidya Sagar
Balinski, Michel Louis
Banagh, Robert Peter
Banks, Dallas O
Barcilon, Victor
Bareiss, Erwin Hans
Barr, George E
Barston, Eugene Myron
Bate, George Lee
Bateson, Lewis E
Bauer, Frances Brand
Bazley, Norman William
Bebernes, Jerrold William
Belford, Geneva Grosz
Belinfante, Johan G F
Bell, Raymond Frank
Bendat, Julius Samuel
Ben-Israel, Adi
Bennett, John Henry
Berger, Neil Everett
Berguist, James William
Bernstein, Barry
Berry, Paul McClellan
Berstein, Barry
Bezdek, James Christian
Bharucha-Reid, Albert Turner
Bishir, John William
Bisshopp, Frederic Edward
Black, Richard H
Blackwell, John Henry
Blaisdell, Basilis Edwin
Bland, Robert Gary
Blankinship, William Aubrey
Bleistein, Norman
Block, Henry David
Blue, James Lawrence
Blum, Lenore Carol
Blum, Marvin
Blythe, Philip Anthony
Boal, Jan List
Boalvarsson, Gunnar
Bobachevsky, Ihor O
Bollinger, Richard Coleman
Bolmarcich, Joseph John
Bolt, Bruce A
Borrelli, Robert L
Bowie, Oscar L
Boyce, William Edward
Bradford, William Henry
Brady, Brian T
Brady, William Gordon

Bramble, James H
Braun, Martin
Brock, Paul
Brockett, Roger Ware
Bromberg, Eleazer
Bronikowski, Thomas Andrew
Bronzo, Joseph Alexander
Brown, Donald Meeker
Brown, Gerald Leonard
Brown, Harold David
Brown, James Ward
Brown, John Lawrence, Jr
Bryan, Joseph Gerard
Bryant, Robert William
Buchal, Robert Norman
Buckmaster, John David
Bucy, Richard Snowden
Burke, James Edward
Burniston, Ernest Edmund
Burns, John Allen
Burstein, Samuel Z
Cade, Ruth Ann
Calabi, Lorenzo
Callahan, Willie Russell
Callender, E David
Carlson, Bille Chandler
Carroll, Benjamin L
Cashwell, Edmond Darrell
Casten, Richard G
Caviness, Bobby Forrester
Celmins, Aivars Karlis Richards
Chai, Winchung A
Chan, Chiu Yeung
Chappelear, John Emerson
Chen, Paul Ear
Chen, Yung Ming
Chi, Donald Nan-Hua
Childress, William Stephen
Chow, Pao Liu
Chow, Tsu-Sen
Christiano, John G
Chu, Chia-Kun
Chu, Kai-Ching
Chu, William Wei-Ling
Clark, Alfred, Jr
Clark, Robert Arthur
Cline, Randall Eugene
Coburn, Richard Karl
Cochran, James Alan
Cohen, Donald Sussman
Cohen, Hirsh G
Cohen, Moses E
Cohoon, David Kent
Coleman, Albert John
Coleman, Norman P, Jr
Comstock, Craig
Concus, Paul
Conroy, Margaret Frances
Cook, Hollis Lee
Cook, Joseph Marion
Cooke, Kenneth Lloyd
Cook-Ioannidis, Leslie Pamela
Cooley, James William
Coulter, Paul (David Todd)
Cowan, Russell (Walter)
Cowell, Wayne Russell
Criminale, William Oliver, Jr
Crooke, Philip Schuyler
Crout, Prescott Durand
Cruise, Donald Richard
Cullum, Jane Kehoe
Cumberbatch, Ellis
Cushing, Jim Michael
Dafermos, Constantine M
Derzko, Nicholas Anthony
Deverall, LaMar Ivan
Dhaliwal, Ranjit S
Diaz, Joaquin Basilio
Dickey, Ronald Wayne
Dickinson, Deanne
Dickson, Lawrence John
Di Prima, Richard Clyde
Dolezal, Vaclav J
Dolph, Charles Laurie
Drenick, Rudolf F
Drew, Bruce Arthur
Dunn, Joseph Charles
Dwyer, Thomas A
Eargle, George Marvin
Eastman, Willard L
Edelen, Dominic Gardiner Bowling
Edelman, Franz
Edelstein, Warren Stanley
Edwards, Arthur L
Edwards, Carol Abe
Eisemann, Kurt
Eisenfeld, Jerome
Elcrat, Alan Ross
Elkins, Judith Molinar

Ellias, Samuel Aaron
Elliott, Sheldon Ellwood
Ellis, Albert Tromly
Elrod, McLowery
Esch, Robin Ernest
Evans, David Hunden
Everstine, Gordon Carl
Fabrey, James Douglas
Fagot, Wilfred Clark
Falb, Peter L
Fan, Hsin Ya
Farlow, Stanley Jerome
Farmer, James
Fattorini, Hector Osvaldo
Faulkner, Frank David
Federowicz, Alexander John
Fein, Alvin Eli
Fettis, Henry Eason
Filippenko, Vladimir I
Finch, John Hatta
Findler, Nicholas Victor
Fink, James Paul
Finkelstein, Abraham Bernard
Fischer, Charlotte Froese
Fisher, George M C
Fitzpatrick, Philip Matthew
Flanders, Harley
Fleishman, Bernard Abraham
Fletcher, Harvey Junior
Fletcher, John Edward
Foote, Joe Reeder
Ford, William Frank
Foschini, Gerard Joseph
Foster, Manus R
Franklin, Joel Nick
Freiberger, Walter Frederick
Friedlander, Susan Jean
Fuller, Franklyn Belmont
Gaal, Steven Alexander
Gaalswyk, Arie
Gabriel, John R
Gadamer, Ernst Oscar
Gajendar, Nandigam
Gardner, Clifford S
Gaskell, Robert Eugene
Gehatia, Matatiahu T
Genensky, Samuel Milton
George, Theodore Samuel
Gersting, John Marshall, Jr
Gessow, Alfred
Getchell, Bassford Case
Gibbs, Alan Gregory
Gilbarg, David
Giuliano, Vincent E
Glaser, Frank
Glasser, M Lawrence
Glauz, Robert Doran
Goering, Orville
Gold, Richard Robert
Goldstein, Allen A
Goldstein, Charles Irwin
Golomski, Rubin
Gomory, Ralph E
Gordon, Sheldon P
Gordon, William John
Grad, Harold
Graff, Samuel M
Grafton, Robert Bruce
Graham, George Alfred Cecil
Granoff, Barry
Greenberg, Herbert Julius
Greenberger, Martin
Greenspan, Harvey Philip
Greiner, John William
Grenander, Ulf
Grimm, Louis John
Gross, George Lloyd
Grossman, Stanley J
Grotte, Jeffrey Harlow
Gunzburger, Max Donald
Gurel, Okan
Gurtin, Morton Edward
Gustavson, Fred Gehrung
Haas, Violet Bushwick
Haaser, Norman Bray
Haeder, Paul Albert
Haight, Frank Avery
Haines, Charles Wills
Halabisky, Lorne Stanley
Hall, William Spencer
Handelman, George Herman
Hanson, Floyd Bliss
Hare, Robert Ritzinger, Jr
Harmon, Sidney M
Harris, Richard Allen
Harrison, Joseph Owens, Jr
Hart, John Francis
Hartman, James Kern
Hartwig, Robert Eduard
Haug, Edward J, Jr
Haussmann, Ulrich Gunther
Hayes, Patrick Louis
Hearn, Dwight D
Hector, David Lawrence
Heflinger, Lee Opert
Heilberg, Ernest
Heineman-Rose, Marian H

Heller, William R
Hemp, Gene Willard
Henry, Myron S
Herbert, Wallace
Herdman, Terry Lee
Hewer, Gary Arthur
Hicks, Darrell Lee
Hills, Norman L
Hitchcock, Daniel Augustus
Ho, Dar-Veig
Ho, Hung-Ta
Hochstadt, Harry
Hoffman, William Charles
Hogarth, Jacke Edwin
Hohn, Franz Edward
Holford, Richard L
Holland, Charles Jordan
Holland, John Henry
Holt, Frederick Sheppard
Holt, Maurice
Holway, Lowell Hoyt, Jr
Honeck, Henry Charles
Hopf, Eberhard
Horak, Martin George
Howard, Bernard Eufinger
Howard, Louis Norberg
Hsu, In-Ding
Hull, David Lee
Hulme, Bernie Lee
Hunt, Leon Gibson
Hunt, Robert Weldon
Hunter, Christopher
Hurt, James Joseph
Hurt, John Tom
Hutcherson, Joseph William
Hutcheson, Paul Henry
Hwang, John Dzen
Ikenberry, Ernest
Infante, Ettore F
Inselberg, Alfred
Isaacs, Rufus Philip
Israel, Jay Elliot
Iverson, Kenneth Eugene
Jarvis, Stephen, Jr
Jarvis, Roger George
Jawa, Manjit S
Jehn, Lawrence A
Jerome, Joseph Walter
Jerri, Abdul J
Johnson, Charles Royal
Johnson, Frederick Carroll
Johnson, Robert Oscar
Jones, Clinton E
Jones, Louise Hinrichsen
Joseph, Roy D
Juncosa, Mario Leon
Justice, James Horace
Kalensher, Bernard Earl
Kalme, Charles Ivars
Kanwal, Ram Prakash
Karal, Frank Charles, Jr
Karp, Samuel Noah
Karreman, Herman Felix
Kay, Irvin (William)
Keller, Edward Lee
Keller, Herbert Bishop
Kellogg, Royal Bruce
Keown, Ernest Ray
Kevorkian, Jirair
Kidder, Ray Edward
King, Richard Frederick
Kirby, Bruce John
Klein, John Sharpless
Kleinman, Ralph Ellis
Kline, Morris
Kloptfenstein, Ralph Walter
Knowles, James Kenyon
Knudsen, John Roland
Kockinos, Constantine Neophytos
Koh, Eusebio Legarda
Komkov, Vadim
Kotchoubey, Andrew
Kranzer, Herbert C
Kratzke, Albert William
Kraus, Lester
Krener, Arthur James
Kulsrud, Helene E
Kummer, Hans Jacob
Kuo, Shan Sun
Kupperman, Robert Harris
Kvarda, Robert Edward
Kydoniefs, Anastasios D
LaChapelle, 5 Benoit Vincent
Lagerstrom, Paco (Axel)
Lai, Peter Chengliang
Lane, Max Herbert
Lange, Charles Gene
Lanzano, Paolo
Lardner, Robin Willmott
Larsen, Kenneth Martin
Lau, Richard Lewis
Lebovitz, Norman Ronald
Lee, Norman K
Lee, William Thad
Leibowitz, Martin Albert
Leimanis, Eugene
Leitman, Marshall J
Leutert, Werner Walter

Levine, Harold
Levine, Lawrence Elliott
Levy, Paul
Lew, John S
Lewis, John Allen
Li, Wen-Hsiung
Liban, Eric
Lick, Dale W
Lick, Wilbert James
Lin, Chia Chiao
Lindquist, Anders Gunnar
Lindquist, Clarence Bernhart
Lindstrom, Fredrick Thomas
Lipshitz, Stanley Paul
Liu, Pan-Tai
Lomen, David Orlando
Loper, David Eric
Loughlin, Timothy Arthur
Lovass-Nagy, Victor
Low, Emmet Francis, Jr
Lucas, William Franklin
Ludford, Geoffrey Stuart Stephen
Luehr, Charles Poling
Luke, Jon Christian
Lundberg, Gustave Harold
Luxenberg, Harold Richard
Lynch, Robert Emmett
Lynn, Roger Yen Shen
Lynn, Yen-Mow
MacDonald, Carolyn Trott
MacGillivray, Archibald Dean
Majumdar, Samir Ranjan
Mamelak, Joseph Simon
Mandelbrot, Benoit
Mandl, Paul
Mangasarian, Olvi Leon
Mann, James Edward, Jr
Mann, William Robert
Maple, Clair George
Maria, Narendra Lal
Marimont, Rosalind Brownstone
Mark, James Wai-Kee
Martin, Joe Alton
Marshall, Clifford Wallace
Martin, Charles John
Martin, Monroe Harnish
Maslen, Stephen Harold
Massey, Fredrick Alan
Mattson, Harold F, Jr
Mazo, James Emery
McBride, Elna Browning
McClamroch, N Harris
McClure, Donald Ernest
McCormick, Clyde Truman
McCulley, William Straight
McElroy, Michael Brendan
McGehee, Ralph Marshall
McIntyre, Robert Gerald
McKnight, Randy Sherwood
McNeary, Samuel Stuart
Meissner, Loren Phillip
Menkes, Joshua
Mertz, Robert Theodore
Metcalf, Frederic Thomas
Metropolis, Nicholas Constantine
Meyer, Gunter Hubert
Michalik, Edmund Richard
Middleton, David
Millar, Robert Fyfe
Miller, Paul George
Miller, Willard, Jr
Millsaps, Knox
Mitchell, A Richard
Miura, Robert Mitsuru
Mohammed, Auyuab
Mood, Alexander McFarlane
Moore, Douglas Houston
Moore, Edward Forrest
Moorti, Varahur R Guru
Moran, James Herbert
Morawetz, Cathleen Synge
Morgan, Richard C
Morrison, Barbara Ann
Morrison, John Allan
Morse, Burt Jules
Mosevich, Jack Walter
Mueller, Raymond Karl
Mufti, Izhar-Ul Haq
Muller, David Eugene
Muller, Karl Frederick
Mullikin, Thomas Wilson
Mysak, Lawrence Alexander
Nachbar, William
Narasimhan, Mysore N L
Nariboli, Gundo A
Nash, David Henry George
Naugle, Norman Wakefield
Naylor, Derek
Neal, Scotty Ray
Newburg, Edward A
Newell, Gordon Frank
Ng, Bartholomew Sung-Hong
Nickel, James Alvin
Niman, John
Norminton, Edward Joseph
Northover, Francis Henry
Norwood, Frederick Reyes

MATHEMATICS

Nowlin, Charles Henry
Oden, Farouk M
Oettinger, Anthony Gervin
Ogg, Frank Chappel, Jr
Olmstead, William Edward
Olver, Frank William John
O'Malley, Robert Edmund, Jr
O'Mathuna, Diarmuid
O'Neil, Elizabeth Jean
O'Neill, John Cornelius
Ornstein, Wilhelm
Opatowski, Izaak
Oser, Hans Joerg
Owen, David R
Painter, Charles Henry
Parker, Francis Dunbar
Patel, Vithalbhai Ambalal
Patrick, Merrell Lee
Pauls, John F
Paxton, K Bradley
Payne, Lawrence Edward
Pell, William Hicks
Perko, Lawrence Marion
Perryman, John Keith
Peterson, David West
Petrie, George Whitefield, III
Phillips, John Richard
Phillips, Ralph Saul
Pierce, John Gregory
Pinski, Gabriel
Pipkin, Allen Compere
Plummer, Otho Raymond
Pohle, Frederick V
Polachek, Harry (Aaron)
Polowy, Henry
Ponzo, Peter James
Porter, Lawrence Delpino
Powers, David Leusch
Price, Harvey Simon
Puri, Pratap
Queen, William Charles
Quesada, Antonio F
Quinn, Dennis Wayne
Raab, Wallace Albert
Radlow, James
Ramanathan, Ganapathiagraharam V
Ranger, Keith Brian
Rautenstrauch, Carl Peter
Ravesloot, John Lowell
Ray, Ajit Kumar
Ray-Chaudhuri, Dwijendra Kumar
Raychowdhury, Pratip Nath
Rebhuhn, Deborah
Reddy, Kapuluro Chandrasekhara
Rehm, Ronald George
Reid, Walter Phillip
Reid, William Hill
Reynolds, David Stephen
Rice, John Rischard
Rice, Peter Milton
Richardson, Richard Laurel
Riesenfeld, Richard F
Riley, John Aswood
Rinzel, John Matthew
Ritt, Robert King
Ritger, Paul David
Ritter, Klaus Guenter
Ritterman, Murray B
Rivlin, Ronald Samuel
Roberts, Jerry Allan
Roberts, Richard Calvin
Roberts, William Woodruff, Jr
Robinson, Stephen Michael
Roe, Glenn Magnus
Roetman, Ernest Levane
Rogers, Edwin Henry
Rohde, Florence Virginia
Rohde, Steve Mark
Rose, Arnold James
Rosen, Arthur Leonard
Rosen, Judah Ben
Rosenblatt, Murray
Ross, Edward William, Jr
Ross, Roderick Alexander
Rotenberg, Aubey
Roth, Charles
Rothenberg, David
Russell, Marvin W
Saffman, Philip Geoffrey
Salton, Gerard
Sanders, Bobby Lee
Sangren, Ward Conrad
Sawyers, Kenneth Norman
Schaefer, Brian Morris
Schaefer, Philip William
Schechter, Samuel
Schiavone, George Joseph
Schmeck, John Frank
Schoenstadt, Arthur Loring
Schryer, Norman Loren
Schuldt, Spencer Burt
Schultz, Nathan
Schwartz, Daniel George
Schweikert, Daniel George
Schwerdski, Ernst Walter
Scott, Meckinley

Scott, Richard Anthony
Sebastian, Richard Lee
Segel, Lee Aaron
Seidman, Thomas Israel
Seifert, Arnold David
Selander, George
Selander, Walter Andrew
Sevian, Walter Nils
Shah, Ghulam M
Shana'a, Joyce A
Shapiro, Claude Elwood
Shapiro, Donald M
Shannon, Charles Bergman, Jr
Shaw, Charles Bergman, Jr
Shaw, Harry, Jr
Shen, Chung Yi
Shen, Mei-Chang
Shenton, Leonard Roy
Shere, Kenneth David
Sherman, Michael
Shnider, Ruth Wolkow
Shnider, Gerald Martin
Shoemaker, Edward Milton
Shonka, Richard Edward A
Shu, Shien Siu
Sibul, Leon Henry
Sida, Derek William
Sigillito, Vincent George
Simmonds, James Gordon
Simons, Roger Mayfield
Sincovec, Richard Frank
Skalafuris, Angelo James
Skinner, Lindsay A
Slatkin, Montgomery Wilson
Sloan, Alan David
Slotnick, Daniel Leonid
Sluyter, Marshall M
Smith, Gerald Francis
Snyder, Herbert Howard
Snyder, Martin Avery
Snygg, John Morrow
Solomon, Alan D
Spahn, Robert Joseph
Spangler, Charles Bishop
Spickerman, William Reed
Spielberg, Kurt
Srivastav, Ram Prasad
Stakgold, Ivar
Stalford, Harold Lenn
Stanions, Victor Adam
Stauffer, Howard Boyer
Steeg, Carl W, Jr
Steinberg, Stanly
Steinmetz, William John
Stenger, Frank
Sternberg, Robert Langley
Stiles, Raeburn Brackett
Straeter, Terry Anthony
Strauss, Charles Michael
Streifer, William
Strumpf, Albert
Stuck, Barton W
Sullivan, Paul Joseph
Surkan, Alvin John
Swann, Dale William
Tai, Clement Leo
Tainiter, Melvin
Tait, Kevin S
Talham, Robert J
Tam, Christopher K W
Tam, Kwok Kuen
Tauber, Selmo
Taylor, Howard Lawrence
Taylor, Howard Milton, III
Teague, David Boyce
Temple, Austin Limiel, (Jr)
Temple, William Benson
Tewarson, Reginald P
Thomas, Donald Henry
Thomas, Larry Emerson
Thompson, Gerald Luther
Thompson, James Lowry
Thorne, Charles Joseph
Thurber, James Kent
Ting, Lu
Toalson, Wilmont
Tompson, Robert Norman
Toomre, Alar
Tory, Elmer Melvin
Traband, Edward Arthur
Trapp, George E, Jr
Troesch, Beat Andreas
Tropf, Cheryl Griffiths
Troy, William Christopher
Tsan, Alice Tung-Hua
Tsokos, Chris Peter
Tu, Yih-o
Tuan, San Fu
Tulenko, James Stanley
Turcheck, Joseph Edward
Tweed, John
Uhlenbrock, Dietrich A
Ullman, Nelly Szabo
Van Norton, Roger Norman
Vayo, Harris Westcott
Ventriglia, Anthony E
Verma, Ghasi Ram
Vogeley, Clyde Eicher, Jr
Von Rohr, Beatrice Louise
Waid, Margaret Cowsar

Walker, Kelsey, Jr
Wan, Frederic Yui-Ming
Wang, Chang-Yi
Warner, William Hamer
Wasserman, Robert H
Weber, Richard Robert
Weill, Georges Gustave
Weinberg, I Jack
Weinberger, Hans Felix
Weiner, Jerome Harris
Weiss, George Herbert
Weitzner, Harold
Wells, William T
Wenzel, Alan Richard
Westbrook, David Rex
Wester, John Walter, Jr
Weston, Vaughan Hatherley
White, George Nichols, Jr
White, Ronald Joseph
Whitham, Gerald Beresford
Whittle, Charles Edward, Jr
Williams, Gareth
Willke, Herbert Louis, Jr
Willman, Warren Walton
Wilson, James Blake
Wilson, Lynn Olson
Wilson, Raymond Hiram, Jr
Wing, George Milton
Winrich, Lonny B
Winter, Donald F
Winthrop, Joel Albert
Witsenhausen, Hans S
Witterholt, Edward John
Wolf, Henry
Wolfe, Philip
Wong, Chak-Kuen
Wu, Lilian Shiao-Yen
Wu, Tai Te
Yang, Chung-Chun
Yanowitch, Michael
Yarmush, David Leon
Yee, Kane Shee-Gong
Yen, David Hsein-Yao
Yett, Fowler Redford
Yeung, Shiu Fong
Yip, Patrick Cheung-Yun
Yorke, James Alan
Zabusky, Norman J
Zachmanoglou, Eleftherios Charalambos
Zatzkis, Henry
Zemanian, Armen Humparsoum
Zitron, Norman Ralph
Zlobec, Sanjo

Biomathematics

Altshuler, Bernard
Berman, Mones
Bernard, Selden Robert
Bertell, Rosalie
Blakley, George Robert
Blumenson, Leslie Eli
Brennerman, Hans J
Bright, Peter Bowman
Britt, Patricia Marie
Brown, Barry W
Cardus, David
Carpenter, Gail Alexandra
Cerimele, Benito Joseph
Cole, Walter Eckle
Conrad, Jane Smiley (Mrs Joseph C Scanlon)
Deysach, Lawrence George
Dixon, Wilfrid Joseph
Evans, John W
Feldman, Marcus William
Francis, Robert Colgate
Garfinkel, David
Gold, Harvey Joseph
Gridgeman, Norman Theodore
Howell, John Robert
Huang, H K
Hutchison, Gerald Andrew
Jacquez, John Alfred
Jansson, Birger
Karrerman, George
Katholi, Charles Robinson
Klipper, Robert William
Kimberling, William J
Koong, Ling-Jung
Landahl, Herbert Daniel
Licko, Vojtech
Maraman, Grady Vancil
Metzler, Carl Maust
Meyers, Philip Robert
Miller, Donald Richard
Niklas, Karl Joseph
Oster, George F
Palatt, Paul Jay
Paloheimo, Jyri Erkki
Peinado, Rolando E
Reiner, John Maximilian
Resigno, Aldo
Richardson, Irvin Whaley
Rosen, Robert
Rudman, Sanford Winton
Saltzberg, Bernard
Schimmel, Herbert

Segel, Lee Aaron
Siler, William MacDowell
Silvers, Abraham
Simon, William
Smeach, Stephen Charles
Spalding, Gary E
Tallarida, Ronald Joseph
Thames, Howard Davis, Jr
Turner, Malcolm Elijah, Jr
Voorhees, Burton Hamilton
Waltman, Paul E
Whitmore, Alice S
Woodbury, Max Atkin
Zimmerman, Stuart O
Zuker, Michael

Combinatorics

Albertson, Michael Owen
Cusick, Thomas William
Di Paola, Jane Walsh
Gardner, Marianne Lepp
Hodel, Margaret Jones
Roberts, Fred Stephen
Shult, Ernest E
Simmons, Gustavus James
Vanderlugt, Donald W

Geometry

Aczel, Janos D
Adelberg, Arnold M
Albu, Evelyn D
Altman, Allen Burchard
Antonelli, Peter Louis
Artzy, Rafael
Banks, John Houston
Barnett, David
Beckenbach, Edwin Ford
Beem, John Kelly
Berg, Gene Arthur
Berton, John Andrew
Blum, Peter
Boehning, Rochelle Lloyd
Borgman, William Martin, Jr
Breen, Marilyn
Cecil, Thomas E
Chavel, Isaac
Chen, Bang-Yen
Chern, Shiing-Shen
Childress, Noel A
Collier, James Bryan
Cox, Milton D
Data, Dilip Kumar
Dennis, Foster Leroy
de Witt, Paul
Di Paola, Jane Walsh
Doyle, William Cletus
Ehrmann, Rita Mae
Feldman, Edgar A
Feldman, Louis A
Flaherty, Francis Joseph
Freese, Raymond Joseph
Fulton, Curtis Maxwell
Gardner, Robert B
Gentry, Frank Cook
Gerber, Leon E
Gold, Sydell Perlmutter
Gray, Alfred
Grunbaum, Branko
Gulliver, Robert David, II
Halsey, Eric Richard
Hausner, Melvin
Hoffman, David Allen
Houston, William Bernard, Jr
Hsu, Chen-Jung
Hughes, Daniel Richard
Jackson, Stanley Bartlett
Johnson, Norman L
Kambayashi, Tatsuji
Kay, David Clifford
Kazdan, Jerry Lawrence
Koch, Richard Moncrief
Kockinos, Constantine Neophytos
Layton, William Isaac
Lewis, George McCormick
Martin, George Edward
Mast, Cecil B
Matsusaka, Teruhisa
Mattuck, Arthur Paul
Millman, Richard Steven
Moolgavkar, Suresh Hiraji
Osserman, Robert
Ozley, Elsie Church
Petti, Richard James
Petty, Clinton Myers
Portnoy, Esther
Sacks, Jonathan
Sawyer, John Wesley
Schell, Joseph Francis
Scherk, Peter
Speiser, Steven Irwin
Thorpe, John Alden
Tanimoto, Taffee Tadashi
Totten, James Edward
Vick, George R
Warner, Frederic Cooper
Wavrik, John J
Weinstein, Alan David
Whitmore, Edward Hugh

Williamson, Robert Emmett
Wrona, Wlodzimierz Stefan
Wylie, Clarence Raymond, Jr
Yaqub, Jill Courtney Donaldson Spencer
Zaustinsky, Eugene Michael

Mathematical Analysis
Aczel, Janos D
Aliprantis, Charalambos Dionisios
Amelin, Charles Francis
Anders, Edward B
Anderson, Clifford Harold
Andrushkiw, Roman Ihor
Anselone, Philip Marshall
Antosiewicz, Henry Albert
Artemiadis, Nicholas
Arterburn, David Roe
Askey, Richard Allen
Atkinson, Harold Russell
Babbitt, Donald George
Baernstein, Albert, II
Bajaj, Prem Nath
Baker, James Dennard
Baker, John Warren
Barnhill, Robert E
Barth, Karl Frederick
Beals, Richard William
Beard, Helen Pearl
Beckenbach, Edwin Ford
Beckenstein, Edward
Beesack, Paul Richard
Bell, Raymond Frank
Bellenot, Steven F
Berberian, Sterling Khazag
Billingheimer, Claude Elias
Bilodeau, Gerald Gustave
Bilyeu, Russell Gene
Bleistein, Norman
Bojanic, Ranko
Bolstein, Arnold Richard
Brakke, Kenneth Allen
Brandstein, Alfred George
Browder, Felix Earl
Brown, Arthur Barton
Brown, James Russell
Buck, Robert Creighton
Buckholtz, James Donnell
Buoni, John J
Campbell, Douglas Michael
Campbell, Robert Calvin
Caradus, Selwyn Ross
Cargo, Gerald Thomas
Carmichael, Richard Dudley
Carroll, Francis W
Cash, Dewey Byron
Cassens, Patrick
Causey, William McLain
Cayford, Afton Herbert
Cesari, Lamberto
Chai, Winchung A
Chan, Chiu Yeung
Chang, I-Lok
Chang, Sun-Yung Alice
Chihara, Theodore Seio
Chimenti, Frank A
Clark, David C
Clark, Douglas Napier
Clark, William Dean
Cohen, Elaine
Cohen, Marion Deutsche
Coleman, Bernard David
Coleman, Courtney (Stafford)
Connett, William C
Conway, Edward Daire, III
Conway, John Bligh
Cook, James Marion
Coppin, Charles Arthur
Crownover, Richard McCranie
Curtiss, John Hamilton
Daly, James Edward
Dawson, David Fleming
Deal, Albert Leonard, III
Deeter, Charles Raymond
Dell, Roger Marcus
Dettman, John Warren
DeVito, Carl Louis
Dickson, Douglas Grassel
Dickson, Lawrence John
Dombrowski, Joanne Marie
Drasin, David
Driver, Rodney David
Eames, William
Easton, Richard J
Elderkin, Richard Howard
Eliason, Stanley B
Elliott, Helen Margaret
Eltze, Ervin Marvin
Embry, Mary Rodriguez
Engert, Martin
Epstein, Harvey Irwin
Erdelyi, Ivan Nicholas
Espelie, Mary Solveig
Fabes, Eugene Barry
Faires, John Douglas
Fast, Henryk
Federer, Herbert

Fefferman, Charles Louis
Feinstein, Irwin K
Filgo, Holland Cleveland
Finlayson, Henry C
Fobes, Melcher Prince
Fornaess, John Erik
Foster, Kent Ellsworth
Fountain, Leonard Du Bois
Frawley, William James
Freedman, Herbert Irving
Freud, Geza
Friedberg, Stephen Howard
Friedman, Harvey Martin
Fuller, Derek Joseph Haggard
Gaer, Marvin Charles
Gasper, George, Jr
Gibson, William Loane
Gilbert, Richard Carl
Gill, John Paul, Jr
Ginsberg, Jonathan I
Goldstein, Jerome A
Gonzales, Mario Octavio
Goodman, Roe William
Gordon, Hugh
Gosselin, Richard Pettengill
Graves, William Howard
Green, William Lohr
Greenhall, Charles August
Greenleaf, Frederick P
Gregory, Michael Baird
Grimmell, William C
Gross, Kenneth Irwin
Gupta, Chaitan Prakash
Gustafson, Grant Bernard
Hagler, James Neil
Hahn, Kyong T
Haimo, Deborah Tepper
Hall, Leon Morris, Jr
Hall, Robert Lester
Hannsgen, Kenneth Bruce
Hansen, Lowell John
Harris, Beverly Howard
Heath, Larry Francis
Heins, Maurice Haskell
Herod, James Victor
Hersh, Reuben
Hiergeist, Franz Xavier
Hillam, Bruce Parks
Hille, Einar
Hilzman, John
Himmelberg, Charles John
Hochstadt, Harry
Hoffman, Stephen Peter, Jr
Holmes, Richard Bruce
Horner, Donald Ray
Horvath, John
Howland, James Lucien
Hubbard, Bertie Earl
Hudson, William Nathaniel
Huff, William Nathan
Hughes, Eugene Morgan
Huneycutt, James Ernest, Jr
Hunt, Robert Weldon
Iha, Franklin Takashi
Ionescu Tulcea Cassius
Isaacs, Godfrey Leonard
Jain, Mahendra Kumar
James, Robert Clarke
Jewett, John William
Johnson, Guy, Jr
Johnson, William Buhmann
Jones, Harold Trainer
Jones, William B
Jurkat, Wolfgang Bernhard
Kalka, Morris
Kallman, Robert Richard
Kalme, Charles Ivars
Kaminker, Jerome Alvin
Kay, Alvin John
Kazdan, Jerry Lawrence
Keener, Marvin Stanford
Kent, James Ronald Fraser
Kerr, Sandria Neidus
Kim, Moon W
Kimes, Thomas Fredric
Kirwan, William English
Klein, Benjamin Garrett
Klepper, Adam
Klipple, Edmund Chester
Knopp, Marvin Isadore
Koosis, Paul
Koren, Charles
Krabbe, Gregers Louis
Laatsch, Richard G
Ladas, Gerasimos E
Laetsch, Theodore Willis
Lahr, Charles Dwight
Lambert, Joseph Michael
Larsen, Ronald John
Lavine, Richard Bengt
Legg, David Alan
Lehner, Joseph
Leibowitz, Gerald Martin
Leipnik, Roy Bergh
Levin, Jacob Joseph
Lindberg, John Albert, Jr
Linscheid, Harold Wilbert

Lippman, Gary Edwin
Littman, Walter
Liu, Tai-Ping
Liukkonen, John Robie
Livingston, Albert Edward
Lonseth, Arvid Turner
Lorentz, George G
Lubin, Arthur Richard
Lubin, Clarence Isaac
Lustfield, Charles Davenport
Luxemburg, Wilhelmus Anthonius Josephus
Machover, Maurice
MacNerney, John Sheridan
Makar, Boshra Halim
Malbrock, Jane C
Maloney, John P
Marik, Jan
Martin, Robert Paul
Mastin, Charles Wayne
May, Sherry Jan
McCarthy, Charles Alan
McClure, John Arthur
McKinney, Alfred Lee
McLain, David Kenneth
McLaughlin, Renate
McLeod, Robert Melvin
McWilliams, Ralph David
Mehlenbacher, Lyle E
Miller, Sanford Stuart
Minda, Carl David
Moore, Marian Alease
Mosak, Richard David
Moskovitz, David
Moyer, Robert Dale
Muckenhoupt, Benjamin
Mugler, Dale H
Mullins, Robert Emmet
Narici, Lawrence Robert
Newton, Tyre Alexander
Nunemacher, Jeffrey Lynn
Nur, Hussain Sayid
Nussbaum, Roger David
Oberle, Richard Alan
Oberlin, Daniel Malcolm
O'Malley, Richard John
Ornstein, Wilhelm
Osserman, Robert
Packel, Edward Wesler
Park, Samuel
Parrish, Herbert Charles
Paschke, William Lindall
Peck, Newton Tenney
Peterson, Gerald E
Phillips, Keith L
Pinney, Edmund Joy
Pipes, Charles Jefferson, Jr
Plakun, Geraldine Taiani
Plunkett, Robert Lee
Podolak, Esther
Pohrer, Robert George
Pool, James C T
Price, Griffith Baley
Pursell, Lyle Eugene
Ragozin, David Lawrence
Raimi, Ralph Alexis
Raphael, Louise Arakelian
Read, Thomas Thornton
Reid, William Thomas
Reneke, James Allen
Richardson, Leonard Frederick
Ripy, Sara Louise
Roach, Francis Aubra
Robinson, R Clark
Roques, Alban Joseph
Rothman, Neal Jules
Royden, Halsey Lawrence
Rung, Donald Charles, Jr
Ryder, Gerald H
Ryff, John V
Saccoman, John Joseph
Saff, Edward Barry
Salehi, Habib
Sardinas, August A
Scanlon, Charles Harris
Schechter, Martin
Schiffer, Menahem Max
Schlesinger, Ernest Carl
Scranton, Bruce Edward
Schweinsberg, Allen Ross
Seeley, Robert T
Shah, Swarupchand Mohanlal
Shapiro, Victor Lenard
Sherbert, Donald R
Shuchat, Alan Howard
Sidney, Stuart Jay
Skoug, David L
Smith, James F
Smith, William Norman
Soni, Kusum
Sours, Richard Eugene
Spikes, Paul Wenton
Stahl, Neil
Stegenga, David Allan
Steib, Michael Lee
Stewart, James Drewry
Stoll, Manfred
Swetharanyam, Lalitha

Swetits, John Joseph
Synowiec, John A
Tam, Kwok-Wai
Taylor, Joseph Lawrence
Thomas, Robert Malcolm
Tompson, Robert Norman
Ton, Bui An
Torchinsky, Alberto
Transue, William Reagle
Utz, Winfield Roy, Jr
Vessey, Theodore Alan
Vest, Marvin Lewis
Walsh, Bertram (John)
Waltman, Paul E
Warfield, Virginia McShane
Webb, Glenn Francis
Weill, Georges Gustave
Weiss, Norman Jay
Wells, Benjamin B, Jr
Wigley, Neil Marchand
Wilde, Carroll Orville
Williams, Lloyd Bayard
Williams, Lynn Dolores
Winslow, Leon E
Wong, Bing Kuen
Wong, James Sai-Wing
Wong, Roderick Sue-Cheung
Wood, James Alan
Worth, Roy Eugene
Yood, Bertram
Young, Eutiquio Chua
Youse, Bevan K
Zafran, Misha
Ziegler, Michael Robert
Zielezny, Zbigniew Henryk
Zimering, Shimshon

Mathematical Biophysics
Macy, Josiah, Jr
Matthysse, Steven William
Newton, Abba Verbeck

Mathematical Physics
Adler, Felix T
Aks, Stanley Olaf
Bargmann, Valentine
Bart, George Raymond
Bender, Carl Martin
Berger, Martin Jacob
Bezuszka, Stanley John
Biggs, Frank
Blank, Albert Abraham
Bleick, Willard Evan
Blue, James Lawrence
Bowden, Robert Lee, Jr
Bragg, Lincoln Ellsworth
Brans, Carl Henry
Brooks, Franklin Coolidge
Burgoyne, Peter Nicholas
Burke, James Edward
Canosa, Jose M
Challifour, John Lee
Childs, Donald Ray
Cohen, E Richard
Critchfield, Charles Louis
Dankel, Thaddeus George, Jr
Davidon, William Cooper
Derr, John Sebastian, Jr
Dory, Robert Allan
Dyson, Freeman John
Emch, Gerard G
Faris, William Guignard
Faulkner, James Earl
Fette, Clarence William
Findley, George Bernard
Fisher, Edward
Fisher, Michael Ellis
Fong, Jeffrey Tse-Wei
Giles, Robin
Granger, Robert A, II
Greenberg, William
Guenther, Ronald Bernard
Habetler, George Joseph
Harvey, George Graham
Herman, Richard Howard
Hershey, Allen Vincent
Hoffmann, Banesh
Humi, Mayer
Hurt, Norman Edward
Jackson, David Phillip
Jackson, Robert Franklin
Jaffe, Arthur Michael
Kaplan, Bernard
Karp, Ira Lawrence
Kelley, Robert Lee
Klapman, Solomon Joel
Klein, Abel
Kohlmayr, Gerhard Franz
Kroll, Norman Myles
Kruskal, Martin David
Kummer, Martin
Kunzle, Hans Peter
Kyrala, Ali
Kyame, Joseph John
Landshoff, Rolf
Lanford, Oscar E, III
Lee, Clarence Edgar
Lee, Kotik Kai

Leith, Cecil Eldon, Jr
Leitner, Alfred
Lenard, Andrew
Lieb, Elliott Hershel
Lomont, John S
Louck, James Donald
Mainster, Martin Aron
Marable, James Holley
Marchand, Jean-Paul
Marcus, Paul Malcolm
Marcuvitz, Nathan
Mark, J Carson
Mayer, Meinhard Edwin
McCoy, Jimmy Jewell
McDonald, James Frederick
Meecham, William Coryell
Miller, Willard, Jr
Min, Kwang-Shik
Moore, Mortimer Norman
Morgan, Samuel Pope
Neuringer, Joseph Louis
Norton, John Leslie
Nussenzveig, Herch Moyses
Ogden, Edwin Burman
Ozsvath, Istvan
Park, Chong Jin
Parker, Lee Ward
Parvulescu, Antares
Pasta, John Robert
Pasternack, Simon
Penico, Anthony Joseph
Plummer, Otho Raymond
Plybon, Benjamin Francis
Prugovecki, Eduard
Radin, Charles Lewis
Rauch, Jeffrey Baron
Reed, Michael Charles
Renken, James Howard
Rindler, Wolfgang
Roman, Paul
Rosenbaum, Marcos
Rubenfeld, Lester A
Saffren, Melvin Michael
Schay, Geza
Schild, Alfred
Schutz, Bernard Frederick
Seidl, Frederick Gabriel Paul
Sen, Dipak Kumar
Siegert, Arnold John Frederick
Simon, Barry Martin
Srivastava, Hari Mohan
Symon, Keith Randolph
Tamburino, Louis Anthony
Tappert, Frederick Drach
Toupin, Richard A
Tross, Ralph G
Tulczyjew, Wlodzimierz Marek
Twersky, Victor
Uhlenbrock, Dietrich A
Van Winter, Clasine
Wagner, Richard John
Warner, David Charles
Warsi, Nazir Ahmed
Weinstock, Robert
Wheeler, Nicholas Allan
Whitten-Wolfe, Barbara L
Wightman, Arthur Strong
Wigner, Eugene Paul
Williams, David Noel
Wilson, Robert Norton
Wolf, Emil
Wong, Maurice King Fan
Wu, Hsin-I
Yeh, Hsin-Yang

Mathematics Education
Allen, Ernest E
Alton, Elaine Vivian
Archer, Cass L
Berenson, Lewis Jay
Biddle, John Charles
Bitter, Gary G
Brieske, Thomas John
Buchanan, O Lexton, Jr
Carroll, Edward Major
Clemens, Kermit Grover
Colharp, Forrest Lee
Crosswhite, F Joe
Deer, George Wendell
Early, Joseph E
Farrell, Margaret Alice
Fitting, Marjorie Ann
Frank, Stanley
Gibb, Elizabeth Glenadine
Grenier, Marie-Anne Cecile
Hall, Wayne Hawkins
Henke, Clarence Henry
Herring, Carey Reuben
Hess, Adrien LeRoy
Hill, Shirley Ann
Hirschi, L Edwin
Hostetler, Robert Paul
Johnson, Louise H
Knodel, Raymond Willard
Lipsey, Sally Irene
Lockley, Jeanette Elaine
Love, William P
McBride, Ralph Book

Miller, William Anton
Moursund, David G
Nelson, Norman Neibuhr
Olsen, Glenn W
Paquette, Gerard Arthur
Prener, Robert
Pruitt, Ralph L
Rappaport, David
Retzer, Kenneth Albert
Riggs, Richard
Robinson, Gertrude Edith
Sams, Emmett Sprinkle
Shana'a, Joyce A
Smith, John Melvin
Steinbrenner, Arthur H
Strickmeier, Henry Bernard, Jr
Stringall, Robert William
Vance, Irvin Elmer
Vervoort, Gerardus
Vest, Floyd Russell
Webb, Leland Frederick
Williams, Vernon
Willoughby, Stephen Schuyler

Number Theory
Adams, William Wells
Ballew, David Wayne
Bernstein, Leon
Borosh, Itshak
Brande, Edward Woodrow
Briggs, William Egbert
Brownawell, Woodrow Dale
Brzenk, Ronald Michael
Bumby, Richard Thomas
Cable, Charles Allen
Calloway, Jean Mitchener
Cayford, Afton Herbert
Cusick, Thomas William
DeLeon, Morris Jack
Denenberg, Charlotte Goryn
Divis, Bohuslav B
Dressler, Robert Eugene
Durst, Lincoln Kearney
Ecklund, Earl Frank, Jr
Eigel, Edwin George, Jr
Entringer, Roger Charles
Freitag, Herta Taussig
Friedlander, John Benjamin
Galambos, Janos
Garrison, Betty Bernhardt
Gelbart, Stephen Samuel
Gerth, Frank Emmett, III
Gioia, Anthony Alfred
Gold, Robert
Goldstein, Larry Joel
Grabois, Neil
Green, Harold Rugby
Hardy, John Thomas
Harris, Stephen Joseph
Harris, Vincent Crockett
Hayes, David R
Hazlewood, Donald Gene
Hensley, Douglas Austin
Hightower, Collin James
Hilliker, David Lee
Hodel, Margaret Jones
Howard, Fredric Timothy
Hsia, John S
Kimble, Gerald Wayne
Kleiman, Howard
Knopp, Marvin Isadore
Kurtzo, Philip C
Lamont, Patrick John Coll
Lardy, Lawrence James
Lefton, Phyllis
Lehner, Joseph
Levit, Robert Jules
Li, Wen-Ching Winnie
Lichtenbaum, Stephen
Lippa, Erik Alexander
Marsh, Donald Charles Burr
Massell, Paul Barry
Mattics, Leon Eugene
Meyer, Franklin Vincent
Muskat, Joseph Baruch
Niedereiter, Harald Guenther
Parry, Charles J
Rabung, John Russell
Robinson, Julia (Bowman)
Smart, John Roderick
Smith, Robert Arnold
Stark, Harold Mead
Steiner, Ray Philip
Sun, Hugo Sui-Hwan
Sunley, Judith S
Sweet, Melvin Millard
Taussky, Olga (Mrs John Todd)
Terras, Audrey Anne
Thanigasalam, Kandiah
Theusch, Colleen Joan
Tsao, Liang-Chi
Vanden Eynden, Charles Lawrence

Numerical Analysis
Ahlberg, John Harold
Andrews, George Harold
Archer, David Anderson
Ascher, Marcia

Atchison, Thomas Andrew
Babuska, Ivo Milan
Benokraitis, Vitalius
Boland, Willard Robert
Bradley, James Henry Stobart
Bramble, James H
Britton, Otha Leon
Bunch, James R
Byrne, George D
Chan, Paula Pui-Ying
Chaney, Robin W
Cody, William James, Jr
Concus, Paul
Conte, Samuel D
Daniel, James Wilson
Dendy, Joel Eugene, Jr
Diesen, Carl Edwin
Dodes, Irving Allen
Dupont, Todd
Eaves, Burchet Curtis
Eberlein, Patricia James
Ehle, Byron Leonard
Ehrlich, Louis William
Eilerts, Charles Kenneth
Fairweather, Graeme
Garder, Arthur Oris, Jr
Gautschi, Walter
Geddes, Keith Oliver
Godfrey, Brendan Berry
Goldstine, Herman Heine
Goodman, Richard Henry
Gregory, Robert Todd
Haber, Seymour
Hager, William Ward
Hall, Charles Allan
Herriot, John George
Hilbert, Stephen Russell
Hoff, John Clifford
Huddleston, Robert E
James, Ralph L
Jameson, William J, Jr
Jefferson, Thomas Hutton, Jr
Johnson, Robert Shepard
Jones, William B
Juncosa, Mario Leon
Kammler, David W
Karon, John Marshall
Kaufman, Linda Carol
Keller, Herbert Bishop
Kemper, Gene Allen
Kimes, Thomas Fredric
Klip, Dorothea A
Konen, Harry P
Kowalik, Janusz Szczesny
Kurowski, Gary John
Kurtz, Lawrence Alfred
Lashley, Gerald Ernest
Leung, Kin-Vinh
Liniger, Werner
Long, Paul Eastwood, Jr
Lonseth, Arvid Turner
Lowell, Sherman Cabot
Lucey, Juliana Margaret
Macon, Nathaniel
Malcolm, Michael Alexander
Manohar, Rampurkar
Mansfield, Lois E
Manteuffel, Thomas Albert
McCarthy, Mary Anne
McCormick, Stephen Fahrney
McGehee, Ralph Marshall
McKinney, Alfred Lee
McKinney, Earl H
McWilliams, Gerald Vernon
Mercer, Robert J
Meyer, Gunter Hubert
Meyer, Harold David
Morris, John Llewelyn
Munro, William Delmar
Naugle, Norman Wakefield
Nazareth, John Lawrence
Niedereiter, Harald Guenther
Orcutt, Bruce Call
Ortega, James M
Petti, Richard James
Pool, James C T
Poole, William George, Jr
Pope, Wendell LaVon
Preiser, Stanley
Prentice, Wilbert Neil
Pruess, Steven Arthur
Rall, Louis Baker
Ramirez, Enrique Humberto
Rheinboldt, Werner Carl
Rigler, Everett C
Rigler, A Kellam
Rodabaugh, David Joseph
Rose, Donald James
Saylor, Paul Edward
Schechter, Samuel
Schutz, David Harold
Shank, George Deane
Shipman, Jerome Saul
Slutz, Ralph Jeffrey
Smith, Brian Thomas
Sobieszczanski, Jaroslaw Eugeniusz
Solomon, Alan D
Segun, Irene Anne

Stephens, Arthur Brooke
Street, Robert Elliott
Sussman, Myron Maurice
Swartz, Blair Kinch
Tapia, Richard
Thacher, Henry Clarke, Jr
Ting, Annsheng Chien
Todd, John
Tornheim, Leonard
Treat, Charles Herbert
Tsan, Alice Tung-Hua
Uhetka, David Jerome
Usmani, Riaz Ahmad
Vastano, Andrew Charles
Voigt, Robert Gary
Ward, Robert Cleveland
Warner, Daniel Douglas
Watson, Velvin Richard
Wheeler, Mary Fanett
Witzgall, Christoph Johann

Operations Research
Abrams, John Werner
Abramson, Stanley L
Balachandran, Kashi Ramamurthi
Bare, Barry Bruce
Barker, William Hamblin, II
Beck, Robert Edward
Beckwith, Richard Edward
Belkin, Barry
Berghofer, Fred G
Bhat, Uggappakodi Narayan
Bialas, Wayne Francis
Billera, Louis Joseph
Bixby, Robert Eugene
Blakemore, George Jefferson, Jr
Bland, Robert Gary
Blumstein, Alfred
Bolmarcich, Joseph John
Bolstein, Arnold Richard
Boodman, David Morris
Borsting, Jack Raymond
Bossard, David Charles
Bottoms, Albert Maitland
Bracken, Jerome
Bradley, Hugh Edward
Brown, Robert Goodell
Bryan, Carroll William
Bryson, Marion Ritchie
Burke, Paul J
Butler, Charles Morgan
Carstens, Allan Matlock
Castater, Robert Dewitt
Caywood, Thomas E
Chaney, Robin W
Chasalow, Ivan G
Chatterjee, Samprit
Cinlar, Erhan
Clementson, Gerhardt C
Cline, James Edward
Clough, Donald J
Coad, L Keith
Coile, Russell Cleven
Conway, Richard Walter
Cook, William H
Cottle, Richard W
Craw, Alexander R
Cullen, Daniel Edward
Cushen, Walter Edward
Davis, Edward Alex
Dawson, Reed
Dean, Burton Victor
De Cicco, Henry
Derman, Cyrus
Desiderio, Anthony Michael
Dockery, John T
Dresch, Francis William
Dreyfus, Stuart Ernest
Drobnies, Saul Isaac
Dunlap, Duane Sherbert
Dwyer, Wendell Arthur
Eaves, Burchet Curtis
Eckler, Albert Ross
Egner, Donald Otto
Eisner, Mark Joseph
Emlet, Harry Elsworth Jr
Emmons, Hamilton
Englund, John Arthur
Enns, Ernest Gerhard
Epstein, Benjamin
Evans, Leonard
Fain, William Wharton
Farquhar, Peter Henry
Fask, Alan S
Fay, Edward Allen
Feibes, Walter
Finkel, Peter William
Finkelstein, Robert
Fishman, George Samuel
Forrest, Robert Neagle
Fromovitz, Stan
Freeman, Raoul James
Gilliam, Ronald Roy
Gladstone, Robert Jay
Golub, Abraham
Goor, Charles G
Grainger, Robert Moore

Greene, Joseph S
Gregory, James McKanna
Grosh, Doris Lloyd
Gross, Donald
Grossling, Bernardo Fruedenburg
Gutjahr, Allan L
Haake, James Walter
Haering, George
Haines, Larry Kent
Halsey, John Joseph
Hance, Paul D
Hannan, Edward Lees
Hanson, Morgan A
Harbus, Fredric Ira
Hardy, William Christopher
Hare, Robert Ritzinger, Jr
Harris, Carl Matthew
Harrison, Don Edward, Jr
Hartman, James Kern
Haussmann, R K Dieter
Hayes, Patrick Louis
Hazelwood, Robert Nichols
Hearne, Horace Clark, Jr
Heilenday, Frank W
Heinberg, Milton
Helbush, Robert Edwin
Helly, Walter S
Hertweck, Gerald
Hightower, James K
Hildebrand, David Kent
Hilldrup, David J
Hiller, Robert Ellis
Hillier, Frederick Stanton
Hirsch, Ernest
Hodge, Donald Ray
Hoffman, Gerald M
Hoffman, Oscar Allen
Hoisington, Laurence Earl
Holz, Betty Weber
Honig, John Gerhart
Hooke, John Allen
Hopkins, Nigel John
Howard, Gilbert Thoreau
Hu, Te Chiang
Hudson, John Leslie
Hull, Bradley Zangerle
Hunter, Kenneth M
Hurt, James Joseph
Iglehart, Donald Lee
Ignall, Edward J
Ignizio, James Paul
Jackson, Barbara Bund
Jacobs, Patricia Anne
Jeroslow, Robert G
Johnson, Ellis Lane
Johnson, Howard Arthur, Sr
Johnson, Russell Dee, Jr
Johnson, Stuart Earl
Johnson, William Lee
Johnsrud, Alan Edwin
Kaplan, David Jeremy
Kapos, Ervin
Kaslow, David Edward
Katcher, David Abraham
Kaufman, Sol
Keeler, Emmett Brown
Keeney, Ralph Lyons
Keilson, Julian
Klein, Morton
Klingman, Darwin Dee
Knapp, Harold Anthony, Jr
Kneece, Roland Royce, Jr
Knight, John Cian
Koenigsberg, Ernest
Kolesar, Peter John
Korsh, James F
Kortanek, Kenneth O
Kreiner, Howard William
Kromer, Ralph Eugene
Kshirsagar, Anant Madhav
Kunce, Henry Warren
Kupperman, Robert Harris
Landry, Preston Myles
Laughlin, James Stanley
Law, Cecil E
Lea, James Dighton
Leckie, Donald Stewart
Leibowitz, Martin Albert
Leight, Walter Gilbert
Levine, Eugene
Lewinson, Victor A
Liang, Tung
Lieberman, Gerald J
Liebman, Judith Stenzel
Lindsey, George Roy
Lipshutz, Nelson Richard
Little, John Dutton Conant
Lodato, Michael W
Lowry, Philip Holt
Lucas, William Franklin
Lyons, Joseph Paul
Macon, Nathaniel
Malcolm, Janet May
Maloney, James Eugene
Marlowe, William Henry
Marshall, Kneale Thomas
Martin, James John, Jr
Martin, LeRoy Brown, Jr

Martino, Joseph Paul
Martz, Harry Franklin, Jr
Massel, Gary Alan
Maxim, Leslie Daniel
Maxwell, William L
May, Donald Curtis, Jr
Mayberry, John Patterson
Mazumdar, Mainak
McCallum, Charles John, Jr
McGee, Edward Arthur
McMasters, Alan Wayne
McMillan, Louis Kelly, Jr
Mihram, George Arthur
Milch, Paul R
Miles, Ralph Fraley, Jr
Mitchell, Theodore
Moore, Bill C
Morey, Richard Carl
Morin, Thomas Lee
Morris, Peter Alan
Morse, Philip McCord
Moses, Edward Joel
Moss, Ronnie Lee
Murad, Emil Moise
Murty, Katta Gopalakrishna
Naddor, Eliezer
Nair, Sreekantan S
Nance, Richard E
Nemhauser, George L
Neuhardt, John Bernard
Neuts, Marcel Fernand
Oberbeck, Thomas Edmond
O'Dell, Jean Marland
Offutt, William Franklin
Orden, Alex
Oren, Shmuel Shimon
Ott, Teunis Jan
Overholt, John Lough
Owen, Joel
Papadopoulos, Alex Spero
Paskert, Paul F
Paxson, Edwin Woolman
Peterson, Elmor Lee
Pewitt, Nelson Douglas
Philipson, Lloyd L
Pilet, Stanford Christian
Pina, Eduardo Isidorio
Pipino, Raymond Joseph
Pitman, George Rubi
Pollock, Stephen M
Poltorak, Andrew Stephen
Potthoff, Richard Frederick
Powell, Bruce Allan
Powers, Joseph Edward
Pugh, George Edgin
Puri, Prem Singh
Qureishi, A Salam
Randolph, Paul Herbert
Rao, Valluru Bhavanarayana
Reardon, William Albert
Reierson, James (Dutton)
Reisman, Arnold
Richards, Francis Russell
Riley, John Astwood
Ritter, Klaus Guenter
Roark, Adelbert Lee
Robinson, Richard Carleton, Jr
Robinson, Stephen Michael
Rockower, Edward Brandt
Rosenthal, Robert Wernick
Roth, Walter John
Rothkopf, Michael H
Rowen, John William
Rowntree, Robert Frederic
Russo, Richard F
Sakaguchi, Dianne Koster
Schaefer, Brian Morris
Schaffer, Marvin Baker
Schaniel, Carl L
Scheuer, Ernest Martin
Schmidt, John Richard
Schneider, Alfred Marcel
Schultz, Hilbert Kenneth
Schwartz, Benjamin L
Schweitzer, Paul Jerome
Scranton, Bruce Edward
Secrest, Everett Leigh
Segers, Richard George
Sengupta, Sailes Kumar
Serfozo, Richard Frank
Sessler, John Charles
Shaw, Leonard G
Shawhan, Gerald L
Shedler, Gerald Stuart
Sheehy, Myles Joseph
Sherbrooke, Craig C
Shere, Kenneth David
Sherman, Gordon R
Sherman, Seymour
Shershin, Anthony Connors
Shibe, Abe Jeffrey
Shimi, Ismail Nabih
Shock, Robert Charles
Shorr, Bernard
Shubert, Bruno Otto
Sielken, Robert Lewis, Jr
Singer, Arnold
Singer, Emanuel

Singpurwalla, Nozer Drabsha
Smallwood, Richard Dale
Smith, Ernest Lee, Jr
Smith, Malcolm
Smith, Stephen Allen
Smith, Wray
Soland, Richard Martin
Spaht, Carlos G, II
Sposito, Vincent Anthony
Steinwachs, Donald Michael
Stewart, William Charles
Stinson, Perri June
Stone, Lawrence David
Stover, James Anderson, Jr
Sutherland, William Harrison
Svilokos, Nikola
Swann, Dale William
Tam, Kwok-Wai
Terry, Herbert
Theusch, Colleen Joan
Thomas, Clayton James
Thomas, Ralph Edward
Thomas, Ronald Emerson
Tolle, Jon Wright
Trauring, Mitchell
Tullier, Peter Marshall, Jr
Turoff, Murray
Tysver, Joseph Bryce
Van Slyke, Richard M
Veinott, Arthur Fales, Jr
Verkhovsky, Boris Samuel
Verkler, Robert Charles
Visco, Eugene Paul
Wagner, Gerald Richard
Wakeley, Jay Townsend
Walker, Terry M
Walker, Warren Elliott
Ware, Glenn Oren
Wasielewski, Paul Francis
Wheeler, Alan Clement
White, William Wallace
Whitt, Ward
Wiener, Howard Lawrence
Wilkinson, William Lyle
Witzgall, Christoph Johann
Woisard, Edwin Lewis
Wolfe, Harry Bernard
Wolman, Eric
Wood, Robert E
Wright, Gordon Pribyl
Yantis, Richard P
Yaspan, Arthur
Young, John Paul
Yu, Greta Y
Yudowitch, Kenneth Louis
Zadeh, Norman
Zindler, Richard Eugene
Zweig, Hans Jacob

Probability
Fabens, Augustus Jerome
Pruitt, William Edwin

Pure Mathematics
Abian, Alexander
Agarwal, Arun Kumar
Alf, Carol Jean
Allard, William Kenneth
Arnold, Bradford Henry
Ash, J Marshall
Ball, Joseph Anthony
Barker, William Hamblin, II
Barratt, Michael George
Billigheimer, Claude Elias
Bose, Anil Kumar
Brace, John Wells
Camillo, Victor Peter
Carlson, James Andrew
Carruth, James Harvey
Ceder, Jack G
Chen, Yuh-Ching
Childress, Denver Ray
Cohen, Joel M
Cross, George Elliot
Davis, Robert Clay
Debnath, Lokenath
Deever, David Livingstone
Deveney, James Kevin
Dixon, John Douglas
Dlab, Vlastimil
Dwinger, Philip
Ellers, Erich Werner
Ellis, James Watson
Faires, Barbara Trader
Fell, James Michael Gardner
Fillmore, Peter Arthur
Fitzpatrick, Patrick Michael
Flanders, Harley
Friedell, John C
Fuchs, Wolfgang Heinrich Johannes
Gandhi, Jeet-Mal
Garner, Cyril Wilbur Luther
Gilpin, Michael James
Goes, Gunther Walter
Gonshor, Harry
Goodman, Victor Wayne
Goodner, Dwight Benjamin
Goodwyn, L Wayne

Hall, Dick Wick
Halmos, Paul Richard
Halperin, Miriam Patricia
Hamstrom, Mary-Elizabeth
Harrell, Ronald Earl
Hart, Lawrence Alan
Heilbronn, Hans Arnold
Henrich, Christopher John
Hopf, Eberhard
Husain, Taqdir
Idowu, Elayne Arrington
Kaul, S K
Koehler, Donald Otto
Laha, Radha Govinda
Larcher, Heinrich
Lick, Dale W
Longyear, Judith Querida
Lowig, Henry Francis Joseph
Luther, Norman Y
Mahler, Kurt
Makar, Boshra Halim
Mauldon, James Grenfell
Metzger, Thomas Andrew
Metzler, Richard Clyde
Meyer, Paul Richard
Morgan, John Clifford, II
Nemitz, William Charles
Novodvorsky, Mark Evgenievich
Ornstein, Donald Samuel
Pearson, Terrance Laverne
Pelletier, Joan Wick
Phillips, Ralph Saul
Protter, Philip Elliott
Puttaswamaiah, Bannikuppe M
Rao, Veldanda Venugopal
Ray-Chaudhuri, Dwijendra Kumar
Riley, James Daniel
Roeder, David William
Sabharwal, Ranjit Singh
Sahney, B N
Saworotnow, Parfeny Pavolich
Shawyer, Bruce L R
Shields, Paul Calvin
Simms, Nathan Frank, Jr
Singer, Isadore Manual
Siu, Yum-Tong
Slye, John Marshall
Speiser, Robert David
Srivastava, Tariq Naseer
Stehney, Ann Kathryn
Stokes, Arnold Paul
Stone, Arthur Harold
Tolimieri, Richard
Torrance, Ellen McCormick
Treves, Jean Francois
Tromba, Anthony Joseph
Wagreich, Philip Donald
Wei, Diana Yun Dee
Wetzig, Calvin Ulysses
Wilken, Donald Rayl
Yang, Chung-Chun
Young, Laurence Chisholm

Symbolic Logic
Addison, John West, Jr
Alton, Donald Alvin
Andrews, Peter Bruce
Applebaum, Charles H
Asenjo, Florencio Gonzalez
Baldwin, John Theodore
Barwise, Kenneth Jon
Blum, Lenore Carol
Boone, William Werner
Bowen, Kenneth Alan
Burton, Charles R
Cannonito, Frank Benjamin
Catlin, Seth
Chang, Chen Chung
Chapin, Edward William, Jr
Clay, Robert Edward
Cogan, Edward J
Comer, Stephen Daniel
Cunningham, Ellen M
Curry, Haskell Brooks
Daigneault, Aubert
Davis, Martin (David)
De Lillo, Nicholas Joseph
Ellentuck, Erik
Enderton, Herbert Bruce
Fabian, Robert John
Ferrante, Jeanne
Friedman, Harvey Martin
Gaal, Ilse Lisl Novak
Gersting, Judith Lee
Gilmore, Paul Carl
Godel, Kurt
Hailperin, Theodore
Halperin, James Daniel
Harrop, Ronald
Hatcher, William S
Hay, Louise (Schmir)
Hechler, Stephen Herman
Henkin, Leon (Albert)
Henson, C Ward
Hill, Victor Ernst
Hodes, Louis
Jacobson, Ned LeRoy
Jeroslow, Robert G

Jockusch, Carl Groos, Jr
Keisler, Howard Jerome
Kino, Akiko
Kleene, Stephen Cole
Kopperman, Ralph David
Kreider, Donald Lester
Kugler, Lawrence Dean
Leggett, Anne Marie
Lercher, Bruce L
Lerman, Manuel
Manaster, Alfred B
Marsh, William Ernest
Mendelson, Elliott
McNulty, George Frank
McCloskey, Teresemarie
Metakides, George
Morley, Michael Darwin
Nelson, David
Mendenhall, Robert Vernon
Olin, Philip
Owings, James Claggett, Jr
Penk, Anna Michaelides
Platek, Richard Alan
Rescher, Nicholas
Robinson, Julia (Bowman)
Rosenstein, Joseph Geoffrey
Schlipf, John Stewart
Schneider, Hubert H
Seldin, Jonathan Paul
Shoft, Alan McKean
Snapper, Ernst
Stark, William Richard
Takeuti, Gaisi
Tamburino, John
Thomason, Steven Karl
Tierney, Myles
Wheeler, William Hollis
Wiebe, Richard Penner
Wolf, Carol Euwema
Wolf, Robert Stanley
Wood, Carol Saunders
Young, Paul Ruel
Zuckerman, Martin Michael

Systems Theory
Adomian, George
Belanger, David Gerald
Benesch, Samuel Eli
Bevis, Jean Harwell
Chu, Kai-Ching
Cohen, Elaine
Goodall, Marcus Campbell
Jameson, William J, Jr
Krener, Arthur James
LaSalle, Joseph Pierre
Linn, John Charles
Marder, Stanley
Midgley, James Eardley
Myers, John Martin
Pattee, Howard Hunt, Jr
Robertshaw, Joseph Earl
Rosenberg, Arnold Leonard
Thornton, Kent W
Wolf, Eric W

Topology
Akasaki, Takeo
Andrew, David Robert
Antonelli, Peter Louis
Armentrout, Steve
Bajaj, Prem Nath
Baildon, John David
Baker, John Warren
Bellamy, David P
Bendersky, Martin
Berard, Anthony D, Jr
Biefko, Robert L
Boes, Eldon C
Born, Alfred Ervin
Bousfield, Aldridge Knight
Brook, Robert B
Broussard, Lois Mary
Browder, William
Brown, Edgar Henry, Jr
Brown, Marianne
Brown, Robert Freeman
Burke, Dennis Keith
Cameron, Douglas Ewan
Cantrell, James Cecil
Cappell, Sylvain Edward
Case, James Hughson
Christenson, Charles O
Christoph, Francis Theodore, Jr
Church, Philip Throop
Cobb, John Iverson
Cochran, Allan Chester
Conrad, Bruce
Cook, David Edwin
Cook, Howard
Craggs, Robert F
Cude, Joe E
Daverman, Robert Jay
Davis, Donald Miller
Davis, Harvey Samuel
Day, Jane Maxwell
Dickman, Raymond F, Jr
Douglas, Roy Rene

Doyle, Patrick H
Dristy, Forrest E
Durham, Harvey Ralph
Duvall, Paul Frazier, Jr
Eisenberg, Murray
Fearnley, Lawrence
Feldman, Louis A
Finney, Ross Lee
Fletcher, Peter
Floyd, Edwin Earl
Frank, David Lewis
Franklin, Stanley Phillip
Freiwald, Ronald Charles
Gemignani, Michael C
Gersten, Stephen M
Glaser, Leslie
Gorda, George Rudolph, Jr
Grace, Edward Everett
Gropen, Arthur Louis
Gugenheim, Victor Kurt Alfred
Haddock, Aubura Glen
Hager, Anthony Wood
Hagopian, Charles Lemuel
Halfar, Edwin
Handel, William Anthony
Hansen, John Kenneth
Harris, William Emery
Haver, William Emery
Hechler, Stephen Herman
Hempel, John Paul
Hildebrant, John A
Himmelberg, Charles John
Horelick, Brindell
Houghton, Charles Joseph
Howell, Leonard Rudolph, Jr
Husch, Lawrence S
Jaco, William Howard
Jaworowski, Jan W
Jerrard, Richard Patterson
Jobe, John M
Jones, Floyd Burton
Jungek, Gerald Frederick
Kaminker, Jerome Alvin
Keesling, James Edgar
Kent, Joseph Francis
Kinoshita, Shin'ichi
Klein, Albert Jonathan
Knill, Ronald John
Kraines, David Paul
Ku, Hsu-Tung
Ku, Mei-Chin Hsiao
Kumpel, Paul Gremminger, Jr
Lambert, Howard W
Lane, Ernest Paul
Lang, George E, Jr
Latch, Dana May
Lea, James Wesley, Jr
Lee, Ronnie
Lehner, Guydo R
Lelek, Andrew Stanislaus
Levine, Jerome Paul
Levy, Ronald Fred
Lin, Shwu-Yeng Tzen
Lin, You-Feng
Malm, Donald E G
Mann, Benjamin Michael
Mann, Larry N
Mansfield, Maynard Joseph
Marty, Roger Henry
McAuley, Patricia Tulley
McCoy, Robert A
McMillan, Daniel Russell, Jr
Meyer, Paul Richard
Meyers, Philip Robert
Meyerson, Mark Daniel
Michael, Ernest Arthur
Michelsohn, Marie-Louise
Mielke, Marvin V
Miller, Gary Glenn
Montgomery, Deane
Mooljavkar, Suresh Hiraji
Moran, Daniel Austin
Morris, Joseph Richard
Murdeshwar, Mangesh Ganesh
Nagata, Jun-Iti
Naimpally, Somashekhar Amrith
Nel, Louis Daniel
Nyikos, Peter Joseph
O'Brien, Thomas V
O'Neill, Edward John
O'Neill, Ronald C
Patty, Clarence Wayne
Pedersen, Katherine L
Peltier, Charles Francis
Penney, David Emory
Pixley, Carl Preston
Plunkett, Robert Lee
Porter, Jack R
Powers, Michael Jerome
Price, Thomas Munro
Puckett, William Thomas, Jr
Rayburn, Marlon Cecil, Jr
Reddy, William L
Rhee, Choon Jai
Rigdon, Robert David
Rogers, Jack Wyndall, Jr

Roitman, Judy
Rosenholtz, Ira N
Rowlett, Russell Johnston, III
Rubin, Leonard Roy
Rushing, Thomas Benny
Ryeburn, David
Sacks, Jonathan
Sanderson, Donald Eugene
Schneider, Harold William
Schori, Richard M
Schultz, Reinhard Edward
Segal, Jack
Seidman, Stephen Benjamin
Sher, Richard B
Shulman, Herbert Byron
Singer, William Merrill
Singh, Sukhjit
Sinkinson, Kathleen
Sloyer, Clifford W, Jr
Smith, Alton Hutchison
Smith, Carol McDonald
Smith, Spurgeon Eugene
Stallings, John Robert, Jr
Stancl, Mildred Luzader
Starbird, Michael Peter
Stern, Ronald John
Stevenson, Nell Elizabeth
Stone, Ellen Rose
Stubblefield, Beauregard
Su, Jin-Chen
Tangora, Martin Charles
Thedford, William Andrew
Thomas, Barbara Smith
Thomas, Paul Emery
Thomeier, Siegfried
Varadarajan, Kalathoor
Vaughan, Jerry Eugene
Venema, Gerard Alan
Verma, Sadanand
Vought, Eldon Jon
Voxman, William L
Wagner, Neal Richard
Wan, Yieh-Hei
Wang, Chin San
Wenner, Bruce Richard
Wicke, Howard Henry
Wilkerson, Clarence Wendell, Jr
Williams, Robert Fones
Williamson, Robert Emmett
Woodruff, Edythe Parker
Wright, Alden Halbert
Wright, David Grant
Wright, Thomas Perrin, Jr
Yang, Jeong Sheng

MECHANICS & FLUID MECHANICS

Mechanics
Allen, Mildred
Antman, Stuart S
Berstein, Barry
Chow, Tsu-Sen
Desai, Rashmi C
De Zeeuw, Carl Henri
Elter, John Frederick
Ferguson, Harry
Foster, Irving Gordon
Gent, Alan Neville
Graham, George Alfred Cecil
Gurtin, Morton Edward
Kolsky, Herbert
Latady, William Robertson
Nicholson, David William
Phillips, Owen M
Sacks, Alvin Howard
Shield, Richard Thorpe
Smith, Gerald Francis
Sternberg, Eli
Stoker, James Johnston
Takagi, Shunsuke
Tasi, James
Washa, George William
Wehausen, John Vrooman
Weil, Nicholas A
Wetenkamp, Harry R
Williams, William Orville

Fluid Mechanics
Agee, Ernest Mason
Anderson, Roland Carl
Bennett, Frederick Dewey
Blythe, Philip Anthony
Bohachevsky, Ihor O
Buckmaster, John David
Burgers, Johannes Martinus
Childress, William Stephen
Coleman, Neil Lloyd
Fox, Ronald Lee
Fultz, Dave
Gans, Roger Frederick
Jacobs, Stanley J
Kim, Yong Wook
Krishnamurti, Ruby Ebisuzaki
Nigam, Lakshmi Narayan

Classical Mechanics
Brackenridge, John Bruce
Dragt, Alexander James

Biomechanics
Becker, Roland Frederick
Chandran, Krishnan Bala
Eickelberg, W Warren B
Farah, Jean William
Hodgson, Voigt R
Kazarian, Leon Edward
Lardner, Thomas Joseph
Liu, Young King
Roberts, David
Rubinow, Sol Isaac
Stern, Jack Tuccur, Jr

Continuum Mechanics
Beatty, Millard Fillmore, Jr
Coleman, Bernard David
Fife, Paul Chase
Huilgol, Raja Ramesh
Jawa, Manjit S
Kearsley, Elliot Armstrong
Kydonieus, Anastasios D
Noll, Walter
Orth, William Albert
Puri, Pratap
Trulio, John George
Wang, Chao-Cheng

Dynamics
Bailey, Cecil Dewitt
Charvonia, David Alan
Duvall, Paul Frazier, Jr
Evvard, John Cooper
Gordon, Joseph Grover, II

Applied Mechanics
Amazigo, John C
Armen, Harry A, Jr
Bahar, Leon Y
Benjamin, Roland John
Burns, Grover Preston
Caughey, Thomas Kirk
Chang, Chang-Sun
Chen, Paul Ear
Chow, Pao Liu
Chu, William Wei-Ling
Conners, Gary Hamilton
Cowin, Stephen Corteen
Danielson, Donald Alfred
Deverall, LaMar Ivan
Davison, Lee Walker
Ehlers, Francis Edward
Erdogan, Fazil
Eringen, Ahmed Cemal
Frasier, John T
Garber, Theodore Bruce
Garnet, Hyman
Goldberg, Martin A
Gurel, Okan
Gurel, Oktay
Hansmann, Douglas R
Haug, Edward J, Jr
Huseyin, Koncay
Jones, Orval Elmer
Junger, Miguel Chapero
Kazakia, Jacob Yakovos
Kehle, Ralph Ottmar
Klingbeil, Werner Walter
Langhaar, Henry Louis
Lechey, Patrick
Lind, Niels Christian
Mallett, Russell Lloyd
Mann, James Edward, Jr
Mansinha, Lalatendu
Mesmer, Gustav
Nemergut, Paul Joseph, Jr
New, John Calhoun
Parkinson, Truman David
Pindera, Jerzy Tadeusz
Prager, William
Schmiedeshoff, Frederick William
Schuerch, Hans
Snowdon, John Colin
Sobieszczanski, Jaroslaw Eugeniusz
Stippes, Marvin C
Strawderman, Wayne Alan
Tai, Clement Leo
Tang, James Juh-Ling
Tiersten, Harry Frank
Tsui, Yeong Ging
Warren, William Ernest
Weigle, Robert Edward

Rahman, Matiur
Reddy, Kapuluro Chandrasekhara
Saffman, Philip Geoffrey
Selwyn, Philip Alan
Somerscales, Euan Francis Cuthbert
Sonnerup, Bengt Ulf Osten
Stegen, Gilbert Rolland
Su, Chau-Hsing
Turcotte, Donald Lawson
Vrebalovich, Thomas
Wallace, James
Waters, Elmer Dale
Weinstock, Jerome

Mallett, Russell Lloyd

Fluid Dynamics
Amsden, Anthony Avery
Arya, Satya Pal Singh
Barcilon, Victor
Barker, Steven Joseph
Barnes, Derek A
Blumen, William
Butler, Thomas Daniel
Cladis, Patricia Elizabeth Ruth
Cline, Michael Castle
Cloutman, Lawrence Dean
Cunningham, Mary Elizabeth
Daly, Bartholomew Joseph
Deem, Gary Spencer
Duff, Russell Earl
Emrich, Raymond Jay
Fohl, Timothy
Folse, Raymond Francis, Jr
Frenkiel, Francois N
Gibson, Carl H
Gilman, Peter A
Harlow, Francis Harvey, Jr
Haussling, Henry Jacob
Hector, David Lawrence
Ho, Hung-Ta
Hoge, Harold James
Howard, Louis Norberg
Hussey, Robert Gregory
Ingard, Karl Uno
Jackson, Roscoe George, II
Jacobs, Sigmund James
Johnson, David Harley
Klein, Milton M
Koschmieder, Ernst Lothar
Kroll, John Ernest
Kuo, Hsiao-Lan
Laurmann, John Alfred
McCartney, Michael Scott
Miksad, Richard Walter
Mjolsness, Raymond C
Mollo-Christensen, Erik Leonard
Mooers, Christopher Northrup Kennard
Ng, Bartholomew Sung-Hong
O'Brien, Vivian
Rehm, Ronald George
Rose, Harvey Arnold
Rosenkilde, Carl Edward
Rossow, Vernon J
Savic, Peter
Scott, Paul Brunson
Seely, Leslie B, Jr
Sheldon, John William
Sirovich, Lawrence
Somerville, Richard Chapin James
Sparks, Cecil Ray
Strehlow, Roger Albert
Suess, Steven Tyler
Travis, John Richard
Van Driest, Edward Reginald
Wearn, Richard Benjamin, Jr
Welge, Henry John
Wyngaard, John C
Yu, James Chun-Ying

Hydrodynamics
Baggett, Lester Marchant
Booker, John Ratcliffe
Browne, Philip Lincoln
Cox, Arthur Nelson
Deal, William E, Jr
Devin, Charles, Jr
Gilman, Peter A
Hunter, James Hardin, Jr
Izzo, Anthony Joseph
Jones, Eric Manning
Kamegai, Minao
Kuo, Albert Yi-Shuong
Langlois, William Edwin
Larsen, Lawrence Harold
Le, Fang An
Le Mehaute, Bernard J
Luecke, Glenn Richard
Marino, Lawrence Louis
McCrory, Robert Lee, Jr
McMaster, William H
Mutschlechner, Joseph Paul
Rattray, Maurice, Jr
Seguin, Fredrick Hampton
Steinberg, Daniel J
Stern, Melvin Ernest
Tachmindji, Alexander John
Torza, Sergio
Witting, James M
Woods, Frank Robert

Hydromechanics
Fang, Ching Seng

Quantum Mechanics
Averill, Frank Wallace
Boyce, Stephen Scott
Brownstein, Kenneth Robert
Corinaldesi, Ernesto
Crater, Horace William
Graham, Roger Neill
Harris, Richard Allan

Innes, Frederick Rush
Kelly, Paul Sherwood
Kikuchi, Tom T
Lamb, Willis Eugene, Jr
Lippmann, Bernard Abram
Macklin, Philip Alan
Moldauer, Peter Arnold
Pack, Russell T
Piziak, Robert
Sabin, John Rogers
Silverman, Jeremiah Nordau

Rheology
Ballman, Richard Lea
Birnboim, Meyer Harold
Byler, Lee Landis, Jr
Cox, David Buchtel
Dodge, James Stanley
Dunning, Robert Lewis
Elliott, John Habersham
Everage, Archie Edward, Jr
Gaskins, Frederick Hudson
Holden, Geoggrey
Hunston, Donald Lee
Huntsberger, James Robert
Kearsley, Elliott Armstrong
Klingbeil, Werner Walter
Kurath, Sheldon Frank
Landel, Robert Franklin
Lin, Otto Chui Chau
Marvin, Robert Sidney
McKinney, John Edward
Mendelson, Robert Allen
Riley, David Waegar
Soussou, Joseph Elias
Weymann, Helmut Dietrich
Wissbrun, Kurt Falke

Solid Mechanics
Abey, Albert Edward
Boyd, William Warren
Burck, Larry Harold
Edelstein, Warren Stanley
Genin, Joseph
Hamstad, Marvin Arnold
Lardner, Robin Willmott
Moon, Francis C
Norwood, Frederick Reyes
Reytblatt, Zinovy V
Senich, Donald
Shu, Larry Steven
Stevens, Alfred Lyman
Stockmann, Volker Erwin
Wan, Frederic Yui-Ming

Statistical Mechanics
Allen, Kenneth Richard
Baker, George Allen, Jr
Barker, John Adair
Bartis, James Thomas
Battle, Ed Len
Bergeron, Kenneth Donald
Bose, Subir Kumar
Bowden, Charles Malcom
Callen, Herbert Bernard
Chandler, David
Coldwell, Robert Lynn
Cooper, Martin Jacob
Corngold, Noel Robert David
Deutch, John Mark
Dorfman, Jay Robert
Dufty, James W
Falk, Harold
Fesciyan, Sezar
Frisch, Harry Lloyd
Garrison, John Carson
Gilmer, George Hudson
Girardeau, Marvin Denham, Jr
Gould, Harvey A
Green, Melville Saul
Gunton, James D
Haggerty, Michael John
Halperin, Bertrand Israel
Harris, Stewart
Hecht, Charles Edward
Helfand, Eugene
Hoegy, Walter R
Huang, Huey Wen
Huerta, Manuel Andres
Jepsen, Donald William
Karo, Douglas Paul
Kikuchi, Ryoichi
Klein, Michael W
Klein, William
Kraichnan, Robert Harry
Krinsky, Samuel
Landau, David Paul
Langer, James Stephen
Leaf, Boris
Mazenko, Gene Francis
McClure, Charles Frederick
McLennan, James Alan, Jr
Millard, Kenneth Young
Montgomery, David Campbell
Moon, Tag Young
Murphy, Thomas James
Nagle, John F
Nichols, William Herbert

Nicoll, Jeffrey Fancher
Pack, Russell T
Pathria, Raj Kumar
Pokrant, Marvin Arthur
Porter, James Colegrove
Rainwater, James Carlton
Ramanathan, Ganapathiagraharam V
Richardson, Robert William
Riedel, Eberhard Karl
Robertson, Harry Stroud
Sak, Joseph
Schieve, William
Servaes, Tahira Minhaj
Shea, Dion Warren Joseph
Siegel, Armand
Siggia, Eric Dean
Stanley, Harry Eugene
Titus, William James
Valleau, John Philip
Vezzetti, David Joseph
Viswanathan, Kadayam Sankaran
Walker, Grayson Howard
Weinberg, Michael C
Wood, William Wayne

Theoretical Mechanics
Cawley, Robert
Denman, Harry Harroun
Deresiewicz, Herbert
Lee, Sung Mook
Liu, Joseph Jeng-Fu
Moss, Marvin Kent
Parry, Myron Gene
Rim, Kwan
Simmons, John Arthur
Smith, Frederick Adair, Jr
Todd, Leonard
Worley, Will J

MEDICINE

Medicine
Abdou, Nabih I
Abell, Murray Richardson
Abelmann, Walter H
Abernathy, Robert Shields
Abildskov, Junior A
Abrams, Archie Adam
Abruzzo, John L
Acheson, George Hawkins
Ackerman, Gustave Adolph, Jr
Adams, John Milton
Adams, John R
Adams, Robert Walker, Jr
Adams Smith, William Nelson
Adelson, Bernard Henry
Aegerter, Ernest E
Agras, William Stewart
Ahrens, Edward Hamblin, Jr
Aisen, Philip
Akutsu, Tetsuzo
Albright, Edwin C
Alexander, Benjamin
Alexander, Carl Stuart
Alexander, Charles Edward, Jr
Alexander, Leslie Luther
Alexander, Ralph William
Allen, Joseph Garrott
Allen, William E, Jr
Allende, Manuel Francisco
Alling, David Wheelock
Allison, Fred, Jr
Alpert, Louis Katz
Alpert, Nelson Leigh
Alpert, Seymour
Amador, Elias
Amidon, Ellsworth Lyman
Anderson, Albert Douglas
Anderson, Donald Grigg
Anderson, Howard Arne
Anderson, Ray Carl
Anderson, Richmond Karl
Andres, Reubin
Andrews, Albert H, Jr
Ankeney, Jay Lloyd
Aponte, Gonzalo Enrique
Aquilina, Joseph Thomas
Archer, John Dale
Archer, Victor Eugene
Arhelger, Roger Boyd
Arias, Irwin Monroe
Arnoldi, Louis B
Aronson, David L
Aronson, Stanley Maynard
Arrick, Myron S
Arrowsmith, William Rankin
Artenstein, Malcolm S
Atwell, Robert James
Auerbach, Oscar
Aufranc, Will H
Auld, Peter A McF
Aurbach, Gerald Donald
Avery, Gordon B
Avery, Mary Ellen
Avioli, Louis
Bacaner, Marvin Bernard

Bach, Sven Aage
Bailey, Richard Elmore
Baisden, Charles Robert
Baker, Robert Norton
Bakke, John Langum
Balint, John Alexander
Ballard, Robert Wilson
Bang, Frederik Barry
Bang, Nils Ulrik
Barach, Richard L
Barclay, William R
Barelare, Bruno
Barger, Abraham Clifford
Barker, Wiley Franklin
Barlow, Charles F
Barnes, Allan Campbell
Barnes, Frederick Walter, Jr
Barnes, William Alexander
Barnett, Eugene Victor
Barnett, Guy Octo
Barnett, Thomas Buchanan
Barnhard, Howard Jerome
Barron, Kevin D
Barry, William Eugene
Barton, Preston Nichols
Bartos, Henry R
Bartter, Frederic Crosby
Baruch, Sulamita B
Bashour, Fouad A
Batchelor, William Henry
Bates, Joseph H
Batsakis, John G
Bauer, Jere Marklee
Baum, Gerald L
Bauman, Norman
Baxter, Donald William
Bayrd, Edwin Dorrance
Beall, Arthur Charles, Jr
Bean, William Bennett
Bearn, Alexander Gordon
Bechtol, Lavon Dee
Becker, David Victor
Becker, Joshua A
Becker, Samuel William, Jr
Beckloff, Gerald Lee
Beecher, Henry Knowles
Beem, John Raymond
Beeson, Paul Bruce
Beierwaites, William Henry
Beigelman, Paul Maurice
Bell, Alfred Lee Loomis, Jr
Bell, Robert Lloyd
Bender, Leonard Franklin
Bender, Merrill Arthur
Benenson, Abram Salmon
Benner, Ernest Jack
Bennett, Hugh Deveraux
Bennison, Bertrand Earl
Benson, Walter Russell
Benton, Joseph George
Berard, Costan William
Berenberg, William
Berger, Eugene Y
Berger, Lawrence
Berger, Sheldon
Bergman, Abraham
Bergman, Russel Theodore
Bergnes, Manuel
Bering, Edgar Andrew, Jr
Berk, Jack Edward
Berkman, James Israel
Berkson, David M
Bernhard, Victor Montwid
Bernstein, Lionel M
Bernstein, Sol
Berry, Leonidas Harris
Berson, Robert Chambliss
Berthrong, Morgan
Bertino, Joseph R
Best, Maurice McDonald, Jr
Bettman, Jerome Wolf
Betts, Henry Brognard
Beutler, Ernest
Bieber, Irving
Bierly, Mahlon Zwingli, Jr
Bierman, Howard Richard
Birt, Arthur Robert
Black, Paul H
Black, Richard Glynn
Bland, John (Hardesty)
Bland, Samuel P W
Bliss, Eugene Lawrence
Blodi, Frederick Christopher
Bloom, Joseph Dutch
Bloomfield, Daniel Kermit
Bloor, Byron Michel
Blum, Leon Leib
Blumberg, Baruch Samuel
Blumenstock, David A
Blumgart, Herrman Ludwig
Blumstein, George I
Boak, Ruth Alice
Boaz, Willard Denton
Bodansky, Oscar
Bodfish, Ralph E
Bogardus, Carl Robert, Jr
Boger, William Pierce

Bolinger, Robert E
Bollet, Alfred Jay
Bolte, Edouard
Bongiovanni, Alfred Marius
Booth, Richard W
Bordley, James, III
Bordley, Richard W
Bornmann, Robert Clare
Bornstein, Paul
Bosomworth, Peter Palliser
Bowling, Franklin Lee
Boyer, Norman Howard
Boyer, Philip A, Jr
Boylan, John W
Boyle, Edwin
Brachman, Philip Sigmund
Bradley, Edward Charles
Bradley, Stanley Edward
Brady, Luther Weldon, Jr
Bragg, David Gordon
Branch, Charles Franklin
Brand, Leonard
Brandaleone, Harold
Brandenburg, Robert O
Braude, Abraham Isaac
Braverman, Irwin Merton
Bray, George A
Brayer, Franklin T
Bream, Charles Anthony
Brem, Thomas Hamilton
Brennan, Michael James
Bressler, Bernard
Brest, Albert N
Brigagman, Robert Alan
Briggs, Arthur Harold
Briggs, Donald K
Brightman, I Jay
Brill, A Bertrand
Brindley, Clyde Owens
Briscoe, William Alexander
Bristow, John David
Brody, Bernard B
Brogden, Byron Gilliam
Bronstein, Eugene L
Brosin, Henry Walter
Brown, David Frederick
Brown, Ellen
Brown, Elmer Burrell
Brown, Harold
Brown, Herbert Eugene
Brown, John Welch
Brown, Josiah
Brown, Robert Calvin
Browning, Torrey Carl
Browning, Robert Hamilton
Bruce, David Lionel
Bruce, Robert Arthur
Brues, Austin M
Brumlik, Joseph V
Bryan, George Terrell
Bryans, Alexander (McKelvey)
Bryant, Lester Richard
Buchan, Ronald Forbes
Bucher, Nancy L R
Buhrow, Charles, Jr
Bulkley, George
Bunce, Paul Leslie
Bunker, John Phillip
Bunnell, Ivan Lee
Burch, Robert Ray
Burchell, Howard Bertram
Burchenal, Joseph Holland
Burke, James Otey
Burko, Henry
Burnell, James McIndoe
Burns, Robert B P
Burr, William Wesley, Jr
Burrell, Craig Donald
Burtis, B Cullen
Burt, Robert C
Buse, Maria F Gordon
Butler, John
Butterworth, Julian Scott
Caceres, Cesar A
Cahill, Kevin M
Camerini-Davalos, Rafael
Cameron, Douglas George
Campbell, James A
Campbell, John Alexander
Canfield, Robert E
Canham, John Edward
Cantarow, Abraham
Cantin, Marc
Capps, Richard Brooks
Capra, J Donald
Carbone, John Vito
Carbone, Thomas E
Cardon, Phillippe Vincent, Jr
Carrasquer, Gasper
Carregal, Enrique Jose Alvarez
Carstens, Herman Paul
Carter, Kenneth
Carter, Paul Richard
Cassady, George
Cassels, Donald Ernest
Catz, Boris
Caulfield, James Benjamin
Chait, Arnold

Chandler, Arthur Cecil, Jr
Chaplin, Hugh, Jr
Chapman, Carleton Burke
Chapman, John S
Charache, Patricia
Charache, Samuel
Chase, Norman E
Chase, Randolph Monteith, Jr
Chasis, Herbert
Chermack, Reuben Mitchell
Childers, Roderick W
Chinn, Austin Brockenbrough
Chirino, Fernando Porfirio
Chisolm, James Julian, Jr
Chiu, G C
Chodos, Robert Bruno
Chown, H Bruce
Christensen, Leonard
Christenson, Paul John
Christopherson, William Martin
Christy, Nicholas Pierson
Chung, Ed Baik
Chung, Choong Wha
Cinotti, Alfonse A
Clark, Charles Malcolm, Jr
Clark, Duncan William
Clark, Lincoln Dutton
Clark, Mervin Leslie
Cleckley, Wallace Henderson, Jr
Clemens, James Jennings
Clemens, Raymond Leopold
Clyde, David F
Cohen, Alan Seymour
Cohen, Sidney
Cole, Harold S
Collen, Morris F
Collins, Vincent J
Compton, Walter Ames
Conan, Neal Joseph, Jr
Conley, Carroll Lockard
Conn, Jerome W
Conn, Rex Boland
Connor, Thomas Byrne
Constantine, Herbert Patrick
Cook, Charles Davenport
Cooke, Charles Robert
Cooper, David Young
Cooper, Gerald Rice
Cooper, Maurice Zealor
Cooper, Robert Arthur, Jr
Cooper, William Clark
Copeland, Bradley Ellsworth
Corden, Brian Joseph
Corder, Clinton Nicholas
Cosby, Richard Sheridan
Cotran, Ramzi S
Cotsonas, Nicholas John, Jr
Cotzias, George Constantin
Courmand, Andre Frederic
Covel, Mitchel Dale
Cowan, Donald William
Craig, Albert Burchfield, Jr
Craig, James William
Craige, Ernest
Crandall, Paul Herbert
Crane, August Reynolds
Crawford, Raymond Bertram
Crawhall, John C
Crepea, Seymour B
Crispell, Kenneth Raymond
Cronkite, Eugene Pitcher
Crosson, John William
Crowley, Leonard Vincent
Cuatrecasas, Pedro
Cucci, Cesare Eleuterio
Culbertson, William Richardson
Cullen, James Henry
Cullen, Stuart Chester
Culo, David Albert
Cummings, Martin Marc
Cunningham, George J
Dack, Simon
Daguillard, Fritz
Dallaire, Louis
Dalmasso, Agustin Pascual
Dalrymple, Glenn Vogt
Dancis, Joseph
Danforth, William H
Danowski, Thaddeus Stanley
Dao, Thomas Ling Yuan
Das Gupta, Tapas Kumar
Daughaday, William Hamilton
Davies, Dean Fletcher
Davies, Richard O
Davignon, Jean
Davis, David A
Davis, Dorland Jones
Davis, Hamilton Seymour
Davis, James Ernest
Davis, Joseph Harrison
Davis, Joyce S
Davis, Robert Paul
Davis, Thomas
Davis, William Duncan, Jr
Day, Stacey Biswas
DeCosse, Jerome J
Deed, Eleanor Polk
De Forest, Ralph Edwin

DeGraff, Arthur Christian
Deisher, Robert William
deJong, Rudolph H
Delabarre, Everett Merrill, Jr
de la Chapelle, Clarence Ewald
De La Huerga, Jesus
Delagi, Edward F
Del Greco, Francesco
DeLor, Camille Joseph
Deming, Quentin Burritt
Demhof, Christopher Harry
Denny-Brown, Derek Ernest
Denson, Judson Samuel
dePeyster, Frederic A
Desforges, Jane Fay
Deuschle, Kurt W
Dexter, Morris W
Diamond, Louis Klein
Dickie, Helen Aird
Dickson, James Francis, III
Dickson, James Gillespie, Jr
Di Ferrante, Nicola Mario
Dillard, Morris, Jr
Dillon, Hugh C, Jr
Doff, Simon David
Dole, Vincent Paul
Donald, William David
Donaldson, James Bowie
Donaldson, Virginia Henrietta
Donnell, Joseph Lawrence
Donnelly, Martin W
Donovan, John Charles
Dorst, John Phillips
Douglas, William Kennedy
Dow, James W
Downey, John A
Dray, Sheldon
Dreifuss, Fritz Emanuel
Dreizen, Paul
Drucker, William Richard
Ducker, Thomas Barbee
Duff, Fratis L
Dumas, Kenneth J
Dunbar, Burdett Sheridan
Duncalf, Deryck
Dunphy, J Englebert
Durkan, James P
Duval, Addison (McGuire)
Dyke, Richard Warren
Dyken, Mark Lewis
Egan, Thomas J
Egan, Richard L
Eggers, George W Nordholtz, Jr
Eiben, Robert Michael
Eich, Robert
Eichna, Ludwig Waldemar
Eickhoff, Theodore C
Eisele, Charles Wesley
Eisenberg, Seymour
Eisenberg, Sidney Edwin
Eitzman, Donald V
Elkinton, Moylan, oseph Russell
Elmore, Stanley Mcdowell
Elsbach, Peter
Elsom, Kendall Adams
Emerson, Charles Phillips
Emerson, Frederick Beauregard, Jr
Emerson, Geraldine Mariellen
Engbring, Norman H
Engel, Eric
Engel, George Libman
Engel, Rudolf
Engle, Ralph Landis, Jr
Engleman, Ephraim Philip
Engler, Harold S
Enneking, William Fisher
Epstein, Frederick Hermon
Epstein, Jeanne Alice
Epstein, Lois Barth
Epstein, William L
Erdman, William James, II
Erecinska, Maria
Erickson, Donald Johan
Erslev, Allan Jacob
Erwin, Chesley Para
Estep, Herschel Leonard
Estes, Edward Harvey, Jr
Estes, Harrison Silas
Evans, John McCallum
Evans, Tommy Nicholas
Evelyn, Kenneth Austin
Everett, Mark Allen
Evers, Carl Gustav
Ewing, William McDaniel
Fairbairn, John F, II
Faludi, Georgina

Fand, Sally Bogolub
Farber, Saul Joseph
Farley, Eugene Shedden, Jr
Fenninger, James Joseph, Jr
Farquhar, John William
Farr, Lee Edward
Farr, Richard Studley
Farrar, John Thruston
Favour, Cutting Broad
Farrow, Joseph Helms
Feder, Walter
Fefter, James Joseph
Feigenbaum, Harvey
Feldman, Daniel Jared
Feldman, Elaine Bossak
Feldman, Clement Alfred
Fine, Samuel
Finch, Bernard Nathan
Finestone, Albert Justin
Finkbeiner, John Aris
Finkel, Asher Joseph
Finkelstein, John Turner
Fisch, Charles
Fischbein, William Nichols
Fischman, Donald A
Fisher, Delbert A
Fisher, John Byron
Fisher, Leonard V
Fisk, Guy Hubert
Fitch, Coy Dean
FitzGerald, Joseph Arthur
Fitzpatrick, Thomas Bernard
Flanagan, Stevenson
Flanigan, William J
Fleeson, William
Fleischer, Clara Joel
Fletcher, Gilbert Hungerford
Flink, Edmund Berney
Flote, C Thomas
Foege, William Herbert
Fogh, Jorgen (Engell)
Foley, William Thomas
Fontana, Vincent J
Ford, Denys Kensington
Fordham, Christopher Columbus, III
Foster, Willis Roy
Fournier, Pierre William
Fox, Arthur Charles
Fox, Samuel Mickle, III
Foye, Laurance V, Jr
Frank, Lawrence
Franke, Frederick Rahde
Franklin, Edward Claus
Fraser, Havelock Frank
Fraser, Robert Gordon
Frawley, Thomas Francis
Frazier, Howard Stanley
Frederickson, Donald Sharp
Freedberg, Abraham Stone
Freedman, Philip
Freeman, Gustave
Frenkel, Ruth Kimmelstiel
Freis, Edward David
Freymann, John Gordon
Friedell, Gilbert H
Friederici, Hartmann H R
Friedland, Fritz
Friedman, Benjamin
Friedman, Irwin
Friedman, Sigmund L
Friesinger, Sydney Murray
Frimpter, Gottlieb Christian
Fritts, Harry Washington, Jr
Frost, John Kingsbury
Fruthaler, George James, Jr
Fuller, Benjamin Franklin, Jr
Furth, Eugene David
Furth, Frank Willard
Fuszard, Barbara
Futcher, Palmer Howard
Gaffney, Paul Cotter
Gasque, Mac Roy
Gastineau, Clifford Felix
Gaudino, Mario
Gault, Neal L, Jr
Gee, J Bernard S
Geer, Jack Charles
Gellis, Sydney Saul
Genton, Edward
Ghadimi, Hossein
Gibson, Count Dillon, Jr
Giesecke, Adolph H
Gifford, Ray Wallace, Jr

Gildenhorn, Hyman L
Gillis, Martin Fern
Gilmore, Hugh Richmond, III
Giovacchini, Rubert Peter
Givler, Robert L
Glaser, Robert Joy
Glaser, Warren
Glass, Leonard
Glezen, William Paul
Glicksman, Arvin Sigmund
Glueck, Helen Iglauer
Goetz, Robert Hans
Gold, Allen
Gold, Martin I
Goldberg, Harry
Goldberg, Irving Hyman
Goldberg, Morton Falk
Goldbloom, Richard B
Goldenberg, Ira Stovin
Goldfien, Alan
Goldsmith, Carl
Goldsmith, Ralph Samuel
Goldsmith, Richard E
Goldstein, Eli
Goldstein, Gerald
Goldstein, Robert
Gollan, Frank
Good, Armin E
Goodgold, Joseph
Goodhart, Robert Stanley
Goodheart, Clyde Raymond
Goodlin, Robert Clair
Gorby, Charles K
Gordan, Gilbert Saul
Gordon, Burgess Lee
Gordon, Douglas Littleton
Gordon, Edgar Stillwell
Gordon, Edward Emanuel
Gordon, Edwin Earl
Gorlin, Richard
Goshen, Charles Ernest
Gottlieb, Abraham Mitchell
Grabill, James Rodney
Gramiak, Raymond
Granger, Carl V
Grant, William Wallace
Grantham, Jared James
Gravenstein, Joachim Stefan
Gray, Frieda Gersh
Gray, Mary Jane
Graybiel, Ashton
Green, Robert Holt
Greenfield, George B
Greenhill, Stanley E
Greenough, William B, III
Grisham, Joe Wheeler
Groel, John Trueman
Grollman, Arthur Patrick
Gromisch, Donald S
Groom, Dale
Groover, Marshall Eugene, Jr
Gross, Jerome
Gross, John Burgess
Gross, Ruth T
Grossman, Jacob
Grover, M Roberts, Jr
Grover, Robert Frederic
Grumbach, Melvin Malcolm
Grynbaum, Bruce B
Gulyassy, Paul Francis
Gum, Oren Berkley
Gunton, Ramsay Willis
Gusberg, Saul Bernard
Gutelius, Margaret Frances
Guthrie, Eugene Harding
Guthrie, Marshall Beck
Guze, Lucien Barry
Gwaltney, Jack Merrit, Jr
Haas, Albert B
Haavik, Arne Goodwin
Habel, Karl
Haft, David Edward
Hagedorn, Albert Berner
Haggard, Mary Ellen
Haig, Pierre Vahe
Hales, Milton Reynolds
Hall, Charles A
Hall, Philip Wells, III
Hall, Thomas Livingston
Hallett, William A
Hallett, Wilbur Y
Halpern, Daniel
Halpern, Myron Herbert
Hamelberg, William
Hammack, William Jack
Hammar, Sherrel L
Hammering, James Solomon

Hanenson, Irwin Boris
Haney, Hance Francis
Hansen, James L
Hanson, John Sherwood
Hanson, Virgil
Hardin, Robert Calvin
Harford, Carl Gayler
Haring, Olga M
Harlan, William R, Jr
Harned, Herbert Spencer, Jr
Harper, Edward O'Neil
Harris, Albert Hall
Harris, Benedict Richard
Harris, Edward Day, Jr
Harris, Henry William
Harris, Herbert H
Harris, Jerome Jones
Harris, Raymond
Harrison, Joan Elizabeth
Harrison, Tinsley Randolph
Hart, Michael Martin
Hartman, James T
Hartz, Jerome
Harvey, Abner McGehee
Harvey, Rejane M
Harvey, Watkins Proctor
Hass, William K
Hasterlik, Robert Joseph
Hastings, Donald Wilson
Hathaway, Beulah M
Haukohl, Robert Somers
Haurani, Farid I
Havel, Richard Joseph
Hayashi, Teruo Terry
Hayes, Guy Scull
Haymovits, Asher
Haynie, Thomas Powell, III
Haywood, Theodore J
Headings, Verle Emery
Heald, Felix Pierpont, Jr
Heinemann, Henry Otto
Heinle, Robert Walter
Heller, Irving Henry
Heller, John Herbert
Heller, John Roderick
Heller, Melvin S
Hellerstein, Herman Kopel
Helpern, Milton
Henderson, Brian E
Henderson, Donald Ainslie
Hendrix, James Paisley
Henkin, Robert I
Henneman, Dorothy
Henneman, Philip Harry
Henry, James Paget
Henry, John Bernard
Henry, Walter Lester, Jr
Hentel, William J
Heptinstall, Robert Hodgson
Herbert, Victor
Herion, John Carroll
Herrmann, John Bellows
Herrmann, Kenneth L
Hewitt, William Lane
Hiatt, Albert Edmund
Hickey, Robert Cornelius
Hiebert, Talmage Gordon
Higgins, Millicent Williams Payne
Hillman, Robert Wright
Hines, Thomas Franklin
Hinkle, Lawrence Earl, Jr
Hinko, Edward N
Hirsch, Jules
Hirsch, Solomon
Hirschowitz, Basil Isaac
Hirst, Albert Edmund
Ho, Monto
Hodges, Fred Jenner, III
Hodges, John Hendricks
Hodgman, Joan Elizabeth
Hoekenga, Mark T
Hoffman, Herbert Saul
Hoffman, Joseph
Hogan, Bartholomew William
Holaday, William J
Holinger, Paul Henry
Holland, Albert Harold, Jr
Holland, James Frederick
Hollenberg, Charles H
Hollenberg, Milton
Hollenhorst, Robert William
Holler, Jacob William
Holley, Howard Lamar
Holling, Herbert Edward
Holman, Charles Nixon
Holmes, Albert William, Jr
Holtkamp, Dorsey Emil
Homburger, Freddy
Hoobler, Sibley Worth
Horlick, Louis
Horn, Henry Joseph
Horner, George John
Horton, Richard
Horwitz, Orville
Hotta, Shoichi Steven
Howard, John Campbell, Jr
Howard, Robert Bruce
Howard, Rufus Oliver

Howell, David Sanders
Howland, William Stapleton
Huang, Kun-Yen
Hubbard, William Neill, Jr
Hubel, Kenneth Andrew
Hudson, James Bloomer
Huestis, Douglas William
Huf, Ernst Gustav
Huffer, Sarah Virginia
Huffines, William Davis
Hughes, Edwin R
Hughes, James Gilliam
Hultgren, Herbert Nils
Humphrey, Arthur Allan
Humphries, J O'Neal
Hunt, George Halsey
Hunt, T E
Hunt, William B, Jr
Hunter, Andrew Tate
Hunter, Oscar Benwood, Jr
Huseby, Robert Arthur
Hussey, Hugh Hudson
Hutaff, Lucile W
Huth, Edward J
Hyatt, George W
Hyde, Henry Van Zile
Hyman, Arthur Bernard
Iber, Frank Lynn
Iezzoni, Domenic G
Iglesias, Rigoberto
Ihrke, Royal Ernest
Imparato, Anthony Michael
Ingbar, Sidney Harold
Irby, William Robert
Irwin, John Wellington
Isard, Harold Joseph
Iseri, Lloyd T
Ishikawa, Sadamu
Isselbacher, Kurt Julius
Jackson, Dudley Pennington
Jackson, Francis Charles
Jackson, Laird G
Jackson, Marvin Alexander
Jacob, Stanley W
Jacobs, Sydney
Jacobson, Leon Orris
Jacobus, David Penman
Jaenike, John Robert
Jaffe, Ernst Richard
James, George Watson, III
James, John Alexander
James, Leonard Stanley
Jameson, Arthur Gregory
Jandl, James Harriman
Janeway, Richard
Janssen, William C
Jarvis, Garth Louray
Jarvis, Jack Reynolds
Javitt, Norman B
Jawetz, ERnest
Jenkins, Daniel Edwards
Jennings, Donald B
Jepson, William W
Johns, Richard James
Johnson, Charles F
Johnson, Fenimore Thomas
Johnson, George, Jr
Johnson, Horton Anton
Johnson, Joseph Richard
Johnson, Murray Leathers
Johnson, Philip Carl
Johnson, Ralph Emil
Johnson, Sture Archie Mansfield
Johnston, Barbara Jane
Jones, Bryant Lee
Jones, Edward Gomer
Jones, Granville Lillard
Jones, John Paul, Jr
Jones, John R
Jones, Oliver William
Jones, Richard Theodore
Jordan, Raymond Ellsworth
Jordan, William Stone, Jr
Jorgens, Joseph
Josephson, Alan S
Joyner, Claude Reuben
Judson, Walter Emery
Julius, Stevo
Juniper, Kerrison, Jr
Kaiser, Raymond Francis
Kallen, Roland Gilbert
Kamner, Mildred Elsie (Mrs Edward M Tolman)
Kane, John Power
Kanof, Abram
Kaplan, Melvin Hyman
Kaplan, Solomon Alexander
Kaplan, Stanley Baruch
Karmen, Arthur
Kasik, John Edward
Kass, Irving
Kattus, Albert Adolph, Jr
Katz, Arnold Martin
Katz, Julius
Katz, Sol
Kauffman, Stuart Alan
Kauffman, Herbert S
Kautz, Harold Douthitt

Kayden, Herbert J
Kaye, Carmen Jimenez
Kaye, Donald
Kayhoe, Donald Ellsworth
Kearns, Owen Austin
Keeler, Martin Harvey
Keliher, Thomas Francis
Kelley, Maurice Leslie, Jr
Kellow, William Francis
Kelly, R Emmet
Kennedy, Jesse Ward
Kennedy, William B
Kerby, Grace Pardridge
Kerr, Andrew, Jr
Kessler, Alexander
Kessler, Richard Howard
Khachadurian, Avedis K
Khalili, Ali A
Kiesewetter, William Burns
Kiley, John Edmund
Kim, Young Shik
Kimbel, Philip
Kimmel, Joe Robert
King, Benton Davis
King, Thomas Creighton
Kirby, William M M
Kirkham, Frederic Theodore, Jr
Kirkland, Richard Horace
Kirkman, Henry Neil
Kirkpatrick, John Arthur, Jr
Kirtzman, Julius
Kissane, John M
Klatzo, Igor
Klavins, Janis Vilberts
Klein, Edmund
Klein, Sidney Wayne
Klemperer, Friedrich Wilhelm
Kline, Irwin Kaven
Klionsky, Bernard Leon
Klipstein, Frederick August
Knies, Phillip Thomas
Knight, Vernon
Knox, Gaylord Shearer
Knox, Walter Eugene
Knudsen, Knud David
Knudson, Alfred George, Jr
Koepke, George Henry
Kohl, Schuyler G
Kohlstaedt, Kenneth George
Kokko, Ukko Pentti
Kolb, Felix Oscar
Kontos, Hermes A
Kopf, Alfred Walter
Korns, Michael Edward
Kory, Ross Conklin
Kossmann, Charles Edward
Koteen, Herbert
Kovacs, Eve Maria
Kozinski, Andrzej
Kraemer, Duane Carl
Krakoff, Irwin Harold
Kramer, Elmer E
Kramer, Philip
Krane, Stephen Martin
Krantz, Kermit Edward
Krasno, Louis Richard
Kraus, Alfred Paul
Krehl, Willard Arthur
Krevans, Julius Richard
Kronenberg, Richard Samuel
Krugman, Saul
Krusen, Edward Montgomery
Kumagai, Lindy Fumio
Kunesh, Jerry Paul
Kunin, Arthur Saul
Kunkel, Henry George
Kurland, Leonard T
Kurnick, Nathaniel Bertrand
Kushner, Irving
Kydd, David Mitchell
Kylstra, Johannes Arnold
Labby, Daniel Harvey
Lacy, William White
Lalezari, Parviz
Lamb, Albert R, Jr
Lamb, Lawrence Edward
Lamdin, Ezra
Lamoureux, Gilles
Landau, Bernard Robert
Landau, Joseph White
Landesman, Robert
Landis, Eugene Markley
Lane, Alexander Z
Langdon, Edward Allen
Langer, Glenn A
Langman, Jan
Langsjoen, Per Harald
Lanthier, Edward Howell
Lanthier, Andre
Lapham, Roger Fulmer
Laragh, John Henry
Laramore, George Ernest
Larson, Frank Clark
Larson, Daniel Lewis
Lashof, Joyce Cohen
Lasser, Elliott Charles
Laster, Leonard
Laszlo, John

Lathem, Willoughby
Lattes, Raffaele
Lau, Francis You King
Lauer, Dolor John
Laupus, William E
Laurenzi, Gustave
Lawrence, Henry Sherwood
Lawrence, John Hundale
Lawrence, Montague Schiele
Laws, E Harold
Lawton, Alexander R, III
Lax, Louis Carl
Lazzara, Ralph
Leach, Robert Ellis
Leavell, Byrd Stuart
Leavitt, Lewis A
Leavitt, Milo David, Jr
Ledsome, John R
Lee, Hyung Mo
Lee, Kyu Taik
Lee, Peter Van Arsdale
Lee, Yien-Hwei
Leer, John Addison, Jr
Lees, Robert S
Leevy, Carroll M
Lefebvre, Rene
Lefkowitz, Lewis Benjamin, Jr
Lefkowitz, Robert Joseph
LeGolvan, Paul Celestin
Lehman, Robert Nathan
Lehman, Roger H
Lehr, David
Leighton, Dorothea Cross
Leiter, Elliot
Lemon, Henry Martyn
Lenkoski, L Douglas
Leon, Robert Leonard
Leonard, Alvin Robert
Leonard, Edward Joseph
Lepper, Mark (Hummer)
le Riche, William Harding
LeRoy, George Veach
Leslie, Stephen Howard
Lester, Roger
Lester, Thomas William, Jr
Levan, Norman E
Levine, Robert Alan
Levine, Bernard Benjamin
Levinsky, Norman George
Levinson, Walter John
Levison, Gilbert E
Levit, Matthew Edmund
Levit, Edithe J
Levy, Harvey Louis
Levy, Robert I
Lewin, Isaac
Lewis, Alan Ervin
Lewis, Charles E
Lewis, Jerry Parker
Lewis, Jessica Helen
Lewis, John Albert
Ley, Allyn Bryson
Leymaster, Glen Ronald
Lezer, Leon Robert
Leznoff, Arthur
Li, Ting Kai
Lichtenstein, Lawrence M
Likoff, William
Lillenfield, Lawrence Spencer
Lillick, Lois Carol
Limarzi, Louis Robert
Lindberg, Howard Avery
Lindsey, Edward Stormont
Linfoot, John Ardis
Linman, James William
Lippmann, Heinz Israel
Lipsitz, Philip Joseph
Lipson, Richard L
Little, James Alexander
Little, Joseph Alexander
Little, Robert Colby
Littlefield, James Beaton
Litman, Armand
Llewellyn-Thomas, Edward
Lockhart, Lillian Hoffman
Loeser, Eugene William
Logothetopoulos, J
Logue, Bruce R
London, S J
London, Irving Myer
Long, Wilmer Newton, Jr
Loop, John Wickwire
Loosli, Clayton G
Lopez-Santolino, Alfredo
Lorber, Stanley H
Lott, James Stewart
Lounsbury, Franklin
Lourenco, Ruy Valentim
Low, James Alexander
Low, Niels Leo
Lowell, Francis Cabot
Lowenstein, Leah Miriam
Lu, Gordon Go
Luessenhop, Alfred John
Luetscher, John Arthur, Jr
Luisada, Aldo Augusto
Lukens, Francis Dring Wetherill

Lumeng, Lawrence
Lupton, Charles Hamilton, Jr
Lynch, Eva Maria
Lynn, Thomas Neil, Jr
Lynn, William Sanford
Lyons, Harold Aloysius
MacDonald, Richard Annis
MacKenzie, Walter Campbell
Mader, Ivan John
Maffly, LeRoy Herrick
Magidson, Oscar
Maha, George Edward
Maher, Frank Thomas
Maher, John Francis
Maier, John G
Majerus, Philip W
Makhlouf, Gabriel Michel
Malkinson, Frederick David
Mallams, John Thomas
Maloney, Mario Salazar
Maloney, William Farlow
Mancini, Robert Eusebio
Manger, William Muir
Manion, Marlow William
Mann, Edward Cullee
Mann, John G
Manning, George William
Manning, Robert Thomas
Marble, Alexander
Marchetta, Frank Carmelo
Marden, Philip Ayer
Margetis, Edward Lambert
Mark, Herbert
Mark, Lester Charles
Marko, Arthur Myroslaw
Marks, Asher
Marks, Charles
Marquart, Philip Butler
Marrack, David
Marston, Robert Quarles
Martin, Carroll James
Martin, Christopher Michael
Martin, Donald Beckwith
Martin, Harold Roland
Martin, James Franklin
Martin, Richard Harvey
Martz, Carl D
Mason, Richard Patrick
Massey, Donald John
Massey, Douglas Gordon
Mastromatteo, Ernest
Mateer, Frank Marion
Matovinovic, Josip
Matsen, John Martin
Mauer, Alvin Marx
Mayer, Florence E
Mayer, William Dixon
McAllister, William H
McAllister, William Barriss, Jr
McCall, Charles B
McClenahan, James Brice
McCloskey, Joseph Francis
McConnell, Jack Baylor
McCoy, Ernest E
McCreary, John Ferguson
McCrumb, Fred Rodgers
McCulloch, Ernest Armstrong
McDermott, Walsh
McDowell, Marion Edward
McDuffie, Frederic Clement
McElfresh, Arthur Edward, Jr
McElin, Thomas (Welsh)
McElroy, Robert C
McEwen, Currier
McGarry, Eleanor E
McGreal, Douglas Anthony
McGuire, Joseph Smith, Jr
McHardy, George Gordon
McIntosh, Hamish William
McIntyre, Patricia Ann
McKusick, Victor Almon
McLaurin, James Walter
McLees, Byron O
McMahon, John Martin
McMaster, Robert H
McMorris, Rex O
McPherson, James C, Jr
McQueen, John Donald
McWhinney, Ian Renwick
McWhorter, Clarence Austin
Meads, Manson
Meador, Clifton Kirkpatrick
Medina, Antonio Samuel
Meeker, C Irving
Meinkejohn, Gordon
Melamed, Myron Roy
Melby, James Christian
Mellin, Gilbert Wylie
Mellinkoff, Sherman Mussoff
Mendelson, Jack H
Menecely, George Rodney
Mengel, Charles E
Menguy, Rene
Merckel, Charles George
Meredith, Jesse Hedgepeth
Merriam, George Rennell, Jr
Merril, Carl R

Merrill, John Putnam
Merrill, Reynold Clurf
Merskey, Clarence
Metcalfe, James
Meyerowitz, Sanford
Meyers, Muriel Charlotte
Mezey, Kalman C
Michelakis, Andrew M
Middlebrook, Gardner
Migliore, Philip Joseph
Mignault, Jean De L
Milam, Denver Franklin
Miles, James S
Millar, Jack William
Miller, Aaron
Miller, Billie Lynn
Miller, George Edward.
Miller, Gerald
Miller, James Roscoe
Miller, Jarrell E
Miller, Jerry Roland
Miller, Max
Miller, Roland Drew
Miller, Roscoe Earl
Miller, William Franklin
Milikan, Clark Harold
Milman, Doris H
Milstoc, Mayer
Minners, Howard Alyn
Mintz, A Aaron
Mitchell, Harold Hugh
Mitchell, John Campbell
Mitchell, Robert Dalton
Mitchell, Roger Sherman
Mithoefer, John Caldwell
Mittelman, Arnold
Moertel, Charles George
Mohler, William C
Moldawer, Marc
Monroe, Russell Ronald
Montalvo, Jose Miguel
Montgomery, Hugh
Montgomery, Max Malcolm
Montreuil, Fernand
Moolten, Sylvan E
Moore, James A
Moore, Thomas Carleton
Moore, Thomas D
Moores, Russell R
Moorhouse, John A
Moran, John J
Moran, Thomas James
Moreno, Esteban
Morgan, Herbert Roy
Morishima, Hisayo Oda
Mork, Byron O
Morley, George W
Morlock, Carl G
Mortimer, Edward Albert, Jr
Moscarello, Mario Antonio
Moser, Kenneth Miles
Moses, Campbell, Jr
Moss, Charles Norman
Mou, Thomas William
Moyer, John Henry
Mudge, Gilbert Horton
Mullaney, Owen Christopher
Mulrow, Patrick J
Mundie, J Ryland
Munnell, Equinn W
Murphy, Edward G
Murphy, Mary Lois
Murphy, Patrick Aidan
Murphy, William Parry, Jr
Murray, Murray J
Mustacchi, Piero Oscar
Myers, David
Myers, Melvi Bertrand, Jr
Nachman, Ralph Louis
Nadell, Judith
Naide, Meyer
Naumann, Dorothy Ethel
Nayak, Ramnath V
Neary, Edward R
Neely, Charles Lea, Jr
Neghme, Amador
Neill, William Alexander
Nelson, Alan R
Nelson, Don Harry
Nelson, Paul Andrew
Nelson, Robert S
Nelson, Sidney W
Nerlich, William Pierrepont, III
Neu, Harold Conrad
Neville, William Edward
Neville, John F, Jr
Newcombe, David S
Newman, Louis Benjamin
Newman, Max Karl
Newton, Carol Marilyn

Ney, Robert Leo
Nghiem, Quang Xuan
Nichols, Buford Lee, Jr
Nichols, John Fraser
Nickson, Margaret Jane
Niebauer, John J
Nitowsky, Harold Martin
Nitz, Robert E
Noble, Nancy Lee
Nolan, James P
Nolan, Floyd (Alvin)
Norton, Edward W D
Nussbaum, Murray
O'Brien, John S
Obrinsky, William
Ochoa, Severo
Oettgen, Herbert Friedrich
O'Keefe, John Joseph
Olansky, Sidney
Olmstead, Edwin Guy
Olson, Kenneth B
Olson, Robert Eugene
Olson, Stanley William
O'Malley, Joseph Paul
Oppenheimer, Henry Ernest
Orkin, Lazarus Allerton
Orloff, Jack
Ortiz, Antonia
Osborn, John Jay
Osborne, John Alan
Oski, Frank
Osmond, Humphry Fortescue
Osserman, Elliott Frederick
Ovalle, William Keith
Owellen, Richard John
Pabico, Rufino C
Page, James A
Page, Robert Griffith
Paine, Thomas Fite, Jr
Pallie, Wazir
Palmer, Darwin L
Palmer, Roger
Pan, Steve Chia-Tung
Parfitt, Gilbert J
Parker, Charles W
Parker, Mary Langston
Parlett, Robert Frederic
Parlett, Robert Carleton
Parmelee, Arthur H, Jr
Parrish, Alvin Edward
Parsons, William Belle, Jr
Partin, John Calvin
Pate, James Wynford
Patek, Arthur Jackson, Jr
Patel, Dali Jehangir
Paton, David
Patrick, Edward Alfred
Patterson, John Ward
Patterson, Marcel
Patton, John Franklin
Pauk, George Lyon
Paul, Jerome Thomas
Paul, Milton Holliday
Paul, Oglesby
Pavlatos, Vytautas
Pearce, Keith Ian
Pearson, David
Pearson, James Gordon
Pearson, Olof Hjalmer
Pechel, Liberto
Peck, Harold Mitchell
Peete, Charles Henry, Jr
Peltier, Hubert Conrad
Penner, Donald Wills
Penneys, Raymond
Percy, John Smith
Perillie, Pasquale E
Per-Lee, John H
Periman, Robert
Permutt, Solbert
Perry, Horace Mitchell, Jr
Pesch, LeRoy Allen
Peskin, Gerald William
Peterdorf, Robert George
Peterson, Harold Oscar
Peterson, Oscar Sylvander, Jr
Petrakis, Nicholas Louis
Petricciani, John C
Petrillo, Charles
Petty, Charles Sutherland
Pevehouse, Byron C
Pfeffer, John B, Jr
Pfeiffer, Mildred Clara Julia
Pfuetze, Karl Hamilton
Phair, John P
Phelan, John T
Phelps, Gerald B
Phillips, John Hunter, Jr
Phillips, Leon A
Phillips, Otto C
Phillips, Samuel
Pick, Alfred
Pingeon, Rene
Pinkerson, Alan L
Pinter, Gabriel George

MEDICINE

Sussman, Howard H
Sussman, Karl Edgar
Swanzey, Eugene Harry
Swenson, Edward N
Swerdlow, Martin A
Sydnor, Katherine Lee
Tabakin, Burton Samuel
Taggart, John Victor
Tait, Columbus Downing, Jr
Talbot, Bernard
Talmage, David Wilson
Talso, Peter Jacob
Tamm, Igor
Tanaka, Kouichi Robert
Tanzer, Radford Chapple
Tapley, Norah duVernet
Taplin, George Vorce
Tarail, Robert
Taschdjian, Claire Louise
Tate, Charles Frank, Jr
Tatelman, Maurice
Tatter, Dorothy
Taussig, Helen Brooke
Taves, Donald R
Taylor, Isaac Montrose
Taylor, Richard Ray
Telfer, Nancy
Templeton, Arch W
Ten Eyck, David Roderick
Teplick, Joseph George
Tepperman, Jay
Terry, Luther Leonidas
Terry, Luther Leonidas, Jr
Teschan, Paul E
Test, Charles Edward
Texon, Meyer
Thayer, Walter Raymond, Jr
Thomas, Clayton Lay
Thomas, E Llewellyn
Thomas, Edward Donnall
Thomas, John
Thompson, William Taliaferro, Jr
Thompson, Wilmer Leigh, Jr
Thomson, Alexander
Thomson, Ashley Edwin
Thornton, William Edgar
Thorup, Oscar Andreas, Jr
Thuline, Horace Crockett
Tierney, Donald Frank
Tishkoff, Garson Harold
Tisza, Veronica Elizabeth Benedek
Tobis, Jerome Sanford
Toigo, Angelo
Tomlinson, Glen E
Toole, James Francis
Toone, Elam Cooksey, Jr
Torres-Rodriguez, Victor M
Tosh, Fred Eugene
Towery, Beverly Todd
Townes, Charles Henry
Triantaphyllopoulos, Eugenie
Trobaugh, Frank Edwin, Jr
Troen, Philip
Trubek, Max
Truby, Charles Paul
Trufant, Samuel Adams
Trunnell, Jack B
Trussell, Ray
Tse, Rose (Lou)
Tucker, Arthur Smith
Tuholski, James Martin
Tumen, Henry Joseph
Tumulty, Philip A
Turino, Gerard Michael
Turkington, Roger W
Turner, Michael D
Twigg, Homer Lee
Tyler, Edward Alton
Tyler, Frank Hill
Tyor, Malcolm Paul
Tzagournis, Manuel
Uhr, Jonathan William
Ulstrom, Robert
Underwood, Rex J
Upjohn, Everett Gifford
Urbach, John Robert
Ureles, Alvin L
Utz, John Philip
Vagnucci, Anthony Hillary
Valberg, Leslie S
Valenta, Lubomir Jan-Vaclav
Valenti, Carlo
Valentine, William Newton
Valloton, William Wise
Valvassori, Galdino E
Van Arsdel, Paul Parr, Jr
Van Italie, Theodore Bertus
Van Metre, Thomas Earle, Jr
Van Middlesworth, Lester
Van Mierop, Lodewyk H S
Van Pernis, Paul Anton
Varon, Myron Izak
Vastola, Edward Francis
Vaughan, John Heath
Vaun, William Stratin
Veech, Richard L
Verzeano, Marcel
Veverbrants, Egils
Vial, Lester Joseph, Jr

Vilter, Richard William
Vinocur, Myron
Visotsky, Harold M
Vogel, Philip James
Vogt, Marguerite Maria Paulette
Volwiler, Wade
Wada, John A
Wadsworth, Joseph Allison Cannon
Wagley, Philip John Franklin
Wakefield, George Earle
Waldman, Richard Trussell
Waldrop, Francis N
Walker, Arthur Earl
Walker, Joe Aaron
Walker, Richard Isley
Wall, Robert Leroy
Wallace, Alexander Cameron
Wallace, Craig Kesting
Wallace, Jackson M
Wallace, Jacques Burton
Wallach, Stanley
Wallerstein, Harry
Walser, Mackenzie
Walter, Richard D
Walton, Kenneth Nelson
Walton, Richard Joseph
Ward, John Robert
Ward, Louis Emmerson
Warner, Willis L
Warren, Robert Fletcher
Warthin, Thomas Alastair
Watanabe, Mamoru
Waterhouse, Christine
Watlington, Charles Oscar
Watson, Cecil James
Weaver, David Dawson
Weaver, Neill Kendall
Weber, Francis John
Webster, David Dyer
Webster, Paul Daniel, III
Wegner, Karl Heinrich
Wegria, Rene
Weigert, Alfred Peter
Weil, Robert W
Weil, Hans
Weiner, Murray
Weinstein, I Bernard
Weisberg, Herbert
Weisman, Russell
Weiss, Thomas E
Weisse, Allen B
Weissgerber, Rudolph E
Weksler, Marc Edward
Welsh, Federico
Wen, Sung-Feng
Wenger, Nanette Kass
Wenner, Herbert Allan
Werkman, Sidney Lee
Werner, Sidney Charles
Wertheim, Arthur Robert
Weser, Elliot
Wessler, Stanford
West, John B
West, Kelly M
Westerberg, Martha Rosalie
Weston, James T
Westphal, Milton C, Jr
Wetstone, Howard J
Whayne, Tom French
Wheeler, Henry Orson
Wheeler, Warren E
Whipple, Gerald Howard
Whisnant, Jack Page
White, Clayton Samuel
White, Denis Naldrett
White, Jack Edward
Whitehorn, William Victor
Whited, Harold Horatio
Wickstrom, Jack
Wiener, Alexander S
Wier, James Arista
Wigger, H Joachim
Wigh, Russell
Wigle, Ernest Douglas
Wigton, Robert Spencer
Wilder, Russell Morse
Wilkinson, Charles Brock
Williams, Christopher P S
Williams, George Rainey
Williams, Robert Frank
Williams, Robert Hardin
Williams, William Joseph
Willis, Park Weed, III
Willmon, Thomas L
Wilson, Charles B
Wilson, Donald Laurence
Wilson, Frank Crane
Wilson, John F
Wilson, Lester A, Jr

Wilson, Miriam Geisendorfer
Winchell, C Paul
Winegrad, Albert Irvin
Winkelstein, Warren, Jr
Winsor, Travis Walter
Winter, Chester Caldwell
Winternitz, William Welch
Witherspoon, Don Meade
Woldich, Francis De Sales
Wolf, George Anthony, Jr
Wolf, James Stuart
Wolf, Julius
Wolfe, Bernard Martin
Wolff, Frederick William
Wolff, Peter Hartwig
Wolfson, Edward A
Wolinsky, Harvey
Wollaeger, Gertraud
Wollschlaeger, Paul Bernhard
Wolma, Fred J
Wood, Francis Clark
Wood, James Edwin, III
Wood, Margaret Gray
Wood, W James Bainter
Woodbury, John F L
Woodruff, Calvin Watts
Woods, James Watson
Woodward, Theodore Englar
Woolsey, Robert S
Worrell, John Mays, Jr
Wrenshall, Gerald Alfred
Wright, Hastings Kemper
Wright, Jane Cooke
Wright, Joseph William, Jr
Wrogemann, Klaus
Wyman, Stanley M
Yablonski, Michael Eugene
Yale, Charles E
Yang, Wen-Kuang
Yankee, Ronald August
Yielding, K Lemone
Yolles, Stanley Faust
Young, Irving
Young, Lawrence Eugene
Young, Reuben B
Yount, Ernest H
Yu, Ts'ai-Fan
Yunis, Edmond
Zaleski, Witold Andrew
Zanecnik, Paul Charles
Zaneveld, Lourens Jan Dirk
Zavon, Mitchell Ralph
Zieler, Kenneth Levie
Zieve, Leslie
Zimmerman, Emery Gilroy
Zumoff, Barnett
Zvaifler, Nathan J
Zweiman, Burton

Aerospace Medicine

Armstrong, George Glaucus, Jr
Barron, Charles Irwin
Barry, William Earl
Baxter, Neal Edward
Beljan, John Richard
Berry, Charles A
Billingham, John
Billings, Charles Edgar, Jr
Campbell, Paul Andrew
Catterson, Allen Duane
Davis, Jefferson C
Dille, John Robert
Douglas, William Kennedy
Downey, Vincent M
Ellingson, Harold Victor
Ewing, Channing Lester
Frese, Frederick Joseph, Jr
Gell, Charles Fredric
Generales, Constantine D John
Gerathewohl, Siegfried Johannes
Hawkins, Willard Royce
Hayes, William Clifton, Jr
Henderson, John Warren
Henzl, Milan Rastislav
Hessberg, Rufus R
Houghton, Karl H
Jones, Geoffrey Melvill
Karstens, Andres Ingver
Kerwin, Joseph Peter
Lancaster, Malcolm
Maio, Domenic Anthony
Mason, William Van Horn
McCally, Michael
McFarland, Ross Armstrong
Mitchell, Hugh Bertron
Mohler, Stanley Ross
Nevison, Thomas Oliver, Jr
Nuttall, James B
Pierson, William R
Rowen, Burt
Schamadan, James Louis
Siegel, Peter Vincent
Snyder, Richard Gerald
Talbot, John Mayo
Von Beck, Harald Johannes
Waggoner, James Norman
Wagner, Henry George

Winget, Charles M
Wolcott, John H
Yerg, Raymond A

Allergy

Baer, Harold
Bardana, Emil John, Jr
Bernton, Harry Saul
Buckley, Charles Edward
Carryer, Haddon McCutchen
Cluff, Leighton Eggertsen
Cohen, Sheldon Gilbert
Crepea, Seymour B
Dice, Stanley Frost
Falliers, Constantine J
Farr, Richard Studley
Feingold, Ben F
Fink, Jordan Norman
Foubert, Edward Louis, Jr
Gregoire, Fernand
Haywood, Theodore J
Heiner, Douglas C
Howell, Charles Maitland
Itkin, Irving Herbert
Kaufman, Herbert S
Kirkpatrick, Charles Harvey
Kniker, William Theodore
Kovacs, Bela A
Layton, Laurence Laird
Leavitt, William Grenfell
Lee, Daisy Si
Levy, David Alfred
Lockey, Richard Funk
Lovell, Robert Gibson
MacLaren, Walter Rogers
Mathews, Kenneth Pine
McGovern, John Phillip
McKee, Kelly Tilson
McLean, James Amos
Nelson, Thomas Lothian
Perlman, Ely
Reed, Charles E
Richerson, Hal Bates
Slavin, Raymond Graham
Spock, Alexander
Swartz, Harry
Szentivanyi, Andor
Vanselow, Neal A
Walton, Charles Hutchinson Acourt
Wilhelm, Rudolf Ernst
Witting, Heinz Joseph
Woodin, William Graves
Wyrick, Ronald Earl
Zeleznick, Lowell D

Anesthesiology

Adriani, John
Aldrete, Jorge Antonio
Alexander, Samuel Craighead, Jr
Alper, Milton H
Amory, David William
Andersen, Thorkild Waino
Artusio, Joseph F, Jr
Ashman, Michael Nathan
Bach, Mary Jean
Baker, Louis Reed
Baker, Mary Rebecca
Balagot, Reuben Castillo
Ballinger, Carter M
Bamforth, Betty Jane
Barash, Paul G
Bennett, Peter Brian
Benson, Donald Warren
Bergman, Norman
Berman, M Lawrence
Black, Richard Glynn
Bonica, John Joseph
Bosomworth, Peter Palliser
Boyd, William Adam
Brand, Leonard
Bromage, Philip R
Brown, Elwyn S
Bruce, David Lionel
Brunner, Edward A
Buckley, Joseph J
Byles, Peter Henry
Cameron, Donald Forbes
Chase, Harold Frederick
Cheney, Frederick Wyman
Coakley, Charles Seymour
Cohen, Ellis N
Collins, Vincent J
Craythorne, N W Brain
Cullen, Bruce F
Deas, Thomas C
deJong, Rudolph H
Dekornfeld, Thomas John
Denson, Judson Samuel
Dent, Sara Jamison
Derrick, William Sheldon
Deutsch, Stanley
Diaz, Pedro Miguel
Dobkin, Allen Benjamin
Dornette, William Henry Lueders
Duflot, Leo Scott
Dunbar, Burdett Sheridan
Duncalf, Deryck

Eastwood, Douglas William
Eckenhoff, James Edward
Edwards, McIver Williamson, Jr
Eggers, George W Nordholtz, Jr
Elam, James O
Elliott, Henry Wood
Etherington, Lorne
Epstein, Robert Marvin
Erickson, James C, III
Esten, Benjamin E
Fabian, Leonard William
Fairley, Henry Barrie Fleming
Faulconer, Albert, Jr
Feingold, Alfred
Fink, Bernard Raymond
Frederickson, Evan Lloyd
Galindo, Anibal H
Galla, Stephen Joseph
Garner, Harold E
Giesecke, Adolph H
Gillies, Alastair J
Gissen, Aaron J
Gleaton, Harriet Elizabeth
Gold, Martin I
Goldberg, Alan Herbert
Graff, Thomas D
Gravenstein, Joachim Stefan
Green, Cloid Darryl
Greene, Nicholas Misplee
Greifenstein, Ferdinand Ernest
Grove, Daniel Dwight
Hall, Kenneth Daland
Hamelberg, William
Hamilton, William Kennon
Hinkley, Robert Edwin, Jr
Holaday, Duncan Asa
Hanks, John M
Hansen, John M
Harbord, Richard P
Harmel, Merel H
Hebert, Clarence Louis
Hedley-Whyte, John
Helrich, Martin
Henschel, Ernest O
Hershey, Solomon George
Hingson, Robert Andrew
Horton, Bennett Franklin
Howland, William Stapleton
Hug, Carl Casimir, Jr
Jacoby, Jay
Jastak, J Theodore
Jenicek, John Andrew
Jenkins, Leonard Cecil
Jenkins, Marion Thomas
Jones, John R
Kafer, Enid Rosemary
Kallus, Frank Theodore
Katz, Ronald Lewis
Kavan, Eva Mary
Keown, Kenneth K
Kim, Kil Chol
Kitz, Richard J
Kretchmer, Henry Edmund
Kripke, Benjamin Joshua
Landmesser, Charles Monroe
Laver, Myron B
Lawrence, Robert Marshall
Liao, Ji-Chia
Lim, Henry S
Linde, Harry Wight
Longnecker, David Eugene
Lorhan, Paul Herman
Lowe, Harry J
Lowenstein, Edward
Macnamara, Thomas E
Marx, Gertie F
Massion, Walter Herbert
Mazzia, Valentino Don Bosco
McCaughey, T J
McCarthy, Frank Martin
Merin, Robert Gillespie
Modell, Jerome Herbert
Moffitt, Emerson Amos
Moore, Daniel Charles
Morch, Ernst Trier
Morris, Lucien Ellis
Morrow, Dean Huston
Motoyama, Etsuro K
Moya, Frank
Moyers, Jack
Mueller, Robert Arthur
Munson, Edwin Sterling
Nagel, Eugene L
Nahrwold, Michael Lange
Ngai, Shih Hsun
North, William Charles
Orkin, Louis R
Pantuck, Eugene Joel
Papper, Emanuel Martin
Parmley, Ray T
Parrish, R Gibson
Parrish, Rob Gene
Patterson, Joseph Flanner, Jr
Patterson, JRichard Westcott
Pavlin, Edward George
Philbin, Daniel Michael
Pittinger, Charles Bernard

Pontoppidan, Henning
Poppers, Paul Jules
Price, Henry Locher
Purkis, Ian Edward
Rackow, Herbert
Rattenborg, Christen C
Redding, Joseph Stafford
Rehder, Kai
Rendell-Baker, Leslie
Reynolds, Robert N
Ridley, Roger W
Rigor, Benjamin Morales, Sr
Roberts, Edwin Bryan
Rushia, Edwin Louis
Rusy, Ben F
Ryan, John F
Sadove, Max Samuel
Safar, Peter
Salanitre, Ernest
Sanner, Fern Rusteberg
Sawyer, Donald C
Schmidt, Kurt F
Sechzer, Philip Haim
Selvin, Beatrice L
Shulman, Morton
Siebecker, Karl LaFollete, Jr
Smith, Jan D
Smith, Norman Ty
Smith, Robert H
Smith, Theodore Craig
Soma, Lawrence R
Spoerel, Wolfgang Eberhart G
Steinhaus, John Edward
Stephen, Charles Ronald
Stoddard, Carl C
Sugioka, Kenneth
Telford, Ruth Jane
Thomas, Lewis Jones, Jr
Turndorf, Herman
Van Bergen, Frederick Hall
Vandam, Leroy David
Vandewater, Stuart Leslie
Volpitto, Perry Paul
Walker, John David
Ward, Richard John
Way, Walter
Webb, Edmund Leslie
Weintraub, Herbert D
Weiskopf, Richard Bruce
Weisman, Harold
Weitzner, Stanley Wallace
White, Joseph Mallie
Willenkin, Robert L
Wilson, Roy D
Winter, Peter Michael
Witherspoon, Samuel McBridge
Wollman, Harry
Wyant, Gordon Michael
Yakaitis, Ronald William
Yeakel, Allen Egger
Zauder, Howard L
Zsigmond, Elemer K

Biomedical Engineering

Albisser, Anthony Michael
Alvi, Zahoor Mohem
Aronow, Saul
Bahner, Carl Tabb
Bales, Paul Dobson
Barr, Roger Coke
Beeler, George W, Jr
Beljan, John Richard
Beretsky, Irwin
Blecharczyk, Walter Joseph
Borgnis, Fritz Edward
Boston, John Robert
Bourne, John Ross
Bruck, Stephen Desiderius
Burgess, Richard Ernest
Chang, Kuo Wei
Cheng, George Chiwo
Childress, Charles Curtis
Childress, Dudley Stephen
Clamann, H Peter
Clark, Howard Garmany
Clarke, Alexander Mallory
Cobbold, R S C
Cohen, Gerald Stanley
Coombs, William, Jr
Covell, James Wachob
Crump, Jesse Franklin
Dallos, Peter John
Daubenspeck, John Andrew
Davis, Stanley David
Deindoerfer, Fred H
Detwiler, John Stephen
Dick, Donald Edward
Doane, Marshall Gordon
Donley, Clark Stephen
Dow, James W
Drossman, Melvyn Miles
Dutton, Robert Edward, Jr
Eberle, Jon William
Edmonds, Peter Derek
Enger, Carl Christian
Falb, Richard D
Fischell, Robert E
Fischer, Grace Mae

Flamboe, Eugene Earl
Francis, Howard Thomas
Freedman, William
Freeman, Donald Chester, Jr
Fromm, Eli
Frommer, Peter Leslie
Fuchs, Albert Frederick
Gall, Donald Alan
Galletti, Pierre Marie
Gardner, David L
Gerson, Raymond E
Gibbons, Donald Frank
Gibbs, Charles Howard
Gordy, Edwin
Gradijan, Jack R
Greatbatch, Wilson
Grodins, Fred Sherman
Grynszpan, Flavio
Haas, Gustav Frederick
Hahn, Allen W
Halleck, Frank Eugene
Hambrecht, Frederick Terry
Hamilton, Leroy Leslie
Hammond, William Edward
Hargest, Thomas Sewell
Harrison, Walter Kirby, Jr
Heinz, John Michael
Hinds, Marvin Harold
Homsy, Charles Albert
Houge, James C
Huang, H K
Huckaba, Charles Edwin
Hughes, G M K
Hulbert, Samuel Foster
Hutchins, Philip Michael
Ingels, Neil Barton, Jr
Jeutter, Dean Curtis
Johnson, Barry Lee
Johnson, Richard Noring
Joyner, Howard Sajon
Kahn, Paul
Katz, J Lawrence
Keller, Edward Lowell
Kelley, Thomas F
King, Paul Harvey
Kline, Jacob
Knox, Francis Stratton, III
Kohn, Michael
Konikoff, John Jacob
Krasner, Jerome Lee
Kreifeldt, John Gene
Krishnan, Silvarama S
Kwatny, Eugene Michael
Lafferty, James Francis
Lam, Chan Fun
Laszlo, Charles A
Lavallee, Marc
Lederman, David Mordechai
Lee, Jen-Shih
Leif, Robert Cary
Leininger, Robert Irvin
Levy, Matthew Nathan
Likuski, Robert Keith
Lisa, Joseph Daniel
Llaurado, Josep G
Macklin, Martin
MacNeill, Arthur Edson
Marsh, Donald Jay
Massey, Joe Thomas
May, Edwin Anthony
McGraw, Charles Patrick
McLeod, William D
McRae, Lorin Post
Meier, Peter M
Mikhail, Adel Ayad
Miller, George Edward
Mishelevich, David Jacob
Moskowitz, Harvey D
Moss, Gerald
Mozley, James Marshall, Jr
Myers, Donald Albin
Myers, George Henry
Negin, Michael
Newell, Jonathan Clark
Newhouse, Vernon Leopold
Norton, Allen C
Nunnally, Huey Neal
Olson, Walter Harold
Pacela, Allan Fred
Pearce, David Harry
Pennock, Bernard Eugene
Peura, Robert Allan
Phillips, Chandler Allen
Pickard, William Freeman
Plantz, Robert Glenn
Plonsey, Robert
Pollak, Victor A
Post, Bernard Saul
Powis, Raymond Leslie
Primiano, Frank Paul, Jr
Purdy, David Lawrence
Ramsey, Maynard, III
Rawlings, Charles Adrian
Reichert, Thomas Andrew
Reinhart, Richard Joseph
Reswick, James Bigelow
Richardson, Peter Damian
Riffenburgh, Roger Harry

Roberge, Fernand Adrien
Roesel, Oscar Fred
Ronel, Samuel Hanan
Rosen, Alan
Roxburgh, James Maxwell
Ruchkin, Daniel S
Rugge, Henry F
Sagawa, Kiichi
Saidel, Gerald Maxwell
Sances, Anthony, Jr
Schmitt, Neil Martin
Schmitt, Otto Herbert
Schrodt, Ariel Gilbert
Schwan, Herman Paul
Scott, Peter Douglas
Searle, John Randolph
Shacter, Bernard
Shapiro, Max
Shapiro, Stephen Irving
Sharpless, Thomas Kite
Sher, Lawrence D
Sherebrin, Marvin Harold
Smith, Robert Elphin
Sodickson, Lester A
Solonche, David Joshua
Spooner, Robert Bruce
Spurlock, Jack Marion
Stewart, George Hamill
Stokely, Ernest Mitchell
Strohbehn, John Walter
Taber, David
Taylor, Bruce Cahill
Taylor, Donald Rudolph, Jr
Terdiman, Joseph Franklin
Thompson, Noel Page
Thurston, George Butte
Thurstone, Frederick Louis
Tichauer, Erwin Rudolph
Timm, Gerald Wayne
Tuthill, Harlan Lloyd
Viernstein, Lawrence J
Villarroel, Fernando
Vogl, Thomas Paul
Vose, George Parlin
Wachtel, Howard
Watson, John Thomas
Weil, Max Harry
Welford, Norman Traviss
Wheeless, Leon Lum, Jr
White, Augustus Aaron, III
Wieland, Bruce Wendell
Wilkins, Michael Gray
Williams, Maryon Johnston, Jr
Williams, Ronald Alvin
Williamson, Donald Elwin
Wilson, James Bruce
Winter, David Arthur
Wolbarsht, Myron Lee
Woo, Kwang Bang
Wooten, Frank Thomas
Worsham, James Essex, Jr
Yamashiro, Stanley Motohiro
Youdin, Myron
Yurek, Gerald G
Ziskin, Marvin Carl

Cancer

Abbott, Betty Jane
Ackerman, Norman Bernard
Anderson, David Eugene
Anderson, Lucy Macdonald
Aszalos, Adorjan
Auersperg, Nelly
Baker, Michael Allen
Banerjee, Satyendra Nath
Barranco, Sam Christopher
Batra, Gopal Krishan
Baum, William Stanhope
Benade, Leonard E
Benjamin, Robert Stephen
Bernstein, William Carl
Bisel, Harry Ferree
Black, Homer Selton
Bloom, Eda Terri
Bond, Howard Edward
Borsos, Tibor
Buck, Clayton Arthur
Buehring, Gertrude Case
Burholt, Dennis Robert
Burke, Arthur Wade, Jr
Burton, Albert Frederick
Cantrell, Elroy Taylor
Carroll, Walter William
Carruthers, Christopher
Cervenka, Jaroslav
Cleaver, James Edward
Cohn, Naomi Kenda
Consigli, Richard Albert
Criss, Wayne Eldon
Davis, William Ellsmore, Jr
Demoise, Charles Francis
Depue, Robert Hemphill
Dipple, Anthony
Do-Van-Quy, Dominic
Dubreuil, Robert
Dunn, Bruce Partridge
Duryee, William Rankin
Elford, Howard Lee

MEDICINE

Elhilali, Mostafa M
Evans, Mary Jo
Evans, Virginia John
Fiala, Silvio Emerich Ivan
Fink, Mary Alexander
Fischinger, Peter John
Freeman, Aaron Eliot
Furth, Arthur
Furth, Jacob
Gailani, Salman
Gehan, Edmund A
Gelhorn, Alfred
Geran, Ruth Iris
Gerschenson, Lazaro E
Gilbertsen, Victor Adolph
Giovanella, Beppino C
Goldschmidt, Bernard Morton
Golumbic, Norma
Goodman, Jay Irwin
Grandjean, Carter Jules
Greenberg, Arnold Harvey
Greenman, David Lewis
Gross, Ludwik
Guerin, Michael Richard
Gunn, Samuel Albert
Gurwara, Sweet K
Hall, Thomas Christopher
Han, Tin
Harder, Harold Cecil
Harewood, Ken Rupert
Hearn, Henry James, Jr
Hellman, Leon
Hill, Donald Lynch
Hillcoat, Brian Leslie
Holper, Jacob Charles
Hosick, Howard Lawrence
Howell, Stephen Barnard
Huggins, Charles Brenton
Ingall, John
Jacobs, Edwin M
Jensen, Keith Edwin
Johnson, Bruce McDougall
Johnson, Irving Stanley
Jones, Bryant Lee
Karasaki, Shuichi
Keefer, Larry Kay
Kelln, Elmer
Kelton, Diane Elizabeth
Kerbel, Robert Stephen
Khwaja, Tasneem Afzal
Kit, Saul
Kline, Ira
Kripke, Margaret Louise (Cook)
Kupchik, Herbert Z
Kurzepa, Henryka Janina
Laham, Souheil
Laishes, Brian Anthony
Lala, Peeyush Kanti
Lamon, Eddie William
Landon, John Campbell
Lemonde, Paul
Levin, Victor Alan
Levine, Philip
Lewis, Martin Gwent
Liebelt, Annabel Glockler
Lieberman, Miriam
Lin, Chi-Wei
Linna, Timo Juhani
Lloyd, Harris Horton
Loeb, Lawrence Arthur
Maher, Veronica Mary
Marquardt, Hans Wilhelm Joe
McCollester, Duncan L
McCormick, J Justin
McCormick, Kenneth James
McCredie, John A
McCulloch, Peter Blair
McQueen, James Lee
Meck, Robert Allen
Mendelsohn, Mortimer Lester
Meredith, Ruby Frances
Mirvish, Sidney Solomon
Mitchen, Joel Ramon
Moolten, Frederick London
Moyer, Rex Carlton
Munroe, Joscelyn Spencer
Murphy, Samuel G
Mushinski, Joseph Frederic
Nagel, Donald Lewis
Nelson, Janet Sue Rasey
Nettesheim, Paul
Newberry, Truman Albert
Novak, Josef Frantisek
O'Brien, Elinor Murray
Ohnuma, Takao
Okey, Allan Bernhardt
Old, Lloyd John
Oppenheimer, Steven Bernard
Pancake, Samuel Joseph
Patterson, Manford Kenneth, Jr
Pettit, George Robert
Pfaffenberger, Carl Dale
Pietras, Richard Joseph
Pilgrim, Hyman Ira
Poirier, Lionel Albert
Pratt, Charles Benton
Purnell, Dallas Michael
Quisenberry, Walter Brown

Rivera, Evelyn Margaret
Ruoslahti, Erkki Ilmari
Rutenberg, Alexander Michael
Sani, Brahma Porinchu
Schmid, Franz Anton
Schmitz, Robert L
Schwartz, Arthur Gerald
Segal, Beatrice Carrier
Segal, Alvin
Selawry, Oleg S
Seon, Byeong Kuk
Sharma, Gopal Chandra
Shimkin, Michael Boris
Shultz, Leonard Donald
Shuster, Joseph
Silagi, Selma
Singer, Robert Mark
Singhal, Ram Pratap
Solter, Davor
Sonstegard, Ronald Arlyn
Souto, Jose
Spackman, Darrel H
Spiegelman, Solomon
Spolsky, Christina Maria
Sporn, Michael Benjamin
Stein, Justin John
Steinfeld, Jesse Leonard
Stern, Kurt
Stevens, John Joseph
Strong, Louise Connally
Sykes, Marguerite Prince
Sykes, John A
Takasugi, Mitsuo
Tan, Charlotte
Tarnowski, George Serge
Teller, Morris N
Terzaghi, Margaret
Thiersch, Johannes Bernhard
Tobey, Robert Allen
Todaro, George Joseph
Tryfiates, George P
Young, Charles William
Yuhas, John M
Yushok, Wasley Donald
Yuspa, Stuart Howard
Zubrod, Charles Gordon

Cardiology

Aagaard, George Nelson
Adams, Forrest Hood
Adams, Wright (Rowe)
Anderson, Milton Winfield
Aranda, Juan Manuel
Baker, Saul Phillip
Bargeron, Lionel Malcolm, Jr
Baum, David
Beamish, Robert Earl
Beard, Owen Wayne
Beller, Barry M
Benson, Herbert
Benzing, George, III
Berliner, Kurt
Berman, Reuben
Bertran, Carlos Enrique
Besch, Henry Roland, Jr
Bicoff, Juan Pedro
Bierenbaum, Marvin L
Bigger, J Thomas, Jr
Blake, Thomas Mathews
Blankenhorn, David Henry
Blomqvist, Carl Gunnar
Bloomfield, Dennis Alexander
Blumenthal, Sidney
Boake, William Charles
Bouchard, Richard Emile
Brandaleone, Harold
Brandtonbrener, Martin
Braunwald, Eugene
Bueher, Martin Stowell
Cady, Lee (De), Jr
Campbell, James A
Carleton, Richard Allyn
Case, Robert B
Ceretti, Elena
Cheitlin, Melvin Donald
Cheng, Tsung O
Childers, Roderick W
Chiong, Miguel Angel
Cho, Young Won
Cochran, Paul Terry
Cohen, Jules
Cohen, Lawrence Sorel
Cohen, Louis
Connell, Walter Ford
Coodley, Eugene Leon
Cox, John William
Daoud, Georges
Davies, David Hywel
Dawber, Thomas Royle
Dean, David Campbell
Demers, Pierre-Paul
Dickhaus, Donald William

Diehl, Antoni Mills
Dimond, Edmunds Grey
Dodge, Harold T
Doherty, James Edward, III
Dolgin, Martin
Downing, Daniel Francis
Doyle, Eugene F
Doyle, Joseph Theobald
Dunean, William Thompson
Dunn, Marvin I
Du Shane, James William
Eddleman, Elvia Etheridge, Jr
Elliot, Robert S
Emmanouilides, George Christos
Engle, Mary Allen English
Escher, Doris Jane Wolf
Falsetti, Herman Leo
Fashena, Gladys Jeannette
Ferencz, Charlotte
Finkelstein, David
Finnerty, Frank Ambrose, Jr
Fletcher, Evan
Franch, Robert H
Frank, Charles Warren
Frank, Martin J
Fraser, Robert Stewart
Friedman, William Foster
Frommer, Peter Leslie
Funk, David Crozier
Ganz, William
Garcia-Palmieri, Mario R
Genecin, Abraham
Genovese, Pasquale Dante
Gershman, Lewis C
Gessner, Ira Harold
Gibson, Thomas Chometon
Gilani, Shamshad H
Gilmore, Hugh Richmond, III
Glancy, David L
Glick, Gerald
Glushien, Arthur Samuel
Goldberg, Henry Peter
Goldmann, Morton Aaron
Goldstein, Sidney
Gosfield, Edward, Jr
Gottlieb, Abraham Mitchell
Gould, Kenneth Lance
Gould, Lawrence A
Grant, Colin
Greene, David Gorham
Groom, Dale
Gubner, Richard S
Guimaraes, Armenio Costa
Gulotta, Stephen Joseph
Gunnar, Rolf McMillan
Guntheroth, Warren G
Hall, Robert Joseph
Hammermeister, Karl E
Han, Jack
Harris, Leonard Crossley
Harris, Willard Samuel
Harrison, Donald C
Hastreiter, Alois Rudolf
Haywood, L Julian
Headley, Robert N
Hellems, Harper Keith
Hejtmancik, Milton R
Helwig, John, Jr
Hill, John Donald
Hoffman, Julien Ivor Ellis
Hohn, Arno R
Hollis, Walter Jesse
Hood, William Boyd, Jr
Hope, Ronald Richmond
Horan, Leo Gallaspy
Hurst, Victor Waldemar, III
January, James O
January, Lewis Edward
Jarmakan, Jay M
Jelliffe, Roger Woodham
Joos, Howard Arthur
Jordan, Jerry Dugger
Kaplan, Samuel
Katus, Albert Adolph, Jr
Kelly, John J
Kennedy, Jesse Ward
Kerwin, Alfred John
Kezdi, Paul
Killip, Thomas, III
Knoebel, Suzanne Buckner
Kraus, William Ludwig
Krovetz, L Jerome
Kuhn, Leslie A
Lancaster, Malcolm
Langendorf, Richard
Langston, Per Harald
Leach, John Kline
Lees, Martin H
Lepeschkin, Eugene
Levin, Aaron R
Levine, Herbert Jerome
Levine, Robert
Levinson, Gilbert E
Levy, Arthur Maurice
Levy, Louis, II
Levy, David Harold
Linde, Leonard M

Lippschutz, Eugene J
Lucas, Russell Vail, Jr
Lukas, Daniel Stanley
Malinow, Manuel R
Manning, James Arthur
Marcus, Frank I
Marcus, Melvin L
Markle, Herbert
Marriott, Henry Joseph Llewellyn
Marshall, Robert James
Martinez-Lopez, Jorge Ignacio
Martt, Jack M
Massie, Edward
Mauck, Henry Page, Jr
Mazzoleni, Alberto
Meadows, William R
Menashe, Victor D
Mesel, Emmanuel
Mignault, Jean De L
Miller, Kay
Miller, William Weaver
Minhas, Kareem
Mitchell, Shiela Craig
Mohiuddin, Syed M
Moore, Earl Neil
Mooring, Paul K
Morgan, Beverly Carver
Morganroth, Joel
Morin, Yves
Morris, James Joseph, Jr
Moser, Marvin
Mudd, J Gerard
Murray, Raymond Harold
Nadas, Alexander Sandor
Nadeau, Reginald Antoine
Naughton, John Patrick
Neill, Catherine Annie
Neill, William Alexander
Noonan, Jacqueline Anne
Nora, James Jackson
Omura, Yoshiaki
Osborne, John Alan
Park, Myung Kun
Parker, Brent M
Parker, John Orval
Parmley, William W
Pearce, Morton Lee
Pensinger, Robert P
Phillips, John Hunter, Jr
Pilz, Clifford G
Pugh, David Milton
Pyle, Robert Lee
Reeves, John T
Regan, Timothy Joseph
Reynolds, Ernest West, Jr
Richardson, David W
Rike, Paul Miller
Ritzmann, Leonard W
Robin, Erwin
Rolett, Ellis Lawrence
Ross, John, Jr
Ross, Richard Starr
Rossall, Richard Edward
Rotem, Chava Eve
Rutenberg, Herbert David
Samet, Philip
Sanders, Charles Addison
Sanyal, Shyamal Kumar
Saslaw, Milton Sibley
Sawyer, C Glenn
Scherr, Lawrence
Schlant, Robert C
Schroeder, John Speer
Scott, Ralph Carmen
Segal, Bernard L
Sellers, Frank Jamieson
Selvester, Ronald H
Setzer, Arthur
Shabetai, Ralph
Shah, Pravin Mangaldas
Shapiro, William
Singer, Donald H
Smith, Eldon Raymond
Smulyan, Harold
Sodhi, Harbhajan S
Soloff, Louis Alexander
Sommer, Leonard Samuel
Spach, Madison Stockton
Spann, James Fletcher, (Jr)
Spitzer, Joseph Maurice
Spodick, David Howard
Stern, Aaron Milton
Strang, Ruth Hancock
Sutman, Leonard Jay
Surawicz, Borys
Talner, Norman Stanley
Taylor, William Jape
Tsagaris, Theofilos John
Tobin, John Robert, Jr
Udall, John Alfred
Vaules, David Wilson
Vlad, Peter
Walsh, John Joseph
Wang, Yang
Weisse, Allen B
Weissler, Arnold M
Wells, Charles Robert Edwin
Wenger, Nanette Kass

Wessel, Hans U
Whittemore, Ruth
Wiener, Leslie
Wigle, Ernest Douglas
Williams, John F, Jr
Willis, Park Weed, III
Witham, Abner Calhoun
Woske, Harry Max
Young, Maurice Durward
Yu, Paul N
Yuceoglu, Yusuf Ziya
Zao, Zang Z
Zinsser, Harry Frederick

Cardiovascular Diseases
Abramson, David Irvin
Amorim, Dalmo de Souza
Askew, Cornelius, Jr
Beckwith, Julian Ruffin
Befeler, Benjamin
Bishop, Sanford Parsons
Brown, Helen Bennett
Bruce, Thomas Allen
Butkus, Antanas
Chobanian, Aram V
Chrysant, Steven George
Cohn, Jay Norman
Conway, Gene Farris
Crawford, Donald W
Didisheim, Paul
Ferrario, Carlos Maria
Flowers, Nancy Carolyn
Fowler, Noble Owen
Frankl, William S
Garner, Harold E
Gerrity, Ross Gordon
Gobel, Frederick L
Goodyer, Allan Victor
Gootman, Norman Lerner
Greenberg, Stanley
Grossman, Laurence Abraham
Haber, Edgar
Harb, Joseph Marshall
Heistad, Donald Dean
Hirsch, Jacob Irwin
Insull, William, Jr
James, Thomas Naum
Kannel, William B
Katz, Murray Alan
Kerber, Richard E
Mandal, Anil Kumar
Mason, Dean Towle
Morgan, William L, Jr
Nelson, Albert Wendell
Nichaman, Milton Z
Oliver, G Charles
Page, Irvine Heinly
Pannami, Motilal Bhagwandas
Pick, Ruth
Prineas, Ronald James
Rakita, Louis
Romack, Frank Eldon
Runge, Thomas Marschall
Ruskin, Arthur
Schatz, Irwin Jacob
Scherf, David
Schertis, Sidney
Schwartz, Stephen Mark
Sheppard, Joseph Jackson, Jr
Simon, Geza
Spittell, John A, Jr
Sullivan, Jay Michael
Swan, Harold James Charles
Threefoot, Sam Abraham
Tuna, Naip
Vahouny, George V
Volicer, Ladislav
Weil, Max Harry
Westlake, Robert Elmer
Westura, Edwin Eugene
Whereat, Arthur Finch
Yeh, Billy Kuo-Jiun
Zatuchni, Jacob
Zugibe, Frederick T
Zukel, William John

Chemotherapy
Actor, Paul
Allen, Lois Brenda
Arnold, John D
Berryhill, Walter Reece
Bonner, Daniel Patrick
Braemer, Allen C
Brody, Gerald
Burrous, Stanley Emerson
Calabresi, Paul
Cantrell, William Fletcher
Chou, Ting-Chao
Cromwell, Norman Henry
Dammin, Gustave John
Davidson, Charles Sprecher
DeLorenzo, William F
Florestano, Herbert Joseph
Fondy, Thomas Paul
French, Frederick Alexis
Froelich, Ernest
Fugmann, Ruth Adele
Furner, Raymond Lynn

Furusawa, Eiichi
Golbey, Robert (Bruce)
Gordee, Robert Stouffer
Graessle, Otto Edward
Grunberg, Emanuel
Heifetz, Carl Louis
Heman-Ackah, Samuel Monie
Hughes, Everett C
Jacobs, Edwin M
Jacobs, Richard Lewis
Khwaja, Tasneem Afzal
Kreis, Willi
Leitner, Felix
Lloyd, Harris Horton
McCoy, John Roger
McKinstry, Donald W
Morgan, Lee Roy, Jr
Moss, Jack N
Price, Kenneth Elbert
Schabel, Frank Milton, Jr
Schlichting, David Arthur
Schwartz, Pauline Mary
Shelton, Robert William
Sidwell, Robert William
Simpson-Herren, Linda
Smith, Frank E
Stock, Charles Chester
Straus, Marc J
Streitfeld, Murray Mark
Waitz, Jay Allan
Warren, George Harry
Welch, Arnold DeMerritt
Wichelhausen, Ruth Hechler
Yarinsky, Allen
Yoshimura, Sei

Clinical Medicine
Aikawa, Jerry Kazuo
Bird, Robert Montgomery
Bloom, Walter Lyon
Boost, Gerhard
Brodman, Keeve
Bukantz, Samuel Charles
Burke, Richard Michael
Castle, William Bosworth
Cate, Thomas Randolph
Daugherty, Guy Wilson
DeGowin, Elmer Louis
Durant, Thomas Morton
Ebert, Richard Vincent
Falk, Abraham
Frachtman, Hirsh Julian
Ginsburg, Isadore Wilcher
Hager, Chester Bradley
Hamilton, Henry Edward
Hamilton, William Kennon
Harris, John William
Hirschboeck, John Stephen
Howlett, Kirby Smith, Jr
Ingram, Marylou
Jelinek, Bohdan
Kay, Calvin Frederick
Kirtley, William Raymond
Klinefelter, Harry Fitch
Klotz, Arthur Paul
Kolff, Willem Johan
Lanzoni, Vincent
Lozner, Eugene Leonard
McAllister, Ronald Eric
McDonald, Roger Koefod
Medway, William
Miller, Charles Phillip
Miller, Joseph Morton
Moersch, Herman John
Myers, Jack Duane
Nichols, Donald Richardson
Noth, Paul Henry
Pellegrino, Edmund Daniel
Reynolds, Leslie Boush, Jr
Rose, Harry Melvin
Samter, Max
Schoenberger, James A
Seeff, Leonard Barry
Segall, Harold Nathan
Segaloff, Albert
Shaffer, James Milton
Sims, John LeRoy
Soffer, Louis Julius
Spaulding, William Bray
Wolf, Stewart George, Jr
Wollaeger, Eric Edwin

Comparative Medicine
Cassell, Gail Houston
Cohen, Bennett J
Easterday, Bernard Carlyle
Hoag, Warren George
Micks, Don Wilfred
Moreland, Alvin Franklin
Palumbo, Nicholas Eugene
Schwartz, Anthony
Whitney, Robert Arthur, Jr

Dermatology
Akers, William Alexander
Anderson, Philip Carr
Arluk, David Jay
Baden, Howard Philip

Baer, Rudolf L
Bagatell, Fillmore Kenneth
Baldini, James Thomas
Becker, Samuel William, Jr
Beerman, Herman
Beirne, Gilbert Arthur
Bereston, Eugene Sydney
Bergstresser, Paul Richard
Blank, Harvey
Blaylock, W Kenneth
Blois, Marsden Scott, Jr
Bluefarb, Samuel M
Braverman, Irwin Merton
Briggaman, Robert Alan
Burnett, Joseph W
Burnham, Thomas K
Callaway, Jasper Lamar
Canizares, Orlando
Caplan, Richard Melvin
Carney, Robert Gibson
Carr, Richard Dean
Carter, David Martin
Cawley, Edward Philip
Clendening, William Edmund
Cooper, Garrett
Couperus, Molleurus
Cripps, Derek J
Crounse, Robert Griffith
Demis, Dermot Joseph
Dobes, William Lamar
Dobson, Richard Lawrence
Eglitis, Irma
Emele, Jane Frances
Epstein, Ervin Harold
Epstein, William L
Esterly, Nancy Burton
Everett, Mark Allen
Farber, Eugene M
Felsher, Zachary
Fleischmajer, Raul
Frank, Samuel B
Fukuyama, Kimie
Fusaro, Ramon Michael
Gigli, Irma
Goltz, Robert W
Griem, Sylvia F
Griesemer, Robert Daniel
Gubersky, Victor R
Guthrie, Marshall Beck
Halprin, Kenneth M
Hambrick, George Walter, Jr
Harber, Leonard C
Hasegawa, Junji
Hashimoto, Ken
Howell, Charles Maitland
Hu, Funan
Hurley, Harry James
Jansen, Gerald Thomas
Johnson, Thomas W
Jordon, Robert Earl
Kersting, David William
King, Lloyd Elijah, Jr
Klaus, Sidney N
Klein, Edmund
Kligman, Albert Montgomery
Knox, John Marshall
Korenyi, Charles
Krupp, Iris M
Landau, Joseph White
Lazarus, Gerald Sylvan
Leavell, Ullin Whitney, Jr
Lerner, Aaron Bunsen
Lerner, Marguerite Rush
Lever, Walter Frederick
Lipkin, George
Livingood, Clarence Swinehart
Loomans, Maurice Edward
Lorincz, Allan Levente
Lowney, Edmund Dillahunty
Lumpkin, Lee Roy
Luscombe, Herbert Alfred
Lynch, Francis Watson
Lynch, Peter John
Maguire, Henry C, Jr
Maibach, Howard H
Mandel, Edward H
Matoltsy, Alexander Gedeon
Melbye, Susanne Warner
Mescon, Herbert
Mihm, Martin C, Jr
Miller, Laurence Herbert
Millikan, Larry Edward
Mitchell, John Campbell
Montes, Leopoldo F
Moore, Morris
Muller, George Heinz
Mullins, J Fred
Nelson, Carl Truman
Nicholas, Leslie
Noojin, Ray O
Norins, Arthur Leonard
Odland, George Fisher
Olansky, Sidney
O'Quinn, Silas Edgar
Osment, Lamar Sutton
Pariser, Harry
Pathak, Madhukar
Perry, Harold Otto

Pinkus, Hermann (Karl Benno)
Pochi, Peter E
Popkin, George Lionel
Potter, Brian
Radcliffe, Christian Elmore
Raskin, Joan
Rees, William James
Reisner, Ronald M
Reiss, Frederick
Reller, Herbert Henry
Robinson, Harry Maximilian, Jr
Rostenberg, Adolph, Jr
Samitz, M H
Sams, Wiley Mitchell, Jr
Schmidt, Otto Ernest Lincoln
Schwartzman, Robert M
Shalita, Alan Remi
Shatin, Harry
Shelley, Walter Brown
Silver, Alene Freudenheim
Smith, Jesse Graham, Jr
Solomon, Lawrence Marvin
Stone, Orville Joseph
Stoughton, Richard Baker
Thomas, Carmen Christine
Traub, Eugene Frederick
Uitto, Jouni Jorma
Urbach, Frederick
Van Scott, Eugene Joseph
Voorhees, John James
Wachs, Gerald N
Weary, Peyton Edwin
Welton, William Arch
Wheeler, Clayton Eugene, Jr
Wildnauer, Richard Harry
Wilgram, George Friederich
Willis, Isaac
Windhorst, Dorothy Baker
Winkelmann, Richard Knisely
Winkler, Norman Walter
Wood, Margaret Gray
Wright, Edwin T
Wuepper, Kirk Dean
Zeligman, Israel

Environmental Medicine
Albert, Roy Ernest
Beck, Jacob Walter
Boyles, James McGregor
Bromberger-Barnea, Baruch (Berthold)
Ferin, Juraj
Frazier, John Melvin
Giel, Bohdan Gielecinski
Goldstein, David
Gunnison, Albert Farrington
Hamilton, Robert William, Jr
Heimann, Harry
Hubbard, Roger W
Jones, LeeRoy George
Nelson, Norton
Orris, Leo
Selikoff, Irving John
Swift, David Leslie
Warren, Douglas Robson
Wilson, John Thomas, Jr
Zaki, Mahfouz H

Epidemiology
Albrecht, Robert Michael
Alexander, Edward Russell
Allen, Alfred Marston
Ames, Wendell Russell
Anderson, Donald Oliver
Asal, Nabih Rafia
Azar, Joseph E
Babbott, Frank Lusk, Jr
Babione, Robert William
Baer, George M
Baker, Susan Pardee
Banta, James E
Benarde, Melvin Albert
Bennett, Peter Howard
Blackburn, Henry Webster, Jr
Brody, Jacob A
Buck, Alfred A
Burch, Thomas Adams
Bures, Milan F
Cassel, John Charles
Chin, Tom Doon Yuen
Choi, Nung Won
Cobb, Sidney
Cohen, Daniel
Comstock, George Wills
Constantine, Denny G
Curnen, Mary G McCrea
Cutler, Sidney Joshua
Dahl, Elmer Vernon
D'Atri, David Albert
Davies, Jack Neville Phillips
Day, Robert Winsor
Dean, Andrew Griswold
Doege, Theodore Charles
Donnell, Henry Denny, Jr
Dunn, Frederick Lester
Ehrlich, S Paul, Jr
Eklund, Carl Milton
Eller, Charles Howe
Emanuel, Irvin

MEDICINE

Evans, Alfred Spring
Fagan, Raymond
Fisher, James Wallvin
Freitag, Julia Louise
Fritz, Roy Fredolin
Gale, James Lyman
Gangarosa, Eugene J
Garfinkel, Lawrence
Geiger, H Jack
Gelfand, Henry Morris
Gezon, Horace Martin
Glass, Robert Loring
Goldstein, Inge F
Golubjatnikov, Rjurk
Goodwin, Melvin Harris, Jr
Gruenberg, Ernest Matsner
Gurgis, Hoda A
Gullen, Warren Hartley
Hambler, Frederick Edwards
Hannon, William McDowell
Hankin, Jean H
Heath, Clark Wright, Jr
Heeren, Ralph Heinrich
Held, Joe N
Henderson, Donald Ainslie
Henderson, Maureen McGrath
Herman, Bertram
Higgins, Ian T
Hirohata, Tomio
Hollister, Arthur Clair, Jr
Hook, Ernest Benjamin
Horsmann, Dorothy Millicent
Horton, Robert John Munro
Houser, Harold Byron
Hutchison, George B
Ibrahim, Michel A
Ingalls, Theodore Hunt
Ingraham, Hollis Steadman
Ipsen, Johannes, Jr
Jamison, Homer Claude
Jekel, James Franklin
Johnson, Benjamin C
Joly, Daniel Jose
Kahn, Harold A
Kahrs, Robert F
Kapikian, Albert Zaven
Kaufmann, Arnold Francis
Keil, Julian Eugene
Keller, Martin David
Kelsey, Jennifer Louise
Kessler, Irving Isar
Kissinger, David George
Kokernot, Robert Hutson
Korns, Robert Fulton
Kraus, Arthur Samuel
Kraus, Jess F
Kurland, Leonard T
Labarthe, Darwin Raymond
Landau, Emanuel
Langmuir, Alexander Duncan
Last, John Murray
Lebowitz, Michael David
Lee, John Alexander Hugh
Leiby, George Martin
Lemon, Frank Raymond
Lennette, Edwin Herman
Lilienfeld, Abraham Morris
Lundin, Frank E, Jr
Maassab, Hunein Fadlo
Macdonald, Eleanor Josephine
MacMahon, Brian
Magoffin, Robert Louis
Martin, Russell James
Masi, Alfonse Thomas
Mason, John
Mausner, Judith S
McCollum, Robert Wayne
McCorkle, Lois Pake
McCroan, John Edgar, Jr
McCusker, Jane
McDonald, Alison Dunstan
McDonald, John Corbett
Milham, Samuel, Jr
Miller, Anthony Bernard
Milmore, Benno Karl
Monk, Mary Alice
Monto, Arnold Simon
Moore, Roscoe Michael, Jr
Morton, William Edwards
Mosley, James W
Mosley, Wiley Henry
Mulvihill, John Joseph
Munan, Louis
Nathanson, Neal
Nelson, Norman Bartram
Newell, Kenneth Wyatt
Newill, Vaun Archie
Oakmann, Margaret Claire
O'Brien, William M
Oleinick, Arthur
Omohundro, Richard E
Omran, Abdel Rahim
Oseasohn, Robert
Ostfeld, Adrian Michael
Ovellen, Richard John
Paffenbarger, Ralph Seal, Jr
Palmer, Alan
Pasternack, Bernard Samuel
Payne, Fred J

Peacock, Peter N B
Peters, James Alexander
Peterson, Donald Richard
Philip, Robert Neil
Portnoy, Bernard
Priester, William Alfred
Purcell, Robert Harry
Purshottam, Natesaier
Ravenholt, Reimert Thorolf
Reed, Dwayne (Milton)
Reeves, William Carlisle
Rose, Gordon Wilson
Rosen, Leon
Rotkin, Isadore David
Saiger, George Lewis
Sartwell, Philip Earl
Scholtens, Robert George
Schreiber, Hans
Schultz, Myron Gilbert
Schultz, Leonard Michael
Schuman, Stanley Harold
Schwabe, Calvin Walter
Schweitzer, Morton David
Scrivani, Robert P
Selby, Lloyd A
Selevan, Sherry Gail
Sheehe, Paul Robert
Shekelle, Richard Barten
Shy, Carl Michael
Siegel, Morris
Silberg, Stanley Louis
Silverman, Charlotte
Slusky, Herbert L
Sohler, Katherine Berridge
Speizer, Frank Erwin
Spiers, Philip Sackville
Stein, Zena
Stewart, Chester Bryant
Susser, Mervyn W
Tesh, Robert Bradfield
Terris, Milton
Thompson, Gordon William
Thorne, Melvyn Charles
Tokuhata, George K
Tosh, Fred Eugene
Trussell, Ray
Vakilzadeh, Javad
Vaughn, John B
Vianna, Nicholas Joseph
Villarejos, Victor Moises
Vobecky, Josef
Voors, Antonie Wouter
Vought, Robert Louis
Wallace, Gordon Dean
Walter, William Arnold, Jr
Watson, Robert Lee
Watt, James
White, Kerr Lachlan
Winsberg, Gwynne Roeseler
Worth, Robert McAlpine
Wyon, John Benjamin

Experimental Medicine
Bain, Barbara
Barka, Tibor
Bond, Victor Potter
Bowdler, Anthony John
Copley, Alfred Lewin
Coutinho, Claude Bernard
Desjardins, Raoul
Dufour, Didier
Evans, Silas McAfee
Ghosh, Nimai Kumar
Grollman, Arthur
Grupp, Gunter
Gunter, Bobby J
Hamilton, Leonard Derwent
Henness, Donald Merle
Iacono, James M
Kolff, Willem Johan
Lepow, Irwin Howard
Levin, Jack
Lis, Martin
Macklem, Peter Tiffany
Maxfield, Myles
Poydock, Mary Eymard
Rabinovitch, Michel Pinkus
Renold, Albert Ernst
Rotem, Chava Eve
Samter, Max
Schmidt, Leon Herbert
Shorter, Roy Gerrard
Sjoerdsma, Albert
Stanbury, John Bruton
Thal, Alan Philip
Tracy, M Joanna
von Kaulla, Kurt Nikolaj
Wescoe, W Clarke
Wexler, Bernard Carl

Family Medicine
Brucker, Paul Charles
Bryan, Thornton Emry
Caley, David William
Carmichael, Lynn Paul
Ciriacy, Edward W
Corley, John Bryson
Davis, Leroy Thomas
Geyman, John Payne
Heck, Robert Skinrood
Hill, Charles Earl
Kowalewski, Edward Joseph
Smith, Robert
Stokes, David Kershaw, Jr
Verby, John E
Wen, Chi-Pang
Wilson, Frederick Surphen
Wilson, Vernon Earl
Worden, Leonard Russell
Wragg, Laurence Edward
Zeschke, Richard Herman

Forensic Medicine
Bigelow, Nolton H
Bucklin, Robert Van Zandt
Di Maio, Vincent J M
Fatteh, Abdullah Valimohammed
Hicks, James Thomas
Hudson, R Page, Jr
Krishnan, Silvarama S
Lehotay, Judith Mona
Lowry, William Thomas
Mills, Don Harper
Perr, Irwin Norman
Spitz, Werner Uri
Zugibe, Frederick T

Gastroenterology
Adibi, Siamak A
Ament, Marvin Earl
Arias, Irwin Monroe
Bacon, Harry Elliott
Baldus, William Philip
Barbero, Giulio J
Barrett, Peter Van Doren
Barrowman, James Adams
Bartholomew, Lloyd Gibson
Beck, Ivan Thomas
Behar, Jose
Benson, John, Alexander, Jr
Biempica, Luis
Binder, Henry Joseph
Bondy, Donald Clarence
Bounous, Gustavo
Boyce, Henry Worth, Jr
Bradford, James Carrow
Bralow, S Philip
Breuhaus, Herbert Charles
Brick, Irving B
Buhac, Ivo
Cain, James Clarence
Calkins, William Graham
Cayer, David
Chalmers, Thomas Clark
Cheng, Hazel Pei-Ling
Christensen, James
Connell, Alastair McCrae
Cook, Michael Arnold
Corcino, Jose Juan
Crump, Malcolm Hart
Dagradi, Angelo E
Davenport, Horace Willard
Davis, William Duncan, Jr
Dietschy, John Maurice
Donaldson, Robert M, Jr
Dower, John Charles
Dworken, Harvey J
Dyck, Walter Peter
Eisenberg, M Michael
Englert, Edwin, Jr
Faloon, William Wassell
Farrar, John Thruston
Finkelstein, James David
Fleshler, Bertram
Folley, Jarrett Harter
Freston, James W
Fromm, Hans
Fuller, Richard Kenneth
Gambescia, Joseph Marion
Garrett, J Marshall
Glass, George B Jerzy
Goldstein, Franz
Gompertz, Michael L
Grace, Norman David
Gray, Gary M
Gregory, Daniel Hayes
Grossman, Morton Irvin
Grossmann, Roberto Jose
Guth, Paul Henry
Hall, William Harvey
Hanson, Wayne Robert
Hanson, Richard Jacob
Henley, Keith Stuart
Herber, Raymond
Hofmann, Alan Frederick
Hugon, Jean S
Hughes, William Stevenson
Janowitz, Henry David
Jeffries, Graham Harry
Joseph, Ramon R
Kalser, Martin
Keller, Reed Theodore
Kern, Fred, Jr
Kimberg, Daniel Victor
Kirsner, Joseph Barnett
Knight, William Allen, Jr
Knill, James Reginald
Koff, Raymond Steven
Kraft, Sumner Charles
Law, David H
Legerton, Clarence W, Jr
Levitan, Ruven
Lipkin, Michael D
Loewenstein, Matthew Samuel
MacDonald, Walter Charlton
McGill, Douglas B
McGuigan, James E
Merritt, Alfred M, II
Moore, Edward Weldon
Morrissey, John F
Nahrwold, David Lange
Nigaglioni, Adan
Olsen, Ward Alan
Ostrow, Jay Donald
Palmer, Eddy Donald
Palmer, Robert Howard
Paustian, Frederick Franz
Paynter, Camen Russell
Pfeiffer, Carl J
Phillips, Sidney Frederick
Pittman, Fred Estes
Plaut, Andrew George
Polish, Edwin
Pope, Charles Edward, II
Powell, Don Watson
Raffensperger, Edward Cowell
Redinger, Richard Norman
Rhodes, James B
Rider, Joseph Alfred
Rodgers, John Barclay, Jr
Rosenberg, Irwin Harold
Rosenthal, William S
Roth, Harold Philmore
Roth, James Luther Aumont
Roy, Claude Charles
Rubin, Cyrus F
Rubin, Walter
Ruffin, Julian Meade
Sabesin, Seymour Marshall
Samloff, I Michael
Sapp, Oscar LeMay, III
Scheig, Robert L
Schenker, Steven
Schoenfield, Leslie Jack
Schofield, Brian
Schwabe, Arthur David
Scott, Norman McLean, Jr
Segal, Harry L
Senior, John Robert
Sevelius, Hilli
Sherlock, Paul
Sifre, Ramon Alberto
Simon, Jerome Barnet
Skoryna, Stanley C
Soergel, Konrad H
Sorrell, Michael Floyd
Sparks, Robert D
Sternlieb, Irmin
Summers, Robert Wendell
Summerskill, William Hedley John
Tavill, Anthony Sydney
Tennant, Bud C
Texter, Elmer Clinton, Jr
Thayer, Walter Raymond, Jr
Torres-Pinedo, Ramon
Trier, Jerry Steven
Vennes, Jack A
Vlahcevic, Zdravko Reno
Watson, William Crawford
Webster, Paul Daniel, III
Weser, Elliot
Williams, Christopher Noel
Wilson, Richard Atwood
Wilson, Edward C
Winawer, Sidney J
Witten, Thomas A
Wong, Patrick Yu-Pei
Zamcheck, Norman
Zfass, Alvin Martin

Geriatrics
Baker, Saul Phillip
Beauchene, Roy E
Grannis, George Franklin
Lang, Calvin Allen
Lawton, Alfred Henry
Miquel, Jaime
Nandy, Kalidas
Rossman, Isadore
Schneider, Edward Lewis
Thompson, Phebe Kirsten

Hematology
Abaidoo, Kodwo-James R
Adler, Solomon Stanley
Alfrey, Clarence P, Jr
Alter, Harvey James
Ambrus, Clara Maria
Ambrus, Julian Lawrence
Asakura, Toshio
Athens, John William
Axelrad, Arthur Aaron
Axelrod, Arnold Raymond

Babior, Bernard M
Baiardi, John Charles
Baker, Michael Allen
Balcerzak, Stanley Paul
Baldini, Mario G
Ballard, Marguerite Candler
Bang, Nils Ulrik
Barker, Jane Ellen
Barnes, Asa, Jr
Barrett, O'Neill, Jr
Bateman, Joseph R
Beck, William Samson
Begg, Charles Frederic
Beizer, Lawrence H
Bell, Warren Napier
Bell, William Robert, Jr
Bennett, John M
Bennett, Michael
Berman, Irwin
Berry, Daisilee H
Bertles, John F
Best, William Robert
Bick, Rodger Lee
Blaisdell, Richard Kekuni
Block, Matthew Harold
Boggs, Dane Ruffner
Bond, William H
Borges, Wayne Howard
Bottomley, Sylvia Stakle
Bove, Joseph Anthony
Bowdler, Anthony John
Bowie, Edward John Walter
Bradley, Thomas Bernard, Jr
Briggs, Donald K
Brody, Jerome Ira
Brown, Audrey Kathleen
Brown, Elmer Burrell
Bruning, Richard Dale
Bullock, William Horace
Burka, Edward Richard
Butler, John Joseph
Calhoun, Mary Lois
Call, Tracey Gillette
Cameron, Bruce Francis
Campbell, Earl William
Carmel, Ralph
Carter, Robert Eldred
Castle, William Bosworth
Cawein, Madison Julius
Chaudhuri, Tuhin
Chernoff, Amoz Immanuel
Chervenick, Paul A
Clarkson, Bayard D
Claveau, Rosario
Clifford, George O
Coleman, Morton
Coltman, Charles Arthur, Jr
Conrad, Marcel E
Cooper, Bernard A
Cooper, George William
Cooper, Miles Robert
Corby, Donald G
Corcino, Jose Juan
Corley, Charles Calhoun, Jr
Corrigan, James John, Jr
Cousineau, Leo
Crookston, John Hamill
Crosby, William Holmes, Jr
Dallman, Peter R
Davis, Richard Bradley
Day, Harvey James
Dugdale, Marion
Dyke, Richard Warren
Eaton, John Wallace
Ebaugh, Franklin G, Jr
Eichner, Edward Randolph
Engle, Ralph Landis, Jr
Epstein, Robert Bernard
Eurenius, Karl
Fairbanks, Virgil
Farnes, Patricia
Fernbach, Donald Joseph
Filip, Donald Joseph
Fisher, Lyman McA
Flexner, John M
Fratantoni, Joseph Charles
Freireich, Emil J
Frenkel, Eugene Philip
Fried, Walter
Fudenberg, H Hugh
Furth, Frank Willard
Gabuzda, Thomas George
Gallagher, Neil Ignatius
Gamble, Jess Franklin
Ganguly, Pankaj
Gesinski, Raymond Marion
Gibbons, Katherine Bond
Giblett, Eloise Rosalie
Gibson, John Graham, II

Gidari, Anthony Salvatore
Gilbert, Harriet S
Giorgio, Anthony Joseph
Glomski, Chester Anthony
Goh, Kong-Oo
Goldwein, Manfred Isaac
Golomb, Harvey Morris
Goswitz, Francis Andrew
Goswitz, Helen Vodopick
Gottfried, Eugene Leslie
Gottlieb, Arlan J
Goulian, Mehran
Green, David
Green, Ralph
Greenwalt, Tibor Jack
Handler, Evelyn Erika
Harkness, Donald R
Harley, John Barker
Harrison, David Ellsworth
Hartmann, John Rudolf
Hartmann, Robert Carl
Hattersley, Paul G
Haut, Arthur
Hayes, Donald M
Hayes, Esther Fincher
Heller, Paul
Heyssel, Robert M
Hinz, Carl Frederick, Jr
Howell, Doris Ahlee
Hrynuik, William
Hutton, John James, Jr
Ihler, Garret Martin
Israels, Lyonel Garry
Jackson, Carl Wayne
Jackson, Dudley Pennington
Jacob, Harry S
Jensen, Wallace Norup
Johnson, Alan J
Josephson, Aaron Mortimer
Kaplan, Manuel E
Kaplan, Sandra Solon
Kashatus, William C
Kazal, Louis Anthony
Keene, Willis Riggs
Kim, Agnes Kyung-Hee
Kingdon, Henry Shannon
Kirby, Edward Paul
Kitchen, Hyram
Klemperer, Martin R
Klingberg, William Gene
Knospe, William H
Kochen, Joseph Abraham
Koepke, John Arthur
Kornfeld, Stuart Arthur
Krantz, Sanford B
Krivit, William
Kurnick, John Edmund
Kwaan, Hau Cheong
LaCelle, Paul (Louis)
Lackner, Henriette
Landaw, Stephen Arthur
Lange, Robert Dale
Langley, G R
Lanzkowsky, Philip
Lee, Glenn Richard
Lee, Stanley L
Levere, Richard David
Levin, Jack
Lewis, Richard John
Lichtman, Marshall A
Lobuglio, Albert Francis
Loh, Jerome Wei-Ping
Lopez, Rafael
Lorber, Mortimer
Louis, John
Lozzio, Bismarck Berto
Lucas, Oscar Nestor
Luhby, Adrian Leonard
Lundblad, Roger Lauren
Lynch, Edward Conover
MacDonald, Ronald Neil Angus
MacKinney, Archie Allen, Jr
Majerus, Philip W
Maldonado, Jorge Eusebio
Marciniak, Ewa
Marcus, Aaron Jacob
Masouredis, Serafeim Panagiotis
Mayer, Klaus
McCurdy, Paul Ranney
McDonald, John William David
McFarland, William
McIntyre, Oswald Ross
McKee, Patrick Allen
McKelvey, Eugene Mowry
McManus, Thomas (Joseph)
McMillan, Campbell White
Meck, Robert Allen
Medici, Paul T
Mendel, Gerald Alan
Mengel, Charles E
Meyer, Leo Martin
Mink, Irving Bernard
Minnich, Virginia
Mizukami, Hiroshi
Molinari, Pietro Filippo
Moloney, William Curry
Monette, Francis C
Moreno, Hernan

Morse, Edward Everett
Muller-Eberhard, Ursula
Nadell, Judith
Nagel, Ronald Lafuente
Neely, Charles Lea, Jr
Nelson, Douglas A
Nemerson, Yale
Newcomb, Thomas F
Niemetz, Julian
Nossel, Hymie L
Noyes, Ward David
Nussbaum, Murray
Osoba, David
Palmer, Jeffress Gary
Payne, Rose Marise
Pearson, Howard Allen
Perkins, Herbert Asa
Perry, Seymour Monroe
Phillips, Louise Lang
Piliero, Sam Joseph
Pinkerton, Peter Harvey
Piomelli, Sergio
Prasad, Ananda S
Price, David C
Rabiner, Saul Frederick
Ragan, Harvey Albert
Rambach, Walter A
Rand, Peter
Ranney, Helen M
Rapaport, Samuel I
Rappaport, Henry
Rath, Charles E
Rauchwerger, Joel M
Rausen, Aaron Reuben
Rebuck, John Walter
Reilly, Emmett B
Rhodes, Robert Shaw
Rieder, Ronald Frederic
Rifkind, Richard A
Rigby, Perry G
Ritzmann, Stephan E
Roberts, Harold R
Rosenfield, Richard Ernest
Rosse, Cornelius
Rothenberg, Sheldon Philip
Rubin, Arnold David
Sachs, John Richard
Sainte-Marie, Guy
Sassa, Shigeru
Saunders, Richard Henry, Jr
Sawitsky, Arthur
Schechter, Brent Allan
Schechter, Geraldine Poppa
Schick, Paul Kenneth
Schiffer, Lewis Martin
Schlossman, Stuart Franklin
Schmukler, Morton
Schnitzer, Bertram
Schorr, Julian
Schreiner, Albert William
Schwartz, Irving Robert
Schwartz, Robert Stewart
Schwartz, Steven Otto
Scott, John Gerald
Scott, Robert Bradley
Sears, David Alan
Shahidi, Nasrollah Thomas
Shanberge, Jacob N
Shanbrom, Edward
Shapiro, Lorne
Shapiro, Sandor Solomon
Sheehy, Thomas W
Shively, John Adrian
Shore, Nomie Abraham
Shulman, Nahum Raphael
Shumway, Clare Nelson, (Jr)
Silber, Robert
Silver, Ruth Kunkle
Simon, Ernest Robert
Simone, Joseph Vincent
Sirridge, Marjorie Spurrier
So, Antero Go
Solomon, Alan
Spivak, Jerry Lepow
Sprague, Charles Cameron
Stengle, James Marshall
Stolc, Viktor
Strauss, Ronald George
Streiff, Richard Reinhart
Stuckey, Walter Jackson, Jr
Sturgeon, Phillip
Stutman, Leonard Jay
Suhrland, Leif George
Swisher, Scott Neil
Tanaka, Kouichi Robert
Taub, Robert Norman
Tavassoli, Mehdi
Tishkoff, Garson Harold
Troup, Stanley Burton
Tsan, Min-Fu
Vietti, Teresa Jane
Vye, Malcolm Vincent
Wallerstein, Harry
Walsh, Peter Newton
Walters, Thomas Richard
Wang, Yeu-Ming Alexander
Waxman, Herbert Sumner
Weed, Robert I

Weintraub, Lewis Robert
Weisman, Russell
Weiss, Gary Bruce
White, Peter
Will, John Junior
Wilner, George Dubar
Wilson, Floyd Dee
Wilson, Henry E
Wimer, Bruce Meade
Wolff, James A
Wright, Claude-Starr
Yachnin, Stanley
Zalusky, Ralph
Zanjani, Esmail Dabaghchian
Zolton, Raymond Peter
Zucker, Marjorie Bass

History of Medicine
Agnew, Leslie Robert Corbet
Atwater, Edward Congdon
Bates, Donald George
Bensley, Edward Horton
Bodemer, Charles William
Brieger, Gert Henry
Brodman, Estelle
Burns, Chester Ray
Divett, Robert Thomas
Howard, Robert Palmer
Keys, Thomas Edward
Lutzker, Edythe
Mallen, Mario Salazar
Mayer, Claudius Francis
Mercer, Sherwood Rock
Nathan, Helmuth M
Necker, Walter Ludwig
Paterson, Garnet Russell
Risse, Guenter Bernhard
Roland, Charles Gordon
Saunders, John Bertrand De Cusance
Morant
Sharpe, William D
Veith, Ilza
Viseltear, Arthur Jack
Wilson, Leonard Gilchrist

Hospital Administration
Drosness, Daniel Leed
Gentry, John Tilmon
Griffith, John Randall
Hall, Esther Jane Wood
Levey, Samuel
Litsky, Bertha Yanis
Meredith, Charles Eymard
Mussells, Francis Lloyd
Nolan, Janiece Simmons
Phillips, Samuel
Rosenkrantz, Jacob Alvin
Simpson, William Stewart
Springall, Arthur Newton
Urbach, Karl Frederic
Weil, Thomas P
Westman, Ragnar Theophile

Industrial Medicine
Braun, Daniel Carl
Cady, Lee (De), Jr
Clyne, Robert Martin
Love, Robert Alexander
Page, Robert Clinton
Schmidt, James L
Thuss, William Getz, Jr

Infectious Diseases
Andriole, Vincent T
Artenstein, Malcolm S
Asher, David Michael
Bass, James W
Battigelli, Mario C
Baum, Stephen Graham
Beisel, William R
Bisno, Alan Lester
Bodel, Phyllis T
Bodey, Gerald Paul, Sr
Brandriss, Michael W
Brooks, George Frank, Jr
Cadigan, Francis C, Jr
Carpenter, Charles C J
Cassell, Gail Houston
Chang, Te Wen
Cherry, James Donald
Clark, Robert Amos
Cluff, Leighton Eggertsen
Clyde, Wallace Alexander, Jr
Cooper, Louis Zucker
Couch, Robert Barnard
Cox, Frederick Eugene
Craig, John Philip
Criswell, Bennie Sue
Crocker, Thomas Timothy
Cromartie, William James
Davis, Starkey D
Dewan, Edmond M
Donta, Sam Theodore
Douglas, Robert Gordon, Jr
Drusin, Lewis Martin
DuPont, Herbert Lancashire
Durr, Friedrich (E)
Echeverria, Peter Durand

MEDICINE

Eckert, Herbert L
Elisberg, Bennett La Dolce
Espana, Carlos
Evans, Hugh E
Farrar, William Edmund, Jr
Feigin, Ralph David
Fekety, F Robert, Jr
Finegold, Sydney Martin
Finland, Maxwell
Fischer, Janet Jordan
Forsyth, Ben Ralph
Foy, Hjordis M
Freiner, Earl Howard
Fricke, Howard Henry
Glezen, William Paul
Gold, Jerome A
Goldstein, Elliot
Gorbach, Sherwood Leslie
Gorzynski, Eugene Arthur
Grieble, Hans G
Gump, Dieter W
Gurwith, Marc Joseph
Hermans, Paul E
Hill, Gale Bartholomew
Hirschman, Shalom Zarach
Holzman, Robert Stephen
Hornick, Richard B
Hudson, Bruce William
Johnson, B Lamar, Jr
Johnson, Joseph Eggleston, III
Joncas, Jean Harry
Kantor, Harvey Sherwin
Kass, Edward Harold
Keller, Margaret Anne
Keusch, Gerald Tilden
Kieff, Elliott Dan
Klainer, Albert S
Klebanoff, Seymour J
Lamb, George Alexander
Liu, Chien
Madoff, Morton A
Majde, Jeannine Adkins
Makley, Torrence Aloysius, Jr
Marks, Melvin Issac
Marr, James Joseph
Martin, Rufus Russell
Mathies, Allen Wray, Jr
Mattern, Carl Frederick Theodore
McCabe, William R
McCall, Charles Emory
McCloskey, Richard Vensel
McCullough, Norman B
McGarrity, Gerard John
Merigan, Thomas Charles, Jr
Meyer, Richard David
Miller, James Nathaniel
Moffet, Hugh L
Montasir, Magda Moustafa
Niblack, John Franklin
Osborn, June Elaine
Page, Malcolm I
Palmer, Darwin I
Paterson, Philip Y
Paterson, Maria Jevitz
Perkins, Robert Louis
Pierce, Nathaniel Field
Postic, Bosko
Powanda, Michael Christopher
Pryles, Charles Victor
Raff, Martin Jay
Reisberg, Boris Elliott
Rifkind, David
Ritter, Harry Woodward
Rohling, Stephen Roy
Ronald, Allan Ross
Russell, Philip King
Sabin, Albert Bruce
Sanford, Jay Philip
Sayed, Hamdy I
Schacher, John Fredrick
Schiff, Gilbert Martin
Seligman, Stephen Jacob
Sellers, Thomas F, Jr
Shepard, Charles Carter
Shirgley, Edward White
Smith, Ian Maclean
Smith, John Kelly
Tamini, Handi Ahmad
Tempest, Bruce Dean
Thornton, George Fred
Thrupp, Lauri David
Togo, Yasushi
Topping, Norman Hawkins
Trifan, Deonisie
Turck, Marvin
Udem, Stephen Alexander
Valentine, Fred Townsend
Vavra, James Joseph
Watanakunakorn, Chatchai
Weinstein, Louis
Weller, Thomas Huckle
White, Arthur C
Wolff, Sheldon Malcolm
Wolinsky, Emanuel

Internal Medicine
Aagaard, George Nelson
Abboud, Francois Mitry

Adhikari, P K
Adolph, Robert J
Albrink, Margaret Joralemon
Aldrich, Franklin Dalton
Alexander, Fred
Alexander, James Kermott
Alexander, Sydenham Benoni
Alger, Elizabeth A
Allen, Herbert Clifton, Jr
Allen, Max Scott
Allen, Milton Winfield
Anderson, Paul Nathaniel
Anderson, Robert Spencer
Anderson, William Evan
Andreoli, Thomas Eugene
Andrews, Charles Edward
Andrews, Gould Arthur
Andriole, Vincent T
Angle, William Dodge
Appelbaum, Emanuel
Aranow, Henry
Archer, Juanita Almeta
Armstrong, John Buchanan
Astwood, Edwin Bennett
Athens, John William
Atkins, Elisha
Atwater, Edward Congdon
Auchincloss, Joseph Howland, Jr
Austen, K Frank
Austrian, Robert
Avad, William Michel, Jr
Ayers, Carlos R
Ayvazian, L Fred
Bacchus, Habeeb
Bainton, Dorothy Ford
Balceerzak, Stanley Paul
Balchum, Oscar Joseph
Balderrama, Francisco E
Baldus, William Phillip
Ball, Gene V
Ballard, Ian Matheson
Ballard, Marguerite Candler
Barbour, Allen Leroy
Barbour, Donald Curtiss
Barker, Earl Stephens
Barrett, Peter Van Doren
Barry, Kevin Gerard
Bartholomew, Lloyd Gibson
Barton, Evan Mansfield
Bauer, Franz Karl
Baum, Stephen Graham
Baum, William Stanhope
Baumeister, Carl Frederick
Baxter, Neal Edward
Baylor, Curtis Horton
Bazzano, Gaetano
Beard, Owen Wayne
Beck, Ivan Thomas
Beck, John Christian
Becker, Ernest Lovell
Becker, Kenneth Louis
Beckwith, Julian Ruffin
Bedell, George Noble
Beebe, Robert Townsend
Behnke, Roy Herbert
Benfield, William Harvey
Beizer, Lawrence H
Bell, William Robert, Jr
Beller, Barry M
Bello, Carmen T
Bennett, John M
Berenson, Gerald Sanders
Berge, Kenneth G
Bergsagel, Daniel Egil
Berliner, Robert William
Berman, Jerome Richard
Bernstein, Stanley H
Berris, Barnet
Berry, Maxwell (Rufus)
Bertles, John F
Berry, Richard John
Billings, Frederic Tremaine
Binder, Henry Joseph
Bing, Richard John
Bisno, Alan Lester
Bissell, Grosvenor Willse
Black, Roger Lewis
Blackard, William Griffith
Blackmon, John R
Blahd, William Henry
Blair, A James, Jr
Blaisdell, Richard Kekuni
Blaufox, Morton D
Bleicher, Sheldon Joseph
Bloom, Joseph Dutch
Bloomfield, Dennis Alexander
Bluemle, Lewis W, Jr
Blythe, William Brevard

Boake, William Charles
Boas, Norman Francis
Bodel, Phyllis T
Bole, Giles G, Jr
Bolt, Robert James
Bond, William H
Bono, Vincent Horace, Jr
Borden, Craig W
Borkon, Eli Leroy
Boshell, Buris Raye
Bottomley, Richard H
Bottomley, Sylvia Stakle
Bouchard, Richard Emile
Bourne, Frederick Munroe
Bowen, Peter
Bowerfind, Edgar Sihler, Jr
Boyce, Henry Worth, Jr
Bozian, Richard Charles
Bradley, Thomas Bernard, Jr
Bralow, S Philip
Bramante, Pietro Ottavio
Brandfonbrener, Martin
Brandt, J Leonard
Braunwald, Eugene
Brenner, Barry Morton
Brettell, Herbert R
Breuhaus, Herbert Charles
Brick, Irving B
Bricker, Neal S
Briehl, Robin Walt
Brien, Francis Staples
Brody, Jerome Ira
Bronsky, David
Brooks, George Frank, Jr
Brown, Walter John
Brown, Robert Wayne
Brown, Robert Stephen
Brown, Paul Wheeler
Brown, John Welch
Brown, Harvey Earl, Jr
Brown, J Malcolm
Brown, Douglas John
Bruce, Thomas Allen
Brubaker, Merlin L
Buchan, Douglas John
Buckley, Charles Edward
Buechner, Howard Albert
Buehler, Martin Stowell
Buhac, Ivo
Bullock, William Horace
Burch, Robert Emmett
Burke, John T
Burns, Thomas Wade
Burrow, Gerard N
Burrows, Benjamin
Buse, John Frederick
Butler, Vincent Paul, Jr
Butler, Hugh Roland
Butterworth, Charles E, Jr
Cade, James Robert
Cain, James Clarence
Caldwell, John R
Calkins, William Graham
Campbell, Eugene Paul
Cander, Leon
Canellos, George P
Capeci, Nicholas Ernest
Carpenter, Charles C J
Carpenter, Robert Raymond
Carr, David Turner
Carr, Edward Albert
Carryer, Haddon McCutchen
Carter, Curtis Harold
Carter, Stephen Keith
Cartwright, George Eastman
Cassan, Stanley Morris
Cassidy, Carl Eugene
Castor, Cecil William
Challoner, David Reynolds
Chalmers, Thomas Clark
Cheitlin, Melvin Donald
Cherniack, Neil S
Chernick, Paul A
Chester, Edward M
Chew, William Hubert, Jr
Chidsey, Charles Augustus
Chinard, Francis Pierre
Chiong, Miguel Angel
Chobanian, Aram V
Choppin, Purnell Whittington
Chopra, Inder Jit
Chosy, Julius J
Chou, Ching-Chung
Claman, Henry Neumann
Clapp, James R
Clapper, Muir
Clark, James Edward
Clarke, W T W
Claveau, Rosario
Clewe, Thomas Hailey
Clifford, George O
Clifton, James Albert
Coburn, Jack Wesley
Coe, Fredric Lawrence
Coffman, Jay D
Cohen, Burton D
Cohen, Jordan J
Cohen, Jules

Cohen, Lawrence Sorel
Cohen, Louis
Cohen, Margo Nita Panush
Cohen, Marvin J
Colsky, Jacob
Colwell, John Amory
Colwell, Jack M
Combes, Burton
Conn, Harold O
Connell, Walter Ford
Connor, William Elliott
Conrad, Marcel E
Conway, Gene Farris
Coodley, Eugene Leon
Corley, Charles Calhoun, Jr
Corman, Lew Andre
Cotter, Edward F
Couch, Robert Barnard
Couture, Roger
Coyne, Veronica E
Crawford, John S
Crede, Robert H
Crofford, Oscar Bledsoe
Cross, Richard James
Crozier, Dan
Crump, Jesse Franklin
Cummings, Nancy Boucot
Cummings, Norman Allen
Cummins, Alvin J
Curry, Francis J
Cutler, Paul
Dagradi, Angelo E
Daly, Walter J
Damon, Albert
Dardas, Terry Jay
Dascomb, Harry Emerson
Daugherty, Robert M, Jr
Davenport, Fred M
Davey, Winthrop Newbury
David, John R
Davidson, John Keay, III
Davidson, John R
Davies, David Hywel
Davis, Harry L
Davis, John Staige, IV
Davis, Richard Bradley
Davis, John Laws
Dawber, Thomas Royle
Dawson, J W
Day, Emerson
Day, Harvey James
Day, J H
Dayton, Seymour
Dean, David Campbell
Deane, Norman
Dearing, William Hill
Decker, John Laws
DeFelice, Eugene Anthony
DeGowin, Richard Louis
Deiss, William Paul, Jr
Deller, John Joseph, Jr
Delp, Mahlon (Henry)
Dennis, Edward Wimberly
Denny, William F
Des Prez, Roger Moister
DeVita, Vincent T, Jr
Dexter, Richard Newman
DeYoung, Willard G
Dickerman, Herbert W
Dickhaus, Donald William
Dickson, Robert Clark
Dietschy, John Maurice
Diosy, Andrew
Dobson, Harold Lawrence
Dodge, Harold T
Doe, Richard P
Doherty, James Edward, III
Doigin, Martin
Doikart, Ralph Elson
Donaldson, Robert M, Jr
Donati, Robert M
Douglas, Robert Gordon, Jr
Dousa, Thomas Patrick
Drucker, William D
Duff, Ivan Francis
Dugdale, Marion
Dujovne, Carlos A
Duncan, Katherine
Duncan, Leroy Lancashire
Dunn, Marvin I
DuPont, Herbert Lancashire
Durant, John Ridgway
Dutra Oliveira, Jose Eduardo
Dutton, Cynthia Baldwin
Dworken, Harvey J
Dyck, Walter Peter
Easterling, Ronald E
Ebbert, Arthur, Jr
Ebert, Robert H
Eckhardt, Richard Dale
Eckman, Elvira (Mrs Arthur Kirschbaum)
Eckstein, John William
Eddleman, Elvia Etheridge, Jr
Egeberg, Roger O
Ehrlich, Edward Norman
Ehrlich, George Edward
Eichenholz, Alfred
Eisenfeld, Arnold Joel
Eisenstein, Albert Bernard

Elisberg, Bennett La Dolce
Engelhardt, Hugo Tristram
Englert, Edwin, Jr
Epstein, Franklin Harold
Epstein, Wallace Victor
Ertel, Norman H
Evans, Robert L
Ezrin, Calvin
Fairbanks, Virgil
Fajans, Stefan Stanislaus
Fanestil, Darrell Dean
Farrar, George Elbert, Jr
Feinstein, Alvan Richard
Fekety, F Robert, Jr
Felig, Philip
Felts, John Harvey
Felts, William Robert
Ferris, Thomas Francis
Fialkow, Philip Jack
Figueroa, William Gutierrez
Finch, Stuart Cecil
Finley, Theodore N
Firstbrook, John Bradshaw
Fischel, Edward Elliot
Fischer, Janet Jordan
Fishman, Sherold
Flanagan, Charles Larkin
Fletcher, Anthony Phillips
Fletcher, Evan
Flexner, John M
Foley, H Thomas
Foley, John F
Folley, Jarrett Harter
Forbes, Allan Louis
Forker, Edson Lee
Forland, Marvin
Forsham, Peter Hugh
Forsyth, Ben Ralph
Foster, Daniel W
Fowler, Noble Owen
Fozzard, Harry A
Frank, Charles Warren
Frank, Martin J
Franks, John Julian
Fraser, Robert Stewart
Fraumeni, Joseph F, Jr
Freedman, Aaron David
Freedman, Lawrence Raphael
Freeman, Irving
Freeman, Richard B
Frei, Emil, III
Freilich, Joseph Kenneth
Frenkel, Norbert
Freireich, Abraham Walter
Freireich, Emil J
Frischer, Henri
Frohlich, Edward David
Frommeyer, Walter Benedict, Jr
Fulop, Milford
Gable, James Jackson, Jr
Gaintner, John Richard
Gallagher, Neil Ignatius
Gambescia, Joseph Marion
Gamble, Jess Franklin
Ganesan, Devaki
Gantt, Clarence Leroy
Garcia-Palmieri, Mario R
Gardner, Kenneth Drake, Jr
Gefter, William Irvin
Gendel, Benjamin Robert
Genecin, Abraham
Generales, Constantine D John
George, Ronald Baylis
Geraci, Joseph E
Gerber, Donald Albert
Gersberg, Herbert
Gertler, Menard M
Gibson, Sam Thompson
Gibson, Thomas Chometon
Gilbert, Elmer Wilhelm
Gilbert, James Alan Longmore
Gilbert, Norman Pettibone
Gilbert, Robert Pettibone
Gilgore, Sheldon G
Ginn, H Earl
Glabman, Sheldon
Gleich, Gerald J
Glick, Gerald
Glushien, Arthur Samuel
Goetz, Frederick Charles
Goger, Pauline Rohm
Gold, Jay Joseph
Gold, Jerome A
Goldberg, Leon Isadore
Goldenberg, Gerald J
Goldman, Ralph
Goldmann, Morton Aaron
Goldner, Martin Gerhard

Goldstein, Franz
Goldstein, Gideon
Goldwein, Manfred Isaac
Gompertz, Michael L
Goodfriend, Theodore L
Goodkind, Morton Jay
Goodman, DeWitt Stetten
Goodwin, Robert Archer, Jr
Gorbach, Sherwood Leslie
Gordon, Paul
Goresky, Carl A
Gorman, Colum A
Gosfield, Edward, Jr
Gottlieb, Arlan J
Gould, Lawrence A
Grace, Norman David
Graef, Irving Philip
Graham, David Tredway
Gray, Frank Davis, Jr
Green, Robert A
Greenberger, Norton Jerald
Gregg, Michael Barrows
Gregory, Raymond (Leslie)
Grieble, Hans G
Grissom, Robert Leslie
Griswold, Herbert Edward
Grob, David
Grossman, Laurence Abraham
Gubner, Richard S
Guerrant, John Lippincott
Guimaraes, Armenio Costa
Gump, Dieter W
Gundersen, Kare
Gunn, Chesterfield Garvin, Jr
Gunnar, Rolf McMillan
Gurney, Clifford W
Gurney, Ramsdell
Guth, Paul Henry
Guze, Samuel Barry
Gwinup, Grant
Hackney, Jack Dean
Hadden, John Winthrop
Haddock, Lillian
Haddy, Francis John
Hahn, Richard David
Hakala, Reino William
Hall, Wendell Howard
Hamerman, David Jay
Hamlin, James T, III
Hammarsten, James Francis
Hammond, James B
Hamolsky, Milton William
Handler, Joseph S
Hanson, Russell Floyd
Harley, John Barker
Harrington, William J
Harris, Barton A
Harvey, John Collins
Harwood, Theodore Henry
Hashim, Sami A
Haut, Arthur
Havens, Walter Paul, Jr
Haverback, Bernard Jacob
Hayes, Donald M
Hays, Marguerite Thompson
Hayward, Oliver Stoddard
Haywood, L Julian
Heaney, Robert Proulx
Heath, Frederick Kriete
Hefner, Lloyd Lee
Hejtmancik, Milton R
Heller, Paul
Helm, Robert Albert
Helwig, John, Jr
Henderson, Edward S
Hendler, Ernesto Danilo
Henig, Philip Edward
Henn, Mary Josephine
Herman, Robert Harold
Hermans, Paul E
Herring, William Benjamin
Hershfield, Earl S
Hershman, Jerome Marshall
Herting, Robert Leslie
Hess, Evelyn V
Hetig, Robert A
Heyman, Albert
Hiatt, Howard Haym
Hickman, Jack Walter
Higgins, James Thomas, Jr
Hildreth, Eugene Augustus
Hill, Samuel Richardson, Jr
Hilton, James Garrett
Hinz, Carl Frederick, Jr
Hirshaut, Yashar
Hoak, John C
Hobson, Lawrence Bennett
Hodges, Robert Edgar
Hoeprich, Paul Daniel
Holland, Paul Vincent
Hollander, Vincent Paul
Hollander, Walter, Jr
Hollander, William
Hollifield, Guy F
Hollingsworth, James W
Hollis, Walter Jesse
Holman, Halsted Reid
Holmes, Joseph Henry

Holmes, Randall Kent
Holub, Donald Arthur
Hood, William Boyd, Jr
Hook, Edward W, Jr
Horan, Leo Gallaspy
Hornick, Richard B
Houston, Charles Snead
Howard, Robert Palmer
Howe, Clifton Dexter
Howie, Donald Lavern
Howland, Joe Wiseman
Huff, Thomas Allen
Huguley, Charles Mason, Jr
Hulet, William Henry
Hull, Edgar
Hunter, Thomas Harrison
Huntley, Robert Ross
Hursh, Laurence M
Hurst, Victor Waldemar, III
Hurxthal, Lewis Marshall
Huston, John Howard
Hutt, Martin P
Hyatt, Robert Eliot
Hyman, Edward Sidney
Ingelfinger, Franz Joseph
Inkley, Scott Russell
Irwin, Glenn Ward, Jr
Isbell, Harris
Jackson, George Gee
Jackson, John Fenwick
Jacob, Harry S
Jacox, Ralph Franklin
Jaffe, Israeli Aaron
Jager, Blair Valdemar
Jarroid, Thomas
Jasin, Hugo E
Jeffries, Graham Harry
Johanson, Waldemar Gustave, Jr
Johns, Varner Jay, Jr
Johnsen, Sherman Edward Jerome
Johnson, B Lamar, Jr
Johnson, Ben Butler
Johnson, Joseph Eggleston, III
Johnson, Kenneth Gerald
Johnson, Tom Milroy
Johnson, William Cone
Johnston, Cyrus Conrad, Jr
Jones, John Evan
Jones, Ralph, Jr
Jourdonais, Leonard Francis
Joy, Robert John Thomas
Joy, Vincent Anthony
Joyner, John T, III
Juergens, John Louis
Kabler, J D
Kagiwanda, Harriet Hatsune
Kalmanson, George Maurice
Kannel, William B
Kaplan, Ervin
Kaplan, Manuel E
Kaplan, Norman
Kaufmann, Leon A
Kaufmann, John Simpson
Keene, Willis Riggs
Kefalides, Nicholas Alexander
Kelleher, Philip Conboy
Kelley, William Nimmons
Kelly, Howard Garfield
Kelly, John J
Kelser, George Archibald
Kempner, Walter
Kennedy, Byrl James
Kern, Fred, Jr
Kessner, David Morton
Kiang, David Teh-Ming
Kilburn, Kaye Hatch
Killip, Thomas, III
Killough, John Harvey
Kimball, Chase Patterson
Kimberg, Daniel Victor
King, Morris Kenton
Kinsella, Ralph A, Jr
Kipnis, David Morris
Kirkendall, Walter Murray
Kirschner, Marvin Abraham
Kirsner, Joseph Barnett
Kissin, Benjamin
Kitay, Julian Israel
Klainer, Albert S
Klapper, Margaret Strange
Klatskin, Gerald
Klevay, Leslie Michael
Kliman, Allan
Klinenberg, James Robert
Klinghoffer, June F
Klumpp, Theodore George
Knoebel, Suzanne Buckner
Knopf, Ralph Fred
Knowles, Harvey C, Jr
Knowles, John Hilton
Koch-Weser, Jan
Koelsche, Giles Alexander

Koff, Raymond Steven
Koide, Samuel Saburo
Kopin, Irwin J
Korst, Donald Richardson
Kosbab, Frederic Paul Gustav
Kottke, Bruce Allen
Kowal, Jerome
Kraft, Sumner Charles
Kraikipanitch, Sompong
Kramer, Norman Clifford
Krantz, Sanford B
Kredich, Nicholas M
Kriss, Joseph P
Kuida, Hiroshi
Kunin, Calvin Murry
Kuo, Peter Te
Kupfer, Sherman
Kwaan, Hau Cheong
Kyser, Franklin A
Ladd, Anthony Thornton
LaDue, John Samuel
Landaw, Stephen Arthur
Landowne, Milton
Lange, Kurt
Langford, Herbert Gaines
Langley, G R
Lansbury, John
Lasseter, Kenneth Carlyle
Lavender, Ardis Ray
Law, David H
Lawrason, F Douglas
Leaf, Alexander
Lee, Glenn Richard
Lee, Philip Randolph
Lee, Stanley L
Leedom, John Milton
Lehnhoff, Henry John, Jr
Leiter, Louis
LeMaistre, Charles Aubrey
Lemieux, Guy
Lennon, Edward Joseph
Leonard, James Joseph
Lerner, Albert Martin
Leser, Ralph Ulrich
Lesser, Gerson Theodore
Levere, Richard David
Levey, Gerald Saul
Levin, Murray Laurence
Levin, Seymour R
Levin, William Cohn
Levine, Harry
Levine, Isidore
Levine, Robert John
Levitan, Ruven
Levitsky, John M
Levitt, Abel
Levitt, Barrie
Lewis, George W
Lewis, Howard Phelps
Li, Min Chiu
Lichtman, Herbert Charles
Lieber, Charles Saul
Lindeman, Robert D
Lindholm, Dale David
Linn, Richard Harry
Lipicky, Raymond John
Lipkin, Mack
Lipsett, Mortimer Broadwin
Lipton, Allan
Lisansky, Ephraim Theodore
Livingston, Virginia
Locke, William
Lockwood, William Rutledge
Loewenstein, Joseph Edward
London, William Thomas
Lorber, Arthur
Loudon, Robert G
Louria, Donald Bruce
Lovell, Robert Gibson
Lubash, Glenn David
Luckey, Egbert Hugh
Lufkin, Edward Gwynne
Lurie, Aron Osher
Luton, Edgar Frank
Lynch, Edward Conover
MacDiarmid, William Donald
Mack, Irving
Mack, Robert Emmet
MacKenzie, Malcolm R
Mackowiak, Robert Carl
Macpherson, Walter Everett
Mahler, Richard Joseph
Malawista, Stephen E
Maloof, Farahe
Mandelstam, Paul
Manning, Phil Richard
Marcus, Aaron Jacob
Marcus, Donald M
Marcus, Frank I
Mardiney, Michael Ralph, Jr
Margolin, Esar Gordon
Marine, William Murphy
Mark, Roger G
Marks, Leon Joseph
Marks, Paul A
Maronde, Robert Francis
Marr, James Joseph
Martin, Helen Eastman

Martin, John Harvey
Martin, Rufus Russell
Martin, Samuel Preston, III
Martinez-Lopez, Jorge Ignacio
Martinez-Maldonado, Manuel
Martt, Jack M
Massey, Robert Unruh
Mathewson, Francis Alexander Lavens
Matz, Robert
Mayer, Klaus
McBain, John Keith
McCann, Margaret Mary
McCarty, Daniel J, Jr
McCollister, Robert John
McCombs, Robert Pratt
McConahey, William McConnell, Jr
McCurdy, Paul Ranney
McDonald, John William David
McFarland, Kay Flowers
McIlroy, Malcolm B
McIntosh, Henry Deane
McKee, Kelly Tilson
McKelvey, Eugene Mowry
McKenzie, John Maxwell
McLean, James Amos
McMahon, Francis Gilbert
McPhaul, John J, Jr
Meadows, William R
Meakin, James William
Meek, Joseph Chester, Jr
Meiselas, Leonard E
Mendel, Gerald Alan
Mendlowitz, Milton
Meroney, William Hyde, III
Merriman, John Edward
Merritt, Arthur Donald
Metz, Robert John Samuel
Michael, Max, Jr
Middleton, Elliot, Jr
Mikkelsen, William Mitchell
Milder, Jack Walter
Miller, Kay
Miller, Myron
Milliken, John Andrew
Mills, Lewis Craig, Jr
Mitchell, Ian Alastair
Mock, David Clinton, Jr
Mogabgab, William Joseph
Mohiuddin, Syed M
Molnar, George D
Morgan, David Zackquill
Morgan, Jean McNeil
Morgan, Thomas Edward, Jr
Morgan, William Keith C
Morgan, William L, Jr
Morganroth, Joel
Morita, Yoshikazu
Morris, James Joseph, Jr
Morrison, Lester Marvin
Moseley, Vince
Moser, Marvin
Mosesson, Michael W
Moss, James Mercer
Motulsky, Arno Gunther
Motz, Robin Owen
Muchmore, Harold Gordon
Mueller, John Frederick
Mulhausen, Robert Oscar
Murdaugh, Herschel Victor, Jr
Murray, John Patrick
Musser, Marc James, (Jr)
Myers, Warren Powers Laird
Myerson, Ralph M
Nadler, Charles Fenger
Nance, Walter Elmore
Naughton, John Patrick
Nedwicki, Edward G
Nelp, Wil B
Neva, Franklin Allen
Newberry, William Marcus
Newcomb, Thomas F
Nichols, George, Jr
Niden, Albert H
Niederman, James Corson
Nigaglioni, Adan
Noback, Richardson K
Noehren, Theodore Henry
Northrop, Gretajo
Noyes, Ward David
Nuetzel, John Arlington
Nunemaker, Charles Arter, Jr
Nuttall, Frank Q
Nyboer, Jan
Oaks, Wilbur W
Oates, John Alexander
O'Donohue, Walter John, Jr
O'Donovan, Cornelius Joseph
Ogden, David Anderson
Ogylzo, Metro Alexander
Ohnuma, Takao
Olsen, Arthur Martin
Olsen, Ward Alan
Oppenheimer, Jack Hans
Osterland, C Kirk
Owen, John Atkinson, Jr

Owens, Albert Henry, Jr
Page, Lot Bates
Page, Malcolm I
Palmer, Jeffress Gary
Palmer, Robert Howard
Pannill, Fitzhugh Carter, Jr
Papac, Rose
Papavasiliou, Paul S
Papper, Solomon
Parker, Brent M
Parker, Donal C
Parker, Richard H
Parson, William
Patterson, James Fulton
Patterson, John Legerwood, Jr
Patterson, Roy
Paulsen, Charles Alvin
Paustian, Frederick Franz
Paynter, Carmen Russell
Peake, Robert Lee
Peckinpaugh, Robert Owen
Periman, Philip
Perkins, Robert Louis
Perkoff, Gerald Thomas
Peter, James Bernard
Peterson, Edward S
Peterson, Richard Elsworth
Petit, Donald William
Phillips, Charles Alan
Pierce, Alan Kraft
Pierce, John Eric
Pierre, Robert V
Pilz, Clifford G
Pines, Kermit L
Pisciotta, Anthony Vito
Pitcairn, Donald M
Pittman, James Allen, Jr
Place, Virgil Alan
Plager, John Everett
Platzer, Richard France
Plotz, Charles M
Plough, Irvin Chaffee
Poffenbarger, Philip Lynn
Polish, Edwin
Pollak, Victor Eugene
Pollard, Herman Marvin
Posen, Gerald
Potter, Jacobus Louw
Potts, John Thomas
Powell, Richard Cinclair
Power, Gordon G
Power, Lawrence
Prasad, Ananda S
Price, John David Ewart
Proger, Samuel
Proper, Robert
Pugh, David Milton
Pullman, Theodore Neil
Puschett, Jules Bernard
Rabiner, Saul Frederick
Rabinovich, Sergio Rospigliosi
Rabinowitz, Murray
Raisz, Lawrence Gideon
Rakita, Louis
Rambach, Walter A
Rammelkamp, Charles Henry
Ramos-Morales, Francisco
Ramsey, Lloyd Hamilton
Randall, Russell E, Jr
Rankin, John
Ranney, Helen M
Rapaport, Elliot
Rapaport, Samuel I
Ratnoff, Oscar Davis
Ray, Clarence Thorpe
Ray, Edward Scott
Reed, Charles E
Reed, Melvin LeRoy
Regan, Timothy Joseph
Reichlin, Seymour
Remington, Jack Samuel
Reynolds, Telfer Barkley
Rhoades, Everett Ronald
Riesselbach, Richard Edgar
Rifkind, David
Rigby, Perry G
Riggs, Byron Lawrence
Rike, Paul Miller
Ritzmann, Leonard W
Riven, Samuel Saul
Roberts, Harold R
Robinson, Roscoe Ross
Robinson, William Dodd
Robinson, William Sidney
Rodgers, John Barclay, Jr
Rogers, David Elliott
Roguska-Kyts, Jadwiga
Rolett, Ellis Lawrence
Romansky, Monroe James
Roof, Betty Sams
Rose, Harold D
Rosen, Saul W
Rosenberg, Saul A
Rosenzweig, David Yates
Ross, Gordon
Ross, Joseph C

Rossall, Richard Edward
Rossen, Roger Downey
Rothenberg, Sheldon Philip
Rothschild, Edmund Otto
Rothschild, Marcus Adolphus
Rowley, Peter Templeton
Rowley, George G
Rubenstein, Edward
Rubin, Albert Louis
Rubin, Samuel H
Rundles, Ralph Wayne
Runyan, John William, Jr
Russe, Henry Paul
Sabesin, Seymour Marshall
Sachs, Frederick Lee
Sachs, Marvin Leonard
Sackett, David Lawrence
Sagan, Leonard A
Said, Sami I
Salvaggio, John Edmond
Samet, Philip
Samson, Donald C
Sandberg, Avery Aba
Sanders, Louis Lee
Sandstead, Harold Hilton
Sapp, Oscar LeMay, III
Saville, Paul D
Saward, Ernest Welton
Sawyer, William D
Schatz, Irwin Jacob
Scheig, Robert L
Schenker, Steven
Schere, Lawrence
Schiff, James Ferdinand
Schiff, Leon
Schlant, Robert C
Schlueter, Donald Paul
Schmid, Rudi
Schmidt, Alexander Mackay
Schnabel, Truman Gross, Jr
Schneckloth, Roland Edmunds
Schneiderman, Lawrence J
Schoch, Henry Kramer
Schoenfield, Leslie Jack
Schreiner, Albert William
Schroder, Jack Spalding
Schrogie, John Joseph
Schtengart, David E
Schulman, John, Jr
Schultz, Alvin Leroy
Schur, Peter Henry
Schwartz, William Benjamin
Scott, John Gerald
Scott, Norman McLean, Jr
Scott, Ralph Carmen
Scott, Robert Bradley
Scott, Virgil Cole
Seabury, John Hollister
Seal, John Ridley
Sears, John Alan
Segal, Bernard L
Segre, Eugene J
Seldin, Donald Wayne
Seneca, Harry
Sessoms, Stuart McGuire
Seto, Jane Mei-Chun
Shanbrom, Edward
Shane, Samuel Jacob
Shapiro, Alvin Philip
Shapiro, Lorne
Shapiro, William
Sharp, John T
Shear, Leroy
Shearn, Martin Alvin
Sheets, Raymond Franklin
Sherlock, Paul
Sherman, Jacques Lawrence, Jr
Sherman, Lawrence
Shields, George Seamon
Shinn, Robert A
Shnider, Bruce I
Shorey, Winston Kinney
Shuhman, Jones Alvin
Shulman, Lawrence Edward
Shuman, Charles Ross
Sidel, Victor William
Siegal, Frederick Paul
Siegler, Peter Emery
Siemsen, Jan Karl
Sifre, Ramon Alberto
Silber, Robert
Silva, Omega Logan
Simmons, Daniel Harold
Simmons, Harry Dady, Jr
Simon, Geza
Simpson, David Gordon
Singewald, Martin Louis
Sinkovics, Joseph
Siperstein, Marvin David
Skoog, William Arthur
Slade, H Clyde
Slavin, Howard Bernard
Slavin, Raymond Granam
Sleeth, Clark Kendall
Slocumb, Charles Henry

Slotkoff, Lawrence M
Smith, Ian Maclean
Smulyan, Harold
Snyder, Louis Michael
Snyderman, Ralph
So, Antero Go
Soergel, Konrad H
Sokal, Joseph Emanuel
Solberg, Lawrence Arthur, Jr
Southern, A Louis
Southworth, Hamilton
Sparks, Robert D
Spickler, J William
Spink, Wesley William
Spittell, John A, Jr
Sprague, Charles Cameron
Sprague, Randall George
Spraragen, Sanford C
Stead, William White
Steel, R Knight
Stengle, James Marshall
Stephens, Charles Arthur Lloyd, Jr
Stewart, Thomas Henry McKenzie
Stokes, Joseph, III
Stolfi, Julius E
Stone, Daniel Boxall
Stonehill, Robert Berrell
Stopford, Woodhall
Stopps, Gordon James
Stout, Landon Clarke, Jr
Straus, Bernard
Streeten, David Henry Palmer
Strong, Cameron Gordon
Stuckey, Walter Jackson, Jr
Sturzman, Leon
Surawicz, Borys
Sutliff, Wheelan Dwight
Sutnick, Alton Ivan
Sweeney, William Michael
Sweet, Herbert C
Sweet, Norman Joseph
Swezey, Robert Leonard
Swisher, Scott Neil
Tamburro, Carlo Horace
Tan, Charles Hua-Min
Tan, Meng Hee
Tanzi, Fausto
Tapley, Donald Fraser
Taranta, Angelo
Tarlov, Alvin R
Taylor, Fletcher Brandon, Jr
Taylor, Robert Mackay
Taylor, William Jape
Terepka, Anthony Raymond
Tesar, Joseph Thomas
Theil, George B
Theilen, Ernest Otto
Theologides, Athanasios
Thomas, George Edward
Thomas, Lewis
Thomas, William Clark, Jr
Thompson, David Duval
Thompson, George Richard
Thompson, John S
Thomson, Andrew
Thomson, Gerald Edmund
Thornton, George Fred
Ting, Er Yi
Tisi, Gennaro Michael
Tobian, Louis
Tobin, John Robert, Jr
Tompsett, Ralph Raymond
Tourtellotte, Charles Dee
Towbin, Eugene Jonas
Townes, Alexander Sloan
Tranquada, Robert Ernest
Travis, David M
Triantaphyllopoulos, Demetrios
Trier, Jerry Steven
Troup, Stanley Burton
Tsagaris, Theofilos John
Tschudy, Donald P
Tucker, William Boose
Tuna, Naip
Tupper, Charles John
Turchi, Joseph J
Turck, Marvin
Turner, John Dean
Tuttle, Elbert P, Jr
Tyan, Marvin L
Udall, John Alfred
Uhl, Henry Stephen Magraw
Ulmer, David D
Unger, Roger Harold
Vaillancourt, De Guise
Vaisrub, Samuel
Valtin, Heinz
Van Dellen, Theodore Robert
Van Woert, Melvin H
Vaules, David Wilson
Vennes, Jack A
Vertes, Victor
Vester, John William

Wagner, Henry N, Jr
Waife, Sholom Omi
Wakim, Khalil Georges
Waldstein, Sheldon Saul
Walker, James Elliot Cabot
Walker, Wilbur Gordon
Wallace, Andrew Grover
Wallace, Tracy I
Walsh, John Richard
Walsh, William J
Wang, Richard I H
Ware, Frederick
Warren, James Vaughn
Wasserman, Albert J
Watanakunakorn, Chatrchai
Watkin, Donald Morgan
Watson, David Werner
Watson, William Crawford
Watts, Malcolm S M
Waugh, William Howard
Wayburn, Edgar
Weidler, Donald John
Weil, John Victor
Weinstein, Louis
Weintraub, Bruce Dale
Weir, Donald Douglas
Weiss, Arthur Jacobs
Weiss, Harvey Jerome
Weissler, Arnold M
Weissmann, Gerald
Weller, John Martin
Wells, Roe
Welsh, George W, III
Wesson, Laurence Goddard, Jr
Westlake, Robert Elmer
Westura, Edwin Eugene
Whang, Robert
Wheby, Munsey S
Whitcomb, Walter Henry
White, Arthur C
Whitelaw, Donald Mackay
Whitsett, Thomas L
Whyte, Michael Peter
Wiener, Stanley L
Wightman, Keith John Roy
Wildenthal, Kern
Williams, John F, Jr
Williams, Marshall Henry, Jr
Williams, Ralph C, Jr
Williams, Thomas Franklin
Wilson, Colon Hayes, Jr
Wilson, Donald Robert
Wilson, Edward C
Wilson, Henry E
Wilson, Jean Donald
Wilson, Sloan Jacob
Wimer, Bruce Meade
Winawer, Sidney J
Wintrobe, Maxwell Myer
Wise, Harold B
Wise, Robert Irby
Witten, Thomas A
Wolf, Stewart George, Jr
Wood, Francis C, Jr
Woolf, C R
Woske, Harry Max
Yachnin, Stanley
Yam, Lung Tsiong
Yoo, Tai-June
Young, Charles William
Young, Daniel Test
Yu, Paul N
Yuceoglu, Yusuf Ziya
Yunis, Adel A
Zalusky, Ralph
Zarafonetis, Chris John Dimiter
Zatuchni, Jacob
Ziff, Morris
Zinneman, Horace Helmut
Ziskind, Morton Moses
Zsoter, Thomas
Zweifler, Andrew J

Laboratory Medicine
Blaker, Robert Gordon
Bove, Joseph Richard
Butts, William Cunningham
Chilcote, Max Eli
Clem, Judy Roberta
Dickman, Albert
Eldridge, John Charles
Felsenfeld, Oscar
Hubbard, Richard W
Hyndman, Lee Allen
Jackson, James Albert
James, Mary Frances
Kumar, Vijay
Lewis, William Weston
McLendon, William Woodard
Monroe, Robert Adams
Orto, Louise A (Mrs Patrick Famighetti)
Shamos, Morris Herbert
Sirridge, Marjorie Spurrier
Werner, Mario
Wichelhausen, Ruth Hechler
Zuckerman, Leo

Laryngology
Canfield, Norton
O'Brien, Joan A
Tucker, Gabriel Frederick, Jr

Leprology
Imaeda, Tamotsu

Malariology
Cabrera, Edelberto Jose
Canfield, Craig Jennings
Cox, Herbert Walton
Jeffrey, Geoffrey Marron
Kessinger, Walter Paul, Jr
Pletsch, Donald James
Rencricca, Nicholas John
Siu, Patrick Mew Lum
Warren, McWilson
Young, Martin Dunaway

Medical Administration
Alberts, Walter Watson
Alexander, C Alex
Armstrong, James G
Asper, Samuel Phillips, Jr
Ausman, Robert K
Beatty, Oren Alexander
Beck, John Edwin
Black, Emilie A
Bluestone, E Michael
Bonnet, Philip D
Brown, Ernest Benton, Jr
Cadmus, Robert R
Campbell, Colin
Clark, William Melvin, Jr
Crawford, Stanley Everett
Dyson, James Everett, Jr
Ebaugh, Franklin G, Jr
Ebert, Robert H
Egeberg, Roger O
Engebretson, Gordon Roy
Gallagher, James Daniel
Gay, William Ingalls
Goldstein, Murray
Green, Robert A
Grover, M Roberts, Jr
Hilgar, Arthur Gilbert
Joyner, Howard Sajon
King, Dannie Hilleary
Korn, Roy Joseph
La Salle, Gerald
Lee, Leavie Edgar, Jr
Levine, Jules Ivan
Martindale, Robert Warren
McCarroll, James Renwick
McLeod, Lionel Everett
Merrill, Joseph Melton
Mitchell, Ian Alastair
Mulhausen, Robert Oscar
Parkin, Robert Charles
Petersen, John Robert
Peterson, Lysle Henry
Pollack, Burton Robert
Porterfield, John Donaldson
Ramsey, Lloyd Hamilton
Reifler, Clifford Bruce
Reinhardt, William Oscar
Rieke, William Oliver
Riemer-Rubenstein, Delilah
Robertson, William O
Rosen, Harry Mark
Rothstein, Robert
Ruhe, Carl Henry William
Slavin, Howard Bernard
Steinmetz, Charles Henry
Sternfeld, Leon
Szumski, Stephen Aloysius
Vanselow, Neal A
Ware, Ray Wilsford
Wildberger, William Campbell
Wilson, Frank MacDonnell
Wilson, Vernon Earl

Medical Anthropology
Abernethy, Virginia Deane
Ablon, Joan
Blumberg, Baruch Samuel
Chrisman, Noel Judson
Hackenberg, Robert Allan
Hochstrasser, Donald Lee
Hughes, Charles Campbell
Jordan, Brigitte
Kennedy, Donald Alexander
Leslie, Charles Miller
Nydegger, Corinne Nemetz
Pearsall, Marion
Protsch, Reiner Robert Rudolf
Romanucci-Ross, Lola
Simeon, George John
Sipes, Richard Grey
Snow, Loudell Fromme
Townsend, John Marshall
Treloar, Alan Edward
Weiss, Charles
Wiese, Helen Jean Coleman
Wood, Corinne Shear
Zarrugh, Laura Hoffman

Medical Bacteriology
Binn, Leonard Norman
Blair, Eugene Baxter
Brewer, John Hanna
Briody, Bernard Aloysius
Bubel, Hans Curt
Calderone, Julius G
Deal, Samuel Joseph
Diamond, Ben Elkan
Eigelsbach, Henry Thomas
Eveland, Warren C
Foster, Robert Scott
Freter, Rolf Gustav
Gaines, Sidney
Gilardi, Gerald Leland
Gilmore, Eleanor La Verne
Going, Dora Henley
Goldenberg, Martin Irwin
Grainger, Thomas Hutcheson, Jr
Holtman, Darlington Frank
Hugh, Rudolph
Jellard, Charles H
Joseph, Sammy William
Kubica, George P
Martineau, Bernard
McCarty, Maclyn
Michelson, Israel David
Milgrom, Felix
Mohn, James Frederic
Moody, Max Dale
Moore, Harold Beveridge
Raffel, Sidney
Ransom, John Paul
Sarber, Raymond William
Shepard, Maurice Charles
Slack, John Madison
Smith, Peter Byrd
Smith, Philip Nixon
Speck, Reinhard Staniford
Stone, Joseph Louis
Syeklocha, Delfa
Toshach, Sheila
Vella, Philip Peter
Walker, Richard V
Weaver, Robert Elwin
Willoughby, Donald S
Wilson, Raphael
Woolridge, Robert Leonard

Medical Biophysics
Brady, Al H
Breckenridge, John Robert
Cohen, Beverly Shapiro
Dicello, John Francis, Jr
Driscoll, Dorothy H
Greenfield, Harvey Stanley
Greenstock, Clive Lewis
Haak, Richard Arlen
Myers, William Graydon
Paterson, Malcolm Cyril
Petkau, Abram
Rainbow, Andrew James
Stoll, Alice Mary
Witcofski, Richard Lou
Wyrobek, Andrew Julius

Medical Ecology
Haddon, William, Jr
Krueger, Albert Paul

Medical Education
Ames, Edward R
Andreoli, Kathleen Gainor
Armistead, Willis William
Bandick, Neal Raymond
Billings, Frederic Tremaine
Bishop, William Richard
Broitman, Selwyn Arthur
Brown, Edwin Wilson, Jr
Bryan, George Thomas
Burford, Hugh Jonathan
Burg, Frederic David
Calma, Victor Charles
Caughey, John Lyon, Jr
Clark, William Melvin, Jr
Cochran, Paul Terry
Coggeshall, Lowell Thelwell
Cooper, William Gregory
Coppola, Edward Dante
Cox, John William
Creditor, Morton C
Darley, Ward
Daw, John Charles
DeBakey, Lois
Dickson, Arthur David
Duffey, Margery
Dyson, James Everett, Jr
Elliott, Paul Russell
Evans, Robert L
Fore, Harry Waugh, Jr
Goldberg, Herbert Sam
Greenberg, Stephen Robert
Grover, William Johnson
Grover, Paul L, Jr
Harrell, George Thomas, Jr
Hegyvary, Sue Thomas
Hess, Joseph W, Jr
Hoffman, Arlene Faun

James, Mary Frances
Kettel, Louis John
Laurenson, Rae Duncan
Lewis, David James
Lindsay, Dale Richard
Mayer, Eugene Stephen
McLeod, Lionel Everett
Meleca, Cosmo Benjamin
Meltzer, Richard S
Millis, John Schoff
Mills, Russell Clarence
Novak, Milan Vaclav
Nystrom, Richard Alan
Overman, Ralph Theodore
Peterson, Malcolm Lee
Reynolds, Richard Clyde
Rinsley, Donald Brendan
Rosinski, Edwin F
Rothman, Arthur I
Schwartz, Peter Larry
Scrafani, Joseph Thomas
Shaffer, James Grant
Shoemaker, Richard Nelson
Solomon, Lawrence Marvin
Sorensen, Andrew Aaron
Spink, Gordon Clayton
Spivey, Bruce Eldon
Stoudt, Howard Webster
Suter, Emanuel
Tacker, Willis Arnold, Jr
Tavano, Donald C
Taylor, John Langdon, Jr
Texter, Elmer Clinton, Jr
Tidball, Charles Stanley
Treat, Donald Fackler
Votaw, Robert Grimm
Waife, Sholom Omi
Walters, Richard Francis
Waugh, Douglas Oliver William
Weintraub, Herbert D
Whitehill, Jules Leonard
Wismar, Beth Louise
Woods, James W

Medical Entomology
Aitken, Thomas Henry Gardiner
Amin, Omar M
Andersen, Dean Martin
Axtell, Richard Charles
Baerg, David Carl
Barnes, William Wayne
Barnett, Herbert Chester
Barr, Allan Ralph
Beadle, Leslie Dewey
Benach, Jorge L
Berry, Richard Lee
Blanton, Franklin Sylvester
Bonnet, David Dudley
Boreham, Melvin Murray
Burden, George Stanley
Butler, Joseph Miles
Chamberlain, Roy William
Christensen, Howard Anthony
Clements, Burle Webster
Collins, William Erle
Darsie, Richard Floyd, Jr
Eads, Richard Bailey
Edman, John David
Elbel, Robert E
Eldridge, Bruce Frederick
Emerson, Kary Cadmus
Evans, Burton Robert
Ezell, William Bruce, Jr
Fairchild, Homer Eaton
Favorite, Frank G, Jr
Fleming, Glenn Allen
Fontaine, Russell Edgar
Fowler, Harland Wade, Jr
Fox, Irving
Garcia, Richard
Gingrich, Richard Earl
Goodwin, William Jennings
Gorham, John Richard
Gouck, Harry (Kydd)
Gould, Douglas Jay
Graham, Owen Hugh
Gratz, Norman G
Grimstad, Paul Robert
Grothaus, Roger Harry
Grundmann, Albert Wendell
Gwadz, Robert Walter
Haines, Thomas Walton
Hall, Donald William
Havertz, David S
Hayes, George Roy, Jr
Hayes, Richard Oliver
Herrin, Charles Selby
Hill, Alfred, Jr
Hilton, Donald Frederick James
Hitchcock, James Carroll, Jr
Hobbs, Jesse H
Holway, Richard Thomas
Hopla, Cluff Earl
Hu, Stephen Moi Kee
Hull, William Ballou
Hurlbut, Herbert Sumner
Jameson, Everett Williams, Jr
Johnson, William E, Jr

Joseph, Stanley Robert
Kappus, Karl Daniel
Keegan, Hugh Lawrence
Keirans, James Edward
Kitzmiller, James Blaine
Kliewer, John Wallace
Klowden, Marc Jeffrey
Kurtz, Frederick Winfield
La Brecque, Germain C
Lacaillade, Charles William
Lane, Robert Sidney
Lee, Vernon Harold
Lewis, Leyburn F
Lingstrom, Leon
McClelland, George Anderson Hugh
McDaniel, Ivan Noel
McLintock, John James Reid
Miller, Albert
Miura, Takeshi
Morgan, Neal O
Mount, Gary A
Mullen, Gary Richard
Murray, William Donald
Nelson, Bernard Clinton
Nelson, Robert LeRoy
Newhouse, Verne Frederic
Newson, Harold Don
Nowell, Wesley Raymond
O'Connor, Charles Timothy
Olson, Jimmy Karl
Ouzts, Johnny Drew
Parrish, Dale Wayne
Patterson, Richard Sheldon
Patterson, William Junior
Peck, John Hubert
Pitts, Charles W
Provost, Maurice Wilfred
Rathburn, Carlisle Baxter, Jr
Reed, Horace Beecher, Jr
Rice, Paul LaVerne
Rigby, Paul T
Roberts, Richard Harris
Rogers, Andrew Jackson
Ryckman, Raymond Edward
Scanlon, John Earl
Siverly, Russell Emmett
Smith, Carroll N
Smith, William Ward
Snoddy, Edward L
Spielman, Andrew
Steelman, Carrol Dayton
Taylor, Robert Tieche
Terwedow, Henry Albert, Jr
Thurber, George A
Thurman-Swartzwelder, Ernestine H
Tonn, Robert J
Traub, Robert
Travis, Bernard Valentine
Trpis, Milan
Vanderberg, Jerome Philip
Wallis, Robert Charles
Ward, Ronald Anthony
Weidhaas, Donald E
Williams, David Francis
Williams, Roger Wright
Wright, James Elbert
Wright, Russell Emery
Zaharis, John Louis
Zimmerman, John Harvey

Medical Ethics

Kass, Leon Richard
Towle, David Walter

Medical Genetics

Altmiller, Dale Henry
Andrews, Lucy Gordon
Bannerman, Robin Mowat
Bartalos, Mihaly
Bernstein, Seldon Edwin
Boyer, Samuel H, IV
Christian, Joe Clark
Comings, David Edward
Cox, Rody Powell
Curran, John Phineas
Donahue, Roger Purtee
Edwards, John Anthony
Eldridge, Roswell
Elsas, Louis Jacob, II
Epstein, Charles Joseph
Erlenmeyer-Kimling, L
Fialkow, Philip Jack
Finley, Sara Crews
Fischman, Harlow Kenneth
Fraser, Frank Clarke
Friedman, Robert David
Garza-Chapa, Raul
Gordon, Hymie
Guthrie, Robert
Hillman, Richard Ephraim
Hodes, Marion Edward
Hollister, David William
Jannback, Hugo Andrew, Jr
Kaufmann, Berwind Norman
Koler, Robert Donald
Lozzio, Carmen Bertucci
Lynch, Henry T
Lyons, Richard Bernard

Medical Microbiology

Ackermann, Hans Wolfgang
Alexander, Aaron D
Allen, William Peter
Arden, Sheldon Bruce
Athar, Mohammed Aqeel
Bacon, Marion
Baker, Herman
Baker, Edgar Eugene, Jr
Barlow, James Lawrence
Barnhart, Donald Delbert
Barr, Fred S
Bartell, Pasquale
Beggs, William Joe
Beasley, William H
Berky, John James
Bibel, David Jan
Bird, Thomas Joseph
Blazevic, Donna Jean
Bojalil, Luis Felipe
Boljpes, Ben Harold
Bolyn, Anthony Edward
Bowman, Bernard Ulysses, Jr
Bowman, Ernest John
Bradshaw, Lawrence Jack
Brancato, Frank Paul
Brockett, Royce Merrett
Brosbe, Edwin Allan
Brown, James Allen, Jr
Brown, Lee Roy, Jr
Brown, Russell Wilfrid
Butas, Constandina
Butt, Elizabeth
Cameron, J A
Campbell, James B
Champlin, William G
Chaudhary, Rabindra Kumar
Chen, Joseph Ke-Chou
Cleary, Paul Patrick
Cleary, Timothy Joseph
Cohen, Jacob Ortlieb
Cohn, Maurice Leon
Collin, William Kent
Cooper, Ronda Fern
Covert, Scott Veasey
Cox, Charles Donald
Coyle, Marie Bridget
Crecelius, Harry Gilbert
Cundy, Kenneth Raymond
Dalzell, Robert Clinton
Deas, Jane Ellen
De Courcy, Samuel Joseph, Jr
DeMeio, Joseph Louis
Dempster, George
Denton, James Fred
DePalma, Philip Anthony
DeVoe, Irving Woodrow
Dickerman, John Melville
Diercks, Fred Herman
Dixon, John Michael Siddons
Dolman, Claude Ernest
Domer, Judith E
Duda, John J
Duncan, I B R
Duncan, James Lowell
Eaves, George Newton
Ethymiou, Constantine John
Elberg, Sanford Samuel
Elliott, Larry P
Ellner, Paul Daniel
Eubanks, Julia Flynt
Evans, Charles Albert
Ewing, William Howell
Fabrizio, Angelina Maria
Fellowes, Oliver Nelson
Felton, Frances Grace
Fishel, Charles Wesley
Foley, George Edward
Forgacs, Joseph
Fox, Eugene N
Francis, Robert Dorl
Friedman, Lorraine
Friend, Charlotte

Medical Microbiology

McAlpine, Phyllis Jean
Mellman, William Jules
Metrakos, Julius Demetrius
Migeon, Barbara Ruben
Mitchell, Donald Folk
Murray, Robert Fulton, Jr
Nightingale, Elena Ottolenghi
Omenn, Gilbert Stanley
Prasad, Rupi
Prescott, Gerald H
Reed, Terry Eugene
Rimoin, David (Lawrence)
Sanger, Warren Glenn
Schaible, Robert Hilton
Shokeir, Mohamed Hassan Kamel
Soltan, Hubert Constantine
Stamatoyannopoulos, George
Steinberg, Arthur Gerald
Thurmon, Theodore Francis
Tishler, Peter Verveer
Williams, Hibbard E
Yamauchi, Toshio
Young, William Irving

Hannan, Charles Kevin
Haque, Riaz-Ul
Harris, Elizabeth Forsyth
Hatten, Betty Arlene
Hawkins, Linda Louise
Hawley, Robert John
Haynes, Ralph Edwards
Hench, Miles Ellsworth
Hendrickson, Donald Allen
Hierholzer, John Charles
Higa, Harry Hiroshi
Hill, Douglas Wayne
Hill, Edward Orson
Hinze, Harry Clifford
Housewright, Riley Dee
Hsu, Hsiu-Sheng
Huber, Thomas Wayne
Hurlburt, Evelyn McClelland
Hurst, Valerie
Isenberg, Henry David
Ivler, Daniel
Jacks, Thomas Mauro
Jackson, Hans Oliver
James, Ann Nixon
Jarolmen, Howard
Jemski, Joseph Victor
Jeter, Marcus Martin
Jensen, Wayburn Stewart
Johnson, Daniel Everett
Johnston, Paul Bruns
Jonsson, Valgard
Joseph, Junior Mehsen
Joys, Terence M
Judefind, Thomas Francis
Jutila, John W
Kaminski, Zigmund Charles
Kaplan, Milton Temkin
Kapral, Frank Albert
Karlson, Alfred Gustav
Kelleher, James Joseph
Kellerman, George D
Kellogg, Douglas Sheldon, Jr
Kern, Earl R
Kettering, James David
Keyser, Peter D
Kilbourn, Joan Priscilla Payne
Kim, Kwang Shin
Kiser, Jackson Sebree
Koh, Won Young
Kotcher, Emil
Kraeger, Spring Juliet
Kuehner, Calvin Charles
Kundsin, Ruth Blumfeld
Kuo, Cho-Chou
Kutner, Leon Jay
Lachapelle, Rene Charles
Lambe, Dwight Wilson, Jr
Larsen, Don Hyrum
Laskowski, Leonard Francis, Jr
Lee, Spencer Hon-Sun
Leers, Wolf-Dietrich
Leibovitz, Albert
Leise, Joshua Melvin
Lewis, William Perry
Lichstein, Herman Carlton
Lief, Florence Suskind
Lindberg, Lois Helen
Lindberg, Robert Benjamin
Livermore, Brian Paul
Lomanitz, Rachel
Lynn, Raymond J
MacFarlane, John O'Donnell
Mackel, Donald Charles
MacLeod, Donald Richard Eason
Mandel, John Herbert
Mann, Dennis Keith
Marrao, Robert V
Marvin, Richard Martin
Mastromarino, Anthony John
McBride, Mollie Elizabeth
McClatchy, Joseph Kenneth
McCracken, Alexander Walker
McFadden, Harry Webber, Jr

Fuller, Vernon Jack
Fulton, Macdonald
Furtado, Dolores
Gadebusch, Hans Henning
Garrison, Robert Gene
Gemski, Peter, Jr
Gibby, Irvin Welch
Glantz, Paul Joseph
Goetchius, George Richard
Goldner, Herman
Goldschmidt, Millicent
Good, Robert Campbell
Gordon, Irving
Gots, Joseph Simon
Greene, Robert Alva
Green, Erika Ana
Green, James Edward
Groschel, Dieter Hans Max
Gruber, Jack
Guckunst, Richard Ralph
Hageage, George John, Jr
Halbert, Seymour Putterman
Halkias, Demetrios
Halstead, Scott Barker
Hampson, C Ross

McLean, Donald Millis
McRipley, Ronald James
McVickar, David Langston
Meyers, Wayne Marvin
Miles, Donald Orval
Milner, Kelsey Charles
Minsavage, Edward Joseph
Miragia, Gennaro J
Moehring, Thomas John
Moore, Walter Edward Cladek
Morello, Josephine A (Mrs Mrs Robert E Burz)
Morter, Raymond Lione
Murphy, Juneann Wadsworth
Murphy, Richard Allan
Nachbar, Martin Stephen
O'Hern, Elizabeth Moot
Ottolenghi, Abramo Cesare
Padgett, Billie Lou
Pai, Chik Hyun
Pappagianis, Demosthenes
Paradise, Lois Jean
Paulissen, Leo John
Peterson, Ward Davis, Jr
Pollard, Morris
Potash, Louis
Puhvel, Sirje Madli
Pumper, Robert William
Puttlitz, Donald Herbert
Qadri, Syed M Hussain
Quarles, John Monroe
Quick, Jacquelin Dunn
Ramsey, Penny Tina
Randall, Eileen Louise
Rasmussen, Aaron Frederick, Jr
Reback, John Frederick
Reddick-Mitchum, Rhoda Anne
Reeves, James Blanchette
Reich, Claude Virgil
Reitman, Morton
Rhodes, Andrew James
Richardson, Harold
Rigney, Mary Margaret
Riley, Perry Stephen
Roantree, Robert Joseph
Robertsen, John Alan
Rogers, Howell Wade
Rogers, Nancy Graham
Rommel, Frederick Allen
Roosa, Robert Andrew
Rosan, Burton
Rossier, Edmond
Royal, George Calvin, Jr
Russell, Catherine Marie
Rustigian, Robert
Sampath, Angus C
Sanders, Ruth Evelyn
Sanders, Christine Culp
Savage, Dwayne Cecil
Schmidt, Jerome P
Scholes, Vernon Eugene
Schuytema, Eunice Chambers
Shadomy, Smith
Shaffer, Morris Frank
Shands, Joseph Walter, Jr
Shaygani, Mehdi
Shechmeister, Isaac Leo
Shotts, Emmett Booker, Jr
Sinha, Shyamal K
Sippel, John Edward
Sliftin, Malcolm
Sloan, Bernard Joseph
Smith, Josephine Reist
Snyder, Irvin S
Solorzano, Robert Francis
Somerson, Norman L
Sonnenwirth, Alexander Coleman
Stevens, Roy White
Steward, John P
Stiffler, Paul W
Stinebring, Warren Richard
Straus, David Conrad
Suling, William John
Summers, William Allen, Sr
Surgalla, Michael Joseph
Sutter, Vera La Verne
Tachibana, Dora K
Theis, Jerold Howard
Thomas, Virginia Lynn
Todd, William McClintock
Tonik, Ellis J
Tully, Joseph George, Jr
Tyeryar, Franklin Joseph, Jr
Vaiturzis, Zigfrdas
Valentine, Bob Leon
Vas, Stephen Istvan
Vera, Harriette Dryden
Verwey, Willard Foster
Vickrey, Herta Miller
Von Graevenitz, Alexander W C
Von Riesen, Victor Lyle
Wang, San-Pin
Ware, Lawrence L, Jr
Webb, Alfred Mohr
Weinberg, Eugene David
Weiss, Emilio
Whang, Sukoo Jack

Whitescarver, Jack Edward
Widmark, Rudolph M
Wilson, Elizabeth
Wilson, Marion Evans
Wiseman, Charles Louis, Jr
Woehler, Michael Edward
Wood, Ronald McFarlane
Wu, Chii-Huei
Wu, William Gay
Wynne, Elmer Staten
Yarzabal, Luis Alberto
Yaverbaum, Sidney
Youngner, Julius Stuart
Yurchenco, John Alfonso
Zajac, Ihor
Zeigler, Joseph Alton
Zubrzycki, Leonard Joseph

Medical Mycology
Ajello, Libero
Al-Doory, Yousef
Bulmer, Glenn Stuart
Campbell, Charlotte Catherine
Cazin, John, Jr
Cozad, George Carmon
Davis, Michael Allan
Erke, Keith Howard
Feldman, Fred
Georg, Lucille Katherine (Mrs W L Pickard)
Gordon, Morris Aaron
Greer, Donald Lee
Halde, Carlyn Jean
Howard, Dexter Herbert
Huppert, Milton
Kaplan, William
Kaufman, Leo
Kozel, Thomas Randall
Kwon-Chung, Kyung Joo
Land, Geoffrey Allison
Larsh, Howard William
Lupan, David Martin
Mackinnon, Juan Enrique
Marshall, James John
Merz, William George
Pore, Robert Scott
Rippon, John Willard
Rogers, Alvin Lee
Romberg, Paul Frederick
Rosenthal, Stanley Arthur
Schmitt, John Arvid, Jr
Schneidau, John Donald, Jr
Sinski, James Thomas
Smith, Andrew George
Stevens, Joseph Alfred
Sun, Sung Huang
Taylor, Robert Lee
Weeks, Robert Joe

Medical Parasitology
Abadie, Stanley Herbert
Beasley, William Joe
Bruckner, David Alan
Chen, David Hou-Chung
Chernin, Eli
Chi, Lois Wong
Contacos, Peter George
Cross, John Henry
Daly, James Joseph
Ferguson, Malcolm Stuart
Fremount, Henry Neil
Harper, Kathleen Lucille
Headlee, William Hamilton
Hunter, George William, III
Ivey, Michael Hamilton
Jacobs, Leon
Larsh, John E, Jr
Lesser, Elliott
Lewert, Robert Murdoch
Little, Maurice Dale
Luttermoser, George William
Malek, Emile Abdel
Mathews, Henry Mabbett
McConnell, Ellicott
McQuay, Russell Michael, Jr
Melvin, Dorothy Mae
Miller, Joseph Henry
Miller, Max Joseph
Orihel, Thomas Charles
Palmer, Timothy Trow
Preisler, Harvey D
Radke, Myron Glen
Ritchie, Lawrence Starr
Ritterson, Albert L
Schiller, Everett L
Sheffield, Harley George
Stirewalt, Margaret Amelia
Sulzer, Alexander Jackson
Swartzwelder, John Clyde
Thorson, Ralph Edward
van der Schalie, Henry
Weatherly, Norman F

Medical Physics
Adams, Paul Louis
Anderson, David Walter
Bagne, Farideh
Baily, Norman Arthur

Blau, Lawrence Martin
Brownell, Gordon Lee
Bruels, Mark Charles
Budinger, Thomas Francis
Bunch, Phillip Carter
Cameron, John Roderick
Carlton, William Herbert
Carson, Paul Langford
Chamberlain, Charles Craig
Chesler, David Alan
Cohen, Montague
Connor, William Gorden
Correia, John Arthur
Cox, Hollace Lawton, Jr
Cradduck, Trevor David
Cunningham, John (Robert)
Doi, Kunio
Douglass, Kenneth Harmon
Eagle, Donald Frohlichstein
Ekstrand, Kenneth Eric
Epp, Edward Rudolph
Erickson, Jon Jay
Fields, Theodore
Flower, Robert Walter
Friedland, Stephen Scholom
Friedmann, Gerhart B
Gledhill, Barton L
Goldstein, Albert
Goodenough, David John
Grant, Roderick M, Jr
Greenfield, Moses A
Groce, David Eiben
Harris, Gale Ion
Haymond, Herman Ralph
Heller, Zindel Herbert
Hendee, William Richard
Henschke, Ulrich K
Hevezi, James M
Hill, David G
Holden, James Edward
Holloway, Arthur F
Hoop, Bernard, Jr
Hoory, Shlome
Howarth, John Lee
Hsieh, Jen-Shu
Hubbard, Lincoln Beals
Huesman, Ronald Henry
Joseph, Peter Maron
Kelsey, Charles Andrew
Kiker, William Edward
Kim, Ki-Hyon
Konneker, Wilfred R
Kortright, James McDougall
Kuchnir, Franca Tabliabue
Lanzi, Lawrence Herman
Laughlin, John Seth
Lerch, Irving A
Liboff, Abraham R
MacIntyre, William James
Malamud, Herbert
McCullough, Edwin Charles
Metz, Charles Edgar
Moore, Richard
Muldawer, Leonard
Nichols, Alexander Vladimir
Nickles, Robert Jerome
Norman, Amos
O'Foghludha, Fearghus Tadhg
Oliver, George Davis, Jr
Orton, Colin George
Orvis, Alan Leroy
Ovadia, Jacques
Palmer, Harvey Earl
Pfalzner, Paul Michael
Reinstein, Lawrence Elliot
Ritter, Rogers C
Robertson, James Sydnor
Robinson, James Eugene
Rosenthal, Donald Jack
Rotenberg, A Daniel
Rothenberg, Lawrence Neil
St Onge, Richard Norbert
Schulz, Robert J
Scott, Mary Jean (Mrs E C H Silk)
Sorenson, James Alfred
Sprawls, Perry, Jr
Starchman, Dale Edward
Stein, Philip
Sternick, Edward Selby
Syed, Ibrahim Bijli
Tai, Douglas L
Taylor, Lauriston Sale
Ter-Pogossian, Michel Mathew
Tran-Manh, Ngo
Waggener, Robert Glenn
Wallace, Robert Henry
Walton, Philip Wilson
Waterman, Frank Melvin
Webb, Robert Howard
Webster, David Alexander
Webster, Edward William
White, William
Winchell, Harry Saul
Wootton, Peter
Wynveen, Robert Allen
Yalow, Rosalyn Sussman
Yang, Kei-Hsiung

Medical Physiology
Alexander, Natalie
Allen, Robert Erwin
Barr, Ronald Edward
Beard, James David
Berne, Robert Matthew
Blatt, Elizabeth Kempske
Bond, Gary Carl
Bricker, Jerome Gough
Brown, Marvin Ross
Chapler, Christopher Keith
Chargois, Deborah Majeau
Claus, Thomas Harrison
Curry, Donald Lawrence
D'Agrosa, Louis Salvatore
Dahlen, Roger W
Dellenback, Robert Joseph
Denniston, Joseph Charles
De Nuccio, David Joseph
Dobrin, Philip Boone
Edwards, Brian Ronald
Ehrhart, Ina C
Ely, Daniel Lee
Emmers, Raimond
Feigen, Larry Philip
Fisher, Martin Joseph
Ford, Lincoln Edmond
Frey, Mary Anne Bassett
Fuhrman, Frederick Alexander
Gaar, Kermit Albert, Jr
Gaugl, John F
Geisler, Fred Harden
Gold, Armand Joel
Gonzalez, Norberto Carlos
Grace, Robert Ambrose
Grubbs, Clinton Julian
Hale, Creighton J
Hamosh, Paul
Hansen, Timothy Ray
Hartman, John Leo, Jr
Heisey, Samuel Richard
Heller, Lois Jane
Hetenyi, Geza Joseph
Hiles, Linda Gayle
Hislop, Helen Jean
Hutchison, Kenneth James
Jensen, David
Johnson, Melvin Andrew, Jr
Kaufman, Albert Irving
Kenney, Richard Alec
Knudsen, James Frederick
Konecci, Eugene B
Korecky, Borivoj
Krieger, Eduardo Moacyr
Kunze, Diana Lee
Laiken, Nora Jane
Lambert, Edward Howard
Larsen, Eleanor Marie
Lebrie, Stephen Joseph
Lee, Robert John
Liu, Maw-Shung
Lo, Chu-Shek
Malindzak, George Steve, Jr
McCutcheon, Ernest P
Miller, David Arthur
Milligan, John Vorley
Millstein, Lloyd Gilbert
Milnor, William Robert
Murphy, John Thomas
Nelson, Eldon Lane, Jr
Neville, James Ryan
Nicholes, Henry Joseph
Ott, Cobern Erwin
Overbeck, Henry West
Pak, Moon Jae
Pflanzer, Richard Gary
Pike, Eileen Halsey
Ponessa, Joseph Thomas
Rall, Jack Alan
Ramsay, David John
Rapela, Carlos Enrique
Redden, David Ray
Reilly, Frank Daniel
Reynolds, Monica
Robertson, William G
Sanborn, Warren Gordon
Sargent, William Quirk
Sernka, Thomas John
Sheng, Hwai-Ping
Sladek, Celia Davis
Snively, William Daniel, Jr
Sparks, Harvey Vise
Stauffer, Edward Keith
Swanson, Robert E
Tidball, Mary Elizabeth
Toivola, Pertti Toivo Kalevi
Trainer, Joseph B
Van Liew, Hugh Davenport
Webb, Paul
Whitten, Elmer Hammond
Wilde, Walter Samuel
Zogg, Carl A

Berger, Frank Milan
Berlin, Nathaniel Isaac
Bonner, R Alan
Borg, Donald Cecil
Brenner, John Francis
Bresler, Emanuel H
Chang, Thomas Ming Swi
Charman, Howard Prentis
Clayton, John Mark
Clewe, Thomas Hailey
Cohen, Morris
Conard, Robert Allen
Courtney, Kenneth Oliver
Crofford, Oscar Bledsoe
Dublin, Thomas David
Eder, Howard Abram
Ewing, Dean Edgar
Fallat, Ronald Walter
Farah, Alfred Emil
Forstner, Janet Ferguson
Fox, Roy Alan
Francis, Marion David
Gorodetzky, Charles W
Hansen, Hobart Raymond
Hawrylewicz, Ervin J
Heller, Milton David
Hilding, Anderson C
Hilgar, Arthur Gilbert
Hirschfeld, William Jacob
Hobson, Lawrence Bennett
Hodgkin, Brian Charles
Hrubec, Zdenek
Hunt, William Lynn
Jones, Joie, Pierce
Kagawa, Charles M
Kayan, Sabih
Klein, LeRoy
Kropp, Helmut
Leon, Arthur Sol
Levine, David Morris
MacFarlane, Malcolm David
Macgregor, Alexander Hamilton
Malik, Vedpal Singh
Manner, Richard John
McKee, Patrick Allen
McLamore, William Merrill
McMillin, Carl Richard
Meryman, Harold Thayer
Miller, Eleanor Marie
Mintz, Sheldon
Murphy, Donald G
Murray, Mary Patricia
Mustafa, Syed Jamal
Nalin, David Robert
Palaia, Frank Lincoln, Jr
Patel, Dhanooprasad Gordhanbhai
Philpott, Delbert E
Pittman, Fred Estes
Ransil, Bernard Jerome
Robbins, Jacob
Rosenfeld, George
Sawchuk, Ronald John
Saxena, Vishv Prakash
Schacter, Brent Allan
Schafer, David Edward
Schechter, Alan Neil
Schiff, Robert
Schimmer, Bernard Paul
Schneider, Henry Peter
Schwartz, Samuel
Schwartz, Anton
Shannon, Wilburn Allen, Jr
Shaw, Stanley Miner
Simpson, Stuart Douglas
Slater, Grant Gay
Smith, Victor Herbert
Snyder, Ann Knabb
Spaulding, Stephen Waasa
Stanley, Evan Richard
Steiner, George
Stibitz, George Robert
Stolz, Hal Fisher
Stone, Irwin
Storrs, Eleanor Emerett
Strahilevitz, Meir
Stuhlman, Robert August
Takesue, Edward I
Tan, Meng Hee
Tobin, Sidney Morris
Traub, Richard Kimberley
Varma, Ranbir S
Voorhees, John James
Wathen, Ronald Larry
Weir, James Henry, III
Whedon, George Donald
Williams, Gerald Albert
Worrall, Paul Michael

Medical Technology
Branch, Garland Marion, Jr
Fillipi, Gordon Michael
Gleich, Carol Sue
Hovde, Ruth Frances
Hunter, Katherine Morton
Long, Mary Jean
Love, Betholene Frances
Scott, Walter L, Jr
Spicher, John L

Medical Research
Arcese, Paul Salvatore
Baxter, James Hubert
Bellville, John Weldon

Vance, Miles Elliott

Neonatology
Abramson, David C
Allen, Alexander Charles
Dickson, Donald Ward
Glass, Leonard
Golden, Alva Morgan
Green, Marvin
Kay, Jacob Lindy
Kleinman, Leonard I
Larks, Saul David
Schwanecke, Rebecca G Pineda
Smith, Barry Thomas Sturt
Tsang, Reginald C

Nephrology
Bank, Norman
Bergeron, Michel
Besarab, Anatole
Brandt, J Leonard
Bresler, Emanuel H
Brown, Robert Stephen
Burg, Maurice B
Coburn, Jack Wesley
Coe, Fredric Lawrence
Cohen, Jordan J
Couture, Roger
Cummings, Nancy Boucot
Czerwinski, Anthony William
Denis, Gustave
Dillard, Mark Gregory
Dirks, John Herbert
Dosseter, John Beamish
Easterling, Ronald E
Fanestil, Darrell Dean
Fellers, Francis Xavier
Freeman, Richard B
Genest, Jacques
Gentile, Dominick E
Glabman, Sheldon
Goldberg, Martin
Goldsmith, Carl
Gonick, Harvey C
Gonzalez, Francisco Manuel
Gottschalk, Carl William
Grantham, Jared James
Gutmann, Ronald D
Heymann, Walter
Hill, Louis Leighton
Hillis, William Daniel
Holland, Nancy H
Hollander, Walter, Jr
Kamm, Donald E
Katz, Murray Alan
Kennedy, Thomas James, Jr
Kraikitpanich, Sompong
Kuchel, Otto George
Lassiter, William Edmund
Lavender, Ardis Ray
LeSher, Dean Allen
Levin, Murray Laurence
Levin, Sidney
Levitt, Marvin Frederick
Lindholm, Dale David
Lubash, Glenn David
Maher, John Francis
Martinez-Maldonado, Manuel
Michael, Alfred Frederick, Jr
Mitch, William Evans
Morita, Yoshikazu
Nash, Franklin D
Nawar, Tewfik
O'Donohue, Walter John, Jr
Ogden, David Anderson
Pabico, Rufino C
Pollak, Victor Eugene
Posen, Gerald
Pryles, Charles Victor
Pullman, Theodore Neil
Purkerson, Mabel Louise
Puschett, Jules Bernard
Rector, Floyd Clinton, Jr
Robinson, Roscoe Ross
Roguske-Kyts, Jadwiga
Rubin, Mitchell Irving
Sebastian, Anthony
Shear, Leroy
Sigler, Miles Harold
Simpson, David Patten
Smith, Fred George, Jr
Stenzel, Kurt Hodgson
Strong, Cameron Gordon
Suk, Wadi Nagib
Swyer, Paul Robert
Thomson, Gerald Edmund
Tisi, Gennaro Michael
Travis, Luther Brisendine
Weller, John Martin
West, Clark Darwin
Whang, Robert

Neurology
Ades, Harlow Whiting
Adler, Alexandra
Aird, Robert Burns
Aita, John Andrew

Ajax, Ernest Theodore
Alter, Milton
Amick, Lawrence Douglas
Anderson, William Westerlin
Appel, Stanley Hersh
Appenzeller, Otto
Arieff, Alex J
Aring, Charles Dair
Armstrong, John Briggs
Arnason, Barry Gilbert Wyatt
Austin, James Henry
Baird, Henry W, III
Baker, Abe Bert
Barbeau, Andre
Barnet, Ann B
Barnett, H J M
Bass, Norman Herbert
Batkin, Stanley
Bell, William Earl
Bender, Morris Boris
Bergen, Donna Catherine
Biehl, Joseph Park
Blaw, Michael Ervin
Bloomberg, Wilfred
Booker, Harold E
Boshes, Benjamin
Brill, Norman Quintus
Broder, Samuel B
Brown, Joe Robert
Brown, Meyer
Bruetman, Martin Edgardo
Buschke, Herman
Calhoun, Calvin L
Calverley, John Robert
Campbell, Berry
Carter, Charles Conrad
Carter, Sidney
Caviness, Verne Strudwick, Jr
Cebalos, Ricardo
Celesia, Gastone G
Chambers, Richard
Chase, Thomas Newell
Churchill, John Alvord
Chusid, Joseph George
Clark, David Barrett
Cohen, Bernard
Cohen, Robert
Cole, Monroe
Conomy, John Paul
Cook, Stuart D
Corbin, Kendall Brooks
Cote, Lucien Joseph
Couch, James Russell, Jr
Currier, Robert David
Curry, Hiram Benjamin
Cutler, Robert W P
Daly, David DeRouen
Davis, Courtland Harwell, Jr
Davis, Floyd Asher
DeJong, Russell Nelson
Dekaban, Anatole S
De Myer, William Erl
De Salva, Salvatore Joseph
Drachman, Daniel Bruce
Drachman, David A
Drew, Arthur Leslie
Duffy, Philip
Duncan, Margaret Caroline
Dunn, Henry George
Duvoisin, Roger C
Dyken, Paul Richard
Eichman, Peter L
Eliasson, Sven Gustav
Elizan, Teresita S
Elliott, Frank A
Engel, William King
Epstein, Arthur William
Ettinger, Milton G
Fahn, Stanley
Farmer, Thomas Wohlsen
Feinberg, Donald Lester
Feldman, Daniel S
Fenichel, Gerald M
Feringa, Earl Robert
Ferrendelli, James Anthony
Ferriss, Gregory Stark
Fields, William Straus
Fisher, Morris Alan
Fishman, Marvin Allen
Fishman, Robert Allen
Foley, Joseph Michael
Forster, Francis Michael
Foster, David Bernard
Fox, Jacob H
Freedman, David Asa
Freeman, John Mark
Freeman, Frank Reed
French, Joseph H
Friedlander, Walter Jay
Gabor, Andrew John
Gallagher, Brian Boru
Gallant, Donald
Gardner, Ernest (Dean)
Geschwind, Norman
Giblin, Denis Richard
Gibson, William Carleton
Gilden, Donald Harvey
Gilman, Sid

Ginsberg, Myron David
Glaser, Gilbert Herbert
Goldberger, Michael Eric
Goldensohn, Eli Samuel
Goldstein, Norman Philip
Gould, Harry J, III
Graf, Carl John
Grossman, Herbert Jules
Growdon, John Herbert
Gutmann, Ludwig
Haase, Gunter R
Hackett, Earl R
Haerer, Armin Friedrich
Halsey, James H, Jr
Harter, Donald Harry
Hass, William K
Heath, Robert Galbraith
Heck, Albert Frank
Heller, Irving Henry
Herndon, Robert McCulloch
Herrman, Christian, Jr
Hinterbuchner, Ladislav Paul
Hochwald, Gerald Martin
Hoffman, Julius
Hogan, Edward L
Hollander, Joshua
Holmes, John Eric
Horenstein, Simon
Hoyt, William F
Itabashi, Hideo Henry
Jabbour, J T
Janeway, Richard
Jarcho, Leonard Wallenstein
Johns, Thomas Richards, II
Johnson, Richard T
Joynt, Robert James
Kaeber, William Walbridge
Kalyanaraman, Krishnaswamy
Kark, Robert Adrian Pieter
Karp, Herbert Rubin
Katzman, Robert
Kennedy, Charles
Kennedy, William Robert
Kinsbourne, Marcel
Koenig, Harold
Korein, Julius
Krieger, Howard Paul
Kurczynski, Thaddeus Walter
Kurtzke, John F
Ladwig, Harold Allen
Lam, Robert Lee
Landau, William M
Lawrence, Donald Gilbert
Lawyer, Tiffany, Jr
Lehrer, Gerard Michael
Lenn, Nicholas Joseph
Lessell, Simmons
Leventhal, Carl M
Levin, Victor Alan
Levi-Montalcini, Rita
Levy, David Edward
Levy, Irwin
Lisak, Robert Philip
Luban, Joseph Anton
MacFadyen, Donald John
Macrae, Donald
Madow, Leo
Magee, Kenneth Raymond
Magladery, John William
Malone, Michael Joseph
Mann, Leslie Bernard
Markesbery, William Ray
Markham, Charles Henry
Marks, Morton
Mason, Beryl Troxell
Mavor, Huntington
McDowell, Fletcher Hughes
McHugh, Paul Rodney
McKhann, Guy Mead
McKinney, William Markley
McMasters, Robert Earl
McQuarrie, Irvine Gray
Menkes, John H
Merlis, Sidney
Mettler, Frederick Albert
Meyer, John Stirling
Miller, James Q
Millhouse, Oliver Eugene
Millichap, Joseph Gordon
Mohr, Jay Preston
Montes, Robert
Moody, Robert Yates
Moossy, John
Morrell, Frank
Morris, Charles Elliot
Morris, Harold H
Mulder, Donald William
Myers, Ronald Elwood
Nachmansohn, David
Nagler, Benedict
Namba, Tatsuji
Nathanson, Neil Marc
Nayyar, Rajinder
Neale, Claude Linwood
Nelson, Erland
Netsky, Martin George

Nevis, Arnold Hastings
Niedermeyer, Ernst F
Norris, Forbes Holten, Jr
North, Richard Ralph
O'Brien, George Sivesind
Obrist, Walter Dorn
O'Doherty, Desmond Sylvester
Oh, Shin Joong
Oldendorf, William Henry
Olsen, Clarence Wilmott
Paddison, Richard Milton
Perez-Borja, Carlos
Petajan, Jack Hougen
Peters, Bruce Harry
Peters, Henry A
Peterson, Donald I
Philippart, Michel Paul
Pierce, Charles Middlebrook
Pincus, Jonathan Henry
Pitner, Samuel Ellis
Plum, Fred
Poirier, Louis
Porter, Huntington
Poskanzer, David Charles
Posner, Jerome B
Prince, David Allan
Prockop, Leon D
Rabe, Edward Frederick
Raichle, Marcus Edward
Randt, Clark Thorp
Rapin, Isabelle (Mrs Harold Oaklander)
Rasmussen, Theodore Brown
Redding, Foster Kinyon
Reinmuth, Oscar McNaughton
Reis, Donald J
Reivich, Martin
Renuart, Adhemar William
Resch, Joseph Anthony
Robb, J Preston
Robert, Emery Dean
Robinson, Bryan Wright
Roos, Raymond Philip
Rose, Arthur L
Rose, Augustus Steele
Ross, Alexander Treloar
Ross, Gilbert Stuart
Rowan, A James
Rowland, Lewis Philip
Sabin, Thomas Daniel
Sabra, Fuad Amin
Satran, Richard
Schain, Richard J
Scheinberg, Labe Charles
Scheinberg, Peritz
Schlagenhauff, Reinhold Eugene
Schlezinger, Nathan Stanley
Schmidt, Richard Penrose
Schneck, Larry
Schoenberg, Bruce Stuart
Schulman, Sidney
Segarra, Joseph M
Seipel, John Howard
Semrad, Elvin Vavrinec
Sherwin, William Richard
Sherwin, Allan Leonard
Shuman, Duane O
Shuter, Eli Ronald
Sibley, William Austin
Siegel, George Jacob
Sickert, Robert George
Silberberg, Donald H
Silverstein, Alexander
Simon, Benjamin
Smith, Bernard H
Smith, Carlton G
Sobkowicz, Hanna Maria
Spehlmann, Rainer
Stevens, Harold
Stevens, Janice R
Stewart, Lever F
Still, Charles Neal
Strobos, Robert Julius
Sullivan, John Francis
Swank, Roy Laver
Swanson, August George
Swanson, Phillip D
Teasdall, Robert Douglas
Teitelbaum, Harry Allen
Thompson, Hartwell Greene, Jr
Thompson, Herbert Stanley
Toglia, Joseph U
Torres, Fernando
Tourtellotte, Wallace William
Towler, Martin Lee
Trufant, Samuel Adams
Twitchell, Thomas Evans
Van Allen, Maurice Wright
Van Den Noort, Stanley
Van Der Meulen, Joseph Pierre
Vastola, Edward Francis
Waggoner, Raymond Walter
Waltz, Arthur G
Watters, Gordon Valentine
Waxman, Stephen George
Webster, Henry deForest
Wehman, Henry Joseph

Stander, Richard Wright
Steer, Charles Melvin
Stepto, Robert Charles
Stewart, David Bradshaw
Stone, Martin L
Swartwout, Joseph Rodolph
Swartz, Donald Percy
Taber, Ben Z
Talbert, Luther M
Talledo, Oscar Eduardo
Tatum, Howard James
Taylor, Edward Stewart
Taylor, Howard Canning, Jr
Taynor, Melvin Lester
Teteris, Nicholas John
Thiede, Henry A
Thompson, John Daniel
Thompson, William Benbow, Jr
Thornton, William Norman, Jr
Thysen, Benjamin
Tobin, Sidney Morris
Trelford, John D
Tricomi, Vincent
Tupper, W R Carl
Twombly, Gray Huntington
Ueland, Kent
Ulfelder, Howard
Vandeviele, Raymond Laurent
Vasicka, Alois
Vollman, Rudolf F
Vorherr, Helmuth Wilhelm
Wallach, Edward E
Walters, Jack Henry
Weed, John Conant
Weingold, Allan Byrne
Wentz, William Budd
White, Charles A, Jr
Whitridge, John, Jr
Wied, George Ludwig
Wilds, Preston Lea
Williams, James Hutchison
Wingate, Martin Bernard
Wolfe, Walter McElhaney
Wolkoff, A Stark
Woodruff, James Donald
Wynn, Ralph Matthew
Yang, Sen-Lian
Zetenik, John Slowko
Zerzavy, Frederick M
Zuspan, Frederick Paul

Oncology

Abell, Creed Wills
Adler, Solomon Stanley
Aisenberg, Alan Clifford
Akamatsu, Yasuyuki
Albert, Samuel
Ambrus, Julian Lawrence
Anderson, Roger Allen
Anderson, Paul Nathaniel
Ansfield, Fred Joseph
Apple, Martin Allen
Arcos, Joseph (Charles)
Ascione, Richard
Auletta, Angela Elaine
Awad, William Michel, Jr
Baluda, Marcel A
Bateman, John Laurens
Bateman, Joseph R
Barrett, O'Neill, Jr
Becker, Frederick F
Bergsagel, Daniel Egil
Bick, Rodger Lee
Biedler, June Lee
Block, Jerome Bernard
Bodey, Gerald Paul, Sr
Bottomley, Richard H
Bradner, William Turnbull

Occupational Medicine

Alvis, Harry J
Baylor, Curtis Horton
Blank, Fritz
Bleier, Hector P
Bowen, Edward H, Jr
Brooks, Stuart Merrill
Bulat, Thomas Joseph
Denne, Edward James
Derryberry, Oscar Merton
deTreville, Robert T P
Doran, William Thomas
Dinman, Bertram David
Felton, Jean Spencer
Goldwater, Leonard John
Howe, Henry Forbush
Lerner, Sidney Isaac
Lieben, Jan
Milby, Thomas Hutchinson
Miller, Alvin Leon
O'Connell, Richard Lee
Osterriter, John Ferdinand
Reinhardt, Charles Francis
Voelz, George Leo
Warren, Douglas Robson
Williams, Norman
Yerg, Raymond A
Zenz, Carl

Brown, J Martin
Burgess, Richard Ray
Burns, Fredric Jay
Burzynski, Stanislaw Rajmund
Byfield, John Eric
Cáceres, Eduardo
Cady, Blake
Cane, Paul E
Cancilos, George P
Capizzi, Robert L
Carbone, Paul P
Carr, David Turner
Carter, Stephen Keith
Chang, Kenneth Shueh-Shen
Charyulu, Komanduri K N
Chuang, Ronald Yan-Li
Clarkson, Bayard D
Coleman, Morton
Collard, William David
Colsky, Jacob
Coltman, Charles Arthur, Jr
Cone, Clarence Donald, Jr
Cooper, Miles Robert
Cooper, Robert Arthur, Jr
Crispens, Charles Gangloff, Jr
Custer, Richard Philip
Daly, James William
Davis, Joseph William
DeOme, Kenneth Benton
Dethlefsen, Lyle A
DiAugustine, Richard Patrick
Dilley, William G
Dodge, William Howard
Dunham, Wolcott Balestier
Dunkel, Virginia Catherine
Durant, John Ridgway
Eaton, Monroe Davis
Ellison, Rose Ruth
Falletta, John Matthew
Ferris, Philip
Filip, Donald Joseph
Finkel, Miriam Posner
Fishman, William Harold
Foley, H Thomas
Frenster, John H
Frigerio, Norman Alfred
Gardner, Murray Briggs
Glazer, Robert Irwin
Goldberg, David Milton
Goldenberg, Gerald J
Golomb, Harvey Morris
Goodman, Louis E
Gould, Thelma Bernice Clark
Groth, Donald Paul
Hall, Thomas Christopher
Hammond, George Denman
Harris, Jules Eli
Harris, Matthew N
Hart, Jacqueline Spoerer
Hartmann, John Rudolf
Hays, Esther Fincher
Healey, John Edward, Jr
Hellstrom, Karl Erik Lennart
Helson, Lawrence
Hennings, Henry
Henderson, Edward S
Hersh, Evan Manuel
Hirshaut, Yashar
Holcenberg, John Stanley
Holroyde, Christopher Peter
Holt, Charlene Poland
Horton, John
Horwitz, Harry
Horwitz, Jerome Philip
Hoshino, Kazumasa
Hrynuk, William
Hughes, Robert Rule
Huguley, Charles Mason, Jr
Hutter, Robert V P
Joachim, Harry L
Jackson, Carlton Darnell
Johnson, Robert O
Jones, Robert F
Kaizer, Herbert
Kaplan, Barry Hubert
Kiang, David Hubert
King, Charles Miller
Klemperer, Martin R
Kolb, Leonard H
Kopf, Alfred Walter
Kubinski, Henry A
Kuo, Eric Yung-Huei
Kucera, Louis S
Kurnick, John Edmund
Lampkin-Asam, Julia McCain
Lasfargues, Etienne Yves
Laux, David Charles
Lavietes, Beverly Blatt
Law, Lloyd William
Leighton, Joseph
Leung, Benjamin Shuet-Kin
Leventhal, Brigid Gray
Lewis, George Campbell, Jr
Li, Min Chiu
Liegner, Leonard M
Lilly, Frank
Lipkin, Martin
Lippman, Marc Estes

Lipton, Allan
Livingston, Virginia
Loeb, Virgil, Jr
Lower, Gerald Malcolm, Jr
MacDonald, Ronald Neil Angus
Martin, Richard Gordon
McFarlane, Ellen Sandra
McGrath, Charles Morris
Medina, Daniel
Meisler, Arnold Irwin
Miller, Edward Godfrey, Jr
Miller, Elizabeth Cavert
Miller, James Alexander
Mizell, Merle
Mitchell, William Marvin
Moldovano, Graziella
Molomut, Norman
Moore, Erin Colleen
Moore, George Eugene
Morris, John Emory
Morton, Donald Lee
Moyer, Mary Pat Sutter
Munson, Benjamin Ray
Murphy, Eduardo S
Murphy, Samuel G
Myers, Warren Powers Laird
Nayak, Debi Prosad
O'Brien, Richard Lee
Oishi, Noboru
Olson, Kenneth B
Padnos, Morton
Patterson, William Bradford
Pearson, John William
Peraino, Carl
Perry, Seymour Monroe
Pienta, Roman Joseph
Pinkus, Hermann (Karl Benno)
Pinsky, Carl Muni
Pollack, Robert Elliot
Potter, John F
Prasad, Kedar N
Price, James Melford
Rusch, Harold Paul
Raam, Shanthi
Ramirez, Guillermo
Reed, Melvin LeRoy
Rees, Earl Douglas
Reiskin, Allan B
Reynolds, Vernon H
Rose, David Peter
Rosenberg, Saul A
Rosenkranz, Herbert S
Rubin, Daniel Justin
Saffiotti, Umberto
Salerno, Ronald Anthony
Samuels, Larry D
Santos, George Wesley
Sarkar, Nurul Haque
Sawitsky, Arthur
Schein, Philip Samuel
Scribner, John David
Selim, Mostafa Ahmed
Shaw, Charles Raymond
Shimaoka, Katsutaro
Shingleton, Hugh Maurice
Shnider, Bruce I
Shubik, Philippe
Simpson, William Loyal
Sinclair, Nicholas Roderick
Smith, Frank E
Smith, Mary Elizabeth
Smith, Ralph Earl
Sokal, Joseph Emanuel
Solomon, Alan
Southam, Chester Milton
Stambaugh, John Edgar, Jr
Starr, Jason Leonard
Steeves, Richard Allison
Stockdale, Frank Edward
Stolbach, Leo Lucien
Straus, Albert Charles
Straus, Marc J
Sutzman, Leon
Suhrland, Leif George
Szybalski, Waclaw
Taylor, Billy G
Taylor, Samuel G, III
Temin, Howard Martin
Theologides, Athanasios
Thomas, Edward Donnall
Thurman, William Gentry
Toth, Bela
Trefford, John D
Trosko, James Edward
Troy, Frederic Arthur
Tucker, William Gough
Ultmann, John Ernest
Urtasun, Raul C
Vaitkevicius, Vainutis K
Vaughn, Clarence Benjamin
Vesselinovich, Stan Dushan
Von Leden, Hans Victor
Webber, Mukta Mala
Weisburger, Elizabeth Kreiser
Welsch, Clifford William, Jr
Wentz, William Budd
Williamson, Charles Elvin
Wodinsky, Isidore

Ophthalmology

Wolberg, William Harvey

Allen, Henry Freeman
Allen, James Harrill
Anderson, Douglas Richard
Anderson, Robert E
Apter, Julia Tutelman
Atkinson, Marshall B
Bagchi, Mihir
Bannon, Robert Edward
Basu, Prasanta Kumar
Beckerman, Barry Lee
Beetham, William Parkes
Behrens, Herbert Charles
Bettman, Jerome Wolf
Biersdorf, William Richard
Brenin, Goodwin Milton
Brown, Stuart Irwin
Buesseler, John Aure
Carr, Ronald E
Chandler, Arthur Cecil, Jr
Cinotti, Alfonse A
Cogan, David Glendenning
Colenbrander, August
Costin, Tullos Oswell
Culver, James F
Dawson, Thomas Amos
Decker, Chandler R
De Margerie, Jean-Marie
DeVoe, Arthur Gerard
Diamond, James Gary
Dohlman, Claes Henrik
Drance, S M
Duane, Thomas David
Eakins, Kenneth E
Edelhauser, Henry F
Eifrig, David Eric
Elliot, Alfred Johnston
Elliott, James, H
Ellis, Philip Paul
Engerman, Ronald Lester
Ernest, John Terry
Fraunfelder, Frederick Theodor
Friedman, Alan Herbert
Friedman, Ephraim
Falls, Harold Francis
Feeney, Mary Lynette
Ferguson, Edward C, III
Ferry, Andrew P
Fine, Ben Sion
Fink, Austin Ira
Fischbarg, Jorge
Foos, Robert Young
Fox, Samuel Louis
Galin, Miles A
Geeraets, Walter J
Gifford, Harold
Girard, Louis Joseph
Goldberg, Morton Falk
Grant, Walter Morton
Grayson, Merrill
Haik, George Michel
Hall, Michael Oakley
Hamasaki, Duco I
Hansen, Axel C
Harley, Robison Dooling
Harris, John Edward
Hart, William Milton
Havener, William H
Hedges, Thomas Reed, Jr
Helmsen, Ralph John
Henderson, John Woodworth
Henkind, Paul
Hepler, Robert S
Hoffert, Jack Russell
Hogan, Michael John
Holland, Monte Gene
Hollenhorst, Robert William
Howard, Rufus Oliver
Hughes, William Franklin, Jr
Ide, Carl Heinz
Jampel, Robert Steven
Jampolsky, Arthur
Jensen, Harlan Ellsworth
Johnson, Samuel Britton
Kambara, George Kiyoshi
Kaufman, Herbert Edward
Kearns, Thomas P
Keeney, Arthur Hail
Kiffney, Gustin Thomas, Jr
Kimura, Samuel Jiro
Knox, David Lalonde
Kolder, Hansjoerg E
Krolman, Gordon M
Kupfer, Carl
Kuwabara, Toichiro
Laibson, Peter R
Laties, Alan M
Leinfelder, Placidus Joseph
Lemoine, Albert N, Jr
Leopold, Irving Henry
Levene, Ralph Zalman
Linksz, Arthur
Linn, Jay George, Jr
Lynn, John R
Macdonald, Roderick, Jr

Mainster, Martin Aron
Maumenee, Alfred Edward
McCaslin, Murray Frew
McClure, Coye Willard
McDonald, James E
McDonald, Phillip Robb
McGavic, John Samuel
McGraw, James Lorenz
McLean, Ian William
Merz, Earl H
Michaels, David D
Morrison, Marcus Eugene
Neufeld, Arthur Harvey
Newell, Frank William
Nicholls, John Van Vliet
O'Connor, George Richard
O'Rourke, James
Ortwerth, Beryl John
Paton, David
Patz, Arnall
Pfeiffer, Raymond L
Phelps, Charles Dexter
Pico, Guillermo
Pilkerton, A Raymond
Pinschmidt, Norman William
Pushkin, Edward A
Rabinovitch, Bernard
Raiford, Morgan B
Reed, Howard Newns
Reinecke, Robert Dale
Richards, Richard Davison
Riley, Michael Verity
Roberts, Rufus Winston
Rockey, John Henry
Rosen, David A
Rosenthal, J William
Rubin, Melvin Lynne
Safir, Aran
Scheie, Harold Glendon
Schultz, Richard Otto
Schwartz, Bernard
Schwartz, John T
Scott, Philip Dell
Sears, Marvin Lloyd
Sexton, Marvin Ross
Shaad, Dorothy Jean
Shearer, Robert Vernon
Sherman, Samuel Murray
Shipley, Thorne
Shoch, David Eugene
Simon, Meredith Ann
Smith, Joseph Lawton
Sowell, John Gregory
Spaeth, George L
Spivey, Bruce Eldon
Swan, Kenneth Carl
Swanson, Arnold Arthur
Tasman, William S
Thomas, Charles I
Thomas, Robert P
Thompson, Herbert Stanley
Thygeson, Phillips
Trotter, Robert Russell
Troutman, Richard Charles
Tso, Mark On-Man
Tyner, George S
Von Noorden, Gunter Konstantin
Wadsworth, Joseph Allison Cannon
Waldman, Joseph
Watzke, Robert Coit
Wirtschafter, Jonathan Dine
Wolter, J Reimer
Wyman, Milton
Young, Morris Nathan
Young, Sue Ellen

Optometry

Afanador, Arthur Joseph
Allen, Merrill James
Baldwin, William Russell
Bannon, Robert Edward
Borish, Irvin Max
Brisbane, William Neely
Carter, John Haas
DeLong, Merrill B
Dowaiby, Margaret Susanne
Everson, Ronald Ward
Flom, Merton Clyde
Gregg, James R
Haffner, Alden Norman
Heath, Gordon Glenn
Hebbard, Frederick Worthman
Hickman, Don Winston
Hirsch, Monroe Jerome
Hofstetter, Henry W
Lewner, John Reuben
Loveter, Gerald Eugene
Ludlam, William Myrton
Lyle, William Montgomery
Madden, Thomas M
Mandell, Robert Burton
Margach, Charles Boyd
Morgan, Meredith Walter
Pheiffer, Chester Harry
Pitts, Donald Graves
Rosenbloom, Alfred A, Jr
Rosenblum, William M
Saladin, Jimmie James

Scheier, Arthur
Schoessler, John Paul
Shaifer, Arthur
Uniacke, Charles Allyn
Wesley, Newton K
Wild, Bradford Williston
Wong, Siu Gum
Woo, George Chi Shing

Oral Medicine

Brightman, Vernon
Buchholz, Robert E
Burman, Louis Robert
Cheraskin, Emanuel
Chilton, Neal Warwick
Cohen, Lawrence
Feltman, Reuben
Guggenheimer, James
Halstead, Charles Lemuel
Kaplan, Herman
Kaslick, Ralph Sidney
Lawson, Benjamin F
Myall, Robert William T
Peters, Henry Buckland
Scopp, Irwin Walter
Ship, Irwin I
Siskin, Milton
Yamane, George M

Orthopedics

Amstutz, Harlan Cabot
Cole, Madison Brooks, Jr
Connolly, John Francis
Dumbleton, John Herbert
Eyring, Edward J
Hudgin, Richard Henry
Jowsey, Jenifer
Miller, Donald Sidney
Schwartz, Edith Richmond

Osteopathy

Denslow, John Stedman
Gross, George Alvin
Jones, Charles M
Marcowitz, Stewart
Sharp, Roland Paul
Townsend, Alexandra A
Ward, Robert C

Otolaryngology

Abramson, Allan Lewis
Alberti, Peter W R M
Alford, Bobby R
Ausband, John R
Bailey, Byron James
Bennett, E Maxine
Bernstein, Leslie
Blatt, Irving Myron
Brandenburg, James H
Brown, David Edward
Brummett, Robert E
Catlin, Francis I
Chandler, J Ryan
Chasin, Werner David
Cody, D Thane
Cohen, Noel Lee
Colton, Raymond H
Daly, John F
DeWeese, David D
Duvall, Arndt John, III
Eglitis, John Arnold
Frazer, John P
Fredrickson, John Murray
Goldstein, Jerome Charles
Gregg, John Bailey
Hallberg, Olav Erik
Hammerling, James Solomon
Hampton, James Wilburn
Harbert, Fred
Harrill, James Albert
Harvey, Joseph Eldon
Hemenway, William Garth
Hetrick, Jacob Adam Werner
Hilding, David Anderson
Honrubia, Vicente
House, Leland Richmond
Hudson, William Rucker
Joseph, Donald J
Kirchner, John Albert
Kohut, Robert Irwin
Lehman, Roger H
Lim, David J
Lindsay, John Ralston
Lore, John M, Jr
Lyons, George D
Marovitz, William F
McCabe, Brian Francis
Montgomery, William Wayne
Montreuil, Fernand
Myers, Eugene Nicholas
Nadol, Bronislaw Joseph, Jr
Nager, George Theodore
Naunton, Ralph Frederick
Ogura, Joseph H
Per-Lee, John H
Petryshyn, Walter A
Pou, Jack Wendell
Proctor, Donald Frederick

Proud, G O'Neil
Reed, George Farrell
Richardson, John Clifford
Ronis, Bernard Joseph
Ronis, Max Lee
Ruben, Robert Joel
Rutledge, Lewis James
Saunders, William H
Schuknecht, Harold Frederick
Senturia, Ben Harlan
Shimizu, Hiroshi
Singleton, George Terrell
Sisson, George Allen
Snow, James Byron, Jr
Sooy, Francis Adrian
Sprinkle, Philip Martin
Stephens, Charlene Barr
Strong, Mervyn Stuart
Torok, Nicholas
Tucker, Harvey Michael
Ward, Paul H
Whitaker, Clay Westerfield
Wolfson, Robert Joseph

Otology

Basek, Milos
Canfield, Norton
Davis, Michael Allan
Donaldson, James Adrian
Duvall, Arndt John, III
Glorig, Aram
Kos, Clair Michael
Outerbridge, John Stuart
Reddy, John Bernard
Sataloff, Joseph
Smith, Catherine Agnes
Wilpizeski, Chester Robert

Otorhinolaryngology

Basek, Milos
Bernstein, David
Johnsson, Lars-Goran
Myers, Phillip Ward
Parkins, Charles Warren
Pastore, Peter Nicholas
Sasaki, Clarence Takashi
Shumrick, Donald A
Toomey, James Michael
Webster, Douglas B
Yarington, Charles Thomas, Jr

Pediatrics

Abildgaard, Charles Frederick
Abramson, David C
Ackerman, Bruce David
Adams, Forrest Hood
Adams, William Curtis
Agre, Karl
Aldrich, Robert Anderson
Allen, Alexander Charles
Amato, R Stephen S
Ambuel, John Philip
Ament, Marvin Earl
Anast, Constantine Spiro
Andrews, Billy Franklin
Arena, Jay M
Arnold, Mary B
Asher, David Michael
Auld, Peter A McF
Avery, Mary Ellen
Ayoub, Elia Moussa
Bacon, George Edgar
Bailie, Michael David
Baker, David H
Barba, William P, II
Barbero, Giulio J
Bargeron, Lionel Malcolm, Jr
Barness, Lewis Abraham
Barnett, Ann B
Barnett, Henry Lewis
Bartram, John Bowman
Bashe, Winslow Jerome, Jr
Bass, James W
Batson, Blair Everett
Batson, Margaret Bailly
Batson, Oscar Randolph
Battaglia, Frederick Camillo
Baublis, Joseph V
Baum, David
Baumann, Thiema Marie Wolf
Baumgartner, Leona
Beckmann, Albert Jules
Behrle, Franklin C
Behrman, Richard Elliot
Beitins, Inese Zinta
Bell, William E
Bellanti, Joseph A
Bender, Norman Charles
Benton, John William, Jr
Benzing, George, III
Bergman, Abraham
Bergstrom, William H
Berlow, Stanley
Berman, Julian L
Berry, Daisilee H
Bessman, Samuel Paul
Beuzeville, Carlos F
Bicoff, Juan Pedro

Birdsong, McLemore
Black, Emilie A
Blaese, Robert Michael
Blattner, Russell John
Blaw, Michael Ervin
Blizzard, Robert M
Blodgett, Frederic Maurice
Bloom, Arthur David
Blumberg, Richard Winston
Boehm, John Joseph
Boelsche, ArrNell
Boley, Scott Jason
Bosma, James Frederick
Boston, Robert Wesley
Boverman, Harold
Bradford, William Dalton
Bradford, William L
Brandt, Ira Kive
Brasel, Jo Anne
Breg, William Roy
Brooke, Clement Eustace
Brown, Audrey Kathleen
Bruck, Erika
Brunell, Philip Alfred
Brusilow, Saul W
Buchanan, Robert Alexander
Buckley, Rebecca Hatcher
Burdi, Alphonse R
Burg, Fredric David
Burke, Edmund C
Calcagno, Philip Louis
Caliguiri, Lawrence Anthony
Calma, Victor Charles
Camacho, Alvro Manuel
Campbell, John Richard
Campbell, Robert A
Capp, Michael Paul
Carey, Benjamin Watson
Carpenter, Gary Grant
Carson, Merl J
Carter, Robert Eldred
Carver, David Harold
Cassady, George
Cassels, Donald Ernest
Catz, Charlotte Schifra
Chamberlin, Harrie Rogers
Chanock, Robert Merritt
Char, Donald F B
Chernick, Victor
Cherry, James Donald
Chicoine, Luc
Christian, Joseph Ralph
Chun, Raymond Wai Mun
Clemmens, Raymond Leopold
Cleveland, William West
Clyde, Wallace Alexander, Jr
Cochrane, William
Cohen, William
Cohen, Flossie
Cohen, Peter
Cohlan, Sidney Quex
Cole, Wilfred Q
Collin, Pierre-Paul
Collins-Williams, Cecil
Collipp, Platon Jack
Colten, Harvey Radin
Cone, Thomas E, Jr
Connor, James D
Cooke, Robert E
Cooper, Herbert Asel
Cooper, Louis Zucker
Cooper, Max Dale
Corby, Donald G
Cornblath, Marvin
Cornfeld, David
Corrigan, James John, Jr
Coursin, David Baird
Cowger, Marilyn L
Cox, Frederick Eugene
Cramblett, Henry G
Crawford, Stanley Everett
Crigler, John F, Jr
Crocker, Allen Carrol
Cropp, Gerd J A
Cunningham, Nicholas
Curnen, Edward Charles, Jr
Curran, John Phineas
Daeschner, Charles William, Jr
Dallman, Peter R
Daniel, William A, Jr
Danis, Peter Godfrey
Davidson, Ronald G
Davis, Sarah Fredericka
Davis, Starkey D
Debuskey, Matthew
Dees, Susan Coons
Deibel, Rudolf
Deisher, Robert William
Demaria, William John Amsterdam
Demers, Pierre-Paul
DeMuth, George Richard
Denny, Floyd Wolfe, Jr
Desmond, Murdina Macfarquhar
Diamond, Louis Klein
Diehl, Antoni Mills
Di George, Angelo Mario
di Sant'Agnese, Paul Emilio Artom
Dodge, Philip Rogers
Dodge, Warren Francis

Done, Alan Kimball
Donnell, George Nino
Dower, John Charles
Doyle, Eugenie F
Drummond, Keith N
Ducharme, Jacques R
Duff, Raymond Stanley
Dungan, Margaret Caroline
Dungan, William Thompson
Dunn, Henry George
Dunphy, Donal
Dupont, Claire Hammel
Du Shane, James William
Dyken, Paul Richard
Eagle, John Frederick
Ebbs, J Harry
Echeverria, Peter Durand
Eckert, Herbert L
Edelmann, Chester M, Jr
Edlin, John Charles
Edwards, Walton
Egan, Thomas J
Ehlers, Hertha
Eichenwald, Heinz Felix
Elders, Minnie Joycelyn
Ellis, Elliot F
Emmanouilides, George Christos
Engle, Mary Allen English
Ensign, Paul Roselle
Enzer, Norbert Beverley
Esterly, Nancy Burton
Etteldorf, James N
Evans, Audrey Elizabeth
Evans, Hugh E
Falkner, Frank Tardrew
Fallers, Constantine J
Fashena, Gladys Jeannette
Feigin, Ralph David
Ferencz, Charlotte
Fernbach, Alfred Leonard
Filer, Lloyd Jackson, Jr
Finberg, Laurence
Fink, Chester Walter
Fink, Donald Lloyd
Fireman, Philip
Firlit, Casimir Francis
Fisher, Delbert A
Fishman, Marvin Allen
Fleisher, Daniel S
Fleming, Arthur William
Floriman, Alfred Leonard
Fogel, Bernard J
Fomon, Samuel Joseph
Fong, Jack Sun-Chik
Forbes, Gilbert Burnett
Forbis, Orie Lester, Jr
Fowler, Richard Edmond
Fraad, Lewis M
Frank, Donald Joseph
Frasier, S Douglas
Freeman, Arnold J
Freeman, John Mark
French, Joseph H
Frick, Oscar L
Friedman, David Belais
Friedman, William Foster
Fujimoto, Atsuko Ono
Fulginiti, Vincent Anthony
Gagliano, Nicholas Charles
Gajdusek, Daniel Carleton
Gardner, Lytt Irvine
Garrard, Sterling Davis
Gaull, Gerald E
Gedgoud, John Leo
Gerald, Park S
Gerrard, John Watson
Gessner, Ira Harold
Ghadimi, Hossein
Giammona, Samuel T
Giannini, Margaret Joan
Gibbs, Gordon Everett
Gibson, William Miles
Githens, John Horace
Gitlin, David
Glasgow, Lowell Alan
Glick, Samuel Shipley
Gluck, Louis
Gold, Eli
Golden, Archie Sidney
Goldman, Armond Samuel
Good, Robert Alan
Good, Thomas Arnold
Gordis, Leon
Gordon, Harry Haskin
Graham, Bruce Douglas
Graham, George G
Graham, William Wallace
Grant, William Wallace
Graven, Stanley N
Graub, Milton
Green, Hubert Gordon
Green, Marvin
Green, Orville
Grewar, David
Gromisch, Donald S
Gross, Ruth T
Grossman, Burton Jay
Grossman, Herbert Jules

Grossman, Milton S
Grossman, Moses
Grulee, Clifford Grosselle, Jr
Grunbach, Melvin Malcolm
Grunn, Jerome Alvin
Guenther, Donna Marie
Guinane, James Edward
Guntheroth, Warren G
Haggerty, Robert Johns
Hamburger, Robert Newfield
Hammar, Sherrel L
Hammond, George Denman
Hammond, Wylda
Hansen, Marc F
Hanshaw, James Barry
Hara, Saburo
Hardgree, Mary Carolyn
Hardwick, David Francis
Harper, Paul Alva
Harris, Jerome Sylvan
Harris, Leonard Crossley
Harris, Ruth Cameron
Harrison, Gunyon M
Harrison, Harold Edward
Hastreiter, Alois Rudolf
Hawke, W A
Haworth, James C
Hayles, Alvin Beasley
Haynes, Ralph Edwards
Hays, Daniel Mauger
Heald, Felix Pierpont, Jr
Hecht, Frederick
Heinlich, Ernest Maurice
Heiner, Douglas C
Hellerstein, Stanley
Helwig, Floy C
Henley, Joseph Owen
Henley, Walter L
Hepner, Ray
Herdman, Roger Cole
Heuser, Eva T
Heymann, Walter
High, Robert Huggins
Hildick-Smith, Gavin
Hill, Harry Raymond
Hill, Louis Leighton
Hillman, Elizabeth S
Hinman, Alanson
Hinton, George Greenough
Hodes, Horace Louis
Hoefnagel, Dick
Hohn, Arno M
Holder, Thomas M
Holland, Nancy H
Holliday, Malcolm A
Hollowell, Joseph Gurney, Jr
Holman, Gerald Hall
Holt, Charlene Poland
Hoffman, Julien Ivor Ellis
Holtzman, Neil Anton
Hong, Richard
Horwitz, Marshall Sydney
Howatt, William Frederick
Howell, Doris Ahlee
Howell, Ralph Rodney
Hsia, Yujen Edward
Hsu, Katherine Han Kuang
Huang, Nancy N
Hug, George
Hughes, James Gilliam
Hughes, Walter T
Hunt, Andrew Dickson, Jr
Huntley, Carolyn Coker
Iob, Vivian
Isaacson, Edward Kenneth
Isom, John B
Israels, Sydney
Izant, Robert James, Jr
Jabbour, J T
Jaco, Nicholas Trevenen
James, Leonard Stanley
Jarmakan, Jay M
Jelliffe, Derrick Brian
Jensen, Gordon D
Jensen, Philip J
Johnston, Richard Boles, Jr
Jones, Barbara
Jones, James David
Joos, Howard Arthur
Jordan, Robert, Jr
Joshi, Vijay Vinayak
Juberg, Richard Caldwell
Kagan, Benjamin M
Kaplan, Selna L
Karzon, David T
Katz, Michael
Katz, Samuel Lawrence
Kaufmann, Herbert Joseph
Kaye, Robert
Keele, Doman Kent
Keitel, Hans George Emil
Kelley, Vincent Charles
Kempe, Charles Henry
Kendall, Norman
Kennedy, Charles
Kennell, John Hawks
Kerr, George R

Kerrigan, Gerald Austin
Kibrick, Sidney
Kimura, Kazuo Kay
King, Katherine Chung-Ho
Kinsbourne, Marcel
Kirkman, Henry Neil
Klein, Robert
Klinberg, William Gene
Knobloch, Hilda
Koch, Richard
Kochen, Joseph Abraham
Kogut, Maurice D
Kohler, Elaine Eloise Humphreys
Kohrman, Arthur Fisher
Kontras, Stella B
Kopel, William Otis
Kornreich, Helen Kass
Kowarski, A Avionoam
Kramar, Jeno Louis
Kretchmer, Norman
Krivit, William
Krovetz, L Jerome
Kugel, Robert Benjamin
Kulczycki, Lucas Luke
Kushnick, Theodore
Lahey, M Eugene
LaMarche, Paul H
Lamb, George Alexander
Lambdin, Morris Arthur
Lanman, Jonathan T
Lanzkowsky, Philip
Lawson, Robert Barrett
Lee, Daisy Si
Lee, Henry Foster
Leer, John Addison, Jr
Lees, Martin H
Lemire, Ronald John
Leonard, Martha Frances
Letarte, Jacques
Leventhal, Brigid Gray
Levin, Aaron B
Levin, Sidney
Levine, Milton Isra
Levine, Robert
Levy, Harvey Louis
Light, Irwin Joseph
Lippard, Vernon William
Lipsitz, Philip Joseph
Lis, Edward Francis
Littlefield, John Walley
Livingston, Samuel
Lobeck, Charles Champlin
Logan, George Bryan
Lopez, Rafael
Lorenzen, Evelyn June
Lorincz, Andrew Endre
Lowe, Charles Upton
Lowrey, George Harrison
Lubchenco, Lula O
Lucas, Russell Vail, Jr
Lucey, Jerold Francis
Luhby, Adrian Leonard
Lurie, Paul Raymond
Luzzatti, Luigi
Mackler, Bruce
MacMillan, Duncan Robert
Macpherson, Roderick Ian
Manning, James Arthur
Marchessault, Victor Henri
Marienfeld, Carl J
Markowitz, Milton
Marks, James Frederic
Marks, Melvin Issac
Martin, John Kenneth
Martin, Lester W
Martin, Malcolm Mencer
Martins da Silva, Mauricio
Matalon, Reuben
Matanoski, Genevieve M
Mathies, Allen Wray, Jr
Mayer, Florence M
McAdams, Arthur James
McAllister, Robert Milton
McBryde, Angus Murdoch
McCormack, William C
McCrory, Wallace Willard
McCue, Carolyn M
McGuinness, Aims Chamberlain
McIntire, Matilda S
McKay, Robert James, Jr
McMillan, Campbell White
McNamara, Dan Goodrich
Medearis, Donald N, Jr
Medovy, Harry
Meehan, Robert John
Melin, Gilbert Wylie
Mellins, Robert B
Meltzer, Richard S
Menashe, Victor D
Menkes, John H
Meredith, Howard Voas
Merritt, Doris Honig
Mesel, Emmanuel
Meyer, Harry Martin, Jr
Meyer, Roger J
Meyer, Ruben
Michael, Alfred Frederick, Jr
Middelkamp, John Neal

Migeon, Claude Jean
Miller, C Arden
Miller, Herbert Chauncey
Miller, John Johnston, III
Miller, Robert Warwick
Miller, William Weaver
Mindlin, Rowland L
Minhas, Kareem
Mirkin, Bernard Leo
Mitchell, Shiela Craig
Mize, Charles Edward
Moffet, Hugh L
Montalvo, Jose Miguel
Montgomery, Theodore Ashton
Mooring, Paul K
Morgan, Beverly Carver
Morishima, Akira
Mosier, H David, Jr
Murphy, Edward G
Nadas, Alexander Sandor
Nadler, Henry Louis
Nahmias, Andre Joseph
Najjar, Victor Assad
Nankervis, George Arthur
Nebert, Daniel Walter
Necheles, Thomas
Neill, Catherine Annie
Neims, Allen Howard
Nelson, John D
Nelson, Thomas Lothian
Nelson, Waldo Emerson
Nemir, Rosa Lee
New, Maria Iandolo
Newcomb, Richard William
Newton, William Allen, Jr
Nghiem, Quang Xuan
Noonan, Jacqueline Anne
Nora, Audrey Hart
Nora, James Jackson
Nyhan, William Leo
O'Brien, Donough
Obrinsky, William
Odell, George Berlage
Oh, William
Ohersen, Henry Biemann, Jr
Oliver, George Murdock
Olmsted, Richard W
Olson, Lloyd Clarence
O'Neill, James A, Jr
Opitz, John Marius
Ordway, Nelson Kneeland
Ortiz, Antonia
Oski, Frank
Owen, George Murdock
Ozere, Rudolph L
Pachman, Daniel James
Page, Arthur R
Palmisano, Paul Anthony
Park, Myung Kun
Parkman, Paul Douglas
Parks, John S
Parmelee, Arthur H, Jr
Partington, Michael W
Patrick, James R
Paulsen, Elsa Proehl
Pearson, Howard Allen
Pearson, Paul (Hammond)
Peebles, Thomas
Perez, Carlos A
Perrin, Eugene Victor
Perry, Thomas Lockwood
Peterson, John Cyril
Peterson, Raymond Dale August
Phibbs, Roderic H
Pickering, Richard Joseph
Piel, Carolyn F
Pierce, Alexander Webster, Jr
Pinkel, Donald Paul
Pious, Donald A
Plotkin, Stanley Alan
Polgar, George
Pollara, Bernard
Porter, Frederick Stanley, Jr
Pradilla, Alberto Gonzalo
Pratt, Charles Benton
Price, Edward Lowell
Pronove, Pacita
Pueschel, Siegfried M
Qazi, Qutubuddin H
Quie, Paul Gerhardt
Rabe, Edward Frederick
Rashkind, William Jacobson
Rasmussen, Lester Paul
Rausen, Aaron Reuben
Ray, Charles George
Read, Charles H
Read, John Hamilton
Reed, George Benson
Reichelderfer, Thomas Elmer
Reiner, Charles Brailove
Reinhart, John Belvin
Rembolt, Raymond Ralph
Rennert, Owen M
Renuart, Adhemar William
Revesz, Clara Rona

Reynolds, John Weston
Ribble, John C
Riley, Conrad Milton
Riley, Harris D, Jr
Robbins, Frederick Chapman
Robertson, Alex F
Robertson, William O
Robinson, Arthur
Roche, Alexander F
Rodgers, Bradley Moreland
Rodriguez-Trias, Helen
Rook, George David
Root, Allen William
Rosenbloom, Arlan Lee
Rosenfield, Robert Lee
Rosenthal, Ira Maurice
Rosin, Sidney
Rothberg, Richard Martin
Roy, Claude Charles
Rubin, Mitchell Irving
Rudolph, Arnold Jack
Russell, Attie Yvonne
Rutledge, Lewis James
Ruttenberg, Herbert David
Ruvalcaba, Rogelio H A
Sanyal, Shyamal Kumar
Scaglione, Peter Robert
Scarpelli, Emile Michael
Schafer, Irwin Arnold
Schain, Richard J
Schiebler, Gerold Ludwig
Schlegel, Robert John
Schmickel, Roy David
Schneck, Larry
Schorr, Julian
Schubert, William K
Schulkind, Martin Lewis
Schulman, Irving
Schulz, Jeanette
Schwanecke, Rebecca G Pineda
Schwartz, Herbert C
Schwartz, James F
Schwartz, Robert
Scott, Roland B
Scriver, Charles Robert
Seeger, Robert Charles
Seely, J Rodman
Segar, William Elias
Seidel, Henry Murray
Sell, Sarah H Wood
Sellers, Frank Jamieson
Senior, Boris
Senn, Milton J E
Sever, John Louis
Shaffer, Thomas Eugene
Shapiro, Donald Lawrence
Shepard, Thomas H
Shih, Vivian EAn
Shinefield, Henry R
Shipley, Thorne
Shirkey, Harry Cameron
Shore, Nomie Abraham
Shulman, Stanford Taylor
Shumway, Clare Nelson, (Jr)
Sidbury, James Buren, Jr
Sifontes, Jose E
Silver, Henry K
Silverman, Frederic Noah
Simone, Joseph Vincent
Simopoulos, Artemis Panageotis
Sinks, Lucius Frederick
Sinnette, Calvin Herman
Sisson, Thomas Randolph Clinton
Slobody, Lawrence Boris
Smith, David S
Smith, David W
Smith, Edna H
Smith, Elizabeth Knapp
Smith, Fred George, Jr
Smith, George F
Smith, Hugo Dunlap
Smith, Joseph Darrel
Smith, Nathan James
Smith, Richard Thomas
Smithwick, Elizabeth Mary
Snyder, Christopher Harrison
Snyderman, Selma Eleanore
Sobel, Edna H
Soifer, Morton Marshall
Solomons, Clive (Charles)
Solomons, Gerald
Solomons, Gerald
Sotos, Juan Fernandez
South, Mary Ann
Spach, Madison Stockton
Speck, William T
Spector, Samuel
Spencer, William Albert
Spitzer, Roger Earl
Spock, Alexander
Spragins, Melchijah
Stafford, George Ewing
Stahlman, Mildred
Stark, Charles R
Stave, Uwe
Steigman, Alex J
Steiner, Morris

Steinschneider, Alfred
Sterling, Harold Melvin
Stern, Aaron Milton
Stevenson, Stuart Shelton
Stiehm, E Richard
Stillerman, Maxwell
Stine, Oscar Cebren
Stoesser, Albert V
Stover, Samuel Landis
Strang, Ruth Hancock
Strauss, Ronald George
Sullivan, Donita B
Sullivan, Margaret P
Summitt, Robert L
Sutherland, James McKenzie
Sweetman, Lawrence
Swyer, Paul Robert
Talbert, James Lewis
Talbot, Nathan Bill
Talner, Norman Stanley
Taylor, Fred M
Taylor, Paul M
Taylor, William Clyne
Thaler, M Michael
Thilenius, Otto G
Thomas, George Howard
Thomas, John Martin
Thorp, Frank Kedzie
Thurman, William Gentry
Tiller, Ralph Earl
Tips, Robert Leonard
Tooley, William Henry
Torres-Pinedo, Ramon
Towers, Bernard
Tragais, Demetrius G
Traisman, Howard Sevin
Travis, Luther Brisendine
Treves, Salvador
Turner, Edward V
Underwood, Louis Edwin
Valdes-Dapena, Marie A
Vallbona, Carlos
Van Leeuwen, Gerard
Vann, Robert Lee
Van Wyk, Judson John
Vaughan, Victor Clarence, III
Vernier, Robert L
Victorica, Benjamin (Eduardo)
Vietti, Teresa Jane
Vogt, Frank Conrad
Voorhess, Mary Louise
Walcher, Dwain N
Walker, Lillie Cutlar
Walters, Thomas Richard
Wannamaker, Lewis William
Wappner, Rebecca Sue
Ward, Richard S
Ward, Robert
Waring, William Winburn
Warkany, Josef
Warner, Robert
Warwick, Warren J
Wasserman, Edward
Watson, David Goulding
Webb, Patricia Ann
Weber, Wendell W
Wedgwood, Ralph Josiah Patrick
Wegman, Myron Ezra
Wehrle, Paul F
Weichsel, Morton E, Jr
Weil, Marvin Lee
Weil, William B, Jr
Weiss, Charles Frederick
Weisskopf, Bernard
Wells, Charles Robert Edwin
Westphal, Milton C, Jr
Whittemore, Ruth
Wieczorowski, Elsie
Williams, Christopher P S
Wilson, John T
Wilson, Miriam Geisendorfer
Winick, Myron
Winter, Jeremy Stephen Drummond
Winter, Robert John
Winters, Robert Wayne
Wishnick, Marcia M
Witte, John Jacob
Wohltmann, Hulda Justine
Wolff, James A
Wong, Paul Wing-Kon
Woody, Norman Cooper
Woolley, Paul Vincent
Work, Henry Harcus
Worthen, Howard George
Wray, Joe D
Wright, David Laverne
Wright, Francis Howell
Wright, Francis Stuart
Wright, Harry Tucker, Jr
Yaffe, Sumner J
Yang, Dorothy Chuan-Ying
Young, Lionel Wesley
Young, Maurice Durward
Yow, Martha Dukes
Zee, Paulus
Zellweger, Hans Ulrich
Zelson, Carl

Ziegra, Sumner Root
Zinkham, William Howard
Zipursky, Alvin
Zuelzer, Wolf W

Physical Medicine & Rehabilitation

Abramson, Arthur Simon
Anderson, Albert Douglas
Anderson, Thomas Page
Archibald, Kenneth C
Athelstan, Gary Thomas
Awad, Essam A
Bennett, Robert Leo
Campbell, Suzann Kay
Christopher, Robert Paul
Chyatte, Samuel Baruch
Dail, Clarence Wilding
Darling, Robert Croly
Demopoulos, James Thomas
Ellwood, Paul M, Jr
Forster, Sigmund
Fowler, William Mayo, Jr
Freed, Murray Monroe
Gersten, Jerome William
Goldfine, Lewis John
Grant, Arthur E
Haskin, Myra Ruth Singer
Healey, John Edward, Jr
Hinterbuchner, Catherine Nicolaides
Hogue, Raymond Ellsworth
Jebsen, Robert H
Kamenetz, Herman Leo
Kirby, William Henry, Jr
Kottke, Frederic James
Lake, Lorraine Frances
Lee, Mathew Hung Mun
Lehmann, Justus Franz
LeVeau, Barney Francis
Liao, Sung Jui
Licht, Sidney
Lowman, Edward Wynne
Martin, Gordon Mather
Melvin, John Lewis
Mennell, John McMillan
Miller, John Melville, III
Mongeau, Maurice
Olson, Don A
Park, Herbert William, III
Paul, Boris Jerome
Payton, Otto D
Rose, Donald L
Rossier, Alain B
Ryan, Allan James
Singleton, Mary Clyde
Spendlove, George Arthur
Sterling, Harold Melvin
Stillwell, George Keith
Swezey, Robert Leonard
Truong, Xuan Thoai
Urbscheit, Nancy Lee
Wellock, Lois Margaret
Worden, Ralph Edwin

Preventive Medicine

Arnold, Richard C
Baker, Timothy D
Barrett, Harold Spencer
Bashe, Winslow Jerome, Jr
Baxley, William Allison
Beard, Rodney Rau
Bergsma, Daniel
Boaz, Thurmond DeWitte, Jr
Breslow, Lester
Brooks, Thomas Joseph, Jr
Buck, Carol Whitlow
Cadmus, Robert R
Catterson, Allen Duane
Chick, Ernest Watson
Christenson, Paul John
Cobb, John Candler
Comer, Ralph Dudley
Cunningham, William John
Dodge, Horace Jackson
Dodge, Warren Francis
Drusin, Lewis Martin
Dull, Harold Bruce
Ellingson, Harold Victor
Farquhar, John William
Ferguson, Ronald James
Finklea, John F
Fisher, Wilton Monroe
Fleming, William Leroy
Flipse, Martin Eugene
Foy, Hjordis M
Freitag, Julia Louise
Fuenning, Samuel Isaiah
Fulmer, Hugh Scott
Gagliano, Nicholas Charles
Gottlieb, Marise Suss
Grayston, J Thomas
Greenberg, Jerome Herbert
Griffith, Jack Dee
Gustafson, David Harold
Hagstrom, Ruth Murray
Hamrick, Joseph Thomas
Henig, Philip Edward
Herbolsheimer, Henrietta
Hicks, Edward James

Hoff, Richard Lee
Howard, Lee Milton
Hume, John Chandler
Hutchison, William Forrest
Imperato, Pascal James
Jones, David John
Kashgarian, Mark
Kessler, Irving Isar
Koch-Weser, Dieter
Kunin, Calvin Murry
Lane, Wallace
Langford, Ernest Robert
Legers, Llewellyn J
Le Seney, Catherine Cosgray
Levin, Morton Loeb
Levitsky, John M
Levy, Leo
Lewis, Charles E
Lichter, Edward A
Lieben, Jan
Lynn, Thomas Neil, Jr
MacDonald, Kenneth
Macdonald, Michael Raymond
Marienfeld, Carl J
Marra, Edward Francis
Marshall, Carter Lee
McNamara, Michael Joseph
McNinch, Joseph Hamilton
Micks, Don Wilfred
Millar, John Donald
Miller, Joseph Morton
Morgan, Donald Pryse
Mott, Frederick Dodge
Mou, Thomas William
Nau, Carl August
Nemuth, Harold I
Northrop, Cedric
Oberman, Albert
Osterud, Harold T
Packer, Henry
Parrish, Henry Mack
Peckinpaugh, Robert Owen
Peterson, Osler Luther
Powers, Leland Earle
Prather, Elbert Charlton
Quinn, Robert (William)
Quinones, Mark A
Quisenberry, Walter Brown
Read, John Hamilton
Rice-Wray, Edris
Richardson, William Perry
Roemer, Milton Irwin
Rogers, Fred Baker
Rogers, Kenneth D
Rutstein, David Davis
Sellers, Thomas F, Jr
Seltser, Raymond
Sheps, Cecil George
Siegel, Morris
Simmons, Robert Lowrey
Simpson, William Loyal
Smith, Donald C
Sox, Ellis Dean
Sparer, Phineas Jack
Spencer, Frederick J
Springall, Arthur Newton
Stamler, Jeremiah
Stave, Uwe
Steele, Robert
Steen, Wilson D
Stiles, William Whitfield
Tabershaw, Irving R
Taylor, Carl Ernest
Thomas, Heriberto Victor
Top, Franklin Henry, Sr
Uzodinma, John E
Van Peenen, Peter Franz Dirk
Van Duser, Arthur L
Vivona, Stefano
Walker, Lillie Cutlar
Webb, Nathaniel Conant, Jr
Wen, Chi-Pang
Wenzel, Richard Louis
Werner, Sanford Benson
Wharton, James Dumont
Wheatley, George Milholland
White, Stanley C
Winter, Phillip E
Witte, John Jacob
Wolfson, Edward A
Wynder, Ernst Ludwig
Zimmer, James Griffith

Proctology

Hill, John Roger

Pulmonary Diseases

Ackerman, Neil Richard
Beatty, Oren Alexander
Berglund, Erik
Brody, Jerome Saul
Brooks, Stuart Merrill
Cassan, Stanley Morris
Cherniak, Louis
Cugell, David Wolf
Des Prez, Roger Moister
Farber, Seymour Morgan
Goldman, Allan Larry

MEDICINE

Hadley, Susan Jane
Hansen, James E
Israel, Harold L
Johanson, Waldemar Gustave, Jr
Kauffman, Leon A
Keim, Lon William
King, Thomas K C
Klocke, Robert Albert
Knowles, John Hilton
Lieberman, Jack
McClement, John Henry
Nedwicki, Edward G
Noehren, Theodore Henry
Petty, Thomas Lee
Ray, Edward Scott
Renzetti, Attilio D, Jr
Rhodes, Mitchell Lee
Ryan, Una Scully
Smith, Jan D
Snider, Gordon Lloyd
Stead, William White
Weiss, William
Woolf, C R

Radiology

Abrams, Herbert L
Adams, Andrew Borden
Alexander, Leslie Luther
Allen, Joseph Hunter
Allen, William E, Jr
Allison, Morris Jonathan Carl
Altman, Kurt Ison
Altschuler, Martin David
Amplatz, Kurt
Anderson, James Howard
Andrews, John Robert
Angleton, George M
Archambeau, John Orin
Ash, Clifford L
Atkins, Harold Lewis
Baker, David H
Baker, Sol Ronald
Ball, Robert P
Barnhard, Howard Jerome
Batley, Frank
Baum, Stanley
Baylin, George Jay
Becker, Joshua A
Bedford, Joel S
Bennett, Leslie R
Berenbaum, Samuel Louis
Berk, Robert Norton
Bilaniuk, Larissa Tetiana
Bilbao, Marcia Kepler
Bogardus, Carl Robert, Jr
Bonakdarpour, Akbar
Bonte, Frederick James
Bookstein, Joseph Jacob
Bosniak, Morton A
Botstein, Charles
Brady, Luther Weldon, Jr
Bragg, David Gordon
Brascho, Donn Joseph
Brodeur, Armand Edward
Brown, Mark
Buhrow, Charles, Jr
Burgener, Francis Andre
Burkell, Charles Craig
Caldwell, William L
Capp, Michael Paul
Carlsson, Erik
Carpender, James Wood Johnson
Chang, Chae Han Joseph
Chang, Chu Huai
Charyulu, Komanduri K N
Chu, Florence Chien-Hwa
Clark, John Whitcomb
Cleary, Stephen Francis
Collins, Lois Cowan
Cooley, Robert Nelson
Craig, Stanley Harold
Crummy, Andrew B
Daves, Marvin Lewis
Dealey, James Bond, Jr
De Carlo, John, Jr
Deed, Eleanor Polk
Del Regato, Juan A
Deslattes, Richard D, Jr
Dewing, Stephen Bronson
Dienes, Domenico
Diner, Wilma Canada
Dobben, Glen D
Doby, Tibor
Dodd, Gerald Dewey, Jr
Don, Conway J
Donner, Martin W
Dorst, John Phillips
Dowdy, Andrew Hunter
Duda, Eugene Edward
Duggan, Hector Ewart
Dunbar, John Scott
Edeiken, Jack
Egan, Robert L
Elkin, Milton
Elliott, Larry Paul
Eschner, Edward George
Eiter, Lewis Elmer
Fabrikant, Jacob I
Fayos, Juan Vallvey
Feder, Bernard Herbert
Felson, Benjamin
Fennessy, John James
Figley, Melvin Morgan
Finby, Nathaniel
Fischer, Harry William
Fox, Eva Fernandez
Freimanis, Atis K
Friedell, Hymer Louis
Friedenberg, Richard M
Friedman, Paul J
Gabrielsen, Trygve O
Gabrielsen, Orlando Frederick
Gannon, William E
Gargano, Fredie Patrick
Gaulden, Mary Esther
Geffen, Abraham
George, Frederick W, III
Gildenhorn, Hyman L
Girolamo, Rita Frances
Good, Clarence Allen
Goodrich, Jack K
Gorson, Robert O
Gottschalk, Alexander
Graham, C Benjamin
Gramiak, Raymond
Grantmyre, Edward Bartlett
Greenfield, George B
Greenspan, Richard H
Griem, Melvin Luther
Grigg, Emanuel Radu Newman
Grossman, Herman
Hall, Ernest Lenard
Hanafee, William Norman
Harle, Thomas Stanley
Haskin, Marvin Edward
Hempelmann, Louis Henry, Jr
Hendrickson, Frank R
Hilal, Sadek K
Hill, B Jay
Hinck, Vincent C
Hodes, Philip J
Hoffer, Paul B
Holiman, B Leonard
Holt, John Floyd
Houston, Clarence Stuart
Howard, William Henry Richard
Howland, Willard J
Hunt, Howard Beeman
Isard, Harold Joseph
Izenstark, Joseph Louis
Jacobson, Arnold P
Jacobson, George
Jacobson, Harold Gordon
James, Alton Everette, Jr
Jessen, Carl Roger
Jing, Bao-Shan
Johnson, Philip Martyn
Jones, Malcolm David
Jorgens, Joseph
Juhl, John Harold
Kanick, Virginia
Kaplan, Henry Seymour
Kartha, Mukund K
Kaufmann, Herbert Joseph
Keats, Theodore Eliot
Keefer, George Prahler
Kent, Henry Peter
Kieffer, Stephen A
Klatte, Eugene
Knox, Gaylord Shearer
Koehler, P Ruben
Kolodny, Gerald Mordecai
Kuhl, David E
Kundel, Harold Louis
Lane, Edwin Lever
Lampe, Isadore
Lang, Erich Karl
Langdon, Edward Allen
Langland, Olaf Elmer
Lasser, Elliott Charles
Latourette, Howard Bennett
Lebel, Jack Lucien
Lehrer, Harold Z
Lester, Richard Garrison
Levin, Bertram
Liebner, Edwin J
Lin, Shu-Ren
Linnemann, Roger E
Lodwick, Gwilym Savage
Loop, John Wickwire
Loring, Marvin F
Lott, James Stewart
Love, Leon
Lowman, Robert Morris
MacEwan, Douglas W
Macpherson, Roderick Ian
Mallams, John Thomas
Manchester, J Stewart
Margulis, Alexander Rafailo
Martel, William
Martin, Charles Louis
Martin, James Franklin
Martinez, Luis Osvaldo
Mathieu, Roger Maurice
McAfee, John Gilmour
McAlister, William H
McDonel, Gerald M
McRae, Donald Lane
Mellins, Harry Zachary
Meschan, Isadore
Mewissen, Dieudonne Jean
Meyers, Harvey I
Meyers, Philip Henry
Miller, Earl Roy
Milne, Eric Campbell
Moos, Walter Sam
Moreton, Robert Dulaney
Morgan, Russell Hedley
Morkovin, Dimitry
Moseley, Robert David, Jr
Myers, Karl Johnson, Sr
Neal, Marcus Pinson, Jr
Newton, Thomas Hans
Nice, Charles Monroe, Jr
Nickson, James Joseph
Nooman, Charles D
Norman, Alex
Ochsner, Seymour F
Oder, Alexander
Oestreich, Alan Emil
O'Loughlin, Bernard James
O'Mara, Robert E
Oppenheim, Bernard Edward
Osmond, Leslie H
Ottoman, Richard Edward
Overfors, Carl-Olof Nils Sten
Palubinskas, Alphonse J
Parker, Robert O
Patterson, Virginia Norrell
Patton, Dennis David
Paul, Robert E, Jr
Pentel, Leon
Perlmutter, Henry Irwin
Perryman, Charles Richard
Peterson, Harold Oscar
Peterson, Oscar Sylvander, Jr
Phillips, Jerry Clyde
Phillips, Leon A
Pinck, Robert Lloyd
Pochaczevsky, Rubem
Pomeroy, Herbert Guy
Ponzanski, Andrew K
Pool, Winford H, (Jr)
Potchen, E James
Potter, Guy Dill
Powell, Clinton Cobb
Poyton, Herbert Guy
Pugh, David Graham
Quinn, James L, III
Rabinowitz, Jack Grant
Ragan, Harvey Albert
Ranniger, Klaus
Rapp, Robert
Ravenos, Antolin, IV
Rejali, Abbas Mostafavi
Reuter, Stewart R
Revesz, George
Reynolds, Jack
Rhodes, Buck Austin
Ridings, Gus Ray
Riemenschneider, Paul Arthur
Rigler, Leo George
Roach, John Faunce
Robbins, Laurence Lamson
Rockoff, Seymour David
Rogers, James Virgil, Jr
Rogoff, Stanley Myron
Rosenberg, Henry Mark
Roth, Robert Earl
Rubenfeld, Sidney
Rubin, Philip
Rugh, Roberts
Ruzicka, Francis Frederick, Jr
Saba, George Peter, II
Saenger, Eugene L
Sagerman, Robert H
Salik, Julian Oswald
Sargent, Nicholas E
Scanlin, James Howard
Schechter, Mannie M
Schlaeger, Ralph
Schreiber, Melvyn Hirsh
Schultz, Everett Hoyle, Jr
Schulz, Milford David
Schwartz, Emanuel Elliot
Schwartz, Solomon Samuel
Schwartz, Gerhart Steven
Scott, Ralph Mason
Seaman, William B
Shapiro, Robert
Shehadi, William Henry
Shin, Myung Soo
Siegel, Barry Alan
Sieniewicz, David James
Silverman, Frederic Noah
Simon, Allan Lester
Skucas, Jovitas
Smith, Edward Herbert
Snipes, Morris Burton
Sobkowski, Frank J
Song, Chang Won
Sooby, Donna Louise
Soulen, Renate Leroi
Southard, Martha Ellen
Staple, Tom Weinberg
Steele, James Patrick
Stein, George Nathan
Steinbach, Howard Lynne
Stickley, Elmer Eugene
Stilson, Walter Leslie
Swischuk, Leonard Edward
Tampas, John Peter
Taplin, George Vorce
Tarleton, Gadson Jack, Jr
Tavares, Juan M
Taybi, Hooshang
Templeton, Arch W
Templeton, Frederic Eastland
Teplick, Joseph George
Thornbury, John R
Torrance, Daniel J
Tripp, Russell Maurice
Tristan, Theodore A
Tucker, Arthur Smith
Tuddenham, William J
Tulsky, Emanuel Goodel
Twigg, Homer Lee
Urtasun, Raul C
Valvassori, Galdino E
Van Buskirk, Frederick William
Vermund, Halvor
Viamonte, Manuel, Jr
Vitek, Jiri Jakub
Waggener, Ronald E
Wallace, Sidney
Wang, Yen
Warnick, Edward George
Webber, Milo M
Wege, William Richard
Whitehouse, Walter Machintre
Whitley, Joseph Efird
Whitley, Nancy O'Neil
Wigh, Russell
Wilson, McClure
Wilson, William Jewell
Wohl, George T
Wolf, Bernard Saul
Wolf, Gerald Lee
Wollaeger, Gertraud
Wollschlaeger, Paul Bernhard
Woloshin, Henry Jacob
Woodruff, John H, Jr
Wyman, Stanley M
Yale, Seymour Hershel
Youker, James Edward
Young, Lionel Wesley
Youngstrom, Karl Arden
Yune, Heun Yung
Zanca, Peter
Zatz, Leslie M
Zboralske, F Frank
Zingesser, Lawrence H

Radiotherapy

Bagshaw, Malcolm A
Baxter, Donald Henry
Bloor, Robert John
Boone, Max L M
Brennan, James Thomas
Byfield, John Eric
Collins, Amy L Tsui
D'Angio, Giulio J
Feder, Bernard Herbert
Fischer, James Joseph
George, Frederick W, III
Hellman, Samuel
Kim, Jae Ho
King, E Richard
Kinzie, Jeannie Jones
Kligerman, Morton M
Kurohara, Samuel S
Levene, Martin Barrack
Levitt, Seymour H
Madoc-Jones, Hywel
Marcial, Victor A
Maruyama, Yosh
Million, Rodney Reiff
Montague, Eleanor D
Order, Stanley Elias
Ozarda, Ahsen T
Parsons, John Andresen
Pearson, James Gordon
Perez-Tamayo, Ruberti
Perry, Harold
Peters, Lester John
Raju, Mudundi Ramakrishna
Seydel, Horst Gunter
Slater, James Munro
Stefani, Stefano
Suit, Herman Day
Tahan, Theodore Wahba
Tefft, Melvin
von Essen, Carl Francois
Wang, Chiu-Chen
Webster, John H
Withers, Hubert Rodney
Wizenberg, Morris Joseph
Woodward, Kent Thomas

Rheumatology
Baum, John
Berney, Steven
Boas, Norman Francis
Cassidy, James T
Castor, Cecil William
Decker, John Laws
Dorwart, Bonnie Brice
Ehrlich, George Edward
Ford, Denys Kensington
Friou, George Jacob
Goldberg, Leonard Stephen
Harris, Edward Day, Jr
Hogan, Robert Steadham
Hollander, Joseph Lee
Kaplan, Stanley Baruch
Klinenberg, James Robert
Krey, Phoebe Regina
Kushner, Irving
Lockshin, Michael Dan
Lorber, Arthur
Martin, John Harvey
Moore, Mary Elizabeth
Mullinax, Perry Franklin
Myers, Ira Lee
Newcombe, David S
Plotz, Charles M
Rankin, Thomas Johnson
Ridolfo, Anthony Sylvester
Steinman, Charles Robert
Thompson, George Richard
Vaillancourt, De Guise
Vignos, Paul Joseph, Jr
Weiss, Joseph Jacob
Wilson, Colon Hayes, Jr
Winfield, John Buckner
Wynne-Roberts, Caroline Rosales

Rickettsial Diseases
Berge, Trygve Obert
Kordova, Nonna
Marchette, Nyven John
Osterman, Joseph Vincent, Jr
Vinson, John William

Syphilology
Izzat, Nawfal Nouri
Nicholas, Leslie
Pariser, Harry

Teratology
Banerjee, Bhola Nath
Barghusen, Herbert Richard
Beck, Sidney L
Blake, David Andrew
Bradford, James Carrow
Catz, Charlotte Schifra
Ceron, Gabriel
Clegg, David John
DeSesso, John Michael
Fisher, Don Lowell
Gale, Thomas Francis
Gilani, Shamshad H
Holson, Joseph Franklin, Jr
Jordan, Robert Lawrence
Kalter, Harold
Kaplan, Stanley
Kernis, Marten Murray
Khera, Kundan Singh
Kimmel, Carole Anne
Lemire, Ronald John
Long, Sally Yates
McCallion, David John
Morrison, Edward Joseph
Mottet, N Karle
Nanda, Ravindra
Overman, Dennis Orton
Pruzansky, Samuel
Reno, Frederick Edmund
Schmidt, Richard Ralph
Schwetz, Bernard Anthony
Scott, William James, Jr
Shenefelt, Ray Eldon
Staples, Robert Edward
Szabo, Kalman Tibor
Thompson, Daniel James
Wilson, James Graves

Therapeutics
Baker, Sol Ronald
Benfey, Bruno Georg
Durr, Friedrich (E)
Freeman, Leon David
Gruber, Charles Michael, Jr
Houde, Raymond Wilfred
Lane, Montague
Nechay, Bohdan Roman
Nickerson, Mark
Parker, William Arthur
Rockwell, Harriet Esther
Smith, Austin Edwards
Weilerstein, Ralph Waldo

Thoracic Diseases
Farber, Seymour Morgan
Lim, Thomas Pyung Kee
Sproule, Brian J

Webster, Burnice Hoyle

Tropical Medicine
Ash, Lawrence Robert
Beaver, Paul Chester
Brady, Frederick Jonathon
Cadigan, Francis C, Jr
Cahill, Kevin M
Coggeshall, Lowell Thelwell
Comer, Ralph Dudley
D'Alessandro-Bacigalupo, Antonio
Frothingham, Thomas Eliot
Imperato, Pascal James
Jung, Rodney Clifton
Kaiser, Robert L
Kean, Benjamin Harrison
Markell, Edward Kingsmill
Martin, Edgar J
Miller, Louis Howard
Most, Harry
Olivier, Louis John
Patterson, Athol James
Rendtorff, Robert Carlisle
Senft, Alfred Walter
Siddiqui, Wasim A
Srinhongse, Sunthorn
Stone, William Spencer
Taylor, Diane
Von Lichtenberg, Franz
Warren, Kenneth S
Warren, McWilson
Weinman, David, II
Weller, Thomas Huckle
Yoeli, Meir

Tuberculosis
Edwards, Phyllis Q
Kendig, Edwin Lawrence, Jr

Urology
Al-Askari, Salah
Ansell, Julian Samuel
Baker, Roger (Carroll), (Jr)
Barelare, Bruno
Barrett, William A
Blackard, Clyde Erhardt
Boyarsky, Saul
Boyce, William Henry
Bulkley, George
Bunge, Raymond George
Chapman, Warren Howe
Cockett, Abraham Timothy K
Collins, William E
Comarr, Avriom Estin
Corriere, Joseph N, Jr
Cox, Clair Edward, II
Culp, Ormond S
Evans, Arthur T
Farrell, James I
Finlayson, Birdwell
Firlit, Casimir Francis
Fraley, Elwin E
Garrett, Robert Austin
Gillenwater, Jay Young
Glenn, James Francis
Goldstein, Abraham M B
Goodwin, Willard E
Govan, Duncan Eban
Grayhack, John Thomas
Greene, Laurence Francis
Herr, Harry Wallace
Hotchkiss, Robert S
Jewett, Hugh Judge
Jordan, Willis Pope, Jr
Karafin, Lester
Kaufman, Joseph J
Kiefer, Joseph Henry
King, Lowell Restell
Koontz, Warren Woodson, Jr
Lapides, Jack
Lattimer, John Kingsley
Leberman, Paul R
Lee, Leroy William
Leiter, Elliot
Lich, Robert, Jr
Litvak, Austin S
Lloyd, Frederick A
Lytton, Bernard
Madsen, Paul O
Magoss, Imre V
Malashock, Edward Marvin
Markland, Alan Colin
McAninch, Lloyd Nealson
McDonald, Donald Fiedler
McDonald, James Hogue
McRoberts, J William
Melick, William F
Mellinger, George T
Miller, George H, Jr
Mobley, Jack Ervin
Mulcahy, John Joseph
Murphy, John Joseph
Newman, Harry
O'Conor, Vincent John, Jr
Parry, William Lockhart
Persky, Lester
Powell, Norborne Berkeley
Prout, George Russell, Jr

Radwin, Howard Martin
Rhamy, Robert Keith
Rieser, Charles Ely
Rohner, Thomas John
Rous, Stephen N
Saleh, Wasfy Seleman
Scott, Russell, Jr
Shirley, Sheridan William
Smith, Donald Ridgeway
Smith, Maurice John Vernon
Stamey, Thomas Alexander
Staubitz, William Joseph
Straffon, Ralph Atwood
Taguchi, Yoshinori
Taylor, Jack Neel
Tessier, Arthur Ned
Uhle, Charles Augustus Woerwag
Urry, Ronald Lee
Veenema, Ralph J
Vermeulen, Cornelius William
Waterhouse, Keith R
Wear, John Brewster, Jr
Weinberg, Sidney R
Weiss, Robert Martin
Woodruff, Marvin Wayne

Venereal Diseases
Bauer, Theodore James
Fiumara, Nicholas J
Furness, Geoffrey
Moody, Eric Edward Marshall
Vinson, John William

METALLURGY

Metallurgy
Ahmad, Iqbal
Coriell, Sam Ray
Delavault, Robert Edmund
Estes, John H
Flinn, Paul Anthony
Franklin, Wilbur Mitchell
Gabrysh, Andrew Francis
Gilman, John Joseph
Habermann, Clarence E
Hallada, Calvin James
Harvey, Walter William
Haseman, Joseph Fish
Kanda, Frank Albert
Lingane, Peter James
Manning, John Randolph
Mueller, Walter A
Peiffer, Howard R
Radford, Kenneth Charles
Stevens, Laurence Guy
Wei, Chuan-Tseng

Chemical Metallurgy
Benner, Blair Richard
Carter, Giles Frederick
Catlett, Duane Stewart
Curtis, Ralph Wendell
El Guindy, Mahmoud Ismail
Englehart, Edwin Thomas, Jr
Felten, Edward J
Foreman, Robert Walter
Glasser, Julian
Goggin, Donald Edward
Hillner, Edward
Hollingshead, Ethan Allen
Illis, Alexander
King, William Robert, Jr
Lynd, Langtry Emmett
Mears, Dana Christopher
Mears, Robert Bruce
Minford, James Dean
Nafziger, Ralph Hamilton
Paine, Robert Madison
Pierson, Hugh Ottho
Richards, Kenneth Julian
Ricksecker, Ralph E
St Cyr, Lewis Alpha
Scharfstein, Lawrence Robert
Shafer, William McKinley
Shores, David Aurth
Staudenmayer, Ralph
Stewart, Robert Daniel
Walsh, Kenneth Albert

Physical Metallurgy
Alfred, Louis Charles Roland
Barton, Richard J
Lynn, Kelvin Gideon
Mulford, Robert Neal Ramsay
Newkirk, John Burt
Thornburg, David Devoe
Wood, Floyd William
Yoakum, Anna Margaret

METEOROLOGY

Meteorology
Acheson, Donald Theodore
Adem, Julian

Adler, Robert Frederick
Alaka, Mikhail A
Alberty, Ronnie Lee
Allen, Louis Pinckney Jr
Almazan, James A
Alpert, Leo
Anthes, Richard Allen
Arya, Satya Pal Singh
Astling, Elford George
Atlas, David
Austin, James Murdoch
Austin, Pauline Morrow
Badgley, Franklin Ilsley
Baer, Ferdinand
Baer, Ledolph
Baker, Kay Dayne
Barnes, Arnold Appleton, Jr
Barnes, James Clarkson
Barrett, Earl Wallace
Battan, Louis Joseph
Batten, Edmund Stanley
Baum, Werne A
Baynton, Harold Wilbert
Behn, Robert Collins
Bellaire, Frank Rolland
Belmont, Arthur David
Benton, George Stock
Beran, Donald Wilmer
Bergman, Kenneth Harris
Berning, Warren Walt
Bernstein, Abram Bernard
Bertholf, Lloyd Bernard
Bierly, Eugene Wendell
Bigler, Stuart Grazier
Bilello, Michael Anthony
Black, James Francis
Blackadar, Alfred Kimball
Bleck, Rainer
Blumen, William D
Bonner, William D
Bornstein, Robert D
Boudreau, Robert Donald
Boville, Byron Walter
Bowling, Sue Ann
Breiland, John Gustavson
Bretherton, Francis P
Brier, Glenn Wilson
Brock, Fred Vincent
Brodrick, Harold James, Jr
Brooks, Edward Morgan
Brown, George Bruce
Brown, John A, Jr
Brown, Rodger Alan
Brundidge, Kenneth Cloud
Bryan, Kirk, (Jr)
Bryson, Reid Allen
Buell, Carleton Eugene
Bunting, Donald Charles
Businger, Joost Alois
Byers, Horace Robert
Campbell, William Joseph
Canfield, Norman L
Cardone, Vincent J
Carlson, Toby Nahum
Carlson, William Samuel
Carlstead, Edward Meredith
Carson, James Estle
Caskey, James Edward, Jr
Chappell, Charles Franklin
Charney, Jule Gregory
Chiu, Wan-Cheng
Clark, Robert Alfred
Clodman, Joseph
Cole, Alan L
Colon, Jose A
Conover, John Hoagland
Cooley, Duane Stuart
Coulson, Kinsell Leroy
Craig, Richard Ansel
Cramer, Harrison Emery
Crane, Robert Kendall
Crawford, Todd V
Cressman, George Parmley
Croom, Herman Lee
Crutcher, Harold L
Culkowski, Walter Martin
Culnan, Robert Neville
Cunningham, Robert M
Cunnold, Derek M
Czapski, Ulrich Hans
Danard, Maurice Beverley
Darkow, Grant Lyle
Darling, Eugene Merrill, Jr
Davidson, Kenneth LaVern
Davies-Jones, Robert Peter
Davis, Paul A
Day, John Arthur
Decker, Wayne Leroy
Degani, Meir Hershtenkorn
De Leonibus, Pasquale S
De Percin, Fernand
Dethier, Bernard Emile
Devereux, Robert Francis
Dewey, Kenneth Frederic
Dickerson, Marvin Hubert
Dickson, Don Robert
Diemer, Edward Devlin
Dingle, Albert Nelson

Djuric, Dusan
Dodd, Arthur V
Doe, Learmont Anstice Earlston
Dorman, Clive Edgar
Doviak, Richard J
Driscoll, Dennis Michael
Droessler, Earl George
Duquet, Robert Theodore
Dutton, John Altnow
East, Conrad
Eddy, George Amos
Edinger, James (G)
Elliott, William Paul
Elliott, Robert Dunshee
Ellis, John Ogborn
Elsberry, Russell Leonard
Engle, Edison Grove, Jr
Epstein, Edward Selig
Erickson, Carl O
Estoque, Mariano A
Evans, William Buell
Faller, Alan Judson
Fankhauser, James Christian
Fawcett, Edwin Babcock
Finger, Frederick George
Fisher, Perry Wright
Fleagle, Robert Guthrie
Fletcher, Robert Dawson
Fogle, Benson Tarrant
Fowlis, William Webster
Fox, Douglas Gary
Franceschini, Guy Arthur
Frank, Sidney Raymond
Fraser, Robert Stuart
Freeman, John Clinton, Jr
Friedman, Don Gene
Frisinger, H Howard, II
Fritschen, Leo J
Fritz, Sigmund
Fujita, Tetsuya T
Fuquay, James Jenkins
Fymat, Alain L
Gage, Kenneth Seaver
Gallagher, James Emerson
Garstang, Michael
Gates, William Lawrence
Geer, Ira W
Gentry, Robert Cecil
Gerhard, John Randolph
Gifford, Franklin Andrew, Jr
Gille, John Charles
Gilman, Donald Lawrence
Glahn, Harry Robert
Gleeson, Thomas Alexander
Godshall, Fredric Allen
Godson, Warren Lehman
Goldman, Joseph L
Gommel, William Raymond
Gossard, Earl Everett
Goyer, Guy Gaston
Grams, Gerald William
Gray, Thomas Ira, Jr
Gray, William Mason
Greenfield, Richard Sherman
Griffiths, John Frederick
Guy, George Anderson
Hadlock, Ronald K
Hage, Keith Donald
Haig, Thomas O
Halitsky, James
Hall, Ferguson
Hall, Freeman Franklin, (Jr)
Hallanger, Norman Lawrence
Halpern, Paul
Haltiner, George Joseph
Hamilton, Harry L, Jr
Haney, Robert Lee
Hanna, Steven Rogers
Haragan, Donald Robert
Hare, Frederick Kenneth
Harrington, James Bishop, Jr
Harris, D Lee
Harris, Miles Fitzgerald
Harris, Russell G
Harrison, Paul Roger
Harrison, Wilks Douglas
Harshbarger, Harold Beck
Hasler, Arthur Frederick
Hastenrath, Stefan Ludwig
Hauwitz, Bernhard
Hauser, Rolland Keith
Havens, Abram Vaughn
Havens, James Meryle
Hedden, Gregory Dexter
Helbush, Robert Edwin
Herman, Benjamin Morris
Herman, Gerald Francis
Hess, Seymour Lester
Hewson, Edgar Wendell
Hiser, Homer Wendell
Hoehne, Walter Elmer
Holl, Manfred Matthias
Holland, Joshua Zalman
Holmes, David W
Horn, Lyle Henry
Hosler, Charles Luther, Jr

Hosler, Charles R
Houghton, David Drew
Houghton, Henry Garrett
Hovermale, John Bruce
Howell, Wallace Egbert
Hsu, Shih-Ang
Huang, Joseph Chi Kan
Hudlow, Michael Dale
Huebsch, Ian O
Hughes, Lawrence Ambrose
Hughes, Patrick E
Humphries, Robert Gordon
Huschke, Ralph Ernest
Hutchinson, Leonard Hugh
Izumi, Yutaka
Jacobs, Woodrow Cooper
James, Ralph Paul
Jensen, Clayton Everett
Jess, Edward Orland
Jiusto, James E
Johnson, Stephen Alton
Johnson, Donald R
Johnson, David Simonds
Johnson, Francis Severin
Johnson, Harry McClure
Jones, David Lloyd
Jordan, Charles Lemuel
Jung, Glenn Harold
Kadlec, Paul William
Kahan, Archie M
Kaimal, Jagadish Chandran
Kao, Shih-Kung
Kaplan, Lewis David
Kasahara, Akira
Katz, Yale H
Kelley, John Joseph, II
Kellogg, William Welch
Kern, Clifford Dalton
Kessler, Edwin, III
Kindle, Earl Clifton
Klein, William H
Klieforth, Harold Ernest
Knox, Joseph Blatt
Koenig, Lloyd Randall
Koschmieder, Ernst Lothar
Kraus, Eric Bradshaw
Kreitzberg, Carl William
Krey, Philip W
Krishnamurti, Tiruvalam N
Kuhn, Peter Mouat
Kung, Ernest Chen-Tsun
Kurihara, Yoshio
Kuschenreuter, Paul Herbert
LaChapelle, Edward Randle
Laevastu, Taivo
Lahey, James Frederick
Lally, Vincent Edward
Landers, Holbrook
La Rue, Jerrold A
La Seur, Noel Edwin
Latham, Don Jay
Lavoie, Ronald Leonard
Laws, Kenneth Lee
Lecher, David Wayne
Lee, Jean T
Lee, John Denis
Lee, Roy
Leighton, Henry George
LeMone, Margaret Anne
Lenschow, Donald Henry
Leovy, Conway B
Lettau, Heinz Helmut
Lewis, Billy M
Lewis, Frank
Li, Ta-Yung
Lilly, Douglas Keith
Lin, Yeong-Jer
Link, Peter K
London, Julius
Long, Paul Eastwood, Jr
Long, Robert Radcliffe
Lorenz, Edward Norton
Lowry, William Prescott
Lyons, Walter Andrew
MacHattie, Leslie Blake
MacKay, Kenneth Pierce, Jr
Malkin, William
Malone, Thomas Francis
Mantis, Homer Theodore
Markham, Charles G
Marlatt, William Edgar
Marshall, John Stewart
Martin, David William
Martin, Donald Efton
Martin, Frank Lionel
Martinelli, Mario, Jr
McCarthy, John
McClain, Ernest Paul
McDaniel, Willard Rich
McDowell, Dawson Clayborn
McFadden, James Douglas
McGovern, Wayne Ernest
McIntyre, Donald Patrick
McTaggart-Cowan, Patrick Duncan
Means, Lynn L
Meyer, James Henry
Miksad, Richard Walter
Miller, Albert

Miller, James E
Miller, John Frederick
Miller, Marvin E
Minzner, Raymond Arthur
Mirabito, John A
Mitchell, John Murray, Jr
Mitchell, Lloyd Vernon
Miyake, Mikio
Molo, William L
Morgenstern, Paul
Mount, Wayne Delano
Moyer, Vance Edwards
Munn, Robert Edward
Murakami, Takio
Murino, Clifford John
Murino, Vincent S
Murray, Francis William
Myer, Glenn Evans
Myers, Robert Frederick
Nagler, Kenneth Malcolm
Namias, Jerome
Neiburger, Morris
Neuberger, Hans Hermann
Neumann, Paul Gerhard
Newell, Reginald Edward
Newstein, Herman
Newton, Chester Whittier
Niemeyer, Lawrence E
Noffsinger, Terrell L
O'Brien, James F
O'Connor, James Francis
Ohring, George
Oliver, Vincent J
Olivero, John Joseph, Jr
Olson, Boyd E
O'Neill, Thomas Hall Robinson
Ooyama, Katsuyuki
Orville, Harold Duvall
Pandolfo, Joseph P
Parry, Hubert Dean
Paulhus, Joseph Louis Honore
Paulson, Clayton Arvid
Perry, John Stephen
Peterson, Ernest W
Pielke, Roger Alvin
Pierce, Edward Thomas
Pierson, Willard James, Jr
Pike, Arthur Clausen
Pitchford, Kenneth Lee
Platzman, George William
Podzimek, Josef
Pooler, Francis, Jr
Pore, Norman Arthur
Porter, John M
Portig, Wilfried Helmut
Price, John Charles
Purdom, James Francis Whitehurst
Pyle, Robert Lawrence
Quinn, William Hewes
Quiroz, Roderick S
Ragotzkie, Robert Austin
Ramage, Colin Stokes
Rao, Gandikota V
Rao, P Krishna
Rapp, R Robert
Rasmusson, Eugene Martin
Rasmussen, James L
Ray, Peter Sawin
Raymond, David James
Read, Robert G
Record, Frank Alaster
Reed, Jack Wilson
Reed, Richard John
Reichelderfer, Francis Wilton
Renard, Robert Joseph
Renne, David Smith
Reynolds, George (Warren)
Richards, Marshall Monroe
Rigby, Malcolm
Robinson, Allan Richard
Robinson, Lewis Howe
Rogers, Roddy R
Rosenthal, Stanley Lawrence
Rubin, Morton Joseph
Ruff, Irwin S
Runnels, Robert Clayton
Russo, John A, Jr
Ryan, Bill Chatten
Sadler, James C
Salomonson, Vincent Victor
Saltzman, Barry
Sanders, Frederick
Sanderson, Alan Nichols
Sartor, James Doyne
Sasamori, Takashi
Saucier, Walter Joseph
Saxena, Vinod Kumar
Schaefer, Joseph Thomas
Schaefer, Vincent Joseph
Schell, Irving Israel
Schleusener, Richard A
Schotland, Richard Morton
Schroeder, Mark Joseph
Scoggins, James R
Seguin, Ward Raymond
Senn, Harry V
SethuRaman, S

Shaw, Roderick Wallace
Sheets, Robert Chester
Sheridan, Laurence Ward
Sherr, Paul Edgar
Shieh, Liau Jang
Shuman, Frederick Gale
Siklar, Dhirendra N
Simpson, Joanne
Simpson, Robert H
Smith, Phillip J
Smith, Theodore Beaton
Snellman, Leonard W
Somerville, Richard Chapin James
Spalding, George Robert
Spar, Jerome
Spengler, Kenneth Clifford
Spilhaus, Athelstan Frederick
Squires, Patrick
Stackpole, John Duke
Staelin, David Hudson
Stearns, Charles R
Steere, Richard C
Stickel, Philip Rice
Stidd, Charles Ketchum
Stinson, James Robert
Stolov, Harold L
Stone, Newton C
Stout, Glenn Emanuel
Stuart, David W
Suchman, David
Suomi, Verner Edward
Summers, Peter William
Swanson, Dwight Wesley
Super, Arlin B
Tang, Chung-Muh
Taylor, George Frederic
Taylor, Ronald Charles
Tepper, Morris
Thomas, Morley Keith
Thomas, Aylmer Henry
Thompson, Jack Coats
Thompson, Owen Edward
Thompson, Philip Duncan
Thompson, Dennis Walter
Thuillier, Richard Howard
Thyer, Norman Harold
Todd, Clement Jameson
Tracton, Martin Steven
Trout, Dennis Alan
Twitchell, Paul F
van der Bijl, Willem
Van den Hoven, Isaac
Vaughan, William Walton
Vernal, Donald L
Vernekar, Anandu Devarao
Vestal, Claude Kendrick
Vincent, Dayton George
Vonder Haar, Thomas Henry
Wahl, Eberhard Wilhelm
Wahl, Edward Robert
Wallace, John Michael
Wanta, Raymond Casimir
Wang, Jen Yu
Wark, David Quentin
Washington, Warren Morton
Wasko, Peter Edmund
Webb, Willis Lee
Weinstein, Alan Ira
Weiss, Edward Leonhardt
Weiss, Richard Raymond
Weller, Gunter Ernst
Wendler, Gerd Dierk
White, Robert M
Widger, William Knowlton, Jr
Wilkins, Eugene Morrill
Willett, Hurd Curtis
Williams, Philip, Jr
Williams, Merlin Charles
Williams, Scott Lansing
Winninghoff, Francis Joseph
Woodbridge, David Davis
Woodridge, Gene Lysle
Woodruff, Rodger King
Worth, James Judson Blackley
Wulf, Oliver Reynolds
Yerg, Donald G
Young, George Anthony
Young, John A

Agricultural Meteorology
Baier, Wolfgang
Bark, Laurence Dean
Brown, Donald Murray
Campbell, Gaylon Sanford
Caprio, Joseph Michael
Gerber, John Francis
Gillespie, Terry James
Gillette, Dale Alan
Marsolf, J David
Pack, Albert Boyd
Robertson, George Wilber
Shaw, Robert Harold
Taylor, Sterling Elwynn
Wesley, Marvin Larry

Atmospheric Chemistry
Bandy, Alan Ray

Berger, Jerry Eugene
Cicerone, Anthony John
Delany, Anthony Charles
de Pena, Rosa G
Duce, Robert Arthur
Farber, Robert James
Finnegan, William George
Friend, James Philip
Gay, Bruce Wallace, Jr
Graedel, Thomas Eldon
Green, William Delap
Harrison, Halstead
Heaney, Robert John
Hudson, Frank Peter
Kolb, Charles Eugene, Jr
Lee, Richard Norman
Levy, Arthur
Levy, Hiram, II
Lodge, James Piatt, Jr
Mason, Allen Smith
McLaren, Eugene Herbert
Newman, Leonard
Pena, Jorge Augusto
Rosinski, Jan
Sedlacek, William Adam
Slanger, Tom George
Stampfer, Joseph Frederick, Jr
Stedman, Donald Hugh
Todd, Edward Payson
Wilson, William Enoch, Jr

Biometeorology
Parton, William Julian, Jr
Ripley, Earl Allison

Climatology
Adam, David Peter
Anstey, Robert L
Arnfield, Anthony John
Bailey, Harry Paul
Barry, Roger Graham
Batten, Edmund Stanley
Baum, Werne A
Bell, Barbara
Bennett, Iven
Bilello, Michael Anthony
Blasing, Terence Jack
Butson, Keith D
Canfield, Norman L
Changnon, Stanley Alcide, Jr
Chapman, Lyman John
Church, Phil Edwards
Court, Arnold
Critchfield, Howard John
Cross, Ralph Donald
Davis, Jerry Mallory
Donn, William L
Durrenberger, Robert W
Fletcher, Roy Jackson
Fuquay, James Jenkins
Furman, Roland William, II
Gallagher, James Weldon
Giovinetto, Mario Bartolome
Haggard, William Henry
Harman, Jay Reginald
Harshbarger, Harold Beck
Hastenrath, Stefan Ludwig
Havens, Abram Vaughn
Havens, James Meryle
Houghton, John G
Howe, George Marvel
Hubbard, John Edward
Justham, Stephen Alton
Kakela, Peter John
Landsberg, Helmut Erich
Lawson, Merlin Paul
Lettau, Katharina
Longley, Richmond Wilberforce
Lydolph, Paul E
Mather, John Russell
Meentemeyer, Vernon George
Miller, David Hewitt
Mitchell, Val Leonard
Moran, Joseph Michael
Muller, Robert Albert
Norwine, James Randolph
Ohtake, Takeshi
Portig, Wilfried Helmut
Potter, Gerald Lee
Rayner, John Norman
Roden, Gunnar Ivo
Rumney, George Richard
Sabbagh, Michael E
Schneider, Stephen Henry
Spuhler, Walter S
Steila, Donald
Thomas, Morley Keith
Tuller, Stanton Ernest
Vestal, Claude Kendrick
Waggoner, Paul Edward
Webb, Thompson, III
Wendland, Wayne Marcel
Wilson, Cynthia
Wilson, Richard Garth

Cloud Physics
Alkezweeny, Abdul Jabbar
Anderson, Bernard Jeffrey

Bartley, David Lauren
Birstein, Seymour J
Byers, Horace Robert
Cannon, Theodore Wiles
Chisholm, Alexander James
Dingle, Albert Nelson
Dye, James Eugene
Fukuta, Norihiko
Gunn, Kenrick Lewis Stuart
Hallett, John
Jayaweera, Lakshman
Kassner, James Lyle, Jr
Katz, Ulrich
Klett, James Dean
Knight, Charles Alfred
Kyle, Thomas Gail
Lamb, Dennis
Long, Alexis Boris
Ohtake, Takeshi
Pitter, Richard Leon
Plummer, Patricia Lynne Moore
Podzimek, Josef
Saxena, Vinod Kumar
Schemenauer, Robert Stuart
Young, Kenneth Christie

Dynamic Meteorology
Agee, Ernest Mason
Alyea, Fred Nelson
Barcilon, Albert I
Bosart, Lance Frank
Bradley, James Henry Stobart
Cavalieri, Donald Joseph
Cho, Han-Ru
Derome, Jacques Florian
Dirks, Richard Allen
Fels, Stephen Brook
Fultz, Dave
Geisler, John Edmund
Hayashi, Yoshikazu
Holloway, John Leith, Jr
Holton, James R
Kousky, Vernon E
Kuo, Hsiao-Lan
Lalas, Demetrius P
Merilees, Philip
Meyer, Walter Davidson
Mudrick, Stephen Edward
Oort, Abraham H
Paulin, Gaston (Ludger)
Pedlosky, Joseph
Peltier, William Richard
Pfeffer, Richard Lawrence
Riegel, Christopher Albert
Rodenhuis, David Roy
Ross, Bruce Brian
Sarachik, Edward S
Sasaki, Yoshi Kazu
Schemm, Charles Edward
Shapiro, Ralph
Smagorinsky, Joseph
Staver, Allen Ernest
Tang, Wen
Tenenbaum, Joel
Vukovich, Fred Matthew
Wagner, Kit Kern
Webster, Peter John
Williams, Roger Terry
Wurtele, Morton Gaither

Forest Meteorology
Armstrong, John Alexander
Federer, C Anthony
Fosberg, Michael Allen
Herrington, Lee Pierce
Reifsnyder, William Edward

Marine Meteorology
Clark, Nathan Edward
Eber, Laurence Elwin
Glenn, Alfred Hill

Microclimatology
Amundsen, Clifford C
Baker, Donald Gardner
Dando, William Arthur
DeWalle, David Russell
Hannell, Francis George
Lee, Richard
Mecklenburg, Roy Albert
Oke, Timothy Richard
Tanner, Champ Bean

Micrometeorology
Barry, P J S
Belt, George Harley, Jr
Boston, Noel Edward James
Covey, Winton Guy, Jr
Droppo, James Garnet, Jr
Hamilton, Harry Lemuel, Jr
Haugen, Duane Arthur
Havard, Jesse Boyd
Hutchinson, Gordon Lee
Langleben, Manuel Phillip
Martin, Hans Carl
Mason, Conrad Jerome
Norman, John Matthew
Panofsky, Hans Arnold

Pennell, William Todd
Portman, Donald James
Radke, Jerry Kieth
Riehl, Herbert
Rosenberg, Norman J
Sinclair, Thomas Russell
Swanson, Robert Nels
Swift, Lloyd Wesley, Jr
Thayer, Scott Dwight
Wagner, Norman Keith
Weber, Allen Howard
Wolf, Marvin Abraham

Physical Meteorology
Anderson, Charles Edward
Chermack, Eugene E A
Davis, Francis Kaye, Jr
Hoffer, Thomas Edward
Lhermitte, Roger M
Magarvey, Raymond Halliday
Stephens, Jesse Jerald
Taylor, Charles Luther
Warner, Charles
Zdunkowski, Wilford G

MICROBIOLOGY

Microbiology
Abernathy, Robert Shields
Abodeely, Robert Assad
Abrahams, Irving
Abram, Dinah
Abramson, Irwin Jerome
Abshire, Claude James
Acker, Robert Flint
Adair, Frank William
Adams, Alfred Birk
Adams, Bruce Gordon
Adams, James Norman
Adams, Jane N
Affronti, Lewis Francis
Ahmed, Saiyed I
Aiston, Stewart Samuel
Ajl, Samuel Jacob
Akers, Thomas Gilbert
Albach, Richard Allen
Albertson, John Newman, Jr
Albright, Lawrence John
Alexander, James King
Allen, Emma Gates
Allen, George Perry
Allen, Mary A Mennes
Allen, Norris Elliott
Allen, Orfin
Allen, Wendall E
Allison, Marvin J
Allison, Milton James
Alms, Thomas H
Amelunxen, Remi Edward
Amerault, Thomas Eugene
Andersen, Kenneth J
Anderson, Dwight Lyman
Anderson, Lucia Lewis
Anderson, Orfin
Anderson, Richard Lee
Anderson, Theodore Edmund
Andrews, Wallace Henry
Angelotti, Robert
Anita, Naval Jamshedji
Appelgren, Walter Phon
Arceneaux, Joseph Lincoln
Argot, Jeanne
Armbruster, Frederick Carl
Aronson, Arthur Ian
Arret, Bernard
Asato, Yukio
Ash, Ronald Joseph
Ashe, Warren (Kelly)
Aspitarte, Thomas (Robert)
Axelrod, David E
Ayengar, Padmasini (Mrs Frederick Aladjem)
Ayers, William Arthur
Bach, John Alfred
Bachman, Marvin Charles
Bahn, Arthur Nathaniel
Baker, Phillip John
Baker, Thomas Irving
Balduzzi, Piero
Baldwin, Jack Norman
Baldwin, William Walter
Balish, Edward
Balows, Albert B
Banman, Elmer Alexander
Baram, Peter
Barber, Thomas Lynwood
Bard, Raymond Camillo
Barden, Ned Thorson
Barile, Michael Frederick
Barkate, John Albert
Barksdale, Lane
Barnes, Isabel Janet
Barnum, Donald Alfred
Barrett, James Thomas
Barry, Arthur Leland
Bartlett, Glenn Wilfred
Basaraba, Joseph

Baseman, Joel Barry
Bassett, Emmett W
Bassin, Robert Harris
Battisto, Jack Richard
Battley, Edwin Hall
Batzing, Barry Lewis
Baum, Robert Harold
Bauman, David Stanley
Baumstark, John Spann
Bausor, Sydney Charles
Bausum, Howard Thomas
Bayan, Aris Paul
Beaman, Blaine Lee
Beardmore, William Boone
Beardsley, Robert Eugene
Beck, Jay Vern
Beck, Raymond Warren
Becker, Benjamin
Becker, Jeffrey Marvin
Becker, Joseph Gerald
Beckhorn, Edward John
Behme, Ronald John
Beining, Paul R
Bell, Gordon Russell
Bell, Samuel Dennis, Jr
Bellamy, Winthrop Dexter
Bellanti, Joseph A
Benedict, Robert Glenn
Beneke, Everett Smith
Benham, Ross Stephen
Benjaminson, Morris Aaron
Bennett, Ralph Edgar
Bennett, Travis H
Benoit, Jean Claude
Bent, Donald Frederick
Benz, Edward John
Berg, Gerald
Berk, Richard Samuel
Berliner, Martha D
Berman, Sanford
Bernal-Llanas, Enrique
Bernheimer, Alan Weyl
Bernstein, Aleck
Bernstein, Sidney
Berquist, Kenneth R
Berry, Levette Joe
Bertani, Lilian Elizabeth
Best, Gary Keith
Beutner, Ernst Herman
Bienvenu, Rene Joseph, Jr
Bilimoria, Minoo Hormasji
Billen, Daniel
Biondo, Frank X
Birdsell, Dale Carl
Birnbaum, Jerome
Bissett, Marjorie Louise
Black, Samuel Harold
Blackmore, Robert Valentine
Blackwood, Allister Clark
Blanchard, Gordon Carlton
Blank, Carl Herbert
Blaustein, Ernest Herman
Blechman, Harry
Bleiweis, Arnold Sheldon
Bliznakov, Emile George
Blouse, Louis E, Jr
Blumenthal, Harold Jay
Boatman, Edwin S
Bobb, Yvonne Dolores
Bocchieri, Samuel Francis
Bockrath, Richard Charles, Jr
Boisvenue, Rudolph Joseph
Bondi, Amedeo
Bonner, Daniel Patrick
Bontempo, John A
Bonventre, Peter Frank
Booth, Sheldon James
Borchardt, Kenneth
Bordt, Dale Emil
Borensztajn, David Zelman
Borg, Alfred Francis
Boring, John Rutledge, III
Borlaug, Norman Ernest
Boros, Dov Lewis
Bose, Henry Robert, Jr
Botan, Edward Allan
Bott, Kenneth F
Bott, Thomas Lee
Bottone, Edward Joseph
Bower, Raymond Kenneth
Bowling, Robert Edward
Boyd, Donald Mitchell
Boyer, Herbert Wayne
Bozeman, F Marilyn
Bradley, Sterling Gaylen
Bradshaw, Willard Henry
Brady, Robert James
Braendle, Donald Harold
Brand, Karl Gerhard
Brawner, Thomas Allan
Brehm, Sylvia Patience
Brennan, Patricia Conlon
Brenneman, Faith Nielsen
Brent, Morgan McKenzie
Bretz, Harold Walter
Brewer, Carl Robert
Brezenski, Francis T
Brierley, James Alan

Brill, Winston J
Briner, William J
Brinton, Charles Chester, Jr
Britt, Eugene Maurice
Brock, Katherine Middleton
Brock, Thomas Dale
Brockman, Ellis R
Broderick, Lynne Sechrist
Broitman, Selwyn Arthur
Brooks, John Bill
Brooks, John Bill
Brown, Arthur
Brown, George L
Brown, Milton Herbert
Brown, Olen Ray
Brown, William E
Brown, William Everett
Brown, William John
Brownell, George H
Brubaker, Robert Robinson
Bruch, Carl William
Brundage, William Gregory
Bruner, Dorsey William
Bryant, James Berry, Jr
Bryner, John Henry
Bryson, Vernon
Buchanan, Bob Branch
Buckelew, Albert Rhoades, Jr
Buckley, Patricia M
Budde, Mary Lawrence
Buddingh, Gerrit John
Budney, Mary Lillian
Buecher, Edward Joseph
Bulich, Anthony Andrew
Burrell, Robert Guthrie
Burton, Alice Jean
Burton, David Norman
Burton, Sheril Dale
Byers, Benjamin Rowe
Cabelli, Victor Jack
Cadmus, Eugene L
Cain, William Aaron
Calabi, Ornella
Caldwell, Daniel R
Calhoun, David H
Calisher, Charles Henry
Callahan, Lynn Thomas, III
Camenzer, Gerald Walter
Caminita, Barbara Hobson
Campbell, James Nicoll
Campbell, Linzy Leon
Campbell, Norman E Ross
Canale-Parola, Ercole
Cantor, Ena D
Cardon, Bartley Pratt
Carlberg, David Marvin
Carlquist, Philip Rich
Carlson, Margaret Joyce
Carp, Richard Irvin
Carpenter, Philip Lewis
Carski, Theodore Robert
Casaz, Geronimo
Casida, Lester Earl, Jr
Cassel, William Alwein
Castenholz, Richard William
Castro, Gilbert Anthony
Catlin, B Wesley
Cazin, John, Jr
Ceglowski, Walter Stanley
Celis, Teodoro F R
Chadwick, June Marie
Chalgren, Steve Dwayne
Chalquest, Richard Ross
Chamberlain, Robert English
Chambers, Velma Catherine
Chambliss, Glenn Hilton
Chan, James C
Chang, Kenneth Shueh-Shen
Chang, Shih Lu
Chang, Yung-Feng
Chaparas, Soitros D
Chaperon, Edward Alfred
Charney, William
Charon, Nyles William
Cheever, Francis Sargent
Cheng, Shu-Sing
Chesbro, William Ronald
Child, Jeffrey James
Childress, Evelyn Tutt
Chipley, John Raymond
Chmura, Norman Walter
Choo, Byung-Ryul
Choman, Bohdan Russell
Chorpenning, Frank Winslow
Christian, Walter
Chu, Yang-Ming
Chung, Choong Wha
Chung, Kwok-Leung
Church, Brooks Davis

Churchill, Bruce Wenzel
Ciegler, Alex
Claridge, Charles Alfred
Clark, A Gavin
Clark, Evelyn Genevieve
Clark, Francis Matthew
Clarke, Junius Manson
Claus, Norman Arthur
Claus, George William
Clesceri, Lenore Stanke
Clewell, Don Bert
Clowes, Royston Courtenay
Cochrane, Vincent Winner
Cody, Reynolds M
Coggin, Joseph Hiram
Coghlan, Anne Eveline
Cohen, Gary H
Cohen, Larry William
Cohen, Melvin Joseph
Cohen, Sidney
Cole, Edward Anthony
Cole, George Christopher
Coleman, Charles Mosby
Coleman, William H
Coles, Embert Harvey, Jr
Colley, Daniel George
Collier, Robert Eugene
Colwell, Rita R
Conner, James D
Conti, Samuel Francis
Contois, David Ely
Cook, David Wilson
Cook, Elizabeth Anne
Cook, Margaret Mary
Cook, Thomas M
Coolbaugh, James Cameron
Cooney, Donald George
Cooney, Joseph Jude
Cooney, Marion Kathleen
Cooper, Billy Howard
Cooper, Murray Sam
Cooper, Robert Chauncey
Cooper, Stephen
Copson, David Arthur
Corbett, Jules John
Corey, Roland Reece, Jr
Corner, Thomas Richard
Coscarelli, Waldimero
Cosderton, J William F
Costlow, Ralph Norman
Costlow, Richard Dale
Cota-Robles, Eugene H
Coy, Vernon Frank
Coulter, Wilson H
Coward, Joe Edwin
Cowell, James Leo
Cowman, Richard Ammon
Cox, Dudley
Cox, Edmond Rudolph, Jr
Cox, Nelson Anthony
Crabtree, Koby Takayashi
Craig, James Morrison
Craig, John Philip
Crane, Anatole
Crane, George Thomas
Crawford, Donald Lee
Crawford, James Gordon
Crawford, Ronald Lyle
Cretz, Joseph Reuel
Crisley, Francis Daniel
Croft, Charles Clayton
Croley, Thomas Edgar
Cromartie, William James
Cronholm, Lois S
Crook, Philip George
Crowle, Alfred John
Cummings, Dennis Paul
Cummins, Cecil Stratford
Curnen, Edward Charles, Jr
Curtis, Paul Robinson
Cutler, Jimmy Edward
Cynkin, Morris Abraham
Czarnecki, Reynold Bernard
Dabbah, Roger
Daily, Lawrence Eugene
Daily, Otis Patrick
Daily, William Allen
Danco-Moore, Loiita
Daoust, Donald Roger
Das, Naba Kishore
Dave, Bhalchandra A
Davies, Helen Jean Conrad
Davies, Warren Lewis
Davis, Bernard David
Davis, Charles Patrick
Davis, Edwin Nathan
Davis, John Barney
Davis, Norman Duane
Davis, Norman Seymour
Davis, William C
Day, Lawrence Eugene
DeBell, Robert Michael
Deeney, Anne O'Connell
de Fiebre, Conrad William
De Goes, Paulo
Dehority, Burk Allyn
Deitz, William Harris

DeLamater, Edward Doane
DeLuca, Patrick John
DeMars, Robert Ivan
DeMoss, John Allen
DeMoss, Ralph Dean
Dennen, Frank Willis, Jr
Dennison, David W
Dimopoullos, George Takis
DePinto, John A
Derow, Matthew Arnold
De Siervo, August Joseph
Detroy, Robert William
Dewar, Norman Ellison
Dewitt, William
Dewitt, Frank Matthew, Jr
Dick, Elliot C
Di Cuolo, C John
DiLiello, Leo Ralph
Dinowitz, Marshall
DiSalvo, Arthur F
Dobrogosz, Walter Jerome
Docherty, John Joseph
Dockstader, Wilmer Beldon
Doenel, William Beldon
Doetsch, Raymond, Jr
Doi, Roy Hiroshi
Donaldson, Paul
Dondero, Norman Carl
Dondershine, Frank Haskin
Dougherty, Robert Marvin
Douglas, Howard Clark
Douglas, Robert John
Douros, John Drenkle
Dowell, Clifton Enders
Dowell, Vitus Raymond, Jr
Drexler, Henry
Drucker, Harvey
Duff, James Thomas
Duffy, Carl Edward
Dugan, Patrick R
Dulaney, Eugene Lambert
Duncan, Bettie
Duncan, Douglas Wallace
Dunkelberg, Wilbur Eugene, Jr
Durachta, Chester William
Durand, Donald P
Dutka, Bernard J
Dworkin, Martin
Dyer, John Kaye
Eagon, Robert Garfield
Easterbrook, Kenneth Brian
Eaton, Norman Ray
Edberg, Stephen Charles
Edgar, Samuel Allen
Edwards, Ogden Frazelle
Edwards, Roselyn Jane
Eiserling, Frederick A
Eisenberg, William Victor
Eklund, Curtis Einar
Eklund, Melvin Wesley
Ekstedt, Richard Dean
Elander, Richard Paul
Elbein, Alan D
Elliker, Paul R
Elliott, Arthur York
Elliott, Lloyd Floren
Ellis, Edwin M
Ellis, Robert J
Ellison, Solon Arthur
Eltz, Robert Walter
Emswiler, Bonnie Sue
Enders, George Leonhard, Jr
Engley, Frank B, Jr
Engley, John Franklin
Ennever, John Joseph
Ennis, Herbert Leo
Enriquez, Nitza M
Erb, Kenneth
Erickson, Raymond C
Ernst, Robert R
Esber, Henry Jemil
Esposito, Vito Michael
Estes, Edna E
Eudy, William Wayne
Evans, George Leonard
Evans, Mary Jo
Evans, Richard Todd
Eveleigh, Douglas Edward
Ezekiel, David Hirsch
Fabrikant, Irene Berger
Falkow, Stanley
Farber, Paul Alan
Farkas-Himsley, Hannah
Farmer, John James, III
Farrand, Stephen Kendall
Faust, Richard Ahlvers
Favero, Martin
Feary, Thomas W

Feit, Ira (Nathan)
Feldman, Lawrence A
Feldman, Louis Israel
Fenje, Paul
Fenters, James Dean
Ferebee, Robert Newton
Ferguson, Dale Vernon
Fernelius, Lloyd C
Fernelius, Albert Lawrence
Ferone, Robert
Ferretti, Joseph Jerome
Field, Marvin Frederick
Fields, Marion Lee
Filippi, Gordon Michael
Fincher, Edward Lester
Fine, Donald Lee
Finegold, Sydney Martin
Fingerhut, Marilyn Ann
Fink, Frederick Charles
Finkelstein, Frances
Finkelstein, Richard Alan
Finney, James William
Finstein, Melvin S
Firstein, William
Fischbach, George
Fischer, Robert George
Fiscus, Alvin G
Fiset, Paul
Fish, Donald C
Fisher, Clark Alan
Fitzgerald, Robert James
Fitzgerald, Thomas James
Fleischman, Alan Isadore
Fleischman, Julian B
Fleming, Henry Fridgen
Florentano, Herbert Joseph
Florestano, Herbert Joseph
Fodor, Andrew Robert
Folds, James Donald
Forbes, Martin
Formal, Samuel Bernard
Forney, John Edgar
Foster, Alvin Garfield
Foster, Billy Glen
Foster, Terry Lynn
Fox, Charles Lewis, Jr
Fox, Sally Ingersoll
Frampton, Elon Wilson
Franklin, Mervyn
Frazier, William Robert
Freeman, Bob A
Freimer, Earl Howard
Friesen, James Donald
Friend, Patric Lee
Friedman, Herman
Friedman, Stephen Burt
Froelich, Ernest
Froman, Seymour
Frost, Bettina Mary
Fry, Kenneth Alvin
Fryer, John Louis
Fuerst, Robert
Fugate, Kearby Joe
Fuhs, George Wolfgang
Fukui, George Massaki
Fukuyama, Thomas T
Fulkerson, John Frederick
Fuller, Rufus Clinton
Fung, Daniel Yee Chak
Funk, Helen Beatrice
Furness, Geoffrey
Fusillo, Matthew Henry
Gabliks, Janis
Gaffney, Peter Edward
Gainor, Charles
Galasso, George John
Gale, George Osborne
Gale, Nord Loran
Galsky, Alan Gary
Gannon, Richard Anthony
Ganaway, James Rives
Garay, Gustav John
Gardner, Earl William, Jr
Gardner, Charles H
Gardner, Frederick Albert
Garfinkle, Barry David
Garro, Anthony Joseph
Gaucher, George Maurice
Gaughran, Eugene Robert Lawrence
Gaumer, Herman Richard
Gavin, John Joseph
Gebhardt, Louis Philipp, Jr
Gee, Lynn LaMarr
Gehrig, Robert Frank
Geldreich, Edwin Emery
Gelzer, Justus
Genghof, Dorothy Schaefer
George, Elmer, Jr
Gerber, Paul
Gerberich, John Barnes
Gerhardt, Philipp
Gerke, John Royal
Gerloff, Robert Kay
Gest, Howard
Gezon, Horace Martin
Ghosh, Chitta Ranjan

Gibbons, Ronald J
Gibson, Audrey Jane
Gifford, George Edwin
Gilfillan, Robert Frederick
Gillis, Robert Edward
Gilmour, Marion Nyholm H
Gilpin, Richard William
Gilpin, Robert Harry
Gingold, James Lehman
Ginoza, Herbert S
Ginsberg, Harold Samuel
Girardi, Anthony Joseph
Girolami, Roland Louis
Gitterman, Charles Oscar
Glasgow, Lowell Alan
Gledhill, William Emerson
Goble, Frans Cleon
Gochnauer, Thomas Alexander
Godzeski, Carl William
Goehler, Brigitte Hanna
Goepfert, John McDonnell
Goldberg, Herbert Sam
Goldberg, Julius
Goldfine, Howard
Goldstein, Gerald
Golubjatnikov, Rjurik
Gooder, Harry
Goodhue, Charles Thomas
Goodman, Harold Stanley
Goodman, Norman L
Gordee, Robert Stouffer
Gorden, Robert Wayne
Gordon, Milton
Goren, Mayer Bear
Gorski, Theodore William
Goss, Robert Charles
Goss, William Albert
Gottlieb, Sheldon F
Grady, Joseph Edward
Graham, Angus Frederick
Granger, Gale A
Grant, Dale Walter
Gray, Clarke Thomas
Gray, Raymond Francis
Grecz, Nicholas
Green, John H
Green, Margaret
Green, Richard H
Greenblatt, Charles Leonard
Greene, Joyce Marie
Greene, Velvl William
Greenfield, Seymour
Greenspan, George
Gregory, Francis Joseph
Gregory, Kenneth Fowler
Grilione, Patricia Louise
Grogan, James Bigbee
Groman, Neal Benjamin
Gronlund, Audrey Florence
Grossberg, Sidney Edward
Groves, David Lynn
Grula, Edward Alan
Gruter, Frederick Herbert
Guerault, Armand
Guidry, Dieu-Donne Joseph
Gunter, Shirley Edna Anne
Gupta, Krishana Chandara
Gutman, Helene Augusta Nathan
Haas, Gerhard Julius
Hadley, Herbert Frank
Hadley, Susan Jane
Hadley, William Keith
Haff, Richard Francis
Hagen, Charles Alfred
Haglund, John Richard
Haight, Roger Dean
Haines, Richard Francis
Hall, Charles Thomas
Hall, William Myron, Jr
Halvorson, Harlyn Odell
Hamdy, Aziz H
Hamilton, Ian Robert
Hamilton, Pat Brooks
Hamilton, Robert Duncan
Hamilton, Thomas Reid
Hammel, Allan H
Hammel, Jay Morris
Hammond, Benjamin Franklin
Hammond, Ray Kenneth
Hampar, Berge
Han, Youn Woo
Hanka, Ladislav James
Hanna, Edgar Ethelbert, Jr
Hanna, Lavelle
Hans, Robert Joseph
Hanson, Hazel Jean
Hanson, Richard Steven
Hanson, Robert Jack
Hardin, Hilliard Frances
Harmon, George Andrew
Harris, Denny Olan
Harris, Geraldine Bender
Harris, Susanna
Harrison, Arthur Pennoyer, Jr
Hart, Joseph L
Hart, Lewis Thomas
Hartman, James Xavier
Hartman, Philip Emil

Hartman, Ronald Earl
Haskins, Reginald Hinton
Hatch, Melvin T
Hatcher, Herbert John
Hatfield, G Wesley
Hargi, John Neal
Hatheway, Charles Louis
Hattman, Stanley
Havas, Helga Francis
Hawirko, Roma Zenovea
Hawthorne, Donald Clair
Hay, Charles Alfred
Haynes, William Clarence
Hearn, Henry James, Jr
Heberlein, Gary T
Heckly, Robert Joseph
Heddleston, Kenneth Luther
Hedrick, Leslie Ray
Hegeman, George D
Hehre, Edward James
Heim, Allen Homer
Heim, Lyle Raymond
Heinis, Julius Leo
Heinze, John Edward
Heise, Eugene Royce
Hempfling, Walter Pahl
Hendin, David
Hendricks, Charles Wendell
Henney, Henry Russell, Jr
Henry, Clay Allen
Henry, Sydney Mark
Henson, Joseph Lawrence
Hentges, David John
Heplar, Joseph Quincy
Herron, David Kent
Herscowitz, Herbert Bernard
Herson, Diane S
Herzberg, Mendel
Heuer, Ann Elizabeth
Heydrick, Fred Painter
Hicks, Garland Fisher, Jr
Hicks, Harold Richard
Higginbotham, Robert David
Higgins, Michael Lee
Hill, Gale Bartholomew
Hill, James Carroll
Himmelfarb, Philip
Hinck, Lawrence Wilson
Hinz, Charles F
Hirschberg, Nell
Hobby, Gladys Lounsbury
Hoch, Sallie O'Neil
Hochstadt, Joy
Hochsteins, Reginald Donald
Hocker, Reginald Orson
Hodgins, Harold Osborne
Hoehn, Marvin Martin
Hoeksema, Walter David
Hoerl, Bryan G
Hoff, John
Hoffee, Patricia Anne
Hoffmann, Conrad Edmund
Hogg, Richard A
Hogg, Robert W
Hoggan, Malcolm David
Hok, Karol Anton
Holder, Ian Alan
Holland, John Joseph
Holmes, Randall Kent
Holt, John Gilbert
Holt, Stanley Carl
Holtermann, Ole A
Homme, Paul John
Hook, William Arthur
Hopps, Hope Elizabeth Byrne
Hoptman, Julian
Horwitz, Marshall Sydney
Howe, Calderon
Howell, Dennis George
Howell, Elmer Virgil, Jr
Howells, John David
Hoyer, Bill Henriksen
Hsu, Wen-Tah
Huang, Alice Shih-Hou
Hubbard, Jerry S
Huber, Floyd Milton
Hubscher, Thomas
Huff, Dennis Karl
Humphrey, Ronald Mack
Hunderfund, Richard C
Hunt, Dale E
Hunt, Fern Ensminger
Hunt, George Albert
Hunt, Lois Turpin
Hunter, Katherine Morton
Hunter, William James
Huntley, Bobby E
Huntley, James Edgar
Hurley, Maureen
Hurst, Andre
Hurwitz, Jerard
Hutchinson, Wesley Gillis
Hutchison, Dorris Jeannette
Hutner, Seymour Herbert
Hutton, William Elmer
Huxsoll, David Leslie
Hyde, Richard Moorehead

Hyman, Richard W
Ibrahim, Adly N
Ingle, Morton Blakeman
Ingledew, William Michael
Ingraham, John Lyman
Inniss, William Edgar
Insalata, Nino F
Iralu, Vichazelhu
Irgens, Roar L
Isa, Abdallah Mohammad
Isokane, Robert Kazuo
Ittensohn, Oswald Ludwig
Iverson, Warren Philip
Iyer, Rajul V
Izzat, Nawtal Nouri
Jackson, Raymond John
Jackson, Sally Womack
Jacobs, Alma Alice
Jahiel, Rene
Jakoby, William Bernard
Jameson, Patricia Madoline
Janicki, Bernard William
Jannasch, Holger Windeklide
Jansons, Vilma Karina
Jaouni, Katherine Cook
Jawetz, ERnest
Jaye, Murray Joseph
Jeffers, Edmund E
Jeffreys, Donald Bearss
Jeffries, Charles Dean
Jeffries, Thomas William
Jenifer, Franklyn G
Jenkin, Howard M
Jennings, Billy Ray
Jensen, Erling Maurice
Jensen, Roy A
Jensen, Thorkil
Jezeski, James John
Johansson, Karl Richard
Johnson, Byron F
Johnson, Charles William
Johnson, Corinne Lessig
Johnson, Donal Dabell
Johnson, Donovan Earl
Johnson, Emmett John
Johnson, Howard M
Johnson, John LeRoy
Johnson, Roy Melvin
Johnson, Russell Clarence
Johnson, Terry Charles
Johnson, Theodore Reynold
Johnson, William
Johnston, Harry Henry
Johnston, Richard Boles, Jr
Johnstone, Donald Boyes
Jones, Daniel L
Jones, Garth Wicks
Jones, Rena Talley
Jones, Theodore Harold Douglas
Jordan, Harold Vernon
Joseph, Sammy William
Joswick, Harry Loren
Jourdian, George William
Joyner, Alan E, Jr
Judge, Leo Francis, Jr
Jurasek, Lubomir
Kachikian, Rouben
Kadis, Solomon
Kadis, Vincent William
Kallestad, Steven Bix
Kalsow, Carolyn Marie
Kane, James Francis
Kaneshiro, Tsuneo
Kang, Tae Wha
Kantor, Harvey Sherwin
Kaplan, Arthur Milton
Kaplan, Louis
Kaplan-Koch, Dora Deborah
Kasel, Julius Albert
Kassira, Edward Naim
Katz, Edward
Kaufmann, Anthony J
Kautter, Donald Albert
Kawatomari, Toshio
Kaye, Saul
Keay, Leonard
Keele, Bernard B, Jr
Keller, Dolores Elaine
Keller, Kenneth F
Kelton, Virginia Crawford
Kelton, William Henry
Kemp, Gordon Arthur
Kempton, Alan George
Kennedy, Harvey Edward
Kenny, George Edward
Kereluk, Karl
Kerr, Sylvia Joann
Kerr, Thomas James
Kessel, John Flenniken
Kessel, Roslyn William Ian
Khan, Abdul Waheed
Khare, Gyaneshwar Prasad
Kiggins, Edward M
Kilbourne, Edwin Dennis
Kim, Juhee
Kim, Kenneth
Kim, Kwang Soo

Kind, Leon Saul
King, Dannie Hilleary
Kingsbury, David Thomas
Kinney, Roland Walter
Kinsel, Norma Ann
Kirber, Maria Wiener
Kirchheimer, Waldemar Franz
Kirchoff, William F
Kirk, Billy Edward
Kirkham, Wayne Wolpert
Kirkland, Jerry J
Kirsch, Edwin Joseph
Kissinger, John Calvin
Kittaka, Robert Shinnosuke
Klasky, Sheldon
Klausmeier, Robert Edward
Klein, Donald Albert
Klemme, Dorothea Elizabeth
Klemmer, Howard Wesley
Klens, Paul Frank
Kleyn, John Gerard
Klinman, Norman Ralph
Klosek, Richard C
Kluepfel, Dieter
Klug, Michael J
Knittel, Martin Dean
Knowles, Roger
Kocan, Richard M
Koerber, Walter Ludwig
Koesterer, Martin George
Koffler, Henry
Kolodziej, Bruno J
Kominek, Leo Aloysius
Kong, Yi-Chi Mei
Koontz, Frank P
Koostra, Walter L
Kopeloff, Lenore Moolten
Koprowski, Hilary
Kornfeld, Joseph M
Kos, Edward Stanley
Krabbenhoft, Kenneth Louis
Krampitz, Lester Orville
Kraskin, Kenneth Stanford
Krause, Richard Michael
Kreier, Julius Peter
Krichevsky, Micah I
Krieg, Noel Roger
Kronish, Donald Paul
Krueger, Russell Francis
Kuchler, Robert Joseph
Kuhn, Daisy Angelika
Kuhnley, Lyle Carlton
Kuo, Mau H
Kupferberg, Alfred Ballen
Kurz, Wolfgang Gebhard Walter
Kurzepa, Henryka Janina
Kutler, Kenneth Latimer
Kvetkas, Marilyn J
Kwapinski, J B George
LaBrie, David Andre
Lacko, Andras Gyorgy
Lacy, Patricia
Laffin, Robert James
Lam, Gow Thue
Lamanna, Carl
LaMotte, Louis Cossitt, Jr
Lampen, J Oliver
Lampidis, Theodore James
Lampky, James Robert
Landau, Burton Joseph
Landau, William
Landry, Marshall Edwin
Landry, Martha Moseley
Langlykke, Asger Funder
Langston, Clarence Walter
Lanni, Yvonne Thery
Lanzilotta, Raymond Philip
Lark, Carl Gordon
Lark, Cynthia Ann
Larsen, Austin Ellis
Larson, Edgar William
Lascelles, June
Lasfargues, Etienne Yves
Lashen, Edward S
Laskin, Allen I
Lasley, Betty Jean
Laurence, Kenneth Allen
Lauro, Gabriel Joseph
Lavender, John Francis
Lawlis, John Frank, Jr
Lazaroff, Norman
Leach, Eddie Dillon
Leadbetter, Edward Renton
Leathen, William Warrick
Leavitt, Richard Irwin
Le Beau, Leon Joseph
Lecce, James Giacomo
Lechevalier, Hubert Arthur
Lechevalier, Mary P
Lechtman, Max Dressler
Lederberg, Esther Miriam
Lee, Chin-Chiu
Lee, Jong Sun
Lee, Sylvan Burton
Lee, Wei Hwa
Lefkowitz, Stanley S
Lein, Joseph

Previte, Joseph James
Prince, Herbert N
Prindle, Bryce
Provost, Philip Joseph
Provost, Philip Joseph
Punch, James Darrell
Putman, Donald Lee
Putman, Hugh D
Puziss, Milton
Pyke, Thomas Richard
Pynes, Gene Dale
Quisno, George L
Rabin, Robert
Rachmeler, Martin
Rado, Thomas A
Ragheb, Hussein S
Rakosky, Joseph, Jr
Ramaley, Robert Folk
Ramsey, William Scott
Randall, Charles Chandler
Randles, Chester
Rapoza, Norbert Pacheco
Rapp, Herbert Joseph
Raska, Karel Frantisek, Jr
Ratliff, Charles Ray
Rauch, Helene Coben
Rauscher, Frank Joseph
Ray, Charles George
Ray, Paul H
Ray, Verne A
Raymond, Richard Laverne
Razzell, Wilfred Edwin
Reardon, John Devereaux
Reddy, Chilecampalli Adinarayana
Reddy, Sunki, Gopal
Redfield, William David
Redmond, William Brinson
Reed, Mary Valedia
Reich, Melvin
Reilly, Hilda Christine
Reilly, Marguerite
Reiner, Abbey M
Reiner-Deutsch, William
Reinhardt, Donald Joseph
Reisner, Gerald Seymour
Remsen, Charles C. III
Retsema, James Allan
Reusser, Fritz
Reynolds, Donald Montgomery
Rheins, Melvin S
Rhoades, Everett Ronald
Ribbons, Douglas William
Rich, Marvin A
Richardson, Lavon Preston
Richardson, Stephen H
Ridgway, George Junior
Riggs, Hammond Greenwald, Jr
Riggs, Stuart
Rights, Fred Lewis
Riley, Vernon Todd
Ringel, Samuel Morris
Ringen, Leif Matt
Ritchey, Thomas William
Rittenberg, Marvin Barry
Ritts, Roy Ellot, Jr
Rizzuto, Anthony Augustine
Rizzuto, Anthony B
Robb, Leslie Allan
Robbins, Mary Louise
Roberge, Marcien Romeo
Roberts, Clifford Evans, Jr
Roberts, Robert Russell
Roberts, Walden Kay
Robertson, Donald Claus
Robertson, Hugh Elburn
Robertson, John Harvey
Robey, Robert Ellis
Robinow, Carl Franz
Robinson, Henry William
Robinson, John Price
Robinson, Robert James
Robinton, Elizabeth Dorothy
Rockwood, Susan Williams
Roessler, William George
Rogers, Charles Graham
Rogers, Palmer, Jr
Rogers, Thomas Earl
Rogers, Thomas Olin
Rogolsky, Marvin
Rohlfing, Stephen Roy
Rohrmann, George Frederick
Roia, Frank Costa, Jr
Rollo, Ian McIntosh
Romano, Antonio Harold
Romano, Paula Josephine
Roon, Robert Jack
Rose, Dyson
Rose, Harry Melvin
Rose, Noel Richard
Rosen, Samuel
Rosenbaum, Manuel
Rosenbaum, Max J
Rosenberg, Fred A
Rosenberg, Steven Loren
Rosenblum, Eugene David
Rosenfeld, Martin Herbert
Rosenkranz, Herbert S
Roslycky, Eugene Bohdan

Rossman, Toby Gale
Roth, Frank J, Jr
Roth, Frieda
Roth, Ivan Lambert
Roth, Norman Gilbert
Rothblat, George H
Rothfield, Lawrence I
Rothman, Sara Weinstein
Rotman, Boris
Roubicek, Rudolph
Routien, John Broderick
Rowland, Ivan W
Rowland, May Eloise
Royt, Paulette Anne
Rozee, Kenneth Roy
Rubin, Benjamin Arnold
Ruchman, Isaac
Rudbach, Jon Anthony
Rude, Theodore Alfred
Rupp, Frank Adolph
Russell, Ruth Lois
Ryden, Fred Ward
Sabath, Leon David
Sabet, Sohair Farid
Sabina, Leslie Robert
Sabiston, Charles Barker, Jr
Sachan, Dileep Singh
Sack, Richard Bradley
Sacks, Lawrence Edgar
Sadoff, Harold Lloyd
Safe, Stephen Harvey
Safferman, Robert S
Sagers, Richard Douglas
Sall, Theodore
Salliman, Bennett
Salser, Josephine See
Salton, Milton Robert James
Salzman, Lois Ann
Sanchez, Gilbert
San Clemente, Charles Leonard
Sanders, Murray
Sanders, W Eugene, Jr
Sandham, Herbert James
Sandine, William Ewald
Sandoval, Howard Kenneth
Sanford, Barbara Ann
Santer, Melvin
Santoro, Thomas
Saperstein, Sidney
Sarachek, Alvin
Sardinas, Joseph Louis
Sarles, William Bowen
Saunders, Allen Perry
Saurino, Vincent Robert
Sauter, Erwin Andrew
Savage, George Mason
Savage, George Roland
Savageau, Michael Antonio
Savard, Edward Victor
Sawyer, William D
Saz, Arthur Kenneth
Scaletti, Joseph Victor
Scarce, LeRoy E
Scarpino, Pasquale Valentine
Schachele, Charles Francis
Schade, Arthur Lincoln
Schaechter, Moselio
Schaefer, Russell William
Schaefler, Sam
Schafer, Thomas Wayne
Schain, Philip
Scheer, Eleanor Ruth
Scheff, George Julius
Scheiner, David M
Scherer, William Franklin
Scherr, David DeLano
Scherr, George Harry
Scherrer, Rene
Schindler, Charles Alvin
Schlech, Barry Arthur
Schlenk, Fritz
Schlesinger, Milton J
Schlesinger, Robert Walter
Schlesinger, Sondra
Schlesinger, David
Schlissel, Harvey Joel
Schloer, Gertrude M
Schlom, Jeffrey
Schmidt, Jean M
Schmidt, Willard Carl
Schmiege, Clement Carl
Schnaitman, Carl A
Schnaper, Edna Stern
Schneierson, S Stanley
Schnell, Gene Wheeler
Schrank, Gordon Dabney
Schricker, Robert Lee
Schuhardt, Vernon Truett
Schuldt, Erich Henry
Schuler, George Albert
Schulze, Karl Ludwig
Schuster, George Sheah
Schutzbach, John Stephen
Schwalb, Marvin N
Schwartz, Benjamin Sam
Schwartz, Leander Joseph
Scott, Marvin Wade
Seaman, Gerald Robert

Seed, John Richard
Seed, Thomas Michael
Seeley, Donald Bernard
Segal, William
Sehgal, Surendra N
Seidler, Ramon John
Seligmann, Edward Baker, Jr
Senitzer, David
Senterfit, Laurence Benfred
Senyk, George
Seshachalam, Dutta
Sever, John Louis
Sevilla-Gardiner, Josefina Zialcita
Shadomy, Helen Jean
Shahidi, Syed Abdus-Salam
Shankel, Delbert Merrill
Shannon, William Michael
Shapiro, James Alan
Shapiro, Martin
Shapiro, Stanley Kallick.
Sharma, Jagdev Mittra
Sharon, Nehama
Sharp, John T
Sharp, William R
Shaw, Eugene
Shefty, Ben Edward
Sheinin, Rose
Shema, Bernard Francis
Sheneman, Jack Marshall
Sherman, Joseph E
Sherris, John C
Shieh, Hang Shan
Shifrine, Moshe
Shipkowitz, Nathan L
Shirling, Elwood Brent
Shively, Jessup MacLean
Shklair, Irving L
Shockman, Gerald David
Shorb, Mary Shaw
Shovlin, Francis Edward
Shrigley, Edward White
Shulls, Wells Alexander
Shuster, Robert C
Sibal, Louis Richard
Siboo, Russell
Siddique, Irtaza H
Siebeling, Ronald Jon
Siegrist, Urban J
Silliker, John Harold
Silva-Hutner, Margarita
Silver, Warren Seymour
Silverman, David J
Silverman, Gerald
Silverman, Melvin Philip
Simard, Ronald E
Siminovitch, Louis
Simmon, Vincent Fowler
Simon, Ellen McMurtrie
Simon, Selwyn
Simons, Daniel J
Simpson, Robert E
Sinclair, Norval A
Singer, Jacques Mauriciu
Singer, Samuel
Singh, Balwant
Singh, Kartar
Sinskey, Anthony J
Sipe, Jerry Eugene
Sipos, Tibor
Siracusano, Vincent C
Sistrom, William R
Sjogren, Robert Erik
Skean, James Dan
Slade, Hutton Davison
Slanetz, Lawrence William
Slankis, Visvaldis
Slepecky, Ralph Andrew
Slotnick, Victor Bernard
Slysh, Anton Roman
Slyter, Leonard L
Small, James David
Smart, Kathryn Marilyn
Smibert, Robert Merrall, II
Smiley, Karl Leroy, Jr
Smith, Andrew George
Smith, Claire Leroy
Smith, Donald Ward
Smith, Dorothy Gordon
Smith, Edith Lucile
Smith, Eugene William
Smith, Harry Logan, Jr
Smith, James Eldon
Smith, James Lee
Smith, John Milton
Smith, Kendall O
Smith, Louis De Spain
Smith, Paul Francis
Smith, Robert Matthews
Smith, Robert William
Smith, William Russell
Snyder, John Crayton
Sobsey, Mark David
Socolofsky, Marion David
Sohler, Arthur
Sokolski, Walter Thomas
Soli, Giorgio
Somers, Kenneth Donald
Sonenshein, Abraham Lincoln

Songer, Joseph Richard
Sonneborn, David R
Sonstein, Stephen Allen
Sorensen, Lloyd J
Spahn, Gerard Joseph
Spangler, William J
Spaulding, Earle Henry
Speck, Marvin Luther
Spence, Hilda Adele
Spence, Kemet Dean
Spence, Leslie Percival
Spencer, John Francis Theodore
Sperber, William Henry
Spitznagel, John Keith
Spizizen, John
Spotts, Charles Russell
Sprott, Gordon Dennis
Spurrier, Elmer R
Squires, Robert Wright
Squires, William Campbell
Sreevalsan, Thazepadath
Srinivasan, Vadake Ram
Srivastava, Kunwar Krishna
Stahly, Donald Paul
Stainer, Dennis William
Stamer, John Richard
Stansly, Philip Gerald
Stapley, Edward Olley
Stark, Egon
Starr, Mortimer Paul
Starr, Patricia Rae
Starr, Theodore Jack
Starzyk, Marvin John
Stechschulte, Agnes Louise
Steel, Robert
Steele, Frances M
Steenbergen, James Franklin
Steers, Edward
Steinberg, Bernard Albert
Steinberg, William
Steinkraus, Keith Hartley
Steinman, Irvin David
Stempen, Henry
Stenback, Wayne Albert
Stephens, James Fred
Stephens, William Leonard
Stern, Joseph Aaron
Stevens, Audrey L
Stevens, William Clark
Stevenson, L Harold
Stevenson, Robert Edwin
Stewart, James Ray
Stock, John Joseph
Stockland, Alan Eugene
Stockton, Jack Jenks
Stojanovic, Borislav Jovan
Stokes, Jacob Leo
Stollar, Victor
Stone, Robert Louis
Stone, Robert William
Stoner, Gary David
Stopkie, Roger John
Storck, Roger Louis
Storz, Johannes
Stoudt, Thomas Henry
Strasdine, George Alfred
Strauss, Robert R
Streckfuss, Joseph Larry
Streips, Uldis Normunds
Streitfeld, Murray Mark
Strom, David Womack
Stuart, David Gordon
Stukus, Philip Eugene
Stumbo, Charles Raymond
Sturgess, Jennifer Mary
Sudo, Sara Zeece
Suggs, Morris Talmage, Jr
Suie, Ted
Suit, Joan C
Sullivan, Julia Christine
Sultzer, Barnet Martin
Sunrall, H Glenn
Suskind, Sigmund Richard
Sussman, Raquel Rotman
Sutton, Donald Dunsmore
Suzuki, Isamu
Suzuki, Tsuneo
Swanson, John Lee
Swatek, Frank Edward
Sweet, Herman Royden
Sword, Christopher Patrick
Sypherd, Paul Starr
Szaniszlo, Paul Joseph
Szybalski, Elizabeth Hunter
Taber, Harry Warren
Taber, Willard Allen
Taber, Morris
Takayama, Kuni
Tanimini, Hamdi Ahmad
Tanenbaum, Stuart William
Tang, Terry Chu
Tanner, Jurate E
Tatini, Sita Ramayya
Taubler, James H
Taussig, Andrew
Taylor, Alton Robert
Taylor, Austin Laurence
Taylor, John Jacob

Taylor, Mary Marshall
Taylor, Milton William
Taylor, Robert Gay
Taylor, Welton Ivan
Tendler, Moses David
Tenenbaum, Saul
Tenney, Raymond Wallace
Tepper, Byron Seymour
Terry, Thomas Milton
Tershak, Daniel R
Testa, Raymond Thomas
Tew, John Garn
Tew, Richard Wilcox
Thanos, Andrew
Thayer, Donald Wayne
Thayer, Philip Standish
Thomason, Donald C
Thomason, Berenice Miller
Thompson, Kenneth David
Thor, Daniel Einar
Thrupp, Lauri David
Tidwell, William Lee
Tilton, Richard C
Timm, Eugene Alvin
Ting, Robert Chin-Yao
Tipper, Donald John
Tischer, Robert George
Tittiger, Franz
Tjostem, John Leander
Todd, Ewen Cameron David
Tokuda, Sei
Tolberg, Adelaide Brokaw
Tomlinson, Geraldine Ann
Tornabene, Thomas Guy
Torres-Blasini, Gladys
Trankle, Robert John
Treadwell, Perry Edward
Treagan, Lucy
Treffers, Henry Peter
Tremaine, Mary M
Trent, Dennis W
Trentham, Jimmy N
Tritz, Gerald Joseph
Troller, John Arthur
Truant, Joseph Paul
Truitt, Robert Lindell
Trust, Trevor John
Tsai, Yuan-Hwang
Tsien, Hsienchyang
Tsui, Kiyoshi
Tu, Chin Ming
Tubiash, Haskell Solomon
Tucci, Anthony Frederick
Tulis, Jerry John
Tunstall, Lucille Hawkins
Turner, Frank Joseph
Turner, Jan Ross
Tuttle, Robert Lewis
Twedt, Robert Madsen
Tyrrell, Elizabeth Ann
Uffen, Robert L
Ugarte, Eduardo
Uhlenhopp, John E
Ukeles, Ravenna
Ulmer, David Heading Bartine, Jr
Ulrich, Arlene Louise
Ulrich, John August
Umanzio, Carl Beeman
Upadhyay, Jagdish M
Updegraff, David Maule
Ushijima, Richard M
Uzodinma, John E
Valdiveso, Dario
Vanaman, Thomas Clark
Vander Wyk, Raymond Winston
Van Dyke, Henry
Van Dyke, Russell Austin
Van Eck, Edward Arthur
Van Tassell, Morgan Howard
Vasington, Paul John
Vaughan, James Roland
Vedros, Neylan Anthony
Vela, Gerard Roland
Veltri, Robert William
Venneman, Martin Ray
Ventura, Arnoldo K
Vercen, Larry Edwin
Vermeulen, Carl William
Vestal, James Robie
Vezina, Claude
Vice, John Leonard
Vicedo, Eusebio
Vick, Jan Tomas
Vincent, Phillip G
Voelz, Herbert Gustav
Vogel, Henry
Vogel, Ralph A
Volcani, Benjamin Elazari
Vold, Barbara Schneider
Volk, Wesley Aaron
Voos, Jane Rhein
Vopicka, Ellen Vandersee
Vorbeck, Marie L
Voss, Jack Goddard
Vredevoe, Donna Lou
Wachsman, Joseph T

Wacker, Waldon Burdette
Wagner, Conrad
Wagner, Robert Roderick
Wagstaff, Paul Arlen
Walker, Duard Lee
Walker, James Roy
Walker, Jerry
Walker, Robert W
Wallace, John Howard
Wallace, Raphael Herman
Wallace, Willie Robert
Walraff, James R
Walraff, Evelyn Bartels
Walter, William Goff
Walters, Curla Sybil
Walton, Robert Bruce
Waner, Joseph Lloyd
Wang, Barbara Kascenko
Wang, Augustine Weisheng
Wannamaker, Lewis William
Ward, Edmund William Beswick
Wargel, Robert Joseph
Warner, George S
Warner, Peter
Warren, George Harry
Washam, Clinton Jay
Watkins, Paul Donald
Watrel, Warren George
Watson, Benjamin Bruce
Watson, Barbara Franklin
Watson, Dennis Wallace
Watson, Joseph Alexander
Watson, Richard White, Jr
Wawszkiewicz, Edward John
Wayne, Lawrence Gershon
Weatherwax, Robert Stanton
Weaver, John Martin
Weaver, Terry L
Webb, Frederick M
Webb, Neil Broyles
Webb, Robert Bradley
Webb, Sydney James
Weber, Allen Thomas
Weber, Morton M
Weber, William Adolph
Wedberg, Stanley Edward
Wegener, Warner Smith
Wehrle, Theodore Mil
Weichlein, Russell George
Weinbaum, George
Weiner, Lawrence Myron
Weiner, Ronald Martin
Weinstein, Marvin Joseph
Weis, Dale Stern
Weslow, Owen Stuart
Wells, David Walter
Welch, Gordon E
Wellerson, Ralph, Jr
Wells, Frank Edward
Welter, C Joseph
Wendt, Theodore Mil
Wentworth, Bertina Brown
Werder, Alvar Arvid
Westlake, Donald William Speck
Whalen, Joseph Wilson
Wheat, Robert Wayne
Wheeler, Harry Ogden
Wheelis, Mark Lewis
Whisler, Howard Clinton
White, David
White, Mack
White, Maurice Leopold
White, Roseann Spicola
Whitehouse, Frank, Jr
Whitehouse, Ronald Leslie S
Whitney, Dorothy McCartney (Mrs
    Ludwik Anigstein)
Whitney, John Glen
Whittington, Melvin Othal, Jr
Widra, Abe
Wiesmeyer, Herbert
Wilcox, Wesley C
Wilder, Martin Stuart
Wiley, Bill Beauford
Wiley, William Rodney
Wilhelm, Alan Roy
Wilhelm, James Maurice
Wilkins, Peter Osborne
Wilkins, Tracy Dale
Wilkinson, Thomas Ross
Wilkoff, Lee Joseph
Willet, Hilda Pope
Willet, Norman P
Williams, Anna Maria
Williams, Jimmy Calvin
Williams, Phletus P
Williams, Ralph Benjamin
Williams, Robert Pierce
Williamson, Clarence Kelly
Willis, Dawn Butler
Wilson, Benjamin James
Wilson, Bobby Eugene
Wilson, Harold Albert
Wilson, Jack Harold
Winkler, Charles Herman, Jr
Winkler, Herbert H
Winter, Charles Ernest
Winter, Jeanette

Winter, Joseph Wolfgang
Winters, Harvey
Wise, Ernest George
Wiseman, Ralph Franklin
Wistreich, George A
Witkin, Evelyn Maisel
Witz, Dennis Fredrick
Wolberg, Gerald
Wolf, Benjamin
Wolf, Kenneth Edward
Wolf, Pierre L
Wolf, Ralph Stoner
Wolfson, Leonard Louis
Wolgamot, Gary
Wolin, Harold Leonard
Wolk, Coleman Peter
Wollum, Arthur George, II
Wolochow, Hyman
Wood, Harland G
Wood, Willis A
Woodburn, Margy Jeanette
Woodruff, Harold Boyd
Woodruff, Theodore Englar
Woolfolk, Clifford Allen
Woolsey, Marion Elmer
Worthen, Leonard Robert
Wright, George Green
Wullstein, Leroy Hugh
Wyatt, Philip Joseph
Wyatt, Roger Dale
Wyckoff, Delphine Grace Rosa
Wyss, Orville
Xavier, K S
Yamamoto, Nobuto
Yamamoto, Tatsuzo
Yamashiroya, Herbert Mitsugi
Yang, Shaw-Ming
Yarbrough, Karen Marguerite
Yaw, Katherine Emily
Yelton, David Baetz
Yip, Lily Chung
Yoch, Duane Charles
Yohn, David Stewart
York, George Kenneth, II
Yotis, William William
Young, Bobby Gene
Young, Frank E
Young, Viola Mae
Zajic, James Edward
Zealey, Marion Edward
Zeikus, J Gregory
Zeltner, Jack
Zimmerer, Robert P
Zimmerman, Robert A
Zimmerman, Sheldon Bernard
Zimmermann, Eugene Robert Charles
Zobell, Claude E
Zsigray, Robert Michael
Zwadyk, Peter, Jr
Zwarun, Andrew Alexander
Zygmunt, Walter A

Agricultural Microbiology

Appleton, George Sanders
Baribo, Lester E
Barran, Leslie Rohit
Bell, Robert Graham
Chandra, Purna
Cheng, Kuo-Joan
Cook, Harold Andrew
Jones, Graham Alfred
Kennedy, Elhart James
Marsden, David Henry
Norstadt, Fred A
Steinke, Paul Karl Willi
Stevenson, Ian Lawrie
Stoller, Benjamin Boris
Volz, Michael George

Clinical Microbiology

Ball, William
Branson, Dorothy Swingle
Capers, Evelyn Lorraine
Chester, Brent
Cohen, Joel Ralph
D'Amato, Richard Frank
Garvin, Donald Frank
Kelly, Michael Thomas
Klein, Dolph
Liberman, Daniel Franklin
Libonati, Joseph Peter
Lindsey, Norma J
Roberts, Glenn Dale
Schwab, Bernard
Silberman, Ronald
Snyder, Merill J
Strenkoski, Leon Francis
Toney, Marcellus E, Jr
Washington, John A, II
Zabransky, Ronald Joseph

Dairy Microbiology

Martin, James Harold
Mikolajcik, Emil Michael

Nath, K Rajinder
Smith, Kenneth Leroy
White, Charles Henry
Willits, Richard Ellis

Food Microbiology

Alford, John Abright
Ashton, David Hugh
Avens, John Stewart
Babel, Frederick John
Banwart, George J
Bell, Wayne Harrel
Beuchat, Larry Ray
Bothast, Rodney Jacob
Brown, William Lewis
Bullerman, Lloyd Bernard
Busta, Francis Frederick
Cameron, Donald Eugene
Clark, David Sedgefield
Collins, Edwin Bruce
Dahiya, Raghunath S
Davidson, Charles Mackenzie
Denny, Cleve B
El-Bisi, Hamed Mohamed
Elliott, James Angus
Fox, Kenneth Ian
Frank, Hilmer Aaron
Gabis, Damien Anthony
Gilliland, Stanley Eugene
Goulet, Jacques
Hartman, Paul Arthur
Heinsohn, Richard Charles
Hickernell, Gary L
Huskey, Glen E
Idziak, Edmund Stefan
King, Alfred Douglas, Jr
Koburger, John Alfred
LaGrange, William Somers
Langlois, Bruce Edward
Larkin, Edward P
Ledford, Richard Allison
Maisch, Weldon Frederick
Marth, Elmer Herman
McAnelly, John Kitchel
Mercuri, Arthur J
Morgan, Bruce Henry
Morita, Toshiko M
Mundt, John Orvin
Naumann, Hugh Donald
Nelson, Frank Eugene
Nickelson, Ranzell, II
O'Leary, Virginia Sawyer
Park, Chong Eel
Patel, Girishchandra Babubhai
Perkins, William Edward
Potter, Norman N
Powers, Edmund Maurice
Rayman, Mohamad Khalil
Rowley, Durwood B
Segner, Wayne Philip
Smittle, Richard Baird
Solberg, Myron
Spira, William Martin
Stevenson, Kenneth Eugene
Stine, James Bryan
Tompkin, Robert Bruce
Vanderzant, Carl
Vaughan, Reese Haskell
Walker, Homer Wayne
Wehrle, Louis, Jr
Westhoff, Dennis Charles
Wilcox, Joseph Clifford
Witter, Lloyd David
Yates, Alfred Randolph

Industrial Microbiology

Ackart, Watson Boudinot
Andreasen, Arthur Albinus
Cayle, Theodore
Cino, Paul Michael
Colasito, Dominic James
Demain, Arnold Lester
Drescher, Robert Frederick
Dunn, Cecil Gordon
Dworschack, Robert George
Engel, Paulinus P
Erickson, Robert Joseph
Garvin, Donald Frank
Goodman, Nelson
Graham, Blanche D
Grundy, Walton Earle
Hahn, Peter Anthony
Hedrick, Harold Gilman
Hodson, Phillip Harvey
Karve, Roger Alan
Kele, Roger Alan
Lakshminarayanan, Krishnaiyer
Larsen, Don Hyrum
Levin, Joseph David
Liberman, Daniel Franklin
Litchfield, John Hyland
Lutey, Richard William
Masurekar, Prakash Sharatchandra
Maul, Stephen Bailey
Palochak, Muriel E
Rogers, Morris Ralph
Rosenman, Sanford Becker

Schuurmans, David Meinte
Sharpley, John Miles
Shieh, Kenneth Kuang-Zen
Sjolander, Newell Oscar
Soller, Arthur
Stark, William Max
Taussig, Steven J
Taylor, Wilbur Spencer
Townsley, Philip McNair
Unowsky, Joel
Ward, Charles Bradley, Jr
Wideburg, Norman Earl
Zorn, Ralph Allan

Marine Microbiology
Ahearn, Donald G
Baross, John Allen
Barvenik, Frank W
Bourquin, Al Willis J
Buck, John David
Claus, George
Faust, Maria Anna
Gundersen, Kaare Reinhardt
Hanson, Roger Brian
Hayasaka, Steven S
Juge, Damian Joseph
Lee, John Joseph
Liguori, Vincent Robert
Litchfield, Carol Darline
Meyers, Samuel Philip
Miget, Russell John
Nealson, Kenneth Henry
Prager, Jan Clement
Sieburth, John McNeill
Taylor, Barrie Frederick
Yonenaka, Hideo H

Medical Microbiology
Ackermann, Hans Wolfgang
Alexander, Aaron D
Allen, William Peter
Arden, Sheldon Bruce
Athar, Mohammed Aqeel
Bacon, Marion
Baer, Herman
Baker, Edgar Eugene, Jr
Barlow, James Lawrence
Barnhart, Donald Delbert
Barr, Fred S
Bartell, Pasquale
Beggs, William H
Berky, John James
Bird, Thomas Joseph
Blazevic, Donna Jean
Bojalil, Luis Felipe
Boltjes, Ben Harold
Bolyn, Anthony Edward
Bowman, Bernard Ulysses, Jr
Bradshaw, Lawrence Jack
Brancato, Frank Paul
Brockett, Royce Merrett
Brosbe, Edwin Allen
Brown, James Allen, Jr
Brown, Lee Roy, Jr
Brown, Russell Wilfrid
Butas, Constandina
Butt, Elizabeth
Cameron, J A
Campbell, James B
Champlin, William G
Chaudhary, Rabindra Kumar
Chen, Joseph Ke-Chou
Cleary, Timothy Joseph
Cohen, Jacob Ortlieb
Cohn, Maurice Leon
Collin, William Kent
Cooper, Ronda Fern
Cords, Carl Ernest, Jr
Covert, Scott Veasey
Cox, Charles Donald
Coyle, Marie Bridget
Crecelius, Harry Gilbert
Cundy, Kenneth Raymond
Dalzell, Robert Clinton
D'Amato, Richard Frank
Deas, Jane Ellen
De Courcy, Samuel Joseph, Jr
DeMeio, Joseph Louis
Dempster, George
Denton, James Fred
DePalma, Philip Anthony
DeVoe, Irving Woodrow
Dickerman, John Melville
Dierckx, Fred Herman
Dixon, John Michael Siddons
Dolman, Claude Ernest
Domer, Judith E
Duda, I B J
Duncan, James Lowell
Eaves, George Newton
Elberg, Sanford Samuel
Elliot, Larry P
Ellner, Paul Daniel
Eubanks, Julia Flynt
Evans, Charles Albert
Ewing, William Howell
Fabrizio, Angelina Maria

Fellowes, Oliver Nelson
Felton, Frances Grace
Fishel, Charles Wesley
Foley, George Edward
Forgacs, Joseph
Fox, Eugene N
Francis, Robert Dorl
Friedman, Lorraine
Friend, Charlotte
Fuller, Vernon Jack
Fulton, Macdonald
Furtado, Dolores
Gadebusch, Hans Henning
Garrison, Robert Gene
Gemski, Peter, Jr
Gibby, Irvin Welch
Glantz, Paul Joseph
Goetchius, George Richard
Goldner, Herman
Goldschmidt, Millicent
Good, Robert Campbell
Gordon, Irving
Gots, Joseph Simon
Green, Erika Ana
Greene, Robert Alva
Greer, James Edward
Groschel, Dieter Hans Max
Gruber, Jack
Gutekunst, Richard Ralph
Hageage, George John, Jr
Halbert, Seymour Putterman
Halkias, Demetrios
Halstead, Scott Barker
Hampson, C Ross
Hannan, Charles Kevin
Haque, Riaz-Ul
Harris, Elizabeth Forsyth
Hatten, Betty Arlene
Hawkins, Linda Louise
Hawley, Robert John
Hench, Miles Ellsworth
Hendrickson, Donald Allen
Hierholzer, John Charles
Higa, Harry Hiroshi
Hill, Douglas Wayne
Hinze, Harry Clifford
Hottle, George Austin
Housewright, Riley Dee
Hsu, Hsiu-Sheng
Huber, Thomas Wayne
Hurlburt, Evelyn McClelland
Hurst, Valerie
Isenberg, Henry David
Ivler, Daniel
Jacks, Thomas Mauro
Jackson, James Oliver
James, Ann Nixon
Jarolmen, Howard
Jemski, Joseph Victor
Jensen, Marcus Martin
Jeter, Wayburn Stewart
Johnston, Daniel Everett
Johnston, Paul Bruns
Joseph, Junior Mehsen
Kaminski, Zigmund Charles
Kaplan, Milton Temkin
Kapral, Frank Albert
Karlson, Alfred Gustav
Kellerman, George D
Kellogg, Douglas Sheldon, Jr
Kern, Earl R
Kettering, James David
Keyser, Peter D
Kilbourn, Joan Priscilla Payne
Kim, Kwang Shin
Kiser, Jackson Sebree
Kline, Bruce Clayton
Koh, Won Young
Kotcher, Emil
Kraeger, Spring Juliet
Kuehner, Calvin Charles
Kundsin, Ruth Blunfeld
Kuo, Cho-Chou
Kutner, Leon Jay
Lachapelle, Rene Charles
Lambe, Dwight Wilson, Jr
Laskowski, Leonard Francis, Jr
Lee, Spencer Hon-Sun
Leers, Wolf-Dietrich
Leibovitz, Albert
Leise, Joshua Melvin
Lewis, William Perry
Lichstein, Herman Carlton
Lief, Florence Suskind
Lindberg, Lois Helen
Lindberg, Robert Benjamin
Livermore, Brian Paul
Lomanitz, Rachel
Lynn, Raymond J
MacFarlane, John O'Donnell
MacLeod, Donald Charles
MacLeod, Donald Richard Eason
Mandel, John Herbert
Mann, Dennis Keith
Marraro, Robert V
Marvin, Richard Martin
Mastromarino, Anthony John

McBride, Mollie Elizabeth
McClatchy, Joseph Kenneth
McCracken, Alexander Walker
McFadden, Harry Webber, Jr
McRipley, Ronald James
McVickar, David Langston
Meyers, Wayne Marvin
Miles, Donald Orval
Millman, Irving
Milner, Kelsey Charles
Minsavage, Edward Joseph
Miraglia, Gennaro J
Moehring, Thomas John
Moore, Walter Edward Cladek
Morello, Josephine A (Mrs Mrs Robert E Butz)
Morter, Raymond Lione
Murphy, Juneann Wadsworth
Murphy, Richard Allan
Nachbar, Martin Stephen
Nahhas, Fuad Michael
O'Hern, Elizabeth Moot
Ottolenghi, Abramo Cesare
Padgett, Billie Lou
Pai, Chik Hyun
Pappagianis, Demosthenes
Paradise, Lois Jean
Paulissen, Leo John
Peterson, Ward Davis, Jr
Pilcher, K Stephen
Pollard, Morris
Potash, Louis
Prince, James T
Puhvel, Sirje Madii
Putlitz, Donald Herbert
Qadri, Syed M Hussain
Quarles, John Monroe
Quick, Jacquelin Dunn
Ramsey, Penny Tina
Randall, Eileen Louise
Rasmussen, Aaron Frederick, Jr
Reback, John Frederick
Reddick-Mitchum, Rhoda Anne
Reeves, James Blanchette
Reich, Claude Virgil
Reitman, Morton
Rhodes, Andrew James
Richardson, Harold
Rigney, Mary Margaret
Riley, Perry Stephen
Roantree, Robert Joseph
Robertsen, John Alan
Rogers, Howell Wade
Rogers, Nancy Graham
Rommel, Frederick Allen
Roosa, Robert Andrew
Rosan, Burton
Rossier, Edmond
Royal, George Calvin, Jr
Russell, Catherine Marie
Rustigian, Robert
Sampath, Angus C
Sanders, Christine Culp
Sanders, Ruth Evelyn
Savage, Dwayne Cecil
Schmidt, Jerome P
Scholes, Vernon Eugene
Schuytema, Eunice Chambers
Severin, Matthew Joseph
Shadomy, Smith
Shaffer, James Grant
Shaffer, Morris Frank
Shands, Joseph Walter, Jr
Shayegani, Mehdi
Shechmeister, Isaac Leo
Shotts, Emmett Booker, Jr
Sinha, Shyamal K
Sippel, John Edward
Slifkin, Malcolm
Sloan, Bernard Joseph
Smith, Josephine Reist
Snyder, Irvin S
Somerson, Norman L
Sonnenwirth, Alexander Coleman
Stevens, Roy White
Steward, John P
Stiffler, Paul W
Stinebring, Warren Richard
Straus, David Conrad
Suling, William John
Surgalla, Michael Joseph
Sutter, Vera La Verne
Tanzer, Charles
Theis, Jerold Howard
Thomas, Leo Alvon
Thomas, Virginia Lynn
Todd, William McClintock
Tonik, Ellis J
Tully, Joseph George, Jr
Tyeryar, Franklin Joseph, Jr
Vaituzis, Zigfridas
Valentine, Bob Leon
Vas, Stephen Istvan
Vera, Harriette Dryden
Verwey, Willard Foster
Von Graevenitz, Alexander W C
Von Riesen, Victor Lyle
Wang, San-Pin

Ware, Lawrence L, Jr
Webb, Alfred Mohr
Weidanz, William P
Weinberg, Eugene David
Weiss, Emilio
Whang, Sukoo Jack
Whitescarver, Jack Edward
Wilson, Elizabeth
Wilson, Marion Evans
Wisseman, Charles Louis, Jr
Wood, Ronald McFarlane
Wu, Chii-Huei
Wu, William Gay
Wynne, Elmer Staten
Yarzabal, Luis Alberto
Yaverbaum, Sidney
Youmans, Anne Stewart
Youngner, Julius Stuart
Yurchenco, John Alfonso
Zajac, Ihor
Zeigler, Joseph Alton
Zubrzycki, Leonard Joseph

Microbial Biochemistry
Aaronson, Sheldon
Adams, Gordon Albert
Aleem, M I Hussain
Andres, William Wolcott
Baptist, James (Noel)
Blakley, Edwin Raymond
Blumenthal, Harold Jay
Brown, Robert George
Bungay, Henry Robert, III
Burg, Richard William
Cort, Winifred Mitchell
Dawson, Peter Stephen Shevyn
Ferguson, Donald Allen, Jr
Fitz-James, Philip Chester
Georgi, Carl Edward
Greasham, Randolph Louis
Gum, Ernest Kemp, Jr
Hanson, Barbara Ann
Hidy, Phil Harter
Ho, Richard I-Fu
Hook, Derek John
Hurley, Laurence Harold
Hutchings, Brian Lamar
Jiu, James
Johnson, Marvin Joyce
Keister, Donald Lee
Killick, Kathleen Ann
Klucas, Robert Vernon
Kronish, Donald Paul
Kushner, Donn Jean
Lueking, Donald Robert
Merkel, Joseph Robert
Murray, Edward Donald
Peruzzotti, George Peter
Repaske, Roy
Roland, John Francis
Sansing, Gerald Allen
Schwartz, Jeffrey Lee
Seitz, Eugene W
Shedlarski, Joseph George, Jr
Singh, Akhand Pratap
Twarog, Robert
Wagman, Gerald Howard
Walter, Richard Webb, Jr
Wise, Edmund Merriman, Jr
Wodzinski, Rudy Joseph

Microbial Ecology
Adams, John Collins
Azam, Farooq
Ballentine, Robert
Bartha, Richard
Belly, Robert T
Bounds, Harold C
Bromel, Mary Cook
Bryant, Marvin Pierce
Burnison, Bryan Kent
Carlucci, Angelo Francis
Cassin, Joseph M
Cullimore, Denis Roy
Doemel, William Naylor
Fletcher, Donald Warren
Gillespie, Paul Albert
Gordon, Ronald Claire
Grula, Mary Muedeking
Guthrie, Rufus Kent
Hoffmann, Harrison Adolph
Koditschek, Leah K
Lange, Willy
Mullins, Jeanette Somerville
Pfaender, Frederic Karl
Pleasants, Julian Randolph
Post, Frederick Just
Pramer, David
Pritchard, Parmely Herbert
Robinson, John Bertram
Staley, James Trotter
States, Jack Sterling
Stotzky, Guenther
Tiedje, James Michael
Tyler, Max Ezra
Vishniac, Helen Simpson
Weston, Charles Richard
Whitt, Dixie Dailey

MICROBIOLOGY

Wolin, Meyer Jerome

Microbial Genetics
Al-Aidroos, Karen Messing
Allen, Marcia Katzman
Atkins, Charles Gilmore
Bachmann, Barbara Joyce
Barratt, Raymond William
Bean, Carl Adams
Beck, Phyllis Dorothy
Bear, Doris Jean
Bernheimer, Harriet P
Bezdicek, David Fred
Bhattacharjee, Jnanendra K
Brescia, Vincent Thomas
Bullas, Leonard Raymond
Campbell, Allan McCulloch
Case, Mary Elizabeth
Chakrabarty, Ananda Mohan
Cole, Michael Allen
Copeland, James Clinton
Crawford, Irving Pope
Cross, Ronald Allan
Curtiss, Roy, III
DeGiovanni-Donnelly, Rosalie F
Diehl, William Paul
Dimmick, Robert Lewellyn
Ely, Berten E, III
Falkinham, Joseph Oliver, III
Fantini, Amedeo Alexander
Felkner, Ira Cecil
Folsome, Clair Edwin
Glatzer, Louis
Goldberg, Ivan D
Goldschmidt, Raul Max
Gorini, Luigi Costantino
Gough, Michael
Gutz, Herbert
Harriman, Philip Darling
Iha, Thomas H
Johnson, Ben Francis
Johnson, Edward Miles, Jr
Jollick, Joseph Darryl
Jones, Elizabeth W
Jones, Lily Ann
Kelly, Beatrice L
Kowalski, John Bernard
Lacy, Ann Matthews
Lancaster, John
Landman, Otto Ernest
Lasure, Linda Lee
Levin, Barbara Chernov
Levinthal, Mark
Liss, Alan
Lo, Theodore Ching-Yang
Marcus, Leon
Marinus, Martin Gerard
McDougall, Kenneth J
Mehta, Bipinchandra Mohanlal
Michalka, Jack
Middleton, Richard B
Miller, Lynn
Miller, Robert Verne
Morrow, Terry Oran
Nakamura, Kazuo
Novick, Richard P
Ogg, James Elvis
Parker, John Hamilton
Pattee, Peter A
Peck, Harry Dowd, Jr
Prasad, Ishwari
Puhalla, John Edward
Rae, Margaret Engel
Raizen, Carol Eileen
Reid, Parlane John
Reilly, Bernard Edward
Rotheim, Minna B
Sanderson, Kenneth Edwin
Schoenhard, Delbert E
Schwinghamer, Erwin A
Scott, June Rothman
Siegel, Eli Charles
Sinha, Raj P
Smith, Hamilton Othanel
Stewart, Charles Ranous
Stine, Gerald James
Stocker, Bruce Arnold Dunbar
Takahashi, Francois Iwao
Thorne, Curtis Blaine
Tien, Weichen
Voll, Mary Jane
Walker, William Stanley
Wax, Richard Gerald
Zahler, Stanley Arnold

Microbial Physiology
Akagi, James Masuji
Albrecht, Alberta Marie
Andrykovitch, George
Bain, William Murray
Barnsley, Eric Arthur
Barrera, Cecilio Richard
Baxter, Ann Webster
Begue, William John
Berger, Leslie Ralph
Beterton, Harry O
Binder, Franklin Lewis
Botsford, James L

Bovell, Carlton Rowland
Boyer, Ernest Wendell
Caltrider, Paul Gene
Carpenter, David Francis
Chan, Eddie Chin Sun
Chang, George Washington
Charlang, Gisela Wohlrab
Clark, James Bennett
Clarke, Gary Anthony
Cobb, Howell Dee, Jr
Davis, James Royce
DeCicco, Benedict Thomas
Dhople, Arvind Madhav
Duerre, John A
Earhart, Charles Franklin, Jr
Eldridge, David Wyatt
Farina, Joseph Peter
Fernandez, Bernal
Fina, Louis R
Finnerty, William Robert
Forney, Frederick Willis
Forsberg, Cecil Wallace
Gary, Norman Dwight
Goldman, Manuel
Goodman, Richard E
Hall, John Bradley
Hechemy, Karim E
Hedden, Kenneth Forsythe
Higerd, Thomas Braden
Hooper, Alan Bacon
Horvath, Raymond S
Humphrey, Ronald DeVere
Ingram, Lonnie O'Neal
Jurstuk, Peter, Jr
Kallio, Reino Emil
Kang, Kenneth S
Ketchum, Paul Abbott
Klein, Harold Paul
Konetzka, Walter Anthony
Konisky, Jordan
Krulwich, Terry Ann
Krywolap, George Nicholas
Larkin, John Montague
Langworthy, Thomas Allan
Lessie, Thomas Guy
Levin, Richard Alexander
MacQuillan, Anthony M
Mahoney, Robert Patrick
Mandels, Gabriel Raphael
Mandels, Mary Hickox
Marquis, Robert E
Marshall, Vincent dePaul
Machett, William H
Matin, Abdul
Matsumura, Philip
Mattingly, Stephen Joseph
McCowen, Sara Moss
McDonald, Ian Johnson
McFeters, Gordon Alwyn
Mindich, Leonard Eugene
Mortlock, Robert Paul
Nash, Claude Hamilton, III
Neilands, John Brian
Nielsen, Allen Madsen
Nierlich, Donald P
Oginsky, Evelyn Lenore
Ondrako, Joanne Marie
Ornston, Leo Nicholas
Parks, Leo Wilburn
Parmegiani, Raulo
Pengra, Robert Monroe
Pitts, Robert Gary
Previc, Edward Paul
Raj, Harkison D
Ranhand, Jon M
Riederer-Henderson, Mary Ann
St John, Ann Carlson
Sebek, Oldrich Karel
Sguros, Peter Louis
Shugart, Lee Raleigh
Silver, Simon David
Simon, Robert David
Sleeper, Bayard Paul
Sojka, Gary Allan
Sperry, Jay Franklin
Tanzer, Jason Michael
Taylor, Mary Lowell Branson
Taylor, Walter Herman, Jr
Theodore, Theodore Spiros
Thomas, Julian Edward, Sr
Treick, Ronald Walter
Trelawny, Gilbert Sterling
Turner, Hal Russell
Tyler, Bonnie Moreland
Urban, James Edward
Van Etten, James L
Villa, Vicente Domingo
Vinopal, Robert Thomas
Wellman, Angela Myra
Wells, Martha Carol
Werth, Jean Marie
Westby, Carl A
White, James Patrick
Whiteley, Helen Riaboff
Wilson, Donald Alan
Yall, Irving
Yousten, Allan A

Microscopic Anatomy
Benes, Elinor Simson
Burns, Edward Robert
Clermont, Yves Wilfred
Cotton, Douglas Howard
Dickson, William Robert
Dietlein, Lawrence Frederick
Dugan, Kimiko Hatta
Elias, Michael Hans
Epling, Glenwood Pershing
Erlandsen, Stanley L
Feleppa, Alfred E, Jr
Golarz-De Bourne, Maria Nelly
Hadek, Robert
Hooker, William Mead
Jersild, Ralph Alvin, Jr
Kendall, Michael Welt
King, Barry Frederick
LeBouton, Albert V
Luchtel, Daniel Lee
Matthews, James Lester
McCallister, Lawrence P
Richer, Claude-Lise
Roosen-Runge, Edward C
Rosenthal, Theodore Bernard
Sedar, Albert William
Shryock, Edwin Harold
Snodgrass, Michael Jens

Oral Microanatomy
Gwinnett, A John

Oral Microbiology
Bowen, William H
Clem, William Henry
Coykendall, Alan Littlefield
Dogon, Leon I
Hoffman, Heiner
Linzer, Rosemary
Mashino, Paul Akira
Miller, Chris H
Orland, Frank J
Socransky, Sigmund Sydney

Petroleum Microbiology
Allred, Raymond Charles
Knecht, Albert T T
Schwartz, Robert David
Wegner, Gene H

Soil Microbiology
Alarie, Albert
Alexander, Martin
Anderson, Guy Richard
Bollag, Jean-Marc
Broadbent, Francis Everett
Cameron, Roy (Eugene)
Chase, Francis Edward
Chilgreen, Donald Ray
Chinn, Stanley H F
Cook, Fred D
Corke, Charles Thomas
Davey, Charles Bingham
Davis, Robert Jaquette, Jr
Doxtader, Kenneth Guy
Dunigan, Edward P
Duryea, William R
Focht, Dennis Douglas
Frederick, Lloyd Randall
Giddens, Joel Edwin
Gilbert, Richard Gene
Hagedorn, Charles
Ham, George Eldon
Harris, John Orville
Hiltbold, Arthur Edward, Jr
Hubbell, David Heuston
Ivarson, Karl C
Jones, Kenneth Lester
Ko, Wen-Hsiung
Loynachan, Thomas Eugene
Lu, Kuo Chin
Lynch, Darrel Luvene
Martin, James Hamilton
Martin, William Paxman
McCalla, Thomas Mark
Parr, James Floyd, Jr
Peterson, Harold LeRoy
Reuszer, Herbert William
Rice, Wendell Alfred
Robson, Hope Howeth
Smith, Jay Hamilton
Sommers, Lee Edwin
Sparrow, Elena Bautista
Stanovick, Richard Paul
Tate, Robert Lee, III
Tennille, Aubrey W
Weaver, Richard Wayne
Weber, Deane Fay
Widden, Paul Rodney
Woicott, Arthur Ripatte
Wolf, Duane Carl

Veterinary Microbiology
Anderson, David Prewitt
Bailie, Wayne E
Berkhoff, German Adolfo
Bittle, James Long

Brown, Albert Loren
Campbell, Charles Haywood
Carmichael, Leland E
Cole, John Rufus, Jr
Collier, John Raymond
Colwell, William Maxwell
Corstvet, Richard E
Cunningham, Charles Henry
Deshmukh, Damodar Ramchandra
Donnermuth, Charles Henry, Jr
Dommert, Arthur Roland
Ferris, Dean Hunter
Firchammer, Burton Deforest
Folkerts, Thomas Mason
Frey, Merwin Lester
Gaskin, Jack Michael
Gyles, C L
Hanson, Lyle Eugene
Harris, Delbert Linn
Hidalgo, Richard Jack
Hirsh, Dwight Charles, III
Hughes, David Edward
Jensen, Wayne Ivan
Kemeny, J Lorant
Killinger, Arden Holmes
King, Nelson Byron
Kleckner, Albert Louis
Kleven, Stanley H
Kohler, Erwin Miller
Kramer, Theodore Tivadar
Krishnamurti, Pullabhotla V
Kunar, Mahesh C
Lambert, George
Lauerman, Lloyd Herman, Jr
Loken, Keith I
Madden, David Larry
Marois, Paul Henri
McCune, Cornelius John
Mengeling, William Lloyd
Meyer, Richard Charles
Moore, Richard Wayne
Morse, Erskine Vance
Newbould, Francis Henry Samuel
O'Berry, Phillip Aaron
Packer, Raymond Allen
Page, Leslie Andrew
Patton, Eugene Claude
Pirtle, Emmett L
Pomeroy, Benjamin Sherwood
Price, Jessie Isabel
Radhakrishnan, Chittur Venkitasubhan
Rosenquist, Bruce David
Ross, Richard Francis
Rubin, Harvey Louis
Saif, Yehia Mohamed
Simpson, Russell Bruce
Smith, Clyde Konrad
Smith, Gail Bevington
Stalheim, Ole Henry
Stunkard, Jim A
Targowski, Stanislaw P
Tripathy, Deoki Nandan
Truscott, Robert Bruce
Waldhalm, Donald George
Wedman, Elwood Edward
White, Franklin Henry
Whitford, Howard Wayne
Wichmann, Robert W
Wood, Richard Lee
Yamamoto, Richard

MINERALOGY

Mineralogy
Alexander, Alexandre Emil
Alper, Allen Myron
Ames, Lloyd Leroy, Jr
Anderson, Orson Lamar
Anthony, John Williams
Asquith, George Benjamin
Bannerman, Harold MacColl
Barczak, Virgil J
Barnard, Walther M
Baskin, Yehuda
Bassett, William Akers
Bates, Thomas Fulcher
Baur, Werner Heinz
Bayliss, Peter
Beard, William Clarence
Beland, Rene
Berkebile, Charles Alan
Berry, Leonard Gascoigne
Biederman, Edwin Williams, Jr
Bieler, Barrie Hill
Bloss, Fred Donald
Boone, Gary M
Boos, Margaret (Fuller)
Borg, Iris Y P
Borst, Roger Lee
Bostrom, Kurt G V
Brindley, George W
Britton, Marvin Gale
Brock, Kenneth Jack
Brophy, Gerald Patrick
Brunton, George Delbert

Brydon, James Emerson
Buerger, Martin Julian
Burnham, Charles Wilson
Burns, Roger George
Burt, Donald McLain
Butler, John C
Butterman, William Charles
Cabri, Louis J
Calver, James Lewis
Carlson, Ernest Howard
Chang, Luke Li-Yu
Christensen, Nikolas Ivan
Christensen, Odin Dale
Cofer, Harland E, Jr
Conrad, Malcolm Alvin
Cook, William R, Jr
Corlett, Mabel Isobel
Crawford, Maria Luisa Buse
Desborough, George A
DeVries, Robert Charles
Dodd, Robert Taylor
Doehler, Robert William
Donnay, Joseph Desire Hubert
Dowty, Eric
Drummond, Paul Linwood
Duesing, Constantin Michael
Dwornik, Edward John
Eades, James L
Ehlers, Ernest George
Ehlmann, Arthur J
Erickson, Edwin Sylvester, Jr
Evans, Howard Tasker, Jr
Faizi, Salih
Fang, Jen-Ho
Faust, George Tobias
Ferguson, Robert Bury
Ferrell, Ray Edward, Jr
Finger, Larry W
Finney, Joseph J
Fisher, Daniel Jerome
Fleischer, Michael
Forbes, Warren C
Foreman, Dennis Walden, Jr
Foster, Wilfrid Raymond
Friedlaender, Carlo Gotthelf Immanuel
Frondel, Clifford
Frondel, Judith W
Frueh, Alfred Joseph, Jr
Frye, Keith
Furlong, Robert B
Gaines, Richard Venable
Gait, Robert I
Garske, David Herman
Garvin, Paul Lawrence
Gheith, Mohamed A
Ghose, Subrata
Giardini, Armando Alfonzo
Gielisse, Peter Jacob Maria
Gibbs, Gerald V
Gin, Thon Too
Glass, Herbert David
Grant, Sheldon Kerry
Gremillion, Louis Ray
Gude, Arthur James, III
Guven, Necip
Hagner, Arthur Feodor
Hall, Henry Thompson
Hamil, Martha M
Hansen, Gary Ralph
Harris, Donald C
Harvey, Richard David
Hatch, Robert Alchin
Hawkins, William Max
Hayes, John Bernard
Hazen, Robert Miller
Henderson, Donald Munro
Henderson, Edward Porter
Herber, Lawrence Justin
Hewins, Roger Herbert
Hewitt, Charles Hayden
Hill, Hamilton Stanton
Hoekstra, Karl Egmond
Holmes, Ralph Jerome
Holser, William Thomas
Hounslow, Arthur William
Howard, Clarence Edward
Hsu, Liang-Chi
Hurlbut, Cornelius Searle, Jr
Insley, Herbert
Isphording, Wayne Carter
Ito, Jun
Jansen, George James
Jensen, David Edward
Johnson, Donald Haskall
Johnson, Eben Lennart
Kamb, Walter Barclay
Kay, Suzanne Mahlburg
Khan, Aijaz Ahmed
Kiessling, Oscar Edward
Klein, Cornelis
Kodama, Hideomi
Koucky, Frank Louis, Jr
Landy, Richard Allen
Langer, Arthur M
Larson, Lawrence T
Leavens, Peter Backus
Ledoux, Robert Louis

Lee, Donald Edward
Leung, Irene Sheung-Ying
Levinson, Alfred Abraham
Liddicoat, Richard Thomas, Jr
Liebling, Richard Stephen
Liese, Homer C
Lyon, Ronald James Pearson
Lyons, Paul Christopher
Mandarino, Joseph Anthony
Marvin, Ursula Bailey
Maxwell, Dwight Thomas
McClellan, Guerry Hamrick
McCormick, George R
Meyer, Henry Oostenwald Albertijn
Meyer, Joachim Dietrich
Mikami, Harry M
Mitchell, Richard Scott
Moll, William Francis, Jr
Monroe, Eugene Alan
Moore, Duane Milton
Moore, Paul Brian
Moore, Raymond Kenworthy
Morris, Elliot Cobia
Morton, Roger David
Mossman, David John
Mumpton, Frederick Albert
Myer, George Henry
Nagle, Frederick, Jr
Nickel, Ernest Henry
Nuffield, Edward Wilfrid
Odom, Ira Edgar
Ohashi, Yoshikazu
Olsen, Donald Ray
Pabst, Adolf
Papike, James Joseph
Parham, Walter Edward
Parker, Ronald Bruce
Parry, William Thomas
Patchett, Joseph Edmund
Peacor, Donald Ralph
Pearl, Richard Maxwell
Perrotta, Anthony Joseph
Phillips, William Revell
Philpotts, Anthony Robert
Piotrowski, Joseph Martin
Pollard, Charles Oscar, Jr
Ponder, Herman
Pough, Frederick Harvey
Powell, Benjamin Neff
Prinz, Martin
Quon, David Shi Haung
Radtke, Arthur Sears
Reasenberg, Julian Robert
Ribbe, Paul Hubert
Riggs, Karl A
Roberson, Herman Ellis
Robie, Richard Allen
Robinson, Stephen Clive
Rockett, Thomas John
Rooney, Thomas Peter
Rosenweig, Abraham
Rossi, Sally Wentworth
Rucklidge, John Christopher
Ruotsala, Albert P
Rutstein, Martin S
Sand, Leonard B
Servos, Kurt
Shade, John William
Simmons, William Bruce, Jr
Simpson, Dale R
Skinner, Helen Catherine W
Smith, Deane Kingsley, Jr
Smith, Dorian Glen Whitney
Smith, Joseph Victor
Smyth, Joseph Richard
Snetsinger, Kenneth George
Snow, Roland B
Speed, Robert Clarke
Stanonis, Francis Leo
Stevenson, John Sinclair
Stevenson, Louise Stevens
Stevenson, Ralph Girard, Jr
Stewart, David Benjamin
Strahl, Erwin Otto
Sturm, Edward
Sun, James Ming-Shan
Switzer, George S
Taylor, Lawrence August
Tettenhorst, Rodney Tampa
Thibault, Newman William
Tien, Thomas Ta-Pin
Valentine, Wilbur Goodrich
Van Cott, Harrison Corbin
Vassiliou, Andreas H
Veit, Jiri Joseph
Venuto, Carmine Joseph
Wadsworth, William Bingham
Wahl, Floyd Michael
Walther, Frank H
Warren, Harry Verney
Warshaw, Charlotte Marsh
Wehrenberg, John P
Weitz, John Hills
Wenden, Henry Edward
Wendland, Wayne Marcel
White, Eugene Wilbert
White, William Arthur
Wicks, Frederick John

Williams, Eugene G
Williams, Sidney Arthur
Winchell, Horace
Winchell, Robert E
Wise, William Stewart
Wittels, Mark C
Woo, Ching Chang
Yatsu, Eiju
Young, Davis Alan
Young, Edward Joseph
Yund, Richard Allen
Zoltai, Tibor

Clay Mineralogy
Austin, Roger Seth
Bailey, Sturges Williams
Bohor, Bruce Forbes
Bundy, Wayne Miley
Burst, John Frederick
Clemency, Charles V
Clementz, David Michael
Douglas, Lowell Arthur
Huang, Wen Hsing
Jacobs, Marian Beckmann
Jonas, Edward Charles
McCaleb, Stanley B
Morris, Horton Harold
Murray, Haydn Herbert
Nelson, Bruce Warren
Prescott, Paul Ithel
Rhoades, James David
Rowland, Richards Atwell
Smith, John M
Snowden, Jesse Otho
Weaver, Robert Michael
Young, Raymond Hinchcliffe, Jr

Gemology
Fryer, Charles W
Gaal, Robert
Scharf, Walter

Soil Mineralogy
Ahlrichs, James Lloyd
Barnhisel, Richard I
Borchardt, Glenn Arnold
Foscolos, Anthony E
Harward, Moyle E
Hawkins, Richard Horace
Johnson, Leon Joseph
Jones, Robert L
Kittrick, James Allen
Kunze, George William
Milford, Murray Hudson
White, Joe Lloyd
Zelazny, Lucian Walter

MORPHOLOGY

Morphology
Adams, Franklin Scott
Alsop, David W
Anderson, Charles Eugene
Basile, Dominick V
Buell, Katherine Mayhew
Carluccio, Leeds Mario
Dalley, Arthur Frederick, II
DeSantis, Mark Edward
Dittmer, Howard James
Fimian, Walter Joseph, Jr
Gaudin, Anthony J
Goldstein, Bernard
Grove, Alvin Russell, Jr
Harris, Joseph Pollard, Jr
Kramer, Alfred William, Jr
Lombard, Richard Eric
Macdonald, Alastair David
McLain, Donald Davis, Jr
Meredith, Howard Voas
Nabrit, Samuel Milton
Oelrich, Thomas Mann
Ramm, Gordon Morley
Ruben, John Alex
Seiden, David
Smith, Cornelia Marschall
Spagnoli, Harriet
Thiruvathukal, Kuriakose V
Thompson, Rufus Henney
Tupa, Dianna Lou Dowden
Wagner, Charles Eugene
Ward, Max
Wetmore, Ralph Hartley
White, Richard Alan
Worthington, Richard Dane
Yasso, Warren E
Youssef, Nabil Naguib
Zug, George R

Experimental Morphology
Baic, Dusan
Dougall, Donald K
Faulkner, Russell Conklin, Jr
Forsyth, David Emory
Hikida, Robert Seiichi
Krupp, Patricia Powers
Neufeld, Daniel Arthur

Tartar, Vance

Insect Morphology
Craig, Douglas Abercrombie M
Dicke, Robert Jerome
Drecktrah, Harold Gene
Helms, Thomas Joseph
Hermann, Henry Remley, Jr
Parsons, Margaret Cranston
Stay, Barbara
Woolever, Patricia S
Zacharuk, R Y

Plant Morphology
Abbott, Rose Marie Savelkoul
Banks, Harlan Parker
Beck, Charles Beverley
Bell, Max Ewart
Bennett, Herald Durward
Berg, Arthur R
Berg, Dwight Hillis
Bierhorst, David William
Breil, David A
Burch, Charles
Canright, James Edward
Cox, Hiden Toy
deLanglade, Ronald Allan
Dickison, William Campbell
Duffy, Regina Maurice
Erspamer, Jack Laverne
Garrison, Rhoda
Giesy, Robert
Graham, Alan Keith
Gray, Lewis Richard
Greyson, Richard Irving
Hansen, Harold Westberg
Haskell, David Andrew
Henry, Robert David
Hewitson, Walter Milton
Hindman, Joseph Lee
Hostetter, Heber P, III
Hotchkiss, Arland Tillotson
Jagels, Richard H
Johnson, Tillman Joseph
Kaplan, Donald Robert
Kaufman, Peter Bishop
Kavaljian, Leroy Gregory
Kelley, Alden Gerard
Lommasson, Robert Curtis
Mallory, Thomas E
Martens, Jacob Louis
McGahan, Merritt Wilson
Meyer, Samuel Lewis
Mickel, John Thomas
Miksche, Jerome Phillip
Mills, Howard Leonard
Morlang, Charles, Jr
Nickerson, Norton Hart
Posluszny, Usher
Prior, Paul Verdayne
Rahn, Joan Elma
Rappleye, Robert Du Bois
Riopel, James L
Rodin, Robert Joseph
Sattler, Rolf
Shutts, Clarence Francis
Siemer, Eugene Glen
Simone, Leo Daniel
Singh, Surendra Pratap
Smith, Bruce Barton
Spurr, Arthur Raymond
Stevenson, Forrest Frederick
Sturrock, Thomas Tracy
Sutherland, Robert Carver
Swift, Lloyd Harrison
Tepfer, Sanford Samuel
Unger, James William
Vieth, Joachim
Whittier, Dean Page
Wilson, Kenneth Allen
Wilson, Thomas Kendrick
Wolfson, Alfred Mortimer
Wyatt, Raymond L

Soil Morphology
Arkley, Rodney John
Arnold, Richard Warren
Berdanier, Charles Reese, Jr
Bidwell, Orville Willard
Bourbeau, Gerard Auguste
Buntley, George Jule
Carlisle, Victor Walter
Cline, Marlin George
Dudas, Marvin Joseph
Fosberg, Maynard Axel
Gersper, Paul Logan
Harlan, Phillip Walker
Holzhey, Charles Steven
Huddleston, James Herbert
Kovar, John Alvis
Lewis, David Thomas
Lund, Lanny Jack
Olson, Gerald Walter
Petersen, Gary Walter
Petry, David Emory
Rieger, Samuel
Ross, Sam Jones, Jr
Rowles, Charles A

MORPHOLOGY

Vertebrate Morphology

Springer, Maxwell Elsworth
Van Ryswyk, Albert Leonard
Wright, William Ray

NEUROSCIENCES

Neurosciences

Bock, Walter Joseph
Boord, Robert Lennis
Callison, George
Clark, Glenn R
Cracraft, Joel Lester
Dapson, Richard W
Delphia, John Maurice
Dinsmore, Charles Earle
Fisher, Harvey Irvin
Goodge, William Russell
Goslow, George E, Jr
Heckerman, Raymond Otto
Henderson, Alex
Hildebrand, Milton
Klemm, Robert David
Lee, Sue Ying
Lehmann, Wilma Helen
Liem, Karel F
Moulton, James Malcolm
Rosenberg, Herbert Irving
Sokol, Otto M
Ward, Robert Porter

Alberts, Walter Watson
Albright, Bruce Calvin
Aronson, Lester Ralph
Bahill, Andrew Terry
Bartlett, John Richard
Berent, Stanley
Bernsohn, Joseph
Bickford, Reginald G
Bockholt, Anton John
Breen, Gail Anne Marie
Brink, John Jerome
Brown, Dawn LaRue
Cassaday, John Herbert
Cavonius, Carl Richard
Chalupa, Leo M
Chan-Palay, Victoria
Charlebois, Clarence Thomas
Chi, Che
Chow, Kao Laing
Clendinen, Martha Anne
Cohen, Maynard
Cohen, Monroe W
Costin, Anatol
Cueto, Cipriano, Jr
Currie, Julia Ruth
Davis, William J
del Cerro, Manuel (Perez)
Denenberg, Victor Hugo
DeWer, Pieter D
Doetsch, Gernot Siegmar
Ebner, Ford Francis
Edds, Mac Vincent
Eisenman, Leonard Max
Elbaum, Charles
Fields, Kay Louise
Forbes, Thomas Eugene
Geyer, Mark Allen
Gibson, Gary Eugene
Goldberg, Louis J
Granda, Allen Manuel
Hafemann, Dennis Reinhold
Hajra, Amiya Kumar
Haller, Ann Cordwell
Halpern, Bruce Peter
Hamilton, Charles Lewis
Hanson, Frank Edwin, Jr
Harth, Erich Martin
Hearney, Elaine Schmidt
Herrup, Karl Franklin
Hirsch, Frederic Geake
Holloway, Frank A
Holloway, Ralph L, Jr
Holtzman, Eric
Hunt, Richard Kevin
Jain, Mahendra Kumar
Jourdikian, Felor Kaloust
Karpiak, Stephen Edward
Klemm, William Robert
Klivington, Kenneth Albert
Korr, Irvin Morris
LaManna, Joseph Charles
Lamotte, Robert Hill
LaVail, Matthew Maurice
Leibovic, K Nicholas
Lennon, Vanda Alice
Levine, Kenneth Albert
Levy, Michael S
Levy, Nelson Louis
Lipton, James Matthew
Marg, Elwin
Marshall, Louise Hanson
McCaman, Marilyn Wales
McElligott, James George
McMasters, Robert Earl
Monjan, Andrew Arthur

Moskowitz, Michael Arthur
Mullen, Richard Joseph
Munson, John Bacon
Murray, Margaret Ransone (Mrs Burton Le Doux)
Nahrwold, Michael Lange
Neff, William Duwayne
Nelson, Phillip Gillard
O'Tanyi, Theodore John, Jr
Patton, Nancy Jane
Pearl, Gary Steven
Perkel, Donald Howard
Pinheiro, Marilyn Lays
Polley, Edward Herman
Popper, Arthur N
Rakic, Pasko
Rall, Wilfrid
Rosenzweig, Mark Richard
St Omer, Vincent Victor
Schoultz, Ture Victor
Sherman, Samuel Murray
Shneour, Elie Alexis
Skinner, James Ernest
Skinner, Robert Dowell
Spector, Novera Herbert
Stevens, Richard Joseph
Stockwell, Charles Warren
Stoney, Samuel David, Jr
Straile, William Edwin
Swiatek, Kenneth Robert
Tamir, Hadassah
Teyler, Timothy James
Thorner, Melvin Wilfred
Tieman, Suzannah Bliss
Valenstein, Elliot Spiro
Vanderwolf, Cornelius Hendrik
Wagner, Henry George
Walker, Don Wesley
Wang, Howard Hao
Wayner, Matthew John
Weiss, Ira Paul
Westfall, Jane Anne
White, Susan Ruth
Wilson, David Louis
Wine, Jeffrey Justus
Worth, William Sutherland
Wright, Dennis Charles
Wyman, Robert J
Young, Simon Nesbitt
Zweig, George

Neuroanatomy

Abplanalp, Paul LeRoy
Aker, Franklin David
Albernaz, Jose Geraldo
Anderson, William John
Angevine, Jay Bernard, Jr
Astruc, Juan A
Augustine, James Robert
Batton, Robert Ralph
Bayer, Shirley Ann
Bennett, Marvin Herbert
Berman, Alvin Leonard
Bernstein, Jerald Jack
Bertram, Ewart George
Bhatnagar, Kunwar Prasad
Bos, Jane
Brightman, Milton Wilfred
Brizzee, Kenneth Raymond
Brownson, Robert Henry
Bueker, Elmer Daniel
Butler, Ann Benedict
Campbell, Carlos Boyd Godfrey
Chambers, Wilbert Franklin
Clark, Ronald Grey
Clemente, Carmine Domenic
Cole, Wilbur Vose
Contu, Paolo
Cookson, Francis Bernard
Courville, Jacques
Cowan, W Maxwell
Cowley, A Ronald
Coyle, Peter
Creps, Elaine Sue
Curtis, Robin Livingstone
Debacker, Hilda Spodheim
Demski, Leo Stanley
DeVito, June Logan
Diamond, Ivan
Dill, Russell Eugene
Ebbesson, Sven O E
Feldman, Martin Leonard
Felten, David L
Finger, Thomas Emanuel
Fox, Clement Alphonsine
Fredman, Steven Leslie
Frontera-Reichard, Jose Guillermo
Fuller, Peter McAfee
Garrett, Frederic Daugherty
Gerstner, Robert
Gfeller, Eduard
Gibson, Kathleen Rita
Gioili, Roland A
Globus, Albert
Glover, Roy Andrew
Gomez, Daniel Guillermo
Goodfellow, Elsie F
Goodman, Donald Charles

Guillery, Rainer Walter
Guth, Lloyd
Hafner, Gary Stuart
Haines, Duane Edwin
Hard, Walter Leon
Harper, Jon William
Hartman, James Francis
Humberston, Albert O, Jr
Hunt, Guy Marion, Jr
Hurst, Edith Marie Maclennan
Hyde, John Baskerville
Jacobson, Stanley
Johnson, John Irwin, Jr
Johnson, Thomas Nick
Johnston, Naomi Lemkey
Karten, Harvey J
Keller, Jeffrey Thomas
Kichler, Ernest Earl, Jr
Kimmel, Donald Loraine
King, James S
Kolb, Helga Ellen Thor
Kriebel, Richard Marvin
Krieg, Wendell Jordan
Kruger, Lawrence
LaMotte, Carole Choate
Larson, Sanford J
LaVail, Jennifer Hart
Lavelle, George Arthur
Leichnetz, George Robert
Leonard, Christiana Morison
Loeser, Charles Nathan
Loewy, Arthur DeCosta
Lu Qin, Ivan James
Magoun, Horace Winchell
Mariscal, Richard North
Martin, George Franklin, Jr
Matzke, Howard Arthur
McClure, Theodore Dean
Mehler, William Raphael
Montgomery, Royce Lee
Moore, Josephine Carroll
Morrison, Adrian Russel
Mugnaini, Enrico
Nandy, Kalidas
Nolan, Michael Francis
Northcut, Richard Glenn
Norvell, John Edmondson, III
Ollerich, Dwayne A
Palay, Sanford Louis
Pearlman, Alan Lee
Peele, Talmage Lee
Penny, Joe Edward
Pentney, Roberta Pierson
Petras, James Minas
Potter, Henson David
Powell, Ervin William
Ralston, Henry James, III
Ramon-Moliner, Enrique
Rennels, Marshall L
Riss, Walter
Rothballer, Alan Burns
Rubinson, Kalman
Russell, Glenn Vinton
Saunders, Richard L
Schnitzlein, Harold Norman
Schroeder, Dolores Margaret
Schwartz, Ilsa Roslow
Sechrist, John William
Severin, Charles Matthew
Shanthaveerappa, Totada Ramaiah
Smith, Diane Elizabeth
Stensaas, Larry J
Stensaas, Suzanne Sperling
Sterling, Peter
Strominger, Norman Lewis
Sullivan, James Michael
Sutin, Jerome
Swigart, Richard Hanawalt
Taslitz, Norman
Thomas, Carolyn Eyster
Tigges, Johannes
Trachtenberg, Michael Carl
Turner, Robert Stuart
Ulinski, Philip Steven
Vaughan, Deborah Whittaker
Vijayan, Vijaya Kumari
Votaw, Charles Lesley
Ware, Carolyn Bogardus
Warr, William Bruce
Wells, Joseph
Werner, Joan Kathleen
White, Edward Lewis
Whitlock, David Graham
Williams, Terence Heaton
Winans, Sarah Schilling
Wong-Riley, Margaret Tze Tung
Woolsey, Thomas Allen
Wright, Charles Gary
Yoss, Robert Eugene
Young, M Wharton
Yu, Mang Chung

Barker, Jeffery Lange
Barmack, Neal Herbert
Bastian, Joseph
Bennett, Edward Leigh
Berry, Robert Wayne
Besso, Joseph Augustus, Jr
Bodian, David
Bondy, Stephen Claude
Bownds, M Deric
Brammer, Jimmie Duane
Brandt, Bruce Losure
Breese, George Richard
Bridgers, William Frank
Brimijoin, William Stephen
Browner, Robert Herman
Bullock, Theodore Holmes
Burgess, Paul Richards
Bystrom, Barbara Gillooly
Campenot, Robert Barry
Capranica, Robert R
Chappell, Richard Lee
Cohen, Adolph Irvin
Cohen, Rochelle Sandra
Coleman, Paul David
D'Adamo, Amedeo Filiberto, Jr
Dahl, Nancy Ann
Davis, George Berthold
Dempsey, Colby Wilson
Dennis, Michael Joseph
Descarries, Laurent
Diamond, Marian C
Disterhoft, John Francis
Doolin, Paul F
Dowling, John Elliott
Dubin, Mark William
Dubner, Ronald
Dusenbery, David Brock
Faber, Donald S
Fambrough, Douglas McIntosh
Fernald, Russell Dawson
Fisher, Leslie John
Fisher, Steven Kay
Frigyesi, Tamas L
Furshpan, Edwin Jean
Gainer, Harold
Geinsman, Yuri
Gerstein, George Leonard
Getting, Peter Alexander
Gibbs, James Gendron, Jr
Glassman, Edward
Glasser, Richard Lee
Gobel, Stephen
Goldman, Leonard Jay
Grabowski, Sandra Reynolds
Graham, Lewis Texada, Jr
Gwilliam, Gilbert Franklin
Haight, John Richard
Hara, Toshiaki
Harris, Charles Leon
Hauschka, Stephen D
Henson, Anna Miriam (Morgan)
Hickey, Terry Lee
Hillman, Dean Elof
Hirsch, Helmut V B
Hubbard, Jack Edward, Jr
Ingram, Walter Robinson
Irwin, Louis Neal
Kandel, Eric Richard
Karlson, Ulf Lennart
Keller, Edward Lowell
Kelly, James P
Kelly, Regis Baker
Kendig, Joan Johnston
Kennedy, Michael Craig
Koenig, Edward
Konopka, Ronald J
Kung, Ching
Landis, Story Cleland
Lang, Frederick
Larimer, James Lynn
Lasek, Raymond J
Ledeen, Robert
Lester, Henry Allen
Levitt, Melvin
Llinas, Rodolfo
Marks, Neville
Maynard, Edith Adele
McEwen, Bruce Sherman
McGaugh, James L
McKelvy, Jeffrey Forrester
McLardy, Turner
McLaughlin, Barbara Jean
McNamara, Mary Colleen
Miselis, Richard Robert
Moran, David Taylor
Murphey, Rodney Keith
Murray, Marion
Nadelhaft, Irving
Nakajima, Yasuko
Nicholson, Charles (Godfrey)
Niklowitz, Werner Johannes
Norden, Jeanette Jean
Novick, Alvin
Nuite, Jo Ann

Neurobiology

Alley, Keith Edward
Anderson, Margaret
Arch, Stephen William
Baker, Peter C
Barker, David Lowell

Oakley, Bruce
Okun, Lawrence M
Oldstone, Michael Beaureguard Alan
Olivo, Richard Francis
O'Steen, Wendall Keith
Ostroy, Sanford Eugene
Pappas, George Demetrios
Paul, Dorothy Hayman
Peek, Frank Willard
Peterson, Richard George
Pollard, Harvey Bruce
Possmayer, Fred
Potter, Lincoln Truslow
Prior, David James
Quarton, Gardner Cowles
Rash, John Edward
Redburn, Dianna Ammons
Roberge, Fernand Adrien
Robertson, Richard Thomas
Ross, Leonard Lester
Rubel, Edwin W
Salpeter, Miriam Mirl
Salzberg, Brian Matthew
Samson, Frederick Eugene, Jr
Scalia, Frank
Schafer, Rollie R
Schlapfer, Werner T
Sechzer, Jeri Altneu
Sharma, Sansar C
Siegel, Allan
Sisken, Betty Florio
Somers, Michael Eugene
Spanis, Curt William
Spirito, Carl Peter
Stefano, George Bogdon
Steinhardt, Richard Antony
Stuart, Ann Elizabeth
Talamo, Barbara Lisann
Turner, James Eldridge
Tweedle, Charles David
Uzman, Betty Geren
Vernadakis, Antonia (Mrs H L Ockerman)
Voneida, Theodore J
Weinreich, Daniel
Wilkens, Lon Allan
Wilson, Barry William
Wood, John Grady
Woodhull, Ann McNeal
Woodson, Paul Bernard
Yu, Riley Chaoping
Zagon, Ian Stuart
Zigmond, Richard Eric
Zomzely-Neurath, Claire Eleanore
Zottoli, Steven Jaynes

Neurochemistry
Abdel-Latif, Ata A
Abramson, Morris Barnet
Albers, Robert Wayne
Aprison, Morris Herman
Archer, Ellen Gleason
Banik, Narendra Lal
Barraco, Robin Anthony
Benjamins, Joyce Ann
Benuck, Myron
Bleecker, Margit
Blume, Arthur Joel
Boulton, Alan Arthur
Brockerhoff, Hans
Brostoff, Steven Warren
Browning, Edward T
Burt, Alvin Miller, III
Chao, Li-Pen
Cheng, Sze-Chuh
Churchill, Lynn
Clendenon, Nancy Ruth
Cohen, Gerald
Cohen, Stephen Robert
Costantino-Ceccarini, Elvira
Daly, John William
Datta, Ranajit Kumar
Deitbarn, Wolf Dietrich
De Vries, George Henry
Dhopeshwarkar, Govind Atmaram
Dreyfus, Pierre Marc
Duffy, Thomas Edward
Dunn, Adrian John
Eiduson, Samuel
Einstein, Elizabeth Roboz
Freedman, Lewis Simon
Gaskin, Felicia
Geison, Ronald Leon
Geller, Edward
Gould, Robert Michael
Greenfield, Seymour
Haubrich, Dean Robert
Held, Irene Rita
Higman, Henry Booth
Hirsch, Hilde Esther Zwirn
Ho, Ing Kang
Holt, Thomas Manning
Horrocks, Lloyd Allen
Hoss, Wayne Paul
Iqbal, Zafar
Kakari, Sophia
Khan, Mozzam Ali

Kies, Marian Wood
Kovachich, Gyula Bertalan
Lapin, Evelyn P
Larrabee, Martin Glover
Lees, Marjorie Berman
Lim, Ramon (Khe Siong)
Lin, Sping
Ling, Alfred Soy Chou
Lowden, J Alexander
Luine, Victoria Nall
Maker, Howard Smith
Margolis, Frank L
McBride, William Joseph
McClure, William Owen
McIlwain, David Lee
Modak, Arvind T
Moscatelli, Ezio Anthony
Nordyke, Ellis Larrimore
Norton, William Thompson
O'Brien, Richard Desmond
Palmer, Eugene Charles
Pappius, Hanna M
Peck, Ernest James, Jr
Peterson, Rudolph Price
Piepho, Robert Walter
Quarles, Richard Hudson
Ramsey, Robert Bruce
Rassin, David Keith
Rivera, Americo, Jr
Roberts, Sidney
Rosenberry, Terrone Lee
Rosenblatt, Dorrie Ellen
Rothenberg, Mortimer Abraham
Samuels, Stanley
Schacht, Jochen Heinrich
Shapira, Raymond
Shaskan, Edward Gregory
Siegel, Laurane Geary
Singh, Vijendra Kumar
Sun, Albert Yung-Kwang
Suzuki, Kunihiko
Swaiman, Kenneth F
Tanaka, Ryo
Teller, David Norton
Varon, Silvio Salomone
Vrba, Rudolf
Wander, Joseph Day
Weinstein, Howard
Weitsen, Howard Arthur
White, Thomas David
Wolfe, Leonhard Scott
Wolfgram, Frederick John
Wood, James Douglas
Yamamura, Henry Ichiro

Neurocytology
Adinolfi, Anthony M
Bennett, Kimberly D
Blanks, Janet Marie (Clarenbach)
Brawer, James Robin
Hattori, Toshiaki
Neale, Elaine Anne
Phelps, Creighton Halstead
Pysh, Joseph John

Neuroembryology
Baird, John Jeffers
Bekoff, Anne Laurens
Das, Gopal Dwarka
Heaton, Marieta Barrow
Lyser, Katherine May (Mrs E Shouby)
Meyer, Ronald Leo
Sohal, Gurkirpal Singh
Woods, Geraldine Pittman

Neuroendocrinology
Alvarez-Buylla, Ramon
Ben-Jonathan, Nira
Borer, Katarina Tomljenovic
Campbell, Gary Thomas
Clemens, James Allen
Corbin, Alan
Cramer, Oneida Morningstar
Curry, Jon Joseph, III
Dunn, Jon D
Fawcett, Colvin Peter
Fowler, Dona Jane
France, Evelyn S (Kalagher)
Gorski, Roger Anthony
Grota, Lee J
Hardin, Carolyn Myrick
Jacobs, John Joseph
Kalra, Satya Paul
Kizer, John Stephen
Knigge, Karl Max
Kozlowski, Gerald P
Krey, Lewis Charles
Lamperti, Albert A
Legan, Sandra Jean
Litteria, Marilyn
Lynch, Harry James
Malven, Paul Vernon
Mason, James Wayne
McCann, Samuel McDonald
Milmore, John Edward
Montemurro, Donald Gilbert
Nishioka, Richard Seiji
Norman, Reid Lynn

Ondo, Jerome G
Passo, Stanley Samuel
Peter, Richard Ector
Quadagno, David Michael
Quadri, Syed Kaleemullah
Quinn, David Lee
Reiter, Russel Joseph
Rhees, Reuben Ward
Rubinstein, Lydia
Santisteban, George Anthony
Saxena, Anjali
Scott, David Evans
Smith, Erla Ring
Sokol, Hilda Weyl
Sorrentino, Sandy, Jr
Spies, Harold Glen
Stumpf, Walter Erich
Turgeon, Judith Lee
Tyrey, Lee
Ulrich, Renee Sandra
Weiner, Richard Ira
Wheaton, Jonathan Edward
Wilbur, Donald Lee

Neurology
Ades, Harlow Whiting
Adler, Alexandra
Aird, Robert Burns
Aita, John Andrew
Ajax, Ernest Theodore
Alexander, Leo
Allen, Norman
Alter, Milton
Amick, Lawrence Douglas
Anderson, William Westerlin
Appel, Stanley Hersh
Appenzeller, Otto
Arieff, Alex J
Aring, Charles Dair
Austin, James Henry
Baird, Henry W, III
Baker, Abe Bert
Barbeau, Andre
Barnett, H J M
Barron, Kevin D
Batkin, Stanley
Bender, Morris Boris
Benton, John William, Jr
Bergen, Donna Catherine
Biehl, Joseph Park
Booker, Harold E
Boshes, Benjamin
Broder, Samuel B
Brown, Joe Robert
Brown, Meyer
Bruetman, Martin Edgardo
Buschke, Herman
Calhoun, Calvin L
Calverley, John Robert
Campbell, Berry
Carter, Charles Conrad
Carter, Sidney
Caviness, Verne Strudwick, Jr
Celesia, Gastone G
Chambers, Richard
Chase, Thomas Newell
Chun, Raymond Wai Mun
Churchill, John Alvord
Chusid, Joseph George
Clark, David Barrett
Cohen, Bernard
Cohn, Robert
Cole, Monroe
Collins, George H
Cook, Stuart D
Corbin, Kendall Brooks
Couch, James Russell, Jr
Currier, Robert David
Cutler, Robert W P
Daly, David DeRouen
Davis, Courtland Harwell, Jr
Davis, Floyd Asher
DeJong, Russell Nelson
Dekaban, Anatole S
De Myer, William Erl
De Salva, Salvatore Joseph
Drachman, Daniel Bruce
Drachman, David A
Drew, Arthur Leslie
Duffy, Philip
Duvoisin, Roger C
Eichman, Peter L
Eliasson, Sven Gustav
Elizan, Teresita S
Elliott, Frank A
Engel, William King
Ettinger, Milton G
Fahn, Stanley
Farmer, Thomas Wohlsen
Feinberg, Donald Lester
Feldman, Daniel S
Fenichel, Gerald M
Feringa, Earl Robert
Ferrendelli, James Anthony
Ferriss, Gregory Stark
Fields, William Straus
Fisher, Morris Alan

Fishman, Robert Allen
Foley, Joseph Michael
Forster, Francis Michael
Forster, David Bernard
Fox, Jacob H
Freedman, David Asa
Freemon, Frank Reed
Friedlander, Walter Jay
Gabor, Andrew John
Gallagher, Brian Boru
Gardner, Ernest (Dean)
Geschwind, Norman
Giblin, Denis Richard
Gibson, William Carleton
Gilden, Donald Harvey
Gilman, Sid
Ginsberg, Myron David
Glaser, Gilbert Herbert
Goldberg, Mark Arthur
Goldberger, Michael Eric
Goldensohn, Eli Samuel
Goldstein, Norman Philip
Gould, Harry J, III
Graf, Carl John
Growdon, John Herbert
Gutmann, Ludwig
Haase, Gunter R
Hackett, Earl R
Haerer, Armin Friedrich
Halsey, James H, Jr
Harter, Donald Harry
Heck, Albert Frank
Herndon, Robert McCulloch
Herrman, Christian, Jr
Heyman, Albert
Hinterbuchner, Ladislav Paul
Hochwald, Gerald Martin
Hogan, Edward L
Hollander, Joshua
Horenstein, Simon
Hoyt, William F
Isom, John B
Itabashi, Hideo Henry
Jarcho, Leonard Wallenstein
Johns, Thomas Richards, II
Johnson, Richard T
Joynt, Robert James
Kaelber, William Walbridge
Kalyanaraman, Krishnaswamy
Kark, Robert Adriaan Pieter
Karp, Herbert Rubin
Katzman, Robert
Kennedy, William Robert
Koenig, Harold
Korein, Julius
Krieger, Howard Paul
Kurtzke, John F
Ladwig, Harold Allen
Lam, Robert Lee
Landau, William M
Lawrence, Donald Gilbert
Lawyer, Tiffany, Jr
Lehrer, Gerard Michael
Lenn, Nicholas Joseph
Lessell, Simmons
Leventhal, Carl M
Levi-Montalcini, Rita
Levy, David Edward
Levy, Irwin
Lisak, Robert Philip
Luhan, Joseph Anton
MacFadyen, Donald John
Macrae, Donald
Magee, Kenneth Raymond
Magladery, John William
Malone, Michael Joseph
Mann, Leslie Bernard
Markesbery, William Ray
Markham, Charles Henry
Marks, Morton
Martin, Herbert Lloyd
Masland, Richard Lambert
Mason, Beryl Troxell
Mavor, Huntington
Mayer, Richard F
McDowell, Fletcher Hughes
McHugh, Paul Rodney
McKhann, Guy Mead
McKinney, William Markley
McQuarrie, Irvine Gray
Mettler, Frederick Albert
Meyer, John Stirling
Mickel, Hubert Sheldon
Miller, James Q
Millhouse, Oliver Eugene
Millichap, Joseph Gordon
Millikan, Clark Harold
Mohr, Jay Preston
Mones, Robert J
Moore, Robert Yates
Moossy, John
Morrell, Frank
Morris, Charles Elliot
Morris, Harold H
Mulder, Donald William
Myers, Ronald Elwood
Nachmansohn, David
Nagler, Benedict

Namba, Tatsuji
Nathanson, Neil Marc
Nayyar, Rajinder
Nelson, Erland
Netsky, Martin George
Nevis, Arnold Hastings
Niedermeyer, Ernst F
Norris, Forbes Holten, Jr
North, Richard Ralph
O'Brien, George Sivesind
O'Doherty, Desmond Sylvester
Oh, Shin Joong
Oldendorf, William Henry
Oliver, John Eoff, Jr
Olsen, Clarence Wilmott
O'Reilly, Sean
Paddison, Richard Milton
Perez-Borja, Carlos
Petajan, Jack Hougen
Peters, Bruce Harry
Peters, Henry A
Peterson, Donald I
Philippart, Michel Paul
Pincus, Jonathan Henry
Pitner, Samuel Ellis
Plum, Fred
Poirier, Louis
Porter, Huntington
Poskanzer, David Charles
Posner, Jerome B
Prince, David Allan
Prockop, Leon D
Raichle, Marcus Edward
Randt, Clark Thorp
Rapin, Isabelle (Mrs Harold Oaklander)
Rasmussen, Theodore Brown
Redding, Foster Kinyon
Reinmuth, Oscar McNaughton
Reis, Donald J
Reivich, Martin
Resch, Joseph Anthony
Richardson, John Clifford
Robb, J Preston
Robert, Emery Dean
Robinson, Bryan Wright
Roos, Raymond Philip
Rose, Arthur L
Rose, Augustus Steele
Ross, Alexander Treloar
Ross, Gilbert Stuart
Rowan, A James
Rowland, Lewis Phillip
Sabin, Thomas Daniel
Sabra, Fuad Amin
Satran, Richard
Scheinberg, Labe Charles
Scheinberg, Peritz
Schlagenhauff, Reinhold Eugene
Schlezinger, Nathan Stanley
Schoenberg, Bruce Stuart
Schulman, Sidney
Schwartz, James F
Segarra, Joseph M
Seipel, John Howard
Senrad, Elvin Vavrinec
Shapiro, William Richard
Sherwin, Allan Leonard
Shuter, Eli Ronald
Sibley, William Austin
Siegel, George Jacob
Siekert, Robert George
Silberberg, Donald H
Smith, Bernard H
Smith, Carlton G
Sobkowicz, Hanna Maria
Spehlmann, Rainer
Stevens, Harold
Stevens, Janice R
Stewart, Lever F
Still, Charles Neal
Strobos, Robert Julius
Sullivan, John Francis
Swank, Roy Laver
Swanson, August George
Swanson, Phillip D
Teasdall, Robert Douglas
Teitelbaum, Harry Allen
Thompson, Hartwell Greene, Jr
Togila, Joseph U
Torres, Fernando
Tourtellotte, Wallace William
Towler, Martin Lee
Twitchell, Thomas Evans
Van Allen, Maurice Wright
Van Den Noort, Stanley
Van Der Meulen, Joseph Pierre
Waltz, Arthur G
Waters, Gordon Valentine
Waxman, Stephen George
Webster, David Dyer
Webster, Henry deForest
Wehman, Henry Joseph
Weitzman, Elliot D
Whisnant, Jack Page
White, Harry Houston
White, Philip Taylor
Whittier, John Rensselaer
Wiederholt, Wigbert C
Wirschafter, Jonathan Dine
Woolsey, Robert S
Yahr, Melvin David
Yang, Dorothy Chuan-Ying
Yatsu, Frank Michio
Yeager, Charles Levant
Young, Gilbert Flowers, Jr
Ziegler, Dewey Kiper

Neuropathology
Adams, Raymond D
Alvord, Ellsworth Chapman, Jr
Anderson, Paul J
Boehme, Diethelm Hartmut
Breazile, James E
Brown, William Jann
Buchan, George Colin
Budzilovich, Gleb Nicholas
Choi, Byung Ho
Cowen, David
Cravioto, Humberto
Davis, Richard LaVerne
Derby, Bennett Marsh
Donahue, Sheila
Dudley, Alden Woodbury, Jr
Earle, Kenneth Martin
Engel, Andrew G
Feigin, Irwin Harris
Friede, Reinhard L
Garcia, Julio H
Gilles, Floyd Harry
Gonzalez-Angulo, Amador
Graham, Doyle Gene
Harkin, James C
Haymaker, Webb Edward
Herman, Mary M
Hirano, Asao
Jones, Margaret Zee
Kaufman, Mavis Anderson
Kim, Seung U
Kirkpatrick, Joel Brian
Kornfeld, Mario O
Krigman, Martin Ross
Lapham, Lowell Winship
Levine, Seymour
Lindenberg, Richard
Lipkin, Lewis Edward
Liss, Leopold
Malamud, Nathan
Mancall, Elliott L
Manuelidis, Elias Emmanuel
Matthews, Murray Albert
McCarty, Paul Anthony
Nelson, James S
Olney, John William
Pope, Alfred
Raine, Cedric Stuart
Rewcastle, Neill Barry
Robertson, David Murray
Roizin, Leon
Rorke, Lucy Balian
Rozdilsky, Bohdan
Rubinstein, Lucien Jules
Schochet, Sydney Sigfried, Jr
Shaw, Cheng-Mei
Shelanski, Michael L
Sidman, Richard Leon
Song, Sun Kyu
Steward, Vincent William
Sung, Joo Ho
Suzuki, Kinuko
Suzuki, Minoru
Terry, Robert Davis
Unterharnscheidt, Friedrich J
Verity, Maurice Anthony
Wagner, John Alfred
Wisniewski, Henryk Miroslaw
Wolf, Abner
Zeman, Wolfgang
Zu Rhein, Gabriele Marie

Neuropharmacology
Aghajanian, George Kevork
Allman, John Morgan
Baker, Walter Wolf
Barnett, Allen
Be Ment, Spencer L
Biber, Margaret Clare Boadle
Bond, Harley William
Boyd, Eleanor H
Brazier, Mary A B
Brezenoff, Henry Evans
Broadie, Larry Lewis
Christensen, Howard Dix
Clay, George A
Clineschmidt, Bradley Van
Colasanti, Brenda Karen
De la Torre, Jack Carlos
Douglas, William Wilton
Ebadi, Manuchair
Edmonds, Harvey Lee, Jr
Fernstrom, John Dickson
Franz, Donald Norbert
Gallagher, Joel Peter
Geller, Herbert M
Glick, Stanley Dennis
Goldman, Harold
Goldstein, Frederick J
Goldstein, Leonide
Haigler, Henry James
Hance, Anthony James
Hanig, Joseph Peter
Harris, Jane Ellen
Harris, Louis Selig
Hess, Helen Hope
Hiller, Jacob Moses
Hillman, Gilbert R
Hosko, Michael J, Jr
Huffman, Ronald Dean
Hui, Ferdinand W
Iturrian, William Ben
Jobe, Phillip Carl
Johnston, John O'Neal
Joy, Robert McKernon
Juorio, Augusto Victor
Kahn, Norman
Kirsten, Edward Bruce
Kosh, Joseph William
Koster, Rudolf
Kuhar, Michael Joseph
Kramer, Stanley Zachary
Lorenzo, Antonio V
Malick, Jeffrey Bevan
Marrazzi, Amedeo S
Marrazzi, Mary Ann
Miyamoto, Michael Dwight
Molinoff, Perry Brown
Parker, Robert Bruce
Patek, David Rushton
Potter, David Dickinson
Quest, John Anthony
Quock, Raymond Mark
Robert, Richard Ross
Rech, Richard Howard
Richter, Judith Anne
Roth, Jerome Allan
Roth, Robert Henry, Jr
Sabelli, Hector C
Schaeppi, Ulrich Hans
Schoener, Eugene Paul
Seeman, Philip
Segal, David S
Share, Norman N
Shinnick-Gallagher, Patricia L
Simon, Marcia Lee Mielke
Smith, Orville Auverne, Jr
Snyder, Solomon H
Spooner, Charles Edward, Jr
Stein, Barry Edward
Suria, Amin
Tuttle, Warren Wilson
Vazquez, Alfredo Jorge
Von Voigtlander, Philip Friedrich
Walker, Charles A
Walker, Don Wesley
Wallach, Marshall Ben
Warnick, Jordan Edward
Waters, Donald Hilton
Welch, Bruce L
Winters, Wallace Dudley
Yarbrough, George Gibbs
Zablocka-Esplin, Barbara
Zigmond, Michael Jonathan

Neurophysiology
Abrahams, Vivian Cecil
Adey, William Ross
Adkins, Ronald James
Agin, Daniel Pierre
Adolph, Alan Robert
Allen, Gary Irving
Altman, Joseph
Amassian, Vahe Eugene
Ames, Adelbert, III
Anderson, David J
Anschel, Steven William
Asanuma, Hiroshi
Ashworth, Robert David
Austin, George M
Bach, L Matthew N
Bach-Y-Rita, Paul
Baldwin, Bernell Elwyn
Barlow, John Sutton
Barlow, Robert Brown, Jr
Barrett, Robert
Batsel, Henry Lewis
Baylor, Denis Aristide
Beaudreau, David R
Beckman, Alexander Lynn
Benevento, Louis Anthony
Benjamin, Robert Myles
Bennett, Michael Vander Laan
Bernard, Rudy Andrew
Bignall, Keith E
Birzis, Lucy
Blankenship, James Emery
Bloedel, James R
Bonkalo, Alexander
Borda, Robert Paul
Boudreau, James Charles
Bowerman, Robert Francis
Boyarsky, Louis Lester
Brooks, Vernon Bernard
Broughton, Roger James
Brown, George Wallace
Brown, Joel Edward
Brust-Carmona, Hector
Buchwald, Jennifer S
Buchwald, Nathaniel Avrom
Burchfiel, James Lee
Burton, Harold
Calvin, William Howard
Camhi, Jeffrey Martin
Carpenter, David O
Carpenter, Richard Chester
Caveness, William Fields
Chapple, William Dismore
Christensen, Burgess Nyles
Clark, Francis John
Clark, George
Coceani, Flavio
Coggeshall, Richard E
Cohen, Alfred Cornell
Cohen, David Harris
Cohen, Lawrence Baruch
Cohen, Morton Irving
Cooke, Ian McLean
Cooper, Gary Pettus
Crain, Stanley M
Cuningham, James Gordon
Dafny, Nachum
Davidoff, Robert Alan
Davidson, Julian M
Davis, Earle Andrew, Jr
Davis, Hallowell
Daw, Nigel Warwick
Del Castillo, Jose
Delgado, Jose Manuel Rodriguez
Dement, William Charles
Demetrescu, M
Dory, Robert William
Drake, Rosemarie Sheila
Dudar, John Douglas
Duffy, Frank Hopkins
Durham, Ross M
Dykes, Robert William
Eagles, Douglas Alan
Easton, Dexter Morgan
Eaton, Douglas Charles
Eckert, Roger Otto
Edinger, Henry Milton
Egger, Maurice David
Eisenman, Joseph Sol
Eldred, Earl
Engel, Jerome, Jr
Epstein, Alan Neil
Erulkar, Solomon David
Evoy, William (Harrington)
Eyzaguirre, Carlos
Fairchild, Mahlon David
Farley, Roger Dean
Fernandez-Guardiola, Augusto
Fertziger, Allen Philip
Fields, Howard Lincoln
Frank, Karl
Frazier, Donald Tha
Fredrickson, John Murray
Freed, Simon
Freeman, John A
Fuortes, Michelangelo Giorgio F
Fuster, Joaquin Maria
Galindo, Cesar
Galeano, Anibal H
Garcia-Rill, Edgar E
Gardner, Esther Polinsky
Garoutte, Bill (Charles)
Gasteiger, Edgar Lionel, Jr
Gebber, Gerard L
Gelperin, Alan
Gesteland, Robert Charles
Gibson, John Michael
Gibson, Mary Morton
Gillary, Howard L
Glantz, Raymon M
Gloor, Pierre
Gold, Richard Michael
Goldberg, Jay M
Gootman, Phyllis Myrna Adler
Grant, Ronald
Grinnell, Alan Dale
Guinan, John Joseph, Jr
Guttman, Rita
Hackett, John Taylor
Hagiwara, Susumu
Hanna, George R
Hancock, Michael B
Happel, Leo Theodore, Jr
Hartline, Daniel Keffer
Hartline, Peter Haldan
Havlicek, Viktor
Henneman, Elwood
Heydemann, Peter Ludwig Martin
Hind, Joseph Edward
Hobson, John Allan
Hockman, Charles Henry
Hoff, Ebbe Curtis
Holloway, James Ashley
Horch, Kenneth William
Horowitz, John M
Horvath, Fred Ernest
Hubel, David Hunter

Hughes, John Russell
Humphrey, Donald R
Hunt, Carlton Cuyler
Jabbur, Suhayl Jibra'il
Jacklet, Jon Willis
Jasper, Herbert Henry
Javel, Eric
John, E Roy
Johnson, Jeffery Lee
Johnson, Kenneth Olafur
Kahan, Linda Beryl
Kashin, Philip
Kater, Stanley B
Katz, George Maxim
Katz, Sidney
Kelly, Dennis D
Kennedy, Duncan Tilly
Kennedy, Thelma Temy
Kiang, Nelson Yuan-Sheng
King, Jerome Stovall
King, Robert Lee
Knapp, Francis Marion
Knott, John Russell
Knox, Charles Kenneth
Koella, Werner Paul
Koopowitz, Harold
Krauhamer, George Michael
Kreisman, Norman Richard
Krnjevic, Kresimir
Kuffler, Stephen William
Kusano, Kiyoshi
Lamarche, Guy
Lange, Gordon David
Laskowski, Michael Bernard
Latimer, Clinton Narath
Lavine, Robert Alan
Lawry, James Voris
LeBlanc, Francis Ernest
Lebovitz, Robert Mark
Lesse, Henry
Letbetter, William Dean
Lettvin, Jerome Y
Levine, Leonard
Li, Choh-Luh
Liberson, Wladimir Theodore
Liu, Chan Nao
Livingston, Robert Burr
Loewenstein, Werner Randolph
Low, Morton David
Lowy, Karl
Lucas, Edgar Arthur
Luco, Joaquin
Lynch, James Carlyle
MacLean, Paul Donald
Magleby, Karl LeGrande
Mailman, Richard Bernard
Mann, Diana Witherspoon
Mann, Michael David
Martin, Alexander Robert
Masserman, Jules Homan
Mayer, David Jonathan
Mayeri, Earl Melchior
Mayor, Stephen Joseph
McArdle, Joseph John
McGraw, Charles Patrick
McMillan, James Alexander
Mellon, DeForest, Jr
Merlis, Jerome K
Merzenich, Michael Matthias
Meyers, H Russell
Michael, Charles Reid
Michael, Joel Allen
Mikiten, Terry Michael
Miller, Arthur Joseph
Miller, Larry O'Dell
Miskimen, George William
Mistretta, Charlotte Mae
Molnar, Charles Edwin
Moore, John Wilson
Moore, Lee E
Morgane, Peter J
Morin, Walter Arthur
Morison, Robert Swain
Moss, Robert L
Mote, Michael Isnardi
Murray, George Cloyd
Nakajima, Shigehiro
Narahashi, Toshio
Nathan, Marc A
Nelson, Margaret Christina
Nicholls, John Graham
Nicoll, Roger Andrew
Nyquist, Judith Kay
Ochs, Sidney
O'Connell, Robert James
O'Leary, Dennis Patrick
Orr, Alfonso
Page, Charles Henry
Paika, John Milan
Palmer, Dwight Miller
Palmer, Larry Alan
Parsons, L Claire
Partridge, Lloyd Donald
Pasik, Pedro
Pasik, Tauba
Peacock, Samuel Moore, Jr
Pearson, Keir Gordon
Perret, George (Edward)

Perryman, James Harvey
Peterson, Barry Wayne
Pfaff, Donald Wells
Phillis, John Whitfield
Pickens, Peter E
Pilar, Guillermo Roman
Pinto, Lawrence Henry
Pomeranz, Bruce Herbert
Poppele, Richard E
Pos, Robert
Poulos, Dennis A
Pozos, Robert Steven
Price, Donald Dennis
Pubols, Benjamin Henry, Jr
Pubols, Lillian Menges
Purple, Richard L
Purpura, Dominick Paul
Randic, Mirjana
Ratliff, Floyd
Redgate, Edward Stewart
Reitan, Ralph Meldahl
Retzlaff, Ernest (Walter)
Rhode, William Stanley
Robbins, Norman
Robinson, David Adair
Robinson, David Lee
Rodin, Martha Kinscher
Roig, Juan Antonio
Rose, Jerzy Edwin
Rosenthal, Fred
Rovainen, Carl (Marx)
Rowe, Edward C
Rowell, Charles Hugh Fraser
Ruch, Theodore Cedric
Rudenberg, Frank Hermann
Rudomin, Pedro
Ruggero, Mario Alfredo
Ryall, Ronald W
St John, Walter McCoy
Sasaki, Clarence Takashi
Sato, Makoto
Saunders, Frank Austin
Saunders, Virginia Fox
Schanberg, Saul M
Scheibel, Arnold Bernard
Schlag, John
Schor, Robert Hyllel
Scott, Donald Jr
Scott, John Wilson
Segundo, Jose Pedro
Selverston, Allen Israel
Sessle, Barry John
Shankland, Daniel Leslie
Shapiro, Bert Irwin
Shepherd, Gordon Murray
Sherman, Robert George
Siminoff, Robert
Sinclair, John C
Smith, James Darrell
Smith, Thomas Graves, Jr
Soechting, John F
Soltysik, Szczesny Stefan
Speck, Louise Barrett
Spray, David Conover
Steinberg, Roy Herbert
Steriade, Mircea
Stevens, Charles F
Streicher, Eugene
Swett, John Emery
Symmes, David
Szumski, Alfred John
Talbot, William Henry
Talbot, Richard Evans
Tang, Pei Chin
Tapper, Daniel Naptali
Tasaki, Ichiji
Taub, Arthur
Tessel, Richard Earl
Thompson, William D
Thorson, John Wells
Tietz, William John, Jr
Trevino, Daniel Louis
Trubatch, Janett
Van Buren, John Miller
Van Sluyters, Richard Charles
Velez, Samuel Jose
Vidoli, Vivian Ann
Villablanca, Jaime Rolando
Von Baumgarten, Rudolf Jury
Wagman, Irving Henry
Walter, Donald Oliver
Webb, George Dayton
Welker, Carol
Welker, Wallace I
Werman, Robert
Wetzel, Allan Brooke
White, Robert J
Whitehorn, David
Whitsel, Barry L
Willis, William Darrell, Jr
Williston, John Stoddard
Willows, Arthur Owen Dennis
Wilson, David Franklin
Wilson, Victor Joseph
Wolfe, James Wallace
Wolin, Lee Roy
Woody, Charles Dillon
Woolsey, Clinton Nathan

Wulff, Verner John
Wurtz, Robert Henry
Wylie, Richard Michael
York, Donald Harold
Zabara, Jacob
Zajac, Felix Edward, III
Zimmerman, Irwin David

Neuropsychiatry

Blass, John P
Boshes, Louis D
Fink, Max
Freeman, Leslie Sherwood
Glusman, Murray
Henn, Fritz Albert
Holloway, Harry Charles
Lal, Samarthji
Moore, Matthew Thibaud
Ng, Lorenz Keng-Yong
Rioch, David McKenzie
Urse, Vladimir George
Weinstein, Edwin Alexander

Neuropsychology

Anderson, Kenneth Verle
Beach, Frank Ambrose
Benton, Arthur Lester
Bernstein, Stephen
Bland, Brian Herbert
Boll, Thomas Jeffrey
Brauh, Steven Earle
Caplan, Paula Joan
Cleeland, Charles Samuel
Clynes, Manfred
Cranford, Jerry L
Creel, Donnell Joseph
Eason, Robert Gaston
Efron, Robert
Fitzhugh-Bell, Kathleen
Flanigan, William Francis, Jr
Frommer, Gabriel Paul
Ganchrow, Donald
Gerken, George Manz
Green, Thomas Kerr
Hall, William Charles
Hartlage, Lawrence Clifton
Hillyard, Steven Allen
Jackson, William James
Junge, Douglas
Justesen, Don Robert
King, Frederick Alexander
Klove, Hallgrim
Knapp, Theodore Martin
Kohn, Herbert Myron
Kupfermann, Irving
Laudenslager, Mark LeRoy
Logan, Cheryl Ann
Lynch, Wesley Clyde
Marsh, Gayle G
Matthews, Charles George
Meier, Manfred John
Mellinger, Melvin Wayne
Mendell, Lorne Michael
Miller, Lyle Herbert
Milner, Brenda (Atkinson)
Paschke, Richard Eugene
Peeler, Dudley Flavus, Jr
Pribram, Karl Harry
Rabe, Ausma
Rhodes, John Marshell
Ritter, Walter Paul
Rosenblum, William I
Sainsbury, Robert Stephen
Schwartz, Melvin Lewis
Settle, Richard Gregg
Singh, Pauline Jirik
Smith, Charles James
Smith, Nicole Schupf
Snyder, Daniel Raphael
Sterman, Maurice B
Storms, Lowell H
Stricker, Edward Michael
Taub, Edward
Thomas, Garth Johnson
Thompson, Robert
Veale, Warren Lorne
Wenzel, Bernice Martha (Mrs Wendell E Jeffrey)
Whitaker, Harry Allen
Witelson, Sandra Freedman
Zaidel, Eran
Zolman, James F

Neurosurgery

Adams, John Edwin
Alexander, Eben, Jr
Allen, Marshall B, Jr
Andy, Orlando Joseph
Arnold, Arthur
Bakay, Louis
Balagura, Saul
Battista, Arthur Francis
Bell, Robert Lloyd
Bering, Edgar Andrew, Jr
Black, Perry
Bloor, Byron Michel
Boop, Warren Clark, Jr
Botterell, Edmund Harry

Brodkey, Jerald Steven
Browder, Eli Jefferson
Bucy, Paul Clancy
Campbell, Robert Louis
Carrea, Raul
Carton, Charles Allan
Chou, Shelley Nien-Chun
Clark, William Kemp
Collins, William Francis, Jr
Cook, Albert William
Cooper, Irving S
Davis, Richard A
DeSaussure, Richard Laurens, Jr
Donaghy, Raymond Madiford Peardon
Dugger, Gordon Shelton
Dunbar, Howard Stanford
Ehni, George (John)
Elvidge, Arthur Roland
Erickson, Theodore Charles
Feindel, William Howard
Ferguson, Gary Gilbert
Fisher, Robert George
Foltz, Eldon Leroy
French, John Douglas
French, Lyle Albert
Furlow, Leonard Thompson
Galbraith, James Garber
Gallo, Anthony Edward, Jr
Garcia-Bengochea, Francisco
Gildenberg, Philip Leon
Golding, Sidney
Grossman, Robert G
Hamlin, Hannibal
Heifetz, Milton David
Heimburger, Robert Francis
Hekmatpanah, Javad
Hetherington, R F
Houseplan, Edgar M
Huertas, Jorge
Hunt, William Edward
Jakoby, Ruth Elizabeth Kerr
Jane, John Anthony
Javid, Manucher J
Kalsbeck, John Edward
Kaplan, Harry Arthur
Kelly, William Albert
Kerr, Frederick William Lawson
Kindt, Glenn W
Kirsch, Wolff M
Kline, David G
Kurze, Theodore
Langfitt, Thomas William
Llewellyn, Raeburn Carson
Long, Donlin Martin
Lourie, Herbert
Luessenhop, Alfred John
MacCarty, Collin Stewart
Malis, Leonard I
Manganiello, Louis O J
McLaurin, Robert L
Mealey, John, Jr
Meyer, Glenn Arthur
Moody, Robert Adams
Morley, Thomas Paterson
Mount, Lester Adran
Mullan, John F
Murtagh, Frederick, Jr
Myers, Paul Walter
Nashold, Blaine S
Nugent, George Robert
O'Connor, Michael John
Odom, Guy Leary
Otenasek, Frank (Joseph)
Owens, Guy
Paine, Kenneth William
Patterson, Russel Hugo, Jr
Perot, Phanor L, Jr
Pollay, Michael
Poppen, James L
Porter, Robert Willis
Pudenz, Robert Harry
Puletti, Flavio
Raimondi, Anthony John
Ransohoff, Joseph
Rayport, Mark
Rhoton, Albert Loren, Jr
Richardson, Donald Edward
Robertson, James Thomas
Rovit, Richard Lee
Ruge, Daniel
Russell, John Robert
Salmon, James Henry
Schemm, George Walker
Schlesinger, Edward Bruce
Schneider, Richard Coy
Schwartz, Henry Gerard
Scott, Michael
Scoville, William Beecher
Shenkin, Henry A
Smith, Kenneth Rupert, Jr
Stein, Bennett M
Story, Jim Lewis
Stratford, Joseph
Sugar, Oscar
Sweet, William Herbert
Sypert, George Walter
Taren, James A

NEUROSCIENCES

Tasker, Ronald Reginald
Tator, Charles Haskell
Thomas, Llywellyn Murray
Thompson, Raymond K
Tindall, George Taylor
Udvarhelyi, George Bela
Voris, Harold Cornelius
Walker, Michael Dirck
Ward, Arthur Allen, Jr
Weissman, William Kent
Westrum, Lesnick Edward
Wetzel, Nicholas
Whisler, Walter William
White, Lowell Elmond, Jr
Wise, Burton Louis
Wolfson, Sidney Kenneth, Jr
Woodhall, Barnes
Yamamoto, Y Lucas
Yashon, David
Youmans, Julian Ray

OCEANOGRAPHY

Oceanography

Abbott, William Harold
Albright, Lawrence John
Alexander, James E
Allen, Louis Pinckney Jr
Anderson, Franz Elner
Ankouchine, William A
Arthenius, Gustaf Olof Svante
Arthur, Robert Siple
Aunderson, Aubrey Lee
Baer, Ledolph
Bainbridge, Arnold Ernest
Barkley, Richard Andrew
Barlow, John Peleg
Barnes, Clifford Adrian
Barrientos, Celso Saquitan
Barlow, Charles Carpenter
Bascom, Willard
Bates, Charles Carpenter
Bennett, John Richard
Benson, Richard Hall
Benson, Bruce Buzzell
Berger, Wolfgang Helmut
Berthof, Lloyd Bernard
Bieri, Robert
Blanchard, Duncan Cromwell
Bowman, James Floyd, II
Boyd, James Emory
Buddemeier, Robert Worth
Burling, Ronald William
Burns, Robert Earle
Burrell, Victor Gregory, Jr
Burt, Wayne Vincent
Bush, James
Byrne, Hugh Michael
Byrne, John Vincent
Byrne, Robert Howard
Byrnes, Bernard Christopher
Caldwell, Douglas Ray
Campbell, William Joseph
Caperon, John
Carr, Jerome Brian
Carritt, Dayton Ernest
Carsola, Alfred James
Chamberlain, Theodore Kiock
Chew, Frank
Churgin, James
Chute, John Lawrence, Jr
Clarke, Thomas Arthur
Coachman, Lawrence Keyes
Coleman, Charles R
Collias, Eugene Evans
Cononos, Tasso John
Corcoran, Eugene Francis
Cox, Charles Shipley
Craig, Harmon
Deffeyes, Kenneth Stover
Dehlinger, Peter
De Leonibus, Pasquale S
Devereux, Robert Francis
Dietz, Robert Sinclair
Doe, Learmont Anstice Earlston
Drummond, Kenneth Herbert
Dubbeldey, Pieter Steven
Duedall, Iver Warren
Dunbar, Maxwell John
Dunsch, Harry Robert
Duxbury, Alyn Crandall
Ebbesmeyer, Curtis Charles
Eliot, Joe Oliver
Ellis, Roy
El Wardani, Sayed Aly
Enns, Theodore
Ewing, Gifford Cochran
Fahlquist, Davis A
Felsher, Murray
Fleming, Richard Howell
Folsom, Theodore Robert
Ford, William Livingstone
Forrester, Warren David
Foster, Theodore Dean
Fowlis, William Webster

Franceschini, Guy Arthur
Fry, John Craig
Fuglister, Frederick Charles
Galvin, Cyril Jerome, Jr
Garrison, Louis Eldred
Garstang, Michael
Gast, James Avery
Gereben, Istvan B
Geyer, Richard Adam
Gibson, Carl H
Giese, Graham Sherwood
Grigg, Richard Wyman
Gross, Meredith Grant, Jr
Hadlock, Ronald K
Haney, Robert Lee
Hansen, George Lyle
Hardy, Wilton Audobon
Harlett, John Charles
Harris, D Lee
Harris, Russell G
Harriss, Robert Curtis
Hayes, David Wayne
Hayes, Miles O
Healy, Michael L
Heath, G Ross
Henry, Vernon James, Jr
Hess, George Dale
Hicks, Steacy Dopp
Hirshman, Julius
Hood, Donald Wilbur
Hsueh, Ya
Huang, Joseph Chi Kan
Hunkins, Kenneth
Hunt, Lee McCaa
Hunter, John Robert
Hyndman, Roy D
Isaacs, John Joseph, II
Jacobs, Marian Beckmann
Jacobs, Woodrow Cooper
Jennings, Charles David
Jennings, Feenan Dee
Johnson, David Ashby
Johnson, Harry McClure
Johnston, James Howard
Jung, Glenn Harold
Kaye, George Thomas
Keen, Michael J
Kelley, James Charles
Kelley, John Charles
Klein, George deVries
Klemas, Vytas
Knauss, John Atkinson
Knox, Cameron
Komar, Paul D
Krause, Dale Curtiss
Kroopnick, Peter Michael
Laevastu, Taivo
La Fond, Eugene Cecil
Lam, Ronald Ka-Wei
Langseth, Marcus G, Jr
Larsen, Lawrence Harold
Laurs, Robert Michael
Lauzier, Louis Marcel
Laws, Edward Allen
Leipper, Dale F
Lewis, Edward Lyn
Lingren, Wesley Earl
Linnenbom, Victor John
Littlewood, William Herbert
Loomis, Alden Albert
Loring, Arthur Paul
Ludeke, Carl Arthur
Mackenzie, Kenneth Victor
Markham, James J
Martin, John Holland
Mathewson, James H
Matthews, John Brian
Maughan, Paul McAlpine
Maxwell, Arthur Eugene
McAlice, Bernard John
McAllister, Raymond Francis
McBlair, William
McFadden, James Douglas
McGeary, David F R
McGill, David A
McLellan, Hugh John
McManus, Dean Alvis
Menzel, David Washington
Menzies, Robert James
Metcalf, William Gerrish
Meyers, Philip Alan
Millman, John D
Miyake, Mikio
Molo-Christensen, Erik Leonard
Molo, William L
Monahan, Edward Charles
Moore, Theodore Carlton, Jr
Morelock, Jack
Morita, Richard Yukio
Munday, John Clingman, Jr
Myer, Glenn Evans
Myers, Thomas DeWitt
Natarajan, Kottayam Viswanathan
Neumann, A Conrad
Neumann, Paul Gerhard
Nixon, Scott West

Noakes, John Edward
Normark, William Raymond
North, Wheeler James
Nutt, David Clark
Nygren, Paul W
O'Brien, James J
Olson, Howard Thomas
Olson, Boyd E
Oostdam, Bernard Lodewijk
Oppenheimer, Carl Henry, Jr
Osterberg, Charles Lamar
Owen, Robert W, Jr
Pandolfo, Joseph P
Park, Paul Kilho
Parr, Albert Eide
Parsons, Timothy Richard
Pasby, Brian
Paulson, Clayton Arvid
Pedlosky, Joseph
Pequegnat, Willis Eugene
Perkins, William Enfield
Phleger, Fred B
Pickard, George Lawson
Pierson, Willard James, Jr
Piper, David Zink
Pirie, Robert Gordon
Platzman, George William
Polcyn, Fabian Casimir
Posner, Gerald Seymour
Powell, Allen LaRue
Price, William Armstrong
Pritchard, Donald William
Proni, John Roberto
Pytkowicz, Ricardo Marcos
Ragotzkie, Robert Austin
Rao, Desraju Bhavanarayana
Rao, P Krishna
Read, Robert G
Redfield, Alfred Clarence
Reid, Joseph Lee
Reid, Robert Osborne
Revelle, Roger (Randall Dougan)
Richards, Adrian Frank
Richards, Francis Asbury
Riley, Gordon Arthur
Rinkel, Murice O
Roach, David Michael
Robinson, Margaret King
Rucker, James Bivin
Ryan, William B F
Salo, Ernest Olavi
Sarachik, Edward S
Sawyer, Constance B
Scafe, Donald William
Schaffel, Simon
Schmitt, Walter R
Sclater, John George
Seckel, Gunter Rudolf
Seibel, Erwin
Seymour, Richard Jones
Shabica, Charles Wright
Shenton, Edward Heriot
Shevenell, Thomas Cortland
Simmons, William Frederick
Smith, Paul Albert
Smith, Stuart D
Spence, Thomas Wayne
Spencer, Derek W
Spilhaus, Athelstan Frederick
Spilhaus, Athelstan Frederick, Jr
Starr, Robert Brewster
Sternberg, Richard
Stevenson, Robert Everett
Stewart, Harris Bates, Jr
Stommel, Henry Melson
Strick, Ellis
Strong, Alan Earl
Stubblefield, Bonnie McGregor
Suhayda, Joseph Nicholas
Sund, Paul N
Swift, Donald J P
Taber, Walter William
Taylor, Walter Rowland
Terry, Richard D
Thiruvathukal, John Varkey
Thompson, Geoffrey
Thompson, Warren Charles
Tibby, Richard Bitner
True, Merrill Allan
Tully, John Patrick
Uchida, Richard Noboru
Van Sciver, Wesley J
Veronis, George
Vetter, Richard C
Von Arx, William Stelling
Waldichuk, Michael
Wallen, Irvin Eugene
Walsh, John Joseph
Wangersky, Peter John
Warsh, Catherine Evelyn
Wearn, Richard Benjamin, Jr
Weihaupt, John George
Weisberg, Joseph Simpson
Weyl, Peter K
Whetten, John T
Wilde, Pat
Wilkniss, Peter Eberhard
Williams, Albert James, III

Williams, David Lee
Williams, Robert Glenn
Windom, Herbert Lynn
Wiseman, William Joseph, Jr
Wollin, Goesta
Woodcock, Alfred Herbert
Wooster, Warren Scriver
Worthington, Lawrence Valentine
Wunsch, Carl Isaac
Yalkovsky, Ralph
Young, David Ross
Zaneveld, Jacques Ronald Victor
Zeiter, Bernard David

Biological Oceanography

Alvarino de Leira, Angeles
Alverson, Dayton L
Anderson, Lars William James
Aron, William James
Ayers, John Carr
Backus, Richard Haven
Banner, Albert Henry
Banse, Karl
Beers, John R
Berner, Leo Dewitte, Jr
Best, Edgar Allan
Bienfang, Paul Kenneth
Bigelow, Maurice Hubbard
Blanton, William George
Blumenthal, Reuben R
Boden, Brian Peter
Boesch, Donald Friedrich
Boyd, Carl M
Bradshaw, John Stratili
Bright, Thomas J
Brinton, Edward
Brooks, Albert Law
Brown, Dail Woodward
Brunhugh, Joe H
Bullis, Harvey Raymond, Jr
Bush, Louise Fulton
Butler, Philip Alan
Byrd, Isaac Burlin
Calder, Dale Ralph
Carey, Andrew Galbraith, Jr
Carlisle, David B
Carlisle, John Griffin, Jr
Carpenter, Edward J
Carriker, Melbourne Romaine
Carter, John C H
Cheng, Paul James
Chew, Kenneth Kendall
Collier, Albert Walker
Coull, Bruce Charles
Courtenay, Walter Rowe, Jr
Crabtree, David Melvin
Crandell, George Frank
Cummings, William Charles
Curl, Herbert (Charles), Jr
Cutler, Edward Bayler
Davis, Curtiss Owen
Davis, David
Deevey, Georgiana Baxter
Demond, Joan
de Sylva, Donald Perrin
Devol, Allan Houston
Dickie, Lloyd Merlin
Digby, Peter Saki Bassett
Di Girolamo, Rudolph Gerard
Dobkin, Sheldon
Dorfman, Donald
Dugdale, Richard Cooper
Dunstan, William Morgan
Eisler, Ronald
Eldredge, Lucius G
Elsner, Robert
English, Thomas Saunders
Epifanio, Charles Edward
Eppley, Richard Wayne
Essaias, Wayne Evor
Fast, Thomas Normand
Fehr, Roger Richard
Feder, Howard Mitchell
Fehlmann, Herman Adair
Fish, James Franklin, Jr
Fish, Marie Poland
Fowler, Scott Wellington
Fox, Denis Llewellyn
Frankenberg, Dirk
Frolander, Herbert Farley
Frost, Bruce Wesley
Fryxell, Greta Albrecht
Gaskin, David Edward
Gilbert, Perry Webster
Gilmartin, Malvern
Glooschenko, Walter Arthur
Glude, John Bryce
Gonzalez, Juan Gerardo
Gordon, Bernard Ludwig
Graham, Jeffrey Brent
Grainger, Edward Henry
Grant, George C
Grassle, John Frederick
Green, John M
Grindis, Richard Byron
Guillard, Robert Russell Louis
Haderlie, Eugene Clinton
Haefner, Paul Aloysius, Jr

Haffner, Rudolph Eric
Hanks, James Elden
Hargis, William Jennings, Jr
Hargraves, Paul E
Hart, Josephine Frances Lavinia
Hartwick, Robert Frank
Hastings, John Woodland
Hayes, Helen Landau
Helfrich, Philip
Herman, Sidney Samuel
Hildebrand, Henry H
Hillis-Colinvaux, Llewellya Williams
Hoese, Hinton Dickson
Holmes, Robert W
Holton, Robert Lawrence
Hopkins, Sewell Hepburn
Hopkins, Thomas Sterling
Hulsemann, Kuni
Humm, Harold Judson
Hunter, John Roe
Ingle, Robert Maurice
James, Bela Michael
Jones, Everet Clyde
Jones, Galen Everts
Jumars, Peter Alfred
Kabata, Zbigniew
Kamykowski, Daniel
Kelly, Mahlon George, Jr
Ketchum, Bostwick Hawley
Kielhorn, William Vineyard
Knowlton, Robert Earle
Kohler, Carl
Kraeuter, John Norman
Lalli, Carol Marie
Larsen, Peter Foster
Leavitt, Benjamin Burton
Lewis, Alan Graham
Lewis, John Bradley
Lippson, Robert Lloyd
Littlepage, Jack Leroy
Loesch, Harold Carl
Loesch, Joseph
Lynch, Maurice Patrick
MacKenzie, Clyde Leonard, Jr
Mansfield, Arthur Walter
Manzer, James Ivan
Marshall, Harold George
Marshall, Nelson
Maturo, Frank Juan Sarno, Jr
McCain, John Charles
McCarthy, Francis Davey
McCarthy, James Joseph
McCauley, James Elias
McConnaughey, Bayard Harlow
McGowan, John Arthur
McLean, Richard Bea
McRoy, C Peter
Menzel, Robert Winston
Merriman, Daniel
Miller, Charles Benedict
Mills, Eric Leonard
Mitsui, Akira
Moeller, Henry William
Mohr, John Luther
Morgan, David William
Moser, H Geoffrey
Mullin, Michael Mahlon
Nace, Paul Foley
Nakamura, Eugene Leroy
Napora, Theodore Alexander
Needler, Alfred Walker Hollinshead
Newman, William Anderson
Nichols-Driscoll, Jean Ann
Nuzzi, Robert
Orth, Robert Joseph
Otsu, Tamio
Otto, David Arthur
Packard, Theodore Train
Pamatmat, Mario M
Pearse, John Stuart
Pequegnat, Linda Lee Haithcock
Perkins, Frank Overton
Perry, Mary Jane
Phelps, Donald Kenneth
Phillips, David William
Pickwell, George Vincent
Pieper, Richard Edward
Pilson, Michael Edward Quinton
Pratt, David Mariotti
Price, Kent Sparks, Jr
Pulley, Thomas Edward
Rae, Kenneth MacFarlane
Randall, John Ernest
Rankin, John Stewart, Jr
Ray, Dixy Lee
Ray, Sammy Mehedy
Raymond, Anne Frances
Rebach, Steve
Renfro, William Charles
Richards, Thomas L
Rodriguez, Gilberto
Roels, Oswald A
Roper, Clyde Forrest Eugene
Ross, June Rosa Pitt
Roth, Ariel A
Rowe, Gilbert Thomas
Ruggieri, George D
Sastry, Akella N

Scarratt, David Johnson
Scheltema, Rudolf S
Schoener, Amy
Sears, Mary
Shabica, Stephen Vale
Sherman, Kenneth
Shih, Chang-Tai
Silver, Mary Wilcox
Sinderman, Carl James
Skud, Bernard Einar
Small, Lawrence Frederick
Smayda, Theodore John
Smith, Roberta Katherine
Smith, Vann Elliott
Soule, Dorothy (Fisher)
Sparks, Albert Kirk
Spies, Robert Bernard
Squires, Donald Fleming
Staiger, Jon Crawford
Stevenson, James Cameron
Strathmann, Richard Ray
Swift, Dorothy Garrison
Swift, Elijah, V
Taft, Jay Leslie
Taylor, Frank John Rupert (Max)
Teal, John Moline
Templeman, Wilfred
Tenore, Kenneth Robert
Thomas, Lowell Phillip
Thomas, William Hewitt
Thompson, Rosemary Ann
Thurberg, Frederick Peter
Tibby, Richard Bitner
Tietjen, John H
Trench, Robert Kent
True, Renate (Schlenz)
Van Dyke, Henry
Van Engel, Willard Abraham
Vernberg, Frank John
Voss, Gilbert Lincoln
Wacasey, Jervis Winn
Wade, Richard Archer
Weaver, Sylvia Short
Welch, Walter Raynes
Wells, John Morgan, Jr
West, Arthur James, II
Westley, Ronald E
Wheeler, Ellsworth Haines, Jr
Wilkie, Donald W
Williams, Francis
Wilson, William Buford
Winn, Howard Elliott
Wulff, Barry Lee
Yang, Won-Tack
Yentsch, Charles Samuel
Young, Richard Edward
Zaneveld, Jacques Simon
Zubkoff, Paul Leon

Chemical Oceanography
Andersen, Neil Richard
Anderson, James Jay
Atkinson, Larry P
Atwood, Donald Keith
Burrell, David Colin
Calder, John Archer
Carpenter, Roy
Chau, Yiu-Kee
Codispoti, Louis Anthony
Cutshall, Norman Hollis
Fitzgerald, William Francis
Gardner, Wayne Stanley
Gordon, Louis Irwin
Green, Edward Jewett
Hochman, Harry
Kester, Dana R
Krause, Daniel, Jr
Lyman, John
MacIntyre, Ferren
Martens, Christopher Sargent
Mattson, James Stewart
Neihof, Rex A
Ostlund, H Gote
Owen, Robert Michael
Phillips, Timothy Dukes
Presley, Bobby Joe
Reeburgh, William Scott
Schink, David R
Sharp, Jonathan Hawley
Shaw, David George
Siegel, Alvin
Stansby, Maurice Earl
Swinnerton, John W
Tan, Francis C
Walton, Alan
Warner, Theodore Baker
Williams, Peter M
Wolfe, Douglas Arthur
Yamamoto, Sachio
Zirino, Alberto
Zuehlke, Richard William

Geological Oceanography
Anderson, John B
Andrews, James Einar
Andrews, Robert Sanborn
Bailey, James Stuart
Behrens, Earl William

Berry, Richard Warren
Biggs, R B
Bonatti, Enrico
Bouma, Arnold Heiko
Bryant, William Richards
Buffington, Edwin Conger
Carlson, Paul Roland
Christofferson, Eric
Cleary, William James
Creager, Joe Scott
Curray, Joseph Ross
Difford, Winthrop Cecil
Dill, Robert Floyd
Dillon, William Patrick
Donohue, John J
Doyle, Larry James
Duane, David Bierlein
Edgar, Norman Terence
Emery, Kenneth Orris
Field, Michael Ehrenhart
Fink, Loyd Kenneth, Jr
Fisher, Robert Lloyd
Fleischer, Peter
Folger, David W
French, William Edwin
Galehouse, Jon Scott
Gardner, James Vincent
Gartner, Stefan, Jr
Glass, Billy Price
Goldsmith, Victor
Gorsline, Donn Sherrin
Grant, Alan Carson
Griffin, George Melvin, Jr
Griggs, Gary B
Grinnell, Robert S, Jr
Hamilton, Edwin Lee
Harding, James Lombard
Holmes, Charles Ward
Hulsemann, Jobst
Hurley, Robert Joseph
Hyne, Norman John
Inderbitzen, Anton Louis, Jr
Johnson, George Leonard
Jones, James I
Keller, George H
Komar, Paul D
Kroenke, Loren William
Kulm, Laverne Duane
Lavelle, John William
Lill, Gordon Grigsby
Lins, Thomas Wesley
Loughridge, Michael Samuel
Ludwick, John Calvin, Jr
Marlowe, James Irvin
Matthews, Jerry Lee
McIntyre, Andrew
Meisburger, Edward Paul
Meyerson, Arthur Lee
Molnia, Bruce Franklin
Moore, David Gillis
Moore, George Thomas
Nayudu, Y Rammohanroy
Nichols, Maynard M
Oetking, Philip
Palmer, Harold Dean
Pelletier, Bernard Roderick
Peter, George
Peterson, Melvin Norman Adolph
Pilkey, Orrin H
Pyle, Thomas Edward
Rea, David Kenerson
Rezak, Richard
Rodolfo, Kelvin S
Rona, Peter Arnold
Ross, David A
Schmalz, Robert Fowler
Schmidt, Volkmar
Schneidermann, Nahum
Schubel, Jerry Robert
Sharma, Ghanshyam D
Shykind, Edwin B
Sly, Peter G
Smith, James Dungan
Smith, Riley Seymour, Jr
Southam, John Ralph
Stanley, Daniel Jean
Sternberg, Richard Walter
Stevens, Richard S
Taft, William H
Thayer, Paul Arthur
Thiede, Jorn
Vallier, Tracy L
Van Andel, Tjeerd Hendrik
Walton, William Ralph
Wang, Frank Feng Hui
White, Stanton M
Wimberley, Stanley
Wright, Frederick Fenning

Marine Botany
Eleuterius, Lionel Numa
Foster, Michael Simmler
Fralick, Richard Allston
Gallagher, John Leslie
Johansen, Hans William
Lee, Robert K S
Liddle, Larry Brook
Menez, Eranani Guingona

Newroth, Peter Russell
Norris, James Newcome, IV
Norris, Richard Earl
Phillips, Ronald Carl
Prezelin, Barbara Berntsen
Prowse, Gerald Albert
Silberhorn, Gene Michael
Sorensen, Lazem Otto
Wynne, Michael

Marine Ecology
Adelson, Lionel Morton
Ainley, David George
Aldrich, David Virgil
Bakus, Gerald Joseph
Bane, Gilbert Winfield
Banus, Mario Douglas
Baylor, Edward Randall
Boynton, Walter Raymond
Bright, Donald Bolton
Brunel, Pierre
Calabrese, Anthony
Carr, William Edward Statter
Cerwonka, Robert Henry
Chestnut, Alphonse F
Cirino, Elizabeth Fahey
Clark, Kerry Bruce
Croker, Robert Arthur
Cronin, Lewis Eugene
Dahl, Arthur Lyon
Dauer, Daniel Martin
Diaz, Robert James
Dragovich, Alexander
Druehl, Louis D
Dupuy, John L
Engstrom, Norman Ardell
Farris, Richard Austin
Flittner, Glenn Arden
Ford, Richard Fiske
Foreman, Ronald Eugene
George, Carl Joseph Winder
Glynn, Peter W
Goering, John James
Gotshall, Daniel Warren
Haines, Evelyn Brown
Hendler, Gordon Lee
Hidu, Herbert
Hollister, Charles Davis
Hurley, Ann Catherine
Hyer, Paul Vincent
Johannes, Robert Earl
Johnson, Samuel Edgar, II
Jones, Gilbert Fred
Kaufman, Donald Wayne
Kelso, Donald Preston
Kolipinski, Milton Charles
Lacroix, Guy
Lee, Welton Lincoln
Levandowsky, Michael
Lowe, Jack Ira
Marsh, James Alexander, Jr
Maurer, Donald Lee
Maynard, Nancy Gray
McAlister, William Bruce
McCormick, Jon Michael
Miller, Robert Joseph
Moore, Johnes Kittelle
Morris, Byron Frederick
Newkirk, Gary Francis
Nybakken, James W
Ogden, John Conrad
Osman, Richard William
Powles, Percival Mount
Rose, Theodore Roosevelt
Rose, Curt D
Scott, Kenneth John, Jr
Setzler, Eileen Marie
Singletary, Robert Lombard
Smith, David Francis
Stickney, Alden Parkhurst
Summers, William Clarke
Sutherland, John Patrick
Taylor, Peter Berkley
Templeton, William Lees
Terry, Orville Whitfield
Thomson, Donald A
Thum, Alan Bradley
Tolderlund, Douglas Stanley
Trott, Lamar Brice
Vadas, Robert Louis
Virnstein, Robert W
Warner, John Edward
Warner, Robert Ronald
Wennemer, Jay
Whitney, David Earle
Wigley, Roland L
Williams, Richard Birge
Willingham, Charles Allen
Woelke, Charles Edward
Wohlschag, Donald Eugene
Wood, Carl Eugene
Woodmansee, Robert Asbury

Marine Geochemistry
Berner, Robert A
Biscaye, Pierre Eginton
Burnett, William Craig
Chambers, Richard Lee

OCEANOGRAPHY

Dayal, Ramesh
Deuser, Werner Georg
Eberhardt, Robert Louis
Farrington, John William
Frakes, Lawrence Austin
Goldberg, Edward D
Huggett, Robert James
Keeling, Charles David
Lerman, Abraham
Loring, Douglas Howard
Man, Eugene Herbert
Mangelsdorf, Paul Christoph, Jr
Moore, Willard S
Ristvet, Byron Leo
Schultz, David Michael
Scott, Martha Richter
Whelan, Thomas, III
Wolgemuth, Kenneth Mark
Wood, Elwyn Devere

Marine Geography
Naidu, Angi Satyanarayan
Ray, John Bernard
Shor, George G, Jr

Marine Geophysics
Anderson, George Boine
Atwater, Tanya Maria
Behrendt, John Charles
Bennett, Lee Cotton, Jr
Frosch, Robert Alan
Geddes, Wilburt Hale
Grim, Paul J
Hammond, Stephen Randolph
Haworth, Richard Thomas
Hussong, Donald MacGregor
Johnson, Rockne Hart
Keen, Charlotte Elizabeth
Loncarevic, Bosko (D)
Ludwig, William Jackson
Luyendyk, Bruce Peter
Riggin, John Dewitt
Taylor, Patrick Timothy
Tiffin, Donald Lloyd
Vogt, Peter Richard
Von Herzen, Richard P
Wold, Richard John

Marine Meteorology
Caputo, Luis R A
Clark, Nathan Edward
Eber, Laurence Elwin
Faust, Maria Anna
Glenn, Alfred Hill
Nealson, Kenneth Henry

Marine Microbiology
Baross, John Allen
Barvenik, Frank W
Bell, Wayne Harrell
Bourquin, Al Willis J
Button, Don K
Gundersen, Kaare Reinhardt
Hanson, Roger Brian
Hayasaka, Steven S
Juge, Damian Marie
Lee, John Joseph
Liguori, Vincent Robert
Litchfield, Carol Darline
Meyers, Samuel Philip
Miget, Russell John
Prager, Jan Clement
Sieburth, John McNeill
Taylor, Barrie Frederick
Yonenaka, Hideo H

Marine Virology
Dearborn, John Holmes

Marine Zoology
Abbott, Marie Bohrn-Lambert
Baker, James Haskell
Banta, William Claude
Barnard, Jerry Laurens
Beeman, Robert D
Bookhout, Cazlyn Green
Castro, Peter
Cheng, Lanna
Costlow, John DeForest
Dexter, Deborah Mary
Fontaine, Arthur Robert
Gardiner, Lion Frederick
Gladfelter, William Bayard
Grice, George Daniel, Jr
Heavers, Barbara Ann
Lamberts, Austin E
Losey, George Spahr, Jr
Michel, Harding B
Miller, John Wesley, Jr
Ostarello, Georgiandra Little
Pawson, David Leo
Rausch, James Peter
Reish, Donald James
Sandifer, Paul Alan
Smith, Edmund Hobart
Smith, Frederick George Walton
Struhsaker, Paul James
Wood, Forrest Glenn

Physical Oceanography
Aagaard, Knut
Allender, James Harry
Apel, John Ralph
Bailey, William Best
Baker, Donald James, Jr
Beardsley, Robert Cruce
Beckerle, John C
Berman, Alan
Blandford, Robert Roy
Blanton, Jackson Orin
Bohlen, Walter Franklin
Boicourt, William Closson
Boston, Noel Edward James
Bowman, Malcolm James
Breeding, J Ernest, Jr
Bretschneider, Charles Leroy
Broida, Saul
Brown, Wendell Stimpson
Bryden, Harry Leonard
Bumpus, Dean Franklin
Callahan, Jeffrey Edwin
Callaway, Richard Joseph
Calman, Jack
Cameron, William Maxwell
Campbell, Neil John
Cannon, Glenn Albert
Caputo, Luis R A
Carder, Kendall L
Carter, Harry Hart
Chen, Davidson Tan-Chuen
Clayton, William Howard
Collins, Curtis Allan
Criminale, William Oliver, Jr
Davis, Russ S
Del Grosso, Vincent Alfred
Dorman, Clive Edgar
Duing, Walter
Elliott, Alan John
Emery, William Jackson
Garvine, Richard William
Godin, Gabriel
Gordon, Arnold L
Gossard, Earl Everett
Green, Theodore, III
Gregg, Michael Charles
Groves, Gordon William
Hacker, Peter Wolfgang
Halpern, David
Hansen, Donald Vernon
Hawkes, H H Bowman
Hickey, Barbara Mary
Hires, Richard Ives
Hueter, Theodor Friedrich
Huyer, Adriana (Jane)
Ichiye, Takashi
Ingham, Merton Charles
Irish, James David
Jefferson, Donald Earl
Johnson, Ronald Ernest
Joy, Joseph Wayne
Keen, Dorothy Jean
Kennedy, Robert E
Kielhorn, William Vineyard
Kirwan, Albert Dennis, Jr
Knowles, Charles Ernest
Knox, Robert Arthur
Korgen, Benjamin Jeffry
Kroll, John Ernest
Kuo, Albert Yi-Shuong
Kupferman, Stuart L
Lambert, Richard Bowles, Jr
LeBlond, Paul Henri
Lee, Fang An
Lee, Jack R
Lovett, Jack R
Luyten, James Reindert
Mader, Charles Lavern
Mandelbaum, Hugo
Mazeika, Paul A
McCartney, Michael Scott
McLean, Max C
Merrell, William John
Mesecar, Roderick Smit
Miller, Arthur R
Mofjeld, Harold Oswald
Montgomery, Raymond Braislin
Mooers, Christopher Northrup Kennard
Mortimer, Clifford Hiley
Muench, Robin Davie
Munk, Walter Heinrich
Murray, Stephen Patrick
Mysak, Lawrence Alexander
Neal, Victor Thomas
Neshyba, Steve
Noble, Vincent Edward
Nowlin, Worth D, Jr
Okubo, Akira
Olson, Franklyn C W
Paquette, Robert George
Paskausky, David Frank
Perkins, Henry Thomas
Philander, Samuel George
Pillsbury, Dale Ronald

Pinkel, Robert
Plant, William James
Pochapsky, Theodore Elias
Pond, George Stephen
Pounder, Elton Roy
Preisendorfer, Rudolph W
Quinn, William Hewes
Rattray, Maurice, Jr
Roden, Gunnar Ivo
Rossby, Thomas
Royer, Thomas Clark
Rufenach, Clifford L
Sanford, Thomas Bayes
Saunders, Kim David
Saur, Jesse Francis Theodore
Schmitz, William Joseph, Jr
Smith, James Dungan
Smith, Ned Philip
Smith, Raymond Calvin
Smith, Robert Lloyd
Stanley, Everett Michael
Steere, Richard C
Stevenson, Merritt Raymond
Sturges, Wilton, III
Swanson, Robert Lawrence
Tabata, Susumu
Taft, Bruce A
Tatro, Peter Richard
Thompson, Rory
Trites, Ronald Wilmot
Tsuchiya, Mizuki
Van Leer, John Cloud
Vastano, Andrew Charles
Verber, James Leonard
Vukovich, Fred Matthew
Walsh, Don
Warren, Bruce Alfred
Warsh, Kenneth Lee
Watts, Dennis Randolph
Webster, Ferris
Weisberg, Robert H
Wert, Richard Thomas
Wilkerson, John Christopher
Williams, Jerome
Williams, Robert Bruce
Witting, James M
Woodruff, Rodger King
Wright, William Redwood
Wyrtki, Klaus
Younce, Gordon Baldwin

OPTICS

Optics
Abramson, Edward
Allen, Richard Ballantine
Allen, Susan Davis
Ameer, George Albert
Andrews, Wayne Jay
Anderson, Ronald Allen
Arsenault, Henri H
Axelrod, Norman Nathan
Baker, James Gilbert
Ballard, Stanley Sumner
Barrett, Harrison Hooker
Bartels, Peter H
Bechtold, Edwin William
Becker, Randolph Armin
Beer, Reinhard
Benjamin, Roland John
Bennett, Harold Earl
Bennett, Jean McPherson
Benton, Stephen Anthony
Berlinghieri, Joel Carl
Bernal G, Enrique
Bernstein, Burton
Besancon, Robert Martin
Bettis, Jerry Ray
Billings, Bruce Hadley
Blaker, John Warren
Borelli, Nicholas Francis
Bottka, Nicholas
Boynton, Robert M
Bradley, Lee Carrington, III
Braithwaite, John Geden North
Brandt, Gerald Bennett
Breckinridge, James Bernard
Brousseau, Nicole
Browder, James Steve
Browell, Edward Vern
Brown, Douglas Edward
Brumley, Corwin Hoyt
Burdette, Ernest Linwood
Burke, James Joseph, Jr
Burris, Charles Andrew, Jr
Bystricky, Karl M
Calvert, James Bowles
Carleton, Herbert Ruck
Carman, Philip Douglas
Carswell, Allan Ian
Carter, William Harold
Chisholm, James Joseph
Christensen, Charles Richard
Chu, William Peter
Church, Charles Henry

Church, Eugene Lent
Close, Donald Henry
Cohen, Fredric Sumner
Condell, William John, Jr
Connolly, John Irving, Jr
Crittenden, Eugene Casson, Jr
Crooker, Peter Peirce
Cupery, Kenneth N
Davidson, Gilbert
Davis, John K
DeBell, Arthur Gerald
Decker, John Alvin, Jr
Delano, Erwin
DePalma, James John
Derderian, George
Dietz, Paul Hamilton
Drews, Udo Wilhelm
Dumas, Herbert M, Jr
Ebersole, John Franklin
Eby, John Edson
Edelberg, Seymour
Emmons, Larrimore Browneller
Errett, Daryl Dale
Eyer, James Arthur
Falconer, David G
Feder, Donald Perry
Feit, Michael Dennis
Fisher, Robert Alan
Foreman, Jesse William, Jr
Forrest, John Wilson
Forsyth, James M
Franken, Peter Alden
Freeman, James R
Fried, David L
Frieden, Bernard Roy
Gagne, Jean-Marie
Gamersfelder, George Royce
Gara, Aaron Delano
Garbuny, Max
Giannino, Peter Dominic
Gimlett, James I
Giordmaine, Joseph Anthony
Giroux, Guy
Givens, Miles Parker
Goble, Alfred Theodore
Gonshery, David
Gordon, Howard R
Gordon, Marvin E
Graham, Robert (Clark)
Green, Eugene L
Green, Robert Wood
Greenebaum, Michael
Grieser, Daniel R
Gruber, John B
Grum, Franc
Gschwendner, Alfred Benedict
Guenther, Bobby Dean
Gunter, Roy Chalmers, Jr
Hager, Nathaniel Ellmaker, Jr
Hanau, Richard
Hannan, W Kelley
Hard, Thomas Michael
Hargrove, Logan Ezral
Harrington, Marshall Cathcart
Harris, Richard Jacob
Harris, Ronald Wilbert
Hartman, Richard Leon
Hass, Georg
Hauesler, Donald Carl
Hayes, John Durham
Heinisch, Roger Paul
Hennes, John Peter
Herzberger, Maximilian Jacob
Hessel, Kenneth Ray
Hett, John Henry
Hildebrand, Bernard Percy
Hills, Robert, Jr
Hochheimer, Bernard Ford
Holst, Gerald Carl
Howe, Dennis George
Howell, Barton John
Huang, Jacob Wen-Kuang
Hubbard, William Marshall
Hudson, Richard Delano, Jr
Ingrao, Hector Carlos
Ivey, Henry Franklin
Janson, Peter Allan
Jeffries, Robert Alan
Jenney, Joe Allen
Johnson, Edgar Gustav
Johnston, Alan Robert
Johnston, George Taylor
Johnston, Lawrence Harding
Joshi, Yoginder N
Jungling, Kenneth Corneal
Jupnik, Helen
Kaprelian, Edward Karnig
Kapany, Narinder Singh
Kennedy, Felix Paul
Kennedy, Larry Zane
Kermisch, Dorian
Kessler, Ernest George, Jr
Khadjavi, Abbas
Kingslake, Rudolf
Kirz, Janos
Knapp, Richard Allen

Knox, Cameron
Koehl, George Martin
Koehler, Wilbert Frederick
Koester, Charles John
Korpel, Adrianus
Kozma, Adam
Krishnan, Kamala Sivasubramaniam
Krueger, George Corwin
Kuppenheimer, John D, Jr
Kurtz, Clark N
Land, Edwin Herbert
Langworthy, Harold Frederick
Latady, William Robertson
Latta, John Neal
Lean, Eric Gung-Hwa
Lee, Pui Kum
Leet, Henry Peter
Leith, Emmett Norman
Liberman, Irving
Linder, Solomon Leon
Lit, John Wai-Yu
Little, Gordon Rice
Loewenstein, Ernest Victor
Lovell, Donald Joseph
Lovette, Maribeth
Malacara, Daniel
Malarkey, Edward Cornelius
Mallory, Willam R
Marathay, Arvind Shankar
Marchand, Erich Watkinson
Martin, Ashley Marvin, III
Mauer, Paul Bernard
Mauro, Jack Anthony
Mavco, George Edward
McKenney, Dean Brinton
McKinley, Harry R
McMahon, Donald Howland
Meier, Rudolf H
Meyer-Arendt, Jurgen Richard
Mielenz, Klaus Dieter
Miller, Herman Lunden
Miller, Robert Charles
Milne, Gordon Gladstone
Mimmack, William Edward
Moeller, Karl Dieter
Montgomery, Anthony John
Moore, Duncan Thomas
Mundie, Lloyd George
Nash, Harry Charles
Nelson, Arthur Robert
Nelson, David Torrison
Nicodemus, Frederick Edwin
Nicolai, Van Olin
Noble, Robert Hamilton
Nyssonen, Diana
Ott, William Roger
Owens, James Carl
Park, Kwangjai
Parke, Edward Charles, Jr
Paul, Frederick William
Pavloopoulos, Theodore G
Pedinoff, Melvin Eli
Pedrotti, Leno Stephano
Pershing, William Raymond
Phelan, Robert J, Jr
Phillips, E Alan
Phillips, Richard Arlan
Piech, Kenneth Robert
Pike, John Nazarian
Pollack, John L
Porter, Terence Lee
Potter, Robert Joseph
Poultney, Sherman King
Pryor, Paul L
Purdom, Emil Garness
Quinn, Jarus William
Ramdas, Anant Krishna
Rao, B Seshagiri
Ravitsky, Charles
Rawcliffe, Robert Douglas
Rawson, Eric Gordon
Reichert, John Douglas
Reichardt, Claude
Riedel, Ernest Paul
Rigler, A Kellam
Rimmer, Matthew Peter
Ring, James George
Rockwell, David Alan
Roesler, Frederick Lewis
Roetling, Paul G
Rosendahl, Gottfried R
Rudolph, Joseph Anthony
Rupp, Wiktor
Rush, Joseph Harold
Sandefur, Kermit Lorain
Sanmann, Everett Eugene
Saunders, Morton Jefferson
Sawatari, Takeo
Schneider, Irwin
Schroeder, Daniel John
Schulman, James Herbert
Schulte, Daniel Herman
Schultz, Rodney Brian
Schweitzer, Walter Gareld, Jr
Scidmore, Wright Harwood
Seagrave, John Dorrington
Seeser, James William
Shack, Roland Vincent

Shannon, Robert Rennie
Shaw, Rodney
Shealy, David Lee
Shenker, Martin
Shumaker, John Benjamin, Jr
Sinclair, Douglas C
Skinner, Thomas Junior
Skolnik, Lyn Howard
Slater, Philip Nicholas
Smiley, Vern Newton
Smith, David Young
Smith, Frederick Dowswell
Smith, Howard Michael
Smith, James Lynn
Smith, Richard Carper
Smith, Winfield Scott
Sollid, Jon Erik
Spiller, Eberhard Adolf
Staebler, David Lloyd
Stamm, Robert Franz
Stavroudis, Orestes Nicholas
Stierwalt, Donald L
Stone, Sidney Norman
Stoner, William Weber
Swindell, William
Tannenwald, Ludwig Max
Thompson, Brian J
Treu, Jesse Isaiah
Tricoles, Gus P
Tropf, William Jacob
Tschunko, Hubert F A
Tubbs, Eldred Frank
Turner, Arthur Francis
Tynes, Arthur Richard
Ulrich, Peter B
Urbach, John C
Vahey, David William
Vande Kieft, Laurence John
Van den Akker, Johannes Archibald
Vanderkooy, John
VanKerkhove, Alan Paul
Van Ligten, Raoul Fredrik
Venable, William Howell, Jr
Verber, Carl Michael
Walker, Jearl Dalton
Walker, Marshall John
Wallace, James
Walther, Adriaan
Weber, Heinz Paul
Weidner, Victor Ray
Weiss, James Paul
Wells, Michael Byron
Wells, Willard H
Wertheimer, Alan Lee
White, Donald Robertson
Whitmer, Romayne Flemming
Wieder, Harold
Wiggins, Thomas Arthur
Wilder, Donald Richard
Wilson, Richard J
Winter, Thomas C, Jr
Wittke, James Pleister
Wolfe, William Louis, Jr
Wyant, James Clair
Yergey, Alfred L, III
Yoder, Paul Rufus, Jr
Young, Charles Gilbert
Young, Russell Dawson
Zechiel, Leon Norris
Zelenka, Jerry Stephen
Zissis, George John

Color Science
Allen, Eugene (Murray)
Demers, Pierre (A E)
Garland, Charles E
Gibson, Gilbert Lewis
Goddard, Murray Cowdery
Goldfinger, George
Gray, Russell Houston
Hoffenberg, Paul Henry
Marcus, Robert Toby
Mimeault, Victor Joseph
Nickerson, Dorothy
Schuler, Mathias John
Shepp, Allan
Simon, Frederick Tyler

Electron Optics
Collier, Robert Jacob
Coltman, John Wesley
Dayton, James Anthony, Jr
Griboval, Paul
Harris, Warren Whitman
Harte, Kenneth J
Herrmannsfeldt, William Bernard
Jacobsen, Edward Hastings
Kuyatt, Chris Ernie Earl
Retsky, Michael Walter
Thomson, Michael George Robert

Electrooptics
Almeida, Silverio Pedro
Anthony, Romuald
Astheimer, Robert W
Barrekette, Euval S
Bartolini, Robert Alfred

Beall, Horace Ansley
Brienza, Michael Joseph
Buckman, Alvin Bruce
Chow, Christopher N
Christensen, Niels Gunnar
Coleman, Howard S
Dougherty, Joseph Patrick
Dressel, Herman Otto
Erdmann, Joachim Christian
Geller, Myer
Han, Ki Sup
Hansen, J Richard
Harvey, George Lloyd
Heckathorn, Harry Mervin, III
Henning, Harley Barry
Honey, Richard Churchill
Hummer, Robert Franklin
Hutter, Edwin Christian
Jablonski, Frank Edward
Jaffe, Hans
Kaminow, Ivan Paul
Kinzly, Robert Edward
Knight, Gordon Raymond
Kornstein, Edward
Kruse, Paul Walters, Jr
Kurtz, Stewart K
Maccabee, Bruce Sargent
Medved, David Bernard
Morris, Glen Jeffs
Nelson, Kyler Fischer
O'Brien, Brian
Penz, P Andrew
Quelle, Fred W, Jr
Richman, Isaac
Riethof, Thomas Robert
Ruderman, Irving Warren
Sauermann, Gerhard Otto
Scifres, Donald Ray
Shackelford, Robert G
Sheridan, Nicholas Keith
Silverstein, Elliot Morton
Sneed, Richard J
Soref, Richard Allan
Spaulding, Richard Alan
Stegelmann, Erich J
Titterton, Paul James
Traub, Alan Cutler
Vance, Dennis William
Volz, William Beckham
Whitney, Colin Gordon
Williamson, Arthur Elridge, Jr
Wunderman, Irwin
Zarem, Abe Mordecai

Lasers
Alcaraz, Ernest Charles
Allan, Barry David
Anderson, Robert Sven
Andringa, Keimpe
Asher, Irvin Mark
Asmus, John Fredrich
Auston, David H
Avizonis, Petras V
Balog, George
Barry, James Dale
Baumann, Winfried
Bean, Brent Leroy
Behringer, Robert Ernest
Belanger, Pierre Andre
Bergmann, Ernest Eisenhardt
Berman, Paul Ronald
Bernal G, Enrique
Bernstein, Burton
Billman, Kenneth William
Binns, Walter Robert
Birnbaum, Milton
Blum, Fred A
Bricks, Bernard Gerard
Campillo, Anthony Joseph
Center, Robert E
Chang, Ren-Fang
Chase, Charles Elroy, Jr
Chin, See Leang
Cohen, Martin Gilbert
Cohn, Daniel Ross
Collins, Carl Baxter, Jr
Corcoran, Vincent John
Crisp, Michael Dennis
Decker, Charles David
Dénariez-Roberge, Marguerite Marie
Denk, Ronald H
Derderian, George
Deutsch, Thomas
Devaney, Joseph James
De Wit, Michiel
Dreyfus, Russell Warren
Eckstrom, Donald James
Erickson, Lynden Edwin
Fahlen, Theodore Stauffer
Fairand, Barry Philip
Faust, Walter Luck
Feldman, Barry Joel
Fetterman, Harold Ralph
Figueira, Joseph Franklin
Fisher, Robert Alan
Fleck, Joseph Amadeus, Jr
Flicker, Herbert
Gelbwachs, Jerry A

Gerardo, James Bernard
Goldberg, Lawrence Spencer
Goldsborough, John Paul
Goldstein, John Cecil
Grams, Gerald William
Greiner, Norman Roy
Gundersen, Martin Adolph
Haueisen, Donald Carl
Heer, Clifford V
Henry, Eugene Michael
Hernqvist, Karl Gerhard
Hessel, Merrill
Hilborn, Robert Clarence
Holt, Helen Keil
Hughes, Richard Swart
Izatt, Jerald Ray
Janiak, Daniel Robert
Janney, Gareth Maynard
Johnson, Eric G, Jr
Johnson, Leo Francis
Judd, O'Dean P
Jungling, Kenneth Corneal
Karras, Thomas William
Kennedy, Chandler James
Kessler, Bernard Von
Kikuchi, Tom T
Killion, Lawrence Eugene
Kim, Boris Fincannon
Koenig, Albert A
Komai, Leo G
Kreuzer, Lloyd Barton
Lee, Haynes A
Leland, Wallace Thompson
Levenson, Marc David
Levine, Barry Franklin
Levy, Richard H
Liberman, Irving
Liebenberg, Donald Henry
Loree, Thomas Robert
Lowke, John James
Lynk, Edgar Thomas
Marcus, Stephen
Massey, Gail Austin
McCorkle, Richard Anthony
McKnight, William Baldwin
McNally, James Henry
Measures, Raymond Massey
Montgomery, George Paul, Jr
Nereson, Norris (George)
Nichols, Davis Betz
Nicolai, Van Olin
Nilson, John Anthony
Offenberger, Allan Anthony
O'Sullivan, William John
Pan, Yu-Li
Parbhakar, Kanwal Jit
Pedrotti, Leno Stephano
Picard, Richard Henry
Piloff, Herschel Sydney
Pitch, Martin Stanley
Pinsley, Edward Allan
Poehler, Theodore O
Quelle, Fred W, Jr
Rao, Devulapalli V G L N
Ready, John Fetsch
Reintjes, John Francis, Jr
Rhodes, Charles Kirkham
Rhodes, George Wyatt
Rice, James Kinsey
Rich, John Charles
Riedel, Ernest Paul
Rinehart, Edgar A
Roberts, Thomas George
Robin, James Edmond
Robinson, C Paul
Rockwood, Stephen Dell
Rosenkrantz, Lawrence Jay
Sattler, Joseph Peter
Schawlow, Arthur Leonard
Schubert, David Crawford
Scully, Charles Norman
Searles, Stuart Kenneth
Simonis, George Jerome
Singh, Shobha
Smith, Archibald William
Smith, William Vick
Stoicheff, Boris Peter
Strome, Forrest C, Jr
Suchard, Steven Norman
Sutton, David George
Taylor, Lyle Herman
Taylor, Raymond L
Telle, John Martin
Temkin, Richard Joel
Thielman, Leroy Oswald
Tuccio, Sam Anthony
Wada, James Yasuo
Walters, Craig Thompson
Watson, Robert Barden
Weber, Marvin John
Werkheiser, Arthur H, Jr
York, George William
Zar, Jacob L

Physical Optics
Avizonis, Petras V
Burge, Dennis Knight
Castle, John Granville, Jr

OPTICS

Conklin, Richard Louis
Cook, Ancel Eugene
Dakss, Mark Ludmer
DeBitetto, Dominick John
Donohue, Robert J
Ehrlich, Morris Joseph
Engh, Robert Oswald
Friedman, Helen Lowenthal
Glasser, Leo George
Hadley, Lawrence Nathan
Hunter, William Ray
King, Marvin
Osborn, Richard Kent
Sherman, George Charles
Starkweather, Gary Keith
Stroke, George W
Strong, Herbert Maxwell

Physiological Optics
Bernard, Gary Dale
Enoch, Jay Martin
Everson, Ronald Ward
Fatt, Irving
Freeman, Ralph David
Frumkes, Thomas Eugene
Fry, Glenn Ansel
Green, Daniel G
Greenspon, Thomas Stephen
Gunkel, Ralph D
Hebbard, Frederick Worthman
Hill, Richard M
Julesz, Bela
Kerr, Kenton E
Kidder, John Newell
Mauro, Jack Anthony
Pease, Paul Lorin
Pratt, Carol Bert
Randle, Robert James
Reading, Rogers W
Remole, Arnulf
Roth, Niles
Sivak, Jacob Gershon
Takahashi, Ellen Shizuko
Webber, Richard Lyle
Westheimer, Gerald

Quantum Optics
Abella, Isaac D
Aggarwal, Roshan Lal
Barrett, Joseph John
Bean, Brent Leroy
Blake, Julian Gaskill
Bloembergen, Nicolaas
Bluemel, Van (Fonken Wilford)
Bowden, Charles Malcom
Brown, Judith
Crisp, Michael Dennis
Cummins, Herman Z
Deck, Robert Thomas
Fairbank, William Martin, Jr
Faries, Dillard Wayne
Feldman, Barry Joel
Fork, Richard Lynn
Fowles, Grant Robert
Franken, Peter Alden
Friedlander, Zitta Zipora
Gelbwachs, Jerry A
Giallorenzi, Thomas Gaetano
Gibbs, Hyatt McDonald
Goldberg, Lawrence Spencer
Gurski, Thomas Richard
Hahn, Yu Hak
Hargis, Philip Joseph, Jr
Heer, Clifford V
Lee, Ching Tsung
Matulic, Ljubomir Francisco
Narducci, Lorenzo M
Nygaard, Kaare Johann
Picard, Richard Henry
Rabin, Herbert
Reintjes, John Francis, Jr
Rogovin, Daniel Noel
Salamo, Gregory Joseph
Sargent, Murray, III
Scarl, Donald B
Shoemaker, Richard Lee
Slusky, Mark Sender
Small, James Graydon
Stoler, David
Stroud, Carlos Ray
Yang, Kei-Hsiung
Zardecki, Andrzej

PALEONTOLOGY

Paleontology
Addicott, Warren O
Bayer, Thomas Norton
Beerbower, James Richard
Benson, Richard Hall
Bishop, Gale Arden
Bork, Kennard Baker
Boucot, Arthur James
Bowen, Zeddie Paul
Brown, George D, Jr
Buchanan, Hugh
Burke, John James
Campbell, Catherine Chase
Caster, Kenneth Edward
Chaffee, Robert Gibson
Chatterton, Brian Douglas Eyre
Church, Clifford Carl
Clark, David Leigh
Clark, Thomas Henry
Clarke, Robert Travis
Clemens, William Alvin
Coates, Anthony George
Coffin, Harold Glen
Collins, Desmond H
Conkin, James E
Coogan, Alan H
Copper, Paul
Cowen, Richard
Curran, Harold Allen
Cygan, Norbert Everett
Davis, Richard Arnold
Davis, William Edwin, Jr
Deboo, Phiroz B
DeMott, Lawrence Lynch
Dresser, Hugh W
Dunbar, Carl Owen
Durden, Christopher John
Easton, William Heyden
Edwards, Richard Archer
Eiler, Eugene Rudolph
Ellis, Charles Howard
Erdtmann, Bernd Dietrich
Etheridge, Richard Emmett
Ethinton, Raymond Lindsay
Faas, Richard William
Farmer, George Thomas, Jr
Ferguson, Laing
Fields, Robert William
Fisher, Donald William
Fisher, James Lawrence
Flessa, Karl Walter
Flower, Rousseau Hayner
Fox, Stephen Knowlton, Jr
Frebold, Hans (Wilhelm Ludwig August Herman)
Fritz, Madeleine Alberta
Fritz, William Harold
Frost, Stanley H
Gale, Hoyt Rodney
Gauri, Kharaiti Lal
Gilmour, Ernest Henry
Gordon, William Anthony
Gould, Stephen Jay
Green, Morton
Greggs, Robert George
Haas, Otto Henry
Hamlin, William Henry
Hansman, Robert Herbert
Harker, Peter
Harrington, Jonathan W
Helfrich, Charles Thomas
Heller, Robert Leo
Hewitt, Philip Cooper
Hinman, Eugene Edward
Hirschfeld, Sue Ellen
Holland, Frank Delno, Jr
Horowitz, Alan Stanley
Hotchkiss, Frederick Hatfield Clark
Huddle, John Warfield
Imlay, Ralph Willard
Johnson, Ralph Gordon
Jones, David Lawrence
Jones, Douglas Epps
Juskevice, John Anthony
Kent, Lois Schoonover
Kesling, Robert Vernon
Kindle, Cecil Haldane
Klapper, Gilbert
Kleinpell, Robert Minnsen
Klement, Karl Walter
Knox, Larry William
Koenig, John Waldo
Kummel, Bernhard
Kurtz, Vincent E
Lane, Harold Richard
Lane, Norman Gary
Langenheim, Ralph Louis, Jr
Levin, Harold Leonard
Levinson, Stuart Alan
Lewis, Standley Eugene
Lillegraven, Jason Arthur
Lowther, John Stewart
Lundin, Robert Folke
Lunsden, William Watt, Jr
Lynts, George Willard
Macomber, Richard Wiltz
Macurda, Donald Bradford, Jr
Mallory, Bob Franklin
Marks, Jay Glenn
McGrew, Paul Orman
McGugan, Alan
McKee, James W
McLaughlin, Robert Everett
McNulty, Charles Lee, Jr
Mello, James Francis
Merrill, Glen Kenton
Meszoely, Charles Aladar Maria
Miller, Halsey Wilkinson, Jr
Mintz, Leigh Wayne
Mosher, Loren Cameron
Moyle, Richard W
Nave, Floyd Roger
Nelson, Katherine Greacen
Nelson, Samuel James
Nichols, Douglas James
Nicol, David
Nodine-Zeller, Doris Eulaia
Norford, Brian Seeley
Norris, Geoffrey
Oliver, William Albert, Jr
Olsen, Rex E
Palmer, Katherine Van Winkle
Pampe, William R
Peck, Joseph Howard
Pestana, Harold Richard
Pitrat, Charles William
Pope, John Keyler
Potter, Franklin Carl
Puri, Harbans Singh
Repenning, Charles Albert
Rhodes, Frank Harold Trevor
Richards, Horace Gardiner
Rickard, Lawrence Vroman
Rigby, J Keith
Ritland, Richard Martin
Riva, John
Robinson, Peter
Robison, Richard Ashby
Roellig, Harold Frederick
Ross, Reuben James, Jr
Russell, Loris Shano
Sandberg, Philip A
Sanders, Nestor John
Sanderson, George Albert
Sandidge, John Roy
Sando, William Jasper
Sarjeant, William Antony Swithin
Sass, Daniel B
Saul, Louella Rankin
Saunders, Jeffrey John
Savage, Donald Elvin
Schumacher, Dietmar
Scott, Alan Johnson
Scott, Robert W
Shrock, Robert Rakes
Sloan, Robert Evan
Smith, Judith Terry
Smith, Raymond N
Smith, Roberta Katherine
Sohl, Norman Frederick
Sohn, Israel Gregory
Spencer, Randall Scott
Spinosa, Claude
Stanley, Steven Mitchell
Stanton, Robert James, Jr
Staplin, Frank Lyons
Stearn, Colin William
Steele, Grant
Steinker, Don Cooper
Stitt, James Harry
Stover, Lewis Eugene
Stratton, James Forrest
Sturgeon, Myron Thomas
Teeter, James Wallis
Teichert, Curt
Tesmer, Irving Howard
Theokritoff, George
Thompson, Marcus Luther
Thoms, Richard Edwin
Thomson, Keith Stewart
Toomey, Donald Francis
Toulmin, Lyman Dorgan, Jr
Trexler, David William
Van Couvering, Judith Anne Harris
Van Den Bold, William Aaldert
Vincent, Jerry William
Warthin, Aldred Scott, Jr
Weart, Richard Claude
Webb, Sawney David
Webers, Gerald F
Webster, Gary Dean
Wells, Dana
Wells, John West
White, John Anderson
Wiggins, Virgil Dale
Wise, Sherwood Willing, Jr
Wolleben, James A
Womardt, Walter William, Jr
Young, Keith Preston
Ziegler, Alfred M
Zingula, Richard Paul

Invertebrate Paleontology
Allison, Richard C
Amsden, Thomas William
Bailey, Richard Hendricks
Batten, Roger Lyman
Berdan, Jean Milton
Beus, Stanley S
Birkhead, Paul Kenneth
Boardman, Richard Stanton
Bolton, Thomas Elwood
Bonem, Rena Mae
Bostwick, David Arthur
Bretsky, Peter William
Bretsky, Sara (Su) Stewart
Brower, James Clinton
Carter, John Lyman
Chamberlain, Charles Kent
Chronic, John
Citelli, Richard
Cloud, Preston E, Jr
Copeland, Charles Wesley, Jr
Cuffey, Roger James
Darby, David G
Diffendal, Robert Francis, Jr
Douglass, Raymond Charles
DuBar, Jules R
Durham, John Wyatt
Erickson, John Mark
Fagerstrom, John Alfred
Feldmann, Rodney Mansfield
Finks, Robert Melvin
Firby, James R
Fischer, William Alfred
Frankel, Larry
Gibson, Thomas George
Glawe, Lloyd Neil
Grant, Richard Evan
Hall, Donald E
Harper, Charles Wood, Jr
Hazel, Joseph Ernest
Heaslip, William Graham
Hoare, Richard David
Imbrie, John
Keen, (Angeline) Myra
Kern, John Philip
Kier, Porter Martin
Kirtley, David Warren
Kolata, Dennis Robert
Lane, Bernard Owen
Laufeld, Sven
Lawrence, David Reed
LeMone, David V
Lesperance, Pierre J
Lipps, Jere Henry
Lochman-Balk, Christina
MacClintock, Copeland
McKinney, Frank Kenneth
Meyer, David Lachlan
Miller, James Frederick
Moore, Reginald George
Nye, Osborne Barr, Jr
Okulitch, Vladimir Joseph
Palmer, Allison Ralph
Pannella, Giorgio
Parsley, Ronald Lee
Perkins, Bobby Frank
Pojeta, John, Jr
Popenoe, Willis Parkison
Raup, David Malcolm
Rowell, Albert John
Scolaro, Reginald Joseph
Shaw, Frederick Carleton
Sprinkle, James (Thomas)
Strimple, Harrell Leroy
Sweet, Walter Clarence
Thayer, Charles Walter
Tischler, Herbert
Titus, Robert Charles
Vokes, Emily Hoskins
Vokes, Harold Ernest
Walter, Thomas Richard
Wanshauer, Steven Michael
Weisbord, Norman Edward
Wilde, Garner Lee
Williams, Michael Eugene
Wilson, Edward Carl
Yochelson, Ellis Leon
Zullo, Victor August

Micropaleontology
Abbott, William Harold
Arnold, Zach McLendon
Baxter, James Watson
Benson, Richard Norman
Berger, Wolfgang Helmut
Berggren, William Alfred
Bock, Wayne Dean
Bordine, Burton W
Braun, Willi Karl
Bukry, John David
Charmatz, Richard
Copeland, Murray John
Cornell, William Crownshield
Cousminer, Harold L
Crouch, Robert Wheeler
Everett, Robert W, Jr
Foreman, Helen Pulver
Forester, Richard Monroe
Frederiksen, Norman Oliver
Frerichs, William Edward
Garret, Julius Benjamin, Jr
Gartner, Stefan, Jr
Gibson, Lee B
Habib, Daniel
Hay, William Winn
Hecht, Alan David
Howe, Robert C
Huff, William J

Ingle, James Chesney, Jr
Kaesler, Roger LeRoy
Kennett, James Peter
Kontrovitz, Mervin
Krutak, Paul Russell
Lamb, George Marion
Lamb, James L
Limper, Karl Esslinger
Ling, Hsin Yi
Loeblich, Alfred Richard Jr
Loeblich, Helen Nina (Tappan)
Maddocks, Rosalie Frances
Masters, Bruce Allen
Medioli, Franco
Miller, James Frederick
Morales, Gustavo Adolfo
Norton, Charles Warren
Parker, Frances Lawrence
Rau, Weldon Willis
Riedel, William Rex
Roberts, Thomas Glasdir
Roth, Peter Hans
Rothwell, William Thomas, Jr
St Jean, Joseph, Jr
Saito, Tsunemasa
Schnitker, Detmar
Seigle, George A
Sen Gupta, Barun Kumar
Smith, Lee Anderson
Smith-Evernden, Roberta Katherine
Vilks, Gustavs
Wall, John Hallett
Webb, Peter Noel
Weiss, Dennis
Wiles, William Walter

Paleobiology
Armstrong, Augustus K
Barghoorn, Elso Sterrenberg
Bell, Bruce McConnell
Berry, William Benjamin Newell
Blake, Daniel Bryan
Bordine, Burton W
Cooper, Gustav Arthur
Cvancara, Alan Milton
Dodson, Peter
Eldredge, Niles
Fisk, Lanny Herbert
Glenister, Brian Frederick
Hickman, Carole Stentz
Kauffman, Erle Galen
Maglio, Vincent Joseph
McAlester, Arcie Lee, Jr
Pirozynski, Krzysztof Andrzej
Schopf, James William
Schopf, Thomas Joseph Morton
Simmonds, Robert T
Stanley, Edward Alex
Sutherland, Patrick Kennedy
Thompson, Ida
West, Ronald Robert

Paleobotany
Abbott, Maxine Langford
Andrews, Henry Nathaniel, Jr
Archangelsky, Sergio
Arnold, Chester Arthur
Basson, Philip Walter
Baxter, Robert Wilson
Becker, Herman Frederick
Beyer, Arthur Frederick
Campbell, John Duncan
Cridland, Arthur A
Darrah, William Culp
Delevoryas, Theodore
Dilcher, David L
Dolph, Gary Edward
Eggert, Donald A
Engelhardt, Donald Wayne
Felix, Charles Jeffrey
Frankenberg, Julian Myron
Fry, Wayne Lyle
Gillette, Norman John
Grierson, James Douglas
Hueber, Francis Maurice
Kasper, Andrew E, Jr
Kosanke, Robert Max
MacGinitie, Harry Dunlap
Mamay, Sergius Harry
Matten, Lawrence Charles
Nitecki, Matthew H
Penny, John Sloyan
Phillips, Tom Lee
Radforth, Norman William
Schopf, James Morton
Scott, Richard Albert
Spackman, William, Jr
Stidd, Benton Maurice
Taylor, Thomas Norwood
Traverse, Alfred
Tschudy, Robert Haydn
Willis, Jeanne Eleanor
Wood, Joseph M

Paleoecology
Alexander, Richard Raymond
Anderson, Edwin J
Baer, James L

Basan, Paul Bradley
Buzas, Martin A
Cameron, Barry Winston
Camp, Mark Jeffrey
Clark, George Richmond, II
Colinvaux, Paul Alfred
Dodd, James Robert
Ekdale, Allan Anton
Erdtmann, Bernd Dietrich
Hoskins, Cortez William
Kennett, James Peter
Kissling, Don Lester
Logan, Alan
McCrone, Alistair William
Nations, Jack Dale
Parks, James Marshall, Jr
Rhoads, Donald Cave
Richards, Richard Peter
Risk, Michael John
Shaak, Graig Dennis
Shabica, Charles Wright
Stevens, Calvin H
Tsukada, Matsuo
Whitehead, Donald Reed
Wright, Herbert Edgar, Jr
Wright, Robert Paul

Paleozoology
Richardson, Eugene Stanley, Jr
Schram, Frederick R
Utgaard, John Edward

Palynology
Adam, David Peter
Baker, Richard Graves
Brenner, Gilbert J
Clarke, Robert Travis
Clendening, John Albert
Cohen, Arthur David
Cornell, William Crownshield
Cropp, Frederick William, III
Darrell, James Harris, II
Drugg, Warren Sowle
Elsik, William Clinton
Fasbender, M Veronica
Fournier, George Richard
Frederiksen, Norman Oliver
Gray, Jane
Grayson, John Francis
Guennel, Gottfried Kurt
Gupta, Sujoy
Habib, Daniel
Hansen, Henry Paul
Hedlund, Richard Warren
Heusser, Linda Olga
Hopkins, William Stephen, Jr
Jarzen, David MacArthur
King, James Edward
Knox, Arthur Stewart
LeBlanc, Arthur Edgar
Lieux, Meredith Hoag
Loeblich, Alfred Richard Jr
Maher, Louis James, Jr
Mathewes, Rolf Walter
McGregor, Duncan Colin
Mehringer, Peter Joseph, Jr
Nichols, Harvey
Norem, Winfred Luther
Nowicke, Joan Weiland
Oltz, Donald Frederick
Peppers, Russel A
Peterson, Earl T
Piel, Kenneth Martin
Pocock, Stanley Albert John
Richard, Glenn Everett
Rouse, Glenn Everett
Sirkin, Leslie A
Stewart, Robert Archie, II
Stover, Lewis Eugene
Sullivan, Herbert J
Terasmae, Jaan
Thompson, Gary Gene
Ting, William Su
Traverse, Alfred
Tynan, Eugene Joseph
Urban, James Bartel
Von Almen, William Frederick I
Walker, Philip Caleb
Warter, Janet Kirchner
Webb, Thompson, III
Wilson, Leonard Richard

Vertebrate Paleontology
Alf, Raymond Manfred
Applegate, Shelton P
Baird, Donald
Bardack, David S
Berman, David S
Bird, Samuel Oscar, II
Bjork, Philip R
Black, Craig C
Carroll, Robert Lynn
Churcher, Charles Stephen
Colbert, Edwin Harris
Coombs, Margery Chalifoux
Cosgriff, John W, Jr
Crompton, Alfred W
Dawson, Mary (Ruth)

DeMar, Robert E
Denison, Robert Howland
Dorr, John Adam, Jr
Downs, Theodore
Dunkle, David Hosbrook
Echols, Joan
Edmund, Alexander Gordon
Estes, Richard
Fox, Richard Carr
Galbreath, Edwin Carter
Gingerich, Philip Dean
Greaves, Walter Stalker
Gregory, Joseph Tracy
Holman, J Alan
Hopson, James A
Howard, Hildegarde (Mrs Henry Anson Wylde)
Howe, John A
Hutchison, John Howard
Jakway, George Elmer
James, Gideon T
Jenkins, Farish Alston, Jr
Jenkins, Floyd Albert
Kitts, David Burlingame
Koerner, Harold Elton
Konizeski, Richard L
Langston, Wann, Jr
Lewis, George Edward
Lindsay, Everett Harold, Jr
Linsley, Robert Martin
Lund, Richard
Lundelius, Ernest Luther, Jr
Macdonald, James Reid
MacIntyre, Giles T
Madsen, James Henry, Jr
Martin, Larry Dean
Martin, Robert Allen
Mawby, John Evans
McKenna, Malcolm Carnegie
Mellett, James Silvan
Melton, William Grover, Jr
Miller, Wade Elliott, II
Olson, Everett Claire
Olson, Storrs Lovejoy
Ostrom, John Harold
Patterson, Bryan
Patton, Thomas Hudson
Rackoff, Jerome S
Ray, Clayton Edward
Reinhart, Roy Herbert
Rich, Patricia Vickers
Richards, Lawrence Phillips
Rollins, Harold Bert
Russell, Dale A
Schaeffer, Bobb
Schultz, Charles Bertrand
Schultz, Gerald Edward
Seltin, Richard James
Simons, Elwyn Laverne
Simpson, George Gaylord
Stephens, John James, III
Stout, Thompson Mylan
Strain, William Samuel
Tedford, Richard Hall
Thurmond, John Tydings
Tullar, Richard Montgomery
Vaughn, Peter Paul
Voorhies, Michael Reginald
Wahlert, John Howard
Warter, Stuart L
Welles, Samuel Paul
West, Robert MacLellan
Wheeler, Walter Hall
Whistler, David Paul
Whitmore, Frank Clifford, Jr
Wilson, Mark Vincent Hardman
Wilson, Robert Warren
Wood, Albert Elmer
Woodburne, Michael O
Zakrzewski, Richard Jerome
Zangerl, Rainer

PATHOLOGY

Pathology
Abell, Murray Richardson
Abrams, Albert Maurice
Abrams, Gerald David
Ackart, Richard Jenks
Ackerman, Lauren Vedder
Adams, Dolph Oliver
Adelson, Lester
Afonsky, Dimitri Aleksandrovich
Akamatsu, Yasuyuki
Albrecht, Paul
Albrink, Wilhelm Stockman
Aleo, Joseph John
Alexander, Robert W
Allen, Arthur Charles
Allen, James R, Jr
Allen, Miller Shannon, Jr
Allen, Raymond A
Allen, Robert Carter
Altshuler, Charles Haskell
Amador, Elias
Amromin, George David

Anderson, Harrison Clarke
Anderson, Robert Edwin
Anderson, William Arnold Douglas
Angevine, Daniel Murray
Archer, Francis L
Arhelger, Roger Boyd
Arquilla, Edward R
Ashley, Charles Allen
Asofsky, Richard Marcy
Aterman, Kurt
Auerbach, Stewart Hart
Ayer, Darrell, (Jr)
Azar, Henry A
Azar, Miguel M
Baba, Nobuhisa
Baggenstoss, Archie Herbert
Bahn, Robert Carlton
Bahr, Gunter F
Bailey, Orville Taylor
Bain, Gordon Orville
Bainborough, Arthur Raymond
Bainton, Dorothy Ford
Baisden, Charles Robert
Baker, Roger Denio
Bakerman, Seymour
Balogh, Karoly
Bangle, Raymond, Jr
Barajas, Luciano
Bardawil, Wadi Antonia
Barget, James Daniel
Barnes, Asa, Jr
Barth, Rolf Frederick
Baserga, Renato
Batsakis, John G
Battifora, Hector A
Bauer, Neinz
Baum, Joseph Herman
Beamer, Parker Reynolds
Bean, Michael Arthur
Becker, Carl George
Becker, Frederick F
Becker, Norwin Howard
Beckfield, William John
Beckwith, John Bruce
Begg, Charles Frederic
Bencosme, Sergio Arturo
Benditt, Earl Philip
Benirschke, Kurt
Bennett, Ivan Loveridge, Jr
Bennington, James Lynne
Benson, Ellis Starbranch
Benson, Walter Russell
Berard, Costan William
Berdjis, Charles Choaib
Berkman, James Israel
Bernstein, Jay
Berton, William Morris
Bevans, Margaret
Bhaskar, Surindar Nath
Bickley, Harmon C
Biempica, Luis
Bigelow, Nolton H
Binford, Chapman Hunter
Black, William Cormack
Blackburn, Will R
Black-Schaffer, Bernard
Blakeslee, George M
Blanc, William Andre
Bloodworth, James Morgan Bartow, Jr
Bloor, Colin Mercer
Blozis, George G
Blum, Leon Leib
Blumberg, Joe Morris
Blumenthal, Herman T
Blundell, George Phelan
Boehme, Diethelm Hartmut
Boggs, Joseph D
Bohrod, Milton George
Bokelman, Delwin Lee
Bolande, Robert Paul
Bolden, Theodore Edward
Bollman, Jesse Louis
Boone, Charles Walter
Borgmann, August Russell
Bostick, Warren Lithgow
Bowden, Drummond Hyde
Bowman, James E
Bowmer, Ernest John
Boyle, Paul Edmund
Bradford, William Dalton
Bratenahl, Charles George
Braunstein, Herbert
Breedis, Charles
Breslow, Alexander
Brinkhous, Kenneth Merle
Briziarelli, Giuliano
Bronk, Theodore Tobias
Brown, Arnold Lanehart, Jr
Brown, Robert Calvin
Browning, Edward T
Brunning, Richard Dale
Bucklin, Robert Van Zandt
Buddingh, Gerrit John
Budzilovich, Gleb Nicholas
Bulger, Ruth Ellen
Bull, Brian S
Burkholder, Peter M
Burt, Robert C

Butler, James Johnson
Calabresi, Massimo
Call, Richard A
Campbell, James Stewart
Campbell, Wallace G, Jr
Cannon, Paul Roberts
Cantero, Antonio
Cardiff, Robert Darrell
Carlson, Arthur Stephen
Carnes, William Henry
Carone, Frank
Carrera, Guillermo Manuel
Carter, James Edward
Carter, John Robert
Casarett, George William
Casorso, Donald Roy
Castleman, Benjamin
Caulfield, James Benjamin
Cawley, Leo Patrick
Ceballos, Ricardo
Chandler, A Bleakley
Chang, Suk Chul
Chang, William Wei-Lien
Changus, George William
Chapman, Arthur Owen
Chason, Jacob Leon
Cheatham, William J
Chester, Clarence Lucian
Chiga, Masahiro
Child, Proctor Louis
Choi, Byung Ho
Christensen, Lauritz Royal
Christian, Howard J
Chu, Elizabeth Wann
Chuang, Hanson Yii-Kuan
Chung, Ed Baik
Chung, Jacob
Cipparrone, Joseph Robert
Clapp, Neal K
Clark, Robert M
Clark, Wallace Henderson, Jr
Clasen, Raymond Adolph
Clemmons, Jackson Joshua Walter
Clum, Floyd Myron
Coalson, Jacqueline Jones
Coe, John Ira
Cohen, Stanley
Collins, George H
Collins, Richard Andrew
Conan, Dale Rex
Conen, Patrick E
Congdon, Charles C
Connor, Daniel Henry
Connor, Nolen Duncan
Constantine, Anthony Benedict
Cook, Robert Thomas
Coon, Robert William
Cooper, Norman S
Cosgrove, Gerald Edward
Costero, Isaac (Tudanca)
Cote, Roger Albert
Cotran, Ramzi S
Cottrell, Thomas S
Coulston, Walter F
Coulston, Frederick
Cowan, Daniel Francis
Cox, Alvin Joseph, Jr
Craig, John Merrill
Craig, Peter Harry
Craighead, John Edward
Crane, August Reynolds
Crass, Gwendolyn
Crawford, William Howard, Jr
Crocker, Diane Winston
Cronin, Michael Thomas Ignatius
Crowley, Leonard Vincent
Crowson, Charles Neville
Cruse, Julius Major, Jr
Culberson, Clyde Gray
Cunningham, George J
Cuppage, Francis Edward
Cutright, Duane Edwin
Czernobilsky, Bernard
D'Agostino, Anthony N
Dahl, Elmer Vernon
Dahlin, David Carl
Daildorf, Frederic Gilbert
Daildorf, Gilbert
Damjanov, Ivan
Damjanov, Gustave John
Dannenberg, Arthur Milton, Jr
Daoud, Assaad S
Daroca, Philip Joseph, Jr
Davidson, Israel
Davis, John Robert
Davis, Joyce S
Davis, Richard LaVerne
Davis, Robert Wilson
Dawe, Clyde Johnson
Dawson, James Robertson, Jr
Dawson, Peter J
Defendi, Vittorio
De La Huerga, Jesus
Denko, John V
Deodhar, Sharad Dinkar
Dhurandhar, Hamida Nina
Diamond, Israel
Diener, Robert Max

Dietert, Scott Edward
D'iang, Arthur H K
Drnochowski, Leon Ludomir
Dockerty, Malcolm Birt
Dolphin, John Michael
Donahue, Sheila
Donohue, William B
Donohue, William Leslie
Dorfman, Howard David
Dowling, Edmund Augustine
Dowling, Shirley Evans
Dublin, Isadore Nathan
Dublin, William Brooks
Dubos, Rene Jules
Duckworth, John Kelly
Dunlap, Charles Edward
Dunn, Thelma Brumfield
Dunn, William Lawrie
Dunning, Wilhelmina Frances
Dutra, Frank Robert
Dutz, Werner
Dvorak, Harold Fisher
Earle, Kenneth Martin
Eckner, Fredrich August Otto
Eddy, Hubert Allen
Edgar, Samuel Allen
Edginton, Thomas S
Edmondson, Hugh Allen
Edwards, Jesse Efren
Edwards, Joshua Leroy
Ehrenreich, Theodore
Eichwald, Ernest J
Eichlepp, Jane G
Elliott, George Alginon
Ellis, John Taylor
Emson, Harry Edmund
Ende, Norman
Endicott, Kenneth Milo
Enterline, Horatio Theodore
Enzinger, Franz Michael
Epstein, Samuel Stanley
Erickson, Cyrus Conrad
Erwin, Chesley Para
Esterly, John Roosevelt
Evers, Carl Gustav
Eyestone, Willard Halsey
Fahimi, Hossein Dariush
Fanger, Herbert
Farber, Emmanuel
Faulconer, Robert Jamieson
Feldman, Joseph David
Fennell, Robert Henry, Jr
Ferry, Andrew P
Fetter, Bernard Frank
Firminger, Harlan Irwin
Fisher, Edwin Ralph
Fisher, Russell Sylvester
Fitch, Frank Wesley
Fitzgerald, Patrick James
Flatt, Ronald Eugene
Flax, Martin Howard
Fletcher, Oscar Jasper, Jr
Fodden, John Henry
Foft, John William
Foos, Robert Young
Foster, Eugene A
Foushee, J Henry Smith, Jr
Fowler, Edward Herbert
Fox, John Gerald
Fox, Karl Richard
Fraley, Elwin E
Francisco, Jerry Thomas
Freeman, Aaron Eliot
Freeman, Robert Glen
Frei, Jaroslav Vaclav
Freiman, David Galland
French, Adam James
Frenkel, Jacob Karl
Fresh, James W
Friedberg, Errol Clive
Friedell, Gilbert H
Friederici, Hartmann H R
Friedman, Nathan Baruch
Friedman, Robert Morris
Fritz, Katherine Elizabeth
Fritz, Thomas Edward
Frost, John Kingsbury
Furth, John J
Gabriell, Elemer Rudolph
Gall, Edward Alfred
Gallager, Harry Stephen
Ganner, George E, Jr
Gardner, Alvin Frederick
Gardner, Murray Briggs
Gareau, Roger
Garneau, Robert (Paul)
Garret, Marta
Garret, Rudolph
Gasic, Gabriel J
Geer, Jack Charles
Geer, Erving Francis
Geevar, Joachim Dieter
Germuth, Frederick George, Jr
Gershon, Richard K
Geruchty, Ronald Mills
Getz, Godfrey S
Gherardi, Gherardo Joseph
Ghidoni, John Joseph

Ghose, Tarunendu
Gibson, John Phillips
Gikas, Paul William
Gilbert, Enid May Fischer
Gilbert, Frederick Emerson, Jr
Gill, Atticus James
Gill, Thomas James, III
Gillespie, Jerry Ray
Gillmore, Hugh Richmond, Jr
Girerd, Rene Jean
Givler, Robert L
Glagov, Seymour
Glenner, George Geiger
Glinski, Ronald P
Glunz, Paul R
Godman, Gabriel C
Godwin, John Thomas
Goepp, Robert August
Goetz, Catherine Gertrude
Goldberg, Burton David
Goldberg, Melvin Leonard
Goldblatt, Harry
Golden, Abner
Golden, Alfred
Goldfischer, Sidney L
Gordon, Benjamin Solomon
Gordon, Gerald Bernard
Gore, Ira
Gottlieb, Leonard Solomon
Goyer, Robert Andrew
Goyings, Lloyd Samuel
Grady, Hugh Gerard
Graef, Irving Philip
Graf, Lisciotte
Graham, John Borden
Grand, Nicholas George
Grauer, Robert Coleman
Gravanis, Michael Basil
Gray, Andrew P
Green, Louis Douglas
Greenberg, Stephen Robert
Grice, Harold C
Grimley, Philip M
Griner, Lynn Adel
Grishman, Joe Wheeler
Gronvall, John Arnold
Gross, Mircea Adrian
Gross, Paul
Gruenwald, Peter
Gruhn, John George
Gueft, Boris
Gulino, Pietro M
Gutman, Paul H
Haber, Meryl H
Hackel, Donald Benjamin
Hackett, Raymond Lewis
Haddad, Jamil Raouf
Hadley, Gilbert George
Hagstrom, Jack Walter Carl
Hales, Milton Reynolds
Hall, Octavia
Hamilton, John Drennan
Hamilton, Thomas Reid
Hamonic, Marcel J
Handforth, Christopher Peter
Handler, Alfred Harris
Hanson, James L
Hanson, Daniel James
Harding-Barlow, Ingeborg
Hardwick, David Francis
Harkin, James C
Harris, Charles
Hartmann, Henrik Anton
Hartmann, William Herman
Hartwell, Ralph Milo
Harvey, Thomas Stoltz
Hass, George Marvin
Hathaway, Beulah M
Haukohl, Robert Somers
Hausman, Robert
Haust, M Daria
Hazard, John Beach
Headington, John Terrence
Hebbel, Robert
Heggtveit, Halvor Alexander
Helpern, Milton
Helwig, Elson Bowman
Henderson, James Stuart
Hendrix, Robert Cowgill
Hennigar, Gordon Ross, Jr
Hentel, William
Hepler, Opal Elsie
Hepinstall, Robert Hodgson
Herbut, Peter Andrew
Herman, Mary M
Hernandez, Juan Antonio
Herrold, Katherine McDermott
Heuser, Eva T
Hicklin, Martin Dale
Hicks, Samuel Pendleton
Higgins, George Kendall
Higginson, John
Highman, Benjamin
Hilberg, Albert William
Hill, Donald P
Hill, Harry Raymond
Hill, Joseph MacGlashan
Hill, Rolla B, Jr

Hineman, Dorin Lee
Ho, Kang-Jey
Hoch-Ligeti, Cornelia
Hoerr, Bryan G
Hoff, Henry Frederick
Hokama, Yoshitsugi
Hollander, David Huizier
Holmes, Robert Hicks
Holmquist, Nelson D
Holyoke, John Bartlett
Hood, Claude Ian
Hoogstraten, Jan
Hopkins, Gordon Bruce
Hopps, Howard Carl
Horn, Robert Chisholm, Jr
Horowitz, Richard E
Howard, Cecil
Howard, Robert Eugene
Hruban, Zdenek
Hubbard, Jesse Donald
Hubbard, John Castleman
Hudson, R Page, Jr
Huestis, Douglas William
Huffines, William Davis
Hukill, Peter Biggs
Hunter, Robert L
Huntington, Robert (Watkinson), Jr
Hussey, Clara Veronica
Hutcheson, James Byron
Hutter, Robert V P
Hyun, Bong Hak
Iannamico, Richard Michael
Ibanez, Michael Louis
Ichinose, Herbert
Igel, Howard Joseph
Imai, Hideshige
Inhorn, Stanley L
Innes, James Robert Maitland
Ioachim, Harry L
Iossifides, Ioulios A
Irey, Nelson Sumner
Iseri, Oscar Akio
Itano, Harvey Akio
Jackson, John Ranicar
Jackson, Marvin Alexander
Jacobs, Myron Samuel
Jacobson, Joan
Jaeschke, Walter Henry
Jaques, William Everett
Jasmin, Gaetan
Javitt, Norman B
Jennings, Frank Lamont
Jennings, Robert Burgess
Jindrak, Karel
Johnson, Frank Bacchus
Johnson, Horton Anton
Johnson, Warren W
Johnston, William Webb
Jones, Albert Moshe
Jones, David B
Jones, Robin Richard
Jones, Russell Stine
Jordan, Scott Wilson
Joshi, Vijay Vinayak
Judefind, Thomas Francis
Karnovsky, Morris John
Kasagarian, Michael
Kashatus, William C
Kaufman, Shirley Louise
Kaufmann, Nathan
Kaufmann, William
Kaye, Gordon I
Kelner, Aaron
Kendrick, Francis Joseph
Kent, Geoffrey
Kent, Sidney Page
Kent, Thomas Hugh
Kepes, John J
Key, Charles R
Kidd, John Graydon
Killingsworth, Lawrence Madison
Kim, Han-Soo
Kim, Harry Hi-Soo
Kimbrough, Renate Dora
King, Donald West, Jr
King, Eileen Brenneman
King, Lester Snow
Kinney, Thomas DeArman
Kipkie, George Frederick
Kirkham, William R
Kirschner, Robert Howard
Kirschstein, Ruth Lillian
Kirsten, Werner H
Klein, Albert William
Kleinerman, Jerome
Kline, Irwin Kaven
Klintworth, Gordon K
Klionsky, Bernard Leon
Klotz, Max Otto
Knabe, George W, Jr
Kniseley, Ralph Marion
Ko, Wen-Hsiung
Kobernick, Sidney D
Koepke, John Arthur
Koerner, Theodore Alfred
Koletsky, Simon
Koobs, Dick Herman
Koprowska, Irena

Korn, David
Kornblum, Roland Norman
Korns, Michael Edward
Korson, Roy
Koss, Leopold George
Kotin, Paul
Kovachevich, Rudy
Kozam, George
Kraft, Lisbeth Martha
Krause, Robert Louis
Krieg, Arthur F
Krikos, George Alexander
Kuhn, Charles, III
Kuhns, William Joseph
Kulka, Johannes Peter
Kuo, Tseng-Tong
Kurtz, Stanley Morton
Kuschner, Marvin
Kuzma, Joseph Francis
Kyriazis, Aikaterini A
Kyriazis, Andreas P
Lacy, Paul Eston
Lagunoff, David
Lalich, Joseph John
Lamm, Michael Emanuel
Lamson, Baldwin Gaylord
Landers, James Walter
Landing, Benjamin Harrison
Lane, Bernard Paul
Langdell, Robert Dana
Langham, Robert Fred
Lanks, Karl William
Laqueur, Gert Ludwig
Latta, Harrison
Lautsch, Elizabeth Virginia
La Via, Mariano Francis
Layton, Jack Malcolm
Layton, William Malloy, Jr
Lazarus, Sydney Simon
Leasure, Elden Emanuel
Leavell, Ullin Whitney, Jr
Lee, Chang Ling
Lee, Joseph Chuen Kwun
Lee, Kyu Taik
Leavie, Leavie Edgar, Jr
Lee, Norman Kunhan
LeGolvan, Paul Celestin
Leighton, Joseph
Leith, John Douglas, Jr
LeQuire, Virgil Shields
LeV, Maurice
Levine, Pincus Philip
Levine, Seymour
Lewis, Franklin Beach
Lewis, Martin Grant
Lichtenberg, Franz Von
Liebow, Averill Abraham
Lillie, Ralph Dougall
Lindberg, Donald Allan Bror
Lindner, Luther Edward
Lindsay, Stuart
Lindsey, James Russell
Lippincott, Stuart Wellington
Lisco, Hermann
Lobdell, David Hill
Lober, Paul Hallam
Loh, Jerome Wei-Ping
Lombard, Louise Scherger
Long, Esmond Ray
Longnecker, Daniel Sidney
Lopushinsky, Theodore
Lord, Geoffrey Haverton
Loring, William Ellsworth
Love, Maurice
Lucas, Fred Vance
Luginbuhl, William Hossfeld
Luke, James Lindsay
Lumb, George Dennett
Lumpkin, Lee Roy
Lundberg, George David
Lunin, Martin
Lupton, Charles Hamilton, Jr
Lurie, Harry I
Lushbaugh, Clarence Chancelum
Lynn, Joseph Alden
MacDonald, Richard Annis
MacFadyen, Douglas Archibald
Mackay, Bruce
MacMahon, Harold Edward
Macpherson, Colin Robertson
Madden, Sidney Clarence
Madrazo, Alfonso A
Maenza, Ronald Morton
Malinin, Theodore I
Malmgren, Richard Axel
Manhold, John Henry, Jr
Manuelidis, Elias Emmanuel
Marchesi, Vincent T
Marcial-Rojas, Raul Armando
Margaretten, William
Margolis, George
Marin-Padilla, Miguel
Marks, Sidney
Marrack, David
Marshall, Richard Blair
Martens, Vernon Edward
Martineau, Perry Cyrus
Maruyama, Koshi

Mason, Earl James
Mason, Reginald G, Jr
Maxwell, Ian David
Mayfield, Ernest Durward, Jr
Maynard, Russell Milton
McAdams, Arthur James
McClellan, Betty Jane
McClellan, James T
McCluskey, Robert Timmons
McCormick, George M, II
McCormick, William F
McDonagh, Jan M
McDonald, John Roland
McDonald, Larry William
McDonnell, William Vincent
McDowell, Elizabeth Mary
McGarry, Paul Anthony
McGavran, Malcolm Howard
McGill, Henry Coleman, Jr
McGrath, John Thomas
McGrew, Elizabeth Anne
McKay, Donald George
McKee, Frank Wray
McKenna, Robert Wilson
McKinney, Ralph Vincent, Jr
McLean, Ian William
McLendon, William Woodard
McManus, Joseph Forde Anthony
Meader, Roland Darrell
Meissner, William Avison
Melamed, Myron Roy
Mellors, Robert Charles
Mendlowitz, Max Gene
Menefee, Max Gene
Meranze, David Raymond
Merin, Robert Gillespie
Merrill, James Allen
Miale, John Buyer
Michaelson, Merle Edward
Mider, George Burroughs
Midgley, A Rees, Jr
Miles, Charles P
Miller, Frank Nelson, Jr
Milliser, Russell Von
Milstoc, Mayer
Milstone, Jacob Haskell
Minckler, Jeff
Minckler, Tate Muldown
Ming, Si-Chun
Minick, Charles Richard
Minkowitz, Stanley
Minowada, Jun
Mitchell, William Marvin
Mitus, Wladyslaw J
Miyai, Katsumi
Mohos, Steven Charles
Molnar, Zelma Villanyi
Monlux, William S
Montasir, Magda Moustafa
Montgomery, Philip O'Bryan, Jr
Moolten, Sylvan E
Moore, Richard Donald
Moran, Thomas James
More, Robert Hall
Morehead, Robert P
Moreno, Esteban
Morgan, Winfield Scott
Mori, Ken
Moritz, Alan Richards
Morris, John Leonard
Morrison, Ashton Byrom
Moss, Leo (David)
Mostofi, Fathollah Keshvar
Mottet, N Karle
Movat, Henry Zoltan
Mowry, Robert Wilbur
Moyer, Dean La Roche
Muirhead, Ernest Eric
Mulcahy, Gabriel Michael
Mulligan, Richard Michael
Murad, Tariq Mohammed
Murphy, Eduardo S
Murphy, James Clair
Mushett, Charles Wilbur
Musser, A Wendell
Mustard, James Fraser
Myers, Jeffery
Myhre, Byron Arnold
Nadel, Eli Maurice
Naeye, Richard L
Nagler, Arnold Leon
Naib, Zuher M
Nakamura, Robert Motoharu
Naumann, Hans Norbert
Neiman, Benjamin H
Nelson, Dallas Leroy
Nettleship, Anderson
Neubecker, Robert Duane
Neustein, Harry (Bernard)
Newberne, Paul M
Newman, William
Newton, Berne Loyst
Newton, William Allen, Jr
Nice, Philip Oliver
Nickerson, Peter Ayers
Nikles, Nelson Robinson
Nopanitaya, Waykin

Nordquist, Robert Ersel
Nordschow, Carleton Deane
Normann, Sigurd Johns
Noto, Thomas Anthony
Nowell, Peter Carey
Nye, Sylvanus William
Ober, William B
Oberman, Harold A
O'Brien, John S
O'Connor, Michael L
O'Conor, Gregory Thomas
O'Donovan, Cornelius Joseph
Oels, Helen C
O'Gara, Roger
Oh, Jang Ok
Oka, Masamichi
Okazaki, Haruo
Olmstead, Edwin Vincent
Olson, Albert Lloyd
O'Neal, Robert Munger
O'Neill, Frank John
Orbison, Carl Andrew
Osborne, Carl Andrew
Osmanski, C Paul
Ozzello, Luciano
Padgett, George Arnold
Pakes, Steven P
Palm, Paul Eugene
Panner, Bernard J
Pape, Brian Eugene
Paronetto, Fiorenzo
Parsons, Robert Jerome
Passmann, John Martin
Past, Wallace Lyle
Patrick, James R
Patten, Stanley Fletcher, Jr
Pechet, Giselle S
Peckham, John Cecil
Peery, Thomas Martin
Peng, Shi-Kaung
Penick, George Dial
Penner, Donald Wills
Perrin, Eugene Victor
Pettit, Robert Eugene
Phillips, Melville James
Phillips, Mildren E
Pickren, John Warren
Pierce, Emmett Coin
Pierce, Gordon Barry
Pierson, K Kendall
Pietra, Giuseppe G
Pirani, Conrad Levi
Pirkle, Hubert Chaille
Pitot, Henry C, III
Pizzolato, Philip
Platt, William Rady
Poel, William Elias
Polet, Herman
Pollak, Otakar Jaroslav
Popper, Hans
Porter, David Dixon
Powers, Robert D
Powsner, Edward R
Pratt, Philip Chase
Pratt-Thomas, Harold Rawling
Prendergast, Robert Anthony
Pretlow, Thomas Garrett, II
Pribor, Hugo Casimer
Price, Harold M
Price, Robert Allen
Prichard, Robert Williams
Prior, John Thompson
Prose, Philip H
Quigley, Herbert Joseph, Jr
Quittner, Howard
Rabin, Erwin R
Rabson, Alan S
Race, George Justice
Ragland, William Lauman, III
Ramsey, Elizabeth Mapelsden
Rapp, John P
Rapp, William Rodger
Rappaport, Henry
Raska, Karel Frantisek, Jr
Raskova, Jana D
Rasmussen, Peter
Rather, Lelland Joseph
Rawnsley, Howard Melody
Rawson, Arnold Joseph
Reagan, James W
Reals, William Joseph
Reddy, Janardan K
Redman, Robert Shelton
Reed, George Benson
Reed, Richard Jay
Reilly, Emmett B
Reimann, Dexter Leroy
Reiner, Charles Brailove
Reinhard, Karl R
Reiskin, Allan B
Reuber, Melvin D
Reyes-Mota, Alfonso
Reynolds, Rolland
Reynoso, Gustavo D
Rice, Walter Gowans
Richardson, Howard Lockhart
Richardson, Mary Elizabeth
Richart, Ralph M

Richter, Goetz Wilfried
Richter, Maxwell
Richter, Ward Robert
Rieksiniece, Emilija Katrina
Rigdon, Raymond Harrison
Ritchie, Alexander Charles
Robb, James Arthur
Robbins, Stanley Leonard
Roberts, James C, Jr
Robertson, Abel Alfred Lazzarini, Jr
Robinson, Hamilton Burrows Greaves
Robinson, Harry L
Rodin, Alvin E
Rodman, Nathaniel Fulford, Jr
Roeckel, Irene E
Rogers, Stanfield
Rohovsky, Michael William
Roma, Doris Spector
Rona, George
Rosan, Robert Carl
Rosenau, Werner
Rosenthal, Sol Roy
Ross, Oscar Alan
Ross, Roderick Clendenning
Roth, Daniel
Roth, Lawrence Max
Roth, Sanford Irwin
Routh, Joseph Isaac
Rowlands, David T, Jr
Rubin, Emanuel
Rubnitz, Myron Ethan
Rude, Theodore Alfred
Ruebner, Boris Henry
Rukstinat, George John
Russell, William Ogburn
Russfield, Agnes Burt
Rywilin, Arkadi Michael
Saccomanno, Geno
Sadek, Salah Eldine
Salazar, Hernando
Salgado, Ernesto D
Sampson, Calvin Coolidge
Sander, Charles H
Santamarina, Enrique
Savory, John
Sayre, George Pomeroy
Schenken, John Rudolph
Schmidt, Robert W
Schnitzer, Bertram
Schnuda, Nasr Danial
Schochet, Sydney Sigfried, Jr
Schor, Norberto Aaron
Schrek, Robert
Schulz, Dale Metherd
Schwartz, Heinz (Georg)
Scully, Robert Edward
Seidman, Irving
Seidman, David
Seligson, David
Sell, Stewart
Selzer, Isidore
Semancik, Joseph Stephen
Senhauser, Donald Albert
Seronde, Joseph, Jr
Seybolt, John Francis
Shanberge, Jacob N
Shankin, Douglas R
Shapiro, John Lawton
Shea, Stephen Michael
Shelburne, John Daniel
Sheldon, Huntington
Sheldon, Walter Herman
Sherman, Frank Edward
Sherrick, Joseph C
Sherwin, Russell P
Shimamura, Tetsuo
Shively, John Adrian
Shivers, Billie Rae
Shnitka, Theodor Khyam
Shubik, Philippe
Sidransky, Herschel
Siegel, Henry
Sikes, Dennis
Silberberg, Ruth
Silver, Malcolm David
Simon, Gerard Theodor
Simon, Morris Arthur
Sinclair, Elizabeth Faye
Sisson, Joseph A
Skinner, Margaret Sheppard
Skinsnes, Olaf Kristian
Slager, Ursula Traugott
Slatkin, Daniel Nathan
Smith, Albert Goodin
Smith, Alice Lorraine
Smith, David Waldo Edward
Smith, George Stahl
Smith, George Thomas
Smith, John Chandler
Smith, Roger Dean
Snell, Katherine Chapin
Sobel, Harold John
Socha, Leslie Howard
Socha, Wladyslaw Wojciech
Sohn, David
Sokoloff, Leon
Solomon, Gordon Charles
Solomon, Robert Douglas

Jurgelsky, William, Jr
Kaley, Gabor
Kamm, Richard Conrad
Kanegis, Leon Abbott
Koch-Weser, Dieter
Kohn, Robert Rothenberg
Lehman, John Michael
Leifer, Calvin
Liebelt, Robert Arthur
Lockwood, William Rutledge
Ludwig, Frederic C
Machado, Emilio Alfredo
Mahaffey, Kathryn Rose
Martin, George Monroe
Mason, James William
Matukas, Victor John
McCully, Kilmer Serjus
McDonagh, Richard Patrick, Jr
McMaster, Philip Robert Bache
Meier, Hans
Messier, Bernard
Middleton, Charles Cheavens
Migliozzi, Joseph Andrew
Minor, Ronald R
Molteni, Agostino
Murphy, Edwin Daniel
Murphy, George Edward
Nakano, James Hiroto
Nichols, Barbara Ann
Nishimura, Edwin Takayasu
Odenheimer, Kurt John Sigmund
Osborne, John Clark
Page, Roy Christopher
Parker, John William
Pearce, Richard Hugh
Porta, Eduardo Angel
Poste, George Henry
Potter, Jacobus Louw
Prehn, Richmond Talbot
Priest, Robert Eugene
Purnell, Dallas Michael
Ram, Madhira Dasaradhi
Ranadive, Narendranath Santuram
Renaud, Serge
Reynolds, Edward Storrs, Jr
Rice, Jerry Mercer
Richters, Arnis
Rifkin, Barry Richard
Robinovitch, Murray R
Robison, Wilbur Gerald, Jr
Rogers, Adrianne Ellefson
Ross, Russell
Rowley, Donald Adams
Rubenstein, Howard Stuart
Russo, Jose
Santos-Buch, Charles A
Sarma, Dittakavi S R
Scarpelli, Dante Giovanni
Smith, William Edgar
Smuckler, Edward Aaron
Stern, Elizabeth
Stevens, Jack Gerald
Studzinski, George P
Sun, Chao Nien
Sunderman, Frederick William, Jr
Swanson, John Lee
Thorbecke, Geertruida Jeanette
Tomashefsky, Philip
Ts'ao, Chung-Hsin
Upton, Arthur Canfield
Weigensberg, Bernard Irvine
Weinstein, Ronald S
Weston, Jean Kendrick
Wiener, Stanley L
Wilson, Floyd Dee
Wilson, Robert Burton
Wolf, Norman Sanford
Yuen, Ted Gim Hing

Fish Pathology
Amend, Donald Ford
Gibson, George G
Landolt, Marsha LaMerle
Lewis, Donald Howard
Post, George
Rogers, Wilmer Alexander
Walker, Roland
Wobeser, Gary Arthur

Forest Pathology
Amburgey, Terry L
Berry, Frederick Hamer
Boyce, John Shaw, Jr
Campana, Richard John
Campbell, William Andrew
Canfield, Elmer Russell
Crosby, Emory Spear
Davidson, Alexander Grant
Davidson, Ross Wallace
Davis, Terry Chaffin
Driver, Charles Henry
Eslyn, Wallace Eugene
Etheridge, David Elliott
French, David Weston
Gilbertson, Robert Lee
Hawksworth, Frank Goode
Henry, Berch Waldo
Hepting, George Henry

Hesterberg, Gene Arthur
Highley, Terry L
Houston, David Royce
Howe, Virgil K
Hunt, Richard Stanley
Kimmey, James William
Krebill, Richard G
Lachance, Denis
Leaphart, Charles Donald
Lightle, Paul Charles
McCracken, Francis Irvin
McGrath, William Thomas
McKenzie, Malcolm Arthur
McNabb, Harold Sanderson, Jr
Merrill, William
Morton, Harrison Leon
Nordin, Vidar John
Paine, Lee Alfred
Patton, Robert Franklin
Phelps, William Robert
Rice, Peter (Franklin)
Ross, Eldon Wayne
Roth, Lewis Franklin
Rowan, Samuel James
Setliff, Edson Carmack
Shaw, Charles Gardner, III
Shea, Keith Raymond
Smith, Richard Barrie
Stambaugh, William James
Tainter, Franklin Hugh
Toole, Eben Richard
Towers, Barry
Whitney, Roy Davidson
Wilcox, Webster Wayne
Zabel, Robert Alger
Zak, Bratislav

Histopathology
Artizzu, Maria
Daoust, Roger
Goltz, Robert W
Harris, Norman Oliver
Jewell, Frederick Forbes
Koepp, Stephen John
Kondo, Edward Shin-Chi
Levy, Barnet M
Ouellette, Guillemond Benoit
Smith, James Knox
Yevich, Paul Peter

Human Pathology
Bensch, Klaus George
Cohen, Martin William
Conway, Kenneth Edward
Cooper, John (Hanwell)
Volk, Thomas Lewis

Immunopathology
Bennett, William Ernest
Chase, William Henry
Duncan, John Robert
Dvorak, Ann Marine-Tompkins
Fierer, Joshua A
Fleetwood, Mildred Kaiser
Hall, Irvin Monroe
Halonen, Marilyn Jean
Hammond, Mary Elizabeth
Haskill, John Stephen
Jones, David Stevens
Koffler, David
Locke, Robert F
Marx, James John, Jr
Perper, Robert J
Pinckard, Robert Neal
Smith, Henry John
Valenzuela, Rafael
White, Thomas George
Whiteside, Theresa L
Yoshida, Takeshi

Insect Pathology
Adams, Jean Ruth
Angus, Thomas Anderson
Bell, Marion Randolph
Briggs, John Dorian
Brooks, Wayne Maurice
Brown, Anthony William Aldridge
Cantwell, George E
Castillo, Jessica Maguila
Clark, Truman Benton
Cunningham, John Castel
Dulmage, Howard Taylor
Goodwin, Ronald Hayse
Harper, James Douglas
Holbrook, Frederick R
Hughes, Kenneth Marion
Jaques, Robert Paul
Kaya, Harry Kazuyoshi
Kramer, John Paul
Kurtti, Timothy John
Maddox, Joseph Vernard
Nordin, Gerald LeRoy
Padhi, Sally Bulpitt
Ramoska, William Allen
Reed, David Kent
Roberts, Donald Wilson
Shimanuki, Hachiro

Stairs, Gordon R
Tamashiro, Minoru
Tanada, Yoshinori
Vaughn, James L
Zimmack, Harold Lincoln

Invertebrate Pathology
Canerday, Thomas Donald
Davidson, Elizabeth West
Drake, Edward Lawson
Federici, Brian Anthony
Granados, Robert R
Hoskin, George Perry
Ignoffo, Carlo Michael
Johnson, Phyllis Truth
Mix, Michael Cary
Reichelderfer, Charles Franklin
Richards, Charles Selwyn
Tripp, Marenes Robert
Wilson, William Thomas

Neuropathology
Adams, Raymond D
Alvord, Ellsworth Chapman, Jr
Anderson, Paul J
Berry, Richard G
Brown, William Jann
Cowen, David
Cravioto, Humberto
Derby, Bennett Marsh
Dudley, Alden Woodbury, Jr
Engel, Andrew G
Feigin, Irwin Harris
Friede, Reinhard L
Garcia, Julio H
Gilles, Floyd Harry
Gonzalez-Angulo, Amador
Graham, Doyle Gene
Haymaker, Webb Edward
Hirano, Asao
Jenkins, Thomas William
Jones, Margaret Zee
Kaufman, Mavis Anderson
Kepes, John J
Kim, Seung U
Kirkpatrick, Joel Brian
Kornfeld, Mario O
Krigman, Martin Ross
Lapham, Lowell Winship
Lindenberg, Richard
Lipkin, Lewis Edward
Liss, Leopold
Malamud, Nathan
Mancall, Elliott L
McCormick, William F
McDonald, Larry William
Nelson, James S
Okazaki, Haruo
Peress, Nancy E
Pope, Alfred
Raine, Cedric Stuart
Rewcastle, Neill Barry
Robertson, David Murray
Roizin, Leon
Rorke, Lucy Balian
Rosenblum, William I
Rozdilsky, Bohdan
Rubinstein, Lucien Jules
Shaw, Cheng-Mei
Shelanski, Michael L
Sidman, Richard Leon
Song, Sun Kyu
Steward, Vincent William
Terry, Robert Davis
Unterharnscheidt, Friedrich J
Wagner, John Alfred
Wisniewski, Henryk Miroslaw
Wolf, Abner
Zeman, Wolfgang

Oral Pathology
Archard, Howell Osborne, Jr
Bartley, Murray Hill, Jr
Bernick, Sol
Bernier, Joseph Leroy
Chaudhry, Anand P
Chen, Sow-Yeh
Cohen, Lawrence
Conroy, Charles William
Crawford, William Howard, Jr
Dick, Henry Marvin
Drinnan, Alan John
Elzay, Richard Paul
Fischman, Stuart L
Friedman, Nathan
Funk, Edward C
Futrell, Maurice Chilton
Gardner, David Godfrey
Garrington, George Everett
Goldman, Henry Maurice
Gorlin, Robert James
Greene, George W, Jr
Halperin, Victor
Halstead, Charles Lemuel
Hammond, Harold Logan
Hanks, Carl Thomas
Hansen, Louis Stephen
Higa, Leslie Hideyasu

Howell, Francis V
Hunter, Harold Alexander
Hurt, William Clarence
Johansen, Erling
Johnson, Clinton Charles
Keene, Harris J
Kelln, Elmer
King, Ordie Herbert, Jr
Krutchkoff, David James
Little, James W
Main, James Hamilton Prentice
Martinez, Mario Guillermo, Jr
McMurchy, Kenneth Allan
Melrose, Raymond John
Medak, Herman
Meyer, Irving
Mezl, Zdenek
Miller, Arthur Simard
Nalbandian, John
Olson, Donald Lee
Porter, Chastain Kendall
Pullon, Peter Akins
Regezi, Joseph Alberts
Rickles, Norman Harold
Rifkin, Barry Richard
Rossi, Edward P
Rovin, Sheldon
Rowe, Nathaniel H
Rubinstein, Alicia Susana
Sabes, William Ruben
Salley, John Jones
Schiess, Alvin V
Schuback, Philip
Shafer, William Gene
Shklar, Gerald
Smith, Roy Martin
Smulow, Jerome B
Spouge, John Douglas
Stahl, S Sigmund
Standish, Samuel Miles
Stanley, Harold Russell
Susi, Frank Robert
Thompson, Charles Calvin
Tomich, Charles Edward
Toto, Patrick D
Waldron, Charles A
Weathers, Dwight Ronald
Wertheimer, Frederick William
Wescott, William B
Witkop, Carl Jacob, Jr
Wood, Norman Kenyon
Woolley, LeGrand H
Wussow, George C
Yamane, George M
Zimmermann, Eugene Robert Charles

Pathobiology
Collier, Robert John
Couch, John Alexander
Cutler, Leslie Stuart
Foley, David Allen
Harshbarger, John Carl, Jr
Jakowska, Sophie (Mrs C L Jeannopoulos)
Rifkin, Erik
St Clair, Richard William
Saunders, George Cherdron
Schultz, Warren Walter
Smith, Albert Carl
Taylor, Donald Fulton
Van Sickle, David C
Vernick, Sanford H
Yang, Tsu-Ju (Thomas)

Pathological Chemistry
Bide, Richard W
Glende, Eric A, Jr
Naumann, Hans Norbert
Nicholson, Thomas Frederick

Pathological Physiology
Burton, Russell Rohan

Phytopathology
Abney, Thomas Scott
Adams, Peter B
Adams, Robert Evans
Agrios, George Nicholas
Aho, Paul E
Aichele, Murit Dean
Aist, James Robert
Alcorn, Stanley Marcus
Alexander, Paul Marion
Allen, Arthur (Silsby)
Allen, Ross Marvin
Allen, Thomas Cort, Jr
Allen, Wayne Robert
Allington, William B
Allison, Joseph Lewis
Allison, Patricia (Lee) (Van Burgh)
Althaus, Ralph Elwood
Altmann, Jack
Alvarez, Anne Maino
Amador, Jose Manuel
Amato, Vincent Alfred
Andersen, Axel Langvad
Anderson, Neil Albert
Anzalone, Louis, Jr
Apple, Jay Lawrence

Kaufman, Harold
Kaufman, Donald DeVere
Kaufmann, Maurice John
Keeling, Bobbie Lee
Keen, Noel Thomas
Keil, Harry Louis
Keller, John Randall
Kelman, Arthur
Kenaga, Clare Burton
Kendrick, Edgar Lohr
Kendrick, James Blair, Jr
Kenknight, Glenn
Kennedy, Bill Wade
Kernkamp, Milton F
Kerr, Eric Donald
Kessler, Kenneth J, Jr
Khan, Sekender Ali
Kiesling, Richard Lorin
King, Thomas Henry
Kingsland, Graydon Chapman
Kingsolver, Charles H
Kirk, Ben Truett
Klarman, William L
Kliejunas, John Thomas
Klisiewicz, John Michael
Klomparens, William
Klos, Edward John
Knaphus, George
Knauss, James Frederick
Knorr, Louis Carl
Knutson, Kenneth Wayne
Knutson, Roger M
Koike, Hideo
Kommedahl, Thor
Konicek, Donald E
Kraft, John M
Kreutzer, William Alexander
Krupa, Sagar
Krupka, Lawrence Ronald
Krusberg, Lorin Ronald
Kucharek, Thomas Albert
Kuhlman, Elmer George
Kuhn, Cedric W
Kulik, Martin Michael
Lacasse, Norman L
Lachance, Rene Onesime
Lacy, Melvyn Leroy
Laemmlen, Franklin
Lai, Ping-Yuen
Lambe, Robert Carl
Lamey, Howard Arthur
Langenberg, Willem G
LaPrade, Jesse Cobb
Larsen, Philip O
Latham, Archie J
Latterell, Frances Meehan
Lautz, William
Laviolette, Francis A
Leach, Charles Morley
Leach, Lysle Douglas
Lear, Bert
Leath, Kenneth T
Lebeau, Jack Bertram
Leben, Curt (Charles)
Leppik, Elmar Emil
Lewis, Frank Herbert
Lewis, Gwynne David
Lewis, Jack A
Lewis, Ralph William
Leyendecker, Philip Jordon
Lim, Sung Man
Lindberg, George Donald
Linderman, Robert G
Lindsey, Donald Leroy
Linn, Manson Bruce
Lippert, Laverne Francis
Littlefield, Larry James
Livingston, Clark Holcomb
Locke, Seth Barton
Lockwood, John LeBaron
Logsdon, Charles Eldon
Lorber, James W
Louie, Raymond
Lucas, George Blanchard
Lucas, Leon Thomas
Ludwig, Ralph Antony
Luepschen, Norman Siegfried
Luke, Herbert Hodges
Lukens, Raymond James
Lukezic, Felix Lee
Lumsden, Robert Douglas
Lutes, Dallas D
Lyda, Stuart D
Maas, John Lewis
MacKenzie, David Robert
MacLachlan, Donald Stuart
MacSwan, Iain Christie
Mader, Erich Otto
Magie, Robert Ogden
Mai, William Frederick
Main, Charles Edward
Maloy, Otis Cleo, Jr
Mankin, Cleon J
Manzer, Franklin Edward
Marchetti, Marco Anthony
Marlatt, Robert Bruce
Martens, John William
Martin, Terry Joe

Martin, Weston Joseph
Martinson, Charlie Anton
Marx, Donald Henry
Marx, Gerald Alvin
Mathre, Donald Eugene
Maxwell, Douglas Paul
Mayhew, Dennis Ed
Mayol, Perpetuo S
McAnelly, Charles William
McCain, Arthur Hamilton
McCallan, Samuel Eugene Alan
McClellan, Wilbur Dwight
McCrum, Richard Caswell
McDonald, Geral Irving
McDonald, William Craik
McDowell, Larry Leon
McElroy, Fred Dee
McFadden, Lorne Austin
McGrew, John Roberts
McGuire, James Marcus
McIntosh, David Livingston
McIntyre, Gary A
McIntyre, John Lee
McKeen, Colin Douglas
McKeen, Wilbert Ezekiel
McLaughlin, James L
McLean, Darrell Marshall
McMeekin, Dorothy
McMillan, Robert Thomas, Jr
McNew, George Lee
Meiners, Jack Pearson
Meister, Charles William
Melching, J Stanley
Meyer, Ronald Warren
Milbrath, Gene McCoy
Miller, Carol Raymond
Miller, Harold James
Miller, Howard Nile
Miller, John Wesley
Miller, Lawrence Ingram
Miller, Patrick Martin
Miller, Paul R
Miller, Paul William
Miller, Robert Ernest
Miller, Robert Walker, Jr
Miller, Thomas
Mills, John T
Mills, Wilford Richard
Mircetich, Srecko M
Mirocha, Chester Joseph
Miskimen, Carmen Rivera
Mitchell, John Edwards
Mitchell, Norman L
Moline, Harold Emil
Moody, Arnold Ralph
Moore, Ermer Leon
Moore, James Frederick, Jr
Moore, John Duain
Moore, John (Newton)
Moore, Larry Wallace
Moore, Laurence Dale
Morehart, Allen L
Morgan, Omar Drennan, Jr
Moseman, John Gustav
Moser, Paul E
Motsinger, Ralph E
Mount, Mark Samuel
Mueller, Walter Carl
Muir, William Howard
Mullin, Robert Spencer
Mumford, David Louis
Munger, George Donald
Munnecke, Donald Edwin
Munro, James
Murakishi, Harry Haruo
Murphy, Douglas Richard
Mussell, Harry W
Nagel, Clatus Martin
Nair, Gangadharan V M
Nakayama, Takao
Neely, Robert Dan
Nelson, Earl Edward
Nelson, Gordon Albert
Nelson, Klayton Edward
Nelson, Merritt Richard
Nelson, Paul Edward
Nelson, Richard Robert
Nemec, Stanley
Newton, H Calvin, Jr
Niblett, Charles Leslie
Nichols, Carl William
Nicholson, Ralph Lester
Niederhauser, John Strong
Nielsen, Jens Juergen
Nielsen, Lowell Wendell
Nigh, Edward Leroy, Jr
Nighswander, James Edward
Novak, Robert Otto
Nugent, Thomas John
Nusbaum, Charles Joseph
Nyland, George
Nyvall, Robert Frederick
Ogawa, Joseph Minoru
Ohms, Richard Earl
Olah, Arthur Frank
Olsen, Carl Mark
Ooka, Jeri Jean
O'Reilly, Henry James

Orellana, Rodrigo Gonzalo
Ortega, Jacobo
Oshima, Nagayoshi
Ostazeski, Stanley A
Oswald, John Wieland
Otta, Jack Duane
Palm, Elmer Thurman
Palmer, John Gilbert
Palmer, Louis Thomas
Panopoulos, Nickolas John
Papavizas, George Constantine
Pappelis, Aristotel John
Parker, Kenneth Gardner
Parmeter, John Richard, Jr
Parrish, David Joe
Partridge, Arthur Dean
Partyka, Robert Edward
Patil, Suresh Siddheshwar
Patrick, Zenon Alexander
Paulsen, Avelina Quiaoit
Paulus, Albert
Paxton, Jack Dunmire
Pecknold, Paul Carson
Peeples, John Lee, Jr
Pelletier, Eugene Neil
Pelletier, Real Lucien
Pepin, Herbert Spencer
Petersen, Donald Harry
Peterson, Glenn Walter
Phipps, Patrick Michael
Pierson, Charles Frederick
Pirone, Pascal Pompey
Pirone, Thomas Pascal
Politis, Demetrios John
Pontis-Videla, Rafael Edmundo
Pool, Robert Alfred Frank
Porter, Clark Alfred
Porter, Daniel Morris
Potter, Howard Spencer
Pound, Glenn Simpson
Powell, Charles Carleton, Jr
Powell, Dwight
Powell, Nathaniel Thomas
Powell, William Morton
Powelson, Robert Loran
Powers, Harry Robert, Jr
Prescott, Jon Michael
Presley, John Thomas
Pring, Daryl Roger
Pristou, Robert
Pryor, Dean Earl
Pueppke, Steven Glenn
Purdy, Laurence Henry
Pyenson, Louis L
Raabe, Robert Donald
Rader, William Ernest
Radewald, John Dale
Rahe, James Edward
Ramsdell, Donald Charles
Ramsey, Richard Harold
Ranney, Carleton David
Raymer, William Bruce
Reed, Howard Ernest
Reiling, Theodore Paul
Reinert, Richard Allyn
Reynolds, James Edward
Rice, William Newell
Rich, Avery Edmund
Rich, Saul
Richards, Bert Lorin, Jr
Richardson, Lloyd Thomas
Ries, Stephen Michael
Riggs, Robert D
Riker, Albert Joyce
Roane, Curtis Woodard
Roberts, Daniel Altman
Rochow, William Frantz
Roelfs, Alan Paul
Rogers, Jack David
Rogers, Wallace Edward
Rohde, Richard Allen
Rohrbach, Kenneth G
Rohringer, Roland
Romanko, Richard Robert
Romig, Robert William McClelland
Roncadori, Ronald Wayne
Rosberg, David William
Rosenberg, Dan Yale
Rosenkranz, Eugen Emil
Rosher, Ronald Mailand
Ross, John Paul
Ross, Robert Gordon
Roth, Jonathan Nicholas
Rowe, Randall Charles
Rowell, John Bartlett
Ruehle, John Leonard
Ruppel, Earl George
Rusden, Philip Lowry
Rush, Milton Charles
Russell, Thomas Edward
Ryker, Truman Clifton
Saari, ...
Sackston, Waldemar Esi
Salch, Richard K
Sanden, Gerald Ernest
Sands, David Chandler
Sauer, David Bruce
Savage, Earl John

Sayre, Richard Martin
Schaad, Norman W
Schafer, John Francis
Scharen, Albert Lois
Scharpf, Robert F
Scheffer, Robert Paul
Schein, Richard David
Schenck, Norman Carl
Schieber, Eugenio
Schipper, Arthur Louis, Jr
Schlegel, David Edward
Schmidt, Donald Peter
Schmitt, Chris George
Schmithenner, August Fredrick
Schnathorst, William Charles
Schneider, Henry
Schneider, Irving Robert
Schoeneweiss, Donald F
Schoulties, Calvin Lee
Schreiber, Lawrence
Schroeder, Harry William
Schroth, Milton Neil
Schultz, Otto Ernst
Schwegmann, Jack Carl
Schwenk, Fred Walter
Scott, Donald Howard
Scott, Howard Allen
Seaman, William Lloyd
Segall, Raphael Herman
Sehgal, Om Parkash
Seliskar, Carl Edward
Semeniuk, George
Sequeira, Luis
Shalla, Thomas Allen
Shaner, Gregory Ellis
Sharp, Eugene Lester
Sharvelle, Eric George
Shaw, Charles Gardner
Shaw, John Gordon
Shaw, Michael
Shay, Junior Ralph
Shepard, James F
Sherf, Arden Frederick
Sherwood, Robert Tinsley
Shigo, Alex, Lloyd
Shortle, Walter Charles
Shriner, David Sylva
Shurtleff, Malcolm C, Jr
Siddiqui, Wasi Mohammad
Siegel, Malcolm Richard
Sievert, Richard Carl
Silbernagel, Matt Joseph
Sill, Webster Harrison, Jr
Silverman, William Bernard
Simons, Marr Dixon
Simpson, William Roy
Sims, Asa C, Jr
Sinclair, James Burton
Sinclair, Wayne A
Sinden, James Whaples
Sinden, Stephen Lee
Singh, Dilbagh
Singh, Jaswant
Sinha, Ramesh Chandra
Sisler, Hugh Delane
Sitterly, Wayne R
Skilling, Darroll Dean
Skoropad, William Peter
Skotland, Calvin B
Slack, Derald Allen
Slack, Steven Allen
Sleeth, Bailey
Slykhuis, John Timothy
Smalley, Eugene Byron
Smiley, Richard Wayne
Smith, Glenn Edward
Smith, Harlan Eugene
Smith, Jeffrey Drew
Smith, Richard S, Jr
Smith, Samuel H
Smith, Thomas Earle
Smith, William Hulse
Smith, Wilson Levering, Jr
Smoot, John Jones
Snow, Gordon Franklin
Snow, Jean Anthony
Snow, Johnnie Park
Snow, Michael Dennis
Snyder, Hugh Donald
Solberg, Richard Allen
Sommer, Noel Frederick
Sonoda, Ronald Masahiro
Southards, Carroll J
Sowell, Grover, Jr
Spalding, Donald Hood
Spencer, James Alphus
Spotts, Robert Allen
Springer, John Kenneth
Sproston, Thomas, Jr
Spurr, Harvey Wesley, Jr
Stace-Smith, Richard
Stadther, Richard James
Staffeldt, Eugene Edward
Stakman, Elvin Charles
Staley, John M
Stall, Robert Eugene
Stanghellini, Michael Eugene
Staples, Richard Cromwell

Johnson, Kenneth Harvey
Jones, Larry Philip
Jones, Thomas Carlyle
Keahey, Kenneth Karl
Kennedy, George Artie
Kennedy, Peter Carleton
Kerr, Kirklyn M
Kluge, John McKain
Kluge, John Paul
Kociba, Richard Joseph
Koestner, Adalbert
Krook, Lennart Per
Kubin, Rosa
Kurtz, Harold John
Liu, Si-Kwang
Long, John Frederick
Lund, John Edward
MacNamee, James K
Mason, Marcus M
Maurer, Fred Dry
McClure, Harold Monroe
McEntee, Kenneth
McGavin, Matthew Donald
McGill, Lawrence David
McKelvie, Douglas H
McKenzie, Basil Everard
Mebus, Charles Albert
Migaki, George
Miller, Janice Margaret
Miller, Lyle Devon
Mills, James Herbert Lawrence
Miner, Merthyr Leilani
Monlux, Andrew W
Moon, Harley W
Moorhead, Philip Darwin
Morehouse, Lawrence G
Moulton, Jack E
Murchison, Thomas Edgar
Nagode, Larry Allen
Newberne, James Wilson
Nielsen, N Ole
Nielsen, Svend Woge
Norcross, Marvin Augustus
Nordin, Robert W
Olander, Harvey Johan
Olson, LeRoy David
Osburn, Bennie Irve
Panciera, Roger J
Payne, Bobby Joe
Perman, Victor
Phemister, Robert David
Pierce, Kenneth Ray
Piper, Richard Carl
Pool, Roy Ransom
Post, John E
Ramsey, Frank K
Ray, James Alton
Rehfield, Carl Ernest
Reynolds, Harry Aaron, Jr
Ribelin, William Eugene
Rich, Lonnie Joe
Riser, Wayne H
Roberts, Edgar D
Robinson, Farrel Richard
Robinson, Virgil Benton
Rooney, James Rowell
Rothenbacher, Hansjakob
Rowsell, Harry Cecil
Sanger, Vance L
Sasmore, Daniel Paul
Saunders, James Robert
Saunders, Leon Z
Sautter, Jay Howard
Schiefer, H Bruno
Schmidt, Donald Arthur
Schmitte, Samuel Conrad
Schmitz, John Albert
Schwartz, William Lewis
Seaton, Vaughn Allen
Seibold, Herman Rudolph
Shields, Robert Pierce
Shively, James Nelson
Shott, Leonard D
Siccardi, Frank John
Simon, Joseph
Simpson, Charles Floyd
Sippel, William Lawrence
Sisk, Dudley Byrd
Sleight, Stuart Duane
Smith, David Lawrence Thomson
Snyder, Stanley Paul
Stevens, Jerry Bruce
Storrs, Ralph Woodrow
Stula, Edwin Francis
Swenberg, James Arthur
Swerczek, Thomas Walter
Szczech, George Marion
Thomassen, Robert William
Thompson, Samuel Wesley, II
Thomson, Reginald George
Trapp, Allan Laverne
Tryphonas, Leander
Tucker, Walter Eugene, Jr
Tyler, David E
Van Kampen, Kent Rigby
Van Vleet, John F
Van Zwieten, Matthew Jacobus
Vlaovic, Milan Stephen

Voelker, Richard William
Waller, Ernest Frederick
Walsh, Alexander Hamilton
Ward, Jerrold Michael
Washko, Floyd Victor
Watrach, Adolf Michael
Waxler, Glenn Lee
Webb, Willis Keith
Webster, Harris Duane
Whiteman, Charles E
Wolf, George L
Wolke, Richard Elwood
Young, Stuart

Wildlife Pathology
Cheatum, Evelyn Leonard

PHARMACOLOGY

Pharmacology
Abernathy, Charles Owen
Aboul-Enein, Hassan Youssef
Abuzzahab, Faruk S, Sr
Aceto, Mario Domenico Giulio
Acosta, Daniel, Jr
Adams, Herbert Jack
Adams, John George
Adams, Max David
Adamson, Richard H
Adler, Martin William
Agrawal, Krishna Chandra
Ahlquist, Raymond Perry
Ahrens, Franklin Alfred
Aiello, Edward Lawrence
Akera, Tai
Aldinger, Earl Edward
Aldred, J Phillip
Alexander, Samuel Craighead, Jr
Allen, Donald Orrie
Allen, Julius Cadden
Alper, Milton H
Alphin, Reevis Stancil
Altshuler, Harold Leon
Altszuler, Norman
Altura, Bella T
Altura, Burton Myron
Alvares, Alvito Peter
Ambrus, Clara Maria
Amer, Mohamed Samir
Amory, David William
Anderson, Archie Duane
Anderson, Edmund George
Anderson, Julius Horne, Jr
Anton, Aaron Harold
Antos, Robert John
Aposhian, Hurair Vasken
Appel, Warren Curtis
Appelt, Glenn David
Archer, John Dale
Armstrong, Robert D
Aronson, A L
Aronson, Carl Edward
Ashmore, James Guy
Askari, Amir
Assaykeen, Tatiana Anna
Aston, Roy
Auyong, Theodore Koon-Hook
Aviado, Domingo Mariano
Babcock, George, Jr
Bachmann, Kenneth Allen
Bachur, Nicholas R, Sr
Back, Kenneth Charles
Back, Nathan
Baer, John Elson
Bagdon, Robert Edward
Bagdon, Walter Joseph
Baggett, Billy
Bagwell, Ervin Eugene
Balek, Richard William
Banziger, Ralph Frederick
Bar, Hans-Peter
Barker, Louis Allen
Barker, Samuel Booth
Barnes, Byron Ashwood
Barrett, Walter Edward
Bass, Allan Delmage
Bass, Paul
Bassett, Arthur Leon
Batra, Karam Vir
Batterman, Robert Coleman
Battista, Sam P
Bauer, Robert Oliver
Beck, Lloyd
Beck, Lyle Vibert
Bederka, John Paul, Jr
Beech, John Alan
Begany, Albert John
Beinfield, William Harvey
Belford, Julius
Bello, Carmen T
Ben, Max
Bender, Allan Douglas
Benfey, Bruno Georg
Bennett, Donald Raymond
Benoit, Philippe Stanislas

Benson, Wilbur Maxwell
Bentley, Patrick E
Bentley, Peter John
Benton, Byrl E
Berger, Frank Milan
Berger, James Edward
Bergman, Hyman Chaim
Berke, Harry L
Berkowitz, Barry Alan
Berman, David Albert
Berman, Eleanor
Bernal-Llanas, Enrique
Berndt, William O
Bernheim, Frederick
Berry, Charles Arthur
Bertino, Joseph R
Besch, Henry Roland, Jr
Besse, John C
Bester, John (Francis)
Bettonville, Paul John
Beuthin, Frederic C
Bevan, John Acton
Beyer, Karl Henry, Jr
Bhagat, Budh Dev
Bhatnagar, Ranbir Krishna
Bhattacharya, Amar Nath
Bianchi, Robert George
Bianchine, Joseph Raymond
Bickerton, Robert Keith
Bierwagen, Max Eugene
Bigger, J Thomas, Jr
Biggs, David Frederick
Bihler, Ivan
Bingel, Audrey Susanna
Black, Jack
Blackmore, William Peter
Blair, Murray Reid, Jr
Blake, David Andrew
Blake, Robert L
Blankenship, James William
Bleiberg, Marvin Jay
Blinks, John Rogers
Blouin, Andre
Blumberg, Harold
Bogdanove, Lasca Hospers
Bogdanski, Donald Frank
Booker, Walter Monroe
Booth, Nicholas Henry
Bordeleau, Jean-Marc
Borella, Luis Enrique
Borgstedt, Harold Heinrich
Borison, Herbert Leon
Borowitz, Joseph Leo
Borzelleca, Joseph Francis
Boshart, Charles Ralph
Bosmann, Harold Bruce
Boulos, Badi Mansour
Bourgault, Priscilla C
Bourke, Anne Rosaleen
Bousquet, William F
Bowen, John Metcalf
Bowman, Douglas Clyde
Bowman, Edward Randolph
Bowman, Faye Johnson
Boxill, Gale Clark
Boyd, Carl Edmund
Boyd, Eugene Stanley
Brahen, Leonard S
Brands, Allen J
Braude, Monique Colsenet
Brauer, Ralph Werner
Brazeau, Paul
Breckenridge, Bruce (McLain)
Brendel, Klaus
Brenner, George Marvin
Briggs, Arthur Harold
Brill, Earl
Brillhart, Russell Edward
Brimijoin, William Stephen
Brodie, Bernard Beryl
Brodie, David Alan
Brody, Alfred Walter
Brody, Michael J
Brody, Theodore Meyer
Brooker, Gary
Brostrom, Charles Otto
Brostrom, Margaret Ann
Brown, Daniel Joseph
Brown, Elise Ann
Brown, John Haynes
Brown, Neal Curtis
Brown, Richard Don
Brown, Theodore Gates, Jr
Brummett, Robert E
Brunner, Edward A
Bryan, George Terrell
Bryant, Gordon Henry
Bryant, Harold Horn
Bryant, Shirley Hills
Buchanan, Robert Alexander
Buckley, Joseph Paul
Buday, Paul Vincent
Buller, Robert Henry
Bunag, Ruben David
Burba, John Vytautas
Burford, Hugh Jonathan
Burgison, Raymond Merritt
Burkhalter, Alan

Burkman, Allan Maurice
Burks, Thomas F
Burns, Richard Henry
Burns, Robert B P
Busch, Harris
Butler, Thomas Cullom
Buttar, Harpal Singh
Buyniski, Joseph P
Byck, Robert
Byrd, Daniel Madison, III
Byrd, William Joseph
Byrne, Jeffrey Edward
Byron, Joseph Winston
Byrum, Woodrow Robert
Cafruny, Edward Joseph
Calabresi, Paul
Caldwell, Robert William
Calesnick, Benjamin
Call, Tracey Gillette
Campbell, Lorne Arthur
Campbell, William Bryson
Campbell, William Howard
Cantrell, Elroy Taylor
Cardoso, Sergio Steiner
Carmichael, Ralph Harry
Carr, Charles Jeleff
Carr, Edward Albert
Carr, Laurence A
Carrano, Richard Alfred
Carrier, Oliver, Jr
Carroll, Marcus Newman, Jr
Carson, Steven
Carter, James Roland
Carter, Mary Kathleen
Caruso, Frank San Carlo
Casselman, Warren Gottlieb Bruce
Castellion, Alan William
Castle, Manford C
Castles, Thomas R
Cattell, McKeen
Cenedella, Richard J
Century, Bernard
Cervoni, Peter
Cha, Sungman
Chakrin, Lawrence William
Chalmers, Robert Kenny
Chambers, John William
Chan, Peter Sinchun
Chan, Wah Yip
Chandler, Michael Lynn
Chang, Yi-Chi
Chapman, John E
Chappell, Elizabeth
Charnock, John S
Chau, Raymond Ying Pui
Chen, Ko Kuei
Chen, Philip Stanley, Jr
Chernick, Warren Sanford
Chernov, Harvey Irwin
Chhabra, Rajendra S
Chi, Christina Hadinata
Chiang, Tzu Sung
Chignell, Colin Francis
Chin, Jane Elizabeth Heng
Chiou, George Chung-Yih
Chiou, Win Loung
Chiu, Andrew Tak-Chau
Chiu, Peter Jiunn-Shyong
Cho, Arthur Kenji
Cho, Jae Yne
Chou, Ting-Chao
Christ, Daryl Dean
Ciancio, Sebastian Gene
Cinti, Dominick Louis
Clark, Byron Bryant
Clark, Carl Heritage
Clark, John Kapp
Clark, Julia Berg
Clark, Wesley Gleason
Clay, Michael M
Cochin, Joseph
Cochran, Kenneth William, Jr
Coffey, Donald Straley
Coffey, Ronald Gibson
Cohen, Marlene Lois
Cohen, Marvin
Cohen, Sanford Ned
Cohen, Steven Donald
Coker, Samuel Terry
Colclough, Norma Vesey
Colella, Donald Francis
Collier, Brian
Collins, Alfred Patterson
Collins, Robert James
Combs, Alan B
Commarato, Michael A
Condouris, George Anthony
Conley, Bernard Edward
Connamacher, Robert Henle
Conney, Allan Howard
Connor, John D
Connor, Nolen Duncan
Conrad, Eugene Anthony
Conroe, Paul F
Conway, Alvin Charles
Conzelman, Gaylord Maurice, Jr
Cook, Donald Latimer

PHARMACOLOGY

Huang, Kee-Chang
Huang, Yung-Chen
Huber, Thomas Lee
Hudak, William John
Hudgins, Patricia Montague
Hudson, Roy D
Huebner, Richard A
Hug, Carl Casimir, Jr
Hughes, Maysie J H
Hui, Ferdinand W
Hulme, Norman Arthur
Hume, Arthur Scott
Hunter, Francis Edmund, Jr
Hupka, Arthur Lee
Hurwitz, David Allan
Huston, Mervyn James
Hutcheon, Duncan Elliot
Hutchings, Donald Edward
Hutsell, Thomas Carlyle
Hwang, Kao
Ichniowski, Casimir Thaddeus
Ignarro, Louis Joseph
Imondi, Anthony Rocco
Inashima, Osami James
Inchiosa, Mario Anthony, Jr
Ingalls, James Warren, Jr
Ingenito, Alphonse J
Innes, Ian Rome
Inscoe, Joseph Kenneth
Inturrisi, Charles E
Iorio, Louis Carmen
Irwin, Richard Leslie
Isom, Gary E
Israel, Yedy
Jaanus, Siret Desiree
Jacobowitz, David
Jacobson, Keith Hazen
Jaffe, Julian Joseph
Jandhyala, Bhagavan S
Jaques, Louis Barker
Jardetzky, Oleg
Jarvik, Murray Elias
Jelliffe, Roger Woodham
Jenden, Donald James
Jenkins, Howard Jones
Jenney, Elizabeth Holden
Jensen, Clyde B
Jensen, Richard Arthur
Jerram, David Carroll
Jessup, Daniel Clifford
Jewett, Robert Elwin
Jhamadas, Khem
Johns, Anthony
Johns, David Garrett
Johnson, Alice Ruffin
Johnson, Carl Lynn
Johnson, Eugene Malcolm, Jr
Johnson, Garland A
Johnson, Gordon E
Johnson, Gordon Lee
Johnson, Henry Douglas
Johnson, Howard (Laurence)
Johnson, John Raymond
Johnson, Melvin Clark
Johnson, Wallace W
Johnson, William Everett
Johnston, Raymond F
Joiner, Paul David
Jondorf, Werner Robert
Jones, Vernon Douglas
Jordin, Marcus Wayne
Juchau, Mont Rawlings
Julien, Robert Michael
Jurgelsky, William, Jr
Kabara, Jon Joseph
Kadar, Dezso
Kaiser, Joseph Anthony
Kalant, Harold
Kalelis, Theodore S
Kalow, Werner
Kaiser, Sarah Chinn
Kandel, Alexander
Kao, Chien Yuan
Kaplan, Harvey Robert
Kappas, Attallah
Karczmar, Alexander George
Karel, Leonard
Kariya, Takashi
Karow, Armand Monfort, Jr
Karr, Gerald William
Kasik, John Edward
Kasirsky, Gilbert
Kasses, Kenneth George
Katz, Norman L
Katzung, Bertram George
Kauffman, Frederick C
Kauker, Michael Lajos
Kaul, Pushkar Nath
Keasling, Hugh Hilary
Keats, Arthur Stanley
Keith, Eaden Francis
Kelleher, Roger Thomson
Kelliher, Gerald James
Kelsey, Frances Oldham
Kelsey, Fremont Ellis
Kemp, John Wilmer

Kempen, Rene Richard
Kenny, Alexander Donovan
Kensler, Charles Joseph
Kerley, Troy Lamar
Keyl, Alexander Charles
Khairallah, Philip Asad
Khazan, Naim
Killam, Richard A
Killam, Ellen Eva King
Killam, Keith Fenton, Jr
Kim, Kil Chol
Kim, Yee Sik
Kimura, Eugene Tatsuru
Kimura, Kazuo Kay
King, John Albert
King, Theodore Oscar
Kinnard, William J, Jr
Kinoshita, Florence Keiko
Kiplinger, Glenn Francis
Kirpekar, Sadashiv M
Kissel, John Walter
Kitzes, George
Klaassen, Curtis Dean
Klavano, Paul Arthur
Klein, Richard Lester
Klingman, Gerda Isolde
Klubes, Philip
Knapp, Daniel Roger
Knoppers, Antonie Theodoor
Knotts, Glenn Richard
Kocsis, James Joseph
Kodama, Jiro Kenneth
Koeferl, Michael Tallyn
Koelle, George Brampton
Kopin, Irwin J
Kosersky, Donald Saadia
Kosman, Mary Ellen
Koss, Michael Campbell
Kovacs, Bela A
Kraatz, Charles Parry
Kraemer, Richard John
Krahl, Maurice Edward
Kramer, Sherman Francis
Kramer, Stanley Phillip
Kraus, Shirley Ruth
Krayer, Otto
Kreutner, William
Krivoy, William Aaron
Kroeger, Donald Charles
Krop, Stephen
Krueger, Hugo Martin
Krzanowski, Joseph John, Jr
Kuhn, William Lloyd
Kunos, George
Kuntzman, Ronald Grover
Kuperman, Albert Sanford
Kupferberg, Harvey J
Kupferman, Allan
Kuo, Jyh-Fa
Kvam, Donald Clarence
LaBella, Frank Sebastian
Ladinsky, Herbert
Lage, Gary Lee
Lal, Harbans
Lamb, Sandra Ina
Lambertsen, Christian James
Lampe, Kenneth Francis
Lanzoni, Vincent
Lapidus, Herbert
Larner, Joseph
Larson, Paul Stanley
Larson, Robert Elof
Lasagna, Louis (Cesare)
Laut, Wilfred Wayne
Lawrence, William Homer
Leaders, Floyd Edwin, Jr
Leake, Chauncey D
Leary, John Sylvester
Lech, John James
Lederis, Karl
Lee, Cheng-Chun
Lee, Kwang Soo
Lee, Lyndon Edmund, Jr
Lee, Peter Van Arsdale
Lee, Tee-Ping
Lee, Yien-Hwei
Lehr, David
Leitz, Frederick Henry
Lenney, James Francis
Leonard, Charles Arthur
Leonard, Robert Meyer
Leong, Basil K J
Leslie, Steven Wayne
Lester, David
Leveque, Phillip Edwin
Levi, Roberto
Levin, Robert Martin
Levin, Sidney Seamore
Levine, Robert Alan
Levine, Ruth R
Levine, Walter (Gerald)
Levinskas, George Joseph
Levitt, Barrie
Levitt, Gerhard
Levy, Joseph Victor
Levy, Louis
Lewis, John Reed

Lewis, Roger Abbott
Liao, Ji-Chia
Light, Amos Ellis
Ligon, Edgar William, Jr
Linde, Harry Wight
Lindsay, Raymond H
Linegar, Charles Ramon
Ling, George M
Lipicky, Raymond John
Lish, Paul Merrill
Little, James Maxwell
Liu, C T
Lomax, Peter
Lombardini, John Barry
Long, John Paul
Loomis, Ted Albert
Lorenzetti, Olfeo J
Lotlikar, Prabhakar Dattaram
Lotz, Frederick
Lowenthal, Julius
Lowry, Oliver Howe
Lu, Frank Chao
Lu, Gordon Go
Lubawy, William Charles
Luckens, Mark Manfred
Ludlum, David Blodgett
Luduena, Froilan Pindaro
Lum, Bert Kwan Buck
Lynch, Vincent De Paul
MacCanon, Donald Moore
MacDonald, William E, Jr
Macko, Edward
Mackowiak, Elaine DeCusatis
Macmillan, William Hooper
Macri, Frank John
Maengwyn-Davies, Gertrude Diane
Magee, Donal Francis
Maha, George Edward
Mahgoub, Ahmed
Maickel, Roger Philip
Maines, Mahin D
Makman, Maynard Harlan
Malanga, Carl Joseph
Maling, Harriet Mylander
Mallov, Samuel
Malone, Marvin Herbert
Mancini, Robert Edward
Mandel, Harold George
Mandel, Lewis Richard
Mandik, Jayant V
Manley, Emmett S
Mann, David Edwin, Jr
Manner, Georg Karl
Mannering, Gilbert James
Manno, Barbara Reynolds
Manno, Joseph Eugene
Mansour, Tag Eldin
Manthei, Roland William
Marczynski, Thaddeus John
Maren, Thomas Hartley
Margolin, Solomon
Margolis, Renee Kleinmann
Margolis, Richard Urdangen
Marier, Guy
Markey, Sanford Philip
Marks, Bernard Herman
Marks, Gerald Samuel
Marois, Robert Leo
Marquardt, Hans Wilhelm Joe
Marshall, Franklin Nick
Martel, Rene R
Martin, Edgar J
Martin, Frank Gene
Martin, George Reilly
Martin, Sarah Smith
Martin, William Robert
Martineau, Perry Cyrus
Marzulli, Francis Nicholas
Mason, Robert C
Mason, Walter Harry
Masuoka, David Takashi
Matheny, James Lafayette
Matsumoto, Charles
Matthews, Richard John, Jr
Mautner, Henry George
Mawhinney, Michael G
Maxwell, Donald Robert
Maxwell, Robert Arthur
Mayer, Steven Edward
Mayner, Everett William
Mazel, Paul
Mazurkiewicz-Kwilecki, Irena Maria
McCaman, Richard Eugene
McCann, William Peter
McCarthy, Duncan Arthur, Jr
McCarty, Leslie Paul
McCawley, Elton Leeman
McClure, David Albert
McColl, John Duncan
McCormack, John Joseph, Jr
McCreesh, Arthur Harold
McCurdy, David Harold
McCutcheon, Rob Stewart
McDougal, David Blean, Jr
McGrath, William Robert
McGuire, John L
McIlreath, Fred J
McIsaac, Robert James

McKinney, Gordon R
McLain, Paul Larimer
McManus, Edward Clayton
McMillan, Donald Edgar
McNamara, Bernard Patrick
McNeill, John Hugh
McNutt, Walter Scott
McPhillips, Joseph John
Meacham, Roger Hening, Jr
Medina, Miguel Angel
Meezan, Elias
Megirian, Robert
Mellett, Lawrence B
Mennear, John Hartley
Menzel, Daniel B
Merker, Philip Charles
Meyers, Donald Bates
Meyers, Frederick Henry
Mezey, Kalman C
Michael, William R
Michaelson, I Arthur
Miech, Ralph Patrick
Mieyal, John Joseph
Milch, Lawrence Jacques
Miller, Jack W
Miller, Kenneth Wayne
Miller, Lowell D
Miller, Ralph English
Miller, Tracy Bertram
Mills, Donald Grant
Mills, Elliott
Milne, George McLean, Jr
Minatoya, Hiroaki
Minsker, David Harry
Mirsky, Joseph Herbert
Misher, Allen
Mitchell, Clifford L
Mittag, Thomas Waldemar
Miya, Tom Saburo
Moawad, Atef H
Mockle, Jerry Auguste
Modell, Walter
Modrak, John Bruce
Moffitt, Robert LeVere
Montgomery, Edward Harry
Moon, Byong Hoon
Moore, Joanne Iweita
Moore, Peter Francis
Moran, Neil Clymer
Morgan, Lee Roy, Jr
Morris, N Ronald
Morris, Ralph William
Morris, Robert Nicholas
Morrison, Robert William
Morrissette, Maurice Corlette
Morrow, Paul Edward
Morse, William Herbert
Morton, Joseph James Pandozzi
Moss, Jack N
Mossberg, Howard E
Motsavage, Vincent Andrew
Mueller, Robert Arthur
Mueller, William H
Mule, Salvatore Joseph
Mundy, Roy Lee
Munn, John Irvin
Munson, Paul Lewis
Murdock, Harold Russell, Jr
Murphy, Robert Carl
Murphy, Sheldon Douglas
Murray, John Randolph
Murthy, Vishnubhakta Shrinivas
Mycek, Mary J
Myers, Howard M
Naeger, Leonard L
Nahas, Gabriel Georges
Nair, Velayudhan
Namm, Donald H
Narrod, Marian Freedman
Nash, Charles William
Nash, Clinton Brooks
Nash, Joe Bert
Nebert, Daniel Walter
Nechay, Bohdan Roman
Neidle, Enid Anne
Nelson, James Arly
Nelson, John White
Nelson, Neal Stanley
Nelson, Robert B
Nelson, Stanley Reid
Nelson, Thomas Eustis, Jr
Neu, Harold Conrad
Neuman, Margaret Wrightington
Newman, Robert Alwin
Newman, Walter Hayes
Niblack, John Franklin
Nichol, Charles Adam
Nichols, William Kenneth
Nicholson, John Angus
Nickander, Rodney Carl
Nickerson, Mark
Nishie, Keica
Noble, John F
North, William Charles
Norton, Stata Elaine
Norton, Ted Raymond
Novick, William Joseph, Jr
Nuite, Jo Ann

Tocco, Dominick Joseph
Tocus, Edward C
Tolman, Edward Laurie
Tompkins, E Crosby
Torchiana, Mary Louise
Torosian, George
Toverud, Svein Utheim
Traber, Daniel Lee
Traitor, Charles Eugene
Tramell, Paul Richard
Trams, Eberhard Georg
Trapold, Joseph Hugh
Travis, David M
Trifaro, Jose Maria
Triggle, David J
Triolo, Anthony J
Truitt, Edward Byrd, Jr
Tsai, Tsui Hsien
Tucker, Geoffrey Thomas
Turkanis, Stuart Allen
Tusing, Thomas William
Tuttle, Richard Suneson
Tuttle, Ronald Ralph
Tye, Arthur
Tyler, David Bernard
Tyler, Tipton Ransom
Ungar, Georges
Unna, Klaus Robert
Upson, Dan W
Urquilla, Pedro Ramon
Ursillo, Richard Carmen
Uyeki, Edwin M
Uyeno, Edward Teiso
Valadares, Joseph R E
Van Arman, Clarence Gordon
Vander Brook, Milton John
Van Deripe, Donald R
Vander Wende, Christina
Van Dyke, Knox
Van Maanen, Evert Florus
Van Petten, Garry R
Van Remoortere, Emile C
Van Stee, Ethard Wendel
Van Tyle, William Kent
Venditti, John M
Ventura, William Paul
Verma, Subhash Chander
Vernier, Vernon George
Vernikos-Danellis, Joan
Vick, Robert Lore
Villar-Palasi, Carlos
Villarreal, Julian Ernesto
Vincenzi, Frank Foster
Vinegar, Ralph
Vogel, Wolfgang Hellmut
Vogh, Betty Pohl
Volle, Robert Leon
Von Hagen, D Stanley
Vore, Mary Edith
Vostal, Jaroslav Joseph
Waddell, William Joseph
Wade, Adelbert Elton
Wagle, Gilmour Lawrence
Wagle, Shreepad R
Waite, Leonard Charles
Wajda, Isabel
Wakade, Arun Ramchandra
Walaszek, Edward Joseph
Wallace, Donald Albin
Wallach, Donald P
Wallin, Richard Franklin
Walsh, Michael Joseph
Walton, Charles Anthony
Walz, Donald Thomas
Wang, Hsueh-Hwa
Wang, Richard I H
Wang, Shih Chun
Ward, Charles O
Ward, John Wesley
Wardell, Joe Russell, Jr
Waterbury, Lowell David
Waters, Irving Wade
Watson, Jack Throck
Watts, Daniel Thomas
Waud, Douglas Russell
Wax, Joan
Way, E Leong
Way, James Leong
Way, Walter
Weaver, Lawrence Clayton
Weber, George
Weber, Lavern J
Weber, Wendell W
Webster, Leslie T, Jr
Webster, Marion Elizabeth
Weeks, James Robert
Wei, Eddie Tak-Fung
Weiner, Irwin M
Weiner, Myron
Weiner, Norman
Weisbrodt, Norman William
Weiss, George B
Weiss, Harvey Richard

Weiss, Lawrence R
Weiss, Robert Martin
Weiss, William P
Welch, Arnold DeMerritt
Wells, Herbert
Wells, Jack Nulk
Wells, Joseph Albert
Wells, Patrick Roland
Welsh, Ronald
Welty, Joseph D
Wendel, Herbert A
Wenger, Herbert Charles
Wenig, Jeffrey
Wenzel, Duane Greve
Werner, Gerhard
Wescoe, W Clarke
West, Bob
West, Fred Ralph, Jr
West, Theodore Clinton
West, William Lionel
Westfall, David Patrick
Westfall, Bertis Alfred
Westfall, Thomas Creed
Wetherell, Herbert Ranson, Jr
Wheeler, Allan Gordon
White, Richard Paul
White, Wallace Fletcher
Wicks, Wesley Doane
Wiebelhaus, Virgil D
Wiegand, Ronald Gay
Wiggins, Jay Ross
Wilcox, Henry G
Wiles, Joseph St Clair
Wilk, Sherwin
Wilkerson, Robert Douglas
Wilkinson, Ann Broom
Wilkinson, Grant Robert
Willard, Paul W
Williams, Betty Jean
Williams, Byron Bennett, Jr
Williams, Patricia Bell
Williams, Ronald Lee
Williamson, Harold Emanuel
Wills, James Henry
Wilson, John T
Wilson, Leslie
Wilson, Marvin Cracraft
Winbury, Martin M
Winek, Charles L
Wingard, Lemuel Bell, Jr
Winkler, Bertram Stanley
Winter, Irwin Clinton
Winters, Ronald Howard
Wit, Andrew Lewis
Withrow, Clarence Dean
Witt, Peter Nikolaus
Wohl, Arnold J
Wolen, Robert Lawrence
Wolf, Harold Herbert
Wolff, Donald John
Wolff, Frederick William
Wolowyk, Michael Walter
Wolpert-DeFilippes, Mary Katherine
Wolters, Robert John
Wong, Stewart
Wood, Charles Donald
Woodard, Geoffrey
Woodbury, Dixon Miles
Woodbury, Robert Arthur
Woodhouse, Bernard Lawrence
Woods, Lauren Albert
Woods, Maribelle
Woodward, James Kenneth
Wooles, Wallace Ralph
Woolley, Dorothy Elizabeth Schumann
Wosilait, Walter Daniel
Wulf, Ronald James
Yaffe, Sumner J
Yard, Allan Stanley
Yeh, Shu-Yuan
Yellin, Tobias O
Yim, George Kwock, Wah
Yong, Man Sen
Zaharko, Daniel Samuel
Zannoni, Vincent G
Zaratzian, Virginia Louis
Zaroslinski, John F
Zedeck, Morris Samuel
Zeitlin, Benjamin Raphael
Ziance, Ronald Joseph
Zimmerman, Ernest Frederick
Zins, Gerald Raymond
Zupko, Arthur George

Biochemical Pharmacology
Agarwal, Ram Prakash
Axelrod, Julius
Bain, James Arthur
Baird, Malcolm Barry
Barboriak, Joseph Jan
Barnard, Eric A
Belleau, Bernard Roland
Bellville, Gail Dianne
Bennett, Leonard Lee, Jr
Besch, Paige Keith
Bhargava, Hemendra Nath

Bloch, Alexander
Buhler, Donald Raymond
Byington, Keith H
Chang, Shaw Fai
Chang, Yi-Han
Chapman, George Herbert
Chello, Paul Larson
Cheng, Yung-Chi
Cicero, Theodore James
Coffey, John Joseph
Collins, Allan Clifford
Conard, Gordon Joseph
Cooke, William Joseph
Cushman, David Wayne
Dalton, Colin
Dannenburg, Warren Nathaniel
Deckert, Fred W
Denber, Herman C B
Diedrich, Donald Frank
Drach, John Charles
Duggan, Daniel Edward
Dus, Karl M
Ecobichon, Donald John
Ellenbogen, Leon
Finger, Kenneth F
Fischer, Lawrence J
Foster, Donald Myers
Freudenthal, Ralph Ira
Galbraith, William
Gallo, Michael Anthony
George, William Jacob
Gibb, James Wooley
Gillette, James Robert
Gordon, Harry William
Green, Donald Eugene
Green, Vernon Albert
Gunther, Jay Kenneth
Hacker, Bruce
Hakala, Maire Tellervo
Harrison, Yvonne E
Hebborn, Peter
Ho, Beng Thong
Ho, Dah-Hsi
Irving, George Washington, Jr
Jacobson, Martin Michael
Judis, Joseph
Kalman, Thomas Ivan
Kamm, Jerome Jr
Kary, Christina Dolores
Khanna, Jatinder Mohan
Killinger, Joanne Marie
Kohn, Leonard David
Krell, Robert Donald
La Du, Bert Nichols, Jr
Langner, Ronald O
Lanman, Robert Charles
Leibman, Kenneth Charles
Levin, Jerome Allen
Lin, Chin-Chung
Lindenmayer, George Earl
Lippmann, Wilbur
Lockridge, Oksana Maslivec
Loh, Horace H
Majchrowicz, Edward
Maragoudakis, Michael E
Matthews, Hazel Benton, Jr
Miller, William Lawrence
Milman, Harry Abraham
Mitoma, Chozo
Muni, Indu A
Muschek, Lawrence David
Nadkarni, Moreshwar Vithal
Nahas, Aly
Nelson, Donald J
Pereira, Michael Alan
Persico, Francis J
Pinson, Rex, Jr
Poulsen, Lawrence Leroy
Randerath, Erika
Rikans, Lora Elizabeth
Roberts, DeWayne
Robison, George Alan
Rosenkrantz, Harris
Roth, Robert Andrew, Jr
Sartorelli, Alan Clayton
Saunders, Priscilla Prince
Schayer, Richard William
Schroeder, David Henry
Schroer, Richard Allen
Schulman, Martin Phillip
Seifried, Harold Edwin
Slaga, Thomas Joseph
Snyder, Robert
Sommer, Kathleen Ruth
Steele, William John
Sterbenz, Francis Joseph
Stone, Deborah Bennett
Trevor, Anthony John
Vandor, Sandor Laszlo
Van Rossum, George Donald Victor
Vesell, Elliot S
Watkins, Mary Louise
Weinstein, Ira
Weisburger, John Hans
Welch, Richard Martin
Wilkens, Hans J
Wiseman, Edward H
Wong, Patrick Yui-Kwong

Wykes, Arthur Albert
Zimmerman, Thomas Paul

Chemical Pharmacology
Friedman, Herbert Alter
Glazko, Anthony Joachim
Hirst, Maurice
Jarboe, Charles Harry
Kohn, Kurt William
Roche, Edward Browning
Rodda, Bruce Edward

Clinical Pharmacology
Agre, Karl
Ambre, John Joseph
Appell, Raynor Norbert
Arnold, John D
Azarnoff, Daniel Lester
Ballard, Ian Matheson
Ballin, John Christian
Bass, Steven William
Beaver, William Thomas
Benforado, Joseph Mark
Benjamin, Robert Stephen
Berman, M Lawrence
Bickerman, Hylan A
Bird, Joseph Gordon
Block, Lawrence Howard
Boullin, David John
Bunde, Carl Albert
Canfield, Craig Jennings
Capizzi, Robert L
Carlozzi, Michael
Cawein, Madison Julius
Chenoweth, Maynard Burton
Chidsey, Charles Augustus
Cho, Young Won
Corman, Lew Andre
Creaven, Patrick Joseph
Crout, John Richard
Davis, Charles Stewart
Desjardins, Raoul
Dietz, Albert J, Jr
Done, Alan Kimball
Doolittle, Charles Herbert, III
Driever, Carl William
Dujovne, Carlos A
Fahrenbach, Marvin Jay
Ferguson, Roger K
Foster, Thomas Scott
Fox, Samuel Louis
Freeman, Leon David
Freund, Jack
Gaut, Zane Noel
Glick, Aaron
Grozier, Michael Lawrence
Harun, Joseph Stanley
Hava, Milos
Herting, Robert Leslie
Holcenberg, John Stanley
Huffman, David H
Imboden, Clarence Alphonse, Jr
Jasinski, Donald Robert
Kaufmann, John Simpson
Kaul, Pushkar Nath
Knill, James Reginald
Koch-Weser, Jan
Kolman, Wilfred Aaron
Kreis, Willi
Kulkarni, Anant Sadashiv
Kunesh, Jerry Paul
Lane, Montague
Larson, Jerry King
Lasseter, Kenneth Carlyle
Lemberger, Louis
Leon, Arthur Sol
LeSher, Dean Allen
Lewis, Richard John
Lietman, Paul Stanley
London, S J
Loo, Ti Li
Louis, John
MacCannell, Keith Leonard
MacFarlane, Malcolm David
Maggio-Cavaliere, Mary
Mahon, William A
Margolius, Harry Stephen
Maronde, Robert Francis
Masaki, Beverly Wong
Maykut, Madelaine Olga
McDonald, Robert H, Jr
McKinstry, Doris Naomi
Melmon, Kenneth Lloyd
Meyer, Marvin Chris
Mirkin, Bernard Leo
Mischler, Terrence Wynn
Mulinos, Michael George
Murad, Ferid
Murphree, Henry Bernard Scott
Nies, Alan Sheffer
Oates, John Alexander
Ogilvie, Richard Ian
Oken, Donald
Olsen, George Duane
Page, Robert Clinton
Palmer, John Davis
Panzer, James David
Perel, James Maurice

PHARMACOLOGY

Picozzi, Anthony
Place, Virgil Alan
Reichgott, Michael Joel
Priddle, Osgood Daniel, Jr
Richards, Richard Kohn
Ridolfo, Anthony Sylvester
Ross, Norton Morris
Sayers, Ross
Schreiber, Bruce David
Schrogie, John Joseph
Segre, Eugene J
Sellers, Edward Moncrieff
Shaw, Jane E
Sjoerdsma, Albert
Stambaugh, John Edgar, Jr
Steelman, Sanford Lewis
Stern, Leo
Taylor, W J Russell
Thompson, Wilmer Leigh, Jr
Tweedale, Martin George
Wagner, William Edward, Jr
Wan, Suk Han
Wardell, William Michael
Wasserman, Albert J
Weisberg, Jerry
Wehky, Irving
White, Albert M
Whitman, Erwin N
Whitsett, Thomas L
Woosley, Raymond Leon
Yeh, Billy Kuo-Jiun
Yu, Ruey Jiin
Zaske, Darwin Erhard

Drug Metabolism
Baer, John Elson
Breault, George Omer
Bruce, Robert Black
Buhs, Rudolf Paul
Chu, Sou Yie
Cone, Edward Jackson
Cosulich, Donna Bernice
Crabtree, Ross Edward
Crew, Malcolm Charles
Dain, Jeremy George
Davidson, Nancy McConnell
Davison, Clarke
DiFazio, Louis T
Doshan, Harold David
Edelson, Jerome
Gaver, Robert Calvin
Gingell, Ralph
Heeg, Joel Francis
Hucker, Howard Benjamin
Intoccia, Alfred Paul
Jansson, Frank Walter
Kimball, Aubrey Pierce
King, Mary Margaret
Knowles, John Appleton, III
Kripalani, Kishin J
Lan, Shih-Jung
Lang, James Frederick
Laurencot, Henry Jules, Jr
Leinweber, Franz Josef
Ludden, Thomas Marcellus
McIlhenny, Hugh M
Mehta, Nariman Bomanshaw
Migdalof, Bruce Howard
Morrison, Joseph Louis
Pittman, Kenneth Arthur
Ranney, Robert Earl
Ruelius, Hans Winfried
Sadee, Wolfgang
Schach Von Wittenau, Manfred
Schwartz, Morton Allen
Shargel, Leon David
Sisenwine, Samuel Fred
Steffens, James Jeffrey
Sullivan, Hugh R
Symchowicz, Samson
Teng, Lina Chen
Vanden Heuvel, William John Adrian, III
Vane, Floie Marie
Vickers, Stanley
Walkenstein, Sidney S
Weinstein, Stephen Henry
Wright, George Joseph
Zacchei, Anthony Gabriel

Molecular Pharmacology
Apple, Martin Allen
Boime, Irving
Brynes, Paul Jeffrey
Cohen, Jonathan Brewer
DiAugustine, Richard Patrick
Horwitz, Susan Band
Hsia, Jen-Chang
Lefkowitz, Robert Joseph
Lindell, Thomas Jay
Martin, Yvonne Connolly
Miller, Jon Philip
Nelson, Eric Loren
Sirotnak, Francis Michael
Talalay, Paul
Van Frank, Richard Mark
Vesell, Elliot S
Wilson, William Ewing

Neuropharmacology
Aghajanian, George Kevork
Baker, Walter Wolf
Barnett, Allen
Biber, Margaret Clare Boadle
Birzis, Lucy
Bond, Harley William
Boyd, Eleanor H
Brazier, Mary A B
Brezenoff, Henry Evans
Brodie, Larry Lewis
Christensen, Howard Dix
Clay, George A
Clineschmidt, Bradley Van
Colasanti, Brenda Karen
De la Torre, Jack Carlos
Douglas, William Wilton
Ebadi, Manuchair
Edmonds, Harvey Lee, Jr
Ellinwood, Everett Hews, Jr
Fernstrom, John Dickson
Franz, Donald Norbert
Gallagher, Joel Peter
Geller, Herbert M
Glick, Stanley Dennis
Goldman, Harold
Goldstein, Frederick J
Goldstein, Leonide
Haigler, Henry James
Halperin, Lawrence Mayer
Hance, Anthony James
Hanig, Joseph Peter
Harris, Jane Ellen
Harris, Louis Selig
Hiller, Jacob Moses
Hillman, Gilbert R
Hosko, Michael J, Jr
Huffman, Ronald Dean
Iturrian, William Ben
Jobe, Phillip Carl
Joy, Robert McKernon
Juorio, Augusto Victor
Kahn, Norman
Kirsten, Edward Bruce
Kosh, Joseph William
Koster, Rudolf
Kramer, Stanley Zachary
Kuhar, Michael Joseph
Lorenzo, Antonio V
Malick, Jeffrey Bevan
Marrazzi, Amedeo S
Marrazzi, Mary Ann
Miyamoto, Michael Dwight
Molinoff, Perry Brown
Parker, Robert Bruce
Patek, David Rushton
Quock, Raymond Mark
Robert, Richard Ross
Rech, Richard Howard
Richter, Judith Anne
Roth, Robert Henry, Jr
Sabelli, Hector C
Schaeppi, Ulrich Hans
Schanberg, Saul M
Schoener, Eugene Paul
Seeman, Philip
Segal, David S
Share, Norman N
Sheard, Michael Henry
Shinnick-Gallagher, Patricia L
Simon, Marcia Lee Mielke
Snyder, Solomon H
Spooner, Charles Edward, Jr
Suria, Amin
Tessel, Richard Earl
Tuttle, Warren Wilson
Vazquez, Alfredo Jorge
Von Voigtlander, Philip Friedrich
Walker, Charles A
Wallach, Marshall Ben
Warnick, Jordan Edward
Waters, Donald Hilton
Weisman, Harvey
Winters, Wallace Dudley
Yarbrough, George Gibbs
Zablocka-Esplin, Barbara

Pharmacodynamics
Greenspan, Kalman
Levy, Rene Hanania
Reynolds, Warren Dudley
Wan, Suk Han
Wilkinson, Paul Kenneth
Young, John Falkner

Pharmacognosy
Awad, Albert T
Babcock, Philip Arnold
Bailey, Harold Edwards
Beal, Jack Lewis
Blomster, Ralph N
Brewer, Willis Ralph
Catalfomo, Philip
Chaubal, Madhukar Gajanan
Cole, Franklin Ruggles
Doorenbos, Norman John
Doskotch, Raymond Walter

Pratt, Robertson
Redman, Kenneth
Robbers, James Earl
Rosazza, John Paul
Rosenberg, Harry
Ruehle, Archie Edwin
Schradie, Joseph
Schramm, Lee Clyde
Schwarting, Arthur Ernest
Sciuchetti, Leo A
Segelman, Alvin Burton
Shimizu, Yuzuru
Shough, Herbert Richard
Sim, Stephen Kahsun
Simonian, Vartkes Hovanes
Staba, Emil John
Stohs, Sidney John
Sullivan, Gerald
Svoboda, Gordon H
Tanner, Noall Stevan
Taylor, Elmore Hector
Tin-Wa, Maung
Turner, Carlton Edgar
Tutupalli, Lohit Venkateswara
Tyler, Varro Eugene
Van Horne, Robert Loren
Walter, Wilbert George
Wier, Jack Knight
Youngken, Heber Wilkinson, Jr

Psychopharmacology
Abel, Ernest Lawrence
Alpern, Herbert P
Ban, Thomas Arthur
Barry, Herbert, III
Bayne, Gilbert M
Beaton, John McCall
Blackwell, Barry M
Bowe, Robert Looby
Brand, Eugene Dew
Carlson, Kristin Rowe
Carroll, Bernard James
Ciccone, Patrick Edwin
Clark, Robert
Cox, Raymond H, Jr
Davidson, Arnold B
D'Encarnacao, Paul S
Elkes, Joel
Evans, Hugh Lloyd
Evans, Wayne Orien
Feldman, Harold Samuel
Flood, James Felix
Flynn, Patrick
Ganzu, Elkan
Gaitpon, Glenn Blaise
Geyer, Mark Allen
Goldstein, Burton Jack
Hanin, Israel
Hartmann, Ernest Louis
Harvey, John Adriance
Holliday, Audrey Rose
Houser, Vincent Paul
Hunt, Howard Francis
Irwin, Samuel
Jaffe, Jerome Herbert
Johanson, Chris Ellyn
Jones, Ben Morgan
Jones, Bill Edson
Kay, David Cyril
Klein, Donald Franklin
Kornetsky, Conan
Korol, Bernard
Ksir, Charles Joseph
Kuczenski, Ronald Thomas
Laties, Victor Gregory

Doughty, Richard Morrison
Duke, Victor Hal
Elsohly, Mahmoud Ahmed
Euler, Kenneth L
Farnsworth, Norman R
Ferguson, Noel Moore
Fong, Harry H S
Gibson, Melvin Roy
Goldner, Karl John
Goodeve, Allan McCoy
Hatfield, George Michael
Hershenson, Benjamin R
Hocking, George Macdonald
Hufford, Charles David
Kelleher, William Joseph
Khalil, Shoukry Khalil Wahba
Kokoski, Robert John
Lamba, Surendar Singh
Lau-Cam, Cesar A
Leonard, Robert Meyer
Leung, Albert Yuk-Sing
Locock, Robert A A
Lopez, Antonio Vincent
Mary, Nouri Y
McLaughlin, Jerry Loren
Medora, Rustem Sohrab
Mika, Edward Stanley
Mockle, Jerry Auguste
O'Connell, Frank Dennis
Pandya, Krishnakant Hariprasad
Paul, Ara Garo
Persinos, Georgia J

Latz, Arje
Leander, John David
Leibowitz, Sarah Fryer
Lin, Reng-Lang
Lytle, Loy Denham
Maitz, Sidney
Manian, Albert Ardashes
McLendon, David Mark
McNair, Douglas McIntosh
Moore, Kenneth Edwin
Nieforth, Karl Allen
Perel, James Maurice
Poschel, Bruno Paul Henry
Potts, Walter Joseph
Rickels, Karl
Robichaud, Roger Charles
Sassenrath, Ethelda Norberg
Schechter, Martin David
Schmidt, Dennis Earl
Scudder, Charles Lee
Seiden, Lewis S
Sepinwall, Jerry
Singh, Jasbir Mahajan
Stein, Larry
Szara, Stephen Istvan
Taber, Robert Irving
Ternes, Joseph Wayne
Thut, Paul Douglas
Uhlenhuth, Eberhard Henry
Usdin, Earl
Van Orden, Lucas Schuyler, III
Watzman, Nathan
Wayner, Matthew John
Weiss, Bernard
Weissman, Albert
Wyatt, Richard J
Zigmond, Michael Jonathan

Veterinary Pharmacology
Adams, Henry Richard
Baggot, John Desmond
Bevill, Richard F, Jr
Borchard, Ronald Eugene
Burrows, George Edward
Carlson, Arthur J F
Coppoc, Gordon Lloyd
Curts, Kephart Maynard
Dickinson, John Otis
Dunlop, Robert Hugh
Hines, James Albert
Jones, L Meyer
Knapp, William Arnold, Jr
Link, Roger Paul
McCurdy, Dennis
Mercer, Henry Dwight
Oliver, Jack Wallace
Powers, Thomas E
Ray, Richard Schell
Reuter, Gerald Louis
St Omer, Vincent Victor
Sisodia, Chaturbhuj Singh
Swenson, Gene Holstrom
Yeary, Roger A

PHARMACY

Pharmacy
Ansel, Howard Carl
Apple, William S
Araujo, Oscar Eduardo
Autian, John
Ballard, Kenneth J
Ballard, Gilbert Stephen
Bardell, Eunice Bonow
Barker, Donald Young
Bartilucci, Andrew J
Baxter, Ross M
Becker, Charles Henry
Belcastro, Patrick Frank
Bender, George Almon
Benica, William Steinhart
Benton, Byrl E
Bergen, John Vanderveer
Bhatia, Vishnu Narain
Billups, Norman Frederick
Blake, Martin Irving
Blaug, Seymour Morton
Blissitt, Charles W
Bone, Jack Norman
Borisenok, Walter A
Born, Gordon Stuart
Bowers, Roy Anderson
Boylan, James Charles
Breckinridge, Charles Edward, Jr
Breese, George Richard
Brockmeyer, Eugene William
Brodie, Donald Crum
Brown, Vivia Jean
Brunett, Emery W
Bruns, Lester George
Burkholder, David Frederick
Burlage, Henry Matthew
Burton, Lloyd Edward
Cadwallader, Donald Elton
Carew, David P
Cataline, Elmon Lamont

Chalmers, Robert Kenny
Chavkin, Leonard Theodore
Ciminera, Joseph Louis
Cohen, Jack
Colaizzi, John Louis
Constantine, George Harmon, Jr
Cooper, Robert Michael
Coran, Aubert Y
Cosgrove, Frank P
Crouthamel, William Guy
Cyr, Gilman Norman
Danian, Michael S
Danti, August Gabriel
Dash, Barry Harold
Dauphinais, Raymond Joseph
Davis, Neil Monas
Dickison, Walter Lee
Dittert, Lewis William
Dodge, Austin Anderson
Drommond, Fred George
Duncan, Gerald R
Eckel, Frederick Monroe
Elowe, Louis N
Entrekin, Durward Neal
Eugene, Edward Joseph
Evanson, Robert Verne
Fenney, Nicholas William
Ferguson, Mary Hobson
Franke, Norman Henry
Gagnon, Leo Paul
Galpin, Donald R
Gans, Eugene Howard
Gerraughty, Robert Joseph
Gjerstad, Gunnar
Gloor, Walter Thomas, Jr
Gold, Gerald
Granatek, Edmund Stanley
Granberg, Charles Boyd
Greco, Salvatore Joseph
Griffenhagen, George Bernard
Grosicki, Thaddeus Stanley
Groves, William G
Guess, Wallace Louis
Guillory, James Keith
Hall, Esther Jane Wood
Hall, William Earl
Hamill, Richard David
Hamlow, Eugene Emanuel
Hammarlund, Edwin Roy
Hammerness, Francis Carl
Hammond, Elmer Lionel
Hamner, Martin
Hanus, Edward Joseph
Havemeyer, Ruth Naomi
Heller, William Mohn
Herd, Allen K, III
Hickman, Eugene
Higgins, Nevada Marie
Hikal, Ahmed Hassan
Hill, Richard A
Hilty, Wayne Woodrow
Hix, Elliott Lee
Holstius, Elvin Albert
Hopkins, Howard
Huber, Raymond C
Hufford, Charles David
Hughes, Francis Norman
Hussar, Daniel Alexander
Hutchinson, Richard Allen
Ice, Rodney D
Illian, Carl Richard
Jackson, Gerald James
Jackson, James Robert
Jensen, Erik Hugo
Johnston, Carl Henry
Johnston, Gene Woods
Julian, Edward A
Jusko, William Joseph
Kapadia, Abbaysingh J
Kapadia, Yash M
Katz, Irwin Alan
Kaufman, Karl Lincoln
Kavula, Michael P, Jr
Kazerovskis, Karlis
Keller, Bernard Gerard, Jr
Kelly, Charles James
Kerr, Wendle Louis
Kessler, Henry A
King, Charles McDonald, Jr
King, James C
King, Louis Delwin
Kinkel, Arlyn Walter
Kirchmeyer, Frederick Joseph
Kirkland, Walter Dean
Knoechel, Charles Joseph
Kokoski, Charles Joseph
Kokoski, Robert John
Koshy, Karyanil Thomas
Kostenbauder, Harry Barr
Krahnke, Harold C
Kramer, Sherman Francis
Lachman, Leon
Lamb, Donald Joseph
Lamba, Surendar Singh
Lamonde, Andre M
Lamy, Peter Paul
Lange, Winthrop Everett
Larose, Roger

Le Blanc, Floyd Joseph
Leeson, Lewis Joseph
Lesshaff, Charles Thomas, Jr
Levine, Philip J
Leyda, James Perkins
Littlejohn, Oliver Marsilius
Lloyd, William Reese
Lordi, Nicholas George
Lowenthal, Werner
Lu, Matthias Chi-Hwa
Luce, Everett N
Ludwig, Walter John
Lueck, Leslie Melvin
MacDonnell, Donald R
Marcus, Arnold David
Marcus, David
Martell, Michael Joseph, Jr
McCarron, Margaret Mary
McCowan, James Robert
Mehta, Himatlal R
Mezei, Catherine
Miller, Howard Anthony
Mittelstaedt, Stanley George
Mlodozeniec, Arthur Roman
Monaco, Anthony L
Montgomery, Kenneth O
Moore, Willis Eugene
Morrison, Robert William
Mossberg, Howard E
Mrtek, Marsha Bedford
Mrtek, Robert George
Mullins, John Dolan
Munden, Bill J
Nairn, John Graham
Nash, Robert Arnold
Nashed, Wilson
Nessel, Robert J
Neuroth, Milton L
Niebergall, Paul J
Nithman, Charles Joseph
O'Brien, Francis Joseph
Oddis, Joseph Anthony
Orr, Jack Edward
Osborne, George Edwin
Parke, Russell Frank
Parker, Martin Dale
Patterson, Harry Robert
Paul, William Larry
Pazienza, Joseph Peter
Peck, Garnet E
Penzotti, Stanley Clare, Jr
Peterson, Charles Fillmore
Plein, Elmer Michael
Plein, Joy Bickmore
Plourde, J Rosaire
Poulsen, Boyd Joseph
Price, James Clarence
Putney, Blake Fuqua
Raff, Allan Maurice
Rasero, Lawrence J
Ravin, Louis Joseph
Reichmann, Keith Wilford
Rhodes, Christopher Thomas
Riedel, Bernard Edward
Riegelman, Sidney
Robinson, Ira Charles
Rodell, Michael Byron
Rodowskas, Christopher A, Jr
Rose, Wayne Burl
Rosenbluth, Sidney Alan
Rowe, Edward John
Rowe, Thomas Dudley
Rowland, Ivan W
Rudolph, Jeffrey Stewart
Ruggiero, John S
Sager, Robert William
Salisbury, Rupert
Saute, Robert E
Scarpone, Anthony John
Schermerhorn, John W
Schuler, Edward Emerson
Schwarz, Thomas Werner
Seugling, Earl William, Jr
Sewell, Emma Winifred
Shaw, Margaret Ann
Sheffield, William Johnson
Sheinaus, Harold
Sherman, Gerald Philip
Sica, Albert Joseph
Silverman, David Norman
Silverman, Harold I
Simon, Thomas H
Sisson, Harriet E
Skauen, Donald M
Smith, Pierre Frank
Sokoloski, Theodore Daniel
Sommers, Ella Blanche
Sorby, Donald Lloyd
Sorenson, Glen Joseph
Sperandio, Glen J
Spiegel, Allen J
Stavchansky, Salomon Ayzenman
Stella, Valentino John
Stempel, Edward
Stenseth, Raymond Eugene
Stolar, Morris Emmanuel
Strauss, Steven
Strianse, Sabbat John

Strickland, William Alexander, Jr
Stuart, David Marshall
Sullivan, Gerald
Sumner, Edward D
Susina, Stanley V
Swafford, William Bryson
Swartz, Harry Sip
Swintosky, Joseph Vincent
Takruri, Harun
Tansey, Robert Paul, Sr
Taylor, William West
Theodore, Joseph M, Jr
Thomasson, Claude Larry
Thompson, Herman O
Tingstad, James Edward
Tkacheff, Joseph, Jr
Torosian, George
Tousignaut, Dwight R
Tranner, Frank
Tucker, Geoffrey Thomas
Turco, Salvatore J
Urdang, Arnold
Van Horne, Robert Loren
Vincent, Muriel C
Visconti, James Andrew
Volicer, Ladislav
Wailes, John Leonard
Walkling, Walter Douglas
Walsh, Robert A
Walton, Charles Anthony
Waters, Kenneth Lee
Webb, Norval Ellsworth, Jr
Whitworth, Clyde W
Willis, Carl Raeburn, Jr
Willits, Lyle Wilmot
Wilson, Edwin E
Wilson, Marvin Cracraft
Winters, Edward Phillip
Wood, James Alexander
Wurster, Dale E
Yakatan, Gerald Joseph
Yanchick, Victor A
Yunker, Martin Henry
Zalucky, Theodore B
Zimmerman, James Joseph

History of Pharmacy

Berman, Alex
Hamarneh, Sami
Mrtek, Robert George

Industrial Pharmacy

Appino, James B
Brochu, William Eugene
Brown, Michael
Hagman, Donald Eric
Hamlin, William Earl
Hecht, Gerald
Larson, Allan Bennett
Manudhane, Krishna Shankar
McKenzie, Walter Lawrence
Mendes, Robert W
Nedich, Ronald Lee
Olsen, James Leroy
Orcutt, Donald Adelbert
Parrott, Eugene Lee
Runkel, Richard A
Schleif, Robert H
Sciarra, John J
Shangraw, Ralph F
Slotsky, Myron Norton
Turi, Paul George
Young, James George

Pharmacy Administration

Barker, Kenneth Neil
Hammond, Elmer Lionel
Kabat, Hugh F
Kern, Joseph Herschel
Lehrman, George Philip
Moore, Robert Conley
Pradhan, Suresh B
Rodowskas, Christopher A, Jr
Segal, Harold Jacob
Shrader, Kenneth Ray
Swartz, Harry Sip

Physical Pharmacy

Aguiar, Armando Joseph
Bahal, Surendra Mohan
Barr, Martin
Bernardo, Peter D
Bighley, Lyle Delevan
Bornstein, Michael
Brochu, William Eugene
Chertkoff, Marvin Joseph
Chrzanowski, Francis Alan
Chun, Alexander Hing Chinn
Conomy, John Paul
Fincher, Julian H
Goldberg, Arthur H
Gupta, Vishnu Das
Hem, Stanley L
Henderson, Norman Leo
Hiestand, Everett Nelson
Hom, Foo Song
Hong, Wen-Hai
Jacobson, Harold

Janssen, Richard William
Kabadi, Balachandra N
Kaspar, Hans Heinrich
Kildsig, Dane Olin
Kornblum, Saul S
Kramer, Paul Alan
Lach, John Louis
Levi, Ralph Sigmund
Lin, Song-Ling
Lindstrom, Richard Edward
Luzzi, Louis A
Mario, Ernest
Maulding, Hawkins Valliant, Jr
Nairn, John Graham
Nash, J Frank
Nedich, Ronald Lee
Patel, Nagin K
Pisano, Frank D
Reinstein, Jerome Alan
Restaino, Frederick A
Rich, Arthur Gilbert
Rogers, James Albert
Rowe, Englebert L
Saad, Hosny Younes A
Sawardeker, Jawahar Sazro
Shah, Devendra Hargovindas
Sheth, Bhogilal
Smith, Harold Linwood
Surpuriya, Vijay B
Thakkar, Arvind Lavji
Wadke, Deodatt Anant
Windheuser, John Joseph
Woller, William Henry

PHYSICS

Physics

Aagard, Roger L
Aamodt, R L
Aamodt, Richard E
Abels, Larry L
Abers, Ernest S
Abraham, Farid Fadlow
Abraham, Marvin Meyer
Abrahams, Elihu
Abramson, Stanley L
Acheson, Willard Phillips
Achor, William Thomas
Adams, Clifford Lowell
Adler, Robert
Agnew, Harold Melvin
Agnew, Lewis Edgar, Jr
Agocs, William Bailey
Agresta, Joseph
Ahmadzadeh, Akbar
Aikman, Edward Percy
Aisenberg, Sol
Aitken, Donald W, Jr
Akers, Lawrence Curtice
Albright, James Curtice
Albright, John Rupp
Alburger, David Elmer
Alexander, Chester, Jr
Alexander, Frank Creighton, Jr
Alexander, Robert Stanley
Allan, David Wayne
Alldredge, Gerald Palmer
Allen, Edward Franklin
Allen, Frank B
Allen, Frank Joseph
Allen, Lew, Jr
Allen, Matthew Arnold
Allen, William Hubert
Alley, Reuben Edward, Jr
Allin, Elizabeth Josephine
Allred, Harry Milburn
Allred, John Caldwell
Alpert, Seymour Samuel
Alpher, Ralph Asher
Alsmiller, Rufard G, Jr
Alstadt, Don Martin
Altman, Joseph Henry
Alvarez, Raymond Angelo, Jr
Alyea, Ethan Davidson, Jr
Ames, Irving
Ammar, Raymond George
Anantha Narayanan, Venkataraman
Ancker-Johnson, Betsy
Anderson, Arthur George
Anderson, Carl David
Anderson, Carl Einar
Anderson, David Leonard
Anderson, Herbert Lawrence
Anderson, J Robert
Anderson, Neal Sample
Anderson, Owen Thomas
Anderson, Robert Lester
Anderson, Roy Stuart
Anderson, Victor Charles
Anderson, Wallace Ervin
Anderson, Weston Arthur
Anderson, William Raymond
Anderson, Wilmer Clayton
Andres, Klaus
Andrews, Charles Luther

Dunlap, William Crawford
Dunn, Cecil Gordon
Dunning, John Ray, Jr
Duntley, Seibert Quimby
Du Pre, Frits Karel
Duret, Maurice Francis
Dutton, David B
Duvall, George Evered
Duvall, Wilbur Irving
Dwyer, Robert Joseph
Dyal, Palmer
Dyke, Walter Payne
Dyke, David L
Eagleson, Halson Vashon
Earl, James Arthur
Earls, Lester Thomas
Early, James M
Easley, James W
Easley, Ronald L
East, Larry Verne
Eastman, Daniel Robert Peden
Eastmond, Elbert John
Eatherly, Walter Pasold
Ebert, Paul J
Eby, Ronald Kraft
Eck, Robert Edwin
Eck, Thomas G
Eckerman, Jerome
Eckert, Hans Ulrich
Eckhardt, Gisela (Marion)
Eckhardt, Wilfried Otto
Eckroth, Charles Angelo
Eckstein, Herbert Philipp
Edelsack, Edgar Allen
Edelson, Sidney
Edelstein, Richard Malvin
Eden, Henry Francis
Edlund, Milton Carl
Edmonds, James D, Jr
Edwards, David Olaf
Edwards, John Elza
Edwards, Martin Hassall
Edwards, Merrill Arthur
Edwards, Palmer Lowell
Edwards, Thomas Harvey
Egner, Donald Otto
Eichelberger, Robert John
Eichholz, Geoffrey Gunther
Einarsson, Alfred W
Einstein, Lloyd Theodore
Eisele, Louis John
Eisen, Fred Henry
Eisenstadt, Maurice
Eisner, Melvin
Eisner, Philip Nathan
Elder, Fred Kingsley, Jr
Elder, James Tait
Elder, Samuel Adams
Eldridge, Francis Reed, Jr
Eliason, Afton Yeates
Elings, Virgil Bruce
Eljon, Herbert Aaron
Elkholy, Hussein A
Elleman, Daniel Draudt
Elleman, Paul M
Elliott, Shelden Douglass, Jr
Elliott, Stuart Bruce
Ellis, Eric Hans
Ellis, Fred E
Ellis, Homer Godsey
Ellis, Jason Arundel
Ellis, Reed Hobart, Jr
Ellis, Robert Anderson, Jr
Ellsworth, Louis Daniel
Elmore, Robert E
Elmore, William Cronk
Elms, James Cornelius
Elsasser, Walter M
Elson, John Merle
Elwell, Albert R
Ely, Robert P, Jr
Emberson, Richard Maury
Emch, George Frederick
Emery, Guy Trask
Emigh, Charles Robert
Emslwiller, Maclellan
Ender, Hans Henry
Enderby, Charles Eldred
Engelhardt, Albert George
England, Alan Coulter
Engle, Paul Randal
Engler, Arnold
Engstrom, Ralph Warren
Enns, John Hermann
Enns, Theodore
Ennulat, Reinhard D
Epstein, Seymour
Erber, Thomas
Erhlich, Robert
Erich, Lester Charles
Erickson, Richard Ames
Erlbach, Erich
Erichson, Herman
Ernst, Martin L
Ernst, Albert R
Eschenfelder, Andrew Herbert
Esquivel, Agerico Liwag
Essig, Gustave Alfred

270

Estabrook, Frank Behle
Estes, Nelson N
Estin, Robert William
Forstat, Harold
Forster, Harriet Herta
Forster, John Heslop
Forsyth, Kurt
Forsyth, Peter Allan
Forward, Robert Lull
Fossan, David B
Foster, John Stuart, Jr
Foster, Theodore Dean
Fowler, Clarence Maxwell
Fowler, Howland Auchincloss
Fowler, Richard Gildart
Fowler, Robert Dudley
Fowler, William Alfred
Fox, Herbert Leon
Fox, John Dana
Francis, Charles E
Frank, Louis Albert
Frank, Wilson James
Frankel, Richard Barry
Franklin, Allan David
Franklin, Philip Jaquins
Franz, Frank Andrew
Franzen, Wolfgang
Fraser, John Stiles
Fraser, Peter Allan
Frazier, Thomas Vernon
Fred, Mark Simon
Frederick, Laurence W
Fredricks, Robert W
Fredkin, Donald Roy
Fredrickson, John E
Freeman, Ira M
Freeman, Mark Simon
Freier, Phyllis S
Fremont, Claude
French, Anthony Philip
French, John Donald
French, Walter Russell, Jr
Fretter, William Bache
Freyre, Raoul Manuel
Frick, Richard Henry
Friedberg, Simeon Adlow
Friedman, Jerome Isaac
Friedman, Marvin Harold
Friedman, William Albert
Fristrom, Robert Maurice
Fritchie, Frank Paul
Fritts, Robert Washburn
Froehlich, Fritz Edgar
Frosle, Harold Milton
Frost, Robert T
Fry, David Lloyd George
Fry, Glenn McKinley, Jr
Fry, Graham Eugene
Frye, Royal Merrill
Fulco, Jose Roque
Fuller, Harold Q
Fuller, Melvin Otis
Fulmer, Clyde Benson
Furman, Walter L
Furry, Wendell Hinkle
Furst, Milton
Furth, Ulrich Richard
Furth, Harold Paul
Fusfeld, Herbert Irving
Futch, Archer Hanmer
Gabrysh, Andrew Francis
Gade, Sandra Ann
Gager, William Ballantine
Gaidos, James A
Gailar, Norman Milton
Gaines, Owen M
Gaines, James R
Galginaitis, Simeon Vitis
Gallagher, Charles Joseph
Gallagher, James J
Gallagher, Lawrence Joseph
Galloway, Richard Thomas
Galloway, William Joyce
Gamble, Francis Trevor
Gammon, Robert Winston
Gano, Hideya
Ganguly, Bishwa Nath
Garber, David Harrison
Garber, Meyer
Garcia-Colin, Leopoldo Scherer
Gardner, Wilford Robert
Gardner, David Arthur
Garfunkel, Myron Paul
Garmire, Elsa Meints
Garrett, Charles Geoffrey Blythe
Garrett, Robert Ogden
Garrick, Isadore Edward
Garrison, Allen K
Garrison, John Dresser
Garvey, Robert Joseph
Garvey, James Emnett
Gates, Halbert Frederick
Gatinger, Richard Larry
Gault, John M
Gautreau, Ronald
Gauvin, Hervey Paul
Gavan, Francis Michael

Geiger, James Stephen
Geiker, Charles Don
Gelernter, Herbert Leo
Geller, Kenneth N
Gelles, Isadore Leo
Gelman, Harry
Gelman, Alan Neville
Gerdes, John William
Gericke, Otto Reinhard
Germain, Lawrence Seymour
Gerritsen, Hendrik Jurjen
Gerson, Harold Arthur
Gerson, Robert
Geschwind, Stanley
Geshwind, Robert Andrew
Gessert, Walter Louis
Gessow, Alfred
Getting, Ivan Alexander
Gettner, Marvin
Ghiorghiu, Paul
Giannini, Gabriel Maria
Giannoni, Umberto Ferdinando
Giauer, Ivar
Gibbs, Peter (Godbe)
Gibbs, William Royal
Gibson, Gilbert Lewis
Gideon, Donald Nason
Gilbert, Barry Jay
Gildart, Lee William
Giles, John Crutchlow
Giles, Peter Cobb
Gille, John Charles
Gillery, Frank Howard
Gillette, Frank Newton
Gillette, Philip Roger
Gilliam, Otis Randolph
Gillis, Murlin Fern
Gilman, John Joseph
Gilmore, Forrest Richard
Gilmore, Robert
Giordmaine, Joseph Anthony
Glauber, Roy Jay
Gleicher, Paul Harry
Glenn, William Henry, Jr
Glickman, Walter A
Gloeckler, George
Gioersen, Per
Glover, Alan Marsh
Glover, Francis Nicholas
Glucksten, Robert Leonard
Godfrey, Charles S
Godfrey, Thomas Nigel King
Goedertier, Kuei-Ling Li
Goedertier, Peter V
Goerz, David Jonathan, Jr
Golay, Marcel Jules Edouard
Gold, Andrew Vick
Gold, Richard Robert
Goldberg, Benjamin
Goldberg, Conrad Stewart
Goldberg, Hyman
Goldberg, Joshua Norman
Goldberg, Norman
Goldberg, Philip A
Goldberg, Stanley
Goldberg, Harold Mark
Goldhaber, Gertrude Scharff
Goldhaber, Maurice
Goldman, Jacob E
Goldman, Peter Carl
Goldmark, George Jason
Goldsmith, Herbert
Goldstein, Jack Stanley
Goldstein, Raymond
Goldwasser, Edwin Leo
Golibersuch, David Clarence
Golowich, Eugene
Goncz, John Henry
Good, Bill Jewel
Good, Myron Lindsay
Good, Wilfred Manly
Goodenough, John Bannister
Goodkind, John M
Goodman, Ralph Abijah
Goody, Richard (Mead)
Goody, Eugene Irving
Gordon, James Power
Gordon, John Ethan
Gordon, Joel Ethan
Gordon, Morton Maurice
Gordon, Richard Lee
Gordy, Walter
Gorog, Istvan
Goslin, Roy Nelson

Goss, Wilbur Hummon
Gossard, Arthur Charles
Gossick, Ben Roger
Gothard, Nicholas
Gott, Preston Frazier
Gottlieb, Melvin Burt
Gottlieb, Milton
Gould, Gordon
Gould, Robert Henderson
Gould, Roy Walter
Gourley, Lloyd Eugene, Jr
Gove, Harry Edmund
Graetzer, Hans Gunther
Graham, Gordon Alexander Robert
Graham, Richard H
Graham, Robert Albert
Graham, Robert Lockhart
Graham, William Rendall
Granato, Andrew Vincent
Grandy, Charles Creed
Granger, John Van Nuys
Grant, Frederick Cyril
Granzow, Kenneth Donald
Graves, Bruce Bannister
Gray, Donald M
Gray, William MacDonald
Green, Alex Edward Samuel
Green, Ben Arthur, Jr
Green, Daniel Thomas
Green, Donald Wayne
Green, Eugene L
Green, John Root
Green, Richard James
Greenberg, Arthur
Greenberg, Jack Sam
Greene, Jack Bruce
Greene, David C
Greenland, Miles Griffith
Greenslade, Thomas Boardman, Jr
Gregg, Earle Covington
Gregory, Christopher
Greifinger, Phyllis Stoliar
Greig, John Henry
Greisen, Kenneth I
Grenning, Daniel A
Greytak, Thomas John
Griem, Hans Rudolf
Griem, Melvin Luther
Griffin, Charles Frank
Griffith, Gordon Lamar
Griffiths, David
Griffiths, David John
Griffiths, Robert Budington
Grigg, Harold R
Grimm, George Walter
Grisaru, Marcus Theodore
Grischkowsky, Daniel Richard
Grismore, Roger
Grissinger, Earl H
Grosch, Chester Enright
Gross, Edward Emanuel
Gross, Thomas Alfred Otto
Grosse, Fred A
Grove, Donald Jones
Grove, George Richard
Groves, Thomas Hoopes
Gruber, Gary Richard
Gruber, Piet C
Guiragossian, Zaven George
Gulkis, Samuel
Gullikson, Charles William
Gundlach, Robert William
Gunther-Mohr, Gerard Robert
Gupta, Neelam
Gurevitch, Mark
Gurr, Graham Edward
Guth, Sylvester Karl
Guthrie, Albert Nelson
Guthrie, Andrew
Gutsche, Graham Denton
Guyer, Edwin Michael
Gwin, Reginald
Gwinn, Joel Alderson
Haar, Lester
Haberecht, Rolf Reinhold
Haber-Schaim, Uri
Habib, Edwin Emile
Hadley, Charles Peleg
Hadley, James Warren
Hafstad, Lawrence Randolph
Hageman, Donald Henry
Hagen, Carl Richard
Hagenlocker, Edward Emerson
Hageseth, Gaylord Terrence
Hahn, Erwin Louis
Hahn, George
Hahn, Thomas Marshall, Jr
Haisley, Waldo Emerson
Hake, Richard Robb

Halbach, Klaus
Hale, Barbara Nelson
Hale, Robert E
Hales, Everett Burton
Hall, David Ballou
Hall, Donald Eugene
Hall, Forrest G
Hall, James Alexander
Hall, John L
Hall, Peter M
Hall, Robert Noel
Hallett, Archibald Cameron Hollis
Hallgren, Richard E
Halpern, Isaac
Halpin, Walter J
Halprin, Arthur
Halverson, Gilbert
Ham, Frank Slagle
Ham, Joe Strother
Ham, William Taylor, Jr
Hamerly, Robert Glenn
Hamil, Charles Norbert
Hanawalt, Joseph Donald
Hamermesh, Bernard
Hamermesh, Morton
Hamilton, William Oliver
Hamm, William Joseph
Hammerle, William Gordon
Hammond, Robert Hugh
Hampel, Viktor Erwin
Hampton, Loyd Donald
Hanawalt, Joseph Donald
Hancock, John Ogden
Handrup, Bernarda
Hannum, William Hamilton
Hansen, Bernard Lyle
Hansen, Carl John
Hansen, Wilford Nels
Hanson, Alvin Walter
Hanson, Fred Sumner, (III)
Hanson, Harold Palmer
Hanson, Howard Grant
Hanson, William Bert
Hanzely, Stephen
Harbold, Mary Leah
Hardie, Robert Howie
Hardin, Clyde D
Harding, Samuel William
Hardy, Robert J
Hardy, Wilton Audobon
Harke, Douglas J
Harker, Kenneth James
Harling, Otto Karl
Harnwell, Gaylord Probasco
Harper, Richard Allan
Harries, Wynford Lewis
Harrington, David Rogers
Harrington, John Vincent
Harris, Donald R, Jr
Harris, Frank Bower, Jr
Harris, Franklin Stewart, Jr
Harris, Sigmund Paul
Harrison, Edward Robert
Harrison, George Russell
Harrison, Mark
Harrison, Melvin Arnold
Harrison, Ralph Joseph
Harrower, George Alexander
Hart, Hiram
Hart, John Birdsall
Hart, Robert John
Hartig, Elmer Otto
Hartle, James Burkett
Hartmann, Gregory Kemenyi
Hartmann, Sven Richard
Hartzler, Alfred James
Hartzler, Harrod Harold
Haruta, Kyoichi
Harvalik, Zaboj Vincent
Harvey, Alexander Louis
Hasty, Turner Elilah
Hatcher, Charles Richard
Hatcher, John Burton
Hatfield, Theo Noel
Hathaway, Charles Edward
Haun, Robert Dee, Jr
Haupt, Curtis Raymond
Hause, Clarence Duane
Hauser, Isidore
Haynes, Sherwood Kimball
Hayn, Carl Hugo
Haynie, William Howard
Hayward, Evans Vaughan
Hazen, Wayne Eskett
Haworth, Leland John
Hay, Donald Ross
Haybron, Ronald M
Hayden, Richard John
Hayes, Charles Franklin
Hayes, Dallas T
Hayes, John Nicholas
Heard, Harry Gordon
Hearst, Joseph R
Heaton, LeRoy
Hebb, Malcolm Hayden
Hebert, Gerard Rosaire
Hecht, Eugene
Heckman, Richard Cooper

Hedgcock, Frederick Thomas
Heid, Roland Leo
Heinberg, Milton
Heinsch, Roger Paul
Heinz, Otto
Helbing, Reinhard Karl Bodo
Heller, Douglas Max
Heller, Gerhard Bernhard
Heller, John Herbert
Heller, John Philip
Heller, Marvin W
Hellman, William S
Hellwarth, Robert Willis
Helmer, John
Heltemes, Eugene Casmir
Hemenway, Curtis Leland
Hemmendinger, Henry
Hendel, Alfred Z
Henderson, Robert Edward
Hendrickson, Thomas James
Hendrie, David Lowery
Henins, Ivars
Henke, Burton Lehman
Henkel, Elmer Thomas
Henminger, Ernest Herman
Henoch, William Weil
Henry, Allan Francis
Henry, Hugh Fort
Henry, Richard Warfield
Henry, Robert Ledyard
Henshaw, Clement Long
Herb, Raymond George
Hereford, Frank Loucks
Herget, William F
Herickhoff, Robert John
Herlin, Melvin Arnold
Herman, Paul Theodore
Heroux, Leon J
Herrey, Erna Miranda Julia
Herrmann, Ulrich Otto
Hershkowitz, Noah
Herwig, Lloyd Otto
Herzfeld, Karl Ferdinand
Herzog, Richard (Franz Karl)
Hess, David Clarence
Hessel, Alexander
Heusinkveld, Myron Ellis
Heverly, John Ross
Hewitt, Frederick George
Hexner, Peter Eugen
Heydemann, Peter Ludwig Martin
Heynick, Louis Norman
Hicks, Bruce Lathan
Hicks, Donald A
Hicks, Harry Frank, Jr
Higgins, George Clinton
High, Marathon Eby
Highland, Virgil Lee
Highsmith, Phillip E
Higinbotham, William Alfred
Hilberry, Norman
Hilker, Harry Van Der Veer, Jr
Hill, Armin John
Hill, Benny Joe
Hill, Dale Eugene
Hill, Freeman Kenneth
Hill, Henry Hunter
Hill, James
Hill, John Joseph
Hill, Otto Herman
Hill, Robert Dickson
Hill, Robert Matteson
Hill, Ronald Ames
Hilliard, Ronnie Lewis
Hillier, James
Hilsenrath, Joseph
Hilt, Richard Leighton
Hilton, Henry H
Hilton, Wallace Atwood
Hinkle, John Marion
Hinkley, Everett David
Hinman, George Wheeler
Hinnov, Einar
Hinrichs, Clarence H
Hintregger, Hans Erich
Hipple, John Alfred
Hirschberg, Joseph Gustav
Hitschfeld, Walter
Hobbie, Russell Klyver
Hobson, Robert Marshall
Hochstim, Adolf R
Hock, Donald Charles
Hodge, Mary Wilma
Hodges, Sidney Edward
Hodgson, Rodney
Hoerlin, Herman William
Hoffman, Herbert Saul
Hoffman, Marvin Morrison
Hoffman, Robert Albert
Hoffmann, Richard Bruce
Hoffmann, Paul Otto
Hoffmann, William Frederick
Hofstadter, Robert
Hogan, William
Hohenemser, Charles Morris
Hohenemser, Christoph
Hohl, Frank
Hoisington, Laurence Earl
Holcomb, Donald Frank

Holdeman, Jonas Tillman, Jr
Holl, Herbert Barthold
Holland, Monte W
Hollinger, James Pippert
Hollyer, Robert Nelson, Jr
Holmstrom, Fred Edward
Holroyd, Louis Vincent
Holstein, Theodore David
Holt, Roland Bell
Holter, Marvin Rosenkrantz
Holter, Norman Jefferis
Holton, Gerald James
Holway, Lowell Hoyt, Jr
Hood, John Mack, Jr
Hood, Richard Fred
Hooper, William John, Jr
Hoover, Gary McClellan
Hopfer, Samuel
Hopkins, Don Carlos
Hopkins, Robert Earl
Horak, Jerry Robert
Hord, Charles W
Horie, Yasuyuki
Horn, William Everett
Horwitz, Nahmin
Houska, Charles Robert
Houston, John Mapes
Houston, Robert Edgar, Jr
Hovis, Louis Samuel
Hovorka, John
Howard, Betty
Howard, John Nelson
Howard, Raydeen Roland
Howard, Richard John
Howard, Robert Adrian
Howe, David Allen
Howe, Stephen Henry
Howell, Everette Irl
Hower, Meade M
Hoy, Gilbert Richard
Hoyer, Wilmer Adolf
Hoyt, Gordon Dunwell
Hoyt, Harry Charles
Hoyt, Rosalie Chase
Hu, Pung Nien
Huang, Wei-Feng
Hubbard, Harmon William
Hubbell, Harry Hopkins, Jr
Hubbs, John Charles
Huberman, Bernardo Abel
Hudson, Cecil Ivan, Jr
Hudson, Donald Edwin
Huffaker, James Neal
Huggett, Richard William, Jr
Hughes, Harold Kenneth
Hughes, Raymond Hargett
Hughes, Thomas Rogers
Hughes, Vernon Willard
Hughes, Victor A
Huish, Howard Paul
Hujer, Karel
Hull, Gordon Ferrie, Jr
Hull, Harvard Leslie
Hull, Robert Joseph
Hults, Malcom E
Hultsch, Roland Arthur
Hume, William
Hummel, Harry Horner
Humphrey, Charles Harve
Hunt, Angus Lamar
Hunt, Earle Raymond
Hunt, James L
Hunt, Robert Harry
Hunter, George Truman
Hunter, Hugh Wylie
Hunter, Joseph Lawrence
Hurlbut, Lloyd Philip
Hurlbut, Franklin Charles
Hurley, James D
Hurst, Donald Geoffrey
Hurt, James Edward
Hutchinson, Thomas Sherret
Hutzenlaub, John F
Huzinaga, Sigeru
Hwang, Chester F
Hyams, Henry C
Hyde, Walter Lewis
Hynek, Walter Joseph
Iberall, Arthur Saul
Ibser, Homer Wesley
Ifft, Edward M
Igo, George (Jerome)
Ikenberry, Dennis L
Ingalls, Robert L
Ingraham, Mark Gordon
Ingram, Peter
Inn, Edward Chang Yul
Intemann, Robert Louis
Irvine, Merle M
Irwin, John Charles
Isaacson, Richard Allen
Isenor, Neil R
Isler, Ralph Charles
Israel, Martin Henry
Itzkan, Irving
Ivey, Donald Glenn
Ivory, John Edward
Jack, Hulan E, Jr

PHYSICS

Jacko, Michael George
Jackson, Edwin Atlee
Jackson, Francis J
Jackson, Prince A, Jr
Jackson, Ray Weldon
Jacobs, Ira
Jacobs, James Albert
Jacobs, Robert Byron
Jacobs, Sigmund James
Jacobs, Stephen Frank
Jacobson, Edward Hastings
Jacobson, Harry C
Jaffe, Bernard Mordecai
Jahns, Monroe Frank
Jain, Piyare Lal
Jamerson, Frank Edward
James, Brian David
James, Dennis Bryan
Jammu, K S
Janacek, Joachim Wilhelm
Janak, James Francis
Janes, George Sargent
Janis, Allen Ira
Janney, Donald Herbert
Janos, William Augustus
Jarrett, Steven Michael
Jarvis, James Gordon
Jarzynski, Jacek
Jasperse, John R
Jastram, Philip Sheldon
Jaswal, Sitaram Singh
Javan, Ali
Jefferson, Donald Earl
Jefimenko, Oleg D
Jekeli, Walter
Jenzano, Anthony Francis
Jeppesen, Myron Alton
Jette, Archelle Norman
Jezek, Edward V
Jin, Rong-Sheng
Joanou, George David
Jobes, Forrest Crossett, Jr
Johannes, Robert
Johns, Harold E
Johnson, Arthur Franklin
Johnson, C Scott
Johnson, Charles Minor
Johnson, DeWayne Carl
Johnson, Elizabeth Briggs
Johnson, Fred Tulloch
Johnson, Larry Claud
Johnson, Larry Don
Johnson, Lloyd Kenneth
Johnson, Montgomery Hunt
Johnson, Paul Sosinski
Johnson, Ralph T, Jr
Johnson, Rolland Paul
Johnson, Walter Heinrick, Jr
Johnson, Walter Richard
Johnson, William Pierre
Johnson, Woodrow Eldred
Johnston, Robert Ward
Jones, Charles Miller, Jr
Jones, Claude Kitchener
Jones, Creighton Clinton
Jones, Dale Robert
Jones, Dallas Wayne
Jones, Douglas Emron
Jones, Elbert Ellery
Jones, Emerson
Jones, Ernest Addison
Jones, Gordon Ervin
Jones, Phillips Russell
Jones, Robert Edward
Jones, Stanley Tanner
Jones, William Barclay
Jordan, Albert Raymond
Jordan, Walter Edwin
Jordan, Walter Harrison
Jory, Farnham Stewart
Joseph, Alexander
Joseph, Alfred S
Joseph, James
Jossem, Edmund Leonard
Joynson, Reuben Edwin, Jr
Judge, Darrell L
Juenker, David W
Julian, Rene Stephen
Jungerman, John (Albert)
Kadaba, Prasad Krishna
Kadesch, Robert R
Kadyk, John Amos
Kagarise, Ronald Eugene
Kahalas, Sheldon Lee
Kahn, Arnold Herbert
Kahn, David
Kahn, Jack Henry
Kahng, Dawon
Kalikstein, Kalman
Kalmbach, Syndey Hobart
Kammer, Erwin William
Kamper, Robert Andrew
Kane, Conrad Gabriel
Kannenberg, Lloyd C
Kanney, Lynna Babs
Kantor, Paul B

Kao, Yi-Han
Kapany, Narinder Singh
Kaplan, Joseph
Kaplon, Morton Fischel
Karas, John Athan
Karle, Isabella Lugoski
Karr, Philip R
Kartha, Gopinath
Kasemir, Richard Allen
Kashnow, Richard Allen
Kasner, William Henry
Kasnitz, Harold Louis
Kastner, Sidney Oscar
Katcher, David Abraham
Katz, Maurice Joseph
Katz, Robert
Katzenstein, Jack
Katzin, Gerald Howard
Kauffman, George Emmett Clarence
Kauffman, Richard L
Kavadas, Alexander D
Kawatra, Mahendra P
Kaye, Brian H
Kaylor, Hoyt McCoy
Kazan, Benjamin
Kearney, Philip Daniel
Keck, Donald Bruce
Keck, James Collyer
Keck, William George
Keck, Winfield
Kedesdy, Horst H
Keeling, Rolland Otis, Jr
Keene, Wayne Hartung
Keesom, Pieter Hendrik
Keevil, Norman Bell
Keezer, Richard Clark
Kehoe, Brandt
Keiffer, David Goforth
Keiber, Charles Norman
Kelland, David Ross
Keller, Harry Bert, III
Kelley, Edward M
Kelly, Elmer Lewis
Kelly, Frederick Miles
Kelly, Henry Charles
Kelly, Hugh P
Kelly, Martin Joseph
Kelly, Robert Emmett
Kelly, Thomas Michael
Kelly, William Clark
Kelton, Gilbert
Kemble, Edwin Crawford
Kendall, Harry White
Kendall, Henry Way
Kendall, James Tyldesley
Kennard, Ralph Brandreth
Kennedy, Jerry Dean
Kennedy, John Edward
Kenney, Vincent Paul
Kenney, George S
Kent, Donald Wetherald, Jr
Kepple, Paul C
Kereiakes, James Gus
Kern, Joseph
Kernaghan, Marie
Kernan, William J, Jr
Kerns, Quentin Alexander
Kershenstein, Karen Weis
Kersta, Lawrence George
Kerth, Leroy Thomas
Kerwin, John Larkin
Kessler, John Otto
Ketterson, John Boyd
Keuper, Jerome Penn
Kevane, Clement Joseph
Keyes, Robert William
Keyes, John David
Khan, Imdad Haque
Khanna, Sardari Lal
Khare, Bishun Narain
Khorana, Brij Mohan
Kicska, Paul A
Kidder, John Newell
Kidder, Ray Edward
Kiehn, Robert Mitchell
Kiess, Edward Marion
Kilgore, William Arlow
Kim, Daniel Y
Kim, Young Bae
Kime, Joseph Martin
King, Allen Lewis
King, David Thane
King, James, Jr
King, John Swinton
King, L D Percival
King, Ronold (Wyeth Percival)
King, William Connor
Kingery, Bernard Troy
Kingston, Robert Hildreth
Kinnison, Gerald L
Kinsella, John J
Kinsey, Bernard Bruno
Kinsey, Kenneth F
Kip, Arthur Frederic
Kirby, Roger D
Kirchgessner, Joseph L
Kirkpatrick, Larry Dale

Kirsch, Lawrence Edward
Kisliuk, Paul
Kiszenick, Walter
Kitazawa, George
Kittel, Charles
Klaiber, George Stanley
Klarmann, Joseph
Kleesattel, Claus
Kleinman, Herbert
Kleinman, Leonard
Kleinman, Chemia Jacob
Kleinsteuber, Tilmann Christoph Werner
Klemens, Paul Gustav
Kleppner, Daniel
Kline, Donald Edgar
Kliwer, James Karl
Klontz, Everett Earl
Kmetko, Edward Andrew
Knapp, Edward Alan
Knecht, Walter Ludwig
Knechtli, Ronald (C)
Knight, Larry V
Knights, John Christopher
Knipp, Julian Knause
Knoblaugh, Armand Frank
Knop, Harry William, Jr
Knop, Robert Edward
Knopoff, Leon
Knowles, Harold Loraine
Knowles, Harold B
Knudsen, John Peter
Kohane, Theodore
Kohler, Roger Patrick
Kohn, Walter
Kohnke, Elton Everett
Kohr, Charles Byron
Kolar, Oscar Clinton
Kolb, Alan Charles
Kolb, Earl Leonard
Kolossvary, Bela Gabriel
Kolsky, Harwood George
Kolsky, Herbert
Kompfner, Rudolf
Konopinski, Emil John
Kopp, Jay Patrick
Koppius, Otto Gustav
Kossler, William Johns
Koster, George Fred
Kostkowski, Henry John
Kouts, Herbert John Cecil
Kozora, Andrew John
Kozanowski, Henry Nikodem
Kraft, David Werner
Kram, William Eric
Krall, Nicholas Anthony
Kramer, Paul Robert
Kranz, Alan Zelig
Krasner, Sol H
Kraus, Alfred Andrew, Jr
Kraus, Olen
Krause, Ernst Henry
Krause, Lucjan
Kraushaar, William Lester
Kraut, Edgar A
Krembheller, Alfred
Kremser, Thurman Rodney
Krewer, Semyon E
Kriminn, Samuel
Krishnamurti, Ruby Ebisuzaki
Kriss, Michael Allen
Kristian, Jerome
Kroemer, Herbert
Kroening, John Leo
Krohn, Victor Eugene, Jr
Kronenberg, Klaus Johannes
Kronenberg, Stanley
Krongelb, Sol
Kropp, William A
Krouse, Howard Roy
Krueger, Robert William
Krueger, Roland Frederick
Kruger, Peter Gerald
Kruglak, Haym
Kruschwitz, Walter Hillis
Kruse, Olan Ernest
Kruse, Ulrich Ernst
Krutter, Harry
Krzywoblocki, Maria Zbigniew von
Kuan, Hsin Min
Kuan, Teh Soong
Kuhns, Ellen Swomley
Kuipers, George Albertus
Kulp, Bernard Andrew

Kunz, Hans Joseph
Kunz, Kaiser Schoen
Kunze, Jay Frederick
Kuper, J B Horner
Kupferberg, Kenneth Maurice
Kurshan, Jerome
Kusch, Polykarp
Kutzscher, Edgar Walter
Kwo, William T
Kyhl, Robert Louis
Labaw, Louis Warne
Lach, Joseph T
Lacke, Robert Edgar
Ladd, John Herbert
Lado, Fred
Lagemann, Robert Theodore
Lagergren, Carl Robert
Lai, Ravindra Behari
Lal, Ravindra Behari
Lalos, George Theodore
Lamarche, J L Gilles
Lamb, George Lawrence, Jr
Lamb, Walter Robert
Lambe, Edward Dixon
Lamberts, Robert L
Landaeur, Joseph K
Landauer, Rolf
Lander, Richard Leon
Landon, Donald Omar
Landrum, Bobby L
Lane, George H
Lang, David (Vern)
Lange, James Neil, Jr
Langevin, Robert Arthur
Langmuir, David Bulkeley
Langmuir, Robert Vose
Langreth, David Chapman
Langstroth, George Forbes Otty
Lanou, Robert Eugene, Jr
Lanza, Giovanni
Lanza, Richard Charles
Lanzl, Lawrence Herman
Lapsley, Alwyn Cowles
Laramore, George Ernest
LaRocca, Anthony Joseph
Larsen, Robert Arthur
Larsen, Karl Davis
Larsen, Ted LeRoy
Larson, Lee Edward
Lashinsky, Herbert
Lashof, Theodore William
Laslett, Annulya Lal
Laslett, Lawrence Jackson
Lathrop, Arthur LaVern
La Tourette, James Thomas
Laufer, Arthur Russell
Laughlin, Robert David
Laukonis, Joseph Vainys
Lavatelli, Leo Silvio
LaVier, Eugene Clark
Law, Harold Bell
Lawrence, George Melvin
Lawrence, James Neville Peed
Lawson, Andrew Werner
Lawson, Bob Leroy
Lawson, James Llewellyn
Lawson, Juan (Otto)
Layton, Richard Gary
Layton, Thomas William
Layzer, Arthur James
Lazar, Norman Henry
Lea, Robert Martin
Leavitt, Christopher Pratt
Leavitt, John Adams
Le Blanc, Marcel A R
Lebo, George Robert
Lebowitz, Joel Louis
Leclaire, Roger
Lederer, Charles Alexander
Lee, David Morris
Lee, Hoong-Chien
Lee, Neville Ka-Shek
Lee, Thomas Henry
Leffel, Claude Spencer, Jr
Legg, Thomas Harry
Legvold, Sam
Leiby, Clare C, Jr
Leidecker, Henning William, Jr
Leinbach, Harold
Leiss, James Elroy
Leiter, Howard Allen
Leivo, William John
LeLacheur, Robert Murray
Lemieux, Paul E
Lemonick, Aaron
Lempicki, Alexander
Lengyel, Bela Adalbert
Lennert, Andrew E
Lents, James Marcellus
Leonard, Byron Peter
Leonard, Bowen Raydo, Jr
Leonard, Frederick Wilhelm
Lepore, Joseph Vernon
Lepoff, Jack H
Lesensky, Leonard
Letcher, John Henry
Letcher, Stephen Vaughan
Levenson, Leonard L
Levin, Eugene (Manuel)
Levinger, Joseph S
Levinstein, Henry

Levinthal, Elliott Charles
Levi Setti, Ricardo
Levit, Lawrence Bruce
Levitas, Alfred Dave
Lewis, Elmer James
Lewis, Francis Hotchkiss, Jr
Lewis, Harold Ralph
Lewis, Harold Warren
Lewis, Henry Rafalsky
Lewis, Margaret Nast
Lewis, Robert Taber
Lewis, Vance De Spain
Lewis, Wilfrid Bennett
Lewis, Willard Deming
Li, Kelvin K
Li, Ming Chiang
Liang, Ching Yu
Libby, Leona Marshall
Licht, Arthur Lewis
Lichten, William Lewis
Lichtenstein, Roland Max
Liebermann, Leonard Norman
Liebes, Sidney, Jr
Liebman, Alan Joel
Liebschutz, Alan Morton
Liebson, Sidney Harold
Liley, Peter Edward
Lilly, Arnys Clifton, Jr
Limperis, Thomas
Lin, Duo-Liang
Lind, Vance Gordon
Linder, Ernest Gustaf
Lindquist, Richard Wallace
Lindsay, Robert Bruce
Lindstrom, Ivar E, Jr
Ling, Samuel Chen-Ying
Link, John Clarence
Liou, Horng Ing
Lipsicas, Max
Lipson, Joseph
Litovitz, Theodore Aaron
Little, William Arthur
Liuima, Francis Aloysius
Livingood, John Jacob
Livingston, Peter Moshchansky
Livingston, Robert Simpson
Lloyd, James Newell
Lobb, Duncan Edward
Lloyd, Donald Edward
Lochstet, William A
Locke, Jack Lambourne
Locke, Stanley
Lockwood, John Alexander
Lodge, Arthur Scott
Lodge, John I
Loeber, Adolph Paul
Loeffler, Frank Joseph
Lofgren, Edward Joseph
Logan, Joseph Granville, Jr
Logan, Robert Kalman
Loh, Hung Yu
Long, Daniel R
Long, George Donald
Long, Howard Charles
Long, Jerome R
Long, John Vincent
Longaker, Perry R
Longini, Richard Leon
Longley, William Warren, Jr
Longmire, Conrad Lee
Lontz, Robert Jan
Loomis, Wheeler
Looney, Duncan Hutchings
Lord, Arthur E, Jr
Lord, Jere Johns
Lorenz, Max Rudolph
Lotspeich, James Fulton
Lott, Layman Austin
Lott, Sam Houston, Jr
Loughborough, Dwight Logan
Louisell, William Henry
Love, Hugh Morrison
Love, William Alfred
Low, William
Lowance, Franklin Elta
Lowe, Robert Peter
Lowell, Sherman Cabot
Lowndes, Robert P
Loyd, Coleman Monroe
Lu, Phillip Kehwa
Luban, Marshall
Lubatti, Henry Joseph
Lubkin, Gloria Becker
Luck, Clarence Frederick, Jr
Luck, David George Croft
Luckey, Paul David, Jr
Lucy, Frank Allen
Lucy, Carl Arthur
Ludeke, Carl Arthur
Luebke, Emmeth August
Lufburrow, Robert Allen
Luhman, Gladys Finney
Lukasik, Stephen Joseph
Lund, Donald S
Lurie, Fred Marcus
Lurio, Allen
Lustig, Claude David
Lustig, Harry
Lutz, Bruce Charles
Lykken, Glenn Irven

Lyman, Ernest McIntosh
Lyman, Ona Rufus
Lynch, Eugene Joseph Michael
Lyon, Gordon Frederick
Lyon, Waldo (Kampmeier)
Lyons, Donald Herbert
Lyons, Harold
Lysiak, Richard John
Ma, Shang-Keng
MacAdam, David Lewis
MacArthur, John Wood
MacDonald, Alexander Daniel
MacDonald, John Campbell Forrester
MacDowell, Samuel Wallace
Macedo, Pedro Buarque de
Macek, Joseph
MacFarlane, John T
MacGregor, Malcolm Herbert
MacHattie, Lloyd Elliot
Machlup, Stefan
MacInnis, Martin Benedict
Mack, Rex Charles
Mack, Stanley Zaner
MacKellar, Alan Douglas
MacKenzie, Innes Keith
MacKenzie, Kenneth Ross
MacKenzie, Kenneth Victor
Mackin, Robert James, Jr
MacNaughton, Earl Bruce
MacNeill, Rupert Heath
Mac Rae, Alfred Urquhart
Madden, Hannibal Hamlin, Jr
Madden, Robert Phyfe
Madigosky, Walter Myron
Maenchen, George
Magarvey, Raymond Halliday
Mahan, Archie Irvin
Maher, James Vincent
Mahlman, George William
Mahoney, Francis Joseph
Maidanik, Gideon
Main, William Francis
Maissel, Leon I
Major, Robert Wayne
Major, Schwab Samuel, Jr
Majumdar, Debaprasad
Majumdar, Ernest Ilya
Maley, Martin Paul
Malhiot, Robert Joseph
Maling, George Croswell, Jr
Malkus, Willem Van Rensselaer
Mallozzi, Philip James
Malmon, Arthur Gerald
Manakkil, Thomas Joseph
Manasse, Roger
Mandelberg, Hirsch I
Mandeville, Charles Earle
Mandula, Jeffrey Ellis
Maninger, Ralph Carroll
Mann, Alfred Kenneth
Mann, Kenneth Clifford
Mann, Leonard Andrew
Mann, Ralph Willard
Mann, Wilfrid Basil
Manning, Walter H, Jr
Manoharan, A Chelvanayakam
Manoogian, Armen
Manring, Edward Raymond
Manson, Donald Joseph
Mantis, Homer Theodore
Mapother, Dillon Edward
Mara, Richard Thomas
Marburger, Richard Eugene
March, Robert Henry
March, Robert Herbert
Marcoux, Jules E
Marcus, Jules Alexander
Marcus, Sanford M
Marcuse, Dietrich
Margolis, Bernard
Marick, Louis
Mariner, Thomas
Marino, Robert Anthony
Mark, Hans Michael
Markham, Jordon Jeptha
Marks, Darrell L
Marks, Luther Whitfield, III
Markworth, Alan John
Marquis, Richard Jack
Marr, Robert B
Marshall, Donald James
Marshall, J Howard, III
Marshall, John Hart
Marshall, John Stewart
Marshall, Lauriston Calvert
Marshall, Samuel Wilson
Marshall, Thomas C
Martin, Albert Byron
Martin, David William
Martin, Gordon Eugene
Martin, James Edwin
Martin, Richard McFadden
Martin, Robert Leonard
Martz, Dowell Edward
Masek, Joseph Walter
Mather, Joseph Walter
Mather, Robert Laurance

Matheson, Willard Edward
Mathieson, Alfred Herman
Mathieu, Roger Maurice
Matlock, Rex Leon
Matsen, Frederick Albert
Matthews, David LeSueur
Matthews, Herbert Maurice
Matthews, John Wauchope
Matthews, Lee Drew
Matthews, Peter Wren
Matthias, Bernd T
Mattox, Donald Moss
Matzner, Richard Alfred
Maudlin, Lloyd Z
Maunsell, Charles Dudley
Maurer, Hans Andreas
Mautz, Charles William
Mavroides, John George
Maxson, Donald Robert
Maxwell, Emanuel
Maxwell, Howard Nicholas
May, John Elliott, Jr
May, Michael Melville
Mayer, Alex
Mayer, Walter Georg
Mayes, Billy Woods, II
Mayfield, Melburn Ross
Mayo, Thomas Tabb, IV
McAllister, Howard Conlee
McAllister, Robert Wallace
McAneny, Laurence Raymond
McAvoy, Bruce Ronald
McBride, John Barton
McCart, Bruce Ronald
McCarthy, Francis Wadsworth
McCarthy, Raymond Lawrence
McCaslin, John Garfield
McCay, Myron Stanley
McClain, John William
McCloskey, James
McClung, Ronald Edwin Dawson
McClure, Gordon Wallace
McCollum, Donald Carruth, Jr
McColm, Douglas Woodruff
McConkey, John William
McConville, George T
McCorkle, William C, Jr
McCormick, William Devlin
McCoubrey, Arthur Orland
McCown, Dean Augustus
McCoy, Jerome Dean
McCracken, Curtis W
McCreary, Ralph Leroy
McCubbin, Thomas King, Jr
McCumber, Dean Everett
McCune, Robert Franklin
McCutchen, Charles Walter
McDaniel, Earl Wadsworth
McDermott, Lillian Christie
McDonald, Allan W
McDonald, Arthur Bruce
McDonald, Frank Bethune
McDonald, William John
McElaney, James H
McElhinney, John
McElroy, Michael Brendan
McEnally, Terence Ernest, Jr
McFarland, Charles Elwood
McFarland, Robert Harold
McFarlane, Ross Alexander
McFarlane, Walter Kenneth
McFee, Raymond Herbert
McGavin, Raymond E
McGill, Thomas Conley, Jr
McGinnis, Carl Leonardt
McGinnis, Eugene A
McGonnagle, Warren James
McGowan, Francis Keith
McGrath, James Russell
McGrath, James Williamson
McGraw, Delford Armstrong
McIlwain, Carl Edwin
McIntire, Sumner Harmon
McIntosh, Bruce Andrew
McIntosh, Harold Leroy
McKay, Alexander Scott
McKee, James Stanley Colton
McKee, John W
McKenna, James
McKeown, Joseph
McKinley, William Albert
McKinney, Chester Meek
McLaughlin, James E
McLean, William Burdette
McLean, William L
McLeroy, Edward Glenn
McManus, Elizabeth Catherine
McMickle, Robert Hawley
McMillan, Daniel Russell
McMillan, Edwin Mattison
McMurry, Earl William
McNutt, Douglas P
McPherson, Ross
McQuistan, Richmond Beckett
McRae, Eion Grant
McWilliams, Donald A
Mead, Sylvester Warren, III

Mechler, Mark Vincent
Mechlin, George Francis, Jr
Medsger, Gerald William
Meek, Jack Henry
Meeker, Ralph Dennis
Megill, Lawrence Rexford
Mehl, James Bernard
Mehran, Farrokh
Meijer, Robert Randal
Meincke, P P M
Meitzler, Allen Henry
Melchior, Jack L
Melkonian, Edward
Mellen, Gilbert Emery
Mellen, Robert Harrison
Melnick, Donald A
Meltzer, Carl Martin
Melvin, Mael Avramy
Memory, Jasper Durham
Mendelsohn, Lawrence Barry
Mendelson, Kenneth Samuel
Menes, Meir
Menius, Arthur Clayton, Jr
Menke, John Roger
Mercer, Robert Allen
Mercereau, James Edgar
Mercure, Ruel Coe, Jr
Merideth, George Thomas
Mermin, N David
Merrill, John Jay
Merritt, Jack
Meshkov, Sydney
Metcalf, Harold
Metzger, Daniel Schaffer
Metzger, Wesley James
Meyer, Donald Irwin
Meyer, Johannes Horst Max
Meyer, Stuart Lloyd
Michael, Paul Andrew
Michael, Bertrand John
Michels, Donald Joseph
Michelson, Louis
Michener, John William
Mikeska, Emory Eugene
Mikkelsen, Harry E
Miles, Kelly George
Milford, Frederick John
Miller, Allen H
Miller, Bertrand John
Miller, Bruce Linn
Miller, David Charles
Miller, Douglas Gordon
Miller, Emery Parker
Miller, Emil C
Miller, Franklin, Jr
Miller, Gabriel Lorimer
Miller, Gerald R
Miller, Glenn Houston
Miller, Harry Galen
Miller, Julian Malcolm
Miller, Roger Edward
Miller, Julius Sumner
Miller, Leonard Robert
Miller, Park Hays, Jr
Miller, Ralph J
Miller, Raymond Edwin
Miller, Robert Bruce
Miller, Robert DuWayne
Miller, William Alfonso
Millett, Walter Elmer
Mills, Frederick Eugene
Mills, James Sidney
Mills, Roger Edward
Milne, Allen Ritchie
Milone, Eugene Frank
Minichino, Camille
Minkoff, John
Minnix, Richard Bryant
Mintz, Esther Uress
Mires, Raymond William
Misener, Austin Donald
Misheloff, Michael Norman
Miskovsky, Nicholas Matthew
Misner, Robert David
Mistry, Nariman Burjor
Mitcham, Donald
Mitchell, Earl Nelson
Mitchell, Ferdinand H
Mitchell, Gary Earl
Mitchell, Henry Rees
Mitchell, John Peter
Mitchell, John Wesley
Mitchell, Michael A
Mitchell, Olga Mary Mracek
Mitchell, Richard Warren
Mitchner, Morton
Mitescu, Catalin Dan
Mitteman, Marvin Harold
Mittler, Arthur
Mo, Luke Wei
Moak, Charles Dexter
Moazed, Cyrus
Mochel, Jack McKinney
Mochel, Virgil Dale
Moe, Chesney Rudolph
Moe, Osborne Kenneth
Mogford, James A
Moldover, Michael Robert
Mollenauer, Linn F

Prinz, Dianne Kasnic
Prior, Michael Herbert
Pripstein, Morris
Proctor, David George
Proctor, Warren George
Prodell, Albert Gerald
Prohammer, Frederick George
Propst, Franklin Moore
Proud, Joseph Mason, Jr
Provencio, Jesus Roberto
Pruett, George Richard
Pruett, John Robert
Pruitt, Roger Arthur
Pryce, Aubrey William
Pryor, Marvin J
Pullan, Harry
Pullman, Ira
Purbrick, Robert Lamburn
Purcell, Edward Mills
Purcell, James Eugene
Purdom, Ray Caldwell
Purdy, William Henry
Pyle, Robert Wendell, Jr
Quarles, Gilford Godfrey
Querry, Marvin Richard
Quinn, Helen Rhoda
Quinn, Warren Eugene
Quirk, Arthur Lincoln
Quisenberry, Karl Spangler, Jr
Rabi, Isidor Isaac
Radcliffe, Alec
Raeuchle, Richard Frank
Rahm, David Charles
Ramsey, Alan T
Rainwater, Clarence Saunders
Rainwater, Leo James
Randall, Charles Hamilton
Randels, Robert Basil
Randhawa, Jagir Singh
Raney, William Perin
Rank, David Herr
Rankin, David
Rao, K V N
Rao, Kandarpa Narahari
Ralph, Waldo
Ralston, H Robert
Ram, Budh
Rambauske, Werner
Ramberg, Edward Granville
Ramsdale, Dan Jerry
Ramsey, Norman Foster, Jr
Raka, Eugene Cd
Rakestraw, James William
Rasband, S Neil
Rassweiler, Merrill (Paul)
Rau, R Ronald
Raub, Harry Lyman, III
Rautenberg, Theodore Herman
Rawlins, Stephen Last
Rawls, William Shelton
Ray, John Robert
Rayburn, Louis Alfred
Rayfield, George W
Raymond, Richard James
Raymond, Richard Collyer
Rayne, John A
Read, Albert James
Read, Floyd M
Read, Philip Lloyd
Read, Richard Bradley
Reader, Joseph
Ream, Donald F
Reardon, Anna Joyce
Reardon, William Albert
Reaves, Harry Lee
Redding, John Lawford
Redding, Rogers Walker
Reddy, Satti Paddi
Reder, Friedrich H
Redhead, Paul Aveling
Rediker, Robert Harmon
Redlich, Martin George
Redmond, Peter John
Regener, Victor H
Reibel, Kurt
Reichertz, Paul Peter
Reid, Charles David
Reidy, James Joseph
Reiffel, Leonard
Reilly, Edwin David, Jr
Reinheimer, Julian
Reisinger, Joseph G
Reisler, Donald Laurence
Reisner, John Henry
Reitz, Robert Alan
Relf, Kenneth E
Remler, Edward A
Remley, Marlin Eugene
Rempfer, Gertrude Fleming
Reneker, Darrell Hyson
Renner, John Wilson
Renzetti, Nicholas A
Reppy, John David
Resnick, Robert

Reudink, Douglas O
Reynolds, Harry Lincoln
Reynolds, John Hamilton
Reynolds, Joseph Melvin
Reynolds, Robert Eugene
Rezanka, Ivan
Rhein, Walter Joseph
Rheinstein, John
Rhodes, Richard Ayer, II
Ribe, Fred Linden
Rich, Wiley Foster
Richards, Charles Marvin
Richards, James Austin, Jr
Richards, Paul Irving
Richards, Paul Linford
Richardson, Charles Bonner
Richardson, John Marshall
Richardson, Robert Esplin
Richmond, James Kenneth
Richter, Burton
Richtmyer, Robert Davis
Rickard, James Alexander
Ricker, Charles William
Rickert, Russell Kenneth
Rider, Ronald Edward
Riedl, H Raymond
Riehl, Jerry A
Riess, Karlem
Riesz, Robert Richard
Riggs, James W, Jr
Rigney, Carl Jennings
Riley, James A
Riley, William Robert
Rinehart, John Sargent
Ring, Harold Francis
Ring, Paul Joseph
Riseberg, Leslie Allen
Risser, Jacob Rutt
Ritson, David Mark
Ritter, Enloe Thomas
Roach, Kenneth Alphonsa
Robbins, Allen Bishop
Robbins, Donald
Robbins, Donald Eugene
Roberds, Wesley Milton
Roberson, John Howard
Roberts, Arthur
Roberts, David Llewellyn
Roberts, Denys Thomas
Roberts, James Richard
Roberts, John England
Roberts, Leonidas Howard
Roberts, Louis Douglas
Roberts, Louis W
Robertson, Randal McGavock
Robertson, Robert Graham Hamish
Robertson, William Woodrow
Robin, James Edmond
Robinson, Clark Shove, Jr
Robinson, David
Robinson, David Zay
Robinson, Edward J
Robinson, Howard Addison
Robinson, Hugh Gettys
Robinson, James H
Robinson, Joseph Dewey
Robinson, Lawrence Baylor
Robinson, Marguerite Moilliet
Robinson, Mark Tabor
Robinson, William Kirley
Robson, John William
Robusto, C Carl
Rodden, Robert Morris
Rodebaugh, George Wayne
Rodriguez, Argelia Velez
Rodriguez, Haydee C
Roe, Byron Paul
Roeder, Robert Charles
Roesch, William Carl
Roessler, David Martyn
Roetling, Paul G
Rogers, Emery Herman
Rogers, Eric Malcolm
Rogers, Howard Gardner
Rogers, Kenneth Cannicott
Rogers, Marguerite Moilliet
Rogers, Peter H
Rogers, Thomas F
Rogers, William Alan
Rogosa, George Leon
Rohr, Robert Charles
Rohrer, Robert Harry
Rojansky, Vladimir
Rol, Pieter Klaas
Roll, Peter Guy
Rollefson, Ragnar
Rollin, Frank David
Rollor, Edward Albert
Rollosson, George William
Rolnick, William Barnett
Romer, Alfred
Romo, William Joseph
Rood, Joseph Lloyd
Rooney, James Arthur
Roothaan, Clemens Carel Johannes
Rorschach, Harold Emil, Jr
Rose, Albert
Rose, Carl Martin, Jr

Rose, David
Rose, Peter Henry
Rosen, Arthur Zelig
Rosen, James Martin
Rosen, Louis
Rosen, Paul
Rosen, Simon Peter
Rosenbaum, Ira Joel
Rosenbaum, Joseph Hans
Rosenberg, Leonard
Rosenberg, Robert
Rosenblum, Bruce
Rosenblum, William M
Rosencwaig, Allan
Rosengren, Jack Whitehead
Rosenthal, Jenny Eugenie (Mrs Arthur Bramley)
Rosenzweig, Walter
Rosin, Seymour
Rosner, Jonathan Lincoln
Rosner, Sheldon David
Ross, Charles Burton
Ross, Donald Alexander
Ross, Sidney
Rosser, Shirley Ewart
Rossi, Bruno B
Rossing, Thomas Dean
Rossmann, Kurt
Rostoker, Norman
Roth, Richard Francis
Rothberg, Joseph Eli
Rothstein, Jerome
Rothwell, William Stanley
Roudebush, William Campbell
Rowe, David John
Rowe, Irving
Rowell, John Martin
Rowell, Neal Pope
Rowland, Sattley Clark
Rowland, Theodore Justin
Rowntree, Robert Frederic
Roys, Paul Allen
Rozsnyai, Balazs
Rubens, Sidney Michel
Rubenstein, Albert Marvin
Rubin, G A
Rubin, Howard Arnold
Rubin, Kenneth
Rubin, Lawrence G
Ruddick, James John
Rudmose, H Wayne
Rudnick, Paul
Ruedenberg, Klaus
Ruegg, Fillmer William
Ruff, George Antony
Ruffine, Richard S
Rugge, Hugo R
Rugheimer, Norman MacGregor
Rundel, Robert Dean
Runyan, Walter R
Rupprecht, Hans S
Russek, Arnold
Russell, Allan Melvin
Russell, George A
Russell, James Edward
Russell, Marvin W
Russell, Ralph Keith
Ruth, Norbert Joseph
Ruvalds, John
Ryan, Alden Hoover
Ryan, Donald F
Ryan, Frederick Merk
Ryder, Martha
Ryge, Gunnar
Sabo, Jesse Jerry, Jr
Saby, John Sanford
Sachs, Allan Maxwell
Sachs, Donald Charles
Sachtleben, Clyde Clinton
Sadowski, Henry
Safford, Richard Whiley
Safko, John Loren
Sagurton, James Reynolds
Salem, Semaan Ibrahim
Salisbury, Winfield Wyman
Salkovitz, Edward Isaac
Salmons, George Beverly
Salzarulo, Leonard Michael
Samara, George Albert
Samios, Nicholas Peter
Sampson, John Laurence
Sampson, William B
Sander, Leonard Michael
Sanders, Steven Gill
Sanders, Theodore Michael, Jr
Sanders, Timothy D
Sanderson, Richard Blodgett
Sandhu, Harbhajan Singh
Sandiford, Peter Johnston
Sandlin, Billy Joe
Sands, Matthew
Sandstrom, Donald Richard
Sandweiss, Jack
Sanford, Edward Richard
Sanford, James R
Saporoschenko, Mykola
Sapp, Richard Cassell
Sarles, Lynn Redmon

Sarwinski, Raymond Edmond
Satterthwaite, Cameron B
Sauder, William Conrad
Sauer, Herbert H
Saunders, Peter Reginald
Saupe, Alfred (Otto)
Sauro, Joseph Pio
Savage, William Ralph
Sawatzky, Erich
Sawyer, Baldwin
Saxl, Erwin Joseph
Sayer, Arthur Robert
Sayer, Michael
Sayvetz, Aaron
Scalapino, Douglas J
Scandrett, John Harvey
Scanlon, Wayne Walter
Scarfone, Leonard Michael
Schacher, Gordon Everett
Schafer, George Edward
Schaffert, Roland Michael
Scharn, Herman Otto Friedrich
Schechter, Daniel
Schechter, Joseph M
Scheel, Nivard
Scheib, Richard, Jr
Scheie, Murray
Scheibe, Paul Olaf
Schelberg, Arthur Daniel
Scherr, Charles W
Schetzina, Jan Frederick
Schick, Jerome David
Schilling, Martin
Schillinger, Edwin Joseph
Schimitschek, Erhard Josef
Schindler, Alan Michael
Schindler, Albert Isadore
Schlecht, Richard Guenther
Schlein, Peter Eli
Schlosser, Jon A
Schmidt, Bruno (Francis)
Schmidt, Fred Henry
Schmidt, George
Schmidt, Helmut
Schmidt, Paul Woodward
Schmieg, Glenn Melwood
Schmugge, Thomas Joseph
Schneider, Harold O
Schneider, James Roy
Schneider, Martin V
Schneider, Sol
Schneiderwent, Myron Otto
Schneps, Jack
Schocken, Klaus
Schoepfle, George Kern
Schofield, Derek
Schone, Harlan Eugene
Schorr, Marvin Gerald
Schrack, Ronald Amundsen
Schramm, Robert William
Schrank, Glen Edward
Schrieffer, John Robert
Schroader, Irvin Homer
Schroeder, Frank, Jr
Schroeder, Manfred Robert
Schroeder, Peter A
Schroer, Dietrich
Schuette, Oswald Francis, Jr
Schulman, Lawrence S
Schultheis, James J
Schultz, Cramer William
Schultz, Frederick Herman Carl
Schultz, Howard Louis
Schultz, Jonas
Schulz, George J
Schulz, Manfred Bruno
Schumaker, Robert Louis
Schuster, Nick August
Schwarcz, Ervin H
Schwartz, Daniel Manning
Schwartz, Herman Meyer
Schwartz, Melvin
Schwarz, Klaus W
Schweinler, Harold Constantine
Schwerdtfeger, Charles Frederick
Schwertz, Frederick Anton
Schwettman, Herbert Dewitt
Schwettman, Harry Alan
Schwinger, Julian Seymour
Schwirzke, Fred
Schwitters, Roy Frederick
Schwoebel, Richard Lynn
Scifres, Donald Ray
Scofield, Dillon Foster
Scott, George David
Scott, George William, Jr
Scott, Paul Brunson
Scott, Peter Leslie
Scott, Robert Wallace
Scott, Thomas A
Scully, Marlan Orvil
Seacord, Daniel Freeman, Jr
Seager, Carleton Hoover
Seagondollar, Lewis Worth
Sears, Raymond W
Sears, Robert F, Jr
Sears, William Clifton

Seay, Glenn Emmett
Seddon, John Carl
Seeley, Paul Ellsworth
Seevers, Delmar Oswell
Segel, Ralph E
Seifert, Howard Stanley
Seitz, Frederick
Sekula, Stanley Ted
Seliger, Howard Harold
Sells, Robert Lee
Seneca, Gerard
Senf, Henry Ruwe
Senftle, Frank Edward
Senitzky, Benjamin
Sense, Karl August
Seppi, Edward Joseph
Serber, Robert
Sessler, John Charles
Sessoms, Faison Thomson
Settles, Ronald Dean
Severance, Dean Charles
Sewell, Duane Campbell
Sewell, Kenneth Glenn
Shaffer, John Clifford
Shaffer, Lawrence Bruce
Shaffer, Wave Henry
Shandley, Paul David
Shankland, Robert Sherwood
Sharma, Gerald White
Sharma, Ram Ratan
Sharp, James Martin
Sharp, Paul Chester
Sharrah, Paul Chester
Shaver, William Walker
Shaw, Charles Bergman, Jr
Shaw, Edgar Albert George
Shaw, John H
Shaw, William Corr
Shawhan, Elbert Neil
Shay, Dennis (John)
Shear, Sidney Kingsbury
Shearin, Paul Edmondson
Sheer, Charles
Sheldon, William Robert
Shelton, John Winthrop
Shelupsky, David I
Shen, Chun-Shan
Shen, Samuel Yi-Wen
Shenfil, Leon
Shenker, Henry
Shepherd, Jimmie George
Sheppard, Charles Wilcox
Sheppard, David W
Sheppard, Douglas Murray
Sher, Arden
Sherman, Charles Henry
Sherman, Christopher
Sherman, Harold
Sherwin, Chalmers William
Shields, Fletcher Douglas
Shipley, Edward Nicholas
Shinners, Carl W
Shire, Norman Steven
Shinn, Donald Leroy
Shih, Hsio Chang
Shilliber, Harry Albert
Shilliday, Theodore Smith
Shilts, James Leonard
Shinkle, Norman Leroy
Shnotin, Louis Marvin
Shockley, William
Shobert, Erle Irwin, II
Shnider, Ruth Wolkow
Shnider, George Aaron
Shotland, Edwin
Shoupp, William Earl
Shrauner, James Ely
Shugart, Howard Alan
Shull, Franklin Buckley
Sia, Richard Mae
Sichel, Enid Keil
Siderowitz, Joshua
Siegel, Lester Aaron
Siegel, Sidney
Siegler, Edourard Horace, Jr
Siem, Robert Arthur
Sikdar, Dhirendra N
Silbar, Richard Robert
Silbernagel, Bernard George
Silcox, John
Silfvast, William Thomas
Sill, Richard Clements
Silsbee, Henry Briggs
Silsbee, Robert Herman
Silver, Robert
Silver, Samuel
Slividi, Anthony Alfred
Simanek, Eugen
Simmons, Ralph Oliver

Simon, Eliot Morton
Simon, Ivan
Simon, Ralph
Simon, Ralph Emanuel
Simon, William
Simpson, Antony Michael
Simpson, James Henry, Jr
Simpson, John Alexander
Sinai, John Arol
Sinai, John Joseph
Sinclair, Rolf Malcolm
Singer, Stanley
Singh, Jag Jeet
Singleton, Edgar Bryson
Singleton, John Byrne
Sinsky, Joel A
Sipson, Roger Fredrick
Sittig, Erhard Karl
Sittig, Orvid Dayle
Six, Norman Frank, Jr
Skadron, George
Skaggs, Lester S
Skaragard, Harvey Milton
Skinner, Richard Emery
Skinner, William Carey
Skoili, Lester L
Skorinko, George
Skrable, Kenneth William
Skramstad, Harold Kenneth
Skudrzyk, Eugen J
Slade, Chaloner Berry
Slater, John Clarke
Slater, Rose Mooney
Slattery, Richard Erick
Slawsky, Zaka Israel
Sleator, William Warner, Jr
Slobodrian, Rodolfo Jose
Slocum, Robert Earle
Sloope, Billy Warren
Slusher, Richard Elliott
Smakula, Alexander
Smart, James Samuel
Smith, Albert Ernest
Smith, Carl Walter
Smith, Charles Sydney, Jr
Smith, Chester Lional
Smith, Daniel Montague
Smith, David English
Smith, Donald Charles
Smith, Edward John
Smith, George Foster
Smith, Harold Duncan
Smith, Harold Edmond
Smith, Harold Glenn
Smith, Harvey Alvin
Smith, Peter William
Smith, Roger M
Smith, Ronald E
Smith, Lloyd P
Smith, Louis Ezra, Jr
Smith, Luther W
Smith, Lyle W
Smith, Malcolm (Kinmonth)
Smith, Neville Vincent
Smith, Newell Hart
Smith, Orville L
Smith, P Scott
Smith, Roy E
Smith, Stephen Judson
Smith, Stephen Roger
Smith, Todd Iverson
Smith, William Conrad
Smith, Willy
Smithson, John Royston
Smits, Friedolf M
Smyth, John Bridges
Smythe, William Rodman
Snavely, Benjamin Lichty
Snider, John William
Snitzer, Elias
Snodgrass, Herschel Roy
Snover, Kurt Albert
Snow, George Abraham
Snow, Joel Alan
Snyder, Conway Wilson
Snyder, Donald DuWayne
Snyder, Howard Arthur
Snyder, James Newton
Snyder, Thoma Mees
Snyder, Wilbert Frank
Sobey, Arthur Edward, Jr
Sobottka, Stanley Earl
Sodickson, Lester A
Soest, Jon Fredrick
Soffer, Bernard Harold
Sogo, Power Bunmei
Sokollu, Adnan
Sokolowski, Henry Alfred
Solem, Anson Donald
Soller, Theodore
Solifrey, William
Sommers, Henry Stern, Jr
Sonder, Edward
Sonett, Charles Philip
Sooy, Walter Richard
Soren, Allan Louis

Sorrows, Howard Earle
Soule, David Elliot
Soules, Jack Arbuthnott
Southgate, Peter David
Southwick, Russell Duty
Soven, Paul
Spalding, Dan Wesley
Spalding, George Robert
Spangler, Charles Bishop
Spangler, George Wesley
Spangler, John David
Sparks, Joseph Theodore
Sparks, Marshall Scott
Spatz, Wilber De Villa Bernhart
Speck, David Ralph
Spence, Robert Dean
Spencer, Domina Eberle (Mrs Parry Moon)
Spencer, Herbert Ward, III
Spencer, William Turner
Sperber, Daniel
Spiegel, Valentin, Jr
Spielberg, Nathan
Spiers, Fred Noel
Spight, Carl
Spilman, George Raymond
Spinner, Theodore
Spinrad, Bernard Israel
Spitzer, William George
Spitznogle, Frank Raymond
Spohr, Daniel Arthur
Sposito, Garrison
Sprague, Basil Sheldon
Sprague, Gale Clifford
Sprague, Newton G
Spremulli, Paul Francis
Springett, Brian E
Sproull, Robert Lamb
Sprouse, Gene Denson
Spruch, Grace Marmor
Sreenivasan, Sreenivasa Ranga
Staats, Percy Anderson
Stacey, John Sydney
Stacey, Larry Milton
Stambaugh, Richard Bulla
Stamer, Peter Eric
Stanbrough, Jess Hedrick, Jr
Standil, Sidney
Stanley, Robert Weir
Stansbury, Edward James
Stanton, Henry Edmund
Stapleton, Harvey James
Stapp, Henry P
Starr, Walter LeRoy
Stauss, George Henry
Stauss, Martin
Stearns, Mary Beth Gorman
Stearns, Robert L
Stebbings, Ronald Frederick
Stebbins, Dean Waldo
Stecher, John D
Stecker, Floyd William
Stefanski, Raymond Joseph
Steiger, Walter Richard
Stein, Frank S
Stein, Seymour
Stein, Talbert Sheldon
Steinberger, Jack
Stekel, Frank D
Stell, George Roger
Stelson, Paul Hugh
Stenger, Victor John
Stenzel, Wolfram G
Stephen, Michael John
Stephens, William Edwards
Stephenson, Francis Creighton
Stephenson, Paul Bernard
Stergis, Christos George
Sternberg, David
Sternglass, Ernest Joachim
Sterzer, Fred
Stevens, Michael Thomas
Stevens, D Richard
Stevenson, Alec Thompson
Stewart, Frank Edwin
Stewart, Hugh Barnes
Stewart, James Lloyd
Stewart, John Westcott
Stewart, Melbourne George
Stewart, Robert William
Steyert, William Albert
Stickley, Elmer Eugene
Stiler, Paul Max
Stiller, Jerry Lee
Stimler, Morton
Stinson, Robert Henry
Stitch, Malcolm Lane
Stockdale, John Alexander Douglas
Stoddard, Alonzo Edwin, Jr
Stoeckly, Robert E
Stolov, Harold L
Stone, Edward Carroll, Jr
Stone, Julian
Stone, Richard Spillane
Stoner, Richard Griffith

Stout, Virgil L
Stover, Raymond Webster
Strahm, Norman Dale
Stranahan, George S
Stranahan, John Philip
Strassenburg, Arnold Adolph
Stratton, Julius Adams
Stratton, William R
Strauss, Wilbur Alexander
Strauss, Walter
Streib, John Fredrick
Stroh, William Richard
Strohmeier, Gustav H
Strong, Ian B
Strong, John (Donovan)
Strong, Robert Norwood
Stuart, Richard David
Stuetzer, Otmar Michael
Stuhlinger, Ernst
Stull, John Leete
Stump, Robert
Stumpf, Folden Burt
Stutz, Conley I
Suhl, Harry
Suits, Chauncey Guy
Suits, James Carr
Suits, James Douglas
Sullivan, Charles Raymond
Sullivan, Harry Morton
Sullivan, James Douglas
Sullivan, Seraphin A
Summerfield, Martin
Sundaram, Swaminatha
Surkan, Alvin John
Surko, Clifford Michael
Sutcliffe, Charles Herbert
Sutcliffe, William George
Sutherland, Bill
Sutter, Philip Henry
Suttle, Jimmie Ray
Sutton, Paul McCullough
Sutton, Paul Porter
Sydoriak, Stephen George
Sykes, Paul Jay, Jr
Symko, Orest George
Szegho, Constantin Stephen
Sztankay, Zoltan Geza
Szydlik, Paul Peter
Taft, Horace Dwight
Taimuty, Samuel Isaac
Takeo, Makoto
Talley, Robert Morrell
Talley, Thurman Lamar
Talman, Richard Michael
Tamarkin, Paul
Tanaka, Katsumi
Tang, Chung Liang
Tang, Kwong-Tin
Tang, Lun Han
Tang, Yau-Chien
Tangherlini, Frank R
Tannenwald, Peter Ernest
Tanner, Earl C
Tantilla, Walter H
Tapp, Charles Millard
Targ, Russell
Tarr, Charles Edwin
Taschek, Richard Ferdinand
Taylor, Charles Joel
Taylor, Edwin Floriman
Taylor, Gerald Reed, Jr
Taylor, Howard S
Taylor, Jack Eldon
Taylor, Jack Howard
Taylor, Jackson Johnson
Taylor, John Gardiner Veitch
Taylor, Merlin Gene
Taylor, Thomas Tallott
Teasdale, John G
Tebo, Edith Janssen
Teegarden, Kenneth James
Teiger, Martin
Teitler, Sidney
Telegdi, Valentine Louis
Telfair, David
Telford, William Murray
Tell, Benjamin
Teller, Edward
Tellinghuisen, Joel Barton
Tenzer, Rudolf Kurt
Terrell, James (Jr)
Tesche, Frederick Rutledge
Tesser, Herbert
Tessler, George
Tessman, Jack Robert

276

Thackeray, Ernest Russel
Thaddeus, Patrick
Thaler, William John
Thatcher, Everett Whiting
Theimer, Otto
Theisen, Wilfred Robert
Thekaekara, Matthew Pothen
Theobald, J Karl
Theriot, Edward Dennis, Jr
Thielking, David H
Thieme, Melvin T
Thiessen, George Jacob
Thiessen, Henry Archer
Thilo, Edward Rudolf
Thomas, Benjamin William
Thomas, Charles Danser
Thomas, Clinton Edward
Thomas, Dan Anderson
Thomas, Edward Carl
Thomas, Garland Leon
Thomas, Harold Albert
Thomas, Henry Coffman
Thomas, James E
Thomas, Keith Skelton
Thomas, Llewellyn Hilleth
Thomas, Montcalm Tom
Thomas, Richard Garland
Thomas, Roy
Thompson, James Chilton
Thompson, John Darrell
Thompson, Richard Scott
Thomson, John Oliver
Thomson, Keith Patrick Bowner
Thorn, Robert Nicol
Thorndike, Alan Moulton
Thorndike, Edward Moulton
Thorngate, John Hill
Thornton, Charles De Wane
Thornton, Robert Lyster
Thornton, William Andrus, Jr
Thourson, Thomas Lawrence
Thrower, Peter Albert
Thurston, Robert Norton
Tibbetts, Gary George
Tiedeman, John Albert
Tierman, Robert Joseph
Tiersten, Martin Stuart
Tiffany, Otho Lyle
Tiller, Calvin Omah
Tilley, Donald E
Tillotson, James Glen
Timusk, Thomas
Tinker, Robert Frederick
Tipsword, Ray Fenton
Tipton, Isabel Hanson
Tisza, Laszlo
Titone, Luke Victor
Tittle, Charles William
Titus, Walter Franklin
Tobin, Marvin Charles
Todd, Hollis N
Todd, Jay, Jr
Toepfer, Alan James
Tohver, Hanno Tiit
Tokita, Noboru
Toller, Louis
Tollestrup, Alvin V
Tomas, Francisco
Tomaselli, Vincent Paul
Tomlinson, Walter John, III
Tompkins, Donald Roy, Jr
Tompson, Clifford Ware
Torrance, Jerry Badgley, Jr
Toth, Robert Allen
Tousey, Richard
Towe, George Coffin
Townes, Charles Hard
Townsend, John Robert
Townsend, John William, Jr
Townsend, Jonathan
Trageser, Milton B
Tralli, Nunzio
Trambarulo, Ralph
Trauger, Donald Byron
Travis, James Roland
Travis, John Richard
Treanor, Charles Edward
Treat, Jay Emery, Jr
Trefny, John Ulric
Treiman, Sam Bard
Trentelman, George Frederick
Trexler, Frederick David
Triebwasser, Sol
Trimmer, John Dezendorf
Tripp, Robert D
Trischka, John Wilson
Troiano, Paul Francis
Trommel, Jan
Tross, Carl Henry
Trost, Walter Raymond
Trousdale, William Latimer
Trout, Edrie Dale
True, William Wadsworth
Trulio, John George
Truxillo, Stanton George
Trytten, Merriam Hartwick

Tsao, Chen-Hsiang
Tsao, Peter
Tschiegg, Carl Emerson
Tschunko, Hubert F A
Tterlikkis, Lambros
Tuchman, Albert
Tuck, James Leslie
Tucker, Edmund Belford
Tuckfield, Ralph George, Jr
Tuman, Vladimir Shlimon
Tunnicliffe, Philip Robert
Tupac, James Daniel
Turchinetz, William Ernest
Turkot, Frank
Turley, Sheldon Gamage
Turner, Arthur Francis
Turner, Clarence Marshall
Turner, Edward Felix, Jr
Turner, Eugene Bonner
Turner, Terry Earle
Turrell, Brian George
Tuul, Johannes
Tuve, Merle Antony
Tuzzolino, Anthony J
Tyler, John Edwards
Uhlenbeck, George Eugene
Ulrich, David Lee
Ulrich, Dale V
Underwood, Newton
Unruh, Henry, Jr
Unterberger, Robert Ruppe
Ushioda, Sukekatsu
Uzgiris, Egidijus E
Vajda, Geza Laszlo
Vajk, Joseph Peter
Valley, George Edward, Jr
Van Allen, James Alfred
Vanden Bout, Paul Adrian
Vander Sluis, Kenneth Leroy
Vander Velde, John Christian
Vanderven, Ned Stuart
van der Ziel, Aldert
Van Dilla, Marvin Albert
Van Domelen, Bruce Harold
Van Ginneken, Andreas J M
Van Heerden, Pieter Jacobus
Van Heuvelen, Alan
Van Horn, John A
Van Lint, Victor Anton Jacobus
Van Sciver, Wesley J
Van Steenbergen, Arie
Vant-Hull, Lorin Lee
Van Vleck, John Hasbrouck
Van Vliet, Carel (Karel) M
Van Wijngaarden, Arie
Vaughn, Michael Thayer
Vawter, Spencer Max
Veazey, Sidney Edwin
Vedam, Kuppuswamy
Vehse, William E
Veigele, William John
Vella-Coleiro, George
Venable, Douglas
Venables, John Duxbury
Vermillion, Robert Everett
Vernon, Russel
Vessot, Robert F C
Vette, James Ira
Via, Giorgio G
Viette, Michael Anthony
Vignos, James Henry
Vilches, Oscar Edgardo
Vineyard, George Hoagland
Vinson, James S
Violet, Charles Earl
Violet, Theodore Dean
Visner, Sidney
Vissat, Peter Louisa
Vogan, Eric Lloyd
Vogt, Rochus E
Vollmer, James
Von Braun, Wernher
Vonbun, Friedrich Otto
Vondrak, Edward Andrew
Von Keszycki, Carl Heinrich
Von Meerwall, Ernst Dieter
Von Weyssenhoff, Hanns
Vook, Frederick Ludwig
Vossen, John Louis
Vought, Robert Howard
Vreeland, John Allen
Vuillemin, Joseph J
Wachtell, George Peter
Wada, Walter W
Waddell, Charles Noel
Waddell, Robert Clinton
Waddington, Cecil Jacob
Wadey, Walter Geoffrey
Wagoner, Glen
Wahlig, Charles F
Wahr, John Cannon
Waine, Martin
Wajda, Edward Stanley
Wakelin, James Henry, Jr
Waksberg, Armand L
Waldron, Stephen
Wales, Walter D
Walker, Charles Edward, Jr

Walker, Christopher Bland
Walker, David Kenneth
Walker, Michael Barry
Walker, Robert Hugh
Walker, Robert Lee
Walker, Ronald Elliot
Wallace, Reuben Henry
Wallick, George Castor
Wallingford, John Stuart
Wallis, Robert L
Walsh, John M
Walsh, Peter
Walske, Max Carl
Walters, Virginia F
Walters, William Le Roy
Walton, Derek
Walton, Richard Bruce
Walton, William Upton
Wang, Charles T P
Wang, Chung Shan
Wang, Frederick E
Wang, Theodore Joseph
Wangler, Thomas P
Wangsness, Roald Klinkenberg
Waniek, Ralph Walter
Ward, Alford L
Ward, John Frank
Ward, Truman L
Ware, Walter Elisha
Waring, Richard C
Waring, Robert Kerr, Jr
Warner, Raymond M, Jr
Warner, Robert Edson
Warren, Holland Douglas
Warren, John Bernard
Warren, Kenneth Lyle
Warren, Richard
Warschauer, Douglas Marvin
Wasielewski, Paul Francis
Waterhouse, Richard (Valentine)
Watermeier, Leland A
Waters, William Edward
Watkins, George Daniels
Watkins, Ivan Warren
Watkins, Sallie Ann
Watrous, Ralph Melvin
Watson, Bernard Bennett
Watson, Edmond Evelyn
Watson, Hugh Alexander
Watson, John H L
Watson, Kenneth Marshall
Watson, Richard Elvis
Watson, Richard William
Watson, Tully Franklin
Wattenberg, Albert
Watts, Terence Leslie
Way, Harold E
Way, Katharine
Weatherly, Thomas Levi
Weaver, Albert Bruce
Weaver, Allen Dale
Webb, Maurice Barnett
Webb, Watt Wetmore
Webber, Donald Salyer
Weber, Alfons
Weber, Heinz Paul
Weber, Joseph
Weber, Robert L
Webster, Harold Frank
Webster, William Merle
Weeks, Dorothy Walcott
Weems, Malcolm Lee Bruce
Weertman, Johannes
Wefel, John Paul
Wehner, Gottfried Karl
Weichel, Hugo
Weidner, Richard Tilghman
Weil, Raoul Bloch
Weimer, Katherine E
Weimer, Paul Kessler
Weinberg, Irving
Weinberg, Irving
Weinreich, Gabriel
Weinrich, Marcel
Weisberg, Howard Louis
Weisberg, Leonard
Weiss, Edward Leonhardt
Weiss, Max Tibor
Weiss, Paul
Weiss, Rainer
Weiss, Richard Jerome
Weissbluth, Mitchel
Weisskopf, Victor Frederick
Weissler, Gerhard Ludwig
Weissman, Herman Benjamin
Weller, Charles Stagg, Jr
Weller, Richard Irwin
Wellner, Marcel
Wells, Daniel R
Wells, John Clarence
Wells, Joseph S
Welsh, John Cannon
Wennersten, Dwight L
Werntz, James Herbert, Jr
Wert, Charles Allen
Werth, Glenn Conrad
Wertwijn, George
Wessel, Gunter Kurt

Westberg, Karl Rogers
Westerfield, Everett Commodore
Westervelt, Donald Ramsey
Westwood, William Dickson
Weyland, Jack Arnold
Whaley, Randall McVay
Whalin, Edwin Ansil, Jr
Wheatley, John Charles
Wheeler, Donald Bingham, Jr
Wheeler, George Willis
Wheeler, Samuel Crane, Jr
Whetten, Nathan Rey
Whitcomb, Stuart Estes
White, Bruce Langton
White, Franklin Estabrook
White, Frederick Andrew
White, Frederick Elmer
White, George Charles, Jr
White, Gifford
White, Harvey Elliott
White, Henry W
White, James Wilson
White, John Underhill
White, Lawrence S
White, Lowell Deane
White, Paul Chapin
White, Robert Lee
White, William Charles
Whitehead, Andrew Bruce
Whitehead, James Rennie
Whitehurst, Robert Neal
Whiteside, Haven
Whitfield, George Danley
Whitmer, Robert Morehouse
Whitmore, Stephen Carr
Whitney, Robert C
Whitney, William Merrill
Whittemore, William Leslie
Whittier, Angus Charles
Whittle, Charles Edward, Jr
Wicher, Enos R
Wick, Gian Carlo
Wickersham, Arthur Frank, Jr
Wieder, Irwin
Wieder, Sol
Wiederhold, Pieter Rijk
Wiens, Jacob Henry
Wieting, Terence James
Wilbarger, Edward Stanley, Jr
Wilbur, Howard Albert
Wilcox, Howard Albert
Wild, Jack William
Wild, John Frederick
Wild, Wayne Grant
Wildenthal, Bryan Hobson
Wiley, Samuel L
Wilkerson, Robert C
Wilkins, Homer Clifton
Wilkinson, David Todd
Wilkinson, Michael Kennerly
Wilks, William Taylor
Willard, Daniel
Willenbrock, Frederick Karl
Williams, Arthur Olney, Jr
Williams, David C
Williams, Earl R
Williams, Ferd Elton
Williams, Joel Quitman
Williams, Philip Sidney
Williams, Robert Leroy
Williams, Robert Walter
Williams, Ross Edward
Williamson, Douglas Bleecker
Williamson, Hugh A
Williamson, Robert Marshall
Williamson, Robert Samuel
Willis, Charles Richard
Willis, French Hoke
Willis, James Stewart, Jr
Willis, William J
Willis, William Russell
Wills, James E, Jr
Wilsey, Neal DAvid
Wilska, Alvar P
Wilson, Charles Woodson, III
Wilson, Jerry Dick
Wilson, John Neville
Wilson, Kenneth Geddes
Wilson, Oscar Bryan, Jr
Wilson, Richard
Wilson, Richard J
Wilson, Robert Rathbun
Winckler, John Randolph
Windham, Pat Morris
Wineland, William Clemard
Wingfield, Edward Christian
Winhold, Edward John
Winkler, Ernst Hans
Winkler, Eva Maria
Winslow, George Harvey
Winston, Arthur William
Winter, Thomas Greeley
Winter, William Kenneth
Winters, Roger
Wiseman, Gordon G
Wisseman, William Rowland
Witteborn, Fred Carl
Witten, Louis

Gibson, Edward George
Glover, Kenneth Merle
Gokhale, Narayan Ramchandra
Goldan, Paul David
Goldberg, Richard Aran
Greenebaum, Michael
Hall, Richard Travis
Hamlin, Daniel Allen
Hanel, Rudolf A
Hardy, Kenneth Reginald
Harper, Robert M
Harris, Kent Karen
Heer, Raymond Robert, Jr
Hernandez, Gonzalo J
Hess, Wilmot Norton
Hicks, Bruce Boundy
Hobbs, Peter Victor
Hoffman, John Harold
Hofmann, David John
Honsaker, John Leonard
Hooke, William Hines
Israel, Gerhard Wilhelm
Jastrow, Robert
Jiusto, James E
Jurica, Gerald Michael
Justus, Carl Gerald
Katsaros, Kristina
Kim, Jai Soo
Kleis, William Delong
Konigsberg, Alvin Stuart
Kraakevik, James Henry
Kyle, Herbert Lee
List, Roland
Little, Charles Gordon
Lorenz, Philip Jack
MacCracken, Michael Calvin
Mandics, Peter Alexander
Markson, Ralph Joseph
Maybank, John
McClatchey, Robert Alan
McCormick, Michael Patrick
McPeters, Richard Douglas
Megaw, William James
Mejia, Gaston Rene
Miller, August
Moe, Mildred Minasian
Mohnen, Volker A
Moore, Charles Bachman, Jr
Murcray, David Guy
Neff, William David
Old, Thomas Eugene
Orgill, Montie M
Orville, Richard Edmonds
Pallmann, Albert J
Penndorf, Rudolf
Peterson, Kendall Robert
Peterson, Lennart Rudolph
Philbrick, Charles Russell
Pike, Julian M
Plooster, Myron Nieveen
Pope, Joseph Horace
Potter, John Fred
Prag, Arthur Barry
Procunier, Richard Werner
Randall, Charles McWilliams
Rao, Ganti Lakshminarayana
Reck, Ruth Annette
Remsberg, Ellis Edward
Reynolds, George William, Jr
Richmond, Arthur Dean
Robinson, George David
Roper, Robert George
Rosenkilde, Carl Edward
Rudd, Millard Eugene
Ruhnke, Lothar Hasso
Ruskin, Robert Edward
Rust, Walter David
Ruttenberg, Stanley
Schwiesow, Ronald Lee
Scott, William Taussig
Semonin, Richard Gerard
Shah, Govindlal M
Shapiro, Mark Howard
Shardanand
Shaw, Glenn Edmond
Shemansky, Donald Eugene
Shettle, Eric Payson
Shimazaki, Tatsuo
Shlanta, Alexis
Silverman, Bernard Allen
Simons, Theodore J
Sinclair, Peter C
Smiley, Vern Newton
Smith, Paul Letton, Jr
Stakutis, Vincent John
Stephens, James Briscoe
Stern, Sidney Charles
Stewart, Richard Willis
Straus, Joe Melvin
Swenson, Gary Russell
Taffe, William John
Tarpley, Jerald Dan
Telford, James Wardrop
Thomas, Gary E
Tisone, Gary C
Turner, Robert Elwood
Vali, Gabor
Veirs, Val Rhodes

Vickers, William W
Volz, Frederic Ernst
Walton, John Joseph
Ward, Gray (Ganesh)
Weickmann, Helmut K
Weinreb, Michael Philip
Welch, Ronald Maurice
Whitney, Cynthia Kolb
Wilkening, Marvin Hubert
Wukelic, George Edward
Yarger, Douglas Neal
Young, Stephen James
Ziauddin, Syed

Chemical Physics

Abramowitz, Stanley
Adams, Wade J
Alder, Berni Julian
Alexander, Michael Norman
Allen, Harry Clay, Jr
Alms, Gregory Russell
Amer, Nabil Mahmoud
Anderson, David Hamel
Anderson, John Howard
Anderson, Robert James
Andrews, Frank Clinton
Anysas, Jurgis Arvydas
Arendale, William Frank
Artman, Joseph Oscar
Ashy, Peter Jawad
Atalla, Rajai Hanna
Aviram, Ari
Axe, John Donald
Aziz, Ronald A
Baer, Tomas
Baetzold, Roger C
Bailey, Thomas L, III
Baldeschwieler, John Dickson
Barker, John Adair
Barnes, Donald George
Barr, Tery Lynn
Bates, John Bryant
Baughman, Ray Henry
Bauman, Robert Poe
Bearman, Richard John
Beaudet, Robert A
Benedict, William Sidney
Benson, Bruce Buzzell
Berlad, Abraham Leon
Berne, Bruce J
Bernheim, Robert A
Bernstein, Elliot R
Berry, Michael James
Bertie, John E
Birely, John H
Birge, Robert Richards
Blais, Normand C
Blint, Richard Joseph
Bloch, Aaron Nixon
Bonham, Russell Aubrey
Boudreaux, Edward A
Bowers, Michael Thomas
Bowman, Robert Clark, Jr
Bragg, John Kendal
Braun, Charles Louis
Bregman, Judith
Brown, Ian David
Brown, Walter Eric
Brus, Louis Eugene
Burnell, Edwin Elliott
Butler, James Ehrich
Cahill, Jerry Edward
Campbell, Edwin Stewart
Cardillo, Mark J
Cartwright, David Chapman
Castle, Peter Myer
Casteman, Albert Welford, Jr
Ceasar, Gerald P
Center, Robert E
Chaiken, Robert Francis
Chandler, David
Charney, Elliot
Chen, Freeman Philip
Chen, Kuo-Mei
Claassen, Howard Hubert
Cladis, Patricia Elizabeth Ruth
Closmann, Philip Joseph
Coffey, Dewitt, Jr
Coffman, Robert Edgar
Cohen, Ronald Bruce
Cole, George David
Cole, Robert Hugh
Cole, Terry
Coleman, Marcia Lepri
Collins, Francis Allen
Collins, Russell Lewis
Colson, Steven Douglas
Combs, Leon Lamar, III
Cook, Charles Falk
Cook, Robert Crossland
Copeland, David Anthony
Copeland, Gary Earl
Coplan, Myron Julius
Cornwell, Charles Daniel
Cross, Jon Byron
Dalton, Larry Raymond
Datz, Sheldon
Davenport, John Eaton

Davidovits, Paul
Davis, Howard Ted
Dawson, Peter Henry
Dehmer, Joseph Leonard
Dehmer, Patricia Moore
Delahay, Paul
DeRocco, Andrew Gabriel
Desai, Rashmi C
Dibeler, Vernon Hamilton
DiCarlo, Ernest Nicholas
Dingle, Raymond
Dodgen, Harold Warren
Doering, John P
Doetschman, David Charles
Dorothy, Robert Glenn
Drost-Hansen, Walter
Drullinger, Robert Eugene
Duff, Russell Earl
Dugan, Charles Hammond
Dumke, Warren Lloyd
Du Pre, Donald Bates
Eaker, Charles William
Eckstrom, Donald James
Ehrlich, Sanford Howard
Eichhorn, Edgar Leo
El Saffar, Zuhair M
Emptage, Michael Rollins
Epstein, Irving Robert
Evans, Glenn Thomas
Ewing, James Joyce
Faber, Roger Jack
Fadley, Charles Sherwood
Farrow, Leonilda Altman
Fayer, Michael David
Feinberg, Melvyn Joel
Felton, Ronald H
Field, Robert Warren
Fink, William Henry
Fitts, Donald Dennis
Fitzpatrick, John Michael
Flygare, Willis H
Flynn, George William
Foner, Samuel Newton
Fong, Francis K
Fraenkel, George Kessler
Franceschetti, Donald Ralph
Francis, Anthony Huston
Freed, Jack H
Freund, Robert Stanley
Fujimoto, Minoru
Fukushima, Eiichi
Fulton, Robert Lester
Gagliardi, L John
Gale, Paula Jane
Gans, Paul Jonathan
Garrett, William Ray
Geacintov, Nicholas
Gelbart, William M
Gelerinter, Edward
Gelfand, Jack Jacob
Gentry, William Ronald
Gerry, Michael Charles Lewis
Giese, Clayton
Ginell, Robert
Goodisman, Jerry
Goodman, Lionel
Gordon, Robert Jay
Gouterman, Martin (Paul)
Gravatt, Claude Carrington, Jr
Greene, Frank T
Greer, William Louis
Grover, James Robb
Gruen, Dieter Martin
Hakala, Reino William
Hamblen, David Philip
Hance, Robert Lee
Hansen, Carl Frederick
Hanson, David M
Harendza-Harinxma, Alfred Josef
Harris, Frank Owen
Harris, Frank Ephraim, Jr
Harris, Robert A
Harrison, Shirley Wanda
Hastie, John William
Haugh, Michael J
Hazzard, George William
Hebert, Alvin Joseph
Heller, Eric Johnson
Herber, Rolfe H
Herschbach, Dudley Robert
Hochstrasser, Robin
Hogan, Patrick Brian
Holloway, Thomas Thornton
Holm, Charles Hawthorne
Hoover, William Graham
Horne, Frederick Herbert
Hornig, Arthur William
Hougen, Jon T
Houston, Paul Lyon
Howgate, David W
Huang, John S
Hubert, Jay Marvin
Hughes, Robert Clark
Hunter, Lawrence Wilbert
Hunziker, Heinrich Erwin
Hutta, Paul John
Ijams, Charles Carroll

Innes, Kenneth Keith
Ivey, Robert Charles
Jackson, William Morgan
Jacobs, Theodore Alan
Jain, Duli Chandra
Jamieson, Alexander MacRae
John, George Swisher
Johnson, Wilbur Vance
Johnston, Don Richard
Jones, Edward Grant
Jung, Hilda Zifle
Kafalas, Peter
Kaiser, Reinhold
Karl, Robert Raymond, Jr
Karo, Arnold Mitchell
Kasha, Michael
Kaufmann, Kenneth James
Keiderling, Timothy Allen
Keizer, Joel Edward
Kell, George Sinclair
Kelley, John Daniel
Kenney-Wallace, Geraldine Anne
Kerstiner, James David
Kestner, Neil R
Kincaid, John Franklin
King, Gerald Wilfrid
King, Gilbert William
Kispert, Lowell Donald
Kivelson, Daniel
Knipe, Richard Hubert
Koeppl, Gerald Walter
Koningstein, Johannes A
Kovac, Jeffrey Dean
Kozak, John Joseph
Kraitchman, Jerome
Kramer, Jerry Martin
Krause, Herbert Francis
Krause, Manfred Otto
Krenos, John Robert
Krisher, Lawrence Charles
Kromhout, Robert Andrew
Kunz, Albert Barry
Kuppermann, Aron
Kwiram, Alvin L
Kyle, Nanse Rector
Labes, Mortimer Milton
Larsen, Russell D
Lassettre, Edwin Nichols
Laudenslager, James Bishop
Laurenzi, Bernard John
Lee, John William
Leffert, Charles Benjamin
Lemberg, Howard Lee
Leone, Stephen Robert
Leroi, George Edgar
Levin, Ira William
Levy, Donald Harris
Lewis, George Kenneth
Lide, David Reynolds, Jr
Light, John Caldwell
Lighty, Paul Elliott
Lin, Sheng Hsien
Lindenberg, Katja Lakatos
Linder, Bruno
Lineberger, William Carl
Livingston, Ralph
Loeb, Arthur Lee
Loebl Ernest Moshe
Long, Charles Anthony
Longmire, Martin Shelling
Longworth, James W
Lossing, Frederick Pettit
Lowitz, David Aaron
Ludwig, Howard C
Lund, Louis Harold
Lutz, Robert William
Lynch, Richard Wallace
Mack, Julius L
Macomber, James Dale
Magde, Douglas
Mahendroo, Prem P
Mahoney, Richard Theodore
Malli, Gulzari Lal
Mann, Joseph Bird, (Jr)
Manson, Steven Trent
Margolis, Jack Selig
Mason, Edward Allen
Massa, Louis
McKnight, Lee Graves
Matlow, Sheldon Leo
Mattoon, Richard Wilbur
Mayer, Joseph Edward
McAfee, Kenneth Bailey, Jr
McConaughy, David Lester
McCourt, Frederick Richard Wayne
McDowell, Harding Keith
McFadden, David Lee
McLachlan, Dan, Jr
McLaughlin, Donald Reed
McMillan, William George
McNeal, Robert Joseph
McTague, John Paul
Meiboom, Saul
Melton, Charles Estel
Menger, Eva L
Merrifield, Richard Ebert
Messmer, Richard Paul

Metzger, Robert Melville
Meyer, Vincent D
Michaels, Ira A L
Michels, Horace Harvey
Miller, Donald Piguet
Miller, Richard Edward
Miller, Terry Alan
Miller, William Hughes
Millikan, Roger Conant
Minden, Henry Thomas
Miyagawa, Ichiro
Monchick, Louis
Montei, George Louis
Moore, James Elton
Mullen, Joseph Matthew
Munn, Robert James
Muschlitz, Earle Eugene, Jr
Myers, Howard
Myers, Ira Thomas
Naylor, Robert Ernest, Jr
Nicol, Malcolm Foertner
Nilsen, Walter Grahn
Noble, Gordon Albert
Norris, Carol Lee
Norris, Wilfred Glen
North, Harper Qua
Ogilvie, John Franklin
Ogilvie, Keith W
Ogle, William Elwood
O'Loane, James Kenneth
Olympia, Pedro Lim, Jr
Onton, Aare
Oosterhuis, William Tenley
Oppenheim, Irwin
O'Reilly, Donald Eugene
Owens, Frank James
Palke, William England
Papaefthymiou, Georgia Christou
Parkinson, David B
Parshall, Clarence Merton
Patterson, Gary David
Pearlstein, Robert Milton
Pedersen, Lee G
Perkins, Willis Drummond
Person, James Carl
Petruska, John Andrew
Phillies, George David Joseph
Phillipson, Paul Edgar
Piskin, William Aaron
Plock, Richard James
Pohl, Herbert Ackland
Poindexter, Edward Haviland
Poland, Helen M
Pollara, Luigi Zummo
Poshusta, Ronald D
Pratt, David W
Present, Richard David
Pullen, Bailey Price
Quirk, James Denis
Raff, Lionel M
Raghunathan, Partha
Ramshaw, John David
Ravache, Harold Joseph
Reddoch, Allan Harvey
Reid, Cyril
Reilly, Charles Austin
Reinhardt, William Parker
Reno, Martin A
Rentzepis, Peter M
Rhee, Jay Jea-yong
Rhodin, Thor Nathaniel, Jr
Rice, James Kinsey
Rieckhoff, Klaus E
Ring, James Walter
Robinson, C Paul
Roeder, Stephen Bernard Walter
Ronn, Avigdor Meir
Rosa, Eugene John
Rose, Timothy Laurence
Rosenberger, Franz
Rothschild, Walter Gustav
Ruffa, Anthony Richard
Runnels, Lynn Kelly
Rutledge, Carl Thomas
Runner, Emile
Sage, Martin Lee
Salwin, Arthur Elliot
Sams, John Robert, Jr
Sanzone, George
Sathoff, H John
Saturno, Antony Fidelas
Schaefer, Dale Wesley
Schalit, Lewis Martin
Schatzki, Thomas Ferdinand
Schempp, Ellory
Schmidt, Parbury Pollen
Schmidt, Thomas William
Schneider, Robert Fournier
Schnepp, Otto
Schnur, Joel Martin
Schulman, Jerome M
Scott, Gary Walter
Sell, Nancy Jean
Selwyn, Philip Alan
Sharma, Ramesh Dutt
Sharp, James H
Sharp, Terry Earl
Sheldon, John William

Shirley, David Arthur
Shoemaker, Richard Lee
Shulman, Robert Gerson
Sibilia, John Philip
Siebert, Donald Robert
Siebrand, Willem
Siegel, Irving
Siegel, Seymour
Silverman, Shirleigh
Siska, Peter Emil
Skofronick, James Gust
Smalley, Richard Errett
Smith, George Wolfram
Snyder, Lawrence Clement
Spain, Ian L
Spangler, Glenn Edward
Springer, John Mervin
Steinhardt, Ralph Gustav, Jr
Stern, Richard Cecil
Stevens, Walter Joseph
Story, Troy Lee
Stout, John Willard
Stuebing, Edward Willis
Swalen, Jerome Douglas
Swets, Don Eugene
Tanczos, Frank I
Taylor, Raymond L
Tauer, Kenneth J
Tiernan, Thomas Orville
Tokuhiro, Tadashi
Trimble, Russell Harold
Truby, Frank Keeler
Tsai, Bilin Paula
Tuan, Debbie Fu-Tai
Tucci, James Vincent
Tully, Frank Paul
Tully, John Charles
Van Duyne, Richard Palmer
Vezzoli, Gary Christopher
Viehland, Larry Alan
Vierina, Teri L
Voelz, Frederick
Vold, Robert Lawrence
Vroom, David Archie
Wagner, Albert Fordyce
Wahnsiedler, Walter Edward
Wallace, William Edward, Jr
Warshel, Arieh
Weeks, John David
Weil, John A
Weiner, John
Weiss, Charles, Jr
Weitz, Eric
Wessel, John Emmit
Westenberg, Arthur Ayer
Wexler, Solomon
Wheeler, John C
White, John Michael
White, Locke, Jr
Wilkins, Roger Lawrence
Willett, Roger
Wilson, Edgar Bright
Wilson, Jack Martin
Wilson, Kent Raymond
Wine, Paul Harris
Winicur, Daniel Henry
Wolken, George, Jr
Wolter, Gerhard Herman
Wrobel, Joseph Jude
Wu, En Shinn
Yao, Shang Jeong
Yates, Kenneth Pidcock
Yeh, George Chiayou
Yeh, Yin
Young, Charles Edward
Young, Ralph Howard
Zare, Richard Neil
Zinning, Lois Jacobs
Zwanzig, Robert Walter

Cryogenics
Ambler, Ernest
Bendt, Philip Joseph
Bohn, Randy G
Brooks, Margaret Hoover
Chau, Cheuk-Kin
Chester, Marvin
Daniel, Michael Roger
Dabbs, John Wilson Thomas
Green, Robert Wood
Gubser, Donald Urban
Holdeman, Louis Brian
Hornung, Erwin William
Kropschot, Richard Henry
Kuchnir, Moyses
Laquer, Henry L
Lax, Edward
McInturff, Alfred D
Miller, Robert Carl
Miranda, Gilbert A
Onn, David Goodwin
Parker, James Henry, Jr
Pigott, Miles Thomas
Roder, Hans Martin
Schermer, Robert Ira
Sherman, Robert Howard
Soulen, Robert J, Jr
Sparks, Larry Leon
Strauss, Bruce Paul
Wilkes, William Roy

Energy Conversion
Andria, George D
Barr, William J
Boer, Karl Wolfgang
Clark, Arnold Franklin
Cohen, Howard David
Cohen, Robert
Cope, David Franklin
Dickinson, William Clarence
Farber, Joseph
Fetkovich, John Gabriel
Firebaugh, Morris W
Gajewski, Ryszard
Goodrich, Robert Bruce
Grannemann, Glenn Niel
Groskreutz, Joseph Charles
Habich, Ernst Rollemann, Jr
Hamilton, Carole Lois
Heldman, Julius David
Holden, John Paul
Huberman, Marshall Norman
Hunt, Thomas Kinzing
Joseph, Bernard William
Kooi, Clarence F
Keeney, Joe
Leffert, Charles Benjamin
Levi, Enrico
Lindena, Siegfried Johannes
Meyer, James Wagner
Miksch, Edmund Stewart
Mitchell, William Cobbey
Morrison, Richard Charles
Moulthrop, Peter Hill
Nozik, Arthur Jack
Parkins, William Edward
Pfeiffer, Heinz Gerhard
Pollard, William Grosvenor
Rabl, Ari
Radosevich, Lee George
Ranken, William Allison
Reed, Thomas Binnington
Ridgway, Stuart L
Rittner, Edmund Sidney
Schiff, Daniel
Schmid, Loren Clark
Shaw, Robert William, Jr
Skrabek, Emanuel Andrew
Speen, Gerald Bruce
Stearns, Brenton Fisk
Stevens, Alfred Lyman
Stirn, Richard J
Telkes, Maria
Tourin, Richard Harold
Trevoy, Donald James
Wahlig, Michael Alexander
Warschauer, Douglas Marvin
White, Ronald Keith
Zauderer, Bert

Engineering Physics
Bacon, Ralph Hoyt
Batdorf, Samuel Burbridge
Benumof, Reuben
Bickmore, John Tarry
Bussard, Robert W
Carpenter, Delma Rae, Jr
Carver, John Guill
Chesnut, Dwight
Chu, Wei-Kan
Connor, Joseph Gerard, Jr
Cox, Everett Franklin
Cruzan, Charles Grant
Gallo, Charles Francis, Jr
Genin, Dennis Joseph
Ginsburgh, Irwin
Gold, Lorne W
Herzfeld, Charles Maria
Hitchie, Douglas Walter
Hoffman, John Raleigh
Huberman, Marshall Norman
Kahl, Ford
Killian, Thomas Joseph
Kuenhold, Kenneth Alan
Lardner, Thomas Joseph
Lawless, Philip Austin
Levine, Melvin Mordecai
Lewis, Lloyd George
Mott, Jack Edward
Nichols, Nathaniel Burgess
Nicol, John
Oleksiuk, Leslie William
Reiling, Gilbert Henry
Reiss, Keith Westcott
Richardson, Jasper E
Rosa, Richard John
Ross, Frederick William
Sank, Victor J
Scheie, Carl Edward
Schumacher, Berthold Walter
Schwartz, Jack
Seldin, Emanuel Judah
Shea, Michael Francis
Sloma, Leonard Vincent
Taylor, Robert Joseph

Thomson, James Alex L
Victor, Andrew C
Wagoner, Earl V, Jr
Walker, Michael Stephen
Waugh, John Blake-Steele
Weigman, Bernard J
Whitman, Walter William
Wilcox, William Jenkins, Jr
Williams, Neal Thomas

Environmental Physics
Barnett, Albert Gerald
Breed, Benny Ray
Brown, George Raymond
Chernosky, Edwin Jasper
Chopra, Kuldip P
Constant, Frank Woodbridge
Crawford, George Wolf
de Latour, Christopher
English, Bruce Vaughan
Fricke, Werner
Garrell, Martin Henry
Helminger, Paul Andrew
Herndon, Roy C
Hyder, Charles Latif
John, Walter
Kester, William Lee
Ku, Peh Sun
Loos, Hendricus G
Ludwig, Claus Berthold
Martin, John H
Matta, Joseph Edward
Milford, Sidney Nevil
Miller, Lyster Keith
Nelson, John William
Olson, Willard Paul
Rae, Stephen
Rosen, Leonard Craig
Seaman, Gregory G
Soberman, Robert K
Swissler, Thomas James
Sydor, Michael
Warren, Mashuri Laird
Watson, Robert Dale
White, Ronald Keith

Fluid Physics
Apel, John Ralph
Aptel, Robert Edmund
Cary, Boyd Balford, Jr
Chahine, Moustafa Toufic
Chang, Ching Ming
Cowan, John Arthur
Crawford, Roy Kent
Crowley, Patrick Arthur
Cutler, Warren Gale
Doolittle, Arthur King
Farber, Joseph
Fickett, Wildon
Fritsch, Klaus
Goldberg, Arnold
Hackett, Colin Edwin
Hartley, Charles LeRoy
Leribaux, Henri Romain
Levanoni, Menachem
Nagamatsu, Henry T
O'Sullivan, William John
Ozawa, Kenneth Susumu
Rivard, William Charles
Sengers, Jan V
Smith, Wesley R
Swinney, Harry Leonard
Thacker, William Carlisle
Wang, H E Frank
Wegener, Peter Paul

Gas Dynamics
Baum, Dennis Willard
Bowman, Craig Thomas
Cassen, Patrick Michael
Gill, Stephen Paschall
Kantrowitz, Arthur (Robert)
King, Creston Alexander, Jr
Mintzer, David
Ozawa, Kenneth Susumu
Watson, Velvin Richard
Wegener, Peter Paul
Weimer, David
Woods, Frank Robert

High Pressure Physics
Krohn, Burton Jay

History of Physics
Bork, Alfred Morton
Gifford, Fay Evan
Goldberg, Stanley
Greenslade, Thomas Boardman, Jr
Klein, Martin Jesse
Rowe, John Edward
Stuewer, Roger Harry
Weiner, Charles
Wise, Matthew Norton

Low Temperature Physics
Adair, Thomas Weymon, III
Adams, Earnest Dwight

Addison, John Rundle
Anderson, Ansel Cochran
Anderson, John Thomas
Anderson, Roger Harris
Beasley, Malcolm Roy
Berman, Barry L
Bernat, Thomas Phillip
Bhatnagar, Anil Kumar
Black, William Carter, Jr
Blackstead, Howard Allan
Boghosian, Charles
Bohn, Randy G
Boorse, Henry Abraham
Bowen, Samuel Philip
Brandt, Richard Gustave
Bretz, Michael
Brickwedde, Ferdinand Graft
Cabrera, Blas
Carroll, Paul Joseph
Chan, Moses Hung-Wai
Chung, David Yih
Clark, Ronald Hershel
Coffey, Howard Thomas
Cole, Milton Walter
Connolly, John Irving, Jr
Constable, James Harris
Corak, William Sydney
Corruccini, Linton Reid
Crooks, Michael John Chamberlain
Daybell, Melvin Drew
Dillinger, Joseph Rollen
Ebner, Charles Arthur
Elwell, David Leslie
Everitt, C W Francis
Fairbank, Henry Alan
Fenichel, Henry
Fietz, William Adolf
Gamota, George
Gasparini, Francis Marino
Geballe, Theodore Henry
Gerritsen, Alexander Nicolaas
Ginsberg, Donald Maurice
Glaberson, William I
Glick, Forrest Irving
Goldman, Allen Marshall
Goldstein, Louis
Goodstein, David Louis
Goree, William Strozier
Graf, Erlend Haakon
Greenberg, Allan S
Gregory, Brooke
Greywall, Dennis Stanley
Grimes, Charles Cullen
Grimsrud, David T
Guenther, Raymond A
Haase, David Glen
Hafstrom, John William
Hallock, Robert B
Halperin, William Paul
Hansma, Paul Kenneth
Hanson, Henry Nicholas
Harris-Lowe, Rodney Frederic Brandon
Hebard, Arthur Foster
Henson, Bob Londes
Herb, John A
Hess, George Burns
Hildebrandt, Alvin Frank
Honig, Arnold
Hudson, Ralph P
Huff, G Bradley
Ihas, Gary Gene
Isaacs, Leslie Laszlo
Jones, Harris Cleve
Kellers, Charles Frederick
Kiersted, Henry Andrew
Koenig, Albert A
Kojima, Haruo
Kuchnir, Moyses
Landau, Judah
Lawson, Dewey Tull
Lindenfeld, Peter
Lipa, John A
Lipschultz, Frederick Phillip
Love, Norman Duane
Lowndes, Douglas H, Jr
Lynton, Ernest Albert
Mangum, Billy Wilson
Markham, Arleigh Holden
Marshall, Billy Jack
Mihalisin, Ted Warren
Moberly, Lawrence Allan
Mota, Ana Celia
Moulton, William G
Mrozinski, Peter Matthew
Mueller, Karl Hugo, Jr
Naugle, Donald
Neidhardt, Walter Jim
Nicol, James
Nisenoff, Martin
Nosanow, Lewis H
Osborne, Darrell Wayne
Palmer, Leigh Hunt
Pandolf, Robert Clay
Parker, William Henry
Paterson, James Lenander
Pipes, Paul Bruce

Pollack, Gerald Leslie
Pribil, Stephen
Prober, Daniel Ethan
Rabinowitz, Mario
Rauckhorst, William H
Reynolds, Claude Lewis, Jr
Richardson, Robert Coleman
Rollefson, Robert John
Romer, Robert Horton
Rosen, Carol Zwick
Rosenshein, Joseph Stanley
Rudnick, Isadore
Salinger, Gerhard Ludwig
Saslow, Wayne Mark
Shen, Sin-Yan
Silverman, Peter Jay
Singer, James Robert
Sites, James Russell
Smith, Charles William, Jr
Smith, Howard John Treweek
Snyder, Nancy Simon
Spitzer, Hermann Josef
Stark, Royal William
Strauss, Bruce Paul
Stromberg, Thorsten Frederick
Strongin, Myron
Struzynski, Raymond Edward
Sullivan, Donald Barrett
Sybert, James Ray
Tanner, David John
Taylor, Barry Norman
Taylor, Raymond Dean
Tedrow, Paul Muller
Templeton, Ian M
Titus, William James
Tough, James Thomas
Turneaure, John Paul
Ulbrich, Carlton Wilbur
Ulrich, Bruce T
van Kann, Frank Joachim
Vernon, Frank Lee, Jr
Wagner, David Loren
Weinstock, Harold
Weyhmann, Walter Victor
Williamson, Samuel Johns
Woerner, Robert Leo
Wolf, Robert Peter
Wollan, John Jerome
Woods, Alfred David Braine
Woolf, Michael A
Worthington, Thomas Kimber
Zimmerman, George Ogurek
Zipfel, Christie Lewis
Zübeck, Robert Bruce

Magnetospheric Physics
Burch, James Leo
Chappell, Charles Richard
Hartz, Theodore Robert
Heacock, Richard Ralph
Katz, Ludwig
Knecht, David Jordan
Ledley, Brian G
McCoy, James Ernest
Radoski, Henry Robert
Reiff, Patricia Hofer
Sharp, Richard Dana
Walt, Martin
Winningham, John David

Mathematical Physics
Abu-Shumays, Ibrahim Khalil
Adler, Felix T
Aks, Stanley Olaf
Bargmann, Valentine
Bart, George Raymond
Berger, Martin Jacob
Bezuszka, Stanley John
Biggs, Frank
Blank, Albert Abraham
Bleick, Willard Evan
Bowden, Robert Lee, Jr
Bragg, Lincoln Ellsworth
Brans, Carl Henry
Brooks, Franklin Coolidge
Burgoyne, Peter Nicholas
Canosa, Jose M
Capri, Anton Zizi
Carhart, Richard Alan
Challifour, John Lee
Chang, Shau-jin
Childs, Donald Ray
Cohen, E Richard
Cook, David Marsden
Critchfield, Charles Louis
Dankel, Thaddeus Cooper, Jr
Davidon, William Cooper
Derr, John Sebastian, Jr
Devaney, Joseph James
Dillard, Margaret Bleick
Dory, Robert Allan
Dyson, Freeman John
Emch, Gerard G
Faris, William Guignard
Faulkner, James Earl
Fette, Clarence William
Findley, George Bernard
Fisher, Edward

Fisher, Michael Ellis
Fong, Jeffrey Tse-Wei
Giles, Robin
Granger, Robert A, II
Greenberg, William
Guenther, Ronald Bernard
Habetler, George Joseph
Hammer, Charles Lawrence
Hanhauser, Martin A
Harvey, George Graham
Herman, Richard Howard
Hershey, Allen Vincent
Herzberger, Maximilian Jacob
Hoffmann, Banesh
Hudgin, Richard Henry
Humi, Mayer
Jackson, David Phillip
Jackson, Robert Franklin
Jaffe, Arthur Michael
Johnson, Porter W
Kalman, Calvin Shea
Kaplan, Bernard
Karp, Ira Lawrence
Kelley, Robert Lee
Klapman, Solomon Joel
Klein, Abel
Kohlmayr, Gerhard Franz
Kroll, Norman Myles
Kruskal, Martin David
Kummer, Martin
Kunzle, Hans Peter
Kyame, Joseph John
Kyrala, Ali
Landshoff, Rolf
Lanford, Oscar E, III
Lee, Clarence Edgar
Lee, Clarence Tsung
Leipnik, Roy Bergh
Leith, Cecil Eldon, Jr
Leitner, Alfred
Lenard, Andrew
Lieb, Elliott Hershel
Lomont, John S
Louck, James Donald
Marable, James Holley
Marchand, Jean-Paul
Marcus, Paul Malcolm
Marcuvitz, Nathan
Mayer, Meinhard Edwin
McCoy, Jimmy Jewell
McDonald, James Frederick
Meecham, William Coryell
Min, Kwang-Shik
Mishoe, Luna l
Moore, Mortimer Norman
Morris, George William
Neuringer, Joseph Louis
Norton, John Leslie
Nussenzveig, Herch Moyses
Ogden, Edwin Burman
Ozsvath, Istvan
Park, Chong Jin
Parker, Lee Ward
Parvulescu, Antares
Pasta, John Robert
Pasternack, Simon
Penico, Anthony Joseph
Plybon, Benjamin Francis
Porter, John Robert
Prugovecki, Eduard
Radin, Charles Lewis
Rauch, Jeffrey Baron
Raychowdhury, Pratip Nath
Reed, Michael Charles
Renken, James Howard
Rindler, Wolfgang
Rodney, Paul Frederick
Rosenbaum, Marcos
Rubenfeld, Lester A
Rund, Hanno
Runnels, Lynn Kelly
Saffren, Melvin Michael
Schay, Geza
Schild, Alfred
Schlitt, Dan Webb
Seidl, Frederick Gabriel Paul
Sen, Dipak Kumar
Siegert, Arnold John Frederick
Simon, Barry Martin
Snyder, Herbert Howard
Symon, Keith Randolph
Tamburino, Louis Anthony
Tappert, Frederick Drach
Toupin, Richard A
Tross, Ralph G
Tulczyjew, Wlodzimierz Marek
Twersky, Victor
Ulam, Stanislaw Marcin
Van Winter, Clasine
Wagner, Richard John
Warner, David Charles
Warsi, Nazir Ahmed
Weinstock, Robert
Wheeler, Nicholas Allan
Whitten-Wolfe, Barbara L
Wightman, Arthur Strong
Wigner, Eugene Paul

Williams, David Noel
Wilson, Robert Norton
Wolf, Emil
Wong, Maurice King Fan
Wu, Hsin-I
Yeh, Hsin-Yang
Zund, Joseph David

Medical Physics
Adams, Paul Louis
Anderson, David Walter
Bagne, Farideh
Baily, Norman Arthur
Barnes, Charles M
Blau, Lawrence Martin
Brownell, Gordon Lee
Budinger, Thomas Francis
Bunch, Phillip Carter
Cameron, John Roderick
Carlton, William Herbert
Carson, Paul Langford
Chamberlain, Charles Craig
Chesler, David Alan
Cohen, Montague
Connor, William Gorden
Correia, John Arthur
Cox, Hollace Lawton, Jr
Cradduck, Trevor David
Cunningham, John (Robert)
Doi, Kunio
Douglass, Kenneth Harmon
Eagle, Donald Frohlichstein
Ekstrand, Kenneth Eric
Epp, Edward Rudolph
Erickson, Jon Jay
Fields, Theodore
Flower, Robert Walter
Friedland, Stephen Scholom
Friedmann, Gerhart B
Gledhill, Barton L
Goldstein, Albert
Goodenough, David John
Gorson, Robert O
Greenfield, Moses A
Groce, David Eiben
Hartman, Roger Duane
Heller, Zindel Herbert
Hendee, William Richard
Henschke, Ulrich K
Hevezi, James M
Hill, David G
Holden, James Edward
Holloway, Arthur F
Hoop, Bernard, Jr
Hoory, Shlome
Howarth, John Lee
Hubbard, Lincoln Beals
Huesman, Ronald Henry
Joseph, Peter Maron
Kelsey, Charles Andrew
Kiker, William Edward
Kim, Ki-Hyon
Konneker, Wilfred R
Kuchnir, Franca Tabliabue
Laughlin, John Seth
Lerch, Irving A
Liboff, Abraham R
Malamud, Herbert
Mandelkern, Mark Alan
McCullough, Edwin Charles
Metz, Charles Edgar
Moore, Richard
Murphy, Paul Henry
Nichols, Alexander Vladimir
Norman, Amos
O'Foghludha, Fearghus Tadhg
Oliver, George Davis, Jr
Orton, Colin George
Orvis, Alan Leroy
Ovadia, Jacques
Pfalzner, Paul Michael
Reinstein, Lawrence Elliot
Robertson, James Sydnor
Robinson, James Eugene
Rogers, William Leslie
Rosenthal, Donald Jack
Rothenberg, A Daniel
Rothenberg, Lawrence Neil
Schulz, Robert J
Scott, Mary Jean (Mrs E C H Silk)
Shapiro, Philip
Sorenson, James Alfred
Sprawls, Perry, Jr
Starchman, Dale Edward
Stein, Philip
Sternick, Edward Selby
Syed, Ibrahim Bijli
Tai, Douglas L
Taylor, Lauriston Sale
Taylor, Morris Chapman
Tepley, Norman
Ter-Pogossian, Michel Mathew
Waggener, Robert Glenn
Wallace, Robert Henry
Walton, Philip Wilson
Waterman, Frank Melvin
Webb, Robert Howard
Weber, David Alexander

PHYSICS

Webster, Edward William
White, William John
Winchell, Harry Saul
Wootton, Peter
Yalow, Rosalyn Sussman

Metal Physics
Baxter, William John
Bose, Shyamalendu M
Buck, Otto
Crozier, Edgar Daryl
Doran, Donald George
Einziger, Robert Emanuel
Gerritsen, Alexander Nicolaas
Ginsberg, Donald Maurice
Green, Walter Verney
Hasegawa, Ryusuke
Hess, George Dale
Jesser, William Augustus
Kissinger, Homer Everett
Koistinen, Donald Peter
Leisure, Robert Glenn
Martin, Douglas Leonard
Meyer, Axel
Nevitt, Michael Vogt
Nine, Harmon D
Peters, James John
Rogers, Douglas Herbert
Rouze, Stanley Rupie
Rowlands, John Alan
Ruff, Arthur William, Jr
Sandenaw, Thomas Arthur
Sartain, Carl Clinton
Templeton, Ian M
Van Horn, David Downing
Williams, David Llewelyn

Metrology
Eagleman, Joe R
Frank, Neil LaVerne
Gregory, John B
Guildner, Leslie Arnold
Hersh, John Franklin
Hitschfeld, Walter
Holzworth, George Charles
Julian, Paul R
Kirk, Wiley Price
Loevinger, Robert
Longley, Richmond Wilberforce
Morris, Derek
Staley, Dean Oden
Taylor, Warren Egbert
Wait, David Francis

Microwave Physics
Anderson, Milo Vernette
Blake, Lamont Vincent
Burkhalter, James Herbert
Dagg, Ian Ralph
Debs, Robert Joseph
Fante, Ronald Louis
Farber, Morton Sheldon
Granatstein, Victor Lawrence
Gutmann, Ronald Jay
Hafner, Theodore
Herskovitz, Sheldon Bernard
Hirschmann, Erwin
Jen, Chih Kung
Knox, Jack Rowles
Larson, Tyrone Ray
McCue, John Joseph Gerald
Moloney, Michael J
Moody, Harry John
Oelke, William C
Recsor, Glyn Edward
Reiss, Keith Westcott
Stewart, Gordon Ervin
Stone, Albert Mordecai
Weiss, Jerald Aubrey
Whicker, Lawrence R

Molecular Physics
Anderson, David Kent
Anderson, Richard Alan
Bender, Charles F
Benesch, William Milton
Brewer, Richard George
Broersma, Sybrand
Browne, James Clayton
Bur, Anthony J
Burke, Anna Mae Walsh
Carlson, Thomas Arthur
Carpenter, Robert Francis
Cartwright, David Chapman
Cederberg, James W
Cho, Chung Won
Christophorou, Loucas Georgiou
Compton, Robert Norman
Coope, John Arthur Robert
Das, Tara Prasad
Decker, John P
Dehmer, Patricia Moore
Delos, John Bernard
De Lucia, Frank Charles
Denison, Arthur B
Doverspike, Lynn D
Dunn, Gordon Harold
Ellis, Donald Edwin

Evenson, Kenneth Melvin
Fischer, C Rutherford
Fishburne, Edward Stokes, III
Foltz, Nevin D
Gangemi, Francis A
Garbuny, Max
Ginter, Marshall L
Green, Thomas Allen
Harris, Richard Allan
Hart, Ronald Wilson
Haugsjaa, Paul O
Herman, Roger Myers
Hessel, Merrill
Hirsh, Merle Norman
Hubbard, Paul Stancyl, Jr
Innes, Kenneth Keith
Inokuti, Mitio
Jason, Andrew John
Jones, Charles E
Kelley, Ralph Edward
Kerr, Donald Philip
Kim, John Jungyu
Klein, Douglas J
Kohin, Barbara Castle
Kyle, Thomas Gail
Larson, Everett Gerald
Leventhal, Jacob J
Lewis, James W L
Lichter, James Joseph
Lipari, Nunzio Ottavio
Mahadevan, Parameswar
Manson, Earle Lowry, Jr
McCourt, Frederick Richard Wayne
McGowan, James William
McLay, David Boyd
Menendez, Manuel Gaspar
Miller, Carl Elmer
Mizushima, Masataka
Moore, John Hays, Jr
Moseley, John Travis
Mulligan, Joseph Francis
Mumma, Michael Jon
Nalley, Samuel Joseph
Neumann, Herschel
Oelke, William C
Parker, Paul Michael
Paske, William Charles
Phelps, Arthur Van Rensselaer
Phillips, Edward
Pilloff, Herschel Sydney
Polo, Santiago Ramos
Pritchard, David Edward
Quade, Charles Richard
Radford, Harrison E
Rodgers, James Earl
Sanders, Frank Clarence, Jr
Santaram, Chilukuri
Sarachman, Theodore N
Schaefer, Albert Russell
Schneider, Barry I
Schoen, Richard Isaac
Sharnoff, Mark
Siegel, Melvin Walter
Sinnott, George
Smith, Wesley R
Snow, William Rosebrook
Snyder, Lewis Emil
Stephens, Timothy Lee
Stephenson, David Allen
Stryland, Jan Cornelis
Tabisz, George Conrad
Tam, Andrew Ching
Temple, Wade Jett
Tipping, Richard H
Toburen, Larry Howard
Toennies, Jan Peter
Trajmar, Sandor
Truhlar, Donald Gene
Vaughan, Alfred Leland
Van Brunt, Richard Joseph
Varney, Robert Nathan
Valley, Leonard Maurice
Vuillet, William George
Walker, Keith Gerald
Warmack, Robert Joseph
Weiss, Andrew W
Welsh, Harry L
Westhaus, Paul Anthony
White, Kevin Joseph
Wing, William Hinshaw
Witriol, Norman Martin
Wong, Shek-Fu
Zinnes, Irving Isadore

Polymer Physics
Ambler, Michael Ray
Andersen, Michael George
Ang, Tjoan-Liem
Avakian, Peter
Ballou, Jack Wayne
Barach, Joseph Leonard
Barker, Robert Edward, Jr
Barnes, John David
Bata, George L
Baughman, Ray Henry
Beatty, Charles Lee

Bersted, Bruce Howard
Blackwell, John
Boye, Charles Andrew, Jr
Brenner, Douglas
Brierre, Roland Theodore, Jr
Burnett, James Joseph, Jr
Canter, Nathan H
Carter, Bruce Burton
Cessna, Lawrence C, Jr
Chu, William How-Jen
Coleman, Marcia Lepri
Conant, Floyd Sanford
Crissman, John Matthews
Curro, John Gillette
Dannis, Mark Libman
Dickens, Elmer Douglas, Jr
Dillon, John Henry
Dudek, Thomas Joseph
Ehlers, Gerhard Friedrich Louis
Erhardt, Peter Franklin
Fanconi, Bruno Mario
Fields, Alfred E
Finegold, Leonard X
Fornes, Raymond Earl
Forster, Michael Jay
Friedman, Emil Martin
Froix, Michael Francis
Fulmer, Glenn Elton
Garber, Charles A
Gaylord, Richard J
Gillham, John K
Gordon, Gerald Arthur
Hammer, Clarence Frederick, Jr
Hardy, George Fisk
Harrison, Ian Roland
Hartmann, Bruce
Hay, Ian Leslie
Heffelfinger, Carl John
Henry, Arnold William
Henry, Arthur Charles
Hobbs, Stanley Young
Holland, Virgil Fortune
Ikeda, Richard Masayoshi
Ivey, Donald Glenn
Jaffe, Michael
Jamieson, Alexander MacRae
Johnson, Robert William
Kaelble, David Hardie
Kambour, Roger Peabody
Kanakkanatt, Sebastian Varghese
Kimmel, Robert Michael
Kliman, Harvey Louis
Knight, Alan Campbell
Kolbeck, Andrew Gerard
Kotliar, Abraham Morris
Kubu, Edward Thomas
Laible, Roy C
Lee, Wylie In-Wei
Lehr, Marvin Harold
Lewis, Thomas Brinley
Longworth, Ruskin
Massa, Dennis Jon
Mazur, Jacob
McCabe, Chester Charles
McPeters, Arnold Lawrence
Meier, Dale Joseph
Mendelson, Robert Allen
Mikell, William Gaillard
Mikes, Peter
Miller, Robert Llewellyn
Miller, William Reynolds, Jr
Moore, Jon Thomas
Morrow, Darrell Roy
Murayama, Takayuki
Nakajima, Nobuyuki
Newman, Seymour
Nielsen, Lawrence Ernie
Olf, Heinz Gunther
O'Malley, James Joseph
O'Reilly, James Michael
Ouano, Augustus Ceniza
Padden, Frank Joseph, Jr
Patterson, Gary David
Peterlin, Anton
Peterson, James Macon
Pochan, John Michael
Pollock, Dorothy Jean
Prest, William Marchant, Jr
Prevorsek, Dusan Ciril
Quynn, Richard Grayson
Rupert, John Paul
Sammak, Emil George
Sanchez, Isaac Cornelius
Sarko, Anatole
Seaman, Richard Eric
Shroff, Ramesh N
Shugart, Cecil G
Simeral, William Goodrich
Slagowski, Eugene Louis
Smith, Jack Carlton
Southern, John Hoyle, II
Stacy, Carl J
Statton, William Osborne
Stefanou, Harry
Steger, Theodore Roosevelt, Jr
Sternstein, Sanford Samuel
Stockmann, Volker Erwin
Sullivan, Peter Kevin

Tonelli, Alan Edward
Van der Hoff, Bernard Maria Euphemius
Ver Strate, Gary William
Wickham, William Terry, Jr
Wilcinsky, Zigmond Walter
Woodbrey, James Calvin
Woodruff, Robert Wilson
Work, Richard Nicholas
Wu, Wen-Li
Yau, Wallace Wen-Chuan

Psychophysics
Bartleson, Christian James
Breneman, Edwin Jay
Desor, Jeannette Ann
Doty, Richard Leroy
Galloway, William Don
Goldstein, E Bruce
Green, Barry George
Gulick, Walter Lawrence
Hartmann, William Morris
Hicks, Robert Eugene
Hirsh, Ira Jean
Johnson, Chris Alan
Jones, George R
Kozak, Wiodzimierz Maciej
Nye, Patrick William
Pangborn, Rose Marie Valdes
Stevens, Joseph Charles
Stevens, John Douglas
Verrillo, Ronald Thomas
Ward, Lawrence McCue
Wheeler, Lawrence
Zwislocki, Jozef John

Quantum Physics
Aronowitz, Frederick
Band, Yehuda Benzion
Bay, Zoltan Lajos
Brewer, Richard George
Clauser, John Francis
Cochran, Andrew Aaron
Cole, Robert Stephen
Gangemi, Francis A
Glaser, Frederic M
Hammer, Jacob Meyer
Hu, Chia-Ren
Joshi, Bhairav Datt
Just, Kurt W
Keller, William Edward
Kobe, Donald Holm
Morris, Derek
Naymik, Daniel Allan
Neidhardt, Walter Jim
Perkins, James Francis
Rauscher, Elizabeth Ann
Reiss, Howard Robert
Servaes, Tahira Minhaj
Tuft, Richard Allan
Vernon, Frank Lee, Jr

Radiation Physics
Baba, Anthony John
Baker, Samuel I
Ball, William Paul
Beique, Rene Alexandre
Bennett, Gerald William
Berkowitz, Harry Leo
Bichsel, Hans
Bryce, Donald Hewitt
Buck, Warren Louis
Burke, Gail De Planque
Cathey, LeConte
Chappel, Samuel Estelle
Chen, Sow-Hsin
Chu, William Tongil
Coulter, Claude Alton
Davis, Thomas Gearue
Duncan, Arthur Gustav, Jr
Ehrlich, Margarete
Grant, David Graham
Haas, Peter Herbert
Harharan, Palghat Venketeswar
Huebner, Russell Henry
Hurley, John Paul
Hurst, George Sam
Kunz, Walter Ernest
Luntz, Myron
MacCallum, Crawford John
McLaughlin, William Lowndes
Miller, John Howard
Moyer, Robert (Findley)
Mozumder, Asokendu
Naber, James Allen
Newton, Carlos E, Jr
O'Brien, Keran
Painter, Linda Robinson
Painter, Richard Carl
Parker, Cleofus Varren, Jr
Petree, Ben
Phelps, Michael Edward
Placious, Robert Charles
Ponnraninge, Gerald C
Rains, Albert Edward
Riley, Richard Charles
Ritchie, Rufus Haynes
Schilliaci, Mario Edward

Schweizer, Felix
Sherman, Norman K
Shupe, Robert Eugene
Spokas, John J
Stinchcomb, Thomas Glenn
Trice, James Buckner
Trombka, Jacob Israel
Trubey, David Keith
Vacirca, Salvatore John
Woodward, Ervin Chapman, Jr
Yuster, Philip Harold
Zucker, Martin Samuel

Radiological Physics
Adams, Gail Dayton, Jr
Agarwal, Suresh Kumar
Alvi, Zahoor Mohem
Arnols, Howard Ira
Attix, Frank Herbert
August, Leon Stanley
Bahr, Gustave Karl
Bates, Lloyd M
Beach, Joseph Lawrence
Benton, Eugene Vladimir
Berger, Harold
Berggren, Michael J
Bjarngard, Bengt E
Braby, Leslie Alan
Braestrup, Carl Bjorn
Bramlet, Roland C
Brookeman, Valerie Ann
Bruels, Mark Charles
Carlson, James C
Chaney, Edward L
Cohen, Norman
Cooper, Philip Harlan
Cruty, Michael Robert
Dauer, Maxwell
Dillard, Margaret Bleick
Dixon, Robert Leland
Doleck, Elwyn Haydn
Elder, Robert Lee
Fairchild, Ralph Grandison
Feldman, Arnold
Finston, Roland A
George, Robert Eugene
Glass, William A
Goodman, Leon Judias
Goodwin, Paul Mewcomb
Gunter, Karlene Klages
Hagee, George Richard
Hale, John
Hoecker, Frank Edward
Horwitz, Norman H
Inatsugu, Seiji
Janney, Clinton Dales
Johnson, Raleigh Francis, Jr
Jones, Ernest Olin
Kelley, John Paul
Khan, Faiz Mohammad
Klein, David Joseph
Lange, Robert Carl
LeBlanc, Adrian David
Loevinger, Robert
Malsky, Stanley Joseph
McCall, Richard C
Meckstroth, George R
Moore, Vaughn Clayton
Nath, Ravinder Katyal
Oddie, Harvey Earl
Palmer, Harvey Earl
Rao, Gopala U V
Rohrig, Norman
Rundo, John
Sargent, Frederick Peter
Sashin, Donald
Shrivastava, Prakash Narayan
Slocum, Robert Richard
Soldat, Joseph Kenneth
Solon, Leonard Raymond
Stoner, William Weber
Suntharalingam, Naglingam
Tanner, Raymond Lewis
Wilson, Robert John
Wilson, Walter Ervin
Wingate, Catharine L
Witherell, Egilda DeAmicis
Wrede, Don Edward
Wright, Ann Elizabeth
Yaniv, Shlomo Stefan
Zelac, Ronald Edward

Radiophysics
Anderson, Lloyd James
Basler, Roy Prentice
Beard, Charles Irvin
Beckmann, Petr
Booker, Henry George
Crane, Robert Kendall
Evans, John V
Fremouw, Edward Joseph
Grossi, Mario Dario
Hagen, Jon Boyd
Hogg, David Clarence
Kelso, John Morris
Kundu, Mukul Ranjan
Mechtly, Eugene A

Nelson, Raymond Adolph
Paghis, Irvine
Sherrill, William Manning
Smith, Ernest Ketcham
Strohbehn, John Walter
Thome, George Durst

Reactor Physics
Ahlfeld, Charles Edward
Bach, David Rudolph
Baird, Quincey Lamar
Ball, Russell Martin
Bell, George Irving
Bennett, Edgar F
Busik, Arthur J
Butler, Daniel Knowles
Carter, Leland LaVelle
Church, John Phillips
Clark, Hugh Kidder
Coats, Richard Lee
Cohen, Leonard David
Corngold, Noel Robert David
Court, Anita
Davis, James Allen
Davis, Ronald Stuart
De Volpi, Alexander
Dingee, David Aaron
Driggers, Frank Edgar
Durant, W S
Ehrlich, Richard
Esch, Louis James
Farr, William Morris
Field, Herbert Cyre
Fox, Thomas Allen
Galey, John Apt
Gerstl, Siegfried Adolf Wilhelm
Graves, Michael Ewing
Green, Ralph Ellis
Gregory, Michael Vladimir
Hardy, Judson, Jr
Harker, Yale Deon
Honeck, Henry Charles
Kato, Walter Yoneo
Kellman, Simon
Kepes, Joseph John
Krumbein, Aaron Davis
Kurey, Thomas John
Lathrop, Kaye Don
Lazareth, Otto William, Jr
Levine, Samuel Harold
McCrosson, F Joseph
McElroy, William Nordell
Mehta, Kishor Kalidas
Millar, Charles Howard
Pennington, Edwin McFarlane
Pettus, William Gower
Phelps, James Parkhurst
Pomraning, Gerald C
Price, Glenn Albert
Purohit, Surendra Nath
Ragan, George Leslie
Redman, William Charles
Roggenkamp, Paul Leonard
Roy, Donald H
Schweitzer, Donald Gerald
Serdula, Kenneth James
Shen, Peter Ko-Chun
Snelgrove, James Lewis
Stacey, Weston Monroe, Jr
Staker, William Paul
Stanford, George Stailing
Thie, Joseph Anthony
Topp, Stephen V
Toppel, Bert Jack
Ulrich, Aaron Jack
Valente, Frank Anthony
Wimett, Thomas Frederick
Woodard, Ralph Emerson
Woodruff, William Lee
Yule, Thomas J

Solar Physics
Acton, Loren Wilber
Altrock, Richard Charles
Altschuler, Martin David
Anderson, Kinsey Amor
Arnquist, Warren Nelson
Baker, Neal Kenton
Balif, Jae R
Bell, Barbara
Bohlin, John David
Brown, Douglas Ross
Burek, Anthony John
Canfield, Richard Charles
Cantrell, Joseph Sires
Castelli, John P
Chapman, Gary Allen
Clark, Arnold Franklin
Cole, Robert Stephen
Curtis, George William
Davis, John Moulton
Dietz, Richard Darby
Dodson, Helen Walter (Mrs Edmond L Prince)
Doschek, George A
Eddy, John Allen
Epstein, Gabriel Leo
Fausey, Norman Ray

Feit, Julius
Feynman, Joan
Firor, John William
Gosling, John Thomas
Hayes, John William
Hewett, Lionel Donnell
Hildebrandt, Alvin Frank
Howard, Russell Alfred
Jackson, Ray Dean
Jones, William Denver
Jordan, Stuart Davis
Keil, Stephen Lesley
Kirk, John Gallatin
Kohl, John Leslie
Kopp, Roger Alan
Krieger, Allen Stephen
Landman, Donald Alan
Leibacher, John William
Lin, Robert Peichung
Lincoln, Jeannette Virginia
Lufkin, Daniel Harlow
MacQueen, Robert Moffat
McDaniels, David K
Meekins, John Fred
Meinel, Aden Baker
Neupert, Werner Martin
Nye, Alan Hall
O'Brien, James Edward
Oertel, Goetz Kuno Heinrich
Peale, Stanton Jerrold
Phillips, Ronald Edward
Pierce, Austin Keith
Plumley, Harold Jamison
Poland, Arthur I
Pollard, William Grosvenor
Querfeld, Charles William
Richardson, Clarence Robert
Roberts, Walter Orr
Sakurai, Kunitomo
Schatten, Kenneth Howard
Silk, John Kevin
Tanner, Champ Bean
Thomas, Roger Jerry
Thompson, Arthur Howard
Underwood, James Henry
Van Hoven, Gerard
Wengel, Raymond William
Widing, Kenneth Gordon
Wilcox, John Marsh
Withbroe, George Lund
Worden, Simon Peter
Zirker, Jack Bernard

Space Physics
Allen, Joseph Percival
Allum, Frank Raymond
Amstutz, Larry Ihrig
Anderson, Hugh Riddell
Anderson, Kinsey Amor
Armstrong, John William
Armstrong, Thomas Peyton
Arnoldy, Roger L
Arthur, Wallace
Asbridge, John Robert
Axford, William Ian
Bame, Samuel Jarvis, Jr
Barry, James Dale
Bauer, Siegfried Josef
Becker, Robert Adolph
Bettinger, Richard Thomas
Blake, J Bernard
Bordoloi, Kiron
Bostrom, Carl Otto
Brecher, Aviva
Bridge, Herbert Sage
Bruner, Elmo Cody, Jr
Burrows, John Ronald
Cahill, Laurence James, Jr
Calio, Anthony John
Campbell, Malcolm John
Carovillano, Robert L
Chen, Abel Jer-Jiunn
Clark, Malcolm A
Coleman, Paul Jerome, Jr
Coon, James Huntington
Coroniti, Ferdinand Vincent
Crooker, Nancy Uss
Cummins, Herman Z
Dalton, Charles Chester
Davis, Leverett, Jr
Deehr, Charles Sterling
De Los Reyes, B William
Despain, Lewis Gail
Dessler, Alexander Jack
Drummond, Andrew Jamieson
Duggal, Shakti Prakash
Dumas, Herbert M, Jr
Erickson, Kenneth Neil
Everitt, C W Francis
Feit, Julius
Feynman, Richard
Fischell, Robert E
Fitz, Harold Carlton, Jr
Fontheim, Ernest Gunter
Francavilla, Thomas Lee
Fraser-Smith, Antony Charles
Freeman, John Wright, Jr
Freyer, Gustav John

Friedman, Richard M
Fu, Jerry Hui Ming
Garrett, Henry Berry
Garriott, Owen Kay
Gibson, Luther Ralph
Goedeke, Arthur Donald
Goodenough, David George
Gordon, Gary Donald
Gosling, John Thomas
Grobecker, Alan J
Gross, Stanley H
Heikkila, Walter John
Hennes, John Peter
Heppner, James P
Hills, Howard Kent
Hoch, Richmond Joel
Hoegy, Walter R
Hoffman, Robert A
Holzer, Robert Edward
Horita, Robert Eiji
Hudson, Mary Katherine
Imhof, William Lowell
Intriligator, Devrie Shapiro
Johnson, Francis Severin
Katz, Ludwig
Kendall, Bruce Reginald Francis
Kennel, Charles Frederick
Kissell, Kenneth Eugene
Kivelson, Margaret Galland
Klumpar, David Michael
Konradi, Andrei
Kostiuk, Theodor
Krimigis, Stamatios Mike
Lanzano, Paolo
Lanzerotti, Louis John
Liemohn, Harold Benjamin
Lin, Robert Peichung
Lin, Wei-Ching
Lind, Arthur Charles
Linsky, Jeffrey L
Liwshitz, Mordehai
Lundquist, Charles Arthur
Luthey, Joe Lee
Machl, Ronald Charles
Maloy, John Owen
Mange, Phillip Warren
Mayfield, Earle Byron
McCarthy, Kathryn Agnes
McDiarmid, Ian Bertrand
McEntire, Richard Willian
McNeal, Robert Joseph
McPherron, Robert Lloyd
McPherson, Donald Attridge
Meredith, Leslie Hugh
Michael, Irving
Michael, William Herbert, Jr
Michel, F Curtis
Modisette, Jerry Lee
Mogro-Campero, Antonio
Morse, Fred A
Mozer, Forrest S
Nash, Douglas B
Nerney, Steven Francis
Ness, Norman Frederick
Newkirk, Lester Leroy
Nordberg, William
Northrop, Theodore George
Oder, Frederic Carl Emil
Ogilvie, Keith W
Oliver, Brian Malcolm
Otterman, Joseph
Page, Larry J
Page, Thornton Leigh
Palmeira, Ricardo Antonio Ribeiro
Papagiannis, Michael D
Parkin, Curtis Willard
Patel, Vithalbhai L
Paulikas, George A
Pavel, Arthur Lawrence
Pazich, Philip Michael
Pieper, George Francis, Jr
Pongratz, Morris Bernard
Preszler, Alan Melvin
Quinn, Robert George
Reidy, William Patrick
Rense, William A
Rhee, John Williams
Roederer, Juan Gualterio
Rosen, Alan
Rosenbaum, Bernard
Rosenberg, Theodore Jay
Rostoker, Gordon
Rothwell, Paul L
Rubin, Allen Gershon
Sagalyn, Rita C
Samir, Uri
Scherb, Frank
Schmerling, Erwin Robert
Schulz, Michael
Semar, Cary Lloyd
Shelley, Edward George
Silbergeld, Mae Driscoll
Siscoe, George L
Smith, Edward John
Snodgrass, Rex Jackson
Sonnerup, Bengt Ulf Osten
Stansberry, Kent Gardner
Stern, David P

PHYSICS

Strong, Stanley Sterling
Su, Shin-Yi
Sugiura, Masahisa
Swift, Daniel W
Taylor, Harold Evans
Thompson, Thomas William
Thorne, Richard Mansergh
Timothy, John Gethyn
Tinsley, Brian Alfred
Tomblin, Fred Fitch
Vampola, Alfred Ludvik
Vande Noord, Edwin Lee
Venkatesan, Doraswamy
Vrebalovich, Thomas
Walker, Arthur Bertram Cuthbert, Jr
Walker, Raymond John
Walker, Robert Mowbray
Walt, Martin
Wax, Robert LeRoy
Weber, Alfred Herman
Weddell, James Blount
Wehrenberg, Paul James
Wende, Charles David
West, Harry Irwin, Jr
Whipple, Fred Lawrence
White, Robert Stephen
Wiggins, John Shearon
Wilkerson, Thomas Delaney
Williams, Donald J
Winter, Thomas C, Jr
Wolf, Richard Alan
Yates, George Kenneth
Yen, Chen-Wan Liu

Surface Physics

Armstrong, Robert A
Arthur, John Read, Jr
Bagchi, Amitabha
Bauer, Ernst Georg
Beck, Donald Edward
Becker, Gordon Edward
Bedell, Louis Robert
Bell, Anthony E
Bellina, Joseph James, Jr
Berkes, John Stephan
Best, Philip Ernest
Biegen, Joseph Robert
Bradshaw, Benjamin Crenshaw
Brady, James Joseph
Brillson, Leonard Jack
Buchholz, Jeffrey Carl
Buck, Thomas M
Burkstrand, James Michael
Callcott, Thomas Anderson
Case, Clinton Meredith
Cho, Alfred Y
Chopra, Dev Raj
Cohen, Philip Ira
Cohen, Samuel Alan
Crowell, Albert Dary
Cunningham, Robert Gail
Current, Michael Ira
Cuthrell, Robert Eugene
Czanderna, Alvin Warren
Dalins, Ilmars
Das, Santosh Kumar
Dash, Jay Gregory
Davison, Sydney George
Denison, Dean R
Dionne, Gerald Francis
Ditman, Richard Henry
Dresner, Joseph
Dresser, Miles Joel
Economou, Eleftherios Nickolas
Ehrlich, Gert
Ekern, Ronald James
Emerson, Lewis Cotesworth
Fabish, Thomas John
Faust, John William, Jr
Fernelius, Nils Conard
Feuchtwang, Thomas Emanuel
Fischer, Traugott Erwin
Fossum, Steve P
Fowler, Alan B
Frankl, Daniel Richard
Furnak, Thomas Elton
Gaines, Gordon Bradford
Goldstein, Bernard
Greenler, Robert George
Haas, George Arthur
Hagstrum, Homer Dupre
Hansma, Paul Kenneth
Harrison, Don Edward, Jr
Hartman, James Keith
Holloway, Dennis Michael
Holden, Stanley J
Houston, Jack E
Huang, John S
Huff, G Bradley
Hurych, Zdenek
Jason, Andrew John
Jenkins, Leslie Hugh
Jepsen, Donald William
Jona, Franco Paul
Jorgenson, Gordon Victor
Kaminsky, Manfred Stephan
Khan, Jhan M
Knotek, Michael Louis

Kollen, Wendell James
Krumbein, Simeon Joseph
Lagally, Max Gunter
Lange, William James
Lee, Ronald Norman
Levie, Harold Walter
Levine, Leonard P
Lichtman, David
Lucke, William Hunter
MacDonald, Noel Charles
Madey, Theodore Eugene
Malo, Salvador Alejandro
Mark, Peter Herman
May, John Walter
McKinney, James T
Moen, Allen LeRoy
Morrison, Stanley Roy
Murday, James Stanley
Nishikawa, Osamu
Norton, Peter Robert
Palmer, Robert Lewis
Peavey, Jerris Hinkins
Porteus, James Oliver
Prince, Robert Harry
Ritz, Victor Henry
Rolfson, Robert John
Schmidt, Paul F
Seki, Hajime
Sinha, Mahendra Kumar
Smith, Joe Nelson, Jr
Smith, Richard James
Stearns, Brenton Fisk
Stevenson, James Rufus
Stewart, Charles Neil
Strozier, John Allen, Jr
Tan, Yen T
Taylor, Thomas Newton
Tomaschke, Harry E
Tracy, Joseph Charles, Jr
Trolan, J Kenneth
Trusty, Josann Watkins
Wachman, Harold Yehuda
Weicher, James F
Wolf, Edward D
Zehner, David Murray

Textile Physics

Barnes, James Crowell
Bernier, Edward Joseph
Buchanan, David Royal
Cushing, James Thomas
Dennison, Richard Wheeler
Gupta, Bhupender Singh
Hardy, Henry Benjamin, Jr
Kobetich, Edward John
Kyame, George John
Lehmicke, David John
Lothrop, Everett Winfred, Jr
Pontrelli, Gene J
Shaffner, Thomas Jackson
Stanley, Harry Eugene
Van Veld, Robert Dale
Weiss, Louis Charles
Williams, Kenneth Roger

Thermal Physics

Arumi, Francisco Noe
Brewer, LeRoy Earl, Jr
Cezarliyan, Ared
Farkass, Imre
Hager, Nathaniel Ellmaker, Jr
Ho, Cho-Yen
Hoge, Harold James
Hsia, Jack Jinn-Goe
Leff, Harvey Sherwin
Lewis, Edwin Augustus Stevens
Martin, Douglas Leonard
Otto, George W
Parker, William James
Quay, Paul Michael
Reese, William
Rutledge, Delbert Leroy
Schooley, James Frederick
Sengers, Johanna M H Levelt

Thermodynamics

Andrews, John Timothy Sawford
Anthrop, Donald F
Armstrong, George Thomson
Ashworth, T
Brown, Billings
Busey, Richard Hoover
Chang, Ren-Fang
Farley, Thomas Albert
Fisher, Robert Amos, Jr
Furukawa, George Tadaharu
Gant, Fred Allan
Garcia-Colin, Leopoldo Scherer
Grilly, Edward Rogers
Halteman, Eber Kington
Holt, Ray James
Holt, Vernon Emerson
Hsia, Jack Jinn-Goe
Kiersted, Henry Andrew
Kreglewski, Alexander
Kurzweg, Hermann Herbert
LaPietra, Richard Andrew

LeFebre, Vernon Glen
Longtin, Bruce
Maccabee, Bruce Sargent
McKinney, John Edward
Northrup, Clyde John Marshall, Jr
Oetting, Franklin Lee
Pirkle, Willis Nathaniel
Pribil, Stephen
Smilters, Juris
Somerscales, Euan Francis Cuthbert
Speen, Gerald Bruce
Spitzer, Hermann Josef
Steinberg, Daniel J
Tang, James Juh-Ling
Tong, Long Sun
Treat, Charles Herbert
Urtiew, Paul Andrew
Van den Akker, Johannes Archibald
Waters, Elmer Dale
Westrum, Edgar Francis, Jr
Wilhoit, Randolph Carroll
Wist, Abund Ottokar
Wu, Yung-Chi
Zimmer, Martin F

PHYSICS, ATOMIC

Atomic Physics

Albridge, Royal
Alrick, Philip Lewis
Anderson, David Kent
Anderson, Louis Wilmer
Anderson, Richard Alan
Armstrong, Baxter Hardin
Arroe, Hack
Atkinson, John Brian
Bagus, Paul Saul
Bainter, Monica Evelyn
Band, Yehuda Benzion
Bardsley, James Norman
Barnes, John Fayette
Bashkin, Stanley
Beck, Donald Richardson
Bengtson, Roger D
Berman, Paul Ronald
Berry, Henry Gordon
Bhalla, Chander P
Bhatia, Anand K
Biondi, Manfred Anthony
Boehm, Felix H
Borst, Walter Ludwig
Brackmann, Richard Theodore
Bradford, John Norman
Brandenberger, John Russell
Branscomb, Lewis McAdory
Bronco, Charles John
Brown, Robert Lamme
Brown, William Arnold
Budick, Burton
Burns, Donal Joseph
Burr, Alexander Fuller
Callan, Edwin Joseph
Carlson, Robert Warner
Cecchi, Joseph Leonard
Chang, Berken
Chaturvedi, Ram Prakash
Chen, James Ralph
Chiad, Tang
Childs, William Jeffries
Christensen, Robert Lee
Christophorou, Loucas Georgiou
Cipolla, Sam J
Compton, Robert Norman
Cook, Charles J
Cooper, John (Jinx)
Cowan, Robert Duane
Crampton, Stuart J B
Crandall, David Hugh
Crasemann, Bernd
Curtis, Lorenzo Jan
Dardis, John G
Darewych, Juraj Wasyl'
Das, Tara Prasad
Datz, Sheldon
Dehmer, Joseph Leonard
Dietrich, Daniel David
Dittner, Peter Fred
Dobson, David A
Dodd, Jack Gordon, (Jr)
Donnally, Bailey Lewis
Doverspike, Lynn D
Doyle, Walter M
Drake, Charles Whitney
Drake, Gordon William Frederic
Driskill, William David
Driver, Richard D
Duckworth, Henry Edmison
Dunn, Gordon Harold
Dunning, Frank Barrymore
Edmonds, Dean Stockett, Jr
Ehlers, Vernon James
Eidson, William Whelan
Elston, Stuart B

Elton, Raymond Carter
Engelke, Charles Edward
Epstein, Gabriel Leo
Essenwanger, Oskar M
Evenson, Kenneth Melvin
Ewbank, Wesley Bruce
Fairchild, Clifford Eugene
Fetzer, Homer D
Fink, Manfred
Flannery, Martin Raymond
Forsman, Earl N
Fox, Russell Elwood
Franz, Judith Rosenbaum
Freerks, Marshall Cornelius
Fromhold, Lothar Werner
Fystrom, Dell O
Gallagher, Alan C
Garbellano, David Wesley
Garcia, Jose Dolores, Jr
Gerardo, James Bernard
Gerjuoy, Edward
Gibbs, Hyatt McDonald
Gibbs, Richard Lynn
Gibson, Henry Clay, Jr
Ginter, Marshall L
Glass, Alexander Jacob
Golden, David E
Green, Thomas Allen
Greenwood, Ivan Anderson
Grissom, John Thomas
Gupta, Rajendra
Hales, J Vern
Hammond, Gordon Leon
Happer, William, Jr
Hardcastle, Donald Lee
Hatfield, Lynn LaMar
Haugsjaa, Paul O
Hayden, Howard Corwin
Heestand, Glenn Martin
Helbig, Herbert Frederick
Henry, Eugene Michael
Henry, Ronald James Whyte
Herman, Roger Myers
Hilborn, Robert Clarence
Hill, Albert Gordon
Hird, Brian
Hirsh, Merle Norman
Holt, Helen Keil
Holzberlein, Thomas M
Hopkins, John Isaac
Hotz, David Franklin
House, Lewis Lundberg
Huebner, Walter F
Hutcherson, Joseph William
Innes, Frederick Rush
Jacobs, Verne Louis
Jaecks, Duane H
Jefferts, Keith Bartlett
Johnson, Charles Edward
Jones, Charles Edward
Judd, Brian Raymond
Junker, Bobby Ray
Kang, Ik-Ju
Kaufman, Victor
Kavanagh, Thomas Murray
Kelly, Ralph Edward
Kelly, Paul Sherwood
Kerr, Donald Philip
Kessler, Karl Gunther
Khadjavi, Abbas
Khan, Jhan M
Kieffer, Lee Joseph
Kim, Dae Mann
Kim, Yong Wook
Kim, Yong-Ki
Kim, Yong Gordon
King, John Gordon
Klose, Jules Zeiser
Knudson, Alvin Richard
Kostroun, Vaclav O
Krause, Manfred Otto
Krause, Herbert Francis
Krotkov, Robert Vladimir
Kuenhold, Kenneth Alan
Kumar, Cidambi Krishna
Kuyatt, Chris Ernie Earl
Lamb, Willis Eugene, Jr
Lambert, Robert Henry
Lambropoulos, Melissa Margaret
Landman, Donald Alan
Lane, Neal F
Larson, Daniel A
Lee, Chia-Ming
Lee, Tong-Nyong
Leventhal, Marvin
Leventhal, Jacob J
Lin, Chii-Dong
Lipeles, Martin
Lipworth, Edgar
Lorents, Donald C
Loyd, David Heron
Lulla, Kotusingh
Luther, Marvin L
Machacek, Milos
MacLennan, Donald Allan
MacVicar-Whelan, Patrick James

Madison, Don Harvey
Magnuson, Gustav Donald
Mahadevan, Parameswar
Malik, Fazley Bary
Mallow, Jeffry Victor
Manson, Steven Trent
Marmet, Paul
Marquet, Louis C
Marrus, Richard
McDermott, Mark Nordman
McGee, James Francis
McGowan, James William
McGuire, Eugene J
McNutt, John Dwight
Melvin, Jonathan David
Menendez, Manuel Gaspar
Merts, Athel Lavelle
Meyerott, Roland Edward
Miller, Frank L
Miller, Thomas Marshall
Mills, Allen Paine, Jr
Mjolsness, Raymond C
Moore, Cornelius Fred
Moore, Edwin Neal
Morgan, Thomas Joseph
Mosburg, Earl R, Jr
Moseley, John Travis
Moses, Herbert A
Mulligan, Joseph Francis
Nalley, Samuel Joseph
Neynaber, Roy Harold
Norcross, David Warren
Nygaard, Kaare Johann
Omidvar, Kazem
Park, John Thornton
Perel, Julius
Peterson, James Ray
Phelps, Arthur Van Rensselaer
Pichanick, Francis Martin
Pollack, Edward
Pomilla, Frank Rocco
Posner, Martin
Pretzer, Donavon Donald
Pritchard, Carroll Adair, Jr
Quarles, Carroll Edward
Rajnak, Katheryn Edmonds
Rao, Pemmaraju Venugopala
Rau, A Ravi Prakash
Ray, Surendra Nath
Rayborn, Grayson Hanks
Rhodes, Charles Kirkham
Ricci, Enzo
Rich, Arthur
Richard, Patrick
Rieser, Leonard M, Jr
Robb, William Derek
Roberts, Thomas D
Rodgers, James Earl
Roesler, Frederick Lewis
Roszman, Larry Joe
Rotenberg, Manuel
Sahni, Viraht
Saloman, Edward Barry
Samson, James Alexander Ross
Sanders, Frank Clarence, Jr
Sands, Richard Hamilton
Sattler, Allan R
Schaefer, Albert Russell
Schearer, Laird D
Schectman, Richard Milton
Schlachter, Alfred Simon
Schmieder, Robert W
Schowengerdt, Franklin Dean
Schuessler, Hans Achim
Sellers, Francis Bachman
Sellin, Ivan Armand
Sen, Sunil Kumar
Shafroth, Stephen Morrison
Sherwood, Jesse Eugene
Shore, Bruce Walter
Smith, Felix Teisseire
Snell, Arthur Hawley
Snider, Joseph Lyons
Snow, William Rosebrook
Soltysik, Edward A
Spence, David
Starace, Anthony Francis
Stauffer, Allan Daniel
Steinhaus, David Walter
Sternheimer, Rudolph Max
Stone, Philip M
Strickler, Thomas David
Strwalley, William Calvin
Sutton, Emmet Albert
Swan, Peter Howard
Synek, Miroslav (Mike)
Tam, Andrew Ching
Temkin, Aaron
Thoe, Robert Steven
Thomas, Clarence Delmar
Thomsen, John Stearns
Thonnard, Norbert
Toburen, Larry Howard
Van Ausdal, Ray Garrison
Van Brunt, Richard Joseph
Van Den Bos, Jan
Vulliet, William George
Waber, James Thomas

Waldrop, Morgan A
Walker, Keith Gerald
Walters, Geoffrey King
Wehrenberg, Paul James
Weinhous, Martin S
Weiss, Andrew W
Welsh, Harry L
Westhaus, Paul Anthony
White, John Arnold
Whitehead, Walter Dexter, Jr
Wiese, Wolfgang Lothar
Williams, William Lee
Williamson, William, Jr
Wilson, Walter Davis
Wing, William Hinshaw
Winters, Loren Mel
Wittkower, Andrew Benedict
Woldseth, Rolf
Wong, Shek-Fu
Woo, Shien-Biau
Wood, Calvin Dale
Woodyard, Jack Ramon
Wright, John Jay
Yarosewick, Stanley J
Zander, Arlen Ray
Zorn, Jens Christian

Atomic Spectroscopy
Andrew, Kenneth L
Brown, Charles Moseley
Cloud, William Max
Conway, John George, Jr
Corliss, Charles Howard
Derby, Stanley Kingdom
Feldman, Uri
George, Simon
Gilbert, David Erwin
Head, Charles Everett
Head, Martha E Moore
Hill, Kenneth Wayne
Holmes, John Richard
Humphreys, Curtis Judson
Kelly, Raymond Leroy
Kielkopf, John F
Kingsbury, Robert Freeman
Lu, Kwang-Tzu
Martin, William Clyde
McCormick, William Wallace
Merritt, Thomas Parker
Phelps, Frederick Martin, III
Pinnington, Eric Henry
Radziemski, Leon Joseph, Jr
Reeves, Edmond Morden
Ross, John Stoner
Shenstone, Allen Goodrich
Smith, Peter Lloyd
Stoner, John Oliver, Jr
Stroke, Hinko Henry
Sugar, Jack
Tomkins, Frank Sargent
Valero, Francisco Pedro Jorge
Wiegand, Roy Vernon

Experimental Atomic Physics
Alton, Gerald Dodd
Amme, Robert Clyde
Bingham, Felton Wells
Brink, Gilbert Oscar
Burns, John Francis
Cano, Gilbert Lucero
Chopra, Dev Raj
Chu, Wei-Kan
Cuderman, Jerry Ferdinand
Dayton, Irving Eugene
de Zafra, Robert Lee
Edwards, Alan Kent
Erickson, Lynden Edwin
Friedlander, Zitta Zipora
Gale, Douglas Shannon, II
Greenebaum, Ben
Inatsugu, Seiji
Jones, Keith Warlow
Kessel, Quentin Cattell
Kohl, John Leslie
Kugel, Henry W
Lockwood, Grant John
Meyerhof, Walter Ernst
Moe, George Wylbur
Mumma, Michael Jon
Mungall, Allan George
Ott, William Roger
Pipkin, Francis Marion
Pitchford, Leanne Carolyn
Redi, Olav
Reynolds, Harlan Kendall
Ries, Richard Ralph
St John, Robert Mahard
Sautter, Chester A
Schneider, Jacob David
Sheridan, John Roger
Simmons, Melvin Kurt
Smith, Winthrop Ware
Stewart, Albert Burns
Suchannek, Rudolf Gerhard
Varghese, Sankoorikal Lonappan
Zitzewitz, Paul William

PHYSICS, EXPERIMENTAL

Experimental Physics
Abella, Isaac D
Addelman, Sidney
Alvarez, Luis Walter
Anderson, Robert Alan
Antal, John Joseph
Arnold, James Tracy
Arnold, William H, Jr
Ascoli, Giulio
Ashkin, Arthur
Bauer, Rudolf Wilhelm
Baum, Paul M
Bederson, Benjamin
Benjamin, Robert Fredric
Besse, Arthur L
Bickel, William Samuel
Blake, Julian Gaskill
Blevins, Gilbert Sanders
Bogle, Robert Worthington
Bradbury, James Norris
Bradner, Hugh
Bullen, Thomas Gerrard
Burkhardt, James Lee
Campbell, Larry Enoch
Chasson, Robert Lee
Cheng, David
Cluxton, David H
Cohn, Charles Erwin
Cole, James A
Collins, Ernest Hobart
Curtis, Roger William
Cutts, David
DeBenedetti, Sergio
DeBitetto, Dominick John
Dehmelt, Hans Georg
Denison, Arthur B
Donovan, Terence M
DuBridge, Lee Alvin
Dunn, Andrew Fletcher
Eppenstein, Walter
Essig, Frederick Charles
Fenstermacher, Charles Alvin
Frey, William Francis
Frost, Robert Hartwig
Fry, William Frederick
Garwin, Richard Lawrence
Gasten, Burt R
Gemmell, Donald Stewart
Gibson, Walter Maxwell
Goodrich, Max
Gunst, Samuel Burton
Hagelberg, Myron Paul
Hamblen, David Gordon
Hartman, Paul Leon
Hebel, Louis Charles
Heflinger, Lee Opert
Helms, Rufus Marshall
Hempstead, Charles Francis
Hickey, Roger
Hirsch, John Michele
Hirvonen, James Karsten
Holloway, John Thomas
Jackson, Jasper Andrew, Jr
Jaffe, Morry
Jahoda, Franz Carl
Jones, Mark Wallon
Kaminsky, Manfred Stephan
Kaufman, Raymond
Keith, Harvey Douglas
Keller, Donald V
Kelly, Harry Charles
Knight, Stephen
Kratz, Howard Russel
Kuswa, Glenn Wesley
Lawson, Kent DeLance
Leder, Lewis Beebe
Lee, Long Chi
Lee, Robert W
Lee-Franzini, Juliet
Lees, Wayne Lowry
Li-Scholz, Angela
Llewellyn, Edward John
Lombard, David Bishop
Long, Dale Donald
Macdonald, Burns
Macdonald, James Robert
Malik, John Stanley
Martin, Donald Clayton
Martin, Frederick Wight
Meissner, Hans Walter
Mink, Lawrence Albright
Moore, Donald Baker
Muller, Richard A
Myer, Jon Harold
Ogier, Walter Thomas
O'Meara, Francis Edmund
Overhage, Carl F J
Page, Lorne Albert
Parks, Ronald Dee
Passenheim, Burr Charles
Perlow, Gilbert Jerome
Pfeiffer, Loren Neil
Phillips, Donald Davis
Plum, William Byrle

Pondrom, Lee Girard
Preston-Thomas, Hugh
Price, Paul Buford, Jr
Reisman, Elias
Richard, Jean-Paul
Richards, Lorenzo Adolph
Ringo, George Roy
Robertson, Merton M
Sandmann, William Henry
Schmidt, Klaus H
Schmidt, Kurt
Schulte, Harry John, Jr
Schumacher, Berthold Walter
Sciulli, Frank J
Shattes, Walter John
Shewell, John Robert
Silver, Ernest Gerard
Stahl, Ralph Henry
Stokes, Richard Hivling
Strickland, James Shive
Styris, David Lee
Sumberg, David Allan
Swigart, John Irvin
Terroux, Ferdinand Richard
Thomas, Bruce Robert
Utterback, Nyle Gene
Vaala, Allen Richard
Vosburgh, Kirby Gannett
Weaver, Harry Edward, Jr
Webb, Julian Pierce
Weber, Louis Russell
Wedding, Brent (M)
Weiffenbach, George Charles
Wesemeyer, Harald
Wheeler, George William
Wiegand, Clyde Edward
Winston, Roland
Zehr, Floyd Joseph
Zei, Dino
Ziegler, George Elliott
Zimmerer, Robert W
Ziock, Klaus Otto H
Zollweg, Robert John

Experimental High Energy Physics
Abrams, Robert Jay
Ankenbrandt, Charles Martin
Ashford, Victor Aaron
Ayres, David Smith
Berge, Jon Peter
Berkelman, Karl
Brenner, Alfred Ephraim
Brown, Charles Nelson
Burnett, Thompson Humphrey
Button-Shafer, Janice
Caldwell, David Orville
Canter, Joseph M
Carithers, William Cornelius, Jr
Chamberlain, Owen
Cohen, Kenneth Joel
Cooper, John Wesley
Dahl, Orin I
Derrick, Malcolm
DiBianca, Frank Anthony
Eartly, David Paul
Ecklund, Stanley Duane
Eisner, Alan Mark
Eisner, Robert Lawrence
Ficenec, John Robert
Francis, William Richard
Friedberg, Carl E
Frisch, Henry Jonathan
Gelfand, Norman Mathew
Gibbard, Bruce Gregory
Gidal, George
Gladding, Gary Earle
Goshaw, Alfred Thomas
Grigorian, Alexander
Groves, Eric Stedman
Hargrove, Clifford Kingston
Harris, Roger Mason
Holloway, Leland Edgar
Humphrey, John William
Hyman, Lloyd George
Jaeger, Klaus Bruno
Kalbach, Robert Michael
Kammerud, Ronald Claire
Kirz, Janos
Koester, Louis Julius, Jr
Kraemer, Stephen Leonard
Kramer, Stephen Leonard
Kraybill, Henry Lawrence
Law, Margaret Elizabeth
Lindenbaum, Seymour Joseph
Littauer, Raphael Max
Loken, Stewart Christian
Louttit, Robert Irving
Lynch, Harvey Lee
Madaras, Ronald John
Margulies, Seymour
McDaniel, Boyce Dawkins
Messner, Robert Lee
Miller, David Harry
Miller, Robert Carl
Miller, Robert Joseph
Mockett, Paul M
Moore, Craig Damon
Moriyasu, Keihachiro

Murphy, Frederick Vernon
Nash, Edward Thomas
Neal, Homer Alfred
Nelson, Jerry Earl
Nodulman, Lawrence Jay
Nussbaum, Mirko
Ogren, Harold Olof
Osborne, Weymar Zack
Oversth, Oliver Enoch, Jr
Pan, Yu-Li
Pawlicki, Anthony Joseph
Peck, Charles William
Perez-Mendez, Victor
Peterson, Vincent Zetterberg
Pewitt, Edward Gale
Phelan, James Joseph
Pipkin, Francis Marion
Protopopescu, Serban Dan
Pruss, Stanley McQuaide
Rey, Charles Albert
Rhines, Don Scott
Romanowski, Thomas Andrew
Rubinstein, Roy
Sanders, Gary Hilton
Sard, Robert Daniel
Schluter, Robert Arvel
Schreiner, Philip Allen
Schultz, Peter Frank
Shephard, William Danks
Shibata, Edward Isamu
Shoemaker, Frank Crawford
Sinclair, Charles Kent
Singer, Richard Alan
Slater, William E
Smith, Arthur John Stewart
Smith, Richard Paul
Stanfield, Kenneth Charles
Steinberg, Phillip Henry
Strovink, Mark William
Sulak, Lawrence Richard
Sutton, Roger Beatty
Swartz, Clifford Edward
Thaler, Jon Jacob
Ticho, Harold Klein
Tompkins, John Carter
Toohig, Timothy E
Trippe, Thomas Gordon
Tycko, Daniel H
Watson, Jerry Mike
Watt, Robert Douglas
West, Edmund Cary
White, David Hywel
White, Milton Grandison
Wicklund, Arthur Barry
Wilkes, Richard Jeffrey
Williams, Hugh Harrison
Yager, Philip Marvin
Yamada, Ryuji
Yelin, Steven Joseph

Experimental Nuclear Physics
Adelberger, Eric George
Albergotti, Jesse Clifton
Andrews, Hugh Robert
Auchampaugh, George Fredrick
Bandel, Kenneth Charles
Beach, Louis Andrew
Beard, George B
Beard, Percy Morris, Jr
Bennett, Elbert White
Berners, Edgar Davis
Blair, John Morris
Blue, Richard Arthur
Bollinger, Lowell Moyer
Bond, Peter Danford
Browne, John C
Burger, John Martin
Burleson, George Robert
Burton, John Williams
Caldwell, John Thomas
Camp, David Conrad
Carter, H Kennon
Carter, Hubert Kennon
Chander, Jagdish
Chapman, Kenneth Reginald
Clegg, Thomas Boykin
Cline, James E
Conzett, Homer Eugene
Cook, Charles William
Cue, Nelson
Daehnick, Wilfried A W
Dayton, Irving Eugene
Duray, John R
Egan, James Joseph
Elwyn, Alexander Joseph
Ferguson, Stephen Mason
Foster, Bruce Parks
Freedman, Melvin Slein
Friesel, Dennis Lane
Garrett, Jerry Dale
Gehrke, Robert James
Glascock, Michael Dean
Gold, Raymond
Goss, John Douglas
Graves, Robert Gage
Gray, Edward Ray
Green, Lawrence
Greiner, Douglas Earl

Hamilton, Joseph H, Jr
Handler, Harry Elias
Hanson, Roger James
Hayward, Thomas Doyle
Helmer, Richard Guy
Herzenberg, Caroline Littlejohn
Holbrow, Charles H
Holland, Charles Emmett
Jain, Mahavir
Jensen, Gary Lee
Jha, Shacheenatha
Jones, William Philip
Kashy, Edwin
Kelly, William H
Keyworth, George A, II
Knapp, Myron William
Krane, Kenneth Saul
Lamaze, George Paul
Laszewski, Ronald M
Lee, David Mallin
Levin, Walter H G
Liebenauer, Paul (Henry)
Lone, Muhammad Aslam
Magisch, Bogdan
Major, John Keene
Mann, Lloyd Godfrey
Mark, Shew-Kuey
Maute, Robert Edgar
May, John Thomas
McNaughton, Michael Walford
Medsker, Larry Robert
Meunier, Jean-Louis
Miller, Daniel Weber
Miller, Walter Charles
Moss, Calvin E
Murdoch, Bruce Thomas
Nathan, Alan Marc
Nerbun, Robert Charles, Jr
Nicholson, Nicholas
O'Fallon, Nancy McCumber
Parker, Vincent Eveland
Parks, Paul Blair
Peek, Neal Frazier
Penner, Samuel
Pinkerton, John Edward
Pitts, Thomas Griffin
Pleasonton, Frances
Putnam, Thomas Milton, Jr
Quinton, Arthur Robert
Ramsayya, Akunuri V
Rapaport, Jacobo
Riley, Peter Julian
Roalsvig, Jan Per
Roberts, James Herbert
Saunders, Bernard Gray
Scott, Hugh Logan, III
SieRKen, Hugh Edward
Simmons, James E
Smither, Robert Karl
Sober, Daniel Isaac
Stinson, Glen Monette
Stoler, Paul
Summers-Gill, Robert George
Sunier, Jules Willy
Tesmer, Joseph Ransdell
Thomas, George E
Thornton, Stephen Thomas
Tollefsrud, Philip Bjorn
Valkovic, Vlado
Vlieks, Arnold Evald
Ward, David
Warner, Ray Allen
Wells, John Calhoun, Jr
Wetzel, Karl Joseph
Wharton, William Raymond
Whetstone, Stanley L, Jr
Whitlock, Lapsley Craig
Witten, Thomas Riner
Young, Lloyd Martin

Experimental Solid State Physics
Abkowitz, Martin Arnold
Akselrad, Aline
Andrew, James F
Ayers, Raymond Dean
Bauer, Robert Steven
Biegelsen, David K
Birchak, James Robert
Birkhoff, Robert D
Blankenship, James Lynn
Boolchand, Punit
Bottom, Virgil Eldon
Boyle, David Joseph
Braunlich, Peter Fritz
Bright, Arthur Aaron
Budzinski, Walter Valerian
Cappelletti, Ronald Louis
Carver, Gary Paul
Cheng, Li-Jen
Chesser, Nancy Jean
Clark, William Gilbert
Cleland, John W
Collings, Edward William
Colwell, Jack Harold
Colwell, Priscilla J
Connelly, John Joseph, Jr
Current, David Harlan
Daniels, William Burton

Deaver, Bascom Sine, Jr
De Mayo, Benjamin
Domb, Ellen Ruth (Colmer)
Dunifer, Gerald Leroy
Dunlap, Bobby David
Engstrom, Herbert Leonard
Fawcett, Eric
Fleming, Phyllis Jane
Friedberg, Charles Bruce
Garland, James C
Gathers, George Roger
Gavenda, John David
Glicksman, Maurice
Golding, Brage, Jr
Gonano, John Roland
Gottlieb, Albert Maxwell
Griffiths, Clifford H
Gustafson, Daniel Ray
Gyorgy, Ernst Michael
Haering, Rudolph Roland
Halpern, Teodoro
Hamburger, Paul David
Hanson, Roland Clements
Huang, Fan-Hsiung Frank
Hudgins, Aubrey C, Jr
Isett, Lawrence C
Jackel, Lawrence David
Jaklevic, Robert C
Kalma, Arne Haerter
Keller, Frederick Jacob
Kepler, Raymond Glen
Kercher, Harold Richard
Khan, Sharif Ahmad
Krebs, James John
Kroger, Harry
Kyser, David Sheldon
Madaci, David Peter
Magerlein, John Harold
Marquardt, Charles Lawrence
Martin, Martin Claude
Mazumdar, Purabi
McBride, Duncan Eldridge
McColl, James Renfrew
McGervey, John Donald
McNiff, Edward J, Jr
Meyer, Stephen Frederick
Miller, William
Morris, Robert Carter
Moulton, Grace Charbonnet
Mozzi, Robert Lewis
Murphy, James Edward
Narayanamurti, Venkatesh
O'Neal, Thomas Norman
Pandey, Surendra, Nath
Payne, James Edward
Philipp, Herbert Reynold
Possin, George Edward
Reed, William Alfred
Rehn, Victor Leonard
Reichert, Jonathon F
Renzema, Theodore Samuel
Risko, John James
Roger, William Alexander
Rogers, James Stewart
Rollins, Roger William
Roper, John Gordon
Rosenblum, Stephen Saul
Rowe, John Edward
Schein, Lawrence Brian
Schumacher, John Christian
Shea, Michael Joseph
Sidles, Paul Howard
Sill, Larry R
Smith, Donald Leonard
Smith, Frederick William
Smith, Robert Owens
Stombler, Milton Philip
Straus, Jozef
Stroud, Jackson Swaverly
Sturge, Michael Dudley
Swanson, Max Lynn
Swenson, Clayton Albert
Tantraporn, Wirojana
Thomas, Gordon Albert
Thorsen, Arthur C
Thygesen, Kenneth Helmer
Tomasch, Walter J
Touger, Jerold Steven
Wagner, David Kendall
Walker, Michael Stephen
Wallace, John Longstreet
Weichman, Frank Ludwig
Weiss, Jonathan David
Wilson, Ronald Harvey
Wohlgemuth, John Harold
Wood, Charles
Yelon, William B
Yu, Peter Yound
Zumsteg, Fredrick C, Jr

Nuclear Physics
Abashian, Alexander
Adair, Robert Kemp

Adams, Harry
Adams, Jerry L
Adelberger, Rexford E
Adkins, Rutherford Hamlet
Agarwal, Som Prakash
Agin, Gary Paul
Aitken, James Henry
Aizenberg-Selove, Fay
Albert, Richard David
Albridge, Royal Gene
Alcock, Norman Zinkan
Alexander, Peter
Alexander, Thomas Kennedy
Alexeff, Igor
Alford, William Lumpkin
Alias, Richard G
Allen, James Sircom
Allen, Joseph Percival
Alonso, Carol Travis
Alonso, Marcelo
Alvager, Torsten Karl Erik
Anderson, Bryon Don
Anderson, John D
Armstrong, Dale Dean
Arthur, Robert George
Arthur, Wallace
Artna-Cohen, Agda
Arya, Atam Parkash
Auerbach, Elliot H
August, Leon Stanley
Aull, Luther Bachman, III
Austin, Samuel Manly
Avignone, Frank Titus, III
Awschalom, Miguel
Axen, David
Bach, David Rudolph
Bachor, Andrew Dow
Baer, Helmut W
Baer, William
Baggett, Lester Marchant
Bainbridge, Kenneth Tompkins
Bair, Joe Keagy
Baker, David Thomas
Baker, Francis Todd
Baker, Paul, Jr
Baker, Robert G
Bakker, Stephen Denio
Bakhru, Hassaram
Balamuth, David
Ball, William Paul
Baltay, Charles
Barber, Walter Carlisle
Bardin, Russell Keith
Bardin, Tsing Tchao
Barnes, Charles Andrew
Barnes, Peter David
Barnothy, Jeno Michael
Barschall, Henry Herman
Bartholomew, Gilbert Alfred
Bartnoff, Shepard
Barton, Richard Donald
Baskin, Stanley
Bassel, Robert Harold
Batay-Csorba, Peter Andrew
Bauman, Norman Paul
Bayman, Benjamin
Beach, Eugene Huff
Beach, Paul L
Becker, John Angus
Becker, Lawrence Charles
Beer, George Atherley
Beghian, Leon E
Bell, Robert Edward
Bemis, Curtis Elliot, Jr
Bendel, Warren Lee
Bender, Roger Stillman
Benenson, Raymond Elliott
Benenson, Walter
Benjamin, Richard Walter
Bennett, Gary Lee
Bennett, Michael J
Bentley, Richard Foster
Benveniste, Jacob
Beres, William Philip
Bergen, Delmar Wesley
Berman, Barry L
Bernabei, Austin M
Bernan, Davy Lee
Bernstein, Eugene Merle
Bertsch, George Frederick
Bevington, Philip Raymond
Beyer, Jan Edgar
Beyer, Louis Martin
Beyster, John R
Bhaduri, Prem Datta
Bharadwaj, Prem Datta
Bhat, Mulki Radhakrishna
Bigham, Clifford Bruce
Bigler, Rodney Errol
Bilaniuk, Olexa-Myron
Bingham, Carol R
Bingham, Robert Lodewijk
Bittner, John William
Blanchard, Converse Herrick
Blatchley, Donald E
Blatt, S Leslie

Blecher, Marvin
Bleuler, Ernst
Block, Robert Charles
Bloom, Stewart Dave
Blosser, Henry Gabriel
Bock, Charles Walter
Bockelman, Charles Kincaid
Bodfish, Ralph E
Boehm, Felix H
Bohannon, George Edmond
Bolen, Lee Napier, Jr
Bolotin, Herbert Howard
Bond, Albert Haskell, Jr
Bondelid, Rollon Oscar
Borchers, Robert Reece
Bowman, Charles D
Bowman, James David
Bowsher, Harry Fred
Boyd, Herman Wayne
Boyd, James Emory
Boye, Robert James
Braden, Charles Hosea
Bradford, John Norman
Brady, Franklin Paul
Braid, Thomas Hamilton
Bramblett, Richard Lee
Brannen, Eric
Brantley, William Henry
Bretscher, Manuel Martin
Brient, Charles E
Britt, Harold Curran
Brockmeier, Richard Taber
Bromley, David Allan
Bronson, Jeff Donaldson
Brown, Hugh Needham
Brown, James Roy
Brown, Louis
Brown, Peter Frank
Browne, Cornelius Payne
Brugger, Robert Melvin
Brunson, Glenn Samuel, Jr
Brynjolfsson, Ari
Buccino, Salvatore George
Buford, William Holmes, Jr
Cahill, Thomas A
Calarco, John Richard
Cameron, John Alexander
Canada, Robert
Canavan, Frederick Louis
Canfield, Eugene H
Cardman, Lawrence Santo
Carlock, Henry Arthur
Carlson, Richard Raymond
Carpenter, Raymond T
Carter, Robert Sague
Casten, Richard Francis
Celitans, Gerard John
Chagnon, Paul Robert
Chalmers, Robert Anton
Chambers, William Hyland
Chasman, Chellis
Chertok, Benson T
Choudhury, Deo C
Choudry, Amar
Chrien, Robert Edward
Church, Eugene Lent
Cipolla, Sam J
Cladis, John Baros
Clark, David Delano
Clarke, John F
Clarke, Robert Lee
Class, Calvin Miller
Clator, Irvin Garrett
Clayton, Donald Delbert
Cleland, Marshall Robert
Cleveland, Bruce Taylor
Cline, Douglas
Close, Donald Alan
Coats, Richard Lee
Cobb, Grover Cleveland, Jr
Cochran, Donald Roy Francis
Cohen, Bernard Leonard
Cohen, Karl (Paley)
Cohen, Leonard David
Cohen, Leslie
Coleman, Ernest
Collins, Warren Eugene
Comfort, Joseph Robert
Congel, Frank Joseph
Conner, Jerry Power
Connolly, Thomas Worthington
Connor, Robert Dickson
Cook, Charles Falk
Cooper, Benjamin Stubbs
Cooper, Larry Russell
Cooperman, Edward Lee
Cope, David Franklin
Corman, Emmett Gary
Corson, Dale Raymond
Cosper, Sammie Wayne
Costrell, Louis

Cothern, Charles Richard
Couch, Jack Gary
Couchell, Gus Perry
Craddock, Michael Kevin
Craig, Donald Spence
Cramer, James D
Cramer, John Gleason
Crandall, Walter Ellis
Crannell, Hall L
Crawley, Gerard Marcus
Creager, Charles Bicknell
Creutz, Edward (Chester)
Crew, John Edwin
Cronin, James Watson
Cross, William Gunn
Crowley, Patrick Arthur
Cujec, Bibijana Dobovisek
Cullen, Dermott Edward
Dabbs, John Wilson Thomas
Dahl, Per Fridtjof
Dally, Edgar B
D'Angelo, Nicola
Dangle, Richard L
Danos, Michael
Darden, Colgate W, III
Darden, Sperry Eugene
Dardis, John G
Davids, Cary Nathan
Davidson, Charles Nelson
Davidson, Melvin G
Davies, Kenneth Thomas Reed
Davis, Harold Larue
Davis, Jay C
Davis, Monte V
Davis, Robert Houser
Davis, Ronald Stuart
Davisson, Charlotte Meaker
Dawson, Wilfred Kenneth
Day, Benjamin Downing
Debevec, Paul Timothy
Dehnhard, Dietrich
Del Bianco, Walter
Depommier, Pierre Henri Maurice
der Mateosian, Edward
De Saussure, Gerard
Dessauer, Gerhard
de Takacsy, Nicholas Benedict
Deutsch, Daniel Harold
Devins, Delbert Wayne
DeVries, Ralph Milton
DeWire, John William
Dick, Charles Edward
Dickerman, Charles Edward
Dickens, Justin Kirk
Dietrich, Frank S
Dietz, George Robert
Dillman, Lowell Thomas
Divadeenam, Mundrathi
Diven, Benjamin Clinton
Dixon, Dwight R
Dobson, Dwight A
Dolan, Kenneth William
Dollhopf, William Edward
Donahue, Douglas James
Donaldson, John Riley
Donhowe, John M
Donoghue, Timothy R
Donovan, John Leo
Dooley, John Raymond, Jr
Dorenbusch, William Edwin
Doub, William Blake
Dowdy, Edward Joseph
Drake, Darrell Melvin
Draper, James Edward
DuBard, James Leroy
Duck, Ian Morley
Dudziak, Donald John
Dudziak, Walter Francis
Duke, Charles Lewis
Duller, Nelson M, Jr
Dunlop, William Henry
Dunnam, Francis Eugene
Durham, Frank Edington
Dvorak, Henry Rudolph
Dworzecka, Maria
Dyer, Frank Falkoner
Dyer, John Norvell
Earle, Eric Davis
Easterday, Harry Tyson
Eby, Frank Shilling
Eccles, Samuel Franklin
Eck, John Stark
Eddy, Jerry Kenneth
Eddy, Nelson Wallace
Edge, Ronald (Dovaston)
Eidson, William Whelan
Eisberg, Robert Martin
Eisenhauer, Charles Martin
Eisenstein, Robert Alan
Ellis, Yurdanur Akovali
Ely, Ralph Lawrence, Jr
Enge, Harald Anton
Engelder, Theodore Carl
Engelke, Charles Edward
Epp, Chirold Delain
Eppling, Frederic John
Erskine, John Robert
Evans, Albert Edwin, Jr

Everling, Friedrich Gustav
Ewan, George T
Ewbank, Wesley Bruce
Ezell, Ronnie Lee
Fagg, Lawrence Wellburn
Falk, Willie Robert
Famularo, Kendall Ferris
Fann, Huoo-Long
Farrell, John A
Faulkner, John Edward
Favale, Anthony John
Feingold, Arnold Moses
Feldman, Lawrence
Feldmeier, Joseph Robert
Ferguson, James Malcolm
Ferry, James A
Fessenden, Peter
Fetzer, Homer D
Fielder, Douglas Stratton
Fields, Reuben Elbert
Finkler, Paul
Finlay, Roger W
Firk, Frank William Kenneth
Fischer, David Lloyd
Fisher, Thornton Roberts
Fivozinsky, Sherman Paul
Flaum, Charles
Fletcher, Neil Russel
Flinner, Jack L
Fluharty, Rex Gilbert
Folk, Robert Thomas
Forstner, James Lee
Fortune, H Terry
Foster, Duncan Graham, Jr
Fox, John David
Fox, John Gaston
Fregeau, Jerome Heyde
Fricke, Edwin Francis
Fricke, Martin Paul
Friedes, Joseph Leonard
Friedman, Jack P
Friesen, Earl Wayne
Frissel, Harry Frederick
Froman, Darol Kenneth
Fulbright, Harry Wilks
Fuller, Everett Gladding
Funk, Emerson Gornflow, Jr
Furr, Aaron Keith
Gabbard, Fletcher
Gaedke, Rudolph Meggs
Gale, Douglas Shannon, II
Galloway, Louie A, III
Galonsky, Aaron Irving
Garber, Donald I
Garg, Jagadish Behari
Garrett, Richard Edward
Garvey, Gerald Thomas
Geiger, Klaus Wilhelm
Genevese, Frank
George, Ted Mason
Georgopulos, Peter Demetrios
Gerhart, James Basil
Giamati, Charles C, Jr
Gibbons, John Howard
Gibson, Benjamin Franklin, V
Gibson, Edward F
Gilbert, Francis Charles
Gillespie, Claude Milton
Givens, Wyatt Wendell
Glasgow, Dale William
Glashausser, Charles Michael
Glass, James Clifford
Glass, Neel Warren
Glaubman, Michael Juda
Glickstein, Stanley S
Goldberg, Eugene
Goldhaber, Gerson
Goldhammer, Paul
Goldman, David Tobias
Goldstein, Norman Phillip
Goloskie, Raymond
Goode, Philip Ranson
Goodjohn, Albert J
Goodman, Albert
Goodman, Clark (Drouillard)
Goodman, Leonard Seymour
Goodwin, Lester Kepner
Gore, Bryan Frank
Gorenstein, Paul
Goss, David
Gossett, Charles Robert
Gottschalk, Charles Max
Goulard, Bernard
Gove, Norwood Babcock
Graber, Harlan Duane
Grabowski, Zbigniew Wojciech
Graetzer, Reinhard
Graves, Glen Atkins
Gray, Walter Steven
Green, Ralph Ellis
Greenwood, Reginald Charles
Griffioen, Roger Duane
Griffiths, George Motley
Grinoch, Paul
Grodzins, Lee
Groseclose, Byron Clark
Grover, George Maurice

Guttmann, Mark
Guy, Reed Augustus
Haag, James Norman
Haas, Francis Xavier, Jr
Haddad, Eugene
Haddock, Roy P
Haeberli, Willy
Hafele, Joseph Carl
Hafemeister, David Walter
Haffner, James Wilson
Hafner, Everett M
Hagee, George Richard
Halbert, Melvyn Leonard
Hall, Hugh Edward, Jr
Hall, James Edison
Hanna, Geoffrey Chalmers
Hanna, Stanley Sweet
Hardekopf, Robert Allen
Hardie, Gerald
Hardy, John Christopher
Hardy, Judson, Jr
Harmon, G Lamar
Harms, Edward Albert
Harris, Gale Ion
Harris, William Macy
Harrison, George H
Harvey, Bernard George
Harvey, John Arthur
Hasan, Mazhar
Haskins, Joseph Richard
Hatch, Eastman Nibley
Hausman, Hershel J
Hausser, Otto Friedrich
Havens, William Westerfield, Jr
Hayward, Raymond Webster
Heath, Russell La Verne
Heckman, Harry Hughes
Hein, Warren Walter
Heindl, Clifford Joseph
Heineman, Robert Edwin
Hellens, Robert Linton
Heller, Leon
Helmick, Herbert Hanna
Hemmendinger, Arthur
Hemsky, Joseph William
Henderson, Billy Joe
Henkel, Richard Luther
Herling, Gary Herbert
Hess, Charles Thomas
Hetrick, David Leroy
Hiebert, John Covell
Hill, David Lawrence
Hill, Lemmuel Leroy
Hill, Max W
Hill, Richard William
Hinderliter, Hilton Fay
Hintz, Norton Mark
Hoffman, Gerald Wayne
Hogg, Benjamin Gregory
Holl, Richard Jacob
Hollandsworth, Clinton E
Holmgren, Harry D
Holt, Roy James
Holzer, Alfred
Hopkins, John Chapman
Hopkins, John Isaac
Hoppes, Dale Du Bois
Horen, Daniel
Hornyak, William Frank
Horoshko, Roger N
Horsley, Robert James
Hotz, David Franklin
Hotz, Henry Palmer
Howell, John Foss
Hu, Chi-Yu
Huddleston, Charles Martin
Hudson, Alvin Maynard
Hudspeth, Emmett Leroy
Hull, McAllister Hobart, Jr
Hungerford, Ed Vernon, III
Hunter, Jack Allen
Hunter, Raymond Eugene
Hurst, Robert Rowe
Hurwitz, Henry, Jr
Huston, Norman Earl
Hutchinson, Donald Raymond
Ijaz, Mujaddid A
Imhof, William Lowell
Ingram, Forrest Duane
Inman, Fred Winston
Irfan, Muhammad
Jackson, Harold E, Jr
Jackson, Herbert Lewis
Jackson, William Roy, Jr
Jacobs, Clarence Gilbert, Jr
Jacobson, Larry A
Jakobson, Mark John
Jalbert, Jeffrey Scott
Jankus, Vytautus Zachary
Jarmie, Nelson
Jenkins, George Lovell
Jennings, Burridge
Jensen, Erling N
Jeong, Tung Hon
Jobst, Joel Edward
John, Joseph
Johns, Martin Wesley

Johnson, Arthur Clark
Johnson, Cleland Howard
Johnson, Frederick Allan
Johnson, John Alan
Johnson, Orland Eugene
Johnson, Ronald Gordon
Johnston, A Sidney
Jones, Garth
Jordan, Willard Clayton
Jovanovich, Jovan Vojislav
Joyce, James Martin
Joyner, Weyland Thomas, Jr
Julian, Glenn Marcenia
Kalbach, Constance
Kane, John Robert
Kane, Walter Reilly
Karr, Hugh James
Kavanagh, Ralph William
Kavanagh, Thomas Murray
Keaton, Posey W, Jr
Kegel, Gunter Heinrich Reinhard
Kepin, George Robert, Jr
Kepes, Joseph John
Kent, James Joseph
Kennett, Terence James
Kerlee, Donald D
Keller, Eldon Lewis
Keller, George Earl
Kelley, Raymond H
Kemper, Kirby Wayne
Kendziorski, Francis Richard
Kenealy, Patrick Francis
Kenefick, Robert Arthur
Kennedy, Edward Francis
Klingensmith, Raymond W
Knapp, David Edwin
Knox, William Jordan
Kocher, Charles William
Kocher, Dale Roland
Koehler, Helmut A
Kowalski, Stanley Benedict
Kowalski, Ludwik
Kozub, Raymond Lee
Kohler, Donald Alvin
Kohler, Sigurd H
Koller, Noemie
Kolstad, George Andrew
Koltun, Daniel S
Koshel, Richard Donald
Kostoff, Morris R
Kostroun, Vaclav O
Kouzes, Richard Thomas
Kovar, Frederick Richard
Krane, Ralph Werner
Kuehner, John Alan
Kull, Lorenz Anthony
Kunselman, Arthur Raymond
Kurey, Thomas John
Kraushaar, Jack Jourdan
Krick, Merlyn Stewart
Krisberg, Nathan Louis
Kinard, Frank Efird
King, James Douglas
Kistner, Otmar Casper
Kiena, Ernest Donald
Khanna, Faqir Chand
Kim, Hee Joong
Kim, Hogil
Lin, David Arthur
Lindenmeier, Charles William
Lindsay, James Gordon, Jr
Lindsay, Richard H
Lipkin, Harry Jeannot
Litherland, Albert Edward
Little, Robert Narvaez, Jr
Liverhant, Solomon Elieser
Livingston, Milton Stanley
Llewellyn, Ralph A
Lodhi, Mohammed Arfin Khan
Logan, Brian Anthony
Loewenstein, Walter B
Long, John Kelley
Loper, Gerald D
Love, William Gary
Loyd, David Heron
Lubitz, Cecil Robert
Ludemann, Carl Arnold
Ludin, Roger Louis
Ludington, Martin A
Ludwig, Edward James
Luther, Marvin L
Lutz, Arthur Leroy
Lutz, Harry Frank
Lycklama, Heinz
MacArthur, John Duncan
MacDonald, William
MacKinney, Arland Lee
Macklin, Richard Lawrence
Maclure, Kenneth Cecil
MacPherson, Herbert Grenfell
Madansky, Leon
Maddox, William Eugene
Madey, Richard
Maienschein, Fred (Conrad)
Makens, Royal Francis
Malanify, John Joseph
Malik, Fazley Bary
Mallary, Eugene Cobb
Mallay, James Francis
Malmberg, Philip Ray
Mancuso, Richard Vincent
Mangelson, Nolan Farrin
Manning, Armin William
Margaziotis, Demetrius John
Marino, Lawrence Louis
Marion, Jerry Baskerville
Marlow, Keith Winton
Marrus, Roscoe Earl
Marrus, Richard
Marsh, Bruce Burton
Marsh, David Paul
Marshak, Harvey
Marshalek, Eugene Richard
Martens, Edward John
Martin, Ashley Marvin, III
Martin, Murray John
Martin, Peter Wilson
Martin, Ronald Lavern
Martinelli, Ernest A
Martin, William Macphail
Mason, Grenville R
Mather, Keith Benson
Mathur, Suresh Chandra
Matthews, James Horace
Matthews, June Lorraine
Mauderli, Walter
Max, Claire Ellen
Mayers, Richard Ralph
McAdams, Robert Eli
McClelland, Clyde Lloyd
McClelland, Wilson Melville, Jr
McClure, Donald Allan
McCullen, John Douglas
McDaniels, David K
McDonald, Frank Alan
McDonald, W John
Meklin, Aram Zareh
Melgard, Rodney
Melkanoff, Michel Allan
Mendelson, Robert Alexander, Jr
Meneghetti, David
Menzel, Joerg H
Meriwether, John R
Meyerhof, Walter Ernst
Meyer-Schutzmeister, Luise
Michael, Irving
Mihelich, John William
Milani, Salvatore
Milburn, Richard Henry
Millburn, George P
Miller, Herman Lunden
Miller, Philip Dixon
Miller, Maurice Max
Miller, Robert Gail
Mills, Robert Laurence
Mills, William Raymond, Jr
Milne, Edmund Alexander
Milton, John Charles Douglas
Min, Kongki
Minehart, Ralph Conrad
Mischke, Richard E
Misra, Sushil
Missimer, John Hertel
Mitler, Henri Emmanuel
Mittelman, Philip Sidney
Mize, Jack Pitts
Mobley, Ralph Claude
Moldauer, Peter Arnold
Mollenauer, James Frederick
Monaro, Sergio
Monard, Joyce Anne
Moore, Benjamin LaBree
Moore, Cornelius Fred
Moore, Harold Arthur
Moore, Kenneth Virgil
Moore, Michael Stanley
Moore, Robert B
Mooring, Francis Paul
Morewitz, Harry Alan
Morgan, Ira Lon
Morgan, James Frederick
Morgan, John Robert, III
Morton, John Allen
Moss, Gerald Allen
Motz, Henry Thomas
Motz, Lloyd
Mozley, Robert Fred
Muether, Herbert Robert
Mughabghab, Said F
Muir, Douglas William
Mukerji, Ambuj
Mulligan, Bernard
Murphy, Elias Smith, Jr
Murray, Kenneth Malcolm, Jr
Murtaugh, Walter A
Mutchler, Gordon Sinclair
Nagatani, Kunio
Nagpal, Tarlok Singh
Nair, Kuttenair Gopinathan
Nall, Julian Clark
Naqvi, Saiyid Ishrat Husain
Nath, Ravinder Katyal
Neeson, John Francis
Neher, Leland K
Neiler, John Henry
Nelson, George Croyden
Nelson, Gerald Clifford
Newman, Eugene
Nichols, Davis Betz
Nicholson, Richard Benjamin
Nickles, Robert Jerome
Nickols, Norris Allan
Nigam, Bishan Perkash
Norbeck, Edwin, Jr
Nordheim, Lothar Wolfgang
Norman, Jay Harold
Northcliffe, Lee Conrad
Nutley, Hugh
Obenshain, Felix Edward
Ober, David Ray
Oberhofer, Edward Samuel
O'Dell, Jean Marland
Ogata, Hisashi
Ohlsen, Gerald G
Okrent, David
Oktay, Erol
Ollerhead, Robin Wemp
Olness, John William
Olsen, Robert James
Olsen, William Charles
Onega, Ronald Joseph
O'Neill, George Francis
Ostrander, Peter Erling
Otteson, Otto Harry
Ouseph, Pullom John
Overley, Jack Castle
Owen, George Ernest
Padgett, Doran William
Palmedo, Philip F
Palmieri, John Nicholas
Palms, John Michael
Park, Jae Young
Parker, Jack Lindsay
Parker, Peter Donald MacDougall
Parkinson, William Charles
Parsignault, Daniel Raymond
Patterson, James Reid
Paul, Peter
Pauls, Franklin Benjamin
Paxton, Hugh Campbell
Peacock, Charles LeRoy
Pearce, R Michael
Pearlstein, Sol
Peastie, David Chase
Peelle, Robert W
Pengra, James G
Penny, Keith
Pepper, Thomas Peter
Perdrisat, Charles F
Perkins, Roger Bruce
Perry, Robert Riley
Perry, Charles William
Peterson, Jack Milton
Peterson, Robert W
Peterson, Rolf Eugene
Peterson, Roy Jerome
Petrovich, Fred
Petry, Robert Franklin
Phillips, E Alan
Phillips, Gary Wilson
Phillips, James A
Phillips, Robert Hastings
Philpott, Richard John
Piccioni, Oreste
Piel, William Frederick
Pieper, George Francis, Jr
Pinajian, John Joseph
Pinkston, William Thomas
Pioway, James Mason
Pisano, Daniel Joseph, Jr
Place, Ralph L
Plasil, Franz
Plassmann, Elizabeth Hebb
Plassmann, Eugene Adolph
Piendl, Hans Siegfried
Ploughe, William D
Podgor, Samuel
Poirier, Charles Philip
Pollock, Robert Elwood
Polson, William Jerry
Pond, Thomas Alexander
Poore, Emery Ray Vaughn
Porges, Karl G
Porter, Leonard Edgar
Posner, Martin
Poth, James Edward
Powers, Darden
Pratt, William Winston
Predom, Barry Mason
Priest, Joseph Roger
Prior, Richard Marion
Prosser, Francis Ware, Jr
Prud'homme, John Thomas
Pryor, Richard J
Pugh, George Edgin
Pugh, Howel Griffith
Pullen, David John
Purohit, Surendra Nath
Purser, Fred O
Raboy, Sol
Raff, Samuel J
Ragan, Charles Ellis, III
Ragan, George Leslie
Raman, Subramanian
Ramavataram, Kilambi
Rana, Abdul R
Rao, Pemmaraju Venugopala
Rasera, Robert Louis
Ratcliff, Keith Frederick
Read, William George
Reber, Jerry D
Reich, Charles William
Reichelt, Walter Herbert
Reierson, James (Dutton)
Riesenfeld, Peter William
Rieser, Leonard M, Jr
Riggs, Roderick D
Rester, Alfred Carl, Jr
Rester, David Hampton
Reynolds, Bruce G
Reynolds, Glenn Myron
Rhodes, Jacob Lester
Rhodes, John Rathbone
Richards, Hugh Taylor
Richards, Walter Bruce
Richardson, Allan Charles Barbour
Richardson, John Reginald
Richert, Anton Stuart
Richert, John Lewis
Rickey, Frank Atkinson, Jr
Rickey, Martin Eugene
Riedinger, Leo Louis
Ritter, Alfred C, Jr
Ritter, James Carroll
Ritter, Rogers C
Robertson, Nathan Russell, Jr
Robertshaw, Joseph Earl
Robinson, Lyle Purmal
Robinson, Berol Lee
Robinson, Edward Lee
Robinson, Russell Lee
Robson, Donald
Robson, John Michael
Rochlin, Robert Sumner
Roderick, Hilliard
Rodney, William Stanley
Rodriguez-Fraga, Andres
Roll, John Donald
Rolland, William Woody
Ronningen, Reginald Martin
Roos, Charles Edwin
Roos, Philip G
Rosenbaum, David Mark
Rosenfeld, Arthur H
Rost, Ernest Stephan
Roush, Marvin Leroy
Roy, Donald H
Roy, Radha Raman
Ruby, Lawrence
Rupaal, Ajit S
Russell, John Lynn, Jr
Russell, Leonard Nelson
Rustgi, Moti Lal
Rutkowski, Robert William

Sachs, Martin William
Sailor, Vance Lewis
St Onge, Richard Norbert
Saladin, Jurg X
Salisbury, Stanley R
Salmi, Ernest William
Sample, John Thomas
Sampson, Thomas Edward
Sandmeier, Henry Armin
Sargent, Bernice Weldon
Sargent, Charles P
Sattler, Allan R
Sawyer, Earl Morrow
Sayer, Royce Orlando
Sayres, Alden R
Schaeffer, Norman Morris
Schamberger, Robert D
Schardt, Alois Wolfgang
Scharenberg, Rolf Paul
Schecter, Larry
Schenter, Robert Earl
Schermer, Robert Ira
Schiffer, John Paul
Schlueter, Donald Jerome
Schmitt, Harold William
Schneid, Edward Joseph
Schneider, Ronald E
Scholz, Wilfried
Schramm, David N
Schreiber, Raemer Edgar
Schrils, Rudolph
Schupp, Guy
Schwandt, Peter
Schwartz, Robert Bernard
Schwarzschild, Arthur Zeiger
Schweitzer, Jeffrey Stewart
Schwensfeir, Robert James, Jr
Scofield, Norman Edward
Scolman, Theodore Thomas
Scoville, John J
Seagrave, John Dorrington
Seamon, Robert Edward
Seeger, Philip Anthony
Seeser, James William
Seglie, Ernest Augustus
Segre, Emilio Gino
Sellers, Francis Bachman
Sen, Sunil Kumar
Serduke, Franklin James David
Seth, Kamal Kishore
Severiens, Johannes Coenraad
Seyler, Richard G
Shakin, Carl M
Shapiro, Mark Howard
Shapiro, Philip
Sharma, Ramesh C
Shelton, Frank Harvey
Shen, Benjamin Shih-Ping
Shera, E Brooks
Sherman, Norman K
Sherr, Rubby
Shin, Yong-Moo
Shon, Frederick John
Shore, Ferdinand John
Shugart, Cecil G
Shwe, Hla
Siems, Norman Edward
Sierk, Arnold John
Silverman, Albert
Silverstein, Edward Allen
Simms, Paul C
Simon, William Georee
Simpson, Wilburn Dwain
Singer, Solomon Elias
Singh, Prithe Paul
Singletary, John Boon
Singletary, Lillian Darlington
Slack, Lewis
Slee, Frederick Watford
Small, Timothy Michael
Smith, Alan B
Smith, Donald Larned
Smith, Gary Leroy
Smith, Hastings Alexander, Jr
Smith, Horace Vernon, Jr
Smith, Raymond Andrew
Souder, Wallace William
Southward, Harold Dean
Spejewski, Eugene Henry
Sperry, Willard Charles
Sprung, Donald Whitfield Loyal
Stallwood, Robert Antony
Stamatelatos, Michael G
Standing, Kenneth Graham
Steffen, Rolf Marcel
Steigert, Frederick Edward
Stein, Nelson
Stein, William Earl
Stelts, Marion Lee
Steuer, Malcolm F

Stolovy, Alexander
Stolzfus, Joseph Christian
Stork, Donald Harvey
Storrs, Charles Lysander
St-Pierre, Claude
Strong, Stanley Sterling
Strottman, Daniel
Struble, Gordon Lee
Sturm, William James
Sun, Chih-Ree
Sund, Raymond Earl
Sunyar, Andrew William
Suttle, Andrew Dillard, Jr
Swann, Charles Paul
Talbert, Willard Lindley, Jr
Talbott, Francis Leo
Tao, Shu-Jen
Tapphorn, Ralph M
Taras, Paul
Tatarczuk, Joseph Richard
Taylor, Harry William
Taylor, Morris Chapman
Temmer, Georges Maxime
Temperley, Judith Kantack
Tendam, Donald Jan
Terreault, Bernard J E J
Terrell, Glen Edward
Thaxton, George Donald
Thieberger, Peter
Thomas, Charles Carlisle, Jr
Thompson, Lewis Chisholm
Thwaites, Thomas Turville
Tickle, Robert Simpson
Tilley, David Ronald
Tipler, Paul A
Tittman, Jay
Tomblin, Fred Fitch
Tombrello, Thomas Anthony, Jr
Tomlinson, Everett Parsons
Tomonto, James R
Toops, Edward Chassell
Toth, Kenneth Stephen
Tracy, James Frueh
Trail, Carroll C
Treado, Paul A
Tribble, Robert Edmond
Tripard, Gerald Edward
Tucker, Allen Brink
Tucker, William Eric
Tulenko, James Stanley
Ullman, Jack Donald
Ungrin, James
Updegraff, William Edward
Urone, Paul Peter
Valente, Frank Anthony
van Oers, Willem Theodorus Hendricus
Van Patter, Douglas Macpherson
Van Putten, James D, Jr
Veeser, Lynn Raymond
Vegors, Stanley H, Jr
Veit, Jiri Joseph
Ventrice, Carl Alfred
Verbinski, Victor V
Vincent, Lloyd Drexell
Vogel, Peter
Vogt, Erich W
Volkov, Anatole Boris
Vourvopoulos, George
Wack, Paul Edward
Waggoner, James Arthur
Waggoner, Margaret Ann
Wagner, Robert Thomas
Wahl, John Schempp
Waldman, Bernard
Walker, J Calvin
Walker, William Delany
Walkiewicz, Thomas Adam
Wall, Nathan Sanders
Walter, Richard L
Waltner, Arthur
Walton, Roddy Burke
Wang, Ken Hsi
Warburton, Ernest Keeling
Warner, Laurance Bliss
Wasson, Oren A
Watt, Bob E
Weber, Alfred Herman
Wegner, Harvey E
Weil, Jesse Leo
Weinberg, Alvin Martin
Weitkamp, William George
Weller, Henry Richard
Wells, Donald O
Welt, Martin A
Werthein, Robert Halley
West, Harry Irwin, Jr
Whaling, Ward
Wheelwright, Earl J
White, Donald Harvey
Whitehead, Marian Nedra
Whitehead, Walter Dexter, Jr
Whitten, Charles A, Jr
Wicker, Everett E
Wiedenbeck, Marcellus Lee
Willard, Harvey Bradford
Williams, Donald J
Williamson, Claude F
Willis, William W

Willmes, Henry
Wills, John G
Wilson, Robert Gray
Wilson, Stephen James
Winsberg, Lester
Winter, Rolf Gerhard
Winters, Ronald Ross
Woldseth, Rolf
Wolfe, Bertram
Wolfe, Paul Jay
Wolfson, Joseph Laurence
Wolicki, Eligius Anthony
Wood, Calvin Dale
Wood, Donald Eugene
Wood, Galen Theodore
Wood, Richard Ellet
Wood, Robert Manning
Wood, Saralue
Woodward, William Mooney
Worsham, Herbert J, Jr
Worth, Donald Calhoun
Wright, Byron Terry
Wright, Kenneth Arthur
Wright, Louis Edgar
Wyckoff, James M
Wyly, Lemuel David, Jr
Yang, Chia Hsiung
Yasko, Richard N
Yearian, Mason Russell
Yergin, Paul Flohr
Yip, Patrick Cheung-Yum
Yoshida, Shiro
Young, Frank Coleman
Young, Gale
Young, James Edward
Young, Phillip Gaffney
Young, Thomas Edward
Youtz, Byron Leroy
Yu, David U L
Yule, Thomas J
Zaffarano, Daniel Joseph
Zafiratos, Chris Dan
Zaidi, Syed Amir Ali
Zamick, Larry
Zander, Arlen Ray
Zatzick, Michael Raymond
Zeidman, Benjamin
Zender, Michael J
Zganjar, Edward F
Ziegler, James Francis
Zimmerman, Peter David
Zolnowski, Dennis Ronald
Zucker, Alexander
Zurmuhle, Robert W
Zweifel, Paul Frederick

Particle Physics

Akerlof, Carl W
Albright, Carl Howard
Anderson, Robert Leonard
Appel, Jeffrey Alan
Auvil, Paul R, Jr
Bader, Henri
Baker, Winslow Furber
Bardeen, William A
Barger, Vernon Duane
Bartlett, David Farnham
Bar-Yam, Zvi H
Basri, Saul Abraham
Behrends, Ralph Eugene
Bensinger, James Robert
Bernstein, Seymour
Birnbaum, David
Blackmon, Maurice Lee
Blaha, Stephen
Blecher, Marvin
Bloom, Elliott D
Bohannon, George Edmond
Booth, Norman E
Brabson, Bennet Bristol
Bridgewater, Albert Louis, Jr
Brolley, John Edward, Jr
Brown, Karl Leslie
Burnstein, Ray A
Burwell, James Robert
Bussian, Alfred Erich
Busza, Wit
Byers, Nina
Capri, Anton Zizi
Carey, David Crockett
Carlstone, Darry Scott
Carrigan, Richard Alfred, Jr
Chang, Shau-jin
Chang-Fang, Chuen-Chuen
Chen, James Ralph
Chertok, Benson T
Chiu, Charles Bin
Chou, Tsu-Teh
Chung, Kuk Pyo
Clark, Robert Beck
Cleland, Wilfred Earl
Cooper, Frederick Michael
Cork, Bruce
Coulson, Larry Vernon
Cox, Bradley Burton
Crater, Horace William
Csonka, Paul L

Cutts, David
Deshpande, Nilendra Ganesh
Dolen, Richard
Domokos, Gabor
Downs, Bertram Wilson, Jr
Eberhard, Philippe Henri
Edwards, Kenneth Westbrook
Einhorn, Martin B
Eisenstein, Robert Alan
Ellis, Stephen Dean
Epstein, Gary Martin
Erickson, Kenneth Neil
Everett, Allen Edward
Feld, Bernard Taub
Ferbel, Thomas
Finkler, Paul
Ford, Richard Lyle
Frauenfelder, Hans (Emil)
Frisken, William Ross
Gaisser, Thomas Korff
Galtieri, Angela Barbaro
Gasiorowicz, Stephen G
Glashow, Sheldon Lee
Glaubman, Michael Juda
Gleeson, Austin M
Goldberg, Howard S
Gollon, Peter J
Greenberg, Oscar Wallace
Greene, Arthur Franklin
Greider, Kenneth Randolph
Griffin, James Edward
Guertin, Ralph Francis
Hackman, Roger H
Hammer, Charles Lawrence
Han, Moo-Young
Harris, Benjamin C
Harrison, W Craig
Hartsough, Walter D
Haymaker, Richard Webb
Hendry, Archibald Wagstaff
Heusch, Clemens August
Hildebrand, Roger Henry
Hoffman, Cyrus Miller
Hones, Michael J
Hulsizer, Robert Inslee, Jr
Hutson, Richard Lee
Huwe, Darrell O
Ijaz, Mujaddid A
Innes, Walter Rundle
Internann, Gerald William
Jaffe, Robert Loren
James, Philip Benjamin
Jensen, Douglas Andrew
Johnson, Marvin Elroy
Jones, Lewis Hammond, IV
Jones, Lorella Margaret
Joseph, David Winram
Kalbfleisch, George Randolph
Kamal, Abdul Naim
Kanofsky, Alvin Sheldon
Kaufmann, William B
Kelly, Robert Lincoln
Kenney, Robert Warner
Key, Anthony W
Keyes, Gary Sylvester
Kimel, Jacob Daniel, Jr
Kinzer, Robert Lee
Kislinger, Mark Brecher
Kistiakowsky, Vera
Kovesi-Domokos, Susan
Kropp, William Rudolph, Jr
Lamb, Richard C
Lande, Kenneth
Lannutti, Joseph Edward
Leipuner, Lawrence Bernard
Lennox, Arlene Judith
Lipkin, Harry Jeannot
Lobkowicz, Frederick
Lowenthal, Dennis David
Luke, Robert A
Macek, Robert James
Mahanthappa, Kalyana T
Mandelkern, Mark Alan
Mansfield, John E
Mast, Terry Steven
McClellan, Gene Elvin
McDaniel, Paul William
McGuire, Austin Dole
Meisner, Gerald Warren
Melissinos, Adrian Constantin
Mellen, Walter Roy
Messner, Robert Lee
Miller, Akeley
Mischke, Richard E
Missimer, John Hertel
Mockett, Paul M
Moe, Michael K
Moneti, Giancarlo
Morrow, Richard Alexander
Murphy, Charles Thornton
Nauenberg, Michael
Nelson, Bruce Philip
Nezrick, Frank Albert
Nigam, Bishan Perkash
Oppo, Giuseppe
Orear, Jay
Osborne, Louis Shreve
Paige, Frank Eaton, Jr

Parke, William C
Peterson, Francis Carl
Picciotto, Charles Edward
Plicher, James Eric
Potter, Douglas Marion
Prepost, Richard
Purser, Fred O
Rarita, William Roland
Reines, Frederick
Repko, Wayne William
Resnick, Lazer
Richardson, Clarence Robert
Richert, Anton Stuart
Rittenberg, Alan
Robinson, Donald Keith
Rockower, Edward Brandt
Rosenberg, Eli Ira
Roth, Benjamin
Rothe, Kenneth Warren
Ruggiero, Alessandro Gabriele
Ryan, David George
Sakitt, Mark
Sakmar, Ismail Aydin
Sannes, Felix Rudolph
Savet, Paul H
Schluter, Robert Arvel
Schroeder, Lee S
Schumann, Thomas Gerald
Sciulli, Frank J
Sechi-Zorn, Bice
Segre, Emilio Gino
Selove, Walter
Shapiro, Gilbert
Shepard, Paul Fenton
Shochet, Melvyn Jay
Shpiz, Joseph M
Siegel, Robert Ted
Silverman, Dennis Joseph
Sobel, Henry Wayne
Steiner, Herbert M
Stevenson, Merion Lynn
Stoler, David
Stowe, Keith S
Strauch, Karl
Sulak, Lawrence Richard
Sundelin, Ronald M
Surko, Pamela Toni
Swallow, Earl Connor
Swanson, Robert Allan
Taylor, Snowden
Thompson, Julia Ann
Thorndike, Edward Harmon
Ting, Samuel C C
Tomljanovich, Nicholas Matthew
Toraskar, Jayashree Ravalnath
Trefil, James S
Trilling, George Henry
Trower, William Peter
Tsai, Yung Su
Tung, Wu-Ki
Uritam, Rein Aarne
Walker, James King
Wehmann, Alan Ahlers
Weil, Francis Alphonse
Wenzel, William Alfred
Westgard, James Blake
White, Donald Harvey
Whitehead, Marian Nedra
Widgoff, Mildred
Willmann, Robert B
Wilson, John Edward
Winston, Roland
Witherell, Michael Stewart
Wolf, Albert Allen
Wright, Jon Alan
Yamin, Samuel Peter
Yeh, Noel Kuei-Eng
Yoder, Levon Lee
Yoder, Neil Richard
Young, James Edward
Young, Pol-Shien
Yount, David Eugene
Yu, David U L
Zemach, Charles

Cosmic Ray Physics

Acosta, Virgilio
Adams, James Hall, Jr
Agarwal, Som Prakash
Ahluwalia, Harjit Singh
Albats, Paul
Allum, Frank Raymond
Annis, Martin
Arens, John Frederic
Balasubrahmanyan, Vriddhachalam K
Besancon, Robert Martin
Binns, Walter Robert
Bland, Clifford J
Blue, Michael Henry
Carmichael, Hugh
Duggal, Shakti Prakash
Fazio, Giovanni Gene
Fenyves, Ervin J
Gaisser, Thomas Korff
Green, Phillip Joseph, II
Groom, Donald Eugene
Hartman, Robert Charles
Hsieh, Ke Chiang

Iona, Mario
Jones, Frank Culver
Jones, William Vernon
Kalbach, Robert Michael
Kasha, Henry
Kasper, Joseph Emil
Kropp, William Rudolph, Jr
Lamb, Richard C
Lambe, Margaret B McClements
Lamport, James Everett
Laster, Howard Joseph
L'Heureux, Jacques (Jean)
Lingenfelter, Richard Emery
Mason, Grant William
McKibben, Robert Bruce
Mendell, Rosalind B
Merker, Milton
Meyer, Peter
Miller, Frank L
O'Donnell, Brian Desmond
Osborne, Weymar Zack
Palmeira, Ricardo Antonio Ribeiro
Parnell, Jack Lindsay
Parnell, Thomas Alfred
Randall, Charles Addison, Jr
Reines, Frederick
Shapiro, Maurice Mandel
Silberberg, Rein
Smoot, George Fitzgerald, III
Sobel, Henry Wayne
Staib, Jon Albert
Wilkes, Richard Jeffrey

Experimental Nuclear Physics

Adelberger, Eric George
Albergotti, Jesse Clifton
Andrews, Hugh Robert
Auchampaugh, George Fredrick
Bandtel, Kenneth Charles
Beach, Louis Andrew
Beard, George B
Beard, Percy Morris, Jr
Bendt, Philip Joseph
Bennett, Elbert White
Berners, Edgar Davis
Blair, John Morris
Blue, Richard Arthur
Bollinger, Lowell Moyer
Bond, Peter Danford
Browne, John C
Burger, John Martin
Burleson, George Robert
Burton, John Williams
Caldwell, John Thomas
Camp, David Conrad
Carter, H Kennon
Carter, Hubert Kennon
Chandler, Jagdish
Chapman, Kenneth Reginald
Clegg, Thomas Boykin
Cline, James E
Connors, Philip Irving
Conzett, Homer Eugene
Cook, Charles William
Cue, Nelson
Daehnick, Wilfried A W
Duray, John R
Egan, James Joseph
Elwyn, Alexander Joseph
Ferguson, Stephen Mason
Foster, Bruce Parks
Freedman, Melvin Stein
Friesel, Dennis Lane
Garrett, Jerry Dale
Gehrke, Robert James
Glascock, Michael Dean
Gold, Raymond
Goodman, Charles David
Goss, John Douglas
Graves, Robert Gage
Gray, Edward Ray
Green, Lawrence
Greiner, Douglas Earl
Hamilton, Joseph H, Jr
Handler, Harry Elias
Hanson, Alfred Olaf
Hanson, Roger James
Hayward, Thomas Doyle
Helmer, Richard Guy
Herzenberg, Caroline Littlejohn
Holbrow, Charles H
Holland, Robert Emmett
Jain, Mahavir
Jensen, Gary Lee
Jha, Shacheenatha
Jones, William Philip
Kashy, Edwin
Kelly, William H
Keyworth, George A, II
Knapp, Myron William
Krane, Kenneth Saul
Lamaze, George Paul
Laszewski, Ronald M
Lee, David Mallin
Lewin, Walter H G
Liebenauer, Paul (Henry)
Lone, Muhammad Aslam
Macek, Robert James

Maglich, Bogdan
Major, John Keene
Mann, Lloyd Godfrey
Mark, Shew-Kuey
Maute, Robert Edgar
May, Jack Thomas
McNaughton, Michael Walford
Medsker, Larry Robert
Meunier, Jean-Louis
Miller, Daniel Weber
Miller, Walter Charles
Moss, Calvin E
Murdoch, Bruce Thomas
Nathan, Alan Marc
Nerbun, Robert Charles, Jr
Nicholson, Nicholas
O'Fallon, Nancy McCumber
Parker, Vincent Eveland
Peck, Neal Frazier
Penner, Samuel
Pinkerton, John Edward
Pitts, Thomas Griffin
Pleasonton, Frances
Putnam, Thomas Milton, Jr
Quinton, Arthur Robert
Ramayya, Akunuri V
Rapaport, Jacobo
Riley, Peter Julian
Roalsvig, Jan Per
Roberts, James Herbert
Saunders, Bernard Gray
Scott, Hugh Logan, III
Siefken, Hugh Edward
Simmons, James E
Smither, Robert Karl
Sober, Daniel Isaac
Stinson, Glen Monette
Stoler, Paul
Summers-Gill, Robert George
Sunier, Jules Willy
Tesmer, Joseph Randell
Thomas, George E
Thornton, Stephen Thomas
Tollestrud, Philip Bjorn
Valkovic, Vlado
Vlieks, Arnold Evald
Ward, David
Warner, Ray Allen
Wells, John Calhoun, Jr
Wetzel, Karl Joseph
Wharton, William Raymond
Whetstone, Stanley L, Jr
Whitlock, Lapsley Craig
Witten, Thomas Riner
Young, Lloyd Martin

Magnetic Resonance

Anderson, Miles Edward
Ashkin, Julius
Blackstead, Howard Allan
Boatner, Lynn Allen
Brown, Ian McLaren
Buckmaster, Harvey Allen
Castle, John Granville, Jr
Clark, William Gilbert
Cleveland, Gregor George
Colpa, Johannes Pieter
Drumheller, John Earl
Gardner, John Hale
Goldman, Stephen Allen
Hampton, Don Allen
Harkavy, Allan Abraham
Hester, Richard Knight
Hsieh, Yu-Nian
Hubbard, Paul Stancyl, Jr
Huisjen, Martin Albert
Kemple, Marvin David
Kissinger, Paul Bertram
McCalley, Roderick Canfield
Miller, John Edward
Miner, George Kenneth
Nelson, William Henry
Raghunathan, Partha
Rollwitz, William Lloyd
Rubins, Roy Selwyn
Sayetta, Thomas C
Schreurs, Jan Willem Herman
Segelken, Warren George
Steiner, Pinckney Alston
Torrey, Henry Cutler
Trifunac, Alexander Dimitrije
Weidner, Richard Tilghman
Wolbarst, Anthony Brinton
Woonton, Garnet Alexander
Yannoni, Costantino Sheldon

Nuclear Chemistry

Ahmad, Irshad
Alexander, Edward Lawson
Alexander, John Macmillan, Jr
Almodovar, Ismael
Andre, Herman William
Apt, Kenneth E
Babitch, Joseph Aaron
Baedecker, Philip A
Bahn, Emil Lawrence, Jr
Ball, James Bryan
Ballou, Nathan Elmer

Batzel, Roger Elwood
Bayhurst, Barbara P
Benjamin, Philip Palamoottil
Bernthal, Frederick Michael
Berreth, Julius R
Bishop, Charles Joseph
Bishop, William P
Bizzell, Oscar McArthur
Blann, H Marshall
Borke, Mitchell Louis
Bremer, Daeg Scott
Brodzinski, Ronald Lee
Broom, Knox McLeod, Jr
Brownlee, James Lawton, Jr
Butler, Gilbert W
Carr, Robert Joseph
Caretto, Albert A, Jr
Cefola, Michael
Cerny, Joseph, III
Chu, Yung Yee
Chulick, Eugene Thomas
Church, Larry B
Colvin, George Hunt
Colvin, Curtis A
Croft, Paul Douglas
Cumming, James Burton
Daly, Patrick Joseph
D'Auria, John Michael
Debiak, Ted Walker
Diamond, Herbert
Diamond, Richard Martin
Douthett, Elwood Moser
Dropesky, Bruce Joseph
Dudey, Norman
Eastwood, Thomas Alexander
Eggers, Donald C
Eichler, Eugene
English, Gerald Alan
Erdal, Bruce Robert
Faler, Kenneth Turner
Fasching, James Le Roy
Feng, Paul Yen-Hsiung
Ferguson, Robert Lynn
Filby, Royston Herbert
Fink, Richard Walter
Finston, Harmon Leo
Fleming, Edward Homer, Jr
Friedlander, Gerhart
Friedman, Abraham Solomon
Fritsch, Arnold Rudolph
Gardner, Donald Glenn
Gatrousis, Christopher
George, Raymond S
Gilmore, John T
Gindler, James Edward
Glendenin, Lawrence Elgin
Godbey, William Givens
Gordon, Barry Maxwell
Grant, Patrick Michael
Gresky, Alan Tolstoy
Griffin, Henry Claude
Grisham, Genevieve Dwyer
Hahn, Richard Leonard
Halperin, Joseph
Harbottle, Garman
Harbour, Robert Myron
Harlan, Ronald A
Harmer, Don Studer
Hastings, James Donald
Haustein, Peter Eugene
Heberlein, Douglas Garavel
Hensley, Walter King
Herzog, Gregory F
Hillman, Manny
Hochel, Robert Charles
Hoff, Richard William
Hofstetter, Kenneth John
Hoffman, Darleane Christian
Hogan, James Joseph
Hollander, Jack Marvin
Hontzeas, S
Hopke, Philip Karl
Horrocks, Donald Leonard
Houck, Frank Scanland
Hower, Charles Oliver
Hudis, Jerome
Huizenga, John Robert
Hulet, Ervin Kenneth
Hummel, John Philip
Hunter, Edwin Thomas
Husain, Liaquat
Hyde, Earl K
Hyder, Monte Lee
Ide, Roger Henry
Jackson, Sydney Vern
Jaffe, Harold
Jaffey, Arthur Harold
Johnson, Noah R
Kaplan, Morton
Kardonsky, Stanley
Karol, Paul Jason
Katcoff, Seymour
Kaufman, Sheldon Bernard
Keenan, Joseph Aloysius
Keith, James Ennis
Kenna, Bernard Thomas
Kiefer, Richard L

Kiley, Leo Austin
Kinderman, Edwin Max
Klobuchar, Richard Louis
Kohman, Truman Paul
Korteling, Ralph Garret
Krohn, Kenneth Albert
Ku, Thomas Hsiu-Heng
Lee, Floyd Denman
Lessler, Richard Marshall
Levin, Robert Warren
Levy, Harris Benjamin
Lindemer, Terrence Bradford
Lindner, Manfred
Lisman, Frederick Louis
Loveland, Walter (David)
Lux, Carl Ray
Macias, Edward S
Mahadeviah, Inally
Mahony, John Daniel
Manning, Winston Marvel
Manuel, Oliver K
Markowitz, Samuel Solomon
Martin, Gerald Charles, Jr
Mausner, Leonard Franklin
McDaniel, Thomas Lee
McHarris, William Charles
McHugh, James Anthony, Jr
Menning, Clarence
Menon, Manchery Prabhakara
Metzger, Albert E
Meyer, Richard Adlin
Miller, Dudley Grant
Milsted, John
Molinski, Victor Joseph
Morrison, David Lee
Morrow, Richard Joseph
Muga, Marvin Luis
Namboodiri, Madassery Neelakantan
Nass, Harold William
Natowitz, Joseph Bernard
Nervik, Walter Edward
Nethaway, David Robert
Niday, James Barker
Norman, Jack C
Norris, Andrew Edward
O'Brien, Harold Aloysious, Jr
Ogard, Allen E
O'Kelley, Grover Davis
Oliver, James Russell
Orr, William Campbell
Orth, Charles Joseph
Pacer, John Charles
Parikh, Sarvabhaum Sohanlal
Passell, Thomas Oliver
Payton, Patrick Herbert
Perkins, William Clopton
Perkons, Auseklis Karlis
Perlman, Isadore
Preiss, Ivor Louis
Perry, Dennis Gordon
Person, Lucy Wu
Pickering, Miles Gilbert
Pierce, Elliot Stearns
Pierce, Timothy Ellis
Pillay, K K Sivasankara
Poggenburg, John Kenneth, Jr
Porile, Norbert Thomas
Poskanzer, Arthur M
Powers, James Allen
Prohaska, Charles Anton
Prussin, Stanley Gerald
Rasmussen, John Oscar, Jr
Rayudu, Garimella V S
Reavis, James Gene
Reed, George W, Jr
Reed, Mary Frances
Reeder, Paul Lorenz
Reedy, Robert Challenger
Remsberg, Louis Philip, Jr
Rengan, Krishnaswamy
Reuland, Donald John
Rightmire, Robert
Roesmer, Josef
Rona, Elizabeth
Root, John Walter
Rubinson, William
Ruddy, Francis Henry
Ruiz, Carl P
Runnalls, Nelva Earline Gross
Russell, Irving James
Ryan, Victor Albert
Sabu, Dwarka Das
Santim, D C
Sarantites, Demetrios George
Schell, William R
Schuman, Robert Paul
Seaborg, Glenn Theodore
Sheline, Raymond Kay
Sher, Alvin Harvey
Shudde, Rex Hawkins
Silva, Robert Joseph
Smith, Francis Marion
Sodd, Vincent J
Stephens, Frank Samuel
Stewart, Robert Francis
Stone, John Austin
Sugarman, Nathan

Sugihara, Thomas Tamotsu
Sun, Kuan-Han
Swanson, David G, Jr
Tewes, Howard Allan
Thomas, Thomas Darrah
Thompson, Ronald Hobart
Townley, Charles William
Troutner, David Elliott
Turkevich, Anthony
Turner, Stanley Eugene
Uhl, Dale Lynden
Unik, John Peter
Vandenbosch, Robert
Viola, Victor E, Jr
Wahl, Arthur Charles
Wahl, Werner Henry
Wai, Chien Moo
Walters, William Ben
Ward, Thomas Edmund
Watson, Rand Lewis
Wikjord, Alfred George
Wild, John Frederick
Wiley, John Robert
Wilhelmy, Jerry Barnard
Williams, David Cary
Williams, Evan Thomas
Williams, Robert Allen
Wing, James
Wischow, Russell P
Wogman, Ned Allen
Wolke, Robert Leslie
Yates, Steven Winfield
Yule, Herbert Phillip
Zoller, William H

Nuclear Magnetic Resonance
Armstrong, Robin L
Barnes, Richard George
Brooker, Hampton Ralph
Buchanan, Gerald Wallace
Chu, Keh-Chang
Cotts, Robert Milo
Cox, Richard Harvey
Current, David Harlan
Doane, Joseph William
Dunell, Basil Anderson
Egan, Richard Stephen
France, Peter William
Giannini, Donald Dominic
Gillen, Kenneth Todd
Harrell, James W, Jr
Hogenboom, David L
Isbrandt, Lester Reinhardt
Jeffrey, Kenneth Robert
Johnson, Leroy Franklin
King, Roy Warbrick
Kurland, Robert John
Lundin, Robert Enor
Miknis, Francis Paul
Millman, Sidney
Miranda, Gilbert A
Parker, George W
Petch, Howard Earle
Pintar, Milan Mik
Raban, Morton
Roeder, Stephen Bernard Walter
Rowan, Raymond, III
Sears, Raymond Eric John
Shapiro, Bernard Lyon
Snodgrass, Rex Jackson
Spielvogel, Bernard Franklin
Stockton, Gerald William
Story, Harold S
Taylor, Charles Emory
Vold, Regitze Rosenorn
Warren, William Willard, Jr
Woessner, Donald Edward

Theoretical Nuclear Physics
Altman, Albert
Austern, Norman
Baltz, Anthony John
Barrett, Bruce Richard
Bassichis, William
Becker, Richard Logan
Bhakar, Balram Singh
Chiao-Yap, Lung Wen
Clark, John Walter
Coker, William Rory
Coulter, Philip W
Cusson, Ronald Yvon
Davis, Alvie Douglas
Donnelly, Thomas William
Dorn, David W
Dover, Carl Bellman
Edwards, Steve
Franco, Victor
Freed, Norman
Fuda, Michael George
Fuller, Richard Clair
Ginocchio, Joseph Natale
Glendenning, Norman Keith
Grube, Geraldine Joyce Terenzoni
Hackman, Roger H
Hadjimichael, Evangelos
Haftel, Michael Ivan
Halbert, Edith Conrad
Harris, Samuel Melvin

Harvey, Malcolm
Henley, Ernest Mark
Ho-Kim, Quang
Jackson, Andrew D, Jr
Kazaks, Peter Alexander
Kim, Yeong Ell
Kirwan, Donald Frazier
Kowalski, Kenneth L
Kurath, Dieter
Lehman, Donald Richard
LeTourneux, Jean
Lim, Teck-Kah
Lock, James Albert
Ma, Chin Wah
Madsen, Victor Arviel
McVoy, Kirk Warren
Millener, David John
Miller, Lewis Dudley
Newton, Victor Joseph
Nix, James Rayford
Payne, Gerald Lew
Pieper, Steven Charles
Pittel, Stuart
Purrington, Robert Daniel
Rawitscher, George Heinrich
Redish, Edward Frederick
Rinker, George Albert, Jr
Rouse, Carl Albert
Saxon, David Stephen
Schick, Lee Henry
Scott, Bruce L
Seaborn, James Byrd
Seki, Ryoichi
Sharon (Schwadron), Yitzhak Yaakov
Sheldon, Eric
Stephenson, Gerard J, Jr
Strobel, George L
Tobocman, William
Tomusiak, Edward Lawrence
Towner, Ian Stuart
Uberall, Herbert Michael
Valk, Henry Snowden
Verhanovitz, Richard Frank
Walker, George Edward
Wallace, Stephen Joseph
Weber, Hans Jurgen
Werntz, Carl H

PHYSICS, PLASMA

Plasma Physics
Ahlborn, Boye
Allis, Willam Phelps
Andrews, Merrill Leroy
Arnold, Kenneth Wiltshire
Askew, Raymond Fike
Auer, Peter Louis
Bachynski, Morrel Paul
Baker, John Cummins
Baldwin, David Ellis
Bandel, Herman William
Barach, John Paul
Barnard, Adam Johannes
Barnes, Derek A
Barr, Thomas Albert, Jr
Beasley, Cloyd O, Jr
Beckner, Everet Hess
Bemmels, William David
Benenson, David Maurice
Bengtson, Roger D
Bennett, Willard Harrison
Bergeron, Kenneth Donald
Bernabei, Stefano
Bevc, Vladislav
Bingham, Robert Lodewijk
Bird, Harvey Harold
Bol, Kees
Boley, Forrest Irving
Book, David Lincoln
Briggs, Richard Julian
Brown, Keith H
Burns, Erskine John Thomas
Callen, James Donald
Case, Carl Tyler
Catto, Peter James
Cecchi, Joseph Leonard
Chanin, Lorne Maxwell
Chen, Robert Long Wen
Cheng, Chiang-Shuei
Choi, Duk-In
Cladis, John Baros
Clauser, Milton John
Clements, Reginald Montgomery
Coburn, John Wyllie
Cohen, Bruce Ira
Cohen, Leonard George
Cohen, Samuel Alan
Cohn, Daniel Ross
Conrad, John Rudolph
Cooper, John (Jinx)
Coroniti, Ferdinand Vincent
Cox, James Lester, Jr
Crawford, Frederick William
Creedon, John E
Crevier, William Francis
Crownfield, Frederic Rudolph, Jr

Damm, Charles Conrad
Davidson, Ronald Crosby
Dawson, John Myrick
Demetriades, Sterge Theodore
DeSilva, Alan W
Dingee, David Aaron
Doggett, Wesley Osborne
Donaldson, Robert Evans
Dove, William Francis
Dreicer, Harry
Drouet, Michel Georges
Dulgeroff, Carl Richard
Dum, Christian Thomas
Ehler, Arthur Wayne
Ehrman, Joachim Benedict
Elton, Raymond Carter
Emmerich, Werner Sigmund
Emmert, Gilbert A
Enoch, Jacob
Estabrook, Kent Gordon
Farley, Donald T Jr
Farr, William Morris
Felber, Franklin Stanton
Ferguson, Joseph Luther, Jr
Flynn, Robert W
Fontheim, Ernest Gunter
Foote, James Herbert
Forrester, Alvin Theodore
Forslund, David Wallace
Freidberg, Jeffrey Philip
Fried, Burton David
Frieman, Edward Allan
Frommhold, Lothar Werner
Fu, Jerry Hui Ming
Fukai, Junichiro
Gajewski, Ryszard
Gardner, Andrew Leroy
Gary, Stephen Peter
Gazdag, Jeno
Gentle, Kenneth W
Gibbs, Richard Lynn
Gleman, Stuart Maxwell
Glowienka, John Clement
Godfrey, Brendan Berry
Goldenbaum, George Charles
Goldman, Leonard Manuel
Goodman, Ronald Keith
Granatstein, Victor Lawrence
Gray, Eoin Wedderburn
Gray, Ernest Paul
Greene, John M
Gregory, Brian Charles
Grieg, J Robert
Grisham, Larry Richard
Gross, Robert Alfred
Grossmann, William
Guernsey, Ralph Lewis
Gula, William Peter
Haas, Gregory Mendel
Haberstich, Albert
Hadley, George Ronald
Hai, Francis
Hall, Laurence Stanford
Halverson, Ward Dean
Hartle, Richard Eastham
Hatch, Albert Jerold
Haught, Alan F
Hawkins, Charles Edward
Hawryluk, Richard Janusz
Hazeltine, Richard Deimel
Heald, Mark Aiken
Hedrick, Clyde Lewis, Jr
Hendel, Hans William
Henderson, Dale Barlow
Hickok, Robert Lyman, Jr
Hilton, John L
Hinton, Frederick Lee
Hirose, Akira
Hirshfield, Jay Leonard
Hitchcock, Daniel Augustus
Hofmann, Gunter August George
Hogan, William Alfred
Holzer, Robert Edward
Hosea, Joel Carlton
Hsuan, Hulbert C S
Hubbard, Edward Leonard
Hudson, Mary Katherine
Huerta, Manuel Andres
Ingraham, John Charles
Isham, Elmer Rex
Jancarik, Jiri
Jarboe, Thomas Richard
Jean, Benoit
Jenkins, Alvin Wilkins, Jr
Jensen, Torkil Hesselberg
Johnson, John Lowell
Johnson, Roy Ragnar
Johnston, George Lawrence
Johnston, Russell Shayne
Johnston, Tudor Wyatt
Jones, William Denver
Joyce, Glenn Russell
Judd, O'Dean P
Kadish, Abraham
Kaiser, Thomas Burton
Kalman, Gabor
Kaplan, Daniel Eliot
Karras, Thomas William

291

PHYSICS, PLASMA

Katz, Ira
Kaufman, Allan N
Kellogg, Paul Jesse
Kennel, Charles Frederick
Raether, Manfred
Kerst, Donald William
Kilb, Ralph Wolfgang
Killeen, John
Kim, Jinchoon
Kim, John Jungyu
Kimblin, Clive William
Kindel, Joseph Martin
Kinsinger, Richard Estyn
Knorr, George E
Kohler, Donald Alvin
Koopman, David Warren
Kovar, Frederick Richard
Kribel, Robert Edward
Kruer, William Leo
Krumbein, Aaron Davis
Ku, Robert Tien-Hung
Kulsrud, Russell Marion
Kunkel, Wulf Bernard
Kusse, Bruce Raymond
Lamborn, Bjorn N A
Lamboi, James Gerald
Lamontboise, James Gerald
Landau, Ronald Wolf
Langdon, Allan Bruce
Lauer, Eugene John
Law, William Brough
Lax, Benjamin
Lee, Anthony
Lee, Tong-Nyong
Leonard, Stanley Lee
Lerche, Ian
Levi, Enrico
Liebenberg, Donald Henry
Liewohn, Harold Benjamin
Lilley, John Richard
Lindemuth, Irvin Raymond, Jr
Lindman, Erick Leroy, Jr
Little, Robert
Liu, Chuan Sheng
Lohr, John Michael
Lowenthal, Dennis David
Lowke, John James
Lubin, Moshe J
Luce, John Sidney
Ludwig, Howard C
Lupton, William Hamilton
Lyons, Peter Bruce
MacLachy, Cyrus Shantz
Macrakis, Michael S
MacVicar-Whelan, Patrick James
Majkowski, Richard Francis
Malmberg, John Holmes
Manheimer, Wallace Milton
Manka, Charles K
Marshall, Theodore
Mason, Rodney Jackson
Massel, Gary Alan
Matsuda, Yoshiyuki
Mawardi, Osman Kamel
Maxon, Marshall Stephen
Mayes, Terrill W
Mayer, Barry Newton
Moore, Richard Lee
Morin, Dorris Clinton
Morris, Glen Jeffs
Morrone, Terry
Motley, Robert W
Nagamatsu, Henry T
Nicholson, Dwight Roy
Nielsen, Philip Edward
McNally, James Rand, Jr
Meade, Dale M
Meyerand, Russell Gilbert, Jr
Miller, Bennett
Montgomery, David Campbell
Moore, Barry Newton
Norwood, Joseph, Jr
Noerdlinger, Peter David
Nugent, Leonard James
Oakes, Melvin Ervin Louis
Offenberger, Allan Anthony
Okabayashi, Michio
Oktay, Erol
Oliver, Brian Malcolm
Osborn, Richard Kent
Osher, John Edward
Ossakow, Sidney Leonard
Paolini, Frank John
Parbhakar, Kanwal Jit
Parkin, Curtis Willard
Patel, Vithalbhai L
Phillips, James A
Porkolab, Miklos
Porter, James Colegrove
Poss, Howard Lionel
Poss, Douglass Edmund, Jr
Poularikas, Alexander D
Price, John Charles
Pye, Robert V

Quinn, Robert George
Radin, Shelden Henry
Radoski, Henry Robert
Rechester, Alexander
Reichelt, Walter Herbert
Rich, Joseph Anthony
Richard, Claude
Ripin, Barrett Howard
Robben, Franklin Arthur
Roberts, Thomas Sheldon
Roberts, Charles Sheldon
Roberts, Harry Stroud
Robertson, Scott Harrison
Robinson, Bruce B
Rode, Daniel Leon
Rosen, Bernard
Ross, David Paul
Roszman, Larry Joe
Rothman, Milton A
Rugge, Henry F
Rutherford, Paul Harding
Rynn, Nathan
Sawyer, George Atanson
Schivel, John Francis
Schmidt, John Allen
Schneider, Richard Theodore
Schulz, Michael
Seidl, Milos
Seto, Yeb Jo
Shearer, James Welles
Sherwood, Arthur Robert
Shikarofsky, Issie Peter
Shohet, Juda Leon
Shrauner, Barbara Abraham
Sigmar, Dieter Joseph
Simon, Albert
Simonen, Thomas Charles
Sinnis, James Constantine
Sloan, David Harold
Soper, Gordon Knowles
Spies, Gunther Otto
Sprott, Julien Clinton
Stallings, Charles Henry
Stamper, John Andrew
Stansfield, Barry Lionel
Stetson, Robert Franklin
Stewart, Gordon Ervin
Stix, Thomas Howard
Stone, Albert Mordecai
Stone, Philip M
Sturrock, Peter Andrew
Su, Chau-Hsing
Sudan, Ravindra Nath
Swanson, Donald G
Sweeney, Mary Ann
Swift, Daniel W
Takahashi, Hironori
Taussig, Robert Trimble
Taylor, Harold Evans
Temkin, Richard Joel
Thomassen, Keith I
Thompson, William Bell
Thomson, Jeffrey John
Thorne, Richard Mansergh
Thornton, John Alexander
Tracy, Philip T
Trivelpiece, Alvin William
Tsang, Kang Too
Turner, James Marshall
Unti, Theodore Wayne Joseph
Vahala, George Martin
Valeo, Ernest John
Van Hoven, Gerard
Van Rij, Willem Idaniel
Vaslow, Dale Franklin
Ventrice, Carl Alfred
Vidmar, Paul Joseph
Vlases, George Charpentier
Wachtel, Jonathan Mark
Wada, James Yasuo
Walsh, John Edmond
Walters, Craig Thompson
Ware, Alan Alfred
Watanabe, Tomiya
Weinstock, Jerome
Weitzner, Harold
Wharton, Charles Benjamin
Wheaton, John Hobson
White, Roscoe Beryl
Wiese, Wolfgang Lothar
Wilkerson, Thomas Delaney
Willett, Joseph Erwin
Wilson, Andrew Robert
Witting, Harald Ludwig
Wolf, Neil Steven
Wong, Alfred
Woodall, David Monroe
Worden, David Gilbert
Wright, Thomas Payne
Yamada, Masaaki
Yarborough, William Walter, Jr
Zucker, Martin Samuel

Electrodynamics

Armoudian, Garabed
Barreto, Ernesto
Chu, En Lung

292

Devine, Robert T
Driver, Rodney David
Heberle, Juergen
Henneberger, Walter Carl
McLennan, Donald Elmore
Miller, Leston Wayne
Papas, Charles Herach
van Roggen, Arend
Vittitoe, Charles Norman

Electrohydrodynamics

Phillips, Perry Edward
Prichard, Benjamin Arnold, Jr

Magnetohydrodynamics

Cassen, Patrick Michael
Felt, Michael Dennis
Holt, James Franklin
Killeen, John
Kribel, Robert Edward
Lempert, Joseph
Loeffler, Albert L, Jr
Loper, David Eric
McNab, Ian Roderick
Nerney, Steven Francis
Redman, William Charles
Smith, Richard James
Stevens, John Charles
Vahala, George Martin
White, Willard Worster, III
Zauderer, Bert

PHYSICS, SOLID STATE

Solid State Physics

Abeles, Benjamin
Adair, Thomas Weymon, III
Adams, Peter David
Adams, Richard Owen
Adda, Lionel Paul
Addiss, Richard Robert, Jr
Adler, David
Adler, Eric
Adler, John G
Ahlers, Guenter
Ahrenkiel, Richard K
Ailion, David Charles
Aitken, John Malcolm
Alben, Richard (Samuel)
Albers, Walter Anthony, Jr
Alexander, Michael Norman
Alexander, Ralph William, Jr
Alexandropoulos, Nikos G
Alfred, Louis Charles Roland
Allen, Roger Casanova
Allen, Frederick Graham
Allen, James Ward
Allen, Roland Emery
Alley, Philip Wayne
Allgaier, Robert Stephen
Alperin, Harvey Albert
Alves, Ronald V
Amer, Nabil Mahmoud
Amith, Avraham
Anastassakis, Evangelos M
Andeen, Carl Gustav
Anderson, Anthony
Anderson, Charles Hammond
Anderson, Elmer E
Anderson, Gerald S
Anderson, James Gerard
Anderson, Robert E
Ansbacher, Theodore Henry
Antcliffe, Gault Anderson
Appelbaum, Joel A
Appleton, B R
Arajs, Sigurds
Arakawa, Edward Takashi
Archer, Robert James
Arko, Aloysius John
Arnold, Emil
Arnold, George W
Arp, Vincent D
Artman, Joseph Oscar
Aschner, Joseph Felix
Ashworth, T
Askill, John
Aston, Duane Ralph
Astrue, Robert William
Atwater, Harry Albert
Aukerman, Lee William
Aven, Manuel
Baer, Adrian Donald
Baer, Donald Ray
Baer, Walter S
Bagchi, Amitabha
Bagley, Brian G
Bailey, Carroll Edward
Bailey, Glenn Charles
Bailey, John Martin
Baker, George Severt
Baldwin, Thomas O
Ballard, Lewis Franklin
Banerjee, R L
Baraff, Gene Allen

Bardasis, Angelo
Barker, A S, Jr
Barnaal, Dennis E
Barnes, Gene A
Barnes, Richard George
Barnoski, Michael K
Baron, Robert
Barrera, Joseph S
Barsch, Gerhard Richard
Bartels, Richard Alfred
Bartlett, Roger James
Bartley, David Lauren
Bartram, Ralph Herbert
Bass, Jack
Bate, Geoffrey
Bates, Clayton Wilson, Jr
Batra, Inder Paul
Bauer, Walter
Baum, Gary Allen
Beattie, Alan Gilbert
Bebb, Herbert Barrington
Becker, Milton
Beeler, Joe R, Jr
Bell, Marvin Drake
Bell, Richard Oman
Bell, Robert John
Benbow, Ralph Lawrence
Bender, Paul A
Bendow, Bernard
Bennett, Alan Jerome
Bensel, John Philip
Berg, James Irving
Berger, Luc
Bermon, Stuart
Bermudez, Victor Manuel
Bernstein, Benjamin Tobias
Berry, Chester Ridlon
Berry, Richard Emerson
Berry, Robert John
Bessey, William Higgins
Best, Philip Ernest
Bevolo, Albert Joseph
Bhalla, Ranbir J R Singh
Bhatnagar, Anil Kumar
Bichard, J W
Bickford, Lawrence Richardson
Bienenstock, Arthur Irwin
Birgeneau, Robert Joseph
Bis, Richard F
Bishop, Thomas Parker
Bitler, William Reynolds
Black, John Earle
Black, Truman D
Blair, McClellan Gordon
Blakemore, John Sydney
Blewitt, Thomas Hugh
Blicher, Adolph
Bloomfield, Philip Earl
Blue, Marts Donald
Blum, Haywood
Blum, Norman Allen
Blum, Robert F
Bode, Donald Edward
Boedicker, Olaf A
Boer, Karl Wolfgang
Boettler, James Leroy
Bogardus, Egbert Hal
Boghosian, Charles
Bogle, Tommy Earl
Bohandy, Joseph
Bohm, Thomas Lynch
Bohn, Henry
Bolef, Dan Isadore
Booth, Bruce L
Borchers, Robert H
Borders, James Alan
Boster, Thomas Arthur
Botka, Nicholas
Bourassa, Ronald Ray
Bowen, Donald Edgar
Boyd, James Robert
Boys, Donald W
Brandt, Richard Charles
Bratina, Woymir John
Bratt, Peter Raymond
Bray, Philip James
Bray, Ralph
Breazeale, Mack Alfred
Breig, Marvin L
Brice, David Kenneth
Brient, Samuel John, Jr
Briggs, Jonathan
Brindley, George W
Brinkman, William F
Broadhurst, Martin Gilbert
Brockhouse, Bertram Neville
Brodie, Don E Farrell
Brodsky, Marc Herbert
Brodsky, Mervyn Berkley
Bron, Walter Ernest
Broshar, Wayne Cecil
Brothers, Alfred Douglas
Brouillette, Walter

Browder, James Steve
Brown, Bruce Stilwell
Brown, Frederick Calvin
Brown, Rodney Duvall, III
Brown, Ronald Alan
Brown, Ronald Franklin
Brown, Walter Lyons
Brumage, William Harry
Brungs, Robert Anthony
Brust, David
Bryant, Paul James
Bube, Richard Howard
Buckman, William Gordon
Buescher, Brent J
Buhrman, Robert Alan
Burch, Thaddeus Joseph
Burdick, David Leo
Burger, Robert M
Burk, David Lawrence
Burkey, Bruce Curtiss
Burnham, Dwight Comber
Burr, Alexander Fuller
Burstein, Elias
Burton, Joseph Ashby
Buschert, Robert Cecil
Busik, Arthur J
Butler, Frank Andrew
Butler, Michael Alfred
Butler, William H
Buyers, William James Leslie
Byer, Norman Ellis
Byvik, Charles Edward
Cable, Joe Wood
Cabrera, Nicolas
Callcott, Thomas Anderson
Callen, Earl Robert
Camp, Paul R
Cape, John Anthony
Cardwell, Alvin Boyd
Carleton, Herbert Ruck
Carlson, David Emil
Carlson, Edward H
Carpenter, Steve Haycock
Carr, Paul Henry
Carter, David L
Carter, Gesina C
Caspari, Max Edward
Caspers, Hubert Henri
Castner, Theodore Grant, Jr
Catchings, Robert Merritt, III
Cathey, William Newton
Celli, Vittorio
Chadwick, George F
Chang, Richard Kounai
Charap, Stanley H
Chatterjee, Ramananda
Chau, Cheuk-Kin
Chen, An-Ban
Chen, Charles Chin-Tse
Chen, Ho Sou
Cherin, Paul
Cherry, Leonard Victor
Chester, Marvin
Chevrier, Jean-Claude Jacques
Child, Harry Ray
Choi, Sang-Il
Chow, Paul C
Choyke, Wolfgang Justus
Chrenko, Richard Michael
Christensen, Charles Michael
Christensen, Stanley Howard
Christmann, Marvin Henry
Chu, Ching-Wu
Chung, David Yih
Chynoweth, Alan Gerald
Claassen, Richard Strong
Clapp, Philip Charles
Clark, Alton Harold
Clark, Arthur Edward
Clark, Bill Pat
Clark, Clifton Bob
Clayman, Bruce Philip
Clogston, Albert McCavour
Coburn, Theodore James
Cochran, John Francis
Cody, George Dewey
Cohen, Marvin Morris
Cohen, Richard Lewis
Coibow, Konrad
Cole, George Rolland
Cole, Milton Walter
Colella, Roberto
Coleman, Charles Clyde
Collins, Francis Allen
Collins, Franklyn
Colliver, Michael Carridine
Coltman, Ralph Read, Jr
Colwell, Joseph F
Compaan, Alvin Dell
Conklin, James Byron, Jr
Connolly, John William Domville
Connor, Donald W
Conradi, Edward Ezra
Conradi, Jan
Constable, James Harris
Conwell, Esther Marly
Cooke, Iain

Cooper, Bernard Richard
Copeland, John Alexander
Corbett, James William
Corliss, Lester Myron
Cowan, David Lawrence
Craig, Richard Anderson
Crane, Langdon Teachout, Jr
Craven, Robert Alan
Crawford, James Homer, Jr
Crittenden, Eugene Casson, Jr
Crowe, George Joseph
Crump, John C, III
Cullen, James Robert
Cunningham, Robert Gail
Curtis, Orlie Lindsey, Jr
Cutler, Melvin
Dacey, George Clement
Dahlquist, Wilbur Lynn
Dalbec, Paul Euclide
Dalven, Richard
Daly, Daniel Francis, Jr
Damon, Richard Winslow
Daniel, Thomas Bruce
Daniels, James Maurice
Danielson, Gordon Charles
Dank, Milton
Darnell, Frederick Jerome
Dasgupta, Sunil Priya
Davis, Hawthorne Antoine
Davis, Jack H
Davis, John Litchfield
Dawes, William Redin, Jr
Day, Stephen Martin
Deaton, Bobby Charles
DeCicco, Peter Donald
Deck, Ronald Joseph
Decker, Daniel Lorenzo
Deegan, Ross Alfred
DeFonzo, Alfred Peter
DeFord, John W
de Graaf, Adriaan M
Deis, Daniel Wayne
Delinger, William Galen
Detrio, John A
Detwiler, Daniel Paul
DeWames, Roger
De Wette, Frederik Willem
deWit, Roland
Dexter, Richard Norman
De Young, Donald Bouwman
Di Bartolo, Baldassare
DiCarlo, James Anthony
Dickey, Dana H
Dickinson, Stanley Key, Jr
DiDomenico, Mauro, Jr
Dienes, George Julian
Dietz, Robert E
Dingle, Raymond
Di Salvo, Francis Joseph
DiStefano, Thomas Herman
Dixon, Carl Eugene
Dixon, George Sumter, Jr
Dixon, Richard Wayne
Dobrov, Wadim (Ivan)
Doherty, Paul Michael
Dolling, Gerald
Domenicali, Charles Angelo
Donahoe, Frank J
Doran, Donald George
Downs, David S
Doyle, William David
Dragoo, Alan Lewis
Dresner, Joseph
Dresselhaus, Gene Frederick
Dresselhaus, Mildred S
Drew, Howard Dennis
Drews, Udo Wilhelm
Dropkin, John Joseph
Drum, Charles Monroe
Dubeck, Leroy W
Duffy, William Thomas, Jr
Dunn, Charles Nord
DuVarney, Raymond Charles
Dy, Kian Seng
Dyment, John Cameron
Dynes, Robert Carr
Eagen, Charles Frederick
Eagleton, Robert Don
Earle, Marshall Delphi
Eastman, Philip Clifford
Ebel, Marvin Emerson
Eby, John Edson
Eck, John Stark
Economou, Eleftherios Nickolas
Edelheit, Lewis S
Edelstein, Alan Shane
Eden, Richard Carl
Edge, Ronald (Dovaston)
Edwards, David Franklin
Edwards, Leon Roger
Egan, Walter George
Ehrstein, James Robert
Eichbaum, Barlane R
Eldridge, Jerome Michael
Ellis, Donald Edwin
Emin, David
Emrick, Roy M
Emtage, Peter Roesch

Engel, Jan Marcin
Engeler, William E
English, Floyd L
Ensign, Thomas Charles
Epstein, Arnold S
Epstein, Arthur Joseph
Erdmann, Joachim Christian
Erickson, Dennis John
Erskine, James Christian, Jr
Erskine, James Lorenzo
Esaki, Leo
Eshbach, John Robert
Esposito, Robert John
Estle, Thomas Leo
Etzel, Howard Wesley
Evans, Allan Robert
Evans, Bruce Douglas
Everett, Glen Exner
Everett, Paul Marvin
Ewaraye, Andrew Oteku
Ewald, Arno Wilfred
Fabish, Thomas John
Fagen, Edward Allen
Fain, Samuel Clark, Jr
Fair, Harry David, Jr
Falicov, Leopoldo Maximo
Fan, Hsu Yun
Faraday, Bruce (John)
Faughnan, Brian Wilfred
Feder, Ralph
Feigl, Frank Joseph
Feingold, Earl
Feinleib, Julius
Feist, Wolfgang Martin
Feldman, Charles
Feldman, Joseph Louis
Fenner, Gunther Erwin
Fenton, Edward Warren
Ferris-Prabhu, Albert Victor Michael
Fetterman, Harold Ralph
Feuchtwang, Thomas Emanuel
Filipovich, George
Fink, Herman Joseph
Finnemore, Douglas K
Fischbach, David Bibb
Fischer, C Rutherford
Fischer, John Edward
Fischer-Colbrie, Erwin
Fiske, Milan Derbyshire
Fletcher, Paul Chipman
Fleury, Paul A
Flippen, Richard Bernard
Flynn, Colin Peter
Foglio, Mario Eusebio
Foner, Simon
Fong, Ching-Yao
Fontanella, John Joseph
Fork, Richard Lynn
Fortin, Emery
Fowler, Alan B
Fowler, Wyman Beall
Fox, Mary Eleanor
France, Peter William
Franceschetti, Donald Ralph
Frank, Robert Carl
Franklin, Alan Douglas
Franklin, Wilbur Mitchell
Franz, Judith Rosenbaum
Frazer, Benjamin Chalmers
Frederikse, Hans Pieter Roetert
Free, John Ulric
Freeman, Arthur Jay
Freeman, James R
Freud, Paul J
Friauf, Robert James
Friedman, Edward Alan
Friedman, Lionel Robert
Fritz, Garold Frederic
Fritzsche, Hellmut
Fromhold, Albert Thomas, Jr
Fuchs, Ronald
Fukushima, Eiichi
Fuller, Richard M
Furdyna, Jacek K
Gabriel, Cedric John
Gaertner, Wolfgang Wilhelm
Galeener, Frank Lee
Galli, John Ronald
Galt, John Kirtland
Gamble, Fred Ridley, Jr
Ganguly, Biswa Nath
Gardner, Edward Eugene
Garito, Anthony Frank
Garland, James W, Jr
Garland, Michael McKee
Garstens, Martin Aaron
Garth, John Campbell
Gaunt, Paul
Gauster, Wilhelm Belrupt
Geiger, Felix Eugene
Genin, Dennis Joseph
Gerstenberg, Dieter
Ghosh, Amal Kumar
Giess, Edward August
Gilbert, Robert L
Gillingham, Robert J
Gilmer, Thomas Edward, Jr
Gilvarry, John James

Girouard, Fernand E
Glasser, M Lawrence
Glover, Rolfe Eldridge, III
Gobeli, Garth William
Goff, James Franklin
Goland, Allen Nathan
Gold, Albert
Goldburg, Walter Isaac
Golin, Stuart
Gonzalo, Julio Antonio
Goodings, David Ambery
Goodrich, Roy Gordon
Gordon, William Livingston
Grant, Paul Michael
Grant, Roderick M, Jr
Gray, Peter Vance
Greene, Alan Campbell
Greene, Michael P
Greene, Richard L
Greenfield, Arthur Judah
Grenier, Claude Georges
Griffin, Peter Allan
Griffing, David Francis
Griffith, William Thomas
Grimes, Charles Cullen
Grimes, Hubert Henry
Grindlay, John
Grobman, Warren David
Gruber, John B
Grunwald, Hubert Peter
Gubernatis, James Edward
Gubser, Donald Urban
Guentert, Otto Johann
Guenther, Raymond A
Guenzer, Charles S P
Guertin, Robert Powell
Gueths, James E
Gundersen, Martin Adolph
Gunsul, Craig J W
Gunter, Thomas E, Jr
Gustafson, John C
Guthrie, George Leslie
Gutman, Lester
Haacke, Gottfried
Haase, Oswald
Hafemeister, David Walter
Hafner, Erich
Hagarman, Vincent A
Hagenlocher, Arno Kurt
Haitz, Roland Hermann
Halder, Narayan Chandra
Hale, Edward Boyd
Hall, John Jay
Halliburton, Larry Eugene
Halperin, Bertrand Israel
Hamill, Dennis W
Hamilton, John Frederick
Hamm, Franklin Albert
Hammer, Jacob Meyer
Handler, Paul
Hanna, Stanley Sweet
Hannah, Eric Cabot
Hansen, Uwe Jens
Hardy, Walter Newbold
Harland, Glen Eugene, Jr
Harris, Arthur
Harris, Arthur Brooks
Harris, Erik Preston
Harris, Richard Elgin
Harrison, Walter Ashley
Hart, Howard Roscoe, Jr
Harte, Kenneth J
Hartke, Jerome L
Hartman, Richard Leon
Hartman, Roger Duane
Hartmann, William Morris
Hasegawa, Ryusuke
Hass, Georg
Hass, Marvin
Hatcher, Robert Douglas
Hauser, Joachim J
Hawkins, Gilbert Allan
Hayes, Timothy Mitchell
Hays, Dan Andrew
Hedgecock, Nigel Edward
Heeger, Alan J
Heiman, Donald Eugene
Heller, Gerald S
Heller, William R
Hempstead, Charles Francis
Hengehold, Robert Leo
Henisch, Heinz Kurt
Henkels, Walter Harvey
Henrich, Victor E
Henry, Charles H
Hensel, John Charles
Hensley, Eugene Benjamin
Herber, Rolfe H
Herman, David S
Herman, John Edward
Herman, Allen Max
Hermanson, John Carl
Herrington, John Peter
Herrington, James Roland
Hetzler, Morris Clifford, Jr
Hewes, Ralph Allan
Hewitt, Robert Russell
Higgins, Richard J

PHYSICS, SOLID STATE

Westbrook, Russell David
Weyhmann, Walter Victor
Weymouth, John Walter
Whippey, Patrick William
White, John Joseph, III
White, Ray Henry
White, Robert Marshall
Wickersheim, Kenneth Alan
Wiegand, Donald Arthur
Wiegand, Philip E
Wield, Robert Lee
Wiley, John Duncan
Wilkenfeld, Jason Michael
Wilkinson, Michael Kennerly
Will, Theodore A
Willard, Harold James, Jr
Willardson, Robert Kent
Williams, Arthur Robert
Williams, Brown F
Williams, Clayton Drews
Williams, Emmett Lewis, Jr
Williams, George Abiah
Williams, Richard Taylor
Williams, Ronald Wendell
Williams, Wendell Sterling
Wilson, John Anthony
Wilson, Timothy M
Winder, Dale Richard
Winter, William Dennis
Witels, Mark C
Wolf, Gordon A
Wolfe, Raymond
Wolff, George Buford
Wolfram, Thomas
Wollan, David Strand
Wong, Jacob Yau-Man
Wood, James C, Jr
Wood, John Herbert
Woods, Alfred David Braine
Woods, Stuart B
Wooten, Frederick (Oliver)
Worden, David Gilbert
Worlock, John M
Wortis, Michael
Wright, George Buford
Wrobel, Joseph Stephen
Wrobel, Theodore Frank
Wronski, Christopher Roman
Yafet, Yako
Yedinak, Peter Demerton
Yelon, Arthur Michael
Yen, William Mao-Shung
Yessik, Michael John
Ying, See Chen
Yip, Kwok Leung
Yoshikawa, Herbert Hiroshi
Young, Ralph Howard
Young, Richard Accipiter
Young, Robert Alan
Zallen, Richard
Zebouni, Nadim H
Zeidler, James Robert
Zimmerman, Walter Bruce
Zitter, Robert Nathan
Zucker, Joseph
Zucker, Melvin Joseph
Zukoynski, Stefan

Experimental Solid State Physics

Abkowitz, Martin Arnold
Akselrad, Aline
Andrew, James F
Ayers, Raymond Dean
Batterman, Boris William
Bauer, Robert Steven
Biegelsen, David K
Birchak, James Robert
Birkhoff, Robert D
Blankenship, James Lynn
Boolchand, Punit
Bottom, Virgil Eldon
Boyle, David Joseph
Braun, Charles Louis
Braunlich, Peter Fritz
Bright, Arthur Aaron
Budzinski, Walter Valerian
Cappelletti, Ronald Louis
Carver, Gary Paul
Cheng, Li-Jen
Chesser, Nancy Jean
Cleland, John W
Collings, Edward William
Colwell, Jack Harold
Colwell, Priscilla J
Connelly, John Joseph, Jr
Daniels, William Burton
Deaver, Bascom Sine, Jr
De Mayo, Benjamin
Domb, Ellen Ruth (Colmer)
Dunifer, Gerald Leroy
Dunlap, Bobby David
Engstrom, Herbert Leonard
Faust, Walter Luck
Fawcett, Eric
Fernelius, Nils Conard
Fleming, Phyllis Jane
Friedberg, Charles Bruce
Garland, James C

Gathers, George Roger
Gavenda, John David
Glicksman, Maurice
Golding, Brage, Jr
Gonano, John Roland
Gottlieb, Albert Maxwell
Griffiths, Clifford H
Gustafson, Daniel Ray
Gyorgy, Ernst Michael
Haering, Rudolph Roland
Halpern, Teodoro
Hambourger, Paul David
Hanson, Roland Clements
Ho, John Ting-Sun
Huang, Fan-Hsiung Frank
Hudgens, Aubrey C, Jr
Isaacs, Leslie Laszlo
Isett, Lawrence C
Jackel, Lawrence David
Jaklevic, Robert C
Kalma, Arne Haerter
Keller, Frederick Jacob
Kepler, Raymond Glen
Kerchner, Harold Richard
Khan, Sharif Ahmad
Krebs, James John
Kyser, David Sheldon
Madaci, David Peter
Magerlein, John Harold
Marquardt, Charles Lawrence
Martin, Martin Claude
Mazumdar, Purabi
McBride, Duncan Eldridge
McColl, James Renfrew
McGervey, John Donald
McNiff, Edward J J, Jr
Meinhard, James Edgar
Meyer, Stephen Frederick
Miller, William
Morehead, Frederick Ferguson, Jr
Morris, Robert Carter
Mozzi, Michael
Murphy, James Joseph
Narayanamurti, Venkatesh
O'Neal, Thomas Norman
Pandey, Surendra, Nath
Payne, James Edward
Philipp, Herbert Reynold
Possin, George Edward
Reed, William Alfred
Rehn, Victor Leonard
Reichert, Jonathon F
Renzema, Theodore Samuel
Reynolds, Claude Lewis, Jr
Ritsko, John James
Roger, William Alexander
Rogers, James Stewart
Rollins, Roger William
Roper, John Gordon
Rosenbaum, Stephen Saul
Rosenkrantz, Lawrence Jay
Schein, Lawrence Brian
Schumacher, John Christian
Shea, Michael Joseph
Sidles, Paul Howard
Steckmann, Everett Frederick
Sill, Larry R
Smith, Donald Leonard
Smith, Frederick William
Smith, Robert Owens
Stombler, Milton Philip
Straus, Jozef
Stroud, Jackson Swaverly
Sturge, Michael Dudley
Swanson, Max Lynn
Tantraporn, Wirojana
Thorsen, Clayton Albert
Thorsen, Gordon Albert
Thygesen, Kenneth Helmer
Tomasch, Arthur C
Tomasch, Walter J
Touger, Jerold Steven
Wagner, David Kendall
Weichman, Frank Ludwig
Weiss, Jonathan David
Wilson, Ronald Harvey
Windecker, Richard Chase
Wohlgemuth, John Harold
Wood, Charles
Yelon, William B
Yu, Peter Yound
Zumsteg, Fredrick C, Jr

Semiconductors

Amick, James Albert
Angelo, Stephen James
Aschner, Joseph Felix
Borrello, Sebastian Ronald
Bruss, David
Burnham, Robert Danner
Buss, Dennis Darcy
Carbajal, Bernard Gonzales, III
Chapman, Richard Alexander
Cho, Alfred Y
Clark, Bill Pat
Comizzoli, Robert Benedict
Crozier, Edgar Daryl

Solid State Electronics

Amelio, Gilbert Frank
Amick, James Albert
Anagnostopoulos, Constantine N
Ashby, Robert Morrell
Bartelink, Dirk Jan
Blum, Fred A
Brophy, James John
Buckbichler, Frederick V
Buss, Dennis Darcy
Carnes, James Edward
Clarke, John Ross
Coleman, Donald James, Jr
Comizzoli, Robert Benedict
Davis, Luther, Jr
Feist, Wolfgang Martin
French, Judson Cull
Gluck, Ronald Monroe
Goldberg, Colman
Goodwin, Charles Arthur
Gummel, Hermann K
Gurbaxani, Shyam Hassomal
Hall, James Alexander
Hayward, John Standish
Hensler, Donald H
Josenhans, James Gross
Kannewurf, Carl Raeside
Kotz, Arthur Rudolph
Kroger, Harry
Lubberts, Gerrit
Malmberg, Paul Rovelstad
Marton, John Peter
McKay, Kenneth Gardiner
Mee, Jack Everett
Nordstrom, Terry Victor
Pearson, Gerald Leondus
Petroff, Pierre Marc
Post, Irving Gilbert
Pugh, Emerson William
Sage, Jay Peter
Saunders, Edward A
Sawyer, David Erickson
Seidel, Thomas Edward
Sites, James Russell
Smith, George Elwood
Steele, Martin Carl
Sullivan, Jerry Stephen
Swinehart, Philip Ross
Teichner, Robert W
Valsamakis, Emmanuel
van de Vaart, Herman
Warfield, George
Willis, William M
Willmorth, John H
Wu, Chung Pao

Solid State Kinetics

Lai, David Ying Fat
McKee, Rodney Allen

Theoretical Solid State Physics

Adawi, Ibrahim (Hasan)
Allender, David William
Altarelli, Massimo
Antoniewicz, Peter R
Antonoff, Marvin M
Bandy, William Robert
Baratoff, Alexis
Barrett, John Harold
Bartel, Lewis Clark
Beissner, Robert Edward
Bergstresser, Thomas Karl
Bishop, Marilyn Frances
Blount, Eugene Irving
Bostock, Judith Louise
Bray, James William
Brown, Edmond
Chui, Siu-Tat
Csavinszky, Peter John

Fuls, Ellis
Gilmer, Thomas Edward, Jr
Gleim, Paul Stanley
Groves, Warren Olley
Haisty, Robert W
Honig, Arnold
Huber, Robert John
Hunter, William Leslie
Kahng, Dawon
Koliwad, Krishna M
Lang, David (Vern)
Larsen, Ted LeRoy
Losee, David Lawrence
Ludwig, Gerald W
Mackey, Harry Michael
McMahon, Thomas Joseph
Montgomery, George Paul, Jr
Palik, Edward Daniel
Pliskin, William Aaron
Pollak, Michael
Prince, Morton Bronenberg
Schmidt, Paul F
Smith, Daniel Montague
Stringfellow, Gerald B
Ure, Roland Walter, Jr
van Roosbroeck, Willy Werner
Waltz, Maynard Carleton

Davis, Harold Lloyd
Dexter, David Lawrence
Duke, Charles Bryan
Einstein, Theodore Lee
English, Philip Stephen
Esterling, Donald M
Euwema, Robert Noel
Evenson, William Edwin
Feibelman, Peter Julian
Fetter, Alexander Lees
Flocken, John W
Frank, Barry
Gadzuk, John William
Gay, Jackson Gilbert
Gersten, Joel Irwin
Gillis, Nelson Slisbee
Glick, Arnold J
Glyde, Henry Russell
Goodman, Bernard
Gora, Thaddeus F, Jr
Gray, Alma Marcus
Green, Barry Adams
Gunther, Leon
Gupta, Rajendra P
Haas, Charles Warren
Hannon, James Patrick
Harrison, Michael Jay
Henkel, John Harmon
Herring, Conyers
Hertz, John Atlee
Hohenberg, Pierre Claude
Hone, Daniel W
Huang, Wei-Tze
Ivey, Jerry Lee
Jacobs, Allan Edward
James, Hubert Maxwell
Jennison, Dwight Richard
Karakashian, Aram Simon
Kenkre, Vasudev Mangesh
Kliewer, Kenneth L
Korenman, Victor
Krieger, Joseph Bernard
Lang, Norton David
Lawrence, Walter Edward
Lax, Melvin
Leplae, Luc A
Lines, Malcolm Ellis
Liu, Samuel Hsi-Peh
Lubensky, Tom C
Lurie, Joan B
Mahan, Gerald Dennis
Mahanti, Subhendra Deb
Maradudin, Alexei Alexei
Massey, Walter Eugene
Mavroyannis, Constantine
Menzel, Wolfgang Paul
Moriarty, John Alan
Morley, Gayle L
Morris, Stanley P
Mullin, William Jesse
Myron, Harold William
Nedoluha, Alfred K
Novaco, Anthony Dominic
O'Dwyer, John J
O'Hare, John Michael
Ortenburger, Irene Beardsley
Overhauser, Albert Warner
Painter, Gayle Stanford
Pantelides, Sokrates Theodore
Pardee, William Joseph
Peterson, Robert Lee
Pincus, Philip A
Pink, David Anthony Herbert
Price, Peter J
Rabii, Sohrab
Rice, Thomas Maurice
Rimbey, Peter Raymond
Rothwarf, Allen
Salahub, Dennis Russell
Schaich, William Lee
Shore, Herbert Barry
Silver, Richard N
Silvert, William Lawrence
Simpson, John Hamilton
Sinha, Om Prakash
Sinha, John Robert
Sorbello, Richard Salvatore
Stern, Frank
Swift, Jack Bernard
Taggart, George Bruce
Tait, William Charles
Thorpe, Michael Fielding
Van Vechten, James Alden
Wang, Ching-Ping Shih
Wolf, Dieter
Woodruff, Truman Owen

PHYSICS, THEORETICAL

Theoretical Physics

Aaron, Ronald
Abarbanel, Henry D I
Abrahams, Elihu
Adler, Carl George
Adler, Ronald John
Aitken, Alfred H

Albers, James Ray
Albertson, James Stanislaus
Allen, James Roy
Altar, William
Amado, Ralph
Amar, Henri
Amster, Harvey Jerome
Andersen, Carl Marius
Anderson, James Leroy
Anderson, Philip Warren
Antonoff, Marvin M
Appelquist, Thomas
Argyres, Petros
Armenti, Angelo, Jr
Armstead, Robert Louis
Arnowitt, Richard Lewis
Arnush, Donald
Aron, Walter Arthur
Aronson, Irving
Asendorf, Robert Harry
Ashby, Neil
Au, Chi-Kwan
Auer, Peter Louis
Avery, Joseph B
Aviles, Joseph B
Bade, William Lemoine
Baker, Howard Crittendon
Bakshi, Pradip M
Balazs, Nandor Laszlo
Bander, Myron
Banos, Alfredo, Jr
Banville, Marcel
Baranger, Elizabeth Urey
Baranger, Michel
Barasch, Murray Leonard
Barnett, Claude C
Barnhill, Maurice Victor, III
Barrie, Robert
Barut, Asim Orhan
Baym, Gordon A
Beard, David Breed
Beckel, Charles Leroy
Becker, Stephen Fraley
Beeman, David Edmund, Jr
Beg, Mirza Abdul Baqi
Bender, Charles F
Berger, Jay Manton
Bergmann, Peter Gabriel
Bergmann, Otto
Berman, Sam Morris
Bernstein, Ira Borah
Bernstein, Jeremy
Betts, Donald Drysdale
Bethe, Hans Albrecht
Bhatia, Avadh Behari
Bierman, Arthur
Bincer, Adam Marian
Bird, George Franklin
Bird, Joseph Francis
Biritz, Helmut
Birman, Joseph Leon
Blankenbecler, Richard
Bloch, Ingram
Bloomfield, Philip Earl
Bludman, Sidney Arnold
Boardman, John
Bodner, Stephen E
Bohm, Arno
Bolsterli, Mark
Bond, John Walter, Jr
Borse, Garold Joseph
Bose, Samir K
Bose, Subir Kumar
Boulware, David G
Bowen, Samuel Philip
Boyer, Timothy Howard
Brachman, Malcolm K
Bradbury, Ted Clay
Brandow, Baird H
Brehme, Robert W
Brennan, James Gerard
Brennan, Robert Owings
Brewer, Harold Reid
Brill, Dieter Rudolf
Brittin, Wesley E
Brode, Harold Leonard
Brodsky, Stanley Jerome
Brown, Gerald Edward
Brown, James T, Jr
Brown, Laurie Mark
Brown, Ronald Alan
Bruch, Ludwig Walter
Brueckner, Keith Allan
Bryan, Ronald Arthur
Burke, Edward Aloysius
Burke, William L
Burns, Grover Preston
Burt, Philip Barnes
Bushkovitch, Alexander Viatcheslav
Buskirk, Fred Ramon
Butkov, Eugene
Byers, Nina
Byers, Ronald Elner
Callaway, Joseph
Calusdian, Richard Frank
Campolattaro, Alfonso
Cantwell, Robert Murray

Capps, Richard H
Carley, David Don
Carmeli, Moshe
Carpenter, Jack William
Carruthers, Peter A
Carter, James Clarence
Casper, Barry Michael
Cawley, Robert
Chakkalakal, Dennis Abraham
Chalmers, Joseph Stephen
Chan, Chia Hwa
Chandler, Colston
Chang, Howard How Chung
Charpie, Robert Alan
Chase, David Marion
Chatelain, Jack Ellis
Chen, Joseph Cheng Yih
Chen, Tuan Wu
Chester, Arthur Noble
Childers, Robert Wayne
Chow, Paul C
Chow, Yutze
Christy, Robert Frederick
Chu, Shu-Yuan
Chun, Kee Won
Ciftan, Mikael
Clapp, Roger Edge
Clark, Edward Aloysius
Clinton, William L
Coffman, Moody Lee
Cohen, Jeffrey M
Cohen, Michael
Cohen, Morrel Herman
Cohen, Robert Sonne
Cohen, Stanley
Cohn, Jack
Coish, Harold Roy
Coleman, James Andrew
Coleman, Sidney Richard
Collins, Royal Eugene
Connolly, John William Domville
Contogouris, Andreas P
Cook, David Marsden
Cook, James Marion
Cooke, James Horton
Cooper, Leon N
Cooper, Ralph Sherman
Cooper, Richard Kent
Coopersmith, Michael Henry
Cooperstock, Fred Isaac
Corben, Herbert Charles
Corman, Emmett Gary
Cornwall, John Michael
Coulter, Byron Leonard
Courant, Ernest David
Craig, Richard Anderson
Creutz, Michael John
Cummings, Frederick W
Dalgarno, Alexander
Darewych, Jurij Wasyl'
Davies, Kenneth Thomas Reed
Davis, Alvin Herbert
Day, Benjamin Downing
DeBar, Roger Bryant
Debney, George Charles, Jr
Dedrick, Kent Gentry
Delaney, Robert Michael
Deser, Stanley
DeVelis, John Bernard
Dewan, Edmond M
DeWitt, Bryce Seligman
DeWitt, Cecile Morette
De Witt, Hugh Edgar
De Wolf, David Alter
Dilley, James Paul
Dohnanyi, Julius S
Dolen, Richard
Domash, Lawrence Harold
Doniach, Sebastian
Dorfman, Jay Robert
Drachman, Richard Jonas
Dreiss, Gerard Julius
Dresden, Max
Drummond, William Eckel
Du Bois, Donald Frank
Duffey, George Henry
Dunn, Christian Thomas
Duncan, Marion M, Jr
Durand, Bernice
Durand, Loyal, III
Durso, John William
Eberly, Joseph Henry
Eger, F Martin
Ehrenreich, Henry
Eisenbud, Leonard
Eisenstein, Julian (Calvert)
Elliott, Richard Amos
Ellis, David Greenhill
Engle, Irene May
Enns, Richard Harvey
Epstein, Saul Theodore
Erickson, Glen Walter
Esposito, F Paul
Evans, Foster
Eyges, Leonard James
Fader, Walter John
Falicov, Leopoldo Maximo

Fano, Ugo
Favro, Lawrence Dale
Federbush, Paul Gerard
Feldman, David
Feldman, Gordon
Fernbach, Sidney
Feynman, Richard Phillips
Fine, Paul Charles
Finkelstein, David
Finley, James Daniel, III
Finn, Edward J
Fishman, Frank J, Jr
Fivel, Daniel I
Flaherty, Edward John, Jr
Flannery, Martin Raymond
Fleming, Gordon N
Foglio, Mario Eusebio
Foldy, Leslie Lawrance
Fong, Peter
Fontana, Peter R
Ford, George Willard
Ford, Joseph
Ford, Kenneth William
Fowler, Michael
Fowler, Thomas Kenneth
Fox, David
Fox, Ronald Forrest
Fradkin, David Milton
Frahm, Charles Peter
Francis, Norman
Francis, William Porter
Frank, Robert Morris
Franklin, Jerrold
Franzak, Edmund George
Frazer, William Robert
Freedman, Daniel Z
Freiser, Marvin Joseph
Freund, Peter George Oliver
Fried, Herbert Martin
Friedberg, Richard Michael
Fuchs, Norman H
Fujita, Shigeji
Fulton, Thomas
Gallaher, Donald Frederick
Gammel, John Ledel
Gandhi, Jeet-Mal
Garrod, Claude
Gartenhaus, Solomon
Gatland, Ian Robert
Gauvin, J N Laurie
Geiger, Joseph M
Gell-Mann, Murray
Geltman, Sydney
Genolio, Raymond Joseph
Gerjuoy, Edward
Geroch, Robert Paul
Gibson, Gordon
Gilbert, Thomas Lewis
Gilinsky, Victor
Gillespie, John
Glaser, Harold
Glass, Edward Nathan
Glassgold, Alfred Emanuel
Goble, David Franklin
Goebel, Charles James
Goldberg, Irwin
Goldhaber, Alfred Scharff
Goldstein, Louis
Goldstein, Rubin
Golestaneh, Ahmad Ali
Good, Roland Hamilton, Jr
Goodman, Ralph Raymond
Gora, Edwin Karl
Goswami, Amit
Gottfried, Kurt
Goulard, Bernard
Gould, Robert Joseph
Graben, Henry Willingham
Grandy, Walter Thomas, Jr
Grant, James J, Jr
Gray, Ernest Paul
Green, Joseph Matthew
Greenberg, Howard
Greenberg, Newton Isaac
Greenberger, Daniel Mordecai
Gregorich, David Tony
Greifinger, Carl
Griffin, James J
Griffing, George Warren
Griffiths, David Jeffery
Griffy, Thomas Alan
Grodsky, Irvin T
Gross, Franz Lucretius
Grotch, Howard
Guier, William Howard
Gunther, Marian W J
Gunton, James D
Gupta, Suraj Narayan
Gursky, Martin Lewis
Gutzwiller, Martin Charles
Hahn, Yukap
Haig, Francis Rawle
Haines, Larry Kent
Hall, George Lincoln
Hall, Harold Hershey
Halley, James Woods, (Jr)
Halpern, Francis Robert

Hanson, Harvey Myron
Hardy, James Edward
Harlow, Francis Harvey, Jr
Harris, Edward Grant
Harris, Isadore
Harrison, Bertrand Kent
Hart, Edward Walter
Hart, Harold Bird
Hart, Robert Warren
Harte, John
Hassoun, Ghazi Qasim
Hatch, Dorian Maurice
Havas, Peter
Hearn, Anthony Clem
Heller, Ralph
Helliwell, Thomas McCaffree
Henderson, Douglas
Herling, Gary Herbert
Herman, Frank
Herman, Robert
Herring, Jackson Rea
Herschman, Arthur
Herzenberg, Arvid
Hestenes, David
Hibbs, Albert Roach
Hill, Robert Nyden
Hilsinger, Harold W
Hirt, Cyril William, Jr
Hiskes, John Robert
Hobey, William David
Hobson, Arthur Stanley
Hodgson, Richard John Wesley
Hoffman, Alexander A J
Hogan, William Alfred
Holland, Dan Howard
Hooper, Charles Frederick, Jr
Hope, Lawrence Latimer
Horrigan, Frank Anthony
Horton, Claude Wendell, Jr
Horwitz, Gerald
Horwitz, Lawrence Paul
Howard, Roger
Howland, Louis Philip
Huang, Justin C
Huang, Kerson
Huber, David Lawrence
Hubisz, John Lawrence, Jr
Huddlestone, Richard H
Huggins, Elisha R
Hull, McAllister Hobart, Jr
Hume, James Nairn Patterson
Hundhausen, Arthur James
Hundley, Richard O'Neil
Hunter, Douglas Lyle
Huschilt, John
Iddings, Carl Kenneth
Inglis, David Rittenhouse
Ingraham, Richard Lee
Inomata, Akira
Israel, Werner
Ivash, Eugene V
Jackiw, Roman Wladimir
Jackson, Henry Woodrow
Jackson, John David
Jacob, Richard John
Jacob, Robert L
Jaynes, Edwin Thompson
Jenkins, Hughes Brantley, Jr
Johnson, Kenneth Alan
Johnson, Vivian Annabelle
Johnston, Robert R
Jokipii, Jack Randolph
Jones, George R
Jones, Robert Clark
Jones, Robert William
Jordan, Thomas Fredrick
Jorna, Seibe
Joyce, William Baxter
Judd, Brian Raymond
Judd, David Lockhart
Kabir, Prabahan Kemal
Kacser, Claude
Kaempffer, Frederick Augustus
Kahn, Peter B
Kalman, Gabor
Kalos, Malvin Howard
Kang, Kyungsik
Kaplan, Harvey
Kaplan, Irving
Karl, Gabriel
Karplus, Robert
Kaskas, James
Kaufman, Charles
Kaus, Peter Edward
Kazes, Emil
Kelly, Donald C
Kemeny, Gabor
Kennedy, Frederick James
Kennedy, Robert E
Kerlick, George David
Kerman, Arthur Kent
Kerner, Edward Haskell
Khuri, Nicola Najib
Kiefer, Harold Milton
Kijewski, Louis Joseph
Kim, Sung Kyu
Kim, Young Nok
Kimel, Jacob Daniel, Jr

Kinnersley, William Morris
Kinzer, Earl T, Jr
Kissinger, Leonard Sol
Kivel, Bennett
Kjeldaas, Terje, Jr
Klarfeld, Joseph
Klauder, John Rider
Klein, Abraham
Klein, Lewis S
Klevans, Edward Harris
Klein, William H
Klotz, Frederick Succop
Knapp, Robert Hazard, Jr
Knight, James Milton
Koga, Toyoki
Komar, Arthur Baraway
Kornacker, Karl
Korringa, Jan
Kounosu, Shigeru
Kovacs, Julius Stephen
Krieger, Stephan Jacques
Krieger, Theodore Joseph
Krizan, John Ernest
Kronminga, Albion Jerome
Krueger, David Allen
Kubis, Joseph John
Kuo, Pao-Kuang
Kuo, Thomas Tzu Szu
Kursunoglu, Behram
Lakin, Wilbur
Lamb, Donald Quincy, Jr
Lamb, Frederick Keithley
Lambropoulos, Peter Poulos
Land, David John
Lande, Alexander
Lane, Eric Trent
Lang, Joseph Edward
Lapidus, Ivan Richard
Larsen, David M
Larsen, Sigurd Yves
Larson, Nancy Marie
Latter, Albert L
Latter, Richard
Lavine, James Philip
Law, Jimmy
Lawson, Kent DeLance
Lawson, Robert Davis
Lazarus, Roger Ben
Leacock, Robert A
Lee, Benjamin W
Lee, Monhe Howard
Lee, Tsung Dao
Leech, John Watson
Legendy, Charles Rudolf
Lehman, Guy Walter
Le Levier, Robert Ernest
Lemos, Anthony M
Leon, Melvin
Lerner, David Evan
Lerner, Edward Clarence
Levine, Arnold David
Levine, Paul Hersh
Lieber, Michael
Lichtenberg, Don Bernett
Lightman, Alan Paige
Lincoln, Charles Albert
Lind, Robert Wayne
Lindenmeier, Charles William
Ling, Daniel Seth, Jr
Litvak, Marvin Mark
Liu, Luke Lokia
Liu-Ger, Tsu-Huei
Loly, Peter Douglas
Lomanitz, Ross
Lomon, Earle Leonard
Longley, Herbert J
Lovelace, Claud William Venton
Lovelock, David
Low, Francis Eugene
Lowdin, Per-Olov
Lublin, Elihu
Lucey, Carol Ann
Luttinger, Joaquin Mazdak
Madan, Rabinder Nath
Maharty, David Edwin
Mahmoud, Hormoz Massoud
Malin, Shimon
Mallory, William R
Malone, Robert Charles
Mandelstam, Stanley
Manning, Irwin
Mano, Koichi
Mansfield, John E
Manz, Bruno Julius
Marburger, John Harmen, III
Marker, David
Markley, Francis Landis
Marshak, Robert Eugene
Martin, Arthur Wesley, III
Martin, Paul Cecil
Martineau, Robert Jean
Mascheroni, P Leonardo

Mast, Cecil B
Mateese, John J
Mathews, Jon
Mayer, Harris Louis
McAfee, Walter Samuel
McClure, Joel William, Jr
McClure, Joseph Andrew, Jr
McColl, Daniel Clyde
McCormick, Philip Thomas
McCoy, Barry
McDonald, Frank Alan
McDonald, Keith Leon
McGinn, William David
McGuire, James B
McIntosh, John Stanton
McIntyre, Robert Gerald
McIvrine, Edward Charles
McKinley, John McKeen
McKnight, John Lacy
McManus, Hugh
McMillan, James Malcolm
Meads, Philip Francis, Jr
Meckler, Alvin
Meiere, Forrest T
Meijer, Paul Herman Ernst
Mellen, Walter Roy
Mendlowitz, Harold
Menne, Thomas Joseph
Merzbacher, Eugen
Metropolis, Nicholas Constantine
Metz, Roger N
Merger, Fritz Walter William
Miller, Arthur I
Miller, Bruce Neil
Miller, Gary Glenn
Miller, Irvin Alexander
Miller, Jack Culbertson
Miller, Stanley Custer, Jr
Mills, Robert Laurence
Minerbo, Gerald N
Misner, Charles William
Mitalas, Romas
Mohling, Franz
Moliow, Benjamin R
Monahan, James Emmett
Montgomery, Charles Gray
Moorhead, William Dean
Moran, James Herbert
Morawitz, Hans
More, Richard Michael
Moroi, David S
Morrison, James Leslie
Morse, Philip McCord
Moses, Harry Elecks
Moszkowski, Steven Alexander
Motil, Ronald Allen
Mould, Richard A
Mountain, Raymond Dale
Mullen, George Henry
Mulligan, Bernard
Nadeau, Gerard
Nambu, Yoichiro
Nedelsky, Leo
Negele, John William
Nelkin, Mark
Nelson, Charles Arnold
Nelson, Terence John
Nesbet, Robert Kenyon
Neufeld, Jacob
Newby, Neal Dow, Jr
Newcomb, William A
Newman, Ezra
Newstein, Maurice
Newton, Roger Gerhard
Newton, Theodore Duddell
Nieto, Michael Martin
Nishimura, Keichi
Nogami, Yukihisa
Noble, Julian Victor
Noble, William John
Nordtvedt, Kenneth L
North, Dwight Olcott
North, Gerald R
Norton, Richard E
Nosanow, Lewis H
Noyes, Henry Pierre
Noz, Marilyn E
Nuttall, John
O'Connell, Robert F
Ochme, Reinhard
Olshansky, Robert
Oneda, Sadao
Onley, David S
Onsager, Lars
Opechowski, Wladyslaw
Oppo, Giuseppe
Pagels, Heinz Rudolf
Painter, Norman Harding
Pais, Abraham
Paldus, Josef
Pandres, Dave, Jr
Paquette, Guy
Park, David Allen
Park, Jae Young
Park, Shim C
Parker, Barry Richard
Parker, Leonard Emanuel
Parks, William Frank

Parren, George Burl, Jr
Parzen, George
Pastine, D John
Pathria, Raj Kumar
Pati, Jogesh Chandra
Patsakos, George
Payne, Marvin Gay
Payne, Wilbur Boswell
Peak, David
Pearle, Philip Mark
Pearlstein, Leon Donald
Pearson, John Michael
Peaslee, Alfred Tredway, Jr
Peaslee, David Chase
Pendleton, Hugh Nelson, III
Penney, Robert Vincent
Percus, Jerome K
Pereira, Carlos Martin
Perez, Joseph Dominique
Peshkin, Murray
Peterson, Gerald A
Pethick, Christopher John
Pettit, John Tanner
Philips, Thomas O
Picker, Harvey Shalom
Pinkston, William Thomas
Pines, David
Poll, Jacobus Daniel
Pondrom, Walter Lewis, Jr
Pooch, Udo Walter
Pope, Noel Kynaston
Porter, William Samuel
Prastein, Solomon Matthew
Prats, Francisco
Pratt, George Woodman, Jr
Preston, Melvin Alexander
Price, Richard Henry
Primack, Joel Robert
Primakoff, Henry
Pryce, Maurice Henry Lecorney
Puff, Robert David
Pugh, Milton Earl
Pugh, Robert E
Puhach, Paul Alexander
Pyite, Agnar
Quinn, John Joseph
Quirk, James Denis
Rafanelli, Kenneth R
Rahman, Anesur
Rajagopal, Attipat Krishnaswamy
Ram, Michael
Raman, Varadaraja Venkata
Raphael, Robert B
Rastall, Peter
Ratcliff, Keith Frederick
Ravenhall, David Geoffrey
Reichert, John Douglas
Reid, Roderick Vincent, Jr
Reitz, John Richard
Renau, Jacques
Rendell, David H
Resnick, Lazer
Richardson, John Mead
Riddell, Robert James, Jr
Roberts, Charles A, Jr
Robertson, Baldwin
Robinson, Ivor
Rohl, Hermann R
Rohrlich, Fritz
Roman, Paul
Rosen, Gerald Harris
Rosen, Nathan
Rosenbluth, Marshall N
Rosenzweig, Norbert
Roskies, Ralph Zvi
Ross, David Ward
Ross, Marc Hanson
Roth, Charles
Rubin, Morton Harold
Ruderman, Malvin Avram
Ruppel, Hans Max
Rush, John Edwin, Jr
Rystephanick, Raymond Gary
Sachs, Lester Marvin
Sachs, Mendel
Sachs, Rainer Kurt
Sachs, Robert Green
Saenz, Albert William
Sakakura, Arthur Yoshikazu
Sakmar, Ismail Aydin
Salter, Lewis Spencer
Salwen, Harold
Salzman, George
Salzman, Freda
Santilli, Ruggero Maria
Saperstein, Alvin Martin
Satchler, George Raymond
Savit, Robert Steven
Sawyer, Raymond Francis
Scadron, Michael David
Scalettar, Richard
Scarf, Frederick Leonard
Scheerbaum, Robert R
Schey, Harry Moritz
Schieve, William
Schiff, Harry
Schiller, Ralph
Schlegel, Richard

Schlessinger, Leonard
Schloemann, Earnst
Schmid, Lawrence Alfred
Schneider, Ronald E
Schrenk, George L
Schult, Roy Louis
Schultz, Theodore David
Schulz, William Donald
Schumacher, Clifford Rodney
Schwartz, Brian B
Schwartz, Charles Leon
Schwartz, Melvin J
Schwartz, Robert Alan
Schwartz, John Henry
Schwebel, Solomon Lawrence
Segal, Irving Ezra
Segall, Benjamin
Segre, Gino C
Sein, John Joan
Sessler, Andrew M
Shaffer, Russell Allen
Sham, Lu Jeu
Shapero, Donald Campbell
Sharp, David Howland
Sharp, Robert Thomas
Shaw, Gordon Lionel
Shea, Dion Warren Joseph
Shelton, Russell D
Sherman, Noah
Shewell, John Robert
Shimamoto, Yoshio
Shimony, Abner
Sieckmann, Everett Frederick
Sigal, Richard Frederick
Signell, Peter Stuart
Skinner, Orville Ray
Slutsky, Mark Sender
Smith, Charles Ray
Smith, Gerrit Joseph
Smith, Jack Howard
Smith, Robert Clinton
Smith, Roger Alan
Sobel, Michael I
Sokoloff, Jack
Soln, Josip Zvonimir
Sommerfield, Charles Michael
Sowle, David H
Spector, Richard M
Speisman, Gerald
Spielberg, Kurt
Spruch, Larry
Stack, John D
Stamper, James Harris
Steinberg, Melvin Sanford
Stephenson, John
Stetler, John Dietrich
Steverding, Bernard
Stratton, Robert
Strecker, Joseph Lawrence
Stuart, George Wallace
Sudarshan, Ennackel Chandy George
Sugar, Robert Louis
Sung, Chi Ching
Swan, Peter Howard
Swihart, James Calvin
Tabakin, Frank
Takahashi, Yasushi
Talman, James Davis
Tambasco, Daniel Joseph
Tamor, Stephen
Tani, Smio
Tanner, James Mervil
Tarter, Curtis Bruce
Tauber, Gerald Erich
Tavel, Morton
Taylor, David Ward
Taylor, John Robert
Teeters, William Dale
Teichmann, Theodor
Teng, Lee Chang-Li
Thaler, Raphael Morton
Thews, Robert Leroy
Thomas, Billy Seay
Thompson, Samuel Lee
Thompson, Sanford P
Thorne, Kip Stephen
Toll, John Sampson
Tomljanovich, Nicholas Matthew
Torgerson, Ronald Thomas
Torrence, Robert James
Towne, Dudley Herbert
Trainor, Lynne E H
Trammell, George Thomas
Trickey, Samuel Baldwin
Trigg, George Lockwood
Troubetzkoy, Eugene Serge
Trubatch, Sheldon L
Tsai, Yung Su
Tuan, San Fu
Tuan, Tai-Fu
Tubis, Arnold
Tucker, Robert H
Tung, Wu-Ki
Turner, Leaf
Tuttle, Elizabeth R
Ullrich, George Werner
Umezawa, Hiroomi

298

Underhill, Glenn
Unruh, William George
Uzes, Charles Alphonse
Vallarta, Manuel Sandoval
Van Kranendonk, Jan
Vasavada, Kashyap V
Vassell, Milton O
Villars, Felix Marc Hermann
Visscher, William M
Vogt, Erich W
Volkoff, George Michael
Von Baeyer, Hans Christian
Von Hippel, Frank
Von Roos, Oldwig
Vosko, Seymour H
Wackman, Peter Husting
Waggoner, Jack Holmes, Jr
Wagner, William Gerard
Wald, Robert Manuel
Walecka, John Dirk
Walker, James Frederick, Jr
Walker, James Joseph
Wallace, Philip Russell
Wang, Shao-Fu
Wannier, Gregory Hugh
Watanabe, Michael Satosi
Weaver, David Leo
Webb, Jerry Glen
Weinberg, Steven
Weiss, Peter-Reifer
Welling, Daniel J
Welton, Theodore Allen
Weneser, Joseph
Wentzel, Gregor
Wesley, Walter Glen
Westervelt, Peter Jocelyn
Weston, Vaughan Hatherley
Wheeler, John Archibald
Wheelon, Albert Dewell
Whitesell, William James
Whitlock, Richard T
Wichmann, Eyvind Hugo
Wilcox, Charles Hamilton
Williams, Clayton Drews
Williams, Harry Thomas
Williams, Stanley A
Wills, John G
Winicour, Jeffrey
Winterberg, Friedwardt
Wist, Abund Ottokar
Witriol, Norman Martin
Witten, Thomas Adams, Jr
Wojtaszek, Joseph Henry
Wojtowicz, Peter Joseph
Woo, Chia-Wei
Woods, Roger David
Wortis, Michael
Woznick, Benjamin Joseph
Wright, Jon Alan
Wright, Louis Edgar
Wright, William Robert
Wu, Alfred Chi-Tai
Wu, Fa Yueh
Wulfman, Carl E
Wylen, Herbert E
Yamauchi, Hiroshi
Yano, Fleur Belle
Yeats, Frank Richard
Yedinak, Peter Demerton
York, James Wesley, Jr
Yos, Jerrold Moore
Young, Francis Lorraine
Young, Hugh David
Young, Robert Carl
Zachariasen, Fredrik
Zapolsky, Harold Saul
Zatzkis, Henry
Zavada, John Michael, Jr
Zeleny, William Bardwell
Zilsel, Paul Rudolph
Zimmerman, Edward John
Zinnes, Irving Isadore
Zwanziger, Daniel
Zweifel, Paul Frederick

Theoretical High Energy Physics
Abbe, Winfield Jonathan
Adler, Stephen L
Auvil, Paul R, Jr
Balazs, Louis A P
Barger, Vernon Duane
Bowling, Arthur Lee, Jr
Cahn, Robert Nathan
Campbell, David Kelly
Carhart, Richard Alan
Chang, Ngee Pong
Chen, Min-Shih
Chew, Geoffrey Foucar
Choudhury, Abdul Latif
Cutkosky, Richard Edwin
DeFacio, W Brian
Dobson, Peter N, Jr
Dolan, Louise Ann
Dragt, Alexander James
Drell, Sidney David
Ebel, Marvin Emerson
Einhorn, Martin B

Feinberg, Gerald
Frautschi, Steven Clark
Gaskell, Robert Weyand
Gilman, Frederick Joseph
Goldberger, Marvin Leonard
Gross, David (Jonathan)
Haller, Kurt
Halpern, Martin B
Hwa, Rudolph Chia-Chao
Isgur, Nathan Gerald
Islam, Muhammad Munirul
Jackson, Shirley Ann
Kaku, Michio
Kayser, Boris Jules
Kim, Chung W
Kinoshita, Toichiro
Lam, Harry Chi-Sing
Li, Ling-Fong
Luthe, John Charles
Machacek, Marie Esther
Mainland, Gordon Bruce
Malenka, Bertram Julian
Muzinich, Ivan J
Narayanaswamy, Padmanabha
Neville, Donald Edward
Oakes, Robert James
Pappademos, John Nicholas
Peierls, Ronald F
Rockmore, Ronald Marshall
Roper, Leon David
St Amand, Joseph
Sakita, Bunji
Sakurai, Jun John
Samuel, Mark Aaron
Schlitt, Dan Webb
Schnitzer, Howard J
Sebastian, Kunnat J
Shapiro, Joel Alan
Shepard, Harvey Kenneth
Simmons, Leonard Micajah, Jr
Slansky, Richard Cyril
Soper, Davison Eugene
Teeters, William Dale
Tiktopoulos, George S
Tomozawa, Yukio
Tryon, Edward Polk
Ward, Bennie Franklin Leon
Warnock, Robert Lee
Weinstein, Marvin
Weis, Joe H
Weisberger, William i
Wilson, John William
Wolfenstein, Lincoln
Wong, Tang-Fong Frank
Yao, York-Peng Edward
Yennie, Donald Robert
Zee, Anthony

Theoretical Mechanics
Denman, Harry Harroun
Deresiewicz, Herbert
Lee, Sung Mook
Parry, Myron Gene
Rim, Kwan
Simmons, John Arthur
Smith, Frederick Adair, Jr
Todd, Leonard
Worley, Will J

Theoretical Nuclear Physics
Altman, Albert
Austern, Norman
Baltz, Anthony John
Barrett, Bruce Richard
Bassichis, William
Becker, Richard Logan
Bhakar, Balram Singh
Chiao-Yap, Lung Wen
Clark, John Walter
Coker, William Rory
Cordes, Davies Marcia A
Coulter, Philip W
Cusson, Ronald Yvon
Davis, Alvie Douglas
Donnelly, Thomas William
Dorn, Carl W
Dover, Carl Bellman
Edwards, Steve
Franco, Victor
Freed, Norman
Fuda, Michael George
Fuller, Richard Clair
Ginocchio, Joseph Natale
Glendenning, Norman Keith
Grube, Geraldine Joyce Terenzoni
Hadjimichael, Evangelos
Haftel, Michael Ivan
Halbert, Edith Conrad
Harris, Samuel Melvin
Harvey, Malcolm
Henley, Ernest Mark
Ho, John Ting-Sum
Ho-Kim, Quang
Jackson, Andrew D, Jr
Kazaks, Peter Alexander
Kim, Yeong Ell
Kowalski, Kenneth L

Kurath, Dieter
Lassila, Kenneth Eino
Lehman, Donald Richard
LeTourneux, Jean
Lim, Teck-Kah
Lock, James Albert
Ma, Chin Wah
Madsen, Victor Arviel
McVoy, Kirk Warren
Millener, David John
Miller, Lewis Dudley
Moomaw, William Renken
Newton, Victor Joseph
Nix, James Rayford
Payne, Gerald Lew
Pieper, Steven Charles
Pittel, Stuart
Purrington, Robert Daniel
Radomski, Mark Stephen
Rawitscher, George Heinrich
Redish, Edward Frederick
Rinker, George Albert, Jr
Saxon, David Stephen
Schick, Lee Henry
Scott, Bruce L
Seaborn, James Byrd
Seki, Ryoichi
Sharon (Schwadron), Yitzhak Yaakov
Sheldon, Eric
Stephenson, Gerard J, Jr
Strobel, George L
Tobocman, William
Tomusiak, Edward Lawrence
Towner, Ian Stuart
Uberall, Herbert Michael
Valk, Henry Snowden
Verhanovitz, Richard Frank
Walker, George Edward
Wallace, Stephen Joseph
Weber, Hans Jurgen
Werntz, Carl H

Theoretical Solid State Physics
Adawi, Ibrahim (Hasan)
Allender, David William
Altarelli, Massimo
Antoniewicz, Peter R
Bandy, William Robert
Baratoff, Alexis
Barrett, John Harold
Bartel, Lewis Clark
Beissner, Robert Edward
Bergstresser, Thomas Karl
Bishop, Marilyn Frances
Blount, Eugene Irving
Bostock, Judith Louise
Bray, James William
Brown, Edmond
Chui, Siu-Tat
Csavinszky, Peter John
Davis, Harold Lloyd
Dexter, David Lawrence
Duke, Charles Bryan
Einstein, Theodore Lee
English, Philip Stephen
Esterling, Donald M
Euwema, Robert Noel
Evenson, William Edwin
Feibelman, Peter Julian
Fetter, Alexander Lees
Flocken, John W
Frank, Barry
Gadzuk, John William
Gay, Jackson Gilbert
Gersten, Joel Irwin
Gillis, Nelson Silsbee
Glyde, Arnold J
Glyde, Henry Russell
Goodman, Bernard
Gora, Thaddeus F, Jr
Gray, Alma Marcus
Green, Barry Adams
Gunther, Leon
Gupta, Rajendra P
Haas, Charles Warren
Hall, George Lincoln
Hannon, James Patrick
Harrison, Michael Jay
Henkel, John Harmon
Herring, Conyers
Hertz, John Atlee
Hohenberg, Pierre Claude
Hone, Daniel W
Ivey, Jerry Lee
Jacobs, Allan Edward
James, Hubert Maxwell
Jennison, Dwight Richard
Karakashian, Aram Simon
Kenkre, Vasudev Mangesh
Kliewer, Kenneth L
Korenman, Victor
Krieger, Joseph Bernard
Lang, Norton David
Lawrence, Walter Edward
Lax, Melvin
Leplae, Luc A
Lines, Malcolm Ellis
Liu, Samuel Hsi-Peh

Lubensky, Tom C
Lurie, Joan B
Mahan, Gerald Dennis
Mahanti, Subhendra Deb
Maradudin, Alexei Alexei
Massey, Walter Eugene
Mavroyannis, Constantine
Menzel, Wolfgang Paul
Moriarty, John Alan
Morley, Gayle L
Morris, Stanley P
Mullin, William Jesse
Myron, Harold William
Nedoluha, Alfred K
Novaco, Anthony Dominic
O'Dwyer, John J
O'Hare, John Michael
Ortenburger, Irene Beardsley
Overhauser, Albert Warner
Painter, Gayle Stanford
Pantelides, Sokrates Theodore
Pardee, William Joseph
Peterson, Robert Lee
Pink, David Anthony Herbert
Price, Peter J
Rabii, Sohrab
Rice, Thomas Maurice
Rimbey, Peter Raymond
Rothwarf, Allen
Salahub, Dennis Russell
Schaich, William Lee
Shore, Herbert Barry
Silver, Richard N
Simpson, John Hamilton
Sinha, Om Prakash
Smith, John Robert
Sorbello, Richard Salvatore
Stern, Frank
Swift, Jack Bernard
Taggart, George Bruce
Tait, William Charles
Thorpe, Michael Fielding
Van Vechten, James Alden
Wang, Ching-Ping Shih
Wolf, Dieter
Woodruff, Truman Owen

PHYSIOLOGY

Physiology
Abaidoo, Kodwo-James R
Abarbanel, Abraham Robert
Abraham, Samuel
Abrams, Herbert L
Abrams, Robert Marlow
Ackles, Kenneth Norman
Adamkiewicz, Vincent Witold
Adams, Herbert Jack
Adams, Thomas
Adams, Walter Church
Adelman, William Joseph, Jr
Agersborg, Helmer Pareli Kjerschow, Jr
Agnew, William Finley
Ahrens, Richard August
Akins, Ervin Loraine
Alarie, Yves
Al-Awqati, Qais
Aldred, J Philip
Alexander, Robert Spence
Al-lami, Fadhil
Allen, Fred William
Allen, George Otis
Allgood, Joseph Patrick
Alpen, Edward Lewis
Alpert, Norman Roland
Alteveer, Robert Jan George
Altland, Paul Daniel
Altura, Bella T
Altura, Burton Myron
Alvarado, Francisco
Ambromovage, Anne Marie
Ames, Smith Whittier
Anderson, James Howard
Anderson, John Denton
Anderson, John Francis
Anderson, John Joseph Baxter
Anderson, Lloyd L
Anderson, Norman Gulack
Anderson, Russell K
Andreoli, Thomas Eugene
Andrew, George McCoubrey
Andrews, Richard Vincent
Angelakos, Evangelos Theodorou
Angelone, Luis
Angerer, Clifford Ackerman
Angstadt, Robert B
Annegers, John Herman
Anthony, Adam
Archdeacon, James William
Arendt, Kenneth Albert
Arimura, Akira
Armstrong, Clay M
Armstrong, David Thomas
Armstrong, George Glaucus, Jr
Arnold, William Archibald
Aronson, David L

Aronson, John Ferguson
Asano, Tomoaki
Ascrinsky, Eugene
Ashburn, Allen David
Ashwin, James Guy
Atherton, Robert W
Atterbom, Hemming A
Atwood, Harold Leslie
Atwood, Mark Wyllie
Aull, Felice
Aviado, Domingo Mariano
Awad, Mohamed Zeinelabideen
Bacaner, Marvin Bernard
Bacus, James William
Badeer, Henry Sarkis
Badger, Donald W
Baecher, Charles Albert
Baecker, Anna Medora
Baginski, Eugene S
Baiardi, John Charles
Baile, Clifton Augustus, III
Bailey, Donald Wycoff
Bailie, Michael David
Bair, Thomas De Pinna
Baird, Malcolm Barry
Bajpai, Praphulla K
Baker, Carleton Harold
Baker, Donald Granville
Baker, Lee Edward
Baker, Robert David
Balchum, Oscar Joseph
Baldwin, Ransom Leland, Jr
Baldwin, William Russell
Balfour, William Mayo
Balke, Bruno
Bancroft, Richard Wolcott
Barcelo, Raymond
Bard, Philip
Barger, Abraham Clifford
Barker, June Northrop
Barker, Samuel Booth
Barlow, George
Barnes, Charles Dee
Barnes, Lester E
Barnhart, Marion Isabel
Barr, Lloyd
Barras, Donald J
Barrera, Frank
Barrow, Emily Mildred Stacy
Bar-Sela, Mildred Elwers
Bartlett, Donald, Jr
Baruch, Sulamita B
Bashir, Nasir Ahmad
Bass, Berl G
Bass, David Eli
Bass, Paul
Bassingthwaighte, James B
Batt, Ellen Rae
Battaglia, Frederick Camillo
Battista, San P
Bauer, Adelia Catherine
Bauer, Mark Henry
Baum, Siegmund Jacob
Baumber, John Scott
Bawden, James Wyatt
Baxter, Claude Frederick
Beames, Calvin G, Jr
Bean, John William
Beard, Elizabeth L
Beaty, Clarissa Hager
Beck, Ronald Richard
Beckman, Edward Louis
Bedford, John Michael
Behnke, Harold R
Behrman, Richard Elliot
Beidler, Lloyd M
Belair, Ernest Joseph
Belknap, Robert Wayne
Bell, Richard Dennis
Beltz, Alex D
Ben, Max
Benjamin, Fred Berthold
Benjamin, Hiram Bernard
Bennett, Arthur Lawrence
Bennett, Peter Brian
Benoit, Philippe Stanislas
Bensadoun, Andre
Bentley, Peter John
Benton, Allen William
Bergen, John Richard
Bergeron, Georges Albert
Bergeron, Michel
Bergman, Emmett Norlin
Bergman, Ronald Arly
Bergofsky, Edward Harold
Berliner, Robert William
Berman, Herbert Joshua
Bernard, Patrick Spialetta
Bernardis, Lee L
Bernstein, Leon
Besarab, Anatole
Besse, John C
Best, Charles Herbert
Best, Jay Boyd
Beyer, Karl Henry, Jr
Beznak, Margaret
Bhagat, Budh Dev
Bhattacharya, Amar Nath

Bianchi, Donald Ernest
Bieliler, Harold Victor
Biggers, John Dennis
Bileau, Claire of the Savior
Bird, John William Clyde
Birdsell, Dale Carl
Birks, Richard Irwin
Bishop, Beverly Peterson
Bishop, David Wakefield
Bishop, Jack Garland
Bishop, Vernon Spilman
Bittar, Evelyn Edward
Bitter, Harold Louis
Blackburn, John Gill
Blackshear, Gertrude Liebl
Blackwell, Leo Herman
Blake, William Dewey
Blamberg, Donald Lee
Blank, Martin
Blanpied, George David
Blasingham, Mary Cynthia
Blatteis, Clark Martin
Blaumanis, Otis Rudolf
Blaustein, Ben Burton
Bloomquist, Eunice
Bloor, Colin Mercer
Blount, Don Houston
Blum, Harold Francis
Blum, Jacob Joseph
Blum, Murray Sheldon
Boatman, Joseph Brasher
Bocage, Albert J
Bogenschutz, Robert Parks
Bogner, Phyllis Holt
Bohannon, Randolph F
Bond, Judith
Bond, Robert Franklin
Bond, Ted P
Book, Steven Arnold
Booth, Nicholas Henry
Borgese, Thomas A
Borle, Andre Bernard
Bornmann, Robert Clare
Bortoff, Alexander
Botello, Stella Tayes
Bottoms, Gerald Doyle
Boullin, David John
Bowen, Thomas Earle, Jr
Bowman, Douglas Clyde
Bowman, Edward Randolph
Boyarsky, Saul
Boyd, William Adam
Boylan, John W
Boyle, Robert William
Bozler, Emil
Bradford, Spencer Graves
Bramante, Pietro Ottavio
Brandt, Charles Lawrence
Brandt, Philip Williams
Braun, Eldon John
Brecher, Gerhard Adolf
Breckenridge, Bruce (McLain)
Bredahl, Edward Arlan
Bredeck, Henry E
Breed, Ernest Spencer
Brett, William John
Briggs, Fred Norman
Briscoe, Anne M
Brobeck, John Raymond
Brock, Mary Anne
Brockway, Barbara Fink
Brodish, Alvin
Brodsky, Alfred Walter
Brody, William Aaron
Bronner, Felix
Brookhart, John Mills
Brooks, Chandler McCuskey
Brooks, Frank Pickering
Brooks, Matilda Moldenhauer
Brown, Arthur Charles
Brown, Arthur Morton
Brown, Herbert Ensign
Brown, Jack Harold Upton
Brown, John R
Brown, Kenneth Taylor
Brown, Richard Edward
Brown, Robert Glenn
Brown, Robert Lee
Brownell, Katharine Anna
Bruce, John Irvin, Jr
Brugge, John F
Brunlieve, Stanley John
Brunson, John Taylor
Brusilow, Saul W
Brust, Manfred
Brust-Carmona, Hector
Bryan, John Kent
Bryan, William Ray
Bryan, Shirley Hills
Bryson, George Gardner
Buchholz, Robert Henry
Buckley, Edward Harland
Buckley, Nancy Margaret
Buckley, Ramon D
Budd, Geoffrey Colin
Buderer, Melvin Charles
Budy, Ann Marie

Buerger, Alfred Arthur
Buffett, Rita Frances
Bull, Leonard Seth
Bullard, Truman Robert
Bullock, Howard R
Bundy, Roy Elton
Bunde, Daryl E
Burdick, Harold Charles
Burdine, Howard William
Burgess, Benjamin Franklin, Jr
Burke, Jack Denning
Burke, Thomas Joseph
Burlington, Harold
Burmester, Mary Alice (Horswill)
Burnell, James McIndoe
Burns, John Mitchell
Burns, Moore J
Buschmann, Robert John
Bush, Ian
Buskirk, Elsworth Robert
Butcher, Reginald William
Butler, Hugh C
Byrne, Jeffrey Edward
Cain, Arthur Samuel, Jr
Cain, Stephen Martin
Calaresu, Franco Romano
Caldwell, Robert William
Calhoun, Thomas Bruce
Callantine, Merritt Reece
Cameron, Colin Robert
Cameron, James N
Campbell, Bonnalie Oetting
Campbell, James
Candia, Oscar A
Carbonneau, Roch
Carlson, Blondel Henry
Carlson, Stanley David
Carpenter, Frank Grant
Carpenter, Malcolm Breckenridge
Carr, Ronald E
Carregal, Enrique Jose Alvarez
Carrier, Oliver, Jr
Carroll, Douglas Gordon
Carroll, Marcus Newman, Jr
Carter, Earl Thomas
Carter, Stefan A
Casady, Robert Barnes
Case, Robert B
Cassin, Sidney
Castro, Gilbert Anthony
Catchpole, Hubert Ralph
Cavanagh, Charles Johnson, Sr
Cavert, Henry Mead
Celander, David Robert
Cerroni, Rose E
Chace, Alfred Bernard
Chaffee, Rowand R J
Chakraborty, Prabir Kumar
Chambers, Alfred Hayes
Chan, Lee-Nien Lillian
Chan, Teh-Sheng
Chang, Franklin
Chang, Min Chueh
Chang, Thomas Ming Swi
Chappell, Guy Lee Monty
Chefurka, William
Chen, Philip Stanley, Jr
Chernick, Victor
Chertok, Robert Joseph
Chevalier, Peter Andrew
Chidsey, Jane Louise
Chien, Shu
Chilgren, John Douglas
Chinoskey, John Edward
Chou, Ching-Chung
Chowdhury, Tushar Kumar
Christ, Daryl Dean
Christensen, Eleanor
Christensen, James
Christianson, Richard Louis
Christian, Ross Edgar
Christian, Paul Clayton
Churchill, Leon
Churney, Leon
Cinti, Dominick Louis
Cipriano, Leonard Francis
Civan, Mortimer M
Cizek, Louis Joseph
Clamann, H Peter
Clancy, Richard L
Clark, C Elmer
Clark, John Kapp
Clark, William Gilbert
Clegg, James S
Clements, John Allen
Clinch, Norman Frederick
Close, George Bruce
Close, Perry
coburn, Ronald F
Cockett, Abraham Timothy K
Code, Charles Frederick
Cody, D Thane
Cody, Charles Arthur Isaac
Cohen, Julius Jay
Cohen, Leonard Arlin
Cohen, Stanton Harry
Cohen, Benjamin Theodore
Cole, Edward Anthony

Coleman, Bernell
Collier, Brian
Collier, Clarence Robert
Collins, Robert James
Colvin, Harry Walter, Jr
Colwell, John Amory
Comroe, Julius Hiram, Jr
Conaway, Howard Herschel
Condon, Robert Edward
Connell, Robert Edward
Connell, Rosemary
Conner, Gabel Henry
Connolly, John Joseph
Conrad, John Terry
Conrad, Margaret C
Consalazio, Carlo Frank
Constantine, Herbert Patrick
Constantine, Jay Winfred
Convey, Edward Michael
Cook, Harry
Cooper, George William
Cooper, Keith Edward
Cooper, Richard Grant
Cooper, Theodore
Cooper, William Anderson
Copp, Alfred Lewin
Copp, Douglas Harold
Coppenger, Claude Jackson
Coppinger, Jack David
Cornelius, Charles Edward
Cornell, Creighton N
Corson, Samuel Abraham
Cortney, Marshall Allen
Cosgrove, William Burnham
Cottle, Merva Kathryn Warren
Cottle, Walter Henry
Coulter, Norman Arthur, Jr
Coursen, Bradner Wood
Covino, Benjamin Gene
Cowan, F Brian M
Cowley, Allen Wilson, Jr
Cox, Allan Clayton
Cox, Beverley Lenore
Cox, Robert Harold
Craig, Albert Burchfield, Jr
Craig, Francis Northrop
Cralley, John Clement
Cramer, Carl Frederick
Crane, Robert Kellogg
Cranefield, Paul Frederic
Crass, Maurice Frederick, III
Crawford, Eugene Carson, Jr
Creasy, Leroy L
Crenshaw, Miles Aubrey
Cismon, Jefferson Martineau
Cristofalo, Vincent Joseph
Critz, Jerry B
Croner, Jerry Haltiwanger
Cronin, Robert Francis Patrick
Cropp, Gerd J A
Crouch, Billy G
Crowell, Jack Wesley
Csaky, Tihamer Zoltan
Csapo, Arpad Istvan
Cserr, Helen F
Cullen, John Knox
Cummings, Edmund George
Cunningham, David A
Cunningham, Dorothy J
Curtin, Terrence M
Curtis, Brian Albert
Cvancara, Victor Alan
Cymerman, Allen
D'Aguanno, William
Dahms, Thomas Edward
D'Alecy, Louis George
Dalton, John Charles
Dandy, James William Trevor
Dane, Benjamin
Danforth, David Newton
Danhof, Ivan Edward
Daniels, Farrington, Jr
Dasler, Adolph Richard
Dasler, William Hoyt
Daubenspeck, John Andrew
Daugherty, Robert M, Jr
Davenport, Horace Willard
Davidson, Floyd Francis
Davidson, Ivan William Frederick
Davidson, John Keay, III
Davies, Philip Wynne
Davies, Robert Ernest
Davis, Audrey Kennon
Davis, Darrell Lawrence
Davis, Elizabeth Young
Davis, George Diament
Davis, James Othello
Davis, Thomas
Davis, William James
Davison, John (Amerpohl)
Davison, Kenneth Lewis
Dawson, David Charles
Deane, Norman
DeBias, Domenic Anthony
De Boer, Jelle
Debons, Albert Frank
De Champlain, Jacques

Deen, William Murray
DeFouw, David O
de Freitas, Anthony S
Dejmal, Roger Kent
Delahayes, Jean
Delaney, John P
Del Guercio, Louis Richard M
De Long, Chester Wallace
Del Pozo, Efren Carlos
DeMaggio, Augustus Edward
DeMartinis, Frederick Daniel
De Mello, W Carlos
Denbo, John Russell
Dennis, Clarence
Dennis, Warren Howard
Depocas, Florent
Desjardins, Claude
Desrosiers, Joseph A Jacques
Detar, Reed L
Detweiler, David Kenneth
Deuben, Roger R
Devoe, Robert
De Weer, Paul Joseph
Dewein, Louis F
Dhalla, Naranjan Singh
Diamond, Jared Mason
Diana, John N
Dickson, James Gillespie, Jr
Diecke, Friedrich Paul Julius
Diermeier, Harold Frederick
Dietrich, John P
Dill, David Bruce
Dill, Russell Eugene
Di Luzio, Nicholas Robert
DiPasquale, Gene
Dirks, John Herbert
Di Salvo, Nicholas Armand
Dixon, John Aldous
Dixon, Wallace Clark, Jr
Dobson, Ernest L
Doemling, Donald Bernard
Doisy, Edward Adelbert
Dolyak, Frank
Domroese, Kenneth Arthur
Dooley, Elmo S
Dorchester, John Edmund Carleton
Dorman, Homer Lee
Dost, Frank Norman
Douglas, Ben Harold
Dousa, Thomas Patrick
Dow, Philip
Dow, Robert Stone
Downey, Harry Fred
Downey, John A
Downie, Harry G
Doyle, Lee Lee
Drees, John Allen
DuBois, Arthur Brooks
Duff, Willard Moyle
Dugal, Louis Paul
Dumont, Allan E
Dunagan, Tommy Tolson
Dungan, Kendrick Webb
Dunham, Jewett
Dunham, Philip Bigelow
Dunn, Arnold Samuel
Dunnigan, Jacques
Dunson, William Albert
Durbin, Richard Paul
Dutton, Robert Edward, Jr
Eagan, Charles J
Eagle, Edward
Easter, Stephen Sherman, (Jr)
Eaton, John Wallace
Ebbs, Jane Cotton
Ebert, Paul Allen
Ecker, Richard Henry
Edelhauser, Henry F
Ederstrom, Helge Ellis
Edinger, Henry Milton
Edwards, Carolyn Trowbridge
Edwards, Gerald Elmo
Edwards, Lawrence Jay
Edwards, Leslie Erroll
Eggena, Patrick
Egle, John Lee, Jr
Eidelberg, Eduardo
Eisenman, Joseph Sol
Eitzman, Donald V
Eldredge, Donald Herbert
Ellert, Martha Schwandt
Elliott, Robert Hare Egerton, Jr
Ellis, Charles Herbert
Ellison, Lois Taylor
Elmadjian, Fred
Elrod, Lloyd Melvin
Elsbach, Peter
Elwell, Leonard Hubert
Elzinga, Marshall
Engen, Richard Lee
Entenman, Cecil
Epstein, Franklin Harold
Erasmus, Beth De Wet
Erickson, Howard Hugh
Ernst, Ralph Ambrose

Evonuk, Eugene
Ewing, Larry Larue
Faber, Jan Job
Fabianek, John
Faiman, Charles
Fairbanks, Gilbert Wayne
Fairchild, Edward Joseph, II
Fajer, Abram Bencjan
Farmanfarmaian, Allahverdi
Farmer, Florence Amelia
Farmer, Donald Sankey
Farnsworth, Patricia Nordstrom
Farrell, James I
Faulkner, John A
Fawaz, George
Fawley, John Philip
Fedor, Edward John
Feigen, George Alexander
Feigl, Eric O
Feist, Dale Daniel
Fell, Colin
Feller, David Douglas
Fenichel, Richard Lee
Ferguson, Frederick Palmer
Ferguson, John Carruthers
Ferguson, John Howard
Ferguson, Marion Lee
Ferguson, Samuel A
Fernandez y Cossio, Hector Rafael
Ferrante, Frank L
Fiala, Silvio Emerich Ivan
Fickess, Douglas Ricardo
Fidone, Salvatore Joseph
Fife, William Paul
Filkins, James P
Filley, Giles Franklin
Finberg, Laurence
Finn, Arthur Leonard
Finster, Mieczyslaw
Fiore, Carl
Fiorica, Vincent
Firriolo, Domenic
Firstbrook, John Bradshaw
Fischbarg, Jorge
Fischer, Grace Mae
Fischer, Alfred Paul
Fishman, Harvey Morton
Fitzgerald, Laurence Rockwell
Fitzgerald, Marie Anton
Fitzgerald, Robert Schaefer
Fitzgerrel, William Wright
Flaim, Kathryn Erskine
Flaim, Stephen Frederick
Flanigan, William P
Fleming, Warren R
Flemister, Launcelot Johnson
Flournoy, Robert Wilson
Foa, Piero Pio
Foglia, Virgilio Gerardo
Folk, George Edgar, Jr
Forbes, William Hathaway
Ford, Donald Herbert
Ford, George Dudley
Foreman, Charles William
Forker, Edson Lee
Forster, Robert E, II
Forster, Roy Philip
Forte, John Gaetano
Fortier, Claude
Fosket, Donald Elston
Foss, Donald C
Foster, Douglas Layne
Fougeron, Myron George
Foulkes, Ernest Charles
Fowler, Ward Scott
Fox, Karl Richard
Fozzard, Harry A
Frank, Fred R
Frank, George Barry
Frank, Morton Howard
Frankel, Harry Meyer
Franklin, Thomas Doyal, Jr
Franz, Gunter Norbert
Frascella, Daniel W
Fraser, Donald
Frayser, Katherine Regina
Frazier, Donald Tha
Frazier, Loy William, Jr
Frederick, Edward C
Fredericks, Christopher M
Free, Michael John
Freedman, William
Freeman, Alan R
Freeman, Paul Joel
Freeman, Walter Jackson, III
Fregly, Melvin James
Frehn, John
French, Leslie Howson
Freund, Matthew J
Freygang, Walter Henry, Jr
Fried, George H
Friedman, Julius Jay
Friedman, Kenneth Joseph
Friedman, Moe Hegby Fred
Friedmann, Naomi
Friesinger, Gottlieb Christian
Fritz, George Richard, Jr

Frohman, Charles Edward
Fromm, Eli
Fronek, Arnost
Fry, Donald Lewis
Fry, Richard Jeremy Michael
Fuchs, Franklin
Fuller, Forst Donald
Fulton, George Pearman
Furry, Donald Edward
Fusco, Madeline M
Gaensler, Edward Arnold
Gage, Tommy Wilton
Gage, Adolf Pharo
Gagnon, Andre
Gailey, Franklin Bryan
Gala, Richard R
Galambos, Robert
Gale, Charles C, Jr
Galey, William Raleigh
Galloway, Raymond Alfred
Gamble, James Lawder, Jr
Gandhi, Sham Sunder
Ganley, Oswald Harold
Gann, Donald Stuart
Ganong, William Francis
Garcia Ramos, Juan
Gardner, David L
Garren, Henry Wilburn
Garth, Richard Edwin
Gartner, Lawrence Mitchel
Gaskell, Peter
Gasser, Raymond Frank
Gaudino, Mario
Geber, William Frederick
Geczik, Ronald Joseph
Geddes, David Darwin
Gee, J Bernard L
Gehrmann, William Henry
Gelinas, Douglas Alfred
Gellert, Ronald J
George, John Caleekal
George, Ronald Baylis
Gerritsen, George Contant
Gerst, Jeffery William
Gerst, Paul Howard
Gibbs, Finley P
Gibor, Aharon
Gibson, Quentin Howieson
Gidari, Anthony Salvatore
Giebisch, Gerhard Hans
Giering, John Edgar
Gifford, Cameron Edward
Gilbert, Daniel Lee
Gilboe, David Dougherty
Gillespie, Jerry Ray
Gillilan, Lois Adell
Gilmore, Joseph Patrick
Ginsburg, Jack Martin
Ginski, John Martin
Gisolfi, Carl Vincent
Gissen, Aaron J
Gladfelter, Wilbert Eugene
Glaser, Edmund M
Glaser, Warren
Glauser, Elinor Mikelberg
Glaviano, Vincent Valentino
Glenn, Eldridge Myles
Glenn, Thomas M
Glick, Bruce
Glocklin, Vera Charlotte
Glomset, John A
Goetz, Kenneth Lee
Goff, Loyal Glenn
Gold, Martin
Goldberg, Alfred L
Goldberg, Erwin
Goldberg, Vivian Joyce
Goldenberg, Marvin M
Goldman, David Eliot
Goldman, Ralph Frederick
Goldner, Andrew M
Goldstein, Bernard
Goldstein, Leon
Gonzalez, Ramon Rafael, Jr
Gonzalez, Richard Rafael
Gonzalez-Fernandez, Jose Maria
Goodall, McChesney
Goode, Lemuel
Goodman, Henry Maurice
Goodman, Joan Wright (Mrs Charles D)
Goodrich, Cecilie Ann
Gordon, Albert Saul
Gordon, Archer Samuel
Gordon, David Buddy
Gordon, Helmut Albert
Gordon, Robert Sirkosky, Jr
Gotshall, Robert William
Gottlieb, Gerald Lane
Gottlieb, Sheldon F
Gottschalk, Carl William
Gouras, Peter
Graf, Gottfried Christian
Grafstein, Bernice (Mrs Howard Shanet)
Graham, Edmund F
Grande, Francisco
Granick, Sam
Grant, Wilson Clark
Graubard, Mark Aaron

Graves, Charles Norman
Graves, William Earl
Gray, John Stephens
Grayson, John
Green, Harold David
Green, James Weston
Green, Keith
Greenbaum, Leon J, Jr
Greenberg, Ruven
Greenberger, Norton Jerald
Greene, Leon Charles
Greenfield, Wilbert
Greenway, Clive Victor
Greenwood, Mary Rita Cooke
Gregg, Donald Eaton
Gregor, Nicholas John
Gregoire, Fernand
Greichus, Algirdas
Greif, Roger Louis
Greisman, Sheldon Edward
Griffin, Edmond Eugene
Griffith, David R
Griffo, Zora Jasincuk
Griggs, Douglas M, Jr
Grim, Eugene
Grimm, Arthur F
Grindeland, Richard Edward
Grodins, Fred Sherman
Grollman, Sigmund
Groom, Alan Clifford
Grossie, James Allen
Grubbs, Robert Custis
Gruener, Raphael P
Grumbach, Leonard
Grundfest, Harry
Grundy, Scott Montgomery
Grupp, Gunter
Guentherman, Robert Henry
Guest, Mary Frances
Guest, Maurice Mason
Guillemin, Roger (Charles Louis)
Gutknecht, John William
Guyton, Arthur Clifton
Gwynn, Robert H
Habicht, Gail Sorem
Hackney, Jack Dean
Haddy, Francis John
Hageman, William E
Hahn, Peter
Haimovici, Henry
Haines, Howard Bodley
Haist, Reginald Evan
Hajdu, Stephen
Hale, Henry Bixby
Hall, Charles Eric
Hall, John Felix, Jr
Hall, Octavia
Hall, Peter
Hall, Peter Francis
Hall, Philip Wells, III
Hall, Robert Henry
Haller, Edwin Wolfgang
Hamilton, Charles Lewis
Hamilton, Clara Eddy
Hamilton, Lyle Howard
Hamilton, Mary Alice
Hanzely, Joseph Bernard
Hardt, Alfred Black
Hardt, James Daniel
Hardy, Lester B, Jr
Hardy, William Lyle
Harley, John Paul
Harley, William Lyle
Harper, Jon William
Harrington, Francis Eugene
Harrington, Joseph D
Harris, Albert Sidney
Harris, Don Navarro
Harris, Patrick Donald
Harris, Stanley Cyril
Harrison, Howard N
Harrison, Lura Ann
Harrison, Robert William
Hart, Jayne Thompson
Hartenstein, Roy
Hartley, Marshall Wendell
Hartline, Haldan Keffer
Haskin, Harold H
Haskins, Caryl Parker
Hassett, George Lee, Jr
Hassett, Charles Clifford
Hast, Malcolm Howard
Hatch, Malcolm Harold
Hatch, William James
Hatcher, James Donald
Hatfield, Efton Everett
Hausberger, Franz X

PHYSIOLOGY

Lincicome, David Richard
Linde, Leonard M
Lindley, Barry Drew
Lindsay, Hugh Alexander
Lindsay, William Germer, Jr
Lindsley, David Ford
Ling, Gilbert Ning
Linkenheimer, Wayne Henry
Linkie, Daniel Michael
Lioy, Franco
Lipsky, Joseph Albin
Lisa, Joseph Daniel
Lisk, Robert Douglas
Little, James Maxwell
Little, John Bertram
Little, Perry L
Little, Robert Colby
Liu, C T
Liu, Frank Tsung Yuan
Llaurado, Josep G
Lloyd, David Pierce Caradoc
Lobo, Luiz Carlos Galvao
LoBue, Joseph
Lockshin, Richard Ansel
Loeb, Earl Randall
Loewenfeld, Irene Elizabeth
Loewy, Ariel Gideon
Logan, Rowland Elizabeth
Loizzi, Robert Francis
Lolley, Richard Newton
Long, Ernest Croft
Long, James Frantz
Longerbeam, Jerrold Kay
Longmuir, Ian Stewart
Longo, Lawrence Daniel
Loofbourrow, Guy Norman
Loose, Leland David
Lorand, Laszlo
Lorber, Mortimer
Lorber, Victor
Lorente De No, Rafael
Lorenz, Frederick Wharton
Lorscheider, Fritz Louis
Lossow, Walter Judah
Lott, James Robert
Lotz, Frederick
Lourenco, Ruy Valentim
Love, Chaille M
Low, Robert Burnham
Lowery, Thomas J
Lucas, Oscar Nestor
Luco, Joaquin
Luick, Jack Roger
Luisada, Aldo Augusto
Lukin, Larissa Skvortsov
Lutt, Carl J
Lynch, G Paul
Lynch, Peter Robin
Lyne, Everett
Lytle, Ivan M
Lytle, James Bert
Maaske, Clarence Alfred
Macdonald, Gordon J
Macey, Robert Irwin
MacFadden, Donald Lee
Machne, Xenia
Macht, Martin Benzyl
MacIntosh, Frank Campbell
MacIntyre, Bruce Alexander
MacLeod, John
MacNeill, Arthur Edson
Madsen, David Christy
Magee, Donal Francis
Magno, Michael Gregory
Maher, John Thomas
Mailman, David Sherwin
Maire, Frederick Wirth
Maitlen, Eldon Gene
Major, Charles Walter
Makhlouf, Gabriel Michel
Makholm, Richard Evelyn Reginald
Malhotra, Om Parkash
Malik, Dharam Dev
Malvin, Richard L
Mammen, Eberhard F
Mandel, Lazaro J
Manery, Jeanne Forest
Manning, Emmett S
Manthey, Arthur Adolph
Marbarger, John Porter
Marc-Aurele, Julien
Marieb, Elaine Nicpon
Markley, Kehl, III
Marquis, Sabath Fred
Marquis, Norman Ronald
Marsh, Donald Jay
Marshall, Jean McElroy
Marshall, Norman Barry
Marsland, Douglas Alfred
Martin, Arthur Wesley, Jr
Martin, Duncan Willis
Martin, John Samuel
Martin, Julio Mario
Martin, Loren Gene
Martin, Richard Harvey

Martinez, Margaret Yarnall
Masken, James Frederick
Mason, Elliott Bernard
Mason, George Robert
Mason, Richard Canfield
Mason, Walter Harry
Masoro, Edward Joseph
Massion, Walter Herbert
Matsumoto, Yorimi
Maturo, Joseph Martin, III
Maurice, David Myer
Maxfield, Mary Evans
Mayer, George Pat
Mayer, Jean
Mayerson, Hymen Samuel
Maynard, Francis Louis
Mays, Charles Edwin
Mazer, Ronald Steven
McAfee, Robert Dixon
McBlair, William
McBroom, Marvin Jack
McCally, Michael
McCandless, Esther Leib
McCann, Frances Veronica
McCarrell, Jane Dinsmore
McCarthy, John Lawrence, Jr
McCauley, Charles Edward
McCauley, William John
McClellan, John Forbes
McClintic, Joseph Robert
McConn, Rita
McCook, Robert Devon
McCormack, Charles Elwin
McCormack, William C
McCoy, Elbert Julius
McCoy, Lowell Eugene
McCrorey, Henry Lawrence
McDonagh, Richard Patrick, Jr
McDowell, Robert E, Jr
McElroy, William Tyndell, Jr
McFadden, Ernest B
McGrady, Angele Vial
McIlreath, Fred J
McKean, Thomas Arthur
McKeever, William Paul
McKenzie, Jess Mack
McLain, Paul Larimer
McLennan, Hugh
McMurray, Virginia M (Vollmer)
McPhail, Murchie Kilburn
Meehan, John Patrick
Mehrle, Paul Martin, Jr
Meier, Albert Henry
Meiss, Richard Alan
Meites, Joseph
Mela, Leena Marja
Melampy, Robert Maurice
Meli, Alberto L G
Melin, Theodore Nelson
Melton, Carlton Earl, Jr
Mendel, Verne Edward
Mendelson, Martin
Mendez, Jose De La Vega
Mendez-Bauer, Carlos
Meng, Raymond Hsien Chang
Mengebier, William Louis
Menhiniak, Edward Fulton
Mennega, Aaldert
Mercer, Paul Frederick
Merrick, Arthur West
Merskey, Clarence
Meschia, Giacomo
Metcoff, Jack
Meyer, Dallas Kremer
Meyer, John Roger
Meyer, Maurice Wesley
Meyer, Ralph A, Jr
Michael, Ernest Denzil, Jr
Michaelson, Solomon M
Michal, Edwin Keith
Michel, Edward L
Middleton, Samuel
Milazzo, Francis Henry
Milburn, Nancy Stafford
Miles, Harry McCauley
Milic-Emili, Joseph
Miller, Augustus Taylor, Jr
Miller, Harvey I
Miller, Inglis, Jr
Miller, James Kincheloe
Miller, Josef Mayer
Miller, Robert James, II
Mills, Elliott
Mills, Kenneth Selby
Minatoya, Hiroaki
Minaker, David
Mines, Allan Howard
Minsker, David Harry
Misher, Allen
Mitchell, Robert A
Mitchell, Robert Bruce
Mitchell, Thomas George
Mitchell, William Hinckley
Mittier, James Carlton
Moberg, Gary Philip
Moe, Gordon Kenneth
Moe, Robert Anthony
Moldenhauer, Ralph Roy

Molnar, George William
Mommaerts, Wilfried
Money, Kenneth Eric
Monkhouse, Frank C
Montalvo, Francisco Emilio
Montemurro, Donald Gilbert
Montgomery, Arthur Vernon
Montgomery, Edward Harry
Mookerjea, Sailen
Moore, Earl Neil
Moore, James Carlton
Moore, Patricia Ann
Moore, Ward Wilfred
Moran, Joseph Francis, Jr
Moran, Walter Harrison, Jr
Morehouse, Laurence Englemohr
Morgan, Howard E
Moriarty, C Michael
Morrill, Callis Gary
Morrissette, Maurice Corlette
Morrissey, J Edward
Mortimore, Glenn Edward
Morton, Martin Lewis
Moses, Henry A
Moses, Alfred Jefferson, Jr
Moss, Samuel
Moss, Woodrow Glen
Mostardi, Richard Albert
Motley, Hurley Lee
Moulton, David Gillman
Mountcastle, Vernon Benjamin
Mouw, David Richard
Mudge, Gilbert Horton
Mueller, Werner Julius
Muir, Larry Allen
Muleri, Bertham Scubon
Muller, Otto Heinrich
Mulvey, Philip Francis, Jr
Munro, Donald W, Jr
Munson, Sam Clark
Murano, Genesio
Murdock, Larry Lee
Murphy, Quilian R, Jr
Murphy, Richard Alan
Murphy, Terence Martin
Murthy, Veeraraghavan Krishna
Musacchia, Xavier Joseph
Musgrave, F Story
Myers, Donald Albin
Myers, James Hurley
Myers, Joseph B
Nabrit, Samuel Milton
Nadel, Eli Maurice
Nagai, Toshio
Nagler, Arnold Leon
Naitove, Arthur
Nance, Francis Carter
Napke, Edward
Nardone, Roland Mario
Nash, Franklin D
Nasset, Edmund Sigurd
Nastuk, William Leo
Nathan, Paul
Natzke, Roger Paul
Navar, Luis Gabriel
Necheles, Thomas
Neff, Robert Jack
Neff, William H
Neidle, Enid Anne
Neill, Jimmy Dyke
Nellor, John Ernest
Nelson, Leonard
Nelson, Ralph A
Nelson, Thomas Eustis, Jr
Neufeld, Arthur Harvey
Neville, Margaret Cobb
Nevison, Thomas Oliver, Jr
Newcomer, Wilbur Stanley
Newell, Jonathan Clark
Newkirk, Robert Franklin
Newman, Howard Abraham Ira
Newton, Wiley Clifford, Jr
Newton, William Morgan
Nicholois, Gregory Ralph
Nicholson, Hayden Coler
Nickerson, John Lester
Nicolai, John Henry, Jr
Nicolette, John Anthony
Nicoll, Charles S
Nicoll, Paul Andrew
Nielson, Read R
Noble, Robert Laing
Nocenti, Mero Raymond
Noe, Frances Elsie
Noell, Werner K
Nolan, Janiece Simmons
Nolan, Richard Arthur
Nolasco, Jesus Bautista
Nooman, Thomas Robert
Norcia, Leonard Nicholas
Nordstrom, Jon Owen
Norman, Roger Atkinson, Jr
Norris, Charles Hamilton
Norton, Allen C
Norton, Virginia Marino
Norwich, Kenneth Howard
Noyes, David Holbrook

Nunn, Arthur Sherman, Jr
Nussbaum, Noel Sidney
Nusser, Wilford Lee
Nye, Robert Eugene, Jr
Nyhus, Lloyd Milton
Nystrom, Richard Alan
O'Brien, Dennis Martin
O'Brien, Larry Joe
O'Brien, Thomas W
O'Connell, Alice L
Odell, Theodore Tellefsen, Jr
Odell, William Douglas
O'Dell-Smith, Roberta Maxine
Ogden, Thomas E
O'Hea, Eugene Kevin
Ohler, Edwin Allen
Olewine, Donald Austin
Oliver, Montague
olman, Robert Alexander
Olmsted, Clinton Albert
Olson, Howard H
Olson, Robert Leroy
Olsson, Ray Andrew
Omachi, Akira
Omid, Ahmad
O'Morchoe, Charles Christopher C
Opdyke, David Franklin
Oppenheim, Elliot
Oppenheimer, Morton Joseph
Orahovats, Peter Dimiter
Orkand, Richard K
Orlans, F Barbara
Orloff, Jack
Oronsky, Arnold Lewis
Osborne, James William
Osborne, Melville
Osborne, Paul James
Ou, Lo-Chang
Outerbridge, John Stuart
Overman, Richard Roll
Overpeck, James Gentry
Oyama, Jiro
Oyler, J Mack
Pace, Caroline S
Pace, Henry Buford
Pace, Nello
Paganelli, Charles Victor
Paine, Robert
Palaty, Vladimir
Palincsar, Edward Emil
Pammani, Motilal Bhagwandas
Paolini, Paul Joseph, Jr
Pappas, George Stephen
Pappenheimer, John Richard
Paradise, Norman Francis
Park, Charles Rawlinson
Parker, Harold R
Parker, Roger Edwin
Parkins, Frederick Milton
Parlow, Albert Francis
Parsons, Rodney Lawrence
Pashley, David Henry
Paterson, Christopher Alexander
Patil, Popat N
Patten, Jimmy Ray
Patterson, John Ward
Patterson, Richard Westcott
Patton, Harry Dickson
Paul, Lawrence Thomas
Paulsen, Gary Melvin
Pautler, Eugene L
Pax, Ralph A
Payton, Brian Wallace
Peach, Michael Joe
Peachey, Lee DeBorde
Pearincott, Joseph V
Pearl, William
Pecora, Louis Joseph
Pederson, Vernon Clayton
Peirce, Edmund Converse, II
Peiss, Clarence Norman
Peissner, Lorraine C
Pengelley, Eric T
Penhos, Juan Carlos
Penicnak, Adrian John
Penrod, Kenneth Earl
Perks, Anthony Manning
Perl, Edward Roy
Perlman, Preston Leonard
Perlmutt, Joseph Hertz
Perry, Harold Tyner, Jr
Peskin, James Charles
Peterjohn, Glenn William
Peterson, Clare Gray
Peterson, Donald Frederick
Peterson, Lysle Henry
Peterson, Stephen Craig
Petta, John M
Pettit, Barbara Jane
Pettit, Robert Eugene
Pew, Weymouth D
Philibert, Robert Lawrence
Phillibert, Hugh Jefferson
Phillips, Joy Burcham
Piacsek, Bela Emery
Pickrell, John A
Pickwell, George Vincent
Pilgeram, Laurence Oscar

PHYSIOLOGY

Plikington, Lou Ann
Pinkerson, Alan L
Pinkston, James Oliver
Pinter, Gabriel George
Piscitelli, Joseph
Pitts, Grover Cleveland
Pitts, Robert Franklin
Planck, Roy Jonathan
Platner, Wesley Stanley
Plotka, Edward Dennis
Podolsky, Richard James
Podleski, Thomas Roger
Poggio, Gian Franco
Poland, James Leroy
Polgar, George
Polimeni, Philip Iniziato
Pollard, James Edward
Polosa, Canio
Pond, Samuel Ernest
Popovic, Vojin
Porter, Charles Warren
Porter, John Charles
Porter, Johnny Ray
Posner, Philip
Post, Robert Lickely
Potter, Gilbert David
Potvin, Pierre
Poutsiaka, John William
Power, Gordon G
Powis, Raymond Leslie
Poznansky, Mark Joab
Pratley, James Nicholas
Pratt, David R
Pray, Judith Dunlap
Prejean, Joe David
Preston, Enos G
Previte, James Benson
Price, Joseph James
Prince, Walter Ray
Pring, Martin
Priola, Donald Victor
Pritchard, John B
Proctor, Jerry Franklin
Proffit, William R
Purple, Richard L
Pursel, Vernon George
Putnam, Serpas Jerome
Quastel, D M J
Quimby, Freeman Henry
Raab, Jacob Lee
Rabinowitz, Lawrence
Racotta, Radu
Radford, Edward Parish, Jr
Raeside, James Inglis
Rahn, Hermann
Rakusan, Karel Josef
Rall, Joseph Edward
Ralph, Charles Leland
Ralston, Henry James
Ramazzotto, Louis John
Ramey, Estelle R
Ramp, Warren Kibby
Rampone, Alfred Joseph
Ramwell, Peter William
Randall, Barbara Feucht
Randall, David Clark
Randall, Howard M
Randall, James Edwin
Randall, Walter Clark
Rankin, John
Rankin, John Horsley Grey
Ranneis, Donald Eugene, Jr
Rappaport, Stanley I
Rappaport, Aron M
Rasch, Robert
Rath, Maurice Monroe
Ratner, Albert
Ratzlaff, Kermit O
Raub, William F
Rautaharju, Pentti M
Rawson, Robert Orrin
Ray, Ovid Malcolm
Rayford, Phillip Leon
Read, Willard Oliver
Recknagel, Richard Otto
Reddan, William Gerald
Redfield, Alfred Clarence
Redick, Thomas Ferguson
Redisch, Walter
Reed, Emerson Aloysius
Reed, Stuart Arthur
Reeve, Ernest Basil
Reeves, Robert Blake
Regen, David Marvin
Rehder, Kai
Rehm, Warren Stacy, Jr
Reichard, Sherwood Marshall
Reichlin, Seymour
Reid, David Mayne
Reid, Ian Andrew
Reid, Marion Adelaide
Reinecke, John Philip
Reinecke, Roger Minske
Reinhart, Ezra Paul
Reininger, Richard Joseph
Reinke, Edward Joseph
Reinke, David Albert

Reissmann, Kurt Rudolph
Reit, Barry
Remington, John Wood
Rendon, Leandro
Rendon, Orlando Roberto
Renkin, Eugene Marshall
Rennick, Barbara Ruth
Rennie, Donald Wesley
Renzetti, Attilio D, Jr
Renzi, Alfred Arthur
Replogle, Clyde R
Resnick, Oscar
Reynolds, David George
Reynolds, Leslie Boush, Jr
Reynolds, Orr Esrey
Reynolds, Robert Williams
Reynolds, Samuel R M
Rhamy, Robert Keith
Rhoade, Edward A, Jr
Rhoades, Rodney A
Richards, Edmund A
Richardson, Alfred Wendel
Richardson, Daniel Ray
Richardson, William Norman
Ricks, Beverly Lee
Ridgway, Ellis Branson
Ridley, Peter Tone
Rieck, Alvin Frank
Riedesel, Marvin LeRoy
Rieselbach, Richard Edgar
Riggi, Stephen Joseph
Rillema, James Alan
Rinard, Gilbert Allen
Ringer, Robert Kosel
Ringle, David Allan
Ripps, Harris
Ritchie, Brenda Rachel
Ritchie, Joseph Murdoch
Rixon, Raymond Harwood
Roberts, John Stephen
Robillard, Eugene
Robin, Eugene Debs
Robinson, George Edward, Jr
Robinson, Roy Garland, Jr
Robinson, Sid
Rockstein, Morris
Rockwood, William Philip
Rodahl, Kaare
Roeder, Martin
Roesel, Oscar Fred
Rogers, Terence Arthur
Roheim, Paul Samuel
Rolf, Doris Barbara
Rolf, Lester Leo, Jr
Romanoff, Elijah Bravman
Ronkin, Raphael Rooser
Roos, Albert
Ross, Gordon
Rosborough, John Paul
Rose, Birgit Loewenstein
Rose, John Charles
Rose, Richard Carrol
Rosenberg, Edith E
Rosenfeld, Leonard M
Rosenfeld, Sheldon
Rosenkrantz, Jens Georg
Rosenthal, Jean
Rosenthal, William S
Ross, Gordon
Rostorfer, Howard Hayes
Roth, Rene Roman
Rothman, Stephen Sutton
Rouse, Thomas C
Rovetto, Michael Julien
Rovick, Allen Asher
Rowell, Loring B
Rowley, George Richard
Ruark, Annette
Rubinstein, Eduardo Hector
Ruch, Theodore Cedric
Rudolph, Abraham Morris
Ruh, Mary Frances
Rulon, Olin
Rulon, Russell Ross
Runey, Gerald Luther
Runser, Richard Henry
Rushton, William A H
Russell, Findlay Ewing
Russell, John McCandless
Russell, Raymond Alvin
Russell, Robert Lee
Rustad, Ronald Cameron
Rutledge, Lester T
Saari, Jack Theodore
Saba, Thomas Maron
Sachs, John Richard
Sadler, Clarence Reagan
Sagawa, Kiichi
Sage, Martin
Salatsky, Bernard P
Saldarini, Ronald John
Sallee, Verney Lee
Salmoiraghi, Gian Carlo
Salzano, John Victor
Samis, Harvey Voorhees, Jr
Sampson, Sanford Robert

Samson, Frederick Eugene, Jr
San Antonio, James Patrick
Sanders, Harvey David
Sanders, Raymond Thomas
Sanders, Rosaltha Hagan
Sanderson, Glen Charles
Sandler, Harold
Santos-Martinez, Jesus
Sar, Madhabananda
Sather, Bryant Thomas
Saunders, Richard Lee
Savery, Marie Schlam
Savitt, Harry P
Sawyer, Richard Leander
Sawyer, Wilbur Henderson
Schachter, Melville
Schacter, Richard C
Schaefer, Karl Ernst
Schapiro, Herbert
Scharrer, Raymond
Schauf, Charles Lawrence
Scheer, Bradley Titus
Scheuer, Allen Myron
Schick, Robert Dean
Schiffrin, Milton Julius
Schilb, Theodore Paul
Schmid, Jack Robert
Schmid, Carl Frederic
Schmidt-Nielsen, Bodil Mimi
Schmidt-Nielsen, Knut
Schneider, Edward Greyer
Schneider, Jurg Adolf
Schneider, Robert Arnold
Schneiweiss, Jeannette W
Schneyer, Charlotte A
Schneyer, Leon Harold
Schnitzler, Ronald Michael
Schoenberg, Mark
Schoepfle, Gordon Marcus
Schofield, Brian
Scholander, Per Fredrik
Scholes, Norman W
Schonbaum, Eduard
Schooler, James M, Jr
Schooley, John C
Schopp, Robert Thomas
Schottelius, Byron Arthur
Schrank, Auline Raymond
Schreiber, Sidney S
Schreiner, George E
Schreider, Bruce David
Schuhmann, Robert Ewald
Schultz, Stanley George
Schwab, Robert G
Schwartz, Bernard
Schwartz, Harold
Schwartz, Irving Leon
Schwartz, Neena Betty
Schwartz, Sidney A
Schwartz, Tobias Louis
Schwassman, Horst Otto
Schweizer, Malvina
Schwinghamer, James M
Scorpio, Ralph M
Scoopy, Robert P
Scott, David Alexander
Scott, George Taylor
Scott, Jerry Benjamin
Scott, John Culberson
Scott, Walter Neil
Scoow, Robert Oliver
Searle, Gilbert Leslie
Searle, Gordon Wentworth
Sears, Dewey Frederick
Seegers, Walter Henry
Seguin, Jerome Joseph
Sekelj, Paul
Selkurt, Ewald Erdman
Sellers, Alvin Louis
Sellner, Ronald George
Selye, Hans
Semple, Robert Evans
Sen, Amar Kumar
Senay, Leo Charles, Jr
Senft, Joseph Philip
Setler, Paulette Elizabeth
Sexton, Alan William
Sha'afi, Ramadan Issa
Shade, Robert Eugene
Shanbour, Linda Livingston
Shannon, James Augustine
Shansky, Michael Steven
Shapiro, Herbert
Shapiro, David Gordon
Share, Leonard
Shaw, David Harold
Shaw, Jane E
Shaw, Joseph Clement
Shea, Philip Joseph
Shell, Lester Crane
Shellenberger, Carl H
Shepard, Robert Stanley
Shephard, Roy Jesse
Shephard, Albert Pitt, Jr
Shepherd, John Thompson

Sheppard, Charles Wilcox
Sherman, James H
Shimizu, C Susan
Shinkman, Paul G
Shirer, Hampton Whiting
Shirley, Barbara Anne
Shock, Nathan Wetherill
Shoemaker, Richard Edward
Shoemaker, Richard Leonard
Shore, Bernard
Shore, Virgie Guinn
Shourd, Melvin Lee
Shrager, Peter George
Siebens, Arthur Alexandre
Sigg, Ernest Beat
Siegman, Marion Joyce
Siegel, Irwin Michael
Siegel, John H
Siegel, Herbert S
Sikand, Rajinder S
Silbergleit, Allen
Silver, Burton Barney
Simmons, Daniel Harold
Simmons, David J
Simmons, Francis Blair
Simpson, Allan Angus
Singh, Sant Parkash
Sinha, Arabinda Kumar
Sinha, Ramananda
Sinkford, Jeanne C
Sirek, Anna
Sirek, Otakar Victor
Sisson, George Maynard
Skahen, Julia Goodsell
Skinner, James Stanford
Skinner, Newman Sheldon, Jr
Skopik, Steven D
Skosey, John Lyle
Skulan, Thomas William
Slack, Jim Marshall
Slack, Samuel Thomas
Slayman, Clifford L
Sleator, William Warner, Jr
Slonim, Arnold Robert
Slotkoff, Lawrence M
Smiley, James Watson
Smith, Amelia Lillian
Smith, Arthur Hamilton
Smith, Charles Clinton, Jr
Smith, Charles Welstead
Smith, Curtis Griffin
Smith, Edwin Lee
Smith, Gerald Nelson, Jr
Smith, Gerard Peter
Smith, James Cecil
Smith, James John
Smith, Jay Alfred
Smith, Joseph Emmitt
Smith, Milton Reynolds
Smith, Ora Kingsley
Smith, Paul Edgar, Jr
Smith, Paul Frederick
Smith, Richard Merrill
Smith, Richard Sidney
Smith, Robert Elphin
Smith, Thomas Caldwell
Smith, Willie White
Smoake, James Alvin
Smulders, Anthony Peter
Snedecor, James George
Sneider, Thomas W
Snider, Ray S
Snipes, Charles Andrew
Sobin, Sidney S
Sobol, Bruce J
Soderling, Thomas Richard
Sokoloff, Louis
Solandt, Omond McKillop
Solberg, Lawrence Arthur, Jr
Solinger, Julian Louis
Solomon, Neil
Solomon, Sidney
Somero, George Nicholls
Somjen, George G
Sonnenschein, Ralph Robert
Sordahl, Louis A
Southrada, Joseph Francis
South, Frank E
Souto, Jose
Spangler, Stanley Gordon
Sparkman, Marjorie Frances
Speakman, Clair Raymond
Spear, Joseph Francis
Speckman, Elwood W
Spencer, William Alden
Sperelakis, Nick
Spickler, J William
Spitzer, John J
Spitzer, Judy A
Spoor, Ryk Peter
Spoerlein, Marie Teresa
Spoor, William Arthur
Spurr, Gerald Baxter
Squires, Bruce Paul
Squires, Russell Dill
Srebro, Richard
Stabenfeldt, George H

Stacy, Ralph Winston
Stagg, Ronald M
Stahl, Philip Damien
Stahlman, Mildred
Stainsby, Wendell Nicholls
Stallworthy, Wilson Burnett
Staple, Peter Hugh
Staples, Albert Franklin
Stark, Paul
Staub, Norman Croft
Stauber, William Taliaferro
Steiman, Henry Robert
Stein, Jerome D, Jr
Stein, Myron
Stein, Philip
Stein, Seymour Nomur
Steinbach, Henry Burr
Steinbeck, Klaus
Stephens, Newman Lloyd
Stephenson, Elizabeth Weiss
Stephenson, William Kay
Stevens, Ernest Donald
Stevens, Vernon Cecil
Stevenson, Nancy Roberta
Stewart, Doris Mae (Mrs Felix Powell)
Stewart, P Brian
Stewart, Peter Arthur
Stewart, William Christopher
Stickney, John Clifford
Stien, Howard M
Stiffler, Daniel F
Stiles, Robert Neal
Stillwell, Edgar Feldman
Stirling, Charles E
Stitt, John Thomas
Stone, Hubert Lowell
Stone, William Ellis
Stopps, Gordon James
Storey, Arthur Thomas
Stratbucker, Robert A
Stratton, Donald Brendan
Strasser, Helen R
Strautz, Robert Lee
Strickholm, Alfred
Stringham, Reed Millington, Jr
Stroud, Robert Church
Strumwasser, Felix
Stuart, Douglas Gordon
Stubbs, Donald William
Stullken, Donald Edward
Sturbaum, Barbara Ann
Sturkie, Paul David
Suberkropp, Keller Francis
Suddick, Richard Phillips
Suga, Nobuo
Sugar, Oscar
Sukowski, Ernest John
Sulakhe, Prakash Vinayak
Sullivan, Lawrence Paul
Sullivan, Walter James
Suter, Daniel B
Sutfin, Duane
Sutherland, Donald James
Sutherland, Gerald Bonar
Suthers, Roderick Atkins
Swan, Algernon Gordon
Swann, Harold James Charles
Sybers, Harley D
Szabo, Gabor
Szerb, John Conrad
Szumski, Alfred John
Tade, William Howard
Taggart, John Victor
Tahmisian, Theodore Newton
Takeda, Yasuhiko
Talesnik, Jaime
Tallitsch, Robert Boyde
Talmage, Roy Van Neste
Tamar, Henry
Tanabe, Tsuneo Y
Tanner, George Albert
Tansy, Martin F
Tanz, Ralph
Tauber, Oscar Earnst
Taurog, Alvin
Taylor, Alan Neil
Taylor, Anna Newman
Taylor, Ardell Nichols
Taylor, Aubrey Elmo
Taylor, Henry Longstreet
Taylor, Paul M
Taylor, Robert Clement
Taylor, Robert E
Taylor, Robert Emerald, Jr
Taylor, Stuart Robert
Tedeschi, Henry
Templeton, Gordon Huffine
Tenney, Stephen Marsh
Terzuolo, Carlo A
Thibodeau, Gary Arthur
Thoa, Nguyen Bich
Thoenes, Lawrence Anthony
Thomas, Kurian K
Thomas, Norman Randall
Thomas, Robert E
Thomas, Sarah Neil
Thompson, Alan Morley
Thompson, James Charles

Thompson, William Donald
Thomson, Duncan Maclaren
Thomson, John David
Thornton, Paul A
Thurber, Robert Eugene
Thurber, Frederick Peter
Tierney, Donald Frank
Tigchelaar, Peter Vernon
Ting, Er Yi
Tipton, Charles M
Tipton, Samuel Ridley
Tobin, Richard Bruce
Todd, Margaret Edna
Tolberg, Adelaide Brokaw
Tolles, Walter Edwin
Tomashefski, Joseph Francis
Tong, Winton
Tonndorf, Juergen
Torchiana, Mary Louise
Tormey, John McDivit
Torre-Bueno, Jose Rollin
Tosteson, Daniel Charles
Totel, Gregory Lee
Toth, Louis Andrew
Towbin, Eugene Jonas
Towe, Arnold Lester
Townsend, Alexandra A
Traber, Daniel Lee
Trank, John W
Trapani, Ignatius Louis
Travis, Hugh Farrant
Travis, Randall Howard
Triantaphyllopoulos, Demetrios
Trimble, Mary Ellen
Trout, David Lynn
Trumbore, Roger H
Truong, Xuan Thoai
Ts'ao, Chung-Hsin
Tschirgi, Robert Donald
Tunik, Bernard D
Turinsky, Jiri
Turner, Manson Don
Tuttle, Elbert P, Jr
Tuttle, Richard Suneson
Tuttle, Ronald Ralph
Twarog, Betty Mack
Twente, Janet
Tyler, David Bernard
Ullrick, William Charles
Ulmer, David Heading Bartine, Jr
Ulvedal, Frode
Upson, Dan W
Upton, G Virginia
Urban, Theodore Joseph
Urquhart, John, III
Usami, Shunichi
Vagnucci, Anthony Hillary
Vallbona, Carlos
Valtin, Heinz
Vanamee, Parker
Vanatta, John Crothers, III
Vandemark, Noland Leroy
Vander, Arthur J
Vanderhoof, Ellen Ruth (Mrs Walter L Peterson)
Van der Kloot, William George
Van Gelder, Nico Michel
Van Harn, Gordon L
Van Harreveld, Anthonie
Van Hassel, Henry John
Van Horn, Diane Lillian
Van Horn, Lester Milton
Van Krey, Harry Peter
Van Liew, Judith Bradford
Van Middlesworth, Lester
van Weel, Pieter Boudewijn
Vargo, Thomas Raymond
Varnell, Thomas Raymond
Vassalle, Mario
Vaughan, Burton Eugene
Veale, Warren Lorne
Vela, Adan Richard
Verses, Christ James
Vesselinovitch, Stan Dushan
Vick, Robert Lore
Vidoli, Vivian Ann
Virkar, Raghunath Atmaram
Visscher, Maurice B
Vogel, Steven
Vogh, Betty Pohl
Vollmer, Erwin Paul
Voogt, James Leonard
Vranic, Mladen
Vroman, Leo
Wachtel, Howard
Waggoner, William Charles
Waitzman, Morton Benjamin
Walker, James Frederick
Walker, James Richard
Walker, Joanne Gillespie
Walker, John Lawrence, Jr
Walker, Sheppard Matthew
Walkowiak, Edmund Francis
Wall, Malcolm Jefferson, Jr
Waller, Hardress Jocelyn
Walloch, Richard Arthur

Walsh, Ralph Thomas
Walsh, Raymond Robert
Walter, Roderich
Walters, Cora Etta
Wang, Hsueh-Hwa
Wang, Lawrence Chia-Huang
Wang, Shih Chun
Wangensteen, Ove Douglas
Ward, Coleman Younger
Ward, John M
Ward, Walter Frederick
Wardell, Joe Russell, Jr
Ware, Frederick
Warner, Alexander Carl
Warnick, Alvin Cropper
Warren, Lionel Gustave
Wasserman, Karlman
Wasserman, Robert Harold
Watkins, Don Wayne
Watlington, Charles Oscar
Watrous, James Joseph
Watson, Lloyd Sherman
Waugh, William Howard
Weatherred, Jackie G
Webber, William A
Weber, Annemarie
Weber, William Adolph
Weeks, Donald Paul
Weiant, Elizabeth Abbott
Weil, Hans
Weimar, Virginia Lee
Weiner, Richard
Weinstein, Hyman Gabriel
Weinstein, Marvin Joseph
Weisbrodt, Norman William
Weiser, Philip Craig
Weiss, Adolph Kurt
Weiss, Harold Samuel
Weiss, Harvey Richard
Weissman, Sherman Morton
Wekstein, David Robert
Welch, Billy Edward
Wells, Charles Henry
Wells, J Gordon
Wells, Jay Byron
Wells, Joseph Albert
Welter, Alphonse Nicholas
Weltman, A Stanley
Welty, Joseph D
Wenger, Herbert Charles
Wenig, Jeffrey
Went, Hans Adriaan
Wesner, Gordon Eugene, Jr
Wessel, Hans U
Wesson, Laurence Goddard, Jr
West, John B
Westfall, Bertis Alfred
Whalen, William James
Wheeler, Allan Gordon
Wheeler, Darrell Deane
Whisler, Kenneth Eugene
White, Clayton Samuel
White, John Francis
White, John Irving
White, Ronald Jerome
White, Stephen Halley
White, Thomas Taylor
Whitehorn, William Victor
Whitfield, Carol Faye
Whitford, Gary M
Whitney, John Edward
Whittenberger, James Laverre
Wickelgren, Warren Otis
Widner, William Richard
Wiebe, John Peter
Wiedeman, Mary Purcell
Wiedmeier, Vernon Thomas
Wiersma, Cornelis Adrianus Gerrit
Wiesel, Torsten Nils
Wiggers, Harold Carl
Wiggins, Earl Lowell
Wiggs, Alfred James
Wilber, Charles Grady
Wilbur, Karl Milton
Wildenthal, Kern
Wilgram, George Friederich
Will, James Arthur
Wilard, Paul W
Willenkin, Robert L
Williams, Charles Melville
Williams, Clyde Michael
Williams, Darrell Dean
Williams, Fred Eugene
Williams, John Andrew
Williams, Marshall Henry, Jr
Williams, Robert Jackson
Williams, Ronald Lee
Williams, Walter Ford
Willis, John Steele
Willis, Judith Horwitz
wilmon, Everett D
Wilmon, Thomas L
Wilson, James Albert
Wilson, Jane Austell
Wilson, John Drennan
Wilson, Kenneth MacKenzie
Wilson, Michael Friend

Wilson, Roy D
Wilson, Thomas Hastings
Wilson, Walter LeRoy
Winbury, Martin M
Windhager, Erich E
Winegrad, Saul
Winkert, John Wynia
Winkler, Barry Steven
Winsmann, Fred Rudolph
Winter, Henry Frank, Jr
Winters, Robert Wayne
Wirts, Charles Wilmer
Wise, William Curtis
Wiser, Cyrus Wymer
Witorsch, Raphael Jay
Wittenberg, Beatrice A
Wittenberg, Jonathan B
Wittle, Lawrence Wayne
Wittner, Murray
Woidich, Francis De Sales
Wolbach, Robert Albert
Wolf, Richard Clarence
Wolf, Robert Lawrence
Wolkoff, A Stark
Wollman, Seymour Horace
Wong, Harry Yuen Chee
Wood, Earl Howard
Wood, Henderson Kingsberry
Wood, Jackie Dale
Wood, Stephen Craig
Wood, William Booth
Woodbury, John Walter
Woodbury, Robert Arthur
Woods, Joseph James
Woods, Mark Winton
Woods, Wendell David
Woodside, Kenneth Hall
Woodward, Donald Jay
Wooley, Wallace Ralph
Wooles, Donald Grant
Woolley, Dorothy Elizabeth Schumann
Woychik, John Henry
Wright, Hastings Kemper
Wu, Szu Hsiao Arthur
Wunder, Charles Cooper
Wykoff, Matthew Henry
Wyssbrod, Herman Robert
Yablonski, Michael Eugene
Yamamoto, William Shigeru
Yang, William Chi Tsu
Yankell, Samuel L
Yanof, Howard Merar
Yates, Francis Eugene
Yatvin, Milton B
Yochim, Jerome M
Yonce, Lloyd Robert
Youmans, William Barton
Young, Allan Charles
Young, David Bruce
Young, Donald Rudolph
Young, Ho Lee
Young, Leona Graff
Young, Margaret Claire
Zabara, Jacob
Zabel, Carroll Wayne
Zadunaisky, Jose Atilio
Zaharko, Daniel Samuel
Zanjani, Esmail Dabaghchian
Zar, Jerrold Howard
Zatzman, Marvin Leon
Zavodni, John J
Zawoiski, Eugene Joseph
Zechman, Frederick William, Jr
Ziegler, Frederick Dixon
Zierler, Kenneth Levie
Zimmerman, Arthur Maurice
Zimmerman, Hyman Joseph
Zipursky, Alvin
Zorzoli, Anita
Zucker, Irving H
Zucker, Marjorie Bass
Zweifach, Benjamin William

Animal Physiology

Abbott, Frank Sidney
Alikhan, Muhammad Akhtar
Anderson, Martin Dean
Andrews, Frederick Newcomb
Aschbacher, Peter William
Bailey, Charles Basil Mansfield
Barker, Hal B
Barone, Milo C
Beasley, Philip Gene
Becker, Donald Eugene
Bellinger, Larry Lee
Black, Craig Patrick
Black, Donald Leighton
Boch, Rolf
Bogdonoff, Philip David, Jr
Bradshaw, William Emmons
Bremel, Robert Duane
Brick, Robert Wayne
Brown, Leonard D
Brown, William Paul
Bruce, David Stewart
Carter, Clint Earl
Caruolo, Edward Vitangelo

Wiederhielm, Curt Arne
Wit, Andrew Lewis
Wolthuis, Roger A
Yellin, Edward L

Comparative Physiology
Ahearn, Gregory Allen
Akers, Thomas Kenny
Baird, James Leroy, Jr
Barkman, Robert Cloyce
Barnwell, Franklin Hershel
Beatty, David Delmar
Beekman, Bruce Edward
Behrisch, Hans Werner
Belkin, Daniel Arthur
Bennett, Miriam Frances
Bintz, Gary Luther
Bishop, Stephen Hurst
Boda, James Marvin
Braun, Eldon John
Brockway, Alan Priest
Campbell, James Wayne
Carnes, David Lee, Jr
Case, James Frederick
Childress, James J
Clark, Mary Eleanor
Clark, Nancy Barnes
Conte, Frank Philip
Crawshaw, Larry Ingram
Crowe, John H
Dawe, Albert Rolke
Dawson, William Ryan
Dehnel, Paul Augustus
DeVillez, Edward Joseph
Diggins, Maureen Rita
Douglas, Donald Sterling
Duerr, Frederick G
Elsner, Robert
Evans, David Hudson
Fairbanks, Michael Bruce
Fingerman, Milton
Forward, Richard Blair, Jr
Fraenkel, Gottfried Samuel
Frings, Hubert Martin
Fromm, Paul Oliver
Galster, William Allen
Gatten, Robert Edward, Jr
Gordon, Malcolm Stephen
Greenberg, Michael John
Greenwald, Lewis
Hagadorn, Irvine R
Hammond, Brian Ralph
Hanegan, James L
Harrison, Florence Louise
Hartman, Herman Bernard
Heath, Alan Gard
Henshaw, Clyde F, II
Herreid, Clyde F, II
Higgins, William Joseph
Hill, Richard William
Hill, Robert Benjamin
Hoar, William Stewart
Hodgson, Edward Shilling
Horne, Francis R
Hoversland, Arthur Stanley
Hudson, Jack William, Jr
Hunter, Frissell Roy
Irving, Laurence
Jegla, Thomas Cyril
Johnson, Leland Gilbert
Josephson, Robert Karl
Kaplan, Martin L
Kooyman, Gerald Lee
Kriebel, Mahlon E
Larsen, James Bouton
Lasker, Reuben
Lawrence, Addison Lee
Lawrence, Jean McVay
Lee, Thomas W
Liles, James Neil
Macdonald, John Alan
Machin, J
Mangum, Charlotte P
Mantel, Linda Habas
Marvin, Daniel Ezra, Jr
Masat, Robert James
McCutcheon, Frederick Harold
McDonald, Harry Sawyer
McMillan, Joseph Patrick
McNabb, F M Anne
McNabb, Roger Allen
McWhinnie, Mary Alice
Menaker, Michael
Mendes, Erasmo Garcia
Miller, Donald Morton
Min, Hong Shik
Moon, Thomas William
Morhardt, J Emil
Morrison, Peter Reed
Murrish, David Earl
Muscatine, Leonard
Ogasawara, Frank X
Oglesby, Larry Calmer
Packer, Randall Kent
Pang, Peter Kai To
Parsons, Robert Hathaway
Penney, David George
Phillips, John Edward

Pierce, Sidney Kendrick
Pritchard, Austin Wyatt
Prosser, Clifford Ladd
Pryor, Marilyn Ann Zirk
Puyear, Robert Louis
Rao, Krothapalli Ranga
Rauch, Josefine Constantia
Redmond, James Ronald
Roberts, John Lewis
Roberts, Shepherd (Knapp de Forest)
Rudy, Paul Passmore, Jr
Sasner, John Joseph, Jr
Schiffman, Robert Harry
Schneider, David Edwin
Senturia, Jerome Basil
Shoemaker, Vaughan Hurst
Sivak, Jacob Gershon
Slater, John Vernon
Smith, Lynwood S
Smith, Michael Joseph
Songdahl, John Harald
Stanley, Jon G
Stokes, Robert Mitchell
Swanson, Curtis James
Taylor, Charles Richard
Testerman, John Kendrick
Tremor, John W
Tucker, John Shepard
Tucker, Vance Alan
Tullis, Richard Eugene
Umminger, Bruce Lynn
Venkataramiah, Amaraneni
Walcott, Benjamin
Waterman, Talbot Howe
Weisbart, Melvin
Weisbroth, Steven H
White, Fred Newton
Whitmore, Donald Herbert, Jr
Wyman, Robert J
Young, Janice Edith

Developmental Physiology
Benson, Katherine Adams
Bradley, Robert Martin
Cons, Jean Marie Abele
Cooke, Helen Joan
Gwinn, John Fredrick
Hunt, Richard Kevin
Koldovsky, Otakar
Kordan, Herbert Allen
Lockshin, Richard Ansel
Mallery, Charles Henry
Morse, Richard Kenneth
Procaccini, Donald J
Rawls, John Marvin
Schwalm, Fritz Ekkehardt
Schweber, Miriam Schurin
Singer, Anita Larks
Strange, John Ruble
Thurmond, William
Timiras, Paola Silvestri
Visscher, Saralee Neumann
Wolfson, Nancy Dolly

Electrophysiology
Ceretti, Elena
De Felice, Louis John
Eisenberg, Robert S
Geduldig, Donald Stanley
Greenspan, Kalman
Konikoff, John Jacob
Maaseidvaag, Frode
MacLeod, Don Putnam
McAlister, Ronald Eric
McCann, Frances Veronica
Omura, Yoshiaki
Rosene, Hilda Florence (Mrs E J Lund)
Runion, Howell Irwin
Schaffer, Abraham Isaac
Siegfried, John Barton
Speros, Perry
Ten Eick, Robert Edwin
Volavka, Jan

Environmental Physiology
Akers, Thomas Kenny
Aleksiuk, Michael
Alliston, Charles Walter
Besch, Emerson Louis
Billings, Charles Edgar, Jr
Bintz, Gary Luther
Blume, Frederick Duane
Burlington, Roy Frederick
Carlson, Gerald Eugene
Chute, Robert Maurice
Coburn, Corbett Benjamin, Jr
Dieter, Michael Phillip
Duman, John Girard
Ekberg, Donald Roy
Farhi, Leon Elie
Ferguson, James Homer
Gelderloos, Orin Glenn
Hadley, Neil F
Hall, Arthur Lee
Hansen, James E
Haschemeyer, Audrey Elizabeth Veazie
Hayward, John S
Heroux, Olivier Joseph Paul

Hertig, Bruce Allerton
Hixson, Floyd Marcus
Holmes, Kenneth Robert
Houston, Arthur Hillier
Hurst, Robert Nelson
Johnson, Donald William
Joyce, Elaine C Elder
Kelly, Herbert Barrett, Jr
Kilgore, Delbert Lyle, Jr
Kreider, Marlin Books
Krizek, Donald Thomas
Leon, Henry A
Lustick, Sheldon Irving
Mackay, William Charles
Minnich, John Edwin
Morse, John Thomas
Moss, Sanford Alexander, III
Nagy, Kenneth Alex
Newsom, Bernard Dean
Paim, Uno
Pandolf, Kent Barry
Panuska, Joseph Allan
Parsons, Lawrence Reed
Payne, Jeremiah Frederick
Percy, Jonathan Arthur
Pereira, Martin Rodrigues
Phillips, Richard Dean
Pough, Frederick Harvey
Ramsey, James Marvin
Reid, Donald House
Richards, Oscar White
Ridgway, Sam H
Robinson, Sumner M
Segal, Earl
Sellers, Cletus Miller, Jr
Senturia, Jerome Basil
Shields, Jimmie Lee
Smiles, Kenneth Albert
Stones, Robert C
Strange, John Ruble
Sudia, Theodore William
Umminger, Bruce Lynn
Vernberg, Winona B
Weathers, Wesley Wayne
Whitten, Bertwell Kneeland
Wilson, Henry R
Wood, Jack Sheehan
Wood, Landley Harriss
Woodring, Jay Porter
Yousef, Mohamed Khalil

Exercise Physiology
Allen, Paul D
Bonner, Hugh Warren
Cafarelli, Enzo Donald
Fox, Edward L
Glaser, Roger Michael
Gollnick, Philip D
Greenleaf, John Edward
Harrison, Aix B
Landry, Fernand
Maksud, Michael George
Mole, Paul Angelo
Nagle, Francis J
Oscai, Lawrence B
Pargman, David
Pollock, Michael L
Robertson, Robert James
Ruhling, Robert Otto
Soule, Roger Gilbert
Spence, Dale William
Vogel, James Alan
Wright, James Edward
Zauner, Christian Walter

Forest Physiology
Barnes, Robert Lloyd
Bilan, M Victor
Dickmann, Donald Irvin
Elam, William Warren
Fitzgerald, Charles H
Gatherum, Gordon Elwood
Larson, Merlyn Milfred
Larson, Philip Rodney
Owston, Peyton Wood
Phares, Robert Eugene
Rietveld, Willis James
Sommer, Harry Edward
Wang, Ben Shih-pin
Whitmore, Frank William
Winton, Lawson Lowell
Woodman, James Nelson
Zaerr, Joe Benjamin

Gerontology
Chesky, Jeffrey Alan
Haensly, William Edward
Holloway, Clarke L
Safier, Gwendolyn
Sandler, Rivka Black
Weg, Ruth Bass

Histophysiology
De Robertis, Eduardo Diego P
Desmond, Alton Harold
Martin, Kathryn Helen
Quay, Wilbur Brooks
Wimsatt, William Abell

Human Physiology
Arnold, Frederic G
Ashworth, Murray Alexander
Bandick, Neal Raymond
Barker, Winona Clinton
Clarke, David Harrison
Costill, David Lee
Davis, Larry Dean
Ellis, Forest Albert
Ferguson, Max B
Friar, Robert Edsel
Fronek, Kitty
Harper, Laura Jane
Harris, Nellie Robbins
Hendler, Edwin
Kaufman, William Carl, Jr
Kidd, Derek John
Lewis, Brian Kreglow
Luft, Ulrich Cameron
Mack, Clinton Olmsted
Mackay, Ian Francis Stuart
Mohammed, Kasheed
Mulvihill, Mary Lou Jolie
Riggs, Dixon L
Smolensky, Michael Hale
Snellen, Jan Willem
Tse, Warren W
Waterhouse, Joseph Stallard
Weld, Charles Beecher
Zacharias, Leona Ruth

Insect Physiology
Adams, Jean Burnham
Adams, Terrance Sturgis
Agee, Herndon Royce
Balboni, Edward Raymond
Ball, Harold James
Barsa-Newton, Mary Claire
Bergtrom, Gerald
Berry, Spencer Julian
Butz, Andrew
Cherbas, Peter Thomas
Chevone, Boris Ivan
Clark, Edgar William
Cochran, Donald Gordon
Cook, Benjamin Jacob
Crowder, Larry A
Dahlman, Douglas Lee
Dasch, Gregory Alan
Denlinger, David Landis
Dethier, Vincent Gaston
Downer, Roger George Hamill
Eaton, John LeRoy
Ewen, Alwyn Bradley
Fain, Margery Jones
Feir, Dorothy Jean
Fisk, Frank Wilbur
Flemings, Milton Baker
Florentine, Gerald Joseph
Friedman, Stanley
Friend, William George
Happ, George Movius
Holman, Grant Mark
House, Howard Leslie
Hsiao, Ting Huan
Jones, Jack Colvard
Jones, Richard Lamar
Judson, Charles LeRoy
Judy, Kenneth James
Kadner, Carl George
Kamal, Adel S
Keeley, Larry Lee
Koeppe, John K
Krysan, James Louis
Larsen, Joseph Reuben
Leahy, Mary Gerald
Locke, Michael
Loschiavo, Samuel Ralph
McFarlane, John Elwood
Mittler, Thomas E
Mullins, Donald Eugene
Mutchmor, John A
Nappi, Anthony Joseph
Nation, James Lamar
Nayar, Jai Krishen
Nijhout, H Frederick
Nijhout, Mary McAllister
Noblet, Raymond
Obenchain, Frederick DeCroes
Pappas, Larry George
Patton, Robert Lee
Rao, Balakrishna Raghavendra
Retnakaran, Arthur
Richards, Albert Glenn
Riddiford, Lynn Moorhead (Mrs James W Truman)
Riegert, Paul William
Robertson, Jacqueline Lee
Roth, Louis Marcus
Rousell, Gerald
Salkfeld, E Helen
Schroeder, Mark Edwin
Shuel, Reginald William
Smyth, Thomas, Jr
Svoboda, James Arvid
Truman, James William
Weaver, Nevin

Wilkens, Jewel L
Wimer, Larry Thomas
Woodruff, Laurence Clark
Wyatt, Gerard Robert

Invertebrate Physiology
Bayne, Christopher Jeffrey
Blanquet, Richard Steven
Bliss, Dorothy Elizabeth
Bowers, William Sigmond
Burnett, Bryan Reeder
Bursey, Charles Robert
Carlson, Albert Dewayne, Jr
Carney, Gordon C
Cramer, Kenneth George
Davey, Kenneth George
Dendinger, James Elmer
Fisher, Frank M, Jr
Fournier, Charles Russell
Fried, Frank Edward
Hammen, Carl Schlee
Hubschman, Jerry Henry
Hyde, Cornelia Tuten
Loeb, Marcia Joan
Pardy, Rosevelt Lawrence
Prusch, Robert Daniel
Puckett, Hugh
Roys, Chester Crosby
Schatzlein, Frank Charles
Seabrook, William Davidson
Shepherd, Julian Granville
Silverthorn, Saidee Unglaub
Spencer, Andrew Nigel
Steel, Colin Geoffrey Hendry
Steele, Vladislava Julie
Sweeney, Daryl Charles
Tombes, Averett Snead
Towle, Albert
Webb, Helen Marguerite
Webb, Rodney A
Wittig, Kenneth Paul
Wolfe, Allan Frederick
Young, Stephen Dean

Kinesiology
Hobart, Donald James
Murray, Mary Patricia
Perry, Jacquelin
Reid, James Gavin

Medical Physiology
Alexander, Natalie
Allen, Robert Erwin
Barr, Ronald Edward
Beard, James David
Berne, Robert Matthew
Blatt, Elizabeth Kempske
Bond, Gary Carl
Bricker, Jerome Gough
Brown, Marvin Ross
Chapler, Christopher Keith
Chargois, Loring Frederick
Claus, Deborah Majeau
Curry, Thomas Harrison
D'Agrosa, Louis Salvatore
Dahlen, Roger W
Dellenback, Robert Joseph
Denniston, Joseph Charles
De Nuccio, David Joseph
Dobrin, Philip Boone
Edwards, Brian Ronald
Ehrhart, Ina C
Ely, Daniel Lee
Emmers, Raimond
Feigen, Larry Philip
Fisher, Martin Joseph
Ford, Lincoln Edmond
Fuhrman, Frederick Alexander
Gaar, Kermit Albert, Jr
Gaugl, John F
Geisler, Fred Harden
Gold, Armand Joel
Gonzalez, Norberto Carlos
Grace, Robert Ambrose
Grubbs, Clinton Julian
Hale, Creighton J
Hamosh, Paul
Hansen, Timothy Ray
Hartman, John Leo, Jr
Heisey, Samuel Richard
Heller, Lois Jane
Hetenyi, Geza Joseph
Hiles, Linda Gayle
Hislop, Helen Jean
Hutchison, Kenneth James
Jensen, David
Johnson, Melvin Andrew, Jr
Kaufman, Albert Irving
Keney, Richard Alec
Knudsen, James Frederick
Konecci, Eugene B
Korecky, Borivoj
Krieger, Eduardo Moacyr
Kunze, Diana Lee
Laiken, Nora Dawn
Lambert, Edward Howard
Larsen, Eleanor Marie
Lebrie, Stephen Joseph
Lee, Robert John
Liu, Maw-Shung
Lo, Chu-Shek
Malindzak, George Steve, Jr
McCutcheon, Ernest P
Miller, David Arthur
Milligan, John Vorley
Millstein, Lloyd Gilbert
Milnor, William Robert
Murphy, John Thomas
Nelson, Eldon Lane, Jr
Neville, Henry Joseph
Nicholes, Henry Ryan
Ott, Cobern Erwin
Overbeck, Henry West
Pak, Moon Jae
Pfanzer, Richard Gary
Pike, Eileen Halsey
Ponessa, Joseph Thomas
Rall, Jack Alan
Ramsay, Robert John
Rapela, Carlos Enrique
Redden, David Ray
Reynolds, Monica
Robertson, William G
Sanborn, Warren Gordon
Sargent, William Quirk
Sernka, Thomas John
Sheng, Hwai-Ping
Sladek, Celia Davis
Snively, William Daniel, Jr
Sparks, Harvey Vise
Stauffer, Edward Keith
Swanson, Robert E
Tidball, Mary Elizabeth
Toivola, Pertti Toivo Kalevi
Trainer, Joseph B
Vail, Edwin George
Van Liew, Hugh Davenport
Webb, Paul
Whitten, Elmer Hammond
Wilde, Walter Samuel
Zogg, Carl A

Metabolism
Adams, William S
Alpers, Joseph Benjamin
Baker, Herman
Barringer, Donald F, Jr
Beisel, William R
Bell, Norman H
Bierman, Edwin Lawrence
Bird, Charles Edward
Bleicher, Sheldon Joseph
Brosnan, John Thomas
Cahill, George Francis, Jr
Canary, John Joseph
Cassidy, James Edward
Cevallos, William Herman
Chance, Ronald E
Coon, Craig Nelson
Crandell, Walter Bain
Crawford, John Douglas
Cunningham, Russell D
Davidson, Mayer B
Delaney, Patrick Francis, Jr
Denis, Gustave
Dillard, Morris, Jr
Eichenholz, Alfred
Elias, Loretta Christine
Fallat, Ronald Walter
Featherston, William Roy
Fielding, Christopher J
Fisher, Waldo Reynolds
Foster, Daniel W
Frisell, Wilhelm Richard
Goto, Antonio Marion, Jr
Hazzard, William Russell
Hill, Jim Tom
Hitchcock, Margaret
Horwitz, David Larry
Jenkins, Melvin Earl
Kalant, Norman
Kaplan, Murray Lee
Kapoor, Inder Prakash
Kappas, Attallah
King, Katherine Chung-Ho
Kleeman, Charles Richard
Kolenbrander, Harold Mark
Krzeminski, Leo Francis
Kydd, David Mitchell
Lee, James B
Little, James Alexander
Locke, Raymond Kenneth
Madison, Leonard Lincoln
Montiel, Francisco
Narahara, Hiromichi Tsuda
Owen, John Atkinson, Jr
Pepper, Solomon
Porte, Daniel, Jr
Pradilla, Alberto Gonzalo
Rash, Jay Justen
Rees, Earl Douglas
Rosen, John Friesner
Rovner, David Richard
Salans, Lester Barry
Shaw, Ralph Arthur
Sherwood, Louis Maier
Sodhi, Harbhajan S
Solomon, Solomon Sidney
Sumner, Darrell Dean
Teekell, Roger Alton
Ting, Irwin Peter
Wilson, Dana E
Wyngaarden, James Barnes
Young, Donald Rudolph
Yu, Ts'ai-Fan
Zimmerman, Hyman Joseph
Zulalian, Jack

Microbial Physiology
Adams, Bruce Gordon
Akagi, James Masuji
Albrecht, Alberta Marie
Altman, John Morgan
Andrykovitch, George
Bain, William Murray
Barnsley, Eric Arthur
Barrera, Cecilio Richard
Baxter, Ann Webster
Begue, William John
Berger, Leslie Ralph
Betterton, Harry O
Binder, Franklin Lewis
Botsford, James L
Bottstein, David
Bovell, Carlton Rowland
Boyer, Ernest Wendell
Caltrider, Paul Gene
Carpenter, David Francis
Chan, Eddie Chin Sun
Chang, George Washington
Charlang, Gisela Wohlrab
Clark, James Bennett
Clarke, Gary Anthony
Davis, James Royce
DeCicco, Benedict Thomas
Dimmick, Robert Lewellyn
Duerre, John A
Eldridge, David Wyatt
Farina, Joseph Peter
Fernandez, Bernal
Fina, Louis R
Finnerty, William Robert
Forney, Frederick Willis
Forsberg, Cecil Wallace
Garcia-Rill, Edgar E
Gary, Norman Dwight
Goldman, Manuel
Hall, John Bradley
Hechemy, Karim E
Hedden, Kenneth Forsythe
Higerd, Thomas Braden
Hooper, Alan Bacon
Horvath, Raymond S
Humphrey, Ronald DeVere
Ingram, Lonnie O'Neal
Jurtshuk, Peter, Jr
Kallio, Reino Emil
Kang, Kenneth S
Ketchum, Paul Abbott
Klein, Harold Paul
Kolodziej, Bruno J
Konetzka, Walter Anthony
Konisky, Jordan
Krywolap, George Nicholas
Langworthy, Thomas Allan
Larkin, John Montague
Lessie, Thomas Guy
Levin, Richard Alexander
MacQuillan, Anthony M
Mahoney, Robert Patrick
Malveaux, Floyd J
Mandels, Gabriel Raphael
Mandels, Mary Hickox
Marquis, Robert E
Marshall, Vincent dePaul
Matchett, William H
Matin, Abdul
Matsumura, Philip
Mattingly, Stephen Joseph
McCowen, Sara Moss
McDonald, Ian Johnson
McFeters, Gordon Alwyn
Mindich, Leonard Eugene
Mortlock, Robert Paul
Nash, Claude Hamilton, III
Nielsen, Allen Madsen
Nierlich, Donald P
Oginsky, Evelyn Lenore
Ondrako, Joanne Marie
Ornston, Leo Nicholas
Parks, Leo Wilburn
Parmegiani, Raulo
Peck, Harry Dowd, Jr
Pengra, Robert Monroe
Previc, Edward Paul
Raj, Harkisan D
Ranhand, Jon M
Revzin, Alvin Morton
Riederer-Henderson, Mary Ann
St John, Ann Carlson
Sebek, Oldrich Karel
Sguros, Peter Louis
Simon, Robert David
Sleeper, Bayard Paul
Sojka, Gary Allan
Sperry, Jay Franklin
Taylor, Mary Lowell Branson
Taylor, Walter Herman, Jr
Theodore, Theodore Spiros
Thomas, Julian Edward, Sr
Treick, Ronald Walter
Trelawny, Gilbert Sterling
Turner, Hal Russell
White, James Patrick
Whiteley, Helen Riaboff
Yall, Irving
Yousten, Allan A

Muscular Physiology
April, Ernest W
Bunch, Wilton Herbert
Chesky, Jeffrey Alan
Dukes-Dobos, Francis N
Goldstein, Margaret Ann
Gordon, Albert McCague
Guthe, Karl Frederick
Lorkovic, Hrvoje Radoslav
Marsh, Benjamin Bruce
Mole, Paul Angelo
Pennick, Suzanne Marie
Robinson, Thomas Frank

Neurophysiology
Abrahams, Vivian Cecil
Adey, William Ross
Adkins, Ronald James
Adolph, Alan Robert
Agin, Daniel Pierre
Ajmone-Marsan, Cosimo
Allen, Gary Irving
Altman, Joseph
Amassian, Vahe Eugene
Ames, Adelbert, III
Anderson, David J
Anschel, Steven William
Asanuma, Hiroshi
Austin, George M
Bach-Y-Rita, Paul
Baldwin, Bernell Elwyn
Barlow, John Sutton
Barlow, Robert Brown, Jr
Barrett, Robert
Batsel, Henry Lewis
Baylor, Denis Aristide
Beckman, Alexander Lynn
Benevento, Louis Anthony
Benjamin, Robert Myles
Bennett, Michael Vander Laan
Bennett, Rudy Andrew
Bignall, Keith E
Binggeli, Richard Lee
Black-Cleworth, Patricia Ann
Blankenship, James Emery
Bloedel, James R
Borda, Robert Paul
Boudreau, James Charles
Bowerman, Robert Francis
Boyarsky, Louis Lester
Brooks, Vernon Bernard
Broughton, Roger James
Brown, George Wallace
Brown, Joel Edward
Buchwald, Jennifer S
Buchwald, Nathaniel Avrom
Burchfiel, James Lee
Burton, Harold
Calvin, William Howard
Camhi, Jeffrey Martin
Carlsen, Richard Chester
Carpenter, David O
Casey, Kenneth L
Caveness, William Fields
Chapple, William Dismore
Christensen, Burgess Nyles
Clark, Francis John
Clark, George
Coats, Alfred Cornell
Coceani, Flavio
Cohen, David Harris
Cohen, Lawrence Baruch
Cohen, Morton Irving
Cooke, Ian McLean
Cooper, Gary Pettus
Coulter, Joe Dan
Crain, Stanley M
Cunningham, James Gordon
Dafny, Nachum
Davidoff, Robert Alan

Davidson, Julian M
Davis, Earle Andrew, Jr
Davis, Hallowell
Daw, Nigel Warwick
Del Castillo, Jose
Delgado, Jose Manuel Rodriguez
Dement, William Charles
Demetrescu, M
Doty, Robert William
Drake, Rosemarie Sheila
Dudar, John Douglas
Duffy, Frank Hopkins
Durham, Ross M
Dykes, Robert William
Eagles, Douglas Alan
Easton, Dexter Morgan
Eaton, Douglas Charles
Eckert, Roger Otto
Egger, Maurice David
Eldred, Earl
Engel, Jerome, Jr
Enroth-Cugell, Christina
Erulkar, Solomon David
Evoy, William (Harrington)
Eyzaguirre, Carlos
Farley, Roger Dean
Fernandez-Guardiola, Augusto
Fertziger, Allen Philip
Fields, Howard Lincoln
Frank, Karl
Freeman, John A
Fuortes, Michelangelo Giorgio F
Fuster, Joaquin Maria
Galeano, Cesar
Gardner, Esther Polinsky
Garoutte, Bill (Charles)
Gasteiger, Edgar Lionel, Jr
Gelperin, Alan
Gesteland, Robert Charles
Gibson, John Michael
Gibson, Mary Morton
Gillary, Howard L
Gloor, Pierre
Gold, Richard Michael
Goldberg, Jay M
Gootman, Phyllis Myrna Adler
Grant, Ronald
Grinnell, Alan Dale
Guinan, John Joseph, Jr
Hackett, John Taylor
Hagiwara, Susumu
Hancock, Michael B
Hanna, George R
Happel, Leo Theodore, Jr
Hartline, Daniel Keffer
Hartline, Peter Haldan
Havlicek, Viktor
Henneman, Elwood
Hind, Joseph Edward
Hockman, Charles Henry
Hoff, Ebbe Curtis
Holloway, James Ashley
Honrubia, Vicente
Horch, Kenneth William
Horowitz, John M
Horvath, Fred Ernest
Hubel, David Hunter
Humphrey, Donald R
Hunt, Carlton Cuyler
Jacklet, Jon Willis
Jabbur, Suhayl Jibra'il
Jasper, Herbert Henry
Javel, Eric
John, E Roy
Johnson, Jeffery Lee
Johnson, Kenneth Olafur
Kahan, Linda Beryl
Kashin, Philip
Kater, Stanley B
Katz, George Maxim
Kelly, Dennis D
Kennedy, Duncan Tilly
Kennedy, Thelma Temy
Kiang, Nelson Yuan-Sheng
King, Jerome Stovall
King, Robert Lee
Knott, John Russell
Knox, Charles Kenneth
Knox, Werner Paul
Koella, Werner Paul
Kohn, Michael
Koopowitz, Harold
Krauthamer, George Michael
Kreisman, Norman Richard
Krnjevic, Kresimir
Kuffler, Stephen William
Kusano, Kiyoshi
Lamarche, Guy
Lange, Gordon David
Laskowski, Michael Bernard
Latimer, Clinton Narath
Lavine, Robert Alan
Lawry, James Voris
LeBlanc, Francis Ernest
Lebovitz, Robert Mark
Letbetter, William Dean
Lettvin, Jerome Y
Levine, Leonard
Li, Choh-Luh

Liberson, Wladimir Theodore
Livingston, Robert Burr
Loewenstein, Werner Randolph
Low, Morton David
Lowy, Karl
Lynch, James Carlyle
MacLean, Paul Donald
Magleby, Karl LeGrande
Mann, Diana Witherspoon
Mann, Michael David
Martin, Alexander Robert
Masserman, Jules Homan
Mayer, David Jonathan
Mayeri, Earl Melchior
Mayor, Stephen Joseph
McArdle, Joseph John
McMillan, James Alexander
Mellon, DeForest, Jr
Merlis, Jerome K
Merzenich, Michael Matthias
Meyers, H Russell
Michael, Charles Reid
Michael, Joel Allen
Mikiten, Terry Michael
Miller, Arthur Joseph
Miskimen, George William
Mistretta, Charlotte Mae
Molnar, Charles Edwin
Moore, John Wilson
Moore, Lee E J
Morgane, Peter J
Morin, Walter Arthur
Morison, Robert Swain
Moss, Robert L
Mote, Michael Isnardi
Murray, George Cloyd
Nakajima, Shigehiro
Narahashi, Toshio
Nathan, Marc A
Nelson, Margaret Christina
Nicholls, John Graham
Nicoll, Roger Andrew
Nudelman, Harvey Banet
Nyquist, Judith Kay
Ochs, Sidney
O'Connell, Robert James
Orr, Alfonso
Page, Charles Henry
Palka, John Milan
Parsons, L Claire
Partridge, Lloyd Donald
Pasik, Pedro
Pasik, Tauba
Peacock, Samuel Moore, Jr
Pearson, Keir Gordon
Perret, George (Edward)
Perryman, James Harvey
Peterson, Barry Wayne
Pfaff, Donald Wells
Phillis, John Whitfield
Pickens, Peter E
Pilar, Guillermo Roman
Pinto, Lawrence Henry
Pomeranz, Bruce Herbert
Poppele, Richard E
Potter, David Dickinson
Poulos, Dennis A
Pozos, Robert Steven
Price, Donald Dennis
Pubols, Benjamin Henry, Jr
Pubols, Lillian Menges
Purpura, Dominick Paul
Randic, Mirjana
Ratliff, Floyd
Redgate, Edward Stewart
Reitan, Ralph Meldahl
Retzlaff, Ernest (Walter)
Rhode, William Stanley
Robbins, Norman
Robinson, David Adair
Robinson, David Lee
Rohner, Mary Christopher
Roig, Juan Antonio
Rose, Jerzy Edwin
Rosenthal, Fred
Rovainen, Carl (Marx)
Rowe, Edward C
Rowell, Charles Hugh Fraser
Rudenberg, Frank Hermann
Rudomin, Pedro
Ruggero, Mario Alfredo
Ryall, Ronald W
St John, Walter McCoy
Sato, Makoto
Saunders, Frank Austin
Saunders, Virginia Fox
Scheibel, Arnold Bernard
Schlag, John
Schor, Robert Hyttel
Scobey, Robert P
Scott, Donald, Jr
Scott, John Wilson
Segundo, Jose Pedro
Selverston, Allen Israel
Sessle, Barry John
Shankland, Daniel Leslie
Shapiro, Bert Irwin
Shepherd, Gordon Murray

Sherman, Robert George
Siminoff, Robert
Sinclair, John C
Smith, James Darrell
Smith, Orville Auverne, Jr
Smith, Thomas Graves, Jr
Soechting, John F
Speck, Louise Barrett
Spray, David Conover
Stein, Barry Edward
Stein, Richard Bernard
Steinberg, Roy Herbert
Steriade, Mircea
Stevens, Charles F
Stitt, John Thomas
Streicher, Eugene
Symmes, David
Takashima, Shiro
Talbot, William Henry
Talbot, Richard Evans
Tang, Pei Chin
Tapper, Daniel Naptali
Tarby, Theodore John
Tasaki, Ichiji
Taub, Arthur
Thompson, William D
Thorson, John Wells
Tietz, William John, Jr
Trevino, Daniel Louis
Trubatch, Janett
Van Buren, John Miller
Velez, Samuel Jose
Villablanca, Jaime Rolando
Von Baumgarten, Rudolf Jury
Wagman, Irving Henry
Walter, Donald Oliver
Webb, George Dayton
Welker, Carol
Welker, Wallace I
Werman, Robert
Wetzel, Allan Brooke
Whitehorn, David
Whitsel, Barry L
Willis, William Darrell, Jr
Williston, John Stoddard
Willows, Arthur Owen Dennis
Wilson, David Franklin
Wilson, Victor Joseph
Wolfe, James Wallace
Wolin, Lee Roy
Woody, Charles Dillon
Woolsey, Clinton Nathan
Wulff, Verner John
Wurtz, Robert Henry
Wyse, Gordon Arthur
York, Donald Harold
Zablow, Leonard
Zajac, Felix Edward, III
Zimmerman, Irwin David

Pathological Physiology
Burton, Russell Rohan
Mettrick, David Francis
Rosenfeld, Irene

Physiological Bacteriology
Arceneaux, Joseph Lincoln
Barnekow, Russell George, Jr
Baugh, Clarence L
Boehms, Charles Nelson
Booth, James Samuel
Burchall, James J
Fisher, Robert John
Fraenkel, Dan Gabriel
Hug, Daniel Hartz
Iandolo, John Joseph
Juni, Elliot
Kory, Mitchell
Martinez, Rafael Juan
McIntosh, Elaine Nelson
Mickelson, Milo Norval
O'Brien, Robert Thomas
Stewart, James Edward
Thompson, Thomas Leo
Torriani Gorini, Annamaria
Traxler, Richard Warwick

Physiological Optics
Afanador, Arthur Joseph
Enoch, Jay Martin
Fatt, Irving
Freeman, Ralph David
Fry, Glenn Ansel
Green, Daniel G
Greenspon, Thomas Stephen
Gunkel, Ralph D
Hill, Richard M
Kerr, Kenton E
Levene, John Reuben
Ludlam, William Myrton
Pease, Paul Lorin
Pitts, Donald Graves
Pratt, Carol Bert
Randle, Robert James
Reading, Rogers W
Remole, Arnulf
Roth, Niles
Takahashi, Ellen Shizuko

Webber, Richard Lyle
Westheimer, Gerald

Poultry Physiology
Cherms, Frank Llewellyn, Jr
Edens, Frank Wesley
Hill, Arthur Thomas
Kuenzel, Wayne John
May, James David
Moreng, Robert Edward
Morgan, George Wallace
Pasvogel, Myron W
Voite, Robert Allen

Psychophysiology
Annau, Zoltan
Arkin, Arthur Malcolm
Bartley, Samuel Howard
Beck, Edward C
Begleiter, Henri
Benuck, Irwin
Black, William
Blass, Elliott Martin
Boynton, Robert M
Brown, Barbara Banker
Brown, Clinton Carl
Brown, John Lott
Brown, Warren Shelburne, Jr
Campbell, James Fulton
Capps, Mary Jayne
Crawford, Morris Lee Jackson
Cutt, Roger Alan
Dawson, William Woodson
de Haan, Henry John
Donchin, Emanuel
Donovick, Peter Joseph
Dykman, Roscoe A
Edelberg, Robert
Eisdorfer, Carl
Erickson, Robert Porter
Essman, Walter Bernard
Fay, Richard Rozzell
Fleming, Donovan Ernest
Fox, Stephen Sorin
Gault, Frederick Paul
Gourevitch, George
Goy, Robert William
Graham, Frances Keesler
Greene, Ernest Gerald
Greenough, William Tallant
Guedry, Frederick Ernest
Halpern, Naomi Mimi
Hamilton, Leonard W
Harlow, Harry F
Horel, James Alan
Hornbuckle, Phyllis Ann
Horne, Edward Porter
Hughes, Kenneth Russell
Isaacson, Robert Lee
Jacobs, Harry Lewis
Jones, Bill Edson
Kaas, Jon Howard
Kallman, William Michael
Keesey, Ulker Tulunay
Klein, Sherwin Jared
Kragt, Clifford Lee
Krauskopf, John
Kryter, Karl David
Lacey, Beatrice Cates
Lacey, John Irving
Leshner, Alan Irvin
Levine, Seymour
Lindsley, Donald B
Lubin, Ardie
Maller, Owen
Masuda, Minoru
McGough, William Edward
McKenney, Joel R
Meier, Gilbert W
Miles, Walter Richard
Miller, Neal Elgar
Mogenson, Gordon James
Moushegian, George
Mulligan, Benjamin Edward
Naitoh, Paul Yoshimasa
Newton, Joseph Emory O'Neal
Oesterreich, Roger Edward
Peacock, Lelon James
Phillips, David
Pinneo, Lawrence Robert
Pinsker, Harold M
Pokorny, Joel
Riggs, Lorrin Andrews
Roberts, Alan H
Rodgers, Charles H
Rubin, Leonard Sidney
Sandel, Thomas Theodore
Satterberg, John Arvid
Schneiderman, Neil
Semmes, Josephine
Sharpless, Seth Kinman
Silverman, Albert Jack
Smith, Vivianne Cameron
Snowdon, Charles Thomas
Sparks, David Lee
Stellar, Eliot
Stroebel, Charles Frederick, III
Sturr, Joseph Francis

Survillo, Walter Wallace
Taub, John Marcus
Teas, Donald Claude
Terando, M Loretta
Teuber, Hans-Lukas
Thorn, Frank
Trehub, Arnold
Trumbull, Richard
Vernon, Jack Allen
Vierck, Charles John, Jr
Vinson, David Berwick
Ward, Ingeborg L
Weiss, Jay M
White, Robert Keller
Whitehead, William Earl
Winer, Cynthia Crosby
Winters, Ray Wyatt
Yellin, Absalom Moses
Yeni-Komshian, Grace Helen
Young, Francis Allan
Zerlin, Stanley

Pulmonary Physiology

Beckman, David Lee
Brigham, Kenneth Larry
Chick, Thomas Wesley
Crandall, Edward D
Fedde, Marion Roger
Hazlett, David Richard
Imbruce, Richard Peter
Kettel, Louis John
King, Thomas K C
Klocke, Robert Albert
Knudson, Ronald Joel
Macklem, Peter Tiffany
Mauderly, Joe L
Nadel, Jay A
Nair, Sreedhar
Nattie, Eugene Edward
Newman, Melvin Micklin
Otis, Arthur Brooks
Pennock, Bernard Eugene
Rosenzweig, David Yates
Sackner, Marvin Arthur
Sedensky, James Andrew
Wagner, Wiltz Walker, Jr
Weber, Kenneth C
Webster, James Randolph, Jr
Weiss, Earle Burton
Wiley, Ronald Lee

Reproductive Physiology

Alexander, Nancy J
Almquist, John Olson
Alsum, Donald James
Amann, Rupert Preynoessl
Anand, Amarjit S
Banik, Upendra K
Barfield, Mary Ashton
Barr, Harry L
Bartke, Andrzej
Bazer, Fuller Warren
Bearden, Henry Joe
Bennett, John Phillip
Berndtson, William Everett
Black, Wallace Gordon
Blend, Michael J
Bolt, Douglas John
Bolte, Edouard
Boone, Merritt Anderson
Boyd, Louis Jefferson
Bronson, Franklin Herbert
Brown, Barry Lee
Brown, Keith Irwin
Burfening, Peter J
Butcher, Roy Lovell
Call, Edward Prior
Campbell, Donald Gray
Casida, Lester Earl
Cecil, Helene Carter
Chang, Chin-Chuan
Chizinsky, Walter
Choudary, Jasti Bhaskararao
Chowdhury, Ajit Kumar
Clark, James Henry
Cochrane, Robert Lowe
Collins, Delwood C
Dahnen, Jerome J
Daniel, Joseph Car, Jr
Dawson, John E
de Alba Martinez Jorge
Dickey, Joseph Freedman
Douglas, Robert Hazard
DuFrain, Russell Jerome
Dukelow, W Richard
Dunn, Thomas Guy
Dyck, Gerald Wayne
Dziuk, Philip J
Eddy, Carlton Anthony
Elliott, Fred Irvine
England, Barry Grant
Espey, Lawrence Lee
Fahning, Melvyn Luverne
Falcon, Carroll James
Faulkner, Lloyd (Clarence)
Feder, Harvey Herman
First, Neal L
Flickinger, George Latimore, Jr
Franklin, Beryl Cletis
Fuchs, Anna-Riitta
Garner, Duane LeKoy
Godke, Robert Alan
Grob, Howard Shea
Grosso, Leonard
Guthrie, Howard David
Hafez, Saad Elsayed
Hall, Kent D
Hahn, DaWon
Hammerstedt, Roy H
Harding, Paul George Richard
Harper, Michael John Kennedy
Harrison, Richard Miller
Hart, Robert Gerald
Hasler, Marilyn Jean
Hawk, Harold W
Hodgson, Barrie John
Hoppe, Peter Christian
Hopwood, Mortimer Lloyd
Hoskins, Dale Douglas
Hoversland, Arthur Stanley
Howarth, Birkett, Jr
Howe, George R
Hughes, Buddy Lee
Inskeep, Emmett Keith
Jochle, Wolfgang
Johnson, Archie Doyle
Johnson, Lawrence Arthur
Johnson, Shirley Mae
Jones Witters, Patricia H
Joshi, Madhusudan Shankarrao
Kaltenbach, Carl Colin
Keller, Dolores Elaine
Kennedy, James J
Kent, Harry Alvin, Jr
Ketchel, Melvin M
Kille, John William
Killian, Gary Joseph
King, Gordon James
Kirkpatrick, Roy Lee
Koering, Marilyn Joan
Kraemer, Duane Carl
Kreider, Jack Leon
Kuo, Ching-Ming
Larson, Larry John
Lasley, Bill Lee
Lauderdale, James W, Jr
Leavitt, Wendell W, Jr
Legan, Sandra Jean
Lewis, Paul Edward
Lisano, Michael Edward
Lobl, Thomas Jay
Lodge, James Robert
Loe, William Carrol
Longley, William Joseph
Mahajan, Satish Chander
Maurer, Ralph Rudolf
Mead, Rodney A
Menge, Alan C
Merker, Jerry Wheeler
Miller, William Wadd, III
Mitchell, J Andrew
Mock, Orin Bailey
Morato, Tomas
Mounib, M Said
Murphy, Bruce Daniel
Murray, Finnie Ardrey, Jr
Nayak, Ramesh Kadhet
Nelson, Leonard
Norwood, James S
O'Brien, Coleman Art
O'Connor, William Brian
Opel, Howard
Pace, Marvin M
Palmer, Robert Alexander
Parks, James John
Pasley, James Neville
Patanelli, Dolores J
Peitz, Betsy
Peterson, Roy Phillip
Pickett, Bill Wayne
Pomerantz, David Kurt
Poston, Hugh Arthur
Preslock, James Peter
Rao, Papineni Seethapathi
Rasweiler, John Jacob, IV
Reeves, Jerry John
Resko, John A
Rexroad, Caird Eugene, Jr
Rich, Royal Allen
Rich, Travis Dean
Riegle, Gail Daniel
Riesen, John William
Robison, Odis Wayne
Rousel, Joseph Donald
Saacke, Richard George
Saididdin, Syed
Sammelwitz, Paul H
Sargent, Frank Dorrance
Scanlon, Patrick Francis
Sconmegna, Antonio
Seeley, Rod R
Seidel, George Elias, Jr
Shelden, Robert Merten
Simmons, Kenneth Rogers
Singleton, Wayne Louis
Slechta, Robert Frank
Slyter, Arthur Lowell
Sorensen, Anton Marinus, Jr
Soupart, Pierre
Spaziani, Eugene
Stolzenberg, Sidney Joseph
Sullivan, John Joseph
Taber, Elsie
Talbot, Prudence
Thatcher, William Watters
Thompson, Leif Harry
Turnipseed, Marvin Roy
Ulberg, Lester Curtiss
Urry, Ronald Lee
Virgo, Bruce Barton
Waite, Albert B
Wallace-Haagens, Mary Jean
Walsh, Scott Wesley
Weisz, Judith
Weitlauf, Harry
Wickersham, Edward Walker
Wildt, David Edwin
Wilson, Henry R
Witters, Weldon L
Woody, Charles Owen, Jr
Younglai, Edward Victor
Zemjanis, Raimunds

Sensory Physiology

Albert, Paul Joseph
Bradley, Robert Martin
Bruno, Merle Sanford
Fishman, Irving Yale
Haas, Gustav Frederick
Henderson, Donald
Kauer, John Stuart
Mozell, Maxwell Mark
Witkovsky, Paul

Tree Physiology

Brown, Gregory Neil
Hart, John Henderson
Hook, Donald
Little, Charles Harrison Anthony
Quirk, John Thomas
Skolmen, Roger Godfrey

Vertebrate Physiology

Ackermann, Uwe
Baker, Mary Ann
Barker, Laren Dee Stacy
Bettice, John Allen
Bolls, Nathan J, Jr
Borei, Hans Georg
Buss, Jack Theodore
Comeau, Roger William
Courtney, Gladys A
Cundiff, Milford Fields
Davis, Kenneth Bruce, Jr
Giere, Frederic Arthur
Heisinger, James Fredrick
Ikenberry, Roy Dewayne
Jensen, Donald Reed
Johnson, Patricia R
Kelley, Fenton Crosland
Marshall, Louise Hanson
Martin, James Henry, III
Meyer, Frederick Richard
Mitchell, Henry Andrew
Orr, James Anthony
Person, Steven John
Pettegrew, Raleigh K
Pober, Zalmon
Saunders, Harry Link
Tharp, Gerald D

Respiratory Physiology

Anthonisen, Nicholas R
Banerjee, Chandra Madhab
Bickerman, Hylan A
Chapin, John Ladner
Clayton, John Wesley, Jr
Colas, Antonio E
Eldridge, Frederic L
Hildebrandt, Jacob
Hyde, Richard Witherington
Jaeger, Marc Jules
Josenhans, William T
Motoyama, Etsuro K
Pearce, David Harry
Pontoppidan, Henning
Urbscheit, Nancy Lee
Weiskopf, Richard Bruce
Wiley, Ronald Lee
Zamel, Noe
Zimmerman, Jay Alan

Visual Physiology

Alpern, Mathew
Bernard, Gary Dale
Biersdorf, William Richard
Blackwell, Harold Richard
Brown, Harold Mack, Jr
Chalupa, Leo M
Corless, Joseph Michael James
Cornsweet, Tom Norman
Crescitelli, Frederick
Darlington, Robert Wells
Dratz, Edward Alexander
Enroth-Cugell, Christina
Gibson, James Jerome
Goldsmith, Timothy Henshaw
Gruber, Samuel Harvey
Guth, Sherman Leon
Harkavy, Allan Abraham
Hofstetter, Henry W
Jones, Arthur Eugene
Julesz, Bela
Kelly, Donald Horton
Knoll, Henry Albert
Kozak, Wlodzimierz Maciej
Lit, Alfred
Loop, Michael Stuart
Lowther, Gerald Eugene
MacLeod, Donald Iain Archibald
Mandell, Robert Burton
Marg, Elwin
Morgan, Meredith Walter
Munz, Frederick Wolf
Oyster, Clyde William
Pak, William Louis
Rodieck, Robert William
Sakitt, Barbara
Saladin, Jimmie James
Schoessler, John Paul
Siegfried, John Barton
Sperling, Harry George
Speros, Perry
Srinivasan, Mandyam Veerambudi
Stell, William Kenyon
Taylor, John Hall
Teller, Davida Young
Tieman, Suzannah Bliss
Tobey, Frank Lindley, Jr
Uniacke, Charles Allyn
Van Sluyters, Richard Charles
Wilson, Graeme Stewart

Veterinary Physiology

Appel, Max J
Beadle, Ralph Eugene
Bisgard, Gerald Edwin
Bowie, Walter C
Brewer, Nathan Ronald
Carter, James M
Chapman, Thomas Everett
Chen, Chao Ling
Colmano, Germille
Coulter, Dwight Bernard
Dale, Homer Eldon
Dickson, William Morris
Dobson, Alan
Dusseau, Jerry William
Fremming, Benjamin DeWitt
Gillette, D Dale
Goetsch, Dennis Donald
Good, A L
Houpt, Katherine Albro
Houpt, Thomas Richard
Hull, Maurice Walter
Kohlmeier, Ronald Harold
Martin, Charles Everett
McCrady, James David
McDonald, Leslie Ernest
Muggenburg, Bruce Al
Nachreiner, Raymond F
Nichols, Roy Elwyn
Palmore, William P
Payne, Loyal Cobb
Phillips, George Douglas
Phillips, Robert Ward
Pinjani, Moti
Redding, Richard William
Robinson, Jerry Allen
Schryver, Herbert Francis
Sellers, Alvin Ferner
Smith, Charles Roger
Stevens, Charles Edward
Swenson, Melvin John
Twardock, Arthur Robert
Veum, Tryve Lauritz
Wettemann, Robert Paul
Whipp, Shannon Carl
Witzel, Donald Andrew

POULTRY SCIENCE

Poultry Science

Aho, William A
Andrews, Daniel Keller
Andrews, Walter Glenn
Arrington, Louis Carroll
Bearse, Gordon Everett
Berry, Joe Gene
Bigbee, Daniel E
Bragg, Denver Dayton
Brown, Keith Irwin
Carson, James Rolland
Coleman, Theo Houghton
Deaton, James Washington

Dobson, Donald C
Ernst, Ralph Ambrose
Essary, Eskel Oren
Froning, Glenn Wesley
Fry, Jack L
Goodman, Billy Lee
Hale, Kirk Kermit, Jr
Hamre, Melvin L
Harper, James Arthur
Harris, Paul Chappell
Heath, James Lee
Henderson, Donald Cedric
Hicks, Floyd W
Holleman, Kendrick Alfred
Holmes, Clayton Ernest
Hughes, Buddy Lee
Johnson, Hugh Swaney
Johnson, William Alexander
Jones, Jack Edenfield
Kohlmeyer, William
Kosin, Igor Leonid
Mellor, David Bridgwood
Mills, William Clearon, Jr
Morehouse, Neal Francis
Mountney, George Joseph
Mueller, Werner Julius
Muir, Forest Vern
Nordskog, Arne William
Orr, Henry Lloyd
Ostrander, Charles Evans
Ott, Walther Henry
Palafox, Anastacio Laida
Peterson, Irvin Leslie
Prasad, Suresh
Quisenberry, John Henry
Reid, Willard Malcolm
Ridlen, Samuel Franklin
Rolfes, Thomas J
Schar, Raymond Dewitt
Schwall, Donald V
Shaffner, Clyne Samuel
Shaklee, William Eugene
Sicer, Joe Wellington
Singh, Suresh Pratap
Singsen, Edwin Pierce
Stephens, James Fred
Stiles, Philip Glenn
Strain, John Henry
Wilgus, Herbert Sedgwick
Winter, Alden Raymond
Wisman, Everett Lee
Wyatt, Roger Dale

Poultry Breeding
Arthur, James Alan
Friars, Gerald W
Gavora, Jan Samuel
Gowe, Robb Shelton
Hawes, Edward George
Jones, Dean Graeme
Kashyap, Tapeshwar S
MacLaury, Donald Wayne
Merritt, Edison S
Rishell, William Arthur
Vint, Larry Francis

Poultry Genetics
Berg, Robert W
Bernier, Paul Emile
Buckland, Roger Basil
Buss, Edward George
Crawford, Roy Douglas
Jones, Dean Alexander
Kondra, Peter Alexander
Law, George Robert John
Moore, Claude Henry
Moore, Jay Winston
Proudfoot, F G
Shoffner, Robert Nurman
Washburn, Kenneth W

Poultry Husbandry
Adams, John Lester
Bletner, James Karl
Brant, Albert Wade
Brunson, Clayton (Cody)
Carlson, Roy David
Francis, David Wesson
Glazener, Edward Walker
Goan, Hugh Charles
Godfrey, George Frelaut
Hammond, John Clarke
Hannah, John Alfred
Hill, Arthur Thomas
Jaap, Robert George
Krueger, Willie Frederick
Newell, George Watts
Parker, Jesse Elmer
Rosenberg, Morton Murray
Rowoth, Olin Arthur
Ryan, Cecil Benjamin
Scott, Harold Martin
Sheppard, Charles Campbell
Skoglund, Winthrop Charles
Stephenson, Alfred Benjamin
Stopper, William W
Taylor, Lewis Walter
Thomas, Charles Hill

Thornton, Erly J
Zindel, Howard Carl

Poultry Nutrition
Abbott, Okra Jones
Adams, Richard Linwood
Allen, Neil Keith
Anderson, Jay Oscar
Arnold, Richard Lane
Arscott, George Henry
Atkinson, Robert Leon
Barnett, Bobby Dale
Begin, John Joseph
Berg, Lawrence Raymond
Biely, Jacob
Bray, Donald James
Britton, Walter Martin
Britzman, Darwin Gene
Clandinin, Donald Robert
Damron, Bobby Leon
Davis, Buster Hall
Day, Elbert Jackson
Dilworth, Benjamin Conroy
Donaldson, William Emmet
Donovan, Gerald Alton
Douglas, Carroll Isaac
Draper, Carroll Isaac
Featherston, William Roy
Flegal, Cal J
Fuller, Henry Lester
Gerry, Richard Woodman
Gledhill, Robert Hamor
Grau, Charles Richard
Halloran, Hobart Rooker
Harms, Robert Henry,
Holder, David Parker
Hunt, John R
Keene, James H
Keene, Owen David
Kelly, Mike
Lee, Kwang
Leong, Kam Choy
Likuski, Henry John
Lillie, Robert Jones
March, Beryl Elizabeth
Mehring, Arnon Lewis, Jr
Middendorf, Donald Floyd
Miller, Byron F
Monson, William Joye
Naber, Edward Carl
Nikolaiczuk, Nikolai
Pasvogel, Myron W
Petersen, Charlie Frederick
Potter, Lawrence Merle
Ringrose, Richard Caig
Robblee, Alexander (Robinson)
Roberson, Robert H
Ross, Ernest
Rowland, Lenton O, Jr
Rowoth, Olin Arthur
Salmon, Raymond Edward
Sanford, Paul Everett
Savage, Jimmie Euel
Shutze, John V
Smith, Daniel Newton, Jr
Snetsinger, David Clarence
Speers, George M
Standlee, William Jasper
Steinke, Frederich H
Sullivan, Thomas Wesley
Sunde, Milton Lester
Thayer, Rollin Harold
Thomas, Owen Pestell
Trammell, Jack Harman, Jr
Waibel, Paul Edward
Ward, James B
Warden, William Kent
Wellenreiter, Rodger Henry
West, John Wyatt
Wharton, Ferdinand Decatur, Jr
Yates, Jerome Douglas

Poultry Pathology
Belding, Ralph Cedric
Bierer, Bert Worman
Brewer, Robert Nelson
Bryan, Thomas Alan
Chang, Timothy Scott
Fabricant, Julius
Gentry, Robert Francis
Harvey, Roger Bruce
Hilbert, Kenneth Franklin
Hitchner, Stephen Ballinger
Morgan, George Wallace
Muniz, Raul A
Narotsky, Saul
Purchase, Harvey Graham
Raggi, Livio Giovanni
Schwartz, Leland Dwight
Tudor, David Cyrus
Witter, Richard L

Poultry Physiology
Boone, Merritt Anderson
Cherms, Frank Llewellyn, Jr
Edens, Frank Wesley

May, James David
Moreng, Robert Edward
Voitle, Robert Allen

PSYCHIATRY

Psychiatry
Abram, Harry Shore
Abse, David Wilfred
Abuzzahab, Faruk S, Sr
Adams, John Evi
Adams, John R
Adams, Paul Lieber
Adamson, John Douglas
Adler, Kurt Alfred
Ainslie, John Durham
Aivazian, Garabed Hagpop
Aldrich, Clarence Knight
Alexander, Leo
Amdur, Millard Jason
Amundson, Mary Jane
Anderson, Richard Hayden
Anderson, William B
Apfelberg, Benjamin
Apter, Nathaniel Stanley
Arieti, Silvano
Arkin, Arthur Malcolm
Arlow, Jacob
Arthur, Ransom James
Ascher, Eduard
Astrachan, Boris Morton
Atcheson, J D
Atkins, Robert W
Auerbach, Arthur Henry
Auerback, Alfred
Babcock, Charlotte Gertrude
Babineau, G Raymond
Baekeland, Frederick
Bakker, Cornelis Bernardus
Ban, Thomas Arthur
Bankier, Robert G
Barondes, Samuel Herbert
Barter, James T
Bartholow, George William
Bartlett, James Williams, Jr
Barton, Walter E
Batson, Margaret 'Bailly
Baum, O Eugene
Beck, Aaron Temkin
Beiser, Morton
Beisser, Arnold Ray
Bell, James Milton
Bellak, Leopold
Belmont, Herman S
Bennett, Ivan Frank
Berblinger, Klaus William
Berlin, Irving Norman
Bernstein, Barbara Elaine
Berry, Gail Wruble
Betz, Barbara Jean
Bever, Christopher Theodore
Bieber, Irving
Billig, Otto
Blachly, Paul H
Blackburn, Archie Barnard
Blain, Daniel
Bliss, Eugene Lawrence
Block, Stanley L
Bloom, Victor
Bloomberg, Wilfred
Blum, Robert Allan
Blume, Sheila Bierman
Blumenthal, Irving Jack
Blumenthal, Monica David
Boag, Thomas Johnson
Bogoch, Samuel
Bojar, Samuel
Bond, Douglas Danford
Bonkalo, Alexander
Bordeleau, Jean-Marc
Boslow, Harold Meyer
Boulanger, Jean Baptiste
Boverman, Harold
Braceland, Francis James
Brady, John Paul
Branch, Charles Henry Hardin
Brand, Eugene Dew
Brands, Alvira Bernice
Brauchi, John Tony
Brenner, Charles
Brickman, Harry Russell
Brill, Norman Quintus
Brodie, Harlow Keith Hammond
Brodsky, Carroll M
Brody, Eugene B
Brooks, George Wilson
Brown, Bertram S
Bruce, E Ivan, Jr
Bruch, Hilde
Burnham, Donald Love
Busse, Ewald William
Butler, Robert N
Byck, Robert
Cadoret, Remi Jere
Call, Justin David
Callaway, Enoch, III

Cameron, Dale Corbin
Cancro, Robert
Cantrell, William Allen
Caplan, Gerald
Carlson, Eric Theodore
Carter, George H
Caveny, Elmer Leonard
Chafetz, Morris Edward
Char, Walter F
Charny, Eugene Joseph
Chassan, Jacob Bernard
Chessick, Richard D
Churchill, Don W
Clancy, John
Clark, Gerald Robert
Clark, Robert Alfred
Clayton, Paula Jean
Cleghorn, Robert Allen
Coburn, Frank Emerson
Cohen, Robert Abraham
Cohen, Sanford I
Cohn, Jay Binswanger
Cole, Nyla J
Coleman, Jules Victor
Colman, Arthur David
Comly, Hunter Hall
Coppolillo, Henry
Cormier, Bruno M
Cornelison, Floyd S, Jr
Crain, Jay Bouton
Cramer, Joseph Benjamin
Crawford, James Weldon
Curtis, George Clifton
Curtis, John Russell
Curtis, Thomas Edwin
Cutts, Robert Irving
Daitzman, Reid Joseph
Daly, Robert Ward
Davis, David
Davis, Elizabeth B
Davis, John Marcell
Davis, Oscar F
Davis, Richard Elden
Davis, Stanley David
Davis, Vernam Terrell
DeBow, Lee Richard
Deckert, Gordon Harmon
DeMyer, Marian Kendall
Denber, Herman C B
Denney, Donald Duane
Detre, Thomas Paul
Dickes, Robert
Dingman, Charles Wesley, II
Dizenhuz, Israel Michael
Doane, Benjamin Knowles
Donahue, Hayden Hackney
Donnelly, John
Dorsey, John Morris
Dovenmuehle, Robert Henry
Duhl, Leonard J
Dunsworth, Francis Alfred
DuPont, Robert L, Jr
Durell, Jack
Duval, Addison (McGuire)
Dyrud, Jarl Edvard
Easton, William McAlpine
Eaton, Merrill Thomas, Jr
Egan, Merritt H
Eisenberg, Leon
Eisdorfer, Carl
Elam, Lloyd Charles
Elkes, Joel
Ellinwood, Everett Hews, Jr
Enelow, Allen Jay
Engel, George Libman
Engelhardt, David Meyer
English, Oliver Spurgeon
Epstein, Arthur William
Epstein, Leon J
Epstein, Nathan Bernic
Ermutlu, Ilhan M
Ervin, Frank (Raymond)
Esser, Aristide Henri
Evans, Harrison Silas
Evarts, Edward Vaughan
Ewalt, Jack R
Ewing, John Alexander
Fabre, Louis Fernand, Jr
Fabrega, Horacio, Jr
Faillace, Louis A
Falk, Marshall Allen
Farnsworth, Dana Lyda
Feather, Ben Wayne
Feinberg, Irwin
Feldman, Harold Samuel
Felix, Robert Hanna
Fellner, Carl Heinz
Ferguson, James Mecham
Ferguson, Shirley Martha
Field, Howard Lawrence
Fieve, Ronald Robert
Filley, John Paton
Finch, Stuart McIntyre
Fisher, Barbara
Fisher, Saul Harrison
FitzGerald, Joseph Arthur
Flach, Frederic Francis
Fleck, Stephen

PSYCHIATRY

Fleeson, William
Forbis, Orie Lester, Jr
Fowler, John Alvis
Frank, Jerome David
Frankel, Fred Harold
Frazier, Shervert Hughes
Freed, Herbert
Freedman, Alfred Mordecai
Freedman, Daniel X
Freedman, Lawrence Zelic
Friedhoff, Arnold Jerome
Friedman, Matthew Joel
Froelich, Robert Earl
Frosch, William Arthur
Gallant, Donald
Gallant, Leonard Jay
Garner, Harry Hyman
Garner, Richard Jay
Gaskill, Herbert Stockton
Gaylin, Willard
Gerber, Carl J
Gericke, Otto Luke
Giffen, Martin Brener
Gill, Merton
Gilliland, Robert McMurtry
Ging, Rosalie J
Ginsberg, Stewart Theodore
Giovacchini, Peter L
Glen, Robert S
Glover, Benjamin Howell
Glueck, Bernard Charles
Godene, Ghislaine D
Goldings, Herbert Jeremy
Goldsmith, Jewett
Goldstein, Burton Jack
Gonda, Thomas Andrew
Goodrich, D Wells
Gottheil, Edward
Gottlieb, Jacques Simon
Gottschalk, Louis August
Greaves, Donald Critchfield
Green, Robert Lee, Jr
Greenberg, Harold Abraham
Greenblatt, Milton
Greenhill, Maurice H
Greenson, Ralph Romeo
Gregory, Ian (Walter De Grave)
Greist, John Howard
Grenell, Robert Gordon
Griffith, John Dorland
Grinker, Roy Richard, Sr
Gross, Milton Michael
Grosser, Bernard Irving
Grosz, Hanus Jiri
Gussen, John
Gutman, Samuel Arnold
Guze, Samuel Barry
Hall, Henry Lee
Halleck, Seymour Leon
Halmi, Katherine A
Hamburg, David Alan
Hamilton, Francis Joseph
Hampson, John L
Hankoff, Leon Dudley
Hansen, Douglas Brayshaw
Hansen, Howard Edward
Harper, Edward O'Neil
Harris, Harold Joseph
Harris, Mervin Robert
Harrison, Saul I
Hartman, Boyd Kent
Hartmann, Ernest Louis
Hartocollis, Peter
Hausman, William
Havens, Leston Laycock
Hawke, W A
Hawkins, David Rollo
Headlee, Raymond
Heath, Robert Galbraith
Hein, Peter Leo, Jr
Heller, Melvin S
Hendrickson, Willard James
Heninger, George Robert
Henisz, Jerzy Emil
Henker, Fred Oswald, III
Henry, Billy Wendell
Herz, Marvin Ira
Heston, Leonard L
Hewitt, Robert T
Hiatt, Harold
Hine, Frederick Roy
Hinko, Edward N
Hirsch, Solomon
Hobbs, George Edgar
Hobson, John Allan
Hoehn-Saric, Rudolf
Hoenig, Julius
Hofing, Charles Kreimer
Hoffman, Julius
Hollender, Marc Hale
Hollister, William Gray
Holmes, Thomas Hall, III
Hornick, Edward J
Horowitz, Mardi J
Huber, Wolfgang Karl
Hudgens, Richard Watts
Huessy, Hans Rosenstock

Huffer, Sarah Virginia
Hughes, Joseph F
Hunter, Robin Cyril Adair
Ikenberry, Lynn David
Imboden, John Baskerville
Itil, Turan M
Jackson, Basil Edgar
Jacobson, Avrohm
Jenkins, Richard Leos
Jensen, Gordon D
Jepson, William W
Jessner, Lucie
Jones, Granville Lillard
Jones, James David
Jones, Reese Tasker
Jones, Robert Orville
Joseph, Edward David
Juhasz, Stephen Eugene
Judd, Lewis Lund
Kaebling, Rudolf
Kane, Francis Joseph, Jr
Kaplan, Harold Irwin
Kaplan, Helen Singer
Kaplan, Stanley Meisel
Kapp, Frederic T
Karkalas, John
Kaufman, Irving Charles
Kaufman, Janice Norton
Kay, David Cyril
Kelley, John Fredric
Kemble, Robert Penn
Kemph, John Patterson
Kepecs, Joseph Goodman
Kernberg, Otto F
Kimball, Chase Patterson
King, Lucy Jane
Kleber, Herbert David
Klee, Gerald D'Arcy
Klein, Donald Franklin
Kline, Nathan Schellenberg
Knapp, Peter Hobart
Knight, Edward Henry
Knight, James Allen
Knopp, Walter
Kohl, Richard Niemes
Kohut, Heinz
Kolb, Lawrence Coleman
Korenyi, Charles
Kosbab, Frederic Paul Gustav
Kraft, Alan M
Kraft, Irvin Alan
Kramer, Milton
Kravitz, Henry
Kriegman, George
Kroger, William Saul
Kulka, Johannes Peter
Kurland, Albert A
Labby, Daniel Harvey
Lambert, William Gordon
Lamberti, Joseph W
Landis, Charles Walter
Landis, Edward Everett
Langner, Enid Asher
Langner, Thomas S
Langsley, Donald Gene
Lauter, Maurice Walter
Lavarry, S G
Laybourne, Paul C
Layman, William Arthur
Lebensohn, Zigmond Meyer
Lehmann, Heinz Edgar
Leiderman, P Herbert
Leighton, Alexander Hamilton
Leighton, Dorothea Cross
Lemkau, Paul Victor
Lenkoski, L Douglas
Leon, Robert Leonard
Leopold, Robert L
Lesse, Henry
Levenson, Alan Ira
Levitt, LeRoy P
Levy, Edwin Z
Levy, Leo
Levy, Norman B
Levy, Robert Isaac
Lewis, David James
Lewis, Thomas Howard
Lewis, William O
Lhamon, William Taylor
Liberman, Robert Paul
Lidz, Theodore
Lieberman, Daniel
Lieberman, Edwin James
Lief, Harold Isaiah
Lifton, Robert Jay
Lindley, Stanley Bryan
Lindon, John Arnold
Linton, Patrick Hugo
Lipinski, Zbigniew J
Lipowski, Zbigniew J
Lipsitt, Don Richard
Lipschutz, Louis Sanderson
Lipkin, Mack
Llewellyn, Charles Elroy, Jr
Loeb, Felix Faust, Jr
Lourie, Reginald Spencer
Lovett-Doust, John William

Lowenbach, Hans
Luby, Elliot Donald
Lynn, Edward Joseph
Maas, James Weldon
Macdonald, John Marshall
MacLeod, Alastair William
Maholick, Leonard Thomas
Malamud, William
Malitz, Sidney
Mallott, I Floyd
Malmquist, Carl Phillip
Mandell, Arnold J
Marcus, Anthony Martin
Marcus, Philip Joseph
Margetts, Edward Lambert
Margolin, Sydney Gerald
Margolis, Philip Marcus
Mariner, Allen Shan
Mariner, Hildegard Rand
Markoff, Elliott Lee
Marmor, Judd
Marotta, Charles Anthony
Martin, Jack
Masland, Richard Lambert
Mason, Aaron S
Mathis, James L
May, Philip Reginald Aldridge
Mazer, Milton
McCabe, Michael S
McClure, James Nathaniel, Jr
McConville, Brian John
McCranie, Erasmus James
McCurdy, Layton
McDanald, Eugene Chester, Jr
McDermott, John F, Jr
McDermott, John Francis, Jr
McDonald, Ian MacLaren
McMillan, Donald Edgar
McNeel, Burdett Harrison
Melges, Frederick Towne
Mellor, Clive Sidney
Meltzer, Herbert Yale
Mendel, Werner Max
Mendell, David
Mendelson, Jack H
Mendelson, Myer
Menninger, Karl Augustus
Meredith, Charles Eymard
Mertis, Sidney
Mesnikoff, Alvin Murray
Metcalfe, Grant E
Meyer, Eugene
Meyer, George G
Meyerowitz, Sanford
Meyersburg, Herman Arnold
Michels, Robert
Miles, Henry Harcourt Waters
Miller, Derek Harry
Miller, Marvin Fred
Miller, Milton H
Miller, Milton Leonard
Milman, Doris H
Minde, Karl Klaus
Modlin, Herbert Charles
Moore, Donald Floyd
Moore, Kenneth Boyd
Morgenstein, Alan Lawrence
Morris, Harold Hollingsworth, Jr
Motto, Jerome (Arthur)
Mueller, Peter Sterling
Muensterberger, Werner
Murphy, George Earl
Murphy, Henry Brian Megget
Myers, Jacob Martin
Myerson, Paul Graves
Nagera, Humberto
Neale, Claude Linwood
Nelson, Herbert Leroy
Nenno, Robert Peter
Newton, Joseph Emory O'Neal
Nichol, Hamish
Nicholson, John Fraser
Noble, Ernest Pascal
Norman, Edward Cobb
Norris, Albert Stanley
Novick, Rudolph G
Noyes, Russell, Jr
Offenkrantz, William Charles
Offer, Daniel
Oken, Donald
Olney, John William
O'Malley, Edward Paul
O'Neal, Patricia L
Opler, Marvin Kaufmann
Orne, Martin Theodore
Orr, William Frederick
Osmond, Humphry Fortescue
Oswald, Peter Frederic
Paredes, Alfonso
Pacella, Bernard Leonard
Parker, Joseph B, Jr
Parker, Seymour
Pasamanick, Benjamin
Pattison, Edward Mansell
Payson, Henry Edwards
Pearce, Keith Ian

Pearson, Manuel Malcolm
Perlin, Seymour
Perr, Irwin Norman
Peters, John Emmett
Pettit, Manson Bowers
Pfeiffer, Eric A
Phillips, Richard Hart
Pierce, Charles Middlebrook
Piker, Philip Edward
Pillard, Richard Colestock
Pokorny, Alex Daniel
Polatin, Phillip
Pollack, Irwon W
Pollack, George Howard
Pollin, William
Pos, Robert
Pottle, Clarence H
Powles, William Earnest
Prange, Arthur Jergen, Jr
Proctor, James Thornton
Prosen, Harry
Pumpian-Mindlin, Eugene
Quarton, Gardner Cowles
Racy, John Cecil
Rae-Grant, Quentin A
Rahe, Richard Henry
Rainer, John David
Rainey, John Marion, Jr
Rakoff, Vivian Morris
Rapaport, Ionel
Rapoport, Judith Livant
Ravaris, Charles Lewis
Reading, Anthony John
Reckless, John B
Reding, Georges Rene
Redlich, Frederick Carl
Reese, William George
Reinhart, John Belvin
Reis, Walter Joseph
Reiser, Morton Francis
Resnik, Harvey Lewis Paul
Rexford, Eveoleen Naomi
Reynolds, Thomas De Witt
Rheingold, Joseph Cyrus
Rhoads, John McFarlane
Richman, Alex
Rickels, Karl
Riggs, Benjamin C
Rinsley, Donald Brendan
Ripley, Herbert Spencer
Riskin, Jules
Rittelmeyer, Louis Frederick, Jr
Ritvo, Edward R
Robbins, Lewis Frederick, Jr
Robbins, William S
Roberts, Leigh M
Robins, Eli
Robinson, David Bancroft
Rodriguez, Alejandro
Roessler, Robert L
Roffwarg, Howard Philip
Rogawski, Alexander S
Rollins, Robert Leroy, Jr
Romano, John
Rome, Howard Phillips
Rosenbaum, Milton
Rosenthal, Saul Haskell
Rosenzweig, Norman
Ross, Mabel
Ross, Mathew
Ross, William Donald
Rothenberg, Albert
Rothrock, Irvin Andrew
Rowland, Vernon
Rubin, Jesse Gallant
Rubin, Robert Terry
Rubin, Sidney
Rudy, Lester Howard
Ruesch, Jurgen
Ruff, George Elson
Russell, Donald Hayes
Sabshin, Melvin
Sackler, Arthur M
Sadock, Benjamin
Sadler, Henry Harrison, Jr
Sager, Clifford J
Salzman, Leon
Sampliner, Robert Bruce
Sander, Louis W
Sandifer, Myron Guy, Jr
Sarwer-Foner, Gerald
Saslow, George
Saul, Leon Joseph
Savage, Charles
Schachter, Joseph
Schacter, Joseph
Scheflen, Albert E
Scher, Jordan Mayer
Schiavi, Raul Constante
Schiele, Burtrum Clarence
Schiller, Maurice Aaron
Schmale, Arthur H, Jr
Schober, Charles Coleman
Schooler, Joseph Clayton
Schorer, Calvin E
Schottstaedt, Mary Gardner

Psychobiology
Burghardt, Gordon Martin
Cheal, MaryLou
Dews, Peter Booth
Ellingson, Robert James
Farel, Paul Bertrand
Gazzaniga, Michael Saunders
Goldschmidt, Leontine
Gottlieb, Gilbert
Grossman, Sebastian Peter
Hauty, George Thomas
Hunsaker, Don, II
Kakolewski, Jan Wiktor
Kety, Seymour Solomon
Kluver, Heinrich
Knight, Walter Rea
Komisaruk, Barry Richard
Maxson, Stephen C
Moorcroft, William Herbert
Myers, Robert Durant
Oppenheim, Ronald William
Richter, Curt Paul
Teng, Evelyn Lee
Wallace, Robert B
Werboff, Jack
Zornetzer, Steven Frank

Psychopharmacology
Abel, Ernest Lawrence
Alpern, Herbert P
Barry, Herbert, III
Bayne, Gilbert M
Beaton, John McCall
Blackwell, Barry M
Bowe, Robert Looby
Carlson, Kristin Rowe
Ciccone, Patrick Edwin
Clark, Robert
Cox, Raymond H, Jr
Davidson, Arnold B
D'Encarnacao, Paul S
Evans, Hugh Lloyd
Evans, Wayne Orien
Flood, James Felix
Flynn, Patrick
Gamzu, Elkan
Gatipon, Glenn Blaise
Griffith, John Dorland
Hanin, Israel
Harvey, John Adriance
Holliday, Audrey Rose
Houser, Vincent Paul
Hunt, Howard Francis
Irwin, Samuel
Jaffe, Jerome Herbert
Johanson, Chris Ellyn
Jones, Ben Morgan
Kline, Nathan Schellenberg
Kornetsky, Conan
Korol, Bernard
Ksir, Charles Joseph
Kuczenski, Ronald Thomas
Laties, Victor Gregory
Latz, Arje
Leander, John David
Leibowitz, Sarah Fryer
Lytle, Loy Denham
Manian, Albert Ardashes
McLendon, David Mark
McNair, Douglas McIntosh
Moore, Kenneth Edwin
Poschel, Bruno Paul Henry
Potts, Walter Joseph
Robichaud, Roger Charles
Schechter, Martin David
Schmidt, Dennis Earl
Scudder, Charles Lee
Seiden, Lewis S
Sepinwall, Jerry
Stein, Larry
Szara, Stephen Istvan
Taber, Robert Irving
Ternes, Joseph Wayne
Thut, Paul Douglas
Usdin, Earl
Watzman, Nathan
Weiss, Bernard
Weissman, Albert
Wolff, Peter Hartwig
Wyatt, Richard J

Psychoacoustics
Butler, Robert Allan
Carterette, Edward Calvin Hayes
Collins, Mary Jane
Elfner, Lloyd F
Elliott, Lois Lawrence
Greenwood, Donald Dean
Guttman, Newman
Harrison, John Michael
Moody, David Burritt
Small, Arnold McCollum, Jr
Soderquist, David Richard
Thompson, Paul O
Tobias, Jerry Vernon
Ward, Wallace Dixon
Wright, Herbert N
Yaniv, Simone Liliane
Yost, William A

Psychophysics
Bartleson, Christian James
Breneman, Edwin Jay
Desor, Jeannette Ann
Galloway, William Don
Gibson, James Jerome
Goldstein, E Bruce
Green, Barry George
Gulick, Walter Lawrence
Hicks, Robert Eugene
Hirsh, Ira Jean
Johnson, Chris Alan
Pangborn, Rose Marie Valdes
Sperling, Harry George
Stevens, Joseph Charles
Stewart, John Douglas

Neuropsychiatry
Blass, John P
Boshes, Louis D
DuPraw, Ernest Joseph
Fink, Max
Freeman, Leslie Sherwood
Glusman, Murray
Holloway, Harry Charles
Lal, Samarthji
Moore, Matthew Thibaud
Ng, Lorenz Keng-Yong
Palmer, Dwight Miller
Riemer-Rubenstein, Delilah
Rioch, David McKenzie
Sattar, Syed Abdus
Urse, Vladimir George

Neuropsychology
Anderson, Kenneth Verle
Beach, Frank Ambrose
Bernstein, Stephen
Boll, Thomas Jeffrey
Brauth, Steven Earle
Caplan, Paula Joan
Cleeland, Charles Samuel
Clynes, Manfred
Cranford, Jerry L
Creel, Donnell Joseph
Eason, Robert Gaston
Efron, Robert
Fitzhugh-Bell, Kathleen
Flanigan, William Francis, Jr
Frommer, Gabriel Paul
Ganchrow, Donald
Gerken, George Manz
Glantz, Raymon M
Green, Thomas Kerr
Hall, William Charles
Hartlage, Lawrence Clifton
Hillyard, Steven Allen
Jackson, William James
Jewett, Don L
Junge, Douglas
Justesen, Don Robert
King, Frederick Alexander
Klove, Hallgrim
Knapp, Theodore Martin
Kohn, Herbert Myron
Kupfermann, Irving
Laudenslager, Mark LeRoy
Lynch, Wesley Clyde
Marsh, Gayle G
Matthews, Charles George
Meier, Manfred John
Mellinger, Melvin Wayne
Mendell, Lorne Michael
Miller, Lyle Herbert
Milner, Brenda (Atkinson)
Palmer, Larry Alan
Paschke, Richard Eugene
Peeler, Dudley Flavus, Jr
Pribram, Karl Harry
Rabe, Ausma
Rhodes, John Marshell
Ritter, Walter, Paul
Sainsbury, Robert Stephen
Schwartz, Melvin Lewis
Settle, Richard Gregg
Singh, Pauline Jirik
Smith, Charles James
Smith, Nicole Schupf
Snyder, Daniel Raphael
Sterman, Maurice B
Storms, Lowell H
Stricker, Edward Michael
Taub, Edward
Thomas, Garth Johnson
Thompson, Robert
Weinstein, Edwin Alexander
Wenzel, Bernice Martha (Mrs Wendell E Jeffrey)
Whitaker, Harry Allen
Wilson, William August, Jr
Witelson, Sandra Freedman
Zaidel, Eran
Zolman, James F

Schubert, Daniel Sven Paul
Schulman, Jerome Lewis
Schultz, Richard E
Schuster, Charles Roberts, Jr
Schuster, Daniel Bradley
Schwab, John J
Schwartz, Allan James
Schwartz, Arthur Harold
Schwartz, Donald Alan
Schwarz, Marvin Lawrence
Selzer, Melvin Lawrence
Senescu, Robert A
Senn, Milton J E
Serafetinides, Eustace Agapios
Shagass, Charles
Shamoian, Charles Anthony
Shands, Harley Cecil
Sharpe, Lawrence
Sheard, Michael Henry
Shelton, Winfred Neil
Sheppe, William Marco, Jr
Sherwin, Duane O
Shore, James Henry
Shore, Miles Frederick
Silbergeld, Sam
Silberman, Henry K
Silberstein, Richard M
Silver, Archie Aaron
Silverman, Albert Jack
Silverstein, Alexander
Simmons, Charles Edward
Simmons, James Quimby, III
Simon, Alexander
Simon, Benjamin
Simpson, George M
Simpson, William Stewart
Sisler, George C
Skobba, Joseph Stanley
Slade, H Clyde
Slipp, Samuel
Sloane, Robert Bruce
Small, Iver Francis
Small, Joyce G
Small, Saul Mouchly
Smith, Charles E
Smith, Gerard Peter
Smith, Jackson Algernon
Smith, Kathleen
Snow, Laurence H
Sonis, Meyer
Southard, Curtis Glenn
Sparer, Phineas Jack
Spiegel, John Paul
Spiro, Herzl Robert
Spradlin, Wilford W
Stancer, Harvey C
Starr, Phillip Henry
Stein, Marvin
Stern, Marvin
Stevenson, Ian
Stewart, Mark Armstrong
Stewart, Robert Lewis
Stinnett, James LeBaron
Stokes, Peter E
Stolk, Jon Martin
Storrow, Hugh Alan
Stotsky, Bernard A
Strahilevitz, Meir
Straumanis, John Janis, Jr
Strauss, John Steaven
Stubblefield, Robert Lee
Stunkard, Albert J
Suess, James Francis
Sugerman, Abraham Arthur
Sussex, James Neil
Swanson, David Wendell
Swartz, Jacob
Szasz, Thomas Stephen
Szurek, Stanislaus (Andrew)
Szyrynski, Victor
Taketomo, Yasuhiko
Tapia, Fernando
Tarjan, George
Taubman, Robert Edward
Taylor, Eugene Emerson
Teicher, Joseph D
Teichner, Victor Jerome
Teja, Jagdish Singh
Thale, Thomas Richard
Thaler, Otto Felix
Thomas, Alexander
Thomas, Claudewell Sidney
Thomas, Robert Eugene
Tidd, Charles Wharton
Tislow, Richard Frederick
Titchener, James Lampton
Tourney, Garfield
Toussieng, Povl Winning
Trosman, Harry
Tsuang, Ming Tso
Tucker, Gary Jay
Tupin, Joe Paul
Turrell, Eugene Snow
Tyce, Francis Anthony
Tyhurst, James Stewart
Tyler, Edward Alton
Uhlenhuth, Eberhard Henry
Ulett, George Andrew

Ullman, Montague
Varsamis, Ioannis
Verwoerdt, Adrian
Vestergaard, Per B
Visotsky, Harold M
Volavka, Jan
Volkan, Vamik
Voth, Harold Moser
Waggoner, Raymond Walter
Waller, William Henry
Wallerstein, Robert Solomon
Wallner, Julius Michael
Walsh, Roger Nugent
Ward, Richard S
Warson, Samuel R
Watkins, Charles
Watson, Andrew Samuel
Wayne, George Jerome
Webb, William Logan
Webster, Thomas G
Weckowicz, Thaddeus Eugene
Weiner, Herbert
Weiss, James Moses Aaron
Weiss, Steven Earle
Weisskopf, Bernard
Wells, Charles Edmon
Wender, Paul H
West, Louis Jolyon
White, Robert B
Whitman, Roy Milton
Whybrow, Peter Charles
Wiesel, Benjamin
Wikler, Abraham
Wildberger, William Campbell
Wilder, Russell Morse
Wilkinson, Charles Brock
Will, Otto Allen, Jr
Williams, Ernest Young
Williams, Robert L
Williams, T Glyne
Wilson, William Preston
Winokur, George
Wittkower, Eric David
Witson, Cecil L
Wolford, Jack Arlington
Wolpe, Joseph
Wolpert, Arthur
Woodruff, Robert Arnold, Jr
Woods, Sherwyn Martin
Worden, Frederic Garfield
Work, Henry Harcus
Wynne, Lyman Carroll
Yamamoto, Joe
Yochelson, Leon
Yolles, Stanley Fausst
Yonge, Keith A
Zitrin, Arthur
Zung, William Wen-Kwai
Zusman, Jack
Zwerling, Israel

Child Psychiatry
Anthony, E James
Beiser, Helen R
Blom, Gaston Eugene
Bolman, William Merton
Brugger, Thomas C
Chapel, James L
Christ, Adolph Ervin
Cohen, Richard Lawrence
Comer, James Pierpont
DeMyer, Marian Kendall
Enzer, Norbert Beverley
Freeman, Roger Dante
Glaser, Kurt
Hagenauer, Fedor
Harrison, Saul I
Hirsch, Jay G
Jackson, Basil Edgar
Judd, Lewis Lund
Langford, William Siddon
Lubin, Gerald I
Lucas, Alexander Ralph
Marcus, Joseph
McDermott, John F, Jr
Mellette, Russell Ramsey, Jr
Noshpitz, Joseph Dove
Offord, David Robert
Rafferty, Frank Thomas
Rapoport, Judith Livant
Reichler, Marshall David
Schechter, Marshall David
Schloemer, Robert Henry
Simmons, James Edwin
Sobel, Raymond
Vennes, John Wesley
Wasserman, Martin S
Wender, Paul H
Westman, Jack Conrad
Wolpert, Arthur

Experimental Psychiatry
Agras, William Stewart
Carroll, Bernard James
Lipton, Morris Abraham
Peters, John Emmett

PSYCHIATRY

Verrillo, Ronald Thomas
Ward, Lawrence McCue
Wheeler, Lawrence
Wright, Anthony Aune
Zwislocki, Jozef John

Psychophysiology
Annau, Zoltan
Bartley, Samuel Howard
Beck, Edward C
Benuck, Irwin
Black, William
Blass, Elliott Martin
Brown, Barbara Banker
Brown, Clinton Carl
Brown, John Lott
Brown, Warren Shelburne, Jr
Bryson, George Gardner
Capps, Mary Jayne
Crawford, Morris Lee Jackson
Cutt, Roger Alan
Dawson, William Woodson
de Haan, Henry John
Donchin, Emanuel
Donovick, Peter Joseph
Dykman, Roscoe A
Edelberg, Robert
Erickson, Robert Porter
Essman, Walter Bernard
Fay, Richard Rozzell
Fleming, Donovan Ernest
Fox, Stephen Sorin
Gault, Frederick Paul
Glickstein, Mitchell
Goldstein, Robert
Gourevitch, George
Goy, Robert William
Graham, Frances Keesler
Greene, Ernest Gerald
Greenough, William Tallant
Guedry, Frederick Ernest
Halpern, Naomi Mimi
Hamilton, Leonard W
Harlow, Harry F
Henderson, Donald
Horel, James Alan
Hornbuckle, Phyllis Ann
Horne, Edward Porter
Hughes, Kenneth Russell
Isaacson, Robert Lee
Jacobs, Harry Lewis
Kaas, Jon Howard
Kallman, William Michael
Keesey, Ulker Tulunay
Klein, Sherwin Jared
Kragt, Clifford Lee
Krauskopf, John
Kryter, Karl David
Lacey, Beatrice Cates
Lacey, John Irving
Leshner, Alan Irvin
Levine, Seymour
Lindsley, Donald B
Lubin, Ardie
Mailer, Owen
Masuda, Minoru
McGough, William Edward
McKenney, Joel R
Meier, Gilbert W
Miles, Walter Richard
Miller, Neal Elgar
Mogenson, Gordon James
Moushegian, George
Mozell, Maxwell Mark
Mulligan, Benjamin Edward
Naitoh, Paul Yoshimasa
Oestreich, Roger Edward
Peacock, Leion James
Phillips, David
Pinneo, Lawrence Robert
Pinsker, Harold M
Pokorny, Joel
Riggs, Lorrin Andrews
Roberts, Alan H
Rodgers, Charles H
Rubin, Leonard Sidney
Sandel, Thomas Theodore
Satterberg, John Arvid
Schneiderman, Neil
Semmes, Josephine
Smith, Vivianne Cameron
Sparks, David Lee
Stellar, Eliot
Stroebel, Charles Frederick, III
Sturr, Joseph Francis
Surwillo, Walter Wallace
Taub, John Marcus
Teas, Donald Claude
Terando, M Loretta
Teuber, Hans-Lukas
Thompson, Richard Frederick
Thorn, Frank
Trehub, Arnold
Trumbull, Richard
Vernon, Jack Allen
Vierck, Charles John, Jr
Vinson, David Berwick
Ward, Ingeborg L

Weiss, Jay M
White, Robert Keller
Whitehead, William Earl
Winers, Cynthia Crosby
Winters, Ray Wyatt
Yellin, Absalom Moses
Yerin-Komshian, Grace Helen
Young, Francis Allan
Zerlin, Stanley

Psychosomatic Medicine
Abram, Harry Shore
Berblinger, Klaus William
Chosy, Julius J
Curtis, George Clifton
Giovacchini, Peter L
Greene, William Allan
Iker, Howard Paul
Johnson, Jean Elaine
Kroger, William Saul
Levy, Norman B
Lisansky, Ephraim Theodore
Purcell, Keneth
Reckless, John B
Reichman, Franz Karl
Rosen, Harold
Schwab, John J
Thies, Roger E
Waldron, Ingrid Lore
Williamson, Penelope Rose

SOIL SCIENCES

Soil Science
Aase, Jan Kristian
Adams, Fred
Ahring, Robert M
Alderfer, Russell Brunner
Allmaras, Raymond Richard
Anderson, Marvin A
Araujo, Jose Emilio Goncalves
Arneman, Harold Frederick
Asleson, Johan Arnold
Austin, Morris Edwin
Auten, John Thompson
Axley, John Harold
Ayers, Alvin Dearing
Bailey, Harry Hudson
Baird, Bruce Lloyd
Baird, Jack Vernon
Banerjee, Sushanta Kumar
Banwart, Wayne Lee
Barker, Allen Vaughan
Bartelli, Lindo Joseph
Bartholomew, William Victor
Bartlett, Frank David
Bates, Thomas Edward
Bauer, Armand
Beatty, Marvin Theodore
Beeks, John Charles
Benson, Nels R
Bentley, Charles Fred
Bernier, Bernard
Bigger, Theodore C
Bishop, Robert Frederick
Blevins, Robert L
Blue, William Guard
Bohannon, Robert Arthur
Boswell, Fred Carlen
Brage, Burton L
Brengle, Kenneth Gordon
Broadfoot, Walter Marion
Bronson, Roy DeBolt
Brown, Arthur Lloyd
Brown, James Melton
Brown, Jerry
Brydon, James Emerson
Buol, Stanley Walter
Burleson, Charles Albertis
Burnett, Earl
Burrows, William Chapel
Burton, David Lee
Cady, John Gilbert
Cain, Charles Columbus
Caldwell, Robert Edward
Carlisle, Frank Jefferson, Jr
Christenson, Donald Robert
Clements, Richard Gerald
Clementz, David Michael
Cole, Vernon C
Coleman, Russell
Conyers, Emery Swinford
Cook, Maurice Gayle
Cook, Ray Lewis
Cooper, George S
Corliss, John Franklin
Cox, Frederick Russell
Crawford, Robert Field
Cressman, Harry Keith
Cummings, George August
Daugherty, Zoel W
Dawson, Murray Drayton
Dion, Henry George
Dixon, Joe Boris
Dowdy, Robert H

Dregne, Harold Ernest
Drew, James Van
Drosdoff, Matthew
Duisberg, Peter Caspar
Duke, Everette Loranza
Eck, Paul
El-Swaify, Samir Aly
Elzam, O E
Eno, Charles Franklin
Erickson, Anton Earl
Essington, Edward Herbert
Fanning, Delvin Seymour
Farnham, Rouse Smith
Ferguson, Wilfred Samuel
Fireman, Milton
Flach, Klaus Werner
Fletcher, Herbert Calvin
Fletcher, Joel Eugene
Folks, Homer Clifton
Fox, Robert Lee
Francis, Chester Wayne
Fredriksen, Richard L
Freeman, Kelly Carey
Fulton, James McCullough
Gessel, Stanley Paul
Gieseking, John Eldon
Gilkeson, Raymond Allen
Gillingham, J T
Glenn, Rollin Copper
Godfrey, Curtis Loveling
Goldston, Eugene Frizzelle
Goss, Don Woodson
Graham, Ellis Ray
Grava, Janis (John)
Gray, Fenton
Grigal, David Francis
Grissinger, Earl H
Guttay, Andrew John Robert
Halvorson, Ardell David
Hammond, Luther Carlisle
Haney, Donald C
Hanway, John Joseph
Harper, Kimball T
Harpstead, Milo I
Harries, Hinrich
Harris, Stuart Arthur
Harter, Robert Duane
Hausenbuiller, Robert Lee
Hedlin, Robert Arthur
Herron, George M
Hill, David Easton
Hobbs, James Arthur
Hoeft, Robert Gene
Hoff, Donald Jerome
Hole, Francis Doan
Holowaychuk, Nicholas
Hood, Joseph Talmadge
Hough, Hugh Walter
Humphrys, Clifford Robertson
Hutchinson, Gordon Lee
Isenee, Allan Robert
Jackson, Marion Leroy
Jackson, Thomas Lloyd
Janke, Wilfred Edwin
Jansen, Ivan John
Jaworski, Casimir A
Johnson, Donal Dabell
Johnson, William Martin
Jorgensen, Jacques R
Kanehiro, Yoshinori
Keefer, Robert Faris
Kellogg, Charles Edwin
Keogh, Joseph Lloyd
Kilmer, Victor James
Klemmedson, James Otto
King, Gerald Fairchild
Klingebiel, Albert Arnold
Knox, Ellis Gilbert
Krantz, Bert Allan
Krashevski, Stefan H
Laird, Reggie James
Lambert, Royce Leone
Larson, William Earl
Lathwell, Douglas J
Laughlin, L M
Lavkulich, Winston Means
Leamer, Ross Wilson
Lee, Gerhard Bjarne
Legg, Joseph Ogden
Leggett, Glen Eugene
Lemon, Edgar Rothwell
Leonard, Chester D
Lewis, Barbara-Ann Gamboa
Lewis, Cornelius Crawford
Leyden, Robert Fullerton
Liebhardt, William C
Lindsay, Willard Lyman
Long, Daryl Clyde
Lowry, Gerald Lafayette
Lucas, Robert Elmer
Luebs, Ralph Edward
Lunin, Jesse
Lynd, Julian Quentin
Lynn, Warren Clark
MacGregor, John Malcolm
Malcolm, John Lowrie
Marriott, Lawrence Frederick
Marsh, Albert William

Marshall, Charles Edmund
Martens, David Charles
Mathers, Aubra Clinton
Mayland, Henry Frederick
McAuliffe, Clayton Doyle
McBeath, Douglas Kay
McCants, Charles Bernard
McClelland, John Edward
McClure, George W, Jr
McCollum, Robert Edward
McCormack, Donald Eugene
McCracken, Ralph Joseph
McCreery, Robert Atkeson
McHenry, John Roger
McKim, Harlan L
McMillan, Neil John
Mellor, Jesse Lynn
Melsted, Sigurd Walter
Meredith, Farris Ray
Messenger, Aubrey Steven
Milford, Murray Hudson
Milford, Clarence James
Miller, Frederick Powell
Miller, Robert Harold
Miller, Willie
Mokma, Delbert Lewis
Munns, Donald Neville
Murdock, John Thomas
Musgrave, Orlo Lynn
Nash, Ralph Glen
Neal, John Lloyd, Jr
Neel, James William
Neher, David Daniel
Nelson, Lewis Bailey
Nelson, Lyle Engmar
Nelson, Wallace Warren
Nettleton, Wiley Dennis
Nielsen, Gerald Alan
Norum, Enoch Betuel
Oertli, Johann Jakob
Ohlrogge, Alvin John
Olness, Alan
Olsen, Robert James
Olson, Robert August
Ott, Billy Joe
Ouellette, Gerard Joseph
Overdahl, Curtis J
Patrick, William H, Jr
Paul, Eldor Alvin
Pawluk, Steve
Peele, Thomas Christopher
Peevy, Walter Jackson
Pesek, John Thomas, Jr
Peterson, Arthur Edwin
Peterson, Gary A
Pitner, John Bruce
Plucknett, Donald L
Pluth, Donald John
Pohlman, George Gordon
Popenoe, Hugh
Powell, Richard Donald
Power, James Francis
Pretty, Kenneth McAlpine
Protz, Richard
Radke, Rodney Owen
Rains, Donald W
Randall, Gyles Wade
Rawlins, Stephen Last
Reeve, Eldrow
Reid, William Shaw
Reisenauer, Hubert Michael
Rennie, Peter John
Rhoades, James David
Rible, John Maurice
Richards, Norval Richard
Rieke, Paul Eugene
Robertson, Lynn Shelby, Jr
Robinson, Daniel Owen
Robinson, Glenn Hugh
Rogers, Howard Topping
Romney, Evan M
Rudgers, Lawrence Alton
Rust, Richard Henry
Sabey, Burns Roy
Saini, Gulshan Rai
St Arnaud, Roland Joseph Odilon
Schafer, George Miles
Schmehl, William Reed
Scholtes, Wayne Henry
Scott, Hubert Donovan
Scott, William
Shaw, Ellsworth
Sheldon, Victor Lawrence
Shrader, William D
Shugars, Jonas P
Siddoway, Francis H
Simmons, Charles Ferdinand
Simonson, Clifford Harry
Simonson, Gerald Herman
Singh, Rabindar Nath
Singh, Surinder Shah
Smith, Hubert Donovan
Smith, Bill Ross
Smith, Richard M
Smith, Robert Edward
Sommerfeldt, Theron G
Soper, Robert Joseph
Sor, Kamil

Southard, Alvin Reid
Spomer, Louis Arthur
Steinbrenner, Eugene Clarence
Stevenson, Frank Jay
Stewart, John Allan
Stewart, John Wray Black
Strauch, Fred
Strickling, Edward
Stroehlein, Jack Lee
Struchtemeyer, Roland August
Survant, William G
Swader, Fred Nicholas
Swan, James Byron
Swenson, Royal Jay
Swindale, Leslie D
Tamura, Tsuneo
Taylor, Howard Melvin
Tedrow, John Charles Fremont
Terry, David Lee
Thomas, Frank Harry
Thompson, Leonard Garnett, Jr
Thompson, Louis Milton
Thompson, Lyell
Thornton, George Daniel
Tidball, Ronald Richard
Tompkins, Gary Alvin
Toogood, John Alfred
Troeh, Frederick Roy
Trogdon, William Oren
Troll, Joseph
Tromble, John M
Tucker, Thomas Curtis
Tyner, Edward Henry
Ugolini, Fiorenzo Cesare
Ullery, Charles Howard
Unger, Paul Walter
Van Eck, Willem Adolph
Van Lear, David Hyde
Viets, Frank Garfield, Jr
Vimmerstedt, John P
Vitosh, Maurice Lee
Voigt, Garth Kenneth
Vomocil, James Arthur
Walker, John Martin
Walker, Laurence Colton
Walker, Rudger Harper
Wall, Gregory John
Walsh, Leo Marcellus
Walton, Grant Fontain
Weaver, Robert Michael
Weeks, Leslie Vernon
Wells, Kenneth Lincoln
Westermann, D T
White, Donald Perry
White, Edwin Henry
Whiteside, Eugene Perry
Wiersma, Daniel
Wilding, Lawrence Paul
Wildung, Raymond Earl
Wilkinson, Stanley R
Williams, Gerald Gordon
Willis, William Hillman
Wittsell, Lawrence Eugene
Wollum, Arthur George, II
Woodhouse, William Walton, Jr
Wright, Bill C
Young, Arthur Wesley
Young, Thomas Wilbur
Younts, Sanford Eugene
Yuan, Tzu-Liang
Zicker, Eldon Louis
Zinke, Paul Joseph
Zubriski, Joseph Cazimer
Zwarun, Andrew Alexander

Soil Biochemistry
Allbritten, Herbert Graves
Bergeaux, Phillip James
Clapp, Charles Edward
Engibous, James Charles
Fine, Lawrence Oliver
Mankau, Reinhold
Skujins, John Janis
Young, J Lowell

Soil Chemistry
Adams, Russell S, Jr
Albregts, Earl Eugene
Allen, Seward Ellery
Allison, Lowell Edward
Anderson, Warren Boyd
Babcock, Kenneth Leslie
Balam, Baxish Singh
Barber, Stanley Arthur
Barrows, Harold Lindsey
Barshad, Isaac
Bartlett, Richmond J
Bennett, Allison Carr
Bensen, David Warren
Bingham, Frank Thomas
Blanchar, Robert W
Bohn, Hinrich Lorenz
Bower, Charles Arthur
Bowman, Bruce Tamblyn
Braids, Olin Capron
Branson, Roy
Brown, Donald A
Brown, John Charl

Bushnell, Vernon Clifford
Caldwell, Alfred Craig
Caldwell, Augustus George
Campbell, Constantine Alberga
Carlson, Robert Marvin
Carter, David LaVere
Cescas, Michel Pierre
Cheng, Cheng-Yin
Cheng, Hwei-Hsien
Chesnin, Leon
Chichester, Frederick Wesley
Chien, Sen Hsiung
Clark, John S
Corey, Richard Boardman
Dantzman, Charles L
Davidson, Thomas, Jr
Dormaar, Johan Frederik
Dubrovin, Kenneth P
Dutt, Gordon Richard
Elgawhary, Salah Mohammad
Ellis, Boyd G
Ellis, Roscoe, Jr
Farmer, Walter Joseph
Fenster, William E
Fiskell, John Garth Austin
Fowler, Eric Beaumont
Frink, Charles Richard
Gammon, Nathan, Jr
Gardner, Bryant Rogers
Gast, Robert Gale
Geist, Jon Michael
Geraldson, Carroll Morton
Gilliam, James Wendell
Graetz, Donald Alvin
Grunes, David Leon
Guenzi, Wayne D
Hall, Jon K
Hashimoto, Isao
Hassett, John J
Heald, Walter Roland
Heddleson, Milford Raynord
Heling, Charles Siver
Hermanson, Harvey Philip
Hill, Archie Clyde
Himes, Frank Lawrence
Ho, Clara Lin
Hortenstine, Charles C
Horton, James Henry, Jr
Hossner, Lloyd Richard
Hourigan, William R
Howe, David Orville
Hutchinson, Frederick Edward
James, Ronald Valdemar
Jenne, Everett A
Jones, Randall Jefferies
Jurinak, Jerome Joseph
Kardos, Louis Thomas
Kelley, Omer Joseph
Khasawneh, Fayez Essa
Kissel, David E
Konrad, John Grey
Kunishi, Harry Mikio
Langdale, George Wilfred
Lavy, Terry Lee
Leonard, Ralph Avery
Lewis, Glenn C
Lisk, Donald James
Logan, Terry James
Long, Franklin Leslie
Low, Philip Funk
Lowe, Lawrence E
MacKay, Donald Cyril
MacKenzie, Angus Finley
Magdoff, Frederick Robin
Matthews, Burton Clare
McIntosh, Jerry Leon
McLean, Eugene Otis
McNeal, Brian Lester
Meek, Burl Dean
Melton, James Ray
Menzel, Ronald George
Michaelson, Neil Elbert
Miller, James Roland
Miller, Raymond Jarvis
Millette, Gerard J F
Mortland, Max Merle
Mortvedt, John Jacob
Mugwira, Luke Makore
Muir, Melvin K
Nakayama, Francis Shigeru
Nash, Victor E
Naylor, Denny Ve
Nelson, Darrell Wayne
Nelson, Wesley Eugene
Newland, Leo Winburne
Nishita, Hideo
O'Connor, George Albert
Olsen, Ralph A
Onken, Arthur Blake
Oster, James Donald
Page, Albert Lee
Page, Norwood Rufus
Pearson, Robert Watt
Pendleton, John Davis
Perkins, Henry Frank
Peterson, Howard Boyd
Pionke, Harry Bernhard

Pope, Alex
Porter, Lynn K
Pratt, Parker Frost
Prevatt, Rubert Waldemar
Prince, Allan Bixby
Ragland, John Leonard
Reed, Lester W
Reed, Marion Guy
Reed, William Edward
Reid, Preston Harding
Reneau, Raymond B, Jr
Rennie, Donald Andrews
Rhoads, Frederick Milton
Robertson, William Kitchener
Routson, Ronald C
Rouse, Roy Dennis
Rubins, Edward J
Russell, Glenn C
Ryan, James Anthony
Saiz Del Rio, Jose Francisco
Sandhu, Shingara Singh
Satchell, Donald Prentice
Scarsbrook, Clarence Edwin
Schnitzer, Morris
Scott, Albert Duncan
Seymour, Keith Goldin
Shannon, Stanton
Shelton, James Edward
Shuman, Larry Myers
Simpson, Daniel Martin Henry
Singh, Daulat
Singleton, Paul C
Smika, Darryl Eugene
Smith, Donald Henry
Smith, R L
Sorensen, Robert Carl
Sowden, Frederick John
Spencer, William F
Spinks, Daniel Owen
Swoboda, Allen Ray
Tamimi, Yusuf Nimr
Thomas, Grant Worthington
Toth, Stephen John
Tullock, Robert Johns
Volk, Bob Garth
Volk, Garth William
Volk, Gaylord Monroe
Volk, Veril Van
Weber, Jerome Bernard
Webster, Gordon Ritchie
Weed, Sterling Barg
Wells, Bobby R
Westermann, D T
Westfall, Dwayne Gene
White, Ronald Paul, Sr
Whittig, Lynn D
Williams, David Emerton
Willis, Guye Henry
Wolf, Benjamin
Wright, James R

Soil Classification
Fly, Claude Lee
Frazee, Charles Joseph
Hajek, Benjamin F
Peterson, Frederick Forney
Pomerening, James Albert
Yahner, Joseph Edward

Soil Conservation
Amemiya, Minoru
Arshad, Muhammad Ahmad
Berg, William Albert
Carlson, Carl Wilburn
Harder, Roger Wehe
Heil, Robert Dean
Holt, Robert F
Isgur, Benjamin
Loughry, Frank Glade
Meyer, Lawrence Donald
Moldenhauer, William C
Nelson, Sheldon Douglas
Palmer, Robert Gerald
Peterson, John William
Ramig, Robert E
Rickman, Ronald Wayne
Rollins, Myron B
Sisson, Donald Ray
Skidmore, Edward Lyman
Smucker, Silas Jonathan
Stallings, James Henry
Taylor, Robert E
Thorp, Eldon Marion
Van Meter, Donald Eugene
Voorhees, Ward Byron
Whitehurst, Sanford Huey
Wischmeier, Walter Henry
Zimmerman, Tommy Lynn

Soil Fertility
Aldrich, Daniel Gaskill, Jr
Anderson, Oscar Emmett
Andreis, Henry Jerome
Baker, Aaron Sidney
Beaton, James Duncan
Beauchamp, Eric G
Bhangoo, Mahendra Singh
Black, Charles Allen

Bouldin, David Ritchey
Bradfield, Richard
Brown, James Richard
Brown, Paul Lawson
Brownell, James Richard
Burns, George Robert
Butler, Patrick Colin
Calvert, David Victor
Carson, Paul LLewellyn
Carter, Robert Leonidas
Case, Vern Wesley
Chapman, Homer Dwight
Cheney, Horace Bellatti
Chowdhury, Ikbalur Rashid
Cook, James Arthur
Cope, John Thomas, Jr
Daigger, Louis A
deMooy, Cornelis Jacobus
Diamond, Ray Byford
Dibb, David Walter
Doll, Eugene Carter
Dumenil, Lloyd C
Eck, Harold Victor
Engelstad, Orvis P
Evans, Clyde Edsel
Everett, Paul Harrison
Fitts, James Walter
Forbes, Richard Brainard
Foy, Charles Daley
Franklin, Ralph E
Fried, Maurice
Frye, Wilbur Wayne
Gardner, E Hugh
Gerard, Cleveland Joseph
Golden, Laron E
Gray, Carl
Gray, Robert Copping
Gupta, Umesh C
Halstead, Ronald Lawrence
Harris, Wallace Wayne
Hawkes, George Rogers
Hensel, Dale Robert
Hensler, Ronald Fred
Hipp, Billy Wayne
Hunter, Albert Sinclair
Hunter, Arvel Hatch
Jackson, William Addison
Johnson, Gordon V
Johnson, Ray Edwin
Jones, J Benton, Jr
Jones, James Preston
Jones, Milton Bennion
Jones, Ulysses Simpson, Jr
Justice, John Keith
Kamprath, Eugene John
Keeney, Dennis Raymond
Koehler, Fred Eugene
Langdale, George Wilfred
Laws, Wilford Derby, Jr
Lessman, Gary M
Ludwick, Albert Earl
Lunt, Owen Raynal
Massee, Truman Winfield
Massey, Herbert Fane, Jr
McCall, Wade Wiley
McCaslin, Bobby Duane
McClung, Andrew Colin
McDole, Robert E
McFee, William Warren
Mikkelsen, Duane Soren
Miller, Murray Henry
Miller, Raymond Woodruff
Miner, Gordon Stanley
Munro, Douglas Cartyle
Munson, Robert Dean
Nicholaides, John J
Nielsen, Kenneth Fred
Nossaman, Norman L
Nuttall, Wesley Ford
O'Grady, Lawrence J
Openshaw, Martin David
Orth, Paul Gerhardt
Parks, Clyde Leonard
Peaslee, Doyle E
Peck, Theodore Richard
Peterson, James Robert
Phillips, Marvin W
Pierre, William Henry
Rateaver, Bargyla
Rauschkolb, Roy Simpson
Ray, Howard Eugene
Rehm, George W
Rendig, Victor Vernon
Richards, Graydon Edward
Robertson, James Alexander
Samuels, George
Sanchez, Pedro Antonio
Scott, Thomas Walter
Seatz, Lloyd Frank
Shickluna, John C
Silva, James Anthony
Simkins, Charles Abraham
Skogley, Earl O
Smith, Floyd W
Soltanpour, Parviz Neil
Sowell, Walter F
Stelly, Matthias

Soil Genesis (continued)

Stewart, Bobby Alton
Stewart, Franklin Burton
Stivers, Russell Kennedy
Sutherland, William Neil
Sutton, Paul
Tennille, Aubrey W
Thorne, David Wynne
Thorup, James Tat
Thorup, Richard M
Usherwood, Noble Ransom
Van Ryswyk, Albert Leonard
Varsa, Edward Charles
Vasey, Edward H
Voss, Regis D
Waddington, Donald Van Pelt
Walker, William M
Walsh, Leo Marcellus
Wang, Li Chuan
Waugh, Donovan Lloyd
Wehunt, Ralph Lee
Wesley, Dean E
White, Richard Anton
Wiese, Richard Calvin
Wilcox, Gerald Eugene
Woltz, Willie Garland
Young, Ralph Alden

Soil Genesis

Allen, Bonnie L
Cunningham, Robert Lester
Daniels, Raymond Bryant
Douglas, Lowell Arthur
Fenton, Thomas E
Foss, John E
Franzmeier, Donald Paul
Gile, Leland Henry
McKeague, Justin Alexander
Parsons, Roger Bruce

Soil Microbiology

Alarie, Albert
Alexander, Martin
Anderson, Guy Richard
Bezdicek, David Fred
Bollag, Jean-Marc
Broadbent, Francis Everett
Cameron, Roy (Eugene)
Chilgreen, Donald Ray
Chinn, Stanley H F
Cook, Fred D
Corke, Charles Thomas
Davey, Charles Bingham
Davis, Robert Jaquette, Jr
Doxtader, Kenneth Guy
Dunigan, Edward P
Focht, Dennis Douglass
Frederick, Lloyd Randall
Giddens, Joel Edwin
Gilbert, Richard Gene
Hagedorn, Charles
Ham, George Eldon
Harris, John Orville
Hiltbold, Arthur Edward, Jr
Hubbell, David Heuston
Ivarson, Karl C
Jones, Kenneth Lester
Loynachan, Thomas Eugene
Lu, Kuo Chin
Martin, James Paxman
Martin, William Paxman
McCalla, Thomas Mark
Parr, James Floyd, Jr
Peterson, Harold LeRoy
Reuszer, Herbert William
Rice, Wendell Alfred
Robson, Hope Howeth
Smith, Jay Hamilton
Sommers, Lee Edwin
Sparrow, Elena Bautista
Stanovick, Richard Paul
Tate, Robert Lee, III
Weaver, Richard Wayne
Weber, Deane Fay
Widden, Paul Rodney
Wolcott, Arthur Ripatte
Wolf, Duane Carl

Soil Mineralogy

Ahlrichs, James Lloyd
Barnhisel, Richard I
Borchardt, Glenn Arnold
Chase, Francis Edward
Foscolos, Anthony E
Harward, Moyle E
Hawkins, Richard Horace
Johnson, Leon Joseph
Jones, Robert L
Kittrick, James Allen
Kunze, George William
White, Joe Lloyd
Zelazny, Lucian Walter

Soil Morphology

Arkley, Rodney John
Arnold, Richard Warren
Berdanier, Charles Reese, Jr
Bidwell, Orville Willard
Bourbeau, Gerard Auguste

Buntley, George Jule
Carlisle, Victor Walter
Daugherty, LeRoy Arthur
Dudas, Marvin Joseph
Feuer, Reeshon
Fosberg, Maynard Axel
Foth, Henry Donald
Gersper, Paul Logan
Harlan, Phillip Walter
Hock, Arthur George
Holzhey, Charles Steven
Huddleston, James Herbert
Kovar, John Alvis
Lund, Lanny Jack
Olson, Gerald Walter
Petersen, Gary Walter
Petry, David Emory
Rieger, Samuel
Ross, Sam Jones, Jr
Rowles, Charles A
Springer, Maxwell Elsworth
Wright, William Ray

Soil Physics

Adams, John Edgar
Anderson, Duwayne Mario
Bertrand, Anson Rabb
Blake, George Rowland
Boelter, Don Howard
Bourget, Sylvio-J
Box, James Ellis, Jr
Brandt, Gerald H
Bruce, Robert Russell
Cannell, Glen H
Cassel, D Keith
Corey, John Charles
Danielson, Robert Eldon
Davidson, James Melvin
Day, Paul Russell
Dirksen, Christiaan
Ekern, Paul Chester
Elrick, David Emerson
Enfield, Carl George
Epstein, Eliot
Ferguson, Albert Hayden
Flocker, William Jack
Forsythe, Warren M
Fritton, Daniel Dale
Gardner, Walter Hale
Gifford, Richard Oliver
Green, Richard E
Grover, Ben Leo
Hanks, Ronald John
Hornsby, Arthur Grady
Jensen, Creighton Randall
Kimball, Bruce Arnold
Kohl, Robert A
Kunze, Raymond J
Lugo-Lopez, Miguel Angel
Mansell, Robert Shirley
Mazurak, Andrew Peter
McCauley, Garry Nathan
Meredith, Harvey L
Miller, Robert Demorest
Nielsen, Donald R
Olson, Tamlin Curtis
Page, John Boyd
Patten, Gaylord Penrod
Peters, Doyle Buren
Powers, William L
Radke, Jerry Kieth
Raney, William Andrew
Ritchie, Joe T
Rogowski, Andrew S
Rolston, Dennis Eugene
Rosenberg, Norman J
Rubin, Jacob
Runkles, Jack Ralph
Shawcroft, Roy Wayne
Shaykewich, Carl Francis
Starr, James LeRoy
Stevenson, David Stuart
Stewart, Gordon LeRoy
Stolzy, Lewis Hal
Stone, John Floyd
Stone, Lloyd Raymond
Swartzendruber, Dale
Tackett, Jesse Lee
Thiel, Thomas J
Tsuji, Gordon Yukio
Warrick, Arthur W
Wendt, Charles William
Whisler, Frank Duane
Whiteley, Eli Lamar
Whitt, Darnell Moses
Wiegand, Craig Loren
Wierenga, Peter J
Willey, Cliff Rufus
Willis, Wayne Owen
Wilson, Clyde Livingston
Wilson, Lorne Graham

SPECTROSCOPY

Spectroscopy

Alexander, Ralph William, Jr

Alpert, Nelson Leigh
Anantha Narayanan, Venkataraman
Anastassakis, Evangelos M
Andermann, George
Andrews, Allen Lester Self
Andrychuk, Dmetro
Armstrong, Andrew Thurman
Aronson, James Ries
Baldwin, Jon Michael
Bales, Howard A
Barr, William Lee
Bazzelle, William Edward
Bell, Robert John
Bernstein, Elliot R
Bickel, William Samuel
Blatt, Joel Herman
Boyd, John Robert
Brannon, Paul J
Bucker, Homer Park, Jr
Caspers, Hubert Henri
Chamberlain, Nugent Francis
Ch'en, Inan
Ch'en, Shang-Yi
Chisholm, James Joseph
Chrenko, Richard Michael
Colthup, Norman Bertram
Crane, Robert Anthony
Cushley, Robert John
Davis, Abram
Decker, John P
Deutsch, John Ludwig
Dolin, Stanley A
Donohue, Robert J
Dowling, Jerome M
Drake, John Edward
Durnick, Thomas Jackson
Evans, Floyd Monte
Field, Byron Dustin
Fleury, Paul A
Foster, Leigh Curtis
Frasco, David Lee
Friedmann, Herman H
Frosch, Robert Peter
Frye, Virginia Brigham
Fuchs, Jacob
Fulmor, William
Gagne, Jean-Marie
Gallaway, William S
Giacchetti, Athos
Giddings, Lorrain Eugene, Jr
Gilby, Anthony Christopher
Golden, David E
Gordon, Howard R
Goudsmit, Samuel Abraham
Grasselli, Jeanette Gecsy
Gregg, Robert Quinly
Griffiths, James Edward
Grynvak, David Alexis
Hagan, Lucy Gay
Hammond, Gordon Leon
Hankins, Bobby Eugene
Hanson, Victor Frederick
Harp, William R., Jr
Harris, Robert Henry
Heidel, Robert Henry
Heller, Stephen Richard
Hirshfeld, Martin Abraham
Hochheimer, Bernard Ford
Hooper, Donald Lloyd
Hopper, Michael James
Hunt, Graham M
Jacobs, Gerald Daniel
Joshi, Yoginder N
Judd, Jane Harter
Kaelble, Emmett Frank
Kalantar, Alfred Husayn
Kasha, Michael
Kendall, David Nelson
Klassen, David Morris
Kline, John Virgil
Koster, David F
Krakow, Burton
Kropp, John Leo
Leibhardt, Edward
Lemmon, Donald H
Levine, Leonard P
Loewenstein, Ernest Victor
Macklin, John Welton
Majkowski, Richard Francis
Manka, Charles K
Margolis, Jack Selig
Marquet, Louis C
Matraw, Harold Claude
McCall, Elizabeth Regina
McDonald, Francis Raymond
McDonald, Robert Skillings
McGinness, James Donald
Melera, Attilio
Mielenz, Klaus Dieter
Moeller, Karl Dieter
Muenter, John Stuart
Neupert, Werner Martin
Nieman, George Carroll
Ofelt, George Sterling
Parkinson, William Hambleton
Pearson, Earl Freeman
Perkins, Willis Drummond
Plane, Robert Allen

Poranski, Chester F, Jr
Rao, B Seshagiri
Rawitch, Allen Barry
Roessler, David Martyn
Sams, Lewis Calhoun, Jr
Sarup, Ram
Schmelzkopf, Arthur L, Jr
Schneider, Richard Theodore
Scott, Richard William
Shaw, Robert Wayne
Shirk, James Siler
Siegel, Marshall Mayer
Smith, Kenneth Edward
Snider, Albert Monroe, Jr
Spitzer, Jeffrey Chandler
Stafsudd, Oscar M, Jr
Stephenson, Danny Lon
Strominger, Dennis Patrick
Suhr, Norman Henry
Tarpley, Anderson Ray, Jr
Tingle, William Herbert
Toeplitz, Barbara Keeler
Tripper, William Mowbray
Varanasi, Prasad
Veillon, Claude
Verma, Ram D
Vilcins, Gunars
Wallace, Thomas Homkowycz
Walters, John P
Weber, Marvin John
Wickersheim, Kenneth Alan
Williamson, Kenneth Lee
Wood, David Roy
Woodriff, Ray Alan
Wozniak, Wayne Theodore
Yarosewick, Stanley J
Yellin, Wilbur
Zandy, Hassan F

Atomic Spectroscopy

Andrew, Kenneth L
Brown, Charles Moseley
Cloud, William Max
Conway, John George, Jr
Corliss, Charles Howard
Derby, Stanley Kingdon
Doschek, George A
Feldman, Uri
George, Simon
Gilbert, David Erwin
Head, Charles Everett
Head, Martha E Moore
Hill, Kenneth Wayne
Holmes, John Richard
Hudson, Robert Douglas
Humphreys, Curtis Judson
Kelly, Raymond Leroy
Kessler, Ernest George, Jr
Kielkopf, John F
Kingsbury, Robert Freeman
Lu, Kwang-Tzu
Martin, William Clyde
Mayes, Terrill W
McCormick, William Wallace
Merritt, Thomas Parker
Newsom, Gerald Higley
Phelps, Frederick Martin, III
Pinnington, Eric Henry
Radziemski, Leon Joseph, Jr
Reeves, Edmond Morden
Ross, John Stoner
Shenstone, Allen Goodrich
Smith, Peter Lloyd
Stoner, John Oliver, Jr
Stroke, Hinko Henry
Sugar, Jack
Tomkins, Frank Sargent
Tubbs, Eldred Frank
Valero, Francisco Pedro Jorge
Wiegand, Kenneth Gordon
Wiggand, Roy Vernon

Molecular Spectroscopy

Abu-Zeid, Molyi Eldin
Anderson, Anthony
Babrov, Harold J
Baldwin, Bernard Arthur
Barrett, Joseph John
Basile, Louis Joseph
Becker, Ralph Sherman
Beers, Linn Yardley
Beeson, Edward Lee, Jr
Benesch, William Milton
Berney, Charles V
Blass, William Errol
Bogan, Denis John
Brinen, Jacob Solomon
Broida, Herbert Philip
Brom, Joseph March, Jr
Brooks, Wendell V F
Cabana, Aldee
Carpenter, Raymond Allison

Electronic Spectroscopy

Costa, Lorenzo F
El-Bayoumi, Mohamed Ashraf
Morabito, Joseph Michael

STATISTICS

Larsen, Russell D
Larsen, Wayne Ammon
Laska, Eugene
Laughlin, James Stanley
Lavender, DeWitt Earl
Lawing, William Dennis
Lawrence, William Earl
Leckie, Donald Stewart
Ledin, George, Jr
Lee, Young Jack
LeRovich, Leonard Philip
Lehoczky, John Paul
Leone, Fred Charles
Lepson, Benjamin
Lever, William Vernon
Lewis, Everett Vernon
Lewis, Peter Adrian Walter
Li, Hung Chiang
Lieberman, Gerald J
Liggett, Walter Stewart, Jr
Lilliefors, Hubert W
Linder, Forrest Edward
Ling, Robert Francis
Lochner, Robert Herman
Locke, Charles Stephen
Lockley, Jeanette Elaine
Loewe, Michel
Lucas, Henry Laurence, Jr
Lundgaard, Robert James
Ma, Cynthia Sanman
Maag, Urs Richard
MacNeil, Ian B
Madansky, Albert
Maisel, Herbert
Makowski, Gary George
Malone, Linda Caron
Maloney, Clifford Joseph
Marchand, Margaret O
Marcus, Allan H
Marquardt, Donald Wesley
Martin, Margaret Elizabeth
Martz, Harry Franklin, Jr
Mason, Robert Edward
Matis, James Henry
Mazumdar, Mainak
McCabe, George Paul, Jr
McCarthy, Philip John
McKay, Michael Darrell
McKean, Harley Ellsworth
McKean, Joseph Walter, Jr
Meeker, William Quackenbush, Jr
Meier, Paul
Melnick, Edward Lawrence
Mendenhall, William, III
Menon, Manavazhi Vijaya Krishna
Mensing, Richard Walter
Merrill, Warner Jay, Jr
Metzler, Carl Maust
Meyer, Charles Frederick
Meyer, John Sigmund
Michalek, Joel Edmund
Michalik, Edmund Richard
Mickey, Max Ray, Jr
Mielke, Paul W, Jr
Mikhail, Wadie F
Milch, Paul R
Miller, Irwin
Miller, James Milton
Miller, Rupert Griel
Milton, Roy Charles
Minton, Paul Dixon
Misra, Rajendra Kumar
Mittal, Yashaswini Deval
Mohanty, Sri Gopal
Monroe, Robert James
Moonan, William Jeane
Moore, David Sheldon
Moore, Edwin Forrest
Moss, Lincoln Ellsworth
Moss, Ronnie Lee
Mullen, Kenneth
Muller, Mervin Edgar
Nair, Sreekantan S
Nelson, Robert John
Neuhardt, John Bernard
Neus, Marcel Fernand
Neyman, Jerzy
Nielson, Howard Curtis
Norton, James Augustus, Jr
Olshen, Richard Allen
O'Meara, Desmond
Orcutt, Bruce Call
O'Shaughnessy, Charles Dennis
Ostle, Dean Bernard
Ott, Ray Lyman, Jr
Owen, Joel
Pabst, William Richard, Jr
Padget, William Jowayne
Pagano, Marcello
Park, Heebok
Park, Won Joon
Parunak, Anita Nowlin
Parzen, Emanuel
Patil, Ganapati P
Paul, Gilbert Ivan
Paulie, M Catherine Therese
Paull, Allan E
Pauls, John F

Pendergrass, Robert Nixon
Peterson, Gary A
Peterson, Raymond Glen
Pfeiffer, George William
Pickands, James, III
Ponnapalli, Ramachandramurty
Poole, Robert Wayne
Pope, Paul Terrell
Powers, William Allen, III
Prabhu, Narahari Umanath
Proschan, Frank
Puffer, Ruth Rice
Purdue, Peter
Puri, Prem Singh
Quade, Dana Edward Anthony
Quesenberry, Charles P
Qureishi, A Salam
Raff, Morton Spencer
Raktoe, B Leo
Randles, Ronald Herman
Rao, Poduri S R S
Redman, Charles Edwin
Remmenga, Elmer Edwin
Richards, Dale Owen
Richter, Donald
Riffenburgh, Roger Harry
Rigney, Jackson Ashcraft
Rinn, Alfred A
Ringer, Larry Joel
Robbins, Naomi Bograd
Roberts, George Gilbert
Robinette, Charles Dennis
Robinson, Lewis Howe
Robison, Norman Glenn
Ross, Louis
Roth, Raymond Edward
Rothenberg, Ronald Isaac
Rubin, Donald Bruce
Rubin, Herman
Rubin, Theodore
Russell, Charles Bradley
Russell, Thomas Solon
Rust, Charles Harry
Ryan, Thomas Arthur, Jr
Sahai, Hardeo
Samuels, Myra Lee
Sansing, Raymond Clayton
Satterthwaite, Franklin Eves
Savage, I Richard
Scarth, Robert Douglas
Scheffe, Henry
Scheuer, Ernest Martin
Schilling, Edward George
Schmee, Josef
Schmidt, Richard Nicholas
Schneiderman, Marvin Arthur
Schutz, Wilfred M
Scott, Mecklinley
Searle, Shayle Robert
Sedransk, Joseph Henry
Seely, Justus Frandsen
Seiden, Esther
Shah, Babubhai Vadilal
Shah, Bhupendra K
Shaman, Paul
Shenton, Leonard Roy
Sherbrooke, Craig C
Sherman, Dorothy Helen
Shimi, Ismail Nabih
Simon, Gary Albert
Sing, Charles F
Singer, Burton Herbert
Singh, Jagbir
Smith, Armand Verne, Jr
Smith, James Douglas
Smith, John Henry
Smith, Lewis Turner
Smith, Robert Elijah
Smith, Robert Paul
Smith, Wray
Smouse, Peter Edgar
Snider, Bill Carl F
Snyder, Mitchell
Sobel, Eugene L
Somerville, Paul Noble
Spivey, Walter Allen
Sposito, Vincent Anthony
Sprowls, Riley Clay
Srivastava, Jagdish Narain
Srivastava, Ramesh C
Stanley, Kenneth Earl
Starks, Thomas Harold
Starling, James Lyne
Stephany, Edward O
Stewart, Leland Taylor
Stier, Howard Livingston
Stigler, Stephen Mack
Stinson, Perri June
Stiteler, William Merle III
Stone, Charles Joel
Storm, Leo Eugene
Straf, Miron L
Strauch, Ralph Eugene
Stroud, Thomas William Felix
Stucker, Robert Evan

Subrahmanian, Kathleen
Suich, Ronald Charles
Sukhatme, Balkrishna Vasudeo
Sullivan, George Allen
Sweet, Leonard
Tan, James Chien-Hua
Tang, Victor Kuang-Tao
Tanis, Elliot Alan
Tarver, Mae-Goodwin
Taylor, Howard Milton, III
Taylor, Robert Lee
Testerman, Jack Duane
Tetreault, Florence G
Thayne, William V
Thomas, Harold Lee
Thomas, Ralph Edward
Thomasian, Aram John
Thompson, James Robert
Tiao, George Ching-Hwuan
Tong, Yung Liang
Trail, Stanley M
Trumbo, Bruce Edward
Tryon, Peter Vincent
Tserpes, Nicolas A
Tsutakawa, Robert K
Tukey, John Wilder
Turner, James William
Urquhart, N Scott
van Belle, Gerald
Vance, Joseph Francis
van der Vaart, Huberius Robert
Van Ness, John Winslow
Van Ryzin, John R
Van Veldhuizen, Philip Androcles
Vaux, James Edward, Jr
Visco, Eugene Paul
Vitale, Richard Albert
Wagner, Gerald Richard
Walker, Strother Holland
Wallace, David Lee
Walls, Robert Clarence
Walpole, Ronald Edgar
Wang, Peter Cheng-Chao
Wani, Jagannath K
Wasan, Madanlal T
Waterman, Michael S
Watkins, William
Wearden, Stanley
Webster, John Thomas
Weisberg, Herbert Ira
Welch, Roy Elmer
Wesler, Oscar
West, Eric Neil
Westlake, Wilfred James
Wheeler, Alan Clement
Whipkey, Kenneth Lee
Whitby, Owen
Whitney, Donald Ransom
Wiener, Howard Lawrence
Williams, James Stanley
Williams, William Howard
Williams, Willie Elbert
Wilson, P David
Wine, Russell Lowell
Winger, Milton Eugene
Wiorkowski, John James
Wittes, Janet Turk
Woods, Jimmie Dale
Wright, Farroll Trim
Wright, Gordon Pribyl
Wu, Sing-Chou
Wurtele, Zivia Syrkin
Yahiku, Paul Y
Yang, Grace L
Yang, Mark Chao-Kuen
Young, John Coleman
Young, John William
Yu, George Chinshih
Zemach, Rita
Zubay, Eli Alan

Agricultural Statistics

Alleman, Ray Starr
Altman, Isidore
Ardley, Harry Mountcastle
Arkin, Herbert
Arnold, Jesse Charles
Arvesen, James Norman
Balachandran, Kashi Ramamurthi
Baltz, Howard Burl
Bartlett, Noel Sloane
Bhargava, Triloki Nath
Blakemore, George Jefferson, Jr
Braverman, Jerome David
Brogan, Donna R
Bullock, Paul David
Carter, Melvin Winsor
Chambers, David Smith
Chen, Chung Wei
Cleland, Richard Cook
Cochran, William Gemmell
Cornell, John Andrew
Crosby, David S
Davis, Wilford Lavern
De Cani, John Stapley
Delate, Edward Joseph
Dolby, James Louis
Dutton, Arthur Morlan
Earickson, Robert James
Ekeblad, Frederick Alfred
Eklund, Darrel Lee
Elashoff, Janet Dixon
Erickson, Edwin E
Essenwanger, Oskar M
Everson, Dale O
Fardo, Robert D
Farthing, Barton Roby
Fisher, Gail Feinster
Fishman, George Samuel
Folks, John Leroy
Franti, Charles Elmer
Fromovitz, Stan
Gaumnitz, Erwin Alfred
Geisel, Martin Simon
Geng, Shu
George, Fredna Stone
Giese, David Lyle
Glocker, Edwin Merriam
Goldsmith, Charles Harry
Goor, Charles G
Gourary, Mina Haskind
Gross, Alan Marvin
Gupta, Rajendra Prasad
Haase, Richard Henry
Hallum, Cecil Ralph
Harkness, William Leonard
Hicks, Charles Robert
Highland, Harold Joseph
Hill, Hubert Mack
Holbert, Donald
Hollander, Myles
Hotchkiss, Donald K
Houston, Samuel Robert
Huberty, Carl J
Irick, Paul Eugene
Iversen, Gudmund R
Jabine, Thomas Boyd
Jain, Aridaman Kumar
Janardan, Konanur G
Jebe, Emil H
Johnson, John Peter
Johnson, Leonard Gustave
Kadane, Joseph Born
Kafka, Fritz
Kana, Alfred Jan
Katti, Shriniwas Keshav
Kettering, Jon Roberts
Kiener, Beat
Kossack, Carl Frederick
Krutchkoff, Richard Gerald
Kullback, Solomon
Landry, Richard Georges
Larntz, Kinley
Lasater, Herbert Alan
Lautenberger, William J
Lawton, William Harvey
Leabo, Dick Albert
Lechner, James Albert
Li, Jane Chiao
Locke, Ben Zion
Mandel, Benjamin J
Mann, Charles Roy
Mann, Jacinta
Margolin, Barry Herbert
Marshall, Clifford Wallace
Marzetti, Lawrence Arthur
Mason, David Dickenson
McCall, Chester Hayden, Jr
McLaughlin, Gerald Wayne
McMahan, Chalmers Alexander
McVay, Francis Edward
Miller, Robert Burnham
Murphy, Ray Bradford
Murray, Donald Shipley
Neter, John
Newcomb, Robert Lewis
Ogilvie, John Charles
Ortenburger, Leigh Natus
Ortiz, Melchor Jr
Oviatt, Charles Dixon
Owen, Donald Bruce
Papadopoulos, Alex Spero

Analytical Statistics

Arkin, Herbert
Carey, George Warren
Curtiss, John Hamilton
Dykstra, Richard Lynn
Folks, John Leroy
Gordon, Tavia
Geller, Harvey
Hellman, Louis Philip
Kana, Alfred Jan
Katti, Shriniwas Keshav
Levine, Eugene
Raich, Abraham Leonard
Tabler, Kenneth Ambrose

Applied Statistics

Gates, Charles Edgar
Hurt, Paul Victor
Littell, Ramon Clarence
Munson, Arvid W
Owings, Addison Davis

Parker, Don Earl
Pegram, Joe D
Perry, Robert Leonard
Pierce, David Alan
Pirie, Walter Ronald
Portnoy, Stephen Lane
Reynolds, David Stephen
Richards, Francis Russell
Rickmers, Albert D
Rigby, Paul Herbert
Roberts, Charles DeWitt
Roberts, Norman Hailstone
Robinson, Richard Carleton, Jr
Rosenbaum, David Mark
Sanathanan, Lalitha P
Scheinberg, Eliyahu
Schmid, John
Schneider, Philip Allen David
Schroeder, Anita Gayle
Schucany, William Roger
Seabaugh, Pyrtle W
Sevacherian, Vahram
Shapiro, Samuel S
Shiskin, Julius
Shloming, Robert
Silverman, Franklin Harold
Simon, Gary Albert
Singer, Donald Allen
Sinha, Snehesh Kumar
Sirken, Monroe Gilbert
Sisson, Donald Victor
Sitgreaves, Rosedith
Solomon, Vasanth Balan
Stephens, Kenneth S
Stephens, Michael A
Stewart, Kirkland Bruce
Stout, Paul Richard
Stoyle, Judith
St-Pierre, Jacques
Stryker, Harry Kane
Tamhane, Ajit Chintaman
Tepping, Benjamin Joseph
Thiebaux, Helen Jean
Thomas, Donald Henry
Thompson, William Oxley, II
Tischendorf, John Allen
Tysver, Joseph Bryce
Ungar, Andrew
Wakeley, Jay Townsend
Waksberg, Joseph
Ware, Glenn Oren
Watts, Donald George
Weinland, Bernard Theodore
Weisberg, Sanford
Welch, Quintin B
Wells, William T
Wheeler, Donald Jefferson
Williams, Anthony Vearncombe
Wright, Henry Albert
Zeigler, Royal Keith
Zelikoff, Steven Barry

Biometrics
Abramson, Norman Jay
Anderson, Paul Sigfried, Jr
Angleton, George M
Bearman, Jacob Eleazer
Bliss, Chester Ittner
Briese, Franklin Wagner
Carmer, Samuel Grant
Carter, Melvin Winsor
Chan, Yick-Kwong
Chapman, Douglas George
Clutter, Jerome Lee
Cox, Charles Philip
Damon, Richard Alan, Jr
Flueck, John A
Forbes, William Frederick
Garber, Morris Joseph
Gill, John Leslie
Goldman, Anne Ipsen
Gosslee, David Gilbert
Gould, A Lawrence
Hardin, Robert Toombs
Hayne, Don William
Heath, Robert Gardner
Homer, Louis David
Hsi, Bartholomew P
Jensen, Chester E
Kolakowski, Donald Louis
Kolman, Wilfred Aaron
Kowalski, Charles Joseph
Lake, Robin Benjamin
Li, Ching Chun
Little, Thomas Morton
Loewenster, RRuth Brandenburger
Lucas, Joseph James
Mackey, Bruce Ernest
Marcus, Leslie F
Mattson, Dale Edward
McCaughran, Donald Alistair
McHugh, Judson Ulery, Jr
Meade, James Horace, Jr
Moon, Thomas Edward
Mosteller, Robert Cobb
Mumm, Robert Franklin
Myers, Max H

Myers, Wayne Lawrence
Nordby, Gordon Lee
Petersen, Roger Gene
Rawlings, John Oren
Reutzel, Lawrence Frederick
Robson, Douglas Sherman
Rohlf, F James
Schneiweiss, Jeannette W
Seif, Robert Dale
Siverston, John Neilos
Sollberger, Arne Rudolph
Solomon, Daniel Lester
Taucci, Enes Barbara
Thomas, John M
Townsend, Edwin C
Urban, Willard Edward, Jr
Walker, Rufus Floyd, Jr
Walker, William M
Weaver, Clyde Richard
White, Colin
Zelen, Marvin
Zubin, Joseph

Biostatistics
Abbey, Helen
Abernathy, James Ralph
Allaway, Norman C
Anello, Charles
Assenzo, Joseph Robert
Auerbach, Harry
Badger, George Franklin
Bailar, John Christian, III
Banghart, Frank W
Bartsch, Glenn Emil
Benedetti, Jacqueline Kay
Best, William Robert
Blot, William James
Boen, James Robert
Brandt, Edward Newman, Jr
Brodsky, Allen
Brown, Byron William, Jr
Buncher, Charles Ralph
Carrier, Steven Theodore
Chang, Potter Chien-Tien
Chase, Gerald Roy
Chase, Helen Christina (Matulic)
Chen, Edwin Hung-Teh
Chern, Ming-Fen Myra
Chiang, Chin Long
Chiazze, Leonard, Jr
Choi, Sung Chil
Chung, Chin Sik
Ciminera, Joseph Louis
Clark, Virginia
Clarkson, Quentin Deane
Cornell, Richard Garth
Crowley, John James
Davis, Kathryn Bullock
Deane, Margaret
Densen, Paul M
Downs, Thomas D
Dunn, Olive Jean
Dyer, Alan Richard
Edelman, David Anthony
Ederer, Fred
Edwards, Brenda Kay
Elkin, William Futter
Ellenberg, Jonas Harold
Elveback, Lillian Rose
Federer, Walter Theodore
Federspiel, Charles Foster
Feigl, Polly Catherine
Feinleib, Manning
Feldman, Joseph Gerald
Fellner, William Henry
Fertig, John William
Fisher, Lloyd D, Jr
Fisher, Pearl Davidowitz
Flora, Jairus Dale, (Jr)
Flora, Roger E
Forsythe, Alan Barry
Frazier, Todd Mearl
Free, Spencer Michael, Jr
Gaffey, William Robert
Gartside, Peter Stuart
Gaylor, David William
Gelber, Richard David
George, Stephen L
Givens, Samuel Virtue
Glasser, Jay Howard
Gleason, Ray Edward
Goldberg, Irving David
Greenberg, Bernard George
Greenberg, Richard Alvin
Grizzle, James Ennis
Gross, Alan John
Haenszel, William Manning
Hagans, James Albert
Hawkins, C Morton
Hebel, John Richard
Hoel, David Gerhard
Holford, Theodore Richard
Hopkins, Carl Edward
Horn, Susan Dadakis
Hucke, Dorothy Marie
Hurley, Frank Leo
Hutcheson, Kermit
Jablon, Seymour

Jacobs, David R, Jr
Jain, Anrudh Kumar
Jessup, Gordon L, Jr
Johnson, Eugene A
Johnston, Dennis Addington
Jones, Paul Kenneth
Jones, Richard Hunn
Keenan, Kathleen Margaret
Kilpatrick, S James, Jr
Kimball, Allyn Winthrop
Kitler, Mary Ellen
Kjelsberg, Marcus Olaf
Klotz, Jerome Hamilton
Knatterud, Genell Lavonne
Knoke, James Dean
Kraemer, Helena Chmura
Krall, John Morton
Kramer, Morton
Kronmal, Richard Aaron
Kuzma, Jan Waldemar
Lachenbruch, Peter Anthony
Lagakos, Stephen William
Lavin, Philip Todd
Levy, Paul Samuel
Leyton, Morley Kamler
Litt, Bertram D
Littell, Arthur Simpson
Loadholt, Claude Boyd
Lu, Kuo Hwa
Lurie, Dan
Lynch, Cornelius James
Manos, Nicholas Emmanuel
Marks, Sidney
Marshall, Rosemarie
Mason, Thomas Joseph
McHenry, Hugh Lansden
Meinert, Curtis Lynea
Mellits, E David
Menduke, Hyman
Merchant, Roland Samuel
Metter, Gerald Edward
Meydrech, Edward F
Mietlowski, William Leonard
Miller, Millage Clinton, III
Mohberg, Noel Ross
Moore, Dan Houston, II
Morrison, Robert Dean
Mosimann, James Emile
Mullooly, John P
Neely, Peter Munro
Norusis, Marija Jurate
Oates, Richard Patrick
Odoroff, Charles Lazar
O'Fallon, William M
Ordille, Carol Maria
Orlando, Anthony Michael
Pell, Sidney
Perrin, Edward Burton
Powers, Jean Hensel
Rafter, John Arthur
Rao, Mamidanna S
Rastogi, Suresh Chandra
Reading, James Cardon
Reinke, William Andrew
Remington, Richard Delleraine
Rider, Rowland Vance
Roberts-Marcus, Helen Miriam
Robinson, Harry
Rockette, Howard Earl, Jr
Rosenberg, Saul H
Royall, Richard Miles
Rustagi, Jagdish S
Ryel, Lawrence Atwell
Schoenfeld, David Alan
Schor, Stanley
Schork, Michael Anthony
Sen, Pranab Kumar
Serfling, Robert Elton
Sharma, Ran S
Shonick, William
Siegel, Carole Ethel
Siguel, Eduardo Nestor
Singer, Arthur Chester
Slack, Nelson Hosking
Smoller, Sylvia Wassertheil
Swallow, William Hutchinson
Sylwester, David Luther
Tan, Wai-Yuan
Tarter, Michael E
Taylor, William F
Teichman, Robert
Thompson, Donovan Jerome
Thomson, Gordon Merle
Tonascia, James A
Tsay, Jia-Yeong
Ullman, Betty M
Ullman, Nelly Szabo
Ury, Hans Konrad
Vander Zwaag, Roger
Varady, John Carl
Vaughn, William King
Wahl, Patricia Walker
Wahl, Francis Joseph
Weckwerth, Vernon Ervin
Weinberg, Roger
Weinstein, Abbott Samson
Weiss, Edward Sebastian
Weiss, William

Wells, Henry Bradley
Wette, Reimut
Williams, George W
Wolff, Albert Eli
Woodbury, Lowell Angus
Wyshak, Grace
Yu, Pao-Lo
Zielezny, Maria Anna
Zippin, Calvin

Engineering Statistics
Breunig, Henry Latham
Gifford, Leon
Herbach, Leon Howard
Hunter, John Stuart
Landers, John Herbert, Jr
Moritz, Roger Homer
Oliver, Morris Albert
Smith, Malcolm
Webb, Stephen Richard

Experimental Statistics
Anderson, Richard L
Andrews, Horace Porter
Behforooz, Ali
Behnken, Donald Washington
Bernier, Gloria A
Cardenas, Manuel
D'Agostino, Ralph B
Feyerherm, Arlin Martin
Finkner, Morris Dale
Ginsburg, Herbert
Grandage, Arnold Herbert Edward
Guzman Foresti, Miguel Angel
Hafley, William LeRoy
Jebe, Emil H
Johnson, Arthur Frederick
Keister, Thomas Dwight
Koonce, Kenneth Lowell
Koop, John C
Kueper, Theodore Vincent
Linnerud, Ardell Chester
Martin, Frank Garland
Miller, Forest Leonard, Jr
Murphy, John Riffe
Ott, Ellis Raymond
Rowe, Kenneth Eugene
Schilling, Prentiss Edwin
Sentz, James Curtis
Suzuki, Akio
Weeks, David Lee
Wooding, William Minor

Mathematical Statistics
Abrams, Israel Jacob
Abramson, Lee Richard
Adensteh, Rolf Karl
Alberda, Willis John
Albert, Arthur Edward
Ali, Mir Maswood
Ailing, David Wheelock
Anderson, Peter Ole
Anderson, Theodore Wilbur
Andrews, Fred Charles
Antle, Charles Edward
Antoniak, Charles Edward
Ardley, Harry Mountcastle
Arnold, Harvey James
Arnold, Jesse Charles
Arnold, Kenneth James
Aroian, Leo Avedis
Arvesen, James Norman
Bahadur, Raghu Raj
Bain, Lee J
Baker, George Allen
Banerjee, Kali Shankar
Bargmann, Rolf Erwin
Barlow, Richard Eugene
Barnett, William Arnold
Barr, David Ross
Barr, Donald R
Bartko, John Jaroslav
Bauer, David Francis
Behara, Minaketan
Bentley, Donald Lyon
Berens, Alan Paul
Berman, Simeon Moses
Bernier, Gloria A
Bessler, Stuart Alan
Beyer, William Hyman
Bhapkar, Vasant Prabhakar
Bhattacharya, Rabindra Nath
Bhattacharyya, Bibhuti Bhushan
Bhattacharyya, Gouri Kanta
Bickel, Peter J
Bland, Richard P
Block, Henry William
Blyth, Colin Ross
Borsting, Jack Raymond
Bowden, David Clark
Bowen, Earl Kenneth
Bowman, Albert Hosmer
Bowman, Kimiko Osada
Bradt, Russell Newton
Brockett, Patrick Lee
Buehler, Robert Joseph
Burdick, Donald Smiley
Butler, Calvin Charles

STATISTICS

Cameron, Joseph Marion
Carlson, Roger
Carol, Bernard
Carr, Raymond Niel
Carter, Richard Leston
Carter, Walter Hansbrough, Jr
Chernoff, Herman
Chew, Victor
Cinlar, Erhan
Cobb, Whitfield
Cohen, Arthur
Corwin, Thomas Lewis
Cote, Louis J
Cox, Edwin Lory
Crosby, David S
D'Agostino, Ralph B
Dale, Douglas Keith
Daly, Joseph Francis
Dantzig, George Bernard
Darling, Donald Allan
Das Gupta, Somesh
David, Herbert Aron
Davis, James Avery
Dawson, Reed
Dean, Charles Edwin
Deely, John Joseph
DeGroot, Morris Herman
Deming, William Edwards
Derman, Cyrus
Diaconis, Persi
Dick, Ronald Stewart
Downing, Darryl Jon
Dudley, Richard Mansfield
Dunn, James Eldon
Durling, Frederick Charles
Dwass, Meyer
Eaton, Morris Leroy
Edge, Orlyn P
Ehrenfeld, Sylvain
Eisenhart, Churchill
Enneking, Eugene A
Enns, Ernest Gerhard
Esary, James Daniel
Farrell, Roger Hamlin
Ferguson, Thomas S
Fields, Raymond Ira
Fisher, Doris M
Franck, Wallace Edmundt
Freund, John Ernst
Frishman, Fred
Gardiner, Donald Andrew
Gastwirth, Joseph L
Geisser, Seymour
Geller, Nancy L
Geng, Thomas Michael
Gessaman, Margaret Palmer
Ghosh, Bhaskar Kumar
Ghosh, Sakti P
Gifford, Leon
Gillis, Catherine Josephine
Giri, Narayan C
Gleser, Leon Jay
Gnanadesikan, Mrudulla
Gordon, Florence S
Gordon, Milton Andrew
Gordon, Sheldon P
Graves, Clayborn Lowell
Graybill, Franklin A
Grenander, Ulf
Gujahr, Allan L
Hader, Robert John
Haley, Kenneth David Cann
Hall, William Jackson
Halperin, Max
Hannan, James Francis
Haq, M Safiul
Harris, Bernard
Harshbarger, Boyd
Harter, Harman Leon
Hartigan, John A
Harvey, James Raymond
Healy, William Carleton, Jr
Henry, Robert Lowell
Hendrickson, Arlo Dennis
Henry, Neil Wylie
Hensler, Gary Lee
Herd, George Ronald
Hodges, Joseph Lawson, Jr
Hoerding, Wassily
Hollander, Myles
Holt, William Robert
Hou, Tien Fang
Howard, William Grady
Hoyt, John Paul
Huddleston, Harold Frank
Hudson, William Nathaniel
Hughes, Harry Meachum
Hulquist, Robert Allan
Iglehart, Donald Lee
Inglis, James
Isaacson, Stanley Leonard

Jackson, James Edward
Jacobs, Walter William
Jacobsen, Robert Leland
Janardan, Konanur G
Jensen, Donald Ray
Johns, Peter Milton Meredith
Johnson, Milton Vernon, Jr
Johnson, Charles Henry
Johnson, Dudley Paul
Jones, Roger Daniel Hewart
Kabe, Dattatraya G
Kadane, Joseph Born
Kalbfleisch, James G
Karlin, Samuel
Karr, Balvant Keshav
Kastenbaum, Marvin Aaron
Katz, Leo
Keilson, Julian
Kennard, Robert Wakely
Kettering, Jon Roberts
Kiefer, Jack C
King, Edgar Pearce
Klimko, Lawrence Andrew
Klingman, Darwin Dee
Knight, Frank B
Koopmans, Lambert Herman
Korin, Basil Peter
Kotz, Samuel
Koul, Hira Lal
Kuebler, Roy Raymond Jr
Kullback, Joseph Henry
Kullback, Solomon
Kupperman, Morton
Larnitz, Kinley
Larson, Harold Joseph
Lasater, Herbert Alan
Laurent, Andre Gilbert (Louis)
Le Cam, Lucien Marie
Lechner, James Albert
Lehmann, Eugene H
Lehman, Erich Leo
Levene, Howard
Lieblein, Julius
Link, Richard Forest
Loftsgaarden, Don Owen
Lott, Fred Wilbur, Jr
Low, Leone Yarborough
Lytle, Ernest James, Jr
Maar, James Richard
Madow, William Gregory
Mallows, Colin Lingwood
Mandel, John
Mann, Charles Roy
Mapleton, Robert Allan
Margolin, Barry Herbert
Massey, Frank Jones, Jr
Matha, Arakaparampil M
Matis, James Henry
McDonald, Alfred Jerome
McDonald, Bruce Jerald
McKeon, Alfred Jerome
McLemore, Benjamin Henry, Jr
McCloskey, John W
Mehrotra, Kishan Gopal
Mihram, George Arthur
Mikulski, Piotr W
Miura, Carole K Masutani
Morgan, John Clifford, II
Morrison, Donald Franklin
Morrison, Nathan
Moser, Joseph M
Mosteller, C Frederick
Mudholkar, Govind S
Muirhead, Robb John
Murphy, Ray Bradford
Myers, Raymond Harold
Myhre, Janet M
Nadas, Arthur Joseph
Nair, K Aiyappan
Nash, Stanley William
Navarro, Joseph Anthony
Nelson, A Carl, Jr
Noether, Gottfried Emanuel
Norick, Nancy Xavier
Norton, Horace Wakeman, III
Oakland, Gail Barker
Odeh, Robert Eugene
Odell, Patrick L
Ogawa, Junjiro
Olkin, Ingram
O'Neill, Anne Frances
Ott, Teunis Jan
Paulson, Edward
Perlman, Michael David
Pierce, David Alan
Pillai, Krishna Chennakadu Sreedharan
Pillai, Raman Narayana
Pina, Eduardo Isidorio
Pirie, Walter Ronald
Platt, Ronald Dean
Pledger, Gordon Wayne
Pollak, Edward
Portnoy, Stephen Lane
Posten, Harry Owen
Potthoff, Richard Frederick
Powell, James Henry
Press, S James
Puri, Madan L

Pyke, Ronald
Qualls, Clifford Ray
Ramsey, Fred Lawrence
Rao, Pejaver Vishwamber
Rao, Vallunu Bhavanarayana
Rastogi, Suresh Chandra
Ray, Rose Marie
Read, Robert Richard
Reinhardt, Howard Earl
Resnikoff, George Joseph
Rhyne, A Leonard
Richards, Winston Ashton
Robbins, Herbert Ellis
Roberts, Charles DeWitt
Robertson, Timothy Joel
Rodine, Robert Henry
Rogers, Gerald Stanley
Rohatgi, Vijay
Romano, Albert
Rosenblatt, David
Rosenblatt, Joan Raup
Rosenblatt, Judah Isser
Rosenblatt, Murray
Roth, Arthur Jason
Roussas, George G
Rust, Velma Irene
Sanathanan, Lalitha P
Santner, Joseph Frank
Saunders, Sam Cundiff
Schaufele, Ronald A
Schneider, Alfred Marcel
Schucany, William Roger
Sclove, Stanley Louis
Scott, Elizabeth Leonard
Sengupta, Sailes Kumar
Serfling, Robert Joseph
Seshadri, Vanamamalai
Sethuraman, Jayaraman
Shantaram, Rajagopal
Siddiqui, Mohammed Moinuddin
Siegmund, David O
Sielken, Robert Lewis, Jr
Sievers, Gerald Lester
Simmons, Walt R
Singh, Rajinder
Singleton, Richard Collom
Siotani, Minoru
Sirken, Monroe Gilbert
Skibinsky, Morris
Smith, Walter Laws
Smith, William Boyce
Smoke, Mary E
Sogliero, Luigina Cianfarani
Solomon, Frederick Allen
Solomon, Herbert
Sornburger, George Clinton
Sprott, David Arthur
Srivastava, Muni Shanker
Stapleton, James H
Starr, Norman
Stein, Arthur
Steinberg, Joseph
Stephens, Kenneth S
Stephens, Michael A
Stoneman, David McNeel
Studden, William John
Subrahmaniam, Kocherlakota
Sudderth, William David
Sukhatme, Shashikala Balkrishna
Sweeny, Hale Caterson
Switzer, Paul
Sylvester, David Luther
Tabler, Kenneth Ambrose
Tamhane, Ajit Chintaman
Tan, Peter Ching-Yao
Taneja, Viday Sagar
Tate, Robert Flemming
Teicher, Henry
Tepping, Benjamin Joseph
Terry, Milton Everett
Thall, Peter Francis
Thomas, Ronald Emerson
Thompson, William Oxley, II
Thompson, William Rae
Tiao, George Ching-Hwuan
Timon, William Edward, Jr
Tourgee, Ronald Alan
Tracy, Derrick Shannon
Trawinski, Benon John
Trawinski, Irene Patricia Monahan
Truax, Donald R
Tsao, Chia Kuei
Uppuluri, V R Rao
Usher, William Mack
Van Ryzin, John R
Vardeman, Stephen Bruce
Vasicek, Oldrich Alfonso
Wahba, Grace
Walker, James Wilson
Waller, Ray Albert
Walner, Arthur H
Ware, James H
Watson, Geoffrey Stuart
Weiner, Howard Jacob
Weiss, Irving

Weiss, Lionel Ira
Welch, Peter D
Welker, Everett Linus
Wheeler, Donald Jefferson
Wheeler, Ruric E
Whitney, Donald Ransom
Whitesey, John Rb
Wiggins, Alvin Dennie
Wijsman, Robert Arthur
Wilkinson, John Wesley
Willke, Thomas Aloys
Williford, William Olin
Wolman, William Wolfgang
Wolock, Fred Walter
Wong, Chi Song
Yen, Elizabeth Hsi
Yin, Barbara Hsin-Hsin
Young, Dennis Lee
Yuan, William Jen Chun
Zacks, Shelemyahu
Zidek, James Victor
Zweig, Hans Jacob

Medical Statistics
Beebe, Gilbert Wheeler
Breslow, Norman Edward
Carrier, Steven Theodore
Chapman, Judith-Anne Williams
Cutler, Sidney Joshua
Garg, Mohan Lal
Gent, Michael
Hallum, Cecil Ralph
Hewitt, David
Holt, Bruce Edward
Ipsen, Johannes, Jr
Keehn, Robert John
Kumbaraci, Turkan Emine
MacCormick, Alasdair John
Miller, Gerson Harry
Perrott, George St John
Phillips, Alexander James
Pogue, Richard Ewert
Rich, Herbert
Rodda, Bruce Edward
Webb, Nathaniel Conant, Jr
Wilcox, Roberta Arlene

Statistical Analysis
Abbott, Robert Classie
Ball, Edwin Lawrence
Berger, Philip Jeffrey
Bischke, Wallace Robert
Brier, Glenn Wilson
Bruckner, Lawrence Adam
Cubitt, John Malcolm
Demskey, Sidney
Dressel, Paul Leroy
Eichhorn-von Wurmb, Heinrich Karl
Evans, David Arthur
Gentleman, Jane Forer
Gibbons, Jean Dickinson
Gnanadesikan, Ramanathan
Kolesar, Peter John
Lohrding, Ronald Keith
Maar, James Richard
Mann, Nancy Robbins
Mather, Robert Eugene
McCune, Duncan Chalmers
Moshman, Jack
Nagin, Daniel Steven
Newton, Howard Joseph
Nyquist, Wyman Ellsworth
Pierson, Ellery Merwin
Srivastava, Tariq Naseer
Steel, Robert George Douglas
Stewart, William Charles
Stoline, Michael Ross
Sykes, Donald Joseph
Thacker, John Charles
Tukey, John Wilder
Turner, David Lee
Zacks, Shelemyahu

Statistical Mechanics
Alder, Berni Julian
Allen, Kenneth Richard
Baker, George Allen, Jr
Bartis, James Thomas
Battle, Ed Len
Callen, Herbert Bernard
Coldwell, Robert Lynn
Duffy, James W
Falk, Harold
Fesciyan, Sezar
Garrison, John Carson
Gilmer, George Hudson
Girardeau, Marvin Denham, Jr
Gould, Harvey A
Green, Melville Saul
Haggerty, Michael John
Harris, Stewart
Hecht, Charles Edward
Karo, Douglas Paul
Kikuchi, Ryoichi
Klein, William
Kraichnan, Robert Harry
Krinsky, Samuel
Langer, James Stephen

320

Leaf, Boris
Mazenko, Gene Francis
McClure, Charles Frederick
McLennan, James Alan, Jr
McQuistan, Richmond Beckett
Millard, Kenneth Young
Murphy, Thomas James
Nichols, William Herbert
Nicoll, Jeffrey Fancher
Pokrant, Marvin Arthur
Rainwater, James Carlton
Richardson, Robert William
Riedel, Eberhard Karl
Sak, Joseph
Siegel, Armand
Siggia, Eric Dean
Valleau, John Philip
Vezzetti, David Joseph
Viswanathan, Kadayam Sankaran
Walker, Grayson Howard
Weinberg, Michael C
Wood, William Wayne

SURGERY

Surgery
Abrams, Jerome Sanford
Ackerman, Norman Bernard
Adler, Richard
Adriani, John
Alexander, James Wesley
Allam, Mark Whittier
Allbritten, Frank F, Jr
Allen, Miller Shannon, Jr
Anderson, Marion C
Anlyan, William George
Arbulu, Agustin
Aries, Leon Judah
Artz, Curtis Price
Atik, Mohammad
Aust, J Bradley
Austen, William Gerald
Bachhuber, Edward A
Baker, Ralph Robinson
Baker, Roger (Carroll), (Jr)
Ballinger, Walter F, II
Barker, Clyde Frederick
Barker, Harold Grant
Barnet, Hendrick Boyer
Barnett, William Oscar
Baron, Shirley Harold
Baue, Arthur Edward
Beal, John Mann
Beard, Joseph Willis
Beattie, Edward J
Beck, William Carl
Bell, Cecil Cooper, Jr
Belzer, Folkert O
Benfield, John R
Benz, Edmund Woodward
Berggren, Ronald B
Bernstein, Eugene F
Birtch, Alan Grant
Black, Benjamin Marden
Blackard, Clyde Erhardt
Blakemore, William Stephen
Block, George E
Blocker, Truman Graves, Jr
Boswick, John A
Bounous, Gustavo
Boyce, Frederick Fitzherbert
Bradham, Gilbert Bowman
Branson, Bruce William
Branigan, Otto Charles
Breed, Ernest Spencer
Breidenbach, Lester
Brigham, M Prince
Brockenbrough, Edwin C
Brooks, John Robinson
Brown, Henry
Brown, Paul Woodrow
Brown, Robert Lee
Brown, Roger E
Buchwald, Henry
Buckwalter, Joseph Addison
Burdette, Walter James
Burdick, Daniel
Burke, John Francis
Burns, Francis John
Butcher, Harvey Raymond, Jr
Butler, Hugh C
Byrne, John Joseph
Cady, Blake
Cain, Arthur Samuel, Jr
Callow, Allan Dana
Campbell, Gilbert Sadler
Cantrell, James R
Carey, Larry Campbell
Caron, Wilfrid M
Carroll, Walter William
Carter, Paul Richard
Castleton, Kenneth Bitner
Chanana, Arjun Dev
Chandler, J Ryan
Child, Charles Gardner, III
Christensen, Leonard

Clark, Randolph Lee
Clarke, James Spencer
Cliffton, Eugene Everett
Cohn, Isidore, Jr
Cole, Warren Henry
Condon, Robert Edward
Conger, Kyril Bailey
Connolly, John E
Cooley, Denton Arthur
Coon, William Warner
Cooper, Donald Russell
Copeland, Murray Marcus
Coppola, Edward Dante
Cox, Clair Edward, II
Crandell, Walter Bain
Crikelair, George F
Crowley, Lawrence Grandjean
Daicoff, George Ronald
Dale, William Andrew
Dammann, John Francis
Daniel, Rollin Augustus, Jr
Davis, Loyal
Davis, William Clayton
DeBakey, Michael Ellis
DeCamp, Paul Trumbull
Delaney, John P
Del Guercio, Louis Richard M
Dennis, Clarence
Dennis, Daniel L
Depalma, Ralph G
Derrick, John Rafter
Deterling, Ralph Alden, Jr
DeWeese, David D
DeWeese, James A
DeWeese, Marion Spencer
Dillard, David Hugh
Dillon, Marcus Lunsford, Jr
Dixon, John Aldous
Doberneck, Raymond C
Dobyns, Brown M
Donohue, William B
Dorsey, John M
Dreiling, David A
Dumont, Allan E
Dunphy, J Englebert
Du Val, Merlin Kearfott
Eckert, Charles
Edwards, W Sterling
Egdahl, Richard H
Ehrenhaft, Johann L
Eiseman, Ben
Eisenberg, M Michael
Elliott, Dan Whitacre
Elliott, Robert Hare Egerton, Jr
Emerson, Ernest Benjamin, Jr
Engler, Harold S
Enquist, Iring Fritiof
Etheredge, Edward Ezekiel
Evans, Arthur T
Everhard, Martin Edward
Farmer, Douglas Alexander
Farrell, John Joseph
Farris, Jack Matthews
Feinman, Max L
Feldhaus, Richard Joseph
Feller, William
Ferguson, Colin C
Ferguson, Donald John
Ferrer, Jose M
Ferris, Deward Olmsted
Fisher, Bernard
Fisher, John Herbert
Fitts, Charles Thomas
Fitts, William Thomas, Jr
Flatt, Adrian Ede
Fletcher, William Sigourney
Flotte, C Thomas
Folkman, Moses Judah
Fonkalsrud, Eric W
Fortner, Joseph Gerald
Foster, James Henry
Foster, John Hoskins
Foster, Roger Sherman, Jr
Fox, Charles Lewis, Jr
Frey, Charles Frederick
Friedmann, Paul
Friesen, Stanley Richard
Fry, William James
Fryer, Minot Packer
Furman, Seymour
Gaensler, Edward Arnold
Gann, Donald Stuart
Gans, Henry
Gardner, Bernard
Gaspar, Max Raymond
German, John Dee
Gerst, Paul Howard
Ghent, William Robert
Gilbertsen, Victor Adolph
Gius, John Armes
Gladstone, Arthur A
Glenn, William Wallace Lumpkin
Gliedman, Marvin
Goldman, Leon
Goldsmith, Edward I
Goldsmith, Harry Sawyer
Goldstein, Louis Arnold
Golomb, Frederick M

Gonzalez, Luis L
Goodman, Louis E
Grage, Theodor B
Grayson, Merrill
Greenlee, Herbert Breckenridge
Griffen, Ward O, Jr
Grimson, Keith Sanford
Gross, Robert Edward
Grossi, Carlo E
Grotzinger, Paul John
Gump, Frank E
Gumport, Stephen Lawrence
Gurd, Fraser Newman
Guralnick, Eugene
Gutelius, John Robert
Haig, Thomas Harrison Brian
Halasz, Nicholas Alexis
Hale, Harry W, Jr
Haley, Harold Bernard
Hall, Albert D
Hall, DeLou Perrin
Hall, John Emmett
Hall, Lynn Raymond
Hallenbeck, George Aaron
Haller, Jacob Alexander, Jr
Hamit, Harold F
Handelsman, Jacob C
Hanlon, Cyril Rollins
Harbison, Samuel Pollock
Hardaway, Robert M, III
Hardin, Creighton A
Hardy, James D
Harper, Paul Vincent
Harris, Matthew N
Harrison, John Hartwell
Harrison, Robert Cameron
Harrison, Timothy Stone
Hartford, Charles Edward
Hartman, Albert William
Hattler, Brack Gillium, Jr
Hayes, Mark Allan
Haynes, Boyd W, Jr
Herrmann, John Bellows
Hershey, Falls Bacon
Herter, Frederic P
Hiatt, Robert Burritt
Hickey, Maurice John
Higgins, George A, Jr
Hightower, Felda
Hines, James R
Hinshaw, David B
Hinshaw, J Raymond
Hitchcock, Claude Raymond
Hodgson, Paul Edmund
Holden, William Douglas
Holman, Cranston William
Hong, Pill Whoon
Hopkins, John West
Horsley, John Shelton, III
Hotchkiss, Robert S
Hovnanian, August P
Howard, John Malone
Hubay, Charles Alfred
Hubbard, T Brannon, Jr
Hudson, William Rucker
Hufnagel, Charles Anthony
Huggins, Charles Edward
Hume, Michael
Hummel, Robert P
Humphrey, Edward William
Humphries, Arthur Lee, Jr
Hunt, Thomas Knight
Hurley, John D
Imamoglu, Kamil H
Ingall, John
Jackson, Benjamin T
Jacob, Stanley W
Jacobson, Myron J
Jannetta, Peter Joseph
Jelenko, Carl, III
Jesseph, John Ervin
Jewell, William R
Johnson, George, Jr
Johnson, Robert O
Johnstone, Frederick Robert Carlyle
Jordan, George Lyman, Jr
Jordan, Paul H, Jr
Jurkiewicz, Maurice J
Kahan, Barry D
Kaminski, Donald Leon
Kantrowitz, Adrian
Karl, Richard C
Karlson, Karl Eugene
Kaufman, Joseph J
Kaye, Michael Peter
Keeley, John L
Kelly, William Daniel
Kendrick, Douglas Blair, Jr
Ketcham, Alfred Schutt
King, Harold
King, Robert (Bainton)
Kingsley, Harry Durwood
Kinney, John Martin
Kittie, Charles Frederick
Klass, Alan Arnold
Klassen, Karl Peter
Kleitsch, William Philip
Klopp, Calvin Trexler

Knock, Frances Engelmann
Kokame, Glenn Megumi
Kolb, Leonard H
Koop, Charles Everett
Kottmeier, Peter Klaus
Krementz, Edward Thomas
Krieger, Harvey
Krippaehne, William W
Kukral, John Charles
Kurzweg, Frank Turner
Kwan-Gett, Clifford Stanley
Landor, John Henry
Large, Alfred McKee
Larson, Duane L
Laufman, Harold
Lawrence, Montague Schiele
Lawrence, Walter, Jr
Lawton, Richard L
Lazaro, Eric Joseph
Lee, Herbert Carl
Lee, Hyung Mo
Lee, Leroy William
Lee, Lyndon Edmund, Jr
Lee, Sun
Lehman, Robert Nathan
Lemieux, Jean-Marie
Lemmer, Kenneth Ellery
Lempke, Robert Everett
Leonard, Arnold S
Lepley, Derward, Jr
Letterman, Gordon Sparks
LeVeen, Harry Henry
Levenson, Stanley Melvin
Levitsky, Sidney
Lewis, Floyd John
Lichty, Richard D
Lillehei, C Walton
Lillehei, Richard Carlton
Lind, James Forest
Lindenauer, S Martin
Lindskog, Gustaf Elmer
Litwak, Robert Seymour
Litwin, Martin Stanley
Localio, S Arthur
Lofgren, Karl Adolph
Longerbeam, Jerrold Kay
Longley, B Jack
Longmire, William Polk, Jr
Lord, Jere Williams, Jr
Lore, John M, Jr
Lynn, Hugh Bailey
Lynn, Ralph Beverley
Macbeth, Robert Alexander
MacDougall, John Taylor
Mackay, Albert George
Mackler, Saul Allen
MacLean, Lloyd Douglas
MacMillan, Bruce Gregg
Madden, John William
Mahoney, Earle Barnes
Majarakis, James Demetrios
Malette, William Graham
Maloney, James Vincent, Jr
Malt, Ronald A
Mandelbaum, Isidore
Mansberger, Arlie Roland, Jr
Marceau, Gilles
Marchioro, Thomas Louis
Markland, Alan Colin
Marshall, Victor Fray
Martin, Daniel S
Martin, Richard Gordon
Mason, Edward Eaton
Mason, George Robert
Maxwell, John Gary
McClelland, Robert Nelson
McCord, Colin Wallace
McCorkle, Horace J
McCorriston, James Roland
McCredie, John A
McDermott, William Vincent, Jr
McDonald, John C
McDowell, Frank
McFee, Arthur Storer
McGarity, William Cecil
Meacham, William Feland
Merendino, K Alvin Aurelius
Mersheimer, Walter Lyon
Metcalf, William
Michel, Marshall Louis, Jr
Miller, Fletcher A
Miller, Leonard David
Minor, George Ridgway
Mittelman, Arnold
Mixter, George, Jr
Mohri, Hitoshi
Mohs, Frederic Edward
Monaco, Anthony Peter

SURGERY

Moncrief, John A
Moody, Frank Gordon
Moore, Condict
Moore, Francis Daniels
Moore, George Eugene
Moore, Wesley Sanford
Moosman, Darvan Albert
Moran, Walter Harrison, Jr
Moretz, William Henry
Morfit, Henry Mason
Morgenstern, Leon
Morris, George Cooper, Jr
Morrison-Cleator, Iain Goosta
Morton, Andrew Glenn
Morton, Donald Lee
Morton, John Henderson
Moscovici, Mauricio
Moss, Charles Norman
Moss, Gerald
Moss, Gerald S
Mountain, Clifton Fletcher
Mrazek, Rudolph G
Mueller, C Barber
Muller, William Henry, Jr
Mulligan, Leo Virgil
Murphy, John Joseph
Musgrave, F Story
Musselman, Merle McNeil
Myers, Richard Thomas
Nachlas, Marvin Morton
Nahrwold, David Lange
Naitove, Arthur
Najarian, John Sarkis
Nance, Francis Carter
Nathan, Helmuth M
Nealon, Thomas F, Jr
Nelsen, Thomas Sloan
Nelson, Norman Crooks
Nelson, Russell Marion
Nemir, Paul, Jr
Neville, William E
Newman, Melvin Micklin
Newsome, James Frederick
Nicholas, James A
Noble, J Arnold
Noer, Rudolf Juul
Nora, Paul Francis
Noya Benitez, Jose Antonio
Nyhus, Lloyd Milton
Oberhelman, Harry Alvin, Jr
Ochsner, John Lockwood
Orkin, Lazarus Allerton
Orloff, Marshall Jerome
Owens, James Cuthbert
Owens, Neal
Pace, William Greenville
Paloyan, Edward
Parrish, Robert A, Jr
Patterson, Hubert C
Patterson, William Bradford
Peacock, Erle Ewart
Peete, William P J
Peirce, Edmund Converse, II
Pennell, Timothy Clinard
Perey, Bernard Jean Francois
Perey, Daniel Yves Emile
Perry, Frank Anthony
Perry, John Francis, Jr
Persky, Lester
Pestana, Carlos
Peters, Paul Conrad
Peters, Richard Morse
Peterson, Clare Gray
Pfaff, William Wallace
Pickett, Lawrence Kimball
Pickrell, Kenneth LeRoy
Pierce, James Clarence
Pierpont, Howard Clemeth
Polk, Hiram Carey
Pories, Walter J
Porter, Milton Reeves
Postelthwait, Raymond Woodrow
Poth, Edgar J
Potter, John F
Powers, Samuel Ralph, Jr
Preston, Frederick Willard
Priestley, James Taggart
Prudden, John Fletcher
Putney, Floyd Johnson
Ragins, Herzl
Ram, Madhira Dasaradhi
Randall, Henry Thomas
Randolph, Judson Graves
Ransdell, Herbert Threlkeld, Jr
Ransom, Henry King
Rapaport, Felix Theodosius
Rapaport, Aron M
Ravitch, Mark Mitchell
Ray, Bronson Sands
Redo, Saverio Frank
Reed, Raymond Charles
Reed, George Elliott
Reemtsma, Keith
Reichle, Frederick Adolph
ReMine, William Hervey
Requarth, William H
Reynolds, Vernon F
Rheinlander, Harold F

Rhoads, Jonathan Evans
Rhode, C Martin
Richards, Ralph Chamberlain
Richards, Victor
Richardson, Lyman King
Ritchie, Wallace Parks, Jr
Rittenbury, Max Sanford
Rob, Charles G
Robinson, David Weaver
Rogers, Lloyd Sloan
Romsdahl, Marvin Magnus
Root, Harlan D
Rosemond, George P
Rosenberg, Irwin Kay
Rosenberg, Jerry C
Rosenkrantz, Jens Georg
Rosenthal, J William
Rosoff, Leonard
Ross, Charles Augustus
Rowe, Edward Barry
Royster, Henry Page
Rudolf, Leslie E
Rush, Benjamin Franklin, Jr
Russell, Paul Snowden
Rutenberg, Alexander Michael
Ryan, Robert F
Sabiston, David Coston, Jr
Sako, Kumao
Sako, Yoshio
Saleh, Wasfy Seleman
Salmon, Peter Alexander
Salzman, Edwin William
Sandusky, William Roberts
Santulli, Thomas V
Sauvage, Lester R
Sawyers, John Lazelle
Schein, Clarence Jacob
Schenk, Worthington G, Jr
Schiebel, Herman Max
Schilling, John Albert
Schloerb, Paul Richard
Schramel, Robert Joseph
Schulte, William John, Jr
Schumer, William
Schwegman, Cletus W
Scott, Henry William, Jr
Scott, William Albert
Scovill, William Albert
Seed, Randolph William
Seigler, Hilliard Foster
Seligman, Arnold Max
Seltzer, Albert Pincus
Serkes, Kenneth Dean
Sertin, Oscar
Shaftan, Gerald Wittes
Shedd, Donald Pomroy
Sherman, Roger Talbot
Shires, George Thomas
Shirley, Sheridan William
Shoemaker, William C
Shumacker, Harris B, Jr
Siegel, John H
Siegel, Bernard
Silbergleit, Allen
Silen, William
Simeone, Fiorindo Anthony
Simmons, Richard Lawrence
Siqueira, Edir Barros
Sirek, Anna
Sistek, Vladimir
Skandalakis, John Elias
Skinner, David Bernt
Slattery, Louis R
Smith, Carl Arthur
Smith, Frederick Williams
Smith, Gardner Watkins
Smith, Louis Livingston
Smyth, Nicholas Patrick Dillon
Snyder, John M
Soroff, Harry S
Southwick, Harry W
Souter, Lamar
Sparkman, Robert Satterfield
Spellman, Mitchell Wright
Spratt, John Stricklin, Jr
Starkloff, Gene B
Starzi, Thomas E
State, David
Steenburg, Richard Wesley
Stein, Irving F, Jr
Stein, John Michael
Stein, Justin John
Stern, W Eugene
Stevens, Lloyd Weakley
Stevenson, Jean Moorhead
Sickel, Delford LeFew
Storer, Edward Hammond
Strandness, Donald Eugene, Jr
Stromberg, LaWayne Roland
Stuckey, Jackson H
Sullivan, William Albert, Jr
Swan, Kenneth G
Swenson, Orvar
Syphax, Burke
Taylor, Billy G
Templeton, John Y, III
Ternberg, Jessie L
Thal, Alan Philip
Thomas, Arthur Norman

Thomas, Colin Gordon, Jr
Thompson, James Charles
Thompson, Jesse Eldon
Thompson, Ralph J, Jr
Thompson, Samuel Alcott
Tidrick, Robert Thompson
Tompkins, Ronald K
Tritschler, Louis George
Tsapogas, Makis Joakim
Tucker, Harvey Michael
Turcotte, Jeremiah G
Turell, Robert
Tyner, George S
Tyson, Ralph Robert
Ulfelder, Howard
Valk, William Lowell
Varco, Richard Lynn
Veith, Frank James
Vredevoe, Lawrence A
Waddell, William Rhoads
Waite, John Henry
Waldhausen, John Anton
Walker, Matthew William
Wallace, Herbert William
Walt, Alexander Jeffrey
Walton, Thomas Peyton, III
Wangensteen, Owen Harding
Wangensteen, Stephen Lightner
Warden, Herbert Edgar
Waterhouse, Keith R
Watkins, Elton, Jr
Watne, Alvin Lloyd
Watson, Thomas Richard, Jr
Watters, Neil Archibald
Way, Lawrence Wellesley
Webb, Watts Rankin
Weeks, Paul Martin
Weidner, Michael George, Jr
Weiss, Jack Allan
Welch, Charles Stuart
Welch, Harold Francis
Wesolowski, S Adam
Whiffen, James Douglass
White, Edgar C
White, Robert J
White, Thomas Taylor
Whitehill, Jules Leonard
Whitfield, Graham Frank
Whitsell, John Crawford, II
Wilkins, Samuel Austell, Jr
Williams, Marion Jack
Willman, Vallee L
Wilson, Harwell
Witherspoon, Don Meade
Wolberg, William Harvey
Wolcott, Mark Walton
Wolf, James Stuart
Wolf, William I
Wolfman, Earl Frank, Jr
Wolma, Fred J
Woods, Alan Churchill, Jr
Woodward, Edward Roy
Wylie, Edwin J
Wylie, Charles E
Yollick, Bernard Lawrence
Young, Morris Nathan
Young, William Paul
Yune, Heun Yung
Ziegler, Hriolfe Read
Ziffren, Sidney Edward
Zimmerman, Jack McKay
Zimmerman, Leo M
Zimmermann, Bernard
Zintel, Harold Albert
Zollinger, Robert Milton
Zuidema, George Dale
Zukoski, Charles Frederick

Cardiovascular Surgery

Absolon, Karel B
Albert, Harold Marcus
Baird, Ronald James
Blair, Emil
Boland, James P
Brown, Ivan Willard, Jr
Carroll, Samuel Edwin
Castaneda, Aldo Ricardo
Clauss, Roy H
Coffin, Laurence Haines
Connar, Richard Grigsby
Danielson, Gordon Kenneth
Davila, Julio C
DeWall, Richard A
Ebert, Paul Allen
Ellison, Robert G
Evans, Geoffrey
Frasher, Wallace G, Jr
Frater, Robert William Mayo
Fuson, Robert L
Gordon, Archer Samuel
Gott, Vincent Lynn
Heimbecker, Raymond Oliver
Hewitt, Robert Lee
Husni, Elias A
Julian, Ormand C
Kahn, Donald R
Kaiser, Gerard Alan

Kennedy, John Hines
Kilman, James William
Kirklin, John W
Lansing, Allan M
Lee, William Morris
Lees, William Harris
Levitsky, Sidney
Long, David Michael
Lower, Richard Rowland
Madden, Robert E
Magovern, George Jerome
Mamiya, Richard T
Moulder, Peter Vincent, Jr
Nicoloff, Demetre M
Nolan, Stanton Peelle
Paton, Bruce Calder
Reed, George Elliott
Reeves, Melvin Mitchell
Replogle, Robert Lee
Rohman, Michael
Rosenberg, Dennis Melville Leo
Rossi, Nicholas Peter
Sayegh, Salem F
Schimert, George
Scott, Stewart Melvin
Silver, Donald
Stephenson, Hugh Edward, Jr
Sugg, Winfred Lindley
Symbas, Panagiotis N
Tice, David Anthony
Vargas, Lester Lambert
Vasko, John Stephen
Wheat, Myron William, Jr
Wilcox, Benson Reid
Wilson, Robert Francis
Winterscheid, Loren Covart

Experimental Surgery

Cole, Jack Westley
Day, Stacey Biswas
Downie, Harry G
Fazekas, Arpad Gyula
Heimbecker, Raymond Oliver
Karagianes, Manuel Tom
Kowalewski, Konstanty Piotr
Malinin, Theodore I
Moss, Gerald S
Pierce, Joseph Elliott
Skoryna, Stanley C
Stefko, Paul Lowell
Vogelfanger, Isaac Joel
Wykoff, Matthew Henry

Genitourinary Surgery

Gittes, Ruben Foster
Lakey, William Hall
Lilien, Otto Michael
Lyon, Edward Spafford
Rubin, Seymour Walter

Maxillofacial Surgery

Ackell, Edmund Ferris
Bernstein, Leslie
Bloom, Herbert Jerome
Kremenak, Charles Robert, Jr
Laskin, Daniel M
Schow, Carl Emil, Jr
Small, Ernest William

Neurosurgery

Adams, John Edwin
Alexander, Eben, Jr
Allen, Marshall B, Jr
Andy, Orlando Joseph
Arnold, Arthur
Bakay, Louis
Balagura, Saul
Battista, Arthur Francis
Black, Perry
Boop, Warren Clark, Jr
Botterell, Edmund Harry
Browder, Eli Jefferson
Bucy, Paul Clancy
Campbell, Robert Louis
Carrea, Raul
Carton, Charles Allan
Chou, Shelley Nien-Chun
Clark, William Kemp
Collins, William Francis, Jr
Cook, Albert William
Cooper, Irving S
Davis, Richard A
DeSaussure, Richard Laurens, Jr
Donaghy, Raymond Madiford Peardon
Dugger, Gordon Shelton
Dunbar, Howard Stanford
Ehni, George (John)
Elvidge, Arthur Roland
Erickson, Theodore Charles
Feindel, William Howard
Ferguson, Gary Gilbert
Fisher, Robert George
Foltz, Eldon Leroy
French, John Douglas
French, Lyle Albert

Furlow, Leonard Thompson
Galbraith, James Garber
Gallo, Anthony Edward, Jr
Garcia-Bengochea, Francisco
Gildenberg, Philip Leon
Goldring, Sidney
Grossman, Robert G
Hamlin, Hannibal
Heifetz, Milton David
Heimburger, Robert Francis
Hekmatpanah, Javad
Hetherington, R F
Housepian, Edgar M
Huertas, Jorge
Hunt, William Edward
Jakoby, Ruth Elizabeth Kerr
Jane, John Anthony
Javid, Manucher J
Kalsbeck, John Edward
Kaplan, Harry Arthur
Kelly, William Albert
Kerr, Frederick William Lawson
Kindt, Glenn W
Kirsch, Wolff M
Kline, David G
Kurze, Theodore
Langfitt, Thomas William
Llewellyn, Raeburn Carson
Long, Donlin Martin
Lourie, Herbert
MacCarty, Collin Stewart
Mais, Leonard I
Manganiello, Louis O J
McClure, Claude
McLaurin, Robert L
Mealey, John, Jr
Meyer, Glenn Arthur
Moody, Robert Adams
Morley, Thomas Paterson
Mount, Lester Adran
Mullan, John F
Murtagh, Frederick, Jr
Myers, Paul Walter
Nashold, Blaine S
Nelson, Stanley Reid
Nugent, George Robert
O'Connor, Michael John
Odom, Guy Leary
Otenasek, Frank (Joseph)
Owens, Guy
Paine, Kenneth William
Patterson, Russel Hugo, Jr
Perot, Phanor L, Jr
Pollay, Michael
Poppen, James L
Porter, Robert Willis
Pudenz, Robert Harry
Puletti, Flavio'
Raimondi, Anthony John
Ransohoff, Joseph
Rayport, Mark
Rhoton, Albert Loren, Jr
Richardson, Donald Edward
Roberts, Theodore S
Robertson, James Thomas
Rovit, Richard Lee
Ruge, Daniel
Russell, John Robert
Salmon, James Henry
Schemm, George Walker
Schlesinger, Edward Bruce
Schneider, Richard Coy
Schwartz, Henry Gerard
Scott, Michael
Scoville, William Beecher
Shenkin, Henry A
Smith, Kenneth Rupert, Jr
Stein, Bennett M
Story, Jim Lewis
Stratford, Joseph
Sweet, William Herbert
Sypert, George Walter
Taren, James A
Tasker, Ronald Reginald
Tator, Charles Haskell
Thomas, Llywellyn Murray
Thompson, Raymond K
Tindall, George Taylor
Udvarhelyi, George Bela
Voris, Harold Cornelius
Walker, Arthur William, Jr
Ward, Arthur Allen, Jr
Weissman, William Kent
Westrum, Lesnick Edward
Wetzel, Nicholas
White, Lowell Elmond, Jr
Wise, Burton Louis
Wolfson, Sidney Kenneth, Jr
Woodhall, Barnes
Yamamoto, Y Lucas
Yashon, David
Youmans, Julian Ray

Ophthalmological Surgery
Anderson, Richard Lee
Doughman, Donald James

Oral Surgery
Alling, Charles Calvin III
Archer, William Harry
Bell, William Harrison
Bentley, Kenneth Chessar
Bloom, Herbert Jerome
Boyer, Harold Edwin
Boyne, Philip John
Calhoun, Noah Robert
Chaikin, Lawrence
Choukas, Nicholas C
Cooksey, Donald Ernest
Costich, Emmett Rand
Doku, Hristo Chris
Epker, Bruce Nelson
Fencl, Robert (Daniel)
Funk, Edward C
Gehrig, John D
Giaroli, John Nello
Gores, Robert James
Gregg, John Marshall
Hall, Hugh David
Hall, John Frank
Harris, Melvyn H
Hayward, James Rogers
Heubsch, Raymond Frank
Hinds, Edward C
Jastak, J Theodore
Kaplan, Herman
Korchin, Leo
Laskin, Daniel M
Lovestedt, Stanley Almer
Lynch, Benjamin Leo
Marble, Howard Bennett, Jr
McCallum, Charles Alexander, Jr
McLeran, James Herbert
Mealey, John, Jr
Meyer, Irving
Morris, Estell E
Murnane, Thomas William
Parnell, Anthony George
Peterson, Larry James
Ping, Ronald Stanley
Reed, Homer Vernon
Robinson, Marsh Edward
Russell, Orville Eugene
Schow, Carl Emil, Jr
Small, Ernest William
Spatz, Sidney S
Staples, Albert Franklin
Stern, Martin
Topazian, Richard G
Troiano, Martin Frank
Wade, George Wesley
Waite, Daniel Elmer
Waldrep, Alfred Carson, Jr
Webster, William Wallace
Welborn, Joseph F
White, Raymond Petrie, Jr
Wussow, George C

Orthopedic Surgery
Adams, John Pletch
Akeson, Wayne Henry
Anderson, Carl Edgar
Anderson, Lewis Daniel
Baker, Lenox Dial
Banks, Henry H
Bassett, Charles Andrew Loockerman
Beals, Rodney K
Becker, Robert O
Beller, Martin Leonard
Blair, John Dennis
Bliven, Floyd E, Jr
Bonfiglio, Michael
Bowker, John
Bradford, David S
Brighton, Carl T
Brower, Thomas Dudley
Brown, Paul Woodrow
Bunch, Wilton Herbert
Clawson, David Kay
Clippinger, Frank Warren, Jr
Compere, Clinton Lee
Convery, F Richard
Cooper, Reginald Rudyard
Coventry, Mark Bingham
Dooley, Wallace T
Elmore, Stanley Mcdowell
Fahey, John James
Ferguson, Albert Barnett
Finder, Jerome Gordon
Fox, Theodore Albert
Frankel, Victor H
Fry, Louis Rummel
Garber, J Neill
Glimcher, Melvin Jacob
Goldner, Joseph Leonard
Griffin, Paul Putnam
Hamsa, William Rudolph
Hark, Fred William
Harkess, James W
Harrington, Paul Randall
Hartman, James T
Harvey, J Paul, Jr
Hayes, John Terrence
Hejna, William Frank
Hendryson, Irvin Edward

Herndon, Charles Harbison
Hoaglund, Franklin Theodore
Ingram, Alvin John
Inman, Verne Thomson
Jacobs, Richard L
Jewett, Don L
Johnson, Einer Wesley, Jr
Kane, William J
Katz, Jacob Feuer
Kelly, Patrick Joseph
Kettelkamp, Donald B
Kuhlman, Robert Eugene
Laing, Patrick Gowans
Laros, Gerald Snyder, II
Larson, Carroll Bernard
Lavine, Leroy S
Leach, Robert Ellis
Lipscomb, Paul Rogers
Lucas, Donald Brooks
Luck, James Vernon
McCauley, John Corran, Jr
McCollum, Donald E
McDonnell, Edmond Joseph
McKay, Douglas William
Meyers, Marvin Harold
Michele, Arthur A
Miller, Wallace E
Mindell, Eugene R
Moe, John
Morris, Harry Dunlap
Murray, William R
Nickel, Vernon L
Omer, George Elbert, Jr
Pankovich, Arsen M
Patterson, Frank Porter
Peltier, Leonard Francis
Perry, Jacquelin
Raney, Richard Beverly
Ray, Robert Durant
Reynolds, Fred C
Rhinelander, Frederic William
Riley, Lee Hunter, Jr
Robinson, Robert Alexander
Rothman, Richard Harrison
Salter, Robert Bruce
Scherr, David DeLano
Schmeisser, Gerhard, Jr
Schmidt, Albert Charles
Scuderi, Carlos S
Shands, Alfred Rives. Jr.
Siffert, Robert S
Singh, Iqbal
Smith, William S
Snell, William E
Soren, Arnold
Southwick, Wayne Orin
Steel, Howard Haldeman
Stein, Arthur Henry, Jr
Stein, Irvin
Stinchfield, Frank E
Stradford, H Todd
Street, Dana Morris
Tucker, Frederick Robert
White, Augustus Aaron, III
Wilson, Frank Crane
Wirka, Herman W
Yablon, Isadore Gerald
Yelton, Chestley Lee

Smith, Leroy
Snyder, Clifford Charles
Spira, Melvin
Stark, Richard B
Zarem, Harvey A

Reconstructive Surgery
Bevin, A Griswold
Brand, Paul W
Brizio-Molteni, Loredana
Chase, Robert A
Entin, Martin A
Furnas, David William
Griffith, B Herold
Krizek, Thomas Joseph

Surgical Pathology
Carr, Malcolm Wallace
Chun, Byungkyu
Fahmy, Aly
Hartmann, William Herman
Kay, Saul
Silverberg, Steven George
Taylor, Herbert Bradley
Terry, Roger
Wolff, Marianne

Thoracic Surgery
Absolon, Karel B
Andrews, Neil Corbly
Arbulu, Agustin
Blair, Emil
Boland, James P
Bougas, James Andrew
Brown, Ivan Willard, Jr
Bryant, Lester Richard
Carroll, Samuel Edwin
Castaneda, Aldo Ricardo
Clagett, Oscar Theron
Codd, John Edward
Coffin, Laurence Haines
Connar, Richard Grigsby
Curreri, Anthony Rudolph
Davila, Julio C
DeWall, Richard A
Edmonds, Louis Henry, Jr
Ellis, Franklin Henry, Jr
Ellison, Robert G
Fineberg, Charles
Frater, Robert William Mayo
Fuson, Robert L
Greenfield, Lazar John
Grimes, Orville Frank
Helmsworth, James Alexander
Hewitt, Robert Lee
Hunter, Samuel W
Jacobson, Myron J
Kahn, Donald R
Kaiser, George C
Kaiser, Gerard Alan
Kennedy, John Hines
Kilman, James William
King, Thomas Creighton
Kokame, Glenn Megumi
Laks, Hillel
Langston, Hiram Thomas
Lee, William Hall, Jr
Lees, William Morris
Long, David Michael
Lower, Richard Rowland
MacKenzie, James W
Madden, Robert E
Magovern, George Jerome
Mamiya, Richard T
Nealon, Thomas F, Jr
Padula, Richard Thomas
Paulson, Donald Lowell
Pennell, Timothy Clinard
Replogle, Robert Lee
Roe, Benson Bertheau
Rohman, Michael
Rosenberg, Dennis Melville Leo
Rosensweig, Jacob
Sawyer, Philip Nicholas
Sayegh, Salem F
Schwartz, Seymour I
Scott, Stewart Melvin
Sealy, Will Camp
Sellers, Robert Douglas
Seybold, William Dempsey
Shields, Thomas William
Silver, Donald
Sloan, Herbert
Starr, Albert
Stemmer, Edward Alan
Stephenson, Hugh Edward, Jr
Sugg, Winfred Lindley
Symbas, Panagiotis N
Thomas, Arthur Norman
Tice, David Anthony
Trummer, Max Joseph
Vasko, John Stephen
Wallace, Robert Bruce
Warden, Herbert Edgar
Wareham, Ellsworth Edwin
Watkins, David Hyder
Weil, Peter H
Wheat, Myron William, Jr

Pediatric Surgery
Boley, Scott Jason
Campbell, John Richard
Collin, Pierre-Paul
Hays, Daniel Mauger
Holder, Thomas M
Izant, Robert James, Jr
Martin, Lester W
O'Neill, James A, Jr
Othersen, Henry Biemann, Jr
Talbert, James Lewis

Plastic Surgery
Ashley, Franklin Longley
Bevin, A Griswold
Brody, Garry Sidney
Converse, John Marquis
Dingman, Reed (Otherbert)
Edgerton, Milton Thomas, Jr
Entin, Martin A
Erich, John B
Frackelton, William Hamilton
Furnas, David William
Georgiade, Nicholas George
Gingrass, Ruedi Peter
Goulian, Dicran
Griffith, B Herold
Harvin, James Shand
Hendrix, James Harvey, Jr
Hoopes, John Eugene
Hugo, Norman Eliot
Krizek, Thomas Joseph
Larson, Duane L
Lewis, Stephen Robert
Lynch, John Brown
Masters, Frank Wynne
McCormack, Robert Morris
McKinney, Peter
Randall, Peter

SURGERY

Wilcox, Benson Reid
Williams, Marion Jack
Wilson, Robert Francis
Winterscheid, Loren Covart
Wolcott, Mark Walton
Young, William Glenn, Jr
Zeppa, Robert

Veterinary Surgery

Brinker, Wade Oberlin
Brodey, Robert S
Clifford, Donald H
Fackelman, Gustave Edward
Frye, Fredric Lee
Gabel, Albert A
Grier, Ronald Lee
Hankes, Gerald H
Hardenbrook, Harry, Jr
Hohn, Ronald Bruce
Horne, Robert D
Hymas, Theo Alfred
Jones, Eric Wynn
Keller, Waldo Frank
Kjar, Harold Anthony
Knecht, Charles Daniel
Leighton, Robert Lyman
Love, John Edward
Parkash Ved
Pearson, Phillip T
Raker, Charles W
Rawlings, Clarence Alvin
Redman, Donald Roger
Rudy, Richard L
Wass, Wallace M
Watts, Raymond Ellsworth
Weirich, Walter Edward
Wilson, George Porter, III

SYSTEMATICS

Systematics

Adams, Robert Philip
Anderson, Steven Clement
Ball, George Eugene
Barton, John Robert
Baum, Bernard R
Belkin, John Nicholas
Birney, Elmer Clea
Bond, James
Brooks, John Langdon
Brown, Clair Alan
Bugbee, Robert Earl
Cecilia, Mary
Cowan, Garry Ian McTaggart
Crovello, Theodore John
Dasch, Clement Eugene
Duncan, Wilbur Howard
Eisenman, Richard L
Emery, William Henry Perry
Garay, Leslie Andrew
Gordh, Gordon
Gould, Sydney Ward
Graham, Shirley Ann
Grear, John Wesley, Jr
Grissell, Edward Eric Fowler
Gunn, Charles Robert
Hall, Marion Trufant
Hart, Dabney Gardner
Hoffmann, Robert Shaw
Hogue, Charles Leonard
Holt, Perry Cecil
Horton, James Heathman
Houk, Richard Duncan
Irwin, Howard Samuel, Jr
Johnston, Richard Fourness
Korf, Richard Paul
Lampton, Robert Koerbel
Larisey, Mary Maxine
Levi, Herbert Walter
Lichtenfels, James Ralph
Lint, Harold L
Magill, Robert Earle
Maysilles, James Howard
Mears, James Austin
Mikula, Bernard C
Mockford, Edward Lee
Montgomery, James Douglas
Mosquin, Theodore
Moss, W Wayne
Newell, Irwin Mayer
Nicolson, Dan Henry
Norman, Eliane Meyer
Nowicke, Joan Weiland
Olexia, Paul Dale
Parker, Kittie Fenley
Peterson, Paul Constant
Phipps, James Bird
Pickford, Grace Evelyn
Pinto, John Darwin
Procaccini, Donald J
Rogers, Claude Marvin
Rowell, Chester Morrison, Jr
Sabrosky, Curtis Williams
Schlinger, Evert Irving
Schmidly, David James
Schuyler, Alfred E
Shechter, Yaakov
Shih, Chang-Tai
Slater, James Alexander
Smith, David Rollins
Snyder, Harold
Sokal, Robert Reuven
Sorensen, Paul Davidsen
Springer, Stewart
Stahnke, Herbert Ludwig
Strandtmann, Russell William
Thompson, Jesse Clay, Jr
Thorne, Robert Folger
Tuff, Donald Wray
Van Haverbeke, David F
Vickery, Vernon Randolph
Vladykov, Vadim Dmitri
Williams, Michael Ledell
Yonke, Thomas Richard
Young, William Donald, Jr

Cytotaxonomy

Dietz, Robert Austin
Love, Askell
Love, Doris
Moore, Raymond John

Systematic Botany

Alex, Jack Franklin
Altschul, Siri von Reis
Anderson, Dennis Elmo
Argus, George William
Averett, John E
Baad, Michael Francis
Baranski, Michael Joseph
Barker, William T
Barkley, Arthur S
Barkley, Theodore Mitchell
Bates, David Martin
Beaman, John Homer
Beard, Luther Stanford
Benson, Lyman David
Blasdell, Robert Ferris
Bobear, Jean B
Bowers, Frank Dana
Bradley, Ted Ray
Brashier, Clyde Kenneth
Broome, Carmen Rose
Brown, Richard McPike
Bucheim, Arno Fritz Gunther
Bunting, George Sydney, Jr
Burk, Carl John
Burk, Derek George
Cabrera, Angel Lulio
Carpenter, Irvin Watson, Jr
Carr, Gerald Dwayne
Castaner, David
Channell, Robert Bennie
Chapman, Carl Joseph
Chuang, Tsan Iang
Clausen, Robert Theodore
Clewell, Andre F
Clovis, Jesse Franklin
Coffey, Janice Carlton
Cooperrider, Tom Smith
Corbett, Gail Rushford
Cowan, Richard Sumner
Crawford, Daniel John
Cronquist, Arthur John
Crow, Garrett Eugene
Cruise, James E
Cutter, Lois Jotter
Davidse, Gerrit
Davidson, John Fraser
DeFilipps, Robert Anthony
DeJong, Diederik Cornelis Dignus
Dempster, Lauramay Tinsley
DeWolf, Gordon Parker, Jr
Drapalik, Donald Joseph
Dress, William John
Dudley, Theodore
Dugle, Janet Mary Rogge
Dunman, Maximilian George
Durkee, LaVerne H
Dwyer, John Duncan
Ediger, Robert I
Ehrle, Elwood Bernhard
Eilers, Lawrence John
Eiser, Arthur L
Elias, Thomas S
Ellis, William Haynes
Emboden, William Allen, Jr
Eshbaugh, William Hardy
Essig, Frederick Burt
Estes, James Russell
Ezell, Wayland Lee
Faircloth, Wayne Reynolds
Fay, Marcus J
Freckmann, Robert W
Freeman, John Daniel
Freeman, Myron L
Fryxell, Paul Arnold
Furlow, John Jacob
Gastony, Gerald Joseph
Gentry, Alwyn Howard
Gillett, John Montague
Glassman, Sidney Frederick
Goldberg, Aaron
Goodman, George Jones
Gordon, Donald
Grable, Albert E
Hale, Mason Ellsworth, Jr
Harms, Vernon Lee
Harriman, Neil Arthur
Harrington, Harold David
Harris, Betty Wolf
Hauke, Richard Louis
Hayden, Mary Victoria
Haynes, Robert Ralph
Hehre, Edward James, Jr
Henrickson, James Solberg
Higgins, Larry Charles
Holmgren, Arthur Herman
Holmgren, Noel Herman
Holmgren, Patricia Kern
Holte, Karl E
Hsi, Eugene Yu-Tseng
Hutchison, Paul Clifford
Huttleston, Donald Grunert
Iltis, Hugh Hellmut
Ingram, John (William, Jr)
James, Charles William
James, Lois Elsie
Jensen, Richard Jorg
Johnson, Miles F
Johnson, Raymond Roy
Johnston, Marshall Conring
Jones, Almut Gitter
Keating, Richard Clark
Kiger, Robert William
Kirkbride, Joseph Harold, Jr
Klein, William McKinley, Jr
Koch, Stephen Douglas
Koelling, Alfred Cornell
Kowal, Robert Raymond
Koyama, Tetsuo
Kral, Robert
Kyhos, Donald William
Lang, Frank Alexander
Langdon, Kenneth R
Lathrop, Earl Wesley
Lawrence, George Hill Mathewson
Ledingham, George Filson
Legault, Albert
Lellinger, David Bruce
Lelong, Michel George
Lilly, Percy Lane
Lindsay, Delbert W
Lindsay, George Edmund
Lloyd, Robert Michael
Logan, Lowell Alvin
Lonard, Robert (Irvin)
Long, Robert William, Jr
Longpre, Edwin Keith
Luteyn, James Leonard
Maguire, Bassett
Mahler, William Fred
Marroquin De La Fuente, Jorge Saul
Martin, William Clarence
Matthews, James Francis
Matuda, Eizi
McClintock, Elizabeth
Meijer, Willem
Miller, Gertrude Nevada
Miller, Kim Irving
Milstead, Wayne Lavine
Mitchell, Richard Sheppard
Mohlenbrock, Robert H, Jr
Monson, Paul Herman
Moran, Reid Venable
Morley, Thomas
Murrell, James Thomas, Jr
Neher, Robert Trostle
Nevling, Lorin Ives, Jr
Ownbey, Gerald Bruce
Packard, Patricia Lois
Parks, James C
Payne, Willard William
Pfeifer, Howard William
Pickering, Jerry L
Pippen, Richard Wayne
Porter, Duncan Macnair
Preece, Sherman Joy, Jr
Ramsey, Gwynn W
Read, Robert William
Reveal, James L
Richards, Charles Davis
Richardson, Annie Louise
Roane, Martha Kotila
Roe, Keith Edward
Roland, Albert Edward
Ronninger, James McDonald
Rouleau, Ernest
Sawyer, Paul Thompson
Schaeffer, Robert L, Jr
Schubert, Bernice Giduz
Schwab, Charlotte Ann
Setchemyer, Kenneth Theodore
Shetler, Stanwyn Gerald
Sieren, David Joseph
Skog, Laurence Edgar
Smith, Claude Earle, Jr
Smith, Dale Metz
Smith, Edwin Burnell
Smith, James Payne, Jr
Smith, Stanley Galen
Sohmer, Seymour H
Spellenberg, Richard (William)
Sperry, John Jerome
Spongberg, Stephen Alan
Stanford, Jack Wayne
Stern, Kingsley Rowland
Stern, Ralph Randles
Steyermark, Julian Alfred
Stocking, Kenneth Morgan
Stolze, Robert Gardner
Stone, Benjamin Clemens, III
Stone, Margaret Hodgman
Stoutamire, Warren Petrie
Stuessy, Tod Falor
Sutherland, David M
Taylor, Charles Arthur, Jr
Taylor, Constance Elaine Southern
Taylor, Raymond John
Taylor, Roy Lewis
Terrell, Edward Everett
Theobald, William L
Thomas, John Hunter
Thomas, Roy Dale
Turner, Billie Lee
Tyrl, Ronald Jay
Urbatsch, Lowell Edward
Van Faasen, Paul
Van Horn, Gene Stanley
Voss, Edward Groesbeck
Ward, Daniel Bertram
Ward, George Henry
Ware, Donna Marie Eggers
Wasshausen, Dieter Carl
Watson, James Ray, Jr
Weber, Wallace Rudolph
Webster, Grady Linder
Wedberg, Hale Levering
Weiler, John Henry, Jr
Welch, Stanley L
Wells, James Ray
Wheeler, Louis Cutter
Wilbur, Robert Lynch
Wilken, Dieter H
Williams, Kenneth Bock
Williams, Louis Otho
Windler, Donald Richard
Yates, Willard F, Jr

Biosystematics

Anderson, Gregory Joseph
Beaudry, Jean Romuald
Bishop, Yvonne M
Bratton, Gerald Roy
Cicchinelli, Alexander L
Futcher, Anthony Graham
Ganders, Fred Russell
Guthrie, Roland L
Haig, Janet
Henderson, Douglass Miles
Huckabay, John Porter
Levy, Morris
Moore, Felix E
Morr, Charles Vernon
Simpson, Beryl Brintnall
Small, Ernest
Sullivan, Victoria I
Vogt, George Britton
Wiens, Delbert

Systematic Entomology

Abbasi, Qamar Ali
Allen, Robert Thomas
Anderson, Donald Morgan
Arnell, J Hal
Ashlock, Peter Dunning
Baumann, Richard William
Benoit, Paul
Blocker, Henry Derrick
Brown, Calvin Hugh
Cazier, Mont Adelbert
Christian, Paul Jackson
Connell, Walter Anthony
Cross, Earle Albright, Jr
Curry, La Verne Leon
Daigleish, Robert Campbell
Duckworth, Walter Donald
Ferguson, Douglas Campbell
Flint, Oliver Simeon, Jr
Freytag, Paul Harold
Gagne, Raymond J
Gill, Gordon Drew
Gorwall, William Harrison, Jr
Gregg, Robert Edmond
Hambleton, Edson Jorge
Hobbs, Kenneth Rollin
Hodges, Ronald William
James, Maurice Theodore
Johnson, Clarence Daniel
Kavanaugh, David Henry
Kim, Ke Chung
Kramer, James Phillip
La Rivers, Ira John
Lindquist, Evert E
Mason, William Richardson Miles
Miller, David Clair
Peterson, Lance George
Priddy, Ralph Banta
Rentz, David Charles
Rozen, Jerome George, Jr

Schuh, Randall Tobias
Shaffer, Jay Charles
Smith, Eric Howard
Stage, Gerald Irving
Thorp, Robbin Walker
Valley, Karl Roy
Wasbauer, Marius Sheridan
White, Richard Earl
Young, David Allan

Systematic Ichthyology
Dooley, James Keith
Douglas, Neil Harrison
Fitch, John Edgar
Haedrich, Richard L
Lachner, Ernest Albert
McAllister, Donald Evan
McKenzie, Joseph Addison
Miller, Robert Victor
Morrow, James Edwin, Jr
Yerger, Ralph William

Systematic Zoology
Bell, Ross Taylor
Boyden, Alan Arthur
Brady, Allen Roy
Edwards, J Gordon
Fauchald, Kristian
Friar, Wayne
Gloyd, Howard Kay
Herrmann, Scott Joseph
Hobbs, Horton Holcombe, Jr
Hoff, Clarence Clayton
Hoffman, Richard Lawrence
Hope, William Duane
Leviton, Alan Edward
McGhee, Charles Robert
Rising, James David
Schwartz, Albert
Sibley, Charles Gald
Williams, Austin Beatty

TEXTILES

Textiles
Broome, Esther Roberts
Causa, Alfredo G
Cleary, Laurence Twomey
Fisher, Calvin L
Galbraith, Ruth Legg
Kennedy, Stephen Jay
Laughlin, Kenneth Clifford
Lund, Lillian O
Martin, William Harry
Mikell, William Gaillard
Miller, Bernard
Nelson, Elton Glen
Newell, William Andrews
Petzel, Florence E
Powers, Edward James
Roberson, Elbert B, Jr
St John, Wayne Lloyd
Schuler, Mathias John
Shealy, Otis Lester
Slater, Keith
Spivak, Steven Mark
Sprague, Basil Sheldon
Steadman, Robert George
Thornton, Daniel McCarty
Tucker, Paul Arthur
Turner, George Robert
Vail, Sidney Lee

Textile Chemistry
Arnold, Luther Bishop, Jr
Aspland, John Richard
Aycock, Benjamin Franklin
Baitinger, William F, Jr
Bannerman, Douglas George
Bannister, Robert Grimshaw
Barish, Leo
Barker, Robert Henry
Bassett, Alton Herman
Baum, Bruton Murry
Baxter, James F
Bercaw, James Robert
Birkinhauer, Robert Joseph
Bixler, Dean A
Browne, Colin Lanfear
Calamari, Timothy A, Jr
Caroselli, Remus Francis
Cates, David Marshall
Chamberlin, Howard Allen
Chase, Vernon Lindsay
Compton, Jack
Cooke, Theodore Frederic
Cooper, Margaret Moore
Cramer, John Joseph
Capilla, Joseph
Davis, Richard Cecil
Donahue, Joseph E
Drelich, Arthur (Herbert)
Dupre, Edmund J
Ebert, Philip E
Elliott, John
Euler, Robert Donald

Fearing, Ralph Burton
Foght, James Loren
Fowler, John Rayford
Franklin, William Elwood
Freeman, Richard Carl
Frishman, Daniel
Galil, Fahmy
Goldstein, Herman Bernard
Goodson, Louie Aubrey, Jr
Greer, James Edward
Griffith, Michael Grey
Guenther, Harry Wilbert
Guion, Thomas Hyman
Hager, Glenn Frederick
Hall, Seymour Gerald
Handy, Carleton Thomas
Hanzel, Robert Stephen
Hershkowitz, Robert L
Horning, Roderick Henry
Hughes, William
Hume, Harold Frederick
Jacoby, Thomas Franklin
Joseph, Marjory L
Katz, Manfred
King, Joseph Clarence
Kissa, Erik
Koenig, Harvey Steven
Koenig, Nathan Hart
LaFleur, Kermit Stillman
Landau, Edward Frederick
Louis, Kwok Toy
Lundgren, Harold Palmer
Machell, Greville
Machlis, Samuel
Magee, John Robert
Mandel, Zoltan
Mirhej, Michael Edward
Mizell, Louis Richard
Morbey, Graham Kenneth
Myers, Clovis D
Needles, Howard Lee
Nuessle, Albert Christian
Olson, Arthur Russell
Ostmann, Bernard George
Panto, Joseph Salvatore
Pemrick, Raymond Edward
Pfeiffer, Gerald Peter
Pizzarello, Roy Aloysius
Plamondon, Joseph Edward
Pretka, John E
Read, Robert E
Reeves, Wilson Alvin
Reid, John David
Reimer, Carl Clayton
Rogers, James Wesley
Saltzman, Max
Sands, Seymour
Sargeant, Peter Barry
Scheve, Bernard Joseph
Scott, Peter John
Sello, Stephen Balthazar
Shoaf, Charles Jefferson
Smith, Betty F
Staples, Milfred Lawson
Stroud, Robert Wayne
Swanson, John Melvin
Tucker, Charles R
Wakelyn, Phillip Jeffrey
Walsh, William K
Walters, Philip Marion
Wasley, William Lingel
Wayland, Rosser Lee, Jr
Webb, Myron Quentin
Weeks, Gregory Paul
Whaley, Wilson Monroe
Wham, George Sims
Williams, Ebenezer David, Jr
Williams, Michael John
Williams, Richard Anderson
Worsham, Walter Castine
Yeh, Kwan-Nan
Zeronian, Sarkis Haig

Textile Physics
Abbott, Norman John
Barnes, James Crowell
Bernier, Edward Joseph
Buchanan, David Royal
Cushing, James Thomas
Dennison, Richard Wheeler
Fornes, Raymond Earl
Gupta, Bhupender Singh
Hardy, Henry Benjamin, Jr
Kobetich, Edward John
Kyame, George John
Lehmicke, David John
Lothrop, Everett Winfred, Jr
Pontrelli, Gene J
Seaman, Richard
Shaffner, Thomas Jackson
Stanley, Harry Eugene
Van Veld, Robert Dale
Weiss, Louis Charles
Williams, Kenneth Roger

Toxicology
Agersborg, Helmer Pareli Kjerschow, Jr
Ahrens, Franklin Alfred
Ainsworth, Earl John
Aldrich, Franklin Dalton
Amdur, Mary Ochsenhirt
Anders, Marion Walter
Anderson, Robert Clarke
Aronson, A L
Arthur, B Wayne
Backer, Ronald Charles
Bagdon, Walter Joseph
Balazs, Tibor
Baldwin, Robert Charles
Banerjee, Bhola Nath
Bastos, Milton Lessa
Becker, Bernard Abraham
Beliles, Robert Pryor
Benson, Wilbur Maxwell
Berg, George G
Berndt, William O
Bishop, Jack Belmont
Bitter, Harold Louis
Blanke, Robert Vernon
Blejer, Hector P
Blend, Michael J
Blumenthal, Herbert
Bonderman, Dean P
Borgmann, August Russell
Borgstedt, Harold Heinrich
Borzelleca, Joseph Francis
Bost, Robert Orion
Bowman, Faye Johnson
Boyd, Carl Edmund
Bradley, William Robinson
Braude, Monique Colsenet
Brown, Daniel Joseph
Brown, John R
Brown, Richard Don
Buehler, Edwin Vernon
Burnett, Clyde Marshall
Buttar, Harpal Singh
Byard, James Leonard
Caplan, Yale Howard
Caplis, Michael E
Carlson, Gary P
Carpenter, Charles Patten
Carrano, Richard Alfred
Casida, John Edward
Chakrin, Alan Leonard
Chambers, Howard Wayne
Chau, Raymond Ying Pui
Chen, Shiu-Chin
Chhabra, Rajendra S
Christensen, Herbert Edward
Christopoulos, George Nick
Clarkson, Thomas William
Clayton, John Wesley, Jr
Clegg, David John
Cohen, Steven Donald
Colclough, Norma Vesey
Coldwell, Blake Burgess
Conley, Bernard Edward
Cornish, Herbert Harry
Cotty, Val Francis
Cramer, Charles Russell
Crane, Charles Russell
Cueto, Cipriano, Jr
Dacre, Jack Craven
Dajani, Esam ZagerZafer
Davidow, Bernard
Davis, Joseph Richard
Decker, Walter Johns
Deichmann, William Bernhard
De La Iglesia, Felix Alberto
Diener, Robert Max
Dilley, James V
Dillingham, Elwood Oliver
Dodson, Vernon N
Doedens, David James
Dost, Frank Norman
Drobeck, Hans Peter
Dubowski, Kurt Max
Eagle, Edward
Earl, Alfred Ellsworth
Eckardt, Robert E
Edwards, Gordon Stuart
Ellison, Christian
Ellenberger, Herman Albert
Ellison, Theodore
Elsea, John Robert
Emmerson, John Lynn
Evenson, Merle Armin
Fabacher, David Lawrence
Fabry, Andras
Faiman, Morris David
Fassett, David Walter
Findlay, Glen Marshall
Fitzloff, John Frederick
Flannagan, John Fullan
Foreman, Ronald Louis
Forney, Robert Burns
Fouts, James Ralph

Fowler, Bruce Andrew
Frawley, John Paul
Frederick, George Leonard
Friedman, Marvin Alan
Furst, Arthur
Gabbert, Paul George
Gabriel, Karl Leonard
Gallo, Michael Anthony
Gandolfi, Allen Jay
Gargus, James L
Garriott, James Clark
Garvin, Paul Joseph, Jr
Gatzy, John T, Jr
Gehring, Perry James
Georghiou, George Paul
Gibson, James Edwin
Gibson, John Phillips
Gilman, Martin Robert
Glasser, Ralph Frederick
Golberg, Leon
Goodman, Samuel
Gordon, Samuel
Grant, Donald Lloyd
Green, Sidney
Greenstein, Edward Theodore
Greichus, Yvonne A
Grice, Harold C
Groblewski, Gerald Eugene
Gross, Stanley Burton
Guzman, Ruben Joseph
Hall, Richard Leland
Hallesy, Duane Wesley
Hamelink, Jerry L.
Hammond, Paul B
Hansen, David John
Hansen, Larry George
Harbison, Raymond D
Hardy, Lester B, Jr
Hartung, Rolf
Hatch, Roger Conant
Hatfield, Garry Kent
Hathcock, John Nathan
Hayes, Andrew Wallace
Hayes, Wayland Jackson, Jr
Henderson, Richard
Hietbrink, Bernard E
Hiles, Richard Allen
Hill, Jim Tom
Hille, Kenneth R
Hillman, Elizabeth S
Hite, Mark
Hodge, Harold Carpenter
Hodgson, Ernest
Hoffman, Donald Bertrand
Hollingworth, Robert Michael
Homan, Elton Richard
Hood, Ronald David
Hottendorf, Girard Harold
Hume, Arthur Scott
Issenberg, Phillip
Jackson, Benjamin
Jacobson, Keith Hazen
Jain, Naresh C
Jasper, Robert Lawrence
Johnson, Howard Ernest
Johnson, Melvin Clark
Kadoum, Ahmed Mohamed
Kaminski, Edward Jozef
Kanegis, Leon Abbott
Kapur, Bhushan M
Kay, Kingsley
Kaye, Sidney
Keller, John George
Kelly, Raymond Crain
Kenaga, Eugene Ellis
Keplinger, Moreno Lavon
Kera, Kathel Bedortha
Khera, Kundan Singh
Kier, Lawrence Charles
Kilgore, Wendell Warren
King, Theodore Oscar
Kinoshita, Florence Keiko
Kivela, Edgar Welton
Klaassen, Curtis Dean
Knaak, James Bruce
Kodama, Jiro Kenneth
Koefferl, Michael Tallyn
Kraemer, Richard John
Kraybill, Herman Fink
Krivanek, Neil Douglas
Krop, Stephen
Laham, Souheil
Lahanna, Carl
Lamar, Jule K
Larson, Robert Elof
Lauer, Dolor John
Lawrence, William Homer
Laws, Edward Raymond, Jr
Leach, Eddie Dillon
Leach, Leonard Joseph
Leary, John Sylvester
Lee, Cheng-Chun
Leong, Basil K J
Levinskas, George Joseph
Levenstein, Irving
Levy, Alan C
Levy, Susanna Agnes
Littlefield, Neil Adair

TOXICOLOGY

Liu, David H W
Long, James Earl
Loomis, Ted Albert
Luckens, Mark Manfred
Lyman, Frank Lewis
MacEwen, James Douglas
MacDonald, William E, Jr
MacFarland, Harold Noble
MacKellar, Donald Gordon
Mailman, Richard Bernard
Mancini, Robert Edward
Manell, William Arnold
Manno, Barbara Reynolds
Manno, Joseph Eugene
Marzulli, Francis Nicholas
Massaro, Edward Joseph
Mattis, Paul Alvin
Mayer, Foster Lee, Jr
Mayer, Richard Thomas
McBay, Arthur John
McNerney, James Murtha
Mehring, Jeffrey Scott
Menn, Julius Joel
Middleton, Edward James
Milby, Thomas Hutchinson
Moorefield, Herbert Hughes
Moreland, Ferrin Bates
Morrison, Alexander Baillie
Morrison, Frank Orville
Morrow, Paul Edward
Munson, Sam Clark
Murphy, Sheldon Douglas
Musselman, Nelson Page
Neal, Robert A
Nelson, Dallas Leroy
Nelson, Gordon Wilfred
Newell, Granville Abraham
Nixon, Joseph Eugene
Nolen, Granville Abraham
Norris, Jessie McGowan
Norton, Stata Elaine
Norvell, Michael Jimmy
Nunez, Loys Joseph
O'Brian, Dennis Martin
O'Leary, Robert Kent
Olson, Kenneth Jean
Olson, William Arthur
Oser, Bernard Levussove
Osterberg, Robert Edward
Ottoboni, Minna Alice
Page, John Gardner
Painter, Ruth Coburn Robbins
Palm, Paul Eugene
Pan, Shih Y
Pape, Brian Eugene
Parent, Richard Alfred
Patel, Narayan Ganesh
Paynter, Orville Eugene
Phillips, Barrie Maurice
Pieper, Gustav Rene
Pittman, Kenneth Arthur
Platonow, Nicolas W
Pollock, John Joseph
Pool, William Robert
Powers, Marcelina Venus
Priddle, Osgood Daniel, Jr
Prince, Herbert N
Rahwan, Ralf George
Rakieten, Nathan
Ramsey, John Charles
Rao, Nutaki Gouri Sankara
Rao, Suryanarayana K
Reagor, John Charles
Recknagel, Richard Otto
Redmond, Ninfa Indacochea
Reed, Warren Douglas
Reeves, Andrew Louis
Rehling, Carl John
Reilly, Joseph F
Reinhardt, Charles Francis
Reno, Frederick Edmund
Rieders, Fredric
Risacher, Robert Louis
Robb, Charles Arlee
Rosenblum, Ira
Rosenfield, Irene
Roslinski, Lawrence Michael
Rowe, Verald Keith
Rubin, Robert Jay
Rutter, Henry Alouis, Jr
Ryan, Charles Furrell
Sadek, Salah Eldine
Salem, Harry
Sandi, Emil
Sandmeyer, Esther E
Saunders, Donald Roy
Scala, Robert Andrew
Schafer, Michael Irving
Scharpf, Lewis George, Jr
Schneider, Philip William, Jr
Schwartz, Wayne Stanley
Schwartz, Edward
Schweda, Paul
Schwetz, Bernard Anthony
Serrone, David M
Shaffer, Charles Boyd
Sharma, Raghubir Prasad
Shellenberger, Thomas E

Sinnhuber, Russell Otto
Slywka, Gerald William Alexander
Smith, Douglas Lee
Smith, Jerry Morgan
Smith, Paul Winston
Smith, Roger Powell
Smith, Thomas Harry Francis
Smith, Walter George
Smyth, Henry Field, Jr
Snyder, Fred Hugh
Sohn, David
Sowell, Wendell L
Spencer, Howard Camac
Spiegl, Charles J
Spolyar, Louis William
Stannard, James Newell
Stavinoha, William Bernard
Stavric, Bozidar
Steinberg, Marshall
Stevenson, George William
Stewart, Richard Donald
Stoewsand, Gilbert Saari
Stokinger, Herbert Ellsworth
Stolman, Abraham
Street, Joseph Curtis
Suarez, Kenneth Alfred
Surak, John Godfrey
Taylor, Jean Marie
Taylor, Steve L
Tepper, Lloyd Barton
Terhaar, Clarence James
Terranova, Andrew Charles
Teske, Richard H
Thomas, Richard Dean
Thomas, Vera
Thompson, Daniel James
Thompson, George Rex
Tompkins, E Crosby
Torkelson, Theodore Ruben
Traina, Vincent Michael
Traitor, Charles Eugene
Valerino, Donald Matthew
Van Stee, Ethard Wendel
Verlangieri, Anthony Joseph
Virgo, Bruce Barton
Visek, Willard James
Vore, Mary Edith
Vostal, Jaroslav Joseph
Waggoner, William Charles
Wagstaff, David Jesse
Wallin, Richard Franklin
Wands, Ralph Clinton
Ward, Charles O
Waritz, Richard Stefen
Way, E Leong
Way, James Leong
Webb, Ryland Edwin
Weeks, Maurice Harold
Wei, Eddie Tak-Fung
Weikel, John Henry, Jr
Weil, Carrol Solomon
Weinberg, Myron Simon
Weir, Robert James, Jr
Weisburger, Elizabeth Kreiser
Weiss, Lawrence R
West, Bob
Wetherell, Herbert Ranson, Jr
Whitford, Gary M
Whitmore, George E
Wiberg, George Stuart
Wiles, Joseph St Clair
Williams, Martin Wesley
Wilson, John Ellis
Winek, Charles L
Witschi, Hanspeter Rudolf
Wogan, Gerald Norman
Wolven, Anne M
Woodhouse, Edward John
Wright, James Francis
Young, John Falkner
Zabik, Matthew John
Zabinski, Rose Marie C
Zaratzian, Virginia Louis
Zeitlin, Benjamin Raphael
Ziller, Stephen A, Jr

Inhalation Toxicology

Ballou, John Edgerton
Birky, Merritt Merle
Coate, William Bleecker
Dilley, James V
McClellan, Roger Orville
McGrath, James Joseph
Mokler, Brian Victor
Palazzolo, Matthew Joseph

Insect Toxicology

Berger, Robert S
Brady, Lewis George, Jr
Cutkomp, Laurence Kremer
Dauterman, Walter Carl
Dorough, Hendley Wyman
Hastings, Felton Leo
Johnson, Richard Emanuel
Kapoor, Inder Prakash
Mehendale, Harihara Mahadeva
Menzer, Robert Everett
Nakatsugawa, Tsutomu
Nigam, Prakash Chandra
Perry, Albert Solomon
Quraishi, Mohammed Sayeed
Roberts, Richard Bruce
Scoggin, John Kyle
Soboczenski, Edward John
Sun, Yun Pei
Tate, Laurence Gray
Turquist, Richard Lee
Weiden, Mathias Herman Joseph

Toxinology

Bernheimer, Alan Weyl
Calandra, Joseph Carl
Eisler, Milton
Hessinger, David Alwyn
Johnson, Bob Duell
Moore, Frederick Wolfgang
Rader, William Austin
Smalley, Harry Edwin

Veterinary Toxicology

Bailey, Everett Murl, Jr
Buck, William Boyd
Dollahite, James Walton
Earl, Francis Lee
Gralla, Edward Joseph
Hobbs, Charles Henry
Hueter, Francis Gordon
Madisson, Harry
McCreesh, Arthur Hugh
Moore, Wellington, Jr
Ochme, Frederick Wolfgang
Osweiler, Gary D

VETERINARY MEDICINE

Veterinary Medicine

Aiken, John M
Allen, Fred Ernest
Amstutz, Harold Emerson
Anderson, Allen Clarence
Anderson, George R
Anderson, Neil Vincent
Anthony, Harry D
Archibald, James
Armistead, Willis William
Arnold, John P
Atkinson, Joe William
Babcock, William Edward
Bankowski, Raymond Adam
Barber, Thomas Lynwood
Barber, Clifford Albert Victor
Barnes, Charles M
Barnum, Donald Alfred
Beck, Clifford C
Bell, Wilson Bryan
Bennett, Dwight G, Jr
Berman, David Theodore
Bidlack, Donald Eugene
Blake, Joseph Thomas
Boley, Loyd Edwin
Bolin, Fonsoe M
Bone, Jesse Franklin
Boney, William Arthur, Jr
Bortree, Alfred Lee
Boulanger, Paul
Brackett, Benjamin Gaylord
Bradley, Richard E
Bree, Max M
Brobst, Duane Franklin
Brodie, Bruce Orr
Brown, Richard Wallace
Brown, William Francis
Bryan, Harold Stever
Buckner, Ralph Gupton
Burch, Clark Wayne
Byrne, Robert Joseph
Callis, Jerry Jackson
Campbell, Clarence L, Jr
Cappucci, Dario Ted, Jr
Case, Arthur Adam
Chalquest, Richard Ross
Cho, Byung-Ryul
Clarke, William James
Clarkson, Merton Robert
Cloyd, Grover David
Conner, Gabel Henry
Constantine, Denny G
Cooperrider, Donald Elmer
Cornelius, Larry Max
Craig, Frank Rankin
Curtin, Terrence M
Davidson, David Edward, Jr
Davis, William Thompson
Decker, Winston M
DeTray, Donald Ervin
Deubler, Mary Josephine
Dickinson, Ernest Milton
Diena, Benito B
Dixon, Joe Maurice
Dobbins, Charles Nelson, Jr
Donovan, Edward Francis
Dougherty, Robert Watson
Downey, Ronald Stuart

Dozsa, Leslie
Dua, Prem Nath
Dubose, Robert Trafton
Dutta, Saradindu
Eberhart, Robert J
Edds, George Tyson
Ellett, Edwin Willard
Engelbrecht, Harlen H
Eskelund, Kenneth H
Fabry, Andras
Faddoul, George Peter
Fahning, Melvyn Luverne
Farrell, Daniel Reese
Finco, Delmar R
Flipo, Jean
Fogleman, Ralph William
Ford, Thomas Matthews
Foster, Henry Louis
Foster, Murray Elwood
Fox, Francis Henry
Fowler, Floyd William
Freeman, Arthur
Frerichs, Wayne Marvin
Friedman, Mark Hirsch
Friend, Jonathon D
Galvin, Thomas Joseph
Ganaway, James Rives
Garlick, Norman Lee
Gelinas, Louis-de-Gonzague
Georgi, Jay R
Gibbons, Walter J
Gillespie, James Howard
Gluckstein, Fritz Paul
Gochenour, William Sylva, Jr
Gordon, Donovan
Gorham, John Richard
Graves, John Henry
Gray, Andrew P
Gregory, Richard Parker, Jr
Griswold, Daniel Pratt, Jr
Guerrero, Raul Jaime
Haas, Kenneth Brooks, (Jr)
Haensly, William Edward
Hahn, Allen W
Hall, Charles E
Hall, Robert Everett
Hamby, Lavern R
Hammer, Aziz H
Hanner, Charles Edward, Jr
Hansen, Mike
Harold, LaVerne Collins
Hatch, Ray Davenport
Hayes, Frank Alfred
Haynes, N Bruce
Herrick, John Berne
Hill, George Neal
Hill, John Donald
Hillman, Robert B
Hinshaw, William Russell
Hofstad, Melvin Sidney
Holzworth, Jean
Hooper, Billy Ernest
Hoopes, Keith Hale
Hostetler, Roy Ivan
Huber, William T
Huber, William George
Huebner, Richard A
Hughes, Frank Alfred
Hulland, Thomas John
Hummer, Robert L
Huxsoll, David Leslie
Hyde, John Landis
Jasper, Donald Edward
Jennings, David Phipps
Jezyk, Peter Franklin
Johnson, Donald W
Jones, James Edward
Jones, William Grover
Jones, William Piner
Kahrs, Robert F
Kalison, Seymour Lincoln
Kammula, Raju G
Karstad, Lars
Keefe, Thomas J
Kelley, Donald Clifford
Kenzy, Sam George
Kersting, Edwin Joseph
Khan, Mohammed Nasrullah
Kilgore, Robert L
Kingman, Harry Ellis, Jr
Kirk, Robert Warren
Kirkham, Wayne Wolpert
Kodras, Rudolph
Koger, Lavon M
Kreier, Julius Peter
Krill, Walter Roland
Kuo, Eric Yung-Huei
Ladson, Thomas Alvin
Lank, Robert Byron
Larsen, Austin Ellis
Larson, Kenneth Allen
Larson, Lester Leroy
Laster, William Russell, Jr

Lay, John Charles
Lee, Arthur Clair
Lee, Robert Jerome
Leibovitz, Louis
Leman, Allen Duane
Liddle, Charles George
Live, Israel
Lumb, William Valjean
Lundvall, Richard
Luoto, Lauri
Lustgarten, Catherine Sue
Macheak, Merlin Edward
Maddy, Keith Thomas
Maestrone, Gianpaolo
Maplesden, Douglas Cecil
Marshall, Robert Reuben
Marshall, Arvle Edward
Mather, Edward Chantry
Mather, George Wells
Maxey, Brian William
McClurkin, Arlan Wilbur
McDonald, John Stoner
McFeely, Richard Aubrey
McGowan, Blaine, Jr
McKinley, Raymond Earl
McMurray, Birch Lee
McVicar, John West
Medway, William
Melby, Edward C, Jr
Merritt, Alfred M, II
Messersmith, Robert E
Michaud, Laurent
Mickelsen, W Duane
Miller, Cecil R
Milne, Frank James
Miniats, Olgerts Pauls
Misra, Hara Prasad
Mitrovic, Milan
Mitruka, Brij Mohan
Mongeau, J Denis
Moore, Earl Neil
Moreland, Alvin Franklin
Morrison, Spencer Horton
Moses, Harold Eugene
Mosier, Jacob Eugene
Mueller, George L
Muller, George Heinz
Mulnix, John Arthur
Murdick, Philip W
Murnane, Thomas George
Narotsky, Saul
Neher, George Martin
Nelson, Robert A
Niemeyer, Kenneth H
Nokes, Richard Francis
Nold, Max M
O'Brien, Joan A
O'Harra, John Lewis
Olds, Durward
Olson, Norman O
Orthoefer, John George
Osborne, Carl Andrew
Ott, Richard L
Page, Edwin Howard
Page, Norbert Paul
Palmer, Jack Sidney
Parkash, Ved
Parker, Richard Langley
Parkhie, Mukund Raghunathrao
Patterson, Donald Floyd
Patterson, William Creigh, Jr
Peardon, David Lee
Pensinger, Robert P
Peterson, C Denis
Peterson, Irvin Leslie
Peterson, Kermit Joseph
Pier, Allan Clark
Piperno, Elliot
Plymale, Harry Hambleton
Pogue, John Parker
Poppensiek, George Charles
Porter, David Bruce
Postle, Donald Sloan
Povar, Morris Leon
Price, Alvin Audis
Price, Donald Albert
Priester, William Alfred
Pritchard, William Roy
Pyle, Robert Lee
Rankin, Alexander David
Rao, Ghanta Nageswara
Ratzlaff, Marc Henry
Reddy, Chilecampalli Adinarayana
Redman, Donald Roger
Reichert, Paul F
Reynolds, William Aden
Rhode, Edward A, Jr
Rhodes, William Harker
Richter, Robert E
Riemann, Hans
Ristic, Miodrag
Robens, Jane Florence
Roberson, Edward Lee
Roenigk, William J
Rosenwald, Arnold Samuel
Ross, James Neil, Jr
Sack, Wolfgang Otto
Sadler, Walter White

Sampson, Gary Robert
Sar, Madhabananda
Sarma, Padman S
Savan, Milton
Schalm, Oscar William
Scheidy, Samuel F
Schiller, Alfred George
Schricker, Robert Lee
Schugel, LaVerne
Scott, George Clifford
Scott, Mack Tommie
Selby, Lloyd A
Sevoian, Martin
Shiffman, Morris A
Shor, Aaron Louis
Shupe, James LeGrande
Siddique, Irtaza H
Siegel, Edward T
Siegmund, Otto Hanns
Siegrist, Jacob C
Slocombe, Joseph Owen Douglas
Small, Erwin
Small, James David
Smith, Dean Harley
Smith, George Dale
Smith, Malcolm Crawford, Jr
Smithcors, James Frederick
Snoeyenbos, Glenn Howard
Sorensen, Dale Kenwood
Spencer, Guy Roger
Spurrell, Francis Arthur
Staples, George Emmett
Steele, James Harlan
Stevens, Alan Douglas
Stocking, Gordon Gary
Stone, Winfield S
Stoner, John Clark
Stowe, Clarence M
Sturdy, Robert Allan
Sweat, Robert Lee
Swift, Brinton L
Szabuniewicz, Michael
Tashjian, Robert John
Tasker, John B
Tate, Charles Luther
Tekeli, Sait
Tennant, Bud C
Teske, Richard H
Tharp, Vernon Lance
Theilen, Gordon H
Thomas, Don Wylie
Thompson, Clarence Henry, Jr
Thurmon, John C
Tjalma, Richard Arlen
Todd, Frank Arnold
Tonelli, George
Trace, James Chalmers
Tritschler, Louis George
Trum, Bernard Francis
Tucker, James
Upham, Roy Walter
Usenik, Edward A
Van Houweling, Cornelius Donald
van Marthens, Edith
Vaughan, John Thomas
Vickers, James Hudson
Walburg, Harry E, Jr
Walker, Donald F
Walker, Jerry
Wampler, Stanley Norman
Wang, Guang Tsan
Wass, Wallace M
Watson, Douglas F
Wayt, Lewis Keith
Webb, Alfreda Johnson
Weide, Kenneth Duane
White, Raymond Gene
Whiteford, Robert Daniel
Whitehair, Leo A
Whitemore, Howard Lloyd
Whitlock, Robert Henry
Whitmore, George E
Williams, Theodore Shields
Willoughby, Russell A
Willson, John Ellis
Wolfe, Thomas Lee
Woods, George Theodore
Wyman, Milton
Yager, Robert H
Yedloutschnig, Ronald John
Yoder, Harry Whitaker, Jr
Young, Robert, Jr
Zarkower, Arian
Zwickey, Robert Earl

Epizootiology
Hanson, Robert Paul
McLean, Robert George
Telford, Sam Rountree, Jr

Laboratory Animal Medicine
Cass, Jules Siland
Clark, James Derrell
Cohen, Bennett J
Conrad, Robert Dean
Conti, Pierre Andre
Flatt, Ronald Eugene
Flynn, Robert James

Glick, Phillip Ray
Grafton, Thurman Stanford
Greenstein, Edward Theodore
Hsu, Chao Kuang
Jonas, Albert Moshe
Jonas, Jimmy Barthel
Lutsky, Irving
McPherson, Charles William
Newton, William Morgan
Pakes, Steven P
Pierce, Joseph Elliott
Potkay, Stephen
Powers, Robert D
Ringler, Daniel Howard
Russell, Robert John
Secord, David Cartwright
Serrano, Louis Joseph
Smith, Alvin Winfred
Smith, Mary Elizabeth
Strandjord, Paul Edphil
Stuhlman, Robert August
Stunkard, Jim A
Valerio, David Allen
Weisbroth, Steven H
Whitney, Robert Arthur, Jr

Theriogenology
Kendrick, John Wesley
Williams, David John, III

Veterinary Anatomy
Bartlett, Lawrence Matthews
Bell, John Thomas, Jr
Bhatnagar, Mahesh Kumar
Chibuzo, Gregory Anenonu
Christensen, George Curtis
Cummings, John Francis
Czarnecki, Caroline Mary Anne
Diesem, Charles D
Dobson, Richard Cecil
Fletcher, Thomas Francis
Habel, Robert Earl
Hare, William Currie Douglas
Hinsman, Edward James
Horowitz, Aaron
Hullinger, Ronald Loral
Julian, Logan M
Kitchell, Ralph Lloyd
Lovell, James Edgeley
Magilton, James Henry
McClure, Robert Charles
McCurdy, Jon Alan
McKibben, John Scott
Meyer, Hermann
Parke, Wesley Wilkin
Pierard, Jean Arthur
Sis, Raymond Francis
Skold, Bernard Harold
Staley, Theodore Earnest Leon
Venable, John Howard
Venzke, Walter George
Westerfield, Clifford
Williams, Raymond Crawford
Worthman, Robert Paul

Veterinary Anesthesiology
Heath, Robert Bruce
Jackson, Larry LaVern

Veterinary Bacteriology
Biberstein, Ernest Ludwig
Fales, William Harold
Kirkbride, Clyde Arnold
Langford, Edgar Verden
Morse, Guy Emery
Wills, Franklin Knight

Veterinary Entomology
Butler, Jerry Frank
Crystal, Maxwell Melvin
Rogoff, William Milton

Veterinary Immunology
Belden, Everett Lee
Buening, Gerald Matthew
Cunliffe, Harry R
Fraser, C E Ovid
Ingram, Donald George
Morgan, Donald O'Quinn
Ruckerbauer, Gerda Margareta

Veterinary Microbiology
Anderson, David Prewitt
Bailie, Wayne E
Batte, Edward G
Berkhoff, German Adolfo
Bittle, James Long
Brown, Albert Loren
Campbell, Charles Haywood
Carmichael, Leland E
Cole, John Rufus, Jr
Collier, John Raymond
Colwell, William Maxwell
Corstvet, Richard E
Cunningham, Charles Henry
Deshmukh, Damodar Ramchandra
Dommermuth, Charles Henry, Jr
Dommert, Arthur Roland

Ferris, Deam Hunter
Firehammer, Burton Deforest
Folkerts, Thomas Mason
Frey, Merwin Lester
Gaskin, Jack Michael
Gyles, C L
Hanson, Lyle Eugene
Harris, Delbert Linn
Hidalgo, Richard Jack
Hirsh, Dwight Charles, III
Hughes, David Edward
Jensen, Wayne Ivan
Kemeny, J Lorant
Killinger, Arden Holmes
Kleckner, Albert Louis
Kleven, Stanley H
Kohler, Erwin Miller
Kramer, Theodore Tivadar
Krishnamurti, Pullabhotla V
Kumar, Mahesh C
Lambert, George
Lauerman, Lloyd Herman, Jr
Loken, Keith I
Madden, David Larry
Mare, Cornelius John
Marois, Paul Henri
McCune, Emmett L
Mengeling, William Lloyd
Meyer, Richard Charles
Moore, Richard Wayne
Morse, Erskine Vance
Newbould, Francis Henry Samuel
O'Berry, Phillip Aaron
Packer, Raymond Allen
Page, Leslie Andrew
Patton, William Henry
Pirtle, Eugene Claude
Pomeroy, Benjamin Sherwood
Price, Jessie Isabel
Radhakrishnan, Chittur Venkitasubhan
Rosenquist, Bruce David
Ross, Richard Francis
Rubin, Harvey Louis
Saif, Yehia Mohamed
Simpson, Russell Bruce
Smith, Clyde Konrad
Smith, Gail Bevington
Stalheim, Ole Henry
Tripathy, Deoki Nandan
Truscott, Robert Bruce
Waldhalm, Donald George
Wedman, Elwood Edward
White, Franklin Henry
Whitford, Howard Wayne
Wichmann, Robert W
Wood, Richard Lee
Yamamoto, Richard

Veterinary Parasitology
Ah, Hyong-Sun
Andrews, John Scott
Andrews, Myron Floyd
Bailey, Wilford Sherrill
Baker, Norman Fletcher
Bergstrom, Robert Charles
Besch, Everett Dickman
Colglazier, Merle Lee
Cruthers, Larry Randall
Dewhirst, Leonard Wesley
Drudge, Junior Harold
Dunlap, Jack Sherwin
Enzie, Frank Dorr
Ewing, Sidney Alton
Ferguson, Donald Leon
Folz, Sylvester D
Forrester, Donald Jason
Gaafar, Sayed Mohammed
Greve, John Henry
Griffiths, Henry Joseph
Hayes, Terence James
Herlich, Harry
Jordan, Helen Elaine
Knapp, Stuart Edward
Kohls, Robert E
Lindquist, William Dexter
Porter, James Armer, Jr
Powers, Kendall Gardner
Rausch, Robert Lloyd
Rubin, Robert
Schlotthauer, John Carl
Shelton, George Calvin
Slater, Robert Lee
Splitter, Earl John
Szanto, Joseph
Theodorides, Vassilios John
Todd, Kenneth S, Jr
Wescott, Richard Breslich
Whitlock, John Hendrick

Veterinary Pathology
Adelaer, Henry Elliot
Allen, Archibald Ferguson
Allen, Anton Markert
Allen, Henry L
Al-Nakeeb, Shaheen Mustafa
Altera, Kenneth P
Barnes, Donald McLeod

Boudreault, Armand
Bowen, James Milton
Bowne, John G
Boyd, Virginia Ann Lewis
Boyle, John Joseph
Bradford, Henry Bernard, Jr
Brailovsky, Carlos Alberto
Brandon, Frank Bayard
Brandt, Carl David
Brandt, Walter Edmund
Branton, Philip Edward
Braune, Maximillian O
Brawner, Thomas Allan
Breidenbach, Gearold Peter
Brockman, William Warner
Bronson, David Lee
Brown, Eric Reeder
Brown, Paul Wheeler
Brunell, Philip Alfred
Buescher, Edward Louis
Burger, Charles L
Burmester, Ben Roy
Burness, Alfred Thomas Henry
Burns, Kenneth Franklin
Burnstein, Theodore
Bussell, Robert Harry
Butel, Janet Susan
Byatt, Pamela Hilda
Cabasso, Victor Jack
Caiguiri, Lawrence Anthony
Came, Paul E
Campbell, Hayward
Cardiff, Robert Darrell
Carlberg, David Marvin
Carlson, Harve J
Carp, Richard Irvin
Casals, Jordi
Castro, Anthony Edward
Cate, Thomas Randolph
Cerini, Costantino Peter
Chagnon, Andre
Chalgren, Steve Dwayne
Chang, Robert Shihman
Chang, Te Wen
Chappell, William Adrian
Chernesky, Max Alexander
Choppin, Purnell Whittington
Christensen, James Roger
Christian, Robert Thomas
Cliver, Dean Otis
Coggin, Joseph Hiram
Cohen, Gary H
Cole, Gerald Alan
Coleman, Philip Hoxie
Collard, William David
Collins, Carolyn Jane
Colon, Julio Ismael
Colter, John Sparby
Compans, Richard W
Conant, Robert M
Considine, Richard George
Consigli, Richard Albert
Cooke, Patricia M
Cords, Carl Ernest, Jr
Counter, Frederick T, Jr
Courtney, Richard James
Coward, Joe Edwin
Cox, Donald Cody
Cox, Herald Rea
Craighead, John Edward
Cramblett, Henry G
Crane, George Thomas
Cremer, Natalie E
Crocker, Thomas Timothy
Crouch, Norman Albert
Crowell, Richard Lane
Culp, Lloyd Anthony
Cusumano, Charles Louis
Cutchins, Ernest Charles
Dales, Samuel
Dalrymple, Joel McKeith
Davis, Eldon Vernon
DeBoer, Carl John
Deforest, Adamadia
Deibel, Rudolf
Deinhardt, Friedrich
Del Villano, Bert Charles
Dhaliwal, Amrik S
Diener, Theodor Otto
Dierks, Richard Ernest
Doane, Frances Whitman
Dobrovolny, Charles George
Docherty, John Joseph
Doerfler, Walter
Dougherty, Robert Malvin
Dowdle, Walter R
Downs, Wilbur George
Drake, John W
Dreesman, Gordon Ronald
Dubose, Del Rose M
Dubos, Robert Trafton
Dubreuil, Robert
Dulbecco, Renato
Durand, Virginia Catherine
Durand, Donald P
East, James Lindsay
Eaton, Bryan Thomas
Eckert, Edward Arthur

Eckhart, Walter
Ehrenfeld, Elvera
Eldredge, Kelly Husbands
Elliott, Arthur York
Elvin-Lewis, Memory P F
Emery, Jerrell Bemis
Emmons, Richard William
Escobar, Mario R
Evenson, Donald Paul
Falk, Lawrence A, Jr
Faras, Anthony James
Farber, Florence Eileen
Faulkner, Peter
Feldman, Lawrence A
Felsenfeld, Ambhan Dasaneyavaja
Feltz, Elmer T
Fenje, Paul
Fenters, James Dean
Fields, Bernard Nathan
Fieldsteel, Arnold Howard
Fiscus, Alvin G
Flanagan, Thomas Donald
Fletcher, Ronald D
Ford, Richard Earl
Fox, John Perrigo
Frankel, Jack William
Franklin, Richard Morris
Frickey, Paul Henry
Friedman, Robert Morris
Frist, Ramsey Hudson
Fujioka, Roger Sadao
Furesz, John
Furuasawa, Eiichi
Fuscaldo, Anthony Alfred
Gabelman, Norman
Gainer, Joseph Henry
Gajdusek, Daniel Carleton
Galasso, George John
Garfinkle, Barry David
Gaush, Charles Richard
Geiduschek, Ernest Peter
Gentry, Glenn Aden
Gerard, Gary Floyd
Gerin, John Louis
Gerone, Peter John
Gibbs, Clarence Joseph, Jr
Gifford, George Edwin
Gillespie, James Howard
Ginsberg, Harold Samuel
Glaser, Ronald
Gold, Eli
Gomatos, Peter John
Goodman, Robert Merwin
Gori, Gio Batta
Goulet, Normand Robert
Graham, Frank Lawson
Granoff, Allan
Gratzek, John B
Gravell, Maneth
Gray, Alan
Green, Irving Joseph
Green, Melvin Howard
Greene, Elias Louis
Greig, Andrew Stephen
Gresser, Ion
Griffin, Diane Edmund
Grossberg, Sidney Edward
Groupe, Vincent
Guss, Maurice Louis
Gyulai, Eugene Jeno
Hackett, Adeline J
Hahon, Nicholas
Hallum, Jules Verne
Hampar, Berge
Hamparian, Vincent
Hankins, William Alfred
Hanshaw, James Barry
Hardy, Frank Merril
Hare, John Donald
Hari, V
Hartley, Janet Wilson
Haselkorn, Robert
Hatanaka, Masakazu
Hatch, Milford Harrison
Hattman, Stanley
Hay, Charles Alfred
Healy, George McNeice
Heberling, Richard Leon
Heck, Fred Carl
Hellman, Alfred
Hennessy, Albert Vincent
Henry, Claudia
Heppner, Gloria Hill
Herrmann, Ernest Carl, Jr
Herrmann, Kenneth L
Hershey, Alfred Day
Herrick, Frank M
Hewetson, John Francis
Hiebert, Ernest
Hightower, Lawrence Edward
Hill, William Francis, Jr
Hilleman, Maurice Ralph
Hillis, William Daniel
Hirsch, Martin Stanley
Hirst, George Keble
Ho, Monto
Ho, Peter Peck Koh
Hoffman, Howard Edgar

Hoffman, William Wheeler
Hollinger, Frederick Blaine
Hollinshead, Ariel Cahill
Holmes, Kathryn Voelker
Homme, Paul John
Horan, Paul Karl
Hotchin, John Elton
Houts, Garnette Edwin
Howe, Martha Morgan
Howell, David McBrier
Hsiung, Gueh Djen
Hsu, Yu-Chih
Huang, Alice Shih-Hou
Huang, Kun-Yen
Huebner, Robert Joseph
Hummeler, Klaus
Hung, Paul P
Igel, Howard Joseph
Iglewski, Wallace
Imagawa, David Tadashi
Incardona, Antonino L
Ittensohn, Oswald Ludwig
Jackson, David Archer
Jameson, Patricia Madoline
Jamison, Richard Melvin
Janssen, Robert (James) J
Jerkofsky, Maryann
Johnson, Christine Margaret
Johnson, F Brent
Johnson, James Carl
Johnson, Roger W
Johnson, Terry Charles
Joklik, Wolfgang Karl
Joncas, Jean Harry
Jorgensen, George Norman
Juo, Pei-Show
Kaighn, Morris Edward
Kaizer, Herbert
Kalter, Seymour Sanford
Kang, Chil-Yong
Kaplan, Albert Sydney
Karzon, David T
Kasel, Julius Albert
Katz, Michael
Katz, Samuel Lawrence
Kempf, John Emerson
Kenyon, Richard H
Kern, Jerome
Khare, Gyaneshwar Prasad
Khoobyarian, Newton
Kieff, Elliott Dan
Kilbourne, Edwin Dennis
Kilham, Lawrence
Kim, Kwang Soo
Kimball, Paul Clark
Kingsbury, David Thomas
Kingsbury, David Wilson
Kirber, Maria Wiener
Kissling, Robert E
Klein, Morton
Klein, Paul Alvin
Korant, Bruce David
Kouroupis, George Michael
Kozloff, Lloyd M
Krueger, Robert George
Kucera, Louis S
Kuns, Merle L
Laipis, Philip James
Landau, Burton Joseph
Landon, John Campbell
Landsberger, Frank Robert
Langridge, William Henry Russell
Lark, Carl Gordon
Larke, Robert Peter Bryce
Larson, David L
Larson, Vivian M
Lavelle, George Cartwright
Lavender, John Francis
Lawrence, William Chase
Ledinko, Nada
Lee, Harold Hon-Kwong
Lee, Kyu Myung
Lefkowitz, Stanley S
Lehman, John Michael
Leong, Jo-Ann Ching
Lerner, Michael Paul
Lesnaw, Judith Alice
Levin, Judith Goldstein
Levine, Alvin Saul
Levine, Arnold Jay
Levine, Myron
Levine, Seymour
Levintow, Leon
Levitt, Neil Hilliard
Levy, Allan Henry
Levy, Hilton Bertram
Lewis, Andrew Morris, Jr
Lewis, Lewis James
Lindsay, Harry Lee
Liu, Chien
Lodmell, Donald Louis
Lohmann, Paul Schoenfeld
Loria, Robert Moshe
Lorenz, Douglas
Louie, Robert Eugene
Luginbuhl, Roy Emil
Lunger, Philip Dupont

Maag, Theodore Augustus
Machlowitz, Roy Alan
Magee, Wayne Edward
Mahdy, Mohamed Sabet
Mak, Stanley
Mandel, Benjamin
Manly, Kenneth Fred
Manning, JaRue Stanley
Maramorosch, Karl
Marchette, Nyven John
Marcus, Philip Irving
Martignoni, Mauro Emilio
Marusyk, Raymond George
Maruyama, Koshi
Marx, Preston August, Jr
Mascoli, Carmine Charles
Matsuoka, Tats
Mayor, Heather Donald
Mayyasi, Sami Ali
McCarter, John Alexander
McCarthy, William John
McClelland, Laurella
McClurkin, Arlan Wilbur
McCollum, William Howard
McConnell, Stewart
McGregor, Sandy
McLaren, Leroy Clarence
McLean, Donald Millis
McLimans, William Fletcher
McMillen, Janis Kay
McSharry, James John
McVicar, John West
Medzon, Edward Lionel
Meinke, William John
Melendez, Luis Vargas
Melnick, Joseph Louis
Merigan, Thomas Charles, Jr
Merriam, Esther Virginia
Metzgar, Don P
Meyer, Harry Martin, Jr
Meyers, Paul
Middleton, Peter James
Midlige, Frederick Horstmann, Jr
Miller, Willwam Robert
Millette, Robert Loomis
Millian, Stephen Jerry
Milligan, Wilbert Harvey, III
Milo, George Edward
Miner, Norman Allen
Minocha, Harish C
Minowada, Jun
Mizutani, Satoshi
Mohanty, Sashi B
Morahan, Page Smith
Morgan, Councilman
Moscovici, Carlo
Moss, Bernard
Moss, L Howard, III
Mountain, Isabel Morgan
Moyer, Mary Pat Sutter
Mulder, Carel
Murphy, Frederick A
Murray, Robert Edward
Nahmias, Andre Joseph
Nathanson, Neal
Nayak, Debi Prosad
Nazerian, Keyvan
Neff, Beverly Jean
Neurath, Alexander Robert
Newbold, John Edward
Newlin, Grodon Ermel
Newman, Franklin Scott
Newton, William Atson, Jr
Nichol, Francis Richard, Jr
Nicholson, Donald Paul
Nomura, Shigeko
Nonoyama, Meihan
Nordquist, Robert Ersel
Noronha, Fernando M Oliveira
North, James A
Northrop, Robert L
Notkins, Abner Louis
Noyes, Wilbur Fiske
Nutter, Robert Leland
Obijeski, John Francis
Oh, Jang Ok
Olsen, Richard George
Opton, Edward Milton
Orsi, Ernest Vinicio
Osborn, June Elaine
Oshiro, Lyndon Satoru
Osterhout, Suydam
Overby, Lacy Rasco
Ozer, Harvey Leon
Padnos, Morton
Palmer, Dan F
Papas, Takis S
Paranchych, William
Parikh, Gokaldas Chandulal
Parker, Dorothy Lundquist
Parker, John Clarence
Parkman, Paul Douglas
Parrott, Robert Harold
Paschke, John Donald
Paucker, Kurt
Payne, Francis Eugene
Penhoet, Edward Etienne
Perez, John Carlos

Peterson, David Allan
Pfau, Charles Julius
Phillips, Bruce A
Phillips, Charles Alan
Pienta, Roman Joseph
Pilcher, K Stephen
Pizarro, Enriqueta
Plagemann, Peter Guenter Wilhelm
Plotkin, Stanley Alan
Pollikoff, Ralph
Pons, Marcel William
Porter, David Dixon
Portner, Allen
Pratt, David
Preble, Olivia Toby
Price, Winston Harvey
Prince, James T
Proffitt, Max Rowland
Pumper, Robert William
Purcell, Robert Harry
Purifoy, Dorothy Jane Martin
Quilligan, James Joseph, Jr
Rabin, Erwin R
Rabinovich, Sergio Rospigliosi
Rafajko, Robert Richard
Rapp, Fred
Rawls, William Edgar
Reagan, Reginald L
Rees, Horace Benner, Jr
Reilly, Christopher Aloysius, Jr
Reisinger, Robert
Renis, Harold E
Rich, Marvin A
Richmond, Jonathan Young
Riggs, John L
Righthand, Vera Fay
Riley, Vernon Todd
Roane, Philip Ransom, Jr
Robb, James Arthur
Roberts, Audrey Nadine
Robinson, Roslyn Quinby
Rochovansky, Olga Maria
Rodriguez, Jose Enrique
Rodriguez-Leiva, Manuel
Roizman, Bernard
Roman, Janet Meincer
Rosenbaum, Manuel
Rosenbaum, Max J
Ross, Martin Russell
Roth, Frieda
Rowan, Dighton Francis
Rozee, Kenneth Roy
Rubin, Harry
Rudnick, Albert
Ruecker, Roland R
Russell, Philip King
Sabin, Albert Bruce
Sabina, Leslie Robert
Sagik, Bernard Phillip
Salzman, Norman Post
Sarma, Padman S
Savan, Milton
Saxton, Romaine Edward
Schaffer, Frederick Leland
Schaffer, Priscilla Ann
Schaub, Stephen Alexander
Schell, Klaus Rainer
Schidlovsky, George
Schieble, Jack H
Schieff, Gilbert Martin
Schleicher, Joseph Bernard
Schlesinger, Sondra
Schuederberg, Ann Elizabeth Snider
Schmidt, Nathalie Joan
Schneider, Nathan Joseph
Schuchardt, Lee Frank
Schultz, Irwin
Schulze, Irene Theresa
Schwerdt, Carlton Everett
Scott, Lawrence Vernon
Scraba, Douglas G
Scrivani, Robert P
Seefried, Adolf Von
Sellers, Margret Irene
Seman, Gabriel
Semancik, Joseph Stephen
Seto, Joseph Tobey
Shadduck, John Allen
Shah, Keerti
Shanmugam, Govindaswamy
Sharma, Jagdev Mittra
Sharon, Nehama
Sharp, Philip Allen
Shatkin, Aaron Jeffrey
Shaw, Eugene
Sheek, Martha Reyburn
Sheffield, Joel Benson
Shelokov, Alexis Ioann
Shepherd, Robert James
Shibley, George P
Shipman, Charles, Jr
Shope, Richard Edwin, Jr
Shope, Robert Ellis
Siegel, Benjamin Vincent
Siegel, Pamela Jean
Sigel, Mola Michael
Siminoff, Paul
Simon, Edward Harvey

Simpson, Robert Wayne
Singer, Irwin I
Singh, Rudra Prasad
Sinkovics, Joseph
Skalka, Anna Marie
Smart, Kathryn Marilyn
Smith, Gary Lee
Smith, Hugh Hollingsworth
Smith, Jerry Warren
Smith, Kendall O
Smith, Ralph Earl
Smith, Roger Dean
Smith, William Roy
Soike, Kenneth F
Sokol, Frantisek
Solorzano, Robert Francis
Soret, Manuel Gonzalez
Southam, Chester Milton
Spalatin, Josip
Spanier, Bonnie Barbara
Spear, Patricia Gail
Spendlove, Rex S
Spetzel, Richard O
Srinhongse, Sunthorn
Sreevalsan, Thazepadath
Steele, Frances M
Steere, Russell Ladd
Steeves, Richard Allison
Steiner, Sheldon
Stevens, David Arthur
Stevens, Jack Gerald
Stevens, Thomas McConnell
Stoemer, Herbert George
Stollar, Victor
Stone, Henry Otto, Jr
Stone, Howard Anderson
Stone, Warren Norton
Storz, Johannes
Strano, Alfonso J
Strauss, Ellen Glowacki
Strauss, James Henry
Stringfellow, Dale Alan
Strohl, William Allen
Stulberg, Cyril Sidney
Sturman, Lawrence Stuart
Sudia, William Daniel
Sweat, Robert Lee
Sweet, Benjamin Hersh
Sydiskis, Robert Joseph
Sykes, John A
Synmington, Janey
Takemoto, Kenneth Kaname
Tamm, Igor
Tan, Yin Hwee
Tankersley, Robert Walker, Jr
Taylor, Gerald C
Temin, Howard Martin
Tennant, Judith R
Tennant, Raymond Wallace
Tershak, Daniel R
Tevethia, Satvir S
Thomas, Donald C
Thormar, Halldor
Togo, Yasushi
Tokumaru, Tadasu
Toler, Robert William
Toy, Stephen Thomas
Traub, Karl Arthur
Trkula, David
Truden, Judith Lucille
Tumilowicz, Joseph J
Turner, Willie
Tzianabos, Theodore
Uden, Stephen Alexander
Underdahl, Norman Russell
Underwood, Gerald Emerson
Ushijima, Richard N
Vande Woude, George
Vasington, Paul John
Ventura, Arnoldo K
Vilcek, Jan Tomas
Vogt, Peter Klaus
Volenec, Frank Jerry
Wachter, Ralph Franklin
Wagner, Robert Roderick
Waite, Marilynn Ransom Fairfax
Walker, Duard Lee
Wallbank, Alfred Mills
Ward, David Christian
Wassermann, Felix Emil
Watkins, Harry Mitchell Sherman
Webb, Patricia Ann
Webb, Joseph
Weber, Michael Joseph
Weber, Robert G
Welkie, George William
Werner, Erna Alture
Wertz, Gail T Williams
Wheelock, Earle Frederick
White, Roberta Jean
Wiktor, Tadeusz Jan
Wilhelm, Alan Roy
Wilson, Dwight Elliott, Jr
Winmer, Eckard
Winters, Wendel Delos
Witkin, Steven S
Witmer, Herman John
Wolff, David A

Wolff, John Shearer, III
Woodhour, Allen F
Woodman, Daniel Ralph
Wu, Jia-Hsi
Yamashiroya, Herbert Mitsugi
Yin, Fay Hoh
York, Charles James
Yoshimura, Sei
Young, Seth Yarbrough, III
Yuill, Thomas Mackay
Zaitin, Milton
Zajac, Barbara Ann
Zebovitz, Eugene
Zee, Yuan Chung
Zeigel, Robert Francis
Zichis, Joseph
Ziegler, Richard James
Zimmerman, Eugene Munro
Zoloter, Laurence Arthur

Animal Virology

Allen, Patton Tolbert
Anderson, Carl William
Bachenheimer, Steven Larry
Blatti, Stanley Parris
Bourgeaux, Pierre
Bower, Raymond Kenneth
Bryans, John Thomas
Chow, Tsu Ling
Fischinger, Peter John
Fleischmann, William Robert, Jr
Gray, Carla Winlund
Handy, Farouk Mohamed
Karu, Alexander Edwin
Lake, Robert Samuel
Noble-Harvey, Jane
O'Callaghan, Dennis John
Panem, Sandra
Paterson, Loyd Thomas
Pearson, George Denton
Rosenberger, John Knox
Steiner, Marion Rothberg
Stewart, Robert Bruce
Wagner, Edward Knapp

Marine Virology

Dearborn, John Holmes
Smith, Alvin Winfred

Plant Virology

Barnett, Ortus Webb, Jr
Beachy, Roger Neil
Bozarth, Robert F
Bradley, Roy Henry Edward
Brakke, Myron Kendall
Bruening, George Emil
Cheo, Pen Ching
Christie, Stephen Rolland
Cochran, George Wilson
Feldman, Jose M
Gill, Clifford Cressey
Gracia, Olga
Gumpf, David John
Halliwell, Robert Stanley
Horst, Ralph Kenneth
Jackson, Andrew Otis
Kimmins, Warwick Charles
Lambom, Calvin Ray
Lane, Leslie Carl
Lee, Peter E
Lister, Richard Malcolm
Mink, Gaylord Ira
Purcifull, Dan Elwood
Resconich, Emil Carl
Siegel, Albert
Timian, Roland Gustav
Tolin, Sue Ann
Tremaine, Jack H
Waterworth, Howard E
Weintraub, Marvin
Wood, Harry Alan
Zitter, Thomas Andrew

Veterinary Virology

Addinger, Hans Karl
Bass, Edmund P
Beard, Charles Walter
Benton, William J
Bohl, Edward Homer
Breese, Sydney Salisbury, Jr
Burroughs, Albert Lawrence
Butterfield, Walter K
Calnek, Bruce Wixson
Coggins, Leroy
Crandell, Robert Allen
Darcel, Colin Le Q
Dardiri, Ahmed Hamed
Derbyshire, John Brian
Eugster, A Konrad
Farrell, Roy Keith
Gale, Charles
Griffin, Thomas Ponton
Gustafson, Donald Pink
Haelterman, Edward Omer
Heuschele, Werner Paul
Johnson, Robert Byron
Kelling, Clayton Lynn

ZOOLOGY

Kolar, Joseph Robert, Jr
Marsh, Richard Floyd
Mattson, Donald Eugene
McKercher, Delbert Grant
Pilchard, Edwin Ivan
Schipper, Ithel Arie
Schultz, Ronald David
Scott, Fredric Winthrop
Thorsen, Jan
Smith, Paul Clay
Van Der Maaten, Martin Junior
Yates, Vance Joseph

Zoology

Abbatiello, Michael James
Abegg, Roland
Abercrombie, Warren Fulton
Abram, James Baker, Jr
Abramoff, Peter
Acholonu, Alexander Dozie
Adams, Charles Henry
Adams, Curtis H
Adams, James Russell
Adelmann, Howard Bernhardt
Ahl, Alwynelle S
Albright, Jack Lawrence
Albright, Joseph Finley
Albright, Raymond Gerard
Alexander, Claude Gordon
Alger, Nelda Elizabeth
Alicino, Nicholas J
Allen, Archie C
Allen, Ashael Lester
Allen, Charles Eugene
Allen, Delorean Matthew
Allen, Kenneth William
Altland, Paul Daniel
Alvarado, Ronald Herbert
Amy, Robert Lewis
Anderson, Bertil Gottfrid
Anderson, Bertin W
Anderson, Daniel Craig
Anderson, David Robert
Anderson, Glenn Arthur
Anderson, Glenn Maxwell
Anderson, Lloyd L
Anderson, Mariowe George
Andrew, Warren
Andrews, Cater Wilson
Andrews, Jay Donald
Andrews, Richard D
Andrus, William DeWitt, Jr
Applegate, Richard Lee
Arant, Frank Selman
Argyris, Bertie
Arnold, John Miller
Arnold, John Ronald
Arora, Harbans Lall
Arthaud, Raymond Louis
Arthaud, Vincent Henry
Arvey, Martin Dale
Aulerich, Richard J
Austin, Joseph Wells
Avolizi, Robert Joseph
Aylesworth, Thomas Gibbons
Bahler, Thomas Gibbons
Bailey, John H
Bailey, Joseph Randle
Bailey, Reeve Maclaren
Bailey, Thomas De Pinna
Baird, Derwood McVey
Baker, Clinton Lyle
Baker, Frank Hamon
Baker, Robert Dale
Baker, Rollin Harold
Baker-Cohen, Katherine France
Bakken, Arnold
Balaban, Martin
Ball, Gordon Harold
Ball, Robert Cragin
Ballard, Neil Brian
Band, Henretta Trent
Band, Rudolph Neal
Bandy, Percy John
Banfield, Alexander William Francis
Banks, David Lee
Banks, Edwin Melvin
Banks, William Michael
Barber, Saul Benjamin
Barber, Albert Arnold, Jr
Barkalow, Frederick Schenck, Jr
Barker, John Grove
Barker, Kenneth Ray
Barker, Shirley Hugh
Barnes, Herbert M
Barnes, Lester E
Barrick, Elliot Roy
Barth, Lucena J
Barth, Robert Hood, Jr
Bartholomew, George Adelbert
Bashir, Nasir Ahmad
Bassett, James Wilbur
Bates, Harold Brennan, Jr

Battin, William T
Bayer, Frederick Merkle
Bazer, Fuller Warren
Beach, Neil William
Beamish, Frederick William Henry
Beams, Harold William
Beatty, Alice Ferguson
Beatty, Kenneth Wilson
Beck, Albert J
Beck, Stanley Dwight
Beckel, William Edwin
Becker, David Alvord
Bedinger, Charles Arthur, Jr
Behrens, Mildred Esther
Belcher, Jane Colburn
Belden, Don Alexander, Jr
Bellis, Edward David
Bellmer, Elizabeth Henry
Belshe, John Francis
Beltran, Enrique
Bennett, Harry Jackson
Bennington, Neville Lynne
Benton, Allen Haydon
Bergstresser, Kenneth A
Bergstrom, David Wallace
Berlind, Allan
Bern, Howard Alan
Bernard, Richard Fernand
Beverley-Burton, Mary
Bibeau, Armand A
Bick, George Herman
Bigelow, Robert Sidney
Birge, Wesley Joe
Bischoff, Harry William
Blackburn, Maurice
Blair, Albert Patrick
Blese, John Carl William
Blomquist, Conrad Alvin
Blosser, Timothy Hobert
Bock, Carl E
Boddy, Dennis Warren
Bodenstein, Dietrich H F A
Boertje, Stanley
Boesel, Marion Waterman
Bogart, Ralph
Bolen, Homer Roscoe
Boliek, Irene
Boling, John Landrum
Bolla, Robert Irving
Bond, Edwin Joshua
Bond, James
Bond, Richard Randolph
Boozer, Reuben Bryan
Borton, Anthony
Boss, Willis Robert
Botkin, Merwin P
Bouwman, Fred Ludwig
Bowers, Darl Eugene
Bowles, John Bedell
Bowman, Thomas Elliot
Boyd, Elizabeth Margaret
Boyd, Leroy Houston
Boyer, Charles Chester
Braddock, James Conger
Brandell, Bruce Reeves
Brandhorst, Carl Theodore
Brattstrom, Bayard Holmes
Breed, Helen Illick
Breitenbach, Robert Peter
Breland, Osmond Philip
Breneman, William Raymond
Brisbin, I Lehr, Jr
Britt, Henry Grady
Britten, Bryan Terrence
Broadbeds, Harold Eugene
Brodkorb, Pierce
Broseghini, Albert L
Browman, Ludvig Gustav
Brown, Claudeous Jethro Dahiels
Brown, Lauren Evans
Brown, Leland Arthur
Brown, Patricia Stocking
Brown, Paul Bruce
Brown, Paul Lopez
Brown, Relis Bastian
Browning, Henry (Charles)
Brunson, Royal Bruce
Brush, Alan Howard
Bryan, Clifford Randall
Brynildson, Oscar Marius
Buchanan-Smith, Jock Gordon
Buck, John Bonner
Budde, Mary Laurence
Buhse, Howar Edward, Jr
Bunde, Daryl E
Bunting, Dewey Lee
Burch, John Bayard
Burdick, George Edgar
Burger, James Wendell
Burkholder, John Henry
Burnham, Kenneth Donald
Burns, John McLauren
Burns, Robert Kyle
Burns, Robert David
Burroughs, Wise
Burton, Paul Ray
Busch, Karl Heinrich Daniel
Buscher, Henry Neil

Bush, Francis M
Buss, Keen
Butcher, John Edward
Butler, Ogbourne Duke
Butterworth, Bernard Bert
Byrd, Mitchell Agee
Cable, Raymond Millard
Cameron, Duncan MacLean, Jr
Cameron, Ivan Lee
Camin, Joseph Harvey
Campbell, Robert Seymour
Campbell, William Cecil
Cargo, David Garrett
Carlton, Robert Austin
Carpenter, Charles Congden
Carpenter, Esther
Carpenter, James Woodford
Carpenter, John Melvin
Carpenter, Roger Edwin
Carpenter, Rose Marie
Carpenter, Zerle Leon
Carr, Archie Fairly, Jr
Carter, Howard Payne
Carter, William Alfred
Cassard, Daniel Waters
Castle, Gordon Benjamin
Cather, James Newton
Chamberlain, James Luther
Chambers, Leslie Addison
Chance, Charles Jackson
Chandler, Clay Morris
Chandler, David Culbertson
Chaney, Allan Harold
Chapman, David MacLean
Charters, Elaine Mary
Cheeke, Peter Robert
Cheng, Tien-Hsi
Cherian, Sebastian K
Chia, Fu-Shiang
Chichester, Lyle Franklin
Chipman, Robert K
Chobotar, Bill
Church, Gilbert
Church, Ronald L
Cullo, Robert Henry
Clampitt, Philip Theodore
Clare, Stewart
Clark, Clarence Floyd
Clark, Eugenie
Clark, Mary Eleanor
Clark, Ralph M
Clarke, Lemuel Floyd
Clarke, Robert Francis
Clemens, Howard Paul
Clench, William James
Cliburn, Joseph William
Clise, Ronald Leo
Coffey, Marvin Dale
Cole, Evelyn
Colias, Lamont Cook
Colias, Elsie Cole (Mrs Nicholas E Collias)
Collins, Nicholas Elias
Collins, Charles Thompson
Collins, Hollie L
Colvard, Dean Wallace
Colwin, Arthur Lentz
Colwin, Laura Hunter
Cominsky, Catherine
Conita, Gabriel William
Connelly, Thomas George
Conner, Robert Louis
Cook, Kenneth Martin
Cook, Marie Mildred
Cook, Nathan Howard
Cooke, Herman Glenn
Cooper, William Anderson
Cooper, William E
Corbet, Philip Steven
Corkum, Kenneth C
Cornman, Ivor
Corrick, James Adam, Jr
Costello, Donald Paul
Costello, William James
Costoff, Allen
Cowan, F Brian M
Cox, Beverley Lenore
Cox, David Frame
Crain, James Larry
Crawford, Eugene Carson, Jr
Crites, John L
Crosman, Arthur Marston
Cross, Frank Bernard
Cross, Hiram Russell
Crouch, Hubert Branch
Crowe, David Burns
Crowell, Robert Merrill
Culley, Dudley Dean, Jr
Curd, Milton Rayburn
Curtis, Stanley Evan
Dahlberg, Michael D
Dahlon, John Charles
Damaskus, Charles William
Dandy, James William Trevor
Daniel, Paul Mason
Daniels, Gert L
Darby, Rollo E

Darlington, James McCown
Davenport, Demorest
David, Paul Rembert
Davis, Betty Schuck
Davis, Frank Roscoe
Davis Frederic Whitlock
Davis, Gordon Richard Fuerst
Davis, John Edward, Jr
Davis, Larry E
Davis, Richard Francis
Davis, Russell Price
Dean, Benjamin T
Dearden, Douglas Morey
DeCoursey, Russell Myles
De Gennaro, Louis D
Dehner, Eugene William
DeLanney, Louis Edgerton
Delzell, David Edgar
Dendy, John Stiles
Denning, Jack
Dennison, Clifford C
Dent, James (Norman)
deRoos, Roger McLean
deRoth, Gerardus Cabble
Derrick, Finnis Ray
Derrickson, Charles M
Desser, Sherwin S
De Terra, Noel
Detwyler, Robert
Devlin, Richard Gerald, Jr
De Witt, Robert Merkle
De Wolf, Robert Abel
Deyrup-Olsen, Ingrith Johnson
Dial, Norman Arnold
Diamond, Louis Stanley
Dickerman, Richard Curtis
Diem, Kenneth Lee
Dietrich, John P
Dilks, Eleanor
Dillard, Joe G
Dillery, Dean George
Dillon, Raymond Donald
Dimond, Marie Therese
Dineen, Clarence Francis
Divelbiss, James Edward
Dobson, William Michael
Dolan, James Michael
Dolnick, Ethel Helen
Domm, Lincoln Valentine
Donker, John D
Donley, David Edward
Donohoo, John T
Dorris, Peggy Rae
Doudoroff, Peter
Drabek, Charles Martin
Dreyer, William Albert
Driscoll, William Thorvald
Dropkin, Victor Harry
Drummond, James
DuBose, Leo Edwin
Duerr, Frederick G
Dulin, William E
Dumont, James Nicholas
Duncan, Stewart
Dunham, Donald West
Dunham, Jewett
Dunn, Mary Catherine
Dunn, Robert Fowler
Dunnebacke-Dixon, Thelma Hudson
Dunson, William Albert
Durrant, Stephen David
Dustman, John Henry
Dutt, Ray Horn
Dyer, Irwin Allen
Dyer, Melvin I
Dynes, J Robert
Eakin, Richard Marshall
Easterling, George Riley
Eastlick, Herbert Leonard
Eaton, Theodore Hildreth, Jr
Ebeling, Alfred W
Eberly, William Robert
Edgar, Arlan Lee
Edney, Eric B
Egge, Alfred Severin
Eggert, Robert Glenn
Eisner, Thomas
Elias, Joel Jesse
Elliott, Alfred Marlyn
Elliott, Alice
Elliott, Rush
Ellis, Leslie Lee, Jr
Elrod, Lloyd Melvin
Ely, Donald Gene
Emerson, Alfred Edwards
Emlen, John Thompson, Jr
Enderson, James H
Engelmann, Franz
Engels, William Louis
Enright, James Thomas
Ensminger, Marion Eugene
Erickson, James George
Eschenberg, Kathryn (Marcella)
Etgen, William M
Etheridge, Albert Louis
Etkin, William
Evans, Francis Gaynor
Evans, George William

Evans, John William
Everingham, John
Falcon, Carroll James
Fanale, Louisa P
Fanslow, Don J
Farner, Donald Sankey
Farrell, Charles Ernest
Faulkner, Russell Conklin, Jr
Feducia, John Alan
Fehon, Jack Harold
Feist, Dale Daniel
Feldballe, Jeanette
Feldmeth, Carl Robert
Ferguson, Max B
Ferguson, Thomas
Fickess, Douglas Ricardo
Filice, Francis P
Filteau, Gabriel
Fincher, John Albert
Finkel, Asher Joseph
Finlay, Peter Stevenson
Fiore, Carl
Fisher, Dorothy A
Fisher, Harold Dean
Fisler, George Frederick
Fitzgerald, Paul Ray
Flaim, Francis Richard
Fleming, Richard Cornwell
Fleming, Theodore Harris
Flynn, Carl Munro
Fooden, Jack
Foor, W Eugene
Ford, Floyd Mallory
Ford, James
Foreman, Charles William
Foreman, Darhl Lois
Franks, Edwin Clark
Frantz, William Lawrence
Franzen, Dorothea Susanna
Fraser, Lemuel Anderson
Frazier, Ralph Paul
Frederick, Edward C
Freeman, John Alderman
Freiburg, Richard Eighme
Fribourgh, James H
Frye, Anne Evans
Frye, Billy Eugene
Gallati, Walter William
Gallicchio, Vincent
Gangwere, Stanley Kenneth
Gans, Carl
Garner, Duane LeRoy
Garrigus, Upson Stanley
Garside, Edward Thomas
Garth, John Shrader
Garthe, William A
Gassie, Edward William
Gatz, Arthur John, Jr
Gaufin, Arden Rupert
Gaunt, Abbot Stott
Geen, Glen Howard
Geist, Valerius
George, John Caleekal
George, William
Gering, Robert Lee
Gersbacher, Willard Marion
Gerst, Jeffery William
Ghiselin, Michael Tenant
Gibley, Charles W, Jr
Gilbert, Charles Russell
Gilbert, Lawrence Irwin
Gillette, Roy (James)
Gilliland, Floyd Ray, Jr
Giltz, Maurice Leroy
Glenn, William Grant
Goforth, W Reid
Goldberg, Erwin
Goldberg, Robert Jack
Good, Don L
Goodchild, Chauncey George
Goodnight, Clarence James
Gorbman, Aubrey
Gould, William Robert, III
Gourley, Eugene Vincent
Graber, Richard Rex
Grabowski, Casimer Thaddeus
Graf, William
Grant, Peter Raymond
Grant, William Chase, Jr
Graves, Artis P
Gray, Faith Harriet
Gray, Peter
Gray, Roy C, Jr
Green, George G
Green, Jonathan P
Green, William Asa
Greenhall, Arthur Merwin
Greenstein, Julius S
Gregg, John Richard
Grewe, Alfred H, Jr
Grier, James William
Grim, John Norman
Grimm, Wilbur Winfield
Grobman, Arnold Brams
Grodner, Robert Maynard
Groody, Thomas Conrad

ZOOLOGY

Grosch, Daniel Swartwood
Groseclose, Nancy Pence
Grosvenor, Clark Edward
Gruchy, David Francis
Gude, Richard Hunter
Guhl, Alphaeus Matthew
Guilford, Harry Garrett
Gunderson, Harvey Lorraine
Gunter, Gordon
Guram, Malkiat Singh
Guthrie, Russell Dale
Gutselman, Sheldon
Guyselman, John Bruce
Haas, Richard
Haefner, Paul Aloysius, Jr
Haggis, Alex John
Hagquist, Carl Waldemar
Haines, Howard Bodley
Hainsworth, Fenwick Reed
Hairston, Nelson George
Hall, Gordon Earl
Hall, John Edgar
Hall, John Sylvester
Hamburger, Viktor
Hamilton, Clara Eddy
Hamilton, William John, Jr
Hammat, Allan H
Hampton, Carolyn Hutchins
Hamrte, Christopher John
Hancock, Hunter McRae
Hanan, Herbert Herrick
Hansen, Bruce Winston
Hansen, Ira Bowers
Hansen, Lester Eugene
Hanson, William Roderick
Harclerode, Jack E
Hardberger, Florian Max
Haresign, Thomas
Harkema, Renard
Harman, Walter James
Harmon, Bud Gene
Harmsworth, Rodney V
Harp, George Lemaul
Harrell, Byron Eugene
Harrington, Joseph B
Harrington, Rodney B
Harrises, Antonio Efthemios
Harrison, Robert William
Harriss, Thomas T
Hartley, Marshall Wendell
Hartley, Richard Thomas
Hartung, Ernest William
Harvey, Elmer Bostwick
Hasler, Arthur Davis
Hatch, Melville Harrison
Haupt, Robert Elliott
Hayes, Murray Lawrence
Hayes, Wilbur Frank
Hays, Horace Albennie
Hays, Rodney Malcolm, Jr
Hayward, Bruce Jolliffe
Hazelwood, Donald Hill
Heath, Alan Gard
Heath, James Edward
Heatwole, Harold Franklin
Heckrotte, Carlton
Heed, William Battles
Hefner, Robert Arthur
Heidenthal, Gertrude Antoinette
Heinrich, Gerd H
Heiss, Herbert Ernest
Hellack, Jenna Jo
Heller, David Harold
Helms, Carl Wilbert
Hempel, Franklin Glenn
Hemphill, Andrew Frederick
Henning, Willard Loren
Henrickson, Robert Lee
Henry, Dora Priaulx
Herber, Elmer Charles
Herman, William S
Herring, Harold Keith
Hertel, Elmer William
Hessler, Robert Raymond
Heuberger, Glen (Louis)
Hiatt, Robert Worth
Hibler, Charles Philip
Hichar, Joseph Kenneth
Hickman, Cleveland Pendleton, Jr
Hicks, Ellis Arden
Hile, Ralph Oscar
Hill, Susan Douglas
Hillemann, Howard Herbert
Hillers, Joe Karl
Hilton, Frederick Kelker
Himmel, Keith LaVern
Hinsch, Gertrude Wilma
Hintz, Howard William
Hisaw, Frederick Lee, Jr
Hitchcock, Harold Bradford
Ho, Ju-Shey
Hochman, Benjamin
Hodgson, Harlow James
Hodgson, James Russell
Hoffman, William F
Hoffman, Mark Peter
Hoffmeister, Donald Frederick

Hofslund, Pershing Benard
Hohenboken, William Daniel
Holling, Crawford Stanley
Holman, Leta Jane
Holmes, John Carl
Honigberg, Bronislaw Mark
Hooper, Frank Fincher
Hopkins, Leon Lorraine, Jr
Hopp, William Beecher
Horner, B Elizabeth
Horowitz, Samuel Boris
Horst, G Roy
Howell, Charles DeWitt
Hoyle, Graham
Huddleston, Mary Anne
Hudson, Frank Alden
Hudson, Larry Wilson
Huggins, Sara Espe
Huizinga, Harry William
Hulet, Clarence Veloid
Humason, Gretchen Lyon
Humes, Paul Edwin
Hunsley, Roger Eugene
Hunt, Burton Poulter
Hunter, Frissell Roy
Hunter, Wanda Sanborn
Hurlbut, Henry Winthrop
Husain, Ansar
Hutchinson, George Evelyn
Hutjens, Michael Francis
Illg, Paul Louis
Ingalls, Jesse Ray
Ingersoll, Edwin Marvin
Irwin, Malcolm Robert
Isler, Gene A
Jackson, Dale Latham
Jacobs, John Allen
Jacobs, Merle Emmor
Jahn, J Russell
James, Thomas William
Janovy, John, Jr
Jeffries, Harry Perry
Jennings, William Lamar
Jessop, Nancy Meyer
Jilek, Anthony Francis
Johansen, Peter Herman
Johansson, Tage Sigvard Kjell
Johnsgard, Paul Austin
Johnson, Clyde Edgar, Jr
Johnson, Ivan M
Johnson, John Christopher, Jr
Johnson, Leland Parrish
Johnson, Leslie Kilham
Johnson, Oscar Walter
Johnson, Raymond Earl
Johnson, Rose Mary
Johnson, Vincent Arnold
Johnson, Willis Hugh
Jones, Claiborne Stribling
Jones, Edmund Ruffin, Jr
Jones, Ira
Jopson, Harry Gorgas Michener
Jorgenson, Edsel Carpenter
Judge, Max David
Jungreis, Arthur Martin
Kahl, Marvin Philip
Kamemoto, Fred Isamu
Kampa, Elizabeth Maitland
Kanatzar, Charles Leplie
Kannowski, Paul Bruno
Kapoor, Narinder N
Kastrinos, William, Jr
Kay, Maire Weir
Keeler, Clyde Edgar
Keller, Roger F, Jr
Kelley, George W, Jr
Kellogg, David Wayne
Kelly, Robert Withers
Kelson, Keith R
Kemp, James Dillon
Kennedy, Donald
Kennerly, Thomas Everton, Jr
Kent, George Cantine, Jr
Kent, John Franklin
Kern, Abraham K
Kern, John Polk
Kerr, John Worth
Kessel, Edward Luther
Kessel, Richard Glen
Ketterer, John Joseph
Kezer, James
Khalaf, Kamel T
Kiddy, Charles Augustus
Kilambi, Varad Raj
Kimple, James B
King, John Arthur
King, Thomas B
Kingston, Newton
Kinsman, Donald Markham
Kinzie, Robert Allen, III
Kirkwood, James Benjamine
Klaus, Ewald Fred, Jr
Kline, Edwin A
Klingener, David John
Klopfer, Peter Hubert
Knudsen, Jens Werner, Jr
Kochersberger, Robert Charles
Koepp, Stephen John

Kohn, Alan Jacobs
Kondo, Yoshio
Kopac, Milan James
Korngay, Ervin Thaddeus
Kott, Edward
Kozloff, Eugene Nicholas
Kreider, Jack Leon
Krekeler, Carl Herman
Krishna, Kumar
Krivanek, Jerome Oldrich
Krumbholz, Louis Augustus
Krutzsch, Philip Henry
Kuehne, Robert Andrew
Kunny, Bartholomew Kenneth
Lai-Fook, Joan Elsa J-Ling
Lagler, Karl Frank
Landry, Stuart Omer, Jr
Lane, Charles Edward
La Pointe, Joseph L
Laprade, Mary Hodge
Larson, Wesley P
Larson, Ingemar W
Lauber, Jean Kautz
Lauff, George Howard
Lavoie, Marcel Elphege
Lawrence, Barbara
Lawrence, George Edwin
Lawrence, Jean McVay
Lawrence, Robert G
Lay, Douglas M
Leathem, William Dolars
Leedy, Daniel Loney
Le Febvre, Eugene Allen
Leffel, Emory Childress
Legge, Thomas Nelson
Legler, John Marshall
Lehman, Grace Church
Lehner, James Patrick
Lehrer, William Peter, Jr
Lemay, Yvan
Lennay, Arthur Byron
Leonard, Herbert Arthur
Leonard, Justin Wilkinson
Leonard, Samuel Leeson
Leonard, Walter Raymond
Leopold, Aldo Starker
Levin, Norman Lewis
Levine, Donald Martin
Levins, Richard
Lewis, Herman William
Licht, Paul
Ligon, James David
Liley, Nicholas Robin
Lindeman, Vertus Frank
Linder, Harris Joseph
Linton, Joe R
Little, Ellis Beecher
Loewenthal, Lois Anne
Lofthus, Orin Mervin
LoGerfo, John J
Lober, Werner J
Long, Charles Alan
Long, James Duncan
Loomis, Robert Henry
Lorand, Joyce Bruner
Louch, Charles Dukes
Love, David S
Lowe, Charles Herbert, Jr
Luckman, Cyril Edmund
Lunger, Philip Dupont
Lutsch, Edward F
Lutz, John Ewald
Lynch, John Douglas
Lynn, William Gardner
Lytle, Charles Franklin
Mackie, George Owen
MacRae, Herber F
MacPhee, Craig
Madison, Caroline Rabb
Magalhaes, Hulda
Mahadeva, Madhu Narayan
Mahrt, Jerome L
Main, Robert Andrew
Majors, Rias Hilton
Males, James Robert
Mangat, Balder Singh
Mapp, Frederick Everett
Marler, Peter
Maroney, Samuel Patterson, Jr
Marsden, Halsey M
Marshall, James Dale
Martin, Robert Lawrence
Martin, Walter Edwin
Martof, Bernard Stephen
Mason, William Hickmon
Matteson, Max Richard
Matthews, Samuel Arthur
Mayer, William Vernon
Mazia, Daniel
McArdle, Eugene W
McArthur, William Henry
McAtee, Loyd Thomas
McAuley, Auley Anderson
McClary, Andrew
McCleave, James David
McClellan, John Forbes
McCloskey, Lawrence Richard
McCollum, Clifford Glenn

McCrone, John David
McDaniel, Susan Griffith
McDowell, Robert E, Jr
McFarland, William Norman
McGaha, Young John
McGraw, James Carmichael
McIntyre, Robert Allen, Jr
McKenzie, Frederick Francis
McKinney, Frank
McKinstry, Charles Albert
McLaughlin, David
McLeod, James Archie
McMahan, Elizabeth Anne
McMillan, Joseph Patrick
McMillin, Charles Winslow
McNeil, Robert J
McNutt, Donald Burley
Meade, Robert J
Meagard, Robert O
Meier, Albert Henry
Meinkoth, Norman August
Melton, Carl Wesley
Mendoza, Guillermo
Mengel, Robert Morrow
Menzies, Robert James
Merchant, Henry Clifton
Merrill, Dorothy
Merwin, Ruth Minerva
Meseth, Earl Herbert
Metcalf, Artie Lou
Meyer, Delbert Eugene
Meyer, Henry
Meyer, John Roger
Meyer, Roland Kenneth
Meyer, William Ellis
Michael, William Earl
Michael, Ted C
Middleton, Alex Lewis Aitken
Millemann, Raymond Eagan
Miller, Dorothea Starbuck
Miller, Edwin Lynn
Miller, Elwood Morton
Miller, Grover Cleveland
Miller, John Allen
Miller, Milton Albert
Miller, Robert Cunningham
Millspaugh, Dick Darwin
Miskimen, Mildred
Mitchell, George Ernest, Jr
Mittler, Sidney
Mohler, James Dawson
Mohrle, Harold Leslie
Momberg, Lawrence Henry
Monaco, Paul Amos
Moody, Allen Murdoch
Moore, Hilary Brooke
Moore, James Marvin
Moore, Thomas Edwin
Moree, Ray
Morejohn, G Victor
Morgan, Juliet
Morgans, Leland Foster
Morrissey, J Edward
Mortensen, Edith (Elizabeth)
Morton, Martin Lewis
Mossman, Archie Stanton
Moulton, John Maxim
Muckenthaler, Florian August
Mueller, Helmut Charles
Mueller, Justus Frederick
Muir, Barry Sinclair
Mulcare, Donald J
Mullen, David Anthony
Mullenax, Charles Howard
Muncy, Robert Jess
Munson, Donald Albert
Munyer, Edward Arnold
Murad, John Louis
Murphy, Joseph Robison
Murphy, Ted Daniel
Murray, Bertram George, Jr
Murray, Joseph James, Jr
Musacchia, Xavier Joseph
Musson, Alfred Lyman
Myers, Raymond J
Myser, Willard C
Nace, Paul Ted
Nadler, Charles Fenger
Nagai, Toshio
Nandi, Satyabrata
Nash, Donald Joseph
Naughten, John Charles
Neff, William H
Negus, Norman Curtiss
Nelson, Harry Gladstone
Nelson, Victor Eugene
Nemenzo, Francisco
Neville, Walter Edward, Jr
Newman, James Raney
Nichols, James Ross
Nickum, John Gerald
Nicol, Joseph Arthur Colin
Noetzel, David Martin
Nordie, Frank Gerald
Norden, Carroll Raymond
Norris, Kenneth Stafford
Norris, William Warren
North, Charles A
Nunnally, David Ambrose

ZOOLOGY

Wallentine, Max V
Walley, Willis Wayne
Walters, Lowell Eugene
Ward, Gerald Madison
Ward, Jack A
Ward, Robert T
Warner, Dwain Willard
Warnock, John Edward
Warwick, Everett James
Wass, Marvin Leroy
Wang, Harry (Hsi)
Watling, Harold
Watson, George E. III
Weaver, Morris Eugene
Weeden, Robert Barton
Weibust, Robert Smith
Weisel, George Ferdinand, Jr
Welbourne, Frank Fitzhugh
Welch, Claude Alton
Weller, Harry
Wellington, George Harvey
Wells, Ouida Carolyn
Welsh, James Francis
Welsh, John Henry
Wemyss, Courtney Titus, Jr
Wenner, Adrian Manley
Wenzel, Rupert Leon
Werner, Robert George
Werntz, Henry Oscar
Werth, Robert George
West, Keith P
West, William Lionel
Westcott, Peter Walter
Wharton, George Willard, Jr
Wheeler, Bernice Marion
Wheeler, Marshall Ralph
White, David Arnold
White, Francis Michael
White, Ronald Jerome
Whitaker, McElwyn D
Whiteley, Arthur Henry
Whiteman, Eldon Eugene
Whitley, Larry Stephen
Whitmore, Mary (Elizabeth) Rowe
Whittinghill, Maurice
Wichterman, Ralph
Wickes, Glenn French
Widmayer, Dorothea Jane
Wiebe, Harold T
Wiggs, Alfred James
Wilbur, Henry Miles
Wilcox, Marion Allen
Wildish, David John
Wilhoft, Daniel C
Wilkins, Orin Perry
Williams, George Christopher
Williams, Henry Warrington
Williams, James Lester
Williams, Jesse Bascom
Williams, John C
Williams, Kenneth L
Williams, Norman Eugene
Williamson, Francis Sidney Lainer
Williamson, James Lawrence
Willis, Edwin Roy
Willis, John Steele
Wills, Irvin Andrews
Wilson, Allan Charles
Wilson, Frances G
Wilson, George Rodger
Wilson, James Lester
Wilson, Jane Austell
Wilson, Lowell L
Wilson, Richard Ferrin
Wilson, Richard Howard
Wilson, William Mark Dunlop
Wilt, Fred H
Wing, Elizabeth S
Wing, Hudson Sumner
Winn, Howard Arthur
Wirtz, William Otis II
Wittig, Gertraude Christa
Wohnus, John Frederick
Wolf, Larry Louis
Wolf, Leonard Nicholas
Wolfson, Albert
Wood, Jackie Dale
Wood, Raymond Arthur
Woodin, William Hartman, III
Woodland, John Turner
Wooley, Tyler Anderson
Wooton, Donald Merchant
Wright, Arthur Gilbert
Wright, Kenneth A
Wright, Margaret Ruth
Wright, Philip Lincoln
Wright, Thomas Dodson
Wydoski, Richard Stanley
Wyse, Gordon Arthur
Wysolmerski, Theresa
Yaden, Senkalong
Yarbrough, Charles Gerald
Yeatman, Harry Clay
Yokley, Paul, Jr
Young, Howard Frederick
Young, Janice Edith
Young, Orson Whitney
Yuhas, Joseph George

Yunker, Conrad Erhardt
Zener, Frederick Neyer
Zenisek, Cyril James
Ziegler, John Henry, Jr
Zimmerman, James Roscoe
Zmolek, William G
Zust, Richard Laurence

Animal Behavior

Adler, Kraig (Kerr)
Altmann, Stuart Allen
Ambrose, Harrison William, III
Angstadt, Robert B
Aronson, Lester Ralph
Aspey, Wayne Peter
Avila, Vernon Lee
Bailey, Edward D
Balling, Jan Walter
Balph, David Finley
Barlow, George Webber
Barry, William James
Beauchamp, Gary Keith
Bekoff, Marc
Bekoff, Daniel Arthur
Berrill, Michael
Biggs, Walter Clark, Jr
Boch, Rolf
Boese, Gilbert Karyle
Bradbury, Jack W
Brown, Judith Adele
Brown, Richard Dean
Brown, Robert Zanes
Buskirk, Ruth Elizabeth
Clark, David Lee
Clarke, Robert Francis
Clayton, Dale Leonard
Coles, Richard Warren
Colvin, Dallas Verne
Cornell, James Morris
Dane, Benjamin
Denniston, Rollin H II
Diakow, Carol
Dingle, Richard Douglas Hugh
Drickamer, Lee Charles
Dunning, Dorothy Covalt
Eaton, Gordon Gray
Emlen, Stephen Thompson
Erickson, Mary Marilla
Esch, Harald Erich
Evans, Roger Malcolm
Falls, James Bruce
Ferguson, Gary Wright
Frings, Mable Ruth
Galusha, Joseph G, Jr
Gauthreaux, Sidney Anthony, Jr
Goodrich, Michael Alan
Gould, Edwin
Graham, William Joseph
Greenspan, Beverly Naomi
Gude, Richard Hunter
Hadley, Wayne Franklin
Hale, Edgar Brewer
Hart, Benjamin Leslie
Hartwick, Robert Frank
Herman, Harry August
Hermkind, William Frank
Hess, Eckhard Heinrich
Hirsch, Jerry
Hopkins, Carl Douglas
Horn, Henry Stainken
Howell, Joseph Corwin
Jacobs, William Wood, Jr
Jenssen, Thomas Alan
Jolly, Alison Bishop
Kanzler, Philip Nelson
Kaufmann, Walter Wilhekm
Keeton, William Tinsley
Keiper, Ronald R
Krekorian, Charles O'Neil
Krieg, David Charles
Lee, Ching-Tse
Lehner, Philip Nelson
Levine, Louis
Lockner, Frederick Russell
Loebich, Karen Elizabeth
Logan, Cheryl Ann
Lynn, Robert Thomas
Madison, Dale Martin
Marcellini, Dale Leroy
Marsh, Michael Pierce
Marshall, Joseph Andrew
Matthews, Robert Wendell
Miller, Maria G
Miller, Verna Jean
Moriarty, Daniel Delmar, Jr
Morris, Glenn Karl
Moynihan, Martin Humphrey
Mueller, Helmut Charles
Muller-Schwarze, Dietland
Mulligan, James Anthony
Muric, Jan O
Muul, Illar
Myrberg, Arthur August, Jr
Nadler, Ronald David
Newton, David C
Olla, Bori Liborio
Peek, Frank Willard

Peters, Roger Paul
Pettijohn, Terry Frank
Phillips, Richard Edward
Pion, Lawrence V
Prince, Harold Hoopes
Quanstrom, Walter Roy
Quertermus, Carl John, Jr
Rabach, Steve
Radabaugh, Dennis Charles
Reese, Ernst S
Roberts, Mervin Francis
Rovner, Jerome Sylvan
Rushforth, Norman B
Salmon, Michael
Shapiro, Lorin James
Shaw, Evelyn (Mrs Fred Wertheim)
Silberglied, Robert Elliot
Silver, Rae
Simmel, Edward Clemens
Smith, Douglas Graham
Smith, Susan May
Smith, Susan Trussell
Smith, W John
Snowdon, Charles Thomas
Soltysik, Seczesny Stefan
Stewart, Anne Marie
Steiner, Andre Louis
Stokes, Allen Woodruff
Sullivan, Daniel Joseph
Tavolga, William Nicolai
Te Paske Everett Russell
Tobach, Ethel
Topoff, Howard Ronald
Vandenbergh, John Garry
Vessey, Stephen H
Vestal, Bedford Mather
Victoria, Janice Krutak
Walther, Fritz R
Ward, Jack A
Waring, George Houstoun
Weigl, Peter Douglas
West Eberhard, Mary Jane
Wiley, Robert Bruce
Williams, Henry Warrington
Williams, Timothy Cheney
Wilson, Richard Howard
Windsor, Donald Montgomery
Wolfe, James Leonard
Wright, Anthony Anne
Wright, John Cushing
Yuhas, Joseph George

Animal Breeding

Ahlschwede, William T
Arave, Clive W
Bailey, Curtiss Merkel
Balch, Donald James
Bennett, James Austin
Benson, Robert Haynes
Benyshek, Larry L
Berger, Philip Jeffrey
Bernard, Camille Stephen
Boylan, William J
Brackelsberg, Paul O
Branton, Cecil
Brokaw, Bryan Edward
Brown, Connell Jean
Burnside, Edward Blair
Carpenter, James Andrew, Jr
Carter, Robert Clifton
Cartwright, Thomas Campbell
Chapman, Arthur Barclay
Charette, Laurent A
Christians, Charles J
Comstock, Ralph Ernest
Conlin, Bernard Joseph
Cowan, William Allen
Crenshaw, David Brooks
Cundiff, Larry Verl
Cunningham, Peter John
Dettmers, Almut Edel
Dickerson, Gordon Edwin
Doane, Ted H
Dollahon, James Clifford
Dunbar, Robert Standish, Jr
England, David James
Fletcher, Jesse Lane
Frahm, Richard R
Franke, Donald Edward
Freeman, Albert Eugene
Gowe, Robb Shelton
Gray, Richard C
Green, Willard Wynn
Gregory, Keith Eugene
Gyles, Nicholas Roy
Haenlein, George Friedrich Wilhelm
Harvey, Walter Robert
Hohenboken, William Daniel
Holland, Lewis
Holtmann, Wilfried
Howell, William Edwin
Jamison, Haley M
Koch, Robert Milton
Koger, Marvin

Kotman, Roy Milton
Kratzer, D Dal
Lasley, John Foster
Lush, Jay Laurence
MacNaughton, William Norman
Magee, William Thomas
Mather, Robert Eugene
Matthews, Doyle Jensen
McAllister, Alan Jackson
McClung, Marvin Richard
McCormick, William Conner
McGilliard, Lon Dee
McGuire, John Albert
Meadows, Clinton Elwood
Miller, Paul Dean
Miller, Kenneth Philip
Moore, John V
Moore, Edward A
Mudge, Joseph William
Musgrave, Stanley Dean
Nagai, Jiro
Nehms, George E
Nelson, Edward A
Nicolai, John Henry, Jr
Nielsen, Merlyn Keith
Norman, Howard Duane
Oliver-Padilla, Fernando Luis
Park, Robert Lynn
Parker, Robert J
Patnish, Otto Floyd
Patterson, Troy B
Patterson, Ronald Earl
Pearson, Hobart Frank
Peters, Raymond Glen
Peterson, Carl Eugene
Pollak, Emil John
Powell, Rex Lynn
Ramsay, John Martin
Rempel, William Ewert
Rennie, James Clarence
Rutledge, Jackie Joe
Scarth, Robert Douglas
Schalles, Robert R
Schoonover, Carroll Owen
Seale, Marvin Ernest
Shelton, James Maurice
Shrode, Robert Ray
Sims, John Albert
Swanson, Vern Bernard
Terrill, Clair Elman
Thayne, William V
Thompson, Carl Eugene
Thrift, Frederick Aaron
Touchberry, Robert Walton
Turner, James William
Vernon, Eugene Haworth
Vesely, John Anthony
Vint, Larry Francis
Vogt, Dale William
Wadell, Lyle H
Walton, Robert Eugene
Wells, Milton Ernest
Whatley, James Arnold
White, John Marvin
Whiteman, Joe V
William, Richard James
Wyatt, Andy Jack

Animal Ecology

Balda, Russell Paul
Boag, David Archibald
Bovbjerg, Richard Viggo
Brand, Raymond Howard
Brewer, Robert Hyde
Bryden, Robert Richmond
Colvin, Dallas Verne
Dimmick, Ralph W
Ellis, James Edgar
Engelmann, Manfred David
Fichter, Edson Harvey
Funderburg, John Broadus, Jr
Hanson, Hugh
Harvey, Michael Joseph
Hayward, Charles Lynn
Hazen, William Eugene
Hespenheide, Henry August, III
Hirth, Harold Frederick
Holmes, Richard Turner
Huey, Raymond Brunson
Istock, Conrad Alan
Jackson, William Bruce
Johnson, Wendel J
Jonas, Robert James
Kennington, Garth Stanford
Laughlin, Harold Emerson
Martin, Robert Eugene
Mertz, David B
Miller, Richard Samuel
Mook, Leonard Jan
Mook, David Jay
Mumford, Russell Eugene
Pearcy, William Gordon
Provo, Marvin Monroe
Quay, Thomas Lavelle
Reeder, William Glase
Robel, Robert Joseph
Rymon, Larry Marine
Salt, George William
Scherer, Robert C
Schmid, William Dale

Schultz, Vincent
Shoemaker, Hurst Hugh
Shontz, Charles Jack
Smith, Jean E
Spencer, Dwight Louis
Spillett, James Juan
Stewart, Robin Kenny
Tanner, James Taylor
Terman, Charles Richard
Verner, Jared
Weller, Milton Webster
Wheat, John David
Wiens, John Anthony

Animal Genetics
Ash, William James
Bennett, Cecil Jackson
Berg, Roy Torgny
Blackwell, Robert Leighton
Bohren, Bernard Benjamin
Brinks, James S
Busch, Robert Edward
Byrne, Barbara Jean McManamy
Byrne, Bruce Campbell
Chambers, Doyle
Chase, Herman Burleigh
Cole, Randall Knight
Collins, Walter Marshall
Crockett, Joe Richard
Crowl, Robert Harold
Dillard, Emmett Urcey
Eastwood, Basil R
Eldridge, Franklin Elmer
Emsley, Alan Burns
Fausch, Hemer David
Fechheimer, Nathan S
Fox, James David
Fredeen, Howard T
Garnett, Ian
Garwood, Vernon Abington
Gaunt, Stanley Newkirk
Harris, Dewey Lynn
Havenstein, Gerald B
Hickman, Charles Garner
Hrubant, Henry Everett
Jerome, Frederick Nelson
King, Steven Clarence
Legates, James Edward
Ludwin, Isadore
Marcum, James Benton
Marks, Henry L
Marlowe, Thomas Johnson
Martin, Truman Glen
McDaniel, Benjamin Thomas
Morgan, Walter Clifford
Quevedo, Walter Cole, Jr
Rasmusen, Benjamin Arthur
Roubicek, Carl Ben
Shelby, Charles Edwin
Shreffler, Carol Kauffman
Somes, Ralph Gilmore, Jr
Swiger, Louis Andre
Towner, Richard Henry
Woodward, James Crawford

Animal Husbandry
Alexander, Robert Allen
Anthony, Wilson Brady
Baker, Bryan, Jr
Baker, Frank Sloan, Jr
Bedell, Thomas Erwin
Bell, Thomas Donald
Bovard, Kenly Paul
Briggs, Hilton Marshall
Buchanan, Marion Lynn
Buck, Charles Frank
Carroll, Floyd Dale
Clark, Jack L
Cole, Clarence Lorraine
Cunha, Tony Joseph
Daniel O'Dell G
Day, Billy Neil
Deans, Robert Jack
Dowe, Thomas Whitfield
Durham, Ralph Marion
Embry, Lawrence Bryan
Evans, Lee E
Gobble, James Lawrence
Griffin, Sumner Albert
Grummer, Robert Henry
Guyer, Paul Quentin
Harris, Lorin E
Hathorn, Fred
Hedrick, Harold Burdette
Heidenreich, Charles John
Henderson, Hugh E
Henneman, Harold Albert
Henry, Wayne E
Hodgson, Charles Worth
Hoefer, Jacob A
Hollandbeck, Richard
Holt, Leroy Henry
Johnson, Elton Loyd
Johnson, George Robert
Johnson, LaDon Jerome
Jones, James Robert
Jordan, Robert Manseau

Kennington, Mack Humpherys
Keyes, Everett A
Klosterman, Earle Wayne
Lewis, John Morgan
Lindley, Charles Edward
Litton, George Washington
Madsen, Milton Andrew
Mattingly, Steele F
Meiske, Jay C
Menzies, Carl Stephen
Miller, John Ivan
Nelson, Ronald Harvey
Newland, Herman William
Paules, Leon H
Pease, Lawrence Honeyman
Peters, John Burl
Ramsey, Clovis Boyd
Ray, Maurice L
Ritchie, Harlan
Rust, Joseph William
Sharma, Udhishtra Deva
Smith, Edgar Fitzhugh
Spalding, Robert Wilber
Squiers, Clifford Dale
Stockbridge, Robert R
Taylor, Jack Crossman
Taysom, Elvin David
Tillman, Allen Douglas
Van Dongen, Cornelis Godefridus
Varney, William York
Wakeman, Donald Lee
Warner, Donald R
Warren, William Michael
Webb, Robert Johnson
Welch, James Alexander
Zinn, Dale Wendel

Animal Nutrition
Abe, Ronald Kuraso
Acker, Duane Calvin
Adams, Albert Whitten
Adams, Frank William
Adams, Holyoke Purinton
Adams, Richard Sanford
Aherne, Francis Xavier
Ainslie, Harry Robert
Albert, Waco W
Albin, Robert Custer
Allen, Neil Keith
Alsmeyer, William Louis
Ames, Stanley Richard
Ammerman, Clarence Bailey
Anderson, Cyrus Vincent
Anderson, Donald Lindsay
Anderson, Gerald Clifton
Anderson, Martin Dean
Anderson, Melvin Joseph
Anderson, Russell K
Apgar, William P
Arrington, Lewis Roberts
Baker, David H
Barnhart, Charles Elmer
Barth, Karl M
Bartley, Erle Edwin (St Clair)
Baumgardner, John Henry
Baumgardt, Billy Ray
Beames, R M
Beardsley, Daniel Waldo
Becker, Donald Eugene
Beeson, William Malcolm
Bell, John Milton
Bell, Marvin Carl
Belzile, Rene
Bentley, Orville George
Bergen, Werner Gerhard
Bertrand, Joseph E
Bird, Herbert Roderick
Black, Alex
Blaylock, Lynn Gail
Bloss, Ronald Edward
Bohman, Verle Rudolph
Bowland, John Patterson
Braemer, Allen C
Brent, Benny Earl
Britzman, Darwin Gene
Brooks, Coy Clifton
Brown, Herbert
Brown, Leonard D
Brown, Lynn Ranney
Brown, William Hedrick
Browning, Charles Benton
Buchanan-Smith, Jock Gordon
Bull, Leonard Seth
Bull, Richard C
Burns, Joseph Charles
Burton, John Heslop
Bush, Leon F
Byers, Floyd Michael
Calhoun, Millard Clayton
Carew, Lyndon Belmont, Jr
Carlson, Charles Wendell
Carr, Scott Bligh
Caskey, Charles (Dirxon), Jr
Chah, Cheong Choo
Chalupa, William Victor
Chamberlain, Charles Calvin
Chapman, Herbert L, Jr
Chappell, Guy Lee Monty

Church, David Calvin
Clanton, Donald Cather
Clawson, Albert J
Cline, Jack Henry
Coalson, James Arthur
Cock, Lorne M
Colby, Robert William
Combs, George Ernest
Combs, Gerald Fuson
Conrad, Joseph H
Cook, Robert Merold
Coppock, Carl Edward
Corbin, James Edward
Cowan, Robert Lee
Cox, Dennis Henry
Cox, James Lee
Cramer, David Alan
Cromwell, Gary Leon
Cullison, Arthur Edison
Cummings, Kenneth Ross
Curtin, Leo Vincent
Danielson, David Murray
Danke, Richard John
Davis, George Kelso
DeGeeter, Melvin Joseph
Dehority, Burk Allyn
De Lay, Roger Lee
Dinusson, William Erling
Dollar, Alexander M
Donefer, Eugene
Dua, Prem Nath
Eaton, Hamilton Dean
Edwards, Robert Lee
Elliot, James I
Elliott, John Murray
Elliott, Ralph Francis
Ellis, William C
Ely, Ray E
Emerick, Royce Jasper
Emery, Roy Saltsman
Erfle, James David
Erickson, Duane Otto
Essig, Henry Werner
Evans, Joseph Liston
Everett, James Peek, Jr
Ewan, Richard Colin
Ewing, Solon Alexander
Farlin, Stanley Dean
Fenner, Heinrich
Fisk, George Raymond
Flatt, William Perry
Fonnesbeck, Paul Vance
Fontenot, Joseph Paul
Forrest, Robert J
Fox, Danny Gene
Froseth, John Allen
Gard, Don Irvin
Gardner, Robert Wayne
Garrett, William Norbert
Germann, Albert Frederick Ottomar, II
Geurin, Hobart Beach
Girouard, Rustum Ernest, Jr
Gleaves, Earl William
Glimp, Hudson A
Goodrich, Richard Douglas
Graber, George
Grandhi, Raja Ratnam
Greathouse, Terrence Ray
Green, George G
Haenlein, George Friedrich Wilhelm
Hale, William Harris
Hall, O Glen
Halloran, Hobart Rooker
Harbaugh, Daniel David
Harbers, Leniel H
Hardison, Wesley Aurel
Harmon, Bud Gene
Harris, Lorin E
Harris, Ralph Rogers
Harrold, Robert Lee
Hatfield, Efton Everett
Haynes, Emmit Howard
Hays, Virgil Wilford
Hazzard, DeWitt George
Heaney, David Paul
Hedde, Richard Duane
Heidebrecht, A Allen
Heinemann, Wilton Walter
Heitman, Hubert, Jr
Hembry, Foster Glen
Henges, James Franklin, Jr
Hershberger, Truman Verne
Heuberger, Glen (Louis)
Hibbs, John William
Hidiroglou, Michael
Hill, Douglas Calvert
Hillier, James Calvin
Hinds, Frank Crossman
Hinners, Scott W
Hintz, Harold Franklin
Hironaka, Robert
Hitchcock, John Paul
Hodson, Harold H, Jr
Hogue, Douglas Emerson
Holck, Gary LeRoy
Holden, Palmer Joseph
Hollis, Gilbert Ray
Horvath, Donald James

Howard, W Terry
Howes, Arol Dean
Huber, John Talmage
Husby, Fredric Martin
Hutcheson, David Paul
Hutchinson, Harold David
Ingle, Donald Lee
Jacobson, Norman Leonard
Jahn, Ernesto
Jensen, Aldon Homan
Jensen, Leo Stanley
Jeter, Max Albert
Johnson, Donald Eugene
Johnson, Paul Edwin
Johnson, Richard James
Johnson, Robert Eugene
Johnson, Ronald Roy
Johnson, William Lawrence
Jordan, Charles Edwin
Katz, Robert Sanford
Keener, Harry Allan
Kercher, Conrad J
Kertz, Alois Francis
Kienholz, Eldon W
Klatte, Fred John, Jr
Klay, Robert Frank
Klein, Richard Gale
Knott, Fred Nelson
Koch, Berl Amos
Krider, Jake Luther
Kromann, Rodney P
Kutches, Alexander Joseph
Lambert, Maurice Reed
Lane, Alfred Glen
Lane, Gary (Thomas)
Lassiter, James William
Leach, Roland Melville, Jr
Leatherwood, James M
Lee, Daniel Dixon, Jr
Leibbrandt, Vernon Dean
Lewis, Roscoe Warfield
Li, Ming Fang
Likuski, Henry John
Lippke, Hagen
Lister, Earl Edward
Little, Charles Oran
Lofgreen, Glen Pehr
Loggins, Phillip Edwards
Lohman, Timothy George
Long, Charles H
Long, Theodore Alfred
Longmire, Dennis B
Loosli, John Kasper
Lowrey, Robert S
Luce, William Glenn
Lyford, Sidney John, Jr
Madsen, Fred Christian
Madsen, Louis Linden
Mahan, Donald Clarence
Martin, Jerry Junior
Martz, Fredric A
Mason, Tim Robert
Matsushima, John K
McCroskey, Jack E
McDowell, Lee Russell
McGilliard, A Dare
McKinley, Raymond Earl
McMillen, Warren Newton
Meade, Thomas Leroy
Menge, Henry
Mente, Glen Allen
Merrill, William George
Mertens, David Roy
Miller, Elwyn Ritter
Miller, James Kincheloe
Miller, Robert Frederick
Miller, William Jack
Monson, William Joye
Moody, Edward Grant
Moore, John Edward
Moore, Talmadge Seab
Morris, James Grant
Morrison, Spencer Horton
Morrison, William D
Moss, Buelon Rexford
Muir, Larry Allen
Murdock, Fenoi R
Musgrave, Stanley Dean
Myers, George Scott, Jr
Neagle, Lyle H
Neathery, Milton White
Nelson, Arnold Bernard
Nelson, D Kent
Nelson, Donald Dewey
Nelson, Talmadge Seab
Nesbit, Arthur Henderson
Neumann, Alvin Ludwig
Newman, Clarence Walter
Nicholson, Hugh Hampson
Nicholson, J W G
Nishimura, John Francis
Noble, Robert Lee
Nockels, Cheryl Ferris
Noland, Paul Robert
O'Connor, Jeremiah Joseph
Oldfield, James Edmund
Otagaki, Kenneth Kengo
Otjen, Robert Raymond
Owen, Bruce Douglas

Owens, Fredric Newell
Palmquist, Donald Leonard
Pate, Findlay Moye
Patterson, Eugene B
Pensack, Joseph Michael
Peo, Ernest Ramy, Jr
Perry, Samuel Cassius
Perry, Tilden Wayne
Peter, Albert P
Pettigrew, James Eugene, Jr
Phillips, William Ernest John
Pigden, Wallace James
Pleasants, Julian Randolph
Pond, Wilson Gideon
Porter, Gilbert Harris
Poston, Hugh Arthur
Pothoven, Marvin Arlo
Potter, Emerson Lucine
Powell, George Wythe
Preston, Rodney LeRoy
Prior, Ronald Leon
Purkhiser, E Dale
Quisenberry, John Henry
Ralston, Allen Thurman
Ramsey, Harold Arch
Rapp, Janet Lorraine Cooper
Rasmussen, Edith Svoboda
Raun, Arthur Phillip
Raun, Ned S
Rechcigl, Miloslav, Jr
Reese, Nathan Allan
Reid, John Thomas
Reiser, Sheldon
Reitnour, Clarence Melvin
Reynolds, Paul Joseph
Rice, Richard W
Richardson, Drayrford
Riggs, John Kamm
Roberts, William Kenneth
Robertson, John Connell
Rogler, John Charles
Ronning, Magnar
Ross, Oscar Burr
Rubin, Max
Rucker, Robert Blain
Rusoff, Louis Leon
Salsbury, Robert Lawrence
Sasser, Lyle Blaine
Satter, Larry Dean
Schugel, LaVerne
Schuh, James Donald
Seerley, Robert Wayne
Sell, Jerry Lee
Sewell, Homer B
Sewell, Raymond F
Shaw, Joseph Clement
Sheppard, Alan Jonathan
Sherrod, Lloyd B
Shirley, Ray Louis
Shor, Aaron Louis
Sleeth, Rhule Bailey
Slen, Sydney Bernard
Slinger, Stanley James
Smart, Lewis Isaac
Smith, Garmond Stanley
Smith, Gary E
Smith, John Douglas
Smith, Keith James
Smith, Sedgewick Edward
Soares, Joseph Henry, Jr
Spining, Arthur Milton, III
Stake, Paul Erik
Staples, George Emmett
Stephens, Noel, Jr
Stephenson, Edward Luther
Stoddard, George Edward
Stoewsand, Gilbert Saari
Stuedemann, John Alfred
Sudweeks, Earl Max
Summers, Charles Eugene
Talley, Spurgeon Morris
Taylor, Terry Mac
Theurer, Clark Brent
Thomas, John William
Thomas, Oscar Otto
Thomas, Roy Orlando
Thompson, James Tipton
Tomlin, Don C
Totusek, Robert
Travis, Hugh Farrant
Tyznik, William John
Udall, Robert Hovey
Ullrey, Duane Earl
Vander Noot, George Ward
Vandersall, John Henry
VanSoest, Peter John
Varner, Larry Weldon
Veum, Trygve Lauritz
Walden, Donald E
Wallace, Harold Dean
Wallenius, Roger Wynn
Wangsness, Paul Jerome
Ward, John K
Warner, Richard G
Waterhouse, Howard N
Watts, Alva Burl

Webb, Kenneth Emerson, Jr
Weichenthal, Burton Arthur
Weir, William Carl
Welch, James Graham
Wesoloski, George D
Weswig, Paul Henry
Wharton, Ferdinand Decatur, Jr
White, Thomas Wayne
Wilbur, Robert Daniel
Wilkening, Marvin C
Wilkinson, James Lawrence
Williamson, William Mark Dunlop
Wing, James Marvin
Wise, George Herman
Wise, Milton Bee
Witwer, Leland S
Woods, Walter Ralph
Young, Jerry Wesley
Young, Robert Kevin
Zimmerman, Dean R

Animal Parasitology
Allen, Rex Wayne
Amrein, Yost Ursus Lucius
Behlow, Robert Frank
Collins, Jeffery Allen
Cuckler, Ashton Clinton
Davis, William S
Eckman, Michael Kent
Esch, Gerald Wisler
Hughins, Ernest Jay
Isely, Duane
Meyer, Marvin Clinton
Mead, Robert Warren
Madin, Stewart Harvey
Reed, Raymond Edgar
Schaefer, Frank William, III
Seidel, Michael Caspar
Roberts, Charles Speer
Sparano, Benjamin Michael
Switzer, William Paul
Vegors, Halsey Hugh
Williams, James Carl
Wilson, Grant Ivins

Animal Pathology
Bolton, Wesson Dudley
Chute, Harold LeRoy
Goss, Leonard Joyce
Gray, Jack Ellsworth
Hoerlein, Alvin Bernard
Locke, Louis Noah

Animal Physiology
Abbott, Frank Sidney
Alikhan, Muhammad Akhtar
Andrews, Frederick Newcomb
Aschbacher, Peter William
Bailey, Charles Basil Mansfield
Barker, Hal B
Barone, Milo C
Beasley, Philip Gene
Bellinger, Larry Lee
Black, Craig Patrick
Black, Donald Leighton
Bogdonoff, Philip David, Jr
Bradshaw, William Emmons
Bremel, Robert Duane
Brick, Robert Wayne
Bruce, David Stewart
Carter, Clint Earl
Caruolo, Edward Vitangelo
Cohenour, Francis D
Covalt-Dunning, Dorothy
Cragle, Raymond George
Cupps, Perry Thomas
Curl, Samuel Everett
Demers, Jean-Marie
DiBenedetto, Frank Edward
Dietz, Thomas Howard
Dinnick, John Frederick
Duby, Robert T
Dziuk, Harold Edmund
Eales, John Geoffrey
Ehlers, Melvin H
Ellington, Earl Franklin
Erb, Ralph Eugene
Estergreen, Victor Line
Evans, James Warren
Federico, Olga Maria
Fletcher, Garth L
Folden, Dewey Bray, Jr
Foote, Robert Hutchinson
Foote, Warren Christopher
Foote, Wilford Darrell
Fowler, Arnold K
Fuguay, John Wade
Gasdorf, Edgar Carl
Goetsch, Gerald D
Graham, Arthur Renfree
Gruszczyk, Jerome Henry
Haning, Quentin C
Hansel, William
Harrveld, Anthonie Van
Harris, Grover Cleveland, Jr
Hartner, William Christopher
Heitmann, Richard Norman

Hill, Donald Louis
Hogg, Edd Coolidge
Horwitz, Barbara Ann
House, Edwin W
Hunter, Alan Graham
Kerstetter, Theodore Harvey
Kirkpatrick, Jay Franklin
Kitts, Warren Dale
Kleerekoper, Herman
Knoll, Jack
Krishnamurti, Cuddalore Rajagopal
Lemay, Jean-Paul
Loy, Robert Graves
Lumb, Roger H
Marple, Dennis Neil
Martin, Elden William
Martin, Jerry Junior
Marx, George Donald
McClain, John A
McKinstry, Donald Michael
McMahon, Brian Robert
Mixner, John Paulding
Moore, John David
Nelson, Darren Melvin
Pandit, Hemchandra M
Patrick, Thomas Everette
Pekas, Jerome Charles
Perrault, Marcel Joseph
Piper, Edgar L
Prange, Henry Davies
Reece, Ralph Parlette
Reed, Randall R
Roller, Michael Harris
Rosen, Jeffrey Kenneth
Russek, Mauricio Berman
Schlough, James Sherwyn
Schroeder, Allen C
Senff, Robert S
Shibley, Gilbert A
Shumway, Richard Phil
Simpson, Robert John
Smith, Clifford James
Smith, Robert Emrie
Spielman, Arless A
Staszak, David John
Stewart, Anne Marie
Stufflebeam, Charles Edward
Sullivan, Charlotte Murdoch
Sutterlin, Arnold M
Swierstra, Ernest Emke
Tilton, James Earl
Titelbaum, Sydney
Toews, Daniel Peter
Tone, James N
Topel, David G
Trelka, Dennis George
Tucker, Herbert Allen
Turner, John K
Van Tienhoven, Ari
Van Ummersen, Claire Ann
Wayman, Oliver
Wegner, Thomas Norman
Willett, Lynn Brunson
Wittbank, James N
Witherspoon, James Donald
Yarns, Dale A
Young, Bruce Arthur

Animal Virology
Allen, Patton Tolbert
Bachenheimer, Steven Larry
Bourgaux, Pierre
Bryans, John Thomas
Chow, Tsu Ling
Fleischmann, William Robert, Jr
Gray, Carla Winlund
Handy, Farouk Mohamed
Lake, Robert Samuel
Noble-Harvey, Jane
O'Callaghan, Dennis John
Panem, Sandra
Pearson, George Denton
Rosenberger, John Knox
Sattar, Syed Abdus
Stewart, Robert Bruce
Wagner, Edward Knapp

Economic Zoology
Baker, Maurice Frank
Shuyler, Harlan R

Ethology
Aspey, Wayne Peter
Baylis, Jeffrey Rowe
Black-Cleworth, Patricia Ann
Brooks, Ronald James
Brown, Jerram L
Burger, Joanna
Burger, Edward Raymond
Carter, Carol Sue
Chase, Ivan Dmitri
Cole, James Edward
DeGhett, Victor John
Eisenberg, John Frederick
Evans, Llewellyn Thomas
Farish, Donald James
Fentress, John Carroll
Ferron, Jean H

Ficken, Millicent Sigler
Ficken, Robert W
Fox, Michael Wilson
Friedl, Ernestine
Fry, Christine L
Gambs, Roger Duane
Geist, Valerius
Gould, James L
Graves, Hannon B
Greenberg, Neil
Haas, Richard
Hailman, Jack Parker
Harrington, Fred Haddox
Hazlett, Brian Arthur
Howard, Walter Egner
Ingold, Donald Alfred
Jacobs, William Wood, Jr
Jenni, Donald Alison
Keenleyside, Miles Hugh Alston
Kleiman, Devra Gail
Kleinhammer, Erich
Krear, Harry Robert
Lloyd, James Armon
Lockard, Robert Bruce
McKenzie, Joseph Addison
McKeown, James Preston
Miller, Helen Carter
Miller, Rudolph J
Moller, Peter
Moore, Nelson Jay
Myton, Becky Ann
Noakes, David Lloyd George
Orcutt, Frederic Scott, Jr
Oring, Lewis Warren
Phillips, Robert Rhodes
Ralls, Katherine Smith
Rosenblum, Leonard Allen
Rowland, William Joseph
Russock, Howard Israel
Sarles, Harvey B
Schein, Martin Warren
Schleidt, Wolfgang Matthias
Schusterman, Ronald Jay
Shapiro, Lorin James
Slobodchikoff, Constantine Nicholas
Southern, William Edward
Svendsen, Gerald Eugene
Taylor, Douglas Hiram
Terman, Max R
Wiley, Richard Haven, Jr
Willis, Edwin O'Neill

Experimental Zoology
Foulkes, Robert Hugh
Goldring, Irene P
Green, Edwin Alfred
Schroeder, Paul Clemens
Stanley, Melissa Sue Miliam

Laboratory Animal Science
Barnes, Raymond D
Finer, Jerry
Fox, Richard Romaine
Myers, David Daniel
Pick, James Raymond
Poiley, Samuel Milton
Poole, Calvin Merten
Prasad, Suresh
Schneider, Henry Peter
Simmonds, Richard Carroll
Stillions, Merle C
Van Hoosier, Gerald L, Jr

Marine Zoology
Abbott, Marie Bohm-Lambert
Baker, James Haskell
Banta, William Claude
Barnard, Jerry Laurens
Beeman, Robert D
Bookhout, Cazlyn Green
Castro, Peter
Cheng, Lanna
Dexter, Deborah Mary
Fontaine, Arthur Robert
Gardiner, Lion Frederick
Grice, George Daniel, Jr
Harrison, Robert Edwin
Heavers, Barbara Ann
Juskevice, John Anthony
Lamberts, Austin E
Losey, George Spahr, Jr
Michel, Harding B
Miller, John Wesley, Jr
Ostarello, Georgiandra Little
Pawson, David Leo
Rausch, James Peter
Reish, Donald James
Sandifer, Paul Alan
Smith, Edmund Hobart
Smith, Frederick George Walton
Stainken, Dennis M
Strubsaker, Paul James
Venkataramiah, Amaraneni
Wells, John West
Wilkes, Stanley Northrup
Wood, Forrest Glenn

Paleozoology
Barghusen, Herbert Richard
Bolt, John Ryan
Richardson, Eugene Stanley, Jr
Schram, Frederick R
Semken, Holmes Alford, Jr

Systematic Zoology
Bell, Ross Taylor
Boyden, Alan Arthur
Brady, Allen Roy
Fauchald, Kristian
Friar, Wayne
Gloyd, Howard Kay
Hobbs, Horton Holcombe, Jr
Hoff, Clarence Clayton
Hoffman, Richard Lawrence
Leviton, Alan Edward
McGhee, Charles Robert
Mercer, Malcolm Clarence
Rising, James David
Schwartz, Albert
Sibley, Charles Gald
Williams, Austin Beatty

Zoogeography
Berra, Tim Martin
Power, Dennis Michael
Reuther, Ronald Theodore
Simoons, Frederick John
Wilson, John William, III

ZOOLOGY, INVERTEBRATE

Invertebrate Zoology
Abbott, Donald Putnam
Aldrich, Frederick Allen
Aldrich, Lewis Eugene, Jr
Allen, Charles Marvin
Allison, Terry C
Apley, Martyn Linn
Armstrong, Howard Wayne
Arrington, Richard, Jr
Baker, Elizabeth McIntosh
Barnes, Robert Drane
Beasley, Clark Wayne
Berger, Jacques
Berrend, Robert E
Birch, Robert Lee
Black, Joe Bernard
Black, William Francis
Blackwelder, Richard Eliot
Blake, James A
Blankespoor, Harvey Dale
Boolootian, Richard Andrew
Boswell, James Louis
Bourne, Neil
Bousfield, Edward Lloyd
Broad, Alfred Carter
Brown, Harley Procter
Brusca, Gary J
Brusca, Richard Charles
Buchsbaum, Ralph
Bushnell, John Horace
Campbell, James L
Carlson, Oscar Verdell
Carter, Richard Thomas
Causey, Nell Bevel
Clark, Kerry Bruce
Clausen, Conrad Duane
Clifford, Hugh Fleming
Collins, Richard Lapointe
Costlow, John DeForest
Couser, Raymond Dowell
Crocker, Denton Winslow
Crowell, (Prince) Sears, (Jr)
Cutress, Charles Ernest
Darlington, Julian Trueheart
Davis, Charles (Carroll)
Davis, Luckett Vanderford
Davis, Marjorie
Dehnel, Paul Augustus
DeMartini, John
Dimock, Ronald Vilroy, Jr
Dudley, Patricia
Durflinger, Elizabeth Ward
Eckelbarger, Kevin Jay
Engemann, Joseph George
Engster, Maryann Sandra
Fairbanks, Laurence Dee
Fasbender, M Veronica
Feldmesser, Julius
Fell, Howard Barraclough
Fennel, William Edward
Ferguson, John Carruthers
Fields, William Gordon
Fish, Arthur Geoffrey
Fleminger, Abraham
Foster, William Burnham
Freeman, Carl Jackson, Jr
Friauf, James Joseph
Fritchman, Harry Kier, II
George, Robert Porter
Gilbert, John Jouett

Gilmore, Claude Raymond
Gilmour, Thomas Henry Johnstone
Gladfelter, William Bayard
Gonor, Jefferson John
Grossman, Herbert H
Hadfield, Michael Gale
Hahnert, William Franklin
Hand, Cadet Hammond, Jr
Hand, William Gordon
Harry, Harold William
Hartman, Willard Daniel
Hermans, Colin Olmsted
Hetzel, Howard Roy
Hillman, Robert Edward
Hirshfield, Henry Israel
Holland, Nicholas Drew
Holt, Perry Cecil
Hotchkiss, Frederick Hatfield Clark
Hughes, Roy Linwood, Jr
Jeffries, William Bowman
Jenkins, Marie Magdalen
Jones, Meredith L
Keiser, Edmund Davis, Jr
Kenk, Roman
Kenk, Vida Carmen
Kinsella, John Michael
Kirsteuer, Ernst
Knowlton, Robert Earle
Kudenov, Jerry David
Landers, Earl James
LaRow, Edward J
Lawrence, James Lester
Lawson, James Everett
Lenhoff, Howard Maer
Lochhead, John Hutchison
Longest, William Douglas
Lutz, Paul E
Lyke, Edward Bonsteel
Mangum, Charlotte P
Manning, Raymond B
McCrary, Anne Bowden
McGavock, Walter Donald
McLean, James H
McLean, Norman, Jr
Meade, Thomas Gerald
Miles, Charles David
Miller, Richard Lee
Morin, James Gunnar
Murdock, Gordon Robert
Muscatine, Leonard
Myers, Thomas DeWitt
Neiland, Kenneth Alfred
Nelson, Diane Roddy
Newberry, Andrew Todd
Oglesby, Larry Calmer
Osborne, Paul James
Ott, Karen Jacobs
Park, Tai Soo
Patton, Wendell Keeler
Pearse, Vicki Buchsbaum
Pennington, Tully Sanford
Pohlo, Ross
Pollock, Leland Wells
Porter, Thomas Wayne
Pugh, Jean Elizabeth
Puglia, Charles Raymond
Rees, John Tonks
Rehder, Harald Alfred
Rennie, Thomas Howard
Rice, Carl Stephen
Rice, Mary Esther
Riser, Nathan Wendell
Roche, Edward Towne
Rosing, Lorraine Morin
Ryan, Edward Parsons
Sacks, Martin
St John, Philip Alan
Sasner, John Joseph, Jr
Sawyer, Roy Thomas
Scarborough, Charles Spurgeon
Shuster, Carl Nathaniel, Jr
Simpson, Leonard
Simpson, Margaret
Sissom, Stanley Lewis
Solem, Alan
Spencer, Larry T
Stamper, Maynard N
Swan, Emery Frederick
Szal, Roger Andrew
Taylor, Dwight Willard
Tomlinson, Jack Trish
Tozloski, Albert Henry
Train, Carl T
Virkar, Raghunath Atmaram
Waffle, Elizabeth Lenora
Wainwright, Stephen Andrew
Walter, Waldemar Melchert
Ward, Diana Vaiela
Watson, Wynnfield Young
Weiss, Mitchell Joseph
West, Warwick Reed, Jr
Westervelt, Clinton Albert, Jr
Williams, Eliot Churchill
Williams, Russell Raymond
Windsor, Donald Arthur
Wise, Charles Davidson
Woodwick, Keith Harris
Wyatt, Ellis Junior

Yarnall, John Lee
Yoder, Wayne Alva
Zimmer, Russel Leonard
Zinn, Donald Joseph
Zischke, James Albert
Zottoli, Robert

Helminthology
Ameel, Donald Jules
Dyer, William Gerald
Freeman, Reino Samuel
Graff, Darrell Jay
James, Hugo A
Kruidenier, Francis Jeremiah
Kuntz, Robert Elroy
Lautenschlager, Edward Walter
Olivier, Louis John
Pearson, John Carwardine
Trainer, John Ezra, Jr
Ulmer, Martin John
Uricchio, William Andrew

Invertebrate Ecology
Berry, James William
Crawford, Clifford Smeed
Driscoll, Egbert Gotzian
Dundee, Dolores Saunders
Gable, Michael
Keith, Donald Edwards
Kraft, Kenneth J
McMahon, Robert Francis, III
Paris, Oscar Hall
Rutherford, James Charles
Ryals, George Lynwood, Jr

Invertebrate Embryology
Fell, Paul Erven
Perkins, Bobby Frank
Potswald, Herbert Eugene

Invertebrate Paleontology
Amsden, Thomas William
Bailey, Richard Hendricks
Batten, Roger Lyman
Berdan, Jean Milton
Beus, Stanley S
Boardman, Richard Stanton
Bolton, Thomas Elwood
Bonem, Rena Mae
Bostwick, David Arthur
Bretsky, Peter William
Bretsky, Sara (Su) Stewart
Brower, James Clinton
Carter, John Lyman
Chronic, John
Cifelli, Richard
Copeland, Murray John
Cramer, Howard Ross
Cuffey, Roger James
Darby, David G
Diffendal, Robert Francis, Jr
Douglass, Raymond Charles
DuBar, Jules R
Durham, John Wyatt
Erickson, John Mark
Fagerstrom, John Alfred
Feldmann, Rodney Mansfield
Finks, Robert Melvin
Firby, James R
Fischer, William Alfred
Frankel, Larry
Gibson, Thomas George
Glawe, Lloyd Neil
Grant, Richard Evans
Hall, Donald D
Harper, Charles Wood, Jr
Hazel, Joseph Ernest
Heaslip, William Graham
Hoare, Richard David
Imbrie, John
Keen, (Angeline) Myra
Kern, John Philip
Kier, Porter Martin
Kirtley, David Warren
Lane, Bernard Owen
Laufeld, Sven
LeMone, David V
Lesperance, Pierre J
Lochman-Balk, Christina
MacClintock, Copeland
McKinney, Frank Kenneth
Meyer, David Lachlan
Moore, Reginald George
Morin, James Gunnar
Nye, Osborne Barr, Jr
Palmer, Allison Ralph
Pannella, Giorgio
Parsley, Ronald Lee
Pojeta, John, Jr
Popenoe, Willis Parkison
Raup, David Malcolm
Rowell, Albert John
St Jean, Joseph, Jr
Scolaro, Reginald Joseph
Shaw, Frederick Carleton
Sprinkle, James (Thomas)
Strimple, Harrell Leroy
Thayer, Charles Walter

Tischler, Herbert
Titus, Robert Charles
Vokes, Emily Hoskins
Waller, Thomas Richard
Warshauer, Steven Michael
Williams, Michael Eugene
Wilson, Edward Carl
Yochelson, Ellis Leon
Zullo, Victor August

Invertebrate Pathology
Allen, George E
Caneday, Thomas Donald
Davidson, Elizabeth West
Drake, Edward Lawson
Federici, Brian Anthony
Granados, Robert R
Johnson, Phyllis Truth
Mix, Michael Cary
Reichelderfer, Charles Franklin
Tripp, Marenes Robert
Wilson, William Thomas

Invertebrate Physiology
Bayne, Christopher Jeffrey
Blanquet, Richard Steven
Bliss, Dorothy Elizabeth
Bowers, William Sigmond
Burnett, Bryan Reeder
Bursey, Charles Robert
Carlson, Albert Dewayne, Jr
Carney, Gordon C
Davey, Kenneth George
Dendinger, James Elmer
Fourtner, Charles Russell
Friedl, Frank Edward
Hammen, Carl Schlee
Hubschman, Jerry Henry
Hyde, Cornelia Tuten
Pardy, Rosevelt Lawrence
Prusch, Robert Daniel
Puckett, Hugh
Rodrick, Gary Eugene
Roys, Chester Crosby
Schatzlein, Frank Charles
Seabrook, William Davidson
Shepherd, Julian Granville
Silverthorn, Saidee Unglaub
Simons, Daniel J
Spencer, Andrew Nigel
Steel, Colin Geoffrey Hendry
Steele, Vladislava Julie
Sweeney, Daryl Charles
Tombes, Averett Snead
Towle, Albert
Webb, Helen Marguerite
Webb, Rodney A
Wittig, Kenneth Paul
Wolfe, Allan Frederick
Young, Stephen Dean

Malacology
Abbott, Robert Tucker
Aguayo, Carlos G
Bickel, Edwin David
Bleakney, John Sherman
Boss, Kenneth Jay
Carriker, Melbourne Romaine
Clarke, Arthur Haddleton, Jr
Coan, Eugene Victor
Davis, George Morgan
Du Pont, John Eleuthere
Emerson, William Keith
Etges, Frank Joseph
Flowers, Ralph Wills
Gilbertson, Donald Edmund
Gugler, Carl Wesley
Harman, Willard Nelson
Heard, William Herman
Houston, Roy Seamands
Kraemer, Louise Russert
Mead, Albert Raymond
Michelson, Edward Harlan
Moore, Donald Richard
Morse, Mary Patricia
Murray, Harold Dixon
Pace, Gary Lee
Radwin, George E
Rawls, Hugh Cecil
Richards, Charles Selwyn
Robertson, Robert
Rosewater, Joseph
Thompson, Fred G
Turner, Ruth Dixon
van der Schalie, Henry
Webb, Glenn R

Nematology
Adamo, Joseph Albert
Betz, Daniel Oliver, Jr
Bird, George W
Brodie, Bill Burl
Buecher, Edward Joseph
Caveness, Fields Earl
Chitwood, May Belle Hutson
Di Sanzo, Carmine Pasqualino
Edwards, Dale Ivan
Ellis, Kenneth Carl

Endo, Burton Yoshiaki
Ferris, John Mason
Ferris, Virginia Rogers
Fielding, Max Jae
Hannon, Chancellor Irving
Harrison, Robert Edwin
Hasbrouck, Edward Ralph
Hirschmann, Hedwig
Hope, William Duane
Jenkins, William Robert
Jensen, Harold James
Johnson, Alva William
Johnson, Edsel Carpenter
Kinloch, Robert Armstrong
Maio, Simon E
Mankau, Reinhold
Mankau, Sarojam Kurudamannil
Mark, Daniel Lee
Marks, Charles Francis
McClure, Michael Allen
Mountain, William Buckingham
Myers, Ronald Fenner
Nickle, William R
O'Bannon, John Horatio
Olthof, Theodorus Hendrikus Antonius
Orbin, David Paul
Paracer, Surindar Mohan
Petrielie, Richard P
Poinar, George O, Jr
Raski, Dewey John
Russell, Charles Clayton
Sasser, Joseph Neal
Sher, Samuel Alexis
Stokes, Donald Eugene
Taylor, Albert Lee
Thames, Walter Hendrix, Jr
Turco, Charles Paul
Viglierchio, David Richard

Protozoology
Angell, Robert Walker
Antipa, Gregory Alexis
Arnold, Zach McLendon
Balamuth, William
Baniforth, Stuart Shoosmith
Barnett, Audrey
Barrett, James Martin
Barrow, James Howell, Jr
Berger, Jacques
Boggs, Nathaniel, Jr
Borror, Arthur Charles
Bradbury, Phyllis Clarke
Burbanck, William Dudley
Buttrey, Benton Wilson
Chacharonis, Peter
Clark, Glen W
Collins, Richard Lapointe
Cooley, Nelson Reede
Corliss, John Ozro
Decker, Joan Elise
Dewey, Virginia Caroline
Diller, William Frey, Jr
Dodd, Everett E
Dollahon, Norman Richard
Downing, William Lawrence
Duke, Eleanor Lyon
Dwyer, Dennis Michael
Ernst, John Verion
Evans, Frederick Read
Francis, David W
Gabel, James Russel
Grassnick, Robert Alan
Hanson, William Lewis
Harmon, Wallace Morrow
Hartig, William John
Hayes, Robert Edward
Herman, Robert
Hibberd, Larry Eugene
Holz, George Gilbert, Jr
Isquith, Irwin R
Jahn, Theodore Louis
Jensen, Emron Alfred
Jones, Edward Eugene
Jones, Justine H
Kantor, Sidney
Keeshan, Margaret M
Kloetzel, John Arthur
Kutler, Kenneth Latimer
Levine, Norman Dion
Lilly, Daniel McQuillan
Lorch, Joan
Lund, Everett Eugene
Marquardt, William Charles
McLaughlin, Roy Earl
McLoughlin, Donald Keith
Meglitsch, Paul Allen
Miller, Brinton Marshall
Napolitano, Joseph J
Nielsen, Peter James
Nigrelli, Ross Franco
Osterud, Kenneth Leland
Outka, Darryll E
Packchanian, Ardzrooony (Arthur)
Panitz, Eric
Penn, James H
Pierson, Bernice Frances
Pokorny, Kathryn Stein

Ranganathan, Vajapeyam S
Reilly, Marguerite
Repak, Arthur Jack
Ridgeway, Bill Tom
Ritter, Edward
Rogers, Frances Arlene
Rosenberg, Lauren Emery
Rudzinska, Maria Anna
Saxe, Leroy Hallowell, Jr
Schuster, Frederick Lee
Simpson, Larry P
Southards, Carroll J
Sprague, Victor
Spuller, Robert L
Stump, Alexander Bell
Sutton, William Wallace
Tamar, Henry
Tamburro, Kathleen O'Connell
Tibbs, John Francisco
Torch, Reuben
Vaughn, Charles Melvin
Vetterling, John Martin
von Zellen, Bruce Walfred
Walker, Glenn Kenneth
Weinshank, Donald Jerome
Weis, Dale Stern
Wessenberg, Harry Sanders
Wilhelm, Walter Eugene
Wilson, David Everett
Wood, Sherwin Francis
Yongue, William Henry

ZOOLOGY, VERTEBRATE

Vertebrate Zoology
Allen, Ted Tipton
Anderson, Sydney
Axtell, Ralph William
Barbour, Roger William
Bartell, Marvin H
Biggs, Walter Clark, Jr
Binford, Laurence Charles
Black, John David
Bolt, John Ryan
Boyer, Don Raymond
Boykins, Ernest Aloysius, Jr
Brand, Leonard Roy
Brandon, Ronald Arthur
Brittan, Martin Ralph
Brockway, Barbara Fink
Brown, Bryce Cardigan
Brown, Herbert Allen
Brown, Robert Harrison
Bushman, John Branson
Caldwell, Melba Carstarphen
Chase, Robert Silmon, Jr
Cliff, Frank Samuel
Cockrum, Elmer Lendell
Cohen, Robert Roy
Cole, James Edward
Collier, Gerald
Cook, David Russell
Cruz, Alexander
Dalquest, Walter Woelberg
Davenport, Leslie Bryan, Jr
Davis, John
Deacon, James Everett
Degenhardt, William George
DeWolfe, Barbara Blanchard Oakeson
Dilger, William C
Dixon, Keith Lee
Dumas, Philip Conrad
Dunlap, Donald Gene
Dusi, Julian Luigi
Echternacht, Arthur Charles
Ely, Charles Adelbert
Erickson, Howard Ralph
Evans, Kenneth Jack
Fabian, Michael William
Ferguson, Denzel Edward
Findley, James Smith
Finley, Robert Bryon, Jr
Forbes, Richard Bryan
Frederickson, Richard William
Freeman, Harry W
Frost, Herbert Hamilton
Garner, Herschel Whitaker
Gauthreaux, Sidney Anthony, Jr
Genelly, Richard Emmett
Goertz, John William
Goodwin, Robert Earle
Guthrie, Daniel Albert
Hall, Eugene Raymond
Hall, Joseph Glenn
Harrington, Edward James
Harris, Arthur Horne
Haugh, John Richard
Hazard, Evan Brandao
Heinsohn, George T
Hendrickson, Herbert T
Hendrickson, John Roscoe
Higgins, Robert Price
Holliman, Dan Clark
Holsinger, John Robert
Hooper, Emmet Thurman, Jr

Houck, Warren Jacob
Howell, Thomas Raymond
Hundley, Louis Reams
James, Ted Ralph
Jameson, Everett Williams, Jr
Johnson, Oliver William
Jones, Duvall Albert
Jones, J Knox, Jr
Kaufmann, John Henry
Kay, Fenton Ray
Kennedy, Michael Lynn
Klein, Harold George
Kluge, Arnold Girard
Koford, Carl Buckingham
Krejsa, Richard Joseph
Krieg, David Charles
Lane, James Dale
Larsen, John Herbert, Jr
Lawlor, Anna Catherine
Leitner, Philip
Lewis, Robert Earl
Linder, Allan David
Lindsay, Hague Leland, Jr
Lutton, Lewis Montfort
MacIntyre, Giles T
MacMahon, James A
Martin, Robert Frederick
McCann, Lester J
McCarley, Wardlow Howard
McCauley, Robert Henry, Jr
McCourt, Robert Perry
McManus, John Joseph
Meacham, William Ross
Miller, Helen Carter
Miller, Richard Gordon
Mount, Robert Hughes
Murie, Martin L
Murray, Joan Baird
Myers, Richard F
New, John G
Norton, Virginia Marino
Oaks, Emily Caywood Jordan
Odell, Daniel Keith
Orr, Orty Edwin
Osgood, David William
Palmer, Ralph Simon
Parsons, Thomas Sturges
Pfeiffer, Egbert Wheeler
Price, John Worthington
Priddy, Ralph Banta
Pyburn, William F
Raitt, Ralph James, Jr
Rand, Austin Loomer
Reuther, Ronald Theodore
Richmond, Milo Eugene
Ritland, Richard Martin
Roberson, Walter Volley
Roecker, Robert Maar
Roest, Aryan Ingomar
Rogers, Frances Arlene
Rosenthal, Marcia White
Rossman, Douglas Athon
Roth, Roland Ray
Ruffer, David G
Ruibal, Rodolfo
Ryan, Richard Alexander
Sanford, L G
Schmitz, Eugene H
Sexton, Owen J
Shadowen, Herbert Edwin
Shellhammer, Howard Stephen
Silliman, Emmanuel I
Smith, Donald Alan
Smith, Hobart Muir
Smith, Philip Wayne
Smith, Ronald Earl
Snyder, Dana Paul
Stallcup, William Blackburn, Jr
Standing, Keith M
Sternberg, Charles Mortram
Stewart, Glenn Raymond
Strawcutter, Richard Edward
Thompson, William Lay
Tinkle, Donald Ward
Tomich, Prosper Quentin
Trost, Charles Henry
Turner, Frederick Brown
Vance, Velma Joyce
Vaughan, Terry Alfred
Vestal, Bedford Mather
Von Bloeker, Jack Christian, Jr
Wade, William Frank
Waring, George Houstoun
Wasserman, Aaron Osias
Waters, Joseph Hemenway
Webb, Robert G
Weigel, Robert David
Weisbrod, Alan Richard
Weston, Henry Griggs, Jr
Wharton, Charles Heizer
White, Clayton M
White, George V S
White, John Anderson
Wilder, Cleo Duke, Jr
Winkelmann, John Roland
Wirtz, John Harold
Wolfe, James Leonard
Woolcott, William Starnold

Wunder, Bruce Arnold

Avian Pathology
Dick, John Walter
Duke, Gary Earl
Dunlop, William Robert
Gross, Walter Burnham
Henderson, Wilson
Hwang, Jen
Kahan, I Howard
Zander, Donald Victor

Avian Physiology
Bailey, R L
Bayer, Robert Clark
Dane, Charles Warren
Durfee, Wayne King
Ferguson, Thomas Morgan
Gessaman, James A
Hughes, Maryanne Robinson
Johnston, David Ware
Kern, Michael Don
Marquardt, Ronald Ralph
McGinnis, Charles Henry, Jr
Mellen, William James
Peterson, Ronald A
Polin, Donald
Scholz, Richard W
Thaxton, James Paul
Wentworth, Bernard C
Wheeler, Robert Stevenson
Wilson, Wilbor Owens
Wolford, John Henry

Herpetology
Anderson, James Donald
Auffenberg, Walter
Banta, Benjamin Harrison
Barton, Alexander James
Bell, Edwin Lewis, II
Black, Jeffrey Howard
Brode, William Edward
Brodie, Edmund Darrell, Jr
Chrapliwy, Peter Stanley
Christiansen, James Learned
Cohen, Nathan Wolf
Conant, Roger
Cupp, Paul Vernon, Jr
Darrow, Thomas D
Dixon, James Ray
Dowling, Herndon Glenn, (Jr)
Duellman, William Edward
Dundee, Harold A
Emsley, Michael Gordon
Ernst, Carl Henry
Etheridge, Richard Emmett
Ferguson, Gary Wright
Feuer, Robert Charles
Flury, Alvin Godfrey
Forester, Donald Charles
Fouquette, Martin John, Jr
Freeman, John Richardson
Gaudin, Anthony J
Gibbons, J Whitfield
Goin, Coleman Jett
Gorman, George Charles
Gottschang, Jack Louis
Greding, Edward J, Jr
Harris, Lester Earle, Jr
Healy, William Ryder
Hensley, Marvin Max
Highton, Richard
Hulsey, James Edward
Inger, Robert Frederick
Jones, Kirkland Lee
Karlstrom, Ernest Leonard
List, James Carl
Loomis, Richard Biggar
Loop, Michael Stuart
Lynch, John Douglas
Mahmoud, Ibrahim Younis
Marcellini, Dale Leroy
Marx, Hymen
Maslin, Thomas Paul
Mays, Charles Edwin
McCoy, Clarence John
McDaniel, Van Rick
McDowell, Sam Booker
McGinnis, Samuel M
Means, Bruce
Mecham, John Stephen
Metter, Dean Edward
Minton, Sherman Anthony
Murphy, George Graham
Myers, Charles William
Netting, Morris Graham
Nussbaum, Ronald Archie
Orejas-Miranda, Braulio
Parker, Richard Bewley
Perill, Stephen Arthur
Presch, William Frederick
Preston, William Burton
Rivero, Juan Arturo
Rubin, David Charles
Russell, Anthony Patrick
Sanders, Ottys E
Savage, Jay Mathers

Scott, Norman Jackson, Jr
Sever, David Michael
Tilley, Stephen George
Valentine, Barry Dean
Voris, Harold K
Walker, James Martin
Weintraub, Joel D
Werner, John Kirwin
Williams, Ernest Edward
Worthington, Richard Dane
Zug, George R
Zweifel, Richard George

Ichthyology

Anderson, William Dewey, Jr
Anderson, William Wyatt
Archibald, Kalman Dale
Baird, Ronald C
Barbour, Clyde D
Barclay, Lee Armstead, (Jr)
Barlow, George Webber
Baylis, Jeffrey Rowe
Beadles, John Kenneth
Behnke, Robert J
Berra, Tim Martin
Birdsong, Ray Stuart
Bohlke, James Erwin
Bond, Carl Eldon
Booke, Henry Edward
Boschung, Herbert Theodore
Bradbury, Margaret G
Branson, Branley Allan
Briggs, John Carmon
Bryan, Charles F
Chen, Lo-Chai
Cichocki, Frederick Paul
Clay, William Marion
Cohen, Daniel Morris
Collette, Bruce Baden
Counselman, C J
Crane, Jules M, Jr
Crawford, Ronald Ward
Crossman, Edwin John
Daiber, Franklin Carl
Davis, Billy J
Dawson, Charles Eric
Deubler, Earl Edward, Jr
De Witt, Hugh Hamilton
Echelle, Anthony Allan
Eschmeyer, William Neil
Etnier, David Allen
Evans, David Hudson
Evans, Robert Ralph
Fahy, William Earl
Feddern, Henry A
Fierstine, Harry Lee
Follett, Wilbur Irving
Foster, Neal Robert
Gibbs, Robert Henry, Jr
Gilbert, Carter Rowell
Gosline, William
Graham, Charles Raymond, Jr
Greenfield, David Wayne
Hill, Loren Gilbert
Hobson, Edmund Schofield, Jr
Hoff, James Gaven
Horn, Michael Hastings
Hoyt, Robert Dan
Hubbs, Carl Leavitt
Hubbs, Clark
Ibara, Richard Mamoru
Johnson, Robert Karl
Kallman, Klaus D
Kiley, Charles Walter
Knapp, Leslie W
Kuo, Ching-Ming
Lagueux, Robert
Lindquist, David Gregory
Lindsey, Casimir Charles
Loyacano, Harold Anthony
Lund, William Albert, Jr
Magnin, Etienne Nicolas
Mather, Frank Jewett, III
McCosker, John E
McInerney, John Edward
McKenney, Thomas William
Mead, Giles Willis
Menzel, Bruce Willard
Migdalski, Edward Charles
Miller, George C
Miller, Robert Rush
Miller, Rudolph J
Minckley, Wendell Lee
Molles, Manuel Carl, Jr
Moyle, Peter Briggs
Musick, John A
Nelson, Gareth Jon
Nelson, Joseph Schieser
Newman, Murray Arthur
Page, Larry Merle
Perlmutter, Alfred
Pflieger, William Leo
Quast, Jay Charles
Raney, Edward Cowden
Reisman, Howard Maurice
Relyea, Kenneth George
Reno, Harley
Richards, William Joseph

Rivas, Luis Rene
Robins, Charles Richard
Robison, Henry Welborn
Rofen, Robert Rees
Rojo, Alfonso
Rosen, Donn Eric
Ross, Robert Donald
Schultz, Leonard Peter
Schultz, Roland Jack
Schwartz, Frank Joseph
Scott, William Beverley
Shipp, Robert Lewis
Simco, Bill Al
Snelson, Franklin F, Jr
Strawn, Robert Kirk
Sutkus, Royal Dallas
Swift, Camm Churchill
Taber, Charles Alec
Taylor, Leighton Robert, Jr
Taylor, William Ralph
Thomerson, Jamie E
Tinker, Spencer Wilkie
Tyler, James Chase
Underhill, Adna Heaton
Walker, Boyd Wallace
Wallace, Charles Ray
Walters, Vladimir
Warner, Edward Nelson
Weitzman, Stanley Howard
Williams, Steven Frank
Woods, Loren Paul

Mammalogy

Ackermann, Uwe
Adams, Clark Edward
Ahl, Alwynelle S
Bailey, Alfred Marshall
Baker, Mary Ann
Banks, Richard Charles
Barrington, Burness Austin, Jr
Bettice, John Allen
Birney, Elmer Clea
Bradshaw, Gordon Van Rensselaer
Brocke, Rainer H
Brown, Larry Nelson
Burt, William Henry
Busch, Robert Stearns
Choate, Jerry Ronald
Christianson, Lee (Edward)
Clothier, Ronald Raymond
Comeau, Roger William
Coolidge, Harold Jefferson
Courtney, Gladys A
Cowan, Ian McTaggart
Cross, Stephen P
Dagg, Anne Innis
Davis, Wayne Harry
D'Eliscu, Peter Neal
Douglas, Charles Leigh
Ehrhart, Llewellyn McDowell
Enders, Robert Kendall
Ernst, Carl Henry
Fay, Francis Hollis
Fenton, M Brock
Fisher, Robert L
Fleharty, Eugene
Forman, G Lawrence
Genoways, Hugh Howard
Giere, Frederic Arthur
Glass, Bryan Pettigrew
Gottschang, Jack Louis
Greenhall, Arthur Merwin
Greer, John Keever
Gunderson, Harvey Lorraine
Handley, Charles Overton, Jr
Hardy, Ross
Harington, Charles Richard
Harry, George Yost
Hatfield, Donald Marshall
Hatt, Robert Torrens
Heidt, Gary A
Hershkovitz, Philip
Hill, Clyde Alfred
Houck, Warren Jacob
Huckaby, David George
Ingersol, Robert Harding
Iverson, Stuart Leroy
Jensen, Donald Reed
Jensen, John Neil
Johnson, Murray Leathers
Johnson, Patricia R
Jones, Clyde Joe
Kelley, Fenton Crosland
Kirkland, Gordon Laidlaw, Jr
Klein, David Robert
Koopman, Karl Friedrich
Lackey, James Alden
Lidicker, William Zander, Jr
Lindsay, Dwight Marsee
Linzey, Alicia Vogt
Lovejoy, David Arnold
Lukens, Paul W, Jr
Martin, Robert Lawrence
McDaniel, Van Rick
McKeever, Sturgis
Meyer, Frederick Richard

Miller, Clarence Allan
Mitchell, Henry Andrew
Mock, Orin Bailey
Moore, Joseph Curtis
Myers, Philip
Novotny, Robert Thomas
O'Farrell, Michael John
O'Farrell, Thomas Paul
Orr, James Anthony
Orr, Robert Thomas
Patton, James Lloyd
Perrin, William Fergus
Person, Steven John
Peterson, Randolph Lee
Pettigrew, Raleigh K
Phillips, Carleton Jaffrey
Pober, Zalmon
Repenning, Charles Albert
Richards, Lawrence Phillips
Rosenzweig, Michael Leo
Russell, Robert Julian, Jr
Saunders, Harry Link
Saunders, Jack K, Jr
Schlitter, Duane A
Schmidly, David James
Seabloom, Robert W
Setzer, Henry Wilfred
Smith, Howard Duane
Snow, Beatrice Lee
Starrett, Andrew
Strecker, Robert Louis
Thomas, Charles S
Trapp, Gene Robert
Tuttle, Merlin Devere
Van De Graaff, Kent Marshall
Van Gelder, Richard George
Wetzel, Ralph Martin
Williams, Daniel Frank
Winkelmann, John Roland
Ziegler, Alan Conrad
Zieman, Joseph Crowe, Jr
Zimmerman, Jay Alan

Ornithology

Adkisson, Curtis Samuel
Alcorn, Gordon Dee
Aldrich, John Warren
Amadon, Dean
Arnold, Keith Alan
Arvey, Martin Dale
Austin, Oliver Luther, Jr
Baepler, Donald H
Bailey, Alfred Marshall
Baird, James
Baldwin, Paul Herbert
Balgooyen, Thomas Gerrit
Banks, Richard Charles
Barlow, Jon Charles
Beecher, William John
Behle, William Harroun
Berger, Andrew John
Berrett, Delwyn Green
Biaggi, Virgilio, Jr
Binford, Laurence Charles
Blake, Emmet Reid
Bond, James
Borror, Donald Joyce
Bowman, Robert Irvin
Brewer, Richard (Dean)
Bruning, Donald Francis
Bull, John
Cade, Thomas Joseph
Carter, William Alfred
Cassell, Joseph Franklin
Caswell, Herbert Hall, Jr
Chipley, Robert MacNeill
Clark, George Alfred, Jr
Clement, Roland Charles
Clench, Mary Heimerdinger
Cogswell, Howard Lyman
Collins, Charles Thompson
Cooch, Frederick Graham
Cox, George W
Crouch, James Ensign
Cuthbert, Nicholas Le Huray
Dickinson, Joshua Clifton, Jr
Dinsmore, James Jay
DuMont, Philip Atkinson
Du Pont, John Eleuthere
Eaton, Stephen Woodman
Edwards, Ernest Preston
Eisenmann, Eugene
Erickson, Mary Marilla
Eyer, Lester Emery
Eyster, Marshall Blackwell
Ford, Norman Lee
Foster, Mercedes S
George, William
Gill, Frank Bennington
Goddard, Stephen
Godfrey, William Earl
Greenlaw, Jon Stanley
Grinnell, Lawrence Irving
Gustafson, John Alfred
Hanebrink, Earl L
Hardy, John William
Heppner, Frank Henry
Hicks, David L

Howe, Marshall Atherton
Hubbard, John Patrick
Humphrey, Philip Strong
Hunt, Lawrence Barrie
Huntington, Charles Ellsworth
Jackson, Jerome Alan
Johnson, Eric Van
Johnson, Ned Keith
Johnson, Richard Evan
Kale, Herbert William II
Kee, David Thomas
Kendeigh, Samuel Charles
Kessel, Brina
Kuenzel, Wayne John
Lancaster, Douglas
Lanyon, Wesley Edwin
Lawrence, Louise de Kiriline
Leck, Charles Frederick
Lowery, George Hines, Jr
Lunk, William Allan
Mahan, Harold Dean
Marshall, Joe Truesdell, Jr
Maxwell, George Ralph, II
Mayfield, Harold Ford
McCoy, John J
McGowan, Robert William
McLean, Edward Bruce
McNeil, Raymond
Mehner, John Frederick
Messersmith, Donald Howard
Mewaldt, Leonard Richard
Monroe, Burt Leavelle, Jr
Monson, Gale (Wendell)
Moore, Nelson Jay
Mosher, James Arthur
Myres, Miles Timothy
Newman, George Allen
Nice, Margaret Morse
Nolan, Val, Jr
Norris, Russell Taplin
North, Charles A
Odum, Eugene Pleasants
Orcutt, Frederic Scott, Jr
Orr, Robert Thomas
Owre, Oscar Theodore
Parker, James Willard
Parkes, Kenneth Carroll
Parmelee, David Freeland
Parnell, James Franklin
Pearson, David Leander
Pennington, Tully Sanford
Petersen, Arnold Jerome
Peterson, Roger Tory
Pettingill, Olin Sewall
Peyton, Leonard James
Phillips, Allan Robert
Pough, Richard Hooper
Prescott, Kenneth Wade
Putnam, Loren Smith
Raikow, Robert Jay
Reilly, Edgar Milton, Jr
Robbins, Chandler Seymour
Rogers, Charles Henry
Russell, Stephen Mims
Rylander, Michael Kent
Schreiber, Ralph Walter
Schwalbe, Paul Wayman
Short, Lester Le Roy, Jr
Smith, Neal Griffith
Smith, Wendell Phillips
Southern, William Edward
Spofford, Sally Hoyt
Staebler, Arthur E
Stein, Robert Carrington
Stettenheim, Peter
Stevenson, Henry Miller
Stewart, Paul Alva
Sutton, George Miksch
Tate, James Leroy, Jr
Taylor, Walter Kingsley
Thompson, Charles Frederick
Thoresen, Asa Clifford
Threlfall, William
Thurber, Walter Arthur
Trainer, John Ezra, Sr
Traylor, Melvin Alvah
Twiest, Gilbert Lee
Tyler, Jack D
Wallace, George John
Warter, Stuart L
Watson, George E, III
Wauer, Roland H
Webster, Jackson Dan
Weise, Charles Martin
West, George Curtiss
Wetmore, Alexander
Willis, Edwin O'Neill
Wolk, Robert George
Woolfenden, Glen Everett
Zimmerman, Dale A

Primatology

Ankel-Simons, Friderun Annursel
Bernstein, Irwin Samuel
Blood, Benjamin Donald
Bramblett, Claud Allen
Buettner-Janusch, John
Chevalier-Skolnikoff, Suzanne

Cooper, Robert Woodrow
Delson, Eric
Harrison, Richard Miller
Helmuth, Herman Siegfried
Hertig, Arthur Tremain
Hylander, William Leroy
Jolly, Clifford J
Kurland, Jeffrey Arnold
Lindburg, Donald Gilson
Mann, Alan Eugene
Maples, William Ross
Nadler, Ronald David
Oppenheimer, John Reed
Pilbeam, David Roger
Porter, James Armer, Jr
Povar, Morris Leon
Ripley, Suzanne
Rosen, Stephen I
Shostak, Stanley
Siegel, Michael Ian
Singh, Ripu Daman
Tappen, Neil Campbell
Tattersall, Ian Michael
Van Horn, Richard Norman
Weiss, Mark Lawrence

Systematic Ichthyology
Dooley, James Keith
Douglas, Neil Harrison
Fitch, John Edgar
Haedrich, Richard L
Lachner, Ernest Albert
McAllister, Donald Evan
Miller, Robert Victor
Morrow, James Edwin, Jr
Yerger, Ralph William

Vertebrate Anatomy
Alexander, A Allan
Atkins, David Lynn
Bentley, Cleo L
Bissaillon, Andre
Chantell, Charles J
Coulombe, Harry N
Delahunta, Alexander
Galton, Peter Malcolm
Hart, Nathan Hoult
Kallen, Frank Clements
Liberti, Alfred Vincent
Ownby, Charlotte Ledbetter
Plakke, Ronald Keith
St. Clair, Lorenz Edward
Stump, John Edward
Waters, James Frederick
Whiting, Anne Margaret

Vertebrate Biology
Blair, William Franklin
Caldwell, David Keller
Cunningham, Harry N, Jr
Hecht, Max Knobler
Hoppe, David Matthew
Jackson, Crawford Gardner, Jr
Layne, James Nathaniel

Mayhew, Wilbur Waldo
Nickerson, Max Allen
Nordan, Harold Cecil
Platt, William Joshua, III
Scudday, James Franklin
Secoy, Diane Marie
Smith, Lorraine Catherine
Snyder, David Hilton
Turner, Larry Webster
Uzzell, Thomas Marshall, Jr
Vial, James Leslie
Wake, Marvalee H
Zimmerman, Earl Graves

Vertebrate Ecology
Alexander, Maurice Myron
Allen, Durward Leon
Anderson, Paul Knight
Barbehenn, Kyle Ray
Batson, Jack David
Beaver, Donald Loyd
Beck, John R
Best, Troy Lee
Blem, Charles R
Bongiorno, Salvatore F
Boylan, Elizabeth Shippee
Boyette, Joseph Greene
Brant, Daniel (Hosmer)
Cook, Robert Sewell
Crowell, Kenneth L
Davis, Richard B
Delco, Exalton Alfonso, Jr
Dole, Jim
Fowle, Charles David
Franca, Edward Nathaniel Lloyd
Frydendall, Merrill J
Fuller, William Albert
Gennaro, Antonio Louis
Hardin, James William
Harris, Van Thomas
Johnson, Donald Ralph
Keast, James Allen
Lewke, Robert Edward
Maher, William J
McMillan, John Frank
Minock, Michael Edward
Moore, Robert Emmett
Nelson, David Herman
Orr, Lowell Preston
Parker, William Skinker
Phinney, George Jay
Porter, Kenneth Raymond
Raun, Gerald George
Seibert, Henri Cleret
Stewart, Margaret McBride
Walimo, Olof Charles
Webb, William Leonard
Wecker, Stanley C
Whitaker, John O, Jr
Williams, Olwen

Vertebrate Embryology
Ballard, William Whitney

Fowler, James A
Goudsmit, Esther Marianne
Heim, Werner George
Johnston, Perry Max
Keys, Charles Everel
Walton, Barbara Ann

Vertebrate Morphology
Bock, Walter Joseph
Boord, Robert Lennis
Cracraft, Joel Lester
Delphia, John Maurice
Fisher, Harvey Irvin
Goslow, George E, Jr
Heckerman, Raymond Otto
Henderson, Alex
Hildebrand, Milton
Klemm, Robert David
Lee, Sue Ying
Lehmann, Wilma Helen
Liem, Karel F
Mawby, John Evans
Moulton, James Malcolm
Rosenberg, Herbert Irving
Sokol, Otto M
Ward, Robert Porter

Vertebrate Paleontology
Alf, Raymond Manfred
Applegate, Shelton P
Baird, Donald
Bardack, David
Berman, David S
Bird, Samuel Oscar, II
Bjork, Philip R
Black, Craig C
Carroll, Robert Lynn
Churcher, Charles Stephen
Colbert, Edwin Harris
Coombs, Margery Chalifoux
Cosgriff, John W, Jr
Crompton, Alfred W
Dawson, Mary (Ruth)
DeMar, Robert E
Denison, Robert Howland
Dorr, John Adam, Jr
Downs, Theodore
Dunkle, David Hosbrook
Echols, Joan
Edmund, Alexander Gordon
Estes, Richard
Fox, Richard Carr
Galbreath, Edwin Carter
Gingerich, Philip Dean
Greaves, Walter Stalker
Gregory, Joseph Tracy
Holman, J Alan
Hopson, James A
Howard, Hildegarde (Mrs Henry Anson Wylde)
Howe, John A
Hutchison, John Howard
Jakway, George Elmer

James, Gideon T
Jenkins, Farish Alston, Jr
Kitts, David Burlingame
Koerner, Harold Elton
Konizeski, Richard L
Langston, Wann, Jr
Lindsay, Everett Harold, Jr
Linsley, Robert Martin
Lund, Richard
Lundelius, Ernest Luther, Jr
Macdonald, James Reid
Martin, Larry Dean
Martin, Robert Allen
McKenna, Malcolm Carnegie
Mellett, James Silvan
Melton, William Grover, Jr
Miller, Wade Elliott, II
Olson, Everett Claire
Ostrom, John Harold
Patterson, Bryan
Patton, Thomas Hudson
Rackoff, Jerome S
Ray, Clayton Edward
Rich, Patricia Vickers
Rollins, Harold Bert
Russell, Dale A
Schultz, Gerald Edward
Seltin, Richard James
Simons, Elwyn Laverne
Simpson, George Gaylord
Stephens, John James, III
Strain, William Samuel
Tedford, Richard Hall
Thurmond, John Tydings
Tullar, Richard Montgomery
Vaughn, Peter Paul
Voorhies, Michael Reginald
Wahlert, John Howard
Welles, Samuel Paul
West, Robert MacLellan
Wheeler, Walter Hall
Whistler, David Paul
Whitmore, Frank Clifford, Jr
Wilson, Mark Vincent Hardman
Wilson, Robert Warren
Wood, Albert Elmer
Woodburne, Michael O
Zakrzewski, Richard Jerome
Zangerl, Rainer

Vertebrate Physiology
Barker, Laren Dee Stacy
Bolls, Nathan J, Jr
Borei, Hans Georg
Buss, Jack Theodore
Contrera, Joseph Fabian
Cundiff, Milford Fields
Davis, Kenneth Bruce, Jr
Heisinger, James Fredrick
Ikenberry, Roy Dewayne
Kundig, Fredericka Dodyk
Martin, James Henry, III
Tharp, Gerald D

Geographic Index

ALABAMA

Stevens, Frank Joseph, chemistry
Stowe, Howard Denison, pathology
Strength, Delphin Ralph, biochemistry
Svacha, Anna Johnson, nutrition, medicine
Taylor, Sterling Elwynn, agricultural meteorology, ecology
Thaxton, George Donald, nuclear physics, solid state physics
Thomasson, Claude Larry, pharmacy
Trouse, Albert Charles, agronomy
Truelove, Bryan, weed science
Van de Mark, Mildred S., nutrition
Vaughan, John Thomas, veterinary medicine
Walker, Donald F., veterinary medicine
Ward, Charlotte Reed, physical chemistry
Ward, Curtis Howard, physical chemistry
Warren, James Clark, hydrogeology
Warren, William Michael, animal husbandry, animal breeding
Waslien, Carol Irene, nutrition, biochemistry
Watson, Jack Ellsworth, genetics, human genetics
Wear, John Ingram, agronomy
Weete, John Donald, plant physiology, plant biochemistry
Wiggins, Earl Lowell, physiology
Wilken, Leon Otto, Jr., pharmaceutics, biopharmaceutics
Wilkinson, Paul Kenneth, biopharmaceutics, pharmacodynamics
Williams, Byron Bennett, Jr., pharmacology
Williams, John Caswell, plant genetics, statistics
Williams, Michael Ledell, entomology, systematics
Wilson, Jane Austell, zoology, physiology
Wilson, Stanley Porter, population genetics
Ziegler, Paul Four, chemistry

BAY MINETTE

Boron, John Lee, marine biology, science education

BIRMINGHAM

Agresti, David George, nuclear spectroscopy, materials science
Alford, Charles Aaron, Jr., virology
Alling, Charles Calvin III, oral surgery, dentistry
Andreoli, Kathleen Gainor, medical education
Andreoli, Thomas Eugene, internal medicine, physiology
Arrington, Richard, Jr., invertebrate zoology
Askew, Harold Cochran, anatomy, dentistry
Bailey, Paul C., cytology
Ball, Gene V., internal medicine
Barclare, Bruno, medicine, urology
Bargeron, Lionel Malcolm, Jr., pediatrics, cardiology
Barker, Samuel Booth, physiology, pharmacology
Barnard, Anthony C L, computer science
Barrett, Jerry Wayne, physical chemistry
Barrett, William Jordan, analytical chemistry
Bauman, Robert Poe, chemical physics
Baxley, William Allison, internal medicine
Bearce, Denny N., geology
Beaton, John McCall, psychopharmacology
Becker, Gerald Leonard, biochemistry
Benington, Frederic, organic chemistry
Bennett, Joe Claude, immunology
Bennett, Leonard Lee, Jr., biochemical pharmacology
Benton, John William, Jr., pediatrics, neurology
Besse, John C., pharmacology, physiology
Bishop, David Hugh Langler, molecular biology, virology
Bishop, Sanford Parsons, pathology, cardiovascular diseases
Boardman, William Jarvis, physics, astronomy
Boshell, Buris Raye, endocrinology, internal medicine
Bradford, Henry Bernard, Jr., virology
Bradley, Edwin Luther, Jr., statistics
Brascho, Donn Joseph, radiology
Bridges, William Frank, academic administration, neurobiology
Brockman, Robert W., biochemistry
Brown, Jerry William, anatomy
Bugg, Charles Edward, physical chemistry
Burks, Robert Elbert, Jr., organic chemistry
Butler, William Thomas, biochemistry
Butterworth, Charles E., Jr., internal medicine
Byrum, Woodrow Robert, pharmacology
Cain, Stephen Malcolm, physiology
Calloway, E Dean, physical chemistry
Campbell, William H., mathematics
Carlson, Gerald Lowell, biological chemistry
Carpenter, Frank Grant, physiology
Cassady, George, medicine, pediatrics
Cassell, Gail Houston, comparative medicine, infectious diseases
Caveny, Elmer Leonard, psychiatry
Ceballos, Ricardo, pathology, neurology
Chastain, Benjamin Burton, inorganic chemistry
Cheraskin, Emanuel, oral medicine
Cheung, Herbert Chiu-Ching, biophysics, physical chemistry
Chopra, Dharam Pal, cell biology
Christian, Samuel Terry, biochemistry, organic chemistry
Cline, George Bruce, physiology, biophysics
Cobb, Charles Madison, dentistry, periodontics
Coburn, William Carl, Jr., physical organic chemistry
Compans, Richard W., virology
Conrad, Marcel E., internal medicine, hematology
Cooper, Max Dale, pediatrics, immunology
Cowsar, Donald Roy, biomaterials, polymer chemistry
Crispens, Charles Gangloff, Jr., oncology
Crittenden, Richard James, mathematics
Cunningham, Russell D., endocrinology, metabolism
Curtis, Roy, III, microbial genetics, molecular genetics
Dagg, Charles Patrick, reproductive biology
Daniel, Robert Eugene, analytical chemistry
Daniel, William A., Jr., pediatrics
Davis, Sarah Fredericka, pediatrics
Denton, Tom Eugene, botany
Dillon, Henry Kenneth, analytical chemistry, biochemistry
Dillon, Hugh C, Jr., medicine
Dismukes, Edward Brock, physical chemistry
Duran, John Ridgway, internal medicine, oncology
Eddleman, Elvia Etheridge, Jr., internal medicine, cardiology
Eden, William Gibbs, economic entomology
Elenburg, Janus Yentsch.
Elliott, Howard Clyde, biochemistry
Elliott, Robert Daryl, organic chemistry
Emerson, Geraldine Mariellen, medical science, biochemistry
Fatig, W Donald, genetics
Feagin, Frederick F., dentistry
Feary, Thomas W., microbiology
Feazel, Charles Elmo, Jr., organic chemistry
Finley, Sara Crews, medical genetics, pediatrics
Finley, Wayne House, medical genetics, biochemistry
Finn, Sidney Bernard, dentistry
Fischer, Theodore E., dentistry
Flowers, Charles E., Jr., obstetrics & gynecology
Foft, John William, pathology
Francis, Robert Dorl, medical microbiology, virology
Friedman, Benjamin, medicine
Frommeyer, Walter Benedict, Jr., internal medicine
Fullmer, Harold Milton, experimental pathology
Furner, Raymond Lynn, pharmacology, chemotherapy
Galbraith, James Garber, neurosurgery
Garrett, J Marshall, gastroenterology
Geer, Jack Charles, medicine, pathology
Gfeller, Edward, neuroanatomy, psychiatry
Glaze, Robert P., biochemistry
Goldschmidt, Raul Max, microbial genetics
Goodall, Marcus Campbell, biophysics, systems theory
Gordon, Kenneth Milton, organic chemistry
Greenspon, Thomas Stephen, physiological optics
Griswold, Daniel Pratt, Jr., veterinary medicine
Habeeb, Ahmed Fathi Sayed Ahmed, protein chemistry, immunology
Haggard, James Herbert, biochemistry
Hall, Leo McAtoon, biochemistry
Halpern, James Daniel, mathematical logic
Halsey, James H, Jr., neurology
Hamel, Earl Gregory, Jr., anatomy
Hammack, William Jack, medicine
Hammons, Paul Edward, dentistry
Hand, George Samuel, Jr., experimental embryology
Hanson, Roger Wayne, pharmacology
Harrison, Tinsley Randolph, medicine
Hartley, Marshall Wendell, zoology, physiology
Hathaway, Beulah M., medicine, pathology
Hazlegrove, Leven Savage, physical chemistry
Hefner, Lloyd Lee, internal medicine
Hickey, Terry Lee, neurobiology
Hill, Donald Lynch, biochemistry, cancer
Hill, Samuel Richardson, Jr., internal medicine
Hiramoto, Raymond Natsuo, bacteriology, immunology
Hirschowitz, Basil Isaac, medicine
Ho, Kang-Jey, pathology
Hoffman, Henry Harland, anatomy
Hogan, Robert Steadham, rheumatology
Holley, Howard Lamar, medicine
Holliman, Dan Clark, vertebrate zoology
Hunt, Charles E., experimental pathology
Hunter, Katherine Morton, nutritional biochemistry
Hurst, David Charles, statistics
Hurtt, Oscar Lee, Jr., analytical chemistry
Hutchison, Gerald Andrew, biomathematics
Hutchison, Jeanne S., mathematics
Jackson, Raymond John, microbiology
Jackson, William Morrison, inorganic chemistry, physical chemistry
James, Thomas Naum, cardiovascular diseases
Jamison, Homer Claude, epidemiology, dentistry
Johnson, Brian John, immunochemistry
Johnson, Elton Loyd, animal nutrition
Johnson, Richard Stebbins, physical organic chemistry
Johnston, Richard Boles, Jr., pediatrics, microbiology
Jones, Daniel David, plant physiology, physiology
Katholi, Charles Robinson, biomathematics
Kaylor, Hoyt McCoy, physics
Keele, Bernard B Jr., biochemistry, microbiology
Keller, Stanley E., dentistry
Kelly, Mike, poultry nutrition
Kent, Sidney Page, pathology
Kiely, Donald Edward, synthetic organic chemistry
Kirklin, John W., cardiovascular surgery
Kiser, Lola Frances, mathematics
Klapper, Clarence Edward, embryology
Klapper, Margaret Strange, internal medicine
Klip, Dorothea A., numerical analysis
Klip, Willem, biophysics
Kochakian, Charles Daniel, endocrinology, biochemistry
Koulourides, Theodore I, dentistry, oral biology
Krannich, Larry Kent, inorganic chemistry
Krumdieck, Carlos L, biochemistry, nutrition
Lamon, Eddie William, cancer, immunology
Langston, James Horace, organic chemistry
Laster, William Russell, Jr., veterinary medicine
Lawton, Alexander R, III., medicine
Lebowitz, Jacob, physical chemistry, biochemistry
Lee, Daisy Si, pediatrics, allergy
Legler, Donald Wayne, immunology, physiology
Levedahl, Blaine Hess, biochemistry
Levene, Ralph Zalman, ophthalmology
Lewis, Danny Harve, polymer chemistry
Lindsay, Raymond H, biochemistry, pharmacology
Lindsey, James Russell, pathology
Linton, Patrick Hugo, psychiatry
Lloyd, Harris Horton, cancer, chemotherapy
Locke, John Franklin, mathematics
Lofton, William Milford, Jr., industrial organic chemistry
Logic, Joseph Richard, cardiovascular physiology, nuclear medicine
Lopez, Hady, nutritional biochemistry, dental epidemiology
Lorincz, Andrew Endre, pediatrics, biochemistry
Lupton, Charles Hamilton, Jr., medicine, pathology
Macy, Josiah, Jr., mathematical biophysics
Manson-Hing, Lincoln Roy, dentistry
Martinez, Mario Guillermo, Jr., oral pathology
Matukas, Victor John, experimental pathology
McCallum, Charles Alexander, Jr., oral surgery
McCann, William Peter, pharmacology, internal medicine
McCulloch, Herbert Alfred, biology
McGhee, Jerry Roger, microbiology, immunology
McKibbin, John Mead, biochemistry
McLaughlin, Ellen Winnie, experimental embryology
McMahon, John Martin, medicine
Mellett, Lawrence B., pharmacology
Mesel, Emmanuel, pediatric cardiology, information science
Mestecky, Jiri, immunology, immunochemistry
Miller, Edward Joseph, biochemistry, radiobiology
Miller, Ernest L, prosthodontics
Miller, Herbert Crawford, analytical chemistry
Miller, John Melville, III, rehabilitation medicine
Misra, Hara Prasad, biochemistry, veterinary medicine
Moller, Palmi, dentistry
Montes, Leopoldo F, dermatology, mycology
Montgomery, John Atterbury, organic chemistry, medicinal chemistry
Moreno, Hernan, pediatrics, hematology
Morgan, Jean McNeil, internal medicine
Morin, Richard Dudley, medicinal chemistry
Morton, Perry Wilkes, Jr., physics
Mowry, Robert Wilbur, pathology
Mundy, Roy Lee, pharmacology
Murad, Tariq Mohammed, pathology
Murray, William Mozley, Jr., analytical chemistry
Nance, Charles Roger, cultural anthropology, archaeology
Navar, Luis Gabriel, physiology, biophysics
Navia, Juan Marcelo, nutrition
Niedermeier, William, biochemistry
Noojin, Ray O., dermatology
Oberman, Albert, preventive medicine, epidemiology
Oh, Shin Joong, neurology
Osmen, Lamar Sutton, dermatology
Ostrom, Carl Alston, preventive dentistry
Oyster, Clyde William, visual physiology
Palmisano, Paul Anthony, pediatrics, pharmacology
Peacock, Peter N B, epidemiology
Pearson, Colin Arthur, physics
Peckham, John Cecil, pathology
Peeples, William Dewey, Jr., mathematics
Peters, Henry Buckland, optometry
Piper, James Robert, organic chemistry
Pittman, James Allen, Jr., internal medicine, endocrinology
Poirier, Gary Raymond, reproductive biology
Polt, Sarah Stephens, clinical pathology
Pontius, Duane Henry, solid state physics
Prejean, Joe David, physiology, biochemistry
Pretlow, Thomas Garrett, II, pathology
Pruitt, Kenneth M., biochemistry
Quarles, Thomas Stephen, cell physiology
Quintarelli, Giuliano, medicine, biology
Rapaka, Rao Sambasiva, pharmaceutical chemistry
Reddy, William John, biochemistry, endocrinology
Rehm, Warren Stacy, Jr., physiology, biophysics
Reilly, Kevin Denis, computer science
Reque, Paul Gerhard, medicine
Ringsdorf, Warren Marshall, Jr., dentistry, nutrition
Risman, George Carl, medicine
Roberds, Wesley Milton, physics
Robinson, Edward Lee, nuclear physics
Robinson, Leonard H, dentistry, academic administration
Roozen, Kenneth James, genetics, molecular biology
Rosen, Lawrence, chemistry
Rosenblum, William M, physics, optometry
Roth, Robert Earl, radiology
Sachs, George, biochemistry

St Pierre, Thomas, chemistry
Sani, Brahma Porinchu, biochemistry, cancer
Schabel, Frank Milton, Jr., chemotherapy
Schmidt, Leon Herbert, experimental medicine
Schnaper, Edna Stern, microbiology
Schnaper, Harold Warren, medicine, cardiovascular physiology
Schneyer, Charlotte A., physiology
Schneyer, Leon Harold, physiology
Schoepfle, Gordon Marcus, physiology
Schrohenloher, Ralph Edward, biochemistry, immunology
Schuessler, Carlos Francis, dentistry
Schutzbach, John Stephen, biochemistry, microbiology
Segal, Arthur Cherny, mathematics
Segrest, Jere Palmer, cell biology, molecular biophysics
Sensenig, Edgar Carl, anatomy
Settine, Robert Louis, organic chemistry
Shannon, William Michael, microbiology
Shealy, David Lee, optics
Shealy, Yoder Fulmer, organic chemistry, medicinal chemistry
Sheehy, Thomas W., medicine, hematology
Sheffield, L Thomas
Shin, Myung Soo, radiology
Shingleton, Hugh Maurice, obstetrics & gynecology, oncology
Shiota, Tetsuo, biochemistry
Shirley, Sheridan William, surgery, urology
Shoemaker, Richard Leonard, physiology
Siegel, Abraham Lazarus, physiology, chemistry
Siler, William MacDowell, biomathematics, theoretical biology
Simpson-Herren, Linda, chemotherapy
Skipper, Howard Earle, biochemistry
Smith, Carol McDonald, topology
Smith, Wallace Britton, applied physics
Sowell, John Gregory, pharmacology, ophthalmology
Spangler, Stanley Gordon, biophysics, physiology
Sparks, David Lee, physiological psychology
Speed, Edwin Maurice, dentistry, anatomy
Spencer, Herbert Ward, III, physics, air pollution
Stocks, Douglas Roscoe, Jr., mathematics
Stover, Samuel Landis, pediatrics
Stroud, Robert Malone, immunology
Struck, Robert Frederick, organic chemistry
Suddath, Fred Leroy, (Jr), biological structure, x-ray crystallography
Suling, William John, medical microbiology, chemotherapy
Summerlin, Lee R., chemistry, science education
Susina, Stanley V., pharmacy, pharmacology
Takahashi, Ellen Shizuko, physiological optics, optometry
Tanquary, Albert Charles, polymer science
Tarwater, Oliver Reed, synthetic organic chemistry, polymer chemistry
Tauxe, Welby Newlon, nuclear medicine, clinical pathology
Taylor, Kenneth Boivin, biochemistry, enzymology
Teague, Robert Sterling, pharmacology, endocrinology
Temple, Carroll Glenn, organic chemistry, medicinal chemistry
Thompson, Jerry Nelson, genetics, biochemistry
Thompson, Wynelle Doggett, biochemistry, organic chemistry
Thorpe, Martha Campbell, physical organic chemistry
Thurmond, John Tydings, vertebrate paleontology, environmental geology
Thuss, William Getz, Jr., industrial medicine
Tiller, Ralph Earl, pediatrics
Tohver, Hanno Tiit, physics
Trawinski, Benon John, mathematical statistics
Trawinski, Irene Patricia Monahan, mathematical statistics
Turner, Malcolm Elijah, Jr., mathematical biology, statistics
Urry, Dan Wesley, molecular biophysics, biochemistry
Vaughan, Loy Otis, Jr., mathematics
Vaughan, Gwenyth Ruth, speech pathology, audiology
Vigee, Gerald S., physical inorganic chemistry
Vitek, Jiri Jakub, radiology
Vittor, Barry Adolph, ecology

Volker, Joseph Francis, biochemistry, dentistry
Warne, Ronson Joseph, mathematics
Watkins, Charles Lee, chemistry
Weatherford, Thomas Waller, III, periodontology
Weller, Edwin Matthew, chemical embryology, developmental physiology
West, Seymour S, biophysics, anatomy
Wheeler, Glynn Pearce, biochemistry
Wheeler, Ruric E, mathematical statistics
Wild, Bradford Williston, optometry
Wilkoff, Lee Joseph, microbiology, biochemistry
Williamson, Arthur Elridge, Jr, electro optics
Wilson, Graeme Stewart, visual physiology
Wingo, William Jacob, biochemistry
Winkler, Charles Herman, Jr, microbiology, health sciences
Wintter, John Ernest, medicinal chemistry
Wolcott, Robert Michael, immunology, biochemistry
Wolff, Albert Eli, biostatistics, epidemiology
Wood, John Edward, III, organic chemistry
Wuehrmann, Arthur H, dentistry
Yang, Chao-Chih, computer science
Yelton, Chestley Lee, orthopedic surgery
Yielding, K Lemone, molecular biology, medicine
Young, John H, physics
Young, Wesley O, dentistry

BOAZ
Whitesell, James Judd, Entomology

BREWTON
Maple, William Robert, forestry

DAUPHIN ISLAND
Huntley, Bobby E, microbiology
Rounsefell, George Armytage, fisheries

DECATUR
Craig, James Porter, Jr, physical chemistry
Galil, Fahmy, textile chemistry, chemical engineering
Parks, Ross Lombard, analytical chemistry
Rannefeld, Clarence Edmund, chemistry
Veazey, Thomas Mabry, organic chemistry
Youtsey, Karl John, physics

DEMOPOLIS
Manley, Lillian C, biology, academic administration

EUFAULA
Hastings, Earl L, mining geology

FAIRHOPE
Keppner, Edwin James, parasitology, ecology
Schulze, Karl Ludwig, microbiology

FLORENCE
Alexander, Kliem, chemistry
Allen, Seward Ellery, plant physiology, soil chemistry
Brackin, Eddy Joe, algebra
Brown, Jack Stanley, freshwater biology
Chien, Sen Hsiung, soil chemistry
Curott, David Richard, physics, astrophysics
Doll, Eugene Carter, soil fertility
Hershey, Arthur L, botany
Isbell, Raymond Eugene, organic chemistry
Johnson, Frank Junior, analytical chemistry
Keys, Charles Everel, vertebrate embryology
Locker, John L, mathematics
Murray, Thomas Pinkney, organic chemistry
Richmond, Charles William, organic chemistry
Thomas, Joseph Calvin, science education, organic chemistry
Wooldridge, Elizabeth Taylor, mathematics
Yokley, Paul, Jr., zoology

FOLEY
Coggeshall, Lowell Thelwell, medical education, tropical medicine

GADSDEN
Head, Robert Berturm, entomology
Rosene, Walter, Jr., wildlife ecology, wildlife management

GULF SHORES
Adams, Wright (Rowe), cardiology

GUNTERSVILLE
Witmer, William Byron, inorganic chemistry, physical chemistry

HUNTSVILLE
Adams, Curtis H, entomology, zoology
Allan, Barry David physical chemistry, lasers
Allen, Robert Erwin, medical physiology, human factors engineering
Anderson, Bernard Jeffrey, cloud physics
Arendale, William Frank, chemical physics
Ayers, Orval Edwin, organic chemistry
Barr, Thomas Albert, Jr., plasma physics
Bishnoi, Udai Ram, agronomy
Bond, Albert F, mathematics
Brandon, Walter Wiley, Jr., applied physics
Brown, Glenn Lamar, physics
Burch, James Leo, magnetospheric physics
Byrd, James Dotson, organic polymer chemistry
Castle, John Granville, Jr, magnetic resonance, physical optics
Chan, Chia Hwa, high energy physics, theoretical physics
Chang, Chang-Sun, applied mechanics
Chappell, Charles Richard, magnetospheric physics
Cook, Frederick Lee, mathematics
Cooper, Emerson Amenhotep, organic chemistry
Crenshaw, Jack Westcott, computer sciences
Dalins, Ilmars, surface physics, seismology
Dalton, Charles Chester, physics
Davis, Jack H, solid state physics
Dodson, Charles Leon, Jr., physical chemistry
Doyle, Frank Lawrence, hydrology, geology
Eckstein, Herbert Philipp, physics
Emerson, Merle T, physical chemistry, analytical chemistry
Essenwanger, Oskar M, atmospheric physics, applied statistics
Fishman, Gerald Jay, astrophysics, x-ray astronomy
Gibson, Peter Murray, mathematics
Heller, Gerhard Bernhard, physics, space sciences
Holl, Herbert Barthold, physics, astronomy
Horner, James M, mathematics
Huskins, Chester Walker, organic chemistry
Johnson, Kenneth Eugene, physical chemistry, organic chemistry
Kennedy, Larry Zane, optical physics
Krause, Helmut G L, astronomy
Kroes, Roger L, solid state physics
Liu, Joseph Jeng-Fu, celestial mechanics, theoretical mechanics
Lovingood, Judson Allison, mathematics, electrical engineering
Lundquist, Charles Arthur, space sciences
McKnight, William Baldwin, lasers
McManus, Samuel Pyler, organic chemistry
Miller, Meredith, physical chemistry
Mookherji, Tripty Kumar, solid state physics
Morelock, James Crutchfield, mathematics
Parnell, Thomas Alfred, physics, cosmic ray physics
Passino, Nicholas Alfred, molecular spectroscopy, infrared physics
Patterson, William Jerry, organic polymer chemistry
Perkins, James Francis, quantum physics
Reiff, Patricia Hofer, magnetospheric physics
Riley, Clyde, physical chemistry
Rosing, Lorraine Morin, limnology, invertebrate zoology

Rowland, May Eloise, microbiology, parasitology
Rush, John Edwin, Jr., theoretical physics
Sanmann, Everett Eugene, optical physics
Sears, Donald Richard, environmental chemistry, crystallography
Shapiro, Norman Malvin, physics
Smalley, Larry L., relativistic astrophysics
Smith, Frederick Williams, surgery
Stephens, James Briscoe, atmospheric physics, space physics
Stephens, William D, organic chemistry
Stettler, John Dietrich, theoretical physics
Stone, Max Wendell, computer sciences
Stuhlinger, Ernst, physics
Sung, Chi Ching, theoretical physics
Vaughan, William Walton, meteorology, aerospace sciences
Witriol, Norman Martin, molecular physics

JACKSONVILLE
Boozer, Reuben Bryan, zoology, biology
Cochis, Thomas, botany, horticulture
Gant, Fred Allan, physical chemistry, thermodynamics
Landers, Kenneth Earl, plant physiology, vertebrate zoology
Sanford, L G, entomology, vertebrate zoology
Sowell, Wendell L, toxicology
Youngblood, Bettye Sue, organic chemistry

LINDEN
Compton, Jack, textile technology

LIVINGSTON
Barr, Alvin Francis, mathematics
Canis, Wayne F, geology
Tucker, Charles Eugene, biology

MARION
Schennum, Wayne Edward, population biology
Williams, Martin Barbour, chemistry

MARSHALL SPACE FLIGHT CTR
O'Dell, Charles Robert, astronomy
Smith, Robert Earl, aeronomy
Swenson, Gary Russell, atmospheric physics

MAXWELL AFB
Harris, Richard Allan, molecular physics, quantum mechanics

MCCALLA
Kendrick, Aaron Baker, physiological chemistry

MCINTOSH
Saul, George Archer, organic chemistry
Taylor, Edward Alan, organic chemistry
Trottier, Claude Henry, organic chemistry

MOBILE
Andrews, Russell S, Jr, pulp chemistry
Baugh, Charles M, biochemistry
Belanger, David Gerald, systems theory
Beyers, Robert John, ecology, aquatic biology
Blackburn, Will R, pathology, tropical medicine
Boyles, James McGregor, medical parasitology
Breithaupt, Lea Joseph, Jr, paper chemistry, cellulose chemistry
Brinkley, Amiel Word, Jr, pulp & paper technology
Callahan, William Paxton, III, anatomy
Campbell, Robert Terry, pulp & paper technology
Cappas, C, physical chemistry, mathematics
Dowling, Edmund Augustine, pathology
Eisele, Louis John, physics
Eyster, Henry Clyde, plant physiology, phycology
Farnell, Daniel Reese, veterinary medicine
Ferguson, Joseph Gantt, research administration
Furman, Walter L, mathematics, physics
Garrett, Arthur Randolph, Jr, ecology
Gibson, David Michael, organic chemistry, analytical chemistry

ALABAMA

Glenn, Thomas M., pharmacology, physiology
Goss, Charles Mayo, anatomy
Heminger, Paul Andrew, molecular spectroscopy, environmental physics
Hemphill, Andrew Frederick, zoology
Hudson, Alice Brandon, physical inorganic chemistry
Huff, William J., micropaleontology, stratigraphy
Huggins, Clyde Griffin, biochemistry
Hughes, Edwin R., chemistry, medicine
Isphording, Wayne Carter, geochemistry, mineralogy
Jackson, Margaret E., inorganic chemistry, physical chemistry
Jackson, Thomas Gerald, organic chemistry
Johnson, Ralph Bernard, mathematics
Jones, Edward Eugene, protozoology
Jordan, Jerry Dugger, cardiology
Kearley, Francis Joseph, Jr., organic chemistry
Lamb, George Marion, micropaleontology, stratigraphy
Lambert, James LeBeau, organic chemistry, biochemistry
Larguier, Everett Henry, mathematics
Laycock, David Gerald, molecular genetics
Lelong, Michel George, plant taxonomy
Linzey, Alicia Vogt, mammalogy
Linzey, Donald Wayne, wildlife biology, mammalogy
Mattics, Leon Eugene, number theory
McGill, Suzanne, mathematics
Mercer, Leonard Preston, II, nutritional biochemistry
Miller, Nathan C., organic chemistry
Mitchell, Ferdinand H., physics
Mitchell, Joseph Christopher, parasitology, entomology
Morgan, Paul Harper, biochemistry
Mustafa, Syed Jamal, medical research
Peterson, Raymond Dale August, physiology, genetics
Phillips, Howard Mitchell, cytology
Rawls, John Marvin, developmental
Robert, Richard Ross, neuropharmacology
Regan, Gerald Thomas, psychopharmacology, ecology
Rines, William John, analytical chemistry
Rowell, Neal Pope, physics
Scholes, Vernon Eugene, medical microbiology, infectious diseases
Shackleford, John Murphy, anatomy
Shipp, Robert Lewis, ichthyology
Tate, Laurence Gray, insect physiology
Tauss, Kurt H., organic chemistry
Turner, Rex Howell, cellulose chemistry
Vinson, Donald C., water pollution, environmental sciences
White, Lowell Elmond, Jr., medical education
Whorton, Rayburn Harlen, paper chemistry
Wiborn, Walter Harrison, human anatomy
Wilkerson, Robert Douglas, pharmacology
Williams, Aaron, Jr., physical geography
Williams, Louis Francis, Jr., computer sciences
Wilson, Eugene Murphey, geography
Workman, William Edward, environmental geology
Yett, Fowler Redford, applied mathematics

MONTEVALLO
Beal, James Burton, Jr., inorganic chemistry
Beasley, Philip Gene, animal physiology, plant physiology
Connell, James Frederick Louis, geology
Eagles, Jan, histology, anatomy
Foreman, Jesse William, Jr., optical physics
Harris, Albert Zeke, analytical chemistry, physical chemistry
Kwon, Tai Hyung, solid state physics
McGuire, Robert Frank, phycology
McMillan, Daniel Russell, physics
Merjanian, Aris, organic chemistry
Sledge, Eugene Bondurant, biology
Turner, Henry Ford, zoology, parasitology

MONTGOMERY
Carpenter, Charles, chemistry
Ellis, Richard Bassett, physical chemistry
Funderburk, Henry Hanly, Jr., plant physiology
Johnson, William E., Jr., medical entomology
Kim, K-ang, mathematics
Knockemus, Ward Wilbur, inorganic chemistry
Lee, Norman Kunhan, pathology
Lieberman, Robert, radiochemistry
Mayfield, John Emory, mycology
Myers, Ira Lee, preventive medicine, public health
Sharma, Udhishtra Deva, animal husbandry, biology
Teggins, John E., inorganic chemistry
Ward, Henry Silas, Jr., botany, ecology
Whitehead, Fred, chemistry

MOUNDVILLE
DeJarnette, David Lloyd, anthropology, archaeology

MUSCLE SHOALS
Barnes, William Wayne, medical entomology, insect toxicology
Diamond, Ray Byford, soil fertility
Edwards, Oscar Wendell, physical chemistry
Engelstad, Orvis P., soil fertility
Gray, Robert Copping, soil fertility
Gremillion, Louis Ray, mineralogy
Hashimoto, Isao, soil chemistry
Hatfield, John Dempsey, physical chemistry
Huffman, Ernest Otto, physical chemistry
Kanipe, Larry Gene, physical chemistry, radiochemistry
Kennedy, Frank Metier, chemistry
Khasawneh, Fayez Essa, soil chemistry, agronomy
Kilmer, Victor James, soils
Kovar, John Alvis, soil morphology
McClellan, Guerry Hamrick, mineralogy, geology
Meagher, James Francis, physical chemistry
Mortvedt, John Jacob, soil chemistry
Nelson, Lewis Bailey, soils
Reynolds, George (Warren), meteorology, environmental sciences
Russel, Darrell Arden, agronomy
Scott, William Caswell, Jr., research administration
Tennessen, Kenneth Joseph, fresh water ecology
Terman, Gilbert Leroy, agronomy
Williams, Gerald Gordon, plant physiology, soil science

NORMAL
Bass, Garland Booker, agronomy
Elgawhary, Salah Mohammad, soil chemistry, soil fertility
Lal, Ravindra Behari, physics
Lee, Ching Tsung, quantum optics, mathematical physics
Mangat, Baldev Singh, entomology, zoology
Manger, Martin C., organic chemistry, environmental sciences
Mugwira, Luke Makore, soil chemistry
Nishimura, John Francis, nutrition
Pinjani, Moti, veterinary physiology, ruminant nutrition
Rao, Ganti Lakshminarayana, atmospheric physics
Reddy, Sunki, Gopal, food science, microbiology
Rice, Barbara Slyder, mathematics
Sapra, Val T., plant cytogenetics
Sharma, Govind C., plant science
Singh, Bharat, plant physiology, food science
Thomas, Winfred, agronomy

NORTHPORT
Nesbitt, Paul Homer, anthropology

OPELIKA
Bennett, Allison Carr, soil chemistry, soil fertility
Capps, Julius Daniel, organic chemistry

REDSTONE ARSENAL
Bowden, Charles Malcom, quantum optics, statistical mechanics
Guenther, Bobby Dean, optical physics
Hartman, Richard Leon, optical physics, solid state physics
Holloman, Miles Edward, physical chemistry
Howgate, David W., chemical physics
McCorkle, William C., Jr., physics, aerospace sciences
Roberts, Thomas George, lasers, plasma physics
Smith, James Lynn, solid state physics, optical physics
Steverding, Bernard, theoretical physics
Werkheiser, Arthur H., Jr., lasers
Wharton, Michael Washington, physical chemistry, ceramics
Yee, Tin Boo, chemistry, ceramics

SARALAND
Wheat, Percy Wayne, industrial organic chemistry

SHEFFIELD
Bullough, Vaughn Lynn, organic chemistry
Sheridan, Richard Collins, chemistry

TALLADEGA
Boetler, James Leroy, solid state physics
Ranganathan, Vaijeyan S., biochemistry, electron microscopy
Simpson, Cohen Thomas, analytical chemistry
Walker, Robert Paul, mathematics

THEODORE
Spinner, Ernest, chemistry

TROY
Barras, Donald J., physiology, biochemistry
Dietz, Robert Austin, cytotaxonomy, ecology
Ingram, Sammy Walker, Jr., organic chemistry
Norman, Billy Ray, science education
Ward, Edward Hilson, inorganic chemistry, analytical chemistry
Widdowson, David Carl, biology
Wilkes, James C., biology
Wilks, William Taylor, physics

TUSCALOOSA
Barkley, Lloyd Blair, organic chemistry
Barr, Ernest Scott, physics
Brandt, Luther Warren, physical chemistry
Cross, Earle Albright, Jr., insect taxonomy, insect ecology
Glover, Elsa Margaret, philosophy of science
Gonzales, Mario Octavio, mathematical analysis
Jacobs, William Donald, analytical chemistry
Lampkin-Asam, Julia McCain, oncology
Lloyd, Nelson Albert, analytical chemistry
Sherer, Sankey, inorganic physical chemistry

TUSKEGEE
Biswas, Prosanto K., horticulture, plant physiology
Bowie, Walter C., veterinary physiology
Kammula, Raju G., veterinary medicine
Karp, Ira Lawrence, mathematical physics, computer science
Pollard, Donald Ray, biochemistry
Siddique, Irtaza H., microbiology, clinical chemistry
Vernon, Eugene Haworth, animal breeding

TUSKEGEE INSTITUTE
Briles, Connally Oran, immunogenetics
Carter, Howard Payne, zoology
Chiburzo, Gregory Anenoni, anatomy
Chung, Ronald Aloysius, food science
Foster, Henry Wendell, gynecology, obstetrics & gynecology
Gill, Piara Singh, physical chemistry, radiation chemistry
Goldsberry, Steve, anatomy, embryology
Henderson, James Henry Meriwether, plant physiology
Koons, Lawrence Franklin, physical chemistry
Ludwick, Adriane Gurak, organic chemistry
Ludwick, Larry Martin, inorganic chemistry
McKenzie, Basil Everard, veterinary pathology
Nelson, Wesley Eugene, soil chemistry
Saini, Rajinder S., entomology
Sapp, Walter J., cell biology
Thomas, Julian Edward, Sr., microbial physiology
Tolbert, Margaret Ellen Mayo, biochemistry
Whatley, Booker Tillman, horticulture, plant physiology
Williams, John Watkins, III, cytogenetics
Williams, Raymond Crawford, veterinary anatomy
Williams, Theodore Shields, veterinary medicine
Yamaguchi, Shogo, plant physiology

UNIVERSITY
Abramovitch, Rudolph Abraham Haim, organic chemistry
Ainsworth, Oscar Richard, mathematics
Alexander, Chester, Jr., physics
Alexander, Sydenham Benoni, internal medicine
Atwood, Jerry Lee, inorganic chemistry
Barksdale, Henry Compton, hydrology
Bartlett, James Holley, Jr., physics
Bishop, Everett Lassiter, Jr., biology
Boone, Peter Augustine, stratigraphy
Boschung, Herbert Theodore, ichthyology
Bramlett, Christopher L., inorganic analysis
Chernock, Ralph Lucien, ecology
Cole, George David, chemical physics
Copeland, Charles Wesley, Jr., stratigraphy, invertebrate paleontology
Coulter, Claude Alton, radiation physics
Coulter, Philip W., theoretical nuclear physics
Darden, William H., Jr., algology
Davis, David Gate, genetics
Deason, Temd R., algology
Drahovzal, James Alan, geology
Dukes, Gary Rinehart, inorganic chemistry, bioinorganic chemistry
Garner, Robert Henry, chemistry
Gibbons, Jean Dickinson, statistical analysis
Gill, John Paul, statistics
Green, Margaret, microbiology
Gunter, Shirley Edna Anne, microbiology
Hand, Clifford Warren, physical chemistry
Hansen, Assel Tanner, anthropology
Hardman, John Kemper, biochemistry, molecular genetics
Harms, Benjamin C., elementary particle physics
Hisey, Alan, clinical biochemistry
Hobby, Charles R., mathematics
Hood, Ronald David, developmental biology, toxicology
Hooks, William Gary, stratigraphy
Hornback, Joseph Hope, mathematics
Howell, James Levert, geology
Hughes, Travis Hubert, geology
Jason, Andrew John, molecular physics, surface physics
Jones, Douglas Epps, geology, paleoecology
Jones, Stanley Tanner, physics
Kispert, Lowell Donald, physics, radiation chemistry
Koch, Walter Ferdinand, geography
LaMoreaux, Philip Elmer, geology, hydrology
Lineback, Neal Gambill, geography
Mego, John L., biochemistry
Mettee, Maurice Ferdinand, aquatic biology
Miyagawa, Ichiro, chemical physics
Moore, Bobby Graham, cell biology
Neathery, Thornton Lee, geology
Neggers, Joseph, mathematics
O'Kelley, Joseph Charles, plant physiology
Palmer, George David, Jr., chemistry
Parr, Albert Clarence, physics
Paudler, William W., organic chemistry
Pittman, Charles U, Jr., organic chemistry
Plunkett, Robert Lee, topology, mathematical analysis
Ponder, Billy Wayne, organic chemistry
Ritchie, Adam Burke, theoretical chemistry
Rogers, David T., Jr., ecology
Sayers, Earl Roger, genetics, plant breeding
Scott, Meckinley, statistics
Seebeck, Charles Louis, Jr., mathematics
Sheeley, Eugene C., audiology

Smith, Claude Earle, Jr., systematic botany
Smith, Donald Foss, physical chemistry
Stow, Stephen Harrington, geochemistry, environmental geology
Taylor, John Dallas, geology
Toffel, George Mathias, chemistry
Van Artsdalen, Ervin Robert, physical chemistry, inorganic chemistry
Walker, William Waldrum, physics
Whitehurst, Robert Neal, physics
Whittle, George Patterson, analytical chemistry
Willard, William Robert, public health
Williams, Louis Gressett, fresh water ecology, algology
Wilson, Walter Lucien, Jr., mathematics
Wochok, Zachary Stephen, plant physiology
Zatko, David A, inorganic chemistry

WILSON DAM
McCullough, John Franklin, inorganic chemistry

ALASKA

ANCHORAGE
Bartonek, James Cloyd, ecology, wildlife management
Bigler, Stuart Grazier, meteorology
Calderwood, Keith Wright, geology
Clark, John Harlan, fisheries
Diemer, Edward Devlin, meteorology
Frohne, William Carrington, entomology
Godbey, William Givens, physical chemistry, nuclear chemistry
Long, William Ellis, stratigraphy, structural geology
Mangus, Marvin D, geology
Michaelson, Neil Elbert, soils
Mowatt, Thomas C, geochemistry, geology
Nauman, Louis William, clinical chemistry, forensic science
Ogle, William Elwood, physics
Ragle, Richard Harrison, geology
Reed, Bruce Loring, geology
Rockwell, Julius, Jr., fisheries
Schaff, Ross G, geology
Schindler, John Frederick, fresh water biology
Schmidt, Ruth A M, geology
Slack, Howard Addison, geophysics
Stewart, Rolland Keith, aquatic biology
Woods, John Price, geophysics

BARROW
Mosher, James Arthur, physiological ecology, ornithology
Underwood, Lawrence Statton, physiological ecology

AUKE BAY
Dahlberg, Michael Lee, fisheries, biometrics
Evans, Robert Ralph, ichthyology
Heard, William R, fish biology
McNeil, William Jacob, fisheries
Pella, Jerome Jacob, fisheries
Rice, Stanley Donald, pollution biology, comparative physiology
Smoker, William Alexander, fisheries, forestry

COLLEGE
Allison, Richard C, invertebrate paleontology, geology
Barsdate, Robert John, geochemistry
Bates, Howard Francis, physics, electrical engineering
Beck, William J, environmental health, bacteriology
Button, Don K, biochemistry, marine microbiology
Davis, Thomas Neil, geophysics
Forbes, Robert Briedwell, geology
Galster, William Allen, comparative physiology, biochemistry
Genaux, Charles Thomas, biochemistry
Goering, John James, marine ecology
Gordon, Ronald Claire, microbial ecology
Hood, Donald Wilbur, oceanography
Hunsucker, Robert Dudley, aeronomy
Jayaweera, Lakshman, cloud physics
Lotspeich, Frederick Benjamin, environmental sciences, fresh water ecology
Lyons, Richard Bernard, cell biology, medical genetics

MacLean, Stephen Frederick, Jr., ecology
McRoy, C Peter, biological oceanography
Reger, Richard David, glacial geology, quaternary geology
Roberts, Thomas D, atomic physics
Shaw, Glenn Edmond, atmospheric physics
Stone, David B, geophysics
Weber, Florence Robinson, geology
Weeden, Robert Barton, zoology

EAGLE RIVER
Francis, Karl Earvil, environmental sciences, resource geography

FAIRBANKS
Akasofu, Syun-Ichi, aeronomy
Alexander, Vera, aquatic ecology
Andresen, Marvin John, geology
Behrisch, Hans Werner, biochemistry, comparative physiology
Belon, Albert Edward, aeronomy, remote sensing
Benson, Carl Sidney, glaciology
Bowling, Sue Ann, meteorology
Brown, Jim McCaslin, structural geology, rock mechanics
Brown, Robert Wallace, mathematics
Burrell, David Colin, chemical oceanography
Cameron, James N, physiology, ecology
Cannon, Philip Jan, geomorphology
Dean, Frederick Chamberlain, wildlife management
Deehr, Charles Sterling, space physics
Degen, Vladimir, aeronomy, astrophysics
Dieterich, Robert Arthur, wildlife diseases
Drury, Horace Featherstone, biochemistry
Elsner, Robert, comparative physiology, marine biology
Fay, Francis Hollis, mammalogy
Feder, Howard Mitchell, marine biology
Feist, Dale Daniel, zoology, physiology
Guthrie, Russell Dale, zoology
Hawkins, Daniel Ballou, geochemistry
Heacock, Richard Ralph, geophysics, magnetospheric physics
Head, Thomas James, mathematics
Hiatt, Robert Worth, zoology
Hippler, Arthur Edwin, cultural anthropology
Hoskins, Charles M, sedimentology
Hoskins, Leo Claron, physical chemistry, animal nutrition
Husby, Fredric Martin, animal nutrition, ruminant nutrition
Irving, Laurence, comparative physiology
Kessel, Brina, ornithology
Klein, David Robert, mammalian ecology
Lando, Barbara Ann, algebra
Laufeld, Sven, invertebrate paleontology
Lent, Peter C, behavioral biology, mammalian ecology
Luick, Jack Roger, physiology, nutrition
Lynch, Donald Francis, geography of Alaska & Northern Lands, resource geography
Mather, Keith Benson, geophysics, nuclear physics
Matthews, John Brian, oceanography
Milan, Frederick Arthur, physical anthropology
Miller, Lyster Keith, environmental physiology, comparative physiology
Morack, John Ludwig, physics
Morrison, Peter Reed, comparative physiology, environmental physiology
Morrow, James Edwin, Jr., systematic ichthyology
Muench, Robin Davie, physical oceanography
Naidu, Angi Satyanarayan, sedimentology, marine geology
Neiland, Bonita, plant ecology, resource management
Neiland, Kenneth Alfred, invertebrate zoology
Norton, David William, physiological ecology
Ohtake, Takeshi, cloud physics, climatology
Osterkamp, Thomas Eugene, glaciology
Payne, Myron William, sedimentology, petroleum geology
Peyton, Leonard James, ornithology
Philip, Kenelm Winslow, radio astronomy
Rae, Kenneth MacFarlane, biological oceanography
Reeburgh, William Scott, chemical oceanography
Rees, Manfred Hugh, aeronomy
Romick, Gerald J, aeronomy
Royer, Thomas Clark, physical oceanography

Sharma, Ghanshyam D, marine geology, geochemistry
Shaw, David George, chemical oceanography, organic chemistry
Sheridan, John Roger, experimental atomic physics
Sivjee, Gulamabas Gulamhusen, aeronomy
Slaughter, Charles Wesley, forest hydrology
Smith, Grant Warren, II, organic chemistry
Sparrow, Elena Bautista, soil microbiology
Stanley, Glenn M, geophysics
Swartz, Leslie Gerard, parasitology
Swift, Daniel W, plasma physics, space physics
Teal, John Jerome, Jr., human ecology
Triplehorn, Don Murray, geology
Turner, Donald Lloyd, geology, geochronology
Van Pelt, Rollo Winslow, Jr., pathology
Van Veldhuizen, Philip Androcles, mathematics, statistics
Viereck, Leslie A, plant ecology, plant taxonomy
Weller, Gunter Ernst, meteorology
Wendler, Gerd Dierk, meteorology
Wentink, Tunis, Jr., physical chemistry
West, George Curtiss, physiological ecology, ornithology
Williams, Darrell Dean, physiology
Wilson, Charles R, geophysics
Wooding, Frank James, agronomy
Workman, William Glenn, agricultural economics

JUNEAU
Akins, Glenn John, resource management
Gard, Richard, fisheries
Merrell, Theodore Reed, Jr., fish biology
Nayudu, Y Rammohanroy, marine geology, petrology
Quast, Jay Charles, ichthyology
Schmiege, Donald Charles, forest entomology
Williams, Ralph Benjamin, public health, microbiology
Williamson, Francis Sidney Lainer, zoology, public health administration
Wright, Frederick Fenning, marine geology, oceanography

PALMER
Brundage, Arthur Lain, dairy husbandry
Laughlin, Winston Means, soil science
Logsdon, Charles Eldon, plant pathology
Loynachan, Thomas Eugene, soil microbiology, soil fertility
Mitchell, William Warren, agronomy, botany
Rieger, Samuel, soil morphology
Taylor, Roscoe L, agronomy
Tomlin, Don C, animal nutrition
Washburn, Richard Hancorne, economic entomology

SEWARD
Neve, Richard Anthony, biochemistry, marine chemistry

SOLDOTNA
Wilson, William Solomon, physical chemistry

ARIZONA

AMADO
Weekes, Trevor Cecil, astrophysics

BENSON
Towle, Louis Wallace, analytical chemistry

DOUGLAS
Williams, Sidney Arthur, mineralogy

DRAGOON
Di Peso, Charles Corradino, anthropology

FLAGSTAFF
Ables, Harold Dwayne, astronomy
Adel, Arthur, astrophysics
Allen, Agnes Morgan, geography
Ambler, John Richard, anthropology

Anderson, Glenn Arthur, zoology
Appelgren, Walter Phon, microbiology, health sciences
Balda, Russell Paul, animal ecology
Barnes, Charles Winfred, petrology, structural geology
Baum, William Alvin, astronomy
Beal, Richard Sidney, Jr., entomology
Bedwell, Thomas Howard, physics
Berlin, Graydon Lennis, physical geography
Berry, Richard Wallace, forestry
Beus, Stanley S, invertebrate paleontology, stratigraphy
Bleibtreu, Hermann Karl, physical anthropology, anthropology
Blinn, Dean Ward, phycology, aquatic ecology
Brathovde, James Robert, environmental management, environmental sciences
Breed, William Joseph, geology
Butchart, John Harvey, mathematics
Campbell, Ralph Edmund, forest soils
Capen, Charles Franklin, Jr., planetary sciences
Caple, Gerald, organic chemistry
Carothers, Steven Warren, ecology
Clary, Warren Powell, plant ecology
Colbert, Edwin Harris, vertebrate paleontology
Cotera, Augustus S, Jr., geology
Dahn, Conard Curtis, astronomy
Danson, Edward Bridge, anthropology
DeKorte, John Martin, inorganic chemistry
Delinger, William Galen, solid state physics
Donovan, Terrence John, geology
Eastwood, Raymond L, geochemistry, petrology
Elston, Donald (Parker), paleomagnetism
English, Darrel Starr, genetics
Giclas, Henry Lee, astronomy
Gilbert, Don Dale, analytical chemistry
Gilbert, Norris W, agronomy, botany
Goin, Coleman Jett, herpetology
Goslow, George E, Jr., vertebrate morphology
Gray, Allan, Jr., mathematics
Griffen, William Bedford, anthropology
Griffith, Charles Ray, anthropology
Grim, John Norman, cell biology, zoology
Gunderson, Hans Magelssen, biochemistry
Hall, Richard Chandler, astronomy
Heaton, Charles Daniel, organic chemistry
Hewitt, Anthony Victor, astronomy
Hildebrandt, Wayne Arthur, electroanalytical chemistry
Hoffman, Charles Andrew, Jr., archaeology, cultural anthropology
Holmgren, Paul, cell biology, electron microscopy
Hoyt, Earle B, Jr., organic chemistry
Huffman, Robert Wesly, organic chemistry
Hughes, Eugene Morgan, mathematical analysis
Ives, Ronald Lorenz, geophysics
Johnsen, Ardith B, bryology
Johnsen, Thomas Norman, Jr., ecology, range management
Johnson, Clarence Daniel, systematic entomology
Johnson, Lee Murphy, mathematics
Johnson, Oliver William, vertebrate zoology, physiology
Jones, John Robert, forestry, plant ecology
Karlstrom, Thor Nels Vincent, geology
Kurmes, Ernest A, forestry
Larson, Frederic Roger, forest management, systems analysis
Layton, Richard Gary, physics
Lipke, William G, plant physiology, plant biochemistry
Little, Charles Edward, mathematics
Lucchitta, Baerbel Koesters, astrogeology
Masursky, Harold, astrogeology
McCauley, John Francis, astrogeology
McDougall, Walter Byron, botany
Micklich, John R, mathematics
Millis, Robert Lowell, astronomy
Minor, Charles Oscar, forestry
Mogensen, Hans Lloyd, plant anatomy, plant morphology
Montgomery, Errol Lee, hydrogeology
Moore, Charles Godat, mathematics
Morris, Elliot Cobia, astrogeology, mineralogy
Nations, Jack Dale, paleoecology
Pearson, Keith Laurence, applied anthropology, social anthropology
Perko, Lawrence Marion, applied mathematics

Pogany, Gilbert Claude, embryology, experimental
Rawson, Richard Ray, geology
Reichman, Omer James, ecology
Rietveld, Willis James, forest physiology
Roddy, David John, geology, geophysics
Rominger, James McDonald, systematic botany
Roth, Eldon Sherwood, geomorphology
Salisbury, Howard Graves, geography
Sanderson, Milton William, entomology
Schaber, Gerald Gene, geology, astrogeology
Sexton, James D., anthropology
Slobodchikoff, Constantine Nicholas, evolutionary biology, ethology
Soller, Theodore, physics
States, Jack Sterling, microbial ecology, mycology
Strobel, John Dixon, Jr, geology
Swann, Gordon Alfred, geology, astrogeology
Swenson, Jack Spencer, organic chemistry
Ulrich, George Erwin, geology
Vaughan, Terry Alfred, zoology, vertebrate
Vrba, Frederick John, astronomy
Walter, Everett L., mathematics
Watson, Robert Dale, environmental physics
Whiting, Alfred Frank, ethnobotany
Wick, James Roy, astronomy,
Wildey, Robert Leroy, astrophysics
Wilkes, Stanley Northrup, parasitology, marine zoology
Willis, William Russell, physics
Wolfe, Edward W., geology
Wood, John Jackson, cultural anthropology, archaeology
Wright, Barton Allen, anthropology

FREDONIA
McCulloch, Clay Young, Jr, ecology

FT HUACHUCA
Schafer, George Edward, physics

GLENDALE
Dale, Jack Kyle, pharmaceutical chemistry
Oberbeck, Thomas Edmond, mathematics, operations research

GRAND CANYON
Euler, Robert Clark, anthropology, archaeology
Johnson, Raymond Roy, systematic botany, vertebrate zoology

GREEN VALLEY
Carson, George Walter, mathematics
Dunning, Gordon Merrill, health physics, science education
Fancher, Otis Earl, chemistry
Fisher, Paul John, bacteriology, chemistry
Hixon, Ralph Malcolm, phytochemistry
Iob, Vivian, organic chemistry, pediatrics
Martin, Donald Stover, bacteriology
Powers, Treval Clifford, chemistry

HUMBOLDT
Ryan, Alden Hoover, physics

KINO SPRINGS
Bates, Robert Wesley, chemistry

LITCHFIELD PARK
Hartig, Elmer Otto, physics

MESA
Casto, Clyde Christy, analytical chemistry
Decker, Jesse Smith, inorganic chemistry
Foster, Robert Edward, II, plant breeding
George, Boyd Winston, entomology
Gerhard, Paul Donald, entomology
Klug, Harlan Lyle, biochemistry
Lisonbee, Lorenzo Kenneth, biology, science education
Melaven, Arthur David, inorganic chemistry
Pew, Weymouth D., physiology
Russell, Thomas Edward, plant pathology
Sharples, George Carroll, horticulture
Way, Harold E., physics

PATAGONIA
Quinlan, James Joseph, geology

PEARCE
Quill, Laurence Larkin, inorganic chemistry

PHOENIX
Allen, John Rybolt, chemistry, biochemistry
Antos, Robert John, pharmacology
Arnold, Joseph Frederick, environmental management, plant ecology
Bariola, Louis Anthony, economic entomology
Bartlett, Alan C., genetics
Baum, William Stanhope, cancer, internal medicine
Bell, Marion Randolph, insect pathology
Bennett, Peter Howard, epidemiology, medicine
Berquist, Kenneth R., microbiology
Bradshaw, Gordon Van Rensselaer, mammalogy
Butler, Byron C., obstetrics & gynecology, biophysics
Butler, George Daniel, Jr, entomology
Butler, Joseph Miles, medical entomology; parasitology
Cain, H Thomas, anthropology
Cherny, Walter B., obstetrics & gynecology
Counts, Jon Milton, public health administration
Creelius, Harry Gilbert, medical microbiology
Curtis, David William, applied physics
Davis, Charles Homer, agronomy
Dobyns, Henry Farmer, anthropology, ethnohistory
Eidelberg, Eduardo, physiology
Favero, Martin, microbiology
Feaster, Carl Vance, plant breeding
Fink, Dwayne Harold, agronomy
Fisher, Daniel Jerome, mineralogy, crystallography
Flint, Hollis Mitchell, entomology
Flock, Eunice Verna, biochemistry
Freeman, Irving, internal medicine
Fry, Kenneth E., plant physiology
Galizioli, Steve, wildlife management
Gilbert, Richard Gene, soil microbiology, plant pathology
Goodwin, Melvin Harris, Jr, epidemiology, parasitology
Guinn, Gene, plant physiology
Hale, Harry W, Jr, surgery
Heck, Joseph Gerard, bacteriology
Heinle, Preston Joseph, organic chemistry
Henneberry, Thomas James, entomology
Hicks, Harold Richard, immunology, microbiology
Hinze, Ray Everald, analytical chemistry, inorganic chemistry
Hull, Hugh Boden, cardiovascular physiology
Hunter, William Leslie, plastics chemistry, semiconductors
Jackson, Ray Dean, soil physics
Jungermann, Eric, organic chemistry
Kelly, John V., obstetrics, gynecology
Kimball, Bruce Arnold, soil physics, micrometeorology
Kleitsch, William Philip, surgery
Korb, Ernest Lloyd, petroleum chemistry
Larson, Noal P., economic entomology
Lawver, Donald Allen, mathematics
Le Seney, Catherine Cosgray, medicine, public health
Manera, Paul Allen, hydrogeology
McDonald, James Hogue, urology
McLean, Katharine Weidman, biochemistry
McNamara, John Edward, physical inorganic chemistry
Moyer, Patricia Helen, organic chemistry
Nakayama, Francis Shigeru, soil
Neumann, Richard K., mathematics, computer science
Parsons, William Belle, Jr, medicine
Peterson, John William, resource management, soil conservation
Radin, John William, plant physiology
Rea, Ronald Howard, meat science
Rippere, Ralph Elliott, physical chemistry
Ritchie, Kim, biochemistry
Roney, James Nevile, economic entomology

Zygmunt, Wendell, anatomy

Russell, Ralph Keith, physics
Rutledge, James Luther, solid state physics, electronics
Sandstedt, Robert Morris, plant physiology; plant nematology
Sax, Ellis Dean, preventive medicine, public health administration
Spendlove, George Arthur, physical medicine & rehabilitation
Steffen, Albert Harry, food chemistry
Straus, Thomas Michael, applied physics
Sun, David Chen Hwa, medicine
Swanholm, Carl E., organic polymer chemistry
Turcotte, Edgar Lewis, plant genetics
Tyagi, Avdhesh Kumar, hydrology
Whitcomb, Donald Leroy, analytical chemistry
Wilson, Frank Douglas, plant genetics
Wilson, Richard Lee, entomology
Woodham, Donald W., analytical chemistry
Yates, Ann Marie, analytical chemistry
Zygmunt, Warren W., geology

SAFFORD
Turner, Fred, Jr, agricultural chemistry, soils

SCOTTSDALE
Alsever, John Bellows, internal medicine, immunohematology
Binkerd, Evan Francis, food science
Buswell, Robert James, analytical chemistry
Cattani, Ray August, agricultural chemistry, soil chemistry
Ekblaw, Sidney Everette, physical geography, economic geography
Eyring, LeRoy, physical chemistry
Fackler, Walter Valentine, Jr, surface chemistry
Fink, Kenneth Howard, food chemistry
Foltz, Thomas Roberts, Jr, protein chemistry
Hancock, Elizabeth Dieckenberger, horticulture, pollution biology
Herring, Harold Keith, animal science, biochemistry
Howell, Alvin Hercules, anthropology, America
Kline, Ralph Willard, food science
Rogers, Alan Barde, chemistry
Rominger, Joseph Franklin, geology
Sleeth, Rhule Bailey, food technology, animal nutrition
Stevenson, Alden, solid state physics
Thomas, Robert Malcolm, organic chemistry, mathematical analysis
Wysocki, Allen John, industrial organic chemistry

SUN CITY
Allen, Clifford V., nuclear medicine
Baker, Gladys Elizabeth, botany
Black, Benjamin Marden, surgery
Fischer, Louis, pharmaceutical chemistry
Fritz, Roy Fredoline, entomology; epidemiology
Goff, Stillman R., chemistry
Hawk, Virgil Brown, agronomy
Jones, Stephen Barr, geography
Lott, Richard Vincent, pomology
Pugh, David Graham, radiology
Reitz, Louis Powers, plant breeding
Spencer, Howard Camac, toxicology, biochemistry
Thomas, Joseph James, biochemistry
Urban, Walter Mathias, physical chemistry; food science
Woods, Mark Winton, cytology, physiology

SUPERIOR
Crosswhite, Carol D., desert ecology, entomology

Crosswhite, Frank Samuel, botany, biogeography

PORTAL
Roth, Vincent Daniel, arachnology

PRESCOTT
Backus, Edward James, mycology
Bohning, John William, agronomy
Brown, Claudeous Jethro Daniels, zoology
Compton, William David, organic chemistry
Lightle, Paul Charles, forest pathology
Martin, Lloyd Milo, entomology

TEMPE
Acker, William James, geography, economics
Ahmadzadeh, Akbar, physics
Aldrich, Frank Thacher, geography
Alvarado, Ronald Herbert, zoology
Anderson, Mary Ruth, industrial engineering, statistics
Archer, Stanley J., immunology
Aronson, Jerome Melville, plant physiology
Barrett, Thomas Wilson, agronomy
Bedient, Jack DeWitt, mathematics
Bieber, Allan Leroy, biochemistry
Birkhofer, James Peter, inorganic chemistry
Bitter, Gary G., mathematics education, computer education
Boyd, George Addison, research administration
Brandt, Elizabeth Anne, anthropology
Brown, Duane, physical chemistry
Brown, Peter, organic chemistry
Bunt, Lucas N H, mathematics, physics
Burgoyne, Edward Eynon, organic chemistry
Burke, William James, polymer chemistry
Burt, Donald McLain, mineralogy, petrology
Buseck, Peter R., geochemistry, economic geology
Cadien, James David, physical anthropology
Canright, James Edward, plant morphology, paleobotany
Castle, Gordon Benjamin, zoology
Cazier, Mont Adelbert, systematic entomology
Chalquest, Richard Ross, veterinary medicine, microbiology
Clark, Geoffrey Anderson, anthropology, archaeology
Clothier, Ronald Raymond, mammalogy
Cole, Gerald Ainsworth, limnology
Collins, James Paul, ecology
Conceaux, Malcolm Louis, cultural geography
Cronin, John Read, biochemistry
Davidson, Elizabeth West, invertebrate pathology
Davis, Edwin Alden, plant physiology
Decker, John Peter, applied synecology
Ditert, Alfred Edward, Jr, anthropology, geology
Durrenberger, Robert W., geography, climatology
Duvall, Vinson Lamar, range ecology
Dycus, Augustus Mahon, plant physiology
Eder, James Farnum, Jr, anthropology
Fouquette, Martin John, Jr, herpetology
Freund, John Ernst, mathematical statistics
Frost, Melvin Jesse, geography of Latin America
Fuchs, Jacob, analytical chemistry, spectroscopy
Gerking, Shelby Delos, ecology, fish biology
Glaunsinger, William Stanley, solid state chemistry, physical chemistry
Goldstein, Myron, mathematics
Grace, Edward Everett, topology
Gust, John Devens, Jr, organic chemistry
Hadley, Neil F., environmental physiology
Hanson, Hugh, animal ecology
Hanson, Roland Clements, experimental solid state physics
Harris, Joseph, biochemistry
Hasbrouck, Frank Flinn, entomology
Hayes, Donald Scott, astronomy
Heede, Burchard Heinrich, watershed management, earth sciences
Henkel, Ray, economic geography
Herald, Delbert Leon, Jr, natural products chemistry
Hestenes, David, theoretical physics
Hibbert, Alden R., forest hydrology
Hilgeman, Robert Harry, horticulture
Holloway, John Requa, petrology, geochemistry
Jacob, Richard John, theoretical physics
Jacobowitz, Ronald, mathematics
Johnson, Roy Melvin, microbiology
Judd, Benjamin Ira, agronomy
Justus, Jerry T., developmental biology
Juvet, Richard Spaulding, Jr, analytical chemistry
Kaufmann, William B., elementary particle physics
Kelly, John Beckwith, mathematics
Kevane, Clement Joseph, physics
Kyrala, Ali, mathematical physics

Landers, Earl James, invertebrate zoology
Larimer, John William, geochemistry
Leathers, Chester Ray, mycology
Leon, Burke, physical chemistry
Lin, Sheng Hsien, chemical kinetics, chemical physics
Liu, Chui Hsun, analytical chemistry, inorganic chemistry
Lounsbury, John Frederick, economic geography, physical geography
Luchsinger, Wayne Wesley, biochemistry
Lundin, Robert Folke, geology, paleontology
Marcus, Melvin Gerald, physical geography
McTaggart, William Donald, geography of Southeast Asia, population geography
Meister, Arnold George, molecular spectroscopy
Merbs, Charles Francis, physical anthropology
Miller, Paul Theodore, geology
Miller, Victor Jay, horticulture
Minckley, Wendell Lee, ichthyology, aquatic ecology.
Mings, Robert Charles, economic geography
Moeller, Therald, inorganic chemistry
Moody, Edward Grant, animal nutrition
Moore, Carlton Bryant, geochemistry, meteorites
Moore, John Douglas, mathematics
Moore, Nadine Hanson, mathematics
Munch, Theodore, bacteriology, science education
Navrotsky, Alexandra, chemistry
Nering, Evar Dare, mathematics
Nigam, Bishan Perkash, particle physics, nuclear physics
Northey, William T, immunology
Ohmart, Robert Dale, zoology
O'Keeffe, Michael, solid state chemistry
Page, John Boyd, Jr, solid state physics
Parsons, Michael L, analytical chemistry
Patten, Duncan Theunissen, plant ecology, environmental biology
Patterson, Robert Allen, zoology
Patton, David Roger, wildlife research
Pettit, George Robert, organic chemistry, cancer
Pewe, Troy Lewis, geomorphology
Pinkava, Donald John, botany
Plog, Fred T, anthropology
Ragan, Donal Mackenzie, structural geology
Ransom, John Paul, medical bacteriology
Rasmussen, David Irvin, evolutionary biology, genetics
Rawls, William Shelton, physics
Reeves, Henry Courtland, biochemistry, bacterial physiology
Richardson, Grant Lee, agronomy
Robinson, Daniel Owen, soil science
Roy, Radha Raman, nuclear physics
Ruppe, Reynold Joseph, anthropology, archaeology
Sanderson, Robert Thomas, inorganic chemistry
Sansone, Fred J, mathematics
Sargent, Charles S, Jr, urban geography
Sauck, William August, geology, geophysics
Savage, Nevin William, mathematics
Schamadan, James Louis, bioengineering, aerospace medicine
Schmidt, Jean M, microbiology
Scott, Walter Tandy, mathematics
Sheridan, Michael Francis, petrology, volcanology
Sherman, Thomas Lawrence, mathematics
Short, Henry Laughton, ecology, wildlife biology
Sinkov, Abraham, mathematics
Smith, Lehi Tingen, mathematics
Snyder, Ernest Elwood, Jr, science education
Sommerfeld, Milton R, phycology
Stahnke, Herbert Ludwig, arachnology, systematics
Stark, Barbara Louise, anthropology
Starrfield, Sumner Grosby, theoretical astrophysics
Stewart, Donald George, mathematics
Stewart, Kenneth Malcolm, anthropology, ethnology
Stiles, Philip Glenn, food technology, poultry science
Stocker, Richard Louis, geophysics
Stoner, Richard Griffith, physics
Stutsman, Paul Snell, organic polymer chemistry, petroleum chemistry
Szarek, Stanley Richard, physiological ecology
Taysom, Elvin David, animal husbandry
Thomson, Tom Radford, physical organic chemistry
Tillery, Bill W, science education
Trelease, Richard Norman, cell biology
Turner, Christy Gentry, II, physical anthropology, dental anthropology
Von Dreele, Robert Bruce, solid state chemistry, crystallography
Wagstaff, H Reid, economic geography
Walker, Charles Thomas, solid state physics
Wexler, Charles, mathematics
Whitehurst, Harry Bernard, physical chemistry
Woolf, Charles Martin, genetics
Work, Richard Nicholas, polymer physics
Yale, Francis Gaymon, science education
Young, Dennis Lee, mathematical statistics

TOLLESON
Brown, Richard Edward, nutrition, physiology
Roberts, William Kenneth, ruminant nutrition

TSAILE
Barreras, Raymond Joseph, organic chemistry

TUBAC
Fletcher, Robert Dawson, meteorology

TUCSON
Abt, Helmut Arthur, astronomy
Alcorn, Stanley Marcus, plant pathology
Allen, Ross Marvin, plant pathology
Angevine, Jay Bernard, Jr, neuroanatomy
Anthony, John Williams, geology, mineralogy
Aposhian, Hurair Vasken, cell biology, pharmacology
Ashby, Carl Toliver, physical chemistry
Babcock, Clarence Lloyd, physics
Babcock, Horace Maxson, hydrology
Bagnara, Joseph Thomas, embryology
Bahr, James Theodore, biophysical chemistry
Bannister, Bryant, dendrochronology, academic administration
Barfield, Michael, physical chemistry
Barker, Roy Jean, entomology
Barrett, Bruce Richard, theoretical nuclear physics
Barrett, Harrison Hooker, medical instrumentation, optics
Bartels, Paul George, plant physiology
Bartels, Peter H, optics, computer science
Bashkin, Stanley, atomic physics, nuclear physics
Basso, Ellen Becker, anthropology
Basso, Keith Hamilton, anthropology
Bates, Robert Brown, organic chemistry
Battan, Louis Joseph, meteorology
Belton, Michael J S, astronomy
Bemis, William Putnam, genetics
Bender, George Almon, pharmacy
Bernstein, Carol, molecular biology
Bernstein, Harris, genetics
Berry, James Wesley, organic chemistry
Bessey, Paul Mack, horticulture
Bhattacharya, Rabindra Nath, mathematical statistics
Bickel, William Samuel, experimental physics, spectroscopy
Bicknell, Edward J, comparative pathology, clinical pathology
Bier, Milan, biophysics
Blasing, Terence Jack, climatology
Blitzer, Leon, celestial mechanics
Bloss, Homer Earl, plant pathology, plant biochemistry
Bohn, Hinrich Lorenz, soil chemistry
Bok, Bart Jan, astronomy
Boone, Max L M, radiotherapy
Bourque, Don Philippe, molecular biology
Bowen, Theodore, physics
Bownds, John Marvin, mathematics
Brady, Frederick Jonathon, tropical medicine
Braun, Eldon John, physiology, comparative physiology
Breckinridge, James Bernard, optical physics
Brendel, Klaus, pharmacology
Brewer, Willis Ralph, pharmacognosy
Briggs, Robert Eugene, agronomy
Broadfoot, Albert Lyle, physics
Brosin, Henry Walter, medicine
Brown, Stuart Graeme, hydrology
Brown, William Hedrick, animal nutrition
Bryan, Douglas Everett, entomology
Bull, William Benham, geomorphology
Burke, James Joseph, Jr, optics
Burke, Michael Francis, analytical chemistry
Burrows, Benjamin, internal medicine
Burton, Lloyd Edward, pharmacy, public health
Butler, Robert Franklin, paleomagnetism
Buxton, Dwayne Revere, crop physiology
Cable, Dwight Raymond, forestry
Calder, William Alexander, III, physiological ecology
Caldwell, Mary Estill, bacteriology
Caldwell, Roger Lee, biochemistry, plant pathology
Call, Reginald Lessey, physics
Capp, Michael Paul, radiology, pediatrics
Cardon, Bartley Pratt, microbiology, nutrition
Carruth, Laurence Adams, entomology
Carter, Herbert Edmund, chemistry, biochemistry
Chang, Nada, endocrinology
Chapman, Clark Russell, planetary sciences
Cheema, Mohindar Singh, mathematics
Chiasson, Robert Breton, comparative anatomy, comparative endocrinology
Chilcott, John Henry, anthropology
Childs, Richard Francis, pharmaceutical chemistry
Christian, Charles Donald, obstetrics & gynecology
Chvapil, Milos, physiological chemistry, experimental pathology
Clay, James Ray, mathematics
Clayton, John Wesley, Jr, toxicology, respiratory physiology
Cluff, Carwin Brent, hydrology
Cockrum, Elmer Lendell, vertebrate zoology
Cohen, Judith Gamora, astrophysics
Cole, Jack Robert, medicinal chemistry
Connor, William Gorden, medical physics
Consroe, Paul F, pharmacology
Corrigan, James John, Jr, hematology, pediatrics
Coyne, George Vincent, astronomy
Crawford, David Livingstone, astronomy
Crowder, Larry A, insect physiology, toxicology
Culbert, Thomas Patrick, anthropology, archaeology
Cusanovich, Michael A, biochemistry
Cushing, Jim Michael, applied mathematics
Cutts, Robert Irving, psychiatry
Damon, Paul Edward, geochronology, geochemistry
Dantzler, William Hoyt, physiology
Davis, George Herbert, structural geology, economic geology
Davis, John Robert, pathology
Davis, Nicholas Falconer, geochemistry, exploration geology
Davis, Russell Price, zoology
Davis, Stanley Nelson, hydrogeology
Day, Arden Dexter, plant breeding
Dean, Jeffrey Stewart, archaeology, dendrochronology
DeBell, Arthur Gerald, optics
DeCook, Kenneth James, geology, hydrology
DeFranco, Ronald James, applied mathematics
Dennis, Robert E, agronomy
Denny, William F, hematology, internal medicine
Deutschman, Archie John, Jr, organic chemistry
DeVito, Carl Louis, mathematical analysis
Dickinson, Robert Eric, urban geography, geography of Western Europe
Diebold, Albert Richard, Jr, anthropology, linguistics
Dietrich, Daniel David, atomic physics
Dinowitz, Marshall, microbiology
Donahue, Douglas James, nuclear physics
Drake, Michael Julian, geochemistry
Dunkelman, Lawrence, physics, astronomy
Dunn, Cecil Gordon, food technology, industrial microbiology
Dutt, Gordon Richard, soil chemistry
Du Val, Merlin Kearfott, surgery
Edwards, Phyllis Q, tuberculosis
Eisa, Hamdy Mahmoud, plant breeding, vegetable crops
Emrick, Roy M, solid state physics
Endrizzi, John Edwin, cytogenetics
Enemark, John Henry, inorganic chemistry
Erickson, Charles John, applied anthropology, system analysis
Eskelson, Cleamond D, biochemistry, organic chemistry
Evans, Daniel Donald, hydrology
Eyer, James Arthur, optics, photography
Faris, William Guignard, mathematical physics
Farr, William Morris, plasma physics, reactor physics
Fazio, Steve, horticulture
Felten, James Edgar, astrophysics
Feltham, Robert Dean, inorganic chemistry, physical chemistry
Ferdon, Edwin Nelson, Jr, anthropology, cultural geography
Ferebee, Robert Newton, microbiology
Ferguson, Charles Wesley, dendrochronology
Fernando, Quintus, analytical chemistry
Ferris, Wayne Robert, cytology
Fife, Paul Chase, mathematics, continuum mechanics
Finch, Stuart McIntyre, psychiatry
Fitch, Walter Stewart, astronomy
Fontana, Bernard Lee, anthropology
Forrester, James Donald, geology
Forster, Leslie Stewart, physical chemistry
Franken, Peter Alden, optical physics, quantum optics
Freiser, Henry, analytical chemistry
Frieden, Bernard Roy, optical physics
Fritts, Harold Clark, plant ecology, dendrochronology
Fulginiti, Vincent Anthony, pediatrics, immunology
Fuller, Wallace Hamilton, biochemistry, bacteriology
Fye, Robert Eaton, entomology
Ganguly, Jibamitra, geology, geochemistry
Garcia, Jose Dolores, Jr, atomic physics
Gay, Lloyd Wesley, watershed management
Gehrels, Tom (Anton Marie Jacob), astronomy
Gerner, Eugene Willard, cell biology, radiobiology
Getty, Harry Thomas, anthropology
Gibson, Lay James, economic geography, urban geography
Gilbertson, Robert Lee, mycology, forest pathology
Glass, Richard Steven, organic chemistry
Gloyd, Howard Kay, systematic zoology
Goldberg, Leo, astrophysics
Gordon, Mark A, radio astronomy
Gould, Laurence McKinley, geology
Greenberg, Richard Joseph, planetary sciences, celestial mechanics
Greenstone, Arthur W, inorganic chemistry, science writing
Groemer, Helmut (Johann), mathematics
Grove, Larry Charles, mathematics
Gruener, Raphael P, physiology
Haase, Edward Francis, botany, plant ecology
Hadley, Mac Eugene, comparative endocrinology
Hale, William Harris, animal nutrition
Hall, Henry Kingston, Jr, physical chemistry, organic chemistry
Hall, Robert Lee, plant genetics, ethnobotany
Halonen, Marilyn Jean, immunopathology
Hanahan, Donald James, biochemistry
Harris, Robert Martin, genetics
Harshbarger, John W, geology, hydrology
Hart, Lyman Herbert, geology
Harvey, John Warren, astrophysics
Hattler, Brack Gillium, Jr, surgery, immunology
Haury, Emil Walter, anthropology, archaeology
Hawkes, Herbert Edwin, Jr, economic geology
Haynes, Caleb Vance, Jr, geology
Hecht, Melvin Edwin, geography of the Southwest United States
Hedge, George Albert, physiology
Heed, William Battles, zoology
Heine, Melvin Wayne, obstetrics & gynecology, endocrinology
Hendrickson, John Roscoe, vertebrate zoology, ecology
Henshaw, Paul Stewart, radiobiology
Herman, Benjamin Morris, meteorology
Hetrick, David Leroy, nuclear science
Heylmun, Edgar B, geology
Hilberry, Norman, physics
Hill, Henry Allen, astrophysics
Hilliard, Ronnie Lewis, physics
Hine, Richard Bates, plant pathology
Hinton, Thomas Benjamin, anthropology
Hintzen, Paul Michael, astronomy
Hirschi, L Edwin, mathematics education
Hoag, Arthur Allen, astronomy
Hoffmann, William Frederick, physics, astronomy
Hogan, Le Moyne, horticulture
Hoshaw, Robert William, phycology

ARIZONA

Hruby, Victor J, bio-organic chemistry
Hsieh, Ke Chiang, cosmic ray physics
Hubbard, William Bogel, Jr, planetary sciences
Huber, Roger Thomas, insect ecology, biometeorology
Huestis, Douglas William, medicine, pathology
Huffman, Donald Ray, solid state physics, astrophysics
Hull, Herbert Mitchell, plant physiology
Hulse, Frederick Seymour, anthropology
Hungerford, Charles Roger, wildlife ecology
Hunten, Donald Mount, planetary sciences
Jacobs, Stephen Frank, physics
Janssen, Robert (James) J, virology, immunology
Jelinek, Arthur J, anthropology
Jenkins, Edgar William, high energy physics
Jensen, Richard Grant, biochemistry
Jeter, Wayburn Stewart, medical microbiology, immunology
Johnson, Gordon V, soil fertility
Johnson, Harold Lester, astronomy
Johnson, Jack Donald, research administration, environmental science
Johnson, Paul Christian, physiology
Jokipii, Jack Randolph, theoretical physics, astrophysics
Jones, Lee Bennett, organic chemistry
Jordan, Gilbert Leroy, range science
Joy, Edward Albert, human anatomy
Just, Kurt W, quantum physics
Kalbach, Robert Michael, experimental high energy physics, cosmic ray physics
Kassander, Arno Richard, Jr, geology
Katterman, Frank Reinald Hugh, plant physiology
Katz, Murray Alan, nephrology, cardiovascular diseases
Kauffeld, Norbert M, entomology, apiculture
Keck, Konrad, biology
Keller, Philip Charles, inorganic chemistry
Kelley, Alec Ervin, organic chemistry
Kellman, Raymond, organic chemistry, polymer chemistry
Kelly, William Henderson, cultural anthropology, applied anthropology
Kemmerer, Arthur Russel, biochemistry
Kendrick, Edgar Lohr, plant pathology
Kessler, John Otto, physics
Kettel, Louis John, pulmonary physiology, medical education
Kilkson, Rein, biophysics
Kim, Hyun Dju, physiology, biochemistry
Kinman, Thomas David, astronomy
Kircher, Henry Winfried, organic chemistry
Kirshner, Robert Paul, astronomy
Klemmedson, James Otto, soil science, range ecology
Kneebone, William Robert, agronomy, plant breeding
Knorr, Philip Noel, forest economics, resource management
Knudson, Ronald Joel, pulmonary physiology
Kohler, Sigurd H, nuclear physics
Krayer, Otto, pharmacology
Kremp, Gerhard Otto Wilhelm, geology
Krutzsch, Philip Henry, anatomy, zoology
Kukolich, Stephen George, physical chemistry, structural chemistry
Laetsch, Theodore Willis, mathematical analysis
LaMarche, Valmore Charles, Jr, geology
Lamb, George Lawrence, Jr, physics
Lamb, Willis Eugene, Jr, quantum mechanics, atomic physics
Lane, Leonard James, hydrology
Larson, Harold Phillip, astronomy
Layton, Jack Malcolm, pathology
Leavitt, John Adams, physics
LeBouton, Albert V, microscopic anatomy, cell biology
Lebowitz, Michael David, epidemiology, pulmonary diseases
Leonard, Alvin Robert, medicine, public health
Leonard, John Lander, mathematics
Levenson, Alan Ira, psychiatry
L'Heureux, Jacques (Jean), cosmic ray physics, astrophysics
Lim, Louise Chin, mathematics
Lindell, Thomas Jay, molecular pharmacology
Lindsay, Everett Harold, Jr, vertebrate paleontology
Livingston, William Charles, astronomy
Lomen, David Orlando, applied mathematics
Lomont, John S, mathematical physics

Long, Austin, geochemistry
Longacre, William Atlas, II, anthropology, archaeology
Loomis, Timothy Patrick, petrology, tectonics
Loper, Gerald Milton, agronomy, biochemistry
Lovell, Stuart Estes, computer science
Lovelock, David, mathematics, theoretical physics
Lovering, Thomas Seward, economic geology
Low, Frank James, solid state physics
Lowe, Charles Herbert, Jr, zoology
Lucas, David Owen, immunology
Ludovici, Peter Paul, bacteriology
Lynch, Peter John, dermatology
Lynds, Beverly T, astronomy
Lynds, Clarence Roger, astronomy
Lytle, Ivan M, physiology
Madden, John William, surgery, biology
Mahar, J Michael, anthropology
Mahmoud, Hormoz Massoud, theoretical physics
Mann, Henry Berthold, mathematics
Marathay, Arwind Shankar, optical physics
Marcus, Frank I, cardiology, internal medicine
Martin, Paul Schultz, ecology
Martin, Samuel Clark, range conservation
Marvel, Carl Shipp, organic polymer chemistry
Mason, Charles Thomas, Jr, botany
Massengale, Martin Andrew, agronomy, crop physiology
Mathews, Christopher King, biochemistry
Mayall, Nicholas Ulrich, astronomy
Mayo, Evans Blakemore, geology
McAlister, Dean Ferdinand, agronomy
McAlister, Harold Alister, astronomy
McCaughey, William Frank, nutritional biochemistry
McCauley, William John, physiology, anatomy
McClure, Michael Allen, nematology
McComb, Andrew Logan, forestry, watershed management
McCullen, John Douglas, nuclear physics
McCullough, Edgar Joseph, Jr, structural geology
McDaniel, Robert Gene, genetics, plant physiology
McGinnies, William Grovenor, plant ecology
McIntyre, Laurence Cook, Jr, nuclear physics
McKelvie, Douglas H, veterinary pathology
McKenney, Dean Brinton, optics
McRae, Lorin Post, biomedical engineering, electronics
Mead, Albert Raymond, malacology
Meezan, Elias, biochemistry, pharmacology
Meinel, Aden Baker, astrophysics, solar energy
Mellor, Robert Sydney, plant physiology
Mendelson, Neil Harland, genetics
Metcalfe, Darrel Seymour, agronomy
Mielke, Eugene Albert, pomology, plant physiology
Milkey, Robert William, astronomy
Miller, Walter Bernard, III, physical chemistry
Mitchell, Lloyd Vernon, meteorology, geography
Moffett, Joseph Orr, entomology
Monson, Gale (Wendell), ornithology, wildlife management
Moore, Leon, entomology
Mortara, Lorne B, high energy physics
Morton, Howard LeRoy, plant physiology
Mount, David William Alexander, genetics, molecular biology
Moyers, Jarvis Lee, chemistry
Mulvaney, James Edward, organic chemistry
Muramoto, Hiroshi, plant breeding, genetics
Myers, Donald Earl, mathematics
Myers, Harold Edwin, agronomy
Nagy, Bartholomew Stephen, organic geochemistry
Nelson, Frank Eugene, food microbiology
Nelson, Merritt Richard, plant pathology
Neuman, Shlomo Peter, hydrology, hydrogeology
Nielson, Mervin William, entomology
Nigh, Edward Leroy, Jr, plant pathology, nematology
Norton, Denis Locklin, geochemistry, geology
Novak, Milan Vaclav, medical education, microbiology
Nudelman, Sol, physics

Nugent, Charles Arter, Jr, internal medicine, endocrinology
Nutting, William Leroy, entomology
Odishaw, Hugh, geophysics
Oekber, Norman Fred, horticulture
Officer, James Eoff, anthropology, academic administration
Ogden, David Anderson, nephrology, internal medicine
O'Leary, James William, plant physiology
Olsen, Stanley John, biological anthropology, archaeology
O'Malley, Robert Edmund, Jr, applied mathematics
Openshaw, Martin David, soil fertility, plant physiology
Osborn, Herbert B, hydrology
Pacholczyk, Andrzej Grzegorz, theoretical astrophysics, radio astronomy
Palmer, John Davis, clinical pharmacology
Parmenter, Robert Haley, solid state physics
Pate, James Bruce, plant breeding, genetics
Patton, Dennis David, radiology, nuclear medicine
Paylore, Patricia Paquita, scientific bibliography, scientific documentation
Peacock, Erle Ewart, surgery
Pearson, Paul Brown, nutrition
Pederson, Leland Roger, geography
Peltier, Leonard Francis, orthopedic surgery
Peng, Yeh-Shan, nutrition
Peters, William Callier, geology
Philip, Thomas, food science
Picchioni, Albert Louis, pharmacology
Pickens, Peter E, neurophysiology
Pierce, Austin Keith, solar physics
Pierce, Richard Scott, algebra
Pinckard, Robert Neal, immunopathology, microbiology
Poole, Hubert Kimberly, air pollution, environmental biology
Post, Roy G, physical inorganic chemistry
Price, Michael J, astrophysics
Price, Ralph Lorin, food science, biochemistry
Qashu, Hasan Khalil, hydrology, water resources
Radabaugh, Robert Eugene, geology
Rasmussen, William Otto, geophysics, computer science
Rathje, William Laurens, anthropology, archaeology
Reed, Raymond Edgar, animal pathology
Reid, Bobby Leroy, biochemistry, nutrition
Reinhard, Karl R, pathology
Resnick, Sol Donald, hydrology
Rhodes, Herbert Dawson, chemistry
Rice, Richard W, animal nutrition, biochemistry
Rifkind, David, internal medicine, infectious diseases
Riker, Albert Joyce, plant pathology
Robinson, William James, dendrochronology
Robson, John William, physics
Roe, Arthur, organic chemistry
Roemer, Elizabeth, astronomy
Rosenzweig, Michael Leo, population ecology, mammalogy
Roubicek, Carl Ben, animal genetics
Rubis, David Daniel, plant breeding, genetics
Rund, Hanno, mathematics, mathematical physics
Rund, John Valentine, inorganic chemistry
Rupley, John Allen, biochemistry
Russell, Diane Haddock, biochemistry, pharmacology
Russell, Stephen Mims, ornithology
Saarinen, Thomas Frederick, resource geography
Salzman, William Ronald, physical chemistry
Sanner, Frederick Charles, astronomy
Sargent, Murray, III, quantum optics
Scadron, Michael David, theoretical physics, high energy physics
Schaffer, William Morris, ecology, evolutionary biology
Schmutz, Ervin Marcell, range management
Schneider, Lawrence Kruse, anatomy
Schonhorst, Melvin Herman, agronomy, plant breeding
Schotland, Richard Morton, meteorology
Schreiber, Joseph Frederick, Jr, geology
Schuh, James Donald, animal nutrition
Schumacher, Dietmar, geology, paleontology
Scully, Marlan Orvil, physics

Sechrist, John William, neuroanatomy, developmental biology
Seeley, Millard Garfield, organic chemistry
Seraphin, Bernhard Otto, solid state physics
Serkowski, Krzysztof, astrophysics
Shack, Roland Vincent, optics
Shannon, Robert Rennie, optics
Shaw, Ellsworth, soil science, agricultural chemistry
Shaw, William Wesley, wildlife conservation, resource management
Shemansky, Donald Eugene, atmospheric physics
Shoemaker, Richard Lee, chemical physics, quantum optics
Sibley, William Austin, neurology
Silverstein, Martin Elliot, medical science, health sciences
Simmons, Woodrow Wilson, geology
Simonian, Vartkes Hovanes, pharmacognosy
Simpson, Eugene Sidney, hydrology
Simpson, George Gaylord, vertebrate paleontology, geology
Sinclair, Norval A, microbiology
Sinski, James Thomas, medical mycology
Slater, Philip Nicholas, optics, remote sensing
Small, James Graydon, quantum optics
Smiley, Terah Leroy, geochronology
Smith, Bradford Adelbert, astronomy
Smith, Edwin Lamar, Jr, range management
Smith, Hugh Hollingsworth, virology
Smith, Norman Sherrill, wildlife ecology
Smith, Watson, anthropology
Smith, Winfield Scott, optics
Snider, Robert Gordon, hydrology, research administration
Solomon, Allen M, plant ecology, palynology
Sonett, Charles Philip, physics
Sowls, Lyle Kenneth, wildlife biology
Spangler, Hayward Gosse, entomology
Spicer, Edward Holland, applied anthropology
Stairs, Gerald Ray, genetics, forestry
Staley, Dean Oden, meteorology
Standifer, Lonnie Nathaniel, entomology, parasitology
Stanghellini, Michael Eugene, plant pathology
Stanislawski, Dan, historical geography, regional geography
Stark, Royal William, solid state physics, low temperature physics
Stavroudis, Orestes Nicholas, mathematics, optics
Steelink, Cornelius, organic chemistry
Steinbrenner, Arthur H, mathematics education
Stephens, Charles Arthur Lloyd, Jr, internal medicine
Stewart, William Thomas, organic chemistry
Stith, Lee S, plant breeding
Stoner, John Oliver, Jr, atomic spectroscopy
Stott, Gerald H, dairy science
Streets, Rubert Burley, plant pathology
Strittmatter, Peter Albert, astronomy
Stroehlein, Jack Lee, soil science
Strom, Robert Gregson, astrogeology
Strom, Stephen, astronomy, astrophysics
Stuart, Douglas Gordon, physiology
Stull, Elisabeth Ann, limnology, phycology
Stull, John Warren, dairy science
Sumner, John Stewart, exploration geophysics, geology
Swihart, Thomas Lee, astrophysics
Swindell, William, optics
Taber, Stephen, III, apiculture
Tanner, Clara Lee, cultural anthropology, archaeology
Tapia, Santiago, astrophysics
Thames, John Long, watershed management
Theurer, Clark Brent, animal nutrition
Thews, Robert Leroy, theoretical physics
Thompson, Anson Ellis, plant breeding, plant genetics
Thompson, Raymond Harris, anthropology, archaeology
Thompson, Richard Bruce, mathematics
Thompson, Rodger Irwin, astrophysics
Thomson, Donald A, marine ecology, ichthyology
Tifft, William Grant, astronomy
Titley, Spencer Rowe, geology, geochemistry
Tollin, Gordon, biophysical chemistry
Tollman, James Perry, pathology
Tomizuka, Carl Tatsuo, solid state physics
Toubassi, Elias Hanna, mathematics

Treat, Jay Emery, Jr, physics
Trifan, Deonisie, applied mathematics
Tucker, Thomas Curtis, soils, plant nutrition
Turner, Arthur Francis, optics, physics
Turner, Raymond Marriner, ecology
Tuttle, O Frank, geology
Ulich, Bobby Lee, radio astronomy
Underwood, Jane Hainline, anthropology
Upchurch, Robert Phillip, plant physiology, weed science
Van Asdall, Willard, plant ecology
Vanselow, Neal A, medical administration, allergy
Van Winkle, Walton, Jr, biochemistry
Vavich, Mitchell George, biochemistry
Vincent, Harold Arthur, analytical chemistry
Voigt, Robert Lee, plant breeding
Vuillemin, Joseph J, physics
Wagle, Robert Fay, forestry, botany
Waller, Gordon David, apiculture
Wallraff, Evelyn Bartels, microbiology, immunology
Wangsness, Roald Klinkenberg, physics
Ward, Oscar Gardien, genetics
Warrick, Arthur W, soil physics, mathematics
Watson, Theo Franklin, entomology
Watts, Raymond Ellsworth, veterinary surgery
Weaver, Albert Bruce, physics, academic administration
Weaver, Thomas, social anthropology, applied anthropology
Weber, Charles Walter, biochemistry, nutrition
Webster, Orrin John, agronomy
Wegner, Thomas Norman, animal physiology, biochemistry
Weistrop, Donna Etta, astronomy
Wells, Michael Arthur, biochemistry
Werner, Floyd Gerald, entomology
West, Robert Elmer, exploration geophysics
Weymann, Ray J, astronomy
Wheeler, Lawrence, psychophysics, psychology
Whitaker, Ewen A, planetary sciences
White, Raymond E, astronomy
Whiting, Frank M, nutrition, biochemistry
Wilkening, Laurel Lynn, meteoritics, planetary sciences
Wilska, Alvar P, physics
Wilson, Andrew Wilkins, geography
Wilson, George Spencer, analytical chemistry
Wilson, Lorne Graham, soil physics, hydrology
Wilson, Richard Fairfield, geology
Wing, William Hinshaw, atomic physics, molecular physics
Wise, Edward Nelson, chemistry, research administration
Wolfe, William Louis, Jr, optics, electrical engineering
Wood, Bruce, mathematics, mechanical engineering
Woodin, William Hartman, III, zoology
Woods, Alexander Hamilton, immunology
Wright, Jerome J, geology, hydrogeology
Wyant, James Clair, optics
Wyckoff, Ralph Walter Graystone, physical chemistry, biophysics
Yakaitis, Ronald William, anesthesiology
Yall, Irving, microbial physiology
Yamamura, Henry Ichiro, neurochemistry, neuropharmacology
Young, Jon Nathan, anthropology, resource management
Young, Kenneth Christie, cloud physics
Young, Richard Accipiter, solid state physics
Younggren, Newell A, biology
Zegura, Stephen Luke, human biology, biological anthropology
Zellner, Benjamin Holmes, astronomy
Zukoski, Charles Frederick, surgery
Zwolinski, Malcolm John, forest hydrology, watershed managenent

WEST SEDONA
Cree, Allan, petroleum geology

WICKENBURG
Northrop, John Howard, biochemistry, biology

WILCOX
Schnell, Jay Heist, ecology

WINKELMAN
Cathey, Everett Henry, science education, geophysics

YUMA
Gardner, Bryant Rogers, soil chemistry, plant nutrition
Jackson, Ernest Baker, agronomy, botany
Rodney, David Ross, horticulture
Ross, Anthony, zoology, entomology
Troutman, Joseph Lawrence, plant pathology
Tuttle, Donald Monroe, entomology

ARKANSAS

ALEXANDER
Bartlett, Frank David, soils

ARKADELPHIA
Basford, Adelphia Meyer, ecology
Dorris, Peggy Rae, zoology
Duncan, Thomas O, fish biology
Everett, Wilbur Wayne, biophysical chemistry
Foster, Mary Lee, mathematics
Gosnell, Aubrey Brewer, analytical chemistry, polymer chemistry
Jones, Haskell Lee, analytical chemistry, science education
McCarty, Clark William, physical chemistry
McMasters, Dennis Wayne, plant physiology
Nisbet, Alex Richard, electroanalytical chemistry
Nix, Joe Franklin, geochemistry, analytical chemistry
Oliver, Kelly Hoyet, Jr, aquatic biology
Oliver, Victor L, zoology, parasitology
Palmer, Bryan D, inorganic chemistry
Singer, James Robert, low temperature physics
Strack, Charles Miller, geography
Wright, Joe Carrol, organic chemistry

BATESVILLE
Handford, Stanley Wing, physiology
Singley, Donald Heath, mathematics

BELLA VISTA
Black, Howard Charles, organic chemistry

BENTON
Ezell, James Ben, Jr, analytical chemistry

CAMDEN
Brooker, Francis Milton, food chemistry

CLARKSVILLE
Bridgman, John Francis, parasitology

CONWAY
Buffaloe, Neal Dollison, biology
Collins, Richard Arlen, fisheries biology
Cooper, Harold Eugene, genetics, physical anthropology
Haggard, Bruce Wayne, genetics
Hudson, Frank M, mathematics
Johnson, Arthur Albin, parasitology
Manion, Jerald Monroe, organic chemistry
Moon, Wilchor David, mathematics
Moore, Jewel Elizabeth, botany
Nichols, James Ross, zoology
Prince, Denver Lee, physics
Robbins, Joseph Graves, acoustics
Shideler, Robert Weaver, biochemistry
Stuckey, John Edmund, physical chemistry, inorganic chemistry
Teague, Marion Warfield, physical inorganic chemistry
Teague, Tommy Kay, algebra

FAYETTEVILLE
Allen, Robert Thomas, systematic entomology
Arnis, Edward Stephen, physical chemistry
Anderson, Richard John, atomic physics
Anderson, Robbin Colyer, physical chemistry

Bailey, Lowell Frederick, plant chemistry, plant physiology
Beasley, Joseph Noble, veterinary pathology
Becker, David Alvord, zoology
Blair, Grace, chemistry
Blyholder, George Donald, physical chemistry
Bower, Oliver Kenneth, mathematics
Bower, Raymond Kenneth, microbiology, animal virology
Boyer, William Paul, entomology
Bradley, George Alexander, horticulture
Brown, Connell Jean, animal breeding
Brown, Donald A, soil chemistry, poultry husbandry
Cairns, William Louis, biochemistry, photobiology
Carroll, Catherine, nutrition
Champlin, William G, medical microbiology
Chenhall, Robert Gene, anthropology, archaeology
Clayton, Frances Elizabeth, genetics
Cochran, Allan Chester, topology
Collins, Frederick Clinton, plant breeding, plant genetics
Cordes, Arthur Wallace, inorganic chemistry
Dahms, Reynold George, economic entomology
Dale, Edward Everett, Jr, plant ecology
Dale, James Lowell, plant pathology
Day, Stephen Martin, solid state physics
Dunn, James Eldon, mathematical statistics
Einert, Alfred Erwin, ornamental horticulture, plant physiology
Evans, William L, cytology
Frans, Robert Earl, agronomy
Fry, Arthur James, organic chemistry
Fuller, Roy Joseph, mathematics
Fulton, Joseph Patton, plant pathology
Fulton, Neil Douglas, plant pathology
Goode, Monroe Jack, phytopathology
Goodwin, Tommy Lee, food science
Guest, William C, genetics
Gyles, Nicholas Roy, genetics, animal breeding
Harris, Grover Cleveland, Jr, animal physiology, environmental physiology
Harris, William M, plant anatomy
Hinkle, Dale Albert, agronomy
Hinton, James Faulk, physical chemistry
Hobson, Arthur Stanley, theoretical physics
Hora, Rajinder Bir, statistics, algebra
Howick, Lester Carl, analytical chemistry
Hughes, Raymond Hargett, physics
Jackson, Kern Chandler, geology
Jenkins, Robert M, fish biology
Johnson, Dale A, physical chemistry, inorganic chemistry
Johnson, George Thomas, botany, bacteriology
Johnston, Perry Max, vertebrate embryology
Jones, John Paul, plant pathology
Kattan, Ahmed A, horticulture, food science
Keesee, John William, mathematics
Keown, Ernest Ray, applied mathematics
Kilambi, Varad Raj, fisheries, zoology
Kilgore, Robert L, veterinary medicine
Kimura, Naoki, mathematics
King, John William, agronomy, crop science
Konig, Ronald H, structural geology, economic geology
Kraemer, Louise Russert, malacology
Kuroda, Paul Kazuo, chemistry
Lancaster, Jesse Leonard, Jr, entomology
Lane, Forrest Eugene, plant physiology, plant biochemistry
Lewis, Paul Kermith, Jr, meat science
Lieber, Michael, theoretical physics
Lincoln, Charles Gatewood, entomology
MacDonald, Harold Carleton, geology
Martin, Duncan Willis, physiology, biophysics
Maxfield, Ollie Orland, geography
McCartney, Allen Papin, archaeology, anthropology
McCollum, John Paschal, horticulture
McFerran, Joe, vegetable crops
McGuire, James Marcus, plant pathology
Meek, James Latham, mathematics
Meyer, Richard Lee, mycology
Meyer, Walter Leslie, organic chemistry
Millett, Francis Spencer, biochemistry
Miner, Floyd Duane, entomology
Money, William Lang, endocrinology
Moore, James Norman, horticulture
Morehouse, Neal Francis, poultry science
Morris, Justin Roy, horticulture, plant physiology
Mueller, Clyde Dewey, genetics

Nelson, Talmadge Seab, animal nutrition
Nettleship, Anderson, pathology
Noland, Paul Robert, animal nutrition
Offutt, Marion Samuel, agronomy
Orton, William R, Jr, mathematics
Ostlund, Neil Sinclair, theoretical chemistry
Patterson, Loyd Thomas, immunology, animal virology
Paulissen, Leo John, medical microbiology, immunology
Phillips, Jacob Robinson, entomology
Piper, Edgar L, animal physiology, reproductive physiology
Porter, James Franklin, mathematics
Quinn, James Harrison, geology
Quirk, Roderic Paul, organometallic chemistry
Rakes, Jerry Max, animal science
Ray, Maurice L, animal husbandry
Richardson, Charles Bonner, physics
Riggs, Robert D, phytopathology
Roberts, Thomas David, organic chemistry
Rolingson, Martha Ann, anthropology
Rom, Roy Curt, horticulture, pomology
Salamo, Gregory Joseph, quantum optics
Schafer, Lothar, physical chemistry, inorganic chemistry
Schmitt, Neil Martin, biomedical engineering
Schmitz, Eugene H, invertebrate zoology, limnology
Schwartz, Herman Meyer, mathematics, physics
Scott, Howard Allen, plant pathology
Scott, Hubert Donovan, soil science
Scott, Thomas William, transportation geography, geography of Europe
Scroggs, James Edward, mathematics
Sealander, John Arthur, Jr, zoology
Sharrah, Paul Chester, physics
Siccardi, Frank John, veterinary pathology
Siegel, Samuel, organic chemistry
Sims, Leslie Berl, physical chemistry
Sistrunk, William Allen, food technology, bacteriology
Slack, Derald Allen, phytopathology, nematology
Smith, Edwin Burnell, plant taxonomy, biosystematics
Spooner, Arthur Elmon, agronomy
Springer, Melvin Dale, mathematical statistics
Stallcup, Odie Talmadge, dairy husbandry
Staudenmayer, Ralph, metallurgical chemistry
Steele, Kenneth F, geology, geochemistry
Stephenson, Edward Luther, animal nutrition
Stutte, Charles Auten, plant physiology, agronomy
Summers, William Hunley, mathematics
Tainter, Franklin Hugh, forest pathology
Talbert, Ronald Edward, agronomy, weed science
Te Beest, David Orien, plant pathology
Templeton, George Earl, plant pathology
Thoma, John Anthony, biochemistry
Thompson, Lyell, soils
Vaile, Joseph Edwin, horticulture
Waddle, Bradford Avon, agronomy, plant breeding
Wagner, George Hoyt, nutrition
Waldrop, Park William, nutrition, biochemistry
Walker, James Martin, herpetology
Walters, Hubert Jack, plant pathology
Warren, Lloyd Oliver, entomology
Wells, Bobby R, soil chemistry, soil fertility
Wickliff, James Leroy, plant physiology, photobiology
Wylie, William Dickey, entomology
Yates, Jerome Douglas, poultry nutrition
Yearian, William C, entomology
York, John Owen, plant breeding, plant genetics
Young, Seth Yarbrough, III, entomology, virology
Zinke, Otto Henry, physics

FOREMAN
Hulsey, Jess Dale, geology

HICKORY RIDGE
Taylor, Ella Richards, analytical chemistry

HORSESHOE BEND
Bullington, Robert Adrian, science education, ecology

ARKANSAS

HOT SPRINGS
Gladstone, William Turnbull, forest genetics, wood science

JEFFERSON
Berky, John James, medical microbiology
Bishop, Jack Belmont, genetics, toxicology
Cameron, Alexander Menzies, veterinary pathology
Casciano, Daniel Anthony, cell biology
Davis, Audrey Kennon, physiology
Emerson, Robert L., bacteriology
Farmer, John Henry, statistics
Gaylor, David William, biostatistics
Greenman, David Lewis, cancer, endocrinology
Haley, Thomas John, pharmacology
Holson, Joseph Franklin, Jr., teratology
Jackson, Carlton Darnell, biochemistry, oncology
King, Charles Miller, biochemistry, oncology
King, Jimmie Ray, analytical chemistry
Lindsay, Dale Richard, administration, medical education
Littlefield, Neil Adair, toxicology
Mohrenweiser, Harvey Walter, toxicology
Norvell, Michael Jimmy, nutrition
Shenefelt, Ray Eldon, teratology
Wolff, George Louis, genetics
Young, John Falkner, pharmacodynamics, toxicology
Zolotor, Laurence Arthur, virology

JONESBORO
Nedrow, Warren Wesley, botany

LITTLE ROCK
Abernathy, Robert Shields, medicine, microbiology
Baker, Max Leslie, radiobiology, biophysics
Barnhard, Howard Jerome, medicine, radiology
Barron, Almen Leo, virology
Bates, Joseph H., medicine, bacteriology
Beard, Owen Wayne, internal medicine, cardiology
Berry, Daisilee H., pediatrics, hematology
Betterton, Harry O., microbial physiology, biochemistry
Bond, Gary Carl, medical physiology
Boop, Warren Clark, Jr., neurosurgery
Bowen, William R., botany, cell biology
Bowker, John, orthopedic surgery
Bowling, Robert Edward, microbiology
Breckinridge, Charles Edward, Jr., pharmacy
Brewster, Marjorie Ann, clinical biochemistry
Broach, Wilson J., physical chemistry
Bruce, Thomas Allen, internal medicine, cardiovascular disease
Burns, Edward Robert, microscopic anatomy, experimental embryology
Caldwell, Fred T., Jr., surgery
Campbell, Gilbert Sadler, surgery
Cave, Mac Donald, anatomy, cell biology
Cernosek, Stanley Frank, Jr., immunochemistry
Christ, Daryl Dean, pharmacology, physiology
Conaway, Howard Herschel, physiology
Cranmer, Morris F., pharmacology, biochemistry
Dalrymple, Glenn Vogt, medicine, radiobiology
Daly, James Joseph, medical parasitology
Dean, Andrew Griswold, epidemiology
Deed, Eleanor Polk, medicine, radiology
De Luca, Donald Carl, organic chemistry, biochemistry
Diner, Wilma Canada, radiology
Doherty, James Edward, III, internal medicine, cardiology
Doyle, Lee Lee, reproductive physiology, endocrinology
Duffy, Carl Edward, microbiology
Dungan, William Thompson, pediatric cardiology
Dykman, Roscoe A., psychophysiology
Elders, Minnie Joycelyn, pediatrics, endocrinology
Engle, Paul Randal, astronomy, physics
Ferguson, Dale Vernon, microbiology

Flacke, Werner Ernst, pharmacology
Flanagan, Stevenson, medicine
Flanigan, William J., medicine, physiology
Fraunfelder, Frederick Theodor, ophthalmology
Fribourgh, James H., zoology, science education
Gilmore, Shirley Ann, anatomy
Ginzel, Karl-Heinz, pharmacology
Green, Hubert Gordon, public health, pediatrics
Grizzell, Roy Ames, Jr., fisheries management, forestry
Grosicki, Thaddeus Stanley, pharmacy
Hanna, Calvin, pharmacology
Hardin, Hilliard Frances, microbiology, mycology
Haut, Arthur, hematology, internal medicine
Hawks, Byron Lovejoy, obstetrics & gynecology
Heidt, Gary A., mammalogy
Henker, Fred Oswald, III, psychiatry
Highman, Benjamin, pathology
Howard, James Michael, geology
Jansen, Gerald Thomas, dermatology
Johnson, Howard James, Jr., clinical chemistry
Jones, Robin Richard, pathology
Jordin, Marcus Wayne, pharmacology
Kehler, Philip Leroy, sedimentology
Koike, Thomas Isao, physiology
Krum, Alvin A., physiology
Lattin, Danny Lee, medicinal chemistry
Lindsey, Julia Page, mycology, plant pathology
Lucas, Edgar Arthur, anatomy, neurophysiology
Marquart, Philip Butler, medicine
Marvin, Horace Newell, anatomy, endocrinology
Mason, Curtis Leonel, microbiology
McCowan, James Robert, pharmacy
Meade, James Horace, Jr., biometrics
Mellor, Jesse Lynn, soils
Mittelstaedt, Stanley George, pharmaceutical chemistry, pharmacy
Molnar, George William, physiology
Morgan, Paul Nolan, microbiology, immunology
Morgans, Leland Foster, histology, zoology
Morris, Manford D., biochemistry, physiology
Moss, Alfred Jefferson, Jr., biophysics, radiation biology
Nagle, William Arthur, radiation biophysics, molecular biology
Nelson, Charles A., biochemistry
North, Charles Mallory, Jr., applied mathematics
Pasley, James Neville, endocrinology, reproductive physiology
Pauly, John Edward, anatomy
Peters, John Emmett, psychiatry, experimental psychology
Powell, Ervin William, neuroanatomy
Prior, Richard Marion, nuclear medicine
Pynes, Gene Dale, biochemistry, microbiology
Quitner, Howard, pathology
Rado, Thomas A., microbiology, molecular biology
Raible, Raymond W., instrumentation
Read, Raymond Charles, surgery
Reese, William George, psychiatry
Rhinelander, Frederic William, orthopedic surgery
Sanders, Louis Lee, internal medicine, biochemistry
Scheving, Lawrence Einar, anatomy, biology
Schoultz, Ture William, neurosciences
Setliff, Frank Lamar, organic chemistry
Sherman, Jerome Kalman, anatomy, cryobiology
Shorey, Winston Kinney, internal medicine
Sinclair, Clarence Bruce, botany
Skinner, Robert Dowell, neurosciences
Smith, Edgar Dumont, surgery
Smith, William Grady, organic chemistry
Soloff, Bernard Leroy, anatomy
Spiers, Robert Sisson, immunobiology, toxicology
Spyhalski, Edward James, entomology
Stead, William White, internal medicine, pulmonary diseases
Stone, Joseph, biochemistry, pharmacology
Sun, Chao Nien, experimental pathology
Texter, Elmer Clinton, Jr., medical education, gastroenterology
Towbin, Eugene Jonas, internal medicine, physiology
Uyeda, Carl Kaoru, anatomy, cytopathology

SEARCY
England, James Donald, organic biochemistry
Mackey, James E., invertebrate zoology
Pryor, Joseph Ehrman, physical chemistry

PINE BLUFF
Ashcraft, Thomas Lee, analytical chemistry
Bentley, Cleo L., vertebrate anatomy, parasitology
Bhangoo, Mahendra Singh, soil fertility, plant nutrition
Burleigh, Joseph Gaynor, insect ecology
Lee, Kwang, poultry nutrition
Porter, Owen Archael, agronomy
Shook, Thomas Eugene, chemistry

RUSSELLVILLE
Bronco, Charles John, atomic physics
Couser, Raymond Dowell, entomology
McMillan, Harlan L., biology, academic admin
Trigg, William Walker, inorganic chemistry
Tucker, Gary Edward, botany
Turnipseed, Glyn D., plant physiology

NORTH LITTLE ROCK
Angel, Charles, biochemistry
Long, Charles H., animal nutrition
Newton, Joseph Emory O'Neal, psychiatry, physiological psychology

MORRILTON
Glimp, Hudson A., animal nutrition

MONTICELLO
Bacon, Edmond James, aquatic ecology
Chamberlin, Henry Howard, forestry
Etheridge, Albert Louis, zoology, developmental biology
Ku, Timothy Tao, forestry
Pearson, Robert Stanley, inorganic chemistry
Shade, Elwood B., forestry, biological sciences
Webb, Jerry Glen, theoretical physics

MAGNOLIA
Adams, Randall Henry, entomology, plant science
Dodson, B. C., inorganic chemistry, science education
Huddlestun, Dwight Leroy, economic geography
Jones, James Ogden, stratigraphy, sedimentology
Loe, William Carrol, reproductive physiology
Logan, Lowell Alvin, taxonomy
Robison, Henry Welborn, ichthyology
Rutledge, Carl Thomas, chemical physics
Wetzig, Calvin Ulysses, pure mathematics

MARIANNA
Keogh, Joseph Lloyd, agronomy, soils

Voldeng, Albert Nelson, medicinal chemistry
Wadkins, Charles Leroy, biochemistry
Wagh, Premanand Vinayak, biochemistry
Walls, Robert Clarence, statistics, mathematics
Watson, Robert Lee, entomology
Wear, James Otto, physical chemistry, biophysics
White, Harold J., pathology, physiology
Whitney, John Edward, physiology
Wiggins, James William, inorganic chemistry
Winter, Charles Gordon, biochemistry
Wold, Donald C., acoustics
Yang, Dominic Tsung-Che, organic chemistry
York, John Lyndal, biochemistry, physical organic chemistry
Young, Ruth Steuart, biology, biochemistry

LOWELL
Gring, John Lukins, chemistry

STATE UNIVERSITY
Barton, Harvey Eugene, economic entomology
Beadles, John Kenneth, ichthyology
Bowman, Robert Hunt, mathematics
Doyle, Miles Lawrence, biochemistry
Gwinup, Paul D., physical chemistry
Hanebrink, Earl L., ornithology, animal ecology
Harp, George Lemaul, zoology, limnology
Hexem, Rodney Orlyn, agronomy, plant science
Hinck, Lawrence Wilson, microbiology, parasitology
Hutchinson, James A., mycology
Johnson, Bob Duell, cytology, toxicology
Keene, James H., poultry nutrition
Linnstaedter, Jerry Leroy, mathematics
McCloud, Hal Emerson, Jr., solid state physics, electrical engineering
McDaniel, Van Rick, mammalogy, herpetology
Mink, Lawrence Albright, experimental physics
Mitchell, Richard Sibley, analytical chemistry
Morse, Dan Franklin, anthropology, archaeology
Nave, Paul Michael, organic chemistry
Rossa, Robert Frank, algebra
Scanlon, Charles Harris, mathematical analysis
Sifford, Dewey H., organic chemistry
Smith, Robert Paul, mathematics, statistics
Stevenson, James Harold, fish biology
Tennille, Aubrey W., soil fertility, soil microbiology
Timmermann, Dan, Jr., botany, crop breeding
Vosburg, David Lee, stratigraphy
Wei, Whua Fu, solid state physics
Wittlake, Eugene Bishop, bryology

STUTTGART
Hastings, Waldon Houston, fisheries
Hoffman, Glenn Lyle, parasitology
Johnston, Theodore Herron, agronomy
Smith, Roy Jefferson, Jr., weed science

Smith, Carroll Ward, biochemistry, environmental health
Williams, William Donald, physical chemistry
Wilson, Edmond Woodrow, Jr., physical chemistry

SPRINGDALE
Moore, Marvin G., mathematics
Spivey, Robert Charles, geology

SILOAM SPRINGS
Williams, Earl R., physics, mathematics
Wills, Irvin Andrews, zoology
Woodland, Dorothy Jane, physical chemistry

CALIFORNIA

AGOURA
Makarem, Anis H., biochemistry

ALAMEDA
Avera, Fitzhugh Lee, food science, chemistry
Carr, Robert Joseph, nuclear chemistry
Hughes, Thomas Rogers, physics
Young, Ho Lee, physiology

ALAMO
Alter, Henry Ward, physical chemistry
Bissell, Eugene Richard, organic chemistry
Henika, Richard Grant, food science
Osher, John Edward, plasma physics
Wilhelmsen, Paul Chadwick, physical chemistry

ALBANY
Bernardin, John Emile, physical biochemistry
Bolin, Harold R., food chemistry, biochemistry
Chan, Bock G., phytochemistry

Corse, Joseph Walters, phytochemistry
Fellers, David Anthony, food science
Fullington, J Garrin, biochemistry
Garcia, Richard, parasitology, medical entomology
Garibaldi, John Attilio, biochemistry
Greene, Frank Clemson, entomology
Hagen, Kenneth Sverre, entomology
Huffaker, Carl Barton, insect ecology
Kelley, Kenneth K, physical inorganic chemistry
King, Alfred Douglas, Jr, food microbiology
Lyon, Cameron Kirby, organic chemistry
Menefee, Emory, physical chemistry
Molyneux, Russell John, organic chemistry
Olson, Alfred C, biochemistry
Palmer, Kenneth James, physical chemistry
Roitman, James Nathaniel, organic chemistry
Sacks, Lawrence Edgar, microbiology
Saunders, Robert Montgomery, food science
Sayre, Robert Newton, agricultural chemistry, histology
Schatzki, Thomas Ferdinant, chemical physics
Stanley, William Lyons, organic chemistry
Stevens, Kenneth Lloyd, natural products chemistry
Thomas, Richard Sanborn, biophysics

ALHAMBRA
Liscombe, Ernest A R, entomology

ALPAUGH
Belkin, Daniel Arthur, comparative physiology, animal behavior

ALTA LOMA
Lubarsky, Robert, mycology

ALTADENA
Becker, Randolph Armin, optics, instrumentation
Channel, Lawrence Edwin, physics
Ensor, David Samuel, air pollution
Green, William Delap, atmospheric chemistry, atmospheric physics
Harrison, Paul Roger, air pollution, meteorology
Hill, Hamilton Stanton, mineralogy
Kalensher, Bernard Earl, applied mathematics, statistics
King, James, Jr, physical chemistry, physics
Koga, Toyoki, theoretical physics
Lass, Harry, mathematics
Miles, Ralph Fraley, Jr, operations research
Salkin, David, medicine
Smith, Edward John, physics, space magnetism
Smith, Theodore Beaton, meteorology
Unti, Theodore Wayne Joseph, plasma physics, optics
White, Warren Humphreys, mathematics, air pollution

Krueger, Roland Frederick, physics
Lax, Edward, cryogenics
Licari, James John, organic chemistry
Manasevit, Harold Murray, physical inorganic chemistry, organometallic chemistry
McCloskey, Allen Lyle, industrial chemistry
McDonald, Allan W, physics
Mee, Jack Everett, inorganic chemistry, solid state electronics
Nicholson, Margie May, physical chemistry
Nies, Nelson Perry, chemistry
Norton, Allen C, physiology, biomedical engineering
Passchier, Arie Anton, physical chemistry, analytical chemistry
Pondrom, Walter Lewis, Jr, applied physics, theoretical physics
Shen, Kelvin Kei-Wei, organic chemistry
Smith, Robert Alan, pesticide chemistry
Sprague, Robert W, inorganic chemistry
Teach, William Charles, space physics
Weddell, James Blount, chemistry
Wilson, Martin, chemistry
Woods, William George, physical organic chemistry

ANGWIN
Anderson, A Keith, mathematics
Anderson, Milo Vernette, microwave physics
Clark, Ervil Delwyn, biology
Eighme, Lloyd Elwyn, entomology, horticulture
Fallon, Joseph Greenleaf, public health
Hemphill, Donald Vincent, biology
Koval, Daniel, mathematics
Martz, Dowell Edward, physics
Tillay, Eldrid Wayne, inorganic chemistry
Trivett, Terrence Lynn, bacteriology
Van Hise, James R, nuclear chemistry, chemistry
Winn, A Vernon, organic chemistry
Woods, Cecil Lamborn, mathematics

APO SAN FRANCISCO
Asai, George Napoleon, plant pathology
Cross, John Henry, medical parasitology
Culver, James F, ophthalmology
Echeverria, Peter Durand, infectious diseases, pediatrics
Gould, Douglas Jay, medical entomology
Hagenmaier, Robert Doller, food chemistry
Marshall, Joe Truesdell, Jr, ornithology
McClure, Howe Elliott, ecology, wildlife management
Nowell, Wesley Raymond, medical entomology
Schneider, Curt Richard, parasitology
Van Peenen, Peter Franz Dirk, preventive medicine

ANAHEIM
Almond, Hy, inorganic chemistry
Ameer, George Albert, optical physics
Ames, Smith Whittier, physiology
Boone, James Lightholder, inorganic chemistry
Brotherton, Robert John, organic chemistry, research administration
Docks, Edward Leon, agricultural chemistry
Dunning, Robert Lewis, physical chemistry, rheology
Farrar, John, physical chemistry
Fisher, Richard Paul, synthetic organic chemistry
Graves, Clayborn Lowell, mathematical statistics
Griffin, Thomas Scott, pesticide chemistry
Gudmundsen, Richard Austin, physics, research administration
Helman, Edith Zak, clinical chemistry, radiochemistry
Kendall, Kenneth Keese, Jr, chemistry, geology
Kennel, John Maurice, applied physics
Kimme, Ernest Godfrey, mathematics, engineering
Kobayashi, Tsutomu, organic chemistry, polymer chemistry

ARCADIA
Banigan, Thomas Franklin, Jr, organic chemistry
Brubaker, Wilson Marcus, physics
Cheo, Pen Ching, plant virology
Demetriades, Sterge Theodore, science administration, plasma physics
Farber, Robert James, atmospheric chemistry
Hanson, George Peter, plant breeding
Lawson, Daniel David, organic chemistry, polymer chemistry
Rider, William B, data processing

ARCATA
Allen, George Herbert, fisheries
Anderson, Dennis Elmo, plant taxonomy
Astrue, Robert William, solid state physics
Barratt, Raymond William, microbial genetics
Becking, Rudolf (Willem), environmental sciences, natural resources
Biles, Charles Morgan, mathematics
Brant, Daniel (Hosmer), vertebrate ecology
Brusca, Gary J, invertebrate zoology, marine ecology
Butler, John Earl, botany
Chinn, Phyllis Zweig, mathematics

Crandell, George Frank, biological oceanography
Cranston, Frederick Pitkin, Jr, physics
Davis, Clyde Edward, inorganic chemistry
DeMartini, John, invertebrate zoology
De Witt, John William, Jr, fisheries, limnology
Freeland, Forrest Dean, Jr, hydrology, forestry
Garlick, George Donald, geochemistry, petrology
Gast, James Avery, oceanography
Genelly, Richard Emmett, vertebrate zoology
Hanson, Mervin Paul, physical chemistry
Harris, Stanley Warren, wildlife management
Hassler, Thomas J, fisheries, fresh water ecology
Hauxwell, Donald Lawrence, forest soils conservation
Hewston, John G, wildlife conservation, fish management
Houck, Warren Jacob, mammalogy, vertebrate zoology
Householder, James Earl, mathematics
Kelly, J Paul Sherwood, atomic physics, quantum mechanics
Kersetter, Theodore Harvey, animal physiology
Kieval, Harry Sears, mathematics
Koplin, James Ray, population ecology
Lang, Kenneth Lyle, freshwater ecology
Largent, David Lee, mycology
Lauck, David R, entomology
Lee, Sue Ying, vertebrate morphology
Lester, William Lewis, microbiology
Longshore, John David, petrology
Lovelace, C James, plant physiology, biochemistry
McCrone, Alistair William, geology, paleoecology
Meredith, Farris Ray, botany, soils
Minckler, Jeff, pathology
Minckler, Tate Muldown, pathology
Mossman, Archie Stanton, zoology
Norris, Daniel Howard, botany
Parke, Edward Charles, Jr, optics, science education
Partain, Gerald Lavern, forest economics
Patel, Vithalbhai Ambalal, applied mathematics
Peithman, Roscoe Edward, physics
Rasmussen, Robert A, phycology
Ridenhour, Richard Lewis, fish biology
Roelofs, Terry Dean, fisheries
Russell, John Blair, inorganic chemistry
Sawyer, John Orvel, Jr, plant ecology
Smith, James Payne, Jr, plant taxonomy, agrostology
Springer, Paul Frederick, wildlife research
Strothmann, Rudolph Otto, forestry
Suryaraman, Maruthuvakudi Gopalasastri, analytical chemistry, physical chemistry
Tang, Victor Kuang-Tao, statistics, mathematics
Thornburgh, Dale A, forest ecology
Tropp, Henry S, mathematics
Tucker, Roy Wilbur, mathematics
Vinyard, William Corwin, botany
Walker, Dennis Kendon, botany
Wallace, Robert Allan, biochemistry, organic chemistry
Waters, James Frederick, vertebrate anatomy
Weiss, Roger Harvey, analytical chemistry
Welsh, James Francis, zoology
Woodin, Terry Stern, biochemistry
Yarnall, John Lee, invertebrate zoology
Yocom, Charles Frederick, wildlife management
Young, John Cannon, geology

ATHERTON
Agoston, Max Karl, mathematics
Douglas, Hugh, natural resources economics
Gill, Stephen Paschall, gas dynamics
Lemmon, Dwight Moulton, geology

AZUSA
Ellis, David Allen, physical chemistry, inorganic chemistry
McCloskey, Chester Martin, organic chemistry
McFarland, James Willis, botany
Rodgers, James Edward, physical organic chemistry
Swanson, Lawrence Ray, physics
Welch, Frank Joseph, organic chemistry
Wright, Frederick Hamilton, physics

BAKERSFIELD
Athelger, Roger Boyd, medicine, pathology
Biddle, John Charles, mathematics education
Blume, Christian James, geology
Blume, Frederick Duane, environmental physiology
Calabrese, Philip G, mathematics, physics
Church, Clifford Carl, geology, paleontology
Coash, John Russell, geology
Cohn, Kim, inorganic chemistry
Cornesky, Robert Andrew, health sciences
Detwiler, Daniel Paul, solid state physics
Fang, Fabian Tien-Hwa, organic chemistry
Gilleland, Martha Jane, bio-organic chemistry
Greene, Alan Campbell, solid state science
Hampson, C Ross, medical microbiology
Hardy, John W, Jr, mathematics, computer science
Hinds, David Stewart, physiological ecology
Horton, James Charles, plant pathology
Hunt, Robert Weldon, mathematical analysis, applied mathematics
Izenstark, Joseph Louis, radiology, nuclear medicine
Johnson, Alan Len, plant breeding
Jones, Daniel John, geology
Kearns, Owen Austin, medicine
Knapp, Joseph Leonce, Jr, agricultural chemistry
Lautenschlager, Herman Kenneth, geology
Lawrence, George Edwin, zoology
Malette, William Graham, surgery
Manning, John Craige, geology
Murphy, Ted Daniel, biology, zoology
Patenaude, Robert Alan, algebra
Silverman, Philip Samuel, cultural anthropology, ethnology
Slodowski, Thomas R, exploration geology
Smith, Marion Bush, Jr, mathematics
Wake, William Henry, geography
Webb, Leland Frederick, mathematics education

BELMONT
Day, Jane Maxwell, topology
Di Girolamo, Rudolph Gerard, marine biology, microbiology
Hart, Herbert Dorlan, organic chemistry
Muraca, Raffaele Francesco, analytical chemistry, physics
Ostarello, Georgiandra Little, marine zoology
Sasmore, Daniel Paul, veterinary pathology, veterinary toxicology

BELVEDERE
Bennett, Ralph Decker, physics
Cullen, Stuart Chester, medicine

BERKELEY
Abrams, Gerald Stanley, high energy physics
Addison, John West, Jr, mathematical logic
Akers, Thomas Gilbert, virology, microbiology
Albertson, James Stanislaus, theoretical physics
Alder, Berni Julian, chemical physics, statistical mechanics
Alderman, De Forest Charles, horticulture
Alfert, Max, cytochemistry
Allen, Robert Edward, chemistry
Allen, William Westhead, economic entomology, acarology
Alpen, Edward Lewis, physiology, radiobiology
Alston-Garnjost, Margaret, high energy physics
Alvarez, Luis Walter, experimental physics
Ambrus, Laszlo, organic chemistry
Amer, Nabil Mahmoud, solid state physics, chemical physics
Ames, Bruce Nathan, biochemical genetics
Amoore, John Ernest, biochemistry
Amster, Harvey Jerome, theoretical physics
Anderson, Henry Walter, hydrology, watershed management
Anderson, James Nelson, anthropology

CALIFORNIA

Anderson, John Richard, medical entomology, parasitology
Andres, Lloyd A, entomology, weed science
Anger, Hal Oscar, instrumentation, nuclear medicine
Antoniak, Charles Edward, mathematical statistics
Arkley, Rodney John, soil morphology
Arnon, Daniel Israel, biochemistry
Arveson, William Barnes, mathematics
Ashworth, Lee Jackson, Jr, plant pathology
Babcock, Kenneth Leslie, soil chemistry
Bade, William George, mathematics
Baez, Albert Vinicio, physics
Baker, Herbert George, evolution, ecology
Baker, James Addison, computer science
Baker, Kenneth Frank, phytopathology, soil microbiology
Baker, Patricia Cooper, developmental biology
Balamuth, William, protozoology, parasitology
Ballou, Clinton Edward, biochemistry
Bandman, Everett, molecular biology
Barker, Horace Albert, biochemistry
Barlow, George Webber, animal behavior, ichthyology
Barlow, Richard Eugene, mathematical statistics
Barshad, Isaac, soil chemistry
Bartholomew, James Collins, cell biology
Bartlett, Neil, inorganic chemistry
Bascom, William Russel, ethnology
Bassham, James Alan, biochemistry
Batterman, Robert Coleman, pharmacology
Bayne, Henry Godwin, bacteriology
Beach, Frank Ambrose, neuropsychology
Bearden, Alan Joyce, biophysics
Bega, Robert V, plant pathology
Belsky, Theodore, public health, analytical chemistry
Bennett, Edward Leigh, biochemistry, neurobiology
Berg, William Eugene, biology
Berkner, Klaus Hans, physics
Berlin, Brent, anthropology
Bern, Howard Alan, endocrinology, zoology
Berreman, Gerald Duane, anthropology
Berry, Frederick Almet Fulghum, geology
Berry, William Benjamin Newell, paleobiology
Bickel, Peter J, mathematical statistics
Bingham, Harry H, Jr, physics
Birge, Raymond Thayer, physics
Birge, Robert Walsh, high energy physics
Bissell, Mina Jahan, cell biology
Bissett, Marjorie Louise, microbiology
Biswell, Harold Hubert, ecology, forestry
Blair, Phyllis Beebe, immunology
Blum, Lenore Carol, mathematical logic, applied mathematics
Blum, Manuel, mathematics
Bolt, Bruce A, seismology, applied mathematics
Bonar, Lee, botany
Borsook, Henry, biochemistry
Bourne, Samuel G, mathematics
Bowker, Albert Hosmer, mathematical statistics
Bowyer, C Stuart, astronomy, space science
Bradfield, Robert B, clinical nutrition, biochemistry
Bremermann, Hans J, mathematical biology, biophysics
Brewer, Leo, physical chemistry
Briggs, George McSpadden, nutrition
Brillinger, David Ross, statistics
Brink, David Liddell, wood chemistry
Brodale, Gary Edward, physical chemistry
Brode, Robert Bigham, physics
Broido, Abraham, physical chemistry
Brooks, Matilda Moldenhauer, physiology
Brown, John Clifford, biochemistry, immunology
Brown, Nancy J, theoretical chemistry
Brown, Robert Reginald, physics
Brown, Spencer Wharton, cytogenetics
Brown, Theodore Llewellyn, analytical chemistry, biochemistry
Broyer, Theodore Clarence, plant nutrition
Buchanan, Bob Branch, biochemistry, microbiology
Budinger, Thomas Francis, medical physics, nuclear medicine
Budnitz, Robert Jay, environmental systems and technology
Buehring, Gertrude Case, cancer
Buffington, Andrew, astrophysics
Burki, Henry John, cell biology, radiation biology
Burlingame, Alma L, physical chemistry
Burnside, Mary Beth, cell biology, anatomy
Burr, Horace Kelsey, physical chemistry, food science
Cabasso, Victor Jack, virology
Calendar, Richard, molecular biology
Calloway, Doris Howes, nutrition
Caltagirone, Leopoldo Enrique, entomology
Calvin, Melvin, organic chemistry
Cape, Ronald Elliot, biochemistry
Carithers, William Cornelius, Jr, experimental high energy physics
Carmichael, Ian Stuart, petrology, geochemistry
Carpenter, Frederick Hiltman, biochemistry
Cartwright, Brian Grant, astrophysics
Casida, John Edward, toxicology
Cason, James, Jr, organic chemistry
Castanera, Esther Goossen, biochemistry, nutrition
Cerny, Joseph, III, nuclear chemistry
Chamberlain, Owen, experimental high energy physics
Chang, George Washington, microbial physiology, microbial genetics
Chemsak, John A, entomology
Chern, Shiing-Shen, geometry
Chew, Geoffrey Foucar, theoretical high energy physics
Chiang, Chin Long, biostatistics
Chiao, Raymond Yu, physics
Chihara, Carol Joyce, developmental genetics
Chinowsky, William, physics
Christensen, Herbert Edward, toxicology, occupational health
Clark, Alvin John, genetics, bacteriology
Clark, John Desmond, archaeology, ethnography
Clarke, John, physics
Clauser, John Francis, quantum mechanics
Clayton, James Oliver, chemistry
Clemens, William Alvin, paleontology
Clinnick, Mansfield, mathematics
Cobb, Fields White, Jr, plant pathology
Cockrell, Robert Alexander, dendrology, wood science & technology
Cohen, Marvin Lou, physics
Cohen, Nathan Wolf, herpetology
Cole, Roger David, protein chemistry
Collins, O'Neil Ray, mycology
Colwell, Robert Knight, ecology, entomology
Colwell, Robert Neil, forest mensuration
Concus, Paul, applied mathematics, numerical analysis
Connick, Robert Elwell, inorganic chemistry
Cons, Jean Marie Abele, developmental physiology
Constance, Lincoln, botany
Conway, John George, Jr, atomic spectroscopy
Conzett, Homer Eugene, experimental nuclear physics
Cooper, Robert Chauncey, microbiology, public health
Cooper, William Clark, medicine
Cooper, William S, information science
Cork, Bruce, particle physics
Cox, H C, entomology
Craig, Paul Palmer, physics, science
Crawford, Frank Stevens, Jr, physics
Cremer, Natalie E, immunology, virology
Crisp, Carl Eugene, plant physiology
Critchfield, William Burke, forest genetics
Crowe, Kenneth Morse, physics
Cruty, Michael Robert, radiological physics
Cudaback, David Dill, radio astronomy
Cunningham, Leland E, celestial mechanics
Curtis, Garniss Hearfield, geology
Curtis, Stanley Bartlett, radiation biophysics
Dahl, Orin I, experimental high energy physics
Dahlsten, Donald L, forest entomology
Dalven, Richard, solid state physics
Daly, Howell Vann, entomology
Daniell, Ellen, molecular biology
Dauben, William Garfield, photochemistry
Davis, Sumner P, physics
Day, Boysie Eugene, plant physiology
Day, Paul Russell, soil physics
Deane, Margaret, biostatistics, epidemiology
de Fremery, Donald, biochemistry
Dekker, Charles Abram, biochemistry
De Mars, Clarence John, Jr, forest entomology, insect ecology
Dempster, Lauramay Tinsley, taxonomic botany
DeOme, Kenneth Benton, oncology
Derenzo, Stephen Edward, nuclear medicine
Diamond, Marian C, neurobiology
Diamond, Martin J, organic chemistry
Diamond, Richard Martin, nuclear chemistry
Dickinson, Fred Eugene, forest products
Dieter-Conklin, Nannielou, radio astronomy
Dismukes, Gerard Charles, physical chemistry
Dobson, Ernest L, physiology, biophysics
Dolhinow, Phyllis Carol, physical anthropology
Donovan, John W, physical chemistry
Dreyfus, Stuart Ernest, operations research
Dubins, Lester Eli, mathematics
Duhl, Leonard J, psychiatry
Dunn, John Edward, Jr, public health
Dunnebacke-Dixon, Thelma Hudson, embryology, zoology
Durbin, Patricia Wallace (Mrs James T Heavey), biophysics
Durham, John Wyatt, invertebrate paleontology
Eakin, Richard Marshall, zoology
Eberhard, Philippe Henri, particle physics
Echols, Robert M, forestry
Elberg, Sanford Samuel, medical microbiology
Elioff, Thomas, high energy physics
Elsasser, Albert B, anthropology
Ely, Robert P, Jr, physics
Emerson, Ralph, mycology
Emmons, Richard William, public health, virology
Entenman, Cecil, physiology, biochemistry
Erman, Don Coutre, aquatic ecology, fisheries
Falcon, Louis A, entomology
Falicov, Leopoldo Maximo, theoretical physics, solid state physics
Farkas, Daniel Frederick, food technology
Fary, Istvan, mathematics
Fatt, Irving, physiological optics
Feldman, Jacob, mathematics
Finkle, Bernard Joseph, plant biochemistry
Finley, John Westcott, food science
Fischer-Colbrie, Erwin, solid state physics, electronic physics
Fisher, Robert Amos, Jr, magnetism, thermodynamics
Flom, Merton Clyde, optometry
Fogel, Seymour, genetics
Forte, John Gaetano, physiology
Foster, Alfred Leon, mathematics
Foster, George McClelland, Jr, anthropology
Foster, Mercedes S, ornithology
Fraenkel-Conrat, Heinz Ludwig, molecular biology
Frazier, Norman Walter, entomology
Freedman, David A, statistics
Freeman, Ralph David, physiological optics, neurophysiology
Freeman, Walter Jackson, III, physiology
Freitag, Julius Herman, entomology
Fretter, William Bache, physics
Friedberg, Carl E, experimental high energy physics
Fristrom, James W, genetics
Fry, Wayne Lyle, paleobotany
Fuller, Glenn, organic chemistry
Furman, Deane Philip, entomology, parasitology
Gaffey, Cornelius Thomas, biophysics
Gaffey, William Robert, biostatistics, biometry
Gale, David, mathematics
Galtieri, Angela Barbaro, particle physics
Garrison, Warren Manford, physical chemistry
Gerhart, John C, biochemistry
Gersper, Paul Logan, soil morphology, soil mineralogy
Ghiorso, Albert, physics
Ghiselin, Michael Tenant, zoology
Giauque, William Francis, physical chemistry
Gidal, George, experimental high energy physics
Gilbert, Charles Merwin, physical geology
Gilbert, William Spencer, nuclear physics
Girton, Raymond Elwood, plant physiology
Glacken, Clarence James, cultural geography
Glaeser, Robert M, biophysics
Glaser, Donald Arthur, physics, molecular biology
Glendenning, Norman Keith, theoretical nuclear physics
Gofman, John William, physical chemistry, nuclear medicine
Gold, Alma Herbert, plant pathology
Goldhaber, Gerson, nuclear physics
Goldsmith, John Rothchild, public health
Goldstein, Hyman, biostatistics
Golueke, Clarence George, environmental sciences
Gordon, Benjamin Edward, analytical chemistry, radiochemistry
Gordon, Harold Thomas, biochemistry
Graburn, Nelson Hayes Henry, cultural anthropology
Grah, Rufolf Ferdinand, forestry
Graham, Susan Lois, computer science
Grassetti, Davide Riccardo, biopharmaceutics
Green, Irving Joseph, virology
Gregory, Joseph Tracy, vertebrate paleontology
Greiner, Douglas Earl, experimental nuclear physics, cosmic ray physics
Groves, Eric Stedman, experimental high energy physics
Grunbaum, Benjamin Wolf, biochemistry
Guirard, Beverly Marie, biochemistry
Guzman, Ruben Joseph, toxicology, pharmacology
Gwinn, William Dulaney, physical chemistry
Hackett, Adeline J, virology, genetics
Haddon, William F, (Jr), analytical chemistry
Hahn, Erwin Louis, physics
Halbach, Klaus, physics
Halpern, Martin B, theoretical high energy physics
Hamlin, Kenneth Eldred, Jr, chemistry, research administration
Hammar, Allan H, microbiology, zoology
Hammel, Eugene Alfred, anthropology
Hancock, Joseph Griscom, Jr, plant pathology
Hansen, Eder Lindsay, pharmacology
Harris, Charles Bonner, physical chemistry
Harris, Morgan, cell biology
Harris, Robert A, theoretical chemistry, chemical physics
Harris, Robert Wilson, forest ecology
Harrison, Michael A, computer science
Harte, John, theoretical physics
Hartmann, Floyd Wellington, bacteriology
Hartsough, Walter D, particle physics
Harvey, Bernard George, nuclear physics
Hatch, Melvin T, microbiology
Hatfield, Donald Marshall, mammalogy
Hay, Richard Le Roy, sedimentary petrology
Hayes, Thomas L, biophysics
Heady, Harold Franklin, plant ecology
Hearst, John Eugene, biophysical chemistry
Heathcock, Clayton Howell, synthetic organic chemistry
Hebert, Alvin Joseph, chemical physics, physical chemistry
Heckard, Lawrence Ray, botany
Heckly, Robert Joseph, microbiology
Heckman, Harry Hughes, nuclear physics
Heftmann, Erich, biochemistry
Heinrich, Bernd, physiology, ecology
Heizer, Robert Fleming, anthropology
Helgeson, Harold Charles, geology, physical chemistry
Helmholz, August Carl, high energy physics
Helms, John Andrew, forestry
Helson, Henry, mathematics
Helwig, Harold Lavern, biochemistry
Hendrie, David Lowery, physics
Henkin, Leon (Albert), mathematical logic
Henry, Claudia, virology, immunology
Hershberger, Lee George, endocrinology
Heslep, John McKay, health physics
Hidalgo, John, pharmacology
Hildebrand, Donald Clair, plant pathology
Hildebrand, Joel Henry, physical chemistry
Hill, Russell John, physical chemistry, high temperature chemistry
Hirsch, Ernest, operations research, physical chemistry
Hirsch, Monroe Jerome, optometry, physiology
Hirsch, Morris William, mathematics
Hiyama, Tetsuo, biochemistry, plant physiology
Hodges, Joseph Lawson, Jr, mathematical statistics
Hok, Karol Anton, microbiology

Holdren, John Paul, energy conversion, environmental physics
Hollander, Jack Marvin, nuclear chemistry
Hollister, Arthur Clair, Jr, epidemiology, public health administration
Hollowell, Craig D, air pollution
Holmquist, Richard, biochemistry, molecular evolution
Holt, Maurice, applied mathematics
Hooson, David John Mahler, geography
Horne, Alexander John, aquatic biology
Hornung, Erwin William, physical chemistry, cryogenics
Howell, Francis Clark, biological anthropology
Hudson, Mary Katherine, space physics, plasma physics
Huebsch, Ian O, geophysics, meteorology
Huenemann, Ruth L, nutrition
Huesman, Ronald Henry, medical physics
Huey, Raymond Brunson, animal ecology
Hurlbut, Franklin Charles, physics
Hutchison, John Howard, vertebrate paleontology
Hyde, Earl K, nuclear chemistry
Jackson, John David, theoretical physics
Jacobson, Louis, plant physiology
Jeffries, Carson Dunning, solid state physics
Jensen, William August, botany
John, Walter, environmental physics
Johnson, Ned Keith, ornithology
Johnston, Harold Sledge, physical chemistry
Johnston, Russell Shayne, plasma physics
Jolly, William Lee, inorganic chemistry
Jones, Francis Tucker, chemistry
Jones, Hardin Blair, physiology
Jones, Russell Lewis, plant physiology
Joslyn, Maynard Alexander, food chemistry
Judd, David Lockhart, theoretical physics
Jukes, Thomas Hughes, biochemistry, nutrition
Jura, George, physical chemistry
Kadyk, John Amos, physics
Kahan, William M, mathematics, computer science
Kahn, Louis B, mathematics, statistics
Kahn, Paul, biomedical engineering
Kalm, Max John, medicinal chemistry
Kammen, Harold Oscar, biochemistry
Kaplan, Donald Robert, plant morphology
Karplus, Robert, theoretical physics
Kasarda, Donald David, protein chemistry
Kato, Tosio, mathematics
Kaufman, Allan N, plasma physics
Kay, Paul, anthropology, anthropological linguistics
Kazarinoff, Michael N, biochemistry
Keefe, Denis, high energy physics
Keller, Edward Lowell, biomedical engineering, neurobiology
Kelley, John Le Roy, mathematics
Kelly, Elmer Lewis, physics
Kelly, Lola Szanto, biophysics
Kelly, Robert Lincoln, elementary particle physics
Kendrick, James Blair, Jr, plant pathology
Kennedy, Barbara Mae, nutrition
Kenney, Robert Warner, high energy physics, particle physics
Kerr, Kenton E, physiological optics
Kerth, Leroy Thomas, physics
Kettering, James David, medical microbiology
King, Ivan Robert, astronomy
King, Janet Carlson, nutrition
Kinloch, Bohun Baker, Jr, genetics, plant pathology
Kip, Arthur Frederic, physics
Kirby, Robion C, mathematics
Kirkland, Walter Dean, pharmacy, chemistry
Kirsch, Jack Frederick, biochemistry
Kittel, Charles, physics
Kland, Mathilde June, environmental chemistry
Kleinpell, Robert Minssen, paleontology, geology
Knaff, David Barry, photobiology
Knight, Claude Arthur, molecular biology
Knight, Walter David, Jr, solid state physics
Kobayashi, Shoshichi, mathematics
Kodama, Arthur Masayoshi, physiology
Koenig, Nathan Hart, textile chemistry
Koenigsberg, Ernest, operations research, management science
Koerber, Thomas William, entomology
Koford, Carl Buckingham, vertebrate zoology

Konrad, Michael Warren, molecular biology
Koshland, Daniel Edward, Jr, biochemistry
Koshland, Marian Elliott, immunology
Kozlowski, Robert H, organic chemistry
Krueger, Albert Paul, medical ecology
Kuhi, Leonard Vello, astrophysics
Kunkel, Wulf Bernard, plasma physics
Kuo, Harng-Shen, chemistry, biochemistry
Kurtzman, Ralph Harold, Jr, biochemistry, mycology
Laetsch, Watson McMillan, botany
Lam, Tsit-Yuen, algebra
Lampton, Michael Logan, x-ray astronomy
Lane, Robert Sidney, medical entomology
Lanford, Oscar E, III, mathematical physics
Lankford, Philip Marlin, urban geography, economic geography
Laslett, Lawrence Jackson, physics
Lawrence, John Hundale, medicine
Lawson, Herbert Blaine, Jr, mathematics
Layton, Laurence Laird, biochemistry, allergy
Le Cam, Lucien Marie, mathematical statistics
Lee, Yuan Tseh, chemistry
Lehman, Russell Sherman, mathematics
Lehmann, Erich Leo, mathematical statistics
Lehmer, Derrick Henry, mathematics
Lemmon, Richard Millington, radiation chemistry
Lennette, Edwin Herman, epidemiology, experimental pathology
Leopold, Aldo Starker, zoology, forestry
Leopold, Luna Bergere, geomorphology
Lepore, Joseph Vernon, physics
Lerke, Peter A, bacteriology
Lerner, I Michael, genetics
Lewis, Edwin Reynolds, bioengineering
Lewis, James Clement, biochemistry
Lewy, Hans, mathematics
Libby, William John, (Jr), forestry, genetics
Licht, Paul, zoology
Lidicker, William Zander, Jr, population biology, mammalogy
Lieberman, Michael Merril, microbiology
Lin, Robert Peichung, space physics, solar physics
Lindgren, Frank Tycko, biophysics
Linfoot, John Ardis, medicine, endocrinology
Linn, Stuart Michael, biochemistry
Linsley, Earle Gorton, entomology
Little, Angela C, food science
Loeve, Michel, mathematics, statistics
Lofgren, Edward Joseph, physics
Loher, Werner J, zoology
Loken, Stewart Christian, experimental high energy physics
Loomis, Albert Geyer, chemistry
Louie, Robert Eugene, virology
Lundin, Robert Enor, nuclear magnetic resonance
Luten, Daniel B, Jr, resource geography
Lyman, John Tompkins, biophysics
Lyman, Richard Lee, nutrition, biochemistry
Lyon, David N, physical chemistry
Lyon, Patricia Jean, anthropology
Lyon, Robert Lyndon, forest entomology, insect toxicology
Macdonald, Burns, experimental physics
Macey, Robert Irwin, physiology, biophysics
MacGinitie, Harry Dunlap, paleobotany, stratigraphy
Machlis, Leonard, plant physiology
Mackey, Bruce Ernest, biometrics, plant breeding
Mackinney, Gordon, agricultural chemistry
MacLeod, Guy Franklin, agriculture, entomology
Madaras, Ronald John, experimental high energy physics
Madin, Stewart Harvey, animal pathology, virology
Maestre, Marcos Francisco, biophysics
Magoffin, Robert Louis, epidemiology, microbiology
Mahan, Bruce Herbert, physical chemistry
Maitra, Shyamal Kumar, biochemistry
Malkin, Richard, biochemistry
Mandell, Robert Burton, optometry, visual physiology
Mandelstam, Stanley, theoretical physics
Manougian, Edward, theoretical biology
Marg, Elwin, vision, neurosciences
Margen, Sheldon, human nutrition

Markowitz, Samuel Solomon, nuclear chemistry
Maron, Melvin Earl, information science
Marrus, Richard, atomic physics, nuclear physics
Marsden, Jerrold Eldon, mathematics
Masri, Merle Sid, agricultural chemistry, mammalian physiology
Mast, Terry Steven, elementary particle physics
Mazia, Daniel, zoology
McCain, Arthur Hamilton, plant pathology
McEvilly, Thomas V, geophysics, seismology
McEwen, William John, anthropology
McKee, Christopher Fulton, astrophysics
McKenzie, Ralph Nelson, mathematics
McKillop, William L M, forest economics
McLaren, Arthur Douglas, biophysics, biology
McMillan, Edwin Mattison, physics
McSwain, Berah Davis, biophysics
Meidav, Tsvi, exploration geophysics
Meissner, Loren Phillip, computer science, applied mathematics
Mel, Howard Charles, biophysics
Mercer, Walter Ashby, microbiology, chemistry
Meschi, David John, high temperature chemistry
Mesirov, Jill Portner, mathematics
Messenger, Powers Slater, entomology
Meyer, Charles, geology
Michelsohn, Marie-Louise, topology
Michener, Harold David, microbiology
Middlekauff, Woodrow Wilson, entomology
Midura, Thaddeus, food technology
Milby, Thomas Hutchinson, occupational medicine, toxicology
Miller, William Hughes, chemical physics
Mittler, Thomas E, insect physiology
Montgomery, Theodore Ashton, public health, pediatrics
Moore, Calvin C, mathematics
Morales, Daniel Richard, clinical biochemistry
Morgan, Meredith Walter, optometry, visual physiology
Morrey, Charles Bradfield, Jr, mathematics
Morrison, Huntly Frank, geophysics
Morse, Anthony Perry, mathematics
Mortimer, Robert Keith, genetics, biophysics
Mozer, Forrest S, space physics
Muller, Richard A, experimental physics
Muller, Rolf Hugo, electrochemistry, chemical engineering
Murai, Mary Miyeko, nutrition
Murphy, Collin Grisseau, developmental biology
Murphy, James L, forest economics, forestry
Mutha, Shantilal Chhotmal, analytical chemistry
Myers, Rollie John, Jr, physical chemistry
Nader, Laura, anthropology
Nandi, Satyabrata, zoology, endocrinology
Neilands, John Brian, bioinorganic chemistry
Nelson, Bernard Clinton, medical entomology
Nelson, Jerry Earl, astrophysics, experimental high energy physics
Nelson, Robert LeRoy, medical entomology
Newell, Gordon Frank, applied mathematics
Newman, John Scott, electrochemistry
Newton, Amos Sylvester, chemistry
Neyman, Jerzy, statistics
Ng, Henry, microbiology, food science
Nichols, Alexander Vladimir, medical physics, biophysics
Nicoll, Charles S, physiology, endocrinology
Nikaido, Hiroshi, microbiology, biochemistry
Nimmo, Charles Colvin, chemistry
Nishioka, Richard Seiji, neuroendocrinology, microscopic anatomy
Nooney, Grove C, mathematics
Norick, Nancy Xavier, forest biometry, mathematical statistics
Noyce, Donald Sterling, organic chemistry
Nutting, Lee, food chemistry
Oace, Susan M, nutrition
Oddone, Piermaria Jorge, high energy physics
O'Konski, Chester Thomas, chemistry, biophysics

O'Malley, Thomas Francis, physics
Orleman, Edwin Franklin, physical chemistry
Ornduff, Robert, botany
Orth, Charles Douglas, high energy physics, astrophysics
Oshiro, Lyndon Satoru, virology, electron microscopy
Oster, George F, mathematical biology
Ostwald, Rosemarie, nutrition, biochemistry
Ottoboni, Minna Alice, toxicology, biochemistry
Pabst, Adolf, mineralogy
Pace, Nello, physiology
Packer, Lester, biochemistry
Paffenbarger, Ralph Seal, Jr, epidemiology
Paine, Lee Alfred, forest pathology
Palm, Risa Ileen, urban geography
Panopoulos, Nickolas John, plant pathology, microbial genetics
Papenfuss, George Frederik, algology
Park, Roderic Bruce, plant physiology
Parker, Sherwood, physics
Parlett, Beresford, mathematics
Parmeter, John Richard, Jr, plant pathology
Parson, William, internal medicine
Parsons, James Jerome, geography of Latin America, biogeography
Patton, James Lloyd, mammalogy
Pavlath, Attila Endre, fluorine chemistry, textile chemistry
Pearson, Oliver Payne, zoology
Peck, Joseph Howard, paleontology
Peisakoff, Melvin Philip, mathematics
Pence, James William, agricultural chemistry, agricultural biochemistry
Penhoet, Edward Etienne, biochemistry, virology
Perez-Mendez, Victor, experimental high energy physics
Peterson, Jack Milton, nuclear physics
Phillips, Norman Edgar, physical chemistry
Philp, Richard Paul, organic geochemistry
Pieper, Gustav Rene, entomology, toxicology
Pillans, Helen Mead, astronomy
Pimentel, George Claude, chemistry
Pinney, Edmund Joy, mathematical analysis
Pipa, Rudolph Louis, entomology
Pirelka, Dorothy Riggs, zoology
Pitelka, Frank Alois, zoology
Pitzer, Kenneth Sanborn, physical chemistry
Poinar, George O, Jr, nematology
Pollack, Louis Rubin, chemistry
Portis, Alan Mark, solid state physics
Poskanzer, Arthur M, nuclear chemistry
Potter, Jack M, anthropology
Poulton, Charles Edgar, ecology
Powell, Clinton Cobb, radiology
Powell, Jerry Alan, entomology
Powell, Richard Edward, physical chemistry
Price, Paul Buford, Jr, experimental physics
Prior, Michael Herbert, physics
Pripstein, Morris, physics
Protter, Murray Harold, mathematics
Prussin, Stanley Gerald, nuclear chemistry, radiochemistry
Purcell, Alexander Holmes, III, entomology, plant pathology
Pyle, Robert V, plasma physics
Raabe, Robert Donald, plant pathology
Rabinowitz, Jesse Charles, biochemistry
Ransley, Derek Leonard, organic chemistry
Rapoport, Henry, organic chemistry
Rarick, George Lawrence, physical education
Rarita, William Roland, elementary particle physics
Rasmussen, John Oscar, Jr, nuclear chemistry
Rauscher, Elizabeth Ann, cosmology, quantum physics
Ray, Rose Marie, mathematical statistics
Raymond, Kenneth Norman, inorganic chemistry, crystallography
Redlich, Martin George, physics
Redlich, Otto, physical chemistry
Reed, Robert Ronald, cultural geography, geography of Southeast Asia
Rees, John Tonks, aquatic biology, invertebrate zoology
Reeve, Roger Mansfield, plant anatomy
Reeves, William Carlisle, epidemiology
Reif, Frederick, physics
Reynolds, John Hamilton, physics
Rhodes, John Lewis, mathematics
Rice, Robert Arnot, science education
Richards, Paul Linford, physics

Richmond, Jonas Edward, biochemistry
Riddell, Robert James, Jr., theoretical physics
Rieffel, Marc A., mathematics
Riffer, Richard, natural products chemistry
Riggs, John L., virology
Rittenberg, Alan, particle physics
Robben, Franklin Arthur, plasma physics
Roberts, Richard Bruce, biophysics
Robertson, Jacqueline Lee, insect toxicology
Robinson, Julia (Bowman), logic; number theory
Robinson, Raphael Mitchel, mathematics
Rosenberg, Lawson Lawrence, physiological chemistry
Rowell, Charles Hugh Fraser, neurophysiology; zoology
Rubin, Harry, cell biology; virology
Ruby, Lawrence, nuclear physics
Rugge, Henry F., medical technology; plasma physics
Sachs, Rainer Kurt, theoretical physics
St Lawrence, Patricia, genetics
Sanazaro, Paul Joseph, medicine
Sangren, Ward Conrad, applied mathematics
Sanui, Hisashi, biophysics, cell physiology
Sarason, Donald Erik, mathematics
Sargent, Thornton William, III, nuclear medicine; biochemical pharmacology
Sarich, Vincent M., physical anthropology
Satir, Peter, zoology
Sauer, Kenneth, biophysical chemistry
Saunders, Bernard Gray, experimental systematics
Savage, Donald Elvin, paleontology
Saxon, David Stephen, theoretical nuclear physics
Schachman, Howard Kapnek, molecular biology; biochemistry
Schaefer, Henry Frederick, III, theoretical chemistry
Schaffer, Frederick Leland, virology; biochemistry
Scharpf, Robert F., plant pathology
Scherer, James R., physical chemistry
Scherer, Rene, microbiology
Schiebe, Jack H., virology
Schlachter, Alfred Simon, atomic physics
Schlegel, David Edward, plant pathology
Schlinger, Evert Irving, entomology; systematics
Schmidt, Nathalie Joan, virology
Schonbeck, Niels Daniel, science education; biochemistry
Schooley, John C., physiology
Schroeder, Duane David, biochemistry
Schroeder, Lee S., elementary particle physics
Schroth, Milton Neil, plant pathology
Schultz, Arnold Max, plant ecology
Schultz, Thomas Henry, organic chemistry
Schulz, William Donald, theoretical physics
Schwartz, Charles Leon, theoretical physics
Schwimmer, Sigmund, enzymology; food biochemistry
Scott, Elizabeth Leonard, statistics, astronomy
Scrivani, Robert P., virology; epidemiology
Seaborg, Glenn Theodore, nuclear chemistry
Segre, Emilio Gino, nuclear physics
Seidenberg, Abraham, mathematics
Serat, William Felkner, biochemistry
Sessler, Andrew M., theoretical physics
Shack, William Alfred, anthropology
Shannon, Stephen Randall, high energy physics
Shapiro, Gilbert, particle physics
Shen, Mitchel C., physical chemistry
Shen, Yuen-Ron, solid state physics
Shirley, David Allen, nuclear chemistry
Shirley, David Arthur, chemical physics
Shugart, Howard Alan, physics
Siebert, Jerome Bernard, agricultural economics
Silva, Paul Claude, botany
Silver, Samuel, physics
Simmons, John Everette, Jr., parasitology

Simmons, Melvin Kurt, experimental atomic physics
Simmons, William Scranton, anthropology
Simpson, Warren Candler, physical chemistry
Singer, Beatrice Adell, molecular biology
Singer, Jerome Ralph, bioengineering, biophysics
Siri, William Emil, biophysics
Smale, Stephen, mathematics
Smart, James Conrad, inorganic chemistry
Smith, Ralph Ingram, zoology
Smith, Ray Fred, entomology
Smith, Richard Harrison, forest entomology
Smoot, George Fitzgerald, III, astrophysics, cosmic ray physics
Snell, Esmond Emerson, biochemistry
Solnitz, Frank Morton, nuclear physics
Solovay, Robert M., mathematics
Somorjai, Gabor Arpad, physical chemistry
Soules, David Edward, immunology
Spanier, Edwin Henry, mathematics
Spieth, Philip Theodore, population genetics
Spinrad, Hyron, astronomy
Srebnik, Herbert Harry, anatomy
Stallings, John Robert, Jr., topology
Stapp, Henry P., physics
Stark, Lawrence, bioengineering
Stebbins, Robert Cyril, zoology
Steiner, Herbert M., particle physics
Steinhardt, Richard Antony, neurobiology; cell biology
Stent, Gunther Siegmund, molecular biology; neurobiology
Stephens, Frank Samuel, nuclear chemistry
Stern, Curt, zoology
Sternberg, Hilgard O'Reilly, geography of Brazil & the humid tropics
Stevenson, Merion Lynn, particle physics
Stites, William Whitfield, medicine, public health
Stohler, Rudolf, zoology
Stokstad, Evan Ludvig Robert, biochemistry
Stone, Edward Curry, plant physiology
Strauch, Fred, soil science, agronomy
Strauss, Herbert L., physical chemistry
Streitwieser, Andrew, Jr., physical organic chemistry
Strohman, Richard Campbell, zoology
Strovink, Mark William, experimental high energy physics
Summers, Charles Geddes, economic entomology
Susskind, Charles, bioengineering, history of technology
Sweet, Benjamin Hersh, virology; immunology
Takahashi, William Noburu, plant pathology
Tanada, Yoshinori, insect pathology
Tarski, Alfred, mathematics
Tarter, Michael E., biostatistics
Taub, Abraham Haskel, mathematics
Tavares, Isabelle Irene, mycology; lichenology
Taylor, Angus Ellis, mathematics
Taylor, Lewis Walter, poultry husbandry
Taylor, Norman Burwell George, population studies
Teeguarden, Dennis Earl, forestry economics
Templeton, Constantine H., immunology
Templeton, David Henry, physical chemistry
Terry, Norman, plant physiology
Terwedow, Henry Albert, Jr., medical entomology
Thomas, Herbert Rex, plant pathology
Thomas, Heriberto Victor, preventive medicine
Thomas, Jerome Francis, air pollution, water pollution
Thomas, Paul Emery, plant physiology
Thomas, Ralph Harold, health physics
Thomasson, Aram John, electrical engineering, statistics
Thompson, Stanley Gerald, pharmacology
Thompson, Charles Raymond, physics
Thompson, James Alex L., engineering physics

Thornton, Robert Lyster, physics
Timiras, Paola Silvestri, physiology; neuroendocrinology
Tinoco, Ignacio, Jr., physical chemistry
Tobias, Charles W., electrochemistry, chemical engineering
Tobias, Cornelius Anthony, biophysics
Tokunaga, Chiyoko, genetics
Tomimatsu, Toshio, physical chemistry
Townes, Charles Hard, physics
Trilling, George Henry, elementary particle physics
Tripp, Robert D., physics
Trippe, Thomas Gordon, high energy physics
Turner, Francis John, petrology
Unger, Andrew, applied statistics
Vance, James Elmon, Jr., geography
van den Bosch, Robert, entomology
Van Sluyters, Richard Charles, physiological optics, neurophysiology
Vaught, Robert L., mathematics
Vaux, Henry James, forestry
Verhoogen, John, geophysics
Vickery, Larry Edward, biophysical chemistry
Vlamis, James, plant nutrition
Wahrhaftig, Clyde (Adolph), geology
Wahlig, Michael Alexander, energy conversion
Wake, David Burton, evolutionary biology
Wake, Marvalee H., vertebrate biology
Wakerling, Raymond Kornelius, mathematics
Walker, Charles Robert, biology
Walker, Howard George, Jr., agricultural chemistry
Wallace, Helen M., public health
Wallace, Joan M., plant biochemistry
Wang, James C., biophysical chemistry
Washburn, Sherwood Larned, anthropology
Waters, William E., forest entomology
Watson, Kenneth Marshall, physics
Weaver, Harold Francis, astronomy
Webber, Irma Eleanor, botany
Wehausen, John Vrooman, mechanics
Wei, Eddie Tak-Fung, toxicology; pharmacology
Weinhold, Albert Raymond, plant pathology
Weinmann, Clarence Jacob, parasitology
Weinstein, Alan David, geometry
Weiss, Lionel Edward, geology
Welch, Graeme P., biophysics
Welch, William John, radio astronomy
Welles, Samuel Paul, vertebrate paleontology
Wenk, Hans-Rudolf, crystallography, structural geology
Wenzel, William Alfred, particle physics
Werben, Frank Simon, bioengineering
Werner, Sanford Benson, preventive medicine
Westheimer, Gerald, physiological optics
Whissel-Buechy, Dorothy Y E., human genetics
White, Harvey Elliott, physics
Whitlock, Gaylord Purcell, agriculture, biochemistry
Wichmann, Eyvind Hugo, theoretical physics
Wiegand, Clyde Edward, experimental physics
Wilde, Pat, oceanography
Wilhelm, Stephen, plant pathology
Williams, Carroll Burns, Jr., forest entomology
Williams, David Emerton, soil chemistry
Williams, Howel, geology
Williams, Mary Ann, nutrition, biochemistry
Williams, Robley Cook, biophysics, molecular biology
Wilson, Allan Charles, biochemistry
Wilson, Carl C., forest management
Wilt, Fred H., zoology
Winkelman, Frederick Charles, high energy physics
Winkelstein, Warren, Jr., medicine, epidemiology
Wofsy, Leon, chemistry; immunology
Wolf, Beverly, microbiology
Wolf, Frantsek, mathematics
Wolf, Joseph Albert, mathematics
Wood, David Lee, forest entomology, insect ecology
Wood, Ronald McFarlane, medical microbiology
Woolever, Patricia S., insect morphology
Wu, Hung-Hsi, mathematics
Yang, Chui-Hsu (Tracy), radiation biophysics
Yarwood, Cecil Edmund, plant pathology

Yoch, Duane Charles, microbiology; biochemistry
Yund, Mary Alice, developmental biology
Zahnley, James Curry, biochemistry
Zezulka, Allison Yates, nutrition
Zinke, Paul Joseph, forestry
Zivnuska, John Arthur, forest economics, academic administration
Zucker, Irving, biological rhythms, neuroendocrinology

BEVERLY HILLS
Bierman, Howard Richard, medicine
Bucklin, Robert Van Zandt, pathology, forensic medicine
Carton, Charles Allan, neurosurgery
Catz, Boris, medicine
Field, John Byron, medicine
Hinrichs, Frederick Woods, geology
Kroger, William Saul, psychiatry
Sampliner, Robert Bruce, psychosomatic medicine
Shapiro, Max, periodontology; biomedical engineering

BIG BEAR LAKE
Johnson, Victor (Einar), physiology

BIG PINE
Hardebeck, Ellen Jean, forensic medicine
Seielstad, George A., radio astronomy

BIGGS
Carnahan, Howard Leon, genetics, plant breeding
Tseng, Shu-Ten, agronomy

BISHOP
Little, Thomas Morton, biometrics

BODEGA BAY
Hand, Cadet Hammond, Jr., invertebrate zoology
Hedgecock, Dennis, genetics

BOLINAS
Ainley, David George, marine ecology; ornithology

BORON
Campbell, George Washington, Jr., industrial chemistry
Cooper, Wilson Wayne, industrial chemistry, chemical engineering

BRAWLEY
Bradford, Willis Warren, agronomy
Ehlig, Carl F., plant physiology; biochemistry
Meek, Burl Dean, soil chemistry

BREA
Alley, Starling Kessler, Jr., petroleum chemistry
Askevold, Robert James, petroleum chemistry
Block, Michael Joseph, organic chemistry; agricultural chemistry
Burdett, Lorenzo Worth, analytical chemistry
Chen, Chih Shan, stratigraphy, sedimentary petrology
Christoffersen, Donald John, analytical chemistry
Copelin, Edward Casimere, geochemistry
Crog, Richard Stanley, chemistry; research administration
Fenton, Donald Mason, petroleum chemistry
Fett, E Reinold, analytical chemistry
Filippone, Walter R., geology; geophysics
Fox, John R., research administration, petroleum geology
Fraser, James Mattison, physical chemistry; analytical chemistry
Goldish, Elihu, physical chemistry
Hansford, Rowland Curtis, physical chemistry
Hendricks, Grant Walstein, petroleum chemistry
Hoskins, Cortez William, geology; paleoecology
Huffman, Hal Charles, chemistry
Mallett, William Robert, analytical chemistry; petroleum chemistry
Marsh, Glenn Anthony, physical chemistry; corrosion

McNall, Lester R, plant nutrition
Nahin, Paul Gilbert, chemistry
Neff, Loren Lee, chemistry
Olivier, Kenneth Leo, organic chemistry
Otte, Carel, Jr, geology
Piel, Kenneth Martin, palynology
Redwine, Lowell E, geology
Scully, Charles Norman, lasers
Smith, Gerould Hammond, chemistry
Tilley, George Levis, physical chemistry
Walker, Joseph, analytical chemistry, organic chemistry
Ward, John William, physical chemistry
Wornardt, Walter William, Jr, paleontology

BURBANK
Alaoglu, Leonidas, mathematics
Baker, Andrew Newton, Jr, physics
Barron, Charles Irwin, aerospace medicine
Carlson, Phillip Richard, physics
Floyd, Acey L, physics
Horowitz, Richard E, pathology
Horton, Robert Eugene, mathematics, statistics
Lanchantin, Gerard Francis, biochemistry
Matsudo, Hitoshi, cell biology
Singh, Hakam, polymer chemistry, surface chemistry

BURLINGAME
Beirne, Gilbert Arthur, dermatology
Berry, Edwin X, atmospheric physics
Eads, Richard Bailey, medical entomology
Jones, Russell Heber Blade, economic geology
Ramsey, Hal Harrison, bacteriology
Redeker, Harry Erwin, chemistry

CALEXICO
Ayala, Reynaldo, geography

CALISTOGA
Sakai, William Shigeru, botany, horticulture

CAMARILLO
Liberman, Robert Paul, psychiatry, clinical psychology
Thorne, Charles Joseph, applied mathematics

CAMBRIA
Huntington, Robert (Watkinson), Jr, pathology

CAMPBELL
Blake, Oliver Duncan, geology
Hall, Thomas Kenneth, organic chemistry

CANOGA PARK
Axworthy, Arthur Edward, Jr, physical chemistry
Christe, Karl Otto, inorganic chemistry, physical chemistry
Curtis, Earl Clifton, Jr, physical chemistry
Eichelberger, Robert Leslie, physical chemistry
English, Gerald Alan, nuclear chemistry, analytical chemistry
Fujikawa, Norma Sutton, analytical chemistry
Galloway, William Joyce, physics
Gehri, Dennis Clark, environmental chemistry
Goggin, Donald Edward, chemical metallurgy
Grant, Louis Russell, Jr, inorganic chemistry
Grantham, Leroy Francis, inorganic chemistry, physical chemistry
Huebner, Albert Louis, applied physics
Kirsch, Milton, chemistry, environmental systems
Kleber, Eugene Lawrence, chemistry
Korst, William Lawrence, inorganic chemistry, physical chemistry
Lo, George Albert, physical chemistry, inorganic chemistry
Lysyj, Ihor, analytical chemistry, environmental technology
Madoff, Milton, organic chemistry
Martin, Albert Byron, physics
McKenzie, Donald Edward, physical chemistry, environmental chemistry
Miller, Leroy Jesse, organic chemistry
Morewitz, Harry Alan, nuclear physics

Nealy, Carson Louis, analytical chemistry, energy conversion
Parkins, William Edward, optical physics
Pedinoff, Melvin Eli, mathematics, data processing
Postley, John Appel,
Recht, Howard Leonard, water chemistry, electrochemistry
Remley, Martin Eugene, physics
Schack, Carl J, inorganic chemistry
Silverman, Jacob, physical chemistry, science administration
Stern, Albert Victor, astronomy, systems engineering
Topol, Leo Eli, physical chemistry
Vernon, Gregory Allen, inorganic chemistry
Wagner, Ross Irving, inorganic chemistry
Yates, Wesley Ross, physical chemistry

CARDIFF
Boden, Brian Peter, marine biology

CARLSBAD
Finn, James Crampton, Jr, plant physiology
Thum, Alan Bradley, marine ecology

CARMEL
Bevelander, Gerrit, histology, anatomy
Farr, Lee Edward, medicine
Fowler, Robert Dudley, physics, chemistry
Fry, Thornton Carl, mathematics
Greenwood, Robert, economic geology, mining geology
Harris, Albert Sidney, physiology
Johnson, Kenneth, research management, organic chemistry
Laurmann, John Alfred, fluid dynamics

CARMEL VALLEY
Barrow, Gordon M, physical chemistry
Campbell, Charles Duncan, geology
Davis, Betty Schuck, zoology
Davis, John, vertebrate zoology
Griffin, James Richard, ecology
Hall, Arthur Lee, environmental physiology

CARMICHAEL
Hewitt, Robert T, psychiatry
Kimsey, Jerry Bruce, fish biology
Mayhew, Dennis Ed, plant pathology, plant virology

CARPINTERIA
Sullwold, Harold H, Jr, geology

CARSON
Goldsmith, Henry Arnold, physical chemistry
Marmor, Solomon, organic chemistry
Silliker, John Harold, microbiology

CASTRO VALLEY
Dutra, Frank Robert, pathology
Fa'arman, Alfred, applied physics
Goldberg, Eugene, nuclear physics
Hanneman, Walter W, analytical chemistry, organic chemistry
Penton, Zelda Eve, analytical biochemistry

CERRITOS
Northington, Dewey Jackson, Jr, organic chemistry
Weiss, Harold Gilbert, inorganic chemistry, physical chemistry
Williams, Colin James, physical chemistry, inorganic chemistry

CHICO
Anthony, Margery Stuart, botany, radiation ecology
Baldy, Marian Wendorf, genetics, enology
Barker, LeRoy N, agronomy, plant breeding
Bechtol, Bruce Emerson, geography
Beck, Albert J, zoology
Burleigh, James Reynolds, plant pathology, genetics
Burnett, Marvin Clifton, organic chemistry
Busch, Robert Edward, animal genetics
Chau, Cheuk-Kin, solid state physics, cryogenics

Chinas, Beverly Newbold, cultural anthropology, ethnology
Cliff, Frank Samuel, vertebrate zoology
Corson, George Edwin, Jr, plant morphogenetics
Curtis, Dwayne H, physiology
Demaree, Richard Spottswood, Jr, cytology, parasitology
Dempsey, Wesley Hugh, genetics, plant breeding
Derr, William Frederick, plant anatomy, ecology
Ediger, Robert I, plant taxonomy, ecology
Erpino, Michael James, endocrinology, histology
Evans, Kenneth Jack, vertebrate zoology, ecology
Franzreb, Kathleen E, ecology
Fredenburg, Robert Love, science education
Gold, Marvin B, inorganic chemistry
Guyton, James W, geology
Hall, Donald D, geology, invertebrate paleontology
Harris, David R, physical chemistry
Hauser, Rolland Keith, meteorology
Hiller, Frederick W, inorganic chemistry, physical chemistry
Huntsinger, Karolyn Regina, limnology, ecology
Hunziker, Rodney William, astronomy
Johnson, Ladd Lind, geography
Karinen, Arthur Eli, cartography, population geography
King, Robbins Sydney, embryology
Kistner, David Harold, entomology
Korte, William David, organic chemistry
Kowalski, Donald T, mycology
Kumli, Karl F, organic chemistry
Lantis, David W, geography
Lofgren, Norman Lowell, chemistry
Luxenberg, Harold Richard, applied mathematics
Mallary, Eugene Cobb, nuclear physics
Martin, Gene Ellis, geography
Maurer, Edward Robert, physical science, history of science
McCready, Thomas Arthur, mathematics
McNairn, Robert Blackwood, plant physiology
Meuter, Ralph F, urban geography, economic geography
Mihalyi, Louis James, geography
Murad, Turhon Allen, biological anthropology
Myers, James Edward, anthropology
Neumann, Fred Robert, geology
Olsen, Robert James, soil science
Oswald, Vernon Harvey, zoology
Pease, Burton Frank, analytical chemistry
Pennington, Frank Cook, organic chemistry
Reese, Floyd Ernest, biochemistry, organic chemistry
Riggle, Everett C, numerical analysis
Seawall, Frank, geography
Smith, Valene Lucy, cultural anthropology, ethnology
Stensrud, Howard Lewis, geochemistry
Stephens, William Leonard, microbiology
Stern, Kingsley Rowland, taxonomic botany
Sutton, Dallas Albert, biology
Thomas, Robert E, physiology, biochemistry
Trussell, Margaret Edith, geography
Van Laan, Gordon James, ornamental horticulture
Vasu, Bangalore Seshacham, biology
Vought, Eldon Jon, topology
Wilhelm, Alan Roy, microbiology, virology
Willis, Grover C, Jr, physical chemistry, electrochemistry
Wootton, Donald Merchant, parasitology, zoology
Zicker, Eldon Louis, soil science

CHINA LAKE
Atkins, Ronald Leroy, organic chemistry
Baer, Adrian Donald, solid state physics
Bennett, Harold Earl, optical physics
Bennett, Jean McPherson, optical physics
Bottka, Nicholas, solid state physics, optical physics
Burdick, David Leon, solid state physics
Burge, Dennis Knight, physical optics
Cordes, Herman Fredrick, physical chemistry
Cruise, Donald Richard, applied mathematics
Dieroff, Jack, physical chemistry
Donovan, Terence M, physics
Elliott, Shelden Dougless, Jr, physics, solar energy
Elson, John Merle, physics

Essig, Frederick Charles, experimental physics
Fay, Edward Allen, operations research
Fine, Dwight Albert, inorganic chemistry
Finnegan, William George, atmospheric chemistry
Fletcher, Aaron Nathaniel, physical chemistry, analytical chemistry
Gryting, Harold Julian, chemistry
Heller, Carl A, physical chemistry
Henry, Ronald Andrew, chemistry
Hewer, Gary Arthur, applied mathematics, systems science
Hughes, Richard Swart, lasers
Hunter, Hugh Wylie, physics
Kaufman, Martin Henry, polymer chemistry
Knipe, Richard Hubert, chemical physics
Kyser, David Sheldon, experimental solid state physics
Leonard, Guy William, analytical chemistry
Lepie, Albert Helmut, physical chemistry
Lind, Charles Douglas, physical chemistry
Mallory, Herbert Dean, physical chemistry
Martin, Eugene Christopher, organic polymer chemistry
McBride, William Robert, inorganic chemistry
McEwan, William Shelley, physical chemistry
McMahon, Thomas Joseph, semiconductors, electrooptics
Nadler, Melvin Philip, physical chemistry
Nielsen, Arnold Thor, organic chemistry
Norris, William Phillip, organic chemistry
Odencrantz, Frederick Kirk, physics
Porteus, James Oliver, surface physics
Prentice, Jack L, physical chemistry, high temperature chemistry
Rehn, Victor Leonard, experimental solid state physics
Rogers, Marguerite Moillet, physics
Rowntree, Robert Frederic, operations research, physics
St Amand, Pierre, geophysics
Schaniel, Carl L, operations research
Shlanta, Alexis, atmospheric physics
Thelen, Charles John, organic chemistry, aerospace technology
Victor, Andrew C, engineering physics
Warschauer, Douglas Marvin, physics, energy conversion
White, William Charles, physics, astronomy
Yee, Tucker Tew, organic chemistry, physical organic chemistry

CHULA VISTA
Beckman, William, inorganic chemistry, physical chemistry
Simpson, Jennie Laura Symons, botany

CITRUS HEIGHTS
Vander Wall, Eugene, physical inorganic chemistry

CITY OF INDUSTRY
Wax, Harry, biochemistry

CLAREMONT
Alf, Raymond Manfred, vertebrate paleontology
Allen, Charles Freeman, organic chemistry, biochemistry
Amrein, Yost Ursus Lucius, animal parasitology
Andrus, William DeWitt, Jr, zoology, cell physiology
Baggerly, Leo L, physics
Baird, Alexander Kennedy, geochemistry, petrology
Baskin, Denis George, comparative endocrinology, biological structure
Beechler, Barbara Jean, mathematics
Beeman, David Edmund, Jr, theoretical physics
Beilby, Alvin Lester, analytical chemistry
Bell, Graydon Dee, physics, astrophysics
Benjamin, Richard Keith, mycology
Benson, Lyman David, plant taxonomy
Bentley, Donald Lyon, mathematical statistics
Berg, Selwyn S, applied physics
Borrelli, Robert L, applied mathematics
Bovard, Freeman Carroll, bio-organic chemistry
Busenberg, Stavros Nicholas, mathematics
Campbell, James Arthur, physical chemistry
Carlquist, Sherwin, botany

355

Erickson, Glen Walter, theoretical physics
Ernst, Ralph Ambrose, poultry science, physiology
Espana, Carlos, infectious diseases
Etzler, Marilynn Edith, immunochemistry, biochemistry
Evans, James Warren, animal physiology
Falk, Richard H, botany, cytology
Faulkin, Leslie J, Jr, anatomy
Feeney, Robert Earl, biochemistry, protein chemistry
Feng, Da-Fei, radiation chemistry
Fink, Herman Joseph, solid state physics
Fink, William Henry, chemical physics, theoretical chemistry
Fisher, Gerald Lionel, environmental chemistry, metabolism
Fisher, Kathleen Mary, genetics, science education
Flocker, William Jack, soil physics, vegetable crops
Fong, Ching-Yao, solid state physics
Fowler, Murray Elwood, veterinary medicine
Franti, Charles Elmer, applied statistics
Freedland, Richard A, physiological chemistry
Friedrich, Edwin Carl, organic chemistry
Fulton, Curtis Maxwell, geometry
Gabor, Andrew John, neurology, neurophysiology
Gall, Graham A E, biochemical genetics
Gardner, Ernest (Dean), neurology, anatomy
Garrett, William Norbert, animal nutrition
Garrod, Claude, theoretical physics
Gary, Norman Erwin, apiculture, entomology
Gatlin, Lila L, theoretical biology, physical chemistry
Gehrmann, John Edward, pharmacology
Geschwind, Irving I, endocrinology, biochemistry
Geyman, John Payne, family medicine
Gifford, Ernest Milton, Jr, botany
Gillespie, Jerry Ray, physiology, pathology
Glauz, Robert Doran, applied mathematics
Globus, Albert, neuroanatomy, neurophysiology
Goheen, Austin Clement, plant pathology
Golder, Thomas Keith, cell biology
Goldman, Charles Remington, limnology
Goldman, Marvin, radiobiology
Goldner, Andrew M, physiology
Goldstein, Elliot, infectious diseases
Grau, Charles Richard, poultry nutrition
Green, Harry Western, II, geophysics, structural geology
Green, Melvin Martin, genetics
Griggs, William Holland, pomology
Grogan, Raymond Gerald, plant pathology
Gulyassy, Paul Francis, medicine
Guymon, James Fuqua, enology
Hackett, Wesley P, horticulture, plant physiology
Hall, Dennis Heeley, phytopathology
Hance, Anthony James, neuropharmacology
Hancock, Kenneth George, organic chemistry, photochemistry
Hansen, Robert John, physiological chemistry, endocrinology
Harding, James A, genetics, horticulture
Harrington, James Foster, olericulture
Harris, Daniel Charles, physical chemistry
Harris, Richard Wilson, horticulture
Hart, Benjamin Leslie, animal behavior
Hart, Winfield Hiram, plant nematology
Hartmann, Hudson Thomas, horticulture
Hattersley, Paul G, hematology, clinical pathology
Hayes, Charles Amos, Jr, mathematics
Hedrick, Jerry Leo, biochemistry
Heitman, Hubert, Jr, animal nutrition
Henderson, Gary Lee, pharmacology
Hendricks, Andrew George, embryology
Hess, Charles, horticulture, plant physiology
Higgins, Charles Graham, physical geology

Hildebrand, Milton, vertebrate morphology
Hill, Fredric William, nutrition
Hills, F Jackson, agronomy, agricultural statistics
Hirsh, Dwight Charles, III, veterinary microbiology
Hodges, Robert Edgar, internal medicine
Hollinger, Mannfred Alan, pharmacology
Hope, Hakon, x-ray crystallography
Horowitz, John M, neurophysiology
Horwitz, Barbara Ann, animal physiology, cell physiology
Howard, Walter Egner, ecology, ethology
Hsiao, Theodore Ching-Teh, plant physiology, biochemistry
Huffaker, Ray C, biochemistry, plant physiology
Hughes, John P, veterinary medicine
Hungate, Robert Edward, biology
Hunter, Robert L, anatomy
Hurley, James P, physics
Hurley, Lucille Shapson (Mrs Kenneth Thompson), nutrition
Ingraham, John Lyman, microbiology
Ingraham, Lloyd Lewis, physical chemistry, organic chemistry
Jacobus, William Edward, biochemistry
Jain, Subodh K, population biology, economic botany
Jameson, Everett Williams, Jr, vertebrate zoology, medical entomology
Jasper, Donald Edward, veterinary medicine, clinical pathology
Jennings, Walter Goodrich, natural products chemistry
Jensen, Gordon D, pediatrics, psychiatry
Jett, Stephen Clinton, geography
Joy, Robert McKernon, neuropharmacology, neurophysiology
Judson, Charles LeRoy, insect physiology
Julian, Logan M, veterinary anatomy
Jungerman, John (Albert), physics
Kado, Clarence Isao, molecular biology, plant pathology
Kahler, Alex L, genetics, plant breeding
Kaneko, Jiro Jerry, physiology
Kawatomari, Toshio, microbiology
Keefer, Raymond Marsh, physical chemistry
Keizer, Joel Edward, physical chemistry, chemical physics
Kendrick, Peter Carleton, veterinary pathology
Kennedy, Peter Carleton, veterinary pathology
Kepner, Richard Edwin, organic chemistry
Kester, Dale Emmert, pomology
Ketellapper, Hendrik Jan, plant physiology
Kiger, John Andrew, Jr, genetics, biophysics
Kilgore, Wendell Warren, toxicology
Killam, Ellen Eva King, pharmacology
Killam, Keith Fenton, Jr, pharmacology
Kitchell, Ralph Lloyd, veterinary anatomy
Klems, Joseph Henry, physics
Kliewer, Walter Mark, biochemistry, plant physiology
Klisiewicz, John Michael, plant pathology
Knight, Allen Warner, aquatic ecology, water pollution
Knowles, Paulden Ford, agronomy
Knox, William Jordan, nuclear physics
Ko, Winston Tai-Kan, high energy physics
Kofranek, Anton Miles, floriculture
Kohl, Harry Charles, Jr, floriculture
Koong, Ling-Jung, biomathematics, animal nutrition
Kosuge, Tsune, plant biochemistry, plant pathology
Kratzer, Frank Howard, nutrition
Kraus, Jess F, epidemiology, environmental health
Krebs, Edwin Gerhard, biochemistry
Kreith, Kurt, mathematics
Krener, Arthur James, applied mathematics, systems theory
Krom, Melven R, mathematics
Kumagai, Lindy Fumio, medicine
Kunkee, Ralph Edward, biochemistry, enology
Kurowski, Gary John, numerical analysis
Kyhos, Donald William, plant taxonomy, plant cytogenetics
Laben, Robert Cochrane, genetics
Laidlaw, Harry Hyde, Jr, apiculture
La Mar, Gerd Neustadter, structural chemistry
Lander, Richard Leon, physics
Lange, Norma Jean, phycology
Lange, William Harry, Jr, entomology
Laude, Horton-Meyer, agronomy, botany
Leach, Lysle Douglas, plant pathology
Lear, Bert, plant pathology

Leighton, Robert Lyman, veterinary surgery
Lenn, Nicholas Joseph, neurology, anatomy
Leung, Philip Min Bun, nutrition, biochemistry
Lewis, Alvin Edward, physiology
Lewis, Jerry Parker, medicine
Lider, Lloyd A, genetics
Lilleland, Omund, pomology
Linz, Peter, computer science
Lipps, Jere Henry, geology, invertebrate paleontology
Lipscomb, Paul Rogers, orthopedic surgery
Lofgreen, Glen Pehr, animal nutrition
Lohse, Carleton Leslie, anatomic pathology
Loomis, Edmond Charles, parasitology, acarology
Loomis, Robert Simpson, plant physiology
Lorenz, Frederick Wharton, physiology
Lorenz, Oscar Anthony, vegetable crops, plant nutrition
Love, Robert Merton, cytogenetics, ecology
Lownsbery, Benjamin Ferris, plant nematology
Lowrey, George Harrison, pediatrics
Luh, Bor Shiun, food science
Lundgren, Harold Palmer, textile chemistry
Lyons, James Martin, plant physiology
MacGregor, Ian Duncan, petrology, geochemistry
MacKenzie, Malcolm R, internal medicine, immunology
Maddy, Keith Thomas, veterinary medicine, public health
Madison, John Herbert, Jr, horticulture
Maggenti, Armand Richard, plant nematology
Major, Jack, plant ecology
Maki, August Harold, biophysical chemistry
Manning, JaKue Stanley, virology, biophysics
Marble, Vern L, agronomy, field crops
Marr, Allen Gerald, microbiology
Martin, George C, pomology
Mason, Dean Towle, cardiovascular diseases
Mazelis, Mendel, plant biochemistry
McClelland, George Anderson Hugh, medical entomology, genetics
McColm, Douglas Woodruff, physics
McGowan, Blaine, Jr, veterinary medicine
McHenry, Henry Malcolm, biological anthropology
McKercher, Delbert Grant, veterinary virology
McLean, Donald Lewis, entomology
Meizel, Stanley, reproductive biology, developmental biology
Mendel, Verne Edward, physiology
Metcalf, Robert Alan, population biology
Meyer, James Henry, nutrition
Mikkelsen, Duane Soren, soil fertility, plant physiology
Miller, Martin Wesley, microbiology
Miller, Milton Albert, zoology
Miller, Russel Bryan, organic chemistry
Mircetich, Srecko M, plant pathology
Moberg, Gary Philip, physiology, neuroendocrinology
Moores, Eldridge Morton, petrology, structural geology
Morgan, Joe Peter, veterinary radiology
Morris, James Grant, animal nutrition
Morris, Leonard Leslie, vegetable crops, plant physiology
Moulton, Jack E, veterinary pathology
Moyle, Peter Briggs, ichthyology, aquatic ecology
Munns, Donald Neville, plant nutrition, soil science
Murphy, Terence Martin, plant biochemistry, physiology
Musker, Warren Kenneth, inorganic chemistry
Nash, Charles Presley, physical chemistry
Needles, Howard Lee, polymer chemistry, textile chemistry
Nelson, Klayton Edward, plant pathology
Neville, Melvin K, anthropology
Nickerson, Thomas Andrew, dairy chemistry, food science
Nielsen, Donald R, soil physics
Noble, Ann Curtis, food science
Nold, Max M, radiation biology, veterinary medicine
Norris, Robert Francis, plant physiology, weed science
Norton, Donald Alan, mathematics
Nyland, George, plant pathology

Ogasawara, Frank X, comparative physiology
Ogawa, Joseph Minoru, plant pathology
Olcott, Harold Saft, food science
Olmo, Harold Paul, viticulture
Osburn, Bennie Irve, veterinary pathology
Osebold, John William, immunology, microbiology
Painter, Edgar Page, organic chemistry
Painter, Ruth Coburn Robbins, nutritional biochemistry, toxicology
Pangborn, Jack, electron microscopy
Pangborn, Rose Marie Valdes, food technology, psychophysics
Pappagianis, Demosthenes, medical microbiology
Parker, Harold R, physiology
Parks, Norris Jim, physical chemistry, nuclear chemistry
Pearcy, Robert Woodwell, plant ecology, plant physiology
Peek, Neal Frazier, experimental nuclear physics
Pellett, David Earl, physics
Peoples, Stuart Anderson, pharmacology
Peterson, Daniel Walter, nutrition, biochemistry
Peterson, Maurice Lewellen, agronomy
Pfeffer, Washek F, mathematics
Phaff, Herman Jan, food technology, microbiology
Phillips, David William, marine biology
Pollak, Emil John, animal breeding
Pool, Roy Ransom, veterinary pathology, radiobiology
Post, Richard Freeman, physics
Pratt, David, genetics, virology
Pratt, Harlan Kelley, plant physiology, horticulture
Preiss, Jack, biochemistry
Pritchard, William Roy, veterinary medicine
Prout, Timothy, genetics
Purdy, Richard Eugene, biochemistry
Qualset, Calvin Odell, genetics, plant breeding
Rabinowitz, Lawrence, physiology
Radosevich, Steven Robert, weed science, botany
Raggi, Livio Giovanni, poultry pathology
Rains, Donald W, plant nutrition, soil science
Rappaport, Lawrence, plant physiology, horticulture
Raski, Dewey John, nematology
Rauschkolb, Roy Simpson, soil fertility
Raveling, Dennis Graff, wildlife biology
Raventos, Antolin, IV, radiology
Reiber, Harold George, organic chemistry
Reid, Roderick Vincent, Jr, theoretical physics
Reisenauer, Hubert Michael, soil science
Rendig, Victor Vernon, soil fertility, plant physiology
Renkin, Eugene Marshall, physiology, pharmacology
Reynolds, Donald Montgomery, microbiology
Rhode, Edward A, Jr, veterinary medicine, physiology
Richerson, Peter James, limnology, human ecology
Rick, Charles Madeira, Jr, cytogenetics, evolution
Riemann, Hans, veterinary medicine
Robinson, Neal Clark, biochemistry
Rock, Peter Alfred, electrochemistry
Rogers, Quinton Ray, nutrition, biochemistry
Rollins, Wade Cuthbert, biology
Rolston, Dennis Eugene, soil physics
Romani, Roger Joseph, plant physiology, biochemistry
Ronning, Magnar, animal nutrition
Root, John Walter, physical chemistry, nuclear chemistry
Rosenberg, Lauren Emery, protozoology
Rosenwald, Arnold Samuel, veterinary medicine
Rost, Thomas Lowell, plant anatomy, plant cytology
Rubatzky, Vincent E, plant physiology, horticulture
Rucker, Robert Blain, animal nutrition
Rudd, Robert L, zoology
Ruebner, Boris Henry, pathology
Russell, Gerald Frederick, food chemistry
Rutger, John Neil, genetics, plant breeding
Ryugo, Kay K, pomology, plant physiology
Sachs, Roy M, plant physiology
Sadler, Walter White, veterinary medicine
Sallee, G Thomas, mathematics
Salt, George William, animal ecology

CALIFORNIA

Sassenrath, Ethelda Norberg, physiology, psychopharmacology, behavioral
Saunders, James Allen, plant physiology, phytochemistry
Schaller, Charles William, agronomy
Schalm, Oscar William, veterinary science
Schmid, Carl William, biophysical chemistry
Schnathorst, William Charles, plant pathology
Schwab, Robert G., physiology
Schwabe, Calvin Walter, epidemiology, parasitology
Schwartz, Jerome Leroy, community health
Schweigert, Bernard Sylvester, food science
Scobey, Robert P., physiology, neurophysiology
Seaman, Donald Edward, weed science
Segel, Irwin Harvey, biochemistry
Seiber, James N., organic chemistry
Shalla, Thomas Allen, plant pathology
Shanks, Wayne C., III, geochemistry
Shapiro, Arthur Maurice, population biology
Shepherd, Robert James, plant pathology, virology
Shifrine, Moshe, microbiology
Shleser, Robert A., genetics
Simmons, Frederick John, zoogeography, medical geography
Sims, William Lynn, vegetable crops
Singleton, Vernon Leroy, natural products chemistry, enology
Slonka, Gerald Francis, parasitology
Smith, Arthur Hamilton, physiology
Smith, Lloyd Muir, food chemistry
Smith, Nathan Elbert, animal science
Smith, Paul Gordon, vegetable crops
Smith, Robert Elphin, physiology, biomedical engineering
Smith, Robert Emric, cell physiology
Snow, Sidney Richard, genetics
Sodhi, Harbhajan S., metabolism, cardiology
Sommer, Leo Harry, organic chemistry
Sommer, Noel Frederick, plant physiology, pathology
Spieth, Herman Theodore, zoology
Spurr, Arthur Richard, plant morphology, endocrinology
Stabenfeldt, George H., physiology
Stanford, Ernest Hall, plant breeding
Stark, Larry Gene, pharmacology
Starr, Mortimer Paul, microbiology, philosophy of biology
Stebbins, George Ledyard, botany
Stein, Sherman Kopald, mathematics
Sterling, Clarence, botany
Sterling, Harold Melvin, medicine, pediatrics
Stern, Judith S., nutrition
Stevens, Merwin Allen, plant genetics
Stewart, George Franklin, food technology
Stocking, Clifford Ralph, plant physiology
Stormont, Clyde J., genetics, immunology
Stowell, Robert Eugene, pathology
Stringall, Robert William, mathematics
Stumpf, Paul Karl, lipid biochemistry
Summers, Francis Marion, entomology
Swenerton, Helene Roupen, nutrition
Swinehart, James Herbert, inorganic chemistry, bioinorganic chemistry
Tanji, Kenneth K., water pollution
Tappel, Aloys Louis, biochemistry
Theilen, Gordon H., veterinary medicine
Theis, Jerold Howard, medical microbiology, parasitology
Thompson, Kenneth, geography
Thornton, Robert Melvin, biology, plant physiology
Thorp, Robbin Walker, insect taxonomy, ecology
Timm, Herman, agronomy
Tinti, Dino S., physical chemistry
Toreson, Wilfred Earl, pathology
Traut, Robert Rush, biochemistry, molecular biology
Trommershausen-Smith, Ann L., genetics
Troy, Frederic Arthur, oncology
True, William Wadsworth, physics
Tsao, Makepeace Uho, chemistry
Tucker, John Maurice, botany
Tully, Edward Joseph, Jr., mathematics
Tupper, Charles John, internal medicine
Turgeon, Judith Lee, neuroendocrinology, reproductive physiology
Twiss, Robert John, geology
Tyler, Walter Steele, anatomy

Uriu, Kiyoto, pomology, plant physiology
Valentine, James William, geology
Valentine, Raymond Carlyle, biochemistry
Vaughn, Reese Haskell, food microbiology
Vermeersch, Joyce Ann, nutrition, community health
Verosub, Kenneth Lee, geophysics
Viglierchio, David Richard, nematology
Vijayan, Vijaya Kumari, human anatomy, neuroanatomy
Vohra, Pran Nath, nutrition, biochemistry
Volman, David H., physical chemistry
Wade, Dale A., wildlife management
Wagman, Irving Henry, neurophysiology, physiology
Wagner, Kit Kern, dynamic meteorology
Walters, Richard Francis, information science, medical education
Waring, Worden, physical chemistry
Washino, Robert K., entomology, public health
Watson, David Werner, internal medicine
Watt, Kenneth Edmund Ferguson, ecology
Weathers, Wesley Wayne, environmental physiology
Weaver, Robert John, plant physiology
Webb, Albert Dinsmoor, enology
Webster, Barbara Donahue, plant morphogenesis
Webster, Grady Linder, plant taxonomy
Webster, Robert K., phytopathology
Weiner, Howard Jacob, mathematical statistics
Weir, William Carl, animal nutrition
Welch, James Edward, genetics
Wells, Kenneth, mycology
West, Theodore Clinton, pharmacology
Wheelis, Mark Lewis, microbiology, genetics
Whitaker, John Robert, biochemistry
White, Lynn D., soil chemistry, soil mineralogy
Wichmann, Robert W., veterinary microbiology
Wiggins, Alvin Dennie, mathematical statistics
Wikman-Coffelt, Joan, biochemistry
Williams, William Arnold, agronomy
Wilson, Barry William, cell biology, neurobiology
Wilson, Floyd Dee, experimental pathology, hematology
Wilson, Lowell D., endocrinology, biological chemistry
Wilson, Wilbor Owens, avian physiology
Winters, Wallace Dudley, neuropharmacology, clinical pharmacology
Wolfe, Stephen Landis, cell biology, electron microscopy
Wolfman, Earl Frank, Jr., surgery
Wong, Ming Ming, parasitology
Woolley, Dorothy Elizabeth Schumann, physiology, pharmacology
Wooten, Frederick (Oliver), solid state physics
Yager, Philip Marvin, experimental high energy physics
Yamaguchi, Masatoshi, plant physiology
Yamamoto, Richard, veterinary microbiology
Yang, Shang-Fa, plant physiology
Yeh, Yin, quantum electronics, chemical physics
Yoder, David Lee, plant pathology, soil microbiology
York, George Kenneth, II, microbiology
Youmans, Julian Ray, neurosurgery
Zee, Yuan Chung, virology
Zeman, Frances June, nutrition
Zeronian, Sarkis Haig, textile chemistry
Zink, Frank W., plant breeding
Zscheile, Frederick Paul, Jr., plant physiology, phytopathology
Zweifel, George, organic chemistry

DEL MAR
Bach, Sven Aage, medicine, biophysics
Chawner, William Donald, geology
Grimaldi, Frank Saverio, inorganic chemistry, analytical chemistry
King, Barry Griffith, physiology
Matheson, Arthur Ralph, inorganic chemistry
Ross, Frederick William, physics

DEEP SPRINGS
Mawby, John Evans, vertebrate paleontology

DUARTE
Amromin, George David, pathology
Bangasser, Susan Andretta, immunochemistry
Beutler, Ernest, medicine
Comings, David Edward, genetics, cell biology
Egan, Marianne Louise, immunochemistry
Engvall, Eva Susanna, immunochemistry

Smith, Joe Nelson, Jr., surface physics, solid state physics
Stillwell, William Duncan, applied physics
White, Jack Lee, materials science

DIAMOND BAR
Bauer, Hans Fred, fuel science
Pacela, Allan Fred, biomedical engineering

DILLON BEACH
Blake, James A., invertebrate zoology, marine biology
Smith, Edmund Hobart, marine zoology
Taylor, Dwight Willard, invertebrate zoology

DIXON
Fleschner, Charles Anthony, entomology

DOMINGUEZ HILLS
Armacost, William L., mathematics
Arora, Harbans Lall, zoology
Book, Stephen Alan, statistics
Chi, Lois Wong, parasitology
Childress, Evelyn Tutt, microbiology, immunology
Colvin, Dallas Verne, animal behavior, animal ecology
Dobyns, Leona Danette, physical chemistry
Evett, Arthur A., physics
Fischer, Robert Blanchard, analytical chemistry, academic administration
Gash, Kenneth Blaine, organic chemistry
Gibson, Lyle Edgar, urban geography
Gould, William E., mathematics
Grabiner, Judith Victor, history of science
Hart, Garry Dewaine, mathematics
Hsiung, Chi-Hua Wu, theoretical chemistry
Johnson, Robert Bethune, geography, research management
Jones, William B., mathematics
Kalland, Gene Arnold, reproductive endocrinology
Larmore, Lawrence Louis, mathematics
Layton, Thomas Nutter, anthropology
Lydon, Carol Guze Konrad, cell biology
McCarthy, Francis Davey, biological oceanography
Miles, Frank Belsley, mathematics
Pope, Polly Holman, cultural anthropology
Stinson, James Robert, meteorology
Wiegmann, Norman Arthur, algebra
Wiley, Samuel L., physics
Wilk, William David, inorganic chemistry

DOWNEY
Babrov, Harold J., molecular spectroscopy
Bauer, Franz Karl, internal medicine
Bessman, Alice Neuman, internal medicine
Brody, Gary Sidney, plastic surgery
Buckley, Ramon D., biochemistry, physiology
Cho, Alfred Chih-Fang, acoustics, aerospace sciences
Christie, Bruce Robert, veterinary pathology, anthropology
Comarr, Avrom Estin, urology
Hackney, Jack Dean, internal medicine, physiology
Hafner, James Wilson, nuclear physics
Hislop, Helen Jean, medical physiology, exercise physiology
Jain, Naresh C., toxicology, analytical chemistry
Perry, Jacquelin, orthopedic surgery, kinesiology
Reswick, James Bigelow, biomedical engineering
Rockenmacher, Morris, biochemistry, immunology
Selvester, Ronald H., cardiology
Tai, Clement Leo, applied mechanics, applied mathematics

Gildenhorn, Hyman L., medicine, radiology
Holden, Joseph Thaddeus, biochemistry
Kan, James Hung-Kei, genetics
Kaplan, William David, genetics
Kessler, Michael J., biochemistry
Klevecz, Robert Raymond, cell biology, molecular biology
Levine, Rachmiel, endocrinology
Lieberman, Jack, pulmonary diseases, enzymology
McCann, Marilyn Wales, neurosciences
McCann, Richard Eugene, pharmacology
Melnyk, John H., cytogenetics
Ohno, Susumu, cytogenetics
Okada, Tadashi A., cytogenetics
Pritchard, David Graham, carbohydrate chemistry, protein chemistry
Riggs, Arthur Dale, biochemistry
Roberts, Eugene, biochemistry
Rouser, George, biochemistry
Ruoslahti, Erkki Ilmari, cancer
Shively, John Ernest, biochemistry
Todd, Charles Wyvill, biochemistry
Trout, William Edgar, III, genetics
Vaughn, James E., Jr., anatomy
Winer, Cynthia Crosby, physiological psychology
Wimer, Richard E., behavioral genetics
Wong, Patrick Tim-Choi, genetics
Yoshida, Akira, biochemistry, genetics

DUBLIN
Bergman, Elliot, organic chemistry
Craig, Theodore Warren, food science
Holmes, David G., food science

EDWARDS
Quinn, Lawrence Paul, inorganic chemistry

EL CAJON
Givens, William Geary, physical chemistry
Schelar, Virginia Mae, inorganic chemistry
Taylor, Merrel Arthur, conservation, environmental biology

EL CENTRO
Hauser, William Joseph, fish biology
Jones, Henry Albert, horticulture
Lehman, William Francis, plant genetics, agronomy

EL CERRITO
Anderson, Kinsey Amor, space physics
Cook, David Lewis, data processing
Dimmick, Robert Llewellyn, microbial physiology
Goldschmidt, Alfred, organic chemistry
Hendrickson, Yngve Gust, organic chemistry
Houston, David Fairchild, cereal chemistry
Katsumoto, Kiyoshi, chemistry
Kemp, Jacob David, physical chemistry
Kohler, George Oscar, biochemistry, biochemical engineering
Robinson, Frank Ernest, agronomy
Worker, George F. Jr., agronomy, botany

EL MACERO
Williams, Tom Vare, plant breeding

EL MONTE
Ashby, Robert Morrell, solid state electronics
Graefe, Allen Frederick, environmental chemistry
King, William Mattern, applied chemistry
Klager, Karl, organic chemistry
McCreary, Ralph Leroy, physics
Mishuck, Eli, physical chemistry
Saltonstall, Clarence William, Jr., organic chemistry, polymer chemistry

EL SEGUNDO
Anderson, Neal Sample, physics
Cook, Gilbert R., physics
Frazier, Edward Nelson, astrophysics
Gault, John M., physics

Getting, Ivan Alexander, physics
Hughes, Gordon Frierson, applied physics
Joanou, George David, physics
Keddy, James Richard, computer sciences
King, Gilbert William, chemical physics
Krause, Ernst Henry, physics
Leonard, Byron Peter, physics
Mahadevan, Parameswar atomic physics, molecular physics
Mayer, Harris Louis, theoretical physics
McPherson, Donald Attridge, space physics
Mearns, Alan John, fisheries, pollution biology
Meyer, Norman Joseph, acoustics
Phillips Roger Winston, surface chemistry
Sakaguchi, Dianne Koster, system analysis
Sashkin, Lawrence, computer sciences
Shemer, Jack Evvard, computer science, electrical engineering
Shimabukuro, Fred Ichiro, radio astronomy
Silvertooth, Ernest W, applied physics
Sitney, Lawrence Raymond, physical chemistry
Sterling, Warren Martin, information science, electronic engineering
Stewart, Gordon Ervin, microwave physics, plasma physics
Suchard, Steven Norman, lasers
Sutton, David George, lasers
Swanson, David G, Jr, nuclear chemistry, physical chemistry
Vampola, Alfred Ludvik, space physics
Wang, H E Frank, fluid physics, systems engineering
Wiener, Sidney, materials science
Wilkins, Roger Lawrence, chemical physics
Young, David Ross, oceanography

EL SOBRANTE
Arnold; Zach McLendon, protozoology, micropaleontology

ELDRIDGE
Peterson, Neal Alfred, biochemistry

EMERYVILLE
Adelson, David E, chemistry
Davis, Peyton Nelson, biochemistry
Fleischer, Allan A, nuclear medicine, nuclear science
Lamb, James Francis, nuclear medicine, nuclear science
Somogyi, Laszlo P, food technology
Winchell, Harry Saul, nuclear medicine

ENCINO
Bilow, Norman, chemistry
McAllister, Cyrus Ray, mathematics
Prater, Arthur Nickolaus, nutrition, food science
Senf, J Henry Ruwe physics
Stewart, Robert Malcolm, applied physics, neuro-technology

ESCONDIDO
Hutchison, Paul Clifford, systematic botany
Welker, Everett Linus, mathmatical statistics
Yanick, Nicholas Samuel, physical chemistry

ETNA
Spencer, Harold Foster, cytology

EUREKA
Belfuss, Erwin Roland, parasitology

FAIRFAX
Greenberg, Sidney Abraham, physical chemistry
Warren, Mashuri Laird, environmental physics, plasma physics

FAIRFIELD
Moore, Donald Baker, experimental physics

FALLBROOK
Langham, Derald G, genetics

FONTANA
Soth, Glenn Carroll, physical chemistry

FOSTER CITY
Cole, Richard, physical chemistry, environmental management
Hammer, Frank E, biochemistry
Hotz, Henry Palmer, nuclear physics
Woldseth, Rolf, atomic physics, nuclear physics

FOUNTAIN VALLEY
Desai, Rajendra G, hematology
Hauk, Peter, physical chemistry
Lou, Kingdon, immunology
Mealey, Edward H, biochemistry
Shen, Yvonne Feng, organic chemistry

FREMONT
Hillendahl, Richard Warren, astrophysics
Kashyap, Tapeshwar S, population genetics, poultry breeding
Warren, Don Cameron, genetics

FRESNO
Arce, Gina, biology
Arnold, Robert Fairbanks, mathematics
Avent, Jon C, geology
Ball, Wilbur Perry, agriculture
Bianchi, William C, groundwater hydrology
Biggerstaff, Warren Richard, organic chemistry
Blackerby, Bruce Alfred, geology
Braun, Donald E, analytical chemistry
Bremner, Raymond Wilson, analytical chemistry
Brown, Robert James Sidford, physics
Brown, Sheldon (Jack), physics
Brownell, James Richard, soils
Burger, Othmar Joseph, agronomy
Burtner, Dale Charles, analytical chemistry
Carr, John Halden, bacteriology
Chesemore, David Lee, wildlife ecology
Ciula, Richard Paul, organic chemistry
Clark, David Ellsworth, organic chemistry
Cohen, Moses E, applied mathematics
Cole, Chester F, geography
Collin, William Kent, medical microbiology
Crosby, John Albert, urban geography, urban sociology
Cserna, Eugene George, structural geology
Curtis, Charles Elliott, entomology
Donaldson, John Riley, nuclear physics
Dowler, Lloyd, agriculture
Eliason, Afton Yeates, physics
Ervin, Roger Edward, geography
Ferguson, David B, plant breeding
Ford, Donald Hoskins, plant pathology
Freckman, Diana Wall, plant nematology
Gigliotti, Helen Jean, biochemistry
Gilmore, James Eugene, entomology
Goldbloom, David Ellis, nutrition, biochemistry
Grubbs, David Edward, physiological ecology
Gump, Barry Hemphill, analytical chemistry
Haas, Richard, ethology, zoology
Harmon, Wallace Morrow, parasitology, protozoology
Harvey, John Marshall, plant pathology
Hawbecker, Albert Claude, biology
Helmer, James Douglas, plant science
Herzenberg, Caroline Littlejohn, experimental nuclear physics
Hewitt, Allan A, plant physiology
Hixson, Floyd Marcus, animal breeding, environmental physiology
Holmes, Donald Eugene, biophysics
Hotz, David Franklin, atomic physics, nuclear physics
Hoversland, Arthur Stanley, reproductive physiology, comparative physiology
Ishimoto, Tom, biology
Judd, Floyd L, high energy physics
Kallo, Robert Max, physical chemistry
Karle, Harry P, plant pathology, viticulture
Kauffman, George Bernard, inorganic chemistry, history of science
Kehoe, Brandt, chemistry
Kipps, Thomas Charles, mathematics
Koch, Gary Marlin, ornamental horticulture
Koller, Ernst Frank, geography of Western Europe
Labarre, Anthony E, Jr, mathematics
Latimer, Howard Leroy, plant genetics, plant ecology

Lewis, Leyburn F, medical entomology
Lipton, Werner Jacob, plant physiology
Liskey, Nathan Eugene, health science
Lundeen, Glen Alfred, food science
Mallory, Thomas E, plant morphology, plant cytology
Mangan, Jerrome, developmental biology, genetics
Markham, Charles G, geography, meteorology
McClellan, Wilbur Dwight, plant pathology, research administration
McClintic, Joseph Robert, physiology
Meyer, Ronald Warren, plant pathology
Miller, William Martin, organic chemistry
Miura, Takeshi, medical entomology
Montgomery, Richard C, geology
Nasse, George Nicholas, geography
Neesby, Torben Emil, clinical chemistry
Nelson, Darren Melvin, animal physiology, endocrinology
Nightingale, Harry Irving, agricultural chemistry
Norsworthy, Stanley Frank, resource geography
Nur, Hussain Sayid, mathematical analysis
Peterson, Lance George, insect physiology, toxicology
Pryor, Dean Earl, plant pathology
Ramming, David Wilbur, plant breeding
Rees, Bryant Eugene, entomology
Rempel, Herman G, chemistry
Ritenour, Gary Lee, agronomy, plant physiology
Rodemeyer, Stephen A, physical organic chemistry
Rogoff, William Milton, veterinary entomology
Rousek, Edwin J, agriculture
Russell, Kenneth Homer, physical chemistry
Schaefer, Charles Herbert, entomology
Shacklett, Robert Lee, physics, academic administration
Smith, Philip Nixon, medical bacteriology, virology
Soderstrom, Edwin Loren, entomology
Spieler, Richard Arno, genetics, zoology
Staebler, Arthur E, ornithology
Standing, Keith M, vertebrate zoology
Stuart, Merrill M, geography
Sun, Hugo Sui-Hwan, algebra, number theory
Thorup, James Tat, agronomy, soil fertility
Toney, Joe David, inorganic chemistry
Tribbey, Bert Allen, zoology, ecology
Van Der Elst, Dirk H, behavioral anthropology, cultural anthropology
Vander Meer, Paul, cultural geography
Van Elswyk, Marinus, Jr, plant breeding, genetics
Van Schaik, Peter Hendrik, plant breeding
Vidoli, Vivian Ann, physiology, neurophysiology
Wade, William Howard, entomology
Wagoner, Ronald Lewis, mathematics
Weiler, John Henry, Jr, plant taxonomy
Whaley, Julian Wendell, plant pathology, plant science
White, Stanton M, sedimentary petrology, marine geology
Wiley, Lorraine, plant physiology
Williamson, Hugh A, physics
Woo, Norman Tzu Teh, mathematics
Woodwick, Keith Harris, invertebrate zoology
Zellmer, David Louis, analytical chemistry
Zender, Michael J, nuclear physics

FT BAKER
Rigby, Paul T, medical entomology

FT ORD
Bryson, Marion Ritchie, operations research
Marchi, Raymond Paul, physical chemistry
Stoesser, Albert V, pediatrics

FULLERTON
Adams, Philip A, entomology
Ames, Dennis Burley, mathematics
Anderson, Robert James, chemical physics, electronic engineering
Ashton, David Hugh, food microbiology
Bailey, David Tiffany, organic chemistry, biochemistry
Beckett, Ralph Lawrence, speech pathology

Beckman, Arnold Orville, chemistry
Belloli, Robert Charles, organic chemistry, organometallic chemistry
Bradshaw, Lawrence Jack, medical microbiology
Brattstrom, Bayard Holmes, zoology, x-ray crystallography
Bryden, John Heilner, x-ray crystallography
Chow, Wen Mou, computer science
Clark, Walter Leighton, III, food science
Cooperman, Edward Lee, nuclear physics
Davenport, Calvin Armstrong, bacteriology, immunology
Donaruma, Lorraine Guy, organic chemistry
Dowalby, Margaret Susanne, optometry
Duneer, Arthur Gustav, Jr, radiation physics
Earick, Arthur David, geography
Gallaway, William S, spectroscopy
Gelfer, Daniel Harold, organic polymer chemistry
Gilbert, Richard Carl, mathematical analysis
Gregg, James R, optometry
Guy, George Anderson, meteorology
Haglund, John Richard, microbiology
Haig, Pierre Vahe, medicine
Hanes, Ted L, plant ecology
Hightower, James K, computer science, operations research
Horn, Michael Hastings, ichthyology, marine biology
Horrocks, Donald Leonard, nuclear chemistry
Hoshizaki, Takashi, plant physiology
Hoyt, Justus, physical chemistry
Jaanus, Siret Desiree, pharmacology, endocrinology
Janota, Harvey Franklin, analytical chemistry
Janus, Alan Robert, solid state science
Joseph, Roger, anthropology
Kehoe, Thomas J, analytical chemistry, inorganic chemistry
Keller, James Lloyd, physical chemistry, organic chemistry
Kelly, Michael James, physical biochemistry
Kent, William L, chemistry
Lambert, Charles Calvin, developmental biology, reproductive biology
Lauro, Gabriel Joseph, food science, microbiology
Leonard, John Edward, organic chemistry, physical chemistry
Liebschutz, Alan Morton, physics
Manning, Robert Joseph, physical chemistry, organic chemistry
Margach, Charles Boyd, optometry
Marley, Gerald C, mathematics
Matovich, Edwin, physical chemistry
Matsuyama, George, electroanalytical chemistry
McCarthy, Miles Duffield, reproductive biology
McWilliams, Donald A, physics
McWilliams, Kenneth Leroy, entomology
Mills, Kenneth Selby, physiology
Montana, Andrew Frederick, organic chemistry
Moshy, Raymond Joseph, food science
Nagel, Glenn M, biochemistry
Nanes, Roger, molecular spectroscopy
Neti, Radhakrishna Murty, physical chemistry
Obremski, Robert John, physical chemistry
Oh, Chan Soo, organic chemistry
Perumal, Alexander, horticulture, agriculture
Prenzlow, Carl Frederick, physical chemistry
Presch, William Frederick, herpetology
Ray, Robert Allen, clinical chemistry
Reith, Gertrude McKean, geography
Roebuck, Albert Henry, physical organic chemistry
Rosenberg, Marvin J, biology, molecular genetics
Ross, Robert Edgar, food science
Rothman, Alvin Harvey, biology
Shapiro, Mark Howard, nuclear physics, atmospheric physics
Shields, Loran Donald, analytical chemistry, academic administration
Sie, Edward Hsien Choh, biochemistry
Spenger, Robert E, organic chemistry
Stiel, Edsel Ford, analytical chemistry
Stoub, Kenneth Paul, analytical chemistry
Sutton, Donald Dunsmore, microbiology
Thomas, Barry, zoology, environmental education
Turner, George Cleveland, ecology, botany
Walkington, David L, botany

CALIFORNIA

Wamser, Carl Christian, organic chemistry, photochemistry
Weber, Bruce Howard, biochemistry, neurochemistry
Wegner, Patrick Andrew, organometallic chemistry
Weinraub, Joel D., herpetology
Weiss, Theodore Joel, lipid chemistry
Weiss, William Van, inorganic chemistry, analytical chemistry
Wodicka, Virgil Orville, food technology
Wong, Dorothy Pan, physical chemistry
Wong, Corinne Shear, physical chemistry
Woyski, Margaret Skillman, geology
Yates, Kenneth Pidcock, chemical physics
Young, Donald C., inorganic chemistry, agricultural chemistry

GARDENA
Steinman, Robert, inorganic chemistry, organic chemistry

GARDEN GROVE
Hydock, Joseph J., chemistry
Maggio, Francis Xavier, chemistry
Scott, Kevin M., geology
Sense, Karl August, physics

GILROY
Davis, Elmo Warren, genetics, horticulture
Silberstein, Otmar Otto, food science
White, Thomas Gailand, plant breeding

GLENDALE
Burgess, Richard Ernest, biochemistry, biomedical engineering
Carruth, Willis Lee, physical chemistry
Hainski, Martha Barrionuevo, biological chemistry
Macpherson, Walter Everett, internal medicine
Newburn, Ray Leon, Jr., astronomy
Nilsson, William A., occupational health
Sproull, Wayne Treber, air pollution
van de Kamp, Peter Cornelis, geology, geochemistry

GLENDORA
McNall, Earl George, biochemistry
Zernow, Louis, physics

GOLETA
Bode, Donald Edward, solid state physics
Bratt, Peter Raymond, solid state physics
Church, Stanley Eugene, geochemistry
Davis, Thomas Pearse, radiation physics, optical physics
Du Pre, Frits Karel, physics
Eck, Robert Edwin, physics
Elliott, Robert Dunshee, meteorology
Errett, Daryl Dale, optics
Frank, Sidney Raymond, meteorology
Franks, Larry Allen, molecular spectroscopy
Kikuchi, Tom T., lasers, quantum electronics
Leipnik, Roy Bergh, mathematical analysis, mathematical physics
Levee, Richard Douglas, data processing
Machacek, Milos, atomic physics
Ovrebo, Paul Johannes, applied physics
Renda, Francis Joseph, solid state physics
Rhoads, William Anderson, radiation ecology
Steinberg, Martin, physical chemistry
Watkins, Robert Arnold, electrooptics, military systems
Wilbarger, Edward Stanley, Jr., physics

GONZALES
Peterson, Richard Grant, agricultural chemistry, enology

GRANADA HILLS
Ashburn, Edward Victor, physics
Gaines, John Franklin, biogeography, physical geography
Menke, Andrew Giedrius, organometallic chemistry

GRASS VALLEY
Pitman, Gary Boyd, entomology

GREENBRAE
Hall, Albert D., surgery
Lindquist, Frank Eugene, chemistry
Loosanoff, Victor L., fisheries

HACIENDA HEIGHTS
Nishibayashi, Masaru, physical chemistry

HARBOR CITY
Torrance, Daniel J., radiology

HAWTHORNE
Clark, William Dempsey, physical chemistry
Curtis, Orlie Lindsey, Jr., physics
Pretzer, Donavon Donald, atomic physics
Robin, James Edmond, physics, lasers
Spiwak, Lazarus, physical chemistry
Tippins, Harry H., Jr., solid state physics, optical physics

HAYWARD
Amemiya, Frances (Louise) Campbell, mathematics
Anderson, Barbara Gallatin, cultural anthropology
Baalman, Robert J., botany
Bates, Robert Ellery, geomorphology
Baum, Dennis Willard, gas dynamics
Benseler, Rolf Wilhelm, dendrology
Birge, Ann Chamberlain, radiation biophysics
Bozak, Richard Edward, organic chemistry
Brooks, Elwood Ralph, geology, petrology
Cadogan, Kevin Denis, physical chemistry
Chauffe, Leroy, organic chemistry
Cogswell, Howard Lyman, ornithology, ecology
Cooper, Richard Kent, electromagnetics, theoretical physics
Crews, Robert Wayne, physics
Cummings, Jon Clark, geology
De Vries, John Edward, chemistry
Eder, Herbert Michael, cultural geography, geography of Latin America
Elkin, Lynne Osman, plant physiology, science education
Fisher, Leon Harold, electron physics, academic administration
Foster, Michael Simmler, marine biology, phycology
Fuller, Milton E., physical chemistry
Giles, John Crutchlow, physics
Goetschel, Charles Thomas, organic chemistry, inorganic chemistry
Good, Robert Howard, high energy physics
Goodrick, Richard Edward, mathematics
Groody, Thomas Conrad, vertebrate zoology, science education
Gross, Phyllis P., biology, science education
Guthrie, Andrew, physics
Heath, Harrison Duane, developmental biology
Heuer, Ann Elizabeth, microbiology, immunology
Hirschfeld, Sue Ellen, paleontology, geology
Holtgrieve, Donald Gordon, geography
Johnson, George Robert, cytogenetics
Keller, Edward Lee, applied mathematics
Kennelly, Robert Andrew, geography
Kilbuck, John Henry, food technology
Lier, John, geography
Lippman, Gary Edwin, mathematical analysis
Luibrand, Richard Thomas, organic chemistry
Lutt, Carl J., anatomy, physiology
Lyke, Edward Bonsteel, cytology, invertebrate zoology
Main, Robert Andrew, zoology, limnology
Manjarrez, Victor M., mathematics
McGinnis, Samuel M., herpetology
McKnight, Robert Kellogg, applied anthropology
Merris, Russell Lloyd, algebra
Mintz, Leigh Wayne, paleontology, stratigraphy
Moiseyev, Alexis N., geochemistry, geology
Monson, Richard Stanley, chemistry
Moser, Louise Elizabeth, mathematics
Nimmo, Harry Arlo, cultural anthropology
Park, Heebok, statistics
Parnell, Dennis Richard, plant science
Perrino, Charles T., solid state chemistry
Peterson, Donald Lee, physical chemistry
Peterson, George Harold, microbiology
Pickett, Patricia Booth, cell biology
Purvis, Colbert Thaxton, mathematics
Rebman, Kenneth Ralph, mathematics
Resnikoff, George Joseph, statistics
Rofen, Robert Rees, aquatic biology, ichthyology
Sabharwal, Ranjit Singh, pure mathematics
Schoenholz, Walter Kurt, immunology, serology
Schusterman, Ronald Jay, ethology
Scudder, Harvey Israel, public health, biology
Shaudys, Vincent Kirkbride, geography
Shelton, John C., organic chemistry
Shiffman, Max, mathematics
Simon, Arthur Bernard, aerodynamics
Smith, Marianne (Ruth) Freundlich, mathematics
Smith, Robin Peggy Piety, biochemistry
Southard, Thomas Hollister, mathematics
Stauffer, Howard Boyer, applied physics, particle physics
Thoman, Richard Samuel, economic geography, urban geography
Thomas, William LeRoy, cultural geography, geography of Southeast Asia
Trumbo, Bruce Edward, statistics
Tullis, Richard Eugene, comparative physiology, comparative endocrinology
Vann, John Herman, geography
Warnke, Detlef Andreas, earth sciences
Weidlich, John Edward, Jr., mathematics
Whitehead, Marian Nedra, nuclear physics
Whitney, Robert C., science education, physics
Winzenread, Marvin Russell, mathematics
Wrona, Wlodzimierz Stefan, geometry
Zhivadinovich, Milka Radoicich, inorganic chemistry

HEMET
Fett, John D., geology, geophysics
Thomas, Robert Eugene, psychiatry

HIGHLAND
Gericke, Otto Luke, psychiatry

HOLLISTER
Ogimachi, Naomi Neil, fluorine chemistry

HOLLYWOOD
Anderson, David W., Jr., food chemistry
Baptist, Victor Harry, physical biochemistry
Conner, Robert Thomas, biochemistry
Kaufmann, Peter John, cosmetic chemistry

HOPLAND
Jones, Milton Bennion, soil fertility
Longhurst, William Murray, wildlife management
Torell, Donald Theodore, animal science

HOLTVILLE
Papazian, Harold Aram, physical chemistry

HUNTINGTON BEACH
Dalrymple, Stephen Harris, computer sciences
Feldman, Fredric J., analytical chemistry, inorganic chemistry
Ginell, William Seaman, physical chemistry
Grossman, Jack Joseph, physical chemistry
Holl, Richard Jacob, nuclear science
Houghton, Karl H., aerospace medicine
Janos, William Augustus, physics
Kelly, Herbert Barrett, Jr., environmental physiology, neurophysiology
Knapp, David Edwin, nuclear physics
McFee, Raymond Herbert, physics
McGuckin, Warren Francis, clinical chemistry, instrumentation
Meier, Rudolf H., optical physics
Olson, Willard Paul, environmental physics
Richman, Isaac, electrooptics
Riha, William E., Jr., food science
Rogers, Bruce Joseph, plant physiology, science writing
Schwartz, Sanford Bernard, computer sciences
Silverstein, Elliot Morton, electrooptics
Vander Weyden, Allen Joseph, chemistry
Webb, Stephen Richard, engineering statistics
Yeiser, Andrew Sturm, computer science

IDYLLWILD
James, Philip Nickerson, organic chemistry

INDIO
Giannini, Gabriel Maria, physics
Nixon, Roy Wesley, horticulture

INDUSTRY
Judson, Charles Morrill, physical chemistry
Week, Friedrich Josef, organic chemistry, physical chemistry

INVERNESS
Wistar, Richard, chemistry

INYOKERN
Austin, Carl Fulton, geology

INGLEWOOD
Allen, Raymond A., pathology
Greene, Robert Alva, medical microbiology
Payne, Robert T., III, solid state physics
Reilly, Emmett B., pathology, hematology

IRVINE
Akasaki, Takeo, algebraic topology
Aldrich, Daniel Gaskill, Jr., soil fertility
Allen, Lois Brenda, virology
Arditti, Joseph, plant physiology, chemotherapy
Arffin, Stuart Michael, biochemistry
Arquilla, Edward R., pathology, immunology
Ball, Ernest, botany
Bander, Myron, theoretical physics
Bardo, Richard Dale, quantum chemistry
Berk, Jack Edward, medicine
Berns, Michael W., developmental biology, cell biology
Bork, Alfred Morton, physics, history of physics
Bostick, Warren Lithgow, pathology
Boughey, Arthur Stanley, ecology
Boyd, John Paul, anthropology, communications
Brown, George William, mathematics
Bryant, Susan Victoria, developmental biology, cell biology
Buerger, Alfred Arthur, physiology, psychology
Cannonito, Frank Benjamin, algebra, mathematical logic
Carpenter, Frances Lynn, ecology, evolution
Carter, Paul Richard, medicine, surgery
Caserio, Marjorie C., organic chemistry
Clark, Jeffrey Lee, biochemistry
Colby, Benjamin N., anthropology
Connolly, John E., surgery
Crocker, Thomas Timothy, infectious diseases, virology
Cunningham, Dennis Dean, cell biology, biochemistry
Danner, Jean, biochemistry
Darling, Donald Allan, mathematical statistics
Davis, Earle Andrew, Jr., neurophysiology
Davis, Rowland Hallowell, genetics
Dearden, Lyle Conway, anatomy
Demetrescu, M., neurophysiology, biomedical engineering

Dixon, Peter Stanley, phycology
Doedens, Robert John, physical chemistry, inorganic chemistry
Donoghue, William F, Jr, mathematics
Doyle, Walter M, atomic physics
Earle, Robert Wallace, pharmacology
Eklof, Paul Christian, mathematics
Englert, Robert D, environmental science, research administration
Eriksen, Stuart P, pharmaceutical chemistry
Fain, Margery Jones, insect physiology, developmental biology
Falk, Darrel Ross, developmental genetics
Feiock, Frank Donald, physics
Finkelstein, Mark, mathematics
Fleischer, Everly B, bioinorganic chemistry
Foltz, Eldon Leroy, neurosurgery
Fosket, Donald Elston, plant physiology, cell biology
Frank, Joan Patricia, radiochemistry
Freeman, Fillmore, physical organic chemistry
Furnas, David William, plastic surgery, reconstructive surgery
Gamo, Hideya, physics
Gentile, Dominick E, nephrology
Giolli, Roland A, neuroanatomy, histochemistry
Gittins, Barbara Tyrrell, algology
Gordon, Charles N, cell biology
Gottschalk, Louis August, psychiatry, psychoanalysis
Granger, Gale A, microbiology
Grant, Edward Robert, chemical kinetics
Greenhouse, Gerald Alan, developmental biology, environmental biology
Guinn, Vincent Perry, radiochemistry
Gwinup, Grant, internal medicine
Hall, Peter Francis, physiology, endocrinology
Hamkalo, Barbara Ann, molecular biology, biochemistry
Hanson, Barbara Ann, microbial biochemistry, microbial physiology
Hart, Michael Martin, pharmacology, medicine
Hatfield, G Wesley, biochemistry, microbiology
Healey, Patrick Leonard, developmental biology
Hill, Mason Lowell, geology
Hillyard, Ira William, pharmacology
Hunt, George Lester, Jr, ecology
Ibsen, Kenneth Howard, biochemistry
Josephson, Robert Karl, comparative physiology
Juberg, Richard Kent, mathematics
Julien, Robert Michael, pharmacology
Justice, Keith Evans, ecology, evolutionary biology
Kalisch, Gerhard Karl, mathematics
Kaplan, Milton Temkin, medical microbiology
Kavan, Eva Mary, anesthesiology
Kaye, Wilbur (Irving), chemistry, instrumentation
Keilin, Bertram, water chemistry
Kingsbury, David Thomas, virology, microbiology
Klein, Abel, mathematical physics
Kohut, Robert Irwin, otolaryngology
Koopowitz, Harold, neurophysiology, invertebrate zoology
Krassner, Stuart M, parasitology
Kropp, William Rudolph, Jr, cosmic ray physics, elementary particle physics
Kunze, Ray A, mathematics
Kurnick, Nathaniel Bertrand, biochemistry, medicine
Leak, John Clay, Jr, organic chemistry
Lee, Edward Kyung Chai, physical chemistry
Lenhoff, Howard Maer, biochemistry, invertebrate zoology
Leopold, Irving Henry, ophthalmology
Littler, Mark Masterton, ecology, phycology
Ludwig, Frederic C, experimental pathology
MacMillen, Richard Edward, physiological ecology, vertebrate biology
Maloy, John Owen, high energy physics, space physics
Mandelkern, Mark Alan, elementary particle physics, medical physics
Manning, Jerry Edsel, biochemistry
Maraudin, Alexei Alexei, theoretical solid state physics
Mayer, Meinhard Edwin, mathematical physics
McClure, David Albert, pharmacology
McClure, James Herbert, obstetrics & gynecology

McGaugh, James L, neurobiology, psychopharmacology
McLaughlin, Calvin Sturgis, biochemistry, genetics
Milburn, Burton Burnett, industrial hygiene
Miller, George E, physical chemistry, radiochemistry
Miller, Jon Philip, biochemistry, molecular pharmacology
Mills, Douglas Leon, solid state physics
Milne, Eric Campbell, radiology
Moe, Michael K, elementary particle physics
Moe, Mildred Minasian, atmospheric physics
Moldave, Kivie, biochemistry
Mood, Alexander McFarlane, applied mathematics, public policy
Moore, Harold W, organic chemistry
Moyed, Harris S, bacteriology, biochemistry
Nelson, Eric Loren, molecular pharmacology
Nelson, Thomas Lothian, pediatrics, allergy
Newcomb, Robert Lewis, applied statistics
O'Brien, Darrell Eugene, chemistry
Overman, Larry Eugene, bio-organic chemistry
Paciorek, Kazimiera J L, polymer chemistry, fluorine chemistry
Pardy, Rosevelt Lawrence, invertebrate physiology
Parker, William Henry, low temperature physics, solid state physics
Pirkle, Hubert Chaille, pathology
Piszkiewicz, Dennis, biochemistry
Porter, Robert Willis, neurosurgery, neurophysiology
Quint, Joseph Freeman, clinical chemistry
Ramirez-Munoz, Juan, analytical chemistry, photometry
Reines, Frederick, elementary particle physics, cosmic ray physics
Robertson, Scott Harrison, plasma physics
Robins, Roland Kenith, organic chemistry
Rommey, Antone Kimball, anthropology
Rostoker, Norman, physics
Rundel, Philip Wilson, plant ecology, lichenology
Rynn, Nathan, plasma physics
Schneidermann, Howard Allen, developmental biology, developmental genetics
Schultz, Jonas, physics
Shaw, Gordon Lionel, theoretical physics
Sidwell, Robert William, research administration, chemotherapy
Silverman, Dennis Joseph, elementary particle physics
Sloane, Howard J, spectrochemistry
Smoke, William Henry, mathematics
Sobel, Henry Wayne, particle physics, cosmic ray physics
Sondhaus, Charles Anderson, biophysics, radiobiology
Stanley, Wendell Meredith, Jr, molecular biology, biochemistry
Starkey, Otis Paul, economic geography, geography of Anglo-America
Stephens, Grover Cleveland, zoology
Stone, Orville Joseph, dermatology
Stout, Mason Gardner, organic chemistry
Sutherland, John Clark, biophysics
Suzuki, Noboru, mathematics
Swedes, Jean Susanne, molecular biology
Sypherd, Paul Starr, microbiology
Taagepera, Mare, physical organic chemistry
Taft, Robert Wheaton, Jr, physical chemistry, organic chemistry
Thompson, Richard Frederick, physiological psychology
Thompson, William Benbow, Jr, obstetrics & gynecology
Thorp, Edward O, mathematics
Tobis, Jerome Sanford, medicine
Tomizawa, Henry Hideo, biochemistry
Trimble, Virginia Louise, astronomy, astrophysics
Tucker, Howard Gregory, mathematics
Udall, John Alfred, internal medicine, cardiology
Ulrich, William Frederick, inorganic chemistry, analytical chemistry
Ushioda, Sukekatsu, physics
Valenta, Lubomir Jan-Vaclav, medicine, endocrinology
Van Den Noort, Stanley, neurology
Van Hoven, Gerard, plasma physics, solar physics
Vermund, Halvor, radiology
Verzeano, Marcel, medicine

Wagner, Edward Knapp, animal virology, biochemistry
Wallis, Richard Fisher, solid state physics
Walwick, Earle Richard, clinical chemistry
Warner, Robert Collett, biochemistry
Werner, Christian, geography
White, Stephen Halley, biophysics, physiology
Williams, Robert Edward, applied chemistry
Witkowski, Joseph Theodore, organic chemistry, medicinal chemistry
Wolsberg, Max, physical chemistry
Woolfolk, Clifford Allen, microbiology
Yeh, James Jui-Tin, mathematics

KENSINGTON
Bacskai, Robert, organic chemistry, polymer chemistry
Flath, Robert Arthur, natural products chemistry
Frye, Fredric Lee, veterinary surgery, herpetology
Milmore, Benno Karl, epidemiology
Ralls, Jack Warner, chemistry
Sylvester, Edward Sanford, entomology
Teranishi, Roy, organic chemistry
Tolberg, Adelaide Brokaw, physiology, microbiology

KERNVILLE
Ellis, Emory Leon, chemistry

LA CANADA
Bergquist, James William, computer sciences, applied mathematics
Craven, Charles Waller, chemistry, biochemistry
Eikrem, Lynwood Olaf, spectrochemistry, instrumentation
Hallanger, Norman Lawrence, meteorology
Moran, William Rodes, geology
Riggs, William McKnight, physical chemistry, analytical chemistry
Sherry, Edwin J, mathematics

LA CRESCENTA
Bean, Robert Taylor, hydrogeology, engineering geology
Snyder, Conway Wilson, physics
Ward, Charles Bradley, Jr, industrial microbiology

LA HABRA
Baker, Paul Eugene, physics
Bass, Manuel N, geology, geochemistry
Bonham, Lawrence Cook, geology
Burtner, Roger Lee, geology
Clementz, Carol A, geology
Clementz, David Michael, clay mineralogy, soil science
Coppel, Claude Peter, physical chemistry
Davis, Bruce W, physical chemistry
Drugg, Warren Sowle, botany, palynology
Gerbacia, William Edward, physical chemistry
Hess, Patrick Henry, organic chemistry
Hill, Donald Gardner, geology, geophysics
Jennings, Harley Young, Jr, chemistry
Jizba, Zdenek Vaclav, exploration geology
Johnson, Carl Emil, Jr, physical chemistry
Jones, Stanley Bennett, geology
Mann, John Francis, Jr, geology
McAuliffe, Clayton Doyle, soil science
Moore, George Thomas, marine geology, stratigraphy
Plumley, William Justin, geology
Redfield, William David, microbiology
Reed, Marion Guy, soil chemistry
Riley, N Allen, earth sciences, research administration
Runge, Richard John, geophysics
Sabins, Floyd F, geology
Schremp, Frederic William, physical chemistry
Seevers, Delmar Oswell, physics
Silverman, Sol Robert, organic geochemistry
Slentz, Loren William, geochemistry
Smalley, Robert Gordon, geology, geochemistry
Stephenson, Lee Palmer, geophysics
Waples, Douglas Wendle, organic geochemistry
Yungul, Sulhi Hasan, exploration geophysics

LA JOLLA
Abelson, John Norman, molecular biology, molecular genetics
Agalides, Eugene, biophysics
Aginsky, Bernard Willard, anthropology, cultural anthropology
Aginsky, Ethel G, anthropology
Ahlstrom, Elbert Halvor, biology
Alden, Richard Allen, x-ray crystallography
Alexander, Edward Cleve, organic chemistry
Allison, William S, protein chemistry
Alvarino de Leira, Angeles, biological oceanography, marine biology
Anderson, Donald Werner, mathematics
Arnold, James Richard, chemistry
Arrhenius, Gustaf Olof Svante, oceanography
Arthur, Robert Siple, oceanography
Asmus, John Fredrich, lasers
Asunmaa, Saara K, physical chemistry
Axford, William Ian, astrophysics, space science
Azam, Farooq, microbial ecology
Backus, George Edward, geophysics
Bada, Jeffrey L, geochemistry, geochronology
Bailey, Frederick George, anthropology
Bainbridge, Arnold Ernest, geophysics, oceanography
Bancroft, Richard Wolcott, physiology
Banks, Peter Morgan, atmospheric physics, aeronomy
Barondes, Samuel Herbert, psychiatry, neurobiology
Barrett, Izadore, fisheries
Bassett, Allen Mordorf, geomology, economic geology
Bayliff, William Henry, fisheries
Beers, John R, biological oceanography
Bender, Edward Anton, mathematics
Benirschke, Kurt, pathology
Berger, Wolfgang Helmut, oceanography, micropaleontology
Beyster, John R, nuclear physics
Bickford, Reginald G, neurosciences
Bien, George Sung-Nien, physical chemistry
Bishop, Errett A, mathematics
Black, William Carter, Jr, low temperature physics
Blackburn, Maurice, zoology
Bloor, Colin Mercer, pathology, physiology
Bond, Clifford Walter, virology
Bond, Frederick Thomas, organic chemistry
Booker, Henry George, radio physics
Bowles, Kenneth Ludlam, physics
Boyd, William Clouser, immunobiology
Boynton, Robert M, psychology, optics
Bradbury, Jack W, animal behavior
Bradner, Hugh, experimental physics, geophysics
Brinton, Edward, biological oceanography
Brody, Stuart, biochemical genetics
Brown, Marvin Ross, medical physiology, internal medicine
Brueckner, Keith Allan, theoretical physics
Byrd, Earl William, Jr, developmental biology
Butler, Warren Lee, biophysics
Cannon, John Burns, organometallic chemistry, bioinorganic chemistry
Bukry, John David, micropaleontology
Bullard, Edward Crisp, geophysics
Bullock, Theodore Holmes, neurobiology
Bunch, James R, numerical analysis
Burnett, Bryan Reeder, invertebrate zoology
Carlucci, Angelo Francis, microbial ecology, marine microbiology
Carpenter, Adelaide Trowbridge Clark, genetics
Chen, Joseph Cheng Yih, theoretical physics, theoretical chemistry
Cheng, Lanna, marine zoology, entomology
Chisholm, Sallie Watson, aquatic ecology
Chivers, Hugh John, physics
Chow, Tsaihwa James, analytical chemistry, geochemistry
Chrispeels, Maarten Jan, plant physiology
Clark, Leigh Bruce, spectrochemistry
Clark, Nathan Edward, marine meteorology, fisheries
Clynes, Manfred, neuropsychology, sentics
Coleman, Philip Lynn, applied physics
Connor, James D, pediatrics, microbiology
Covell, James Wachob, cardiovascular physiology, biomedical engineering

CALIFORNIA

Cox, Charles Shipley, oceanography
Craig, Harmon, geochemistry, oceanography
Crawford, Irving Pope, microbial genetics
Croft, Paul Douglas, nuclear chemistry
Crosby, William Holmes, Jr, hematology
Curray, Joseph Ross, marine geology
Dahlberg, Richard Craig, physics
Dandliker, Walter Beach, immunochemistry
D'Andrade, Roy G, anthropology
David, Gary Samuel, immunochemistry
Davis, Russ E, physical oceanography
de Hoffmann, Frederic, physics
Dennis, Edward A, biochemistry, physical organic chemistry
Dixon, Frank James, immunology
Doolittle, Russell F, biochemistry
Dorman, Leroy Myron, geophysics
Drell, William, biochemistry
Duff, Russell Earl, fluid dynamics, chemical physics
Duntley, Seibert Quimby, physics
Dutton, Richard W, cell biology, immunology
Eber, Laurence Elwin, marine meteorology
Edgar, Norman Terence, marine geology, geophysics
Edgington, Thomas S, pathology, immunology
Ellis, Albert Tromly, applied mechanics
El Wardani, Sayed Aly, oceanography, geochemistry
Engel, Albert Edward John, geology, geochemistry
Enns, Theodore, physics, oceanography
Enright, James Thomas, zoology
Epel, David, developmental biology, cell biology
Eppley, Richard Wayne, biological oceanography
Esser, Alfred F, biological chemistry, immunochemistry
Evans, John W, mathematical biology
Fager, Edward William, ecology
Fahey, Robert C, organic chemistry, biochemistry
Faulkner, D John, organic chemistry
Feher, George, physics, biophysics
Fejer, Jules A, atmospheric physics, plasma physics
Feldman, Joseph David, pathology
Ferguson, James Mecham, psychiatry, behavioral biology
Fisher, Robert Lloyd, marine geology
Fleminger, Abraham, invertebrate zoology
Folsom, Theodore Robert, physics, oceanography
Foster, Margaret C, biophysics
Foster, Theodore Dean, physics, oceanography
Fox, Denis Llewellyn, marine biochemistry, marine biology
Fox, William Walter, Jr, fisheries
Francis, Robert Colgate, biomathematics
Frankel, Theodore Thomas, mathematics
Frautschy, Jeffery Dean, geology
Frazer, William Robert, theoretical physics
Fredkin, Donald Roy, physics
Freer, Stephan T, biochemistry, physical chemistry
Fricke, Martin Paul, nuclear physics
Friedkin, Morris Enton, biochemistry
Friedman, Hannah, cell biology
Friedman, Paul J, radiology
Fronek, Arnost, physiology
Fronek, Kitty, human physiology
Galambos, Robert, physiology, psychology
Gantzel, Peter Kellogg, physical chemistry
Garsia, Adriano Mario, mathematics
Gealy, Elizabeth Lee, geology
Geiduschek, Ernest Peter, molecular biology, virology
Getoor, Ronald Kay, mathematics
Geyer, Mark Allen, neurosciences, psychopharmacology
Gibson, Carl H, fluid dynamics, oceanography
Gilbert, James Freeman, geophysics, seismology
Gill, Gordon Nelson, endocrinology
Gluck, Louis, biochemistry, pediatrics
Goldberg, Edward D, marine geochemistry
Goodkind, John M, physics
Goodman, Murray, organic chemistry
Gordon, Alvin S, physical chemistry
Gould, Robert Joseph, theoretical physics, astrophysics
Gould-Somero, Meredith, developmental biology
Goulian, Mehran, hematology
Gragg, William Bryant, Jr, mathematics
Green, Melvin Howard, biochemistry, virology
Green, Ralph, hematology
Griffin, John Henry, biochemistry, experimental pathology
Grine, Donald Reaville, geophysics
Grismore, Roger, physics
Grobstein, Clifford, biology, academic administration
Groce, David Eiben, medical physics, nuclear physics
Gross, Stephen Richard, pharmacology, biochemistry
Habel, Karl, medicine
Halkin, Hubert, mathematics
Hallenbeck, George Aaron, surgery
Halpern, Francis Robert, theoretical physics
Hamburger, Robert Newfield, pediatrics
Hamlin, Daniel Allen, atmospheric physics
Hammel, Harold Theodore, physiology
Hanstein, Walter Georg, biochemistry, physical organic chemistry
Harper, Elvin, biochemistry
Hartline, Daniel Keffer, neurophysiology, biophysics
Hasterlik, Robert Joseph, medicine
Hastings, Albert Baird, biochemistry
Hatefi, Youssef, biochemistry
Haubrich, Richard August, geophysics
Hawkins, James Wilbur, Jr, geology
Haxo, Francis Theodore, plant physiology
Helinski, Donald Raymond, biochemistry, genetics
Helstrom, Carl Wilhelm, applied physics
Helton, John W, mathematics
Henson, Peter Michell, immunology, cell biology
Hessler, Robert Raymond, zoology
Hille, Einar, mathematical analysis
Hillyard, Steven Allen, neuropsychology
Hoch, James Alfred, genetics, biochemistry
Hoch, Sallie O'Neil, microbiology, biochemistry
Holland, John Joseph, microbiology
Holland, Nicholas Drew, invertebrate zoology
Holland, Robert Joseph, organic chemistry, biochemistry
Holliday, Audrey Rose, psychopharmacology, research administration
Holm-Hansen, Osmund, plant physiology
Holter, Norman Jefferis, physics
Hopkins, Gordon Bruce, pathology, nuclear medicine
Hotta, Yasuo, cell biology, genetics
Hougie, Cecil, pathology
Howell, Doris Ahlee, pediatrics, hematology
Howell, Francis V, oral pathology
Howell, Stephen Herbert, cell biology
Hu, Te Chiang, computer science, operations research
Hubbs, Carl Leavitt, ichthyology
Hudson, Cecil Ivan, Jr, physics
Huennekens, Frank Matthew, Jr, biochemistry
Hugli, Tony Edward, biochemistry, protein chemistry
Hulsemann, Jobst, marine geology, sedimentology
Hulsemann, Kuni, marine biology
Hunter, John Roe, marine biology, ichthyology
Hurley, Ann Catherine, marine ecology
Ingwall, Joanne S, molecular biology
Inman, Douglas Lamar, geology
Isaacs, John Dove, oceanography
Itano, Harvey Akio, pathology
Jacobsen, Donald Weldon, biochemistry, chemistry
Jones, Oliver William, medicine, biochemistry
Jordan, David K, cultural anthropology, social anthropology
Jordan, Thomas Hillman, geophysics
Jorna, Seibe, theoretical physics
Joseph, James, marine biology
Judd, Lewis Lund, psychiatry, child psychiatry
Kahler, Richard Lee, cardiovascular physiology
Kampa, Elizabeth Maitland, zoology
Kaplan, Nathan Oram, biochemistry
Katz, Ira, plasma physics
Kaye, Samuel, organic chemistry
Kearns, David R, physical chemistry
Keeling, Charles David, physical chemistry, marine geochemistry
Kenagy, George James, ecology, behavioral physiology
Klawe, Witold L, fish biology
Knox, Robert Arthur, physical oceanography
Kohn, Walter, physics
Kooyman, Gerald Lee, comparative physiology
Korevaar, Jacob, mathematics
Kratz, Howard Russel, experimental physics
Kraut, Joseph, physical biochemistry
Kroll, Norman Myles, mathematical physics
Kull, Lorenz Anthony, nuclear physics
Kyte, Jack Ernst, biochemistry
Laiken, Nora Dawn, medical physiology, medical education
Lal, Devendra, nuclear physics, geochemistry
Lang, Joseph Herman, biochemistry
Lasker, Reuben, comparative physiology
Laudenslager, Mark LeRoy, neuropsychology
Laurs, Robert Michael, oceanography, fisheries
Lee, Sun, surgery
Leffert, Hyam Lerner, cell biology
Lein, Allen, physiology
Lerner, Richard Alan, immunology
Levine, Howard Bernard, physical chemistry
Levy, Robert Isaac, anthropology, psychiatry
Lewin, Ralph Arnold, phycology
Lewis, Urban James, endocrinology
Liebermann, Leonard Norman, physics
Linck, Robert George, inorganic chemistry
Lindenberg, Katja Lakatos, physical chemistry, chemical physics
Lindsley, Dan Leslie, Jr, genetics
Livingston, Robert Burr, neurophysiology, neuroanatomy
Loeb, Felix Faust, Jr, psychiatry, psychoanalysis
Loeffler, Harold Julius, food science, science writing
Loomis, William Farnsworth, Jr, developmental biology
Lux, John Herbert, chemistry
Ma, Shang-Keng, physics
Macdougall, John Douglas, geochemistry, meteoritics
MacLeod, Donald Iain Archibald, vision
Magde, Douglas, chemical physics
Malmberg, John Holmes, plasma physics
Manaster, Alfred B, mathematical logic
Mandell, Arnold J, psychiatry, neurochemistry
Mandell, Paul Irving, economic geography, economic development
Marti, Kurt, cosmochemistry
Masek, George Edward, physics
Masouredis, Serafeim Panogiotis, hematology, immunohematology
Matsumura, Philip, microbial physiology
Matthews, David Allan, biophysical chemistry
Matthews, Jerry Lee, marine geology
Matthias, Bernd T, physics
Mayer, Joseph Edward, chemical physics
Mayer, Steven Edward, pharmacology, biochemistry
McBride, John Barton, physics
McElroy, William David, biology, biochemistry
McGowan, John Arthur, biological oceanography
McIlwain, Carl Edwin, physics
McMillan, Robert, immunohematology
Meeker, Michael Elliott, cultural anthropology
Meinke, Geraldine Chciuk, immunology
Meinke, William John, microbiology, virology
Menard, Henry William, Jr, geology
Miles, John Wilder, geophysics
Miller, Stanley Lloyd, chemistry
Miyai, Katsumi, pathology
Moore, Robert Yates, neurology, pediatric neurology
Mork, Byron O, medicine
Morrison, David Campbell, immunology
Moser, H Geoffrey, marine biology, ichthyology
Mudie, John David, geophysics
Muller-Eberhard, Hans Joachim, immunology, biochemistry
Muller-Eberhard, Ursula, hematology, biochemistry
Mullin, Michael Mahlon, biological oceanography, ecology
Munk, Walter Heinrich, physical oceanography
Murray, George Graham, Jr, mathematics
Mysels, Estella Katzenellenbogen, chemistry
Nachbar, William, applied mathematics
Nakamura, Robert Motoharu, pathology, immunology
Namias, Jerome, meteorology
Nealson, Kenneth Henry, marine microbiology
Nelles, Maurice, physical chemistry
Newman, William Anderson, marine biology
Nguyen-Huu, Xuong, crystallography, biophysics
Nierenberg, William Aaron, physics
Nyhan, William Leo, pediatrics, biochemistry
O'Brien, John S, pathology, medicine
Oesterreicher, Hans, solid state chemistry
Oldstone, Michael Beaureguard Alan, experimental biology, neurobiology
Olshen, Richard Allen, statistics, mathematics
Ondricek, Anatola, biology
Overmyer, Robert Franklin, solid state physics
Owen, Robert W, Jr, oceanography
Parker, Frances Lawrence, micropaleontology
Parks, Donald E, physics
Patrick, Lyle A, physics
Perrin, Charles Lee, organic chemistry
Perrin, William Fergus, mammalogy
Petersen, Carl Frank, geophysics
Peterson, Laurence E, x-ray astronomy
Phleger, Fred B, oceanography
Piccioni, Oreste, nuclear physics
Pollard, John K, Jr, plant physiology
Pomraning, Gerald C, reactor physics, radiation physics
Powers, Joseph Edward, fisheries management, operations research
Price, Paul Arms, biochemistry
Printz, Morton Philip, biochemistry
Raitt, Russell Watson, geophysics
Ranney, Helen M, internal medicine, hematology
Reid, Joseph Lee, oceanography
Reisfeld, Ralph Alfred, immunochemistry, biochemistry
Reynolds, Glenn Myron, nuclear physics
Riedel, William Rex, micropaleontology
Robb, James Arthur, pathology, virology
Roberts, David Hall, theoretical astrophysics
Robinson, Margaret King, oceanography
Rodin, Burton, mathematics
Rogovin, Daniel Noel, solid state physics, quantum optics
Rohrl, Helmut, mathematics
Romanucci-Ross, Lola, anthropology
Rosenblatt, Murray, applied mathematics, mathematical statistics
Rosenblatt, Richard Heinrich, zoology
Ross, Donald, acoustics
Ross, John, Jr, medicine, cardiology
Rotenberg, Manuel, atomic physics, biophysics
Roth, Walter, physical chemistry
Rothschild, Brian James, fish biology, aquatic ecology
Rumsey, Victor Henry, applied physics
Russell, Percy J, biochemistry
Sakagawa, Gary Toshio, fisheries
Saltman, Paul David, biochemistry
Sanchez, Robert A, bio-organic chemistry, organic chemistry
Sato, Gordon Hisashi, biology
Saur, Jesse Francis Theodore, physical oceanography
Savitch, Walter John, computer science, mathematics
Scheffler, Immo Erich, molecular biology
Schmitt Walter R, oceanography
Schneiderman, Lawrence J, internal medicine
Scholander, Per Fredrik, physiology
Schrauzer, Gerhard N, inorganic chemistry
Schultz, Sheldon, solid state physics
Seay, Glenn Emmett, physics
Seegmiller, Jarvis Edwin, biochemistry
Segal, David S, neuropharmacology, neuropsychology
Sell, Stewart, pathology, immunology
Selverston, Allen Israel, neurophysiology, comparative physiology
Seymour, Richard Jones, oceanography
Sham, Lu Jeu, theoretical physics, solid state physics
Shantz, Edgar Moore, biochemistry, plant physiology
Sharp, Gary Duane, population genetics, physiological ecology
Sharpe, Michael John, mathematics
Shelton, John Sewall, geology
Shimkin, Michael Boris, cancer
Shneour, Elie Alexis, biochemistry, neurosciences
Shor, George G, Jr, marine geophysics
Shuler, Kurt Egon, theoretical chemistry
Simon, Harold J, medicine
Singer, Seymour Jonathan, biology
Sinha, Yagya Nand, endocrinology
Siu, Chi-Hung, developmental biology

Smith, David Allan, genetics
Smith, Donald Ray, mathematics
Smith, Douglas Wemp, molecular biology, biophysics
Smith, Paul Edward, ecology, fish biology
Smith, Raymond Calvin, physical oceanography
Somero, George Nicholls, biochemistry, physiology
Soule, Michael E, population biology
Spangler, Charles Bishop, applied mathematics, physics
Spiess, Fred Noel, physics
Spiro, Melford Elliot, cultural anthropology
Spizizen, John, microbiology, biochemistry
Spooner, Charles Edward, Jr, neuropharmacology, neurophysiology
Stauffer, Gary Dean, fisheries
Stein, Wayne Alfred, astrophysics
Steinberg, Daniel, biochemistry
Stern, Herbert, cell biology
Stevenson, Merritt Raymond, physical oceanography
Stevenson, Robert Everett, oceanography
Stidd, Charles Ketchum, meteorology
Stokes, Joseph, III, internal medicine
Stoner, Gary David, microbiology
Storms, Lowell H, neuropsychology
Stoughton, Richard Baker, dermatology
Stuart, George Wallace, theoretical physics
Stutz, Frederick Paul, urban geography, economic geography
Suess, Hans Eduard, chemistry
Suhl, Harry, physics
Swanson, Robert Allan, elementary particle physics
Swartz, Marc Jerome, anthropology
Sweetman, Lawrence, biochemistry, pediatrics
Tamor, Stephen, theoretical physics
Tan, Eng M, immunology
Tavassoli, Mehdi, hematology
Taylor, John Hall, vision
Taylor, Palmer William, pharmacology
Taylor, Susan Serota, biochemistry
Terras, Audrey Anne, number theory
Theiss, Jeffrey Charles, pharmacology
Thomas, William Hewitt, biological oceanography
Thompson, William Bell, plasma physics
Tisi, Gennaro Michael, pulmonary diseases, internal medicine
Traylor, Teddy G, organic chemistry
Tschirgi, Robert Donald, physiology
Tsuchiya, Mizuki, physical oceanography
Turner, John Dean, medicine
Tyler, John Edwards, physics
Urey, Harold Clayton, chemistry
Vacquier, Victor, geophysics
Vanderlaan, Willard Parker, endocrinology
Varon, Silvio Salomone, neurochemistry, neurobiology
Vaughan, John Heath, immunology, medicine
Vazquez, Jacinto Joseph, pathology
Volcani, Benjamin Elazari, microbiology, biochemistry
Vold, Barbara Schneider, biochemistry, microbiology
Vold, Regitze Rosenorn, nuclear magnetic resonance
Warschawski, Stefan Emanuel, mathematics
Wavrik, John J, geometry
Weare, John H, geometry
Weber, Thomas Byrnes, biochemistry
Weigle, William O, immunology, cell biology
Wert, Richard Thomas, physical oceanography
West, John B, physiology, medicine
Westall, Frederick Charles, biochemistry
Wheatley, John Charles, physics
Wheeler, Henry Orson, medicine
Wheeler, John C, theoretical chemistry, chemical physics
Whitaker, Thomas Wallace, genetics
Whitehill, Jules Leonard, surgery, medical education
Wilkie, Donald W, marine biology
Williams, Peter M chemical oceanography
Williamson, Stanley Gill, mathematics
Wills, Christopher J, genetics, biology
Wilson, Andrew Robert, plasma physics
Wilson, Doris Burda, anatomy, embryology
Wilson, Katherine Woods, air pollution
Wilson, Kent Raymond, chemical physics, environmental sciences
Wong, David Yue, physics
Woodson, Paul Bernard, neurobiology

Wright, John Marlin, chemical instrumentation, magnetic resonance
Wulbert, Daniel Ray, mathematics
Yen, Samuel Show-Chih, reproductive endocrinology
Yguerabide, Juan, biochemistry, biophysics
York, Charles James, virology, bacteriology
York, Herbert Frank, physics, science policy
Zetler, Bernard David, oceanography
Ziccardi, Robert John, biochemistry, immunology
Zimm, Bruno Hasbrouck, biophysical chemistry, polymer chemistry
Zobell, Claude E, microbiology
Zweifach, Benjamin William, physiology, bioengineering

LA MESA
Domingo, Wayne Elwin, crop breeding
Haupt, Curtis Raymond, physics
Rennilson, Justin J, photometry
Smyth, John Bridges, physics
Vomhof, Daniel William, forensic science, chemistry

LA MIRADA
Crawford, Robert Field, agronomy, soil science

LA VERNE
Boardman, William Walter, Jr, physical inorganic chemistry
Clague, William Donald, science education
Compton, Leslie Ellwyn, physical chemistry
Dolbear, Geoffrey Emerson, fuel science
Longanbach, James Robert, chemistry
Neher, Robert Trostle, plant taxonomy, plant science
Rhee, Jay Jea-yong, chemical physics, physical chemistry
Shutts, Clarence Francis, plant morphology, ecology
Stewart, Robert Daniel, physical chemistry, metallurgical chemistry
Tipton, Ann Baugh, physical chemistry, fuel science
Tyson, George Noblit, Jr, inorganic chemistry

LAFAYETTE
Langlois, Gordon Ellerby, physical chemistry
Marquis, David Maley, industrial organic chemistry
Newey, Herbert Alfred, organic polymer chemistry
Sampson, William Wilson, entomology
Thomas, Walter Dill, Jr, phytopathology

LAGUNA BEACH
Farrington, William Benford, structural geology
Goddard, Edwin Newell, structural geology
Loos, Hendricus G, environmental physics

LAGUNA HILLS
Ast, David Bernard, dentistry, public health administration
Brown, Morden Grant, biophysics
Brown, Ronald Frederick, organic chemistry
DuBridge, Lee Alvin, experimental physics
McCarty, Harold Hull, economic geography, geography of the United States
Papp, Cornelius Alfred, high temperature chemistry
Young, William Gould, chemistry

LAGUNA NIGUEL
Morgan, Charles O, hydrology

LAKE SAN MARCOS
Grossman, William Lewis, geology

LAKEWOOD
Assony, Steven James, organic chemistry

LAWNDALE
Thanos, Andrew, microbiology, plant pathology
Vajda, Geza Laszlo, physics, engineerings

LEMOORE
Hygh, Earl Hampton, water chemistry, pollution chemistry

LINDSAY
Webster, John Robert, food science

LIVERMORE
Abey, Albert Edward, solid mechanics
Alonso, Carol Travis, nuclear physics
Alvarez, Raymond Angelo, Jr, physics
Anderson, John D, nuclear physics
Anderson, Roger E, physical chemistry, computer sciences
Anspaugh, Lynn Richard, environmental sciences
Baldwin, David Ellis, plasma physics
Bandtel, Kenneth Charles, experimental nuclear physics, research administration
Barr, William Lee, spectroscopy
Barton, George Wendell, Jr, analytical chemistry, radiochemistry
Batzel, Roger Elwood, nuclear chemistry
Bauer, Rudolf Wilhelm, experimental physics
Bauer, Walter, solid state physics
Bender, Charles F, theoretical atomic physics, molecular physics
Berger, Beverly Jane, population biology
Berman, Barry L, nuclear physics
Bing, George Franklin, theoretical physics
Bloom, Stewart Dave, nuclear physics
Bogart, Elliot, biophysics
Bonner, Norman Andrew, chemistry
Borg, Iris Y P, earth sciences, mineralogy
Borg, Richard John, physical chemistry
Boster, Thomas Arthur, solid state physics
Boyle, Walter Gordon, Jr, analytical chemistry
Branscomb, Elbert Warren, molecular biology
Braun, Robert Leore, crystallography, physical chemistry
Briggs, Richard Julian, plasma physics
Browne, John C, experimental nuclear physics
Brownlee, James Lawton, Jr, nuclear chemistry, radiochemistry
Burginyon, Gary Alfred, nuclear physics
Bystroff, Roman Ivan, analytical chemistry
Camp, David Conrad, experimental nuclear physics
Campbell, John Hyde, physical chemistry
Canfield, Eugene H, nuclear physics
Carothers, James (Edward), physics
Carrano, Anthony Vito, cytogenetics, biophysics
Clark, Arnold Franklin, energy conversion, solar physics
Clarkson, Jack E, analytical chemistry, nuclear chemistry
Coensgen, Frederic Harley, physics
Colmenares, Carlos Adolfo, physical chemistry, chemical engineering
Condit, Ralph Howell, solid state chemistry
Cook, Thomas Bratton, Jr, physics
Corman, Emmett Gary, theoretical physics
Costantino, Marc Shaw, high pressure physics, nuclear physics
Coyne, Patrick Ivan, physiological ecology
Cullen, Dermott Edward, nuclear physics, computer science
Culler, Vaughn Edgar, physics
Cunningham, Mary Elizabeth, fluid dynamics
Damm, Charles Conrad, plasma physics
Davis, Jay C, nuclear physics
Dean, Phillip Nolan, physics, biophysics
DeBar, Roger Bryant, theoretical physics
De Witt, Hugh Edgar, theoretical physics
Dickerson, George Fielden, applied physics
Dickerson, Marvin Hubert, meteorology
Dickinson, William Clarence, energy conversion
Dietrich, Frank S, nuclear physics
Donaldson, Robert Evans, plasma physics, atmospheric physics
Dorn, David W, theoretical nuclear physics, research administration

Dorough, Gus Downs, Jr, physical chemistry
Dunlop, William Henry, nuclear physics
Ebert, Paul J, physics
Eby, Frank Shilling, nuclear physics
Eccles, Samuel Franklin, nuclear physics
Edwards, Arthur L, chemical engineering, applied mathematics
Elson, Robert Emanuel, inorganic chemistry
Eltgroth, Peter George, astrophysics
Emerson, Donald Orville, geology, petrology
Estabrook, Kent Gordon, plasma physics
Estill, Wesley Boyd, analytical chemistry
Feit, Michael Dennis, optics, magnetohydrodynamics
Ferguson, James Malcolm, nuclear physics
Fernbach, Sidney, theoretical physics, computer science
Fleck, Joseph Amadeus, Jr, lasers
Fleming, Edward Homer, Jr, nuclear chemistry, science administration
Fletcher, John George, computer science
Foote, James Herbert, plasma physics
Fowler, Thomas Kenneth, theoretical physics
Frank, Wilson James, chemistry
Frazer, Jack Winfield, chemistry
Futch, Archer Hamner, physics
Gardner, Donald Glenn, nuclear chemistry
Garrison, John Carson, statistical mechanics, quantum optics
Gasten, Burt R, experimental physics
Gathers, George Roger, experimental solid state physics
Gatrousis, Christopher, nuclear chemistry, radiochemistry
Germain, Lawrence Seymour, physics
Giles, Peter Cobb, physics
Gledhill, Barton L, medical physics, biophysics
Goishi, Wataru, radiochemistry
Gold, Sydell Perlmutter, geometry, system analysis
Goodman, Ronald Keith, plasma physics
Gordon, David Buddy, physiology
Grabske, Robert Jerold, biochemistry
Gray, Joe William, biophysics
Groseclose, Byron Clark, nuclear physics
Gunn, Stuart Richard, physical chemistry
Gunnink, Raymond, physical chemistry
Gustavson, Marvin Ronald, physical chemistry, research administration
Hadley, James Warren, plasma physics
Hall, Laurence Stanford, plasma physics
Hampel, Viktor Erwin, physics, computer science
Hamstad, Marvin Arnold, acoustics, solid mechanics
Hannon, Willard James, Jr, geophysics
Harrar, JacksonElwood, analytical chemistry
Harrison, Florence Louise, comparative physiology, marine biology
Harrison, Melvin Arnold, physics
Hicks, Harry Gross, radiochemistry
Higgins, Gary Hoyt, earth science
Hill, Richard William, nuclear physics, genetics
Haugen, Gilbert R, physical chemistry
Heard, Hugh Corey, rock mechanics, high pressure physics
Hearst, Joseph R, physics
Henry, Eugene Michael, atomic physics, lasers
Herman, Paul Theodore, physics
Heusinkveld, Myron Ellis, physics
Hiskes, John Robert, theoretical physics
Hogan, William John, applied physics
Holzer, Alfred, nuclear physics
Hoover, William Graham, chemical physics
Hornig, Howard Chester, physical chemistry, inorganic chemistry
Howard, John Hall, geology
Hrubesh, Lawrence Wayne, chemical instrumentation
Huddleston, Robert E, numerical analysis
Hudgins, Arthur Judson, applied physics
Hulet, Ervin Kenneth, nuclear chemistry, experimental nuclear physics
Hunt, Angus Lamar, physics
Ide, Roger Henry, nuclear chemistry, research administration
Jefferson, Thomas Hutton, Jr, numerical analysis, computer science
Jeffries, Thomas William, microbiology
Jensen, Ronald Harry, biophysical chemistry
Johnson, Montgomery Hunt, physics
Johnson, Quintin C, crystallography
Kahn, James Steven, earth sciences
Kaiser, Thomas Burton, plasma physics

CALIFORNIA

Kamegai, Minao, hydrodynamics
Karo, Arnold Mitchell, chemical physics, solid state physics
Kavanagh, Thomas Murray, nuclear physics, atomic physics
Khan, Jhan M, atomic physics, surface physics
Kidder, Ray Edward, applied mathematics, physics
Killeen, John, plasma physics, magnetohydrodynamics
Knapp, Myron William, experimental nuclear physics
Knox, Joseph Blatt, meteorology, applied physics
Koehler, Helmut A, nuclear physics, solid state physics
Kolar, Oscar Clinton, physics
Kovar, Frederick Richard, nuclear physics, plasma physics
Krikorian, Oscar Harold, high temperature chemistry
Kruer, William Leo, plasma physics
Krupke, William F, solid state physics
Kury, John William, physical chemistry, inorganic chemistry
Lai, David Ying Fat, physical chemistry, solid state kinetics
Landauer, Joseph K, physics
Langdon, Allan Bruce, plasma physics
Lanier, Robert George, nuclear physics
Lauer, Eugene John, plasma physics
Leider, Herman R, solid state chemistry
Lessler, Richard Marshall, nuclear chemistry
Levie, Harold Walter, surface physics
Levy, Harris Benjamin, nuclear chemistry
Lewis, Francis Hotchkiss, Jr, physics, mechanical engineering
Lindemuth, Irvin Raymond, Jr, plasma physics
Lindner, Manfred, nuclear chemistry
Loewe, William Edward, applied physics
Lorensen, Lyman Edward, organic polymer chemistry
Luce, John Sidney, plasma physics
Lutz, Harry Frank, nuclear physics
MacCracken, Michael Calvin, atmospheric physics, air pollution
MacGregor, Malcolm Herbert, physics
Maenchen, George, physics
Maninger, Ralph Carroll, physics
Mann, Lloyd Godfrey, experimental nuclear physics
Marino, Lawrence Louis, hydrodynamics, nuclear physics
Max, Claire Ellen, plasma physics, astrophysics
Maxon, Marshall Stephen, plasma physics
May, Michael Melville, physics
Mayall, Brian Holden, cell biology, biophysics
McClelland, Wilson Melville, Jr, nuclear physics
McKague, Herbert Lawrence, geology, geochemistry
McMaster, William H, hydrodynamics
Mead, Sylvester Warren, III, physics
Mendelsohn, Mortimer Lester, biophysics, cancer
Meyer, Richard Adlin, nuclear chemistry & physics
Milanovich, Fred Paul, biophysics
Miller, Donald Gabriel, physical chemistry
Miskel, John Albert, chemistry
Mode, Vincent Alan, inorganic chemistry, computer science
Moore, Dan Houston, II, biostatistics
Morton, John Robert, III, nuclear science
Murphey, Byron Freeze, physics
Nervik, Walter Edward, nuclear chemistry
Nethaway, David Robert, nuclear chemistry
Newcomb, William A, theoretical physics
Newkirk, Herbert William, inorganic chemistry, physical chemistry
Newton, John Chester, analytical chemistry
Nishimura, Keiichi, theoretical physics
Norem, Winfred Luther, palynology
Nuckolls, John Hopkins, applied physics
O'Dell, Jean Marland, nuclear science, operations research
Olness, Dolores Urquiza, solid state physics
Olness, Robert James, nuclear physics
Pan, Yu-Li, experimental high energy physics, lasers
Pearlstein, Leon Donald, theoretical physics
Peterson, Kendall Robert, atmospheric physics
Peterson, Otis G, solid state physics
Pipes, Gayle Woody bionucleonics, nuclear medicine
Potter, Gerald Lee, climatology
Pyper, James William, physical chemistry
Radosevich, Lee George, energy conversion, solid state physics
Ragaini, Richard Charles, environmental chemistry
Raley, John Howard, physical organic chemistry
Ralston, H Robert, physics
Ramsey, William James, inorganic chemistry
Ramspott, Lawrence Dewey, geology
Ree, Francis H, theoretical physics
Reynolds, Harry Lincoln, physics
Richardson, Jeffery Howard, physical chemistry
Rinde, James A, polymer chemistry
Ross, Marvin, high pressure physics
Rozsnyai, Balazs, physics, mathematics
Sanborn, Russell Hobart, physical chemistry
Schmieder, Robert W, atomic physics
Schock, Robert Norman, geophysics
Selig, Walter, analytical chemistry
Sewell, Duane Campbell, physics
Shearer, James Welles, plasma physics
Shore, Bernard, biochemistry, physiology
Shore, Bruce Walter, atomic physics
Shore, Virgie Guinn, biochemistry, physiology
Simonen, Thomas Charles, plasma physics
Smith, Charles Francis, Jr, radiochemistry, environmental chemistry
Smith, Gordon Stuart, x-ray crystallography
Sparks, Joseph Theodore, physics
Speck, David Ralph, physics
Spies, Robert Bernard, marine biology, marine ecology
Steinberg, Daniel J, thermodynamics, hydrodynamics
Stern, Richard Cecil, chemical physics
Stevens, John Charles, magnetohydrodynamics, x-ray astronomy
Stevenson, Peter Cooper, radiochemistry
Struble, Gordon Lee, nuclear chemistry
Stuart, Richard Norwood, physics
Sutcliffe, William George, physics
Tarter, Curtis Bruce, astrophysics, theoretical physics
Taylor, Charles Joel, physics
Taylor, Robert Thomas, biochemistry
Tewes, Howard Allan, nuclear chemistry
Tewhey, John David, petrology
Thompson, Lawrence Hadley, cell biology
Thomson, Jeffrey John, plasma physics
Timourian, Hector, developmental biology
Tompkins, Gary Alvin, soil science, environmental chemistry
Tracy, James Frueh, nuclear physics
Urtiew, Paul Andrew, thermodynamics, aeronautical engineering
Vajk, Joseph Peter, physics
Valeo, Ernest John, plasma physics
Van Dilla, Marvin Albert, physics
Violet, Charles Earl, physics
von Holdt, Richard Elton, mathematics
Walton, John Joseph, atmospheric physics
Ward, Raymond Leland, physical biochemistry
Weber, Marvin John, lasers, spectroscopy
Weed, Homer Clyde, physical chemistry
Weihe, Joseph William, mathematics
Werth, Glenn Conrad, physics
West, Harry Irwin, Jr, space physics, nuclear physics
Wild, John Frederick, nuclear chemistry
Wilson, James R, theoretical astrophysics
Wilson, William Dennis, solid state physics
Wood, Calvin Dale, nuclear physics, atomic physics
Woodward, Ervin Chapman, Jr, radiation physics
Worden, Earl Freemont, Jr, physical chemistry
Wyrobek, Andrew Julius, medical biophysics, animal genetics
Young, David A, physical chemistry

LODI
Cooke, Ron Charles, plant science

LOLETA
Niles, Doris Kildale, botany, geology

LOMA LINDA
Arendt, Kenneth Albert, physiology
Austin, George M, neurophysiology
Baldwin, Bernell Elwyn, neurophysiology
Beltz, Richard Edward, biochemistry
Bergman, Russel Theodore, medicine
Blankenship, James W, biochemistry
Boche, Robert DeVore, genetics, virology
Botimer, Laurence Wallace, organic chemistry
Brand, Leonard Roy, vertebrate zoology
Branson, Bruce William, surgery
Bull, Brian S, pathology
Bullas, Leonard Raymond, microbial genetics
Case, Norman Mondell, human anatomy
Chu, William Tongil, radiation physics
Clausen, Conrad Duane, invertebrate zoology
Couperus, Molleurus, dermatology
Crawford, Raymond Bertram, medicine
Dail, Clarence Wilding, physical medicine & rehabilitation
Dalgleish, Arthur E, anatomy
Dysinger, Paul William, tropical public health, public health
Ehlers, Hertha, pediatrics
Elick, John William, cultural anthropology
Evans, Harrison Silas, medicine, psychiatry
Fraser, Ian McLennan, pharmacology
Gonzalez, Ramon Rafael, Jr, physiology
Hadley, Gilbert Gordon, pathology
Hall, Raymond G, Jr, cell physiology
Hardinge, Mervyn Gilbert, pharmacology
Heifetz, Milton David, neurosurgery
Herber, Raymond, gastroenterology, biochemistry
Hinshaw, David B, surgery
Hirst, Albert Edmund, medicine
Ho, Yuk Lin, microbiology
Holmes, Ivan Gregory, analytical chemistry, history of science
Hooker, William Mead, microscopic anatomy
Hubbard, Richard W, laboratory medicine, biochemistry
Hunt, Guy Marion, Jr, neuroanatomy
Johns, Varner Jay, Jr, internal medicine
Jolley, Weldon Bosen, physiology
Judefind, Thomas Francis, pathology, medical microbiology
Judkins, Melvin P, cardiovascular radiology
Kelln, Elmer, oral pathology, cancer
Kissinger, David George, epidemiology
Koobs, Dick Herman, pathology, biological chemistry
Kuzma, Jan Waldemar, biostatistics
Leech, William Dale, chemistry
Leonora, John, endocrinology
Lonergan, Lester Harold, health education
Longo, Lawrence Daniel, physiology
Magie, Allan Rupert, public health
Mathisen, Maurice Earl, analytical chemistry
McMillan, Paul Junior, biochemistry, histochemistry
Mitchell, Robert Dalton, medicine
Mortensen, Raymond Archie, chemistry
Neilsen, Ivan Robert, physics
Neufeld, Berney Roy, molecular biology
Nickel, Vernon L, orthopedic surgery
Nutter, Robert Leland, biophysics, virology
Olsen, Clarence Wilmott, neurology
Peters, Marvin Arthur, biochemical pharmacology
Peterson, Donald I, neurology, pharmacology
Peterson, John Eric, internal medicine
Power, Gordon G, physiology, internal medicine
Quilligan, James Joseph, Jr, virology
Register, Ulma Doyle, biochemistry, nutrition
Ridley, Roger W, anesthesiology
Roberts, Walter Herbert B, anatomy
Roth, Ariel A, biological oceanography
Rowley, Rodney Ray, audiology, psychoacoustics
Ryckman, Raymond Edward, medical entomology, parasitology
Sanchez, Albert, nutrition, biochemistry
Schultz, Robert Lowell, anatomy
Shearer, Robert Vernon, ophthalmology
Shryock, Edwin Harold, microscopic anatomy
Slater, James Munro, radiotherapy
Slattery, Charles Wilbur, physical chemistry
Smith, Louis Livingston, surgery
Steinman, Ralph R, dentistry
Stilson, Walter Leslie, radiology
Stirling, James Heber, anthropology
Strother, Allen, pharmacology, biochemistry
Thompson, Ralph J, Jr, surgery
Tilton, Bernard Ellsworth, pharmacology
Vogel, Philip James, medicine
Wagner, Edward D, parasitology
Wareham, Ellsworth Edwin, thoracic surgery
Wat, Bo Ying, medicine
Widmer, Elmer Andreas, helminthology, medical parasitology
Wilbur, David Wesley, biophysics
Wilcox, Ronald Bruce, biochemistry, endocrinology
Winter, Charles Ernest, microbiology
Yahiku, Paul Y, statistics
Zimmerman, C Duane, computer science
Zolber, Kathleen Keen, nutrition
Zuccarelli, Anthony Joseph, molecular biology, molecular genetics

LONG BEACH
Albert, Eugene, mathematics
Anfinson, Olaf P, physics
Appleton, George Ludwig, solid state physics
Austin, Charles Ward, mathematics
Ayers, Raymond Dean, experimental solid state physics
Bascom, Willard, oceanography
Batsel, Henry Lewis, neurophysiology
Bauer, Roger Duane, biochemistry
Baumgartner, Werner Andreas, physical chemistry, biophysical chemistry
Becker, Edwin Norbert, physical chemistry
Beekman, Bruce Edward, comparative physiology
Biedebach, Mark Conrad, biophysics
Bodfish, Ralph E, medicine, nuclear physics
Bright, Donald Bolton, marine ecology
Brosbe, Edwin Allan, medical microbiology
Callison, George, evolutionary biology, vertebrate morphology
Carlberg, David Marvin, microbiology, virology
Carlisle, John Griffin, Jr, marine biology
Carpenter, Bruce H, plant physiology
Chow, Richard H, physics
Civen, Morton, biochemistry
Collins, Charles Thompson, zoology, ornithology
Conrey, Bert L, geology
Cox, Hiden Toy, plant morphology, plant anatomy
Dagradi, Angelo E, internal medicine, gastroenterology
Dennis, John Gordon, structural geology, tectonics
Dixon, Keith Alan, anthropology
Ehrreich, Albert LeRoy, geology
Ericksen, Sheldon Danielsen, geography
Erickson, John Otto, electron microscopy
Fairchild, Mahlon David, pharmacology, neurophysiology
Feldman, Daniel Jared, rehabilitation medicine
Fischer, Imre A, clinical biochemistry
Florsheim, Warner Hans, endocrinology
Fredrickson, John E, physics
Fung, Henry C, Jr, immunology
Gaspar, Max Raymond, surgery
Gast, Joseph Henry, clinical chemistry
George, Simon, atomic spectroscopy
Gittleman, Arthur P, mathematics
Goldish, Dorothy May (Bowman), organic chemistry
Gosselin, Edward Alberic, history of science, science education
Greiner, John William, applied mathematics
Hardy, Ross, mammalogy, ecology
Harman, Robert Charles, cultural anthropology
Harris, Edwin Randall, organic chemistry
Henderson, Robert Burr, organic chemistry
Ho, Ju-Shey, zoology, parasitology
Hochberg, Melvin, physical chemistry
Hrubant, Henry Everett, animal genetics
Hu, Chi-Yu, nuclear physics
Huckaby, David George, mammalogy
Huey, Arthur S, petroleum geology
Hunt, Richard Lee, inorganic chemistry
Huppert, Milton, medical mycology
Jenkins, Kenneth Dunning, developmental biology
Jensen, James Leslie, physical organic chemistry
Johnson, Kenneth LeRoy, physiology
Jones, Ira, zoology, parasitology
Karabenick, Edward, geography
Kierbow, Julie Van Note Parker, physical chemistry, radiochemistry
Kim, Juhee, microbiology
Kluss, Byron Curtis, cell biology
Kroman, Ronald Avron, genetics
Larr, Alfred Louis, speech pathology

Leamy, Larry Jackson, quantitative genetics
Legg, Kenneth Deardorff, analytical chemistry
Lerner, Lawrence S, solid state physics, history of science
Libby, Dorothy, anthropology
Lietze, Arthur, immunochemistry, allergy
Lincoln, Richard G, plant physiology
Loomis, Richard Biggar, acarology, herpetology
Lumsden, William Watt, Jr, geology, paleontology
Manheim, Jerome Henry, mathematics
Mardellis, Anthony, mathematics
Margulies, William George, mathematics
Maricich, Tom John, organic chemistry
Marsi, Kenneth Larue, physical organic chemistry
Martin, Roger Charles, geology
Maxwell, Kenneth Eugene, entomology
Mayfield, Darwin Lyell, organic chemistry
McCone, Robert Clyde, cultural anthropology, anthropological linguistics
McCorkle, Thomas, anthropology
McGaughey, Charles Gilbert, biochemistry, dental research
McLeod, Edward Blake, mathematics
Menees, James H, histology, embryology
Mills, Don Harper, forensic medicine
Mosher, Robert E, mathematics
Mosier, H David, Jr, pediatrics, endocrinology
Munsee, Jack Howard, physics
Naish, John Michael, physics
Osborne, Douglas, anthropology, archaeology
Perlgut, Louis E, biochemistry
Petty, Milton Andrew, Jr, microbiology, mycology
Pitesky, Isadore, biochemistry, immunology
Po, Henry Ng, physical chemistry
Raff, Allan Maurice, pharmacy
Raj, Harkisan D, microbial physiology, microscopic anatomy
Reish, Donald James, marine zoology
Roberts, Charles A, Jr, theoretical physics
Rodriquez, Mildred Shepherd, nutrition, biochemistry
Ross, Mathew, psychiatry
Rubin, Louis, chemistry
Russell, Ruth Lois, microbiology
Salem, Semaan Ibrahim, physics
Scalettar, Richard, theoretical physics
Scantling, Frederick Holland, cultural geography
Schatzlein, Frank Charles, invertebrate physiology
Schechter, Daniel, physics
Schindler, Guenter Martin, mathematics
Schultz, Cramer William, physics
Schwarz, Klaus, biochemistry, bioinorganic chemistry
Scott, Bruce L, theoretical nuclear physics
Sekhon, Sant Singh, zoology, cytology
Senozan, Nail Mehmet, physical chemistry
Shlemon, Roy J, geography, geomorphology
Shu, Ping, biochemistry
Simonsen, Donald Howard, biochemistry
Sleeper, Elbert Launee, entomology
Smith, Alton Hutchison, topology
Smoke, Mary E, mathematical statistics
Spallholz, Julian Ernest, biochemistry
Stein, James DeWitt, Jr, mathematics
Stein, Justin John, surgery, cancer
Steiner, Rodney, physical geography
Stemmer, Edward Alan, thoracic surgery
Stephens, Lee Bishop, Jr, embryology
Stern, John Hanus, physical chemistry
Stinson, Perri June, operations research, statistics
Stockton, William Denis, entomology
Stowell, Ellery Cory, (Jr), biochemistry
Sun, Sung Huang, medical mycology
Swatek, Frank Edward, microbiology, mycology
Swingle, Karl Frederick, radiobiology
Szasz, Stephen E, physical chemistry
Tharp, A G, inorganic chemistry
Trubatch, Sheldon L, theoretical physics, biophysics
Verdina, Joseph, mathematics
Warter, Janet Kirchner, palynology, paleobotany
Warter, Stuart L, ornithology, vertebrate paleontology
Wayne, Lawrence Gershon, microbiology
Wenjen, Chien, mathematics
Wharton, Marion Agnes, nutrition
Williams, Albert Doran, nuclear medicine
Wilson, James Newton, geography
Wilson, William Jewell, radiology
Winchell, Robert E, mineralogy, crystallography
Wood, Eunice Marjorie, cell biology, electron microscopy
Wynston, Leslie K, biochemistry

LOS ALAMITOS

Atkins, Don Carlos, Jr, industrial chemistry

LOS ALAMOS

Hall, David Ballou, physics
Harris, Donald R, Jr, physics
Jones, Wesley Morris, physical chemistry
Krick, Merlyn Stewart, nuclear science

LOS ALTOS

Altera, Kenneth P, veterinary pathology
Bates, David James, applied physics
Chang, Howard How Chung, theoretical physics
Ehrmantraut, Harry Charles, instrumentation
Harris, Kent Karren, atmospheric physics, nuclear physics
Jordan, Charles, high energy physics
Lewis, Arthur Edward, earth science
Moss, Lloyd Kent, biochemistry
Newkirk, Lester Leroy, space physics
Noble, Paul, Jr, physical chemistry
Parker, John Abel, chemistry
Raphael, Bertram, computer science, mathematics
Roney, James G, Jr, anthropology, medicine
Rossow, Vernon J, fluid dynamics
Schoellhamer, Jack Edward, geology
Sherman, John Edwin, information science
Singer, Solomon Elias, chemistry, nuclear physics
Smith, Newell Hart, physics
Sobon, Leon Edward, solid state physics, inorganic chemistry
Wehrli, Robert L, instrumentation
Weilmuenster, Earl Adam, organic chemistry
Wiesendanger, Hans Ulrich David, physical chemistry

LOS ALTOS HILLS

Chatland, Harold, mathematics
Dyal, Palmer, physics, astrophysics
Johnson, Richard D, chemistry
Macdonald, James Reid, vertebrate paleontology, geology
Melchor, Jack L, physics
Meyerott, Roland Edward, astrophysics, atomic physics
Nelson, Richard Burton, physics
Pierce, William G, structural geology
Sacks, Alvin Howard, mechanics, bioengineering
Schechter, Samuel, numerical analysis, applied mathematics
Selover, James Carroll, organic chemistry
Stutz, Robert Eugene, biochemistry
Szekely, Ivan J, resource management
Walt, Martin, magnetospheric physics, space physics
Whitmore, William Francis, mathematics

LOS ANGELES

Abarbanel, Abraham Robert, physiology
Abbott, Bernard C, biophysics
Abell, George Ogden, astronomy
Abers, Ernest S, physics
Abrams, Albert Maurice, pathology
Ackell, Edmund Ferris, medical administration, maxillofacial surgery
Adams, Forrest Hood, pediatric cardiology
Adams, John Milton, medicine
Adams, William S, metabolism
Adamson, Arthur Wilson, physical chemistry
Adey, William Ross, neurophysiology
Adinolfi, Anthony M, neurocytology
Aftergood, Lilla, biochemistry, nutrition
Agnew, Leslie Robert Corbet, history of medicine, experimental pathology
Ahmad, Nazir, histology, reproductive endocrinology
Akawie, Richard Isidore, organic chemistry, polymer chemistry
Aklonis, John Joseph, physical chemistry, polymer chemistry
Aladjem, Frederick, immunochemistry, biophysics
Alberte, Randall Sheldon, plant physiology, biochemistry
Alexander, Natalie, medical physiology
Alfin-Slater, Roslyn Bernice (Mrs Grant G Slater), biochemistry
Allan, Benjamin Wilson, inorganic chemistry, physical chemistry
Allen, Arnold Oral, mathematics
Allen, Frederick Graham, solid state physics
Allen, Richard K, entomology
Aller, Lawrence Hugh, astrophysics
Allerton, Samuel E, biophysical chemistry, medical science
Altar, William, theoretical physics, electronics
Amador, Elias, medicine, pathology
Ament, Marvin Earl, pediatrics, gastroenterology
Amey, Ralph Leonard, physical chemistry
Amstutz, Harlan Cabot, orthopedics, bioengineering
Anderson, Clifford Harold, mathematical analysis
Anderson, Orson Lamar, geophysics, mineralogy
Anderson, Peter Ole, mathematical statistics
Andreoli, Anthony Joseph, bacteriology, biochemistry
Anet, Frank Adrien Louis, organic chemistry
Ansari, Ali, clinical chemistry, lipid chemistry
Antosiewicz, Henry Albert, mathematical analysis
Applegate, Shelton P, vertebrate paleontology, ichthyology
Appleman, M Michael, biochemistry
Appleman, Milo Don, food bacteriology
Arens, Richard Friederich, mathematics
Arnell, J Hal, systematic entomology
Arnquist, Warren Nelson, solar physics
Arthur, Ransom James, psychiatry, epidemiology
Ash, Lawrence Robert, parasitology, tropical medicine
Ashley, Franklin Longley, plastic surgery
Atkinson, Daniel Edward, biochemistry
Aukerman, Lee William, solid state physics
Ayengar, Padmasini (Mrs Frederick Aladjem), biochemistry, microbiology
Babbitt, Donald George, mathematical analysis
Backus, John (Graham), physics
Bair, Thomas De Pinna, zoology, physiology
Baird, John Jeffers, neuroembryology
Baker, Kirby Alan, mathematics
Baker, Mary Ann, mammalian physiology
Baker, Neal Kenton, solar physics
Baker, Nome, biochemistry
Baker, Richard Freligh, biophysics
Baker, Robert Frank, molecular biology, genetics
Baker, Sol Ronald, therapeutics, radiology
Bakus, Gerald Joseph, marine ecology
Balchum, Oscar Joseph, internal medicine, physiology
Ball, Gordon Harold, zoology, parasitology
Ballard, Kathryn Wise, cardiovascular physiology
Ballard, Kenneth J, pharmacy
Baluda, Marcel A, virology, oncology
Banos, Alfredo, Jr, theoretical physics
Baptista, Luis Felipe, bioacoustics, behavioral genetics
Barber, Albert Alcide, cell biology
Barber, Thomas King, dentistry
Barker, Steven Joseph, fluid dynamics
Barker, Wiley Franklin, medicine
Barnes, John Landes, mathematics, electrical engineering
Barnett, Eugene Victor, biology, medicine
Barr, Allan Ralph, medical entomology
Barrett, Peter Van Doren, internal medicine, gastroenterology
Barrett, Robert, neurophysiology
Barry, Don Cary, astronomy
Barry, James Dale, space physics, lasers
Bartholomew, George Adelbert, zoology
Bartholomew, James William, bacteriology
Barwise, Kenneth Jon, mathematical logic
Batdorf, Samuel Burbridge, physics, engineering mechanics
Bateman, Joseph R, hematology, oncology
Bau, Robert, inorganic chemistry
Bauer, Robert Oliver, pharmacology
Baur, Mario Elliott, physical chemistry
Bavetta, Lucien Andrew, biochemistry, nutrition
Bayes, Kyle D, physical chemistry
Beals, Ralph Leon, anthropology, cultural anthropology
Beaudet, Robert A, chemical physics, physical chemistry
Beckenbach, Edwin Ford, mathematical analysis, geometry
Becker, Milton, solid state physics
Becker, Robert Adolph, space physics
Beenken, May Margaret, mathematics
Behrens, Herbert Charles, ophthalmology
Beigelman, Paul Maurice, medicine
Beisser, Arnold Ray, psychiatry
Belkin, John Nicholas, entomology, taxonomy
Bellman, Richard Ernest, mathematics
Bellville, John Weldon, medical research
Bendat, Julius Samuel, random data analysis, applied mathematics
Benedetti, Jacqueline Kay, biostatistics
Bengelsdorf, Irving Swem, science writing
Bennett, Charles Franklin, 1 cultural geography
Bennett, Leslie R, radiology
Ben-Zvi, Ephraim, physical chemistry
Berger, Rainer, anthropology, geography
Bergman, Hyman Chaim, pharmacology
Bergren, William Raymond, bio-organic chemistry
Berman, David Albert, pharmacology
Bernick, Sol, anatomy, oral pathology
Bernstein, Gerald Sanford, reproductive biology, obstetrics & gynecology
Bernstein, Sol, medicine
Bernstein, Stephen, neuropsychology
Bessman, Samuel Paul, pediatrics, biochemistry
Bethune, John Edmund, endocrinology
Betz, Barbara Jean, psychiatry
Bevan, John Acton, pharmacology
Biles, John Alexander, pharmaceutical chemistry
Billig, Franklin A, organic chemistry, science education
Bills, Robert F, cytology, electron microscopy
Binggeli, Richard Lee, neurophysiology
Birdsell, Joseph Benjamin, physical anthropology
Birman, Joseph Harold, geology
Bishop, John William, petroleum chemistry
Blackburn, John Francis, physics
Black-Cleworth, Patricia Ann, ethology, neurophysiology
Blahd, William Henry, nuclear medicine, internal medicine
Blake, J Bernard, astrophysics, space physics
Blanc, Robert Parmelee, geology
Blankenhorn, David Henry, cardiology
Blass, John P, neuropsychiatry, biochemistry
Blischke, Wallace Robert, statistical analysis
Bloom, Eda Terri, immunology, cancer
Blum, Edward K, mathematics
Blum, Marvin, applied mathematics, system analysis
Boak, Ruth Alice, medicine
Bock, Robert Oliver, physics, navigation
Boice, Lu Belle, molecular genetics
Bok, P Dean, anatomy, cell biology
Bollman, Vernon Leroy, physics
Boolootian, Richard Andrew, invertebrate zoology
Boyer, Paul Delos, biochemistry
Boynton, William Vandegrift, geochemistry, cosmochemistry
Bradbury, Ted Clay, theoretical physics
Bradley, Edward Charles, medicine
Brady, Allan Jordan, biophysics
Bragin, Joseph, physical chemistry
Braunstein, Rubin, physics
Brazier, Mary A B, neurophysiology
Breisacher, Peter, physical chemistry
Brem, Thomas Hamilton, medicine
Breslow, Lester, public health, preventive medicine
Brewer, Gary David, fish biology
Brickman, Harry Russell, psychiatry, psychoanalysis
Bright, Peter Bowman, mathematical biology
Brill, Norman Quintus, psychiatry, neurology
Brisbane, William Neely, optometry, public health
Britt, Patricia Marie, computer science, biomathematics
Brodie, Arnold Frank, bacteriology, biochemistry
Brown, Douglas Markham, biochemistry
Brown, Harold Victor, environmental health, industrial hygiene
Brown, Josiah, medicine
Brown, Robert Freeman, topology
Brown, Thomas Andrew, mathematics

Brown, Warren Shelburne, Jr., psychophysiology
Brown, William Jann, neuropathology
Bruck, Peter, physical organic chemistry
Bruckner, David Alan, medical parasitology
Bruman, Henry John, cultural geography, geography of Latin America
Brunk, Clifford Franklin, biophysics
Brusca, Richard Charles, invertebrate zoology, marine ecology
Buchwald, Jennifer S., neurophysiology
Buchwald, Nathaniel Avrom, neurophysiology, neuroanatomy
Buggs, Charles Wesley, bacteriology
Bundy, Hallie Flowers, biochemistry
Burg, Anton Behne, inorganic chemistry
Burg, Fred Paul, physics
Byatt, Pamela Hilda, virology
Byers, Nina, particle physics, theoretical physics
Byron, William Glenn, geography
Bystrom, Barbara Gillooly, neurobiology
Callender, E David, applied mathematics, computer science
Campbell, John Howland, mathematics, computer sciences
Caplin, Samuel Milton, botany
Capon, Brian, botany
Carlisle, Donald, economic geology
Carlson, Robert Warner, aeronomy, atomic physics
Carmel, Ralph, hematology
Carnes, William Henry, pathology
Carpenter, Roland LeRoy, astronomy, astrophysics
Carr, Robert H., physics
Carregal, Enrique Jose Alvarez, medicine, physiology
Carroll, June Starr, physical geography, geography of Western United States
Carroll, Walter William, cancer, surgery
Carsten, Mary E., biochemistry
Carterete, Edward Calvin Hayes, psychoacoustics, neuropsychology
Casanova, Joseph, organic chemistry
Cascarano, Joseph, cell physiology
Cassan, Stanley Morris, pulmonary diseases, internal medicine
Cavey, Michael John, biological structure, embryology
Chamberlin, Thomas Leland, geology
Chan, Leland, environmental health
Chang, Berken, atomic physics
Chang, Chen Chung, mathematical logic
Chang, Francis F., physics
Chang, Potter Chien-Tien, biostatistics
Chang, Sun-Yung Alice, mathematical analysis
Chao, Jowett, parasitology
Chao, Li-Pen, neurochemistry
Chew, Robert Marshall, ecology
Chiu, Yam-Tsi, atmospheric physics
Cho, Arthur Kenji, pharmacology, organic chemistry
Chock, Ernest Phaynan, physical inorganic chemistry, polymer chemistry
Chopra, Inder Jit, endocrinology, internal medicine
Christensen, Niels Gunnar, structural geology, electron microscopy
Christie, John McDougall, structural geology
Chu, Edith Ju-Hwa, organic chemistry
Church, Alonzo, mathematics
Clark, Charles Lester, mathematics
Clark, Malcolm A., space physics, aerospace technology
Clark, Virginia, biostatistics
Clark, William Arthur Valentine, urban geography
Clark, William Gilbert, magnetic resonance, experimental solid state physics
Clark, William Richmond, biochemistry
Cleland, James Spencer, surgery
Clement, George Horace, organic chemistry
Clement, Carmine Domenic, neuroanatomy
Coburn, Jack Wesley, internal medicine, nephrology
Coddington, Earl Alexander, mathematics
Code, Charles Frederick, physiology
Cohen, Morris, cell biology, medical research

Cohen, Natalie Shulman, cell physiology, biochemistry
Cohen, Norman, physical chemistry
Cohen, Ronald Bruce, chemical physics
Coburn, Ivan Paul, structural geology, stratigraphy
Cole, Robert Kiev, physics
Coleman, Charles Clyde, solid state physics
Coleman, Paul Jerome, Jr., space physics
Colchman, Eugene Louis, physical chemistry
Collias, Elsie Cole (Mrs Nicholas E Collias), zoology
Collias, Nicholas Elias, zoology
Collier, Clarence Robert, physiology
Collier, James Bryan, geometry
Collier, Robert John, biochemistry
Cook-Ioannidis, Leslie Pamela, mathematics
Cooksey, Donald Ernest, oral surgery
Cooper, Edwin Lowell, immunology, biology
Cooper, Maurice Zealot, medicine
Cornwall, John Michael, theoretical physics
Coroniti, Ferdinand Vincent, space physics, plasma physics
Cosby, Richard Sheridan, medicine
Costea, Nicolas V., immunohematology
Costin, Anatol, neurosciences
Coulson, Walter F., pathology
Covel, Michel Dale, medicine
Cram, Donald James, organic chemistry
Crandall, Paul H., medicine
Crawford, Donald W., cardiovascular diseases
Crawford, William Howard, Jr., pathology, oral pathology
Crescitelli, Frederick, visual physiology
Crocker, Diane Winston, pathology, data processing
Crooker, Nancy Uss, space physics
Cross, Ronald Allan, microbial genetics
Crowell, Wilfred J., pharmaceutical chemistry
Crutchfield, Charlie, analytical chemistry, inorganic chemistry
Culley, Benjamin Hays, mathematics
Cummings, David, geology
Cunningham, William Glenn, geography
Currell, Douglas Leo, organic chemistry
Curtis, Philip Chadsey, Jr., mathematics
Davidson, Christopher, plant anatomy
Davidson, Mayer B., endocrinology
Davis, Gregory Arlen, structural geology, tectonics
Davis, Lily Herlinda, plant pathology
Davis, Richard LaVerne, pathology, neuropathology
Dawson, John Myrick, plasma physics
Daybell, Melvin Drew, low temperature physics
Dayton, Seymour, internal medicine
Dea, Phoebe Kin-Kin, physical chemistry
DeGarmo, Glen Dean, anthropology
De Haan, Frank P., physical chemistry, inorganic chemistry
DeLand, Edward Charles, mathematics
DeLange, Robert J., biochemistry, protein chemistry
Demond, Joan, marine biology
Denson, Judson Samuel, medicine, anesthesiology
Deonier, Richard Charles, biophysical chemistry, molecular biology
De Sapio, Rodolfo Vittorio, mathematics
De Shazer, Larry Grant, physics
Deutsch, Daniel Harold, organic chemistry, nuclear chemistry
Devor, Kenneth Arthur, biochemistry
Dhopeshwarkar, Govind Atmaram, neurochemistry
Diamond, Jared Mason, physiology, ecology
Dignam, William Joseph, obstetrics & gynecology
Dimsdale, Bernard, mathematics
Dinge, Otelia Villa, nutrition
Dirksen, Ellen Roter, cell biology, developmental biology
Dixon, Andrew Derart, anatomy
Dixon, James Francis Peter, experimental ichthyology
Dixon, Wilfrid Joseph, biomathematics
Dolen, Richard, theoretical physics
Donnan, Christopher B., anthropology, archaeology
Donnell, George Nino, pediatrics
Dorman, Leon M., physical chemistry
Dougherty, Harry L., anatomy, orthodontics
Douglas, Robert L., speech pathology
Dowdy, Andrew Hunter, radiology

Dowling, Jerome M., physics, spectroscopy
Downs, Theodore, vertebrate paleontology
Dows, David Alan, physical chemistry
Dubbs, Clyde Andrew, biochemistry
Dugundji, James, mathematics
Dukes, Peter Paul, biochemistry
Dummet, Clifton Orrin, periodontology
Dunbar, Gary Seamans, geography
Dunn, Arnold Samuel, physiology
Dunn, Doris Frankel, biochemistry
Dunn, Olive Jean, biostatistics
Dunn, Robert Fowler, zoology, electron microscopy
Dye, Henry Abel, mathematics
Earle, Marshall Delph, solid state physics
Easton, Walter Heyden, paleontology
Eberhart, Hal, cultural anthropology, archaeology
Eckert, Roger Orto, neurophysiology
Edgerton, Robert Breckenridge, cultural anthropology
Edinger, James (G), meteorology
Edmondson, Hugh Allen, pathology
Edney, Eric B., zoology
Ehlig, Perry Lawrence, petrology, geology
Eidusen, Samuel, neurochemistry
Eisenberg, David, biophysical chemistry
Eisenberg, Robert S., electrophysiology, biophysics
Eisenman, George, biophysics
Eiserling, Frederick A., microbiology
Elashoff, Janet Dixon, applied statistics
Eldred, Earl, neurophysiology
Elison, Christian, pharmacology, toxicology
Elliott, Stuart Bruce, physics
El-Sayed, Mostafa Amr, physical chemistry
Emerson, Gladys Anderson, nutrition, chemistry
Endahl, Gerald Leroy, biochemistry
Enderton, Herbert Bruce, mathematical logic
Engelmann, Franz, zoology
Epps, Harland Warren, astronomy
Epstein, David George, cultural anthropology
Epstein, Eugene Eshan, astronomy
Epstein, George, plastics chemistry, materials science
Ernst, Wallace Gary, petrology, geochemistry
Ervin, Frank (Raymond), psychiatry
Espoy, Henry Marti, analytical chemistry
Ewald, Robert Harold, anthropology
Fahey, John Leslie, immunology
Fajans, Edgar W., industrial chemistry
Fallscheer, Herman O., agricultural chemistry, food chemistry
Farber, Sergio Julio, organic chemistry, clinical pathology
Farrington, Paul Stephen, analytical chemistry
Fattorini, Hector Osvaldo, applied mathematics
Fauchald, Kristian, systematic zoology
Feder, Bernard Herbert, radiotherapy
Felber, Franklin Stanton, plasma physics, theoretical physics
Feliciano-Dodonoff, Manuel, mathematics
Felton, Jean Spencer, occupational medicine
Ferguson, Lloyd Noel, chemistry
Ferguson, Thomas S., mathematical statistics
Fernandez y Cossio, Hector Rafael, marine biology
Fessler, John Hans, molecular biology
Figueroa, William Gutierrez, internal medicine
Finegold, Sydney Martin, infectious diseases, microbiology
Fink, Kathryn Ferguson, biochemistry
Fink, Robert Morgan, biochemistry
Finkelstein, Robert Jay, physics
Fish, Barbara, psychiatry
Fitch, John Edgar, systematic ichthyology
Flambee, Eugene Earl, biomedical engineering
Fleischauer, Paul Dell, physical inorganic chemistry, photochemistry
Flood, James Felix, psychopharmacology
Flood, Thomas Charles, organic chemistry, organometallic chemistry
Fonkalsrud, Eric W., surgery
Foos, Robert Young, ophthalmology, pathology
Foote, Christopher S., organic chemistry
Ford, Holland Cole, astronomy

Forester, Alvin Theodore, plasma physics
Forster, Harriet Herta, physics
Forster, Kurt, physics
Forsythe, Alan Barry, biostatistics
Fowler, Audree Vernee, protein chemistry
Francis, Charles E., physics
Frasher, Wallace G. Jr., cardioscular physiology
Frasier, S Douglas, pediatric endocrinology
Fratiello, Anthony, physical chemistry
Freeman, Raoul James, operations research
Frelinger, Jeffrey, immunogenetics
French, John Douglas, neurosurgery
Frey, Celeste, microbiology
Fried, Burton David, plasma physics
Friedman, David Belais, pediatrics
Friedman, Nathan, periodontology, oral pathology
Friedman, Nathan Baruch, pathology
Friou, George Jacob, immunology
Fryer, Charles W., geology
Fujimoto, Atsuko Ono, pediatrics
Fukuyama, Thomas T., microbiology, biochemistry
Fulco, Armand J., biochemistry
Furstman, Lawrence L., orthodontics, anatomy
Fuster, Joaquin Maria, neurophysiology
Gale, Robert Peter, immunobiology, medicine
Ganguly, Biswa Nath, solid state physics
Ganz, William, cardiology
Garcia-Rill, Edgar E., neurophysiology
Gardner, Murray Briggs, pathology, oncology
Garmire, Elsa Meints, physics
Garth, John Shrader, zoology
Garvey, James Emmett, entomology
Garwood, Victor Paul, audiology
Geiger, Paul Jerome, biochemistry
Gelbart, William M., chemical physics
Gelbwachs, Jerry A., quantum optics, lasers
Geller, Edward, neurochemistry
George, Frederick W. III, radiology, radiotherapy
George, Robert, pharmacology
Gerschenson, Lazaro E., molecular biology, cancer
Gerstein, Melvin, chemistry
Gibson, Edward George, physics, engineering
Gibson, Gary Eugene, neurosciences
Gill, Ayesha Elenin, genetics, population biology
Gillard, Baiba Kurins, chemical kinetics, enzymology
Gilchriest, William Clarence, biochemistry, molecular biology
Gillman, David, mathematics
Giorgio, Anthony Joseph, hematology
Given, Robert R., environmental biology
Gladysz, John A., organometallic chemistry, organic chemistry
Glitz, Dohn George, biology
Goguen, Joseph A. Jr., computer science, information science
Goldberg, Leonard Stephen, immunology, rheumatology
Goldberg, Louis J., neurosciences, dental research
Goldberg, Mark Arthur, pharmacology, neurology
Goldman, Ralph, internal medicine
Goldschmidt, Walter Rochs, cultural anthropology, social anthropology
Goldstein, Abraham M B, urology, histology
Goldwhite, Harold, inorganic chemistry
Golomb, Solomon Wolf, mathematics
Golub, Orville Joseph, bacteriology
Golub, Sidney Harris, immunology
Gonick, Harvey C., nephrology
Gonzalez, Elma, cell biology
Goodwin, Willard E., urology
Gordon, Irving, medical microbiology
Gordon, Malcolm Stephen, comparative physiology, marine biology
Gordon, Robert Julian, environmental chemistry
Gorman, George Charles, herpetology
Gorski, Roger Anthony, neuroendocrinology
Gorsline, Joseph Sherrin, marine geology
Gralla, Jay Douglas, biochemistry
Graves, Glenn William, mathematics
Green, Agnes Ann, physical inorganic chemistry

Green, John Willie, mathematics
Greene, Ernest Gerald, physiological psychology
Greenfield, Moses A, medical physics
Greenson, Ralph Romeo, psychiatry
Greenstadt, Melvin, inorganic chemistry, science education
Gregorich, David Tony, theoretical physics
Greibach, Sheila Adele, computer science, systems science
Grew, Edward Sturgis, petrology
Grew, Priscilla Croswell Perkins, geology
Griesel, Wesley Otto, botany
Griffith, Donal Louis, cell physiology
Grimes, Carol Jane Galles, bioinorganic chemistry
Grinna, Lynn Sharon, cell biology, biochemistry
Grinnell, Alan Dale, neurophysiology
Grodins, Fred Sherman, physiology, biomedical engineering
Grossman, Morton Irvin, gastroenterology
Grossman, Nathaniel, mathematics
Grossman, Richard C, orthodontics
Gussen, John, psychiatry
Guth, Paul Henry, internal medicine, gastroenterology
Guthrie, Donald, statistics
Guze, Lucien Barry, medicine
Haddock, Roy P, nuclear physics
Hagiwara, Susumu, neurophysiology
Hai, Francis, plasma physics
Haig, Janet, biosystematics
Hale, Gerry Alwyn, cultural anthropology, geography of the Sudan
Hales, Alfred Washington, mathematics
Hall, Clarence Albert, Jr, geology
Hall, Ernest Lenard, bioengineering, radiology
Hall, Michael Oakley, biochemistry, ophthalmology
Hall, Richard Travis, atmospheric physics, particle physics
Hallett, Wilbur Y, medicine
Hammond, George Denman, pediatrics, oncology
Hammond, Wylda, pediatrics
Hamner, Karl Clemens, plant physiology
Hamor, Glenn Herbert, pharmaceutical chemistry
Han, Yuri Wha-Yul, organic chemistry, polymer chemistry
Hanafee, William Norman, radiology
Hand, William Gordon, invertebrate zoology, physiology
Handy, Lyman Lee, physical chemistry
Hansen, Howard Edward, psychiatry, psychoanalysis
Hansen, James E, pulmonary diseases, environmental physiology
Hansmann, Douglas R, bioengineering, applied mechanics
Hanson, Virgil, medicine
Hanson, William Roderick, zoology
Harary, Isaac, biochemistry
Harding, Boyd W, biochemistry, endocrinology
Hardwick, Eugene Russell, physical chemistry
Harris, Theodore Edward, mathematics
Harvey, J Paul, Jr, orthopedic surgery
Haun, Charles Kenneth, anatomy
Haverback, Bernard Jacob, internal medicine, gastroenterology
Hawthorne, Marion Frederick, inorganic chemistry
Hayes, Keith James, behavioral biology
Hayes, Robert Mayo, mathematics, information science
Haymond, Herman Ralph, biophysics, medical physics
Hays, Daniel Mauger, pediatric surgery
Hays, Esther Fincher, oncology, hematology
Haywood, L Julian, internal medicine, cardiology
Hazelwood, Robert Nichols, biophysics, operations research
Heiman, Donald Eugene, solid state physics
Heller, Eric Johnson, chemical physics
Hellwarth, Robert Willis, physics
Helmer, Olaf, mathematics
Hemenway, William Garth, otolaryngology
Henderson, Brian E, medicine
Henley, William Ballentine, public administration, medical jurisprudence
Henrickson, James Solberg, plant taxonomy
Henry, James Paget, medicine
Hepler, Robert S, ophthalmology
Herrman, Christian, Jr, clinical neurology
Herschman, Harvey R, biological chemistry, cell biology
Hershey, Alan Unger, organic polymer chemistry
Hershman, Jerome Marshall, internal medicine, endocrinology
Herzberg, Fred, anatomy, dentistry
Hespenheide, Henry August, III, animal ecology, taxonomy
Hestenes, Magnus Rudolph, mathematics
Heuser, Eva T, pediatrics, pathology
Hewitt, William Lane, medicine
Hicks, Donald A, physics, science administration
Hill, James Newlin, anthropology
Hilton, Henry H, physics
Hirsch, Hilde Esther Zwirn, neurochemistry
Ho, Genevieve Po-Ai, foods, nutrition
Hochstein, Paul Eugene, pharmacology, biochemistry
Hodgman, Joan Elizabeth, medicine
Hoel, Paul Gerhard, mathematics
Hogue, Charles Leonard, entomology, taxonomy
Holland, Robert Campbell, anatomy
Holmes, John Eric, physiology, clinical neurology
Holstein, Theodore David, physics
Holzer, Robert Edward, space physics, plasma physics
Homsher, Earl Edwin, II, physiology
Hon, Edward Harry Gee, obstetrics & gynecology
Honrubia, Vicente, otolaryngology, neurophysiology
Hoover, George Nollner, physiology
Hopkins, Carl Edward, biostatistics, public health
Horn, Alfred, algebra
Horn, Diane, cell biology
Horowitz, Ellis, computer science
Horowitz, Sylvia Teich, bio-organic chemistry
Horton, Richard, medicine, physiology
Horwitz, Joseph, biophysics
Hoskins, Raymond Howard, solid state physics
House, Leland Richmond, otolaryngology
Houston, Roy Seamands, malacology, marine ecology
Howard, Bruce David, biochemistry, molecular biology
Howard, Dexter Herbert, medical mycology
Howard, Hildegarde (Mrs Henry Anson Wylde), vertebrate paleontology
Howell, Thomas Raymond, vertebrate zoology
Howton, David Ronald, biochemistry, radiation chemistry
Hu, Chia-Ren, quantum physics
Hu, Sze-Tsen, mathematics
Huddlestone, Richard H, theoretical physics
Hudson, Alvin Maynard, nuclear physics
Hudson, Donald Edwin, physics
Hughes, Everett C, chemotherapy, bioengineering
Hungerford, Gerald Fred, anatomy
Hurrell, John Patrick, solid state physics
Hurwitz, Alexander, computer sciences
Hyers, Donald Holmes, mathematics
Igo, George (Jerome), physics
Inselberg, Alfred, applied mathematics, biomathematics
Intriligator, Devrie Shapiro, space physics
Ivler, Daniel, medical microbiology, infectious diseases
Jackson, David Diether, geophysics
Jacobs, Eugene Howard, mathematics, computer science
Jacobs, Thomas Lloyd, organic chemistry
Jacobson, George, radiology
Jaffe, Sigmund, physical chemistry
Jahn, Theodore Louis, physiology, protozoology
Jakway, George Elmer, vertebrate paleontology
James, John Alexander, medicine
James, Thomas William, zoology
Jarmakan, Jay M, pediatric cardiology
Jarvik, Lissy F, human genetics
Jarvik, Murray Elias, pharmacology
Jelliffe, Derrick Brian, nutrition, pediatrics
Jelliffe, Roger Woodham, cardiology, pharmacology
Jenden, Donald James, pharmacology
Jenkins, Floyd Albert, comparative anatomy, vertebrate paleontology
Jenner, David Charles, astronomy
Jennrich, Ellen Coutlee, animal behavior, ecology
Jennrich, Robert I, statistics
Johnson, Allen Willard, cultural anthropology
Johnson, B Lamar, Jr, internal medicine, infectious diseases
Johnson, Paul Bennett, mathematics
Johnston, Stewart Archibald, physical chemistry
Jones, Edward Gomer, medicine
Jones, Gilbert Fred, marine ecology, histology
Jones, Mary Ellen, biochemistry
Jones, Peter Frank, physical chemistry, forensic science
Jorgens, Joseph, medicine, radiology
Judge, Darrell L, physics
Juge, Damian Marie, marine microbiology, molecular biology
Jung, Michael Ernest, synthetic organic chemistry
Junge, Douglas, neurophysiology
Jura, Michael Alan, astrophysics
Kadlec, Paul William, meteorology
Kadner, Carl George, insect physiology
Kaesz, Herbert David, organometallic chemistry
Kagan, Benjamin M, pediatrics
Kalmanson, George Maurice, internal medicine
Kalra, Vijay Kumar, biochemistry
Kambara, George Kiyoshi, ophthalmology
Kamen, Martin David, physical chemistry
Kanarik, Rosella, mathematics
Kanter, Helmut, electron physics
Kaplan, David Gilbert, physical chemistry
Kaplan, Issac R, geochemistry
Kaplan, Joseph, physics, geophysics
Kaplan, Solomon Alexander, medicine
Kapur, Krishan Kishore, physiology, prosthodontics
Kark, Robert Adriaan Pieter, neurology, neurochemistry
Kataja, Eva I, molecular biology, community health
Kattus, Albert Adolph, Jr, medicine, cardiology
Katz, Joseph, biochemistry
Katz, Ronald Lewis, anesthesiology
Kaufman, Joseph J, urology, surgery
Kaula, William Mason, geophysics
Kavanau, Julian Lee, ethology
Kay, Robert Woodbury, geochemistry
Kay, Suzanne Mahlburg, petrology, mineralogy
Kehl, William Brunner, computer science
Kelly, Douglas Elliott, developmental anatomy, microscopic anatomy
Kennedy, Chandler James, lasers
Kennedy, George Clayton, geophysics
Kennel, Charles Frederick, plasma physics, space physics
Kernan, Keith Thomas, anthropology
Kertesz, Jean Constance, pharmaceutical chemistry
Kessel, John Flenniken, microbiology
Kesterson, Clinton Joe, pesticide chemistry
Keys, Richard Taylor, physical chemistry
Keyzer, Hendrik, physical chemistry, solid state chemistry
Kharasch, Norman, chemistry
Khare, Gyaneshwar Prasad, microbiology, virology
Khwaja, Tasneem Afzal, cancer, chemotherapy
Kieffer, Hugh Hartman, planetary science
Kieffer, Susan Werner, geology
Kilday, Warren D, organic chemistry
Kim, Young Bae, physics
Kissler, Gerald Ray, institutional research
Kivelson, Daniel, chemical physics
Kivelson, Margaret Galland, space physics
Klapman, Solomon Joel, mathematical physics
Kleeman, Charles Richard, physiology, metabolism
Klein, David Joseph, radiological physics
Kline, Frank Menefee, psychiatry
Kline, Nathan Schellenberg, psychiatry, psychopharmacology
Klinenberg, James Robert, internal medicine, rheumatology
Knobler, Charles Martin, physical chemistry
Knopoff, Leon, geophysics, physics
Knutson, John William, public health
Koda, Robert T, pharmaceutical chemistry
Kogut, Maurice D, pediatrics, endocrinology
Kolin, Alexander, biophysics
Koosis, Paul, mathematical analysis
Kornberg, Thomas B, cell biology, developmental biology
Kornreich, Helen Kass, pediatrics
Kostanick, Huey Louis, geography of Europe, geography of Soviet Union
Kratochvil, Frank James, prosthodontics
Krieg, David Ronald, genetics
Krueger, Robert William, physics
Kruger, Lawrence, neuroanatomy, neurophysiology
Ku, Teh-Lung, geochemistry, geochronology
Kuper, Hilda Beemer, anthropology, ethnology
Kurohara, Samuel S, radiobiology, radiotherapy
Kurze, Theodore, neurosurgery
Lamb, Sandra Ina, organic chemistry, pharmacology
Lambert, Frank Lewis, organic chemistry
Lambropoulos, Peter Poulos, theoretical physics
Lamson, Baldwin Gaylord, pathology
Landesman, Herbert, inorganic chemistry
Landing, Benjamin Harrison, pathology
Langdon, Edward Allen, medicine, radiology
Lange, Charles Gene, applied mathematics
Langer, Glenn A, medicine
Langford, Robert Bruce, organic chemistry
Langness, Lewis L, anthropology
Lanski, Charles Philip, mathematics
Lascelles, June, microbiology, biochemistry
Laties, George Glushanok, plant physiology
Latta, Harrison, pathology, biophysics
Lau, Francis You King, medicine
Layne, Ennis C, biochemistry, physcial chemistry
Leahy, Mary Gerald, insect physiology, acarology
Lee, Long Chi, experimental physics
Lee, Peter Van Arsdale, pharmacology, medicine
Lee, Young Chang, cell biology, radiation biology
Leedom, John Milton, internal medicine
Lefever, Robert Allen, solid state chemistry, materials science
Lehrer, William Peter, Jr, animal science
Leitner, Nathan Barr, biochemistry
Leone, Stephen Robert, chemical physics
Leong, Kam Choy, biochemistry, poultry nutrition
Lessa, William Armand, anthropology
Lesse, Henry, psychiatry, neurophysiology
Levan, Norman E, medicine
Levenson, Marc David, lasers, quantum electronics
Levin, Seymour R, internal medicine
Levine, Michael S, neurosciences
Levy, Daniel, biochemistry, protein chemistry
Levy, Louis, pharmacology
Lewis, Charles E, medicine, preventive medicine
Lewis, Frank Harlan, botany
Lewis, Jane Sanford, nutrition
Lewis, William Perry, medical microbiology
Lian, Harold Maynard, geology
Libby, Leona Marshall, physics
Libby, Willard Frank, chemistry
Liddicoat, Richard Thomas, Jr, gemology, mineralogy
Lieb, Margaret, genetics
Lien, Eric Jung-Chi, pharmaceutical chemistry
Liggett, Thomas Milton, mathematics
Lindburg, Donald Gilson, physical anthropology, primatology
Linde, Leonard M, cardiology, physiology
Lindon, John Arnold, psychiatry, psychoanalysis
Lindsley, David Ford, physiology
Lindsley, Donald B, psychophysiology
Lindstedt-Siva, K June, biology
Lingenfelter, Richard Emery, astrophysics, cosmic ray physics
Linnell, Robert Hartley, academic administration, institutional research
Littleton, C Scott, cultural anthropology
Loeblich, Alfred Richard Jr, micropaleontology, palynology
Loeblich, Helen Nina (Tappan), micropaleontology, paleoecology
Logan, Joseph Granville, Jr, physics
Logan, Richard Fink, geography, arid lands
Lomax, Peter, pharmacology
Longmire, William Polk, Jr, surgery
Loosli, Clayton G, medicine
Lorber, Arthur, internal medicine, rheumatology
Lorhan, Paul Herman, anesthesiology
Louisell, William Henry, physics
Lowe, Harry J, anesthesiology
Lowe, Orville G, organic chemistry
Luck, James Vernon, orthopedic surgery
Lundberg, George David, pathology
Luner, Stephen Jay, biophysics

Lunt, Owen Raynal, soil fertility
Lurie, Paul Raymond, pediatrics
MacDonald, Norman Scott, biological chemistry
MacFadden, Clifford Herbert, geography
Mach, Martin Henry, physical organic chemistry
MacInnis, Austin J., parasitology, biochemistry
Mack, Rex Charles, physics
MacKenzie, Kenneth Ross, physics
MacLaren, Walter Rogers, allergy
Madden, Sidney Clarence, pathology
Magdison, Oscar, medicine
Mäh, Robert A., microbiology
Maloney, James Vincent, Jr, surgery
Manning, Phil Richard, internal medicine
Manning, Leslie Bernard, neurology
Maquet, Jacques, cultural anthropology, social anthropology
Marburger, John Harmen, III, theoretical physics
Marcus, Carol Silber, radiation biology, biophysics
Marcus, Leon, microbial genetics, molecular biology
Margaziotis, Demetrios John, nuclear physics
Markham, Charles Henry, neurology, neurophysiology
Markland, Francis Swaby, Jr, biochemistry
Markoff, Elliott Lee, psychiatry, psychoanalysis
Martin, Walter Edwin, parasitology, zoology
Martinez, Rafael Juan, bacterial physiology
Martinson, Harold Gerhard, molecular biology
Marx, Walter, biochemistry
Maronde, Robert Francis, internal medicine, clinical pharmacology
Masaki, Beverly Wong, pharmaceutical chemistry, clinical pharmacology
Marsh, Donald Jay, physiology, biomedical engineering
Massey, Frank Jones, Jr, mathematical statistics
Marsh, Gayle G., neuropsychology
Marshall, Louise Hanson, neurosciences, mammalian physiology
Mathias, Mildred Esther (Mrs Gerald L Hassler), botany
Mathies, Allen Wray, Jr, pediatrics, infectious diseases
Matioli, Gastone, cell biology
Maxwell, David Samuel, anatomy
May, Philip Reginald Aldridge, psychiatry
Mayer, Bromley Morgan, microbiology
Mayer, Stanley Wallace, physical chemistry
McAllister, Robert Milton, pediatrics
McAnally, John Sackett, analytical biochemistry
McCall, Chester Hayden, Jr, applied statistics
McCarroll, James Renwick, medical administration
McCarron, Margaret Mary, internal medicine, pharmacy
McCarthy, Frank Martin, dentistry, anesthesiology
McCarthy, Mary Anne, numerical analysis, system analysis
McClure, William Owen, neurochemistry
McCullough, James Douglas, structural chemistry, inorganic chemistry
McDonnel, Gerald M, radiology, radiobiology
McGinnis, Carl Leonardt, physics
McKee, Ralph Wendell, biological chemistry
McKenna, Charles Edward, bio-organic chemistry
McKnight, Tom, geography
McLean, James H, invertebrate zoology
McMenamin, John William, developmental biology
McMillan, William George, chemical physics
McNeal, Robert Joseph, chemical physics, space physics
McPherron, Robert Lloyd, geophysics, space physics
McTague, John Paul, chemical physics
McVickar, David Langston, medical microbiology
McWilliams, Margaret Ann, nutrition, food
Mead, Giles Willis, ichthyology
Mead, James Franklyn, biochemistry
Meade, Melinda Sue, geography
Meacham, William Coryell, mathematical physics, classical physics
Meehan, John Patrick, physiology

Meerbaum, Samuel, bioengineering
Meighan, Clement Woodward, anthropology, archaeology
Melkanoff, Michel Allan, nuclear physics
Mellinkoff, Sherman Mussoff, medicine
Melrose, Raymond John, oral pathology
Mendel, Werner Max, psychiatry
Mensah, Patricia Lucas, neuroanatomy, neurosciences
Mercer, Robert J., numerical analysis
Merriam, Esther Virginia, virology
Merriam, John Roger, genetics
Merriam, Richard Holmes, geology
Meyer, Richard David, infectious diseases
Meyer, Walter Davidson, dynamic meteorology
Meyers, Harvey I., radiology
Meyers, Marvin Harold, orthopedic surgery
Mickey, Max Ray, Jr., statistics
Miech, Ronald Joseph, mathematics
Miller, James Nathaniel, infectious diseases
Miller, Orville H., pharmaceutical chemistry
Miller, Stanley Johnson, organic chemistry
Mintz, Yale, meteorology
Mishkin, Frederick Seymore, nuclear medicine
Mitchell-Kernan, Claudia I., anthropology
Mitwer, Tod Edwin, bacteriology
Moerman, Michael, anthropological linguistics
Montagni, Magda Moustafa, infectious diseases, pathology
Montgomery, Susan, algebra
Moore, Cecilia Louise, physical organic chemistry
Moore, John George, Jr, obstetrics & gynecology
Mohr, John Luther, marine biology, protozoology
Mommaerts, Wilfried, physiology
Momparler, Richard Lewis, pharmacology, biochemistry
Monroe, Barbara Samson Granger, anatomy
Morgan, Jasper Eugene, physics
Morgenstern, Leon, surgery
Morin, James Gunnar, invertebrate zoology, invertebrate physiology
Mornigo, Fernando Bernardino, physics
Morris, William Joseph, geology
Morse, Fred A., space physics
Morton, Donald Lee, surgery, oncology
Morton, Martin Lewis, zoology, physiology
Moschovakis, Joan Rand, mathematics
Moschovakis, Yiannis N., mathematics
Mosley, James W., epidemiology
Moss, Charles Norman, medicine, surgery
Mosteller, Raymond Dee, biochemistry
Moszkowski, Steven Alexander, theoretical physics
Motley, Hurley Lee, physiology
Moye, Anthony Joseph, academic administration, physical organic chemistry
Moyer, Dean La Roche, pathology
Muessig, Siegfried James, economic geology
Murad, Emil Moise, operations research
Murdoch, Joseph Richard, organic chemistry
Muscatine, Leonard, invertebrate zoology, comparative physiology
Myers, Lawrence Stanley, Jr. radiation biophysics, environmental sciences
Nafpaktitis, Basil G., biology
Nagy, Kenneth Alex, environmental physiology
Nakada, Yoshinao, chemistry
Nakamura, Robert Masao, reproductive biology, biophysics
Nayak, Debi Prosad, virology, oncology
Neiburger, Morris, meteorology
Nelson, Eldred (Carlyle), physics
Nelson, Howard Joseph, geography
Nelson, Rex Roland, acoustics
Nerlich, William Edward, medicine
Neu, Ernest Ludwig, chemistry
Neustein, Harry (Bernard), pathology
Nevenzel, Judd Cuthbert, lipid chemistry
Newman, Bertha L., anatomy
Newman, Philip Lee, anthropology, ethnology

Newton, Carol Marilyn, medicine, computer science
Nichols, Nathaniel Burgess, engineering
Nicol, Malcolm Foertner, chemical physics
Nicolaides, Nicholas, biochemistry, organic chemistry
Nicolson, Margery O'Neal, biochemistry
Niden, Albert H, internal medicine
Niederreiter, Harald Guenther, number theory, numerical analysis
Niehoff, Arthur Herman, cultural anthropology, applied anthropology
Nierlich, Donald P., microbial physiology, molecular biology
Nimni, Marcel Efraim, biochemistry, nutrition
Nishihara, Mutsuko, microbiology, biochemistry
Nishita, Hideo, soil chemistry, plant nutrition
Nobel, Park S., plant physiology, ecology
Nodvik, John S., physics
Norman, Amos, medical physics
Norton, Richard E., theoretical physics
Nyc, Joseph Frank, biological chemistry, nuclear medicine
Obert, Jessie C., nutrition
O'Brien, Richard Lee, oncology, cell biology
O'Connor, John Dennis, zoology,
Oertel, Gerhard Friedrich, structural geology
Oltz, Donald Frederick, palynology
Onak, Thomas Philip, organic chemistry,
Orbach, Raymond Lee, solid state physics
Osborne, Robert Howard, stratigraphy
Oldham, Susan Banks, endocrinology
Olitzky, Irving, bacteriology
Oliver, Richard Charles, periodontology
Olson, Albert Lloyd, pathology
Olson, Everett Claire, vertebrate paleontology
Ogden, Thomas E., physiology
Okrent, David, nuclear physics
Okun, Ronald, clinical pharmacology
Oldendorf, William Henry, neurology, experimental pathology
Parker, John William, organic chemistry
Parks, Joel Harris, physics
Parmelee, Arthur H, Jr, medicine, pediatrics
Patek, Paul R., anatomy
Pathak, Keshav Dattatray, biochemistry
Pattabhiraman, Tammanur R., chemistry, marine biology
Paule, Wendelin Joseph, anatomy
Paulikas, George A., space physics
Paulson, Donald Robert, organic chemistry
Pearce, Morton Lee, cardiology
Pearson, Carl M., medicine
Pease, Daniel Chapin, electron microscopy
Perkins, William Hughes, speech pathology
Peter, James Bernard, internal medicine,
Peters, Geraldine Joan, astronomy
Petit, Donald William, internal medicine
Petruska, John Andrew, chemical physics
Phelan, Nelson Flagge, physical inorganic chemistry
Philippart, Michel Paul, neurology, neurochemistry
Philipson, Lloyd L., mathematics, operations research
Phillies, George David Joseph, physics, spectroscopy
Phillips, Edward, molecular physics
Phinney, Bernard Orrin, botany
Pickett, Morris John, bacteriology
Pieper, Richard Edward, biological oceanography
Pierce, John Gregory, applied mathematics
Pierce, John Grissim, biochemistry
Pierson, William R., aerospace medicine
Pietras, Richard Joseph, endocrinology, cancer
Pincus, Philip A., theoretical solid state physics
Pine, Stanley H., organic chemistry
Pipkin, Bernard Wallace, environmental geology
Pitts, Thomas Dennis, parasitology
Plavec, Miroslav (Mirek) Josef, astrophysics
Poc, Norman Dean, nuclear medicine
Pokras, Harold Herbert, chemistry
Poliack, Seymour, psychiatry

Popenoe, Willis Parkison, invertebrate paleontology
Popjak, George Joseph, biochemistry
Popper, Daniel Magnes, astronomy
Port, Sidney C., mathematics
Porter, David Dixon, virology, pathology
Portnoy, Bernard, epidemiology, pediatrics
Porto, Sergio P S., physics
Prag, Arthur Barry, atmospheric physics, ionospheric physics
Pressman, Thomas Richard, botany
Price, Charles Morton, mathematics
Pridmore-Brown, David Clifford, physics
Prockow, John James, medicine
Puckett, William Thomas, Jr, topology
Puhvel, Sirje Madli, medical microbiology
Pumpian-Mindlin, Eugene, psychiatry
Quinlan, Edward, obstetrics & gynecology
Quinton, Paul Marquis, cell physiology
Radell, David, geography, urban studies
Raghunathan, Lalitha, parasitology, cytogenetics
Ralston, James Vickroy, Jr., mathematics
Rapaport, Samuel I., internal medicine, hematology
Rasmussen, Aaron Frederick, Jr, medical microbiology, immunology
Rawcliffe, Robert Douglas, optical physics
Ray, Dan S., molecular biology,
Rees, John David, biogeography, cultural geography
Reeves, Robert Lloyd, dentistry
Rehman, Irving, anatomy
Reinheimer, Julian, physics
Reinkober, Fred M, anthropology
Reisner, Ronald M, dermatology
Reiss, Howard, physical chemistry
Renau, Jacques, theoretical physics
Reynolds, Don Rupert, mycology
Reynolds, Telfer Barkley, internal medicine
Rice, Leslie Irene, biochemistry
Richardson, John Reginald, nuclear physics
Richters, Arnis, experimental pathology
Rigler, Leo George, radiology
Riley, Richard Fowble, physiological chemistry
Rinderknecht, Heinrich, biochemistry
Rittenberg, Sydney Charles, bacteriology
Ritvo, Edward R, psychiatry
Roberts, Carmel Montgomery, pharmacology
Roberts, Sidney, biological chemistry,
Robertson, Thomas N, mathematics
Robinson, Gerald Dean, Jr, nuclear chemistry
Robinson, Hamilton Burrows Greaves, pathology, histology
Robinson, James McOmber, clinical biochemistry
Robinson, Lawrence Baylor, physics
Robinson, Marsh Edward, oral surgery
Robinson, William John, petroleum geology, geophysics
Roemer, Milton Irwin, public health, preventive medicine
Rogawski, Alexander S., psychiatry
Rogers, Jack Cree, mathematics
Rokaw, Stanley N., medicine
Rolett, Ellis Lawrence, internal medicine, cardiology
Roman, Janet Meincer, immunology,
Romig, William Robert, bacteriology
Romney, Evan M., plant nutrition, soils
Rose, Augustus Steele, clinical neurology
Rosenfeld, John L., geology
Rosenfeld, Sheldon, physiology
Rosenkrantz, Jens Georg, surgery
Rosin, Sidney, pediatrics
Rosoff, Leonard, surgery
Ross, Gordon, physiology, internal medicine
Ross, Joseph Foster, medicine
Rothschild, Bruce Lee, mathematics
Rove, Louis Claude, Jr, geology
Roy-Burman, Pradip, biochemistry, molecular biology
Rubin, Milton Emanuel, medicine
Rubinstein, Eduardo Hector, physiology

Rubinstein, Lydia, endocrinology, neuroendocrinology
Rudnick, Isadore, acoustics, low temperature physics
Rugge, Hugo R, physics
Russell, Christopher Thomas, geophysics
Russell, Findlay Ewing, physiology, toxinology
Russell, John Albert, astronomy
Russell, Mercer P, entomology
Rydelek, Paul Anthony, exploration geophysics
Sacher, Joseph Albert, plant physiology
Sakurai, Jun John, theoretical high energy physics
Salot, Stuart Edwin, inorganic chemistry
Salovey, Ronald, physical chemistry, polymer chemistry
Salser, Winston Albert, molecular biology
Salter, Christopher Lord, cultural geography, geography of China
Saltzman, Max, textile technology
Sanborn, Warren Gordon, medical physiology
Sanchez, David A, mathematics
Sanders, Timothy D, physics
Sargent, Nicholas E, radiology
Sario, Leo Reino, mathematics
Satten, Robert A, solid state physics
Sauer, Jonathon Deininger, biogeography
Saul, Louella Rankin, paleontology
Savage, Jay Mathers, herpetology
Sawyer, Charles Henry, anatomy, neuroendocrinology
Saxton, Romaine Edward, virology
Schacher, John Fredrick, parasitology, infectious diseases
Schain, Richard J, pediatric neurology
Schechter, Joel Ernest, cell biology, electron microscopy
Scheibel, Arnold Bernard, neurophysiology, psychophysiology
Scherer, Kirby Vaughn, Jr, organic chemistry
Schiffman, Sandra, biochemistry
Schlag, John, neurophysiology
Schlegel, Robert John, pediatrics, genetics
Schlein, Peter Eli, physics
Schlesinger, Stewart Irwin, computer science, mathematics
Schneider, Friedemann W, physical chemistry
Schnepp, Otto, physical chemistry, chemical physics
Schoenfield, Leslie Jack, gastroenterology, internal medicine
Schopf, James William, paleobiology, organic geochemistry
Schotz, Michael C, biochemistry
Schroeder, Charles Arthur, botany
Schubert, Gerald, geophysics, planetary sciences
Schulz, Michael, space physics, plasma physics
Schumaker, Verne Norman, physical biochemistry
Schwabe, Arthur David, medicine, gastroenterology
Schwartz, Alice Griffin, molecular genetics
Schwartz, Donald Alan, psychiatry
Schwartz, Joseph Robert, physical chemistry, organic chemistry
Schwartz, Pauline Mary, chemotherapy
Schwinger, Julian Seymour, physics
Scott, Robert Lane, physical chemistry
Seeger, Robert Charles, immunology, pediatrics
Seekins, Charles William, mathematics
Segal, Gerald A, theoretical chemistry
Segundo, Jose Pedro, neurophysiology
Sellers, Alvin Louis, physiology
Sellers, Margret Irene, virology
Senum, Tellef, biochemistry
Serafetinides, Eustace Agapios, psychiatry
Sercarz, Eli, immunology, cell biology
Seto, Joseph Tobey, virology
Shader, Melvin A, mathematics
Shanbrom, Edward, internal medicine, hematology
Shani, Jashovam, pharmacology, radiobiology
Shapiro, Isadore, chemistry, materials science
Sharp, Kenneth George, inorganic chemistry
Shaw, Kenneth Noel Francis, biochemistry
Sheppard, Asher R, biophysics
Sheppard, Joseph Jackson, Jr, cardiovascular diseases
Sherwin, Russell P, pathology
Shimizu, C Susan, biochemistry, physiology
Shome, Basudev, biochemistry, endocrinology
Shonick, William, biostatistics, public health
Shore, Nomie Abraham, pediatrics, hematology
Shreve, Ronald Lee, geomorphology, glaciology
Shugarman, Peter Melvin, plant physiology, biochemistry
Siegel, Lawrence Sheldon, medicine
Siegel, Richard Weil, genetics
Siegel, Seymour, chemical physics
Siemsen, Jan Karl, nuclear & internal medicine
Sigman, David Stephan, biochemistry
Silva, George Douglas, dentistry, medicine
Silver, Arnold Herbert, solid state physics
Simkin, Benjamin, endocrinology
Simmons, Daniel Harold, internal medicine, physiology
Simmons, James Quimby, III, psychiatry
Simon, Marcia Lee Mielke, neuropharmacology
Simpson, Larry P, cell biology, protozoology
Sinder, Riley Monroe, theoretical chemistry
Singer, Anita Larks, developmental physiology, enzymology
Singer, Lawrence Alan, organic chemistry
Siscoe, George L, space physics
Sjostrand, Fritiof S, molecular biology, neuroanatomy
Skiles, Durward DeWitt, geophysics
Skoog, William Arthur, internal medicine
Slager, Ursula Traugott, pathology
Slater, Grant Gay, biochemistry, medical research
Slater, William E, experimental high energy physics
Slavin, Bernard Geoffrey, anatomy
Sloane, Robert Bruce, psychiatry
Slotnick, Irving James, bacteriology
Smit, Jan, solid state physics
Smith, Emil L, biochemistry, biophysics
Smith, George Stahl, pathology, immunology
Smith, Roberts Angus, biochemistry
Smith, Wayne Earl, mathematics
Smith, William R, biophysics, mathematics
Smulders, Anthony Peter, physiology
Snoke, John Edward, biochemistry
Sobin, Sidney S, physiology
Sognnaes, Reidar Fauske, dentistry, biology
Solomon, David Harris, endocrinology
Soltysik, Szczesny Stefan, animal behavior, neurophysiology
Sonnenschein, Ralph Robert, physiology
Sorgenfrey, Robert Henry, mathematics
Soule, Dorothy (Fisher), marine biology
Soule, John Dutcher, zoology
Sparkes, Robert Stanley, medicine, human genetics
Spellman, Mitchell Wright, surgery
Spence, Mary Anne, human genetics
Spencer, Joseph Earle, cultural geography, geography of East & Southeast Asia
Spielman, John Russel, inorganic chemistry
Spitzer, William George, physics
Sprowls, Riley Clay, statistics
Stafsudd, Oscar M, Jr, solid state spectroscopy
Stager, Kenneth Earl, zoology
Stahl, Frieda Axelrod, solid state physics
Stansberry, Kent Gardner, space physics
Stapp, John Paul, biophysics
Stear, Adrian N, inorganic chemistry, management science
Steen, Stephen N, medicine
Steinberg, Howard, organic chemistry
Steinberg, Robert, mathematics
Stell, William Kenyon, visual physiology, neurobiology
Stello, Phyllis Greene, solid state physics
Stellwagen, Robert Harwood, biochemistry
Stephens, John Stewart, Jr, marine biology, fish biology
Stephens, Philip J, theoretical chemistry
Sterling, Rex Elliott, biochemistry
Stern, Elizabeth, experimental pathology, epidemiology
Stern, Richard, acoustics
Stern, W Eugene, surgery
Stevens, Jack Gerald, virology, experimental pathology
Stiehm, E Richard, pediatrics, immunology
Stone, Charles Joel, mathematics, statistics
Stone, Richard L, anthropology
Stone, Richard O'Neill, geomorphology
Stork, Donald Harvey, nuclear physics
Stout, Martin Lindy, geology
Straatsma, Bradley Ralph, medicine
Straughan, Isdale (Dale) Margaret, biology, ecology
Straus, Ernst Gabor, mathematics
Straus, Joe Melvin, atmospheric physics
Strauss, Bernard, medicine
Straw, Richard Myron, population biology, systematic botany
Strehler, Bernard Louis, biology, biochemistry
Strickland, Erasmus Hardin, biophysics
Strong, Stanley Sterling, nuclear physics, space physics
Strouse, Charles Earl, chemistry
Stupian, Gary Wendell, solid state physics, surface physics
Sturgeon, Phillip, hematology
Su, Che, pharmacology
Sutter, Vera La Verne, medical microbiology, infectious diseases
Swan, Harold James Charles, physiology, cardiovascular diseases
Swanson, Virginia Lee, pathology
Sweet, Robert Mahlon, molecular biology
Swendseid, Marian Edna, biochemistry
Swezey, Robert Leonard, internal medicine, physical medicine & rehabilitation
Swift, Camm Churchill, ichthyology
Swift, Charles James, computer sciences
Swift, Jonathan Dean, mathematics
Sykes, John A, virology, cancer
Szabo, Arlene Slogoff, cell biology
Szego, Clara Marian, cell biology
Takasugi, Mitsuo, cancer, immunology
Takesaki, Masamichi, mathematics
Tanzi, Fausto, internal medicine
Taplin, George Vorce, medicine, radiology
Tarjan, George, psychiatry
Tarr, Betty R, chemistry
Tatter, Dorothy, medicine
Taylor, Dermot Brownrigg, pharmacology
Taylor, Howard S, chemistry, physics
Taylor, Thomas Tallott, physics
Teicher, Joseph D, psychiatry
Telfer, Nancy, medicine, nuclear medicine
Templeton, McCormick anatomy
Teng, Evelyn Lee, psychobiology
Teng, Ta-Liang, geophysics
Terasaki, Paul Ichiro, immunology
Terjung, Werner Heinrich, geography
Terry, Roger, surgical pathology
Thacker, John Charles, statistical analysis
Thomas, Benjamin Earl, geography
Thomas, Dudley Watson, biochemistry
Thomas, Lyell Jay, Jr, pharmacology
Thomas, Tracy Yerkes, mathematics
Thompson, Henry Joseph, biology
Thompson, Richard Scott, physics
Thornber, James Philip, plant biochemistry
Thorne, Richard Mansergh, space physics, plasma physics
Tibby, Richard Bitner, oceanography, marine biology
Ticho, Harold Klein, experimental high energy physics
Tierney, Donald Frank, physiology, medicine
Tiktopoulos, George S, theoretical high energy physics
Ting, William Su, geology, palynology
Tintner, Gerhard, mathematics
Tobin, Allan Joshua, developmental biology, molecular biology
Tobin, Elaine Munsey, plant development
Tokes, Zoltan Andras, biochemistry, developmental biology
Tompkins, Ronald K, surgery
Topping, Norman Hawkins, infectious diseases
Tormey, John McDivit, physiology, cell biology
Tourtellotte, Wallace William, neurology
Towers, Bernard, pediatrics, developmental anatomy
Towner, Howard Frost, biology, ecology
Tranquada, Robert Ernest, internal medicine
Trauring, Mitchell, operations research
Troesch, Beat Andreas, applied mathematics
Trosper, Terry Louise, biophysics
Trueblood, Kenneth Nyitray, chemistry
Trulio, John George, continuum mechanics, physics
Truxal, Fred Stone, entomology
Tsuji, Frederick Ichiro, biochemistry
Tubis, Manuel, nuclear medicine
Turner, Eugene Bonner, physics
Turner, Frederick Brown, vertebrate zoology
Ulmer, David D, internal medicine, physical chemistry
Underwood, James Henry, solar physics, x-ray astronomy
Valencich, Trina J, physical chemistry
Valentine, Frederick Albert, mathematics
Valentine, William Newton, medicine
Vance, Velma Joyce, vertebrate zoology
Van Der Meulen, Joseph Pierre, neurology, neurophysiology
Van Driest, Edward Reginald, fluid dynamics
Van Lancker, Julien L, pathology
van Marthens, Edith, veterinary medicine, nutrition
Vaughn, Peter Paul, vertebrate paleontology
Venit, Stewart Mark, mathematics
Verity, Maurice Anthony, pathology, neuropathology
Vernon, Frank Lee, Jr, low temperature physics, quantum physics
Villablanca, Jaime Rolando, neurophysiology, experimental neurology
Villarejo, Merna, biochemistry
Visser, Donald Willis, biochemistry
Voge, Marietta, parasitology
Vogl, Richard J, botany, ecology
Vogt, Peter Klaus, biology, virology
Vold, Marjorie Jean, colloid chemistry
Von Bloeker, Jack Christian, Jr, vertebrate zoology
Von Leden, Hans Victor, otolaryngology
Vredevoe, Donna Lou, immunology, microbiology
Wada, James Yasuo, lasers, plasma physics
Waddell, Charles Noel, physics
Wadley, Margil Warren, inorganic chemistry
Waggoner, James Norman, aerospace medicine
Wagner, William Gerard, theoretical physics, quantum electronics
Waisman, Jerry, pathology
Walberg, Clifford Bennett, clinical chemistry
Walford, Roy Lee, Jr, pathology
Walker, Boyd Wallace, ichthyology
Wall, Thomas Randolph, molecular biology, immunology
Wallace, Arthur, plant nutrition, plant physiology
Wallen, Clarence Joseph, mathematics
Walsh, Don, physical oceanography, ocean engineering
Walter, Donald Oliver, neurophysiology
Walter, Hartmut, biogeography, wildlife management
Walter, Richard D, medicine
Walters, Vladimir, ichthyology
Wan, Suk Han, pharmacodynamics, clinical pharmacology
Ward, John F, radiation chemistry, radiation biochemistry
Ward, Paul H, otolaryngology
Ward, Robert, pediatrics
Ware, Arnold Grassel, biochemistry
Warf, James Curren, inorganic chemistry
Warner, Nancy Elizabeth, pathology
Warshel, Arieh, chemical physics, molecular biology
Wasserman, Martin S, child psychiatry
Wasson, John Taylor, cosmochemistry, planetology
Watson, Kenneth De Pencier, geology
Waxman, Alan David, nuclear medicine
Wayne, George Jerome, psychiatry, psychoanalysis
Wayne, Lowell Grant, air pollution, environmental health
Webber, Milo M, radiology, nuclear medicine
Weber, Heather Ross, biochemistry
Weg, Ruth Bass, gerontology
Wehrle, Paul F, pediatrics, microbiology
Weil, Max Harry, cardiovascular diseases, biomedical engineering
Weimer, Henry Eben, immunochemistry
Weiner, Myron, pharmacology
Weinstock, Alfred, cell biology, periodontology
Weisman, Harold, anesthesiology
Weiss, Max Tibor, physics
Weiss, Richard Louis, biochemistry
Weissler, Gerhard Ludwig, physics
Welch, Lloyd Richard, mathematics
Wells, Patrick Harrington, biology
Welmers, Everett Thomas, mathematics
Wenzel, Bernice Martha (Mrs Wendell E Jeffrey), neuropsychology, medical education
Wessel, John Emmit, chemical physics, forensic science
West, Charles Allen, biochemistry
West, Charles David, analytical chemistry

West, Louis Jolyon, psychiatry, neurology
Westberg, Karl Rogers, chemistry, physics
Westcott, Wayne Leslie, biochemistry
Wettstein, Felix O., molecular biology
Whang, Sukoo Jack, microbiology, immunology
Wheeler, Louis Cutter, systematic botany
Wheelon, Albert Dewell, theoretical physics
Whistler, David Paul, vertebrate paleontology
Whitaker, Clay Westerfield, otolaryngology
Whitaker, Leslie A., analytical chemistry
White, Fred Newton, cardiovascular physiology, comparative physiology
White, Maurice Leopold, microbiology
White, Paul A., mathematics
Whiten, Albert Leon, nuclear physics
Whitenm, Charles A. Jr., nuclear physics, intermediate energy physics
Wicker, Berthold Robert, biochemistry
Wickham, Donald G., inorganic chemistry
Wilbert, Johannes, anthropology
Wilcox, Charles Hamilton, theoretical physics, engineering management
Wilcox, Gary Lynn, biochemical genetics
Williams, Bobby Joe, anthropology, population genetics
Williams, Harold Hamilton, ornamental horticulture
Willis, William W., nuclear physics, solid state electronics
Wisnieski, Bernadine Joann, physical biochemistry
Wistreich, George A., microbiology, state electronics
Wilson, Edward Carl, invertebrate paleontology
Wilson, Frances G., zoology, biology
Wilson, James Woodrow, chemistry
Wilson, Miriam Geisendorfer, medicine, pediatrics
Wolf, Walter, radiochemistry, radiopharmacy
Wolfram, Frederick John, neurochemistry
Wolinsky, Albert, physics, mathematics
Wong, Alfred, plasma physics
Wong, Nancy Elizabeth, speech pathology
Wood, Richard Lyman, cytology
Wood, Sherwin Francis, protozoology, medical entomology
Wood, Spencer Hoffman, geology
Woods, Clyde M., anthropology
Woods, Sherwyn Martin, psychiatry, psychoanalysis
Woody, Charles Dillon, neurophysiology
Worden, Ralph Edwin, physical medicine
Work, Telford Hindley, biology, epidemiology
Wright, Byron Terry, nuclear physics
Wright, Edwin T., dermatology
Wright, Harry Tucker, Jr., pediatrics, electrical engineering
Wurtele, Morton Gaither, dynamic meteorology
Yamamoto, Joe, psychiatry
Yamashita, Stanley Motohiro, engineering
Yang, William Chi Tsu, physiology
Yano, Fleur Belle, theoretical physics
Yates, Francis Eugene, physiology
Yen, Teh Fu, environmental science, biochemical engineering
Yin, Barbara Hsin-Hsin, statistics
Young, Richard Wain, anatomy
Young, Stephen James, atmospheric physics
Yu, David U L, nuclear physics, particle physics
Yu, Grace Wei-Chi Hu, plant physiology, cell biology
Zabin, Irving, biochemistry, molecular biology
Zachary, Rizkalla, human anatomy
Zamenhof, Patrice Joy, molecular genetics
Zamenhof, Stephen, biochemistry, genetics
Zarem, Abe Mordecai, electrooptics, engineering management
Zarem, Harvey A., plastic surgery
Zeldis, Louis Jenrette, pathology
Zighelboim, Jacob, immunobiology
Zimmer, Russel Leonard, invertebrate biology

Zimmerman, Emery Gilroy, anatomy, medicine

LOS GATOS
Barasch, Werner, plastics chemistry
Dorman, Stephen Charles, biochemistry, pesticide chemistry
Einarsson, Alfred W., physics
Johnson, Duane Edward, organic polymer chemistry
Johnson, Ralph E., physical chemistry
Ruggles, Ivan Dale, mathematics
Sussman, Irving, mathematics
Thompson, James Oliver, physical chemistry

LOS NIETOS
Duvall, Jacque L., physical chemistry, organic chemistry

LOS OSOS
Crouch, Robert Wheeler, micropaleontology, paleoecology
Wiese, John Herbert, geology

MALIBU
Abrams, Richard Lee, applied physics
Allen, Susan Davis, optical physics
Arnold, Kenneth Wiltshire, plasma physics
Barnoski, Michael K., solid state physics
Baron, Robert, solid state physics
Chester, Arthur Noble, theoretical physics
Close, Donald Henry, optics
Crandall, Walter Ellis, nuclear physics
Dulgeroff, Carl Richard, plasma physics
Eckhardt, Gisela (Marion), physics
Eckhardt, Wilfried Otto, physics
Forward, Robert Lull, physics
Frashier, Loyd Dola, physical chemistry
Gentile, Anthony L., materials science
Hess, LaVerne Derryl, physical chemistry
Hofmann, Gunter August George, plasma physics
Householder, Alston Scott, mathematics
Hughes, Norman, developmental genetics
Janney, Gareth Maynard, lasers
Kikuchi, Ryoichi, statistical mechanics
Kyle, Nanse Rector, chemical physics
Lotspeich, James Fulton, physics
Smith, George Foster, physics, electronics
Soffer, Bernard Harold, physics
Taylor, Charlotte Clarke, chemistry
Wilks, Charles Edward, mathematics
Wilson, Robert Gray, electronics, nuclear physics
Wolf, Edward D., surface physics
Wong, Shi-Yin, organic chemistry
Yamagishi, Frederick George, organic chemistry

MANHATTAN BEACH
Anthony, Romuald, electrooptics
Greenman, Norman Nathan, geology
Schell, William John, inorganic chemistry
Wax, Robert LeRoy, space physics
Whitmer, Robert Morehouse, electrical engineering

MARINA DEL REY
Alvi, Zahoor Mohem, radiological physics, biomedical engineering
Baak, Tryggve, physical chemistry, inorganic chemistry
Biehl, Arthur Trew, physics
Brode, Harold Leonard, theoretical physics
Dee, Diana, physical chemistry
Fisher, George Phillip, applied physics
Gilmore, Forrest Richard, physics
Greifinger, Carl, theoretical physics
Greifinger, Phyllis Stoliar, physics
Hubbard, Harmon William, physics
Hundley, Richard O'Neil, theoretical physics
Latter, Albert L., theoretical physics
Le Levier, Robert Ernest, theoretical physics
Martinelli, Ernest A., nuclear physics
Mitchell, Harold Hugh, medicine
Rasiewicz, Casimir E., bacteriology
Ridgway, Stuart L., energy conversion
Schatter, Marvin Baker, chemical engineering, operations research
Turco, Richard Peter, aeronomy

MARTINEZ
Dublin, William Brooks, pathology

MCFARLAND
Davis, Larry Alan, crop physiology

MENLO PARK
Aberth, William H., mass spectrometry
Ablow, Clarence Maurice, mathematics
Acton, Edward McIntosh, bio-organic chemistry
Adam, David Peter, palynology, climatology
Addicott, Warren O., paleontology
Anbar, Michael, physical inorganic chemistry
Ashley, Roger Parkmand, economic geology
Bailey, Edgar Herbert, geology
Banks, Norman Guy, geology
Barker, John Roger, chemical kinetics
Barnes, David Fitz, geophysics
Barnes, Ivan, geochemistry
Basler, Roy Prentice, radiophysics
Bateman, Paul Charles, structural geology, petrology
Baxter, John Lewis, fisheries
Benitez, Allen, organic chemistry
Benson, Sidney William, physical chemistry
Berg, Henry Clay, geology
Bischoff, James Louden, geochemistry
Black, Graham, physical chemistry
Blake, Milton Clark, Jr., geology
Bloom, Arnold Lapin, physics
Bloom, Elliot D., particle physics
Boardman, Robert Leland, economic geology
Bohonos, Nestor, biochemistry
Bonilla, Manuel George, geology
Boucher, Gary Wynn, geophysics
Brabb, Earl Edward, geology
Brauman, Sharon Kruse, physical organic chemistry
Brew, David Alan, geology
Brown, Walter Creighton, biology
Buck, Paul Andrews, plant physiology, food technology
Bufe, Charles Glenn, seismology
Buffington, Edwin Conger, marine tectonics
Byerlee, James Douglas, geophysics
Campbell, Catherine Chase, paleontology, geomorphology
Campbell, Russell Harper, earth science, engineering geology
Carlson, Paul Roland, marine geology
Carr, Michael H., astrogeology
Case, James Edward, geology, geophysics
Chapman, Robert Mills, geology
Chesnut, Walter G., engineering physics
Christ, Charles Louis, physical chemistry
Christiansen, Robert Lorenz, geology
Christie, Joseph Herman, analytical chemistry
Churkin, Michael, Jr., geology
Clark, Malcolm Mallory, geology
Clifton, Hugh Edward, geology
Coats, Robert Roy, mining geology
Cobb, Edward Huntington, geology
Coffey, Howard Thomas, low temperature physics
Coleman, Robert Griffin, geology
Collis, Ronald Thomas, atmospheric physics
Colwell, William Tracy, pharmaceutical chemistry, synthetic organic chemistry
Conomos, Tasso John, oceanography, geochemistry
Cook, Charles J., atomic physics, electrical engineering
Cook, Harry E. III, geology
Cook, Lawrence Harvey, chemistry
Cornwall, Henry Rowland, economic geology, petrology
Cory, Michael, medicinal chemistry
Creasey, Savill Cyrus, geology
Crittenden, Max Dermont, Jr., environmental geology
Cubicciotti, Daniel David, Jr., high temperature chemistry, physical inorganic chemistry
Curran, Donald Robert, physics

Efron, Robert, neuropsychology, neurophysiology
Finley, Theodore N., internal medicine
Koopmann, Henry Ferdinand, physical chemistry
McFarland, William Horace, hematology
Mennell, John McMillan, physical medicine
Rao, Ananda G., biochemistry
Schwarzer, Carl G., organic chemistry

Dabberdt, Walter F., air pollution
Dalrymple, Gary Brent, geology
Davenport, John Eaton, physical chemistry, chemical physics
Davis, Paul A., meteorology
Davis, Willard E., geophysics
Davis, William Elsmore, Jr., cancer chemistry
Dawson, Marcia Illton, bio-organic chemistry
Dedrick, Kent Gentry, theoretical physics
Degraw, Joseph Irving, Jr., medicinal chemistry
Denson, Donald D., organic chemistry
Dilley, James V., toxicology, inhalation toxicology
Dimen, William Horace, geophysics
DoAmaral, Jefferson Ribeiro, organic chemistry, mass spectrometry
Dodge, Franklin C W., geology
Doell, Richard Rayman, geophysics
Dresch, Francis William, statistics, operations research
Duffield, Wendell Arthur, petrology, structural geology
Durrum, Emmett Leigh, biochemistry
Eaton, Jerry Paul, seismology
Eckstrom, Donald James, chemical physics, lasers
Everndem, Jack Foord, geophysics
Eberlein, George Donald, geology
Falconer, David G., optical physics, radiophysics
Gardner, James Vincent, marine geology
Ferrians, Oscar John, Jr., geology
Feth, John Henry, geology
Field, Michael Ehrenhart, marine geology
Fieldstead, Arnold Howard, quaternary geology
Fisher, Philip Chapin, physics, astronomy
Ford, Arthur B., geology, petrology
Foster, Helen Laura, geology
Fournier, Robert Orville, geochemistry
Freeman, Gustave, medicine
Fremouw, Edward Joseph, aeronomy, radiophysics
Gillette, James Robert, pharmacology
Glover, Leon Conrad, Jr., organic chemistry, polymer chemistry
Goben, Lola Coleman, operations research
Golden, David Mark, chemical kinetics
Gower, Howard Dale, geology
Grantz, Arthur, geology
Greene, Gordon William, geophysics
Greene, Robert Carl, geology
Griggs, Allan Bingham, geology
Griscom, Andrew, geophysics
Gromme, Charles Sherman, geology
Gryc, George, geology
Guibranden, Robert Allen, geology
Hall, Wayne Everett, geology
Hamilton, Thomas Dudley, quaternary geology, glacial geology
Heard, Harry Gordon, physics
Heller, Jorge, organic chemistry
Hem, John David, geochemistry
Hendry, Dale Glenn, physical organic chemistry
Henry, David Weston, medicinal chemistry
Heynick, Louis Norman, physics, mathematics
Hietanen-Makela, Anna (Martta), geology
Hill, David Paul, geophysics
Hill, Mary Rae, geomorphology
Hill, Robert Mateson, physics
Holzer, Thomas Lequear, geology
Honey, Richard Churchill, electrooptics
Hopkins, David Moody, geology
Horsma, David August, physical chemistry
Hotz, Preston Enslow, geology
Huber, Norman King, geology
Hunter, Ralph Eugene, geology
Inman, Robert Eugene, plant pathology, botany
Iyer, Harihariyer Mahadeva, geophysics, seismology
Jackson, Everett Dale, geology
James, Ronald Valdemar, soil chemistry
Jenne, Everett A., soil chemistry
Jensen, Richard Arthur, pharmacology
Johanson, Robert Gail, analytical chemistry
Johnson, Howard (Laurence), medicinal chemistry, pharmacology
Johnson, Oscar Hugo, organic chemistry
Jones, David Charles Lloyd, geology, physiology
Jones, David Lawrence, geology, paleontology
Joyner, William B., geophysics
Kelly, David Horton, vision
Kenley, Richard Allen, organic chemistry
Kennedy, Vance Clifford, geochemistry, hydrology

Kharaka, Yousif Khoshu, geochemistry, hydrogeology
Kinderman, Edwin Max, nuclear chemistry
King, Chi-Yu, geophysics
King, Philip Burke, geology
Kistler, Ronald Wayne, geology
Kopelman, Jay B, environmental sciences
Kopf, Rudolph William, stratigraphy, structural geology
Krebs, John S, biophysics, radiation biology
Krieger, Medora Hooper, geology
Krishnan, Kamala Sivasubramaniam, optical physics, solid state science
Kryter, Karl David, psychophysiology, psychoacoustics
Kuhlmann, Karl Frederick, biophysical chemistry
Kunkel, Fred, hydrology
Kvenvolden, Keith Arthur, organic geochemistry
Lachenbruch, Arthur Herold, geophysics
Lahr, John Clark, seismology
Lanphere, Marvin Adler, geology, geochemistry
Laughlin, William Sceva, physical anthropology
Lee, William Hung Kan, geophysics
Lee, William Wei, organic chemistry
Leonard, Robert Stuart, geophysics, aeronomy
Liu, David H W, toxicology
Loney, Robert Ahlberg, structural geology, petrology
McCulloch, David Sears, geology, geomorphology
McKee, Edwin H, geology
Merritt, Thomas Parker, atomic spectroscopy
Mitoma, Chozo, biochemical pharmacology
Molnia, Bruce Franklin, marine geology
Mooney, John Bernard, analytical chemistry, materials research
Moore, George William, geology
Moore, J Strother, computer science
Moore, James Gregory, geology
Morris, Hal Tryon, geology
Morrison, Stanley Roy, surface chemistry, surface physics
Moseley, John Travis, atomic physics, molecular physics
Muffler, Leroy John Patrick, geology, geochemistry
Nason, Robert Dohrmann, seismology
Nelson, Carlton Hans, geology
Nelson, Raymond Adolph, radiophysics
Newell, Gordon Wilfred, toxicology
Nielsen, Norman Russell, computer science, information science
Normark, William Raymond, oceanography, marine geology
Nyberg, David Dolph, polymer chemistry
Olmsted, Franklin Howard, geology
O'Neil, James R, geochemistry, physical chemistry
Orcutt, Harold George, fish biology
Ovenshine, A Thomas, geology
Page, Norman J, petrology
Page, Robert Alan, Jr, geophysics
Paskert, Paul F, operations research
Patton, William Wallace, Jr, geology
Pauling, Linus Carl, chemistry, physics
Pease, Marshall Carleton, III, physics, mathematics
Peselnick, Louis, solid state physics
Peters, Howard McDowell, organic chemistry
Peters, John Henry, biochemistry, biochemical pharmacology
Peterson, Donald William, geology
Peterson, James Ray, atomic physics
Pettijohn, Richard Robert, radiochemistry
Pfluke, John H, geophysics, seismology
Pierce, Edward Thomas, meteorology, physics
Pike, Richard Joseph, Jr, geology, physical geography

Pincus, Jack Howard, biochemistry, immunochemistry
Piper, David Zink, oceanography
Rabinowitz, Mario, low temperature physics, electrical engineering
Radbruch-Hall, Dorothy Hill, geology
Radtke, Arthur Sears, economic geology, mineralogy
Raleigh, Cecil Baring, geophysics
Redlich, Dorothy Von, organic chemistry, biochemistry
Reist, Elmer Joseph, organic chemistry
Repenning, Charles Albert, paleontology, mammalogy
Reyes, Zoila, organic chemistry
Rhodes, Charles Kirkham, atomic physics, lasers
Rice, Philip Joseph, Jr, electronic physics
Robbins, Robert Crowell, physical chemistry
Roberson, Charles Elmer, geochemistry
Roberts, Ralph Jackson, geology
Robinson, Arthur Brouhard, medicinal chemistry
Robinson, Gershon Duvall, geology
Robinson, Stephen Clive, mineralogy
Rodden, Robert Morris, physics, biology
Rollosson, George William, physics
Ross, David Samuel, physical organic chemistry
Ross, Donald Clarence, geology
Ross, Donald Lewis, organic chemistry
Rubin, Jacob, soil physics, hydrology
Sancier, Kenneth Martin, physical chemistry
Sass, John Harvey, geophysics
Savage, James Crampton, geophysics
Scholl, David William, geology
Scott, Edward W, geology
Seely, Leslie B, Jr, fluid dynamics
Seiders, Victor Mann, geology
Servos, Kurt, mineralogy
Shreve, George Wilcox, physical chemistry
Silberling, Norman John, stratigraphy
Silberman, Miles Louis, geochemistry
Simmon, Vincent Fowler, microbiology
Sims, John David, sedimentology, quaternary geology
Singleton, Richard Collom, mathematical statistics, information science
Skinner, Wilfred Aubrey, Jr, organic chemistry
Slack, Keith Vollmer, limnology
Slanger, Tom George, chemical kinetics, atmospheric chemistry
Smith, Carl Walter, physics
Smith, Felix Teisseire, atomic physics
Smith, George Irving, quaternary geology
Smith, Judith Terry, paleontology
Snavely, Parke Detweiler, Jr, geology
Snyder, Charles Theodore, hydrology
Sonnenberg, Joseph, organic chemistry
Sovish, Richard Charles, organic chemistry, polymer chemistry
Spieker, Andrew Maute, hydrology
Stager, Harold Keith, geology
Stein, Stephen Ellery, physical chemistry, chemical kinetics
Stewart, John Harris, geology
Stolzenberg, Sidney Joseph, reproductive physiology, endocrinology
Stuart-Alexander, Desiree Elizabeth, geology
Swanson, Donald Alan, geology
Tabor, Rowland Whitney, geology
Tanabe, Masato, organic chemistry
Targ, Russell, physics
Thatcher, Wayne Raymond, seismology
Theodore, Ted George, economic geology
Tobey, Arthur Robert, applied physics
Torbit, Charles Allen, Jr, developmental biology
Truesdell, Alfred Hemingway, geology, chemistry
Unger, John Duey, geophysics, seismology
Uyeno, Edward Teiso, pharmacology
Vallier, Tracy L, marine geology
Von Huene, Roland, geology
Waldinger, Richard J, computer science
Wallace, Graham Franklin, computer science
Wallace, Robert Earl, geology
Walton, Barry Louis, solid state physics, operations analysis
Wang, Frank Feng Hui, marine geology
Ward, Peter Langdon, seismology, volcanology
Warren, David Henry, geophysics, seismology
Wentworth, Carl M, Jr, environmental geology, sedimentology
White, Donald Edward, geology
White, Ronald Keith, energy conversion, environmental physics
Wilshire, Howard Gordon, petrology

Wilson, L Kenneth, exploration geology
Wise, Henry, physical chemistry
Woodriff, Roger L, mathematics
Wrucke, Chester Theodore, Jr, geology
Yadavalli, Sriramamurti Venkata, physics, electrical engineering
Yeend, Warren Ernest, geology
Zand, Siavosh Marc, hydrology, environmental sciences

MERCED
Ainslie, John Durham, psychiatry

MILL VALLEY
Baer, Robert M, mathematics
Elftman, Alice G, anatomy
Elftman, Herbert (Oliver), anatomy

MILLBRAE
Bowen, Myles Foster, entomology

MISSION VIEJO
Maxum, Bernard J, electromagnetism, energy systems

MODESTO
Boyer, Alvin C, biochemistry, organic chemistry
Carr, John B, medicinal chemistry
Carter, Gerald Bate, analytical chemistry
Collins, Jeffery Allen, animal parasitology
Corey, Robert Arden, entomology
Cunningham, Virgil Dwayne, entomology, plant pathology
Fan, Hsing Yun, agricultural biochemistry
Favour, Cutting Broad, medicine
Halloran, Hobart Rooker, poultry nutrition, animal nutrition
Hass, D Kendall, parasitology
Haynes, George Rufus, synthetic organic chemistry
Huston, Charles K, analytical chemistry, biological chemistry
Jackson, Earl Kenneth, plant physiology
Joyner, Alan E, Jr, microbiology
Keasling, Hugh Hilary, pharmacology
Loeffler, Erwin Stanley, agricultural chemistry
Loeffler, Josef Ernst, agricultural chemistry
McKinney, William Jan, analytical chemistry
Mersmann, Harry John, biochemistry
Morenzoni, Richard Anthony, microbiology
Payne, George Bernson, organic chemistry
Porter, Paul Edward, physical chemistry
Potter, John Clarkson, radiochemistry, analytical chemistry
Prudhomme, Ronald Edward, food technology
Rader, William Ernest, plant pathology
Reed, Walter T, entomology
Schroeder, Mark Edwin, insect physiology
Siers, David Gard, population genetics
Simkover, Harold George, entomology
Skelsey, James Jeremiah, entomology
Soloway, Samuel Barney, organic chemistry
Stanton, Hubert Coleman, pharmacology
Stoutamire, Donald Wesley, organic chemistry
Sun, Yun Pei, insect toxicology
Sundelin, Kurt Gustav Ragnar, organic chemistry
Tieman, Charles Henry, Jr, organic chemistry
Wittsell, Lawrence Eugene, soil science
Young, Robert, Jr, veterinary medicine
Youngman, Edward August, organic chemistry, polymer chemistry

MOFFETT FIELD
Altman, Robert Leon, physical chemistry
Bader, Michel, research administration
Billingham, John, aerospace medicine, exobiology
Billings, Charles Edgar, Jr, aerospace medicine, environmental physiology
Billman, Kenneth William, lasers
Black, David Charles, theoretical astrophysics, meteorites
Borucki, William Joseph, atmospheric physics
Buchler, Alfred, physical chemistry, high temperature chemistry
Bunch, Theodore Eugene, geology
Caroff, Lawrence John, astrophysics

Cassen, Patrick Michael, gas dynamics, magnetohydrodynamics
Chackerian, Charles, Jr, physical chemistry, molecular spectroscopy
Debs, Robert Joseph, microwave physics
DeVincenzi, Donald Louis, exobiology, biochemistry
Erickson, Edwin Francis, astrophysics
Feller, David Douglas, physiology
Fohlen, George Marcel, organic chemistry, polymer chemistry
Gault, Donald E, astrogeology, planetary sciences
Gilbreath, William Pollock, technological forecasting, surface chemistry
Ginoza, Herbert S, biochemistry, microbiology
Golub, Morton Allan, physical chemistry, polymer chemistry
Goorvitch, David, astrophysics
Greeley, Ronald, geology
Greenleaf, John Edward, exercise physiology, environmental physiology
Grindeland, Richard Edward, physiology
Hansen, Carl Frederick, chemical physics
Harding-Barlow, Ingeborg, chemistry, pathology
Haymaker, Webb Edward, neuropathology
Heinrich, Milton Rollin, biochemistry
Hochstein, Lawrence I, bacteriology
Khan, Imdad Haque, physics
Klein, Harold Paul, microbial physiology, biochemistry
Kostiw, Luba Liszczynska, microbiology
Lanyi, Janos K, biochemistry
Lawless, James George, analytical chemistry
Lea, Susan Maureen, astrophysics
Lerner, Narcinda Reynolds, polymer chemistry
Malich, Charles Wilson, biophysics
Mark, Hans Michael, physics
McCutcheon, Ernest P, medical physiology, biomedical engineering
Mehler, William Raphael, neuroanatomy
Miquel, Jaime, gerontology
Nerney, Steven Francis, space physics, magnetohydrodynamics
Newsom, Bernard Dean, environmental physiology
Oyama, Jiro, physiology, biochemistry
Oyama, Vance I, biochemistry, organic chemistry
Philpott, Delbert E, molecular biology, medical research
Poppoff, Ilia George, aeronomy
Reinisch, Ronald Fabian, chemistry
Reynolds, Ray Thomas, planetary sciences
Rosser, Robert William, fluorine chemistry, polymer chemistry
Sandler, Harold, medicine, physiology
Scargle, Jeffrey D, astronomy
Schultz, Peter Hewlett, astrogeology
Silverman, Melvin Philip, microbiology, geochemistry
Simmonds, Richard Carroll, laboratory animal science, research administration
Snetsinger, Kenneth George, geochemistry, mineralogy
Sridhar, Champa Guha, solid state physics
Starr, Walter LeRoy, physics
Stein, Seymour Norman, physiology
Stewart, John Douglas, psychophysics
Tremor, John W, comparative physiology, environmental biology
Valero, Francisco Pedro Jorge, atomic spectroscopy, molecular spectroscopy
Vedder, James Forrest, planetary sciences
Vernikos-Danellis, Joan, pharmacology, endocrinology
Watson, Velvin Richard, gas dynamics, numerical analysis
Whitten, Robert Craig, Jr, aeronomy
Winget, Charles M, aerospace sciences
Witteborn, Fred Carl, physics, astrophysics
Wydeven, Theodore, physical chemistry
Young, Donald Rudolph, physiology, metabolism
Young, Richard Evans, dynamic meteorology, geophysics
Young, Wei, biophysics
Zill, Leonard Peter, biochemistry

MONROVIA
Chait, Edward Martin, analytical chemistry, mass spectrometry
Dave, Bhalchandra A, microbiology, food science
Farber, Milton, physical chemistry
Fletcher, Peter C, physical chemistry, ceramics
Gimlett, James I, optics, geophysics

CALIFORNIA

Jelinek, Bohdan, biochemistry, clinical medicine
Odiorne, Truman J., physical chemistry
Zatzick, Michael Raymond, nuclear physics

MONTE SERENO
Reas, William Harry, chemistry
Sutcliffe, Charles Herbert, physics

MONTEREY
Andrews, Robert Sanborn, geological oceanography, marine geophysics
Armstead, Robert Louis, theoretical physics
Barr, Donald R., mathematical statistics
Bleich, Willard Evan, mathematical physics
Borsting, Jack Raymond, mathematical statistics, operations research
Buskirk, Fred Ramon, physics
Cooper, John Niessink, physics
Crittenden, Eugene Casson, Jr., solid state physics, optics
Cunningham, William Peyton, physics
Dahl, Harvey A., physics
Daily, Edgar B., high energy physics, nuclear physics
Davidson, Kenneth LaVern, meteorology
Diamond, Ainsley Herbert, mathematics
Dyer, John Norvell, nuclear physics
Ebert, Earl Ernest, fish biology
Elsberry, Russell Leonard, meteorology
Esary, James Daniel, mathematical statistics
Faulkner, Frank David, applied mathematics
Forrest, Robert Neagle, operations research, operations analysis
Fossum, Robert, mathematics, statistics
Gaskell, Robert Eugene, applied mathematics
Gaver, Donald Paul, mathematics
Gotshall, Daniel Warren, marine ecology
Haderlie, Eugene Clinton, biological oceanography
Haltiner, George Joseph, meteorology
Handler, Harry Elias, experimental nuclear physics
Haney, Robert Lee, meteorology, oceanography
Harrison, Don Edward, Jr., surface physics, operations analysis
Hartman, James Kern, operations research, applied mathematics
Heinz, Otto, physics
Holl, Manfred Matthias, meteorology
Howard, Gilbert Thoreau, operations research
Johnson, James Howard, oceanography, fisheries
Jung, Glenn Harold, oceanography, meteorology
Kalmbach, Syndey Hobart, physics
Kelly, Raymond Leroy, atomic spectroscopy
Kinney, Gilbert Ford, physical chemistry
Kodres, Uno Robert, statistics
Koehler, Wilbert Frederick, optical physics
Laevastu, Taivo, oceanography, meteorology
Larson, Harold Joseph, mathematical statistics
Leipper, Dale F., oceanography
Lewis, Peter Adrian Walter, statistics
Lindmayer, Joseph, solid state physics
Lockhart, Brooks Javins, mathematics
Marshall, Kneale Thomas, operations research
Martin, Frank Lionel, meteorology
Martin, Michael, pollution biology
McMasters, Alan Wayne, operations research
Medwin, Herman, underwater acoustics
Milch, Paul R., operations research, statistics
Milne, Edmund Alexander, nuclear physics
Morris, George William, mathematics, mathematical statistics
Neighbours, John Robert, solid state physics
Olsen, Leonard Oliver, solid state physics, molecular spectroscopy
Paquette, Robert George, physical oceanography, ocean engineering
Pulliam, Francis McConnell, mathematics
Read, Robert Richard, mathematical statistics
Reese, William, thermal physics
Reinhard, Richard Alan, inorganic chemistry
Renard, Robert Joseph, meteorology
Reynolds, Melvin Ferguson, physical

Richards, Francis Russell, operations research, applied statistics
Riggin, John Dewitt, marine geophysics
Rinehart, Robert Fross, mathematics
Rodeback, George Wayne, physics
Rowell, Charles Frederick, physical organic chemistry
Sanders, James Vincent, acoustics
Schacher, Gordon Everett, physics
Schoenstadt, Arthur Loring, applied mathematics
Schultz, John Wilfred, physical chemistry
Schwirzke, Fred, physics
Seckel, Gunter Rudolf, oceanography
Shorb, Alan McKean, oceanographical logic
Shubert, Bruno Otto, mathematics, operations research
Shudde, Rex Hawkins, nuclear chemistry
Smith, Raymond James, civil engineering, geology
Stewart, Robert Everett, physics
Taylor, Charles Luther, physical meteorology
Thompson, Warren Charles, oceanography
Tolles, William Marshall, physical chemistry
Torrance, Esther McCormick, mathematics
Trahan, Donald Herbert, mathematics
Tysver, Joseph Bryce, applied statistics, operations research
van der Bijl, Willem, meteorology
Wang, Peter Cheng-Chao, statistics
Wilde, Carroll Orville, mathematics
Williams, Roger Terry, dynamic meteorology
Wilson, Oscar Bryan, Jr., physics
Woehler, Karlheinz Edgar, physics
Zeleny, William Bardwell, physics
Zweig, Hans Jacob, mathematical statistics, operations research

MONTEREY PARK
Welsky, Norman, organic chemistry, immunochemistry

MORAGA
Barney, James Earl, II, analytical chemistry
Dienst, Carl Sedgwick, food science
Dodd, Everett E., protozoology, cytology
Friedman, Mendel, organic chemistry
Hoch, Paul Edwin, organic chemistry
Leitner, Philip, vertebrate zoology
Rockland, Louis B., agricultural chemistry, food science
Wiebe, Richard Penner, mathematical logic

MOSS LANDING
Arnal, Robert Emile, geology
Martin, John Holland, oceanography, pollution biology
Nybakken, James W., marine ecology

MOUNTAIN VIEW
Armstrong, Donald B., acoustics, crystallography
Beyer, Edgar Herman, plant genetics
Blachman, Nelson Merle, physics
Braemer, Allen C., chemotherapy, animal nutrition
Decker, Charles David, lasers, quantum electronics
Fahlen, Theodore Stauffer, lasers
Fettis, Henry Eason, applied mathematics
Gardner, Phillip John, physical chemistry
Goldsborough, John Paul, lasers
Goree, William Strozier, low temperature physics
Gray, Reed Alden, biochemistry
Harris, Frank Bower, Jr., physics
Hedde, Richard Duane, animal nutrition
Hogan, Clarence Lester, applied physics
Huber, Wolfgang, biochemistry
James, Brian David, physics
Jarrett, Steven Michael, physics
Johnson, Leroy Franklin, nuclear magnetic resonance
Kieff, John A., physical chemistry, inorganic chemistry
Daly, Kevin Richard, genetics
Davison, Walter Francis, mathematics
Dole, Jim, vertebrate ecology
Dulkin, Sol I., clinical chemistry
Elterson, Melvin Elroy, food science
Emboden, William Allen, Jr., systematic botany
Ervin, Guy Jr., physical chemistry, ceramics
Filippenko, Vladimir I., applied mathematics

Schalit, Lewis Martin, chemical physics, research administration
Schnell, Jerome Vincent, biochemistry
Shimazaki, Tatsuo, atmospheric physics, space physics
Siew, Chakwan, pharmacology
Titterton, Paul James, electro optics
Williams, Lewis David, chemistry
Wunderman, Irwin, electrooptics, electronic instrumentation

NEWARK
Foerster, Donald Ray, environmental chemistry

NEWBURY PARK
Cher, Mark, physical chemistry
Parry, Edward Petterson, analytical chemistry, environmental chemistry
Richards, Lorenzo Willard, physical chemistry

NEWHALL
Gerber, Louis P., biochemistry
Howe, George Franklin, plant physiology

NEWPORT BEACH
Bean, Ross Coleman, biochemistry
de Langre, John Paul, organic chemistry
Elms, James Cornelius, physics
English, Floyd L., solid state physics
Garfin, Louis, actuarial science
Gee, Allen, physical chemistry
Glasky, Alvin Jerald, biochemistry
Goodwin, Lester Kepner, nuclear science
Grynnak, David Alexis, spectroscopy, physics
Hance, Paul D., organic chemistry, operations research
Hilker, Harry Van Der Veer, Jr., physics
Hsieh, Paul Yao Tong, surface chemistry, chemical engineering
Kay, Robert Eugene, biochemistry
Kratzer, Reinhold, polymer chemistry
Lee, Yat-Shir, inorganic chemistry
Martin, John Eimslie, inorganic chemistry
Mason, James Willard, synthetic organic chemistry
Nichol, Francis Richard, Jr., virology
Phoenix, David A., geology, hydrology
Potter, Norman D., physical chemistry, inorganic chemistry
Reisman, Elias, experimental physics
Rogers, Howard H., chemistry, electronics
Sutton, Paul McCullogh, physics
Tin-Wa, Maung, pharmacognosy, natural products chemistry

NORTH HOLLYWOOD
Andersen, Wilford Hoyt, physical chemistry
Fernandez, Alberto Antonio, biochemistry, clinical chemistry
Keltz, Alan, biophysical chemistry, clinical chemistry
Lechtman, Max Dressler, microbiology

NORTHRIDGE
Abrash, Henry I., organic chemistry, biochemistry
Allen, James Paul, geography
Astrachan, Max, mathematics, statistics
Bates, Barney Leroy, biophysics
Barber, Mary Lee, entomology
Bellinger, Peter F., entomology
Bianchi, Donald Ernest, physiology, mycology
Biriuk, George, mathematics
Breckenridge, Robert George, physics
Cantor, Marvin H., cell physiology
Chow, Paul C., solid state physics
Collins, Lorence Gene, photogeology
Corcoran, Mary Ritzel, genetics, plant physiology
Court, Arnold, climatology
Cronbein, Georg Erich, biochemistry, pharmacology
Cunningham, Samuel Preston, physics

Fisler, George Frederick, biology, zoology
Foster, Lorraine L., mathematics
Furumoto, Warren Akira, plant science
Gaudin, Anthony J., herpetology, morphology
Grigori, Artur, algebra
Gutstadt, Allan Morton, geology, stratigraphy
Hardcastle, Kenneth Irvin, inorganic chemistry, physical chemistry
Harris, Francis Laurie, Jr., organic chemistry
Heimbold, Robert Lawson, mathematics
Jones, Kenneth Charles, algology
Joseph, Marjory L., textile chemistry
Karush, William, mathematics
Klinedinst, Paul Edward, Jr., organic chemistry
Kuhn, Daisy Angelika, microbiology
Lefevre, George, Jr., genetics
Lengyel, Bela Adalbert, physics
Lewthwaite, Gordon Rowland, historical geography, geography of the Pacific
Liang, Ching Yu, physics
Lin, Tung-Po, mathematics
Marx, Paul Christian, physical chemistry
Maxwell, Joyce Bennett, genetics
McIntire, Elliot Gregor, cultural geography, historical geography
Michaelson, Evalyn Jacobson, anthropology
Moore, Mortimer Norman, mathematics
Nathanson, Weston Irwin, mathematics
Nazarian, Girair Mihran, physical chemistry
Nyquist, Harlan LeRoy, organic chemistry
Olsen, Carl John, organic chemistry
Olson, Roy E., solid state physics
Oppenheimer, Steven Bernard, developmental biology, cancer
Pearlman, Harry, physical chemistry
Pohlo, Ross, invertebrate zoology
Pollock, Edward G., developmental biology, cell biology
Potter, Franklin Carl, paleontology
Poiter, Richard Lyle, biochemistry
Potts, Donald Harry, mathematics
Ravicz, Robert S., anthropology
Reichman, Sandor, physical chemistry
Romagnoli, Robert Joseph, electromagnetics
Rudd, Velva Elaine, botany
Sandhu, Harbhajan Singh, physics
Saavedra, Lydia Goodman, chemistry
Scheuer, Ernest Martin, statistics, operations research
Schiffman, Robert Harry, comparative physiology
Segal, Earl, environmental physiology
Seki, Ryoichi, theoretical nuclear physics, high energy physics
Silva, Ricardo, physical chemistry
Spots, Charles Russell, microbiology
Starrett, Andrew, mammalogy
Wang, I-Shou, population geography, urban geography
Watkins, William, mathematics, statistics
Weston, Charles Richard, developmental biology, microbial ecology
Wilson, Kenneth Allen, plant morphology, systematic botany
Wren, Frank Lynwood, mathematics
Wright, Elisabeth Muriel Jane, mathematics
Zeitin, Joel Loeb, mathematics

NORWALK
Crane, Jules M., Jr., ichthyology, marine biology
Garcia, Eugene N., biochemistry

OAKLAND
Abraham, Samuel, physiology, biochemistry
Ahuja, Jagan N., biochemistry, clinical chemistry
Anderson, Robert Thomas, cultural anthropology
Barany, Ronald, physical chemistry
Barnum, Emmett Raymond, organic chemistry
Beasley, William Joe, medical microbiology, immunology
Bowers, Darl Eugene, zoology, ecology
Boyer, Ruth McDonald, anthropology
Brust, David, solid state physics, semiconductors
Byerly, Perry, seismology
Collen, Morris F., medicine
Constantine, Denny G., veterinary medicine, epidemiology
Eastman, John W., clinical chemistry
Eisler, Daniel M., bacteriology

Fleisher, Daniel S, pediatrics
Gibbons, Mathew Gerald, atmospheric physics, nuclear physics
Guenther, Donna Marie, pediatrics, immunology
Haxo, Henry Emile, Jr, rubber chemistry, plastics chemistry
Hayward, Oliver Stoddard, internal medicine
Hubbs, John Charles, physics
Hundley, James Manson, public health, nutrition
Johnston, Muriel Evelyn, immunobiology
Jory, Farnham Stewart, physics
Kasapligil, Baki, botany
Kido, George Seiji, entomology
Levine, Hillel Benjamin, bacteriology
Liauw, Koei-Liang, organic chemistry
Mango, Frank Donald, organic chemistry, inorganic chemistry
Markell, Edward Kingsmill, parasitology, tropical medicine
Meads, Philip Francis, Jr, theoretical physics, computer science
Moretti, Richard Leo, developmental biology
Nasset, Edmund Sigurd, physiology
Nelson-Rees, Walter Anthony, genetics, cytology
Oakeshott, Gordon B, geology
O'Donnell, Gordon James, chemistry
Parsons, Robert Jerome, pathology
Ploux, Marie Denise Madeleine, plant physiology
Purshottam, Natesaier, epidemiology, nutrition
Reeve, Marian E, plant anatomy
Rosenfeld, George, biochemistry, medical research
Rostler, Fritz S, chemistry
Schwegmann, Jack Carl, plant pathology
Shearn, Martin Alvin, internal medicine
Smith, Elbert George, chemistry
Taybi, Hooshang, radiology
Terdiman, Joseph Franklin, biomedical nutrition
Tocher, Don, seismology
Ury, Hans Konrad, biostatistics, mathematical statistics
Vedros, Neylan Anthony, microbiology, immunology
Vegotsky, Allen, biological chemistry
Wolochow, Hyman, microbiology
York, Carl Monroe, Jr, physics

OCEANSIDE
Pearson, Edward Pillsbury, chemistry
Schumacher, John Christian, experimental solid state physics, industrial chemistry
Schumacher, Joseph Charles, industrial chemistry

ONTARIO
Beisel, Clifford Gordon, food science
Harding, Paul Raymond, Jr, mycology, plant pathology
Oser, Willem, physical chemistry, electrochemistry
Palmer, Grant H, biochemistry, nutrition

ORANGE
Carver, John Guill, physics engineering
Cohn, Jay Binswanger, psychiatry
Elliott, Henry Wood, pharmacology, anesthesiology
Flint, Arthur Emerson, economic geology
Fradkin, Cheng-Mei Wang, genetics
Heinz, David Murray, solid state chemistry
Hollingshead, Anne Huston, resource geography
Hull, Inez Mary, genetics
Kakis, Frederic Jacob, physical organic chemistry, air pollution
Kortright, James McDougall, medical chemistry
Miyada, Don Shuso, biochemistry
O'Loughlin, Bernard James, radiology
Pattison, Edward Mansell, psychiatry, applied anthropology
Quinlivan, William Leslie G, obstetrics & gynecology

Shand, Edwin William, organic chemistry
Thrupp, Lauri David, infectious diseases, microbiology
Westervelt, Clinton Albert, Jr, invertebrate zoology, parasitology
Whipple, Gerald Howard, medicine
Wimberley, Stanley, marine geology, oceanography

ORINDA
Baker, Don Robert, organic chemistry
Graf, Peter Emil, physical chemistry
Kothny, Evaldo Luis, air pollution, geochemistry
Lapporte, Seymour Jerome, organic chemistry, petroleum chemistry
MacIver, Donald Stuart, industrial chemistry
Miskus, Raymond P, organic chemistry, analytical chemistry
O'Meara, John Pierce, food technology
Ottke, Robert Crittenden, organic chemistry, biochemistry
Rust, Frederick Farlow, organic chemistry

OXNARD
Gordon, Archer Samuel, cardiovascular surgery, physiology
Hottle, George Austin, bacteriology, medical microbiology
Richards, William Robert, biophysics

PACIFIC GROVE
Abbott, Donald Putnam, invertebrate zoology
Abbott, Isabella Aiona, phycology, marine biology
Blinks, Lawrence Rogers, plant physiology
Buchsbaum, Ralph, invertebrate zoology, ecology
Coppens, Alan Berchard, physics, acoustics
Patchick, Paul Francis, earth science
Phillips, John Howell, Jr, biochemistry, immunology
Soffer, Louis M, chemistry, economics
Van Niel, Cornelis Bernardus, biochemistry

PACIFIC PALISADES
Csendes, Ernest, organic chemistry
Farmer, Donald Jackson, physics
Jacobs, S Lawrence, clinical chemistry
Katz, Yale H, meteorology
Knox, Cameron, oceanography, optics
Lyons, Harold, physics
Quade, Edward Schaumberg, mathematics, policy sciences
Siegel, Sidney, physics
Slichter, Louis Byrne, geophysics, seismology
Steinberg, Bernhard, pathology
Van Atta, John R, biophysics
Weisz, Robert Stephen, inorganic chemistry
Winston, Harvey, physical chemistry

PALM DESERT
Fischer, Craig Leland, anatomic pathology, laboratory medicine
Julian, Ormand C, cardiovascular surgery
Mahler, Richard Joseph, internal medicine, endocrinology

PALM SPRINGS
Stone, Newton C, meteorology, atmospheric sciences

PALO ALTO
Acton, Loren Wilber, solar physics
Adams, George Baker, Jr, physical chemistry
Adkins, Benjamin Jefferson, biochemistry
Alksne, Alberta Yearian, mathematics, geophysics
Allen, Ronald V, solid state physics
Alves, Ronald V, solid state physics, quantum electronics
Amelio, Gilbert Frank, solid state electronics
Amen, Ronald Joseph, food science, nutrition
Amkraut, Alfred A, immunology, immunochemistry
Anderson, Weston Arthur, physics
Archer, Robert James, solid state physics
Armstrong, Baxter Hardin, atomic physics

Arnold, James Tracy, experimental physics
Aron, Walter Arthur, theoretical physics
Asendorf, Robert Harry, theoretical physics, information science
Ayres, Wesley P, physics
Bailin, Lionel J, physical inorganic chemistry
Bandel, Herman William, plasma physics
Banik, Narendra Lal, neurochemistry
Banovitz, Jay Bernard, immunochemistry
Bardin, Russell Keith, nuclear physics
Bardin, Tsing Tchao, nuclear physics
Barrera, Joseph S, solid state physics
Bartelink, Dirk Jan, solid state electronics
Bauer, Robert Steven, experimental solid state physics, electronic spectroscopy
Baum, John William, organic chemistry
Becker, John Angus, nuclear physics
Beigler, Myron Arnold, nutrition, research administration
Bell, John Perkins, biochemistry, biology
Bennett, John Phillip, reproductive physiology, endocrinology
Benson, Harriet, organic chemistry, scientific bibliography
Bessler, Stuart Alan, mathematical statistics
Bidlack, Donald Eugene, veterinary medicine
Biegelsen, David K, experimental solid state physics
Birzis, Lucy, neurophysiology
Board, Robert Dennis, analytical chemistry, computer science
Bobb, Yvonne Dolores, microbiology, biochemistry
Bobrow, Daniel G, computer science
Boost, Gerhard, clinical medicine
Brown, Harmon W, Jr, chemistry
Brown, William Arnold, atomic physics
Buchanan, Robert Ambrose, physics, semiconductors
Burnham, Robert Danner, pharmacology
Canosa, Jose M, mathematical physics, applied mathematics
Caren, Robert Poston, physics
Carter, Kenneth, medicine, pharmaceutical chemistry
Castle, Robert O, geology, tectonics
Catura, Richard Clarence, physics, x-ray astronomy
Chalmers, Robert Anton, nuclear physics
Chang, Jaw-Kang, biochemistry
Chase, Lloyd Fremont, Jr, physics
Chen, Tu, magnetism, materials science
Cheng, David, experimental physics
Chervenka, Charles Henry, biochemistry
Chiang, Anne, physical chemistry, display technology
Chin, Jane Elizabeth Heng, organic chemistry, protein chemistry
Choong, Hsia Shaw-Lwan, organic chemistry, protein chemistry
Chow, Kao Laing, neurosciences, neuroanatomy
Cladis, John Baros, nuclear physics, plasma physics
Clauss, James K, analytical chemistry, physical chemistry
Clover, Richmond Bennett, magnetism
Codrington, Robert Smith, physics
Cohen, Jack, pharmacy, analytical chemistry
Cohen, Karl (Paley), nuclear science
Cohen, Paul Joseph, mathematics
Comar, Cyril Lewis, environmental health
Conway, Lynn Ann, computer science, electrical engineering
Crepea, Seymour B, medicine, allergy
Crismon, Jefferson Martineau, physiology
Cross, Alexander Dennis, organic chemistry
Cusumano, James A, catalysis
Cutler, Leonard Samuel, physics, mathematics
Czamanske, Gerald Kent, geochemistry
Dalla Betta, Ralph A, surface chemistry
Dawson, Daniel Joseph, organic polymer chemistry
Deal, Bruce Elmer, physical chemistry
Denison, Dean R, surface physics
Deutsch, Laurence Peter, information science
Dobrov, Wadim (Ivan), solid state physics
Dorfman, Ralph Isadore, biochemistry
Dougherty, Jean Hay, hematology
Dumas, Kenneth J, medicine
Duncan, Cecil Eugene, mathematics, information science
Dyck, Rudolph Henry, physical chemistry
Early, James M, physics
Ebaugh, Franklin G, Jr, medical administration, hematology

Elkind, Jerome I, computer sciences
Enderby, Charles Eldred, physics
Eng, Lawrence F, biochemistry
Engelmann, Reinhart Wolfgang H, electronic physics
English, William Kirk, information science
Evans, John Ellis, physics
Fisher, John Crocker, physics, natural science
Fisher, Thornton Roberts, nuclear physics
Fitch, William Lawrence, bio-organic chemistry
Flatt, Horace Perry, mathematics
Flegal, Robert Melvin, computer science
Forchielli, Enrico Henry, biochemistry
Forrest, Irene Stephanie, biochemistry
Foster, Robert L, geology
Fried, John, organic chemistry, medicinal chemistry
Fritchie, Frank Paul, physics
Frye, William Emerson, space physics
Galeener, Frank Lee, solid state physics
Gazdag, Jeno, plasma physics
Geschke, Charles Matthew, computer science
Ginzton, Edward Leonard, physics
Glaser, Robert Joy, medicine
Goldstein, Avram, pharmacology
Gordon, Manuel Joe, genetics, cell biology
Gottlieb, Abraham Mitchell, medicine, cardiology
Green, Donald Eugene, biochemical pharmacology, analytical chemistry
Greene, Paul E, chemistry
Griffith, Owen Malcolm, biochemistry, biophysics
Hahn, Harold Thomas, physical inorganic chemistry
Haitz, Roland Hermann, solid state physics
Hall, Harold Hershey, theoretical physics
Halleny, Duane Wesley, toxicology
Halpern, Paul, meteorology
Hamburg, David Alan, psychiatry
Hamilton, Douglas Holmes, geology
Hannah, Eric Cabot, solid state science
Havemeyer, Ruth Naomi, pharmacy, research administration
Hayes, Timothy Mitchell, solid state physics
Healy, John H, geophysics
Hebel, Louis Charles, experimental physics
Helmer, John, physics, electronics
Henrick, Clive Arthur, organic chemistry
Henzl, Milan Rastislav, medicine, endocrinology
Herd, Allen K, III, pharmacy, physical chemistry
Hill, George Richard, chemistry, fuel science
Hill, Marion Elzie, organic chemistry
Hill, Robert, physiology, biochemistry
Hirsch, Peter M, mathematics
Hiskes, Ronald, solid state physical chemistry
Hoffman, Charles John, inorganic chemistry
Hofstee, Barend Hendrik Jan, biochemistry
Hollister, Leo E, pharmacology
Huberman, Bernardo Abel, physics
Hultgren, Herbert Nils, medicine
Hurd, Paul DeHart, biology, science education
Hyams, Henry C, physics, electronics
Imhof, William Lowell, space physics, nuclear physics
Ingels, Neil Barton, Jr, biomedical engineering, cardiovascular physiology
Jamison, Steven Lyle, mathematics, computer sciences
Jenkins, Ted R, mathematics
Jochle, Wolfgang, reproductive physiology
Johnson, Ben Francis, microbial genetics
Johnson, Hugh Marvin, astronomy
Johnson, Vard Hayes, geology
Johnston, Robert R, theoretical physics
Jones, Gordon Henry, organic chemistry
Jordan, James A, Jr, applied mathematics
Jordan, Willard Clayton, nuclear physics
Judy, Kenneth James, insect physiology
Junga, Frank Arthur, solid state physics
Kahn, Frederic Jay, solid state physics, electrooptics
Kapany, Narinder Singh, physics, optics
Kaplan, Daniel Eliot, solid state physics, plasma physics
Kaplan, Henry Seymour, radiology
Kaplan, Ronald M, computer science
Karasek, Marvin A, biochemistry
Katan, Theodore, physical chemistry
Katz, Martin, pharmaceutics
Kazan, Benjamin, physics

CALIFORNIA

Goetz, Alexander Franklin Hermann, geophysics, geology
Goldreich, Peter, astrophysics
Goldstein, Raymond, physics
Goodstein, David Louis, low temperature physics
Gould, Roy Walter, physics, electrical engineering
Gray, Harry B, inorganic chemistry
Green, Richard H, microbiology
Greenhall, Charles August, mathematical analysis
Greenstein, Jesse Leonard, astrophysics
Grunthaner, Frank John, physical inorganic chemistry, solid state chemistry
Gulkis, Samuel, physics, aeronautical engineering
Gunn, James Edward, astrophysics
Gupta, Amitava, photochemistry, physical organic chemistry
Haagen-Smit, Arie Jan, biochemistry
Hagen, Charles Alfred, microbiology
Hall, Marshall, Jr, algebra
Hamilton, Carole Lois, energy conversion, systems analysis
Hamilton, Charles R., behavioral biology
Harkrider, David Garrison, seismology
Harreveld, Anthonie Van, animal physiology
Hartman, Richard Eugene, biophysics
Hartman, Roberta Smith, biophysics
Hartoog, Mark Richard, astronomy
Hasegawa, Shin, food science
Hayman, Ernest Paul, agricultural biochemistry
Heindl, Clifford Joseph, nuclear physics
Herb, John A, low temperature physics
Hibbs, Albert Roach, theoretical physics
Hoffman, Paul Fajvel, geology
Hood, Leroy E, immunogenetics, genetics
Horowitz, Norman Harold, genetics, biochemistry
Horowitz, Robert Miller, organic chemistry
Howard, Robert Franklin, astronomy
Hughes, Edward Wesley, chemistry, x-ray crystallography
Humphrey, Floyd Bernard, magnetism
Huntress, Wesley Theodore, Jr, chemical physics
Ingersoll, Andrew Perry, planetary atmospheres
Ireland, Robert Ellsworth, organic chemistry
Isokane, Robert Kazuo, microbiology
Jaffe, Leonard David, planetary sciences, aerospace engineering
Janssen, Michael Allen, radio astronomy, planetary sciences
Johnson, Torrence Vaino, planetary sciences, astronomy
Johnston, Alan Robert, optical chemistry
Juster, Norman Joel, organic chemistry
Kahle, Anne Bettine, geophysics
Kaighn, Morris Edward, embryology, virology
Kalfayan, Sarkis Hagop, polymer chemistry
Kamb, Walter Barclay, mineralogy, glaciology
Kavanagh, Ralph William, nuclear physics
Keller, Herbert Bishop, applied mathematics, numerical analysis
Klein, Michael John, radio astronomy
Knapp, Gillian Revill, radio astronomy
Knapp, Stephen Laurence, astronomy, instrumentation
Knowles, James Kenyon, applied mathematics
Koepfli, Joseph Blake, chemistry
Koliwad, Krishna M, microelectronics, semiconductors
Konopka, Ronald J, neurobiology
Kristian, Jerome, physics, astronomy
Kuppermann, Aron, chemical physics
Lagerstrom, Paco (Axel), applied mathematics
Landel, Aurora Mamaug, biochemistry
Landel, Robert Franklin, physical chemistry, rheology
Langmuir, Robert Vose, physics
Lau, Richard Lewis, applied mathematics
Laudenslager, James Bishop, chemical physics
Laufer, Arthur Russell, physics, science administration
Lebofsky, Larry Allen, planetary sciences
Leiga, Algird George, physical chemistry
Leighton, Robert Benjamin, astrophysics
Le Mehaute, Bernard J, hydrodynamics
Lester, Henry Allen, neurobiology, biophysics
Lewis, Edward B, biology
Lindena, Siegfried Johannes, energy conversion, magnetism

Lo, Kwok-Yung, radio astronomy, astrophysics
Locanthi, Dorothy Davis, astrophysics, scientific bibliography
Loefer, John B, biology
Loomis, Alden Albert, geology, oceanography
Loranger, William Farrand, xeroradiography
Lowenstam, Heinz Adolf, ecology
Lowy, Peter Herman, organic chemistry
Luthey, Joe Lee, space physics
Luxemburg, Wilhelmus Anthonius Josephus, mathematical analysis
MacCready, Paul Beattie, Jr, air pollution
Macdoran, Peter Frank, geodesy
Mackin, Robert James, Jr, physics
Manat, Stanley L, organic chemistry
Mann, Kenneth Clifford, physics
Marcus, Rudolph Julius, physical chemistry
Margolis, Jack Selig, chemical physics, spectroscopy
Marrs, Roscoe Earl, nuclear physics
Marsh, Richard Edward, physical chemistry
Marshall, J Howard, III, physics, electronics
Martin, Helen Eastman, internal medicine
Maserjian, Joseph, semiconductors, microelectronics
Mathews, Jon, theoretical physics
Matson, Dennis Ludwig, planetary sciences
McCann, Gilbert Donald, Jr, information science
McGill, Thomas Conley, Jr, physics, electrical engineering
McKoy, Basil Vincent, theoretical chemistry
McLaughlin, William Irving, celestial mechanics
McMahon, Daniel Stanton, biochemistry
Melvin, Jonathan David, atomic physics
Mercereau, James Edgar, physics
Meyer, Ronald Leo, neuroembryology
Mitchell, Herschel Kenworthy, biochemistry
Moffet, Alan Theodore, radio astronomy
Morgan, James John, environmental science, water chemistry
Morris, Mark Root, radio astronomy
Munch, Guido, astrophysics, astronomy
Murray, Bruce C, astronomy, geology
Nash, Douglas B, geology, space physics
Nickerson, Robert Fletcher, computer sciences
Niell, Arthur Edwin, radio astronomy
North, Wheeler James, oceanography
Ohnuki, Yasushi, cytogenetics
Oke, John Beverley, astronomy
Olsen, Edward Tait, radio astronomy, planetary atmospheres
Ong, Kwok Maw, physics
Orton, Glenn Scott, atmospheres
Owen, Ray David, genetics, immunology
Papas, Charles Herach, electrodynamics
Patterson, Claire Cameron, geochemistry, environmental chemistry
Peck, Charles William, experimental high energy physics
Perel, Julius, atomic physics
Persson, Sven Eric, astronomy
Pettit, John Tanner, theoretical physics
Philipson, Joseph, polymer chemistry
Pine, Jerome, physics
Plesset, Milton Spinoza, physics
Pomeroy, Richard Durant, chemistry
Porter, Lawrence Delpino, geophysics
Posner, Edward Charles, mathematics, engineering
Powers, Richard James, physics
Poynter, Robert Louis, molecular spectroscopy
Pudenz, Robert Harry, neurosurgery
Rea, Donald George, planetary sciences
Read, Richard Bradley, physics, radio astronomy
Revel, Jean Paul, cell biology
Rhein, Robert Alden, inorganic chemistry, polymer chemistry
Rho, Joon H, biochemistry
Richards, John Hall, biochemistry
Richstone, Douglas Orange, astronomy
Richter, Charles Francis, seismology
Roberts, John D, organic chemistry
Robinson, George Wilse, photochemistry
Rounds, Donald Edwin, embryology
Russell, Richard Lawson, genetics, neurobiology
Ryason, Porter Raymond, physical chemistry
Ryser, Herbert John, mathematics
Sackmann, I Juliana, astrophysics

Saffman, Philip Geoffrey, applied mathematics, fluid mechanics
Saffren, Melvin Michael, mathematical physics, fluid dynamics
Sandage, Allan Rex, astronomy
Sandefur, Kermit Lorain, optics
Sargent, Wallace Leslie William, astrophysics
Schaefer, William Palzer, inorganic chemistry
Schlipf, John Stewart, mathematical logic
Schmidt, Maarten, astronomy
Schroeder, Walter Adolph, protein chemistry
Schwarz, John Henry, theoretical physics, high energy physics
Sciulli, Frank J, experimental physics, elementary particle physics
Sharp, Robert Phillip, geomorphology
Shoemaker, Eugene Merle, geology
Sierk, Arnold John, nuclear physics
Silver, Leon Theodore, petrology, geochemistry
Simpson, Ruth DeEtte, anthropology, archaeology
Singer, Stanley, chemistry, physics
Sinsheimer, Robert Louis, biochemistry, biophysics
Slade, Martin Alphonse, III, celestial mechanics, radio astronomy
Smith, Darryl Lyle, solid state physics
Somoano, Robert Bonner, solid state physics
Standish, E Myles, Jr, astronomy
Stehsel, Melvin Louis, organic chemistry, plant physiology
Stenger, William, mathematics
Sternberg, Eli, mechanics
Stirn, Richard J, energy conversion
Stone, Alexander Glattstein, mathematics
Stone, Edward Carroll, Jr, physics
Strauss, Ellen Glowacki, molecular genetics, virology
Strauss, James Henry, biochemistry, virology
Strumwasser, Felix, physiology, neurobiology
Sulentic, Jack William, astronomy
Swanson, Paul N, radio astronomy
Swift, Ernest Haywood, analytical chemistry
Swope, Henrietta Hill, astronomy
Tajima, Yuji Alexis, plastics chemistry
Taussky, Olga (Mrs John Todd), mathematics, number theory
Taylor, Donald Stinson, chemistry
Taylor, Hugh P, Jr, petrology, geochemistry
Thompson, Thomas William, space physics
Thorne, Kip Stephen, astrophysics, theoretical physics
Thourson, Thomas Lawrence, physics
Thuan, Trinh Xuan, astrophysics
Todd, John, numerical analysis
Tollestrup, Alvin V, physics
Tombrello, Thomas Anthony, Jr, nuclear physics
Toth, Robert Allen, physics
Trajmar, Sandor, molecular physics
Tschoegl, Nicholas William, physical chemistry
Van Harreveld, Anthonie, physiology
Vaughan, Jack William, astronomy
Vaughan, Arthur Harris, Jr, astronomy
Vaughan, Francis Edward, physical chemistry
Vinograd, Jerome, molecular biology, biological chemistry
Vogel, Peter, nuclear physics
Vogt, Rochus E, physics
Von Roos, Oldwig, theoretical physics
Wales, David Bertram, mathematics
Walker, Robert Lee, physics
Waser, Jurg, chemistry
Wasserburg, Gerald Joseph, geology, geophysics
Wegst, Walter F, Jr, radiological health
Wells, Willard H, optical physics
Werner, Michael Wolock, astrophysics
Westphal, James Adolph, planetary sciences
Whaling, Ward, nuclear physics
Whitehead, Andrew Bruce, planetology, physics
Whitham, Gerald Beresford, applied mathematics
Whitney, William Merrill, physics
Wiersma, Cornelis Adrianus Gerrit, physiology
Wilson, Olin Chaddock, astronomy, astrophysics
Winnail, Douglas Samuel, biology
Winnett, Howard, biophysics
Wood, William Barry, III, biochemistry, genetics
Wulf, Oliver Reynolds, meteorology, physics

Wyckoff, Robert Cushman, physics, biophysics
Yen, Chen-Wan Liu, aerospace science
Yeomans, Donald Keith, astronomy
Yost, Don M, inorganic chemistry
Yuen, Ted Gim Hing, experimental pathology
Zachariasen, Fredrik, theoretical physics
Zaidel, Eran, neuropsychology
Zinn, Robert James, astronomy
Zirin, Harold, astronomy
Zweig, George, high energy physics, neurosciences

PAUMA VALLEY
Rateaver, Bargyla, conservation, soil fertility

PEBBLE BEACH
Comstock, Craig, applied mathematics
Cooper, Alfred William Madison, physics
Elliott, Robert Paul, food bacteriology

PITTSBURG
Beamer, William Howard, physical chemistry, environmental sciences
Bublitz, Donald Edward, organic chemistry
Dalman, Gary, organic chemistry
Edamura, Fred Y, organic chemistry
Little, John Clayton, organic chemistry
Love, Jim, organic chemistry

PLACENTIA
Hoffman, James Tracy, electronics, systems engineering
Neuman, Charles Herbert, physics

PLAYA DEL REY
Nuttall, James B, aerospace medicine

PLEASANT HILL
Buell, George Christopher, biochemistry, organic chemistry
Whitfield, Robert Edward, chemistry

PLEASANTON
Augood, Derek Raymond, physical organic chemistry, chemical engineering
Baker, Bernard Ray, surface chemistry
Bolmer, Perce W, electrochemistry, surface chemistry
Bushey, Albert Henry, physical chemistry, analytical chemistry
Calkins, Russel Crosby, analytical chemistry
Caputi, Roger William, physical chemistry
El-Shimi, Ahmed Fayez, physical chemistry
Faulkner, James Earl, mathematical physics
Hutchin, Maxine E, physiological science, forensic science
Jones, Samuel Stimpson, physical chemistry
King, William Robert, Jr, chemical metallurgy
Martin, Gerald Charles, Jr, nuclear chemistry
Meyers, Gene Howard, physical chemistry, computer sciences
Mikami, Harry M, mineralogy, ceramics
Miles, Charles Burke, chemistry
Moulthrop, Peter Hill, energy conversion
Murphy, James Francis, physical chemistry
Niday, James Barker, nuclear chemistry
Palmer, Thomas Adolph, analytical chemistry
Pearson, Robert Melvin, physical chemistry, analytical chemistry
Rider, Benjamin Franklin, analytical chemistry
Ruiz, Carl P, nuclear chemistry
Seim, Henry Jerome, analytical chemistry
Sobolev, Igor, organic chemistry
Strahl, Erwin Otto, mineralogy, petrology

POINT MUGU
Milne, Walter Leroy, health physics, radiological health
Reid, Donald House, environmental physiology

POINT REYES
Brown, Richard McPike, plant taxonomy, plant ecology

CALIFORNIA

POINT RICHMOND
Garbellano, David Wesley, atomic physics

POMONA
Abernethy, John Leo, bio-organic chemistry
Abrams, Marvin Colin, physical chemistry
Amelin, Charles Francis, mathematical analysis
Ames, Ralph Wolfley, phytopathology
Batchelor, Oliver A., horticulture
Birnbaum, Sidney, mathematics
Blakely, Lawrence Mace, plant physiology
Bowe, Arthur Frederick, analytical chemistry
Bowen, Charles E., biochemistry
Bowen, Ruth Justice, inorganic chemistry, physical chemistry
Brizolis, Demetrios, algebra
Brown, Howard S., biology, genetics
Brown, Keith H., plasma physics, high pressure physics
Campbell, David Paul, genetics
Castro, Peter, marine zoology, parasitology
Celik, Hasan Ali, algebra
Cullen, Theodore John, mathematics
Cunha, Tony Joseph, animal husbandry, nutrition
Dev, Vasu, organic chemistry, medicinal chemistry
Dinniman, Jerome Eugene, plant pathology
Dutra, Ramiro Carvalho, organic chemistry, food chemistry
Eagleton, Robert John, solid state physics
Erspamer, Jack Laverne, plant morphology, plant anatomy
Fan, Hsin Ya, applied mathematics
Fausch, Homer David, animal genetics
Ferris, Horace Garfield, physics
Fluhary, Arvan Lawrence, biochemistry
Force, Don Clement, entomology
Geller, Irwin, fuel technology
Giancoli, Douglas Charles, history of science, biophysics
Gius, John Armes, surgery
Glaser, Frank, applied mathematics
Goehler, Brigitte Hanna, microbiology
Green, Simon, mathematics
Harthill, Marion Paul, plant ecology
Henderson, Gerald Vernon, geology
Herber, Lawrence Justin, mineralogy, engineering geology
Herzog, Emil Rudolph, astronomy
Hesse, Walter Herman, agronomy
Hiemenz, Paul C., physical chemistry, colloid chemistry
Hillam, Bruce Parks, mathematical analysis
Hobbs, Kenneth Rollin, systematic entomology, botany
Hsia, Yu-Ping, physical chemistry
Jackson, James Oliver, medical microbiology, immunology
Keating, Eugene Kneeland, agricultural biochemistry, nutrition
Kelly, Edward M., physics
Kennington, Mack Humpherys, animal husbandry
Kihara, Hayato, biochemistry
Knill, Lamar M., radiobiology
Kronenberg, Klaus Johannes, physics
Lane, Bernard Owen, invertebrate paleontology
Leffler, Esther Barbara, physical chemistry
Lint, Harold L., taxonomy
Mahiman, George William, physics
Martinek, George William, genetics
Maya, Walter, organic chemistry, inorganic chemistry
McElhoe, Forrest Lester, Jr., physical geography, economic geography
Mercer, Edward King, fresh water biology
Moyer, Rudolph Henry, biochemistry
Nelson, Edward A., animal breeding
Palatnick, Barton, physics
Parker, Vincent Eveland, experimental nuclear physics
Patten, Gaylord Penrod, soil physics
Pomerening, James Albert, soil classification
Pye, Earl Louis, physical chemistry, corrosion engineering
Radnitz, Alan, mathematics
Rice, Elmer Harold, biochemistry
Riznyk, Raymond Zenon, botany
Roche, Edward Towne, invertebrate zoology

Rygg, George Leonard, plant physiology
Schafroth, Don W., structural geology, stratigraphy
Schmitz, George William, agronomy
Schweizer, Felix, radiation physics, theoretical physics
Sibbett, Donald Joseph, physical chemistry, biophysics
Simmons, Harold Franklyn, mathematics
Simpson, John Ernest, organic chemistry, enology
Smith, Velma Merriline, algebra
Sneed, Richard J., electrooptics, solid state science
Stanley, Emilo Joseph, environmental studies, geography
Stewart, Glenn Raymond, vertebrate zoology, natural history
Stiffler, Daniel F., physiology, zoology
Stoner, Martin Franklin, plant pathology, mycology
Tuai, Johannes, physics
Verkler, Robert Charles, operations research
Vollmar, Arnulf R., organic chemistry
Wu, Jia-Hsi, plant physiology, virology

PORT HUENEME
Drisko, Richard Warren, organic chemistry
Haynes, Willis Stuart, chemistry, research administration
Hearst, Peter Jacob, organic chemistry
Vind, Harold Pennington, environmental chemistry

PORTERVILLE
Corkins, Jack Philips, entomology, applied physiology
Gilbert, Elmer Wilhelm, internal medicine

PORTOLA VALLEY
Hurd, Cuthbert Corwin, mathematics
Hutchisson, Elmer, applied physics
Rothenberg, Stephen, quantum chemistry, computer systems
Seppi, Edward Joseph, mathematics
Urbach, John C., optics
Wilson, Elwood Justin, Jr., organic chemistry

POWAY
Vuillet, William George, atomic physics, molecular physics

PRESIDIO OF SAN FRANCISCO
Bikle, Daniel David, internal medicine, biochemistry
Canham, John Edward, nutrition
Consolazio, Carlo Frank, nutrition, physiology
Luzzio, Anthony Joseph, immunology
Marshall, John Dean, Jr., microbiology
Sauberlich, Howerde Edwin, biochemistry
Zuck, Thomas Frank, pathology

RANCHO BERNARDO
Enns, John Hermann, physics

RANCHO PALOS VERDES
Binder, Daniel, physics
Gold, Richard Robert, physics, applied mathematics
Singletary, Lillian Darlington, nuclear physics
Wells, J Gordon, physiology

RANCHO SANTA FE
Frederick, Donald Sherwood, polymer chemistry
Rosenthal, Sol Roy, pathology, bacteriology
Steiner, Arnold Byron, organic polymer chemistry
Zichis, Joseph, biochemistry, virology

REDDING
Roy, Douglas Fielding, forestry

REDLANDS
Baty, Roger M., anthropology
Blatchley, Donald E., nuclear physics
Carlson, Richard Frederick, physics

RESEDA
Tannenbaum, Irving Robert physical

REDWOOD CITY
Abrams, Irving Melvin, industrial chemistry
Anderson, Robert Emra, water chemistry
Haag, Robert Marlay, physical chemistry
Lincoln, Kenneth Arnold, high temperature chemistry
Lindsay, Stuart, pathology
McCurdy, Orville L., polymer chemistry, organic chemistry
Millar, John Robert, organic polymer chemistry
Perry, Robert Hood, Jr., organic chemistry, research administration
Tarail, Robert, medicine, community health

RICHMOND
Abbott, Andrew Doyle, physical chemistry
Abell, Jared, organic chemistry
Adams, John Quincy, physical chemistry
Altgelt, Klaus H., physical chemistry, petroleum chemistry
Anderson, Arthur Bernhardt, organic chemistry, forest products

Contento, Isobel Corneli, immunology, nutrition
Corneli, Paul Hampton, physical chemistry
Dana, Stephen Winchester, geology
Gates, Gerald Otis, ecology
Hodson, William Myron, plant ecology
Hollenberg, J Leland, physical chemistry
Howell, Charles DeWitt, zoology
Ifft, James Brown, biophysical chemistry
Krantz, Reinhold John, organic chemistry
Plock, Richard James, chemical physics
Roberts, Julian Lee, Jr., analytical chemistry
Rohrer, William Glen, astronomy, physics
Sanderson, Judson, mathematics
Trolan, J Kenneth, surface physics
Williams, Floyd James, geology

REDONDO BEACH
Ackerman, Donald Godfrey, Jr., analytical chemistry
Adelson, Harold Ely, resource management
Alexander, Madeline J., mathematics, resource management
Andres, John Milton, microwave electronics
Arnush, Donald, theoretical physics
Ball, William Paul, nuclear physics, radiation science
Benesch, Samuel Eli, systems theory
Benveniste, Jacob, nuclear physics
Brown, Frederick S., biochemistry
Day, Robert James, analytical chemistry, organic geochemistry
Doolittle, Robert Frederick, II, physics
Fredericks, Robert W., physics, geophysics
Friedman, Richard M., space physics
Honnold, Vincent Richard, solid state physics
Huberman, Marshall Norman, engineering physics, energy conversion
Kagiwada, Reynold Shigeru, solid state physics, acoustics
Karr, Philip R., physics
Kropp, John Leo, physical chemistry, spectroscopy
Kurz, Richard J., high energy physics, instrumentation
Meyers, Robert Allen, air pollution, organic chemistry
Molmud, Paul, physics
Nordin, Paul, applied physics
Rogers, John Langley, solid state physics
Rosen, Alan, space physics, biomedical engineering
Scarf, Frederick Leonard, theoretical physics
Seo, Eddie Tatsu, analytical chemistry, electrochemistry
Silverman, Herbert Philip, electrochemistry
Smith, Oliver King, mathematics, computer science
Wagner, Richard John, mathematical physics
Walker, Kelsey, Jr., theoretical gas dynamics, applied mathematics
Wiggins, John Shearon, space physics
Wuerker, Ralph Frederick, physics

Anderson, Robert Griffin, organic chemistry
Ballinger, Peter Richard, organic chemistry
Barusch, Maurice R., chemistry
Beach, John Youngs, physical chemistry
Benoit, George Julien, Jr., biochemistry
Bolt, Robert O'Connor, organic chemistry
Brookman, David Joseph, analytical chemistry
Brown, Stuart Houston, organic chemistry
Burrous, Mervyn Lee, organic chemistry
Byrne, Hugh Desmond, entomology
Christofferson, Glen Davis, physical chemistry
Condit, Paul Carr, petroleum chemistry
Csicsery, Sigmund Maria, surface chemistry, petroleum chemistry
Erdman, Timothy Robert, organic chemistry
Etzler, Dort Homer, chemistry
Guffy, Joseph Claude, analytical chemistry
Hall, Kenneth Lynn, petroleum chemistry
Hansberger, Hugh Francis, physical inorganic chemistry
Harrison, Jonas P., physical chemistry, chemical engineering
Hawkes, George Rogers, soil fertility, agronomy
Hickson, Donald Andrew, surface chemistry, petroleum chemistry
Hotten, Bruce Walter, surface chemistry
Houston, Robert John, petroleum chemistry
Hubert, Jay Marvin, chemical physics
Ja, William Yin, analytical chemistry
King, John Mathews, organic chemistry
Kiskis, Ronald Clements, organic chemistry
Kluksdahl, Harris Eudell, inorganic chemistry
Komoto, Robert Gordon, industrial organic chemistry, inorganic chemistry
Kray, Louis Robert, petroleum chemistry
Kurkov, Victor Peter, organic chemistry
LeTourneau, Robert Louis, chemistry
Leventhal, Leon, radiochemistry
Lewis, Robert Allen, petroleum chemical engineering
Lewis, Robert Allen, petroleum chemistry
Lewis, Robert Taber, physics
Liddicoet, Thomas Herbert, pesticide chemistry
Lindquist, Robert Henry, physical chemistry
Lockley, Jeanette Elaine, statistics, mathematics education
Lowe, Warren, organic chemistry
Lukens, Raymond James, plant pathology
Magee, Philip Stewart, physical organic chemistry
Mason, Harold Frederick, physical chemistry
McClellan, Aubrey Lester, physical chemistry
Meader, Arthur Lloyd, Jr., petroleum chemistry
Mehmedbasich, Enver, organic chemistry
Melgard, Rodney, radiochemistry, nuclear physics
Mertens, Edward William, organic chemistry, marine biology
Michlmayr, Manfred, inorganic chemistry, catalysis
Mihailovski, Alexander, organic chemistry
Nathans, Marcel Willem, physical chemistry
Oppenstein, Joseph Nils, organic chemistry
Pallos, Ferenc M., organic chemistry, pesticide chemistry
Phillips, Lee Vern, organic chemistry
Prudhomme, Edward Louis, food science
Quisenberry, Benson F., entomology
Ranmer, Irwyn Alden, chemistry
Rhodes, David R., analytical chemistry, electrochemistry
Richardson, Wallace Lloyd, organic chemistry

Rosenthal, Joel William, fuel science, petroleum chemistry
Schlatter, Maurice Jay, petroleum chemistry
Schneider, Ronald Alan, organic chemistry
Schniewind, Arno Peter, forest products
Seifert, Wolfgang K, organic chemistry
Sinkovic, Jelena, electrochemistry
Spiegler, Kurt Samuel, chemistry
Stanton, Garth Michael, organic chemistry
Stelzer, Lorin Roy, research administration, entomology
Stonebraker, Peter Michael, chemistry
Straus, Alan Edward, organic chemistry
Sullivan, Richard Frederick, physical chemistry
Suzuki, Shigeto, organic chemistry
Sweeney, William Alan, petroleum chemistry
Teach, Eugene Gordon, organic chemistry
Teeter, Richard Malcolm, organic chemistry
Thomas, John Richard, physical chemistry
Thompson, Robert Kruger, entomology
Tilles, Harry, organic chemistry
Tornheim, Leonard, numerical analysis
Tseng, Chien Kuei, organic chemistry
Walker, Francis H, organic chemistry
Wall, Robert Gene, organic chemistry
Wallace, Edwin Garfield, organic chemistry
Weamer, George L, petroleum chemistry
Whipp, Arthur Andrew, entomology
Whittemore, Irville Merrill, petroleum chemistry
Wilcox, Webster Wayne, forest pathology, forest products
Wilgus, Donovan Ray, organic chemistry
Wilkes, John Barker, petroleum chemistry
Woo, Gar Lok, organic chemistry
Young, Janis Dillaha, biochemistry, immunochemistry
Zabin, Burton Allen, inorganic chemistry, analytical chemistry
Zavarin, Eugene, organic chemistry

RIDGECREST

Cleaves, Duncan Worster, physical chemistry
Evans, Allison Bickle, industrial chemistry
Kruse, Howard Wendell, inorganic chemistry
Leet, Henry Peter, optics
Merrow, Raymond Theodore, organic chemistry
Plain, Gilbert John, physics
Stevens, Lewis Axtell, biophysics

RIVERSIDE

Allen, William Merle, organic chemistry
Allen, William Ross, developmental biology
Anderson, Eugene N, Jr, anthropology
Anderson, Howard T, environmental geology, exploration geology
Anderson, Lauren Davis, entomology
Armstrong, Paul Douglas, organic chemistry, medicinal chemistry
Aschmann, H Homer, geography
Bacchus, Habeeb, internal medicine, endocrinology
Bailey, Harry Paul, climatology
Bald, John Grieve, plant pathology
Barnes, Martin McRae, economic entomology
Bartlett, Blair Ralph, entomology
Bartnicki-Garcia, Salomon, biochemistry
Baum, Peter Joseph, physics
Beals, Alan Robin, cultural anthropology
Beasley, C A (Bud), plant physiology, biochemistry
Beaver, Robert John, statistics
Belser, William Luther, Jr, genetics
Bergh, Berthold Orphie, genetics
Biehler, Shawn, geophysics
Bingham, Frank Thomas, soil chemistry
Birge, Robert Richards, chemical physics
Bitters, Willard Paul, horticulture
Block, Richard Earl, mathematics
Bovell, Carlton Rowland, microbial physiology
Bowden, Leonard Walter, geography
Branson, Roy, soil chemistry
Brown, Leland Ralph, entomology
Bruner, Leon James, biophysics
Calavan, Edmond, plant pathology
Cameron, James Wagner, plant genetics
Cannell, Glen H, soil physics
Carman, Glenn Elwin, entomology
Carpelan, Lars Hjalmar, ecology
Castro, Charles E, organic chemistry
Chapman, Homer Dwight, soil fertility
Coggins, Charles William, Jr, plant physiology
Cohen, Lewis H, geochemistry
Cooper, Kenneth Willard, cytogenetics, entomology
Cummings, Frederick W, theoretical physics, mathematics
Darley, Ellis Fleck, air pollution, plant pathology
David, Florence N, statistics, mathematics
De Bach, Paul (Hevener), entomology, insect ecology
de Pillis, John, mathematics
Desai, Bipin Ratilal, high energy physics
Desjardins, Paul Roy, plant pathology
Dirksen, Christiaan, soil physics
Dugger, Willie Mack, Jr, plant physiology
Dunbar, Dennis Monroe, entomology
Dunn, Michael F, biochemistry, enzymology
Eaks, Irving Leslie, plant physiology, biochemistry
Ebeling, Walter, entomology
Eckert, Joseph Webster, plant pathology
Elders, Wilfred Allan, geology
Ely, Daniel Lee, medical physiology
Embleton, Tom William, horticulture, plant nutrition
Endo, Robert Minoru, plant pathology
Erickson, Louis Carl, plant physiology
Erwin, Donald C, phytopathology
Eskew, David Lewis, plant physiology
Evard, Rene, biochemistry
Everett, Glen Exner, solid state physics
Ewart, William Howard, economic entomology
Farley, Roger Dean, neurophysiology
Farmer, Walter Joseph, soil chemistry
Federici, Brian Anthony, invertebrate pathology, virology
Ferguson, Le Baron O, mathematics
Fetter, Neil Ross, analytical chemistry
Fisher, Theodore William, entomology
Focht, Dennis Douglass, soil microbiology
Fukuto, Tetsuo Roy, organic chemistry
Furlong, Clement Eugene, biochemistry
Furuta, Tokuji, environmental horticulture
Galaway, Ronald Alvin, bio-organic chemistry
Garber, Morris Joseph, biometry, computer science
Gaston, Lyle Kenneth, organic chemistry
Georghiou, George Paul, entomology, toxicology
Gibeault, Victor Andrew, ornamental horticulture
Gibian, Morton J, physical organic chemistry, enzymology
Gill, Harmonhindar Singh, plant pathology, mycology
Gill, Robert Wager, ecology
Gillett, George Willson, botany
Goeden, Richard Dean, entomology, weed science
Gokhale, Dattaprabhakar V, statistics
Goodman, Victor Herke, plant anatomy, phycology
Gordon, Samuel Robert, mathematics
Gray, Cliffton Herschel, Jr, geology
Green, Lisle Royal, range management
Gretsky, Neil E, mathematics
Gumpf, David John, plant virology
Gunther, Francis Alan, chemistry
Halberg, Charles John August, Jr, mathematics
Hall, Anthony Elmitt, crop physiology
Hall, Irvin Monroe, insect pathology
Hansen, Carl Lough, geography, geomorphology
Harper, Lawrence Hueston, mathematics
Heath, Robert Louis, biophysics, biochemistry
Helmkamp, George Kenneth, organic chemistry
Hewitt, Robert Russell, solid state physics
Holten, Darold Duane, biochemistry, enzymology
Holten, Virginia Zewe, biochemistry
Houck, Laurie Gerald, plant pathology, plant physiology
Huang, Nai Li, solid state physics
Humphreys, Curtis Judson, atomic spectroscopy
Isom, William Howard, agronomy
Jefferson, Roland Newton, entomology
Jeppson, Lee Ralph, entomology
Johnson, Bertil Lennart, genetics
Johnson, Harry William, Jr, organic chemistry
Johnson, Hyrum Bennett, botany
Johnson, Nancy Sue, statistics
Jones, Floyd Burton, topology
Jones, Gary Edward, genetics, cell biology
Jones, Winston William, horticulture
Jordan, Lowell Stephen, plant physiology
Kaloostian, George H, entomology
Karu, Alexander Edwin, biochemistry, animal virology
Kaufmann, Merrill R, plant physiology, physiological ecology
Kaus, Peter Edward, theoretical physics
Keen, Noel Thomas, plant pathology
Kirkpatrick, John David horticulture
Kolbezen, Martin (Joseph), pesticide chemistry
Kramer, Vernon A, mathematics
Kumamoto, Junji, physical organic chemistry
Labanauskas, Charles Kazys, plant physiology
Lathrop, Earl Wesley, systematic botany
Lawson, Andrew Werner, physics
Leary, John Vincent, genetics
Lee, Donald E, science education
Lee, Tien-Chang, geophysics
Legner, E Fred, entomology, ecology
Leonard, Robert Thomas, plant physiology
Letey, John, Jr, biophysics
Lewis, Lowell N, plant physiology
Lindgren, David Leonard, entomology
Lippert, Laverne Francis, plant pathology
Longerbeam, Jerrold Kay, surgery, physiology
Lund, Lanny Jack, soil morphology
Lutz, Kenneth Russell, speech pathology, audiology
Maas, Eugene Vernon, plant physiology
MacLaughlin, Douglas Earl, solid state physics
Mankau, Reinhold, nematology, soil biochemistry
March, Ralph Burton, entomology
Marsh, Albert William, soil science, irrigation
Martin, James Paxman, soil microbiology
Mayhew, Wilbur Waldo, vertebrate biology, desert ecology
McCollum, Donald Carruth, Jr, physics
McKinney, Ted Meredith, analytical chemistry
McMurtry, James A, entomology
Metcalf, Frederic Thomas, applied mathematics
Midland, Michael Mark, organic chemistry, organometallic chemistry
Miller, Thomas Albert, entomology
Mitchell, Norman L, plant pathology
Moore, Betty Clark, embryology
Moore, John Alexander, evolutionary biology
Mudd, John Brian, biochemistry
Mulla, Mir S, entomology
Munnecke, Donald Edwin, phytopathology
Murashige, Toshio, plant physiology
Murphy, Michael A, geology, stratigraphy
Nettleton, Wiley Dennis, soil science, geology
Neuman, Robert C, Jr, physical organic chemistry
Newell, Irwin Mayer, acarology, systematics
Nieman, Richard Hovey, plant physiology
Noltmann, Ernst August, biochemistry, enzymology
Norman, Anthony Westcott, biochemistry
Oatman, Earl R, entomology
Oddson, John Keith, mathematics
Okamura, William H, synthetic organic chemistry, medicinal chemistry
Oldfield, George Newton, acarology
Olsen, Richard William, biochemistry
Orttung, William Herbert, physical chemistry
Oster, James Donald, soil chemistry
Page, Albert Lee, soil chemistry
Painter, Page Robbins, physics, mathematics
Pauling, Edward Grellin, molecular biology
Paulus, Albert, plant pathology
Pease, Robert Wright, physical geography
Pengelley, Eric T, physiology
Philpot, Charles Walter, forestry
Pinto, John Darwin, entomology, systematics
Pitts, James Ninde, Jr, physical chemistry, photochemistry
Platzer, Anna (Colville), developmental biology
Platzer, Edward George, parasitology, nutrition
Plumb, Timothy Roy, Jr, forest management, plant physiology
Pollak, Michael, semiconductors
Pon, Ning Gin, biochemistry, photobiology
Pratt, Parker Frost, soil chemistry, water chemistry
Preszler, Alan Melvin, space physics
Radewald, John Dale, plant pathology
Rao, Malempati Madhusudana, mathematics
Ratliff, Louis Jackson, Jr, mathematics
Rawlins, Stephen Last, soils, physics
Reed, David Kent, insect pathology
Reid, Brian Robert, biochemistry
Reuther, Walter, horticulture
Reynolds, Harold Truman, entomology
Rhoades, James David, soil science, clay mineralogy
Rible, John Maurice, soil science
Richards, Lorenzo Adolph, experimental physics
Riehl, Louis Adam, entomology
Riggs, James W, Jr, physics
Robinson, Paul Thornton, geology
Ruibal, Rodolfo, vertebrate zoology
Rush, David Eugene, mathematics
Ryan, Bill Chatten, meteorology
Sawyer, Donald Turner, Jr, analytical chemistry, bioinorganic chemistry
Schlanger, Seymour Oscar, geology
Schmidt, Hartland H, physical chemistry
Schneider, Henry, plant pathology & anatomy
Schroeder, Mark Joseph, meteorology
Scora, Rainer W, botany
Scott, Gary Walter, chemical physics
Semancik, Joseph Stephen, pathology, virology
Sevacherian, Vahram, entomology, applied statistics
Shannon, Leland Marion, plant biochemistry
Shannon, Michael Carlyle, plant genetics
Shapiro, Victor Lenard, mathematical analysis
Shaw, John Gilbert, entomology
Shelden, Harold Raymond, II, organic chemistry
Sher, Samuel Alexis, nematology
Sherman, Irwin William, zoology, parasitology
Shoemaker, Vaughan Hurst, comparative physiology
Shorey, Harry Haslam, entomology
Shortridge, Robert Glenn, Jr, chemical kinetics
Simanek, Eugen, physics
Sims, James Joseph, organic chemistry
Sinclair, Walton B, biochemistry
Slota, Peter John, Jr, chemistry
Smith, Albert Ernest, physics
Smith, Charles Aloysius, analytical chemistry, pesticide chemistry
Smith, Malcolm, operations research, engineering statistics
Soost, Robert Kenneth, genetics
Spencer, William F, soil chemistry
Sposito, Garrison, physics
Stephens, Edgar Ray, physical chemistry
Stern, Vernon Mark, entomology
Stolzy, Lewis Hal, soil physics
Storey, Theodore George, forestry
Stralka, Albert R, mathematics
Strong, Rudolph Greer, entomology
Sudmeier, James Lee, analytical chemistry
Taylor, Charles Ellett, population genetics
Taylor, Oliver Clifton, horticulture
Taylor, Royal Ervin, Jr, anthropology
Testerman, John Kendrick, comparative physiology
Thomason, Ivan J, plant nematology, plant pathology
Thompson, Chester Ray, biochemistry
Thompson, Lewis Chisholm, nuclear physics
Thomson, William Walter, botany, cytology
Ting, Irwin Peter, plant physiology, metabolism
Traugh, Jolinda Ann, biochemistry, molecular biology
Tsao, Pamela Wen-Chau Wang, plant pathology
Tsao, Peter Hsing-Tsuen, plant pathology
Turrell, Franklin Marion, plant physiology
Van Der Woude, William Jan, plant cytology, plant physiology
Van Gundy, Seymour Dean, plant pathology, nematology
Vasek, Frank Charles, botany
Wallace, James Merrill, plant pathology
Wallihan, Ellis Flower, plant nutrition
Weathers, Lewis Glen, plant pathology
Weber, John R, plant physiology
Webster, Clyde Leroy, Jr, inorganic chemistry, geochemistry

Wedding, Randolph Townsend, biochemistry
Weeks, Leslie Vernon, soil science
Weinberg, Ralph, bacteriology
White, Robert Stephen, astrophysics, space physics
Wild, Robert Lee, solid state physics
Woodburne, Michael O., vertebrate paleontology, stratigraphy
Yermanos, Demetrios M., genetics, plant breeding
Young, Roy E., plant physiology
Youngner, Victor Bernard, agronomy, plant ecology
Zaugg, Wayne E., physical chemistry
Zentmyer, George Aubrey, (Jr), phytopathology
Zych, Allen Dale, astrophysics

ROHNERT PARK
Arnold, John Ronald, zoology
Blitz, Ruth R., bacteriology, immunology
Brunbaugh, Joe H., marine biology
Clothier, Galen Edward, cell biology, developmental biology
Dickerman, Mildred, cultural anthropology
Duncan, Donald Gordon, mathematics
Dunning, John Ray, Jr., physics
Ebert, Wesley W., genetics, botany
Eck, David Lowell, organic chemistry
Frazer, William James, geography
Hermans, Colin Olmsted, invertebrate zoology
Hoagland, Vincent DeForest, Jr., biochemistry
Johnston, George Lawrence, physics, astrophysics, plasma
Kientz, Marvin L., biochemistry
Kjelsen, Chris Kelvin, phycology
Lockner, Frederick Russell animal behavior
Luttman, Frederick William, mathematics
Marshall, Donald D., inorganic chemistry, analytical chemistry
Nichols, Ambrose Reuben, Jr., physical chemistry
Porter, Thomas Reginald, botany
Rustad, Douglas Scott, inorganic chemistry, physical chemistry
Schaumberg, Gene David, organometallic chemistry
Sherman, Robert James, botany, ecology
Stanek, Jean Chan, mathematics
Trowbridge, Dale Brian, organic chemistry
Wright, William Herbert, III, structural geology

ROLLING HILLS
Layton, Thomas William, physics
Roberts, James C., Jr., pathology

ROLLING HILLS ESTATES
Dyer, Denzel Leroy, biochemistry
Millburn, George P., nuclear physics
Servis, Kenneth L., physical chemistry, organic chemistry

ROSEMEAD
Davis, Ward B., biochemistry
Strachan, Alec Ronald, environmental biology
Swinney, Chauncey Melvin, economic geology, petrology

ROSEVILLE
Gerdel, Robert Wallace, geophysics

ROCKLIN
Underhill, Raymond Alden, biology

SACRAMENTO
Alfors, John Theodore, geology
Arnold, Brigham Alicen, geography
Aston, Duane Ralph, solid state physics
Ayoub, Sadek M., plant nematology, plant pathology
Baad, Michael Francis, plant taxonomy, ecology
Barnes, Gene A., solid state physics
Barry, Arthur Leland, microbiology, bacteriology
Barter, James T., psychiatry
Beeson, William Jean, anthropology
Bencs, Elinor Simson, microscopic anatomy, vertebrate zoology
Benner, Ernest Jack, medicine
Bieler, Barrie Hill, mineralogy

Blanc, Francis Louis, entomology
Bolar, Martin L., plant physiology
Bolt, Robert James, internal medicine
Bressler, David Wilson, mathematics
Brittan, Martin Ralph, vertebrate zoology, ecology
Brown, John Welch, medicine, preventive medicine
Bruemner, Rolf Sylvester, applied chemistry
Burcham, Levi Turner, ecology, geography
Burnett, John Laurence, geology
Calderone, Julius G., medical bacteriology, medical mycology
Chow, Tseng Yeh, mathematics
Christopher, John, mathematics
Cohen, Joseph, inorganic chemistry
Darby, Rollo E., entomology, zoology
Davis, Stanley David, psychiatry, biomedical engineering
Decious, Daniel, physical chemistry
Delisle, Albert Lorenzo, botany, genetics
DeFree, David Otte, microbiology
DiGiorgio, Joseph Brun, physical organic chemistry
Di Milo, Anthony J., polymer chemistry
Draper, Roy Douglas, biochemistry
Dunton, Marguerite Elizabeth, mathematics
Eldredge, Kelly Husbands, virology
Evans, James R., economic geology
Fish, Richard Wayne, organic chemistry
Forkey, David Medrick, organic chemistry
Fowler, William Mayo, Jr., physical medicine & rehabilitation
French, Alexander Murdoch, plant pathology
Fuller, Melvin Otis, physics
Fuller, Thomas Charles, weed science
Gibson, Edward F., nuclear physics
Gold, Eli, virology, pediatrics
Grant, Robert Charles S., physical chemistry
Hagopian, Charles Lemuel, topology
Hall, Donald Eugene, physics, astrophysics
Hamel, Edward E., organic chemistry
Hasbrouck, Edward Ralph, nematology, ecology
Hayashi, Shuki, biophysics
Helm, Donald Cairney, groundwater hydrology
Hoeprich, Paul Daniel, internal medicine
Holdeman, Quintin Lee, plant pathology
Huff, Dennis Karl, microbiology, parasitology
Hughart, Stanley Parlett, mathematics
Hughes, Eldon Parker, fisheries management
Hurley, Carl Robert, inorganic chemistry
Ibser, Homer Wesley, physics
Jackson, Sharon Wesley, genetics, environmental sciences
Jaffa, Robert E., mathematics
Jakob, Fredi, analytical chemistry
James, Laurence Beresford, geology
Janke, Norman C., environmental geology
Jones, Morris Val, speech pathology, audiology
Kantz, Paul Thomas, Jr., phycology
Kavalian, Leroy Gregory, plant morphology
Kim, Chung Sul, organic chemistry, polymer chemistry
Koch, Richard, pediatrics
Konicek, Donald E., plant pathology
Krakowski, Fred, mathematics
Krohn, Kenneth Albert, nuclear chemistry
Ludwig, Carl Edward, biology, entomology
Love, Challie M., physiology
Langsley, Donald Gene, psychiatry, psychoanalysis
Lillick, Lois Carol, medicine
Livezey, Robert Lee, biology
Loomis, Robert Henry, zoology, limnology
McGeary, David F R., geology
Messersmith, James David, fisheries management
Metcalf, Robert Harker, microbiology
Morse, John Thomas, environmental physiology
Moser, Charles R., developmental biology
Nichols, Carl William, plant pathology
Nussenbaum, Siegfried, biochemistry, organic chemistry
Payne, Holland I., research administration, natural science
Peak, Wilferd Warner, geology, engineering geology
Plummer, Charles Carlton, petrology
Poland, Joseph Fairfield, hydrogeology

Pool, Robert Alfred Frank, plant pathology; plant physiology
Rauf, Mohammad A., anthropology
Richter, Raymond C., geology
Roche, George William, forensic science
Rosenberg, Dan Yale, phytopathology
Rosenberg, Sanders David, fuels science, organometallic chemistry
Rowland, Richard Lloyd, physical chemistry
Russell, John George, organic chemistry
Schmidt, Willfred G., chemical kinetics
Schwenck, Julius Rae, science education
Sellas, James Thomas, organic chemistry
Shea, Michael Joseph, experimental solid state physics
Sherifi, Leon, physics
Sime, Rodney J., physical chemistry
Sinne, Ruth Lewin, plant pathology
Skinner, John Eugene, fish biology
Smith, Charles Henry, forensic science
Snow, Gordon Franklin, plant pathology
Snyder, Warren Arthur, anthropology
Trapp, Gene Robert, mammalogy
Trelford, John D., obstetrics & gynecology; oncology
Troxel, Bennie Wyatt, geology
Tupin, Joe Paul, psychiatry
Udvardy, Miklos Dezso Ferenc, zoology
Urone, Paul Peter, nuclear physics
Vanicek, C David, fisheries management
Veigel, Jon Michael, research administration, environmental sciences
Vreeland, John Allen, physics
Wagnon, Harvey Keith, plant pathology
Wasbauer, Marius Sheridan, insect taxonomy
Weilerstein, Ralph Waldo, therapeutics, public health
Weiss, Melford Stephen, cultural anthropology
Wiedman, Harold W, phytopathology
Wilson, Jerry Lee, biochemistry
Wulff, John Leland, mathematics
Yates, Robert Edmunds, physical chemistry

ST HELENA
Amerine, Maynard Andrew, agriculture

SALINAS
Bennett, Carlyle Wilson, plant pathology
Darst, Philip High, economic entomology
Duffus, James Edward, plant pathology, plant virology
Haines, Robert Gordon, entomology
Harry, John Boyer, phytopathology
Hoefert, Lynn Lucretia, botany
Jenkins, Burton Charles, cytogenetics
Kovach, Ladis Daniel, mathematics
Lewellen, Robert Thomas, genetics, plant breeding
Marshall, Ernest (Roy), horticulture
McFarlane, John Spencer, genetics
Ririe, David, agronomy
Ryder, Edward Jonas, genetics, plant breeding
Savitsky, Helen, genetics
Thurman, Duane Edward, entomology, breeding
Whitney, Elvin Dale, plant breeding, plant pathology,
Woodruff, Richard Earl, horticulture, plant physiology
Yu, Ming-Hung, plant cytogenetics

SAN BERNARDINO
Ackerman, William Vaughn, geography of Latin America, economic geography
Anderson, Clyde Lee, chemistry
Braunstein, Herbert, pathology
Carson, George Stephen, mathematics
Craig, John Horace, organic chemistry
Crum, James Davidson, organic chemistry
Dennemeyer, Rene Felix, mathematics
DeRemer, Russell Jay, physics
Dulock, Victor A, Jr., applied physics
Eggs, Alfred Severin, zoology
Goodman, Richard E., molecular biology, microbial physiology
Hafstrom, John Edward, mathematics
Harrington, Dalton, botany, phycology
Harris, Arlo Dean, inorganic chemistry
Ikenberry, Dennis L., physics
Kellers, Charles Frederick, low temperature physics
Liu, Fook Fah, high energy physics
Mankau, Sarojam Kurudamannil, parasitology, nematology
Mantei, Kenneth Alan, physical chemistry
Murphy, James Lee, mathematics

Petrucci, Ralph Herbert, physical chemistry, academic administration
Rowland, Richard Hugh, population geography, geography of the Soviet Union
Scherba, Gerald Marron, zoology
Simmons, Michael Patrick, physical anthropology
Sokoloff, Alexander, genetics
White, Kenneth L, Sr., geography
Woods, Roger David, theoretical physics, computer science

SAN CARLOS
Daehler, Max, Jr., physics
Wickersham, Arthur Frank, Jr., physics
Williams, John Wharton, geology

SAN CLEMENTE
Mehl, John Wilbur, biochemistry
Terry, Richard D., oceanography, space sciences

SAN DIEGO
Abbott, Mitchel Theodore, biochemistry
Abbott, Patrick Leon, sedimentology
Akeson, Wayne Henry, orthopedic surgery
Albert, Jerry David, clinical biochemistry
Alexander, Nicholas Michael, biochemistry
Anderson, Arthur James Outram, anthropology
Anderson, Victor Charles, physics
Anderson, Zoe Estelle, nutrition
Ashburn, William Lee, nuclear medicine
Aurand, Henry Spiese, Jr., underwater acoustics
Avila, Vernon Lee, comparative endocrinology; animal behavior
Awbrey, Frank Thomas, biology
Baer, Adela Dee, genetics
Baily, Norman Arthur, medical physics, radiology
Banta, Benjamin Harrison, evolutionary biology; herpetology
Bartlet, Grant Rogers, biochemistry
Batzler, William Emmet, underwater acoustics
Baxter, William Leroy, virology
Becker, Gerald Anthony, mathematics
Bennett, Larry E., inorganic chemistry
Benson, Andrew Alm, biochemistry, plant physiology
Benson, Peter Howard, ecology, environmental biology
Bernstein, Eugene F., surgery
Berry, Richard Warren, marine geology; marine geochemistry
Blanch, Gertrude, mathematics
Blick, James Donald, geography
Bloom, Joseph Dutich, medicine, internal medicine
Bohnsack, Kurt K., ecology; acarology
Bookstein, Joseph Jacob, radiology
Booth, Newell Ormond, underwater acoustics, ocean engineering
Botger, Gilbert Ted, economic entomology
Bradshaw, John Stratili, biological oceanography
Bramblett, Richard Lee, applied physics, nuclear physics
Brandt, Bruce Losure, neurobiology, electrophysiology
Brandt, Charles Lawrence, physiology
Braude, Abraham Isaac, medicine
Broadhead, Gordon Clifford, fish biology, marine science
Brown, James Roy, nuclear physics
Bucker, Homer Park, Jr., acoustics
Burgus, Roger Cecil, biochemistry, neuroendocrinology
Burke, Richard Lerda, organic chemistry
Burton, Charles R., mathematical logic
Byfield, John Eric, radiotherapy
Byrd, William Joseph, immunology, oncology
Caldwell, Loren Thomas, geology, pharmacology
Cameron, Dale Corbin, psychiatry
Cameron, Sidney Herbert, horticulture
Caprioglio, Giovanni, electrochemistry
Carpenter, Roger Edwin, zoology
Carsola, Alfred James, oceanography, marine geology
Case, Charles Calvin, anthropology
Caspers, Hubert Henri, solid state physics, spectroscopy
Cassidy, Samuel H. mathematics
Chaney, Charles Lester, analytical chemistry
Chen, Lo-Chai, ichthyology
Christensen, Eleanor, physiology

Clark, Mary Eleanor, zoology, comparative physiology
Clark, Orrin H, physics
Cobble, James Wikle, physical chemistry, inorganic chemistry
Coffey, Dewitt, Jr, chemical physics
Cohn, Melvin, immunochemistry, biochemistry
Collier, Boyd David, population ecology
Collier, Gerald, vertebrate zoology, animal behavior
Colwell, Joseph F, solid state physics
Comer, Ralph Dudley, preventive medicine, tropical medicine
Convery, F Richard, orthopedic surgery
Cooper, Charles F, ecology
Cooper, Eugene Perry, physics
Cooper, Robert Woodrow, primate biology
Cottrell, Ian William, carbohydrate chemistry
Courtney, Kenneth Oliver, medical research
Cox, George W, ecology, ornithology
Crawford, Ronald Ward, ichthyology, limnology
Crouch, James Ensign, ornithology
Cummings, William Charles, marine biology, bioacoustics
Dahms, Arthur Stephen, Jr, biochemistry
Daub, Clarence Theodore, Jr, astrophysics
Davis, Craig H, molecular biology
Deaton, Edmund Ike, mathematics
DeLuca, Marlene, biochemistry
Dennis, Martha Greenberg, computer science
Dessel, Norman F, physics
Dexter, Deborah Mary, marine zoology, ecology
Diehl, William Paul, microbial genetics
Dolan, James Michael, zoology
Dorman, Clive Edgar, physical oceanography, meteorology
Drobnies, Saul Isaac, mathematics, operations research
Dukepoo, Frank Charles, human genetics
Dunn, Howard J, food chemistry
Earnest, Sue W, speech pathology
Eberhardt, Robert Louis, marine ecology
Ebert, Thomas A, ecology
Eckhart, Walter, molecular biology, virology
Ehler, Kenneth Walter, bio-organic chemistry
Epstein, Barry D, electrochemistry, analytical chemistry
Estes, Richard, vertebrate paleontology, herpetology
Etheridge, Richard Emmett, herpetology, paleontology
Ezell, Paul Howard, anthropology
Fanestil, Darrell Dean, internal medicine, nephrology
Farris, David Allen, fisheries
Farris, Jack Matthews, surgery
Feher, Elsa, science education, physics
Fish, James Franklin, Jr, biological oceanography, animal behavior
Fisher, Frederick Hendrick, physics
Flanigan, William Francis, Jr, neuropsychology, animal behavior
Fletcher, Paul Chipman, solid state physics
Flittner, Glenn Arden, marine ecology, meteorology
Ford, Richard Fiske, marine ecology, water pollution
Fountain, Leonard Du Bois, mathematical analysis
Frederiksen, Norman Oliver, palynology, micropaleontology
Friedman, William Foster, pediatric cardiology
Furry, Donald Edward, anatomy, physiology
Futch, David Gardner, genetics, evolution
Gabriel, Cedric John, solid state physics
Gales, Robert Sydney, acoustics
Gallup, Avery Houseley, plant physiology
Garrison, Betty Bernhardt, number theory
Garrison, John Dresser, physics
Gastil, R Gordon, geology
Geller, Myer, electrooptics
Gittes, Ruben Foster, genitourinary surgery, endocrinology
Glushien, Arthur Samuel, cardiology, internal medicine
Gochman, Nathan, biochemistry
Goldkind, Victor, cultural anthropology
Goodjohn, Albert J, reactor physics
Gosnell, Rex Beach, organic chemistry, polymer chemistry
Goss, Robert Nichols, mathematics
Grace, Odis Donn, acoustics
Graham, Richard H, physics
Green, Barry Adams, theoretical solid state physics
Green, Charles E, underwater acoustics, electroacoustics
Greenfeld, Philip John, cultural anthropology, anthropological linguistics
Greenwood, Ned H, geography
Griner, Lynn Adel, pathology, bacteriology
Grubbs, Edward, physical organic chemistry
Grundy, Scott Montgomery, physiology, biochemistry
Guillemin, Roger (Charles Louis), physiology, neuroendocrinology
Gunning, Barbara E, nutrition, biochemistry
Hageman, Donald Henry, physics, electrical engineering
Halasz, Nicholas Alexis, surgery
Halley, Robert, underwater acoustics
Hamilton, Edwin Lee, marine geology
Handschuh, Gerald Jay, clinical chemistry
Harris, Vincent Crockett, number theory
Harvey, Albert Raymond, mathematics
Hays, Robert Leroy, physiological ecology
Hazen, William Eugene, animal ecology
Heiges, Harvey Eric, transportation geography, urban geography
Hill, Clyde Alfred, mammalogy
Himes, Ronald Stewart, cultural anthropology
Ho, Hung-Ta, applied mathematics, fluid dynamics
Hobbs, Billy Frankel, mathematics
Holcomb, Walter Floyd, organic chemistry
Holley, Robert William, biochemistry
Holliday, Dale Vance, underwater acoustics
Holmes, Calvin Virgil, mathematics
Hood, John Mack, Jr, physics
Hopkins, George Robert, applied physics, nuclear engineering
Hubbard, Edward Leonard, plasma physics
Huffman, Edward Wight, entomology
Hunsaker, Don, II, psychobiology, ecology
Hunt, George Halsey, medicine
Hunter, George William, III, medical parasitology, tropical medicine
Hunter, Jack Allen, nuclear physics
Hurlbert, Stuart Hartley, ecology, limnology
Hurley, John Paul, radiation physics
Isensee, Robert William, organic chemistry
Jackson, Crawford Gardner, Jr, vertebrate biology, paleontology
Jacobs, Diane Margaret, immunology
Jensen, Torkil Hesselberg, plasma physics
John, Joseph, nuclear science, instrumentation
Johnson, Albert W, plant ecology
Johnson, C Scott, physics, biophysics
Johnson, Kenneth Duane, plant physiology
Johnson, Warren Arthur, geography
Joy, Joseph Wayne, physical oceanography
Kalma, Arne Haerter, experimental solid state physics
Kang, Kenneth S, microbial physiology
Kask, John Laurence, fisheries
Kaston, Benjamin Julian, biology
Kaye, George Thomas, oceanography
Keary, Thomas Joseph, electromagnetism
Keen, Elmer A, geography of East Asia, resource geography
Kelleher, Raymond Joseph, Jr, genetics, biochemistry
Kelly, Beatrice L, microbial genetics
Kern, John Philip, invertebrate paleontology
Kim, Ki Hong, analytical chemistry
Kinnison, Gerald L, physics
Kirkpatrick, Robert James, geology, petrology
Kolb, Alan Charles, physics
Kopp, Harriet Green, audiology, clinical psychology
Kosiba, Walter Louis, physical chemistry
Krekorian, Charles O'Neil, animal behavior
Krummenacher, Daniel, geology, chemistry
Kuhns, Ellen Swomley, physics, research administration
Kushinsky, Stanley, biochemistry
Kutner, Leon Jay, medical microbiology
Kvarda, Robert Edward, applied mathematics
LaBahn, Raymond Willis, ionospheric physics
La Fond, Eugene Cecil, oceanography
Landis, Vincent J, inorganic chemistry
Lange, Gordon David, neurophysiology, theoretical biology
Langer, Sidney, physical chemistry
Lasley, Bill Lee, reproductive physiology
Lasser, Elliott Charles, medicine, radiology
Leach, Larry Lamont, anthropology, archaeology
Lebherz, Herbert G, biochemistry
Lee, Raymond Curtis, physical chemistry
Lennon, Vanda Alice, neuroimmunology
Lennox, Edwin Samuel, biochemistry
Lesley, Frank David, mathematics
Levine, Paul Hersh, theoretical physcis, applied physics
Liebow, Averill Abraham, pathology
Lindner, Elek, analytical biochemistry, marine biology
Liska, Kenneth J, chemistry, ecology
Livingston, Virginia oncology, internal medicine
Lohrmann, Rolf, bio-organic chemistry
Long, David Michael, cardiovascular surgery, thoracic surgery
Long, John Vincent, physics
Longley-Cook, Mark T, acoustics
Lovett, Jack R, acoustics, physical oceanography
Lowenstein, Carl David, applied physics
Lubin, Ardie, psychophysiology, biostatistics
Ludwig, Claus Berthold, molecular spectroscopy, environmental physics
Lyon, Waldo (Kampmeier), physics
Magnuson, Gustav Donald, solid state physics, atomic physics
Malik, Jim Gorden, physical chemistry, inorganic chemistry
Marchand, E Roger, anatomy
Martin, Gordon Eugene, physics, engineering
Mather, Robert Laurance, physics
Mathewson, James H, bio-organic chemistry, oceanography
Maudlin, Lloyd Z, physics
Mauriello, David Anthony, ecology
Mautz, Charles William, physics
McArthur, David Samuel, physical geography
McBlair, William, physiology, oceanography
McDermott, John Patrick, organic chemistry
McEuen, Robert Blair, geophysics
McLaughlin, Charles Albert, zoology
McLean, Norman, Jr, invertebrate zoology
McLean, William Burdette, physics
McNeely, William Harold, chemistry, microbiology
Merz, Paul Louis, rubber chemistry, polymer chemistry
Merzbacher, Claude F, natural history, chemical engineering
Messmer, Trudy Ottilia, cell biology
Metzger, Robert P, biochemistry
Miller, Park Hays, Jr, physics
Miller, Philip Clement, ecology
Moe, Chesney Rudolph, physics
Monroe, Ronald Eugene, entomology
Moonan, William Jeane, statistics
Moore, David Gillis, marine geology, marine geophysics
Moore, Harold Beveridge, medical bacteriology
Moran, Reid Venable, plant taxonomy
Moriarty, Daniel Delmar, Jr, animal behavior
Morris, Halcyon Ellen McNeil, mathematics
Morris, Richard Herbert, electromagnetics
Morse, Garth Edwin, physics
Mosen, Arthur Walter, analytical chemistry
Moser, Joseph M, mathematical statistics
Moser, Kenneth Miles, medicine
Myers, Benjamin Franklin, Jr, physical chemistry
Mysels, Karol Joseph, colloid chemistry, surface chemistry
Naber, James Allen, radiation physics, solid state physics
Naitoh, Paul Yoshimasa, psychophysiology
Nedoluha, Alfred K, theoretical solid state physics
Neel, James William, soil science, botany
Nelson, Burt, astronomy
Neu, John Ternay, physical chemistry
Neubert, Jerome Arthur, physics
Neynaber, Roy Harold, atomic physics
Nilsen-Hamilton, Marit, biochemistry
Nordheim, Lothar Wolfgang, nuclear science
Northrop, John, geophysics
Nower, Leon, mathematics
Nyquist, Judith Kay, neurophysiology
O'Keefe, Dennis Robert, physics
Olson, Andrew Clarence, Jr, parasitology
O'Neal, Harry E, physical chemistry
Orgel, Leslie E, chemistry
Orloff, Marshall Jerome, surgery
Orona, Angelo Raymond, ethnology
Owens, Alvin Jewel, applied mathematics, computer science
Palmer, Robert Lewis, surface physics
Paolini, Paul Joseph, Jr, biophysics, physiology
Park, Chong Jin, mathematical statistics
Parker, Donal C, endocrinology, internal medicine
Parsons, John Arthur, cell biology
Passenheim, Burr Charles, experimental physics
Pavlopoulos, Theodore G, optical physics
Peskin, Gerald William, medicine
Peters, Richard Morse, surgery
Peterson, Donald Bruce, physical chemistry
Peterson, Gary Lee, geology
Peterson, Melvin Norman Adolph, marine geology
Pettit, David J, carbohydrate chemistry
Phelps, Leroy Nash, microbiology, immunology
Phillips, Richard P, earth sciences
Pickwell, George Vincent, physiology, marine biology
Pinkel, Robert, physical oceanography
Piserchio, Robert J, physics
Plymale, Harry Hambleton, zoology, veterinary medicine
Porter, John T, physical chemistry
Preston, Dudley A, botany
Pryde, Philip Rust, geography
Ptacek, Anton D, geology
Radwin, George E, malacology, invertebrate ecology
Rahe, Richard Henry, psychiatry
Rankin, Thomas Johnson, rheumatology
Rast, Howard Eugene, Jr, solid state science
Ratty, Frank John, Jr, genetics
Rayle, David Lee, plant physiology
Richardson, Robert William, geography
Richardson, William Harry, organic chemistry
Ridgway, Sam H, environmental physiology
Riedman, Richard M, audiology
Rigsby, George Pierce, geology
Riley, Robert Lee, polymer chemistry, organic chemistry
Rinehart, Robert R, genetics, radiation biology
Ring, Morey Abraham, inorganic chemistry
Robinson, Dudley Hugh, chemistry
Roeder, Stephen Bernard Walter, nuclear magnetic resonance, chemical physics
Rogers, Spencer Lee, physical anthropology, ethnology
Rohrl, Vivian, anthropology
Rohy, David Alan, solid state physics
Romano, Albert, mathematical statistics
Ross, Bernd, solid state physics, solid state electronics
Rouse, Carl Albert, theoretical astrophysics, theoretical nuclear physics
Rowe, Robert David, chemistry
Rubin, Theodore, statistics
Russell, John Lynn, Jr, nuclear physics
Salk, Jonas Edward, medicine, immunology
Saltz, Daniel, mathematics
Sarwinski, Raymond Edmond, physics
Schapiro, Harriette Charlotte, biochemistry
Scharber, Samuel Robert, Jr, physical chemistry
Schimitschek, Erhard Josef, physical chemistry, physics
Schlapfer, Werner T, neurobiology
Schock, Richard Unger, Jr, organic chemistry
Schopp, John David, astronomy
Segal, Charles Lewis, polymer chemistry, materials engineering
Sendroy, Julius, Jr, physiological chemistry
Shaffer, Patricia Marie, biochemistry
Sharts, Clay Marcus, fluorine chemistry, organic chemistry
Sheehy, Myles Joseph, operations research
Shepard, David C, biology
Sherwin, Chalmers William, physics
Shier, Wayne Thomas, biochemistry
Shipman, William H, radiochemistry

CALIFORNIA

Shore, Herbert Barry, theoretical solid state physics
Short, Donald Ray, Jr, mathematics
Shull, Charles Morell, Jr, physical chemistry, analytical chemistry
Skolil, Lester L, physics
Sloan, William Cooper, biology
Smith, Louis Ezra, Jr, physics
Smith, Norman Ty, anesthesiology, pharmacology
Snodgrass, Herschel Roy, physics
Snowden, Donald Philip, solid state physics
Spangler, John Allen, chemistry
Spanis, Curt William, neurobiology, cell biochemistry
Stahl, Ralph Henry, experimental physics
Steenbergen, James Franklin, microbiology
Sterling, Stewart Allen, solid state physics
Stetson, Alvin Rae, physical chemistry
Stevens, William George, electrochemistry, analytical chemistry
Stewart, Charles Jack, biochemistry
Stewart, Hugh Barnes, physics
Stewart, James Lloyd, physics
Stierwalt, Donald L, optics
Stratopoulos, George, mathematics
Strecker, Robert Louis, ecology, mammalogy
Stull, James Travis, pharmacology, biochemistry
Sund, Raymond Earl, nuclear physics
Taylor, James William, geography
Taylor, Kenneth Monroe, genetics
Teasdale, John G, physics
Thatcher, Everett Whiting, physics
Thomas, Blakemore Ewing, economic geology
Thomas, Harold Albert, physics
Thompson, Paul O, psychoacoustics, bioacoustics
Threet, Richard Lowell, geology
Thwaites, William Mueller, genetics
Traylor, Patricia Shizuko, chemistry, biochemistry
Tricoles, Gus P, optics, electrical engineering
Trummer, Max Joseph, thoracic surgery
Tuckfield, Ralph George, Jr, physics
Van de Wetering, Richard Lee, mathematics
Van Lint, Victor Anton Jacobus, physics
Vaslow, Dale Franklin, plasma physics
Vent, Robert Joseph, underwater acoustics
Verbinski, Victor V, nuclear physics
Victoria, Janice Krutak, animal behavior
Vogt, Marguerite Maria Paulette, medicine
Vold, Robert Donald, colloid chemistry
Vold, Robert Lawrence, colloid chemistry
von Essen, Carl Francois, radiotherapy
Vroom, David Archie, chemical physics
Wade, Robert Harold, organic chemistry, polymer chemistry
Wagner, Richard Vernon, anthropology
Walawender, Michael John, petrology, geology
Walba, Harold, organic chemistry
Walch, Henry Andrew, Jr, mycology
Warn, George Frederick, geology, geography
Watson, Donald Pickett, horticulture
Watson, Lawrence Craig, ethnology, cultural anthropology
Wedberg, Hale Levering, plant taxonomy, evolution
Wedberg, Stanley Edward, microbiology
Weiss, Herbert V, chemistry
Westerfield, Everett Commodore, physics
White, Ray Henry, solid state physics
White, Richard Wallace, underwater acoustics
Whitney, Daniel DeWayne, anthropology
Whittemore, William Leslie, physics
Wiederholt, Wigbert C, neurology, neurophysiology
Wilcox, Howard Albert, physics, environmental management
Wilkenfeld, Jason Michael, solid state physics
Willerding, Margaret Frances, mathematics
Wilson, Wilfred J, embryology
Wilts, James Reed, electronic physics
Wolf, Paul Leon, pathology
Wolter, Gerhard Herman, chemical physics
Wood, Forrest Glenn, marine zoology
Woodson, John Hodges, physical chemistry, theoretical chemistry
Wrasidlo, Wolfgang Johann, polymer chemistry
Wrobel, Theodore Frank, solid state physics
Yahr, Charles Corbin, cultural geography, geography of South Asia
Yamamoto, Sachio, marine chemistry, physical chemistry
Young, Arthur, astronomy
Young, Robert William, physics, acoustics
Zao, Zang Z, cardiology
Zedler, Joy Buswell, ecology
Zedler, Paul Hugo, ecology
Zeidler, James Robert, solid state physics
Zettner, Alfred, clinical pathology
Zirino, Alberto, chemical oceanography, polarography
Zvaifler, Nathan J, medicine, immunology

SAN FERNANDO

Sallo, Jerome Stanley, organic polymer chemistry

SAN FRANCISCO

Ablon, Joan, anthropology, medical anthropology
Adams, Jane N, microbiology, genetics
Adams, John Edwin, neurosurgery
Ainsworth, Cameron, organic chemistry
Aird, Robert Burns, neurology
Akers, William Alexander, dermatology
Albergotti, Jesse Clifton, experimental nuclear physics
Alexander, Claude Gordon, zoology
Allen, Alfred Marston, epidemiology, preventive medicine
Allende, Manuel Francisco, medicine
Ames, David Wason, cultural anthropology
Anderson, Carl Edgar, orthopedic surgery
Anderson, James Arthur, economic geology, resource management
Anderson, William Westerlin, neurology
Anthony, Luean Evangeline, biochemistry, nutrition
Apple, Martin Allen, molecular pharmacology, oncology
Araki, George Shoichi, biology
Archibald, Kenneth C, physical medicine
Ardley, Harry Mountcastle, mathematical statistics, applied statistics
Armstrong, Rosa Mae, histology, embryology
Arnaud, Paul Henri, Jr, entomology
Arrick, Myron S, medicine
Asling, Clarence Willet, anatomy
Astromoff, Andrew, mathematics
Atkinson, Marshall B, ophthalmology
Auerback, Alfred, psychiatry
Aufranc, Will H, medicine
Bach-Y-Rita, Paul, neurophysiology, rehabilitation
Bainton, Dorothy Ford, pathology, internal medicine
Baker, Donald Granville, physiology
Baker, William Laird, geophysics
Barbat, William Franklin, geology
Barbour, Donald Curtiss, internal medicine
Baron, Shirley Harold, surgery
Barton, John Selby, pulp chemistry, paper chemistry
Bassler, Gerald Clayton, organic chemistry
Batchelder, Alan Coleman, chemistry
Beck, John Christian, internal medicine
Beeman, Robert D, marine zoology, invertebrate zoology
Belzer, Folkert O, surgery
Bendix, Selina (Weinbaum), environmental management
Benet, Leslie Z, biopharmaceutics
Bennington, James Lynne, pathology
Benton, Eugene Vladimir, radiological physics
Berblinger, Klaus William, psychiatry, psychosomatic medicine
Berman, Louis, astronomy
Berrend, Robert E, invertebrate zoology
Bertin, Henry John, Jr, physical organic chemistry
Bhatnagar, Rajendra Sahai, biochemistry
Bibel, David Jan, medical microbiology, microbial ecology
Biglieri, Edward George, internal medicine, endocrinology
Binford, Laurence Charles, vertebrate zoology, ornithology
Bishop, John Michael, virology, biochemistry
Blake, James J, organic chemistry, biochemistry
Blois, Marsden Scott, Jr, dermatology
Borchardt, Glenn Arnold, soil mineralogy
Borchardt, Kenneth, microbiology
Bowen, Sarane Thompson, genetics
Bowman, Edgar Cornell, geology
Bowman, Robert Irvin, ornithology
Boyer, Herbert Wayne, microbiology
Bradbury, Margaret G, ichthyology
Bradley, Thomas Bernard, Jr, internal medicine, hematology
Brecher, George, clinical pathology, hematology
Brieger, Gert Henry, history medicine
Brochmann-Hanssen, Einar, pharmaceutical chemistry
Brodie, Donald Crum, pharmacy
Brodsky, Carroll M, psychiatry, anthropology
Brown, Ellen, medicine
Brown, Kenneth Taylor, physiology
Bruhns, Karen Olsen, anthropology
Bucci, Thomas Joseph, comparative pathology
Burbank, Burr Gamaliel, physics
Burkhalter, Alan, pharmacology
Byers, Sanford Oscar, biochemistry
Byers, Vera Steinberger, immunobiology
Callaway, Enoch, III, psychiatry
Campbell, Ian, environmental geology
Cappucci, Dario Ted, Jr, veterinary medicine
Carbone, John Vito, medicine
Carlsson, Erik, radiology
Castagnoli, Neal, Jr, organic chemistry, medicinal chemistry
Cavalieri, Ralph R, endocrinology, nuclear medicine
Chamberlain, Jack G, developmental anatomy
Chang, Hsing-Tze Ruan, protein chemistry, nutritional biochemistry
Chao, Fu-chuan, biochemistry
Cheitlin, Melvin Donald, internal medicine, cardiology
Chierici, George J, prosthodontics
Clark, Mary Margaret, cultural anthropology
Cleaver, James Edward, cancer
Clements, John Allen, physiology
Close, Perry, genetics, physiology
Cloud, William K, seismology, mechanical engineering
Cluff, Lloyd Sterling, geology
Coan, Eugene Victor, environmental sciences, malacology
Cohen, Peter, pediatrics
Colenbrander, August, ophthalmology
Colley, Fredrick Christensen, parasitology, public health
Collins, Carter Compton, biophysics
Colman, Arthur David, psychiatry
Comroe, Julius Hiram, Jr, physiology
Cooke, Roger, biophysics
Coppenger, Claude Jackson, physiology
Corbascio, Aldo Nicola, pharmacology
Coyle, Bernard Andrew, inorganic chemistry, crystallography
Craig, John Cymerman, organic chemistry
Crede, Robert H, internal medicine
Crippen, Gordon Marvin, biophysical chemistry, theoretical chemistry
Curry, Francis J, internal medicine
Dallman, Peter R, pediatrics, hematology
Daniels, Troy Cook, medicinal chemistry
Dawson, Chandler R, ophthalmology
Deen, William Murray, physiology, chemical engineering
De Groot, Jack, anatomy, endocrinology
Deller, John Joseph, Jr, internal medicine, endocrinology
De Ment, Jack (Donovan), agronomy
Dennis, Michael Joseph, neurobiology
Detert, Francis Lawrence, organic chemistry
DeVenuto, Frank, organic chemistry, biological chemistry
Diamond, Ivan, neuroanatomy
Diamond, Louis Klein, medicine, pediatrics
Dickinson, Wade, biophysics
Doell, Ruth Gertrude, biology
Douglas, Robert James, mathematics
Dower, John Charles, pediatrics, gastroenterology
Dreyer, David, organic chemistry
Dreyer, Robert Marx, economic geology
Duke, Joseph A, physical biochemistry
Duncan, James Thayer, developmental biology
Dunicz, Boleslaw Ludwik, physical chemistry
Dunn, Frederick Lester, epidemiology, anthropology
Dunphy, J Englebert, medicine, surgery
Durbin, Richard Paul, physiology, biophysics
Edelman, Isidore Samuel, medicine
Edmondson, Dale Edward, biochemistry
Eiler, John Joseph, biochemistry
Einstein, Elizabeth Roboz, neurochemistry
Elashoff, Robert M, statistics
Elias, Joel Jesse, zoology, anatomy
Ellinwood, Howard Lyman, geology
Ellman, George Leon, biochemistry
Enelow, Allen Jay, psychiatry
Engleman, Ephraim Philip, medicine
Epstein, Charles Joseph, medical genetics, developmental biology
Epstein, Ervin Harold, dermatology
Epstein, Leon J, psychiatry
Epstein, Lois Barth, medical science, health science
Epstein, Wallace Victor, internal medicine
Epstein, William L, medicine, dermatology
Eschmeyer, William Neil, ichthyology
Estermann, Eva Frances, plant physiology
Eydal, Astvaldur, geography
Fairley, Henry Barrie Fleming, anesthesiology
Farber, Seymour Morgan, thoracic diseases, pulmo nary diseases
Farmer, Susan Walker, endocrinology
Farrell, Edward Joseph, mathematics
Feinberg, Irwin, psychiatry
Feingold, Ben F, allergy, immunology
Fielding, Christopher J, metabolism, biochemistry
Fields, Howard Lincoln, neurophysiology
Filice, Francis P, zoology
Fineberg, Richard Arnold, biochemistry
Fink, Donald Lloyd, pediatrics
Fischer, John Eugene, mathematics
Fisher, Newman, mathematics
Fishman, Robert Allen, neurology
Fishman, Sherman Sampson, pharmacology, ultrasound
Fletcher, Donald Warren, microbial ecology
Follett, Wilbur Irving, ichthyology
Forsham, Peter Hugh, internal medicine, endocrinology
Fowler, James Lee, veterinary public health, food microbiology
Fraser, Gordon Simon, sedimentology
French, Frederick Alexis, chemotherapy
Frick, Oscar L, immunology, pediatrics
Friend, Daniel S, experimental pathology, cell biology
Friesen, Earl Wayne, nuclear physics
Fukuyama, Kimie, dermatology
Furst, Arthur, cancer, toxicology
Gabel, James Russel, protozoology
Gabelman, John Warren, geology
Galehouse, Jon Scott, marine geology, sedimentology
Gallo, Robert Vincent, endocrinology
Ganong, William Francis, physiology
Garfin, David Edward, molecular biology
Garoutte, Bill (Charles), neurophysiology
Gates, George Oscar, geology
Geiger, Jacob Casson, public health
Genolio, Raymond Joseph, theoretical physics
Giammona, Samuel T, pediatrics
Gillies, George Alexander, organic chemistry
Glass, Laurel Ellen, developmental biology, anatomy
Gobuty, Allan Harvey, chemistry, medicinal chemistry
Goerke, Rudolph Jon, biophysics, surface chemistry
Goerz, David Jonathan, Jr, physics
Gohr, Frank August, environmental sciences
Goldberg, Estelle Maxine, mathematics
Goldberg, Melvin Leonard, biochemistry, pathology
Goldfien, Alan, medicine
Goldman, Leon, surgery
Goldstein, Bernard, physiology, morphology
Goodman, Howard Michael, biochemistry
Goodman, Joel Warren, immunology
Goodman, Joseph Robert, electron microscopy
Goodson, Jo Max, pharmacology, dental research
Gordan, Gilbert Saul, medicine
Gordon, Burton LeRoy, biogeography
Gordon, Samuel Morris, chemistry
Gorman, Melville, inorganic chemistry, history of science
Goyan, Frank Mayer, physical chemistry
Goyan, Jere Edwin, pharmaceutical chemistry
Greenberg, David Morris, biochemistry
Greenberg, Louis Donald, biochemistry
Grimes, Orville Frank, thoracic surgery
Grodsky, Gerold Morton, biological chemistry, endocrinology
Grossman, Moses, pediatrics
Gruhn, Thomas Albin, physical chemistry

Grumbach, Melvin Malcolm, medicine, pediatrics
Gurchot, Charles, chemistry
Gustafson, Joel Frank, entomology, ecology
Guthrie, Christine, molecular biology, biochemistry
Guttman, Paul H, pathology, radiobiology
Haag, James Norman, computer science, nuclear physics
Hacker, Walter Rudolf, geography
Hadley, William Keith, clinical pathology, microbiology
Haimes, Florence Catherine, chemistry, history of chemistry
Halde, Carlyn Jean, medical mycology
Hall, Joseph Glenn, vertebrate zoology
Hamilton, William Kennon, clinical medicine, anesthesiology
Hanes, Deanne Meredith, biology, immunology
Hanna, Lavelle, microbiology
Hannon, John Patrick, physiology
Hansen, Louis Stephen, oral pathology
Harper, Harold Anthony, biochemistry
Harris, John Wayne, cell biology, radiation biology
Harris, Mervin Robert, psychiatry
Harvold, Egil, dentistry
Havel, Richard Joseph, medicine
Hayashida, Tetsuo, anatomy, endocrinology
Henshaw, Paul Carrington, economic geology
Hensill, John Samuel, biology, physiology
Herman, Robert Harold, internal medicine
Hetzner, Howard Paul, petroleum chemistry
Hoffer, Paul B, radiology, nuclear medicine
Hoffman, Arlene Faun, medical education, physiology
Hoffman, Julien Ivor Ellis, pediatric cardiology
Hogan, Michael John, ophthalmology
Hohenthal, William Dalton, Jr, ethnography, ethnology
Hollahan, John Ronald, physical chemistry
Hollenberg, Milton, medicine, cardiovascular physiology
Holliday, Malcolm A, pediatrics
Holt, Ray James, thermodynamics
Holzer, Walter Frank, chemistry
Hopkins, Lemac, economic entomology
Horowitz, Mardi J, psychiatry, psychoanalysis
Hovey, Richard Dean, geophysics, geology
Hoyt, William F, neurology, ophthalmology
Hughes, Robert Alan, environmental chemistry, limnology
Hunderfund, Richard C, microbiology, parasitology
Hunt, Thomas Knight, surgery
Hurst, Valerie, medical microbiology
Inman, Verne Thomson, anatomy, orthopedic surgery
Jacobs, Edwin M, cancer, chemotherapy
Jaffe, Robert B, endocrinology, obstetrics & gynecology
James, Thomas Larry, biophysical chemistry
Jampolsky, Arthur, ophthalmology
Jawetz, ERnest, microbiology, medicine
Jayne, Jerrold Clarence, analytical chemistry, inorganic chemistry
Jewett, Don L, orthopedic surgery, neurophysiology
Johnson, Russell Tingey, plant genetics, agronomy
Jones, Reese Tasker, psychiatry
Jones, Theodore Harold Douglas, biochemistry, microbiology
Jorgensen, Eugene Clifford, organic chemistry, medicinal chemistry
Judd, Stanley H, environmental health
Kaiser, Raymond Francis, medicine
Kane, John Power, medicine, biochemistry
Kaplan, Herman, oral surgery, oral medicine
Kaplan, Selna L, pediatrics, endocrinology
Kapp, Leon Neal, molecular biology, cell biology
Kaska, Harold Victor, geology

Katzung, Bertram George, pharmacology
Kaufman, Herbert S, medicine, allergy
Kavanaugh, David Henry, systematic entomology
Kearney, Edna Beatrice, biochemistry
Keefe, James Richard, organic chemistry
Kelley, James Charles, oceanography
Kellogg, Ralph Henderson, physiology
Kelly, John J, internal medicine, cardiology
Kelly, Regis Baker, neurobiology
Kenney, William Clark, biochemistry
Kent, James Woodward, physical chemistry
Ketcham, Roger, organic chemistry
Kiefer, Christie Weber, anthropology
Kilgore, Bruce Moody, forest ecology, wildlife ecology
Kim, Young Shik, medicine
Kimura, Samuel Jiro, ophthalmology
King, Eileen Brenneman, pathology, cytology
King, Mary-Claire, genetics
Klatte, Fred John, Jr, animal nutrition
Kleist, John Raymond, geology
Kolb, Felix Oscar, medicine
Kolipinski, Milton Charles, marine ecology
Kollman, Peter Andrew, quantum chemistry
Korenbrot, Juan Igal, biophysics
Kraft, Lisbeth Martha, pathology, radiobiology
Krasno, Louis Richard, medicine
Krevans, James Richard, medicine
Krol, Arthur J, dentistry
Kumler, Warren Donald, chemistry
Kun, Ernest, biochemistry, pharmacology
Kuntz, Irwin Douglas, Jr, physical chemistry
Landahl, Herbert Daniel, mathematical biology
Lawry, James Voris, neurophysiology, neuroanatomy
Leake, Chauncey D, pharmacology, history of science
Ledin, George, Jr, computer science, statistics
Lee, Alfred Tze-Hau, analytical chemistry, organic chemistry
Lee, Kwan-Hua, pharmaceutical chemistry
Lee, Nancy Zee-Nee Ma, biochemistry
Lee, Philip Randolph, internal medicine
Lee, Welton Lincoln, marine ecology, physiological ecology
Leicester, Henry Marshall, biochemistry, cell biology
Lemanski, Larry Fredrick, developmental biology
Levin, Victor Alan, cancer, neurology
Levinson, Warren E, microbiology, cell biology
Levit, Robert Jules, number theory
Leviton, Alan Edward, systematic zoology, zoogeography
Levy, Joseph Victor, physiology, pharmacology
Levy, Louis, pharmacology
Lewin, Ellen, anthropology
Lewis, Richard John, hematology, clinical pharmacology
Li, Choh Hao, biochemistry
Libet, Benjamin, physiology
Licko, Vojtech, mathematical biology
Linde, Peter Franz, physical chemistry
Lindquist, Robert Nels, biochemistry, organic chemistry
Lindsay, George Edmund, plant taxonomy
Linscott, William Dean, immunology
Loh, Horace H, biochemistry, biochemical pharmacology
Long, John Arthur, cytology, anatomy
Loveless, Loyal E, clinical chemistry
Lowenstein, Jerold Marvin, nuclear medicine
Lucas, Donald Brooks, orthopedic surgery
Ludowieg, Julio, biochemistry
Lugassy, Armand Amram, materials science, prosthodontics
Luhman, Gladys Finney, physics
Lukin, Larissa Skvortsov, physiology
Mackey, James P, biology
Macrae, Donald, neurology
Maibach, Howard I, dermatology
Majmundar, Hasmukhrai Hiralal, geochemistry, geology
Malamud, Nathan, neuropathology
Maloney, Mary Adelaide, biology
Mandra, York T, geology
Margaretten, William, pathology
Margulis, Alexander Rafailo, radiology
Maroney, William, chemistry
Mathews, J Rodney, orthodontics
Mayeri, Earl Melchior, neurophysiology

McCarthy, Brian John, microbiology, molecular biology
McCasland, Gifford Ewing, organic chemistry, computer science
McClintock, Elizabeth, systematic botany
McCorkle, Horace J, surgery
McCosker, John E, aquatic biology, ichthyology
McEwen, William Kirk, biochemistry
McIlroy, Malcolm B, internal medicine
McKay, Donald George, pathology
Meagher, Richard Brian, molecular genetics, enzymology
Meggison, David Laurence, food technology
Melmon, Kenneth Lloyd, clinical pharmacology
Mendelson, Robert Alexander, Jr, nuclear physics, molecular biology
Merckel, Charles George, medicine
Merzenich, Michael Matthias, neurophysiology, neuroanatomy
Meyer, Rich Bakke, Jr, medicinal chemistry, organic chemistry
Meyers, Frederick Henry, pharmacology
Michaeli, Dov, immunochemistry, biochemistry
Miller, Arthur Joseph, neurophysiology, physiology
Miller, Earl Roy, radiology
Miller, Malcolm Ray, anatomy
Miller, Philip Allen, inorganic chemistry
Miller, Robert Cunningham, zoology
Milne, David Bayard, nutritional biochemistry, bioinorganic chemistry
Mines, Allan Howard, physiology
Mitchell, Robert A, physiology
Monie, Ian Whitelaw, anatomy
Moore, Wesley Sanford, surgery
Morales, Manuel Frank, biophysics
Mori, Raymond I, organic chemistry
Motto, Jerome (Arthur), psychiatry
Mueller, Manfred Ernst, physical chemistry
Mullen, David Anthony, zoology
Murphy, Alexander James, biochemistry
Murray, John Frederic, internal medicine
Murray, William R, orthopedic surgery
Mustacchi, Piero Oscar, medicine
Nadel, Jay A, pulmonary physiology, cardiovascular physiology
Nasser, DeLill, microbiology, microbial genetics
Neff, William Medina, embryology
Nelson, Arthur Hansen, biology
Neustadter, Siegfried Friedrich, mathematics
Neville, James Ryan, medical physiology
Newbrun, Ernest, oral biology, biochemistry
Newton, Thomas Hans, radiology
Nichols, Barbara Ann, cell biology, experimental pathology
Niebauer, John J, medicine
Nitecki, Danute Emilija, biochemistry
Nolan, Janiece Simmons, hospital administration, physiology
Noonan, Charles D, radiology
Norris, Forbes Holten, Jr, neurology, neurophysiology
Nydegger, Corinne Nemetz, medical anthropology
Nygreen, Paul W, geology, oceanography
Oberlander, George T, botany
O'Connor, George Richard, ophthalmology
O'Donnell, Ashton Jay, physics
Oh, Jang Ok, virology, pathology
Oppenheim, Joseph Harold, algebra
Oppenheimer, Frank, physics
Oppenheimer, Norman Joseph, biochemistry
Orr, Robert Thomas, mammalogy, ornithology
Ortiz De Montellano, Paul R, bio-organic chemistry
Osborn, John Jay, medicine
Ostwald, Peter Frederic, psychiatry, communication sciences
Ottoboni, Fred Lawrence, Jr, occupational health
Ovenfors, Carl-Olof Nils Sten, radiology
Painter, Robert Blair, radiobiology
Palubinskas, Alphonse J, radiology
Papkoff, Harold, biochemistry
Parmley, William W, cardiology
Patt, Harvey Milton, radiobiology, biophysics
Pavone, Ben W, dentistry
Peddicord, Richard G, computer science
Pedersen, Roger Arnold, developmental genetics, radiobiology
Peller, Leonard, physical chemistry
Perkins, Herbert Asa, hematology
Pestrong, Raymond, environmental geology, geomorphology
Petrakis, Nicholas Louis, medicine
Pevehouse, Byron C, medicine

Phelps, Pharo A, physics, civil engineering
Phibbs, Roderic H, pediatrics, biophysics
Philip, Cornelius Becker, entomology
Phillips, Theodore Locke, medicine, radiology
Picker, Robert, geography, environmental management
Piel, Carolyn F, pediatrics
Piper, Walter Nelson, biochemistry, pharmacology
Plaat, Otto, mathematics
Plantz, Robert Glenn, biomedical engineering
Pollycove, Myron, medicine, biophysics
Posin, Daniel Q, physics
Post, Douglas Manners, botany
Powell, Thomas Mabrey, chemistry
Pratt, Robertson, pharmacognosy, microbiology
Price, David C, nuclear medicine, hematology
Printz, Richard H, anatomy, obstetrics & gynecology
Rainwater, Clarence Saunders, physics
Ralston, Henry James, physiology
Ralston, Henry James, III, neuroanatomy, electron microscopy
Ramachandran, Janakiraman, endocrinology, biochemistry
Ramsey, David John, medical physiology
Ramsey, Brian Gaines, physical organic chemistry, molecular spectroscopy
Rapaport, Elliot, internal medicine
Rector, Floyd Clinton, Jr, nephrology
Reid, Ian Andrew, physiology
Reinhardt, William Oscar, medical administration, anatomy
Rentz, David Charles, systematic entomology
Reuther, Ronald Theodore, vertebrate zoology, zoogeography
Richards, Victor, surgery
Rider, Joseph Alfred, gastroenterology
Riegelman, Sidney, pharmacy
Riley, Danny Arthur, anatomy
Robinson, Lewis Howe, meteorology, statistics
Rodda, Peter Ulisse, geology
Roe, Benson Bertheau, thoracic surgery
Rollo, Frank David, physics, medicine
Romberg, Paul Frederick, medical mycology, taxonomy
Rosen, Steven David, cell biology
Rosenau, Werner, pathology, immunology
Rosenman, Ray Harold, medicine
Rosenstein, Ludwig, chemistry
Rosenthal, Fred, neurophysiology
Rosinski, Edwin F, medical education
Ross, Edward Shearman, entomology, biological photography
Rothman, Stephen Sutton, physiology
Rudnick, Albert, virology
Rudolph, Abraham Morris, physiology
Ruesch, Jurgen, psychiatry
Ruffini, Julio Lawrence, anthropology
Rutter, William J, biochemistry
Ryge, Gunnar, dentistry, physics
Sadee, Wolfgang, drug metabolism, analytical biochemistry
Sadler, Henry Harrison, Jr, psychiatry
Sampson, Sanford Robert, physiology, pharmacology
Sandman, Robert Paul, biochemistry
Santi, Daniel V, organic chemistry, biochemistry
Saunders, Frank Austin, psychology, neurophysiology
Saunders, John Bertrand De Cusance Morant, history of medicine, human anatomy
Saunders, Virginia Fox, neuropsychology, neurosciences
Schachter, Julius, bacteriology
Schmid, Peter, biology, biochemistry
Schmid, Rudi, internal medicine, biochemistry
Schofield, Edmund Acton, Jr, environmental sciences, conservation
Schooley, Robert, anatomy
Schuchard, Alfred, dentistry
Schwarz, Thomas Werner, pharmaceutical chemistry, pharmacy
Scott, Norman McLean, Jr, internal medicine, gastroenterology
Scott, Robert Clyde, groundwater geology
Scrivener, Charles Arthur, public health
Searle, Gilbert Leslie, physiology
Sebastian, Anthony, medicine
Selzer, Arthur, medicine, cardiology
Senyk, George, immunology, microbiology
Severinghaus, John Wendell, medicine

381

CALIFORNIA

Shafer, Richard Howard, biophysical chemistry
Shah, Shantilal Nathubhai, biochemistry, neurochemistry
Shames, David Marshall, nuclear medicine
Shapiro, Charles Saul, physics
Shapiro, Edwin Seymour, mathematics
Shaw, Evelyn (Mrs Fred Wertheim), animal behavior
Sheline, Glenn Elmer, medicine
Shetlar, Martin David, photochemistry
Shinefield, Henry R, medicine, pediatrics
Shnider, Sol M, medicine
Simon, Alexander, psychiatry
Singer, Thomas Peter, biochemistry
Siperstein, Marvin David, biochemistry, internal medicine
Siteri, Pentti Kasper, biochemistry
Skala, James Herbert, analytical biochemistry, nutrition
Sleisenger, Marvin Herbert, medicine
Small, John, Jr, geology, geophysics
Smith, Allyn Goodwin, biology
Smith, Arthur R, economic geology
Smith, Donald Ridgeway, medicine, urology
Smith, James Thomas, mathematics
Smith, Lloyd Hollingworth, Jr, medicine
Smith, Robert H, anesthesiology
Sokolow, Maurice, medicine
Solomon, Malcolm David, organic chemistry
Sooy, Francis Adrian, otolaryngology
Sopp, Samuel William, inorganic chemistry, physical chemistry
Speck, Reinhard Staniford, medical bacteriology
Spence, John A, organic chemistry, environmental health
Spencer, E Martin, medicine
Spivey, Bruce Eldon, ophthalmology, medical education
Spotts, John Hugh, geology
Spudich, James Anthony, biochemistry
Stark, Marvin Michael, dentistry
Staub, Norman Croft, physiology
Steinbach, Howard Lynne, radiology
Steinberg, Roy Herbert, neurophysiology
Stewart, Alva Theodore, Jr, organic chemistry, science administration
Stifel, Frederick Benton, biochemistry, nutrition
Stoeckenius, Walther, cytology
Stollberg, Robert, science education
Stone, Deborah Bennett, physical biochemistry, biochemical pharmacology
Strain, Jerome Chamberlain, dentistry
Stubbs, John Dorton, molecular biology, biochemistry
Sullivan, Raymond, geology
Sutherland, Violette Cutter, pharmacology
Swan, Lawrence Wesley, biology
Swanson, Robert Nels, micrometeorology
Sweet, Norman Joseph, internal medicine
Szurek, Stanislaus (Andrew), psychiatry, psychoanalysis
Tabatabaian, Ali Mohammad, mathematics
Tachibana, Dora K, immunology, medical microbiology
Tamimi, Hamdi Ahmad, microbiology, infectious diseases
Tang, Alfred Sho-Yu, algebra
Tappeiner, John Cummings, II, forest ecology, silviculture
Tarver, Harold, chemistry
Taylor, Steve L, food science, toxicology
Taylor, William H, II, solid state physics
Tesche, Frederick Rutledge, physics
Thaler, M Michael, pediatrics, developmental biology
Thiers, Harry Delbert, mycology
Thomas, Arthur Norman, surgery, thoracic surgery
Thomas, Gerald Andrew, radiochemistry
Thomassen, Paul R, Jr, bacteriology
Thorup, Richard M, agronomy, soil fertility
Thuillier, Richard Howard, meteorology, air pollution
Thygeson, Phillips, ophthalmology
Tidd, Charles Wharton, psychiatry
Tieman, Suzannah Bliss, neurosciences, vision
Todd, Harry Flynn, Jr, anthropology, medical anthropology
Toland, William Gridley, Jr, industrial chemistry
Tomlinson, Jack Trish, invertebrate zoology
Tooley, William Henry, pediatrics
Towle, Albert, invertebrate physiology, science education
Tozer, Thomas Nelson, pharmaceutical chemistry
Treagan, Lucy, microbiology
Trevor, Anthony John, biochemical pharmacology, neurochemistry
Tuck, Leo Dallas, physical chemistry
Tyberg, John Victor, cardiovascular physiology
Tyler, John Howard, geology
Urbach, Karl Frederic, hospital administration
Van Beckum, William George, chemistry
Veith, Ilza, history of medicine
Vore, Mary Edith, toxicology, pharmacology
Vyas, Girish Narmadashankar, immunology, genetics
Wakerlin, George Earle, medicine
Wallerstein, Robert Solomon, psychiatry, psychoanalysis
Wallis, Orthello Langworthy, ecology, natural science
Waltz, Arthur G, neurology, cardiovascular diseases
Wang, Harry (Hsi), zoology
Watanabe, Shizuo, biochemistry
Watson, John Alfred biochemistry
Watts, Malcolm S M, internal medicine
Way, E Leong, pharmacology, toxicology
Way, Lawrence Wellesley, surgery
Way, Walter, anesthesiology, pharmacology
Wayburn, Edgar, internal medicine
Webb, William Paul, organic chemistry
Weiner, Richard Ira, neuroendocrinology
Weinkam, Robert Joseph, medicinal chemistry
Weiskopf, Richard Bruce, respiratory physiology, anesthesiology
Wellington, John Sessions, pathology
Werdegar, David, community health
Wessenberg, Harry Sanders, protozoology, parasitology
Westfall, John Edward, historical geography
Wheeler, Kenneth Theodore, Jr, biophysics, radiation biology
Wiggins, Virgil Dale, palynology
Wiley, Lynn M, developmental biology
Williams, George Zur, clinical pathology
Williams, Herbert H, anthropology
Williams, Hibbard E, medical genetics
Williams, John Andrew, physiology
Williams, Stanley Clark, ecology
Williston, John Stoddard, neurophysiology
Wilson, Charles B, medicine
Wilson, Christine Shearer, nutrition
Wise, Burton Louis, neurosurgery
Wolff, Manfred Ernst, medicinal chemistry
Wolff, Sheldon, cytogenetics, radiobiology
Wong, Walter Mun-Fay, dentistry
Wong-Riley, Margaret Tze Tung, neuroanatomy, anatomy
Wood, David Alvra, pathology
Wu, William Gay, medical microbiology, immunology
Wycoff, Samuel John, public health, dentistry
Wylie, Edwin J, surgery
Yamamoto, Keith Robert, molecular biology
Yang, Jen Tsi, biophysical chemistry
Yeager, Charles Levant, neurology, psychiatry
Yonenaka, Hideo H, marine microbiology
Young, John Kiger, dentistry
Zarrugh, Laura Hoffman, medical anthropology
Zavortink, Thomas James, entomology, biology
Zipp, Adam Peter, biophysical chemistry, biophysics
Zippin, Calvin, biostatistics, epidemiology

SAN GABRIEL

Kutches, Alexander Joseph, animal nutrition

SAN JOSE

Abraham, Farid Fadlow, physics
Acrivos, Juana Luisa Vivo, physical chemistry
Aitken, Donald W, Jr, physics, astrophysics
Anderson, Arthur George, physics
Anderson, Herbert Rudolph, Jr, physical chemistry
Anderson, Rodney Ebon, mathematics
Anthrop, Donald F, thermodynamics, geography
Bagus, Paul Saul, atomic physics, quantum chemistry
Balgooyen, Thomas Gerrit, ornithology
Ballard, Ralph Campbell, entomology
Bargon, Joachim, physical chemistry
Barker, John Adair, chemical physics, statistical mechanics
Barlow, Irmela Christiane, physical chemistry, electron microscopy
Barrall, Edward Martin, II, analytical chemistry
Batra, Inder Paul, solid state physics
Beard, Jean, science education
Bell, Charles W, plant physiology
Bennett, John Francis, science education, evolutionary biology
Bickford, Lawrence Richardson, solid state physics
Bird, Marion Taylor, mathematics
Bornstein, Robert D, meteorology
Brewer, Richard George, quantum physics, molecular physics
Brooke, John Percival, geophysics, geochemistry
Calhoun, Bertram Allen, magnetism
Cantow, Manfred Josef Richard, physical chemistry, polymer physics
Carter, David, physics
Castro, Albert Joseph, organic chemistry
Castro, George, research administration, physical chemistry
Chen, Jane Lee, plant physiology
Chen, Tien Chi, computer science
Chu, William How-Jen, polymer physics
Coburn, John Wyllie, plasma physics
Craig, James Morrison, microbiology
Creely, Robert Scott, geology
Croll, Ian Murray, physical chemistry
Dahlstrom, Robert V, biochemistry
David, Lore Rose, biology
DeHollander, William Roger, physical chemistry
Diaz, Arthur Fred, organic chemistry
Dimeff, John, physics
Dolby, James Louis, applied statistics
Dolloff, Norman, geology, geochemistry
Duggan, Michael J, physics
Economy, James, organic chemistry
Edwards, J Gordon, entomology, systematic zoology
Eldridge, Jerome Michael, surface chemistry, solid state physics
Ellis, Forest Albert, human physiology
Embree, Harland Dumond, organic chemistry
Engel, Jan Marcin, solid state physics
Eschenfelder, Andrew Herbert, physics
Feldman, Leonard, mathematics
Ferguson, William E, entomology
Fischer, David Lloyd, nuclear physics
Fitting, Marjorie Ann, mathematics education
Foote, John K, physical chemistry
Foster, Robert John, geology
Fowler, Kenneth Arthur, mathematics
Freeman, Paul Joel, physiology
Frenster, John H, internal medicine, oncology
Friedman, Henry David, mathematics
Fuller, Franklyn Belmont, applied mathematics
Geiss, Roy Howard, electron microscopy
Ghosh, Sakti P, mathematical statistics, computer science
Gladney, Henry M, computer science, research administration
Goodman, Nelson, industrial microbiology
Gordon, Joseph Grover, II, inorganic chemistry, dynamics
Graf, William, zoology
Grant, Barbara Dianne, synthetic organic chemistry
Grant, Paul Michael, solid state physics
Greene, Richard L, solid state physics
Greer, Edison, mathematics
Grilione, Patricia Louise, microbiology
Gritter, Roy John, organic chemistry
Haight, Roger Dean, microbiology, biochemistry
Harris, Hubert Andrew, phytopathology
Hartesveldt, Richard J, ecology, forestry
Harvey, Harry Thomas, biology
Henderson, Douglas, theoretical physics, theoretical chemistry
Hendricks, Lawrence Joseph, limnology, fisheries
Herman, Frank, theoretical physics
Hester, Joseph Aaron, Jr, anthropology
Hewett, William Ainslie, polymer chemistry
Hiraoka, Hiroyuki, physical chemistry
Hoggatt, Verner Emil, Jr, mathematics
Holmstrom, Fred Edward, physics
Hunziker, Heinrich Erwin, chemical physics
Hutton, Kenneth Earl, physiology
Jaffe, Annette Bronkesh, physical organic chemistry, electroanalytical chemistry
James, Dean B, physical inorganic chemistry
Jamison, Herman Free, mathematics
Jurmain, Robert Douglas, physical anthropology
Kay, Eric, inorganic chemistry, physical chemistry
Kelley, Leon A, biochemistry
Kenk, Vida Carmen, invertebrate zoology
Koehler, Thomas Richard, physics
Koenig, Inge Rabes, physical chemistry
Kramer, Max, mathematics
Kutilek, Michael Joseph, wildlife ecology
Lange, Lester Henry, mathematics
Langlois, William Edwin, hydrodynamics
Lanzilotta, Raymond Philip, microbiology
Larrabee, Richard Brian, organic chemistry
Larsen, Charles McLoud, mathematics
Lee, Kenneth, solid state physics
Lester, William Alexander, Jr, physical chemistry
Levy, Morton Frank, organic chemistry
Lindberg, Lois Helen, medical microbiology
Lorenz, Max Rudolph, physical chemistry, physics
Lovaglia, Anthony Richard, mathematics
Macfarlane, Roger Morton, solid state physics
MacKay, Kenneth Pierce, Jr, meteorology, air pollution
Maddock, Marshall, geology
Marrello, Vincent, solid state physics
Matlow, Sheldon Leo, chemical physics
McCallum, George Alexander, genetics
McIntyre, Michael Perry, physical geography, geography of the Pacific Basin
Mewaldt, Leonard Richard, ornithology
Michael, William Alexander, mathematics
Miller, Albert, meteorology
Miller, Robert Dennis, organic chemistry
Minkiewicz, Vincent Joseph, solid state physics
Mitchem, John Alan, mathematics
Monard, Joyce Anne, nuclear physics, nuclear engineering
Morawitz, Hans, theoretical physics
Morejohn, G Victor, zoology, genetics
Murphy, Henry D, histology, anatomy
Myers, William Howard, mathematics
Nakagawa, T William, organic chemistry, polymer chemistry
Naylor, Benjamin Franklin, chemistry
Nebenzahl, Linda Levine, physical organic chemistry, surface chemistry
Nelson, Norman Bartram, epidemiology
Neptune, John Addison, inorganic chemistry
Nesbet, Robert Kenyon, theoretical physics
Noyes, Robert Wallace, obstetrics & gynecology
Olds, Carl Douglas, mathematics
Onton, Aare, physics, electrical engineering
Ortenburger, Irene Beardsley, theoretical solid state physics
Ouano, Augustus Ceniza, chemical engineering, polymer physics
Parrish, William, x-ray crystallography, materials science
Patterson, Harry Robert, pharmacy, bacteriology
Pisano, Rocci George, biology, agriculture
Porter, Charles Warren, zoology, physiology
Posey, Leroy Raadell, Jr, physics
Pratley, James Nicholas, physiology
Preston, Gerald Cowles, mathematics
Raimondi, Donald Louis, chemistry
Read, Robert G, meteorology, oceanography
Replogle, Lanny Lee, organic chemistry
Richter, Robert E, veterinary medicine
Riegel, Christopher Albert, dynamic meteorology
Rissanen, Jorma Johannes, mathematics
Robinson, Henry William, tropical public health, microbiology
Rose, Robert Leon, geology
Rudge, William Edwin, solid state physics
Russell, Attie Yvonne, pediatrics
Sawatzky, Erich, physics
Schein, Arnold Harold, biochemistry
Schmidt, Clifford LeRoy, botany, ecology
Schuck, Robert, chemistry
Seki, Hajime, surface physics, electrochemistry
Selter, Gerald A, organic chemistry
Sharsmith, Carl William, botany
Shedler, Gerald Stuart, operations research, computer science
Shellhammer, Howard Stephen, vertebrate zoology
Simons, Roger Mayfield, applied mathematics

Smith, Richard Avery, geology, science education
Smith, Thor Lowe, polymer science
Snyder, Thoma Mees, physics
Solar, Samuel Louis, organic chemistry
Soller, Arthur, industrial microbiology
Spitze, LeRoy Alvin, chemistry
Sporer, Alfred Herbert, physical chemistry
Sprain, Wilbur, science education
Sprokel, Gerard J, physical chemistry
Stanley, Raymond Wallace, geography of Latin America, political geography
Stevens, Calvin H, paleoecology, stratigraphy
Stone, Irwin, biochemistry, medical research
Suits, James Carr, physics
Swalen, Jerome Douglas, chemical physics
Swann, Howard Story Gray, mathematics
Thaw, Richard Franklin, botany, natural history
Thompson, Jack Coats, meteorology
Tidwell, William Lee, microbiology
Tilden, James Wilson, entomology
Towse, Donald Frederick, geology
Tu, Yih-o, applied mathematics
Tucker, Allen Brink, nuclear physics
Underwood, Francis Wenrich, anthropology
Valletta, Robert M, physical chemistry
Van Alten, Lloyd, inorganic chemistry
Vessel, Matthew F, science education, botany
Wang, Jen Yu, meteorology
Watanabe, Ronald S, biochemistry
Weaver, Ellen Cleminshaw, genetics
West, Donald Markham, analytical chemistry
Weston, Henry Griggs, Jr, ecology, vertebrate zoology
Wieder, Harold, optical physics
Wolfe, Bertram, nuclear physics, nuclear engineering
Wrede, Robert C, Jr, mathematics
Yaffe, Ruth Powers, radiochemistry
Yannoni, Costantino Sheldon, magnetic resonance
Yoon, Do Yeung, polymer chemistry
Young, Joseph Hardie, biology

SAN JUAN BAUTISTA
Nichols, Courtland Geoffrey, plant breeding
Thompson, David J, genetics, plant breeding

SAN LEANDRO
Blair, John Dennis, orthopedic surgery
Carter, Robert Duncan, entomology
Ford, Franklin C, physics
Gillette, Frank Newton, physics
Godfrey, Charles S, physics
Hagan, William Leonard, plant breeding, phytopathology
Hooks, James A, agronomy, plant breeding
Likuski, Robert Keith, biomedical engineering
Moore, Edgar Tilden, Jr, applied physics
Sloan, David Harold, plasma physics
Stallings, Charles Henry, plasma physics
Towner, George Rutherford, information science

SAN LUIS OBISPO
Anderson, Russell K, animal nutrition, physiology
Atwood, Bruce, astrophysics
Atwood, Linda, bio-organic chemistry
Balthaser, Lawrence Harold, geology
Booth, James Samuel, bacterial physiology
Boroughs, Howard, biochemistry
Bowls, Woodford Eugene, physics
Brown, Harold Franklin, solid state physics
Bucy, LaVerne, nutrition, biology
Buschman, William Owen, mathematics
Call, Tracey Gillette, pharmacology, hematology
Cary, Arthur Simmons, high energy physics
Cichowski, Robert Stanley, physical chemistry, ceramics
Dills, Charles E, physical chemistry, organic chemistry
Eatough, Norman L, physical chemistry
Endres, Leland Sander, organic chemistry, physical chemistry
Epstein, Gary Martin, particle physics
Fierstine, Harry Lee, comparative anatomy, ichthyology
Finch, Harry C, plant pathology

Frey, Thomas G, organic chemistry
Frost, Robert Hartwig, experimental physics
Gambs, Roger Duane, ethology, animal behavior
Gupta, Neelam, physics
Hafemeister, David Walter, solid state & nuclear physics
Haskell, Charles Thomson, mathematics
Hawley, Lewis Burton, Jr, organic chemistry, bio-organic chemistry
Houk, Alva Leroy, organic chemistry
Hsu, John Y, computer science
Hughes, Luther Bertram, Jr, agronomy, biochemistry
Jacobson, Gail M, biochemistry
Jacobson, Ralph Allen, biochemistry
Johnson, Corwin McGillivray, agronomy
Johnson, Eric Van, ornithology, wildlife biology
Karpinski, Robert Whitcomb, geology
Katekaru, James, analytical chemistry, nuclear chemistry
Kellerman, Martin, physical chemistry
Kennelly, Bruce, biochemistry
Krejsa, Richard Joseph, vertebrate zoology, resource management
Lambert, Royce Leone, soils, agronomy
Langworthy, William Clayton, organic chemistry, environmental chemistry
Lewis, George McCormick, geometry
Lewis, Vance De Spain, physics
Lukes, Thomas Mark, food science
Mach, George Robert, mathematics
Maksoudian, Y Leon, mathematics
Noble, Glenn Arthur, parasitology
Ozawa, Kenneth Susumu, fluid physics
Pendse, Pratapsinha C, cytogenetics, botany
Perryman, Elizabeth Kay, endocrinology, cytology
Peters, James Milton, biochemistry
Pimentel, Richard A, vertebrate zoology
Richards, Thomas L, marine biology, zoology
Roach, David Michael, oceanography
Rodin, Robert Joseph, plant morphology, taxonomy
Roest, Aryan Ingomar, vertebrate zoology
Rosen, Arthur Zelig, physics
Schumann, Thomas Gerald, elementary particle physics
Siegel, Irving, solid state physics, chemical physics
Sparling, Shirley, phycology
Stansfield, William D, genetics
Stowe, Keith S, elementary particle physics
Strickmeier, Henry Bernard, Jr, mathematics education
Terry, Raymond Douglas, mathematics
Thurmond, William, developmental physiology
Tice, Russell L, physical chemistry
Venerable, Grant Delbert, chemistry, science education
Walker, Howard David, biochemistry
Warren, Ralph Martin, mathematics
Watson, Harold John, poultry nutrition
West, John Wyatt, poultry nutrition
Westover, James Donald, organic chemistry
Wight, Hewitt Glenn, synthetic organic chemistry
Williamson, David Gadsby, chemical kinetics
Wilson, Walter Davis, atomic physics
Wolf, Robert Stanley, mathematical logic
Wu, Sing-Chou, statistics, econometrics

Schmidt, Otto Ernest Lincoln, dermatology
Shnider, Ruth Wolkow, applied mathematics, physics
Wiens, Jacob Henry, physics

SAN PEDRO
Buhrow, Charles, Jr, medicine, radiology
Chang, Han-Chuan Liu, molecular spectroscopy
Michaels, David D, ophthalmology
Ramsey, O C, Jr, mathematics
Valentekovich, Marija Nikoletic, chemistry
Wagoner, Earl V, Jr, biophysics, engineering physics

SAN RAFAEL
Adams, John Howard, organic chemistry
Blakeslee, Theodore Edwin, entomology
Colucci, Anthony Vito, environmental health, toxicology
Denison, George Haigh, chemistry
Goodrich, Judson Earl, organic chemistry
Greenfield, Stanley Marshall, environmental science, environmental management
Nimer, Edward Lee, physical chemistry
Stanek, Peter, mathematics
Warner, Willis L, medicine
Wilgus, Herbert Sedgwick, poultry science

SAN RAMON
Bishop, Jack Lynn, entomology
Kodama, Jiro Kenneth, pharmacology, toxicology
Sloan, Miner Joe, entomology
Whetstone, Richard Roy, pesticide chemistry
Wolven, Anne M, toxicology

SANTA ANA
Anderson, LeRay J, medicinal chemistry, organic chemistry
Beall, Horace Ansley, electrooptics, electronic engineering
Bernard, Joseph Lionel, analytical chemistry, instrumentation
Brake, Jon Michael, biochemistry, organic chemistry
Brimm, Eugene Oskar, inorganic chemistry
Carson, Merl J, pediatrics
Erikson, Jay Arthur, surface chemistry
Forbes, James Franklin, dental materials
Guttentag, Jack David, corrosion, electrochemistry
Hilliard, Jessamine, anatomy
Kesting, Robert E, polymer chemistry
Matkin, Oris Arthur, horticulture
McDonald, John E, physical chemistry, inorganic chemistry
Meinhard, James Edgar, applied chemistry, experimental solid state physics
Moe, George, physical chemistry, inorganic chemistry
O'Connell, John Joseph, Sr, materials science
Palchak, Robert Joseph Francis, organic chemistry
Rankin, Alexander Donald, veterinary medicine
Ryan, Jack A, geophysics
Westman, Thomas Louis, organic chemistry

SANTA BARBARA
Anderson, Curtis Benjamin, organic chemistry
Anikouchine, William A, oceanography, geology
Archer, Douglas Harley, physics
Aue, Donald Henry, organic chemistry
Baldwin, John Arnold, Jr, physics
Barrett, Paul Henry, physics
Bate, George Lee, physics, applied mathematics
Bickerdike, Ernest Lawrence, analytical chemistry, inorganic chemistry
Bischoff, Fritz Emil, organic chemistry, biochemistry
Bohannan, Paul James, anthropology
Boles, James Richard, sedimentary petrology
Bowers, Michael Thomas, chemical kinetics, chemical physics
Branch, Charles Henry Hardin, psychiatry
Broida, Herbert Philip, molecular spectroscopy

Brooks, Franklin Coolidge, mathematical physics
Bruckner, Andrew M, mathematics
Bruce, Thomas Charles, bio-organic chemistry
Bryson, George Gardner, physiology, psychophysiology
Burton, Clifford A, organic chemistry
Byers, Horace Robert, meteorology, cloud physics
Caldwell, David Orville, experimental high energy physics
Carbon, John Anthony, biochemistry
Case, James Frederick, comparative physiology
Ceder, Jack G, pure mathematics
Chaffee, Rowand R J, physiology, biochemistry
Chandler, Donald Ernest, physics
Cheadle, Vernon Irvin, botany
Childress, James J, comparative physiology, biological oceanography
Cloud, Preston E, Jr, geology, invertebrate paleontology
Connell, Joseph H, population biology
Crevier, William Francis, plasma physics, electrodynamics
Cronshaw, James, botany, cell biology
Crowell, John Chambers, geology
Cushing, John (Eldridge), Jr, biology
Davenport, Demorest, zoology, entomology
Davies, Robert Milton, aquatic ecology, water chemistry
DeWolfe, Barbara Blanchard Oakeson, vertebrate zoology
Dewolfe, Robert Hill, organic chemistry
Domaille, Peter John, molecular spectroscopy
Dout, Richard Leroy, entomology, environmental sciences
Dudley, James Robert, water chemistry
Dudziak, Walter Francis, computer science, nuclear science
Ebeling, Alfred W, zoology
Edelson, Sidney, mathematics, physics
Eisberg, Robert Martin, nuclear physics
Eisner, Alan Mark, experimental high energy physics
Elings, Virgil Bruce, physics, instrumentation
Englesberg, Ellis, genetics
Ensign, Stewart Ellery, genetics
Erasmus, Charles John, anthropology
Erickson, Mary Marilla, ornithology, animal behavior
Ernest, John Arthur, mathematics
Esau, Katherine, botany
Everett, Lorne Gordon, hydrology, limnology
Fan, Ky, mathematics
Fisher, Richard Virgil, geology
Fisher, Steven Kay, neurobiology
Ford, Lester Randolph, Jr, mathematics
Ford, Peter Campbell, inorganic chemistry
Fulco, Jose Roque, physics
Gerig, John Thomas, bio-organic chemistry
Gerstein, Larry J, mathematics
Gibor, Aharon, physiology
Goss, Wilbur Hummon, physics
Haase, Ynez Durnford, geography
Hansma, Paul Kenneth, low temperature physics, surface physics
Hardin, Garrett (James), biology
Harris, David Owen, chemical physics, molecular spectroscopy
Hartle, James Burkett, physics
Hatch, Elvin James, history of anthropology, ethnology
Hester, Frank J, environmental sciences, fisheries
Hill, Robert Dickson, physics
Holland, Dan Howard, theoretical physics
Holmes, Robert W, biological oceanography, ecology
Holmes, William Neil, endocrinology
Hone, Daniel W, theoretical solid state physics
Hooker, Thomas M, Jr, biophysical chemistry
Hopson, Clifford Andrae, geology
Horvath, Dee Travis, anthropology
Hudson, Robert Franklin, electrooptics
Hummer, Robert Franklin, electrooptics
Hurst, Richard William, geology, geochemistry
Jaccarino, Vincent, solid state physics
Johnsen, Eugene Carlyle, algebra
Kaska, William Charles, chemistry
Keller, Edward Anthony, geomorphology
Kelly, Paul J, mathematics
Kennedy, John Harvey, electrochemistry, analytical chemistry
Kilb, Ralph Wolfgang, plasma physics, ionospheric physics

King, Jack Lester, genetics
Kirtman, Bernard, physical chemistry
Kohl, David Martin, developmental biology; biochemistry
Ladner, Jane Ellen Crawford, physical chemistry; molecular biology
Lambropoulos, Melissa Margaret, atomic physics
Laris, Philip Charles, physiology
Levin, Louis, biochemistry
Lewis, Harold Warren, physics
Little, Raymond Daniel, organic chemistry
Longley, Herbert J., theoretical physics
Longmire, Conrad Lee, physics
Lowance, Franklin Elta, physics
Luck, David George Croft, physics
Luyendyk, Bruce Peter, marine geophysics
MacIntyre, Ferren, chemical oceanography
Madsen, William, anthropology
Mahall, Brian Elliott, plant ecology
Manasse, Roger, physics
Marcus, Marvin, algebra
Martin, Marilyn Kay, anthropology
Martin, Richard McKelvy, physical chemistry
Mattinson, James Meikle, geochronology, petrology
McKee, John W., physics, biology
Michael, Ernest Denzil Jr., physiology, ergonomics
Miller, Glenn Harry, chemistry
Millett, Kenneth Cary, mathematics
Milliken, Roger Conant, chemical physics
Minc, Henryk, algebra
Mines, Mattison, anthropology
Moore, John Douglas, mathematics
Morhardt, J Emil, comparative physiology; ecology
Morhardt, Sylvia Staehle, ecology, environmental management
Morrison, Rollin John, high energy physics
Moseley, Maynard Fowle, Jr., plant anatomy
Mullen, Robert Keech, physiological ecology; radiation ecology
Muller, Cornelius Herman, plant ecology
Muller, Walter Henry, plant physiology
Murdoch, William W., population biology; ecology
Murphy, Frederick Vernon, experimental high energy physics
Muses, Charles Arthur, mathematics, cybernetics
Nakada, Henry Isao, biochemistry
Neushul, Michael, Jr., botany
Noble, Elmer Ray, parasitology
Norris, Robert Matheson, geology
Offen, Henry William, physical chemistry
Old, Thomas Eugene, atmospheric physics
Ostrand, Phillip Arthur, mathematics
Outcalt, David L., mathematics
Paike, William England, chemical physics
Peale, Stanton Jerrold, solar physics, astrophysics
Pennington, Ralph Hugh, computer science
Prezelin, Barbara Berntsen, biochemistry
Percival, Frank William, marine phycology; photobiology
Philbrick, Ralph Nowell, botany
Phillips, David T., physics
Power, Dennis Michael, biology; zoogeography
Presnell, Alexander Koehne, biochemistry
Pritchard, Glyn O., physical chemistry
Rawlings, Floyd, Jr., chemistry
Recsei, Andrew A., organic chemistry
Redmond, Peter John, physics
Reynolds, Robert Williams, physiology; psychology
Rickborn, Bruce Frederick, organic chemistry
Riemenschneider, Paul Arthur, medicine, radiology
Robertson, James Byron, mathematics
Rosenfeld, Melvin, mathematics
Ross, Ian Kenneth, mycology; cell biology
Sage, Orrin Grant, Jr., environmental geology; environmental management
Sahyun, Melville, physics
Sapperfield, Dale S., physical chemistry
Sawyer, Raymond Francis, theoretical physics
Scalapino, Douglas J., physics
Scheibe, Murray, physics
Schrank, Glen Edward, physics
Schuerch, Hans, applied mechanics
Service, Elman Rogers, cultural anthropology
Simons, Stephen, mathematics
Smith, Dale Metz, systematic botany

Smithcors, James Frederick, veterinary medicine
Sowie, David H., theoretical physics
Spaulding, Albert Clanton, anthropology; archaeology
Sprecher, David A., mathematics
Stephens, Timothy Lee, molecular physics; atmospheric physics
Stewart, William Sheldon, plant physiology
Stoeckly, Robert E., physics, astronomy
Stull, Vincent Robert, applied physics, systems analysis
Sugar, Robert Louis, theoretical physics
Sweeney, Beatrice Marcy, biological rhythms
Taborsky, George, biochemistry
Talley, Robert Morrell, physics
Thompson, Robert Charles, mathematics
Thompson, Rosemary Ann, marine biology
Tilton, George Robert, geochemistry
Triplett, Edward Lee, zoology
Tuinstra, Kenneth Eugene, ecology; limnology
Uterback, Nyle Gene, experimental physics
Voorhies, Barbara, anthropology; archaeology
Walker, William Charles, solid state physics
Walters, James Lee, cytogenetics
Walters, Marta Sherman, cytogenetics
Walton, Lewis F., mathematics
Warner, Robert Ronald, marine ecology
Webb, Robert Wesley, geology
Weiss, Max Leslie, mathematics
Wenner, Adrian Manley, zoology
White, Willard Worster, III, neotohydrodynamics; solid state physics
Wider, Raymond Louis, mathematics
Wilson, Leslie, pharmacology; biology
Wilson, Robert Norton, mathematical physics, atmospheric physics
Wise, William Stewart, mineralogy, petrology
Wooldridge, Dean Everett, physics
Wright, James Edward, exercise physiology
Wyatt, Philip Joseph, biophysics, microbiology
Yaqub, Adil Mohamed, algebra
Yellin, Steven Joseph, experimental high energy physics
Zelmanowitz, Julius Martin, algebra

SANTA CLARA

Alexanderson, Gerald Lee, mathematics
Barker, William Alfred, physics
Carter, Melvin K., physical chemistry; environmental chemistry
Deck, Joseph Francis, chemistry
Deindoerfer, Fred H., clinical chemistry; biomedical engineering
D'Eliscu, Peter Neal, marine ecology; parasitology
Drobot, Vladimir, mathematics
Duffy, William Thomas, Jr., solid state physics
Fast, Thomas Normand, marine biology
Flaim, Francis Richard, zoology
Fraser, Grant Adam, mathematics
Grenning, Daniel A., physics
Hayn, Carl Hugo, physics
Hobbs, M Floyd, physical chemistry; organic chemistry
Huertas, Jorge, neurosurgery
Mansfield, Tom, bacteriology
Markle, Gerald E., zoology
McCormick, Philip Thomas, theoretical physics
McCoy, Thomas Aylesbury, biochemistry
Mei, Alexis Italo, seismology
Moore, Gordon Earle, physical chemistry
Mooring, John Stuart, botany
Mugler, Dale H., mathematical analysis
Muray, Julius J., physics
Nathan, Lawrence Charles, inorganic chemistry
Noyce, Robert Norton, physics
Packwood, Donald Lee, solid state physics
Parkin, Curtis Willard, space physics, plasma physics
Phillips, Veri LeRoy, algebra
Procunier, Richard Werner, atmospheric physics
Sheehan, William Francis, physical chemistry
Steinberg, Gunther, surface chemistry; physical chemistry
Sweeney, Michael Anthony, physical chemistry
Tomlinson, Geraldine Ann, microbiology; biochemistry

White, David Halbert, physical organic chemistry; exobiology

SANTA CRUZ

Abraham, Ralph Herman, mathematics
Anderson, Charles Alfred, geology
Anderson, Richard Hayden, psychiatry
Andrews, Roger W., physical chemistry
Andrews, Frank Clinton, chemical physics
Beavers, Harry, plant physiology; biochemistry
Bernasconi, Claude Francois, physical organic chemistry
Bodenheimer, Peter Herman, astrophysics
Bunnett, Joseph Frederick, organic physics
Burgoyne, Peter Nicholas, mathematical physics
Burke, William L., theoretical physics
Christensen, Mark Newell, geology
Coe, Robert Stephen, geophysics; geochemistry
Cota-Robles, Eugene H. microbiology; cytology
Crews, Philip O., organic chemistry
Daniel, Charles Waller, biology
Danskin, John Moffatt, Jr., mathematics
Davern, Cedric I., molecular genetics
Davis, William J., neurosciences
Diaz, May Nordquist, anthropology
Dice, J Fred, cell biology; biochemistry
Doyle, William T., bryology
Dratz, Edward Alexander, biochemistry; vision
Edgar, Robert Stuart, molecular genetics
Eveleth, Earl Mansfield, theoretical chemistry; organic chemistry
Faber, Sandra Moore, astronomy
Farrell, Bryan Henry, geography of the South Pacific
Feldman, Jerry F., genetics
Fink, Anthony Lawrence, bio-organic chemistry
Fritz, John Merwin, anthropology, archaeology
Garrison, Robert Edward, geology
Goff, Lynda June, phycology
Greenberg, Marvin Jay, mathematics
Griggs, Gary B., marine geology; environmental geology
Hammond, George Simms, organic chemistry
Herbig, George Howard, astronomy
Heusch, Clemens August, elementary particle physics
Hodges, Helen Leslie, bioinorganic chemistry
Kelley, Allen Frederick, Jr., mathematics; forestry
Klemola, Arnold R., astronomy
Kliger, David Saul, physical chemistry
Kraft, Robert Paul, astronomy
Landesman, Edward Milton, mathematics
Langenheim, Jean Harmon, plant ecology; paleobotany
Laporte, Leo Frederic, geology
McMurry, John Edward, synthetic organic chemistry
Menger, Eva L., chemical physics
Nauenberg, Michael, elementary particle physics
Newberry, Andrew Todd, invertebrate zoology
Noller, Harry Francis, Jr., biochemistry
Norris, Kenneth Stafford, zoology
Osterbrock, Donald Edward, astronomy
Pearse, John Stuart, marine biology
Pearse, Vicki Buchsbaum, invertebrate zoology; phycology
Pensinger, Robert P., veterinary medicine; cardiology
Primack, Joel Robert, theoretical physics
Rohlen, Thomas Payne, anthropology
Rosenblum, Bruce, physics
Ruby, Ronald Henry, biophysics
Sands, Matthew, physics
Schlegel, Stuart Allen, anthropology
Schlech, Thomas W., biochemistry; molecular biology
Scott, Peter Leslie, physics
Shane, Charles Donald, astronomy
Sickel, Sharon Rae, computer science
Silver, Mary Wilcox, oceanography; marine biology
Smith-Evernden, Roberta Katherine, micropaleontology
Stoller, Benjamin Boris, agricultural microbiology; biochemistry
Switkes, Eugene, quantum chemistry
Taiz, Lincoln, plant physiology
Thimann, Kenneth Vivian, plant physiology
Tobisch, Othmar Tardin, structural geology

Tromba, Anthony Joseph, pure mathematics
Vasilevskis, Stanislaus, astronomy
Walker, Merle F., astronomy
Wampler, E Joseph, astronomy
Wang, Howard Hao, neurosciences, biochemical pharmacology
Waters, Aaron Clement, geology
Whitford, Albert Edward, astrophysics
Widom, Harold, mathematics
Williamson, Stanley Morris, chemistry
Wipke, Will Todd, chemistry
Zihlman, Adrienne Louella, physical anthropology

SANTA FE SPRINGS

Basinski, John Edward, organic chemistry
Blair, Charles Melvin, Jr., physical chemistry; organic chemistry
Caserio, Frederick F, Jr., physical organic chemistry
Holzman, Richard Thomas, inorganic chemistry; physical chemistry
Kissel, Charles Louis, industrial organic chemistry
Lipson, Melvin Alan, organic chemistry
Rex, Robert Walter, exploration geology
Rohrback, Gilson Henry, physical chemistry; inorganic chemistry
Schulze, William Eugene, synthetic organic chemistry
Speen, Gerald Bruce, heat transfer; energy conversion

SANTA MONICA

Baer, Walter S., solid state physics, telecommunications
Bates, Philip Knight, research administration
Batten, Edmund Stanley, meteorology; climatology
Bick, Rodger Lee, hematology; oncology
Brault, Robert George, organic chemistry
Brennan, Lawrence Edward, electronics
Chaiken, Jan Michael, mathematics
Connors, Theodore Thomas, petroleum, geography
Conrad, Herbert M, biochemistry; nutrition
D'Attorre, Leonardo, physics
Davies, Merton Edward, planetary sciences, photogrammetry
Deirmendjian, Diran, atmospheric physics
Dews, Edmund, atmospheric physics; systems analysis
Dillon, John Thomas, structural geology
Frick, Richard Henry, physics
Gaal, Robert, gemology; geology
Garber, Theodore Bruce, applied mechanics, astronautics
Garfunkel, Irving Minturn, mathematics
Gates, William Lawrence, meteorology
Geisler, Murray Aaron, mathematics
Gladstone, Robert Jay, operations research
Graham, Gordon Alexander Robert, physics
Gilinsky, Victor, theoretical physics
Gill, Robert F, Jr., physical chemistry; chemical engineering
Gilvarry, John James, solid state science, physics
Haverty, John Patrick, computer science, aerospace sciences
Henning, Harley Barry, electrooptics, chemical engineering
Hult, John Luther, electronics, physics engineering
Humphrey, Charles Harve, physics
Johnson, Selmer Martin, mathematics
Juncosa, Mario Leon, applied mathematics, numerical analysis
Kagiwada, Harriet Hatsune, applied mathematics, systems analysis
Keeler, Emmett Brown, systems analysis
Kisliuk, Paul, physics
Koenig, Lloyd Randall, meteorology
Krieger, Firmin Joseph, physical chemistry
Lamar, Donald Lee, geology; geophysics
Landau, Joseph White, medicine, dermatology
Langmuir, David Bulkeley, physics
Larks, Saul David, neonatology
Lee, Charles Albert, economic geology
Leifer, Herbert Norman, solid state physics
Magoun, Horace Winchell, neuroanatomy
Medved, David Bernard, electrooptics
Merifield, Paul M., geology
Mundie, Lloyd George, optics
Murray, Francis William, meteorology

Neufer, John E., physical chemistry
Ory, Horace Anthony, physical chemistry
Paxson, Edwin Woolman, system analysis
Pitchford, Kenneth Lee, meteorology
Rapp, R Robert, meteorology
Schilling, Gerhard Friedrich, geophysics
Schlessinger, Leonard, theoretical physics
Sealander, Carl Elof, mathematics, computer science
Shapiro, Norman Zalman, mathematics, computer science
Shapley, Lloyd Stowell, mathematics, mathematical economics
Sherman, Michael, applied mathematics
Solfrey, William, physics
Specht, Robert Dickerson, mathematics
Springer, Bernard G., solid state physics
Stevenson, George William, toxicology
Strauch, Ralph Eugene, mathematics, statistics
Swerling, Peter, mathematics
Thornton, John Alexander, solid state physics, plasma physics
Vredevoe, Lawrence A., surgery
Webb, Irving D., organic chemistry
Winninghoff, Francis Joseph, meteorology
Wurtele, Zivia Syrkin, statistics, econometrics
Young, Bruce C., geography

SANTA ROSA
Church, Ronald L, zoology, pollution biology
Stocking, Kenneth Morgan, plant taxonomy, ecology
Tye, Arthur, pharmacology

SANTA YNEZ
Amacher, Peter, information science
Busemann, Herbert, mathematics
Kroc, Robert Louis, physiology

SARATOGA
Burke, James Edward, applied mathematics, mathematical physics
Clark, Crosman Jay, mathematics
Kolsky, Harwood George, physics, computer science
Lianides, Sylvia Panagos, physiology, biochemistry
Lueck, Roger Hawks, research administration, corrosion
Phipps, Peter Beverley Powell, solid state chemistry, corrosion
Schaffert, Roland Michael, physics
Tripp, Russell Maurice, radiology, geophysics

SCOTTS VALLEY
Bailey, Stanley Fuller, entomology

SEAL BEACH
Ragland, Ruth Hines, dental hygiene

SEBASTOPOL
Jentoft, Ralph Eugene, Jr, physical chemistry, analytical chemistry

SEPULVEDA
Baxter, Claude Frederick, physiology, biochemistry
Brown, Barbara Banker, psychophysiology
Cherkin, Arthur, biochemistry
Clark, William Gilbert, physiology, biochemistry
Greenblatt, Milton, psychiatry
Korenman, Stanley G, endocrinology, biochemistry
Lolley, Richard Newton, physiology, biochemistry
Lutwak, Leo, endocrinology, nutrition
Masuoka, David Takashi, pharmacology
Sterman, Maurice B, neuropsychology

SHAFTER
Garber, Richard Hammerle, plant pathology
Hyer, Angus Hillyard, crop breeding, genetics
Jorgenson, Edsel Carpenter, zoology, nematology
Leigh, Thomas Francis, entomology

SHERMAN OAKS
Furst, Ulrich Richard, physics
Kutzscher, Edgar Walter, physics

Marantz, Laurence Boyd, organic chemistry
Slosson, James E., geology
Webber, Donald Salyer, physics

SIERRA MADRE
Gale, Hoyt Rodney, paleontology, anthropology
Metzger, Albert E., nuclear chemistry
Notley, Norman Thomas, polymer chemistry, physical chemistry
Verbiscar, Anthony James, organic chemistry
Wadsworth, Donald van Zelm, geophysics, telecommunications

SIMI VALLEY
Joffre, Stephen Paul, chemistry

SOLANA BEACH
Johnson, Wallace E., computer science
Northcraft, Richard Dunn, plant physiology
Poppendiek, Heinz Frank, physics

SOLVANG
Hays, Edwin Everett, biochemistry
Soli, Giorgio, microbiology

SONOMA
Cherms, Frank Llewellyn, Jr, poultry physiology
Holmes, Robert Edward, physical chemistry
Shultz, Fred Townsend, genetics

SOUTH EL MONTE
Fuller, Martin Emil, II, physical chemistry
Orlowski, Jan Alexander, chemistry, chemical engineering

SOUTH PASADENA
Benedict, Harris Miller, pollution biology
Estabrook, Frank Behle, physics
Henson, Carl P., clinical chemistry
Jacobs, Theodore Alan, chemical physics
Robinson, Ross Utley, analytical biochemistry
Wilson, Charles Oren, clinical chemistry

SOUTH SAN FRANCISCO
Levine, Ralph Manuel, polymer chemistry

SPRECKELS
Schulke, James Darrell, plant genetics, agronomy

STANFORD
Agras, William Stewart, medicine, experimental psychiatry
Allen, Joseph Garrott, medicine
Allen, Marcia Katzman, microbial genetics
Allen, Matthew Arnold, physics
Andersen, Hans Christian, physical chemistry
Anderson, John Thomas, low temperature physics
Anderson, Theodore Wilbur, mathematical statistics, econometrics
Bacon, Harold Maile, mathematics
Bagshaw, Malcolm A., radiotherapy, radiobiology
Baldwin, Robert Lesh, physical biochemistry
Ballam, Joseph, physics
Barbour, Allen Babcock, internal medicine
Basch, Paul Frederick, parasitology, public health
Bates, Clayton Wilson, Jr, solid state physics
Baum, David, pediatric cardiology, biochemistry
Baylor, Denis Aristide, neurophysiology
Beard, Rodney Rau, preventive medicine, environmental health
Beasley, Malcolm Roy, low temperature physics
Befu, Harumi, anthropology
Begle, Edward Griffith, mathematics
Bensch, Klaus George, human pathology
Berg, Paul, biochemistry
Berg, Paul Walter, mathematics
Bergman, Stefan, mathematics
Berman, Sam Morris, theoretical physics
Bershader, Daniel, aerophysics
Bettman, Jerome Wolf, medicine, ophthalmology

Bienenstock, Arthur Irwin, solid state physics, x-ray crystallography
Bjorkman, Olle, physiological ecology, photobiology
Blankenbecler, Richard, theoretical physics
Bloch, Felix, physics
Bonner, William Andrew, chemistry
Boudart, Michel, physical chemistry
Bracewell, Ronald Newbold, radio astronomy
Brauman, John I., chemistry
Briggs, Winslow Russell, plant physiology
Brodsky, Stanley Jerome, high energy physics, theoretical physics
Brown, Byron William, Jr, biostatistics
Brown, Harold David, applied mathematics, computer sciences
Brown, J Martin, radiobiology, oncology
Brown, Jeanette Snyder, photobiology
Brown, Karl Leslie, particle physics
Brutlag, Douglas Lee, molecular biology
Buchanan, Bruce G., computer science
Bunker, John Phillip, medicine
Cabrera, Blas, low temperature physics
Calarco, John Richard, nuclear physics
Cambray, Joseph, physical organic chemistry, bio-organic chemistry
Campbell, Alice Del Campillo, biochemistry
Campbell, Allan McCulloch, microbial genetics
Center, Elizabeth M, genetics
Chevalier-Skolnikoff, Suzanne, physical anthropology, primatology
Chodorow, Marvin, acoustics, electronics
Chu, En Lung, electrodynamics
Chung, Kai Lai, mathematics
Clayton, Raymond Brazenor, biochemistry, endocrinology
Cohen, Ellis N, anesthesiology
Collman, James Paddock, inorganic chemistry
Compton, Robert Ross, geology
Conover, Woodrow Wilson, physical biochemistry
Cottle, Richard W., mathematics, operations research
Cowan, W Maxwell, neuroanatomy
Cox, Allan Verne, geophysics
Cox, Alvin Joseph, Jr, pathology
Crawford, Frederick William, plasma physics
Dantzig, George Bernard, mathematical statistics
Davidson, Julian M, neurophysiology
Davis, Ronald Wayne, molecular biology
DeLeeuw, Karel, mathematics
Dement, William Charles, neurophysiology
De Pangher, John, physics
Diaconis, Persi, mathematical statistics
Dickinson, William Richard, geology
Dickson, Frank Wilson, geology
Djerassi, Carl, organic chemistry
Dodge, Alice Hribal, anatomy, cell biology
Doering, Charles Henry, biochemistry, endocrinology
Doniach, Sebastian, theoretical physics
Donnelly, Thomas William, theoretical nuclear physics
Downey, Vincent M., aerospace medicine
Drell, Sidney David, theoretical high energy physics
DuPraw, Ernest Joseph, biology, neuropsychiatry
Durham, Lois Jean, organic chemistry
Eastman, Richard Hallenbeck, organic chemistry
Eaton, Monroe Davis, immunology, oncology
Eaves, Burchet Curtis, operations research, numerical analysis
Ehrlich, Paul Ralph, biology
Einaudi, Marco Tullio, economic geology
Eisenson, Jon, speech & hearing sciences
Eltherington, Lorne, anesthesiology, pharmacology
Everitt, C W Francis, space physics, low temperature physics
Evitt, William Robert, geology
Fairbank, William Martin, physics
Faith, Ray Edwin, statistics
Farber, Eugene M., dermatology
Farquhar, John William, medicine, preventive medicine
Fayer, Michael David, chemical physics
Feigen, George Alexander, physiology, immunochemistry
Feldman, Marcus William, mathematical biology
Fessenden, Peter, nuclear physics
Fetter, Alexander Lees, theoretical solid state physics

Finston, Roland A., health physics, radiological physics
Fischer, Gerhard Emil, high energy physics
Flory, Paul John, physical chemistry
Floyd, Robert W., computer science
Ford, Richard Lyle, particle physics
Fork, David Charles, botany
Frake, Charles Oliver, anthropology
Fraser-Smith, Antony Charles, space physics, geophysics
French, Charles Stacy, plant physiology
Friedberg, Errol Clive, biochemistry, pathology
Fuhrman, Frederick Alexander, medical physiology, toxinology
Ganesan, Adayapalam T., molecular biology, genetics
Ganesan, Ann K., molecular genetics
Garwin, Edward Lee, applied physics
Geballe, Theodore Henry, low temperature physics
Gerow, Bert Alfred, anthropology
Getting, Peter Alexander, neurobiology
Gibbs, James Lowell, Jr, anthropology
Gibson, Count Dillon, Jr, medicine
Giese, Arthur Charles, biology
Gilbarg, David, applied mathematics
Gilman, Frederick Joseph, theoretical high energy physics
Glick, David, histochemistry
Goldstein, Dora Benedict, pharmacology
Golub, Gene H, mathematics
Gonda, Thomas Andrew, psychiatry
Goodlin, Robert Clair, medicine, genetics
Gould, Robert Gordon, biochemistry
Govan, Duncan Eban, urology
Grant, Ronald, neurophysiology
Gray, Donald James, anatomy
Gray, Gary M., gastroenterology
Green, Claude Cordell, computer science
Green, Paul Barnett, botany
Greenberg, Joseph Harold, anthropology, linguistics
Grigorian, Alexander, experimental high energy physics
Gross, Ruth T., medicine, pediatrics
Guiragossian, Zaven George, physics
Hahn, George, geology
Hallet, Bernard, geology, glaciology
Hammond, Robert Hugh, physics
Hanawalt, Philip Courtland, molecular biology
Hanna, Stanley Sweet, nuclear physics, solid state physics
Harbaugh, John Warvelle, economic geology
Harker, Kenneth James, physics
Harrison, Donald C., cardiology, cardiovascular physiology
Harrison, Walter Ashley, solid state physics
Hawley, Newton Seymour, Jr, mathematics
Hayflick, Leonard, cell biology
Hefferman, Laurel Grace, molecular physics
Helm, Richard H, high energy particle accelerator physics
Hepler, Peter Klock, cell biology
Herman, Mary M, pathology, neuropathology
Herriot, John George, computer science, numerical analysis
Herrmannsfeldt, William Bernard, electron optics
Herzenberg, Leonard Arthur, immunology, genetics
Hilfiker, Frederick Stanton, operations research
Hodgson, Keith Owen, bioinorganic chemistry, structural chemistry
Hofstadter, Robert, physics
Hogness, David Swenson, molecular genetics
Holm, Richard William, biology
Holman, Halsted Reid, internal medicine, immunology
Hoots, Harold William, geology
Howard, Arthur David, geology
Hutchinson, Eric, colloid chemistry
Iglehart, Donald Lee, operations research, mathematical statistics
Ingle, James Chesney, Jr, micropaleontology, geology
Jacobs, Patricia Anne, operations research
Jahns, Richard Henry, geology
Jardetzky, Oleg, molecular biology, pharmacology
Johns, Milton Vernon, Jr, mathematical statistics
Johnson, William B, high energy physics
Johnson, William Summer, organic chemistry
Kaiser, Armin Dale, biochemistry, genetics

CALIFORNIA

Kallman, Robert Friend, radiobiology
Kalman, Sumner Myron, biological chemistry
Karlin, Samuel, mathematical statistics, statistics
Keen, (Angeline) Myra, invertebrate paleontology
Kendig, Joan Johnston, neurobiology
Kennedy, Donald, physiology; zoology
Kessler, Seymour, genetics
Kirkman, Hadley, anatomy
Knuth, Donald Ervin, computer science, mathematics
Korn, David, pathology; molecular biology
Kornberg, Arthur, biochemistry
Kraemer, Helena Chmura, biostatistics
Krahl, Ardis June Lostroh, physiology
Krahl, Maurice Edward, biochemistry, pharmacology
Krauskopf, Konrad Bates, geochemistry
Kruger, Fredrick Christian, economic geology
Lederberg, Esther Miriam, microbiology
Lederberg, Joshua, genetics
Lederberg, Gerald J., operations research, statistics
Lieberman, Miriam, cancer
Liebes, Sidney, Jr., physics
Leith, David W G S, high energy physics
Lehman, Israel Robert, biochemistry
Lemons, Ross Alan, applied physics
Levine, Harold, applied mathematics
Levine, Seymour, psychophysiology
Levinthal, Elliott Charles, physics
Lieberman, Arthur David, high energy physics
Liu, Su-Chin Chang, biochemistry
Lipa, John A., low temperature physics
Liou, John G., geology
Lindau, Evert Ingolf, solid state physics
Little, William Arthur, physics
Loew, Gilda M Harris, theoretical biology, biophysics
Luckham, David Compton, computer science
Lung, Ben, cell biology, histology
Luth, William Clair, geology, geochemistry
Luzzatti, Luigi, pediatrics
Lynch, Harvey Lee, experimental high energy physics
Lyon, Ronald James Pearson, geology, mineralogy
Maffly, LeRoy Herrick, medicine
Mallett, Russell Lloyd, mechanics, dynamics
Mallory, Kenneth Brandt, applied physics
Mansour, Tag Eldin, pharmacology; microbial ecology
Matin, Abdul, microbial physiology; microbiology
Maurice, David Myer, physiology
McCall, Richard C., radiological physics
McCarthy, John, computer science
McClenahan, James Brice, medicine
McConnell, Harden Marsden, biophysical
McDowell, Frank, surgery
McLennan, Charles Ewart, obstetrics & gynecology
Merigan, Thomas Charles, Jr., infectious diseases, virology
Meyerhof, Walter Ernst, experimental physics, nuclear physics
Milgram, Richard James, mathematics
Miller, John Johnston, III, pediatrics, immunology
Miller, Rupert Griel, statistics
Miller, William Frederick, computer science, academic administration
Mitchner, Morton, physics
Mittal, Yashaswini Deval, statistics
Mortensen, Otto Axel, anatomy
Moses, Lincoln Ellsworth, statistics
Mosher, Carol Walker, bio-organic chemistry
Mosher, Harry Stone, organic chemistry
Mozley, Robert Fred, nuclear physics
Muller, George Heinz, veterinary dermatology
Neal, Richard B., physics
Nelsen, Thomas Sloan, surgery
Nelson, Walter Ralph, health physics
Newmeyer, Dorothy, genetics
Nichols, John Graham, neurophysiology
Nichols, Kathryn Marion, stratigraphy, sedimentary petrology
Noduluman, Lawrence Jay, experimental high energy physics
Noyes, Henry Pierre, theoretical physics
Nur, Amos M, geophysics
Oberhelman, Harry Alvin, Jr., surgery
Oberste-Lehn, Deane, geology
Olkin, Ingram, mathematical statistics
Ornstein, Donald Samuel, pure mathematics

Osserman, Robert, geometry, mathematical analysis
Page, Benjamin Markham, geology
Panofsky, Wolfgang K H., physics
Parlee, Norman Allen Devine, physical chemistry
Payne, Rose Marise, immunology; hematology
Pearson, Gerald Leondus, solid state electronics
Pecci, Roberto (Daniele), physics
Pecora, Robert, physical chemistry
Perkel, Donald Howard, theoretical biology; neurosciences
Perkins, David Dexter, biology
Perl, Martin Lewis, physics
Phillips, Ralph Saul, pure mathematics, applied mathematics
Pittendrigh, Colin Stephenson, biology
Polya, George, mathematics
Powell, Ronald Allan, solid state physics
Prince, David Allan, neurology; neurophysiology
Raffel, Sidney, medical bacteriology; immunology
Ray, Peter Martin, plant physiology
Rees, John Robert, high energy physics
Regnery, David Cook, biology
Remson, Irwin, hydrology, environmental geology
Resnick, Sidney I., mathematics
Rich, Clayton, medicine, endocrinology
Rich, Ernest I., geology
Richter, Burton, physics
Ritson, David Mark, physics
Roantree, Robert Joseph, medical microbiology
Robertson, William Van Bogaert, biochemistry
Robin, Eugene Debs, physiology; medicine, molecular biology
Robinson, William Sidney, internal medicine
Royden, Halsey Lawrence, mathematical analysis
Rosenberg, Leon T., immunology
Rosenberg, Saul A., internal medicine, oncology
Rosenquist, Grace Link, immunology
Roughgarden, Jonathan David, population biology
Rubenstein, Edward, internal medicine
Rubinstein, Lucien Jules, neuropathology
Ryland, David A., medicine
Sakitt, Barbara, vision
Salzberg, David Aaron, biochemistry
Samelson, Hans, mathematics
Samuel, Arthur Lee, computer science
Sandberg, Eugene Carl, obstetrics & gynecology
Schawlow, Arthur Leonard, lasers
Schinke, Robert T., biochemistry
Schrier, Stanley Leonard, medicine
Schroeder, John Speer, cardiology
Schroth, Mary Dolores, mathematics
Schulman, Irving, medicine, pediatrics
Schwartz, Herbert C., medicine, pediatrics
Schwartz, Melvin, physics
Schwettman, Harry Alan, physics
Schweitzers, Roy Frederick, physics, virology
Seifert, Howard Stanley, physics
Shockley, William, physics
Shooter, Eric Manvers, biochemistry
Simoni, Robert Dario, biochemistry
Simmons, Francis Blair, physiology
Sinclair, Charles Kent, experimental high energy physics
Sigreaves, Rosedith, applied statistics
Skinner, George William, cultural anthropology
Skogg, Douglas Arvid, chemistry
Smith, Charles Allen, molecular biology
Smith, Erla Ring, neuroendocrinology; anatomy
Smith, Kendric Charles, biochemistry, photobiology
Smith, Todd Iversen, physics
Smith, Ward Conwell, geology
Sokol, Otto M, vertebrate morphology
Solomon, Herbert, mathematical statistics
Spicer, William Edward, solid state physics
Spindler, Louise Schaubel, cultural anthropology
Staney, Thomas Alexander, urology
Stark, George Robert, biochemistry
Stebler, Alan James, plant physiology
Stevenson, David Austin, materials science
Steward, John P., medical microbiology, immunology
Stilwell, Donald Lonson, anatomy
Stockdale, Frank Edward, developmental biology; oncology
Stocker, Bruce Arnold Dunbar, microbial genetics, medical microbiology
Stunkard, Albert J., psychiatry

Sturrock, Peter Andrew, astrophysics, plasma physics
Sussman, Howard H., biochemistry, medicine
Swinyard, Chester Allan, anatomy
Switzer, Paul, mathematical statistics, geology
Taube, Henry, inorganic chemistry
Teller, Edward, physics
Textor, Robert Bayard, cultural anthropology
Thomas, John Hunter, plant systematics
Thompson, George Albert, geophysics, geology
Tsai, Yung Su, theoretical physics, elementary particle physics
Tsuboi, Kenneth Kaz, biochemistry
Turneaure, John Paul, low temperature physics
Turner, Robert Stuart, neuroanatomy
van Kann, Frank Joachim, low temperature physics
van Tamelen, Eugene Earl, chemistry
Veinot, Arthur Fales, Jr., operations research
Wagoner, Robert Vernon, Jr., theoretical astrophysics
Wahl, Geoffrey Myles, molecular biology
Walecka, John Dirk, theoretical physics
Walker, Arthur Bertram Cuthbert, Jr., space physics, astronomy
Walsh, Roger Nugent, psychiatry, psychobiology
Watt, Robert Douglas, experimental high energy physics, physics engineering
Watt, Ward Belfield, evolutionary biology
Weinstein, Marvin, theoretical high energy physics
Weissbluth, Mitchel, physics
Wessells, Norman Keith, biology
White, Robert Lee, physics
Whitemore, Alice S., biomathematics; biostatistics
Wilcox, John Marsh, solar physics
Wilkinson, David Ian, biochemistry
Will, Clifford Martin, theoretical astrophysics
Wilson, Perry Baker, high energy physics
Wine, Jeffrey Justus, neurosciences
Winick, Herman, high energy physics
Winograd, Terry Allen, computer science
Wojcicki, Stanley G., high energy physics
Wong, Ming Dak, biochemistry
Wood, Peter Douglas, biology
Woodward, Dow Owen, molecular biology
Yanofsky, Charles, molecular biology
Yau, Shing-Tung, mathematics
Yearian, Mason Russell, nuclear physics, physics
Young, Michael Warren, genetics
Zafran, Misha, mathematical analysis
Zborealke, F Frank, radiology, low temperature physics
Zubeck, Robert Bruce, physics, applied physics

STANTON
Stout, Charles Allison, organic chemistry

STOCKTON
Anderson, Robert Leonard, elementary particle physics
Anderson, Steven Clement, systematics, ecology
Barker, Donald Young, pharmacy
Brown, John Kennedy, phytochemistry
Buss, Charles Delevan, food science, food technology
Carson, J David, David, genetics
Chadwick, Harold King, fisheries
Chaubal, Madhukar Gajanan, pharmacognosy, chemistry
Christianson, Lee (Edward), mammalogy, ecology
Cobb, Emerson Gillmore, organic chemistry
Di Franco, Roland B, algebra
Dodge, Richard Patrick, physical chemistry
Frankel, John Martin, dentistry
Fries, David Samuel, medicinal chemistry
Frye, Herschel Gordon, analytical chemistry
Gentry, Frank Cook, geometry
Gross, Paul Hans, organic chemistry, physical chemistry
Helton, Floyd Franklin, mathematics
Hunter, Alice S (Baker), embryology
Hunter, Francis Robert, physiology
King, James C., pharmacy
Lark, Neil LaVern, nuclear physics
Malone, Marvin Herbert, pharmacology

Matuszak, Charles A., physical organic chemistry
McNeal, Dale William, Jr., botany
Minch, Michael Joseph, physical organic chemistry
Nahhas, Fuad Michael, parasitology; medical microbiology
Newberry, Truman Albert, cancer
Pace, Donald M., cell physiology
Perry, Richard Lee, physics
Potts, John Calvin, inorganic chemistry, physical chemistry
Prend, Joseph, horticulture
Quock, Raymond Mark, neuropharmacology
Riedesel, Carl Clement, pharmacology
Rodriguez-Fraga, Andres, nuclear physics, radiochemistry
Roscoe, Charles William, pharmaceutical chemistry
Rowland, Ivan W., pharmacy, microbiology
Runion, Howell Irwin, electrophysiology
Sayre, Francis Warren, biochemistry
Tocchini, John Joseph, dentistry
Topp, William Robert, mathematics
Tucker, John Shepard, comparative physiology, aquatic biology
Tutupalli, Lohit Venkateswara, pharmacognosy, phytochemistry
Volbrecht, Stanley Gordon, mining geology
Wadman, W Hugh, chemistry, biochemistry
West, Clair Alexander, food chemistry
Wedegaertner, Donald K., organic chemistry
Widmer, Carl, biochemistry
Wulfman, Carl E., theoretical physics, theoretical chemistry

STUDIO CITY
Dearing, Le Roy Mathew, chemistry

SUMMERLAND
Kent, Richard, physical oceanography

SUN CITY
Bishop, Frederic Lendall, physics

SUNNYVALE
Altman, David, chemistry
Armstrong, Augustus K., paleobiology, stratigraphy
Breiner, Sheldon, geophysics
Bunyard, George B., nuclear physics
Chaikin, Saul William, chemistry
Chicks, Charles Hampton, chemistry
Churchill, Dewey Ross, Jr., physics
Cope, Oswald James, polymer chemistry
Cureton, Glen Lee, pharmaceutics
Cutforth, Howard Glen, chemistry, aerospace sciences
Deverall, LaMar Ivan, applied mathematics, applied mechanics
Drew, Daniel L., statistics, computer science
Ernster, Arthur F., analytical chemistry
Friedman, Morris David, mathematics
Gilliam, Ronald Roy, operations research
Harris, William Macy, nuclear physics
Hart, Philip James, electromagnetism
Kaspar, Hans Heinrich, physical chemistry
Keil, Lanny Charles, physiology; endocrinology
Kindsvater, Howard Maxwell, chemistry
Kleitman, David, solid state physics
Lamb, Walter Robert, physics
Linsk, Jack, organic chemistry
Lopata, Eugene Stephen, polymer chemistry
MacLaren, Richard Oliver, physical chemistry
Main, William Francis, physics
Marley, James Aloysius, inorganic chemistry, engineering management
Mayer, Alex, geophysics, space sciences
Miatech, Gerald James, sciences
Miller, Maurice Max, nuclear physics
Mott, Jack Edward, engineering physics
Oder, Frederic Carl Emil, aerospace sciences
Page, Larry J., space physics
Passell, Thomas Oliver, nuclear chemistry
Perloff, David Steven, physics, materials science
Perry, William James, mathematics
Raby, Bruce Alan, analytical chemistry
Randle, Robert James, physiological optics
Romig, Paul William, pharmaceutics

Rubenstein, Kenneth E, organic chemistry, biochemistry
Rudy, Thomas Philip, organic chemistry
Smart, Wilson Harvey, electrochemistry
Snow, Edward Hunter, solid state physics
Sterling, Robert Fillmore, organic polymer chemistry
Stosick, Arthur James, chemistry
Stuart, Derald Archie, solid state physics
Taylor, George Frederic, computer science, meteorology
Teresi, Joseph Dominic, biochemistry, health physics
Tetenbaum, Sidney Joseph, electromagnetism
Thielman, Leroy Oswald, lasers
Toy, Madeline Shen, polymer & organic chemistry
Weissbart, Joseph, electrochemistry

SUSANVILLE
Lossow, Walter Judah, physiology

SYLMAR
Camp, Leon (W), physics
Froman, Seymour, microbiology
Taylor, William Edwin, materials science

TARZANA
Buford, William Holmes, Jr., nucleonics
Maurer, Hans Andreas, physics
Plaut, Herman, chemistry
Woodbury, Eric John, physics

TERMINAL ISLAND
Khayat, Ali, agricultural chemistry, food science

TERRA BELLA
Kusserow, Gerhard William, environmental chemistry

THOUSAND OAKS
Birnbaum, George I, structural chemistry
Blum, Fred A, solid state electronics, lasers
Buck, Otto, metal physics
Cannon, Peter, physical chemistry
Cape, John Anthony, solid state physics
Cohen, E Richard, mathematical physics, reactor physics
Collins, Barbara Jane, botany
DeWames, Roger, solid state physics
Eden, Richard Carl, solid state physics
Edmund, Rudolph William, geology
Eisen, Fred Henry, physics
Evensen, James Millard, geology
Goldberg, Ira Barry, physical chemistry
Housley, Robert Melvin, planetology
Jenney, Joe Allen, optical physics
Joseph, Alfred S., physics
Kaelble, David Hardie, physical chemistry, polymer physics
Kraut, Edgar A, physics
Lauer, George, physical chemistry
Lind, Maurice David, physical chemistry, x-ray crystallography
Lipeles, Martin, atomic physics, atmospheric chemistry
Longo, Joseph Thomas, solid state physics
Mann, Nancy Robbins, statistical analysis
Maxwell, Thomas James, anthropology, archaeology
Morin, Francis Joseph, solid state physics
Muir, Arthur H, Jr, solid state science
Nichols, Robert Ted, physics
Nickel, Phillip Arnold, parasitology, acarology
Pardee, William Joseph, theoretical solid state physics
Raleigh, Douglas Overholt, electrochemistry
Richardson, John Mead, theoretical physics, information sciences
Shaw, Charles Bergman, Jr, physics, applied mathematics
Tennant, William Emerson, solid state physics
Thompson, Donald Oscar, solid state physics
Thorsen, Arthur C, experimental solid state physics
Tittmann, Bernhard R, solid state physics, acoustics
Walz, Alvin Eugene, physical chemistry, analytical chemistry
Wiley, Michael David, organic chemistry

THREE RIVERS
Parsons, David Jerome, plant ecology, environmental management

TIBURON
Abramson, Norman Jay, fisheries, biometrics
Hobson, Edmund Schofield, Jr, ichthyology
Struhsaker, Jeannette Adair Whipple, fish biology, environmental physiology
Sund, Paul N, oceanography

TORRANCE
Barajas, Luciano, pathology
Benfield, John R, surgery
Bhiwandker, Nutan C, inorganic chemistry, ceramics
Blair, George Richard, inorganic chemistry
Bragonier, John Robert, obstetrics & gynecology
Bray, George A, medicine
Burgeson, Robert Eugene, molecular biology, medical genetics
Byfield, Patricia E, immunology
Clark, Brian Roger, biochemistry
Coling, Forrest L, computer sciences
Colodny, Paul Charles, physical chemistry, analytical chemistry
Ehrlich, Morris Joseph, physical optics
Emmanouilides, George Christos, pediatrics, cardiology
Escoffery, Charles Alexander, physical chemistry
Fisher, Delbert A, medicine, pediatrics
Grogan, Michael John, physical chemistry
Heflinger, Lee Opert, applied mathematics, experimental physics
Heiner, Douglas C, pediatrics, allergy
Holden, Geoggrey, physical chemistry, rheology
Hollister, David William, medical genetics, biochemical genetics
Huisjen, Martin Albert, magnetic resonance, microwave engineering
Imagawa, David Tadashi, virology
Itabashi, Hideo Henry, neurology, neuropathology
Kallman, Burton Jay, biochemistry
Karp, Laurence Edward, obstetric & gynecology, medical genetics
Keller, Margaret Anne, infectious diseases
Koyal, Sankar Nath, physiology, nutrition
Long, Robert William, physical chemistry
Libram, Myer Michael, clinical pathology
Marshall, John Romney, obstetrics & gynecology
Mayfield, Earle Byron, space physics
Miller, Julius Sumner, physics
Mohandas, Thuluvancheri, human genetics, cytogenetics
Moore, Thomas Carleton, medicine
Myhre, Byron Arnold, pathology, immunohematology
Nagel, Eugene L, anesthesiology
Natowsky, Sheldon, industrial organic
Oddie, Thomas Harold, radiological physics
Odell, William Douglas, endocrinology, physiology
Parlow, Albert Francis, biology, physiology
Randall, Charles McWilliams, atmospheric physics
Rimoin, David (Lawrence), medical genetics
Rubin, Robert Terry, psychiatry
Samloff, I Michael, medicine, gastroenterology
Shapiro, Irving, audiology
Shapiro, Larry Jay, human genetics
Shoemaker, William C, surgery
State, David, surgery
Stewart, Ray Edward, pedodontics, medical genetics
Strautz, Robert Lee, histology, physiology
Street, Dana Morris, orthopedic surgery
Tanaka, Kouichi Robert, medicine, hematology
Wasserman, Karlman, physiology, medicine
Weichsel, Morton E, Jr, pediatrics, medicine
Weil, Marvin Lee, pediatric neurology, neurology
Winch, Bradley Louis, organic chemistry, virology
Zamboni, Luciano, pathology, electron microscopy

TREASURE ISLAND
Byall, Elliott Bruce, forensic science

TULELAKE
Puri, Yesh Paul, plant breeding, genetics

TURLOCK
Brown, Judith Adele, animal behavior, biological rhythms
Canby, Joel Shackelford, cultural anthropology, ethnology
Chow, Tai-Low, astrophysics
Durbin, Thomas Edmond, anthropology, archaeology
Feldman, Louis A, topology, geometry
Gotelli, David M, botany, mycology
Grillos, Steve John, botany, plant anatomy
Hackwell, Glenn Alfred, entomology
Hamilton, Hobart Gordon, Jr, inorganic chemistry
Hanson, James Charles, cell biology, developmental biology
Hinkson, Jimmy Wilford, biochemistry
James, Ralph L, mathematics, numerical analysis
Javaher, James N, mathematics
Maria, Narendra Lal, applied mathematics
Mayol, Perpetus S, plant pathology, microbiology
Napton, Lewis Kyle, archaeology, physical anthropology
Olson, Walter, geography
Pandell, Alexander Jerry, organic chemistry
Pierce, Wayne Stanley, plant physiology
Roe, Pamela, zoology
Schwarcz, Ervin H, physics
Thompson, Evan M, physical organic chemistry
Tordoff, Walter, III, population biology
Tuman, Vladimir Shlimon, physics, geophysics
Williams, Daniel Frank, mammalogy

TUSTIN
Edwards, Raymond Richard, environmental chemistry
Hamersma, J Warren, environmental chemistry
Newsom, Herbert Charles, organic chemistry, pesticide chemistry

TWAIN HARTE
Westlake, William Ellis, pesticide chemistry

UPLAND
Innes, William Beveridge, physical chemistry
Swisher, Horton Edward, food chemistry

VACAVILLE
Barham, Warren Sandusky, plant breeding, horticulture
Grill, Herman, Jr, dairy science, biochemistry
Sheneman, Jack Marshall, microbiology
Weber, Francis John, medicine

VALENCIA
Miller, Robert Vance, applied physics

VALLEY CENTER
Olson, Ronald Leroy, anthropology

VAN NUYS
Beardslee, Ronald Allen, clinical chemistry
Berkman, Sam, bacteriology
Blaker, Robert Gordon, laboratory medicine, immunology
Blattler, Delbert Paul, clinical biochemistry
Campbell, James L, invertebrate zoology
Campbell, Lorne Arthur, pharmacology
Capers, Evelyn Lorraine, clinical microbiology
Claus, Wilbur Scheirich, chemistry
Demetriou, James A, clinical chemistry, biochemistry
Dickie, John Peter, biochemistry, organic chemistry
Drewes, Patricia Ann, clinical chemistry
Eilers, Russell Jay, clinical pathology
Henry, Richard Joseph, biochemistry
Horwitt, Benjamin Norman, biochemistry
Howard, Raydeen Roland, physics
La Ganga, Thomas S, endocrinology, clinical chemistry
Lee, Norman David, biochemistry, chemistry
Mason, William Burkett, clinical chemistry
McGrath, William Patrick, clinical chemistry
McIntire, Junius Merlin, biochemistry
Morgan, Harry Clark, physics
Nordmann, Joseph Behrens, chemistry
Ottoman, Richard Edward, radiology
Petrowski, Gary E, food chemistry, surface chemistry
Pileggi, Vincent Joseph, clinical chemistry
Roberts, Martin, biochemistry
Saute, Robert E, pharmacy, pharmacology
Scott, Paul Brunson, fluid dynamics, physics
Smith, Henry John, immunopathology
Sparks, Marshall Scott, physics
Ten Eyck, David Roderick, medicine, public health
Weissman, Norman, clinical chemistry
Whatley, Thomas Alvah, physical chemistry, instrumentation
Williams, Marion Porter, food science
Windsor, Emanuel, biochemistry

VANDENBERG AFB
Battle, Ed Len, statistical mechanics, plasma physics
Ewing, Clair Eugene, geodesy
Lichter, James Joseph, molecular physics, computer science
Lippitt, Louis, geophysics

VENICE
Fay, Rimmon C, biochemistry, bacteriology

VENTURA
Bailey, Thomas Laval, geology
Dember, Alexis Berthold, physics
McConnell, John Earl Willard, food technology
Miller, Larry O'Dell, developmental biology, neurophysiology
O'Neill, Thomas Brendan, microbiology, botany
Plum, William Byrle, experimental physics

VERNON
Jules, Leonard Herbert, organic chemistry

VILLA PARK
Byrd, Norman Robert, organic polymer chemistry

VISALIA
Murray, William Donald, medical entomology
Volk, Thomas Lewis, human pathology, endocrinology

WALNUT
Jones, Kenneth La Mar, biological sciences

WALNUT CREEK
Abbott, Seth R, chemical instrumentation, spectroscopy
Aldrich, Robert Clement, forest mensuration
Allison, William Earl, entomology
Arabian, Karekin Gaspar, petroleum, physical chemistry
Arrington, Jack Phillip, organic chemistry
Bailes, Richard Hazel, physical chemistry, organic chemistry
Baumann, Frederick, analytical chemistry
Beck, Henry Nelson, polymer chemistry
Conyers, Emery Swinford, soil science
Cram, Stuart Proud, analytical chemistry
Dance, Eldred Leroy, chemistry
Dowell, Frank Herbert, parasitology, medical entomology
Driver, Harold Edson, cultural anthropology, ethnology
Feay, Darrell Charles, polymer chemistry
Fischback, Bryant C, organic chemistry
Giacobbe, Thomas Joseph, organic chemistry, bio-organic chemistry
Graham, Dee McDonald, food science

CALIFORNIA

Hamaker, John Warren, agricultural chemistry
Hanson, Ronald Gordon, plant breeding
Harris, Guy H., organic chemistry
Hilton, John L., plasma physics
Holmsen, Theodore Waage, applied physiology
Jaye, Murray Joseph, food science, microbiology
Kurihara, Norman Hiromu, chemistry
Lawson, Chester Alvin, biology, psychology
MacWilliams, Dalton Carson, polymer chemistry
Malhotra, Sudarshan Kumar, organic chemistry, pesticide chemistry
Meikle, Richard William, organic chemistry, biochemistry
Molau, Gunther Erich, polymer chemistry
Niven, Charles Franklin, Jr., microbiology, food science
Omid, Ahmad, agronomy, physiology
Reifschneider, Walter, organic chemistry
Rosenblatt, Leon Saul, genetics
Ruetman, Sven Helmuth, organic chemistry
Son, Chung Hyun, food science, technology
Sternberg, Heinz Walter, chemistry
Stevenson, Robert Lovell, analytical chemistry
Taylor, Richard Moreland, public health
Thompson, Clifford Francis, organic polymer chemistry
Tong, Yulan Chang, organic chemistry
Vocks, John Forrest, physical chemistry
Wheaton, Robert Miller, industrial chemistry

WILMINGTON
Battles, Willis Ralph, petroleum chemistry

WHITTIER
Armstrong, Don Leigh, inorganic chemistry
Farmer, Malcolm French, anthropology, communications science
Goldberg, Stephen Robert, ecology
Gutschow, Nathan Robert, physical chemistry
James, Lois Elsie, plant taxonomy
Leighton, Freeman Beach, geology
Newsom, Will Roy, analytical chemistry
Pyle, Henry Randolph, mathematics
Sarachman, Theodore N., molecular physics
Tennyson, Marilyn Elizabeth, sedimentary petrology, tectonics
Wadsworth, William Bingham, mineralogy

WESTLAKE VILLAGE
Copenhaver, John William, organic chemistry, research administration
Hidy, George Martel, physical chemistry, chemical engineering
Lodato, Michael W., operations research

WEST LOS ANGELES
Menkes, John H., pediatrics, neurology
Oswalt, Wendell Hillman, cultural anthropology, archaeology
Ulrich, Renee Sandra, neuroendocrinology

WEST SACRAMENTO
Williams, Harold Edward, plant pathology

WASCO
Colasito, Dominic James, industrial microbiology

WATSONVILLE
Nelson, Richard Douglas, plant pathology

WEED
Beatty, Kenneth Wilson, zoology, limnology

WEST COVINA
Moses, Alfred James, forensic science
Sweny, Keith Holcomb, physical chemistry

WOODLAND HILLS
Ashby, Val Jean, applied physics, electrooptics
Bangle, Raymond, Jr., pathology
Christian, Walter, microbiology, immunology
Darnell, Alfred Jerome, physical inorganic chemistry
Dwyer, Robert Joseph, physics
Friedland, Melvyn, biochemistry
Green, Joseph Matthew, theoretical physics
Harris, Sigmund Paul, physics
Knechtli, Ronald (C), physics
Kuljian, Ernest Sam, analytical chemistry
Montgomery, Christine Anne, anthropology, computer science
Myer, Jon Harold, experimental physics
Notrica, Solomon, biochemistry
Picus, Gerald Sherman, solid state science
Tuffly, Bartholomew Louis, physical chemistry, analytical chemistry
Tullar, Richard Montgomery, vertebrate paleontology, population genetics
Weissman, Earl Bernard, clinical chemistry
Wolten, Gerard Martin, physical chemistry, forensic science
Yosim, Samuel Jack, physical chemistry
Yuwiler, Arthur, biochemistry
Zimmerman, Elmer Leroy, physics

WOODSIDE
Huggins, Maurice Loyal, physical chemistry, polymer chemistry
Smith, Lloyd P., physics, engineering

YORBA LINDA
Foscante, Raymond Eugene, organic chemistry
Fried, David L., optical physics

YOSEMITE NAT PARK
Van Wagtendonk, Jan Willem, forest ecology

YOUNTVILLE
Harris, William Sidney, electrochemistry

WOODLAND
Crill, Pat, plant breeding, plant pathology
Thomas, Paul Clarence, plant breeding, vegetable crops
Watterson, Jon Craig, plant pathology
Wyatt, Colen Charles, horticulture
Yu, Albert Tzeng-Tyng, plant breeding

BERTHOUD
Robertson, Larry Dee, plant breeding

AURORA
Blair, Eugene Baxter, medical bacteriology
Ziporin, Zigmund Zangwill, biochemistry

COLORADO

AKRON
Hinze, Greg Otto, plant breeding
Shawcroft, Roy Wayne, soil physics, micrometeorology
Smika, Darryl Eugene, soil chemistry

ALAMOSA
Burroughs, Richard Lee, petroleum geology
Craft, James Harvey, botany
Dick, Herbert William, anthropology, archaeology
Keen, Veryl F., biology
Lowenstein, Michael Zimmer, physical chemistry
Moore, Frank Archer, biochemistry
Morris, Maurice F., applied physics
Mueller, Theodore Arnold, biophysics
Peterson, Richard Charles, geology
Watkins, Kay Orville, inorganic chemistry, physical chemistry

ARVADA
Ferris, Clinton S, Jr., exploration geology
Pilkington, Harold Dean, geology, petrology

ASPEN
Balke, Bruno, physiology
Hall, James Louis, inorganic chemistry

BOULDER
Aamodt, Richard E., physics
Albaum, Melvin, geography
Albersheim, Peter, plant biochemistry
Albert, Harrison Bernard, chemical instrumentation, computer sciences
Allan, David Wayne, mathematics
Allee, Paul Andrew, atmospheric physics
Alpern, Herbert P., psychopharmacology, neurosciences
Altschuler, Helmut Martin, electrophysics
Altschuler, Martin David, solar physics, radiology
Andrews, John Thomas, geomorphology
Appel, Glenn David, pharmacology
Armitage, John Denton, Jr., communications science, optical physics
Arp, Vincent D., solid state physics
Ashby, Neil, theoretical physics
Atlas, David, meteorology, geophysics
Baggett, Larry W., mathematics
Bailey, Dana Kavanagh, systematic botany
Barnes, James Allen, physics
Barrett, Earl Wallace, meteorology
Barrett, Harold Whilbert, biochemistry
Barry, Roger Graham, climatology
Barth, Charles Adolph, physics
Bartlett, Albert Allen, physics
Bartlett, David Farnham, elementary particle physics, electromagnetism
Barut, Asim Orhan, theoretical physics
Batay-Csorba, Peter Andrew, nuclear physics
Bate, Geoffrey, solid state physics
Baynton, Harold Wilbert, meteorology
Beaty, Earl Claude, physics
Bebernes, Jerrold William, applied mathematics
Beckmann, Petr, radiophysics
Beehler, Roger Earl, physics
Beers, Linn Yardley, molecular spectroscopy
Bekoff, Marc, animal behavior
Bender, Peter Leopold, physics
Benton, Edward Rowell, geophysics
Beran, Donald Wilmer, meteorology
Berg, Howard Curtis, biophysics
Berry, George Willard, geology
Billings, Donald Earl, astrophysics
Bills, Daniel Granville, physics
Birkeland, Peter Wessel, geology, soil science
Blackmon, Maurice Lee, elementary particle physics
Blizard-Cox, Jane Berggren, physics
Blumen, William, fluid dynamics, meteorology
Bock, Carl E., zoology, ecology
Bock, Jane Haskett, botany
Bonde, Erik Kauffmann, plant physiology
Bonneville, Mary Agnes, cell biology
Braddock, William A., geology
Bradley, William Crane, geology
Branson, Farrel Allen, plant ecology
Breternitz, David Alan, anthropology
Bretherton, Francis P., meteorology
Briggs, William Egbert, number theory, academic administration
Britin, Wesley E., theoretical physics
Brock, Fred Vincent, meteorology
Brown, Gordon Elliott, mathematics
Brues, Alice Mossie, physical anthropology
Bruner, Elmo Cody, Jr., space physics
Burt, William Henry, mammalogy
Bushnell, John Horace, invertebrate zoology
Bushnell, Robert Hempstead, atmospheric physics
Bussey, Howard Emerson, electromagnetism
Cadle, Richard Dunbar, physical chemistry
Caine, T Nelson, geomorphology, hydrology
Calfee, Robert Francis, atmospheric physics
Calvert, Wynne, atmospheric physics
Cannon, Theodore Wiles, cloud physics
Castleman, Albert Welford, Jr., chemical physics
Castor, John I., astrophysics
Cavalieri, Donald Joseph, dynamic meteorology
Chappell, Charles Franklin, meteorology, atmospheric physics
Chronic, John, invertebrate paleontology, stratigraphy
Clements, George Francis, mathematics
Coleman, William Earl, organic chemistry
Collins, Allan Clifford, biochemical pharmacology, behavioral genetics
Cooper, John (Jinx), atomic physics, plasma physics
Cox, John Paul, theoretical astrophysics
Crawford, James Worthington, organic chemistry
Cristol, Stanley Jerome, chemistry
Crow, Edwin Louis, statistics, mathematics
Crumpacker, David Wilson, genetics, agronomy
Cruz, Alexander, ecology, vertebrate zoology
Culnan, Robert Neville, meteorology
Cundiff, Milford Fields, vertebrate physiology
Curl, Herbert (Charles) Jr., biological oceanography
Curtis, Bruce Franklin, geology
Curtis, George William, astrophysics, solar physics
Davies, Kenneth, ionospheric physics
Davis, Milford Hall, physics
DeFries, John Clarence, behavioral genetics
de Heer, Joseph, physical chemistry
Delany, Anthony Charles, atmospheric chemistry, cosmochemistry
De Puy, Charles Herbert, organic chemistry
Derr, Vernon Ellsworth, atmospheric physics
Downing, Mancourt, biochemistry
Downs, Bertram Wilson, Jr., elementary particle physics
Drommond, Fred George, pharmacy, pharmaceutical chemistry
Drullinger, Robert Eugene, chemical physics
Dubin, Mark William, neurobiology
Dulk, George A., radio astronomy, astrophysics
Dunn, Gordon Harold, atomic physics, molecular physics
Dye, James Eugene, cloud physics, atmospheric sciences
Easton, Robert Walter, mathematics
Eddy, John Allen, astrophysics, solar physics
Eicher, Don Lauren, geology
Ellis, Homer Godsey, mathematics
Engdahl, Eric Robert, geophysics
Epp, John George, analytical chemistry
Erwin, Virgil Gene, pharmacology, biochemistry
Evenson, Kenneth Melvin, atomic physics, molecular physics
Faller, James E., physics, astrophysics
Fankhauser, James Christian, meteorology
Ferguson, Eldon Earl, aeronomy
Feynman, Joan, solar physics, space science
Fickett, Frederick Roland, physics
Firor, John William, solar physics, radio astronomy
Fischer, Irwin, mathematics
Fotino, Mircea, biophysics, cytology
Franklin, Allan David, physics
Fritz, Richard Blair, atmospheric physics
Froelich, Harry Curt, ionospheric physics
Frye, John Chapman, geology
Fulks, Watson, mathematics
Gallagher, Alan C., atomic physics
Garstang, Roy Henry, astrophysics, atomic physics
Gary, John Mitchell, computer science
Gebbie, Katharine Blodgett, astrophysics
Geltman, Sydney, theoretical physics
Gill, Stanley Jensen, physical chemistry
Gille, John Charles, meteorology, physics
Gillette, Dale Alan, agricultural meteorology
Gilman, Peter A., fluid dynamics, hydrodynamics
Glock, Waldo Sumner, ecology, geology
Goldan, Paul David, atmospheric physics
Goldstein, Lester, atmospheric chemistry
Goodrich, Robert Kent, mathematics
Gossard, Earl Everett, physical oceanography, radio meteorology

Grams, Gerald William, meteorology, lasers
Grant, Michael Clarence, population biology, botany
Greene, David Lee, physical anthropology, dental anthropology
Greenway, John, anthropology, criminology
Gregg, Robert Edmond, systematic entomology
Grim, Paul J, marine geophysics
Gruzensky, Paul M, solid state chemistry
Gustafson, Karl Edwin, mathematics
Hackenberg, Robert Allan, medical anthropology, applied anthropology
Hall, Freeman Franklin, (Jr), atmospheric sciences, remote sensing
Hall, John L, physics
Ham, Richard George, cell biology
Hammerness, Francis Carl, pharmacy
Hanley, Howard James Mason, physical chemistry
Hanna, Melvin Wesley, physical chemistry
Hansen, Carl John, astrophysics, physics
Hansen, Richard (Thomas), astronomy
Harris, Richard Elgin, solid state physics, low temperature physics
Harrison, John Christopher, geophysics
Hassner, Alfred, organic chemistry
Haugen, Duane Arthur, micrometeorology
Heberlein, Douglas Garavel, nuclear chemistry, industrial chemistry
Heim, Harold Clifford, biochemistry, pharmacology
Helburn, Nicholas, geography
Hennig, Arnold John, pharmaceutical chemistry
Hermes, Henry Gustav, mathematics
Hernandez, Gonzalo J, atmospheric physics
Herring, Jackson Rea, theoretical physics
Hess, Wilmot Norton, atmospheric physics, oceanography
Hessel, Merrill, molecular physics, lasers
Hessler, Victor Peter, geophysics, electrical engineering
Hester, James J, anthropology, archaeology
Hewes, Gordon Winant, anthropology
Hill, Alton David, Jr, geography
Hine, Gerald John, health sciences, nuclear medicine
Hodges, John Herbert, mathematics
Hogan, Patrick Brian, chemical physics
Holley, Richard Andrew, mathematics
Hooke, William Hines, atmospheric physics
Hord, Charles W, physics, atmospheric physics
House, Lewis Lundberg, astrophysics, atomic physics
Hufford, George (Allen), mathematics, electromagnetism
Hultquist, Martin Everett, organic chemistry
Hummer, David Graybill, theoretical astrophysics
Hundhausen, Arthur James, theoretical physics, astrophysics
Hurt, James Edward, physics
Iddings, Carl Kenneth, theoretical physics
Ives, John David, physical geography
Jarvis, Stephen, Jr, applied mathematics
Jennings, Donald Alfred, solid state physics
Johnson, Eric G, Jr, lasers
Jones, Burton Wadsworth, mathematics
Jones, Hal Joseph, geophysics
Jones, Richard Evan, comparative endocrinology, reproductive biology
Jones, William B, numerical analysis, mathematical analysis
Julian, Paul R, meteorology
Kaimal, Jagadish Chandran, meteorology
Kamper, Robert Andrew, physics
Kasahara, Akira, meteorology
Kaschube, Dorothea Vedral, anthropology, anthropological linguistics
Kasemir, Heinz, physics
Keammerer, Warren Roy, plant ecology
Keller, Raymond Nevoy, inorganic chemistry
Kellogg, William Welch, meteorology
Kerns, David Marlow, electromagnetics
King, Edward Louis, inorganic chemistry
Kisslinger, Carl, geophysics
Kleis, William Delong, atmospheric physics
Knight, Charles Alfred, cloud physics
Knollenberg, Robert George, instrumentation
Koch, Tad H, organic chemistry
Koerner, Harold Elton, vertebrate paleontology

Kopp, Roger Alan, solar physics
Kraemer, Louise Margaret, physical chemistry, biochemistry
Kraemer, Richard John, pharmacology, toxicology
Kraushaar, Jack Jourdan, nuclear physics
Kroemer, Herbert, physics
Kropschot, Richard Henry, cryogenics, solid state physics
Kuhn, Peter Mouat, meteorology
Kyle, Thomas Gail, cloud physics, molecular physics
Lacher, John Robert, physical chemistry
Lally, Vincent Edward, meteorology, electronics
Lander, James French, geophysics, seismology
Lang, Gottfried Otto, anthropology
Lanham, Urless Norton, / entomology
Larson, Edwin E, geophysics, geology
Larson, Tyrone Ray, microwave physics
Lawrence, George Melvin, physics
Leinbach, Harold, physics
Leith, Cecil Eldon, Jr, mathematical physics
LeMone, Margaret Anne, meteorology
Lenschow, Donald Henry, meteorology
Lillie, Charles Frederick, astrophysics
Lilly, Douglas Keith, meteorology
Lincoln, Jeannette Virginia, geophysics, solar physics
Lind, David Arthur, nuclear physics
Lindquist, Robert Marion, organic chemistry, photographic chemistry
Lineberger, William Carl, chemical physics
Linhart, Yan Bohumil, evolution, genetics
Linsky, Jeffrey L, space physics
Little, Charles Gordon, atmospheric physics, remote sensing
Lodge, James Piatt, Jr, atmospheric chemistry
London, Julius, meteorology
Long, Alexis Boris, cloud physics
Longley, William Warren, geology
Lore, Askell, cytotaxonomy, cytogenetics
Love, Doris, cytotaxonomy
Love, William F, solid state physics
Lund, Donald S, physics
Lundell, Albert Thomas, mathematics
MacPhail, Donald Dougald, geography
MacQueen, Robert Moffat, solar physics
MacRae, Robert E, mathematics
Mahanthappa, Kalyana T, elementary particle physics
Mahler, Robert John, solid state physics
Malville, John McKim, astrophysics
Mandics, Peter Alexander, atmospheric physics
Mankin, William Gray, astrophysics
Marr, John Winton, botany
Martell, Edward A, radiochemistry, nuclear geochemistry
Maslin, Thomas Paul, herpetology
Matsushita, Sadami, geophysics
Matthews, Gary Joseph, organic chemistry
Maybee, John Stanley, mathematics
Mayer, William Vernon, zoology
McCray, Richard Alan, astrophysics
McEntee, Thomas Edwin, synthetic organic chemistry
McFarland, Mack, air pollution
McGavin, Raymond E, physics
McIntosh, John Richard, cell biology
McKeehan, Wallace Lee, biochemistry
Meek, John Sawyers, organic chemistry
Menhusen, Bernadette Remus, botany, ecology
Mercure, Ruel Coe, Jr, physics
Metzger, H Peter, biochemistry, science writing
Meyer, Burnett Chandler, mathematics
Meyers, Herbert, geophysics
Mihalas, Dimitri, astrophysics
Miller, Stanley Custer, Jr, theoretical physics
Mills, George Scott, applied physics
Miner, Frend John, inorganic chemistry, soil chemistry
Mitton, Jeffry Bond, population genetics
Mizushima, Masataka, molecular physics
Mohling, Franz, theoretical physics
Monk, James Donald, mathematics
Morrison, Charles Freeman, Jr, analytical chemistry, medical technology
Morrison, Nancy Dunlap, astronomy
Morse, Robert Malcolm, high energy physics
Mosburg, Earl R, Jr, atomic physics
Munoz, James Loomis, geochemistry
Mycielski, Jan, mathematics
Neff, William David, atmospheric physics
Newkirk, Gordon Allen, astrophysics
Newton, Chester Whittier, meteorology
Nichols, Harvey, palynology
Nicholson, Dwight Roy, plasma physics
Norcross, David Warren, atomic physics

Norman, Arlan Dale, inorganic chemistry
Norris, Charles Hamilton, physiology
Norris, David Otto, comparative endocrinology
Norton, Karl Kenneth, mathematics
Noxon, John Franklin, physics
Oster, Ludwig Friedrich, astrophysics
Ostrow, Sidney Maurice, ionospheric physics
O'Sullivan, William John, lasers, fluid physics
Park, Joseph Dal, chemistry
Peacock, Roy Norman, instrumentation
Pennak, Robert William, zoology
Pennell, William Todd, micrometeorology
Peter, George, geophysics, marine geology
Peterson, Robert Lee, theoretical solid state physics
Peterson, Roy Jerome, nuclear physics
Phelan, Robert J, Jr, optical physics
Phelps, Arthur Van Rensselaer, atomic physics, molecular physics
Phillipson, Paul Edgar, chemical physics, biophysics
Pickart, Don Edward, organic chemistry
Pickett-Heaps, Jeremy David, cell biology, phycology
Pierpont, Cortlandt Godwin, inorganic chemistry
Pike, Julian M, atmospheric physics
Plooster, Myron Nieveen, atmospheric physics
Pneuman, Gerald W, astrophysics
Poland, Arthur I, astrophysics, solar physics
Pollock, Bruce McFarland, plant physiology
Pope, Joseph Horace, atmospheric physics, instrumentation
Porter, Keith Roberts, zoology, cell biology
Powell, Robert Lee, solid state physics
Pueschel, Rudolf Franz, atmospheric physics, atmospheric chemistry
Querfeld, Charles William, solar physics
Quick, Horace Floyd, population geography, population ecology
Radebaugh, Ray, cryogenics
Ramsay, Arlan (Bruce), mathematics
Rearick, David F, mathematics
Reinhardt, William Nelson, mathematics
Reinhardt, William Parker, chemical physics
Rense, William A, space physics
Richardson, William Norman, phycology, physiology
Richmond, Arthur Dean, atmospheric physics
Richtmyer, Robert Davis, physics
Roberts, Walter Orr, solar physics
Robinson, Peter, paleontology, geology
Roder, Hans Martin, cryogenics
Rogers, David James, economic botany
Rogers, Robert N, solid state physics
Rosinski, Jan, atmospheric chemistry
Rost, Ernest Stephan, nuclear physics
Roth, Richard Lewis, mathematics
Runnells, Donald DeMar, geochemistry, hydrogeology
Runner, Meredith Noftzger, embryology
Rush, Joseph Harold, optics
Rust, Walter David, atmospheric physics
Ruttenberg, Stanley, atmospheric physics, meteorology
Sangster, Raymond Charles, technological forecasting, inorganic chemistry
Sartor, James Doyne, meteorology
Sasamori, Takashi, meteorology
Sather, Duane Paul, mathematics
Sauer, Herbert H, physics
Sayvetz, Aaron, physics
Schlesinger, Kurt, behavioral genetics
Schmeltekopf, Arthur L, Jr, spectroscopy
Schmidt, Wolfgang M, mathematics
Schneider, Stephen Henry, climatology
Schwiesow, Ronald Lee, atomospheric physics, physical optics
Scott, James Floyd, solid state physics
Segal, William, microbiology
Shankman, Paul Andrew, economic anthropology
Shapiro, Robert Howard, organic chemistry
Shapley, Alan Horace, geophysics
Sharda, Satish Chander, polymer science, polymer engineering
Sharpless, Seth Kinman, pharmacology, physiological psychology
Shulls, Wells Alexander, microbiology
Shushan, Sam, lichenology, mycology
Siegwarth, James David, solid state physics
Sievers, Robert Eugene, analytical chemistry, inorganic chemistry
Silberger, Donald Morison, mathematics
Skumanich, Andrew, astrophysics

Sliker, Todd Richard, solid state physics
Slutz, Ralph Jeffrey, numerical analysis
Smith, Albert W, geography
Smith, Dean Francis, astrophysics
Smith, Ernest Ketcham, radio physics, telecommunications
Smith, Hobart Muir, vertebrate zoology
Smith, James Miller, Jr, organic chemistry
Smith, James Taylor, nuclear physics
Smith, Stephen Judson, physics
Smythe, William Rodman, physics
Snyder, Howard Arthur, physics, geophysics
Snyder, Judith Armstrong, cell biology
Snyder, Nancy Simon, low temperature physics
Snyder, Wilbert Frank, physics
Somerville, Richard Chapin James, meteorology, fluid dynamics
Spangenberg, Dorothy Breslin, zoology, biochemistry
Sparks, Larry Leon, cryogenics, solid state physics
Speiser, Theodore Wesley, astrophysics, geophysics
Spetzler, Hartmut August Werner, rock mechanics
Staehelin, Lucas Andrew, cell biology, electron microscopy
Stein, Gretchen Herpel, cell biology
Stevens, Walter Joseph, theoretical chemistry, chemical physics
Stewart, Omer Call, applied anthropology, ethnohistory
Strickler, Stewart Jeffery, physical chemistry
Struik, Ruth Rebekka, mathematics
Sueoka, Noboru, genetics
Suess, Steven Tyler, fluid dynamics
Sullivan, Donald Barrett, low temperature physics
Sweet, David Paul, analytical chemistry
Tanttila, Walter H, physics
Taylor, John Robert, theoretical physics
Taylor, Walter Fuller, mathematics
Taylor, William L, meteorology, electromagnetism
Thieme, Frederick Patton, anthropology
Thomas, Gary E, atmospheric physics
Thomas, Richard Nelson, astrophysics
Thompson, Milton Avery, environmental management
Thompson, Philip Duncan, meteorology
Thorne, Oakleigh II, biology, ecology
Thron, Wolfgang Joseph, mathematics
Tolbert, Bert Mills, biochemistry, nutrition
Troeger, Gary Leslie, solid state physics
Tryon, Peter Vincent, statistics
Ulam, Stanislaw Marcin, mathematics, mathematical physics
Van Couvering, John Anthony, stratigraphy, geochronology
Van Couvering, Judith Anne Harris, paleontology
Vandenberg, Steven Gerritjan, behavioral genetics, psychology
Vande Noord, Edwin Lee, space physics
Van Vorous, Ted, analytical chemistry
Van Zandt, Thomas Edward, aeronomy, atmospheric physics
Verschuur, Gerrit L, astronomy
Wacker, Paul Frederick, electromagnetics
Wahlstrom, Ernest Eugene, petrology
Wait, David Francis, metrology
Wait, James Richard, geoenvironmental science
Walker, Deward Edgar, Jr, anthropology
Walker, Theodore Roscoe, geology
Walton, Harold Frederic, analytical chemistry
Warner, Lawrence Allen, geology
Warwick, James Walter, radio astronomy
Washington, Warren Morton, meteorology
Webber, Patrick John, ecology, biology
Weber, William Alfred, botany
Weickmann, Helmut K, atmospheric physics, meteorology
Weinstock, Jerome, plasma physics, fluid mechanics
Weiss, Irving, mathematical statistics
Wheat, Joe Ben, anthropology, archaeology
White, Gilbert Fowler, resource geography
Williams, Donald J, space physics, nuclear physics
Williams, Olwen, vertebrate ecology
Wilson, F Wesley Jr, mathematics
Wilson, Irwin B, biochemistry
Wilson, James Russell, behavioral biology
Wilson, Lynn Harold, polymer chemistry
Windell, John Thomas, aquatic biology
Wingeleth, Dale Clifford, inorganic chemistry, clinical chemistry
Winston, Paul Wolf, biology

Wolfe, Douglas Arthur, marine chemistry, pollution biology
Wyngaard, John C., fluid dynamics
Wyss, Max, seismology
Wyss, Walter, mathematics, physics
Yarus, Michael J., molecular biology, biochemistry
Zafiratos, Chris Dan, nuclear physics
Ziebarth, Timothy Dean, organic chemistry, physical organic chemistry
Zimmerer, Robert W., experimental physics, instrumentation
Zirakzadeh, Aboulghassem, mathematics
Zundel, Steven Stanford, inorganic chemistry

CANON CITY
Middlemiss, Ross Raymond, mathematics

BROOMFIELD
Brown, William Francis, veterinary medicine
Henderson, Leland Clifford, dentistry

COLORADO SPRINGS
Beidleman, Richard Gooch, ecology, history of science
Beyer, Jacquelyn L., geography
Berthrong, Morgan, medicine
Bordner, Charles Albert, Jr, physics
Bradley, Richard Crane, physics
Brock, George William, physics
Brundin, Robert H., analytical chemistry
Bryce, Donald Hewit, radiation chemistry
Buell, Carleton Eugene, mathematics, meteorology
Carter, Jack Lee, botany
Champion, William (Clare), organic chemistry
Connolly, Thomas Worthington, nuclear physics
Dickenson, Donald Dwight, crop breeding, agronomy
Dowling, Patrick J., physics
Eley, James H., biochemistry, plant physiology
Anderson, James H., zoology
Fischer, William Alfred, invertebrate paleontology
Gateley, Wilson York, mathematics
Grainger, Robert Ball, nutrition, biochemistry
Gullikson, Charles William, physics
Hamilton, Mary Alice, physiology
Hansman, Margaret Mary, mathematics
Hathaway, Ronald Philip, embryology, human genetics
Heim, Werner George, vertebrate zoology
Hilt, Richard Leighton, physics
Hitchcock, Eldon Titus, analytical chemistry
Hoffman, John Raleigh, engineering
Hostetler, John Davison, physical chemistry
Huebert, Barry Joe, physical chemistry, analytical chemistry
Irwin, Robert Cook, mathematics
Jones, Harold Lester, organic chemistry, numerical analysis
Karon, Harold Lester, physical anthropology
Kelso, Alec John, physical anthropology
Kendall, James Tyldesley, physics
Kunsche, Paul, cultural anthropology
Landry, Preston Myles, operations research
Lane, Max Herbert, applied mathematics
Lewis, John Hubbard, geology
Michel, Lester Allen, chemistry
Novak, Michael, anthropology
Paine, Richard Bradford, computer sciences
Pearl, Richard Maxwell, mineralogy
Reich, Hans, organic chemistry
Rich, Wiley Foster, physics
Roeder, David William, pure mathematics
Sachs, Donald Charles, physics
Schoffstall, Allen M., organic chemistry
Schwartz, Robert Alan, theoretical physics, astrophysics
Shelton, Frank Harvey, nuclear physics
Simmons, George Finlay, mathematics
Stabler, Robert Miller, zoology
Sterling, Daniel J., mathematics
Sterrett, John Kenneth, mathematics
Stoddard, Dan Warren, mathematics
Taber, Richard Lawrence, chemistry
Tracy, Philip T., plasma physics, electromagnetics
Van Horn, Donald H., ecology
Vegele, William John, physics
Veirs, Val Rhodes, atmospheric physics
Vogel, Richard E., computer science

Wackernagel, Hans Beat, astrodynamics, data processing
Ware, Walter Elisha, physics
Windholz, Walter M., physics
Wood, Donald Eugene, nuclear physics
Wood, James Thornton, mathematics
Wright, Wilbur Herbert, physics

CONIFER
Hoffman, Dale A., limnology

DENVER
Abrams, Adolph, biochemistry
Ahlbrandt, Thomas Stuart, geology, sedimentology
Aikawa, Jerry Kazuo, clinical medicine
Aldrete, Jorge Antonio, anesthesiology
Aldrich, Robert Anderson, pediatrics
Algermissen, Sylvester Theodore, geophysics
Aldredge, Leroy Romney, geophysics
Alpha, Andrew Gray, economic geology
Alpiner, Jerome Gerald, speech pathology
Anne, Robert Clyde, experimental atomic physics, atmospheric physics
Anderson, Roger Arthur, botany, biology
Angell, Robert Walker, protozoology, cytology
Austin, James Henry, neurology
Averitt, Paul, economic geology
Bailey, Alfred Marshall, ornithology, mammalogy
Bailey, Robert Vernon, economic geology
Balch, Alfred Hudson, geophysics
Ball, William David, developmental biology
Balser, Donald S., wildlife management
Barber, Thomas Lynwood, medicine, veterinary medicine, microbiology
Bardwell, George, mathematics, statistics
Barker, Fred, geology
Barnes, Harley, geology
Barnes, Ralph Craig, public entomology
Barrett, Charles Sanborn, physics
Barrett, Dennis, developmental biology, biochemical genetics
Battaglia, Frederick Camillo, physiology, pediatrics
Bekoff, Anne Laurens, neuroembryology
Belden, Don Alexander, Jr, zoophysiology
Beren, Sheldon Kuciel, chemistry
Bergendahl, Maximilian Hilmar, economic geology
Best, LaVar, bioacoustics
Betz, George, obstetrics & gynecology, endocrinology
Black, William Cormack, pathology
Blager, Florence Berman, speech pathology
Blair, Emil, thoracic surgery, cardiovascular surgery
Blair, Robert William, petroleum geology
Bleistein, Norman, applied mathematics, mathematical analysis
Block, Matthew Harold, hematology
Bohannon, Robert Gary, geology
Bohor, Bruce Forbes, geology, clay mineralogy
Bondy, Stephen Claude, neurobiology
Boos, Margaret (Fuller), mineralogy
Borek, Ernest, biochemistry
Boswick, John G., surgery
Bowne, John G., virology, cytology
Brandon, William Franklin, developmental biology, cytogenetics
Brettell, Herbert R., internal medicine
Briese, Franklin Wagner, biometry
Brockway, Barbara Fink, vertebrate zoology, physiology
Bromfield, Calvin Stanton, geology
Brown, George L., microbiology
Brown, Jerry, immunology
Brown, Jerry L., biochemistry
Bryant, Bruce Hazelton, regional geology
Bublitz, Clark, biochemistry
Buckman, Robert Campbell, geology
Buecher, Edward Joseph, microbiology, nematology
Burke, Thomas Joseph, physiology
Burrows, Leslie Raymond, anatomy, research administration
Bush, Alfred Lerner, economic geology, environmental geology
Byers, Frank Milton, Jr, geology
Byers, Virginia Pratt, geology
Cady, Wallace Martin, geology
Calvert, James Bowles, optical physics
Campbell, John Arthur, petrology
Campbell, Priscilla Ann, immunology

Campbell, Wallace Hall, atmospheric physics, geomagnetism
Cann, John Rusweiler, biophysical chemistry, molecular biophysics
Canney, Frank Cogswell, geology
Carpenter, Helen Leighton, geochemistry
Carpenter, Steve Haycock, solid state physics
Carr, Ronald Irving, immunology, biochemistry
Carson, Paul Langford, medical physics, ultrasonic research
Chaffee, Maurice Ahlborn, economic geology, geochemistry
Chao, Tsun Tien, geochemistry
Chasson, Robert Lee, experimental physics
Chidsey, Charles Augustus, internal medicine, clinical pharmacology
Church, Brooks Davis, microbiology
Ciriacks, Kenneth W., geology
Claman, Henry Neumann, internal medicine, immunology
Colton, Roger Burnham, environmental geology
Conner, Jack Michael, chemistry, physical chemistry
Conroy, Margaret Frances, applied mathematics
Cooper, William Gregory, medical education
Corby, Donald G., pediatrics, hematology
Corchary, George Sutter, geology
Costa, John Emil, geomorphology, environmental geology
Cygan, Norbert Everett, geology, paleontology
Dahl, Arthur Richard, geology, civil engineering
Dahlberg, Frances Murray, anthropology
Daily, Frederick Thomas, mathematics
Darley, Ward, medical education
Davies, Marvin Lewis, radiology
David, George Berthold, biophysics, neurobiology
Davies, David Hywel, cardiology, internal medicine
Daviess, Steven Norman, geology
Deitrich, Richard Adam, biochemistry, pharmacology
Desborough, George A., geology, mineralogy
Dewey, Fred McAlpin, organic chemistry
Dick, Donald Edward, biomedical engineering, electrical engineering
Dickey, Dayton Deibert, geology
Dickinson, Kendell A., geology, sedimentary petrology
Dixon, Helen Roberta, geology
Dixon, Linda Kay, behavioral genetics
Dobrovolny, Ernest, geology
Doe, Bruce, geology
Dooley, John Raymond, Jr, nuclear physics
Dorn, William S., mathematics
Drewes, Harald D., geology
Driscoll, William Thorvald, zoology
Dumke, Walter Henry, physical chemistry
Dunlop, Stuart George, bacteriology
Duvall, Wilbur Irving, physics
Earley, James William, geochemistry
Eaton, Gareth Richard, inorganic chemistry
Eckel, Edwin Butt, geology
Eickhoff, Theodore C., medicine
Eisele, Charles Wesley, medicine
Eiseman, Ben, surgery
Ellis, Philip Paul, ophthalmology
Elsey, Margaret Grace, mathematics
Emerson, Frederick Beauregard, Jr, medicine
Engelhardt, Donald Wayne, paleobotany
Erdmann, David E., analytical chemistry, inorganic chemistry
Erickson, Ralph Leroy, geology
Erickson, Raymond Leo, molecular biology
Evans, James Bowen, radiochemistry

Everhart, Edgar, celestial mechanics, astronomy
Falliers, Constantine J., allergy, pediatrics
Fairs, Richard Studley, medicine, allergy
Feinsinger, Peter, ecology
Fennell, Robert Henry, Jr, pathology
Fennell, Arthur Thomas, geology
Feucht, James Roger, horticulture, botany
Filley, Giles Franklin, physiology
Finch, Warren Irvin, geology
Firminger, Harlan Irvin, pathology
Frischknecht, Frank C., geophysics
Fishman, Marvin Joseph, water chemistry
Fleming, Robert William, geology, engineering geology
Flores, Romeo M., sedimentology
Foster, Norman Holland, geology
Fox, James Ellison, geology
Franks, John Julian, internal medicine
Frezon, Sherwood Earl, economic geology
Friedman, Irving, geochemistry
Friedman, Jules Daniel, geology
Fukuta, Norihiko, cloud physics, atmospheric sciences
Fullerton, David Stanley, quaternary geology
Garb, Solomon, pharmacology
Garbarini, George S., geology
Gaskill, Herbert Stockton, psychiatry
Genton, Edward, medicine
Gersten, Jerome William, physical medicine & rehabilitation
Githens, John Horace, pediatrics
Goren, Mayer Bear, microbiology, organic chemistry
Gottschall, Carl, radiation chemistry
Gramera, Robert Eugene, biochemistry, organic chemistry
Graul, Walter Dale, behavioral biology, vertebrate ecology
Greenberg, Herbert Julius, applied mathematics
Greenwood, William R., geology
Grey, Howard M., immunochemistry, pathology
Griffiths, Thomas Melvin, physical geography
Griffitts, Wallace Rush, economic geology
Grover, Robert Frederic, medicine
Gudder, Stanley Phillip, mathematics
Gunter, Bobby J., analytical chemistry, experimental medicine
Hagerman, Dwain Douglas, biochemistry, obstetrics & gynecology
Hahn, William Eugene, molecular biology, cell biology
Hamilton, Warren (Bell), geology, tectonics
Hansen, Alan Ray, geology
Harner, Carol Frances Hodgson, photobiology
Harold, Franklin Marcel, biochemistry
Harshman, Elbert Nelson, economic geology
Hartman, Emily Lou, botany
Hayes, Bridget Ann, developmental biology
Hays, William Henry, geology
Hazlett, David Richard, pulmonary physiology
Hector, David Lawrence, applied mathematics, fluid dynamics
Hedlund, David C., geology
Heidel, Robert Henry, spectroscopy
Hendee, William Richard, medical physics
Herold, Laurance Carter, cultural geography, urban geography
Hersman, Marion Frank, research administration
Heyl, Allen Van, Jr, geology
Hiestand, Thomas Cleon, exploration geology, mineral economics
Hightower, Collin James, number theory
Hills, Francis Allan, geochronology
Hobbs, Samuel Warren, geology
Hoffman, Ruth I., mathematics
Holmes, Joseph Henry, internal medicine, clinical pathology
Holyoke, John Bartlett, pathology
Hoover, Donald Brunton, geophysics, electrical engineering
Hornbein, Joseph Michael, organic chemistry
Howell, Wallace Egbert, meteorology
Hull, Joseph Poyer Deyo, Jr, petroleum geology
Hunt, Graham R., physical chemistry, spectroscopy
Huseby, Robert Arthur, medicine
Huss, Glenn I., meteoritics

Hutt, Martin P, internal medicine
Hutto, Francis Baird, Jr, inorganic chemistry
Iona, Mario, cosmic ray physics
Jacob, Arthur Frank, sedimentary petrology, economic geology
Jensen, David, medical physiology
Johnson, Ernest Walter, physiology
Johnson, Frederick Arthur, Jr, geology
Jones, Carol A, biophysics
Jones, Richard Hunn, biostatistics, computer science
Jordan, James N, earth science
Kadey, Frederic L, Jr, economic geology, exploration geology
Kane, Martin Francis, geophysics
Kanizay, Stephen Peter, geology
Kao, Fa-Ten, genetics
Katsh, Seymour, biology
Katz, Fred H, internal medicine, endocrinology
Kaufman, Irving Charles, psychiatry
Kaufman, Janice Norton, psychiatry
Keefer, William Richard, geology
Keighin, Charles William, geochemistry, economic geology
Keith, James Oliver, wildlife ecology
Kempe, Charles Henry, pediatrics
Kemper, William Alexander, environmental chemistry, ballistics
Kent, Kate Peck, anthropology
Kern, Fred, Jr, internal medicine, gastroenterology
Kerr, Sylvia Jean, biochemistry
Kier, Lawrence Charles, toxicology, forensic science
Kim, Daniel Y, geophysics, physics
Kimberling, William J, human genetics, biomathematics
Kinsinger, Floyd Elton, range ecology
Kirsch, Wolff M, neurosurgery, biochemistry
Kleinkopf, Merlin Dean, geophysics
Kosanke, Robert Max, paleobotany
Kotin, Paul, pathology
Kozloff, Lloyd M, virology, molecular biology
Krikos, George Alexander, pathology
Kubo, Ralph Teruo, immunology
Kuntz, Mel Anton, geology, petrology
Kurnick, John Edmund, hematology, oncology
Lambert, Paul Wayne, geomorphology, quaternary geology
Langan, Thomas Augustine, biochemistry
Leaver, Frederick Wilson, biochemistry
Lee, Donald Edward, mineralogy
Lehman, John Michael, experimental pathology, virology
Leineweber, James Peter, physical chemistry
LeMasurier, Wesley Ernest, geology
Leonard, Benjamin Franklin, geology
Leopold, Estella (Bergere), botany
Lesh, Janet Rountree, astrophysics
Levings, William Stephen, geology
Lewis, George Edward, stratigraphy, vetebrate paleontology
Liechty, Richard D, surgery, endocrinology
Linn, Richard Harry, internal medicine
Lipman, Peter Waldman, geology
Lohman, Stanley William, geology
Lubchenco, Lula O, pediatrics
Luebs, Ralph Edward, soils
Maaske, Clarence Alfred, physiology
Macdonald, John Marshall, psychiatry, criminology
Mackenzie, Cosmo Glenn, biochemistry
Mackenzie, Julia Buzz, biochemistry
Mackinney, Herbert William, polymer chemistry, environmental chemistry
MacLachlan, James Crawford, geology
Magee, Charles Brian, physical chemistry, inorganic chemistry
Makowski, Edgar Leonard, obstetrics & gynecology
Malde, Harold Edwin, geology
Male, Carolyn Joan, microbiology
Mallory, William Wyman, geology
Manes, Cole, developmental biology
Mann, John Allen, geology
March, Andrew Lee, geography
Marchand, Jean-Paul, mathematical physics
Margolin, Sydney Gerald, psychiatry
Marranzino, Albert Pasquale, geochemistry
Martin, Alexander Robert, neurophysiology
Martin, John Lee, biochemistry
Mathies, James Crosby, biochemistry
Maughan, Edwin Kelly, geology
Maxwell, Charles Henry, geology
Mazzia, Valentino Don Bosco, anesthesiology
McAtee, Patricia Rooney, community health
McClatchy, Joseph Kenneth, medical microbiology
McCracken, Robert Dale, anthropology, psychology
McDowell, Marion Edward, medicine, nutrition
McGill, John Thomas, geology
McIntire, Floyd Cottam, biochemistry
McIntyre, David H, geology
McKee, Edwin Dinwiddie, geology
McLaughlin, Thad Gerald, geology
Meade, Robert Heber, Jr, geology
Mehta, Himatlal R, pharmacy
Meiklejohn, Gordon, medicine
Meschia, Giacomo, physiology
Meyer, Harvey John, geology, computer science
Meyer, Richard Thomas, physical chemistry
Middleton, Elliot, Jr, internal medicine
Miesch, Alfred Thomas, geology
Miles, James S, medicine
Millard, Hugh Thompson, Jr, radiochemistry, geochemistry
Miller, Betty M (Tinklepaugh), petroleum geology
Miller, Clifford Daniel, quaternary geology, volcanology
Miller, William Theodore, organic chemistry, biological chemistry
Millette, Robert Loomis, biochemistry, virology
Milligan, Merle Wallace, mathematics
Mitchell, Roger Sherman, medicine
Moench, Robert Hadley, geology
Molinoff, Perry Brown, neuropharmacology
Moore, George Eugene, surgery, oncology
Moore, John Ezra, hydrology, groundwater geology
Moran, David Taylor, neurobiology, cell biology
Morfit, Henry Mason, surgery
Morgenthaler, George William, mathematics
Morrill, Callis Gary, physiology
Morris, Robert Hamilton, geology
Morrison, Roger Barron, geology
Morse, Melvin Laurance, genetics
Mueller, John Frederick, internal medicine
Mulligan, Richard Michael, pathology
Mullineaux, Donal Ray, geology
Murcray, David Guy, atmospheric physics
Murphy, Daniel Lawson, geology
Murphy, Robert Carl, organic chemistry, pharmacology
Muscari, Joseph A, physics
Mytton, James W, geology
Naeser, Charles Wilbur, geology
Nakane, Paul K, histochemistry, electron microscopy
Nash, John Thomas, geology, geochemistry
Naughton, Michael A, biophysics, obstetrics & gynecology
Netzel, Richard G, physics, academic administration
Neumann, Herschel, molecular physics, science education
Neville, Margaret Cobb, physiology
Newcomb, Richard William, pediatrics, immunology
Newkirk, John Burt, physical metallurgy, bioengineering
Newman, Melvin Micklin, surgery, pulmonary physiology
Newton, William Donald, cytology
Nichols, Douglas James, paleontology
Nielsen, G Howard, mathematics
Nielsen, Larry Dennis, biochemistry
Nora, Audrey Hart, pediatrics
Nora, James Jackson, pediatric cardiology, genetics
Northern, Jerry Lee, audiology
Norton, Daniel Remsen, analytical chemistry
Norton, James Jennings, geology
Obradovich, John Dinko, geophysics
O'Brien, Donough, pediatrics
Offield, Terry Watson, geology
Olson, Alan Peter, archaeology, anthropology
Olson, Jerry Chipman, geology
Olson, John Richard, physics
Oriel, Steven S, structural geology
Owens, James Cuthbert, surgery
Ozog, Francis Joseph, physical chemistry, organic chemistry
Pabst, Michael John, biochemistry
Pace, Norman R, biochemistry
Palacas, James George, organic geochemistry
Pannbacker, Richard George, biochemistry, pharmacology
Parker, Raymond Laurence, geology
Parks, Daniel K, mathematics
Parks, James John, obstetrics & gynecology, reproductive physiology
Patel, Vithalbhai L, plasma physics, space physics
Paterson, Christopher Alexander, physiology
Paton, Bruce Calder, cardiovascular surgery
Patterson, David, biochemical genetics, cancer
Patton, Priscilla Candace, economic geology
Pauley, James Donald, bioengineering
Pearlman, Sholom, dentistry
Pearson, John Richard, biochemistry, endocrinology
Perkins, John Phillip, pharmacology, biochemistry
Peterson, Earl T, geology, palynology
Peterson, Fred, geology
Pettijohn, David E, molecular biology
Petty, Thomas Lee, pulmonary diseases
Pierce, Gordon Barry, pathology
Pierce, Kenneth Lee, quaternary geology, geomorphology
Pinckney, Darrell Mayne, geology
Pipiringos, George Nicholas, stratigraphy
Platt, James Earl, comparative endocrinology
Porter, Kenneth Raymond, vertebrate ecology
Prasad, Kedar N, radiobiology, oncology
Pratt, Walden Penfield, geology
Prichard, George Edwards, geology
Prostka, Harold J, geology
Puck, Theodore Thomas, biophysics
Pundsack, Frederick Leigh, inorganic chemistry
Purcell, Kenneth, clinical psychology
Quiatt, Duane D, anthropology
Raab, Joseph A, mathematics
Raup, Omer Beaver, geology, geochemistry
Raup, Robert Bruce, Jr, geology
Rechard, Ottis William, computer science, mathematics
Redman, Robert Shelton, oral biology, pathology
Redmond, John Lynn, petroleum geology
Reeve, Ernest Basil, physiology
Reeves, John T, cardiology
Reinbold, George W, bacteriology, dairy industry
Reiss, Oscar Kully, biochemistry
Richmond, Gerald Martin, quaternary geology
Rickenberg, Howard V, developmental biology
Riley, Conrad Milton, pediatrics
Riter, John Randolph, Jr, physical chemistry
Roberts, Walden Kay, microbiology, biochemistry
Robinson, Arthur, pediatrics, genetics
Rodeck, Hugo George, zoology, museology
Roederer, Juan Gualterio, space physics, psychoacoustics
Roehl, Perry Owen, petroleum geology
Rogers, Frank Bradway, medicine
Rose, Peter R, petroleum geology, stratigraphy
Rosenblum, Sam, geology
Ross, Reuben James, Jr, paleontology
Rowley, Peter Dewitt, geology
Rudd, Robert Dean, physical geography, remote sensing
Ruppel, Edward Thompson, geology
Ryder, Robert Thomas, stratigraphy, petroleum geology
Rye, Robert O, geochemistry
Ryniker, Charles, geology
Sadler, John, biochemistry, molecular biology
Sams, Wiley Mitchell, Jr, dermatology
Sanborn, Albert Francis, geology
Sands, Howard, pharmacology, biochemistry
Sargent, Kenneth Albert, geology
Saunders, George Cherdron, immunology, pathobiology
Saxena, Vinod Kumar, meteorology, cloud physics
Scarborough, Gene Allen, biochemistry
Schallhorn, Robert George, periodontics, biochemistry
Schleicher, David Lawrence, environmental geology
Schmeer, Arline Catherine, cell biology
Schmidt, Dwight Lyman, geology
Schmidt-Collerus, Josef Johannes, physical organic chemistry, biochemistry
Schultz, Leonard G, geology
Schultz, Phyllis W, developmental biology
Schwartz, Charles, medicinal chemistry
Schwendinger, Richard B, biology, chemistry
Scott, James Henry, geophysics, geology
Scott, Richard Albert, paleobotany, plant morphology
Scott, Thomas George, wildlife research
Scott, William Edward, geology
Seeds, Nicholas Warren, biochemistry
Seeland, David Arthur, geology
Severson, Ronald Charles, geochemistry
Sexton, Alan William, physiology
Shacklette, Hansford Threlkeld, botany
Shaklee, Alfred Barral, behavioral biology, psychology
Shaw, Alan Bosworth, geology
Sheppard, Richard A, geology
Shonle, John Irwin, musical acoustics
Shubert, Moras L, ecology
Siemens, George John, biology
Silver, Henry K, pediatrics
Silverberg, Steven George, surgical pathology
Silverman, Bernard Allen, atmospheric physics, meteorology
Simon, Ruth B, geophysics
Simons, Frank Stanton, geology
Simpson, Howard Edwin, geology
Sims, Paul Kibler, geology
Smith, Bruce Dyfrig, geophysics
Smith, Dwight Morrell, physical chemistry, analytical chemistry
Smith, Joe Fred, Jr, geology
Smith, Stuart Werner, anatomy
Smyth, Charley Johnson, medicine
Solomons, Clive (Charles), biochemistry, pediatrics
Souhrada, Joseph Francis, physiology, pathology
Spencer, Joseph Walter, geophysics
Spickler, J William, physiology, internal medicine
Staatz, Mortimer Hay, economic geology
Stacey, John Sydney, physics
Stark, Philip Herald, geology
Starr, Robert I, chemistry, environmental sciences
Starzi, Thomas E, surgery
Steen-McIntyre, Virginia Carol, quaternary geology, tephrochronology
Steven, Thomas August, geology
Stevens, Richard Edward, geography
Stewart, John Morrow, biochemistry, pharmacology
Stickler, William Carl, organic chemistry
Stone, Charles Porter, wildlife research, wildlife ecology
Stone, Gordon Emory, cell biology
Sunderwirth, Stanley George, organic chemistry
Sussman, Karl Edgar, medicine, endocrinology
Sutfin, Duane, physiology
Swanson, Rowena Weiss, information science
Swett, John Emery, anatomy, neurophysiology
Szabo, Barney Julius, geochemistry
Takeda, Yasuhiko, physiology, clinical pathology
Talmage, David Wilson, medicine
Tarby, Theodore John, anatomy, neurophysiology
Tate, James Leroy, Jr, ornithology
Tatsumoto, Mitsunobu, geochemistry, geochronology
Taylor, Austin Laurence, microbiology
Taylor, Edward Stewart, obstetrics & gynecology
Thomsen, Harry Ludwig, geology
Thorman, Charles Hadley, geology
Tidball, Ronald Richard, soil science
Timblin, Lloyd O, Jr, applied physics, physical science
Tobin, Charles Emil, anatomy, embryology
Todd, Clement Jameson, meteorology
Todd, James Hopkins, geology
Tong, James Lun, microbiology
Tosh, Fred Eugene, medicine, epidemiology
Tourtelot, Harry Allison, geology
Toy, Terrence Joseph, geomorphology, geography
Tschanz, Charles McFarland, geology
Tschudy, Robert Haydn, paleobotany
Turner, James Howard, analytical chemistry
Tuttle, Elizabeth R, theoretical physics
Twenhofel, William Stephens, geology
Tweto, Ogden, economic geology
Varnes, David Joseph, engineering geology
Vernadakis, Antonia (Mrs H L Ockerman), developmental neurobiology
Waddell, William Rhoads, surgery
Wagner, Wiltz Walker, Jr, pulmonary physiology
Waldren, Charles Allen, biophysics

Walker, Strother Holland, mathematics, statistics
Wallace, Chester Alan, stratigraphy, sedimentary petrology
Walters, Curla Sybil, immunology, microbiology
Warren, Peter, mathematics, biomathematics
Watson, Kenneth, geophysics
Wayman, Cooper H., environmental management
Weakly, Ward Fredrick, anthropology
Webber, Mukta Mala, oncology, cell biology
Weil, John Victor, internal medicine
Weiner, Eugene Robert, physical chemistry
Weiner, Norman, pharmacology
Weiser, Philip Craig, physiology
Welder, Frank A., geomorphology, stratigraphy
Wells, Joseph S., physics
Wenger, David Arthur, biochemical genetics, pediatrics
Wenrich-Verbeek, Karen Jane, geology
Wentland, Stephen Henry, biochemistry, organic chemistry
Werkman, Sidney Lee, medicine
Wershaw, Robert Lawrence, hydrology, geochemistry
Westphal, Warren Henry, exploration geophysics
White, Franklin Estabrook, physics
White, Richard William, geology
Whitlock, John Stuart, biochemistry
Whittemore, Charles Alan, organic chemistry, neurophysiology
Wickelgren, Warren Otis, physiology, psychology
Wicks, Wesley Doane, biochemistry, pharmacology
Wickstrom, Eric, biophysical chemistry
Wier, James Arista, medicine
Wilcox, Ray Everett, geology
Wilkes, John Stuart, biochemistry
Williams, David Lee, geophysics, oceanography
Williams, Denis R., theoretical chemistry, academic administration
Wilson, John Tucker, anatomy
Wilson, David George, range conservation, research administration
Withers, Arnold Moore, anthropology
Witkind, Irving Jerome, economic geology
Witt, Shirley Hill, biological anthropology
Witten, Thomas A., internal medicine, gastroenterology
Worth, William Sutherland, neurosciences, nutrition
York, Sheldon Stafford, biophysical chemistry
Young, Edward Joseph, geochemistry, mineralogy
Zartman, Robert Eugene, geochronology
Zeiner, Frederick Neyer, zoology
Zeiner, Helen Marsh, botany, ecology

EMPIRE
Kamilli, Robert Joseph, geology

ENGLEWOOD
Bowling, Franklin Lee, economic geology
Chico, Raymundo Jose, mining geology
Dole, Hollis Mathews, economic geology
Hsu, I-Chi, exploration geophysics
Landon, Robert E., geology
Marks, Jay Glenn, geology, paleontology

DURANGO
Baars, Donald Lee, stratigraphy, sedimentary petrology
Craig, Roy Phillip, environmental management, science writing
Dever, John E., Jr., plant physiology,
Erickson, James George, biochemistry
Harrison, Edward Merle, chemistry
Marquiss, Robert W., range science
Mills, James Wilson, physical chemistry
Owen, Herbert Elmer, Jr., plant biochemistry
Peters, Roger Paul, animal behavior
Ritchey, John Michael, chemistry, analytical chemistry
Schwab, Charlotte Ann, plant anatomy
Spencer, Albert William, zoology
Steinhoff, Harold William, wildlife biology

Pizzolato, Philip Joseph, inorganic chemistry
Polhemus, John Thomas, entomology
Sittner, Weldon Rexer, solid state physics, biophysics
Thompson, Raymond Melvin, geology

ESTES PARK
Russell, Richard Dana, sedimentology
Yost, William Jacque, physics

EVERGREEN
Durkee, Edward Fleming, petroleum geology, economic geology
Nieraber, James Henry, geology

FT COLLINS
Abel, John H., Jr., cell biology
Adams, Robert Philip, evolutionary biology, taxonomy
Alexander, Archibald Ferguson, veterinary pathology
Allgower, Eugene L., mathematics
Altman, Jack, plant physiology
Angleton, George M., radiation biology, biometrics
Avens, John Stewart, food microbiology
Azari, Parviz, biochemistry
Bagby, John R., Jr., public health, biology
Bailey, James Allen, wildlife ecology
Baker, R Ralph, plant pathology, mycology
Baldwin, Paul Herbert, ornithology, vertebrate ecology
Bamburg, James Robert, biochemistry
Banks, William Joseph, Jr., veterinary histology, cytology
Barnes, Allan Marion, medical entomology
Barney, Charles Wesley, forest ecology, silviculture
Bashan, Charles W., horticulture, plant physiology
Basri, Saul Abraham, elementary particle physics
Bass, Louis Nelson, plant physiology
Batchelder, Arthur Roland, agronomy
Behnke, Robert J., ichthyology
Bennett, Dwight G., Jr., veterinary medicine
Berg, William Albert, soil conservation
Berge, Trygve Obert, virology, rickettsial diseases
Berndtson, William Everett, reproductive physiology
Bernstein, Elliot R., chemical physics, spectroscopy
Best, Jay Boyd, physiology; biophysics
Billenstein, Dorothy Corinne, anatomy
Bodie, Jozsef, wood science, wood technology
Bonham, Charles D., plant ecology,
Bose, Raj Chandra, statistics
Bowden, David Clark, statistics, biometry
Boyd, Josephine Watson, water pollution
Boyd, William Lee, bacteriology
Bragonier, Wendell Hughell, botany
Braun, Clait E., wildlife biology
Brengle, Kenneth Gordon, soil science
Brewer, Jesse Wayne, entomology, plant pathology
Brier, Glenn Wilson, statistical analysis, meteorology
Brink, Kenneth Maurice, horticulture
Brinks, James S., animal genetics
Burke, Michael John, biophysics
Butler, Jackie Dean, horticulture
Butler, Walter Cassius, mathematics
Caisher, Charles Henry, microbiology
Carlson, Clarence Albert, Jr., fish biology
Carpenter, James Andrew, Jr., animal breeding
Caughey, Winslow Spaulding, biochemistry
Chamberlain, Theodore Klock, geology, oceanography
Charkey, Lowell William, nutritional biochemistry
Charney, Michael, physical anthropology
Chow, Fu Ho Chen, biochemistry, physical chemistry
Chow, Tsu Ling, animal virology
Cole, Vernon C., soil science
Coleman, David Cowan, ecology
Collier, John Raymond, microbiology
Cook, Charles Wayne, range science
Cook, William Boyd, organic chemistry
Cooper, Tommye, agriculture, statistics
Corrin, Myron Lee, physical chemistry

Costello, David Francis, environmental biology
Coulombe, Harry N., vertebrate biology, environmental management
Cox, Stephen Kent, atmospheric physics
Cramer, David Alan, animal nutrition
Cuany, Robin Louis, plant genetics
Culver, Roger Bruce, astronomy
Curtis, Byrd Collins, plant breeding, plant genetics
Danielson, Robert Eldon, soil physics
Danst, Richard B., mathematics
Daugherty, Ned Arthur, inorganic chemistry
Davidson, Ross Wallace, forest pathology
Davis, Lloyd Edward, pharmacology
Davis, Robert Wilson, anatomy, pathology
Deal, Ervin R., mathematics
DeBruin, Kenneth Edward, organic chemistry
DeMeyer, Frank R., mathematics
deMooy, Cornelis Jacobus, soil fertility, plant nutrition
Derbyshire, William Davis, radiobiology
Dewey, Wilson Cornet, radiobiology
Dickens, Lester Emert, plant pathology
Diis, Robert Earl, academic administration, resource management
Dittberner, Phillip Lynn, range ecology,
Dodd, Jerrold Lowell, range ecology
Doehring, Donald O., geomorphology,
Dorzenko, Alexander Daniel, geoenvironmental science
Doxtader, Kenneth Guy, soil microbiology
Driscoll, Richard Stark, range management, remote sensing
Dupont, Jacqueline (Louise), nutrition
Dyer, Melvin I., zoology
Edwards, Harry Wallace, physical chemistry
Ellis, James Edgar, animal ecology, animal behavior
Eling, Glenwood Pershing, microscopic anatomy
Ethridge, Frank Guide, sedimentology
Evans, Howard Ensign, entomology
Fahrney, David Emory, biochemistry
Fairbank, William Martin, Jr., quantum optics
Farnell, Albert Bennett, mathematics
Faulkner, Lloyd (Clarence), reproductive physiology
Fechner, Gilbert Henry, forest genetics
Ferguson, William Sidney, analytical chemistry
Finley, Robert Byron, Jr., vertebrate zoology
Fletcher, Herbert Calvin, soil science
Flickinger, Stephen Albert, fish biology
Fly, Claude Lee, soil chemistry, plant physiology
Fosberg, Michael Allen, forest meteorology
Fox, Douglas Gary, meteorology, air pollution
Frandson, Roven Dale, anatomy
French, Norman Roger, ecology
Frisinger, H Howard, II, mathematics,
Fronk, William Don, economic entomology
Furman, Roland William, II, climatology
Gaines, Robert Earl, mathematics
Gibson, James H., environmental sciences
Gilbert, Douglas L., ecology
Gillespie, John Paul, organic chemistry,
Gillette, Edward LeRoy, veterinary radiology, radiobiology
Gillis, Nelson Silsbee, theoretical solid state physics
Glover, Fred Arthur, wildlife
Gorell, Thomas Andrew, endocrinology, developmental biology
Gorthy, Willis Charles, cytopathology
Grant, Dale Walter, histology
Gray, William Mason, meteorology
Graybill, Franklin A., mathematical statistics
Greathouse, Terrence Ray, animal nutrition
Greenberg, Allan S., low temperature physics
Greene, Kenneth Titsworth, physical chemistry
Guenzi, Wayne D., soil chemistry
Hadley, Lawrence Nathan, physical optics
Hagen, Harold Kolstoe, fish biology
Haise, Howard Ross, agronomy, agricultural engineering
Hall, Peter, physiology; pharmacology

Hamar, Dwayne Walter, biochemistry
Hanan, Joe John, horticulture, plant physiology
Hanchey, Penelope Jane, phytopathology, cytology
Hansen, Richard M., zoology
Happ, George Movius, insect physiology, developmental biology
Hartill, Inez Kemble, nutrition, biochemistry
Harrington, Harold David, plant taxonomy
Harris, David Vernon, geology
Harrison, Monty De Vert, plant pathology
Hasler, Marilyn Jean, reproductive physiology
Hauptman, Bernhard, meteorology
Haus, Thilo Enoch, agronomy
Hawksworth, Frank Goode, forest pathology
Hayes, Richard Oliver, medical entomology
Heath, Robert Bruce, veterinary anesthesiology
Hecker, Richard Jacob, plant breeding
Hegedus, Louis Stevenson, organic chemistry, organometallic chemistry
Heil, Robert Dean, soil conservation
Heil, Dale Arthur, wildlife management,
Heller, Marvin W., physics
Hendrix, John Edwin, plant physiology
Herin, Reginald Augustus, physiology,
Herman, Frederick Joseph, botany
Hess, Archie Davila, public health,
Hess, Frederick Dana, cell biology, plant biology
Hibler, Charles Phillip, zoology,
Hoerlein, Alvin Bernard, animal
Holley, Winfred Davis, horticulture
Hopwood, Mortimer Lloyd, reproductive physiology
Horak, Donald L., fisheries
Houston, Walter Randolph, plant ecology; land reclamation
Hsu, David Kuei-Yu, solid state physics
Hudson, Bruce William, immunology,
Hunt, Henry William, ecology,
Hutchinson, Gordon Lee, soil science,
Hyder, Donald N., range science
Jameson, Donald Albert, ecology
Jansen, Gustav Richard, nutrition
Jennings, Calvin Hunt, archaeology
Jensen, Rue, veterinary pathology
Job, Robert Charles, inorganic chemistry
Johnsen, Richard Emanuel, insect toxicology
Johnson, Donal Dabell, soils,
Johnson, Donald Eugene, animal nutrition
Johnson, James Edward, radiation biophysics
Johnson, Robert Britten, geology
Jones, Everett Bruce, hydrology water resources engineering
Jordan, John Patrick, biochemistry
Kamal, Adel S., insect physiology; insect taxonomy
Kanney, Lynna Babs, physics
Kano, Adeline Kyoko, chemistry, academic administration
Kearney, Philip Daniel, physics
Keim, Wayne Franklin, crop breeding
Kelman, Robert Bernard, mathematics,
Kemp, Graham Elmore, veterinary public health
Kemper, William Doral, agronomy
Kienholz, Eldon W., animal nutrition
Klein, Donald Albert, microbiology
Klopfenstein, Kenneth F., mathematics
Klute, Arnold, agronomy
Knutson, Kenneth Wayne, plant pathology
Koyner, Jacob L., forest biometry
Kozlowski, Gerald P., neuroendocrinology
Kreutzer, William Alexander, plant pathology
Krueger, David Allen, theoretical physics
Larsen, Arnold Lewis, botany, agronomy
Larson, Kenneth Allen, immunology, veterinary medicine
Lauenroth, William Karl, plant ecology
Lauerman, Lloyd Herman, Jr., veterinary microbiology, veterinary immunology
Law, George Robert John, genetics
Laycock, William Anthony, plant ecology

Lebel, Jack Lucien, radiology, radiation biology
Lee, Arthur Clair, veterinary medicine, radiological health
Lee, Virginia Ann, biochemistry
Lehman, Joe Junior, organic chemistry
Lehner, Philip Nelson, animal behavior, ecology
Leisure, Robert Glenn, metal physics
Lett, John Terence, biophysics, radiation biology
Levinger, Bernard Werner, mathematics
Lewis, Gordon Depew, forest economics, marketing
Lindsay, Willard Lyman, soil science
Livingston, Clark Holcomb, plant pathology
Lorenz, Klaus J, cereal chemistry
Ludwick, Albert Earl, soil fertility
Lumb, William Valjean, veterinary medicine
Luoto, Lauri, veterinary medicine, epidemiology
Maag, Dale D, biochemistry
Maciel, Gary Emmet, physical chemistry
Maga, Joseph Andrew, food science,
Magnus, Arne, mathematics
Mahoney, Charles Lindbergh, environmental sciences
Marlatt, William Edgar, atmospheric science
Marquardt, William Charles, protozoology, parasitology
Martin, Stephen George, behavioral ecology
Martin, Susan Scott, plant chemistry
Martinelli, Mario, Jr, meteorology, forestry
Masken, James Frederick, physiology, biochemistry
Matsushima, John K, animal nutrition
Mautz, William Ward, wildlife research, wildlife ecology
Meiman, James R, watershed management
McCallum, Malcolm E, geology
McClellan, John Forbes, zoology
McCormick, Stephen Fahrney, numerical analysis
McGinnies, William Joseph, range science, plant ecology
McIntyre, Gary A, plant pathology
McLean, Robert George, epizootiology, vertebrate ecology
Meyers, Albert Irving, organic chemistry
Mielke, Paul W, Jr, statistics, mathematics
Miller, Byron F, poultry nutrition
Miller, Charles William, biology
Miller, Larry Lee, organic chemistry
Mogren, Edwin Walfred, forestry
Mohlner, David Morris, electrochemistry, analytical chemistry
Moore, Frank Devitt, III, horticulture
Moore, Russell Thomas, plant ecology
Moreng, Robert Edward, poultry physiology
Morrison, John Albert, wildlife management, ecology
Mosier, Arvin Ray, agricultural chemistry
Mulnix, John Arthur, veterinary medicine
Musgrave, Ted Russell, inorganic chemistry
Nabors, Murray Wayne, plant physiology, plant breeding
Nagy, Julius G, nutrition, bacteriology
Nash, Donald Joseph, genetics, zoology
Nelson, Albert Wendell, cardiovascular diseases
Niemann, Ralph Henry, mathematics
Niswender, Gordon Dean, reproductive endocrinology
Nockels, Cheryl Ferris, animal nutrition
Nornes, Howard Onsgaard, neurobiology, developmental biology
Nordin, Robert W, veterinary pathology
Norstadt, Fred A, agricultural chemistry
Ogg, James Elvis, microbial genetics
O'Keefe, Kelly Ray, analytical chemistry
Oshima, Nagayoshi, plant pathology
Osteryoung, Janet G, electroanalytical chemistry
Osteryoung, Robert Allen, analytical chemistry
Owen, William Bert, entomology
Packard, Gary Claire, physiological ecology, zoology
Page, Rector Lee, mathematics, computer science
Painter, Kent, bionucleonics
Painter, Richard J, mathematics
Parton, William Julian, D, biometeorology, mathematical biology

Patton, Alva Rae, chemistry
Patton, Carl E, solid state physics
Pautler, Eugene L, physiology, biophysics
Payne, Merle G, biochemistry
Petit, Michael Geoffrey, physical chemistry, organic geochemistry
Pettus, David, vertebrate zoology
Phemister, Robert David, veterinary pathology
Phillips, Robert Ward, veterinary physiology
Pickett, Bill Wayne, reproductive physiology
Poland, Helen M, chemical physics, atmospheric chemistry
Poland, Jack Dean, medicine
Porter, Lynn K, soil chemistry
Post, George, fish pathology, wildlife pathology
Pressel, Esther Joan, anthropology
Puleston, Harry Samuel, chemistry
Pyle, Robert Lee, veterinary medicine, cardiology
Ragin, James Frederick, genetics, molecular biology
Raich, John Carl, solid state physics
Raleigh, Robert Franklin, fish biology
Ralph, Charles Leland, physiology
Reed, Edward Brandt, limnology, ecology
Reeves, Fontaine, Jr, mycology
Reid, Charles Phillip Patrick, forest ecology, physiology
Reiter, Elmar Rudolf, meteorology
Remmenga, Elmer Edwin, statistics
Rich, Lonnie Joe, veterinary pathology
Riehl, Herbert, meteorology
Robertson, Jerold C, physical organic chemistry
Ronco, Frank, Jr, forest management
Ross, Cleon Walter, plant physiology, plant biochemistry
Rubin, Robert, veterinary parasitology
Ruppel, Earl George, plant pathology, virology
Ryder, Ronald Arch, wildlife management
Sabey, Burns Roy, soils
Savage, Eldon P, public health
Schiager, Keith Jerome, environmental health, health physics
Schmehl, Willard Reed, soils
Schmidt, John Lancaster, wildlife ecology, wildlife management
Schnuelle, Gary Wayne, theoretical chemistry
Schroeder, Herbert August, organic chemistry
Schumann, Stanley Alfred, geomorphology, watershed management
Schweizer, Edward E, plant physiology, agronomy
Scott, James Allan, ecology, entomology
Seidel, George Elias, Jr, reproductive physiology
She, Chiao-Yao, quantum electronics, solid state physics
Siddiqui, Mohammed Moinuddin, mathematical statistics
Simpson, Robert Gene, entomology
Sims, Philip Leon, range science
Sinclair, Peter C, atmospheric physics
Sites, James Russell, solid state physics
Skogerboe, Rodney K, analytical chemistry
Slade, Larry Malcom, animal science
Smith, Vearl Robert, agriculture
Sneider, Thomas W, biochemistry, physiology
Snyder, Stanley Paul, veterinary pathology, oncology
Solie, Thomas Norman, physical chemistry, biophysics
Solomon, Gordon Charles, anatomy, pathology
Soltanpour, Parviz Neil, soil fertility, agronomy
Sommerfeld, Richard Arthur, geochemistry, geology
Splittgerber, George H, inorganic chemistry
Squire, Phil George, biophysical chemistry
Srivastava, Jagdish Narain, statistics, mathematics
Stack, Stephen M, cytology
Staley, John M, plant pathology, forestry
States, James Bruce, ecology
Stein, Frederick Max, mathematics
Stermitz, Frank, organic chemistry
Stevens, Robert E, forest entomology
Storz, Johannes, virology, microbiology
Streeter, Robert Glen, wildlife conservation, wildlife research
Striffler, William D, forest hydrology
Swanson, Gustav Adolph, wildlife ecology
Swanson, Vern Bernard, animal breeding

Teller, Leo, watershed management, environmental science
Tengerdy, Robert Paul, immunochemistry, microbiology
Terwilliger, Charles, Jr, plant ecology, range science
Theodoratus, Robert James, cultural anthropology, ethnology
Thomas, James William, mathematics
Thomas, William Robb, food science, dairy bacteriology
Thomassen, Robert William, veterinary pathology
Thompson, Tommy Burt, economic geology, petrology
Tietz, William John, Jr, neurophysiology
Tornabene, Thomas Guy, microbiology
Townsend, Charley E, plant breeding, genetics
Tracy, C Richard, ecology
Trent, Dennis W, microbiology,
Troxell, Harry Emerson, Jr, wood science & technology
Tsuchiya, Takumi, plant cytology, plant genetics
Tu, Anthony T, biochemistry
Udall, Robert Hovey, animal nutrition
Van Dyne, George M, ecology, nutrition
Vaughan, John Dixon, physical chemistry
Venable, John Howard, veterinary anatomy, histology
Viets, Frank Garfield, Jr, soil science
Vonder Haar, Thomas Henry, meteorology, space science
Walimo, Olof Charles, vertebrate ecology
Wangaard, Frederick Field, forest products
Ward, Gerald Madison, animal science
Ward, Richard Theodore, plant ecology
Warren, Clarence Gerald, inorganic chemistry
Wasser, Clinton Howard, range ecology
Watkins, Kenneth Walter, physical chemistry
Weber, Louis Russell, experimental physics
Weitz, Joseph Leonard, geology
Welch, Ronald Maurice, atmospheric physics, meteorology
Welsh, James Ralph, plant breeding, plant genetics
Whicker, Floyd Ward, radiation biology, ecology
Whiteman, Charles E, veterinary pathology
Wilber, Charles Grady, physiology
Wilken, Dieter H, systematic botany
Wilken, Gene C, geography
Williams, James Stanley, statistics
Wilson, Alma McDonald, plant physiology
Winder, Dale Richard, solid state physics
Wood, Donald Roy, agronomy
Woodmansee, Robert George, range ecology, forest ecology
Woody, A-Young Moon, chemistry, biochemistry
Woody, Robert Wayne, physical chemistry, biochemistry
Woolley, Tyler Anderson, zoology
Workman, Milton, plant physiology
Wunder, Bruce Arnold, physiological ecology, vertebrate zoology
Young, Stuart, veterinary pathology
Youngman, Vern E, agronomy
Zelle, Max Romaine, genetics
Zimdahl, Robert Lawrence, agronomy, weed science
Zimmerman, Robert A, microbiology, epidemiology

GLENWOOD SPRINGS

Baker, M Michelle, fresh water biology, ecology
Trapani, Ignatius Louis, physiology, immunology

GOLDEN

Adams, Richard Owen, solid state physics
April, Robert Wayne, analytical chemistry, physical chemistry
Atwood, Mark Trevor, industrial organic chemistry
Bisque, Ramon Edward, geochemistry
Boes, Ardel J, mathematics
Bowersox, Ralph B, physics
Brady, Brian T, materials science, applied mathematics
Brown, Austin Robert, Jr, mathematics
Brown, James T, Jr, theoretical physics
Burnett, Jerrold J, physics
Cathcart, James Bachelder, geology
Cleveland, Jesse Marvin Jr, inorganic chemistry

Cope, Oliver Brewer, fishery biology
Davidson, Darwin Ervin, botany, mycology
Dickerhoof, Dean W, inorganic chemistry
Dickinson, Deanne, applied mathematics
Dover, James H, structural geology, metamorphic petrology
Edwards, Kenneth Ward, analytical chemistry, physical chemistry
England, Anthony W, geophysics, astrogeology
Epis, Rudy Charles, geology
Finney, Joseph J, geology, mineralogy
Goens, Duane N, physical chemistry, electrochemistry
Gorton, Kenneth Arnold, petroleum geology
Grose, Lucius Trowbridge, geology
Guss, Cyrus Omar, organic chemistry
Hadzeriga, Pablo, inorganic chemistry
Hall, David Warren, organic chemistry
Harlan, Ronald A, nuclear chemistry
Haun, John Daniel, petroleum geology
Hazen, Wayne Colby, inorganic chemistry
Hilchie, Douglas Walter, engineering science, geophysics
Hiltrop, Carl L, geochemistry
Holmer, Ralph Carrol, exploration geophysics
Houghton, Augustus Sherrill, chemistry
Hounslow, Arthur William, mineralogy, petrography
Hundhausen, Joan Rohrer, mathematics
Hutchinson, Robert Maskicll, geology
Hyatt, David Ernest, inorganic chemistry, organometallic chemistry
Jansen, George James, mineralogy
Jordan, Albert Raymond, physics
Kazanjian, Armen Roupen, physical chemistry
Kennedy, George Hunt, surface chemistry
Kent, Harry Christison, stratigraphy
Kline, John Virgil, spectroscopy
Klusman, Ronald William, geochemistry
Kork, John O, mathematics, statistics
Kuhn, Truman Howard, geology
La Fehr, Thomas Robert, geophysics
Law, William Brough, plasma physics
Learned, Robert Eugene, economic geology, geochemistry
Lee, Keenan, geology
Lucas, George Bond, organic geochemistry
Marsh, Donald Charles Burr, number theory
McAllister, Robert Wallace, physics, instrumentation
Merideth, George Thomas, physics
Moore, Fred Edward, geology
Mueller, Raymond Karl, applied mathematics
Navratil, James Dale, industrial chemistry
Newman, Karl Robert, geology
Oetting, Franklin Lee, physical chemistry, thermodynamics
Olson, Richard Hubbell, economic geology
Pickett, George R, geophysics, engineering
Ponder, Herman, mineralogy
Price, Richard Walter, cell biology, biophysics
Putzier, Edward Anthony, health physics
Romberger, Samuel B, geochemistry, economic geology
Romig, Phillip Richardson, geophysics, seismology
Rouse, George Elverton, geochemistry
Sakakura, Arthur Yoshikazu, theoretical physics
Scharon, Harry Leroy, geophysics
Schowengerdt, Franklin Dean, atomic physics
Shendrikar, Arun D, environmental chemistry
Skougstad, Marvin Wilmer, analytical chemistry
Slaughter, Maynard, crystallography
Terada, Kazuji, inorganic chemistry
Trexler, David William, geology, paleontology
Vejvoda, Edward, industrial chemistry
Wallace, Stewart Raynor, geology
Weimer, Robert J, geology
Werkema, Marilyn S, physical chemistry
Whitman, Walter William, mathematics, engineering physics
Wildeman, Thomas Raymond, analytical chemistry, geochemistry
Williams, John T, physical physics
Wiltse, Milton Adair, Jr, geology, geochemistry
Witters, Robert Dale, physical chemistry

COLORADO

Yeats, Frank Richard, theoretical physics

LA JUNTA
Hess, Dexter Winfield, botany
Moore, Donald Clark, physics

JAMESTOWN
Bussian, Alfred Erich, particle physics

GRAND JUNCTION
Alexander, Peter, nuclear physics
Anderson, Paul S., geology
Bell, Wallace G., gynecology
Bowyer, Ben, economic geology
Chenoweth, William Lyman, geology
Duray, John R., experimental nuclear physics
Fischer, Richard Philip, economic geology
Grutt, Eugene Wadsworth, Jr., exploration geology; natural resources
Luepschen, Norman Siegfried, plant pathology
Saccomanno, Geno, pathology
Shannon, Spencer Sweet, Jr., economic geology

GREELEY
Beel, John Addis, organic chemistry
Bond, Richard Randolph, academic administration, zoology
Cavanagh, Timothy D., mathematics
Dietz, Richard Darby, solar physics
Fay, George Emory, archaeology; ethnology
Fields, Clark Leroy, inorganic chemistry
Fish, Forest Norland, mathematics
Fuelberth, John Douglas, algebra
Glidden, Harley Fremont, science education
Hamerly, Robert Glenn, physics
Heiny, Robert Lowell, mathematical statistics
Houston, Samuel Robert, applied statistics
James, MarLynn Rees, physical chemistry
Kearns, Kevin Corrigan, geography
Koch, William George, physical physics
Lehrer, Paul Lindner, physical geography
Lindauer, Ivo Eugene, plant ecology
Moinat, Arthur David, plant physiology
Neal, Louise Adelaide, biology; science education
Peeples, Earle Edward, developmental genetics
Plakke, Ronald Keith, vertebrate anatomy, physiology
Popejoy, William Dean, mathematics
Rehfeld, Carl Ernest, veterinary pathology
Rich, Royal Allen, reproductive physiology
Richards, Edmund A., pharmacology
Schmid, John, applied statistics, educational psychology
Schmidt, Gerald D., zoology; parasitology
Schreck, James Otto, organic chemistry
Selberg, Edith Marie, biology
Stamper, Maynard N., invertebrate zoology
Sund, Robert B., science education
Thomas, Bert O., zoology
Thorpe, Bert Duane, immunology, wildlife diseases
Tomasi, Gordon Ernest, biochemistry, organic chemistry
Trowbridge, Leslie Walter, science education
Winchester, Albert McCombs, biology
Woerner, Dale Earl, analytical chemistry

GUNNISON
DeBoer, Kenneth F., biology
Dorgan, William E., mathematics
Ferchau, Hugo Alfred, botany
Harriss, Thomas T., zoology; entomology
Lawrence, Aubrey Wilford, chemistry
Longpre, Edwin Keith, systematic botany
Mobley, Harold Morton, plant physiology
Prather, Thomas Leigh, geology
Rumburg, Charles Buddy, agronomy
Siemer, Eugene Glen, plant morphology, agronomy
Violett, Theodore Dean, physics

LAKEWOOD
Adams, John Wagstaff, economic geology
Cadigan, Robert Allen, geology
Cain, Joseph Carter, geophysics
Connor, Jon James, geology
Dean, Walter E., Jr., geochemistry, sedimentology
Dickinson, Robert Gerald, geology
Gilluly, James, geology
Gordon, Ellis Davis, hydrology; hydrogeology
Gude, Arthur James, III, mineralogy
Hadley, Richard Frederick, geomorphology
Harrison, Jack Edward, regional geology
Hayes, John Robert, geology; geophysics
Hendricks, Thomas Andrews, geology
Hill, Walter Edward, Jr., geochemistry
Johnson, Ross Byron, environmental geology
Kahan, Archie M., meteorology
Miller, Mary H., geology
Miller, William Robert, geochemistry
Osterwald, Frank William, physical geology
Peterman, Zell Edwin, geology
Quinlan, William David, geology
Sable, Edward George, geology
Shawe, Daniel Reeves, geology
Simmons, George Clarke, geology
Smith, Ralph Emerson, groundwater geology
Spencer, Charles Winthrop, geology
Steele, Timothy Doak, hydrology; resource management
Sutherland, Angus Johnston, chemistry
Swenson, Frank Albert, hydrology
Visher, Frank N., hydrology; geology
Whatley, Alfred T., physical chemistry
Wolfbauer, Claudia Ann, sedimentology

LARKSPUR
Zirkle, Raymond Elliott, biophysics

LITTLETON
Argabright, Perry A., chemistry
Bruce, Charles Robert, physics
Burdge, David Newman, petroleum chemistry
Choquette, Philip Wheeler, geology
Clark, Benton C., geochemistry; biophysics
Clementson, Gerhardt C., computer sciences, operations research
Craig, Dexter Hildreth, geology
Douglas, Larry Joe, physical chemistry; electrochemistry
Duke, Roy Burt, Jr., petroleum chemistry
Ellis, Charles Howard, paleontology
Erskine, Christopher Forbes, geology
Gee, Susan McMillin, geophysics, mineralogy
Gibbons, Louis Charles, petroleum
Goldburg, Arnold, fluid physics
Guewa, Paul Ramon, geology
Guenzel, Gottfried Kurt, palynology
Harms, John Conrad, geology
Hayes, John Bernard, geology
Healy, William Carleton, Jr., mathematical statistics
Knight, Fred G., earth sciences
Lubeck, Axel John, analytical chemistry
MacKenzie, David Brindley, geology
McCarty, Billy Dean, analytical chemistry
McConnell, Andrew Pollock, Jr., petroleum geology
McCubbin, Donald Gene, geology; sedimentology
McKnight, Randy Sherwood, applied mathematics
Norton, Charles J., organic chemistry, anthropology
Nosal, Eugene Adam, geophysics
Patton, James Winton, organic chemistry, geochemistry
Peterson, Alan Herbert, chemistry
Presley, Cecil Travis, physical chemistry
Reitsema, Robert Harold, geochemistry
Rice, Robert Bruce, geophysics, mathematics
Ronzio, Anthony Rose, synthetic organic chemistry
Schalge, Alvin Laverne, analytical chemistry
Shank, George Deane, numerical analysis
Spearing, Darwin Robert, geology
Tackett, James Edwin, Jr., analytical chemistry
Thompson, Loren Edward, Jr., geophysics

Trembly, Lynn Dale, exploration geophysics
Trusell, Fred Charles, analytical chemistry
Viksne, Andy, geophysics
Wells, Frederick Joseph, seismology
Wray, John Lee, geology
Wygant, Noel Darwin, forestry

LONGMONT
Akeson, Walter Roy, biochemistry, agronomy
Dubrovin, Kenneth P., soil chemistry, plant physiology
Litzenberger, Samuel Cameron, agronomy; field crops
Oldemeyer, Robert King, plant breeding, agronomy
Sullivan, Edward Francis, agronomy
Suzuki, Akio, plant breeding, experimental statistics
Svendsen, Kendall Lorraine, geophysics
Widner, Jimmy Newton, crop breeding
Wood, Reed Ralph, agronomy
Yun, Young Mok, entomology

LOVELAND
Linden, James Carl, plant biochemistry

MANCOS
Bement, Robert Earl, range science

MONUMENT
Rhoads, William Denham, analytical chemistry, biochemistry

MORRISON
Lowell, James Diller, petroleum geology; structural geology

PUEBLO
Allen, Ernest E., mathematics education
Bard, Eugene Dwight, science education
Bartlett, Thomas Jefferson, astronomy
Gill, John Paul, Jr., mathematical analysis
Gordon, Robert Wayne, microbiology; ecology
Hammer, Charles Rankin, organic chemistry
Hermann, Scott Joseph, limnology, freshwater invertebrate systematics
Janes, Donald Wallace, bacteriology
Lavelle, James W., limnology
Li, Hung Chiang, statistics, analytical mathematics
Linam, Jay H., entomology
Mahan, Kent Ira, physical chemistry; radiochemistry
McCown, Dean Augustus, physics
Meredith, Charles Eymard, administration, psychiatry, hospital
Mikkelsen, Harry E., physics
Nesbit, Lyle Edwin, physical chemistry
Phillips, David Lowell, mathematics
Raich, Abraham Leonard, analytical statistics, metals
Seilheimer, Jack Arthur, zoology; limnology
Smith, John Elvans, analytical chemistry
Swanson, Clarence A E., mathematics
Watkins, Sallie Ann, physics

SEDALIA
Pakiser, Louis Charles, Jr., geophysics

SPIVAK
Jacobs, Barbara B., endocrinology; immunology
Sharma, Gopal Chandra, pharmacology; cancer
Tripathi, Kamala Kant, biochemistry, analytical chemistry

STERLING
Martin, Jack E., nutrition, biochemistry

URAVAN
Nichols, Chester Encell, geology

US AIR FORCE ACADEMY
Erbacher, John Kornel, physical inorganic chemistry
Hussey, Charles Logan, analytical chemistry
Jendrek, John Paul, Jr., organic chemistry, polymer chemistry

King, Lowell Alvin, physical chemistry
Lamb, Robert W., organic chemistry, physical chemistry
May, John Thomas, experimental nuclear physics
Moran, Michael J., pharmaceutical chemistry
Orth, William Albert, continuum mechanics
Seegmiller, David W., electrochemistry
Shackelford, Scott Addison, organic chemistry; fluorine chemistry
Woodyard, William T., organic chemistry

WHEAT RIDGE
Butler, Arthur Pierce, Jr., economic geology
Huffman, Edward William Dickson, microchemistry
Huffman, Edward William Dickson, Jr., analytical chemistry
Maire, Frederick Wirth, physiology
Post, Edwin Van Horn, economic geology; geochemistry

WINDSOR
Conant, Dale Holdrege, photographic chemistry

WOODY CREEK
Strahalan, George S., physics

CONNECTICUT

BETHANY
Ho, Chong Cheong, polymer science
Karabinos, Joseph Vincent, organic chemistry
Matthews, Demetreos Nestor, organic chemistry
Yiannios, Christ Nicholas, synthetic organic chemistry

BLOOMFIELD
Gomez, Ildefonso Luis, organic chemistry
Steingiser, Samuel, chemistry

BRANFORD
Chitteden, Fayette Dudley, chemistry
Krause, Leonard Anthony, biochemistry

BRIDGEPORT
Banks, Harold Douglas, organic chemistry
Brown, Paul Woodrow, surgery; orthopedic surgery
Calhoun, Gordon Maxwell, physical chemistry
Chih, Chung-Ying, physics
Clarkson, Donald R., mathematics
Davis, Clarence Daniel, obstetrics & gynecology
Denyes, Helen Arliss, environmental biology
Ekeblad, Frederick Alfred, applied statistics, management sciences
Elliott, Helen Margaret, mathematical analysis
Galton, Peter Malcolm, vertebrate paleontology
Ho, Grace Ping-Poo, mathematics
James, Hugo A., parasitology; helminthology
Kim, Byong M., internal medicine
Larsen, Karl Davis, physics, acoustics
Lobdell, David Hill, pathology
Mayer, Stuart Allan, inorganic chemistry
McCarroll, Bruce, physical chemistry
Mellor, John, physical chemistry
Meyer, Goldye M., science education
Moran, Joseph Francis, Jr., physiology
Paul, Jeddeo, biochemistry, analytical chemistry
Perillie, Pasquale E., medicine
Pistey, Warren R., zoology; medicine
Poliubovich, John Jacob, physiological ecology
Prober, Maurice, organic chemistry, polymer chemistry
Raghuvir, Nuggehalli Narayana, entomology
Rao, Valluru Bhavanarayana, mathematical statistics, operations research
Sethi, Dhanwant S., physical chemistry
Singletary, Robert Lombard, marine ecology

Somers, Michael Eugene, neurobiology, histology
Spiltoir, Charles Francis, Jr., mycology
Tucci, James Vincent, chemical physics
Tucker, Edmund Belford, physics
Zandy, Hassan F., physics, spectroscopy
Zuehlke, Richard William, physical chemistry, marine chemistry

BROOKFIELD
Denison, M Carl, organic chemistry, physical chemistry

BYRAM
Allen, Duff Shederic, Jr., organic chemistry

CENTERBROOK
Aikman, Edward Percy, physics

CHESHIRE
Barrante, James Richard, physical chemistry
Beard, Raimon Lewis, entomology
Klanica, Andrew Joseph, industrial chemistry
Roscoe, John Stanley, Jr., physical inorganic chemistry

CLINTON
Ellis, John George, bacteriology
Hunt, Richard Henry, chemistry
Markland, William R., cosmetic chemistry

COLLINSVILLE
Warner, Frederic William, anthropology, archaeology

COS COB
McPherson, James Beverley, Jr., organic chemistry

DANBURY
Adler, Alan David, molecular biology, biophysical chemistry
Beck, Paul W., physical chemistry
Brant, Arthur Albert, geophysics
Brunell, Gloria Florette, mathematics
Denison, Ruth Corbet, chemistry
Dye, Frank J., cytology
Groff, Donald William, geoenvironmental science
Hines, Paul Steward, organic chemistry
Kendziorski, Francis Richard, nuclear physics
Kreizinger, Jean Dolloff, genetics
Landskroener, Peter Armstrong, physical chemistry
LeMay, Charlotte Zihlman, solid state physics
Lewis, Otis Griffin, polymer chemistry
Lu, Phillip Kehwa, astronomy, physics
Moreland, Parker Elbert, Jr., applied physics, engineering management
Nabighian, Misac N., geophysics
Nelligan, William Bryon, geophysics
Turk, Amos, organic chemistry
Vijayendran, Bheema R., physical chemistry
Wasson, Robert Gordon, ethnomycology

DARIEN
McDonough, Everett Goodrich, organic chemistry
Mertz, Robert Theodore, applied mathematics
Reichertz, Paul Peter, physics
Salce, Ludwig, organic chemistry
Saunders, Kenneth Worden, chemistry
Schuster, Nick August, physics, electronics
Thomas, Walter Moreland, chemistry
Wall, Robert Allen, physical chemistry
Webb, Richard Lansing, polymer chemistry

EAST HAMPTON
Hamblen, David Gordon, experimental physics
Sommerman, Kathryn Martha, entomology

EAST HARTFORD
Ard, William Bryant, Jr., physics
Bowman, Craig Thomas, chemical kinetics, gas dynamics

Bronfin, Barry Robert, research administration, physical chemistry
Brooks, Clyde S., physical chemistry
Brown, Charles Thomas, physical chemistry
Fader, Walter John, theoretical physics
Galasso, Francis Salvatore, solid state chemistry, inorganic chemistry
Glenn, William Henry, Jr., physics
Golden, Gerald Seymour, analytical chemistry
Grubin, Harold Lewis, physics
Haught, Alan F., plasma physics
Hobson, Melvin Clay, Jr., physical chemistry
Kellner, Jordan David, physical chemistry
Kohlmayr, Gerhard Franz, mathematical physics
Kolker, Harold Jerrold, theoretical chemistry
Kriege, Owen Hobbs, applied chemistry
Meyerand, Russell Gilbert, Jr., plasma physics
Michels, Horace Harvey, chemical physics
Nilson, Edwin Norman, mathematics
Otocka, Edward Paul, polymer chemistry
Otter, Fred August, Jr., solid state physics
Peterson, Gerald A., theoretical physics, solid state physics
Pike, Roscoe Adams, organic chemistry
Seery, Daniel J., chemical kinetics, fuel
Shuskus, Alexander J., solid state physics
Sogliero, Luigina Cianfarani, mathematical statistics
Solomon, Peter R., solid state physics
Wingfield, Edward Christian, physics, science administration
Yntema, George Busey, physics

EAST LYME
Larrabee, Clifford Everett, organic chemistry

ENFIELD
Baum, Bernard, polymer chemistry
Nickerson, Mortimer Henderson, organic chemistry, research administration

FAIRFIELD
Barone, John A., organic chemistry
Barone, Milo C., animal physiology
Barske, Philip, wildlife management
Boggio, Joseph E., physical chemistry
Bongiorno, Salvatore F., vertebrate ecology
Elder, John William, organic chemistry
Hadjimichael, Evangelos, theoretical nuclear physics
Harms, Edward Albert, nuclear physics
Khadjavi, Abbas, atomic physics, optics
Lang, George E., Jr., topology
Lazaruk, William, horticulture, biology
Lee, Thomas Henry, physics
MacDonald, John Chisholm, analytical chemistry
McElaney, James H., physics
Merrin, Seymour, physical chemistry, geochemistry
Newton, Victor Joseph, theoretical physics
O'Connell, Edmond J, Jr., organic chemistry, photochemistry
Olah, Arthur Frank, plant physiology, phytopathology
O'Neill, Edward John, topology
Pulito, Aldo Martin, chemistry
Rice, Frank J., genetics, biostatistics
Ross, Donald Joseph, physiological chemistry
Shaffer, Dorothy Browne, mathematics
Skutnik, Bolesh Joseph, physical chemistry
Vanderslice, Thomas Aquinas, physical chemistry
Verses, Christ James, physiology, molecular biology
Wong, Maurice King Fan, mathematical physics

FARMINGTON
Becker, Elmer Lewis, immunology
Blechner, Jack Norman, obstetrics & gynecology
Bleich, Hermann Ewald, physical biochemistry
Bronner, Felix, physiology, biophysics
Burstone, Charles Justin, orthodontics
Cancro, Robert, psychiatry
Catalanotto, Frank Alfred, dental research, pedodontics
Chan, Lee-Nien Lillian, developmental biology, physiology

Chan, Teh-Sheng, human genetics, physiology
Chang, Lucy Ming-Shih, biochemistry
Cinti, Dominick Louis, physiology, pharmacology
Cohen, Stanley, immunology, pathology
Cooke, Peter Hayman, cytology
Cooperstein, Sherwin Jerome, anatomy, cell physiology
Cutler, Leslie Stuart, developmental biology, pathobiology
D'Amato, Donald Paul, physics
Damjanov, Ivan, pathology
Deutscher, Murray Paul, biochemistry, molecular biology
Fabrikant, Irene Berger, microbiology
Feinstein, Maurice B., pharmacology
Felsenfeld, Herbert William, biochemistry, pharmacology
Fleeson, William, medicine, psychiatry
Foulds, John Douglas, biochemistry, molecular biology
Gaintner, John Richard, internal medicine
Gay, Thomas John, oral biology, acoustics
Glasel, Jay Arthur, biochemistry, physical chemistry
Golub, Ellis Eckstein, biochemistry
Henderson, Edward George, biophysics, pharmacology
Hinz, Carl Frederick, Jr., internal medicine, hematology
Hoffman, Herbert Saul, physics, medicine
Hohnadel, David Charles, clinical chemistry, biochemistry
Jungas, Robert Leando, biochemistry, physiology
Kalt, Marvin Robert, anatomy, cell biology
Kolakowski, Donald Louis, psychometrics, quantitative genetics
Kollar, Edward James, oral biology, embryology
Kosher, Robert Andrew, developmental biology
Krutchkoff, David James, oral pathology, dentistry
Langeland, Kaare, dental materials, experimental pathology
Lepow, Irwin Howard, immunology, experimental medicine
Levine, Philip Theodore, protein chemistry, oral biology
Lobl, Richard Tolstoi, anatomy, endocrine physiology
Loeser, Charles Nathan, cytology, neuroanatomy
Maenza, Ronald Morton, pathology, electron microscopy
Maher, John Francis, medicine, nephrology
Mann, Lewis Theodore, Jr., clinical biochemistry, immunochemistry
Markowitz, Milton, pediatrics
Massey, Robert Unruh, internal medicine
Miyamoto, Michael Dwight, neuropharmacology
Morse, Edward Everett, medicine, hematology
Mumford, George, prosthodontics
Nalbandian, John, oral pathology, pedodontics
Nanda, Ravindra, orthodontics
Oliver, Janet Mary, cell physiology
O'Rourke, James, ophthalmology
Osborn, Mary Jane, molecular biology, biochemistry
Owens, Guy, neurosurgery, anatomy
Pappano, Achilles Gus, Jr., pharmacology
Patterson, John Ward, physiology, medicine
Peterson, Larry James, oral surgery
Pfeiffer, Steven Eugene, cell biology, neurosciences
Phelps, Creighton Halstead, neurocytology
Poole, Andrew E., human genetics, biochemistry
Poyton, Robert Oliver, molecular biology, microbiology
Raisz, Lawrence Gideon, internal medicine, endocrinology
Reid, Parliane John, biochemistry, microbial genetics
Reiskin, Allan B., experimental pathology, oncology
Ressler, Charlotte, organic chemistry, biochemistry
Rothenberg, Albert, psychiatry
Rothfield, Lawrence I., microbiology, biochemistry
Rothfield, Naomi Fox, medicine
Seewald, David Allan, biophysical chemistry
Setlow, Peter, biochemistry
Sha'afi, Ramadan Issa, biophysics, physiology

Shaskan, Edward Gregory, neurochemistry, neuropharmacology
Sheetz, Michael Patrick, biophysical chemistry
Solonche, David Joshua, biomedical engineering, orthodontics
Spencer, Richard Paul, nuclear medicine, biochemistry
Spitznagle, Larry Allen, bionucleonics
Sunderman, Frederick William, Jr., clinical pathology, experimental pathology
Tanzer, Marvin Lawrence, biochemistry
Taubman, Sheldon Bailey, immunology, pathology
Topazian, Richard G., oral surgery
Venham, Larry Lee, pedodontics, psychology
Volle, Robert Leon, pharmacology
Votaw, Robert Grimm, medical education
Walker, James Elliot Cabot, internal medicine
Wampler, Donald Eugene, biochemistry
Ward, Peter A., pathology, immunology
Watkins, Dudley T., anatomy
Weinstein, Sam, orthodontics
Woronick, Charles Louis, biochemistry, clinical chemistry
Wu, Henry Chi-Ping, biochemistry
Yaeger, James Amos, histology
Yoshida, Takeshi, immunopathology

GALES FERRY
Lees, Thomas Masson, biochemistry
Miles, Walter Richard, physiological psychology

GLASTONBURY
Bertinuson, Torvald Arthur, agronomy
Burlew, John Swaim, physical chemistry
Carroll, James Barr, science administration, science policy
DeSanto, Robert Spilka, environmental management, ecology
McCune, Robert Franklin, physics
Rishell, William Arthur, poultry breeding
Scola, Daniel Anthony, chemistry, materials science
Ultee, Casper Jan, molecular spectroscopy, physical chemistry

GREENWICH
Anderson, Wilmer Clayton, physics
Blodgett, Robert Bell, chemistry
Boehm, Robert Louis, paper chemistry, pulp chemistry
Colclough, Norma Vesey, pharmacology, toxicology
Conolly, John R., geology, natural resources
Fehr, Robert O., acoustics
Hagner, Arthur Feodor, geology, mineralogy
Hall, James Roger, chemistry
Hunter, George Truman, physics
Johnson, Frank Walker, geology
Knot, Donald MacMillan, industrial organic chemistry
Kohn, Gaston G., food technology, agronomy
Lohuis, Delmont John, chemistry
McClenachan, Ellsworth, organic chemistry
McKirahan, Richard Duncan, organic chemistry, environmental management
Rusden, Philip Lowry, plant pathology
Simon, Dorothy Martin, physical chemistry
Smith, Rodger Chapman, inorganic chemistry
Tschudi, Wilbur James, inorganic chemistry
Wendricks, Roland N., physical chemistry
Wier, Charles Eugene, economic geology

GROTON
Ackerman, Neil Richard, pulmonary diseases, immunology
Aguiar, Armando Joseph, physical pharmacy
Althuis, Thomas Henry, organic chemistry, medicinal chemistry
Baird, James Leroy, Jr., comparative physiology
Bereboom, John Joseph, organic chemistry
Belleire, John Lewis, medicinal chemistry
Bigham, Eric Cleveland, medicinal chemistry
Bindra, Jasjit Singh, organic chemistry
Blackwood, Robert Keith, organic chemistry

Bloom, Barry Malcolm, organic chemistry, medicinal chemistry
Bohlen, Walter Franklin, physical oceanography
Borden, George Wayne, organic chemistry, chemical engineering
Broida, Saul, physical oceanography
Buckley, Jay Selleck, Jr., information science
Capotosto, Augustine, Jr., chemistry
Celmer, Walter Daniel, bio-organic chemistry
Chang, Yi-Han, biochemical pharmacology, immunology
Chiburis, Edward Frank, geophysics
Chmurny, Alan Bruce, organic chemistry
Collins, Alfred Patterson, pharmaceutical chemistry
Constantine, Jay Winfred, physiology
Cooke, John Cooper, mycology
Crawford, Thomas Charles, organic chemistry
Cronin, Timothy H., organic chemistry
Curley, James Edward, analytical chemistry
Czuba, Leonard J., medicinal chemistry
Dominy, Beryl W., organic chemistry, information science
Dooley, Joseph Francis, clinical chemistry
Doshan, Harold David, drug metabolism, organic chemistry
Dowling, John J., geophysics
Eastwood, DeLyle, physical chemistry
English, Arthur Robert, bacteriology
Faubl, Hermann, organic chemistry
Figdor, Sanford Kermit, organic chemistry
Fitzgerald, William Francis, chemical oceanography, marine geochemistry
Forcier, George Arthur, analytical chemistry
Garvine, Richard William, physical oceanography
Gasteyer, Charles Earl, astronomy
Gauthier, George James, pharmaceutical
Gell, Charles Fredric, aerospace medicine
Gordon, Philip Newton, plant science
Gupta, Shyam Kirti, medicinal chemistry
Hamsher, James J., bio-organic chemistry
Harber, Charles A., organic chemistry, medicinal chemistry
Hess, Hans-Jurgen Ernst, organic chemistry
Hobbs, Donald Clifford, biochemistry
Hoffman, William Wheeler, virology
Hohnke, Lyle Ashby, physiology
Holland, Gerald Fagan, medicinal chemistry
Howes, Harold, Jr., parasitology
Izzo, Anthony Joseph, hydroacoustics
Johnson, Michael Ross, organic chemistry
Jolly, Ramesh C., food science, dairy science
Kasubick, Robert Valentine, organic chemistry
King, Theodore Oscar, toxicology; pharmacology
Kirchmeier, Robert Lynn, fluorine chemistry, analytical chemistry
Kita, Donald Albert, biochemistry
Koch, Richard Carl, medicinal chemistry
Koe, B Kenneth, organic chemistry
Korst, James Joseph, organic chemistry
Larson, Jerry King, clinical pharmacology
Lipinski, Christopher Andrew, organic chemistry
Lynch, John Edward, bacteriology, parasitology
McFarland, James William, medicinal chemistry
McIlhenny, Hugh M, drug metabolism
McLamore, William Merrill, medical research
Milne, George McLean, Jr., medicinal chemistry, pharmacology
Moore, Peter Francis, pharmacology
Moppet, Charles Edward, organic chemistry
Moreland, Walter Thomas, Jr., medicinal chemistry
Murai, Kotaro, organic chemistry
Nelson, Roger Peter, organic chemistry
Niblack, John Franklin, pharmacology, infectious diseases
Pan, Shih Y., pharmacology, toxicology
Pereira, Joseph, biochemistry
Pinson, Rex, Jr., biochemical pharmacology
Proctor, Alan Ray, genetics
Rash, Jay Justen, metabolism
Ray, Verne A., microbiology, biochemistry
Rennhard, Hans Heinrich, organic chemistry
Retsema, James Allan, biochemistry, microbiology
Rosati, Robert Louis, medicinal chemistry
Routien, John Broderick, microbiology
Sardinas, Joseph Louis, microbiology, biochemistry
Sarges, Reinhard, organic chemistry
Scanio, Charles John Vincent, organic chemistry
Schaaf, Thomas Ken, medicinal chemistry
Schach Von Wittenau, Manfred, organic chemistry, drug metabolism
Schaefer, Karl Ernst, physiology
Schnur, Rodney Caughren, medicinal chemistry, synthetic organic chemistry
Seeley, Donald Bernard, microbiology
Seo, John S., bacteriology, mycology
Tappan, Donald Vester, biochemistry
Tate, Bryce Eugene, organic chemistry
Taylor, James A., Jr., biochemistry
Thomas, Paul David, organic chemistry
Tien, Weichen, microbial genetics, industrial microbiology
Tretter, James Ray, organic chemistry
Truesdail, Susan Jane, molecular biology
Upton, Ronald P., analytical chemistry
Walsh, Alexander Hamilton, veterinary pathology
Weissman, Albert, psychopharmacology
Welch, Willard McKowan, Jr., organic chemistry
Whipple, Earl Bennett, physical chemistry
Wilcox, Roberta Arlene, medical statistics
Wiseman, Edward H., biochemical pharmacology

GUILFORD
Drinkard, Russell Drew, physical chemistry
Drobnyk, John Wendel, geology
Kenson, Robert Earl, physical chemistry
Richards, Frederic Middlebrook, protein chemistry
Stuart, Alfred Herbert, photographic chemistry

HADDAM
Houston, Walter Scott, astronomy, meteoritics

HAMDEN
Beitch, Irwin, embryology, cytology
Bernard, Gary Dale, vision, physiological optics
Bernstein, Richard Fernand, zoology
Borst, Daryll C., limnology
Coppola, Elia Domenico, analytical chemistry
Darling, George Bapst, public health
Doede, Dorothy Ruth, bacteriology
Fenney, Nicholas William, pharmacy
Garratt, George Alfred, forestry
Harris, Lawrence, academic administration
Houston, David Royce, forest pathology
Hoy, Marjorie Ann, insect ecology, acarology
Jacobson, Samuel, chemistry
Johnson, Doris, nutrition
Knollmueller, Karl Otto, organic chemistry
Levine, Harvey Robert, parasitology, entomology
MacMullen, Clinton William, organic chemistry
Mazzone, Horace M., biochemistry
McArthur, Richard Edward, organic chemistry
Milford, Alan Hackney, industrial chemistry
Nigam, Lakshmi Narayan, mechanics, mathematics
Parker, Johnson, plant physiology
Parr, Albert Eide, zoology, oceanography
Payne, Irving John, microbiology
Repat, Arthur Jack, protozoology
Ritchie, Brenda Rachel, physiology
Schroeder, Hansjuergen Alfred, organic chemistry
Tilton, Richard C., microbiology
Toomey, James Michael, otorhinolaryngology
Van Stone, James Morril, zoology
Walde, Ralph Eldon, mathematics
Weina, Cecilia, mathematics
Wetstone, Howard J., internal medicine, biochemistry
Wiesel, Benjamin, psychiatry, neurology
Wilt, John Robert, physical chemistry

HARTFORD
Barrett, Harold Spencer, preventive medicine, public health
Bavisotto, Vincent, chemistry
Bennett, John Henry, applied mathematics
Bobko, Edward, organic chemistry
Braceland, Francis James, psychiatry
Brewer, Robert Hyde, animal ecology, invertebrate zoology
Burnett, Robert Walter, clinical chemistry, analytical chemistry
Castaldi, Cosmo Raymond, dentistry
Chaturvedi, Rama Kant, bio-organic chemistry
Child, Frank Malcolm, cell biology
Clayton-Hopkins, Judith Ann, endocrinology
Collins, Michael Frederick, developmental biology
Constant, Frank Woodbridge, environmental physics
Crawford, Richard Bradway, biochemistry
Day, James Meikle, environmental chemistry
DePhillips, Henry Alfred, Jr., physical chemistry
Donnelly, John, psychiatry
Ellis, Robert Homer, research administration, academic administration
Foster, James Henry, surgery
Freymann, John Gordon, medicine
Friedman, Don Gene, meteorology
Galbraith, Donald Barrett, developmental genetics
Ganis, Frank Michael Gangarosa, biochemistry
Glueck, Bernard Charles, psychiatry
Gregory, Brooke, low temperature physics
Haffner, Rudolph Eric, biological oceanography
Hanson, Henry Nicholas, low temperature physics
Harris, Russell G., oceanography, meteorology
Heeren, James Kenneth, organic chemistry
Honeyman, Merton Seymour, genetics
Hukill, Peter Biggs, pathology
Klimczak, Walter John, mathematics
Klock, Peter Alan, molecular genetics
Kornfeld, Joseph M., microbiology
Lewis, George W., internal medicine, microbiology
Lindsay, Robert, magnetism
Lloyd, Douglas Seward, public health
Moyer, Ralph Owen, Jr., inorganic chemistry
Natarajan, Kottayam Viswanathan, oceanography, microbiology
Noll, Clifford Raymond, Jr., biochemistry
Pandolfo, Joseph P., meteorology
Picker, Harvey Shalom, theoretical physics
Poliferno, Mario Joseph, mathematics
Renfro, William Charles, biological oceanography
Robinson, George David, atmospheric physics
Ross, Martin Russell, virology, immunology
Schlessinger, Gert Gustav, water chemistry
Schneider, Craig William, phycology
Scoville, William Beecher, neurosurgery
Servadio, Gildo Joseph, food science
Shorr, Bernard, operations research
Skopek, Jerry, bacteriology
Smelie, Robert Henderson, Jr., physical chemistry
Steele, Frances M., virology, microbiology
Stewart, Robert Clarence, mathematics
Stolman, Abraham, toxicology
Stroebel, Charles Frederick, III, psychophysiology, neurophysiology
Sulavik, Stephen B., medicine
Tanzer, Jason Michael, dentistry, microbial physiology
Thiruvengada, Seshan, chemistry, physical chemistry
Tokita, Noboru, physics
Whelan, William Paul, Jr., organic chemistry

HAZARDVILLE
White, Leroy Albert, physical chemistry

IVORYTON
Bongiorni, Domenic Frank, physical organic chemistry

LAKEVILLE
Bodel, John Knox, physical anthropology

LEBANON
Jahoda, William John, biology

MANCHESTER
Bandes, Herbert, electrochemistry
Klock, Peter Allan, molecular genetics

MANSFIELD CENTER
Joranson, Philip Nathaniel, conservation
Starke, Albert Carl, Jr., organic chemistry

MERIDEN
East, Larry Verne, physics

MIDDLEBURY
Andersen, Paul George, polymer science
Carr, Clide Isom, physical chemistry, polymer chemistry
Davison, John Alden, polymer chemistry, chemical engineering
DeDecker, Hendrik Kamiel Johannes, physical chemistry
Johnson, Arnold Nathaniel, microscopy
Klingbeil, Werner Walter, applied mechanics, rheology
Kreis, Ronald W., physical medicine
Liao, Sung Jui, physical medicine
McGuinness, James Anthony, organic chemistry
Nebb, Robert Gilman, organic chemistry, organic polymer
O'Shea, Francis Xavier, organic polymer chemistry
Peascoe, Warren Joseph, organic chemistry, polymer chemistry
Reid, William John, polymer chemistry, physical chemistry
Relyea, Douglas Irving, organic chemistry
Rim, Yong Sung, organic chemistry
Rinehart, Robert Eugene, organic chemistry
Roberts, George P., polymer chemistry
Skewis, John David, physical chemistry
Smith, Wendell Vandervort, physical chemistry
Stoner, Allan Wilbur, physical chemistry
Strunk, Richard John, organic chemistry
Sundholm, Norman Karl, polymer chemistry

MIDDLETOWN
Allewelt, Norma Mary, biology
Baierlein, Ralph Frederick, cosmology
Baker, Jeffrey John Wheeler, developmental biology, science writing
Bannerman, Harold MacColl, geology, mineralogy
Berlind, Allan, zoology
Berry, Spencer Julian, insect physiology
Bett, John Alexander Stuart, physical chemistry
Burford, Mortimer Gilbert, physical chemistry
Cochrane, Vincent Winner, microbiology
Comfort, William Wistar, mathematics
Crampton, Earle Wilcox, nutrition
DeBoer, Jelle, geophysics
Donady, John James, developmental genetics
Felten, Edward J., chemical metallurgy
Firshein, William, microbiology
Fry, Albert Joseph, organic chemistry
Gomez-Ibanez, Jose Daniel, chemistry
Gortner, Ross Aiken, Jr., biochemistry
Gottschalk, Walter Helbig, mathematics
Green, Robert Holt, medicine
Haake, Paul C., chemistry
Hager, Anthony Wood, topology
Hall, William A., medicine, environmental health
Hanson, Earl Dorchester, cell biology, evolutionary biology
Horne, Gregory Stuart, marine geology, stratigraphy
Infante, Anthony A., biochemistry

Jacobi, Peter Alan, synthetic organic chemistry
Kiefer, Barry Irwin, developmental biology
Kinoshita, Kimio, electrochemistry
Kohler, Bryan Earl, physical chemistry
Larsen, Ronald John, mathematical analysis
Lindquist, Fred E J, mathematics
Linton, Fred G J, physical chemistry
Lowry, Eric G., mathematics
Lukens, Lewis Nelson, biochemistry
Lynch, Carol Becker, behavioral genetics
McAllester, David Park, anthropology
McIntosh, John Stanton, theoretical physics
Morgan, Thomas Joseph, atomic physics
Pawlowski, Philip John, cell biology
Pessl, Fred, Jr., glacial geology, environmental geology
Pringle, Wallace C Jr., physical chemistry
Reddy, William I., topology
Reid, James Dolan, mathematics
Rollefson, Robert John, low temperature physics, surface physics
Rosenbaum, Robert Abraham, mathematics
Sease, John William, electrochemistry
Singleton, Robert Richmond, mathematics
Stanton, Charles Madison, analytical mathematics
Tishler, Max, organic chemistry
Todd, Harold David, chemistry
Trefny, John Ulric, physics
Trousdale, William Latimer, physics
Turner, Robert Scott, Jr, developmental biology
Upgren, Arthur Reinhold, Jr, astronomy
Walker, Willard Brewster, ethnology, applied anthropology
Weissberger, Edward, inorganic chemistry, organometallic chemistry
Wharton, Peter Stanley, organic chemistry
Wood, Carol Saunders, mathematical logic

MILFORD
Brown, Robert Lawrence, geophysics, astronomy
Calabrese, Anthony, marine ecology
Desor, Jeannette Ann, psychophysics
Dodd, Charles Gardner, physical chemistry, materials science
Graikoski, John T., environmental microbiology
Hanks, James Elden, marine biology
Lisman, Frederick Louis, nuclear chemistry
Longwell, Arlene Crosby (Mazzone), genetics, cytogenetics
Rosenman, Irwin David, organic chemistry
Thurberg, Frederick Peter, physiology, marine biology
Ukeles, Ravenna, microbiology

MONTVILLE
Schwensfeir, Robert James, Jr., nuclear science

MT CARMEL
Davis, Kenneth Pickett, forest management
Haring, Robert Clinton, chemistry

NAUGATUCK
Abeling, Edwin John, bacteriology
Amidon, Roger Welton, organic chemistry
Ariyan, Zaven S., organic chemistry
Baorewicz, Wadim, organic chemistry
Bergen, Robert Ludlum, Jr., polymer chemistry
Brett, Thomas Joseph, Jr, polymer chemistry
Brown, James Allen, Jr, medical microbiology, pharmacology
Brown, Robert Walter, chemistry
Cornell, Robert Joseph, polymer chemistry
Covey, Rupert Alden, organic chemistry
Easterbrook, Eliot Knights, polymer chemistry, organic chemistry
Greenfield, Harold, organic chemistry
Hawley, Thomas G, Jr, organic chemistry
Klender, Gerald, plastics chemistry
Knapp, Robert Lester, organic chemistry
McCleary, Charles David, chemistry
Morris, Harris Lee, physical chemistry, chemical engineering
Nektutin, Vadim Constantin, organic chemistry, polymer chemistry

O'Brien, John Terence, organic polymer chemistry
Plant, Howard Leon, organic chemistry
Smith, Allen Elston, pesticide chemistry
von Schmeling, Bogislav G, agriculture

NEW BRITAIN
Anderson, Herbert Godwin, Jr, biological sciences
Bissett, Orville R, plant physiology, forestry
Buckwold, Sidney Joshua, inorganic chemistry, analytical chemistry
Carluccio, Leeds Mario, botany, morphology
Cash, Rowley Vincent, history of chemistry
Chichester, Lyle Franklin, zoology
De Nuccio, David Joseph, medical physiology
Douville, Phillip Raoul, physical chemistry
Eisenberg, Sidney Edwin, medicine
Fiore, Carl, physiology, zoology
Fu, Wei-ming, cytogenetics
Goodstein, Madeline P, science education, chemistry
Gorski, Leon John, ecology, biology
Groth, Joyce Lorraine, analytical chemistry
Groth, Richard Henry, environmental chemistry
Howard, Rufus Oliver, medicine, ophthalmology
Krause, Adrienne Wickenden, inorganic chemistry
Lee, Thomas W, comparative physiology
Newton, David C, apiculture, animal behavior
Novoa, William Brewster, biochemistry
Ostrander, Darl Reed, limnology
Pereyda, Andrew Daniel, physical geography, geography of the Soviet Union
Salamon, Richard Joseph, science education
Shine, Timothy D, organic chemistry
Slotnick, Herbert, chemical engineering, physical chemistry
Tozloski, Albert Henry, economic entomology, invertebrate zoology

NEW CANAAN
Hanford, William Edward, chemistry
Harrison, Thomas Southworth, physical chemistry
Megrue, George Henry, geochemistry, research administration
Price, Whitfield, physical chemistry
Stubblefield, Robert Lee, psychiatry
Voorhies, John Davidson, surface chemistry

NEW HAVEN
Aaboe, Asger (Hartvig), mathematics
Adair, Robert Kemp, nuclear physics
Adelberg, Edward Allen, genetics
Aghajanian, George Kevork, neuropharmacology
Agrawal, Krishna Chandra, medicinal chemistry, pharmacology
Aitken, Thomas Henry Gardiner, medical entomology, parasitology
Alben, Richard (Samuel), solid state physics
Altman, Sidney, molecular biology
Anderson, John Fredric, entomology
Anderson, Michael Peter, algebra
Andrews, Stephen Brian, cell biology, biophysical chemistry
Andriole, Vincent T., internal medicine, infectious diseases
Ankel-Simons, Friderun Annursel, primatology, comparative anatomy
Anscombe, Francis John, statistics
Atkins, Elisha, internal medicine
Auer, Lawrence H, astrophysics
Babice, John Stanley, Jr, industrial chemistry, research administration
Bachmann, Barbara Joyce, microbial genetics
Barash, Paul G, anesthesiology
Bardeen, James Maxwell, theoretical astrophysics
Barnett, Russell Joffree, anatomy
Barthold, Stephen William, veterinary pathology
Baue, Arthur Edward, surgery, cardiovascular physiology
Baumgarten, Alexander, immunology
Bausher, Larry Paul, organic chemistry

Behrman, Harold R, physiology, biochemistry
Bennett, William Ralph, Jr, physics
Beringer, Robert, physics
Berkowitz, Steven Arlen, biophysical chemistry
Berliner, Robert William, physiology, internal medicine
Berlyn, Graeme Pierce, plant anatomy, tree physiology
Berlyn, Mary Berry, genetics
Berner, Robert A., marine geochemistry
Bernstein, Ira Borah, theoretical physics
Berson, Jerome Abraham, organic chemistry
Bertino, Joseph R, pharmacology, medicine
Bevill, Rardon Dixon, III, biochemistry, clinical chemistry
Binder, Henry Joseph, internal medicine, gastroenterology
Black, Francis Lee, virology
Bliss, Chester Ittner, biometrics
Bloodgood, Robert Alan, cell biology
Bockelman, Charles Kincaid, nuclear physics
Bodel, Phyllis T, internal medicine, infectious diseases
Boell, Edgar John, biology
Bormann, Frederick Herbert, ecology
Bouhuys, Arend, air pollution, pneumology
Bove, Joseph Richard, hematology, laboratory medicine
Braverman, Irwin Merton, medicine, dermatology
Brenneman, Faith Nielsen, microbiology
Bromley, David Allan, nuclear physics
Bromley, Mary Elizabeth, botany
Byck, Robert, pharmacology, psychiatry
Cafarelli, Enzo Donald, exercise physiology
Calabresi, Massimo, pathology
Canellakis, Evangelo S, biochemistry
Canellakis, Zoe Nakos, biochemistry
Capizzi, Robert L, oncology, clinical pharmacology
Carter, David Martin, dermatology
Casals, Jordi, virology
Chang, Kwang-Chih, anthropology, archaeology
Chang, Pauline (Wuai) Kimm, organic chemistry
Chang, Richard Kounai, solid state physics, quantum electronics
Cheney, Charles Brooker, obstetrics & gynecology
Chonacky, Norman J, physics
Chupka, William Andrew, physical chemistry
Clark, Sydney P, Jr, geophysics
Clutter, Mary Elizabeth, botany
Coe, Michael Douglas, anthropology, archaeology
Cohart, Edward Maurice, public health
Cohen, Lawrence Baruch, neurophysiology
Cohen, Lawrence Sorel, internal medicine, cardiology
Cohen, Melvin Joseph, microbiology
Cole, Jack Westley, experimental surgery
Coleman, Joseph Emory, biochemistry, biophysics
Coleman, Jules Victor, psychiatry
Collins, Stephen, ecology
Collins, William Francis, Jr, neurosurgery
Colson, Steven Douglas, chemical physics
Comer, James Pierpont, child psychiatry, public health administration
Cone, Robert Edward, immunology
Conklin, Harold Colyer, anthropology, ethnography
Conn, Harold O, internal medicine
Connolly, John Joseph, physiology, immunology
Cooper, Franklin Seaney, speech, communication science
Cooper, Jack Ross, pharmacology
Cosenza, Benjamin John, bacteriology
Coward, James Kenderdine, bio-organic chemistry, medicinal chemistry
Crawshaw, Larry Ingram, comparative physiology
Creasey, William Alfred, biochemistry, pharmacology
Crelin, Edmund Slocum, human anatomy, human development
Cronan, John Emerson, Jr, biochemistry, molecular biology
Cronin, Michael Thomas Ignatius, pathology
Crothers, Donald M, biophysical chemistry
Csejka, David Andrew, physical chemistry, analytical chemistry
Cullen, Mary Urban, genetics
D'Atri, David Albert, epidemiology
Day, Peter Rodney, genetics

Delgado, Jose Manuel Rodriguez, neurophysiology
Demarque, Pierre, astrophysics
Dillard, Morris, Jr., metabolism, internal medicine
Di Pasquale, Albert Martin, cell biology
Doane, Charles Chesley, entomology
Doane, Winifred Walsh, developmental genetics
Douglas, William Wilton, neuropharmacology, neuroendocrinology
Downing, Shirley Evans, pathology
Downs, Wilbur George, virology
DuBois, Arthur Brooks, physiology
Duff, Raymond Stanley, pediatrics, sociology
Ebbert, Arthur, Jr., internal medicine
Edwards, Kathryn Louise, plant physiology
Eisenfeld, Arnold Joel, pharmacology, internal medicine
Eisenstadt, Jerome Melvin, biochemistry, human genetics
Engelman, Donald Max, biophysics, biochemistry
Ernst-Fonberg, Marylou, biochemistry, epidemiology
Evans, Alfred Spring, epidemiology, internal medicine
Evans, Glenn Thomas, chemical physics
Falk, Isidore Sydney, public health
Faller, John William, inorganic chemistry, organometallic chemistry
Farmer, Douglas Alexander, surgery
Farquhar, Marilyn Gist, cell biology, experimental pathology
Faust, John Philip, water chemistry
Feinstein, Alvan Richard, internal medicine
Feit, Sidnie Marilyn, mathematics
Feit, Walter, mathematics
Felig, Philip, internal medicine, endocrinology
Fenn, John Bennett, chemistry
Finch, Stuart Cecil, internal medicine
Firk, Frank William Kenneth, nuclear physics
Fischer, Diana Bradbury, statistics, epidemiology
Fischer, James Joseph, radiobiology, radiotherapy
Fixman, Marshall, theoretical chemistry
Fleck, Stephen, psychiatry
Fleischer, Joseph, physical chemistry
Fletcher, Paul Litton, Jr., protein chemistry
Forbes, Thomas Rogers, anatomy
Frink, Charles Richard, soil chemistry
Fruton, Joseph Stewart, biochemistry
Fuoss, Raymond Matthew, physical chemistry
Furnival, George Mason, forestry
Gagge, Adolf Pharo, biophysics, physiology
Gale, Paula Jane, chemical physics
Gall, Joseph Grafton, cell biology
Galston, Arthur William, plant physiology
Garen, Alan, developmental biology
Garfinkel, Boris, astronomy
Garland, Howard, mathematics
Gavin, David Francis, synthetic organic chemistry
Gee, J Bernard L, medicine, physiology
Gerit, John Alan, biochemistry
Gershon, Richard K, pathology, immunology
Giebisch, Gerhard Hans, physiology
Gilliam, James Melvin, biophysics, cell biology
Gillis, Charles Norman, pharmacology
Gilman, Alfred, pharmacology
Ginocchio, Joseph Natale, theoretical nuclear physics
Glaser, Gilbert Herbert, neurology
Glassner, Martin Ira, geography
Glenn, William Wallace Lumpkin, surgery
Godson, Godfrey Nigel, molecular biology, molecular genetics
Goldenberg, Ira Stovin, medicine
Goldsmith, Mary Helen Martin, biology, plant physiology
Goldsmith, Timothy Henshaw, visual physiology
Gonzalez, Richard Rafael, physiology
Goodyer, Allan Victor, cardiovascular diseases
Gordon, Robert Boyd, geophysics, engineering
Gottschalk, Alexander, radiology, nuclear medicine
Gould, Sydney Ward, taxonomy
Green, Barry George, psychophysics
Greenberg, Jack Sam, physics
Greene, Nicholas Misplee, anesthesiology
Greengard, Paul, biochemistry
Greenspan, Richard H, radiology

Guries, Raymond Paul, forest genetics
Haakonsen, Harry Olav, science education
Halaban, Ruth, genetics, developmental biology
Haller, Gary Lee, physical chemistry
Handschumacher, Robert Edmund, pharmacology
Hankin, Lester, biochemistry
Hanna, Joseph Gordon, analytical chemistry
Hanson, Kenneth Ralph, biochemistry
Hanson, Robert Bruce, astronomy
Harris, Benedict Richard, medicine
Harris, William Edgar, astronomy
Hartigan, John A, mathematical statistics
Hartman, Willard Daniel, invertebrate zoology
Havir, Evelyn A, biochemistry
Hawkins, Morris, Jr, human genetics
Hayes, Mark Allan, surgery
Hayslett, John P, physiology
Hedlund, Gustav Arnold, mathematics
Heichel, Gary Harold, plant physiology
Henderson, Louis E, biochemistry
Henderson, Richard, toxicology
Hendler, Ernesto Danilo, internal medicine
Heninger, George Robert, psychiatry
Henisz, Jerzy Emil, psychiatry
Herzenberg, Arvid, theoretical physics
Higgins, Joan Amy, biochemistry, cell biology
Hill, David Easton, soil science
Hirshfield, Jay Leonard, plasma physics
Hitchcock, Margaret, pharmacology, metabolism
Hoffleit, Ellen Dorrit, astronomy
Hoffman, Joseph Frederick, physiology
Hofrichter, Charles Henry, chemistry
Holeman, George Robert, health physics
Holford, Theodore Richard, biostatistics
Horrigan, Philip Archibald, inorganic chemistry
Horsfall, James Gordon, plant pathology
Horstmann, Dorothy Millicent, epidemiology, pediatrics
Horvath, Csaba Gyula, physical chemistry, chemical engineering
Howard-Flanders, Paul, biophysics
Howe, Roger, mathematics
Howlett, Kirby Smith, Jr, clinical medicine
Hoyt, Joseph Bixby, geography
Hsia, Yujen Edward, pediatrics, medical genetics
Hubbard, Ann Louise, cell biology
Hughes, Vernon Willard, physics
Hunter, David Emanuel, cultural anthropology, anthropological linguistics
Huszar, Gabor, biochemistry, biology
Hutchinson, Franklin, biophysics, biochemistry
Hutchinson, George Evelyn, zoology, ecology
Hutchinson, Robert Lynn, biology
Jacobson, Florence Dorfman, mathematics
Jacobson, Nathan, mathematics
Jacoby, Robert Ottinger, comparative pathology, immunology
Jamieson, James Douglas, cell biology
Jaynes, Richard Andrus, plant breeding
Jekel, James Franklin, epidemiology, public health
Jonas, Albert Moshe, pathology, laboratory animal medicine
Kakutani, Shizuo, mathematics
Kankel, Douglas Ray, developmental biology, neurobiology
Kantor, Fred Stuart, immunology
Kasgarian, Michael, pathology
Kasha, Henry, high energy physics, cosmic ray physics
Kauer, John Stuart, sensory physiology
Kaya, Harry Kazuyoshi, insect pathology
Kelsey, Jennifer Louise, epidemiology
Keyes, Thomas Francis, theoretical chemistry
Kidd, Kenneth Kay, population genetics, behavioral genetics
Kirchner, John Albert, otolaryngology
Klatskin, Gerald, internal medicine
Klaus, Sidney N, dermatology
Kleber, Herbert David, psychiatry
Klein, Martin Jesse, history of physics
Konigsberg, William Henry, biochemistry
Kraybill, Henry Lawrence, experimental high energy physics
Kring, James Burton, entomology
Krizek, Thomas Joseph, plastic surgery, reconstructive surgery
Kuslan, Louis Isaac, history of science
Ladanyi, Branka Maria, theoretical chemistry
Lande, Saul, biochemistry, organic chemistry

Lange, Robert Carl, nuclear medicine, radiological physics
Larson, Richard Bondo, astrophysics
Ledig, F Thomas, genetics
Lee, Ronnie, topology
Lengyel, Peter, biochemistry
Leonard, Martha Frances, pediatrics
Lepowsky, James Ivan, mathematics
Lerner, Aaron Bunsen, dermatology, biochemistry
Lerner, Marguerite Rush, dermatology
Levine, Robert John, internal medicine, pharmacology
Lichten, William Lewis, physics
Lidz, Theodore, psychiatry
Lifton, Robert Jay, psychiatry
Lindskog, Gustaf Elmer, surgery
Lippard, Vernon William, pediatrics
Lipsky, Seymour Richard, biochemistry, physical chemistry
Lounsbury, Floyd Glenn, ethnology
Low, Kenneth Brooks, Jr, genetics
Lowman, Robert Morris, radiology
Lynch, Wesley Clyde, neuropsychology
Lyons, Philip Augustine, physical chemistry
Lytton, Bernard, urology
Maas, James Weldon, psychiatry
MacClintock, Copeland, invertebrate paleontology
MacDowell, Samuel Wallace, physics
Macnab, Robert Marshall, biophysics
Magee, Paul Terry, genetics, biochemistry
Magnarelli, Louis Anthony, entomology
Maizell, Robert Edward, industrial chemistry
Malawista, Stephen E, internal medicine
Manuelidis, Elias Emmanuel, pathology, neuropathology
Margolin, Barry Herbert, mathematical statistics, applied statistics
Markert, Clement Lawrence, developmental genetics, enzymology
Martin, Geoffrey John, geograpmy
Massey, William S, mathematics
Matthews, Lee Drew, physics
McAllister, William Barriss, Jr, medicine
McBride, James Michael, physical organic chemistry
McClure, Mark Stephen, entomology, ecology
McClure, Robert D, astronomy
McClymont, John Wilbur, botany
McColl, James Renfrew, experimental solid state physics
McCollum, Robert Wayne, epidemiology
McGuire, Joseph Smith, Jr, medicine
McIntyre, John Lee, plant pathology
McKaye, Kenneth Robert, ecology, behavioral biology
McMurray, Walter Joseph, biochemistry
Mengel, John Geist, astronomy
Mergen, Francois, forest genetics
Merriman, Daniel, biological oceanography
Messing, Simon D, cultural anthropology, medical anthropology
Michael, Charles Reid, neurophysiology
Migdalski, Edward Charles, ichthyology
Miller, Patrick Martin, plant pathology
Miller, Richard Samuel, animal ecology
Miller, William Henry, biology
Milstone, Jacob Haskell, biochemistry, pathology
Mitchell, Kenneth John, forest mensuration
Moellmann, Gisela E Bielitz, cell biology, cytochemistry
Moore, Frank William, cultural anthropology
Moore, Peter Bartlett, biochemistry
Morowitz, Harold Joseph, biophysics, biology
Morris, John McLean, medicine
Mostow, George Daniel, mathematics
Motoyama, Etsuro K, anesthesiology, respiratory physiology
Mroczkowski, Stanley, solid state chemistry
Muirhead, Robb John, mathematical statistics
Nadel, Ethan Richard, physiology
Nath, Ravinder Katyal, nuclear physics, radiological physics
Nelson, Vernon A, entomology
Neufeld, Arthur Harvey, physiology, ophthalmology
Newman, Paul, anthropology
Niederman, James Corson, internal medicine, epidemiology
Noack, Manfred Gerhard, inorganic chemistry, industrial chemistry
Novick, Alvin, behavioral physiology, neurobiology
Nye, Patrick William, biophysics, psychophysics
O'Hayre, Arthur Paul, forest hydrology

Ondrako, Joanne Marie, microbial physiology
Opton, Edward Milton, virology
Ornston, Leo Nicholas, microbial physiology, biochemistry
Orville, Philip Moore, geology
Ostfeld, Adrian Michael, epidemiology, public health
Ostrom, John Harold, vertebrate paleontology
Owens, John Harold, biology, science education
Palade, George E, cell biology, histology
Panicci, Ronald J, organic chemistry
Papac, Rose, internal medicine
Parker, Peter Donald MacDougall, nuclear physics
Parlange, Jean-Yves, environmental sciences
Patterson, Andrew, Jr, physical chemistry
Patton, Curtis Leverne, parasitology, immunology
Pawelek, John Mason, biochemistry, genetics
Pearson, Howard Allen, pediatrics, hematology
Pellegrino, Edmund Daniel, clinical medicine
Pelzer, Karl Josef, geography
Perlis, Alan Jay, mathematics
Petrillo, Charles, medicine
Pickett, Lawrence Kimball, surgery
Pilbeam, David Roger, physical anthropology, primatology
Pincus, Jonathan Henry, neurology
Piotrowski, Joseph Martin, petrology, mineralogy
Pitlick, Frances Ann, biochemistry, cell biology
Poincelot, Raymond Paul, Jr, biochemistry
Porter, William Samuel, theoretical physics
Poulson, Donald Frederick, genetics
Preston, Frank James, chemistry
Prober, Daniel Ethan, low temperature physics, solid state physics
Provasoli, Luigi, microbiology
Prusoff, William Herman, biochemistry, pharmacology
Racusen, Richard Harry, plant physiology
Radding, Charles Meyer, biochemistry, genetics
Rae, Margaret Engel, microbial genetics
Rae, Peter Murdoch McPhail, cell biology
Ramus, Joseph, botany
Rawson, Robert Orrin, physiology
Redlich, Frederick Carl, psychiatry
Reid, Ted Warren, biochemistry
Reifsnyder, William Edward, forest meteorology
Reiser, Morton Francis, psychiatry
Remington, Charles Lee, biology
Rhoads, Donald Cave, paleoecology
Ricciuti, Florence Christine, human genetics
Rich, Saul, phytopathology
Richards, Frank Frederick, biochemistry, medicine
Rickart, Charles Earl, mathematics
Riley, Stephen James, physical chemistry
Ritchie, Joseph Murdoch, pharmacology, physiology
Rodgers, John, geology
Rooney, Seamus Augustine, lipid chemistry
Rose, Michael Dudley, physical anthropology
Rosen, George, public health
Rosenberg, Leon Emanuel, genetics
Rosenstein, Robert William, biochemistry, immunology
Rosenthal, Jean, physiology
Roth, Jerome Allan, biochemistry, neuropharmacology
Roth, Robert Andrew, Jr, biochemical pharmacology, environmental medicine
Roth, Robert Henry, Jr, neuropharmacology
Rouse, Irving, anthropology, archaeology
Rubel, Edwin W, neurobiology
Ruddle, Francis Hugh, genetics, cell biology
Ruddle, Nancy Hartman, immunology
Rudnick, Dorothea, embryology
Rupp, W Dean, molecular genetics
Sachs, Frederick Lee, internal medicine
Sachs, Martin William, nuclear physics
Sakalowsky, Peter Paul, Jr, physical geography
Saltzman, Barry, meteorology
Sands, David Chandler, plant pathology, bacteriology
Sandweiss, Jack, physics
Sartorelli, Alan Clayton, biochemical pharmacology

Sasaki, Clarence Takashi, otorhinolaryngology, neurophysiology
Satter, Ruth, developmental biology, plant physiology
Saunders, Martin, organic chemistry
Savage, I Richard, statistics
Scardera, Michael, industrial organic chemistry
Schenkman, John Boris, biochemistry, pharmacology
Schluederberg, Ann Elizabeth Snider, virology, immunology
Schmir, Gaston L, biochemistry, organic chemistry
Schnabel, Wilhelm J, organic chemistry
Schrader, Dorothy Virginia, mathematics
Schultz, Howard Louis, physics
Schultz, Martin H, computer science
Schulz, George J, physics
Schulz, Robert J, medical physics
Schwartz, Anthony, comparative medicine, immunology
Schwartz, Solomon Samuel, radiology
Sciarini, Louis John, organic chemistry
Scott, Alastair Ian, organic chemistry
Scott, Robert Neal, industrial chemistry
Sears, Marvin Lloyd, ophthalmology
Seglie, Ernest Augustus, nuclear physics
Seligman, George Benham, mathematics
Seligson, David, pathology, biochemistry
Shapiro, Donald Lawrence, pediatrics
Shapiro, Robert, radiology
Sheard, Michael Henry, psychiatry, neuropharmacology
Shepherd, Gordon Murray, neurophysiology
Shope, Robert Ellis, virology
Sibley, Charles Gald, systematic zoology, ornithology
Siccama, Thomas G, ecology
Sieckhaus, John Francis, industrial chemistry
Sikand, Rajinder S, physiology, medicine
Silver, George Albert, medicine
Simmonds, Sofia, biochemistry
Simon, Allan Lester, medicine, radiology
Simons, Elwyn Laverne, vertebrate paleontology, primatology
Sinanoglu, Oktay, theoretical chemistry
Sinkinson, Kathleen, topology
Siu, Yum-Tong, pure mathematics
Skinner, Brian John, geochemistry, economic geology
Skinner, Helen Catherine W, mineralogy, bioinorganic chemistry
Slayman, Carolyn Walch, genetics
Slayman, Clifford L, physiology
Smith, David Martyn, silviculture, forest ecology
Smith, Dwight Glenn, population ecology
Smith, Ora Kingsley, physiology, medicine
Smith, William Hulse, plant pathology
Snyder, Daniel Raphael, neuropsychology, primatology
Soll, Dieter Gerhard, molecular biology, organic chemistry
Sommerfield, Charles Michael, theoretical physics
Southwick, Wayne Orin, orthopedic surgery
Spaulding, Stephen Waasa, endocrinology, medical research
Spiro, Howard Marget, medicine
Srinivasan, Mandyam Veerambudi, vision
Starr, James LeRoy, soil physics, microbiology
Staugaard, Burton Christian, embryology
Steigman, Gary, astrophysics
Steinmetz, Charles, Jr, zoology
Steitz, Joan Argetsinger, biochemistry, molecular biology
Steitz, Thomas Arthur, molecular biology
Stevens, Charles F, neurophysiology
Stevens, Joseph Charles, experimental psychology
Stevens, Rosemary Anne, public health
Stevenson, Harlan Quinn, cytogenetics, radiobiology
Stitt, John Thomas, physiology, neurophysiology
Stolwijk, Jan Adrianus Jozef, biophysics
Stowe, Bruce Bernot, plant physiology, biochemistry
Stryer, Lubert, biochemistry, biophysics
Sturtevant, Julian Munson, biophysical chemistry
Summers, William Cofield, molecular biology, biochemistry
Summers, Wilma Poos, biochemical genetics
Sussex, Ian Mitchell, botany
Sweeting, Orville John, organic chemistry
Szczarba, Robert Henry, topology
Taft, Horace Dwight, physics
Talner, Norman Stanley, pediatrics, cardiology

Taub, Arthur, neurophysiology, neurology
Thomson, Keith Stewart, zoology, paleontology
Thorpe, Michael Fielding, theoretical physics
Tinsley, Beatrice Muriel, theoretical astrophysics, cosmology
Tomlinson, Harley, plant pathology
Treffers, Henry Peter, microbiology
Trench, Robert Kent, marine biology, biochemistry
Trinkaus, John Philip, developmental biology
Triolo, Victor Anthony, information science
Trotz, Samuel Isaac, chemistry
Turekian, Karl Karekin, geochemistry
Vaisnys, Juozas Rimvydas, physical
Van Altena, William F, astronomy
Van Wyk, Rodney, mathematics
Veronis, George, oceanography
Vickery, Hubert Bradford, biochemistry, history of medicine
Viseltear, Arthur Jack, public health, history of medicine
Voigt, Garth Kenneth, soils, plant nutrition
Volz, Michael George, agricultural microbiology, plant physiology
Von Graevenitz, Alexander W C, medical microbiology
Waage, Karl Mensch, geology
Waggoner, Paul Edward, climatology
Waksman, Byron Halstead, immunology
Wallace, Douglas Cecil, genetics
Wallis, Robert Charles, medical entomology, parasitology
Walton, Gerald Steven, plant pathology
Ward, David Christian, virology, biochemistry
Wasserman, Harry H, organic chemistry
Waterman, Talbot Howe, comparative physiology
Waters, Levin Lyttleton, pathology
Wegener, Peter Paul, fluid physics, gas dynamics
Weinman, David, II, tropical medicine, parasitology
Weiss, Robert Martin, urology, pharmacology
Weissman, Sherman Morton, physiology, biochemistry
Weseloh, Ronald Mack, entomology
White, Augustus Aaron, III, orthopedic surgery, biomedical engineering
Whitt, Ward, operations research
Whittemore, Ruth, pediatrics, cardiology
Wiberg, Kenneth Berle, organic chemistry
Willenkin, Robert L, anesthesiology, physiology
Williams, Robley Cook, Jr, physical biochemistry
Willis, William J, physics
Wilson, Christopher Paul, astronomy
Winchell, Horace, mineralogy, geology
Wojtowicz, John Alfred, industrial pharmacology
Wolf, Werner Paul, physics
Wong, Shek-Fu, atomic physics, molecular physics
Worrell, Albert Cadwallader, forest economics, resource economics
Wright, Hastings Kemper, medicine, physiology
Wyckoff, Harold Winfield, molecular biophysics
Wyman, Robert J, neurosciences, comparative physiology
Wyshak, Grace, biostatistics
Yu, Robert Kuan-Jen, biochemistry, neurochemistry
Zelitch, Israel, biochemistry, plant physiology
Zeller, Michael Edward, high energy physics
Ziegler, Frederick Edward, organic chemistry

NEW LONDON
Barry, William James, animal behavior
Bell, Thaddeus Gibson, acoustics
Brooks, Albert Law, biological oceanography
Brown, Oliver Leonard Inman, physical chemistry
Cobb, Jewel Plummer, cell biology
Dinapoli, Frederick Richard, acoustics
Eby, Edward Stuart, Sr, mathematics
Einstein, Lloyd Theodore, physics
Fell, Paul Erven, invertebrate embryology, marine biology
Fenton, David George, physics
Goodwin, Richard Hale, botany
Green, Eugene L, physics, optics
Green, Milton, applied physics

Hanrahan, John J, underwater acoustics
Harris, John Donald, experimental psychology, audiology
Hasse, Raymond William, Jr, underwater acoustics
Hostinsky, Lois Aileen, algebra
Kent, John Franklin, zoology
Macklin, June, anthropology
McGill, David A, oceanography
McKeon, Mary Gertrude, electrochemistry
Meyers, Allan Richard, anthropology
Nash, Harold Earl, physics
Niering, William Albert, plant ecology
Prokesch, Jeanne Chase, immunology, endocrinology
Rubega, Robert A, acoustics
Schlesinger, Ernest Carl, mathematical analysis
Schmidt, James L, pharmacology, industrial medicine
Sherman, Charles Henry, physics
Smith, Trudy Enzer, physical chemistry
Stark, Charles R, pediatrics, epidemiology
Strawderman, Wayne Alan, applied mathematics
Thomson, Betty Flanders, botany
Tolderlund, Douglas Stanley, marine ecology
Viccione, Daniel Michael, acoustics
Warren, Richard Scott, plant physiology
Wehman, Anthony Theodore, organic chemistry, organometallic chemistry
Wheeler, Bernice Marion, zoology
Woollett, Ralph Storer, acoustics

NEW MILFORD
Howard, Hartley Wolle, chemistry
Weaver, Warren, mathematics

NEWINGTON
Coykendall, Alan Littlefield, oral microbiology
Elliott, John Raymond, organic polymer chemistry, engineering management
Frauenglass, Elliott, organic chemistry
Hauser, Martin, polymer chemistry
Malofsky, Bernard Miles, organic polymer chemistry
Morris, Richard Knowles, anthropology
Rich, Richard Douglas, organic chemistry
Russo, John A, Jr, meteorology
Scheig, Robert L, internal medicine, gastroenterology
Schmitz, John Vincent, organic chemistry
Wyckoff, Delaphine Grace Rosa, microbiology, bacteriology

NIANTIC
Lombardino, Joseph George, organic chemistry

NOANK
Buck, John David, bacteriology, marine microbiology
Feng, Sung Yen, physiological ecology
Sternberg, Robert Langley, applied mathematics, ocean engineering

NORFOLK
Egler, Frank Edwin, plant ecology

NOROTON
Wohnsiedler, Henry Peter, plastics chemistry

NORTH BRANFORD
Gralla, Edward Joseph, veterinary toxicology

NORTH HAVEN
Heying, Theodore Louis, organic chemistry, research administration
Humphrey, Bingham Johnson, industrial organic chemistry
Lipka, Benjamin, organic chemistry
Lyons, Margaret S, physical chemistry
Nadeau, Herbert Gerard, analytical chemistry, physical chemistry
Richter, Reinhard Hans, organic chemistry
Rose, James Stephenson, organic chemistry
Sayigh, Adnan Abdul Rida, organic chemistry
Smith, Curtis Page, organic chemistry
Stuber, Fred A, physical organic chemistry
Ulrich, Henri, organic polymer chemistry

NORTH STONINGTON
Man, Evelyn Brower, clinical chemistry

NORTH WOODSTOCK
O'Brien, Brian, electrooptics

NORTHFORD
Farrissey, William Joseph, Jr, organic polymer chemistry
Raymond, Maurice A, organic chemistry, polymer chemistry

NORWALK
Boas, Norman Francis, internal medicine, rheumatology
Brienza, Michael Joseph, electrooptics, lasers
Bystricky, Karl M, optics
Conrad, Eugene Anthony, pharmacology
Ettre, Leslie Stephen, analytical chemistry
Fine, Leonard W, organic chemistry
Goedert, Michel G, analytical chemistry
Hartigan, Martin Joseph, analytical chemistry
Imbruce, Richard Peter, pulmonary physiology
Kambli, Vijaykant Bhagwan, biochemistry, clinical chemistry
Karsten, Kenneth Stephen, plant physiology, chemistry
Knapp, Richard Allen, solid state physics, optics
Lange, Winthrop Everett, pharmacy
Locke, Stanley, mathematics, physics
Nair, Sreedhar, pulmonary physiology, pharmacology
Poultney, Sherman King, applied physics, optical physics
Ranalli, Anthony William, food technology
Rosenkrantz, Lawrence Jay, lasers, experimental solid state physics
Savitzky, Abraham, physical chemistry, computer science
Scott, Roderic MacDonald, astronomy
Siegler, Edouard Horace, Jr, physics
Smith, Donald Leonard, experimental solid state physics
Takesue, Edward I, pharmacology, medical research
Taylor, Wilbur Spencer, industrial microbiology

NORWICH
Gordon, Malcolm Wofsy, biochemistry
Heotis, James Peter, agricultural biochemistry, analytical biochemistry
Kavarnos, George James, clinical chemistry
Sabatini, Joseph Francis, physical chemistry

OLD GREENWICH
Lumb, George Dennett, pathology
Strauss, Roger William, paper technology

OLD LYME
McManus, James Michael, organic chemistry
Mellen, Robert Harrison, physics
Peterson, Roger Tory, ornithology
Roberts, Mervin Francis, animal behavior

PAWCATUCK
Casey, John Edward, Jr, organic chemistry
Gunther, Ronald George, electrochemistry
Petrocelli, Americo W, physical inorganic chemistry
Seiger, Harvey Norman, physical chemistry

Schweitzer, Jeffrey Stewart, nuclear physics
Sherman, Harold, physics
Smith, Frederick Albert, physical chemistry, organic chemistry
Tapphorn, Ralph M, nuclear physics
Tobin, Marvin Charles, physical chemistry, physics
Wahl, John Schempp, nuclear physics
Witterholt, Edward John, applied mathematics
Zelson, Carl, pediatrics

RIVERSIDE
Rove, Olaf Norberg, mining geology

ROCKY HILL
Hilst, Glenn Rudolph, meteorology

ROWAYTON
Klemme, Hugh Douglas, petroleum geology
Larson, Allan Bennett, industrial pharmacy, cosmetic chemistry

SHERMAN
Bristol, Melvin Lee, botany

SIMSBURY
Schwarz, Helmut Julius, applied physics, electron physics
Ulmer, Richard Clyde, physical chemistry

SOUTH MERIDEN
Reardon, Joseph Daniel, physical chemistry

SOUTH NORWALK
Gilby, Anthony Christopher, physical chemistry, spectroscopy
Truett, William Lawrence, chemistry

SOUTH WINDSOR
Powers, Joseph, materials science, polymer science

SOUTHBURY
Breg, William Roy, pediatrics, cytogenetics
Senn, Milton J E, pediatrics, psychiatry

SOUTHPORT
Bruno, Arthur John, organic chemistry
Hill, David Lawrence, nuclear physics, instrumentation
Martin, William Sonderman, solid state physics
Taylor, Howard Canning, Jr, obstetrics & gynecology

STAFFORDVILLE
Herrington, William Charles, fisheries management

STAMFORD
Alpert, Nelson Leigh, medical technology, spectroscopy
Arlt, Herbert George, Jr, organic chemistry, polymer chemistry
Armstrong, William David, physical chemistry
Astheimer, Robert W, electrooptics
Barber, William Austin, physical chemistry
Bard, John William, chemistry
Barnes, Carl Edmund, organic chemistry
Barnes, Robert Bowling, physics
Begala, Arthur James, physical chemistry
Behnken, Donald Washington, experimental statistics
Berets, Donald Joseph, physical chemistry
Berry, John William, analytical chemistry
Bil, Milos Sidney, organic chemistry, inorganic chemistry
Breivik, Orville Nolan, food technology
Brinen, Jacob Solomon, molecular spectroscopy, surface chemistry
Brown, Keith Charles, physical organic chemistry
Bruesch, John F, organic chemistry
Burnett, Clyde Marshall, toxicology
Calbo, Leonard Joseph, industrial organic chemistry
Carbonell, Robert Joseph, food science

CONNECTICUT

Carter, Irving Doyle, chemistry
Chrepta, Stephen John, analytical chemistry
Clarke, George, analytical chemistry
Coleman, Denis, organic chemistry
Colthup, Norman Bertram, spectroscopy
Connell, Richard Allen, physics
Cooke, Theodore Frederic, textile chemistry
Corbett, John Frank, organic chemistry
Coscia, Anthony Thomas, polymer chemistry
Capilla, Joseph, organic chemistry, science
Cunningham, Frederick William, physics
Davenport, Lee Losee, physics, telecommunications
Dobay, Donald Gene, polymer chemistry, chemical engineering
Dolian, Frank Eugene, chemistry
Dolin, Stanley A., spectroscopy, computer science
Doolittle, Howard Daniel, physics
Dreyfus, Marc George, physics
Drucker, Arnold, polymer chemistry, organic chemistry
Duda, Edward John, entomology
Ebel, Robert Henry, applied chemistry
Engel, Paulinus P., industrial microbiology
Evans, Charles P., organic chemistry, administration
Feinland, Raymond, analytical chemistry
Ferguson, John Allen, industrial chemistry
Fiore, Joseph Vincent, biochemistry
Firth, William Charles, Jr., organic polymer chemistry
Fischer, Robert George, Jr., inorganic chemistry, industrial organic chemistry
Frank, Simon, industrial chemistry
Freeman, Mark Phillips, chemistry
Fusfeld, Herbert Irving, physics, research administration
Gaertner, Wolfgang Wilhelm, solid state electronics
Gallivan, James Bernard, physical chemistry
Garcia, Mario Leopoldo, physical chemistry, cosmetic chemistry
Gasque, Mac Roy, medicine
Gefter, William Irvin, internal medicine
Gingold, Kurt, chemistry
Goebel, James Christopher, polymer chemistry
Gold, Daniel Howard, industrial chemistry
Goldman, Jacob E., physics
Goldmark, Peter Carl, physics
Goldring, Lionel Solomon, physical chemistry, research administration
Goldstein, Marvin Sherwood, petroleum chemistry
Haacke, Gottfried, solid state science
Haines, George Shuler, chemistry
Halverson, Frederick, chemistry
Hannan, Roy Barton, Jr., analytical chemistry
Henderson, William Arthur, Jr., organic chemistry
Hinz, Charles F., microbiology, biochemistry
Hoffman, Oscar Allen, operations research
Hoffmann, Arthur Kentaro, organic chemistry
Huffman, Kenneth Robert, organic chemistry
Irani, Riyad Rida, physical chemistry
Itter, Stuart, biochemistry
Jones, Daniel Elven, physical chemistry, computer science
Kaufman, Ernest D., physical chemistry
Kennerly, George Warren, industrial chemistry
Klokholm, Erick, solid state physics
Kuck, Julius Anson, organic chemistry
Lahr, Roy (Jeremy), research administration
Lancaster, John Edgar, physical chemistry
Lazo-Wasem, Edgar Arthur, pharmaceutics
Leuter, Werner Walter, applied mathematics
Liebson, Sidney Harold, physics
Linke, William Finan, physical chemistry
Lustig, Bernard, biochemistry
Magrane, John Kearns, Jr., organic chemistry
Matsuda, Ken, organic chemistry
McIrvine, Edward Charles, theoretical physics, information science
McNally, John G., analytical chemistry
Mead, Thomas Edward, organic chemistry, mass spectrometry

Menkart, John, cosmetic chemistry, research administration
Meriwether, Lewis Smith, physical chemistry, organic chemistry
Miller, Arnold, physical chemistry
Miller, Jerry K., analytical chemistry
Montgomery, Anthony John, optics physics
Mulhern, Thomas Patrick, mathematics
Nachtigall, Guenter Willi, organic chemistry
Noland, James Sterling, organic polymer chemistry
Norris, Max Valentine, analytical chemistry
Novak, Robert William, organic chemistry
O'Connell, Richard Lee, occupational medicine
Okaya, Akira, quantum electronics, physical electronics
Panzer, Hans Peter, organic polymer chemistry
Pellon, Joseph, organic chemistry
Pietsch, Gerhard Josef, organic chemistry
Platko, Frank Edward, physical chemistry
Pry, Robert Henry, materials science, research administration
Raghu, Sivaraman, synthetic organic chemistry
Reddy, Thomas Bradley, electrochemistry
Redfern, Sutton, cereal chemistry
Riker, John A., food science
Rose, Stuart Alan, analytical chemistry, photochemistry
Rosenberg, Ira Edward, physical organic chemistry
Schaefer, Frederic Charles, organic chemistry
Schlegel, John C., plastics chemistry
Schmitt, Joseph Lawrence, Jr., physical chemistry
Schmitt, Joseph Michael, polymer chemistry
Sedlak, John Andrew, organic chemistry
Shaffer, Lloyd Hamilton, physical chemistry
Siegel, Lester Aaron, physics
Stamm, Robert Franz, physical chemistry, optical physics
Strazdins, Edward, physical chemistry
Streuli, Carl Arthur, analytical chemistry
Sussman, Sidney, water chemistry, corrosion
Sveda, Michael, chemistry
Swanson, Donald Leroy, analytical chemistry, physical chemistry
Torrance, Ellen McCormick, pure mathematics, actuarial mathematics
Updegraff, Ivor Heberling, organic chemistry
Wasser, Richard Barkman, paper chemistry, physical chemistry
Weir, John Marshall, public health
Welcher, Richard Parke, chemistry
Werneke, Michael Francis, inorganic chemistry
West, Bob, pharmacology, toxicology
White, John Underhill, physics
Whittaker, Mack Page, inorganic chemistry
Witschonke, Charles Richard, environmental chemistry
Wong, David C., semiconductors
Wystrach, Vernon Paul, organic chemistry
Yerg, Raymond A., aerospace medicine, occupational medicine
Young, George Jamison, physical chemistry
Zavisza, Daniel Maximillian, physical organic chemistry
Ziegler, Theresa Frances, chemistry
Zweg, Arnold, physical organic chemistry

STONINGTON
De Zeeuw, John Robert, biochemistry
Woods, Jimmie Dale, mathematics, statistics

STEVENSON
Denues, Arthur Russell Taylor, physical biochemistry, biochemical engineering

STORRS
Aho, William A., poultry science
Aigner, Jean Stephanie, anthropology
Aitken, Janet Mora, geology
Allen, John Logan, historical geography, cultural geography
Allen, Lindsay Helen, nutrition
Amundsen, Lawrence Hardin, organic chemistry

Anderson, Gregory Joseph, biosystematics, pollination biology
Ashley, Richard Allan, weed science
Azaroff, Leonid Vladimirovich, x-ray crystallography
Bailey, William Francis, organic chemistry
Bartram, Ralph Herbert, solid state physics
Bee, Robert L., anthropology
Benson, Robert Haynes, animal breeding
Berg, Claire M., genetics
Best, Philip Ernest, solid state physics
Black, Robert Foster, geology
Bobbit, James McCue, chemistry
Bohn, Robert K., physical chemistry
Braswell, Emory Harold, biophysics, biophysical chemistry
Brown, Lynn Ranney, animal nutrition
Brush, Alan Howard, zoology
Budnick, Joseph Ignatius, physics
Burch, Thaddeus Joseph, solid state physics
Cameron, J A., medical microbiology
Carey, Bernard Joseph, computer science
Carpenter, Edwin David, horticulture, ornamental horticulture
Caswell, Hal, ecology, mathematical biology
Chamberland, Bertrand Leo, inorganic chemistry
Chance, Norman Allee, cultural anthropology
Chapple, William Dismore, neurophysiology, comparative physiology
Chovnick, Arthur, genetics
Clark, George Alfred, Jr., ornithology
Clark, Nancy Barnes, comparative endocrinology, comparative physiology
Cohen, Steven Donald, toxicology
Collins, Ralph Porter, botany
Cowan, William Allen, animal breeding
Crepet, William Louis, paleobotany, evolutionary biology
Damman, Antoni Willem Hermanus, plant ecology
Danon, Dwight Hills, physics
David, Carl Wolfgang, physical chemistry
Davis, Norman Thomas, entomology
DeCoursey, Russell Myles, zoology
Dehlinger, Peter, geophysics, oceanography
Denenberg, Victor Hugo, neurosciences
DiCapua, Richard Anthony, immunochemistry
Doeg, Kenneth Albert, biochemistry
Eaton, Hamilton Dean, animal nutrition
Faris, James Chester, anthropology
Fitch, Robert McLellan, polymer chemistry, colloid chemistry
Frankel, Larry, invertebrate paleontology
Fredrickson, Torgny Norman, veterinary pathology
Frier, Henry Ira, nutritional biochemistry
Fruch, Alfred Joseph, Jr., crystallography, mineralogy
Gaunya, William Stephen, dairy husbandry
Gilliam, Otis Randolph, physics
Ginsburg, Benson Earl, genetics
Goetinck, Paul Firmin, genetics
Gosselin, Richard Pettengill, mathematical analysis
Gould, Steven James, bio-organic chemistry
Greenblatt, Irwin M., genetics, botany
Gutay, Andrew John Robert, soils, plant science
Hahn, Yukap, theoretical physics
Hall, Kenneth Noble, food science, nutrition
Haller, Kurt, theoretical high energy physics
Hayden, Howard Corwin, atomic physics
Henry, Charles Stuart, evolutionary biology, entomology
Herrmann, Heinz, chemical embryology
Hewitt, Harold George, pharmaceutical chemistry
Heywood, Stuart Mackenzie, molecular biology, biochemistry
Highower, Lawrence Edward, virology
Hite, Gilbert J., medicinal chemistry
Huang, Samuel J., organic chemistry, polymer chemistry
Hurley, James Frederick, high energy physics
Islam, Muhammad Munirul, algebra
Janes, Byron Everett, plant physiology
Jahnke, Paul Joseph, pharmaceutical chemistry
Jayne, Edgar Pleasant, anatomy, physiology

Jensen, Robert Gordon, biochemistry
Johnson, Bruce McK., statistics
Johnson, Julian Frank, polymer chemistry
Judd, Roy Whitlock, Jr., plant pathology
Kappers, Lawrence Allen, solid state physics
Katz, Lewis, physical chemistry
Kegeles, Gerson, biophysical chemistry
Kelleher, William Joseph, biochemistry, pharmacognosy
Kennard, William Crawford, plant physiology, horticulture
Kersting, Edwin Joseph, veterinary medicine
Kessel, Quentin Cattell, experimental atomic physics
Khairallah, Edward A., endocrinology, nutritional biochemistry
Kim, Soon-Kyu, mathematics
Kind, Charles Albert, biochemistry
Kinsman, Donald Markham, animal science
Klemens, Paul Gustav, physics
Knox, Kirvin L., nutrition
Koontz, Harold Vivien, plant physiology
Kostiner, Edward S., solid state chemistry
Koths, Jay Sanford, floriculture
Krause, Ronald Alfred, inorganic chemistry
Langrer, Ronald O., biochemical pharmacology
Laufer, Hans, developmental biology
Leacock, Seth, anthropology
Leibowitz, Gerald Martin, mathematical analysis
Lerman, Manuel, mathematical logic
Liese, Homer C., mineralogy, petrology
Lindstrom, Richard Edward, physical pharmacy
Lipschultz, Frederick Phillip, low temperature physics
Liss, Alan, microbial genetics, virology
Lucas, Joseph James, biometrics, genetics
Lucas-Lenard, Jean Marian, molecular biology
Maxson, Stephen C., behavior genetics
Mehlquist, Gustav Arthur Leonard, horticulture, plant breeding
Micheher, Bryan Paul, anthropology
Mittering, Lloyd Alfred, pomology
Moeller, Carl William, Jr., inorganic chemistry
Monahan, Audrey Small, organic chemistry
Montgomery, John C., mathematics
Moran, Thomas Irving, physics
Mugnaini, Enrico, neuroanatomy
Mundkur, Balaji, cytology
Neuwirth, Jerome H., mathematics
Newcomer, Earl Holland, cytology, genetics
Nieforth, Karl Allen, medicinal chemistry, psychopharmacology
Nielsen, Svend Woge, veterinary pathology
Nightingale, Charles Henry, biopharmaceutics
Noether, Gottfried Emanuel, mathematical statistics
Orr, William Campbell, nuclear chemistry
Pelto, Pertti Juho, anthropology
Penner, Lawrence Raymond, parasitology
Peterson, Cynthia Wyeth, solid state physics
Pfeifer, Howard William, plant taxonomy
Phillips, Alvah H., physiological chemistry
Philpotts, Anthony Robert, petrology, mineralogy
Piero, Louis John, biology
Pilar, Guillermo Roman, neurophysiology
Pollack, Edward, atomic physics
Posten, Harry Owen, mathematical statistics
Pudelkiewicz, Walter Joseph, biochemistry
Purves, William Kirkwood, plant physiology
Raney, George Neal, mathematics
Rankin, John Stewart, Jr., marine biology
Rawitscher, George Heinrich, theoretical nuclear physics
Reffner, John A., physical chemistry, crystallography
Reschovsky, Helene J., mathematics
Rettenmeyer, Carl William, entomology, ecology

Rich, Peter Hamilton, limnology
Richelle, Leon Joseph, oral biology
Riesen, John William, reproductive physiology
Romano, Antonio Harold, microbiology
Rosenberg, Philip, pharmacology
Roth, Jay Sanford, biochemistry
Rother, Ana, pharmacognosy
Rousseau, Joseph Edward, Jr., nutrition
Rubins, Edward J, soil chemistry
Rumney, George Richard, geography, climatology
Russek, Arnold, physics
Ryff, John V, mathematical analysis
Sachs, Benjamin David, behavioral biology
Samulski, Edward Thaddeus, physical chemistry, polymer chemistry
Savos, Milton George, entomology
Schaefer, Carl W, II, entomology
Schor, Robert, biophysics, solid state physics
Schramm, Robert Johnson, Jr, plant nutrition
Schultz, Roland Jack, population biology, ichthyology
Schuster, Todd Mervyn, biophysical chemistry
Schwarting, Arthur Ernest, pharmacognosy
Schwartz, Tobias Louis, biophysics, physiology
Schwinck, Ilse, developmental genetics
Scott, Joseph Lybrand, endocrinology
Shelly, Eugene Paul, mathematics
Sholl, Howard Alfred, computer sciences
Sidney, Stuart Jay, mathematical analysis
Simonelli, Anthony Peter, pharmaceutics, chemical kinetics
Singsen, Edwin Pierce, poultry science
Skauen, Donald M, pharmacy
Slater, James Alexander, entomology, systematics
Smith, Arnold Chauncey, dairy industry
Smith, Sidney Ruven, physical chemistry
Smith, Winthrop Ware, experimental atomic physics
Snodgrass, Rex Jackson, solid state physics, nuclear magnetic resonance
Somes, Ralph Gilmore, Jr, animal genetics, nutrition
Spencer, Domina Eberle (Mrs Parry Moon), mathematics, physics
Spiegel, Eugene, mathematics
Stage, Gerald Irving, systematic entomology, ecology
Stake, Paul Erik, animal nutrition
Steigert, Frederick Edward, nuclear physics
Stock, John Thomas, analytical chemistry
Stratford, Eugene Scott, medicinal chemistry
Streams, Frederick Arthur, insect ecology
Strittmatter, Philipp, biochemistry, enzymology
Sze, Paul Yi Ling, biochemistry, neurobiology
Tanaka, John, inorganic chemistry
Terry, Thomas Milton, biophysics, microbiology
Tolimieri, Richard, pure mathematics
Tollefson, Jeffrey L., mathematics
Tourtellotte, Mark Eton, phycology
Trainor, Francis Rice, phycology
Vasington, Frank D, biochemistry
Vaughan, Wyman Ristine, chemistry
Vinopal, Robert Thomas, microbial physiology
Vinsonhaler, Charles I, mathematics
Wachman, Murray, mathematics
Wachtel, Allen W, cell biology, cytology
Walker, Marshall John, optics
Waring, Charles Emmett, physical chemistry
Washko, Walter William, agronomy
Wassmundt, Frederick William, organic chemistry
Waxman, Sidney, ornamental horticulture
Webster, Terry R, botany, plant morphology
Wengel, Raymond William, soil physics
Werboff, Jack, psychobiology
Wetherell, Donald Francis, plant physiology
Wetzel, Ralph Martin, mammalogy
White, John Robert, computer sciences, information sciences
Whitworth, Walter Richard, aquatic biology
Wilson, William August, Jr, neuropsychology
Wolk, Elliott Samuel, mathematics
Wood, David Eldon, physical chemistry
Woody, Charles Owen, Jr, reproductive physiology
Yang, Tsu-Ju (Thomas), pathobiology, immunobiology
Young, Charles Gilbert, optics
Yphantis, David Andrew, biophysics, biochemistry

STRATFORD
Horrocks, Robert H, organic chemistry
Magenheimer, John Joseph, surface chemistry, chemical kinetics

TORRINGTON
Vidone, Romeo Albert, pathology

TRUMBULL
Berdick, Murray, polymer chemistry
Berube, Gene Roland, physical biochemistry
Goldberg, Morris H, biochemistry
Haas, Ward John, biochemistry
Husband, Robert Murray, organic chemistry, chemical engineering
Mulligan, James Edward, Jr, mathematics
Roye, Gerald Stephen, analytical chemistry, computer science
Sawers, James Richard, Jr, solid state physics, instrumentation
Tamorria, Christopher Richard, medicinal chemistry, communication science
Tranner, Frank, pharmacy, chemistry
Tripathi, Uma Prasad, inorganic chemistry, cosmetic chemistry
Weaver, J Ritner, analytical chemistry

WALLINGFORD
Hines, Thomas Franklin, medicine
Padbury, John James, organic polymer chemistry

WATERBURY
Camp, Eldridge Kimbel, electrochemistry
Hurst, Victor Waldemar, III, internal medicine, cardiology
Shilling, Paul R, biology, genetics
Thornton, George Fred, internal medicine, infectious diseases

WATERFORD
Fried, John H, microbiology, chemistry
Wennemer, Jay, marine ecology

WEST CORNWALL
Hechenbleikner, Ingenuin Albin, organic chemistry

WEST HARTFORD
Allen, Richard Ballantine, optical physics
Bogucki, Raymond Francis, inorganic chemistry
Bryan, Joseph Gerard, applied mathematics
Burger, James Wendell, zoology
Cheo, Peter K, physics
Coleman, William H, microbiology
Eisenberg, Sheldon Merven, mathematics
Gardner, Fred Marvin, electron physics
Genevese, Frank, nuclear physics, mathematics
Gwynn, Robert H, physiology
Johnson, Dorothy Dole, histology
Kagan, Joel (David), mathematics
Landsman, Douglas Anderson, physical chemistry
Maguder, Theodore Leo, Jr, wildlife biology, ecology
Markham, Elizabeth Mary, mathematics
Markham, M Clare, physical chemistry
Murphy, Mary Teresa Joseph, geochemistry
Simmons, Donald C, cultural anthropology
Simpson, Tracy L, biology
Staker, William Paul, reactor physics
Stevens, Malcolm Peter, organic polymer chemistry
Swain, Elisabeth Ramsay, zoology
Wallace, Robert B, psychobiology

WEST HAVEN
Behar, Jose, gastroenterology
Brownfield, Robert Bruce, analytical chemistry, organic chemistry
Chun, Kee Won, theoretical physics
Cummings, Dennis Paul, microbiology
Desio, Peter John, organometallic chemistry
Donaldson, Robert M, Jr, internal medicine, gastroenterology
Groszmann, Roberto Jose, gastroenterology
Haley, Edward Everett, biochemistry
Hsiung, Gueh Djen, virology
Lee, Henry C, forensic science, biochemistry
Lemaire, Henry, organic chemistry
Ma, Wai-Sai, immunochemistry
Morrison, Richard Charles, energy conversion
Schafer, David Edward, medical research
Storer, Edward Hammond, surgery
Wright, Herbert Fessenden, medicinal chemistry
Yesner, Raymond, pathology

WESTON
Rees, Roberts M, medicine
Sells, Jean Thurber, mathematics

WESTPORT
Bartholomew, William Holden, engineering
Belford, John F, ultrasound, biomedical engineering
Davis, Robert Irving, economic geology, mining geology
Galloway, Ethan Charles, organic chemistry
Gold, Richard Frank, inorganic chemistry, environmental management
Graham, Jack Bennett, physical geology
Herz, Jack L, physical organic chemistry
Hwa, Jesse Chia Hsi, polymer chemistry
Jackel, Simon Samuel, biochemistry
Levy, Gabor Bela, chemistry
Lieberman, James, public health
Murphy, Douglas Richard, plant pathology
Omohundro, Allen Llewellyn, chemistry
Perkins, Willis Drummond, chemical physics, spectroscopy
Roberts, Elliott John, physical inorganic chemistry
Stamm, Walter, chemistry
Terry, Herbert, operations research
Tressler, Donald Kiteley, food chemistry
Ward, Samuel Abner, electronic physics

WETHERSFIELD
Hunter, Ruth Macmillan, embryology

WILLIMANTIC
Amdur, Millard Jason, psychiatry
Gable, Michael, invertebrate ecology
Meyer, Delbert Eugene, zoology, biochemistry
Roos, Henry, anatomy, physiology
Shapiro, Nathan, genetics
Smith, Raymond N, paleontology
Wright, Alan Carl, chemistry
Wulff, Barry Lee, ecology, marine biology

WILTON
DaVanzo, John Paul, biology
Ettre, Kitty, physical chemistry, ceramics
Ford, Clinton Banker, astronomy
Leyda, James Perkins, pharmacy, pharmaceutical chemistry
Murray, Francis Joseph, bacteriology
Tao, Shu-Jen, nuclear science, physical chemistry
Weissgerber, Rudolph E, medicine
Yoder, Paul Rufus, Jr, optics

WINDSOR
Ahrens, John Frederick, weed science
Goldstein, Rubin, theoretical physics, applied mathematics
Hellens, Robert Linton, nuclear science
Lichtenberger, Harold V, nuclear physics
Storrs, Charles Lysander, nuclear physics
Taylor, Gordon Stevens, plant pathology

WINSTED
Duffy, Regina Maurice, plant morphology

WOODBRIDGE
Bruson, Herman Alexander, organic chemistry
Hardy, James Daniel, physiology
Hershenson, Herbert Malcolm, chemistry
Hunter, Byron Alexander, industrial organic chemistry, rubber chemistry
Lewis, Franklin Beach, pathology, medical entomology

WOODBURY
Carter, William Caswell, mathematics

WOODSTOCK
Bernier, Edward Joseph, textile physics, applied physics
Hyde, Walter Lewis, physics, academic administration

DELAWARE

BRIDGEVILLE
Hammond, John Clarke, poultry husbandry

CLAYMONT
Brodoway, Nicolas, organic chemistry
Crecely, Roger William, analytical chemistry
Le Maistre, John Wesley, chemistry
Paris, Jean Philip, analytical chemistry, physical chemistry
Snyder, Jack Austin, biochemistry
Tryon, Sager, organic chemistry

DAGSBORO
Fogg, Donald Ernest, veterinary pathology

DOVER
Bodola, Anthony, hydrobiology
Brown, Robert Raymond, organic chemistry, polymer chemistry
Cameron, Donald Eugene, food microbiology
Dill, Norman Hudson, botany, plant ecology
Ferguson, Thomas, zoology
Hassler, William Woods, analytical chemistry, organic chemistry
Hill, James Aubrey, organic chemistry
Jones, Edward Raymond, agronomy
LeClaire, Claire Dean, organic chemistry
Mishoe, Luna I, mathematics, mathematical physics
Mitchell, Donald Gilman, food technology
Pollak, Oakar Jaroslav, biochemistry, pathology
Sammak, Emil George, polymer research
Tierkel, Ernest Shalom, veterinary public health, epidemiology

EDGEMOOR
Blumenberg, Karl Edward, industrial chemistry
Dickinson, John G, inorganic chemistry
Linton, Howard Richard, inorganic chemistry
Marshall, William Joseph, physical chemistry

GREENVILLE
Abbott, Robert Tucker, malacology
Dessauer, Rolf, organic chemistry
Du Pont, John Eleuthere, ornithology, malacology
Flook, William Mowat, Jr, physics

HENRY CLAY
Lewis, George Leoutsacos, chemistry

HOCKESSIN
Blunt, Harry William, polymer chemistry
Drysdale, John Jay, organic chemistry
Euston, Charles B, physical chemistry
Gilbert, Walter Wilson, industrial chemistry
Jackson, Harold Leonard, polymer chemistry, fluorine chemistry
Lorenz, John Clark, organic chemistry
Lovejoy, Elwyn Raymond, polymer chemistry
Mastrangelo, Sebastian Vito Rocco, physical chemistry
Montague, Barbara Ann, information science
Ojakaar, Leo, organic chemistry
Sauer, John Carl, organic chemistry
Singh, Gurdial, organometallic chemistry
Sterniski, Michael Andrew, organic chemistry
Welldon, Paul Burke, organic chemistry
Williams, Richard Anderson, textile chemistry

LEWES
Bolton, Ellis Truesdale, biophysics

DELAWARE

Carriker, Melbourne Romaine, malacology, marine biology
Epifanio, Charles Edward, marine biology
Gibbs, Ronald John, geochemistry
Inderbitzen, Anton Louis, Jr., marine geology
Maurer, Donald Leo, marine ecology, pollution biology
Mooers, Christopher Northrup Kennard, physical oceanography, fluid dynamics
Price, Kent Sparks, Jr., marine biology
Sharp, Jonathan Hawley, chemical oceanography, biological oceanography
Sick, Lowell Victor, organometallic chemistry

NEW CASTLE
Bartholomew, Eleanor Rachel, chemistry
Dean, Robert Reed, chemistry
Nebel, Richard Wilson, chemistry
Proops, William Robert, organic chemistry

MILLSBORO
Wills, Franklin Knight, veterinary bacteriology, pathology

NEWARK
Alfieri, Charles C., organic chemistry, inorganic chemistry
Amith, Avraham, solid state physics
Angeler, George William, entomology
Angell, Thomas Strong, mathematics
Austin, Paul Rolland, organic chemistry
Banerjee, Kali Shankar, mathematical statistics
Barnes, Stephen Noble, vision
Barnhill, Maurice Victor, III, theoretical physics
Batra, Karam Vir, pharmacology
Batt, William George, biochemistry
Baxter, Willard Ellis, mathematics
Beachell, Harold Charles, physical chemistry
Bellamy, David P., topology
Benson, Richard Norman, micropaleontology
Benton, William J., veterinary virology
Biddle, Richard Albert, inorganic chemistry
Biebuyck, Daniel P., anthropology
Biggs, R B, marine geology
Birchenall, Charles Ernest, chemistry
Brill, Thomas Barton, inorganic chemistry, physical chemistry
Burbutis, Paul Philip, entomology
Burmeister, John Luther, inorganic chemistry
Boord, Robert Lennis, vertebrate morphology, neuroanatomy
Boyce, Richard Joseph, rubber chemistry
Brasher, Eugene Paul, horticulture
Bray, Dale Frank, entomology
Boer, Karl Wolfgang, solid state physics, energy conversion
Buxbaum, Edwin Clarence, anthropology
Campbell, Linzy Leon, microbiology, biochemistry
Clark, Arnold M, genetics
Clark, Roberta F, psychopharmacology
Colman, Roberta F, biochemistry, protein chemistry
Connell, Walter Anthony, insect taxonomy
Cooper, Charles Burleigh, physics
Cornell, Howard Vernon, biogeography, ecology
Cramer, Francis Barnard, organic chemistry
Crittenden, Henry William, plant pathology
Crossan, Donald Franklin, plant pathology
Daiber, Franklin Carl, ichthyology, ecology
Dalrymple, David Lawrence, organic chemistry
Daniels, William Burton, experimental solid state physics
Davies, Warren Lewis, microbiology
Davis, William H., economic entomology
Dennis, Don, biochemistry
Dickinson, Clifford Lee, Jr, organic chemistry
Dorothy, Robert Glenn, chemical physics
Dunham, Charles W., horticulture
Dyer, Elizabeth, organic chemistry
Dysart, Richard James, entomology
Ehrlich, Robert Stark, biophysical chemistry
Eisenberg, Robert Michael, ecology
Eissner, Robert M, statistics
Elliott, John Habersham, rheology, polymer chemistry
Ewing, Richard Dwight, physics

Fagen, Edward Allen, solid state physics
Fieldhouse, Donald John, horticulture
Francis, David W., protozoology, developmental biology
Gaer, Marvin Charles, mathematical analysis
Gelb, Leonard Louis, organic polymer chemistry
Giese, John H, mathematics
Gilbert, Robert Pertsch, mathematics
Glass, Billy Price, marine geology, astrogeology
Goddin, Avery Howe, entomology
Goodwin, Bruce Edward, mathematics
Gore, Wilbert Lee, physical chemistry
Gorski, Robert Alexander, physical organic chemistry
Gould, Adair Brasted, genetics
Granda, Allen Manuel, neurosciences
Gretak, Robert Paul, pharmacology
Haenlein, George Friedrich Wilhelm, animal nutrition, animal breeding
Hall, Robert Turner, analytical chemistry, pesticide chemistry
Halprin, Arthur, physics
Hauty, George Thomas, psychobiology
Hayman, Selma, enzymology
Hendrickson, Robert Mark, Jr., entomology
Herr, Richard Baessler, astronomy
Herson, Diane S., microbiology
Hesseltine, Wilbur R., dairy husbandry
Heuberger, John William, plant pathology
Hickman, James Joseph, photochemistry
Hill, Richard William, comparative physiology, environmental biology
Hill, Robert Nyden, theoretical physics
Hodson, Robert Cleaves, plant physiology, phycology
Hoffman, Howard Edgar, virology
Holtzen, Dwight Alan, electrochemistry
Hurd, Lawrence Edward, ecology
Jain, Mahendra Kumar, biophysics, neurosciences
Jones, Louise Hinrichsen, applied mathematics, computer science
Jones, Richard Hamilton, physical oceanography
Jordan, Robert R., geology
Kelsey, Lewis Preston, entomology
Kennard, Robert Wakely, mathematical statistics
Kerner, Edward Haskell, theoretical physics, theoretical biology
Kleinman, Ralph Ellis, applied mathematics
Klemas, Vytas, marine sciences, remote sensing
Kraft, John Christian, geology, stratigraphy
Krause, James Barber, embryology
Krivanek, Neil Douglas, toxicology
Kupferman, Stuart L., physical oceanography
Kwart, Harold, physical chemistry, organic chemistry
Leavens, Peter Backus, mineralogy
Leismeier, Ronald Newell, polymer chemistry
Lerner, Harry, electrochemistry
Libera, Richard Joseph, mathematics
Liebhardt, William C., soils, plant physiology
Lighty, Richard William, horticulture
Lin, Kang, organic chemistry
Ling, Hubert, biochemical genetics
Lippert, Arnold Leroy, organic chemistry, mycology
Livingston, Albert Edward, mathematical analysis
Lunger, Philip Dupont, zoology, virology
Lutz, Bruce Charles, physics
MacCreary, Donald, entomology
Maciag, William John, Jr., microbiology
Marascia, Frank Joseph, organic chemistry
Martin, John Robert, analytical chemistry
Mason, Charles Eugene, insect ecology, apiculture
Mather, John Russell, climatology
McCullough, Roy Lynn, physical chemistry
McCurdy, Wallace Hutchinson, Jr, analytical chemistry
Mehl, James Bernard, physics
Miller, John Henry, III, electronic physics
Minn, James, organic chemistry
Mitchell, William H., agronomy, botany
Moore, James Alexander, organic chemistry
Morehart, Allen L., phytopathology, medical mycology
Munson, Burnaby, physical chemistry, analytical chemistry
Murphey, Frank J., biology

Murray, Richard Bennett, solid state physics
Myers, Thomas DeWitt, invertebrate zoology, oceanography
Nelson, Jerry Allen, organic chemistry
Noble-Harvey, Jane, animal virology
Noggle, Joseph Henry, physical chemistry
Nyce, Jack Leland, polymer chemistry, rubber chemistry
Onn, David Goodwin, cryogenics, solid state physics
O'Rourke, F L Steve, horticulture
Pellicciaro, Edward Joseph, mathematics
Pence, Jacques Jean, microbiology
Pickett, Thomas Ernest, geology
Preiss, John William, physics
Price, William Alrich, Jr., pharmacology
Reid, Donald Eugene, organic chemistry
Reinhardt, Charles Francis, occupational medicine, toxicology
Reitnour, Clarence Melvin, animal nutrition
Remage, Russell, Jr., mathematics
Ridge, Douglas Poll, physical chemistry
Rosenberg, Hans Reinhard, organic chemistry
Rosenberger, John Knox, animal virology, avian pathology
Roth, Roland Ray, ecology, vertebrate biology
Rothwarf, Allen, theoretical solid state physics
Rust, Richard W, insect ecology
Salsbury, Robert Lawrence, ruminant nutrition, microbial biochemistry
Sammelwitz, Paul H., reproductive physiology
Sarner, Stanley Frederick, physical chemistry
Schneider, Philip William, Jr., fisheries, toxicology
Schweizer, Edward Ernest, organic chemistry
Sharnoff, Mark, molecular physics, solid state physics
Sharpe, Thomas Ray, medicinal chemistry
Sheppard, David E., biochemical genetics
Sheridan, Robert E., geology
Skopik, Steven D., physiology
Sloyer, Clifford W., Jr., topology
Somers, George Fredrick, Jr., plant physiology
Staagold, Ivar, applied mathematics
Stegner, Robert W., biology
Stetson, Milton H., reproductive endocrinology, biological rhythms
Stevens, Evelyn Victoria, enzymology
Stopps, Gordon James, internal medicine, toxicology
Stula, Edwin Francis, veterinary pathology
Stump, John M, pharmacology
Svec, Leroy Vernon, agronomy, plant physiology
Teikes, Maria, energy conversion
Thompson, Allan M, petrology
Trabant, Edward Arthur, applied mathematics
Tripp, Marenes Robert, invertebrate pathology
Trofimenko, Swiatoslaw, organic chemistry
Trunbore, Conrad Noble, physical chemistry
Vassiliou, Eustathios, physical chemistry, plastics
Vernier, Vernon George, pharmacology
Vincent, Walter Sampson, Jr., zoology
von Frankenberg, Carl Alexander, physical chemistry
Wagner, Roger Curtis, cell biology
Waid, Margaret Cowsar, applied mathematics
Warfield, George, solid state electronics
Weaver, Jeremiah William, organic chemistry
Weinacht, Richard Jay, mathematics
Wenger, Ronald Harold, mathematics
West, William Alvin, organic chemistry
Wetlaufer, Donald Burton, biochemistry
Whitney, Charles Candee, Jr., drug metabolism, biopharmaceutics
Williams, Ferd Elton, physics
Wlech, Raymond Lee, organic chemistry
Wolfe, Stephen James, mathematics
Woo, Shien-Biau, atomic physics, molecular physics
Wood, Robert Hemsley, physical chemistry
Wood, Thomas Ross, biochemistry, health sciences
Woodhouse, John Crawford, chemistry

Wriston, John Clarence, Jr., biochemistry
Yolles, Seymour, chemistry
Zehrung, Winfield Scott, III, organic chemistry
Zikakis, John Philip, biological sciences, biochemistry

NEWPORT
Clarke, John Frederick Gates, Jr., analytical chemistry
Jaffe, Edward E., organic chemistry
Johnson, Roger Alvin, industrial organic chemistry

SEAFORD
Ahrens, Rolland William, physical chemistry
Houghland, Geoffrey Van Clief, plant physiology
Kittila, Allan Benona, physical chemistry
Norton, Lilburn Lafayette, organic chemistry
Prohaska, Charles Anton, nuclear chemistry
Sauerbrunn, Robert Dewey, analytical chemistry
Shuler, Woodfin Epps, physical chemistry
Stone, Robert Marion, physical chemistry

WILMINGTON
Abbiss, Joseph William, clinical pathology
Abernathy, Henry Herman, polymer chemistry
Abrahamson, Earl Arthur, analytical chemistry
Abrams, Lloyd, physical chemistry
Adelman, Robert Leonard, organic chemistry
Ahramjian, Leo, organic chemistry
Ainbinder, Zarah, organic chemistry
Akeley, David Francis, polymer chemistry
Aldrich, Paul E., organic chemistry
Alexander, James Ernest, organic polymer chemistry, rubber chemistry
Allen, Nelson, plastics chemistry
Alvarez, Vincent Edward, chemistry
Andersen, Donald Edward, physical chemistry, research administration
Anderson, Arthur William, polymer chemistry
Anderson, Burton Carl, organic polymer chemistry
Anderson, Lewis L., physical biochemistry
Angelo, Rudolph J., organic chemistry
Angerer, John David, organic chemistry, polymer chemistry
Ansul, Gerald R., organic chemistry
Apotheker, David, polymer chemistry
Applegate, Lynn E., water chemistry
Arhart, Richard James, organic polymer chemistry
Armstrong, Robert Krick, organic chemistry
Arimoto, Fred Shunji, organic chemistry
Arrington, Charles Hammond, Jr., physical chemistry
Armbrecht, Frank Maurice, Jr., organometallic chemistry
Arthur, Paul, Jr., chemistry
Athey, Robert Jackson, organic chemistry
Armitage, John Brian, organic polymer chemistry
Audermarsh, Carl Albert, Jr., organic chemistry
Auspos, Lawrence Arthur, chemistry
Autenrieth, John Stork, organic chemistry
Avakian, Peter, polymer physics
Baer, Donald Robert, research administration
Baidins, Andrejs, physical chemistry
Bair, Thomas Irvin, polymer chemistry
Baird, Richard Leroy, physical organic chemistry
Baker, Bertsil Burgess, analytical chemistry
Baker, Harris Mitchell Jr., analytical chemistry
Ballou, Jack Wayne, polymer physics
Bankert, Ralph Allen, organic chemistry
Bannerman, Douglas George, textile chemistry
Bannister, Robert Grimshaw, textile chemistry
Bare, Paul Orville, organic chemistry
Barker, Harold Clinton, organic chemistry, polymer chemistry
Barnes, David Kennedy, organic chemistry

Barney, Arthur Livingston, polymer chemistry
Barrier, George Edgar, plant physiology
Barron, Eugene Roy, organic chemistry
Bartels, George William, Jr, organic chemistry
Barth, Howard Gordon, analytical chemistry
Barton, Randolph, Jr, physical chemistry, crystallography
Bartron, Lester Ray, organic chemistry
Bauchwitz, Peter Siegbert, organic chemistry, physical chemistry
Bauer, Albert Webb, organic chemistry
Baum, Arthur Aloysius, organic chemistry
Baylor, Charles, Jr, organic chemistry, photographic chemistry
Beare, Steven Douglas, organic chemistry, physical chemistry
Becher, Paul, physical chemistry
Bechtold, Max Fredrick, physical chemistry
Beckerbauer, Richard, organic chemistry
Bellina, Russell Frank, organic chemistry
Bellis, Harold E, physical chemistry, mathematics
Benson, Frederic Rupert, organic chemistry, information science
Benson, Richard Edward, analytical chemistry
Bercaw, James Robert, textile products
Beresniewcz, Aleksander, physical chemistry
Bergna, Horacio Enrique, physical chemistry, colloid chemistry
Berkheimer, Henry Edward, rubber chemistry
Berta, Dominic Andrew, plastic chemistry
Beyer, Elmo Monroe, Jr, plant physiology, plant chemistry
Bingham, Richard Charles, organic chemistry
Birkenhauer, Robert Joseph, textile chemistry
Bissot, Thomas Charles, inorganic chemistry
Bither, Tom Allen, Jr, chemistry, physical chemistry
Bjornson, August Sven, organic
Black, Carl (Ellsworth), polymer science
Blaker, Robert Hockman, physical chemistry, information science
Blankenstein, William E, organic chemistry
Bonner, Willard Hallam, Jr, organic chemistry
Booth, Bruce L, solid state physics, lasers
Borchardt, Hans J, fluorine chemistry
Borchardt, John Keith, synthetic organic chemistry, physical organic chemistry
Boswell, George A, Jr, organic chemistry
Boughton, John Harland, inorganic chemistry
Bowers, George Henry, III, organic chemistry
Boyd, Samuel Neil, Jr, organic chemistry
Boylen, Joyce Beatrice, biochemistry
Brame, Edward Grant, Jr, analytical chemistry
Brandner, John David, industrial chemistry
Braun, Juergen Hans, inorganic chemistry, physical chemistry
Brehm, Warren John, chemistry
Bremer, Keith George, organic chemistry
Brennan, Gerald L, inorganic chemistry
Breslow, David Samuel, organic chemistry, polymer chemistry
Briden, Roger Clarence, clinical chemistry
Briggs, Paul Clayton, Jr, organic chemistry
Brill, Harold Clifford, chemistry
Brinker, Keith Clark, organic chemistry
Bristowe, William Warren, physical chemistry
Brittelli, David Ross, organic chemistry, medicinal chemistry
Brixner, Lothar Heinrich, inorganic chemistry
Brown, Charles Julian, Jr, physical chemistry
Brown, Morton, polymer chemistry
Brown, Robert G, physics

Brown, Stewart Cliff, plastics chemistry, polymer chemistry
Bruce, John MacMillan, Jr, organic chemistry
Buchanan, James Balfour, organic chemistry
Buchta, Raymond Charles, analytical chemistry
Buck, Warren Howard, physical chemistry, polymer chemistry
Burdick, Charles Lalor, chemistry
Burg, Marion, organic chemistry
Burns, Richard Charles, biochemistry
Burton, Louis Lasseter, physical chemistry
Buser, Kenneth Rene, organic chemistry
Bushey, William Raymond, inorganic chemistry, chemical kinetics
Butler, Robert Westbrook, polymer chemistry, organic chemistry
Butterman, William Charles, mineralogy, geology
Cairncross, Allan, organic chemistry
Cairns, Theodore L, organic chemistry
Calkins, William Harold, industrial chemistry, polymer chemistry
Carboni, Rudolph A, organic chemistry
Carlson, Bruce Arne, organic chemistry, organometalic chemistry
Carlson, Norman Arthur, organic chemistry
Carnahan, James Elliot, entomology, environmental biology
Carrano, Richard Alfred, pharmacology, toxicology
Castle, John Edwards, applied chemistry
Cates, Harry Louis, Jr, plastics chemistry
Caywood, Stanley William, Jr, organic chemistry
Cella, Richard Joseph, Jr, polymer chemistry, physical chemistry
Cessna, Lawrence C, Jr, polymer physics, chemical engineering
Chambers, Vaughan Crandall, (Jr), organic chemistry
Chang, Catherine Teh-Lin, photochemistry, photographic chemistry
Chang, Yu-Wei, organic chemistry
Chaudhari, Bipin Bhudharlal, medicinal chemistry
Cheng, Fred Fa Wu, organic chemistry, analytical chemistry
Cheng, Lawrence Kar-Hiu, pharmaceutics
Cherkofsky, Saul Carl, organic chemistry
Chevrier, Jean-Claude Jacques, solid state physics
Chiu, Jen, analytical chemistry, polymer chemistry
Clement, Robert Alton, industrial organic chemistry
Cline, Edward Terry, organic chemistry
Cluff, Edward Fuller, organic chemistry
Cohen, Gordon Mark, physical organic chemistry
Citron, Joel David, organic chemistry, polymer chemistry
Class, Jay Bernard, organic chemistry
Clayton, Anthony Broxholme, organic chemistry
Cleary, Laurence Twomey, textiles, chemistry
Chollet, Raymond, plant physiology
Chromey, Fred Carl, applied physics
Chu, Victor Fu Hua, physical chemistry
Cohen, Martin Allen, inorganic chemistry, textile technology
Cole, George Rolland, solid state science
Coleman, Marcia Lepri, chemical chemistry
Collette, John Wilfred, organic chemistry
Comen, Alan Lee, organic chemistry
Conner, Albert Z, analytical chemistry
Cook, Donald Bowker, physics
Cook, Gordon Smith, organic chemistry
Cooney, John Leo, physical chemistry
Cooper, Terence Alfred, polymer chemistry
Copelin, Harry B, organic chemistry
Coraor, George Robert, organic chemistry
Cords, Donald Philip, organic chemistry
Corner, James Oliver, organic polymer chemistry
Coulson, Dale Robert, organic chemistry, organometallic chemistry
Coyner, Eugene Casper, chemistry
Craig, Alan Daniel, industrial chemistry, pharmaceutical chemistry
Cramer, Richard (David), organometallic chemistry
Crary, James Walter, organic chemistry
Craven, James Milton, polymer chemistry
Crean, Patrick J, organic chemistry
Cripps, Harry Norman, organic chemistry
Criswell, Jerome Glenn, crop physiology

Croach, Jesse William, Jr, physics, research administration
Cummins, Earl Wesley, organic chemistry
Cupery, Willis Eli, pesticide chemistry
Dahl, Alton, physical chemistry
Daly, John Joseph, Jr, organic chemistry
Darby, Robert Albert, organic chemistry
Darnell, Frederick Jerome, solid state physics
Dasgupta, Sunil Priya, solid state physics, organic chemistry
Dautlick, Joseph X, clinical biochemistry
David, Israel A, organic polymer chemistry
Davidson, William John, physical chemistry
Davies, Robert Dillwyn, physics
Davis, Vernam Terrell, psychiatry
Davison, Robert Wilder, physical chemistry
Dawson, Robert Louis, polymer chemistry
Day, Bruce Frederick, organic chemistry
DeBrunner, Marjorie R, organic chemistry
DeDominicis, Alex John, physical chemistry
Delate, Edward Joseph, applied statistics
Delmonte, David William, organic chemistry, polymer chemistry
Delp, Charles Joseph, plant pathology
Dennison, Richard Wheeler, textile chemistry
Dettre, Robert Harold, physical chemistry
Di Giacomo, Armand, polymer chemistry
Dills, William Leonard, chemistry, spectrochemistry
Dippel, William Alan, analytical chemistry
Dively, William Russell, organic chemistry
Domanski, John Joseph, Jr, plant physiology, weed science
Donahue, Joseph E, chemistry, textile chemistry
Donnan, Alvan, organic chemistry
Donohue, Paul Christopher, solid state chemistry, inorganic chemistry
Downing, Joseph Richard, industrial chemistry, spectrochemistry
Downing, Ralph Churchman, organic chemistry
Drinkard, William Charles, Jr, inorganic chemistry
Droucas, John, organic chemistry, polymer chemistry
Drukker, Alexander Emanuel, chemistry, information science
Dunlop, Edward Clarence, chemistry
Dunworth, William Paul, organic chemistry
Durandetta, Donald W, organic polymer chemistry, clinical chemistry
Durbin, Ronald Priestley, analytical chemistry
Dyson, Ian Fraser, organic chemistry
Earle, Ralph Hervey, Jr, science administration, industrial chemistry
Easley, Warren C, physical chemistry
Eaton, David Fielder, physical organic chemistry
Ebert, Philip E, textile chemistry
Edwards, Walter Murray, polymer chemistry
Ehrich, Felix Frederick, organic chemistry
Eichel, Herman Joseph, biochemistry, organic chemistry
Eichman, Martin L, pharmaceutics
Eleuterio, Herbert Sousa, organic chemistry
Elia, Raymond J, organic chemistry
Ellingboe, Ellsworth Knowlton, organic chemistry
Elliott, Ralph Benjamin, physical chemistry
Emmick, Robert D, organic chemistry
Engelhardt, Vaughn Arthur, organic chemistry
Eños, Herman Isaac, Jr, organic chemistry
Erdman, John Paul, rubber chemistry
Ernsberger, Maurice Leon, organic chemistry
Ernst, Richard Edward, industrial chemistry
Esayian, Manuel, organic chemistry
Espy, Herbert Hastings, physical organic chemistry
Evans, Franklin James, Jr, organic chemistry
Eyler, Robert Wilson, analytical chemistry
Fahl, Roy Jackson, Jr, industrial chemistry
Fahsel, Michael John, analytical chemistry

Fassnacht, John Hartwell, organic chemistry
Fawcett, Mark Stanley, organic chemistry
Feltzin, Joseph, organic chemistry
Fenoglio, Richard Andrew, organic chemistry
Ferguson, Raymond Craig, physical chemistry
Fielding, Max Jae, nematology
Fields, Melvin, organic chemistry
Finlay, Joseph Burton, organic chemistry
Fitzgerald, Emerson Blanchard, polymer chemistry
Fleming, Richard Allan, plastics chemistry
Fleming, Sydney Winn, physical chemistry
Flexman, Edmund A, Jr, polymer chemistry
Flippen, Richard Bernard, solid state physics
Flournoy, Philip Alexander, applied physics
Fogiel, Adolf W, polymer chemistry, rubber chemistry
Foldi, Andrew Peter, organic chemistry, polymer chemistry
Ford, Thomas Aven, chemistry
Forshey, William Osmond, Jr, organic chemistry
Foss, Robert Paul, physical organic chemistry, polymer chemistry
Foster, Robert Everett, polymer chemistry, research administration
Fowler, John Rayford, inorganic chemistry, polymer chemistry
Frankenburg, Peter Edgar, organic chemistry
Franta, William Alfred, polymer chemistry, plastics engineering
Frawley, John Paul, toxicology
Freedman, Henry Hillel, immunology
French, James Edwin, inorganic chemistry, polymer chemistry
Frensdorff, Hans Karl, physical chemistry, polymer chemistry
Frey, Harold Joseph, physical chemistry
Fuchs, Julius Jakob, organic chemistry
Fullhart, Lawrence, Jr, chemistry
Funck, Dennis Light, polymer chemistry
Gagliano, Louis John, inorganic chemistry
Gailey, Joseph A, analytical chemistry
Gakenheimer, Walter Christian pharmaceutical chemistry
Gall, Walter George, organic chemistry
Gano, Robert Daniel, organic chemistry
Gardiner, John Alden, pesticide chemistry, analytical chemistry
Garrett, Robert Roth, physical chemistry
Garrison, William Emmett, Jr, organic chemistry
Gay, Frank P, polymer science
Geer, Richard P, polymer chemistry, organic chemistry
Genge, Colin Arthur, physical chemistry
George, Daniel Eugene, organic chemistry
Gerlach, Howard G, Jr, clinical chemistry
Gibbs, Hugh Harper, organic chemistry
Gierke, Timothy Dee, physical chemistry
Gilbert, Arthur Donald, industrial chemistry
Gillow, Edward William, physical chemistry, inorganic chemistry
Gladding, Elinor Hartnell, chemistry
Glaeser, Hans Hellmut, inorganic chemistry
Glasebrook, Arthur Lawrence, industrial chemistry
Glasser, Leo George, physical optics
Glenn, Furman Eugene, organic polymer chemistry
Gloor, Walter Ervin, chemistry
Goddu, Robert Fenno, polymer chemistry
Golike, Ralph Crosby, physical chemistry
Gonick, Ely, inorganic chemistry
Goodman, Alan Lawrence, organic chemistry
Goodman, Albert, polymer chemistry, organic chemistry
Gosser, Lawrence Wayne, physical chemistry, organometallic chemistry
Gould, Charles Webster, information science
Graham, Augustus Washington, inorganic chemistry
Graham, Boynton, chemistry
Graham, James Carl, agronomy, botany
Greiner, Richard William, physical chemistry, organic chemistry
Gribbins, Myers Floyd, biochemistry
Griffith, Lewis John, bacteriology
Griffiths, David, physics

Griswold, Paul Hulet, Jr., organic chemistry
Grogan, Robert Mann, economic geology
Gros, Walther Gustav Fredrich, organic polymer chemistry, electrochemistry
Gruber, Wilhelm F., organic chemistry
Grunert, Rudolf Richard, biochemistry
Guggenberger, Lloyd Joseph, organometallic chemistry, polymer chemistry
Guillotte, John Edward, mathematical statistics
Gulrich, Leslie William, Jr., physical chemistry
Gumprecht, William Henry, organic chemistry
Haas, Charles Warren, theoretical solid state physics
Hagen, Richard Martin, physical chemistry
Hager, Glenn Frederick, textile chemistry
Haglid, Frank Runar, pesticide chemistry
Hall, Edward Duncan, physical chemistry
Hamilton, Jefferson Merritt, Jr., fluorine chemistry
Hamilton, Paul Barnard, biochemistry
Hammer, Clarence Frederick, Jr., polymer chemistry
Han, Jerry C-Y., pesticide chemistry
Handy, Carleton Thomas, textile chemistry
Hanzel, Robert Stephen, organic chemistry, polymer chemistry
Haq, Mohammad Zamir-ul, analytical chemistry, organic chemistry
Hardham, William Morgan, photographic chemistry
Hardy, Henry Benjamin, Jr., textile physics
Hardy, Ralph Wilbur Frederick, biochemistry
Hargreaves, Chester Arthur II, organic chemistry
Harmuth, Charles Moore, organic chemistry
Harrell, Jerald Rice, polymer chemistry
Harris, George Christie, organic chemistry
Harris, John Ferguson, Jr., organic chemistry
Hartzler, Harris Dale, organic chemistry
Harvey, John, Jr., organic chemistry
Hasek, William Robert, organic chemistry
Hasty, Noel Marion, Jr., organic chemistry
Hatchard, William Reginald, polymer chemistry
Haubein, Albert Howard, organic chemistry
Haugh, Eugene (Frederick), physical chemistry
Hause, Norman Laurance, organic chemistry
Hauser, Paul Matthew, organic chemistry
Havelka, Ulysses D., agronomy
Hayek, Mason, organic chemistry
Hays, John Thomas, organic chemistry, information science
Heaton, Richard Clawson, analytical chemistry
Heberling, Jack Waugh, Jr., organic chemistry
Heldt, Walter Z., organic chemistry
Herbin, William Fitts, biochemistry, pharmacology
Herglotz, Heribert Karl Josef, physical chemistry
Herkes, Frank Edward, industrial organic chemistry
Hershkowitz, Robert L., textile chemistry
Herskovitz, Thomas, organometallic chemistry, bioinorganic chemistry
Hertier, Walter Raymond, organic chemistry
Hess, Richard William, chemistry
Hess, William Wilson, organic chemistry
Heytler, Peter George, biochemistry
Hicks, Elija Maxie, Jr., organic polymer chemistry
Higgins, Norton Allen, chemistry
Hilfiker, Franklin Roberts, organic chemistry
Hill, Frederick Burns, Jr., fluorine chemistry, polymer chemistry
Hill, Gideon D., agronomy
Hill, James Theo, polymer chemistry
Hiller, Julian Werner, chemistry
Hirsy, Sylvain Max, physical chemistry
Hirwe, Ashalata Shyamsunder, synthetic organic chemistry
Ho, Floyd Fong-Lok, analytical chemistry

Hobgood, Richard Troy, Jr., physical chemistry
Hock, Charles William, microscopy
Hoegger, Erhard Fritz, organic chemistry
Hoehn, Harvey Herbert, organic polymer chemistry
Hoeschele, Guenther Kurt, organic polymer chemistry
Hoh, George Lok Kwong, organic chemistry
Hoiness, Connie Marquez, organic chemistry
Hoiness, David Eldon, organic chemistry
Holfeld, Winfried Thomas, organic chemistry
Holland, Russell Sedgwick, photographic science
Holmes, David Willis, organic chemistry
Holmes, Richard, organic chemistry
Holmquist, Howard Emil, organic chemistry
Holsten, Richard David, plant physiology, biochemistry
Honsberg, Wolfgang, organic chemistry, polymer chemistry
Hood, Horace Edward, organic chemistry
Hoover, Fred Wayne, organic chemistry
Hopkins, Elbert Erskine, physical chemistry
Hoppenjans, Donald William, physical inorganic chemistry
Howe, Donald Eugene, analytical chemistry
Howe, King Lau, physical organic chemistry, polymer chemistry
Hubben, Klaus, veterinary pathology, toxicology
Hull, Prentice Roy, organic chemistry
Hume, Harold Frederick, textile chemistry
Hummel, Donald George, organic chemistry
Hunter, Frank Ray, organic chemistry
Huntsberger, James Robert, surface chemistry, rheology
Huppe, Francis Frowin, solid state physics
Hurford, Thomas Rowland, polymer chemistry
Hyson, Archibald Miller, chemistry
Ihde, Keith Desmond, entomology
Ikeda, Richard Masayoshi, polymer chemistry
Iler, Ralph Kingsley, chemistry
Ingersoll, Henry Gilbert, physical chemistry
Inskip, Harold Kirkwood, organic chemistry
Ireland, Carol Beard, physical chemistry
Ivett, Reginald William, industrial chemistry
Jabloner, Harold, polymer chemistry, organic chemistry
Jackson, Julius, inorganic chemistry
Jackson, Thomas Edwin, medicinal chemistry
Jacobs, Emmett S., air pollution chemistry, catalysis
James, Daniel Shaw, organic chemistry
Jamison, Joel Dexter, organic chemistry
Jankowski, Stanley John, analytical chemistry
Janson, Peter Allan, optical physics
Jarrett, Howard Starke, Jr., magnetism, solid state physics
Jeischo, Wolfgang K., chemistry, crystallography
Jelinek, Arthur Gilbert, synthetic organic chemistry
Jenkins, William A., physical chemistry
Jenkins, Wilmer Atkinson, II, physical chemistry, inorganic chemistry
Jenner, Edward Levant, agricultural biochemistry
Jepson, Carl Henry, polymer chemistry, plastics engineering
Jewell, Richard A., organic chemistry
John, Andrew, photochemistry, surface chemistry
Johnson, Alexander Lawrence, organic chemistry
Johnson, Donald Richard, analytical chemistry
Johnson, Kenneth Earl, chemistry
Johnson, Melvin Clark, toxicology, pharmacology
Johnson, Paul Robert, rubber chemistry, polymer chemistry

Johnson, Rayner Selby, organic chemistry
Johnson, Robert William, polymer chemistry
Johnson, Rulon Edward, Jr., physical chemistry
Johnson, Vandliff, solid state chemistry
Johnston, Frederick Lewis, organic chemistry
Jolley, John Eric, physical chemistry
Jordan, Kenneth Gary, physical chemistry
Jordan, Walter Edwin, physics
Joyce, Robert Michael, organic chemistry
Julin, Bruce Gustav, analytical chemistry, pesticide chemistry
Kahn, A. Clark, biochemistry
Kaminski, Louis Alfred, clinical chemistry, instrumentation
Kaplan, Ralph Benjamin, organic chemistry
Kasowski, Robert V., solid state physics
Kassal, Robert James, organic chemistry
Katz, Manfred, polymer chemistry, textile chemistry
Kauer, James Charles, organic chemistry
Kay, Peter Steven, clinical chemistry
Kealy, Thomas Joseph, organic chemistry
Keating, James T., organic chemistry, electrochemistry
Kegelman, Matthew Roland, chemistry
Keidel, Frederick Andrew, physical chemistry
Keim, Gerald Inman, paper chemistry
Kelemen, Denis George, physical chemistry
Keller, Philip Joseph, physical chemistry
Kendrick, Lawrence W., Jr., organic chemistry
Keown, Robert William, organic chemistry
Khan, Ausat Ali, organic chemistry, polymer chemistry
Killian, Frederick Luther, polymer chemistry
Kinsinger, William George, microscopy
Kirkland, Joseph Jack, analytical chemistry
Kissa, Erik, colloid chemistry, textile chemistry
Kitson, Robert Edward, polymer chemistry
Kittila, Richard Sulo, organic chemistry
Klacsmann, John Anthony, organic chemistry
Kleinschuster, Jacob John, organic chemistry
Klopping, Hein Louis, organic chemistry, polymer chemistry
Knesley, Joseph Wayne, industrial chemistry
Knipmeyer, Hubert Elmer, organic chemistry, research administration
Knobloch, Fred William, organic chemistry, polymer chemistry
Knop, Harry William, Jr., physics
Knoth, Walter Henry, Jr., organometallic chemistry, inorganic chemistry
Knowles, Joseph N., organic chemistry
Knox, Andrew Gibson, physical chemistry
Kobetich, Edward John, textile physics, materials science
Koch, Theodore Augur, physical chemistry
Kogon, Irving Charles, organic chemistry
Kohan, Melvin Ira, organic polymer chemistry
Kolb, Harry John, chemistry
Koller, Charles Richard, organic chemistry
Konizer, George Burr, physical organic chemistry
Korant, Bruce David, virology
Kosak, John R., industrial chemistry
Krackow, Mark Harry, organic chemistry
Kramer, Brian Dale, physical organic chemistry
Krause, Robert Louis, pathology
Krespan, Carl George, organic chemistry
Kropp, William A., physics
Kruse, Walter, inorganic chemistry, physical chemistry
Krusic, Paul Joseph, physical chemistry
Krysiak, Henry R., organic chemistry
Kung, Harold Hing Chuen, physical chemistry
Kuratle, Henry III, biology, agriculture
Kutner, Abraham, organic polymer chemistry, photochemistry
Kwok, Wo Kong, organic chemistry
Kwolek, Stephanie Louise, polymer chemistry
Landert, Harold Paul, organic chemistry
Landoll, Leo Michael, polymer chemistry

Langkammerer, Carl Martin, organic chemistry
Langsdorf, William Philip, physical organic chemistry
Lann, Joseph Sidney, organic chemistry
Lanzl, George Frank, physical chemistry
Larson, Lester Mikkel, chemistry
Lauderback, Sanford Keith, physical chemistry
Lautenberger, William J., applied chemistry, textile chemistry
Lavery, Bernard James, physical chemistry
Law, Amy Stauber, clinical biochemistry
Lawless, Gregory Benedict, pharmaceutical chemistry
Lazaridis, Christina Nicholson, organic chemistry, photographic chemistry
LeBlanc, Norman Francis, analytical chemistry, research administration
LeBleu, Ronald Eugene, organic chemistry
Lee, Shung-Yan Luke, surface chemistry
Leibu, Henry J., chemistry
Leitch, Robert Edgar, Jr., analytical chemistry
Lenher, Samuel, chemistry
Leser, Ernst Eugene, organic chemistry
Levitt, George Kenneth, chemical physics
Levy, Paul F., analytical chemistry, instrumentation
Lew, Baak Wai, organic chemistry
Lewis, Ernest Eugene, plastics chemistry
Lewis, George Kenneth, organic chemistry
Lewis, John Raymond, polymer chemistry
Little, Ernest Lewis, Jr., chemistry
Lockart, Royce Zeno, Jr., microbiology
Lockwood, William Howard, chemistry
Logullo, Francis Mark, organic polymer chemistry
Lonberg-Holm, Knud Karl, biochemistry
Londergan, Martin Christoper, physical chemistry
Long, James Delbert, horticulture, plant physiology
Longworth, Ruskin, polymer chemistry
Lorenz, Carl Edward, organic chemistry
Lou, Marjorie Feng, biochemistry
Lowen, Warren Kealoha, analytical chemistry
Luckenbaugh, Raymond Wilson, agricultural chemistry
Ludwig, Richard Eli, organic chemistry
Lukach, Carl Andrew, polymer chemistry
Lund, John Turner, physical chemistry
Lupton, John Madison, physical chemistry
Lyon, Donald Wilkinson, inorganic chemistry
Lyons, Peter Francis, physical chemistry, polymer science
MacDonald, Robert Neal, organic chemistry
MacLachlan, James Daniel, polymer chemistry, organic chemistry
Maerov, Sidney Benjamin, organic chemistry
Magat, Eugene Edward, chemistry
Mahler, Walter, inorganic chemistry, organic chemistry
Mahlman, Bert H., polymer chemistry, materials science
Mair, Robert Dixon, physical chemistry
Malbica, Joseph Orazio, biochemistry
Malick, Jeffrey Bevan, neuropharmacology
Malone, Creighton Paul, physical chemistry
Maloney, Daniel Edwin, polymer chemistry
Malya, Govinda P A., clinical biochemistry, pathological chemistry
Mandel, Zoltan, textile chemistry
Mannis, Fred, physical chemistry
Manzer, Leo Ernest, organometallic chemistry
Marcus, Sanford M., physics
Markievitz, Kenneth Helmut, polymer chemistry
Marquard, Donald Wesley, statistics
Marshall, Edwin Randolph, chemistry
Marshall, Thomas Ball, polymer chemistry
Martin, Arthur Francis, chemistry

Mathre, Owen Bertwell, analytical chemistry
Matlack, Albert Shelton, organic polymer chemistry
Mattair, Robert, inorganic chemistry, environmental systems & technology
Maury, Lucien Garnett, physical chemistry, organic chemistry
Maxfield, Mary Evans, physiology
May, Ralph Forrest, agricultural chemistry
Maynard, Carl Wesley, Jr, organic chemistry
Maynard, John Thomas, organic chemistry
McBride, Edward Francis, organic chemistry
McBride, John Joseph, physical chemistry
McBurney, Lane Fordyce, organic chemistry
McCabe, Chester Charles, polymer science
McCane, Donald Irwin, organic chemistry
McCarthy, Raymond Lawrence, physics
McCartney, John Richard, physical chemistry
McClelland, Alan Lindsey, inorganic chemistry, research administration
McClure, George Richard, organic chemistry, physical chemistry
McClure, John Hibbert, analytical chemistry
McCormack, William Brewster, organic chemistry
McCoy, V Eugene, Jr, physical organic chemistry
McCurdy, David Harold, pharmacology
McDevit, William Ferris, physical chemistry
McDonald, Charles Cameron, physical chemistry
McEwen, Charles Nehemiah, mass spectrometry
McGahen, Joe Winfield, microbiology
McGinnis, William Joseph, inorganic chemistry, physical chemistry
McGirk, Richard Heath, organic polymer chemistry
McKinney, Charles Dana, Jr, physical chemistry
McMillian, Frank Lebarron, organic chemistry
Mckler, Arlen B, organic chemistry
Melby, Lester Russell, organic chemistry
Memeger, Wesley, Jr, organic chemistry, analytical chemistry
Merrifield, Richard Ebert, chemical physics
Merrill, John Richard, physical chemistry
Messer, Wayne Ronald, organic chemistry, photochemistry
Metzger, James Douglas, organic chemistry, agricultural chemistry
Meyer, Gregory Carl, organic chemistry
Meyer, James Melvin, polymer chemistry
Meyer, William Paul, organic chemistry
Michel, Rudolph Henry, organic chemistry
Mighton, Charles Joseph, organic chemistry
Mighton, Harold Russell, chemistry
Mikell, William Gaillard, polymer physics, textiles
Miles, James Lowell, chemistry, biochemistry
Milian, Alwin S, Jr, fluorine chemistry
Miller, Ivan Keith, polymer chemistry
Mitchell, John, Jr, analytical chemistry
Mochel, Walter Edwin, physical chemistry, organic chemistry
Monagle, Daniel J, organic polymer chemistry
Moncure, Henry, Jr, organic polymer chemistry
Monroe, Bruce Malcolm, organic chemistry
Monroe, Elizabeth McLeister, organic chemistry
Moody, Frank Baldwin, biochemistry
Moore, Earl Philip, industrial chemistry
Moore, Ralph Bishop, industrial chemistry, analytical chemistry
Moran, Edward Francis, Jr, inorganic chemistry, organic chemistry
Morgan, Paul Winthrop, chemistry
Morgan, Robert Lee, organic chemistry
Mori, Peter Taketoshi, organic chemistry
Morrison, William Harvey, Jr, inorganic chemistry
Moses, Francis Guy, physical organic chemistry

Mrowca, Joseph J, organometallic chemistry
Mrozinski, Peter Matthew, applied physics, low temperature physics
Munn, George Edward, organic chemistry
Murray, Robert Marie, rubber chemistry
Myoda, Toshio Timothy, microbiology, biochemistry
Nader, Allan E, organic chemistry
Narvaez, Richard, polymer chemistry
Nash, James Lewis, Jr, polymer chemistry
Naylor, Marcus A, Jr, organic chemistry
Naylor, Robert Ernest, Jr, chemical physics
Neal, Thomas Edward, analytical chemistry, textile chemistry
Neil, Donald E, physical chemistry
Nersasian, Arthur, organic chemistry
Newby, William Edward, physical chemistry, organic chemistry
Newitt, Edward James, physical organic chemistry
Nichols, James Randall, physical chemistry
Nicolas, Jesus, science education
Niedzielski, Edmund Luke, petroleum chemistry
Nielsen, Norman Arnold, chemistry
Nolin, Joseph Arthur Benoit, polymer chemistry
Norling, Parry McWhinnie, polymer chemistry
O'Dell, Durward George, organic chemistry
Olesen, John Allen, organic chemistry
Olson, Carl Marcus, chemistry
Orphanides, Gus George, organic polymer chemistry
Osborn, Robert Henry, applied physics
Ostmann, Bernard George, textile technology
Overall, Derick William, physical chemistry
Overman, Joseph DeWitt, physical chemistry
Page, James A, medicine
Pailthorp, John Raymond, organic chemistry
Palmer, Alan Blakeslee, physical chemistry
Panar, Manuel, physical organic chemistry, polymer science
Pappas, Nicholas, organic chemistry, polymer chemistry
Pariser, Rudolph, physical chemistry
Park, Chung Ho, organic chemistry
Parkins, John Alexander, physical chemistry
Parrish, Robert G, physical chemistry
Parshall, George William, chemistry
Patel, Narayan Ganesh, toxicology
Patterson, George Harold, organic chemistry, fluorine chemistry
Patterson, Gordon Derby, Jr, analytical chemistry, polymer science
Paulshock, Marvin, organic chemistry
Pavlic, Albert Alan, organic chemistry
Peeples, John Lee, Jr, plant pathology, microbiology
Pegg, Philip John, clinical pathology
Pell, Sidney, biostatistics, epidemiology
Pelosi, Lorenzo Fred, organic polymer chemistry
Pensak, David Alan, chemistry, computer science
Percival, William Colony, organic chemistry
Perri, Joseph Mark, colloid chemistry
Pettit, Paul Herschel, Jr, industrial chemistry
Pfeiffer, Gerald Peter, textile chemistry
Pfeiffer, Joseph George, physical organic chemistry
Phillips, Brian Ross, physical chemistry
Phillips, William Dale, biophysics
Pierce, Marion Armbruster, physical chemistry
Pierce, Robert Henry Horace, Jr, chemistry
Pierrard, John Martin, air pollution, environmental sciences
Pieski, Edwin Thomas, physical chemistry
Plambeck, Louis, Jr, organic chemistry
Plunkett, Roy Joseph, organic chemistry
Podlas, Thomas Joseph, chemistry
Pontrelli, Gene J, textile physics
Porter, Hardin Kibbe, organic chemistry
Portnoy, Robert Charles, pesticide chemistry, synthetic organic chemistry
Potrafke, Earl Mark, environmental chemistry
Powelson, Dorothy May, microbiology
Prather, Charles Wayne, biochemistry
Pretka, John E, applied chemistry, textile chemistry

Price, Edward Hector, organic chemistry
Prichard, William W, organic chemistry
Proctor, James Simpson, chemistry
Prosser, Robert M, organic chemistry
Prosser, Thomas John, organic chemistry
Pruckmayr, Gerfried, organic chemistry
Pugh, Thomas L, analytical chemistry
Purdon, William Andrew Bowie, physical chemistry
Putnam, Stearns Tyler, paper chemistry, polymer chemistry
Quebedeaux, Bruno, Jr, plant physiology, crop science
Quisenberry, Richard Keith, organic chemistry
Raasch, Maynard Stanley, organic chemistry
Rainard, Leo Walter, chemistry
Ramler, Edward Otto, organic chemistry
Ranson, William Wade, organic chemistry
Rave, Terence William, polymer chemistry, organic chemistry
Raymond, Richard Laverne, microbiology
Raynolds, Stuart, organic chemistry, polymer chemistry
Read, Robert E, textile chemistry
Reap, James John, organic chemistry
Reardon, Joseph Edward, organic chemistry, polymer chemistry
Reddy, Gade Subbarami, physical chemistry
Reed, Donald Eugene, biochemistry
Reilly, Edward Leo, organic chemistry
Reimer, Carl Clayton, textile chemistry
Reitz, Carl Rex, organic chemistry
Remington, William Roscoe, organic chemistry
Repka, Benjamin C, Jr, polymer chemistry

Resnick, Paul R, fluorine chemistry
Restaino, Alfred Joseph, physical chemistry, polymer chemistry
Rexford, Dean R, fluorine chemistry
Reynard, John William, inorganic chemistry
Rhodes, Robert Carl, organic chemistry, organic chemistry
Richards, Bert Lorin, Jr, plant pathology
Richards, John Cadwallader, polymer chemistry
Richardson, Graham McGavock, organic chemistry
Richardson, Paul Noel, organic chemistry, polymer chemistry
Riches, Wesley William, physical chemistry
Richwine, John Robert, rubber chemistry, organic chemistry
Riggleman, James Dale, horticulture, plant physiology
Ring, Harold Francis, physics
Ringwald, Owen Edward, physical chemistry
Ripka, William Charles, organic chemistry, pharmaceutical chemistry
Roberson, Elbert B, Jr, organic chemistry, textiles
Robertson, James A, chemistry
Robinson, Ivan Maxwell, organic chemistry
Roder, Thomas Michael, organic chemistry
Rodowskas, Edward Laurence, chemistry
Rogers, James Wesley, physical chemistry, textile chemistry
Rogers, Tommie Gene, organic chemistry
Rolston, Charles Hopkins, industrial organic chemistry
Rondestvedt, Christian Scriver, Jr, industrial organic chemistry
Ropp, Walter Shade, chemistry
Rothrock, George Moore, chemistry
Rothrock, Henry Shirley, chemistry
Roussel, Philip Andrew, chemistry
Rudkin, George Osborne, physical organic chemistry
Rueggeberg, Walter Herman Carl, industrial chemistry
Rutledge, Thomas Franklin, organic chemistry
Ryker, Truman Clifton, plant pathology
Sadler, Monroe Scharff, physical chemistry
Saffer, Henry Walker, physical chemistry, textile chemistry
St John, Daniel Shelton, physical chemistry
Samuels, Robert Joel, polymer chemistry
Sandberg, Robert Gustave, clinical biochemistry
Sandell, Lionel Samuel, colloid chemistry
Sands, Seymour, textile chemistry
Sarasohn, Ilya M, physical organic chemistry
Sauers, Richard Frank, organic chemistry
Sausen, George Neil, organic chemistry

Schadt, Frank Leonard, III, organic chemistry
Schaefgen, John Raymond, polymer chemistry
Schappell, Frederick George, organic chemistry
Scheiber, David Hitz, polymer chemistry
Scheiderbauer, Robert Albert, chemistry
Schenker, Henry Hans, analytical chemistry
Scheve, Bernard Joseph, textile chemistry
Schexnayder, Mary Anne, organic chemistry, photochemistry
Schlatter, Rudolph, chemistry
Schmidt, Francis Henry, physical organic chemistry
Schmidt, Webster Raymond, chemistry
Schmiegel, Walter Werner, rubber chemistry
Schmude, Keith E, physical chemistry
Schneider, Jurg Adolf, physiology, pharmacology
Schreyer, Ralph Courtenay, organic polymer chemistry
Schroeder, Herman Elbert, organic chemistry
Schrof, William Ernst John, organic polymer chemistry
Schuler, Mathias John, textile technology, color science
Schulley, John Damian, organic chemistry
Schultz, John Lawrence, information science
Schunn, Robert Allen, inorganic chemistry
Schuurmans, Hendrik J L, polymer chemistry
Schwartz, Jerome Lawrence, clinical chemistry
Schweiger, James W, dentistry, prosthodontics
Schweitzer, Carl Earle, organic chemistry
Scofield, Dillon Foster, physics
Scoggin, John Kyle, insect toxicology
Scott, Samuel LeRoy, organic chemistry
Seaman, Richard Eric, polymer physics, textile physics
Searle, Norman Edward, organic chemistry, plant biochemistry
Senkler, George Henry, Jr, physical organic chemistry
Setterquist, Robert Alton, polymer chemistry, organometallic chemistry
Seufert, Ludwig E, organic chemistry, inorganic chemistry
Shaffer, Ralph Gunter, geography, political science
Shain, Albert Leopold, physical science
Shands, Alfred Rives, Jr, orthopedic surgery
Shanks, John Amos, organic chemistry
Shannon, Robert Day, solid state chemistry, inorganic chemistry
Sharkey, William Henry, polymer chemistry, organic chemistry
Sharp, Alvin George, organic chemistry
Sharp, Silas, entomology
Shealy, Otis Lester, organic chemistry, textiles
Sheer, Maxine Lana, physical chemistry, analytical chemistry
Sheeran, Patrick Jerome, pesticide chemistry
Shellenbarger, Robert Martin, organic chemistry
Sheppard, William Arthur, organic chemistry
Sherbeck, L Adair, polymer chemistry
Sherman, Albert Herman, organic chemistry
Shih, Hsiang, polymer chemistry
Shivers, Joseph Clois, Jr, polymer chemistry
Shoaf, Charles Jefferson, textile chemistry
Shozda, Raymond John, organic chemistry, inorganic chemistry
Siedschlag, Karl Glenn, Jr, organic chemistry
Sienicki, Edward Alexander, industrial chemistry, inorganic chemistry
Silverman, Howard L, instrumentation
Simeral, William Goodrich, polymer physics
Simmons, Howard Ensign, Jr, organic chemistry
Simms, John Alvin, organic polymer chemistry
Simpson, Clifford Carlton, Jr, physical chemistry
Simpson, David Alexander, physical organic chemistry
Skolnik, Herman, organic chemistry, information science
Slade, Arthur Laird, polymer chemistry, electrochemistry

DELAWARE

Sleight, Arthur William, solid state chemistry
Sloan, Gilbert Jacob, physical chemistry
Sloan, Martin Frank, industrial organic chemistry
Slutsky, Joel, organic chemistry
Smart, Bruce Edmund, physical organic chemistry
Smat, Robert Joseph, organic chemistry
Smiley, Robert Arthur, organic chemistry
Smith, Albert Faris, analytical chemistry
Smith, Carolyn Jean, chemistry
Smith, Clabourne Davis, organic chemistry
Smith, Howard Leroy, organic chemistry
Smith, John Frederick, polymer chemistry
Smith, Kenneth McGregor, physical chemistry
Smith, Ronald W., chemical engineering
Smith, Thomas David, physical chemistry
Smook, Malcolm Andrew, organic chemistry
Smoot, Charles Richard, physical chemistry
Smullin, Charles Frederick, analytical chemistry, chemical engineering
Snyder, Harold Herbert, chemistry
Soboczenski, Edward John, insect toxicology
Solenberger, John Carl, industrial chemistry
Sonnichsen, George Carl, chemistry
Sowards, Donald Maurice, physical chemistry, inorganic chemistry
Speck, Rhoads McClellan, organic chemistry
Speck, Stanley Brooke, chemistry
Spence, Gavin Gary, organic chemistry
Spurlin, Harold Morton, physical chemistry
Squire, Edward Noonan, chemistry
Scrog, Cyrus Efrem, organic chemistry
Stacey, Francis Wilfred, organic chemistry
Stahl, Roland Edgar, organic chemistry
Staikos, Dimitri Nickolas, electrochemistry
Stanley, Harry Eugene, textile physics
Stanton, William Alexander, organic chemistry
Starkweather, Howard Warner, Jr., physical chemistry
Steller, Kenneth Eugene, organic chemistry
Sterling, John Deco, chemistry
Stevens, Sandra, synthetic organic chemistry
Stevenson, Arthur Charles, rubber chemistry
Stevenson, Irone Edmund, Jr., biochemistry
Stewart, Charles Winfield, theoretical chemistry
Stockburger, George Joseph, industrial organic chemistry
Stoptic, Roger John, microbiology, biochemistry
Straw, Harry Arthur, organic chemistry
Strobach, Donald Roy, organic chemistry, biochemistry
Sturgis, Bernard Miller, petroleum chemistry
Suarez, Thomas H., resource management
Subramanian, Pallatheri Manackal, organic chemistry
Summers, John Clifford, pesticide chemistry
Sundet, Sherman Archie, polymer chemistry
Swalheim, Donald'Arthur, chemistry
Swaner, Frederic Wurl, organic chemistry
Swank, Howard Wigton, analytical chemistry
Swanson, John Melvin, textile chemistry
Sweet, Arthur Thomas, Jr., polymer chemistry
Sweetser, Philip Bliss, analytical chemistry
Swingle, Robert Shelton, II, analytical chemistry
Swoboda, Thomas James, physical chemistry
Tabb, David Leo, polymer science
Tabbian, Richard, polymer chemistry
Takeshita, Tsuneichi, physical organic chemistry
Talvorian, Kenneth Bedrose, polymer chemistry, textile chemistry
Tams, William P., industrial organic chemistry
Tan, Henry Harry, organic chemistry
Tanikella, Murty Sundara Sitarama, physical chemistry, polymer chemistry
Tanner, David, polymer chemistry

Tarney, Robert Edward, organic chemistry
Tatum, William Earl, organic chemistry
Taufen, Harvey James, organic chemistry
Taves, Milton Arthur, organic chemistry
Taylor, Barry Edward, solid state chemistry
Taylor, Robert Burns, Jr., organic chemistry
Taylor, Stephen Keith, organic chemistry
Teichman, Robert, biostatistics
Temple, Stanley, organic chemistry
Tennent, Howard Gordon, organometallic chemistry
Terss, Robert H., organic chemistry
Thamm, Richard C., Jr., organic chemistry
Theobald, Clement Walter, organic chemistry
Thomas, Walter William, organic chemistry
Thompson, Douglas Stuart, physical chemistry
Thompson, Kenneth Roy, physical chemistry, analytical chemistry
Thompson, Robert Gene, physical chemistry
Thornton, Daniel McCarty, textiles
Thornton, Roger Lea, organic chemistry
Tillmanns, Emma-June H, information science
Tillson, Henry Charles, rubber chemistry
Tolman, Chadwick Alma, physical chemistry, inorganic chemistry
Tomic, Ernst Alois, inorganic chemistry, physical chemistry
Tordella, John P, chemistry
Toy, Stephen Thomas, immunology, virology
Traumann, Klaus Friedrich, organic chemistry
Tripper, William Mowbray, spectroscopy
Trost, Henry Biggs, organic chemistry
Truemper, Joseph Tucker, chemistry
Tubbs, Robert Kenneth, colloid chemistry, surface chemistry
Tuites, Donald Edgar, polymer chemistry
Tullio, Victor, chemistry
Tullock, Charles William, organic chemistry
Turner, George Robert, textiles, chemistry
Turner, Robert Lawrence, organic polymer chemistry
Turner, Vernon Lee, Jr., chemistry
Turner, William Richard, analytical chemistry
Tyler, Chaplin, chemistry
Un, Howard Ho-Wei, organic polymer chemistry
Upchurch, Donald Gene, inorganic chemistry
Upson, Robert William, organic chemistry, photographic chemistry
Vaala, Gordon Theodore, chemistry
Valdsaar, Herbert, high temperature chemistry
Vandenberg, Edwin James, polymer chemistry
Van Dyk, John William, physical chemistry
Van Fossen, Paul, organic chemistry
Van Gulick, Norman Martin, organic chemistry
van Roggen, Arend, electrodynamics, computer science
Varner, Reed William, plant pathology, forestry
Vassallo, Donald Arthur, polymer chemistry
Venkatachalam, Taracad Krishnan, chemistry
Verbanc, John Joseph, organic chemistry
Victorius, Claus, organic polymer chemistry
Vignes, Robert Paul, organic chemistry
Vosburgh, William George, organic chemistry
Voter, Roger Conant, chemistry
Waddell, James, nutrition
Wagner, Klaus Peter, organic polymer chemistry
Wagner, Richard Lloyd, polymer chemistry, photochemistry
Wagner, Robert Edwin, organic chemistry
Wagner, Romeo Barrick, organic chemistry
Wahlig, Charles F., physics
Wahlberger, Frederick Theodore, organic polymer chemistry
Waller, Francis Joseph, organic chemistry
Walsh, Raymond Anthony, chemistry
Walsh, Robert Michael, physical chemistry
Walter, Henry Clement, organic chemistry
Walters, Philip Marion, textile chemistry

Wang, Victor Kai-Kuo, physical chemistry, industrial chemistry
Ward, George A., analytical chemistry
Ward, Richard Bernard, organic chemistry
Waring, Derek Morris Holt, organic chemistry
Waring, Robert Kerr, Jr., physics
Wartz, Richard Stefen, toxicology
Wat, Edward Koon Wah, organic chemistry
Watkins, Spencer Hunt, organic chemistry
Wayne, Winston Joe, chemistry
Wayrynen, Robert Ellis, photographic chemistry
Webb, Myron Quentin, chemistry
Weber, Arthur George, chemistry
Weber, Vincent Joseph, organic chemistry
Webster, Owen Wright, organic chemistry
Wehter, James F., surface chemistry
Weise, Jurgen Karl, polymer chemistry
Weisgerber, Cyrus Aaron, organic chemistry
Weiss, Douglas Eugene, organic polymer chemistry
Wellings, Ian, organic chemistry, medicinal chemistry
Wells, Adoniram Judson, physical chemistry
Wells, Martha Carol, microbial physiology
Wendt, Robert Charles, surface chemistry
Wermis, Gerald R., biochemistry, clinical chemistry
Werny, Frank, natural products chemistry, textile technology
West, Richard Lowell, organic chemistry
Wetteral, Frank P., analytical chemistry
Wetzel, Franklin Huff, chemistry
Wheeler, Allan Gordon, pharmacology, physiology
Whiteley, Norman McKee, clinical biochemistry
Whitman, Gerald Messner, organic chemistry
Whitney, Joel Gayton, organic chemistry
Wich, Grosvenor Searles, organic chemistry
Wiest, Emil Gabriel, chemistry
Wiley, Douglas Walker, organic chemistry
Wiley, William Lee, organic chemistry
Willong, Robert Edward, chemistry
Williams, Ebenezer David, Jr., textile chemistry
Williams, Harry Douglas, organic chemistry
Williams, Kenneth Roger, textile physics
Williams, Reed Chester, analytical chemistry
Wittbecker, Emerson Laverne, polymer chemistry, textile chemistry
Wittenbach, Vernon Arie, plant physiology
Wolf, Dale E., agronomy
Wolfe, James Richard, Jr., organic chemistry
Wolfe, William Ray, Jr., physical chemistry
Wolter, Frederick John, chemistry
Wonmack, Joel Benjamin, Jr., agricultural chemistry
Woodbury, Elton Norris, entomology
Woods, Thomas Stephen, organic chemistry
Wopschall, Robert Harold, physical chemistry
Wriede, Peter Artur, organic chemistry
Wright, Everett James, organic chemistry
Wright, Leon Wendell, physical chemistry
Wu, Souheng, physical chemistry
Wu, Ting Kai, chemistry
Yau, Wallace Wen-Chuan, analytical chemistry, polymer physics
Yellin, Tobias O, biochemistry, pharmacology
Yin, Fay Hoh, virology
Young, Charles Albert, industrial organic chemistry
Young, Edmond Grove, fluorine chemistry
Young, William Anthony, physical chemistry, engineering management
Yourtee, John Ashby, chemistry
Zapp, John Adam, Jr., biochemistry
Ziegel, Kenneth David, physical chemistry
Zimmerman, Joseph, polymer chemistry
Zumsteg, Fredrick C., Jr., experimental solid state physics

WINTERTHUR

Hanson, Victor Frederick, x-ray spectroscopy

WOODSIDE

Fisher, Norman Gail, organic chemistry

DISTRICT OF COLUMBIA

BOLLING AFB

Haffner, Richard William, surface chemistry

WASHINGTON

Abashian, Alexander, nuclear physics
Abdel-Gawad, Monem, geology
Abel, Robert Berger, science administration
Abelson, Philip Hauge, physical chemistry
Aberhart, Donald John, organic chemistry
Abernathy, Charles Owen, pharmacology
Abramowitz, Stanley, chemical physics
Abramson, David C., pediatrics, neonatology
Abramson, Lee Richard, mathematical statistics
Absolon, Karel B., thoracic & cardiovascular surgery
Acheson, Donald Theodore, meteorology, instrumentation
Achhammer, Bernard George, polymer chemistry
Acker, Robert Flint, microbiology
Adams, James Hall, Jr., cosmic ray physics
Adams, John George, pharmacology
Adams, John Pietch, orthopedic surgery
Adams, Otis William, physical organic chemistry
Adams, Robert John, electronics
Adamson, Lucile Frances, nutritional biochemistry, environmental sciences
Addamiano, Arrigo, inorganic chemistry, physical chemistry
Adkinson, Burton Wilbur, information science
Adler, Hans Henry, geology
Affronti, Lewis Francis, microbiology, immunology
Agins, Barnett Robert, mathematics
Agnew, Allen Francis, geology
Ahearne, John Francis, resource management
Ahluwalia, Balwant Singh, endocrinology
Aitken, Alfred H., theoretical physics
Albert, Ernest Narinder, anatomy, cell biology
Albertson, John Newman, Jr., microbiology, laboratory management
Al-Doory, Yousef, medical mycology
Aldrich, Lyman Thomas, geophysics
Aldridge, Mary Hennen, organic chemistry, biochemistry
Alexander, Charles Edward Jr., medicine, public health
Alexander, Thomas Goodwin, analytical chemistry
Ali, Mahamed Asgar, physical chemistry, quantum chemistry
Allan, Frank Duane, anatomy, embryology
Allas, Richard G., nuclear physics
Allen, Louis Pinckney Jr., meteorology, oceanography
Allen, William Hubert, physics
Alleva, Frederic Remo, perinatal biology
Alleva, John J., biology
Almazan, James A., meteorology
Alonso, Marcelo, nuclear physics
Alpert, Louis Katz, medicine
Alpert, Seymour, medicine
Alter, Harvey, physical chemistry, environmental chemistry
Altschul, Aaron Mayer, nutrition
Ambler, Ernest, cryogenics, science administration
Ambrose, John Russell, chemistry, corrosion
Amster, Adolph Bernard, physical chemistry
Ancker-Johnson, Betsy, physics
Anderle, Richard, mathematics
Andersen, Neil Richard, chemical oceanography
Anderson, Donald Morgan, systematic entomology
Anderson, Floyd Edmond, organic chemistry

Anderson, Lars William James, phycology, biological oceanography
Andrews, George William, geology
Andrews, John Robert, radiology
Andrews, Ronald Allen, lasers, optics
Andrews, Wallace Henry, microbiology
Angel, John Lawrence, physical anthropology
Angelotti, Robert, microbiology
Angevine, Daniel Murray, pathology
Apple, William S., pharmacy
Appleman, Daniel Everett, geology, crystallography
Archer, Alford, population geography, cartography
Archer, Ellen Gleason, neurochemistry
Archer, Juanita Almeta, internal medicine
Artna-Cohen, Agda, information science, nuclear physics
Ashe, Warren (Kelly), microbiology, biochemistry
Assousa, George Elias, astrophysics, science policy
Atkins, David Lynn, vertebrate anatomy, neuroanatomy
Attaway, David Henry, biochemistry
Attix, Frank Herbert, radiological physics
Aufenkamp, Don, mathematics, computer science
Augustine, Robertson J., health physics
Avery, Gordon B., medicine, embryology
Aviles, Joseph B., theoretical physics
Bahr, Gunter F., biophysics, pathology
Bailar, Barbara Ann, statistics
Banks, Harvey Washington, astrophysics
Banks, Richard Charles, ornithology
Banks, William Michael, zoology
Banta, William Claude, marine zoology
Barbehenn, Kyle Ray, vertebrate ecology
Bardon, Marcel, physics
Barker, Winona Clinton, human physiology
Barnard, Jerry Laurens, marine zoology
Barnes, Burton B., exploration geophysics, marine geology
Barnes, John David, polymer physics
Barnes, John Maurice, plant pathology
Barnet, Ann B., pediatric neurology
Baron, Louis Sol, bacteriology
Barr, Nathaniel Frank, radiation chemistry
Barrows, Harold Lindsey, soil chemistry
Barry, Guy Thomas, organic chemistry
Barry, William Earl, aerospace medicine, bioengineering
Bartels, William Charles Joseph, physics
Bartley, William Call, science policy, science administration
Barton, Paul Booth, Jr., geology
Bascom, Willard D., physical chemistry
Bass, Arnold Marvin, physics
Bass, James W., pediatrics, infectious diseases
Bass, Virginia Carvel, analytical chemistry
Bassel, Robert Harold, nuclear physics
Bassler, Richard Albert, information science
Bastiaans, Glenn John, analytical chemistry
Bates, Charles Carpenter, oceanography
Bates, Richard Doane, Jr., physical chemistry
Bauer, Mark Henry, physiology
Bauer, Neinz, pathology
Baumiller, Robert Cahill, genetics
Bautz, Laura Patricia, astronomy, science administration
Bay, Zoltan Lajos, quantum physics
Beach, Louis Andrew, experimental nuclear physics
Bean, Vern Ellis, high pressure physics
Beard, Charles Irvin, radiophysics
Beasley, Edward Evans, physics
Beaver, William Thomas, clinical pharmacology
Beck, Glenn Hans, dairy husbandry

Beck, Roland Arthur, inorganic chemistry, physical chemistry
Becker, Kenneth Louis, internal medicine, endocrinology
Beebe, Gilbert Wheeler, medical statistics
Beer, Charles, agricultural economics, business management
Bell, Peter M., geochemistry, geophysics
Bell, Rosemond Kay, analytical chem
Bellanti, Joseph A., pediatrics, microbiology
Bellmer, Elizabeth Henry, botany, zoology
Bendel, Warren Lee, nuclear physics
Bender, James Arthur, applied physics
Benjamin, Fred Berthold, physiology
Bennett, Gary Lee, nuclear physics
Bennett, Herbert Stanton, physics, materials science
Bennett, Lawrence Herman, physics, materials science
Benson, Richard Hall, paleontology, oceanography
Benson, Walter Roderick, pharmaceutical chemistry, analytical chemistry
Benson, William Edward Barnes, geology
Berdan, Jean Milton, invertebrate paleontology
Berdjis, Charles Choaib, pathology, radiobiology
Berendzen, Richard Earl, astronomy, history of science
Berg, Joseph Wilbur, Jr., geophysics
Berg, Winfred Emil, science administration
Berge, Truman Kent, physics, computer science
Berger, Harold, radiological physics
Berger, Martin Jacob, mathematical physics
Berger, William J., mathematics, computer science
Bergman, Kenneth Harris, meteorology
Bering, Edgar Andrew, Jr., medicine, neurosurgery
Berkson, Harold, environmental biology
Berley, David, physics
Berlowitz, Laurence Jack, developmental biology, cell biology
Berman, Alan, physics, physical oceanography
Berman, Horace Aaron, applied chemistry
Berman, Sanford, microbiology
Bermudez, Victor Manuel, solid state science
Bernier, Joseph Leroy, oral pathology
Berning, Warren Walt, physics, meteorology
Bernstein, Abram Bernard, meteorology
Bernstein, Leon, physiology, public health administration
Bernstein, Lionel M., medicine
Bernton, Harry Saul, allergy
Berryman, Jack Holmes, resource management, fish & game management
Bertaut, J. Edgard Francis, computer science
Bertholf, Lloyd Bernard, meteorology, oceanography
Bever, Arley Tunis, (Jr), biochemistry
Bever, Christopher Theodore, psychiatry
Bhaskar, Surindar Nath, pathology
Bhussry, Baldev Raj, anatomy, dentistry
Bierly, Eugene Wendell, meteorology, science administration
Billings, Bruce Hadley, optics
Binford, Chapman Hunter, pathology
Binn, Leonard Norman, medical bacteriology
Birdsall, John J., food technology, biochemistry
Birks, LaVerne Stanfield, physics
Birky, Merritt Merle, physical chemistry, inhalation toxicology
Bishop, William P., nuclear science
Bissell, Robert, theoretical chemistry, organic chemistry
Blake, Doris Holmes, entomology
Blake, Francis Gilman, research administration
Blanpied, William Antoine, science policy, history of science
Blanquet, Richard Steven, invertebrate physiology, biochemistry
Blanton, Jackson Orin, physical oceanography, limnology
Blaunstein, Robert P., research administration
Blecher, Melvin, biochemistry
Bleiberg, Marvin Jay, pharmacology
Block, Stanley, physical chemistry, crystallography
Blum, Robert Allan, psychiatry, psychoanalysis
Blumberg, Joe Morris, pathology

Blumenthal, Herbert, biochemistry, toxicology
Blundell, George Phelan, pathology
Blunt, Robert F., solid state physics
Boardman, Richard Stanton, invertebrate paleontology
Boaz, Thurmond DeWitte, Jr., preventive medicine
Bobo, Edwin Ray, mathematics
Bodine, John James, anthropology
Bodner, Stephen E., theoretical physics
Boehne, John William, nutrition
Boek, Walter Erwin, anthropology, economics
Bogan, Denis John, chemical kinetics, molecular spectroscopy
Bogdan, Victor Michael, mathematics
Boggess, Nancy Weber, astronomy
Bogle, Robert Worthington, experimental physics
Bohlin, John David, solar physics
Bond, James Oliver, public health
Bondelid, Rollon Oscar, nuclear physics
Bonham, Lawrence Douglas, geology
Bonnell, David William, physical inorganic chemistry, high temperature chemistry
Bonner, Tom Ivan, molecular biology
Bonner, William D., meteorology
Book, David Lincoln, plasma physics
Booker, Walter Monroe, pharmacology
Boosman, Jaap Wim, science administration, sedimentology
Borg, Alfred Francis, microbiology
Botts, Truman Arthur, mathematics
Bouchard, Raymond William, fresh water biology
Bowen, David Hywel Michael, chemistry, chemical engineering
Bowling, Lloyd Spencer, Sr, audiology, speech pathology
Bowman, Charles D., nuclear physics
Bowman, Thomas Elliot, zoology
Bowman, Wallace Deal, environmental management, research administration
Boyce, Peter Bradford, astronomy
Boyd, Francis R., geology
Boyd, James, geology
Brady, Edward Lewis, physical chemistry
Brady, Robert Frederick, Jr, organic chemistry
Brandt, Carl David, virology
Brandt, Robert William, forestry
Brandt, Walter Edmund, virology
Brauer, Gerhard Max, chemistry, dental research
Braungart, Dale Carl, cytogenetics
Breder, Charles Vincent, organic polymer chemistry, analytical chemistry
Bregman, Jacob Israel, environmental management
Breiter, Jerome John, organic chemistry
Brennan, James Gerard, theoretical physics
Brenner, Fivel Cecil, cellulose chemistry
Breslow, Alexander, pathology
Breyere, Edward Joseph, immunogenetics
Brick, Irving B., internal medicine, gastroenterology
Bridgewater, Albert Louis, Jr., elementary particle physics
Briggs, Norman Theodore, parasitology
Bright, Harold Frederick, statistics
Britt, A D., physical chemistry
Broadhurst, Martin Gilbert, solid state physics
Broderick, Grace Nolan, geology
Brodrick, Harold James, Jr, meteorology
Brodsky, Allen, health physics, biostatistics
Brooks, John Langdon, ecology, systematics
Brooks, Walter Lyda, physics, astronomy
Brosseau, George Emile, Jr genetics
Brown, Albert, physics
Brown, Barry Lee, endocrinology, reproductive physiology
Brown, Charles Moseley, atomic spectroscopy, molecular spectroscopy
Brown, Dail Woodward, biological oceanography
Brown, Harry Esmond, forest hydrology
Brown, Louis, nuclear physics
Brown, Myrtle Laurestine, nutrition
Brown, Norman Louis, physical chemistry
Brown, Walter Eric, chemical physics
Brubaker, Merlin L, internal medicine, public health
Brunelle, Richard Leon, forensic science
Brunk, William Edward, astronomy
Bruns, Paul Donald, obstetrics & gynecology
Bryson, Robert Pearne, geology
Buccino, Alphonse, science administration, algebra
Buchal, Robert Norman, applied mathematics

Buchanan, John Donald, radiochemistry, health physics
Buckley, John Leo, ecology
Buescher, Edward Louis, virology
Bullard, Ervin Trowbridge, horticulture
Bullis, William Murray, physics
Bullock, William Horace, internal medicine, hematology
Bultman, John D, chemistry, marine biology
Bur, Anthony J., molecular physics
Burk, Dean, biochemistry
Burk, Jerry Alan, pesticide chemistry
Burkhalter, James Herbert, microwave spectroscopy
Burley, Gordon, physical chemistry, science administration
Burnett, John Lambe, inorganic chemistry, physical chemistry
Burnette, Mahlon Admire, III, food science, nutrition
Burns, Denver Peeper, research administration
Burr, William Wesley, Jr, biochemistry, medicine
Burton, David Lee, horticulture, soils
Burton, James Samuel, physical chemistry
Bushman, John Branson, vertebrate zoology
Butler, Ann Benedict, neuroanatomy
Butler, James Ehrich, chemical physics
Butler, Robert N., psychiatry
Buzzelli, Donald Edward, science policy, philosophy of science
Byrnes, Bernard Christopher, oceanography
Cadigan, Francis C, Jr., infectious diseases, tropical medicine
Cain, Arthur Samuel, Jr, physiology, surgery
Cairns, Robert William, physical chemistry
Calabi, Ornella, microbiology
Calcagno, Philip Louis, pediatrics
Calhoun, Noah Robert, oral surgery, dental research
Calio, Anthony John, space physics
Callaham, Robert Zina, forest ecology, genetics
Callahan, Jeffrey Edwin, physical oceanography
Callanan, Margaret Joan, physical chemistry
Callen, Earl Robert, solid state physics
Calvert, Allen Fisher, biochemical genetics
Cameron, Joseph Marion, mathematical statistics
Campbell, Francis James, physical chemistry, physics
Campbell, Frank Leslie, entomology
Campbell, Paul Gilbert, organic chemistry
Campbell, William Jackson, clinical biochemistry
Canary, John Joseph, metabolism, endocrinology
Canfield, Craig Jennings, clinical pharmacology, malariology
Cantrell, Thomas Samuel, organic chemistry
Cantu, Antonio Arnold, forensic science, quantum chemistry
Caponio, Joseph Francis, biochemistry
Caress, Edward Alan, organic chemistry
Carlson, Carl Wilburn, soil conservation
Carlson, William Theodore, information science
Carnell, Paul Herbert, petroleum chemistry
Carpenter, Richard A, science administration
Carrigan, Richard Alfred, environmental chemistry
Carroll, Gerald V, geology
Carson, Frederick Wallace, organic chemistry, biochemistry
Carson, Theophilus Roosevelt, biology
Carstea, Dumitru Dumitru, environmental sciences, soil chemistry
Carter, Gesina C, solid state physics
Carter, William Harold, optical physics
Cartwright, Oscar Ling, entomology
Casey, Harold W, veterinary pathology
Cass, Jules Siland, laboratory animal medicine
Caswell, James Martin, dental materials
Caswell, Randall Smith, physics
Caswell, Robert Little, pesticide chemistry
Catchings, Robert Merritt, III, solid state physics
Cezariiryan, Ared, thermal physics, materials science
Chace, Fenner Albert, Jr, zoology
Chakrabarti, Siba Gopal, biochemistry

Challinor, David, forest ecology
Chamberlain, David Leroy, Jr., organic chemistry, physical chemistry
Chambers, Charles McKay, Jr., academic administration, mathematics
Chan, An Soo, histology, embryology
Chang, I-Lok, mathematical analysis
Chang, Shu-Sing, physical chemistry
Chapin, Douglas Scott, physical chemistry
Chapline, William Ridgely, (Jr.), range management, forestry
Chapman, Charles R, ecology
Chapman, George Bunker, cytology
Chappel, Samuel Estelle, radiation physics
Charlton, Gordon Randolph, high energy physics, research administration
Charvonia, David Alan, dynamics
Chaves, Felix, petrology
Cheaves, Thomas Henry, industrial chemistry, petroleum chemistry
Chen, Davidson Tah-Chuen, physical oceanography / remote sensing
Cheng, Tsung O., cardiology
Chertok, George, physics
Chertok, Benson T. nuclear physics, particle physics
Chesser, Nancy Jean, experimental solid state physics
Chiao-Yap, Lung Wen, theoretical nuclear physics
Chiazze, Leonard, Jr., biostatistics, epidemiology
Chinn, Herman Isaac, biochemistry
Chirikjian, Jack G., biological chemistry, chemistry, atomic physics
Chiu, Lue-Yung Chow, quantum chemistry, atomic physics
Chiu, Ying-Nan, physical chemistry
Chow, Laurence Chung-Lung, physical chemistry, dental research
Chu, Yang-Ming, microbiology, immunology
Chubb, Talbot Albert, geophysics, astrophysics
Chun, Byungkyu, surgical pathology
Chung, Choong Wha, microbiology, medicine
Chung, David Yih, solid state physics, low temperature physics
Chung, Ed Baik, medicine, pathology
Church, Charles Henry, optical physics
Church, Lloyd Eugene, anatomy
Churgin, James, oceanography
Cifelli, Richard, invertebrate paleontology
Cisin, Ira Hubert, statistics
Clark, Ernst M, physical chemistry
Clark, Eloise Elizabeth, biochemistry, biophysics
Clarke, Roy Slayton, Jr., geochemistry
Clarke, John Frederick Gates, entomology
Clark, Joan Robinson, crystallography
Coates, Anthony George, paleontology, stratigraphy
Coakley, Charles Seymour, anesthesiology
Cobb, William Montague, anatomy
Cobee, George Vincent, petroleum geology
Cohen, Alex, organic chemistry, clinical chemistry
Cohen, Daniel Morris, ichthyology
Cohen, Julius, physics
Cohen, Leslie, nuclear physics
Cohen, Lucy M, anthropology
Cohen, Robert, energy conversion
Cohn, Ernst M, physical chemistry
Cohn, Robert, neurology
Cohn, Victor Hugo, biochemical pharmacology
Coleman, Ernest, nuclear physics, high energy physics
Coleman, John Sherrard, physics
Coleman, Russell, soils
Collat, Justin White, analytical chemistry
Collette, Bruce Baden, ichthyology
Collins, Curtis Allan, physical oceanography
Collins, Henry B, archaeology
Collins, William Carridine, solid state physics
Colvin, Burton Houston, mathematics
Colwell, Jack Harold, experimental solid state physics
Comerford, John J, analytical chemistry
Commerford, John D, organic chemistry
Conant, Louis Cowles, geology
Congel, Frank Joseph, nuclear physics
Connor, Daniel Henry, pathology
Conrad, Geoffrey Wentworth, anthropology
Cook, Ancel Eugene, physical optics, microelectronics

Cook, Charles William, experimental nuclear physics
Cook, Richard Kaufman, acoustics, geophysics
Cook, Robert Carter, genetics, demography
Cooper, Benjamin Stubbs, nuclear physics
Cooper, Glenn Adair, Jr., wood technology
Cooper, Gustav Arthur, paleobiology
Cooper, John Allen Dicks, biochemistry
Cooper, Martin Jacob, statistical mechanics, science administration
Cooper, Raymond David, physics
Cooper, Theodore, physiology, pharmacology
Copeland, Edmund Sargent, biophysical chemistry, magnetic resonance
Corbin, Thomas Elbert, astronomy
Corden, Pierce Stephen, physics
Coriell, Sam Ray, physical chemistry, physical metallurgy
Corliss, Charles Howard, atomic spectroscopy
Corliss, Edith Lou Royner, physics
Cornell, Neal William, biochemistry
Cornell, Samuel Douglas, science administration, academic administration
Cornely, Paul Bertau, public health
Corsaro, Robert Dominic, physical chemistry, acoustics
Cosby, Lynwood Anthony, electronics
Costa, Erminio, pharmacology
Costello, Leslie Carl, cell physiology, endocrinology
Costrell, Louis, nuclear physics, nuclear science
Cotruvo, Joseph Alfred, organic chemistry
Councell, Clara Elizabeth, public health, statistics
Cowan, Richard Sumner, plant taxonomy
Cowser, Kenneth Emery, health physics, environmental engineering
Coyle, Thomas Davidson, inorganic chemistry, organometallic chemistry
Cramer, William Smith, physics
Crampton, Theodore Henry Miller, mathematics, radiobiology
Crannell, Hall L., nuclear physics
Crary, Albert Paddock, geophysics
Crentz, William Luther, chemistry
Cressey, Roger F, parasitology
Creutz, Edward (Chester), nuclear physics
Crisler, Joseph Presley, microchemistry
Crisp, Thomas Mitchell, Jr, anatomy, endocrinology
Criss, Wayne Eldon, cancer, endocrinology
Crissman, John Matthews, polymer science
Crist, DeLanson Ross, physical organic chemistry
Crocker, William Henry, cultural anthropology
Crosby, David S, applied statistics, mathematical statistics
Crosby, Gayle Marcella, developmental biology
Cruddace, Raymond Gibson, xray astronomy
Crum, John Kistler, inorganic chemistry, science administration
Cullen, William Charles, organic chemistry
Cummings, Joseph Gerard, pesticide chemistry
Cummings, Nancy Boucot, nephrology, internal medicine
Currie, Lloyd Arthur, physical chemistry, radiochemistry
Curry, Thomas Harvey, organic chemistry, chemical engineering
Cutchins, Ernest Charles, bacteriology, virology
Cuthill, Elizabeth, mathematics
Cutright, Duane Edwin, pathology
Da Costa, William A, biochemistry
Daechsel, Mark, physics
D'Aguanno, William, physiology, pharmacology
Dailey, Robert Engle, biochemistry, pharmacology
Dalmat, Herbert Theodore, public health administration
Dalrymple, Joel McKeith, virology
Daniel, John Harrison, physics
Danos, Michael, nuclear physics
D'Antonio, Peter, structural chemistry
Darwent, Basil de Baskerville, physical chemistry
Daugherty, David M, entomology
Davey, John Edmund, physics

Davidson, David Edward, Jr., veterinary medicine, radiobiology
Davidson, Robert A., botany
Davis, Charles Mitchell, Jr, underwater acoustics
Davis, Donald Ray, entomology
Davis, Elizabeth Young, nutrition, physiology
Davis, Jack, physics
Davis, James Allen, underwater acoustics, reactor physics
Davis, Ruth Margaret, computer sciences
Davis, William Spencer, fish biology
Davisson, Charlotte Meeker, nuclear science
Daza, Carlos Hernan, nutrition, public health
DeCicco, Benedict Thomas, microbial physiology
DeCicco, Peter Donald, solid state physics
DeFillipps, Robert Anthony, plant taxonomy
DeGiovanni-Donnelly, Rosalie F, microbial genetics, biochemistry
Dehl, Ronald, physical chemistry
Deitz, Victor Reuel, surface chemistry
DeLevie, Robert, electrochemistry, biophysics
Del Grosso, Vincent Alfred, physical oceanography
De Long, Chester Wallace, biochemistry, physiology
De Marco, Ronald Anthony, fluorine chemistry
Demaree, Gale E, pharmacology
Denning, William Edwards, mathematical statistics
Denny, Cleve B, food microbiology
DePaoli, Alexander, veterinary pathology
De Percin, Fernand, physical geography, meteorology
DeRisi, Mary Christine, dentistry
DeSantis, Mark Edward, morphology, neurobiology
De Simone, Daniel Van, science policy
Deslattes, Richard D, Jr, radiology, physics
Desmond, Alton Harold, histophysiology, forensic sciences
Desrosiers, Russell, plant pathology
Dessouky, Ahmad, anatomy, electron microscopy
Deutsch, Mike John, biochemistry
Devereux, Robert Francis, oceanography, meteorology
Devine, Robert T, electrodynamics
deWit, Roland, solid state physics
De Witt, Warren Peyton, mathematics
Diamond, Jacob Joseph, physical chemistry
Dibeler, Vernon Hamilton, chemical physics
Dick, Charles Edward, nuclear physics
Dick, Ronald Stewart, mathematical statistics
Dickens, Brian, polymer chemistry
Dickens, Charles Henderson, science policy
Dickson, George, dental materials
Dickson, James Francis, III, medicine, electrical engineering
Dillard, Martin Gregory, nephrology
Dillon, Wilton Sterling, applied anthropology, behavioral anthropology
DiMarzio, Edmund Armand, physics
Dimond, Marie Therese, zoology
Dockstader, Wilmer Beldon, microbiology
Doctor, Bhupendra Pannalal, biochemistry, nutrition
Dodge, William R, physics
Dodgen, Durward F, chemistry
Donaldson, James E, physics
Doolittle, Warren Truman, mathematics
Doran, William Thomas, occupational medicine, public health
Dorscher, Kenneth Peter, agronomy
Doschek, George A, solar physics, atomic spectroscopy
Doschek, Wardella Woford, research administration
Douglass, Raymond Charles, invertebrate paleontology
Doumani, George Alexander, geology, stratigraphy
Doyle, Thomas Daniel, pharmaceutical chemistry, organic chemistry
Dragoo, Alan Lewis, solid state physics
Dresen, David W, veterinary public health
Dretchen, Kenneth Lewis, pharmacology
Drew, Russell Cooper, physics, science policy
Drewes, Wolfram Ulrich, geography
Dubin, Maurice, physics, astrophysics

Duckworth, Walter Donald, systematic entomology
Dudley, Theodore, plant taxonomy
DuMont, Philip Atkinson, ornithology
Dunbar, Burdett Sheridan, medicine, anesthesiology
Duncombe, Raynor Lockwood, astronomy
Dunning, Kenneth Laverne, atomic physics
Durell, Jack, psychiatry, biochemistry
Durst, Richard Allen, analytical chemistry
Dutro, John Thomas, Jr, geology
Dutta, Sisir Kamal, genetics
Dyer, Donald Ray, population geography, economic geography
Dyer, Randolph H, organic chemistry
Eagles, Douglas Alan, neurophysiology
Eckerle, Kenneth Lee, applied physics
Eagleson, Halson Vashon, physics
Eden, Henry Francis, physics
Edmunds, Charles W, genetics
Edmunds, Lafe Rees, entomology
Edson, Quentin A, fisheries management
Edwards, Cecile Hoover, nutrition, biochemistry
Edwards, Gordon Stuart, toxicology
Edwards, Lydia Bowman, medicine
Earley, Joseph Emmett, physical inorganic chemistry
Eby, Charles J, organic chemistry
Eby, Ronald Kraft, physics
Edington, Charles W, genetics
Earl, Francis Lee, veterinary toxicology
Earle, Kenneth Martin, pathology, neuropathology
Efron, Herman Yale, health sciences
Egan, Howard L, mathematics
Egeberg, Roger O, internal medicine, medical administration
Ehrlich, Margarete, radiation physics
Ehrlich, S Paul Jr, epidemiology, public health
Ehrenstein, James Robert, solid state physics
Eisenberg, John Frederick, ethology
Eisenberg, William Victor, microbiology
Eisenhart, Churchill, mathematical statistics
Eisenstein, Julian (Calvert), theoretical physics
Eldridge, Bruce Frederick, medical entomology
Eldridge, Marie Delaney, statistics
Elkins, Earleen Feldman, audiology
Ellet, William H, radiological health
Elliot, Eric Charles, cardiovascular physiology
Elliot, Joe Oliver, oceanography
Elliott, David Duncan, science policy
Ellis, Frank Russell, clinical pathology
Ellis, John Ogborn, meteorology
Elton, Raymond Carter, atomic physics, plasma physics
Emerson, Kary Cadmus, medical entomology, parasitology
Engler, Reto Arnold, organic chemistry
English, Spofford Grady, chemistry, research administration
Ennulat, Reinhard D, physics
Enzinger, Franz Michael, pathology
Epple, Robert (Paul), solid state physics, high temperature chemistry
Epstein, Arnold S, solid state physics
Erickson, Carl O, meteorology
Erickson, Duane Gordon, parasitology
Espelie, Mary Solveig, mathematical analysis
Estep-Barnes, Patricia Anne, spectrochemistry
Estrin, Norman Frederick, organic chemistry, science administration
Etzel, Howard Wesley, solid state physics, science administration
Evans, Bruce Douglas, solid state physics
Evans, Clifford, anthropology, archaeology
Evans, John McCallum, medicine
Evans, William Harrington, physical chemistry
Everett, Ardell Gordon, geology
Ewers, John Canfield, ethnology
Eyde, Richard Husted, plant anatomy, paleobotany
Fabro, Sergio, pharmacology, obstetrics & gynecology
Fagg, Lawrence Wellburn, nuclear physics
Fairchild, Homer Eaton, entomology
Fanconi, Bruno Mario, polymer physics
Faraday, Bruce (John), solid state physics

Faust, Walter Luck, lasers, experimental solid state physics
Favorite, Frank G, Jr, medical entomology
Fearn, James Ernest, physical organic chemistry, polymer chemistry
Feffer, James Joseph, medicine
Fehlmann, Herman Adair, marine biology
Felch, Richard Elroy, agronomy
Feldman, Alfred Philip, organic chemistry
Feldman, Charles, solid state physics
Feldman, Joseph Louis, solid state physics
Feldman, Martin Robert, organic chemistry
Feldman, Uri, atomic spectroscopy, solar physics
Feldmann, Edward George, pharmaceutical chemistry
Feller, William, surgery
Fellows, Robert Francis, planetary atmospheres
Felsher, Murray, geology, oceanography
Felts, William Robert, internal medicine
Ferguson, Douglas Campbell, insect taxonomy
Ferguson, Mary Hobson, communication science
Ferrigno, Peter D, dentistry
Fiala, Alan Dale, astronomy
Fiege, Herbert Reynold, Jr, research administration
Field, Herbert Cyre, reactor physics
Fike, Harold Lester, organic chemistry
Filipescu, Nicolae, organic chemistry, physical chemistry
Fine, Ben Sion, ophthalmology
Fine, Paul Charles, theoretical physics
Finger, Frederick George, meteorology
Finger, Larry W, mineralogy, crystallography
Finke, Herman Louis, physical chemistry
Finkel, Peter William, operations research
Finkelstein, James David, gastroenterology, biochemistry
Finkner, Alva Leroy, statistics
Finn, Edward J, theoretical physics
Finnerty, Frank Ambrose, Jr, cardiology
Finney, William Jetton, physics
Fireman, Milton, soil science
Fischbach, Henry, chemistry
Fischer, Ronald Howard, petroleum chemistry, chemical engineering
Fischmann, Eugene J, electrocardiology
Fishbein, William Nichols, biochemistry, medicine
Fisher, William Lawrence, geology, paleontology
Fisher, Wilton Monroe, public health, preventive medicine
Fitzgerald, James W, atmospheric physics
Fitzhugh, William W, anthropology
Flach, Klaus Werner, soil science
Flajser, Steven Henry, interdisciplinary sciences, science policy
Flake, John C, dairy science
Flannery, Regina (Mrs Karl F Herzfeld), anthropology
Flax, Lawrence, acoustics
Fleischer, Robert, astronomy, science administration
Fletcher, James Chipman, physics
Flinn, Edward Ambrose, geophysics
Flint, Oliver Simeon, Jr, systematic entomology
Florin, Roland Eric, polymer chemistry
Florio, Lloyd Joseph, public health
Flynn, Joseph Henry, polymer science, thermal sciences
Fogle, Benson Tarrant, aeronomy, meteorology
Folen, Vincent James, magnetism
Foley, H Thomas, internal medicine, oncology
Foley, Robert Thomas, electrochemistry
Fong, Jeffrey Tse-Wei, mathematical physics, continuum mechanics
Forbes, Allan Louis, internal medicine
Forester, Donald Wayne, magnetism
Forester, Richard Monroe, micropaleontology, paleoecology
Formal, Samuel Bernard, microbiology
Forster, William Owen, analytical chemistry, oceanography
Forziati, Alphonse Frank, physical chemistry
Foster, John Stuart, Jr, physics
Foster, Willis Roy, medical science, information science
Fowler, Emil Eugene, chemistry
Fowler, Howland Auchincloss, physics
Fowlis, William Webster, meteorology, oceanography
Fox, Mattie Rae Spivey, nutritional biochemistry
Fox, Samuel Mickle, III, medicine
Fox, William B, physical inorganic chemistry
Foye, Laurance V, Jr, medicine
Francavilla, Thomas Lee, solid state physics
Frank, Martin, cell physiology, cardiovascular physiology
Frank, Richard Stephen, forensic science, science administration
Franklin, Philip Jaquins, physics
Franklin, Ralph E, environmental sciences, soil fertility
Frederikse, Hans Pieter Roetert, solid state physics
Fredrikson, Kurt A, geochemistry
Freeman, Jacob Joachim, physics
Freeman, Martin, computer sciences
Freis, Edward David, medicine
French, Judson Cull, solid state electronics
Frenkiel, Francois N, fluid dynamics
Fricken, Raymond Lee, high energy physics, research administration
Friedberg, Felix, biochemistry
Friedman, Abraham Solomon, nuclear science
Friedman, Leonard, biochemistry
Frishman, Fred, mathematical statistics
Fritz, Gilbert Geiger, x-ray astronomy
Fritz, James Clarence, nutritional biochemistry
Frodyma, Michael Mitchell, analytical chemistry
Froeschner, Richard Charles, entomology
Frohnsdorff, Geoffrey James Carl, physical chemistry, materials science
Fry, John Craig, oceanography, science policy
Fudali, Robert F, geochemistry, petrology
Fulkerson, John Frederick, microbiology, plant pathology
Furfine, Charles Stuart, biological chemistry
Gabriel, R Othmar, biochemistry
Gadzuk, John William, theoretical solid state physics, surface physics
Gagne, Raymond J, systematic entomology
Gajan, Raymond Joseph, analytical chemistry, organic chemistry
Galagan, Donald J, preventive dentistry
Gallagher, Brian Boru, neurology
Galloway, Kenneth Franklin, high energy physics
Gammon, William Howard, information science
Gangstad, Edward Otis, environmental sciences
Ganley, Oswald Harold, bacteriology, physiology
Gann, Richard George, physical chemistry, chemical kinetics
Gardner, Alvin Frederick, pathology
Gardner, William Milton, anthropology
Garner, Daniel Dee, forensic science
Garner, Richard Gordon, food science
Garvin, David, physical chemistry
Gastwirth, Joseph L, mathematical statistics
Geckler, Robert Payne, environmental sciences
Geddes, Wilburt Hale, marine geophysics
Geiger, Felix Eugene, solid state physics
Gemski, Peter, Jr, medical microbiology
Gerathewohl, Siegfried Johannes, aerospace medicine
Gessow, Alfred, physics, applied mathematics
Gevantman, Lewis Herman, physical chemistry
Giacchetti, Athos, spectroscopy
Giallorenzi, Thomas Gaetano, quantum optics
Gibbs, Robert Henry, Jr, ichthyology, biological oceanography
Gibbs, Robert John, physical chemistry, science administration
Gibson, Gordon Davis, anthropology
Gibson, Thomas George, invertebrate paleontology
Giel, Bohdan Gielecinski, environmental medicine, internal medicine
Gilardi, Richard Dean, structural chemistry
Gilbert, Francis Charles, nuclear physics
Gilbert, Robert Arthur, environmental management
Gilfrich, John Valentine, analytical chemistry
Gill, Jocelyn Ruth, astronomy
Gillespie, Robert L, veterinary medicine
Gillespie, Walter Lee, science education
Gillis, James E, Jr, geology, cartography
Gillis, Richard A, pharmacology
Gilman, Donald Lawrence, meteorology
Gilmore, Eleanor La Verne, medical bacteriology
Ginther, Robert J, inorganic chemistry
Gist, Lewis Alexander, Jr, organic chemistry, science administration
Gittes, Hyman Raphael, biochemistry
Glasgow, Augustus Rossell, Jr, pesticide chemistry
Glinos, Andre Dimitri, cell physiology
Glocklin, Vera Charlotte, physiology, biochemistry
Goddard, David Rockwell, botany, plant physiology
Goeringer, Gerald Conrad, cell biology
Goff, James Franklin, solid state physics
Goff, Loyal Glenn, physiology
Gold, Armand Joel, medical physiology
Goldberg, Aaron, taxonomic botany
Goldberg, Lawrence Spencer, quantum optics, lasers
Goldberg, Robert Nathan, physical chemistry
Goldberg, Vivian Joyce, physiology
Goldenberg, Neal, physical chemistry
Goldman, David Tobias, science administration, nuclear physics
Golla, Victor Karl, anthropology
Golub, Abraham, operations research, science administration
Golumbic, Calvin, biochemistry
Gonzalez, Nancie L, cultural anthropology, social anthropology
Goodenough, David John, medical physics, radiological physics
Goodman, Ralph Raymond, theoretical physics
Goodrich, Robert Bruce, energy conversion
Goor, Charles G, applied statistics, operations research
Gordh, Gordon, entomology, systematics
Gordon, George Selbie, physical chemistry, environmental engineering
Gordon, Nathan, biochemistry
Gordon, William Bernard, mathematics
Gorham, John Richard, medical entomology
Gossett, Charles Robert, nuclear physics
Gottschalk, Charles Max, information science, nuclear science
Gottschalk, John Simison, conservation
Gould, Jack Richard, organic chemistry
Graham, Joseph H, organic chemistry
Graminski, Edmond Leonard, physical organic chemistry, rheology
Granatstein, Victor Lawrence, plasma physics, microwave physics
Granger, John Van Nuys, physics
Grant, Richard Evans, invertebrate paleontology
Grant, Warren Herbert, physical chemistry, polymer chemistry
Graumann, Hugo Oswalt, agronomy
Gravatt, Claude Carrington, Jr, chemical physics
Graves, Glen Atkins, nuclear physics
Gray, Irving, biochemistry, biophysics
Gray, Mary Wheat, mathematics
Gray, Thomas Ira, Jr, meteorology
Green, Ben Arthur, Jr, physics, science education
Green, Richard James, physics
Green, Robert Eugene, mathematics, electronics
Green, Sidney, toxicology
Greenberg, Louis, bacteriology, immunology
Greenblatt, Jayson Herschel, earth sciences
Greene, Albert Godfrey, Jr, plant physiology
Greenewalt, David, geophysics
Greenfield, Richard Sherman, meteorology
Greenhall, Arthur Merwin, mammalogy
Greenhouse, Samuel William, statistics
Greenwalt, Tibor Jack, medicine, hematology
Gregg, Donald Eaton, physiology
Grieg, J Robert, plasma physics
Griffenhagen, George Bernard, pharmacy
Griffin, Joe Lee, cell biology
Griffin, Kathleen (Mary), speech pathology, audiology
Griffiths, Norman Henry Campbell, dentistry
Grigsby, Margaret Elizabeth, medicine
Grisamore, Nelson Thomas, science administration
Grobecker, Alan J, space physics, geophysics
Gross, Donald, operations research
Gross, Mircea Adrian, pathology
Grosvenor, Melville Bell, geography
Gruntfest, Irving James, physical chemistry
Gubser, Donald Urban, solid state physics, cryogenics
Guenzer, Charles S P, solid state physics
Guildner, Leslie Arnold, metrology
Gunn, John William, Jr, forensic science, science administration
Gupta, Om Prakash, dentistry
Gurney, Ashley Buell, entomology
Gurney, Margaret, mathematical statistics
Gusman, Samuel, science policy, physical chemistry
Gutelius, Margaret Frances, medicine
Guthe, Otto Emmor, research administration
Guthrie, Albert Nelson, physics
Guttman, Charles M, polymer science
Haar, Lester, physics
Haas, George Arthur, surface physics
Haas, Gustav Frederick, sensory physiology, biomedical engineering
Haas, Peter Herbert, radiation physics
Haber, Seymour, numerical analysis
Haddon, William, Jr, medical ecology
Hadermann, Albert Felix, physical chemistry, polymer chemistry
Haenni, Edward Otto, analytical chemistry
Hafstad, Lawrence Randolph, physics
Haftel, Michael Ivan, theoretical nuclear physics
Hagan, Lucy Gay, physical chemistry, spectroscopy
Hagans, James Albert, biostatistics, therapeutics
Hagler, James Neil, mathematical analysis
Hahn, Fred Ernst, molecular biology, molecular pharmacology
Hale, Mason Ellsworth, Jr, botany
Haller, Wolfgang Karl, physical chemistry, glass technology
Halpern, Katherine Spencer, anthropology, applied anthropology
Halvorson, Lloyd Chester, agricultural economics
Hamarneh, Sami, history of pharmacy
Hambleton, Edson Jorge, systematic entomology
Hamer, Walter Jay, physical chemistry, electrochemistry
Hammer, Carl, computer sciences
Hammer, Charles F, organic chemistry, chemical instrumentation
Hammond, Peter (Boyd), anthropology
Hamosh, Paul, medical physiology
Hampson, Robert F, Jr, physical chemistry
Han, Charles Chih-Chao, polymer chemistry
Handler, Philip, biochemistry
Handley, Charles Overton, Jr, mammalogy
Hanig, Joseph Peter, neuropharmacology, toxicology
Hannah, John Alfred, poultry husbandry
Hannum, William Hamilton, reactor physics
Hansen, James L, medicine, pathology
Hanssen, George Lyle, oceanography
Haque, Rizwanul, environmental chemistry
Harbour, Jerry, geology
Harder, Harold Cecil, pharmacology, cancer
Hardison, Wesley Aurel, animal nutrition
Hare, Peter Edgar, organic chemistry
Harkins, Robert W, nutrition, physiological chemistry
Harland, Barbara Ferguson, nutrition
Harney, Brian Michael, physical chemistry, fuel science
Harrington, John Vincent, physics
Harrington, Robert Sutton, astronomy
Harris, Clare I, plant science
Harris, George Lawrence, cultural anthropology
Harris, Saul Joseph, public health administration
Harrison, Joseph Owens, Jr, applied mathematics
Harrison, Mark, physics
Harshbarger, Harold Beck, meteorology, climatology
Harshbarger, John Carl, Jr, pathobiology
Hart, Charles Willard, Jr, aquatic biology, science writing
Hart, Pembroke J, geophysics, seismology
Hartstein, Arthur M, physical chemistry, analytical chemistry
Hartzler, Alfred James, physics
Harvey, Albert Bigelow, physical chemistry
Harvey, Douglas G, chemistry
Harvey, Elmer Bostwick, embryology, zoology
Harvey, George Lloyd, electrooptics
Harvey, John Collins, internal medicine
Harvey, Watkins Proctor, medicine
Harwood, Clare Theresa, pharmacology
Hashmall, Joseph Alan, quantum chemistry, physical chemistry

Haskins, Caryl Parker, genetics, physiology
Haskins, Edna Ferrell, chemistry
Hass, Marvin, solid state physics
Hastie, John William, inorganic chemistry, chemical physics
Harziolos, Basil Constantine, veterinary pathology
Haubach, Walter Jennings, Jr., physical chemistry
Hawley, Robert John, medical microbiology
Hawthorne, Edward William, physiology
Hayashi, Fumihiko, plant physiology
Hayes, Helen Landau, aquatic biology, biological oceanography
Hayes, John Nicholas, physics
Hayes, Marguerite Thompson, nuclear medicine, internal medicine
Hazel, Joseph Ernest, invertebrate paleontology
Hazen, Robert Miller, mineralogy, history of science
Hazlett, Robert Neil, physical organic chemistry
Headings, Verle Emery, medicine, human genetics
Heath, Robert Gardner, biometrics
Hehir, Robert M., chemistry
Heifer, Melvin Harold, physiology
Heilman, Dorothy Henderson, bacteriology
Heineman, Robert Edwin, nuclear science
Helgesen, Andre E., obstetrics & gynecology
Heller, Douglas Max, physics
Hellman, Louis M., obstetrics & gynecology
Helmann, Max, science administration
Helwig, Elson Bowman, pathology
Heman-Ackah, Samuel Monie, chemotherapy, biopharmaceutics
Hemily, Philip Wright, crystallography
Henderson, Madeline M Berry, information science, computer science
Hendrickson, Wayne Arthur, molecular biophysics
Henkin, Robert I., medicine
Henney, Dagmar Renate, mathematics
Henry, Joseph L., anatomy, dentistry
Henry, Walter Lester, Jr., medicine, endocrinology
Henschke, Ulrich K., medical physics
Hensel, Gustav, mathematics
Henze, Robert Edgerton, biochemistry
Herling, Gary Herbert, nuclear physics
Herman, Eugene H., pharmacology
Herndon, J Loritts, III, polymer chemistry
Herr, Frank Leaman, Jr., geochemistry, physical chemistry
Herring, Jon Lamar, entomology
Herschman, Arthur, theoretical physics, information science
Herscowitz, Herbert Bernard, immunology, microbiology
Hersey, David Floyd, information science
Hershner, Ivan Raymond, Jr., mathematics
Hertz, Hans Georg, celestial mechanics
Hertz, Roy, physiology
Herwig, Lloyd Otto, physics
Herzfeld, Karl Ferdinand, physics
Hess, Walter Cohen, biochemistry
Hessberg, Rufus R., aerospace medicine
Hesser, Leon Francis, agricultural economics
Heydemann, Peter Ludwig Martin, physics, neurophysiology
Heyer, William Ronald, biology
Hickey, Leo Joseph, paleobotany
Hicks, Grady Thomas, geophysics
Hiebert, Gordon Lee, physical chemistry
Higgins, George A., Jr., surgery
Higgins, Robert Price, zoology
Higgs, Robert Hughes, geophysics
Hilberg, Albert William, pathology
Hildebrandt, Paul Knud, veterinary pathology
Hillig, Fred, food chemistry
Himmelsbach, Clifton Keck, pharmacology
Hinckley, Alden Dexter, ecology, entomology
Hinners, Noel W, geochemistry, geology
Hirvonen, James Karsten, experimental physics
Hobbs, Charles Henry, veterinary toxicology
Hobbs, Herman Hedberg, solid state physics
Hobbs, Horton Holcombe, Jr., systematic zoology

Hobson, Lawrence Bennett, medical research, internal medicine
Hodges, Ronald William, systematic entomology
Hoffman, Allan Richard, solid state physics
Hoffman, Jerry Irwin, research administration, dental research
Hoffman, Julius, psychiatry, neurology
Hofmann, Albrecht Werner, geochemistry, geochronology
Hogben, David, statistics
Holdeman, Louis Brian, cryogenics, solid state physics
Hollaender, Alexander, biophysics
Holland, Joshua Zaliman, meteorology
Hollinger, James Pippert, physics, radio astronomy
Hollinshead, Ariel Cahill, pharmacology, virology
Holloway, Harry Charles, neuropsychiatry
Holloway, James Ashley, neurophysiology
Holloway, John Thomas, experimental physics
Holmes, George Edward, biology
Holt, Helen Keil, atomic physics, lasers
Holt, James Allen, biochemistry
Hood, Thomas Robin, biochemistry
Hooper, Edward Temple, Jr., physics
Hoover, Samuel Randolph, food science
Hope, William Duane, nematology
Hopper, Grace Murray, systematic zoology
Hoppman, Julian, microbiology, environmental physiology
Horner, William Harry, biochemistry
Hornstein, Irwin, nutritional biochemistry
Horowitz, Joel Lawrence, environmental management, transportation
Horton, John Edward, immunology, periodontology
Horton, William Sheldon, physical chemistry
Houck, Frank Scanland, nuclear chemistry, operations research
Houck, John Candee, immunology, cell biology
Hougen, Jon T., chemical physics
Hough, Walter Andrew, forestry
Housewright, Riley Dee, microbiology
Howard, John William, analytical chemistry
Howard, Lee Million, preventive medicine, public health administration
Howard, Russell Alfred, solar physics
Howe, Marshall Atherton, ornithology
Howell, Barbara Fennema, physical chemistry, radiochemistry
Hoyer, Bill Henriksen, microbiology
Hrubec, Zdenek, medical research, epidemiology
Hsia, Jack Jinn-Goe, thermal physics, heat transfer
Huang, H K, biomedical engineering, biomathematics
Huang, Kun-Yen, virology, medicine
Hubbard, Harold Benson, public health
Hubbard, Willard Dwight, analytical chemistry
Hubler, Graham Kelder, Jr., applied physics
Huckaba, Charles Edwin, biomedical engineering, chemical engineering
Huddle, John Warfield, geology, paleontology
Hudlow, Michael Dale, meteorology, hydrology
Hudson, Ralph P., low temperature physics
Hueber, Francis Maurice, paleobotany, plant morphology
Hufnagel, Charles Anthony, surgery
Huget, Eugene F., dental materials, organic chemistry
Huggett, Clayton (McKenna), chemistry
Hugh, Rudolph, medical bacteriology
Hughes, Patrick E., journalism
Huh, Oscar Karl, geology
Huie, Robert Elliott, chemical kinetics
Humphrey, Robert Lee, Jr., anthropology, archaeology
Hunston, Donald Lee, chemistry
Hunt, John Baker, inorganic chemistry
Hunt, Lee McCaa, oceanography, geology
Hunt, Lois Turpin, cytology, microbiology
Hunter, James Murry, geography
Hunter, Oscar Benwood, Jr., medicine

Hunter, William Ray, physical optics
Huntley, Robert Ross, internal medicine, health administration
Hurd, Paul David, Jr., entomology
Hurley, Frank Leo, biostatistics
Hussain, Mehdi Hajivani, pharmaceutical chemistry, biochemistry
Huxsoll, David Leslie, veterinary medicine
Hyatt, George W., medicine
Iacono, James M., experimental medicine, biological chemistry
Iampietro, Patsy F., physiology
Ifft, Edward M., physics
Imlay, Ralph Willard, paleontology
Ingentio, Frank Leo, underwater acoustics
Ingersoll, Jasper, anthropology
Ingle, George William, physical chemistry, research administration
Ingle, John Ide, dentistry
Ingram, Glenn R., computer science
Insley, Herbert, mineralogy
Irey, Nelson Sumner, pathology
Irick, Paul Eugene, applied statistics, information science
Irvine, T Neil, geology
Isbell, Horace Smith, carbohydrate chemistry
Iverson, Warren Philip, microbiology, corrosion
Jabine, Thomas Boyd, applied statistics
Jablon, Seymour, biostatistics, epidemiology
Jackson, Dudley Pennington, medicine, hematology
Jackson, Edward Soth, astronomy
Jackson, Francis Charles, medicine
Jackson, George John, parasitology
Jackson, Marvin Alexander, medicine, pathology
Jackson, Michael J., physiology
Jackson, Robert Howard, food science
Jackson, William Morgan, photochemistry, chemical physics
Jacobs, George Joseph, biology
Jacobs, Walter William, mathematical statistics
Jacobs, Marilyn Esther, physical chemistry
Jakoby, Ruth Elizabeth Kerr, neurosurgery
James, David Evan, geophysics
James, Frances Crews, ecology
Janiczek, Paul Michael, astronomy, navigation
Jarvis, Garth Louray, medicine
Jarvis, Neldon Lynn, surface chemistry
Jarzynski, Jacek, physics
Jelinek, Charles Frank, organic chemistry
Jenkins, Dale Wilson, ecology
Jenkins, David Bruce, gross anatomy
Jenkins, Melvin Earl, metabolism
Jennings, Allen Lee, biochemistry, enzymology
Jennings, Feenan Dee, oceanography
Jennings, Robert Kimmel, biochemistry
Jessner, Lucie, psychiatry
Johannesen, Rolf Bradford, inorganic chemistry
Johnson, David Simonds, meteorology
Johnson, Donald Rex, molecular spectroscopy, radio astronomy
Johnson, Edward A., forestry, watershed management
Johnson, Elgy Sibley, mathematics
Johnson, Frank Bacchus, pathology
Johnson, Frederick Carroll, applied mathematics, resource management
Johnson, Harold Hunt, mathematics
Johnson, John Peter, applied statistics
Johnson, Kenneth Delford, air pollution, industrial hygiene
Johnson, Leonard Evans, geophysics
Johnson, Lloyd Kenneth, physics, science education
Johnson, Paul Edwin, animal nutrition
Johnson, Raymond Earl, zoology
Johnson, Stuart Earl, operations research
Johnson, Thomas Nick, neuroanatomy
Johnson, William Martin, soil science
Johnsrud, Alan Edwin, operations research
Jones, Meredith L., invertebrate zoology
Jones, Philip Robert, biology
Joy, Robert John Thomas, internal medicine, physiology
Julienne, Paul Sebastian, quantum chemistry, atmospheric chemistry

Kabler, Milton Norris, solid state physics
Kadish, Abraham, mathematics, plasma physics
Kagarise, Ronald Eugene, physics
Kahn, Arnold Herbert, physics
Kahn, Jack Henry, physics
Kahn, Samuel George, nutrition
Kaiser, Quentin C., solid state physics
Kallander, John William, computer science
Kamenetz, Herman Leo, physical medicine, rehabilitation
Kammer, Erwin William, physics
Kandel, Richard Joshua, physical chemistry
Kapadia, Govind J., pharmaceutical chemistry
Kaplan, David Jeremy, operations research, systems analysis
Kaplan, Nathan, electrochemistry
Kaplan, Raphael, solid state physics
Kapur, Shakti Prakash, human anatomy, histology
Karklins, Olgerts Longins, geology, biology
Karle, Isabella Lugoski, physics
Karle, Jerome, crystallography
Karstadt, Myra Leonore, environmental sciences, occupational health
Kashelar, Dinkar Kashinath, physiology, biochemistry
Kass, Leon Richard, biochemistry, medical ethics
Kassira, Edward Naim, public health, microbiology
Kastenbaum, Marvin Aaron, mathematical statistics
Katz, Edward, microbiology, biochemistry
Katz, Irving, mathematics
Katz, Maurice Joseph, physics
Katz, Sol, medicine
Kauffman, Erle Galen, paleobiology, paleoecology
Kaufman, Jay Victor Richard, inorganic chemistry, explosives
Kaufman, Victor, atomic physics
Kautter, Donald Albert, microbiology
Kayser, Boris Jules, theoretical high energy physics
Kearsley, Elliot Armstrong, rheology, continuum mechanics
Keehn, Robert John, medical statistics
Keeler, Roger Norris, high pressure physics
Keen, Dorothy Jean, physical oceanography
Keenan, Thomas Aquinas, computer science
Kegeles, Lawrence Steven, theoretical astrophysics
Keitt, George Wannamaker, Jr., plant physiology
Kelber, Charles Norman, physics
Kelber, Thomas Francis, medicine
Keller, Richard Alan, physical chemistry
Kellett, James Clarence, Jr., medicinal chemistry
Kelley, John Joseph, II, oceanography, meteorology
Kelly, Henry Charles, physics, science policy
Kelly, William Clark, physics
Kelser, George Archibald, internal medicine
Kelsey, Fremont Ellis, pharmacology
Kelson, Keith R, zoology
Kemple, Marvin David, magnetic resonance
Kempter, Charles Prentiss, chemistry, materials science
Kenk, Roman, invertebrate zoology
Kennard, Ralph Brandreth, bacteriology
Kennedy, Eugene Richard, bacteriology
Kennedy, Thomas James, Jr., physiology, nephrology
Kenney, Richard Alec, medical physiology
Kenyon, Hewitt, mathematics
Kepple, Paul C., physics
Kershenstein, Karen Weis, physics
Kertesz, Dennis Jay, organic chemistry
Kessler, Karl Gunther, atomic spectroscopy
Kessler, Ernest George, Jr., optical physics, atomic spectroscopy
Khanna, Krishan L., phytochemistry, pharmaceutical chemistry
Kieffer, Lee Joseph, atomic physics
Kier, Porter Martin, invertebrate paleontology
Kies, Marian Wood, biochemistry, neurochemistry
Kiess, Norman Halvor, molecular spectroscopy
Killion, Lawrence Eugene, lasers, computer science
Kind, Phyllis Dawn, immunology

King, Donald Roy, entomology
King, Michael M, organic chemistry
Kingman, Harry Ellis, Jr, veterinary medicine
Kingsolver, John Mark, entomology
Kinnamon, Kenneth Ellis, physiology, radiobiology
Kinney, Edward Coyle, Jr, fish biology
Kinney, Roland Walter, microbiology, biochemistry
Kinney, Terry B, Jr, population genetics, research administration
Kinsinger, Jack Burl, physical chemistry
Kinzer, Robert Lee, astrophysics, elementary particle physics
Kirchhoff, William Hayes, physical chemistry
Kirkbride, Joseph Harold, Jr, taxonomic botany
Kirkien-Rzeszotarski, Alicia M, physical chemistry
Kittrell, Flemmie P, nutrition
Klayman, Daniel Leslie, medicinal chemistry, organic chemistry
Kleiman, Devra Gail, ethology, reproductive biology
Klein, Lewis S, theoretical physics
Klein, Philipp Hillel, physical inorganic chemistry
Klein, Ralph, physical chemistry
Klich, Clifford C, solid state physics
Kline, Jerry Robert, analytical chemistry, soil chemistry
Klock, Benny LeRoy, astronomy
Klopp, Calvin Trexler, surgery
Klose, Jules Zeiser, atomic physics
Klove, Robert Charles, geography, cartography
Klubes, Philip, pharmacology
Klute, Charles Henry, physical chemistry
Knapp, Leslie W, ichthyology
Kneece, Roland Royce, Jr, mathematics, operations research
Knez, Eugene Irving, anthropology, ethnology
Knowles, Stephen H, radio astronomy
Knowlton, Robert Earle, invertebrate zoology, marine biology
Knox, Arthur Stewart, palynology, cartography
Knudson, Alvin Richard, atomic physics
Koehl, George Martin, optics
Koering, Marilyn Jean, anatomy, reproductive physiology
Kokoski, Charles Joseph, pharmacy
Kolb, Edward John, research administration, ionospheric physics
Komm, Horace, mathematics
Konnerup, Nels Millard, veterinary public health
Korin, Basil Peter, mathematical statistics
Kornicker, Louis Sampson, geology
Kostkowski, Henry John, physics
Kot, Peter Aloysius, cardiovascular physiology
Kothe, Kenneth Ralph, geophysics
Kowkabany, George Norman, organic chemistry
Kramer, James Phillip, systematic entomology
Kramer, Julian, biology
Kramer, Norman Clifford, internal medicine, immunology
Kramer, Stanley Phillip, medicinal chemistry, pharmacology
Krashevski, Stefan H, soil science
Krause, Ralph M, mathematics
Krebill, Richard G, forest pathology
Krebs, James John, experimental solid state physics
Kreysa, Frank Joseph, organic chemistry, resource management
Krogh, Thomas Edvard, geochemistry
Krombein, Karl Von Vorse, entomology
Krop, Stephen, pharmacology, toxicology
Kruger, Jerome, physical chemistry, corrosion
Krushensky, Richard D, geology, volcanology
Kulczycki, Lucas Luke, pediatrics
Kullback, Joseph Henry, mathematical statistics
Kullback, Solomon, mathematical statistics, applied statistics
Kumar, Cidambi Krishna, astrophysics, atomic physics
Kumar, Soma, biochemistry
Kuperman, William Aaron, acoustics, marine geophysics
Kupperman, Robert Harris, applied mathematics, operations research
Kurfess, James Daniel, astrophysics
Kuriyama, Masao, solid state physics, crystallography
Kurtzke, John F, neurology, epidemiology
Kushner, Lawrence Maurice, physical chemistry
Kuswa, Glenn Wesley, experimental physics
Kutz, Frederick Winfield, ecology, medical entomology
Kuyatt, Chris Ernie Earl, electron optics, atomic physics
Kvenberg, John Eide, entomology
Lachmann, Alfred, physical chemistry
Lachner, Ernest Albert, systematic ichthyology
Ladd, Harry Stephen, geology
Lafferty, Walter J, physical chemistry
Lagnese, John Edward, mathematics
Laird, Wilson Morrow, geology
Lamanna, Carl, microbiology, toxicology
Lamaze, George Paul, experimental nuclear physics
Lambert, James Morrison, nuclear physics
Lambert, Richard Bowles, Jr, physical oceanography
Lance, John Franklin, geology
Landau, Emanuel, epidemiology, biostatistics
Landman, Otto Ernest, microbial genetics
Landman, Ruth Hallo, anthropology
Lanzano, Paolo, applied mathematics, space physics
Lashof, Theodore William, physics
Lasser, Marvin Elliott, solid state physics
Laufer, Allan Henry, photochemistry, chemical kinetics
Laughlin, Robert Moody, anthropology
LaVilla, Robert E, physical chemistry
Lavine, Robert Alan, neurophysiology, psychology
Lazen, Alvin Gordon, biochemistry, research administration
Leachman, Robert Briggs, nuclear physics
Leak, Lee Virn, cell biology, electron microscopy
Leap, William L, anthropology
Leary, Joseph Aloysius, physical chemistry, chemical engineering
LeBaron, Robert (Francis), chemistry
Lebensohn, Zigmond Meyer, psychiatry
Lechner, James Albert, mathematical statistics, applied statistics
Ledley, Robert Steven, biophysics, mathematics
Lee, James A, human ecology, public health
Lee, Kah-Hock, inorganic chemistry
Lee, Lyndon Edmund, Jr, surgery, pharmacology
Lee, Tong-Nyong, plasma physics, atomic physics
Lee, Wei Hwa, microbiology, food technology
Leedy, Daniel Loney, zoology, ecology
LeGolvan, Paul Celestin, medicine, pathology
Legters, Llewellyn J, preventive medicine, tropical public health
Lehman, Donald Richard, theoretical nuclear physics
Lehman, Richard Lawrence, biophysics
Leibowitz, Jack Richard, solid state physics
Leidecker, Henning William, Jr, physics
Leise, Joshua Melvin, medical microbiology
Leiss, James Elroy, physics
Lellinger, David Bruce, taxonomic botany
Leng, Earl Reece, plant genetics, plant breeding
Leonard, Frederic Adams, physiology
Leonard, Joseph Thomas, fuel science
Leone, Fred Charles, statistics
Lepp, Albert, biochemistry
Lepson, Benjamin, mathematics, statistics
Lerner, Melvin, analytical chemistry
Lessoff, Howard, solid state science
Letterman, Gordon Sparks, surgery
Leung, Woot-Tsuen Wu, nutrition
Leverton, Ruth Mandeville, nutrition
Levy, Joseph Benjamin, physical chemistry, organic chemistry
Lewin, Lawrence M, biochemistry, microbiology
Lewis, Herman William, genetics, zoology
Lewis, James Eldon, economic geography, transportation geography
Lewis, Ralph Kepler, anthropology
Lewis, Thomas Howard, psychiatry
Lide, David Reynolds, Jr, chemical physics
Lieberman, Edwin James, psychiatry, social psychology
Liebman, Samuel, polymer chemistry, physical chemistry
Light, Amos Ellis, pharmacology, nutrition
Lilienfield, Lawrence Spencer, medicine, physiology
Lilliefors, Hubert W, statistics
Linde, Alan Trevor, geophysics
Lindgren, Richard Arthur, nuclear physics
Lindholm, Roy Charles, geology
Lindquist, Clarence Bernhart, applied mathematics
Linduska, Joseph Paul, biology
Link, John Clarence, physics
Link, William B, analytical chemistry, organic chemistry
Linnenbom, Victor John, oceanography
Linzer, Melvin, physical chemistry, ultrasound
Litovitz, Theodore Aaron, physics
Little, Elbert Luther, Jr, botany, dendrology
Littlewood, William Herbert, science policy, oceanography
Liverman, James Leslie, plant physiology, biochemistry
Liverman, Thomas Phillip George, mathematics
Livermore, Arthur Hamilton, biochemistry, science education
Livingston, Peter Moshchansky, physics, theoretical chemistry
Lloyd, Joel Joseph, geology, mathematics
Lloyd, Ruth Smith, anatomy
Locke, Krystyna Kopaczyk, biochemistry, enzymology
Locke, Raymond Kenneth, biochemistry, metabolism
Loeb, Marcia Joan, invertebrate physiology
Loeb, Marilyn Rosenthal, biochemistry
Loebenstein, William Vaille, physical chemistry, dental research
Loebner, Egon Ezriel, solid state physics, research administration
Loevinger, Robert, radiological physics, metrology
Lohman, Kenneth Elmo, geology
Lombard, David Bishop, experimental physics
Lombardo, Pasquale, analytical chemistry
Long, John Kelley, nuclear physics, nuclear engineering
Longrie, Dean Paul, wildlife ecology
Lorber, Mortimer, physiology, hematology
Loughridge, Michael Samuel, marine geology
Loustalot, Arnaud Joseph, plant physiology
Lovas, Francis John, molecular spectroscopy, radio astronomy
Lovvorn, Roy Lee, agronomy, botany
Lowenthal, Joseph Philip, microbiology, immunology
Lucas, John Paul, microbiology
Lucke, Robert Lancaster, astrophysics
Lucke, William Hunter, surface physics
Luebke, Emmeth August, physics
Luessenhop, Alfred John, medicine, neurosurgery
Luke, James Lindsay, pathology
Lupton, William Hamilton, plasma physics
Lutz, George John, physical chemistry
Lynn, William Gardner, zoology
Mabie, Curtis Parsons, dental materials, microscopy
MacArthur, Donald M, environmental management, health sciences
Maccini, John Andrew, geology
MacDonald, Rosemary A, solid state physics
Macedo, Pedro Buarque de, physics
MacElroy, Robert David, exobiology
Mack, Julius L, chemical physics
Mackenthun, Kenneth Marsh, aquatic biology
Mackenzie, Kenneth Victor, physics, oceanography
Macnamara, Thomas E, anesthesiology
Macon, Nathaniel, numerical analysis, systems science
Madden, Robert Phyfe, physics
Madey, Theodore Eugene, surface physics, surface chemistry
Madow, William Gregory, mathematical statistics
Maengwyn-Davies, Gertrude Diane, pharmacology
Magin, George Benedict, Jr, inorganic chemistry, geochemistry
Mahaffey, Kathryn Rose, nutrition, experimental pathology
Maienthal, E June, chemistry
Maienthal, Millard, organic chemistry
Maier, John G, medicine, radiobiology
Maio, Domenic Anthony, physiology, aerospace medicine
Maisch, William George, physical chemistry
Maisel, Herbert, computer science, statistics
Majchrowicz, Edward, biochemistry, biochemical pharmacology
Maki, Arthur George, Jr, physical chemistry
Malloy, Alfred Marcus, physical chemistry, electrochemistry
Malmberg, Philip Ray, nuclear physics, solid state physics
Malmgren, Richard Axel, pathology
Malone, Charles R, ecology
Malveaux, Floyd J, microbiology, microbial physiology
Mamay, Sergius Harry, paleobotany
Mandel, Benjamin J, applied statistics
Mandel, Harold George, pharmacology
Mangan, George Francis, Jr, biochemistry
Mange, Phillip Warren, planetary atmospheres, space physics
Manheimer, Wallace Milton, plasma physics
Mann, Charles Roy, mathematical statistics, applied statistics
Mann, Wilfrid Basil, physics
Manning, Irwin, theoretical physics
Manning, John Randolph, solid state physics, metallurgy
Manning, Raymond B, invertebrate zoology, marine biology
Manos, Nicholas Emmanuel, mathematics
Mao, Ho-Kwang, geochemistry
Marcellini, Dale Leroy, herpetology, animal behavior
Markham, Arleigh Holden, low temperature physics, urban research & development
Marks, Sidney, pathology, biostatistics
Marlow, Keith Winton, nuclear physics
Marlow, William Henry, operations research, mathematics
Marquardt, Charles Lawrence, experimental solid state physics
Marrone, Michael Joseph, solid state physics
Marsh, Paul Malcolm, entomology
Marshak, Harvey, nuclear physics
Marshall, Charles Louis, chemistry
Marshall, Lawrence Marcellus, biochemistry
Marshall, Samuel Wilson, physics
Martens, Vernon Edward, pathology
Marth-Snader, Ella Carolyn, mathematics
Martin, Malcolm Mencer, pediatric endocrinology
Martin, Margaret Elizabeth, statistics
Martin, William Clyde, atomic spectroscopy, atomic physics
Martins da Silva, Mauricio, pediatrics
Martire, Daniel Edward, physical chemistry
Marvin, Henry Howard, Jr, physical chemistry
Marzetti, Lawrence Arthur, applied statistics
Mason, Brian Harold, geochemistry
Mason, John, epidemiology, public health
Mason, John Wayne, neuroendocrinology
Massaro, Donald John, medicine
Mathew, Mathai, x-ray crystallography, inorganic chemistry
Mattern, Kenneth Lawrence, inorganic chemistry
Mattson, James Stewart, chemical oceanography, surface chemistry
Matuszko, Anthony Joseph, organic chemistry, inorganic chemistry
Maughan, Paul McAlpine, marine sciences, remote sensing
Maune, David Francis, photogrammetry, research administration
Mavrodineanu, Radu, physical chemistry
Maxwell, Fowden Gene, entomology
May, Leopold, physical biochemistry
Maycock, Paul Dean, solid state physics, science administration
Mayer, Claudius Francis, history of medicine
Mayer, Cornell Henry, astronomy
Mayer, Vernon William, Jr, microbiology, genetics
Mayer, Walter Georg, physics
Mayo, James Wellington, solid state physics
Mayo, Santos, microelectronics
Mazeika, Paul A, physical oceanography
Mazel, Paul, pharmacology, biochemistry
Mazur, Jacob, polymer physics
McArdle, Richard Edwin, forestry
McBryde, Felix Webster, geography, anthropology
McCafferty, Edward, chemistry
McCally, Michael, physiology, aerospace medicine
McCann, James Alwyn, fisheries, biometrics
McCarthy, Dennis Dean, astronomy
McCauley, Charles Edward, physiology

McClelland, John Edward, soils
McClure, Frank James, nutrition
McClure, Joseph Andrew, Jr., theoretical physics
McCormack, Donald Eugene, soil science
McCoubrey, Arthur Orland, physics, science administration
McCoy, William Harrison, physical chemistry
McCracken, Ralph Joseph, soil science
McCue, Edmund Bradley, physics
McCullough, James Matthew, botany, environmental biology
McCurdy, Paul Ranney, internal medicine, hematology
McDonald, Jimmie Reed, physical chemistry, molecular spectroscopy
McDonald, Roger Koefod, clinical medicine
McDowell, Hershel, physical chemistry
McElhinney, John, physics
McGrath, James Russell, physics
McGuire, John R., forest economics
McIntyre, Wallace Edward, economic geography, administrative sciences
McKay, Douglas William, orthopedic surgery
McKay, Ruth Blumenfeld, anthropology
McKeon, Alfred Jerome, mathematical statistics
McKinney, John Edward, thermodynamics, rheology
McKnight, Melvin Edward, entomology
McLaughlin, David, zoology
McLaughlin, Joseph, Jr., organic chemistry
McLaughlin, William Lowndes, radiation chemistry
McLean, Edgar Alexander, plasma physics
McLean, Ian William, pathology, ophthalmology
McNesby, James Robert, physical chemistry
McNett, Charles William, Jr., anthropology, archaeology
McNutt, Douglas P., physics
McRae, Vincent Vernon, science policy, information science
McVey, William Henry, chemistry
McWright, Cornelius Glen, immunology, microbiology
Medz, Robert B., environmental management
Meekins, John Fred, solar physics, x-ray astronomy
Meggers, Betty Jane, anthropology
Megregian, Stephen, analytical chemistry
Meier, Michael McDaniel, nuclear physics
Meier, Robert R., aeronomy
Meijer, Paul Herman Ernst, theoretical physics
Meinke, Arnold Stephen Ernst, entomology
Meisburger, Edward Paul, marine geology
Melendez, Luis Vargas, virology
Mello, James Francis, paleontology
Melson, William Gerald, petrology, geochemistry
Meltzer, Richard S., physics
Mencher, Alan George, science policy education
Mendel, Julius Louis, clinical chemistry
Mendlowitz, Harold, theoretical physics
Menez, Ernani Guingona, marine phycology
Menke, William John, physical oceanography
Menkes, Joshua, applied mathematics
Menzel, Joerg H., nuclear science
Merchant, Henry Clifton, ecology, zoology
Merrell, William John, physical oceanography
Merrill, Malcolm Hendricks, public health
Meshkov, Sydney, physics
Metzger, William Henry, Jr., electrochemistry, physical chemistry
Meyer, Alvin F., Jr., environmental health
Meyer, Edward Raymond, environmental sciences
Meyer, Frederick Gustav, botany
Meyer, Gerald, geology

Meyer, Morton A., geography
Meyer, Ralph O., solid state physics, fuel technology
Meyers, Wayne Marvin, medical microbiology
Michejda, Christopher Jan, physical organic chemistry, bio-organic chemistry
Michels, Donald Joseph, physics
Mielke, James Edward, geochemistry
Mies, Frederick Henry, quantum chemistry
Migaki, George, veterinary pathology, comparative pathology
Millar, Jack William, medicine
Miller, A. Eugene, mathematics, research administration
Miller, Bennett, plasma physics, science administration
Miller, Daniel Robert, chemistry, research administration
Miller, David Jacob, analytical chemistry
Miller, Donald Fletcher, nutrition
Miller, Frank Nelson, Jr., pathology
Miller, George Edward, biomedical engineering
Miller, Robert Victor, systematic ichthyology, research administration
Miller, William Lawrence, metallurgy, geology
Mills, George Alexander, physical chemistry, fuel science
Milner, Max, food science
Milton, Albert Fenner, solid state physics
Mink, John F., hydrology, geology
Minthorn, Martin Lloyd, Jr., biochemistry
Mislivec, Philip Brian, mycology, plant pathology
Misner, Robert David, physics
Misra, Dwarika Nath, physical chemistry, surface chemistry
Mitchell, Dean Lewis, solid state physics
Mitchell, Ian Alastair, internal medicine, medical administration
Mitchell, Thomas George, nuclear medicine
Mitz, Milton Aaron, biochemistry
Modderman, John Philip, analytical chemistry, food chemistry
Mohler, Stanley Ross, aerospace medicine
Moldover, Michael Robert, physics
Moller, Raymond William, mathematics
Moniz, William Bettencourt, physical chemistry
Monius, William S., pathology, microbiology
Montrose, Charles Joseph, Jr., physics
Moody, John Robert, chemistry
Moore, Harvey Cleaver, anthropology, cultural anthropology
Moore, Jerry Arnold, wildlife ecology, pesticide management
Moore, Thomas D., bacteriology
Moore, William Earl, protein chemistry
Mopsik, Frederick Israel, physical chemistry

Morgan, Thomas Edward, Jr., internal medicine, biochemistry
Morrell, William Egbert, chemistry, science education
Morrill, Clyde Arthur, physics, electrical engineering
Morris, Alvin Leonard, dentistry
Morris, Harold Paul, nutrition, biochemistry
Morris, Joseph Burton, analytical chemistry
Morris, Kelso Bronson, physical inorganic chemistry
Morris, Robert Gemmill, solid state physics
Morrison, Clyde Arthur, physics, low temperature physics
Morrissey, Arthur Charles, science policy
Morrissey, Bruce William, physical chemistry
Mortensen, Edith (Elizabeth), zoology
Moshman, Jack, statistical analysis
Moss, James Mercer, internal medicine
Mostofi, Fathollah Keshvar, pathology
Motz, Joseph William, physics
Mountain, Raymond Dale, theoretical physics
Mountney, George Joseph, poultry science, food technology
Mourshed, Farouk Ali, dental radiology
Moynihan, Cornelius Timothy, chemistry, glass technology
Mozer, Bernard, physics, material science
Mueller, Herbert J., physics
Muench, John, Jr., forestry, forest economics
Mukherjee, Tapan Kumar, organic chemistry, solid state science

Muller, Mervin Edgar, computer science, crystallography, mineralogy
Munson, Ronald Alfred, physical chemistry
Munson, Sam Clark, physiology, toxicology
Murday, James Stanley, surface physics
Murdock, Harold Russell, Jr., pharmacology
Murnane, Thomas George, veterinary medicine, resource management
Murphy, Elizabeth Wilcox, nutrition, food chemistry
Murray, Kenneth Malcolm, Jr., nuclear physics
Murray, Robert Fulton, Jr., medical genetics
Myers, Donald Albin, physiology, biomedical engineering
Myers, Phillip Ward, otolaryngology, audiology
Myers, Vernon Work, physics
Naeser, Charles Rudolph, inorganic chemistry
Nandedkar, Arvindkumar Narhari, biochemistry, clinical chemistry
Nardone, Roland Mario, physiology
Nash, Murray L., physical chemistry, chemical engineering
Nash, Philleo, applied anthropology
Natoli, Salvatore John, geography
Naves, Renee G., organic chemistry
Neihof, Rex A., marine chemistry, microbial biochemistry
Nelson, David, mathematical logic
Nelson, Edward R., organic chemistry
Nelson, Neal Stanley, pharmacology, radiobiology
Nesheim, Stanley, analytical chemistry
Nethery, Sidney J., physics
Neubauer, Werner George, physics
Neufeld, Daniel Arthur, human anatomy
Neuman, Morris, mathematics
Neuman, Robert Ballin, geology
Newcomb, Edward Lindsay, economic geology
Newcomb, Thomas F., internal medicine, hematology
Newkirk, Richard Albert Michael, acarology, scientific bibliography
Newman, Morris, mathematics
Ney, Wilbert Roger, physics

Ng, George, inorganic chemistry
Nicholson, Dorothy, color science
Nicholson, Richard Selindh, chemistry
Nickerson, John F., physics
Nicodemus, Frederick Edwin, optical physics
Nicoll, Roger Andrew, neuropharmacology
Nicolson, Dan Henry, botany, taxonomy
Nightingale, Elena Ottolenghi, genetics, medical genetics
Niniger, Robert D., economic geology
Nisenoff, Martin, low temperature physics
Noble, Robert Vernon, academic administration
Noble, Vincent Edward, physical oceanography
Nodvik, Paul Edgard Rudolph, Jr., inorganic chemistry, analytical chemistry
Norman, Wesley P., developmental biology, neuroanatomy
Norr, Sigmund Carl, developmental biology
Norris, James Newcome, IV, marine phycology
Norton, Matthew Frank, geology
Norvell, John Charles, molecular biophysics, crystallography
Nosanow, Lewis H., theoretical physics, low temperature physics
Noshpitz, Joseph Dove, child psychiatry
Novotny, Donald Bob, chemistry
Nowicke, Joan Weiland, palynology, systematics
Noyes, Howard Ellis, bacteriology
Nute, Jo Ann, pharmacology, neurobiology
Nuttall, Ralph Leslie, physical chemistry
Nystrom, J Warren, economic geography, political geography
Nyssonen, Diana, optics
Oakley, David Charles, science administration, nuclear science
O'Brien, John Aloysius, cell biology
O'Connell, Jesse Elbert, botany
O'Dell, Roger Gene, nutritional biochemistry
O'Doherty, Desmond Sylvester, neurology
Oertel, Goetz Kuno Heinrich, solar physics
Ogden, Ingram Wesley, dentistry

Ohashi, Yoshikazu, mineralogy, crystallography
Okabe, Hideo, physical chemistry
Okrend, Harold, microbiology
Oktay, Erol, nuclear engineering, plasma physics
O'Leary, Brian Todd, astronomy
Olenick, John George, molecular biology
Oliver, Jane Chasnoff, mathematics, computer science
Oliver, Lawrence Hampton, science education, computer science
Oliver, William Albert, Jr., paleontology
Olson, Boyd E., oceanography, meteorology
Olson, Joseph Carl, Jr., bacteriology
Olson, Storrs Lovejoy, ornithology
Olson, William Bruce, physical chemistry
Olson, Ray Andrew, physiology
Olympia, Pedro Lim, Jr., chemistry, chemical physics
Ondik, Helen Margaret, crystallography, chemical physics
Oosterhuis, William Tenley, chemical physics
Opal, Chet Brian, aeronomy, astrophysics
Opp, Albert Geelmuyden, astrophysics
Orcutt, Bruce Call, statistics, numerical analysis
O'Reilly, Sean, neurology
Orejas-Miranda, Braulio, herpetology
Orlin, Hyman, geophysics, geodesy
Oroshnik, Jesse, physics
Ortner, Donald John, physical anthropology
Osborn, Elburt Franklin, geology
Oser, Hans Joerg, applied mathematics
Osgood, Charles Freeman, mathematics
Ossakow, Sidney Leonard, plasma physics
Ostenso, Grace Laudon, nutrition
Osterberg, Charles Lamar, oceanography
Osterberg, Robert Edward, pharmacology, toxicology
Osterman, Joseph Vincent, Jr., rickettsial diseases
Ott, William Roger, experimental atomic physics, optical physics
Overpeck, James Gentry, physiology
Oyster, Dale Eugene, physics
Paabo, Maya, analytical chemistry
Pabst, William Richard, Jr., statistics
Packer, Donald MacGregor, physics
Packer, Randall Kent, comparative physiology
Padgett, Alice Adams, physical chemistry
Paffenbarger, George Corbly, dentistry
Page, John Boyd, soil physics
Pallansch, Michael J., biochemistry
Pallett, David Stephen, acoustics
Palmadesso, Peter Joseph, physics
Palmer, Samuel Joseph, cancer
Pannu, Sardul S., analytical chemistry
Panuska, Joseph Allan, environmental physiology
Parke, William C., elementary particle physics
Parker, Dean Roberts, genetics
Parker, James Evans, physics
Parker, Kittie Fenley, taxonomy, botany
Parker, Richard H., internal medicine, microbiology
Parker, Robert Louis, solid state physics
Parker, Vivian, physical chemistry
Parker, William James, thermal sciences
Parkin, Robert Charles, medical administration
Parks, Albert Fielding, analytical chemistry, international trade
Parrish, Alvin Edward, medicine
Parrot, Robert Harold, virology
Partington, Carl Ralph, mathematics, physics
Pascu, Dan, astronomy, celestial mechanics
Passaglia, Elio, solid state physics
Passer, Moses, chemical education
Pasta, John Robert, mathematical physics, computer science
Patten, Raymond Alex, physics
Patten, Joseph Rhea, physics
Paulsen, Paul, analytical chemistry
Pawson, David Leo, marine zoology
Payne, Wilbur Boswell, theoretical physics
Paynter, Orville Eugene, toxicology
Peadon, Annie Mae, zoology
Pearson, Henry Alexander, range science
Peiser, Herbert Steffen, crystallography, international relations
Pell, William Hicks, applied mathematics
Pelletier, Charles A., health physics
Penhos, Juan Carlos, physiology, endocrinology
Penner, Samuel, experimental nuclear physics
Pennington, Arthur Wesley, mathematics

Perez-Farfante, Isabel Cristina, zoology
Perhac, Ralph Matthew, geochemistry
Periman, Phillip, internal medicine, immunology
Perkinson, Jesse Dean, Jr, biochemistry
Perlin, Seymour, psychiatry
Perloff, Alvin, physical chemistry
Perros, Theodore Peter, fluorine chemistry, forensic sciences
Perry, Rufus Patterson, organic chemistry
Peterlin, Anton, polymer physics
Peterson, Miller Harrell, electrochemistry, corrosion
Petras, James Minas, anatomy, neuroanatomy
Petrie, William Leo, geology
Pettibone, Marian Hope, zoology
Peverley, J Roger, physics
Phair, George, petrology, geochemistry
Phares, Robert Eugene, forest physiology, silviculture
Phelps, Harriette Longacre, ecology
Phifer, Kenneth Oscar, parasitology
Phillips, Don Irwin, science education, institutional research
Phillips, Gary Wilson, nuclear science, radiation physics
Phillips, John Richard, computer sciences, applied mathematics
Phillips, Lyle Winston, physics
Phillips, William George, entomology
Phillips, Willie Edward, applied physics, electrical engineering
Pierce, David Alan, mathematical statistics, applied statistics
Pierce, Elliot Stearns, nuclear chemistry
Pierce, Jack Warren, geology
Pierce, John Cleve, Jr, animal science, meat science
Piermarini, Gasper J, physical chemistry, crystallography
Pierpont, Howard Clemeth, surgery
Pilchard, Edwin Ivan, veterinary virology, veterinary immunology
Pilkerton, A Raymond, ophthalmology
Pillsbury, Harold C, biochemistry
Pipberger, Hubert V, medicine
Piper, William Stephen, mathematics
Pipkin, Sarah Bedichek, genetics
Pitman, George Rubi, operations research, systems analysis
Place, John Louis, physical geography
Placious, Robert Charles, radiation physics
Plant, William James, physical oceanography
Platt, Lois Irene, medicine
Plucknett, Donald L, agronomy, soil science
Podgor, Samuel, nuclear physics
Pojeta, John, Jr, invertebrate paleontology, taxonomy
Polishuk, Paul, physics, management
Polivanov, Sergey, genetics
Pomerantz, Irwin Herman, organic chemistry
Pomeroy, John Howard, chemistry
Poon, Bing Toy, organic chemistry
Pope, Michael Thor, inorganic chemistry
Poranski, Chester F, Jr, physical chemistry, spectroscopy
Porter, Terence Lee, optical physics
Post, Boyd Wallace, forestry
Potter, John F, surgery, oncology
Powell, Francis X, chemistry
Powers, James Allen, nuclear chemistry, physical chemistry
Prats, Francisco, theoretical physics
Preer, James Randolph, inorganic chemistry
Price, Elton, physical chemistry
Prince, Edward, solid state physics
Prince, Morton Bronenberg, semiconductors
Prindle, Richard Alan, medicine
Prinz, Dianne Kasnic, physics
Prinz, Gary A, solid state physics
Prosky, Leon, nutritional biochemistry
Puffer, Ruth Rice, statistics
Pummer, Walter John, organic chemistry
Quaife, Mary Louise, biochemistry
Quarles, Gilford Godfrey, physics
Quimby, Freeman Henry, physiology
Quinn, Jarus William, optics
Quiroz, Roderick S, meteorology
Rabin, Herbert, quantum optics, solid state physics
Rabin, Robert, microbiology, research administration
Rabson, Robert, plant physiology
Rado, George Tibor, solid state physics
Raff, Morton Spencer, statistics
Rains, Theodore Conrad, analytical chemistry
Raizen, Senta Amon, physical chemistry, research administration
Ralls, Katherine Smith, ethology
Ralston, Noel Printiss, dairy science
Ramaker, David Ellis, quantum chemistry, surface chemistry
Rambaut, Paul Christopher, nutrition, biochemistry
Ramey, Estelle R, physiology, endocrinology
Ramirez, Enrique Humberto, numerical analysis
Ramsdale, Dan Jerry, physics
Ramsey, Elizabeth Mapelsden, placentology, pathology
Ramsey, Jerry Warren, research administration
Ramwell, Peter William, pharmacology, physiology
Randall, Richard Rainier, geography
Randolph, Judson Graves, surgery
Rankin, Joseph Eugene, psychiatry
Rao, Mamidanna S, biostatistics
Rapoport, Judith Livant, child psychiatry, psychiatry
Ratchford, Joseph Thomas, solid state physics
Rath, Charles E, hematology
Raveche, Harold Joseph, chemical physics
Ravenholt, Reimert Thorolf, epidemiology, public health
Ray, Clayton Edward, vertebrate paleontology
Ray, David Tobias, genetics
Ray, Howard Eugene, communication science, soil fertility
Read, Ralston Baker, Jr, bacteriology
Read, Robert William, taxonomic botany
Reader, Joseph, physics
Reba, Richard Charney, medicine, nuclear medicine
Recant, Lillian, medicine
Reed, John Calvin, geology
Rehder, Harald Alfred, invertebrate zoology
Reich, Melvin, microbiology, biochemistry
Reichardt, Charles Henry, physical chemistry
Reichelderfer, Francis Wilton, meteorology
Reilly, Joseph F, pharmacology, toxicology
Reinemund, John Adam, geology
Reinert, John Francis, entomology
Reintjes, John Francis, Jr, lasers, quantum optics
Reisa, James Joseph, Jr, environmental biology, science policy
Reiss, Howard Robert, quantum physics, electrodynamics
Reneker, Darrell Hyson, physics
Resing, Henry Anton, physical chemistry
Revoile, Sally Gates, audiology
Reynolds, David George, physiology
Reynolds, Thomas De Witt, psychiatry
Rhodes, George Wyatt, lasers
Rhyne, James Jennings, solid state physics
Rhynsburger, Robert Whitman, astronomy
Rice, Frederick Anders Hudson, organic chemistry
Rice, Mary Esther, invertebrate zoology
Rice, Nancy Reed, molecular biology
Rice, Randall Glenn, information science
Rice, William Edward, physical chemistry, inorganic chemistry
Richards, Clyde Rich, dairy husbandry
Richards, Paul Bland, mathematics
Richardson, Clarence Robert, elementary particle physics, solar physics
Richardson, Earl Leroy, cosmetic chemistry
Richardson, Howard Lockhart, pathology
Richardson, John Marshall, physics, telecommunications
Richardson, Mary Elizabeth, pathology, electron microscopy
Riegel, Kurt Wetherhold, research administration, astronomy
Riemer, William John, biology
Ries, Richard Ralph, experimental atomic physics, science administration
Riesenberg, Saul Herbert, anthropology, ethnology
Riley, Robert C, entomology
Ripin, Barrett Howard, plasma physics
Ripley, Sidney Dillon, II, zoology
Rittelmeyer, Louis Frederick, Jr, psychiatry
Ritter, Enloe Thomas, physics
Ritter, James Carroll, nuclear physics
Ritz, Victor Henry, surface physics, radiation physics
Riva, Joseph Peter, Jr, geology
Roane, Philip Ransom, Jr, virology, immunology
Robbins, Mary Louise, microbiology
Roberts, Charles DeWitt, mathematical statistics, applied statistics
Roberts, Howard Richard, public health administration
Roberts, James Richard, physics
Roberts, Richard Brooke, biophysics
Roberts, Richard William, physical chemistry
Robertson, Baldwin, theoretical physics
Robertson, William, IV, science policy, environmental management
Robinette, Charles Dennis, radiobiology, statistics
Robinson, David Mason, cell biology, cryobiology
Robinson, Glenn Hugh, soils
Robinson, Harold Ernest, botany, entomology
Robinson, Ira Charles, pharmacy
Rochlin, Robert Sumner, nuclear physics
Rockoff, Seymour David, radiology
Roessler, William George, microbiology
Rogosa, George Leon, physics
Roller, Paul S, chemistry
Roman, Nancy Grace, astronomy
Romanoff, Elijah Bravman, physiology
Romans, James Bond, physical chemistry
Romansky, Monroe James, internal medicine
Ronkin, Raphael Rooser, physiology
Ronne, Finn, geography
Rook, Harry Lorenz, analytical chemistry
Roper, Clyde Forrest Eugene, biological oceanography, systematic zoology
Roscher, Nina Matheny, physical organic chemistry
Rose, John Charles, physiology
Rose, Milton Edward, mathematics
Rosen, William G, mathematics
Rosenberg, Edith E, physiology
Rosenblatt, David, mathematical statistics
Rosenblatt, Joan Raup, mathematical statistics
Rosenstock, Herbert Bernhard, solid state physics
Rosenwald, Albert John, bacteriology
Rosenwasser, Hyman B, physical chemistry
Rosewater, Joseph, malacology
Rosier, Ronald Crosby, mathematics
Ross, Daphne Riska, geology
Ross, Donald Morris, industrial hygiene
Ross, Philip, ecology
Rossmassler, Stephen Atwater, physical chemistry, science administration
Roszman, Larry Joe, atomic physics, plasma physics
Rotariu, George Julian, physical chemisty, instrumentation
Roth, Robert S, geology
Rothman, Sam, physical chemistry, research administration
Routly, Paul McRae, astrophysics
Row, Clark, forest economics
Rowe, John Michael, solid state physics
Rowley, David Alton, physical inorganic chemistry
Royal, George Calvin, Jr, medical microbiology
Rozman, Robert Sanford, pharmacology
Rubin, Bernard, physical chemistry, research administration
Rubin, Jesse Gallant, psychiatry
Rubin, Martin Israel, biochemistry
Rubin, Robert Joshua, physical chemistry
Rubin, Vera Cooper, astronomy
Rubinstein, Mark, solid state physics
Ruegg, Fillmer William, physics
Ruff, Arthur William, Jr, metal physics
Ruffa, Anthony Richard, chemical physics, solid state physics
Ruffine, Richard S, physics
Ruggiero, John S, pharmacy
Ruhle, George Cornelius, physical chemistry
Ruhnke, Lothar Hasso, atmospheric physics
Rupp, Nelson Woodward, dental research
Rush, John Joseph, solid state physics, physical chemistry
Ruskin, Robert Edward, atmospheric physics
Russell, Philip King, infectious diseases, virology
Russell, Robert John, laboratory animal medicine
Rutford, Robert Hoxie, geology
Ryan, Frank Beall, mathematics, computer science
Saalfeld, Fred Eric, physical chemistry, inorganic chemistry
Sabrosky, Curtis Williams, entomology, taxonomy
Sabshin, Melvin, psychiatry
Sacks, I Selwyn, geophysics
Sadun, Elvio Herbert, parasitology
Saenz, Albert William, theoretical physics
St Hoyme, Lucile Eleanor, physical anthropology
Saiz Del Rio, Jose Francisco, soil chemistry, soil fertility
Saloman, Edward Barry, atomic physics
Salwin, Harold, food chemistry
Sampson, Calvin Coolidge, medicine, pathology
Sanders, William Albert, theoretical chemistry
Sando, William Jasper, stratigraphy, paleontology
Sangree, Anne Coates, geology, geography
Santamour, Frank Shalvey, Jr, plant genetics, biochemistry
Sarber, Raymond William, medical bacteriology
Saslaw, Leonard David, biochemistry
Savitz, Maxine Lazarus, organic chemistry, electrochemistry
Saworotnow, Parfeny Pavolich, pure mathematics
Sawyer, David Erickson, solid state electronics
Saz, Arthur Kenneth, microbiology
Schaefer, Albert Russell, atomic physics, molecular physics
Schaffer, Robert, analytical chemistry, bio-organic chemistry
Schamberger, Robert D, nuclear physics
Schardt, Alois Wolfgang, nuclear physics, astrophysics
Schattner, Robert I, pharmaceutical chemistry, pesticide chemistry
Schechter, Geraldine Poppa, hematology
Scheel, Nivard, physics, academic administration
Scheele, Leonard Andrew, public health, medicine
Scheer, Milton David, physical chemistry
Schein, Philip Samuel, pharmacology, oncology
Schiff, Stefan Otto, cytology, radiation biology
Schindler, Albert Isadore, physics
Schmerling, Erwin Robert, space physics
Schmidt, William Edward, analytical chemistry
Schmitt, Waldo LaSalle, zoology
Schneider, Imogene Pauline, genetics
Schneider, Irwin, solid state physics, optics
Schnur, Joel Martin, chemical physics
Schoen, Richard Isaac, science administration, molecular physics
Schofield, James Roy, academic administration
Schombert, John Leonard, astronomy
Schoning, Robert Whitney, fisheries
Schooley, James Frederick, thermal physics
Schrack, Ronald Amundsen, physics
Schreiber, Manuel, biochemistry, pharmacology
Schreiner, George E, medicine, physiology
Schriempf, John Thomas, solid state physics
Schroeder, LeRoy William, physical chemistry
Schrum, Mary Irene Knoller, biochemistry, information science
Schubert, Leo, analytical chemistry, inorganic chemistry
Schwartz, Robert Bernard, nuclear physics
Schwartz, Robert Saul, analytical chemistry
Schwartz, Sorell Lee, pharmacology
Schweitzer, Walter Gareld, Jr, optical physics
Sciuchetti, Leo A, pharmacognosy
Scott, Kenneth Richard, medicinal chemistry, analytical chemistry
Scott, Ralph Asa, Jr, analytical chemistry, radiochemistry
Scott, Roland B, pediatrics
Scoville, John J, nuclear physics, environmental management
Scribner, Bourdon Francis, physical chemistry
Seaman, Elwood Armstrong, aquatic biology
Searles, Stuart Kenneth, lasers
Seebold, Robert Elvin, chemistry
Seeff, Leonard Barry, clinical medicine
Seguin, Ward Raymond, meteorology
Seidelmann, Paul Kenneth, astronomy, celestial mechanics
Sengers, Johanna M H Levelt, thermal physics
Senich, Donald, soil mechanics
Sentz, James Curtis, plant science, experimental statistics
Senzel, Alan Joseph, analytical chemistry
Setty, Laurel Raymond, anatomy
Setzer, Henry Wilfred, mammalogy
Shackelford, James Marshall, organic chemistry

Shaklee, William Eugene, poultry science, genetics
Shank, Fred Ross, nutrition
Shanklin, John Ferguson, forestry
Shapero, Donald Campbell, theoretical physics, science policy
Shapiro, Maurice Mandel, cosmic ray physics, astrophysics
Shapiro, Philip, nuclear physics, medical physics
Shapiro, Raymond E., chemistry
Share, Gerald Harvey, x-ray astronomy
Shea, Keith Raymond, forest pathology, entomology
Shein, Eric Benjamin, pharmaceutical chemistry
Sheinson, Ronald Swiren, physical chemistry
Sheldon, Richard P., geology
Shelesnyak, Moses Chiam, biodynamics
Shenker, Henry, physics
Shepherd, George Robbins, biological chemistry
Sheppard, Alan Jonathan, nutritional biochemistry, animal nutrition
Sher, Alvin Harvey, nuclear chemistry, semiconductor physics
Sherburne, James Auril, ecology
Sherman, John Foord, pharmacology
Shetler, Stanwyn Gerald, plant taxonomy
Shibko, Samuel Issac, biochemistry
Shiskin, Julius, economic statistics, applied statistics
Shivanandan, Kandiah, astrophysics, cosmology
Shnider, Bruce I., internal medicine, oncology
Shon, Frederick John, nuclear physics
Shotkin, Louis Marvin, physical chemistry
Showell, John Sheldon, organic chemistry
Shuler, Robert Lee, surface chemistry
Shulman, Seth David, x-ray astronomy
Shumaker, John Benjamin, Jr, optical physics
Shuman, Frederick Gale, meteorology
Shykind, Edwin B., marine geology endocrinology
Sieck, L Wayne, physical chemistry
Siegel, Frederic Richard, geology petrology
Silber, Robert L., biochemistry, science education
Silberberg, Rein, cosmic ray physics
Silbergitt, Richard Stephen, solid state physics
Silva, Omega Logan, internal medicine, endocrinology
Silverman, Shirleigh, physics
Simkin, Thomas Edward, geology, administration
Simopoulos, Artemis Panageotis, pediatrics, endocrinology
Simpson, Beryl Brintnall, biosystematics
Simpson, John Arol, physics, science
Skog, Laurence Edgar, plant taxonomy
Slavin, Howard Bernard, medical administration, internal medicine
Sleeman, Harry Kenneth, biochemistry
Slepian, Paul, mathematics
Slotkoff, Lawrence M., physiology, internal medicine
Smardzewski, Richard Roman, physical inorganic chemistry
Smiley, Seymour Howard, chemistry
Smith, Benjamin Williams, biochemistry
Smith, Bruce H., medicine
Smith, Clayton Albert, Jr, astronomy
Smith, David Rollins, systematics, entomology
Smith, Edward, pharmaceutical chemistry research
Smith, Ernest Lee, Jr, biochemistry
Smith, Harlan Eugene, plant pathology
Smith, James Cecil, physiology, biochemistry
Smith, John Eldrid, medicine
Smith, John Henry, statistics
Smith, Leslie E., physical chemistry

Smith, Paul Albert, geodesy, oceanography
Smith, Richard Elbridge, geology, geochemistry
Smith, Willis Dean, solid state physics
Smith, Wray, operations research, statistics
Smith-Gill, Sandra Joyce, developmental genetics, endocrinology
Smulson, Mark Elliott, biochemistry
Smyth, Nicholas Patrick Dillon, surgery
Smythe, Richard Vincent, entomology
Snead, Claybourne C., physical chemistry
Snell, Richard Saxon, anatomy
Snoke, J Arthur, seismology
Snyder, John L., physical geology
Snyder, Joel Alan, physics
Snow, Daniel Isaac, experimental nuclear physics, elementary particle physics
Sofia, Sabatino, astrophysics
Sohl, Norman Frederick, paleontology
Sohn, Israel Gregory, paleontology
Sohns, Ernest Reeves, botany
Sokoloff, Vladimir P., geochemistry
Sollers-Riedel, Helen, entomology
Soper, Gordon Knowles, plasma physics
Sorows, Howard Earle, physics
Soto, Gerardo H., agronomy, soil classification
Souten, Robert J., Jr, cryogenic physics
Spangler, Paul Junior, entomology
Spearing, Cecilia W., biochemistry
Spector, Novera Herbt, neuroscience, biophysics
Speer, Donald Arthur, science administration
Speidel, John Joseph, population biology, public health
Sperling, Frederick, pharmacology
Spiegel, Valentin, Jr, physics
Spilhaus, Athelstan Frederick, meteorology, oceanography
Spilhaus, Athelstan Frederick, Jr, geophysics, oceanography
Spohr, Daniel Arthur, physics
Spotz, Ellen Mae Lackey, physical chemistry
Sprague, Lucian Matthew, genetics
Springer, Victor Gruschka, biology
Spurlock, Langley Augustine, organic chemistry
Sreevalsan, Thazepadath, microbiology, virology
Stafford, Fred E., physical inorganic chemistry
Stahly, Eldon Everett, organic chemistry
Stalford, Harold Lenn, applied mathematical chemistry
Stallings, James Henry, soil conservation
Stamper, John Andrew, plasma physics
Standaert, Frank George, pharmacology
Stanley, Daniel Jean, marine geology
Staples, Bert Roland, physical inorganic chemistry
Stapleton, John F., medicine
Stauss, George Henry, physics
Steele, Edgar Alfred, organic chemistry
Steele, John A., organic chemistry, analytical chemistry
Segun, Irene Anne, numerical analysis
Stein, Marjorie Leiter, mathematics
Steinbach, Wayne Robert, molecular spectroscopy
Steinberg, Marshall, pharmacology, toxicology
Steiner, Bruce, physical chemistry
Steinhardt, Jacinto, physical chemistry
Steinman, Irvin David, biochemistry
Stephens, John James, III, vertebrate paleontology
Stern, Kurt Heinz, physical inorganic chemistry
Stevens, Donald Keith, solid state physics
Stevens, Harold, neurology
Stevens, Russell Bradford, plant pathology
Stever, H Guyford, aeronautics, astronautics
Stewart, Sarah Elizabeth, bacteriology
Stiehler, Robert Daniel, chemistry
Stokes, Arnold Paul, pure mathematics
Stolovy, Alexander, nuclear physics
Stolz, Hal Fisher, medical research
Stone, Philip M., atomic physics, plasma physics
Storm, Carlyle Bell, bioinorganic chemistry
Story, Troy Lee, chemical physics
Stovall, Miron L., statistics
Strand, Kaj Aage, astronomy
Strasburg, Donald Wishart, fish biology
Strassenburg, Arnold Adolph, physics
Straw, James Ashley, pharmacology
Stringfield, Victor Timothy, hydrogeology
Strobel, Darrell Fred, planetary sciences
Stromberg, Robert Renson, polymer science
Stroud, Richard Hamilton, zoology

Sturtevant, William Curtis, anthropology
Sudia, Theodore William, environmental physiology
Sugar, Jack, atomic spectroscopy
Sullivan, John Dennis, forestry
Sunley, Judith S., number theory
Suria, Amin, neuropharmacology
Suter, Emanuel, medical education, immunology
Sutter, David Franklin, instrumentation
Sutter, John Ritter, physical chemistry
Suzuki, George, mathematics
Swanson, August George, neurology
Swanson, Dwight Wesley, meteorology, data processing
Swartzendruber, Lydon James, physics
Sweeney, Thomas Richard, medicinal chemistry
Swenson, Herbert Alfred, hydrology
Swinehead, Jeff, zoology
Switzer, George S., mineralogy
Syphax, Burke, surgery
Tabler, Kenneth Ambrose, mathematical statistics, analytical statistics
Tackle, David, silviculture
Talbert, Preston Tidball, organic chemistry
Tanczos, Frank I., chemical physics
Tanner, James Thomas, radiochemistry
Tape, Gerald Frederick, science administration
Taragin, Morton Frank, solid state physics
Tarpley, Jerald Dan, atmospheric physics
Tarpley, Barry Norton, solid state physics, low temperature physics
Taylor, Jean Marie, toxicology, pharmacology
Taylor, John Keenan, physical chemistry, analytical chemistry
Taylor, Marie Clark, botany
Taylor, Moddie Daniel, inorganic chemistry
Taylor, Patrick Timothy, marine geophysics
Taylor, Richard Ray, medicine
Taylor, William Ralph, ichthyology
Teitler, Sidney, physics
Telang, Vasant G., medicinal chemistry
Teleki, Geza, anatomy
Telford, Ira Rockwood, anatomy
Terant, Seldon W., chemistry
Thaler, Alvin Isaac, algebra
Thomas, Patricia Zeis, biochemistry, information science
Thompson, Frederic Christian, endocrinology
Thompson, Joseph Kyle, entomology, evolutionary biology
Thompson, Warren Elwin, physical chemistry
Thonnard, Norbert, astrophysics, atomic physics
Thoren, Conrad Joseph, geography, cartography
Thorington, Richard Wainwright, Jr, biology
Thurston, William, geology
Tidball, Charles Stanley, computer science, medical education
Tidball, Mary Elizabeth, medical physiology, institutional research
Timmons, Richard B., physical chemistry
Todd, Edward Payson, atmospheric physics, science administration
Tombaugh, Larry William, forest economics
Tompkins, Daniel Reuben, plant physiology, horticulture
Tousey, Richard, physics
Towe, Kenneth McCarn, geology
Towle, Laird C., high pressure physics, solid state physics
Tracton, Martin Steven, meteorology
Treado, Paul A., nuclear physics
Treadwell, Carleton Raymond, biochemistry
Treichler, Ray, agricultural chemistry, biological chemistry
Trevino, Samuel Francisco, solid state physics
Trott, Lamar Brice, marine ecology, ichthyology
Trott, Winfield James, underwater acoustics
Truitt, Edward Byrd, Jr, pharmacology
Tryon, Max, instrumentation, materials science
Tryten, Meriam Hartwick, physiology
Tsang, Tung, physical chemistry
Tsao, Chen-Hsiang, physics
Tschiegg, Carl Emerson, physics
Tseng, Hsiang Len, pathology

Tso, Mark On-Man, ophthalmology, pathology
Tucker, Allen Brown, mathematics
Tucker, William Eric, nuclear physics
Tuckson, Coleman Reed, Jr, dentistry
Turner, Kenneth Clyde, radio astronomy
Turner, Noel Hinton, physical chemistry
Turner, Willie, virology, immunology
Turner, Merle Antony, physics
Twigg, Homer Lee, medicine, radiology
Tyler, James Chase, ichthyology
Ubelaker, Douglas Henry, physical anthropology
Uberall, Herbert Michael, theoretical nuclear physics, acoustics
Ulrich, Peter B., optical physics
Ulsamer, Andrew George, Jr, biochemistry
Ulvedal, Frode, physiology
Umberger, Ernest Joy, biochemistry
Upholt, William Martin, science administration
Utton, Donald Brian, solid state physics
Utz, John Philip, medicine
Vahouny, George V., cardiovascular diseases
Vaiuzis, Zigfridas, medical microbiology
Van Buren, Arnie Lee, acoustics
Van Olphen, Hendrik, colloid chemistry
Vander Hart, David Lloyd, physical chemistry
Vanderveen, John Edward, nutrition, chemistry
Van Flandern, Thomas Charles, clestial mechanics
Venable, William Howell, Jr, optical medicine
Venezky, David Lester, inorganic chemistry
Verdier, Peter Howard, physical chemistry
Verell, Ruth Ann, organic chemistry
Vervey, Willard Foster, medical microbiology
Vetter, Richard C., oceanography
Vietmeyer, Noel Duncan, economic biology, science writing
Vilakazi, Absolom, anthropology
Vincent, Robert Corbin, analytical chemistry
Viola, Herman Joseph, anthropology
Viola, John Thomas, physical chemistry
Visco, Eugene Paul, operations research, statistics
Vittoria, Carmine, magnetism
Vogl, Thomas Paul, biomedical engineering
Vrebalovich, Thomas, space physics, fluid mechanics
Wachholz, Bruce William, radiation biology
Wachman, John Bryan, Jr, solid state science
Wagman, Donald David, chemistry
Wagner, Herman Leon, polymer science
Wakelin, James Henry, Jr, physics
Waldon, Edgar F., audiology
Walker, Charles R., biochemistry, fish biology
Walker, Glenn Anthony, biochemistry
Wall, Robert Ecki, geophysics
Wallenmeyer, William Anton, high energy physics
Waller, Thomas Richard, invertebrate paleontology, geology
Walsh, John Lawrence, solid state physics
Walter, Mable Ruth, genetics
Walton, Theodore Ross, organic chemistry
Wands, Ralph Clinton, toxicology
Wang, Francis Wei-Yu, polymer science
Wang, David Justin, plant genetics
Ward, James Audley, mathematics, computer science
Ward, Keith Bolen, Jr, biophysics
Ward, Max, botany, morphology
Ward, Ronald Anthony, medical entomology
Ware, Lawrence L., Jr, medical microbiology
Ware, Stanton James, environmental management
Wark, David Quentin, meteorology, astrophysics
Warner, Alexander Carl, ecology
Warner, Laurance Bliss, nuclear physics
Warner, Theodore Baker, physical chemistry, chemical oceanography

Warren, Charles Reynolds, geology
Wasshausen, Dieter Carl, systematic botany
Wasson, Oren A, nuclear physics
Waterhouse, Richard (Valentine), physics
Watkins, Don Wayne, physiology
Watkins, Mark Hanna, anthropology
Watson, George E, III, zoology, ornithology
Watson, Robert Barden, acoustics, lasers
Watson, Robert C, organic chemistry
Watson, Robert Francis, chemistry, science education
Watters, Robert Lisle, radiochemistry, health physics
Weaver, Neill Kendall, medicine
Webb, Alan Wendell, physical inorganic chemistry
Webb, Arthur Harper, bacteriology
Weber, John Donald, pharmaceutical chemistry
Webster, Thomas G, psychiatry, academic administration
Weidman, Donald Robert, mathematics
Weidner, Victor Ray, optical physics
Weil-Malherbe, Hans, biochemistry
Weingold, Allan Byrne, obstetrics & gynecology
Weintraub, Herbert D, anesthesiology, medical education
Weintraub, Robert Louis, cell physiology, biochemistry
Weisberg, Samuel Myer, biochemistry
Weiss, Andrew W, atomic physics, molecular physics
Weiss, Charles, Jr, science policy, chemical physics
Weiss, Ira Paul, neurosciences
Weiss, Lawrence R, pharmacology, toxicology
Weiss, Peter Joseph, pharmaceutical chemistry
Weiss, Richard Gerald, photochemistry, physical organic chemistry
Weiss, William P, pharmacology
Weissler, Alfred, chemistry
Weitzman, Stanley Howard, ichthyology
Welbeck, Paa-Bekoe Henry, communications science, public health education
Welch, Billy Edward, physiology, biochemistry
Weller, Charles Stagg, Jr, physics
Wells, Eugene Ernest, Jr, organic chemistry, electrochemistry
Welt, Isaac Davidson, information science, documentation
Wenmersten, Dwight L, physics
Werdel, Judith Ann, information science, science policy
Werner, Mario, laboratory medicine
Werntz, Carl H, theoretical nuclear physics, astrophysics
Wertheim, Robert Halley, nuclear physics
West, Fred Ralph, Jr, pharmacology
West, William Lionel, zoology, pharmacology
Westbrook, Fred Emerson, agronomy, plant science
Weston, Jean Kendrick, experimental pathology
Wetmore, Alexander, ornithology
Wetherill, George West, geophysics
Wetzel, Lewis Bernard, applied physics
Wheeler, James William, Jr, organic chemistry
Whetstone, Stanley L, Jr, experimental nuclear physics
Whicker, Lawrence R, microwave physics, electromagnetism
White, Briggs Johnston, chemistry
White, David C, pathology, radiobiology
White, David Gover, inorganic chemistry
White, Edward Austin, biochemistry, nutrition
White, Jack Edward, medicine
White, John Arnold, atomic physics
White, Mack, microbiology
White, Richard Earl, systematic entomology
White, Robert M, meteorology
White, Stanley C, preventive medicine
White, William Calvin, soil fertility, plant physiology
Whiteker, Roy Archie, analytical chemistry
Whiteside, Haven, physics
Whitmire, Carrie Ella, bacteriology
Whitmore, Frank Clifford, Jr, vertebrate paleontology
Whitted, Harold Horatio, medicine, public health
Whittemore, Frederick Winsor, entomology
Widing, Kenneth Gordon, solar physics, atomic spectroscopy
Wiener, Howard Lawrence, statistics, operations research

Wiese, Wolfgang Lothar, atomic physics, plasma physics
Wiesnet, Donald Richard, geology, hydrology
Wieting, Terence James, physics
Wild, Jack William, physics
Wile, Howard P, research administration
Wilkerson, John Christopher, physical oceanography
Wilkins, J Ernest, Jr, mathematics
Wilkniss, Peter Eberhard, oceanography, radiochemistry
Willard, Daniel, physics
Willcox, Alfred Burton, mathematics
Willenbrock, Frederick Karl, physics
Williams, Austin Beatty, systematic zoology
Williams, Ernest Young, neurology, psychiatry
Williams, Floyd James, phytopathology
Williams, John Covington, chemistry
Williams, Oren Francis, inorganic chemistry
Williams, Owen Wingate, geodesy, geophysics
Williams, Richard Birge, marine ecology
Williams, Richard Taylor, solid state physics
Williams, Robert E, range management, ecology
Williams, Robert Glenn, oceanography
Williams, Sidney, analytical chemistry
Willis, Eric Herbert, geophysics, geochemistry
Willman, Warren Walton, applied mathematics
Wilner, Jerome, dairy industry
Wilsey, Neal David, physics
Wilson, Eugene M, plant pathology
Wilson, James Bruce, computer science, biomedical engineering
Wilson, Mathew Kent, physical chemistry
Wilson, William Mark Dunlop, animal science, ruminant nutrition
Winchester, Clarence Floyd, nutrition
Wing, James, nuclear chemistry
Winkert, John Wynia, physiology, chemistry
Winstead, Jack Alan, biochemistry
Winter, David Leon, research administration
Winter, Phillip E, preventive medicine, epidemiology
Winter, Thomas C, Jr, space physics, optics
Wintercorn, Eleanor Stiegler, audiology, speech pathology
Wintermoyer, John Paul, applied chemistry
Wirth, Willis Wagner, entomology
Wischow, Russell P, nuclear chemistry
Withrow, Alice Phillips, biological chemistry, photobiology
Wittels, Mark C, mineralogy, solid state physics
Witting, James M, physical oceanography, hydrodynamics
Witzgall, Christoph Johann, operations research, numerical analysis
Wohlhieter, John Andrew, biophysics
Woidich, Francis De Sales, medicine, physiology
Wolcott, John H, aerospace medicine
Wolf, Stephen Noll, acoustics
Wolff, Frederick William, pharmacology, medicine
Wolicki, Eligius Anthony, nuclear physics
Wollan, David Strand, solid state physics
Wolman, William Wolfgang, mathematical statistics
Wong, Donald Tai On, immunology
Wong, Harry Yuen Chee, physiology, endocrinology
Wood, Burrell Lusha, Jr, organic chemistry
Wood, Garnett Elmer, biochemistry
Wood, George William, acoustics, research administration
Wood, Lawrence Arnell, physics
Wood, Leonard Alton, geology
Wood, Reuben Esselstyn, physical chemistry
Wood, Robert Winfield, radiation biophysics
Woodbury, Nathalie Ferris Sampson, cultural anthropology, ethnography
Wooding, Wendell Phillips, geology
Woodward, Stephen Cotter, pathology
Woolf, William Blauvelt, mathematics
Woolley, George Walter, genetics
Work, Henry Harcus, pediatrics, psychiatry
Wrathall, Jean Rew, genetics, cell biology
Wright, Arthur Gilbert, zoology
Wright, James Roscoe, organic chemistry
Wright, Robert James, geology

Wright, Robert Raymond, science policy
Wright, Rufus William, physics
Wright, William Wynn, chemistry
Wu, Yung-Chi, thermodynamics
Wurdack, John J, botany
Wyatt, Jeffrey Renner, chemistry
Wyatt, Richard J, psychopharmacology
Wyckoff, James M, nuclear physics
Wyckoff, Dale Emerson, physical oceanography
Wylie, Aubrey Evans, forestry
Wylie, Richard Michael, biology, neurophysiology
Yager, Robert H, veterinary medicine
Yamamoto, William Shigeru, physiology, computer science
Yang, Chung-Chun, pure mathematics, applied mathematics
Yang, David Chih-Hsin, biochemistry
Yaniv, Shlomo Stefan, health physics, radiological physics
Yaniv, Simone Liliane, psychoacoustics, bioacoustics
Yap, William Tan, physical chemistry
Yates, John Thomas, Jr, physical chemistry, surface chemistry
Yip, George, biochemistry, food technology
Yochelson, Ellis Leon, invertebrate paleontology
Yochelson, Leon, psychiatry
Yoder, Hatten Schuyler, petrology
Yohe, Daniel Charles, physical chemistry, science education
Young, Elizabeth Bell, speech pathology, audiology
Young, Harold Henry, analytical chemistry
Young, Richard S, exobiology, developmental biology
Young, Russell Dawson, optics, surface science
Youtcheff, John Sheldon, astrophysics
Zaborsky, Oskar Rudolf, biological chemistry
Zahl, Paul Arthur, natural science, experimental biology
Zalubas, Romuald, astrophysics
Zalucky, Theodore B, pharmacy, pharmaceutical chemistry
Zemach, Charles, elementary particle physics
Zimmerman, Lorenz Eugene, pathology
Zon, Gerald, organic chemistry
Zuchelli, Artley Joseph, physics
Zug, George R, herpetology, morphology
Zusi, Richard Laurence, zoology
Zwolenik, James Joseph, physical chemistry, science policy

FLORIDA

ALACHUA
Yorton, Joan Bannister, organic chemistry

APALACHICOLA
Ingle, Robert Maurice, marine biology

APOPKA
Conover, Charles Albert, ornamental horticulture
Hamlen, Ronald Alan, entomology, plant nematology
Knauss, James Frederick, plant pathology
Poole, Richard Turk, Jr, ornamental horticulture

ATLANTIS
James, Preston Everett, geography

BARTOW
Baumann, Arthur Nicholas, inorganic chemistry, analytical chemistry

BAY PINES
Davis, Robert Leo, biochemistry
Dexter, Morris W, medicine
Friedlander, Jackson Harrison, internal medicine
Hsu, Jeng Mein, biochemistry, nutrition
Lawton, Alfred Henry, geriatrics
Samis, Harvey Voorhees, Jr, biochemistry, physiology
Winslow, Donald J, pathology

BELLE GLADE
Allen, Robert John, Jr, agronomy
Beardsley, Daniel Waldo, animal nutrition
Burdine, Howard William, plant nutrition, physiology
Conley, Cecil, biochemistry, agriculture
Crockett, Joe Richard, animal genetics
Gascho, Gary John, agronomy
Guzman, Victor Lionel, horticulture
Janes, Melvin Joseph, entomology
Kidder, Gerald, agronomy
Orsenigo, Joseph Reuter, weed science
Pate, Findlay Moye, animal nutrition
Snyder, George Heft, agronomy, soil
Tate, Robert Lee, III, soil microbiology
Zitter, Thomas Andrew, plant virology

BOCA RATON
Banter, John C, physical chemistry
Bates, Francis Leslie, physical chemistry
Bieber, Theodore Immanuel, organic chemistry, biochemistry
Blakemore, John Sydney, solid state physics
Boss, Manley Leon, plant physiology
Burnett, Clyde Ray, physics
Burris-Meyer, Harold, acoustics
Clark, Samuel Friend, organic chemistry
Coulter, Neal Stanley, computer sciences, information science
Courtenay, Walter Rowe, Jr, marine biology, ichthyology
Cox, Joseph Robert, physics
Craig, Alan Knowlton, geography, anthropology
DeLamater, Edward Doane, microbiology, electron microscopy
DeLeon, Morris Jack, number theory
Di Paola, Jane Walsh, combinatorics, geometry
Dobkin, Sheldon, biological oceanography, invertebrate zoology
Early, John Drennan, applied anthropology
Fellowes, Oliver Nelson, medical microbiology, immunology
Grimm, Robert Blair, plant pathology
Hartman, James Xavier, microbiology, phytopathology
Hoffmann, Frederick, mathematics
Hoffmann, Harrison Adolph, microbial ecology
Hogan, William Alfred, plasma physics, theoretical physics
Iverson, Ray Mads, biology
Jackson, Carey Birdsong, inorganic chemistry
Kennedy, William Jerald, anthropology
Lamborn, Bjorn N A, plasma physics
Latham, James Parker, geography
Lee, David Raymond, cultural geography, geography of Africa
Lemon, Roy Richard Henry, geology
Levow, Roy Bruce, mathematics, computer science
Livingood, John N B, mathematics
Lombardo, Anthony, organic chemistry
McAllister, Raymond Francis, oceanography, ocean engineering
McGuire, James B, theoretical physics
Melnick, Daniel, biochemistry
Mueller, George, organic geochemistry, cosmochemistry
Nanz, Robert Augustus Rollins, food science
Perumareddi, Jayarama Reddi, inorganic chemistry, physical chemistry
Saurino, Vincent Robert, microbiology
Sax, Newton Irving, occupational health, environmental health
Schultz, Franklin Alfred, analytical chemistry, electrochemistry
Schultz, Ronald Richard, economic geography, urban geography
Sears, William Hulse, anthropology, archaeology
Sguros, Peter Louis, microbial physiology
Stetson, Robert Franklin, plasma physics
Stewart, Herbert, science education, biology
Sturrock, Thomas Tracy, horticulture, plant morphology
Sublett, Audrey J, physical anthropology
Tata, Robert Joseph, geography of Latin America, economic geography
Warburton, David Lewis, geochemistry
Weiss, Gerald, anthropology

BONITA SPRINGS
Caldwell, John Richard, polymer chemistry

BOYNTON BEACH
Linder, Ernest Gustaf, physics
Saslaw, Milton Sibley, cardiology
Westveld, Ruthford Henry, forestry

FLORIDA

BRADENTON
Binnell, James Monroe, organic chemistry; chemical engineering
Engelhard, Arthur William, plant pathology
Geraldson, Carroll Morton, soil chemistry
Harbaugh, Brent Kalen, ornamental horticulture
Hett, John Henry, optics, astronomy
Jones, John Paul, plant pathology
Kelsheimer, Eugene Gillespie, entomology
Magie, Robert Ogden, plant pathology
Marousky, Francis John, horticulture
Overman, Amegda Jack, plant nematology
Pickett, Lucy Weston, physical chemistry
Schuster, DAvid J., entomology
Waters, Willie Estel, horticulture
Wilfret, Gary Joe, plant breeding, genetics
Woltz, Shreve Simpson, horticulture

CANAL POINT
Dean, Jack Lemuel, phytopathology
James, Norman Ivan, agronomy
Lyrene, Paul Magnus, plant breeding
Summers, Thomas Eugene, entomology, plant pathology

CAPE CORAL
Ayers, Alvin Dearing, soil science, plant physiology
Bowers, Alston Gordon, pesticide
Jensen, Clayton Everett, meteorology, computer science
Nickson, Margaret Jane, medicine

CLEARWATER
Baldwin, William John, analytical chemistry
Biver, Carl John, Jr, applied physics, materials science
Jennings, William Lamar, ecology, zoology
Patrick, George Robert, organic chemistry
Steiner, Loren Franklin, entomology
Turner, Stanley Eugene, nuclear chemistry

CLERMONT
Boyd, John Mann, chemistry
Koppius, Otto Gustav, physics

CLEWISTON
Andreis, Henry Jerome, soil fertility
Durham, James Ivey, plant physiology
Holder, David Gordon, plant breeding
Rife, David Cecil, genetics
Todd, Edwin Harkness, plant pathology

COCOA BEACH
Brown, James Wilson, plant physiology
Frese, Frederick Joseph, Jr, aerospace medicine
Kuns, Merle L., virology, ecology

CORAL GABLES
Adams, William Curtis, pediatrics
Alexander, Taylor Richard, botany
Bader, Henri, glaciology, particle physics
Badgley, Wilfrid John, chemistry
Bagley, Robert Waller, mathematics
Bayer, Frederick Merkle, zoology
Bensen, Jack F, speech & hearing sciences
Bergstresser, Paul Richard, dermatology
Bleck, Rainer, meteorology
Bostrom, Kurt G V, mineralogy, geochemistry
Burdick, Everette Marshall, chemistry
Butson, Alton Thomas, mathematics
Clark, Robert M, pathology
Clegg, James S, physiology, biochemistry
Compton, Kenneth Gordon, electrochemistry
Criss, Cecil M, physical chemistry
Dauer, Maxwell, radiological physics
De Bilj, Harm Jan, geography
Deichmann, William Bernhard, pharmacology, toxicology
Doepker, Richard DuMont, physical chemistry
Drost-Hansen, Walter, chemical physics
Duda, Edwin, mathematics
Duke, Douglas, astronomy, astrophysics
Estoque, Mariano A, meteorology
Evans, David Hudson, comparative physiology, ichthyology
Evoy, William (Harrington), neurophysiology
Faber, Shepard Mazor, science education
Flipse, Martin Eugene, preventive medicine
Fox, Sidney Walter, organic polymer chemistry
Frank, Neil LaVerne, meteorology
Garmus, Ralph David, entomology
Geisler, John Edmund, dynamic meteorology
Gilman, Lauren Cundiff, genetics
Goodman, Richard Henry, numerical analysis, computer science
Gordon, Howard R, optics, spectroscopy
Gould, Thelma Bernice Clark, oncology
Grabowski, Casimer Thaddeus, zoology, embryology
Grant, Wilson Clark, physiology
Greenfield, Leonard Julian, biochemistry
Greer, Sheldon, molecular genetics
Gropp, Armin Henry, physical chemistry, analytical chemistry
Hare, Curtis R, inorganic chemistry, physical chemistry
Herbert, Thomas James, biophysics
Hertzig, David, mathematics
Hinsch, Gertrude Wilma, zoology, embryology
Hirschberg, Joseph Gustav, physics
Hiser, Homer Wendell, meteorology, anthropology
Hsu, Laura Hwei-Nien Ling, evolution
Huerta, Manuel Andres, statistical mechanics, plasma physics
Hunt, Burton Poulter, zoology
Hutchinson, Harry William, cultural anthropology
Keenan, Arthur George, physical chemistry
Kelley, Robert Lee, mathematical physics
Kline, Jacob, biomedical engineering
Kraus, Eric Bradshaw, meteorology
Kreske, Richard Daniel, geography
Kursunoglu, Behram, theoretical physics
Latham, Don Jay, atmospheric physics
Leigh, Walter Henry, parasitology
Lhermitte, Roger M, physical meteorology
Licht, Sidney, physical medicine
Loewenstein, Werner Randolph, neurophysiology, biophysics
Lopez, Diana Montes De Oca, microbiology
Luykx, Peter (Van Oosterzee), cytology
Mallery, Charles Henry, developmental physiology, biochemistry
Man, Eugene Herbert, marine geochemistry
McDougle, Paul E, mathematics
McKnight, James Dawson, Jr, mathematics
Metz, Charles Baker, developmental biology
Meyer, Herman, mathematics
Miale, John Buyer, pathology
Miecke, Marvin V, topology
Miller, Elwood Morton, zoology
Mills, Alfred Preston, physical chemistry
Mustard, Margaret Jean, horticulture
Nakashima, Tadayoshi, biochemistry
Norwood, Joseph, Jr, plasma physics
Nuber, John Arthur, mathematics
Onsager, Lars, theoretical physics
Owre, Oscar Theodore, ornithology
Pardo, William Bermudez, physics
Perlmutter, Arnold, physics
Rich, Earl Robert, ecology
Robertson, Harry Stroud, plasma physics, statistical mechanics
Roth, Frank J, Jr, microbiology
Salter, Paul Sanford, urban geography
Saylor, Richard Lynn, mathematics
Schneiderman, Neil, psychophysiology
Schultz, Harry Pershing, organic chemistry
Senn, Harry V, meteorology
Sheets, Robert Chester, meteorology
Sickels, Jackson Pyburn, organic chemistry
Simplicio, Jon, biochemistry, inorganic chemistry
Skramstad, Harold Kenneth, physics
Smolens, Joseph, immunology
Snyder, Carl Henry, organic chemistry
Sparks, William Joseph, chemistry
Strohecker, Henry Frederick, biology
Stuckwisch, Clarence George, chemistry
Teas, Howard Jones, genetics
Wells, Daniel R, physics
Williams, Robert Haworth, plant physiology
Winters, Ray Wyatt, physiological psychology
Wojaszek, Joseph Henry, theoretical physics
Zame, Alan, mathematics

CRYSTAL RIVER
Radhakrishnan, Chittur Venkitasubhan, veterinary microbiology

DADE CITY
Donohoe, Heber Clark, entomology, ecology
Furlow, Leonard Thompson, neurosurgery

DANIA
Markowitz, William, astronomy

DAYTONA BEACH
Copeland, Richard Franklin, physical chemistry; organometallic chemistry
McCloskey, James, physics

DE LAND
Beiler, Theodore Wiseman, organic chemistry
Blair, James Stuart, chemistry
Cannon, Raymond Joseph, Jr, mathematics
Chauvin, Robert S, geology, geography
Coolidge, Edwin Channing, analytical chemistry
DeLap, James Harve, physical chemistry
Everett, Kenneth Gary, inorganic chemistry
Fuller, Dorothy Langford, biology
Hansen, Keith Leyton, biology
Jenkins, George Lovell, nuclear physics
Knapp, Francis Marion, cardiovascular physiology, neurophysiology
Margarian, Elizabeth Ann, algebra
Medlin, Gene Woodard, mathematics
Stock, David Allen, microbiology, genetics
Thwing, Henry Warren, algebra
Williams, Gareth, applied mathematics
Wolfe, James Alvis, botany, ecology

DELRAY BEACH
Ozaki, Henry Yoshio, vegetable crops
Sanders, Murray, microbiology
Stone, William Jack Hanson, plant pathology; weed science

DELTONA
Murphy, Raymond Edward, urban geography

DOVER
Albregts, Earl Eugene, soil chemistry
Howard, Charles Marion, plant pathology

DUNEDIN
Childs, William Henry, pomology
Dunbar, Carl Owen, paleontology
Zinn, Walter Henry, physics

EGLIN AFB
Arpke, Charles Kenneth, science administration
Findley, George Bernard, mathematical physics
Johnson, Howard Arthur, Sr, mathematics, operations research
Patrick, Michael Andrew, microbiology
Zimmer, Martin F, thermodynamics, explosives

ENGLEWOOD
Duisberg, Peter Caspar, natural resources

FT LAUDERDALE
Blum, Alvin Seymour, biophysics
Burke, Anna Mae Walsh, molecular physics
Burt, Evert Oakley, agronomy
Carter, Albert Smith, chemistry
Dasher, Paul James, chemistry
Fateh, Abdullah Valimohammed, forensic medicine
Ludeke, Carl Arthur, physics, oceanography
MacFarlane, Malcolm David, clinical pharmacology; medical research
Menzies, Robert Allen, biochemistry
Miller, John Allen, zoology
Neel, Percy Landreth, ornamental horticulture
Nichols, Philip Ray, entomology
Reinert, James Arnold, entomology
Rupp, Wiktor, optics
Sprinkle, H D, mathematics
Steward, Kerry Kalan, plant physiology
Tamers, Murry Allen, physical chemistry; environmental chemistry
Warren, Joel, cancer research
Wolf, Benjamin, soil chemistry

FT MYERS
Del Fosse, Ernest Sheridan, entomology, aquatic ecology
Freeman, Kenneth Alfrey, organic
Hursh, Charles Raymond, environmental sciences
Kane, Howard L, organic chemistry
Kerr, Kathel Bedortha, parasitology; toxicology
Ramer, Luther Grimm, acoustics
Ramoska, William Allen, insect pathology

FT MYERS BEACH
Barber, Franklin Weston, food science

FT PIERCE
Brolmann, John Bernardus, plant breeding
Bullock, Robert Crossley, entomology
Calvert, David Victor, soil fertility
Eckelbarger, Kevin Jay, invertebrate zoology
Kretschmer, Albert Emil, Jr, agrostology
Sonoda, Ronald Masahiro, plant pathology
Wallen, Irvin Eugene, oceanography

FT WALTON BEACH
Pierce, Harold Hunter, animal science

GAINESVILLE
Abbott, Thomas B, speech pathology
Abrams, Robert Marlow, physiology
Achey, Phillip M, radiation biophysics
Adams, Earnest Dwight, low temperature physics
Adams, John Evi, psychiatry
Agee, Herndon Royce, insect physiology
Ahmed, Esam Mahmoud, food science
Aldrich, Henry Carl, botany, mycology
Allen, Charles Marshall, Jr, biochemistry
Allen, Don Lee, dentistry, periodontology
Allen, George E, entomology
Ammerman, Clarence Bailey, animal nutrition
Andersen, Thorkild Waino, anesthesiology
Anderson, John Francis, fluid mechanics
Anderson, Roland Carl, astrophysics, physiology
Andresen, Brian Dean, mass spectrometry, organic chemistry
Antonini, Gustavo Arthur, geography; resource management
Appledorf, Howard, nutrition, food science
Araujo, Oscar Eduardo, pharmacy
Arrington, Lewis Roberts, animal nutrition
Auffenberg, Walter, herpetology
Austin, Oliver Luther, Jr, ornithology
Ayoub, Elia Moussa, pediatrics
Baer, Herman, medical microbiology, pediatrics
Baig, Mirza Mansoor, biochemistry
Bailey, Henry Carl, biochemistry
Bailey, Donald Leroy, chemical physics
Ballard, Thomas, III, chemical physics
Ballard, Stanley Sumner, physics, optics
Banks, William Alden, entomology
Barney, Duane Lowell, inorganic chemistry
Barron, Donald Henry, anatomy
Bartz, Jerry A, plant pathology
Bassett, Mark Julian, plant breeding
Bates, Roger Gordon, analytical chemistry
Battiste, Merle Andrew, organic chemistry
Baxter, John Franklin, chemistry
Bazer, Fuller Warren, animal science, reproductive physiology

Becker, Charles Henry, pharmacy
Bednarek, Alexander R., mathematics
Bennett, Carroll G., dentistry
Berger, Richard Donald, plant pathology
Berner, Lewis, entomology
Bernstein, Jerald Jack, neuroanatomy, experimental neurology
Besch, Emerson Louis, environmental physiology
Biggs, Robert Hilton, plant physiology
Bingham, Nelson Eldred, biology
Birdsell, Dale Carl, microbiology, physiology
Bitton, Gabriel, environmental biology
Blake, Robert George, mathematics
Blanchard, Frank Nelson, geology
Blanton, Franklin Sylvester, medical entomology
Bleiweis, Arnold Sheldon, microbiology, immunology
Blue, Richard Arthur, experimental nuclear physics
Blue, William Guard, soils
Blythe, Rudolph Hamma, pharmaceutical chemistry
Boote, Kenneth Jay, crop physiology
Boswell, Thomas D., geography
Boyce, Richard P., biophysics
Boyd, Frederick Tilghman, agronomy, soil fertility
Bradfield, Richard, agronomy, soil fertility
Bradley, Richard E., parasitology, veterinary medicine
Brechner, Beverly Lorraine, mathematics
Breland, Herman Leroy, agronomy
Brey, Wallace Siegfried, Jr., physical chemistry
Brezonik, Patrick Lee, water chemistry, limnology
Brodkorb, Pierce, zoology
Brookbank, John Warren, developmental biology
Brookeman, Valerie Ann, radiological physics
Brooks, Harold Kelly, geology
Brooks, James Keith, mathematics
Brown, Henry Clay, III, organic chemistry
Brown, William Lewis, food microbiology
Brown, William Samuel, Jr., speech pathology
Browning, Charles Benton, animal nutrition
Broyles, Arthur Augustus, physics
Bullen, Adelaide Kendall, anthropology
Bunting, Donald Charles, meteorology
Burden, George Stanley, medical entomology
Callahan, Philip Serna, entomology
Campbell, Howard Wallace, environmental sciences, population biology
Cantliffe, Daniel James, plant physiology, vegetable crops
Carlisle, Victor Walter, soil morphology
Carpenter, James Woodford, animal science
Carr, Archie Fairly, Jr., zoology
Carr, Thomas Deaderick, physics, astronomy
Carr, William Edward Statter, marine ecology
Carter, William Earl, ethnology, applied anthropology
Carver, William Angus, plant breeding
Cassin, Sidney, physiology
Centifanto, Ysolina M., bacteriology
Cerutti, Peter A., biochemistry
Chai, An-Ti, molecular spectroscopy
Chambers, Derrell Lynn, entomology
Chew, Victor, mathematical statistics
Chiou, George Chung-Yih, pharmacology, biochemistry
Christie, Stephen Rolland, plant virology
Chun, Paul W., physical biochemistry
Clem, Lester William, immunology
Cohen, Howard Lionel, astronomy
Cohen, Robert Jay, biophysics
Coldwell, Robert Lynn, statistical mechanics
Colgate, Samuel Oran, physical chemistry
Combs, George Ernest, animal nutrition
Conklin, James Byron, Jr., solid state physics, academic administration

Connolly, John William Domville, theoretical physics, solid state physics
Conover, Robert Armine, botany
Conrad, Joseph H., animal nutrition, biochemistry
Conway, Kenneth Edward, mycology, plant pathology
Cook, Allyn Austin, plant pathology
Cooper, Robert Warren, forestry
Cornelius, Charles Edward, pathology
Cornell, John Andrew, applied statistics
Cornwell, George William, wildlife ecology, environmental sciences
Couch, Margaret Wheland, organic chemistry
Crandall, Richard B., parasitology, immunology
Creighton, John Thomas, entomology
Cromroy, Harvey Leonard, radiobiology, entomology
Cross, Clark Irwin, geography, physical science
Cusumano, Charles Louis, virology, immunology
Daicoff, George Ronald, surgery
Daly, James William, obstetrics & gynecology, oncology
Dame, David Allan, entomology
Damron, Bobby Leon, poultry nutrition
Davidson, James Melvin, soil physics
Davis, George Kelso, animal nutrition
Dawson, William Woodson, physiological psychology, biophysics
Dean, Charles Edgar, plant breeding, plant genetics
Decker, Phares, plant pathology
Deevey, Edward Smith, Jr., biology
Deevey, Georgiana Baxter, biological oceanography
Dekle, George Wallace, entomology
Denmark, Harold Anderson, acarology
Dennison, Raymond Alexander, food science
De Witt, James Merkle, zoology
Deyrup, James Alden, organic chemistry
Dickinson, Joshua Clifton, Jr., ornithology
Dickson, Donald Ward, nematology, plant pathology
Dolbier, William Read, Jr., organic chemistry
Donivan, Frank Forbes, Jr., radio astronomy
Doughty, Paul Larrabee, applied anthropology
Douglas, Derek Brian, physics
Dove, Derek Brian, physics
Drake, David Allyn, mathematics
Dresdner, Richard David, chemistry
Dudeck, Albert Eugene, agronomy
Duffield, Robert Brokaw, chemistry
Dufty, James W., statistical mechanics
Duggan, Dennis E, bacterial genetics
Dunn, Adrian John, neurochemistry, biochemistry
Dunn, Ben Monroe, bio-organic chemistry, physical biochemistry
Dunnam, Francis Eugene, nuclear physics
du Toit, Brian Murray, anthropology
Dykstra, Michael Jack, mycology, cell biology
Eades, James L., mineralogy
Edds, George Tyson, pharmacology, veterinary medicine
Edwards, Richard Archer, paleontology
Edwardson, John Richard, genetics
Eitzman, Donald V., medicine, physiology
Elliott, Larry Paul, radiology
Emmel, Thomas C., population biology, genetics
Enneking, William Fisher, medicine
Eno, Charles Franklin, soil science
Enoch, Jay Martin, physiological optics, vision
Eoff, Kay M., physics
Eyler, John Robert, physical chemistry
Fairchild, Graham Bell, entomology
Feaster, John Pipkin, biological chemistry
Feldherr, Carl M., anatomy
Fifield, Willard Merwin, agriculture
Finger, Kenneth F., biochemical pharmacology
Finlayson, Birdwell, urology, biophysics
Fischischweiger, Werner, electron microscopy
Fisher, Dale John, analytical chemistry, instrumentation
Fisher, Martin Joseph, medical physiology
Fisher, Waldo Reynolds, biochemistry, metabolism
Fiskell, John Garth Austin, soil chemistry
Flowers, John Wilson, physics
Fontaine, Thomas Davis, biochemistry
Ford, Ernest Sidney, botany

Forrester, Donald Jason, veterinary parasitology
Fox, Jackson Leland, environmental biology
Fox, Lauretta Ewing, pharmacology
Franke, Donald Edward, animal breeding, animal genetics
Freeman, Thomas Edward, plant pathology
Fregly, Melvin James, physiology
French, Rowland Barnes, biochemistry
Freund, Gerhard, internal medicine, endocrinology
Fried, Melvin, biochemistry
Fritz, George John, plant physiology
Fry, Jack L., academic administration, poultry science
Gabbay, Edmond J, biochemistry, biophysics
Gammon, Nathan, Jr., soil chemistry
Ganguly, Rama, biochemistry
Garcia-Bengochea, Francisco, neurosurgery
Garg, Lal Chand, pharmacology
Garrett, Edward Robert, pharmaceutical chemistry, biopharmaceutics
Garrett, Richard Edward, nuclear physics
Garrington, George Everett, dentistry, oral pathology
Gaskin, Jack Michael, veterinary microbiology
Gaskins, Murray Hendricks, horticulture
George, Theodore Samuel, applied mathematics
Gerber, John Francis, agricultural meteorology
Gessner, Ira Harold, pediatric cardiology
Gibbs, Charles Howard, dental research, biomedical engineering
Giesel, James Theodore, population biology, ecology
Gifford, George Edwin, virology, microbiology
Gilbert, Carter Rowell, ichthyology
Goddard, Ray Everett, forest genetics
Going, Robert Ernest, dentistry, histology
Goldberg, Eugene P., organic chemistry
Gottesman, Stephen T., radio astronomy
Gouck, Harry (Kydd), medical entomology
Graetz, Donald Alvin, soil chemistry, water chemistry
Grainger, David A., restorative dentistry
Gramling, Lea Gene, pharmaceutical chemistry
Gray, John Lewis, forest economics, forest management
Green, Alex Edward Samuel, physics
Green, Victor Eugene, Jr., field crops
Gregg, James Henderson, developmental biology
Griffin, Dana Gove, III, bryology
Griffin, George Melvin, Jr., marine geology, sedimentology
Grissell, Edward Eric Fowler, entomology, taxonomy
Grosenbaugh, Lewis Randolph, forest mensuration
Gull, Dwain D., vegetable crops
Habeck, Dale Herbert, entomology
Hackett, Raymond Lewis, pathology
Hadlock, Edwin Harold, mathematics
Hall, Chesley Barker, plant physiology
Hall, Donald William, medical entomology
Hamilton, Eugene W., entomology, electrical engineering
Hammer, Lowell Clarke, speech pathology, audiology
Hammer, Richard Hartman, pharmaceutical chemistry, organic chemistry
Hammond, Luther Carlisle, soils
Hanrahan, Robert Joseph, physical chemistry
Hanson, Harold Palmer, physics, academic administration
Hardy, John Milton, ornithology
Hariharan, Pallat Venketeswar, biochemistry, radiation physics
Harms, Robert Henry, poultry nutrition
Harper, Verne Lester, forest management, research administration
Harris, Barney, Jr., dairy science
Head, H Herbert, dairy science
Heaton, Marieta Barrow, neuroembryology
Helling, John Frederic, organometallic chemistry
Hemmings, E Thomas, archaeology, anthropology
Hemp, Gene Willard, applied mathematics
Heniges, James Franklin, Jr, ruminant nutrition
Hetrick, Lawrence Andrew, entomology, forestry

Hiebert, Ernest, virology, plant pathology
Hill, Clement Joseph, pedodontics
Himes, James Albert, veterinary pharmacology, veterinary physiology
Hinson, Kuell, genetics, plant breeding
Hoffmann, Edward Marker, immunology
Hogen-esch, Theo Eltjo, organic polymer chemistry, physical organic chemistry
Hollien, Harry, experimental phonetics
Hooper, Charles Frederick, Jr., theoretical physics
Horne, Edward Porter, psychophysiology
Horner, Earl Stewart, plant breeding
Hortenstine, Charles C., soil chemistry
Hubbell, David Heuston, soil microbiology
Huffman, Jacob Brainard, forestry, wood science technology
Humphreys, Thomas Elder, plant biochemistry
Ingram, Lonnie O'Neal, microbial physiology, cell biology
Isaacson, Robert Lee, physiological psychology
Isler, Ralph Charles, physics
Jaeger, Marc Jules, respiratory physiology
Johnson, Carl Henry, pharmacy
Johnson, Charles Robert, ornamental horticulture
Johnson, Chris Alan, psychophysics, physiological optics
Johnson, F Clifford, genetics, ecology
Johnston, David Ware, avian physiology, avian ecology
Joiner, Jasper Newton, horticulture
Jones, Edmund Ruffin, Jr., zoology
Jones, William Maurice, organic chemistry
Kallenbach, Ernst Adolf Theodor, histology, electron microscopy
Kallman, Robert Richard, mathematical analysis
Kalman, Rudolf Emil, engineering, mathematics
Kalra, Satya Paul, neuroendocrinology, endocrinology
Kaufman, Clemens Marcus, silviculture, forest ecology
Kaufman, Harold Loraine, physics
Kaufmann, Herbert Edward, ophthalmology
Kaufmann, John Henry, vertebrate zoology
Kay, Carol Ann, insect ecology
Kay, Fenton Ray, physiological ecology, vertebrate zoology
Keene, Willis Riggs, internal medicine, hematology
Keesling, James Edgar, topology
Kelly, John Francis, horticulture, olericulture
Kerr, Stratton H., entomology
Kilby, John Davis, ecology
Killinger, Gordon Beverly, agronomy
Kimball, Solon T, anthropology
Kimbrough, James W., mycology
King, Frederick Alexander, neuropsychology, neurophysiology
King, Robert Lee, neurophysiology
King, Roy Warbrick, mass spectrometry, nuclear magnetic resonance
Klein, Paul Alvin, virology, immunology
Knapp, Frederick Whiton, food chemistry
Knowles, Harold Loraine, physics
Koburger, John Alfred, food microbiology
Koger, Marvin, animal breeding, genetics
Kramer, Sol, biology
Kroger, Hanns H. inorganic chemistry, electrochemistry
Kucharek, Thomas Albert, plant pathology
Kuitert, Louis Cornelius, entomology
La Brecque, Germain C, medical entomology
Laipis, Philip James, molecular biology, virology
Laitinen, Herbert August, analytical chemistry
Lanciani, Carmine Andrew, ecology
Langdon, Kenneth R, plant taxonomy, nematology
LaPointe, Leonard Lyell, speech pathology
Larkin, Lynn Haydock, anatomy, reproductive biology
Lawrence, Fred Parker, horticulture, entomology
Leacock, Robert Jay, astronomy
Leavitt, Benjamin Burton, marine biology
Lebo, George Robert, physics, radio astronomy
Leibman, Kenneth Charles, biochemical pharmacology
Leonard, Christiana Morison, neuroanatomy, psychology
Levin, Sidney, pediatrics, nephrology

FLORIDA

Lewis, Daniel Ralph, mathematics
Li, Kuang-Pang, analytical chemistry
Littell, Ramon Clarence, agricultural statistics, mathematical statistics
Lloyd, James Edward, evolutionary biology
Locascio, Salvador J, horticulture
Lofgren, Clifford Swanson, entomology
Loggins, Phillip Edwards, animal nutrition
Loosli, John Kasper, animal nutrition
Lorz, Albert Protus, plant cytogenetics
Lowdin, Per-Olov, theoretical physics, quantum biology
Lowe, Ronald Edsel, entomology
Lucansky, Terry Wayne, botany
Luehr, Charles Poling, applied mathematics
Lugo, Ariel Emilio, ecology
Luke, Herbert Hodges, plant pathology
Mackenzie, Richard Stanley, dentistry
Mans, Rusty Jay, biochemistry, enzymology
Mansell, Robert Shirley, soil physics
Maples, William Ross, physical anthropology, primatology
Marcus, Robert Brown, physical geography
Maren, Thomas Hartley, pharmacology
Margolis, Maxine Luanna, anthropology
Marlowe, George Albert, Jr, horticulture
Marshall, Sidney Paul, dairy husbandry
Marston, Robert Quarles, medicine
Martin, Frank Garland, experimental statistics
Marvel, Mason E, horticulture, plant pathology
Matthews, Richard Finis, food chemistry, biochemistry
Maturo, Frank Juan Sarno, Jr, marine biology
Mauderli, Walter, nuclear physics
Mauldin, Richard Daniel, mathematics
Mayer, Marion Sidney, entomology, biochemistry
McCloud, Darell Edison, agronomy, agriculture
McConnell, Dennis Brooks, ornamental horticulture
McCune, Shannon, geography
McDowell, Lee Russell, animal nutrition
McElwee, Edgar Warren, ornamental horticulture
McGuigan, James E, gastroenterology, immunology
McKerns, Kenneth (Wilshire), biochemistry
McPeters, Richard Douglas, atmospheric physics
Mead, Frank Waldreth, entomology
Medina, Jose Enrique, dentistry
Mendenhall, William, III, statistics
Merrill, John Ellsworth, astronomy
Micha, David Allan, chemical physics
Mifflin, Martin David, hydrogeology
Milanich, Jerald Thomas, anthropology
Miller, Billie Lynn, medicine
Miller, George H, Jr, urology
Miller, Howard Nile, plant pathology
Miller, John Wesley, plant pathology
Million, Rodney Reiff, radiotherapy
Millsaps, Knox, applied mathematics
Minnick, Danny Richard, entomology
Mitchell, Everett Royal, entomology
Modell, Jerome Herbert, anesthiosology
Mohammed, M Hamdi A, dental materials, prosthodontics
Montelaro, James, horticulture
Moore, G Alexander, Jr, anthropology
Moore, John Edward, ruminant nutrition
Moore, Theral Orvis, mathematics
Moreland, Alvin Franklin, veterinary medicine, comparative medicine
Moscovici, Carlo, virology
Mott, Gerald O, agronomy
Moulder, Peter Vincent, Jr, cardiovascular surgery
Mount, Gary A, medical entomology
Moye, Hugh Anson, analytical chemistry
Mueller, Paul Allen, geochemistry
Muga, Marvin Luis, nuclear chemistry
Mull, Leon Edmund, bacteriology
Mullin, Robert Spencer, plant pathology
Mullins, John Thomas, botany
Mundy, Belvey Washington, physical chemistry
Munson, Edwin Sterling, anesthesiology
Munson, John Bacon, neurosciences
Munyer, Edward Arnold, zoology, science education
Murphey, Milledge, entomology
Muschlitz, Earle Eugene, Jr, physical chemistry, chemical physics
Musgrave, Carol Ann, entomology
Myers, Gardiner Hubbard, physical chemistry
Nation, James Lamar, insect physiology
Neilson, John Taylor McLaren, immunology, parasitology
Nettles, Victor Fleetwood, vegetable crops
Nevis, Arnold Hastings, neurology, biophysics
Newton, James Henry, physical chemistry
Nichols, Wilmer Wayne, cardiovascular physiology
Nicol, David, peleontology
Niddrie, David Lawrence, geography
Niles, James Alfred, agricultural economics
Noonan, Kenneth Daniel, biochemistry, cell biology
Norden, Allan James, plant breeding, genetics
Nordlie, Frank Gerald, zoology
Normann, Sigurd Johns, pathology
Noyes, Ward David, internal medicine, hematology
Nunez, Theron A, Jr, anthropology
Oberlander, Herbert, biology
O'Brien, Thomas W, physiology, biochemistry
Odum, Howard Thomas, ecology, oceanography
Ohrn, Nils Yngve, chemical physics, quantum chemistry
Oliver, John Parker, astronomy
Olson, Kenneth B, medicine, oncology
Olsson, Carl Niels, radio astronomy
Omer, Guy Clifton, Jr, physics
Otis, Arthur Brooks, pulmonary physiology
Ott, Ray Lyman, Jr, statistics
Overman, Allen Ray, agronomy
Palenik, Gus J, crystallography
Palmore, William P, veterinary physiology
Parkinson, Michael Thaddeus, high energy physics
Patterson, Richard Sheldon, medical entomology
Patton, Thomas Hudson, vertebrate paleontology
Payne, Willard William, plant taxonomy
Pepinsky, Raymond, physics, biology
Perry, Vernon G, plant nematology
Person, Willis Bagley, physical chemistry
Peterson, Lennart Rudolph, atmospheric physics, atomic physics
Pfaff, William Wallace, surgery
Pfahler, Paul Leighton, genetics
Phillips, Richard Lee, horticulture, plant physiology
Pierce, Robert William, geology
Pierson, K Kendall, pathology
Pirenian, Zareh Meguerditch, mathematics
Pirkle, Earl Conly, economic geology
Poe, Sidney LaMarr, entomology
Pokrant, Marvin Arthur, statistical mechanics
Popenoe, Hugh, soil science
Pop-Stojanovic, Zoran Rista, mathematics
Porter, Hayden Samuel, Jr, applied physics, aeronomy
Pradhan, Tapas Kumar, biochemistry
Prange, Henry Davies, animal physiology
Prasad, Sheo Shanker, physics
Preston, James Faulkner, III, biochemistry, microbiology
Previc, Edward Paul, microbial physiology
Prine, Gordon Madison, agronomy
Pring, Daryl Roger, plant pathology
Pritchett, William Lawrence, forest soils
Purcifull, Dan Elwood, plant virology
Purdy, Barbara Ann, anthropology, archaeology
Purdy, Laurence Henry, plant pathology
Putnam, Hugh D, microbiology
Quesenberry, Kenneth Hays, agronomy, plant breeding
Randazzo, Anthony Frank, geology
Rao, Pejaver Vishwamber, mathematical statistics
Rappenecker, Caspar, geology, geography
Reddish, Robert Lee, animal science
Reid, Charles Edward, physical chemistry
Reiskind, Jonathan, evolutionary biology, arachnology
Reith, Edward John, anatomy
Rennert, Owen M, pediatrics, biochemistry
Reynolds, Richard Clyde, medical education
Rhoton, Albert Loren, Jr, neurosurgery
Robbins, Ralph Compton, nutrition, physiology
Roberts, Daniel Altman, plant pathology
Roberts, Leonidas Howard, physical sciences, astronomy
Roberts, Richard Harris, medical entomology, veterinary entomology
Roberts, Robert Michael, biochemistry, plant physiology
Robertson, William Kitchener, soil chemistry
Rodgers, Bradley Moreland, pediatric surgery
Rodgers, Earl Gilbert, agronomy, weed science
Roessler, Charles Ervin, health physics, environmental engineering
Roland, David Alfred, Sr, nutritional biochemistry
Romrell, Lynn John, genetics, human anatomy
Rosenbloom, Arlan Lee, pediatrics, endocrinology
Rosenshein, Joseph Stanley, low temperature physics, biophysics
Ross, Michael H, anatomy, cell biology
Rothman, Howard Barry, speech & hearing science
Rowe, Mary Budd, science education
Rubin, Melvin Lynne, medicine, ophthalmology
Ruelke, Otto Charles, agronomy, plant physiology
Ryschkewitsch, George Eugene, inorganic chemistry
Sabin, John Rogers, quantum mechanics, biophysics
Sailer, Reece Ivan, entomology
Sander, Eugene G, biochemistry
Sawyer, Earl Morrow, nuclear physics
Schank, Stanley Cox, cytogenetics, plant breeding
Scheaffer, Richard Lewis, statistics
Schenck, Norman Carl, plant pathology
Schiebler, Gerold Ludwig, medicine, pediatric cardiology
Schmid, Gerhard Martin, electrochemistry
Schneider, Richard Theodore, plasma physics, spectroscopy
Schoulties, Calvin Lee, plant pathology
Schroder, Vincent Nils, plant physiology
Schulkind, Martin Lewis, pediatrics, immunology
Schulman, Stephen Gregory, analytical chemistry, photochemistry
Schwassman, Horst Otto, biology, physiology
Scott, Thomas A, physics
Seawright, Jack Arlyn, entomology, genetics
Selfridge, Ralph Gordon, mathematics
Selman, Kelly, cell biology, reproductive biology
Shaak, Graig Dennis, paleoecology
Shah, Dinesh Ochhavlal, biophysics
Shands, Joseph Walter, Jr, medical microbiology
Shanor, Leland, mycology
Sharpe, Ralph Harold, horticulture
Sheehan, Thomas John, ornamental horticulture
Sherman, Wayne Bush, plant breeding, horticulture
Shields, Robert Pierce, veterinary pathology, biochemistry
Shirley, Ray Louis, animal nutrition
Showalter, Robert Kenneth, horticulture, food science
Shulman, Stanford Taylor, pediatrics, immunology
Sigmon, Kermit Neal, mathematics
Silhacek, Donald Le Roy, biochemistry, entomology
Silverman, David Norman, pharmacology, biophysics
Simpson, Charles Floyd, veterinary pathology
Singleton, George Terrell, otolaryngology
Singley, John Edward, water chemistry
Sites, John Wilbur, horticulture
Slater, John Clarke, physics
Slater, Rose Mooney, physics
Small, Parker Adams, Jr, immunology, immunochemistry
Smart, Grover Cleveland, Jr, plant nematology
Smith, Alexander Goudy, astrophysics
Smith, Carroll N, medical entomology
Smith, Douglas Lee, geology
Smith, Kenneth Leroy, dairy microbiology
Smith, Paul Howard, bacteriology
Smith, Rex L, plant genetics, plant breeding
Smith, Richard Clark, plant physiology
Smith, Richard Thomas, pediatrics
Smith, Wayne H, forestry
Smith, William Ward, medical entomology, economic entomology
Smittle, Burrell Joe, entomology
Snedaker, Samuel Curry, ecology
Soule, James, horticulture
Spangler, Daniel Patrick, geology
Spellacy, William Nelson, obstetrics & gynecology
Spinks, Daniel Owen, soil chemistry
Stainsby, Wendell Nicholls, physiology
Stall, Robert Eugene, plant pathology
Stanley, Harold Russell, oral pathology
Stearns, Thomas W, biochemistry
Stein, Gary S, biochemistry, cell biology
Stetson, Chandler Alton, medicine
Stevens, Ann Rebecca (Larkin), biochemistry, cell biology
Stokes, Donald Eugene, nematology
Stoufer, Robert Carl, inorganic chemistry
Streiff, Richard Reinhart, medicine, hematology
Stump, Eugene Curtis, Jr, organic chemistry, fluorine chemistry
Su, Stanley Y W, computer science
Sullivan, Edward T, forest economics
Surak, John Godfrey, toxicology
Suzuki, Howard Kazuro, anatomy
Swinford, Kenneth Roberts, forestry
Sypert, George Walter, neurosurgery, neurophysiology
Szinai, Stephen Slomo, chemistry, pharmacology
Talbert, James Lewis, pediatric surgery, thoracic surgery
Tarrant, Paul, chemistry
Taylor, Albert Lee, nematology
Taylor, William Jape, internal medicine, cardiology
Teague, Perry Owen, immunology
Teas, Donald Claude, physiological psychology
Telford, Sam Rountree, Jr, epizootiology
Teply, Mark Lawrence, algebra
Thatcher, William Watters, reproductive physiology, reproduction endocrinology
Thomas, Billy Seay, theoretical physics
Thomas, William Clark, Jr, internal medicine, endocrinology
Thompson, Buford Dale, horticulture, vegetable crops
Thompson, Fred G, malacology
Thompson, Leonard Garnett, Jr, soils
Thompson, Neal Philip, plant physiology, plant anatomy
Tobey, Frank Lindley, Jr, vision, photometry
Torosian, George, pharmacy, pharmacology
Travis, David M, internal medicine, pharmacology
Trickey, Samuel Baldwin, theoretical physics, solid state physics
Tsibris, John Constantine Michael, biochemistry, biophysics
Tu, Chingkuang, biochemistry, pharmacology
Tucker, William Boose, internal medicine
Tyler, Max Ezra, microbial ecology
Urone, Paul, chemistry
Vala, Martin Thorvald, Jr, physical chemistry, spectrochemistry
Vanderwerf, Calvin Anthony, organic chemistry
Van Horn, Harold H, Jr, dairying
Van Mierop, Lodewyk H S, medicine
Varma, Arun Kumar, mathematics
Vasil, Indra Kumar, botany
Victorica, Benjamin (Eduardo), pediatrics
Vierck, Charles John, Jr, psychophysiology
Vogh, Betty Pohl, pharmacology, physiology
Voitle, Robert Allen, poultry physiology
Volk, Bob Garth, soil chemistry
Volk, Gaylord Monroe, soil chemistry
Von Mering, Otto O, anthropology
Wagley, Charles W, anthropology, ethnology
Wahl, Floyd Michael, mineralogy
Waites, Robert Ellsworth, entomology
Wakeman, Donald Lee, animal husbandry
Waldman, Robert H, immunology
Walker, Don Wesley, neuroscience, neuropharmacology
Walker, Thomas Jefferson, entomology
Wallace, Harold Dean, animal nutrition
Wallbrunn, Henry Maurice, zoology
Ward, Coleman Younger, agronomy, physiology
Ward, Daniel Bertram, plant taxonomy
Ward, Gray (Ganesh), atmospheric physics
Ward, Ronald Wayne, agricultural economics, econometrics
Warmke, Harry Earl, plant cytology, plant genetics
Warnick, Alvin Cropper, physiology
Warson, Samuel R, psychiatry
Webb, Sawney David, paleontology, anatomy
Weems, Howard Vincent, Jr, entomology
Weidhaas, Donald E, medical entomology

Weller, Henry Richard, nuclear physics
Weltner, William, Jr., physical chemistry
West, Sherlie Hill, plant physiology, agronomy
Westfall, Minter Jackson, Jr., biology
Wheeler, Willis Boly, biochemistry
Whitcomb, Willard Hall, entomology
White, Franklin Henry, veterinary microbiology
White, Larry Dale, range management
Whitney, Ellsworth Dow, physical chemistry
Whitty, Elmo Benjamin, agronomy
Wilcox, Charles Julian, dairy science
Wilcox, Merrill, plant physiology
Wilkinson, Robert Cleveland, Jr., entomology, ecology
Wilkowske, Howard Hugo, dairy bacteriology
Williams, Clyde Michael, physiology
Williams, David Tyndale, applied physics
Williams, William Norman, speech pathology
Wilson, Henry R., reproductive physiology, environmental physiology
Witbank, William Joseph, horticulture, plant physiology
Winefordner, James D., analytical chemistry
Wing, Elizabeth S, zoology
Wing, James Marvin, animal nutrition
Wittig, Heinz Joseph, allergy, immunology
Wood, Frank Bradshaw, astronomy
Woodard, James Carroll, nutrition, comparative pathology
Woodruff, Robert Eugene, entomology
Woodward, Edward Roy, surgery
Yang, Mark Chao-Kuen, statistics
Yost, William A, psychoacoustics, psychophysics
Young, Martin Dunaway, parasitology, malariology
Yuan, Tzu-Liang, soils
Zam, Stephen G, III, parasitology
Zauner, Christian Walter, exercise physiology, pulmonary physiology
Zettler, Francis William, plant pathology, entomology
Ziegler, Louis William, horticulture
Zoltewicz, John A, organic chemistry
Zornetzer, Steven Frank, psychobiology, neurosciences

GULF BREEZE
Bourquin, Al Willis J, marine microbiology, microbial ecology
Butler, Philip Alan, marine biology
Cooley, Nelson Reede, protozoology
Couch, John Alexander, pathobiology, protozoology
Davies, Tudor T, environmental sciences, geochemistry
Dobinson, Frank, organic chemistry, polymer chemistry
Hansen, David John, toxicology
Lowe, Jack Ira, marine ecology, toxicology
Nimmo, Del Wayne Roy, environmental biology
Schoor, W Peter, biophysical chemistry, comparative biochemistry

HAINES CITY
Hannon, Chancellor Irving, nematology, plant pathology

HALLANDALE
Ender, Hans Henry, organic chemistry, physics

HASTINGS
Hensel, Dale Robert, soil fertility
Weingartner, David Peter, phytopathology, plant nematology
Workman, Ralph Burns, economic entomology, vegetable crops

HIALEAH
Bucolo, Giovanni, organic chemistry, biochemistry
Fox, Gerald, industrial chemistry
Madappally, Mathew Mathai, nutritional biochemistry

HOLIDAY
Morrow, David Clarence, mathematics

HOLLYWOOD
Luhan, Joseph Anton, neurology

HOMESTEAD
Baranowski, Richard Matthew, entomology
Bryan, Herbert Harris, horticulture, plant physiology
Buckwalter, Howard McWilliams, organic chemistry
Campbell, Carl Walter, plant physiology, horticulture
Duncan, Andrew A., horticulture
John, Charles Alfred, plant breeding
Malo, Simon E., horticulture, nematology
Marliatt, Robert Bruce, plant pathology
McMillan, Robert Thomas, Jr., plant pathology
Orth, Paul Gerhardt, soil fertility
Volin, Raymond Bradford, plant pathology, plant breeding
Waddill, Van Hulen, entomology
Wolfenbarger, Daniel Otis, entomology
Young, Thomas Wilbur, horticulture, soils

IMMOKALEE
Blazquez Y Servin, Carlos Humberto, plant pathology
Everett, Paul Harrison, soil fertility, vegetable crops

INDIALANTIC
Johns, Roman Karol Chrzaszcz, geophysics
Manning, Walter H, Jr., physics

INDIAN HARBOUR BEACH
Relf, Kenneth E, physics

JACKSONVILLE
Allen, Ted Tipton, vertebrate zoology
Ash, Willard Osborne, statistics
Bledsoe, James O, Jr., organic chemistry
Boehnke, David Neal, physical chemistry, theoretical chemistry
Bowman, Ray Douglas, physical chemistry, molecular biology
Browder, James Steve, solid state physics, optics
Cardenas, Carlos Guillermo, organic chemistry
Carlson, Willard Emmett, paper chemistry
Councill, Richard J, petrography, sedimentology
DeMort, Carole Lyle, phycology, ecology
Derfer, John Mentzer, industrial organic chemistry
Doff, Simon David, medicine, public health
Dorion, George Henry, organic chemistry
Fleek, James Burton, physical chemistry
Fresh, James W, pathology
Gager, William Ballantine, physics
Groover, Marshall Eugene, Jr., medicine
Hanley, James Richard, Jr., organic chemistry
Hernandez, Juan Antonio, pathology
Huebner, Jay Stanley, biophysics
Kane, Bernard James, organic chemistry
Kelley, George Greene, biological chemistry
Lewis, George Edwin, organic chemistry, internal medicine
Michael, Max, Jr., internal medicine
Miller, Kim Irving, plant taxonomy
Mitch, Frank Allan, organic chemistry
Porter, Lee Albert, natural products chemistry
Prather, Elbert Charlton, preventive medicine, public health
Queen, William Charles, applied mathematics
Relyea, Kenneth George, ichthyology, ecology
Ridings, Gus Ray, radiology
Schneider, Nathan Joseph, virology
Siegel, Peter Vincent, aerospace medicine
Sowder, Wilson Thomas, medicine
Stephenson, Samuel Edward, Jr, medicine, organic chemistry
Stine, Gerald James, microbial genetics
Thomas, Dan Anderson, physics
Timmer, Kathleen Mae, algebra
Trainer, John Ezra, Jr, helminthology, biological chemistry
Weiss, Charles Frederick, pediatrics, pharmacology
West, Felicia Emminger, science education
Winton, Charles Newton, mathematics

JAY
Bertrand, Joseph E, animal nutrition
Dunavin, Leonard Sypret, Jr., agronomy
Kinloch, Robert Armstrong, nematology
Lutrick, Monroe Cornealous, agronomy, soil chemistry
Peacock, Hugh Anthony, agronomy

KENNEDY SPACE CTR
Gayle, John Ben, physical organic chemistry, mathematical statistics

KEY BISCAYNE
Bullis, Harvey Raymond, Jr., marine biology
Hazard, John Beach, pathology

KEY LARGO
Feddern, Henry A, ichthyology

KISSIMMEE
Cooperrider, Donald Elmer, veterinary medicine

KISSIONONEE
Hajna, Anthony Alphonse, bacteriology

LA BELLE
Hebert, Leo Placide, agronomy

LAKE ALFRED
Albrigo, Leo Gene, horticulture, plant physiology
Brooks, Robert Franklin, entomology
Brown, George Eldon, plant pathology, plant genetics
DuCharme, Ernst Peter, plant pathology
Feldman, Albert William, plant pathology
Grierson, William, pomology
Hanks, Robert William, plant chemistry, phytopathology
Koo, Robert Chung Jen, pomology
Leonard, Chester D, soils
McCornack, Andrew Adams, horticulture
McCoy, Clayton William, entomology
Moore, Edwin Lewis, food science
Nigg, Herbert Nicholas, entomology
Pieringer, Arthur Paul, plant breeding
Reese, Robert Lewis, horticulture, weed science
Reitz, Herman J., horticulture
Simanton, William Aldrich, entomology
Stewart, Ivan, plant chemistry
Tarjan, Armen Charles, plant nematology
Ting, Sik Vung, horticulture
Tucker, David Patrick Hislop, horticulture
Wander, Irvin Woodrow, horticulture
Wardowski, Wilfred Francis, II, horticulture
Wheaton, Thomas Adair, plant physiology, horticulture
Whiteside, Jack Oliver, plant pathology
Wilson, William Curtis, plant physiology

LAKE CITY
Hedrick, Glen Willard, organic chemistry
Newsom, William S, Jr., agricultural chemistry

LAKE PLACID
Layne, James Nathaniel, vertebrate biology
Rand, Austin Loomer, vertebrate zoology

LAKE WORTH
Farrell, John Joseph, surgery
Halverson, Gilbert, physics
Kline, Gordon Mabey, plastics chemistry, polymer chemistry
Lindeman, Vertus Frank, zoology
Truchelut, George Burnett, plant physiology

LAKELAND
Attaway, John Allen, organic chemistry, plant biochemistry
Brown, Ivan Willard, Jr., thoracic surgery, cardiovascular surgery
Dinsmore, Howard Livingstone, analytical chemistry
Gerwe, Raymond Daniel, chemistry
Gilbert, Margaret Lois, botany, ecology
Graves, Howard Bradley, Jr., geology

Leaders, William M, inorganic chemistry, chemical engineering
Moore, Joseph Curtis, mammalogy
Prevatt, Rubert Waldemar, soil chemistry, horticulture
Stamper, James Harris, theoretical physics
Taylor, Edward Wyllys, chemistry, psychology
Willard, Thomas Maxwell, inorganic chemistry, analytical chemistry

LAND O'LAKES
Karst, Otto Jacob, mathematics

LARGO
Nelson, Philip R, fisheries, fish biology
Wheat, Myron William, Jr., thoracic surgery, cardiovascular surgery

LAUDERHILL
Stearn, Joseph Leonard, geodesy

LEESBURG
Adlerz, Warren Clifford, entomology
Branham, Joseph Morhart, developmental biology, ecology
Corwin, James Fay, chemistry
Crall, James Monroe, plant breeding, plant pathology
Elmstrom, Gary William, horticulture
Hopkins, Donald Lee, plant pathology
Mortensen, John Alan, horticulture, genetics

LEHIGH ACRES
McCloskey, Kenneth Emory, physical chemistry, food science

LIVE OAK
Rubin, Harvey Louis, veterinary microbiology, pathology

MAITLAND
Hannegan, John Michael, chemistry
Norman, Eliane Meyer, botany, taxonomy
O'Bannon, John Horatio, nematology

MARATHON SHORES
Lane, Richard Dale, forest management

MARIANNA
Gorbet, Daniel Wayne, agronomy, plant breeding
Hebb, Edwin Atkins, forest ecology
Leibbrandt, Vernon Dean, animal nutrition
Lewis, Clifford Eugene, range management, botany
Vipperman, Posey Elmer, Jr., nutrition

MELBOURNE
Blatt, Albert Harold, organic chemistry
Blatt, Joel Herman, spectroscopy, instrumentation
Bosch, Warren Luther, organic chemistry, mathematics
Burns, Jay, III., physics
Clark, Kerry Bruce, invertebrate zoology, marine ecology
Clutterham, David Robert, mathematics
De Fazio, Sally Ruth, immunology
DeSua, Frank Crispin, mathematics
Dubbelday, Pieter Steven, oceanography
Gross, John Howard, physical chemistry
Jin, Rong-Sheng, physics
Keuper, Jerome Penn, physics
Miller, John Edward, magnetic resonance
Mounts, Richard Duane, analytical chemistry
Nevin, Thomas Andrew, bacteriology
Potter, James Gregor, physics
Thomas, Garland Leon, physics
Webster, George Calvin, biochemistry
Woodbridge, David Davis, physics, meteorology
Young, John William, mathematics, computer science
Zeigler, Joseph Alton, medical microbiology, electron microscopy

MELBOURNE BEACH
von Fischer, William, chemistry

419

MELROSE
Teller, Morton Herman, physics

MERRITT ISLAND
McCampbell, John Caldwell, geology

MIAMI
Anderson, David Gordon, biochemistry
Anderson, Douglas Richard, ophthalmology
Anderson, William Arnold Douglas, pathology
Apel, John Ralph, fluid physics, physical oceanography
Aranda, Juan Manuel, cardiology
Awad, William Michel, Jr., internal medicine, oncology
Azarnia, Rooblk, cell physiology, biophysics
Ball, Mahlon M., geology, geophysics
Barrera, Frank, physiology
Bassett, Arthur Leon, cardiovascular physiology, pharmacology
Beardsley, Grant Lindley, Jr., fish biology
Beck, Jacob Walter, medical parasitology
Beeler, Benjamin, cardiovascular disease
Berk, Toby Steven, anatomy, hematology
Berman, Irwin, anatomy, hematology
Bernstein, Stanley H. internal medicine, immunology
Birdsey, Monroe Roberts, botany
Blank, Harvey, dermatology
Block, Ronald Edward, physical chemistry, biophysical chemistry
Blumenthal, Sidney, cardiology
Bock, Wayne Dean, micropaleontology
Bonati, Enrico, marine geology
Bradshaw, Claire Margaret, immunology, immunochemistry
Brady, Al H., medical biophysics
Brill, Earl, organic chemistry, pharmacology
Brown, Harvey Earl, Jr., internal medicine
Burditt, Arthur Kendall, Jr., economic entomology
Burg, Stanley (Paul), plant physiology
Byrne, Hugh Michael, oceanography
Cahill, Donald R., anatomy
Cameron, Bruce Francis, physical biochemistry, hematology
Campbell, Hallock Cowles, chemistry
Carmichael, Lynn Paul, family medicine
Carter, Bettina Bush (Mrs Daniel F Jackson), immunobiology,
Carter, Claude Francis, physics
Castro, Alberto, biochemistry, endocrinology
Chambers, Edward Lucas, cell physiology
Chambliss, Keith Wayne, immunochemistry
Chandler, J Ryan, surgery, otolaryngology
Charyulu, Komanduri K N, radiology
Chesky, Jeffrey Alan, muscular physiology, gerontology
Chester, Brent, clinical microbiology
Chew, Frank, oceanography
Chiu, Andrew Tak-Chau, pharmacology
Claflin, Alice J., immunobiology,
Clark, Ronald Grey, neuroanatomy
Cleary, Timothy Joseph, medical microbiology
Cleveland, William West, pediatrics
Collins, Galen Franklin, clinical biochemistry,
Colsky, Jacob, internal medicine,
Colwin, Arthur Lentz, zoology
Colwin, Laura Hunter, zoology
Copenhaver, Wilfred Monroe, anatomy
Corcoran, Eugene Francis, oceanography
Cowman, Richard Ammon, microbiology, biochemistry
Crayhorne, N W Brain, anesthesiology
Curtiss, John Hamilton, mathematical analysis, analytical statistics
Damhaugh, Luella N., geography
Dash, Bibhu Prasad, geophysics
Dash, Harriman Harvey, biochemistry,
Davidoff, Robert Alan, neuropharmacology
Davis, Joseph Harrison, forensic medicine
DeFerrari, Harry Austin, underwater acoustics
de Sylva, Donald Perrin, biological oceanography, ichthyology
Diaz, Pedro Miguel, anesthesiology
Dickson, David Ross, speech pathology

Dietz, Robert Sinclair, geology, oceanography
Dombro, Roy S., biochemistry
Dragovich, Alexander, marine ecology
Duany, Luis F, Jr., dental epidemiology, preventive dentistry
Dunning, Walter, physical oceanography
Dunning, Wilhelmina Frances, pathology
Feingold, Alfred, anesthesiology
Field, Henry, physical anthropology, archaeology
Fisher, David J., geochronology, meteoritics
Fisher, George Harold, Jr., organic biochemistry
Fisher, Jack Bernard, botany
Fisher, Robert Charles, mathematics
Fitzgerald, Dorothea Babbitt, biochemistry
Fitzgerald, Robert James, microbiology, oral biology
Fletcher, Mary Ann, immunology
Fogel, Bernard J., immunology, pediatrics
Fox, Alfred Earl, immunochemistry, immunohematology
Fuller, Laphalle, immunology
Gargano, Fredie Patrick, radiology
Giegel, Joseph Lester, clinical biochemistry
Gilmore, Hugh Richmond, III., medicine, cardiology
Gilson, Albert Jack, nuclear medicine
Gold, Martin I., medicine, anesthesiology
Goldsmith, Carl, internal medicine, nephrology
Goldstein, Burton Jack, psychiatry, psychopharmacology
Goll, Robert John, physical chemistry
Gollan, Frank, medicine
Gotterer, Malcolm Harold, information science, computer science
Gratzner, Howard G, genetics
Griffiths, James John, medicine
Gross, Michael Ralph, computer sciences
Gruber, Samuel Harvey, behavioral physiology, vision
Gunn, Samuel Albert, cancer, experimental pathology
Halbert, Seymour Putterman, medical microbiology, immunology
Halprin, Kenneth M, dermatology
Hamasaki, Duco I., ophthalmology
Harkness, David R., hematology
Harrington, William J., internal medicine
Harrison, Christopher George Alick, geophysics
Harrison, Robert J., audiology, speech pathology
Haukohl, Robert Somers, medicine, pathology
Hay, William Winn, micropaleontology, marine geology
Heald, Eric James, ecology, fish biology
Healey, John Edward, Jr., oncology, rehabilitation medicine
Heckerman, Raymond Otto, vertebrate morphology, cytology
Heller, Zindel Herbert, medical physics, instrumentation
Herriott, Arthur W, organic chemistry
Higman, James B., fisheries
Hodes, Philip J., radiology
Holaday, Duncan Asa, anesthesiology
Hope, Ronald Richmond, cardiology
Houde, Edward Donald, fisheries
Howell, David Sanders, medicine, physiology
Hsia, Sung Lan, biochemistry, genetics
Huijing, Frans, biochemistry
Hurley, Robert Joseph, marine geology, oceanography
Immon, W Byron, obstetrics & gynecology
Iversen, Edwin Severin, fish biology
Jablon, James Martin, bacteriology
Jacobson, Marcus, physiology, biophysics
Jensen, Joerg, immunology, transplantation biology
Joensuu, Oiva I., geology
Jones, Albert Cleveland, fisheries
Jungbauer, Mary Ann, inorganic chemistry
Kaiser, Gerard Alan, thoracic surgery, cardiovascular surgery
Keller, Martin, gastroenterology
Keller, George H., marine geology
Ketcham, Alfred Schutt, surgery
Kunce, Henry Warren, operations research
Lampe, Kenneth Francis, pharmacology
Landowne, David, physiology, biophysics
Laramore, George Ernest, physics, medicine
Lasseter, Kenneth Carlyle, clinical pharmacology, internal medicine

Lavelle, John William, geological oceanography
Lawson, Robert Barrett, pediatrics
Lazzara, Ralph, medicine
Leif, Robert Cary, immunohematology, biomedical engineering
Leighton, Morris Wellman, exploration geology
Lemberg, Louis, physiology
Levey, Gerald Saul, internal medicine, endocrinology
Lewis, Billy M, meteorology
Lillien, Irving, organic chemistry
Lin, Tsue-Ming, immunology, microbiology
Little, William Asa, obstetrics & gynecology
Magleby, Karl LeGrande, neurophysiology, biophysics
Malinin, Theodore I., pathology
Marks, Asher, medicine
Marlborough, David Ian, biochemistry
Marsh, John MacClenahan, biochemistry
Marshall, James John, biochemistry, medical research
Mason, Allen Smith, atmospheric chemistry
Maue-Dickson, Wilma, developmental anatomy
McFadden, James Douglas, meteorology, oceanography
Mende, Thomas Julius, biochemistry
Mendell, Jay Stanley, technological forecasting
Michel, Harding B., marine zoology
Michel, George C., ichthyology
Miller, Kent D., biochemistry
Miller, Wallace E., orthopedic surgery
Millero, Frank Joseph, Jr., physical chemistry
Mitsui, Akira, biological oceanography
Mock, Gene Vernon, organic chemistry
Mofjeld, Harold Oswald, physical mineralogy
Montes De Oca, Hector, cell biology
Moore, Donald Richard, malacology
Moore, Hilary Brooke, zoology
Morejon, Clara Baez, chemistry
Moya, Frank, anesthesiology
Muench, Karl Hugo, biochemistry, genetics
Munson, Gerald Leonard, cell biology, developmental biology
Murphy, William Parry, Jr., medicine
Myrberg, Arthur August, Jr., animal behavior, marine biology
Nagle, Frederick, Jr., petrology
Noble, Nancy Lee, biochemistry,
Norton, Edward W D, medicine
Noto, Thomas Anthony, pathology
Odell, Daniel Keith, vertebrate zoology
Orgell, Wallace Herman, plant physiology
Ostlund, H Gote, marine chemistry, atmospheric chemistry
Paff, George Hugo, histology
Palmer, Roger, pharmacology, medicine
Pappas, Anthony John, chemistry
Papper, Emanuel Martin, anesthesiology
Parker, John Hilliard, physical chemistry,
Perkins, Henry Thomas, physical oceanography
Peterson, Ernest A., acoustics
Poitras, Adrian William, mycology
Popenoe, John, horticulture
Porter, James Armer, Jr., primatology
Potter, Lincoln Truslow, neurobiology
Pressman, Berton Charles, biochemistry, biophysics
Proni, John Roberto, underwater acoustics, oceanography
Radomski, Jack London, pharmacology
Rauch, Nancy, developmental biology
Reinmuth, Oscar McNaughton, neurology, internal medicine
Reiss, Eric, medicine
Richards, Marvin Sherrill, organic chemistry
Richards, William Joseph, ichthyology
Rivas, Luis Rene, ichthyology
Roach, Don, analytical chemistry
Robins, Charles Richard, ichthyology
Rockstein, Morris, physiology
Roessler, Martin A., fish biology
Rona, Peter Arnold, marine geology, geophysics
Rose, Birgit Loewenstein, physiology, cell physiology
Rosenberg, Benjamin, underwater acoustics
Rosenthal, Stanley Lawrence, meteorology

Rosomoff, Hubert Lawrence, medicine
Rufenach, Clifford L., ionospheric physics, physical oceanography
Ryan, James Walter, medicine, biochemistry
Ryan, Una Scully, cell biology, pulmonary diseases
Sakhnovsky, Alexander Alexandrovitch, physical chemistry
Sallman, Bennett, microbiology
Sandoval, Howard Kenneth, microbiology
Sapolsky, Asher Isadore, biochemistry
Savard, Francis Gerald Kenneth, biochemistry, endocrinology
Sawyer, Constance B., oceanography, astrophysics
Schalberger, George Elmer, ecology
Scheinberg, Peritz, neurology
Schell, Klaus Rainer, virology, immunology
Scherlag, Benjamin J, cardiovascular physiology
Schiff, Leon, internal medicine
Schneider, Arthur Lee, protein chemistry
Schultz, Duane Robert, immunology, protein chemistry
Schultz, Julius, biochemistry
Schwartz, Albert, systematic zoology
Selawry, Oleg S., cancer
Sheldon, Eric Ernest, cell biology
Smith, David Spencer, cell biology
Smith, Frederick George Walton, marine zoology
Shershin, Anthony Connors, operations research, mathematics
Shinn, Eugene Allen, biology, geology
Shipley, Thorne, ophthalmology
Shook, Edwin Martin, anthropology
Sigel, Mola Michael, virology
Skinner, Margaret Sheppard, pathology
Slotta, Karl Heinrich, biochemistry
Smith, Joseph Lawton, ophthalmology
Smith, Nathan Lewis, III, enzymology
So, Antero Go, internal medicine, immunochemistry
Socolar, Sidney Joseph, biophysics, cell physiology
Soldo, Anthony Thomas, biochemistry
Somani, Pitambar, pharmacology
Sommer, Leonard Samuel, cardiology
Soto, Aida R., organic chemistry
Southam, John Ralph, marine geology
Spalding, Donald Hood, plant pathology
Staiger, Jon Crawford, biological oceanography, ichthyology
Starr, Robert Brewster, oceanography
Stave, Uwe, pediatrics, preventive medicine
Stechschulte, Agnes Louise, biology,
Stein, Abraham Morton, biochemistry
Steinberg, John Christian, underwater acoustics
Sterner, John, physics
Stewart, Harris Bates, Jr., oceanography
Streitfeld, Murray Mark, microbiology, chemotherapy
Stubblefield, Bonnie McGregor, microbiology
Sussex, James Neil, psychiatry
Swenson, Orvar, surgery
Swift, Donald J P, sedimentology
Tate, Charles Frank, Jr., medicine
Taylor, Barrie Frederick, marine microbiology
Templer, David Allen, polymer chemistry
Teng, Nelson N H, cell biology,
Tershakovec, George Andrew, biochemistry
Thacker, William Carlisle, fluid dynamics
Thomas, Lowell Phillip, biological oceanography
Thomas, Vera, chemistry, toxicology
Thorhaug, Anitra L., biophysics, algology
Tocci, Paul M, biochemistry
Tomono, James R., nuclear physics
Tucker, Gail Susan, experimental embryology
Van Landingham, John W, environmental chemistry
Van Leer, John Cloud, physical oceanography, ocean engineering
Ventura, Arnoldo K., microbiology, virology
Villemure, M Paul James, mathematics
Voigt, Walter, biochemistry, endocrinology

Voss, Gilbert Lincoln, biological oceanography, systematic zoology
Wanless, Harold Rogers, sedimentology, marine geology
Warren, Richard Joseph, human genetics, cytogenetics
Warren, Robert Holmes, cell biology
Weber, William Adolph, microbiology, physiology
Werner, Rudolf, chemistry
Wheelock, Mark Carroll, pathology
Whelan, William Joseph, biochemistry
Whitney, Philip Lawrence, biochemistry
Williams, Francis, fisheries, biological oceanography
Williams, Willie Elbert, mathematics, statistics
Williamson, Donald Elwin, instrumentation, biomedical engineering
Wilson, David Louis, neurosciences, molecular biology
Wisby, Warren Jensen, fishery biology
Wise, Gary E, cell biology
Woessner, Jacob Frederick, Jr, biochemistry
Wooster, Warren Scriver, oceanography
Wu, Ming-Chi, biochemistry
Yakaitis-Surbis, Albina Ann, anatomy
Yang, Won-Tack, marine biology
Yeh, Billy Kuo-Jiun, cardiovascular diseases, clinical pharmacology
Yunis, Adel A, internal medicine
Zaharis, John Louis, medical entomology
Zarco, Romeo Morales, immunochemistry, public health
Zeppa, Robert, thoracic surgery
Ziboh, Vincent Azubike, biochemistry
Zinner, Doran David, dentistry
Zubrod, Charles Gordon, cancer microbiology
Zucker, Robert Martin, biophysics, biology

MIAMI BEACH
Bennett, Harry, industrial chemistry
Boyle, Edwin, medicine
Cassel, Hans Maurice, surface chemistry
Finn, Ronald Dennet, radiochemistry
Ginsburg, Robert Nathan, sedimentology
Grossman, Milton S, pediatric endocrinology
Halley, Robert Bruce, sedimentology
Hupf, Homer Benjamin, nuclear medicine
Martinez, Luis Osvaldo, radiology
Poiley, Samuel Milton, laboratory animal breeding
Rywirn, Arkadi Michael, medicine, pathology
Sackner, Marvin Arthur, pulmonary physiology
Samet, Philip, internal medicine, cardiology
Schenberg, Samuel, science education, chemical engineering
Viamonte, Manuel, Jr, radiology

NAPLES
Burgess, Hovey Mann, food science, nutrition
Chandler, David Culbertson, zoology
Elliott, Alfred Marlyn, zoology
Kahl, Marvin Philip, zoology
Wilde, Walter Samuel, medical physiology

NEW SMYRNA BEACH
McNary, Robert Reed, biochemistry

NICEVILLE
Bundy, Roy Elton, physiology

NORTH MIAMI
Dougherty, Patrick Henry, chemistry

NORTH MIAMI BEACH
Lavin, George Israel, chemistry
Liberson, Wladimir Theodore, neurophysiology
Rubin, Seymour Walter, genitourinary surgery

NORTH PALM BEACH
Miale, Joseph Peter, pharmaceutical chemistry, organic chemistry

OCALA
McMillen, Warren Newton, animal nutrition

OLUSTEE
Fatzinger, Carl Warren, forest entomology
Franklin, Edward Carlyle, forest genetics
Merkel, Edward Paul, forest entomology
Roberts, Donald Ray, plant physiology, weed science
Squillace, Anthony Eugene, plant genetics

ONA
Chapman, Herbert L, Jr, animal nutrition
Dantzman, Charles L, soil chemistry, soil fertility
Mislevy, Paul, agronomy

ORANGE PARK
Webb, Robert Lee, organic chemistry

ORLANDO
Baker, Graeme Levo, biochemistry
Bobber, Robert John, acoustics
Bolemon, Jay S, physics
Bolte, John R, physics
Brennan, John Joseph, chemistry
Chamelin, Isidor Marie, chemistry
Childs, James Fielding Lewis, plant pathology
Clausen, Chris Anthony, inorganic chemistry
Cooper, William Cecil, plant physiology
Cunningham, Glenn N, biochemistry, nutrition
DeNavarre, Maison Gabriel, cosmetic chemistry
Derderian, George, optics, lasers
Dutton, Arthur Morlan, applied statistics
Ehrhart, Llewellyn McDowell, mammalogy
Ellis, Leslie Lee, Jr, zoology
Falconer, David Ross, computer science
Frazier, Stephen Earl, inorganic chemistry
Garnsey, Stephen Michael, plant pathology, plant virology
Getting, Vlado Andrew, public health
Gheorghiu, Paul, physics, instrumentation
Goldstein, Ernst Moritz, organic chemistry
Grimm, Gordon Ralph, plant pathology
Groves, Ivor Durham, Jr, underwater acoustics
Hales, Everett Burton, physics
Harmon, G Lamar, nuclear physics
Hatton, Thurman Timbrook, Jr, horticulture
Hearn, Charles Jackson, plant breeding
Henderson, Billy Joe, nuclear physics
Henriquez, Theodore Aurelio, acoustics
Hertel, George Robert, physical chemistry
Hobbs, Kenneth Burkett, academic administration, research administration
Idoux, John Paul, bio-organic chemistry, organic chemistry
Ingwalson, Raymond Wesley, industrial organic chemistry
Jones, Roy Carl, Jr, mathematics
Katzin, Joel Carl, science education, radio astronomy
Koevenig, James L, botany, biology
Kuhn, David Truman, genetics
Kujawa, Frank B, geochemistry
Landrum, Bobby L, physics, mathematics
Mann, David Jacob, organic chemistry
Mattson, Guy C, organic chemistry
McGee, William Walter, analytical chemistry
Miller, Harvey Alfred, bryology
Murchison, Thomas Edgar, veterinary pathology
Nemec, Stanley, plant pathology
Norman, Edward, molecular physics, microwave physics
Ostle, Dean Bernard, statistics
Pettofrezzo, Anthony J, mathematics
Pfeiffer, Carroll Athey, anatomy
Rasmussen, Gordon Keith, plant physiology
Rautenstrauch, Carl Peter, applied mathematics
Rhein, Walter Joseph, physics
Risacher, Robert Louis, toxicology, clinical chemistry
Rogers, Peter H, physics
Rosendahl, Gottfried R, optics

Sabin, Gerald Abrodos, underwater acoustics
Schaefer, Arthur Edward, organic chemistry
Segall, Raphael Herman, plant pathology
Smith, Paul Frederick, physiology
Smoot, John Jones, plant pathology
Snelson, Franklin F, Jr, ichthyology
Somerville, Paul Noble, statistics
Stout, Isaac Jack, ecology
Sweeney, Michael Joseph, immunology, immunochemistry
Sweet, Haven C, plant physiology
Taylor, Michael Dee, mathematics
Taylor, Walter Kingsley, ornithology, vertebrate ecology
Timme, Robert William, applied physics
Tweed, Paul Basset, chemistry
Vickers, David Hyle, entomology, biochemistry
White, Roseann Spicola, biochemistry, microbiology
Whittier, Henry O, botany, bryology
Wodzinski, Rudy Joseph, microbial biochemistry
Yelenosky, George, plant physiology

ORMOND BEACH
Coke, Chauncey Eugene, polymer chemistry

PALATKA
Frank, Stanley, mathematics education

PALM BAY
Montgomery, John Richard, speech pathology

PALM BEACH
Bush, James, oceanography
Heggie, Robert, organic chemistry

PANAMA CITY
Allen, Kenneth Richard, statistical mechanics
Breeding, J Ernest, Jr, physical oceanography
Carroll, Paul Joseph, low temperature physics
Clark, Ronald Hershel, electromagnetics, low temperature physics
Clements, Burie Webster, medical entomology
Crecelius, Samuel Brown, biochemistry
Galloway, Richard Thomas, physics
Hogge, Ernest, physical chemistry
McBride, Joseph James, Jr, organic chemistry
McLeroy, Edward Glenn, physics
Nakamura, Eugene Leroy, biological oceanography, fish biology
Olson, Franklyn C W, physical oceanography
Rathburn, Carlisle Baxter, Jr, medical physics
Richards, Charles Marvin, military systems
Rogers, Andrew Jackson, medical entomology
Russell, James, polymer chemistry, physical chemistry
Vail, Edwin George, medical physiology, bioengineering

PATRICK AFB
Dryden, Warren Arnold, computer sciences
Leies, Gerard M, nuclear physics
Morrow, Richard Joseph, nuclear chemistry, physical chemistry
O'Connor, John Joseph, physics

PENSACOLA
Bach, Hartwig C, polymer chemistry
Ballman, Richard Lea, rheology
Baylis, John Robert, Jr, genetics
Beindorff, Arthur Baker, polymer chemistry
Beischer, Dietrich Eberhard, physical chemistry
Birdwhistell, Ralph Kenton, physical inorganic chemistry
Black, William Bruce, organic chemistry, polymer chemistry
Chaet, Alfred Bernard, physiology
Chang, Clifford Wah Jun, organic chemistry
Chapman, Richard David, organic chemistry
Doerr, Arthur Harry, geography of Asia, academic administration

Dunbar, Richard Alan, polymer chemistry
Edwards, Palmer Lowell, physics
Everage, Archie Edward, Jr, rheology
Foster, Virginia, phytopathology
Graybiel, Ashton, medicine
Guedry, Frederick Ernest, psychophysiology
Gurst, Jerome E, organic chemistry
Havard, Jesse Boyd, micrometeorology
Heidner, Robert Hubbard, analytical chemistry
Hieserman, Clarence Edward, chemistry, organic chemistry
Holmer, Donald A, organic chemistry
Hopkins, Thomas Sterling, marine biology
Jenkins, Lloyd Theodore, biological chemistry, organic chemistry
Johnson, Robert, chemistry
Kerr, John Polk, zoology, aquatic biology
Knepton, James Cannie, Jr, physiology
Leonard, Reid Hayward, chemistry
Lundy, Talmage E, biology
Magee, John Robert, textile chemistry
Moshiri, Gerald Alexander, limnology, physiological ecology
Novosad, Robert S, mathematics
Rao, Krothapalli Ranga, comparative physiology, comparative endocrinology
Reid, Roger Delbert, bacteriology
Ridgway, James Stratman, polymer chemistry
Riehm, John P, biochemistry
Royals, Edwin Earl, organic chemistry
Saunders, James Henry, polymer chemistry, organic chemistry
Schaefer, Hermann Joseph, biophysics
Silver, Frank Morris, polymer chemistry, organic chemistry
Simpson, Paul Gravis, physical chemistry
Slade, Philip Earl, Jr, polymer chemistry
Smith, Richard Carper, optics
Spiegelhalter, Roland Robert, analytical chemistry
Squier, Donald Platte, mathematics
Tolbert, Tommy Lyle, organic chemistry, polymer chemistry
Ucci, Pompelio Angelo, physical chemistry, mathematics
Wilson, Glenn Rhodes, organic chemistry, biochemistry
Wu, Wen-Li, polymer physics
Zaukelies, David Aaron, physical chemistry

PIERCE
Haseman, Joseph Fish, metallurgy

PLACIDA
Springer, Stewart, taxonomy, vertebrate ecology

PLANT CITY
Haskett, William Courtney, plant pathology

POMPANO BEACH
Gleyzal, Andre, mathematics

PONTE VEDRA BEACH
Hall, Lawrence Babcock, public health, engineering

PORT CHARLOTTE
Bigelow, Maurice Hubbard, marine biology
Ingraham, Raymond Clifford, physiology, chemistry

PORT EVERGLADES
Hirshman, Julius, geophysics, oceanography

PORT ST LUCIE
Clark, John Francis Bullock, clinical chemistry, analytical chemistry

PUNTA GORDA
Moyle, Clarence Llewellyn, organic chemistry
Vassel, Bruno, biochemistry

QUINCY
Baker, Frank Sloan, Jr, animal husbandry, animal nutrition
Barnett, Ronald David, agronomy, plant genetics
Chapman, Willis Harleston, agronomy

Greene, Gerald L., entomology
Jilek, Anthony Francis, animal science
Rhoads, Frederick Milton, soil chemistry, soil physics
Sanden, Gerald Ernest, plant pathology
Stanley, Robert Lee, Jr., agronomy

SAFETY HARBOR
Gubner, Richard S., internal medicine, cardiology

ST AUGUSTINE
Caldwell, David Keller, vertebrate biology
Caldwell, Melba Carstarphen, vertebrate zoology, communications science
Dobrovsky, Todor Manoloff, entomology
Gurin, Samuel, biochemistry

ST AUGUSTINE SHORES
Stevenson, Heber John Richards, biophysics, science administration

ST LEO
Adisesh, Setty Ravanappa, physical chemistry, inorganic chemistry
Peterson, Robert Hampton, organic chemistry

ST PETERSBURG
Alexander, James E., oceanography
Baird, Ronald C., ichthyology, marine ecology
Beard, Joseph Willis, surgery
Bergs, Victor Visvaldis, virology
Bowen, Cornelius Monroe, public health administration
Byrd, Isaac Burlin, fisheries management, marine biology
Carder, Kendall L., physical oceanography
Ciskowski, Joseph M., organic chemistry
Dillard, Beverly Mincey, industrial chemistry
Doyle, Larry James, marine geology
Farrow, Wendall Moore, microbiology
Ferguson, John Carruthers, invertebrate zoology, physiology
Ferguson, Philip Rex, organic chemistry
Foster, Irving Gordon, mechanics
Frankel, Jack William, virology
Groupe, Vincent, virology
Hannom, William McDowell, epidemiology, virology
Hatala, Robert John, physical chemistry
Holloway, Dennis Michael, surface physics
Houts, Garnette Edwin, industrial chemistry
Howell, John Foss, electronics, nuclear physics
Humm, Harold Judson, marine biology
Hurst, Thomas Felix, environmental sciences
Ivey, Marvin, physical geology
Jefferson, Carol Annette, plant ecology
Julien, Hiram Paul, physical chemistry
Kelley, John Ernest, mathematics
Kiley, Leo Austin, nuclear chemistry
Kreider, Henry Royer, physical chemistry, organic chemistry
Lofquist, George W., mathematics
Lowstuter, William Robert, organic chemistry
Maddox, Billy Hoyte, mathematics
Manheim, Frank T., geochemistry
Marriott, Henry Joseph Llewellyn, cardiology
Meacham, Robert Colegrove, mathematics
Miller, Hillard Craig, applied physics
Mitchen, Joel Ramon, immunology, cancer
Moss, Ernest Kent, physical organic chemistry, polymer chemistry
Munk, Miner Nelson, applied physics
Neithamer, Richard Walter, inorganic chemistry
Odland, Russell Kent, analytical chemistry
Pyle, Thomas Edward, marine geology
Reid, George Kell, aquatic ecology
Rhodes, Richard Ayer, II, physics
Rinkel, Murice O., oceanography
Roess, William B., genetics, molecular biology
Schultz, Everett Hoyle, Jr., medicine, radiology
Scribner, Leonard, chemistry
Seibert, Florence Barbara, biochemistry
Serfass, Earl James, analytical chemistry
Smith, Mary Elizabeth, laboratory animal medicine, oncology

Stevenson, William Henry, fisheries management
Tsiapalis, Chris Milton, biochemistry
Tyler, George William, mathematics

ST PETERSBURG BEACH
Ugarte, Eduardo, biochemistry, microbiology

SANFORD
Darby, John Feaster, plant pathology
Forbes, Richard Brainard, soil fertility, vegetable crops
Rhoades, Harlan Leon, plant nematology
Scudder, Walter Tredwell, weed science

SANIBEL
Gaffron, Hans, plant physiology, photobiology
Klein, Myron William, physics

SARASOTA
Bralow, S Philip, gastroenterology, internal medicine
Brown, Henry, chemistry
Burr, Peter Frederick, genetics
Corbin, Kendall Brooks, neurology, medical administration
Dunbrook, Raymond Frederick, polymer chemistry
Gilbert, Perry Webster, marine biology, vertebrate morphology
Hoagland, Alan D., geology
Kahlenberg, Eilhard Nash, organic chemistry
Kazaks, Peter Alexander, theoretical nuclear physics
Kothe, Herbert John, biochemistry
Laudani, Hamilton, economic entomology
Marlowe, James Irvin, marine geology
McCaffrey, Joseph Clifford, microbiology
Morrill, John Barstow, Jr., developmental biology
Riflin, Erik, pathobiology
Sayles, Everett Duane, zoology
Smith, Earl Westley, organic chemistry
Smith, William K., mathematics
Starcher, George William, mathematics
Vedamuthu, Ebenezer Rajkumar, food science, food technology
Weber, Alfred Herman, nuclear physics, space physics

SATELLITE BEACH
Bush, Norman, statistics

SEMINOLE
Terry, Daniel Hetfield, organic chemistry

SOUTH MIAMI
Tator, Benjamin Almon, geology

SOUTH PASADENA
Duncan, James Francis, physics

STUART
Cowan, Frederick Pierce, health physics
Croom, Herman Lee, science administration, meteorology
Ference, Michael, Jr., physics

SUGARLOAF SHORES
Walton, Charles William, chemistry

SUN CITY CENTER
Neuberger, Hans Hermann, meteorology
Peyton, Floyd Avery, dentistry

TALLAHASSEE
Albright, John Rupp, physics
Anderson, Loran C., botany
Baker, Theodore Paul, computer science
Banghart, Frank W., biostatistics
Barcilon, Albert I., dynamic meteorology
Basu, Debabrata, statistics
Bayfield, Edward Geoffrey, cereal chemistry
Beidler, Lloyd M., physiology, biophysics
Bellenoi, Steven F., mathematical analysis
Bennison, Bertrand Earl, medicine
Boggs, Nathaniel, Jr., protozoology, cytology
Bradley, Ralph Allan, statistics
Brown, Dawn LaRue, neurosciences

Brueckheimer, William Rogers, resource geography
Brundage, Robert Earl, computer science
Bryant, John Logan, mathematics
Buie, Bennett Frank, economic geology
Burkman, Ernest, science education
Calder, John Archer, chemical oceanography, organic geochemistry
Campbell, Clarence L, Jr., veterinary medicine
Chan, Chiu Yeung, mathematical analysis, applied mathematics
Chapman, Kenneth Reginald, experimental nuclear physics
Chase, Harrison Vernon, geography
Choppin, Gregory Robert, inorganic chemistry, nuclear chemistry
Clark, Ronald Jene, inorganic chemistry
Clewell, Andre F., systematic botany
Collier, Albert Walker, marine biology, oceanography, organic geochemistry
Craig, Richard Ansel, meteorology
Davis, Robert Houser, neurophysiology
Deagan, Kathleen A., anthropology
DeBusk, Aron Gib, genetics
DeKloet, Siwo R., biochemistry, molecular genetics
Desloge, Edward Augustine, physics
DeTar, DeLos Fletcher, organic chemistry
DeVore, George Warren, geology
Dougherty, Ralph C., physical organic chemistry
Easton, Dexter Morgan, neurophysiology
Edwards, Steve, theoretical nuclear physics
Eichinger, Jack Waldo, Jr., inorganic chemistry
Elfner, Lloyd F., psychoacoustics, experimental psychology
Elias, Loretta Christine, bacterial metabolism
Elliott, Paul Russell, medical education
Erdman, Anne Marie, nutrition
Evans, Lee E., animal husbandry
Faust, Richard Deane, anthropology
Fisher, James Robert, biochemistry
Fitzgerald, Thomas James, medicinal chemistry, pharmacology
Fletcher, Neil Russel, nuclear physics
Flowers, Ralph Wills, entomology
Ford, Clinita Arnsby, nutrition
Fox, John David, nuclear physics
Freeman, Marc Edward, reproductive endocrinology
Frieden, Earl, biochemistry
Friedmann, Emerich Imre, phycology
Frye, Ozro Earle, Jr., wildlife biology
Fulton, Robert Lester, chemical physics
Garrett, Barry B., physical chemistry, inorganic chemistry
Gilmer, Robert William, Jr., algebra
Gleeson, Thomas Alexander, meteorology
Glick, Richard Edwin, physical chemistry
Godfrey, Robert Kenneth, botany
Goodner, Dwight Benjamin, pure mathematics
Greenberg, Michael John, comparative physiology
Griffith, H C, mathematics
Grindal, Bruce Theodore, cultural anthropology
Gulick, Wilson M, Jr., analytical chemistry
Hagopian, Vasken, high energy physics
Hallam, Thomas Guy, mathematics
Hanson, Morgan A., operations research, statistics
Harris, Nathlyn Dalton, food science
Harris, Robert Curtis, marine sciences
Harrold, Orville Goodwin, Jr., mathematics
Heard, William Herman, malacology
Heerema, Nickolas, mathematics
Heil, Wolfgang Heinrich, mathematics
Hendry, Julius Leo, microbiology
Hendry, Charles Walter, Jr., geology
Herndon, Roy C., environmental physics
Hermkind, William Frank, animal behavior, marine biology
Herz, Werner, organic chemistry
Hess, Seymour Lester, meteorology, planetary atmospheres
Hill, Paul Daniel, mathematics
Ho, Ting-Jui, anthropology
Hofer, Kurt Gabriel, cell biology, radiation biology
Hollander, Myles, mathematical statistics, applied statistics
Homann, Peter H., plant physiology, biochemistry
Hood, Mary Noka, bacteriology
Hsueh, Ya, oceanography
Hunt, Robert Harry, physics

Hunter, Christopher, applied mathematics
Jahoda, Gerald, information science
Johnsen, Russell Harold, radiation chemistry, academic administration
Johnson, Walter Lee, geology
Jones, James I, marine geology, micropaleontology
Jordan, Charles Lemuel, meteorology
Kalin, Michael, mathematics
Kasha, Michael, chemical physics, spectroscopy
Keirs, Russell John, analytical chemistry
Kemper, Kirby Wayne, nuclear physics
Kimel, Jacob Daniel, Jr., particle physics, theoretical physics
Kinoshita, Shin'ichi, topology
Knop, Robert Edward, physics
Kreimer, Herbert Frederick, Jr., mathematics
Krishnamurti, Ruby Ebisuzaki, physics
Krishnamurti, Tiruvalam N., meteorology
Kromhout, Robert Andrew, chemical physics
Lamba, Surendar Singh, pharmacy, pharmacognosy
Lannutti, Joseph Edward, particle physics
La Seur, Noel Edwin, meteorology
Linder, Bruno, theoretical chemistry
Leffler, John Edward, chemistry
Levitz, Hilbert, mathematics
Levy, George Charles, physical organic chemistry
Lipner, Harry Joel, reproductive endocrinology
Loper, David Eric, magnetohydrodynamics, applied mathematics
Leysieffer, Frederick Walter, mathematics
Mandelkern, Leo, polymer chemistry, biophysics
Mann, Charles Kenneth, analytical chemistry
Mariscal, Richard North, marine biology, invertebrate zoology
Marziluff, William Frank, Jr., biochemistry
McArthur, Charles Wilson, mathematics
McWilliams, Ralph David, mathematics
Means, Bruce, ecology, herpetology
Medsker, Larry Robert, experimental nuclear physics
Mellon, Edward Knox, Jr., inorganic chemistry, science education
Menzel, Margaret Young, cytogenetics
Menzel, Robert Winston, marine biology
Menzies, Robert James, oceanography
Merrill, John Raymond, solid state physics
Miles, Ernest Percy, Jr., mathematics
Montgomery, David Carey, physics, institutional research
Morris, Robert Carter, experimental solid state physics
Morse, John Wilbur, geochemistry
Mott, Joe Leonard, algebra
Moulton, Grace Charbonnet, biophysics, experimental solid state physics
Moulton, William G, solid state physics, low temperature physics
Nelson, John William, environmental physics, experimental nuclear physics
Nichols, Eugene Douglas, mathematics
Oberlin, Daniel Malcolm, mathematical analysis
O'Brien, James J., meteorology, oceanography
Ocampo-Friedmann, Roseli C., ecology, phycology
Osmond, John Kenneth, geology
Owens, Clarence Burgess, plant physiology
Paredes, James Anthony, anthropology
Pargman, David, exercise physiology, nutrition
Pates, Anne Louise, bacteriology
Paton, Robert Frederick, physics
Paton, Donald John, geography
Penrod, Kenneth Earl, physiology
Peters, William Lee, aquatic entomology
Petrovich, Fred, nuclear physics
Pfeffer, Richard Lawrence, dynamic meteorology
Phillips, William Baars, solid state physics
Philpott, Richard John, nuclear physics
Plendl, Hans Siegfried, nuclear physics
Poore, Jesse H, Jr., computer science
Proschan, Frank, statistics
Puri, Harbans Singh, stratigraphy, paleontology

Quagliano, James Vincent, inorganic chemistry
Rhodes, William Clifford, theoretical chemistry
Rill, Randolph Lynn, physical biochemistry
Robinson, Bryan Wright, neurology, psychology
Robson, Donald, nuclear physics
Roeder, Martin, physiology, biochemistry
Rushton, William A H, physiology
Safron, Sanford Alan, physical chemistry
Saltiel, Jack, organic chemistry
Schwartz, Martin Alan, organic chemistry
Seals, Rupert Grant, dairy industry
Serfling, Robert Joseph, mathematical statistics
Sethuraman, Jayaram, mathematical statistics
Shah, Devendra Hargovindas, physical pharmacy
Sheline, Raymond Kay, nuclear chemistry
Shelton, Wilford Neil, electron physics
Shimi, Ismail Nabih, statistics, operations research
Short, Robert Brown, parasitology
Simberloff, Daniel S., ecology, mathematical biology
Skofronick, James Gust, chemical physics
Smith, Hale Gilliam, archaeology, cultural anthropology
Snover, James Edward, mathematics
Sparkman, Marjorie Frances, physiology
Speisman, Gerald, theoretical physics
Stephens, Jesse Jerald, physical meteorology
Stevenson, Henry Miller, ornithology
Stiles, Wilbur J, mathematics
Stuart, David W, meteorology
Sturges, Wilton, III, physical oceanography
Stuy, Johan Harrie, bacteriology
Tam, Christopher K W, acoustics, applied mathematics
Tanner, William Francis, Jr, geology
Taylor, James Herbert, biology
Tolstoy, Ivan, geophysics
Toulmin, Lyman Dorgan, Jr, paleontology, geology
Tterlikkis, Lambros, physics
Tucker, Don, biophysics
Turner, Ralph Waldo, physical chemistry, mathematics
Vanderhill, Burke Gordon, geography
Van Middelem, Charles Henry, biochemistry
Vickers, Thomas J, analytical chemistry
Vourvopoulos, George, nuclear physics
Wade, Thomas Leonard, Jr, mathematics
Wakefield, Lucille Marion, nutrition
Walborsky, Harry M, organic chemistry
Walker, Charles A, neuropharmacology
Walker, Courtney Emery, agriculture
Walters, Cora Etta, physiology
Wang, Yung-Li, solid state physics
Weber, Neal Albert, entomology, ecology
Weisbord, Norman Edward, geology, invertebrate paleontology
White, David Cleaveland, biochemistry
Whitney, Eleanor Noss, nutrition
Williams, Theodore P, biophysics
Winchester, John Widmer, physical chemistry
Winsberg, Morton Daniel, geography
Winters, Stephen Samuel, geology
Wise, Sherwood Willing, Jr, geology, paleontology
Wright, Thomas Perrin, Jr, topology
Yerger, Ralph William, systematic ichthyology
Young, Eutiquio Chua, mathematical analysis
Young, Jay Alfred, physical chemistry

TAMPA
Adair, Winston Lee, Jr, biochemistry
Akins, Daniel L, physical chemistry
Ashford, Theodore Askounes, organic chemistry
Aubel, Joseph Lee, physics
Azar, Henry A, pathology
Baker, Carleton Harold, nutrition, physiology
Barness, Lewis Abraham, pediatrics
Barrett, O'Neill, Jr, hematology, oncology
Behnke, Roy Herbert, internal medicine
Bermes, Boris John, hydrogeology
Betz, John Vianney, bacteriology, virology
Binford, Jesse Stone, Jr, physical chemistry
Bloch, Sylvan C, physics

Bloom, Sherman, experimental pathology, cell biology
Boler, Reginald Keith, anatomy, experimental pathology
Boyce, Henry Worth, Jr, internal medicine, gastroenterology
Braman, Robert Steven, analytical chemistry, instrumentation
Brashears, Maurice Lyman, geology
Briggs, John Carmon, ichthyology, zoogeography
Britton, Jack Rolf, mathematics
Brooker, Hampton Ralph, nuclear magnetic resonance
Brown, Larry Nelson, mammalogy, ecology
Bukantz, Samuel Charles, clinical medicine
Burch, Derek George, systematic botany
Clapp, Roger Williams, Jr, physics
Clark, William Edwin, mathematics
Cleaver, Frank L, mathematics
Connar, Richard Grigsby, thoracic surgery, cardiovascular surgery
Cory, Joseph G, biochemistry
Cowell, Bruce Craig, limnology
Davis, Darrell Lawrence, physiology
Davis, Jefferson Clark, Jr, physical chemistry
Davis, Richard Albert, Jr, geology
Dawes, Clinton John, cytology, phycology
Deeds, Joseph Bird, mathematics
Del Regato, Juan A, radiology
Dudley, Frank Mayo, science education
Dwornik, Julian Jonathan, anatomy
Eichhorn-von Wurmb, Heinrich Karl, astronomy, statistical analysis
Ellison, Marlon L, botany
Engebretson, Gordon Roy, physical chemistry, medical administration
Essig, Frederick Burt, systematic botany
Fallon, Frederick Walter, astronomy
Fernandez, Jack Eugene, organic chemistry
Fishel, Charles Wesley, medical microbiology
Flynn, Robert W, plasma physics
Forman, Guy, physics
Fredricks, David Alan, mathematics
Friedl, Frank Edward, invertebrate physiology, parasitology
Fuson, Robert Henderson, geography
Geraghty, James Joseph, geology
Gilmore, Robert, physics
Goldman, Allan Larry, pulmonary diseases
Goodman, Adolph Winkler, mathematics
Grange, Roger T, Jr, anthropology, archaeology
Gude, Richard Hunter, zoology, animal behavior
Hager, William Ward, numerical analysis
Halder, Narayan Chandra, solid state physics
Halkias, Demetrios, medical microbiology
Hartley, Miles C, mathematics
Hartmann, Robert Carl, hematology
Hessinger, David Alwyn, cell biology, toxinology
Hickman, Jack Walter, internal medicine
Higgins, James Jacob, statistics
Holt, Thomas Manning, neurochemistry
Huang, Wen Hsing, clay mineralogy, geochemistry
Hunter, James Hardin, Jr, astronomy, hydrodynamics
Hutcheson, James Byron, pathology
Jackson, George Frederick, III, physical chemistry, analytical chemistry
Jackson, George Richard, chemistry
Johnston, Milton Dwynell, Jr, physical chemistry, molecular spectroscopy
Jones, William Denver, plasma physics, solar physics
Jurch, George Richard, Jr, physical organic chemistry
Kendall, Harry White, physics
Kessler, Evelyn Seinfeld, anthropology
King, Charles Everett, population biology
Kory, Ross Conklin, medicine, physiology
Krivanek, Jerome Oldrich, zoology
Kruschwitz, Walter Hillis, physics
Krzanowski, Joseph John, Jr, pharmacology, physiology
Kushner, Gilbert, applied anthropology
Lawrence, John M, physiology
Lee, Anthony, plasma physics
Liang, Joseph Jen-Yin, mathematics
Lin, Shwu-Yeng Tzen, topology
Lin, You-Feng, topology
Linton, Joe R, zoology
Lockey, Richard Funk, allergy, immunology
Long, Robert William, Jr, systematic botany

Manougian, Manoug N, mathematics
Martin, Dean Frederick, inorganic chemistry
Maybury, Paul Calvin, physical chemistry
McClung, Norvel Malcolm, microbiology
McDiarmid, Roy Wallace, biology, vertebrate ecology
Mitchell, Richard Warren, physics
Monaloy, Stephen Emanuel, genetics
Mukherjea, Arunava, mathematics
Narske, Richard Martin, organic chemistry
Nelson, Gideon Edmund, Jr, biology
Nicholosi, Gregory Ralph, physiology
Niklas, Wilfrid F, physics
Noer, Rudolf Juul, surgery
Nolan, Michael Francis, neuroanatomy
O'Donnell, Edward, geology
Oleson, Norman Lee, physics
Olsen, Eugene Donald, analytical chemistry, clinical chemistry
Owen, Terence Cunliffe, organic chemistry, radiation chemistry
Paradise, Lois Jean, medical microbiology
Polson, James Bernard, pharmacology
Prockop, Leon D, neurology
Raber, Douglas John, chemistry
Ramsey, Maynard, III, biomedical engineering
Rath, Maurice Monroe, pharmacology, physiology
Ratti, Joginder Singh, analytical mathematics
Ray, James Davis, Jr, botany
Riggs, Carl Daniel, zoology
Robinson, Gerald Garland, zoology, endocrinology
Root, Allen William, pediatric endocrinology
Rose, Donald Clayton, mathematics
Rothwell, Stuart Clark, geography
Saff, Edward Barry, mathematical analysis
Schimmel, Steven David, biochemistry
Schmidt, Carl Frederic, physiology, pharmacology
Schmidt, Paul Joseph, medicine, clinical pathology
Schneller, Stewart Wright, organic chemistry
Schnitzlein, Harold Norman, neuroanatomy
Schreiber, Ralph Walter, ornithology, marine ecology
Sellers, John William, organic chemistry, chemical engineering
Sherman, Roger Talbot, surgery
Shiloh, Ailon, cultural anthropology
Sidransky, Herschel, pathology
Silver, Warren Seymour, microbiology
Simon, Joseph Leslie, zoology
Smeach, Stephen Charles, biomathematics
Smith, Donn Leroy, pharmacology
Smith, Edward Byron, clinical pathology, hematology
Smith, Haywood Clark, Jr, astronomy
Snider, Arthur David, mathematics
Solomons, Thomas William Graham, organic chemistry
Solomonson, Larry Paul, biochemistry
Stanko, Joseph Anthony, inorganic chemistry
Stevens, Brian, photochemistry
Stevenson, Ralph Girard, Jr, mineralogy, petrology
Szentirvanyi, Andor, pharmacology, allergy
Taft, William H, marine geology
Tipton, Henry C, nutrition, biochemistry
Truxillo, Stanton George, physics
Tserpes, Nicolas A, mathematics
Tsokos, Chris Peter, applied mathematics, statistics
Tsokos, Janice Oseth, biochemistry
Tyler, David Bernard, physiology, pharmacology
Upchurch, Sam Bayliss, environmental geology, sedimentology
Webb, Sydney James, biophysics, microbiology
Wenzinger, George Robert, chemistry
Whitaker, Robert Dallas, inorganic chemistry
Wiener, Curtis Wakefield, physical chemistry
Wiggins, Jay Ross, pharmacology
Williams, Carol Ann, astronomy, celestial mechanics
Williams, James Raymond, anthropology, archaeology
Wilson, Robert E, astrophysics, x-ray astronomy
Wilson, William Edward, III, hydrogeology
Wolfe, Alvin William, anthropology, social anthropology
Woolfenden, Glen Everett, ornithology

Worrell, Jay H, physical inorganic chemistry
Zerla, Fredric James, mathematics

TIERRA VERDE
Wilbur, Donald Aldrich, physics

VALPARAISO
Head, Ronald Alan, chemistry

VENICE
Seekircher, Richard, petroleum chemistry
Thornton, George Daniel, soils

VERO BEACH
Bidlingmayer, William Lester, entomology
Borovsky, Dov, chemistry
Campbell, William Robert, entomology
Counselman, C J, entomology, ichthyology
Davenport, O Malcolm, forestry
Dow, Richard Phelps, entomology
Frank, John Howard, insect ecology
Hensley, Jerry Ray, plant physiology
Kale, Herbert William II, ornithology, conservation
Kitzmiller, James Blaine, genetics, medical entomology
Lane, Charles Edward, zoology
Linley, John Roger, entomology, insect physiology
Nayar, Jai Krishen, insect physiology, medical entomology
O'Meara, George Francis, genetics
Pelletier, Eugene Neil, plant pathology
Provost, Maurice Wilfred, medical entomology
Simons, John Norton, entomology
Skye, George Eri, II, biochemistry
Van Handel, Emile, organic chemistry
Woofter, Harvey Darrell, weed science

VIRGINIA KEY
Hansen, Donald Vernon, physical oceanography

WARRINGTON
Bond, Bernard Batson, chemistry

WEST PALM BEACH
Alter, Abraham, chemistry
Cohen, Martin Joseph, electronic physics
Fox, Lewis, dentistry
Gibson, Henry Clay, Jr, atomic physics
Hudson, Alice Peterson, surface chemistry
Lisanke, Robert John, Sr, organic polymer chemistry
McCormick, Clyde Truman, applied mathematics
Pinsley, Edward Allan, lasers, engineering management
Smith, Riley Seymour, Jr, marine geology
Woodward, Fred Erskine, surface chemistry, organic polymer chemistry

WHITE SPRINGS
Hirko, Ronald John, physical chemistry

WINTER HAVEN
Berry, Robert Eddy, biochemistry
Cox, Everett Franklin, engineering physics
Hayward, Frederick Warren, food chemistry
Landry, Martha Moseley, microbiology, genetics
McKinnis, Ronald Bishop, chemistry, food technology
Nordby, Harold Edwin, organic chemistry, biochemistry
Shaw, Philip Eugene, organic chemistry
Swift, Lyle James, agricultural chemistry
Westbrook, George Franklin, food chemistry

WINTER PARK
Bowen, Charles Verne, chemistry
Cochran, George Thomas, inorganic chemistry, analytical chemistry
Fluno, John Arthur, entomology
Hellwege, Herbert Elmore, inorganic chemistry, clinical chemistry
Hock, Donald Charles, physics
Lytle, Ernest James, Jr, mathematical statistics
Mulson, Joseph F, electron physics

Ridgway, Robert Worrell, organic chemistry
Riederer-Henderson, Mary Ann, microbial physiology
Robinson, Aaron Zed, Jr., electronics, acoustics
Ross, John Stoner, atomic spectroscopy
Roth, Raymond Edward, statistics
Sandstrom, Carl John, zoology
Vestal, Paul Anthony, botany
Ziegler, George Elliott, physics

GEORGIA

ALBANY
Bates, Harold Brennan, Jr., zoology
Black, Billy C. II, biochemistry,
Green, Edwin Alfred, experimental zoology
Husain, Ansar, zoology, parasitology
Pandey, Surendra, Nath, experimental solid state physics
Steele, Jack, inorganic chemistry, physical chemistry

AMERICUS
Arden, Daniel Douglas, exploration geology
Cofer, Harland E., Jr., economic geology, mineralogy
Counts, Wayne Boyd, organic chemistry
Henderson, Arnold Richard, geology
Mathews, Walter Kelly, organic chemistry, biochemistry
McKinney, Max Terral, mathematics
Powders, Vernon Neil, parasitology
Rhyne, Claude Little, Jr., genetics, plant breeding
Tietjen, William Leighton, entomology, zoology

ATHENS
Abbe, Winfield Jonathan, theoretical high energy physics
Adomian, George, applied mathematics, systems theory
Agosin, Moises, biochemistry, parasitology
Ah, Hyong-Sun, veterinary parasitology, acarology
Allard, Giles Oliver, geology, economic geology
Allinger, Norman Louis, organic chemistry
Anderson, David Prewitt, veterinary microbiology
Ansel, Howard Carl, pharmacy
Ashley, Doyle Allen, agronomy, applied physiology
Atyeo, Warren Thomas, entomology
Ayres, John Clifton, food science
Bailey, George William, water pollution
Bailey, Wilfred Charles, anthropology
Baker, Francis Todd, nuclear physics
Baldwin, Jack Norman, microbiology
Baldwin, Winfield Morgan, Jr., organic chemistry
Ball, Billy Joe, mathematics
Bargmann, Rolf Erwin, statistics
Beard, Charles Walter, veterinary virology
Beaty, Elvis Roy, agronomy
Bell, John Thomas, Jr., veterinary anatomy
Bennett, Frederick William, bacteriology
Benyshek, Larry L., animal breeding
Bergeaux, Phillip James, agronomy, soil chemistry
Bernstein, Irwin Samuel, primatology
Berry, Charles Richard, plant pathology
Black, Clanton Candler, Jr., biochemistry
Blair, Charles Barkley, Jr., cytology, comparative anatomy
Blanton, Charles DeWitt, Jr., organic chemistry, medicinal chemistry
Blomquist, Richard Frederick, organic chemistry
Blum, Murray Sheldon, entomology, physiology
Booth, Nicholas Henry, physiology, pharmacology
Bouldin, Richard Hindman, mathematics
Bowen, John Metcalf, pharmacology, biophysics
Boyd, George Edward, chemistry
Boyd, Louis Jefferson, reproductive physiology

Brady, Ullman Eugene, Jr., insect toxicology, physiology
Brahana, Thomas Roy, mathematics
Brandau, Betty Lee, inorganic chemistry, nuclear chemistry
Brewer, John Michael, biochemistry
Britton, Walter Martin, poultry nutrition
Brown, Acton Richard, plant breeding
Brown, Ronald Harold, agronomy, plant physiology
Bryan, John Henry Donald, cell biology
Burdick, Donald George, biochemistry
Burrill, Robert Meredith, biogeography,
Butts, David, biology, science education
Cadwallader, Donald Elton, pharmacy,
Campbell, William Andrew, forest pathology
Canerday, Thomas Donald, invertebrate pathology, economic entomology
Cantrell, James Cecil, topology
Carlson, Jon Frederick, algebra
Carlton, Bruce Charles, genetics
Carmon, James Lavern, statistics,
Carpenter, Robert Halstead, geology
Carreira, Lionel Andrade, physical chemistry, molecular spectroscopy
Carver, Robert E., sedimentary petrology, natural resources
Case, Mary Elizabeth, microbial genetics
Cassen, Thomas Joseph, physical chemistry
Caster, William Oviatt, nutrition
Chah, Cheong Choo, animal nutrition
Chakrabarty, Rameswar Prasad, statistics
Champney, William Scott, biochemical genetics
Chapman, Willie Lasco, Jr., comparative pathology
Cheatum, Evelyn Leonard, wildlife pathology
Cheng, George Chiwo, biomedical engineering, electrical engineering
Chin, Edward, biology
Chou, Tsu-Teh, elementary particle physics, high energy physics
Clark, Douglas Napier, mathematical analysis
Clark, James Derrell, laboratory animal medicine
Clifton, Carl Moore, dairy husbandry
Clutter, Jerome Lee, biometrics, operations research
Cohen, Alonzo Clifford, Jr., statistics
Cole, Ronald Sinclair, molecular biology, radiobiology
Cooper, Charles Dewey, physics
Cormier, Milton Joseph, biochemistry
Cornelius, Larry Max, veterinary medicine
Cosgrove, William Burnham, physiology
Coulter, Dwight Bernard, veterinary physiology
Coward, Stuart Jess, developmental biology
Cox, Nelson Anthony, microbiology
Cox, Richard Harvey, nuclear magnetic resonance, organic chemistry
Crossley, DeRyee Ashton, Jr., radiation ecology
Cullison, Arthur Edison, animal nutrition
Curtis, John Russell, psychiatry
Dallmeyer, R David, geochronology
Damian, Raymond T., parasitology, immunology
Daniel, O'Dell G., animal husbandry
Davison, Frederick Corbet, veterinary pathology, physiology
De Camp, Wilson Hamilton, physical organic chemistry
Dekazos, Elias Demetrios, plant physiology, food science
De Sa, Richard John, enzymology
Devarianan, Daniel Vartan, biochemistry
DeVorsey, Louis Jr., historical geography
Dietz, Alfred, entomology, apiculture
Dobbins, Charles Nelson, Jr., veterinary medicine
Douglas, Charles Herbert, research administration
Dull, Gerald G., plant biochemistry
Duncan, Marion M. Jr., theoretical physics
Duncan, Wilbur Howard, taxonomy
Dure, Leon S. III, biochemistry
Dwinell, Lew David, plant pathology
Dyson, Peter John, forest economics
Eagon, Robert Garfield, microbiology
Edwards, Alan Kent, experimental atomic physics
Edwards, Hardy Malcolm, Jr., nutritional biochemistry

Eitenmiller, Ronald Ray, food science
Entrekin, Durward Neal, pharmacy
Evans, Burton Robert, medical entomology
Farrell, Robert Lawrence, veterinary medicine
Finco, Delmar R., veterinary medicine
Finnery, William Robert, microbial physiology, biochemistry
Fisher, Donald B., botany
Fishman, Marshall Lewis, polymer chemistry
Fitzgerald, Charles H., forest physiology, pesticide chemistry
Flatt, William Perry, animal nutrition
Fleming, Attie Anderson, plant genetics
Fletcher, Oscar Jasper, Jr., pathology
Fosgate, Olin Tracy, dairy science
Fuller, Henry Lester, poultry nutrition
Fuller, Melvin Stuart, botany, mycology
Garren, Henry Wilburn, physiology, endocrinology
Garrison, Arthur Wayne, environmental chemistry
Garst, John Fredric, physical chemistry, organic chemistry
Giardini, Armando Alfonzo, mineralogy
Giddens, Joel Edwin, soil microbiology
Giles, Norman Henry, genetics
Goetsch, Dennis Donald, veterinary physiology
Golley, Frank Benjamin, ecology
Gonzales, Serge, geology
Goss, Dixie J, biophysical chemistry
Gratzek, John B., virology
Greene, Barbara E., food science,
Greene, James T., forestry
Hale, Kirk Kermit, Jr., poultry science, food science
Hally, David Judson, anthropology, archaeology
Hamdy, Mostafa Kamal, bacteriology
Hamm, Douglas, food science
Hanlin, Richard Thomas, mycology
Hanson, William Lewis, parasitology,
Hargreaves, Leon Abraham, Jr., forestry
Harmer, Don Stuler, nuclear chemistry,
Harris, Emmett Dewitt, Jr., economic entomology
Hatchell, Glyndon Elbert, forest soils, silviculture
Hautala, Judith Ann, molecular biology
Hautala, Richard Roy, organic chemistry,
Hayes, Frank Alfred, veterinary medicine, wildlife diseases
Henderson, Charles Henry, Jr., horticulture
Hendrix, Floyd Fuller, Jr., plant pathology, soil microbiology
Henkel, John Harmon, theoretical solid state physics, exploration geophysics
Hercules, David Michael, analytical chemistry
Heric, Eugene Leroy, physical chemistry
Herrman, Henry Remley, Jr., insect morphology
Herrick, Allyn Marsh, forestry
Herz, Norman, geology
Hewlett, John David, hydrology, ecology
Hill, Richard Keith, organic chemistry
Himel, Chester Mora, biophysics,
Hollingsworth, John Gressett, mathematics
Honigberg, Irwin Leon, medicinal chemistry
Hoover, Thomas Burdett, environmental chemistry, electroanalytical chemistry
Horne, James Grady, Jr., mathematics
Howard, James Hatten, III, geochemistry, astrogeology
Howarth, Birkett, Jr., reproductive physiology
Howe, Henry Branch, Jr., genetics
Hoy, Don Roger, geography of Latin America, economic geography
Huber, Thomas Lee, physiology,
Huberty, Carl J., applied statistics, educational psychology
Huebner, Richard A., veterinary medicine, pharmacology
Huff, Gerald Boone, mathematics
Humphreys, Walter James, biological structure
Hunter, Preston Eugene, entomology, acarology
Hurst, Vernon James, geology, crystallography
Hussey, Richard Sommers, nematology
Huston, Till Monroe, animal physiology
Hutcheson, Kermit, biostatistics

Ike, Albert Francis, forest soils, academic administration
Inman, Franklin Pope, Jr., immunochemistry, biochemistry
Isaac, Robert A., analytical chemistry
Iturrian, William Ben, neuropharmacology
Jacobson, James William, molecular genetics
James, Charles William, systematic botany
Jenkins, James Hobart, wildlife management
Jensen, Leo Stanley, animal nutrition
Johannes, Robert Earl, marine ecology
Johnson, Albert Sydney, III, wildlife biology, wildlife management
Johnson, Archie Doyle, reproduction physiology, biochemistry
Johnson, Henry Douglas, pharmacology
Johnston, Francis J. physical chemistry
Johnstone, Francis Elliott, Jr., horticulture, plant breeding
Jones, J Benton, Jr., soil fertility, plant nutrition
Jones, Lois Marilyn, geochemistry
Jones, Roger Daniel Hewart, mathematical statistics, ecology
Jones, Samuel B Jr., botany
Jordan, Carl Frederick, ecology
Jordan, Cedric Roy, entomology
Junker, Bobby Ray, atomic physics
Kadis, Solomon, microbiology
Kaikofen, Ulrich Paul, parasitology
Karickhoff, Samuel Woodford, physical chemistry
Keith, Lawrence H., pollution chemistry
Kelly, Arthur Randolph, anthropology, archaeology
Kent, Harry Alvin, Jr., reproductive physiology
Kenyon, James Byron, geography
Kerr, Thomas James, microbiology
Key, Joe Lynne, plant physiology
Key, Alexander Charles, pharmacology
King, Robert Bruce, inorganic chemistry
Kleckner, Albert Louis, veterinary
Kleven, Stanley H, microbiology
Koch, George Schneider, Jr., geology, statistics
Kochert, Gary Dean, botany
Koehler, Philip Edward, food science
Koelsche, Charles L., inorganic chemistry, analytical chemistry
Kossack, Carl Frederick, applied statistics
Kuhn, Cedric W., plant pathology
Kushner, Sidney Ralph, molecular genetics, enzymology
Kutal, Charles Ronald, inorganic chemistry
Landau, David Paul, magnetism, statistical mechanics
La Rocca, Joseph Paul, medicinal chemistry
Lassiter, James William, animal nutrition
Lea, Arden Otterbein, entomology
Lee, John William, chemical physics, biophysics
Lee, Monhe Howard, theoretical physics
Leeper, George Frederick, botany
Leyden, Donald E., analytical chemistry
Lillard, Dorris Alton, food chemistry
Lindsay, David Taylor, developmental biology
Ljungdahl, Lars Gerhard, biochemistry, microbiology
Loewenstein, Morrison, dairy chemistry, nutrition
Love, William Gary, nuclear physics
Lovins, Robert E., mass spectrometry, protein biochemistry
Lowrey, Robert S., animal nutrition
Lucas, Myron Cran, biochemical genetics
Lutert, Phil Dean, microbiology
Lund, Horace Odin, entomology
Luttrell, Everett Stanley, mycology, plant pathology
Luzzi, Louis A., physical pharmacy, colloid chemistry
Marchinton, Robert Larry, wildlife ecology, ethology
Marks, Henry L., animal genetics
Marx, Donald Henry, plant pathology, soil microbiology
Matthews, Robert Wendell, entomology
Mattingly, Mary Ellen, biology
May, Jack Truett, forestry
McAlister, Robert Hardy, forest products
McCarter, States Marion, plant pathology, plant breeding
McCartney, Morley Gordon, genetics
McCleary, Stephen Hill, algebra

McCreery, Robert Atkeson, agronomy, soils
McDonald, Leslie Ernest, physiology
McGhee, Robert Barclay, parasitology
McGuire, John Murray, analytical chemistry
McKinney, Leonard Laurence, food chemistry
McRorie, Robert Anderson, biochemistry
Meentemeyer, Vernon George, climatology
Melton, Charles Estel, chemical physics
Melvin, Ernest Eugene, urban geography
Mendicino, Joseph Frank, biochemistry
Menendez, Manuel Gaspar, atomic physics, molecular physics
Mercuri, Arthur J., food microbiology, food science
Mertens, David Roy, ruminant nutrition, dairy science
Michel, Burlyn Everett, plant physiology
Miller, Thomas, plant pathology
Miller, William Jack, animal nutrition
Mills, Harry Arvin, vegetable crops
Mokler, Corwin Morris, cardiovascular physiology
Monk, Carl Douglas, plant ecology
Moore, Robert Conley, pharmacy administration, public health administration
Morris, Harold Donald, agronomy
Morrison, Wiley Herbert, III, organic chemistry
Motsinger, Ralph E., plant pathology
Mulligan, Benjamin Edward, sensory psychology
Murray, Calvin Clyde, agronomy
Nance, Jon Roland, physics
Neathery, Milton White, animal nutrition
Needham, Thomas E, Jr., pharmaceutics
Neter, John, applied statistics
Neufeld, Cornelius Herman Harry, organic chemistry
Nicholson, Harry Page, entomology
Nishie, Keica, pharmacology
Noakes, John Edward, geochemistry, oceanography
Nord, John C, forest entomology
Odum, Eugene Pleasants, ecology, ornithology
Olien, Michael David, anthropology
Oliver, John Eoff, Jr., comparative neurology
Oyler, J Mack, physiology
Pannell, Clifton Wyndham, geography
Papa, Kenneth E, genetics
Parkman, Sammie Bell, agronomy
Patel, Gordhan, molecular biology
Patten, Bernard Clarence, ecology
Patterson, Archie Edgar, forest management
Patterson, William Creigh, Jr., veterinary medicine
Payne, William Jackson, microbiology
Peacock, Leion James, physiological psychology, comparative psychology
Peck, Harry Dowd, Jr., microbial physiology, enzymology
Peifer, James J, biochemistry
Pelletier, S William, organic chemistry
Penney, David Emory, topology
Perkins, Henry Frank, soil chemistry
Pienaar, Leon Visser, forest biometry, population dynamics
Plummer, Gayther Lynn, plant ecology
Pokorny, Franklin Albert, horticulture
Pomerance, Carl, mathematics
Pomeroy, Lawrence Richards, ecology
Porter, David, botany
Powell, William Morton, plant pathology
Powers, Harry Robert, Jr, plant pathology
Powers, John Joseph, food technology
Pressey, Russell, biochemistry, plant physiology
Price, James Clarence, pharmacy
Provost, Ernest Edmund, zoology
Prunty, Merle Charles, geography
Ragland, William Lauman, III, pathology, biochemistry
Rawlings, Clarence Alvin, veterinary surgery, cardiopulmonary physiology
Reagan, James Oliver, meat sciences
Reed, John Fielding, agricultural chemistry
Reid, Willard Malcolm, parasitology, poultry science
Reines, Mervin, botany
Rice, James Thomas, wood technology
Rice, Peter Milton, applied mathematics
Rich, Mark, geology
Ritter, Hope Thomas Martin, Jr, cell biology
Rives, John Edgar, solid state physics
Roberson, Edward Lee, parasitology, veterinary medicine
Robertson, James Aldred, chemistry

Robinson, Gertrude Edith, mathematics education
Rogers, Lockhart Burgess, analytical chemistry
Roncadori, Ronald Wayne, plant pathology
Ross, Herbert Holdsworth, entomology, ecology
Roth, Ivan Lambert, microbiology
Rowan, Samuel James, forest pathology, plant physiology
Ruehle, John Leonard, plant pathology
Ruff, John K., inorganic chemistry
Ruff, Michael David, parasitology
Saade, John Marshall, mathematics
Sanford, Malcolm Thomas, apiculture
Sansing, Norman Glenn, biochemistry, plant physiology
Schelly, Zoltan Andrew, physical chemistry
Schepartz, Abner Irwin, biochemistry
Schindler, James Edward, zoology
Schramm, Lee Clyde, pharmacognosy
Schuler, George Albert, food science, microbiology
Scott, Donald Charles, limnology, fish biology
Sears, William Clifton, physics
Seerley, Robert Wayne, animal science, animal nutrition
Seitz, William Rudolf, analytical chemistry
Sen Gupta, Barun Kumar, micropaleontology, marine geology
Shackelford, Walter McDonald, analytical chemistry
Shaw, James Scott, astronomy
Shear, James Algan, geography
Shenton, Leonard Roy, statistics, applied mathematics
Shotts, Emmett Booker, Jr., medical microbiology, veterinary microbiology
Shrum, John W., geology, science education
Snyder, Willard Monroe, hydrology
Sommer, Harry Edward, forest physiology
Sparks, Darrell, plant physiology, horticulture
Speirs, Mary, nutrition
Srivastava, Prakash Narain, reproductive biology
Stammer, Charles Hugh, organic chemistry
Stanley, Edward 'Alex, paleobiology
Steinbeck, Klaus, forestry, physiology
Steuer, Malcolm F., nuclear physics
Stewart, James T., pharmaceutical chemistry
Stone, Kirk Haskin, geography
Stormer, John Charles, Jr, petrology, geochemistry
Strobel, George L, theoretical nuclear physics, optics
Tan, Kim H., tropical agriculture
Taras, Michael Andrew, forest products, wood technology
Taylor, Jack, plant pathology
Taylor, Robert Clement, physiology, zoology
Thompson, Bobby Blackburn, medicinal chemistry, organic chemistry
Thompson, Frederick Nimrod, Jr, reproductive endocrinology
Thompson, Peter Ervin, genetics
Tinga, Jacob Hinnes, horticulture, plant physiology
Travis, James, biochemistry
Tritz, Gerald Joseph, microbiology, genetics
Tsao, Ching Hsi, entomology
Tyler, David E, veterinary pathology
Uzes, Charles Alphonse, theoretical physics
Van Eseltine, William Parker, bacteriology
Van Fleet, Dick Scott, botany
Vick, Charles Booker, forest products
Vines, Herbert Max, plant physiology
Wade, Adelbert Elton, pharmacology, biochemistry
Waggoner, William Horace, inorganic chemistry
Walker, Alma Toevs, botany
Walker, Dan B, plant anatomy
Wallace, James Bruce, entomology, hydrobiology
Walters, Douglas Bruce, analytical chemistry

Ware, Glenn Oren, applied statistics, operations research
Washam, Clinton Jay, microbiology
Washburn, Kenneth W., poultry genetics
Waters, Kenneth Lee, pharmacy, medicinal chemistry
Weathersby, Augustus Burns, entomology
Weaver, James Bode, Jr., genetics, agronomy
Welch, Roy Allen, geography, photogrammetry
Wenner, David Bruce, geochemistry, geology
Westerfield, Clifford, veterinary anatomy
Westfall, Jonathan Jackson, botany
Wheeler, James Orton, transportation geography, economic geography
Wheeler, Robert Stevenson, avian physiology
White, Charles Henry, dairy microbiology
Whitehead, Thomas Hillyer, analytical chemistry
Whitten, Kenneth Wayne, inorganic chemistry
Whitworth, Clyde W., pharmacy
Wiegert, Richard G., ecology
Williams, David John, III, theriogenology
Williams, Joy Elizabeth P, bacteriology, biochemistry
Williams, William Lawrence, biochemistry
Williford, William Olin, mathematical statistics
Witherspoon, Don Meade, medicine, surgery
Wood, Robert Manning, nuclear physics
Woodruff, James F, geography
Wyatt, Roger Dale, microbiology, poultry science
Wynn, Willard Kendall, Jr, plant pathology
Yates, Harry Orbell, III, forest entomology
Yoder, Harry Whitaker, Jr, veterinary medicine
Young, Louis Lee, food science
Younts, Sanford Eugene, soil science, agronomy
Ziance, Ronald Joseph, pharmacology

ATLANTA

Ahearn, Donald G, mycology, marine microbiology
Ahrens, Rudolf Martin (Tino), physics
Ajello, Libero, medical mycology
Allen, Arthur T, Jr, geology
Ambrose, John Augustine, nutritional biochemistry, biochemical genetics
Ames, William F, applied mathematics
Anderson, Gloria Long, organic chemistry
Anderson, Kenneth Verle, neuropsychology, neuroanatomy
Anschel, Steven William, neurophysiology
Ash, Ronald Joseph, microbiology, virology
Ashby, Eugene Christopher, organic chemistry, inorganic chemistry
Atwood, Sanford Soverhill, cytology, genetics
Ayer, Darrell, (Jr), pathology
Badre, Albert Nasib, information science
Bailey, Gordon Burgess, biochemistry
Bain, James Arthur, biochemical pharmacology
Balachandran, Kashi Ramamurthi, operations research, applied statistics
Ball, John Miller, geography
Ballard, Marguerite Candler, internal medicine, hematology
Balows, Albert B, microbiology
Bartholomai, C W, entomology
Basmajian, John V, anatomy
Batra, Gopal Krishan, virology, cancer chemistry
Beadle, Leslie Dewey, medical entomology
Bean, Brent Leroy, lasers, quantum optics
Bederman, Sanford Harold, geography
Belinfante, Johan G F, applied mathematics, physics
Benoit, Peter Wells, dentistry, anatomy
Berry, Maxwell (Rufus), internal medicine
Bertrand, Joseph Aaron, inorganic chemistry
Bevis, Jean Harwell, algebra, systems theory
Binkley, Francis, biochemistry, immunochemistry
Biritz, Helmut, theoretical physics
Black, Jessie Kate, developmental biology
Blank, Carl Herbert, public health, microbiology
Blitch, Lee Wesley, chemistry

Bloom, Walter Lyon, clinical medicine
Blue, Marts Donald, solid state physics
Blumberg, Richard Winston, pediatrics
Boal, Jan List, applied mathematics
Boring, John Rutledge, III, microbiology
Bornstein, Leopold Frey, plastics chemistry
Bos, Jane, neuroanatomy
Bourne, Geoffrey Howard, experimental pathology, anatomy
Boutwell, Joseph Haskell, clinical biochemistry
Boyd, James Emory, nuclear physics, oceanography
Boykin, David Withers, Jr, organic chemistry
Brachman, Philip Sigmund, medicine
Braden, Charles Hosea, nuclear physics
Braley, James Alexander, organic chemistry
Breeland, Samuel Glover, entomology
Brent, J Allen, science administration
Brewer, Harold Reid, theoretical physics
Brieske, Thomas John, mathematics education
Brody, Aaron Leo, food technology
Brogan, Donna R, applied statistics
Brooke, Marion Murphy, parasitology
Brooks, John Bill, biochemistry, microbiology
Brown, George Raymond, environmental physics, air pollution
Brown, Herbert Eugene, medicine
Brown, Robert Lee, surgery
Browne, John McDonald, development biology
Bryan, Frank Leon, bacteriology, food microbiology
Burbanck, Madeline Palmer, botany, cytology
Burbanck, William Dudley, protozoology, animal ecology
Burgess, Edward Meredith, organic chemistry
Burrows, Walter Herbert, industrial chemistry
Butt, Elizabeth, medical microbiology
Cain, George Lee, Jr, mathematics
Caine, Drury Sullivan, III, organic chemistry
Campbell, Wallace G, Jr, pathology
Caruthers, John Quincy, bacteriology
Casey, Helen Liles, bacteriology, immunology
Cassel, William Alwein, microbiology
Chamberlain, Roy William, medical entomology, virology
Chan, Yick-Kwong, biometry
Chandler, Francis Woodrow, Jr, veterinary pathology
Chappell, William Adrian, virology
Chen, Andrew Tat-Leng, cytogenetics, genetics
Chen, Robert Long Wen, plasma physics
Cherniak, Robert, biochemistry
Cherry, William Bailey, bacteriology
Choi, Keewhan, statistics
Chyatte, Samuel Baruch, physical medicine
Clark, Allen Varden, food science
Clement, Anthony Calhoun, embryology
Clever, Henry Lawrence, physical chemistry
Cohen, Jacob Ortlieb, medical microbiology
Cole, Thomas Winston, Jr, organic chemistry
Collins, Delwood C, biochemistry, reproductive physiology
Collins, William Erle, medical entomology, economic entomology
Constant, Clinton, inorganic chemistry, chemical engineering
Contacos, Peter George, medical parasitology, tropical medicine
Cooper, Gerald Rice, physical chemistry, medicine
Corley, Charles Calhoun, Jr, hematology, internal medicine
Corliss, John Franklin, soil science
Cox, Dennis Henry, animal nutrition
Craft, Thomas Fisher, environmental sciences, nuclear science
Cramer, Ardis Lahann, parasitology, invertebrate physiology
Cramer, Howard Ross, stratigraphy, invertebrate paleontology
Crawford, Vernon, physics
Crenshaw, John Walden, Jr, population genetics
Cross, Hansell Flynn, acarology
Crow-Baste, Claudia Adkison, anatomy
Dale, Edwin, endocrinology
Danner, Dean Jay, biochemistry, medical genetics
Darsie, Richard Floyd, Jr, medical entomology

Davey, Winthrop Newbury, internal medicine
Davidson, John Keay, III, internal medicine, physiology
Davis, Monte V., nuclear physics
Day, Reuben Alexander, Jr, analytical chemistry
Dayton, Peter Gustav, pharmacology
De Felice, Louis John, biophysics, electrophysiology
DeHaan, Robert Lawrence, biology, embryology
Dever, G E Alan, public health administration
Dienhart, Charlotte Marie, anatomy
Dobes, William Lamar, dermatology
Dowdle, Walter R., virology
Dowell, Vulus Raymond, Jr, microbiology, bacteriology
Drucker, Bertram Morris, mathematics
Drummond, Margaret Crawford, biochemistry
Duke, Richard Alter, mathematics
Dull, Harold Bruce, preventive medicine, public health
Dusenbery, David Brock, biophysics, neurobiology
DuVarney, Raymond Charles, solid state physics
Eames, Wilmer B., dentistry
Eberhardt, William Henry, physical chemistry
Edwards, Bety F., anatomy
Egan, Robert L., radiology
Eichholz, Geoffrey Gunther, physics, nuclear engineering
Ellis, Robert J., bacteriology
Elmer, William Arthur, developmental genetics
Elsas, Louis Jacob, II, medical genetics
Eubanks, Julia Flynt, medical microbiology
Evans, Trevor, mathematics
Evans, William Buell, mathematics, meteorology
Falconer, Etta Zuber, algebra
Falek, Arthur, human genetics
Fales, Frank Weck, biochemistry, clinical chemistry
Farmer, John James, III, microbiology
Feeley, John Cornelius, immunology
Felner, Susan K., nephrology
Felton, Ronald H., chemical physics
Fetner, Robert Henry, biology
Fincher, Edward Lester, microbiology
Fink, Richard Walter, nuclear chemistry, physics
Flannery, Martin Raymond, atomic physics
Flaschka, Hermenegild Arved, analytical chemistry
Fluker, Sam Spruill, entomology
Flynn, Arthur Davis, entomology
Fodor, Andrew Robert, microbiology
Foege, William Herbert, medicine, epidemiology
Fong, Peter, theoretical physics, molecular biology
Fontaine, Russell Edgar, medical entomology
Ford, David A., mathematics
Ford, Joseph, theoretical physics
Forney, John Edgar, microbiology, public health
Foster, John Wallace, bacteriology
Fox, Ronald Forrest, theoretical physics
Franch, Robert H., cardiology, cardiovascular physiology
Franke, Norman Henry, pharmacy
Frederick, Lafayette, plant pathology, mycology
Frederickson, Evan Lloyd, anesthesiology
Friedman, Harold Bertrand, physical chemistry, industrial chemistry
Fritz, Michael E., periodontology, cell physiology
Fuller, Ellen Oneil, cardiovascular physiology
Gaffney, Peter Edward, microbiology
Galambos, John Thomas, medicine
Gallagher, James J., physics
Gallaher, Lawrence Joseph, physics, computer science
Gangarosa, Eugene J., epidemiology
Garrison, Allen K., physics
Gatland, Ian Robert, theoretical physics
Gayles, Joseph Nathan, Jr, physical chemistry
Gersch, Harold Arthur, physics
Gianturco, Maurizio, organic chemistry
Gilbert, Frederick Emerson, Jr, pathology, clinical chemistry
Ginsberg, Stewart Theodore, psychiatry
Glazer, Robert Irwin, pharmacology, oncology
Godwin, John Thomas, pathology

Golarz-De Bourne, Maria Nelly, microscopic anatomy, histochemistry
Goldsmith, David Jonathan, synthetic organic chemistry
Goldstein, Jacob Herman, physical chemistry
Goodchild, Chauncey George, zoology
Goslin, Roy Nelson, physics, mathematics
Grant, Willard H., geology, chemistry
Gravanis, Michael Basil, pathology
Gray, Stephen Wood, anatomy
Greenberg, Joseph, biology
Gregg, Michael Barrows, internal medicine
Griffith, William Kirk, agronomy
Groth, Donald Paul, oncology
Grovenstein, Erling, Jr, organic chemistry
Hagler, Henry James, neuropharmacology
Haineline, Adrian, Jr, biochemistry
Hall, Charles Thomas, microbiology, immunology
Hall, Henry Lee, psychiatry
Hall, John Henry, Jr, physical chemistry
Halstead, Charles Lemuel, oral pathology, oral medicine
Harrell, William Knox, bacteriology
Harris, Jane Ellen, neuropharmacology, pathology
Harris, Jesse Max, entomology
Hart, Raymond Kenneth, electron microscopy
Hartshorn, Truman Asa, geography, urban geography
Hatch, Milford Harrison, virology
Hathaway, Charles Louis, microbiology, immunochemistry
Havlicek, Stephen, organic chemistry, water chemistry
Hawkins, Theo M., public health
Healy, George Richard, parasitology
Heath, Clark Wright, Jr, epidemiology, internal medicine
Heise, John J., biology, biophysics
Henneike, Henry Fred, inorganic chemistry
Herndon, John Francis, biochemistry, nutrition
Herod, James Victor, mathematical analysis
Herrmann, Kenneth L., medicine, virology
Hersey, Stephen J., physiology
Hicklin, Martin Dale, pathology
Hicks, Donald Gail, analytical chemistry, inorganic chemistry
Hierholzer, John Charles, medical microbiology, biochemistry
Hill, Edward Orson, medical microbiology
Ho, Dar-Veig, applied mathematics
Hobart, Oscar F, Jr, phytopathology
Hobson, Richard David, geology
Hopkins, Harry P, Jr, physical chemistry
House, Herbert Otis, organic chemistry
Hsu, Frank Hsiao-Hua, solid state physics
Hubbard, Jerry S., microbiology
Hug, Carl Casimir, Jr, pharmacology, anesthesiology
Huguley, Charles Mason, Jr, medicine, oncology
Humphrey, Donald R., neurophysiology, biomedical engineering
Humphries, Asa Alan, Jr, cell biology, embryology
Hunt, Dale E., microbiology
Hunt, Ernest Lowell, comparative endocrinology
Hunt, Harold Russell, Jr, inorganic chemistry
Hunt, Lindsay McLaurin, Jr, oral biology
Hunt, Sara McClanahan, nutrition, biochemistry
Hunter, George L K, organic chemistry, analytical chemistry
Hunter, Roy, Jr, developmental biology
Husted, John E., geology, natural resources
Hutchinson, Leonard Hugh, meteorology
Iacobucci, Guillermo Arturo, bio-organic chemistry
Ibrahim, Adly N., microbiology
Iglarsh, Harvey Jerome, mathematics
Immel, Eric Robert, mathematics
Ingols, Robert Smalley, chemistry
Jackson, Richard Thomas, chemistry
Jeffrey, Geoffrey Marron, malariology
Johnson, Donald Ross, entomology
Johnson, James Dean, cell biology, reproductive biology
Johnson, Richard Clayton, applied physics
Johnson, Roger D., Jr, mathematics
Johnson, Ronald Carl, inorganic chemistry

Jones, Rena Talley, microbiology
Jones, Ronald Goldin, organic chemistry
Jones, Wilbur Douglas, Jr, bacteriology
Jordan, Helen Berry, zoology
Jurkiewicz, Maurice J., surgery
Justus, Carl Gerald, atmospheric physics
Kagan, Irving George, parasitology, immunology
Kahn, Bernd, radiochemistry
Kaiser, Robert L., tropical medicine, epidemiology
Kanmerer, William John, mathematics
Kaplan, William John, mathematics
Kappus, Karl Daniel, medical entomology, environmental health
Karp, Herbert Rubin, neurology
Kasriel, Robert H., mathematics
Kissling, Robert E., virology, comparative pathology
Knight, James Albert, Jr, organic chemistry
Kokko, Ukko Pentti, medicine
Kostyo, Jack Lawrence, physiology, endocrinology
Kreitzman, Stephen Neil, biochemistry, nutrition
Kuchmak, Myron, biochemistry, plant nutrition
Kuck, John Frederick Read, Jr, biochemistry
Kuo, Jyh-Fa, biochemistry, pharmacology
Kimbrough, Renate Dora, pathology
Kirkwood, James Benjamine, zoology, radiation biology
La Via, Mariano Francis, pathology
Lavroff, Viacheslav V., mathematics
LeFlore, William B., parasitology
Leonard, William Wilson, mathematics
Letbetter, William Dean, neurophysiology, neuroanatomy
Levine, Raphael Berg, biophysics
Line, John Paul, mathematics
Lisella, Frank Scott, public health
Littlejohn, Oliver Marsilius, pharmacy
Liu, Fred Wei Jui, physical chemistry
Lockhart, Ernest Earl, food science
Logue, Bruce R., medicine
Lomax, Eddie, organic chemistry
Long, Earl Ellsworth, public health
Long, Leland Timothy, geophysics, seismology
Long, Wilmer Newton, Jr, medicine, obstetrics & gynecology
Lopez, Antonio Vincent, pharmaceutical chemistry, pharmacognosy
Lowell, Robert Paul, geophysics
Lumb, Judith Rae, immunology
Lyon, John Blakeslee, Jr, biochemistry
Mackel, Donald Charles, medical microbiology
Maddison, Shirley Eunice, immunology, parasitology
Mahavier, William S., mathematics
Maholick, Leonard Thomas, psychiatry
Majors, Rias Hilton, animal science
Malaspina, Alex, nutrition, food technology
Mallison, George Franklin, environmental health, public health
Mandell, Leon, organic chemistry
Manning, John W., physiology
Manocha, Sohan Lall, histochemistry
Manson, Steven Trent, atomic physics
Mantueffel, Thomas Albert, numerical analysis
Mapp, Frederick Everett, zoology
Marine, William Murphy, preventive medicine, internal medicine
Martin, Charles K., mathematics
Martin, David Willis, physics
Massey, Fredrick Alan, applied mathematics
Mather, Alan, clinical biochemistry
Mather, Jane H., biochemistry
Mathews, Henry Mabbett, medical parasitology
Masumomo, Yorimi, physiology
Mathews, Hewitt William, pharmaceutical chemistry
May, Sheldon William, biochemistry
McBay, Henry Cecil, chemistry
McClure, Donald Allan, nuclear physics

McClure, Harold Monroe, veterinary pathology
McCroan, John Edgar, Jr, epidemiology
McDaniel, Earl Wadsworth, physics
McFarlin, Richard Francis, inorganic chemistry
McGarity, William Cecil, surgery
McLeod, William D., biomedical engineering
McNinch, Joseph Hamilton, preventive medicine
Melvin, Dorothy Mae, medical parasitology
Menger, Fred M., organic chemistry
Meridenth, Charles Waymond, physical inorganic chemistry
Meyer, Gunter Hubert, numerical analysis, applied mathematics
Miles, James William, pesticide chemistry
Millar, John Donald, preventive medicine, epidemiology
Miller, George Alford, physical chemistry
Miller, William Boynton, Jr, academic administration, radiation physics
Mills, John Blakely, III, biochemistry
Moncrief, John William, chemistry
Moore, Robert Vernon, radiochemistry
Moran, Neil Clymer, pharmacology
Moran, Thomas Francis, physical chemistry
Morgan, Ashley Grantham, Jr, science education
Morgan, Karl Ziegler, health physics
Moss, Claude Wayne, microbiology
Mosteller, Robert Cobb, biometrics
Moulton, George Herbert, dentistry
Murdy, William Henry, botany
Murphy, Frederick A., virology
Murray, Joseph Buford, geology
Murray, Malcolm Arthur, geography
Myers, Dirck V., biochemistry
Myers, Joseph B., physiology
Nabb, Dale Preston, biochemistry
Nabrit, Samuel Milton, morphology, physiology
Nadler, Ronald David, animal behavior, primatology
Nahmias, Andre Joseph, virology, pediatrics
Naib, Zuher M., pathology
Nakano, James Hiroto, microbiology
Nashed, Mohammed Zuhair Zaki, mathematics
Nave, Carl R., physics
Neff, John David, mathematics
Neff, Mary Muskof, mathematics
Neill, Jimmy Dyke, physiology, endocrinology
Nelson, William Henry, magnetic resonance
Nethercut, Philip Edwin, organic chemistry
Neuberger, John William, mathematics
Neumann, Henry Matthew, inorganic chemistry
Neville, Gwen Kennedy, anthropology
Newhouse, Verne Frederic, medical entomology
Nichaman, Milton Z., cardiovascular diseases
Nickerson, John David, physical chemistry
Noe, Bryan Dale, anatomy, cell biology
Norris, Thomas Elfred, biochemistry
Nunnally, Huey Neal, biomedical engineering, electrical engineering
Objeski, John Francis, microbiology, virology
Ogren, David Ernest, geology
O'Hara, James, ecology
Olansky, Sidney, medicine, dermatology
Olkowski, Zbigniew, cytochemistry
Osborn, James Maxwell, analytical mathematics
O'Shea, Daniel Charles, molecular spectroscopy
O'Steen, Wendall Keith, anatomy, neurobiology
Palmer, Dan F., virology, immunology
Palmer, Richard Carl, radiation chemistry, radiation physics
Palms, John Michael, nuclear physics, applied physics
Papageorge, Evangeline (Thomas), biochemistry
Parker, Richard Langley, veterinary medicine
Parrish, Fred Kenneth, biology
Patronis, Eugene Thayer, Jr, physics
Patterson, Rosalyn Mitchell, cytogenetics
Pearl, Gary Steven, neurosciences

Penn, James H, protozoology, parasitology
Pentz, Ella Irene, biochemistry
Perkowitz, Sidney, solid state physics
Per-Lee, John H, medicine
Perlin, Irwin Earl, mathematics
Peters, Charles Frederick, microbiology, chemistry
Petitt, Gus A, physics
Phelps, William Robert, forest pathology
Pierotti, Robert Amadeo, physical chemistry
Pine, Leo, biochemistry, microbiology
Platt, Robert Baxter, ecology
Pollard, Charles Oscar, Jr, geochemistry, mineralogy
Pooler, John Preston, biophysics, photobiology
Popovic, Vojin, physiology
Power, Walter Robert, geology
Powers, James Cecil, organic biochemistry
Pratt, Harry Davis, entomology
Preedy, John Robert Knowlton, medicine
Priest, Robert Eugene, experimental pathology, cell biology
Prothro, Johnnie W, foods, nutrition
Pruitt, Albert Wesley, pharmacology
Purcell, James Eugene, physics
Puri, Om Parkash, solid state physics
Pyron, Raymond Scott, organic chemistry
Radford, Terence, organic chemistry
Radin, Nathan, clinical chemistry
Raiford, Morgan B, ophthalmology
Ramaswamy, H N, inorganic chemistry, analytical chemistry
Ramos, Harold Smith, medicine
Rao, Pemmaraju Venugopala, nuclear physics, atomic physics
Raphael, Louise Arakelian, mathematical analysis
Raphael, Robert B, theoretical physics
Ray, Charles, Jr, genetics
Redmond, William Brinson, microbiology
Reich, George Arthur, medicine, epidemiology
Reichert, Leo E, Jr, biochemistry, endocrinology
Reimer, Charles Blaisdell, immunology
Reinhardt, Donald Joseph, microbiology, mycology
Remington, John Wood, physiology
Rester, Alfred Carl, Jr, nuclear physics
Reuter, Johannes Helmut, organic geochemistry
Richardson, Arthur Pawley, pharmacology
Rieser, Charles Ely, urology
Riley, Edward Eddy, Jr, cell biology, histophysiology
Riley, Perry Stephen, medical microbiology
Rinard, Gilbert Allen, physiology, endocrinology
Roberts, Carlyle Jones, health physics
Robinson, Daniel Alfred, algebra
Robinson, Roslyn Quinby, bacteriology, virology
Rogers, James Virgil, Jr, radiology
Rohrer, Robert Harry, physics
Roper, Robert George, atmospheric physics
Ross, David Paul, institutional research, plasma physics
Ross, Hubert Barnes, cultural anthropology
Royer, Donald Jack, inorganic chemistry
Rudman, Daniel, internal medicine
Rutledge, Dorothy Stallworth, mathematics
Scales, Roy William, immunology
Schantz, Peter Mullineaux, veterinary public health, parasitology
Schlant, Robert C, internal medicine, cardiology
Scholtens, Robert George, epidemiology, parasitology
Schroder, Jack Spalding, internal medicine
Schultz, Myron Gilbert, epidemiology
Schwartz, James F, pediatrics, neurology
Scott, John Watts, Jr, anatomy, psychology
Scott, June Rothman, microbial genetics
Scott, Walter L, Jr, radiobiology, medical technology
Sears, Curtis Thornton, Jr, inorganic chemistry, organometallic chemistry
Seery, Virginia Lee, biochemistry
Seiferle, Edwin James, industrial chemistry
Sellers, Thomas F, Jr, preventive medicine, infectious diseases
Sgoutas, Demetrios Spiros, biochemistry, clinical chemistry
Shackelford, Robert G, electrooptics
Shanthaveerappa, Totada Ramaiah, neuroanatomy, histochemistry
Shapira, Raymond, biochemistry, neurochemistry
Sharp, Henry, Jr, mathematics
Shear, Charles Robert, anatomy, neurobiology
Shepard, Charles Carter, infectious diseases
Sherry, Peter Burum, physical chemistry
Shonkwiler, Ronald Wesley, mathematics
Shulman, Jones Alvin, internal medicine
Shuster, Robert C, biochemistry, microbiology
Simmons, James Wood, molecular spectroscopy
Simmons, Samuel William, public health, biology
Sinha, Om Prakash, theoretical solid state physics
Size, William Bachtrup, geology
Skandalakis, John Elias, surgery
Skobba, Joseph Stanley, psychiatry
Sledd, Marvin Banks, mathematics
Sloan, Alan David, applied mathematics
Smith, Barnett Frissell, parasitology
Smith, James Victor, public health
Smith, Paul Dennis, genetics
Smith, Peter Byrd, medical bacteriology
Smith, William Allen, mathematics
Sophianopoulos, Alkis John, biochemistry, biophysics
Spicer, William Monroe, physical chemistry
Spight, Carl, physics
Sprawls, Perry, Jr, medical physics, biomedical engineering
Spriggs, Alfred Samuel, organic chemistry
Spurlock, Jack Marion, chemical engineering, biomedical engineering
Stanfield, James Aarmond, organic chemistry
Stanford, Augustus Lamar, Jr, solid state physics
Starrett, Austin Leroy, mathematics
Steinhaus, John Edward, anesthesiology
Stephens, Kenneth S, applied statistics, mathematical statistics
Stevens, Charles David, biochemistry
Stevenson, Enola L, plant physiology
Stevenson, James Rufus, surface physics
Straley, H W, III, geology, geophysics
Strange, John Ruble, developmental physiology, environmental physiology
Sturrock, Peter Earle, analytical chemistry, electrochemistry
Sudarsanan, Kesavan, x-ray crystallography, computer science
Sudia, William Daniel, entomology, virology
Suggs, Morris Talmage, Jr, microbiology
Sulzer, Alexander Jackson, medical parasitology
Sutin, Jerome, neuroanatomy, neurophysiology
Sweeny, James Gilbert, natural products chemistry
Symbas, Panagiotis N, thoracic surgery, cardiovascular surgery
Tager, Morris, microbiology
Tait, Columbus Downing, Jr, medicine, psychoanalysis
Tanner, James Mervil, theoretical physics
Taylor, Boyd Eugene, genetics
Taylor, Gerald C, bacteriology, virology
Taylor, James Lester, organic chemistry
Taylor, Robert Tieche, medical entomology
Techo, Robert, information science
Thedford, Roosevelt, organic biochemistry, molecular biology
Thomas, Edward Wilfrid, physics
Thomas, Frank Henry, resource geography, urban geography
Thomas, William Andrew, geology
Thomason, Berenice Miller, microbiology
Thompson, John Daniel, obstetrics & gynecology
Thomson, Junius Richard, biology
Tigges, Johannes, neuroanatomy
Tincher, Wayne Coleman, physical chemistry
Tindall, George Taylor, neurosurgery
Tonne, Philip Charles, mathematics
Topp, Allan Crickington, physical chemistry
Trawick, William George, physical chemistry clinical chemistry
Treadwell, Perry Edward, microbiology
Tunstall, Lucille Hawkins, microbiology, immunology
Turner, James Marshall, plasma physics
Tuttle, Elbert P, Jr, physiology, internal medicine
Tzianabos, Theodore, virology
Underwood, Arthur Louis, biochemistry
Usherwood, Noble Ransom, soil fertility, plant nutrition
Valk, Henry Snowden, theoretical nuclear physics
Venable, John Heinz, public health
Vogel, Ralph A, microbiology
Waag, Charles Joseph, geology
Wagner, Robert Earl, agronomy
Waitzman, Morton Benjamin, physiology, biochemistry
Waldron, Charles A, oral pathology
Walker, Grayson Howard, statistical mechanics, chemical physics
Walker, James Wilson, mathematical statistics
Walker, John J, organic chemistry
Wall, Donald Dines, mathematics
Walls, Kenneth W, immunology, parasitology
Walls, Nancy Williams, bacteriology
Walton, Kenneth Nelson, medicine
Wampler, Jesse Marion, geochemistry, geochronology
Ward, Richard S, psychiatry, pediatrics
Warren, McWilson, malariology, tropical medicine
Warsi, Nazir Ahmed, mathematical physics
Wartell, Roger Martin, biophysical chemistry
Weatherly, Thomas Levi, physics
Weathers, Dwight Ronald, oral pathology
Weaver, Charles Edward, geochemistry, sedimentology
Weaver, Robert Elwin, medical bacteriology
Weinstein, Allan, physical chemistry
Wenger, Nanette Kass, medicine, cardiology
Wharton, Charles Heizer, vertebrate zoology, ecology
White, John Francis, physiology
White, Lendell Aaron, bacteriology
Whitehead, Marvin Delbert, phytopathology, mycology
Wichelhausen, Ruth Hechler, laboratory medicine, chemotherapy
Wilhelmi, Alfred Ellis, biochemistry, endocrinology
Wilkins, Samuel Austell, Jr, surgery
Williams, Joel Quitman, physics
Wilson, Colon Hayes, Jr, rheumatology, internal medicine
Witte, John Jacob, preventive medicine, pediatrics
Wood, Robert E, operations research
Woods, Wendell David, biochemistry, physiology
Woodward, Leroy Albert, physics
Worth, Roy Eugene, functional analysis
Wray, Joe Willie, mathematics
Wyly, Lemuel David, Jr, nuclear physics
Wyse, Frank Oliver, mathematics
Yeargers, Edward Klingensmith, biophysics
Young, Leona Graff, physiology
Young, Robert Alan, solid state physics, crystallography
Youse, Bevan K, mathematical analysis
Yu, Nai-Teng, biophysical chemistry
Zalkow, Leon Harry, organic chemistry
Zubay, Eli Alan, statistics, actuarial science
Zvejnieks, Andrejs, organic chemistry

ATTAPULGUS
Horton, Norman Hagood, organic chemistry

AUGUSTA
Abdel-Latif, Ata A, biochemistry, neurochemistry
Abraham, Edathara Chacko, biochemistry
Ahlquist, Raymond Perry, pharmacology
Akamatsu, Yasuyuki, pathology, oncology
Allen, Lane, anatomy
Allen, Marshall B, Jr, neurosurgery
Bard, Raymond Camillo, microbiology
Beaudreau, David E, restorative dentistry, neurophysiology
Best, Gary Keith, microbiology
Bhalla, Vinod Kumar, endocrinology
Bhargava, Hemendra Nath, biochemical pharmacology
Black, John B, endocrinology, anatomy
Bliven, Floyd E, Jr, orthopedic surgery
Bockman, Dale Edward, anatomy
Bompart, Billy Earl, mathematics
Bowsher, Harry Fred, nuclear physics
Bransome, Edwin D, Jr, endocrinology, biochemistry
Braselton, Webb Emmett, Jr, biochemistry, endocrinology
Bresnick, Edward, biochemistry
Brown, Mark, radiology
Brownell, George H, microbiology, bacterial genetics
Bryans, Charles Iverson, Jr, obstetrics & gynecology
Burnett, George Wesley, microbiology, dentistry
Bustos-Valdes. Sergio Enrique, biochemistry
Byrd, J Rogers, cytogenetics, zoology
Carlton, William Herbert, medical physics
Carter, Curtis Harold, internal medicine
Chandler, A Bleakley, pathology
Chew, William Hubert, Jr, internal medicine, infectious diseases
Chiang, Tzu Sung, pharmacology
Christenberry, George Andrew, mycology, academic administration
Colborn, Gene Louis, anatomy
Coryell, Margaret E, biological chemistry
Costoff, Allen, zoology, endocrinology
Davis, Carl O, dentistry, educational measurement
Davis, Frank Roscoe, zoology
Delahayes, Jean, physiology
Denton, James Fred, medical microbiology, zoology
Dinwiddie, Joseph Gray, Jr, organic chemistry
Dirksen, Thomas Reed, dentistry, biochemistry
Doetsch, Gernot Siegmar, neuroscience
Dow, Philip, physiology
Dyken, Paul Richard, pediatrics, neurology
Ehrhart, Ina C, medical physiology
Ellison, Lois Taylor, physiology
Ellison, Robert G, cardiovascular surgery, thoracic surgery
Engler, Harold S, medicine, surgery
Ezell, Ronnie Lee, nuclear physics
Feldman, Daniel S, neurology, neurophysiology
Feldman, Elaine Bossak, medicine
Frank, Martin J, internal medicine, cardiology
Gangarosa, Louis Paul, Sr, pharmacology
Geber, William Frederick, physiology, pharmacology
Gelfant, Seymour, cell biology
Ginsburg, Jack Martin, physiology
Green, Keith, physiology, biophysics
Greenblatt, Robert Benjamin, endocrinology
Gullen, Warren Hartley, epidemiology, public health
Hall, Walter Knowlton, biochemistry
Hartlage, Lawrence Clifton, neuropsychology, pediatrics
Hawkins, Isaac Kinney, restorative dentistry, human anatomy
Hayes, John Thompson, biology, ecology
Hendrich, Chester Eugene, physiology, endocrinology
Hobbs, Milford Leroy, clinical pathology
Hofman, Wendell Fey, physiology
Howard, Eugene Frank, cell biology
Howard, John Charles, organic chemistry, biochemistry
Hudson, James Bloomer, medicine
Huff, Thomas Allen, internal medicine
Huisman, Titus Hendrik Jan, biochemistry
Humphries, Arthur Lee, Jr, surgery
Jackson, William James, neuropsychology
Jelenko, Carl, III, surgery
Jerram, David Carroll, pharmacology
Karow, Armand Monfort, Jr, pharmacology, cryobiology
Karp, Warren B, biochemistry
Kinzer, Robert Leroy, dentistry
Kolbeck, Ralph Carl, physiology, biochemistry
Lambert, Frank (Edwin), Jr, science education
Leibach, Fredrick Hartmut, biochemistry, endocrinology
Lewis, Jasper Phelps, chemistry, biochemistry
Lewis, Silas Davis, organic chemistry
Little, Robert Colby, physiology, medicine
Liu, Paul Ishen, clinical pathology
Mahesh, Virendra B, organic chemistry, endocrinology
Manganiello, Louis O J, neurosurgery
Mansberger, Arlie Roland, Jr, surgery
Marble, Howard Bennett, Jr, dentistry, oral surgery
Matheny, James Lafayette, pharmacology
McCranie, Erasmus James, psychiatry
McFarland, Kay Flowers, internal medicine, endocrinology
McKenney, Joel R, psychophysiology
McKenzie, John Ward, histology, embryology
McKinney, Ralph Vincent, Jr, pathology, cell biology

GEORGIA

McPherson, James C, Jr, biochemistry, medicine
Meyer, Leon Herbert, physical chemistry, chemical engineering
Miller, David Arthur, medical physiology
Mills, Thomas Marshall, reproductive endocrinology
Mills, William Carlos, food technology
Moores, Russell R, medicine
Moretz, William Henry, surgery
Muldoon, Thomas George, biochemistry
Nelson, George Humphry, biochemistry, obstetrics & gynecology
Ogle, Thomas Frank, reproductive endocrinology
O'Neal, Floyd Breland, analytical chemistry
Page, Malcolm I, internal medicine, infectious diseases
Parrish, Robert A, Jr, surgery
Pashley, David Henry, physiology
Patten, Jimmy Ray, physiology
Plowman, Kent Milton, biochemistry
Pogue, Richard Ewert, medical statistics, computer science
Pool, Winford H, (Jr), radiology
Powers, Walter Lee, physics
Prescott, Glenn Carleton, Jr, hydrology
Puchtler, Holde, histochemistry, pathology
Reichard, Sherwood Marshall, radiology, physiology
Rhode, C Martin, surgery
Rice, Walter Gowans, pathology
Robertson, Alex F, pediatrics, gynecology
Roesel, Catherine Elizabeth, biochemistry
Schuster, George Sheah, microbiology, cell biology
Scoggin, William Allen, obstetrics & gynecology
Scott, David Frederick, biochemistry
Sharawy, Mohamed, anatomy
Smith, Linda Lou, physical biochemistry
Sobel, Robert Edward, clinical chemistry
Sohal, Gurkirpal Singh, neuroembryology
Stoddard, Leland Douglas, pathology
Stoney, Samuel David Jr, neuroscience
Sutherland, James Henry Richardson, pharmacology
Talledo, Oscar Eduardo, obstetrics & gynecology
Teabeaut, James Robert, II, pathology
Thomas, Robert P, ophthalmology
Threefoot, Sam Abraham, cardiovascular diseases
Turner, Janice Butler, molecular spectroscopy
Tyler, Jean Mary, endocrinology, immunochemistry
Urbanek, Vincent Edward, prosthodontics, dentistry
Vargo, Robert Allen, physiology
Volpito, Perry Paul, anesthesiology
Walecka, Jerrold Alberts, organic chemistry
Weatherred, Jackie G, physiology
Webber, Brooke Bland, genetics
Webster, Paul Daniel, III, medicine, gastroenterology
Wege, William Richard, radiology
Welband, Wilbur A, anatomy, neurology
Welter, Dave Allen, anatomy, cytogenetics
Whitford, Gary M, toxicology, physiology
Wiedmeier, Vernon Thomas, physiology
Wilds, Preston Lea, obstetrics & gynecology
Williams, Maryon Johnston, Jr, biomedical engineering, electrical engineering
Witham, Abner Calhoun, cardiology
Wolf, Paul A, biochemistry
Wright, Claude-Starr, hematology
Wycoff, Harland Dewitt, biochemistry
Zwemer, Thomas J, orthodontics

BOWDON
Garrett, Robert Lee, mathematics

BRUNSWICK
Iannicelli, Joseph, organic chemistry, natural resources
Reimold, Robert J, ecology

BYRON
Horton, Billy D, plant physiology, horticulture
Hunter, Richard Edmund, phytopathology
Kirkpatrick, Hugh Charles, plant pathology
Payne, Jerry Allen, entomology, ecology
Thompson, James Marion, plant breeding, plant genetics
White, Andrew Wilson, soil conservation, soil fertility

CARROLLTON
Boyd, Herman Wayne, nuclear physics
Dangle, Richard L, nuclear physics
De Mayo, Benjamin, experimental solid state physics
Duquette, Alfred L, mathematics
Eiss, Albert Frank, science education
Esslinger, William Glenn, organic chemistry
Gardner, Arthur Wendel, genetics
Garmon, Lucille Burnett, science education, history of science
Gilbert, Edward E, entomology
Hahn, Hwa Suk, mathematics
Hecht, Alan David, micropaleontology
Keller, George Earl, nuclear physics
Klee, Lucille Holljes, biochemistry
Lampton, Robert Koerbel, botany
Larson, Lewis Henry, Jr, anthropology
Lockhart, William Lafayette, inorganic chemistry
Long, Clarence Sumner, Jr, geology
Maples, William Paul, parasitology
Pittman, Chatty Roger, mathematics
Poort, Jon Michael, geology
Powell, Bobby Earl, solid state physics
Quertermus, Carl John, Jr, animal behavior, aquatic ecology
Sanders, Richard Pat, geology
Sharp, Thomas Joseph, algebra
Stokes, Jimmy Cleveland, science education
Taylor, Howard Edward, taxonomy
Welch, Robert McClam, genetics, cell biology
Zander, Vernon Emil, mathematics

CARTERSVILLE
George, Lucille Katherine (Mrs W L Pickard), medical mycology

CLARKSTON
Hardy, Flournoy Lane, mathematics
Murray, Joan Baird, vertebrate zoology

COCHRAN
Alderman, A Louis Cleveland, Jr, biology, parasitology
Comeau, Roger William, mammalian physiology
Crider, Fretwell Goer, physical chemistry
DeLorenzo, Ronald Anthony, physical inorganic chemistry
Husa, William John, Jr, analytical chemistry
Rhodes, Robert Allen, pharmaceutical chemistry

COLQUITT
Small, Howard G, Jr, agronomy

COLUMBUS
Cash, Dewey Byron, mathematical analysis
Clark, Flora Mae, genetics
Davis, Clyde William, Jr, algebra
LeNoir, William Cannon, Jr, biology
Lytle, James Bert, physiology, radiation biology
Pyle, John Tillman, analytical chemistry
Ratliff, Charles Ray, biochemistry
Rigsby, Ernest Duward, microbiology
Stanton, George Edwin, aquatic ecology, invertebrate ecology

DAHLONEGA
Callaham, Mac A, wildlife biology, fisheries
Pandres, Dave, Jr, theoretical physics
Wicht, Marion Cammack, mathematics

DALTON
Wiman, Robert Edgar, organic chemistry

DECATUR
Baxter, Gene Francis, polymer chemistry
Bess, John Clifford, audiology, speech pathology
Bridgeman, Anna Josephine, biology
Brown, William E, microbiology, food science
Carlquist, Philip Rich, microbiology
Clark, Marion Thomas, chemistry
Cleek, George Kime, chemistry
Cunningham, Alice Jeanne, electrochemistry, polarography
Dobrovolny, Charles George, virology, public health
Dratz, Arthur Frederick, nuclear medicine
Ewing, William Howell, medical microbiology
Frierson, William Joe, analytical chemistry
Gary, Julia Thomas, analytical chemistry
Groseclose, Nancy Pence, zoology
Herrick, Stephen (Marion), geology
Jarvis, Jack Reynolds, medicine
Lester, Charles Turner, organic chemistry
McBride, Clifford Hoyt, clinical chemistry
Morin, Leo Gregory, clinical biochemistry
Rice, Paul LaVerne, medical entomology
Ripy, Sara Louise, mathematical analysis
Rollor, Edward Albert, physics
White, Alan Jonathon, inorganic chemistry
Willis, Isaac, dermatology
Woodall, William Robert, Jr, aquatic ecology, environmental biology
Yodaiken, Ralph Emile, pathology, endocrinology

DEMOREST
Iloff, Philip Murray, Jr, organic chemistry, inorganic chemistry

DORAVILLE
Pollard, Lin Davis, geochemistry

DRY BRANCH
Smith, John M, economic geology, clay mineralogy

DUNWOODY
Kirchner, Justus George, organic chemistry

EAST POINT
Weipert, Eugene Allen, industrial organic chemistry

EXPERIMENT
Anderson, Oscar Emmett, agronomy, soil fertility
Baird, Derwood McVey, animal science, nutrition
Beuchat, Larry Ray, food microbiology
Boswell, Fred Carlen, agronomy, soils
Bough, Wayne Arnold, chemistry
Burns, Robert Emmett, plant physiology
Cecil, Sam Reber, food science
Cordia, Honorico, parasitology
Cummins, David Gray, agronomy
Daniell, Jeff Walter, horticulture
Dempsey, Alvin Hugh, horticulture
Flora, Lewis Franklin, food science, food chemistry
Hardcastle, Willis Sanford, weed science
Jackson, Curtis Rukes, plant pathology
Jellum, Milton Delbert, agronomy, plant breeding
Landes, Doelas Randy, nutrition
Lane, Ronald Paton, horticulture, plant breeding
Langford, Walter Robert, agronomy
Massey, John Hubert, agronomy
McCullough, Marshall Edward, dairy nutrition
Nelson, Lloyd Russel, plant breeding
Ohki, Kenneth, plant physiology
Purcell, Joseph Carroll, agricultural economics
Savage, Earl Frederick, pomology
Schaad, Norman W, plant pathology
Shewfelt, Albert Lorne, food science
Shuman, Larry Myers, soil chemistry
Smith, Albert Ernest, plant physiology
Sowell, Grover, Jr, plant pathology
Sudweeks, Earl Max, animal nutrition, dairy nutrition
Walker, Jerry Tyler, plant pathology
Wilkinson, Robert Eugene, plant physiology, weed science
Woodroof, Jasper Guy, horticulture
Worthington, Robert Earl, lipid chemistry, food science
Young, Clyde Thomas, food science

FORSYTH
Archer, Allan Frost, arachnology, anthropology
Dunkelberg, Wilbur Eugene, Jr, microbiology
Patterson, William Junior, medical entomology, parasitology

FT MCPHERSON
Cretz, Joseph Reuel, microbiology

FT VALLEY
Abe, Ronald Kuraso, animal nutrition
Blanchet, Waldo W E, science education
Jacklin, Stanley William, entomology
Juenge, Eric Carl, organometallic chemistry
Kromann, Paul Roger, physical chemistry
Moorehead, William Douglas, biology, science education
Phillips, Barry Allen, organic chemistry
Snapp, Oliver Irvin, Sr, economic entomology

GAINESVILLE
Andrews, Charles Lawrence, wildlife ecology, parasitology
Andrews, Lucy Gordon, medical genetics
Benuck, Irwin, psychophysiology, genetics
Maag, Theodore Augustus, virology, immunology

GORDON
Austin, Roger Seth, geology, clay mineralogy
Prescott, Paul Ithel, research administration, clay mineralogy
Young, Raymond Hinchcliffe, Jr, clay mineralogy

GRIFFIN
Cochran, Hulon Lilley, horticulture

LA GRANGE
Hendrix, James Easton, polymer chemistry, organic chemistry
Hicks, Arthur M, organic chemistry
Machell, Greville, organic chemistry
Shibley, John Luke, zoology

LAWRENCEVILLE
Baer, George M, epidemiology, virology

LITHONIA
Marion, James Edsel, food science, nutrition

MACON
Brender, Ernst Victor, forest management
Bush, Powell Daniel, Jr, physics
Caminita, Barbara Hobson, microbiology
Dever, David Francis, physical chemistry
Diboll, Alfred, developmental anatomy, plant anatomy
Furse, Clare Taylor, analytical chemistry
Harrison, James Ostelle, entomology
James, Franklin Ward, analytical chemistry
Kraus, John Franklyn, forest genetics
La Farge, Timothy, forest genetics
Marquart, John R, physical chemistry
Morris, Horton Harold, clay mineralogy, pulp & paper technology
Sluder, Earl Ray, forestry, genetics
Smith, Nat E, medicine

MADISON
Rigdon, Raymond Harrison, pathology

MARIETTA
Davis, Herbert L, Jr, cytogenetics, radiation biology
Laemmle, Joseph Thomas, organometallic chemistry
Stanitski, Conrad Leon, inorganic chemistry
Straley, Tina, mathematics
Weiner, Robert Samuel, physical chemistry

Wolke, Sara Richardson, analytical chemistry

MILLEDGEVILLE
Aliff, John Vincent, parasitology, environmental sciences
Baarda, David Gene, organic chemistry, environmental chemistry
Batson, Jack David, vertebrate ecology, fresh water ecology
Chesnut, Thomas Lloyd, entomology, zoology, ecology
Cotter, David James, plant ecology
Curtis, Jerry Leon, biochemistry
Daniel, Charles Pack, biology
George, Dick Leon, mathematics
Keeler, Clyde Edgar, zoology
Staszak, David John, animal physiology, biochemistry
Vincent, Joseph Francis, biochemistry

MORROW
Aust, Catherine Cowan, algebra
Daniel, Leonard Rupert, computer sciences
Nail, Billy Ray, algebra

MT BERRY
Hancock, Kenneth Farrell, biology
Kiser, Roy Stone, biology
McDowell, John Willis, parasitology
Pirkle, Willis Nathaniel, thermodynamics, physical chemistry
Rhoades, James Lawrence, biochemistry
Shand, Julian Bonham, Jr, solid state physics

NORCROSS
Broyde, Barret, physical chemistry
Gesner, Bruce D, organic polymer chemistry
Marshall, Donald Irving, plastics chemistry
Sabia, Raffaele, polymer chemistry
Saunders, Morton Jefferson, optics
Snyder, Hugh Donald, plant pathology

OXFORD
Ali, Monica McCarthy, chemistry
Landt, James Frederick, biology
Sharp, Homer Franklin, Jr, zoology, ecology

ROBINS AFB
Maraman, Grady Vancil, biomathematics, physiology

ROME
Allee, Marshall Craig, entomology, zoology
Grear, Philip French-Carson, ecology, geoenvironmental science
Lipps, Emma Lewis, plant ecology
Peakes, Lawson Vernon, organic chemistry
Rollinson, Samuel Milton, wood technology

ROSWELL
Wood, Roy Kellum, wildlife management

ST SIMONS ISLAND
Anderson, William Wyatt, ichthyology
Elder, Joseph Denison, science writing
King, Gladys Smith, cell physiology, histology

SAPELO ISLAND
Gallagher, John Leslie, marine botany
Haines, Evelyn Brown, marine ecology
Hanson, Roger Brian, marine microbiology
Wheeler, John Russell, ecology, environmental chemistry
Whitney, David Earle, marine ecology

SAVANNAH
Adamson, William Charles, plant breeding
Anantha Narayanan, Venkataraman, physics, spectroscopy
Arbogast, Richard Terrance, entomology
Atkinson, Larry P, chemical oceanography
Beitz, Alex D, physiology
Brewer, John Gilbert, analytical chemistry
Brower, John Harold, entomology, radiation ecology
Burke, Roger E, industrial organic chemistry
Chandra, Kailash, physics
Chaney, Edward L, radiological physics, health physics
Clark, Charles Kittredge, natural products chemistry
Clemmons, John B, mathematics, physics
Davenport, Leslie Bryan, Jr, vertebrate zoology, ecology
Davis, Robert, insect ecology
Dennis, Norman McLeod, entomology
Dunstan, William Morgan, biological oceanography
Ennor, Kenneth Stafford, organic chemistry
Gardner, Wayne Stanley, water chemistry, marine chemistry
Ghuman, Gian Singh, geoenvironmental science
Guerrant, Gordon Owen, analytical chemistry
Hampton, Burt Laurent, organic chemistry
Harding, James Lombard, geological oceanography
Harris, J Henry Earl, organic chemistry, biochemistry
Henry, Vernon James, Jr, oceanography, geology
Highland, Henry Arthur, economic entomology
Howard, James Dolan, geology
Hudson, Anne Lester, mathematics
Hudson, Sigmund Nyrop, mathematics
Hunter, Frissell Roy, zoology, comparative physiology
Jackson, Prince A, Jr, physics, mathematics
Jen, Yun, organic chemistry
Johnson, Robert William, Jr, organic chemistry
Kimble, Glenn Curry, chemistry
Krishnamurti, Pullabhotla V, veterinary microbiology
LeCato, George Leonard, III, entomology
Lee, Richard Fayao, environmental chemistry, biological oceanography
Lukat, Robert Timon, chemistry
Lum, Patrick Tung Moon, entomology
McKinney, Roger Minor, organic chemistry, immunochemistry
Menon, Manchery Prabhakara, nuclear chemistry, radiochemistry
Menzel, David Washington, oceanography
Nambiar, Govindan Kuppadakkath, genetics
Oertel, George Frederick, sedimentology
Paulsen, Grover Cleveland, Jr, chemistry
Raut, Kamalakar Balkrishna, organic chemistry
Redlinger, Leonard Maurice, entomology
Robbins, Paul Edward, organic chemistry
Shelby, Charles Edwin, animal genetics
Singh, Harpal P, inorganic chemistry
Stratton, Cedric, analytical chemistry
Su, Helen Chien-Fan, organic chemistry
Summerville, Richard Marion, mathematics
Tenore, Kenneth Robert, biological oceanography
Tucker, Willie George, organic chemistry
Windom, Herbert Lynn, oceanography
Wofford, Irvin Mirle, agronomy
Woodhouse, Bernard Lawrence, pharmacology

SMYRNA
Buchanan, O Lexton, Jr, mathematics education

SNELLVILLE
Webb, Patricia Ann, virology, pediatrics

STATESBORO
Bishop, Gale Arden, paleontology
Bishop, Thomas Parker, solid state physics
Boole, John Allen, Jr, plant morphology, plant physiology
Boxer, Robert Jacob, organic chemistry
Bozeman, John Russell, plant ecology, resource management
Bryant, Carroll William, physics
Colvin, Clair Ivan, physical chemistry
Curtis, Henry L, plant physiology
Darrell, James Harris, II, geology, palynology
Drapalik, Donald Joseph, plant taxonomy
Fitzwater, Robert N, chemistry
French, Frank Elwood, Jr, entomology, parasitology
Hanson, Hiram Stanley, geology, geochemistry
Hartberg, Warren Keith, entomology, genetics
Hibbs, Edwin Thompson, entomology
Hyde, Cornelia Tuten, invertebrate physiology
Kellogg, Craig Kent, organic chemistry
Lavender, DeWitt Earl, mathematics, statistics
Marshall, Rosemarie, bacteriology, biostatistics
Mayfield, Harold Gordon, inorganic chemistry
McKeever, Sturgis, ecology, mammalogy
Meadows, Jerriane Kujie Stafford, nutrition
Mobley, Harris W, anthropology
Nelson, Robert Norton, physical chemistry
Obenchain, Frederick DeCroes, acarology, insect physiology
Olewine, Donald Austin, physiology
Oliver, James Henry, Jr, acarology, cytogenetics
Pennington, Tully Sanford, invertebrate zoology, ornithology
Sparks, Arthur Godwin, mathematics
Stone, David Ross, algebra

STONE MOUNTAIN
Nevins, Arthur James, computer science
Weeks, Robert Joe, mycology, medical microbiology

SYLVANIA
Cotten, Marion deVeaux, pharmacology

THOMASVILLE
Komarek, Edwin Vaclav, biology
Pettit, Manson Bowers, psychiatry

TIFTON
Arnett, James DeLos, Jr, plant pathology
Austin, Max E, horticulture
Barry, Robert Merritt, entomology
Burton, Glenn Willard, agronomy
Carter, Robert Leonidas, soil fertility
Chalfant, Richard Bruce, entomology
Cole, John Rufus, Jr, veterinary microbiology
Cutler, Horace Garnett, plant physiology, plant nematology
Dogger, James Russell, entomology
Douglas, Charles Francis, agronomy
Dowler, Clyde Cecil, weed science, agronomy
Forbes, Ian, agronomy, plant breeding
Gaines, Tinsley Powell, analytical biochemistry
Gill, Denzell Leigh, plant pathology
Hammons, Ray Otto, genetics, agronomy
Hanna, Wayne William, plant genetics
Harmon, Silas Albert, horticulture
Hauser, Ellis W, weed science
Jaworski, Casimir A, soil science, horticulture
Johnson, Alva William, nematology, plant pathology
Jones, Richard Lamar, insect physiology
Lewis, Wallace Joe, entomology
McCain, Francis Saxon, plant breeding
McCormick, William Conner, animal breeding
McMillian, William Wallard, entomology
Minton, Norman A, plant nematology
Mixon, Aubrey Clifton, agronomy, plant breeding
Monson, Warren Glenn, agronomy
Morey, Darrell Dorr, agronomy
Neville, Walter Edward, Jr, animal science
Phatak, Sharad Chintaman, horticulture, plant physiology
Powell, George Wythe, animal nutrition
Segars, William Isaac, agronomy
Smittle, Doyle Allen, olericulture, plant physiology
Snoddy, Edward L, medical entomology, ecology
Sparks, Alton Neal, entomology
Stewart, Thomas Bonner, zoology, parasitology
Stone, William Morgan, parasitology
Sumner, Donald Ray, plant pathology
Tai, Peter Yao-Po, plant breeding
Thomas, Frank Harry, soils, inorganic chemistry
Thompson, Samuel Stanley, Jr, phytopathology
Todd, James Wyatt, entomology
Utley, Philip Ray, animal science
Waller, Ernest Frederick, veterinary pathology
Wells, Homer Douglas, plant pathology
Wesley, William Keith, agronomy
Widstrom, Neil Wayne, genetics, plant breeding
Wiseman, Billy Ray, entomology, horticulture
Womack, Herbert, entomology
Worley, Ray Edward, horticulture, agronomy

VALDOSTA
Brown, Calvin Hugh, systematic entomology, insect ecology
Duncan, Donald Lee, mathematics
Duvall, Harry Marean, organic chemistry
Faircloth, Wayne Reynolds, systematic botany
Howell, Leonard Rudolph, Jr, topology
Jenkins, Hughes Brantley, Jr, theoretical physics
Lindauer, Maurice William, analytical chemistry, physical chemistry
Little, Robert Lewis, geology
Marks, Dennis William, astrophysics
Martin, James Edwin, physics
Sumerford, Wooten Taylor, pharmaceutical chemistry, organic chemistry
Wall, James Graham, mathematics

WARM SPRINGS
Bennett, Robert Leo, physical medicine
Schultz, Donald Paul, pesticide chemistry

WATKINSVILLE
Box, James Ellis, Jr, soil physics
Bruce, Robert Russell, soil physics
Jackson, William Andrew, analytical chemistry
Langdale, George Wilfred, soil fertility, soil chemistry
Leonard, Ralph Avery, soil chemistry
Pallas, James Edward, Jr, plant physiology
Steudemann, John Alfred, animal nutrition
Wilkinson, Stanley R, agronomy, soil science

WEST POINT
Waddle, Howard Meffert, organic chemistry

WINTERVILLE
Stoehr, Henry Arthur, forestry

HAWAII

AIEA
Buren, Lawrence Lamont, crop physiology, plant physiology
Comstock, Jack Charles, plant pathology
Cushing, Robert Leavitt, agronomy
Hilton, H Wayne, organic chemistry
Mongelard, Joseph Cyril, plant physiology, plant ecology
Osgood, Robert Vernon, weed science
Richards, George Manning, plant biochemistry
Roberts, Robert Russell, microbiology

CAPTAIN COOK
Bower, Charles Arthur, soil chemistry

COCONUT ISLAND
Bardach, John E, aquatic ecology, fish biology

HAWAII NATIONAL PARK
Banko, Winston Edgar, wildlife ecology
Eaton, Gordon Pryor, physical geology
Lockwood, John Paul, geology
Tilling, Robert Ingersoll, geology

HILO
Cunningham, Roy Thomas, entomology
Dority, Guy Hiram, entomology
Downs, James Francis, cultural anthropology, applied anthropology
Finlayson, James Bruce, analytical chemistry, geochemistry
Fullerton, Charles Michael, atmospheric physics

HAWAII

Halbig, Joseph Benjamin, geochemistry, environmental geology
Hemmes, Don E., cell biology, mycology
Hildemann, William Henry, genetics, immunology
Howell, Richard Wesley, anthropology
Ito, Philip J., horticulture, plant breeding
Kliejunas, John Thomas, plant pathology
Ko, Wen-Hsiung, plant pathology, soil microbiology
Little, Harold Franklin, entomology
Miura, Carole K Masutani, mathematical statistics
Noda, Kaoru, parasitology
Reimer, Diedrich, genetics, animal science
Sogo, Power Bunmei, physics
Sood, Satya P., physical chemistry, polymer chemistry
Tamimi, Yusuf Nimr, soil chemistry
Thompson, John R., agronomy

HONOKAA
Tomich, Prosper Quentin, vertebrate zoology, animal ecology

HONOLULU
Adams, Bruce Gordon, microbiology, microbial physiology
Adams, William Mansfield, geophysics
Ahearn, Jayne Newton, developmental genetics
Akamine, Ernest Kisei, plant physiology
Alicata, Joseph Everett, parasitology
Allen, Richard Dean, cell biology
Alvarez, Anne Maino, plant pathology
Amundson, Mary Jane, psychiatric nursing
Andermann, George, spectroscopy
Andrews, James Einar, geological oceanography
Ansberry, Merle, speech pathology, audiology
Apt, Walter James, plant pathology, nematology
Aragaki, Minoru, plant pathology
Armstrong, R Warwick, geography
Arnold, John Miller, zoology, embryology
Ashton, Geoffrey C., genetics
Awada, Minoru, plant physiology
Baker, Harold Lawrence, forest economics
Bandermann, Lothar W., physics, astrophysics
Barkley, Richard Andrew, oceanography
Bartholomew, Duane P., plant physiology
Batkin, Stanley, neurology, neurosurgery
Bear, Herbert S, Jr., mathematics
Beardsley, John Wyman, Jr., entomology
Benedict, Albert Alfred, immunochemistry
Berg, Eduard, geophysics
Berger, Andrew John, anatomy, ornithology
Berger, Leslie Ralph, microbial physiology
Bess, Henry Alver, entomology
Bhagavan, Nadhipuram V, clinical biochemistry
Blaisdell, Richard Kekuni, internal medicine, hematology
Boesgaard, Ann Merchant, astronomy
Boggs, Stephen Taylor, anthropology
Bolman, William Merton, child psychiatry
Bonsack, Walter Karl, astrophysics
Bopp, Thomas Theodore, physical chemistry
Bowers, Neal Monroe, geography
Brantley, Lee Reed, physical chemistry
Bretschneider, Charles Leroy, physical oceanography, ocean engineering
Brewbaker, James Lynn, plant genetics
Bridges, Kent Wentworth, ecology
Brooks, Coy Clifton, animal nutrition
Brown, Thomas Townsend, physics, biophysics
Bryan, Edwin Horace, Jr., natural history
Buddemeier, Robert Worth, oceanography, hydrology
Budy, Ann Marie, physiology, pharmacology
Bullock, Richard Melvin, horticulture
Burch, Thomas Adams, epidemiology
Burr, George Oswald, biochemistry, plant physiology
Burreson, Burton Jay, natural products chemistry
Caperon, John, ecology, oceanography
Carlstead, Edward Meredith, meteorology, computer science
Carr, Gerald Dwayne, evolution
Carson, Hampton Lawrence, biology

Cence, Robert J., physics
Chan, Harvey Thomas, Jr., food science
Chang, Franklin, physiology, entomology
Chang, Jen Hu, geography
Chang, Sen-Dou, geography of China, urban geography
Char, Donald F B., public health, pediatrics
Char, Walter F., psychiatry
Chave, Keith Ernest, geology
Chiu, Wan-Cheng, meteorology
Chung, Chin Sik, human genetics, biostatistics
Contois, David Ely, microbiology
Cooil, Bruce James, plant physiology
Cooke, Ian McLean, neurophysiology, comparative physiology
Cox, Doak Carey, environmental geology
Cramer, Roger Earl, inorganic chemistry
Crooker, Peter Peirce, optical physics
De Feo, Vincent Joseph, endocrinology, reproductive biology
Demanche, Edna Louise, science education, plant physiology
Desowitz, Robert, parasitology
Diamond, Milton, reproductive biology immunology
Dizon, Andrew Edward, fish biology
Dobson, Peter N, Jr., theoretical high energy physics
Dollar, Alexander M, animal nutrition, food biochemistry
Doty, Maxwell Stanford, phycology
Ekern, Paul Chester, soil physics, hydrology
El-Swaify, Samir Aly, soil science, soil chemistry
Emory, Kenneth Pike, anthropology
Fadley, Charles Sherwood, chemical physics
Fan, Pow-Foong, geology
Finney, Ben R, anthropology
Folsome, Clair Edwin, microbial genetics, exobiology
Force, Roland Wynfield, anthropology
Fox, Robert Lee, soils
Frank, Hilmer Aaron, food microbiology
Fuchs, Roland John, economic geography, geography of the Soviet Union
Fujioka, Roger Sadao, virology, water pollution
Fuller, Gary Albert, population geography, cultural geography
Furumoto, Augustine S, seismology
Furusawa, Eiichi, chemotherapy, virology
Gaines, Sidney, medical bacteriology
Gallagher, Brent S, physical oceanography
Gibbons, Ian Read, molecular biology
Gibbons, Barbara Hollingworth, biochemistry
Gilbert, James Carl, plant breeding
Gilje, John, inorganic chemistry
Gillary, Howard L., neurophysiology
Goto, Shosuke, agriculture
Gould, Richard Allan, cultural anthropology, archaeology
Green, Richard E, soil physics
Greenwood, Frederick C, biochemistry, endocrinology
Gregory, Christopher, physics
Gressitt, Judson Linsley, biogeography, entomology
Griffith, Jeffrey Knowles, developmental biology
Groves, Gordon William, oceanography
Guillory, Richard, biochemistry
Gundersen, Kaare Reinhardt, microbiology
Hadfield, Michael Gale, invertebrate zoology
Haley, Samuel Randolph, developmental anatomy
Hall, John Bradley, microbial physiology
Halstead, Scott Barker, medical microbiology, tropical medicine
Hamilton, Richard Airth, agriculture
Hammar, Sherrel L., pediatrics, medicine
Hammond, Stephen Randolph, marine geophysics
Hankin, Jean H, nutrition, epidemiology
Hanna, Joel Michael, physical anthropology
Haramoto, Frank H, entomology
Hardy, D Elmo, entomology
Hardy, Wilton Audobon, physics, oceanography
Harris, Ernest James, entomology
Hartmann, Richard W, plant breeding
Hartroft, Walter Stanley, experimental pathology, nutrition
Hayes, Charles Franklin, physics
Heinz, Don J, plant breeding, plant genetics

Helfrich, Philip, marine biology
Henke, Burton Lehman, physics
Hepton, Anthony, vegetable crops
Herzberg, Mendel, rickettsial diseases
Higa, Harry Hiroshi, medical microbiology
Hilker, Doris M, nutrition, biochemistry
Hirohata, Tomio, epidemiology, cancer biology
Hiu, Dawes Nyukeu, organic chemistry
Hokama, Yoshitsugi, immunology, pathology
Holmes, John Richard, atomic spectroscopy
Holtzman, Oliver Vincent, plant pathology
Hong, Pill Whoon, surgery
Howard, Alan, cultural anthropology
Hsiao, Sidney Chihti, developmental biology
Hubbard, Arthur T, analytical chemistry, electrochemistry
Humphreys, Susie Hunt, cell biology
Humphreys, Tom Daniel, developmental biology
Hunt, John A, molecular genetics
Hussong, Donald MacGregor, marine geophysics
Hylin, John Walter, plant biochemistry
Iha, Franklin Takashi, mathematical analysis
Ihrig, Judson La Moure, physical chemistry
Inatsugu, Seiji, experimental atomic physics, radiological physics
Inskeep, Richard Guy, physical chemistry
Ishii, Mamoru, plant pathology
Itoga, Stephen Yokio, computer science
Jacobs, Virgil Leon, anatomy
Jeffries, John Trevor, astrophysics
Johnson, James Stephen, mathematics
Johnson, Jerry Michael, environmental health
Johnson, Rockne Hart, marine geophysics
Kamemoto, Fred Isamu, zoology
Kamemoto, Haruyuki, horticulture
Kane, Robert Edward, cell biology
Kanehiro, Yoshinori, soil science
Kaneshiro, Kenneth Yoshimitsu, evolutionary biology
Kay, Elizabeth Alison, biology
Kefford, Noel Price, plant physiology
Khan, Mohammad Asad, geodesy
Kiefer, Edgar Francis, organic chemistry
Kinzie, Robert Allen, III, zoology
Kirch, Patrick Vinton, anthropology
Kleinfeld, Ruth Gratman, cell biology, reproductive biology
Klemmer, Howard Wesley, microbiology
Kloetzel, Milton Carl, organic chemistry
Kokame, Glenn Megumi, surgery, thoracic surgery
Kondo, Yoshio, zoology
Kornhauser, David Henry, geography of Japan, urban geography
Kortschak, Hugo Peter, plant chemistry
Krauss, Beatrice Hilmer, plant physiology, ethnobotany
Kroenke, Loren William, marine geology
Kroopnick, Peter Michael, oceanography, geochemistry
Lamoureux, Charles Harrington, botany
Landman, Donald Alan, atomic physics, solar physics
LaPlante, Albert Aurel, Jr., economic entomology
Larson, Harold Olaf, organic chemistry
Laurila, Simo Heikki, geodesy
Laws, Edward Allen, oceanography
Lebra, Takie Sugiyama, anthropology
Lee, William Philip, anthropology
Lenney, James Francis, pharmacology
Liang, Tung, operations research, bioengineering
Lichton, Ira Jay, physiology
Lieban, Richard Warren, cultural anthropology
Lin, Yu-Chong, physiology
Liu, Robert Shing-Hei, organic chemistry
Loh, Philip Choo-Seng, virology
Loomis, Harold George, mathematics
Losey, George Spahr, Jr., marine zoology, ethology
Lum, Bert Kwan Buck, pharmacology
Luomala, Katharine, cultural anthropology, ethnography
Macdonald, Gordon Andrew, geology
Maciolek, John A, limnology
Mader, Adolf algebra
Malecha, Spencer R, genetics
Mamiya, Richard T, thoracic surgery, cardiovascular surgery
Mandel, Morton, molecular biology
Manghnani, Murli Hukumal, geophysics, geochemistry

Manoharan, Arthur, occupational health, public health
Marchette, Nyen John, virology
Maretzki, Andrew, biochemistry
Mason, Leonard Edward, anthropology
McDonald, Ray Locke, physical chemistry
McCain, John Charles, marine biology
McCall, Wade Wiley, soil fertility
McConnell, Bruce, biochemistry
McDermott, John F, Jr., psychiatry, child psychiatry
Massey, Douglas Gordon, medicine
Matsumoto, Hiromu, agricultural biochemistry
McKay, Robert Harvey, biochemistry
McNamara, Joseph Judson, surgery
McPherson, Donald Frank, audiology
Merriam, Robert Arnold, forest hydrology
Mi, Ming-Pi, genetics
Mitchell, Wallace Clark, entomology
Moberly, Ralph M, geology
Montalvo, Francisco Emilio, geophysics
Moore, Richard E, organic chemistry
Morisugu, Toshio, sugar chemistry
Morris, Gerald Brooks, geophysics
Morrison, David Douglas, astronomy
Morton, Bruce Eldine, biochemistry, reproductive biology
Morton, Newton Ennis, population genetics
Moser, Roy Edgar, food technology
Mower, Howard Frederick, organic chemistry
Mueller-Dombois, Dieter, plant ecology
Muenow, David W, physical chemistry
Murakami, Takio, meteorology
Murdoch, Charles Loraine, horticulture
Murphy, Garth Ivor, fish biology
Nakane, Henry Yoshiki, horticulture
Nakata, Shigeru, plant physiology
Nakayama, Tommy, food science
Namba, Ryoji, entomology
Naughton, John Joseph, chemistry
Nelson, Marita Lee, human anatomy, endocrinology
Nelson, Robert Eldon, forestry
Newhouse, W Jan, phycology, science education
Nishimoto, Roy Katsuto, weed science, vegetable crops
Nishimura, Edwin Takayasu, experimental pathology
Nitz, Robert E, medicine
Nobusawa, Nobuo, algebra
Norton, Ted Raymond, organic chemistry, pharmacology
Odegard, Mark Eric, seismology, marine geophysics
Oishi, Noboru, oncology, immunohematology
Olbrich, Steven Emil, diary science
Orrall, Frank Quimby, astrophysics
Ortner, Mary Joanne, pharmacology
Ota, Asher Kenhachiro, entomology
Otagaki, Kenneth Kengo, animal nutrition
Pager, David, computer science, mathematics
Pakvasa, Sandip, physics
Palafox, Anastacio Laida, nutrition, poultry science
Palumbo, Nicholas Eugene, comparative medicine
Pankiwskyj, Kost Andrij, geology
Parvulescu, Antares, mathematical physics, acoustics
Patil, Suresh Siddheshwar, plant pathology
Pecsok, Robert Louis, analytical chemistry
Peters, Michael Wood, physics
Peterson, Frank Lynn, geology, hydrology, groundwater
Pietruszewsky, Michael, Jr., physical anthropology
Piette, Lawrence Hector, biophysics
Pike, Arthur Clausen, meteorology
Pilcher, Carl Bernard, planetary sciences
Pion, Ronald Joseph, obstetrics & gynecology
Pirie, Peter Nigel Douglas, population geography
Pitcher, Tom Stephen, mathematics
Pitts, Forrest Ralph, geography
Pong, William, physics
Popper, Arthur N, zoology, neurosciences

Porta, Eduardo Angel, experimental pathology
Preisendorfer, Rudolph W., physical oceanography
Prowse, Gerald Albert, marine phycology
Putman, Edison Walker, plant biochemistry
Pyle, Terence Lawrence, meteorology
Quisenberry, Walter Brown, preventive medicine, cancer
Radovsky, Frank Jay, acarology, medical entomology
Ramage, Colin Stokes, meteorology
Randall, John Ernest, marine biology
Rauch, Fred D., horticulture, plant physiology
Read, George Wesley, pharmacology
Reed, Stuart Arthur, zoology, physiology
Reese, Ernst S., animal behavior, ecology
Rogers, Beverly Jane, biochemistry, reproductive biology
Rogers, Terence Arthur, physiology
Rohrbach, Kenneth G., plant pathology
Rose, John Creighton, geophysics
Rosen, Leon, epidemiology
Ross, Ernest, poultry nutrition
Rotar, Peter P., crop breeding, cytogenetics
Rubin, Joan, anthropology
Sadler, James C., meteorology
Sagawa, Yoneo, cytogenetics
St John, Harold, botany
Sanford, Wallace Gordon, plant physiology
Schaeger, Larry L., organic chemistry
Scheuer, Paul Josef, organic chemistry
Schwind, Paul Jackson, urban geography
Seff, Karl, physical chemistry, crystallography
Seo, Stanley Toshio, chemistry
Shaklee, James Brooker, fish biology, biochemical genetics
Sherman, Martin, entomology
Shibata, Shoji, pharmacology
Shomura, Richard Sunao, fish biology
Siddiqui, Wasim A., parasitology, tropical medicine
Siegel, Barbara Zenz, biology
Siegel, Sanford Marvin, biochemistry, environmental chemistry
Silva, James Anthony, soil fertility
Simeon, George John, medical anthropology
Sinoto, Yoshiko, anthropology
Sinton, William Merz, astronomy
Skillman, Robert Allen, fisheries management, population ecology
Skinsnes, Olaf Kristian, pathology
Skolmen, Roger Godfrey, tree physiology, forest products
Smith, Albert Carl, pathobiology, pathology
Smith, Richard Merrill, physiology
Smith, Richard S, Jr, plant pathology
Solheim, Wilhelm Gerhard, II, anthropology
Standal, Bluebell R S'Yiem, nutrition
Stanley, Richard W., nutrition, biochemistry
Steffan, Wallace Allan, entomology
Steiger, Walter Richard, physics
Steiner, Gary W., plant pathology
Stemmermann, Grant N., pathology
Stenger, Victor John, physics
Street, John Malcolm, biogeography
Struhsaker, Paul James, marine zoology, fisheries
Sutton, George Harry, seismology, geophysics
Swindale, Leslie D., soil science, research administration
Tamashiro, Minoru, insect pathology
Tang, Chung-Shih, plant chemistry
Taussig, Steven J., industrial microbiology
Taylor, Diane, tropical medicine
Taylor, Leighton Robert, Jr, ichthyology
Tesh, Robert Bradfield, epidemiology
Tinker, Spencer Wilkie, ichthyology
Townsley, Sidney Joseph, radiobiology
Trujillo, Eduardo E, plant pathology
Tsuji, Gordon Yukio, soil physics
Tu, Chen Chuan, chemistry
Tuan, San Fu, theoretical physics, applied mathematics
Uchida, Richard Noboru, marine sciences
Vann, Douglas Carroll, immunobiology
Van Reen, Robert, biochemistry
Van Royen, Pieter, biogeography, plant taxonomy
van Weel, Pieter Boudewijn, physiology
Vitousek, Martin J., geophysics
Vogt, Dale William, animal breeding, animal genetics
Voulgaropoulos, Emmanuel, public health
Walker, Daniel Alvin, seismology
Wallace, Gordon Dean, epidemiology
Warner, John Northrup, plant breeding, agronomy

Warner, Robert. Malcolm, ecology, plant physiology
Wasfi, Sadiq Hassan, inorganic chemistry
Watanabe, Michael Satosi, theoretical physics, information science
Waugh, John Lodovick Thomson, physical inorganic chemistry
Wayman, Oliver, animal physiology, animal nutrition
Wells, Benjamin B, Jr, mathematical analysis
Wenska, Tom Marion, mathematics
Wester, Lyndon Leonard, geography
Whittow, George Causey, physiology
Wolff, Richard James, astrophysics
Wolff, Sidney Carne, astrophysics
Wolstencroft, Ramon David, astrophysics
Woodcock, Alfred Herbert, oceanography
Woollard, George Prior, geophysics
Worth, Robert McAlpine, epidemiology
Wyrtki, Klaus, physical oceanography
Yamamoto, Harry Y., biochemistry, food technology
Yamauchi, Hiroshi, theoretical physics
Yanagimachi, Ryuzo, reproductive biology
Yasunobu, Kerry T., biochemistry
Young, Franklin, nutrition, biochemistry
Young, Hong Yip, agricultural chemistry
Young, Richard Edward, biological oceanography
Yount, David Eugene, elementary particle physics
Ziegler, Alan Conrad, mammalogy
Zirker, Jack Bernard, solar physics

KAILUA
Bonnet, David Dudley, medical entomology
Otsu, Tamio, marine biology
Peterson, Vincent Zetterberg, experimental high energy physics
Shallenberger, Robert Jenkins, zoology
Weinbaum, Carl Mattin, mathematics

KAILUA KONA
Alexander, Earl Glynn, resource management, food science

KANAI
Theobald, William L, systematic botany, horticulture

KAPAA
Ooka, Jeri Jean, phytopathology

KULA
Carter, William Eugene, geodesy
Mickey, Donald Lee, astrophysics
Parvin, Philip Eugene, floriculture
Statton, William Osborne, polymer physics
Whitney, Arthur Sheldon, agronomy

LAIE
Andersen, Dean Martin, medical entomology, environmental biology
Berrett, Delwyn Green, ornithology
Coburn, Richard Karl, applied mathematics
Dalton, Patrick Daly, ecology, botany
Nicholes, Henry Joseph, medical physiology
Wrathall, Jay W., inorganic chemistry

LIHUE
Kikuchi, William Kenji, anthropology, archaeology

MAKAWAO
Williams, David Douglas F, horticulture

MANOA
Hoffmann, Joan Carol, physiology

WAHIAWA
Manly, Philip James, health physics

WAIALUA
Degener, Otto, botany

WAIANAE
Ellis, Nathan Kent, horticulture

WAIMANALO
Bienfang, Paul Kenneth, biological oceanography
Hanson, Joe A., resource management, science writing
Kuo, Ching-Ming, reproductive physiology, ichthyology
Nash, Colin Edward, resource management
Steinbach, Henry Burr, zoology, physiology

IDAHO

ABERDEEN
Corsini, Dennis Lee, agricultural biochemistry
Davis, James Robert, plant pathology
Douglas, Dexter Richard, plant pathology, botany
McDole, Robert E, soil fertility, soil genesis
Pavek, Joseph John, plant genetics, plant breeding
Sparks, Walter Chappel, horticulture
Stallknecht, Gilbert Franklin, plant physiology, plant biochemistry
Wesenberg, Darrell, agronomy, genetics

BLACKFOOT
Huber, Clayton Shirl, food science
Polzer, Wilfred L, geochemistry, environmental sciences

BOISE
Applegate, James Keith, exploration geophysics, geotechnical engineering
Banks, Richard C., organic chemistry
Bush, David Clair, organic chemistry
Bushnell, Vernon Clifford, soil chemistry
Carter, Loren Sheldon, physical chemistry, inorganic chemistry
Dalton, Jack L, biochemistry, organic chemistry
Duke, Victor Hal, pharmacology, pharmacognosy
Ferguson, David John, mathematics
Fritchman, Harry Kerr, II, invertebrate zoology
Fuller, Eugene George, embryology, histology
Hibbs, Robert A., chemistry, bacteriology
Hollenbaugh, Kenneth Malcolm, environmental geology
Jones, Leo Edward, plant morphogenesis
Juola, Robert C., statistics, mathematics
Kelley, Fenton Crosland, mammalian physiology, environmental physiology
Luke, Robert A., particle physics
Maloof, Giles Wilson, mathematics, geophysics
Meade, James Montgomery, resource geography, population geography
Mech, William Paul, mathematics
Obee, Donald Jennings, botany
Platts, William Sidney, fisheries
Smith, Wendell Eugene, fish biology
Spinosa, Claude, paleontology, geology
Stokes, Lee W. Boley, water pollution
Tracy, Joseph Walter, inorganic chemistry, meat science
Valcarce, Arland Casiano, entomology
Ward, Frederick Roger, mathematics
Warner, Mont Marcellus, geology
Wilson, Monte Dale, geology
Wyllie, Gilbert Alexander, biology, ecology

CALDWELL
Baird, Craig Riska, entomology
Blood, Franklin Harvey, inorganic chemistry
Bratz, Robert Davis, biology, ecology
Henry, Boyd (Herbert), mathematics

Higdem, Roger Leon, mathematics
Howes, Arol Dean, animal nutrition
Hunt, Gilbert John, chemistry
Marshall, James Dale, zoology, entomology
Packard, Patricia Lois, systematic botany
Stanford, Lyle Morris, zoology, botany
Waldhalm, Donald George, veterinary microbiology
Willmorth, John H., solid state electronics

COEUR D'ALENE
Wright, Kenneth James, physical inorganic chemistry, environmental sciences

DUBOIS
Hulet, Clarence Veloid, animal science

EMMETT
Knisely, Ralph Marion, pathology

HOPE
Stearns, Harold Thornton, geology

IDAHO FALLS
Baldwin, Jon Michael, analytical chemistry, spectroscopy
Berreth, Julius R., nuclear chemistry, physical chemistry
Bills, Charles Wayne, inorganic chemistry, organic chemistry
Booman, Glenn Lawrence, analytical chemistry
Brunson, Glenn Samuel, Jr, nuclear physics
Burgus, Warren Harold, radiochemistry
Dahl, Adrian Hilman, biophysics
Delmastro, Ann Mary, analytical chemistry
Delmastro, Joseph Raymond, analytical chemistry
Dixon, Carl Eugene, solid state physics
Einziger, Robert Emanuel, metal physics
Gehrke, Robert James, experimental nuclear physics
Greenwood, Reginald Charles, nuclear physics
Hammer, Robert Russell, physical inorganic chemistry
Harker, Yale Deon, reactor physics
Heath, Russell La Verne, nuclear physics
Helmer, Richard Guy, experimental nuclear physics
Kunze, Jay Frederick, physics
Lewis, Leroy Crawford, physical chemistry, inorganic chemistry
Lott, Layman Austin, physics
Makens, Royal Francis, nuclear science
McCaslin, John Weaver, health physics, safety engineering
McClure, John Arthur, mathematical analysis
Moore, Kenneth Virgil, nuclear science
Nyer, Warren Edwin, physics
Philipson, Joseph Bion, physics, economics
Reich, Charles William, nuclear physics
Rhodes, Donald Walter, applied chemistry
Rohde, Kenneth Lincoln, organic chemistry
Rutledge, Gene Preston, physical chemistry, research administration
Shank, Ralph Chalmer, analytical chemistry
Sill, Claude Woodrow, analytical chemistry
Slansky, Cyril Method, chemistry
Thalgott, Fred William, mathematics
Walker, Donald I, analytical chemistry, chemical microscopy
Wheeler, Gilbert Vernon, chemistry
Willard, Harold James, Jr, solid state physics, nuclear engineering
Wood, Richard Ellet, nuclear physics
Young, Robert Carl, theoretical physics, computer sciences
Young, Thomas Edward, nuclear physics
Zack, Neil Richard, inorganic chemistry, fluorine chemistry

JEROME
Massee, Truman Winfield, soil fertility

KELLOGG
Weiss, Michael Karl, chemistry

IDAHO

KIMBERLY
Carpenter, Gene Paul, economic entomology
Carter, David LaVere, soil chemistry
Carter, John Newton, agronomy
Dean, Leslie L., plant pathology, plant breeding
Kleinkopf, Gale Eugene, plant physiology
Kolar, John Joseph, plant breeding
LeBaron, Marshall John, field crops
Leggett, Glen Eugene, soil science
Mayland, Henry Frederick, soil science
Smith, Jay Hamilton, soil microbiology
Westermann, D T, soil science, soil chemistry
Wright, James Louis, agronomy, agricultural meteorology

KUNA
Neunauter, Lloyd Joseph, veterinary public health, laboratory animal medicine

MALAD
Jones, Lewis William, bacteriology

LEWISTON
Blair, Roger L., forest genetics
Brown, John Henry, wood science, wood technology
Laval, William Norris, geology
McKean, Herbert Baldwin, wood technology
Sutton, John Curtis, organic chemistry, analytical chemistry

MOSCOW
Ables, Ernest D., wildlife ecology, zoology
Anderegg, Doyle Edward, lichenology
Anderson, Guy Richard, soil microbiology
Ardrey, William Boyle, bacteriology
Augustin, Jorg A L, food science, plant biochemistry
Baker, William Hudson, botany
Barnhart, John Lorce, dairy science
Barr, William Frederick, entomology
Beck, Sidney M., bacteriology
Bell, Thomas Donald, animal husbandry
Bell, George Harley, Jr, micrometeorology, forest hydrology
Bethlahmy, Nedavia, forest hydrology
Bishop, Guy William, entomology
Bobisud, Larry Eugene, mathematics
Boe, Arthur Amos, horticulture
Bond, John Gilbert, stratigraphy, regional geology
Browne, Michael Edwin, physics
Brusven, Merlyn Ardel, entomology
Bull, Richard C., animal nutrition
Burrows, George Edward, veterinary pharmacology, veterinary toxicology
Calpouzos, Lucas, plant pathology
Campbell, Howard Ernest, mathematics
Canfield, Elmer Russell, forest pathology, mycology
Christenson, Charles O., topology
Christian, Ross Edgar, physiology, genetics
Cobb, John Iverson, topology
Cooley, James Hollis, organic chemistry
Dahmen, Jerome J., genetics, reproductive physiology
Davis, Lawrence William, Jr, physics
Davis, Steven Lewis, endocrinology
Day, Richard Lorey, geography
Deitschman, Glenn Howard, forest management
Ehrenreich, John Helmuth, forestry, range management
Ensign, Ronald D., agronomy, crop breeding
Erickson, Lambert Cornelius, weed science, agronomy
Eroschenko, Victor Paul, human anatomy
Everson, Dale O., statistics, applied statistics
Fenwick, Harry, microbiology
Ferguson, James Homer, environmental physiology
Finley, Arthur Marion, plant pathology
Flinders, Jerran Trueman, wildlife ecology, range ecology
Forbes, Oliver Clifford, genetics
Fosberg, Maynard Axel, soil morphology
Frank, Floyd William, veterinary medicine, veterinary microbiology
Garrard, Verl Grady, physical chemistry, analytical chemistry
Gilmour, Campbell Morrison, bacteriology
Gittins, Arthur Richard, entomology
Grahn, Edgar Howard, inorganic chemistry
Grieb, Merland William, inorganic chemistry
Gustafson, Donald Arvid, chemistry
Guthrie, James Warren, phytopathology
Hall, William Bartlett, geology
Hartung, Ernest William, zoology, developmental genetics
Heimsch, Richard Charles, food microbiology
Helton, Audus Winzle, plant pathology
Henderson, Douglass Miles, biosystematics, botany
Hodgson, Charles Worth, animal husbandry
Howe, John Prentice, forestry
Hungerford, Kenneth Eugene, wildlife ecology, management
Jacobs, John Allen, animal science
Johnson, Donald Ralph, vertebrate ecology
Johnson, Frederic Duane, forest ecology
Johnston, Lawrence Harding, optical physics
Jones, James Preston, soil fertility, plant nutrition
Jones, Robert William, geology
Kearney, Robert James, solid state physics
Kluetz, Michael David, biophysical chemistry
Knudson, Ruthann, anthropology, archaeology
Kraus, James Ellsworth, horticulture
Leaphart, Charles Donald, forest pathology
Lee, Gary Albert, weed science
LeTourneau, Duane John, biochemistry
Lewis, Glenn C., soil chemistry
Lingg, Al Joseph, microbiology
Loewenstein, Howard, forestry
MacPhee, Craig, zoology
McCroskey, Jack E., animal nutrition
McDonald, Geral Irving, plant pathology
McKean, Thomas Arthur, physiology
McMullen, John Lloyd, botany
Mead, Rodney A., reproductive physiology, endocrinology
Meserve, Peter Lambert, ecology
Miller, Raymond Jarvis, soil chemistry, physical chemistry
Montoure, John Ernest, dairy science, agricultural chemistry
Moslemi, Ali A., forest products
Mullins, Auttis Marr, meat sciences, food technology
Muneta, Paul, food science
Murray, Glen A., crop physiology
Naskali, Richard John, botany
Naylor, Denny Ve, agronomy
O'Keeffe, Lawrence Eugene, entomology
Partridge, Arthur Dean, phytopathology
Patsakos, George, theoretical physics
Peck, Edson Ruther, physics
Peck, James Merrell, wildlife ecology
Petersen, Charlie Frederick, poultry nutrition
Pope, Warren Kirkpatrick, agronomy
Porter, Richard A., physical chemistry
Powell, James Daniel, geology
Raunio, Elmer Kauno, geology
Reid, Rolland Ramsay, geology
Renfrew, Malcolm MacKenzie, physical chemistry
Rice, David Gordon, archaeology
Roberts, Lorin Watson, plant physiology
Ross, Richard Henry, dairy science
Rourke, Arthur W., cell physiology
Ruboston, George M., organic chemistry
Sauter, Erwin Andrew, food science, microbiology
Savage, Carleton Norman, geology
Schell, Stewart Claude, parasitology
Schenk, John Albright, forest entomology
Scripter, Morton W., geography
Seale, Robert Henry, forest economics
Seely, Clarence Ivan, agronomy
Sharp, Lee Ajax, range management
Shreeve, Jean'ne Marie, inorganic chemistry
Sieckmann, Everett Frederick, theoretical physics, experimental solid state physics
Siems, Peter Laurence, geology
Smiley, Charles Jack, geology, paleobotany
Smith, Howard Weedon, zoology, paleontology
Sponer, George Guy, plant physiology, ecology
Sprague, Roderick, anthropology, archaeology
Stark, Ronald William, forest entomology
Teresa, George Washington, bacteriology
Tisdale, Edwin William, range science
Tyutki, Edmund Eugene, mycology
Voxman, William L., topology
Wai, Chien Moo, nuclear chemistry, geochemistry
Wang, Chi-Wu, forest genetics
Wang, Ya-Yen Lee, mathematics, computer science
Watson, Roscoe Derrick, plant pathology
Wicker, Ed Franklin, plant pathology, forestry
Wiese, Alvin Carl, biochemistry
Willmes, George Arthur, geology, engineering
Williams, Roy Edward, hydrogeology
Williams, Henry, nuclear physics
Wohletz, Ernest W., forestry
Zaehringer, Mary Veronica, foods

NEW MEADOWS
Smith, George Dale, veterinary medicine

NAMPA
Marks, Darrell L., physics
Tillotson, Donald Bearse, mathematics

PARMA
Franklin, DeLance Flournoy, horticulture
Kochan, Walter J., plant physiology
Romanko, Richard Robert, plant pathology
Simpson, William Roy, plant pathology
Walz, Arthur Joseph, economic entomology
Waters, Norman Dale, entomology

POCATELLO
Anderson, Robert Curtis, entomology
Arcand, George Myron, analytical chemistry, inorganic chemistry
Benson, Ernest Phillip, Jr, inorganic chemistry
Beuthin, Frederic C., pharmacology
Bigelow, Melvin Jerome, organic chemistry
Bowner, Richard Glenn, plant physiology
Braun, Loren L., organic chemistry
Bunde, Daryl E., zoology, physiology
Cole, Franklin Ruggles, pharmacognosy
Cosgrove, Frank P., pharmacy
Faler, Kenneth Turner, nuclear chemistry, environmental chemistry
Fichter, Edson Harvey, animal ecology, mammalogy
Fleischmann, William Robert, Jr, animal virology
Fontenelle, Lydia Julia, biochemistry
Goettsch, Robert Wayne, pharmaceutics, biopharmaceutics
Heckler, George Earl, physical chemistry
Higgins, Nevada Marie, pharmacy
Hilzman, John, mathematical analysis
Holte, Karl E., nuclear chemistry, environmental chemistry
House, Edwin W., animal physiology
Isaacson, Eugene I., pharmaceutical chemistry, organic chemistry
Ison, Gary E., pharmacology
Jarvis, Francis George, bacteriology
Johnson, Donald William, environmental physiology, fish biology
Laughlin, James Stanley, operations research
Linder, Allan David, vertebrate zoology
McCune, Mary Joan Huxley, microbiology
McCune, Ronald William, biochemistry
Minshall, Gerry Wayne, aquatic biology, aquatic ecology
Ore, Henry Thomas, sedimentology
Parker, Barry Richard, theoretical physics
Price, Joseph Earl, physics
Ronald, Bruce Pender, organic chemistry
Rose, Frederick Louis, Jr, aquatic biology, botany
Seeley, Rod R., reproductive physiology
Taylor, Albert Edward, analytical chemistry
Thompson, Joseph Lippard, radiochemistry, physical chemistry
Trost, Charles Henry, physiological ecology
Tullis, James Earl, genetics
Vegors, Stanley H. Jr, nuclear physics
Watson, Ralph A., exploration geol, mineral economics
White, John Anderson, vertebrate zoology, paleontology
Wiegand, Gayl, organic chemistry

ARGO
Armbruster, Frederick Carl, microbiology, biochemistry
Cleveland, Elonza Alexander, Jr, chemistry
Danzig, Morris Judah, organic chemistry, inorganic chemistry
Harjes, Clarence Frank, organic chemistry
Kite, Francis Ervin, cereal chemistry
Kooi, Earl Robert, biophysical chemistry
Molotsky, Hyman Max, carbohydrate chemistry
Parmerter, Stanley Marshall, organic chemistry
Paschall, Eugene F., carbohydrate chemistry
Pomes, Adrian Francis, sugar chemistry
Turner, Hal Russell, microbial physiology
Watson, Stanley Arthur, biochemistry
Zobel, Henry Freeman, physical chemistry, analytical chemistry

ARGONNE
Abraham, Bernard M., chemistry
Ackermann, Raymond J., high temperature chemistry
Adams, Richard Melverne, physical chemistry
Ahmad, Irshad, nuclear chemistry
Ainsworth, Earl John, toxicology, radiobiology
Allender, James Harry, physical oceanography
Altenberger-Siczek, Aldona, theoretical chemistry
Appleman, Evan Hugh, inorganic chemistry
Arko, Aloysius John, solid state physics
Atoji, Masao, physical chemistry
Avery, Robert, theoretical physics
Ayres, David Smith, experimental high energy physics
Baker, Louis, Jr., physical chemistry
Band, Yehuda Benzion, atomic physics, quantum physics
Basile, Louis Joseph, molecular spectroscopy, inorganic chemistry
Beck, William Nelson, mathematics, physics
Beitinger, Thomas Lee, physiological ecology
Benioff, Paul, physics
Bennett, Edgar F., reactor physics
Berger, Edmond Louis, high energy physics
Berkowitz, Joseph, physical chemistry
Bhattacharyya, Maryka Horsting, biochemistry
Blackburn, Paul Edward, high temperature chemistry
Blander, Milton, physical chemistry
Blewitt, Thomas Hugh, solid state physics
Bollinger, Lowell Moyer, nuclear physics
Boyle, James Martin, computer sciences

REXBURG
Biddulph, Lowell George, geology
Hibbert, Larry Eugene, parasitology, protozoology
Hoggan, Roger D., geology
Lindsay, Delbert W., systematic botany
Winkel, Cleve R., biochemistry, organic chemistry

TWIN FALLS
Butler, Calvin Charles, mathematical statistics
Lambern, Calvin Ray, plant virology, plant breeding
Morris, John Leonard, plant breeding, plant pathology
Moser, Paul E., plant breeding, plant pathology
Ohms, Richard Earl, plant pathology
Schweitzer, Leland Ray, agronomy, vegetable crops
Stoltz, Robert Lewis, entomology

WEISER
Holt, Charlene Poland, pediatrics

ILLINOIS

ALTON
Bahn, Arthur Nathaniel, microbiology, dental research
Kadis, Barney Morris, biochemistry

Braid, Thomas Hamilton, nuclear physics
Brennan, Patricia Conlon, microbiology, immunology
Bretscher, Manuel Martin, nuclear physics
Brodsky, Merwyn Berkley, solid state physics
Brown, Bruce Stilwell, solid state physics
Brues, Austin M, medicine, radiobiology
Buck, Warren Louis, radiation physics
Buffington, John Douglas, ecology, entomology
Butler, Margaret K, mathematics
Cameron, Roy (Eugene), soil microbiology, ecology
Carroll, Kenneth Girard, physics
Carson, James Estle, meteorology
Chasanov, Martin Gerson, physical chemistry
Cheever, Charles Lyle, industrial hygiene, environmental health
Chellew, Norman Raymond, inorganic chemistry
Childs, William Jeffries, atomic physics
Cody, William James, Jr, numerical analysis
Coester, Fritz, physics
Cohen, Donald, radiochemistry
Cohen, Stanley, theoretical physics
Cohn, Charles Erwin, experimental physics
Connor, Donald W, solid state physics
Cook, Joseph Marion, applied mathematics
Cowell, Wayne Russell, applied mathematics
Crespi, Henry Lewis, physical chemistry
Crosbie, Edwin Alexander, physics
Crosswhite, Henry Milton, Jr, physics
Crouse, David Austin, radiobiology
Cunningham, Paul Thomas, physical chemistry
Daniels, Edward William, cell biology
Das, Santosh Kumar, surface physics
Davids, Cary Nathan, nuclear physics, astrophysics
Day, Benjamin Downing, theoretical physics, nuclear physics
Day, Paul Palmer, physics
Dehmer, Joseph Leonard, chemical physics, atomic physics
Dehmer, Patricia Moore, chemical physics, molecular physics
Delbecq, Charles Jarchow, solid state chemistry
Derrick, Malcolm, experimental high energy physics
De Volpi, Alexander, reactor physics
Diamond, Herbert, nuclear chemistry
Dickerman, Charles Edward, nuclear science
Diebold, Robert Ernest, high energy physics
Draley, Joseph Edward, chemistry
Dudey, Norman, nuclear chemistry
Dunlap, Bobby David, experimental solid state physics
Eberhart, James G, surface chemistry
Edmundson, Allen B, biochemistry
Eggenberger, Delbert Norgaard, applied physics
Ekern, Ronald James, surface physics
Elkind, Mortimer M, biophysics, radiobiology
Elwyn, Alexander Joseph, experimental nuclear physics
Epstein, Leo Francis, physical chemistry
Erskine, John Robert, nuclear physics
Evans, Allan Robert, solid state physics
Failla, Patricia McClement, biophysics
Fairman, William Duane, analytical chemistry, radiological health
Falco, Charles Maurice, physics
Feinstein, Robert Norman, enzymology, biochemistry
Ferraro, John Ralph, molecular spectroscopy
Finkel, Miriam Posner, oncology, radiobiology
Fischer, Albert Karl, physical inorganic chemistry
Flynn, Robert James, laboratory animal medicine
Foss, Martyn (Henry), physics
Foster, Melvin S, physical chemistry
Fred, Mark Simon, physics
Freedman, Melvin Slein, experimental nuclear physics
Frigerio, Norman Alfred, oncology, environmental health
Fritz, Thomas Edward, pathology, radiobiology
Fry, Richard Jeremy Michael, physiology, radiobiology
Gabelnick, Stephen David, physical chemistry
Gabriel, John R, applied mathematics, computer science
Gemmell, Donald Stewart, experimental physics
Gilbert, Thomas Lewis, theoretical physics
Gindler, James Edward, nuclear chemistry
Givens, Wallace, (Jr), mathematics
Glendenin, Lawrence Elgin, nuclear chemistry
Golchert, Norbert William, environmental radiochemistry
Goodman, Leonard Seymour, nuclear physics
Gordon, Sheffield, radiation chemistry
Grahn, Douglas, genetics, radiobiology
Green, David William, physical chemistry
Gruen, Dieter Martin, chemical physics, surface chemistry
Guilmette, Raymond Alfred, radiological health
Gustafson, Philip Felix, environmental sciences
Guttman, Lester, solid state physics
Hafstrom, John William, low temperature physics
Hanson, Wayne Robert, radiobiology, gastroenterology
Harrison, Wyman, geoenvironmental science
Hatch, Albert Jerold, plasma physics
Hayatsu, Ryoichi, organic chemistry
Hess, David Clarence, physics
Hess, George Dale, meteorology, oceanography
Hicks, Bruce Boundy, atmospheric physics
Hinchman, Ray Richard, plant ecology, environmental biology
Hinks, David George, physical chemistry
Hoegerman, Stanton Fred, genetics, radiobiology
Hoekstra, Henry Raymond, chemistry
Holland, Robert Emmett, experimental nuclear physics
Holt, Ben Dance, analytical chemistry
Holt, Roy James, nuclear physics
Holtzman, Richard Beves, radiochemistry, radiobiology
Horwitz, Earl Philip, inorganic chemistry
Hubble, Billy Ray, physical chemistry
Huebner, Russell Henry, electron physics, radiation physics
Hummel, Harry Horner, physics
Hyman, Lloyd George, experimental high energy physics
Inokuti, Mitio, molecular physics
Jackson, Harold E, Jr, nuclear physics
Jaeger, Klaus Bruno, experimental high energy physics
Jaffey, Arthur Harold, nuclear chemistry
Jankus, Vytautus Zachary, nuclear science
Janson, Thomas Ralph, physical chemistry, physical biochemistry
Jaroslow, Bernard Norman, immunology
Johnson, Irving, physical chemistry
Jonah, Charles D, radiation chemistry
Kaminsky, Manfred Stephan, experimental physics, surface physics
Katz, Joseph J, physical chemistry
Katzin, Leonard Isaac, physical inorganic chemistry
Kaufman, Sheldon Bernard, nuclear chemistry
Kierstead, Henry Andrew, low temperature physics, thermodynamics
Kim, Yong-Ki, atomic physics
King, Richard Frederick, applied mathematics
Koelling, Dale Dean, solid state science
Kramer, Stephen Leonard, experimental high energy physics
Krisch, Robert Earle, biophysics
Krohn, Victor Eugene, Jr, physics
Kubitschek, Herbert Ernest, biophysics
Kurath, Dieter, theoretical nuclear physics
Kyger, Jack Adolphus, physical chemistry, nuclear engineering
Land, Robert H, nuclear physics
Langsdorf, Alexander, Jr, nuclear physics
Larsen, Robert Peter, physical chemistry
Laszewski, Ronald M, experimental nuclear physics
Lawson, Robert Davis, theoretical physics
Lee, Tsung-Shung Harry, nuclear physics
Leibowitz, Leonard, physical chemistry
Lewis, Barbara-Ann Gamboa, environmental sciences, soil science
Lewis, Lloyd George, nuclear physics
Lindenbaum, Arthur, biochemistry, radiobiology
Livingood, John Jacob, physics
Lloyd, Elizabeth Luke, biophysics, microbiology
Lombard, Louise Scherger, pathology
Lu, Kwang-Tzu, atomic spectroscopy
Luner, Charles, physical chemistry
Lyon, William Graham, physical chemistry
Maccoss, Malcolm, bio-organic chemistry
Manning, Winston Marvel, nuclear chemistry
Marshall, Jack Stanton, limnology
Marshall, John Hart, physics
Marshall, Samson A, solid state physics
Martin, John H, environmental physics
Martin, Ronald Lavern, nuclear physics
Matheson, Max Smith, radiation chemistry
Matsushita, Tatsuo, biochemical genetics
McDaniel, Paul William, particle physics
Meadows, James Wallace, Jr, physical chemistry
Meisel, Dan, physical chemistry, radiation chemistry
Meneghetti, David, nuclear physics
Meyer-Schutzmeister, Luise, nuclear physics
Miller, Charles Everett, Jr, radiochemistry, analytical chemistry
Miller, John Robert, physical chemistry, radiation chemistry
Miller, Joseph Edwin, plant biochemistry, plant physiology
Miller, Robert Carl, experimental high energy physics, cryogenics
Miller, Robert Joseph, experimental high energy physics
Milsted, John, nuclear chemistry
Moldauer, Peter Arnold, nuclear physics, quantum mechanics
Monahan, James Emmett, theoretical physics
Mooring, Francis Paul, nuclear physics
Mueller, Fred Michael, solid state physics
Muller, Robert Neil, plant ecology
Nazareth, John Lawrence, numerical analysis, computer science
Norris, William Penrod, radiobiology
O'Connor, Timothy Edmond, organic chemistry, biochemistry
O'Fallon, Nancy McCumber, experimental nuclear physics, instrumentation
O'Hare, Patrick, physical chemistry
O'Reilly, Donald Eugene, chemical physics
Osborne, Darrell Wayne, physical chemistry, low temperature physics
Paddock, Robert Alton, physics
Parks, Eric K, physical chemistry
Pawlicki, Anthony Joseph, experimental high energy physics
Pennington, Edwin McFarlane, reactor physics
Peppard, Donald Francis, inorganic chemistry
Peraino, Carl, biochemistry, oncology
Perlow, Gilbert Jerome, experimental physics
Persiani, Paul J, physics
Person, James Carl, chemical physics
Person, Lucy Wu, nuclear chemistry, physics
Peshkin, Murray, theoretical physics
Peterson, Selmer Wilfred, physical chemistry
Pewitt, Edward Gale, experimental high energy physics, engineering management
Phelan, James Joseph, experimental high energy physics
Pieper, Steven Charles, theoretical nuclear physics
Poenitz, Wofgang P, physics
Pool, James C T, mathematical analysis, numerical analysis
Poole, Calvin Merten, laboratory animal science
Porges, Karl G, nuclear physics
Prastein, Solomon Matthew, theoretical physics
Price, David Long, solid state physics
Primak, William Leo, chemistry, physics
Prohammer, Frederick George, physics
Rabl, Ari, energy conversion
Rahman, Aneesur, theoretical physics
Rahman, Yueh Erh, medicine, health sciences
Redman, William Charles, reactor physics, magnetohydrodynamics
Reed, George W, Jr, nuclear chemistry, geochemistry
Reilly, Christopher Aloysius, Jr, virology
Reis, Arthur Henry, Jr, inorganic chemistry
Renner, Terrence Alan, physical chemistry
Reynolds, Bruce G, nuclear physics
Romanowski, Thomas Andrew, experimental high energy physics
Rosenthal, Marcia White, vertebrate zoology, radiation biology
Rothman, Alan Bernard, physical chemistry, nuclear engineering
Routbort, Jules Lazar, materials science
Rowland, Robert Edmund, radiation biophysics
Rundo, John, radiological physics, radiobiology
Sacher, George Alban, Jr, biology
Sachs, Robert Green, research administration, theoretical physics
Sauer, Myran Charles, Jr, radiation chemistry
Saunders, Barbara Gail Breidenbach, chemical kinetics
Saunders, Kim David, physical oceanography
Schiffer, John Paul, nuclear physics
Schiffer, Marianne Tsuk, x-ray crystallography
Schlenker, Robert Alison, radiation physics
Schmidt, Klaus H, radiation chemistry, experimental physics
Schnizlein, John Glenn, physical inorganic chemistry
Schreiner, Felix, physical chemistry, inorganic chemistry
Schreiner, Philip Allen, experimental high energy physics
Schultz, Harvey Albert, radiobiology
Schultz, Peter Frank, experimental high energy physics
Schwarz, Ricardo, solid state physics
Sedlet, Jacob, physical chemistry
Seed, Thomas Michael, microbiology
Serduke, Franklin James David, nuclear physics
Shaftman, David Harry, mathematics
Shen-Miller, Jane, plant physiology, horticulture
Shipman, Lester Lynn, theoretical chemistry
Siegel, Stanley, crystallography
Simpson, Oliver Cecil, solid state science
Sinclair, Warren Keith, biophysics
Singer, Richard Alan, experimental high energy physics
Smith, Alan B, nuclear physics
Smith, Brian Thomas, numerical analysis
Smith, David Young, solid state physics, optical physics
Smith, Donald Larned, nuclear physics
Smith, Richard Paul, experimental high energy physics
Smither, Robert Karl, experimental nuclear physics
Snelgrove, James Lewis, reactor physics
Spence, David, atomic physics
Stacey, Weston Monroe, Jr, reactor physics, applied physics
Stanford, George Stailing, reactor physics
Stearner, Sigrid Phyllis, cardiovascular physiology, radiobiology
Stehney, Andrew Frank, radiochemistry, radiobiology
Stein, Lawrence, inorganic chemistry
Steinberg, Ellis Philip, chemistry
Steindler, Martin Joseph, inorganic chemistry, research administration
Steunenberg, Robert Keppel, inorganic chemistry
Stevenson, Charles Edward, organic chemistry
Stewart, Donald Charles, radiochemistry
Strecok, Anthony J, mathematics
Studier, Martin Herman, chemistry
Sturm, William James, nuclear physics, applied physics
Tahmisian, Theodore Newton, physiology
Tetenbaum, Marvin, physical chemistry
Thomas, George E, experimental nuclear physics
Thomson, John Ferguson, pharmacology, biochemistry
Throw, Francis Edward, research administration
Tisue, George Thomas, environmental chemistry, analytical chemistry
Tomkins, Frank Sargent, atomic spectroscopy
Toppel, Bert Jack, reactor physics
Trevorrow, Laverne Everett, inorganic chemistry, physical chemistry
Trifunac, Alexander Dimitrije, physical chemistry, magnetic resonance
Ulrich, Aaron Jack, reactor physics
Unik, John Peter, nuclear chemistry
Veal, Boyd William, Jr, solid state physics
Veleckis, Ewald, chemistry
Wagner, Albert Fordyce, chemical physics
Wagner, Lawrence Carl, high temperature chemistry
Wahlgren, Morris A, environmental chemistry

Wangler, Thomas P., physics
Watson, Jerry Mike, energy physics
Webb, Robert Bradley, microbiology, genetics
Wesley, Marvin Larry, agricultural meteorology
Wexler, Solomon, chemical physics
Wicklund, Arthur Barry, experimental high energy physics
Williams, Jack Marvin, structural chemistry, inorganic chemistry
Winslow, George Harvey, physics
Woodruff, William Lee, reactor physics
Yntema, Jan Lambertus, physics
Yokosawa, Akihiko, high energy physics
Young, Charles Edward, chemical physics
Yule, Thomas J., nuclear physics, reactor physics
Zeidman, Benjamin, nuclear physics
Zielen, Albin John, physical chemistry, inorganic chemistry
Zwerdling, Solomon, physics, physical chemistry

ARLINGTON HEIGHTS
Anderson, Bror Ernest, paper chemistry
Atwater, Norman Willis, organic chemistry
Braun, Otto Godfrey, inorganic chemistry
Gray, Lewis Richard, plant morphology
Knodt, Cloy Bernard, dairy science
Lakshminarayanan, Krishnaiyer, biochemistry, industrial microbiology
Paynter, Camen Russell, internal medicine, gastroenterology
Schick, Margery Leone, industrial chemistry, organic polymer chemistry
Wahl, Werner Henry, nuclear chemistry, radiochemistry

ASHTON
Kemp, Albert Raymond, food science

AUBURN
Huck, Morris Glen, plant science, system analysis
Land, James Edward, chemistry
Smith, Robert C., biochemistry

AURORA
Alexander, Robert W., pathology
Hannum, Steven Earl, physical chemistry
Lay, Steven R., mathematics
Burkwall, Morris Paton Jr., biochemistry, food chemistry
Cheng, Shu-Sing, organic chemistry, microbiology
Crane, Anatole, microbiology
Dunlop, Andrew P., chemistry
Elson, William O., organic chemistry
Germino, Felix Joseph, food chemistry
Gould, Max Randall, food sciences
Hardy, Paul Wilson, chemistry
Haupt, Arthur Wing, botany
Heydanek, Menard George, Jr., food chemistry
Kincs, Frank Raymond, food chemistry
Kumar, Surinder, food science
Lebermann, Kenneth Wayne, food sciences
Lillwitz, Lawrence Dale, industrial organic chemistry
Lockhart, Haines Boots, nutrition, biochemistry
Lusas, Edmund W., food science, food technology
Min, David Byong, food chemistry
Moeller, Theodore William, food science
Nesheim, Robert Olaf, nutrition
Newton, Stephen Bruington, microbiology, food science
O'Mahony, John Patrick, food science
Rentmeester, Kenneth R., organic chemistry

BARRINGTON
Beese, Ronald Elroy, physical chemistry, inorganic chemistry
Bujake, John Edward, Jr., physical chemistry, food science

BATAVIA
Abarbanel, Henry D I., theoretical physics
Ankenbrandt, Charles Martin, experimental high energy physics
Appel, Jeffrey Alan, elementary particle physics
Ashford, Victor Aaron, experimental high energy physics
Atac, Muzaffer, physics
Awschalom, Miguel, nuclear physics
Baker, Samuel I., radiation physics
Baker, Winslow Furber, high energy physics, elementary particle physics
Bardeen, William A., particle physics
Berge, Jon Peter, experimental high energy physics
Brenner, Alfred Ephraim, physics
Brown, Bruce Claire, high energy physics
Brown, Charles Nelson, experimental high energy physics
Carey, David Crockett, high energy physics, particle physics
Carrigan, Richard Alfred, Jr., elementary particle physics
Cole, Francis Talmage, physics
Coulson, Larry Vernon, particle physics, radiation physics
Cox, Bradley Burton, elementary particle physics
Curtis, Cyril Dean, physics
DiBianca, Frank Anthony, experimental high energy physics
Eartly, David Paul, experimental high energy physics
Ecklund, Stanley Duane, experimental high energy physics
Einhorn, Martin B, theoretical high energy physics, elementary particle physics
Fast, Ronald Walter, physics
Fisk, Henry Eugene, high energy physics
Goldwasser, Edwin Leo, physics
Gollon, Peter J., particle physics, health physics
Gray, Edward Ray, experimental nuclear physics
Greene, Arthur Franklin, elementary particle physics
Griffin, James Edward, elementary particle physics
Groves, Thomas Hoopes, physics
Harris, Roger Mason, experimental high energy physics
Huson, Frederick Russell, high energy physics
Innes, Walter Rundle, particle physics
Jackson, Shirley Ann, theoretical high energy physics
Johnson, David Edwin, high energy physics
Johnson, Marvin Elroy, particle physics
Johnson, Rolland Paul, physics
Jovanovitch, Drasko D, high energy physics
Kammerud, Ronald Claire, experimental high energy physics
Kerns, Quentin Alexander, physics
Kuchnir, Moyses, low temperature physics, cryogenics
Lach, Joseph T., physics
Lee, Benjamin W., theoretical physics
Lennox, Arlene Judith, elementary particle physics
Lundy, Richard Alan, high energy physics
Lys, Jeremy Eion Alleyne, high energy physics
MacLachlan, James Angell, Jr., high energy physics
Malamud, Ernest Ilya, physics
Mikenberg, Giora, high energy physics
Mills, Frederick Eugene, physics
Moore, Craig Damon, health physics, experimental high energy physics
Murphy, Charles Thornton, elementary particle physics
Nash, Edward Thomas, experimental high energy physics
Nezrick, Frank Albert, elementary particle physics
Oleksiuk, Leslie William, engineering physics
Peoples, John, Jr., high energy physics
Peters, Robert Edward, electromagnetics
Potter, Douglas Marion, elementary particle physics
Pruss, Stanley McQuaide, experimental physics
Raja, Rajendran, high energy physics
Roberts, Arthur, physics
Rubinstein, Roy, experimental high energy physics
Ruggiero, Alessandro Gabriele, particle physics
Rupp, Eldor Gustav, agricultural chemistry
Sanford, James R., physics
Savit, Robert Steven, high energy physics, theoretical physics
Shannon, Edward Leo, food science
Shapiro, Rubin, chemistry
Shea, Michael Francis, engineering physics
Sherman, Edward, research administration, organic chemistry
Smart, Wesley Mitchell, high energy physics
Smith, Robert Ewing, nutrition
Stefanski, Raymond Joseph, physics
Stiening, Rae Frank, high energy physics
Stout, Paul Richard, applied statistics, particle physics
Strand, Robert Charles, organic polymer chemistry
Strauss, Bruce Paul, cryogenics, low temperature physics
Teng, Lee Chang-Li, theoretical physics
Theriot, Edward Dennis, Jr., physics
Thie, Joseph Anthony, reactor physics
Tomonaso, Hideo, organic chemistry
Tompkins, John Carter, experimental high energy physics
Toohig, Timothy E., experimental high energy physics
Tortorello, Anthony Joseph, synthetic organic chem
Turkot, Frank, physics
Van Ginneken, Andreas J M., physics
Vosti, Donald Curtis, organic chemistry, food technology
Voyvodic, Louis, high energy physics
Walker, James King, particle physics
Wehmann, Alan Ahlers, particle physics
Wilkinson, Raleigh James, food science, food chemistry
Wilson, Robert Rathbun, physics
Wiseblatt, Lazare, chemistry
Wuskell, Joseph P., chemistry
Yamada, Ryuji, physics
Yamanouchi, Taiji, physics
Young, Donald Edward, high energy physics
Yovanovitch, Drasko D., high energy physics

BAYLIS
Seybold, Mary Anice, mathematics

BENTON
Allinson, Morris Jonathan Carl, radiology

BERWYN
Kukral, John Charles, vascular surgery, general surgery
Mrazek, Rudolph G, surgery

BLOOMINGTON
Arteman, Robert Lloyd, plant pathology
Berthof, Lloyd Millard, apiculture
Brown, Robert Irwin, plant genetics, agronomy
Briggs, Robert Wilbur, genetics, biology
Briggs, Robert William, embryology
Craig, William F., agronomy, plant genetics
Criley, Bruce, experimental embryology
Evans, George Harlowe, physical chemistry
Englert, Du Wayne Cleveland, genetics
Frank, Forrest Jay, organic chemistry
Franzen, Dorothea Susanna, zoology
Hess, Wendell Wayne, inorganic chemistry
Jump, Lorin Keith, genetics, plant breeding
Kaufman, Thomas Charles, genetics
Laible, Charles A., genetics
McNeill, Michael John, genetics, plant breeding
Steele, Leon, plant breeding
Wantland, Evelyn Kendrick, mathematics
Wilcox, Wesley Crain, agronomy, botany

BOLINGBROOK
Phillips, Harold Bruce, applied physics

BROOKFIELD
Boese, Gilbert Karyle, animal behavior
Chamot, Walter M, polymer chemistry, colloid chemistry
Rabb, George Bernard, zoology

BUFFALO GROVE
Layton, Roger, organic chemistry

CAHOKIA
Roberts, John England, physics

CALUMET CITY
Willis, Victor Max, organic chemistry
Winner, Bernard Mark, organic chemistry, biochemistry

CARBONDALE
Altschuler, Milton, anthropology
Arnold, Richard Thomas, physical organic chemistry
Artemiadis, Nicholas, mathematical analysis
Ashby, William Clark, physiological ecology
Bailey, James Michael, physical biochemistry
Banerjee, Chandra Madhab, respiratory physiology; cardiovascular physiology
Baumann, Duane Dennis, geography, environmental management
BeMiller, James Noble, biochemistry, carbohydrate chemistry
Beyler, Roger Eldon, organic chemistry
Biesterfeldt, Herman John, Jr., mathematics
Blackwelder, Richard Eliot, zoology
Bolen, David Wayne, physical biochemistry
Borkon, Eli Leroy, internal medicine
Borst, Walter Ludwig, atomic physics, atmospheric physics
Bose, Subir Kumar, theoretical physics, statistical mechanics
Bouwsma, Ward D., mathematics
Brandon, Ronald Arthur, vertebrate zoology
Brown, George Earl, organic chemistry
Burton, Theodore Allen, mathematics
Caskey, Albert Leroy, analytical chemistry
Chang-Fang, Chuen-Chuen, particle physics
Christensen, David Emin, academic administration, geography
Cook, Edwin Aubrey, anthropology
Coorts, Gerald Duane, ornamental horticulture
Cox, James Allan, analytical chemistry, electrochemistry
Cutnell, John Daniel, physical chemistry
Dark, Philip John C., ethnology
Dunagan, Tommy Tolson, physiology
Dutcher, Russell Richardson, geology
Dyer, William Gerald, helminthology
Elkins, Donald Marcum, agronomy
Emptage, Michael Rollins, chemical physics
Fang, Jen-Ho, mineralogy
Fisher, Harvey Irvin, vertebrate zoology
Foland, Neal Eugene, mathematics
Foote, Florence Martindale, anatomy, morphology
Fraunfelter, George H., geology
Funk, David Truman, forest genetics
Galbreath, Edwin Carter, vertebrate paleontology, geology
Garoian, George, parasitology
Gass, George Hiram, endocrinology, pharmacology
Gates, Leslie Dean, Jr., mathematics
George, William, ornithology; zoology
Gibbard, H Frank, Jr., physical chemistry
Goodman, Billy Lee, poultry science
Gunerman, George John, III, anthropology
Guyon, John Carl, analytical chemistry
Haas, Hermann Josef, developmental biology
Hadley, Herbert Isaac, biochemistry
Hadley, Elbert Hamilton, chemistry
Hall, J Herbert, organic chemistry
Handler, Jerome Sidney, ethnology
Hardin, James William, vertebrate ecology, animal behavior
Hargrave, Paul Allan, biochemistry
Harris, Stanley Edwards, Jr., geology
Henneberger, Walter Carl, electrodynamics
Hillyer, Irvin George, horticulture
Hinckley, Conrad Cutler, physical chemistry
Hinners, Scott W., animal nutrition
Hodson, Harold H, Jr., animal nutrition
Hood, William Calvin, geology
Horton, Frank Elba, academic administration, urban geography
Hsu, Yu-Sheng, mathematics, statistics
Huang, Wei-Tze, biophysics, theoretical solid state physics

Hunter, William Sam, physiology
Jones, David Lloyd, earth sciences, meteorology
Kammler, David W., numerical analysis
Kaplan, Harold M, physiology
Kelley, John Charles, anthropology
Klimstra, Willard David, wildlife management, vertebrate ecology
Koch, Charles Frederick, mathematics
Konishi, Frank, nutrition
Koster, David F, physical chemistry, spectroscopy
Kuipers, Lauwerens, mathematics
Langenhop, Carl Eric, mathematics
Lee, Daniel Dixon, Jr., animal nutrition
Le Febvre, Eugene Allen, zoology
Lewis, William Madison, fisheries
Lit, Alfred, vision, experimental psychology
Males, James Robert, animal science
Marshall, Lauriston Calvert, physics
Martan, Jan, reproductive biology
Matten, Lawrence Charles, paleobotany
Maxwell, Charles Neville, mathematics
McClary, Daniel Otho, microbiology
McDaniel, Wilbur Charles, mathematics
McPherson, John Edwin, entomology
Meyers, Cal Yale, organic chemistry
Miller, Donald Morton, comparative physiology
Millman, Richard Steven, geometry
Mohlenbrock, Robert H, Jr, systematic botany
Moore, Robert Alonzo, mathematics
Mowry, James B., horticulture
Muller, Jon David, anthropology, archaeology
Myers, Charles Christopher, forestry, physiology
Myers, James Hurley, physiology
Myers, Oval, Jr., plant genetics
Nathanson, Melvyn Bernard, mathematics
Nickell, William Everett, physics
Ogur, Maurice, biochemistry
Olmsted, John Meigs Hubbell, mathematics
Olson, Farrel John, agronomy
Olson, Howard H, dairy science, physiology
Pappelis, Aristotel John, plant pathology
Parke, Wesley Wilkin, vertebrate anatomy, embryology
Parsons, John David, limnology
Payne, Irene R, nutrition
Pedersen, Franklin D, mathematics
Pedersen, Katherine L, topology
Petersen, Bruce Wallace, zoology
Portz, Herbert Lester, plant physiology
Rands, Robert Lawrence, anthropology, archaeology
Rawlings, Charles Adrian, biomedical engineering
Reed, Alex, animal science
Richardson, Alfred Wendel, physiology, biophysics
Ritter, Dale Franklin, geology
Robertson, Philip Alan, plant ecology
Slocum, Donald Warren, organometallic chemistry, organic chemistry
Smith, Gerard Vinton, physical organic chemistry
Snyder, Herbert Howard, applied mathematics, mathematical physics
Sollberger, Arne Rudolph, biometry
Stacy, Ralph Winston, physiology, bioengineering
Stahl, John Benton, limnology
Stains, Howard James, zoology
Starks, Thomas Harold, statistics
Sundberg, Walter James, mycology
Taylor, George Thomas, developmental biology, zoology
Trimble, Russell Fay, inorganic chemistry
Tweedy, James Arthur, horticulture
Tyrrell, James, theoretical chemistry
Ufgaard, John Edward, geology, paleozoology
Van Lente, Kenneth Anthony, physical chemistry

Varsa, Edward Charles, soil fertility
Verguin, Jacob, plant physiology
Voigt, John Wilbur, plant ecology
Wade, David Robert, biochemistry
Waring, George Houstoun, animal behavior, vertebrate zoology
Watson, Richard Elvis, physics
Weaver, George Thomas, forest ecology, silviculture
Welton, Richard Frederick, agricultural education
Wolff, Robert L, agriculture
Wotiz, John Henry, organic chemistry
Yambert, Paul Abt, conservation
Young, Otis Bigelow, physics
Zitter, Robert Nathan, solid state physics, quantum electronics

CARLINVILLE
Campbell, Jack Allen, physical chemistry
Itschner, Kenneth Frank, biochemistry
Singh, Dilbagh, plant pathology
Werner, William Ernest, Jr., ecology

CAROL STREAM
Maltenfort, George Gunther, chemistry
Trout, Paul Eugene, chemistry

CHAMPAIGN
Bever, Wayne Melville, plant pathology
Brown, Theodore Lawrence, inorganic chemistry
Buddemeier, Wilbur Dahl, agricultural economics
Carter, Carol Sue, ethology
Clark, Francis Matthew, microbiology
Donchin, Emanuel, physiological psychology
Drago, Russell Stephen, inorganic chemistry
Forsberg, Junius Leonard, plant pathology
Frankenberg, Julian Myron, paleobotany, plant morphology
Getz, Lowell Lee, ecology
Greenough, William Tallant, physiological psychology, neurobiology
Heaton, LeRoy, physics
Hirsch, Jerry, animal behavior, behavioral genetics
Jugenheimer, Robert William, plant genetics
Kendeigh, Samuel Charles, ornithology
Kruger, Peter Gerald, physics
Mapother, Dillon Edward, academic administration, physics
Mitchell, George Weston, biochemistry, electronics
Peters, Joseph William, mathematics
Powell, Dwight, plant pathology
Rose, William Cumming, biochemistry
Salmon, Michael, animal behavior
Semonin, Richard Gerard, atmospheric physics
Shoemaker, Hurst Hugh, animal ecology, ichthyology
Smith, Janice Minerva, nutrition
Smith, Leslie Garrett, geophysics
Stout, William F, mathematics
Tuckey, Stewart Lawrence, dairying
Vander Wyk, James Colby, research
Willson, Mary Frances, ecology, evolution
Young, Lloyd Martin, experimental nuclear physics
Zemlin, Willard R, speech & hearing sciences

CHARLESTON
Amos, Dewey Harold, petrology
Andrews, Richard D., zoology
Atkins, Ferrel, mathematics
Bailey, Zeno Earl, botany, genetics
Baker, Weldon Nicholas, analytical chemistry
Balbach, Harold Edward, phytogeography, ecology
Baumgardner, Kandy Diane, genetics
Becker, Steven Allan, plant anatomy, morphology
Breig, Marvin L, solid state physics
Buchanan, David Hamilton, chemistry
Butler, William Albert, physics
Cloud, William Max, atomic spectroscopy
Coon, Lewis Hulbert, mathematics
Cunningham, George Lewis, Jr, physical chemistry
Davis, Alvie Douglas, theoretical nuclear physics
Duffett, Walter N, physical geography, cartography

Durham, Leonard, aquatic ecology, fisheries management
Ebdon, David William, physical chemistry
Ebinger, John Edwin, botany, taxonomy
Ellis, Jerry William, organic chemistry
Ferguson, Karen Anne, biochemistry, lipid chemistry
Ferguson, Max B, zoology, human physiology
Foote, Carlton Dan, biochemistry, organic chemistry
Ford, John Philip, stratigraphy, geology
Goodrich, Michael Alan, entomology, animal behavior
Hamerski, Julian Joseph, inorganic chemistry
Henderson, Giles Lee, molecular spectroscopy, physical chemistry
Hsu, Nai-Chao, mathematics
Hunt, Lawrence Barrie, ornithology, vertebrate ecology
Karraker, Robert Harreld, inorganic chemistry
Keiter, Richard Lee, inorganic chemistry
Keppler, William J, genetics, evolution
Kniskern, Verne Burton, parasitology
Krehbiel, Eugene B, embryology, endocrinology
Laible, Jon Morse, algebra
Meyer, Douglas Kermit, cultural geography
Nanda, Jagdish L, mathematics
Palmer, James E, geology, geophysics
Price, Dalias Adolph, geography
Rawls, Hugh Cecil, malacology
Read, Mary Jo, geography of Latin America
Ridgeway, Bill Tom, parasitology, protozoology
Riegel, Garland Tavner, entomology
Ringenberg, Lawrence Albert, mathematics
Schram, Frederick R, invertebrate paleobiology
Scott, William Wallace, mycology
Smith, P Scott, physics
Smith, Richard Lawrence, phycology
Smith, Robert Johnson, organic chemistry
Steele, Sidney Russell, chemistry
Stratton, James Forrest, paleontology
Waddell, Robert Clinton, physics
Wallace, Ronald Gary, geomorphology
Weidner, Terry Mohr, plant physiology
Weiler, William Alexander, bacteriology, microbial ecology
Whalin, Edwin Ansil, Jr, physics
Whiteside, Wesley C, botany
Whitley, Larry Stephen, zoology

CHICAGO
Abbasi, Qamar Ali, systematic entomology
Abella, Isaac D, experimental physics, quantum optics
Abels, Larry L, physics
Abrams, Edward, research administration
Abrams, Israel Jacob, mathematical statistics
Adams, Robert Jay, experimental high energy physics
Adams, John R, medicine, psychiatry
Adams, Robert McCormick,
Adams, William L, geology
Adelson, Bernard Henry, medicine, oncology
Adler, Robert, physics, electronics
Adler, Solomon Stanley, hematology, oncology
Aduss, Howard, orthodontics
Agin, Daniel Pierre, neurophysiology, biophysics
Ainis, Herman, bacteriology, immunology
Akbar, Abulfatah Maksood, endocrinology
Aks, Stanley Olaf, mathematical physics
Albach, Richard Allen, cell physiology, microbiology
Albrecht, Robert H, chemistry
Albrecht, William Lloyd, physical chemistry
Alexander, Aaron D, medical microbiology
Alexander, Benjamin H, organic chemistry
Alivisatos, Spyridon Gerasimos Anastasios, biochemistry
Allgood, Jonathan L, mathematics
Alperin, Jonathan Lazarus, mathematics
Altmann, Stuart Allen, animal behavior
Amarose, Anthony Philip, biology, cytology
Ames, Peter L, environmental sciences
Anders, Edward, cosmochemistry, radiochemistry

Anderson, Alfred Titus, Jr, petrology, volcanology
Anderson, Burton, internal medicine, immunology
Anderson, Byron, biochemistry, immunology
Anderson, Edmund George, pharmacology
Anderson, Herbert Lawrence, physics
Anderson, Louise Eleanor, biochemistry
Anderson, Ralph F, pesticide chemistry, industrial chemistry
Anderson, William Raymond, physics
Andres, Cal L, food science
Andrews, Albert H, Jr, medicine
Andrews, Eugene Raymond, organic chemistry
Anker, Herbert S, biochemistry
Annegers, John Herman, physiology
Anyxas, Jurgis Arvydas, chemical physics
Apter, Nathaniel Stanley, psychiatry
Archer, Francis L, pathology
Archer, John Dale, medicine, pharmacology
Aries, Leon Judah, surgery
Arieff, Alex J, neurology, psychiatry
Arnold, Arthur, neurosurgery
Arora, Kasturi Lal, biochemistry
Ash, J Marshall, pure mathematics
Ashenhurst, Robert Lovett, applied mathematics, computer sciences
Askew, Cornelius, Jr, cardiovascular diseases, epidemiology
Auerbach, Earl, parasitology, food science
Auerbach, Harry, biostatistics
Ayuso, Katharine, food science
Babler, James Harold, synthetic organic chemistry
Bachrach, Joseph, carbohydrate chemistry
Bacus, James William bioengineering, physiology
Bahadur, Raghu Raj, mathematical statistics
Bailey, Edward Thomas, organic chemistry, polymer chemistry
Bailey, Orville Taylor, pathology
Baily, Walter Lewis, Jr, mathematics
Baker, David Warren, structural geology
Baker, William Kaufman, genetics
Bakshy, Stanley, biochemistry
Balagot, Reuben Castillo, anesthesiology
Baldwin, John Theodore, mathematical logic
Ball, Clayton Garrett, geology
Ballin, John Christian, clinical pharmacology
Baram, Peter, immunology, microbiology
Barany, Michael, biochemistry
Baratz, Robert Sears, developmental biology
Barcilon, Victor, applied mathematics, fluid dynamics
Barclay, William R, medicine
Bardack, David, vertebrate paleontology, ichthyology
Barden, Howard Stavers, biological anthropology
Barghusen, Herbert Richard, paleozoology, anatomy
Barret, Harold Jay, biology
Barston, Eugene Myron, applied mathematics
Bart, George Raymond, mathematical physics, elementary particle physics
Barton, Ambrose Donald, biochemistry
Barton, Evan Mansfield, internal medicine
Battifora, Hector A, pathology
Bauer, Ludwig, organic chemistry, medicinal chemistry
Baumeister, Carl Frederick, internal medicine
Baumgart, John Keppler, mathematics
Baumgarten, Ronald J, organic chemistry, environmental chemistry
Baur, Werner Heinz, mineralogy, crystallography
Beach, George Winchester, chemistry
Beadle, George Wells, plant genetics
Beal, John Mann, surgery
Beals, Richard William, mathematical analysis
Beamer, Parker Reynolds, pathology
Beck, Robert Nason, nuclear medicine
Becker, Milton J, biochemistry, biophysics
Becker, Samuel William, Jr, medicine, dermatology
Bederka, John Paul, Jr, pharmacology, organic chemistry
Beecher, William John, ornithology
Beem, Marc O, medicine
Behof, Anthony F, Jr, physics

Behrman, Abraham Sidney, water chemistry, industrial chemistry
Beiser, Helen R., psychoanalysis, child psychiatry
Bell, Clara G., immunology.
Belner, Robert Joseph, physical chemistry, polymer chemistry
Benevento, Louis Anthony, neurophysiology, neuroanatomy
Benjamin, Roland John, applied mechanics, optics
Bennet, Kimberly D. neurocytology
Benoit, Philippe Stanislas, pharmacology, physiology
Benson, Donald Warren, anesthesiology
Beraha, Louis, plant pathology
Berger, Donna Catherine, neurology
Berger, Neil Everett, applied mathematics
Berger, Sheldon, medicine, endocrinology
Bergstrom, Clarence George, organic chemistry
Berkson, David M. medicine
Berlin, Byron Sanford, virology
Berlin, Nathaniel Isaac, medical research
Berman, Eleanor, pharmacology, biochemistry
Berman, Joel David, mathematics
Bernard, Julian L., pediatrics, genetics
Bernhardt, Robert L. III, mathematics
Bernstein, Barry, applied mathematics
Bernstein, Leon, number theory
Bernstein, Seymour, particle physics
Berry, Charles Arthur, pharmacology
Berry, Henry Gordon, atomic physics
Berry, Leonidas Harris, medicine
Berry, Richard Stephen, physical chemistry
Berry, Robert Wayne, neurobiology
Berryman, George Hugh, nutrition, research administration
Berstein, Barry, applied mathematics, mechanics
Besic, Frank Charles, dentistry
Best, William Robert, hematology, biostatistics
Betts, Henry Brognard, rehabilitation medicine
Betz, Robert F., biochemistry, ecology
Bezkorovainy, Anatoly, biochemistry
Bianchi, Robert George, pharmacology
Bibbo, Marluce, cytopathology
Bienarz, Joseph, obstetrics & gynecology
Billingley, Patrick Paul, mathematics
Bingel, Audrey Susanna, pharmacology, reproductive endocrinology
Birnbaumer, Lutz, biochemistry, endocrinology
Bishop, William Richard, medical education
Bixby, William Ellis, physics
Blake, Emmet Reid, ornithology
Blake, Martin Irving, pharmacy
Blaut, James Morris, agricultural geography
Blivaiss, Ben Burton, physiology
Block, George E., surgery
Blomquist, Conrad Alvin, zoology
Bloom, Jack, food chemistry
Bluefarb, Samuel M. dermatology
Blumberg, Avrom Aaron, physical chemistry
Bodley, Herbert Daniel, II, anatomy
Boehm, John Joseph, pediatrics
Bogdanove, Lasca Hospers, pharmacology
Boggs, Joseph D. pathology
Bohannan, Laura, anthropology
Boit, John Ryan, vertebrate zoology, paleozoology
Bond, James Arthur, II, ecology
Bondareff, William, anatomy
Bongard, Steven J., anatomy, embryology
Bonstein, Abraham, information science
Booth, David Layton, pesticide chemistry
Borden, Craig W., internal medicine
Borges, Wayne Howard, hematology
Bosen, Sidney Frederick, forensic sciences, analytical chemistry
Boshart, Gregory Lew, organic chemistry
Boshes, Benjamin, neurology
Boshes, Louis D., neuropsychiatry
Bouck, G Benjamin, botany
Boulos, Badi Mansour, pharmacology
Bourguial, Priscilla C., pharmacology
Bousfield, Aldridge Knight, topology
Bowman, James E., pathology, human genetics
Bowman, Joel Mark, theoretical chemistry
Boyer, Joseph Henry, organic chemistry
Braham, Roscoe Riley, Jr., cloud physics
Braidwood, Robert J., anthropology, archaeology
Bramante, Pietro Ottavio, physiology, internal medicine

Brandfonbrener, Martin, internal medicine, cardiology
Breitbeil, Fred W., III, organic chemistry
Brenniman, Gary Russell, environmental health
Bretz, Harold Walter, microbiology
Breuhaus, Herbert Charles, gastroenterology, internal medicine
Brewer, John Isaac, obstetrics & gynecology
Brewer, Nathan Ronald, veterinary physiology
Brink, Marion Francis, nutrition
Brissey, Ruben Marion, physical chemistry
Britan, Norman, applied anthropology
Broder, Samuel B., neurology, psychiatry
Bronsky, David, internal medicine, endocrinology
Brophy, James John, solid state electronics
Browder, Felix Earl, mathematical analysis
Brown, Eric Reeder, chemistry, virology
Brown, Kenneth Howard, organic polymer chemistry
Brown, Meyer, neurology, psychiatry
Brown, Weldon Grant, chemistry
Brubaker, George Randell, inorganic chemistry, bioinorganic chemistry
Bruce, David Lionel, anesthesiology, medicine
Bruetman, Martin Edgardo, neurology
Brunner, Edward A., pharmacology, anesthesiology
Budrys, Rimgauda S., surface chemistry
Buhse, Howar Edward Jr., zoology
Bulkley, George, medicine, urology
Buntinas, Martin George, mathematics
Burgauer, Paul David, organic chemistry, pharmaceutical chemistry
Burney, Donald Eugene, industrial organic chemistry
Burns, Richard Price, physical chemistry, inorganic chemistry
Burnstein, Ray A., elementary particle physics
Burrell, Elliott Joseph, Jr., physical chemistry
Burrill, Dan Y., dentistry
Burris, B Cullen, medicine
Burrows, William, bacteriology
Buschman, Robert John, physiology, electron microscopy
Bush, C Allen, biophysical chemistry
Bussert, Jack Francis, organic chemistry
Butler, Robert Allan, psychoacoustics
Butzer, Karl Wilhelm, physical geography, archaeology
Calandra, Joseph Carl, toxicology
Callighan, Owen Hugh, immunochemistry
Cammarata, Peter S., biochemistry, physical organic chemistry
Campbell, James A., medicine, cardiology
Campbell, John Alexander, chemical engineering
Camras, Marvin, magnetism, electronics
Capps, Richard Brooks, medicine
Carhart, Richard Alan, theoretical high energy physics, mathematical physics
Carlin, Richard Lewis, inorganic chemistry
Carlson, Eric Dungan, geography, archaeology
Carnow, Bertram Warren, environmental health, thoracic diseases
Carone, Frank, pathology
Carstens, Herman Paul, internal medicine
Carton, Robert Wells, internal medicine
Cassaretto, Frank Philip, chemistry
Cassels, Donald Ernest, medicine, pediatrics
Catchpole, Hubert Ralph, physiology
Catrambone, Joseph Anthony, Sr., mathematics, data processing
Caviness, Bobby Forester, computer science, applied mathematics
Caywood, Thomas E., operations research
Cecilia, Mary, taxonomy
Cedar, Warren Richard, dentistry
Ceithaml, Joseph James, biochemistry
Century, Bernard, pharmacology
Cerefice, Steven A., organic chemistry, organometallic chemistry
Chakrin, Alan Leonard, clinical chemistry, toxicology
Chamberlain, Joseph Miles, astronomy
Chandran, Satish Raman, entomology, parasitology
Chandrasekhar, Subrahmanyan, astronomy, astrophysics
Changus, George William, / pathology
Charlier, Roger Henri, geology, geography
Chatterton, Robert Treat, Jr., endocrinology, biochemistry
Chen, Chiadao, biochemistry

Chen, Edwin Hung-Teh, biostatistics
Chen, Shepley S., biology, plant physiology
Cheng, Sze-Chuh, neurochemistry, biology
Cheston, Warren Bruce, physics
Chiakulas, John James, anatomy
Chiang, Kwen-Sheng, molecular biology
Childers, Roderick W., medicine, cardiology
Childress, Dudley Stephen, biomedical engineering
Chiou, Win Loung, pharmacology
Chorvat, Robert John, medicinal chemistry
Christopoulos, George Nick, toxicology
Chung, Jiwhey, biochemistry, food science
Cibils, Luis Angel, obstetrics & gynecology
Cifonelli, Joseph Anthony, carbohydrate chemistry
Clark, Harlan Eugene, physical chemistry, analytical chemistry
Clark, John Whitcomb, radiology
Clasen, Raymond Adolph, pathology
Clay, George A., neuropharmacology
Clayton, Robert Norman, geochemistry
Clem, William Henry, endodontics, oral microbiology
Close, Warren James, pharmaceutical chemistry
Closs, Gerhard Ludwig, organic chemistry
Clough, Robert Ragan, mathematics
Cochrane, Chappelle Cecil, organic chemistry
Coe, Elmon Lee, biochemistry
Coe, Fredric Lawrence, internal medicine, nephrology
Cohen, Maynard, neurosciences
Cohen, Morrel Herman, theoretical physics
Cohen, Carl, immunogenetics
Cohen, Harry, chemistry
Cohen, Lawrence, oral medicine, oral pathology
Cohen, Sidney, medicine, microbiology
Cohen, Gerald Edward, molecular biophysics
Colbert, Marvin J., internal medicine
Cole, Edmond Ray, biochemistry
Coleman, Bernell, physiology
Coleman, Harold Mitchell, physical chemistry
Coley, Ronald Frank, physical chemistry
Collier, Donald, anthropology, archaeology
Collins, Paul Waddell, medicinal chemistry
Collins, Vincent J., medicine, anesthesiology
Combs, Clarence Murphy, anatomy
Compere, Clinton Lee, orthopedic surgery
Condoulis, William V., immunology, developmental biology
Cook, Donald Latimer, pharmacology
Coolidge, Thomas Buckingham, biochemistry
Corbett, Jules John, microbiology
Corcoran, John William, biochemistry
Cordes, William Charles, plant physiology, cytochemistry
Coulter, Charles L., structural chemistry, biophysical chemistry
Cowan, Jack David, biophysics, theoretical biology
Cozzarelli, Nicholas Robert, biochemistry
Cramer, Archie Barrett, food chemistry
Cream, Joseph Gaylord, plant physiology, plant biochemistry
Crewe, Albert Victor, physics
Crompton, Charles Edward, physical chemistry
Cronin, James Watson, nuclear physics
Crovetti, Aldo Joseph, organic chemistry
Crumrine, David Shafer, physical organic chemistry
Cudd, Herschel Herbert, chemistry
Cugell, David Wolf, pulmonary diseases
Cummings, Michael R., cytology, developmental genetics
Currie, Bruce LaMonte, medicinal chemistry, organic chemistry
Curtis, Herbert John, mathematics
Cutler, Irving Herbert, geography

Cutler, Robert W P., neurology, neurochemistry
Cutshall, Alden Denzel, geography
Czesler, Jeffrey Lance, biophysical chemistry
Czerlinski, George Heinrich, biophysics, biochemistry
Dahlberg, Albert A., dental anthropology
Damle, Suresh B., organic chemistry, biochemistry
Dan, Teruki Clark, biochemistry, psychology
Daniel, David Newton, physiology
Danforth, William (Frank), fresh water ecology
Daniel, Jon Cameron, developmental biology, cell biology
Daniels, Ralph, organic chemistry
Das Gupta, Tapas Kumar, medicine
Dasler, Waldemar, biochemistry
Davidson, Israel, pathology
Davis, Floyd Asher, neurology, neurosciences
Davis, James Ernest, medicine
Davis, John Marcell, psychiatry
Davis, Loyal, surgery
Davis, Oscar F., psychiatry, pharmacology
Dawe, Albert Rolke, comparative physiology
Dawson, Glyn, biochemistry
Day, Emerson, internal medicine
Dean, Richard Raymond, pharmacology
Dean, Robert Waters, food science
DeBoer, Frank Edward, physical chemistry
De Bruyn, Peter Paul Henry, histology
De Cicco, John, mathematics
DeCosta, Edwin J., obstetrics & gynecology
de Figueiredo, Mario P., food science, food technology
De Forest, Ralph Edwin, medicine
DeGroot, Leslie Jacob, endocrinology
Dierickx, Friedrich, virology
Dekirmenjian, Haroutune, biochemistry
De la Torre, Jack Carlos, neuropharmacology, neurosurgery
Del Greco, Francesco, medicine
Deliyannis, Platon Constantine, mathematics
DeMar, Robert E., vertebrate paleontology
Denning, Jack, biology, zoology
dePeyster, Frederic A., medicine
DeSombre, Eugene Robert, biochemistry, endocrinology
DeYoung, Edwin Lawson, organometallic chemistry
DeYoung, Willard G., internal medicine
Dhaliwal, Amrik S., horticulture, virology
Diab, Ihsan M., pharmacology, biochemistry
Diamond, Jack, pharmacology
Dierickx, Charles Wallace, geography
Dinerstein, Robert Alvin, chemistry
Dinkines, Flora, mathematics
Disterhoft, John Francis, neurobiology
Dmowski, W Paul, obstetrics & gynecology
Dodge, James Stanley, polymer chemistry, rheology
Doede, John Henry, resource management
Doege, Theodore Charles, epidemiology, public health
Doehler, Robert William, mineralogy
Doerschuk, Albert Peter, organic chemistry, economics
Doi, Kunio, medical physics, optics
Dolkart, Ralph Elson, internal medicine
Domsky, Irving Isaac, analytical chemistry
Donaldson, Alan Weston, parasitology, public health administration
Dordick, Isadore L., geography
Dorfman, Albert, biochemistry
Doughty, Clyde Carl, biochemistry
Douglas, Bruce L., dentistry
Dow, James W., medicine, biomedical engineering
Dowling, Forrest Leroy, geophysics
Dowling, Stephen Ward, cell biology
Doyle, William Lewis, cell biology
Drachman, David A., neurology
Dray, Sheldon, medicine, immunochemistry
Drill, Victor Alexander, pharmacology
Driscoll, Richard James, analytical chemistry
Dubin, Alvin, biochemistry
DuBrul, E Lloyd, anatomy
Duda, Eugene Edward, radiology
Dudley, Eugene F., ecology
Dudley, Horace Chester, physics, radiobiology

Duncan, James Lowell, medical microbiology
Dunn, Dorothy Fay, public health
Dunn, William Joseph, medicinal chemistry
Dupont, Todd, numerical analysis
Dwinger, Philip, pure mathematics
Dybalski, Jack Norbert, industrial organic chemistry, highway engineering
Dyer, Alan Richard, biostatistics
Dyrud, Jarl Edvard, psychiatry
Earle, David Prince, Jr, medicine
Eaton, Philip Eugene, organic chemistry
Ecanow, Bernard, pharmaceutical sciences
Eckenhoff, James Edward, anesthesiology
Eckner, Fredrich August Otto, pathology
Economou, Steven George, medicine
Edelstein, Alan Shane, solid state physics
Edelstein, Warren Stanley, applied mathematics, solid mechanics
Egan, Richard L, medicine
Egermeier, Edward R, dairy science, bacteriology
Eggan, Fred R, anthropology, social anthropology
Eggert, Donald A, paleobotany
Ehrlich, Richard, environmental health, infectious diseases
Eich, Stephen Joseph, biochemistry
Eigsti, Orie Jacob, botany
Ekstedt, Richard Dean, microbiology
Elam, James O, anesthesiology
Elias, Michael Hans, microscopic anatomy
Ellinger, Rudolph H, food science, food microbiology
Emrick, Edwin Roy, analytical chemistry
Engel, Milton Baer, dentistry
Ensor, Elwood Henderson, organic chemistry
Epstein, Robert Bernard, hematology
Epstein, Wolfgang, biochemistry
Erber, Thomas, physics
Erby, William Arthur, organic chemistry
Erickson, Wallace Alfred, chemistry
Ermler, Walter Carl, theoretical chemistry
Ernest, John Terry, ophthalmology
Erskine, James Christian, Jr, soild state physics
Esposito, Michael Salvatore, genetics, biochemistry
Esposito, Rochelle E, genetics
Esterly, John Roosevelt, pathology
Esterly, Nancy Burton, pediatrics, dermatology
Estes, Reedus Ray, chemistry
Evans, Earl Alison, Jr, biochemistry
Evans, William John, inorganic chemistry
Everingham, John, zoology, anatomy
Failey, Crawford Fairbanks, chemistry
Falk, Lawrence A, Jr, virology, immunology
Falk, Marshall Allen, psychiatry
Fano, Ugo, theoretical physics
Fanslow, Don J, zoology, endocrinology
Fanta, Paul Edward, organic chemistry
Farbman, Albert Irving, histology, cytology
Farnsworth, Norman R, pharmacognosy, phytochemistry
Fay, Richard Rozzell, psychophysiology, neurosciences
Fedor, Edward John, physiology
Feinberg, Harold, pharmacology
Feinstein, Irwin K, mathematical analysis
Felsher, Zachary, dermatology
Fencl, Robert (Daniel), oral surgery
Fennessy, John James, radiology
Fenninger, Leonard Davis, medicine
Fenters, James Dean, virology, microbiology
Ferguson, Donald John, surgery
Fields, Theodore, medical physics
Filler, Robert, organic chemistry
Finder, Jerome Gordon, orthopedic surgery
Finkel, Asher Joseph, medicine, zoology
Finney, Mildred Irene, geography
Firlit, Casimir Francis, pediatrics, urology
Fischman, Donald A, medicine
Fisher, Morris Alan, neurology
Fitch, Frank Wesley, pathology
Fitko, Chester Walter, organic polymer chemistry
Fitzloff, John Frederick, medicinal chemistry, toxicology
Flanagan, Charles Larkin, internal medicine
Fletcher, Dean Charles, biochemistry, pharmacology
Flores, Samson Sol, prosthodontics
Flouret, George R, biochemistry, endocrinology
Fong, Harry H S, pharmacognosy
Fooden, Jack, zoology
Forbes, Warren C, mineralogy
Ford, Lincoln Edmond, medical physiology
Ford, Virginia, anthropology
Foreman, Ronald Louis, pharmacology, toxicology
Formanek, Edward William, algebra
Forrette, John Elmer, analytical chemistry
Foster, Leigh Curtis, spectroscopy, research administration
Foulser, David A, mathematics
Fox, Eugene N, medical microbiology
Fox, Jacob H, neurology
Fox, Theodore Albert, orthopedic surgery
Fozzard, Harry A, physiology, internal medicine
Francis, Howard Thomas, biomedical engineering
Frechette, Arthur Roy, dentistry
Freed, Karl F, theoretical chemistry, physical chemistry
Freedman, Daniel X, psychiatry
Freedman, Lawrence Zelic, psychiatry
Freedman, Philip, medicine
Freeman, Susan Tax, anthropology, ethnology
Freeman, Wade Austin, inorganic chemistry
Freilich, Joseph Kenneth, internal medicine
Freinkel, Norbert, internal medicine, endocrinology
Freinkel, Ruth Kimmelstiel, medicine
Freund, Peter George Oliver, theoretical physics
Fried, Josef, organic chemistry
Fried, Walter, internal medicine, hematology
Friedlander, Susan Jean, applied mathematics
Friedman, Bernard Samuel, petroleum chemistry
Friedmann, Herbert Claus, biochemistry
Frisch, Henry Jonathan, experimental high energy physics
Frischer, Henri, internal medicine, genetics
Frisque, Alvin Joseph, analytical chemistry, physical chemistry
Fritz, Carol S, mathematics
Fritzsche, Hellmut, solid state physics
Frohman, Lawrence Asher, endocrinology
Fry, Christine L, anthropology, ethnology
Fuerholzer, James J, polymer chemistry, organic chemistry
Fujii, Atsushi, bio-organic chemistry
Fujita, Tetsuya T, meteorology
Fultz, Dave, dynamic meteorology, fluid mechanics
Garbarino, Merwyn Stephens, anthropology
Garber, Beatrice B, developmental biology
Garber, Edward David, cytogenetics
Garcia-Munoz, Moises, astrophysics
Garland, Robert Bruce, organic chemistry
Garner, Harry Hyman, clinical psychiatry
Gassman, Merrill Loren, plant physiology
Gavlin, Gilbert, organic chemistry, chemical engineering
Gavrilovic, John, microchemistry, mineral engineering
Gearien, James Edward, medicinal chemistry, organic chemistry
Gehrie, Mark Joshua, psychological anthropology, psychoanalysis
Gehring, Harvey Thomas, organic polymer chemistry
Geinisman, Yuri, neurobiology
Gelfand, Henry Morris, epidemiology
Gelfand, Norman Mathew, experimental high energy physics
Gelperin, Abraham, public health
Georgakis, Constantine, mathematics, statistics
Gerbie, Albert B, obstetrics & gynecology
Geroch, Robert Paul, theoretical physics
Gershbein, Leon Lee, organic chemistry, biochemistry
Getz, Godfrey S, biochemistry, pathology
Giacomoni, Dario, cell biology
Gilbert, Eugene Charles, synthetic organic chemistry, organic polymer chemistry
Gilbert, Robert L, solid state physics
Gill, Merton, psychoanalysis
Gillette, Roy (James), zoology
Ginsburg, Norton Sydney, geography of Asia, political geography
Giori, Claudio, polymer chemistry
Giovacchini, Peter L, psychiatry, psychosomatic medicine
Gislason, Eric Arni, physical chemistry
Gladfelter, Bruce G, geography
Glagov, Seymour, pathology
Glass, Howard George, pharmacology
Glassman, Sidney Frederick, plant taxonomy
Glauberman, George Isaac, mathematics
Glaviano, Vincent Valentino, physiology
Glendening, Norman Willard, applied chemistry
Glenister, Paul Robson, food science
Glick, Gerald, internal medicine, cardiology
Goedken, Virgil Linus, inorganic chemistry
Goepp, Robert August, dentistry, pathology
Goes, Gunther Walter, pure mathematics
Goheen, Peter George, urban geography, historical geography
Gold, Jay Joseph, internal medicine, endocrinology
Goldberg, Howard S, particle physics
Goldberg, Jay M, neurophysiology, neuroanatomy
Goldberg, Julius, microbiology, public health
Goldberg, Leon Isadore, pharmacology, internal medicine
Goldberg, Morton Falk, medicine, ophthalmology
Goldberg, Robert Jack, zoology
Goldie, Mark, developmental biology
Goldin, Milton, bacteriology
Goldman, Jack Leslie, philosophy of science
Goldman, Manuel, microbial physiology, metabolism
Goldmann, Morton Aaron, internal medicine, cardiology
Goldsmith, Jewett, psychiatry
Goldsmith, Julian Royce, geology, geochemistry
Goldwasser, Eugene, biochemistry
Golomb, Harvey Morris, hematology, oncology
Golomski, William Arthur, applied mathematics, statistics
Gomer, Robert, physical chemistry
Gonzales, Federico, cell biology
Goodman, Harold Stanley, microbiology
Goran, Morris, chemistry
Gordon, Burgess Lee, medicine
Gordon, Gerald Arthur, polymer science
Gordon, Milton Andrew, mathematical statistics
Gordon, Robert Jay, physical chemistry, chemical physics
Gorecki, Donna, cytology, developmental biology
Gorenstein, David George, bio-organic chemistry
Gorman, Susan B, organic chemistry
Gotoff, Samuel P, immunology
Gottlieb, Gerald Lane, electrical engineering, physiology
Gourse, Jerome Allen, organic chemistry
Graber, Touro Mor, orthodontics
Grand, Nicholas George, pathology
Graves, Robert Lawrence, mathematics
Gray, Allan P, organic chemistry
Gray, Brayton, mathematics
Gray, John Stephens, physiology
Grayhack, John Thomas, urology
Greaves, Walter Stalker, vertebrate paleontology
Grecz, Nicholas, microbiology
Green, David, hematology
Green, Orville, pediatrics, endocrinology
Greenberg, Bernard, entomology, environmental biology
Greenberg, Ruven, physiology
Greenberg, Stephen Robert, pathology, medical education
Greenfield, George B, radiology, medicine
Greenstein, David Snellenburg, mathematics
Grieble, Hans G, internal medicine, infectious diseases
Griem, Melvin Luther, radiology, physics
Griem, Sylvia F, medicine, dermatology
Griesbach, Robert Anthony, cytology, genetics
Griffith, B Herold, plastic surgery, reconstructive surgery
Griffith, James H, polymer chemistry
Grigg, Emanuel Radu Newman, radiology, nuclear medicine
Grimm, Arthur F, physiology, histology
Grinker, Roy Richard, Sr, psychiatry
Grollig, Francis Xavier, anthropology, archaeology
Groskopf, William R, biochemistry
Grossman, Burton Jay, pediatrics
Grossman, Herbert Jules, pediatric neurology
Grossman, Lawrence, cosmochemistry, geochemistry
Grossman, Sebastian Peter, psychobiology
Grossweiner, Leonard Irwin, biophysics
Grove, William Johnson, medical education
Grubbs, Clinton Julian, medical physiology, cancer
Gugenheim, Victor Kurt Alfred, topology
Gurney, Benjamin Franklin, endodontics
Gurney, Clifford W, internal medicine
Guthmann, Walter Sigmund, chemistry
Gutman, David, physical chemistry
Haberman, Shelby Joel, statistics
Hac, Lucile (R), biochemistry
Hadley, Elmer Burton, botany, plant ecology
Hahn, Paul Gene, anthropology
Hall, Robert Leonard, anthropology, archaeology
Halm, James Maurice, organometallic chemistry, physical organic chemistry
Halpern, Bernard, bacteriology
Halpern, Jack, inorganic chemistry, organometallic chemistry
Hamilton, Byron Bruce, pharmacology
Hamilton, Robert W, entomology
Hammond, James B, internal medicine
Hamp, Eric Pratt, anthropology
Hanley, Joseph Wall, food science
Hanlon, Cyril Rollins, surgery
Hanlon, Mary Sue, biochemistry
Hanly, W Carey, biochemistry, immunology
Hansen, Bruce Winston, animal science
Hansen, Donald Willis, Jr, medicinal chemistry, organic chemistry
Hansen, Timothy Ray, medical physiology
Hanson, Floyd Bliss, applied mathematics, rarefied gas dynamics
Haque, Riaz-Ul, medical microbiology
Harbord, Richard P, anesthesiology
Haring, Olga M, medicine
Hark, Fred William, orthopedic surgery
Harper, Paul Vincent, surgery
Harris, Arthur, solid state physics
Harris, Chauncy Dennison, geography, academic administration
Harris, Ronald Wilbert, optics
Harris, Stanley Cyril, physiology, pharmacology
Harris, Willard Samuel, cardiology
Harrison, William Henry, biochemistry
Hart, Alice Goodson, mathematics
Harter, Donald Harry, neurology, virology
Hartmann, James Francis, neuroanatomy
Hartsuch, Paul Jackson, chemistry
Harvey, Ronald Gilbert, organic chemistry
Hasegawa, Junji, dermatology, histochemistry
Haselkorn, Robert, physical biochemistry, virology
Hass, George Marvin, pathology
Hast, Malcolm Howard, physiology
Hastreiter, Alois Rudolf, pediatric cardiology
Hauser, Isidore, physics
Haussmann, R K Dieter, system analysis, research administration
Hawley, Philip Lines, physiology
Hawrylewicz, Ervin J, biochemistry, medical research
Hay, Louise (Schmir), mathematical logic
Hayashi, James Akira, biochemistry
Hayashi, Teru, physiology
Hayes, Alice Bourke, botany
Healy, Michael James, plant pathology
Hechter, Oscar Milton, physiology
Heck, Robert Skinrood, family medicine
Hedayat, Abdossamad, mathematics, statistics
Hefferren, John James, chemistry
Hefley, Alta Jean, analytical chemistry, food chemistry
Hegyvary, Csaba, physiology
Hegyvary, Sue Thomas, medical education
Hejna, William Frank, orthopedic surgery
Hekmatpanah, Javad, neurosurgery, neurology
Heller, Alfred, pharmacology
Heller, David Harold, academic administration, zoology
Heller, Paul, internal medicine, hematology
Helms, Mary Wallace, cultural anthropology, ethnology
Helwig, Floy C, pediatrics
Henderson, Richard Elliott Lee, medicinal chemistry
Henderson, Thomas Otis, bacteriology, biochemistry
Hendrickson, Frank R, radiology
Henkin, Hyman, physical chemistry
Henry, Robert Edwin, chemistry
Hepler, Opal Elsie, pathology
Herbolsheimer, Henrietta, preventive medicine, public health

Herlinger, Albert William, inorganic chemistry
Herring, Susan Weller, anatomy
Hersh, Herbert N., physical chemistry
Hershkovitz, Philip, mammalogy
Herstein, Israel Nathan, mathematics
Hertz, John Atlee, theoretical solid state physics
Herzfeld, Simon Herman, physical chemistry
Hess, Eckhard Heinrich, animal behavior
Heyd, Charles E., organic chemistry
Higgins, Walter Mayo, pharmaceutical chemistry
Hildebrand, Roger Henry, elementary particle physics, astronomy, biological rhythms
Hiles, Linda Gayle, medical physiology
Hill, B Jay, radiology
Hines, James R., radiology
Hinkley, Robert Edwin, Jr, cell biology, anesthesiology
Hirsch, Jay G., child psychiatry
Hirschfeld, William Jacob, medical research, biomathematics
Hoerman, Kirk Conklin, biochemistry
Hoersch, Theodore Matthew, food science
Hoff, Gloria Thelma (Alburne), high energy physics
Hoffer, Abraham, earth sciences, geophysics
Hoffmann, Gerald M., operations research
Hoffmann, Philip Craig, pharmacology
Hofmann, Lorenz M., physiology
Hokama, Takeo, organic chemistry
Holinger, Paul Henry, medicine
Hollenberg, Paul Frederick, biochemistry, enzymology
Hollifield, Guy F., internal medicine
Holmes, Albert William, Jr, medicine
Hopson, James A., vertebrate paleontology
Horwitz, David Larry, endocrinology, metabolism
Hoskin, Francis Clifford George, biochemistry
Houk, Richard J., geography of Europe, geography of Africa
Hovde, Christian Arneson, anatomy
Howe, Henry Forbush, occupational medicine
Hruban, Zdenek, pathology
Huffman, George Wallen, biochemistry, organic chemistry
Huggins, Charles Brenton, cancer chemistry
Hughes, John Russell, electroencephalography, neurophysiology
Hughes, William Franklin, Jr, ophthalmology
Hsu, Wen-Tah, microbiology
Hubbard, Lincoln Beals, medical physics
Hubby, John L., biological sciences, genetics
Hugo, Norman Eliot, plastic surgery
Huilgol, Raja Ramesh, continuum mechanics
Humphrey, Gordon Laird, physiology, psychology
Humphrey, William Elliott, geology
Hunt, Charles Kellogg, organic chemistry
Hunt, William Lynn, medical research
Hunter, Robert L., pathology, immunology
Hunter, Wood E., organic chemistry
Hupert, Julius Jan Marian, electrophysics
Hurd, Richard Nelson, organic chemistry
Hussey, Hugh Hudson, medicine
Huston, John Lewis, physical chemistry, inorganic chemistry
Hutchens, John Oliver, physiology
Hutchison, Richard Allen, astrophysics
Hutchison, Clyde Allen, Jr, physical chemistry
Hutsell, Thomas Carlyle, biochemistry, pharmacology
Ihrke, Royal Ernest, medicine
Inger, Robert Frederick, herpetology
Ingham, Mark Gordon, physics
Ingle, James Davis, food science
Ipser, James Reid, theoretical astrophysics
Irvin, Howard H., polymer chemistry
Isaacson, Edward Kenneth, pediatrics
Isaacson, Michael Saul, electron physics, electron microscopy
Isaacson, Richard Allen, physics
Ito, Jun, mineralogy, analytical chemistry
Ito, Noboru, mathematics
Ivory, John Edward, physics
Jackson, George Gee, internal medicine
Jackson, Edmund, physiology
Jacobson, Leon Orris, medicine

Jacobson, Myron J., surgery, thoracic surgery
Jameson, A Keith, physical chemistry
Jamieson, John Calhoun, geophysics, high pressure physics
Janota, Rudolph Benjamin, chemistry
Jaselskis, Bruno, analytical chemistry
Jaskoski, Benedict Jacob, parasitology
Jasper, Donald Kohen, biology
Jeffay, Henry, biochemistry
Jensen, Elwood Vernon, biochemistry
Jerome, Joseph Benedict, organic chemistry
Jiu, James, microbial biochemistry
Joffe, Morris H., chemistry
Johansen, Chris Ellyn, psychopharmacology
Johns, William Francis, organic chemistry, medicinal chemistry
Johnson, Michael Evan, biophysics
Johnson, Porter W., high energy physics, mathematical physics
Johnson, Ralph Gordon, paleontology
Johnson, Robert Karl, ichthyology
Johnson, Terry Charles, microbiology, virology
Johnston, A Sidney, nuclear physics
Johnston, Naomi Lemkey, neuroanatomy, anatomy
Jones, David A, Jr, organic chemistry
Jones, Donald Wade, economic geography
Jordan, Steven Lee, mathematics
Joslin, Robert Scott, pharmaceutical chemistry
Jungmann, Richard A., organic chemistry
Justice, Parvin, biochemistry
Kachmar, John Frederick, biochemistry
Kaeck, Jack A., solid state physics
Kagan, Jacques, organic chemistry, biological chemistry
Kahan, Barry D., immunology, surgery
Kaiser, Emil Thomas, physical organic chemistry
Kaistha, Krishan K., pharmaceutical chemistry
Kaltenbach, John Paul, biochemistry
Kaminski, Edward Jozef, biochemistry, toxicology
Kane, William J., orthopedic surgery
Kantorovitz, Shmuel, mathematics
Kaplan, Ephraim Henry, analytical chemistry
Kaplansky, Irving, mathematics
Karim, Aziz, pharmaceutical chemistry
Kaplan, Julius Frank, organic chemistry
Kaplan, Maurice, psychiatry
Kaplan, Morris Aaron, biochemistry, immunology
Kaplan, Lewis David, meteorology
Kaplan-Koch Dora Deborah, microbiology, biochemistry
Kasner, Fred E., physical chemistry
Kaspin, Ben Louis, applied chemistry
Kasprow, Barbara Ann, anatomy, reproductive biology
Kass, Guss Sigmund, cosmetic chemistry
Kassner, Richard J., biochemistry
Kathan, Ralph Herman, biochemistry
Katz, Adrian I., internal medicine, physiology
Katz, Norman L., pharmacology
Katz, Sidney, physical chemistry
Kaufman, Richard Gilbert, physical physics
Keidering, Timothy Allen, chemical physics
Keller, Robert, microbiology
Kelly, Robert Edward, anatomy, cytology
Kelso, Albert Frederick, physiology
Kempt, John Emerson, virology
Kennedy, Elhart James, agricultural chemistry
Kennelly, Mary Marina, microbiology, botany
Kent, Geoffrey, organic chemistry
Kernis, Marten Murray, anatomy
Kessler, Richard Howard, medicine, teratology
Kethley, John Bryan, entomology
Kezdy, Ferenc J., biochemistry
Khalili, Ali A., rehabilitation medicine
Khan, Aijaz Ahmed, mineralogy
Khan, Mohammed Abdul Quddus, insect physiology, insect toxicology
Khan, Mohammed Nasrullah, physiology, veterinary science
Khoobyarian, Newton, virology
Kiang, Ying Cheng, urban geography, geography of China
Kiefer, Helen Chilton, enzymology
Kiefer, John Harold, physical chemistry

Kiefer, Joseph Henry, urology
Kieff, Elliott Dan, infectious diseases, virology
Kim, Byung Suk, immunology
Kimball, Chase Patterson, psychiatry, internal medicine
King, Lester Snow, pathology
King, Lowell Restell, urology
Kinzie, Jeannie Jones, radiotherapy
Kirk, Ralph Gary, physiology, molecular biology
Kirschner, Robert Howard, pathology
Kirsner, Joseph Barnett, internal medicine, gastroenterology
Kirsten, Werner H., pathology
Kislinger, Mark Brecher, elementary particle physics
Kittaka, Robert Shinnosuke, food science, microbiology
Kittie, James Frederick, surgery
Klaas, Rosalind Amelia (Mrs G Weber Schimpff), biochemistry
Klass, Donald Leroy, organic chemistry
Klavan, Bennett, dentistry, periodontology
Kleiman, Morton, organic chemistry
Klein, Morton Joseph, inorganic chemistry
Klein, Peter Douglas, biochemistry
Klein, Richard G., anthropology
Kleppa, Ole Jakob, physical chemistry
Klimstra, Paul D., medicinal chemistry
Klinger, Lawrence Edward, bacteriology
Klowden, Marc Jeffrey, medical entomology
Kluver, Heinrich, biological psychology
Kniebes, Duane Van, analytical chemistry
Knight, Katherine Lathrop, biochemistry, immunology
Knock, Frances Engelmann, surgery, organic chemistry
Knopp, Marvin Isadore, number theory, mathematical analysis
Knospe, William H., hematology
Kobick, Daniel Cecil, physiology
Koch, Elizabeth Anne, biochemical genetics, electron microscopy
Koehler, Henry Max, scientific information
Koenig, Harold, neurology
Kohler, Heinz, protein chemistry, immunology
Kohut, Heinz, psychoanalysis, psychiatry
Kokoris, Louis A., algebra
Kolb, Leonard H., surgery, oncology
Korn, Roy Joseph, medical administration
Kornel, Ludwig, endocrinology
Korpel, Adrianus, optics, acoustics
Kosman, Mary Ellen, pharmacology
Koster Van Groos, August Ferdinand, geochemistry
Kotin, Leonard, physical chemistry
Koushanpour, Esmail, physiology, biophysics
Kouvel, James Spyros, magnetism
Kozak, Edward Joseph, prosthodontics
Kraft, Sumner Charles, internal medicine, gastroenterology
Krasner, Sol H., physics
Kraus, Frank Joseph, chemistry
Krawetz, Arthur Altshuler, analytical chemistry, physical chemistry
Kraychy, Stephen, organic chemistry, nuclear medicine
Krebiehl, Robert Henry, anatomy
Kreider, Eunice S., medicinal chemistry
Krieg, Wendell Jordan, neuroanatomy
Krieger, Stephan Jacques, theoretical physics
Krimmel, Peter, medicinal chemistry, parasitology, immunology
Kruskal, William Henry, statistics
Kuceski, Vincent Paul, organic chemistry
Kuchnir, Franca Tablabue, medical physics, nuclear medicine
Kuettner, Klaus, biochemistry
Kukla, Michael Joseph, pharmaceutical chemistry
Kulkarni, Anant Sadashiv, clinical pharmacology, immunology
Kuo, Hsiao-Lan, dynamic meteorology, fluid dynamics
Kuramitsu, Howard Kikuo, biochemistry
Kusano, Kiyoshi, neurophysiology
Kvetkas, Marilyn J., microbiology
Kwaan, Hau Cheong, physics, nuclear physics, internal medicine, hematology
Kyriazis, Aikaterini A., pathology
Kyriazis, Andreas P., pathology, immunology
Lamp, Herbert F., plant ecology
Lamport, James Everett, cosmic ray physics
Landau, Richard Louis, endocrinology
Landau, William, microbiology
Landing, James Edward, cultural geography, historical geography

Lane, Ardelle Catherine, physiology
Langendorf, Richard, cardiology
Langston, Hiram Thomas, thoracic surgery
Lanzi, Lawrence Herman, physics, medical physics
Laros, Gerald Snyder, II, orthopedic surgery
Larson, Albert Jo, cultural geography, historical geography
Larson, Richard Gustavus, mathematics
Lash, Abraham Fae, obstetrics & gynecology
Lasher, Sim, mathematics
Lashof, Joyce Cohen, medicine
Lashof, Richard Kenneth, mathematics
Laskin, Daniel M., oral surgery, maxillofacial surgery
Lavelle, George Arthur, neuroanatomy
Lavenda, Nathan, physiology
Lathrop, Katherine Austin, endocrinology
Latimer, Donald Andrew, physical organic chemistry, biochemistry
Lautenschlager, Eugene Paul, biomaterials
Law, John Harold, biochemistry
Lawton, Irene Elizabeth, physiology
Lazda, Velta Abuls, anatomy
Le Beau, Leon Joseph, molecular biology
Lebovitz, Norman Ronald, applied mathematics, astrophysics
Lee, Chang Ling, immunology, pathology
Lee, Chung, reproductive endocrinology, nutrition
Lee, Martin J G, solid state physics
Lee, Warren G., physical organic chemistry
Left, Harvey Sherwin, thermal physics
Lehman, Dennis Dale, inorganic chemistry, organometallic chemistry
Lehmann, Wilma Helen, vertebrate morphology
Lemberger, August Paul, pharmaceutics
Lemon, Harry Degener, physiology
LeRoy, George Veach, medicine
Lester, George Ronald, physical chemistry, petroleum chemistry
Lester, Thomas William, Jr, medicine
Lev, Maurice, pathology
Levin, Bertram, radiology
Lerche, Ian, astrophysics, plasma physics
Levin, Kathryn J., solid state physics
Levin, Murray Laurence, internal medicine, nephrology
Levin, Samuel Joseph, biochemistry
Levi Setti, Riccardo, physics
Levitan, Ruven, internal medicine, gastroenterology
Levitsky, John M., internal medicine, preventive medicine
Levitsky, Sidney, cardiovascular surgery, surgery
Levitt, LeRoy P., psychiatry
Levy, Donald Harris, chemical physics
Levy, Leo, psychology, preventive medicine
Levy, Roy Stephen, meteoritics
Levy, Paul Samuel, biostatistics, epidemiology
Lewert, Robert Murdoch, medical parasitology, immunology
Lewis, Floyd John, surgery
Lewis, John Reed, pharmacology
Lewis, Phillip H., anthropology, ethnology
Likins, Robert Campbell, biochemistry
Lim, Ramon (Khe Siong), neurochemistry
Limarzi, Louis Robert, medicine
Lin, Reng-Lang, biochemistry
Lin, Sue Chin, mathematics
Lindberg, Howard Avery, medicine
Linde, Harry Wight, pharmacology, anesthesiology
Lindsay, John Ralston, otolaryngology
Lis, Edward Francis, pediatrics

Pogrund, Robert Seymour, environmental health, physiology
Pokorny, Joel, psychophysiology
Polen, Percy B., agricultural chemistry
Polimeni, Philip Iniziato, physiology
Pollack, Emanuel Davis, zoology, embryology
Polley, Edward Herman, neurosciences
Pollock, George Howard, psychiatry
Porterfield, John Donaldson, public health, medical administration
Poskozim, Paul Stanley, inorganic chemistry
Postman, Clarence, Jr., inorganic chemistry
Posvic, Harvey Walter, organic chemistry
Potempa, Sylvester Joseph, chemistry
Potter, Elizabeth Vaughan, medicine
Potts, Walter Joseph, psychopharmacology
Powell, Eugene Loren, medical microbiology
Prabu, Venkataray G., pharmacology
Prasad, Rameshwar, immunology
Preston, Frederick Willard, surgery
Prout, Franklin Sinclair, organic chemistry
Pruzansky, Jacob Julius, biochemistry
Pruzansky, Samuel, teratology, orthodontics
Pscheidt, Gordon Robert, biochemistry
Pullman, Theodore Neil, internal medicine, nephrology
Pumper, Robert William, medical microbiology
Pushkin, Edward A., ophthalmology
Puski, Gabor, food science, biochemistry
Putnam, Alfred Lunt, mathematics
Pyler, Richard Ernst, biochemistry, organic chemistry
Pysh, Joseph John, neurocytology, electron microscopy
Quanstrom, Walter Roy, animal behavior
Queen, Daniel, electronic engineering, acoustics
Quinn, James L, III, radiology, nuclear medicine
Qutub, Musa Y., water resources, geology
Rabideau, Glenn Sylvester, botany
Rabinowitz, Murray, internal medicine, biochemistry
Rachmeler, Martin, microbiology, genetics
Radzialowski, Frederick M., pharmacology, biochemistry
Rafelson, Max Emanuel, Jr., biochemistry
Rafferty, Frank Thomas, child psychiatry
Rafferty, Keen Alexander, Jr., embryology
Rafferty, Nancy S., cell biology, anatomy
Raghavan, Thirukkannamangai E S., mathematics
Raimondi, Anthony John, neurosurgery
Raisen, Elliott, inorganic chemistry, physical chemistry
Ramanathan, Ganapathiagraharam V., applied mathematics, statistical mechanics
Rambach, Walter A., internal medicine, hematology
Ranhotra, Gurbachan Singh, nutritional biochemistry
Ranney, David Francis, immunobiology, oncology
Ranney, Robert Earl, drug metabolism enzymology / genetics
Rao, Suryanarayana K., toxicology
Rao, Yedavalli Shyamsunder, organic chemistry
Rapoza, Norbert Pacheco, microbiology
Rappaport, David, mathematics education
Rappaport, Henry, pathology, hematology
Rattenborg, Christen C., anesthesiology
Rau, Herbert Lawrence, Jr., economic geography
Ravin, Arnold Warren, genetics
Ray, Robert Durant, orthopedic surgery
Rayudu, Garimella V S., nuclear chemistry, nuclear medicine
Reddy, Gunda, entomology
Reed, Charles Allen, anthropology, zoology
Reed, John Francis, physical chemistry
Rees, Thomas Charles, organic chemistry
Rees, William James, medicine, dermatology
Reid, William Hill, applied mathematics
Reiffel, Leonard, physics
Reilly, Richard W., medicine, cell biology
Reingold, Haim, mathematics
Reisberg, Boris Elliot, medicine, infectious diseases
Reisel, Robert Benedict, mathematics
Renfroe, Earl Wiley, orthodontics

Replogle, Robert Lee, cardiovascular surgery, thoracic surgery
Requarth, William H., surgery
Resnekov, Leon, medicine
Ressler, Newton, biochemistry
Reynolds, Wynetka Ann King, embryology
Reyblatt, Zinovy V., mathematics, solid mechanics
Rhines, Ruth, anatomy
Rice, Stuart Alan, physical chemistry
Richardson, Eugene Stanley, Jr., paleozoology, paleoecology
Richter, Sidney Bernard, organic chemistry
Richter, Ward Robert, pathology, electron microscopy
Ricker, Neil William, mathematics
Ries, Herman Elkan, Jr., physical chemistry
Rietz, Edward Gustave, chemistry
Riley, Reed Farrar, industrial chemistry
Rippon, John William, medicine, nephrology
Robbins, Kenneth Carl, biochemistry
Robert, Emery Dean, neurology, radiology
Roberts, Willard Lewis, nutritional biochemistry
Robin, Burton Howard, organic chemistry
Robinson, John E., Jr., dentistry
Rocek, Jan, physical organic chemistry
Roden, Carl Nils Lennart, biochemistry
Rodgers, Charles H., physiological psychology
Rodolfo, Kelvin S., marine geology
Roguska-Kyts, Jadwiga, internal medicine, nephrology
Roizman, Bernard, virology
Roll, John Donald, nuclear physics
Rommel, Marjorie Ann, analytical chemistry
Roothaan, Clemens Carel Johannes, physics
Rosen, Arthur Leonard, biophysics, applied mathematics
Rosenberg, Eli Ira, elementary particle physics
Rosenberg, Henry Mark, radiology, periodontology
Rosenberg, Irwin Harold, gastroenterology, nutrition
Rosenberg, Saul H., biostatistics
Rosenbloom, Alfred A, Jr., optometry
Rosenfeld, Robert Lee, pediatric endocrinology
Rosenstein, Sheldon William, orthodontics
Rosenthal, Gerson Max, Jr., zoology
Rosenthal, Ira Maurice, pediatrics
Rosner, Lawrence, biochemistry
Ross, Camilla Brems, organic chemistry
Ross, Mabel, psychiatry, public health
Rossi, Ennio Claudio, medicine
Rossmann, Kurt, physics
Rosenberg, Adolph, Jr., dermatology
Rotermund, Albert J, Jr., cell biology, physiology
Roth, Lloyd Joseph, pharmacology
Roth, Robert Mark, biology, genetics
Rothberg, Richard Martin, immunology, pediatrics
Rothenberg, Melvin G., mathematics
Rotkin, Isadore David, cancer
Rouffa, Albert Stanley, botany
Roush, Allan Herbert, biochemistry
Rovick, Allen Asher, physiology
Rowley, Donald Adams, experimental pathology, immunology
Rowley, Janet Davison, cytogenetics
Rubenstein, Arthur Harold, medicine, endocrinology
Rubin, Howard Arnold, physics
Ruddat, Manfred, plant physiology
Rudy, Lester Howard, psychiatry
Ruge, Daniel, neurosurgery
Ruhe, Carl Henry William, medical administration
Rukstinat, George John, pathology
Russe, Henry Paul, internal medicine, immunology
Rust, Charles Harry, mathematics, statistics
Rust, John Howard, radiobiology, biophysics
Sabelli, Hector C., neuropharmacology
Sabelli, Nora Hojvat, theoretical chemistry, computer science
Sacco, Louis Joseph, Jr., organic chemistry
Sachs, Jerome Michael, mathematics
Sadove, Max Samuel, medicine, anesthesiology

Sadowski, Anthony James, applied chemistry
Safier, Gwendolyn, gerontology, applied anthropology
Sahlins, Marshall David, anthropology
Saidel, Leo James, biochemistry
Sair, Louis, chemistry
Salamon, Ivan Istvan, organic chemistry
Saleem, Zubair A., groundwater hydrology, water resources
Samter, Max, clinical medicine, experimental medicine
Sanathanan, Lalitha P., applied statistics, mathematical statistics
Sander, Nestor John, geology, paleontology
Sangster, William, zoology
Sanner, Fern Rusteberg, anesthesiology
Sanner, John Harper, pharmacology
Sarkar, Francis Joseph, endocrinology
Sause, Henry William, organic chemistry
Savitz, Jan, limnology
Saxton, Augustus Donovan, analytical chemistry, pesticide chemistry
Scanu, Angelo, medicine, biochemistry
Scapino, Robert Peter, anatomy, dentistry
Scarce, LeRoy E., microbiology
Schaffer, Michael Irving, toxicology, pharmacology
Scharf, Arthur Alfred, biology, biophysics
Schaut, Charles Lawrence, physiology, biophysics
Scheff, George Julius, microbiology
Schensul, Stephen Lewis, anthropology
Scher, Jordan Mayer, psychiatry
Scherberg, Neal Harvey, molecular biology, biochemistry
Schiff, Sheldon K., psychiatry
Schiller, Maurice Aaron, physics
Schillinger, Edwin Joseph, physics, science education
Schlaeger, Albert Joseph, industrial chemistry
Schlenk, Fritz, biochemistry
Schlissfeld, Louis Harold, biochemistry, microbiology
Schmid, Frank Richard, medicine
Schmidt, Anthony John, embryology
Schmiege, Clement Carl, food science, microbiology
Schmitt, Kenneth Frederick, dentistry
Schmitz, Robert L., cancer
Schneider, David M., cultural anthropology
Schneider, Harold William, topology
Schoenberger, James A., clinical medicine
Scholz, Robert George, analytical chemistry
Schopf, Thomas Joseph Morton, paleobiology
Schrage, Samuel, physical chemistry, history of science
Schramm, David N., theoretical astrophysics, nuclear physics
Schreiber, David Seyfarth, solid state physics
Schreiber, Hans, oncology, cytopathology
Schreiber, Bruce David, clinical pharmacology, physiology
Schrotenboer, Gordon Harvey, organic chemistry
Schug, Kenneth, inorganic chemistry
Schulman, Jerome Lewis, psychiatry
Schulman, Martin Philip, biochemical pharmacology
Schulman, Sidney, neurology
Schulz, Jeanette, pediatrics
Schumacher, Gebhard Friedrich B., biochemistry, obstetrics & gynecology
Schuster, Charles Roberts, Jr., psychology, pharmacology
Schuytema, Eunice Chambers, medical microbiology
Schwab, Joseph Jackson, genetics
Schwartz, Robert Nelson, physical statistics
Scommegna, Antonio, reproductive physiology
Scott, Richard William, spectroscopy, analytical chemistry
Scuderi, Carlos S., orthopedic surgery
Seale, Raymond Ulric, anatomy, experimental embryology
Seaman, Gerald Robert, microbiology
Seed, Randolph William, surgery
Segner, Wayne Philip, food microbiology
Seiden, Lewis S., psychopharmacology

Seiling, Alfred William, plastics chemistry
Semrad, Joseph Edward, zoology
Shabica, Charles Wright, marine sciences, paleoecology
Shaffer, James Grant, medical education, medical microbiology
Shank, Max Carleton, ecology
Shanklin, Douglas R., pathology
Shansky, Michael Steven, visual physiology, psychology
Shapiro, James Alan, microbiology, molecular genetics
Shapiro, Stanley Kallick, biochemistry
Sharma, Ram Ratan, physics
Shaw, Richard Franklin, genetics
Shekelle, Richard Barten, epidemiology
Shekelton, Joseph, organic chemistry
Shepherd, Herndon Guinn, Jr., clinical chemistry, biochemistry
Sherman, Warren V., photochemistry, radiation chemistry
Sherrerd, Joseph C., organic chemistry, research administration
Sherrod, Theodore Roosevelt, pharmacology
Sherwood, Louis Maier, endocrinology, metabolism
Shields, Thomas William, thoracic surgery
Shih, Hsio Chang, physics
Shiner, Edward Arnold, organic chemistry
Shirk, James Siler, physical chemistry, spectroscopy
Shochet, Melvin Jay, elementary particle physics
Shoemaker, Clarence Jay, ophthalmology
Shoemaker, David Eugene, zoology
Shone, Robert L., organic chemistry, medicinal chemistry
Shornay, David, organic chemistry
Sibley, Hiram, public health administration
Sigel, Bernard, surgery
Sigler, Paul Benjamin, biochemistry
Simon, Donald A., biochemistry
Silverman, Peter Jay, solid state physics, low temperature physics
Silverstein, Edward Allen, nuclear medicine, nuclear physics
Simmons, Eric Leslie, biology
Simon, Lee Will, astronomy
Simon, Selwyn, microbiology
Simpson, Donald Ray, botany
Simpson, John Alexander, physics
Simpson, Donald H., cardiology
Singer, Ira, parasitology
Singer, Milton, anthropology
Singer, Rolf, mycology
Singer, Ronald, anatomy, physical anthropology
Singh, Eric John, biochemistry
Siqueira, Edir Barros, surgery
Sisson, George Allen, otolaryngology
Skaggs, Lester S., physics
Skinner, David Berni, surgery
Sklar, Abraham, mathematics
Skosey, John Lyle, medicine, physiology
Sky-Peck, Howard H., biochemistry
Slade, Hutton Davison, microbiology
Slatkin, Montgomery Wilson, applied mathematics, population biology
Slutsky, Harold L., epidemiology
Smith, David Waldo Edward, pathology, molecular biology
Smith, Eric Howard, systematic entomology
Smith, George F., pediatrics, human genetics
Smith, Vivianne Cameron, psychophysiology
Smith, Jay Alfred, physiology
Smith, Joseph Victor, mineralogy
Smith, Margo Lane, ethnography
Smith, Norman Dwight, sedimentology
Smith, Patricia Anne, developmental genetics
Snarr, John Frederic, physiology
Soare, Robert I., mathematics
Soffer, Alfred, internal medicine
Solem, Alan, invertebrate zoology
Solin, Stuart Allan, solid state physics
Solomon, Irvine Jerome, organic chemistry, inorganic chemistry
Solomon, Julius, high energy physics
Solomon, Lawrence Marvin, dermatology, medical education
Solomon, Robert Douglas, pathology
Sommers, Herbert M., pathology
Sonek, Mojmir G., cytology, obstetrics & gynecology
Sood, Manmohan K., petrology, geochemistry
Sorensen, Leif Boge, medicine, biochemistry

Sorensen, Ralph Albrecht, developmental biology
Southwick, Harry W, surgery
Spaeth, Ralph, medicine
Spannuth, Hiram Troutman, organic chemistry, nutrition
Spargo, Benjamin H, pathology
Spear, Patricia Gail, virology, cell biology
Specht, William Harold, mathematics
Spector, Harold Norman, solid state physics
Spector, Samuel, pediatrics
Spehlmann, Rainer, neurology, neurophysiology
Spiegl, Charles J, chemistry, toxicology
Spiess, Eliot Bruce, genetics
Spirakis, Charles N, pharmacology, nutrition
Spiroff, Boris E N, biology, embryology
Spitzer, Robert Harry, biochemistry
Spitzer, William Carl, organic polymer chemistry
Spofford, Janice Brogue, genetics
Srinivasan, B, cosmochemistry
Stamler, Jeremiah, preventive medicine, public health
Stanford, John W, organic chemistry, biochemistry
Steck, Theodore Lyle, biochemistry
Steffek, Anthony J, dentistry, pharmacology
Steigmann, Frederick, medicine
Steiner, Donald Frederick, biochemistry, endocrinology
Stejskal, Rudolf, pathology
Stenn, Frederick, medicine
Stepto, Robert Charles, obstetrics & gynecology, pathology
Stern, Paula Helene, pharmacology
Stevens, Joseph Alfred, medical mycology, medical microbiology
Stevens, Michael Thomas, physics
Stevenson, George Franklin, pathology
Steward, Vincent William, neuropathology
Stiffler, Paul W, medical microbiology
Stinchcomb, Thomas Glenn, radiation physics, nuclear physics
Stine, James Bryan, food microbiology
Stock, Leon M, organic chemistry
Stolze, Robert Gardner, systematic botany
Stoolmiller, Allen Charles, biological chemistry
Stotz, Robert William, research administration
Stout, John Willard, physical chemistry, chemical physics
Stover, Leon Eugene, cultural anthropology
Strahm, Norman Dale, physics, electrical engineering
Straus, Francis Howe, II, pathology
Straus, Helen Lorna Puttkammer, anatomy, biology
Straus, Werner, biochemistry
Strauss, Bernard S, molecular biology, cell biology
Strohmeier, Edwin Frank, physics
Sturtevant, Frank Milton, pharmacology
Su, Cheh-Jen, polymer chemistry, paper chemistry
Suarez, Kenneth Alfred, pharmacology, toxicology
Sugar, Oscar, physiology, neurosurgery
Sugarman, Meyer Louis, Jr, research administration
Sugarman, Nathan, nuclear chemistry
Suker, Jacob Robert, internal medicine
Sukowski, Ernest John, physiology, pharmacology
Sullivan, Daniel Richard, organic chemistry, lipid chemistry
Sullivan, Michael Joseph, Jr, mathematics
Sundaram, Swaminatha, physics
Swallow, Earl Connor, elementary particle physics, experimental physics
Swan, Richard Gordon, mathematics
Swanson, Don R, information science
Swartwout, Joseph Rodolph, obstetrics & gynecology
Swerdlow, Martin A, medicine, pathology
Swiatek, Kenneth Robert, neurosciences
Swift, Hewson Hoyt, cytology
Swinford, David Charles, cell biology
Szegho, Constantin Stephen, physics
Szuhaj, Bernard F, biochemistry, lipid chemistry
Tang, Pei Chin, neurophysiology, neuroanatomy
Tangora, Martin Charles, topology
Tarlov, Alvin R, internal medicine, biochemistry
Tarver, Mac-Goodwin, statistics
Tax, Sol, cultural anthropology, applied anthropology

Taylor, Ardell Nichols, physiology
Taylor, Edwin William, biophysics
Taylor, Elmore Hector, pharmacognosy, phytochemistry
Taylor, Samuel G, III, oncology
Taylor, Welton Ivan, microbiology
Teeri, James Arthur, ecology, polar biology
Teeters, William Dale, theoretical high energy physics, theoretical physics
Telegdi, Valentine Louis, physics
Telser, Alvin Gilbert, biochemistry, cell biology
Tenczar, Francis J, pathology
Ten Eick, Robert Edwin, electrophysiology, pharmacology
Tesar, Joseph Thomas, internal medicine, immunology
Tesk, John A, dental research
Teuscher, George William, dentistry
Thaemert, Jona Carl, anatomy
Thielen, Lawrence Eugene, organic chemistry
Thilenius, Otto G, pediatric cardiology, physiology
Thoennes, Lawrence Anthony, physiology, pharmacology
Thomas, Carolyn Eyster, neuroanatomy, histology
Thomas, William Arthur, ecology
Thommes, Robert Charles, zoology
Thompson, Emmanuel Bandele, pharmacology
Thompson, Phebe Kirsten, endocrinology, geriatrics
Thompson, Vinton Newbold, industrial hygiene
Thomson, Andrew, internal medicine, immunology
Thorp, Frank Kedzie, biochemistry, pediatrics
Throckmorton, Lynn Hiram, zoology
Tiecke, Richard William, pathology
Tiedemann, Clifford E, geography
Titchener, Edward Bradford, biochemistry
Titelbaum, Sydney, animal physiology, evolutionary biology
Titman, Paul Wilson, botany
Tobin, John Robert, Jr, internal medicine, cardiology
Tom, Baldwin Heng, transplantation immunology
Tomisek, Arthur John, biochemistry
Tomlinson, Glen E, medicine
Torok, Nicholas, otolaryngology
Tosteson, Daniel Charles, physiology, biophysics
Towns, Clarence, Jr, pathology, histology
Townsend, Alexandra A, physiology, osteopathy
Traise, Thornton, organic chemistry
Traisman, Howard Sevin, pediatrics
Traxler, James Theodore, synthetic organic chemistry, pesticide chemistry
Traylor, Melvin Alvah, ornithology
Treptow, Richard S, inorganic chemistry
Trobaugh, Frank Edwin, Jr, medicine
Trosman, Harry, psychiatry, psychoanalysis
Ts'ao, Chung-Hsin, physiology, experimental pathology
Tse, Warren W, human physiology
Tung, Wu-Ki, theoretical physics, elementary particle physics
Turkevich, Anthony, nuclear chemistry, space chemistry
Turner, Fred Allen, organic chemistry
Tuttle, Russell H, anthropology, primatology
Tuzar, Jaroslav, mathematics
Tweit, Robert Christopher, medicinal chemistry
Twersky, Victor, mathematical physics
Tybor, Philip Thomas, food science
Tyler, Edward Alton, medicine, psychiatry
Uhlenhuth, Eberhard Henry, psychiatry, psychopharmacology
Ulinski, Philip Steven, neuroanatomy
Ultmann, John Ernest, hematology, oncology
Unger, Lloyd George, physical chemistry
Unna, Klaus Robert, pharmacology
Upholt, William Boyce, molecular biology
Uretz, Robert Benjamin, biophysics
Urry, Wilbert Herbert, chemistry
Urse, Vladimir George, neuropsychiatry
Vaisrub, Samuel, internal medicine
Valsassori, Galdino E, medicine, radiology
Van Alten, Pierson Jay, embryology, immunology
Van Dellen, Theodore Robert, internal medicine

Van der Kloot, Albert Peter, food chemistry
Vandervoort, Peter Oliver, astronomy
Van Lanen, Robert Jerome, bio-organic chemistry, physical organic chemistry
Van Ostenburg, Donald Ora, solid state physics
Van Pernis, Paul Anton, medicine
Van Valen, Leigh, evolutionary biology
Vassiliades, Anthony E, physical chemistry, polymer chemistry
Vatuk, Sylvia Dutra, anthropology
Vazquez, Alfredo Jorge, neuropharmacology, electrophysiology
Veis, Arthur, biochemistry, physical chemistry
Vermeulen, Cornelius William, urology
Vesselinovich, Stan Dushan, physiology, oncology
Vezzetti, David Joseph, statistical mechanics
Vice, John Leonard, microbiology, biochemistry
Vicher, Edward Ernest, bacteriology
Visotsky, Harold M, medicine, psychiatry
Voedisch, Robert W, organic chemistry
Von Koeppen, Andreas, pulp technology, paper technology
Voris, Harold Cornelius, neurosurgery
Voris, Harold K, herpetology
Vorres, Karl S, physical chemistry
Vygantas, Auste Marija, biochemistry, organic chemistry
Wagner, Hans, organic chemistry
Wagreich, Philip Donald, pure mathematics
Wald, Robert Manuel, theoretical physics
Waldstein, Sheldon Saul, internal medicine
Wallace, David Lee, statistics
Wallace, Donald Albin, pharmacology, chemistry
Walter, Robert Irving, physical organic chemistry
Walter, Roderich, chemistry, physiology
Walton, William Ralph, geological oceanography
Ward, Paul J, dairy bacteriology
Warnecke, Melvin Oscar, food science
Warnick, Edward George, radiology
Warnock, Martha L, pathology
Warnock, Robert Lee, theoretical high energy physics
Warren, Charles Preston, biological anthropology
Washington, Elmer L, physical chemistry
Wassersug, Richard Joel, evolutionary biology
Waterhouse, John P, pathology, histology
Waterman, Frank Melvin, medical physics
Watson, David Livingston, entomology
Wawszkiewicz, Edward John, microbiology, biochemistry
Webber, Gayle Milton, organic chemistry
Webster, Dale Arroyo, biochemistry
Webster, James Randolph, Jr, pulmonary physiology
Webster, Leslie T, Jr, pharmacology
Wefel, John Paul, physics, astrophysics
Weier, Richard Mathias, organic chemistry
Weinstein, Ronald S, experimental pathology, electron microscopy
Weinstock, Harold, solid state physics, low temperature physics
Weinzweig, Avrum Israel, mathematics
Weiss, Jack Allan, surgery
Weiss, Marvin B, dentistry
Weiss, Samuel Bernard, biochemistry
Weissmann, Herman Benjamin, physics
Weissmann, Bernard, biochemistry
Welborn, Joseph F, oral surgery
Weller, Glenn Peter, mathematics
Weller, James Marvin, geology
Wells, Warren F, biochemistry
Wenzel, Rupert Leon, zoology
Wesley, Newton K, optometry
Westley, John Leonard, enzymology, physical biochemistry
Wetzel, Allan Brooke, neuropsychology
Wetzel, Nicholas, neurosurgery
Wharton, Lennard, physical chemistry
Whisler, Walter William, biochemistry, neurosurgery
Whitacre, David Martin, environmental sciences
Widra, Abe, microbiology
Wieczorowski, Elsie, pediatrics
Wied, George Ludwig, obstetrics & gynecology
Wiercinski, Floyd Joseph, cell physiology
Wilcox, Lee Roy, algebra, science education
Wilks, Louis Phillip, chemistry
Willey, Robert Bruce, animal behavior
Willey, Ruth Lippitt, entomology

Williams, Gerald Albert, medical research, endocrinology
Williams, Leamon Dale, food science, chemistry
Williams, Lesley Lattin, physical chemistry
Williams, Louis Otho, systematic botany
Williams-Ashman, Howard Guy, biochemistry
Wilson, Donald Alan, biochemistry, microbial physiology
Witt, James William, organic chemistry
Winsberg, Gwynne Roeseler, epidemiology
Winsberg, Lester, nuclear science
Winslow, John Durfee, geology
Winston, Roland, experimental physics, particle physics
Winter, Irwin Clinton, pharmacology
Winter, Robert John, pediatric endocrinology
Wirszup, Izaak, mathematics
Wissler, Robert William, pathology
Wistreich, Hugo Eryk, food science
Witmer, Hernan John, virology
Witt, John, Jr, organic chemistry
Wojcik, Anthony Stephen, computer science, algebra
Wolfe, Lauren Gene, pathology
Wolff, Arthur Harold, environmental health
Wong, Paul Wing-Kon, pediatrics, biochemical genetics
Wong, Ruth (Lau), pathology
Wong, Yuen-Fat, mathematics
Woodard, John, biology
Woodland, Bertram George, structural geology
Woods, James E, endocrinology, histochemistry
Woods, Loren Paul, ichthyology
Woodward, David Alfred, cartography
Wool, Ira Goodwin, biochemistry
Wozniak, Wayne Theodore, chemistry, spectroscopy
Yokoyama, Mitsuo, immunology, immunohematology
Wright, Francis Howell, pediatrics
Wright, Sydney Courtenay, physics
Wylie, Peter John, geology, geochemistry
Wynn, Ralph Matthew, obstetrics & gynecology
Yachnin, Stanley, internal medicine, hematology
Yale, Seymour Hershel, radiology
Yamashiroya, Herbert Mitsugi, microbiology, virology
Yang, Nien-Chu, chemistry
Yang, Sen-Lian, obstetrics & gynecology
Yogore, Mariano G, Jr, parasitology, public health
Youmans, Anne Stewart, medical microbiology, tuberculosis
Youmans, Guy Parry, bacteriology
Young, Harold, biochemistry
Young, James George, industrial pharmacy
Zabinski, Rose Marie C, clinical chemistry, toxicology
Zachariasen, Fredrik William Houlder, physics
Zadrozny, Mitchell G, geography
Zaki, Abd El-Moneim Emam, histology, oral biology
Zaneveld, Lourens Jan Dirk, biochemistry, medicine
Zangerl, Rainer, vertebrate paleontology
Zawadzki, Joseph Francis, organic chemistry
Zerlin, Stanley, psychophysiology, electrophysiology
Zerzavy, Frederick M, obstetrics & gynecology, public health administration
Ziegler, Alfred M, paleontology, stratigraphy
Zimmer, Albert Michael, nuclear medicine, bionucleonics
Zimmerman, Leo M, surgery
Zintel, Harold Albert, surgery
Zygmund, Antoni, mathematics

CHICAGO HEIGHTS
Gabis, Damien Anthony, food microbiology
Warden, William Kent, poultry nutrition

CICERO
Peterson, James Robert, soil fertility

CLARENDON HILLS
Ecker, Richard Eugene, physiology
Giggard, Earl David, food technology

ILLINOIS

Kretchmer, Richard Allan, synthetic organic chemistry
Ryznar, John William, chemistry

COLLINSVILLE
Horner, Henry John, analytical chemistry

CRYSTAL LAKE
Brown, Lloyd H., industrial organic chemistry
Foster, Frederick Calvin, polymer chemistry

COLUMBIA
Davis, Paul A., food science

DANVILLE
Cole, Walter Earl, organic chemistry
Pilkington, Dwain H., food science
Stefanini, Mario, pathology

DARIEN
Urbas, Branko, organic chemistry

DE KALB
Albright, Carl Howard, particle physics
Angotti, Rodney, mathematics
Beach, James Wilson, mathematics
Becker, Jerry Page, mathematics
Bennett, Cecil Jackson, animal genetics, behavioral genetics
Bower, John Edwin, physical chemistry
Briles, Worthie Elwood, immunogenetics, poultry genetics
Broussard, Lois Mary, topology
Brower, James E., physiological ecology
Bushnell, David L., physics
Casella, Clarence J., geology
Christiano, John G., applied mathematics
Cole, Alan L., meteorology, physics
Cunico, Robert Frederick, organic chemistry
Dahlberg, Richard Ernest, cartography
Dillman, Charles Daniel, geography of Latin America
Dwyer, Thomas Aloysius Walsh, III, mathematics
Erman, James Edwin, physical biochemistry
Fisher, Cletus G., speech pathology, audiology
Gravel, Pierre Bertez, cultural anthropology
Greenfield, David Wayne, ichthyology
Groskiggs, James Henry, mycology
Guest, Buddy Ross, resource geography
Gupta, Chaitan Prakash, mathematical analysis
Hampel, Arnold E., biochemistry, molecular biology
Haning, Quentin C., animal physiology
Hanzely, Laszlo, botany, cytology
Hardgrove, Clarence Ethel, mathematics
Hart, Donn Vorhis, cultural anthropology, ethnology
Hassan, Mazhar, nuclear physics
Holtzman, Stephen Ford, physical anthropology, history of anthropology
Honea, Kenneth Howard, anthropology, archaeology
Hurych, Zdenek, solid state physics, surface physics
Jollie, Malcolm Thomas, comparative anatomy
Jones, Dean Graeme, poultry breeding
Kambayashi, Tatsuji, algebra, geometry
Kevill, Dennis Neil, physical organic chemistry
Kimball, Clyde William, solid state physics
Kresheck, Gordon C., physical biochemistry
Kuller, Robert G., mathematics
Lange, Charles Henry, cultural anthropology, applied anthropology
Ledwitz-Rigby, Florence Ina, reproductive endocrinology
Leonard, Henry Siggins, Jr., mathematics
Lindbeck, Wendell Arthur, organic chemistry

Lindsey, Marvin Frederick, plant breeding
Lynch, Darrel Luvene, soil microbiology
McAlister, Donald Beaton, mathematics
Mason, W Roy, III, inorganic chemistry
McCleary, James A., botany
McFadden, Robert, mathematics
McIlrath, Wayne Jackson, plant physiology
Messenger, Aubrey Steven, soil science, forest ecology
Meyer, Axel, metal physics
McGinnis, Lyle David, geophysics
Miller, Francis Marion, organic chemistry
Mitchell, John Laurin Amos, cell physiology, biological rhythms
Mitler, Sidney, genetics, zoology
Morris, Robert Clarence, geology
Nortog, Knut Jonson, marine biology
Odom, Ira Edgar, mineralogy, petrology
Osman, Richard William, marine ecology, evolutionary biology
Otten, Charlotte Marie, biological anthropology, serology
Perry, Eugene Carleton, Jr., geochemistry
Piatak, David Michael, organic chemistry
Powers, Michael Jerome, topology
Prahlad, Kadaba V., developmental biology
Preston, Richard Swain, physics
Provencher, Ronald, anthropology
Reinemann, Martin, anthropology
Reynolds, Rosalie Dean (Sibert), organic chemistry
Rodine, Robert Henry, mathematical statistics
Rohde, Charles John, Jr., acarology, ecology
Rolf, Frederick William, analytical chemistry
Rossing, Thomas Dean, physics
Rubel, Daniel Nicholas, petrology, volcanology
Russell, Morley Egerton, physical chemistry
Schilt, Alfred Ayars, analytical chemistry
Schjeide, Ole Arne, cell biology, radiobiology
Shaffer, John Clifford, physics
Shapiro, Harvey Lee, mathematics, psychology
Shearer, William McCague, speech & hearing science
Sill, Larry R., experimental solid state physics, magnetism
Simons, Allan Barnard, agronomy
Simonson, Clifford Harry, soil science, environmental sciences
Skok, John, plant physiology
Sons, Linda Ruth, mathematics
Sorensen, Paul Davidsen, botany, taxonomy
Southern, William Edward, ornithology, ethology
Spangler, Charles William, physical organic chemistry
Starzyk, Marvin John, microbiology
Staver, Allen Ernest, dynamic meteorology
Stevens, George Putnam, economic geography
Trail, Stanley M., statistics
Trott, Charles Eugene, urban geography
Tsao, Peter, physics, geophysics
Vaughn, Joe Warren, inorganic chemistry, physical chemistry
Villmow, Jack R., geography
Vint, Larry Francis, animal breeding, poultry breeding
Voelker, Alan Morris, science education
von Zellen, Bruce Walfred, parasitology, protozoology
Voth, Paul Dirks, botany
Weaver, Allen Dale, physics
Webb, Peter Noel, geology, micropaleontology
Weiss, Malcolm Pickett, geology
Williams, Eddie Robert, mathematics
Wilson, Robert Steven, physical chemistry
Wong, How Kin, geophysics, physics
Wood, Charles, experimental solid state physics
Wunderlich, Marvin C., mathematics
Zar, Jerrold Howard, ecology, physiology
Zetti, Anton J., mathematics

DECATUR
Askill, John, solid state physics, metallurgy
Balgley, Ely, chemistry
Brobst, Kenneth Martin, analytical chemistry
Bulich, Anthony Andrew, microbiology

Campbell, Michael Floyd, food science
Cole, Morton S., food technology
Drenan, James Warner, physical chemistry
Duke, Jodie Lee, Jr, carbohydrate chemistry
Empen, Joseph A., organic polymer chemistry
Forbes, Malcolm Holloway, organic chemistry
Hahn, Richard Ray, food science
Horan, Francis E, physical organic chemistry
Hurst, Thomas Lighthall, food chemistry
Larson, Roy Fred, carbohydrate chemistry
McMurray, Birch Lee, veterinary medicine, agriculture
Moser, Kenneth Bruce, carbohydrate chemistry
Pour-El, Akiva, biochemistry
Schanefelt, Robert Von, food science, cereal chemistry
Seidman, Martin, carbohydrate chemistry
Shell, Lester Crane, embryology, physiology
Shelton, Ronald M., mathematics
Short, Rolland William Phillip, carbohydrate chemistry
Silver, Samuel Lewis, chemistry
Vander Burgh, Leonard F., physical organic chemistry
Verbanac, Frank, organic chemistry
Walton, Henry Miller, organic chemistry
Weatherbee, Carl, organic chemistry
Young, Austin Harry, physical chemistry

DEERFIELD
Bellamy, David, physics
Benn, Walter R., organic chemistry
Fisher, Earl Eugene, organic chemistry, research administration
Ginger, Leonard George, organic chemistry
Goldstein, Maurice Sabin, endocrinology
Hakewill, Henry, Jr., organic chemistry, inorganic chemistry
Hamill, Richard David, pharmacy
Howard, Kenneth Leon, organic chemistry
Josephson, Aaron Mortimer, hematology, genetics
Mazur, Robert Henry, organic chemistry
Stipanovic, Bozidar J., organic chemistry
Townsley, William W, Jr, plant pathology & physiology
Wessel, Hans U., cardiology, physiology

DES PLAINES
Barr, Tery Lynn, chemical physics, surface chemistry
Bloch, Herman Samuel, petroleum chemistry
Boehme, Werner Richard, organic chemistry
Burch, Wendell Dale, inorganic chemistry
Burroughs, James Edward, analytical chemistry
Chang, Franklin Shih Chuan, polymer chemistry
deRosset, Armand John, chemistry
Dyck, Arnold Wolff Jan, chemistry, organic chemistry, agricultural chemistry
Eby, Lawrence Thornton, organic chemistry
Falk, John Carl, organic chemistry
Fish, Ferol F, Jr., applied physics
Flagg, John Ferard, analytical chemistry, research administration
Frame, Robert Roy, industrial organic chemistry
Haensel, Vladimir, organic chemistry
Hoeg, Donald Francis, physical organic chemistry
Homeier, Edwin H Jr., physical inorganic chemistry
Illingworth, George Ernest, physical organic chemistry, industrial chemistry
Johnson, Robert William, polymer chemistry
Joy, George Cecil, III, inorganic chemistry
Kahn, Alfred Jerome, physiology
Kuehner, Richard Louis, bacteriology
Kutik, Leon, organic chemistry
Lane, William James, analytical chemistry
Lanterman, Elma, analytical chemistry
Levy, Joseph, organic chemistry
Lira, Emil Patrick, organic chemistry
McLaughlin, Robert Lawrence, research administration, organic chemistry
Monroe, Robert Adams, laboratory medicine
Neuzil, Richard William, chemistry

Nichols, George Morrill, physical chemistry
Novick, Rudolph G, psychiatry
Padria, Frank George, organic chemistry,
Palmer, Jay, physical chemistry
Pollitzer, Ernest Leo, petroleum chemistry, surface chemistry
Pollock, James Percy, geology
Reily, William Singer, organic chemistry
Schiefelbein, Benedict, inorganic chemistry, electron microscopy
Schmeling, Louis, organic chemistry
Skala, Hertha, chemistry
Sloma, Leonard Vincent, engineering physics
Stevens, Lawrence Guy, inorganic chemistry, extractive metallurgy
Strobel, Charles William, polymer chemistry
Tuomi, Donald, physical chemistry
Welsh, Lawrence B., solid state physics
White, William, medical physics, research administration
Wolf, Richard Eugene, organic chemistry, polymer chemistry

DIVERNON
Frazee, Charles Joseph, soil classification

DOWNERS GROVE
Bodmer, Arnold R., physics
Cosper, David Russell, organic chemistry, polymer chemistry
Endres, Joseph George, food science
Erickson, David R., agricultural chemistry
Fritzsche, Herbert William, physical chemistry
Garland, James W, Jr., solid state physics
Gibbs, James Albert, organic chemistry
Goodman, Gordon Louis, information science
Ketterson, John Boyd, physics
Lee, Richard Jui-Fu, organic chemistry
Montet, George Louis, chemical physics
Muzyczko, Thaddeus Marion, polymer chemistry
Perlow, Mina Rea Jones, inorganic chemistry, radiochemistry
Phifer, Harold Edwin, physical chemistry
Ring, James George, solid state physics, optics
Rose, David, physics
Yuster, Philip Harold, radiation physics

DOWNEY
Breen, Moira, biochemistry
Gaballah, Saeed S., biochemistry, molecular biology
Kantor, Harvey Sherwin, infectious diseases, microbiology
Kautz, Harold Douthitt, medicine
Litteria, Marilyn, neuroendocrinology
Schumer, William, surgery, biochemistry
Singh, Sant Parkash, endocrinology, physiology
Weinstein, Hyman Gabriel, physiology

DUNDEE
Burger, George Vanderkarr, wildlife conservation
Keyser, William Lacy, food science

DUNLAP
Brigham, Nelson Allen, mathematics

EAST MOLINE
Powell, Lanny C., economic geography

EAST ALTON
Kozicky, Edward Louis, wildlife management

EAST PEORIA
Hafele, Joseph Carl, nuclear physics
Kauffman, Harry Frey, organic chemistry
Kolb, Doris Kasey (Mrs K E Kolb), organic chemistry

EDWARDSVILLE
Argos, Patrick, crystallography, molecular biophysics
Aschenbrenner, Joyce Cathryn, anthropology
Axtell, Ralph William, vertebrate zoology
Baich, Annette, biochemistry
Bain, Ralph Lee, chemistry

Baker, William Bryan, geography
Baldwin, Thomas O, solid state physics
Bardolph, Marinus Peter, organic chemistry
Boedecker, Richard Roy, physics
Bouman, Thomas David, theoretical chemistry
Braundmeier, Arthur John, Jr, physics
Broadbooks, Harold Eugene, zoology
Clemans, Kermit Grover, mathematical statistics
Collier, James Eli, geography
Coy, Richard Eugene, dentistry
Custer, Frederic, dentistry
Davis, Norman Seymour, microbiology, biochemistry
Firsching, Ferdinand Henry, analytical chemistry
Frisbie, Charlotte Johnson, cultural anthropology, ethnology
Garder, Arthur Oris, Jr, numerical analysis
Gore, Dorothy J, geology, geography
Hall, Stephen Kenneth, inorganic chemistry, environmental chemistry
Hattemer, Jimmie Ray, mathematics
Hazen, Stanley P, dentistry
Hess, Charles F, geography
Ho, Chung-Wu, mathematics
Hoffman, Alan Bruce, physical inorganic chemistry
Jason, Emil Fred, chemistry
Kang, Ik-Ju, atomic physics
Kazeck, Melvin E, geography
Keating, Richard Clark, systematic botany
King, Ordie Herbert, Jr, oral pathology
Kircher, Harry Bertram, economic geography
Kotiah, Thoddi Chandrasekara, mathematics, statistics
Kulfinski, Frank Benjamin, plant physiology, cell biology
Kumler, Marion Lawrence, plant ecology, plant physiology
Kurth, Rudolf, mathematics
Levy, Michael R, cellular biology
Linder, Louis Jacob, analytical chemistry
Lindstrum, Andrew O, Jr, mathematics
Maloney, Thomas J, cultural anthropology
Matta, Michael Stanley, biological chemistry, organic chemistry
McAneny, Laurence Raymond, physics
Miller, Halsey Wilkinson, Jr, geology, paleontology
Milligan, Wilbert Harvey, III, virology
Peterson, Roy Phillip, endocrinology, reproductive physiology
Ratzlaff, Kermit O, physiology, zoology
Rutledge, Robert B, mathematics, electrical engineering
Nelson, Thomas Eustis, Jr, pharmacology, physiology
Oursler, Clellie Curtis, mathematics
Parker, Nancy Johanne Rentner, developmental biology
Parker, Richard Bewley, ecology, herpetology
Parrill, Irwin Homer, physical chemistry, agricultural chemistry
Patrick, Timothy Benson, organic chemistry
Pendergrass, Robert Nixon, statistics
Sanders, Steven Gill, physics
Schmidt, James Robert, prosthodontics
Schopp, Robert Thomas, physiology
Schusky, Ernest Lester, anthropology
Shaw, William Corr, physics
Smith, Raymond Dale, anatomy
Sobkowski, Frank J, radiology, radiobiology
Sturley, Eric Avern, mathematics
Thomerson, Jamie E, ichthyology, zoology
Vasileff, Vasil, dentistry
Voget, Fred W, cultural anthropology, ethnology
Walford, Lionel K, solid state physics
White, Jesse Edmund, physical inorganic chemistry
Wilson, Howell Kenneth, mathematics
Wittig, Gertraude Christa, zoology, insect pathology
Zahalsky, Arthur C, biochemistry, parasitology

ELGIN
Averill, Frank Wallace, quantum mechanics, applied mathematics
Heinicke, Herbert Raymond, nutrition
Janssen, Arthur Gray, polymer chemistry

Juergensmeyer, Elizabeth B, cell biology, genetics
Monson, William Joye, poultry nutrition, animal nutrition
Orcutt, Donald Adelbert, biochemistry, industrial pharmacy
Simon, Wilbur, chemistry

ELK GROVE VILLAGE
Gordon, Edward Emanuel, medicine
Radanovics, Charles, food science
Scott, Don, biochemistry, food science

ELMHURST
Allen, Frank B, mathematics, physics
Beck, Keith Russell, synthetic organic chemistry
Ganchoff, John Christopher, analytical chemistry, inorganic chemistry
Ginn, Martin E, physical chemistry, analytical chemistry
Glogovsky, Robert L, physical chemistry, biochemistry
Gorsic, Joseph, plant genetics
Iskenderian, Haig Parnag, magnetism
Jump, John Austin, mycology, plant pathology
Meseth, Earl Herbert, zoology
Sweeney, Robert Milton, developmental biology, zoology

ELMWOOD PARK
Berg, Dana B, science writing

ELSAH
Cornell, David Allan, physics
Holzberlein, Thomas M, atomic physics
Robertson, Forbes, economic geology

EUREKA
Binkley, Stephen Bennett, biochemistry
Snyder, Jack Willard, biology

EVANSTON
Albert, Ethel Mary, anthropology
Allred, Albert Louis, inorganic chemistry
Ambuel, John Philip, pediatrics
Austin, Donald Guy, mathematics
Auvil, Paul R, Jr, elementary particle physics, theoretical high energy physics
Bahng, John Deuck Ryong, astronomy
Bailyn, Martin H, physics
Baker, Robert Henry, organic chemistry
Bareiss, Erwin Hans, applied mathematics
Barnothy, Jeno Michael, nuclear physics, astrophysics
Barnothy, Madeleine Forro, physics
Barratt, Michael George, pure mathematics
Bartlett, Glenn Wilfred, microbiology
Basolo, Fred, inorganic chemistry
Bender, Myron Lee, chemistry
Ben-Israel, Adi, applied mathematics
Bicoff, Juan Pedro, pediatric cardiology
Birchfield, Gene Edward, atmospheric physics, geophysics
Block, Martin M, physics
Boas, Mary Layne, physics
Boas, Ralph Philip, Jr, mathematics
Borchers, Curtis Edward, physical chemistry
Bordwell, Frederick George, organic chemistry
Brown, Frank Arthur, Jr, biology
Brown, Laurie Mark, theoretical physics
Brownrigg, Leslie Ann, applied anthropology
Buikstra, Jane Ellen, biological anthropology, archaeology
Burch, Benjamin Clay, mathematics
Burnett, Allison L, biology
Burwell, Robert Lemmon, Jr, catalysis
Buscombe, William, astrophysics
Carpenter, Carolyn Virus, biochemistry
Cashman, Robert Joseph, physics
Cember, Herman, radiobiology, health physics
Chessick, Richard D, psychiatry
Chin, Yeh-Hao, computer sciences
Cinlar, Erhan, mathematical statistics, operations research
Cohen, Ronald, anthropology, political science
Colton, Frank Benjamin, chemistry
Crawford, James Weldon, psychiatry, experimental psychology
Crawford, Susan N, information science
Crist, Buckley, Jr, polymer science
Culver, David Clair, ecology
Dacey, Michael Francis, geography, statistics

Dallos, Peter John, biomedical engineering, biophysics
Dapples, Edward Charles, geology
DeFord, Donald Dale, analytical chemistry
De La Huerga, Jesus, medicine, biochemistry
Devinatz, Allen, mathematics
Deysach, Lawrence George, mathematical biology
Dobbs, Frank W, physical chemistry
Dorsey, John M, surgery
Dumas, Lawrence Bernard, molecular biology, biochemistry
Dwass, Meyer, mathematical statistics
Eagle, Edward, physiology, toxicology
Elliott, Lois Lawrence, psychoacoustics, audiology
Enroth-Cugell, Christina, vision, neurophysiology
Espenshade, Edward Bowman, Jr, physical geography, cartography
Evens, Leonard, mathematics
Ewald, Arno Wilfred, solid state physics
Fahey, John James, orthopedic surgery
Farquhar, Peter Henry, operations research, statistics
Fisher, Stephen D, mathematics
Forman, Donald T, biochemistry, analytical chemistry
Frank, Evelyn, mathematics
Freeman, Arthur Jay, solid state physics
Friederici, Hartmann H R, medicine, pathology
Friedman, Avner, mathematics
Frost, Arthur Atwater, physical chemistry
Gantt, Clarence Leroy, internal medicine
Garrels, Robert Minard, geochemistry
Gasper, George, Jr, mathematical analysis
Gesteland, Robert Charles, neurophysiology, electrical engineering
Gilbert, Lawrence Irwin, zoology
Goldberg, Erwin, zoology, physiology
Goldberg, Colin C, mathematics
Grau, Albert A, mathematics, computer science
Grober, Samuel, forestry
Gupta, Rajendra P, theoretical solid state physics
Hall, Edward Twitchell, anthropology
Halperin, William Paul, low temperature physics
Henschen, Lawrence Joseph, computer science
Hines, Roderick Ludlow, solid state physics
Hoffman, Brian Mark, physical chemistry
Howard, William Michael, astrophysics
Howland, Arthur Lloyd, petrology
Hsu, Francis Lang-Kwang, cultural anthropology, psychological anthropology
Huang, Su-Shu, astrophysics
Huntzicker, Harry Noble, physical chemistry, electrochemistry
Hussey, Allen Sanborn, chemistry
Hynek, Joseph Allen, astrophysics
Ibers, James Arthur, structural chemistry
Ionescu Tulcea, Alexandra, mathematics
Ionescu Tulcea, Cassius, mathematical analysis
Jackson, Roscoe George, II, sedimentology, fluid dynamics
Jerome, Joseph Walter, applied mathematics
Jourdonais, Leonard Francis, bacteriology, internal medicine
Kahn, Daniel Stephen, mathematics
Kannewurf, Carl Raeside, solid state electronics
Kauffman, John W, solid state physics
Kaye, Saul, microbiology
Keren, Joseph, physics
Kille, John William, developmental biology, reproductive physiology
Killip, Thomas, III, internal medicine, cardiology
King, Lafayette Carroll, organic chemistry
King, Robert Charles, genetics
Klotz, Irving Myron, physical biochemistry
Krumbein, William Christian, geology
Kucera, Thomas J, organic medicine
Kyser, Franklin A, internal medicine
Lambert, Joseph B, organic chemistry
Lambert, Mary Pulliam, biochemistry
Lerman, Abraham, marine geochemistry, limnology
Letsinger, Robert Lewis, organic chemistry
Lewis, Frederick D, photochemistry
Lewis, Marvin Burton, theoretical physics
Leymaster, Glen Ronald, medicine

Lippincott, Barbara Barnes, microbiology, plant physiology
Lippincott, James Andrew, plant physiology
Liu, Liu, solid state physics
Loach, Paul A, biochemistry, physical biochemistry
Lorand, Joyce Bruner, zoology
Lorand, Laszlo, biochemistry, physiology
Lucchesi, Claude A, analytical chemistry, physical chemistry
Mackenzie, Frederick Theodore, geochemistry, sedimentology
Mahowald, Mark Edward, mathematics
Marcus, Jules Alexander, physics
Marcus, Michael Barry, mathematics
Margoliash, Emanuel, molecular biology, protein chemistry
Marks, Tobin Jay, chemistry
Marshall, James Arthur, organic chemistry
Matlis, Eben, mathematics
McElin, Thomas (Welsh), medicine
McKeever, William Paul, physiology
Mendel, Gerald Alan, internal medicine, hematology
Meyer, Edwin F, physical chemistry
Meyer, Roger J, pediatrics
Meyer, Stuart Lloyd, physics, telecommunications
Miller, James Roscoe, medicine
Mintzer, David, gas dynamics, underwater acoustics
Morin, Thomas Lee, operations research
Mount, Kenneth R, mathematics
Neuhaus, Francis Clemens, biochemistry
Nobles, Laurence Hewit, glaciology
Noll, Hans, molecular biology
Norton, David L, biological rhythms
Novales, Ronald Richards, comparative endocrinology, cell biology
Oakes, Robert James, theoretical high energy physics
Offner, Franklin Faller, biophysics
Olmsted, William Edward, applied mathematics
Olmsted, Richard W, pediatrics
Pearson, Ralph Gottfrid, chemistry
Peterson, Elmor Lee, mathematics, operations research
Pines, Herman, organic chemistry
Pinsky, Mark A, mathematics
Priddy, Stewart Beauregard, mathematics
Randall, Eileen Louise, medical microbiology
Ratner, Mark A, physical chemistry
Robinson, R Clark, mathematical analysis
Rosenthal, Robert Wernick, operations research, economics
Rulon, Olin, physiology
Saari, Donald Fene, mathematics, celestial mechanics
Sacks, Jerome, mathematics
Schaefer, Brian Morris, operations research, applied mathematics
Schluter, Robert Arvel, elementary particle physics, experimental high energy physics
Schwartz, Neena Betty, physiology
Schwartz, Theodore Benoni, medicine
Segel, Ralph E, physics
Seth, Kamal Kishore, nuclear physics
Sharon, Nehama, microbiology, virology
Shemin, David, biochemistry
Shen, Sin-Yan, low temperature physics
Shriver, Duward F, inorganic chemistry
Siegert, Arnold John Frederick, mathematical physics
Simon, Norman M, medicine
Simpson, Sidney Burgess, Jr, developmental biology
Sloss, Laurence Louis, geology
Slotter, Richard Arden, inorganic chemistry
Smith, Donald E, physical chemistry, analytical chemistry
Speed, Robert Clarke, geology, mineralogy
Springer, Georg F, immunochemistry
Stein, Michael Roger, mathematics
Strueuer, Stuart, anthropology, archaeology
Sudborough, Ivan Hal, information science
Swanson, Leonard William, mathematics
Tamhane, Ajit Chintaman, applied statistics, mathematical statistics
Turek, Fred William, reproductive endocrinology, photobiology
Van Duyne, Richard Palmer, analytical chemistry, chemical physics
Vye, Malcolm Vincent, hematology, pathology
Waber, James Thomas, atomic physics, solid state physics
Wachtel, Carl, biochemistry
Wagner, James Bruce, Jr, physical chemistry

ILLINOIS

Wagner, John Alexander, entomology
Weertman, Johannes, physics
Weertman, Julia Randall, physics
Weitz, Eric, chemical physics, physical physics
Welker, Neil Ernest, biochemistry
Welland, Robert Roy, mathematics
Werner, Oswald, anthropology; linguistics
Whitten, Eric Harold Timothy, geology
Williams, Robert Fones, topology
Wolfson, Albert, zoology
Woo, Chia-Wei, theoretical physics
Wu, Tai Te, biophysics, applied mathematics
Zelinsky, Daniel, algebra

GLENCOE
Darsow, William Frank, mathematics
Pincus, Irving, organic chemistry
Savit, Joseph, electrophotography

GLEN ELLYN
Brasfield, Travis Winford, plant physiology, mycology,
Groszos, Stephen Joseph, academic administration, industrial chemistry,
Kaufman, Priscilla C., organic chemistry, radiation chemistry
Nicholson, Hayden Coler, physiology
Roberson, John Howard, physics
Sherman, William Cyrus, nutritional biochemistry

EVANSVILLE
Ellis, Donald Edwin, solid state physics, molecular physics

EVERGREEN PARK
D'Ouville, Edmond Lawrence, chemistry
Talso, Peter Jacob, medicine

FAIRBURY
Alsmeyer, William Louis, animal nutrition, biochemistry

FLOSSMOOR
Machinger, Lawrence Arnold, mathematics, education

FOREST PARK
Johnson, Calvin Keith, organic chemistry
Merdinger, Emanuel, biochemistry

FORREST
Godfrey, George Frelaut, poultry husbandry

FOX LAKE
Whymark, Roy R., acoustics

FRANKLIN PARK
Vetter, James Louis, food technology

FREEPORT
Coleman, Joseph Johnston, physics
Glidden, Kenneth Eugene, chemistry

GALESBURG
Boyd, John William, physics
DeMott, Lawrence Lynch, geology, paleontology
Geer, Billy W., genetics, nutrition
Green, Donald Wayne, physics
Harris, Leland, physical organic chemistry
Hourston, Rogvvald Charles Nicol, mathematics
Johnson, Robert Eugene, physiology, animal nutrition
Kooser, Robert Galen, physical chemistry
Mark, Daniel Lee, parasitology, nematology
Moore, Duane Milton, mineralogy, geochemistry
Neumiller, Harry Jacob, Jr, organic chemistry
Perry, Eugene Arthur, microbiology, mycology
Salter, Lewis Spencer, theoretical physics
Sutton, Russell Paul, chemistry
Ward, George Henry, systematic botany
Young, Frank Hood, mathematics

GLENVIEW
Applewhite, Thomas Hood, organic chemistry, petroleum chemistry
Baldwin, Heber Ross, biochemistry, food technology
Berkman, Michael G., chemistry
Chan, Ming Sui Michael, food technology, biochemistry
Dework, Frank Matthew, Jr., microbiology
Gloyer, Stewart Edward, organic chemistry, industrial
Gordon, Arthur Leonard, agricultural biochemistry
Holcomb, David Nelson, physical chemistry
Huffman, Clarence W., organic chemistry
Hynes, John Francis, food biochemistry,
Jackson, Harold Woodworth, analytical chemistry
Jonas, John Joseph, organic chemistry,
Krishnamurthy, Ramanathapur Gundachar, food chemistry, biochemistry
Kunkel, Reinold Walter, dairy science
Loeb, Melvin Lester, organic chemistry
Lowrie, Harman Smith, organic chemistry
Lucas, Glennard Ralph, organic polymer chemistry
Lushbough, Channing Harden, nutrition, biochemistry
Meisner, Donald F., cereal chemistry
Nath, K Rajinder, dairy chemistry
Norris, Frank Arthur, food science
Proctor, Jerry Franklin, nutrition, physiology
Roland, John Francis, microbial biochemistry
Schnell, Gene Wheeler, microbiology, biochemistry
Tyner, David Anson, organic chemistry
Wargel, Robert Joseph, microbiology
Williams, Joseph Lee, lipid chemistry
Matson, Howard John, organic chemistry, petroleum chemistry
Nowak, Anthony Victor, analytical chemistry
Snow, Adolph Isaac, chemistry
Turnquest, Byron W., organic chemistry
Voelz, Frederick, chemical physics
Wulfers, Thomas Frederick, organic chemistry

GLENWOOD
Capone, James Joseph, immunochemistry; bacteriology

GODFREY
Chacharonis, Peter, protozoology
Jones, Richard Evan, Jr., physical chemistry; inorganic chemistry
Sokolowski, Danny Hale, physics

GOLF
Johnson, Robert Ivar, microscopy
Willis, Clifford Leon, geology

GREAT LAKES
Devine, Leonard Francis, bacteriology
Keene, Harris J., oral pathology
Miller, Ralph J., physics
Richards, R Ronald, physical chemistry
Siefken, Hugh Edward, experimental nuclear physics
Stromberg, Verner L., Jr, organic chemistry
Tao, Robert Chi-Mei, nutrition
Tomaschke, Harry E., surface physics

GREENVILLE
Hodson, Adrian Zachariah, nutritional biochemistry
McMullen, Warren Anthony, organic chemistry

GURNEE
Berdahl, James Maynard, organic chemistry
Nolan, Chris, biochemistry
Peckinpaugh, Robert Owen, preventive medicine, internal medicine

HARVEY
Gallagher, James Patrick, chemistry
Jaecker, John Alvin, inorganic chemistry
Johnson, Marvin Francis Linton, physical chemistry
Knecht, Albert T., petroleum microbiology; biochemistry

HINSDALE
Ashley, Warren Cotton, organic chemistry
Elkins, Robert Hiatt, chemistry
Freeman, Arthur, veterinary medicine
Lichtenwalter, Glen, organic chemistry
McElroy, Donald L., dentistry
Mesrobian, Robert Benjamin, chemistry
Ringo, George Roy, experimental physics
Tevebaugh, Arthur David, physical chemistry
Witzel, Everet Wayne, anatomy

HOFFMAN ESTATES
Lax, Louis Carl, physiology; medicine

HOMEWOOD
Crews, Lowell Thomas, petroleum chemistry
Jacobs, Robert Byron, physics
Muggli, Robert Zeno, bio-organic chemistry; chemical microscopy
Wester, John Walter, Jr, applied mathematics

HUNTLEY
Venerable, James Thomas, organic chemistry

JACKSONVILLE
Ecker, Edwin D., mathematics
Evans, Robert John, organic chemistry
Filson, Don P., physical chemistry
Franz, Edgar Arthur, mathematics
Freiburg, Richard Eighme, zoology; ecology
Kanatzar, Charles Leplie, zoology
Kohlbecker, Eugene Edmund, mathematics
Leland, Frances Elbridge, physical chemistry
McCollough, Fred, Jr, inorganic chemistry

HINES
Bernsohn, Joseph, neurosciences
Bird, Thomas Joseph, medical microbiology
Dietz, Albert Arnold Clarence, biochemistry
Greenlee, Herbert Breckenridge, surgery
Held, Irene Rita, neurochemistry
Hiebert, Talmage Gordon, anesthesiology
Johnson, Arthur Frederick, experimental medicine
Kanabrocki, Eugene Ladislaus, biochemistry
Kaplan, Ervin, internal medicine, nuclear medicine
Littman, Armand, medicine
Meadows, William R., cardiology
Molnar, Zelma Villanyi, pathology
Oester, Yvo Thomas, pharmacology
Rubnitz, Myron Ethan, pathology
Schrek, Robert, pathology
Sharp, John Turner, internal medicine
Stanley, Malcolm McClain, medicine
Stefani, Stefano, radiotherapy

HIGHLAND PARK
Baskin, Aaron David, plant pathology
Bernstein, Benjamin Tobias, solid state physics, physical chemistry
Bernstein, Elaine Katz, biochemistry,
Coleman, Richard J., organic chemistry, food technology
Curtice, Jay Stephen, physical organic chemistry, pharmaceutical chemistry
Nysted, Leonard Norman, pharmaceutical chemistry
Rahn, Joan Elma, plant morphology
Shulman, John Morton, anesthesiology
Stein, Irving F, Jr., surgery

HAVANA
Sparks, Richard Edward, aquatic biology

KANKAKEE
Aldred, J Phillip, physiology, pharmacology
Bastian, James W., endocrinology
Clements, Gerald Richard, experimental pathology
Dailey, Joseph Patrick, organic chemistry
Feldman, Fred, biochemistry; medical research
Ferren, Larry Gene, biochemistry
Grothaus, Clarence (Edward), organic chemistry
Hanson, John Elbert, inorganic chemistry
Hines, Wallis Gartside, organic chemistry
Hughes, John Lawrence, organic chemistry
Kaiser, Emil, chemistry
Reams, Max Warren, geomorphology, sedimentary petrology
Schlueter, Robert John, biochemistry
Skibbe, Martin Otto, pharmaceutical chemistry
Strickler, Dwight Johnston, genetics
Westfall, Robert Judson, biochemistry

KEWANEE
Van Riper, Gordon Everett, agronomy

KINMUNDY
Buck, David Homer, fisheries

LA GRANGE
Carnall, William Thomas, physical chemistry
Damaskus, Charles William, biochemistry; zoology
Tiemstra, Peter J, food chemistry

LA GRANGE PARK
Gilbert, Francis Evalo, chemistry
Vissat, Peter Louisa, physics
Wood, Scott Emerson, physical chemistry

LA SALLE
Moore, Eunice Martha, physical chemistry
Reidies, Arno H, inorganic chemistry

LAKE BLUFF
Beck, Karl Maurice, organic chemistry
Frederick, Kenneth Jacob, physical chemistry
Leffler, Martin Templeton, organic chemistry
Thorner, Melvin Wilfred, neurosciences

LAKE FOREST
Couts, John Wallace, physical chemistry
Dames, Charlotte A., physical chemistry
Donnally, Bailey Lewis, atomic physics
Dunn, William Lewis, analytical chemistry
Faber, Roger Jack, chemical physics
Giere, Frederic Arthur, mammalian physiology
Gross, Herbert Michael, pharmaceutical chemistry
Jeong, Tung Hon, nuclear physics
John, Lucille, inorganic chemistry
Long, Charles Anthony, chemical physics
Louch, Charles Dukes, zoology
Packel, Edward Wesler, mathematics
Runge, Richard R, genetics
Shively, Ralph Leland, mathematics
Spiess, Luretta Davis, developmental biology
Thompson, Martin Leroy, inorganic chemistry
Troyer, Robert James, mathematics

LAKE VILLA
Rymer, Harry, astronomy; mathematics

JOLIET
Foster, Raymond Orrville, clinical biochemistry, reproductive biology
Iveson, Herbert Todd, chemistry
Mead, John Marcus, organic chemistry
Rolth, Robert J., organic chemistry
Rudzitis, Edgars, inorganic chemistry, forensic science
Zeller, Mary Claudia, mathematics

Rainbolt, Mary Louise, biology

LAKE ZURICH
Malloy, Thomas Patrick, organic chemistry
Salutsky, Murrell Leon, water chemistry

LAWRENCEVILLE
Moore, Marian Alease, mathematical analysis

LEBANON
Jones, R E Douglas, mathematics

LIBERTYVILLE
Ballmann, Donald Lawrence, geology
Bogner, Phyllis Holt, physiology
Brown, Lindsay Dietrich, horticulture, plant physiology
Case, Vern Wesley, soil fertility
Everhart, Donald Lough, economic geology
Ferrara, Louis W, analytical chemistry, biochemistry
Friedland, Waldo Charles, chemistry
Hamer, Martin, organic chemistry
Holik, Melville James, organic polymer chemistry, synthetic organic chemistry
Kraemer, John Francis, organic chemistry, polymer chemistry
Lane, Alfred Glen, animal nutrition
Neagle, Lyle H, animal nutrition, biochemistry
Palaia, Frank Lincoln, Jr, medical research
Peeler, Herbert Tremble, nutrition, biochemistry
Roderick, William Rodney, organic chemistry, medicinal chemistry
Sandvik, Peter Olaf, geology
Stewart, John Allan, soil science
Teague, Kefton Harding, economic geology

LINCOLNWOOD
Taber, David, biomedical engineering, organic chemistry
Tuzzolino, Anthony J, physics, solid state physics

LISLE
Bowe, Joseph Charles, physics
Carney, Rose Agnes, physics
Etter, Alfred Gordon, ecology
Hall, Marion Trufant, taxonomy, cytogenetics
Hazdra, James Joseph, analytical chemistry
McMillan, Clara Albertina, physical chemistry
Meeker, Ralph Dennis, physics
Rausch, David John, organic chemistry
Shonka, Richard Edward A, applied mathematics
Spokas, John J, radiation physics
Thiruvathukal, Kuriakose V, zoology, morphology

LOCKPORT
Dreska, Noel, molecular spectroscopy
Hogan, Philip, organic chemistry
Meyer, Eugene Frank Jr, physical chemistry
Ware, George Henry, plant ecology

LOMBARD
Awad, Mohamed Zeinelabideen, physiology
Bachop, William Earl, developmental biology, gross anatomy
Himmel, Keith LaVern, biology, zoology
Jourdikian, Felor Kaloust, neurosciences
Kuehner, Calvin Charles, medical microbiology, public health
Poppe, Wassily, physical chemistry
Wertwijn, George, physics, chemistry

MACOMB
Archbold, Norbert L, geology
Bean, Gerritt Post, organic chemistry
Bergen, John Victor, geography
Bundschuh, James Edward, physical chemistry
Chu, Keh-Chang, nuclear magnetic resonance, radiation physics
Crall, Howard William, biology
Dove, Lewis Dunbar, plant physiology
Edwards, Harold Herbert, plant physiology, biochemistry

El-Awady, Abbas Abbas, physical inorganic chemistry
Fink, Rodney James, agronomy, weed science
Franks, Edwin Clark, zoology
Gabler, Robert Earl, physical geography
Gandhi, Jeet-Mal, pure mathematics, theoretical physics
Gardner, Franklin Pierce, agronomy
Goeckner, Norbert Anthony, organic chemistry
Griffin, Donald William, geography
Hardin, Richard Lynn, biochemistry
Harrod, Scott Boynton, mathematics
Hart, Harold Bird, theoretical physics
Henry, Robert David, plant morphology
Hess, David Filbert, geology, petrology
Holmes, Edward Bruce, comparative anatomy
Howe, Virgil K, forest pathology
Hughes, Benjamin G, inorganic chemistry, analytical chemistry
Hurren, Weiler R, solid state physics
Jahn, Lawrence A, aquatic biology
Jones, Reece Alexander, geography
Juskevice, John Anthony, paleontology, marine zoology
Keller, Allen S, stratigraphy, structural geology
Kirkpatrick, James W, analytical chemistry
Kreiling, Daryl, mathematics
Kurjack, Edward Barna, anthropology
Larkin, Jeanne Holden, developmental biology
Lathrop, Arthur LaVern, physics
Ma, Te Hsiu, cytogenetics
Martin, Kenneth Robert, geography
McVickar, John S, agronomy
Mock, Gordon Duane, mathematics
Morey, Robert V, cultural anthropology, ethnology
Morris, Everett Franklin, botany
Mumik, Mary Rengo, genetics
Myers, Roy Maurice, botany
Neas, Robert Edwin, analytical chemistry
Nelson, Ronald Eugene, geography
Nielsen, Peter James, cell physiology, protozoology
Noble, John Dale, physics
Nollen, Paul Marion, parasitology
O'Flaherty, Larrance Michael Arthur, algology
Palmer, Robert Gerald, soil conservation
Pederson, Vernon Clayton, physiology, endocrinology
Rawlinson, David John, organic chemistry
Sather, John Henry, zoology
Sedman, Yale S, entomology
Sheldon, Victor Lawrence, soils
Shelton, Robert Wayne, organic chemistry
Shryock, A Jerry, mathematics
Singer, Samuel, microbiology
Soule, David Elliot, physics
Stidd, Benton Maurice, paleobotany
Stipanowich, Joseph J, mathematics
Synovitz, Robert J, health science
Taneja, Viday Sagar, mathematical statistics
Thurow, Gordon Ray, zoology, anatomy
Turner, John K, animal physiology
Walter, Waldemar Melchert, invertebrate zoology, ecology
Warnock, John Edward, zoology
Weller, Paul Franklin, inorganic chemistry
Wendt, Arnold, mathematics
Wesley, Dean E, soil fertility
White, David Arnold, zoology, aquatic ecology
Wingard, Norman Edward, geology
Wylie, Douglas Wilson, physics

MAHOMET
Bryan, Hugh D, pharmaceutical chemistry

MANSFIELD
Undeen, Albert Harold, parasitology

MAPLE PARK
Gorenz, August Mark, plant pathology

MARION
Sesco, Jerry Anthony, forestry, economics

MAYWOOD
Aktipis, Stelios, biochemistry, biophysics
Bermes, Edward William, Jr, clinical chemistry

Bloor, Byron Michel, medicine, neurosurgery
Blumenthal, Harold Jay, microbiology, microbial biochemistry
Bowman, Douglas Clyde, physiology, pharmacology
Bunch, Wilton Herbert, orthopedic surgery, muscular physiology
Choukas, Nicholas C, oral surgery
Cole, Madison Brooks, Jr, cell biology, orthopedics
Davis, Joseph Richard, toxicology, oncology
Dobrin, Philip Boone, medical physiology, cardiovascular physiology
Doemling, Donald Bernard, physiology
Domm, Lincoln Valentine, anatomy, zoology
Doolin, Paul F, neurobiology
Dunn, Jon D, neuroendocrinology
Farrand, Stephen Kendall, microbiology, molecular biology
Filkins, James P, physiology
Friedman, Alexander Herbert, pharmacology
Gerhard, Rinert J, dentistry
Goaz, Paul William, dentistry
Gowgiel, Joseph Michael, dentistry, anatomy
Grandel, Eugene Robert, dentistry
Gunnar, Rolf McMillan, cardiology, internal medicine
Hadek, Robert, microscopic anatomy
Juhasz, Stephen Eugene, psychiatry
Karczmar, Alexander George, pharmacology, physiology
Keeley, John L, surgery
Keis, Adelbert Ferdinand Richard, human anatomy, histology
Keresztes-Nagy, Steven, physical chemistry, biochemistry
Kiely, Michael Lawrence, anatomy
Kulkarni, Bidy D, reproductive endocrinology, maternal & child health
Lange, Charles Ford, Jr, biochemistry, immunology
LaVelle, Faith Wilson, histology, neuroembryology
Lees, William Morris, thoracic surgery, cardiovascular surgery
L'Heureux, Maurice Victor, biochemistry
Love, Leon, radiology
Masterson, John G, obstetrics & gynecology
McDonald, Hugh Joseph, biochemistry, physical chemistry
Melchior, Norten Cass, organic chemistry, biological chemistry
O'Morchoe, Charles Christopher C, anatomy, physiology
O'Morchoe, Patricia Jean, histology, cytology
Paloyan, Edward, surgery
Peiss, Clarence Norman, physiology
Perez-Tamayo, Ruheri, radiotherapy
Pollock, Robert J, Jr, oral biology
Randall, Walter Clark, physiology
Rapp, Gustav Victor, biochemistry
Rubinstein, Alicia Susana, oral pathology
Sandrik, James Leslie, dental materials
Schmidt, Robert Sherwood, behavioral physiology
Schmitt, Allen F, chemistry, biology
Schoen, William P, dentistry, dental materials
Schultz, Richard Michael, biochemistry
Scudder, Charles Lee, psychopharmacology
Smith, Jackson Algernon, psychiatry
Soper, Edward Henry, dentistry, anatomy
Sturtevant, Ruthann Patterson, gross anatomy, biological rhythms
Toto, Patrick D, oral pathology
Velardo, Joseph Thomas, anatomy
Wells, Joseph Albert, pharmacology, physiology
Wood, Norman Kenyon, oral pathology
Yotis, William William, microbiology

MCCOOK
Abramitis, Walter William, agricultural chemistry
Castro, Anthony J, organic polymer chemistry
Gray, Linsley Shepard, Jr, analytical chemistry
Jakubiec, Robert Joseph, analytical chemistry
Metcalfe, Lincoln Douglas, analytical chemistry
Shapiro, Sydney Harold, organic chemistry
Wan, Kwok Ming, organic biochemistry, business

MCHENRY
Barker, George Ernest, organic chemistry

MELROSE PARK
Alsberg, Henry, polymer chemistry, industrial chemistry
Berger, Daniel Richard, organic chemistry
Dasher, George Franklin, Jr, physical chemistry, surface chemistry
Forest, Harvey, physical chemistry
Nelson, Arthur Kendall, chemistry
Poulos, Nicholas A, organic chemistry
Retsky, Michael Walter, electron optics, electron microscopy
Shaneyfelt, Duane L, polymer chemistry, applied chemistry

MILFORD
Mumm, Walter John, genetics

MOLINE
Burrows, William Chapel, soil science, agronomy
Pauli, Arland Walter, plant physiology
Stickler, Fred Charles, agronomy, crop ecology
Wiedenmann, Lynn G, polymer chemistry, organic chemistry

MONMOUTH
Allison, David C, plant genetics, cytology
Boswell, Rupert Dean, Jr, mathematics
Buchholz, Robert Henry, physiology
Gebauer, Peter Anthony, organic chemistry
Johnson, Arthur Franklin, physics
Jones, Berwyn E, analytical chemistry
Ketterer, John Joseph, zoology
Kieft, Richard Leonard, inorganic chemistry, analytical chemistry
Nagel, Terry Marvin, inorganic chemistry, physical chemistry
Skov, Charles E, solid state physics
Williams, Lyman O, structural geology
Wills, Donald L, geology

MONTICELLO
Collins, Vernon Kirkpatrick, biochemistry

MORRIS
Fettes, Edward Mackay, polymer chemistry
Hovnanian, August P, surgery

MORTON GROVE
Ausman, Robert K, medical administration, research administration
Berger, Arthur, organic chemistry
Brochu, William Eugene, industrial pharmacy, physical pharmacy
Chinn, Leland Jew, medicinal chemistry
Darby, Thomas Dillard, pharmacology
Feldman, Louis Israel, microbiology
Garvin, Paul Joseph, Jr, toxicology
Ginsburgh, Irwin, engineering physics
Jaffe, Philip Monlane, inorganic chemistry
Koefert, Michael Tallyn, toxicology, pharmacology
Mascoli, Carmine Charles, virology
Mather, Adaline Nicoles, biochemistry
Nandan, Rajiva, microbiology, biochemistry
Nedich, Ronald Lee, physical pharmacy, industrial pharmacy
Rakosky, Joseph, Jr, microbiology, food science
Serkes, Kenneth Dean, surgery
Stern, Ivan J, biochemistry, bacteriology
Tucker, Robert Gene, nutrition
Wallin, Richard Franklin, toxicology, pharmacology
Weary, Marlys E, anatomy, histology

MT PROSPECT
Borgman, Robert John, medicinal chemistry
Gardella, Libero Anthony, pharmaceutical chemistry
Morris, Robert Nicholas, pharmacology
Possley, Leroy Henry, pharmaceutical chemistry
Zaroslinski, John F, pharmacology, biochemistry

MT VERNON
Rue, Edward Evans, exploration geology, resource management

MUNCIE
Van Atta, Robert Ernest, analytical chemistry

NAPERVILLE
Adams, Max Dwain, inorganic chemistry
Allen, John Kay, polymer chemistry
Banas, Emil Mike, physics
Bersted, Bruce Howard, polymer physics
Bertolacini, Ralph James, inorganic chemistry
Blaha, Eli William, organic chemistry
Bolton, B A, organic chemistry
Brennan, Harry Michael, physical chemistry
Brown, John Stewart, chemistry
Buntrock, Robert Edward, organic chemistry
Cengel, John Anthony, physical organic chemistry
Chen, Paul Ear, applied mechanics, applied mathematics
Chipman, Gary Russell, polymer chemistry
Clardy, LeRoy, physical chemistry, polymer chemistry
Conway, Hertsell S, information science, chemical literature
Denenberg, Charlotte Goryn, number theory, computer science
Dietrich, Verne Eugene, physics
Dobry, Alan (Mora), physical chemistry
Ebner, Herman George, polymer chemistry, organic chemistry
Feinstein, Allen Irwin, organic chemistry
Fenoglio, David John, physical organic chemistry
Golinkin, Herbert Sheldon, physical chemistry
Goretta, Louis Alexander, organic chemistry
Guterliet, Louis Charles, organic chemistry
Guttman, Newman, psychoacoustics
Hanson, Robert Burton, polymer chemistry
Hanson, Russell, pharmacology
Harper, Jon Jay, industrial organic chemistry
Harris, Samuel William, petroleum chemistry
Hopkins, Paul Donald, inorganic chemistry
Howsmon, John Arthur, chemistry, research administration
Hughes, Robert David, polymer chemistry
Hunt, Russell Aubrey, Jr, organic chemistry
James, David Eugene, organic chemistry, research administration
Jezl, James Louis, organic chemistry, chemistry
Johnson, Carl Edwin, organic chemistry
Johnson, Donald Elwood, mathematics
Jones, Thomas Hubbard, physical chemistry
Kalinowski, Mathew Lawrence, petroleum chemistry
Kapff, Six Frederick, organic chemistry
Karayannis, Nicholas M, inorganic chemistry
Karll, Robert E, organic chemistry
Kissel, William John, polymer chemistry
Knobloch, James Otis, organic chemistry
Knox, Jack Rowles, polymer science
Kuhlmann, George Edward, organic chemistry
Lee, Robert James, industrial organic chemistry
Lindberg, Steven Edward, chemistry, physical chemistry
Little, Randel Quincy, Jr, organic chemistry
Marcus, Sheldon H, research administration
Marsh, Terrence George, environmental biology
Martin, Ronald LeRoy, analytical chemistry
McCollum, John David, chemistry
Meguerian, Garbis H, physical organic chemistry
Meyer, Delbert Henry, organic chemistry
Meyerson, Seymour, chemistry, mass spectrometry
Mhatre, NageshShamrao, biochemistry, enzymology
Morello, Edwin Francis, organic chemistry
Myerholtz, Ralph W Jr, polymer chemistry

Nevitt, Thomas D, organic chemistry, petroleum chemistry
Paschke, Edward Ernest, polymer chemistry
Peri, John Bayard, physical chemistry
Peters, Edwin Francis, chemistry
Piehl, Frank John, organic chemistry
Poel, Russell J, organic chemistry
Pohlmann, Hans Peter, inorganic chemistry, physical chemistry
Radford, Herschel Donald, organic chemistry
Rakowsky, Frederick William, physical chemistry
Rife, William C, organic chemistry
Rogan, John B, organic chemistry, polymer chemistry
Schaap, Luke Anthony, organic chemistry
Schwartz, Michael Muni, industrial chemistry
Sherren, Anne Terry, analytical chemistry
Sisko, Arthur William, physical chemistry
Siana, Francis J, chemistry
Stanley, William Gordon, physical chemistry
Stephens, James Regis, organic polymer chemistry
Stofler, Robert Llewellyn, industrial hygiene, analytical chemistry
Swakon, Edward Antone, organic chemistry
Towle, Philip Hamilton, organic chemistry
Trent, John Ellsworth, organic chemistry
Trevillyan, Alvin Earl, organic chemistry
Tucker, Marie, embryology, biology
Udelhofen, John Henry, petroleum chemistry
Vander Haar, Roy William, analytical chemistry
Van Strien, Richard Edward, organic chemistry
Warne, Thomas Martin, organic chemistry, petroleum chemistry
Weil, Thomas Andre, inorganic chemistry
Wheeler, George Willis, physics
White, Philip Cleaver, chemistry
Wolff, William Francis, organic chemistry
Zimmerscheid, William John, chemistry
Zietz, Alex, physical chemistry, organic chemistry

NILES
Pankratz, Ronald Ernest, analytical chemistry
Pavkovic, Stephen F, inorganic chemistry

NEOGA
Nance, James Francis, plant physiology

NEW WINDSOR
Samuels, Larry D, oncology, radiation biology

NORMAL
Berk, Kenneth N, mathematics
Birkenholz, Dale Eugene, biology
Born, Harold Joseph, physics
Bristol, Benton Keith, agriculture
Brockman, Herman E, genetics
Brown, Francis Robert, mathematics
Brown, Lauren Evans, zoology
Brown, Walter Howard, plant ecology, physiology
Bunting, Roger Kent, inorganic chemistry
Cain, Jerome Richard, algology
Calef, Wesley Carr, resource geography
Chasson, Robert Morton, plant physiology
Chuang, Tsan Iang, systematic botany
Clemens, Stanley Ray, mathematics
Cralley, John Clement, human anatomy, physiology
Crew, John Edwin, nuclear physics
Crumley, Richard D, mathematics
Dilks, Eleanor, zoology
Duey, Robert C, organic chemistry, physical chemistry
Ecklund, Earl Frank, Jr, number theory, computer science
Edwards, Thomas F, science education
Eggan, Lawrence Carl, mathematics
Fenshol, Dorothy Eunice, phycology
Fitch, Kenneth Leonard, anatomy
Frahm, Charles Peter, theoretical physics
Frehn, John, physiology
Friedberg, Stephen Howard, mathematical analysis
Fuess, Frederick William, III, agronomy

Hansen, John Frederick, organic chemistry
Hart, Richard Royce, geology, stratigraphy
Hetzel, Howard Roy, invertebrate zoology
House, James Evan, Jr, physical inorganic chemistry
Huizinga, Harry William, zoology, parasitology
Ichniowski, Thaddeus Casimir, chemistry
Insel, Arnold J, mathematics
Jacobson, Alvin Raymond, environmental health
Jensen, Donald Reed, mammalian morphology
Jesse, Kenneth Edward, solid state physics
Johnson, Eric Shepherd, geography of Latin America, economic geography
Kurz, Michael E, organic chemistry
Liberta, Anthony E, mycology
Luther, Marvin L, atomic physics
Mentzer, Loren Willis, ecology
Miller, Edith Joan Wilson, cultural geography, geography of the United States
Miller, John Grier, mathematics
Mockford, Edward Lee, entomology
Moore, Clarence L, dairy science
Neville, Melvin Edward, microbiology
Otto, Albert Dean, mathematics
Parr, James Theodore, algebra
Parr, Phyllis Graham, mathematics
Patterson, James Edward, physical chemistry
Retzer, Kenneth Albert, mathematics education
Rhymer, Ione, bacteriology
Richardson, Arlan Gilbert, biochemistry, organic chemistry
Rilett, Robert Omar, biology
Ritt, Robert King, applied mathematics
Ryder, Bernard Leroy, organic chemistry
Sagebiel, Joe Alfred, animal science
Schroeer, Juergen Max, solid state physics, mass spectrometry
Schwalm, Fritz Ekkehardt, zoology, developmental physiology
Searight, Thomas Kay, geology
Seligman, Isaac Morris, insect physiology
Shulman, Sol, organic chemistry, polymer chemistry
Speiser, Robert David, pure mathematics
Sublett, Michael Dean, geography
Thompson, James Tipton, animal science, animal nutrition
Tone, James N, animal physiology
Trotter, John Ellis, geography
Vanden Eynden, Charles Lawrence, number theory
Verner, Jared, animal ecology
Waage, Edward Vern, physical chemistry
Ward, Jack A, zoology, animal behavior
Weber, David Frederick, cytogenetics
Weigel, Robert David, vertebrate zoology
West, Douglas Xavier, inorganic chemistry
Willis, Edwin Roy, zoology, entomology

NORTH CHICAGO
Andres, William Wolcott, microbial biochemistry
Appell, Raynor Norbert, clinical pharmacology
Appleton, George Sanders, agricultural microbiology
Barlow, Grant Harold, physical biochemistry
Becker, Bernard Abraham, toxicology
Biel, John Hans, chemistry
Braendle, Donald Harold, microbiology
Brodie, David Alan, pharmacology
Chadde, Frank Ernest, analytical chemistry
Chappell, Elizabeth, pharmacology
Chu, Sou Yie, drug metabolism
Cole, Wayne, natural products chemistry
Couch, Edward Hing Loy, chemistry
Couch, Terry Lee, entomology
Decker, Richard H, biochemistry
Denison, Frank Willis, Jr, microbiology
DeRose, Anthony Francis, medicinal chemistry, pharmaceutical chemistry
Dodge, Patrick William, pharmacology
Egan, Richard Stephen, structural chemistry, nuclear magnetic resonance

Estep, Charles Blackburn, entomology, pharmacology
Ford, Thomas Matthews, veterinary medicine
Fricke, Howard Henry, infectious diseases, biochemistry
Garven, Floyd Charles, organic chemistry
Girolami, Roland Louis, microbiology
Grundy, Walton Earle, bacteriology, industrial microbiology
Hasbrouck, Richard Berend, organic chemistry
Herting, Robert Leslie, internal medicine, clinical pharmacology
Hirata, Arthur Atsunobu, immunology
Holleman, William H, biochemistry
Hung, Paul P, biochemistry, virology
Hwang, Kao, pharmacology
Jones, Peter Hadley, organic chemistry, cardiovascular diseases
Jones, Ralph William, pharmaceutical chemistry
Kimura, Eugene Tatsuru, pharmacology
Kirchmeyer, Frederick Joseph, pharmacy
Krimen, Lewis Irwin, organic chemistry
Kurath, Paul, organic chemistry
Lambert, Glenn Frederick, biochemistry, medicinal chemistry
Lee, Cheuk Man, organic chemistry
Ling, Chung-Mei, biochemistry
Lynch, Don Murl, organic chemistry
Mao, James Chieh Hsia, biochemistry, medicinal chemistry, molecular pharmacology
Martin, Yvonne George, chemistry
Mattoon, Richard Wilbur, chemical physics
McAlpine, James Bruce, organic chemistry
Minard, Frederick Nelson, biochemistry
Nutting, Leighton Adams, biology
Levenberg, Milton Irwin, mass spectrometry, computer science
Otto, Robert H, bacteriology
Overby, Lacy Rasco, biochemistry
Patel, Dhanooprasad Gordhanbhai, medical research, pharmacology
Perun, Thomas John, organic chemistry
Peterson, Merlin Henry, biochemistry
Sievert, Herman William, biochemistry
Singiser, Robert Eugene, research administration, pharmaceutics
Schoepke, Hollis George, pharmacology
Shipkowitz, Nathan L, microbiology
Schenck, Jay Ruffner, biochemistry
Scheicher, Joseph Bernard, cell biology, virology
Saunders, Allen Perry, microbiology
Ranade, Vinayak Vasudeo, medicinal chemistry
Plotnikoff, Nicholas Peter, pharmacology
Price, James Melford, oncology, physiology
Ringler, Ira, biochemistry
Rivett, Robert Wyman, biochemistry
Rosenbrook, William, Jr, organic chemistry
Smith, Irvin Darrow, biochemistry
Snyder, Ann Knabb, medical research
Sommers, Armiger Henry, chemistry
Stein, Herman H, biochemistry
Stein, Robert George, agricultural chemistry, medicinal chemistry
Taylor, Julius David, biochemistry
Tekeli, Sait, veterinary medicine, pathology
Thomas, Alford Mitchell, chemistry
Thomas, Elizabeth Wadsworth, analytical chemistry
Thompson, George Rex, toxicology, pharmacology
Washburn, William H, analytical chemistry
Weatherwax, Robert Stanton, microbiology
Weston, Arthur Walter, organic chemistry
Wideburg, Norman Earl, microbiology
Wiegand, Ronald Gay, biochemistry
Wimer, David Carlisle, analytical chemistry, organic chemistry
Winfield, Arnold Harris, chemistry
Winn, Martin, medicinal chemistry
Winters, George Philip, pharmacy
Woroch, Eugene Leo, medicinal chemistry
Yunker, Martin Henry, pharmacy
Zaugg, Harold Elmer, organic chemistry

NORTH RIVERSIDE
Duke, Phillip S, experimental pathology, biochemistry

NORTHBROOK
Cox, Billy Joe, environmental sciences
Cushing, Vincent Jerome, applied physics
Goodheart, Clyde Raymond, biology, medicine
Gordon, Donovan, veterinary medicine, veterinary pathology
Hime, William Gene, analytical chemistry
Keplinger, Moreno Lavon, toxicology
Kinoshita, Florence Keiko, toxicology, pharmacology
Litman, Irving Isaac, food technology
Lloyd, Frederick A., urology
Perrine, Eugene Louis, physics
Robertson, Reed S., water chemistry, chemical engineering
Wingender, Ronald John, analytical chemistry

NORTHFIELD
Knaggs, Edward Andrew, organic chemistry
Magnus, George, plastics chemistry

OAK BROOK
Beuk, Jack Frank, biochemistry, enzymology
Bowers, Raymond Harold, analytical chemistry
Davies, Donald Leslie, food science
Donnelly, Thomas Henry, biophysical chemistry
Drews, Udo Wilhelm, solid state physics, optics
Duxbury, Dean David, food science
Freedman, Arthur Jacob, chemistry
Greenberg, Richard Aaron, bacteriology
Heuberger, Glen (Louis), animal science, animal nutrition
Holty, David Webster, organic chemistry, finance
Johnston, Richard S., chemistry
Kang, Changhee Kim, food science
Kueper, Theodore Vincent, experimental statistics
Lewis, Morton, organic chemistry
McIntyre, George Francis, applied chemistry
Murphy, Robert Emmett, meat sciences
Pavey, Robert Louis, food science
Poling, Clyde Edward, nutrition
Rice, Eldon Emerson, biochemistry, nutrition
Schuytema, Carl G, food science
Taylor, James Robert, analytical chemistry
Tompkin, Robert Bruce, food microbiology
Trelease, Richard Davis, food chemistry
Wilcox, Joseph Clifford, food science, food microbiology

ORLAND PARK
Casaz, Geronimo, microbiology, clinical pathology

PALATINE
Albers, Robert Jay, biophysical chemistry
Burkhard, Mahlon Daniel, acoustics
Hoff, Raymond E, polymer chemistry
Kokalis, Soter George, inorganic chemistry
Krueger, Robert Harold, physical chemistry
Laurin, Pushpamala, electromagnetism
Mavity, Julian Maris, organic chemistry
Mulvihill, Mary Lou Jolie, human physiology
Stucker, Joseph Bernard, chemistry

PALOS HEIGHTS
Cook, Harry, endocrinology, physiology

PALOS HILLS
Vasiliauskas, Edmund, organic chemistry

PALOS PARK
Marcowitz, Stewart, osteopathy

PARK FOREST
Babcock, Robert Frederick, analytical chemistry
Berry, Brian Joe Lobley, urban geography, urban research & development
Eaton, James Edmonds, environmental chemistry
Fox, Kenneth Ian, food microbiology, food science
Greenberg, Elliott, physical chemistry, inorganic chemistry
Kleinschmidt, Albert Willoughby, organic chemistry
Lustig, Stanley, physical chemistry
Scherr, George Harry, microbiology
Shaw, Norman Yon-Shong, biochemistry, immunochemistry
Stock, Werner, biochemistry
Wells, Jane Frances, mathematics

PARK FOREST SOUTH
Andrews, Theodore Francis, ecology
Casagrande, Daniel Joseph, organic geochemistry
Chambers, John Edward, phycology, electron microscopy
Douglas, Donald Sterling, comparative physiology
Fenner, Peter, geology, science education
Rocke, David M, mathematics

PARK RIDGE
Fisher, Perry Wright, meteorology
Gupta, Indra Narayan, geophysics, seismology
Stern, Joel R, biochemistry

PEKIN
Terando, M Loretta, psychophysiology, transactional analysis

PEORIA
Abbott, Thomas Paul, polymer chemistry
Anderson, Robert Lewis, biochemistry, protein chemistry
Bagby, Marvin Orville, organic chemistry
Bietz, Jerold Allen, protein chemistry
Bjorklund, Richard Guy, ecology, conservation
Bothast, Rodney Jacob, food microbiology
Brown, Joseph Ross, mathematics
Burmeister, Harland Reno, microbiology
Chang, Shu-Pei, polymer chemistry, industrial organic chemistry
Christianson, Donald Duane, plant biochemistry
Ciegler, Alex, microbiology
Cotsonas, Nicholas John, Jr., medicine
Cowan, John C, organic chemistry
Cummings, Thomas Fulton, physical chemistry, organic chemistry
Davis, Edwin Nathan, microbiology, biochemistry
Detroy, Robert William, biochemistry, microbiology
Dick, William Edwin, Jr, chemistry
Doane, William M, organic chemistry
Dutton, Herbert Jasper, chemistry
Earle, Fontaine Richard, chemistry
Eissler, Robert L, physical chemistry
Emken, Edward Allen, lipid chemistry, organic biochemistry
Evans, William Paul, physics

Fanta, George Frederick, organic polymer chemistry
Feldman, Arnold, radiological physics
Finnerty, James Lawrence, biochemistry, information science
Fitzgerald, Paul Jackson, plant breeding, plant pathology
Frankel, Edwin N, biochemistry, organic chemistry
Friedrich, John Philip, synthetic organic chemistry
Galsky, Alan Gary, plant physiology, microbiology
Gardner, Harold Wayne, biochemistry, lipid chemistry
Gasdorf, Edgar Carl, animal physiology
Gast, Lyle Everett, organic chemistry
Glover, Allen Donald, organic chemistry
Glover, Earl Robert, agricultural marketing, agricultural education
Grimm, Wilbur Winfield, zoology
Grove, Michael Dean, natural products chemistry
Grundbacher, Frederick John, human genetics
Hammond, William Marion, communications, computer systems
Haynes, William Clarence, microbiology
Herrmann, Ernest Carl, Jr., virology
Hesseltine, Clifford William, mycology
Ho, Andrew K S, pharmacology, cell biology
Hodge, John Edward, carbohydrate chemistry
Hofreiter, Bernard T, organic chemistry, paper technology
Honig, David Herman, food chemistry
Inglett, George Everett, agricultural chemistry
Johnson, Donovan Earl, microbiology, biochemistry
Jungck, Gerald Frederick, topology
Kaneshiro, Tsuneo, microbiology
Knutson, Clarence Arthur, Jr, organic chemistry
Kohlhase, William Lawrence, organic chemistry, polymer chemistry
Kolb, Kenneth Emil, organic chemistry
Koritala, Sanbasivaroa, lipid chemistry
Kurtzman, Cletus Paul, mycology
Kwolek, William F, entomology, statistics
Lillehoj, Eivind B, plant physiology, biochemistry
Maher, George Garrison, bio-organic chemistry
Maisch, Weldon Frederick, food microbiology, industrial microbiology
Martin, Loren Gene, physiology, biochemistry
Mathis, Billy John, limnology
McGaughey, Albert Wayne, mathematics
McMorris, Rex O, medicine
Migliozzi, Joseph Andrew, experimental pathology
Mikolajczak, Kenneth Lee, natural products chemistry
Miller, William Riedel, organic chemistry
Mills, Frank D, organic chemistry
Miwa, Thomas Kanji, organic chemistry, biochemistry
Monoson, Herbert L, mycology
Moore, Harold Arthur, nuclear physics
Morris, Herbert Allen, mathematics
Mounts, Timothy Lee, agricultural chemistry
Mullendore, James Myers, speech pathology, audiology
Nielsen, Harald Christian, biochemistry
Niffenegger, Daniel Arvid, botany, agronomy
Otey, Felix Harold, organic chemistry
Princen, Lambertus Henricus, physical chemistry
Pryde, Everett Hilton, lipid chemistry, industrial organic chemistry
Rackis, Joseph John, biochemistry, food science
Rakoff, Henry, organic chemistry
Rankin, John Carter, cereal chemistry
Rohwedder, William Kenneth, mass spectrometry
Rothfus, John Arden, biochemistry
Russell, Charles Richard, organic polymer chemistry
Sandford, Paul A, carbohydrate chemistry, microbial chemistry
Sathoff, H John, chemical chemistry
Schaefer, Wilbur Carls, agricultural chemistry
Schwab, Arthur William, organic chemistry
Sessa, David Joseph, organic biochemistry
Shasha, Baruch, agricultural chemistry
Shoop, George Jerome, agronomy
Shotwell, Odette Louise, organic chemistry
Sinclair, Henry Beall, organic chemistry

Sisson, Donald Ray, soil conservation, agricultural engineering
Slodki, Morey Eli, biochemistry
Smiley, Karl L, bacteriology
Smith, Cecil Randolph, Jr, natural products chemistry, lipid chemistry
Stout, Edward Irvin, organic chemistry
Strandberg, Gerald William, bacteriology, biochemistry
Stubblefield, Robert Douglas, analytical chemistry
Stutz, Conley I, physics
Szeto, George, mathematics
Tallent, William Hugh, organic chemistry, biochemistry
Teeter, Howard Maple, organic chemistry
Thiel, Thomas J, soil physics, hydrology
Tookey, Harvey Llewellyn, biochemistry
Truong, Xuan Thoai, physical medicine & rehabilitation, physiology
Van Cleve, John Woodbridge, carbohydrate chemistry
Van Etten, Cecil Herman, natural products chemistry
Wall, Joseph Sennen, biochemistry
Wallen, Lowell Lawrence, bio-organic chemistry
Wang, Hwa Lih, biochemistry
Wang, Li Chuan, biochemistry, soil fertility
Wang, Wun-Cheng, water pollution
Wing, Robert Edward, chemistry
Wolf, Michael Joseph, chemistry
Wolf, Walter J, biochemistry
Wu, Ying Victor, physical chemistry
Yates, Shelly Gene, chemistry

PEOTONE
Mehring, Jeffrey Scott, nutrition, toxicology

QUINCY
Campbell, Robert Samuel, botany
French, Allen Lee, entomology, animal parasitology
Gasser, William, organic chemistry
Hutchinson, Harold David, animal nutrition
Joshi, Madan Mohan, plant pathology, soil microbiology
Klay, Robert Frank, animal nutrition
Lang, Robert Phillip, physical chemistry
Natalini, John Joseph, biological rhythms, vertebrate biology
Nesbit, Arthur Henderson, animal nutrition
Ostdiek, John L, ecology
Peter, Albert P, ruminant nutrition
Petrigrew, James Eugene, Jr, animal nutrition
Pothoven, Marvin Arlo, animal nutrition
Siefker, Joseph Alphonse, industrial chemistry
Siems, Norman Edward, nuclear science
Sturdy, Robert Allan, veterinary medicine
Velesz, Dunstan George, mathematics
Windolph, Joseph R, mathematics

RIVER FOREST
Allard, Nona Mary, mathematics
Bartell, Marvin H, vertebrate zoology, endocrinology
Domroese, Kenneth Arthur, physiology
El Saffar, Zuhair M, chemical physics
Fields, Ellis Kirby, organic chemistry
Gross, Herbert Henry, physical geography
Just, Mary Agnes, geology, geography
Koetke, Donald D, high energy physics
McSweeney, Jean, organic chemistry
O'Malley, Mary Therese, mathematics
Steele, Mary Philip, mathematics
Woods, Mary, inorganic chemistry

RIVERSIDE
Hinrichs, Marie Agnes, public health education

ROBINSON
Hughes, Raymond Hadley, analytical chemistry

ROCHESTER
Martin, Russell James, epidemiology

ROCK ISLAND
Anderson, Richard Charles, geology
Berntsen, Robert Andyv, inorganic chemistry

ILLINOIS

Borrong, Bernard John, chemistry, military systems
Coleman, Norman P., Jr., mathematics
DeArmon, Ira Alexander, Jr., applied mathematics
Eliason, Morton A., physical chemistry
Frank, Robert Carl, statistics
Fryxell, Fritiof Melvin, geology
Hanning, Edward, geography
Johnson, Robert Leroy, mathematics
Larson, Ingemar W., zoology, parasitology
McCart, Bruce Ronald, physics
Moline, Norman Theodore, cultural geography, resource geography
Moore, Richard Lee, plasma physics
Muffley, Harry Chilton, biology, organic chemistry
Neely, Florence Elizabeth, botany
Nelson, Harry Ernest, mathematics, astronomy
Peterson, Melbert Eugene, organic chemistry
Renneke, David Richard, physics, computer science
Rennie, Thomas Howard, aquatic ecology, invertebrate zoology
Sundelius, Harold Wesley, geology
Talitsch, Robert Boyde, physiology
Troll, Ralph, zoology, botany
Turnquist, Richard Lee, insect toxicology, biochemistry

ROCKFORD
Block, Douglas Alfred, geology, stratigraphy
Fell, George Brady, ecology
Forman, G Lawrence, mammalogy, comparative anatomy
Grulee, Clifford Grosselle, Jr., pediatrics
Hetrick, Alan Henry, dairy technology
Hutchcroft, Alan Charles, organic medicine
Muck, George A., dairy science, biochemistry
Muller, Donald Edward, analytical chemistry
Schumaker, John Abraham, mathematics
Wesner, Gordon Eugene, Jr., physiology

ROLLING MEADOWS
Douglas, David Lewis, physical chemistry
Oxley, James Edward, electrochemistry
Pullukat, Thomas Joseph, chemistry
Schmuler, Seymour, organic chemistry
Shida, Mitsuzo, chemistry
Shroff, Ramesh N, polymer physics
Tisone, Thomas C., materials science

ROSEMONT
Grove, Ewart Lester, analytical chemistry, physical chemistry
Menz, William Wolfgang, organic chemistry, chemical literature
Speckmann, Elwood W., physiology, nutrition

ROUND LAKE
Stith, William Joseph, biochemistry
Williams, Ronald Alvin, biomedical engineering, electrical engineering

St CHARLES
Carlborg, Frank William, statistics
Hewson, William Bell, organic chemistry
Horst, Albert W., underwater acoustics

SAUGET
Lynch, Dan K., industrial chemistry
Palmer, John Frank, Jr., industrial organic chemistry

SCHAUMBURG
Ames, Edward R., medical education
Corliss, Glenn Arthur, food science
Decker, Winston M., veterinary medicine
Price, Donald Albert, veterinary medicine

SCHILLER PARK
Hansen, William Anthony, topology, computer science

SIMPSON
Faix, James Jacob, agronomy
Lewis, John Morgan, animal husbandry
Webb, Robert Johnson, animal husbandry

SKOKIE
Agre, Karl, clinical pharmacology, pediatrics
Arenson, Donald Lewis, mathematics
Berman, Lawrence Uretz, organic chemistry
Buzard, James Albert, biochemistry
Cohen, Gloria, information science
Cook, Harold Dale, physics
Duelgen, Ronald Rex, information science
Ewen, Edward Francis, analytical chemistry
Farrell, James I., physiology, urology
Gruhn, John George, pathology
Hampel, Clifford Allen, chemistry
Kantor, David Leon, physical chemistry
Kusak, Lloyd James, mathematics, computer sciences
Le Von, Ernest Franklin, organic chemistry
Mayron, Lewis Walter, biological chemistry, nuclear medicine
McTreath, Fred J, physiology, pharmacology
Mihina, Joseph Stephen, organic chemistry
Scheie, Carl Edward, engineering physics
Shapiro, Howard Maurice, medicine, bioengineering
Sherman, Joseph E, microbiology, food technology
Smith, William Roy, virology

SOUTH HOLLAND
Dobbem, Glen D., radiology
Rieke, Carol Anger, astronomy
Sample, James Halverson, organic polymer chemistry

SPRINGFIELD
Ahler, Stanley Albert, anthropology
Arnott, Robert A., air pollution, analytical chemistry
Barnes, Isabel Janet, microbiology
Beineke, Thomas Andrew, physical chemistry
Birtch, Alan Grant, surgery
Bloemer, William Louis, physical chemistry, chemical instrumentation
Campbell, Charlotte Catherine, medical mycology
Casella, Alexander Joseph, biophysics
Haynes, Robert C., limnology
Janardan, Konanur G, mathematical statistics, applied statistics
Juniper, Kerrison, Jr., medicine
Kabisch, William Thomas, anatomy
Kakela, Peter John, climatology, hydrology
King, James Edward, palynology
Koeling, Alfred Cornell, plant taxonomy
Koppa, F Russell, science education
Markanus, Peter Charles, analytical chemistry
McConnachie, Peter Ross, immunobiology
McMillan, R Bruce, archaeology, biogeography
Moulton, Wilbur Norton, organic chemistry
Murov, Steven Lee, photochemistry
Myers, Walter Loy, immunology
Norris, Albert Stanley, psychiatry
Paul, John R., zoology
Rabinovich, Sergio Rospigliosi, internal medicine, virology
Rauchhorst, William H., low temperature physics
Reininger, Edward Joseph, physiology
Roddick, John William, Jr., obstetrics & gynecology
Rollins, Earl Arthur, developmental biology
Rowan, Dighton Francis, medical virology
Salmon, James Henry, neurosurgery
Sames, Richard William, bacteriology
Saunders, Jeffrey John, paleontology
Schaefter, David Joseph, environmental management
Sebelius, Carl Louis, public health
Somani, Satu M., pharmacology, biochemical pharmacology
Strahlevitz, Meir, psychiatry, medical research
Strano, Alfonso J, virology, pathology
Taylor, D Dax, anatomic pathology, clinical pathology
Thompson, Milton D., zoology
Upham, Roy Walter, veterinary medicine, food technology
Van Fossan, Donald Duane, biochemistry

Wright, Richard Donald, cell biology
Wynne-Roberts, Caroline Rosales, rheumatology, electron microscopy
Yntema, Mary Katherine, mathematics

SYCAMORE
Mack, Clinton Olmsted, human physiology, developmental biology

THOMASBORO
Johnson, Glenn Richard, plant breeding

URBANA
Abikoff, William, mathematics
Acker, Geraldine Enod, foods, nutrition
Ades, Harlow Whiting, experimental neurology
Adler, Felix T., mathematical physics
Albert, Waco W., animal nutrition
Aldrich, Samuel Roy, agronomy
Alexander, Charles Stevenson, geography, geomorphology
Alexander, Denton Eugene, plant breeding, cytogenetics
Alger, Nelda Elizabeth, zoology
Altarelli, Massimo, theoretical solid state physics
Ambrose, Harrison William, III, ecology, animal behavior
Anderson, Ansel Cochran, low temperature physics
Anderson, John Denton, physiology
Anderson, Thomas Frank, geochemistry
Anderson, William Leno, wildlife biology
Appel, Kenneth I., mathematics
Appley, James E., entomology
Applequist, Douglas Einar, organic chemistry
Argoudelis, Chris J., organic chemistry, biochemistry
Armstrong, James Walter, Jr., mathematics
Arnold, Charles Yesbra, horticulture
Ascoli, Giulio, experimental physics
Ash, Robert B., mathematics
Atherton, Elwood, geology
Axel, Peter, physics
Bailar, John Christian, Jr., inorganic chemistry
Baker, David H., animal nutrition
Baldwin, Thomas Oakley, biochemistry, evolutionary biology
Bank, Steven Barry, mathematics
Banks, Edwin Melvin, biology, zoology
Banwart, Wayne Lee, soils
Bardeen, John, physics
Barefield, Edward Kent, inorganic chemistry, organometallic chemistry
Barr, Lloyd, physiology
Bartle, Robert Gardner, mathematics
Bateman, Felice Davidson, mathematics
Bateman, Paul Trevier, mathematics
Baxter, James Watson, economic geology, micropaleontology
Baym, Gordon A., theoretical physics
Bazzaz, Maarib Bakri, plant physiology
Beak, Peter, organic chemistry
Beamer, Paul Donald, veterinary pathology
Beavers, Alvin Herman, agronomy
Becker, Donald Eugene, animal nutrition
Belford, Geneva Grosz, applied mathematics
Belford, Rue Linn, physical chemistry, inorganic chemistry
Bell, Roma Raines, nutrition
Bentley, Orville George, nutritional biochemistry, animal nutrition
Benzinger, Harold Edward, Jr., mathematics
Berg, Ira David, mathematics
Bergstrom, Robert Edward, geology
Berkson, Earl Robert, mathematics
Bernard, Richard Lawson, plant breeding
Berndt, Bruce Carl, mathematics
Bevill, Richard F., Jr., veterinary pharmacology
Birkeland, Charles John, horticulture
Bishop, Richard Lawrence, geometry
Blair, Charles Eugene, operations research
Blake, Daniel Bryan, paleobiology
Blint, Richard Joseph, chemical physics
Bloomfield, Daniel Kermit, medicine, biochemistry
Boggess, Samuel Forest, plant physiology
Bohmer, Heinrich Everhard, physics
Bohrer, Robert Edward, mathematics, statistics
Boley, Loyd Edwin, veterinary medicine
Bond, Donald C., physical chemistry
Boone, William Werner, mathematical logic, algebra

Booth, Alfred Whaley, geography
Bowhill, Sidney Allan, aeronomy
Boyer, John Strickland, plant physiology
Bradbury, James Clifford, geology
Braunfeld, Peter George, mathematics
Bray, Donald James, poultry nutrition
Brodie, Bruce Orr, veterinary medicine
Brown, Charles Myers, plant breeding
Brown, Frederick Calvin, solid state physics
Brown, John Wesley, mathematics
Brown, Richard Maurice, computer science
Bruner, Edward M., cultural anthropology
Brussel, Morton Kremen, physics
Bryan, Harold Stever, veterinary medicine
Bryant, Marvin Pierce, microbial ecology
Buckmaster, Dennis Clifford, fluid mechanics, applied mathematics
Buetow, Dennis Edward, cell biology
Burger, Ambrose William, agronomy science education
Burkholder, Donald Lyman, mathematics
Buschelder, Thomas Charles, geology
Butterworth, Douglas Stanley, cultural anthropology
Cahn, Julius Hofeller, astrophysics
Cain, Charles Alan, bioengineering
Cairns, Stewart Scott, mathematics
Cardman, Lawrence Santo, nuclear physics
Carl, Philip Louis, molecular genetics
Carmer, Samuel Grant, biometrics
Carothers, Zane Bland, botany
Carozzi, Albert Victor, geology
Carroll, Robert Wayne, mathematics
Cartwright, Keros, hydrogeology
Casagrande, Joseph Bartholomew, anthropology
Chamberlain, Donald William, plant pathology
Chandler, David, chemical physics, statistical mechanics
Chang, Shau-jin, elementary particle physics, mathematical physics
Changnon, Stanley Alcide, Jr., climatology
Chapman, Carleton Abramson, geology
Chen, Kuo-Tsai, mathematics
Cheng, Hazel Pei-Ling, gastroenterology, cell biology
Clark, Howard Selby, chemistry
Clark, John Magruder, Jr., biochemistry
Claussen, Walter Frederick, physical chemistry
Coates, Robert Mercer, organic chemistry
Cole, Michael Allen, microbial genetics, soil microbiology
Collinson, Charles William, geology
Conrad, Harry Edward, biochemistry
Conroy, John Wesley, veterinary pathology
Cooper, John Wesley, experimental high energy physics
Cooper, Richard Lee, plant breeding, plant genetics
Corbin, James Edward, animal nutrition
Craggs, Robert F., topology
Crandell, Robert Allen, veterinary virology
Crane, Joseph Leland, mycology
Crawford, Roy Kent, materials science, fluid physics
Creditor, Morton C., medical education
Culbert, John Robert, horticulture
Cureton, Thomas (Kirk), applied physiology
Curtin, David Yarrow, organic chemistry
Curtis, Stanley Evan, animal science, environmental physiology
Damberger, Heinz Heinrich, geology
Daniel, William L., human genetics
Davis, Carl Lee, nutrition, biochemistry
Davis, Robert Benjamin, mathematics
Day, Mahlon Marsh, mathematics
Dayton, Daniel Francis, horticulture
Debrunner, Peter Georg, physics
DeMoss, Ralph Dean, microbiology
DeWet, Jan M J., microbiology
Diamond, Harold George, mathematics
Dickel, Helene Ramseyer, astronomy
Dickinson, David Budd, plant physiology
Dirr, Michael Albert, ornamental horticulture
Donath, Fred Arthur, geology
Doob, Joseph Leo, mathematics
Dor, Leonard Eliezer, mathematics
Dornhoff, Larry Lee, mathematics
Drake, John W., genetics, virology
Drickamer, Harry George, chemical engineering
Dudley, John Wesley, plant breeding
DuMontelle, Paul Bertrand, geology
Dinn, Floyd, biophysics, bioengineering

Dunsing, Marilyn Magdalene, human ecology
Dziuk, Philip J, reproductive physiology, endocrinology
Earley, Ernest Benton, agronomy
Edwards, Dale Ivan, nematology, plant pathology
Ehrlich, Gert, surface physics
Eisenstein, Bob I, high energy physics
Erdman, John Wilson, Jr, food science
Erskine, James Lorenzo, solid state physics
Evers, Robert August, botany
Faulkner, Larry Ray, chemistry
Fauntleroy, Amassa Courtney, algebra
Fellmann, Jerome Donald, geography
Ferguson, William Allen, mathematics
Ferrier, Leslie Kenneth, food science, enzymology
Finney, Ross Lee, topology
Fitzgerald, Paul Ray, zoology, parasitology
Flygare, Willis H, chemical physics
Flynn, Colin Peter, solid state physics
Forbes, Richard Mather, nutrition
Ford, Richard Earl, virology, plant pathology
Fossum, Robert Merle, mathematics
Foster, Fred William, geography of the Middle East
Fraenkel, Gottfried Samuel, entomology, comparative physiology
Francis, Bettina Magnus, genetics
Francis, George Konrad, mathematics
Francis, William Richard, experimental high energy physics
Frauenfelder, Hans (Emil), particle physics, biophysics
Friedman, Stanley, insect physiology
Friesen, Donald Kent, algebra
Fruth, Lester Sylvester, Jr, sedimentology
Fryman, Leo Ray, dairy science
Futrelle, Robert Peel, developmental biology, computer sciences
Garland, John Henry, geography
Garrigus, Upson Stanley, animal science
Gartner, John Bernard, floriculture, ornamental horticulture
Gaylord, Richard J, polymer science
Gear, Charles William, computer science
Gennis, Robert Bennett, biophysical chemistry
Gerdemann, James Wessel, plant pathology
Ghent, Arthur W, ecology, biometrics
Gieseking, John Eldon, soil science
Giles, Eugene, physical anthropology
Gilmore, Alvan Ray, forestry
Ginsberg, Donald Maurice, low temperature physics, metal physics
Gladding, Gary Earle, experimental high energy physics
Glass, Herbert David, mineralogy
Gluskoter, Harold Jay, geology
Goldberg, Samuel I, mathematics
Goodman, Robert Merwin, virology, plant pathology
Gottlieb, David, plant pathology
Gould, Harold A, cultural anthropology, academic administration
Govindjee, biophysics, plant physiology
Graber, Richard Rex, zoology
Graf, Donald Lee, geochemistry
Graffis, Don Warren, agronomy
Granato, Andrew Vincent, physics
Graves, Charles Norman, physiology, biochemistry
Gray, John Walker, mathematics
Green, Edward Lewis, algebra
Griffith, Phillip A, mathematics
Grim, Ralph Early, geology
Gross, David Lee, geology
Grossman, Michael, population genetics, quantitative genetics
Grove, David Cliff, anthropology, archaeology
Guiher, John Kenneth, forest products
Gumport, Richard I, biochemistry, enzymology
Gunsalus, Irwin Clyde, biochemistry
Gutowsky, Herbert Sander, physical chemistry
Hadley, Henry Hultman, genetics, plant breeding
Hageman, Richard Harry, plant physiology
Hager, Lowell Paul, biochemistry
Haight, Gilbert Pierce, Jr, inorganic chemistry, science education
Haken, Wolfgang, mathematics
Hamstrom, Mary-Elizabeth, pure mathematics
Handler, Paul, solid state physics
Haney, Alan William, plant ecology
Hansen, Larry George, toxicology
Hanser, Donald Frary, fisheries
Hanson, Alfred Olaf, experimental nuclear physics
Hanson, John Bernard, plant physiology
Hanson, Lyle Eugene, veterinary microbiology
Hardenbrook, Harry, Jr, veterinary surgery
Harlan, Jack Rodney, agronomy, genetics
Harmon, Bud Gene, animal science, animal nutrition
Harper, James Eugene, plant physiology, agronomy
Harshbarger, Kenneth E, dairy science
Hartley, Arnold Manchester, analytical chemistry
Hartline, Peter Haldan, neurophysiology, animal behavior
Hartstirn, Walter, plant pathology
Harvey, Richard David, mineralogy, petrology
Haskell, Betty Echternach, nutritional biochemistry
Hassett, John J, soil chemistry
Hatch, Ray Davenport, veterinary medicine
Hatfield, Efton Everett, animal nutrition, physiology
Hays, Ray Leroy, physiology
Heath, James Edward, zoology, physiology
Heigold, Paul C, groundwater geology
Helms, Lester LaVerne, mathematics
Henderson, Donald Munro, mineralogy
Hensley, Douglas Austin, number theory
Henson, C Ward, mathematical logic
Herdt, Robert William, agricultural economics
Hershberger, Charles Lee, molecular biology
Hertig, Bruce Allerton, environmental physiology
Hicks, Bruce Lathan, physics
Hieronymus, Thomas Applegate, agricultural economics
Hiltibran, Robert Comegys, biochemistry
Himelick, Eugene Bryson, plant pathology
Himoe, Albert, biochemistry
Hinds, Frank Crossman, animal nutrition
Hittle, Carl Nelson, crop breeding
Hoeft, Robert Gene, soil science, agronomy
Hoffman, Larry Ronald, phycology
Hoffmeister, Donald Frederick, zoology
Hohn, Franz Edward, applied mathematics
Holland, Israel Irving, forest economics
Holloway, Leland Edgar, experimental high energy physics
Holmes, Kenneth Robert, environmental physiology
Hooker, Arthur Lee, genetics, phypathology
Hoover, Paul Swegman, astrophysics
Hopen, Herbert, horticulture
Hopke, Philip Karl, nuclear chemistry, environmental chemistry
Hopkins, M E, geology
Horsfall, William Robert, entomology
Howe, Wayne Lamoyne, entomology
Howell, Robert Wayne, plant physiology
Hummel, John Philip, nuclear chemistry
Hursh, Laurence M, internal medicine, nutrition
Hymowitz, Theodore, genetics, plant breeding
Iben, Icko, Jr, astrophysics
Jackobs, Joseph Alden, agronomy
Jackson, Edwin Atlee, physics
Jackson, Gary Loucks, endocrinology, physiology
Jacobsen, Barry James, plant pathology
Jansen, Ivan John, agronomy, soil science
Janusz, Gerald Joseph, mathematics
Jaycox, Elbert Ralph, entomology
Jedlinski, Henryk, plant pathology
Jennison, Dwight Richard, theoretical solid state physics
Jensen, Aldon Homan, animal nutrition
Jerrard, Richard Patterson, topology
Jockusch, Carl Groos, Jr, mathematical logic
Johnson, Donald Lee, geography
Johnson, Hugh Swaney, poultry science
Johnson, Richard Ray, agronomy
Johnson, William Hilton, quaternary geology
Jokela, Jalmer John, forest genetics
Jones, Almut Gitter, plant taxonomy
Jones, L Meyer, veterinary pharmacology
Jones, Lorella Margaret, elementary particle physics
Jones, Robert L, soil mineralogy
Kaler, James Bailey, astronomy
Kallio, Reino Emil, microbial physiology
Katzenellenbogen, Benita Schulman, reproductive endocrinology
Katzenellenbogen, John Albert, bio-organic chemistry, synthetic organic chemistry
Kaufmann, Kenneth James, chemical physics
Keller, Charles M, anthropology
Kempton, John Paul, groundwater geology
Kent, Lois Schoonover, geology, paleontology
Kicliter, Ernest Earl, Jr, neuroanatomy
Kidd, Richard Wayne, physical chemistry
Killinger, Arden Holmes, veterinary microbiology
Kirk, Thomas Bernard Walter, high energy physics
Klein, Barbara P, food chemistry
Klein, George deVries, sedimentology, oceanography
Klein, Miles Vincent, physics
Klepinger, Linda Lehman, biological anthropology
Knake, Ellery Louis, weed science, agronomy
Knight, Frank B, mathematics, mathematical statistics
Koehler, James Stark, physics
Koeppe, David Edward, plant biochemistry, plant physiology
Koester, Louis Julius, Jr, experimental high energy physics
Kolata, Dennis Robert, stratigraphy, invertebrate paleontology
Konisky, Jordan, microbial physiology
Krasnow, Marvin Ellman, physical chemistry
Kruidenier, Francis Jeremiah, helminthology
Kruse, Ulrich Ernst, physics
Kummerow, Fred August, food science
Kunz, Albert Barry, chemical physics, solid state science
Kurtz, Lester Touby, agronomy
LaBerge, Wallace E, entomology
Labisky, Ronald Frank, wildlife biology
Lamb, Donald Quincy, Jr, theoretical physics
Lamb, Frederick Keithley, theoretical physics
Lambert, Robert John, plant genetics
Langebartel, Ray Gartner, mathematics, astronomy
Langenheim, Ralph Louis, Jr, paleontology, stratigraphy
Langhaar, Henry Louis, applied mechanics
Lardner, Thomas Joseph, engineering mechanics, biomechanics
Larimore, Richard Weldon, fish biology
Larsen, Joseph Reuben, insect physiology
Larson, Bruce Linder, biochemistry
Larson, Thurston E, chemistry
Lathrap, Donald Ward, anthropology
Laughnan, John Raphael, genetics
Lavatelli, Leo Silvio, physics
Lazarus, David, solid state physics
Lee, Merlin Raymond, evolutionary biology
Leonard, Nelson Jordan, organic biochemistry
Levine, Norman Dion, parasitology, protozoology
Levy, Allan Henry, computer sciences, virology
Levy, Harry, mathematics
Liebman, Judith Stenzel, operations research
Lim, Sung Man, plant pathology
Lin, Lily, photobiology
Lineback, Jerry Alvin, geology
Link, Roger Paul, veterinary pharmacology
Linn, Manson Bruce, plant pathology
Liu, Chung Laung, computer science
Livant, Peter David, physical organic chemistry
Lodge, James Robert, reproductive physiology
Loeb, Peter Albert, mathematics
Lohman, Timothy George, animal nutrition
Loomis, Wheeler, physics
Loop, Michael Stuart, vision, herpetology
Lorenz, Ralph William, forestry
Lovell, James Edgeley, veterinary anatomy
Luckmann, William Henry, entomology
Lueking, Donald Robert, microbial biochemistry
Lyman, Ernest McIntosh, physics
MacLeod, Ellis Gilmore, evolutionary biology, entomology
Maddox, Joseph Vernard, insect pathology
Malek, Richard Barry, plant nematology
Malmstadt, Howard Vincent, chemistry
Mann, Christian John, geology
Marcus, Rudolph Arthur, physical chemistry
Markowitz, Melvin Myron, phycology
Marsh, Richard Riley, human biology, food science
Marshall, Arvle Edward, veterinary medicine, veterinary neurology
Martin, James Cullen, organic chemistry
Matteson, Max Richard, zoology
Maurer, Robert Joseph, solid state physics, research administration
McGlamery, Marshal Dean, agronomy, weed science
McLaughlin, Thomas G, mathematics
McLinden, Lynn, mathematics
Mechtly, Eugene A, radiophysics
Mehta, Tara, nutritional biochemistry
Meins, Frederick, Jr, developmental biology, biochemistry
Melsted, Sigurd Walter, agronomy, soils
Messner, Robert Lee, experimental high energy physics, elementary particle physics
Metcalf, Robert Lee, entomology
Meyer, Martin Marinus, Jr, ornamental horticulture, plant physiology
Meyer, Richard Charles, veterinary microbiology, animal virology
Meyer, Ronald Harmon, economic entomology
Meyer, Stephen Frederick, experimental solid state physics, nuclear magnetic resonance
Meyerson, Mark Daniel, topology
Michalski, Ryszard Stanislaw, computer sciences
Milbrath, Gene McCoy, plant pathology
Miles, Henry Jarvis, mathematics
Miles, Joseph Belsley, mathematics
Miller, Darrell Alvin, genetics, plant breeding
Milner, John Austin, nutrition
Milner, Reid Thompson, food science
Mistry, Sorab Pirozshah, biochemistry
Mochel, Jack McKinney, physics
Moore, Stevenson, III, entomology
Moran, Joseph Michael, climatology
Moreno, Carlos Julio, mathematics
Morgan, Donald O'Quinn, veterinary immunology
Muller, David Eugene, applied mathematics
Muroga, Saburo, computer science
Nalbandov, Andrew Vladimir, genetics
Nanney, David Ledbetter, genetics
Neely, Robert Dan, plant pathology
Nelson, Alvin I, food technology
Nelson, Bruce Philip, elementary particle physics
Neumann, Alvin Ludwig, animal nutrition
Newton, William Morgan, laboratory animal medicine, physiology
Nicoli, Miriam Ziegler, protein chemistry, enzymology
Nieman, Timothy Alan, analytical chemistry
Nievergelt, Jurg, computer science
Nishida, Toshiro, food chemistry
Normandin, Diane Kilbourne, electron microscopy
Norton, Horace Wakeman, III, mathematical statistics
Nyikos, Peter Joseph, topology
Nystrom, Robert Forrest, organic chemistry, radiochemistry
O'Brien, William Daniel, Jr, bioacoustics, bioengineering
Ogren, William Lewis, plant physiology
O'Halloran, Thomas A, physics
Oldfield, Eric, molecular biology, biophysical chemistry
Olson, Edward Cooper, astrophysics
Olson, Walter Harold, biomedical engineering
Ordal, Zakarias John, microbiology
Ormiston, Emmett Ezekiel, dairy science
Osborn, Howard Ashley, mathematics
Page, Larry Merle, ichtyhology, invertebrate zoology
Paley, Hiram, algebra
Pang, Chan Yueh, high energy physics
Pappas, Larry George, insect physiology
Parker, Ernest Tilden, mathematics
Parker, Helen Meister, physical biochemistry
Patterson, Earl Byron, genetics
Paul, Iain C, physical chemistry, biological chemistry
Paul, Pauline Constance, food chemistry
Pausch, Robert Dale, economic entomology, insect physiology
Paxton, Jack Dunmire, plant pathology, biochemistry
Peck, Newton Tenney, mathematical analysis
Peck, Theodore Richard, soil fertility, soil chemistry
Peppers, Russel A, geology, palynology

Peressini, Anthony L., analytical mathematics
Perkins, Edward George, organic chemistry
Peters, Doyle Buren, soil physics
Pethica, Christopher John, theoretical physics
Petty, Howard B., entomology
Philipp, Walter V., mathematics
Phillips, Tom Lee, paleobotany, plant morphology
Picciano, Mary Frances Ann, nutrition
Pines, David, theoretical physics
Pirkle, William H., organic chemistry
Porta, Horacio A., mathematics
Portnoy, Esther, geometry
Portnoy, Stephen Lane, mathematical statistics, applied statistics
Price, Peter Wilfrid, ecology, entomology
Propst, Franklin Moore, physics
Prosser, Clifford Ladd, comparative physiology
Raether, Manfred, plasma physics
Rasmussen, Benjamin Arthur, animal genetics
Ravenhall, David Geoffrey, theoretical physics
Reichmann, Manfred Eliezer, biochemistry
Reiner, Irma Moses, mathematics
Reiner, Irving, mathematics
Reinertsen, David Louis, geology
Reingold, Edward Martin, computer science
Remels, Robert Gossett, forest entomology
Reynolds, Claude Lewis, Jr., low temperature physics, experimental solid state physics
Reynolds, Harry Aaron, Jr., veterinary pathology
Rhodes, Ashby Marshall, plant breeding
Ricketts, Gary Eugene, animal science
Ridlen, Samuel Franklin, poultry science
Ries, Stephen Michael, plant pathology
Riley, Thomas Joseph, anthropology, archaeology
Rinehart, Kenneth Lloyd, Jr., organic chemistry
Rinne, Robert W., plant physiology
Ristic, Miodrag, veterinary medicine
Robinson, Clark Shove, Jr., physics
Robinson, Derek Scott, mathematics
Robinson, James Lawrence, biochemistry
Roepke, Howard George, economic geography, resource geography
Rogers, Donald Philip, botany
Romack, Frank Eldon, cardiovascular diseases, animal physiology
Romans, John Richard, animal science, biochemistry
Rosales-Sharp, Maria Consolacion, developmental biology, tissue culture
Rosen, Sidney, history of science
Ross, Harold Marion, anthropology, ethnology
Rothman, Neal Jules, mathematical analysis
Rotman, Joseph Jonah, mathematics
Rowland, Theodore Justin, physics
Rubel, Lee Albert, mathematics
Ruch, Rodney R., analytical chemistry
Russell, George A., physics
Russell, Joseph Albert, geography
Russell, Morell Belote, agronomy
Sacks, Jonathan, geometry, topology
Safanie, Alvin H., anatomy, histology
St. Clair, Lorenz Edward, veterinary anatomy
Salamon, Myron B., solid state physics
Salisbury, Glenn Wade, dairy science
Sandberg, Philip A., geology, paleontology
Sanderson, Glen Charles, zoology
Sard, Robert Daniel, experimental high energy physics
Sargent, Malcolm Lee, biochemical genetics: biological rhythms
Satterthwaite, Cameron B., physical chemistry, physics
Savage, Dwayne Cecil, medical microbiology
Saylor, Paul Edward, numerical analysis
Schiller, Alfred George, veterinary medicine
Schleicher, John Anthony, analytical chemistry
Schmid, Glenn Roy, meat sciences
Schmid, Nancy Jeanne, cultural anthropology
Schmidt, Paul Gardner, biophysical chemistry
Schoeneweiss, Donald F., plant pathology
Schubert, Jewell Emma, mathematics
Schulz, Roy Louis, theoretical physics
Schulz, Arthur Jay, inorganic chemistry

Schupp, Paul Eugene, mathematics
Schuster, Gary Benjamin, photochemistry
Scott, Edward Joseph, mathematics
Scott, Harold Martin, poultry husbandry
Scott, Walter O'Daniel, agronomy
Sechriest, Ralph Earl, economic entomology
Sechrist, Chalmers Franklin, Jr., aeronomy, ionospheric physics
Secrest, Donald H., physical chemistry
Segre, Diego, immunology
Segre, Mariangela Bertani, immunology
Seif, Robert Dale, biometry
Seigler, David Stanley, organic chemistry; botany
Seitz, Wesley Donald, agricultural economics, environmental management
Selander, Richard Brent, entomology
Shapiro, David Jordan, biochemistry
Shaw, Paul Dale, biochemistry
Sherbert, Donald R., mathematical analysis
Shield, Richard Thorpe, mechanics
Shiley, Robert Wilmer, veterinary pathology
Shimkin, Demitri Boris, applied anthropology, cultural geography
Shimp, Neil Frederick, analytical chemistry
Shurtleff, Malcolm C., Jr., phytopathology
Siedler, Arthur James, food science, nutrition
Sleator, William Warner, Jr., physics
Slichter, Charles Pence, solid state physics
Slife, Fred Warren, agronomy
Sligar, Stephen Gary, biochemistry
Slotnick, Daniel Leonid, applied mathematics, electrical engineering
Small, Erwin, veterinary medicine
Smith, James Hammond, nuclear physics
Smith, Kenneth Edward, analytical chemistry, spectroscopy
Smith, Philip Wayne, vertebrate zoology
Smith, Stanley Glen, organic chemistry
Smith, William Calhoun, geology
Snyder, Harold Ray, organic chemistry
Snyder, James Newton, physics, computer science
Snyder, Lewis Emil, astrophysics, molecular physics
Spahr, Sidney Louis, agriculture, dairy science
Sperber, Steven Irwin, geometry
Spittstoesser, Walter E., plant physiology
Spomer, Louis Arthur, plant physiology;
Sprague, George Frederick, agronomy
Sprugel, George, Jr., zoology, ecology
Stack, John D., theoretical physics
Stannard, Lewis Judson, Jr., entomology
Stapleton, Harvey James, physics
Steffensen, Dale Marriott, genetics
Steinberg, Marvin Phillip, food technology
Sternburg, James Gordon, entomology
Stevenson, Frank Jay, soils
Stipes, Marvin C., applied mechanics
Stolarsky, Kenneth B., mathematics
Stoller, Edward W., plant physiology; weed science
Stolpe, Stanley George, endocrinology
Storm, Daniel Ralph, biochemistry
Stout, Glenn Emanuel, meteorology
Strehlow, Roger Albert, physical chemistry, fluid dynamics
Stucky, Galen Dean, inorganic chemistry
Swann, Sherlock, Jr., electrochemistry
Sweeney, Daryl Charles, invertebrate physiology; neurochemistry
Switzer, Robert Lee, biochemistry
Takeuti, Gaisi, mathematical logic
Taylor, Albert Cecil, biology
Taylor, Aubrey Bryant, zoology
Thomas, Josephus, Jr., analytical chemistry
Thompson, John, historical geography, geography of Latin America
Thompson, Marcus Luther, geology, paleontology
Thornberry, Halbert Houston, plant pathology
Thorne, Marlowe Driggs, agronomy
Thurman, John C., veterinary anesthesiology
Ting, Tsuan Wu, mathematics
Titus, John S., pomology

Tobias, Joseph, dairy science, food science
Todd, Kenneth S., Jr., veterinary parasitology
Tondeur, Philippe, mathematics
Tripathy, Deoki Nandan, veterinary microbiology
Truran, James Wellington, Jr., astrophysics
Tsao, Liang-Chi, number theory
Turgeon, Alfred J., weed science, ecology
Turgeon, Robert Williams, genetics, botany
Twardock, Arthur Robert, veterinary physiology
Tyler, Tipton Ransom, pharmacology
Tyner, Edward Henry, soils
Uhl, John Jerry, Jr., mathematics
Uhlenbeck, Karen K., mathematics
Uhlenbeck, Olke Cornelis, biophysical chemistry
Ullom, Stephen Virgil, mathematics
Ulrich, Stephen Edgar, chemistry
Unzicker, John Duane, arachnology, entomology
Van Duyne, Frances Olivia, food chemistry
Vaughan, Herbert Edward, mathematics
Visek, Willard James, nutrition, toxicology
Voss, Edward William, Jr., immunochemistry, microbiology
Wachsman, Joseph T., microbiology, biochemistry
Wagstaff, Samuel Standfield, Jr., mathematics
Waldbauer, Gilbert Peter, entomology
Walker, William M., soil fertility, biometrics
Walter, John Harris, algebra
Walters, Charles Sebastian, forestry
Watanabe, Daniel Seishi, computer science
Watrach, Adolf Michael, veterinary pathology
Watson, William Douglas, astrophysics
Wattenberg, Albert, physics
Watterson, Ray Leighton, embryology
Webb, Donald Wayne, entomology
Weber, Evelyn Joyce, lipid chemistry
Weber, Gregorio, biochemistry, biophysics
Weber, Michael Joseph, virology, cell biology
Wei, Lun-Shin, food science
Weichenthal, Burton Arthur, animal nutrition
Weichsel, Paul M., mathematics
Weinberg, Elliot Carl, mathematics
Wenk, Charles Allen, physics
Wetenkamp, Harry R., mechanics
Wetzel, John Edwin, mathematics
Whigham, David Keith, agronomy
White, Donald Glenn, plant pathology
White, George Willard, geology
White, William Arthur, mineralogy, petrology
Whitney, Robert McLaughlin, food
Whitt, Dixie Dailey, microbial ecology, microbial physiology
Whitt, Gregory Sidney, developmental genetics, biochemical genetics
Whitten, Norman Earl, Jr., anthropology
Widholm, Jack Milton, plant physiology, genetics
Wijsman, Robert Arthur, mathematical statistics
Williams, Wendell Sterling, physics
Willis, John Steele, physiology, zoology
Willis, Judith Horwitz, physiology, developmental biology
Willman, Harold Bowen, geology
Wilson, Curtis Marshall, plant biochemistry
Wilson, Linda S Whaley, chemistry; research administration
Witter, Lloyd David, food science
Wirz, Klaus Gerhard, mathematics
Wolfe, Ralph Stoner, microbiology
Wolfowitz, Jacob, mathematics
Woods, George Theodore, veterinary medicine, public health
Wooley, Joseph Tarbet, plant physiology
Worley, Will J., theoretical mechanics; applied mechanics
Wortis, Michael, solid state physics
Wraight, Colin Allen, biophysics
Wright, Jon Alan, theoretical physics, particle physics

Wyatt, Stanley Porter, Jr., astronomy
Wyld, Henry William, Jr., physics
Yankwich, Peter Ewald, physical chemistry
Yardley, James Thomas, III, physical chemistry
Yohe, Gail Robert, organic chemistry
Yoss, Kenneth M., astronomy
Zaring, Wilson Miles, mathematics
Zych, Chester Charles, horticulture

VILLA PARK
Matz, Samuel Adam, food science

WAUKEGAN
Chun, Alexander Hing Chinn, physical pharmacy
Glaser, Milton Arthur, organic polymer chemistry
Holstein, Arthur G., biochemistry
Karnattu, Joseph J., clinical biochemistry
Schleif, Robert H., industrial pharmacy
Smart, William Donald, organic chemistry
Taylor, Donald Francis, inorganic chemistry

WEST CHICAGO
Boggs, Dallas Ervin, biochemistry
Brannen, Cecil Gray, organic chemistry
Rasmussen, Oscar Gustav, nutrition, biochemistry
Silverman, Walter Lawrence, industrial chemistry

WESTERN SPRINGS
Hull, Harvard Leslie, physics
Kaiser, Edward William, organic chemistry
Rosenwald, Robert Henry, petroleum chemistry
Tiefenthal, Harlan E., organic chemistry
Young, Harland Harry, organic chemistry

WESTMONT
Young, Marvin Kendall, Jr., biochemistry

WHEATON
Alm, Robert M., organic chemistry
Boardman, Donald Chapin, geology
Brace, Neal Orin, organic chemistry; fluorine chemistry
Brand, Raymond Howard, animal ecology
Classen, Howard Hubert, chemical physics
Faries, Dillard Wayne, quantum optics
Fiess, Harold Alvin, chemistry
Funck, Larry Lehman, inorganic chemistry
Green, Frank Orville, chemistry
Haddock, Gerald Hugh, geology
Kraakevik, James Henry, atmospheric physics
Luckman, Cyril Edmund, zoology
Mixter, Russell Lowell, anatomy
Nelson, Bernard Andrew, organic chemistry
Nevitt, Michael Vogt, metal physics
Price, David Thomas, algebra
Seven, Raymond Peter, chemistry
Wright, Paul McCoy, physical chemistry

WHEELING
Johnson, Dale Waldo, food science
Krajewski, John J., synthetic organic chemistry, polymer chemistry
Loire, Norman Paul, organic chemistry

WILMETTE
Brown, John Max, medicinal chemistry, science writing
Fish, Harold Somers, anatomy
Hildebrandt, Emanuel Henry Carl, mathematics
Jenkins, William Wesley, pharmaceutical chemistry
Jones, Haydn, industrial chemistry
Maselli, John Anthony, food chemistry
McFee, Donald Ray, industrial health, mechanical engineering
Odell, Clarence Burt, geography; cartography
Sager, William Frederick, organic chemistry
Sollman, Paul Benjamin, organic chemistry
Weiner, Louis Max, algebra

WINNETKA
Crosson, John William, medicine
Miner, Carl Shelley, Jr, organic chemistry
Tenney, Robert Imboden, microbiology

WOOD RIVER
Ryan, Julian Gilbert, petroleum chemistry

WOODSTOCK
Baldoni, Andrew Ateleo, organic chemistry
Bare, Thomas M, synthetic organic chemistry
Decker, Kenneth Harold, analytical chemistry
Frank, Robert Loeffler, nutrition
Frey, David Allen, organic chemistry, polymer chemistry
Isbister, Roger John, polymer chemistry
Jenny, Neil Allan, pesticide chemistry
Katsaros, Constantine, pesticide chemistry
Langler, James Edward, food science
Leibhardt, Edward, astronomy, spectroscopy
Martin, William C, agronomy
Newcome, Marshall Millar, analytical chemistry
Staetz, Charles Alan, economic entomology
Turner, Robert James, organic chemistry, polymer chemistry
Weyna, Philip Leo, organic chemistry, polymer chemistry

YORKVILLE
Cannon, Paul Roberts, pathology

INDIANA

ANDERSON
Cook, Kenneth Emery, organic chemistry
Cruikshank, Donald Burgoyne, Jr, acoustics
Goodman, John David, parasitology
Mayo, Marie Joiner, embryology, endocrinology
Reed, A Thomas, inorganic chemistry
Shaffer, Lawrence Bruce, physics
Sipe, Jerry Eugene, biochemistry, microbiology
Stephens, Stanley LaVerne, mathematics

ANGOLA
Leiter, Howard Allen, physics
Moulder, Jerry Wright, physics
Pinkham, Chester Allen, III, physical chemistry
Young, Ralph Waldo, mathematics

BLOOMINGTON
Afanador, Arthur Joseph, optometry, physiological optics
Al-lami, Fadhil, anatomy, physiology
Allerhand, Adam, physical chemistry, biophysical chemistry
Alyea, Ethan Davidson, Jr, physics
Atkinson, Robert d'Escourt, astronomy
Auchmuty, James Francis Giles, applied mathematics
Aufderheide, Karl John, cell biology, developmental biology
Austin, Mary Lellah, genetics
Azumaya, Goro, algebra
Bacher, Andrew Dow, nuclear physics
Bair, Edward Jay, physical chemistry, analytical chemistry
Baxter, Neal Edward, internal medicine, aerospace medicine
Beck, Lyle Vibert, pharmacology
Bennett, Don C, geography
Bent, Robert Demo, physics
Billman, John Henry, organic chemistry
Black, William Richard, transportation geography
Black-Schaffer, Bernard, pathology
Blakely, Robert Fraser, geophysics
Bleuer, Ned Kermit, glacial geology
Bonham, Russell Aubrey, chemical physics
Borish, Irving Max, optometry
Bosin, Talmage R, optometry
Brabson, Bennet Bristol, elementary particle physics
Breneman, William Raymond, zoology
Bron, Walter Ernest, solid state physics

Brothers, John Edwin, mathematics
Brown, Arlen, mathematics
Burkhead, Martin Samuel, astrophysics
Campaigne, Ernest Edwin, medicinal chemistry
Carmack, Marvin, organic chemistry
Carr, Donald Dean, geology
Challifour, John Lee, mathematical physics
Chase, Lloyd Lee, physics
Chu, Shu-Yuan, theoretical physics
Clevenger, Sarah, plant biochemistry
Comfort, Joseph Robert, nuclear physics
Conway, John Bligh, mathematical analysis
Cordes, Eugene H, biochemistry, organic chemistry
Crandall, Jack Kenneth, organic chemistry
Crittenden, Ray Ryland, physics
Crowell, (Prince) Sears, (Jr), elopmental biology, invertebrate zoology
Davis, Jerry Mallory, climatology
Day, Harry Gilbert, nutritional biochemistry
Debevec, Paul Timothy, nuclear physics
Devins, Delbert Wayne, nuclear physics
Dilcher, David L, paleobotany
Dippell, Ruth Virginia, cell biology, genetics
Dodd, James Robert, paleoecology
Droste, John Brown, sedimentology
Edmondson, Frank Kelley, astronomy
Emery, Guy Trask, physics
Emlen, John Merritt, ecology
Epstein, Aubrey, audiology, speech pathology
Epstein, George, mathematics, computer science
Everson, Ronald Ward, optometry, physiological optics
Ewing, George Edward, physical chemistry
Floyd, Alton David, anatomy
Franz, Frank Andrew, physics
Franz, Judith Rosenbaum, solid state physics, atomic physics
Fraser, William Dean, molecular biology
Frey, David Grover, limnology
Friesel, Dennis Lane, experimental nuclear physics
Frommer, Gabriel Paul, neuropsychology
Gajewski, Joseph J, organic chemistry
Gastony, Gerald Joseph, systematic botany
Gest, Howard, microbiology
Goodman, Victor Wayne, pure mathematics
Gray, Henry Hamilton, geology
Gurd, Frank Ross Newman, biochemistry
Gustafson, William Howard, algebra
Gustin, William, mathematics
Guth, Gary Stuart, neuroanatomy, neurocytology
Hafner, Charles William, Jr, academic administration, plant science
Hagstrom, Stanley Alan, physical chemistry
Hake, Richard Robb, physics
Halmos, Paul Richard, pure mathematics
Hartwell, George E, Jr, inorganic chemistry, organometallic chemistry
Hattin, Donald Edward, geology
Haurowitz, Felix (Michael), biochemistry, immunochemistry
Hayes, John Michael, organic geochemistry, mass spectrometry
Haymore, Barry Lant, inorganic chemistry, organometallic chemistry
Heath, Gordon Glenn, optometry, physiology
Hegeman, George D, microbiology, biochemistry
Heinz, Richard Meade, high energy physics
Heiser, Charles Bixler, Jr, botany
Hendrix, Thomas Eugene, geology
Hendry, Archibald Wagstaff, elementary particle physics
Herreid, Ernest Oliver, dairy science
Hidore, John J, physical geography
Hietfje, Gary Martin, analytical chemistry
Hippensteele, James Robert, cardiovascular physiology, biophysics
Hirs, Christophe Henri Werner, organic chemistry
Hofstetter, Henry W, visual physiology, optometry
Holland, James Philip, endocrinology
Hopf, Eberhard, applied mathematics, pure mathematics
Horowitz, Alan Stanley, paleontology, geology
Hudock, George Anthony, biochemical genetics

Humphrey, Rufus R, histology, experimental embryology
Hurt, Wesley Robert, archaeology, ethnology
Jacobson, Richard Martin, synthetic organic chemistry
Jacubs, John Francis, urban geography
Jaworowski, Jan W, topology
Jenkins, Winborne Terry, biological chemistry, enzymology
Johnson, Hollis Ralph, astrophysics
Jones, Arthur Eugene, vision
Jones, Harris Cleve, solid state physics, low temperature physics
Jones, William Philip, experimental nuclear physics
Kaslow, Christian Edward, organic chemistry
Kellar, James Harley, anthropology, archaeology
Kerr, Donald R, mathematics
Kerr, Ralph Oliver, organic chemistry
Klein, Cornelis, mineralogy, petrology
Klinge, Paul E, biology, science education
Koch, Arthur Louis, theoretical biology
Kochi, Jay Kazuo, organometallic chemistry, physical organic chemistry
Konetzka, Walter Anthony, microbial physiology
Konopinski, Emil John, physics
Kouzes, Richard Thomas, nuclear physics
Lane, Norman Gary, paleontology
Langer, Lawrence Marvin, nuclear physics
Langhoff, Peter Wolfgang, theoretical chemistry
Lenard, Andrew, mathematical physics
Lichtenberg, Don Bernett, theoretical physics
Lowell, Wayne Russell, economic geology
Lowengrub, Morton, mathematics
Lurie, Fred Marcus, physics
MacFadyen, Douglas Archibald, biochemistry, pathology
MacKenzie, Robert Earl, mathematics
Madden, Thomas M, optometry
Mahan, Gerald Dennis, theoretical solid state physics
Mahlberg, Paul Gordon, plant anatomy, cell biology
Mahler, Henry Ralph, biochemistry
Mahowald, Anthony P, developmental biology
Maickel, Roger Philip, biochemistry, pharmacology
Malacinksi, George M, developmental biology, biochemistry
Malik, Fazley Bary, atomic physics, nuclear physics
Martin, Hugh Jack, Jr, high energy physics
McClung, Leland Swint, microbiology
McMasters, Donald L, analytical chemistry
McQuarrie, Donald Allan, theoretical chemistry
Mead, Judson, geophysics
Meinschein, Warren G, organic chemistry
Merriam, Alan Parkhurst, anthropology
Merritt, Lynne Lionel, Jr, analytical chemistry
Miller, Carlos Oakley, plant physiology
Miller, Daniel Weber, experimental nuclear physics
Minty, George James, Jr, mathematics
Mintz, Jerome Richard, anthropology
Mizell, Sherwin, gross anatomy
Montgomery, Lawrence Kernan, organic chemistry, physical chemistry
Moolgavkar, Suresh Hiraji, geometry, topology
Moran, Emilio Federico, applied anthropology
Morrel, Bernard Baldwin, mathematics
Murray, Haydn Herbert, clay mineralogy, economic geology
Murray, Raymond Gorbold, histology
Mutschlecner, Joseph Paul, astrophysics, hydrodynamics
Neal, Homer Alfred, experimental high energy physics
Nebergall, William Harrison, inorganic chemistry
Neff, William Duwayne, neurosciences
Nelson, Craig Eugene, ecology, evolutionary biology
Newton, Roger Gerhard, theoretical physics
Nicoll, Paul Andrew, physiology
Nolan, Val, Jr, ornithology, ecology
Odland, John, urban geography
Ogren, Harold Olof, experimental high energy physics
Ortoleva, Peter Joseph, biological rhythms, biophysics

Parkhurst, David Frank, ecology, plant physiology
Parmenter, Charles Stedman, physical chemistry
Pataki, Louis Peter, Jr, astronomy
Patton, John Barratt, geology
Peery, Benjamin Franklin, Jr, astronomy, astrophysics
Peters, Dennis Gail, analytical chemistry, electrochemistry
Pietsch, Paul Andrew, anatomy, molecular biology
Pollock, Robert Elwood, nuclear physics
Potter, Henson David, neuroanatomy, neurophysiology
Pounds, Norman John Greville, geography, history
Preer, John Randolph, Jr, zoology
Puri, Madan L, mathematical statistics
Putnam, Frank William, biochemistry
Randall, Barbara Feucht, physiology
Randall, James Edwin, biophysics, physiology
Randolph, James Collier, ecology
Randolph, Polley Ann, ecology
Reading, Rogers W, physiological optics, optometry
Rexroad, Carl Buckner, geology
Rhoades, Billy Eugene, mathematics
Rhoades, Marcus Morton, genetics
Richardson, John Paul, biochemistry
Rickey, Martin Eugene, nuclear physics, musical acoustics
Robinson, Sid, physiology
Rostorfer, Howard Hayes, physiology
Rowland, William Joseph, ethology
Rudman, Albert Julius, geophysics
Ruesink, Albert William, plant physiology
Ruhe, Robert Victory, geology
Schaap, Ward Beecher, inorganic chemistry, analytical chemistry
Schaich, William Lee, theoretical solid state physics
Scheffe, Henry, statistics
Schlesinger, Barry Michael, astrophysics
Schmidt, Ingeborg, medicine
Schober, Glenn E, mathematics
Schochet, Claude Lewis, mathematics
Schroeder, Dolores Margaret, neuroanatomy
Schulman, Lawrence S, physics
Schwandt, Peter, nuclear physics
Schwartz, Drew, genetics
Schwartz, Ilsa Roslow, neuroanatomy
Sebeok, Thomas Albert, psycholinguistics, anthropology
Seifert, Ralph Louis, physical chemistry
Shalucha, Barbara, horticulture
Shaver, Robert Harold, geology
Sherman, Seymour, mathematics
Shiner, Vernon Jack, Jr, physical organic chemistry
Shull, Harrison, quantum chemistry, academic administration
Sinclair, John Henry, cell biology
Singh, Prithe Paul, nuclear physics
Smith, Hastings Alexander, Jr, nuclear physics
Sojka, Gary Allan, microbial physiology, genetics
Sonneborn, Tracy Morton, genetics
Springer, George, mathematics
Starr, Richard Cawthon, phycology
Stegenga, David Allan, mathematical analysis
Streib, William E, physical chemistry
Strickholm, Alfred, physiology, biophysics
Suthers, Roderick Atkins, physiology
Suttner, Lee Joseph, geology
Swihart, James Calvin, theoretical physics
Szabo, Attila, theoretical biophysical chemistry
Taaffe, Robert Norman, economic geography
Tansey, Michael Richard, mycology
Taylor, Milton William, microbiology, genetics
Thompson, Maynard, mathematics
Todd, Lee John, inorganic chemistry
Togasaki, Robert K, plant physiology, cell biology
Torchinsky, Alberto, mathematical analysis
Torrey, Theodore Willett, developmental anatomy
Towell, David Garrett, geochemistry
Vaughan, James Herbert, Jr, anthropology
Vitaliano, Charles Joseph, geology
Vitaliano, Dorothy Brauneck, geology
Vitousek, Peter Morrison, geophysics
Voegelin, Charles Frederick, anthropology
Walker, George Edward, theoretical nuclear physics

INDIANA

Ward, Thomas Edmund, nuclear chemistry, nuclear physics
Weinberg, Eugene David, medical microbiology
Wentworth, Rupert A D, inorganic chemistry
Wheeler, William Hollis, mathematical logic
White, David, microbiology
Whitehead, Donald Reed, botany, paleoecology
Williams, Gene R, plant physiology, plant biochemistry
Wills, John G, theoretical physics, nuclear physics
Wilson, Stephen Ross, synthetic organic chemistry
Wohlenberg, Ernest Harold, economic geography
Wonenburger, Maria Josefa, mathematics
Wu, Jang-Mei Gloria, analytical mathematics
Young, Frank Nelson, Jr, biology
Zeller, Frank Jacob, reproductive endocrinology
Ziemer, William P, mathematics

BROOKSTON
Patil, Sakharan Karsan, cereal chemistry, plant genetics
Ghosh, Saroj Bandhu, physical chemistry, analytical biochemistry
Thomas, Johnny Ray, agronomy

BUTLERVILLE
Culley, William James, biochemistry

CARMEL
Evanega, George R, organic chemistry
Nielsen, Kaj Leo, mathematics

CLARKSVILLE
Trott, Gene F, rubber chemistry

CLINTON
Caltrider, Paul Gene, microbial physiology
Hodson, Phillip Harvey, industrial microbiology

CRANE
Klausmeier, Robert Edward, microbiology
Tanner, John Eyer, Jr, physical chemistry

CRAWFORDSVILLE
Brooks, Austin Edward, phycology
Carter, James Cedric, plant pathology
Cole, Thomas A, biochemistry, genetics
Cooley, Robert Lee, mathematics
Doemel, William Naylor, microbiology, microbial ecology
Dollhopf, William Edward, physics
Haenisch, Edward Lauth, chemistry
Henry, Robert Ledyard, physics
Johnson, Willis Hugh, zoology
Lenox, Ronald Sheaffer, organic chemistry
McKinney, Paul Caylor, physical chemistry
Mielke, Paul Theodore, mathematics
Petty, Robert Owen, ecology
Swift, William Clement, mathematics
Williams, Eliot Churchill, invertebrate zoology, animal ecology
Wilson, David E, mathematics
Zimmerman, John F, mathematics

CULVER
Baker, David Thomas, science education

DECATUR
Childs, George Richard, nutrition
Coalson, James Arthur, animal nutrition
Gledhill, Robert Hamor, poultry nutrition, biochemistry
Longmire, Dennis B, ruminant nutrition
Middendorf, Donald Floyd, poultry nutrition, biochemistry
Waterhouse, Howard N, animal nutrition
Wesoloski, George D, animal nutrition

EAST CHICAGO
Holoway, Michael O, inorganic chemistry
Oliver, Montague, genetics, physiology

ELKHART
Adams, Ernest Clarence, biochemistry, immunochemistry
Alpert, Morton, cytochemistry, research administration
Alter, John Emanuel, physical chemistry
Anderson, Theodore Edmund, microbiology, biochemistry
Batti, Mario Alex, botany
Bauer, Robert, clinical biochemistry
Beers, Roland Frank Jr, biochemistry
Bluhm, Leslie, food science, food technology
Borchert, Peter Jochen, organic chemistry
Borglum, Gerald Baltzer, biochemistry
Boyer, J M, biochemistry, microbial physiology, microbial genetics
Burd, John Frederick, biochemistry
Carrico, Robert Joseph, biochemistry
Casey, James Patrick, chemistry
Chang, Peter Hon, pharmaceutics
Clemens, Anton Hubert, bioinstrumentation
Colbourn, Joseph Leason, biochemistry, pharmaceutics
Compton, Walter Ames, medicine, pharmacology
Croxall, Willard (Joseph), chemistry
Duvall, Ronald Nash, pharmaceutical chemistry
Erickson, Robert Joseph, industrial microbiology
Foster, George A Jr, biochemistry
Free, Alfred Henry, biochemistry
Gardner, David Arnold, cell biology, biochemistry
Gavin, John Joseph, microbiology
Gold, Gerald, pharmacy
Grabill James Rodney, pharmacology
Greyson, Jerome, physical chemistry, clinical chemistry
Gunter, Claude Ray, biochemistry
Hager, Chester Bradley, clinical medicine
Hendershot, William Fred, biochemistry
Himmelsbach, William Anthony, occupational health
Ingle, Morton Blakeman, microbiology, biochemistry
Jackson, James Albert, biochemistry
Janik, Borek, biochemistry, molecular biology
Johnston, Katharine Gentry, industrial biology
Kotick, Michael Paul, organic chemistry
Kreiser, Thomas Harry, biochemistry
Kuo, Mau H, microbiology, biochemistry
Kurchacova, Elva S, organic chemistry
Lasure, Linda Lee, microbial genetics
Leeling, Jerry L, biochemistry
Lowery, Charles E Jr, microbiology
Macgregor, Alexander Hamilton, research
Miller (Gilbert), Carol Ann, microbiology, immunology
Mirza, John, organic chemistry
Muni, Indu A, biochemical pharmacology
O'Donovan, Cornelius Joseph, internal medicine
Osuch, Mary Ann V, biochemistry
Parker, John Hamilton, microbial genetics
Phillips, Barrie Maurice, toxicology
Platt, Ronald Dean, statistics
Rand, Phillip Gordon, biochemistry
Rosenfield, Daniel, food technology
Safdy, Max Errol, organic chemistry, medicinal chemistry
Sastry, Sindgi Dattu, natural products chemistry
Schroeder, Hartmut Richard, biochemistry
Schut, Robert N, medicinal chemistry
Sherman, Leslie, pharmaceutical chemistry
Skarstedt, Mark Teofil, biochemistry
Smith, Richard Scott, bacteriology
Spradlin, Joseph Edward, enzymology
Sternberg, Moshe, food science
Strenkoski, Leon Francis, clinical microbiology
Surula, Chester Louis, physical chemistry, clinical chemistry
Trimmer, Robert Whitfield, industrial organic chemistry
Van Dyke, John William, Jr, organic chemistry
Vogelhut, Paul Otto, biophysics
Ward, Frederick Edmund, medicinal chemistry
Weiss, Ronald, food science
Wishinsky, Henry, bio-organic chemistry
Yoder, John Menly, endocrinology, immunochemistry
Zienty, Mitchell Frank, organic chemistry

ENGLISH
Komp, Richard Joseph, solid state physics, physical chemistry

EVANSVILLE
Amer, Mohamed Samir, pharmacology, biochemistry
Baranowsky, Paul E, biochemistry
Boenigk, John William, chemistry
Brenneman, James Alden, pharmaceutical science
Browder, Henry Polk, Jr, plant physiology
Bryant, Rhys, organic chemistry
Buesking, Clarence W, pharmacology
Byrne, Jeffrey Edward, physiology
Callantine, Merritt Reece, endocrinology
Campbell, Barbara Knapp, medicinal chemistry
Campbell, Kenneth Neilsen, chemistry
Carnahan, Robert Edward, organic chemistry
Claybaugh, Glenn Alan, bacteriology
Coleman, Ralph H, mathematics, statistics
Comer, William Timmey, organic chemistry
Cook, David Allan, nutritional biochemistry
Corrigan, John Raymond, organic chemistry
Cox, Raymond H, Jr, psychopharmacology
Dansby, Doris, hematology
Dungan, Kendrick Webb, pharmacology, physiology
Dunham, Donald West, organic chemistry
Dunn, Howard Eugene, zoology
Duracha, Chester William, microbiology
Dyer, Rolla McIntyre, Jr, analytical chemistry
Dykstra, Stanley John, organic chemistry
Ellingson, Rudolph Conrad, organic chemistry
Fairless, Billy J, physical chemistry
Ferguson, Hugh Carson, pharmacology
Gallo, Duane Gordon, biochemistry
Gomoll, Allen W, reproductive biology
Govier, William Miller, pharmacology
Gundersen, Larry Edward, research administration, information science
Gunther, Jay Kenneth, biochemistry
Guthrie, George Drake, biochemistry
Hamlow, Eugene Emanuel, pharmacy
Hastings, Carl Wayne, food science
Hayes, Donald Charles, biochemistry
Henness, Donald Merle, experimental medicine
Hubregs, William Henry, physiology, science administration
Kaplan, Allan Steven, pharmaceutics
Kinsey, David Webster, mathematics
Kinsey, Philip A, physical chemistry
Kreighbaum, William Eugene, medicinal chemistry
Larsen, Aubrey Arnold, medicinal chemistry
Lawson, John Edward, medicinal chemistry
Ludwig, Walter John, pharmacy
Majewski, Robert Francis, organic chemistry
Marquis, Norman Ronald, biochemistry
Martin, Tellis Alexander, organic chemistry
Mayol, Robert Francis, immunochemistry, biochemistry
McKinney, Gordon R, pharmacology, endocrinology
Mikolasek, Dougals Gene, medicinal chemistry
Miles, Kelly George, chemistry
Moffit, Robert LeVere, pharmacology
Morrison, William Alfred, inorganic chemistry
Morrow, Duane Francis, pharmaceutical chemistry
Mueller, Arthur Jacob, biochemistry
Mueller, Wayne Paul, environmental sciences
Nicholson-Guthrie, Catherine Shirley, genetics
Nielsen, Eldon Denzel, biochemistry
Nunes, Mathews Anthony, chemistry
Ogilvy, Winston Stowell, food science
Orlando, Anthony Michael, biostatistics
Ott, Karen Jacobs, parasitology, invertebrate zoology
Perbach, James Lawrence, Jr, pharmacology
Pohl, Victoria Mary, algebra
Robertson, Charles William, biological structure
Rogers, Howell Wade, medical microbiology, virology
Ryan, Richard Patrick, organic chemistry, medical science
Saret, Herbert Paul, biochemistry
Schmitt, Anthony Paul, inorganic chemistry, electrochemistry
Schneider, Donald Louis, biochemistry, nutrition
Seidehamel, Richard Joseph, pharmacology
Stevenson, Eugene Hamilton, nutrition
Stevenson, William Campbell, pharmacology
Shaw, Vernon Reed, analytical chemistry
Siegman, Fred Stephen, biochemistry
Sisson, George Maynard, physiology, pharmacology
Temple, Davis Litteton, Jr, medicinal chemistry
Tennyson, Richard Harvey, organic chemistry
Theuer, Richard Charles, nutrition
Timma, Donald Lee, chemistry
Tompkins, E Crosby, toxicology
Tuholski, James Martin, medicine
Uloth, Robert Henry, organic chemistry
Wallander, Jerome F, food science
Weikel, John Henry, Jr, toxicology
Weller, Lowell Ernest, organic chemistry, biochemistry
Wheeler, Keith Wilson, information science
Wintenheimer, P Louis, botany
Worrall, Paul Michael, medical research
Wu, Yao Hua, organic chemistry
Yevich, Joseph Paul, organic chemistry, medicinal chemistry
Zygmunt, Walter A, microbiology, biochemistry

Simonson, Donald Raymond, organic chemistry
Trankle, Robert John, botany

FRANKLIN
Cowan, Raymond, physics

FT WAYNE
Becker, Benjamin, microbiology
Beineke, Lowell Wayne, mathematics
Blair, Robert Paul, organic chemistry
Chowdhury, Dipak Kumar, geophysics
Coburn, Stephen Putnam, biochemistry
Cranton, Thomas James, mathematics
Cravens, William Windsor, nutrition, research administration
Davies, Harold William, (Jr), aquatic biology, phycology
Erdmann, Bernd Dietrich, paleoecology
Felger, Maurice Monroe, chemistry
Finco, Arthur A, mathematics
Flynn, John Joseph, Jr, organic chemistry
Friedel, Arthur W, inorganic chemistry, history of science
Glidewell, Marvin Elmer, industrial chemistry
Gottlieb, Sheldon F, physiology
Hatke, Mary Agnes, mathematics
Healy, Robert Michael, physical chemistry, inorganic chemistry
Hoffman, Warren E, organic chemistry
Legg, David Alan, mathematical analysis
Longroy, Allen Leroy, organic chemistry
Mansfield, Maynard Joseph, topology
Martin, M Celine, ecology, biology

Mathews, Geoffrey William, petrology, geology
McCarthy, Douglas Robert, mathematics
Muhler, Joseph Charles, biochemistry
Onwood, David P., physical chemistry
Pacer, Richard A., analytical chemistry
Pippert, Raymond Elmer, mathematics
Ringeisen, Richard Delose, mathematics
Sallay, Stephen, organic chemistry
Scheiber, Donald Joseph, underwater acoustics
Schwartz, Donald, chemistry
Steeg, Carl W, Jr, applied mathematics
Stevenson, Kenneth Lee, photochemistry
Suberkropp, Keller Francis, mycology, physiology
Summers, Charles Eugene, animal nutrition
Svoboda, Rudy George, mathematics
Walle, Oscar Theodore, zoology
Zilz, Melvin Leonard, cell biology, biochemistry

GARY
Baldwin, William Walter, microbiology, biochemistry
Bhattacharya, Pradeep Kumar, plant physiology, plant pathology
Brock, Kenneth Jack, mineralogy
Caplis, Michael E, biochemistry, toxicology
DiSalvo, Joseph, information science
Dustman, John Henry, zoology, endocrinology
Hanks, George D., genetics
Lorentzen, Keith Eden, physical organic chemistry
Mason, Earl James, pathology, microbiology
Peterson, Harold Leroy, mathematics
Randall, Rogers Ellis, Sr, inorganic chemistry, science education
Reshkin, Mark, geomorphology
Richardson, F C, botany
Stabler, Timothy Allen, developmental biology, endocrinology
Synowiec, John A, mathematical analysis

GOSHEN
Bishop, Charles Franklin, plant pathology
Buschert, Robert Cecil, solid state physics
Grove, Stanley Neal, botany, cell biology
Jacobs, Merle Emmor, zoology
Miller, Glen Russel, organic chemistry
Roth, Jonathan Nicholas, plant pathology, marine biology
Smucker, Arthur Allan, biochemistry
Smucker, Silas Jonathan, soil conservation
Weaver, Henry D, Jr, physical chemistry
Zimmerman, Lester J, agronomy, mathematics

GRANGER
Pleasants, Julian Randolph, animal nutrition, microbial ecology

GREENCASTLE
Adams, Preston, botany
Burkett, Howard (Benton), organic chemistry
Cook, Donald Jack, organic chemistry
Dudley, Underwood, mathematics, history of science
Fuller, Forst Donald, physiology
Gammon, James Robert, biology, ecology
Gass, Clinton Burke, mathematics
George, James E, inorganic chemistry
Henninger, Ernest Herman, physics
Henry, Hugh Fort, physics
Kissinger, Paul Bertram, magnetic resonance
Loring, Robert David, geography, geography of Latin America
Madison, James Ambrose, geology
Mays, Charles Edwin, physiology, herpetology
McFarland, John William, organic chemistry
Morrill, John Elliott, mathematics
Reynolds, Albert Eugene, zoology
Ricketts, John Adrian, physical chemistry
Sullivan, Dan Allen, geology
Thomas, Robert Jay, computer sciences
Welch, Winona Hazel, botany
Youse, Howard Ray, botany

GREENFIELD
Alder, Edwin Francis, plant physiology
Amundson, Merle E, analytical chemistry, pharmaceutical chemistry

Begue, William John, microbial physiology, analytical biochemistry
Bennett, Travis H., microbiology
Boisvenue, Rudolph Joseph, microbiology
Bowen, Richard Eli, parasitology
Brown, Herbert, animal nutrition, animal husbandry
Burow, Kenneth Wayne, Jr, synthetic organic chemistry
Campau, Edward Junior, entomology
Cohen, Harold Karl, veterinary pathology
Counter, Frederick T, Jr, virology
Day, Edgar William, Jr, analytical chemistry
Dettwiler, Herman Andrew, bacteriology
Donoho, Alvin Leroy, biochemistry
Emmerson, John Lynn, pharmacology, toxicology
Emmick, Thomas Lynn, organic chemistry
Froyd, James Donald, plant pathology
Gale, Charles, veterinary virology
Gard, Don Irvin, animal nutrition
Gibson, William Raymond, pharmacology
Gramlich, James Vandle, plant physiology
Gregory, Richard Parker, Jr, veterinary medicine
Griffing, William James, comparative pathology, veterinary toxicology
Hamelink, Jerry L, limnology, toxicology
Harper, Richard Waltz, synthetic organic chemistry
Jordan, Charles Edwin, animal nutrition
Klingman, Glenn Charles, weed science
Mann, Robert Leslie, agricultural chemistry
Matsuoka, Tats, virology
McCowen, Max Creager, parasitology
McDougald, Larry Robert, parasitology
Melliere, Alvin L, nutrition
Meyers, Donald Bates, pharmacology
Ose, Earl Eugene, microbiology
Page, John Gardner, pharmacology, toxicology
Perkins, Alvin Thomas, agronomy
Pierce, Emmett Coin, pathology
Potter, Emerson Lucine, animal nutrition
Probst, Gerald William, biochemistry
Raun, Arthur Phillip, animal nutrition
Sampson, Gary Robert, veterinary medicine
Schwer, Joseph Francis, agronomy, plant physiology
Shumard, Raymond Fred, parasitology
Smith, James William, bacteriology
Thayer, Paul Loyd, plant pathology
Thibault, Thomas Delor, organic chemistry
Tschabold, Edward Evertt, plant physiology
Waldrep, Thomas William, plant physiology
Wellenreiter, Rodger Henry, poultry nutrition
Wright, William Leland, weed science

HAMMOND
Apter, Julia Tutelman, ophthalmology, mathematical biology
Bates, Charles, food science, food technology
Bechtel, Robert D, mathematics
Chihara, Theodore Seio, mathematical analysis
Constant, Marc Duncan, analytical chemistry
Coolidge, Ardath Anders, nutrition
Demkovich, Paul Andrew, petroleum chemistry, analytical chemistry
Fayle, Harlan Downing, biochemistry, pharmacology
Forbes, Jack Edwin, mathematics
Heydegger, Helmut Roland, physical chemistry, analytical chemistry
Hill, Robert Joe, mathematics
Hullinger, Clifford, biochemistry
Jarabak, Joseph R, physiology
Miller, Charles Ellsworth, bionucleonics
Neely, John Charles, genetics
Nelson, Nils Keith, organic chemistry
Phillips, Travis J, inorganic chemistry
Phillips, Travis J, physical chemistry
Roebuck, Alan Kittson, zoology
Shoup, Jane Rearick, zoology
Troy, Daniel Joseph, mathematics
Tseng, Charles C, plant anatomy, morphology
Wermuth, Jerome Francis, developmental biology
Werth, Robert Joseph, zoology
Wilson, Kenneth Sheridan, mycology, plant pathology
Yates, Richard Lee, mathematics

HANOVER
Cassidy, Harold Gomes, organic chemistry, science writing
Conklin, Richard Louis, physical optics
Ellefsen, Paul, analytical chemistry
Maysilles, James Howard, taxonomy, plant ecology
Pray, Enos G, physiology
Seifert, Ralph Louis, Jr, mathematics
Totten, Stanley Martin, geology
Webster, Jackson Dan, parasitology, ornithology
White, Harold Keith, organic chemistry, biochemistry
Yarnelle, John E, mathematics

HUNTINGTON
Bergdall, Irene Floy, mathematics
Hale, Robert E, physics

INDIANAPOLIS
Aliprantis, Charalambos Dionisios, mathematical analysis
Allen, Norris Elliott, microbiology
Allmann, David William, biochemistry
Alton, Elaine Vivian, mathematics education
Anderson, Robert Clarke, toxicology
Andrew, Warren, zoology, anatomy
Aprison, Morris Herman, neurochemistry, biophysics
Arbogast, John Lynn, clinical pathology
Archer, Robert Allen, organic chemistry
Armstrong, James G, medical administration
Ashmore, James, pharmacology
Babayan, Vigen Khachig, food science, lipid chemistry
Bailey, Thomas Daniel, organic chemistry
Bang, Nils Ulrik, medicine, hematology
Barnett, Charles Jackson, organic chemistry
Barnhart, James William, biochemistry
Bauer, Dietrich Charles, immunology
Bechtol, Lavon Dee, medicine, chemistry
Becker, John W, pharmaceutical chemistry
Behrens, Otto Karl, biochemistry
Bell, Norman H, endocrinology, metabolism
Ben-Jonathan, Nira, neuroendocrinology, reproductive physiology
Bennett, Ivan Frank, psychiatry
Berger, James Edward, pharmacology
Bernard, Marie (Witte), biology
Berry, James William, invertebrate ecology
Besch, Henry Roland, Jr, pharmacology, cardiology
Bessey, William Higgins, solid state physics
Bhatti, Waqar Hamid, phytochemistry, hematology
Bidney, David, history of anthropology, ethnology
Bikin, Henry, pharmaceutical chemistry
Bishara, Rafik Hanna, analytical chemistry
Bixler, David, genetics, dentistry
Black, William, physiological psychology
Blair, Paul V, biochemistry
Bland, Hester Beth, health sciences
Blasingham, Mary Cynthia, physiology
Blevins, Charles Edward, anatomy
Blickenstaff, Robert Theron, organic chemistry
Boaz, Patricia Anne, physical chemistry, analytical chemistry
Bobbit, Jesse Leroy, biochemistry
Bockrath, Richard Charles, Jr, radiation biophysics, microbiology
Bogan, Robert L, dentistry
Bohlen, Harold Glenn, cardiovascular physiology
Bond, William H, internal medicine, hematology
Bonderman, Dean P, clinical pathology, toxicology
Bonner, John Franklin, Jr, biochemistry
Booe, James Marvin, electrochemistry
Borden, Kenneth Duane, physical chemistry, nuclear chemistry
Bornstein, Michael, physical pharmacy, analytical chemistry
Boschmann, Erwin, bioinorganic chemistry
Bottorff, Edmond Milton, organic chemistry
Bowman, Donald Edwin, biochemistry
Boyer, Charles Chester, zoology
Boylan, James Charles, pharmacy
Brandt, Ira Kive, pediatrics
Brannon, Donald Ray, organic chemistry

Bray, Malcolm Davonne, pharmaceutical chemistry
Brennan, David Michael, endocrine physiology
Breunig, Henry Latham, engineering statistics
Bromer, William Wallis, biochemistry
Brooker, Robert Munro, organic chemistry
Brooks, George Frank, Jr, internal medicine, infectious diseases
Brown, Daniel Joseph, pharmacology, toxicology
Brown, David Edward, otolaryngology
Brown, Edwin Wilson, Jr, community health, medical education
Brown, William Francis, information science
Bryan, William Phelan, biochemistry
Burck, Philip John, biochemistry
Campbell, Hayward, virology
Campbell, Robert Louis, neurosurgery
Cannon, William Nathaniel, organic chemistry
Carmichael, Ralph Harry, pharmacology
Carter, James Edward, obstetrics & gynecology, pathology
Cerimele, Benito Joseph, biomathematics, computer sciences
Chamberlin, James Wesley, organic chemistry
Chance, Ronald E, biochemistry, metabolism
Chen, Ko Kuei, pharmacology
Childers, Ray Fleetwood, pharmaceutical chemistry, analytical chemistry
Chiu, G C, medicine
Christian, Joe Clark, medical genetics
Churchill, Don W, psychiatry
Cislak, Francis Edward, organic chemistry
Clark, Charles Malcolm, Jr, medicine
Clark, Julia Berg, biochemistry, pharmacology
Clemens, James Allen, neuroendocrinology, neurophysiology
Cochrane, Robert Lowe, reproductive physiology, endocrinology
Cohen, Marlene Lois, pharmacology
Comer, Jack Payne, analytical chemistry
Conine, James William, pharmaceutical chemistry
Cooper, Robin D G, organic chemistry
Corum, Cyril Joseph, mycology
Cox, Martha, physics
Crabtree, Ross Edward, drug metabolism
Culbertson, Clyde Gray, pathology
Cutshall, Theodore Wayne, organic chemistry
Daily, Fay Kenoyer (Mrs William A Daily), botany
Daily, William Allen, microbiology
Daily, Walter J, internal medicine
Davis, Eldred Jack, biochemistry
Davisson, Edwin Orlando, physical chemistry
Day, Lawrence Eugene, microbiology
Debono, Manuel, organic chemistry
Delfin, Eliseo Dais, biology, entomology
DeMyer, Marian Kendall, child psychiatry, psychiatry
De Myer, William Erl, neurology
Dennen, David W, biochemistry, microbiology
Dexter, Richard Newman, internal medicine
DiGiovanna, Charles V, organic chemistry
Diller, Erold Ray, biochemistry, pharmacology
Dinner, Alan, organic chemistry
Dixon, Henry Marshall, III, physics
Doedens, David James, toxicology
Dominianni, Samuel James, organic chemistry
Dorman, Douglas Earl, spectrochemistry
Douthart, Richard James, biophysical chemistry
Drew, Arthur Leslie, neurology, pediatrics
Driesens, Robert James, bacteriology
Durflinger, Elizabeth Ward, invertebrate zoology
Dusenberry, William Earl, statistics
Dyke, Richard Warren, medicine, hematology
Dyken, Mark Lewis, medicine, neurology
Easton, Nelson Roy, organic chemistry
Eble, John Nelson, biochemistry
Edmundowicz, John Michael, biochemistry
Edwards, Joshua Leroy, pathology, immunology
Eggleton, Reginald Charles, ultrasound, medical science
Elliott, Robert A, bacteriology
Emerson, James L, veterinary pathology

Farkas, Eugene, organic chemistry, medicinal chemistry
Fasola, Alfred Francis, cardiopulmonary physiology
Feigenbaum, Harvey, medicine
Felten, David L., neuroanatomy
Fife, Wilmer Krafft, bio-organic chemistry
Fisch, Charles, medicine
Fisher, George M.C., applied mathematics, telecommunications engineering
FitzGerald, Joseph Arthur, medicine, psychiatry
Fitzhugh-Bell, Kathleen, neuropsychology
Fix, James D., anatomy, physical anthropology
Flaugh, Michael Edward, synthetic organic chemistry
Fleisch, Jerome Herbert, pharmacology
Floreancig, Herbert Joseph, microbiology, chemotherapy
Folkers, Thomas Mason, veterinary microbiology
Fornefeld, Eugene Joseph, organic chemistry
Forney, Robert Burns, toxicology
Frank, Bruce Hill, biochemistry, biophysics
Franklin, Thomas Doyal, Jr., physiology
Fraser, Havelock Frank, medicine
Fricke, Gordon Hugh, analytical chemistry
Fuller, Roy W. biochemistry, research
Furman, Robert Howard, administration
Gainer, Frank Edward, analytical chemistry
Garber, J Neill, orthopedic surgery
Garbrecht, William Lee, organic chemistry
Garner, LaForrest D., orthodontics
Garrett, Robert Austin, urology
Gemignani, Michael C., topology
Genovese, Pasquale Dante, cardiology
Gersting, John Marshall, Jr., applied mathematics, computer science
Gersting, Judith Lee, mathematical logic
Gibson, David Mark, biochemistry
Godzeski, Carl William, microbiology
Goe, Gerald Lee, organic chemistry
Golab, Tomasz, agricultural biochemistry
Gommel, William Raymond, meteorology
Gordee, Robert Stouffer, microbiology, pharmacology, therapeutics
Gutowski, Gerald Edward, bio-organic chemistry
Gorman, Marvin, bio-organic chemistry
Grant, Ernest Walter, pharmaceutical chemistry
Grayson, Merrill, ophthalmology, surgery
Greist, John Howard, psychiatry
Grosz, Hanus Jiri, psychiatry
Gruber, Charles Michael, Jr., pharmacology
Haak, Richard Arlen, medical biophysics, radiation biophysics
Hadd, Harry Earle, endocrinology
Hall, David Alfred, electrochemistry
Hall, John Frank, oral surgery
Hamill, Robert L., natural products chemistry
Hamilton, Jean A. x-ray crystallography, biochemistry
Harper, Edwin T. organic chemistry
Harper, Kathleen Lucille, medical parasitology, medical microbiology
Harris, Robert Allison, biochemistry
Headlee, William Hugh, medical parasitology, tropical medicine
Heimburger, Robert Francis, neurosurgery
Henderson, Robert Edward, physics, engineering
Hennon, David Kent, dentistry, pedodontics
Henry, Harry James, biochemistry
Herr, Earl Binkley, Jr., biochemistry
Herrmann, Roy G., pharmacology
Herron, David Kent, synthetic organic chemistry, microbiology
Hicks, Edward James, immunology, preventive medicine
Higgins, Harvey (M), Jr., biochemistry
Higgins, James Thomas, Jr., internal medicine, physiology
Hiley, Wayne Woodrow, pharmacy
Hine, Maynard Kiplinger, dentistry
Hinman, Charles Wiley, organic chemistry
Ho, Peter Peck Koh, biochemistry, virology
Hodes, Marion Edward, medical genetics, biochemistry
Hoehn, Marvin Martin, microbiology

Holtz, David, research administration, chemistry
Hopper, Samuel Hersey, bacteriology, public health
Hosley, Robert James, science administration
Hubbard, Jesse Donald, pathology
Huber, Floyd Milton, microbiology
Hull, Robert Normond, cell biology
Indelicato, Joseph Michael, organic chemistry
Ingraham, Joseph Sterling, immunology
Iqbal, Zafar, neurochemistry, biochemistry
Irwin, Glenn Ward, Jr., endocrinology, internal medicine
Ishaq, Mohammed, anatomy
Jackson, Richard Lee, biochemistry, pharmacy
Jarvis, Albert E., pharmaceutical chemistry
Jersild, Ralph Alvin, Jr., microscopic anatomy, cell biology
Jesseph, John Ervin, surgery
Jeter, Max Albert, animal nutrition, bacteriology
Johnsen, Sherman Edward Jerome, applied mathematics
Johnson, Cyrus Conrad, Jr., internal medicine, endocrinology
Johnson, Ronald Doyle, natural products chemistry
Johnson, Charles F. medicine, electron microscopy
Johnson, Irving Stanley, cancer, research administration
Johnson, Ernest Raymond, mathematics
Jones, Noel Duane, crystallography
Judson, Walter Emery, medicine
Kalsbeck, John Edward, neurosurgery
Kaminker, Jerome Alvin, topology, mathematical analysis
Kammeraad, Adrian, experimental biology, research administration
Kaplan, Jerome I., solid state physics
Karnatz, Frank Albert, organic chemistry
Katner, Allen Samuel, organic chemistry
Katzber, Allan Alfred, anatomy
Kaufman, Karl Lincoln, pharmacy, medicinal chemistry
Kavanagh, Frederick, plant physiology
Kavula, Michael P., Jr., pharmacy, radiochemistry
Kebler, Richard William, applied physics
Kelly, Edward Joseph, physical chemistry
Kelly-Fry, Elizabeth, bioacoustics, science administration
Kennedy, Edward Earl, analytical chemistry
Kenney, Michael Thomas, immunology
Kern, Raymond Adrian, physical chemistry
Kettelkamp, Donald B., orthopedic surgery
Kisheimer, Sidney Arthur, organic chemistry
Kim, Kil Choi, anesthesiology, pharmacology
King, Edgar Pearce, mathematical statistics
King, Harold, surgery
Kinsel, Norma Ann, microbiology
Kiplinger, Glenn Francis, pharmacology
Kirsch, Joseph Lawrence, Jr., physical chemistry
Kirst, Herbert Andrew, organic chemistry
Kirtley, William Raymond, clinical medicine
Klatte, Eugene, radiology
Kleber, John William, biopharmaceutics
Kleinschmidt, Walter John, biochemistry
Knobel, Leon Kenneth, physiology
Knobel, Suzanne Buckner, internal medicine, cardiology
Koch, Kay Frances, organic chemistry
Kohlstaedt, Kenneth George, medicine
Koppel, Gary Allen, organic chemistry
Kornfeld, Edmund Carl, medicinal chemistry
Kory, Mitchell, physiological bacteriology
Kossov, Aaron David, analytical chemistry, organic chemistry
Kranzfelder, Arthur Leonard, organic chemistry
Kress, Thomas Joseph, organic chemistry
Kuczkowski, Joseph Edward, algebra
Kuzel, Norbert R. analytical chemistry, instrumentation
Lappas, Lewis Christopher, pharmaceutical chemistry
Lavagnino, Edward Ralph, organic chemistry
Lavender, John Francis, microbiology, virology
Laverell, William David, mathematics
LeFor, William Mathew, immunology
Lemberger, Louis, clinical pharmacology
Lempke, Robert Everett, surgery

Leser, Ralph Ulrich, internal medicine
Li, Ting Kai, medicine, biochemistry
Lin, Tsung-Min, physiology
Lively, David Harryman, microbiology
Lobb, Barry Lee, mathematics
Luke, Jon Christian, applied mathematics
Lukemeyer, Jack Warren, microbiology, biochemistry
Luneng, Lawrence, medicine, biochemistry
Lund, Melvin Robert, dentistry
Maciak, George M., microchemistry
Mallet, Gordon Edward, microbiology
Malone, Thomas Francis, meteorology
Mandelbaum, Isidore, surgery
Manion, Marlow William, biochemistry
Manthey, John August, plant physiology
Marconi, Gary G., natural products chemistry
Marsh, Max Martin, physical chemistry, analytical chemistry
Marshall, Frederick J., organic chemistry
Marshall, Winston Stanley, medicinal chemistry
Martz, Bill L., medicine
Martz, Carl D., medicine
Mason, Norman Ronald, biochemistry
Massey, Eddie H., pharmaceutical chemistry
Matsumoto, Charles, pharmacology, biochemistry
Matsumoto, Ken, organic chemistry
McKeehan, Charles Wayne, pharmaceutical chemistry, research administration
McKinney, Myron William, pharmaceutical chemistry
Mealey, John, Jr., neurosurgery
Megraw, Robert Ellis, clinical chemistry
Meierc, Forrest T., theoretical physics
Meiss, Richard Alan, physiology
Merritt, Arthur Donald, internal medicine
Merritt, Doris Honig, research administration, pediatrics
Metz, Clyde, physical chemistry
Michel, Karl Heinz, bio-organic chemistry
Miesel, John Louis, organic chemistry
Millar, Wayne Norval, microbiology
Miller, Chris H., oral microbiology
Miller, Emery Parker, physics
Miller, James Franklin, analytical chemistry
Miller, Jerry Roland, medicine
Miller, Richard Pressly, biochemistry
Miller, Roscoe Earl, medicine
Minassian, Donald Paul, mathematics
Minton, Sherman Anthony, herpetology, microbiology
Mirsky, Arthur, environmental geology, stratigraphy
Mitchell, David Farrar, dentistry
Montieth, Richard Voorhees, chemistry
Moore, Donald Floyd, psychiatry
Moore, Ward Wilfred, physiology
Moorehead, Wells Rufus, clinical chemistry, biochemistry
Morgan, Carl Robert, anatomy
Morris, Estell E., oral surgery
Munden, Bill J., pharmacy
Munsick, Robert Alliot, obstetrics & gynecology
Murphy, Charles Franklin, medicinal chemistry
Murphy, Hubert William, pharmaceutical chemistry
Murphy, John Riffe, experimental statistics
Murphy, Patrick Joseph, biochemistry
Murray, Raymond Harold, cardiology
Myers, Franklin Guy, mathematics
Nash, Claude Hamilton, III, microbial physiology
Nash, Franklin D., nephrology
Nash, J Frank, pharmaceutical chemistry
Nevill, William Albert, organic chemistry
Nevin, Robert Stephen, organic chemistry, polymer chemistry
Ng, Bartholomew Sung-Hong, applied mathematics, fluid dynamics
Nickander, Rodney Carl, pharmacology
Niederpruem, Donald J., microbiology, biochemistry

Niklowitz, Werner Johannes, neurobiology, electron microscopy
Niss, Hamilton Frederick, microbiology
Nordschow, Carleton Deane, pathology, biochemistry
Norins, Arthur Leonard, dermatology
Norman, Richard D., dentistry, analytical chemistry
Norton, James Augustus, Jr., statistics
Nunn, Arthur Sherman, Jr., physiology
Nurnberger, John Ignatius, cytology, chemistry
Occolowitz, John Lewis, mass spectrometry
Ochs, Sidney, neurophysiology, medical biophysics
O'Connor, Brian Lee, human biology, physical anthropology
Ong, John Tjoan Ho, pharmaceutics
Osborne, John William, dental research
Osgood, David William, vertebrate zoology
Ott, John Lewis, bacteriology
Palmer, Catherine Gardella, cytogenetics
Paradise, Raymond R., pharmacology
Parka, Stanley John, weed science
Parke, Russell Frank, pharmacy
Patrick, Edward Alfred, computer science, medicine
Patterson, Samuel S., dentistry
Pelton, John Forrester, plant ecology
Perkins, Lois Claire, anatomy
Perrill, Stephen Arthur, herpetology
Peters, Lynn Randolph, industrial organic chemistry
Pettinga, Cornelius Wesley, biochemistry
Pfanzer, Richard Gary, medical physiology
Phillips, Ralph W., dental materials
Pikal, Michael Jon, physical chemistry, pharmaceutics
Ping, Ronald Stanley, dentistry, oral surgery
Pioch, Richard Paul, medicinal chemistry
Pittenger, Robert Carlton, microbiology
Pohland, Albert, organic chemistry
Porter, Herschel Donovan, organic chemistry
Potter, Rosario H Yap, genetics, dental research
Powell, Richard Cinclair, internal medicine, endocrinology
Proksch, Gary J., biochemistry, clinical chemistry
Quay, John Ferguson, biophysics
Quener, Sherry Fream, biochemistry
Quinney, Paul Reed, analytical chemistry
Rabideau, Peter W., organic chemistry
Reed, Terry Eugene, medical genetics
Rhodes, Mitchell Lee, pulmonary diseases
Richards, Alice B., pharmacology
Richter, Judith Anne, neuropharmacology, neurochemistry
Rickard, Eugene Clark, analytical chemistry
Ridge, William Clayton, mathematics
Ridolfo, Anthony Sylvester, rheumatology, clinical pharmacology
Rieger, William Holley, organic chemistry
Rigdon, Robert David, topology
Rivers, Paul Michael, organic chemistry
Robey, Roger Lewis, synthetic organic chemistry
Rodda, Bruce Edward, medical statistics, clinical pharmacology
Roeske, Roger William, bio-organic chemistry
Rohn, Robert Jones, medicine
Root, Mary Avery, pharmacology
Ross, Alexander Treloar, neurology
Roth, Lawrence Max, pathology, electron microscopy
Rothe, Carl Frederick, physiology
Rowe, Edward John, physiology
Rowe, James Lincoln, organic chemistry
Rubin, Alan, pharmacology
Russell, John Robert, invertebrate zoology
St John, Philip Alan, zoology
Salerni, Oreste Leroy, organic chemistry, medicinal chemistry
Samuels, Robert, zoology
Sargent, Roger N., zoology
Schloemer, Robert Henry, virology
Schmedtje, John Frederick, anatomy
Schramm, Jacob Richard, botany
Sawyer, William D., microbiology
Schaible, Robert Hilton, medical genetics, animal genetics
Schulz, Arthur R., biochemistry, nutrition
Schulz, Dale Metherd, pathology

Schwarz, Anton, medical research
Selkurt, Ewald Erdman, physiology
Seymour, Keith Morton, organic chemistry
Shafer, William Gene, oral pathology
Shanks, James Clements, Jr, speech pathology
Sharp, Louis James, IV, polymer chemistry
Shaw, Margaret Ann, pharmacy
Shaw, Walter Norman, biological chemistry
Shellhamer, Robert Howard, anatomy, histochemistry
Shepard, Edwin Reed, organic chemistry
Shields, James Edwin, biochemistry
Shrigley, Edward White, microbiology, infectious diseases
Shumacker, Harris B, Jr, surgery
Shupe, Robert Eugene, radiation physics, radiobiology
Siakotos, Aristotle N., biochemistry
Simmons, James Edwin, child psychiatry
Skadron, Peter, solid state physics
Slater, Irwin Holzman, pharmacology
Small, Joyce G, psychiatry
Small, Iver Francis, psychiatry
Smith, Gerald Floyd, bio-organic chemistry, molecular biol
Smith, James Warren, clinical pathology
Smith, Milton Reynolds, biochemistry, physiology
Smith, Ronald Earl, vertebrate zoology
Smits, Stephen Edward, pharmacology
Snively, William Daniel, Jr, medical physiology
Sonday, Francis Llewellyn, audiology
Soper, Quentin Francis, chemistry
Sowers, Edward Eugene, industrial organic chemistry
Spencer, John Lawrence, organic chemistry
Spining, Arthur Milton, III, animal nutrition, biochemistry
Spolyar, Louis William, toxicology
Squires, Robert Wright, microbiology
Stamper, Martha C, organic chemistry
Standish, Samuel Miles, oral pathology
Stark, Paul, pharmacology, physiology
Stark, William Max, industrial microbiology
Starkey, Paul Edward, pedodontics, dentistry
Steinberg, Mitchell Irwin, pharmacology
Steinrauf, Larry King, biochemistry, physical chemistry
Stockton, Mary Rose, organic chemistry
Stoelting, Vergil Kenneth, medicine
Stone, Robert Louis, microbiology
Stonehill, Robert Berrell, internal medicine
Stookey, George K, dentistry
Storvick, Waldemar O, biochemistry
Su, Kenneth Shyan-Eli, pharmaceutics
Sullivan, Hugh R, drug metabolism
Summers, William Allen, Sr, parasitology, medical microbiology
Svoboda, Gordon H, pharmacognosy
Swartz, Howard, radiobiology
Swartz, Marjorie Louise, inorganic chemistry
Swenson, Henry Maurice, dentistry
Takruri, Harun, pharmacy
Tanner, George Albert, physiology
Taylor, Harold Leland, biochemistry
Tensmeyer, Lowell George, physical chemistry
Test, Charles Edward, medicine
Thakkar, Arvind Layji, physical pharmacy
Thompkins, Leon, pharmaceutical chemistry
Thompson, Daniel James, teratology, toxicology
Thompson, Gerald Lee, organic chemistry
Thompson, Richard Michael, biochemistry, analytical chemistry
Todd, Glen Cory, pathology
Tomich, Charles Edward, dentistry, oral pathology
Toomey, Joseph Edward, synthetic organic chemistry, structural chemistry
Troxell, Terry Charles, biophysical chemistry
Turner, Jan Ross, microbiology
Turrell, Eugene Snow, psychiatry
Tuttle, Ronald Ralph, pharmacology, physiology
Van Frank, Richard Mark, cell biology, molecular pharmacology
Van Heyningen, Earle Marvin, organic chemistry
Van Tyle, William Kent, pharmacology
Vasavada, Kashyap V, theoretical physics
Vawter, Spencer Max, physics
Vondrak, Edward Andrew, physics

Von Schuching, Susanne, organic chemistry, biochemistry
Wagle, Shreepad R, pharmacology
Waife, Sholom Omi, internal medicine, medical education
Wallace, Gerald Wayne, analytical chemistry, physical chemistry
Walsh, Patrick Noel, high temperature chemistry
Wappner, Rebecca Sue, pediatrics, biochemical genetics
Watson, Ronald Ross, immunology, nutrition
Webber, John Alan, medicinal chemistry
Weber, George, pharmacology, biochemistry
Weber, Janet Crosby, biochemistry, computer science
Webster, Rex N, botany
Webster, Richard Curtis, histology, cytology
Wegener, Warner Smith, microbiology, biochemistry
Weihaupt, John George, astrogeology, oceanography
Weist, William Godfrey, Jr, geology, hydrology
Whang, Robert, internal medicine, nephrology
Wheeler, Frank Carlisle, pharmaceutical chemistry
Wheeler, William Joe, medicinal chemistry
Wheeler, William Raleigh, organic chemistry, chemical engineering
White, Arthur C, internal medicine, infectious diseases
White, Halbert Constantine, organic chemistry
White, Lowell Deane, physics
Whitney, John Glen, microbiology
Wigh, Russell, medicine, radiology
Wild, Gene Muriel, biochemistry
Wilson, Claude E, analytical chemistry, electrochemistry
Wolen, Robert Lawrence, pharmacology
Wong, David Taiwai, biochemistry
Wren, Henry K, organic chemistry
Wright, Ian Glaisby, synthetic organic chemistry
Wright, Joseph William, Jr, medicine
Wright, Peter Hedley, pharmacology, endocrinology
Wright, Walter Eugene, biopharmaceutics
Wyma, Richard J, inorganic chemistry
Yates, Willard F, Jr, plant taxonomy, cytogenetics
Yen, Terence Tsin Tsu, biochemistry, genetics
Yoho, Robert Oscar, public health education
Young, Peter Chun Man, biochemistry
Yu, Pao-Lo, genetics, biostatistics
Yune, Heun Yung, radiology, surgery
Zeman, Wolfgang, neuropathology
Zimmerman, Sarah E, immunology, biochemistry

JEFFERSONVILLE
Putnam, Hamilton Wallace, chemistry, food technology

KOKOMO
Boneham, Roger Frederick, geology
Dolph, Gary Edward, paleobotany
Haffley, Philip Gene, organic chemistry
Hanig, Ruth Belle Cohn, biochemistry, chemistry
Harland, Glen Eugene, Jr, solid state physics
Jaumot, Frank Edward, Jr, solid state physics
McLafferty, John J, analytical chemistry
Stein, Frank S, physics

LA PORTE
Miller, Kenneth Melvin, fresh water ecology, food science

LAFAYETTE
Adams, Richard Linwood, poultry nutrition
Allen, Durward Leon, vertebrate ecology
Altman, Joseph, neuropsychology
Axtell, John David, genetics, plant breeding
Bancroft, John Basil, biology
Barr, Rita, plant physiology
Baxter, Glen Earl, mathematics
Bohren, Bernard Benjamin, animal genetics
Brandt, Karl Garet, biochemistry
Buchi, Julius Richard, mathematics
Burnstein, Theodore, virology

Carmony, Donald Duane, high energy physics
Chiscon, J Alfred, genetics, evolution
Diestler, Dennis Jon, theoretical chemistry
Drasin, David, mathematical analysis
Drazin, Michael Peter, mathematics
Edgell, Walter Francis, chemistry
Fan, Hsu Yun, solid state physics
Featherston, William Roy, poultry nutrition, metabolism
Filmer, David Lee, biochemistry, biophysics
Floss, Heinz G, biochemistry
Foster, Joseph Franklin, biochemistry
Fowler, Earle Cabell, high energy physics
Fry, Cleota Gage, mathematics
Fry, Robert E, anthropology
Gambill, Robert Arnold, mathematics
Garfinkel, Arthur Frederick, high energy physics
Golub, Edward S, immunology, cell biology
Gottlieb, Daniel Henry, mathematics
Green, Ralph J, Jr, plant pathology
Harrington, Rodney B, animal science, statistics
Harris, Samuel Melvin, theoretical nuclear physics
Hartl, Daniel L, developmental genetics, population genetics
Hayden, Richard Amherst, horticulture, pomology
Hennen, Joe Fleetwood, botany, mycology
Herrmann, Klaus Manfred, biochemistry, molecular biology
Herron, James Dudley, science education, chemistry
Hicks, John W, III, agricultural economics
Hinze, William James, geophysics
Hollandbeck, Richard, animal husbandry
Howe, Herbert James, stratigraphy
Keesom, Pieter Hendrik, physics
Keller, Marion Wiles, mathematics
Kildsig, Dane Olin, physical pharmacy
Kivioja, Lassi A, geodesy
Klinghammer, Erich, ethology, psychology
Krabbe, Gregers Louis, mathematical analysis
Laskowski, Michael Jr, biochemistry
Levandowski, Donald William, geology
Levy, Morris, evolutionary biology, biosystematics
Lindsey, Alton Anthony, plant ecology
Lobstein, Otto Ervin, clinical biochemistry
Loeffler, Frank Joseph, physics
McLaughlin, Jerry Loren, pharmacognosy
Mennear, John Hartley, pharmacology
Meyer, Henry Oostenwald Albertijn, mineralogy, geochemistry
Moldenhauer, William C, soil conservation
Moore, David Sheldon, statistics
Morre, D James, biochemistry
Morse, Erskine Vance, veterinary microbiology, veterinary public health
Mueller, Charles Richard, theoretical chemistry, physical chemistry
Mullen, James G, solid state physics
Muller, Norbert, physical chemistry
Mumford, Russell Eugene, animal ecology
Myers, Robert Durant, psychobiology, neurobiology
Nakajima, Shigehiro, neurophysiology
Page, Edwin Howard, veterinary medicine
Patterson, Fred La Vern, plant breeding, genetics
Pinto, Lawrence Henry, neurophysiology, bioengineering
Pollard, Harry, mathematics
Porile, Norbert Thomas, nuclear chemistry
Puri, Prem Singh, statistics, operations research
Ramdas, Anant Krishna, solid state physics, optics
Ray, William Jackson, Jr, organic chemistry
Regnier, Frederick Eugene, biochemistry
Remers, William Alan, organic chemistry, medicinal chemistry
Rickey, Frank Atkinson, Jr, nuclear physics
Robinson, William Robert, inorganic chemistry
Rodriguez, Victor William, solid state physics
Rodwell, Victor William, biochemistry
Roesel, Oscar Fred, physiology, biomedical engineering
Rossmann, Michael G, crystallography
Roy, Robert Francis, geophysics

Schlueter, Donald Jerome, nuclear physics
Scott, Donald Howard, phytopathology
Shaw, Stanley Miner, bionucleonics, medical research
Silverman, Edward, mathematics
Simms, Paul C, nuclear physics
Simon, Edward Harvey, virology
Singleton, Wayne Louis, animal science, reproductive physiology
Smolen, Victor Frank, pharmacology, bioengineering
Studden, William John, mathematical statistics
Tessman, Irwin, biophysics, molecular biology
Thurber, James Kent, applied mathematics
Tobias, Russell Stuart, bioinorganic chemistry, organometallic chemistry
Tyler, Varro Eugene, pharmacognosy
Van Etten, Robert Lee, chemistry
Wetsman, Allen William, mathematics
Wildfeuer, Marvin Emanuel, biochemistry
Willmann, Robert B, particle physics
Wollan, Gerhard Norval, mathematics
Wolszon, John Donald, analytical chemistry
Yim, George Kwock Wah, pharmacology
Young, Paul Ruel, computer sciences, mathematical logic
Zachmanoglou, Eleftherios Charalambos, applied mathematics
Zalkin, Howard, biochemistry
Zink, Robert Edwin, mathematics
Zitron, Norman Ralph, applied mathematics

LAWRENCEBURG
Higson, Harold George, analytical chemistry

MARION
Ansbacher, Stefan, biochemistry
Chilgreen, Donald Ray, soil microbiology
Hinds, Marvin Harold, physiology, biomedical engineering
Klinger, William Russell, mathematics
Lee, James K, physical chemistry, analytical chemistry
Porter, Donald Henry, mathematics
Werking, Robert Junior, science education

MARTINSVILLE
Van Abeele, Frederick Richard, biochemistry

MEMPHIS
Ginski, John Martin, physiology

MICHIGAN CITY
Johnston, John, rubber chemistry
Potter, Brian, dermatology

MISHAWAKA
Kroon, James Lee, electrochemistry

MOORESVILLE
Trischler, Floyd D, organic chemistry, polymer chemistry

MT VERNON
Bostick, Edgar E, physical organic chemistry
Campbell, Allen James, organic chemistry, polymer chemistry
Jaquiss, Donald B G, organic chemistry
Liberti, Frank Nunzio, polymer chemistry, physical chemistry
Mackinney, Arland Lee, nuclear physics, research administration
Mark, Victor, organic chemistry

MUNCIE
Beatty, George Franklin, geography of Asia
Beekman, John Alfred, mathematics
Byers, Stanley A, ceramics
Carmin, Robert Leighton, geography
Costill, David Lee, human physiology
Crankshaw, William Bliss, forest ecology
Dillon, Lowell Ivan, geography
Doeden, Gerald Ennen, analytics[1]
Eiser, Arthur L, plant taxonomy, plant ecology

Engstrom, Lee Edward, developmental genetics
Flores-Meiser, Enya P., anthropology, sociology
Gaiser, Romey Arthur, inorganic chemistry
Hendrickson, Donald Allen, medical microbiology, microbial ecology
Henzlik, Raymond Eugene, physiology, health sciences
Higgins, William Brouillard, mathematics
Hutts, Malcom E., physics
Joyner, Ralph Deimer, inorganic chemistry
Justham, Stephen Alton, climatology, meteorology
Kallman, Ralph Arthur, mathematics
Kane, Henry Edward, geology, geography
Kennedy, Duncan Tilly, neurophysiology
Kirkpatrick, Ralph Donald, wildlife ecology
Lawrence, Richard Manley, structural chemistry
Lee, Norman K., applied mathematics
Lesh, Thomas Allan, physiology
List, James Carl, herpetology
Ludwig, Hubert Joseph, statistics, computer science
Ma, Cynthia Sanman, science
Ma, Pang-Fai, biochemistry
Martinson, Tom L., economic geography, geography of Latin America
McCormick, Roy L., mathematics
McElhinney, Margaret M (Cocklin), science education
McKinney, Earl H., numerical analysis
Meiser, John H., chemistry
Mertens, Thomas Robert, genetics, science education
Mosbo, John Alvin, inorganic chemistry
Nisbet, Jerry J., science education, plant anatomy
Norton, Norma J., botany
Ober, David Ray, nuclear physics
Oliver, Jeanette Clements, plant anatomy, plant taxonomy
Orr, Robert William, geology
Paschall, Homer Donald, cell physiology
Place, Ralph L., nuclear physics
Robold, Alice Ilene, mathematics
Roepke, Harlan Hugh, sedimentary petrology
Sannella, Joseph L., research administration
Schoen, Meera, cultural geography, geography of South Asia
Siverly, Russell Emmett, medical entomology
Smith, Charles Edward, Jr., limnology, physiology
Smith, Shelby Dean, mathematics
Sprague, Newton G., physics, astronomy
Stevenson, Forrest Frederick, plant morphology
Storhoff, Bruce Norman, inorganic chemistry
Van Meter, Donald Eugene, soil conservation
Wagner, Eugene Stephen, biological chemistry, physical chemistry
Welker, George W., parasitology, bacteriology
Wise, Charles Davidson, zoology, limnology
Wiseman, Park Allen, organic chemistry
Zimmack, Harold Lincoln, insect pathology

NEW CASTLE
Van Meter, John Connell, biochemistry

NEW HAVEN
Davis, Joseph Anthony, inorganic chemistry, solid state chemistry

NORTH MANCHESTER
Berry, Dwight Beecher, physics

NEW ALBANY
Atwood, Kenton, physical chemistry
Jansing, Jo Ann, analytical chemistry, physical chemistry

MUNSTER
Basila, Michael Robert, physical chemistry
Schwoegler, Edward John, organic chemistry

NASHVILLE
Erickson, Jon Jay, medical physics

NOTRE DAME
Asano, Tomoaki, physiology
Balka, Don Stephen, mathematics
Bamberek, Mark A., analytical chemistry
Basu, Subhash Chandra, biochemistry
Bellina, Joseph James, Jr., surface physics
Bender, Harvey Alan, genetics
Benton, Francis Lee, organic chemistry
Berners, Edgar Davis, experimental nuclear physics
Betchov, Robert, physics
Bick, George Herman, entomology, zoology
Bishop, James Martin, physics
Biswas, Nripendra Nath, physics
Blackstead, Howard Allan, magnetic resonance, low temperature physics
Bose, Samir K., theoretical physics
Bottei, Rudolph Santo, analytical chemistry, inorganic chemistry
Brennecke, Kenneth Raymond, geography
Bretnauer, Roger R., biochemistry
Brown, Ferdinand Louis, mathematics
Browne, Cornelius Payne, nuclear physics
Burton, Milton, chemistry
Cason, Neal M., high energy physics
Castellino, Francis Joseph, biochemistry
Chagnon, Paul Robert, nuclear physics
Chapin, Edward William, Jr., mathematical logic
Clay, Robert Edward, mathematical logic
Cooney, Miriam Patrick, mathematics
Craig, George Brownlee, Jr., entomology
Crovello, Theodore John, systematics
Curran, Columba, inorganic chemistry
Cushing, James Thomas, theoretical physics
Danehy, James Philip, organic chemistry
Darden, Sperry Eugene, nuclear physics
DeCelles, Paul C., physics
Despres, Leo Arthur, social anthropology, anthropology
Dineen, Clarence Francis, zoology, ecology
Dull, Martin Homer, mathematics
Duman, John Girard, environmental physiology
Esch, Harald Erich, animal behavior
Fairley, William Merle, geology
Fehlner, Thomas Patrick, physical chemistry
Feigl, Dorothy Marie, organic chemistry
Freeman, Jeremiah Patrick, organic chemistry
Fuchs, Morton S., biochemical genetics
Funk, Emerson Gornflow, Jr., nuclear physics
Goetz, Abraham, mathematics
Gordon, Robert Edward, ecology
Greene, Richard Wallace, invertebrate biology, plant physiology
Grimstad, Paul Robert, medical entomology, epizootiology
Gutschick, Raymond Charles, geology
Haaser, Norman Bray, applied mathematics
Hamill, William Henry, physical chemistry
Hayes, Robert Green, physical chemistry
Helman, William Philip, physical chemistry
Hennion, George Felix, organic chemistry
Hickey, William August, entomology, genetics
Hofman, Emil Thomas, inorganic chemistry

Eberly, William Robert, zoology, limnology
Farringer, Leland Dwight, physics
Lutz, Wilson Boyd, physics
McBride, Ralph Book, mathematics education
Miller, Edward George, organic chemistry
Orput, Philip Arvid, mycology

Huckleberry, Alan Trinler, mathematics
Johnson, Walter Richard, physics
Kenney, Vincent Paul, physics
Khorana, Brij Mohan, physics
Kozak, John Joseph, chemical physics, biophysical chemistry
Kronstein, Karl Martin, algebra
Madsen, David Christy, physiology, genetics
Magee, John Lafayette, physical chemistry
Marshalek, Eugene Richard, nuclear physics
Martinez-Carrion, Marino, biochemistry
Mast, Cecil B., geometry, theoretical physics
McGlinn, William David, theoretical physics

McGrath, James J., plant anatomy
McIntosh, Robert Patrick, plant ecology
Miller, Walter Charles, experimental nuclear physics
Monga, David William, biological oceanography, aquatic biology
Mozumder, Asokendu, radiation chemistry, radiation physics
Mrowca, Adalbert, physics
Murphy, Michael Joseph, physical geology
O'Meara, O Timothy, mathematics
O'Neil, Carl William, anthropology
Otter, Richard Robert, mathematics
Pasto, Daniel Jerome, organic chemistry
Peltier, Charles Francis, topology
Petrauskas, Alexander Anselmus, physics
Phan, Kok-Wee, algebra
Pierce, Louis, physical chemistry
Pilger, Richard Christian, Jr., physical chemistry
Poirier, John Anthony, physics
Pollak, Barth, algebra
Pollard, Morris, medical microbiology
Rai, Karamjit Singh, cytogenetics
Rey, Charles Albert, experimental high energy physics
Rhines, Don Scott, experimental high energy physics
Riget, James Aloysius, rock mechanics
Sasaki, Tom Taketo, anthropology, sociology
Saz, Howard Jay, biochemistry, parasitology
Schaefer, Frank William, III, animal parasitology
Scheidt, Walter Robert, chemistry, x-ray crystallography
Schwartz, Maurice Edward, theoretical chemistry
Sever, David Michael, herpetology
Sheng, Shan-Jen, physical chemistry
Shephard, William Danks, experimental high energy physics
Shilts, James Leonard, physics, astronomy
Smith, Peter David, mathematics, computer science
Stoll, Wilhelm, mathematics
Strieder, William, chemical engineering
Susalla, Anne A., plant anatomy
Thomas, John Kerry, physical chemistry
Thorson, Ralph Edward, medical parasitology, veterinary parasitology
Tihen, Joseph Anton, zoology
Tomasch, Walter J., experimental solid state physics
Trozzolo, Anthony Marion, organic chemistry, photochemistry
Tweedell, Kenyon Stanley, zoology
Vuckovic, Vladeta, mathematics
Wagner, Morris, bacteriology, immunology
Waldman, Bernard, nuclear physics
Walter, Joseph L., inorganic chemistry
Webb, Phyllis Marie, immunology
Weinstein, Paul P., parasitology
Williams, Daniel Charles, cell biology, electron microscopy
Winicur, Daniel Henry, chemical physics
Winkler, Erhard Mario, geology
Wong, Warren James, algebra
Wostmann, Bernard Stephan, biochemistry, nutrition

RENSSELAER
Davis, Michael Edward, geology
Jones, Duvall Albert, vertebrate zoology
Kramer, William J., organic chemistry, analytical chemistry
Rodia, Jacob Stephen, chemistry
Rueve, Charles Richard, mathematics
Siegrist, Urban J., microbiology
Sleeman, Lyle Herman, Jr., geology

RICHMOND
Alexander, Howard Wright, mathematics, statistics
Bakker, Gerald Robert, organic chemistry
Camp, Mark Jeffrey, paleoecology, malacology
Fishback, William Thompson, mathematics
Gooding, Ansel Miller, geology
Hanes, Harold, mathematics
Martin, Charles Wellington, Jr., physiology
Rhoads, Paul Spottswood, medicine
Scherer, George Allen, chemistry
Stephenson, William Kay, physiology
Stratton, Wilmer Joseph, chemistry
Strong, Laurence Edward, physical chemistry
Telfair, David, physics

Ward, Gertrude Luckhardt, entomology
Whitcomb, Stuart Estes, physics

ROCKPORT
Rothwell, Frederick Mirvan, mycology, physiological ecology

ST MARY-OF-THE-WOODS
Cunningham, Ellen M., mathematical logic

SEYMOUR
Montgomery, Kenneth O., pharmacy, pharmaceutical chemistry

ST MEINRAD
Ostdick, Thomas, inorganic chemistry, academic administration
Schmelz, Damian Vincent, ecology

SOUTH BEND
Coomes, Edward Arthur, physics
Critz, Jerry B., physiology
D'Aelio, Gaetano Francis, polymer chemistry
Garber, Lawrence L., inorganic chemistry
Haines, Larry Kent, theoretical physics
Huemmer, Thomas Francis, polymer chemistry
Huitink, Geraldine M., analytical chemistry
Lamont, Patrick John Coll., algebra, number theory
Marcus, Philip Selmar, mathematics
Mead, Darwin James, chemistry
Meagher, Ralph Ernest, applied physics, computer science
Metcalfe, Grant E., psychiatry
Mihelich, John William, nuclear physics
Nazaroff, George Vasily, theoretical chemistry
Pioway, James Mason, nuclear physics
Rice, Francis Owen, chemistry
Ross, Alberta B., chemistry, information science
Ross, Joseph Hansbro, organic chemistry
Savage, Earl John, plant pathology, mycology
Snyder, Donald DuWayne, physics
Specht, Edward John, mathematics
Teumac, Fred N., chemistry, science
Uznanski-Botter, Rita Marlene, analytical chemistry
Williams, Lynn Roy, mathematics
Williamson, Frederick Dale, polymer chemistry
Winicur, Sandra, cell physiology
Zimmerman, Walter Bruce, solid state physics

SOUTH WHITLEY
German, Albert Frederick Ottomar, II, animal nutrition

SYRACUSE
Herrold, Katherine McDermott, pathology

TERRE HAUTE
Albright, Joseph Finley, zoology
Alvager, Torsten Karl Erik, biophysics, nuclear physics
Anderson, William John, neuroanatomy
Babcock, William Edward, veterinary medicine
Baca, Glenn, physical inorganic chemistry, electrochemistry
Bachman, Marvin Charles, microbiology
Bailey, Herbert R., mathematics
Bakken, George Stewart, ecology, biophysics
Barton, Byron Kurtz, geography
Baxter-Gabbard, Karen Lee, molecular biology
Benjaminov, Benjamin S., organic medicine
Blum, Leon Leib, medicine, pathology
Brehm, Sylva Patience, microbiology
Brett, William John, physiology
Bunger, William Boone, physical chemistry
Chamberlain, William Maynard, aquatic biology
Cleveland, John H., economic geology
Corrigan, John Joseph, biochemistry
Crawford, James Gordon, microbiology
De Korvin, Andre, mathematics
Dial, Norman Arnold, zoology

Diekhans, Herbert Henry, mathematics
Di Lavore, Philip, III, physics
Druelinger, Melvin L, organic chemistry, photochemistry
Drummond, Robert Roland, urban geography, physical geography
Dusanic, Donald G, parasitology
Easton, Richard J, mathematical analysis
Eckler, Paul Eugene, synthetic organic chemistry
Eversole, Wilburn John, endocrinology
Greasham, Randolph Louis, microbial biochemistry
Greenspan, Kalman, electrophysiology, pharmacodynamics
Gross, Jean Alvah, cell biology
Guernsey, James Lee, resource geography, urban geography
Guthrie, Frank Albert, analytical chemistry
Hansen, Uwe Jens, solid state physics
Hasan, Khwaja Arif, cultural anthropology
Hidy, Phil Harter, microbial biochemistry
Hill, Roy Dean, inorganic chemistry
Hodge, Edward Butler, chemistry
Hook, John Clinton, physical geography
Hopp, William Beecher, zoology
Howe, David Orville, soil chemistry, plant physiology
Howe, Robert C, micropaleontology
Hull, Clarence Joseph, organic chemistry
Hunsucker, Jerry H, organic chemistry
Jackson, Marion T, plant ecology
Jacobs, Martin John, synthetic organic chemistry
Johnson, David Franklin, biology, bacteriology
Kessel, William George, organic chemistry
Klein, Richard Gale, animal nutrition
Knudsen, Oran Milton, physical chemistry
Levine, Alvin Saul, virology
Lewis, Dennis Allen, organic chemistry
Llewellyn, Ralph A, nuclear physics
Mahoney, Joan Munroe, biochemistry
Martin, Jerome, chemistry
Matz, John J, Jr, nutrition, biochemistry
Mausel, Paul Warner, physical geography
McGregor, John Robert, economic geography
McMillan, Graham Watson, chemistry
Meeks, Wilkison (Winfield), acoustics
Method, Peter Francis, physical chemistry
Miescher, Guido, microbiology
Miller, Victor Charles, geomorphology
Moloney, Michael J, microwave physics
Moulton, Benjamin, geography, geology
Murphy, Robert Carl, anatomy
Oliver, John Edward, physical geography
Oster, Mark Otho, biochemistry, microbiology
Owen, Donald Eugene, geology
Palmer, Theodore Paine, mathematics
Pang, Hildegard Elisabeth, ethnography, archaeology
Parrish, Clyde Franklin, physical chemistry, radiation chemistry
Poorman, Lawrence Eugene, physics
Quist, Raymond Willard, speech pathology
Reuland, Donald John, nuclear chemistry
Rhee, John Williams, space physics
Robertson, Donald Edwin, organic chemistry
Sakano, Theodore K, physical chemistry
Sansing, Gerald Allen, microbial biochemistry
Sartain, Carl Clinton, metals physics
Schoknecht, Jean Donze, mycology
Sherman, Gary Joseph, mathematics
Shipchandler, Mohammed Tyebji, medicinal chemistry, organic chemistry
Shriner, Walter Owen, mathematics
Siddiqi, Akhtar Husain, economic geography
Siefker, Joseph Roy, analytical chemistry, inorganic chemistry
Smith, Armand Verne, Jr, mathematics, statistics
Smith, Earl Cooper, analytical chemistry
Smith, Leonard Charles, biochemistry
Summers, William Allen, Jr, industrial organic chemistry, physical organic chemistry
Swanson, Lynn Allen, analytical chemistry
Swez, John Adam, biophysics
Sword, Christopher Patrick, microbiology
Tamar, Henry, protozoology, physiology
Tatum, James Patrick, physical chemistry
Thrasher, George W, animal nutrition
Trinler, William A, organic chemistry
Uhlhorn, Kenneth W, science education
Wakim, Khalil Georges, internal medicine
Webb, George Willis, economic geography, cultural geography
Wehrmeister, Herbert Louis, chemistry
Westgard, James Blake, elementary particle physics
Whitaker, John O, Jr, vertebrate ecology, mammalogy
Wilcox, Frank H, genetics
Zwick, Earl J, mathematics

TIPTON
Hoffbeck, Loren John, agronomy

UPLAND
Burden, Stanley Lee, Jr, chemical instrumentation, analytical chemistry
Neuhouser, David Lee, mathematics
Nussbaum, Elmer, biophysics
Snyder, Harold, taxonomy, conservation

VALPARAISO
Bloom, William Whiley, biology
Buls, Erwin Julius, geography
Carlson, Kermit Howard, mathematics
Cook, Addison Gilbert, organic chemistry
Deters, John Frederick, inorganic chemistry
Gunther, Waldemar Carl, biology
Hallerberg, Arthur Edward, mathematics
Hansen, Waldemar Conrad, chemistry
Hanson, Robert Jack, microbiology
Hicks, Garland Fisher, Jr, microbiology
Kallay, Ferencz Paul, geography
Krekeler, Carl Herman, zoology, entomology
Larson, Raymond George, chemistry
Leoschke, William Leroy, biochemistry
Manning, Armin William, nuclear physics
Marks, Gayton Carl, botany
Meyer, Alfred Herman Ludwig, geography
Meyer, Frederick Richard, mammalian physiology
Nagel, Edgar Herbert, analytical chemistry
Nichols, Kenneth E, plant physiology
Schwan, Theodore Carl, organic chemistry, inorganic chemistry
Shankland, Rodney Veeder, chemistry
Shirer, Donald Leroy, physics
Smith, Lewis Oliver, Jr, organic chemistry
Strietelmeier, John Henry, cultural geography, geography of Great Britain
Tan, James Chien-Hua, genetics, statistics

WABASH
Connell, Balfour, rubber chemistry

WARSAW
Fuson, Robert L, thoracic surgery, cardiovascular surgery

WEST LAFAYETTE
Aberle, Elton D, meat sciences, food science
Abernathy, Richard Paul, nutrition
Abhyankar, Shreeram, mathematics
Abney, Thomas Scott, plant pathology, mycology
Agee, Ernest Mason, dynamic meteorology, fluid mechanics
Albright, Jack Lawrence, animal science
Alliston, Charles Walter, environmental physiology, reproductive physiology
Alvey, David Dale, agronomy
Amstutz, Harold Emerson, veterinary medicine
Amy, Jonathan Weekes, chemistry
Anderson, Virgil Lee, mathematics, statistics
Andre, Herman William, nuclear chemistry
Andrew, Kenneth L, atomic spectroscopy
Andrews, Frederick Newcomb, animal physiology
Angell, C A, physical chemistry
Arnott, Struther, molecular biology
Aronson, Arthur Ian, microbiology
Athow, Kirk Leland, plant pathology
Axelrod, Bernard, biochemistry
Babel, Frederick John, food microbiology
Bachman, Gustave Bryant, chemistry
Badenhop, Arthur Fredrick, food science, biochemistry
Balazs, Louis A P, theoretical high energy physics
Banker, Gilbert Stephen, pharmacy
Barber, Stanley Arthur, soil chemistry, soil fertility
Barnes, Virgil Everett, II, high energy physics
Bauer, Marvin E, agronomy, remote sensing
Bauman, Loyal Frederick, plant breeding
Baumgardner, Marion F, agronomy
Bayer, Shirley Ann, neuroanatomy
Beers, Thomas Wesley, forest mensuration
Beeson, William Malcolm, animal nutrition
Beineke, Walter Frank, forest genetics
Belcastro, Patrick Frank, pharmacy
Belinfante, Frederik J, physics
Bell, Audra Earl, genetics
Benkeser, Robert Anthony, organic chemistry
Bergeson, Glenn Bernard, plant nematology
Berkhoff, German Adolfo, veterinary microbiology
Berkovitz, Leonard David, mathematics
Berry, Joe Gene, food science, poultry science
Bishop, Marilyn Frances, theoretical solid state physics
Bittinger, Marvin Lowell, mathematics
Blair, Byron Oliver, crop ecology
Born, Gordon Stuart, bionucleonics, pharmacy
Borowitz, Joseph Leo, pharmacology
Bottoms, Gerald Doyle, physiology, endocrino logy
Bousquet, William F, pharmacology, biochemistry
Bracker, Charles E, Jr, plant pathology
Brakke, Kenneth Allen, mathematical analysis
Bramble, William Clark, forestry
Branges, Louis De, mathematics
Bray, Ralph, solid state physics
Brewster, James Henry, organic chemistry
Broersma, Delmar B, entomology
Bronson, Roy DeBolt, soil science
Brown, Herbert Charles, inorganic chemistry, organic chemistry
Brundage, Roy Charles, forestry
Buker, Robert Joe, plant breeding
Bula, Raymond J, agronomy
Butler, Larry G, biochemistry
Byrn, Stephen Robert, biophysical chemistry, solid state chemistry
Byrnes, William Richard, forestry
Cable, Raymond Millard, zoology
Capps, Richard H, theoretical physics
Carlson, Gary P, toxicology, pharmacology
Carlton, William Walter, veterinary pathology, veterinary toxicology
Carson, James Rolland, poultry science
Carter, James M, veterinary physiology
Carter, Mason Carlton, plant physiology, forestry
Cassady, John Mac, organic chemistry, medicinal chemistry
Casten, Richard G, applied mathematics
Chalmers, Robert Kenny, pharmacy, pharmacology
Chambers, James Vernon, food science
Chandler, Leland, entomology
Chaney, William R, forestry, plant physiology
Chang, Ching-Jer, natural products chemistry, medicinal chemistry
Cherry, Joe H, plant physiology, biochemistry
Chiscon, Martha Oakley, immunobiology
Chrisman, Charles Larry, cytogenetics
Christmas, Ellsworth P, agronomy
Chylek, Petr, atmospheric physics, optics
Claflin, Robert Malden, veterinary pathology
Clark, Allan H, mathematics
Clark, Helen Edith, nutrition
Colella, Roberto, solid state physics
Conte, Samuel D, numerical analysis
Coppoc, Gordon Lloyd, pharmaculogy, veterinary pharmacology
Cote, Louis J, mathematical statistics
Cramer, William Anthony, biophysics
Crane, Frederick Loring, biochemistry
Crane, Paul Levi, plant breeding, agronomy
Cumberbatch, Ellis, applied mathematics
Curtis, Roy Walter, plant physiology
Cushman, Mark, organic chemistry, medicinal chemistry
Cwalina, Gustav Edward, medicinal chemistry
Dalby, Arthur, biochemical genetics
Daly, Patrick Joseph, nuclear chemistry
Daniel, William Hugh, agronomy
Das, Gopal Dwarka, neuroembryology, neuroanatomy
Data, John Batiste, pharmaceutical chemistry
Davenport, Derek Alfred, inorganic chemistry
Davis, Ralph Lanier, genetics, plant breeding
Denning, Dorothy Elizabeth Robling, information science
Dilley, Richard Alan, biochemistry, plant physiology
Dixon, Jack Edward, biochemistry
Dobson, Richard Cecil, veterinary entomology, economic entomology
Dolch, William Lee, physical chemistry
Doolittle, Donald Preston, genetics
Dostal, Herbert C, plant physiology, horticulture
Eckelman, Carl A, wood science, structural engineering
Emerson, Frank Henry, horticulture, plant pathology
Erb, Ralph Eugene, animal physiology
Erickson, Homer Theodore, horticulture
Evanson, Robert Verne, pharmacy
Ferris, John Mason, nematology
Ferris, Virginia Rogers, nematology
Feuer, Henry, organic chemistry
Feuer, Paula Berger, physics
Fischang, William John, entomology
Flint, Harrison Leigh, ornamental horticulture
Fong, Francis K, chemical physics, biophysics
Forrest, John Charles, meat science
Foster, John Edward, entomology, plant breeding
Franzmeier, Donald Paul, soil genesis, soil classification
Freeman, Max James, immunobiology, immunopathology
Freeman, Verne Crawford, agriculture
Fuchs, Norman H, theoretical physics
Fujii, Koichi, population ecology
Fuller, William Richard, mathematics
Fuqua, Mary Elizabeth, nutrition
Furdyna, Jacek K solid state physics
Gaafar, Sayed Mohammed, veterinary parasitology
Gaidos, James A, physics
Gailar, Owen H, physics
Galloway, Harry M, agronomy
Gallun, Robert Louis, entomology
Garman, John Andrew, organic chemistry, research administration
Gartenhaus, Solomon, theoretical physics
Garwood, Vernon Abington, animal genetics
Gautschi, Walter, numerical analysis
Geddes, Leslie Alexander, cardiovascular physiology, biomedical engineering
Geib, Irving George, x-ray crystallography
Geller, Susan Carol, algebra
Genin, Joseph, solid mechanics
George, Robert Eugene, radiological physics, health physics
Gerns, Fred Rudolph, organic chemistry
Gerritsen, Alexander Nicolaas, metal physics, low temperature physics
Gerritsen, Jacqueline Koolhaas, information science
Gilham, Peter Thomas, organic chemistry, biochemistry
Glazer, Mark, anthropology
Gleser, Leon Jay, mathematical statistics, statistics
Glover, David Val, cytogenetics
Goetsch, Gerald D, animal physiology
Golomb, Michael, mathematics
Goonewardene, Hilary Felix, entomology, plant pathology
Gould, George Edwin, entomology
Grabowski, Sandra Raynolds, neurobiology
Grabowski, Zbigniew Wojciech, nuclear physics
Graham, Dale Elliott, molecular biology
Grimley, Robert Thomas, high temperature chemistry
Grutzner, John Brandon, organic chemistry
Gupta, Shanti Swarup, mathematical statistics
Gustafson, Donald Pink, veterinary virology, pathobiology
Haas, Felix, mathematics
Haas, Violet Bushwick, applied mathematics, electrical engineering
Haelterman, Edward Omer, veterinary virology
Hall, Randall Clark, analytical chemistry
Halstead, Maurice Howard, computer science
Hancock, John Ogden, physics
Harrington, Daniel Dale, veterinary pathology
Harris, Dewey Lynn, animal genetics, statistics
Heinstein, Peter, biochemistry
Hem, Stanley L, physical pharmacy

Hicks, Charles Robert, applied statistics
Hill, Donald Louis, animal physiology
Hilst, Arvin Rudolph, agronomy
Hinsman, Edward James, veterinary anatomy, comparative neurology
Ho, Cho-Yen, thermal physics, research administration
Hodges, Thomas Kent, plant physiology
Hoff, Johan Eduard, food science, analytical chemistry
Holland, Charles Jordan, applied mathematics
Hollingworth, Robert Michael, toxicology, pesticide chemistry
Holmes, Richard Bruce, mathematical analysis, systems analysis
Holt, Donald Alexander, agronomy, plant physiology
Holt, Harvey Allen, forest ecology
Hooper, Billy Ernest, veterinary medicine, veterinary pathology
Hornemann, Ulfert, organic chemistry, biochemistry
Housley, Thomas Lee, plant physiology
Huber, Don Morgan, plant pathology, soil microbiology
Hughes, Howard Kibble, mathematics
Hulinger, Ronald Loral, veterinary anatomy, microscopic anatomy
Hunsley, Roger Eugene, animal science
Hunt, Michael O'Leary, forest products
Hurst, Robert Nelson, environmental physiology, science education
Jackson, Andrew Otis, plant virology
Jaffe, Lionel, developmental physiology, biophysics
Jaffe, Miriam Walther, astronomy
James, Hubert Maxwell, theoretical solid state physics
Janick, Jules, plant genetics, plant breeding
Jardine, Ian, mass spectrometry, bio-organic chemistry
Jenkins, Glenn Llewellyn, pharmaceutical chemistry
Jennings, Ted Vernon, structural geology, petrology
Jerison, Meyer, mathematics
Johnson, Orland Eugene, nuclear physics
Johnson, Vivian Annabelle, theoretical physics
Jones, Hobart Wayne, animal breeding
Jones, Russell K., veterinary public health
Judge, Max David, animal science, food science
Kenaga, Clare Burton, plant pathology
Kent, Claudia Marie, biochemistry
Kessler, Wayne Vincent, bionucleonics
Kim, Yeong Ell, theoretical nuclear physics
Kirkham, Wayne Wolpert, veterinary science, microbiology
Kirkpatrick, Charles Milton, wildlife ecology
Kirksey, Avanelle, nutrition, biochemistry
Kirsch, Edwin Joseph, microbiology
Klontz, Everett Earl, physics
Knecht, Charles Daniel, veterinary surgery, neurology
Knevel, Adelbert Michael, medicinal chemistry
Knudson, Douglas Marvin, forest economics, silviculture
Kochhaw, Stanley Oscar, mathematics
Kohlhaw, Gunter B., biochemistry
Kohnke, Helmut, agronomy
Kornblum, Nathan, organic chemistry
Kramer, Paul Alan, physical pharmacy
Krider, Jake Luther, animal nutrition
Krishnamurthy, Sundaram, organic chemistry
Krogmann, David William, biology
Kullerud, Gunnar, geochemistry
Kuo, Tzee-Ke, high energy physics
Lam, Shue-Lock, cytogenetics, plant physiology
Landolt, Robert Raymond, bionucleonics
Laviolette, Francis A., plant pathology
Lechtenberg, Victor Louis, agronomy
Lessman, Koert J., plant breeding, plant genetics
Levinthal, Mark, microbial genetics
Light, Albert, biochemistry
Liley, Peter Edward, physics, chemical engineering
Lippa, Erik Alexander, mathematics
Lipman, Joseph, mathematics, number theory

Lipschutz, Michael Elazar, physical chemistry, cosmochemistry
Liska, Bernard Joseph, food technology
Lister, Richard Malcolm, plant virology
Livingston, Robert Louis, physical chemistry
Lovett, James Satterthwaite, botany
Low, Philip Funk, soil chemistry
Lundquist, Norman Stanley, dairy science
Lynch, Robert Emmett, applied mathematics
Lytle, Fred Edward, chemistry
MacKay, John Warwick, solid state physics
Maiven, Paul Vernon, neuroendocrinology, reproductive physiology
Mannering, Jerry Vincent, agronomy
Magerum, Dale William, inorganic chemistry, analytical chemistry
Markley, John Lute, physical biochemistry, protein chemistry
Martin, Truman Glen, animal genetics
McCabe, George Paul, Jr., statistics
McFee, William Warren, forest soils, soil fertility
McIlvain, Jay Edward, neurobiology
Miller, David Harry, experimental high energy physics
Miller, William Lloyd, agricultural economics
Mittenthal, Jay Edward, neurobiology
Miya, Tom Saburo, pharmacology
Montague, Fredrick Howard, Jr., wildlife ecology
Morrison, Harry, organic chemistry
Mortenson, Leonard Earl, biochemistry
Morter, Raymond Lione, medical microbiology, immunopathology
Moser, Bruno Carl, horticulture
Moser, John William, Jr., forest biometry
Moses, Harold Eugene, veterinary medicine
Moskowitz, Mervin, microbiology, immunochemistry
Mullen, Russell Edward, crop physiology
Mullikin, Thomas Wilson, mathematics
Nakajima, Yasuko, anatomy, neurobiology
Neher, George Martin, veterinary science
Neie, Van Elroy, physics education
Nelson, Darrell Wayne, soil chemistry, soil microbiology
Nelson, Werner Lu Lind, agronomy
Neugebauer, Christoph Johannes, mathematics
Neus, Marcel Fernand, statistics, operations research
Newhouse, Vernon Leopold, biomedical engineering
Newman, James Edward, agronomy
Nicholson, Ralph Lester, phytopathology
Nichols, David Earl, medicinal chemistry
Novodvorsky, Mark Evgenievich, pure mathematics
Nyquist, Wyman Ellsworth, statistical analysis, quantitative genetics
Ohlrogge, Alvin John, soils
Olander, Harvey Johan, veterinary pathology
Olson, John Bennet, zoology, science education
Ortman, Eldon Emil, entomology
Osmun, John Vincent, entomology
Ostroy, Sanford Eugene, neurobiology
Outhouse, James Burton, animal science
Overhauser, Albert Warner, theoretical solid state physics
Pak, William Louis, visual physiology, neurobiology
Palfrey, Thomas Rossman, Jr., physics
Pardue, Harry L., analytical chemistry
Parker, George Ralph, ecology, forestry
Parker, Herbert Edmund, biochemistry
Parmelee, Carlton Edwin, food science
Paschke, John Donald, entomology, virology
Pearlman, Norman, solid state physics
Peck, Garnet E., pharmacy
Pecknold, Paul Carson, plant pathology
Perlis, Sam, algebra
Perone, Samuel Patrick, analytical chemistry
Perry, Tilden Wayne, animal nutrition
Peterson, John Booth, agronomy
Phillips, Marvin W., soil fertility
Pickett, Robert Cooper, plant breeding
Pillai, Krishna Chennakadu Sreedharan, mathematical statistics

Plumlee, Millard P., Jr., nutrition
Postlethwait, Samuel Noel, botany
Pratt, Dan Edwin, food chemistry, nutrition
Probst, Albert Henry, agronomy
Prohofsky, Earl William, solid state physics
Pullen, Milton William, Jr., geology
Quackenbush, Forrest Ward, biochemistry
Ragheb, Hussein S., microbiology
Ranieri, Richard Leo, organic chemistry, natural products chemistry
Rao, Koneru Venkata Rajeswara, mathematics
Reiss, William Dean, agronomy
Reuszer, Herbert William, soil microbiology
Rhykerd, Charles Loren, plant physiology
Rice, John Richard, applied mathematics, computer science
Richardson, James Wyman, quantum chemistry
Robbers, James Earl, pharmacognosy
Robinson, Farrel Richard, veterinary pathology, toxicology
Robinson, Kenneth Ronald, developmental biology
Robitaille, Henry Arthur, horticulture
Rogier, John Charles, animal nutrition
Rosen, Saul, mathematics
Rosen, Simon Peter, physics
Ross, Merrill Arthur, Jr., weed science
Roth, Charles Barron, agronomy, soil science
Rubin, Herman, statistics, mathematics
Rubin, Jean E., mathematics
Russock, Howard Israel, ethology
Samuels, Myra Lee, statistics
Sato, Hiroshi, solid state physics
Schaffert, Robert Eugene, plant breeding, genetics
Schall, Elwyn DeLaurel, analytical chemistry
Scharenberg, Rolf Paul, nuclear physics
Schenkman, Eugene Victor, algebra
Schreiber, Marvin Mandel, agronomy, plant physiology
Schuder, Donald Lloyd, entomology
Schultz, Reinhard Edward, topology, mathematics
Senft, John Franklin, forest products
Shands, Henry Lee, plant genetics, plant breeding
Shaner, Gregory Ellis, plant pathology
Shankland, Daniel Leslie, neurophysiology
Shanks, Merrill Edward, mathematics
Shen, Chun-Shan, physics, astrophysics
Shibata, Edward Isamu, experimental high energy physics
Shu, Shien Siu, applied mathematics
Sicer, Joe Wellington, poultry science
Sinclair, Annette, mathematics
Sladek, Ronald John, solid state physics
Smith, Ned Myron, geology
Smith, Phillip J., meteorology
Smith, Robert William, microbiology, biochemistry
Smith, William Harold, animal science
Sneen, Richard Allen, organic chemistry
Somerville, Ronald Lamont, biochemistry
Sommers, Lee Edwin, soil microbiology
Spacie, Anne, limnology, pollution biology
Sperandio, Glen Joseph, clinical pharmacy
Stadelman, William Jacob, food science
Stanfield, Kenneth Charles, experimental sciences
Stanley, Robert Weir, physics
Stark, Eric Walter, wood technology
Steer, Max David, speech & hearing sciences
Steffen, Rolf Marcel, nuclear physics
Stiller, Mary Louise, plant physiology
Stivers, Russell Kennedy, soil fertility
Stob, Martin, animal science
Stockton, Jack Jenks, microbiology, parasitology
Stromberg, Melvin Willard, gross anatomy, neuroanatomy
Stump, John Edward, veterinary anatomy
Suddarth, Stanley Kendrick, forestry
Svanes, Torgny, algebra
Swartzendruber, Dale, soil physics
Swearingin, Marvin Laverne, agronomy
Tacker, Martha McClelland, biochemistry
Tacker, Willis Arnold, Jr., cardiovascular physiology, medical education
Tendam, Donald Jan, nuclear physics
Tigchelaar, Edward Clarence, genetics, plant breeding
Tomes, Mark Louis, genetics

Truce, William Everett, chemistry
Tsai, Chia-Yin, genetics, biochemistry
Tsai, Min-Shen Chen, carbohydrate biochemistry
Tubis, Arnold, theoretical physics
Tuite, John F., phytopathology
Turpin, Frank Thomas, economic entomology
Vanable, Joseph William, Jr., developmental biology
Van Sickle, David C., histology, pathology
Van Vleet, John F., veterinary pathology
Van Zandt, Lonnie L., solid state physics
Vardeman, Stephen Bruce, mathematical statistics
Vetter, Richard J., health physics, radiobiology
Vincent, Dayton George, meteorology
Vorst, James J., agronomy
Ward, Bennie Franklin Leon, theoretical high energy physics
Warren, George Frederick, weed science
Warren, Herman Lecil, plant pathology
Webster, Merritt Samuel, mathematics
Weeks, Harmon Patrick, Jr., wildlife ecology
Weinberg, Bernd, speech pathology
Weiner, Henry, organic chemistry, biochemistry
Weinstein, Bernard Allen, solid state physics
Weirich, Walter Edward, veterinary physics
Weismiller, Richard A., agronomy
Welch, Zara D., organic chemistry
Wernimont, Grant (Theodore), chemistry
Weston, Vaughan Hatherley, applied mathematics, theoretical physics
Whistler, Roy Lester, chemistry
White, Joe Lloyd, soil mineralogy, soil chemistry
Wiebers, Joyce Adams, analytical biochemistry
Wiersma, Daniel, soil science
Wilcox, Gerald Eugene, soil fertility
Wilcox, James Raymond, plant breeding, genetics
Williams, Edwin Bruce, phytopathology
Williams, James Lovon, Jr., weed science, plant physiology
Williams, Luther Steward, molecular biology
Wilson, Karl A., biochemistry
Wilson, Mark Curtis, entomology
Winfree, Arthur T., biology, biophysics
Winograd, Nicholas, analytical chemistry
Wischmeier, Walter Henry, soil conservation
Wolinsky, Joseph, organic chemistry
Woodruff, David Scott, ecology, medical parasitology
Woods, Walter Ralph, animal nutrition
Wott, John Arthur, horticulture
Wright, Gordon Pribyl, operations research, statistics
Yackel, James W., mathematics
Yahner, Joseph Edward, soil classification
Young, Nancy Lizotte, biochemistry
Ziemer, Paul L., health physics

WEST TERRE HAUTE
Husain, Syd S., microbiology
Regan, Francis, mathematics
Schultz, Fred Henry, Jr., pharmacology

WESTVILLE
Brill, Wilfred G., physics
Countryman, James Joseph, algebra
Hawthorne, Robert Montgomery, Jr., organic chemistry, history of chemistry
Porter, Clyde L. Jr., ecology

WHITING
Porsche, Francis William, chemistry
Tom, Theodore Benton, organic chemistry

WINDFALL
Nanda, Devender (Dave) Kumar, plant breeding, genetics

WINONA LAKE
De Young, Donald Bouwman, science
Humberd, Jesse David, science education
Tanner, Gary Dale, entomology, biology

ZIONSVILLE

Milch, Lawrence Jacques, pharmacology, biochemistry
Robinson, Virgil Benton, veterinary pathology, comparative pathology
Strycker, Stanley Julian, organic chemistry
Tedeschi, Ralph Earl, pharmacology
Wang, Samuel S M, medicinal chemistry

IOWA

AMES

Abian, Alexander, pure mathematics
Ahrens, Franklin Alfred, pharmacology, toxicology
Allen, Edward Switzer, mathematics
Allison, Milton James, microbiology
Amemiya, Minoru, soil conservation, agronomy
Anderson, Irvin Charles, plant physiology
Anderson, Lloyd L., animal science, physiology
Angelici, Marvin A, soils
Angelici, Robert Joe, inorganic chemistry
Applequist, Jon Barr, biophysical chemistry
Arnrich, Lotte, nutrition
Atherly, Alan G, genetics, molecular biology
Atkins, Richard Elton, agronomy
Bachmann, Roger Werner, limnology
Baker, Albert L, clinical chemistry
Baetz, Durwood L, veterinary medicine
Bancroft, Theodore Alfonso, statistics
Barnes, Richard George, solid state physics, nuclear magnetic resonance
Barnes, Wilfred E, mathematics
Barton, Thomas J, organic chemistry
Basart, John Philip, radio astronomy
Beavers, Willet I, astrophysics
Beitz, Donald Clarence, nutritional biochemistry
Bensend, Dwight Winfred, wood science, wood technology
Beran, George Wesley, veterinary public health
Berger, Philip Jeffrey, animal breeding, statistical analysis
Berkey, Dennis Alan, seed physiology
Bevolo, Albert Joseph, solid state physics
Biggs, Donald Lee, geology
Bird, Emerson Wheat, clinical chemistry
Black, Charles Allen, soil fertility
Boney, William Arthur, Jr, veterinary medicine, immunology
Bowen, Charles Clark, cell biology, botany
Bowen, George Hamilton, Jr, physics
Brackelsberg, Paul O, animal breeding
Bremner, John McColl, chemistry
Brewer, Wilma Denell, nutrition
Brown, Richard Wallace, veterinary medicine
Browning, John Artie, plant pathology
Bryner, John Henry, microbiology
Buck, Griffith J, horticulture
Buck, William Boyd, veterinary toxicology
Bulkley, Ross Vivian, fisheries
Burris, Joseph Stephen, plant physiology
Burroughs, Wise, animal science
Buttrey, Benton Wilson, protozoology
Carithers, Jeanine Rutherford, anatomy, endocrinology
Carlander, Kenneth Dixon, fisheries
Carlin, Frances, food chemistry
Carlson, Bille Chandler, applied mathematics
Carlson, Irving Theodore, plant breeding
Carlson, Richard Eugene, agronomy
Chapman, Orville Lamar, organic chemistry
Chen, Kuo-Mei, chemical physics
Cheville, Norman F, veterinary pathology
Chiotti, Premo, physical chemistry
Christensen, George Curtis, veterinary anatomy
Clardy, Jon Christel, structural chemistry
Clark, Raymond Loyd, plant pathology
Clem, John Richard, physics
Cook, Barnett C, physics
Corbett, John Dudley, inorganic chemistry
Cornette, James L, mathematics
Countryman, David Wayne, forest management
Cox, Charles Philip, statistics, biometrics
Cox, David Frame, statistics, animal science

Cox, George Stanley, biochemistry, molecular biology
Crump, Malcolm Hart, gastroenterology
Cutlip, Randall Curry, veterinary pathology
Dahiya, Rajbir Singh, mathematics
Dahlgren, Robert Bernard, wildlife research
Dahm, Paul Adolph, entomology, biochemistry
Danielson, Gordon Charles, solid state physics
Danks, Maureen Lee, plant ecology
David, Herbert Aron, mathematical statistics
Davis, Craig Brian, ecology
Dean, Nathan Wesley, physics
Dellmann, Horst-Dieter, histology, neuroendocrinology
Denisen, Ervin Loren, horticulture
Dicke, Ferdinand Frederick, entomology
Dickson, Spencer E, mathematics
Diehl, Harvey Clarence, analytical chemistry
Dinsmore, James Jay, ornithology
Dodd, John Durrance, botany
Dougherty, Robert Watson, veterinary medicine
Dumenil, Lloyd C, soil fertility
Dunham, Jewett, zoology, physiology
Dunleavy, John M, plant pathology
Durand, Donald P, microbiology, virology
Dyer, Donald Chester, pharmacology
Dyer, James Arthur, mathematics
Earls, Lester Thomas, physics
Eastwood, Basil R, animal genetics, dairy science
Eberhart, Steve A, genetics, statistics
Ellinghausen, Herman Charles, Jr, bacteriology
Ellis, Edwin M, microbiology
Engen, Richard Lee, physiology, biomedical engineering
Espenson, James Henry, inorganic chemistry
Everson, Leroy Everett, agronomy
Ewan, Richard Colin, animal nutrition, biochemistry
Ewing, Solon Alexander, animal nutrition
Fassel, Velmer Arthur, physical chemistry, analytical chemistry
Fenton, Thomas E, soil genesis, soil classification
Fink, Arlington M, mathematics
Finnemore, Douglas K, solid state physics
Firestone, Alexander, high energy physics
Flatt, Ronald Eugene, pathology, laboratory animal medicine
Floyd, Denis Ragan, algebra
Foley, Dean Carroll, phytopathology
Foreman, Charles Frederick, dairy science
Franke, Robert G, botany
Franzen, Hugo Friedrich, physical chemistry
Frederick, Lloyd Randall, soil microbiology, soil fertility
Freeman, Albert Eugene, animal breeding
French, Dexter, biochemistry
Frey, Kenneth John, plant breeding
Fritz, James Sherwood, analytical chemistry
Fromm, Herbert Jerome, biochemistry
Fuchs, Ronald, solid state physics
Fuller, Wayne Arthur, agricultural economics, statistics
Furtak, Thomas Elton, surface physics
Garcia, Pilar A, nutrition
Gentle, James Eddie, statistics
George, John Ronald, plant breeding
Gerstein, Bernard Clemence, physical chemistry
Ghoshal, Nani Gopal, comparative anatomy, neuroanatomy
Gillette, D Dale, veterinary physiology
Gilman, Henry, organic chemistry, organometallic chemistry
Glenn-Lewin, David Carl, plant ecology
Goetz, Charles Albert, analytical chemistry
Goll, Darrel Eugene, molecular biology, biochemistry
Gordon, John C, plant physiology, silviculture
Gough, Patricia Marie, biochemistry, immunology
Gradwohl, David Mayer, anthropology
Grant, John Gray, hygiene
Graves, Donald J, biochemistry
Green, Detroy Edward, agronomy
Greve, John Henry, veterinary parasitology
Grier, Ronald Lee, veterinary surgery, physiology, endocrinology

Grossman, Allen S, astrophysics
Hall, Charles Virdus, vegetable crops, horticulture
Hall, Richard Brian, forest genetics, silviculture
Hallauer, Arnel Roy, plant breeding
Hammer, Charles Lawrence, elementary partical physics, mathematical physics
Hammond, Earl Gullette, food chemistry, biochemistry
Han, Chien-Pae, statistics
Hansen, Robert Suttle, physical chemistry
Hanway, John Joseph, soils
Harris, Delbert Linn, veterinary microbiology
Hart, Elwood Roy, entomology
Hartman, Paul Arthur, food microbiology
Harville, David Arthur, statistics
Hathcock, John Nathan, nutrition, toxicology
Haupt, Robert Elliott, zoology
Haynes, Emmit Howard, animal nutrition
Hearn, Walter Russell, biochemistry
Heddleston, Kenneth Luther, microbiology
Heintz, Roger Lewis, biochemistry
Hembrough, Frederick B, cardiovascular physiology
Herm, Ronald Richard, physical chemistry
Herrick, John Berne, veterinary medicine
Hickman, Roy D, statistics
Hicks, Ellis Arden, zoology
Hinrichsen, John James Luett, mathematics
Hinz, Paul Norman, statistics
Hodges, Clinton Frederick, plant pathology
Hodges, Laurent, solid state physics
Hoffman, Mark Peter, animal science
Hofstad, Melvin Sidney, veterinary medicine
Holden, Palmer Joseph, animal nutrition, animal husbandry
Hollander, Willard Fisher, genetics
Holt, John Gilbert, microbiology
Homer, Roger Harry, mathematics
Hopkins, Frederick Sherman, Jr, forestry
Horner, Harry Theodore, Jr, botany
Horowitz, Jack, biochemistry
Hotchkiss, Donald K, applied statistics
Hughes, David Edward, veterinary microbiology
Huntsberger, David Vernon, statistics
Hussey, Keith Morgan, geomorphology
Hutchcroft, Charles Dennett, agronomy
Hutton, Wilbert, Jr, inorganic chemistry, science education
Imsande, John, biochemistry, molecular biology
Isely, Duane, plant taxonomy, economic botany
Jackson, Larry LaVern, veterinary anesthesiology
Jacobson, Norman Leonard, animal nutrition
Jacobson, Robert Andrew, physical chemistry
Jensen, Erling N, nuclear physics
Jeska, Edward Lawrence, immunology
Keller, Roy Fred, mathematics, computer science
Kelley, James Durrett, ornamental horticulture
Kemeny, J Lorant, veterinary microbiology
Kempthorne, Oscar, statistics, mathematical biology
Kernan, William J, Jr, physics
King, Walter Bernard, inorganic chemistry
Kirkham, Don, geophysics
Klaas, Erwin Eugene, wildlife biology
Kliewer, Kenneth L, theoretical solid state physics
Kline, Edwin A, animal science
Kluge, John Paul, veterinary pathology, comparative pathology
Klun, Jerome Anthony, entomology
Knaphus, George, plant pathology
Knight, Harry Hazelton, entomology
Kniseley, Richard Newman, spectrochemistry
Kraft, Allen Abraham, food technology
Kreider, Orlando Clark, mathematics
Kunesh, Jerry Paul, clinical pharmacology, medicine
LaGrange, William Somers, food microbiology, dairy bacteriology
Lamb, Richard C, cosmic ray physics, elementary particle physics
Lambert, George, veterinary microbiology
Lambert, Robert J, mathematics
Lamotte, Clifford Elton, plant physiology
Landers, Roger Q, Jr, plant ecology, range management

Larock, Richard Craig, organic chemistry
Lassila, Kenneth Eino, high energy physics, theoretical nuclear physics
Laveglia, James Gary, entomology
Leacock, Robert A, theoretical physics
Legvold, Sam, physics
Lemish, John, economic geology, geochemistry
Lersten, Nels R, botany
Lewis, Robert Earl, entomology, vertebrate zoology
Lindahl, Clarence Homer, mathematics
Liu, Samuel Hsi-Peh, theoretical solid state physics
Lockhart, William Raymond, bacteriology
Luecke, Glenn Richard, mathematical analysis
Lundvall, Richard, veterinary medicine
Lush, Jay Laurence, animal breeding, animal genetics
Lynch, David William, solid state physics
Macheak, Merlin Edward, veterinary medicine, microbiology
Magilton, James Henry, veterinary anatomy
Mahlstede, John Peter, ornamental horticulture
Maple, Clair George, applied mathematics
Mare, Cornelius John, veterinary microbiology
Marion, William W, food science
Martin, Don Stanley, Jr, physical inorganic chemistry
Martinson, Charlie Anton, plant pathology
Mathews, Jerold Chase, mathematics
McCarley, Robert Eugene, inorganic chemistry
McClurkin, Arlan Wilbur, veterinary medicine, virology
McCorkle, Willard Homer, physics
McCormack, William C, pediatrics, physiology
McDonald, John Stoner, veterinary medicine, microbiology
McGill, John Joseph, entomology
McGilliard, A Dare, animal nutrition
McNabb, Harold Sanderson, Jr, forest pathology
Meeker, William Quackenbush, Jr, statistics
Melampy, Robert Maurice, physiology
Mengeling, William Lloyd, veterinary pathology
Mensing, Richard Walter, statistics
Menzel, Bruce Willard, ichthyology
Merkal, Richard Sterling, microbiology
Metzler, David Everett, biochemistry
Mickelson, Milo Norval, physiological bacteriology
Miller, Janice Margaret, veterinary pathology
Miller, Lyle Devon, veterinary pathology
Miller, Richard Keith, mathematics
Miller, Wilmer Jay, immunology, genetics
Mitchell, Lawrence Gustave, animal parasitology, protozoology
Mock, James Joseph, agronomy, plant breeding
Moon, Harley W, veterinary pathology
Moon, Jay Ryong, veterinary public health
Moorman, Robert Bruce, fisheries
Muncy, Robert Jess, zoology
Muthmor, John A, insect physiology
Nariboli, Gundo A, applied mathematics
Nelson, D Kent, animal nutrition
Nevins, Donald James, plant physiology
Nielsen, Verner Henry, food science
Nordlie, Bert Edward, geochemistry, geology
Nordskog, Arne William, poultry science
Norton, Don Carlos, plant nematology
Nyvall, Robert Frederick, plant pathology
O'Berry, Phillip Aaron, veterinary microbiology
Olson, James Allen, nutritional biochemistry
Outka, Darryll E, protozoology, cell biology
Pacer, John Charles, nuclear chemistry
Packer, Raymond Allen, veterinary microbiology
Page, Leslie Andrew, veterinary microbiology, wildlife diseases
Palmer, Reid G, plant breeding, genetics
Palmquist, Robert Clarence, geomorphology, environmental geology
Parrish, Frederick Charles, Jr, animal science, biochemistry
Pattee, Peter A, microbial genetics
Peake, Edmund James, Jr, algebra
Pearce, Robert Brent, agronomy, plant physiology

Pearson, Phillip T., veterinary surgery, biomedical engineering

Pedigo, Larry Preston, entomology

Peet, Louise Jenison, entomology

Peglar, George W., mathematics

Pesek, John Thomas, Jr., soils

Peterson, Peter Andrew, genetics

Peterson, Francis Carl, elementary particle physics

Phillips, Marshall, biochemistry, organic chemistry

Pier, Allan Clark, veterinary medicine

Pierce, William Henry, agronomy, soil fertility

Pirtle, Eugene Claude, veterinary microbiology

Pohl, Richard Walter, botany

Pollak, Edward, mathematical statistics

Porter, Arthur R., dairy science

Powell, Jack Edward, physical chemistry, inorganic chemistry

Prestemon, Dean R., forest products

Quinn, Loyd Yost, bacteriology

Rahman, Mushtaq-Ur, geography, cultural anthropology

Ramsey, Frank K., veterinary pathology

Randic, Mirjana, neurophysiology

Randolph, Paul Herbert, operations research

Rebers, Paul Armand, immunochemistry

Redmond, James Ronald, comparative physiology

Richard, John Lee, mycology

Rimbey, Peter Raymond, theoretical solid state physics

Ross, Dennis Kent, astrophysics

Ross, Richard Francis, veterinary microbiology

Rougvie, Malcolm Arnold, biophysics

Rowley, Wayne A., entomology

Roy, Chalmer John, geology

Robertson, Donald Sage, genetics

Robinson, Joseph Lee, agronomy, animal science

Robson, Richard Morris, biochemistry, animal science

Ruedenberg, Klaus, theoretical chemistry, physics

Robyt, John F., biochemistry

Roderuck, Charlotte Elizabeth, nutrition

Rosauer, Elmer Augustine, materials science

Russell, Glen Allan, organic chemistry

Russell, Wilbert Ambrick, plant breeding, genetics

Runyan, Thora J., nutrition

Runyan, William Scottie, veterinary

Sadanaga, Kiyoshi, genetics

Sanderson, Donald Eugene, topology

Schafer, John William, Jr., soil science

Schaller, Frank Willard, agronomy

Schilletter, Julian Claude, horticulture

Schillinger, John Andrew, Jr., agronomy, crop breeding

Scholes, Wayne Henry, soil science

Scott, Albert Duncan, soil chemistry

Seaton, Vaughn Allen, veterinary pathology

Sebranek, Joseph George, meat sciences

Seifert, George, applied mathematics

Seifert, Karl Earl, geology

Self, Hazzle Layfette, animal science

Sendelin, Lyle V A, geological engineering, geophysics

Shaw, Kenneth C, zoology

Shaw, Robert Harold, agricultural meteorology

Shelton, Winfred Neil, psychiatry, psychology

Sherwood, Charles H, horticulture

Shibles, Richard Marwood, agronomy

Shrader, William D, soils

Sidles, Paul Howard, experimental solid state physics, solar physics

Simons, Marr Dixon, plant pathology

Sinha, Sunil K, solid state physics

Skold, Bernard Harold, veterinary anatomy, pathology

Skrdla, Willis Howard, agronomy

Smith, Frederick George, biochemistry

Smith, Paul Clay, veterinary virology

Smith, Stuart Newton, genetics, crop breeding

Snyder, Harry E, food science

Songer, Joseph Richard, biochemistry

Spedding, Frank Harold, physical chemistry

Speer, Vaughn C, nutrition, biochemistry research

Sposito, Vincent Anthony, operations research, statistics

Sprague, Richard Howard, mathematics

Stadther, Leon Gregory, bioinorganic chemistry

Stahelin, Ole Henry, veterinary microbiology

Staniforth, David William, agronomy

Steiner, Anne Kercheval, mathematics

Steiner, Eugene Francis, mathematics

Stewart, Cecil R, plant physiology

Stewart, Robert Murray, Jr., computer science

Stockdale, Harold James, economic entomology

Stone, Stanley S, biochemistry, immunochemistry

Stritzel, Joseph Andrew, agronomy

Stromer, Marvin Henry, cell biology, biochemistry

Sukhatme, Balkrishna Vasudeo, statistics

Sukhatme, Shashikala Balkrishna, mathematical statistics

Summers, Dennis Brian, plant breeding

Svec, Harry John, physical chemistry, analytical chemistry

Swenson, Clayton Albert, experimental solid state physics

Swenson, Melvin John, veterinary physiology

Switzer, William Paul, animal pathology, microbiology

Sylwester, Erhard Paul, botany

Tachibana, Hideo, plant pathology

Talbert, Willard Lindley, Jr., nuclear physics

Tauber, Oscar Earnst, physiology

Taylor, Howard Melvin, soil science

Thomas, Byron Henry, nutrition, biochemistry

Thomas, James Arthur, biochemistry

Thompson, Harvey K, agronomy

Thompson, Louis Milton, soils

Thomson, George Willis, forestry

Thurston, John Robert, bacteriology, immunology

Tiffany, Lois Hattery, plant pathology, mycology

Tipton, Carl Lee, biochemistry

Tondra, Richard John, mathematics

Topel, David G, animal physiology

Trahanovsky, Walter Samuel, organic chemistry

Trenkle, Allen H, nutrition, biochemistry

Troeh, Frederick Roy, soil science

Ulmer, Martin John, helminthology

Van Der Maaten, Martin Junior, veterinary virology

Verkade, John George, bioinorganic chemistry, organometallic chemistry

Vetter, Richard L, animal science, nutrition

Vinograde, Bernard, mathematics

Voigt, Adolf Frank, radiochemistry

Volz, Emil Conrad, horticulture

Vondra, Carl Frank, geology

Voss, Regis D, soil fertility

Wagner, William Charles, reproductive endocrinology

Walker, Homer Wayne, food microbiology

Warner, Carol Miller, biochemistry, immunobiology

Warner, Donald R, animal nutrition

Warner, Philip Mark, organic chemistry

Wass, Wallace M, veterinary medicine, veterinary surgery

Webb, John Raymond, agronomy

Wedin, Walter F, agronomy

Weigle, Jack LeRoy, plant breeding

Welshons, William John, genetics

Wessman, Garner Elmer, bacteriology

Whipp, Shannon Carl, veterinary physiology, veterinary surgery

White, Bernard J, biochemistry

Wiggers, Kenneth Dale, nutrition

Widman, William Cooper, organic biology

Widman, Ruth Bowman, aquatic biology

Wilhelm, Harley Alney, physical chemistry

Willham, Richard Lewis, animal breeding

Williams, Fred Devoe, bacteriology

Williams, Stanley A, theoretical physics

Willson, Lee Anne Mordy, astronomy

Wing, Larry Dean, wildlife ecology

Wolf, Carol Euwema, mathematical logic

Wood, Richard Lee, veterinary microbiology

Woolley, Donald Grant, agronomy, physiology

Wright, Fred Marion, mathematics

Wunder, William W, population genetics, biometrics

Yarger, Douglas Neal, atmospheric physics, meteorology

Young, Bing-Lin, high energy physics

Young, Jerry Wesley, animal nutrition

Zaffarano, Daniel Joseph, nuclear physics

Zimmerman, Dean R, animal nutrition

Zimmerman, William John, parasitology

Zmolek, William G, animal science

ANKENY

Guthrie, Wilbur Dean, entomology

BETTENDORF

Bulat, Thomas Joseph, biomedical engineering, acoustics

Hurt, James Joseph, operations research, applied mathematics

Vogel, Ronald Frank, acoustics

BURLINGTON

Vanderhoof, Ellen Ruth (Mrs Walter L Peterson), physiology

CEDAR FALLS

Allegre, Charles Frederick, biology

Anderson, Wayne I, geology

Chang, James C, physical chemistry, inorganic chemistry

Dowell, Virgil Eugene, fish biology

Downey, John Charles, entomology

Eilers, Lawrence John, plant taxonomy, phytogeography

Goss, Robert Charles, microbiology

Hamilton, Elbert W, physics

Hanson, Robert W, science education, chemistry

Hanson, Roger James, experimental nuclear physics

Kercheval, James William, organic chemistry

Lott, Fred Wilbur, Jr., mathematics, mathematical statistics

McCollum, Clifford Glenn, zoology

Olson, Dale Wilson, magnetism

Orr, Alan R, botany, cell biology

Poppy, Willard Joseph, physics

Richter, Erwin (William), biochemistry

Rider, Paul Edward, physical chemistry

Riggs, Dixon L, human physiology

Sauer, Pauline Louise, natural history, conservation

Schurrer, Augusta, mathematics

Schwartz, Ralph Jerome, speech pathology, audiology

Simpson, Robert John, animal physiology

Te Paske, Everett Russell, biology, animal behavior

Wilkinson, Jack Dale, mathematics, education

Wilson, Leland Leslie, chemistry

Wilson, Nixon Albert, acarology

CEDAR RAPIDS

Carr, Duane Tucker, chemistry

Cook, Kenneth Martin, zoology

Drexler, Robert Virgil, bryology

Goellner, Karl Eugene, vertebrate zoology

Halva, Carroll J, geochemistry, analytical

Jackobs, John Joseph, physical chemistry, x-ray crystallography

Jameson, William J, Jr., numerical mathematics, systems analysis

Kasper, Joseph Emil, cosmic ray physics

Keiser, Jeffrey E, organic chemistry

Kirby, Kenneth William, biochemistry

Lawrence, Montague Schiele, medicine, surgery

Lindsay, Charles McCown, mathematics

McManus, Margaret Ann (Mary Annunciata), biology

Rapp, Waldean G, biochemistry, organic chemistry

Watkins, Stanley Read, analytical chemistry, environmental chemistry

Weir, Donald Douglas, internal medicine

Witherell, Donald Ray, organic chemistry

CHARLES CITY

Baron, Robert Richard, parasitology

Green, Harry Edward, chemistry

Salsbury, John Greensmith, pharmaceutical chemistry

Welch, Dean Earl, organic chemistry

CLARINDA

Showalter, Donald Lee, radiochemistry

CLARION

Baker, Maurice Frank, economic zoology

CLARKSVILLE

Priepke, Rudolf Julius, physical chemistry

CLINTON

Dworschack, Robert George, industrial microbiology

Ewan, Maurice Albertson, organic chemistry

Lindsey, William B, applied mathematics

Lloyd, Norman Edward, biochemistry

Meier, John Warren, polymer chemistry

Newton, John Marshall, carbohydrate chemistry

DALLAS CENTER

Welter, C Joseph, microbiology, immunology

DAVENPORT

Brinkman, John Allen, solid state physics

Haug, Edward J, Jr., applied mechanics, applied mathematics

Masat, Robert James, comparative physiology, biochemistry

Miller, Bertrand John, physics

Paulson, Ivan Wunder, plant genetics

Rice, Carl Stephen, invertebrate zoology

Vinje, Mary M, biology

DECORAH

Barnal, Dennis E, solid state physics

Docken, Adrian (Merwin), synthetic organic chemistry

Eckblad, James Wilbur, aquatic ecology

Heins, Walden Leo, inorganic chemistry

Knudson, George E, chemistry

Knutson, Roger M, plant pathology

Miller, Emil C, physics

Moorcroft, William Herbert, psychobiology

Mottley, Carolyn, physical chemistry

Nelson, David Torrison, optics

Pilgrim, Donald, mathematics

Reitan, Phillip Jennings, zoology

Ruton, Russell Ross, physiology

Tjostem, John Leander, microbiology

Trytten, George Norman, mathematics

DENISON

Howard, Glenn Willard, Jr., molecular biology

DES MOINES

Akin, Wallace Elmus, physical geography

Atwood, Mark Wyllie, entomology

Benton, Byrl E, pharmacology, pharmacy physiology

Bordt, Dale Emil, microbiology

Brown, William Lacy, genetics, plant breeding

Canfield, Earle Lloyd, mathematics, statistics

Celander, David Robert, biochemistry, physiology

Celander, Evelyn Faun, biochemistry

Christiansen, James Learned, herpetology

Cooper, Richard Grant, pharmacology, chemistry

Coppock, William Homer, organic chemistry

Gillam, Basil Early, mathematics

Granberg, Charles Boyd, pharmacy

Heeren, Ralph Heinrich, epidemiology, public health

Huff, George Charles, biology

Isaacson, Stanley Leonard, mathematical statistics

Jacob, Fielden Emmitt, analytical chemistry, physical chemistry

Johnson, Leland Parrish, zoology

Kleiner, Alexander F, Jr., mathematics

Kodama, Robert Makoto, cell physiology

Leveque, Phillip Edwin, pharmacology

Levine, Phillip J, pharmacy

Lindberg, James George, organic chemistry

Lucas, Gene Allan, genetics

Lutz, Robert William, chemical physics, computer science

Madison, Don Harvey, atomic physics

Megitsch, Paul Allen, protozoology

Merkley, Wayne Bingham, pollution biology, marine aquatic ecology

Newcomb, Harvey Russell, microbiology

Newlin, Owen Jay, agronomy

O'Brien, Dennis Craig, geology

Orcutt, James Addison, chemistry

Oxley, Theron D, Jr., mathematics

Pipkin, George Ervin, biochemistry, nutrition

Riggs, Philip Shaefer, astronomy

Robinson, David, physics

Rogers, Frances Arlene, protozoology, vertebrate zoology
Rogers, Rodney Albert, biology
Skadron, George, physics, astrophysics
Solomon, Vasanth Balan, applied statistics
Song, Joseph, pathology
Southard, Wendell Homer, biochemistry
Stratton, Donald Brendan, physiology
Swanson, Harold Dueker, cell biology
Teppert, William Allan, Sr., pharmacology
Thomson, Duncan Maclaren, physiology
Watkins, David Hyder, thoracic surgery
Wildberger, William Campbell, medical administration, psychiatry
Wilson, Shirley Lane, plant physiology, plant pathology
Woods, Joe Darst, inorganic chemistry
Woodworth, Wayne Leon, mathematics

DUBUQUE
Bamrick, John Francis, biology, genetics
Binz, Carl Michael, analytical chemistry
Cawley, Richard T, plant ecology
Ernsdorff, Louis Edward, mathematics
Friedell, John C, pure mathematics
Guest, Mary Frances, physiology
Hart, Lawrence Alan, pure mathematics
Hoekstra, John Junior, physical chemistry
Hutchinson, Donald Raymond, nuclear physics
Kapler, Joseph Edward, biology
Kaufmann, Gerald Wayne, animal behavior, ecology
Keller, Mary Kenneth, computer science, mathematics
Kopp, Jay Patrick, physics
Kraus, Kenneth Wayne, organic chemistry
Maruyama, George Masao, chemistry
Miller, Francis Michael, astronomy
Neumann, Marguerite, organic chemistry, medicinal chemistry
Nye, Warren Edward, biology
Osuch, Carl, organic chemistry
Reuland, Robert John, radiochemistry
Rothlisberger, Hazel Marie, mathematics
Ryder, Martha, physics
Schaefer, Joseph Albert, solid state physics
Schulte, George Nicholas, chemistry
Zusy, Dennis, ecology, philosophy of science

FAIRFIELD
Domash, Lawrence Harold, theoretical physics
Kapiloff, Paul Louis, science education, physiology
Wagoner, Dale E, genetics, biology
Weinless, Michael Howard, mathematics

FAYETTE
Coleman, Richard Walter, ecology, biology
Longley, William Warren, Jr, physics
Naylor, Ernst E, botany

FT DODGE
Engelbrecht, Harlen J, veterinary medicine
Kalton, Robert Rankin, crop breeding
Sweat, Robert Lee, veterinary medicine, virology
Trace, James Chalmers, veterinary medicine

FT MADISON
Hansman, Robert Herbert, paleontology, geology
Schmidt, Reese Boise, analytical chemistry

GRINNELL
Adelberg, Arnold M, geometry, algebra
Andelson, Jonathan Gary, ethnology, physical anthropology
Bays, James Philip, bio-organic chemistry, organic chemistry
Christiansen, Kenneth Allen, evolutionary biology, speleology
Clotfelter, Beryl Edward, physics
Danforth, Joseph Davis, physical chemistry
Delong, Karl Thomas, ecology
De Long, Sharon Koepcke, mycology
Denbo, John Russell, physiology
Denny, Wayne Belding, physiology
Duke, Charles Lewis, nuclear physics
Durkee, La Verne H, plant taxonomy
Erickson, Luther E, physical chemistry

Fishman, Irving Yale, sensory physiology
Graham, Benjamin Franklin, Jr, botany, plant ecology
Herman, Eugene Alexander, mathematics
Kurtz, Ronald Joseph, ethnology
Luebben, Ralph A, anthropology
Martinek, John Joel, anatomy
Mendoza, Guillermo, zoology
Oelke, William C, physical chemistry
Walker, Waldo Sylvester, botany
Wubbels, Gene Gerald, chemistry

HAMPTON
Moore, John V, animal breeding

HUDSON
Smith, Keith James, animal nutrition

INDIANOLA
Alberding, Herbert, earth sciences
Considine, Judith Mayberry, biological chemistry
De Lisle, Donald Gordon, botany
Meints, Clifford Leroy, biochemistry
Myers, Clovis D, textile chemistry
Watson, Margaret Liebe, genetics

IOWA CITY
Abadi, Djahanguir M, clinical chemistry
Abboud, Francois Mitry, internal medicine
Al-Awqati, Qais, physiology, biophysics
Alton, Donald Alvin, computer sciences, mathematical logic
Ambre, John Joseph, clinical pharmacology
Anderson, Charles V, audiology
Anderson, Richard Lee, ophthalmological surgery
Anderson, Thomas Alexander, biochemistry, nutrition
Appleby, Ralph Carson, dentistry
Arnone, Arthur Richard, biochemistry
Atkinson, Kendall E, mathematics
Autor, Anne Pomeroy, biochemistry
Avcin, Matthew John, Jr, economic geology, stratigraphy
Baenziger, Norman Charles, physical chemistry
Baker, Richard Graves, palynology, quaternary geology
Beams, Harold William, zoology
Bedell, George Noble, internal medicine
Bell, William E, pediatric neurology
Bennett, William Earl, inorganic chemistry
Benton, Arthur Lester, neuropsychology
Berg, Clarence Peter, biochemistry
Bhatnagar, Ranbir Krishna, pharmacology
Bickley, Harmon C, pathology, dentistry
Bighley, Lyle Delevan, physical pharmacy
Bishara, Samir Edward, orthodontics
Bjorndal, Arne Magne, dentistry
Blakley, Raymond L, biochemistry
Blodi, Frederick Christopher, medicine
Bonfiglio, Michael, biology, orthopedic surgery
Bovbjerg, Richard Viggo, animal ecology
Bradbury, James Thomas, endocrinology
Brody, Michael J, pharmacology
Buchanan, Edward Bracy, Jr, analytical chemistry
Buckles, Robert Edwin, physical organic chemistry
Bull, Henry Bolivar, physical biochemistry, protein chemistry
Bunge, Raymond George, urology
Burton, Donald Joseph, organic chemistry
Butler, John Edward, immunology
Cadoret, Remi Jere, psychiatry
Cain, George D, parasitology
Camillo, Victor Peter, pure mathematics
Cannon, Joseph G, medicinal chemistry
Caplan, Richard Melvin, dermatology
Carew, David P, pharmacy
Carlson, Richard Raymond, nuclear physics
Carlson, Wayne R, cytogenetics
Carman, John Homer, petrology, geochemistry
Carney, Robert Gibson, dermatology
Carpenter, Raymond T, nuclear physics
Cater, Earle David, physical chemistry
Cazin, John, Jr, microbiology, medical mycology
Chalkley, G Roger, biochemistry
Chan, Kai Chiu, dentistry
Chaudhuri, Tuhin, nuclear medicine, hematology
Cheng, Frank Hsieh Fu, biochemistry, immunochemistry

Chipman, Daniel Myron, quantum chemistry
Christensen, James, physiology, gastroenterology
Clancy, John, psychiatry
Clark, Kenneth Frederick, economic geology
Clifton, James Albert, internal medicine
Coffman, Robert Edgar, chemical physics
Comly, Hunter Hall, psychiatry
Conway, Thomas William, biochemistry
Cooke, Helen Joan, developmental physiology
Cooper, Reginald Rudyard, orthopedic surgery
Cortney, Marshall Allen, physiology
Coucouvanis, Dimitri N, inorganic chemistry, crystallography
Crouch, Norman Albert, virology
Cruden, Robert William, ecology, evolution
Cryer, Jonathan D, statistics
Culo, David Albert, medicine
Davis, Leodis, biochemistry
Dawson, David Charles, physiology
Dean, Henry Lee, botany
DeGowin, Elmer Louis, clinical medicine
DeGowin, Richard Louis, internal medicine
Diamond, James Gary, ophthalmology
Diana, John N, physiology
Dingle, Richard Douglas Hugh, animal behavior, ecology
Dolch, John Parker, computer science
Donelson, John Everett, biochemistry
Donta, Sam Theodore, infectious diseases
Dorheim, Fredrick Houge, economic geology
Doyle, John Robert, organometallic chemistry
Drake, Lon David, geology
Dryer, Robert Leonard, biochemistry
Dueker, Kenneth John, urban geography, regional planning
Duke, Frederick Robert, physical chemistry
Eckhardt, Richard Dale, internal medicine
Eckstein, John William, internal medicine
Ehrenhaft, Johann L, surgery
Embree, Robert William, botany
Eyman, Darrell Paul, inorganic chemistry
Falsetti, Herman Leo, cardiology
Feldbush, Thomas Lee, immunobiology
Fellows, Robert Ellis, Jr, endocrinology
Filer, Lloyd Jackson, Jr, pediatrics, nutrition
Fischer, Lawrence J, biochemical pharmacology, toxicology
Fix, John Dekle, astrophysics
Flatt, Adrian Ede, surgery
Fleck, Arthur C, computer science
Folk, George Edgar, Jr, physiology
Fomon, Samuel Joseph, pediatrics, nutrition
Forker, Edson Lee, physiology, internal medicine
Fox, Stephen Sorin, physiological psychology
Frank, Louis Albert, physics, astronomy
Frankel, Joseph, developmental genetics
Friedrich, Bruce H, physical chemistry
Fuller, Kent Ralph, mathematics
Funk, David Crozier, cardiology
Furnish, William Madison, Jr, geology
Geraghty, Michael A, mathematics
Gisolfi, Carl Vincent, physiology
Gleich, Carol Sue, medical technology
Glenister, Brian Frederick, paleobiology
Goldberg, Richard Robinson, mathematics
Goplerud, Clifford P, obstetrics & gynecology
Graf, Carl John, neurology
Grant, Stanley Cameron, geology, science administration
Grigsby, William Redman, dentistry, biochemistry
Guillory, James Keith, pharmacy
Gussin, Gary Nathaniel, molecular biology
Hahne, Rolf Mathieu August, environmental management, physical chemistry
Halmi, Katherine A, psychiatry
Halmi, Nicholas Stephen, endocrinology
Hamilton, Henry Edward, clinical medicine
Hammond, Harold Logan, oral pathology
Hardin, Robert Calvin, medicine
Hardy, James C, speech pathology
Hartford, Charles Edward, surgery
Harvey, John Adriance, psychopharmacology
Hausler, William John, Jr, public health administration
Hayden, Jess, Jr, anatomy, dentistry
Heckel, Philip Henry, geology

Hegmann, Joseph Paul, quantitative genetics
Heidger, Paul McClay, Jr, anatomy
Heistad, Donald Dean, cardiovascular diseases, cardiovascular physiology
Helm, June, ethnology, anthropology
Henn, Fritz Albert, neuropsychiatry, neurochemistry
Hermsmeyer, Ralph Kent, physiology
Hershey, Howard Garland, geology
Hershkowitz, Noah, physics
Higa, Leslie Hideyasu, oral pathology
Hoak, John C, internal medicine
Hochstrasser, Donald Lee, medical anthropology, community health
Hoffmann, Louis Gerhard, immunology
Hogben, Charles Adrian Michael, physiology
Hogenkamp, Henricus Petrus C, biochemistry
Hogg, Robert Vincent, Jr, statistics
Hoppin, Richard Arthur, geology
Hsu, Hsi Fan, parasitology, immunology
Hsu, Shu Ying Li, parasitology, immunology
Hubel, Kenneth Andrew, medicine, physiology
Hug, Daniel Hartz, physiological bacteriology, photobiology
Hulbary, Robert Louis, botany
Imig, Charles Joseph, physiology
Ingram, Forrest Duane, nuclear physics, biophysics
Ingram, Walter Robinson, neurobiology
Jackson, Herbert Lewis, nuclear physics
Jacobs, Richard M, orthodontics, anatomy
January, Lewis Edward, cardiology
Jenkins, Richard Leos, psychiatry
Johnson, Alan Kim, behavioral biology
Johnson, Eugene W, algebra
Johnson, Leslie Kilham, zoology
Johnson, Norman L, geometry
Johnson, Wallace W, pharmacology, dentistry
Johnson, William, microbiology
Joyce, Glenn Russell, plasma physics
Kaelber, William Walbridge, neurology
Kalnitsky, George, biochemistry
Karlsson, Ulf Lennart, anatomy, neurobiology
Kater, Stanley B, neurophysiology
Keettel, William Charles, obstetrics & gynecology
Keim, Lon William, pulmonary disease
Kennedy, Robert Alan, plant physiology
Kent, Thomas Hugh, pathology
Kerber, Richard E, cardiovascular diseases
Kerr, Wendle Louis, pharmacy
Kessel, Richard Glen, zoology, anatomy
Khurana, Surjit Singh, mathematics
Kirk, William Arthur, mathematics
Klapper, Gilbert, paleontology
Kleinfeld, Erwin, algebra
Kleinfeld, Margaret Humm, mathematics
Klink, William H, theoretical physics
Knorr, George E, plasma physics
Knowler, Lloyd A, actuarial science, statistics
Koch, Donald Leroy, stratigraphy, groundwater geology
Koepke, John Arthur, pathology, hematology
Kohn, Clyde Frederick, geography
Kolder, Hansjoerg E, ophthalmology, physiology
Kollros, Jerry John, embryology, cell biology
Koontz, Frank P, microbiology
Kos, Clair Michael, otology
Kosier, Frank J, mathematics
Kremenak, Charles Robert, Jr, maxillofacial prosthetics, prosthodontics
Kuiper, Logan Keith, groundwater hydrology, geophysics
Kuo, Chao-Ying, immunochemistry
Kutzko, Philip C, number theory
Lach, John Louis, physical pharmacy
Lambert, Howard W, topology
Lara-Braud, Carolyn Weathersbee, biochemistry
Larson, Carroll Bernard, orthopedic surgery
Lata, Gene Frederick, biochemistry
Latourette, Howard Bennett, radiology
Lawton, Richard L, surgery
Lediaev, John P, mathematics
Lee, Agnes C, biochemistry
Leinfelder, Placidus Joseph, ophthalmology
Lewis, Lewis James, virology, bacteriology
Lilly, David J, audiology
Lin, Bor-Luh, mathematics
Lindberg, James Beckwith, geography
Long, John Paul, pharmacology
Long, Keith Royce, environmental health

461

IOWA

Lorkovic, Hrvoje Radoslav, muscular physiology
Lucas, James Robert, geology
MacDonald, Kenneth, hygiene, preventive medicine
Marcus, Melvin L., cardiology
Markovetz, Allen John, microbiology, biochemistry
Mason, Edward Eaton, surgery
McCabe, Brian Francis, otolaryngology
McCabe, Michael S., psychiatry
McCormick, George R., mineralogy, petrology
McCrone, John David, zoology, administrative sciences
McKusick, Marshall Bassford, anthropology, archaeology
McLean, James Herbert, dentistry, oral surgery
McNulty, Michael Leigh, geography
Menninger, John Robert, molecular biology
Milkman, Roger Dawson, genetics
Mohler, James Dawson, zoology
Moll, Kenneth Leon, speech pathology
Montgomery, David Campbell, plasma physics, statistical mechanics
Montgomery, Rex, biochemistry
Morris, Hughlett Lewis, speech pathology
Morris, Robert Lyle, chemistry
Moyers, Jack, anesthesiology
Muir, Robert Mathew, plant physiology
Nair, Vasu, organic chemistry, bio-organic chemistry
Neff, John S., astronomy, astrophysics
Nelson, Edward Bryant, physics
Nelson, Herbert Leroy, psychiatry
Nicholson, Donald Paul, microbiology, virology
Nolf, Luther Owen, parasitology
Norbeck, Edwin, Jr., nuclear physics
Nowak, Arthur John, pedodontics
Noyes, Russell, Jr., psychiatry
Nuki, Klaus, histochemistry, periodontology
Oaks, John Adams, cell biology, parasitology
Oberg, Edwin Nathaniel, mathematics
O'Connor, Michael L., pathology
Oehnke, Robert H., mathematics
Olin, William (Harold), orthodontics
Oliver, Denis Richard, biochemistry
Osborn, Margaret Olive, nutrition
Osborne, James William, physiology, radiobiology
Osman, Elizabeth Mary, organic chemistry
Parkins, Frederick Milton, pedodontics
Parrish, Dale Wayne, medical physiology
Peterson, Richard Elsworth, internal medicine, nuclear medicine
Pfluum, Ronald Trenda, analytical chemistry
Phelps, Charles Dexter, ophthalmology
Pietryk, Donald John, analytical chemistry
Pitkin, Roy Macbeth, obstetrics & gynecology
Plapp, Bryce Vernon, biochemistry
Platt, William Joshua, III, ecology, vertebrate biology
Ponseti, Ignacio Vives, medicine
Porter, John Roger, microbiology
Powell, Robin Dale, medicine
Price, Thomas Munro, topology
Radcliffe, Christian Elmore, dermatology
Ramsey, James Carroll, cell biology, molecular biology
Randels, Ronald Herman, statistics
Read, Charles H., pediatrics, endocrinology
Rembolt, Raymond Ralph, pediatrics
Reynolds, David Reid, geography
Richardson, Hal Bates, allergy, immunology
Riley, Edgar Francis, radiobiology
Rim, Kwan, theoretical mechanics, applied mechanics
Robertson, Timothy Joel, mathematical statistics
Rodriguez, Jose Enrique, virology
Rosazza, John Paul, pharmacognosy, bio-organic chemistry
Rose, Earl Forrest, pathology
Roskoski, Robert, Jr., biochemistry
Rossi, Nicholas Peter, cardiovascular surgery

Routh, Joseph Isaac, clinical biochemistry, pathology
Rushton, Gerard, economic geography, urban geography
Sabiston, Charles Barker, Jr., microbiology, dentistry
Salisbury, Neil Elliot, geography
Sando, Kenneth Martin, theoretical chemistry
Savage, William Ralph, physics
Schedl, Harold Paul, medicine
Schochet, Sydney Sigfried, Jr., pathology, neuropathology
Schotelius, Byron Arthur, physiology
Schotelius, Dorothy Dickey, biochemistry
Schweitzer, John William, solid state physics
Scott, James Raymond, obstetrics & gynecology, reproductive biology
Searle, Gordon Wentworth, physiology
Searls, James Collier, anatomy
Seebohm, Paul Minor, medicine
Semken, Holmes Alford, Jr., quaternary geology, paleozoology
Sheets, Raymond Franklin, internal medicine
Sherman, Dorothy Helen, speech pathology, statistics
Shih, Ching-Yuan, plant physiology, crop breeding
Shires, Thomas Kay, cell biology
Shuler, Richard Jr., anthropology
Six, Erich Walther, biophysics
Sjolund, Richard David, botany
Small, Arnold McCollum, Jr., psychoacoustics
Smith, Fred George, Jr., pediatrics, nephrology
Smith, Ian Maclean, infectious diseases, internal medicine
Smith, Jan D., pulmonary diseases, anesthesiology
Snider, Bill Carl F., statistics
Snyder, Irvin S., medical microbiology
Soil, David Richard, developmental biology
Solomons, Gerald, pediatrics, child growth
Solursh, Michael, zoology, developmental biology
Sooby, Donna Louise, radiation biology
Soper, Robert Tunnicliff, medicine
Spaziani, Eugene, endocrinology, reproductive physiology
Spector, Arthur Abraham, biochemistry, internal medicine
Spratt, James Leo, pharmacology
Stahly, Donald Paul, microbiology
Staley, Robert Newton, orthodontics, physical anthropology
Stamler, Frederic Leroy, pathology
Stauber, William Taliaferro, physiology
Stay, Barbara, insect morphology
Steele, William John, biochemical pharmacology
Stegink, Lewis D., biological chemistry
Stellwagen, Earle C., biochemistry
Stewart, Mark Armstrong, psychiatry
Stille, John Kenneth, organic chemistry, polymer chemistry
Strimple, Harrell Leroy, invertebrate paleontology, stratigraphy
Stwalley, William Calvin, physical chemistry, atomic physics
Summers, Robert Wendell, gastroenterology
Sunshine, Melvin Gilbert, bacterial genetics
Swenson, Charles Allyn, physical chemistry
Swett, Keene, geology
Tade, William Howard, pathology, physiology
Tephly, Thomas R., pharmacology
Tester, Allen Crawford, geology
Thayer, Keith Evans, dentistry
Theilen, Ernest Otto, internal medicine
Thompson, Herbert Stanley, ophthalmology, neurology
Thompson, John S., internal medicine, immunology
Thompson, John David, physiology
Tipton, Charles M., physiology
Tsuang, Ming Tso, psychiatry
Tuthil, Samuel James, geology, paleoecology
Tuttle, Sherwood Dodge, geomorphology
Van Allen, James Alfred, physics
Van Allen, Maurice Wright, neurology
Van Demark, Duane R., speech pathology
Van Orden, Lucas Schuyler, III, neurobiology, psychopharmacology
Vesting, Carl Swensson, biochemistry
Voots, Richard Joseph, acoustics
Waltman, Paul E., mathematical analysis, biomathematics

Warner, Emory Dean, pathology
Watzke, Robert Coit, ophthalmology
Wawzonek, Stanley, organic chemistry
Weeg, Gerard Paul, computer science
Wei, Stephen Hon Yin, pedodontics, histology
Wernick, Joel, audiology, speech pathology
Whitehead, Floy Eugenia, nutrition
Williams, Dean E., speech pathology, audiology
Williams, Norman Eugene, zoology
Williams, Terence Heaton, neuroanatomy, electron microscopy
Williamson, Harold Emanuel, pharmacology
Wilson, Elizabeth, medical microbiology
Winokur, George, psychiatry
Witte, David L., biochemistry, pathology
Wong, James Sai-Wing, mathematical analysis
Wright, Farroll Tim, statistics, mathematics
Wunder, Charles Cooper, physiology, biophysics
Wurster, Dale E., pharmacy
Yager, Robert Eugene, science education, plant physiology
Yoder, John L., dentistry
Yoo, Tai-June, immunology, internal medicine
Zellweger, Hans Ulrich, pediatrics
Ziffren, Sidney Edward, surgery

JOHNSTON
Arthur, James Alan, poultry breeding
Duvick, Donald Nelson, genetics, plant breeding
Frey, Nicholas Martin, crop physiology
Hawes, Robert Oscar, poultry breeding
Wilkinson, Daniel R., plant pathology, plant breeding

KNOXVILLE
Ralston, Furman Paul, Jr., botany, ecology

LAMONI
Graybill, Bruce Myron, physical organic chemistry
Jacobson, Ned LeRoy, symbolic logic

LE MARS
Cram, Sheldon Lewis, physics
Divelbiss, James Edward, zoology
Franklin, Robert Louis, biochemistry, organic chemistry
Kells, Lyman Francis, chemistry
Marty, Wayne George, parasitology
Rebstock, Theodore Lynn, plant biochemistry
Ulrich, Mervyn Gene, zoology

MARION
Bywaters, James Humphreys, genetics

MARSHALLTOWN
Harbaugh, Daniel David, animal nutrition
McInroy, Elmer Eastwood, agricultural chemistry

MT PLEASANT
Graf, Dolores Irma, ecology, plant morphology
Rila, Charles Clinton, inorganic chemistry, organic chemistry

MT VERNON
Ault, Addison, organic chemistry
Bailey, Donald Forest, mathematics
Barthel, William Frederick, agricultural chemistry
Christiansen, Paul Arthur, botany, plant ecology
Dam, Cecil Frederick, physics
Davis, Watson M., mathematics
Deskin, William Arna, inorganic chemistry
Garvin, Paul Lawrence, mineralogy
Graber, Harlan Duane, nuclear physics
Hendriks, Herbert Edward, geology
Hill, Edward T., mathematics
Hinman, Eugene George, paleontology
Jacob, Richard L., theoretical physics
Jordan, Truman H., physical chemistry
Meyer, John Sigmund, statistics
Pray, Francis Adams, zoology
Rogers, Thomas Edwin, zoology

MUSCATINE
Carlson, Emil Herbert, organic chemistry
Garbutt, John Thomas, biochemistry
Gardner, Charles H., microbiology
Hanson, Austin Moe, bacteriology
Kiser, Donald Lee, analytical chemistry
Lewis, Charles J., animal science
Maizann, Ronald C., bacteriology
Mente, Glen Allen, animal nutrition
Morehouse, Alpha L., biochemistry
Rohwer, Robert G., microbiology
Smith, Claire Leroy, microbiology
Smith, Daniel Newton, Jr., poultry nutrition
Zorn, Ralph Allan, industrial microbiology, analytical chemistry

OAKDALE
Berry, Clyde Marvin, industrial hygiene, chemical engineering
Kasik, John Edward, medicine, pharmacology
Llinas, Rodolfo, neurobiology, electrophysiology
Morgan, Donald Pryse, preventive medicine
Nicholson, Charles (Godfrey), neurobiology
Top, Franklin Henry, Sr., preventive medicine, epidemiology

ORANGE CITY
Hammerstrom, Harold Elmore, analytical chemistry, inorganic chemistry
Van Eck, Edward Arthur, microbiology

OSKALOOSA
Gygi, Francis Richard, bacteriology
Stoltzfus, William Bryan, entomology

PELLA
Bosch, Arthur James, biochemistry
Bowles, John Bedell, zoology
Byers, Ronald Elner, theoretical physics
Crichton, David, analytical chemistry
Graber, Leland D., mathematics
Huffman, Donald Marion, mycology
Kolenbrander, Harold Mark, metabolism, enzymology
Lau, Kenneth Kwok-Kwan, mathematics
Moen, Allen LeRoy, surface physics
Ogren, Paul Joseph, physical chemistry
Wilson, David Everett, cell physiology

SERGEANT BLUFF
Homan, Joseph M., inorganic chemistry, analytical chemistry

SHENANDOAH
Anderzhon, Mamie Louise, geography

SIOUX CENTER
Alberda, Willis John, mathematical statistics
De Young, Marvin, physics
Geels, Edwin James, organic chemistry
Maatman, Russell Wayne, physical chemistry
Mennega, Aalder, anatomy, physiology

SIOUX CITY
Dible, William Trotter, Jr., chemistry
Green, Robert Wood, cryogenics, optics
Johnson, Robert P., mathematics
Muller, Elsie, mathematics
Privett, James E., organic chemistry
Rundell, Harold Lee, zoology
Webb, Morgan Chofield, III, entomology

SOLON
Brown, George Wallace, neurophysiology

SPENCER
Holck, Gary LeRoy, animal nutrition

STORM LAKE
Borgman, Robert P., plant ecology
Christiansen, James Brackney, biological chemistry
Mayeux, Jerry Vincent, research administration
Moenter, Richard Luther, mathematics

WASHINGTON
Bauer, Clifford David, biochemistry
Thorne, John Carl, plant breeding, genetics

WAVERLY
Azbell, William, physics
Bridgman, George Henry, mathematics
Eiben, Galen J, entomology
Hampton, David Clark, organic chemistry, inorganic chemistry
Hertel, Elmer William, zoology, genetics
Main, Stephen Paul, botany, aquatic ecology
Petri, Leo Henry, parasitology
Waltmann, William Lee, mathematics
Zemke, Warren T, physical chemistry, quantum chemistry

WEST BRANCH
Heins, Conrad F, organic polymer chemistry

WILLIAMSBURG
Reynolds, James Edward, plant pathology

KANSAS

ATCHISON
Bassi, Sukh D, developmental genetics
Baumgartner, George Julius, organic chemistry
Brom, Joseph March, Jr, physical chemistry, molecular spectroscopy
Brothers, Alfred Douglas, solid state physics
Dehner, Eugene William, zoology
Senecal, Gerard, physics

BALDWIN CITY
Boyd, Ivan Louis, biology
Foreman, Calvin, mathematics
Nelson, Victor Eugene, zoology
White, R Milford, physical chemistry

CUNNINGHAM
Allbritten, Frank F, Jr, surgery

DODGE CITY
Paulie, M Catherine Therese, mathematics, statistics
Rodrick, Gary Eugene, parasitology, invertebrate physiology

EMPORIA
Boles, Robert Joe, aquatic biology
Breukelman, John (William), ecology
Bridge, Thomas E, geology
Bruyr, Donald Lee, mathematics
Burger, John Martin, experimental nuclear physics
Carlson, John W, mathematics
Clarke, Robert Francis, zoology, animal behavior
Cram, Stanford Winston, physics
Creager, Charles Bicknell, nuclear physics
Durst, Harold Everett, environmental biology
Eddy, Thomas A, entomology, wildlife management
Emerson, Marion Preston, algebra
Ericson, Alfred (Theodore), biochemistry
Gimple, Glenn Edward, inorganic chemistry
Greenlief, Charles M, surface chemistry
Hill, James Wagy, organic chemistry
Homman, Guy Burger, science education, inorganic chemistry
Keeling, Richard Paire, mycology
Le Fever, Hermon Michael, genetics, electron microscopy
Leisman, Gilbert Arthur, botany
McElree, Helen, immunobiology
Neufeld, Gaylen Jay, cell physiology
Peterson, John Edward, Jr, mycology, microbiology
Poole, George Douglas, mathematics
Prophet, Carl Wright, aquatic ecology
Purcell, Terry, solid state physics
Rowe, Edward C, neurophysiology
Smalley, Robert Lee, biochemistry
Spencer, Dwight Louis, animal ecology
Ulrich, Arlene Louise, microbiology
Wilson, James Stewart, botany

Witten, Gerald Lee, science education

GARDEN CITY
Herron, George M, soil science
Kyle, Jack Hiram, horticulture
Norwood, Charles Arthur, agronomy
Stone, Jay D, entomology

GREAT BEND
Hausler, Rudolf H, corrosion, petroleum chemistry

HAYS
Beougher, Elton Earl, mathematics
Choate, Jerry Ronald, mammalogy, evolutionary biology
Choguill, Harold Samuel, chemistry
Dressler, Robert Louis, organic chemistry
Eltze, Ervin Marvin, mathematical analysis
Ely, Charles Adelbert, vertebrate zoology
Fleharty, Eugene, mammalian ecology
Harris, Wallace Wayne, agronomy, soil fertility
Harvey, Thomas Larkin, entomology
Launchbaugh, John L, Jr, plant ecology
Livers, Ronald Wilson, plant breeding
Marshall, Delbert Allan, analytical chemistry
Martin, Terry Joe, plant pathology
Pierson, David W, science education, conservation
Pruitt, Roger Arthur, physics
Rice, Jimmy Marshall, mathematics
Rumpel, Max Leonard, inorganic chemistry
Shearer, Edmund Cook, physical chemistry
Toalson, Wilmont, applied mathematics
Tomanek, Gerald Wayne, botany
Votaw, Charles Isac, mathematics
Walker, Neil Allan, acarology
Witten, Maurice Haden, physics
Zakrzewski, Richard Jerome, vertebrate paleontology

HILLSBORO
Johnson, William Jacob, inorganic chemistry
Terman, Max R, ecology, ethology

HUGOTON
Wagoner, John Allen, carbohydrate chemistry

HUTCHINSON
Henry, Wayne E, food science, animal husbandry

KANSAS CITY
Abdou, Nabih I, medicine, immunology
Allen, Max Scott, internal medicine
Amelunxen, Remi Edward, microbiology
Arnold, Richard Lane, poultry nutrition
Arnold, Wilfred Niels, biochemistry
Azarnoff, Daniel Lester, clinical pharmacology
Barth, Rolf Frederick, pathology, immunology
Behbehani, Abbas M, virology
Bolinger, Robert E, medicine
Brown, Ernest Benton, Jr, medical administration
Brown, Robert Wayne, internal medicine
Bunag, Ruben David, cardiovascular physiology, pharmacology
Carr, Daniel Oscar, biochemistry
Chang, Chae Han Joseph, radiology
Chapman, Albert Lee, anatomy
Chiga, Masahiro, pathology
Chin, Tom Doon Yuen, epidemiology
Clancy, Richard L, physiology
Couch, James Russell, Jr, neurology, neuropharmacology
Cuppage, Francis Edward, pathology
Danzer, Laurence Alfred, clinical chemistry, physical chemistry
Delp, Mahlon (Henry), medical medicine
Diehl, Antoni Mills, pediatrics, cardiology
Doull, John, pharmacology
Duffey, Margery, medical education
Dujovne, Carlos A, clinical pharmacology, internal medicine
Dunn, Marvin I, internal medicine, cardiology
Ebner, Kurt E, biochemistry
Fainstat, Theodore, obstetrics & gynecology
Foltz, Floyd Mathew, anatomy
Frakes, Elizabeth (McCune), nutrition

Frenkel, Jacob Karl, pathology
Friesen, Stanley Richard, surgery
Fullmer, Curtis Sheridan, physical biology, biochemistry
Furtado, Dolores, medical microbiology
Glick, John Henry, Jr, clinical chemistry
Goetzinger, Cornelius Peter, audiology, psychology
Goldberg, Ivan D, microbial genetics
Goldstein, Albert, medical physics
Gonzalez, Norberto Carlos, medical physiology
Grady, Harold James, clinical biochemistry
Grantham, Jared James, medicine, nephrology
Greaves, Donald Critchfield, psychiatry
Greenberger, Norton Jerald, internal medicine, physiology
Greene, Frank T, chemical physics
Greenwald, Gilbert Saul, anatomy
Grisolia, Santiago, biochemistry
Haas, Herbert Frank, microbiology
Hardin, Creighton A, surgery
Harris, Lewis Philip, biochemistry
Heuschele, Werner Paul, veterinary virology, immunology
Hoff, Richard Lee, immunology, preventive medicine
Hollowell, Joseph Gurney, Jr, pediatric endocrinology, genetics
Hudson, Billy Gerald, biochemistry
Huffman, David H, clinical pharmacology
Hung, Kuen-Shan, anatomy, histology
Jerome, Norge Winifred, nutrition, public health
Jewell, William R, surgery
Johnson, Donald Charles, endocrinology
Kepes, John J, pathology, neuropathology
Kimmel, Joe Robert, biochemistry, medicine
Kinsey, John Aaron, Jr, genetics
Klaassen, Curtis Dean, pharmacology, toxicology
Klotz, Arthur Paul, clinical medicine
Krantz, Kermit Edward, medicine
Langston, Clarence Walter, microbiology
Laybourne, Paul C, psychiatry
Lemoine, Albert N, Jr, ophthalmology
Lindsey, Norma J, clinical microbiology
Liu, Chien, infectious diseases, virology
Loofbourrow, Guy Norman, physiology
Masters, Frank Wynne, plastic surgery
McMillen, Janis Kay, virology
Meek, Joseph Chester, Jr, internal medicine, endocrinology
Menninger, Karl Augustus, psychiatry
Miller, Herbert Chauncey, pediatrics
Milne, Thomas Anderson, physical chemistry
Mohn, Melvin P, anatomy, histology
Molteni, Agostino, experimental pathology
Mulford, Dwight James, biochemistry
Murdock, Archie Lee, biochemistry
Nelson, Stanley Reid, pharmacology, neurosurgery
Noelken, Milton Edward, physical chemistry
Norton, Stata Elaine, pharmacology, toxicology
Nurge, Ethel, cultural anthropology
Parmley, Ray T, anesthesiology
Poisner, Alan Mark, pharmacology, cell physiology
Poorman, Douglas Harold, anatomy
Poretz, Ronald David, biochemistry, immunochemistry
Porter, Chastain Kendall, dentistry, oral pathology
Pottinger, M Aelred, biology
Proud, G O'Neil, otolaryngology
Pugh, David Milton, internal medicine, cardiology
Rawitch, Allen Barry, biochemistry, spectroscopy
Reddy, Janardan K, pathology
Redford, John W B, rehabilitation medicine
Reissmann, Kurt Rudolph, medicine, physiology
Rhodes, James B, gastroenterology
Rieke, William Oliver, medical administration, anatomy
Riley, Richard Charles, radiation biophysics
Ringle, Stata Norton, pharmacology
Rising, Jesse David, medicine, pharmacology
Robinson, David Weaver, surgery
Rockwell, Wayne L, obstetrics & gynecology
Rose, Donald L, medicine, physical medicine
Rothrock, Irvin Andrew, psychiatry

Samson, Frederick Eugene, Jr, physiology, neurobiology
Scarpelli, Dante Giovanni, experimental pathology
Schloerb, Paul Richard, surgery
Shaad, Dorothy Jean, ophthalmology, psychology
Shadbolt, C Allan, horticulture, plant physiology
Sheek, Martha Reyburn, virology, biochemistry
Shellenberger, Melvin Kent, neuropharmacology
Shoeman, Don Walter, pharmacology
Silverstein, Richard, biochemistry
Smith, Kenneth E, audiology
Sullivan, Lawrence Paul, physiology
Suzuki, Tsuneo, microbiology, biochemistry
Tansey, Robert Paul, Sr, pharmacy
Thal, Alan Philip, surgery, experimental medicine
Thompson, Alan Morley, physiology
Trank, John W, physiology, electrical engineering
Uyeki, Edwin M, pharmacology, radiobiology
Valk, William Lowell, surgery
Walaszek, Edward Joseph, pharmacology
Welling, Daniel J, theoretical physics, biophysics
Werder, Alvar Arvid, microbiology
Wilson, Sloan Jacob, internal medicine
Wolkoff, A Stark, obstetrics & gynecology, physiology
Wong, Kin-Ping, biochemistry, biophysical chemistry
Wood, Jackie Dale, physiology, zoology
Youngstom, Karl Arden, radiology, anatomy
Ziegler, Dewey Kiper, neurology

LAWRENCE
Adams, Ralph Norman, analytical chemistry
Adams, Robert D, mathematics
Ahshapanek, Don Colesto, plant ecology
Akagi, James Masuji, microbial physiology
Ammar, Raymond George, physics
Angino, Ernest Edward, geochemistry
Argersinger, William John, Jr, physical chemistry
Armitage, Kenneth Barclay, ecology
Armstrong, Thomas Peyton, space physics
Aronszajn, Nachman, mathematics
Ashlock, Peter Dunning, taxonomic entomology
Augelli, John Pat, cultural geography, geography of Latin America
Balfour, William Mayo, physiology
Baptist, Jeremy Eduard, biophysics
Barkhurst, Rodney Charles, organic chemistry
Baxter, Robert Wilson, paleobotany
Beach, Joseph Lawrence, radiological physics
Beard, David Breed, theoretical physics
Bearse, Robert Carleton, applied physics
Beer, Robert Edward, entomology
Bickford, Marion Eugene, Jr, petrology, geochemistry
Bodor, Nicolae Stefan, medicinal chemistry
Boguslawski, George, genetics, biochemistry
Borchert, Rolf, plant physiology
Bradt, Russell Newton, mathematical statistics
Brady, Lawrence Lee, geology, economic geology
Brewer, James W, algebra
Bricker, Clark Eugene, chemistry
Brown, Robert Dillon, mathematics
Buller, Clarence S, microbiology
Burchill, Brower Rene, cell biology
Burgstahler, Albert William, organic chemistry
Burton, Paul Ray, zoology, cell biology
Bussell, Robert Harry, virology
Byers, George William, entomology
Camin, Joseph Harvey, zoology
Carlson, Robert Gideon, organic chemistry
Christoffersen, Ralph Earl, physical chemistry
Coil, William Herschell, parasitology
Conrad, Paul, mathematics
Crawford, Michael H, anthropology
Creese, Thomas Morton, mathematics
Culhouse, Jack Wayne, zoology
Cunningham, Robert Stephen, mathematics
Dahl, Nancy Ann, neurobiology
Davidson, John Pirnie, physics

Davis, John Clements, geology
Davis, Robin Eden Pierre, high energy physics
Deddens, James Albert, mathematics
Detlwig, Louis Field, regional geology, remote sensing
Dienes, Leslie Dennis, geography
Dorr, Wakefield, Jr., quaternary geology, geomorphology
Draper, Laurence Rene, immunology
Duellman, William Edward, herpetology
Eagleman, Joe R., meteorology, soil physics
Eaton, Theodore Hildreth, Jr., zoology
Enoch, Jacob, plasma physics
Erickson, Carlton Kuehl, pharmacology
Faiman, Morris David, pharmacology
Fitch, Henry Sheldon, ecology
Foley, Frank Clingan, groundwater geology
Fox, J Eugene, plant physiology, plant biochemistry
Friauf, Robert James, solid state physics
Friesen, Benjamin S., plant biology
Gaines, Michael Stephen, population biology
Gilles, Paul Wilson, physical chemistry
Givens, Michael Spencer, organic chemistry
Goldhammer, Paul, inorganic chemistry
Griswold, Ernest, inorganic chemistry
Grunewald, Gary Lawrence, medicinal chemistry, bio-organic chemistry
Hall, Eugene Raymond, vertebrate zoology
Hambleton, William Weldon, geology, geophysics
Hamrick, James Lewis, population biology
Hanna, Martin Slafter, mathematics
Hanzlik, Robert Paul, bio-organic chemistry, bio
Harmony, Martin D., physical chemistry
Haslam, John Lee, pharmaceutical chemistry
Hedrick, Philip William, genetics
Hersh, Robert Tweed, biophysics
Hierl, Peter Marston, chemical kinetics
Higuchi, Takeru, physical chemistry, organic chemistry
Himes, Richard H., biochemistry
Himmelberg, Charles John, topology, mathematical analysis
Hoecker, Frank Edward, radiological physics, radiation biophysics
Hoffmann, Robert Shaw, systematics, ecology
Houston, L L., biochemistry
Humphrey, Philip Strong, ornithology
Huyser, Earl Stanley, organic chemistry
Iwamoto, Reynold Toshiaki, analytical chemistry
Jenks, George Frederick, geography
Jensen, Thorkil, microbiology
Johnson, Alfred Edwin, anthropology
Johnson, Richard Fourness, systematics, ecology
Kaesler, Roger LeRoy, micropaleontology
Kaminski, James Joseph, organic chemistry
Kitos, Paul Alan, biochemistry
Kleinberg Jacob, inorganic chemistry
Kollmorgen, Walter Martin, agricultural geography
Krone, Ralph Werner, nuclear physics
Kucher, August Wilhelm, phytogeography, ecology
Kuo, Norman Yu-Neng, organic chemistry, pharmaceutical chemistry
Kwak, Nowhan, high energy physics
Landgrebe, John A., organic chemistry
Lee, Byungkook, x-ray crystallography, biological structure
Leonard, Arthur Byron, zoology
Lerner, David Evan, theoretical physics
Lichwardt, Robert William, mycology
Lien, Yeong-Chung Edmund, computer sciences
Lindenbaum, Siegfried, physical chemistry
Ling, Daniel Seth, Jr., theoretical physics
Lundsgaarde, Henry P., anthropology
Maggiora, Gerald M., molecular biophysics
Maher, Michael John, comparative endocrinology
Mansfield, Lois E., numerical analysis
Martin, Frank Gene, pharmacology
Martin, Larry Dean, vertebrate paleontology
McCarthy, Paul Joseph, mathematics
McChesney, James Dewey, plant science, bio-organic chemistry
McClendon, James Fred, mathematics
McColl, Robert William, geography of Asia, political geography

McGregor, Ronald Leighton, botany
Mengel, Robert Morrow, zoology
Merrill, William Meredith, geology
Mertes, Mathias Peter, medicinal chemistry
Michener, Charles Duncan, entomology
Middaugh, Richard Lowe, inorganic chemistry
Mitscher, Lester Allan, bio-organic chemistry
Moos, Felix, anthropology
Mossberg, Howard E., pharmacy
Moser, Paul Stallings, pharmacy
Moyer, Robert Dale, mathematical analysis
Munczek, Herman J., physics
Neely, Peter Munro, statistical biology
Newmark, Marjorie Zeiger, biochemistry
Nodine-Zeller, Doris Eulaia, paleontology, stratigraphy
Nunley, Robert E., geography
O'Brien, William John, aquatic ecology, limnology
Oros, Margaret Oiava (Erickson), petroleum geology
Orr, James Anthony, mammalian physiology, cardiopulmonary physiology
Paretsky, David, microbiology
Paschke, William Lindall, mathematical analysis
Patton, Thomas Floyd, biopharmaceutics analysis
Peoples, James A., Jr., geophysics
Porter, Jack R., topology
Price, Griffith Baley, mathematical analysis
Prosser, Francis Ware, Jr., nuclear physics
Quadagno, David Michael, physiology
Renich, Paul William, physical chemistry
Reynolds, Charles Albert, chemistry
Robertson, Donald Claus, biochemistry, microbiology
Robison, Richard Ashby, geology, paleontology
Roofe, Paul Gibbons, anatomy
Rosenshein, Joseph Samuel, hydrology, hydrogeology
Rowell, Albert John, invertebrate paleontology
Rutledge, Charles O., pharmacology
Rutter, Edgar A Jr., mathematics
Ryting, Joseph Howard, physical chemistry, pharmaceutics
Sanders, Robert B, biological chemistry
Sapp, Richard Cassell, physics
Schlager, Gunther, genetics
Schowen, Richard Lyle, physical organic chemistry
Schumacher, Clifford Rodney, theoretical physics
Schweppe, Earl Justin, computer science, mathematics
Sedelow, Sally Yeates, linguistics
Sedelow, Walter Alfred, Jr., computer science, sociology
Shankel, Delbert Merrill, microbiology
Shaw, Edward Irwin, radiation biology
Shirer, Hampton Whiting, physiology
Slade, Norman Andrew, population ecology
Smith, Carlyle Shreeve, anthropology, archaeology
Smith, Thomas Russell, economic geography
Steeples, Donald Wallace, geophysics
Stella, Valentino John, pharmacy
Stini, William Arthur, human biology, physical anthropology
Stockhammer, Karl Adolf, zoology, entomology
Stone, Henry Orto, Jr., virology
Stump, Robert, physics
Szepycki, Pawel, mathematics
Teichert, Curt, geology, paleontology
Tessel, Richard Earl, neuropharmacology, psychopharmacology
Thompson, Rufus Henney, morphology
Torres, Andrew Marion, botany
Ulmer, Gilbert, physics
Van Schmus, William Randall, geology, meteoritics
Van Vleck, Fred Scott, mathematics
Wallace, Victor Lew, computer science
Weaver, Robert F., biochemistry
Weir, John Arnold, genetics
Wells, Philip Vincent, botany
Wenzel, Duane Greve, pharmacology
White, Anta M., anthropology
Wiley, Robert A, medicinal chemistry
Willner, Dorothy, cultural anthropology
Windhueser, John Joseph, physical pharmacy
Wiseman, Gordon G, physics

Wolfe, Herbert Glenn, developmental genetics
Wong, Kai-Wai, physics
Woodruff, Laurence Clark, insect physiology
Wytenbach, Charles Richard, developmental biology
Yarger, Harold Lee, exploration geophysics
Yochim, Jerome M, endocrinology, physiology
Zeller, Edward Jacob, geochemistry
Zimbrick, John David, radiation biophysics, radiation chemistry

LEAVENWORTH
Johnston, Andrea, mathematics
Miller, Jesse William, Jr, physical geography, geomorphology

LEAWOOD
Bancroft, George Herbert, physics
Burch, Clark Wayne, veterinary medicine

LENEXA
Volence, Frank Jerry, virology,

LINDSBORG
Bellah, Robert Glenn, environmental biology
England, Charles R., organic chemistry
Hermanson, Joseph Leonard, organic chemistry
Lungstrom, Leon, medical entomology
Swenson, Christina N. physics

MANHATTAN
Acker, Duane Calvin, animal nutrition
Adams, Albert Whitten, animal nutrition
Allen, Deloran Matthew, animal science, meat science
Ameel, Donald Jules, helminthology
Anderson, Neil Vincent, veterinary medicine
Andrews, Arthur Clinton, physical chemistry
Anthony, Harry D., veterinary medicine
Bailie, Wayne E., veterinary microbiology
Bark, Laurence Dean, agricultural meteorology
Barkley, Theodore Mitchell, systematic botany
Bartley, Erle Edwin (St Clair), animal nutrition
Bassette, Richard, dairy science
Bechtle, Robert M., food science
Beck, Henry V., quaternary geology, hydrogeology
Bhalla, Chander P., atomic physics
Bidwell, Orville Willard, soil classification
Bieberly, Frank Gearhart, agronomy
Blocker, Henry Derrick, systematic entomology
Bode, Vernon Cecil, biochemical genetics
Bohannon, Robert Arthur, soil science
Bowers, Jane Ann (Raymond), food science
Brent, Benny Earl, animal nutrition
Browder, Lewis Eugene, plant pathology
Bulla, Lee Austin, Jr., microbiology, biochemistry
Burckel, Robert Bruce, mathematics
Burkhard, Raymond Kenneth, biochemistry
Burroughs, Albert Lawrence, veterinary virology
Butler, Hugh C., surgery, physiology
Call, Edward Prior, reproductive physiology, agriculture
Campbell, Ronald Wayne, horticulture
Cardwell, Alvin Boyd, solid state physics
Casady, Alfred Jackson, agronomy, plant breeding
Caul, Jean Frances, food science
Chapman, Thomas Everett, veterinary physiology
Chelikowsky, Joseph R, geology
Chen, Chao Ling, veterinary physiology, endocrinology
Chung, Okkyung Kin, cereal chemistry
Clarenburg, Rudolf, physiological chemistry
Clayberg, Carl Dudley, genetics
Clegg, Robert Edward, biochemistry
Coles, Embert Harvey, Jr., clinical pathology, microbiology
Compan, Alvin Dell, solid state physics

Conrad, Gary Warren, developmental biology
Conrow, Kenneth, computer science
Consigli, Richard Albert, virology, cancer
Cook, James Ellsworth, veterinary pathology, comparative pathology
Copeland, James Lewis, physical chemistry
Cox, David Jackson, biochemistry
Currutte, Basil, Jr., physics
Dale, Ernest Brook, physics
Danen, Wayne C., organic chemistry
Davis, Lawrence Clark, biochemistry, molecular biology
Davis, Marjorie, developmental biology, invertebrate zoology
Denell, Robin Ernest, genetics
DesMarteau, Darryl D., inorganic chemistry
Deyoe, Charles W., biochemistry, nutrition
Dickenson, Ottie J., plant pathology
Dixon, Lyle Junior, mathematics
Dragsdorf, Russell Dean, solid state physics
Dressler, Robert Eugene, number theory
Ealy, Robert Phillip, ornamental horticulture
Eck, John Stark, nuclear physics, solid state physics
Edmunds, Leon K., plant pathology
Ellis, Roscoe, Jr., soil chemistry
Ellsworth, Louis Daniel, physics
Eltzinga, Richard John, entomology
Evans, John C., astrophysics
Fateley, William Gene, structural chemistry
Faulkner, Lindsey Ralph, plant nematology
Fedde, Marion Roger, pulmonary physiology
Feyerherm, Arlin Martin, experimental statistics
Filinger, George Albert, horticulture
Fina, Louis R., microbial physiology
Finkelstein, Beatrice, nutrition
Finnegan, Michael, physical anthropology
Finney, Karl Frederick, chemistry
Folland, Nathan Orlando, physics
Fretwell, Steve D., population ecology
Fryer, Elsie Beth, nutrition
Fryer, Holly Claire, biological statistics
Fuller, Leonard Eugene, mathematics
Gier, Herschel Thomas, cytology, embryology
Good, Don L., animal science
Goss, James Arthur, botany
Gray, Andrew P., veterinary medicine, pharmacology
Greechie, Richard Joseph, mathematics
Greenaway, Walter Thomas, agricultural biochemistry, statistics
Greig, James Kibler, Jr., horticulture
Grosh, Doris Lloyd, statistics, operations research
Guhl, Alpheus Matthew, zoology
Hackerot, Harold Leroy, genetics, plant breeding
Halazon, George Christ, biology
Hammaker, Geneva Sinquefield, inorganic chemistry
Hammaker, Robert Michael, physical chemistry
Hansen, Merle Frederick, parasitology
Hansing, Earl Dahl, plant pathology
Harbers, Leniel H., animal nutrition
Harris, John Orville, soil microbiology
Harrison, Dorothy Lucile, food science
Hatchett, Jimmy Howell, entomology
Hathaway, Charles Edward, physics
Hawley, Merle Dale, analytical chemistry
Hedgcoth, Charlie, Jr., biochemistry
Hess, Carroll V., agricultural economics
Heyne, Elmer George, agronomy
Hobbs, James Arthur, agronomy, soils
Hoover, William Jay, food science
Hopkins, Theodore Louis, entomology
Horber, Ernst Konrad, entomology
Hosency, Russell Carl, cereal chemistry
Hsu, Chen-Jung, geometry
Hulbert, Lloyd Clair, plant ecology
Iandolo, John Joseph, bacterial physiology, food microbiology
Jack, Hulan E., Jr., physics
Jacobs, Hyde Spencer, agronomy
Johnson, George Dana, chemistry
Johnson, John Alexander, biochemistry
Johnson, Lowell Boyden, plant pathology, plant physiology
Johnson, Michael Paul, plant ecology

Kadoum, Ahmed Mohamed, entomology, toxicology
Kammer, Ann Emma, neurobiology
Kanemasu, Edward Tsukasa, agriculture
Keen, Ray Albert, ornamental horticulture
Kelley, Donald Clifford, veterinary medicine
Kemp, Kenneth E, statistics
Kennedy, George Arlie, veterinary pathology
Klaassen, Harold Eugene, ecology, fish biology
Klemm, Robert David, vertebrate morphology
Klopfenstein, William Elmer, biochemistry
Knutson, Herbert Claus, entomology
Koch, Berl Amos, animal nutrition
Kramer, Charles Lawrence, mycology
Kromm, David Elwyn, resource geography
Kropf, Donald Harris, meat science, animal husbandry
Kruh, Robert Frank, physical chemistry
Labhsetwar, Anant Pandurang, reproductive endocrinology
Lambert, Jack Leeper, analytical chemistry, inorganic chemistry
Lanning, Francis Chowing, chemistry
Larson, Vernon C, agriculture, academic administration
Leasure, Elden Emanuel, pathology, physiology
Lee, Ronald S, solid state physics
Legg, James C, nuclear physics
Leipold, Horst Wilhelm, anatomic pathology
Leland, Stanley Edward, Jr., parasitology
Lenhert, Anne Gerhardt, organic chemistry
Liang, George H L, plant genetics, plant breeding
Lindquist, William Dexter, veterinary parasitology
Lineback, David R, carbohydrate chemistry
Macdonald, James Robert, experimental physics
Manney, Thomas Richard, biophysics
Marchin, George Leonard, molecular biology
Marr, John Maurice, mathematics
Marzolf, George Richard, limnology
Maxfield, John Edward, algebra
Maxfield, Margaret Waugh, mathematics
McDonald, Richard Norman, organic chemistry
McGaughey, William Horton, physiology
McGavin, Matthew Donald, veterinary pathology
Meloan, Clifton E, analytical chemistry
Miles, Neil Wayne, horticulture, plant physiology
Mills, Robert Barney, entomology
Minocha, Harish C, biochemical virology
Mitchell, Howard Lee, agricultural biochemistry
Mitchell, Roger L, agronomy, crop physiology
Morril, James Lawrence, Jr, dairy science
Moser, Herbert Charles, physical chemistry
Mosier, Jacob Eugene, veterinary medicine
Mueller, Delbert Dean, physical biochemistry
Murphy, Larry S, agronomy, plant physiology
Newell, Kathleen, nutrition
Niblett, Charles Leslie, plant pathology, virology
Nickell, Cecil D, plant genetics, biometry
Nilson, Erick Bogseth, agronomy
Nordin, Philip, biochemistry
Norton, Charles Lawrence, dairy science
Oehme, Frederick Wolfgang, veterinary toxicology, comparative medicine
Olson, Raymond Verlin, agronomy
Ottenheimer, Harriet Joseph, anthropology
Ottenheimer, Martin, anthropology
Owensby, Clenton Edgar, range science, ecology
Pady, Stuart McGregor, botany
Parker, Sidney Thomas, mathematics
Parker, Willard Albert, mathematics
Parrish, Donald Baker, nutrition, biochemistry
Partida, Gregory John, Jr., economic entomology
Paske, William Charles, molecular physics, science education
Paulsen, Avelina Quiaoit, plant pathology
Paulsen, Gary Melvin, physiology

Petersen, John David, inorganic chemistry
Pickett, William Francis, horticulture
Pittenger, Thad Heckle, Jr., genetics
Pitts, Charles W., medical entomology, insect physiology
Pomeranz, Yeshajahu, biochemistry, cereal chemistry
Ponte, Joseph G, Jr., food science
Posler, Gerry Lynn, agronomy, plant breeding
Poston, Freddie Lee, Jr., entomology
Powers, William L, soil physics
Purcell, Keith Frederick, inorganic chemistry
Reeck, Gerald Russell, biochemistry
Richardson, Draytford, animal nutrition
Robel, Robert Joseph, animal ecology
Robinson, Robert James, microbiology, biochemistry
Roche, Thomas Edward, biochemistry
Rodkey, Leo Scott, immunology
Rosenkilde, Carl Edward, fluid dynamics, atmospheric physics
Roth, Linwood Evans, zoology
Roufa, Donald Jay, biochemistry, genetics
Rulliffson, Willard Sloan, biochemistry
Sanford, Paul Everett, poultry nutrition
Sauer, David Bruce, plant pathology
Schalles, Robert R, animal breeding, population genetics
Schrenk, William George, chemistry
Schwenk, Fred Walter, plant pathology, virology
Seaman, Gregory G, environmental physics
Seib, Paul A, organic chemistry, biochemistry
Seitz, Larry Max, analytical biochemistry, cereal chemistry
Setser, Carole Sue, food science
Setser, Donald W., physical chemistry
Shellenberger, John Alfred, cereal chemistry
Shenkel, Claude W, Jr., geology
Shreve, Loy William, forestry
Shult, Ernest E, algebra, combinatorics
Siddall, William Richard, transportation geography
Sincovec, Richard Frank, applied mathematics
Siotani, Minoru, mathematical statistics
Skelton, Marilyn Mae, food science, biochemistry
Skidmore, Edward Lyman, soil conservation, agronomy
Sloat, Floyd Brooksher, mathematics
Smith, Christopher Carlisle, evolutionary biology
Smith, Edgar Fitzhugh, animal husbandry
Smith, Floyd W, soil fertility, soil chemistry
Smith, Joseph Emmitt, clinical pathology, physiology
Sorensen, Edgar Lavell, plant breeding
Spangler, John David, physics
Spooner, Brian Sandford, developmental biology, cell biology
Stamey, William Lee, mathematics
Stone, Loyd Raymond, soil physics
Stover, Stephen Leech, geography
Strafuss, Albert Charles, comparative pathology, oncology
Strecker, George Edison, mathematics
Stromberg, Karl Robert, mathematics
Stuteville, Donald Lee, plant pathology
Tatschl, Annehara Kathleen, botany
Taylor, Robert Bartley, cultural anthropology
Teare, Iwan Dale, crop physiology
Thien, Stephen John, agronomy
Thompson, Hugh Erwin, entomology
Tinklin, Gwendolyn L, foods
Tsen, Cho Ching, biochemistry, nutrition
Tuma, Harold J, meat science, food science
Twiss, Page Charles, geology
Upson, Dan W, pharmacology, physiology
Urban, James Edward, microbial physiology
Vanderlip, Richard L, agronomy
Van Swaay, Maarten, analytical chemistry
Varghese, Sankoorikal Lonappan, experimental atomic physics
Walters, Charles Philip, astrogeology
Ward, Arlin Bruce, food science
Ward, George Merrill, dairying, nutrition
Wassom, Clyde E, agronomy
Weber, Richard Gerald, entomology
Weis, Jerry Samuel, biology

West, Ronald Robert, paleobiology, paleoecology
Westfall, Jane Anne, neurosciences
Wheat, John David, animal genetics
Whittemore, Donald Osgood, geochemistry
Wilde, Gerald Eldon, entomology
Williams, Dudley, molecular spectroscopy, planetary atmospheres
Williams, Larry Gale, genetics
Williams, Robert Elvin, mathematics
Wilson, Freddie Elton, endocrinology
Yee, Kane Shee-Gong, applied mathematics
Young, Paul McClure, mathematics
Zimmerman, John Lester, ecology

MCPHERSON
Burkholder, John Henry, zoology
Ikenberry, Gifford John, Jr., plant anatomy, plant development
Krehbiel, Jesse D., mathematics

MERRIAM
Cahoy, Roger Paul, organic chemistry
Curless, William Toole, environmental sciences, pesticide chemistry
Doyle, William Carter, Jr., pesticide chemistry
Gehrig, Neil Edward, inorganic chemistry, organic polymer chemistry
Harwell, Kenneth Elzer, industrial organic chemistry, polymer chemistry
Hedrich, Loren Wesley, medicinal chemistry, agricultural chemistry
Hoffmann, Otto Louis, plant physiology
Kirkpatrick, Joel Lee, organic chemistry, pesticide chemistry
Riden, Joseph Robert, Jr., biochemistry
Rutter, Jerry L., organic chemistry
Schroeder, Robert Samuel, pesticide chemistry, organic biochemistry
Schwartzbeck, Richard Arthur, weed science
Wright, Charles Hubert, analytical chemistry
Young, Dale W., plant physiology

NORTH NEWTON
Platt, Dwight Rich, zoology
Schmidt, Robert W., biochemistry
Wedel, Arnold Marion, mathematics

OLATHE
Childress, Charles Curtis, biochemistry, biomedical engineering
Kruse, Carl William, organic chemistry
Lawrence, Robert G, zoology
Nelson, Dallas Leroy, toxicology, pathology
Roth, Elmer Alfred, botany

OTTAWA
Bacon, John Alvin, entomology
Bemmels, William David, plasma physics
Chaney, George L, mathematics
Diaz, Justo A., physics
Flusser, Peter R., mathematics
Morrissey, J Edward, zoology, physiology

OVERLAND PARK
Matulis, Raymond M, analytical chemistry

PITTSBURG
Bass, J Carl, biology, science education
Bednekoff, Alexander G, biochemistry
Coltharp, Forrest Lee, mathematics education
Crandall, Elbert Williams, organic chemistry
Daniel, Thomas Bruce, solid state physics
Davis, Elwyn H, algebra
Duncan, Bettie, microbiology
Haggard, J D., mathematics
Hays, Horace Wayne, mathematics
Hight, Donald Wayne, mathematics
Johnson, John Christopher, Jr., zoology, parasitology
Keller, Leland Edward, anatomy
Kelting, Ralph Walter, plant ecology
Kriegsman, Helen, mathematics
Pauley, James L., physical chemistry
Potts, Melvin Lester, physical inorganic chemistry
Riches, Ralph Harvard, genetics, biology
Sperry, Theodore Melrose, botany, ecology
Thomas, Harold Lee, mathematics, statistics

Thomas, James E., physics
Walker, Joe M., analytical chemistry

PRAIRIE VILLAGE
McLeod, William Stirling, entomology

SALINA
Buser, Mary Paul, mathematics
Owen, Bernard Lawton, entomology
Voth, Orville Lester, biochemistry

SHAWNEE
Carlson, Arthur, Jr., veterinary pharmacology
Leaton, John Roger, analytical chemistry, pharmaceutical chemistry
Matzke, Howard Arthur, neuroanatomy
Rose, Wayne Burl, physical chemistry, pharmacy

SHAWNEE MISSION
Curts, Kephart Maynard, veterinary pharmacology
Hearne, Horace Clark, Jr., mathematics, operations research
Mills, Russell Clarence, medical education
Parks, Terry Everett, organic chemistry, science education
Stutz, Robert L, organic chemistry, pharmaceutical chemistry
Von Rumker, Rosmarie, agricultural chemistry

STERLING
Crosen, Robert Glenn, analytical chemistry, science education
McAllister, Stuart Allan, physical chemistry
Powers, Daniel D, organic chemistry
Taylor, Dale L, ecology

TOPEKA
Alexander, Robert Stanley, astronomy, physics
Boyer, Don Raymond, vertebrate zoology
Cohen, Sheldon H, inorganic chemistry
Dunphy, James Francis, polymer chemistry
Eberhart, Paul, mathematics
Foster, David Bernard, neurology
Glazier, Robert Henry, organic chemistry
Hartocollis, Peter, psychiatry, psychoanalysis
Hickox, John Ekstrom, geology
Kopper, Paul Heinz, bacteriology
Lamb, Frank Bruce, forestry
Levy, Edwin Z, psychiatry, psychoanalysis
McAdam, Terry Donald, mathematics
Modlin, Herbert Charles, psychiatry
Morrison, Garrett Louis, geochemistry
Nelson, Woodrow Ensign, food technology, microbiology
Rabin, Erwin R, pathology, virology
Rinsley, Donald Brendan, psychiatry, medical education
Simpson, William Stewart, psychiatry, hospital administration
Thompson, Robert Harry, mathematics, computer science
Virmani, Yash Paul, physical chemistry
Voth, Harold Moser, psychiatry
Wolf, Thomas Michael, genetics, environmental biology

WICHITA
Allen, Anneke S., physical chemistry
Arteaga, Lucio, mathematics, statistics
Axman, Mary Claudine, biology
Bajaj, Prem Nath, mathematical analysis, topology
Beard, William Quinby, Jr., industrial chemistry
Berg, John Robert, geology, chemistry
Blythe, Jack Gordon, geology
Bosley, Elizabeth Caswell, speech pathology
Bradley, Doris P, speech pathology, audiology
Brady, Stephen W., mathematics
Buess, Charles Merlyn, organic chemistry
Carper, William Robert, enzymology
Cawley, Leo Patrick, pathology, immunochemistry
Chopra, Dharam-Vir, statistics
Christena, Ray Clifford, industrial chemistry
Christian, Robert Vernon, Jr., chemistry

465

KANSAS

Eichler, Victor B., experimental embryology
Elcrat, Alan Ross, applied mathematics
Fenrick, Maureen Helen, algebra
Gries, John Charles, structural geology
Gundersen, James Novotny, economic geology
Gusenius, Edwin Mauritz, inorganic chemistry
Harvey, Thomas Stoltz, pathology
Hirschmann, Robert P., applied chemistry
Ho, James Chien Ming, physical chemistry, solid state physics
Holmes, Lowell Don, anthropology, history of anthropology
Johnson, John Webster, Jr., organic chemistry
Joyner, Howard Sajon, medical administration, biomedical engineering
Lakin, Wilbur, theoretical physics
Linscheid, Harold Wilbert, mathematical analysis
Loper, Gerald D., nuclear physics
McCroskey, Robert Lee, audiology, speech pathology
Mellinger, George T., urology
Miller, Glendon Richard, microbiology
Penner, Siegfried Edmund, organic chemistry
Perel, William Morris, mathematics
Reals, William Joseph, pathology
Roberts, Daniel Keith, obstetrics & gynecology
Rodenberg, Sidney Dan, cell physiology
Rohn, Arthur Henry, Jr., anthropology
Sarachek, Alvin, microbiology
Simons, Gary, quantum chemistry
Singh, Surendra Pratap, plant morphology
Singhal, Ram Pratap, biochemistry, cancer
Strecker, Joseph Lawrence, theoretical physics
Sweet, George H., immunology
Talaty, Erach R., organic chemistry, electrochemistry
Unruh, Henry, Jr., physics
Wahlbeck, Phillip Glenn, high temperature chemistry, surface chemistry
Watson, Tully Franklin, physics
Wilson, James Alexander, plant breeding
Youngman, Arthur L., botany
Zandler, Melvin E., physical chemistry

WINFIELD
Laws, Leonard Stewart, mathematics

KENTUCKY

ANCHORAGE
Landis, Edward Everett, psychiatry

ASHLAND
Kovach, Stephen Michael, petroleum chemistry

BEREA
Baker, Howard Crittenden, theoretical physics
Barnes, Richard N., ecology
Boyce, Stephen Scott, mathematics, quantum mechanics
Gailey, Franklin Bryan, physiology
Gentry, Claude Edwin, agronomy
Hogg, Edd Coolidge, animal physiology
Levey, Gerrit, physical chemistry
Powell, Smith Thompson, III, physics
Ramsay, John Martin, animal breeding
Roberts, George Gilbert, mathematics, statistics
Shugars, Jonas P., plant science, soil science
Stephens, Noel, Jr., animal nutrition
Strickler, Thomas David, atomic physics
Vogel, Willis Gene, range science
Wright, John Cushing, animal behavior

BARBOURVILLE
Cohenour, Francis D., animal physiology, microbiology
Gilbert, Frank Albert, biology
Jankovics, Lawrence Robert, analytical chemistry, physical chemistry
Myers, Dale Kamerer, organic chemistry
Riesz, Robert Richard, physics
Santaram, Chitikuri, molecular physics

COVINGTON
Boyle, David Joseph, experimental solid state physics
Buzzee, David H., biochemistry
Fox, Mary Eleanor, solid state physics
Landay, Marshall Edwin, microbiology
Miner, George Kenneth, magnetic resonance
Rose (Rauen), Mary, mathematics
Zembrodt, Anthony Raymond, analytical chemistry, physical chemistry

CAMPBELLSVILLE
Roberts, Noble, zoology, cytology

CALVERT CITY
Reitz, John Marsteller, organic chemistry, chemical engineering

BRANDENBURG
Gilmore, Forrest Cubley, physical chemistry

KEVIL
Magruder, Samuel Rossington, human anatomy

LEWISPORT
Clinton, Charles Anthony, ethnography

LEXINGTON
Abbott, Susan, anthropology
Abramson, Fredric David, community health, epidemiology
Adams, William Yewdale, cultural anthropology

JEFFERSONTOWN
Masters, John Edward, organic chemistry
Wamsley, Robert Alan, paper chemistry

HIGHLAND HEIGHTS
Kearns, Thomas J., algebra
Smith, Joe K., mathematics
Thieret, John William, botany

GOLDEN POND
Nall, Raymond Willett, botany

HENDERSON
Stanonis, Francis Leo, petrology

GEORGETOWN
Clark, Evelyn Genevieve, microbiology
Glass, Dudley Brewer, organic chemistry
Hanely, Wayne Stewart, chemistry
Lindsay, Dwight Marsee, mammalogy
Seay, Thomas Nash, entomology
Senter, James Parker, biophysics

FT THOMAS
Donnelly, Joseph Lawrence, medicine

FT MITCHELL
Budde, Mary Laurence, zoology, microbiology
Hieber, Thomas Eugene, biochemistry,
Humphreys, Wallace F., biochemistry,
Lang, Joseph Edward, theoretical physics
Mueller, Mary Casimira, physical chemistry, theoretical chemistry
Reed, Kenneth Paul, analytical chemistry
Stevens, Alan Douglas, occupational health, veterinary medicine

FRANKFORT
Clark, Minor E., fisheries
Hill, Carl McClellan, organic chemistry
Letton, James Carey, pharmaceutical chemistry
Scott, Linda Preston, physical education, human development

FLORENCE
Wise, Hugh Edward, Jr., organic chemistry

ERLANGER
Luft, Stanley Jeremie, geology

DANVILLE
Cook, Eugene Wilbur, Jr., biology
Ellis, Roy, oceanography
Feese, Bennie Taylor, molecular biology
Hammond, Ray Kenneth, biochemistry, microbiology
Hanson, Harold Nelson, inorganic chemistry, physical chemistry
Marsh, Michael Pierce, ecology, animal behavior
Moyle, Susan Mary, bryology
Pizlak, Robert, mathematics, quantum mechanics
Robinson, Wilbur Judson, mathematics
Sagar, William Clayton, synthetic organic chemistry
Walkup, John Harper, physical chemistry
Whittle, Charles Edward, Jr., physics, applied mathematics
Wilt, Paxton Marshall, molecular spectroscopy

BOWLING GREEN
Ahsan, S Reza, geography
Bailey, Donald Wycoff, physiology
Barksdale, James Bryan, Jr., mathematics
Beal, Ernest O., aquatic biology
Brown, Leonard D., animal nutrition, physiology
Buckman, William Gordon, solid state physics
Chamberlin, John Macmullen, inorganic chemistry, physical chemistry
Coohill, Thomas Patrick, biophysics
Crumb, Glenn Howard, academic administration
Csallany, Sandor Csergo, hydrology, water resources
Davis, Chester L., computer science, applied mathematics
Davis, James Leslie, geography
Dillard, Gary Eugene, phycology, aquatic ecology
Elliott, Larry P., medical microbiology
Farina, Robert Donald, bioinorganic chemistry
Feibes, Walter, statistics, operations research
Foster, Robert H., biogeography
Gildersleeve, Benjamin, geology
Gleason, Larry Neil, parasitology
Gray, Elmer, plant breeding
Hartman, David Robert, biochemistry, nutrition
Hegen, Edmund Eduard, geography, resource geography
Hoffman, Wayne Larry, urban geography
Holy, Norman Lee, organic chemistry
Hourigan, William R., soil chemistry
Hoyt, Robert Dan, ichthyology
Hunter, Norman W., chemistry
Jenkins, Jeff Harlin, plant pathology
Johnson, Ray Edwin, soil fertility
Longmire, Martin Shelling, chemical physics, physical chemistry
Lowman, Bertha Pauline, mathematics
Nicely, Kenneth Aubrey, botany
Parks, James Edgar, physics
Pearson, Earl Freeman, physical chemistry, spectroscopy
Prins, Rudolph, aquatic ecology, zoology
Puckett, Hugh, invertebrate endocrinology
Reasoner, John W., photochemistry
Riley, John Thomas, inorganic chemistry
Russell, Marvin W., physics, applied mathematics
Seeger, Charles Ronald, structural geology, geophysics
Shadowen, Herbert Edwin, vertebrate zoology
Shank, Lowell William, chemistry
Six, Norman Frank, Jr., physics
Skean, James Dan, microbiology
Stephens, Henry LeRoy, plant physiology
Stokes, Joseph Franklin, mathematics
Stroube, William Hugh, plant science
Taylor, James Woodall, physical geography
Toman, Frank R., biochemistry
Toups, Polly Anticich, anthropology, archaeology
Wallace, Kyle David, mathematics
Watson, Martha F., mathematics
Wilkins, Curtis C., physical chemistry
Wilson, Gordon, Jr., organic chemistry
Winstead, Joe Everett, botany, ecology
Yungbluth, Thomas Alan, genetics, plant breeding

Aleem, M I Hussain, microbial biochemistry
Allen, David Mitchell, statistics
Alter, Ronald, mathematics, computer science
Ambrose, Charles T., immunology
Andersen, Roger Allen, biochemistry, oncology
Anderson, Richard L., experimental statistics
Archdeacon, James William, physiology
Attig, Thomas George, inorganic chemistry, organometallic chemistry
Bailey, Harry Hudson, soil science
Barbour, Roger William, vertebrate zoology
Barnhart, Charles Elmer, animal nutrition
Barnhisel, Richard I., soil mineralogy, soil chemistry
Baskin, Jerry Mack, plant ecology
Bauer, Henry Hermann, electrochemistry, analytical chemistry
Bauman, David Stanley, microbiology
Beatty, Millard Fillmore, Jr., continuum mechanics
Begin, John Joseph, poultry nutrition
Beidleman, James C., algebra
Benenson, Abram Salmon, medicine
Benton, Robert S., anatomy
Bixby, Robert Eugene, operations research
Black, Rodney Elmer, physical chemistry
Blackburn, William Howard, petrology
Blevins, Robert L., soil science
Bollum, Frederick James, biochemistry
Boone, Donald Joe, clinical chemistry,
Bortner, Charles Eugene, agronomy
Bosomworth, Peter Palliser, medicine, anesthesiology
Boyarsky, Lila Harriet, biology
Boyarsky, Louis Lester, neurophysiology
Brandenberger, Jerry D., physics
Brock, Carolyn Pratt, structural chemistry
Broida, Theodore Ray, research administration
Brower, Thomas Dudley, orthopedic surgery
Brown, Ellis Vincent, organic chemistry
Brown, Leland Arthur, zoology
Brown, William Randall, geology
Bryans, John Thomas, animal virology
Buck, Charles Frank, animal husbandry
Buckholtz, James Donnell, mathematical analysis
Buckner, Robert Cecil, plant genetics
Bull, Leonard Seth, animal nutrition, physiology
Bush, Lowell Palmer, plant physiology
Butler, Frank Andrew, solid state physics
Butterfield, David Allan, biophysical chemistry
Campbell, Lois Jeannette, geology
Carpenter, John Melvin, zoology
Chan, Shung Kai, biochemistry
Chapman, Richard Alexander, plant pathology
Chappell, Guy Lee Monty, ruminant nutrition, physiology
Chen, Linda Li-Yueh Huang, biochemistry, nutrition
Cheniae, George Maurice, plant biochemistry
Chick, Ernest Watson, mycology, preventive medicine
Christensen, Christian Martin, economic entomology
Clark, David Barrett, neurology, neuropathology
Clark, Jimmy Dorral, mycology
Cochran, Lewis Wellington, physics
Coleman, Donald Brooks, mathematics
Collins, Glenn Burton, plant genetics
Coltharp, George B., forest hydrology, water pollution
Companion, Audrey (Lee), quantum chemistry
Cone, Edward Jackson, drug metabolism
Conti, Samuel Francis, microbiology
Costich, Emmett Rand, oral surgery
Cotter, William Bryan, Jr., genetics, anatomy
Cox, Raymond H., anatomy
Crawford, Eugene Carson, Jr., physiology, zoology
Cromwell, Gary Leon, animal nutrition
Csaky, Tihamer Zoltan, physiology, pharmacology

Curtz, Thaddeus Bankson, mathematics
Dahlman, Douglas Lee, insect physiology
Das, Nirmal Kanti, cell biology
Davis, Daniel Layten, plant physiology, plant genetics
Davis, Wayne Harry, mammalogy
Dean, Benjamin T, animal science
DeLand, Frank H, nuclear medicine
De Luca, Patrick Phillip, pharmaceutics
DeMarcus, Wendell Carden, physics
Dennen, William Henry, geology, geochemistry
Diachun, Stephen, plant pathology
Diamond, Louis, pharmacology
Diedrich, Donald Frank, biochemical pharmacology
Digenis, George A, medicinal chemistry, nuclear medicine
Dillon, Marcus Lunsford, Jr, surgery
Dittert, Lewis William, pharmacy
Donnelly, Grace Marie, cell biology
Dorough, Hendley Wyman, insect toxicology, entomology
Doughty, Richard Morrison, pharmacognosy
Drucker, Philip, archaeology, cultural anthropology
Drudge, Junior Harold, veterinary parasitology
Duffield, Lathel Flay, anthropology, archaeology
Duhring, John Lewis, obstetrics & gynecology
Duncan, William Graham, plant physiology
Dutt, Ray Horn, animal science
Edwards, Ogden Frazelle, microbiology
Egli, Dennis B, agronomy, crop physiology
Ehmann, William Donald, radiochemistry, geochemistry
Elwood, William K, dentistry
Ely, Donald Gene, animal science
Engelberg, Joseph, biophysics
Enochs, Edgar Earle, mathematics
Essene, Frank J, anthropology
Everett, Paul Marvin, solid state physics
Fairweather, Graeme, numerical analysis
Field, Thomas Parry, cultural geography, geography of Australia
Fisher, Irving Sanborn, geology
Flesher, James Wendell, pharmacology
Fordham, Joseph Raymond, nutritional biochemistry
Foster, Thomas Scott, biopharmaceutics, clinical pharmacology
Fowler, Ira, experimental embryology
Frazier, Donald Tha, physiology, neurophysiology
Freeman, Theodore Russell, dairying
Freytag, Paul Harold, systematic entomology
Frye, Wilbur Wayne, soil fertility
Furcolow, Michael L, public health
Gabbard, Fletcher, nuclear physics
Gallaher, Art, Jr, anthropology, applied anthropology
Garrigus, Wesley Patterson, agriculture
Gillilan, Lois Adell, anatomy, physiology
Gochenour, William Sylva, Jr, veterinary medicine
Golden, Abner, pathology
Goldenberg, David Milton, oncology, experimental pathology
Goodman, Norman L, microbiology, biochemistry
Goodwill, Robert, quantitative genetics, population genetics
Goodwyn, L Wayne, pure mathematics
Gordon, Helmut Albert, physiology, pharmacology
Gorodetzky, Charles W, pharmacology, medical research
Gossick, Ben Roger, physics
Govindarajulu, Zakkula, mathematics, statistics
Green, William Warden, audiology
Greene, John W, Jr, obstetrics & gynecology
Gregory, Wesley Wright, Jr, economic entomology
Griffen, Ward O, Jr, surgery
Griffith, John Dorland, psychiatry, psychopharmacology
Griffith, Robert Bell, plant physiology, biochemistry
Grunwald, Claus Hans, plant physiology
Guthrie, Robert D, physical organic chemistry
Hagan, Wallace Woodrow, geology
Hallberg, George Robert, quaternary geology
Hamilton, James Lewis, plant physiology
Hammaker, Ellwood Meacham, analytical chemistry
Hanau, Richard, optics
Harris, Denny Olan, phycology, microbiology
Hayden, Thomas Lee, mathematics
Hays, Rodney Malcolm, Jr, zoology
Hays, Virgil Wilford, animal nutrition
Hemken, Roger Wayne, dairy nutrition
Hendrix, James William, plant pathology, mycology
Herrick, Clifford Ernest, Jr, chemistry
Herron, James Watt, botany
Hiatt, Andrew Jackson, plant physiology, agronomy
Hirsch, Henry Richard, biophysics, gerontology
Hofstetter, Kenneth John, nuclear chemistry, physical chemistry
Holland, Nancy H, pediatrics, nephrology
Hollingsworth, James W, internal medicine
Hook, Donald, forestry, tree physiology
Hopkins, Howard, pharmacy, pharmaceutical chemistry
Howard, Henry Cobourn, mathematics
Hu, Alfred Soy Lan, biochemistry
Humphries, James Charles, bacteriology
Hurley, Laurence Harold, microbial biochemistry
Hutton, John James, Jr, hematology, biochemistry
Isbell, Harris, internal medicine
Jacobson, Don Richard, nutrition, biochemistry
James, Mary Frances, medical education, laboratory medicine
Jasinski, Donald Robert, clinical pharmacology
Jokl, Ernst, physiology
Jones, Bill Edson, experimental psychology
Jordan, William Stone, Jr, medicine
Kadaba, Pankaja Kooveli, organic chemistry, biochemistry
Kadaba, Prasad Krishna, physics, electronics
Karan, Pradyumna P, economic geography, cultural geography
Kasperbauer, Michael J, plant physiology
Kay, David Cyril, psychiatry, psychopharmacology
Keller, George Randy, Jr, seismology
Kellerman, George D, medical microbiology
Kemp, James Dillon, animal science
Kemp, Thomas Rogers, structural chemistry
Keogh, Frank Richard, mathematics
Kern, Bernard Donald, nuclear physics
Kiser, Robert Wayne, inorganic chemistry
Knapp, Fred William, entomology
Knavel, Dean Edgar, horticulture
Kornet, Milton Joseph, pharmaceutical chemistry, organic chemistry
Kostenbauder, Harry Barr, pharmacy
Kratzer, D Dal, animal breeding, statistics
Krivoy, William Aaron, pharmacology
Krogdahl, Wasley Sven, cosmology
Kuc, Joseph, plant biochemistry, plant pathology
Kuehne, Robert Andrew, zoology
Lafferty, James Francis, biomedical engineering, mechanical engineering
Laine, Roger Allan, biochemistry
Lambson, Roger O, anatomy
Langlois, Bruce Edward, food microbiology
Lantis, Margaret Lydia, anthropology
Lasheen, Aly M, plant physiology, horticulture
Leavell, Ullin Whitney, Jr, dermatology, pathology
Leggett, James Everett, plant physiology
Lehman, Guy Walter, theoretical physics
Lemon, Frank Raymond, epidemiology
Lesnaw, Judith Alice, virology
Lesshafft, Charles Thomas, Jr, pharmacy
Lester, Robert Leonard, biochemistry
Lillich, Thomas Tyler, microbiology, cell biology
Lindquist, Anders Gunnar, applied mathematics
Little, Charles Oran, animal nutrition, agriculture
Little, James W, oral pathology
Litvak, Austin S, urology
Lloyd, William Gilbert, organic chemistry, applied chemistry
Lockard, Raymond G, plant physiology
Lowe, Richie Howard, plant physiology, biochemistry
Loy, Robert Graves, animal physiology
Lubawy, William Charles, pharmacology
Luckens, Mark Manfred, pharmacology, toxicology
Lyons, Erwin John, geology
Lyons, Eugene T, parasitology
MacKellar, Alan Douglas, physics
MacLaury, Donald Wayne, poultry breeding
MacQuown, William Charles, Jr, geology
Mandelstam, Paul, biochemistry, internal medicine
Marciniak, Ewa, hematology
Markesbery, William Ray, neurology, neuropathology
Marlatt, Abby Lindsey, nutrition
Martin, William David, anatomy
Martin, William Robert, pharmacology
Maruyama, Yosh, radiotherapy, radiobiology
Mason, Aaron S, psychiatry, health administration
Massey, Herbert Fane, Jr, agronomy, soil fertility
Matulionis, Daniel H, anatomy, embryology
Mazzoleni, Alberto, cardiology
McClellan, James T, pathology
McCollum, William Howard, virology
McCook, Robert Devon, physiology
McCord, Michael Campbell, information science
McDowell, Robert Carter, geology
McEllistrem, Marcus Thomas, nuclear physics
McKean, Harlley Ellsworth, mathematics, statistics
McRoberts, J William, urology
Meck, Robert Allen, cancer, hematology
Meijer, Willem, plant taxonomy
Miller, Ralph English, pharmacology
Mink, John R, dentistry
Mitchell, George Ernest, Jr, animal science
Mohr, Hubert Charles, horticulture
Moore, Kenneth Boyd, psychiatry
Morrow, Dean Huston, anesthesiology, cardiovascular physiology
Mulcahy, John Joseph, urology
Munjal, Devidayal, immunochemistry
Munson, James William, analytical chemistry
Musser, A Wendell, pathology
Naae, Douglas Gene, fluorine chemistry, solid state chemistry
Neelakantan, Lakshmanan, organic chemistry, biochemistry
Nelson, Vincent Edward, geology
Newbery, A Chris, mathematics
Nicolai, John Henry, Jr, physiology, animal breeding
Niedenzu, Kurt, inorganic chemistry
Nikitovitch-Winer, Miroslava B, anatomy, endocrinology
Noonan, Jacqueline Anne, pediatric cardiology
Nordin, Gerald LeRoy, insect pathology
Olds, Durward, dairy science, veterinary medicine
Olive, Wilds Williamson, physical geology
Ordman, Edward Thorne, mathematics
O'Reilly, James Emil, analytical chemistry
Othmer, Ekkehard, psychiatry
Ott, Cobern Erwin, medical physiology, nephrology
Overman, Timothy Lloyd, clinical microbiology
Packett, Leonard Vasco, biochemistry
Papka, Raymond Edward, anatomy
Parker, Martin Dale, pharmacy, pharmaceutics
Parks, Harold Francis, anatomy, cytology
Pass, Bobby Clifton, economic entomology, insect ecology
Pattengill, Merle Dean, physical chemistry
Patterson, John Miles, organic chemistry
Pearsall, Marion, medical anthropology, applied anthropology
Peaslee, Doyle E, soil fertility
Phillips, Ronald Edward, soil physics, agronomy
Pirone, Thomas Pascal, plant pathology
Piziak, Veronica Kelly, biochemistry
Plucknett, William Kennedy, chemistry
Poneleit, Charles Gustav, plant genetics, plant breeding
Pratt, Judith Dunlap, physiology
Prior, David James, neurobiology
Purdue, Peter, mathematics, statistics
Ragland, John Leonard, soil chemistry, plant nutrition
Raitz, Karl Bennett, cultural geography, geography of the United States
Randall, David Clark, physiology
Raney, Harley Gene, entomology
Rebagay, Teofila Velasco, pollution chemistry, surface chemistry
Reed, Mary Frances, nuclear chemistry, medical physics
Rees, Earl Douglas, oncology, metabolism
Richards, Dean Boyd, forestry
Richardson, Daniel Ray, physiology
Riley, Herbert Parkes, cytogenetics
Rishel, Raymond Warren, mathematics
Roberts, Clarence Richard, horticulture, plant physiology
Roberts, Thomas Glasdir, micropaleontology, stratigraphy
Robertson, John Connell, animal nutrition
Robinson, George Waller, biochemistry
Rodriquez, Juan Guadalupe, entomology
Roeckel, Irene E, pathology
Rosenbaum, Harold Dennis, medicine
Rosenthal, Gerald A, plant physiology, biochemistry
Rovin, Sheldon, oral pathology
Royster, Wimberly Calvin, mathematics
Ruchman, Isaac, microbiology
Sabes, William Ruben, oral pathology
Sabharwal, Pritam Singh, developmental biology
Sandifer, Myron Guy, Jr, psychiatry
Sands, Donald Edgar, physical chemistry
Saxe, Stanley Richard, periodontology, dentistry
Sayeg, Joseph A, biophysics
Scheibner, Rudolph A, entomology
Schelling, Gerald Thomas, nutrition
Schneider, George William, pomology
Schrils, Rudolph, nuclear physics
Schwert, George William, biochemistry
Sears, Paul Gregory, physical chemistry
Shabetai, Ralph, cardiology, medicine
Shah, Swarupchand Mohanlal, mathematical analysis
Shaw, John Gordon, plant pathology
Sheen, Shuh-Ji, plant genetics
Sherman, Gerald Philip, pharmacology, pharmacy
Shivers, Billie Rae, immunology, pathology
Siegel, Malcolm Richard, plant pathology, toxicology
Sigafus, Roy Edward, agriculture
Sills, Joe Fred, public health
Simmons, Guy Held, nuclear medicine
Simpson, James Edward, mathematics
Sims, John Leonidas, agronomy
Sisk, Dudley Byrd, veterinary pathology
Sisken, Betty Florio, neurobiology
Sisken, Jesse Ernest, cell biology
Slagel, Donald E, biochemistry, neurobiology
Smiley, Jones Hazelwood, agronomy, plant pathology
Smith, Gilbert Edwin, geology
Smith, Stanford Lee, organic chemistry
Smith, Stephen D, anatomy, embryology
Smith, Walter Thomas, Jr, organic chemistry
Spedding, Robert H, dentistry
Stokes, Granville Woolman, crop breeding, plant pathology
Stoltz, Leonard Paul, horticulture
Storrow, Hugh Alan, psychiatry
Surawicz, Borys, internal medicine, cardiology
Survant, William G, soil science
Swerczek, Thomas Walter, veterinary pathology
Swintosky, Joseph Vincent, pharmacy
Sydnor, Katherine Lee, medicine
Tai, Douglas L, medical physics
Tang, Ruen Chiu, forest products
Taylor, Norman Linn, agronomy
Taylor, Timothy H, agronomy
Templeton, William Chelcy, Jr, agronomy
Thacher, Henry Clarke, Jr, computer science, numerical analysis
Thio, Alan Poo-An, medicinal chemistry, pesticide chemistry
Thomas, Grant Worthington, soil chemistry
Thompson, William Oxley, II, mathematical statistics, applied statistics
Thornton, Paul A, physiology, nutrition
Thrailkill, John Vernon, geology, geochemistry
Thrift, Frederick Aaron, animal breeding
Thurston, Richard, entomology
Tietz, Norbert W, clinical chemistry
Townsend, Lee Hill, entomology
Traurig, Harold Henry, anatomy, endocrinology
Tucker, Ray Edwin, animal science
Uglem, Gary Lee, parasitology, zoology
Vandiviere, H Mac, public health
Van Willigen, John Gilbert, applied anthropology, ethnography
Van Winter, Clasine, mathematical physics
Varney, William York, animal husbandry
Waddell, William Joseph, pharmacology
Wagner, William Frederick, chemistry
Wallace, Roberts Manning, economic geology

Walton, Charles Anthony, pharmacy, pharmacology
Wang, Chung Chian, computer science
Ware, Ray Wilsford, medical administration, bioengineering
Wasson, John R., physical chemistry, inorganic chemistry
Webster, Gilbert Theodore, agronomy
Weil, Jesse Leo, nuclear physics
Weinstock, Barnet Mordecai, mathematics
Weiss, Daniel Leigh, pathology
Wekstein, David Robert, physiology
Wells, James Howard, mathematics
Wells, Kenneth Lincoln, soil science, agronomy
Wesley, Robert Cook, dentistry
Wesley, Roger J B., mathematics
Whayne, Tom French, medicine
Wheeler, Harry Ernest, phytopathology
Wheeler, Warren E., medicine
Whicker, McElwyn D., animal science
Wiese, Helen Jean Coleman, medical physics
Wiseman, Ralph Franklin, microbiology
Withington, William Adriance, geography of Southeast Asia
Wrede, Don Edward, radiological physics
Wyatt, John Poyner, pathology
Yates, Steven Winfield, nuclear chemistry, radiochemistry
Yeh, Shu-Yuan, pharmaceutical chemistry, pharmacology
Yost, Francis Lorraine, theoretical physics
Zechman, Frederick William, Jr., physiology
Zolman, James F., neuropsychology, psychopharmacology

LOUISVILLE
Adams, Paul Lieber, psychiatry, sociology
Alexander, Lloyd Ephraim, embryology
Andreasen, Arthur Albinus, industrial microbiology
Andrews, Billy Franklin, pediatrics
Atik, Mohammad, surgery
Beatty, Oren Alexander, medical administration, pulmonary diseases
Bedoit, William Clarence, Jr., polymer chemistry, physical organic chemistry
Berger, Frank Milan, medical research, pharmacology
Best, Maurice McDonald, Jr., medicine
Bhatnagar, Kunwar Prasad, anatomy, neuroanatomy
Bierman, Don Edward, transportation geography, political geography
Blake, James Neal, speech & hearing sciences
Bordoloi, Kiron, solid state physics
Bos, William G., inorganic chemistry, solid state chemistry
Boyer, Harold Edwin, oral surgery
Bronsky, Albert J., biochemistry
Brown, John Wesley, biological chemistry, reproductive endocrinology
Burton, Robert McMahon, immunology
Burzynski, Norbert J., experimental pathology
Calhoun, Thomas Bruce, physiology
Camp, Frank Rudolph, Jr., immunology
Campbell, Ferrell Rulon, anatomy
Carr, Laurence A., pharmacology
Carrasquer, Gaspar, medicine
Chalamalasetty, Venkateswara Rao, reproductive endocrinology
Chalmers, Joseph Stephen, theoretical physics
Christopherson, William Martin, medicine
Clay, William Marion, ichthyology, herpetology
Conkin, James E., geology, paleontology
Cooke, Samuel Leonard, Jr., physical chemistry, analytical chemistry
Covell, Charles Van Orden, Jr., entomology
Crawford, Thomas H., chemistry
Cronholm, Lois S., microbiology
Cummings, Norman Allen, internal medicine
Dagirmanjian, Rose, pharmacology
Dallam, Richard Duncan, biochemistry
Daly, John Matthew, inorganic chemistry

Davis, William S., plant taxonomy, plant cytogenetics
Davitt, Richard Michael, mathematics
Deck, Joseph Charles, physical chemistry
Denber, Herman C B., psychiatry, biochemical pharmacology
Dillon, John Andrew, Jr., physics
Doderer, George Charles, chemistry
Du, Julie (Yi-Fang) Tsai, biochemistry
DuBard, James Leroy, nuclear physics
Du Pre, Donald Bates, chemical physics
Dyar, James Joseph, plant physiology
Edwards, Walton, pediatrics
Emery, Philip Anthony, hydrogeology, geology
Ewing, William McDaniel, medicine
Fields, Raymond Ira, mathematical statistics
Flowers, Nancy Carolyn, cardiovascular diseases
Fonda, Margaret Lee, biochemistry
Fontaine, Julia Clare, anatomy, biology
France, Peter William, organic polymer chemistry
Fuller, Peter McAfee, neuroanatomy
Gauri, Kharaiti Lal, geology
Geeslin, Roger Harold, mathematics
Gibson, Dorothy Hinds, organic chemistry, organometallic chemistry
Granger, Joseph Edward, anthropology
Graver, Richard Byrd, organic polymer chemistry
Gray, Robert Dee, biochemistry
Green, Charles David, organic polymer chemistry
Greenberg, Richard Alvin, epidemiology
Gregg, Robert Vincent, anatomy
Gwinn, Joel Alderson, physics
Hagan, Ralph S., polymer chemistry
Hall, DeLou Perrin, surgery
Harkess, James W., orthopedic surgery
Hayden, Mary Victoria, systematic botany
Haynes, Douglas Martin, obstetrics & gynecology
Heitkamp, Norman Denis, research administration, applied statistics
Herbener, George Henry, electron microscopy
Hicks, Frederic Noble, anthropology
Higginbotham, Robert David, microbiology
Hilton, Frederick Kelker, zoology
Hilton, Mary Anderson, biochemistry
Holt, Joseph Paynter, physiology
Horan, Leo Gallaspy, internal medicine, cardiology
Hotchkiss, Arland Tillotson, plant morphology
Huang, Kee-Chang, pharmacology, physiology
Huang, Wei-Feng, physics
Hubbard, Jack Edward, Jr., neurobiology, anatomy
Hunt, Graham Hugh, economic geology
Huttenlocher, Dietrich F., organic chemistry, lubrication engineering
Jarboe, Charles Harry, chemical pharmacology
Johnson, Robert Reiner, organic chemistry
Johnston, Paul Bruns, medical microbiology
Jordan, Thomas Earl, industrial organic chemistry
Justus, David Eldon, immunology, parasitology
Kalsow, Carolyn Marie, microbiology
Kargl, Thomas E., biochemistry
Keefe, John Richard, biology, cell biology
Keely, William Martin, physical chemistry
Keeney, Arthur Hail, ophthalmology
Keller, Kenneth F., microbiology
Kennedy, John Elmo, Jr., organic chemistry, organic chemistry
Kielkopf, John F., atomic spectroscopy, chemistry
Kleinhenz, Margie Joyce, human anatomy
Knopf, Daniel Peter, chemistry
Kovachevich, Rudy, biochemistry, pathology
Krumholz, Louis Augustus, zoology
Kwo, William T., physics
Lambert, Roger Gayle, plant physiology
Lang, Calvin Allen, gerontology
Lansing, Allan M., cardiovascular surgery
Letter, Charles William, physiological ecology
Levy, Robert Sigmund, biochemistry, polymer nutrition
Lipscomb, Nathan Thornton, chemistry

Liu, Pinghui Victor, medical microbiology
Loeb, Leopold, physical chemistry
Logothetis, Anestis Leonidas, polymer chemistry
Longley, James Baird, histochemistry
Lukes, Robert Michael, oranic chemistry
Macdonald, Roderick, Jr., ophthalmology
MacMillan, Duncan Robert, pediatric endocrinology
Mansfield, John Michael, immunology, microbiology
Mapes, William Henry, analytical chemistry
McBain, John Keith, nuclear medicine, internal medicine
McConnell, Kenneth Paul, biochemistry
McGeachin, Robert Lorimer, biochemistry
Meyer, John Roger, physiology, zoology
Miller, Kenneth Leron, polymer chemistry
Mills, Roger Edward, biology
Minhas, Kareem, organic chemistry, pharmacology
Monroe, Burt Leavelle, Jr., ornithology
Moore, Condict, surgery
Moore, James Carlton, physiology
Murray, Marvin, clinical pathology
Naake, Hans Joachim, acoustics, solid state physics
Neff, Stuart Edmund, biology
Nelson, Lloyd Steadman, mathematics
Nicholson, John Angus, organic chemistry, pharmacology
Noland, Jerre Lancaster, biochemistry
Ouseph, Pullom John, nuclear physics
Past, Wallace Lyle, pathology
Pearson, William Dean, fisheries
Phillips, John Perrow, analytical chemistry
Plummer, Louisa Greenleaf, analytical chemistry
Polk, Hiram Carey, surgery
Potts, Albert Mintz, biochemistry
Price, Martin Burton, organic chemistry
Queenan, John T., obstetrics & gynecology
Racuglia, Giovanni, medicine
Raff, Martin Jay, infectious diseases
Ransdell, Herbert Threlkeld, Jr., surgery
Reazin, George Harvey, Jr., plant physiology, biochemistry
Richter, Edward Eugene, organic chemistry, inorganic chemistry
Rink, Richard Donald, human anatomy, operations research, systems analysis
Roark, Adelbert Lee, operations research, systems analysis
Robinson, Thane Sparks, ecology
Roseman, Ephraim, medicine
Rosenberg, Alexander F., analytical chemistry
Rosene, Clarence James, organic chemistry, analytical chemistry
Sabel, Clara Ann, mathematics
Sanford, Robert Alois, organic chemistry
Scharff, Thomas G., pharmacology
Schwab, John J., psychiatry, psychosomatic medicine
Scott, George William, organic chemistry
Scott, Leland Latham, mathematics
Scott, Ralph Mason, radiology
Segal, Edwin Stanley, cultural anthropology
Seibert, Mary Angelice, biochemistry
Shoemaker, Gradus Lawrence, organic chemistry
Short, James Harold, pharmacology
Silver, Burton Barney, physiology, electron microscopy
Sinai, John Joseph, physics
Smith, Richard Petri, pharmacology
Smothers, James Llewellyn, animal physiology
Spatola, Arno F., bio-organic chemistry
Spragens, William Henry, Jr., mathematics
Stewart, Arthur Van, dentistry
Streips, Uldis Normunds, microbiology
Strunk, Duane H., analytical chemistry
Surwillo, Walter Wallace, psychophysiology
Swartz, Frank Joseph, anatomy
Sweeny, Daniel Michael, inorganic chemistry
Swigart, Richard Hanawalt, neuroanatomy
Tamburro, Kathleen O'Connell, protozoology
Taylor, John Fuller, biochemistry
Taylor, Kenneth Grant, organic chemistry
Teller, David Norton, neurochemistry, psychopharmacology
Towery, Beverly Todd, medicine
Trapp, Charles Anthony, physical chemistry

Tucker, Irwin William, organic chemistry, environmental engineering
Turner, Ella Victoria, immunology
Vance, Lawrence Joseph, food science
Voogt, James Leonard, physiology, neuroendocrinology
Wacker, Waldon Burdette, immunology
Wagner, Charles Eugene, morphology
Waite, Leonard Charles, pharmacology
Walker, Sheppard Matthew, physiology, biophysics
Wallace, John Howard, microbiology, immunology
Waller, Mary Concetta, inorganic chemistry
Wead, William Badertscher, cardiovascular physiology
Weber, Frederick, Jr., microbiology, public health
Weisskopf, Bernard, pediatrics, psychiatry
Westphal, Ulrich Friedrich, biochemistry
Whitaker, Frederick Horace, parasitology
Wiedeman, Varley Earl, botany, ecology
Williams, Donald Elmer, x-ray crystallography, physical chemistry
Wittwer, John William, periodontology
Wolfe, Walter McIlhaney, obstetrics & gynecology
Yam, Lung Tsiong, internal medicine
Yankeelov, John Allen, Jr., biochemistry, organic chemistry
Zavist, Algerd Frank, chemistry

MOREHEAD
Clark, William Thomas, Jr., geography of Latin America
Derrickson, Charles M., animal science
DuBar, Jules R., invertebrate paleontology
Falls, William Randolph, Sr., science education
Gould, Robert Barris, geography
Heaslip, Margaret Barkley, plant ecology, plant genetics
Philley, John Calvin, geology
Simon, Verne A., physical chemistry
Spears, James Richard, reproductive endocrinology, experimental embryology

MURRAY
Beyer, Louis Martin, nuclear physics
Cantrell, Grady Leon, mathematics
Clark, Armin Lee, geology
Clark, Howell R., inorganic chemistry, analytical chemistry
Cole, Evelyn, zoology
Conley, Harry Lee, Jr., physical chemistry
Duncan, Don Darryl, physics
Eversmeyer, Harold Edwin, plant pathology, botany
Fuller, Marian Jane, plant taxonomy, genetics
Gordon, Annette Waters, organic chemistry, biochemistry
Gordon, Marshall, physical organic chemistry
Hancock, Hunter McRae, zoology
Hendon, Joseph D., organic chemistry
Henley, Melvin Brent, Jr., physical chemistry
Hussung, Karl Frederick, organic chemistry
Klein, William Richard, acoustics
Kline, James Martin, solid state physics
Maddox, William Eugene, nuclear physics
McClellan, Bobby Ewing, analytical chemistry
McHugh, William Paul, anthropology, archaeology
Mikulcik, John D., agronomy
Panzera, Pete, organic chemistry
Read, William James, nuclear physics
Spann, Liza Agnes, biology
Vandegrift, Vaughn, biochemistry
Visher, Halene Hatcher, geography, resource geography
Whaley, Peter Walter, geology
Wilder, Cleo Duke, Jr., vertebrate zoology
Wolfson, Alfred Mortimer, plant morphology

NAZARETH
Heines, Virginia, chemistry
Juhasz, Roderick, analytical chemistry, biochemistry

NICHOLASVILLE
Gray, Roy C. Jr., animal science, biology

OWENSBORO
Applin, Paul Livingston, geology
Armendarez, Peter X., physical chemistry
Dalziel, Robert Clinton, medical microbiology
Flachskam, Robert Louis, Jr., organic chemistry
Gerteisen, Thomas Jacob, organic chemistry
Magnuson, Winifred Lane, inorganic chemistry
Purdom, Ray Caldwell, physics

PADUCAH
Levin, Robert Warren, nuclear chemistry
Plebuch, Raymond Otto, hydrology

PEMBROKE
Bradshaw, Benjamin Crenshaw, surface physics

PIKEVILLE
Bailey, Maurice Eugene, science education, mining technology
Darlage, Larry James, organic chemistry
Williams, Martin Wesley, toxicology

PRINCETON
Trace, Robert Denny, geology

PROSPECT
Lich, Robert, Jr., urology

RICHMOND
Barkley, Dwight G., horticulture
Bendall, Victor Ivor, organic chemistry
Branson, Branley Allan, ichthyology, malacology
Britt, Danny Gilbert, dairy nutrition
Byrn, Ernest Edward, physical chemistry, analytical chemistry
Cupp, Paul Vernon, Jr., physiological ecology, herpetology
Davidson, John Edwin, analytical chemistry, inorganic chemistry
Dixon, Wallace Clark, Jr., physiology
George, Ted Mason, nuclear physics, electronics
Haney, Donald C, geology, soils
Harley, John Paul, physiology, parasitology
Helfrich, Charles Thomas, paleontology, stratigraphy
Hess, Edwin A., biology, physiology
Hill, Roberta Bleiler, nutrition, biochemistry
Hoge, Harry Porter, sedimentary petrology
Householder, William Allen, agriculture
Howard, Aughtum Smith, mathematics
Jones, Sanford L, reproductive endocrinology
Keefe, Thomas Leeven, botany, biology
King, Amy P., mathematics
Kubiak, Timothy James, urban geography, resource development
LaFuze, Henry Harvey, botany
Laird, Christopher Eli, nuclear physics
Lane, Bennie Ray, mathematics
Lassetter, John Stuart, botany
LeVan, Marijo Marie, economic geography, planning
Luts, Heino Alfred, medicinal chemistry
Martin, William Haywood, III, plant ecology, forest ecology
McGlasson, Alvin Garnett, mathematics
Meisenheimer, John Long, organic chemistry
Otero, Raymond B, microbiology, biochemistry
Powell, Howard B., inorganic chemistry
Rudersdorf, Ward J, wildlife management, ecology
Salyer, Darnell, analytical chemistry
Schulz, William, analytical chemistry
Schwendeman, Joseph Raymond, Jr., physical geography, resource geography
Scorsone, Francesco G., mathematics
Sebor, Milos Marie, economic geography, planning
Smiley, Harry M, physical chemistry
Taylor, Morris D., science education, physical chemistry
Thompson, Marvin Pete, Jr., wildlife ecology, mammalogy
Thompson, Ralph J., physical chemistry
Williams, John C, zoology

WILLIAMSBURG
Early, Joseph E, mathematics education
Wilder, William Baylor, botany, genetics

WILMORE
Hamann, Cecil Boyce, biology
Higgins, Jerry Mitchell, speech pathology
Howell, Henry Howze, aquatic biology, aquatic ecology
Morris, Edward C, chemistry
Ray, Jesse Paul, analytical chemistry
Smith, John Milton, microbiology, biochemistry

WINCHESTER
Marsh, Robert Cecil, molecular biology
Perozzi, Edmund Frank, organic chemistry

LOUISIANA

ALEXANDRIA
Britt, Henry Grady, zoology
McMillin, Charles W., wood science & technology
Wells, Darthon Vernon, organic chemistry

ARCADIA
Ozley, Elsie Church, geometry

BATON ROUGE
Aaron, Charles Sidney, molecular genetics, mutagenesis
Adams, Sam, science education
Addison, Leslie Mandeville, organic chemistry
Allen, Robert Scott, biochemistry
Andersen, Harold Veral, geology
Anderson, Richard Davis, mathematics
Anzalone, Louis, Jr., phytopathology
Avault, James W, Jr., biology
Baham, Arnold, dairy nutrition
Bailey, George William, analytical chemistry
Baker, John Bee, plant physiology
Barrett, Robert Earl, physical chemistry, polymer chemistry
Barrios, Earl P., horticulture
Beadle, Ralph Eugene, veterinary physiology
Berg, Eugene Walter, analytical chemistry
Bennett, Harry Jackson, zoology
Bernat, Thomas Phillip, low temperature physics
Besch, Everett Dickman, veterinary parasitology, veterinary public health
Bigger, Cynthia Anita Hopwood, molecular biology
Birchfield, Wray, plant pathology
Black, Lowell Lynn, plant pathology
Bond, Howard Emerson, astronomy
Boozer, Charles (Eugene), rubber chemistry
Boudreaux, Henry Bruce, entomology
Boulware, Ralph Frederick, genetics
Bowden, Joe Allen, biochemistry, solar energy
Branton, Cecil, animal breeding
Braymer, Hugh Douglas, molecular biology
Brown, Clair Alan, taxonomy, plant ecology
Brown, Jerome Engel, industrial chemistry
Brown, Paul Bruce, animal science
Bryant, James Berry, Jr., microbiology
Bryan, Charles F., ichthyology, limnology
Bullock, Howard R., entomology, physiology
Burnett, William Thomas, Jr., radiochemistry
Burns, Edward Columbus, entomology
Burns, Paul Yoder, forestry
Bursh, Talmage Poutrau, physical chemistry
Burt, William Enos, chemistry
Butts, Hubert S, mathematics
Caffey, Horace Rouse, agronomy, plant breeding
Caldwell, Augustus George, soil chemistry
Callaway, Joseph, theoretical physics
Carpenter, Dewey Kenneth, physical chemistry, polymer chemistry
Carpenter, Paul Gershom, organic chemistry
Cartledge, Frank, organic chemistry
Causey, Nell Bevel, invertebrate zoology
Chabreck, Robert Henry, wildlife management, ecology
Chakkalakal, Dennis Abraham, theoretical physics
Chambers, Doyle, animal genetics, animal breeding
Chanmugam, Ganesar, theoretical astrophysics
Chapman, Russell Leonard, phycology
Chardon, Roland E., cultural geography
Chilton, St John Poindexter, plant pathology
Choong, Elvin T., wood technology, forestry
Clarke, Wilbur Bancroft, organic chemistry
Clower, Dan Frederic, entomology
Coleman, James Malcolm, geomorphology
Collins, Heron Sherwood, mathematics
Collins, Richard Lapointe, invertebrate zoology, protozoology
Colmer, Arthur Russell, bacteriology
Conner, Pierre Euclide, Jr., mathematics
Conrad, Franklin, industrial chemistry
Cook, Shirl Eldon, chemistry
Cooper, Douglas Elhoff, organic chemistry
Cooper, Ronda Fern, medical microbiology
Corkum, Kenneth C., parasitology, invertebrate zoology
Cragg, Hoyt J, organic chemistry
Crow, Alonzo Bigler, forest ecology
Cull, Neville, inorganic chemistry
Culley, Dudley Dean, Jr., zoology
Cutler, Janice Zemanek, algebra
Daly, William Howard, organic chemistry, polymer chemistry
Damann, Kenneth Eugene, Jr., plant pathology
Davenport, Tom Forest, Jr., organic chemistry, pharmaceutical chemistry
Davis, Buster Hall, poultry nutrition
Day, Marion Clyde, Jr., inorganic chemistry, physical chemistry
Dietz, Thomas Howard, animal physiology
Dixon, Joe Maurice, veterinary science
Dixon, Louis H., geology
Dommert, Arthur Roland, veterinary microbiology
Dorroh, James Robert, mathematics
Drushel, Harry (Vernon), analytical chemistry
Dunigan, Edward P., soil microbiology
Durham, Clarence Orson, Jr., geology
Earle, Norman Williston, entomology
Edwards, Warrick Rigeley, Jr., organic chemistry
Eisenbraun, Allan Alfred, organic chemistry
Farthing, Barton Roby, applied statistics
Ferrell, Ray Edward, Jr., mineralogy, geochemistry
Fischer, Nikolaus Hartmut, natural products chemistry, synthetic organic chemistry
Floyd, Ernest Hazel, entomology
Fogg, Peter John, wood science, wood technology
Forbes, Irvin L., botany
Foster, Walter Edward, industrial organic chemistry, research administration
French, Wilbur Lile, cytogenetics
Frey, Frederick Wolff, Jr., inorganic chemistry
Frye, Jennings Bryan, Jr., dairy science
Gagliano, Sherwood Moneer, physical geography, archaeology
Gandour, Richard David, physical organic chemistry, bio-organic chemistry
Gassie, Edward William, animal science, information science
Giamalva, Mike J., horticulture, plant pathology
Gladrow, Elroy Merle, physical chemistry
Glasgow, Leslie Lloyd, wildlife management
Godke, Robert Alan, reproductive physiology
Golden, Laron E., soil fertility, plant nutrition
Goodrich, Max, experimental physics
Goodrich, Roy Gordon, solid state physics
Gordon, Douglas Littleton, medicine
Gosselink, James G, plant ecology, marine sciences
Graves, Jerry Brook, entomology
Grenchik, Raymond Thomas, astrophysics
Grenier, Claude Georges, solid state physics
Griffin, Ernest Lyle, mathematics
Grodner, Mary Laslie, embryology, developmental biology
Grodner, Robert Maynard, food science, zoology
Haag, William George, anthropology
Hamilton, Robert Bruce, wildlife ecology, ornithology
Hamilton, William Oliver, physics
Hammond, Abner M, Jr., entomology
Hanchey, Richard Howard, floriculture
Harbo, John Russell, apiculture
Harman, Walter James, zoology
Hart, Lewis Thomas, microbiology, biochemistry
Harvey, Clarence Charles, (Jr), chemistry
Hawthorne, Percy Lynnwood, horticulture
Haymaker, Richard Webb, particle physics
Hembry, Foster Glen, animal nutrition
Henderson, Merlin Theodore, agronomy
Hendrick, Rodney Douglas, insect ecology
Henry, Ronald James Whyte, atomic physics
Hensley, Sess D, entomology
Hernandez, Teme P., horticulture, plant breeding
Hildebrant, John A., algebra, topology
Hilliard, Sam Bowers, geography
Ho, Clara Lin, soil chemistry, marine chemistry
Hoff, Bert John, plant breeding
Holcomb, Gordon Ernest, plant pathology
Hollis, John Percy, Jr., plant pathology
Horn, Norman Louis, Jr., plant pathology
Hornbaker, Edwin Dale, organic chemistry
Hoskins, Frederick Hall, food science, nutrition
Houk, Kendall Newcomb, organic chemistry
Howie, Milham Salem, organic chemistry
Hsu, Shih-Ang, meteorology
Hubbert, William T., veterinary medicine, immunology
Huggett, Richard William, Jr., physics
Humes, Paul Edwin, animal science, genetics
Hussey, Robert Gregory, fluid dynamics
Iddings, Frank Allen, nuclear science, analytical chemistry
Impastato, Fred John, organic chemistry
Jackson, Blanche Ellen, physiology
Jackson, Hezekiah, plant physiology
James, William Holden, food science
Jamison, Robert Edward, mathematics
Jefferson, Jack Howard, analytical chemistry
Johnson, Clyde Edgar, Jr, physiology, zoology
Johnson, James Augustus, Jr., petroleum chemistry, chemical engineering
Johnson, William Alexander, poultry science
Johnston, James Douglas, organic chemistry
Jones, Jack Earl, agronomy
Jones, Lloyd George, horticulture
Jones, Paul Hastings, geology
Jones, William Vernon, cosmic ray physics
Karnes, Houston Thurman, mathematics
Keisler, James Edwin, mathematics
Keister, Thomas Dwight, forestry, experimental statistics
Kent, George Cantine, Jr., zoology
Kesel, Richard Herman, geography
Kestner, Neil R, chemical physics
Kewish, Ralph Wallace, chemistry
Kirsch, Warren Bernard, inorganic chemistry
Kniffen, Fred Bowerman, cultural geography, ethnography
Koch, Robert Jacob, mathematics
Koenig, Paul Edward, organic chemistry
Koonce, Kenneth Lowell, experimental statistics
Kreider, Jack Leon, animal science, reproductive physiology
Kupfer, Donald Harry, structural geology
Lambremont, Edward Nelson, entomology, nuclear science
Landolt, Arlo Udell, astronomy
Lank, Robert Byron, veterinary medicine
Larkin, John Montague, microbial physiology, microbial ecology
Lawson, Jimmie Don, mathematics
Lee, Jordan Grey, Jr., chemistry
Lee, Pang-Kai, physical chemistry
Lee, Paul D., astronomy, astrophysics
Lee, William Roscoe, genetics
Lesniak, Linda Marie, mathematics
Lewis, Anthony James, physical geography
Lewis, Harrye Fleming, nutrition
Li, Hsueh Ming, polymer chemistry
Lieux, Meredith Hoag, palynology, botany
Lindberg, George Donald, plant pathology

LOUISIANA

Lindsay, Kenneth Lawson, organic chemistry
Linnartz, Norvin Eugene, forest soils
Loesch, Harold Carl, biological oceanography
Lowe, Donald Ray, sedimentology
Lowry, George Hines, Jr, ornithology
Luizzo, Joseph Anthony, food science
Lyles, George Robert, analytical chemistry
Macomber, James Dale, chemical physics
Madison, Bernard L., mathematics
Mangham, Jesse Roger, organic chemistry
Mann, Edward Cullee, medicine
Martin, Clifton Boyd, forest economics
Martin, Freddie Anthony, plant physiology; crop physiology
Martin, Julia Mae, biochemistry
Martin, Weston Joseph, plant pathology
Martinez, Joseph Didier, environmental geology; environmental engineering
Mattice, Wayne Lee, physical biochemistry
Mayes, McKinley, agronomy
McDermid, Robert Wesson, forest engineering; forest economics
McGehee, Oscar Carruth, mathematics
McGlynn, Sean Patrick, physical chemistry
McIntire, William Grant, geography
McKnight, William Carl, organic chemistry
McLellan, Crawford Reid, organic chemistry
McNary, Sidney A Jr, biochemistry, organic chemistry
Meier, Albert Henry, zoology; physiology
Merrill, Howard Emerson, petroleum chemistry
Metevia, Louis Anthony, cell biology, electron microscopy
Meyers, Samuel Philip, marine microbiology
Miller, Russell Lee, crop science
Mitchell, Benjamin Evans, mathematics
Moore, Clyde H, Jr, geology
Morgan, James Plummer, geology
Morrissette, Maurice Corlette, chemistry
Morrow, Norman Louis, physical chemistry, analytical chemistry
Novak, Arthur Francis, chemistry
Nyman, Dale James, groundwater geology
Owens, Edward Henry, geomorphology, sedimentology
O'Connell, Robert F, theoretical astrophysics, theoretical physics
Ohm, Jack Elton, mathematics
Oliver, Abe D, Jr, entomology; forestry
O'Rourke, Edmund Newton, Jr, horticulture
Pahl, Gordon, mathematics
Patrick, Thomas Everette, animal physiology
Patrick, William H J, Jr, agronomy; soils
Pearson, Tilmon Henry, organic chemistry
Peevy, Walter Jackson, soils
Perry, Charles Lewis, astronomy
Pesson, Lynn L, agricultural education
Piller, Herbert, solid state physics
Pine, Lloyd A, organic chemistry
Plonsker, Larry, organic chemistry
Porter, Lee Grant, physics
Pryor, William Austin, organic chemistry; biological chemistry
Rajagopal, Atirpat Krishnaswamy, theoretical physics
Ralph, Dorr C, physics
Rao, Ramachandra M R, food science & technology
Rau, A Ravi Prakash, atomic physics
Rees, Paul Klein, mathematics
Reid, Kenneth Brooks, mathematics
Retherford, James Ronald, mathematics
Reynolds, Joseph Melvin, physics
Richardson, Leonard Frederick, mathematical analysis
Richardson, Miles Edward, cultural anthropology

Rickey, Frank Atkinson, mathematics
Riddick, John Allen, analytical chemistry
Rinderer, Thomas Earl, genetics, insect pathology
Risinger, Gerald E., chemistry
Roberts, Edgar D., veterinary pathology
Roberts, Harry Heil, geology
Roberts, John Harvey, entomology
Robertson, George Leven, animal science
Robinson, Gene Conrad, organic chemistry
Robinson, George Edward, Jr, physiology
Robinson, James William, analytical chemistry
Robinson, Press L., physical chemistry
Robson, Harry Edwin, physical chemistry
Rolston, Lawrence H, entomology
Rossman, Douglas Athon, vertebrate zoology
Roussel, John S, entomology; agronomy
Roussel, Joseph Donald, reproductive physiology
Rudd, Walter Greyson, chemistry
Ruffin, Spaulding Merrick, biological chemistry
Runnels, Lynn Kelly, chemical physics
Rush, Milton Charles, phytopathology
Rusoff, Louis Leon, animal nutrition
Sachder, Sham L., chemistry
Sanders, Robert N, physical chemistry; inorganic chemistry
Schilling, Prentiss Edwin, statistics
Scholz, Dan Robert, mathematics
Schori, Richard M., topology
Schrodt, Ariel Gilbert, physical chemistry; biomedical engineering
Scott, Louis, zoology
Sedberry, Joseph E, Jr, agronomy; soil chemistry
Sethin, Joel, inorganic chemistry
Sen, Buddhadev, physical inorganic chemistry
Shapiro, Hymin, organic chemistry
Short, Charles Robert, pharmacology
Siebeling, Ronald Jon, microbiology
Smalley, Arnold Winfred, organic chemistry
Smart, Lewis Isaac, animal nutrition
Smith, Isaac Litton, analytical chemistry
Smith, Martin Bristow, physical chemistry
Smith, Robert Leonard, analytical chemistry; mass spectrometry
Snow, Johnnie Park, plant pathology
Socolofsky, Marion David, microbiology
Spink, William T, entomology
Srinivasan, Vadake Ram, biochemistry; microbiology
Stadther, Richard James, plant physiology; plant breeding
Standifer, Leonides Calmet, Jr, plant physiology
Steelman, Carrol Dayton, medical entomology; veterinary entomology
Steib, Rene J, plant pathology
Suhayda, Joseph Nicholas, oceanography
Teekell, Roger Alton, metabolism, biochemistry
Thielges, Bart A., forest genetics
Thompson, Robert, neuropsychology
Thrasher, Donald Miller, animal science, animal husbandry
Tipton, Kenneth Warren, plant genetics, agronomy
Titkemeyer, Charles William, anatomy
Traynham, James Gibson, organic chemistry
Tsai, Tom Chung Hsiung, polymer chemistry; colloid chemistry
Tucker, Kenneth Wilburn, apiculture
Tucker, Shirley Cotter, botany
Van Den Bold, Willem Aaldert, paleontology
Van Lopik, Jack Richard, research administration; marine sciences
Vermeer, Donald E., geography
Vick, Maurice M, analytical chemistry
Vidaurreta, Luis E, analytical chemistry
Von Bodungen, George Anthony, physical chemistry
Wade, Luther Irwin, mathematics
Wagner, Raymond Lee, theoretical astrophysics, astronomy
Waite, John Henry, surgery
Walker, Harley Jesse, geography
Walker, John Robert, entomology; research administration
Walter, Thomas James, organic chemistry
Wang, Ching-Ping Shih, theoretical solid state physics
Watkins, Steven F, structural chemistry; crystallography
Watts, Alva Burl, animal nutrition
Wayt, Lewis Keith, veterinary medicine

Werner, Henry James, histology; cytology
West, Philip William, chemistry
West, Robert Cooper, geography of Latin America, historical geography
Wharton, James Henry, physical chemistry
Wheelock, Kenneth Steven, petroleum chemistry; inorganic chemistry
Whelan, Thomas, III, marine geochemistry
White, Lewis L., embryology
Wilie, John Jacob, Jr, cell biology
Williams, Curtis, agronomy; plant breeding
Williams, George, Jr, botany
Williams, Hulen Brown, physical chemistry
Williams, James Carl, animal parasitology
Williams, Lynn Dolores, mathematics
Williams, Vernon, mathematics, education
Willis, Guye Henry, soil chemistry
Willis, William Hillman, soils
Wiseman, William Joseph, Jr, oceanography
Wood, James Manley, Jr, physical chemistry
Woodring, Jay Porter, environmental physiology
Yadav, Raghunath P, entomology
Yang, Chia Hsiung, nuclear physics
Younathan, Ezzat Saad, biochemistry
Younathan, Margaret Tims, food science; nutrition
Zagar, Walter T., physical chemistry; polymer chemistry
Zavon, Mitchell Ralph, medicine
Zebouni, Nadim H., solid state physics
Zganjar, Edward F., nuclear physics
Zimmerman, Peter David, nuclear physics, meteorics

BERWICK
Connell, Frank Herman, parasitology

BELLE CHASSE
Rose, Sylvan Meryl, embryology
Wieviorowski, Tadeusz Karol, chemistry

CALHOUN
Woodward, Ralph Stanley, horticulture

CARVILLE
Brand, Paul W., reconstructive surgery; rehabilitation
Kirchheimer, Waldemar Franz, microbiology

CHASE
Hernandez, Travis Paul, plant genetics

CORBIN
Wascom, Earl Ray, ecology

COVINGTON
Brizzee, Kenneth Raymond, neuroanatomy; neuropathology
Felsenfeld, Oscar, laboratory medicine; virology
Felsenfeld, Amihan Dasaneyavaja, virology
Gerone, Peter John, virology
Hanson, Alvin Maddison, stratigraphy
Harrison, Richard Miller, reproductive physiology; primatology
Hofer, Helmut Otto, anatomy; comparative anatomy
Li, Su-Chen, biochemistry
Martin, Louis Norbert, immunology

CROWLEY
McIlrath, William Oliver, plant breeding; genetics
White, Thomas Wayne, ruminant nutrition

EUNICE
Couvillion, John Lee, organic chemistry
Lembeck, William Jacobs, microbiology; academic administration
Roques, Alban Joseph, mathematical analysis

GEISMAR
La Rochelle, John Hart, physical chemistry

GRAMBLING
Agarwal, Arun Kumar, pure mathematics
Bailey, R L, avian physiology; veterinary anatomy
Gajendar, Nandigam, applied mathematics, computer science
Hill, Franklin D, biochemistry
Howell, Monticello Jefferson, horticulture
Moorti, Varahur R Guru, physical chemistry; applied mathematics
Perkins, Lee E, cultural geography; history
Sabu, Dwarka Das, nuclear chemistry; cosmochemistry

GRETNA
Simonoff, Robert, organic chemistry

HAMMOND
Bond, William Payton, plant pathology
Brown, Robert C, mathematics
Corkern, Walter Harold, organic chemistry
Crain, Charles Larry, zoology
Hayes, Donald H, physiology; genetics
Kirk, Ben Truett, plant pathology
Munchausen, Linda Lou, organic chemistry
Nelson, Harvard G, agronomy; agricultural statistics
Pullen, Bailey Price, chemical physics
Settoon, Patrick Delano, biochemistry; food chemistry
Shepherd, David Preston, radiation biology; zoology
Wallace, Willie Robert, microbiology; dairy science
Watson, James Arthur, Jr, chemistry
White, Zebulon Waters, forestry

HAUGHTON
Pope, Robert Eugene, veterinary public health

HOUMA
Benda, Gerd Thomas Alfred, plant physiology
Irvine, James Estill, plant physiology
Jackson, Robert Dewey, entomology
Koike, Hideo, phytopathology
Sanford, James Walker, economic entomology
Yang, Shaw-Ming, microbiology

JEANERETTE
Davis, Johnny Henry, agronomy

KAPLAN
Girouard, Rustum Ernest, Jr, animal nutrition; animal physiology

LA PLACE
Maskornick, Michael J, physical organic chemistry

LAFAYETTE
Andrew, David Robert, topology
Authement, Ray Paul, applied mathematics
Batson, Lewis E., applied mathematics
Bernard, Davy Lee, nuclear physics
Cain, Charles Columbus, soil science
Chow, Bryant, mathematics, statistics
Cosper, Sammie Wayne, nuclear physics
Crisler, Robert Morris, geography
Dakin, Matt Eitel, entomology
DeLaitsch, Dale M, organic chemistry
Dickinson, Peter Charles, statistics
Durio, Walter O'Neal, parasitology
Edwards, Joseph D Jr, organic chemistry
Erbe, Lawrence Wayne, botany
Eyster, Marshall Blackwell, ornithology
Fitzpatrick, Jimmie Doile, organic chemistry; analytical chemistry
Flannery, William Louis, bacteriology
Fletcher, William Ellis, horticulture
Foret, James A., ornamental horticulture; weed science
Forney, Frederick Willis, microbial physiology
Gimbrede, Louis de Agramonte, physiology
Graham, Lewis Texada, entomology
Grimsal, Edward George, acoustics

Hamlin, William Henry, geology, paleontology
Hathorn, Fred, animal husbandry
Hearn, Dwight D., applied mathematics
Heatherly, Henry Edward, mathematics
Hoese, Hinton Dickson, marine biology
Keeley, Dean Francis, inorganic chemistry, radiochemistry
Keiser, Edmund Davis, Jr., vertebrate zoology, wildlife ecology
Kessinger, Walter Paul, Jr., micropaleontology
Lafleur, Louis Dwynn, solid state physics
Lanoux, Sigred Boyd, inorganic chemistry
Leutze, Willard Parker, geology
Ligh, Steve, mathematics
Matese, John J., theoretical physics
Meriwether, John R., nuclear physics, computer science
Oliver, James Russell, physical chemistry, nuclear chemistry
Parker, Robert Davis, chemistry
Pease, Lawrence Honeyman, animal husbandry, veterinary medicine
Perkins, Richard Scott, electrochemistry
Reese, William Dean, biology
Reilly, Timothy Frank, geography
Schroeder, Rudolph Alrud, physical chemistry
Shriver, Bruce Douglas, computer science
Sidman, Robert David, mathematics
Stallings, Emmett Francis, physical geography
Stueben, Edmund Bruno, parasitology
Sullivan, Victoria I., biosystematics
Testerman, Jack Duane, statistics
Walker, Terry M., operations research, computer science
Walton, Thomas Peyton, III, surgery
Webb, Robert MacHardy, physical geography
White, Ronald Joseph, theoretical chemistry, applied mathematics
Wiechelman, Karen Janice, biophysical chemistry
Wood, William Hulbert, geology, mathematics

LAKE CHARLES

Black, Joe Bernard, invertebrate zoology
Bogle, Tommy Earl, solid state physics
Bromberg, Milton Jay industrial organic
Bryant, Robert L., forestry
Chapman, Harold Clyde, entomology
Cobb, Glenn Wayne, plant morphogenesis, plant pathology
Dahlquist, Wilbur Lynn, solid state physics
Ford, Patrick Lang, mathematics
Hankins, Bobby Eugene, analytical chemistry, spectroscopy
Ieyoub, Kalil Phillip, organic chemistry
Iglinsky, William, entomology
Lane, James Dale, vertebrate zoology
Monsour, Victor, microbiology
Petersen, James J, entomology
Robinson, Roy Garland, Jr., physiology, histology
Stilley, Jerry Lee, physics
Swetharanyam, Lalitha, mathematical analysis
White, George V S, vertebrate zoology
Young, John Coleman, statistics

MANDEVILLE

MacDougall, Robert Douglas, petroleum geology

METAIRIE

Bailiff, Ralph Norman, cytology
Becnel, Irwin Joseph, entomology
Berni, Ralph John, analytical chemistry
Brown, Lawrence Eldon, microchemistry
Franklin, George Joseph, geology
Hoy, Robert C, solid state physics, crystallography
Hughes, James Perry, chemistry
Jacks, Thomas Jerome, biochemistry, plant physiology
McAndrews, Harry, geology
Owens, Neal, surgery
Reeves, Wilson Alvin, textile chemistry
Singh, Jasbir Mahajan, pharmacology, psychopharmacology
Wallace, Alexander Doniphan, mathematics
Welch, Clark Moore, organic chemistry

MONROE

Baker, Earl Wayne, organic chemistry
Barrett, Elton Ray, plant pathology, breeding
Baum, Lawrence Stephen, plant physiology, cell physiology
Bedell, Louis Robert, surface physics
Bennett, Lonnie Truman, mathematics, statistics
Biersmith, Edward L, organic chemistry, bio-organic chemistry
Bounds, Harold C, microbial ecology
Boyd, Frank McCalla, bacteriology
Byrd, David Shelton, organic chemistry, physical chemistry
Cason, James Lee, dairy husbandry, nutrition
Cauther, Sally Eugenia, biochemistry
Danti, August Gabriel, pharmacy, chemistry
DePoe, Charles Edward, botany, fresh water ecology
Douglas, Neil Harrison, systematic ichthyology
Dupree, Daniel Edward, mathematics
Eickholt, Theodore Henry, pharmacology
Ferguson, Gary Gene, pharmacology
Franklin, Beryl Cletis, reproductive physiology
Fritsche, Paul Frank, mathematics
Geiger, Paul Frank, pharmacology
Glawe, Lloyd Neil, invertebrate paleontology
Hammons, Jasper Glen, agronomy
Holt, Robert Louis, organic chemistry
Johnson, Larry Don, physics
Kee, David Thomas, ornithology
Kern, Joseph Herschel, pharmacy administration
Miller, Kenneth Jay, physical chemistry
Miller, William Wadd, III, reproductive physiology
Norris, William Warren, zoology
Overton, Edward Beardslee, analytical chemistry, inorganic chemistry
Prince, Buford Earl, parasitology
Ricks, Beverly Lee, physiology, zoology
Saenz, Reynaldo V, pharmaceutical chemistry
Shrader, Kenneth Ray, pharmacy administration
Shugart, Cecil G, nuclear physics, polymer physics
Smith, Charles R, mathematics
Smith, Ronald E, physics
Stevenson, Dennis A, biophysics
Thomas, Roy Dale, plant taxonomy
Vingiello, Frank Anthony, organic chemistry
Wright, Oscar Lewis, physical chemistry
Young, Leonard M, geology

NATCHITOCHES

Allen, Arthur (Silsby), plant pathology, mycology
Anders, Edward B, mathematical analysis
Baumgardner, Ray K, limnology
Bienvenu, Rene Joseph, Jr., microbiology
Bissell, Charles Lynn, physical chemistry
Daugherty, Zoel W, soil science, plant physiology
Knipmeyer, William B, geography
Lin, James C H, genetics, cell physiology
Martin, Frank Winstead, botany
Outland, Roderick Henderson, cytology
Ryan, Donald Edwin, mathematics, chemistry
Stokes, George Alwin, geography
Temple, Austin Limiel, (Jr), applied mathematics
Williams, Kenneth L, zoology

NEW IBERIA

Bernal, Ernesto, veterinary pathology
Burchfield, Harry P, biochemistry
Price, Keith Clayton, ecology
Storrs, Eleanor Emerett, biochemistry, medical research
Wheeler, Ralph John, analytical chemistry

NEW ORLEANS

Abadie, Stanley Herbert, medical parasitology, medical education
Adriani, John, surgery, anesthesiology
Adrouny, George Adour (Kuyumjian), biochemistry
Alam, Syed Qamar, nutritional biochemistry, dental research
Albert, Harold Marcus, cardiovascular surgery
Aldinger, Earl Edward, pharmacology
Aldous, Duane Leo, organic chemistry, biochemistry
Allen, Emory Raworth, anatomy, cell biology
Allen, Gary Curtiss, geochemistry, petrology
Allen, James Harrill, ophthalmology, bacteriology
Allison, Fred, Jr., medicine
Alworth, William Lee, bio-organic chemistry
Anderson, Walter Clinton, forest economics
Andrus, Jan Frederick, applied mathematics
Anex, Basil Gideon, physical chemistry
Arcos, Joseph (Charles), biochemistry, oncology
Argus, Mary Frances, biochemistry
Arimura, Akira, endocrinology, physiology
Arquembourg, Pierre Charles, immunology, immunochemistry
Arrowsmith, William Rankin, medicine
Arthur, Jett Clinton, physical chemistry
Bains, Malkiat Singh, physical inorganic chemistry
Bamforth, Stuart Shoosmith, protozoology
Banta, James E, epidemiology
Barber, John Clark, forest genetics
Barber, John Threlfall, plant physiology
Baril, Albert, Jr., research administration
Bartell, Clelmer Kay, cell physiology, invertebrate physiology
Beacham, Woodard Davis, obstetrics & gynecology
Beard, Elizabeth L, physiology
Beaver, Paul Chester, parasitology, tropical medicine
Beckford, Philip Rains, public health administration
Beckwith, Richard Edward, statistics, operations research
Beeler, Myrton Freeman, clinical pathology, clinical chemistry
Beeson, Edward Lee, Jr., molecular spectroscopy
Bell, Charles Bernard, Jr., mathematics, statistics
Benard, Mark, algebra
Benedetto, Frank Aristide, physics
Benerito, Ruth Rogan, physical chemistry
Bennett, Joan Wennstrom, genetics, mycology
Berardi, Leah Castillon, protein chemistry
Berenson, Gerald Sanders, internal medicine
Bergeron, Clyde J, Jr., physics
Berlin, Charles I, audiology
Bernofsky, Carl, biochemistry
Berri, Manuel Phillip, mathematics
Bertoniere, Noelie Rita, organic chemistry
Betterton, Jesse Oatman, Jr., physics
Birtel, Frank T, mathematics
Bitzer, Carl Wilfrid, mathematics
Blackard, William Griffith, internal medicine, endocrinology
Blouin, Florine Alice, organic chemistry
Bocage, Albert J, physiology, biochemistry
Boertje, Stanley, zoology, parasitology
Bornside, George Harry, bacteriology
Boudreaux, Richard A, inorganic chemistry, chemical physics
Boyce, Frederick Fitzherbert, surgery
Brans, Carl Henry, mathematical physics
Braunstein, Jules, stratigraphy
Brazda, Fred George, biochemistry
Bresler, Emanuel H, medical research, nephrology
Bresnick, Gerald Irwin, limnology
Bricker, Victoria Reifler, ethnology
Brockett, Patrick Lee, mathematical statistics
Brown, John Haynes, pharmacology
Bryan, Clifford Randall, genetics, zoology
Bryan, Sara E, biochemistry
Bryant, Lester Richard, medicine, thoracic surgery
Buccino, Salvatore George, nuclear physics
Buddingh, Gerrit John, microbiology, pathology
Buechner, Howard Albert, internal medicine
Burch, Robert Ray, medicine
Burns, Kenneth Franklin, virology
Byers, Larry Douglas, biochemistry
Cabasso, Israel, polymer chemistry
Calamari, Timothy A, Jr., textile chemistry
Carmichael, J W, Jr., physical chemistry, inorganic chemistry
Carrera, Guillermo Manuel, pathology
Carter, James Clarence, theoretical physics
Carter, Mary Eddie, organic chemistry
Carter, Mary Kathleen, pharmacology
Chandran, Krishnan Bala, biomechanics, mechanical engineering
Chargois, Deborah Majeau, medical physiology
Cherry, John Paul, food biochemistry, agricultural biochemistry
Chirino, Fernando Porfirio, medicine
Cho, Young Won, clinical pharmacology, cardiology
Churney, Leon, physiology
Clifford, Alfred Hoblitzelle, mathematics
Clyde, David F, medicine
Cohen, William biochemistry
Cohn, Isidore, Jr, surgery
Collins, Jason Haydel, obstetrics & gynecology
Conway, Edward Daire, III, mathematical analysis
Copeland, Donald Eugene, biology
Coughlin, John W, dentistry
Coulson, Roland Armstrong, biochemistry
Coward, Joe Edwin, virology, microbiology
Cowin, Stephen Corteen, applied mechanics, engineering mechanics
Coy, David Howard, endocrinology
Craig, William Warren, geology
Croley, Thomas Edgar, anatomy, microbiology
Cullen, John Knox, physiology, electrical engineering
Cusachs, Louis Chopin, computer science, quantum chemistry
Daigle, Donald J, chemistry
Daigle, Josephine Siragusa, pharmaceutics
Daigneault, Ernest Albert, pharmacology
D'Alessandro-Bacigalupo, Antonio, parasitology, tropical medicine
Darensbourg, Donald Jude, inorganic chemistry, organometallic chemistry
Darensbourg, Marcetta York, inorganic chemistry, organometallic chemistry
Daroca, Philip Joseph, Jr., pathology
Dascomb, Harry Emerson, internal medicine
Daul, Carolyn Virginia Beach, immunology
Dauns, John, algebra
Davis, Donald G, analytical chemistry
Davis, George Diament, physiology
Davis, William Duncan, Jr, medicine, gastroenterology
Deas, Jane Ellen, medical microbiology, parasitology
DeCamp, Paul Trumbull, surgery
Dechary, Joseph Martin, organic chemistry
Deck, Ronald Joseph, solid state physics
Dell, Tommy Ray, forest biometry
Dessauder, Herbert Clay, biochemistry
Dhuandhar, Hamida Nina, pathology
Dickey, Richard Palmer, obstetrics & gynecology
Diem, John Edwin, mathematics
Di Luzio, Nicholas Robert, physiology
DiMaggio, Anthony, III, biochemistry
Dollear, Frank Gilbert, chemistry
Domer, Floyd Ray, pharmacology
Domer, Judith E, medical microbiology, mycology
Domingue, Gerald James, bacteriology, immunology
Duncan, Margaret Caroline, pediatric neurology
Dundee, Dolores Saunders, invertebrate ecology
Dundee, Harold A, herpetology
Dunlap, Charles Edward, pathology
Dupuy, Harold Paul, biochemistry
Durham, Frank Edington, nuclear physics
Dyer, Robert Frank, anatomy, cell biology
Earle, Thomas Theron, botany
Easson, William McAlpine, psychiatry
Edmonson, Munro Sterling, anthropology
Eggen, Douglas Ambrose, biophysics
Ehrlich, Melanie, biochemistry, microbiology
Eisenstatt, Phillip, geology
Ellgaard, Erik G, genetics, developmental biology
Ellis, James Watson, pure mathematics
Ellzey, Samuel Edward, Jr, organic chemistry
Englande, Andrew Joseph, environmental sciences
Englert, Mary Elizabeth, chemistry
Epps, Anna Cherrie, immunology
Epstein, Arthur William, psychiatry, neurology
Esslinger, Jack Houston, parasitology
Everett, Robert W, Jr., micropaleontology
Evilia, Ronald Frank, analytical chemistry

Ewan, Joseph (Andorfer), botany, history of biology
Ewing, Channing Lester, bioengineering, aerospace medicine
Fagley, Thomas Fisher, physical chemistry
Fairbanks, Laurence Dee, invertebrate zoology
Feigen, Larry Philip, medical physiology
Ferriss, Gregory Stark, neurology
Finerty, John Charles, anatomy
Fingerman, Milton, comparative physiology
Fingerman, Sue Whitsell, fisheries
Fisher, James W., pharmacology
Flurry, Robert Luther, Jr., theoretical chemistry
Foote, Joe Reeder, applied mathematics
Fort, Arthur Tomlinson, obstetrics & gynecology
Fowler, Richard Edmond, pediatrics
Fox, Marjorie Hopkins, anatomy
Frampton, Vernon Lachenous, chemistry
Franklin, William Elwood, textile chemistry, polymer chemistry
Frey, Maurice G., structural geology, petroleum geology
Friedman, Lorraine, microbiology
Fruthaler, George James, Jr., medicine
Fuchs, Laszlo, mathematics
Gagliano, Nicholas Charles, pediatrics, preventive medicine
Gallant, Donald, psychiatry, neurology
Gasser, Raymond Frank, anatomy, physiology
Gaumer, Herman Richard, microbiology, immunology
George, William Jacob, biochemical pharmacology
Gilbert, Norman Sutcliffe, internal medicine
Glancy, David L., cardiology
Glenn, Alfred Hill, marine meteorology, physical oceanography
Goheen, Gilbert Earl, organic chemistry
Goldberg, Stanley Irwin, bio-organic chemistry, organic chemistry
Goldblatt, Leo Arthur, organic chemistry
Goldstein, Jerome A., mathematical analysis
Goldthwaite, Duncan, biochemistry
Gonzales, Elwood John, microbiology
Gonzalez, Francisco Manuel, nephrology
Good, Bill Jewel, physics
Good, Mary Lowe, inorganic chemistry, radiochemistry
Gottlieb, A Arthur, immunology, biochemistry
Gottlieb, Marise Suss, preventive medicine, epidemiology
Griffin, Gary Walter, organic chemistry
Guidry, Dieu-Donne Joseph, neurosciences
Gunn, Oren Berkley, medicine
Gunning, Gerald Eugene, fish biology
Guth, Paul Spencer, pharmacology
Guthrie, John Daulton, biochemistry
Hack, Marvin Howard, histochemistry
Hackett, Earl R., neurology
Haik, George Michel, ophthalmology
Haller, Ann Cordwell, human anatomy, neurosciences
Hallum, Cecil Ralph, applied statistics, medical statistics
Halperin, Victor, oral pathology
Hamer, Jan, chemistry
Hamilton, James T. III, internal medicine
Hanori, Eugene, biochemistry, physical chemistry
Hamrick, Joseph Thomas, preventive medicine
Hancock, John Charles, pharmacology
Happel, Leo Theodore, Jr., neurophysiology
Harb, Joseph Marshall, cell biology, cardiovascular diseases
Hargis, Larry G., analytical chemistry
Harkin, James C., pathology, neuropathology
Harper, Jon William, physiology, neuroanatomy
Hartman, James Austin, geology
Hartwell, Ralph Milo, pathology
Hatgi, John Neal, microbiology, serology
Hayes, George Roy, Jr., medical entomology
Head, Charles Everett, atomic spectroscopy, molecular spectroscopy
Heath, Robert Galbraith, psychiatry, neurology
Heneghan, James Beyer, physiology
Herbert, Jack Durnin, biochemistry, nutrition
Hermann, Allen Max, solid state physics
Hernandez, Thomas, pharmacology

Herzberger, Maximilian Jacob, optics, mathematical physics
Hess, Melvin, endocrinology
Hewitt, Robert Lee, thoracic surgery, cardiovascular surgery
Heyn, Anton Nicolaas Johannes, biology, biophysics
Hofmann, Karl Heinrich, mathematics
Holland, Monte Gene, ophthalmology
Hollis, Walter Jesse, internal medicine, cardiology
Holmquist, Jan Olle Valter, dental
Holmsted, Nelson D., pathology
Hood, Marion Winifred, bacteriology, pathology
Hornung, Maria O'C, immunology
Howe, Calderon, microbiology
Hulbert, Samuel Foster, biomedical engineering, materials engineering
Hurley, Maureen, microbiology, cell biology
Hyde, Paul Martin, biochemistry
Hyman, Edward Sidney, internal medicine, physiology
Ibanez, Manuel Luis, biochemistry, microbiology
Ichinose, Herbert, pathology
Ignarro, Louis Joseph, pharmacology, cell biology
Istre, Clifton O., Jr., audiology
Jackson, Johnny, plant physiology, biochemistry
Jacobs, John Joseph, neuroendocrinology
Jacobs, Sydney, medicine
Johnson, Emmett John, microbiology
Johnson, Hamilton McKee, geology, geophysics
Johnson, Horton Anton, medicine, pathology
Johnson, Mary Knettles, bacteriology
Johnston, James Baker, marine sciences, science education
Joiner, Paul David, pharmacology
Jonassen, Hans Boegh, inorganic chemistry
Jung, Hilda Zirle, physical chemistry, chemical physics
Jung, Rodney Clifton, tropical medicine
Kasten, Frederick H., cell biology
Keffer, David Goforth, physics
Kelly, Charles James, pharmacy
Kern, Ralph Donald, Jr., physical chemistry
Khalaf, Kamel T., zoology, medical entomology
King, Arden Ross, anthropology
King, Creston Alexander, Jr., low temperature physics
Kirgis, Homer Dale, anatomy
Klein, Elias, physical chemistry
Kline, David G., neurosurgery
Kloepfer, Henry Warner, human genetics
Knight, Edward Henry, psychiatry, psychoanalysis
Knill, Ronald John, topology
Kocurko, Michael John, geology, sedimentology
Kokatnur, Mohan Gundo, biochemistry, nutrition
Koppa, Vasantha, bioinorganic chemistry
Kreisman, Norman Richard, neurophysiology
Krementz, Edward Thomas, surgery
Krupp, Iris M., parasitology, dermatology
Kuck, James Chester, agricultural biochemistry, organic chemistry
Kujala, Robert Oakes, mathematics
Kyame, George John, textile physics, textile engineering
Kyane, Joseph John, mathematical physics
Laguaite, Jeannette Katherine, speech pathology
Lai, Ping-Yuen, plant pathology, biology
Lartigue, Donald Joseph, biochemistry, enzymology
Laseter, John Luther, biochemistry
Lassen, Laurence E., forestry
Lehmann, Hermann Peter, clinical biochemistry
Lentz, Patrick Edmund, physiology
Lesseps, Roland Joseph, developmental biology
Levy, Louis, II, cardiology
Li, Yu-Teh, biochemistry
Lilies, Samuel Lee, physiology
Lillie, Ralph Dougall, pathology, histochemistry
Lindberg, David Seaman, Sr., academic administration, health sciences
Lindsey, Edward Stormont, medicine
Little, Charles Eugene, speech pathology
Little, Maurice Dale, medical parasitology
Litwin, Martin Stanley, surgery
Liu, Maw-Shung, medical physiology

Liu, Young King, biomechanics, biomedical engineering
Liukkonen, John Robie, mathematical analysis
Llewellyn, Raeburn Carson, neurosurgery
Locke, William, internal medicine
Lockmann, Ronald Frederick, geography
Longenecker, Herbert Eugene, biological chemistry
Lopez-Santolino, Alfredo, medicine, physiology
Lowe, Robert Franklin, Jr., cardiovascular biochemistry
Lumsden, Richard, cell biology
Lyons, George D., otolaryngology
Mague, Joel Tabor, inorganic chemistry, organometallic chemistry
Makielski, Sally Kimball, population biology, environmental health
Malek, Emile Abdel, medical parasitology
Marks, Charles, surgery
Martinez-Lopez, Jorge Ignacio, internal medicine, cardiology
Matthews, Murray Albert, anatomy, neuropathology
May, Paul David, polymer chemistry
Mayerson, Hymen Samuel, physiology
Mazzeno, Laurence William, organic chemistry
McAfee, Robert Dixon, physiology
McCall, Elizabeth Regina, biophysics
McClugage, Samuel Gardner, Jr., anatomy
McConnell, Virginia Fenner, organic chemistry
McDonald, John C., surgery, immunology
McDowell, John Parmelee, sedimentary petrology
McGarry, Paul Anthony, pathology
McHale, John T., botany, cytology
McHardy, George Gordon, medicine
McKinnon, William Mitchell Patrick, surgery
McLaurin, James Walter, medicine, surgery
McLean, Robert T., algebra
McMahan, Chalmers Alexander, applied statistics, demography
McMahon, Francis Gilbert, internal medicine
McMillen, Larry Byron, prosthodontics
Meckstroth, George R., radiological physics
Meyer, Joachim Dietrich, mineralogy, petrology
Meyers, Philip Henry, radiology
Michael, Marshall Louis, Jr., surgery
Mickal, Abe, obstetrics & gynecology
Miles, Henry Harcourt Waters, psychiatry, psychoanalysis
Miller, Albert, medical entomology
Miller, Harvey I., physiology
Miller, James Albert, Jr., anatomy
Miller, Joseph Henry, medical parasitology
Miller, Marvin Fred, psychiatry
Miller, Max Joseph, medical parasitology
Mislove, Michael William, mathematics
Mitcham, Donald, physics
Mitchell, Wilbur Leonard, mathematics
Mizell, Merle, developmental genetics, oncology
Mogabgab, William Joseph, internal medicine
Mokrasch, Lewis Carl, biochemistry
Mole, Paul Angelo, muscular physiology, exercise physiology
Montalvo, Joseph G., analytical chemistry
Moore, Walter Guy, aquatic biology, ecology
Morgan, Lee Roy, Jr., pharmacology, chemotherapy
Morris, Cletus Eugene, organic chemistry
Morris, Harry Dunlap, orthopedic surgery
Mulcrone, Thomas Francis, mathematics
Mule, James Gaspare, obstetrics & gynecology
Mullahy, John Henry, biology, phycology
Myers, Melvi Bertrand Jr., medicine
Nance, Francis Carter, surgery
Nelson, Charles Edward, geography
Nelson, Mary Lockett (Mrs John D Guthrie), cellulose chemistry
Newman, Wiley Clifford, Jr., physiology
Nice, Charles Monroe, Jr., radiology
Nico, William Raymond, mathematics
Nissly, Charles Martin, geography of Latin America, cultural geography
Norman, Edward Cobb, public health education, psychiatry

Nugent, Maurice Joseph, Jr., organic chemistry
Nussmann, David George, geology, geochemistry
Oalmann, Margaret Claire, epidemiology
Ochsner, John Lockwood, surgery
Ochsner, Seymour F., radiology
O'Dell-Smith, Roberta Maxine, physiology
Odenheimer, Kurt John Sigmund, clinical pathology, experimental pathology
Olmsted, Clinton Albert, physiology
O'Quinn, Silas Edgar, dermatology
Orihel, Thomas Charles, medical parasitology
Ory, Robert Louis, biochemistry
Paddison, Richard Milton, neurology
Page, Clayton R. III, cell biology, biology
Pargaonkar, Padmaker Shankar, biochemistry
Parker, Calvin Alfred, geology
Parsley, Ronald Lee, invertebrate paleontology, geology
Paterson, Athol James, public health, tropical medicine
Peacock, Charles LeRoy, nuclear physics
Peebles, Edward McCrady, anatomy, cell biology
Pekarthy, James Maurice, anatomy, cell biology
Pepperman, Armand Bennett, Jr., organic chemistry
Peppler, Richard Douglas, anatomy
Petterson, Robert Carlyle, organic chemistry
Phillips, John Hunter, Jr., internal medicine, cardiology
Pierce, William Arthur, Jr., microbiology
Pinschmidt, Norman William, ophthalmology
Pisano, Joseph Carmen, cell physiology, immunology
Pizzolato, Philip, pathology
Plessy, Boake Lucien, physical chemistry, polymer chemistry
Politzer, Ieva Ruks, organic chemistry
Politzer, Peter Andrew, theoretical chemistry
Pons, Walter A, Jr., analytical chemistry, food chemistry
Ponthier, Roy Leonce, Jr., endocrinology
Porter, Johnny Ray, physiology
Price, Leonard, organic chemistry
Puri, Pratap, applied mathematics, continuum mechanics
Purrington, Robert Daniel, theoretical nuclear physics
Puyau, Francis A., medicine
Quigley, Frank Douglas, mathematics
Radhakrishnamurthy, Bhandaru, biochemistry
Randall, Howard M., physiology
Ratchford, Robert James, physical chemistry
Rawls, Henry Ralph, physical chemistry, dental research
Ray, Clarence Thorpe, internal medicine
Rayson, Jack Henry, dentistry
Reed, Richard Jay, medicine, pathology
Rees, Charles Sparks, mathematics
Reeves, Richard Edwin, organic chemistry
Reid, John David, textile chemistry
Reinhardt, Robert Milton, organic chemistry
Richardson, Donald Edward, neurosurgery
Riess, Karlem, physics
Roberts, Earl John, chemistry
Rodriguez, Mario Santos, biochemistry, dental materials
Rogers, James Ted, Jr., mathematics
Ronstrom, George Nelson, anatomy
Rosenberg, Dennis Melville Leo, thoracic surgery, cardiovascular surgery
Rosencrans, Steven I., mathematics
Rosenthal, J William, ophthalmology, surgery
Rothschild, Henry, molecular biology, medicine
Rowland, Stanley Paul, cellulose chemistry, polymer chemistry
Ruby, John Robert, anatomy
Russell, Raymond Alvin, physiology
Rutledge, Lewis James, otolaryngology, pediatrics
Ryan, Robert F., surgery
Sacco, Paul, plant physiology
Salvaggio, John Edmond, internal medicine, immunology
Sappenfield, Robert W., medicine
Sayegh, Salem F., thoracic surgery, cardiovascular surgery
Schally, Andrew Victor, endocrinology

Schilb, Theodore Paul, biophysics, physiology
Schimek, Robert Alfred, medicine
Schlegel, Jorgen Ulrik, medicine
Schmidt, Raymond LeRoy, physical chemistry
Schneidau, John Donald, Jr, medical mycology
Schoellmann, Guenther, biochemistry, organic chemistry
Schor, Norberto Aaron, pathology
Schramel, Robert Joseph, surgery
Scott, Harold George, entomology
Seabury, John Hollister, internal medicine
Sears, Dewey Frederick, physiology
Seed, John Richard, microbiology, parasitology
Segal, Leon, chemistry
Segaloff, Albert, endocrinology, clinical medicine
Senn, Vincent John, biochemistry
Serfling, Robert Elton, biostatistics, demography
Seto, Jane Mei-Chun, internal medicine
Seto, Yeb Jo, electronics, plasma physics
Sevenair, John P., bio-organic chemistry
Shedlarski, Joseph George, Jr, microbial biochemistry
Shenkel, J Richard, anthropology, archaeology
Shilstone, Cecil Maxwell, analytical chemistry, chemical engineering
Shirkey, Harry Cameron, pediatrics
Siddall, Thomas Henry, III, chemistry
Siegel, William Carl, resource economics
Simmons, Robert Lowrey, public health, preventive medicine
Simmons, William Bruce, Jr, mineralogy, petrology
Sims, Asa C, Jr, plant pathology
Skau, Evald Laurids, physical organic chemistry
Skinner, Hubert Clayton, geology
Sly, Ridge Michael, medicine
Smalley, Alfred Evans, ecology
Smith, Diane Elizabeth, neuroanatomy
Smith, Jerry Warren, virology
Smith, Joshua Daniel, organic chemistry, immunology
Snowden, Jesse Otho, clay mineralogy
Snyder, Christopher Harrison, pediatrics
Spence, Hilda Adele, microbiology
Spires, Morris Albert, chemistry, pharmacology
Spitzer, John J, physiology
Spitzer, Judy A., physiology, biochemistry
Srinivasan, Sathanur Ramachandran, chemistry, biochemistry
Stakis, Andris A., organic chemistry
Stanfield, Manie K, organic chemistry, biochemistry
Stanonis, David Joseph, organic chemistry
Stansbury, Mack Fulton, agricultural chemistry
Steele, Richard Harold, biochemistry
Stephens, Maynard Moody, petroleum engineering, economic geology
Stephens, Raymond Weathers, Jr, petroleum geology
Sternberg, William Howard, pathology
Stewart, William Huffman, medicine
Stjernholm, Rune Leonard, biochemistry
Stocker, Jack Hubert, organic chemistry
Stone, Doris Zemurray, anthropology, archaeology
Stowell, John Charles, organic chemistry
Strong, Jack Perry, pathology
Stuckey, Walter Jackson, Jr, internal medicine, hematology
Sumrell, Gene, organic chemistry
Suttkus, Royal Dallas, ichthyology, fish biology
Sutton, William Wallace, protozoology, cell biology
Swanson, Charles Richard, plant physiology
Swartzwelder, John Clyde, medical parasitology, tropical public health
Taylor, Billy G, oncology, surgery
Thomas, Dempsey Lee, biological sciences
Thomas, Estes Centennial, III, physical chemistry
Thomson, Alan, geology
Thurber, George A., medical entomology
Thurman-Swartzwelder, Ernestine H, medical entomology, research administration
Thurmon, Theodore Francis, medical genetics
Toth, Louis Andrew, physiology
Tracy, Richard E, pathology
Trapido, Harold, biology
Trefonas, Louis Marco, structural chemistry, physical biochemistry
True, Merrill Allan, marine biology

True, Renate (Schlenz), biological oceanography
Tucker, Woodson Coleman, Jr, academic administration, physical chemistry
Tyrrell, Willis W, Jr, geology
Unterharnscheidt, Friedrich J, neuropathology, neurology
Upadhyay, Jagdish M, microbiology, biochemistry
Vail, Sidney Lee, organic chemistry, textiles
Vaughn, John B, epidemiology
Vaupel, Martin Robert, anatomy, embryology
Vela, Adan Richard, physiology
Vial, Lester Joseph, Jr, medicine, pathology
Vigo, Tyrone Lawrence, polymer chemistry, textile chemistry
Villarejos, Victor Moises, epidemiology, tropical medicine
Vokes, Emily Hoskins, invertebrate paleontology, malacology
Vokes, Harold Ernest, stratigraphy, invertebrate paleontology
Volpe, Erminio Peter, zoology
Von Almen, William Frederick I, palynology, geology
Voors, Antonie Wouter, epidemiology
Walia, Jasjit Singh, organic chemistry
Walker, Leon Bryan, Jr, anatomy
Walsh, John Joseph, cardiology
Ward, Martha Coonfield, cultural anthropology
Ward, Truman L, physics, physical chemistry
Ward, William Cruse, sedimentary petrology
Waring, William Winburn, pediatrics
Warren, Lionel Gustave, parasitology, physiology
Webster, Douglas B, otorhinolaryngology
Webster, Richard Milroy, anatomy
Weed, John Conant, gynecology, infertility
Weidie, Alfred Edward, geology
Weill, Hans, medicine, physiology
Weinberg, Roger, biostatistics
Weiss, Louis Charles, textile physics
Weiss, Thomas E, medicine
Weitsen, Howard Arthur, neurochemistry, anatomy
Welden, Arthur Luna, mycology
Welsh, Ronald, pathology
Wendel, William Bean, biochemistry
Wendt, Richard P, physical chemistry
Wheeler, Philip Ridgly, forestry
Wickstrom, Jack, medicine
Williams, Marion Jack, thoracic surgery, surgery
Williams, Ronald Lee, pharmacology, physiology
Wolleben, James A, paleontology, stratigraphy
Wood, Elwyn Devere, marine geochemistry
Woody, Norman Cooper, pediatrics
Wright, David Laverne, pediatrics
Yaeger, Robert George, parasitology
Yates, Robert Doyle, cytology
Yatsu, Lawrence Y., plant physiology
Yeadon, David Allou, chemistry
Zame, William Robin, mathematics
Zamikoff, Irving Ira, dentistry, prosthodontics
Zimny, Marilyn Lucile, anatomy
Ziskind, Morton Moses, internal medicine

PEARL RIVER
Harris, James A, analytical chemistry, radiochemistry

PINEVILLE
Barnett, James P, plant physiology, forestry
Barras, Stanley J, forest entomology
Cavanaugh, Charles Johnson, Sr, physiology
Grelen, Harold Eugene, range science
Koch, Peter, wood science & technology
Lorio, Peter Leonce, Jr, forest soils, forest ecology
Manwiller, Floyd George, wood technology
McGraw, Gerald Wayne, biochemistry, organic chemistry
McKee, William Henry, Jr, agronomy
McLemore, Bobbie Frank, plant physiology
Moser, John C, acarology
Simmons, David Rae, mathematics
Thatcher, Robert Clifford, forest entomology
Van Sambeek, Jerome William, plant physiology
Watson, Dennis Ronald, chemistry

PLAQUEMINE
Roberts, Reginald Francis, physical chemistry
Rosene, Robert Bernard, organic chemistry

RUSTON
Abegg, Roland, zoology
Atteberry, Billy Joe, mathematics
Barker, Hal B, animal physiology, animal nutrition
Bernard, William Hickman, physics
Brumage, William Harry, solid state physics
Christian, James A, botany, genetics
Crump, Kenny Sherman, mathematics, statistics
Davis, Billy J, ichthyology, water pollution
Flournoy, Robert Wilson, physiology
Freasier, Ben Forest, physical chemistry
Garner, Jackie Bass, mathematics
Gibbs, Richard Lynn, atomic physics, plasma physics
Gilbert, Jimmie Dale, algebra
Goertz, John William, vertebrate zoology
Hackbarth, Winston (Philip), physiological ecology
Hedrick, Harold Gilman, industrial microbiology
Herbert, Wallace, applied mathematics
Herrmann, Leo Anthony, geology
Jewell, Frederick Forbes, histopathology
Kelly, Edgar Preston, Jr, mathematics
Lutes, Dallas D, plant pathology
Miller, Agnes Chambless, nutrition, foods
Moseley, Harry Edward, chemistry
Moseley, Patterson B, chromatography
Murad, John Louis, zoology
Patton, Selma Hicks, clinical biochemistry, clinical chemistry
Ramsey, Paul Roger, population biology, genetics
Roberts, Donald Duane, physical organic chemistry
Smith, Charles Hooper, chemistry
Stephenson, Paul Bernard, physics
Stewart, Gordon Arnold, dairy science
Temple, William Benson, applied mathematics
Thompson, Ronald Hobart, nuclear chemistry
Trisler, John Charles, organic chemistry
White, James Clarence, plant pathology
Wright, Johnie Algie, horticulture

ST GABRIEL
Babb, Robert Massey, organic chemistry

ST JOSEPH
Rogers, Robert Larry, plant physiology, biochemistry

SHREVEPORT
Baker, Glenn Jackson, geology
Battarbee, Harold Douglas, endocrinology, cardiovascular physiology
Brashier, Gary Kermit, physical chemistry
Brauchi, John Tony, psychiatry
Brown, Richard Don, pharmacology, toxicology
Carlton, Virginia, mathematics
Clawson, Robert Charles, histology, embryology
Corley, Glyn Jackson, mathematics
Cowley, A Ronald, neuroanatomy, neurosciences
Dailey, John William, pharmacology
Deutel, Robert, bacteriology
Donaldson, Paul, microbiology, immunology
Eichner, Edward Randolph, hematology
Forgotson, James Morris, Jr, petroleum geology, petroleum engineering
Gaar, Kermit Albert, Jr, medical physiology
Galloway, Louie A, III, nuclear physics
George, Ronald Baylis, internal medicine, physiology
Graham, Lewis Texada, Jr, biochemistry, neurobiology
Griswold, Kenneth Edwin, Jr, clinical biochemistry, population genetics
Hall, John Whiting, geography
Hanson, Marvin Wayne, organic chemistry
Haynes, Robert Ralph, plant taxonomy
Henderson, Ralph Joseph, Jr, biochemistry
Hibbs, Richard Guythal, histology
Hood, Robert Luther, inorganic chemistry

Jamison, Richard Melvin, virology
Jobe, Philip Carl, neuropharmacology
Juberg, Richard Caldwell, human genetics, pediatrics
Kalinsky, Robert George, phycology, environmental biology
Kamm, Richard Conrad, experimental pathology
Kenknight, Glenn, plant pathology
Knox, Francis Stratton, III, physiology, biomedical engineering
Koerner, Theodore Alfred, pathology
Kotcher, Emil, medical microbiology
Kurzweg, Frank Turner, surgery
Lang, Erich Karl, radiology
Little, Joseph Alexander, medicine
Loewenstein, Joseph Edward, endocrinology, internal medicine
Lowrey, Charles Boyce, physical organic chemistry
Manno, Barbara Reynolds, pharmacology, toxicology
Manno, Joseph Eugene, toxicology, pharmacology
Matlock, Rex Leon, physics
McCormick, George M, II, pathology, endocrinology
McElroy, William Tyndell, Jr, physiology
McKinney, Alfred Lee, mathematical analysis, numerical analysis
Menely, George Rodney, medicine
Olson, Anita Cora, biochemistry,
Penny, Joe Edward, neuroanatomy, human anatomy
Pou, Jack Wendell, otolaryngology
Redetzki, Helmut M, pharmacology
Reed, Adrian Faragher, medicine, anatomy
Rudolph, Guilford George, biochemistry
Sardisco, John Baptist, physical chemistry
Schwartz, Robert Donald, analytical chemistry
Serna, Thomas John, medical physiology, medical biophysics
Shaw, Nolan Gail, geology
Silberman, Ronald, clinical microbiology, immunology
Smith, Albert Goodin, pathology
Smith, Robert Lewis, biochemistry
Spaht, Carlos G, II, mathematics, operations research
Steinmetz, Walter Edmund, organic chemistry
Uzman, Betty Geren, neurobiology, pathology
Walker, Rufus Floyd, Jr, biometrics, biophysics
Warters, Mary, genetics
Wilkes, Ann Broom, pharmacology
Wilkins, Orin Perry, zoology, botany
Wood, Charles Donald, pharmacology

THIBODAUX
Avonda, Frank Peter, organic chemistry
Ayo, Donald Joseph, agronomy, horticulture
Blatt, Irving Myron, otolaryngology
Falcon, Carroll James, reproductive physiology
Field, Jack Everett, inorganic chemistry
Hardberger, Florian Max, zoology, botany
Harris, Alva H, parasitology, marine biology
Long, William Henry, entomology
Ohmer, Merlin Maurice, mathematics
Rathle, Pierre, physical chemistry
Sachs, Jules Barry, geology
Veith, Daniel A, solid state physics
Webert, Henry S, botany

WEST MONROE
Somsen, Roger Alan, pulp chemistry
Zentner, Thomas Glenn, pulp chemistry, paper chemistry

MAINE

AUGUSTA
Bond, Lyndon Herrick, fish biology
Bourque, Bruce Joseph, anthropology, archaeology
Lind, Howard Eric, bacteriology
Nash, Robley Wilson, entomology
Norton, Cynthia Friend, microbiology
Page, Robert Leroy, mathematics, physical science
Pasvogel, Myron W, poultry nutrition, poultry physiology

MAINE

BANGOR
Hsu, Yu Kao, mathematics, aeronautics
LaMarche, Paul H., genetics, pediatrics
Lindgren, Robert M., physical chemistry
Warner, Kendall, fisheries

BAR HARBOR
Bailey, Donald Wayne, immunogenetics
Bernstein, Seldon Edwin, medical genetics
Blake, Robert L., biochemistry, pharmacology
Chai, Chen Kang, genetics
Chen, Harry Wu-Shiong, biochemistry
Cherry, Marianna, immunogenetics
Eicher, Eva Mae, genetics
Eleftheriou, Basil E., endocrinology, biochemistry
Fox, Richard Romaine, laboratory animal science, genetics
Green, Earl Leroy, genetics
Green, Margaret Creighton, genetics
Harrison, David Ellsworth, immunohematology
Holway, Richard Thomas, medical entomology
Hoppe, Peter Christian, reproductive physiology
Kaliss, Nathan, immunology
Kandusch, Andrew August, biochemistry
Keighley, Geoffrey Lorrimer, physiology
Leiter, Edward Henry, cell biology
Les, Edwin Paul, genetics
Meier, Hans, experimental pathology
Mobraaten, Larry Edward, immunogenetics
Murphy, Edwin Daniel, experimental pathology
Myers, David Daniel, laboratory animal science, animal pathology
Olday, Frederick Combs, plant nutrition
Roderick, Thomas Huston, genetics
Russell, Elizabeth Shull, genetics
Shultz, Leonard Donald, cancer immunogenetics
Snell, George Davis, genetics
Stevens, Leroy Carlton, Jr., developmental biology
Waymouth, Charity, cell biology
Womack, James E., genetics

BAR MILLS
Mattor, John Alan, synthetic organic chemistry

BETHEL
Willard, John Jay, organic chemistry

BIDDEFORD
Maryan, David Richard, biochemistry
Yuhas, Joseph George, animal behavior, zoology

BOOTHBAY
Welsh, John Henry, zoology

BOOTHBAY HARBOR
Stickney, Alden Parkhurst, marine ecology

BROOKSVILLE
Bancroft, Dennison, physics

BRUNSWICK
Anderson, George Robert, physical chemistry
Bohan, Thomas Lynch, biophysics, solid state physics
Butcher, Samuel Shipp, physical chemistry
Chittim, Richard Leigh, mathematics
Greenspan, Beverly Naomi, animal behavior
Gustafson, Alton Herman, cytology, algology
Howland, John LaFollette, biochemistry
Hughes, William Taylor, biophysics
Huntington, Charles Ellsworth, ornithology
Hussey, Arthur M. II., geology
Jeppesen, Myron Alton, physics
Johnson, Robert Wells, mathematics
LaCasce, Elroy Osborne, Jr., acoustics
Mayo, Dana Walker, organic chemistry
Mooz, Elizabeth Dodd, biochemistry
Moulton, James Malcolm, vertebrate morphology
Page, David Sanborn, bio-organic chemistry, marine chemistry
Settlemire, Carl Thomas, biochemistry
Silver, Murray, mathematics
Turner, James Henry, solid state physics
Ward, James Edward, III, mathematics

CAMDEN
Mueller, George Peter, organic chemistry

CUMBERLAND MILLS
Eames, Arnold C., paper chemistry

DIXFIELD
Pollister, Priscilla Frew, biology

DRYDEN
Heinrich, Gerd H., zoology, biology

EAST WINTHROP
Pond, Samuel Ernest, physiology

ELLSWORTH
Lambdin, Morris Arthur, pediatrics
Simnock, Pomeroy, human genetics

FALMOUTH
Cartier, George Thomas, applied chemistry
Loring, William Ellsworth, pathology

FARMINGTON
Aft, Harvey, organic chemistry
Doran, Peter Cobb, health sciences
Easter, Thomas Edward, environmental geology
Lange, Gail Laura, algebra

FRIENDSHIP
Trefethen, Joseph Muzzy, geology

FT KENT
Gillis, John Ericsen, evolutionary biology

GORHAM
Clough, Philip James, physical inorganic chemistry
Whitten, Maurice Mason, analytical chemistry; history of science

KENNEBUNK
Fassett, David Walter, toxicology

KENNEBUNKPORT
Frink, Aline, mathematics
Lothrop, Warren Craig, organic chemistry

LEWISTON
Boyles, James Glenn, physical chemistry
Branch, Charles Franklin, medicine
Briggs, Richard M., organic chemistry
Chute, Robert Maurice, environmental physiology; limnology
Farnsworth, Roy Lothrop, geology
Fellers, Francis Xavier, nephrology; medicine
Haines, David Clark, mathematics
Hoffman, John Stephen, Jr., mathematical analysis
Kingsbury, Robert Freeman, atomic spectroscopy
Minkoff, Eli Cooperman, evolutionary biology; comparative anatomy
Pribram, John Karl, solid state physics
Ruff, George Antony, physics
Sprowles, Jolyon Charles, inorganic chemistry
Stauffer, Charles Henry, physical chemistry
Thomas, William Benjamin, organic chemistry
Vierina, Teri L., chemical physics

LISBON FALLS
Bowie, Harold E., mathematics

LUBEC
Pierce, Arleen Cecilia, organic chemistry

MACHIAS
Fraenkel-Conrat, Jane E., biochemistry

MANSET
Evans, Jacqueline P., mathematics

MILBRIDGE
Bradley, Wilmot Hyde, geology
Mapleton, Robert Allan, mathematical physics

MILLINOCKET
Mattson, Victor Frank, chemistry

MONMOUTH
Stiles, Warren Cryder, horticulture

NEW SHARON
Martin, Robert Lawrence, mammalogy, zoology

NORWAY
Dewing, Stephen Bronson, radiology

ORONO
Allen, Kenneth William, zoology
Apgar, William P., animal nutrition
Ashley, Marshall Douglas, forest biometry; remote sensing
Bain, William Murray, microbial physiology
Barden, Albert Arnold, Jr., zoology
Bayer, Robert Clark, avian physiology
Beamesderfer, John William, physical chemistry
Bennett, Clarence Edwin, physics
Bentley, Michael David, organic chemistry
Bird, Francis Howe, nutrition
Blamberg, Donald Lee, nutrition
Blum, Barton Morrill, silviculture
Borns, Harold William, Jr., quaternary geology
Brownstein, Kenneth Robert, quantum mechanics
Bryan, Thomas Alan, poultry pathology
Buck, Charles Elon, bacteriology
Camp, Paul R., solid state physics
Campana, Richard John, forest pathology
Canavera, David Stephen, forest genetics
Carpenter, Paul Nathaniel, agronomy
Carr, Edward Frank, physics
Chute, Harold LeRoy, animal pathology
Clark, Alton Harold, solid state physics
Cook, James Richard, cell physiology
Cooper, George Raymond, plant ecology; plant physiology
Coulter, Malcolm Wilford, wildlife ecology; forest ecology
Crawford, Hewlette Spencer, Jr., wildlife ecology; forest ecology
Csavinszky, Peter John, theoretical solid state physics
Dearborn, John Holmes, marine ecology
De Haas, Herman, biological chemistry
Dickey, Howard Chester, dairy husbandry
Dimond, John Barnet, entomology; ecology
Douglass, Irwin Bruce, environmental chemistry
Dunlap, Robert D., physical chemistry
Eggert, Franklin Paul, horticulture
Emerick, Richard Gibbs, anthropology
Eves, Howard (Whitley), mathematics
Farlow, Stanley Jerome, applied mathematics
Fink, Loyd Kenneth, Jr., marine geology
Flynn, Carl Munro, zoology
Forsythe, Howard Yost, Jr., entomology; plant pathology
Frank, James Anthony, plant pathology
Gelinas, Douglas Alfred, botany; physiology
Gerry, Richard Woodman, poultry nutrition
Gershman, Melvin, bacteriology
Gibbs, Harold Cuthbert, parasitology
Glenn, Rollin Copper, soils
Goodfriend, Paul Louis, physical chemistry
Gregory, Richard Wallace, fisheries
Harris, Paul Chappell, poultry science
Hatch, Richard Wallace, fish biology
Hepler, Paul Raymond, horticulture
Hess, Charles Thomas, nuclear physics
Highlands, Matthew Edward, bacteriology, food science
Hilton, Merle Tyson, plant pathology
Hogan, John Mathew, virology, molecular genetics
Holbrook, Frederick R., insect pathology
Hooper, Henry Olcott, solid state physics
Howd, Frank Hawver, economic geology, environmental geology
Hutchinson, Frederick Edward, soil chemistry
Jerkofsky, Maryann, virology, molecular genetics
Knight, Fred Barrows, forest entomology
Krueger, George Corwin, optics
Kuschia, Norman Paul, forest products
Laber, Larry Jackson, plant physiology
Langford, Eric Siddon, mathematics
Langille, Alan Ralph, crop physiology
Lennon, Robert Earl, fish biology
Leonard, David E., entomology
Leonard, Herbert Arthur, animal science
Lerner, Joseph, biochemistry
Locke, Philip M., mathematics
Mairhaber, John Carl, mathematics
Major, Charles Walter, physiology
Manzer, Franklin Edward, plant pathology
McCleave, James David, zoology
McCrum, Richard Caswell, plant pathology
McDaniel, Ivan Noel, medical entomology
Mendall, Howard Lewis, wildlife biology, ornithology
Meyer, Marvin Clinton, parasitology
Morrow, Richard Alexander, particle physics
Muir, Forest Vern, poultry science
Mun, Alton M., developmental biology
Murphy, Grattan Patrick, mathematics
Musgrave, Stanley Dean, animal nutrition
Neubauer, Benedict Francis, plant anatomy
Nicholson, Bruce Lee, microbiology
Northam, Edward Stafford, mathematics
Norton, Stephen Allen, geology
Olson, Robert Edward, forest entomology
Osberg, Philip Henry, geology
Owen, Ray Bucklin, Jr., wildlife ecology
Patterson, Howard Hugh, physical chemistry
Pogorzelski, Henry Andrew, mathematics
Pratt, Darrell Bradford, bacteriology
Radke, Frederick Herbert, biochemistry
Rassiah, Jayendran C., physical chemistry
Richards, Charles Davis, plant taxonomy
Richens, Voit B., wildlife ecology
Ringo, John Moyer, behavioral genetics
Roberts, Franklin Lewis, cytology; genetics
Rooney, James Arthur, physics
Roxby, Robert, biochemistry
Russ, Charles Roger, inorganic chemistry
Sanger, David, anthropology
Shottafer, James Edward, materials science, wood technology
Shuler, Craig Edward, forest products, wood science & technology
Simard, Gerald Lionel, physical chemistry
Simmons, Gary Adair, forest entomology
Simpson, Geddes Wilson, entomology
Slabyj, Bohdan M., food science
Smith, Charles William, Jr., low temperature physics
Specher, Benjamin Robert, zoology
Stanley, Jon G., comparative physiology
Storch, Richard Harry, entomology
Struchtemeyer, Roland August, soils
Summers, Robert Gentry, Jr., zoology
Tarr, Charles Edwin, physics
Thompson, Edward Valentine, physical chemistry
Vadas, Robert Louis, marine ecology
Viette, Michael Anthony, physiology
Wave, Herbert Edwin, entomology; plant pathology
Weigang, Oscar Emil, Jr., physical chemistry
Whitehill, Alvin Richard, bacteriology
Whittaker, James Curtiss, forestry
Wilson, John M., food science
Wolfhagen, James Langdon, organic chemistry
Wolford, John Henry, avian physiology, poultry science
Young, Harold Edle, forestry
Zabel, Lowell Wallace, physics
Zollweg, John Allman, physical chemistry

OXFORD
Nutting, Albert Deane, forestry

PORTLAND

Bonner, R Alan, medical research
Doby, Tibor, radiology, nuclear medicine
Hodgkin, Brian Charles, medical research
Kern, Abraham K, botany, zoology
Mazer, Ronald Steven, endocrinology, physiology
Najarian, Haig Hagop, parasitology
Nelson, Clifford Vincent, cardiovascular physiology
Novak, Irwin Daniel, geomorphology
O'Brien, Katharine Elizabeth, mathematics
Rand, Peter, cardiovascular physiology, hematology
Rundell, Clark Ace, clinical chemistry
Shenton, Edward Heriot, oceanography
Sottery, Theodore Walter, physical organic chemistry
Walking, Robert Adolph, acoustics

POWNAL

Hoffman, John Leslie, anthropology

ROCKLAND

Bixler, Harris Jacob, physics, chemistry
Guiseley, Kenneth B, carbohydrate chemistry
Renn, Donald Walter, natural products chemistry, immunochemistry
Stanley, Norman Francis, chemistry

SALISBURY COVE

Kinter, William Boardman, physiology
Schmidt-Nielsen, Bodil Mimi, physiology

SOUTH HARPSWELL

McEwen, Currier, medicine

SOUTH PARIS

Carroll, Robert Baker, organic chemistry, plant pathology

SOUTH PORTLAND

Horton, Donald Bion, fisheries, marine biology

SOUTH WINDHAM

Thurlow, John Frank, biochemistry

SOUTHWEST HARBOR

Dunham, Wolcott Balestier, oncology

SPRINGVALE

Ciullo, Robert Henry, zoology
Gilmore, Claude Raymond, invertebrate zoology, ecology
La Prade, Marie Douglas, structural chemistry, inorganic chemistry

STILLWATER

Green, Brian, organic chemistry

SWANS ISLAND

Bailey, Norman Sprague, biology

TOGUS

Wise, Robert Irby, bacteriology, internal medicine

WALPOLE

Dean, David, biological oceanography
De Witt, Hugh Hamilton, ichthyology, marine biology
Green, Edward Jewett, chemical oceanography
Hidu, Herbert, marine ecology
McAlice, Bernard John, oceanography
Schnitker, Detmar, micropaleontology, ecology

WATERVILLE

Bennett, Miriam Frances, biological rhythms, comparative physiology
Briggs, Jonathan, solid state physics
Champlin, Arthur Kingsley, reproductive biology, genetics
Combellack, Wilfred James, mathematics
Dudley, John Minot, physics
Easton, Thomas W, cytology, evolution
Eskelund, Kenneth H, veterinary medicine

Koons, Donaldson, geomorphology
Machemer, Paul Ewers, analytical chemistry
Maier, George D, biochemistry
Metz, Roger N, theoretical physics
Pestana, Harold Richard, paleontology
Ray, Wendell Augustus, physical chemistry
Reid, Evans Burton, organic chemistry
Scott, Allan Charles, zoology
Small, Donald Bridgham, mathematics
Smith, Wayne Lee, inorganic chemistry, physical chemistry
Terry, Robert Lee, cell physiology
Zukowski, Lucille Pinette, mathematics

WAYNE

Pettingill, Olin Sewall, ornithology

WEST BOOTHBAY HARBOR

Dugdale, Richard Cooper, biological oceanography, limnology
Graham, Joseph James, population biology
Larsen, Peter Foster, biological oceanography
Packard, Theodore Train, biological oceanography
Welch, Walter Raynes, marine biology

WESTBROOK

Allen, Roger Baker, physical chemistry
Gezon, Horace Martin, microbiology, epidemiology
La Vallee, William Alfred, pulp & paper technology
Millard, Ben, physical chemistry
Wheatland, David Alan, inorganic chemistry

WINSLOW

Haug, Arthur John, chemistry

MARYLAND

ABERDEEN

Gion, Edmund, physics

ABERDEEN PROVING GROUND

Allen, Frank Joseph, physics
Barrows, Austin Willard, Jr, applied physics
Benokraitis, Vitalius, numerical analysis
Boldt, Roger Earl, biochemistry
Callahan, John Joseph, physical chemistry
Celmins, Aivars Karlis Richards, applied mathematics
Coates, Arthur Donwell, physical chemistry, fuel science
Crosley, David Risdon, physical chemistry
Daasch, Lester William, mass spectrometry, spectrochemistry
Derr, John Sebastian, Jr, mathematical physics, biophysics
Dietz, Paul Hamilton, optics
Dubin, Henry Charles, electromagnetics
Eccleshall, Donald, research administration, applied physics
Eichelberger, Robert John, physics
Elder, Alexander Stowell, mathematics
Eng, Leslie, physical organic chemistry, organic chemistry
Epstein, Joseph, physical chemistry
Frasier, John T, applied mechanics
Freedman, Eli (Hansell), high temperature chemistry
Hedden, Kenneth Forsythe, microbial physiology, environmental sciences
Hillstrom, Warren W, surface chemistry, organic chemistry
Hollandsworth, Clinton E, nuclear physics
Klein, Nathan, physical chemistry
Leser, Tadeusz, mathematics
Lyman, Ona Rufus, physics
Masaitis, Cestovas, mathematics
McCreesh, Arthur Hugh, pharmacology, veterinary toxicology
McNamara, Bernard Patrick, pharmacology
Odell, Floyd Adams, biophysics
Olson, Wendell Clarence, organic chemistry
Poziomek, Edward John, organic chemistry, research administration
Rainis, Albert Edward, radiation physics
Shelton, Russell D, theoretical physics
Sliney, David H, health physics

Strenzwilk, Denis Frank, solid state physics
Tannenbaum, Harvey, physical chemistry
Temperley, Judith Kantack, nuclear physics
Vande Kieft, Laurence John, solid state physics, optical physics
Varney, Robert Nathan, molecular physics
Wangemann, Robert Theodore, biophysics
Ward, Joseph Richard, chemistry
Watermeier, Leland A, physical chemistry, physics
Witten, Benjamin, organic chemistry
Yockey, Hubert Palmer, theoretical biology

ABINGDON

Yurow, Harvey Warren, analytical chemistry

ADELPHI

Affens, Wilbur Allen, chemistry
Baba, Anthony John, radiation physics
Beard, David Franklin, plant breeding
Brandstein, Alfred George, mathematical analysis
Colglazier, Merle Lee, veterinary parasitology
Conrad, Edward Ezra, solid state physics, radiation chemistry
Hausner, Arthur, computer science
Marx, Egon, electromagnetics
McLean, Flynn Brevard, solid state physics
Michalowicz, Joseph Victor, mathematics
Miller, Robert Harold, botany
Ravitsky, Charles, applied physics, optics
St John, Peter Alan, analytical chemistry
Sattler, Joseph Peter, lasers
Saylor, Charles Hamilton Proffer, microscopy
Silbergeld, Sam, psychiatry
Simonis, George Jerome, lasers
Soln, Josip Zvonimir, theoretical physics
Sztankay, Zoltan Geza, physics
Ward, Alford L, physics
Yeatman, John Newton, food science

ANDREWS AFB

Kulp, Bernard Andrew, physics

ANNAPOLIS

Acosta, Virgilio, cosmic ray physics
Alley, Reuben Edward, Jr, physics
Betz, Ebon Elbert, mathematics
Blankinship, William Aubrey, applied mathematics
Calame, Gerald Paul, astrophysics, physics
Clark, Trevor H, physics
Clarke, Frank Eldridge, hydrology
Corey, Roland Reece, Jr, microbiology
Crum, Lawrence Arthur, acoustics
Currier, Albert (Eldred), mathematics
Elder, Samuel Adams, physics
Fischer, Eugene Charles, research administration, marine biology
Fontanella, John Joseph, solid state physics
Godshall, Fredric Allen, meteorology
Goodwin, Ralph Abijah, physics
Gorman, John Richard, mathematics
Granger, Robert A, II, physics, fluid dynamics
Gutsche, Graham Denton, physics
Ho, Louis T, acoustics
Jones, Owen Lloyd, physical chemistry, radiochemistry
Kalme, John S, mathematics
Kelley, John Francis, Jr, biochemistry
Klinkhammer, Michael Dennis, academic administration
Koubek, Edward, inorganic chemistry
Ladson, Thomas Alvin, veterinary medicine
Lagally, Paul, polymer chemistry
Massie, Samuel Proctor, chemistry
Matin, Shaikh Moizul, high energy physics
McQuaid, Richard William, fuel science, fuel technology
Milkman, Joseph, mathematics
Miller, Frank L, atomic physics, cosmic ray physics
Morgan, Bruce Harry, physics
Moulis, Edward Jean, Jr, mathematics
Palmer, Harold Dean, marine geology
Panzer, Richard Earl, physical inorganic chemistry
Pinkston, Earl Roland, physics

Rector, Charles Willson, solid state physics
Rollins, Orville Woodrow, inorganic chemistry, analytical chemistry
Saslaw, Samuel, mathematics
Schneider, Carl Stanley, solid state physics
Schrader, Malcolm Elliot, physical chemistry, surface chemistry
Sheets, Donald Guy, chemistry
Skinner, Dale Dean, underwater acoustics, electrical engineering
Smedley, William Michael, organic chemistry
Smithson, John Royston, physics
Stanley, Everett Michael, physical oceanography
Strohl, George Ralph, Jr, mathematics
Tierney, John A, mathematics
Tullier, Peter Marshall, Jr, operations research, systems analysis
Wardlaw, William Patterson, mathematics
Wiebush, Joseph Roy, analytical chemistry, marine chemistry
Willey, Cliff Rufus, soil physics
Williams, Jerome, physical oceanography
Wittler, Ruth Graeser, microbiology
Wolfe, Carvel Stewart, mathematics
Zimmerman, John Gordon, physical chemistry, inorganic chemistry

ARNOLD

Berrier, John Vincent, synthetic organic chemistry
Sagan, Leon Francis, mathematics
Schnappinger, Melvin Gerhardt, Jr, agronomy

ASHTON

Rice, Rip G, chemistry

BALTIMORE

Abbey, Helen, biostatistics
Adams, Elijah, biochemistry
Adams, James Miller, biochemistry, analytical chemistry
Albinak, Marvin Joseph, inorganic chemistry
Allen, Willard M, obstetrics & gynecology, endocrinology
Anderson, Mauritz Gunnar, mycology
Anderson, Paul Nathaniel, internal medicine, oncology
Anderson, Robert I, serology, parasitology
Andres, Reubin, medicine, gerontology
Annau, Zoltan, psychophysiology
Anthony, Ronald Lewis, immunology, serology
Aranow, Ruth Lee Horwitz, physical chemistry
Armstrong, Lloyd, Jr, atomic physics
Ascher, Eduard, psychiatry
Ashman, Michael Nathan, anesthesiology, computer sciences
Aurelian, Laure, virology
Bachur, Nicholas R, Sr, biochemistry, pharmacology
Bacon, Hilary Edwin, water chemistry, pollution chemistry
Baetjer, Anna Medora, physiology
Baisden, Charles Robert, medicine, pathology
Baker, Percy Hayes, genetics
Baker, Ralph Robinson, surgery
Baker, Richard H, genetics
Baker, Susan Pardee, epidemiology
Baker, Timothy D, preventive medicine
Ballentine, Robert, biochemistry, microbial ecology
Banerjee, Subrata, analytical chemistry
Bang, Frederik Barry, medicine, pathobiology
Bard, Philip, physiology
Barnett, Herbert Chester, medical entomology
Barraclough, Charles Arthur, endocrinology
Barrows, Charles Harry, Jr, biochemistry
Barry, Sue-ning C, cell physiology
Bartels, Richard Harold, computer science
Bearden, Joyce Alvin, physics
Beauge, Luis Alberto, biophysics
Beck, David Paul, biochemistry
Beck, Jeanne Crawford, biochemistry
Beer, Michael, molecular biology
Beiler, Adam Clarke, physics
Beishlag, George Albert, geography
Bell, William Robert, Jr, internal medicine, hematology
Benton, George Stock, meteorology
Bereston, Eugene Sydney, dermatology, mycology

MARYLAND

Berger, Hillard, molecular genetics
Berns, Kenneth, biochemistry
Bernstein, Dorothy Lewis, mathematics
Benson, Robert Chambliss, medicine
Bessman, Maurice Jules, biochemistry
Cohen, Bernice Hirschhorn, human genetics, epidemiology
Bhalla, Nam Parshad, mathematics
Bias, Wilma B., human genetics
Biltonen, Rodney Lincoln, biophysical chemistry, biophysics
Black, Perry, neurosurgery,
Blake, David Andrew, pharmacology, neurophysiology
Blake, William Dewey, physiology
Blankenship, Floyd Allen, physical chemistry
Blass, Elliott Martin, psychophysiology, animal behavior
Blaumanis, Otis Rudolf, physiology,
Bloch, Aaron Nixon, chemical physics, solid state chemistry
Block, Jerome Bernard, oncology, immunology
Blomster, Ralph N., pharmacognosy
Boardman, John Michael, mathematics
Bodian, David, neurobiology
Boicourt, William Closson, physical oceanography
Bonnet, Philip D., medical administration
Borgaonkar, Digamber Shankarrao, genetics, cytology
Boslow, Harold Meyer, psychiatry
Boye, Robert James, nuclear physics
Boyer, Samuel H. IV, medical genetics
Brand, Ludwig, biochemistry
Branigan, Otto Charles, surgery
Bricker, Owen P. III, geochemistry
Brinley, Floyd John, Jr., biophysics
Brock, Mary Anne, biology, physiology
Brody, Eugene B., psychiatry, psychoanalysis
Bromberger-Barnea, Baruch (Berthold), environmental medicine, physiology
Brown, Clinton Carl, psychophysiology
Brown, Donald D., developmental biology
Brown, Neal Curtis, biochemistry, pharmacology
Brown, Torrey Carl, medicine
Brush, Grace Somers, forest ecology,
Brusilow, Saul W., pediatrics, physiology
Bucci, Enrico, physical biochemistry
Bueding, Ernest, biochemistry
Burgison, Raymond Merritt, pharmacology
Burke, Morris, protein chemistry, organic chemistry
Burnett, Joseph W., dermatology
Burns, Hugh Donald, organic chemistry, nuclear medicine
Butcher, Henry Clay, IV, plant physiology, biochemistry
Butzow, James J., biochemistry
Byer, Norman Ellis, solid state physics
Byron, Joseph Winston, pharmacology
Cady, John Gilbert, soils
Callahan, Mary Vincent, chemistry, instrumentation
Campolattaro, Alfonso, theoretical physics
Caplan, Yale Howard, toxicology
Carlson, Francis Dewey, biophysics
Carroll, Douglas Gordon, physiology
Carter, Harry Hart, physical oceanography
Carver, David Harold, pediatrics, infectious diseases
Cash, Richard Alan, tropical public health
Cebra, John Joseph, immunochemistry, immunology
Chace, Gary Andrew, medicine
Charache, Samuel, medicine
Chang, John Wan-Yun, biochemistry
Chang, Yung-Feng, biochemistry, microbiology
Channing, Cornelia Post, endocrinology
Chappelle, Emmett W., biochemistry, photobiology
Charache, Patricia, medicine
Chase, Gary Andrew, human genetics, statistics
Chen, Chung-Ho, biochemistry
Chi, Kuo-Ruey, crop breeding, genetics
Chien, Chih-Yung, high energy physics
Chisolm, James Julian, Jr., medicine
Cincotta, Joseph John, analytical chemistry
Clark, Carl Cyrus, biophysics
Cleaves, Emery Taylor, geomorphology, environmental geology
Clemmens, Raymond Leopold, medicine, pediatrics

Coffey, Donald Straley, pharmacology, biochemistry
Cohan, Leonard Hecht, colloid chemistry
Cohen, Alan Mathew, biology, embryology
Cohen, Bernice Hirschhorn, human genetics, epidemiology
Cohen, Howard Joseph, chemistry
Cohn, Leslie, mathematics
Cole, Gerald Alan, virology, immunology
Comstock, George Wills, epidemiology
Cone, Richard Allen, biophysics
Conley, Carroll Lockard, medicine
Conn, Rex Boland, medicine, clinical pathology
Connor, Thomas Byrne, medicine
Cooke, Charles Robert, medicine
Corak, William Sydney, materials
Cornblath, Marvin, pediatrics, biochemistry
Corwin, Alsoph Henry, organic chemistry
Cotter, Edward F., internal medicine
Cowan, Dwaine O., organic chemistry
Craig, Nessly Coile, cell biology
Craig, Susan Walker, immunobiology
Crouthamel, William Guy, pharmacy, pharmaceutics
Crowley, William Patrick, geology
Dannenberg, Arthur Milton, Jr., pathology, immunology
Davidsen, Arthur Falnes, astrophysics, x-ray astronomy
Davies, Philip Wynne, physiology, biophysics
Davis, Leroy Thomas, family medicine
DeBow, Lee Richard, psychiatry, organic chemistry
Debuskey, Matthew, pediatrics
Debye, Nordulf Wiking Gerud, physical inorganic chemistry, physical chemistry
De Carlo, John, Jr., radiology
De Lorenzo, Anthony John, anatomy, cell biology
Devoe, Robert, physiology, biophysics
Dhople, Arvind Madhav, clinical biochemistry, microbial physiology
Diamond, Earl Louis, statistics
Dicke, Sally Harrison, astronomy
Dietz, Albert J., Jr., clinical pharmacology, biochemistry
Dintzis, Howard Marvin, biophysics
Doering, John P., chemical physics
Domokos, Gabor, elementary particle physics
Donati, Edward Joseph, anatomy
Donnelly, Charles Joseph, dental epidemiology
Donner, Martin W., medicine, radiology
Dorst, John Phillips, medicine, radiology
Douglass, Kenneth Harmon, medical physics
Dowling, Marie Augustine, mathematics
Drachman, Daniel Bruce, neurology, muscular physiology
Dubin, Norman H., reproductive endocrinology
Ducker, Thomas Barbee, medical sciences, health sciences
Duncan, Acheson Johnston, applied statistics
Duncan, David Beattie, statistics
Dunning, Henry Armitt Brown, Jr., chemistry
Durkan, James P., medicine, obstetrics & gynecology
Earickson, Robert James, urban geography, applied statistics
Ebert, James David, embryology
Eby, Denise, chemistry
Eddin, Michael Aaron, embryology, immunology
Edwards, Jonathan, Jr., structural geology
Ehrlich, Walter, cardiovascular physiology
Eichorn, Gunther Louis, bioinorganic chemistry
Elkes, Joel, psychiatry, psychopharmacology
Elliott, Howard Ralph, vertebrate zoology
Eugster, Hans Peter, geochemistry, sedimentology
Ewing, Larry Larue, endocrinology
Embree, Earl Owen, mathematics
Englund, Paul Theodore, biochemistry
Eppler, Richard A., ceramics, inorganic chemistry
Erickson, Howard Ralph, physical chemistry
Fajer, Abram Benjanin, physiology, endocrinology
Falker, William Alexander, Jr., immunology

Fambrough, Douglas McIntosh, neurobiology
Fastie, William George, astronomy
Feild, Frank Joseph, biochemistry
Fein, Alvin Eli, applied mathematics
Feldman, Gordon, theoretical physics
Feldman, Paul Donald, atmospheric physics, aeronomy
Fenselau, Allan Herman, organic chemistry
Fenselau, Catherine Clarke, pharmacology, mass spectrometry
Ferguson, Gordon John, geophysics
Fertziger, Allen Philip, neurophysiology, neurobiology
Fioto, George Anthony, cosmetic chemistry
Firman, David, geography
Fiset, Paul, microbiology
Fisher, George Wescott, geology
Fisher, Russell Sylvester, pathology
Fitch, Steven Joseph, industrial chemistry
Fitzgerald, Edwin Roger, physics
Fitzgerald, Robert Schaefer, physiology
Fleischer, Clara Joel, medicine, surgery
Flotte, C Thomas, medicine, surgery
Forbes, Michael Shepard, biological structure, cell biology
Forester, Donald Charles, behavioral biology, herpetology
Foster, Girard Vernam, endocrinology
Fox, Samuel Louis, ophthalmology, clinical pharmacology
Frank, Jerome David, psychiatry
Frank, Leonard Harold, biochemistry
Fraser, Thomas Petigru, science education, science writing
Frazier, John Melvin, environmental medicine
Freeman, John Mark, pediatric neurology
Froehlich, Jeffrey Paul, biophysics
Frost, John Kingsbury, medicine, pathology
Fulton, Thomas, theoretical physics
Funk, Helen Beatrice, microbiology
Gathey, Betty Jean, biophysical chemistry
Gallant, Leonard Jay, psychiatry
Gallier, Sidney Roland, ecology, limnology
Gamble, James Lawder, Jr., physiology, biochemistry
Gann, Donald Stuart, physiology, surgery
Garcia, Julio H., neuropathology, electron microscopy
Gartner, Leslie Paul, histology, dental research
Genecin, Abraham, internal medicine, cardiology
Gerberg, Eugene Jordan, entomology
Gertz, Samuel David, anatomic pathology
Gillespie, Elizabeth, biochemistry, pharmacology
Glaser, Edmund M, computer science, physiology
Glaser, Kurt, child psychiatry
Glick, Samuel Shipley, pediatrics
Godene, Ghislaine D., psychiatry
Goldberg, Alan Marvin, pharmacology
Goldberg, Benjamin, histochemistry
Golden, Archie Sidney, pediatrics
Goldenson, Jerome, chemistry
Goldfine, Lewis John, physical medicine & rehabilitation
Goldheim, Samuel Lewis, chemistry
Goldman, Moise Herbert, Jr., biophysics, electrical engineering
Goldstein, Jeannine, mathematics
Goodman, Louis E., surgery, oncology
Gordis, Leon, pediatrics, epidemiology
Gornic, Fred, physical chemistry
Gott, Vincent Lynn, cardiovascular surgery
Goterer, Gerald S., biochemistry
Gould, Edwin, ecology, animal behavior
Graff, Thomas D., anesthesiology
Graham, George G., pediatrics, nutrition
Granick, Jeannine, mathematics
Grant, David Graham, radiation physics
Greeley, James H., prosthodontics
Green, Lawrence Winter, public health education
Greenberger, Martin, applied mathematics
Greenhouse, Harold Mitchell, physical inorganic chemistry
Greenhouse, Walter Van Vleck, biochemistry
Greenough, William B. III, medicine
Gregerman, Robert Isaac, gerontology
Greisman, Sheldon Edward, physiology, medicine
Grenell, Robert Gordon, anatomy, psychiatry

Grewe, John Mitchell, orthodontics, anatomy
Griffin, Diane Edmund, immunology, virology
Grollman, Sigmund, physiology
Gross, Fred, mathematics
Gross, James Harrison, anatomy, neuroanatomy
Gross, Meredith Grant, Jr., oceanography
Grossman, Lawrence, biochemistry
Gruenberg, Ernest Matsner, epidemiology, psychiatry
Gryder, John William, physical chemistry, inorganic chemistry
Guchhait, Ras Bihari, biochemistry
Guth, Lloyd, neuroanatomy
Hacker, Peter Wolfgang, physical oceanography
Hahn, Richard David, internal medicine
Haig, Frances Rawle, theoretical physics
Hall, James Alexander, physics, solid state electronics
Haller, Jacob Alexander, Jr., surgery
Hambrick, George Walter, Jr., medicine, dermatology
Handelsman, Jacob C., surgery
Hanks, John Harold, bacteriology, immunology
Harper, Paul Alva, pediatrics
Hardy, William George, audiology, speech pathology
Harrington, William Fields, molecular biology
Harrison, Harold Edward, pediatrics
Harrison, Helen Coplan, biochemistry
Harrison, Walter Kirby, Jr., biomedical engineering
Hartman, Philip, mathematics
Hartman, Philip Emil, microbiology
Hartz, Jerome, anesthesiology, evolutionary
Harvey, Abner McGehee, medicine
Haskins, Arthur L, Jr., obstetrics & gynecology
Hathaway, Wilfred Bostock, biology
Heald, Felix Pierpont, Jr., medicine, pediatrics
Hebel, John Richard, biostatistics
Heck, Albert Frank, neurology
Hedman, Fritz Algot, physical chemistry
Heinz, John Michael, biomedical engineering
Helrich, Martin, anesthesiology
Heltne, Paul Gregory, evolutionary biology, gross anatomy
Hendrickson, Constance McRight, biophysical chemistry, pesticide chemistry
Henney, Christopher Scot, immunology
Henry, Richard Conn, astrophysics
Hepner, Ray, pediatrics
Hepinstall, Robert Hodgson, pathology
Herman, David S., solid state physics
Herman, Robert McCulloch, neurology
Herriott, Roger Moss, biochemistry
Heyssel, Robert M., hematology; nuclear medicine
Hill, Charles Earl, family medicine
Hillis, William Daniel, virology
Hinton, David Earl, comparative pathology
Hobart, Donald James, kinesiology
Hodges, Fred Jenner, III, medicine
Hoehn-Saric, Rudolf, psychiatry
Hollander, David Hutzler, pathology
Hollis, Donald Pierce, biophysics
Holtzman, Neil Anton, pediatrics, medical genetics
Hoopes, John Eugene, plastic surgery
Horak, Martin George, mathematics
Horelick, Brindell, mathematics
Houk, James Charles, physiology, engineering
Horn, Lyle William, biophysics
Horn, Roger Alan, mathematics
Horn, Susan Dadakis, biostatistics
Horn, William Everett, physics
Hornick, Richard B., internal medicine, infectious diseases
Hoster, Donald Paul, organic chemistry
Hsu, Chao Kuang, laboratory animal medicine, public health
Hsu, Susan Hu, immunogenetics
Hsu, Yu-Chih, developmental biology, virology
Hu, Stephen Moi Kee, entomology
Huang, Jacob Wen-Kuang, optical physics, solid state physics

Huang, Pien-Chien, biochemistry, genetics
Huang, Ru-Chih Chow, molecular biology
Hubbard, T Brannon, Jr, surgery
Huddleston, Mary Anne, zoology
Huffer, Sarah Virginia, medicine, psychiatry
Hume, John Chandler, preventive medicine, public health
Humphries, J O'Neal, medicine
Hunt, Richard Kevin, developmental physiology, neurosciences
Hunter, John Robert, oceanography
Hurley, Forrest Reyburn, inorganic chemistry
Hybl, Albert, biophysics, crystallography
Iber, Frank Lynn, medicine
Ichniowski, Casimir Thaddeus, pharmacology
Imboden, John Baskerville, psychiatry
Isaacs, Rufus Philip, applied mathematics
Iseri, Oscar Akio, pathology
Ishizaka, Kimishige, immunology, immunochemistry
Jaffee, Harris Alexander, algebra
Jandorf, Bernard Joseph, biochemistry
Jensen, Arthur Seigfried, electronic physics
Jewett, Hugh Judge, urology
Johns, Richard James, medicine
Johnson, Richard T, neurology, virology
Jonas, Leonard Abraham, physical chemistry
Joseph, Junior Mehsen, medical microbiology
Joseph, Richard Isaac, solid state physics
Judd, Brian Raymond, theoretical physics, atomic physics
Judson, Horace Augustus, organic chemistry
Jurf, Amin N, physiology
Kahn, Harold A, epidemiology
Kaizer, Herbert, oncology, virology
Kaltreider, D Frank, obstetrics
Kan, Lou Sing, physical chemistry, biophysics
Kaplan, Emanuel, biochemistry
Kauffman, Frederick C, biochemistry, pharmacology
Kaufman, Joyce J, quantum chemistry, psychopharmacology
Kayser, Robert Helmut, bio-organic chemistry
Kessel, Rosslyn William Ian, immunology, microbiology
Kessler, Irving Isar, preventive medicine, epidemiology
Khazan, Naim, pharmacology
Kidd, Bernard Sean Langford, cardiovascular physiology
Kidder, George Wallace, III, physiology, biophysics
Kim, Chung W, theoretical high energy physics, theoretical nuclear physics
Kimball, Allyn Winthrop, biostatistics
Kincaid, John Franklin, chemical physics
King, John Wesley, biology
King, Theodore Matthew, obstetrics & gynecology, physiology
Kinnard, William J, Jr, pharmacology
Kirtley, Mary Elizabeth, biochemistry
Kistenmacher, Thomas John, inorganic chemistry, crystallography
Klee, Gerald D'Arcy, psychiatry, medicine
Klein, Albert William, anatomy, pathology
Klemm, Waldemar Arthur, Jr, silicate chemistry
Klimt, Christian Robert, public health
Klinefelter, Harry Fitch, clinical medicine
Knatterud, Genell Lavonne, biostatistics
Knoblock, Edward C, biochemistry
Knox, David Lalonde, ophthalmology
Knox, Gaylord Shearer, medicine, radiology
Knudsen, James Frederick, medical physiology
Koblin, Abraham, analytical chemistry
Kodras, Rudolph, veterinary medicine, biochemistry
Kok, Bessel, plant physiology
Kokoski, Robert John, pharmacy, pharmacognosy
Kornblum, Roland Norman, pathology
Koski, Walter S, physical chemistry
Kovesi-Domokos, Susan, elementary particle physics
Kowalewski, Edward Joseph, family medicine
Kowarski, A Avionoam, pediatrics, endocrinology
Kowarski, Chana Rose, pharmaceutics
Krahl, Vernon Edward, anatomy
Kramer, Norman, bacteriology
Krikorian, Samuel Edward, Jr, analytical chemistry
Kruse, Cornelius Wolfram, public health, sanitary engineering
Krywolap, George Nicholas, microbial physiology
Kuhar, Michael Joseph, neuropharmacology, neurobiology
Kuhn, Albin Owings, agronomy
Kumada, Mamoru, physiology
Kundig, Fredericka Dodyk, cell physiology, vertebrate physiology
Kurland, Albert A, psychiatry
LaBrosse, Elwood Henry, biochemistry
Laden, Hyman Nathaniel, mathematics, electronics
Lafon, Guy Michel, geochemistry
Lambooy, John Peter, biochemistry
LaMotte, Carole Choate, neuroanatomy, neurophysiology
LaMotte, Robert Hill, neurosciences
Lamy, Peter Paul, biopharmaceutics, clinical pharmacy
Lane, Malcolm Daniel, biochemistry
Larkin, David, electrochemistry
Larrabee, Martin Glover, neurochemistry, neurophysiology
Laws, Edward Raymond, Jr, toxicology
Lee, Frederick Strube, physical chemistry
Lee, Yuan Chuan, biochemistry
Lee, Yung-Keun, nuclear physics
Lehninger, Albert Lester, biochemistry
Lemkau, Paul Victor, psychiatry, public health
Lennarz, William Joseph, biochemistry
Leonard, Charles Brown, Jr, biochemistry
Leons, Madeline Barbara, anthropology, cultural anthropology
Lesinski, John Silvester, obstetrics & gynecology
Lesko, Stephen Albert, biochemistry
Leslie, James, biophysical chemistry
Levin, Isador, physical chemistry
Levin, Jack, internal medicine, hematology
Levin, Morton Loeb, preventive medicine, public health
Levin, Nathan, pharmaceutical chemistry
Levine, David Morris, public health, medical research
Levy, David Alfred, immunology, allergy
Levy, Robert, biochemistry, clinical chemistry
Libonati, Joseph Peter, clinical microbiology
Lichtenstein, Lawrence M, medicine, immunology
Liebman, Joel Fredric, theoretical chemistry
Lietman, Paul Stanley, clinical pharmacology
Lilienfeld, Abraham Morris, epidemiology, biostatistics
Lim, Henry S, anesthesiology
Lin, Shin, biochemistry, biophysics
Lindenberg, Richard, neuropathology
Linder, Seymour Martin, industrial organic chemistry
Lisansky, Ephraim Theodore, internal medicine, psychosomatic medicine
Little, Robert Greenwood, inorganic chemistry, x-ray crystallography
Littlefield, John Walley, pediatrics
Livingston, Samuel, pediatrics
Lo, Chu-Shek, medical physiology
Logue, Marshall Woford, chemistry
Long, Donlin Martin, neurosurgery, electron microscopy
Long, Robert Radcliffe, meteorology
Lonky, Martin Leonard, electronic physics, solid state physics
Love, Warner Edwards, biophysics
Ludlum, David Blodgett, pharmacology
Lunin, Martin, pathology
Lustgarten, Jack Abraham, clinical chemistry
Lye, Robert Glen, solid state physics
Lynch, James Carlyle, neurophysiology
Lynn, Yen-Mow, applied mathematics
Madansky, Leon, nuclear physics
Magladery, John William, neurology
Maher, Philip Kenerick, physical chemistry
Makinodan, Takashi, immunology
Malarkey, Edward Cornelius, optical physics
Mandelberg, Hirsch I, physics
Mandell, Wallace, psychology, public health
Marcus, Allan H, statistics, environmental sciences
Mardiney, Michael Ralph, Jr, immunology, internal medicine
Margolis, Simeon, biochemistry
Marsh, Bruce David, geology
Marsh, David George, immunogenetics, biochemistry
Marsho, Thomas V, plant biochemistry
Maslen, Stephen Harold, applied mathematics
Mason, George Robert, surgery, physiology
Matanoski, Genevieve M pediatrics, epidemiology
Mattoon, James Richard, biochemistry
Maumenee, Alfred Edward, ophthalmology
Max, Stephen Richard, biochemistry
Mayer, Manfred Martin, immunology, biochemistry
Mayer, Richard F, neurology
Mazumder, Bibhuti R, surface chemistry, solid state chemistry
McCarthy, Vincent Cormac, parasitology, microbiology
McCarty, Maclyn, Jr, physical chemistry
McCoart, Richard F, Jr, mathematics
McCord, Colin Wallace, public health, surgery
McCrumb, Fred Rodgers, medicine
McDonnell, Edmond Joseph, orthopedic surgery
McDowell, Elizabeth Mary, pathology
McGuire, Francis Joseph, organic chemistry, analytical chemistry
McHugh, Paul Rodney, neurology, psychiatry
McIntyre, Patricia Ann, medicine
McKee, James Robert, organic chemistry
McKhann, Guy Mead, neurology, neurochemistry
McKusick, Victor Almon, medicine
McMillion, C Robert, pharmaceutical chemistry, organic chemistry
McQueen, John Donald, medicine
Meckler, Alvin, theoretical physics
Meinert, Curtis Lynea, biostatistics
Mellits, E David, biostatistics
Merlis, Jerome K, neurophysiology
Merz, William George, medical mycology, immunology
Meszler, Richard M, cell biology, neuroanatomy
Meyer, Eugene, psychiatry
Meyer, Jean-Pierre, mathematics
Middlebrook, Gardner, medicine
Migeon, Barbara Ruben, medical genetics
Migeon, Claude Jean, pediatrics, endocrinology
Miller, Melvin P, physical chemistry
Miller, Paul Scott, bio-organic chemistry
Milnor, William Robert, medical physiology
Minkowski, Jan Michael, solid state physics
Mitch, William Evans, nephrology
Mithun, Jacqueline Stearns, applied anthropology
Moment, Gairdner Bostwick, developmental biology
Monjan, Andrew Arthur, neurosciences, immunopathology
Monroe, Russell Ronald, medicine
Montgomery, Raymond Braislin, physical oceanography
Montgomery, Stewart Robert, industrial chemistry
Moos, Henry Warren, physics
Morgan, Russell Hedley, radiology
Moskowski, Erica F, obstetrics & gynecology, endocrinology
Mosley, Wiley Henry, epidemiology
Mountcastle, Vernon Benjamin, physiology
Mulligan, Joseph Francis, atomic physics, molecular physics
Mullins, Lorin John, biophysics
Murphy, Patrick Aidan, microbiology, medicine
Murr, Brown L, Jr, physical organic chemistry
Musselman, Nelson Page, toxicology, analytical chemistry
Nachlas, Marvin Morton, surgery
Naddor, Eliezer, operations research
Nager, George Theodore, otolaryngology
Nair, Padmanabhan Padmanabhan, biochemistry
Nalin, David Robert, medical research
Nason, Alvin, enzymology
Nathans, Daniel, microbiology, molecular biology
Nathanson, Neal, virology, epidemiology
Nauman, Robert Karl, microbiology
Neill, Catherine Annie, pediatrics, cardiology
Nelson, Erland, neurology, neuropathology
Newburger, Sylvan Henry, cosmetic chemistry
Newman, John Brown, physics
Nickon, Alex, organic chemistry
Niedermeyer, Ernst F, neurology
Nipper, Henry Carmack, clinical chemistry
Nordin, Albert Andrew, immunology
Norman, Philip Sidney, immunology, biochemistry
Odell, George Berlage, pediatrics
Odell, Lois Dorothea, biology, science education
Olson, Donald Lee, dentistry, oral pathology
Oppenheimer, John Reed, primatology, animal ecology
Osman, Jack, pharmaceutical chemistry, medicinal chemistry
Otenasek, Frank (Joseph), neurosurgery
Otto, Robert George, aquatic ecology
Owellen, Richard John, oncology, medicine
Owen, George Ernest, nuclear physics
Owens, Albert Henry, Jr, internal medicine, pharmacology
Ozol, Michael Arvid, geology
Pagano, Richard Emil, biophysics
Parikh, Indu, biochemistry
Parker, Robert Tarbert, internal medicine
Parker, Rodger D, mathematics, mathematical biology
Patz, Arnall, ophthalmology
Pedersen, Peter L, biochemistry, molecular biology
Permutt, Solbert, medicine
Petersen, Kyle W, cytogenetics
Petersen, Raymond Carl, physical chemistry
Peterson, Malcolm Lee, medical education
Pettijohn, Francis John, geology
Pevsner, Aihud, physics
Phillips, Owen M, mechanics
Phizackerley, Richard Paul, x-ray crystallography, applied physics
Piavis, George Walter, zoology
Pierce, Nathaniel Field, infectious diseases
Pinter, Gabriel George, medicine, physiology
Pipino, Raymond Joseph, operations research, systems analysis
Pitha, Josef, bio-organic chemistry
Pitha, Paula Marie, biochemistry, molecular biology
Pittenger, Arthur O, mathematics
Plapinger, Robert Edwin, organic chemistry
Poggio, Gian Franco, physiology
Pollack, Ralph Martin, bio-organic chemistry, physical organic chemistry
Pomerantz, Seymour Herbert, biochemistry
Portnoy, Joseph, immunology
Posner, Gary Herbert, organic chemistry
Prendergast, Robert Anthony, immunology, pathology
Price, David Edgar, public health administration
Price, Richard Swee, ethnology, ethnography
Price, Winston Harvey, biochemistry, virology
Pritchard, Donald William, oceanography
Proctor, Donald Frederick, otolaryngology, environmental medicine
Prout, Thaddeus Edmund, medicine
Provenza, Dominic Vincent, zoology
Radcliffe, Alec, physics, systems analysis
Radford, Edward Parish, Jr, physiology, environmental health
Ramsay, Frederick J, anatomy
Rao, Gopala U V, radiological physics
Rasera, Robert Louis, solid state physics, nuclear physics
Rash, John Edward, neurobiology
Raskin, Betty Lou, plastics chemistry, psychology
Raskin, Howard F, medicine
Raskin, Joan, medicine, dermatology
Rasmussen, Peter, pathology
Ray, Carleton, zoology
Reading, Anthony John, psychiatry
Reback, John Frederick, medical microbiology
Rednor, Abraham, analytical chemistry
Reed, Clyde F, botany
Reed, Warren Douglas, biochemistry, toxicology
Reeder, Ronald Howard, molecular biology
Rehak, Matthew Joseph, clinical chemistry
Reich, Claude Virgil, medical microbiology, leprology
Reichelderfer, Thomas Elmer, pediatrics, public health
Reimann, Dexter Leroy, pathology
Reinke, William Andrew, biostatistics
Reissig, Magdalena, electron microscopy, parasitology
Rennels, Marshall L, neuroanatomy
Revell, Samuel Thompson Redgrave, Jr, medicine
Rhodes, Buck Austin, radiology, pharmacology

MARYLAND

Richards, Joseph Dudley, industrial chemistry
Richards, Richard Davison, ophthalmology
Richardson, Edward Henderson, Jr., obstetrics & gynecology
Richardson, Paul Frederick, medicine
Richter, Curt Paul, psychobiology
Rider, Rowland Vance, biostatistics
Ridley, Esther Joanne, plant physiology
Rieksniece, Emilija Katrina, pathology, endocrinology
Rifkind, Joseph Moses, physical biochemistry
Riley, Lee Hunter, Jr., orthopedic surgery
Ritter, Richard Arnold, medicine
Roberts, Fredric Calvin, applied mathematics
Robinson, Cecil Howard, organic chemistry, pharmacology
Robinson, David Adair, neurophysiology
Robinson, Dean Wentworth, physical chemistry
Robinson, Harry Maximilian, Jr., dermatology
Robinson, James Eugene, medical physics
Robinson, Robert Alexander, orthopedic surgery
Rodriguez, Alejandro, psychiatry
Roos, Raymond Philip, neurology, virology
Roseman, Saul, biochemistry
Rosen, Barry Philip, biochemistry
Rosen, Harold, psychosomatic medicine
Ross, Richard Starr, cardiology
Roswell, David Frederick, chemistry
Roth, Stephen (Allen), developmental biology
Roth, Thomas Frederic, cell biology, developmental biology
Rothstein, Frances Marilyn, ethnology
Rowe, Elizabeth Snow, physical biochemistry
Royall, Richard Miles, biostatistics
Rozeboom, Lloyd Eugene, medical entomology
Rubin, Morton Harold, theoretical physics
Rubin, Robert Jay, toxicology, biochemistry
Ruchkin, Daniel S., biomedical engineering
Rudo, Frieda Galindo, pharmacology
Rush, Cecil Archer, microchemistry
Saba, George Peter, II, radiology
Sachs, Lester Marvin, theoretical physics, information science
Sack, Richard Bradley, medicine, microbiology
Sacktor, Bertram, biochemistry
Sagawa, Kiichi, physiology; biomedical engineering
Sakai, Richard Kazuichi, genetics
Salik, Julian Oswald, radiology
Sampson, Joseph Harold, mathematics
Santos, George Wesley, oncology, immunology
Savage, Charles, psychiatry
Schefris, Sidney, cardiovascular diseases
Schiller, Everett L., medical parasitology
Schirmer, Horst K A., cell physiology
Schmeisser, Gerhard Jr., orthopedic surgery
Schmukler, Morton, hematology, biochemistry
Schneider, Edward Lewis, human genetics
Schon, Miguel Antonio, anatomy, physical anthropology
Schubert, David Crawford, lasers
Schuetz, Allen W., endocrinology, cell biology
Schuler, Frances Pierce, anatomy, physical anthropology
Schwartz, Martin, plant physiology
Scocca, John Joseph, biochemistry, molecular biology
Scott, William Wallace, physiology, photobiology
Seidel, Henry Murray, pediatrics
Seidman, Thomas Israel, applied mathematics
Seiger, Howard Harold, physics, photobiology
Seligman, Arnold Max, surgery
Seltser, Raymond, preventive medicine, public health
Selvin, Beatrice L., anesthesiology
Shafer, William McKinley, metallurgical chemistry
Shah, Keerti, virology
Shangraw, Ralph F., industrial pharmacy
Shapiro, George, mathematics

Shapiro, Sam, research administration, epidemiology
Shay, Donald Emerson, bacteriology
Shearn, Allen Barnard, developmental genetics
Sheldon, Walter Herman, pathology
Shepard, Richard Hance, medicine
Shiffman, Bernard, mathematics
Shimizu, Hiroshi, otolaryngology, audiology
Shin, Yong Ae Im, inorganic chemistry, gerontology
Shock, Nathan Wetherill, physiology; gerontology
Shuger, Leroy Woodrow, chemistry
Shulman, Lawrence Edward, internal medicine
Siebens, Arthur Alexandre, physiology; rehabilitation medicine
Silva, Joseph A., mathematics
Silver, Seymour David, chemistry
Silverman, David J., microbiology; cell biology
Silversmith, Ernest Frank, organic chemistry
Silverstone, Harris Julian, quantum chemistry
Simpson, David Gordon, internal medicine
Singewald, Martin Louis, internal medicine
Sisca, Rodger Franklin, histology; restorative dentistry
Siwinski, Arthur George, medicine
Sjodin, Raymond Andrew, biophysics
Skainy, Jan Peter, silicate chemistry
Sladen, William Joseph Lambart, medicine, ecology
Smith, Andrew George, microbiology; medical mycology
Smith, Gardner Watkins, surgery
Smith, Hamilton Othanel, microbial genetics
Snope, Andrew John, genetics, cytogenetics
Snyder, Merrill J., clinical microbiology
Snyder, Solomon H., neuropharmacology
Sofer, William Harold, developmental biology
Solomon, Neil, physiology
Southwick, Charles Henry, zoology
Spear, Gerald Sanford, pathology
Speros, Perry, visual physiology; electrophysiology
Spivak, Jerry Lepow, hematology
Spragins, Melchiah, pediatrics
Squire, Robert Alfred, comparative pathology
Stanley, Steven Mitchell, paleontology, evolution
Starfield, Barbara Holtzman, pediatrics
Starr, John Thornton, Jr., transportation geography; urban geography
Steckler, Robert, chemistry
Steinwachs, Donald Michael, operations research; public health administration
Stewart, Doris Mae (Mrs Felix Powell), zoology; physiology
Stine, Oscar Cebren, pediatrics, public health
Stokowski, Stanley E., solid state physics
Stone, William Spencer, tropical medicine
Stubbs, Ulysses Simpson Jr., chemistry
Stuelpnagel, John Clay, mathematics
Suskind, Sigmund Richard, microbiology
Sweeting, Linda Marie, molecular pharmacology
Swift, David Leslie, environmental medicine, physiology
Sysiskis, Robert Joseph, virology
Tadros, Maher Ebeid, surface chemistry
Taft, Jay Leslie, biological oceanography
Talalay, Paul, molecular pharmacology
Talamo, Barbara Lisann, neurobiology, biochemistry
Talbot, Walter Richard, mathematics
Talbot, William Henry, neurophysiology
Taussig, Helen Brooke, medicine
Tayback, Matthew, public health
Taylor, Carl Ernest, preventive medicine, epidemiology
Taylor, Julius Henry, solid state physics
Taylor, Walter Rowland, oceanography
Teitelbaum, Harry Allen, neurology, psychiatry
Tepper, Byron Seymour, microbiology
Tesar, Charles, chemical kinetics
Thomas, George Howard, pediatrics
Thompson, Raymond K., mathematics
Thomsen, John Stearns, atomic physics
Thorne, Richard Mark, immunology
Thorne, Melvyn Charles, epidemiology
Thut, Paul Douglas, psychopharmacology
Tigertt, William David, pathology
Tildon, J Tyson, biochemistry
Togo, Yasushi, infectious diseases, virology
Tonascia, James A., biostatistics

Traub, Robert, medical entomology
Treffner, Walter Sebastian, inorganic chemistry, physical chemistry
Trips, Milan, medical entomology
Truesdell, Clifford Ambrose, III, mathematics
Trump, Benjamin Franklin, pathology, cell biology
Tsan, Min-Fu, hematology, nuclear medicine
Ts'o, Paul On Pong, biophysical chemistry
Tsong, Tian Yow, biophysical chemistry, biochemistry
Tumulty, Philip A., medicine
Turner, Thomas Bourne, microbiology
Tyson, Greta E. zoology
Udvarhelyi, George Bela, neurosurgery
Vance, Hugh Gordon, analytical chemistry, biochemistry
Vandegaer, Jan Edmond, physical chemistry; organic chemistry
van der Hoeven, Theo A., biochemistry
Van Metre, Thomas Earle, Jr., medicine
Varnhorn, Mary Catherine, physics
Venables, John Duxbury, physics
Vincent, James Sidney, physical chemistry
Vitullo, Victor Patrick, organic chemistry
Von Korff, Richard Walter, biochemistry, enzymology
Wadsworth, Gladys Elizabeth, anatomy
Wagley, Philip Franklin, medicine
Wagner, Henry N. Jr., internal medicine, nuclear medicine
Wagner, John Alfred, neuropathology
Walker, Donald Gregory, anatomy
Walker, J Calvin, nuclear physics, solid state physics
Walker, Michael Dirck, neurosurgery
Walker, Wilbur Gordon, internal medicine
Walser, Mackenzie, medicine
Wang, Shih Yi, organic chemistry
Wang, Virginia Li, public health
Warner, George S., microbiology; immunology
Warnick, Jordan Edward, neuropharmacology
Watt, John Yin Chieh, parasitology
Weaver, Kenneth Newcomer, geology
Webb, James L A. organic chemistry
Weber, Leon, surface chemistry, physical chemistry
Weigman, Bernard J., engineering physics
Weinreich, Daniel, neurobiology
Weiss, Leon, anatomy
Weiss, Michael Stephen, speech & hearing sciences, speech pathology
Welch, Bruce L., ecology; neuropharmacology
White, Emil Henry, organic chemistry
White, John Irving, physiology
White, Kerr Lachlan, epidemiology
Whitridge, John, Jr., obstetrics & gynecology
Wiggins, James Wendell, biophysics
Williamson, Penelope Rose, psychosomatic medicine, human ecology
Wilson, John Phillips, mathematics
Wilson, P David, statistics
Windler, Donald Richard, plant taxonomy
Wingrove, Alan Smith, organic chemistry
Wisseman, Charles Louis, Jr., medical microbiology
Wiswell, John Gordon, endocrinology
Wizenberg, Morris Joseph, radiotherapy
Wolman, Markley Gordon, geology
Wood, Colin, pathology
Woodroff, James Donald, obstetrics & gynecology
Woods, Alan Churchill, Jr., surgery
Woodward, Theodore Englar, microbiology; medicine
Wright, Norbert Dean, public health
Wu, En Shinn, chemical physics
Yarbrough, Arthur C Jr., organic chemistry; science education
Yardley, John Howard, pathology
Yeni-Komshian, Grace Helen, psychophysiology
Yergey, Alfred L. III, ion optics
Young, James Howard, entomology
Young, John Paul, operations research
Young, M Wharton, neuroanatomy
Young, Viola Mae, microbiology
Zaczek, Norbert Marion, organic chemistry
Zankel, Kenneth L., biophysics
Zdanis, Richard Albert, physics
Zeligman, Israel, dermatology
Zenker, Nicolas, pharmaceutical chemistry

Zielke, Horst Ronald, biochemistry, cell biology
Zierler, Kenneth Levie, medicine, physiology
Zimmerman, Burke Kisling, molecular biology
Zimmerman, Jack McKay, surgery
Zinkham, William Howard, pediatrics
Zuidema, George Dale, surgery

BEL AIR

Bennett, Frederick Dewey, fluid mechanics
Brady, Howard Shaub, organic chemistry
Carion, Hugh Robert, atmospheric physics
Dronamraju, Krishna Rao, evolutionary biology; genetics
Horton, Richard Greenfield, physiology
Ostrom, Thomas Ross, health physics, sanitary engineering
Simmons, Thomas Carl, organic chemistry
Van Antwerp, Walter Robert, solid state physics
Waine, Martin, physics
Weeks, Maurice Harold, toxicology
Williamson, Charles Elvin, bio-organic chemistry; oncology
Wood, William Otto, cell biology; biophysics

BELTSVILLE

Abdul-Baki, Aref Asad, plant physiology
Adams, Jean Ruth, insect pathology
Adams, Peter B., plant pathology
Adler, Victor Eugene, entomology
Adolph, Horst Guenter, organic chemistry
Alford, John Abright, food microbiology
Ambler, John Edward, plant nutrition
Amerault, Thomas Eugene, microbiology, immunochemistry
Anderson, James David, plant physiology
Anderson, Raymond Edward, horticulture
Argauer, Robert John, analytical chemistry
Armbrecht, Bernard Henry, biochemistry
Asen, Sam, plant physiology
Ayers, William Arthur, microbiology
Baker, Doris, analytical chemistry, cereal chemistry
Baker, Edward William, entomology
Baker, James Earl, plant physiology
Barclay, Arthur S., plant taxonomy
Barksdale, Thomas Henry, plant pathology
Barnes, Robert F., agronomy
Batra, Lekh Raj, mycology
Batra, Suzanne Wellington Tubby, entomology
Beecher, Gary Richard, nutritional biochemistry
Bennett, Jesse Harland, plant physiology
Berlin, Elliott, biochemistry
Beroza, Morton, analytical chemistry
Bitman, Joel, biochemistry
Blosser, Timothy Hobert, dairy science, animal science
Bodwell, Clarence Eugene, nutrition
Bolt, Douglas John, reproductive physiology; endocrinology
Bond, James, animal science
Borkovec, Alexej B., organic chemistry; biochemistry
Briggle, Leland Wilson, plant genetics
Brown, John Charl., plant physiology; soil sciences
Burge, Wylie D., microbiology
Burks, Barnard De Witt, entomology
Cantelo, William Wesley, entomology
Cantwell, George E. insect pathology
Carlson, Gerald Eugene, environmental physiology; agronomy
Caro, Joseph Henry, environmental chemistry
Cecil, Helene Carter, reproductive physiology
Chandra, G Ram, biochemistry
Chang, Mei Ling (Wu), nutritional biochemistry
Chang, Shen Chin, entomology, biochemistry
Christiansen, Meryl Naeve, plant physiology
Civerolo, Edwin Louis, plant virology
Clark, Truman Benton, insect pathology
Coe, Gerald Edwin, genetics, cytology
Coffman, Charles Benjamin, agronomy

Coleman, Robert E., plant physiology
Conner, Ray M., bacteriology
Coulson, Jack Richard, entomology
Cox, Edwin Lory, mathematical statistics
Cross, Hiram Russell, food science, animal science
Curtis, Charles R., plant pathology
Danielson, Loran Leroy, plant physiology
Davis, Robert Edward, plant pathology, microbiology
Deahl, Kenneth Luvere, phytopathology
Dermen, Haig, botany
Devine, Thomas Edward, plant genetics, plant breeding
Dickinson, Frank N., genetics, statistics
Diener, Theodor Otto, virology
Dolnick, Ethel Helen, zoology
Doran, David James, parasitology
Dosier, Larry Waddell, plant morphogenetics, crop breeding
Douvres, Frank William, educ: univ md, bs, 48, ms
Dowler, William Minor, phytopathology
Duke, James A., botany
Emswiler, Bonnie Sue, microbiology
Endo, Burton Yoshiaki, nematology
England, Charles Bennett, watershed management
Enzie, Frank Dorr, veterinary parasitology
Epstein, Eliot, soil physics
Faust, Miklos, pomology
Feinstein, Louis, biochemistry
Feldmesser, Julius, plant nematology, invertebrate zoology
Ferretti, Aldo, organic chemistry
Finegold, Harold, physical chemistry, organic chemistry
Fogle, Harold Warman, horticulture
Foote, Richard Herbert, entomology
Foy, Charles Daley, soil fertility, plant nutrition
Freeman, Horatio Putnam, inorganic chemistry
Frerichs, Wayne Marvin, veterinary science, biochemistry
Frobish, Lowell T., animal nutrition, biochemistry
Gamble, Dean Franklin, organic chemistry, information science
Gentner, Walter Andrew, plant pathology, cell biology
Gortner, Willis Alway, nutrition
Goth, Robert W., plant pathology
Graham, Joseph Harry, phytopathology
Gunn, Charles Robert, taxonomy, botany
Guthrie, Howard David, reproductive physiology
Gutierrez, Jose, parasitology
Haeskaylo, Edward, botany
Hanson, Angus Alexander, agronomy
Hardenburg, Robert Earle, horticulture
Harvey, Paul Henry, crop science, plant breeding
Hawk, Harold W., reproductive physiology
Hayes, Dora Kruse, biochemistry
Hays, Harry Witherow, pharmacology
Heggestad, Howard Edwin, phytopathology
Heimpel, Arthur MacLeod, entomology
Helling, Charles Siver, soil chemistry, pesticide chemistry
Herlich, Harry, veterinary parasitology
Hill, Kenneth Richard, pesticide chemistry
Hilton, James Lee, pesticide chemistry
Hopkins, Homer Thawley, Jr., plant physiology, agriculture
Hruschka, Howard Wilbur, horticulture, plant physiology
Hwang, Joseph Cen, parasitology, public health
Hyde, John Landis, veterinary medicine, clinical chemistry
Inscoe, May Nilson, organic chemistry
Jasper, Robert Lawrence, endocrinology, toxicology
Johnson, Lawrence Arthur, reproductive physiology
Johnson, Robert Morton, chemistry
Jones, Quentin, economic botany
Kaper, Jacobus M., biochemistry, molecular biology
Kaplanis, John Nicholas, entomology
Kaufman, Donald DeVere, pathology, soil microbiology
Kearney, Philip C, biochemistry, agriculture

Keil, Harry Louis, plant pathology
Kelsay, June Lavelle, nutrition
Kiddy, Charles Augustus, animal science
Kilpatrick, R A., plant pathology
King, Steven Clarence, animal genetics
Klassen, Waldemar, entomology, genetics
Klingman, Dayton L., agronomy
Knight, Robert Arthur, parasitology
Knipling, Edward Fred, entomology
Knutson, Lloyd Vernon, entomology
Kotula, Anthony W., food science
Krizek, Donald Thomas, environmental physiology
Kunishi, Harry Mikio, soil chemistry
Lakshmanan, Florence Lazicki, biochemistry
Landgrebe, Albert R., chemistry
Leese, Bernard M, plant physiology
Leffel, Robert Cecil, agronomy, plant breeding
Legg, Joseph Ogden, soil science
Lentz, Paul Lewis, mycology
Leppik, Elmar Emil, plant pathology
Levander, Orville Arvid, biochemistry
Levin, Marshall David, entomology, apiculture
Lewis, Frank, meteorology
Lewis, Jack A., plant pathology, soil microbiology
Lieberman, Morris, plant physiology
Lillie, Robert Jones, poultry nutrition
Little, Ruby Rice, botany
Longenecker, William Hilton, organic chemistry
Louloudes, Spiro James, entomology
Lumsden, Robert Douglas, plant pathology
Lund, Everett Eugene, parasitology, protozoology
Lunin, Jesse, soil science
Lynch, G Paul, physiology
Maas, John Lewis, plant pathology
Mageau, Richard Paul, microbiology, immunology
Mandava, Nagabhushanam, bio-organic chemistry
Marsh, Paul Bruce, plant physiology
Martin, Ethelbert Cowley, entomology
McClain, Philip Edwin, nutritional biochemistry
McCollum, Gilbert Dewey, Jr, plant cytogenetics
McGovern, Terrence Philip, organic chemistry
McGrew, John Roberts, phytopathology, viticulture
McGuire, Judson Ulery, Jr., biometrics
McHaffey, David George, entomology, plant pathology
McLoughlin, Donald Keith, protozoology, parasitology
Meiners, Jack Pearson, phytopathology
Menge, Henry, animal nutrition
Menzies, James David, microbiology
Mercer, Henry Dwight, veterinary pharmacology, veterinary toxicology
Mertz, Walter, nutrition, biochemistry
Meudt, Werner J., plant physiology
Miller, Douglass Ross, entomology
Miller, Robert Hoover, genetics
Mitchell, John William, plant physiology
Moats, William Alden, food biochemistry
Mohanty, Sashi B., microbiology, virology
Moline, Harold Emil, phytopathology, plant virology
Moore, Ermer Leon, plant pathology, genetics
Morgan, Neal O., medical entomology, ecology
Morris, Eugene Ray, nutrition, physiological chemistry
Moseman, John Gustav, plant pathology, agronomy
Nash, Ralph Glen, soil science, chemistry
Neal, John William, Jr., entomology
Nickle, William R., nematology
Norcross, Marvin Augustus, veterinary pathology
Norman, Howard Duane, animal breeding
Oltjen, Robert Raymond, ruminant nutrition
Opel, Howard, reproductive physiology, endocrinology
Orellana, Rodrigo Gonzalo, plant pathology
Ostazeski, Stanley A, plant pathology, mycology
Owens, Lowell Davis, plant physiology
Papavizas, George Constantine, plant pathology
Parencia, Charles R., entomology
Parr, James Floyd, Jr., soil microbiology, plant physiology
Pearson, Ronald Earl, dairy science, animal breeding

Perdue, Robert Edward, Jr., economic botany
Peterson, Irvin Leslie, veterinary medicine, poultry science
Plimmer, Jack Reynolds, organic chemistry
Powell, Jerrel B., genetics, agronomy
Powell, Rex Lynn, animal breeding
Purchase, Harvey Graham, poultry virology
Pursel, Vernon George, physiology, animal science
Raney, William Andrew, soil physics
Redfern, Robert Earl, entomology
Reiser, Sheldon, biochemistry, animal nutrition
Rexroad, Caird Eugene, Jr, reproductive physiology
Reynolds, Paul Joseph, animal nutrition, biochemistry
Reynolds, Robert David, biological chemistry
Ridgway, Richard L., entomology
Robbins, William E, entomology
Romanowski, Robert David, biochemistry, plant physiology
Romberger, John Albert, plant physiology
Rose, Joseph Edward, biochemistry
Russell, Louise May, entomology
Ruth, John Moore, physical chemistry
St John, Judith Brook, biochemistry, physiology
San Antonio, James Patrick, horticulture, physiology
Sarvella, Patricia Ann, cytogenetics
Sayre, Richard Martin, plant pathology
Schaeffer, Gideon W., cell physiology, molecular genetics
Schalk, James Maximillian, entomology
Schar, Raymond Dewitt, poultry science
Schechter, Milton Seymour, organic chemistry, agricultural chemistry
Schneider, Bernard Arnold, plant physiology, agronomy
Schneider, Irving Robert, plant pathology
Schwab, Bernard, clinical microbiology
Schwartz, Paul Henry, entomology
Schwarz, Meyer, organic chemistry
Shaw, Warren Cleaton, agronomy, plant physiology
Shear, Cornelius Barrett, horticulture, physiology
Shimanuki, Hachiro, insect pathology, apiculture
Simpson, Marion Emma, plant physiology, botany
Sinden, Stephen Lee, plant pathology, plant physiology
Sloger, Charles, plant physiology
Slyter, Leonard L., microbiology, nutrition
Smith, James W., dairy science
Smith, Wilson Levering, Jr., plant pathology
Snyder, Freeman Woodrow, plant physiology
Sonnet, Philip E, organic chemistry
Souders, Helen Jeanette, nutrition
Stavely, Joseph Rennie, plant pathology
Steere, Russell Ladd, virology, biophysics
Steffens, George Louis, plant physiology
Stewart, Robert N., phytochemistry
Stoetzel, Manya Brooke, entomology
Stoner, Allan K., horticulture
Stringfellow, Frank, zoology
Svoboda, James Arvid, insect physiology
Szepesi, Bela, nutrition, molecular biology
Taylorson, Raymond Brierly, plant physiology, horticulture
Terrell, Edward Everett, plant taxonomy
Terrill, Clair Elman, animal breeding
Terry, Paul H., organic chemistry
Tester, Cecil Fred, biochemistry, environmental management
Thompson, Malcolm J., organic chemistry
Tromba, Francis Gabriel, parasitology
Trout, David Lynn, physiology
Tso, Tien Chioh, phytochemistry
Tyrrell, Henry Flansburg, nutrition, biometry
Uecker, Francis August, mycology
Vincent, Phillip G, biochemistry
van der Zwet, Tom, plant pathology
Vaughan, David Arthur, nutrition
Vaughn, James L, insect physiology
Vegors, Halsey Hugh, animal parasitology
Vetterling, John Martin, protozoology, parasitology
Wagstaff, David Jesse, toxicology
Warthen, John David, Jr., natural product chemistry
Warwick, Everett James, animal science
Watada, Alley E., horticulture, plant physiology
Weber, Deane Fay, soil microbiology

Weinland, Bernard Theodore, applied statistics
Weir, C Edith, nutrition, food technology
Wergin, William Peter, cytology
Wester, Robert Emerson, horticulture
White, George Albert, agronomy, botany
Whitmore, George E., veterinary medicine, toxicology
Windlan, Harold Milton, dairy chemistry, biochemistry
Wolf, Wayne Robert, analytical chemistry, nutrition
Womack, Madelyn, nutrition
Woods, Charles William, organic chemistry
Woodstock, Lowell Willard, plant physiology, biochemistry
Woolson, Edwin Albert, pesticide chemistry
Worley, Joseph Francis, plant pathology
Zimmerman, Richard Hale, plant physiology

BETHESDA
Aaslestad, Halvor Gunerius, virology
Abbott, Betty Jane, cancer
Abrell, John William, bioorganic chemistry
Adamson, Richard H, pharmacology, biochemistry
Ajmone-Marsan, Cosimo, neurophysiology
Akers, Robert Preston, public health administration
Albers, Robert Wayne, neurochemistry
Alberts, Walter Watson, neurosciences, medical administration
Albrecht, Paul, virology, pathology
Allen, Anton Markert, veterinary pathology
Allen, Ernest Mason, public health administration
Allen, Gordon, human genetics
Allen, William Peter, medical microbiology
Alling, David Wheelock, mathematical statistics, medicine
Allison, Richard Gall, nutritional biochemistry
Alter, Harvey James, internal medicine, hematology
Altland, Paul Daniel, zoology, physiology
Altman, Philip Lawrence, information science, communication science
Anderson, W French, biochemistry, genetics
Anfinsen, Christian Boehmer, biochemistry
Angelone, Luis, physiology
Appella, Ettore, immunology
Archard, Howell Osborne, Jr., oral pathology
Armstrong, Wallace David, biochemistry
Aronson, David L, medicine, physiology
Ascione, Richard, oncology, biochemistry
Asher, David Michael, infectious diseases, pediatrics
Ashwell, G Gilbert, biochemistry
Asofsky, Richard Marcy, immunology, pathology
Astin, Allen Varley, physics
Atchley, Floyd Owen, public health
Atkinson, Joe William, veterinary medicine
Auletta, Angela Elaine, oncology, cell biology
Aurbach, Gerald Donald, medicine
Avigan, Joel, biochemistry
Axelrod, Julius, biochemical pharmacology
Baer, Harold, allergy, immunology
Bailar, John Christian, III, research administration, biostatistics
Baker, Phillip John, microbiology, immunology
Banfield, William Gethin, experimental pathology
Barban, Stanley, biochemistry
Barbehenn, Elizabeth Kern, biochemistry
Barker, Jeffery Lange, neurobiology, neuropharmacology
Bartko, John Jaroslav, mathematical statistics
Bartter, Frederic Crosby, medicine
Bassin, Robert Harris, microbiology
Batchelor, William Henry, medicine, physical chemistry
Batlin, Alexander, bacteriology
Bauer, Adelia Catherine, physiology
Bauer, Ernest, atmospheric physics
Baum, Siegmund Jacob, physiology, radiobiology

Bawden, Monte Paul, parasitology, immunology

Baxter, James Hubert, medical research

Beall, Robert Joseph, biochemistry

Beaven, Michael Anthony, pharmacology, biochemistry

Becker, Edwin Demuth, physical chemistry

Beisler, John Albert, medicinal chemistry

Benedict, Jean Davidson, biochemistry

Bennett, William Ernest, immunopathology

Berard, Costan William, pathology

Berger, Robert Lewis, biophysics, instrumentation

Berger, Shelby Louise, biochemistry, cell biology

Bergman, Fred Heinz, genetics

Berman, Mones, biomathematics

Bessey, Otto Arthur, biochemistry

Bieri, John Genther, biochemistry, nutrition

Blood, Benjamin Donald, primatology, science administration

Blot, William James, biostatistics, epidemiology

Blount, Don Houston, physiology

Blye, Richard Perry, reproductive biology

Bogdanski, Donald Frank, pharmacology

Bond, Howard Edward, biophysics, cancer

Bono, Vincent Horace, Jr., internal medicine, biochemistry

Boone, Charles Walter, pathology, pediatrics

Bornmann, Robert Clare, medicine, physiology

Borsos, Tibor, cancer, immunology

Bosma, James Frederick, pediatrics

Boss, Willis Robert, zoology, public health administration

Bourke, Anne Rosaleen, pharmacology

Bowen, William H., oral microbiology, biochemistry

Bowers, Mary Blair, cell biology

Bowery, Thomas Glenn, health sciences

Bowman, Robert Lewis, instrumentation

Bozeman, F Marilyn, microbiology

Bracken, Marilyn C., information science

Brady, Roscoe Owen, biochemistry

Breitman, Theodore Ronald, biochemistry

Brenner, Stephen Louis, biophysics

Brewer, Carl Robert, microbiology

Bridgman, Charles Floyd, anatomy

Brightman, Milton Wilfred, neuroanatomy

Brody, Jacob A., epidemiology, virology

Brown, Elise Ann, pharmacology

Brown, Paul Wheeler, virology, internal medicine

Bruck, Stephen Desiderius, polymer chemistry, biomedical engineering

Buck, John Bonner, zoology

Burg, Maurice B., nephrology

Burnham, Donald Love, psychiatry

Byrne, Robert Joseph, veterinary science

Cabib, Enrico, biochemistry, enzymology

Cahnmann, Hans Julius, organic chemistry

Cain, Dennis Francis, biochemistry

Calhoun, John Bumpass, ecology

Callahan, Lynn Thomas, III, microbiology

Canby, Henry Fawcett, dentistry

Cannon, Donald Charles, clinical pathology, immunology

Cantarow, Abraham, biochemistry, medicine

Cardon, Phillipe Vincent, Jr., medicine

Carlson, Harve J., bacteriology, virology

Carlson, Margaret Joyce, microbiology

Carpenter, David O., neurophysiology, biophysics

Carr, Charles Jelleff, pharmacology, chemistry

Carroll, William Robert, biochemistry

Carter, Robert Emerson, physics

Carter, Stephen Keith, internal medicine, oncology

Cashel, Michael, biochemistry, genetics

Castle, Manford C., pharmacology

Catignani, George Louis, biochemistry, nutritional biochemistry

Catravas, George Nicholas, organic chemistry, biochemistry

Catz, Charlotte Schifra, pediatrics, teratology

Caveness, William Fields, neurophysiology, clinical neurology

Chader, Gerald Joseph, biochemistry

Chalkley, Donald Thomas, embryology, science policy

Chang, Kenneth Shueh-Shen, microbiology, oncology

Chanock, Robert Merritt, pediatrics, bacteriology

Chaparas, Sotiros D., microbiology, immunology

Charney, Elliot, chemical physics

Chase, Thomas Newell, neurology, neuropharmacology

Chassy, Bruce Matthew, biochemistry, organic chemistry

Chen, David Hou-Chung, medical parasitology, immunology

Chen, Philip Stanley, Jr., pharmacology, physiology

Chen, Yi-Der, theoretical chemistry, pathology

Chu, Elizabeth Wann, cytology, pathology

Chu, Sherwood Cheng-Wu, mathematics

Chused, Thomas Morton, immunology

Clark, Byron Bryant, pharmacology

Clum, Floyd Myron, pathology, mycology

Cogan, David Glendening, ophthalmology

Cohen, Gerson H., physical chemistry

Cohen, Jack Sidney, physical biochemistry

Cohen, Louis Arthur, physical organic chemistry, bio-organic chemistry

Cohen, Pinya, biochemistry, immunology

Cohen, Robert Abraham, psychiatry

Cohen, Sheldon Gilbert, immunology, allergy

Colburn, Nancy Hall, biochemistry

Colburn, Robert Warren, biochemistry

Cole, Kenneth Stewart, biophysics

Coleman, William Gilmore, Jr., molecular biology

Combs, Gerald Fuson, animal nutrition

Conant, Robert M., virology, immunology

Condliffe, Peter George, biochemistry, research administration

Constantopoulos, George, biochemistry

Cook, Ellsworth Barrett, pharmacology

Coolbaugh, James Cameron, microbiology

Corfman, Philip Albert, obstetrics & gynecology

Corning, Mary Elizabeth, chemistry

Costlow, Richard Dale, microbiology

Cotton, William Robert, microscopic anatomy, oral microanatomy

Coulombre, Alfred Joseph, embryology

Cox, John William, medical education, cardiology

Creveling, Cyrus Robbins, pharmacology

Crouch, Madge Louise, public health administration

Crowson, Henry L., mathematics

Crump, Stuart Faulkner, physics, research administration

Cueto, Cipriano, Jr., pharmacology, toxicology

Cummings, Martin Marc, medicine

Curtis, Roger William, experimental physics

Cushen, Walter Edward, operations research

Dahlen, Roger W., medical physiology, science administration

Daily, Otis Patrick, microbiology

Dalton, Albert Joseph, biology

Dalton, John Charles, zoology, physiology

Daly, John William, neurochemistry, natural products chemistry

Darby, Eleanor Muriel Kapp, biochemistry

Dasch, Gregory Alan, insect physiology, microbiology

Dasler, Adolph Richard, physiology

Datta, Padma Rag, biochemistry, ecology

Davidson, Harold Michael, biochemistry

Davies, David R., biophysics, crystallography

Davis, Dorland Jones, medicine

Davis, Richard Henry, Jr. biochemistry

Dawe, Clyde Johnson, pathology

Day, Paul Louis, biochemistry

DeBell, Robert Michael, microbiology

Decker, John Laws, rheumatology, internal medicine

Dekaban, Anatole S., neurology

Delappe, Irving Pierce, science administration

De Lisi, Charles, biophysics

De Luca, Luigi M., biochemistry

Demoise, Charles Francis, cancer

Depue, Robert Hemphill, cancer

Devin, Charles, Jr., acoustics, hydrodynamics

DeVita, Vincent T., Jr., internal medicine, pharmacology

Dhyse, Frederick George, biochemistry, information science

Diamond, Louis Stanley, zoology

DiPaolo, Joseph Amadeo, genetics

di Sant'Agnese, Paul Emilio Artom, pediatrics

Doan, David Bentley, earth science

Dockery, John T., physics, operations research

Domanski, Thaddeus John, research administration

Dorr, John Van Nostrand, II, geology

Doty, Stephen Bruce, biology

Douglas, Carl Dean, biochemistry

Doukas, Harry Michael, organic chemistry, biochemistry

Douros, John Drenkle, microbiology

Driscoll, John Stanford, medicinal chemistry

Duane, David Bierlein, marine geology, coastal engineering

Dublin, Thomas David, medical research

Dubner, Ronald, neurobiology

Duff, James Thomas, microbiology

Duncan, Katherine, internal medicine

Duncan, Leroy Edward, Jr., internal medicine

Dunkel, Virginia Catherine, virology, oncology

Dunn, Thelma Brumfield, pathology

Eagles, Eldon Lewis, medicine, public health administration

Eanes, Edward David, biophysical chemistry

Eaton, William Allen, biophysics, physical biochemistry

Eaves, George Newton, medical microbiology

Eddy, Bernice Elaine, bacteriology

Ederer, Fred, biostatistics

Ehrenstein, Gerald, biophysics

Eisenberg, Frank, Jr., biochemistry

Eldridge, Roswell, medical genetics, neurology

Elsberg, Bennett La Dolce, infectious diseases, internal medicine

Elkin, William Futter, biostatistics

Eilenberg, Jonas Harold, biostatistics

Engel, William King, neurology

Englander, Harold Robert, dentistry

Engle, Virginia John, cell biology, cancer

Evans, Edward Vaughan, psychiatry, neurology

Evarts, Ritva Poukka, experimental pathology

Evenden, Fred George, wildlife conservation

Everstine, Gordon Carl, applied mathematics

Falk, Charles Eugene, physics, science policy

Feinberg, Richard, vision, human development

Feinleib, Manning, epidemiology

Felsenfeld, Gary, biophysics

Ferguson, Frederick Palmer, physiology

Ferguson, Malcolm Stuart, medical parasitology, communication science

Fink, Mary Alexander, cancer

Finlayson, John Sylvester, biochemistry

Fischinger, Peter John, cancer, animal virology

Fisher, Kenneth D., plant pathology

Foltz, Calvin Martin, organic chemistry, medicinal chemistry

Fournelle, Harold John, bacteriology

Frank, Karl, neurophysiology

Frank, Michael M., immunology

Fratantoni, Joseph Charles, hematology

Fraumeni, Joseph F Jr., internal medicine, epidemiology

Fredrickson, Donald Sharp, medicine

Freese, Ernst, molecular biology

Freygang, Walter Henry, Jr., physiology

Friedman, Mischa Elliot, bacteriology

Friedman, Robert Morris, pathology, virology

Friess, Seymour Louis, physical chemistry, organic chemistry

Frommer, Peter Leslie, cardiology, biomedical engineering

Fry, Donald Lewis, physiology

Fuhr, Irvin, biophysical chemistry

Fuller, Everett Gladding, nuclear physics

Fuller, Vernon Jack, medical microbiology

Fuortes, Michelangelo Giorgio F., neurophysiology

Furano, Anthony, biochemistry

Fuson, Roger Baker, immunology

Galasso, George John, microbiology, virology

Gallo, Robert C., cell biology

Ganaway, James Rives, veterinary medicine, microbiology

Ganz, Aaron, research administration

Garber, David Harrison, physics, research administration

Gardner, Earl William, Jr., microbiology

Gardner, Sara A., pharmacology, neurochemistry

Garruto, Ralph Michael, biological anthropology, medical anthropology

Gart, John Jacob, statistics

Gary, Norman Dwight, microbial physiology

Gay, William Ingalls, medical

Gelboin, Harry Victor, biochemistry

Geller, Harvey, analytical statistics, research management

Gellert, Martin Frank, physical biochemistry

Geran, Ruth Iris, cancer

Gerring, Irving, environmental sciences, public health administration

Gerwin, Brenda Isen, biochemistry

Giacometti, Luigi, biology

Gladner, Jules A., biochemistry

Gleissner, Gene Heiden, mathematics, computer science

Gibbs, Clarence Joseph, Jr., virology

Gibson, Sam Thompson, internal medicine

Gilbert, Daniel Lee, physiology

Gilford, Leon, mathematical statistics, administration

Gillette, James Robert, biochemical pharmacology

Ginsburg, Victor, biochemistry

Giermer, George Geiger, pathology

Gluckstein, Fritz Paul, veterinary medicine

Gobel, Stephen, cytology, neurobiology

Goggins, John Francis, dental research

Goldberg, Irving David, biostatistics, public health

Goldberger, Robert Frank, biochemistry

Goldin, Abraham, pharmacology

Goldstein, Murray, medical administration

Goldwater, William Henry, research administration

Golumbic, Norma, cancer

Gonzalez-Fernandez, Jose Maria, mathematics, physiology

Goodman, Lester, bioengineering

Gouras, Peter, physiology

Goor, Ronald Stephen, biochemistry, molecular biology

Gordon, Milton, microbiology

Gori, Gio Batta, virology, environmental health

Gottlieb, Melvin Harvey, biophysics

Goulet, Normand Robert, virology

Graff, Morris Morse, endocrinology

Gram, Theodore Edward, pharmacology

Greenbaum, Leon J Jr., physiology

Greenberg, Harold Abraham, psychiatry

Greenstein, Edward Theodore, laboratory animal medicine, toxicology

Greulich, Richard Curtice, anatomy, oral biology

Griffo, Zora Jasincuk, physiology

Gross, Erhard, organic chemistry

Gross, Noel Harden, biochemistry

Gruber, Jack, medical microbiology

Guarino, Anthony Michael, biochemical pharmacology, toxicology

Gullino, Pietro M., pathology

Gulyas, Bela Janos, cell biology, developmental biology

Gunkel, Ralph D. physiological optics

Guss, Maurice Louis, virology
Gutter, Frederick Jay, biochemistry, science administration
Guttman, Helene Augusta Nathan, microbiology, biochemistry
Gwadz, Robert Walter, medical entomology, parasitology
Haenszel, William Manning, biostatistics, epidemiology
Haggerty, James Francis, biochemistry
Hajdu, Stephen, physiology
Hall, William Thomas, cytology
Halperin, Max, mathematical statistics
Hambrecht, Frederick Terry, biomedical engineering
Hamilton, Clara Eddy, zoology, physiology
Hampar, Berge, microbiology, virology
Hampp, Edward Gottlieb, bacteriology
Hand, Arthur Ralph, cell biology, cytochemistry
Handler, Joseph S, internal medicine
Hanna, Edgar Ethelbert, Jr, microbiology, immunology
Hansen, Carl Tams, genetics
Hansen, Ira Bowers, zoology
Hansen, Morris Howard, statistics
Hao, Yu-Lee, biochemistry
Hardegree, Mary Carolyn, pediatrics, immunology
Harding, Fann, anatomy
Harris, Maureen Isabelle, public health administration
Harris, Rudolph, cell biology
Harte, Robert Adolph, biochemistry, information science
Hartley, Janet Wilson, virology
Hartley, Robert William, Jr, biophysics
Hartwell, Jonathan Lutton, chemistry
Hascall, Gretchen Katharine, cell biology
Hascall, Vincent Charles, Jr, biochemistry
Hasenclever, Herbert Frederick, bacteriology
Hassell, John Robert, developmental biology, biochemistry
Hasselmeyer, Eileen Grace, perinatal biology
Hatanaka, Masakazu, biochemistry, virology
Hatchett, Stephen Pinckney, public health administration
Hatfield, Dolph Lee, molecular biology
Haussling, Henry Jacob, fluid dynamics
Hayden, George A, biochemistry
Hayes, Richard Lloyd, dentistry
Hayward, Evans Vaughan, physics
Hayward, Raymond Webster, nuclear physics
Hearing, Vincent Joseph, Jr, biochemistry
Hebert, Clarence Louis, anesthesiology
Heim, Allen Homer, microbiology
Heine, Ursula Ingrid, cytology, biology
Heinrich, Max Alfred, Jr, pharmacology
Held, Joe R, parasitology, epidemiology
Held, Victor Maxwell, science administration
Heller, John Roderick, medicine
Heller, Stephen Richard, spectroscopy, computer science
Hellman, Alfred, virology, cell physiology
Hellman, Louis Philip, analytical statistics, management sciences
Helmsen, Ralph John, biochemistry, ophthalmology
Helvig, Raymond Jennings, public health
Heming, Arthur Edward, biochemistry
Hendler, Richard Wallace, biochemistry
Henkert, Pierre, biochemistry
Henley, Catherine, embryology, cytology
Hennings, Henry, oncology
Herman, Lloyd George, bacteriology
Hess, Eugene Lyle, biochemistry
Hess, Helen Hope, biochemistry, neuropathology
Heston, Walter Enoch, genetics
Heumann, Karl Fredrich, organic chemistry
Heydrick, Fred Painter, microbiology, science administration
Hickson, John LeFever, sugar chemistry
Highet, Robert John, organic chemistry
Hill, James Carroll, microbiology, bacteriology
Hill, Robert Towner, anatomy
Hill, Terrell Leslie, biophysics, physical biochemistry
Hisaoka, Kenichi Kenneth, embryology
Hobbs, Robert Boyd, chemistry
Hochstein, Herbert Donald, microbiology
Hodgen, Gary Dean, endocrinology
Hoffman, Harold A, genetics
Hoggan, Malcolm David, microbiology
Holland, Paul Vincent, internal medicine
Homer, Louis David, biometrics, physiology
Honig, John Gerhart, operations research, management
Hook, William Arthur, microbiology, immunology
Hopps, Hope Elizabeth Byrne, microbiology
Horenstein, Evelyn Anne, developmental biology, cell biology
Horner, Theodore Wright, statistics, genetics
Horton, Richard E, immunology
Howell, David McBrier, virology
Huebner, Robert Joseph, virology
Hummel, Donald Joseph, information science, cytology
Hunter, Jehu Callis, cell physiology
Hurd, Suzanne Sheldon, biochemistry
Hutchinson, George Allen, algebra
Hyatt, Asher Angel, science administration, organic chemistry
Hyde, Henry Van Zile, medicine
Ihndris, Raymond Will, organic chemistry
Ingham, Kenneth Culver, physical biochemistry
Ingram, Robert Lee, biochemistry
Inman, John Keith, protein chemistry
Innes, James Robert Maitland, pathology
Irreverre, Filadelfo, biochemistry
Irving, George Washington, Jr, biochemical pharmacology
Irwin, Richard Leslie, pharmacology
Jablonski, Frank Edward, electrooptics
Jacobowitz, David, pharmacology
Jacobs, Leon, medical parasitology
Jacobs, Richard Lewis, chemotherapy
Jacobson, Arthur E, medicinal chemistry
Jakoby, William Bernard, biochemistry, microbiology
Jakus, Marie A, biology
James, John Cary, organic chemistry
Jamieson, Graham Archibald, biochemistry
Janicki, Bernard William, microbiology, immunology
Jaouni, Katherine Cook, microbiology
Jaouni, Taysir M, organic chemistry
Jefferson, David Kenoss, computer science
Jenerick, Howard Peter, physiology
Jennings, William Harney, Jr, biophysics
Jerina, Donald M, organic chemistry, biochemistry
Jernigan, Robert Lee, physical chemistry, polymer chemistry
Joftes, David Lion, physiology, developmental biology
Johns, David Garrett, pharmacology, biochemistry
Johnson, David Freeman, biochemistry
Johnson, Paul Sosinski, physics
Johnson, Ralph Emil, medicine, experimental biology
Jones, Morris Thompson, bacteriology
Jones, Thomas Oswell, chemistry
Joseph, Sammy William, microbiology, medical bacteriology
Josephs, Melvin Jay, plant physiology
Kafka, Marian Stern, physiology
Kaiser, Joseph Anthony, pharmacology
Kakehashi, Samuel, dentistry
Kalberer, John Theodore, Jr, physiology, biology
Kales, Morris L, mathematics
Kalser, Sarah Chinn, pharmacology
Kapikian, Albert Zaven, epidemiology, virology
Kaplan, Ann Esther, biochemistry, physiology
Karel, Leonard, pharmacology
Karten, Marvin J, organic chemistry
Kaufman, Ann Anderson, research administration
Kaufman, Bernard Tobias, biochemistry
Kaufman, Seymour, biochemistry
Kayhoe, Donald Ellsworth, medicine
Keefer, Larry Kay, organic chemistry, cancer
Kempner, Ellis Stanley, biophysics
Kendrick, Francis Joseph, pathology, dentistry
Kennedy, Charles, pediatric neurology
Kenneke, Albert Patrick, health physics
Kern, Milton, biochemistry
Ketley, Jeanne Nelson, enzymology, developmental biology
Kielley, William Wayne, biochemistry
King, Cecil Thomas G, physiology, endocrinology
King, Thomas Joseph, embryology
Kinoshita, Jin Harold, biological chemistry
Kirkpatrick, Charles Harvey, allergy, immunology
Kirschstein, Ruth Lillian, pathology
Kissman, Henry Marcel, organic chemistry
Kitzes, George, biochemistry, pharmacology
Klatzo, Igor, medicine
Klein, Michael, anatomy
Kline, Ira, cancer
Kline, Oral Lee, biochemistry
Kohn, Kurt William, chemical pharmacology
Kohn, Leonard David, biochemical pharmacology
Kolb, Helga Ellen Thor, neuroanatomy
Kolbye, Albert Christian, Jr, public health
Kominz, David Richard, protein chemistry
Kopin, Irwin J, internal medicine, pharmacology
Korn, Edward David, biochemistry, cell biology
Kornfeld, Lottie, immunobiology
Krakowski, Martin, mathematics
Krause, Richard Michael, microbiology, immunology
Kreiner, Howard William, operations research
Kreshover, Seymour J, dental pathology
Kretchmer, Norman, pediatrics, biochemistry
Krichevsky, Micah I, microbiology, biochemistry
Kroll, Bernard Hilton, statistics, systems science
Krueger, Keatha Kathrine, biochemistry
Kuether, Carl Albert, biochemistry
Kuff, Edward Louis, molecular biology
Kulik, Martin Michael, plant pathology
Kulwich, Roman, biochemistry
Kupfer, Carl, ophthalmology
Kupferberg, Harvey J, pharmacology
Kusiak, John Warren, biochemistry
Kuwabara, Toichiro, ophthalmology
Kwon-Chung, Kyung Joo, medical mycology
Kyle, Wendell Henry, genetics
Labaw, Louis Warne, physics
Lambert, Paul Dudley, public health administration
Land, Charles Even, statistics
Land, William Everett, physical chemistry
Lane, John Edward, public health administration
Langone, John Joseph, bio-organic chemistry
Laqueur, Gert Ludwig, pathology
Larson, Clarence Edward, physical chemistry
Larson, Rachel Harris (Mrs John Watson Henry), biochemistry, dental research
LaVier, Eugene Clark, physics
Law, Lloyd William, oncology
Leach, Berton Joe, biology, science administration
Leavitt, Milo David, Jr, medicine, science administration
Lecar, Harold, biophysics
Leder, Irwin Gordon, biochemistry
Ledney, George David, biology
Leight, Walter Gilbert, operations research, meteorology
Leiserson, Lee, organic chemistry, physical chemistry
Leiter, Joseph, biochemistry
Leive, Loretta, biochemistry, microbiology
Lenfant, Claude J M, physiology
Leonard, Edward Joseph, medicine
Leonard, Fred, chemistry
Levenbook, Leo, biochemistry
Leventhal, Brigid Gray, pediatrics, oncology
Levin, Barbara Chernov, microbial genetics, molecular biology
Levin, Ira William, chemical physics
Levin, Judith Goldstein, biochemistry, virology
Levy, Hilton Bertram, virology
Levy, Robert I, medicine, biochemistry
Lewis, Andrew Morris, Jr, virology
Lewis, Marc Simon, biochemistry, biophysics
Li, Choh-Luh, neurophysiology, neurosurgery
Ligon, Edgar William, Jr, pharmacology
Lipkin, Lewis Edward, neuropathology, computer sciences
Lippel, Kenneth, biochemistry
Lippman, Marc Estes, endocrinology, oncology
Lipsett, Marie Nieft, biological chemistry
Liu, Teh-Yung, biochemistry
London, Jack P, bacteriology, biochemistry
Lorenz, Douglas, microbiology, virology
Love, Robert, pathology
Lovenberg, Walter McKay, biochemistry
Lowe, Charles Upton, pediatrics
Luborsky, Samuel William, physical chemistry
Luttermoser, George William, medical parasitology
Lymn, Richard Wesley, biophysics
MacCanon, Donald Moore, cardiovascular physiology, pharmacology
MacLean, Paul Donald, neurophysiology
MacNamee, James K, veterinary pathology
Macri, Frank John, pharmacology
Madden, David Larry, veterinary microbiology
Mage, Michael Gordon, immunology
Mahoney, Francis Joseph, physics, biology
Maidanik, Gideon, physics
Maling, Harriet Mylander, pharmacology
Malmon, Arthur Gerald, physics
Malone, Winfred Francis, environmental health
Maloney, Clifford Joseph, statistics
Manaker, Robert Anthony, microbiology
Manclark, Charles Robert, microbiology, immunology
Marchesi, Vincent T, biochemistry, pathology
Marciani, Dante Juan, biochemistry
Margolis, Sam Aaron, biochemistry, molecular biology
Marimont, Rosalind Brownstone, applied mathematics
Markey, Sanford Philip, organic chemistry, pharmacology
Markley, Kehl, III, physiology
Marsden, Halsey M, zoology, research administration
Martin, George Reilly, pharmacology
Marx, Stephen John, endocrinology
Marzulli, Francis Nicholas, pharmacology, toxicology
Mason, Karl Ernest, anatomy, nutrition
Mason, Thomas Joseph, biostatistics, epidemiology
Mathis, John Buell, science administration
Mattern, Carl Frederick Theodore, infectious diseases
Matthews, Ruth Hastings, food science, nutrition
Maxwell, Elizabeth Starbuck, biochemistry, molecular biology
May, Everette Lee, medicinal chemistry
Mayer, Florence E, medicine, pediatrics
Mazzur, Scott Ruigh, microbiology, medical anthropology
McCauley, Robert Henry, Jr, vertebrate zoology
McCutchen, Charles Walter, physics, biology
McCutcheon, Rob Stewart, pharmacology
McIntire, Kenneth Robert, biology, immunology
McKigney, John Ignatius, nutrition, biochemistry
McMaster, Philip Robert Bache, immunology, experimental pathology
McPherson, Charles William, laboratory animal medicine
McQuarrie, Irvine Gray, experimental neurology
McQueen, James Lee, cancer, epidemiology
Mell, Leroy Dayton, Jr, analytical chemistry
Melville, Robert Seaman, clinical biochemistry
Meredith, Orsell Montgomery, radiological health, nuclear medicine
Mergenhagen, Stephan Edward, immunology, microbiology
Merrick, Arthur West, physiology
Merril, Carl R, molecular biology, medicine
Merrill, Warner Jay, Jr, statistics
Merwin, Ruth Minerva, zoology
Meryman, Harold Thayer, cryobiology, medical research
Metzger, Henry, biochemistry
Mider, George Burroughs, pathology
Mihalyi, Elemer, biochemistry
Millar, David Bosie-Seurs, III, physical biochemistry
Miller, Dorothea Starbuck, zoology
Miller, Laurence Herbert, dermatology
Miller, Louis Howard, tropical medicine, parasitology
Miller, Robert Warwick, pediatrics, epidemiology
Milman, Harry Abraham, biochemical pharmacology
Milne, George William Anthony, chemistry
Milstien, Julie Block, biochemistry
Milton, Roy Charles, statistics
Minners, Howard Alyn, medicine
Minton, Allen Paul, biophysical chemistry
Mironescu, Stefan Gheorghe Dan, cell biology

Mitchell, Shiela Craig, pediatric cardiology, epidemiology
Mittal, Kamal Kant, immunogenetics, transplantation immunology
Mohler, William C., computer science, medicine
Moloney, John Bromley, biology
Monaellia, Vincent Joseph, hydrodynamics
Monnier, Dwight Chapin, academic administration
Moore, John Criswell, chemistry
Morris, Humbert, physical chemistry
Morris, Joseph Anthony, bacteriology
Morris, Rosemary Shull, science administration, nutrition
Morrow, Andrew Glenn, surgery
Morse, Herbert Carpenter, III, immunology
Mosimann, James Emile, biostatistics
Moss, Bernard, biochemistry, virology
Moss, Samuel, physiology
Moss, Woodrow Glen, physiology, science administration
Mudd, Stuart Harvey, biochemistry
Mueller, Helmut, biochemistry
Mulders, Gerard Francis William, astrophysics
Mullinix, Kathleen Patricia, biochemistry
Mulvihill, John Joseph, epidemiology, pediatrics
Murayama, Makio, biochemistry, cell biology
Murphy, Donald G., medical research, pharmacology
Murphy, James Gilbert, organic chemistry
Murray, George Cloyd, neurophysiology, physics
Murray, Margaret Ransone (Mrs Burton Le Doux), neurosciences
Murrell, Kenneth Darwin, parasitology, immunology
Mushinski, Joseph Frederic, biochemistry, cancer
Myers, Max H., biometrics, statistics
Myers, Ronald Elwood, neurology
Myrianthopoulos, Ninos, human genetics
Nadkarni, Moreshwar Vithal, biochemical pharmacology
Naff, Marion Benton, organic chemistry
Nathanson, Neil Marc, neurobiology
Navarro, Joseph Anthony, mathematical statistics
Naylor, Alfred F., genetics
Neale, Elaine Anne, electron microscopy, neurocytology
Nebert, Daniel Walter, pediatrics, pharmacology
Nelson, Elton Glen, textiles
Nelson, Gary Joe, biophysics, biochemistry
Nelson, Phillip Gillard, neurosciences, cell biology
Ness, Arthur Thomas, physics
Ness, Robert Kiracofe, carbohydrate chemistry
Neufeld, Elizabeth Fondal, biochemistry, human genetics
Neva, Franklin Allen, microbiology, internal medicine
Neville, David Michael, Jr, molecular biology
Newton, Walter Lloyd, parasitology
Nijhout, H Frederick, insect physiology
Nijhout, Mary McAllister, insect physiology
Nimeroff, Isadore, physics
Nirenberg, Marshall Warren, biochemistry
Niswander, Jerry David, human genetics
Nomura, Shigeko, microbiology, virology
Nossal, Nancy, biochemistry
Nossal, Ralph J., biophysics, physics
Notkins, Abner Louis, virology, immunology
Nusser, Wilford Lee, parasitology, physiology
Nylen, Marie Ussing, dental research, electron microscopy
O'Brien, Paul J., biochemistry
O'Brien, Stephen James, genetics
O'Conor, Gregory Thomas, pathology, research administration
Oddis, Joseph Anthony, pharmacy
Odle, John William, mathematics
O'Donnell, James Francis, biochemistry
O'Gara, Roger, pathology
Ohe, Keiji, biochemistry
O'Hern, Elizabeth Moot, medical microbiology
Oka, Takami, developmental biology, pharmacology
Oliver, Eugene Joseph, microbiology, biochemistry

Olivier, Louis John, helminthology, tropical medicine
Olman, Robert Alexander, physiology
O'Malley, Joseph Paul, medicine
Omata, Robert Rokuro, bacteriology
Orians, F Barbara, physiology
Orloff, Jack, physiology, medicine
Orr, Henry Clayton, biology
Osbahr, Albert J, Jr, biochemistry, physical chemistry
Osborn, David Gordon, economic geography
Otani, Theodore Toshiro, biochemistry
Oxman, Michael Allan, information science, research administration
Padian, Eduardo Agustin, biophysics
Page, Norbert Paul, veterinary medicine, radiation biology
Pages, Robert Alex, bio-organic chemistry
Palmer, Timothy Trow, medical parasitology
Papadopoulos, Nicholas M., biochemistry
Papas, Takis S., biochemistry, virology
Parker, John Clarence, virology, parasitology
Parkman, Paul Douglas, virology, pediatrics
Parsegian, Vozken Adrian, biophysics
Passonneau, Janet Vivian, biochemistry
Pastan, Ira Harry, molecular biology, endocrinology
Patanelli, Dolores J., reproductive physiology
Patel, Dali Jehangir, medicine, biophysics
Pauk, George Lyon, medicine, endocrinology
Paul, William Erwin, immunology
Pavlovskis, Olgerts Raimonds, microbiology
Payne, Fred J., epidemiology
Payne, William Harold, information science
Pazoles, Christopher James, molecular biology
Peacock, Andrew Clinton, biochemistry
Pearson, John William, microbiology, oncology
Peebles, Paul Thomas, pediatrics, microbiology
Petier, Louis Cook, geography, geology
Pennington, Robert Miles, microbiology
Perrott, George St John, medical statistics
Perry, Seymour Monroe, oncology, hematology
Peterkofsky, Alan, biochemistry
Peterkofsky, Beverly, biochemistry
Peters, James Alexander, epidemiology
Peters, Elbert Axel, biochemistry
Peterson, John Ivan, analytical chemistry
Petricciani, John C., physical chemistry, medicine
Phillips, Leo Augustus, molecular biology, biophysical chemistry
Piatigorsky, Joram Paul, developmental biology, biophysics
Pierce, Jack Vincent, biochemistry
Pierce, Joseph Elliot, laboratory animal medicine, experimental surgery
Pierson, Bernice Frances, protozoology
Piez, Karl Anton, biochemistry
Pinkerson, Alan L., medicine, physiology
Pisano, John Joseph, biochemistry
Pitcairn, Donald M., internal medicine
Pittman, Margaret, bacteriology
Pledger, Richard Alfred, microbiology
Plough, Irvin Chaffee, internal medicine, research administration
Podolsky, Richard James, biophysics
Poirier, Lionel Albert, cancer biology, biochemistry
Pollard, Harvey Bruce, neurobiology
Pollok, Nicholas Lewis, III, microbiology
Pomerance, William, obstetrics & gynecology
Pomeroy, Thomas Charles, radiology
Portugal, Franklin H., biochemistry
Postow, Elliot, biophysics
Powers, Kendall Gardner, parasitology
Pratt, Arnold Warburton, biophysics
Prescott, Benjamin, biochemistry
Price, Paul Jay, cell biology
Price, Samuel, genetics
Priester, William Edward, biochemistry
Priester, William Alfred, veterinary medicine, epidemiology
Pronove, Pacita, pediatrics
Purcell, Robert Harry, virology
Putman, Donald Lee, microbiology
Quarles, Richard Hudson, neurochemistry
Quinn, Frank Russell, medicinal chemistry, information science

Rabinovitz, Marco, biochemistry
Rabson, Alan S., pathology
Rall, Joseph Edward, physiology
Rall, Wilfrid, biophysics, neurophysiology
Ram, J Sri, biochemistry, research administration
Ramsay, William Charles, resource management
Ranhand, Jon M., microbial physiology
Rapoport, Stanley I., medicine, physiology
Rapp, Herbert Joseph, microbiology
Rastogi, Suresh Chandra, biochemistry
Raub, William F, physiology, computer science, biostatistics
Rauscher, Frank Joseph, microbiology
Ray, Ovid Malcolm, physiology
Read, Merrill Stafford, biochemistry, nutrition
Reagan, Reginald L., virology
Reid, James Cutler, organic chemistry
Reifenstein, George Henry, medicine
Reisinger, Robert, virology, immunology
Reitman, Morton, medical microbiology, immunology
Repaske, Roy, microbial biochemistry
Resigno, Aldo, mathematical biology
Resnik, Robert Alan, physiology
Rice, Jerry Mercer, experimental pathology, biochemistry
Rice, Kenner Craile, synthetic organic chemistry, medicinal chemistry
Richards, Charles Selwyn, malacology
Richardson, Allan Charles Barbour, nuclear physics, radiological health
Riesz, Peter, physical chemistry, radiation biology
Ringler, Robert L., biochemistry
Rinzel, John Matthew, applied mathematics
Risser, Nancy M., biochemistry
Rivera, Americo, Jr., biochemistry
Rizzo, Anthony Augustine, dentistry, microbiology
Robbins, Jacob, endocrinology, medical research
Robbins, Jay Howard, cell biology, medicine
Robens, Jane Florence, veterinary medicine
Robinson, David Lee, neurophysiology
Robinson, Wilbur Gerald, Jr., cell biology, experimental pathology
Rodbell, Martin, biochemistry
Rodkey, Frederick Lee, biochemistry
Rodowskas, Christopher A, Jr., pharmacy, pharmacy administration
Rogers, William Edward, Jr., pharmacology, nutrition
Roller, Peter Paul, organic chemistry
Romano, Paula Josephine, immunology, microbiology
Roscoe, Henry George, biochemistry, science administration
Rosen, Saul W., internal medicine, endocrinology
Rosenberger, Charles Rupley, entomology
Rosenblatt, Dorrie Ellen, neurochemistry
Rosenfield, Joan Samour, physical chemistry
Rosner, Judah Leon, molecular biology
Ross, Ronald Burns, organic chemistry, lipid chemistry
Roth, Harold Philmore, gastroenterology
Roth, Jesse, endocrinology
Rotherham, Jean, biochemistry
Royt, Paulette Anne, microbiology
Rubin, Daniel Justin, immunology
Sabol, Steven Layne, molecular biology
Safer, Brian, molecular biology, oncology
Saffiotti, Umberto, oncology
Salzman, Lois Ann, microbiology
Salzman, Norman Post, virology
Sandberg, Ann Linnea, immunology
Sanford, Barbara Hendrick, genetics, immunology
Sanford, Jay Philip, infectious diseases
Sanford, Katherine Koontz (Mrs Charles F R Mifflin), zoology
Sansone, William Robert, nutritional biochemistry
Sarin, Prem S., biochemistry, cell biology
Sarma, Padman S., virology, veterinary medicine
Saroff, Harry Arthur, organic chemistry
Saunders, Joseph Francis, biochemistry
Savchuck, William Basil, physiological chemistry
Schade, Arthur Lincoln, microbiology
Schaefer, Morris, medicine, microbiology

Scharff, Raymond, physiology, biochemistry
Schechter, Alan Neil, medical research, protein chemistry
Schepartz, Saul Alexander, biochemistry
Schaffino, Silvio Stephen, research administration
Schiffmann, Elliot, biochemistry, organic chemistry
Schlamm, Norbert Arnold, zoology, bacteriology
Schlom, Jeffrey, microbiology
Schmidt, Edward Matthews, bioengineering
Schmidt, Jack Russell, research
Schneider, Walter Carl, chemistry
Schneiderman, Marvin Arthur, statistics
Schoenberg, Bruce Stuart, neurology, epidemiology
Schoenberg, Mark, biochemistry
Schoolman, Harold M., internal medicine
Schrecker, Anthony Wolfgang, biochemistry
Schubert, John Rockwell, nutritional biochemistry
Schacter, Bernard, biochemistry, biomedical engineering
Shaffer, Paul Raymond, geology
Shanks, Daniel, mathematics
Shapiro, Marvin Benjamin, computer science
Sharpless, Norman Edward, physical chemistry
Shaw, Robert William, Jr., solid state physics, energy conversion
Shean, Gerald Michael, Jr., electrochemistry
Sherbrooke, Craig C., operations research, statistics
Shichi, Hitoshi, biochemistry, biophysics
Shields, Jimmie Lee, environmental physiology
Shih, Thomas Yutzong, biochemistry, molecular biology
Shilling, Charles Wesley, medicine
Shulman, Nahum Raphael, hematology, immunology
Shumway, Norman Price, medicine
Sibal, Louis Richard, microbiology
Sidbury, James Buren, Jr, pediatrics
Sieber-Fabro, Susan M., pharmacology
Silverton, James Vincent, physical chemistry, crystallography
Simkins, Ronald Allen, biochemistry
Simon, Ernest Robert, medicine, hematology
Simonds, Josephine Abigail, biochemical genetics, microbiology
Simpson, Robert Todd, biochemistry
Singer, Maxine Frank, biochemistry
Sippel, John Edward, medical microbiology
Skolnick, Phil, pharmacology
Slawsky, Zaka Israel, physics
Small, James David, veterinary medicine, microbiology
Smith, Lester, biochemistry
Smith, Thomas Graves, Jr, neurophysiology
Smith, Willie White, physiology
Smith, Mary Elizabeth, developmental biology
Smyrniotis, Pauline Zoe, biochemistry
Snell, Katherine Chapin, pathology
Snyder, Stephen Laurie, biological chemistry
Sokoloff, Louis, physiology, biochemistry
Soliner, Karl, chemistry
Soliner-Webb, Barbara Thea, molecular biology
Solomon, Joel Martin, immunology
Solovey, Mathilde, immunohematology
Spande, Thomas Frederick, organic chemistry

Sporn, Michael Benjamin, cancer, biochemistry
Stadman, Earl Reece, biochemistry
Stadman, Thressa Campbell, biochemistry, enzymology
Stahl, Charles Jay, III, pathology
Stansly, Philip Gerald, biochemistry, microbiology
Stanton, Mearl Fredrick, medicine
Steers, Edward, Jr, biological chemistry
Steinman, Harry Gordon, organic chemistry
Stengle, James Marshall, internal medicine, hematology
Stephenson, Elizabeth Weiss, physiology
Stephenson, John Leslie, biophysics
Stern, Robert, medicine, biochemistry
Stetten, DeWitt, Jr, medicine
Stetten, Marjorie Roloff, biochemistry
Still, Edwin Tanner, radiobiology
Stillman, Irving Mayer, biophysics, medicine
Stimler, Suzanne Stokes, physical chemistry
Stone, Sanford Herbert, immunology
Strasberg, Murray, acoustics
Straube, Robert Leonard, radiobiology
Streicher, Eugene, neurophysiology
Stromberg, Kurt, pathology
Stromberg, LaWayne Roland, surgery, nuclear medicine
Stroud, Robert Church, physiology
Surrey, Kenneth, phytochemistry
Symmes, David, neurophysiology, psychology
Szu, Shousun Chen, molecular biophysics
Tabor, Celia White, pharmacology
Tabor, Herbert, pharmacology, biochemistry
Takemoto, Kenneth Kaname, virology
Talbot, John Mayo, aerospace medicine, radiobiology
Tasaki, Ichiji, neurophysiology
Taylor, Dorothy Jane, biology
Taylor, Lauriston Sale, medical physics
Taylor, Robert E, physiology, biophysics
Termine, John David, biochemistry
Terry, William David, immunology
Theilheimer, Feodor, mathematics
Theodore, Theodore Spiros, microbial physiology
Therriault, Donald G, biochemistry
Thoa, Nguyen Bich, pharmacology, physiology
Thomas, Louis Barton, pathology
Thompson, Edward Ivins Bradbridge, molecular biology, cell biology
Thompson, William D, neurophysiology
Tietze, Frank, biochemistry
Titus, Elwood Owen, organic chemistry, veterinary medicine
Tjalma, Richard Arlen, veterinary medicine
Tjio, Joe Hin, cytogenetics
Tobie, John Edwin, immunology
Todaro, George Joseph, cancer
Topper, Yale Jerome, biochemistry
Torchia, Dennis Anthony, biophysics
Tousignaut, Dwight R, pharmacy administration
Tower, Donald Bayley, research
Trams, Eberhard Georg, biochemistry, pharmacology
Triantaphyllopoulos, Demetrios, physiology, internal medicine
Triche, Timothy Junius, pathology
Tross, Carl Henry, physics, mathematics
Trueblood, Emily Walcott Emmart, cytology
Tsai, Lin, organic chemistry
Tsan, Alice Tung-Hua, numerical analysis, applied mathematics
Tschudy, Donald P, internal medicine,
Tullner, William W, endocrinology
Tully, Joseph George, Jr, medical microbiology
Turbyfill, Charles Lewis, research administration
Turner, James Henry, parasitology
Tyeryar, Franklin Joseph, Jr, medical microbiology
Uphoff, Delta Emma, genetics
Van Buren, John Miller, neurophysiology, neuroanatomy
Vanderryn, Jack, physical chemistry, research administration
Vande Woude, George, biochemistry, virology
Vannier, Wilton Emile, immunochemistry, biochemistry
Varon, Myron Izak, radiobiology, medicine
Vasta, Bruno Morreale, information science, biochemistry
Vaughan, Martha, biochemistry
Venditti, John M, pharmacology, biochemistry
Vernon, Mina Lee, anatomy

Villarroel, Fernando, biomedical engineering, chemical engineering
Vlahakis, George, genetics
Vollman, Rudolf F, obstetrics & gynecology
Vollmer, Erwin Paul, physiology
Von Euler, Leo Hans, pathology
Vought, Robert Louis, epidemiology, preventive medicine
Vreeland, Herbert Harold, III, anthropology
Waalkes, T Phillip, public health
Wagner, Henry George, neurosciences, aerospace medicine
Wahl, Sharon Knudson, immunology
Waksberg, Joseph, applied statistics
Waldmann, Thomas A, medicine, immunology
Walske, Max Carl, physics
Walter, William Arnold, Jr, epidemiology, biochemistry
Waravdekar, Vaman Shivram, biochemistry
Ward, Jerrold Michael, veterinary pathology
Ware, James H, mathematical statistics
Waterman, Allyn Jay, embryology
Waters, James Augustus, organic chemistry
Watzman, Nathan, psychopharmacology, pharmacology
Waxdal, Myron John, biochemistry, physiology
Webb, Alfred Mohr, medical microbiology
Webber, Richard Lyle, physiological optics, dentistry
Webster, Henry deForest, neurology, neuropathology
Webster, Marion Elizabeth, biochemistry, pharmacology
Weiger, Robert W, medicine
Weinbach, Eugene Clayton, biochemistry
Weinstein, Constance de Courcy, biochemistry
Weinstein, Howard, neurochemistry, cell physiology
Weintraub, Bruce Dale, endocrinology, internal medicine
Weisburger, Elizabeth Kreiser, oncology, toxicology
Weisgraber, Karl Heinrich, organic chemistry, biochemistry
Weisiger, James Richard, biochemistry
Weiss, Emilio, medical microbiology
Weiss, George Herbert, applied mathematics
Weiss, Ulrich, organic chemistry
Weiss, William, biostatistics, epidemiology
Weissberg, Alfred, mathematics, science administration
Whedon, George Donald, medical research administration, aerospace medicine
Whitehart, David Ralph, biochemistry
White, Florence Roy, biochemistry
White, Frederick Howard, Jr, protein chemistry
White, Howard Julian, Jr, science administration
Whitehair, Leo A, veterinary medicine, food science
Whitney, Robert Arthur, Jr, laboratory animal medicine, comparative medicine
Wickerhauser, Milan, biochemistry
Wilcox, Marion Allen, zoology
Williams, Jimmy Calvin, microbiology, biochemistry
Williams, Robert Jackson, physiology, ecology
Willis, George Mirron, plant pathology
Wilson, Katherine Schmitkons, genetics
Windhorst, Dorothy Baker, dermatology, immunology
Windmueller, Herbert George, biochemistry
Wineman, Robert Judson, bio-organic chemistry, biomaterials
Winestock, Claire Hummel, organic chemistry
Witkop, Bernhard, organic chemistry
Wittenberger, Charles Louis, bacteriology
Wolf, Robert Oliver, oral biology
Wolff, Jan, endocrinology
Wolff, John Bruno, biophysics
Wolff, Sheldon Malcolm, infectious diseases, immunology
Wolfe, Thomas Lee, veterinary medicine, animal behavior
Wollman, Seymour Horace, physiology
Woo, Kwang Bang, biomedical engineering, electrical engineering
Wood, David Belden, astronomy
Wood, Harry Burgess, Jr, organic chemistry
Woodman, Daniel Ralph, virology
Woodman, Richard J, biochemistry

Woods, Geraldine Pittman, neuroembryology
Woodside, Gilbert Llewellyn, developmental biology
Wooldridge, Robert Leonard, medical bacteriology, immunology
Wooster, Harold Abbott, information science
Wortman, Bernard, biochemistry
Wurtz, Robert Henry, neurophysiology, neuropsychology
Wykes, Arthur Albert, biochemical pharmacology, toxicology
Yagi, Haruhiko, organic chemistry, electrochemistry
Yamamoto, Richard Susumu, biochemistry
Yeh, Herman Jia-Chain, physical chemistry
Yellin, Herbert, health sciences, experimental neurology
Young, David Marshall, pathology
Young, Donald Stirling, clinical pathology
Yu, Leepo Cheng, biophysics
Yurek, Gerald G, biomedical engineering, electrical engineering
Zaharko, Daniel Samuel, pharmacology, physiology
Zatz, Marion M, immunology
Zebovitz, Eugene, virology
Zelenka, Peggy Sue, developmental biology
Ziffer, Herman, organic chemistry
Zimmerman, Daniel Hill, biochemistry, immunology
Zimmerman, Hyman Joseph, physiology, metabolism
Zimmerman, Steven B, biochemistry
Zukel, William John, cardiovascular diseases
Zwanzig, Frances Ryder, chemistry

BOWIE
Bernstein, Barbara Elaine, mathematics, psychiatry
Jones, Elbert Ellery, physics
Koch, J Frederick, solid state physics
Lichtenfels, James Ralph, parasitology, taxonomy
Saba, William George, physical chemistry
Schultze, Walter Donald, dairy bacteriology
Stunkard, Jim A, laboratory animal medicine, veterinary microbiology
Wortman, Roger Matthew, physics
Young, Ulysses Simpson, anthropology

BROOKLANDVILLE
Bryant, Harold Horn, pharmacology

BROOKMONT
Gazin, Charles Lewis, geology

BRUNSWICK
Kao, Kung-Ying Tang, biochemistry

BURTONSVILLE
Bettinger, Richard Thomas, space physics

CABIN JOHN
Sewell, Emma Winifred, information science, pharmacy

CAMBRIDGE
Cronin, Lewis Eugene, marine ecology, zoology
Savage, John Edward, environmental sciences
Tatro, Mahlon Charles, environmental sciences
Traub, Eugene Frederick, dermatology

CAMP SPRINGS
Davis, Kent J, veterinary pathology
Fawcett, Edwin Babcock, meteorology
O'Connor, Mary Jo, physical chemistry
Strauss, James Francis, meteorology
Strauss, Simon Wolf, chemistry
Witwer, Bruce Donald, resource management

CATONSVILLE
Burchard, Robert P, bacteriology
Fischer, Roland, biochemistry
Freimuth, Henry Charles, chemistry, forensic science
Gethmann, Richard Charles, genetics

Hanson, Frank Edwin, Jr, neurosciences, behavioral physiology
Kloetzel, John Arthur, cell biology, protozoology
Kung, Shain-Dow, molecular biology
Lovett, Paul Scott, microbiology
Platt, Austin Pickard, ecology, population biology
Ruskin, Bernice Heyman, medicinal chemistry, organic chemistry
Schamp, Homer Ward, Jr, high pressure physics
Simon, Stephen Wistar, biology
Wardell, William Lewis, plant physiology
Williams, T Glyne, psychiatry

CHESAPEAKE BEACH
Moore, Willard S, marine geochemistry
Vogt, Peter Richard, marine geophysics

CHESTERTOWN
Brown, Richard Harland, mathematics
Gwynn, Edgar Percival, cytogenetics
McLain, Joseph Howard, physics, plasma physics
Yaw, Katherine Emily, microbiology

CHEVY CHASE
Bates, Lloyd M, radiological physics
Bingham, Robert Lodewijk, nuclear physics, research administration
Blanche, Ernest Evred, mathematics, statistics
Brenner, Abner, electrochemistry
Burrill, Meredith Frederic, geography
Campbell, Eugene Paul, public health, internal medicine
Causey, George Donald, audiology, speech pathology
Chovitz, Bernard H, geodesy, cartography
Crane, Langdon Teachout, Jr, solid state physics, research administration
Crosby, Percy, petrology
French, Bevan Meredith, geochemistry, astrogeology
Gibson, Gilbert Lewis, physics, color science
Goldman, Alan Joseph, mathematics
Hunt, Charles Maxwell, environmental chemistry
Kielhorn, William Vineyard, physical oceanography, biological oceanography
Knox, William Tyndall, organic chemistry
Korgen, Reinhard Lunde, mathematics
Lourie, Reginald Spencer, psychiatry
Manger, George Edward, geology
McCabe, Louis Cordell, geology
Naugle, John Earl, physics
Newman, Sanford Bernhart, materials science
Pohrer, Robert George, mathematical analysis
Price, Donna, physical chemistry, explosives
Rader, William Austin, veterinary toxicology
Raff, Samuel J, nuclear physics
Rubenstein, Albert Marvin, physics
Segel, Edward, chemistry
Smith, Charlotte Damron, biochemistry
Smith, Jack Carlton, polymer physics
Vincent, Monroe Mortimer, cell biology, immunobiology
Wang, Theodore Joseph, physics

CLARKSBURG
Revesz, Akos George, solid state chemistry
Rittner, Edmund Sidney, applied physics, energy conversion

CLARKSVILLE
Albers, Edwin Wolf, physical chemistry
Block, Jacob, water chemistry
Duecker, Heyman Clarke, inorganic chemistry
Fulmer, Glenn Elton, polymer physics, polymer chemistry
Guthrie, James Leverette, organic chemistry
Janney, John Hall, Jr, public health
Lard, Edwin Webster, analytical chemistry
Maselli, James Michael, inorganic chemistry
Vaughan, David Evan William, chemistry, geochemistry
Wood, Louis L, organic chemistry

COCKEYSVILLE
Carr, Edward Mark, analytical chemistry

Carski, Theodore Robert, immunology, microbiology
Esposito, Vito Michael, immunology, microbiology
Evans, George Leonard, microbiology
Galetto, William George. food chemistry
Glenn, William Grant, zoology

COLLEGE PARK

Adams, William Wells, number theory
Agrawala, Ashok Kumar, computer science
Ahnert, Frank, physical geography, geography of America
Ahrens, Richard August, nutrition, physiology
Alexander, James Crew, mathematics
Alexander, Millard Henry, theoretical chemistry
Allan, J David, ecology
Ammon, Herman L, organic chemistry
Anastos, George, acarology
Anderson, J Robert, physics
Angell, Frederick Franklyn, plant breeding, genetics
Antman, Stuart S, applied mathematics, mechanics
Arbuckle, Wendell Sherwood, dairy science
Arisumi, Toru, plant genetics
Atchison, William Franklin, mathematics, computer science
Auslander, Joseph, mathematics
Austin, Morris Edwin, soils
Axley, John Harold, soils
Aycock, Marvin Kenneth, Jr, plant breeding, plant genetics
Babuska, Ivo Milan, numerical analysis, applied mathematics
Bagchi, Amitabha, solid state physics, surface physics
Bailey, William John, organic chemistry
Baker, Robert Lewis, horticulture, botany
Bandel, Vernon Allan, agronomy, plant physiology
Bardasis, Angelo, solid state physics
Barnett, Audrey, genetics, protozoology
Barnett, Neal Mason, plant physiology
Barry, Kevin Gerard, internal medicine
Bean, George A, plant pathology
Beaton, John Rogerson, nutrition
Bell, Roger Alistair, astrophysics
Benedetto, John, analytical mathematics
Benedict, William Sidney, chemical physics
Benesch, William Milton, molecular physics, molecular spectroscopy
Berk, Abraham Albert, chemistry
Bernstein, Allen Richard, mathematics
Bhagat, Satindar M, physics
Bickley, William Elbert, entomology
Bigbee, Daniel E, poultry science
Blevins, Dale Glenn, plant nutrition
Bottino, Paul James, plant genetics
Bouwkamp, John C, horticulture, plant breeding
Boyd, Alfred Colton, Jr, inorganic chemistry
Brace, John Wells, pure mathematics
Brauth, Steven Earle, neuropsychology
Brill, Dieter Rudolf, theoretical physics
Brinkley, Howard J, reproductive endocrinology
Broome, Carmen Rose, systematic botany
Brown, Kenneth E, mathematics
Brown, Russell Guy, botany
Brush, Stephen George, history of science
Buchler, Edward Raymond, ethology
Buck, Raymond Wilbur, Jr, cytogenetics
Burgers, Johannes Martinus, physics, fluid dynamics
Burt, Gordon Willis, agronomy
Caceres, Cesar A, medicine, computer science
Cairns, Gordon Mann, dairy husbandry
Caron, Dewey Maurice, entomology, apiculture
Carroll, Edward James, Jr, developmental biology, biochemistry
Castellan, Gilbert William, physical chemistry
Chance, Charles Marion, dairy nutrition
Chang, Ren-Fang, thermodynamics, lasers
Chapin, John Ladner, respiratory physiology
Chen, Abel Jer-Jiunn, space physics, aeronomy
Cheung, Augustine Y, radiological health, microwave engineering
Clark, Eugenie, zoology
Clark, Neri Anthony, agronomy
Clarke, David Harrison, human physiology
Cohen, Joel M, pure mathematics
Cohen, Leon Warren, mathematics
Cohen, Philip Ira, surface physics
Colwell, Rita R, microbiology
Contrera, Joseph Fabian, vertebrate physiology, neuropharmacology
Cook, Clarence Harlan, mathematics
Cook, Thomas M, microbiology
Coplan, Michael Alan, physical chemistry
Corbett, M Kenneth, plant pathology
Corliss, John Ozro, protozoology
Davidson, John Angus, entomology
Davidson, Ronald Crosby, plasma physics
Davis, Richard Francis, animal science, dairy science
Day, Thomas Brennock, high energy physics
Deal, Elwyn Ernest, agronomy
Decker, Alvin Morris, Jr, agronomy
DeRocco, Andrew Gabriel, chemical physics, biophysics
De Saint-Jean, Alain Y, ethnology, economic anthropology
Deshler, Walter W, cultural geography, geography of Africa
DeSilva, Alan W, plasma physics
DeVoe, Howard Josselyn, physical chemistry
Doetsch, Raymond Nicholas, bacteriology
Dorfman, Jay Robert, theoretical physics, statistical mechanics
Douglis, Avron, mathematics
Dragt, Alexander James, theoretical high energy physics, classical mechanics
Drew, Howard Dennis, solid state physics
Dworzecka, Maria, nuclear physics
Earl, James Arthur, physics
Edmundson, Harold Parkins, mathematics
Einstein, Theodore Lee, theoretical solid state physics, surface physics
Ellis, Robert L, mathematics
Elsasser, Walter M, physics, geophysics
Erickson, William Clarence, astronomy
Faller, Alan Judson, meteorology
Fanning, Delvin Seymour, soil science
Fitzpatrick, Patrick Michael, pure mathematics
Fivel, Daniel I, theoretical physics
Flyger, Vagn Folkmann, wildlife ecology
Fonaroff, Leonard Schuyler, biogeography
Foss, John E, soil genesis, soil classification
Freeman, David Haines, physical chemistry, analytical chemistry
Fromovitz, Stan, operations research, applied statistics
Galloway, Raymond Alfred, physiology
Gammon, Robert Winston, physics
Gardner, Marjorie Hyer, science education, chemistry
Gass, Saul Irving, mathematics
Ginter, Marshall L, molecular physics, atomic physics
Glascock, Michael Dean, experimental nuclear physics
Glasser, Robert Gene, physics
Glick, Arnold J, theoretical solid state physics
Gloeckler, George, physics
Glover, Rolfe Eldridge, III, solid state physics
Gluckstern, Robert Leonard, physics, academic administration
Goldberg, Seymour, mathematics
Goldenbaum, George Charles, plasma physics
Goldhaber, Jacob Kopel, algebra
Goldsby, Richard Allen, biochemistry, cell biology
Goldstein, Larry, Joel, number theory
Good, Richard Albert, algebra
Goode, Melvyn Dennis, cell biology, biophysics
Gouin, Francis R, plant physiology
Gray, Alfred, geometry
Green, John H, microbiology, biochemistry
Green, Willard Wynn, animal breeding
Greenberg, Oscar Wallace, elementary particle physics, high energy physics
Griem, Hans Rudolf, physics
Griffin, James J, theoretical physics
Grim, Samuel Oram, inorganic chemistry, organometallic chemistry
Guernsey, Ralph Lewis, plasma physics
Gulick, Sidney L, III, mathematics
Haley, Albert James, parasitology
Hansen, John Norman, biochemistry, molecular biology
Harding, Wallace Charles, Jr, entomology, botany
Harper, Robert Alexander, urban geography, economic geography
Harrington, James Patrick, astrophysics
Harrison, George Keithley, plant physiology
Haut, Irvin Charles, horticulture
Heath, James Lee, poultry science, food science
Heilprin, Laurence Bedford, information science
Heins, Maurice Haskell, mathematical analysis
Helz, George Rudolph, geochemistry
Helzer, Garry A, algebra
Henery-Logan, Kenneth Robert, organic chemistry
Henkelman, James Henry, mathematics
Hetrick, Frank M, virology
Higgins, William Joseph, comparative physiology
Highton, Richard, herpetology, population genetics
Hodgson, Ralph Edward, dairy husbandry
Holmgren, Harry D, nuclear physics
Holmlund, Chester Eric, biochemistry
Hornyak, William Frank, nuclear physics
Horvath, John, mathematical analysis
Howell, Barton John, optics
Hu, Charles Y, geography
Hubbard, Bertie Earl, mathematical analysis
Huheey, James Edward, inorganic chemistry, herpetology
Hummel, James Alexander, mathematics
Imberski, Richard Bernard, developmental genetics
Israel, Gerhard Wilhelm, atmospheric physics, air pollution control
Jachowski, Leo Albert, Jr, parasitology
Jackson, Stanley Bartlett, geometry
Jaquith, Richard Herbert, inorganic chemist ry
Jarvis, Bruce B, organic chemistry
Jensen, Philip J, pediatrics
Johnson, Charles Royal, algebra, applied mathematics
Johnson, Jerry Wayne, agronomy
Johnson, Raymond Lewis, mathematics
Johnson, Robert Byron, veterinary virology
Johnson, William Pierre, physics
Jones, Grover Stephen, mathematics
Jones, Jack Colvard, insect physiology
Jones, Lewis Hammond, IV, elementary particle physics
Joseph, Stanley Robert, medical entomology
Kacser, Claude, theoretical physics
Kalil, Ford, engineering physics
Kanal, Laveen Nanik, computer science, statistics
Karlander, Edward P, algology
Kasler, Franz Johann, analytical chemistry
Keeney, Mark, dairy science
Kellogg, Royal Bruce, applied mathematics
Kerley, Ellis R, physical anthropology, forensic science
Kerr, Frank John, radio astronomy
Kim, Hogil, nuclear physics, applied physics
King, Raymond Leroy, food science
Kirby, Karen Marie, statistics, population genetics
Kirwan, William English, mathematical analysis
Klarman, William L, phytopathology
Kleppner, Adam, mathematical analysis
Koopman, David Warren, plasma physics
Korenman, Victor, theoretical solid state physics
Krall, Nicholas Anthony, physics
Kramer, Amihud, food science
Krisher, Lawrence Charles, chemical physics
Krusberg, Lorin Ronald, plant pathology
Kueker, David William, mathematics
Kuenzel, Wayne John, poultry physiology, ornithology
Kundu, Mukul Ranjan, radiophysics, electronics
Laffer, Norman Callender, bacteriology
Landsberg, Helmut Erich, climatology
Langford, George Shealy, entomology
Larson, Jerome Valjean, geophysics, electrical engineering
Lashinsky, Herbert, physics
Laster, Howard Joseph, cosmic ray physics
Lay, David Clark, mathematics
Lee, Robert Jerome, veterinary medicine
Lee, Young Jack, statistics
Leffel, Emory Childress, animal science
Lehner, Guydo R, topology
Leone, Mark Paul, anthropology, archaeology
Li, Ta-Yung, meteorology
Linder, Harris Joseph, zoology
Link, Conrad Barnett, horticulture
Liu, Chuan Sheng, theoretical astrophysics, plasma physics
Liu, Tai-Ping, mathematical analysis
Lockard, J David, botany, science education
MacDonald, William, nuclear physics
MacQuillan, Anthony M, microbial physiology
Maehl, Ronald Charles, space physics
Marion, Jerry Baskerville, nuclear physics
Marr, Harold Everett, III, physical chemistry
Martin, David Lee, biochemistry
Martin, Frederick Wight, experimental physics
Matthews, David LeSueur, physics
Mattick, Joseph Francis, biochemistry, bacteriology
Mayor, John Roberts, mathematics
Mazzocchi, Paul Henry, organic chemistry
McClellan, Gene Elvin, elementary particle physics
Menzer, Robert Everett, insect toxicology
Messersmith, Donald Howard, entomology, ornithology
Mikulski, Piotr W, mathematical statistics
Miller, Frederick Powell, soil science, agronomy
Miller, Gerald Ray, physical chemistry
Miller, James Roland, soil chemistry, agronomy
Minker, Jack, mathematics
Misner, Charles William, theoretical physics, cosmology
Mitchell, Robert Davis, geography
Moore, John Hays, Jr, physical chemistry, molecular physics
Morgan, Delbert Thomas, Jr, botany
Morse, Douglass Hathaway, ecology
Motta, Jerome J, mycology
Mulchi, Charles Lee, agronomy, plant physiology
Munn, Robert James, chemical physics
Murdoch, Wallace Pierce, entomology
Murphy, Thomas James, statistical mechanics
Myers, Ralph Duane, physics
Neri, Umberto, mathematics
Newby, Hayes Augustus, speech pathology
O'Gallagher, Joseph James, physics
O'Haver, Thomas Calvin, analytical chemistry
Olver, Frank William John, applied mathematics
Oneda, Sadao, theoretical physics
Opik, Ernest Julius, astronomy
Osborn, John Edward, mathematics
Otto, Gilbert Fred, parasitology
Owings, James Claggett, Jr, mathematical logic
Park, Robert L, physics
Parochetti, James V, agronomy
Pati, Jogesh Chandra, theoretical physics
Patterson, Glenn Wayne, plant physiology, plant biochemistry
Pearl, Martin Herbert, mathematics
Pereira, Carlos Martin, theoretical physics, mathematics
Pickard, Hugh Brown, physical chemistry
Pierce, Sidney Kendrick, comparative physiology
Plowman, Ronald Dean, dairy husbandry, research administration
Ponnamperuma, Cyril Andrew, bio-organic chemistry
Potter, Jane Huntington, zoology
Prange, Richard E, physics
Prather, Mary Elizabeth Sturkie, nutrition
Pugh, Howel Griffith, nuclear physics, high energy physics
Purdy, William Crossley, analytical chemistry
Ragsdale, Nancy Nealy, pesticide chemistry, cell physiology
Ramm, Gordon Morley, embryology, morphology
Rappleye, Robert Du Bois, plant morphology
Rasekh, Jamshid G, food science
Redish, Edward Frederick, theoretical nuclear physics
Rees, Colin Pritchard, zoology
Reeve, Edward Wilkins, organic chemistry
Reichelderfer, Charles Franklin, invertebrate pathology, entomology
Reinhart, Bruce Lloyd, mathematics
Resnik, Harvey Lewis Paul, psychiatry, psychoanalysis
Reveal, James L, plant taxonomy
Reynolds, Charles William, horticulture
Rheinboldt, Werner Carl, numerical analysis
Richard, Jean-Paul, experimental physics

Rodenhuis, David Roy, dynamic meteorology
Rogers, Benjamin Lanham, horticulture
Rollinson, Carl Linden, chemistry
Roos, Philip G., nuclear physics
Rose, William K., astrophysics
Rosen, Stephen I, physical anthropology, primatology
Rosenfeld, Azriel, computer science
Roush, Marvin Leroy, nuclear physics
Rowan, Robert, III, nuclear magnetic resonance
Rubin, Max, animal nutrition, biochemistry
Sampugna, Joseph, biochemistry, neurochemistry
Schafer, James A., mathematics
Scheinberg, Sam Louis, genetics
Schemm, Charles Edward, dynamic meteorology
Schleidt, Wolfgang Matthias, ethology, bioacoustics
Sechi-Zorn, Bice, elementary particle physics
Segovia, Antonio, geology
Sengers, Jan V, fluid physics
Shaffner, Clyne Samuel, poultry science
Shalter, Michael David, ethology
Shanks, James Bates, horticulture
Shapere, Dudley, philosophy of science, history of science
Sherburne, Russell Knight, physical chemistry
Shorb, Mary Shaw, microbiology
Sidwell, Virginia DeCecco, fisheries, nutrition
Silverman, Joseph, radiation chemistry, polymer chemistry
Simonson, Simon Christian, III, astronomy
Sisler, Hugh Delane, plant pathology
Smith, Betty F, textile chemistry, carbohydrate chemistry
Smith, Elske Van Panhuys, astronomy
Snow, George Abraham, physics
Soares, Joseph Henry, Jr, animal nutrition, biochemistry
Sommer, Sheldon E, geochemistry
Sorokin, Constantine Alexis, plant physiology
Spain, Ian L, chemical physics
Spivak, Steven Mark, textiles, textile engineering
Staley, Stuart Warner, organic chemistry
Stark, Francis C, Jr, horticulture
Steinberg, Phillip Henry, experimental high energy physics
Steinhauer, Allen Laurence, entomology, ecology
Stellmacher, Karl L, analytical mathematics
Stern, William Louis, plant anatomy
Stifel, Peter Beekman, geology
Strauss, Aaron Solomon, mathematics
Stuntz, Calvin Frederick, chemistry
Suppe, Frederick (Roy), philosophy of science, history of science
Syski, Ryszard, mathematics
Taylor, Ronald Charles, meteorology
Thomas, Owen Pestell, poultry nutrition
Thompson, Arthur Howard, pomology
Thompson, Owen Edward, meteorology, atmospheric physics
Trivelpiece, Alvin William, plasma physics
Twigg, Bernard Alvin, horticulture, food science
Vandersall, John Henry, animal nutrition
Vanderslice, Joseph Thomas, physical chemistry
Veitch, Fletcher Pearre, biochemistry, evolutionary biology, biogeography
Vernekar, Anandu Devarao, meteorology
Vermeij, Geerat Jacobus, evolutionary biology, biogeography
Viola, Victor E, Jr, nuclear chemistry
Voll, Mary Jane, microbial genetics
Wabeck, Charles J, food science
Wall, Nathan Sanders, nuclear physics
Wallace, Stephen Joseph, theoretical nuclear physics
Walters, William Ben, nuclear chemistry, physical chemistry
Warner, Charles Robert, mathematics
Weaver, Leslie O, plant pathology
Weber, Joseph, physics
Weidner, Jerry R, geochemistry
Weiner, Ronald Martin, microbiology
Wentzel, Donat Gotthard, astrophysics
Westerhout, Gart, astronomy
Westhoff, Dennis Charles, food microbiology
Wiley, Robert Craig, food science, horticulture
Wilkerson, Thomas Delaney, plasma physics, space physics
Williams, Aubrey Willis, Jr, anthropology, ethnography

Williams, Eleanor Ruth, nutrition, foods
Williams, Walter Ford, physiology
Wolf, Duane Carl, soil microbiology
Wood, Francis Eugene, entomology
Worthington, John Thomas, III, horticulture
Yang, Grace L, statistics
Yeh, Kwan-Nan, textile chemistry
Yoder, Neil Richard, particle physics
Yodh, Gaurang Bhaskar, physics
Yorke, James Alan, applied mathematics, biomathematics
Young, Bobby Gene, microbiology, genetics
Zajac, Felix Edward, III, neurophysiology, biomedical engineering
Zedek, Mishael, mathematics
Zoller, William H, nuclear chemistry
Zorn, Bice Sechi, high energy physics
Zorn, Gus Tom, high energy physics
Zuckerman, Benjamin Michael, astronomy
Zwanzig, Robert Walter, chemical physics

COLUMBIA
Baird, James Haythorn, analytical chemistry
Bass, Jonathan Langer, physical chemistry
Brown, Douglas Edward, optical physics
Bush, Richard Wayne, polymer chemistry
Cogliano, Joseph Albert, organic chemistry
Das, Naba Kishore, microbiology
De Matte, Michael L, organic chemistry
Federighi, Enrico Thomas, algebra
Feinberg, Stewart Carl, organic polymer chemistry
Ferington, Thomas Edwin, polymer chemistry
Fisher, Elton, industrial chemistry
Frank, Victor Samuel, industrial organic chemistry, research administration
Gaidis, James Michael, organometallic chemistry
Gin, Jerry Ben, chemistry
Hill, David G, medical physics, high energy physics
Kelso, John Morris, radio physics
Ketley, Arthur Donald, photochemistry, polymer chemistry
Kiker, William Edward, medical physics, radiological health
Knox, Ellis Gilbert, soil science
McDaniel, Carl Vance, physical chemistry
McKinney, Robert Wesley, analytical chemistry
Morgan, Charles Robert, physical organic chemistry
Moyer, Joseph Donald, organic chemistry
Murch, Robert Matthews, inorganic chemistry, organometallic chemistry
Nickell, Louis G, plant physiology
Reuber, Melvin D, pathology, medicine
Rie, John E, polymer chemistry, photochemistry
Smith, Jean Gillen, physical inorganic chemistry, colloid chemistry
Spohn, William Gideon, Jr, mathematics
Veltman, Preston Leonard, chemistry
Weiss, Edward Sebastian, biostatistics
Whitney, Ambrose Grunhagen, organic chemistry

COOKSVILLE
Magee, John Storey, Jr, inorganic chemistry, physical chemistry

CROFTON
Veazey, Sidney Edwin, physics
Vestal, Claude Kendrick, meteorology, climatology

CUMBERLAND
Brown, Donald Meeker, applied mathematics
Gibson, James Donald, organic chemistry
Hartman, Kenneth Owen, physical chemistry
Jacobs, Albert Michael, applied chemistry
Miller, Roy Richard, physical chemistry
Pawlowski, Anthony T, physical chemistry
Pratt, Thomas Herring, Jr, physical chemistry
Preckel, Ralph Frederick, chemistry
Simon, John Antony, clinical biochemistry
Simpson, Myron Lee, parasitology

Starr, Duane Frank, physical chemistry

DAMASCUS
Bridgers, Bernard Thomas, botany, horticulture

DENTON
Martin, Monroe Harnish, applied mathematics

DERWOOD
Brinckman, Frederick Edward, Jr, inorganic chemistry
Folk, John Edard, biochemistry
Sinclair, Edward Elliot, chemistry

DISTRICT HEIGHTS
Campbell, William Joseph, physical chemistry
Stackpole, John Duke, meteorology

EASTON
Grove, Donald Cooper, pharmaceutical chemistry
Guthrie, Eugene Harding, medicine, public health

EDGEWATER
Faust, Maria Anna, marine microbiology, marine phycology

EDGEWOOD
Hinton, Dennis Melvin, immunology, biophysics

EDGEWOOD ARSENAL
Aaron, Herbert Samuel, organic chemistry
Armstrong, Robert D, pharmacology
Bales, Paul Dobson, biomedical engineering
Berkowitz, Lewis Maurice, organic chemistry
Broomfield, Clarence A, biochemistry
Buckles, Marjorie Fox, micro-chemistry
Crabtree, Eleanor Voorhees, organic chemistry, analytical chemistry
Craig, Francis Northrop, physiology
Crook, James Washington, pharmacology
Cummings, Edmund George, physiology
Davis, George Thomas, organic chemistry
Ellin, Robert Isadore, pharmaceutical chemistry
Gaskins, Frederick Hudson, rheology
Gibby, Irvin Welch, medical microbiology
McClure, Claude, biochemistry, neurosurgery
O'Leary, John Francis, pharmacology
Papirmeister, Bruno, biochemistry
Parent, Paul Andrew, chemistry
Sarver, Emory William, environmental chemistry
Sass, Samuel, analytical chemistry
Sternberger, Ludwig Amadeus, medicine
Traub, Richard Kimberley, medical research
Vocci, Frank Joseph, biochemistry
Wiles, Joseph St Clair, pharmacology, toxicology
Wilson, Kenneth MacKenzie, physiology, chemistry

ELKRIDGE
Legal, Casimer Claudius, Jr, inorganic chemistry

ELKTON
Vriesen, Calvin W, organic polymer chemistry, synthetic organic chemistry
Ward, Laird Gordon Lindsay, inorganic chemistry

ELLICOTT CITY
Hanson, James Edward, mathematics
Steiner, Robert Frank, physical biochemistry

EMMITSBURG
Meredith, William G, ecology
Richards, John Watson, physical chemistry
Stephens, Arthur Brooke, numerical analysis

FLINTSTONE
Harris, Van Thomas, vertebrate ecology

FREDERICK
Abeles, Ann Lindstrom, biophysics
Aszalos, Adorjan, bio-organic chemistry, cancer
Bausum, Howard Thomas, microbiology
Beisel, William R, infectious diseases, metabolism
Bonde, Morris Reiner, plant pathology
Boyd, Virginia Ann Lewis, virology
Bromfield, Kenneth Raymond, plant pathology
Buzzell, Anne, biophysics
Canonico, Peter Guy, cell biology
Cherrington, Ernest Hurst, Jr, astronomy
Cowen, William Frank, water chemistry
Craw, Alexander R, mathematics, operations research
Crozier, Dan, internal medicine
Damsteegt, Vernon Dale, plant pathology
Daniels, William Fowler, biochemistry
Darrow, Robert Arthur, plant physiology, plant ecology
Dipple, Anthony, cancer
Dorrell, William Woodrow, bacteriology
Eigelsbach, Henry Thomas, medical bacteriology
Eisman, Leon Philip, bacteriology, public health
Emge, Robert George, plant pathology
Fine, Donald Lee, microbiology
Fish, Donald C, microbiology
Fowler, Arnold K, animal physiology, genetics
French, Richard Collins, plant biochemistry
Glassman, Harold Nelson, biochemistry
Graf, Lloyd Herbert, biochemistry
Graham, Charles Lee, entomology, parasitology
Hanna, Michael G, Jr, immunobiology
Hearn, Henry James, Jr, microbiology, cancer
Hegyeli, Andrew Francis, biochemistry, experimental pathology
Hinshaw, William Russell, veterinary microbiology
Hurtt, Woodland, plant physiology
Issaq, Haleem Jeries, analytical chemistry
Janini, George Musa, analytical chemistry, physical chemistry
Jemski, Joseph Victor, medical microbiology
Johnson, Roger W, virology
Kenyon, Richard H, virology, immunology
Kingsolver, Charles H, plant pathology
Kleinspehn, George Gehret, organic chemistry
Kripke, Margaret Louise (Cook), cancer, immunobiology
Langlykke, Asger Funder, microbiology
Larson, Edgar William, microbiology
Latterell, Frances Meehan, plant pathology
Liu, C T, physiology, pharmacology
Lufkin, Daniel Harlow, solar physics
Marchetti, Marco Anthony, plant pathology
McCarrell, Jane Dinsmore, physiology
McLellan, William L, Jr, biochemistry, microbiology
Meier, Eugene Paul, environmental chemistry, analytical chemistry
Melching, J Stanley, plant pathology
Munson, Donald Albert, zoology, parasitology
Muul, Illar, ecology, animal behavior
Neufeld, Harold Alex, biochemistry
Oshiro, Yuki, biochemistry
Palmer, Winifred G, biochemistry
Payne, William Walker, public health
Pienta, Roman Joseph, virology, oncology
Pitillo, Robert Francis, bacteriology
Powanda, Michael Christopher, infectious diseases
Schaub, Stephen Alexander, virology, environmental health
Schidlovsky, George, virology, electron microscopy
Schmit, Chris George, phytopathology
Shibley, George P, virology, immunology
Silverman, Sidney Joseph, bacteriology
Slein, Milton Wilbur, biochemistry
Smith, Dorothy Gordon, microbiology
Smith, Frank Orvil, economic botany
Smith, Sharron Williams, biochemistry
Spahn, Gerard Joseph, microbiology
Spero, Leonard, biochemistry
Spertzel, Richard O, virology, radiobiology
Sterrett, John Paul, plant physiology

Stevenson, Robert Edwin, microbiology, science administration
Vlaovic, Milan Stephen, veterinary pathology
Wachter, Ralph Franklin, biochemistry, virology
Wade, Clarence W R, organic chemistry
Walker, Jerry, veterinary medicine, microbiology
Wannemacher, Robert, Jr, biochemistry, nutrition
Warner, Mark Clayton, aquatic biology
Warshowsky, Benjamin, analytical chemistry
Weislow, Owen Stuart, immunology, microbiology
White, John David, bacteriology
Willis, Phylida Maye, physical chemistry
Zielinski, Walter L, Jr, analytical chemistry

FREELAND
Scott, Charles Covert, pharmacology

FRIENDSHIP
Ream, Donald F, physics

FROSTBURG
Cotton, James V, geography
Gilpin, Robert Harry, microbiology
Hunt, Paul Payson, physical chemistry
Jones, John Paul, mathematics
Redick, Thomas Ferguson, physiology
Schrock, Alta, plant ecology
Snyder, Jack Russell, zoology
Yoder, Wayne Alva, invertebrate zoology, entomology

FT DETRICK
Burrows, William Dickinson, environmental chemistry
Dacre, Jack Craven, toxicology, biochemistry
Houston, William Eddie, immunobiology
Levitt, Neil Hilliard, virology, immunology
Rosenblatt, David Hirsch, chemistry

FT HOWARD
Tonik, Ellis J, medical microbiology

FT GEORGE G MEADE
Abercrombie, Jay, entomology
Allen, Lew, Jr, physics
Bandy, William Robert, theoretical solid state physics
Maar, James Richard, mathematical statistics, statistical analysis

FULTON
Chitwood, May Belle Hutson, parasitology, nematology
Rand, Robert Collom, environmental chemistry
Sulzbacher, William Louis, food science

GAITHERSBURG
Bizzell, Oscar McArthur, nuclear chemistry
Bloomfield, Philip Earl, solid state physics, theoretical physics
Bowen, Rafael Lee, dentistry, polymer chemistry
Bryan, William Ray, physiology
Coriell, Kathleen Patricia, computer sciences
Culver, William Howard, physics
Davis, George Thomas, polymer chemistry
Devoe, James Rollo, radiochemistry
Franklin, Alan Douglas, solid state physics
Frazer, Lowell Keith, mathematics
Gould, Gordon, physics
Greer, Sandra Charlene, physical chemistry
Herron, John Thomas, physical chemistry
Kinnie, Irvin Gray, research management, information science
Krimgold, Dov B, hydrology
McAlister, Archie Joseph, solid state physics
Miles, Wyndham Davies, history of chemistry
Milly, George Harwood, earth science
Rehm, Ronald George, applied mathematics, fluid dynamics
Rowen, John William, operations research, physical chemistry
Shibe, Abe Jeffrey, operations research
Shipman, Jerome Saul, numerical analysis

Thayer, Scott Dwight, air pollution, micrometeorology
Tong, Long Sun, heat transfer
Wiggin, Edwin Albert, chemistry

GARRETT PARK
Hartman, Gregory Kemenyi, physics
Yeandle, Stephen Safford, biophysics

GERMANTOWN
Griffo, Joseph Salvatore, physical chemistry, inorganic chemistry
Hamburger, Richard, geology
Knapp, Harold Anthony, Jr, operations research
Patel, Prafull Raojibhai, dental research
Uhl, Edward George, applied physics
Von Braun, Wernher, physics, aerospace engineering
Yoder, Robert E, health physics, industrial hygiene

GLEN ECHO
Cannon, Edward Whitney, mathematics

GLENELG
Glocker, Edwin Merrian, applied statistics

GLENN DALE
Darrow, George McMillan, horticulture
Preston, William Henry, Jr, horticulture
Waterworth, Howard E, plant virology

GLENWOOD
Johnson, Chester Curtis, solid state physics

GREENBELT
Abbas, Mian Mohammad, infrared astronomy, space physics
Adler, Robert Frederick, meteorology
Arens, John Frederic, cosmic ray physics
Arking, Albert, physics
Balasubrahmanyan, Vriddhachalam K, cosmic ray physics
Barker, John L, Jr, earth sciences, remote sensing
Bauer, Siegfried Josef, space physics
Benson, Robert Frederick, ionospheric physics
Bhatia, Anand K, atomic physics
Blaine, Lamdin Roden, physics
Boggess, Albert, III, astrophysics
Bohlin, Ralph Charles, astrophysics
Boldt, Elihu (Aaron), physics
Bonavita, Nino Louis, physics, magnetism
Brace, Larry Harold, ionospheric physics
Brandt, John Conrad, astronomy
Brown, Douglas Ross, astrophysics
Buhl, David, radio astronomy
Burlaga, Leonard F, physics
Cameron, Winifred Sawtell, astronomy
Chapman, Robert DeWitt, astrophysics
Clark, John Fulmer, planetary sciences
Clark, Thomas Arvid, radio astronomy
Cline, Thomas L, physics
Coffman, John W, physics
Crannell, Carol Jo Argus, astrophysics
DeGasparis, Aurelio Alfonso A, geophysics, meteoritics
Dolan, Joseph Francis, astronomy
Donn, Bertram (David), astrophysics
Drachman, Richard Jonas, theoretical physics
Eckerman, Jerome, physics
Endal, Andrew Samson, astrophysics
Epstein, Gabriel Leo, solar physics
Fainberg, Joseph, radio astronomy
Feibelman, Walter A, astronomy, astrophysics
Fichtel, Carl Edwin, astrophysics
Fischel, David, astrophysics
Fleming, James Joseph, computer science
Foster, James Russell, entomology
Fraser, Robert Stewart, meteorology
Freden, Stanley Charles, science administration, remote sensing
Gloersen, Per, physics
Goldberg, Richard Aran, atmospheric physics
Hanel, Rudolf A, atmospheric physics
Harris, Isadore, theoretical physics
Harrison, Floyd Perry, economic entomology

Hartle, Richard Eastham, astrophysics, plasma physics
Hartman, Robert Charles, cosmic ray physics
Hasler, Arthur Frederick, meteorology
Hauser, Michael George, physics
Heap, Sara Ridgway, astronomy
Heppner, James P, space physics, geophysics
Hirschmann, Ervin, microwave physics
Hoegy, Walter R, space physics
Hoffman, Robert A, space physics
Holt, Stephen S, astrophysics
Jennings, Donald Edward, molecular spectroscopy
Jones, Frank Culver, cosmic ray physics, theoretical astrophysics
Jones, George R, theoretical physics, psychophysics
Jordan, Stuart Davis, solar physics
Kastner, Sidney Oscar, physics
Kniffen, Donald Avery, astrophysics
Kostiuk, Theodor, space physics
Kyle, Herbert Lee, atmospheric physics
Langel, Robert Allan, geophysics
Leckrone, David Stanley, astrophysics
Ledley, Brian G, magnetospheric physics
Lee, Yuen San, foods, biochemistry
Lipton, Samuel Harry, biochemistry, organic chemistry
Lowman, Paul Daniel, Jr, astrogeology, photogeology
Maier, Eugene Jacob Rudolph, geophysics
Malitson, Harriet Hutzler, astronomy
Maran, Stephen Paul, astrophysics
McCracken, Curtis W, astronomy, physics
McDonald, Frank Bethune, physics
Mead, Gilbert Dunbar, geophysics
Mead, Jaylee Montague, astronomy
Meredith, Leslie Hugh, space physics
Minzner, Raymond Arthur, aeronomy, meteorology
Mumma, Michael Jon, experimental atomic physics, molecular physics
Musen, Peter, astronomy
Nagy, Theresa Ann, astronomy
Nakada, Minoru Paul, physics
Ness, Norman Frederick, space physics
Neupert, Werner Martin, spectroscopy, solar physics
New, John Calhoun, applied mechanics
Nordberg, William, space physics
Northrop, Theodore George, space physics
Oglivie, Keith W, physics, space science
O'Keefe, John Aloysius, astronomy
Omidvar, Kazem, atomic physics
Pellerin, Charles James, Jr, astrophysics
Philpotts, John Aldwyn, geochemistry
Piccioni, Grace Lee, cell physiology
Pieper, George Francis, Jr, nuclear physics, space sciences
Price, John Charles, plasma physics, meteorology
Ramaty, Reuven, astrophysics
Rango, Albert, astrophysics, watershed management, remote sensing
Rosenbaum, Bernard, aerospace sciences
Rudnick, Paul, physics
Sakurai, Kunitomo, solar physics, radio astronomy
Salomonson, Vincent Victor, meteorology, hydrology
Schmid, Lawrence Alfred, theoretical physics
Schnugge, Thomas Joseph, physics
Short, Nicholas Martin, geology
Siry, Joseph William, mathematics, astrophysics
Stecker, Floyd William, physics
Stecker, Theodore P, astronomy
Stern, David P, space physics
Stief, Louis J, photochemistry
Sugiura, Masahisa, space physics
Temkin, Aaron, atomic physics
Thekaekara, Matthew Pothen, physics
Thomas, Ann P, organic chemistry
Thomas, Herman Holt, geochemistry
Thomas, Roger Jerry, solar physics
Thompson, David John, astrophysics
Trombka, Jacob Israel, radiation physics, space physics
Tschunko, Hubert F A, physics, optics
Underhill, Anne Barbara, astrophysics
Vette, James Ira, physics
Vonbun, Friedrich Otto, physics, mathematics
Walter, Louis S, geochemistry

Weber, Richard Rand, radio astronomy, instrumentation
Webster, William John, Jr, planetary sciences
Wende, Charles David, space physics
West, Donald K, astrophysics
Woodgate, Bruce Edward, astrophysics
Yap, Fung Yen, computer sciences, science education

HANCOCK
Krestensen, Elroy R, entomology

HAGERSTOWN
Maruyama, Hitoshi, biochemistry
Rogers, Wallace Edward, plant pathology
Taylor, Harold Nathaniel, chemistry

HAVRE DE GRACE
Grubbs, Frank Ephraim, mathematical statistics, operations research
Wason, Satish Kumar, physical inorganic chemistry

HUNT VALLEY
Abramson, Irwin Jerome, microbiology
Hall, Richard Leland, food science, toxicology
Pareles, Stephen Ronald, chemical instrumentation
Stahl, William Herbert, chemistry
Zeger, William Nathaniel, natural products chemistry

HYATTSVILLE
Abercrombie, Warren Fulton, zoology
Allen, J Frances, ecology
Barnett, Douglas Elton, entomology
Benjamin, Chester Ray, mycology
Bertrand, Kenneth John, physical geography, geography of Polar Regions
Bram, Ralph A, entomology
Brazzel, James Roland, entomology
Brown, Joshua Robert Calloway, cell biology
Carlisle, Frank Jefferson, Jr, soil science
Carter, Edward Pendleton, botany
Casady, Robert Barnes, physiology
Daum, Richard J, agricultural economics, insect toxicology
Drechsler, Charles, botany
Ehrlich, Gertrude, mathematics
Embody, Daniel Robert, statistics
French, Leslie Howson, economic geology
Hanley, John Bernard, agricultural chemistry
Hardesty, John Oliver, agricultural chemistry
Heer, Raymond Robert, Jr, atmospheric physics
Hoffman, Clarence Howard, entomology
Imle, Ernest Paul, plant pathology
Kahn, Robert Philip, plant pathology
Kates, Kenneth Casper, Sr, parasitology
Kellogg, Charles Edwin, soil science
Klemme, Dorothea Elizabeth, microbiology, chemistry
Morgan, Omar Drennan, Jr, plant pathology
Omohundro, Richard E, epidemiology, poultry pathology
Page, Louise, nutrition
Patterson, Charles Meade, geology
Pringer, Albert Aloysius, Jr, horticulture
Ramsay, Maynard Jack, entomology
Richardson, Martha, nutrition
Schricker, Robert Lee, veterinary medicine, microbiology
Strickling, Edward, soils
Thompson, Clarence Henry, Jr, veterinary medicine

LIAMSVILLE
Vickers, James Hudson, veterinary medicine, pathology

INDIAN HEAD
Browning, Joe Leon, chemical engineering
Fauth, Mae Irene, analytical chemistry
Malotky, Lyle Oscar, polymer chemistry
Nayak, Ramesh Kadbet, reproductive physiology
Rochlin, Phillip, chemistry, information science
Tompa, Albert S, analytical chemistry, physical chemistry
Wilmot, George Barwick, physical chemistry

JOPPA

Mednick, Morton L, organic chemistry

White, Kevin Joseph, molecular physics

KENNEDYVILLE

Smith, Paul Edgar, Jr, biophysics, physiology

KENSINGTON

Aronson, Casper Jacob, physics, explosives

Bethke, Philip Martin, geology

Bruner, Harry Davis, radiobiology, research administration

Cockrell, Beverly Yvonne, veterinary pathology, electron microscopy

Eng, Chee Ping, immunology

Fitzhugh, Oscar Garth, pharmacology

Goodwin, William Jennings, medical entomology

Gordon, Robert Sirkosky, Jr, physiology

Gull, Cloyd Dake, information science

Hart, Earl Ross, pharmacology

Hayes, Joseph Edward, Jr, biochemistry, research administration

Hoisington, Laurence Earl, physics, operations research

Holper, Jacob Charles, cancer, immunology

Kafka, Fritz, applied statistics

Klement, Alfred William, Jr, health physics, environmental sciences

Lill, Gordon Grigsby, marine geology

Matheson, Harry, immunology

Maurer, Bruce Anthony, immunobiology

Meyersburg, Herman Arnold, psychiatry

Mielenz, Klaus Dieter, optics, spectroscopy

Moulton, James Frank, Jr, research administration

Robinson, Richard Carleton, Jr, operations research, applied statistics

Schoonover, Irl Corley, inorganic chemistry

Seibold, Herman Rudolph, veterinary pathology

Slocum, Glenn Gerald, bacteriology

Spealman, Clair Raymond, physiology

Weir, Robert James, Jr, toxicology

Zimmerman, Eugene Munro, virology

KINGSVILLE

Buckles, Lawrence Calvin, physical organic chemistry

LA VALE

Harman, Dan M, forest entomology

Orlick, Charles Alex, physical chemistry

LANHAM

Wadleigh, Cecil Herbert, plant physiology

LAUREL

Adrian, Frank John, physical chemistry

Bader, Frank, physics

Bird, Joseph Francis, theoretical physics

Blau, Edmund Justin, information science, scientific bibliography

Blevins, Gilbert Sanders, experimental physics

Blum, Norman Allen, solid state science

Bohandy, Joseph, solid state physics

Bostrom, Carl Otto, space physics

Cavagna, Giancarlo Antonio, organic chemistry, paper chemistry

Clark, Donald Ray, Jr, pollution biology

Clark, Paul Enoch, physical chemistry

Cornett, Richard Orrin, physics, communications science

Crissey, Walter Ford, biology

Dieter, Michael Phillip, environmental physiology

Ehrlich, Louis William, numerical analysis

Elrod, McLowery, applied mathematics

Emch, George Frederick, physics

Erickson, Ray Charles, wildlife research

Farrell, Richard Alfred, biophysics

Flower, Robert Walter, medical physics

Follin, James Wightman, Jr, physics

Foner, Samuel Newton, chemical physics, mass spectrometry

Friedman, Morton Harold, biophysics

Fristrom, Robert Maurice, chemistry, physics

Geis, Aelred Dean, biology

Getchell, Bassford Case, applied mathematics

Gibson, Ralph Edward, physical chemistry

Gilbert, Myron B, communications science, science administration

Gray, Ernest Paul, plasma physics, theoretical physics

Guier, William Howard, theoretical physics

Hart, Robert Warren, theoretical physics

Hensler, Gary Lee, mathematical statistics

Hill, Freeman Kenneth, physics

Hochheimer, Bernard Ford, optics, spectroscopy, lasers

Hunger, Gunther K, organic chemistry

Hunter, Lawrence Wilbert, chemical physics

Ikeda, George J, biochemistry

Isensee, Allan Robert, soil science, plant physiology

Jen, Chih Kung, microwave physics

Jette, Archelle Norman, physics

Kershner, Richard Brandon, mathematics

Kim, Boris Fincannon, molecular spectroscopy, lasers

Krimigis, Stamatios Mike, space physics

Lawson, Mildred Wiker, mathematics, computer sciences

Leffel, Claude Spencer, Jr, physics

Ludke, James Larry, environmental biology

Marton, Joseph, physical organic chemistry, wood chemistry

McEntire, Richard Willian, space physics

Meyer, Charles Frederick, mathematics, statistics

Murphy, John Cornelius, physics

Newton, Robert Russell, physics

O'Brien, Vivian, fluid dynamics

Ohlendorf, Harry Max, wildlife research

Poehler, Theodore O, solid state physics, lasers

Reichel, William Louis, environmental chemistry

Rich, Robert Peter, mathematics, computer sciences

Robbins, Chandler Seymour, ornithology

Rogers, John Patrick, zoology, wildlife ecology

Satkiewicz, Frank George, physical chemistry

Schroeder, Irvin Homer, physics

Serafin, John Augustus, nutrition

Shaw, Harry, Jr, applied mathematics

Silver, David Martin, theoretical chemistry

Sleight, Thomas Perry, computer science, spectrochemistry

Smith, Gary Leroy, nuclear physics, systems analysis

Stainer, Howard Martin, computer sciences

Stendell, Rey Carl, pollution biology

Stickel, Lucille Farrier, zoology

Stickel, William Henson, research, pollution biology

Stone, Albert Mordecai, plasma physics, microwave physics

Taylor, Robert Joseph, engineering physics

Thompson, Robert John, Jr, physical chemistry

Tropf, Cheryl Griffiths, applied mathematics

Trotter, Gordon Trumbull, computer sciences

Viernstein, Lawrence J, biomedical engineering

Walter, John Fitler, applied physics

Warsh, Kenneth Lee, physical oceanography

Westenberg, Arthur Ayer, chemical physics

LAYTONSVILLE

Kolstad, George Andrew, nuclear physics, earth sciences

MARLOW HEIGHTS

La Rue, Jerrold A, meteorology

Ruff, Irwin S, meteorology

MILLINGTON

Trimmer, John Dezendorf, physics

MT AIRY

Carpenter, Raymond Allison, molecular spectroscopy

Darneal, Robert Lee, earth sciences, chemical microscopy

MT RAINIER

Greene, Michael P, solid state physics, research administration

NEW CARROLLTON

Daly, Joseph Francis, mathematical statistics

Robinson, Malcom Emerson, geography

ODENTON

Hudson, Robert Leslie, organic polymer chemistry

OLNEY

Pelleu, George Bernard, Jr, microbiology

Valega, Thomas Michael, organic chemistry

OXFORD

Dalldorf, Gilbert, pathology

Johnson, Phyllis Truth, invertebrate pathology

Meigs, Frederick Madison, chemistry

Tubiash, Haskell Solomon, microbiology

OXON HILL

Barnes, Charlie James, physical chemistry, organic chemistry

Cox, Anna Lucile, physical chemistry

Drummeter, Louis Franklin, Jr, optics

Fernstrom, John Richard, geography

Fusillo, Matthew Henry, microbiology

North, Harper Qua, physics

Palik, Edward Daniel, semiconductors

Schuldiner, Sigmund, physical chemistry

Stauch, John Edward, bacteriology

Walter, Dean Irving, chemistry

Williams, Frederick Wallace, analytical chemistry

Young, Frank Coleman, nuclear physics

PARKVILLE

White, Walter Finch, hydrology

PERRYVILLE

Van Der Heem, Peter, physical chemistry, chemical engineering

PHOENIX

Vennos, Mary Susannah, chemistry

PIKESVILLE

Graham, Charles Raymond, Jr, ichthyology, physiology

POOLESVILLE

Greenberg, Neil, ethology

Potkay, Stephen, laboratory animal medicine

PORT REPUBLIC

Hollowell, Eugene A, agronomy

Schultz, Leonard Peter, ichthyology

POTOMAC

Alsmeyer, Richard Harvey, meat science, biochemistry

Avena, Remedios M, biochemistry, lipid chemistry

Carter, Robert Sague, nuclear physics

Casella, Russell Carl, physics

Cohn, David Lionel, physics, theoretical biology

Dingman, Charles Wesley, II, cell biology, psychiatry

Engle, Robert Rufus, organic chemistry

Gelles, Isadore Leo, physics

Glick, J Leslie, health sciences

Holman, John Ervin, Jr, veterinary pathology

Jaffe, Harold, nuclear science

Kirshbaum, Amiel, bacteriology

Kratchman, Jack, geology

Lakein, Richard Bruce, mathematics

Landon, John Campbell, virology, cancer

Levine, Eugene, analytical statistics, operations research

Marvin, Robert Sidney, physical chemistry, rheology

Mohler, Irvin C, Jr, bacteriology

Murphy, John Francis, geology

Pecora, Louis Joseph, physiology

Pomerantz, Reuben, food chemistry

Reardon, John Devereaux, microbiology

PRINCE FREDERICK

Boynton, Walter Raymond, marine ecology

Setzler, Eileen Marie, marine ecology, fish biology

PRINCESS ANNE

Hopkins, Thomas Franklin, reproductive endocrinology

Quinn, Caroline Elisabeth, biochemistry

Rebach, Steve, animal behavior, biological oceanography

Smith, F Harrell, animal science, agricultural mechanics

Vaughn, Moses William, food technology

RIVERDALE

Chang, Charles Yu-Chun, pharmaceutical chemistry

Wells, William T, applied statistics, applied mathematics

ROCK HALL

Parsons, James Bayard, inorganic chemistry

ROCKVILLE

Alleman, Ray Starr, underwater acoustics, applied statistics

Andersen, Frank Alan, radiation biophysics

Anello, Charles, biostatistics, epidemiology

Asher, Irvin Mark, biophysics, lasers

Baer, Ledolph, oceanography, meteorology

Baier, Edward John, industrial hygiene

Baker, Carl Gwin, research administration

Baker, Charles Albert, research administration, biological chemistry

Baker, Leonard Samuel, geodesy

Banes, Daniel, pharmaceutical chemistry

Barile, Michael Frederick, microbiology

Barnstein, Charles Hansen, pharmaceutical chemistry

Barron, Charlie Nelms, veterinary pathology

Bensen, David Warren, soil chemistry

Bernstein, Emil Oscar, cell biology

Bester, John (Francis), pharmacology

Bestul, Alden Beecher, physical chemistry

Black, Roger Lewis, internal medicine

Bledsoe, John D, geophysics, engineering

Bockstahler, Larry Earl, virology, biophysics

Boss, Bruce David, physical chemistry

Bovee, Harley Howard, analytical chemistry, occupational health

Brackett, Frederick Sumner, physics

Brands, Allen J, pharmacology

Brands, Alvira Bernice, psychiatric nursing

Braude, Monique Colsenet, pharmacology, toxicology

Bricker, Jerome Gough, medical physiology

Brown, Bertram S, psychiatry

Brown, John A, Jr, meteorology

Brown, Robert Francis, biology, electron microscopy

Bruch, Carl William, microbiology

Bruckner, Benjamin Harry, biochemistry

Buras, Edmund Maurice, organic chemistry

Canfield, Norman L, meteorology, climatology

Casola, Armand Ralph, pharmaceutical chemistry, medicinal chemistry

Chafetz, Morris Edward, psychiatry

Charman, Howard Prentis, medical research

Chernov, Harvey Irwin, pharmacology

Cleland, Charles Frederick, plant physiology

Cline, James E, experimental nuclear physics

Cohen, Gerald Stanley, biomedical engineering

Cohen, Marvin Morris, solid state physics

Conklin, Glenn Ernest, physics

Rodney, William Stanley, nuclear physics, astrophysics

Schulkin, Morris, acoustics

Sharp, Gerald White, physics

Skidmore, Wesley Dean, biochemistry, radiobiology

Smith, Robert Leland, volcanology

Tabershaw, Irving R, preventive medicine

Conley, Bernard Edward, pharmacology, toxicology
Conley, Veronica Lucey, public health
Cosnides, George James, pharmacology
Cressman, George Parmley, meteorology
Crosby, Lon Owen, nutrition, nutritional biochemistry
Crout, John Richard, clinical pharmacology; public health administration
Cyr, Wilbur Howard, radiation genetics
De Leonibus, Pasquale S, oceanography, meteorology
DeLong, Merill B, optometry
Diesen, Carl Edwin, numerical analysis
Dighe, Shrikant Vishwanath, organic chemistry, biopharmaceutics
Dill, Robert Floyd, marine geology
Donovick, Richard, bacteriology
Duggar, Benjamin Charles, public health
DuPont, Robert L, Jr, psychiatry
Eisenhauer, Charles Martin, nuclear science
Elder, Robert Lee, radiological physics
Elmadjian, Fred, physiology
Endicott, Kenneth Milo, pathology
Epstein, Edward Selig, meteorology
Evans, Richard Castleman, electrochemistry
Fairweather, William Ross, statistics
Feith, Kenneth Edwin, applied statistics, acoustics
Finkelstein, Robert, preventive medicine, research administration
Finklea, John F, preventive medicine, research administration
Fisher, Gail Feinster, applied statistics, demography
Fivozinsky, Sherman Paul, nuclear physics
Flick, Donald Franklin, biochemistry
Flowers, Earl Shederick, environmental health, chemistry
Frieders, Robert B, zoology, endocrinology
Frost, Harold Maurice, III, acoustics, biophysics
Furukawa, George Tadaharu, thermodynamics
Gainer, Joseph Henry, virology
Galloway, William Don, psychophysics
Gantt, Elisabeth, plant physiology, biological structure
Gelberg, Alan, organic chemistry, information science
Gerber, Paul, microbiology
Gereben, Istvan B, oceanography
Gerin, John Louis, virology; biochemistry
Gilden, Raymond Victor, immunology, genetics
Giovacchini, Rubert Peter, medicine
Gottwald, Jimmy Thorne, acoustics
Graber, George, animal nutrition
Grady, Lee Timothy, analytical chemistry, pharmaceutical chemistry
Griffin, Thomas Ponton, veterinary virology
Griffith, Jack Dee, preventive medicine
Gross, Michael Alan, biochemistry
Grove, George Richard, physics
Gryder, Rosa Meyersburg, biochemistry
Hall, Ferguson, meteorology
Hamilton, Leroy Leslie, biomedical engineering, electrical engineering
Hanlon, John Joseph, public health
Hardin, Clyde D, physics
Hardy, Lester B, Jr, physiology; toxicology
Hay, Robert J, cell physiology; developmental biology
Heinrich, Kurt Francis Joseph, chemistry
Heller, William Mohn, pharmacy
Herner, Albert Erwin, clinical chemistry, radiochemistry
Hicks, Steacy Dopp, oceanography
Hollies, Norman Robert Stanley, physical chemistry
Hopkins, Thomas A, physical biochemistry
Hoppes, Dale Du Bois, nuclear physics
Houk, Albert Edward Hennessee, biochemistry, biophysics
Howie, Donald Lavern, internal medicine, research administration
Hull, William Ballou, medical entomology
Hutchison, Marilyn Kay, occupational health
Hwang, Shuh-Wei, mycology
Idyll, Clarence Purvis, fish biology
Inscoe, Joseph Kenneth, pharmacology, biochemistry
Israel, Jay Elliot, computer sciences, applied mathematics
Jacobs, Woodrow Cooper, meteorology
Jacobson, Keith Hazen, pharmacology, toxicology

Jensen, Erling Maurice, microbiology
Jessup, Gordon L, Jr, biostatistics
John, George Swisher, chemical physics, chemical engineering
Johnson, Fatima Nunes, organic chemistry
Jones, Bryant Lee, medicine, cancer
Jong, Shung-Chang, mycology
Kagan, Benjamin, organic chemistry
Kappe, David Syme, physical chemistry, radiochemistry
Katcher, David Abraham, physics, operations research
Kazper, Meyer, biophysics, computer science
Kawin, Bergene, physiology
Kelsey, Frances Oldham, pharmacology
Kerr, Jerome, virology
Kilgore, William Arlow, physics
King, R Maurice, Jr, mathematics, statistics
Klein, William Herbert, plant physiology
Knapp, William Arnold, Jr, pharmacology; toxicology
Knobl, George Martin, Jr, organic chemistry
Kramer, Morton, biostatistics, epidemiology
Kruse, Jurgen M, chemistry
Kunkuman, Charles Simon, medicinal chemistry
Laden, Karl, biochemistry
Lamar, Jule K, toxicology
Lavappa, Kantharajapura S, biology
Lavoie, Ronald Leonard, meteorology
Leach, William Matthew, cell biology; radiobiology
Lepovetsky, Barney Charles, science administration
Leventhal, Carl M, neurology; neuropathology
Levin, Gilbert Victor, environmental health, engineering
Lieblein, Julius, mathematical statistics
Litt, Bertram D, biostatistics
Locke, Ben Zion, applied statistics, epidemiology
Long, Cedric William, biochemistry, cell biology
Love, Daniel Lindsley, radiochemistry
Luginbyhl, Thomas Terence, occupational health, information science
Lukasik, Stephen Joseph, physics
Lundin, Frank E, Jr, epidemiology
Lynch, Cornelius James, biostatistics, operations research
Lytle, Carl David, biophysics
MacDonald, William E, Jr, pharmacology; toxicology
Mangum, Billy Wilson, physics
Malmberg, Marjorie Schooley, physical chemistry
Manian, Albert Ardashes, psychopharmacology
Margulies, Maurice, biochemistry, plant physiology
Martin, Edgar J, tropical medicine, pharmacology
Maxfield, Myles, experimental medicine, biophysics
Maycock, John Norman, solid state physics, chemical physics
McGovern, Wayne Ernest, meteorology
McKinstry, Donald W, microbiology, chemotherapy
Means, Lynn L, meteorology, physical science
Meyer, Harry Martin, Jr, virology; pediatrics
Meyer, Richard Irwin, food science, food technology
Miller, John Frederick, meteorology
Mirabito, John A, meteorology
Moore, Roscoe Michael, Jr, epidemiology
Moriyama, Iwao Milton, public health
Morton, John Dudley, environmental health
Morton, Joseph James Pandozzi, toxicology
Mowry, David Thomas, industrial chemistry
Murray, William Sparrow, science administration
Ng, Lorenz Keng-Yong, neuropsychiatry
Noble, Ernest Pascal, psychiatry, biochemistry
Offutt, William Franklin, operations research
O'Meara, Francis Edmund, experimental physics
Oroszlan, Stephen, immunochemistry, biochemical pharmacology
Otley, Kurt O, inorganic chemistry
Pahl, Herbert Bowen, public health administration
Palazzolo, Matthew Joseph, inhalation toxicology

Paras, Peter, radiological health, nuclear medicine
Parkhie, Mukund Raghunathrao, veterinary medicine, endocrinology
Paule, Robert Charles, physical chemistry
Persinos, Georgia J, pharmacognosy
Polachek, Harry (Aaron), applied mathematics
Pollin, William, psychiatry, psychoanalysis
Potash, Louis, medical microbiology
Powell, Charles Herbert, industrial hygiene
Purdom, James Francis Whitehurst, meteorology
Putterman, Gerald Joseph, biochemistry, protein chemistry
Puziss, Milton, microbiology
Quest, John Anthony, neuropharmacology
Rader, Charles Allen, surface chemistry, organic chemistry
Rafajko, Robert Richard, virology
Rasmussen, Eugene Martin, meteorology
Rasmussen, James L, meteorology
Ray, Richard G, geology
Rechcigl, Miloslav, Jr, nutritional biochemistry, animal nutrition
Reif, Van Dale, pharmaceutical chemistry
Rockoff, Maxine Lieberman, mathematics
Roddy, Martin Thomas, cell biology
Roedel, John Morgan, biology
Rucker, James Bivin, geology, oceanography
Rugh, Roberts, zoology, radiology
Ruskin, Arthur, cardiovascular diseases
Sassoon, Humphrey Frederick, nutrition
Schmidt, Alexander Mackay, internal medicine
Schroeder, Anita Gayle, applied statistics
Schwartz, Anthony Max, surface chemistry
Semmes, Josephine, physiological psychology
Seubold, Frank Henry, Jr, organic chemistry; public health administration
Shaheen, Donald G, analytical chemistry
Sherrod, John, mathematics, information science
Shroff, Arvin Pranlal, pharmaceutical chemistry
Shropshire, Walter, Jr, biophysics, plant physiology
Siguel, Eduardo Nestor, biostatistics, operations research
Silverman, Charlotte, epidemiology
Simons, Daniel J, microbiology, invertebrate physiology
Sirken, Monroe Gilbert, statistics, applied statistics
Skelly, Jerome Philip, Sr, biopharmaceutics
Smith, Ieuan Trevor, chemistry
Smith, Roland F, zoology
Sokol, Philip Edward, polymer chemistry
Southard, Curtis Glenn, psychiatry
Steere, Richard C, meteorology; physical oceanography
Stein, Harvey Philip, organic chemistry
Stevens, David Arthur, bacteriology; virology
Stirewalt, Margaret Amelia, medical parasitology
Straat, Patricia Ann, biochemistry, enzymology
Stratmeyer, Melvin Edward, radiobiology, biochemistry
Strier, Murray Paul, physical chemistry
Szara, Stephen Istvan, psychopharmacology
Taber, Robert William, oceanography, marine geology
Taylor, Eugene Alfred, geodesy
Taylor, Jack Crossman, animal husbandry
Tepper, Lloyd Barton, toxicology, occupational health
Teske, Richard H, veterinary medicine, toxicology
Thompson, Donald Leroy, health physics
Ting, Robert Chin-Yao, microbiology
Tocus, Edward C, pharmacology
Tolgyesi, Eva, organic chemistry, polymer chemistry
Tolgyesi, William Steven, organic chemistry
Townsend, John William, Jr, physics
Trapani, Robert-John, immunology
Triantaphyllopoulos, Eugenie, medicine, biochemistry
Tropf, William Jacob, optics, military systems
Usdin, Earl, psychopharmacology, pharmacology
Usdin, Vera Rudin, biochemistry
Valerino, Donald Matthew, pharmacology, toxicology

Van Arsdel, William Campbell, III, pharmacology
Van Dreal, Paul Arthur, biochemistry, cytology
Van Houweling, Cornelius Donald, veterinary medicine
Varadi, Peter Ferencz, physical chemistry
Wade, George Wesley, oral surgery
Waldrop, Francis N, medicine
Warsh, Catherine Evelyn, oceanography
Weiss, Martin, geology
Weistrop, Jessie Syd, plant physiology, cell biology
Wells, John Morgan, III, marine biology
Wessel, Carl John, biochemistry
White, Julius, organic chemistry
White, William Victor, physiology, medicine
Willette, Robert Edmond, medicinal chemistry, organic chemistry
Williams, Scott Lansing, meteorology
Wills, James Henry, pharmacology
Wolters, Robert John, chemistry
Wolford, Richard Kenneth, chemistry, pharmaceutical chemistry
Wood, Fergus James, geophysics
Wu, Roy Shih-Shyong, cell biology
Wyckoff, Harold Orville, physics
Yao, Kenneth Tsoong-Sieu, genetics
Yule, Herbert Phillip, nuclear chemistry, computer science
Zaratzian, Virginia Louis, pharmacology, toxicology

ROYAL OAK
Fife, Earl Hanson, Jr, immunology

ST MARY'S CITY
Novotny, Robert Thomas, ecology; paleoecology

SALISBURY
Carte, Ira F, genetics
Craig, Frank Rankin, veterinary medicine
Estes, Edna E, botany, microbiology
Fusaro, Bernard A, mathematics
Hackert, Raymond A, analytical chemistry
Kantzes, James (George), plant pathology
Kernaghan, Roy Peter, biology, genetics
Van Breemen, Verne Leroy, histology; cytology

SANDY SPRING
Sowder, Arthur Merrill, forestry

SEABROOK
Bryant, Robert William, applied mathematics

SILVER SPRING
Ablard, James Elbert, chemistry
Alaka, Mikhail A, meteorology, mathematics
Allgaier, Robert Stephen, solid state physics
Alpern, Harvey Albert, solid state physics
Andrews, John Scott, veterinary parasitology
Avery, William Hinckley, physical chemistry, physics
Baker, Roger Denio, pathology
Barasch, Murray Leonard, theoretical physics
Barrientos, Celso Saquinan, oceanography
Bartelli, Lindo Joseph, soil science, soil conservation
Beach, Eugene Huff, nuclear physics, engineering management
Belson, Henry S, physics
Berch, Julian, chemistry
Bernecker, Richard Rudolph, physical chemistry
Bis, Richard F, solid state physics
Blaistein, Ira M, underwater acoustics
Blake, Lamont Vincent, microwave physics, radio engineering
Brown, Frank Burkhead, physics
Brown, Robert Eugene, physical chemistry
Bryant, Jay Clark, biochemistry
Carver, Gary Paul, experimental solid state physics
Cawley, Robert, theoretical mechanics
Cha, Moon Hwa, physics
Christian, Ermine America, theoretical physics
Clark, Arthur Edward, solid state physics

Clark, Robert Alfred, meteorology, civil engineering
Cleveland, Merrill L, entomology
Cook, William H, operations research
Cullen, James Robert, solid state physics
Dacons, Joseph Carl, organic chemistry
Dahlstrom, Robert Kirchner, physics
Darwin, James T, Jr, mathematics
Davis, Edward Alex, operations research
Davis, Henry McRay, analytical chemistry, physical chemistry
Davis, Joseph Berry, research administration
Dawson, Reed, mathematical statistics, operations research
Dayhoff, Edward Samuel, physics, military systems
Dayhoff, Margaret Oakley, biochemistry
deMonsabert, Winston Russel, chemistry, research administration
Dietemann, Allan B, chemistry
Dixon, Jack Richard, physics
Dribin, Daniel Maccabaeus, algebra
Elliott, William Paul, meteorology
Enig, Julius William, mathematics
Farr, Marion Margaret, parasitology
Fischell, Robert E, space physics, biomedical engineering
Fishbein, Morris, food bacteriology
Fisher, Pearl Davidowitz, biostatistics
Fowler, Harland Wade, Jr, medical entomology, invertebrate zoology
Fox, David William, mathematics
Gary, Robert, science administration, communications science
Glahn, Harry Robert, meteorology
Glaser, Harold, theoretical physics
Goering, Orville, applied mathematics, computer science
Goldberg, Benjamin, physics
Gordon, Daniel Israel, electromagnetism
Gordon, Tavia, analytical statistics
Greenspan, Martin, acoustics
Hallgren, Richard E, meteorology, physics
Hammond, Gordon Leon, atomic physics, spectroscopy
Hartmann, Bruce, polymer physics, acoustics
Haupt, Ralph Freeman, astronomy
Haynes, James Lester, agronomy
Hazzard, DeWitt George, animal nutrition, biochemistry
Houston, Bland Bryan, Jr, solid state physics
Huddleston, Charles Martin, nuclear physics
Hudson, George Elbert, applied physics, mathematics
Hurlburt, Evelyn McClelland, medical microbiology
Hodgson, Harlow James, agronomy, animal science
Johnson, Edward Miles, Jr, microbial genetics
Johnson, Russell Dee, Jr, operations research
Hoffsommer, John C, organic chemistry
Holmes, David W, geology
Hooker, Marjorie, geology
Horowitz, Emanuel, polymer chemistry
Horwitz, William, food chemistry
Jacobs, Sigmund James, physics, fluid dynamics
James, Ralph Paul, meteorology
Jefferson, Donald Earl, physics, physical oceanography
Johnson, David Harley, applied physics, fluid dynamics
Kamlet, Mortimer Jacob, physical organic chemistry
Kaplan, Lloyd Allan, physical organic chemistry
Katz, Sidney Marco, data processing, computer sciences
Kaufman, Samuel, physical chemistry
Kehr, Clifton Leroy, organic chemistry, polymer chemistry
Kessler, Bernard Von, lasers
Kirk, John Gallatin, solar physics, aerospace sciences
Klein, William H, meteorology
Klingebiel, Albert Arnold, soil science
Kneale, Samuel George, mathematics
Kossiakoff, Alexander, physical chemistry
Kurzweg, Hermann Herbert, aerodynamics, thermodynamics
Land, David John, theoretical physics
Lee, Ronald Norman, surface physics
Leonard, Robert Meyer, pharmacognosy, pharmacology
Leventhal, Edwin, solid state physics
Libelo, Louis Francis, theoretical physics
Maccabee, Bruce Sargent, thermodynamics, electrooptics
Madigosky, Walter Myron, physics
Mahan, Archie Irvin, physics
Mandel, John, mathematical statistics
Marans, Nelson Samuel, chemistry
Markley, Francis Landis, theoretical physics
Martin, James Milton, geophysics
Martin, Margaret Eileen, biochemistry
Massey, Joe Thomas, biomedical engineering
McClure, Charles Frederick, statistical mechanics, operations research
Mead, Marshall Walter, analytical chemistry
Melnick, Donald A, physics
Meyer, James Henry, meteorology
Miller, C David, analytical chemistry
Miller, Paul R, plant pathology
Milton, Charles, geology
Mitchell, John Murray, Jr, meteorology
Mitchell, Michael A, solid state physics
Montalvo, Ramiro A, physics,
Moon, Milton Lewis, environmental sciences
Moorjani, Kishin, physics
Moseley, Donald Stillman, underwater acoustics
Moses, Edward Joel, underwater acoustics, operations research
Mullen, Joseph Matthew, chemical physics, astrophysics
Murino, Vincent S, meteorology, research administration
Myers, Peter Briggs, physics
Nagler, Kenneth Malcolm, meteorology
Noffsinger, Terrell L, meteorology, agriculture
Page, Chester Hall, physics
Paulhus, Joseph Louis Honore, meteorology
Peck, Eugene Lincoln, hydrology, meteorology
Petree, Ben, radiation physics
Pollorak, Andrew Stephen, operations research, physics
Pomerantz, Jacob, physics, aerospace engineering
Pore, Norman Arthur, meteorology
Potema, Thomas Andrew, physics
Pugh, Milton Earl, theoretical physics
Reid, Walter Phillip, applied mathematics
Reinhart, Frank Walter, plastics chemistry
Rozanski, George, geology
Salwin, Arthur Elliott, chemical physics, computer science
Scanlon, Wayne Walter, physics
Schonfeld, Hyman Kolman, public health
Schroeder, Frank, Jr, physics
Schulz, Alvin George, nuclear medicine,
Schwartz, John T, ophthalmology, epidemiology
Scofield, Norman Edward, nuclear physics
Seifer, Arnold David, applied mathematics, underwater acoustics
Sessoms, Faison Thomson, physics
Sheldon, Donald Russell, science administration
Shleien, Bernard, health physics
Shotland, Edwin, physics
Sigillito, Vincent George, applied computer science
Silbergeld, Mae Driscoll, space physics, computer sciences
Singewald, Quentin Dreyer, economic geology
Smith, Floyd Franklin, entomology
Smith, Ralph Grafton, pharmacology
Snavely, Benjamin Lichty, physics
Splitter, Earl John, veterinary parasitology
Stanberson, Bradford Roy, chemistry
Steinberg, Joseph, mathematical statistics
Stewart, Kent Kallam, biochemistry
Stimler, Morton, physics, engineering
Stoloff, Leonard, biochemistry
Strauser, Wilbur Alexander, physics, mathematics
Sutherland, William Harrison, operations research
Tanada, Takuma, plant physiology
Tarshis, Irvin Barry, parasitology, zoology
Taub, Edward, neuropsychology
Tepper, Morris, meteorology, science administration
Tepping, Benjamin Joseph, mathematical statistics, applied statistics
Travis, Dorothy Frances, cell biology
Tumilowicz, Joseph J, virology, immunology
Turner, Mortimer Darling, economic geology, engineering geology
Tuve, Richard Larsen, physical chemistry, inorganic chemistry
Urick, Robert Joseph, underwater acoustics
Van der Hoven, Isaac, meteorology
Van Tuyl, Andrew Heuer, mathematics
Waddell, Mathews Cary, mathematics
Wadey, Walter Geoffrey, physics
Walker, Ronald Elliot, physics, physical chemistry
Walner, Arthur H, mathematical statistics
Wang, Frederick E, chemistry, physics
Weinstein, Alexander, mathematics
Weinstein, Marvin Stanley, acoustics
Wertz, James Richard, theoretical astrophysics, astronautics
Westman, Ragnar Theophile, hospital administration, public health
White, Blanche Babette, chemistry
Wineland, William Clemard, physics, research administration
Wolpert-DeFilippes, Mary Katherine, pharmacology
Young, George Anthony, meteorology
Zerbe, John Irwin, wood science, wood technology
Zisman, William Albert, physical chemistry
Zusman, James, Fred Selwyn, mathematics

SOLOMONS

Cargo, David Garrett, zoology
Koo, Ted Swei-Yen, fish biology
Rose, Curt D, marine ecology, fisheries
Sprague, Victor, protozoology
Ulanowicz, Robert Edward, theoretical biology

SPARROWS POINT

Reichenbach, George Suter, Jr, industrial hygiene

SUITLAND

Coleman, Charles R, oceanography
Fritz, Sigmund, meteorology
Hanson, Robert Harold, statistics, mathematics
Johnson, Harry McClure, meteorology, oceanography
McClain, Ernest Paul, meteorology
Morse, Burt Jules, applied mathematics
Oliver, Vincent J, meteorology
Porter, John M, meteorology, engineering
Rao, P Krishna, meteorology,
Strong, Alan Earl, oceanography
Weinreb, Michael Philip, atmospheric physics

TAKOMA PARK

Brace, Kirkland, radiobiology
Dixon, Peggy A, physics
Fatiadi, Alexander Johann, organic chemistry
Fischer, Irene Kaminka, geodesy
Ford, Richard Alan, physical organic chemistry
Futcher, Anthony Graham, biosystematics, natural history
Gentry, Robert Vance, geochemistry, nuclear geophysics
Harris, Lester Earle, Jr, herpetology
Hoffman, William Martin, pesticide chemistry
Lincicome, David Richard, parasitology, physiology
Menefee, Robert William, science education
Naibert, Zane Elvin, science education, chemistry
Pratt, Ernest Fay, organic chemistry
Pratt, Yolanda Tota, organic chemistry
Scorpio, Ralph M, biochemistry, physiology
Sickels, Margaret Hines, zoology

TIMONIUM

Gabbert, Paul George, biochemistry, toxicology
Harrell, George Thomas, Jr, medical education
Kirby, William Henry, Jr, physical medicine, industrial engineering
Rudolph, Ray Ronald, mathematics
Skrabek, Emanuel Andrew, energy conversion, physical chemistry

TOWSON

Berlinrood, Martin, developmental biology
Carret, Robert Laurent, organic chemistry, environmental chemistry
Caskey, Charles (Dixon), Jr, animal nutrition
Collagan, Robert Bruce, geology, astronomy
Coon, Geraldine Alma, mathematics
Cox, Louis Thomas, Jr, science education
Habermann, Helen Margaret, plant physiology
Hassett, Charles Clifford, physiology
Huffington, Jesse Marion, horticulture
Igusa, Jun-Ichi, mathematics
Kask, Uno, inorganic chemistry
Koppelman, Elaine, mathematics
Lacy, Ann Matthews, microbial genetics
Michel, Harry Oscar, biochemistry
Rodewald, Lynn B, organic chemistry
Salmon, James F, inorganic chemistry, physical chemistry
Siegel, Martha J, mathematics
Stroh, William Richard, analytical chemistry
Topping, Joseph John, analytical chemistry
Webb, Helen Marguerite, invertebrate physiology

UPPER MARLBORO

Cuniff, Patricia A, physical chemistry
McAtee, Lloyd Thomas, cell biology, zoology
McKee, Claude Gibbons, agronomy
Turner, Anne Halligan, analytical chemistry

WALDORF

Root, Charles Brian, physical chemistry, inorganic chemistry
Tiller, Richard Edward, environmental health

WASHINGTON GROVE

Gordon, Gary Donald, space physics

WEST HYATTSVILLE

Pierson, Dolores Lehmann, zoology, human ecology

WESTMINSTER

Achor, William Thomas, physics
Brown, Michael Mathison, plant physiology, cytology
Cross, David Ralston, organic chemistry, physical chemistry
Jones, Donald Eugene, analytical chemistry
Kerschner, Jean, genetics
Lightner, James Edward, mathematics, education
Reed, Mary Valedia, microbiology, cytology
Sturdivant, Harwell Presley, zoology
Yedinak, Peter Demerton, theoretical physics, solid state physics

WHEATON

Alvarez, Robert, analytical chemistry, physical chemistry
Gibson, Colvin Lee, parasitology
Levin, Irvin, physical chemistry
Snow, Milton Leonard, physical chemistry

WHITE MARSH

Egner, Donald Otto, physics, operations research

WHITE OAK

Burlinson, Nicholas Edward, physical organic chemistry
Connor, Joseph Gerard, Jr, engineering physics

Holden, James Richard, physical chemistry
James, Stanley D., electrochemistry
Kubose, Don Akeru, physical chemistry,
Lalos, George Theodore, explosives
O'Keeffe, David John, physics
Pastine, D John, theoretical physics
Phipps, Thomas Erwin, Jr., physics
Warfield, Robert Welmore, polymer chemistry
Winkler, Ernst Hans, physics
Winkler, Eva Maria, physics

WYE MILLS
Paul, Frederick William, optical physics
Shah, Shirish Kalyanbhai, radiation chemistry, physical chemistry

MASSACHUSETTS

ABINGTON
Waller, William Henry, psychiatry

ACTON
Crane, Robert Kendall, radiophysics, meteorology
Wechsler, Harry C., polymer chemistry
Westcott, Donald Elvin, food technology

AMHERST
Abbott, Herschel George, forestry
Agrios, George Nicholas, plant pathology, virology
Allen, Stephen Ives, mathematics
Anderson, Donald Lindsay, animal nutrition
Arbib, Michael A., computer sciences
Archer, Robert Raymond, applied mathematics, engineering mechanics
Archer, Ronald Dean, inorganic chemistry
Armacost, David Lee, mathematics
Armelagos, George John, biological anthropology
Arny, Thomas Travis, astrophysics
Atkinson, Lenette Rogers, plant cytology,
Bailey, Duane W., mathematics
Barker, Allen Vaughan, plant physiology, soil science
Barnes, Ramon M, analytical chemistry
Bartlett, Lawrence Matthews, vertebrate anatomy, ornithology
Bar-Zev, Asher, molecular biology
Beal, Virginia Asta, nutrition
Becker, William Bernard, entomology
Beebe, Ralph Alonzo, surface chemistry
Belt, Edward Scudder, sedimentology
Bennett, Emmett, biochemistry
Bennett, Mary Katherine, mathematics
Benson, Bruce Buzzell, chemical physics,
Bert, Mark Henry, nutritional biochemistry
Bierhorst, David William, plant morphology
Bigelow, Howard Elson, botany
Bigelow, Margaret Elizabeth Barr, mycology
Black, Donald Leighton, animal physiology
Black, Wallace Gordon, reproductive physiology
Bond, Robert Sumner, forest economics, resource economics
Borrego, Joseph Thomas, mathematics
Borton, Anthony, animal science
Bramlage, William Joseph, horticulture
Brandts, John Frederick, physical chemistry, biochemistry
Brehm, John Joseph, physics
Breusch, Robert Hermann, mathematics
Bromery, Randolph Wilson, geology, geophysics
Brophy, Gerald Patrick, mineralogy
Brower, Lincoln Pierson, ecology
Bruno, Merle Sanford, sensory physiology, science education
Buck, Ernest Mauro, meat science
Burke, Terence, geography
Button-Shafer, Janice, experimental high energy physics
Camerino, Pat William, biochemistry, academic administration
Canale-Parola, Ercole, microbiology
Cannon, George Wesley, organic chemistry
Carpino, Louis Albert, chemistry
Carritt, Dayton Ernest, chemistry, oceanography

Catlin, Donald E., mathematics
Chen, Yu Why, mathematics
Chien, James C W., physical chemistry
Clydesdale, Fergus Macdonald, chemistry, food science
Cohen, Haskell, mathematics
Cole, Charles Franklyn, fish biology
Cole, John Wallace, cultural anthropology, economic anthropology
Cole, Johnnetta B., anthropology
Coler, Robert A., limnology
Collins, Frances Wilmoth, organic chemistry
Cook, LeRoy Franklin, (Jr), physics
Cook, Thurlow Adrean, mathematics
Coombs, Margery Chalifoux, paleontology, vertebrate
Coppinger, Raymond Parke, biology, ecology
Cordes, James Martin, radio astronomy
Cox, Charles Donald, medical microbiology
Craker, Lyle E., plant physiology
Cullen, Helen Frances, mathematics
Curran, David James, analytical chemistry
Czarnecki, Reynold Bernard, microbiology
Damon, Richard Alan, Jr., biometrics, genetics
Davidson, Robert Bellamy, theoretical chemistry
Davis, Edward Lyon, botany
Decker, Joan Elise, protozoology, comparative biochemistry
Dempsey, Colby Wilson, physics, neurobiology
Dethier, Vincent Gaston, insect physiology
Dewey, Virginia Caroline, protozoology,
Dhar, Sachidula, high energy physics
Dickinson, David (James), mathematics
Dickinson, Leonard Charles, physical chemistry, biophysical chemistry
Dowell, Clifton Enders, microbiology
Duby, Robert T., animal physiology, reproductive physiology
Edman, John David, medical entomology
Edwards, Dallas Craig, population ecology
Edwards, Lawrence Jay, physiology,
Eisenberg, Murray, topology
Ellis, Richard Steven, mathematics
Esselen, William B, food science
Fagerson, Irving Seymour, food science
Farquhar, Oswald Cornell, economic geology
Feder, William Adolph, pollution biology
Fenner, Heinrich, animal nutrition
Fink, Richard David, physical chemistry, nuclear chemistry
Fischer, Mark Samuel, biology, biochemistry
Fisher, Stuart Gordon, ecology, limnology
Fogarty, John Charde, algebra
Foose, Richard Martin, geology
Ford, Norman Cornell, Jr., biophysics
Fortier, David Harry, anthropology
Foster, John McGaw, mathematics
Foulis, David James, mathematics
Fournier, Maurille Joseph, Jr., molecular biology, biochemistry
Fox, Thomas Walton, animal genetics,
Francis, Frederick John, food science
Fraser, Thomas Mott, Jr., anthropology
Fuller, Rufus Clinton, biochemistry
Gatsick, Harold Bailey, wood technology
Gaunt, Stanley Newkirk, animal genetics
Gawienowski, Anthony Michael, biochemistry
Gay, David Lawrence, physical chemistry
George, John Warren, inorganic chemistry
Gilgut, Constantine Joseph, plant pathology, horticulture
Godchaux, Walter, III, molecular biology
Gold, Richard Michael, neuropsychology
Goldberg, Stanley, physics, history of physics
Goldenberg, Harold Mark, physics
Goldhor, Susan, developmental biology, science education
Golowich, Eugene, physics
Gordon, Courtney Parks, astronomy
Gordon, Joel Ethan, physics
Gordon, Kurtiss Jay, astronomy
Goulock, Edward Vernon, Jr., physical chemistry, polymer chemistry
Greeley, Frederick, biology, wildlife management
Gross, Alan John, mathematical statistics, biological statistics

Gunner, Haim Bernard, environmental biology
Hadlock, Charles Robert, mathematics
Hafner, Everett M, nuclear physics
Hafner, James Allan, cultural geography, economic geography
Hall, Leo M, geology
Hallock, Robert B, low temperature physics
Halpern, Joel Martin, cultural anthropology, applied anthropology
Hankinson, Denzel J., food science
Hanson, John Francis, entomology
Harrison, Edward Robert, physics, astrophysics
Hartshorn, Joseph Harold, glacial geology
Havis, John Ralph, horticulture
Hayes, David R., number theory
Hayes, Kirby Maxwell, food technology
Hertz, Douglas Nelson, mathematics
Hexter, William Michael, biology, genetics
Hoadley, Robert Bruce, wood science & technology
Hoffman, David Allen, geometry
Holland, Samuel S, Jr., mathematics
Holmes, Francis William, plant pathology
Holmes, Robert Richard, inorganic chemistry
Holt, Stanley Carl, microbiology
Honigberg, Bronislaw Mark, zoology
Howe, George R., reproductive physiology
Hubert, John Frederick, geology
Hudson, Alfred Bacon, anthropology
Huguenin, George Richard, astronomy, astrophysics
Hultin, Herbert Oscar, food science
Hurt, Norman Edward, mathematical physics
Inglis, David Rittenhouse, theoretical physics
Irvine, William Michael, planetary atmospheres, radio astronomy
Isgur, Benjamin, soil conservation
Ives, Philip Truman, genetics
Jackson, Darrell R., solid state physics
Jacob, Henry George, Jr, mathematics
Jaffe, Howard William, geology
Janowitz, Melvin Fiva, algebra
Jennings, Paul Harry, plant physiology
Johnson, James Edward, fish biology,
Jones, Phillips Russell, physics
Karasz, Frank Erwin, polymer science,
Karfunkel, Perry, biophysical chemistry, developmental biology
Kennelly, James J., reproductive physiology
Kidder, George Wallace, biology,
Kilham, Eleanor, biochemistry
Killam, Mary Beth, plant physiology
Klekowski, Edward Joseph, Jr, botany,
Klingener, David John, zoology
Koller, Richard Robert, physics
Krause, Daniel Jr., chemical oceanography
Kreisler, Michael Norman, high energy physics
Kropf, Allen, biophysical chemistry
Krolow, Robert Vladimir, atomic physics
Ku, Hsu-Tung, topology
Ku, Mei-Chin Hsiao, topology
Kunkert, Esayas G., algebra
Lachman, William Henry, Jr, plant breeding
Langley, Kenneth Hall, biophysics
Larson, Joseph Stanley, wildlife biology
Lavalee, Lorraine Doris, mathematics
Leadbetter, Edward Renton, microbiology
Lenz, Robert William, polymer chemistry
Lessie, Thomas Guy, microbial physiology
Levin, Robert E., microbiology, food science
Lilly, John Henry, entomology
Lillya, Clifford Peter, organic chemistry
Lister, Bradford Carlton, population biology
Litsky, Bertha Yanis, microbiology, hospital administration
Litsky, Warren, bacteriology
Little, Henry Nelson, biochemistry
Livingston, Robert Blair, botany
Lockhart, James Arthur, plant physiology
Lockwood, Linda Gail, environmental biology, science education
Lord, William John, pomology
Lowry, Nancy, physical organic chemistry
Lyford, Sidney John, Jr., animal nutrition, biochemistry
MacConnell, William Preston, forestry
Mac Knight, William John, physical chemistry

Mader, Donald Lewis, forest soils, forest hydrology
Mange, Arthur P., genetics
Marcum, James Benton, animal genetics,
Marra, Allan A., wood technology
Marsh, William Ernest, mathematical logic
Mathieson, Alfred Herman, physics
Matthews, Herbert Maurice, physics
Mauldin, James Grenfell, pure mathematics
Maynard, Donald Nelson, plant physiology
McEwen, William Edwin, organic chemistry
McGill, George Emmert, geology
McWhorter, Earl James, organic chemistry
Mellen, William James, avian physiology
Miller, Bernard, organic chemistry
Miller, Lynn, microbial genetics
Moir, David Ross, botany
Moner, John George, cell physiology,
Morcock, Robert Edward, physiological ecology
Morse, Stearns Anthony, petrology
Mortlock, Robert Paul, microbial biology,
Mosher, Harold Elwood, horticulture
Motts, Ward Sundt, geology
Mount, Mark Samuel, plant pathology,
Mueller, William Samuel, microbiology
Mulcahy, David Louis, ecology,
Mullin, William Jesse, theoretical solid state physics
Munn, Nancy D., anthropology
Naegele, John Adam, environmental biology, entomology
Navon, David H., electronic physics, microelectronics
Nawar, Wassef W., food chemistry
Noden, Drew M., experimental embryology, neuroembryology
Nolan, Linda Lee, biological chemistry
Nordin, John Hoffman, biochemistry
Nutting, William Brown, acarology
Oakland, Gail Barker, mathematical statistics
O'Connor, William Brian, reproductive physiology
Oliver, John Walter, analytical chemistry
Oyewole, Saundra Herndon, microbiology
Palmer, John Derry, biological rhythms
Pease, Lila Gierasch, biophysical
Penchina, Claude Michel, solid state physics
Peters, Howard August, environmental health
Peters, Thomas Michael, entomology
Peterson, Gerald Alvin, physics
Peterson, Jerome S., public health
Pichanick, Francis Martin, atomic physics
Pitkin, Donald Stevenson, anthropology
Pitrat, Charles William, paleontology
Porter, Roger Stephen, physical
Potswald, Herbert Eugene, invertebrate embryology, cytology
Potter, Frank Elwood, food science, dairy
Price, Fraser Pierpont, chemistry
Prouix, Donald Allen, anthropology, archaeology
Quinton, Arthur Robert, experimental nuclear physics
Ragle, John Linn, physical chemistry
Randall, Charles Hamilton, mathematics,
Rausch, Harold, zoology
Rausch, Marvin D., organometallic chemistry, organic chemistry
Reed, Ellen Elizabeth, mathematics
Reed, Roger J., fish biology
Reiner, Albey W., microbiology, genetics
Rhodes, Arnold Densmore, forestry
Ricci, Benjamin, applied physiology
Rice, William Newell, plant pathology
Richason, George R, Jr., inorganic chemistry
Roberts, John Edwin, chemistry
Roberts, John Lewis, comparative physiology
Roberts, Larry Spurgeon, zoology, parasitology
Robinson, Peter, structural geology, petrology
Robinson, Trevor, biochemistry
Robison, Hope Howeth, soil microbiology
Rohde, Richard Allen, plant pathology

Rollason, Grace Saunders, embryology
Rollason, Herbert Duncan, histology
Romer, Robert Horton, low temperature physics
Rosen, Philip, biophysics
Rosenau, William Allison, plant nutrition
Rowell, Robert Lee, physical chemistry
Rudvalis, Arunas, algebra
Salzmann, Zdenek, ethnography
Sanchez, Isaac Cornelius, polymer physics
Sargent, Theodore David, zoology
Sawyer, Frederick Miles, food technology, nutrition
Schneider, Donald Leonard, biochemistry
Schuster, Rudolf Mathias, botany
Schwartz, Ira, biochemistry
Schweizer, Berthold, mathematics
Searcy, Dennis Grant, cell physiology
Sevoian, Martin, veterinary medicine
Shapiro, Seymour, botany
Shaw, Frank Robert, entomology
Siggia, Sidney, analytical chemistry
Silver, Marc Stamm, bio-organic chemistry
Simmons, Emory Guy, mycology
Skibinsky, Morris, mathematical statistics
Smith, Albert Charles, botany
Smith, Glenn Harold, physical chemistry, inorganic chemistry
Smith, Marion Estelle, entomology
Snedecor, James George, physiology
Snoeyenbos, Glenn Howard, veterinary medicine
Snyder, Dana Paul, vertebrate zoology
Soltysik, Edward A, atomic physics
Southwick, Franklin Walburg, pomology
Spielman, Arless A, animal physiology
Starr, Norton, mathematics
Stein, Otto Ludwig, plant morphogenesis
Stein, Richard Stephen, polymer chemistry
Stengle, Thomas Richard, physical chemistry
Stern, Arthur Irving, plant physiology, photobiology
Stewart, Gordon LeRoy, soil physics, hydrology
Stidham, Howard Donathan, physical chemistry
Stockton, Doris S, mathematics
Stoffolano, John George, Jr, entomology, neurobiology
Stone, Marshall Harvey, mathematics
Strong, John (Donovan), physics
Strother, Wayman L, mathematics
Stumbo, Charles Raymond, microbiology, food science
Su, Jin-Chen, topology
Swanson, Carl Pontius, cytogenetics
Swedlund, Alan Charles, biological anthropology
Swift, Arthur Reynders, physics
Taylor, Joseph Hooton, Jr, astronomy
Thomson, Cecil Lyman, agronomy
Vengris, Jonas, agronomy
Vietor, Donald Melvin, crop physiology
Vogl, Otto, polymer chemistry
Waggoner, Alan Stuart, biophysical chemistry
Wagner, Robert Wanner, mathematics
Walker, James Frederick, Jr, theoretical physics
Walker, James Willard, evolutionary biology
Walker, Robert W, microbiology
Wang, Ju-Kwei, mathematics
Wattenberg, Franklin Arvey, mathematics
Webb, Gregory Worthington, geology
Weeks, Martin Edward, agronomy
Westhead, Edward William, Jr, biochemistry
Whaley, Ross Samuel, forest economics
Whitney, Robert Byron, organic chemistry
Wilce, Robert Thayer, botany
Wilcox, Louis Van Inwegen, Jr, ecology
Wilder, Martin Stuart, microbiology
Wilkie, Richard W, geography
Wilson, Brayton F, botany, forestry
Wise, Donald U, geology
Wise, William Bernard, physical chemistry
Wood, John Stanley, chemistry

Woodbury, Richard Benjamin, anthropology
Woodcock, Christopher Leonard Frank, cell biology, botany
Woodhull, Ann McNeal, biophysics, neurobiology
Wun, Chun Kwun, biology
Wyse, Gordon Arthur, zoology, neurophysiology
Yost, Henry Thomas, Jr, radiation genetics
Zimmerman, William Frederick, cell biology
Zimmermann, Robert Alan, molecular biology, biochemistry
Zuckerman, Bert Merton, plant pathology

ANDOVER
Alexander, Edward Lawson, nuclear chemistry, radiation chemistry
Crabtree, Douglas Everett, mathematics
Frishman, Daniel, textile chemistry
Hempstead, Charles Francis, experimental physics, solid state physics
Minne, Ronn N, inorganic chemistry
Pike, Ronald Marston, organic chemistry

ARLINGTON
Castelli, John P, radio astronomy, solar physics
Dunning, James Morse, dentistry, public health
Goldberg, Gershon Morton, photographic chemistry
Haas, Howard Clyde, polymer chemistry
Mano, Koichi, theoretical physics
Rickter, Donald Oscar, organic chemistry
Russo, Richard F, operations research
Sonnichsen, Harold Marvin, organic chemistry
Widder, David Vernon, mathematics

ASHFIELD
Markle, Carrolle Anderson, plant anatomy
Woolf, Michael A, low temperature physics

ASHLAND
O'Connor, Thomas Lee, colloid chemistry

ATTLEBORO
Baboian, Robert, electrochemistry
Jost, Ernest, physical chemistry
Vaudo, Anthony Frank, physical chemistry

AUBURN
McQuarrie, Bruce Cale, algebra

BABSON PARK
Bowen, Earl Kenneth, mathematical statistics
Prindle, Bryce, microbiology

BEDFORD
Aarons, Jules, ionospheric physics
Armington, Alton, physical chemistry, fuel technology
Baird, Walter Scott, physics
Bedinger, John Franklin, physics
Bedo, Donald Eiro, physics
Bradford, John Norman, atomic physics, nuclear physics
Cabaniss, Gerry Henderson, geophysics, tectonics
Carr, Paul Henry, solid state physics
Comer, Joseph John, inorganic chemistry, electron microscopy
Conley, Thomas Daniel, physics
Cunningham, Robert M, meteorology
Curtis, Harold Ormand, physics
Dandekar, Balkrishna S, geophysics, aeronomy
Deutsch, Marshall Emanuel, clinical chemistry, nuclear medicine
Dewan, Edmond M, theoretical physics, applied mathematics
Dickinson, Stanley Key, Jr, geochemistry, solid state science
Drane, Charles Joseph, Jr, electromagnetism
Dybwad, Jens Peter, physical chemistry
Ebersole, John Franklin, optical physics
Eckhardt, Donald Henry, geophysics
Engelman, Arthur, atmospheric physics, meteorology

Eyges, Leonard James, theoretical physics
Fishman, Robert Sumner, mathematics
Fitzgerald, Donald Ray, atmospheric physics
Fougere, Paul Francis, geophysics
Frimper, Michael Howard, geology, hydrology
Gallagher, James Emerson, meteorology
Garing, John Seymour, molecular spectroscopy
Garrett, Henry Berry, space physics
Gauvin, Hervey Paul, physics
Gelman, Harry, physics
Gerlach, Alan Meyer, geophysics
Gianino, Peter Dominic, optical physics
Glover, Kenneth Merle, atmospheric physics, electronics
Golubovic, Aleksandar, chemistry
Greeley, Richard Stiles, physical chemistry
Hagenlocher, Arno Kurt, solid state physics
Hanson, Douglas MacArthur, biochemistry, molecular biology
Heroux, Leon J, physics
Hinteregger, Hans Erich, physics
Hornig, Arthur William, chemical physics
Huffman, Robert Eugene, physical chemistry
Innes, Frederick Rush, atomic physics, quantum mechanics
Izumi, Yutaka, meteorology
Jasperse, John R, physics
Kaiser, Christopher B, philosophy of science, computer science
Katayama, Daniel Hideo, molecular spectroscopy
Kohlberg, Ira, electromagnetism
Kolb, Charles Eugene, Jr, chemical kinetics, atmospheric chemistry
Legat, Wilhelm Hans, solid state physics
Loewenstein, Ernest Victor, optics, spectroscopy
Ludman, Jacques Ernest, solid state physics
Mack, Richard Bruce, applied physics
Marsh, Howard Stephen, applied physics
McClatchey, Robert Alan, atmospheric physics
Michael, Irving, space physics, nuclear physics
Miranda, Henry A, Jr, geoenvironmental science
Moses, Harry Elecks, theoretical physics
Mudrick, Stephen Edward, dynamic meteorology
Muller, Karl Frederick, applied mathematics
Murad, Edmond, physical chemistry
Nandy, Kalidas, geriatrics, neuroanatomy
Narcisi, Rocco S, aeronomy
Newburgh, Ronald Gerald, electromagnetism
Paine, Clair Maynard, biochemistry
Paulsen, Duane E, physical chemistry
Paulson, John Frederick, physical chemistry
Pazich, Philip Michael, space physics
Philbrick, Charles Russell, atmospheric physics, solid s tate physics
Picard, Richard Henry, lasers, quantum optics
Quinlan, Kenneth Paul, inorganic chemistry
Radoski, Henry Robert, plasma physics, magnetospheric physics
Rambauske, Werner, physics, astronomy
Rao, K V N, physics, electrical engineering
Ratney, Ronald Steven, industrial hygiene
Read, Philip Lloyd, physics, instrumentation
Record, Frank Alaster, air pollution, meteorology
Riecker, Robert E, geology, geophysics
Rubin, Allen Gershon, physics
Rush, Charles Merle, ionospheric physics
Safford, Richard Whiley, physics, systems analysis
Sagalyn, Rita C, space physics, ionospheric physics
Salisbury, John William, Jr, geology
Sauermann, Gerhard Otto, electrooptics
Schneider, Frederick Howard, pharmacology
Scholberg, Harold Milton, chemistry
Sherman, Christopher, physics
Shettle, Eric Payson, atmospheric physics
Silverman, Sam M, geophysics, physical chemistry
Smiltens, Juris, thermodynamics
Stakutis, Vincent John, atmospheric physics, optics
Stergis, Christos George, physics

Stidworthy, George H, biochemistry
Straub, Wolf Deter, solid state physics
Swider, William, Jr, aeronomy
Thomson, Ker Clive, geophysics, seismology
Thun, Rudolf Eduard, solid state physics
Toman, Kurt, ionospheric physics
Volz, Frederic Ernst, atmospheric physics
Walter, Charlton M, information science
Wanta, Raymond Casimir, meteorology, environmental sciences
Yannoni, Nicholas, physical chemistry
Yates, George Kenneth, space physics

BELCHERTOWN
Page, Alan Cameron, forest management
Page, Joanna R Ziegler, phycology

BELMONT
Apt, Charles Maurice, chemistry
Chan, David Siupoon, biochemistry
Douglass, Raymond Donald, mathematics
Ellingboe, James, biochemistry
Engel, Frank August, Jr, computer science
Folch-Pi, Jordi, biochemistry
Frazier, Shervert Hughes, psychiatry, psychoanalysis
Frondel, Clifford, mineralogy
Frondel, Judith W, geology, mineralogy
Garrard, Sterling Davis, pediatrics
Harvey, George Graham, mathematical physics
Hauser, George, biochemistry
Lees, Marjorie Berman, neurochemistry
Lex, Barbara Wendy, cultural anthropology
Lilley, Arthur Edward, astronomy
Macrakis, Michael S, electromagnetics, plasma physics
Mendelson, Jack H, medicine, psychiatry
Orowan, Egon, physics
Parkes, Alan Schofield, instrumentation
Pope, Alfred, neuropathology
Shahrik, H Arto, cytology, oral biology
Street, Jabez Curry, high energy physics
Trachtenberg, Michael Carl, neuroanatomy, neurophysiology
Whipple, Fred Lawrence, astronomy, space physics
Zacharias, Jerrold Reinach, physics, science education

BEVERLY
Adams, David Lawrence, science education
Boyer, William Montgomery, organic chemistry
Coolidge, Harold Jefferson, mammalogy, conservation
Harris, Miles Fitzgerald, meteorology, science writing
Hoover, M Frederick, organic polymer chemistry, paper chemistry
Meloni, Edward George, inorganic chemistry
Sullivan, Edward Augustine, inorganic chemistry
White, Frederick Elmer, physics

BILLERICA
Atkins, Jaspard Harvey, physical chemistry, analytical chemistry
Boonstra, Bram B, electrochemistry
Cotten, George Richard, polymer chemistry, physical chemistry
Friedenstein, Hanna, chemistry, information science
Heckman, Francis Austin, electron microscopy, microscopy
Medalia, Avrom Izak, colloid chemistry, rubber chemistry
Rivin, Donald, surface chemistry
Rozelle, Lee Theodore, physical chemistry
Stearns, Charles Edward, geology
Troiano, Paul Francis, physical chemistry

BOSTON
Aaron, Ronald, theoretical physics
Abelmann, Walter H, medicine
Abrams, Archie Adam, medicine
Abrams, Herbert L, radiology, physiology
Adams, Raymond D, neuropathology, neurology
Adelstein, Stanely James, radiation biophysics, nuclear medicine
Adolph, Alan Robert, neurophysiology
Afonsky, Dimitri Aleksandrovich, pathology

491

MASSACHUSETTS

Aguilera, Francisco Enrique, cultural anthropology
Aisenberg, Alan Clifford, oncology
Albert, Arthur Edward, mathematical statistics, applied mathematics
Albright, John T., biology dentistry
Albin, Henry Freeman, ophthalmology
Allen, James Richard, geomorphology
Allen, John Ernest, statistics
Alper, Milton H., anesthesiology, pharmacology
Alpers, Joseph Benjamin, biological chemistry, metabolism
Ames, Adelbert, III, neurophysiology, neurochemistry
Amos, Harold, bacteriology
Anastassakis, Evangelos M, solid state physics, spectroscopy
Anselme, Jean-Pierre L M, organic chemistry
Antoniades, Harry Nicholas, biochemistry
Argyres, Petros, theoretical physics
Arnason, Barry Gilbert Wyatt, immunology, neurology
Arnison, Paul Grenville, plant physiology
Arnowitt, Richard Lewis, theoretical physics
Aronow, Saul, biomedical engineering
Auld, David Stuart, organic chemistry, biological chemistry
Auslander, Bernice Liberman, mathematics
Austen, K Frank, internal medicine, immunology
Austen, William Gerald, surgery
Avery, Mary Ellen, medicine, pediatrics
Ayres, Barbara Chartier, cultural anthropology
Azpetia, Alfonso Gil, mathematics
Babior, Bernard M., biochemistry, hematology
Baden, Howard Philip, dermatology
Badger, Alison Mary, immunobiology
Bailey, Richard Hendricks, invertebrate paleontology
Baker, Edgar Eugene, Jr., medical microbiology
Baldwin, William Russell, optometry, physiology
Bandes, Dean, mathematics
Banks, Henry H., orthopedic surgery
Barger, Abraham Clifford, physiology, medicine
Barlow, Charles F., medicine
Barlow, John Sutton, neurophysiology, biophysics
Barnett, Guy Octo, medicine
Barosh, Patrick James, geology
Batchelder, Robert Bruce, geography
Beardsley, George Peter, bio-organic chemistry
Beck, Mae Lucille, organic chemistry endocrinology
Beck, William Samson, biology
Belamarich, Frank Alexander, cell biology
Bell, Jerry Alan, physical chemistry
Bell, Warner Duane, anatomy, cytology
Benacerraf, Baruj, immunology, experimental pathology
Bennett, Michael, experimental pathology, hematology
Benson, Herbert, cardiology
Berenberg, William, medicine
Berkowitz, Joan B, physical chemistry
Berliner, Martha D., microbiology, mycology
Berman, Elliot, organic chemistry
Berman, Herbert Joshua, physiology
Best, Alan C G, geography
Best, Troy Lee, vertebrate ecology, systematic zoology
Bethune, John Lemuel, physical chemistry, biophysics
Biggers, John Dennis, physiology
Bing, David H., immunology, protein chemistry
Birstein, Seymour J., surface chemistry, cloud physics
Bishop, Yvonne M., biostatistics
Bittman, Loran R., applied physics
Bjamgard, Bengt E., radiological physics
Black, Paul H., medicine, virology
Blackett, Donald Watson, mathematics
Blake, Jules, organic chemistry
Blanchard, Gordon Carlton, mathematics
Blaustein, Ernest Herman, microbiology
Bloomquist, Eunice, physiology
Blout, Elkin Rogers, biochemistry
Blumenstein, Michael, biochemistry

Bogoch, Samuel, biochemistry, psychiatry
Bojar, Samuel, psychiatry
Bolker, Ethan D., mathematics
Bonaventura, Maria Migliorini, chemistry
Cori, Carl Ferdinand, pharmacology, biochemistry
Borysenko, Myrin, immunobiology
Bougas, James Andrew, thoracic surgery, cardiopulmonary physiology
Bowen, Edward H Jr, occupational medicine
Bowers, Peter George, physical chemistry
Boyer, Norman Howard, medicine
Brantley, John Calvin, radiological health, radiochemistry
Braunwald, Eugene, internal medicine, cardiology
Brawerman, George, biochemistry
Brecher, Peter I., biochemistry, endocrinology
Brenner, Barry Morton, internal medicine
Brenner, John Francis, medical research medicine, nephrology
Broitman, Selwyn Arthur, microbiology, medical education
Brooks, John Robinson, surgery
Brosephini, Albert L, zoology
Brown, Henry, surgery
Brown, Robert Stephen, internal medicine, nephrology
Brody, Jerome Saul, pulmonary diseases
Brownlow, Arthur Hume, geology
Brownlee, Gordon Lee, medical physics
Brudevold, Finn, dentistry
Bucher, Nancy L R, medicine
Burchfiel, James Lee, neurophysiology
Burke, John Francis, surgery
Butler, Patrick Colin, soil fertility
Byrne, John Joseph, surgery
Cady, Blake, surgery, oncology
Cahill, George Francis, Jr., metabolism
Caldwell, Dabney Withers, geology
Callard, Gloria Vincz, comparative endocrinology
Callow, Allan Dana, surgery
Cameron, Barry Winston, paleoecology
Campenot, Robert Barry, neurobiology
Canellos, George P., internal medicine, oncology
Cannon, Donald Joseph, biochemistry
Carter, John Haas, optometry
Caskey, James Edward, Jr., meteorology
Cass, William Emerson, organic polymer chemistry
Castaneda, Aldo Ricardo, thoracic surgery, cardiovascular surgery
Castleman, Benjamin, pathology
Cathou, Renata Egone, immunology
Caulfield, James Benjamin, medicine, pathology
Cavazos, Lauro Fred, anatomy
Chang, Chin-Hai, cancer, immunochemistry
Chang, Te Wen, infectious diseases, virology
Chan-Palay, Victoria, neurosciences
Charpie, Robert Alan, theoretical physics
Chase, David Marion, theoretical physics, applied physics
Chesler, David Alan, medical physics
Chobanian, Aram V, cardiovascular diseases, internal medicine
Christensen, Thomas Gash, experimental biology
Chaykovsky, Michael, organic chemistry, endocrinology
Christian, Howard J., pathology
Chrzanowski, Francis Alan, physical pharmacy
Cheever, Francis Sargent, microbiology
Chen, Ching-Chih, information science, scientific bibliography
Chu, William Wei-Ling, applied mechanics, applied mathematics
Clarke, Richard Henry, physical chemistry
Cochin, Joseph, pharmacology
Codington, John F., biochemistry
Coffman, Jay D., internal medicine
Coghlan, Anne Eveline, microbiology
Coghlin, Alan Seymour, medicine
Colten, Harvey Radin, pediatrics
Cone, Thomas E, Jr., pediatrics
Conley, Patrick, electromagnetics
Coons, Albert Hewett, immunology

Cooperband, Sidney R, immunology, cell biology
Copeland, Bradley Ellsworth, clinical pathology
Crigler, John F, Jr., pediatrics
Crisley, Francis Daniel, microbiology, bacteriology
Crocker, Allen Carrol, pediatrics
Cromer, Alan H., science writing, physics
Curtis, Brian Albert, physiology
Cynkin, Morris Abraham, biochemistry
D'Agostino, Ralph B., mathematical statistics, experimental statistics
Daggy, Richard Henry, entomology, public health
Dammin, Gustave John, medicine, pathology
David, John R., internal medicine, immunology
Davidson, Richard Laurence, genetics
Davies, Geoffrey, inorganic chemistry, analytical chemistry
Davis, Bernard David, biochemistry, microbiology
Davis, Elizabeth Allaway, developmental biology
Davis, Michael Allan, nuclear medicine, medical research
Davis, William Edwin, Jr., geology, paleontology
Davison, Peter Fitzgerald, physical chemistry
Dawber, Thomas Royle, internal medicine, cardiology
Dealey, James Bond, Jr., radiology
Delabarre, Everett Merrill, Jr., medicine
Densen, Matthew Arnold, microbiology
Derow, Matthew Arnold, biostatistics
Desforges, Jane Fay, medicine
Detering, Ralph Aiden, Jr., surgery
Dews, Peter Booth, psychobiology
Dinsmore, Charles Earle, developmental biology, vertebrate morphology
Ditmer, John Edward, developmental biology
Doane, Marshall Gordon, biophysics
Doellgast, George John, biochemistry
Dogon, Leon I., dentistry, oral microbiology
Dohlman, Claes Henrik, ophthalmology
Doku, Hristo Chris, oral surgery
Dorf, Martin Edward, immunogenetics
Downing, Donald Talbot, biochemistry
Drinker, Philip Aldrich, bioengineering
Drysdale, James Wallace, biochemistry
Duffy, Frank Hopkins, neurophysiology, neurology
Duncan, Stewart, zoology
Dunn, Gary Raymond, genetics
Dvorak, Ann Marine-Tompkins, immunopathology
Dvorak, Harold Fisher, pathology
Ebert, Robert H., internal medicine, medical administration
Edds, Louise Luckenbill, developmental biology
Edmonds, Dean Stockett, Jr., atomic physics
Egdahl, Richard H., surgery
Ehrich, Robert William, anthropology, archaeology
Eipper, Alfred Ward, fish biology, conservation
Eisenberg, Leon, psychiatry
El Ghamry, Mohamed Tawfik, analytical chemistry, inorganic chemistry
Ellis, Franklin Henry, Jr., thoracic surgery
Elzinga, Marshall, biochemistry
Emerson, Charles Phillips, medicine
Enders, John Franklin, microbiology
Engel, Lewis Libman, biochemistry
Epstein, Franklin Harold, internal medicine, physiology
Erhardt, Paul William, medicinal chemistry
Eriks, Klaas, structural chemistry
Everett, Marylee Sicklin, plant physiology
Ewalt, Jack R., psychiatry
Faissler, William L., physics
Fang, So-Fei, mathematics
Farnsworth, Dana Lyda, psychiatry
Fawcett, Don Wayne, anatomy, histology
Federman, Micheline, cell biology, electron microscopy

Feldman, Julius, organic chemistry
Feldman, Martin Leonard, neuroanatomy
Ferris, Benjamin Greeley, Jr., environmental health, pulmonary diseases
Fields, Bernard Nathan, medicine, virology
Filgo, Holland Cleveland, mathematical analysis
Fillios, Louis Charles, biochemistry, nutrition
Fine, Richard Eliot, cell biology
Fine, Samuel, medicine, biophysics
Finland, Maxwell, infectious diseases
Fischbein, Irwin William, biophysics
Fischer, John Herbert, surgery
Fisher, Robert John, bacterial physiology
Fishman, William Harold, biochemistry, oncology
Fitzpatrick, Thomas Bernard, medicine
Fiumara, Nicholas J, public health, venereal diseases
Flax, Martin Howard, pathology
Foley, George Edward, medical microbiology
Folkman, Moses Judah, surgery
Forbes, William Hathaway, physiology
Fossel, Eric Thor, biophysical chemistry, enzymology
Foye, William Owen, medicinal chemistry
Fraenkel, Dan Gabriel, bacterial physiology
Frankel, Fred Harold, psychiatry
Franzblau, Carl, biochemistry
Franzen, Wolfgang, physics
Frazier, Howard Stanley, medicine
Frazier, Todd Mearl, biostatistics
Frechette, Alfred Leo, public health
Freeberg, John Arthur, botany
Freed, Murray Monroe, rehabilitation medicine
Freedman, Marvin I., mathematics
Frei, Emil, III., internal medicine
Freilich, Morris, cultural anthropology
Freeman, David Galland, pathology
Freyre, Raoul Manuel, physics, mathematics
Friedland, Fritz, medicine
Friedman, Emanuel A., gynecology
Friedman, Marvin Harold, physics
Furshpan, Edwin Jean, neurobiology
Gabbiks, Janis, microbiology
Gaensler, Edward Arnold, surgery, physiology
Gainor, Charles, microbiology
Gallop, Paul Myron, protein chemistry
Galmarino, Alberto Raul, mathematics
Gamst, Frederick Charles, ethnology
Garelick, David Arthur, physics
Garg, Hari Gopal, biological chemistry
Gellis, Sydney Saul, medicine
Genes, Andrew Nicholas, geology
Gensler, Walter Joseph, organic chemistry
George, Harvey, biochemistry, clinical pathology
Gerald, Park S., pediatrics
Gergely, John, biochemistry
Gerhardt, John Randolph, meteorology
Gershman, Louis Leo, analytical chemistry
Gershoff, Stanley Norton, biochemistry
Gettner, Marvin, physics
Geyer, Robert Pershing, biochemistry
Ghetti, Mohamed A., geology
Gherardi, Gherardo Joseph, pathology
Gianelly, Anthony Alfred, orthodontics
Gianelly, Michael Francis, Jr., biological anthropology, anatomy
Gibbons, Ronald J., microbiology
Gibson, John Graham, II., hematology
Giering, Warren Percival, organometallic chemistry
Gilbert, Thomas Rexford, analytical chemistry
Gilfillan, Robert Frederick, microbiology
Gilles, Floyd Harry, neurology
Gillespie, John, theoretical physics
Gillis, Catherine Josephine, mathematical statistics
Gilmore, Maurice Eugene, mathematics
Ginsberg, Edward S., high energy physics
Gissen, Aaron J., anesthesiology
Glaubman, Michael Juda, particle physics, nuclear physics
Gleason, Ray Edward, biostatistics
Glimcher, Melvin Jacob, orthopedic surgery, biophysics
Gold, Norman Irving, biochemistry

Goldberg, Alan Herbert, anesthesiology, cardiovascular physiology
Goldberg, Alfred L., physiology, biochemistry
Goldberg, Edward B., molecular biology
Goldberg, Hyman, physics
Goldberg, Irving Hyman, medicine, pharmacology
Goldhaber, Paul, dentistry
Goldman, Henry Maurice, oral pathology, periodontology
Goldman, Peter, biochemistry, pharmacology
Golubic, Stephan, ecology, phycology
Goodenough, Daniel Adino, cell biology
Goolsby, Charles Martel, cell biology
Gorbach, Sherwood Leslie, infectious diseases, internal medicine
Gordon, Bernard Ludwig, marine biology
Gorini, Luigi Costantino, microbial genetics
Gottlieb, Leonard Solomon, pathology
Gottschalk, Bernard, high energy physics
Granchelli, Felix Edward, medicinal chemistry
Granger, Carl V., medicine
Granoff, Barry, applied mathematics
Grant, Walter Morton, ophthalmology
Grasso, Joseph Anthony, anatomy, cytology
Greenfield, Seymour, microbiology, biochemistry
Greengard, Olga, biochemistry
Greep, Roy Orval, anatomy
Griesemer, Robert Daniel, dermatology
Gron, Poul, dentistry
Gross, Jerome, biology, medicine
Gross, Robert Edward, surgery
Guinane, James Edward, pediatrics
Gustafson, Alvar Walter, histology, reproductive biology
Guterman, Sonia Kosow, molecular biology
Habener, Joel Francis, endocrinology, neuroendocrinology
Haber, Edgar, immunology, cardiovascular diseases
Haber-Schaim, Uri, physics, science education
Haggerty, Robert Johns, pediatrics
Hajian, Arshag B., mathematics
Hall, John Emmett, orthopedic surgery
Halpern, Arthur Merrill, physical chemistry, photochemistry
Hamilton, David Whitman, anatomy
Hamlin, Hannibal, neurosurgery
Hammond, Mary Elizabeth, immunopathology
Hardy, William Lyle, biophysics, physiology
Hargis, Betty Jean, biology, immunology
Harlett, John Charles, oceanography
Harris, Melvyn H, dentistry, oral surgery
Harris, Wayne G, bionucleonics, biochemistry
Harrison, Bettina Hall, cytology, immunology
Harrison, John Hartwell, surgery
Harrison, John Michael, psychoacoustics, psychophysics
Hartkopf, Arleigh Van, analytical chemistry
Hartman, Iclal Sirel, biochemistry, physiology
Hartman, Standish Chard, biochemistry
Hartmann, Ernest Louis, psychiatry, psychopharmacology
Havens, Leston Laycock, psychiatry
Hawkins, Thomas William, Jr., mathematics, history of science
Hay, Elizabeth Dexter, anatomy
Hayes, Kenneth Cronise, nutritional pathology
Hedley-Whyte, John, anesthesiology, physiology
Hegsted, David Mark, nutrition, biochemistry
Hein, John William, dentistry
Hellman, Samuel, radiotherapy
Hellman, William S., physics
Henneman, Elwood, neurophysiology
Herda, Hans-Heinrich Wolfgang, mathematics
Herrmann, Robert Lawrence, biochemistry
Herrup, Karl Franklin, neuroscience
Herscovics, Annette Antoinette, biochemistry
Hershenson, Benjamin R., pharmacognosy, academic administration
Hershey, John William Baker, biochemistry, organic chemistry
Herzfeld, Judith, biophysics
Hexner, Peter Eugen, physics

Heyn, Arno Harry Albert, analytical chemistry
Hiatt, Howard Haym, internal medicine, public health
Higgins, Dorothy, inorganic chemistry
Hildebrand, John Grant, III, biochemistry, entomology
Hinds, James Wadsworth, anatomy
Hirsch, Martin Stanley, immunology, virology
Ho, Richard I-Fu, microbial biochemistry, medicinal chemistry
Hobson, John Allan, psychiatry, neurophysiology
Hoffman, Morton Z., physical inorganic chemistry
Hogan, Guy T., mathematics
Hollander, William, internal medicine
Holman, B Leonard, nuclear medicine, radiology
Holmquist, Barton, biological chemistry
Holzworth, Jean, veterinary medicine
Hood, William Boyd, Jr, internal medicine, cardiology
Hoop, Bernard, Jr, medical physics, radiological physics
Howell, David Moore, organic chemistry
Howell, Stephen Barnard, cancer, immunology
Hoy, Gilbert Richard, physics
Huang, Alice Shih-Hou, microbiology, virology
Hubel, David Hunter, neurophysiology
Huggins, Charles Edward, surgery, cryobiology
Hughes, Roy Linwood, Jr, invertebrate, zoology
Hughes, Walter Lee, Jr, biochemistry
Hume, Michael, surgery
Hunt, Ann Hampton, medicine, physiology
Hunt, M Eva, anthropology
Hurxthal, Lewis Marshall, internal medicine
Ito, Susumu, electron microscopy, cytology
Jackson, Barbara Bund, operations research
Jackson, Benjamin T, surgery, physiology
Jacobs, Alma Alice, microbiology, immunology
Jacobson, Bernard Jerome, biochemistry
Jacobson, Stanley, neuroanatomy
Jandl, James Harriman, medicine
Jankowski, Conrad M., analytical chemistry, chemical instrumentation
Jeanloz, Roger William, biochemistry
Jenkins, Howard Jones, pharmacology
Johannessen, Leif Bertram, dentistry, histology
Jones, Elmer Everett, organic chemistry
Jones, Guilford, II, organic chemistry, photochemistry
Jones, James Patrick, urban geography
Jordan, Harold Vernon, microbiology
Kagan, Herbert Marcus, biochemistry
Kalckar, Herman Moritz, biochemistry
Kaminer, Benjamin, physiology
Kamowitz, Herbert M., mathematics
Kaplan, James, mathematics
Kaplan, Lawrence, botany
Karger, Barry Lloyd, analytical chemistry
Karnovsky, Manfred L., biochemistry
Karnovsky, Morris John, pathology
Kashket, Eva Ruth, bacteriology, biochemistry
Kashket, Shelby, biochemistry
Kass, Edward Harold, infectious diseases
Kass, Seymour, algebra
Kassabian, Garabet Haroutian, analytical chemistry
Kayne, Herbert Lawrence, physiology
Kazmaier, Harold Eugene, environmental sciences, environmental management
Kelleher, Roger Thomson, pharmacology
Kennedy, Eugene P., biochemistry
Kerr, George R., pediatrics
Kety, Seymour Solomon, physiology, psychobiology
Khudairi, Abdul Karim, plant physiology
Kiang, Nelson Yuan-Sheng, physiology
Kibrick, Sidney, pediatrics
Kier, Lemont Burwell, medicinal chemistry
Killick, Kathleen Ann, microbial biochemistry

Kim, Agnes Kyung-Hee, clinical pathology, hematology
Kishimoto, Yasuo, biological chemistry, pharmaceutical chemistry
Kisliuk, Roy Louis, biochemistry
Kitz, Richard J, anesthesiology, enzymology
Klein, Robert, pediatrics
Kliman, Allan, internal medicine
Knapp, Peter Hobart, psychiatry, psychoanalysis
Knott, John Russell, neurophysiology
Knox, Walter Eugene, medicine
Knutgen, Howard G, applied physiology
Kobayashi, Yutaka, biochemistry
Koch-Weser, Dieter, experimental pathology, preventive medicine
Koff, Raymond Steven, internal medicine, gastroenterology
Kohn, Henry Irving, radiobiology
Kolodner, Richard David, molecular biology
Kolodny, Gerald Mordecai, radiology
Konigsbacher, Kurt S, organic chemistry, biochemistry
Kornberg, Roger David, biochemistry
Kornetsky, Conan, psychology
Kosasky, Harold Jack, obstetrics & gynecology
Kosersky, Donald Saadia, pharmacology
Kramer, Gerald M, periodontology, dentistry
Kramer, Philip, medicine
Krane, Stephen Martin, medicine, biochemistry
Krasner, Jerome Lee, biomedical engineering, cardiovascular physiology
Kravitz, Edward Arthur, biochemistry, neurochemistry
Krinsky, Norman Irving, biochemistry
Kripke, Benjamin Joshua, anesthesiology
Kronman, Joseph Henry, anatomy, dentistry
Kuffler, Stephen William, neurophysiology
Kumar, Ajit, cell biology
Kundsin, Ruth Blumfeld, medical microbiology
Kupchik, Herbert Z., cancer
Kupferman, Allan, pharmacology, ophthalmology
Kushmerick, Martin Joseph, physiology
Laiderman, Donald D., chemistry
Laing, Ronald Albert, biophysics
Lamb, George Alexander, pediatrics, infectious diseases
Lambert, Helen Haynes, endocrinology
Lampidis, Theodore James, microbiology, cell biology
Lancaster, Jane, transportation geography
Landis, Story Cleland, neurobiology
Landy, David, anthropology
Lange, Yvonne, biophysics
Langmuir, Alexander Duncan, epidemiology
Lanza, Giovanni, physics
LaSala, Edward Francis, pharmaceutical chemistry, organic chemistry
Latz, Arje, psychopharmacology
Laufer, Daniel A, organic chemistry
Laursen, Richard Allan, bio-organic chemistry
LaVail, Jennifer Hart, neuroanatomy, neuroembryology
LaVail, Matthew Maurice, neurosciences, cell biology
Laver, Myron B, anesthesiology
Leach, Robert Ellis, medicine, orthopedic surgery
Leaf, Alexander, internal medicine
Leary, John Dennis, biochemistry
Lee, Michael John, biochemistry
Leeds, Anthony, anthropology
Leeman, Susan Epstein, physiology
Lees, Robert S., medicine, biochemistry
Lehrer, Sherwin Samuel, biochemistry
Leibowitz, Lila, behavioral anthropology
Lembo, Nicholas J, analytical chemistry
Le Quesne, Philip William, organic chemistry
Leskowitz, Sidney, immunology
Lessell, Simmons, neurology
Levene, Martin Barrack, radiotherapy
Lever, Walter Frederick, dermatology
Levering, Dale Franklin, Jr, botany, plant ecology
Levine, Herbert Jerome, cardiology
Levine, Leo, immunology, microbiology
Levine, Oscar, physical chemistry
Levine, Ruth R, pharmacology
Levins, Richard, population biology, mathematical biology
Levinsky, Norman George, medicine
Levy, Charles Kingsley, radiation ecology
Levy, Harvey Louis, medicine, pediatrics
Levy, Michael Green, parasitology

Lewin, Isaac, medicine
Lewis, George Knowlton, geography
Lichtenberg, Franz Von, pathology
Lin, Chi-Wei, cancer, biochemistry
Lin, Edmund Chi Chien, biochemistry
Linck, Richard Wayne, cell biology
Lionetti, Fabian Joseph, biochemistry
Lipke, Herbert, biochemistry
Lisco, Hermann, pathology
Litt, Mortimer, immunology
Little, John Bertram, physiology, radiation biology
Lobene, Ralph Rufino, dentistry, periodontology
Loewenstein, Matthew Samuel, gastroenterology
Lombroso, Cesare Thomas, medicine
Lorenzo, Antonio V, neuropharmacology, neurochemistry
Lowell, Francis Cabot, medicine
Lowenstein, Edward, anesthesiology, cardiopulmonary physiology
Lowenstein, Leah Miriam, medicine, biochemistry
Lowey, Susan, protein chemistry, physical chemistry
Lowndes, Robert P, physics
Lukas, Joan Donaldson, mathematics
Lyons, Donald Herbert, physics
Lyons, Paul Christopher, mineralogy, petrology
Lutts, John A, mathematics
Lynton, Ernest Albert, academic administration, low temperature physics
Macchi, I Alden, endocrinology
MacDonald, Alex Bruce, immunology, endocrinology
Mackay, Ralph Stuart, biophysics, electronics
MacMahon, Brian, epidemiology
Maker, Philip T, mathematics
Malamud, Daniel F, cell biology
Malamud, William, psychiatry
Malenka, Bertram Julian, theoretical high energy physics
Malkiel, Saul, immunochemistry, allergy
Malone, Michael Joseph, neurology, neurochemistry
Maloof, Farahe, internal medicine, endocrinology
Mait, Ronald A, surgery, molecular biology
Mandula, Barbara Blumenstein, biology
Mann, James, medicine
Marchant, Douglas J, obstetrics & gynecology
Margulis, Lynn, cell biology, evolution
Margulis, Thomas N, structural chemistry
Mariani, Henry A, biochemistry, physical chemistry
Marks, Leon Joseph, internal medicine, endocrinology
Marotta, Charles Anthony, molecular biology, psychiatry
Marshall, Theodore, plasma physics
Martin, Arthur Wesley, III, theoretical physics
Martin, Donald Beckwith, medicine
Masih, Shabir Zahoor, biopharmaceutics
Mason, Marion, nutrition
Massler, Maury, dentistry
Matoltsy, Alexander Gedeon, dermatology
Matthysse, Steven William, mathematical biophysics, psychiatry
Mautner, Henry George, biochemistry, pharmacology
Mayer, Jean, physiology, nutrition
Mayfield, Robert Charles, cultural geography, geography of South Asia
McArthur, Janet W, endocrinology
McCabe, William R, infectious diseases, microbiology
McCall, Daniel Francis, anthropology
McCluskey, Robert Timmons, pathology
McCombs, Robert Pratt, internal medicine
McCully, Kilmer Serjus, experimental pathology
McDermott, William Vincent, Jr, surgery
McFarland, Ross Armstrong, physiological psychology, aerospace medicine
McLardy, Turner, neurobiology, neuropsychiatry
McLeod, Guy Collingwood, plant physiology
McMahan, Uel Jackson, II, anatomy
McNair, Douglas McIntosh, psychopharmacology, psychology
McNary, William Francis, Jr, anatomy
McNutt, Walter Scott, biochemistry, pharmacology
Meader, Ralph Gibson, anatomy

Meienhofer, Johannes Arnold, biochemistry, organic chemistry
Messner, William Avison, pathology
Melby, James Christian, medicine
Melbye, Susanne Warner, biochemistry, dermatology
Mellins, Harry Zachary, radiology
Metvold, Roger Wayne, genetics
Menge, Bruce Allan, ecology
Menges, Michael George, Jr., geography
Menzin, Margaret Schoenberg, mathematics
Merker, Philip Charles, pharmacology
Merrill, John Putnam, medicine
Mescon, Herbert, dermatology
Meszoely, Charles Aladar Maria, paleontology, parasitology
Michelson, Edward Harlan, malacology, public health
Mickel, Hubert Sheldon, neurology, neurochemistry
Mickles, James, medicinal chemistry
Mihm, Martin C, Jr., dermatology, pathology
Milburn, Ronald McRae, inorganic chemistry
Miller, Aaron, medicine
Miller, Joseph Morton, clinical medicine, preventive medicine
Millette, Clarke Francis, reproductive biology, immunology
Mindel, Joseph, history of science, science education
Minichino, Camille, physics
Mitchell, George W, Jr, obstetrics & gynecology
Mitus, Wladyslaw J, pathology, hematology
Mohr, Jay Preston, neurology
Mohr, Scott Chalmers, biochemistry
Mollow, Benjamin R, theoretical physics
Moloney, William Curry, hematology
Monaco, Anthony Peter, surgery, immunology
Monette, Francis C, hematology, cell physiology
Mongini, Patricia Katherine Ann, immunology
Montgomery, William Wayne, otolaryngology
Moolten, Frederick London, cancer
Moore, Francis Daniels, surgery
Moorrees, Coenraad Frans August, orthodontics
Morest, Donald Kent, anatomy
Morse, William Herbert, pharmacology
Mullaney, Owen Christopher, experimental pathology
Mullen, Richard Joseph, developmental genetics, neurosciences
Mulvey, Philip Francis, Jr., radiobiology, physiology
Murnane, Thomas William, oral surgery, anatomy
Murphy, Sheldon Douglas, toxicology
Murthy, Krishna A S, histochemistry, pharmacology
Myers, John Martin, applied physics, systems theory
Myerson, Paul Graves, psychiatry
Nadas, Alexander Sandor, pediatric cardiology
Nadol, Bronislaw Joseph, Jr., otolaryngology
Nagy, Bela Ferenc, biochemistry
Naidus, Harold, chemistry
Najjar, Victor Assad, pediatrics
Nasir-Ud-Din, Nasir, biochemistry, carbohydrate chemistry
Naylor, Richard Stevens, geology, geochemistry
Nechtes, Thomas, pediatrics, physiology
Nelson, Margaret Christina, neurophysiology, ethology
Neumeyer, John L, medicinal chemistry
Neurath, Peter Wolfgang, information science
Newell, William Andrews, textiles
Newman, William Alexander, glacial geology
Nichols, George, Jr, biochemistry
Nichols, Roger Loyd, internal medicine
Nizel, Abraham Edward, preventive dentistry, nutrition
O'Connell, Alice L, anatomy, physiology
O'Neil, Elizabeth Jean, applied mathematics
O'Neil, Patrick Eugene, computer science
Order, Stanley Elias, radiotherapy, immunology
Owades, Joseph Lawrence, biochemistry
Palay, Sanford Louis, neuroanatomy, neurocytology
Pan, Steve Chia-Tung, medicine

Papagiannis, Michael D, astrophysics, space physics
Pappenheimer, John Richard, physiology
Paquette, Gerard Arthur, mathematics education
Pardee, Arthur Beck, biochemistry
Park, James Theodore, microbiology, biochemistry
Parrott, Stephen Kinsley, medicine
Patek, Arthur Jackson, Jr., medicine
Pathak, Madhukar, biochemistry, dermatology
Patt, Donald Irving, biology
Patterson, James Fulton, internal medicine
Paul, Benoy Bhushan, biochemistry, radiobiology
Paul, Robert E, Jr, radiology
Paz, Mercedes Aurora, biochemistry
Pearincott, Joseph V, physiology, zoology
Pease, Paul Lorin, physiological optics
Pecci, Joseph, biochemistry
Peczon, Benigno David, biochemistry
Pelavin, Lawrence, synthetic organic chemistry
Pelikan, Edward Warren, pharmacology
Pennell, Robert Brown, biochemistry
Perlman, Robert, biochemistry, immunology
Perry, Clive Howe, solid state physics
Peters, Alan, anatomy
Peters, Stefan, mathematics
Peterson, Osler Luther, preventive medicine
Pharo, Richard Levers, biochemistry
Philbin, Daniel Michael, anesthesiology
Pillard, Richard Colestock, psychiatry
Pinkus, Jack Leon, organic chemistry
Piper, James Underhill, organic chemistry
Plank, Stephen J, public health, population biology
Plate, Janet Margaret, immunology
Plaut, Andrew George, biochemistry, gastroenterology
Pochi, Peter E, medicine, dermatology
Polet, Herman, pathology, cell biology
Pollard, Thomas Dean, cell biology, medicine
Pontoppidan, Henning, anesthesiology, respiratory physiology
Poppen, James L, neurosurgery
Porter, Huntington, neurology, neurochemistry
Poskanzer, David Charles, neurology, preventive medicine
Posner, Martin, nuclear physics, atomic physics
Posner, Tamar Beatrice, physical organic chemistry
Potter, David Dickinson, neurophysiology
Potts, John Thomas, endocrinology, medicine
Powell, Arnet L, physical chemistry
Prager, Gerald David, internal medicine
Procaccini, Donald J, economic geology
Prock, Alfred, physical chemistry
Proffitt, Max Rowland, immunology, physiology, systematics
Proger, Samuel, internal medicine
Prout, George Russell, Jr, urology
Quelle, Fred W, Jr, lasers, electrooptics
Raacke, Ilse Dorothea, biology, insect physiology
Rabe, Edward Frederick, evolutionary biology
Raben, Maurice Solomon, neurology
Raffauf, Robert Francis, medicine
Raisbeck, Barbara, organic chemistry
Rakic, Pasko, developmental neurosciences, developmental biology
Ramras, Mark Bernard, mathematics
Ransil, Bernard Jerome, physical chemistry, medical research
Rao, Devulapalli V G L N, solid state physics, lasers
Rapaport, Eliezer, biochemistry
Read, Kenneth Richard Hodgson, biochemistry
Refojo, Miguel Fernandez, biomaterials, polymer chemistry
Reif, Arnold E, immunology
Reiff, William Michael, inorganic chemistry
Reif-Lehrer, Liane, molecular biology
Reinhard, John Frederick, pharmacology
Reit, Barry, endocrinology, physiology
Rexford, Eveoleen Naomi, psychiatry, psychoanalysis
Reynolds, Robert N, anesthesiology
Rheingold, Joseph Cyrus, psychiatry
Rheinlander, Harold F, surgery
Richardson, Charles Clifton, biochemistry

Richardson, George S, endocrinology, obstetrics & gynecology
Richman, Justin Lewis, medicine
Richmond, Julius Benjamin, medicine
Richmond, Martha Ellis, biochemistry
Rieder, Sidney Victor, biochemistry
Riordan, James F, biological chemistry
Robbins, Laurence Lamson, radiology
Robbins, Stanley Leonard, pathology
Roberts, Francis Donald, organic chemistry
Roebber, John Leonard, physical chemistry
Roman, Paul, theoretical physics, mathematical physics
Romayne, Michael Richard, physical chemistry
Rosen, Walter George, plant physiology
Rosenberg, Fred A, microbiology
Rosner, Anthony Leopold, enzymology, endocrinology
Rosowsky, Andre, organic chemistry
Roth, Sanford Irwin, pathology, electron microscopy
Rothman, Sara Weinstein, microbiology
Ruben, Morris P, periodontology, oral biology
Ruber, Ernest, ecology
Rule, Allyn H, immunochemistry, biochemistry
Russell, Donald Hayes, psychiatry
Russell, Paul Snowden, surgery, immunology
Rutenberg, Alexander Michael, surgery, cancer
Ryan, John F, anesthesiology
Ryan, Kenneth John, endocrinology, biochemistry
Ryser, Hugues Jean-Paul, cell biology, pharmacology
Sabin, Thomas Daniel, neurology
Sacks, Marie Luisetti, developmental biology
Sager, Ruth, genetics
Salhanick, Hilton Aaron, obstetrics & gynecology
Salzman, Edwin William, surgery
Salzman, George, theoretical physics
Sanadi, D Rao, biochemistry
Sanders, Daniel Charles, botany
Sandler, Louis W, psychiatry
Sandler, Sheldon Samuel, cardiology
Sandson, John I, medicine
Saravis, Calvin, immunology
Sbarra, Anthony J, bacteriology
Schaechter, Moselio, microbiology, molecular biology
Schay, Geza, mathematical physics
Scheid, Francis, mathematics
Scherer, Daniel Charles, botany
Scheuplein, Robert J, physical chemistry, biophysics
Schilder, Herbert, dentistry
Schilling, Albert, medicine
Schimmel, Elihu Myron, medicine
Schlossman, Stuart Franklin, immunology, hematology
Schmid, Karl, biochemistry
Schmid, Gerhard, biochemistry
Schmid, Kurt F, anesthesiology
Schmidt, William Morris, medicine
Schmitt, Francis Otto, biology
Schmitter, Ruth Elizabeth, cell biology, physiology
Schorr, Marvin Gerald, physics
Schuknecht, Harold Frederick, otolaryngology
Schultz, Martin C, speech pathology, audiology
Schulz, Milford David, radiology
Schur, Peter Henry, internal medicine, immunology
Schwartz, Bernard, ophthalmology, physiology
Schwartz, Lowell Melvin, physical chemistry
Schwartz, Robert Stewart, hematology
Schwartz, William Benjamin, internal medicine
Schweber, Miriam Schurin, cell biology, developmental biology
Scott, Jesse Friend, biochemistry, biophysics
Scully, Robert Edward, pathology
Seaman, Edna, biology
Seeley, Robert T, mathematical analysis
Seidel, John Charles, biochemistry
Seltzer, Carl Coleman, physical anthropology
Senrad, Elvin Vavrinec, psychiatry
Sengupta, Sisir K, organic chemistry, cancer
Senior, Boris, pediatrics, endocrinology
Shane, John Richard, magnetism
Shargel, Leon David, pharmacology, drug metabolism

Shashoua, Victor E, biochemistry
Shaw, James Headon, biochemistry
Shepard, Robert Andrews, organic chemistry
Shepro, David, cardiovascular physiology
Sherman, Thomas Oakley, mathematics
Shiere, Frederic Roland, dentistry
Shiffman, Carl Abraham, physics
Shih, Vivian E-an, pediatrics
Shimony, Abner, theoretical physics, philosophy of science
Shipley, George Graham, biophysics
Shore, Miles Frederic, psychiatry
Shuster, Louis, pharmacology
Shwachman, Harry, medicine
Siciliano, Arthur Anthony, cosmetic chemistry
Sidman, Richard Leon, neuropathology
Sidman, Bernard, pharmaceutics
Siegel, Armand, statistical mechanics, biophysics
Sifneos, Peter E, medicine
Silen, William, surgery
Silverman, Harold I, pharmacy
Simon, Benjamin, psychiatry, neurology
Simon, Meredith Ann, immunology, ophthalmology
Simons, Elizabeth Reiman, biophysical chemistry
Smithson, Janet Eleanor, histology
Smulow, Jerome B, oral pathology, cytology
Sinex, Francis Marott, biochemistry
Singer, Robert Mark, radiology
Slack, Warner Vincent, neurology
Slechta, Robert Frank, reproductive physiology
Smith, Daniel James, immunology, oral biology
Smith, Edward Herbert, radiology
Smith, Gerald Nelson, Jr, developmental biology
Smith, Grahame J C, insect ecology
Smith, Pierre Frank, pharmacy, medicinal chemistry
Snider, Gordon Lloyd, pulmonary diseases
Snow, Beatrice Lee, mammalian genetics
Snyder, John Crayton, microbiology, population biology
Socransky, Sigmund Sydney, oral microbiology
Sokoloff, Jeffrey Bruce, solid state physics
Solomon, Arthur Kaskel, biophysics
Soloway, Albert Herman, medicinal chemistry
Soltzberg, Leonard Jay, physical chemistry, crystallography
Sonenshein, Abraham Lincoln, microbiology, molecular biology
Sorensen, Andrew Aaron, medical education
Sorokin, Sergei Pitirimovitch, histology, embryology
Soule, Roger Gilbert, exercise physiology
Spayne, Robert William, geography
Spector, Elliot, pharmacology
Spetzer, Frank A, biochemistry
Spengler, Kenneth Clifford, meteorology
Spielman, Andrew, medical entomology
Spiro, Robert Gunter, biochemistry
Spiro, Mary Jane, biochemistry
Spodick, David Howard, cardiology
Stare, Fredrick John, nutrition
Stark, Betty Salzberg, algebra
Stein, Bennett M, neurosurgery
Stephenson, Mary Louise, molecular biology
Stern, Irving B, dentistry, cell biology
Stier, Howard Livingston, research administration
Stoeckle, John Duane, medicine
Stolberg, Marvin Arnold, organic chemistry
Stollar, Bernard David, immunology
Stolzberg, Richard Jay, analytical chemistry
Stone, Frederick Logan, biology, public health
Stosky, Bernard A, psychiatry
Strauss, Marc J, oncology, chemotherapy
Strauss, John Steinert, medicine
Strauss, Phyllis R, cell physiology
Strong, Mervyn Stuart, otolaryngology
Struthers, Robert Claflin, comparative anatomy, developmental anatomy
Stuart, Ann Elizabeth, neurobiology
Suit, Herman Day, radiotherapy
Sullivan, John Francis, neurology
Sullivan, Louis Wade, internal medicine

Sullivan, Thomas Frederick, organic chemistry, nuclear chemistry, oral pathology
Susi, Frank Robert, anatomy, oral pathology
Swartz, Jacob, psychiatry, psychoanalysis
Sweet, William Herbert, neurosurgery
Sze, Heven, plant physiology
Taft, Edgar Breck, pathology
Talbot, Nathan Bill, pediatrics
Tamarin, Robert Harvey, genetics, ecology
Tanimoto, Taffee Tadashi, mathematics, geometry
Tashjian, Armen H, Jr, endocrinology, pharmacology
Taubman, Martin Arnold, immunology, oral biology
Taveras, Juan M, radiology
Taymor, Melvin Lester, obstetrics & gynecology
Terner, Charles, biochemistry
Terres, Geronimo, immunobiology
Tetvethia, Mary Judith (Robinson), molecular biology, genetics
Tetvethia, Satvir S, virology, immunology
Thenen, Shirley Warnock, nutrition, biochemistry
Thomas, Charles Allen, Jr, biophysical chemistry
Thomas, David Dale, molecular biophysics
Tishler, Peter Verveer, medical genetics
Tisza, Veronica Elizabeth Benedek, medicine
Tomlinson, Michael Bangs, mathematics
Traggis, Demetrius G, pediatrics
Trelstad, Robert Laurence, embryology, pathology
Treves, Salvador, nuclear medicine, pediatrics
Trier, Jerry Steven, internal medicine, gastroenterology
Troxler, Robert Fulton, plant physiology
Tullis, James Lyman, biochemistry
Turesky, Samuel Saul, dentistry
Twitchell, Paul F, meteorology
Twitchell, Thomas Evans, neurology
Ulfelder, Howard, obstetrics & gynecology, surgery
Ulrich, William Charles, physiology
Ulrick, Frank, cell physiology
Umans, Robert Scott, biophysical chemistry
Unanue, Emil R, immunology
Vaillant, Henry Winchester, population biology
Vallee, Bert L, biochemistry, biophysics
Vandam, Leroy David, anesthesiology
Vander Wyk, Raymond Winston, microbiology
Van Meter, David, applied physics, information sciences
Van Ummersen, Claire Ann, developmental biology, animal physiology
Van Zwieten, Matthew Jacobus, veterinary pathology
Vaughan, Deborah Whittaker, neuroanatomy
Vaughn, Michael Thayer, physics
Veillon, Claude, analytical chemistry, spectroscopy
Vernon, Robert Carey, solid state physics
Villee, Claude Alvin, Jr, biochemistry
Villee, Dorothy, endocrinology, biochemistry
Vinson, John William, rickettsial diseases, venereal diseases
Viola, Alfred, organic chemistry
Vitale, Joseph John, nutrition, biochemistry
Volicer, Ladislav, pharmacology, cardiovascular diseases
Von Goeler, Eberhard, high energy physics
Von Lichtenberg, Franz, pathology, tropical medicine
Vouros, Paul, analytical chemistry, organic chemistry
Walker, Eugene Hoffman, economic geology
Walker-Nasir, Evelyne, carbohydrate chemistry
Walkowiak, Edmund Francis, physiology
Wallace, Thomas Homkowycz, spectroscopy
Walsh, Michael Patrick, cytology
Walsh, Robert A, pharmacy, biology
Walters, David Royal, cell biology, developmental biology
Waner, Joseph Lloyd, microbiology
Wang, Chi-Hua, organic chemistry
Wang, Chiu-Chen, radiotherapy
Wang, Nancy Yang, organic chemistry
Ward, Samuel, biochemistry
Warga, Jack, mathematics
Warner, Ann Marie, clinical chemistry

Warner, Victor Duane, medicinal chemistry
Warr, William Bruce, neuroanatomy
Warren, Christopher David, carbohydrate chemistry
Warren, Shields, pathology
Washburn, Henry Bradford, Jr, geography
Watkins, Elton, Jr, surgery, cancer
Watson, Barbara Kascenko, bacteriology, microbiology
Waxman, Stephen George, neurology
Weaver, Nevin, insect physiology
Webster, Edward William, medical physics
Weiant, Elizabeth Abbott, physiology
Weibrecht, Walter Eugene, inorganic chemistry
Weinstein, Louis, internal medicine, infectious diseases
Weintraub, Lewis Robert, hematology
Weiss, Edwin, algebra
Weiss, Karl H, physical chemistry
Weiss, Robert Jerome, psychiatry
Weller, Thomas Huckle, tropical medicine, infectious diseases
Wells, Herbert, pharmacology, orthodontics
Wells, Roe, internal medicine
Wellum, Glyn Richard, chemistry
Wentz, Henry Oscar, zoology
Werthessen, Nicholas Theodore, endocrinology
West, Arthur James, II, parasitology, marine biology
White, Edward Lewis, neuroanatomy
Whitescarver, Jack Edward, medical microbiology
Whiteside, Roberta Emerson, immunology
Wiener, Robert Newman, chemistry
Wiesel, Torsten Nils, physiology
Wikswo, Muriel Anastasia, cell biology
Wildasin, Harry Lewis, biochemistry
Wilgram, George Friederich, physiology, dermatology
Wilkes, Hilbert Garrison, Jr, plant genetics
Williams, David Allen, medicinal chemistry
Willis, Charles Richard, physics
Wilson, Thomas Hastings, physiology
Winn, Henry Joseph, immunology
Wohlrab, Hartmut, biochemistry
Wolff, Peter Hartwig, medicine, psychology
Woodland, John Turner, zoology
Wotiz, Herbert Henry, organic chemistry, biochemistry
Woznick, Benjamin Joseph, theoretical physics, computer science
Wright, Andrew, molecular biology
Wright, Barbara Evelyn, biochemistry
Wright, George Green, immunology, microbiology
Wu, Fa Yueh, theoretical physics
Wyon, John Benjamin, epidemiology, demography
Yablon, Isadore Gerald, orthopedic surgery
Yankee, Ronald August, medicine
Yarbrough, Lynn Douglas, computer science
Yen, Peter Kai Jen, dentistry, biophysics
Yerby, Alonzo Smythe, public health, health administration
Yerganian, George, biology
Yoshinaga, Koji, endocrinology
Young, Delano Victor, biochemistry
Young, Richard L, organic chemistry
Yurkstas, A Albert, dentistry
Zacharias, Leona Ruth, human development
Zamcheck, Norman, gastroenterology, clinical pathology
Zamecnik, Paul Charles, medicine
Zigmond, Richard Eric, neurobiology, neuroendocrinology
Zimmerman, George Ogurek, low temperature physics
Zompa, Leverett Joseph, inorganic chemistry
Zurawski, Vincent Richard, Jr, biochemistry, immunology
Zwaan, Johan, developmental biology, human anatomy

BOXFORD
Bogdonoff, Philip David, Jr, animal physiology, animal nutrition

BOYLSTON
Sornburger, George Clinton, mathematics, statistics

BREWSTER
Pyle, Robert Wendell, zoology
Sheldon, William Gulliver, ecology

BRIDGEWATER
Boutilier, Robert Francis, geology
Brennan, James Robert, plant anatomy, cytology
Caluscian, Richard Frank, theoretical physics
Chipman, Wilmon B, organic chemistry, biochemistry
Cirino, Elizabeth Fahey, marine ecology
D'Alarcao, Hugo T, algebra
Daley, Henry Owen, Jr, physical chemistry
Furlong, Ira E, geology
Hewitson, Walter Milton, plant morphology
Hiferty, Frank Joseph, bryology
Horner, George Roland, anthropology
Howe, Kenneth Jesse, plant physiology
Jahoda, John C, ecology, mammalogy
Maier, Emanuel, physical geography
Marganian, Vahe Mardiros, inorganic chemistry
Mish, Lawrence Bronislaw, botany
Morin, Walter Arthur, neurophysiology
Muckenthaler, Florian August, zoology
Palubinskas, Felix Stanley, biophysics
Sethares, George C, mathematics
Summer, Kenneth, bioinorganic chemistry
Wall, William James, entomology

BRIGHTON
Carroll, Thomas Joseph, physics

BROCKTON
Claff, Chester Eliot, Jr, organic chemistry
Gabay, Sabit, biochemistry, pharmacology
Leith, John Douglas, Jr, pathology, cell biology
Rustigian, Robert, medical microbiology

BROOKLINE
Altschule, Mark David, medicine
Castle, William Bosworth, clinical medicine, hematology
Cohen, Saul Israel, nutrition
Craig, John Merrill, pathology
Curby, William Adolph, physical biology
Davidson, Betty, biochemistry
Friedman, Orrie Max, research administration, bio-organic chemistry
Friedman, Robert Bernard, biochemistry, carbohydrate chemistry
Glazman, Yuli, surface chemistry, colloid chemistry
Shklar, Gerald, oral pathology
Smith, Olive Watkins, biochemistry, endocrinology

BURLINGTON
Argue, Gary R, inorganic chemistry, electrochemistry
Hanson, Per Roland, electronics, electrical engineering
Lees, Wayne Lowry, experimental physics
Liang, Charles Chi, physical chemistry, analytical chemistry
Reidy, William Patrick, space physics
Seeley, Paul Ellsworth, physics
Sutton, Emmet Albert, atomic physics
Tait, Kevin S, applied mathematics
Tomjanovich, Nicholas Matthew, theoretical physics, elementary particle physics
Winston, Arthur William, physics, mathematics
Wolsky, Sumner Paul, physical chemistry

BUZZARDS BAY
Wright, David Franklin, physical chemistry

CAMBRIDGE
Adler, David, solid state physics
Adler, Norman, analytical chemistry, physical chemistry
Aggarwal, Roshan Lal, quantum optics
Ahearn, James Joseph, Jr, analytical chemistry
Ahlfors, Alfred Valerian, mathematics
Ajami, Alfred Michel, pesticide chemistry, pharmaceutical chemistry
Aki, Keiiti, seismology
Aksnes, Kaare, astronomy

Alberty, Robert Arnold, chemistry
Aldrich, Franklin Dalton, toxicology, internal medicine
Allen, Frank Lluberas, information science
Allis, William Phelps, plasma physics, electron ph ysics
Alyea, Fred Nelson, dynamic meteorology
Amdur, Mary Ochsenhirt, toxicology
Anderson, Donald Gordon Marcus, applied mathematics
Angeline, John Frederick, organic chemistry
Annis, Martin, cosmic ray physics
Apgar, Edward G, physics
Arsenault, Guy Pierre, bio-organic chemistry
Atkinson, Edward Redmond, organic chemistry
Atwater, Tanya Maria, marine geophysics, paleomagnetism
Ausubel, Frederick Michael, molecular genetics
Avrett, Eugene Hinton, astrophysics
Bader, Henry, organic chemistry
Bainbridge, Kenneth Tompkins, nuclear physics
Baker, Raymond Milton, genetics, cell biology
Baldwin, Jack Edward, organic chemistry
Baltimore, David, virology, biochemistry
Bamberger, Joan Thatcher, anthropology
Bangerter, Benedict W, physical chemistry
Baranger, Michel, theoretical physics
Barber, Walter Carlisle, nuclear physics
Barger, James Edwin, acoustics
Barghoorn, Elso Sterrenberg, paleobiology
Barrett, Alan H, radio astronomy
Bartels-Kieth, James Richard, organic chemistry
Battista, Sam P, pharmacology, physiology
Baumgartner, Leona, pediatrics, public health
Becker, Ulrich J, high energy physics
Bekefi, George, physics
Bell, Barbara, solar physics, climatology
Bell, Eugene, developmental biology
Bell, Samuel Dennis, Jr, microbiology
Bender, Carl Martin, high energy physics, mathematical physics
Benedek, George Bernard, physics
Bennett, John Richard, oceanography
Bennett, Stewart, physics
Benton, Stephen Anthony, optical physics
Berchtold, Glenn Allen, organic chemistry
Berger, Steven Barry, physics
Berney, Charles V, molecular spectroscopy
Bernfeld, Peter, enzymology, biochemistry
Bertozzi, William, physics
Biemann, Klaus, organic chemistry
Bigelow, Nolton H, pathology, forensic medicine
Billings, Marland Pratt, geology
Birch, Albert Francis, geophysics
Birgeneau, Robert Joseph, solid state physics
Birkhoff, Garrett, mathematics
Bjorkholm, Paul J, astrophysics
Blake, Julian Gaskill, quantum optics, experimental physics
Blakemore, George Jefferson, Jr, applied statistics, operations research
Bloch, Konrad Emil, biological chemistry
Bloembergen, Nicolaas, quantum optics
Bloomberg, Wilfred, psychiatry, neurology
Blumgart, Herrman Ludwig, medicine
Bogorad, Lawrence, plant physiology
Boodman, David Morris, systems science
Borror, Alan L, organic chemistry
Boss, Kenneth Jay, malacology
Bostock, Judith Louise, theoretical solid state physics
Bothwell, Alfred Lester Meador, molecular biology
Botstein, David, molecular genetics, microbial physiology
Bott, Raoul H, mathematics
Brace, William Francis, structural geology
Bradt, Hale Van Dorn, physics
Branton, Daniel, cell biology, botany
Brauer, Richard Dagobert, mathematics
Brecher, Aviva, space physics, paleomagnetism
Brecher, Kenneth, theoretical astrophysics
Bridge, Herbert Sage, space physics

MASSACHUSETTS

Brockett, Roger Ware, applied mathematics
Brooks, Harvey, physics
Brown, Gene Monte, biochemistry
Buchi, George, organic chemistry
Buechner, William Weber, physics
Buerger, Martin Julian, mineralogy, crystallography
Burg, Alan Walter, biochemistry
Burke, Bernard Flood, physics
Burnham, Charles Wilson, mineralogy
Burns, John McLauren, zoology, evolution
Burns, Roger George, geochemistry, mineralogy
Burton, Charles Jewell, physics
Busby, William Fisher, Jr, biochemistry
Busza, Wit, elementary particle physics
Butler, James Newton, physical chemistry, environmental sciences
Button, Kenneth J, physics
Cairncross, Stanley Everett, chemistry
Calman, Jack, physical oceanography
Canizares, Claude Roger, x-ray astronomy
Caplan, Gerald, psychiatry, psychoanalysis
Carleton, Nathaniel Phillips, astrophysics
Carpenter, Frank Morton, entomology
Carpenter, Gail Alexandra, biomathematics
Carter, George H, psychiatry
Catsimpoolas, Nicholas, biophysics
Cerankowski, Leon Dennis, physical chemistry
Chalmers, Bruce, applied physics
Charkoudian, John Charles, inorganic chemistry, photographic chemistry
Charney, Jule Gregory, meteorology
Chen, Sow-Hsin, experimental fluid physics, radiation physics
Cherbas, Lucy Fuchsman, developmental biology
Cherbas, Peter Thomas, developmental biology, insect physiology
Chernoff, Herman, mathematical statistics
Chinnery, Michael Alistair, geophysics, seismology
Clapp, Roger Edge, theoretical physics
Clark, George Whipple, physics
Cline, James Edward, operations research
Clow, James Rodgers, physics
Cobb, Carolus M, physical chemistry
Cochran, William Gemmell, applied statistics
Coffey, John Joseph, biochemistry, biochemical pharmacology
Coffin, Perley Andrews, rubber chemistry, polymer chemistry
Cohen, David, magnetism
Cohen, Fredric Sumner, polymer chemistry, optics
Cohen, Howard David, chemical physics, energy con
Cohen, I Bernard, history of scienceversion
Cohn, Daniel Ross, lasers, plasma physics
Coleman, Sidney Richard, theoretical physics
Comba, Paul Gustavo, computer science
Cook, Allan Fairchild, II, astrophysics
Coppi, Bruno, physics
Corbato, Fernando Jose, physics
Corey, Elias James, organic chemistry
Costello, Catherine E, mass spectrometry, organic biochemistry
Cotton, Frank Albert, chemistry
Counselman, Charles Claude, III, planetary sciences, radio astronomy
Crompton, Alfred W, vertebrate paleontology
Crout, Prescott Durand, applied mathematics
Cunnold, Derek M, meteorology
Dainty, Anton Michael, seismology
Dalgarno, Alexander, theoretical physics, astrophysics
Damon, Albert, internal medicine, physical anthropology
Darling, Eugene Merrill, Jr, mathematics, meteorology
Darlington, Philip Jackson, Jr, entomology
Davidson, Charles Sprecher, clinical medicine
Davidson, Samuel James, biochemistry, cell physiology
Davis, Charles Freeman, Jr, physics
Davis, John Moulton, solar physics, x-ray astronomy
Davis, Marc, astrophysics
Davis, Robert James, astrophysics
Davis, Thomas, medicine, physiology
Davison, Alan, inorganic chemistry
Delvaille, John Paul, physics
Demain, Arnold Lester, industrial microbiology
De Member, John Raymond, organic chemistry
Demos, Peter Theodore, physics
Dempster, Arthur Pentland, statistics
Denison, Robert Howland, vertebrate paleontology
Denlinger, David Landis, insect physiology
Denny-Brown, Derek Ernest, medicine
Deutch, John Mark, physical chemistry, statistical mechanics
Deutsch, Martin, physics
DeVore, Boyd Irven, anthropology
Dickey, John Sloan, Jr, petrology
Dickinson, Dale Flint, physics
Dobbs, Gregory Melville, physical chemistry
Dodson, Vance Hayden, Jr, inorganic chemistry
Doerfler, Thomas Eugene, statistics
Doering, William Von Eggers, organic chemistry
Dolan, Louise Ann, theoretical high energy physics
Doty, Paul Mead, biochemistry
Dowling, John Elliott, neurobiology
Downey, John Francis, Jr, photographic chemistry
Dresselhaus, Mildred S, solid state physics
Dressler, David, biochemistry, molecular biology
Driver, Richard D, atomic physics
Du Bois, Cora, anthropology
Dudley, Richard Mansfield, mathematical statistics
Dupree, Andrea K, astrophysics
Dyer, Ira, acoustics
Dziewonski, Adam Marian, seismology, geomagnetism
Eagleson, Peter Sturges, hydrology
Edsall, John Tileston, biochemistry
Ehrenreich, Henry, theoretical physics
Eisen, Herman Nathaniel, immunology
Ekman, Carl Frederick W, inorganic chemistry, solid state chemistry
Elgin, Sarah Carlisle Roberts, molecular genetics
Elion, Herbert Aaron, physics
Enge, Harald Anton, nuclear physics
Eppling, Frederic John, nuclear physics
Erickson, Alan Eric, embryology
Ernst, Martin L, physics
Evans, Robley Dunglison, physics
Exarhos, Gregory James, physical chemistry, solid state chemistry
Fairbairn, Harold Williams, petrology
Fazio, Giovanni Gene, astrophysics, cosmic ray physics
Feld, Bernard Taub, elementary particle physics, high energy physics
Feld, Michael S, physics
Fell, Howard Barraclough, invertebrate zoology
Fernstrom, John Dickson, neuropharmacology, neuroendocrinology
Feshbach, Herman, physics
Field, George Brooks, theoretical astrophysics
Field, Robert Warren, chemical physics
Fieser, Louis Frederick, organic chemistry
Fireman, Edward Leonard, physics
First, Melvin William, public health
Flynn, George Patrick, physical chemistry
Foner, Simon, solid state physics
Forman, Earl Julian, analytical chemistry
Fox, Herbert Leon, physics, economics
Fox, Maurice Stanford, genetics
Frankel, Richard Barry, physics, chemistry
Franklin, Fred Aldrich, astronomy
Frant, Martin S, analytical chemistry
Freeman, Harold Adolph, statistics
French, Anthony Philip, physics
Frey, Frederick August, geochemistry
Friedlaender, Jonathan Scott, physical anthropology
Friedlander, John Benjamin, number theory
Friedman, Jerome Isaac, physics
Frishkopf, Lawrence Samuel, biophysics
Furry, Wendell Hinkle, physics
Gajewski, Ryszard, plasma physics, energy conversion
Garay, Leslie Andrew, taxonomy
Garland, Carl Wesley, physical chemistry
Gefter, Malcolm Lawrence, molecular biology, biochemistry
Geller, Margaret Joan, theoretical astrophysics, cosmology
Gerassimenko, Michel, astrophysics
Giacconi, Riccardo, astrophysics
Gilbert, Walter, molecular biology
Gill, David Michael, biochemistry, toxinology
Gingerich, Owen (Jay), astronomy
Giotta, Gregory John, biochemistry
Glashow, Sheldon Lee, elementary particle physics
Glass, Robert Loring, epidemiology, dentistry
Glauber, Roy Jay, physics
Gleason, Andrew Mattei, mathematics
Goetze, Christopher, geophysics
Goldberg, Paul, physical chemistry
Goldblith, Samuel Abraham, food science
Gole, James Leslie, high temperature chemistry
Goodenough, Ursula Wiltshire, cell biology, genetics
Goody, Richard (Mead), physics
Gorenstein, Paul, astrophysics, nuclear physics
Gould, Bernard Sidney, biochemistry
Gould, Stephen Jay, paleontology, evolutionary biology
Grasshoff, Jurgen Michael, polymer chemistry, organic chemistry
Green, Howard, cell biology
Greene, Curtis, mathematics
Greene, Frederick Davis, II, organic chemistry
Greenspan, Harvey Philip, applied mathematics
Greenwald, Richard B, organic chemistry
Greytak, Thomas John, physics
Griffin, Robert Guy, physical chemistry, biophysics
Grindlay, Jonathan Ellis, astrophysics
Grodzins, Lee, nuclear physics
Grossi, Mario Dario, radio physics, engineering
Growdon, John Herbert, neurology, neuropharmacology
Guidotti, Guido, biochemistry
Guinan, John Joseph, Jr, neurophysiology
Guralnick, Eugene, surgery
Gursky, Herbert, astronomy
Haff, Lawrence Allen, molecular biology
Haller, John, structural geology
Halperin, Bertrand Israel, solid state physics, statistical mechanics
Halverson, Ward Dean, plasma physics
Handrick, George Richard, organic chemistry
Hard, Thomas Michael, physical chemistry, optics
Harris, Allen Joel, astronomy
Harrison, Stephen Coplan, biophysical chemistry
Hart, Stanley Robert, geology, geochemistry
Hastings, John Woodland, marine biology, biological rhythms
Hays, James Fred, petrology, geophysics
Hearn, David Russell, x-ray astronomy
Hecht, Sidney Michael, chemistry, biological chemistry
Helgason, Sigurdur, mathematics
Henderson, Ellen Jane, biochemistry
Henry, Allan Francis, physics
Henry, Joseph Patrick, astrophysics
Herschbach, Dudley Robert, chemical physics
Hester, Richard Knight, magnetic resonance
Hiebert, Erwin Nick, history of science
Hildebrand, Francis Begnaud, mathematics
Hill, Albert Gordon, atomic physics
Hirschfeld, Tomas Beno, analytical chemistry, molecular spectroscopy
Hnatowich, Donald John, nuclear medicine
Hoaglin, David Caster, statistics
Hobbs, John Robert, physical chemistry, analytical chemistry
Hodes, William, organic chemistry, polymer chemistry
Hoffman, Jeffrey Alan, x-ray astronomy
Hoffman, Kenneth Myron, mathematics
Holland, Heinrich Dieter, geochemistry
Holldobler, Berthold Karl, behavioral biology
Hollister, Charlotte Ann, quantum chemistry
Holt, Charles Edward, molecular biology
Holton, Gerald James, physics
Homburger, Freddy, medicine
Hook, Edwin Oscar, organic chemistry
Hopkins, Esther Arvilla Harrison, physical chemistry
Horne, Ralph Albert, environmental chemistry
Houghton, Henry Garrett, meteorology
Howard, Louis Norberg, applied mathematics, fluid dynamics
Howells, William White, physical anthropology
Hsui, Albert Tong-Kwan, geophysics
Hu, Shiu-Ying Hsu, botany
Huang, Kerson, theoretical physics
Hubbard, Ruth, biology
Hulsizer, Robert Inslee, Jr, elementary particle physics
Hume, David Newton, chemistry
Hunt, Albert Melvin, analytical chemistry
Hurlbut, Cornelius Searle, Jr, mineralogy, crystallography
Hurley, Patrick Mason, geochronology
Hutchison, George B, epidemiology
Hutzenlaub, John F, physics, mathematics
Idelson, Martin, organic chemistry
Ingard, Karl Uno, fluid dynamics
Ingram, Vernon Martin, biochemistry
Irvine, John Withers, Jr, radiochemistry
Irwin, John Wellington, medicine
Jacchia, Luigi Giuseppe, astronomy
Jackiw, Roman Wladimir, theoretical physics, high energy physics
Jacobson, Bruce Shell, biological chemistry
Jaffe, Arthur Michael, mathematical physics
Jaffe, Robert Loren, elementary particle physics
Javan, Ali, physics
Jenkins, Farish Alston, Jr, vertebrate paleontology, anatomy
Johnson, Keith Huber, quantum chemistry
Johnson, Kenneth Alan, theoretical physics
Jones, Richard Victor, solid state physics
Jones, Robert Clark, theoretical physics
Jones, Roger Alan, organic chemistry
Jordan, John Emmett, physical chemistry
Josefowicz, Jack Yitzhak, biophysics
Joss, Paul Christopher, theoretical astrophysics, x-ray astronomy
Junger, Miguel Chapero, acoustics, applied mechanics
Kafatos, Fotis C, developmental biology
Kahn, David, physics
Kalkofen, Wolfgang, astrophysics
Kamilli, Diana Chapman, geochemistry, petrology
Kaplan, Irving, theoretical physics
Karel, Marcus, food science
Karplus, Martin, physical chemistry
Kasman, Sidney, photographic chemistry
Kasnitz, Harold Louis, physics
Keck, James Collyer, physics
Keeney, Ralph Lyons, operations research
Kelland, David Ross, physics
Kellogg, Edwin M, x-ray astronomy
Kemble, Edwin Crawford, physics
Kemp, Daniel Schaeffer, organic chemistry
Kendall, Henry Way, physics
Kensler, Charles Joseph, pharmacology, biochemistry
Kerman, Arthur Kent, theoretical physics
Kerwin, Edward Michael, Jr, acoustics
Kerzman, Norberto Luis, mathematics
Kessin, Richard Harry, developmental genetics
Kessler, Myer M, information science
Ketteringham, John M, physical chemistry
Khorana, Har Gobind, organic chemistry
King, John Gordon, atomic physics
King, Jonathan Alan, molecular biology, developmental genetics
King, Robert Wilson, Jr, geodesy
Kinsey, James Lloyd, physical chemistry
Kistiakowsky, George Bogdan, physical chemistry
Kistiakowsky, Vera, elementary particle physics
Klainer, Stanley M, instrumentation, physical chemistry
Kleitman, Daniel J, mathematics
Klemperer, William, physical chemistry
Kleppner, Daniel, physics
Klotz, Lynn Charles, physical biochemistry
Kohl, John Leslie, experimental atomic physics, solar physics
Kolm, Henry Herbert, magnetism
Koltun, Walter Lang, biophysical chemistry
Kopito, Louis Eliezer, nutrition
Kostant, Bertram, mathematics
Koster, George Fred, physics
Kovac, Jeffrey Dean, chemical physics
Kowalski, Stanley Benedict, nuclear physics
Krieger, Allen Stephen, solar physics
Krieger, Jeanne Kann, physical organic chemistry
Kroll, John Ernest, fluid dynamics, physical oceanography
Kummel, Bernhard, paleontology
Kyhl, Robert Louis, physics
Labate, Samuel, acoustics
Lagow, Richard James, chemistry

Solomon, Edward I., physical inorganic chemistry
Solomon, Sean Carl, geophysics
Sousson, Joseph Elias, rheology
Southard, John Brelsford, geology
Southworth, Richard Boynton, astronomy
Spongberg, Stephen Alan, systematic botany
Staelin, David Hudson, radio astronomy, meteorology
Staley, Ralph Horton, physical chemistry
Stanbury, John Bruton, experimental medicine
Stanley, Harry Eugene, statistical mechanics, solid state physics
Stanley, Richard Peter, algebra
Stark, Harold Mead, number theory
Steele, Warren Cavanaugh, physical chemistry
Steiner, Lisa Amelia, immunology
Steinfeld, Jeffrey Irwin, physical chemistry
Steminski, John Roman, organic chemistry
Stephenson, Clark Conkling, physical chemistry
Stevens, Raymond Sawtell, research administration
Stevenson, Donald Thomas, solid state physics
Stommel, Henry Melson, oceanography
Stone, Richard Spillane, physics
Strandberg, Malcom Woodrow Pershing, chemistry
Strang, William Gilbert, mathematics
Stratton, Julius Adams, physics
Strauch, Karl, particle physics
Suchanek, Rudolf Gerhard, experimental atomic physics
Suit, Joan C., microbiology
Sulak, Lawrence Richard, elementary particle physics, experimental high energy physics
Sullivan, James Douglas, physics, space science
Swain, Charles Gardner, chemistry
Szonyi, Geza, computer science, information science
Taylor, Edwin Florisman, physics
Taylor, Lloyd David, chemistry
Tedrow, Paul Muller, low temperature physics
Temkin, Richard Joel, lasers, plasma physics
Teubler, Hans-Lukas, psychophysiology
Teyler, Timothy James, neurosciences
Thayer, Philip Standish, microbiology
Thomas, George Brinton, Jr, mathematics
Thompson, Alan Bruce, petrology
Thompson, James Burleigh, Jr, petrology, geochemistry
Timothy, John Gethyn, space physics
Ting, Samuel C.C., particle physics
Tinkham, Michael, solid state physics
Tisza, Laszlo, physics
Toksoz, Mehmet Nafi, geophysics
Toomre, Alar, astronomy, applied mathematics
Torriani Gorini, Annamaria, bacterial physiology
Towle, Margaret Ashley, ethnobotany
Townley, Judy Ann, information science
Traficante, Daniel Dominick, structural chemistry
Trageser, Milton B., physics
Traub, Wesley Arthur, astrophysics, planetary atmospheres
Trinkaus, Erik, physical anthropology
Trum, Bernard Francis, veterinary medicine
Tryon, Rolla Milton, Jr, botany
Turchinetz, William Ernest, physics
Turnbull, David, physical chemistry
Turner, Ruth Dixon, malacology
Tyler, Bonnie Moreland, microbial physiology, molecular biology
Ulmer, Melville Paul, x-ray astronomy
Underwood, Donald Lee, physical chemistry
Van Dongen, Cornelis Godefridus, animal husbandry, reproductive physiology
Vanelli, Ronald Edward, organic chemistry
Van Heerden, Pieter Jacobus, physics
VanSpeybroeck, Leon Paul, x-ray astronomy
Van Vleck, John Hasbrouck, physics
Vessot, Robert F.C., physics

Villars, Felix Marc Hermann, theoretical physics
Vinti, John Pascal, celestial mechanics
Vogt, Evon Zartman, anthropology
Wachman, Harold Yehuda, surface physics
Wadsworth, George Proctor, mathematics
Waff, Harve S., geophysics
Wald, George, biochemistry
Walker, David, petrology
Wallace, Frederic Andrew, physical chemistry; analytical chemistry
Waller, David Percival, organic chemistry; photographic chemistry
Walsh, Christopher Thomas, biochemistry
Walsh, Joseph Broughton, geology, geophysics
Ward, William Roger, planetary sciences
Watson, Fletcher Guard, science education
Waugh, David Floyd, biology
Waugh, John Stewart, physical chemistry
Webb, Nathaniel Conant, Jr, preventive medicine, medical statistics
Webb, Robert Howard, medical physics
Weiffenbach, George Charles, experimental physics
Weinberg, Steven, theoretical physics
Weiner, Charles, history of physics, history of biology
Weinstein, Alexander, genetics, history of science
Weisberg, Herbert Ira, statistics
Weiss, Rainer, physics
Weisskopf, Victor Frederick, physics
Welsch, Roy Elmer, statistics
West, Stanley A., applied anthropology
Westheimer, Frank Henry, organic chemistry; enzymology
Wetmore, Ralph Hartley, botany, morphology
Whitehead, George William, mathematics
Whiteside, George McClelland, organic chemistry
Whitney, Charles Allen, astrophysics
Whitney, Cynthia Kolb, atmospheric physics, mathematical physics
Whittenberger, James Laverre, physiology
Wiley, Don Craig, biophysics
Willet, Hurd Curtis, meteorology
Williams, Carroll Milton, biology
Williams, Ernest Edward, herpetology, ecology
Williams, Stephen, anthropology; archaeology
Williamson, Claude F., nuclear physics
Wilke, Herbert Louis, Jr, applied mathematics
Wilson, Edgar Bright, chemical physics
Wilson, Edward Osborne, behavioral biology; evolutionary biology
Wilson, Richard, physics
Winbroe, George Lund, astrophysics, solar physics
Wodinsky, Isidore, oncology
Wogan, Gerald Norman, toxicology
Wolf, George, nutrition, biochemistry
Wolfe, Harry Bernard, operations research
Wolff, Peter Adalbert, physics
Wolfson, James, physics
Wood, Carroll E., Jr, botany
Wood, John Armsted, Jr, geochemistry
Woodward, Robert Burns, organic chemistry
Worden, Frederic Garfield, psychiatry
Wright, Frances Woodworth, astronomy
Wrighton, Mark Stephen, photochemistry; inorganic chemistry
Wu, Tai Tsun, physics
Wunsch, Carl Isaac, oceanography
Wurman, Judith Joy, nutrition
Wurtman, Richard Jay, endocrinology; neurobiology
Wyman, Stanley M., medicine, radiology
Yamamoto, Richard Kumeo, physics
Yang, Julie Chi-Sun, inorganic chemistry
Yesair, David Wayne, biochemistry
Young, James Edward, particle physics, nuclear physics
Young, Vernon Robert, nutrition, biochemistry
Zariski, Oscar, mathematics
Zeldin, Michael Hermen, cell biology; plant physiology
Zombeck, Martin Vincent, physics

CANTON
Bluhm, Aaron Leo, organic chemistry
Rowe, Paul E., physical organic chemistry

CARLISLE
Calabi, Lorenzo

Fohl, Timothy, fluid dynamics

CHATHAM
Tomlinson, Everett Parsons, nuclear physics, high energy physics

CHELMSFORD
Cochrane, Hector, physical chemistry
Rubinstein, Harry, organic chemistry
Skrable, Kenneth William, physics, radiological health

CHERRY VALLEY
Campbell, William (Aloysius), developmental anatomy, electron microscopy

CHESTERFIELD
Sayre, Geneva, botany

CHESTNUT HILL
Bade, Maria Leipelt, biochemistry
Bakshi, Pradip M., theoretical physics
Bennett, Ovell Francis, organic chemistry
Bezuszka, Stanley John, mathematical physics
Billo, Edward Joseph, inorganic chemistry; analytical chemistry
Boger, Eliahu, organic chemistry, enzymology
Bornstein, Joseph, synthetic organic chemistry
Brown, Edward Morgan, meteorology
Brown, George D. Jr, geology; paleontology
Cardullo, Maria Ann, cell physiology; bacteriology
Carovillano, Robert L., space physics, astrophysics
Chen, Joseph H., astrophysics
Davidovits, Paul, chemical physics
de Bethune, Andre Jacques, chemistry
Di Bartolo, Baldassare, solid state physics
Epstein, Harvey Irwin, mathematical analysis
Faber, Richard Leon, mathematics
Fabens, Augustus Jerome, probability
Finian, Walter Joseph, Jr, morphology, astrophysics
Gilroy, James Joseph, bacteriology
Goldsmith, George Jason, radiology
Greenberg, Orrin, dental materials, anatomy
Kalman, Gabor, plasma physics, theoretical physics
Kelly, Thomas Ross, organic chemistry
Lin, Jeong-Long, physical chemistry
Liss, Maurice, biochemistry; biology
Liuima, Francis Aloysius, physics
Marble, Alexander, medicine
Marcou, Rene Joseph, medicine
Maynard, Francis Louis, physiology
McCaffrey, Francis, solid state physics
McCaffrey, Timothy Edward, biochemistry
McFadden, David Lee, chemical physics
O'Brien, Elinor Murray, cancer, electron microscopy
O'Malley, Robert Francis, fluorine chemistry
Orlando, Joseph Alexander, biological chemistry
Pan, Yuh Kang, theoretical chemistry
Plocke, Donald J, biophysical chemistry
Ring-Carroll, Rose, mathematics
Russell, Irving James, nuclear chemistry
Sardella, Dennis Joseph, chemistry
Schubert, Clarence Joseph, chemistry
Schwebel, Solomon Lawrence, theoretical physics
Stetson, Richard Pratt, medicine
Sullivan, Joseph Arthur, mathematics
Sullivan, William Daniel, biochemistry
Ting, Yu-Chen, plant cytology, plant genetics
Uritam, Rein Aarne, elementary particle physics
Vogel, George, organic chemistry
Yoon, Chai Hyun, genetics
Zeiger, Herbert J., physics

CHICOPEE
Klouda, Mary Ann Aberle, physiology
Schneider, Maxine Dorothy, physical chemistry
Wright, Mary Lou, developmental biology

CLINTON
Dhami, Kewal Singh, organic chemistry, polymer chemistry

COHASSET
Buddington, Arthur Francis, geology
Mullen, James A., applied physics

CONCORD
Austin, James Murdoch, meteorology
Austin, Pauline Morrow, meteorology
Barbour, William E. Jr, applied physics, electrical engineering
Barnes, James Clarkson, meteorology, physics
Blau, Henry Hess, Jr, atmospheric physics
Clapp, Philip Charles, solid state physics
Damon, Richard Winslow, solid state physics
Decker, John Alvin Jr, optics
Dinneen, Gerald Paul, mathematics
Dion, Andre R., physics
Esch, Robin Ernest, applied mathematics
George, James Henry Bryn, physical chemistry, chemical engineering
Gurski, Thomas Richard, quantum optics, electrooptics
Hanselman, Raymond Bush, analytical chemistry
Hardy, Kenneth Reginald, atmospheric physics
Harrison, George Russell, physics
Hersh, John Franklin, meteorology
Kissmeyer-Nielsen, Erik, food science
Long, Louis, Jr, organic chemistry
Murphy, Brian Logan, air pollution, science policy
Parker, Lee Ward, mathematical physics
Pittelli, Ernest, physics
Reed, Thomas Binnington, energy conversion
Sherr, Paul Edgar, meteorology
Souter, Lamar, surgery
Stern, Ernest, solid state physics, acoustics
Thomas, Martha Jane Bergin, analytical chemistry; physical chemistry
Valley, George Edward, Jr, physics
Waymouth, John Francis, applied physics

DANVERS
Bragole, Robert Anthony, organic chemistry
Cadmus, Eugene L., microbiology
Eby, John Edson, solid state physics, optics
Lovette, Maribeth, optics
Meyer, Vincent D., physical chemistry, chemical physics
Sanford, Robert Alan, physical chemistry, electrochemistry
Shinn, Dennis Burton, inorganic chemistry
Strem, Michael Edward, organic chemistry

DEDHAM
Abbott, Norman John, textile physics
Barish, Leo, microscopy; textile chemistry
Cohen, Harvey Martin, inorganic chemistry
Conover, John Hoagland, meteorology
Davis, Robert Bernard, chemistry; polymer chemistry
McCarthy, Francis Wadsworth, physics
Panto, Joseph Salvatore, textile chemistry
Quynn, Richard Grayson, polymer physics

DORCHESTER
Alper, Joseph Seth, theoretical chemistry
Clench, William James, zoology

DOVER
Nicol, James, engineering physics, low temperature physics
Russell, Henry D., ecology

DRACUT
Dunlap, Albert Atkinson, plant pathology; plant physiology

DUDLEY
Songdahl, John Harald, comparative physiology; marine biology

Stancl, Mildred Luzader, topology

DUXBURY
Hillman, Robert Edward, invertebrate zoology, marine biology
Townley, Charles William, nuclear chemistry, research administration
Willingham, Charles Allen, marine ecology, marine biology

EAST WALPOLE
Knudson, Harold William, paper chemistry

EAST WAREHAM
Cross, Chester Ellsworth, botany
Deubert, Karl Heinz, environmental biology
Devlin, Robert Martin, plant physiology

EDGARTOWN
Atwood, Francis Clarke, chemistry
Mazer, Milton, psychiatry

EVERETT
Ewing, James Joyce, chemical physics
Herzlinger, George Arthur, biophysics, bioengineering
Itzkan, Irving, physics, electrical engineering
Janes, George Sargent, physics
Kantrowitz, Arthur (Robert), gas dynamics, health physics
Karo, Douglas Paul, statistical mechanics, optics
Kivel, Bennett, theoretical physics
Lederman, David Mordechai, biomedical engineering, biophysics
Nyilas, Emery, organic chemistry
Peng, Fred Ming-Sheng, polymer chemistry
Petschek, Harry E, physics
Zar, Jacob L, lasers

FALMOUTH
Stanbrough, Jess Hedrick, Jr, physics
Zinn, Donald Joseph, invertebrate zoology

FEEDING HILLS
Marieb, Elaine Nicpon, anatomy, physiology

FITCHBURG
Condike, George Francis, inorganic chemistry
Davis, Frederic Whitlock, zoology, wildlife manage ment
Fuhrman, Albert William, organic chemistry
Jablonski, Werner Louis, organic chemistry
Larson, Leslie L, organic chemistry
Martens, Edward John, nuclear physics
McNaney, John A, inorganic chemistry
Strong, Robert Stanley, analytical chemistry
Vignale, Michael Joseph, organic chemistry
Wolf, Frank E, biology
Zottoli, Robert, invertebrate zoology

FOXBORO
Fletcher, Kenneth Steele, III, analytical chemistry
McCrea, Peter Frederick, instrumentation
Vignos, James Henry, physics

FRAMINGHAM
Adams, Herbert Jack, pharmacology, physiology
Anstey, Robert L, geography, climatology
Bhatia, Sushil, polymer chemistry
Brack, Karl, organic polymer chemistry, photochemistry
Bustead, Ronald Lorima, Jr, food technology, food chemistry
Castagna, Frank, mathematics
Covino, Benjamin Gene, physiology
Haight, Thomas H, botany
Hirt, Howard Franklin, geography of Southern Asia, urban geography
Ingalls, Theodore Hunt, epidemiology, nutrition
Jost, Dana Nelson, phycology

Jost, Paul Douglas, analytical chemistry
Kannel, William B, cardiovascular diseases, internal medicine
Keough, Allen Henry, organic chemistry
Martin, Thomas George, III, health physics
McPhillips, Joseph John, pharmacology
Previte, Joseph James, microbiology, physiology
Prindle, William Roscoe, ceramics
Rankin, Joel Sender, obstetrics & gynecology
Rosenberg, Isadore Nathan, medicine
Rosman, Bernard Harvey, mathematics, computer science
Sopka, John J, Jr, mathematics
Spence, Sidney Charles, botany
Stern, Willard Lewis, atmospheric physics, environmental sciences
Traub, Alan Cutler, electro-optics
Zapsalis, Charles, food science, chemistry

FRAMINGHAM CENTRE
Polanyi, Thomas Gabriel, physics
Schiff, Daniel, resource management, energy conversion

FRANKLIN
Wendt, Theodore Mil, microbiology, chemistry

GLOUCESTER
Coon, Carleton Stevens, anthropology
Hanks, Robert William, fisheries management, research administration
King, Frederick Jessop, food science
Rathjen, Warren Francis, fisheries biology
Rose, Peter Henry, physics
Witkower, Andrew Benedict, atomic physics
Yentsch, Charles Samuel, marine biology

GROTON
Webber, Harold Haskell, aquatic biology

HADLEY
Smith, Elinor Van Dorn, bacteriology

HAMPDEN
Bump, Charles Kilbourne, chemistry
Phillips, Yorke Peter, polymer chemistry

HANSCOM AFB
Barnes, Arnold Appleton, Jr, meteorology
Bendow, Bernard, solid state physics
Garth, John Campbell, solid state physics
Guidice, Donald Anthony, radio astronomy
Hayes, Dallas T, physics, mathematics
Hunt, Mahlon Seymour, geodesy
Herskovitz, Sheldon Bernard, microwave physics
Katz, Ludwig, space physics, magnetospheric physics
Knecht, David Jordan, magnetospheric physics
Leiby, Clare C, Jr, physics
Pavel, Arthur Lawrence, space physics
Rooney, Thomas Peter, mineralogy
Skolnik, Lyn Howard, optics
Van Tassel, Roger A, aeronomy
Weinstein, Alan Ira, meteorology

HANSON
Tillotson, James E, food science

HARVARD
Bliss, Arthur Dean, organic chemistry
Saxl, Erwin Joseph, physics

HARWICH PORT
Lucke, John Becker, geology

HAVERHILL
Hill, Henry Aaron, organic chemistry

HINGHAM
Latady, William Robertson, optics, mechanics
Mayburg, Sumner, solid state physics

HOLDEN
Lemaire, Minnie Ethel, geography

HOLYOKE
Dunny, Stanley, organometallic chemistry
Fitzgerald, Marie Anton, physiology
Hodgins, George Raymond, photographic chemistry
Walter, Henry Alexander, organic chemistry

HYANNIS
Gyure, William Louis, clinical biochemistry
Kattsoff, Louis Osgood, mathematics, philosophy of science

INDIAN ORCHARD
Auerbach, Michael Howard, analytical chemistry
Baer, Massimo, organic chemistry
Barkhuff, Raymond Addison, Jr, polymer chemistry
Brodsky, Philip Hyman, polymer chemistry
Cairns, Robert Edward, organic chemistry
Corey, Albert Eugene, organic chemistry
Drumm, Manuel Felix, organic chemistry
Fitzhugh, Andrew Fyfe, chemistry
Gardner, Donald Murray, polymer chemistry
Hardwicke, Norman Lawson, polymer chemistry
Le Blanc, John Roger, organic polymer chemistry
Lee, Yoon Chai, chemistry
Magin, Ralph Walter, organic chemistry, polymer chemistry
Martin, Richard Hadley, Jr, polymer chemistry
Mendelson, Robert Allen, polymer science, rheology
Mont, George Edward, polymer chemistry
Nemeth, Ronald Louis, physical chemistry
Rademacher, Leo Edward, organic chemistry
Reisman, Abraham Joseph, polymer chemistry
Serlin, Irving, polymer chemistry
Shapras, Peter, analytical chemistry
Snelgrove, James Arthur, physical chemistry
Southern, John Hoyle, II, polymer science
Terry, Stuart Lee, polymer chemistry

IPSWICH
Hanley, Peter Ronald, applied physics
Wade, Robert Charles, inorganic chemistry

JAMAICA PLAIN
DeWolf, Gordon Parker, Jr, taxonomic botany
Edds, Mac Vincent, developmental biology, neurosciences
Freedberg, Abraham Stone, medicine
Girard, Kenneth Francis, bacteriology, immunology
Howard, Richard Alden, botany
Madoff, Morton A, infectious diseases, immunology
Ofner, Peter, organic chemistry, andrology
Raam, Shanthi, oncology
Segarra, Joseph M, neurology

LAWRENCE
Botan, Edward Allan, microbiology, biochemistry
Holt, Roland Bell, physics
Stone, Herman, organic chemistry
Winstrom, Leon Oscar, physical chemistry, textile chemistry

LEE
Mattina, Charles Frederick, Jr, physical chemistry

LEOMINSTER
Biletch, Harry, organic polymer chemistry
Cantrill, James Egbert, organic chemistry
Divis, Roy Richard, polymer chemistry
Gibbs, William Eugene, physical chemistry
Hamilton, Charles William, chemistry
Ikeda, Tatsuya, molecular spectroscopy
Millington, James E, organic chemistry

Phillips, Richard Arlan, optics, applied physics

LEXINGTON
Animalu, Alexander Obiefoka Enukora, physics, mathematics
Antonoff, Marvin M, theoretical physics, solid state physics
Ash, Michael Edward, applied mathematics
Bass, Arthur, air pollution
Bradley, Lee Carrington, III, optical physics
Brundage, Robert Scott, biophysical chemistry
Buchanan, John Machlin, biochemistry
Cathles, Lawrence MacLagan, III, geophysics
Champion, Kenneth Stanley Warner, physics
Chase, Fred Leroy, chemistry, research administration
Chiu, Tai-Woo, physical chemistry
Clagett, Donald Carl, organic chemistry
Cook, Glenn Melvin, physical chemistry
Daly, Richard Farrell, physics
De Lorenzo, Eugene Joseph, physical chemistry
Deutsch, Thomas, lasers
Dickey, Dana H, solid state physics
Dionne, Gerald Francis, magnetism, surface physics
Dresselhaus, Gene Frederick, solid state physics
Edelberg, Seymour, optical physics
Enzmann, Robert D, geology, electrical engineering
Evans, John V, radio physics
Evans, Thomas George, mathematics, computer science
Fan, John Chin Chiang, applied physics
Farber, Morton Sheldon, microwave physics
Feinleib, Julius, solid state physics
Fetterman, Harold Ralph, lasers, experimental solid state physics
Ford, Peter Wilbraham, applied physics
Freed, Aubyn, mathematics
Gaposchkin, Edward Michael, geodesy
Gerdes, John William, physics, electronic engineering
Giuliano, Vincent E, applied mathematics
Goodenough, John Bannister, physics
Goodman, Philip, physical chemistry, chemical instrumentation
Green, Jerome Joseph, magnetism
Gschwendner, Alfred Benedict, optics
Harte, Kenneth J, electron optics, solid state physics
Henrich, Victor E, solid state physics
Herlin, Melvin Arnold, physics
Hills, Robert, Jr, optics
Hinkley, Everett David, physics
Hong, Henry Yao-Pen, solid state chemistry
Hudgin, Richard Henry, mathematical physics, atmospheric physics
Hull, Robert Joseph, physics
Jackson, Earl Graves, physical inorganic chemistry
Kafalas, James A, physical inorganic chemistry
Kafalas, Peter, chemical physics
Kelley, Thomas F, biomedical engineering, clinical chemistry
Kildal, Helge, applied physics
Kingston, Robert Hildreth, physics
Kleiman, Herbert, physics
Klein, Milton M, fluid dynamics
Kofsky, Irving Louis, physics
Krag, William Eric, physics
Ku, Robert Tien-Hung, plasma physics, molecular spectroscopy
Kust, Roger Nayland, inorganic chemistry, physical chemistry
Larsen, David M, theoretical physics
Leipziger, Fredric Douglas, analytical chemistry
Lell, Eberhard, photochemistry
Lewis, Henry Rafalsky, physics
Lewis, William Weston, laboratory medicine
Light, Truman S, analytical chemistry
Lingane, Peter James, electrochemistry, metallurgy
Longaker, Perry R, physics
Lowder, J Elbert, applied physics
Mack, Charles Lawrence, Jr, applied physics
Madden, Stephen James, Jr, computer science, geodesy
Malenfant, Arthur Lewis, analytical chemistry
Manning, William Joseph, microbiology, plant pathology
Marcus, Stephen, lasers
Marquet, Louis C, atomic physics, spectroscopy

Martineau, Robert Jean, theoretical physics, electronics
Mavroides, John George, physics
McCue, John Joseph Gerald, microwave physics
Mehgahii, John, applied physics
Menashi, Jameel, inorganic chemistry, physical chemistry
Nedzel, V Alexander, physics
Nesbeda, Paul, mathematics
Park, Won Choon, economic geology
Pian, Charles Hsueh Chien, organic chemistry
Pippert, Glen Francis, electromagnetics
Raisbeck, Gordon, mathematics, electrical engineering
Rediker, Robert Harmon, physics
Rheinstein, John, physics
Rimmer, Matthew Peter, optics
Rindner, Wilhelm, solid state physics
Rockstad, Howard Kent, solid state physics
Rosato, Frank Joseph, applied physics
Sampson, John Laurence, physics
Sasiela, Richard, physics
Schilling, Martin, lasers
Schulte, Daniel Herman, astronomy, optics
Schultz, Herman Solomon, polymer chemistry
Scott, Peter Hamilton, chemistry
Shapiro, Irwin Ira, physics, astronomy
Shrock, Robert Rakes, paleontology, sedimentology
Siegel, Pamela Jean, molecular biology, virology
Slade, Chaloner Berry, physics
Slattery, Richard Erick, physics
Smith, Donald Oscar, solid state physics
Spears, David Lewis, solid state physics
Speliotis, Dennis Elias, solid state physics, magnetism
Spencer, William Turner, physics
Spergel, Philip, instrumentation, electrical engineering
Tannenwald, Peter Ernest, physics
Temin, Samuel Cantor, polymer physics
Wand, Ronald Herbert, ionospheric physics
Weiner, Stephen Douglas, applied physics
Wexler, Arthur Samuel, analytical chemistry
Wyman, John E., chemistry
Zisson, James, organic chemistry

LINCOLN
Birkett, James Davis, physical chemistry
Bolt, Richard Henry, physics, acoustics
Buckler, Sheldon A., organic chemistry
Drury, William Holland, ecology
Friel, Patrick Joseph, physical chemistry
Gerson, Nathaniel Charles, ionospheric physics, electromagnetism
Gross, Thomas Alfred Otto, physics
Koopman, Bernard Osgood, mathematics
Luft, Ludwig, physical chemistry
McMahon, Howard Oldford, physical chemistry
Mount, Wayne Delano, meteorology, environmental management
Smith, Alan Bradford, applied physics

LINCOLN CENTER
Fullerton, Albert L, Jr., data processing
Terrell, John Hart, applied physics

LITTLETON
Baird, James, ornithology

LONGMEADOW
Basdekis, Costas H, polymer chemistry
Ezrin, Myer, analytical chemistry, plastics chemistry
McNeil, Harry Daniel, Jr, organic chemistry, polymer chemistry
Milani, Victor John, microbiology
Platzer, Norbert, polymer chemistry
Pollard, Robert Eugene, polymer chemistry
Schnitzler, Ronald Michael, physiology

LOWELL
Altman, Albert, theoretical nuclear physics, solid state physics
Baker, Adolph, physics
Bannister, William Warren, physical organic chemistry, marine chemistry
Barnes, Barry Keith, physics
Beghian, Leon E, nuclear physics
Blumstein, Alexandre, polymer chemistry
Bruce, John Irvin, Jr., parasitology, physiology
Carr, George Leroy, physics, science education
Clough, Stuart Benjamin, physical chemistry, polymer chemistry
Cole, Charles Daniel, physics
Coleman, Robert Marshall, parasitology, immunology
Couchell, Gus Perry, nuclear physics
Deann, Rudolph D., polymer chemistry
Egan, James Joseph, experimental nuclear physics
Farina, Joseph Peter, microbial physiology, health sciences
Griffin, George Robert, organic chemistry
Herman, John R., ionospheric physics
Isaks, Martin, physical organic chemistry
Israel, Stanley C., polymer chemistry, physical organic chemistry
James, Ernest P, analytical chemistry
Kamien, Ethel N, plant physiology, human biology
Kannenberg, Lloyd C., physics
Karakashian, Aram Simon, theoretical solid state physics
Kegel, Gunter Heinrich Reinhard, nuclear physics
Kowalak, Albert Douglas, physical inorganic chemistry
Lee, Siu-Lam, insect ecology, evolution
Macdonald, Timothy, genetics, biochemistry
Mathur, Suresh Chandra, nuclear chemistry
Mellen, Walter Roy, elementary particle physics, theoretical physics
Miller, Arthur I., history of science, theoretical high energy physics
Mittler, Arthur, physics
Peireti, Robert John, chemistry
Phelps, James Parkhurst, reactor physics
Pierce, James Bruce, electrochemistry
Port, William Solomon, organic chemistry
Pullen, David John, nuclear physics
Pyun, Chong Wha, physical chemistry
Rencricca, Nicholas John, hematology, malariology
Ring, Paul Joseph, physics
Salamone, Joseph C., organic chemistry, polymer chemistry
Sebastian, Kunnat J., theoretical high energy physics
Sheldon, Eric, theoretical nuclear physics
Tang, Wen, dynamic meteorology
Watterson, Arthur C, Jr, organic chemistry, polymer chemistry
Weinberg, I Jack, computer science, applied mathematics
Winslow, Richard Edward, mathematics
Wong, Chuen, physics
Worrell, Francis Toussaint, physics

LYNN
Clemson, Harry C., biochemistry, clinical chemistry
Kalmes, Otto Julius, pulp & paper technology

LYNNFIELD
Newkirk, Terry Franklin, ceramics, applied physics

MALDEN
Mijal, Chester Francis, physical chemistry, organic chemistry
Willwerth, Lawrence James, chemistry

MANCHESTER
Karas, John Athan, physics
Klein, Gerhart Paul, physical chemistry, electrochemistry
Simon, Edward, solid state science

MARBLEHEAD
Butler, Keith Huestis, chemistry
Cohen, Merrill, organic chemistry
Most, David S., paper chemistry
Pritchard, David Edward, atomic physics
Quinn, Daniel James, high energy physics
Reichlin, Seymour, internal medicine, physiology
Reuss, Robert L., geology
Reynolds, William Francis, algebra
Roeder, Kenneth David, biology
Roys, Chester Crosby, invertebrate physiology
Sames, George L., zoology
Sample, Howard H, solid state physics
Schlesinger, James William, mathematics
Schneps, Jack, physics

MARLBOROUGH
Frey, John H, environmental management, geophysics

MARSHFIELD
Butler, John Louis, physics

MATTAPOISETT
Goetz, Rudolph W, organic chemistry, analytical chemistry
Hoff, James Gaven, ecology, ichthyology

MAYNARD
Bell, James Richard, computer science, research administration
Brender, Ronald Franklin, information science
Payne, Mary Hewlett, mathematics
Reid, Marion Adelaide, physiology

MEDFORD
Berk, Harold, histology
Berman, Kenneth Sidney, oral biology
Bresler, Jack Barry, research administration, human biology
Canter, Joseph M, experimental high energy physics
Chew, Frances Sze-Ling, biology
Cooper, James William, chemical instrumentation, organometallic chemistry
Dane, Benjamin, animal behavior, physiology
Dewald, Robert Reinhold, physical chemistry
Eddy, Robert Devereux, inorganic chemistry
Evans, Gordon Goodwin, organic chemistry
Everett, Allen Edward, elementary particle physics
Feinleib, Mary Ella (Harman), plant physiology
Feldberg, Ross Sheldon, biochemistry
Ferrante, Jeanne, mathematical logic, computer science
Frommer, Jack, anatomy
Fulton, Dawson Gerald, mathematics
Georgian, Vlasios, organic chemistry
Gibb, Thomas Robinson Pirie, Jr., analytical chemistry, marine chemistry
Grace, Norman David, gastroenterology, internal medicine
Greenwood, Fred Laurel, organic chemistry
Guerin, Robert Powell, solid state physics, magnetism
Gunther, Leon, theoretical solid state physics
Guterman, Martin Mayr, algebra
Haas, Terry Evans, inorganic chemistry
Halpern, Miriam Patricia, pure mathematics
Hecht, Norman B, cell biology
Hodgson, Edward Shilling, comparative physiology
Hoge, Harold James, thermal sciences, fluid mechanics
Holt, Frederick Sheppard, applied mathematics
Hume, James David, geology
Illinger, Karl Heinz, physical chemistry
Isles, David Frederick, mathematics
Kaczmarczyk, Alexander, inorganic chemistry
Kirtzman, Julius, medicine
Klema, Ernest Donald, nuclear physics
Knipp, Julian Knause, physics
Kreifeldt, John Gene, biomedical engineering, human factors engineering
Maloney, William Farlow, medicine
McCarthy, Kathryn Agnes, solid state physics
Messer, Charles Edward, physical chemistry
Milburn, Nancy Stafford, physiology, electron microscopy
Milburn, Richard Henry, nuclear physics
Munford, George Saltonstall, III, astronomy
Nichols, Robert Leslie, geology
Nickerson, Norton Hart, plant morphology, plant anatomy
Page, Lot Bates, internal medicine, molecular physics
Siegel, Eli Charles, microbial genetics
Slapikoff, Saul Abraham, biochemistry
Stolow, Robert David, organic chemistry
Sweet, Herman Royden, botany
Szerenyi, Peter, physical chemistry
Tessman, Jack Robert, physics
Urry, Grant Wayne, inorganic chemistry
Weaver, David Leo, theoretical physics
Sullivan, Charles Irving, organic polymer chemistry
Wu, Tse Cheng, organic chemistry, polymer chemistry
Williams, Philip, prosthodontics
Winter, Stephen Samuel, science education

MEDWAY
Seacord, Daniel Freeman, Jr., physics

MELROSE
Moriarty, John Henry, food science
Orphanos, Demetrius George, organic chemistry

METHUEN
Saad, Farida M, biochemistry

MIDDLETON
Rossito, Conrad, polymer & organic chemistry

MILFORD
Bidlingmeyer, Brian Arthur, analytical chemistry
Little, James Noel, analytical chemistry
Vivilecchia, Richard, analytical chemistry

MILTON
Esten, Benjamin E, anesthesiology
Evans, Hiram John, embryology
Holloway, John Robert, comparative endocrinology
Hovorka, John, physics
Jeffries, David, applied physics
Touger, Jerold Steven, experimental solid state physics
Tramondozzi, John Edmund, organic chemistry

MONUMENT BEACH
Bottoms, Albert Maitland, operations research, systems analysis

NAHANT
Morse, Mary Patricia, malacology
Riser, Nathan Wendell, invertebrate zoology

NANTUCKET
Mahoney, Earle Barnes, surgery

NATICK
Allen, Paul D, exercise physiology
Anderson, Edward Everett, food science
Anellis, Abe, bacteriology
Angelini, Pio, food chemistry
Bailey, Milton, industrial chemistry
Ball, Derek Harry, carbohydrate chemistry
Bazinet, Maurice L., chemistry
Breckenridge, John Robert, biophysics
Brockmann, Maxwell Curtis, food science
Brynjolfsson, Ari, nuclear physics, geophysics
Calhoun, William Kenneth, nutrition
Carpenter, David Francis, microbial physiology
Coplan, Myron Julius, chemical physics
Cymerman, Allen, physiology
Denniston, Joseph Charles, medical physiology
Evans, Wayne Orien, psychopharmacology
Fix, Richard Conrad, environmental chemistry
Florentine, Gerald Joseph, insect physiology
Francesconi, Ralph P, biological chemistry
Goldman, Ralph Frederick, physiology
Heidelbaugh, Norman Dale, food technology, public health

Hertweck, Gerald, operations research, systems analysis
Herz, Matthew Lawrence, physical organic chemistry
Hubbard, Roger W., environmental medicine
Jacobs, Harry Lewis, psychophysiology, nutrition
Johnson, Karl Robert, food science
Jones, LeeRoy George, environmental medicine, internal medicine
Kaplan, Arthur Milton, microbiology
Kapsalis, John George, food science, biochemistry
Laible, Roy C, polymer physics
Lampi, Rauno Andrew, food technology
Levinson, Hillel Salmon, microbiology
Mabrouk, Ahmed Fahmy, agricultural chemistry
Macnair, Richard Nelson, organic chemistry
Mager, Milton, biochemistry
Maher, John Thomas, physiology
Mandels, Gabriel Raphael, microbial physiology, plant physiology
Mandels, Mary Hickox, microbial physiology
McCormick, Neil Glenn, microbiology
Merritt, Charles, Jr., analytical chemistry
Narayan, Krishamurthi Ananth, nutritional biochemistry, food science
Pandolf, Kent Barry, environmental physiology, exercise physiology
Pober, Zalmon, mammalian physiology
Porter, William L., biochemistry
Powers, Edmund Maurice, food microbiology
Pratt, John Jacob, Jr, entomology
Ramsley, Alvin Olsen, physical chemistry
Reese, Elwyn Thomas, mycology
Richmond, Robert Chaffee, radiobiology
Robinson, Sumner M, environmental physiology
Rogers, Morris Ralph, industrial microbiology
Ross, Edward William, Jr, mathematics, applied mathematics
Roth, Louis Marcus, insect physiology
Rowley, Durwood B, food microbiology
Secrist, John Leonard, food science
Sieling, Dale Harold, research administration
Silverman, Gerald, microbiology, food science
Spira, William Martin, food microbiology, microbial physiology
Thomas, Miriam Mason Higgins, nutrition
Vogel, James Alan, exercise physiology, environmental medicine
White, Robert Manson, anthropometrics, physical anthropology
Wierbicki, Eugen, biochemistry, agricultural chemistry
Wilson, Angus, rubber chemistry
Winsmann, Fred Rudolph, physiology, biology

NEEDHAM
Chase, Charles Elroy, Jr, lasers
Harris, Sheldon Richard, biochemistry
Lowen, Jack, analytical chemistry
Oman, Henry James, electron physics
Riblet, Henry James, mathematics
Yonda, Alfred William, mathematics

NEEDHAM HEIGHTS
Brusca, Donald Richard, clinical chemistry
Kennedy, Stephen Jay, textiles
Myers, Robert Frederick, chemical engineering

NEW BEDFORD
Friedman, Lawrence Boyd, inorganic chemistry
Porter, Raymond P, physical chemistry
Sauro, Joseph Pio, physics

NEWTON
Albert, Mary Roberts Forbes (Day), cytology, biochemistry
Brummer, S Barry, physical chemistry
Copley, Lawrence Gordon, applied physics
Cunningham, Frances, cell biology
Harvey, Walter William, physical chemistry, extractive metallurgy
Josephson, Edward Samuel, food science, research administration
Ludwin, Isadore, animal genetics, physiology
Marten, James Frederick, biochemistry
Michelson, Louis, physics

Offenhartz, Peter O'Donnell, physical chemistry
Rauh, Robert David, Jr, electrochemistry, photochemistry
Walton, William Upton, physics, science education

NEWTON CENTER
Abeles, Robert Heinz, biochemistry
Davidson, Gilbert, science administration, optics
Green, Milton, photographic chemistry
Kalnajs, Janis Arvids, physical chemistry, inorganic chemistry
Lichtin, Norman Nahum, physical organic chemistry
Schuler, Robert Frederick, organic chemistry
Smith, Frederick Dowswell, optics

NEWTON HIGHLANDS
Howard, John Nelson, physics
Pandorf, Robert Clay, low temperature physics
Solinger, Julian Louis, physiology

NEWTON LOWER FALLS
Laurenzi, Gustave, medicine
Witherell, Egilda DeAmicis, radiological physics, nuclear medicine

NEWTON UPPER FALLS
Kahalas, Sheldon Lee, physics

NEWTONVILLE
Bilodeau, Gerald Gustave, mathematical analysis
Gordy, Edwin, biomedical engineering
Loew, Earl Randall, physiology
Sheingold, Leonard Sumner, applied physics
Todd, Neil Bowman, evolutionary biology, genetics

NORTH ACTON
Marshall, Donald James, physics

NORTH ADAMS
Bernard, Walter Joseph, physical chemistry
Burn, Ian, ceramics
Engel, Adolph James, electroanalytical chemistry
Finkelstein, Manuel, organic chemistry
Fresia, Elmo James, electrochemistry
Hess, John Monroe Converse, physical chemistry
Millard, Richard James, electrochemistry
Randall, John J, Jr, physical chemistry, inorganic chemistry
Ross, Sidney David, chemistry
Shirn, George Aaron, physics
Sleeman, Richard Alexander, chemistry
Summers, Selby Edward, chemistry

NORTH AMHERST
McKenzie, Malcolm Arthur, forest pathology

NORTH ANDOVER
Apley, Martyn Linn, invertebrate zoology, physiological ecology
Costello, Ernest F, Jr, physics
DeVelis, John Bernard, theoretical physics
Emshwiller, Maclellan, physics, communications engineering
Fann, Huoo-Long, nuclear physics
Hart, Joseph L, microbiology
Kearns, Donald Allen, mathematics
Ozimkoski, Raymond Edward, mathematics
Singleton, John Byrne, physics
Tambasco, Daniel Joseph, theoretical physics
Thomas, Aubrey Stephen, Jr, plant physiology

NORTH BILLERICA
Estin, Robert William, physics, science education
Lebowitz, Elliot, nuclear medicine, nuclear chemistry
Rocco, Gregory Gabriel, radiochemistry
Tripodi, Daniel, immunochemistry, microbiology

NORTH DARTMOUTH
Asato, Yukio, microbiology
Baker, Dwight Lynds, chemistry
Bar-Yam, Zvi H, elementary particle physics, high energy physics
Bessette, Russell Romulus, analytical chemistry
Conrad, Walter Edmund, organic chemistry
De Pagter, James Keith, physics
Dowd, John P, high energy physics
Dupre, Edmund J, textile chemistry, organic polymer chemistry
Freier, Jerome Bernard, mathematics
Hooper, Robert John, inorganic chemistry, molecular spectroscopy
Ibara, Richard Mamoru, ichthyology, ecology
Kern, Wolfhard, high energy physics
McCabe, Robert Lyden, mathematics
Moss, Sanford Alexander, III, environmental physiology, marine biology
Mowery, Dwight Fay, Jr, carbohydrate chemistry, chemical kinetics
Mulcare, Donald J, zoology, developmental biology
Nichols-Driscoll, Jean Ann, biological oceanography
Reardon, John Joseph, ecology
St Amand, Joseph, theoretical high energy physics, science education
Stauder, Jack, cultural anthropology
Stone, Samuel Arthur, mathematics
Szal, Roger Andrew, invertebrate zoology
Tykodi, Ralph John, physical chemistry
Wechter, Margaret Ann, analytical chemistry, radiochemistry
Wolock, Fred Walter, mathematical statistics

NORTH EASTON
Hurley, Francis Joseph, parasitology
Moore, Maryalice Conley, organic chemistry

NORTH READING
Riemer-Rubenstein, Delilah, medical administration, neuropsychiatry
Smith, Malcolm (Kinmonth), physics, instrumentation

NORTHAMPTON
Adams, Helen Elizabeth, mathematics
Albertson, Michael Owen, combinatorial mathematics
Anderson, Margaret, neurobiology
Burger, Henry Robert, III, structural geology
Burk, Carl John, plant taxonomy, plant ecology
Carpenter, Esther, zoology
Cohen, David Warren, mathematics
Curran, Harold Allen, paleontology, paleoecology
Davis, John Dunning, environmental sciences
de Villafranca, George Warren, comparative biochemistry
Dickinson, Alice B, mathematics
Durham, George Stone, physical chemistry
Fleck, George Morrison, physical chemistry
Greene, Joyce Marie, immunology, microbiology
Haskell, David Andrew, plant morphology
Hawkins, Bruce, musical acoustics
Hedlund, James Howard, mathematics
Hellman, Kenneth P, biochemistry
Horner, B Elizabeth, zoology
Ivey, Elizabeth Spencer, acoustics
Josephs, Jess J, musical acoustics
Kemble, Robert Penn, psychiatry
Laprade, Mary Hodge, zoology
Lowry, Thomas Hastings, organic chemistry
Mendelson, Bert, mathematics
Merritt, Robert Buell, population genetics
Olivo, Richard Francis, neurobiology
Powell, Jeanne Adele, developmental biology
Reid, Philip Dean, plant physiology
Schalk, Marshall, geology
Soffer, Milton David, synthetic organic chemistry, natural products chemistry
Steinberg, Melvin Sanford, theoretical physics
Tilley, Stephen George, population biology, herpetology
Tyrrell, Elizabeth Ann, microbiology
White, Brian, geology

NORTON
Beck, Sidney L, developmental genetics, teratology
Chidsey, Jane Louise, physiology
Jennings, Bojan Hamlin, organic chemistry
Kricher, John C, ecology
Marshall, Maud Alice, organic chemistry
O'Neill, Anne Frances, mathematical statistics
Pearson, Myrna Schmidt, organic chemistry
Spanier, Bonnie Barbara, virology, molecular biology
Spearing, Ann Marie, plant physiology, plant ecology
White, Elizabeth Lloyd, experimental embryology, molecular biology

NORWELL
Crandlemere, Robert Wayne, applied chemistry, forensic science
Dugan, Gary Edwin, pharmaceutical chemistry
MacCoy, Clinton Viles, ecology

NORWOOD
Cabell, Pamela Whiting, plastics chemistry
Madore, Bernadette, botany, bacteriology
MacDonald, Richard Annis, medicine, pathology
Malin, Murray Edward, physical chemistry
Markstein, George Henry, applied physics
Warthin, Thomas Angell, medicine

OXFORD
Lippson, Robert Lloyd, marine biology

PALMER
Thomas, Clayton Lay, medicine

PAXTON
Cole, Edward Anthony, microbiology, physiology
Socquet, Irene, chemistry

PEABODY
Greif, Mortimer, surface chemistry, colloid chemistry

PEPPERELL
Light, Donald Willis, chemistry

PETERSHAM
Gould, Ernest Morton, Jr, forest economics, resource management
Langenauer, Haviva Dolgin, botany
Posluszny, Usher, plant morphology
Raup, Hugh Miller, botany
Tomlinson, Philip Barry, botany
Torrey, John Gordon, plant physiology
Zimmermann, Martin Huldrych, plant anatomy, plant physiology

PITTSFIELD
Flowers, Ralph Grant, physical chemistry
Fox, Daniel Wayne, plastic chemistry, polymer chemistry
Goossens, John Charles, organic polymer chemistry
Nelson, John Daniel, plastics chemistry
Osthoff, Robert Charles, inorganic chemistry
Peters, Till Justus Nathan, inorganic chemistry
Pitha, John Joseph, inorganic chemistry

PRIDES CROSSING
Vickers, William W, atmospheric physics

PROVINCETOWN
Giese, Graham Sherwood, oceanography

QUINCY
Caley, Wendell J, Jr, physics
Dallos, Andras, electronic physics
Ernest, Leland C, plant physiology
Free, John Ulric, solid state physics, acoustics
Gleichauf, Paul Harry, physics
Hall, Lowell Headley, II, physical chemistry
Naylor, Jasper Ross, mathematics

MASSACHUSETTS

Stark, James Cornelius, organic chemistry, biochemistry

READING
Chiu, Tin-Ho, surface chemistry
Fante, Ronald Louis, microwave physics
Neuringer, Joseph Louis, mathematical physics

ROCKPORT
Hull, Gordon Ferrie, Jr., physics
Norris, Russell Taplin, ornithology
Robinson, Elizabeth Dorothy, microbiology; public health

ROYALSTON
Frye, Royal Merrill, physics
Frye, Virginia Brigham, spectroscopy

SALEM
Briney, Robert Edward, mathematics
Centorino, James Joseph, physical geography
DePalma, Philip Anthony, medical microbiology
Engelke, John Leland, physical chemistry
George, John C., urban geography
Harrises, Antonio Ethremos, parasitology; zoology
Moore, Johnes Kittelle, marine ecology
Roberts, Edith Adelaide, botany
Ryan, Thomas I., biology
Salley, Paul V., geography
Shahin, Jamal Khalil, mathematics
Thompson, Gary Gene, geology; palynology
Wolfe, Caleb Wroe, geology

SANDWICH
Korgen, Benjamin Jeffry, physical oceanography

SCITUATE
Emslie, Alfred George, applied physics

SEEKONK
Doolittle, Charles Herbert, III, clinical pharmacology

SHARON
Elfbaum, Stanley Goodman, clinical chemistry
Kornetsky, Aaron, food science

SHREWSBURY
Bartke, Andrzej, reproductive physiology; biochemistry
Bergen, John Richard, endocrinology
Brodie, Angela (Hartley), physiology
Brodie, Harry Joseph, biochemistry, chemistry
Caspi, Eliahu, organic chemistry, biochemistry
Chang, Min Chueh, physiology
Greenberg, Jay R., cell biology; molecular biology
Hele, Priscilla, biochemistry
Hoagland, Hudson, physiology
Hoagland, Mahlon Bush, molecular biology; research administration
Hochstadt, Joy, biochemistry; microbiology
Klaiber, Edward L., endocrinology
Knight, Robert Hallowell, molecular biology; cell biology
Kupfer, David, biochemical pharmacology; drug metabolism
Luftig, Ronald Bernard, microbiology; biophysics
Mason, Marcus M., veterinary pathology
McCracken, John Aitken, endocrinology
McNiven, Neal Lindsay, organic chemistry
Morgane, Peter J., neurophysiology; psychopharmacology
Ozer, Harvey Leon, virology; cell biology
Resnick, Oscar, physiology; pharmacology
Roberts, John Stephen, physiology; endocrinology
Skarnes, Robert C., biochemistry; immunology
Welsh, Federico, molecular biology; medicine

SOMERSET
Gartner, Edward A., organic chemistry

SOUTH DARTMOUTH
Whitaker, Ellis Hobart, plant physiology

SOUTH HADLEY
Allen, Mildred, mechanics
Bates, Grace Elizabeth, mathematics
Beeman, Curt Pletcher, analytical chemistry
Beeman, Elizabeth Ann, behavioral biology
Boyd, Elizabeth Margaret, zoology
Campbell, Mary Kathryn, biophysical chemistry
Clancy, Edward Philbrook, physics
D'Amato, Richard John, physical chemistry
Delton, Mary Helen, organic chemistry
Dennis, Tom Ross, astronomy
Durfee, William Hetherington, mathematics
Durso, John William, theoretical physics
Enggass, Peter Maurice, historical geography; economic geography
Eschenberg, Kathryn (Marcella), embryology
Godchaux, Martha Miller, petrology
Griffiths, David Jeffery, theoretical physics
Hall, George E., organic chemistry
Harrison, Anna Jane, physical chemistry
Hixson, Susan Harvill, biochemistry
Jones, Theodore Charles, genetics; biochemistry
Kaltenbach, Jane Couffer, histology
Moore, Thomas Warner, solid state physics
Muus-Jytte Marie, biochemistry
Nicholson, Howard White, Jr., high energy physics
Pollatsek, Harriet Suzanne, mathematics
Pryor, Marilyn Ann Zirk, comparative physiology
Senechal, Lester John, mathematics
Smith, Curtis Griffin, physiology
Sprague, Isabelle Baird, biology
Tavakolian, Bahram Mehdi, ethnology
Weaver, Edwin Snell, physical chemistry
Wick, Emily Lippincott, organic chemistry; academic administration
Williamson, Kenneth Lee, organic chemistry; spectroscopy

SOUTHAMPTON
McKinley, Harry R., optics

SOUTH LANCASTER
Kryger, Roy George, physical organic chemistry
Nyirady, Stephen Arnold, microbiology

SOUTH ORLEANS
Anthony, David Henry, dental materials
Berry, Chester Ridlon, solid state physics

SOUTHBOROUGH
Fraser, C E Ovid, veterinary immunology; veterinary microbiology
Herd, James Alan, physiology
Hertig, Arthur Tremain, embryology; primatology
Hunt, Ronald Duncan, comparative pathology
Jones, Thomas Carlyle, veterinary pathology; comparative pathology
Ma, Nancy Shui Fong, cytogenetics
Stipe, John Gordon, Jr., planetary sciences

SOUTHBRIDGE
Begun, Fred P., acoustics, physics
Brandt, Neill Matteson, physics
Hannan, W Kelley, optics
Hoffman, Donald Oliver, industrial chemistry
Jones, Florence Shirley (Patterson), astronomy
Smith, Luther W., physics, mathematics
Snitzer, Elias, physics

SPRINGFIELD
Ahlberg, Henry David, biology
Aplington, Henry Webster, Jr., anatomy
Barkman, Robert Cloyce, comparative physiology
Brainerd, John Whiting, biology
Cassidy, Carl Eugene, endocrinology; internal medicine
Chapin, Earl Cone, organic chemistry
Cohen, Isadore, botany

Cohen, Joel Ralph, clinical microbiology
Cohen, Saul Mark, organic chemistry
Coleman, James Andrew, theoretical physics
Deets, Gary Lee, polymer chemistry
Demko, Donald, geography
Dickinson, Alan Charles, physical chemistry
Ford, Emory A., organic polymer chemistry
Frank, Jean Ann, organic chemistry
Friedmann, Paul, surgery
Kay, Maire Weir, zoology
Keeney, Clifford Emerson, physiology
Locke, Frederic John, chemistry
Maniscalco, Ignatius Anthony, organic chemistry; biochemistry
Meyer, Irving, oral surgery; oral pathology
Nemphos, Speros Peter, polymer chemistry
O'Connor, Michael Gerald, food technology
Pease, Roger Waterman, Jr., evolutionary biology
Santer, James Owen, organic chemistry
Scheuchenzuber, H Joseph, biomechanics, physical education
Shear, Leroy, internal medicine; nephrology
Syed, Ibrahim Bijli, medical physics, radiological health
Taggart, William Paul, polymer chemistry
Torre, Frank John, physical chemistry

STOCKBRIDGE
Will, Otto Allen, Jr., psychiatry

STOUGHTON
Condit, Carlton, geology

STOW
Lovell, Donald Joseph, optics
Seronde, Joseph, Jr., pathology

STURBRIDGE
Hovey, Richard John, physical chemistry

SUDBURY
Boyle, Paul Edmund, pathology
Eastman, Willard L., applied mathematics
Hartke, Jerome L., solid state physics
Keene, Wayne Hartung, physics
Kestigian, Michael, inorganic chemistry
Kroger, Harry, solid state electronics; experimental solid state physics
Langmuir, Margaret Elizabeth Lang, physical chemistry; photochemistry
Lustig, Claude David, physics
Maloney, William Thomas, magnetism
McMahon, Donald Howland, optical physics
Minden, Henry Thomas, chemical physics
Nelson, Arthur Robert, optics
Rafuse, Mary Jane Lounsbury, chemistry
Newman, Roger, chemistry
Seavey, Marden Homer, Jr., solid state physics
Sewell, Frank Anderson, Jr., electronic physics
Soile, Leland Peter, acoustics
Soref, Richard Allan, electrooptics
van de Vaart, Herman, solid state electronics
Wallace, James, optics, fluid mechanics
Wegener, Horst Albrecht Richard, solid state science
Whitney, Colin Gordon, electrooptics

TEMPLETON
Chandler, Robert Flint, Jr., agronomy

WABAN
Garrison, Rhoda, plant morphology
Sodickson, Lester A., physics, biomedical engineering

WAKEFIELD
Addiss, Richard Robert, Jr., solid state physics
Holst, William Frederick, computer sciences
Marmo, Frederick Francis, aeronomy; environmental chemistry
Richards, Paul Irving, physics
Taylor, Raymond L., lasers, chemical physics

Veidis, Mikelis Valdis, chemistry

WALPOLE
Fadner, Thomas Alan, polymer chemistry
Olson, Arthur Russell, textile chemistry
Roth, Roy William, polymer chemistry
Stolbach, Leo Lucien, oncology; internal medicine

WALTHAM
Adler, Alice Joan, biophysical chemistry
Aisenberg, Sol, physics
Allemand, Charly D., applied physics
Auborn, James John, physical chemistry
Auslander, Maurice, mathematics
Baglio, Joseph Anthony, physical chemistry
Bair, Kenneth Walter, synthetic organic chemistry
Baird, Donald Heston, physical chemistry
Bell, Richard Oman, solid state physics
Bensinger, James Robert, elementary particle physics
Birnbaum, Sanford Milton, biochemistry
Bowness, Colin, physics
Bracco, Donato John, physical chemistry; analytical chemistry
Brackett, John Washburn, computer science, system analysis
Brecher, Charles, physical chemistry
Brown, Edgar Henry, Jr., topology
Buchsbaum, David Alvin, mathematics
Carnevale, Edmund Henry, physics
Caviness, Verne Strudwick, Jr., neurology; neuropathology
Chang, Kuo Wei, biomedical engineering
Chernosky, Edwin Jasper, environmental physics; space physics
Chretien, Max, physics
Codere, Helen, anthropology
Cohen, Carolyn, biophysics
Cohen, Saul G., organic chemistry
Cowgill, George L., anthropology
Cukor, Peter, analytical chemistry
Dakss, Mark Ludmer, physical optics
Davis, Luther, Jr., solid state electronics
Debye, Peter Paul Ruprecht, physics
DeRosier, David J., biophysics, molecular biology
Deser, Stanley, theoretical physics
Dorain, Paul Brendel, physical chemistry
Dressel, Herman Otto, electrooptics
Dulaney, John Thornton, biochemistry
Eisenbud, David, algebra
Epstein, Herman Theodore, biophysics
Epstein, Irving Robert, chemical physics, theoretical chemistry
Ewing, Sheila Pauline, physical organic chemistry
Faddoul, George Peter, veterinary medicine, public health
Fasman, Gerald David, biochemistry; biophysics
Feist, Wolfgang Martin, solid state physics, solid state electronics
Fong, Franklin, cell biology
Freifelder, David, molecular biology
French, Kenneth William, applied chemistry
Fulton, Chandler Montgomery, developmental biology; cell biology
Galinat, Walton Clarence, genetics, plant morphology
Gentile, Adrian George, economic entomology
Gibbs, Martin, biology
Golden, Sidney, physical chemistry; theoretical chemistry
Goldstein, Jack Stanley, physics
Gorbunoff, Marina J., organic chemistry
Grisaru, Marcus Theodore, physics
Grunwald, Ernest Max, physical organic chemistry
Guenter, Otto Johann, solid state physics
Gustafson, John C., solid state physics
Haber, James Edward, molecular genetics
Halvorson, Harlyn Odell, microbiology
Haugejaa, Paul O., atomic physics, molecular physics
Henchman, Michael J., physical chemistry
Hendrickson, James Briggs, synthetic organic chemistry, research administration
Holland, Melvin Gerald, solid state physics
Hollocher, Thomas Clyde, Jr., biochemistry
Holway, Lowell Hoyt, Jr., applied mathematics
Hope, Lawrence Latimer, physics
Horr, David Agee, anthropology
Horrigan, Frank Anthony, theoretical physics
Jencks, William Platt, biochemistry

Job, Donald Dexter, environmental sciences, photobiology
Johnson, Earnest J., solid state physics
Jones, Chase Breese, biochemistry
Jordan, Peter C H., theoretical chemistry
Jungalwala, Firoze Bamanshaw, biochemistry, neurochemistry
Kadakde, Prakash Gopal, plant physiology, plant biochemistry
Kaplan, David, anthropology
Kelner, Albert, biology
Kirsch, Lawrence Edward, physics
Klein, Attila Otto, plant physiology
Klein, Claude A., physics
Kliem, Peter O., analytical chemistry, photographic chemistry
Kohane, Theodore, physics
Kolbeck, Andrew Gerard, polymer physics, glass technology
Kramer, Jerry Martin, chemical physics
Kubin, Rosa, biochemistry, veterinary pathology
Kustin, Kenneth, inorganic chemistry, physical chemistry
Lapuck, Jack Lester, food chemistry, bacteriology
Lattman, Eaton Edward, molecular biophysics
Lauer, Robert B., materials science, solid state science
Lempicki, Alexander, physics
Lesensky, Leonard, physics
Lester, Joseph Eugene, physical chemistry
Levine, Harold Irving, mathematics
Levine, Jerome Paul, topology
Levine, Lawrence, immunochemistry
Linschitz, Henry, physical chemistry
Lipworth, Edgar, atomic physics
Lowenstein, John Martin, biochemistry
Lublin, Paul, physical chemistry
Luck, Clarence Frederick, Jr., physics
Manners, Robert Alan, anthropology
Matsusaka, Teruhisha, geometry
Mattison, Roland Lees, computer sciences
McCluer, Robert Hampton, biochemistry
McKinzie, Howard Lee, surface chemistry, solid state chemistry
Mlavsky, Abraham Isaac, physical chemistry, materials science
Mozzi, Robert Lewis, experimental solid state physics
Nelson, William Frank, physics
Nisonoff, Alfred, immunochemistry, immunobiology
Olsen, Gjerding, endocrinology
Osepchuk, John M., physics
Palais, Richard Sheldon, mathematics
Palilla, Frank C., inorganic chemistry
Pappalardo, Romano Giuseppe, solid state physics
Parry, Ronald John, bio-organic chemistry, natural products chemistry
Pendleton, Hugh Nelson, III, theoretical physics
Phillips, Walter Charles, physics
Preble, Norman Alexander, biology
Propst-Reciuti, Catherine Lamb, molecular biology
Redfield, Alfred Guillou, physical biochemistry
Reid, F Joseph, solid state physics
Riseberg, Leslie Allen, physics
Rosenblum, Myron, organic chemistry
Ross, Ronald D., chemistry
Sage, Jay Peter, solid state electronics
Schiff, Jerome A., plant physiology
Schlaikjer, Carl Roger, chemistry
Schleif, Robert Ferber, molecular biology, biochemistry
Schnitzer, Howard J, theoretical high energy physics
Schulz, Manfred Bruno, physics
Schweber, Silvan Samuel, theoretical physics
Seibert, Michael, photobiology, plant physiology
Sellers, Francis Bachman, atomic physics, nuclear physics
Servaes, Tahira Minhaj, quantum physics, statistical mechanics
Simpson, William Henry, solid state chemistry, photographic chemistry
Soodak, Morris, biochemistry
Spiegel, John Paul, psychiatry, sociology
Statz, Hermann, physics
Steel, Colin, physical chemistry
Stein, Robert Foster, astrophysics
Stein, Seymour, physics
Stoffyn, Pierre Jules, chemistry, biochemistry
Stoner, William Weber, optical physics, radiological physics
Swank, Thomas Francis, colloid chemistry, photographic chemistry
Swartz, John Croucher, physics, materials science

Szent-Gyorgyi, Andrew Gabriel, biochemistry
Taunton-Rigby, Alison, chemistry, biochemistry
Tiernan, Robert Joseph, physics, ceramics
Timasheff, Serge Nicholas, physical biochemistry
Tuttle, Thomas R, Jr., physical chemistry
Van Vunakis, Helen, biochemistry
Vassell, Milton O, theoretical physics
Waack, Richard, physical chemistry, polymer chemistry
Wallace, Robert William, science education, environmental chemistry
Wardle, John Francis Carleton, radio astronomy
Wasserman, Moe Stanley, physical chemistry
Woehler, Michael Edward, immunology, medical microbiology
Wong, Jacob Yau-Man, solid state physics, bionics
Young, Robert Ellsworth, horticulture
Zemon, Stanley Alan, physics
Zucker, Joseph, solid state physics
Zuckerman, Bernard, photographic chemistry

WATERTOWN

Alexander, Michael Norman, solid state physics, chemical physics
Antal, John Joseph, experimental physics
Band, Hans Eduard, physics, materials science
Beres, John Joseph, physical organic chemistry
Bowie, Oscar L., applied mathematics
Burkhardt, James Lee, experimental physics
Chipman, David Randolph, x-ray crystallography
Croft, William Joseph, crystallography
Desper, Clyde Richard, polymer science
Dutta, Sunil, ceramics
Gericke, Otto Reinhard, physics
Haase, Kurt Harald, mathematics
Harrison, Ralph Joseph, physics
Hodgdon, Russell Bates, Jr., organic chemistry
Hynes, Thomas Vincent, solid state physics, materials science
Isserow, Saul, materials science, materials engineering
Jennings, Laurence Duane, solid state physics
Kauffman, Joel Mervin, organic chemistry
Lin, Sin-Shong, high temperature chemistry
McCauley, James Weymann, materials science, crystallography
McRae, Wayne Alan, physical chemistry
Perkins, Janet Sanford, high temperature chemistry, organic polymer chemistry
Priest, Homer Farnum, chemistry
Sagalyn, Paul Leon, solid state physics
Shuford, Richard Joseph, organic polymer chemistry
Singer, Robert Edward, polymer chemistry
Staknis, Victor Richard, mathematics
Tauer, Kenneth J, physical chemistry, chemical physics
Tewksbury, Charles Isaac, rubber chemistry
Thomas, George Richard, organic chemistry
Walker, Christopher Bland, physics
Weiss, Richard Jerome, physics
Wentworth, Stanley Earl, organic polymer chemistry
Wiederhold, Pieter Rijk, physics
Wilde, Anthony Flory, physical chemistry

WAYLAND

Au, Andrew Taichiu, organic chemistry
Boardway, Nancy Louise, organic chemistry
Clark, Melville, physics, electrical engineering
Freedman, Harold Hersh, organic chemistry
Hedlund, Donald A, ionospheric physics, radio engineering
Holland, Andrew Brian, solid state physics, photographic chemistry
Kreidl, Ekkehard Ludwig, physical chemistry
Langer, Horst Gunter, inorganic chemistry
Thome, George Durst, radiophysics, geophysics
Windsor, Robert Beach, applied physics

WELLESLEY

Adler, Stephen Miller, astronomy
Allen, Mary A Mennes, microbiology
Balmer, Clifford Earl, organic chemistry
Birney, Dion Scott, Jr., astronomy
Brauner, Phyllis Ambler, analytical chemistry, inorganic chemistry
Brown, Judith, quantum optics
Carr, Jerome Brian, oceanography, geophysics
Coyne, Mary Frances D, endocrinology
Crawford, Jean Veghte, organic chemistry
Creighton, Harriet Baldwin, botany
Dinger, Ann St Clair, astrophysics
Dobbins, David Ross, botany
Donohue, David Arthur Timothy, petroleum
Fiske, Virginia Mayo, endocrinology
Fleming, Phyllis Jane, experimental solid state physics
Gauthier, Geraldine Florence, cell biology
Guernsey, Janet Brown, physics
Hicks, Sonja Elaine, biochemistry
Hill, Sarah Jeannette, astronomy
Kolodny, Nancy Harrison, physical chemistry
Lester, William Wright, applied physics
Little-Marenin, Irene Renate, astrophysics
Loehlin, James Herbert, physical chemistry
Mehrich, Ferdinand Paul, food science
Norvig, Torsten, mathematics
Padykula, Helen Ann, cell biology
Proud, John Mason, Jr., physics
Rock, Elizabeth Jane, physical chemistry
Roitman, Judy, topology, mathematical logic
Rosenberg, Joseph, organic chemistry
Schafer, Alice Turner, algebra
Shimony, Annemarie Anrod, anthropology
Shuchat, Alan Howard, mathematical analysis
Stehney, Ann Kathryn, pure mathematics
Webb, Andrew Clive, developmental biology
Webster, Eleanor Rudd, organic chemistry, history of science
Weeks, Dorothy Walcott, physics
Widmayer, Dorothea Jane, zoology, microbial genetics
Wilcox, Howard Joseph, mathematics

WELLESLEY HILLS

Ingrao, Hector Carlos, astronomy, optics
Penndorf, Rudolf, atmospheric physics
Satterthwaite, Franklin Eves, statistics

WENHAM

Camp, Russell R, plant pathology, electron microscopy
Dent, Thomas Curtis, botany, plant taxonomy
Haas, John William, Jr, physical chemistry
Wright, Richard T, aquatic ecology

WEST BARNSTABLE

French, Berlin Carson, organic chemistry

WEST HARWICH

Minarik, Charles Edwin, plant physiology

WEST LYNN

Jones, Claude Kitchener, physics

WEST NEWTON

Alexander, Leo, psychiatry, neurology
Dunlap, William Crawford, physics
Goldings, Herbert Jeremy, psychiatry
Simons, Harold Lee, physical chemistry

WEST ROXBURY

Rossier, Alain B, physical medicine, rehabilitation
Watkin, Donald Morgan, internal medicine, public health

WESTBOROUGH

Malone, Joseph James, algebra
Rotenberg, Don Harris, polymer chemistry

WESTFIELD

Harris, Betty Wolf, systematic botany, plant genetics
Lovejoy, David Arnold, mammalian ecology
Majumder, Sanat Kumer, botany, ecology
McGuigan, Robert Alister, Jr, mathematics
Taylor, James Kenneth, ecology

WESTFORD

Meeks, Marion Littleton, radio astronomy
Rogers, Alan Ernest Exel, radio astronomy
Thorstensen, Thomas Clayton, chemistry

WESTON

Ashbrook, Joseph, astronomy
Besse, Arthur L, experimental physics
Burke, Leonarda, mathematics, data processing
Evans, Robert L, internal medicine, medical education
Floyd, William Beckwith, computer science
Funkhouser, John Tower, analytical chemistry
Johnson, Clark E, Jr, magnetism
Landowne, Milton, internal medicine
Lawrence (Roche), Mary Anna, biology
McGarry, Margaret, inorganic chemistry, analytical chemistry
Mulrennan, Cecilia Agnes, biology
Ritt, Paul Edward, Jr, chemistry
Schloemann, Earnst, theoretical physics
Skehan, James William, geology, tectonics
Striming, Walter Eugene, mathematics, electrical engineering
Williamson, Susan, mathematics
Wyman, Donald, horticulture

WESTPORT

Kennison, Lawrence Sanford, mathematics

WESTWOOD

Ebert, Andrew Gabriel, pharmacology, food science
Manly, Marian LeFevre, dental research
Manly, Richard Samuel, biochemistry
Shoup, Charles Samuel, Jr, physical chemistry

WILBRAHAM

Norris, Forrest Harvey, polymer chemistry

WILLIAMSTOWN

Art, Henry Warren, forest ecology, environmental biology
Brown, Fielding, physics
Burgess, Thomas Edward, inorganic chemistry
Chandler, Dean Wesley, physical chemistry, analytical chemistry
Chang, Raymond, physical chemistry
Compton, Charles (Daniel), chemistry
Copeland, Frederick Cleveland, biology
Crampton, Stuart J B, atomic physics
Dewitt, William, microbiology, biochemistry
Drickamer, Lee Charles, animal behavior, ecology
Foote, Freeman, geology
Fox, William Templeton, geology
Grabois, Neil, algebra, number theory
Grant, William Chase, Jr, zoology
Green, William Lohr, mathematical analysis
Hill, Victor Ernst, algebra, mathematical logic
Jacobsen, Robert Thomas, colloid chemistry
Jewett, Sandra Lynne, biochemistry, bio-organic chemistry
Kaplan, Lawrence Jay, chemistry
Kleier, Daniel Anthony, physical chemistry, theoretical chemistry
Koepnick, Richard Borland, sedimentary petrology
Koppenheffer, Thomas Lynn, biochemistry, immunology
Kozelka, Robert M, statistics
MacFadyen, John Archibald, Jr, geology
Markgraf, John Hodge, organic chemistry
Matthews, Samuel Arthur, zoology
Moomaw, William Renken, molecular spectroscopy, environmental sciences

Oliver, Henry William, mathematics
Park, David Allen, theoretical physics
Pasachoff, Jay Myron, astronomy, physics
Pierce, Camden Ballard, solid state physics
Roosenraad, Cris Thomas, mathematics
Semon, John H., science education
Shelton, John Winthrop, physics
Skinner, James F., physical inorganic chemistry
Spencer, Guilford Lawson, II, mathematics
Tuchman, Albert, physics
Vankin, George Lawrence, cell biology
Warren, Harold Hubbard, organic chemistry
Wobus, Reinhard Arthur, geology

WILMINGTON
Bade, William Lemoine, theoretical physics
Bush, Martin Bruce, analytical chemistry, clinical chemistry
Foster, Henry Louis, veterinary medicine
Merken, Henry, polymer chemistry
Schneider, Harold O., physics

WINCHESTER
Aronson, James Ries, spectroscopy
Baker, James Gilbert, optical physics, astronomy
Baratta, Edmond John, analytical chemistry, radiochemistry
Beranek, Leo Leroy, acoustics, communications
Carpenter, Russell Le Grand, radiobiology, histology
Kadesch, Richard Gilmore, organic chemistry
King, Ronold (Wyeth Percival), physics, electrical engineering
Morse, Philip McCord, theoretical physics, operations research
Rogers, Hartley J., mathematics
Schultz, John Russell, geology
Terzaghi, Ruth Doggett, geology
Zerweich, Charles Ezra, Jr., organic chemistry

WINTHROP
Moses, Ronald Elliot, organic chemistry
Vasilos, Thomas, ceramics, chemistry

WOBURN
Carpenter, Jack William, theoretical physics
Ives, Robert Southwick, chemistry
Kuist, Charles Howard, physical chemistry, polymer chemistry

WOLLASTON
Babcock, William James Verner, science education
MacKenzie, Donald Hershey, inorganic chemistry
Wrigley, Walter, physics

WOODS HOLE
Abbott, Marie Bohm-Lambert, zoology
Acheson, George Hawkins, medicine
Adelman, William Joseph, Jr., physiology, biophysics
Backus, Richard Haven, marine biology
Ball, Eric Glendinning, biochemistry
Barth, Lucena J., zoology
Beardsley, Robert Cruce, physical oceanography
Beckerle, John C., physical oceanography
Behrendt, John Charles, marine geophysics
Berggren, William Alfred, geology, micropaleontology
Blumer, Max, geochemistry
Botkin, Daniel Benjamin, ecology
Bowen, Vaughan Tabor, biochemistry
Bowin, Carl Otto, geology
Brown, Bradford E., fish biology, ecology
Bryan, Wilfred Bottrill, petrology, marine geology
Bumpus, Dean Franklin, physical oceanography
Bunce, Elizabeth Thompson, geophysics
Cornman, Ivor, zoology
Deuser, Werner Georg, marine geochemistry
Dick, Henry Jonathan Biddle, petrology

Dillon, William Patrick, geological oceanography
Edwards, Robert Lomas, ecology
Emery, Kenneth Orris, marine geology
Ewing, Gifford Cochran, oceanography
Farrington, John William, marine geochemistry
Folger, David W., geological oceanography
Frosch, Robert Alan, marine geophysics
Fuglister, Frederick Charles, oceanography
Fye, Paul McDonald, physical chemistry
Gifford, Cameron Edward, ecology, physiology
Gillespie, Paul Albert, microbial ecology
Grassle, John Frederick, ecology, biological oceanography
Grassle, Judith Payne, population genetics
Grice, George Daniel, Jr., marine zoology
Grosslein, Marvin Darrel, fish biology
Guillard, Robert Russell Louis, biological oceanography, microbiology
Haedrich, Richard L., systematic ichthyology, biological oceanography
Harvey, George Ranson, organic chemistry
Heirtzler, James Ransom, geophysics
Heyerdahl, Eugene Gerhardt, biology
Hobbie, John Eyres, ecology
Hollister, Charles Davis, marine geology
Hoskins, Hartley, geophysics
Humes, Arthur Grover, parasitology
Hunt, John Meacham, organic chemistry, geochemistry
Jannasch, Holger Windekilde, microbiology
Johnson, David Ashby, oceanography
Keosian, John, exobiology
Ketchum, Bostwick Hawley, ecology, biological oceanography
Knott, Sydney T, seismology, marine geophysics
Lang, Frederick, neurobiology, invertebrate zoology
Luyten, James Reindert, physical oceanography
MacNichol, Edward Ford, Jr., biophysics, physiology
Marsland, Douglas Alfred, oceanography
Mather, Frank Jewett, III, ichthyology
Maxwell, Arthur Eugene, oceanography
McCartney, Michael Scott, physical oceanography, fluid dynamics
Metcalf, William Gerrish, oceanography
Miller, Arthur R., physical oceanography
Milliman, John D., oceanography, geology
Monahan, Edward Charles, oceanography
Morse, Robert Warren, physics
Phillips, Joseph D., geophysics
Redfield, Alfred Clarence, oceanography, physiology
Ridgway, George Junior, microbiology, biochemistry
Rosenfeld, Melvin Arthur, sedimentology, academic administration
Ross, David A., geological oceanography
Rowe, Gilbert Thomas, biological oceanography
Sabo, Dennis John, nutrition, biochemistry
Salmon, Edward Dickinson, cell biology
Sanford, Thomas Bayes, physical oceanography
Sayles, Frederick Livermore, geochemistry
Schelema, Rudolf S., marine biology
Schlee, John Stevens, geology
Schmitz, William Joseph, Jr., physical oceanography
Sears, Mary, marine biology
Sichel, Elsa Keil, zoology
Spencer, Derek W., oceanography, marine geology
Spindel, Robert Charles, underwater acoustics, electrical engineering
Stephens, Raymond Edward, protein chemistry
Szent-Gyorgyi, Albert, biochemistry
Teal, John Moline, marine biology, ecology
Thompson, Geoffrey, geochemistry, oceanography
Thompson, Rory, physical oceanography
Tiffney, Wesley Newell, mycology, marine biology
Uchupi, Elazar, geology
Von Arx, William Stelling, oceanography
Von Herzen, Richard P., marine geophysics
Warren, Bruce Alfred, physical oceanography
Watson, Stanley W., bacteriology
Webster, Ferris, physical oceanography
Wells, Jay Byron, physiology

Whelan, Barbara Jean King, organic chemistry
Whiting, Anna Rachel, genetics
Wichterman, Ralph, zoology
Wigley, Roland L., marine ecology
Williams, Albert James, III, oceanography, ocean engineering
Woo, Ching Chang, petrology, mineralogy
Woodwell, George Masters, ecology, botany
Worthington, Lawrence Valentine, oceanography
Wright, William Redwood, physical oceanography

WORCESTER
Ahmadjian, Vernon, botany
Allen, Harry Clay, Jr., chemical physics
Anderson, Roy Stuart, physics
Babich, George Leon, developmental biology
Baker, John Richard, pathology, histochemistry
Balogh, Karoly, pathology, histochemistry
Bergstresser, Thomas Karl, theoretical solid state physics
Berry, Leonard, physical geography, resource management
Beyea, Jan Edgar, nuclear physics
Blacklow, Neil Richard, virology
Blair, Murray Reid, Jr., pharmacology
Blake, Louis Harvey, mathematics
Bluemel, Van (Fonken Wilford), quantum optics
Bogden, Arthur Eugene, immunology
Brenner, Daeg Scott, nuclear chemistry
Bridgman, Wilbur Benjamin, physical chemistry
Brink, John Jerome, neurosciences
Buell, Elliott Lyndon, mathematics
Bulger, Ruth Ellen, anatomy, pathology
Bures, Milan F., epidemiology, public health
Bushweller, Charles Hackett, physical organic chemistry, structural chemistry
Butcher, Reginald William, biochemistry, physiology
Buttimer, Anne, urban geography, urban sociology
Byrnes, Eugene William, organic chemistry
Camougis, George, research administration
Campbell, Douglas Arthur, molecular genetics
Clark, Sam Lillard, Jr., anatomy
Clark, Edward Nielsen, research administration
Cohen, Saul Bernard, geography
Cooke, William Joseph, biochemical pharmacology
Crusberg, Theodore Clifford, biochemistry
Curtis, Joseph C., cell biology, biophysics
D'Amato, Henry Edward, pharmacology
Danielli, James F., biology
Davis, Sherman Gilbert, chemistry, food technology
Erickson, Karen Louise, organic chemistry
Esber, Henry Jemil, immunology, microbiology
Flavin, John William, cytology, pathology
Fleischman, Robert Werder, veterinary pathology
Friedell, Gilbert H., medicine, pathology
Fulmer, Hugh Scott, preventive medicine
Girard, James Emery, organic chemistry
Glassbrenner, Charles J., physics
Goloskie, Raymond, nuclear physics
Goodman, Henry Maurice, endocrinology
Gottlieb, Albert Maxwell, experimental solid state physics, biophysics
Gould, Harvey A., statistical mechanics
Graham, Terry Edward, physiological ecology
Greenfield, Robert Edman, Jr., biochemistry
Gunter, Roy Chalmers, Jr., optics, electronics
Gut, Marcel, organic chemistry
Hagopian, Miasnig, organic chemistry, biological chemistry
Halkerston, Ian D K, biochemistry, endocrinology
Handler, Alfred Harris, pathology
Hardell, William John, mathematics
Hazzard, George William, mathematics, physics
Healy, William Ryder, population ecology, herpetology
Heller, John Ralph, theoretical physics, medicine, surgery

Hilgar, Arthur Gilbert, medical research, medical administration
Hilsinger, Harold W., theoretical physics
Hobey, William David, theoretical chemistry
Hohenemser, Christoph, physics
Holle, Paul August, biology
Hopper, Anita Klein, genetics, molecular biology
Hopper, James Ernest, cell biology, developmental biology
Humi, Mayer, mathematical physics
Humphreys, Robert Edward, immunochemistry
Inglefield, Paul T., physical chemistry
Jasperson, Stephen Newell, solid state physics
Johansen, Hans William, marine phycology
Johnson, John Clark, applied physics
Johnson, John Samuel Edgar, II, marine ecology
Johnston, Michael Adair, biochemistry
Jones, Alan A., polymer chemistry
Kaplan, Melvin Hyman, medicine, immunology
Kaseta, Francis William, solid state physics
Kasperson, Roger Eugene, geography, environmental management
Kates, Robert William, resource geography
Keil, Thomas H., solid state physics
Kelton, Diane Elizabeth, genetics, cancer
Kennedy, Edward Francis, nuclear physics
Kennison, John Frederick, mathematics
Kessner, David Morton, internal medicine
Knadler, George Arthur, geography of Soviet Union
Kohin, Barbara Castle, molecular physics
Kohin, Roger Patrick, physics
Kreider, Martin Books, environmental physiology
Kuo, Eric Yung-Huei, oncology, veterinary medicine
Lanyon, Hubert Peter David, solid state physics
Leonard, Edward H., analytical chemistry
Lewis, Lawrence A., geography
Lingappa, Banadakoppa Thimmappa, microbiology
Lingappa, Yamuna, microbiology
Lyerla, Timothy Arden, developmental genetics
MacDonnell, John Joseph, mathematics
Manchester, Kenneth Edward, surface physics
Marinus, Martin Gerard, microbial genetics
Marks, Sandy Cole, Jr., anatomy
McBrien, Vincent Owen, mathematics
McClintock, Michael, quantum electronics
McGrath, Michael Glennon, organic chemistry
McGuire, Robert Francis, biophysics
McMaster, Paul D., biochemistry, organic chemistry
Menninger, Florian Francis, Jr., immunology, immunochemistry
Merken, Melvin, chemistry, science education
Merrithew, Paul Burton, chemistry
Miller, Tracy Bertram, pharmacology
Miller, William Brunner, mathematics
Molinari, Pietro Filippo, endocrinology, hematology
Morris, Rita, M L., geography
Morton, Donald John, information science
Morton, Richard Freeman, physics
Mulder, Carel, virology, molecular biology
Mulroy, Michael Joseph, anatomy
Narducci, Lorenzo M., quantum optics
Nelson, Donald John, biochemistry
Noonan, James Waring, mathematics
Nunnemacher, Rudolph Fink, histology
O'Neill, Edward Leo, physics
Paciorek, Joseph Walter, mathematics
Paracer, Surindar Mohan, nematology
Parker, Allan Elwood, physics
Pechet, Giselle S., pathology
Pechet, Liberto, medicine
Peet, Richard, geography
Perkins, Peter, mathematics
Perry, Robert F. Jr., geography
Peura, Robert Allan, biomedical engineering, electrical engineering
Peusner, Leonardo, chemistry
Plumb, Robert Charles, physical chemistry
Reisert, Patricia, mycology
Reynolds, John Theodore, bacteriology

Ricci, Robert William, physical organic chemistry
Rich, Stanley R, environmental sciences
Roberts, Thomas L, microbial biochemistry, genetics
Rosenkrantz, Harris, biochemical pharmacology
Rothmeier, Jeffrey, medicine, computer science
Rusfield, Agnes Burt, pathology
Sand, Leonard B, mineralogy
Sarup, Ram, solid state physics, spectroscopy
Saunders, Richard Henry, Jr, hematology
Scala, Alfred Anthony, photochemistry
Schaeppi, Ulrich Hans, neurophysiology, pharmacology
Schoen, Kenneth, mathematics
Shanahan, Patrick, mathematics
Shaw, Earl Bennett, economic geography
Sherman, Robert George, neurophysiology, biology
Smith, Edgar Eugene, biochemistry
Smith, Emil Richard, pharmacology
Snyder, Louis Michael, internal medicine
Srinivasan, Bhama, mathematics
Stubbe, John Sunapee, mathematics
Sulski, Leonard C, mathematics
Takman, Bertil Herbert, chemistry
Tangherlini, Frank R, physics
Tashjian, Robert John, veterinary medicine
Thibault, Newman William, mineralogy
Tipper, Donald John, microbiology, molecular biology
Todd, David, chemistry
Trachtenberg, Edward Norman, organic chemistry
Tuft, Richard Allan, quantum optics
Van Alstyne, John Pruyn, mathematics
Van Hook, Andrew, physical chemistry
Vellaccio, Frank, bio-organic chemistry
Vidulich, George A, physical chemistry
Walther, Adriaan, optics
Waud, Douglas Russell, pharmacology
Weininger, Stephen Joel, organic chemistry
Weiss, Earle Burton, pulmonary physiology
Weiss, Jerald Aubrey, microwave physics
Wen, Wen-Yang, physical chemistry
Wild, John Frederick, physics
Wooten, Benjamin Allen, physics
Wright, George Edward, medicinal chemistry
Yankauer, Alfred, public health
Zimmer, William Frederick, Jr, organic polymer chemistry

MICHIGAN

ADA
Schaafsma, Bernard Richard, organic chemistry

ADRIAN
Anderson, Amos Robert, organic chemistry
Bruner, Leonard Bretz, Jr, chemistry
Craft, Willard Leahman, Jr, chemistry
Duggan, Helen Ann, chemistry
Fishman, Frank J, Jr, theoretical physics
Geipel, Lothar Ernst, organic chemistry
Husband, Robert W, entomology
Latham, Ross, Jr, inorganic chemistry
Lewis, Richard Newton, polymer chemistry
Marchand, Margaret O, mathematics, statistics
Miller, Robert Clay, organic chemistry
Rintamaa, David Lee, ecology
Wu, Ching Kuei, biology, genetics
Xavier, K S, botany, microbiology
Yoder, Levon Lee, elementary particle physics

ALBION
Armstrong, Robert Lee, biochemistry
Cook, Paul Laverne, organic chemistry
Crump, John William, organic chemistry
Dillery, Dean George, zoology
Fryxell, Ronald C, mathematics
Gaswick, Dennis C, inorganic chemistry
Gilbert, William James, algology
Guyselman, John Bruce, zoology
Ludington, Martin A, nuclear physics
Moore, Warren Keith, mathematics
Pettersen, Howard Eugene, solid state physics
Ricker, Charles William, physics
Stowell, Ewell Addison, plant pathology
Taylor, Lawrence Dow, glaciology, geomorphology

Williams, John Albert, astronomy

ALLEGAN
Curlin, Lemuel Calvert, pharmaceutical chemistry
Lutz, Harold John, forestry

ALLEN PARK
Moore, Willis Eugene, pharmacy

ALLENDALE
Bailey, Rodney Albert, environmental chemistry
Bajema, Carl J, human ecology, biological anthropology
Hammer, Preston Clarence, computer science, mathematics
Hunter, Kenneth M, operations research
Knop, Charles Philip, inorganic chemistry
Koppel, Sheldon Jerome, history of science
Lefebvre, Richard Harold, geology
Lewis, Brian Kreglow, human physiology
MacTavish, John N, environmental sciences, geology
MacVicar-Whelan, Patrick James, atomic physics, plasma physics
Martin, Abram Venable, mathematics
Meloy, Carl Ridge, organic chemistry
Merrill, Dorothy, zoology
Miles, Donald Orval, medical microbiology
Neal, William Joseph, sedimentary petrology
Northup, Melvin Lee, environmental sciences
Paschke, Richard Eugene, neuropsychology
Shontz, John Paul, plant ecology
Shontz, Nancy Nickerson, plant ecology
Stein, Howard Jay, plant physiology
Strickland, James Shive, experimental physics
Ten Brink, Norman Wayne, quaternary geology
Toft, Robert Jens, academic administration
Vanderlagt, Donald W, combinatorics
Warren, John Stanley, geology
Weldon, John William, chemistry

ALMA
Beaumont, Randolph Campbell, inorganic chemistry
Bremer, Robert F, physical chemistry
DeYoung, Jacob J, organic chemistry
Edgar, Arlan Lee, zoology
Edison, Larry Alvin, mathematics
Eyer, Lester Emery, ornithology
Hutchison, James Robert, physical inorganic chemistry
Kapp, Ronald Ormond, botany
Potter, Howard A, organic chemistry
Toller, Louis, physics
Wilson, Paul Robert, mathematics
Wittle, Lawrence Wayne, physiology

ALPENA
Miller, David Charles, physics

ANN ARBOR
Abbrecht, Peter H, physiology, bioengineering
Abell, Murray Richardson, medicine, pathology
Abrams, Gerald David, pathology
Adams, Julian Philip, population genetics
Agranoff, Bernard William, biochemistry
Akerlof, Carl W, elementary particle physics
Albright, James Andrew, organic chemistry
Alexander, Richard Dale, zoology
Allen, John Morgan, cell biology
Allen, Sally Lyman, genetics
Aller, Margo Friedel, astronomy
Alpern, Mathew, visual physiology
Aminoff, David, biochemistry, immunogenetics
Anderson, David G, obstetrics & gynecology
Anderson, David J, neurophysiology, bioengineering
Anver, Miriam R, comparative pathology
Arnold, Chester Arthur, paleobotany
Asgar, Kamal, bioengineering
Ash, Major McKinley, Jr, periodontics
Ashe, Arthur James, III, organic chemistry
Atreya, Sushil Kumar, atmospheric sciences, planetary atmospheres

Aufdemberge, Theodore Paul, physical geography
Avery, James Knuckey, embryology, anatomy
Axelrod, Solomon Jacob, public health
Ayers, John Carr, biological oceanography
Bach, David Rudolph, nuclear physics, reactor physics
Bacon, George Edgar, pediatric endocrinology
Baer, Ferdinand, meteorology
Baic, Dusan, experimental morphology
Bailey, Reeve Maclaren, zoology
Baker, Burton Lowell, anatomy
Baker-Blocker, Anita Linda, geoenvironmental science, atmospheric chemistry
Ballert, Albert George, geography of the Great Lakes, economic geography
Ballou, David Penfield, biochemistry, enzymology
Bardwick, John, III, physics
Barnes, Burton Verne, forestry, botany
Barnes, James Milton, physics
Bartell, Lawrence Sims, physical chemistry
Bartels, Robert Christian Frank, mathematics
Batsakis, John G, medicine, pathology
Baublis, Joseph V, pediatrics, virology
Bauer, Jere Marklee, medicine
Bean, John William, physiology
Beardsley, Richard King, cultural anthropology, ethnography
Beaudoin, Allan Roger, experimental embryology
Beck, Charles Beverley, plant morphology
Beck, Clifford C, veterinary medicine, pharmaceutics
Behrents, Rolf Gordon, human development, orthodontics
Behrman, Samuel J, obstetrics & gynecology
Beierwaltes, William Henry, medicine
Be Ment, Spencer L, electronics, neurophysiology
Benninghoff, William Shiffer, plant ecology
Bernstein, Isadore A, biochemistry
Beyer, Robert Edward, biochemistry
Bigelow, Wilbur Charles, physical chemistry
Bixter, A L M, paper chemistry
Black, Martin Luther, synthetic organic chemistry
Blankespoor, Harvey Dale, invertebrate zoology, parasitology
Blankley, Clifton John, medicinal chemistry
Blinder, Seymour Michael, theoretical chemistry
Block, Walter David, biochemistry, nutrition
Bloomer, Henry Harlan, speech pathology
Blouin, Leonard Thomas, cardiovascular physiology, analytical chemistry
Blumenthal, Monica David, psychiatry
Boettner, Edward Alvin, physics
Bohr, David Francis, cardiovascular physiology
Bole, Giles G, Jr, internal medicine, biochemistry
Boorman, Evelyn Hutterer, algebra
Borer, Katarina Tomljenovic, neuroendocrinology, comparative psychology
Bornemeier, Dwight D, physics
Brace, C Loring, physical anthropology
Bradley, Robert Martin, sensory physiology, developmental physiology
Braithwaite, John Geden North, optics
Bree, Max M, veterinary medicine
Bretz, Michael, low temperature physics
Briggs, Darinka Zigic, stratigraphy, information science
Briggs, Louis Isaac, Jr, petrology
Brinkley, Linda Lee, developmental biology
Britt, Eugene Maurice, microbiology
Brockman, William Warner, molecular biology, virology
Brockway, Lawrence Olin, physical chemistry
Brown, Morton, mathematics
Brunfiel, Charles, mathematics
Brysk, Miriam Mason, biological chemistry
Buchanan, Robert Alexander, pediatrics, pharmacology
Buchholz, Robert E, dental radiology, oral medicine
Burch, John Bayard, zoology
Burckhalter, Joseph Harold, medicinal chemistry

Burdi, Alphonse R, dental research, child growth
Burkel, William E, anatomy
Burling, Robbins, anthropology
Burroughs, Robert Eli, physics
Burt, Brian Aubrey, dental epidemiology
Butler, Donald Eugene, organic chemistry
Butsch, Robert Stearns, mammalogy
Byers, Dohrman Harold, industrial hygiene
Cain, Stanley Adair, botany
Campbell, Alfred, organic chemistry
Campbell, Colin, obstetrics & gynecology, medical administration
Cantrall, Irving James, entomology
Capps, David Bridgman, medicinal chemistry
Carlson, Bruce Martin, anatomy
Carlson, David Sten, anthropology, dental research
Carow, John, forestry
Carpenter, Robert Raymond, internal medicine
Carr, John Frank, fisheries
Carroll, Bernard James, experimental psychiatry, psychopharmacology
Carroll, Vern, anthropology
Casey, Kenneth L, neurophysiology, neurology
Cassidy, James T, rheumatology
Castelli, Walter Andrew, anatomy
Castor, Cecil William, internal medicine, rheumatology
Cather, James Newton, zoology
Cesari, Lamberto, mathematical analyses
Chamberlain, John Paul, developmental biology
Chambers, Richard Lee, marine geochemistry, sedimentology
Chameides, William Lloyd, aeronomy, atmospheric chemistry
Charbeneau, Gerald T, dentistry
Cheal, MaryLou, psychobiology
Chen, Min-Shih, theoretical high energy physics
Child, Charles Gardner, III, surgery
Christensen, Halvor Niels, biochemistry, biophysics
Christman, Adam A, physiological chemistry
Chu, Ernest Hsiao-Ying, genetics
Cicerone, Ralph John, aeronomy, atmospheric chemistry
Clark, Bruce R, geology
Clarkson, James David, geography
Clewell, Don Bert, biochemistry, microbiology
Clifton, Yeaton Hopley, mathematics
Cloke, Paul LeRoy, geochemistry
Cochran, Kenneth William, Jr, pharmacology
Cohen, Bennett J, laboratory animal medicine, comparative medicine
Collins, Carolyn Jane, molecular biology, virology
Conn, Jerome W, medicine
Connelly, Thomas George, anatomy, zoology
Conrad, Michael, biophysics, biomathematics
Coon, Minor J, biochemistry, pharmacology
Coon, William Warner, surgery
Cooper, Gerald Paul, fish biology
Cooper, Stephen, microbiology, genetics
Cornell, Richard Garth, biostatistics
Cornish, Herbert Harry, toxicology
Counsell, Raymond Ernest, medicinal chemistry, pharmacology
Cowan, Archibald B, wildlife diseases, wildlife management
Cowley, Anne Pyne, astronomy
Cowley, Charles Ramsay, astronomy
Cox, David Buchtel, rheology, organic chemistry
Coyle, Peter, neuroanatomy, neurophysiology
Craig, Cecil Calvert, mathematics
Craig, Robert George, dental research
Crane, Horace Richard, physics
Crary, Douglas Dunham, geography
Creger, Paul LeRoy, organic chemistry
Crooks, Harry Means, Jr, organic chemistry
Crum, Howard Alvin, bryology
Curtis, George Clifton, psychiatry, psychosomatic medicine
Curtis, Myron David, inorganic chemistry, organometallic chemistry
D'Alecy, Louis George, physiology
Datta, Prasanta, biochemistry, molecular biology
Davenport, Fred M, internal medicine, epidemiology
Davenport, Horace Willard, physiology, gastroenterology
David, Chelladurai S, immunogenetics

MICHIGAN

Davis, Charles Hargis, information science
Davis, Curtiss Owen, biological oceanography
Davis, Roger (Edward), behavioral biology
Dawson, William Ryan, comparative physiology
Deninger, Rolf A., civil engineering, environmental health
DeJong, Russell Nelson, neurology
Dekker, Eugene Earl, biochemistry
Dekornfeld, Thomas John, anesthesiology
DeMuth, George Richard, pediatrics
Dermody, William Christian, reproductive endocrinology
Detwyler, Thomas Robert, geography
DeWald, Horace Albert, organic chemistry
DeWeese, Marion Spencer, surgery
Diamond, Norma Joyce, cultural anthropology
Dice, John Raymond, medicinal chemistry
Dickson, Douglas Grassel, mathematical analysis
Dingle, Albert Nelson, meteorology, cloud physics
Dingman, Reed (Othelbert), plastic surgery
Dodge, Horace Jackson, preventive medicine, public health
Dolin, Morton Irwin, biochemistry
Dolph, Charles Laurier, applied mathematics
Domino, Edward Felix, pharmacology
Donabedian, Avedis, public health
Donaldson, Wayne, biochemistry
Dorr, John Adam, Jr., vertebrate paleontology
Douthit, Harry Anderson, Jr., molecular biology
Dowson, John, dentistry
Drach, John Charles, biochemical pharmacology
Drayson, Sydney Roland, atmospheric physics
Duderstadt, James Johnson, applied physics
Duell, Elizabeth Ann, biochemistry, molecular biology
Duff, Ivan Francis, internal medicine
Dunn, Thomas M., physical chemistry
Duren, Peter Larkin, mathematics
Dziewiatkowski, Dominic Donald, biochemistry
Easter, Stephen Sherman, (Jr), physiology, biophysics
Easterling, Ronald E., internal medicine, nephrology
Eckert, Edward Arthur, virology
Edgren, Richard Arthur, endocrinology
Ege, Seyhan Nurettin, organic chemistry
Elderfield, Robert Cooley, organic chemistry
Ellis, Wade, mathematics
Elslager, Edward Faith, organic chemistry
Elving, Philip Juliber, analytical chemistry
England, Barry Grant, endocrinology, reproductive physiology
Epstein, Frederick Hermon, medicine
Ericson, William Arnold, statistics
Eschman, Donald Frazier, geomorphology
Eschmeyer, Paul Henry, biology
Essene, Eric J., petrology
Eveland, Warren C., medical bacteriology
Evans, Billy Joe, solid state chemistry, physics
Evans, Francis Cope, ecology
Evans, Francis Gaynor, anatomy, zoology
Farah, Jean William, genetics
Fairfax, Sally Kirk, forestry
Fajans, Stefan Stanislaus, internal medicine
Falls, Harold Francis, ophthalmology
Farrand, William Richard, quaternary geology
Faulkner, John A., physiology
Faust, Homer Edward, dentistry
Fayos, Juan Valivey, radiology
Federbush, Paul Gerard, theoretical physics
Fee, James Arthur, physical biochemistry
Fekety, F Robert, Jr., internal medicine, infectious diseases
Feldman, Chester, mathematics
Feringa, Earl Robert, neurology
Ferrell, William James, biochemistry, organic chemistry
Fetterolf, Carlos De La Mesa, Jr., fisheries, pollution biology

Fischer, Theodore Vernon, human anatomy
Fisher, Don Lowell, embryology, teratology
Fisher, Leslie John, neurobiology
Fitzgerald, James Edward, veterinary pathology, toxicology
Flannery, Kent Vaughn, anthropology, archaeology
Fleming, Arthur William, pediatrics
Fleming, Robert Willerton, medicinal chemistry
Flood, Merrill Meeks, management science
Flora, Jairus Dale, (Jr), biostatistics, medical statistics
Folk, William Robert, biochemical genetics
Fontheim, Ernest Gunter, plasma physics, space physics
Ford, George Willard, theoretical physics
Ford, Richard Irving, anthropology, ethnology
Forman, Shepard L., anthropology
Foster, Douglas Layne, reproductive endocrinology, physiology
Foster, Morris, immunogenetics, oncology
Francis, Anthony Huston, chemical physics
Frantz, Roberto, physical anthropology
French, Adam James, pathology
French, Arthur Bancroft, internal medicine
French, James C., organic chemistry
Freter, Rolf Gustav, medical bacteriology, immunology
Frey, Charles Frederick, surgery
Friedman, Joyce Barbara, computer sciences
Frisancho, Roberto, physical anthropology
Frohlich, Moses Michael, psychiatry
Fry, William James, surgery
Frye, Billy Eugene, developmental biology
Frye, Helen, zoology
Gabrielsen, Trygve O., radiology
Galler, Bernard Aaron, computer science, mathematics
Gans, Carl, zoology
Garn, Stanley Marion, physical anthropology
Gates, David Murray, biophysics
Gay, Helen, biology
Geen, Henry Cory, chemistry
Gehring, Frederick William, mathematics
Gelman, Charles, chemistry
Gerlach, Eberhard, mathematics
Gershowitz, Henry, human genetics
Gikas, Paul William, pathology
Ging, Rosalie J., psychiatry
Gingerich, Philip Dean, vertebrate paleontology
Glazko, Anthony Joachim, chemical pharmacology
Glinski, Ronald P., bio-organic chemistry, pathology
Glover, Roy Andrew, neuroanatomy, microscopic anatomy
Gloyd, Leonora Katherine, entomology
Good, Armin E., medicine
Gordus, Adon Alden, analytical chemistry
Gosine, William, ichthyology
Gosling, John Roderick Gwynne, obstetrics & gynecology
Gosling, Lee Anthony Peter, geography
Graber, Lee Winn, orthodontics, human development
Gray, Robert Howard, environmental health, cell biology
Gray, Walter Steven, nuclear physics
Green, Daniel G., physiological optics
Green, Robert A., medical administration, internal medicine
Greenberg, Goodwin Robert, biochemistry
Griess, Robert Louis, Jr., algebra
Griffin, Henry Claude, nuclear chemistry
Griffin, James Bennett, anthropology, archaeology
Griffith, John Randall, hospital administration
Gronvall, John Arnold, pathology
Groves, John Taylor, III, organic chemistry
Gupta, Arjun Kumar, mathematical statistics
Guthe, Karl Frederick, muscular physiology
Haddock, Frederick Theodore, Jr., radio astronomy
Haines, Richard Francis, microbiology
Hajra, Amiya Kumar, biochemistry, neurosciences
Hallada, Calvin James, industrial chemistry, metallurgy
Han, Seong S., cell biology, science education

Hanawalt, Joseph Donald, physics
Hanks, Carl Thomas, oral pathology
Harary, Frank, mathematics
Harlan, William R. Jr., medicine
Harris, James Edward, orthodontics, genetics
Harrison, Saul I., psychiatry, child psychiatry
Hartsook, Joseph Thurman, dentistry
Hartung, Rolf, toxicology
Haschke, John Maurice, solid state chemistry, high temperature chemistry
Haskell, Theodore Herbert, Jr., natural products chemistry
Hatfield, George Michael, pharmacognosy
Hawkins, Joseph Elmer, Jr., physiology, acoustics
Hay, George Edward, mathematics
Hayward, James Rogers, oral surgery
Hazen, Wayne Eskett, physics
Hazlett, Brian Arthur, ethology
Headington, John Terrence, pathology
Heederks, William John, bioengineering
Hegyi, Dennis, astrophysics
Heinrich, Eberhardt William, geology
Heins, Albert Edward, mathematics
Helling, Robert Bruce, genetics
Hendal, John Frederick, acoustics
Hendel, Alfred Z, physics
Henderson, John Woodworth, ophthalmology
Hendrickson, Willard James, psychiatry
Hendrix, Robert Cowgill, pathology
Henley, Keith Stuart, gastroenterology
Henness, Albert Vincent, virology
Hess, Lloyd William, ecology, environmental education
Hicks, Noel J., mathematics
Hicks, Samuel Pendleton, pathology
Higgins, Ian T., epidemiology
Higgins, Millicent Williams Payne, medicine, epidemiology
High, Robert Huggins, pediatrics
Higman, Donald Gordon, mathematics
Higuchi, William Iyeo, pharmaceutical chemistry
Hilbert, Morton S., environmental health, public health
Hile, Ralph Oscar, zoology
Hill, Bruce M., statistics
Hills, Jack Gilbert, theoretical astrophysics
Hiltner, William Albert, astrophysics
Hiltunen, Jarl Kalervo, fresh water ecology
Hinerman, Dorin Lee, pathology
Hoch, Frederic Louis, biochemistry, endocrinology
Hoefle, Milton Louis, organic chemistry
Holden, Joseph August, biological chemistry
Holland, John Henry, applied mathematics, computer science
Holt, John Floyd, medicine
Holter, Marvin Rosenkrantz, physics, mathematics
Hoobler, Icie Macy, nutritional biochemistry
Hoobler, Sibley Worth, medicine
Hooper, Emmet Thurman, Jr., vertebrate zoology
Hooper, Frank Fincher, zoology
Horvath, William John, biophysics
Horwitz, Harold M., mathematics
Houk, Nancy (Mia), astronomy
Howat, William Frederick, pediatrics, physiology
Huang, Joseph Chi Kan, oceanography, meteorology
Hubbell, Theodore Huntington, entomology
Huelke, Donald Fred, anatomy
Hughes, Byron Orville, human genetics
Hull, Robert Lee, medicinal chemistry
Hultquist, Donald Elliott, biochemistry
Hunt, Thomas Kintzing, energy conversion
Hutterer, Karl Leopold, anthropology
Ice, Rodney D., pharmacy, nuclear medicine
Ikuma, Hiroshi, plant physiology
Imboden, Clarence Alphonse, Jr., public health, clinical pharmacology
Jackson, David Archer, molecular biology, virology
Jackson, Ethel Noland, molecular biology
Jackson, Philip Landis, geology
Jacobs, Stanley J., geophysics, fluid mechanics
Jacobson, Arnold P., radiation biology
Jacquez, John Alfred, physiology, biomathematics
Janecke, Joachim Wilhelm, physics
Janzen, Daniel Hunt, ecology, evolution
Jebe, Emil H, experimental statistics, applied statistics

Jochim, Kenneth Erwin, physiology
Johnson, Arthur Gilbert, immunology
Johnson, Benjamin C., epidemiology
Johnsson, Lars-Goran, otorhinolaryngology
Jones, Donald Akers, actuarial science
Jones, Eldon Melton, medicinal chemistry
Jones, Garth Wicks, microbiology
Jones, George Henry, biochemistry, molecular biology
Jones, Kenneth Lester, soil microbiology
Jones, Lawrence William, high energy physics
Jones, Philip Sanford, mathematics
Jourdian, George William, biochemistry, microbiology
Julius, Stevo, medicine
Juni, Elliot, bacterial physiology
Kahn, Raymond Henry, endocrinology, histology
Kalme, Charles Ivars, applied mathematics, mathematical analysis
Kaltenborn, James S., organic chemistry
Kane, Gordon Leon, high energy physics
Kaplan, Wilfred, mathematics
Karsch, Fred Joseph, reproductive endocrinology
Katz, Ernst, solid state physics
Kaufman, Peter Bishop, plant morphology
Kelley, William Nimmons, biochemistry
Kelly, William Crowley, economic geology
Kemp, George Norman Everett, developmental biology
Kesling, Robert Vernon, paleontology
Keyes, Paul Landis, reproductive endocrinology, physiology
Kilham, Peter, limnology, geochemistry
Kincaid, Wilfred Macdonald, mathematics
Kindt, Glenn W., neurosurgery
King, John Swinton, physics
Kish, George, geography
Kister, James Milton, mathematics
Kleinsmith, Lewis Joel, cell biology, biochemistry
Kluger, Arnold Girard, vertebrate zoology
Kluger, Matthew Jay, physiology
Knopf, Ralph Fred, internal medicine
Kochen, Manfred, information science
Koepke, George Henry, medicine
Kolars, John F., cultural geography, geography of Turkey
Kooi, Kenneth Ashley, electroencephalography
Kopelman, Raoul, physical chemistry
Kottak, Conrad Philip, anthropology
Kowalski, Charles Joseph, statistics, biometrics
Kozma, Adam, optical physics, electrooptics
Kramer, Theodore Christian, anatomy
Krause, Eugene Franklin, mathematics
Krimm, Samuel, physics, biophysics
Krisch, Alan David, high energy physics
Kubis, Joseph John, theoretical physics
Kuczkowski, Robert Louis, physical inorganic chemistry
Kuhn, William R., planetary atmospheres
Kurtz, Stanley Morton, pathology
Kutkuhn, Joseph Henry, ecology, fisheries
La Du, Bert Nichols, Jr., biochemical pharmacology
Lagler, Karl Frank, fisheries, zoology
Lampe, Isadore, radiology
Lands, William Edward Mitchell, biochemistry
Lapides, Jack, urology
Larimore, Ann Evans, geography
Larkin, Frances Ann, nutrition
LaRocca, Anthony Joseph, physics, electrical engineering
Latta, John Neal, optics, electrical engineering
Latta, William Carl, fish biology
Lauer, Edward Willard, anatomy
Lawrence, Merle, physiology
Lawton, Richard G., synthetic organic chemistry, bio-organic chemistry
Leabo, Dick Albert, applied statistics, economic statistics
Lee, Chung N., mathematics
Legan, Sandra Jean, reproductive physiology, neuroendocrinology
Lehman, Grace Church, zoology, endocrinology
Leinonen, Ellen A., anatomy, physiology
Leisenring, Kenneth Baylis, mathematics
Leith, Emmett Norman, optics
Leonard, John Wilkinson, zoology
Levengood, William Camburn, biophysics
Levine, Myron, genetics, virology
Lewis, Donald John, mathematics

Lewis, Robert Richards, Jr., theoretical physics
Liao, Shu-Chung, physical-analytical chemistry
Lillie, John Howard, anatomy, dentistry
Limperis, Thomas, physics
Lin, Chang Kwei, phycology
Lindenauer, S Martin, surgery
Livingstone, Frank Brown, physical anthropology
Lockridge, Oksana Maslivec, biochemical pharmacology, biochemical genetics
Lockwood, William Grover, cultural anthropology
Loewenthal, Lois Anne, zoology
Lohr, Lawrence Luther, Jr., theoretical chemistry
Lomax, Margaret Irene, molecular genetics
Longo, Michael Joseph, high energy physics
Longone, Daniel Thomas, organic chemistry
Louis, Lawrence Hua-Hsien, biochemistry
Lovell, Robert Gibson, internal medicine, allergy
Low, Bobbi Stiers, evolutionary biology, ecology
Lowry, Robert James, botany
Ludwig, Martha Louise, biochemistry
Lunk, William Allan, ornithology
Luthe, John Charles, theoretical high energy physics
Lyndon, Roger Conant, mathematics
Maassab, Hunein Fadlo, epidemiology, virology
Macalpine, Gordon Madeira, astrophysics
MacCallum, Donald Kenneth, anatomy, histology
Macurda, Donald Bradford, Jr., paleontology, marine biology
Magee, Kenneth Raymond, neurology
Magnuson, Harold Joseph, occupational medicine
Malvin, Richard L, physiology
Malvitz, Dolores Marie, public health, dental hygiene
Mann, William Richard, dentistry
Manny, Bruce Andrew, limnology, aquatic ecology
Marcus, Joyce, anthropology
Margolis, Philip Marcus, psychiatry
Marino, Joseph Paul, organic chemistry
Martel, James Edward, polymer chemistry
Martel, William, radiology
Martin, Michael McCulloch, biological chemistry
Martin, Sarah Smith, pharmacology, biochemistry
Mason, Conrad Jerome, micrometeorology
Mason, Merle, biochemistry
Massey, Vincent, biochemistry
Mathews, Kenneth Pine, allergy
Mathews, Rowena Green, biochemistry, protein chemistry
Matovinovic, Josip, medicine
Maxwell, Richard Elmore, biochemistry
McBride, Gordon E, phycology
McCarthy, Duncan Arthur, Jr., pharmacology
McClamroch, N Harris, applied mathematics
McClelland, Nina Irene, environmental health
McCord, Carey Pratt, occupational health
McCormick, William Wallace, atomic spectroscopy
McCullough, Dale Richard, wildlife management, ecology
McFadden, James Thompson, population ecology
McLaughlin, Jack Enloe, mathematics
McLean, James Amos, internal medicine, allergy
McLean, John Robert, biochemistry
McMaster, Marvin Clayton, Jr., biochemistry
McNamara, James Alyn, Jr, orthodontics, anatomy
McVaugh, Rogers, botany
Medzihradsky, Fedor, biochemistry
Meinke, William Wayne, analytical chemistry
Menge, Alan C, reproductive physiology
Meyer, Donald Irwin, physics
Meyer, Robert F, pharmaceutical chemistry
Meyer, Ruben, pediatrics, public health
Meyers, Muriel Charlotte, medicine
Meyers, Philip Alan, organic geochemistry, oceanography
Mich, Thomas Frederick, organic chemistry
Midgley, A Rees, Jr, endocrinology, pathology

Mikkelsen, William Mitchell, internal medicine
Millard, Herbert Dean, dentistry
Miller, Derek Harry, psychiatry, psychoanalysis
Miller, Freeman Devold, astronomy
Miller, Lila, biochemistry
Miller, Robert Rush, ichthyology
Milone, Nicholas Arthur, public health
Mistretta, Charlotte Mae, neurophysiology, developmental physiology
Moersch, George William, organic chemistry
Mohler, Orren (Cuthbert), astronomy
Moll, Russell Addison, limnology
Monto, Arnold Simon, epidemiology, infectious diseases
Moody, David Burritt, psychoacoustics
Moore, Alexander Mazyck, organic chemistry, information science
Moore, Felix E, biostatistics
Moore, Thomas Edwin, zoology, entomology
Moosman, Darvan Albert, anatomy, surgery
Morawa, Arnold Peter, pedodontics, oral biology
Morley, Dourossoff Edmund, speech pathology
Morley, George W, medicine
Morris, Michael D, analytical chemistry
Morton, Harrison Leon, forest pathology
Mouw, David Richard, physiology
Moyer, Carl Edward, physiological chemistry
Moyers, Robert Edison, orthodontics, human development
Mozley, Samuel Clifford, aquatic ecology
Murphy, William Henry, Jr., microbiology
Murty, Katta Gopalakrishna, operations research
Myers, Philip, mammalogy
Nace, George William, developmental biology
Nagera, Humberto, psychiatry, psychoanalysis
Nagy, Andrew F, aeronomy, atmospheric physics
Nasjleti, Carlos Eduardo, cytogenetics, teratology
Neel, James Van Gundia, genetics
Neidhardt, Frederick Carl, microbiology, biochemistry
Nesbit, Cecil James, actuarial mathematics
Nicolaides, Ernest D, organic chemistry
Nooden, Larry Donald, plant physiology, biochemistry
Nordby, Gordon Lee, biochemistry, biometry
Nordin, Ivan Conrad, organic chemistry
Nordman, Christer Eric, physical chemistry
Norman, Arthur Geoffrey, plant biochemistry
Northcutt, Richard Glenn, neuroanatomy
Nussbaum, Ronald Archie, herpetology
Nystuen, John David, geography
Oakley, Bruce, neurobiology
Oakley, Burks, II, bioengineering
Oberman, Harold A, medicine, pathology
O'Brien, William Joseph, dental materials, surface chemistry
Oelrich, Thomas Mann, morphology
Okerholm, Richard Arthur, biochemistry
Oliver, William J, pediatrics
Olsen, Ronald H, microbiology
Olsson, Gunnar, economic geography
Oncley, John Lawrence, molecular biophysics
O'Neal, Russell D, physics
Osborn, Richard Kent, plasma physics, physical optics
Outcalt, Samuel Irvine, physical geography
Overberger, Charles Gilbert, polymer chemistry
Overseth, Oliver Enoch, Jr., experimental high energy physics
Owen, Robert Michael, chemical oceanography, marine geochemistry
Oxender, Dale LaVern, biochemistry
Palmore, Julian Ivanhoe, III, mathematics, celestial mechanics
Parcell, Robert Ford, organic chemistry
Parker, Robert Bruce, neuropharmacology
Parkinson, William Charles, nuclear physics
Patriarche, Mercer Harding, fish biology
Patterson, Richard L, resource management
Paul, Ara Garo, pharmacognosy
Payne, Anita H, biochemistry, endocrinology
Payne, Francis Eugene, virology
Payne, Robert B, zoology

Peacock, Keith, astronomy, physics
Peacor, Donald Ralph, mineralogy
Pearcy, Carl Mark, Jr., mathematics
Peebles, Christopher Spalding, anthropology, archaeology
Piranian, George, mathematics
Platt, John Radar, biophysics
Pliske, Edward Carl, anatomy
Polcyn, Fabian Casimir, physics, oceanography
Pollack, Henry Nathan, geophysics
Pollard, Herman Marvin, internal medicine
Pollock, Stephen M, operations research
Porter, Richard Janvier, parasitology, immunology
Portman, Donald James, micrometeorology
Poschel, Bruno Paul Henry, psychopharmacology
Powers, John Michael, dental materials
Poznanski, Andrew K, radiology
Prentice, Virgina Lee, geography
Preston, Stephen Boylan, forest products
Price, Alan Roger, biochemistry
Quarton, Gardner Cowles, psychiatry, neurobiology
Radin, Norman Samuel, biochemistry
Ramanujan, Melapalayam Srinivasan, mathematics
Ramford, Sigurd, dentistry, periodontology
Ransom, Henry King, surgery
Rao, Desiraju Bhavanarayana, oceanography
Rapp, Robert, radiology
Rappaport, Ray A, anthropology
Rasmussen, Paul, physical inorganic chemistry
Rauch, Jeffrey Baron, mathematical physics
Rea, David Kenerson, geological oceanography
Reade, Maxwell Ossian, mathematics
Rebstock, Mildred Catherine, organic chemistry
Reel, Jerry Royce, endocrinology
Regezi, Joseph Alberts, oral pathology
Reiher, Harold Frederick, acoustics
Reuter, Stewart R, radiology
Reynolds, Herbert McGaughey, physical anthropology, anthropometrics
Rhodes, Frank Harold Trevor, geology, paleontology
Rhodin, Johannes Arne Gosta, anatomy
Rich, Arthur, astrophysics, atomic physics
Richards, Albert Gustav, dental radiology
Riggs, Thomas Rowland, biochemistry, nutrition
Rikans, Lora Elizabeth, biochemical pharmacology
Ringler, Daniel Howard, laboratory animal medicine
Rittenhouse, Harry Gcogre, biological chemistry
Rizki, Tahir Mirza, developmental genetics, cell biology
Robertson, Andrew, aquatic ecology
Robinson, William Dodd, medicine, internal medicine
Rodney, Gertrude, biochemistry
Roe, Byron Paul, physics
Rogers, William Leslie, medicine, medical physics
Rootare, Hillar Muidar, dental materials, surface chemistry
Rosen, Jeffrey Kenneth, animal physiology
Rosen, Ronald Haiam, mathematics
Ross, Marc Hanson, theoretical physics
Rowe, Nathaniel H, oral pathology
Rowe, Thomas Dudley, pharmacy
Rucknagel, Donald Louis, human genetics
Ruddon, Raymond Walter, Jr, pharmacology
Rudolph, Ralph W, inorganic chemistry
Rulfs, Charles Leslie, analytical chemistry, inorganic chemistry
Rupp, Ralph Russell, audiology, speech pathology
Rusch, Wilbert, Sr, biology, geology
Russell, Albert Lee, dental epidemiology
Ruthruff, Robert Freeborn, organic chemistry
Rutledge, Lester T, physiology
St Maurice, Jean-Pierre, ionospheric physics
Samir, Uri, space physics
Sander, Leonard Michael, physics
Sanders, Theodore Michael, Jr, physics
Sands, Richard Hamilton, atomic physics, biophysics
Savageau, Michael Antonio, microbiology, systems science
Schacht, Jochen Heinrich, neurochemistry

Schafer, Rollie R, neurobiology
Schatz, Irwin Jacob, internal medicine, cardiovascular diseases
Scheider, Walter, biophysics
Schelske, Claire L, limnology
Schmickel, Roy David, pediatrics, genetics
Schmidt, Robert W, pathology
Schneider, Richard Coy, neurosurgery
Schnitzer, Bertram, hematology, pathology
Schoch, Henry Kramer, internal medicine
Schorger, William Davison, cultural anthropology
Schork, Michael Anthony, biostatistics
Schriber, Thomas J, computer science, mathematics
Schteingart, David E, internal medicine
Schultz, Jane Schwartz, immunogenetics
Scott, Richard Anthony, applied mathematics
Sears, Richard Langley, astrophysics
Seevers, Maurice Harrison, pharmacology
Seibel, Erwin, oceanography
Sell, John Edward, biochemistry, immunology
Sellinger, Otto Zivko, biochemistry
Selzer, Melvin Lawrence, psychiatry
Senior, Thomas Bryan Alexander, electromagnetics
Shafer, Jules Alan, physical organic chemistry, biochemistry
Shaffer, Robert Lynn, mycology
Shappirio, David Gordon, physiology, biochemistry
Sharf, Donald Jack, speech & hearing sciences
Sharp, Robert Richard, physical chemistry
Sherman, James H, physiology
Sherman, Noah, theoretical physics
Shields, Allen Lowell, mathematics
Shipman, Charles, Jr., virology
Short, Franklin Willard, medicinal chemistry
Sicko-Goad, Linda May, cell biology, algology
Siegel, George Jacob, neurology, neurobiology
Silverman, Albert Jack, psychiatry, psychophysiology
Simon, Carl Paul, mathematics
Sing, Charles F, human genetics, statistics
Sinsheimer, Joseph Eugene, medicinal chemistry, pharmaceutical chemistry
Sippel, Theodore Otto, anatomy
Sloan, Herbert, thoracic surgery
Slutsky, Mark Sender, theoretical physics, quantum optics
Smith, Alexander Hanchett, mycology
Smith, Charles Isaac, geology
Smith, Donald C, preventive medicine, maternal & child health
Smith, Edwin Mark, medicine
Smith, Gerald Ray, zoology
Smith, Newbern, geophysics, electromagnetism
Smith, Peter Alan Somervail, organic chemistry
Smith, Ralph G, industrial hygiene, analytical chemistry
Smith, Ronald Duane, pharmacology
Smith, Stanford Henry, fish biology
Smith, Thomas Charles, pharmacology
Smith, William S, orthopedic surgery
Smoller, Joel A, mathematics
Smouse, Peter Edgar, human genetics, statistics
Snyder, Richard Gerald, aerospace medicine, anthropology
Solomon, David Eugene, research administration, materials science
Somers, Joseph Henry, environmental chemistry
Spang, Arthur William, microchemistry
Sparks, Harvey Vise, medical physiology
Sparrow, Frederick Kroeber, Jr.
Spivey, Walter Allen, statistics
Stapp, William B, conservation
Starr, Norman, mathematical statistics
Stebbins, William Cooper, bioacoustics
Stedman, Donald Hugh, atmospheric chemistry
Steiner, Erich E, genetics
Stern, Aaron Milton, pediatrics, cardiology
Stewart, Ralph Randles, systematic botany
Stiles, Martin, organic chemistry
Stoermer, Eugene F, algology, limnology
Stone, Robert Edward, Jr, speech & hearing sciences, speech pathology
Storer, Robert Winthrop, zoology
Storer, Thomas, mathematics
Strachan, Donald Stewart, histology, oral biology

Strang, Ruth Hancock, pediatrics, cardiology
Stratton, Charlotte Dianne, organic chemistry
Striffler, David Frank, public health, dentistry
Stucki, William Paul, biochemistry
Sullivan, Donita B., pediatrics
Sun, Nai Chau, genetics, cell biology
Sussman, Alfred Sheppard, mycology
Swain, Henry Huntington, pharmacology
Swartz, Walter H., dentistry
Tamres, Milton, physical inorganic chemistry
Tanner, David John, low temperature physics
Tarapchak, Stephen J., limnology
Taren, James A., neurosurgery
Tashian, Richard Earl, biochemical genetics
Taylor, Robert Cooper, physical chemistry
Taylor, William Randolph, botany
Teichroew, Theodore, mathematics
Terry, Samuel Matthew, chemistry
Terwilliger, Kent Melville, high energy physics
Teske, Richard Glenn, astronomy
Test, Frederick Harold, zoology
Thornbury, John R., radiology
Tickle, Robert Simpson, nuclear physics
Tiffany, Mary Lois, biophysics
Tiffany, Otho Lyle, physics
Tinkle, Donald Ward, vertebrate zoology
Tinney, Francis John, medicinal chemistry
Titiev, Mischa, ethnology
Titus, Charles Joseph, mathematics
Tobler, Waldo Rudolph, geography, cartography
Tocher, Stewart Ross, forestry
Tomozawa, Yukio, theoretical high energy physics
Toth, Robert S., solid state physics
Townes, Henry Keith, Jr., entomology
Townes, Marjorie Chapman, entomology
Truden, Judith Lucille, virology
Tsai, Alan Chung-Hong, nutrition
Tsigdinos, George Andrew, inorganic chemistry
Turcotte, Jeremiah G., surgery
Turner, Robert Elwood, atmospheric physics, theoretical astrophysics
Turpening, Roger Munson, seismology
Valenstein, Elliot Spiro, neurosciences, psychology
van der Schalie, Henry, malacology
Vandenbelt, John Melvin, physical chemistry
Vander, Arthur J., physiology
Vandermeer, John H., population biology, aquatic ecology
Vander Velde, John Christian, physics
van der Voo, Rob, geology, geophysics
Vinh, Nguyen Xuan, celestial mechanics
Voorhees, John James, dermatology, medical research
Voss, Edward Groesbeck, taxonomic botany
Votaw, Charles Lesley, neuroanatomy
Waggoner, Raymond Walter, psychiatry, neurology
Wagner, Florence Signaigo, botany
Wagner, John Garnet, pharmaceutical chemistry, organic chemistry
Wagner, Warren Herbert, Jr., botany
Walker, Charles Frederic, zoology
Walker, Glenn Kenneth, cell biology, protozoology
Wallner, Julius Michael, psychiatry
Ward, John Frank, physics
Wasserman, Arthur Gabriel, mathematics
Watson, Andrew Samuel, psychiatry
Wax, Joan, pharmacology
Weber, Wendell W., pharmacology, pediatrics
Wegman, Myron Ezra, public health, pediatrics
Weidler, Donald John, internal medicine
Wenberger, Peter Jay, mathematics, computer science
Weinhold, Paul Allen, biochemistry
Weinreich, Gabriel, physics
Weiss, David Steven, organic chemistry, photochemistry
Weller, John Martin, internal medicine, nephrology
Wellock, Lois Margaret, physical medicine
Wendel, James Gutwillig, mathematics
Werbel, Leslie Morton, organic chemistry
Westerberg, Martha Rosalie, neurology
Westrum, Edgar Francis, Jr., physical chemistry, thermodynamics
Whallon, Robert, Jr., anthropology, archaeology
Wheeler, Albert Harold, bacteriology
Whipple, George Hoyt, Jr., radiological health
Whitehouse, Frank, Jr., microbiology, immunology
Whitehouse, Walter MacIntire, radiology
Whitfield, Carolyn Dickson, biochemistry
Whitfield, Harvey James, Jr., molecular biology
Wiedenbeck, Marcellus Lee, nuclear physics
Wiley, John Herbert, speech pathology, audiology
Wilkinson, Bruce H., sedimentology
Williams, Charles Haddon, Jr., biochemistry
Williams, David Noel, mathematical physics
Williams, George W., biostatistics
Williams, William Lee, atomic physics
Willis, Park Weed, III, internal medicine, cardiovascular diseases
Wilmsen, Edwin Norman, anthropology
Wilson, James Tinley, geophysics
Winans, Sarah Schilling, neuroanatomy, neuropsychology
Winter, David John, mathematics
Winter, Harry Clark, biochemistry
Winter, William Phillips, biochemistry
Wiseman, John R., organic chemistry
Witten, Thomas Adams, Jr., theoretical physics
Witter, John Allen, forest entomology, insect ecology
Wolpoff, Milford Howell, physical anthropology
Wolter, J Reimer, ophthalmology
Wong, Victor Kenneth, physics
Woodburne, Russell Thomas, anatomy
Worden, Leonard Russell, organic chemistry, family medicine
Wu, Alfred Chi-Tai, theoretical physics
Wu, Chung, biochemistry
Yao, York-Peng Edward, theoretical high energy physics
Yengoyan, Aram A., anthropology
Yocum, Charles Fredrick, biochemistry
Yocum, Conrad Schatte, plant physiology
Yu, Terry Ta-Jen, organic chemistry, medicinal chemistry
Zand, Robert, biophysical chemistry, neurochemistry
Zannoni, Vincent G., pharmacology
Zarafonetis, Chris John Dimiter, internal medicine
Zelenka, Jerry Stephen, electronic engineering, optics
Zissis, George John, optics
Zorn, Jens Christian, atomic physics
Zsigmond, Elemer K., anesthesiology
Zweifler, Andrew J., internal medicine

BATTLE CREEK

Carlotti, Ronald John, nutritional biochemistry
Collins, Richard Andrew, pathology
Fulger, Charles Von, food technology
Hopper, John Henry, food technology
Humphrey, Arthur Allan, nutrition
John, Ralph Alfred, organic chemistry
Mafaractisti, Boaz Amon, food technology
Schaller, Daryl Richard, nutrition
Vondel, Richard M., food science

BENTON HARBOR

Blachford, John Kerslake, organic chemistry
Einset, Eystein, food science
Grace, Robert Ambrose, medical physiology
Miranda, Thomas Joseph, organic polymer chemistry
Nelson, John Arthur, chemistry, chemical engineering
Peterson, Edward Charles, physics
Rutkowski, Beverly Jean, physical chemistry
Stright, Paul Leonard, organic chemistry

BERKLEY

Mehlenbacher, Lyle E., mathematical analysis

BERRIEN SPRINGS

Brown, Robert Henry, geophysics
Chobotar, Bill, zoology, animal parasitology
Coffin, Harold Glen, paleontology
Ford, Dewain, chemistry
Hare, Leonard N., plant physiology
Jones, Harold Trainer, mathematical analysis
Marsh, Alice Garrett, nutrition, foods
Marsh, Frank Lewis, ecology
Minesinger, Richard Rockwell, organic chemistry
Mutch, George William, chemical kinetics, physical chemistry
Ritland, Richard Martin, vertebrate zoology, paleontology
Rowland, Sattley Clark, materials science, paleontology
Stout, John Frederic, neurobiology
Thoresen, Asa Clifford, ornithology
Wong, Peter Alexander, chemical instrumentation

BIG RAPIDS

Buss, Jack Theodore, vertebrate physiology
Chulski, Thomas, analytical chemistry
Friar, Robert Edsel, human physiology
Hoeksema, Walter David, microbiology, biochemistry
Holcomb, George N., medicinal chemistry
Jacoby, Ronald Lee, medicinal chemistry
Kazerovskis, Karlis, pharmacy
Larson, Gustav Olof, organic chemistry
Lehnert, James Patrick, zoology
Lofquist, Marvin John, inorganic
Nash, Edmund Garrett, organic inorganic
Nienhouse, Everett J., organic chemistry
Oldfield, Thomas Edward, physiological ecology
Poland, Lloyd Orville, inorganic chemistry, analytical chemistry
Reitz, Henry Charles, protein chemistry
Rue, Sigurd Oscar, analytical chemistry
Slywka, Gerald William Alexander, analytical chemistry, toxicology
Swartz, Harry Sip, pharmacy administration, pharmacy
Wagner, Kenneth A., botany

BIRMINGHAM

Faggan, Joseph Edward, organic chemistry

BLOOMFIELD HILLS

Butterworth, Allen Virgil, physical chemistry
Clampitt, Philip Theodore, zoology, animal ecology
Hollyer, Robert Nelson, Jr., physics
Liboff, Abraham R., medical physics, biophysics

BRIDGMAN

Vanderburg, Vance Dilks, high energy physics

BUCHANAN

Jarvis, Lactance Aubrey, organic polymer chemistry

CALUMET

Wege, Randall James, geology

CARO

Karpovich, John, physical chemistry

CASSOPOLIS

Purkhiser, E Dale, animal nutrition

DEARBORN

Adams, Franklin Scott, botany, morphology
Agresti, William W., computer sciences
Anderson, James E., physical chemistry
Asik, Joseph R., applied physics
Block, Duane Llewellyn, occupational medicine
Borchers, Robert H., solid state physics
Brown, James Ward, applied mathematics
Chang, Tai Yup, air pollution, quantum chemistry
Chattha, Mohinder Singh, polymer chemistry
Chetsanga, Christopher J., molecular biology, biochemistry
Cole, Terry, chemical physics
Compton, Walter Dale, physics
Davis, Lloyd Craig, physics
DeCamp, Mark Rutledge, organic chemistry
Dickie, Ray Alexander, polymer chemistry, physical chemistry
Dzieciuch, Matthew Andrew, electrochemistry
Eagen, Charles Frederick, solid state physics
Emerson, David Winthrop, organic chemistry
Gelderloos, Orin Glenn, environmental physiology
Hagenlocker, Edward Emerson, physics
Heller, Hanan Chonon, physical chemistry
Hertzler, Emanuel Cassel, physiology
Jaklevic, Robert C., experimental solid state physics
Kaiser, Edward William, Jr., physical chemistry
Kim, John Jungun, molecular physics, plasma physics
Kleinman, Roberta Wilma, organic chemistry, biochemistry
Kummer, Joseph T., solid state chemistry
Kushida, Toshimoto, solid state physics
Labana, Santokh Singh, organic chemistry, polymer science
Laurance, Neal L., computer science
Ludwig, Frank Arno, electrochemistry
Lyjak, Robert Fred, mathematics
Meitzler, Allen Henry, physics
Mencik, Zdenek, physical chemistry, x-ray crystallography
Milberg, Morton Edwin, solid state chemistry
Newman, Seymour, polymer science, plastics
Oei, Djong-Gie, physical chemistry, inorganic chemistry
Oene, Henk Van, physical chemistry
O'Shea, Timothy Allan, environmental chemistry, analytical chemistry
Otto, Klaus, environmental chemistry
Pierson, William Roy, physical chemistry
Potts, John Earl, solid state physics
Potts, Richard Allen, inorganic chemistry
Reitz, John Richard, theoretical physics
Robertson, Richard Earl, physical chemistry, polymer science
Roslinski, Lawrence Michael, environmental health
Rothschild, Walter Gustav, chemical physics
Ruof, Clarence Herman, chemistry
Saillant, Roger Barry, inorganic chemistry
Schneider, Charles Louis, obstetrics & gynecology
Schneider, Michael J., plant physiology, photobiology
Sickafus, Edward N., solid state physics
Stearns, Mary Beth Gorman, physics
Truex, Timothy Jay, inorganic chemistry
Ullman, Robert, polymer chemistry
Verhey, Roger Frank, mathematics
Weaver, Ervin Eugene, environmental chemistry
Weber, Willes Henry, solid state physics
Weiner, Steven Allan, physical organic chemistry
Weinstock, Bernard, physical chemistry
Wilhelm, Rudolf Ernst, allergy, immunology
Wu, Ching-Hsong, chemical kinetics
Yao, Yung Fang Yu, physical chemistry
Zitzewitz, Paul William, experimental atomic physics

DETROIT

Abramson, Hanley N., medicinal chemistry
Albert, Samuel, oncology
Albright, Raymond Gerard, zoology
Alcala, Jose Ramon, anatomy
Alegnani, William Charles, bacteriology
Anderson, Harlan Dwight, biochemistry
Antipa, Gregory Alexis, protozoology, cell biology
Arbulu, Agustin, surgery, thoracic surgery

Arking, Robert, developmental genetics
Arnold, William James, cell biology
Ash, Arthur Burr, organic chemistry
Aston, Roy, pharmacology
Axelrod, Arnold Raymond, hematology
Bachelis, Gregory Frank, mathematics
Bacon, Larry Dean, genetics, immunogenetics
Baechler, Charles Albert, physiology
Bagchi, Mihir, cell biology, ophthalmology
Bagshaw, Joseph Charles, molecular biology, biochemistry
Bailey, Harold Edwards, pharmacognosy
Bailey, Virginia Long, botany
Balek, Richard William, pharmacology
Barnhart, Marion Isabel, physiology
Barr, Martin, physical pharmacy
Barraco, Robin Anthony, neurochemistry
Bartlett, Paul Devere, biochemistry
Barton, M Xaveria, cultural geography
Basinski, Daniel Henry, clinical chemistry
Bates, John Bertram, colloid chemistry
Beal, John Anthony, anatomy, neuroanatomy
Beard, George B, experimental nuclear physics
Becker, Charles Edward, biochemistry
Beher, William Tyers, biochemistry
Bender, Leonard Franklin, medicine
Benjamins, Joyce Ann, neurochemistry
Beres, William Philip, nuclear physics
Berk, Richard Samuel, microbiology, biochemistry
Berke, Harry L, biochemistry, pharmacology
Bernstein, Maurice Harry, anatomy, cell biology
Bharucha-Reid, Albert Turner, applied mathematics
Blass, Gerhard Alois, theoretical physics
Blend, Michael J, reproductive physiology, toxicology
Bloom, Herbert Jerome, oral surgery, maxillofacial surgery
Bloom, Victor, psychiatry
Bohm, Henry, solid state physics
Bollinger, Robert Otto, cell biology
Boltz, David Ferdinand, analytical chemistry
Bond, Harley William, neuropharmacology, pharmacy
Borgman, William Martin, Jr, geometry
Boros, Dov Lewis, microbiology, immunology
Boving, Bent Giede, anatomy, reproductive biology
Boyce, Charles, obstetrics & gynecology
Brackett, Robert Giles, immunology
Brandon, Frank Bayard, biophysics, virology
Brandt, Manuel, analytical chemistry
Brennan, Michael James, medicine
Briggs, Charles Francis, mathematics
Brooks, Samuel Carroll, biochemistry, endocrinology
Brown, Ray Kent, protein chemistry
Brown, William John, microbiology
Burnham, Thomas K, dermatology, immunology
Caldwell, John R, internal medicine
Carlson, Gustaf Harry, organic chemistry
Chason, Jacob Leon, pathology
Chavin, Walter, endocrinology, radiobiology
Chen, Ming Chih, organic chemistry
Choe, Byung-Kil, microbiology, cell biology
Chow, Pao Liu, applied mechanics, applied mathematics
Christensen, James Boyd, anthropology
Churchill, John Alvord, neurology, pediatric neurology
Churchill, Paul Clayton, physiology
Clapper, Muir, internal medicine
Cohen, Flossie, immunology, pediatrics
Cohen, Margo Nita Panush, internal medicine, endocrinology
Cohen, Sanford Ned, pharmacology, developmental biology
Cole, George Christopher, microbiology
Connor, Nolen Duncan, pharmacology, pathology
Cook, David Russell, invertebrate zoology, entomology
Cooper, Margaret Hardesty, anatomy
Corrigan, Kenneth Edwin, physics
Cosgriff, John W, Jr, vertebrate paleontology
Coye, Robert Dudley, anatomic pathology
Curtin, Thomas J, entomology, research administration
Damusis, Adolfas, chemistry
Dauphinais, Raymond Joseph, pharmacy
Davila, Julio C, cardiovascular surgery, thoracic surgery

Debich, Danica, biochemistry
De Giusti, Dominic Lawrence, parasitology
de Graaf, Adriaan M, solid state physics
Denman, Harry Harroun, theoretical mechanics, mathematical physics
Dohrs, Fred E, geography
Done, Alan Kimball, pediatrics, clinical pharmacology
Dorenbusch, William Edwin, nuclear physics
Dorsey, John Morris, psychiatry
Doscher, Marilyn Scott, biochemistry
Driscoll, Egbert Gotzian, invertebrate ecology, invertebrate paleontology
Dunifer, Gerald Leroy, experimental solid state physics
Dunker, Melvin Frederick William, pharmaceutical chemistry
Dutta, Saradindu, pharmacology, veterinary medicine
Ebbing, Darrell Delmar, physical chemistry
Ehrlich, John, microbiology
Eisenstadt, Bertram Joseph, mathematics
Endicott, John F, inorganic chemistry
Engel, John Hal, Jr, organic chemistry, polymer chemistry
Ervin, Robert Francis, bacteriology
Essner, Edward Stanley, cell biology, electron microscopy
Ettinger, Anna Marie Conway, anatomy
Evans, Tommy Nicholas, medicine
Fand, Sally Bogolub, medicine, endocrinology
Fast, Henryk, mathematical analysis
Favro, Lawrence Dale, theoretical physics
Filbey, Allen Howard, organic chemistry
Fischer, George A, biochemistry, clinical chemistry
Fisher, Mike W, bacteriology
Fleming, Suzanne M, inorganic chemistry
Foa, Piero Pio, physiology
Folley, Karl Wilmot, mathematics
Foor, W Eugene, zoology, parasitology
Foreman, Robert Walter, metallurgical chemistry
Forsthoefel, Paulinus Frederick, genetics
Fox, Clement Alphonsine, neuroanatomy
Fradkin, David Milton, theoretical physics
Francis, Dawn Elizabeth, bioinorganic chemistry, clinical chemistry
Fredrick, William George, industrial hygiene
Frisch, Kurt Charles, chemistry
Frohman, Charles Edward, biochemistry, physiology
Furlong, Robert B, mineralogy, geochemistry
Gaebler, Oliver Henry, biochemistry, clinical chemistry
Gaeth, John Henry, audiology
Gala, Richard R, physiology, endocrinology
Gangwere, Stanley Kenneth, zoology
Gayer, Karl Herman, chemistry
Gemant, Andrew, physical chemistry, biophysical chemistry
Glass, Alexander Jacob, atomic physics
Glick, Milton Don, physical chemistry, inorganic chemistry
Goel, Om Prakash, organic chemistry, medicinal chemistry
Goldfarb, Abraham Robert, biochemistry
Goldman, Harold, neuropharmacology
Goldstein, Sidney, cardiology
Goliber, Edward William, chemistry
Goodman, Morris, immunology
Goodman, Robert Joseph, geography of India
Goodwin, Jesse Francis, clinical biochemistry
Goodwin, Stefan Cornelius, anthropology
Gorelic, Lester Sylvan, biochemistry, photobiology
Gottlieb, Jacques Simon, psychiatry
Graham, Blanche D, industrial microbiology
Greer, James Edward, medical microbiology
Grosscup, Gordon Leonard, anthropology, archaeology
Gupta, Suraj Narayan, theoretical physics
Gustafson, Daniel Ray, experimental solid state physics
Hafez, Saad Elsayed, reproductive physiology
Hagarman, Vincent A, solid state physics
Haggis, Alex John, zoology
Hahn, Peter Anthony, industrial microbiology
Hahn, Richard Balser, chemistry
Hall, Richard Coumains, organic chemistry
Halvorson, Herbert Russell, biophysical chemistry

Handel, David, topology
Hans, Robert Joseph, microbiology
Hansen, Lowell John, mathematical analysis
Harding, Clifford Vincent, Jr, cell physiology
Hari, V, virology, molecular biology
Harmison, Charles Rice, protein chemistry, biophysics
Hause, Helen E, anthropology
Heifetz, Carl Louis, bacteriology, chemotherapy
Heller, Wilfried, chemistry
Henry, Leonard Francis, III, materials science, polymer science
Henry, Raymond Leo, anatomy, physiology
Hess, Joseph W, Jr, medical education
Hill, Jane Hassler, anthropology, anthropological linguistics
Hinkamp, James Benjamin, chemistry
Hitchcock, Dorothy Jean, biology
Hochstim, Adolf R, physics
Hodapp, Eugene Lorenz, physiology
Hodari, A Alberto, obstetrics & gynecology
Hodgson, Voigt R, biomechanics
Holcomb, Ira James, analytical chemistry
Hom, Foo Song, physical pharmacy
Hong, Wen-Hai, physical pharmacy
Horn, Robert Chisholm, Jr, pathology
Horowitz, Samuel Boris, cell physiology, zoology
Horwitz, Jerome Philip, organic chemistry, oncology
Hough, Richard Anton, limnology
Houh, Chorng Shi, mathematics
Howells, John David, microbiology
Huang, Chian Li, pharmacology, medicinal chemistry
Hutchinson, Kenneth A, analytical chemistry, chemical engineering
Ignatia (Frye), Mary, mathematics
Ihrman, Kryn George, organic chemistry
Imamoglu, Kamil H, surgery
Ionescu, Lavinel G, physical chemistry, biochemistry
Irwin, John McCormick, mathematics
Irwin, Louis Neal, neurobiology, neurochemistry
Jampel, Robert Steven, ophthalmology, anatomy
Jay, James Monroe, bacteriology, microbial ecology
Jeffries, Charles Dean, microbiology
Johnson, Carl Randolph, organic chemistry
Johnson, Jean Elaine, psychosomatic medicine
Johnson, Paul Hickok, biochemistry, molecular biology
Johnson, Robert Michael, biochemistry
Jollick, Joseph Darryl, microbial genetics, medical microbiology
Jones, Lily Ann, microbial genetics, molecular biology
Kaldor, George, physiology
Kaplan, Bernice Antoville (Mrs Gabriel Lasker), applied anthropology
Kaskas, James, theoretical physics
Kedzie, Robert Walter, solid state physics
Kenealy, Patrick Francis, nuclear physics
Kenney, Donald J, physical chemistry
Kessel, David Harry, biochemistry
Kevan, Larry, physical chemistry
Kienle, Robert Nelson, organic chemistry
Kimura, Tokuji, enzymology, biophysics
Kinkel, Arlyn Walter, pharmacy
Kirschner, Stanley, inorganic chemistry
Klempner, Daniel, physical chemistry, polymer science
Kobernick, Sidney D, pathology
Kong, Yi-Chi Mei, immunology, microbiology
Kuo, Pao-Kuang, theoretical physics
Kurtz, George Wilbur, dairy science
LaCroix, Joseph Donald, botany
Laframboise, Marc Alexander, mathematics
Lalas, Demetrius P, dynamic meteorology
Landers, James Walter, pathology
Large, Alfred McKee, surgery
Lasker, Gabriel (Ward), human anatomy, biological anthropology
Laurent, Andre Gilbert (Louis), mathematical statistics
LeBel, Norman Albert, organic chemistry
Lee, Chuan-Pu, biochemistry, physical chemistry
Leffert, Charles Benjamin, energy conversion, chemical physics
Lelek, Andrew Stanislaus, topology
Leon, Myron A, immunology
Lerner, Albert Martin, internal medicine
Levine, Laurence, cell biology
Levine, Seymour, virology

Lewis, Benjamin Marzluff, physiology
Lightbody, James James, immunology, biochemistry
Lim, Edward C, physical chemistry
Lintvedt, Richard Lowell, physical inorganic chemistry
Livingood, Clarence Swinehart, dermatology
Loewenfeld, Irene Elizabeth, physiology
Long, Loren Marlin, chemistry
Longyear, Judith Querida, pure mathematics
Luby, Elliot Donald, psychiatry, law
Lucey, Juliana Margaret, numerical analysis
Lueck, Leslie Melvin, pharmacy
Lu Qui, Ivan James, neuroanatomy
MacDonald, Roderick Patterson, biochemistry
Machamer, Harold Eugene, microbiology
Mack, Robert Emmet, internal medicine
Madden, Hannibal Hamlin, Jr, physics
Maddox, V Harold, Jr, organic chemistry
Maher, Veronica Mary, cancer, molecular biology
Malkin, Leonard Isadore, biochemistry
Mammen, Eberhard F, physiology, pharmacology
Mansour, Agnes Mary, biochemistry
Mantik, David Wayne, biophysics
Marks, Bernard Herman, pharmacology, biochemistry
Marrazzi, Mary Ann, neuropharmacology, neurochemistry
Mathews, Willis Woodrow, embryology
Mattman, Lida Holmes, bacteriology
Mayeda, Kazutoshi, genetics
McAllister, Harmon Carlyle, Jr, biochemistry
McCormick, J Justin, cancer, molecular biology
McCoy, Lowell Eugene, physiology, biochemistry
McCrum, Wilbur Ross, anatomy
McDonald, John Roland, pathology
McGinness, James Donald, analytical chemistry, spectroscopy
McGrath, Charles Morris, oncology
McLean, John A, Jr, inorganic chemistry
Mergentime, Max, food chemistry
Meyer, David Bernard, anatomy
Miceli, Angelo Sylvestro, physical chemistry
Miller, Charles Alexis, organic chemistry, research administration
Miskel, John J, (Sr), organic chemistry, pharmaceutical chemistry
Mitchell, J Andrew, reproductive physiology, neuroendocrinology
Mitchell, Robert Alexander, biochemistry
Mizeres, Nicholas James, anatomy
Mizukami, Hiroshi, biophysics, hematology
Moazed, Cyrus, physics
Moghissi, Kamran S, obstetrics & gynecology
Mooney, Thomas Faulkner, Jr, analytical chemistry, industrial hygiene
Moore, William Samuel, population biology
Morse, Philip Dexter, II, molecular biology
Moss, Leonard Wallace, anthropology
Mozola, Andrew John, geology
Murano, Genesio, physiology, biochemistry
Nedwicki, Edward G, pulmonary diseases, internal medicine
Nemeth, Abraham, mathematics
Newman, Max Karl, medicine
Nichols, Parks Montgomery, rubber chemistry
Nims, John Buchanan, physics
Noe, Eric Arden, organic chemistry
Noe, Frances Elsie, physiology
Noonan, Sharon Mariella, electron microscopy, cell physiology
Nordstrom, J David, organic polymer chemistry
Nunez, William J, III, immunology, microbiology
Nyboer, Jan, internal medicine
Oliver, John Preston, inorganic chemistry
O'Neill, John Dacey, mathematics
Orten, James M, biochemistry
Owens, Glynn, mathematics
Owens, James Samuel, physics
Paine, Philip Lowell, cell physiology
Palaszek, Mary De Paul, chemistry
Papadakis, Emmanuel Philippos, acoustics, solid state physics
Parke, Hervey Cushman, pharmaceutical chemistry
Parker, Charles J, Jr, biochemistry
Parshall, Clarence Merton, physics
Parsons, Willard Hall, geology
Pence, Leland Hadley, bioorganic chemistry

MICHIGAN

Perez-Borja, Carlos, neurology
Perle, Eugene Daniel, urban geography, regional science
Perrin, Eugene Victor, pediatrics, pathology
Perry, Harold, radiotherapy
Peterson, Ward Davis, Jr, medical microbiology
Petsko, Gregory Anthony, biophysics, biochemistry
Philips, Judson Christopher, organic chemistry
Philip, Michael J., genetics
Pilling, Arnold Remington, anthropology, ethnohistory
Pinkus, Hermann (Karl Benno), dermatology
Polgar, George, pediatrics, physiology
Pourcho, Roberta Grace, anatomy, cell biology
Power, Lawrence, internal medicine
Powsner, Wendell Holmes, biochemistry
Powsner, Edward R., pathology, nuclear medicine
Prasad, Ananda S., internal medicine, hematology
Prychodko, William Wasyl, ecology
Putney, James Wiley, Jr, pharmacology
Raban, Morton, organic chemistry, nuclear magnetic resonance
Rafols, Jose Antonio, anatomy
Rainey, John Marion, Jr, psychiatry, biochemistry
Randinitis, Edward J., biopharmaceutics
Ransford, George Henry, carbohydrate chemistry, medicinal chemistry
Rauch, Helene Coben, immunology, microbiology
Rebuck, John Walter, hematology
Reck, Gene Paul, physical chemistry
Redding, Foster Kinyon, neurology, neurophysiology
Reed, Melvin LeRoy, internal medicine, oncology
Reegen, Sidney Lloyd, polymer chemistry
Reeves, Andrew Louis, toxicology
Rhee, Choon Jai, topology
Rich, Marvin A., microbiology, virology
Richards, George Martin, organic biochemistry
Righthand, Vera Fay, virology, microbial biochemistry
Rights, Fred Lewis, microbiology
Rillema, James Alan, endocrinology, physiology
Roberts, Joan Marie, radiobiology
Robin, Erwin, cardiology
Rodin, Martha Kinscher, anatomy
Rol, Pieter Klaas, physics
Rohnick, William Barnett, physics
Ronca, Luciano Bruno, geology
Rorabacher, David Bruce, analytical chemistry, inorganic chemistry
Rose, Gordon Wilson, epidemiology, clinical microbiology
Rose, Noel Richard, microbiology, immunology
Rosenbaum, Manuel, microbiology, virology
Rosenberg, Irwin Kay, surgery
Rosenberg, Jerry C., surgery, transplantation immunology
Rosenzweig, Norman, psychiatry
Rossmoore, Harold W., bacteriology
Rozhin, Jurij, biochemistry
Ruffner, James Alan, history of science, philosophy of science
Russo, Jose, experimental pathology
Saperstein, Alvin Martin, theoretical physics
Sardesai, Vishwanath M., biochemistry, clinical chemistry
Schaap, Arthur Paul, organic chemistry
Schaeffer, Joseph Negley, medicine
Schatz, Leo, anatomy, histology
Scheerer, Eugene Paul, neuropsychology, clinical psychology
Schorer, Calvin E., psychiatry
Schuler, Edward Emerson, pharmacy
Schwartz, Melvin Lewis, neuropsychology, clinical psychology
Sedensky, James Andrew, physiology, pulmonary physiology
Segers, Walter Henry, physiology, biochemistry
Seidel, Wladimir, mathematics
Seizer, Isidore, pathology
Serafini, Angela, bacteriology
Shafer, A William, medicine
Shepard, Robert Stanley, physiology, pharmacology

Sherman, Alfred Isaac, obstetrics & gynecology
Shore, Joseph D., enzymology, physical biochemistry
Siebert, Karl Joseph, biochemistry
Siegel, Albert, molecular genetics, plant virology
Silbergleit, Allen, surgery, physiology
Simpson, William Loyal, oncology, preventive medicine
Sinclair, Robert, geography
Skaff, Michael Samuel, mathematics
Slaby, Harold Theodore, algebra
Sloan, Bernard Joseph, medical microbiology
Smith, Richard Watson, Jr, endocrinology
Smith, Robert Jay, zoology
Spector, Richard M., theoretical physics
Sperley, Richard Jon, organic chemistry
Spitz, Werner Uri, pathology, forensic medicine
Stanley, John Pearson, mathematics
Stearns, Martin, physics
Steiman, Henry Robert, physiology, dentistry
Stein, Talbert Sheldon, physics
Stephens, John Firth, industrial hygiene
Stevens, Calvin Lee, chemistry
Stewart, Melbourne George, physics
Stulberg, Cyril Sidney, tissue culture, virology
Sundick, Roy, immunology
Swanborg, Robert Harry, immunology, immunochemistry
Szuka, Anton, analytical chemistry
Tang, Lun Han, nuclear science, engineering
Swanson, Curtis James, comparative physiology, biochemistry
Swartz, Robert David, medicine
Swisher, Joseph Vincent, organic chemistry
Syner, Frank N., biochemistry
Szmant, Herman Harry, organic chemistry
Tatelman, Maurice, medicine
Taylor, Alton Robert, microbiology
Taylor, John Dirk, developmental biology
Tchen, Tche Tsing, biochemistry, industrial
Teague, Dwight Maxwell, industrial chemistry, automotive engineering
Teasdall, Robert Douglas, neurology
Teodoro, Rosario Reyes, bacteriology
Ter Haar, Gary L., inorganic chemistry
Tetreault, Florence G., statistics
Thomas, Llywellyn Murray, neurosurgery
Thomas, Robert L., solid state physics
Thompson, William Lay, vertebrate zoology
Timberlake, William Edward, mycology, developmental biology
Timm, Eugene Alvin, microbiology
Tokuhiro, Tadashi, physical chemistry, chemical physics
Tourney, Garfield, psychiatry
Trivich, Dan, physical chemistry
Trix, Phelps, organic chemistry
Tsao, Chia Kuei, mathematical statistics
Turner, Almon George, Jr, theoretical chemistry, inorganic chemistry
Vaitkevicius, Vainutis K., oncology
Vinogradov, Serge, biochemistry
Waggoner, Philip Ray, developmental biology
Walsh, Ralph Thomas, physiology
Walt, Alexander Jeffrey, surgery
Ward, Richard Floyd, geology
Watson, John H L., physics, biophysics
Wechsler, Martin H., mathematics
Weidig, Charles Ferdinand, inorganic chemistry, bioinorganic chemistry
Weiner, Lawrence Myron, microbiology, immunology
Weinsieder, Allan, cell biology
Weinstart, Melvin, comparative endocrinology, comparative physical anatomy
Weiss, Mark Lawrence, physical anthropology, primatology
Weiss, Paul, physics
Weissler, Arnold M., cardiology, internal medicine
Wertheimer, Frederick William, periodontology, oral pathology
Williams, Allan, Rawson, organic chemistry
Williams, Mary Louise Monica Fritts, anatomy, embryology
Willson, Philip James, high temperature chemistry, ceramics
Wilson, George Donald, biochemistry
Wilson, Robert Francis, thoracic surgery, cardiovascular surgery
Wojtowicz, Wesley Joseph, physical chemistry
Wolfson, Seymour J., computer science
Wollensak, John Charles, organic chemistry

Wollschlaeger, Gertraud, medicine, radiology
Wollschlaeger, Paul Bernhard, medicine, radiology
Wood, Pauline J., developmental biology, histology
Woodyard, James Robert, physics
Woolley, Paul Vincent, pediatrics
Wormser, Henry C., pharmaceutical chemistry
Wright, Robert Crunn, urban geography, education
Yanari, Sam Satomi, immunology, biochemistry
Zak, Bennie, clinical biochemistry
Zemlicka, Jiri, organic chemistry, biochemistry
Zuelzer, Wolf W., pathology, pediatrics

DEXTER
Chapman, Robert Earl, Jr, biophysics
Maffett, Andrew L., mathematics
Taube, Clarence Martin, fisheries

DOLLAR BAY
Baldwin, Keith Malcolm, physics, electronics

DOWAGIAC
Nicklow, Clark W., horticulture

EAST LANSING
Abolins, Maris Arvids, high energy physics
Abou-El-Seoud, Mohamed Osman, mycology
Adams, Maurice Wayne, plant breeding
Adams, Thomas, physiology
Adney, Joseph Elliott, Jr, mathematics
Ahl, Alwynelle S., zoology, mammalogy physiology
Akera, Tai, pharmacology
Anderson, Richard J., animal science statistics
Aust, Steven Douglas, biochemistry
Anderson, Glen Douglas, mathematics
Austin, Samuel Manly, nuclear physics
Anderson, Richard Lee, biochemistry, microbiology
Arata, Dorothy, nutrition
Arnold, Kenneth James, mathematical statistics
Aulenich, Richard J., animal science statistics
Balaban, Martin, zoology
Baker, Rollin Harold, zoology
Ball, Robert Cragin, zoology
Band, Henrietta Trent, genetics, zoology
Band, Rudolph Neal, zoology
Bandurski, Robert Stanley, plant biochemistry
Barker, Robert, biochemistry
Barrett, Paul Howard, limnology
Bartley, Samuel Howard, psychophysiology
Bass, Jack, solid state physics
Bath, James Edmond, entomology, plant pathology
Beaman, John Homer, systematic botany
Beaver, Donald Loyd, vertebrate ecology
Becker, Roland Frederick, biomechanics, anatomy
Bedford, Clifford Levi, food science
Behr, Eldon August, wood science, wood technology
Belding, Ralph Cedric, poultry pathology
Beneke, Everett Smith, botany, microbiology
Benenson, Walter, nuclear physics
Bergen, Werner Gerhard, animal nutrition
Bernard, Rudy Andrew, neurophysiology
Bernthal, Frederick Michael, nuclear chemistry, nuclear physics
Bertsch, George Frederick, nuclear physics
Bieber, Loran Lamoine, biochemistry
Bird, George W., nematology, plant pathology
Blatt, Frank Joachim, physics
Blinn, Walter Craig, science education, natural science
Blom, Gaston Eugene, child psychiatry, educational psychology
Blosser, Henry Gabriel, nuclear physics
Blyth, Mary Isobel, mathematics
Boezi, John A., biochemistry, molecular biology
Bond, Jenny Taylor, nutrition
Borgstrom, Georg Arne, food science
Bowdler, Anthony John, hematology
Braddock, James Conger, zoology

Bradley, Robert Lincoln, inorganic chemistry
Brandou, Julian Robert, science education
Bredeck, Henry E., physiology
Brierley, Wade Oberlin, veterinary surgery
Brody, Theodore Meyer, pharmacology
Bromley, Stephen C., developmental biology
Brown, Anthony William Aldridge, insect physiology, toxicology
Brubaker, Robert Robinson, microbiology
Brunn, Stanley David, geography
Brunner, Jay Robert, food science, dairy chemistry
Brunnschweiler, Dieter Heinz, geography
Bukovac, Martin John, horticulture, plant physiology
Burmeister, Ben Roy, virology
Burmeister, Mary Alice (Horswill), physiology
Burness, James Hubert, bioinorganic chemistry
Burnett, Jean Bullard, biological chemistry
Burton, Thomas Maxie, aquatic ecology
Butcher, James Walter, entomology, ecology
Byerrum, Richard Uglow, biochemistry
Calhoun, Mary Lois, histology, anatomy
Campbell, Earl William, hematology
Cantino, Edward Charles, mycology
Cantlon, John Edward, ecology
Carde, Ring Richard Tomlinson, entomology, behavioral biology
Carew, John, horticulture
Carlson, Edward H., solid state physics
Carlson, Robert Fritz, horticulture
Carlson, William H., horticulture
Carmichael, Robert Stewart, geophysics, geology
Carpenter, William John, floriculture
Cederquist, Dena Caroline, nutrition
Chang, Timothy Scott, bacteriology, poultry pathology
Chappelle, Daniel Eugene, forest economics
Chen, Bang-Yen, geometry
Chou, Ching-Chung, physiology, internal medicine
Christenson, Donald Robert, soil science, plant physiology
Chubb, Michael, resource geography
Cleland, Charles Edward, archaeology, cultural anthropology
Cook, Robert Merold, animal nutrition
Cooper, William E., zoology
Copeland, Lawrence O., crop science
Coppola, Edward Dante, surgery, medical science
Collings, William Doyne, biology
Colwell, Priscilla J., experimental solid state physics, quantum electronics
Conklin, James L., anatomy
Conner, Gabel Henry, veterinary medicine, physiology
Convey, Edward Michael, physiology, endocrinology
Corcos, Alain Francois, genetics
Cortier, Thomas Richard, microbiology
Costilow, Ralph Norman, microbiology
Cowen, Jerry Arnold, physics
Cox, Herbert Walton, malariology
Crawley, Gerard Marcus, nuclear physics
Cress, Charles Edwin, statistics, quantitative genetics
Cress, Donald Chauncey, entomology
Crittenden, Lyman Butler, genetics
Crouch, Stanley Ross, analytical chemistry
Cukier, Robert Isaac, theoretical chemistry
Cunningham, Charles Henry, veterinary microbiology
Cunningham, James Gordon, neurophysiology, neurology
Dardas, Terry Jay, immunology, internal medicine
Darden, Joe Turner, urban geography
Daugherty, Robert M., Jr, physiology, internal medicine
Davidson, Harold, landscape horticulture
Davis, Harvey Samuel, topology
Dawson, Lawrence E., food science
Deal, William Cecil, Jr, physical biochemistry
Deans, Robert Jack, animal husbandry

De Hertogh, August Albert, plant physiology, horticulture
Delmer, Deborah P, plant physiology, biochemistry
Dennis, Frank George, Jr, pomology, plant physiology
Dewey, Donald Henry, horticulture, pomology
deZeeuw, Donald John, plant pathology
Dickmann, Donald Irvin, forest physiology, silviculture
Dilley, David Ross, plant physiology
D'itri, Frank M, analytical chemistry, water chemistry
Douglas, Robert Hazard, reproductive physiology
Doyle, Patrick H, topology
Dressel, Paul Leroy, statistical analysis
Drew, William Brooks, plant ecology
Dugan, LeRoy, Jr, organic chemistry
Dukelow, W Richard, reproductive physiology
Dutton, Frederic Booth, chemistry, academic administration
Dye, James Louis, physical chemistry
Edwards, Thomas Harvey, physics
Eick, Harry Arthur, physical chemistry, inorganic chemistry
El-Bayoumi, Mohamed Ashraf, physical chemistry, electronic spectroscopy
Ellingboe, Albert Harlan, plant pathology, genetics
Elliott, Fred Craig, field crops
Elliott, James McFarland, science education
Elliott, Jane Elizabeth Inch, geology
Ellis, Boyd G, soil chemistry
Emerson, Thomas Edward, Jr, cardiovascular physiology
Emery, Roy Saltsman, animal nutrition, biochemistry
Engelmann, Manfred David, animal ecology
Enke, Christie George, electrochemistry, analytical chemistry
Enochs, Nettie Jean, developmental biology, botany
Enzer, Norbert Beverley, child psychiatry, pediatrics
Erickson, Anton Earl, soil science
Evans, Robert John, biochemistry, nutrition
Everson, Everett Henry, genetics, plant breeding
Fabrega, Horacio, Jr, psychiatry, anthropology
Fairley, James Lafayette, Jr, biochemistry
Farnum, Donald G, organic chemistry
Fennell, Richard Adams, cytochemistry
Ferguson, Roger K, clinical pharmacology, pharmacology
Filner, Philip, biochemistry
Fischer, Roland Lee, entomology
Fisher, James Harold, petroleum geology
Flegal, Cal J, poultry nutrition
Foiles, Carl Luther, physics
Forstat, Harold, physics
Foth, Henry Donald, agronomy, soil morphology
Fox, Danny Gene, ruminant nutrition
Fox, Martin, statistics
Fraker, Pamela Jean, immunobiology
Frame, James Sutherland, mathematics
Frantz, William Lawrence, zoology
Fromm, Paul Oliver, comparative physiology
Fuentes, Ricardo, Jr, bioinorganic chemistry
Gallin, Bernard, cultural anthropology
Galonsky, Aaron Irving, nuclear physics
Gebber, Gerard L, pharmacology, neurophysiology
Gerhardt, Philipp, microbiology
Gibbs, Carter B, forestry
Gibson, James Edwin, pharmacology, toxicology
Gill, John Leslie, biometrics
Gilliland, Dennis Crippen, statistics, mathematics
Goatley, James Leon, biochemistry
Good, Norman Everett, plant physiology
Goodman, Jay Irwin, pharmacology, cancer
Gordon, Morton Maurice, physics
Grafius, John Edward, genetics, plant breeding
Gross, George Alvin, osteopathy
Grubbs, Robert Howard, organic chemistry, inorganic chemistry
Gruhn, Charles R, physics
Guile, Ralph Lawrence, organic chemistry
Guyer, Gordon Earl, entomology
Gysel, Leslie William, forestry, wildlife management
Hackel, Emanuel, human genetics, transplantation immunology

Haddy, Francis John, physiology, internal medicine
Hafs, Harold David, reproductive physiology, endocrinology
Hamilton, James Beclone, inorganic chemistry
Hammer, Robert Nelson, inorganic chemistry
Hannan, James Francis, mathematical statistics
Hanover, James W, forest genetics
Harman, Jay Reginald, geography, climatology
Harmon, Laurence George, food science
Harpstead, Dale D, genetics, plant breeding
Harris, Gale Ion, nuclear physics, medical physics
Harrison, James Francis, theoretical chemistry, quantum chemistry
Harrison, Michael Jay, theoretical solid state physics
Hart, Harold, organic chemistry, photochemistry
Hart, John Henderson, tree physiology, wood chemistry
Hartmann, William Morris, solid state physics, psychophysics
Hause, Clarence Duane, physics
Haynes, Dean L, forest entomology
Haynes, Sherwood Kimball, physics
Hedrick, Theodore Isaac, food science
Heisey, Samuel Richard, medical physiology
Henneman, Harold Albert, animal husbandry
Hensley, Marvin Max, herpetology
Herzog, Fritz, mathematics
Higgins, James Victor, human genetics
Hill, James Leslie, behavioral biology
Hill, Stephen James, astrophysics
Hill, Susan Douglas, zoology
Hillman, Donald, dairy nutrition
Hills, Norman L, applied mathematics
Hiscoe, Helen Brush, reproductive biology
Hocking, John Gilbert, mathematics
Hoefer, Jacob A, animal husbandry
Hoffert, Jack Russell, comparative physiology, ophthalmology
Hoffman, Julius R, entomology
Hogaboam, George Joseph, plant breeding
Hollensen, Raymond Hans, bryology
Holman, J Alan, vertebrate paleontology, herpetology
Honma, Shigemi, horticulture
Hook, Jerry Bruce, pharmacology
Hooker, William James, plant pathology
Hooper, Gerald Ray, plant pathology, cytology
Hoopingarner, Roger A, entomology
Horne, Frederick Herbert, physical chemistry, chemical physics
Howell, Gordon Stanley, Jr, horticulture, plant physiology
Howitt, Angus Joseph, entomology
Huber, John Talmage, animal nutrition, dairy science
Hull, Jerome, Jr, agriculture, horticulture
Humphrey, Harold Edward Burton, Jr, environmental health
Humphrys, Clifford Robertson, soils
Hunt, Andrew Dickson, Jr, pediatrics
Ishino, Iwao, anthropology
Jacobs, Allen Wayne, anatomy
Jacobson, Daniel, geography
James, Lee Morton, forestry
Jenkins, Thomas William, anatomy, neuropathology
Johnson, Eric Robert, analytical chemistry, chemical instrumentation
Johnson, Howard Ernest, fresh water ecology, toxicology
Johnson, John Irwin, Jr, neuroanatomy
Johnson, Shirley Mae, reproductive physiology
Johnston, Raymond F, physiology, pharmacology
Johnston, Taylor Jimmie, agronomy, plant physiology
Jones, Alan Lee, plant pathology
Jones, Charles M, osteopathy
Jones, Margaret Zee, neuropathology, pathology
Jordan, Brigitte, medical anthropology
Kabara, Jon Joseph, pharmacology, organic chemistry
Kamrin, Michael Arnold, science education
Kaplan, Thomas Abraham, solid state physics
Karabatsos, Gerasimos J, organic chemistry
Kashy, Edwin, experimental nuclear physics
Katz, Leo, mathematical statistics

Keahey, Kenneth Karl, veterinary pathology
Keck, William George, physics
Keller, Waldo Frank, veterinary surgery
Kelly, Leroy Milton, mathematics
Kelly, William H, experimental nuclear physics
Kemeny, Gabor, theoretical physics
Kende, Hans Janos, plant physiology
Kennedy, Maldon Keith, entomology
Kenworthy, Alvin Lawrence, horticulture
Kessler, George Morton, horticulture
Kevern, Niles Russell, aquatic biology, ecology
Kindel, Paul Kurt, biochemistry
King, Darrell Lee, limnology
King, Herman (Lee), entomology
King, John Arthur, zoology
Kirschbaum, Thomas H, obstetrics & gynecology
Kivilaan, Aleksander, plant physiology
Klos, Edward John, plant pathology
Knobloch, Irving William, botany
Koelling, Melvin R, forestry
Kohrman, Arthur Fisher, pediatrics
Korr, Irvin Morris, physiology, neurosciences
Koul, Hira Lal, mathematical statistics
Kovacs, Julius Stephen, theoretical physics
Kowert, Bruce Arthur, physical chemistry
Krupka, Lawrence Ronald, plant pathology
Krzywoblocki, Maria Zbigniew von, fluid mechanics, quantum mechanics
Kuntz, Eloise, cell biology
Kunze, Raymond J, soil physics
Kwun, Kyung Whan, mathematics
Lacy, Melvyn Leroy, plant pathology, soil microbiology
Laemmlen, Franklin, plant pathology
Lamport, Derek Thomas Anthony, biochemistry
Lang, Anton, developmental plant biology
Langham, Robert Fred, pathology
Lassiter, Charles Albert, dairy science
Lawton, Kirkpatrick, agronomy
Leader, Robert Wardell, comparative pathology
LeGoff, Eugene, organic chemistry
Leroi, George Edgar, chemical physics
Leveille, Gilbert Antonio, nutrition, biochemistry
Lewis, Clarence E, horticulture
Lewis, Ralph William, plant pathology
Liang, Edison Park-Tak, astrophysics, cosmology
Lillevik, Hans Andreas, biochemistry
Lockwood, John LeBaron plant pathology
Loomis, Robert Morgan, forestry
Lopushinsky, Theodore, ecology, pathology
Lucas, Robert Elmer, soil science
Ludden, Gerald D, mathematics
Luecke, Richard William, biochemistry, nutrition
Ma, Zee-Ming, high energy physics
Mack, Walter Noel, bacteriology
Magee, William Thomas, animal breeding
Mahanti, Subhendra Deb, theoretical solid state physics, biophysics
Mallmann, Virginia H, microbiology
Manthy, Robert Sigmund, forest economics
Marik, Jan, mathematical analysis
Markakis, Pericles, food science
Marty, Robert Joseph, forest economics
Mathews, A L, chemistry
Matley, Ian Murray, geography
Maxwell, Moreau Sanford, anthropology
McCarty, Charles Norman, chemistry
McClary, Andrew, zoology, anthropology
McConnell, David Graham, biophysics, biochemistry
McCoy, Thomas LaRue, mathematics
McCullough, Norman B, microbiology, infectious diseases
McGilliard, Lon Dee, animal breeding, dairy science
McHarris, William Charles, nuclear chemistry
McManus, Hugh, theoretical physics
McMeekin, Dorothy, plant pathology
McNabb, Clarence Duncan, Jr, limnology
Meadows, Clinton Elwood, animal breeding
Mecklenburg, Roy Albert, plant physiology, microclimatology
Meggitt, William Fredric, agronomy
Meites, Joseph, physiology
Mericle, Leo Willis, genetics, radiation biology
Mericle, Rae Phelps, developmental genetics
Merkel, Robert Anthony, meat science
Merritt, Richard William, entomology

Michelakis, Andrew M, medicine
Mickelsen, Olaf, biochemistry
Miller, Elwyn Ritter, animal nutrition
Miller, Harold Charles, microbiology, immunology
Miller, Maynard Malcolm, geology
Minkel, Clarence Wilbert, geography of Latin America, economic geography
Mokma, Delbert Lewis, soil science
Montgomery, Donald Joseph, physics
Moon, Robert John, microbiology
Moore, John (Newton), plant pathology
Moore, Kenneth Edwin, psychopharmacology
Moran, Daniel Austin, topology
Morris, Allan J, biochemistry
Morrow, David Austin, theriogenology
Mortland, Max Merle, soil chemistry
Moulton, James Edward, botany
Mouser, Gilbert Warren, biology
Mullins, John A, chemistry
Murakishi, Harry Haruo, plant pathology
Murphy, Peter George, ecology, botany
Myers, Wayne Lawrence, forestry, biometry
Nadler, Kenneth David, plant physiology, biochemistry
Nazerian, Keyvan, virology, electron microscopy
Nellor, John Ernest, physiology, endocrinology
Nelson, Clarence Herbert, biology
Nelson, Leyton Vincent, agronomy
Nelson, Ronald Harvey, animal husbandry
Newson, Harold Don, medical entomology
Nicholas, Richard Carpenter, food science
Noerdlinger, Peter David, astrophysics, plasma physics
Nordhaus, Edward Alfred, mathematics
O'Neill, Ronald C, topology
Overbeck, Henry West, medical physiology, internal medicine
Palmer, Edgar M, mathematics
Pamnani, Motilal Bhagwandas, physiology, cardiovascular diseases
Paris, Clark Davis, horticulture
Parke, Donna, biochemistry
Parker, Paul Michael, molecular physics
Parsons, George E, dairy husbandry
Patrick, Richard Allen, immunology, biochemistry
Patterson, Maria Jevitz, infectious diseases
Patterson, Ronald James, immunobiology
Pax, Ralph A, physiology
Payne, Kenyon Thomas, agronomy
Peabody, Frank Robert, microbiology
Pearson, Albert Marchant, food science
Pedrey, Charles P, speech pathology, audiology
Peebles, Charles Robert, parasitology
Penner, Donald P, weed science
Petrides, George Athan, ecology, wildlife management
Phillips, Richard E, Jr, mathematics
Pinnavaia, Thomas J, inorganic chemistry
Pittman, Robert Preston, cardiovascular physiology
Plotkin, Jacob Manuel, mathematics
Polin, Donald, avian physiology, avian nutrition
Pollack, Gerald Leslie, low temperature physics
Popov, Alexander Ivan, analytical chemistry, inorganic chemistry
Porter, Thomas Wayne, invertebrate zoology
Porzio, Michael Anthony, protein chemistry
Potchen, E James, radiology, nuclear medicine
Potter, Howard Spencer, plant pathology, entomology
Price, Hugh Criswell, horticulture
Price, James F, food science
Prince, Harold Hoopes, behavioral ecology
Prouty, Chilton Eaton, geology
Racie, Fred Arnold, natural science
Radomski, Mark Stephen, theoretical nuclear physics
Ramsdell, Donald Charles, plant pathology
Rasmussen, Harry Paul, plant physiology, plant nutrition
Rathke, Michael William, organic chemistry
Ratzlaff, Marc Henry, anatomy, veterinary medicine
Rech, Richard Howard, neuropharmacology, psychopharmacology
Reddy, Aluru Raghuram, mathematics
Reddy, Chilecampalli Adinarayana, veterinary medicine, microbiology

Reid, Donald J., agronomy, field crops
Reineke, Ezra Paul, physiology
Reinke, David Albert, pharmacology, physiology
Repko, Wayne William, elementary particle physics
Retzlaff, Ernest (Walter), neurophysiology, medical physiology
Reusch, William Henry, synthetic organic chemistry
Riegle, Gail Daniel, endocrinology, reproductive physiology
Rieke, Paul Eugene, soil science
Ries, Stanley K., horticulture, plant physiology
Ringer, Robert Kosel, physiology
Rippen, Alvin Leonard, dairy science
Ritchie, Harlan, animal husbandry
Rivera, Evelyn Margaret, endocrinology, cancer
Robbins, Leonard Gilbert, genetics
Robertson, Lynn Shelby, Jr., soil science
Robertson, Robert Graham Hamish, physics
Roelofs, Eugene Woodrow, aquatic biology
Rogers, Alvin Lee, medical mycology, medical microbiology
Rogers, Max Tofield, physical chemistry
Romsos, Dale Richard, nutrition
Ronzio, Robert A., biochemistry, cell biology
Ross, William T., anthropology
Rossman, Elmer Chris, plant breeding
Rottman, Fritz M., biochemistry
Rous, Stephen N., urology
Rovner, David Richard, endocrinology, metabolism
Rubel, Arthur Joseph, cultural anthropology
Rudolph, Victor J., forest management
Ruppel, Robert Frank, entomology
Sadoff, Harold Lloyd, microbiology, biochemistry
Saettler, Alfred William, plant pathology
Salehi, Habib, mathematical analysis
San Clemente, Charles Leonard, microbiology
Sander, Charles H., medicine, pathology
Sanders, Ruth Evelyn, medical microbiology
Sanger, Vance L., veterinary pathology
Sauer, Richard John, entomology
Sawyer, Donald C., anesthesiology
Scarborough, Charles Spurgeon, invertebrate zoology, acarology
Scheffer, Robert Paul, plant pathology
Schemmel, Rachel A., nutrition
Schlegel, Richard, theoretical physics, philosophy of science
Seil, Harold Melvin, vertebrate paleontology
Shapiro, Lee Tobey, astronomy
Sharma, Jagdev Mittra, microbiology, virology
Shaver, John Rodney, embryology
Sheppard, Charles Campbell, poultry husbandry, agricultural economics
Shicluna, John C., soil fertility
Sibley, Duncan Fawcett, sedimentary petrology
Signell, Peter Stuart, theoretical physics
Simkin, Susan Marguerite, astronomy
Singh, Daulat, soil chemistry, analytical chemistry
Sinha, Indranand, mathematics
Sink, Kenneth C., Jr., ornamental horticulture
Slatis, Herman Moses, genetics
Sledd, William T., mathematics
Sleight, Stuart Duane, veterinary pathology
Sliker, Alan, wood technology
Smith, David Harrison, Jr., plant breeding
Smith, Eugene Joseph, biochemistry
Smith, Gerald A., high energy physics
Smith, William Lee, biochemistry
Snow, Loudell Fromme, medical anthropology
Snyder, Loren Russell, molecular biology
Solomon, John Junior, bacteriology
Sommers, Lawrence Melvin, geography

Sonneborn, Lee Meyers, mathematics
Speck, John Clarence, Jr., biochemistry
Spees, Steven Tremble, Jr., inorganic chemistry
Spence, Robert Dean, physics
Spielberg, Joseph, cultural anthropology
Spink, Gordon Clayton, medical education, cytology
Spira, Robert Samuel, mathematics
Sprafka, Robert J., health physics
Stapleton, James H., mathematical statistics
Stehr, Frederick William, entomology
Stelson, Hugh Eugene, mathematics
Stephenson, Stephen Neil, botany, ecology
Stevenson, Kenneth Eugene, food microbiology
Stickney, Janice Lee, pharmacology, cardiovascular physiology
Stine, Charles Maxwell, food science
Stoeckley, Thomas Robert, astronomy
Stone, Ellen Rose, topology
Stone, Howard Anderson, genetics, virology
Stonehouse, Harold Bertram, geochemistry
Stoudt, Howard Webster, medical education, public health
Strawbridge, Dennis Winslow, ecology
Stuart, Sarah Elizabeth, molecular biology
Stuhl, John Neal, wildlife diseases
Suchsland, Otto, wood technology
Suelter, Clarence Henry, enzymology
Suhrland, Leif George, hematology, oncology
Sweeley, Charles Crawford, biochemistry, organic chemistry
Swisher, Scott Neil, internal medicine, hematology
Tack, Peter Isaac, fish biology
Tai, William, cytogenetics
Tavano, Donald C., public health education, medical education
Taylor, James Lee, horticulture
Tesar, Milo, agronomy
Thomas, John William, animal nutrition
Thomas, Robert Nelson, geography, population studies
Thompson, Norman Robert, plant breeding, genetics
Tiedje, James Michael, microbial ecology, soil microbiology
Tien, Hsin Ti, biophysics
Timmick, Andrew, analytical chemistry
Tolbert, Nathan Edward, biochemistry
Tomber, Marvin L., algebra
Tosi, Oscar I., audiology, acoustics
Trapp, Allan Laverne, veterinary pathology
Trosko, James Edward, genetics, oncology
Trow, James, geology
Tsai, Chester E., algebra
Tschirley, Fred Harold, ecology
Tucker, Herbert Allen, animal physiology
Tulinsky, Alexander, structural chemistry
Tully, Frank Paul, physical chemistry
Tweedle, Charles David, neurobiology
Twohy, Donald Wilfred, parasitology
Ullrey, Duane Earl, animal nutrition
Useem, John, anthropology
Vargas, Joseph Martin, Jr., plant pathology
Vest, Hyrum Grant, Jr., plant genetics
Vitosh, Maurice Lee, agronomy, soil science
Vogel, Thomas A., geology
Wagner, Peter J., physical organic chemistry, photochemistry
Walker, Bruce Edward, anatomy
Walker, John Martin, water pollution
Walner, William E., entomology, plant pathology
Wang, Chang-Yi, applied mathematics
Ward, Robert C., osteopathy
Warner, Ray Allen, experimental nuclear physics
Wasserman, Robert H., applied mathematics
Waxler, Glenn Lee, veterinary pathology
Webster, James Allan, entomology
Wei, Chuan-Tseng, metallurgy
Wei, Guang-Jong Jason, physical biochemistry
Weil, Clifford Edward, mathematics
Weil, William B., Jr., pediatrics
Weinshank, Donald Jerome, morphology
DeGraef, Donald Earl, physics
Kammeyer, Carl William, physical chemistry
King, Larry Michael, mathematics
Konarny, Julius Michael, physical chemistry, organic chemistry

Welsch, Clifford William, Jr., physiological chemistry, oncology
Wen, Chi-Pang, preventive medicine, family medicine
Weymouth, Patricia Perkins, natural science, history of science
White, Donald Perry, forestry, soil science
Whitehair, Charles Kenneth, nutrition, pathology
Whiteside, Eugene Perry, soil science
Whittier, Herbert Lincoln, anthropology
Wiese, Maurice Victor, plant pathology
Wildenthal, Bryan Hobson, physics
Williams, Jeffrey F., parasitology, immunology
Williams, Wells Eldon, fisheries
Wilson, John Edward, neurochemistry, enzymology
Wilson, Louis Frederick, entomology
Wilson, Ronald Wayne, botany, mycology
Winters, Harold Abraham, geography, geology
Wisnretsky, Theodore, food technology
Witter, Richard L., poultry pathology
Witwer, Sylvan Harold, horticulture
Wolcott, Arthur Ripatte, soil microbiology
Wolk, Coleman Peter, developmental physiology
Wolternik, Lester Floyd, biophysics
Wong, Pui Kei, mathematics
Wood, Willis A., microbiology, biochemistry
Woodby, Lauren G., mathematics
Woodruff, Truman Owen, theoretical solid state physics
Wright, David Grant, topology
Yen, David Hsein-Yao, applied mathematics, engineering mechanics
Zabik, Matthew John, chemistry, toxicology
Zeevaart, Jan Adriaan Dingenis, plant physiology
Zimring, Lois Jacobs, chemical physics
Zindel, Howard Carl, poultry husbandry

ELOISE
Blair, A James, Jr., internal medicine, endocrinology
Joseph, Ramon R., gastroenterology
McCann, Daisy S., biochemistry, immunology
Thompson, George Richard, internal medicine, rheumatology
Vickrey, Herta Miller, immunology, medical microbiology
Weiss, Joseph Jacob, rheumatology

ESCANABA
Campbell, DeWayne E., fisheries

FARMINGTON
Wynn, Charles Martin, Sr., organic chemistry, academic administration

FERNDALE
Bartleson, John David, chemistry
Kardos, Otto, chemistry
Kehoe, Lawrence Joseph, organic chemistry
Knapp, Gordon Grayson, organic chemistry
Perlstein, Warren Louis, applied chemistry
Seyb, Edgar John, Jr., electrochemistry
Shin, Kiu Hi, chemistry
Shubkin, Ronald Lee, organometallic chemistry
Sistrunk, Thomas Olloise, inorganic chemistry
Zaweski, Edward F., organic chemistry

FLINT
Blecker, Harry Herman, organic chemistry
Bolander, Richard, mathematics, physics
Boys, Donald W., solid state physics
Caldwell, William V., mathematics
Caraway, Wendell Thomas, biochemistry
Cope, Virgil W., inorganic chemistry
Courtney, Cameron B., mathematics
Dapson, Richard W., ecology, vertebrate anatomy
Snyder, Robert Harvey, organic chemistry
Thompson, James Lowry, anatomy
Tillitson, Edward Walter, dental materials, chemical engineering
Wheeler, Larry Meade, pharmaceutical chemistry

Kugler, Lawrence Dean, mathematical logic
McLaughlin, Renate, mathematical analysis
Merkle, John, ecology
Otero, Joseph Guillermo, parasitology
Pace, Gary Lee, malacology
Roobol, Norman R., organic chemistry, bacteriology
Purvis, George Allen, food science, nutrition
Stewart, Robert A., nutrition

GALESBURG
Hiestand, Everett Nelson, physical pharmacy, pharmaceutics

GRAND HAVEN
Christenson, Paul John, medicine, preventive medicine

GRAND LEDGE
Lucas, Alfred Martin, anatomy

GRAND RAPIDS
Baldwin, Ralph Belknap, astrophysics
Bouwman, Fred Ludwig, zoology
Bratt, Albertus Dirk, entomology
Broene, Herman Henry, chemistry
Cogan, Harold Louis, physical chemistry
Dirkse, Thedford Preston, chemistry
Ehlers, Vernon James, atomic physics
Estes, Timothy King, paper chemistry
Gebben, Alan Irwin, plant ecology
Griffioen, Roger Duane, nuclear physics,
Higgins, James Joseph, paper chemistry
Johnson, Tom Milroy, internal medicine
Johnson, Warren Charles, chemistry
Kromminga, Albion Jerome, theoretical physics
Kuipers, Jack, mathematics
Lambers, Austin E., marine zoology,
Mattano, Leonard August, organic neurosurgery
Menninga, Clarence, nuclear chemistry, physical chemistry
Servis, Robert Eugene, bio-organic chemistry
Sinke, Carl, mathematics
Smith, Eugene William, botany, microbiology
Sutherland, Jeffrey C., geochemistry, hydrology
Tasker, Clinton Waldorf, chemistry
Tenbroek, Bernard John, zoology
Tigchelaar, Peter Vernon, physiology
Tilmann, Jean Paul, geography
Van Doorne, William, inorganic chemistry
Van Harn, Gordon L., physiology
Van Till, Howard Jay, astronomy
Van Zwalenberg, George, mathematics
Van Zytveld, John Bos, solid state physics
Zwier, Paul J., mathematics

GRAYLING
Wallace, George John, ornithology

GROSSE ILE
Kaufman, Daniel, organic chemistry
Lee, Kwang Woo, chemistry

GROSSE POINTE
Controulis, John, organic chemistry
Marrazzi, Amedeo S., neuropharmacology, neurophysiology
Mohammed, Clive Imram, dentistry

FREMONT
Billerbeck, Fred William, Jr., food science
Frodey, Ray Charles, food science
Motawi, Kamal El-Din Hussein, food science
Shantaram, Rajagopal, mathematical statistics
Studier, Eugene H., physiological ecology

HARTFORD
Owen, Frank William, pomology

HICKORY CORNERS
Cummins, Kenneth William, limnology
Klug, Michael J., microbiology, ecology
Lauff, George Howard, limnology
zoology
Wetzel, Robert George, limnology

HILLSDALE
Albaugh, A Henry, mathematics, statistics
Catenhusen, John Alfons, plant ecology
Herbener, Roland Eugene, organic chemistry
Johnson, Ray Leland, physical chemistry, environmental chemistry
Peters, James John, metal physics
Toczek, Donald Richard, entomology, botany

HOLLAND
Bonem, Rena Mae, invertebrate paleontology, paleoecology
Brady, Allen Roy, systematic zoology
Brockmeier, Richard Taber, nuclear physics
Dershem, Herbert L., computer science
Doyle, Michael P., physical organic chemistry
Dusseau, Jerry William, vertebrate physiology, endocrinology
Folkert, Jay Ernest, mathematics
Frissel, Harry Frederick, nuclear physics
Hu, Clyde Kuen-Hua, organic polymer chemistry
Iyengar, Doreeswamy Raghavachar,
Jekel, Eugene Carl, inorganic chemistry
Klein, David Henry, environmental chemistry
Lee, Shui Lung, organic chemistry
Lyznicki, Edward Peter, Jr., organic chemistry
Malchick, Sherwin Paul, nuclear physics, optics
Marker, David, theoretical physics
Moser, Frank Hans, pollution chemistry
Mungall, William Stewart, bio-organic chemistry
Ockerse, Ralph, plant physiology, biochemistry
Ouderkirk, John Thomas, organic chemistry, physical chemistry
Pepoy, Louis John, organic chemistry
Petfield, Robert Joseph, organic polymer chemistry
Reinking, Robert Louis, geology
Rieck, Norman Wilbur, anatomy
Seeser, James William, nuclear physics, optics
Tanis, Elliot Alan, mathematics, statistics
Tharin, James Cotter, geology
Van Faasen, Paul, plant taxonomy
Van Putten, James D, Jr., nuclear physics
Van Schaack, Eva Blanche, botany
Webb, Philip Gilbert, industrial organic chemistry
Wettack, F Sheldon, physical chemistry
Williams, Donald Howard, inorganic chemistry

HOUGHTON
Agin, Gary Paul, nuclear physics
Allison, Betzabe M., cell physiology
Allison, John Arthur Charles, organic chemistry, inorganic chemistry
Anderson, Howard Benjamin, applied mathematics
Berry, Myron Garland, physical chemistry
Brown, Robert Thorson, plant ecology
Byers, Gordon Cleaves, mathematics
Coffman, Michael S., forest ecology, plant physiology
Crowther, C Richard, forestry
Dawson, Gladys Quinty, inorganic chemistry
El Khademi, Hassan S., organic chemistry, carbohydrate chemistry
Erbisch, Frederic H., botany, cytology
Francis, William Porter, theoretical physics, statistical mechanics
Frantti, Gordon Earl, seismology
Funkenbusch, Walter William, mathematics
Garland, Hereford, forest products
Gilpin, Michael James, pure mathematics
Glime, Janice Mildred, ecology, botany
Heuvers, Konrad John, mathematics
Hunzeker, Hubert LaVon, mathematical statistics
Jacobsen, Robert Leland, mathematical statistics

Janke, Robert A., plant ecology, physics
Julien, Larry Marlin, physical chemistry
Kalliokoski, Jorma Osmo Kalervo, economic geology
Keeling, Rolland Otis, Jr., physics
Kenny, David Herman, organic chemistry
Kraft, Kenneth J., invertebrate ecology
Krear, Harry Robert, ethology, ecology
Lai, Yuan-Zong, wood chemistry
Lee, Sung Mook, theoretical mechanics, solid state physics
Leifer, Leslie, physical chemistry
Linn, Robert, plant ecology, resource management
Longacre, William Atlas, geophysics
Lowther, John Lincoln, computer science
Luehrs, Dean C., inorganic chemistry
Mandeville, Charles Earle, physics
McMillin, Kenneth M., mathematics
Mericle, R Bruce, mathematics
Miller, Roswell Kenfield, forest management
Mitchell, Henry Rees, physics
Motteler, Zane Clinton, mathematics, computer science
Nordeng, Stephan C., stratigraphy
Peterson, Rolf Olin, wildlife ecology
Potnis, Vasant Raghunath, physics
Remondini, David Joseph, genetics
Reynolds, Garth Fredric, physical chemistry
Rose, William Ingersoll, Jr., volcanology
Ruotsala, Albert P., mineralogy
Sandel, Vernon Ralph, physical organic chemistry
Shandley, Paul David, physics
Sloan, Norman F., forest entomology, wildlife ecology
Spahn, Robert Joseph, applied mathematics
Spain, James Dorris, Jr., biochemistry
Stebbins, Dean Waldo, physics, academic administration
Stones, Robert C., environmental physiology
Sun, Bernard Ching-Huey, forest products, pulp chemistry
Vichich, Thomas E., mathematics
Warrington, Terrell L., physical chemistry, biochemistry
Whitten, Bertwell Kneeland, environmental physiology, comparative physiology
Williams, Fredrick David, polymer chemistry, physical chemistry
Wright, Thomas Dodson, zoology, limnology
Wyble, D O., geophysics
Yerg, Donald G, meteorology

HOWELL
Bianco, Donald R., physics

HUNTINGTON WOODS
Chang, Charles C., physical chemistry, chemical microscopy, crystallography
Krc, John, Jr., chemical microscopy, optics
Lippmann, Seymour A., applied physics

INKSTER
Bunting, Albert L, organic chemistry

JACKSON
Cruse, Julius Major, Jr., immunology, pathology
Leonard, Charles Grant, astronomy
Lyons, Don Chalmers, bacteriology
Miller, Herman Lunden, nucleonics, optics
Riggs, Roderick D., nuclear physics
Weitzner, Stanley, pathology

Bartoo, Harriette Valletta (Krick), economic botany, paleobotany
Batts, Henry Lewis, Jr., ecology
Beal, Philip Franklin, III, organic chemistry
Beeler, Fred A., mathematics
Berndt, Donald Carl, physical organic chemistry
Berneis, Hans Ludwig, physical chemistry, organic chemistry
Bernstein, Eugene Merle, nuclear physics
Bhuyan, Bijoy Kumar, biochemistry
Blefko, Robert L., topology
Bloss, Ronald Edward, animal nutrition
Booth, Roger Elwood, pharmaceutical chemistry
Boyack, Gerald (Arthur), organic chemistry
Bradfield, Stillman, anthropology
Bradley, George Edgar, physics
Bradley, Hugh Edward, operations research
Brewer, Richard (Dean), ecology, ornithology
Brodasky, Thomas Francis, physical chemistry, organic chemistry
Brown, Donald John, inorganic chemistry
Bryan, Jack T., pharmaceutical chemistry
Buckley, Joseph Thaddeus, mathematics
Bundy, Gordon Leonard, organic chemistry
Byrnes, William Winfield, endocrinology
Calloway, Jean Mitchener, number theory
Carley, David Don, theoretical physics
Carlson, Robert George, veterinary pathology
Chang, Albert Yen, biochemistry
Chartrand, Gary, mathematics
Churchill, Bruce Wenzel, microbiology
Clark, John Jefferson, veterinary pathology
Clarke, Allen Bruce, mathematics
Clarke, Richard Penfield, physical chemistry educ: princeton univ, ab, 41, ma, 48, phd(phys che
Cohen, Eckford, algebra
Coleman, Lester Lyman, biochemistry
Colingsworth, Donald Rudolph, chemistry, bacteriology
Collins, Robert James, physiology, pharmacology
Cook, Richard James, physical organic chemistry
Cooke, Dean William, inorganic chemistry
Cronk, Caspar, exploration geophysics, chemistry
Deal, Ralph Macgill, physical chemistry
DeGeeter, Melvin Joseph, animal nutrition
Derby, Stanley Kingdon, atomic spectroscopy
Dickason, David Gordon, transportation geography, geography of Asia
Drennan, Ollin Junior, history of science
Duchamp, David James, physical chemistry, x-ray crystallography
DuCharme, Donald Walter, pharmacology
Dulin, William E, zoology
Eble, Thomas Eugene, biochemistry
Eichenlaub, Val L, geography
Eisenberg, Robert C, microbiology
Elliott, George Algimon, pathology
Engemann, Joseph George, invertebrate zoology
Erhart, Rainer R, physical geography
Erlandson, Arvid Leonard, bacteriology
Evans, David Arnold, entomology
Evans, John Stone, endocrinology
Feenstra, Ernest Star, veterinary pathology
Ferguson, Stephen Mason, experimental nuclear physics
Ficsor, Gyula, genetics
Fitzpatrick, Francis Anthony, analytical biochemistry
Floksta, John Hilbert, clinical chemistry
Folz, Sylvester D, veterinary parasitology, medical parasitology
Fonken, Gunther Siegfried, research administration
Foote, Joel Lindsley, biochemistry
Ford, Jared Hewes, organic biochemistry
Forist, Arlington Ardeane, biochemistry
Forsblad, Ingemar Bjorn, organic chemistry
Fowler, Dona Jane, neuroendocrinology
Frank, Fred R., physiology, biochemistry
Freyburger, Walter Alfred, pharmacology
Friedman, Stephen Burt, microbiology
Gall, Martin, medicinal chemistry, organic chemistry
Garland, William, cultural anthropology
Gault, Frederick Paul, physiological psychology
Geng, Shu, applied statistics

Gerritsen, George Contant, physiology
Gilbertson, Terry Joel, clinical biochemistry, clinical chemistry
Gioia, Anthony Alfred, number theory
Glenn, Eldridge Myles, physiology
Glenn, Marldon Weldon, veterinary pathology
Goldsmith, Donald Leon, mathematics
Goodnight, Clarence James, zoology
Goodnight, Marie Louise, biology
Gorman, Robert Roland, biochemistry
Goyings, Lloyd Samuel, pathology, biochemistry
Grady, Joseph Edward, microbiology, biochemistry
Graham, Walter Robert, veterinary pathology
Gray, Gary D, biochemistry, immunology
Gray, Jack Ellsworth, animal pathology, toxicology
Green, Ernestene Leverne, anthropology, archaeology
Gregg, David Henry, pharmaceutical chemistry, organic chemistry
Grostic, Margaret Elizabeth, pharmacology
Grostic, Marvin Ford, organic chemistry
Haas, Kenneth Brooks, (Jr), veterinary medicine
Hall, Charles Mack, organic chemistry
Hamdy, Aziz H., veterinary medicine, microbiology
Hamlin, William Earl, industrial pharmacy
Hanka, Ladislav James, microbiology
Hannon, Herbert Harold, mathematics
Hardie, Gerald, nuclear physics
Hardwidge, Edward Albert, pharmaceutical chemistry
Harmon, Robert E, organic chemistry, medicinal chemistry
Hawker, Charles Davis, biochemistry
Heather, James Brian, synthetic organic chemistry
Heinle, Robert Walter, medicine
Heinzelman, Richard Voorhees, medicinal chemistry
Heller, Charles Frederick, Jr., geography
Herman, John Edward, solid state physics
Herr, Ross Robert, organic chemistry
Hester, Jackson Boling, Jr., organic chemistry, pharmaceutical chemistry
Heyd, William Ernst, medicinal chemistry, organic chemistry
Hinman, Jack Wiley, biochemistry, research administration
Hoeksema, Herman, Jr., organic chemistry
Holkeboer, Paul Edward, analytical chemistry
Holysz, Roman Paul, organic chemistry
Horst, Oscar Heinz, geography of Latin America, economic geography
Houser, Thomas J., physical chemistry
Howey, William Jeffrey, organic chemistry, computer sciences
Howell, James Arnold, analytical chemistry
Hsi, Richard S P., organic chemistry
Hsieh, Po-Fang (Philip), mathematics
Hubbard, William Neill, Jr., medicine
Huber, Joel E., organic chemistry
Hurst, Elaine H., phycology, aquatic ecology
Hylton, Thomas Anthony, organic chemistry
Iffland, Don Charles, organic chemistry
Inselberg, Edgar, plant physiology
Jackman, Albert Havens, physical geography, geography of North American Arctic & Sub-Arctic
Jaeger, Herbert Karl, organic chemistry
Jensen, Erik Hugo, pharmacy
Johnson, Byron Andrew, pharmaceutical chemistry
Johnson, Fenimore Thomas, medicine
Johnson, Garland A, biochemistry, pharmacology
Johnson, James Leslie, chemistry
Johnson, Richard Allen, physical chemistry
Josten, John James, bacteriology
Kagan, Fred, medicinal chemistry, organic chemistry
Kaiser, David Gilbert, pharmaceutical chemistry
Kana'an, Adli Sadeq, physical chemistry
Kanamueller, Joseph M, inorganic chemistry
Kapoor, S F, mathematics
Kaufman, Kurt Dunn, organic chemistry
Kelly, Robert Charles, organic chemistry
Kimball, Frances Adrienne, reproductive endocrinology

Kirchner, Eugene Carl, urban geography, geography of Sub-Saharan Africa
Klomparens, William, plant pathology
Knoebel, Edwin Lewis, pharmacy
Ko, Howard Wha Kee, biophysics
Kommnek, Leo Aloysius, microbiology, biochemistry
Koshy, Karyanil Thomas, pharmacy, pharmaceutical chemistry
Kragt, Clifford Lee, psychophysiology
Kramer, Sherman Francis, pharmacy, pharmacology
Kruglak, Haym, physics
Krzeminski, Leo Francis, analytical biochemistry, metabolism,
Kuenzi, W David, geology
Kukolich, Stephen Irvin, chemistry
Kupiecki, Floyd Peter, biochemistry
Kurtz, Richard Robert, synthetic organic chemistry
Lamb, Donald Joseph, pharmacy
Lauderdale, James W., Jr., reproductive physiology, endocrinology
Lawrence, Jean McVay, comparative physiology, zoology
Ledhicer, Daniel, organic chemistry
Leja, Stanislaw, mathematics
Li, Li-Hsieng, biochemistry
Lick, Don R., mathematics
Littner, Carl John, Jr., pharmaceutical chemistry
Lobl, Thomas Jay, pharmaceutical chemistry, reproductive physiology
Loeblich, Karen Elizabeth, animal behavior, entomology
Loughman, Barbara Ellen Evers, immunobiology
Lowry, George Gordon, physical chemistry
Lustgarten, Ronald Krisses, chemistry
Magerlein, Barney John, medicinal chemistry
Maher, Robert Francis, anthropology
Malik, Vedpal Singh, medical research
Mallinson, George Greisen, science administration, science education
Mann, Kingsley M., science writing
Marschke, Charles Keith, biochemistry, microbiology
Merritt, Margaret Virginia, analytical chemistry
Metzler, Carl Maust, statistics
Meyer, Heinz Friedrich, biomathematics
Marshall, Norman Barry, physiology
Marshall, Vincent dePaul, microbial physiology
Mason, Donald Joseph, microbiology
Maxey, Brian William, veterinary medicine
McCarville, Michael Edward, enzymology
Maxon, William Densmore, biochemistry
Miller, Curtis C., genetics, animal breeding
Miller, Robert Bruce, physics
Miller, Thomas Lee, biochemistry, microbiology
Miller, William Louis, biochemistry
Moffett, Robert Bruce, organic chemistry, medicinal chemistry
Mohberg, Noel Ross, biostatistics
Moisides-Hines, Lydia Elizabeth, medicinal chemistry, industrial chemistry
Morozovich, Walter, medicinal chemistry
Murray, Herbert Charles, medicine
Myers, Donald Royal, organic chemistry
Nagler, Robert Carlton, geography
Nathan, Alan Hart, chemistry
Neff, Alven William, biochemistry
Neil, Gary Lawrence, biochemical pharmacology
Nelson, James Donald, chemistry
Nelson, Norman Allan, pharmaceutical chemistry
Nichols, Nathan Lankford, physics
Nielson, Morris Lowell, chemistry
Nielson, George Marius, mathematics
Nishizawa, Edward Eichi, biochemistry, biology
O'Connell, Paul William, chemistry
Ogilvie, Marvin Lee, physiological chemistry
Olexia, Paul Dale, mycology, taxonomy
Owen, Stanley Paul, biochemistry
Pals, Donald Theodore, pharmacology
Panzer, James David, clinical pharmacology
Parcells, Alan Jerome, biochemistry
Parikh, Jekishan R., steroid chemistry

Peery, Clifford Young, industrial organic chemistry, medicinal chemistry
Peterson, Durcy Harold, biochemistry
Petro, John William, algebra
Petzold, Edgar, biochemistry
Phillips, Richard Peter, physiology
Pippen, Richard Wayne, systematic botany, plant ecology
Poel, Robert Herman, science education
Powell, James Henry, mathematical statistics
Punch, James Darrell, microbiology
Pyke, Thomas Richard, microbiology, biochemistry
Rajnak, Katheryn Edmonds, atomic physics
Rajnak, Stanley L., mathematics
Raup, Henry Armstrong, resource geography
Reid, William Bradley, pharmaceuticals
Renis, Harold E., virology, biochemistry
Reusser, Fritz, molecular biology, microbiology
Robert, Andre, endocrinology
Robertson, John Harvey, analytical chemistry, microbiology
Rogers, John Sinclair, agronomy
Roseman, Theodore Jonas, pharmaceutical chemistry
Rowe, Englebert L., physical pharmacy
Rudzik, Allan D., pharmacology
Salmond, William Glover, synthetic organic chemistry, steroid chemistry
Schumann, Edward Lewis, organic chemistry, medicinal chemistry
Sebek, Oldrich Karel, microbial physiology, industrial microbiology
Seber, Robert Charles, mathematics
Shah, Ashok Chandulal, pharmaceutics
Sievers, Gerald Lester, mathematical statistics
Schlagel, Carl Alvin, pharmacology
Schmaltz, Lloyd John, geomorphology, glacial geology
Schneider, William Paul, organic chemistry
Schreiner, Erik Andrew, mathematics
Schultz, Ida Beth, science education
Schumann, Edward Lewis, organic chemistry
Sims, Bernard, pharmaceutical chemistry, physical chemistry
Sinkula, Anthony Arthur, pharmaceutical chemistry, medicinal chemistry
Slomp, George, physical chemistry
Smith, Robert Jack, microbiology
Smith, Robert Matthews, microbiology
Smith, Thomas Jefferson, geophysics, mathematics
Sokoloski, Walter Thomas, microbiology, virology
Soret, Manuel Gonzalez, pathology
Spilman, Charles Hadley, reproductive biology
Stafford, Robert Oppen, endocrinology
Steen, Edwin Benzel, parasitology
Steenhaus, Ralph K., analytical chemistry
Stenesh, Jochanan, biochemistry
Stocking, Gordon Gary, veterinary medicine
Stoline, Michael Ross, statistical analysis
Straw, Robert Niccolls, pharmacology
Stringfellow, Dale Alan, virology
Stroman, David Womack, microbiology
Stromsta, Courtney Paul, speech pathology, audiology
Struck, William Anthony, analytical chemistry
Stucki, Jacob Calvin, endocrinology, research administration
Sud, Gian Chand, zoology, biochemistry
Sun, Frank F., biochemistry
Swenberg, James Arthur, veterinary pathology
Swenson, Gene Holstrom, pharmacology
Szczech, George Marion, pathology, teratology
Tang, Andrew H., pharmacology
Taraszka, Anthony John, analytical chemistry, pharmaceutical chemistry
Taraszka, Mildred J., physical chemistry
Taylor, Betty, foods, nutrition
Tidd, Joseph Shepard, botany
Tiffany, Burris Dwight, organic chemistry
Trimitsis, George B., organic chemistry
Trotter, Allen Richard, plant breeding
Tsuji, Kiyoshi, microbiology, analytical chemistry
Tucker, William Gough, oncology
Underwood, Gerald Emerson, virology
Upjohn, Everett Gifford, medicine

Ursprung, Joseph John, organic chemistry, medicinal chemistry
Valvani, Shri Chand, pharmaceutical chemistry, pharmaceutics
Vanderbeek, Leo Cornelis, plant physiology
Van Deventer, William Carl, biology
Van Rheenen, Verlan H., organic chemistry
Vavra, James Joseph, infectious diseases
Verseput, Herman Ward, paper technology
Von Voigtlander, Philip Friedrich, neuropharmacology
Vuicich, George, urban geography
Walker, Jerry Arnold, chemistry
Wallach, Donald P., biochemistry, pharmacology
Warner, Donald Theodore, biochemistry
Weber, Dennis Joseph, analytical chemistry, physical chemistry
Webster, Harris Duane, veterinary pathology
Wechter, William Julius, organic chemistry
Weeks, James Robert, pharmacology
Weisblat, David Irvin, chemistry
Wells, Franklin Burnham, organic chemistry
Whitfield, George Buckmaster, Jr., biochemistry
Wickrema Sinha, Asoka J., biochemistry, pharmacology
Wierenga, Wendell, organic chemistry
Whaley, Howard Arnold, chemistry
Wiley, Paul Fears, organic chemistry
Wilks, John William, reproductive endocrinology
Wilson, Laurence Edward, inorganic chemistry
Witz, Dennis Fredrick, microbiology
Wood, Jack Sheehan, environmental physiology
White, Arthur Thomas II, mathematics
White, David Raymond, organic chemistry
White, Gordon Justice, immunochemistry, biochemistry
Wright, Alden Halbert, topology
Wright, John Brenton, medicinal chemistry
Wright, Wayne Mitchell, acoustics
Yalkowsky, Samuel Hyman, pharmaceutical chemistry
Yankee, Ernest Warren, organic chemistry, medicinal chemistry
Youngdale, Gilbert Arthur, organic chemistry
Zietlow, James Philip, physics
Zimbelman, Robert George, reproductive endocrinology
Zins, Gerald Raymond, pharmacology

L'ANSE

Bourdo, Eric A., Jr., forest management
Meteer, James William, forestry, computer science

LAINGSBURG

Purselove, Laurence Albert, science writing, computer science

LAKE LEELANAU

Herbst, Robert Max, organic chemistry

LAKE LINDEN

Hesterberg, Gene Arthur, forest pathology

LAKESIDE

Jackson, John Mathews, food science

LANSING

Anderson, George R., veterinary medicine, virology
Becker, Maurice Edwin, bacteriology
Bedford, James William, water chemistry
Blouch, Ralph Irving, wildlife management
Borgeson, David P., fish biology
Chang, Ik-Chin, immunology
Cipparone, Joseph Robert, pathology
Cook, Ray Lewis, soils
Foy, Robert Bastian, soils
Gottshall, Russell Y., bacteriology
Hyndman, Lee Allen, biochemistry, laboratory medicine

Isleib, Donald Richard, plant physiology
Jenkins, David H., wildlife management
Kivela, Edgar Welton, toxicology
Kuwahara, Steven Sadao, biochemistry
Lievense, Stanley James, fisheries
Lorch, Steven Kalman, forensic science
McCall, Keith Bradley, biochemistry
Mitchell, John Richard, microbiology, public health
Muenterer, Donald Arthur, bacteriology
Olson, Birger Henry, microbiology
Portwood, Lucile Mitchell, bacteriology
Price, Harold Anthony, organic chemistry
Ruby, John L., forest genetics
Ryel, Lawrence Atwell, biostatistics, wildlife research
Schuurmans, David Meinte, industrial microbiology
Stewart, Bonnie Madison, mathematics
Tishkoff, Garson Harold, hematology, medicine
Tung, Fred Fu, biochemistry
Wentworth, Bertina Brown, microbiology
Wetherell, Herbert Ranson, Jr., pharmacology, toxicology
Zemach, Rita, statistics

LIVONIA

Dennis, Mary, inorganic chemistry
Mayer, Raymond Parm, organic chemistry
Miller, Donald Elbert, biology

LUDINGTON

Maskal, John, physical chemistry
Ryan, William Alexander, Jr., geology
Smith, Arthur Gerald, corrosion, organic polymer chemistry
Yessik, Michael John, solid state physics

MADISON HEIGHTS

Howell, John Kenneth, Jr., industrial chemistry
Lacy, Robert M., surface chemistry, electrochemistry
Symons, Philip Charles, electrochemistry

MARQUETTE

Allenstein, Richard Van, analytical chemistry
Barry, Roger Donald, organic chemistry, biochemistry
Bowers, Maynard C., botany
Earney, Fillmore Christy Fidelis, resource geography
Frey, John Erhart, inorganic chemistry
Froiland, Thomas Gordon, developmental genetics
Gill, Gordon Drew, insect taxonomy
Griffith, Thomas, biochemistry
Heath, Roy Elmer, inorganic chemistry
Heikkinen, Donald D., mathematics
Heimonen, Henry Samuel, geography, geology
Hughes, John Derek, geomorphology
Jacobs, Gerald Daniel, physical chemistry, spectroscopy
Jamrich, John Xavier, academic administration, statistics
Johnson, Whitney Larsen, computer science
Macalady, Donald Lee, physical inorganic chemistry, environmental chemistry
McNeill, Robert Bradley, mathematics
Merry, William James, botany
Niemi, Alfred Otto, conservation
Parejko, Ronald Anthony, environmental sciences, microbiology
Peters, Lewis, parasitology
Reynolds, Orland Bruce, biochemistry
Robinson, William Laughlin, wildlife ecology
Roth, Jerome A., organic chemistry
Snitgen, Donald Albert, biology, science education
Stauffer, Thomas Miel, fish biology
Stortz, Clarence B., mathematics
Trentelman, George Frederick, physics
Vande Berg, Warren James, phycology
Verley, Frank A., genetics
Wagner, Robert Thomas, nuclear physics
Werner, John Kirwin, herpetology

MASON

Beil, Gary Milton, plant breeding
Stribley, Rexford Carl, organic chemistry

514

MATTAWAN
Goldenthal, Edwin Ira, pharmacology
Jessup, Daniel Clifford, physiology, pharmacology

MENDON
Estes, Frances Lorraine, environmental chemistry

MIDLAND
Alfrey, Turner, Jr., polymer chemistry
Anders, Oswald Ulrich, radiochemistry
Archer, Wesley Lea, industrial organic chemistry
Arends, Charles Bradford, plastics chemistry
Armentrout, David Noel, analytical chemistry
Asperger, Robert George, chemistry
Atwell, William Henry, organic chemistry
Bajzer, William Xavier, organic chemistry, medicinal chemistry
Baney, Ronald Howard, industrial chemistry
Barnes, Garrett Henry, Jr., organic chemistry
Barry, Arthur John, organic chemistry, physical chemistry
Bass, Shailer Linwood, silicon chemistry
Bauman, William Carrel, chemistry
Bauriedel, Wallace Robert, biochemistry
Bennett, Donald Raymond, pharmacology
Betso, Stephen Richard, analytical chemistry, polymer chemistry
Bjork, Carl Kenneth, Sr, agricultural chemistry
Blair, Ercyl Howell, organic chemistry
Blanchard, Fred Ayres, biophysics
Boundy, Ray Harold, chemistry
Bowman, Carlos Morales, information science, computer science
Boyer, Raymond Foster, physics
Braley, Silas Alonzo, Jr, industrial chemistry
Brandt, Gerald H, soil physics
Bredeweg, Robert Allen, analytical chemistry
Bremmer, Bart J, organic chemistry, polymer chemistry
Brower, Frank M, environmental management
Bruss, Harry Francis, organic chemistry agricultural
Budde, Paul Bernard, agricultural chemistry
Burgert, Bill E, organic chemistry
Casorso, Donald Ray, pathology
Chamberlain, Malcolm, organic chemistry
Chandler, Michael Lynn, pharmacology, biochemistry
Chao, Mou Shu, electrochemistry, analytical chemistry
Chenoweth, Maynard Burton, clinical pharmacology
Clark, Harold Arthur, polymer chemistry
Cobler, John George, organic polymer chemistry
Colby, Robert William, animal nutrition, biochemistry
Coleman, Anna M, information science
Coulter, Llewellyn Legrande, weed science, wildlife management
Crable, George Francis, physics
Crummett, Warren B, analytical chemistry
Currutt, Jerry Lee, physical chemistry
Davis, Ralph Anderson, chemistry
Dennis, Kent Seddens, physical chemistry
Diesen, Ronald W, physical chemistry
Dilling, Wendell Lee, organic chemistry
Dorenbos, Harold E, organic chemistry
Dorman, Linneaus Cuthbert, medicinal chemistry
Drake, Stevens Stewart, polymer chemistry
Dunbar, Joseph Edward, pharmaceutical chemistry
Edmonds, James William, crystallography
Elias, Hans Georg, polymer chemistry
Fairless, Charles Michael Haskel, analytical chemistry
Farber, Hugh Arthur, organic chemistry
Fearon, Frederick William Gordon, chemistry
Finley, Arlington Levart, solid state chemistry, inorganic chemistry
Fletcher, ed Walker, entomology
Flynn, James Patrick, physical chemistry
Frevel, Ludo Karl, inorganic chemistry
Frye, Cecil Leonard, organic chemistry
Gantz, Ralph Lee, weed science

Gehring, Perry James, toxicology, pharmacology
Gentry, Willard Max, Jr., organic chemistry
Getzendaner, Milton Edmond, pesticide chemistry
Gilkey, John Woodbury, chemistry
Grabiel, Charles Edward, organic chemistry
Gray, Henry Emil, entomology
Greene, Bettye Washington, physical chemistry, colloid chemistry
Greminger, George King, Jr., pulp chemistry, paper chemistry
Habermann, Clarence E, physical chemistry
Hall, Richard Harold, organic chemistry, chemical engineering
Hansen, Robert Douglas, physical chemistry
Harmer, David Edward, radiation chemistry, polymer chemistry
Heeg, Joel Francis, drug metabolism, bionucleonics
Heeschen, Jerry Parker, physical chemistry
Hennis, Henry Emil, organic chemistry
Hinkamp, Paul Eugene, (II), organic chemistry
Hoerger, Fred Donald, research administration, environmental sciences
Houtman, Thomas, Jr., organic chemistry
Howe, Robert George, entomology
Hymas, Theo Alfred, veterinary surgery
Irish, Don DeLance, biochemistry
Jersey, George Carl, veterinary pathology
Johnson, Julius Earl, biochemistry
Jones, Giffin Denison, organic chemistry
Kagel, Ronald Oliver, analytical chemistry
Kenaga, Duane Leroy, wood technology
Kenaga, Eugene Ellis, toxicology
Keskkula, Henno, chemistry
Kim, Yungki, organic chemistry, polymer chemistry
King, Stanley Shih-Tung, physical chemistry, analytical chemistry
Koch, Melvin Vernon, pharmaceutical chemistry
Kocher, Charles William, nuclear physics
Kociba, Richard Joseph, veterinary pathology, toxicology
Koerker, Frederick William, inorganic chemistry
Kolat, Robert S, analytical chemistry
Konkle, George Melvin, chemistry
Kostelnik, Robert J, chemistry
Koster, Robert Allen, organic chemistry
Kurfman, Virgil Benson, physical chemistry
LaBarge, Robert Gordon, applied chemistry, industrial chemistry
Lane, George Ashel, physical chemistry, applied chemistry
Langley, Neal Roger, polymer chemistry
Langner, Ralph Rolland, environmental chemistry
Larsen, Eric Russell, organic chemistry
Leathers, Joel Monroe, chemistry
Leavitt, Frederick Carlton, polymer chemistry
Ledbetter, Harvey Don, organic chemistry
Leddy, James Jerome, industrial chemistry
Lee, Wei-Ming, physical chemistry, polymer chemistry
Legrow, Gary Edward, chemistry
Lehman, Duane Stanley, inorganic chemistry
Leng, Marguerite Lambert, agricultural biochemistry, analytical biochemistry
Lentz, Charles Wesley, chemistry
Lewenz, George F, organic chemistry
Lipowitz, Jonathan, organometallic chemistry, polymer chemistry
Maasberg, Albert Thomas, chemistry, chemical engineering
Mahle, Nels H, analytical chemistry
Mani, Inder, organic chemistry
Markley, Lowell Dean, organic chemistry
Marquardt, Roland Paul, analytical chemistry, organic chemistry
Marshall, Franklin Nick, pharmacology
McCarthy, James Ray, Jr., organic chemistry
McCarty, Leslie Paul, pharmacology
McGregor, Stanley Dane, organic chemistry
McKeever, L Dennis, polymer chemistry, physical chemistry
McLean, James Dennis, analytical chemistry, polarography
Meier, Dale Joseph, polymer physics
Messing, Sheldon Harold, organic chemistry
Miller, Frederick Arnold, physical chemistry

Miller, Robert Llewellyn, polymer science
Mills, Jack F, organic chemistry
Moll, Harold Westbrook, organic chemistry, chemical engineering
Monroe, Ezra, pharmaceutical chemistry
Moolenaar, Robert John, physical chemistry, environmental sciences
Moon, Tag Young, physical chemistry, statistical mechanics
Moore, Carl, polymer chemistry, colloid chemistry
Moore, Eugene Roger, polymer chemistry, chemical engineering
Morehouse, Donald S, Jr, industrial chemistry
Morris, Leo Raymond, industrial organic chemistry
Moss, Rodney Dale, organic chemistry, analytical chemistry
Moyer, John Raymond, inorganic chemistry
Mullison, Wendell Roxby, plant physiology
Murchison, Pamela W, physical chemistry
Neely, Brock Wesley, biochemistry, physical chemistry
Neumann, Fred William, organic chemistry, analytical chemistry
Norris, Jessie McGowan, toxicology, parasitology
Norris, Mark Gilbert, Jr., organic chemistry
Nowak, Robert Michael, chemistry
Nummy, William Ralph, organic chemistry
Nyquist, Richard Allen, molecular spectroscopy
Olson, Kenneth Jean, toxicology
Oriel, Patrick John, physical chemistry
Orvik, Jon Anthony, physical organic chemistry
Osborne, David Wendell, organic chemistry
Peet, Norton Paul, organic chemistry, medicinal chemistry
Peters, James, colloid chemistry, polymer chemistry
Petersen, Donald Ralph, physical chemistry, computer science
Petrella, Ronald Vincent, physical chemistry
Pews, Richard Garth, organic chemistry
Pierce, James Kenneth, chemistry
Pierce, Ogden Ross, chemistry
Platt, Alan Edward, physical organic chemistry
Plueddemann, Edwin Paul, polymer chemistry
Polmanteer, Keith Earl, physics, chemistry
Pruitt, Malcolm E, chemistry
Rabold, Gary Paul, industrial chemistry
Raich, William Judd, polymer chemistry
Rainey, Mary Louise, analytical chemistry
Rakshys, Joseph W, Jr, polymer chemistry
Raley, Charles Francis, Jr, organic chemistry
Ramsey, John Charles, toxicology
Rausch, Douglas Alfred, organic chemistry
Reguiski, Thomas Walter, polymer chemistry
Reinecke, Charles Everett, organic chemistry
Renzi, Alfred Arthur, physiology, endocrinology
Rieke, James Kirk, physical chemistry
Roberts, Charles Brockway, inorganic chemistry, analytical chemistry
Rogers, Richard Brewer, agricultural chemistry
Rowe, Verald Keith, toxicology, industrial hygiene
Ryan, John William, organic chemistry
Salinger, Rudolf Michael, organometallic chemistry, chemical engineering
Sarge, Theodore William, physical organic chemistry
Sarkar, Nitis, physical chemistry, mineral engineering
Saunders, Frank Linwood, physical chemistry
Scheidt, Francis Matthew, organic chemistry
Schmidt, Donald Dean, physical inorganic chemistry
Schmidt, Donald L, organic polymer chemistry
Schmitt, John Aloysius, physical chemistry
Schneider, John Arthur, organic chemistry, polymer chemistry
Schwetz, Bernard Anthony, toxicology, teratology
Scott, Richard Lynn, analytical chemistry

Selby, Theodore W, physical chemistry
Seymour, Keith Goldin, soil chemistry
Shadoff, Lewis Allan, analytical chemistry, mass spectrometry
Shaver, Robert John, zoology, parasitology
Shea, Philip Joseph, pharmacology, physiology
Sheetz, David P, physical chemistry
Skelcey, James Stanley, inorganic chemistry
Skelly, Norman Edward, analytical chemistry, physical chemistry
Skochdopole, Richard E, physical chemistry, polymer chemistry
Smith, Albert Lee, physical chemistry
Smith, Harry Andrew, organic chemistry, polymer chemistry
Snyder, Robert Gene, molecular spectroscopy
Solc, Karel, physical chemistry
Southwick, Lawrence, agronomy, horticulture
Speier, John Leo, Jr., chemistry
Spencer, Robert Shirley, physical chemistry, management science
Stark, Forrest Otto, chemistry
Stenger, Vernon Arthur, analytical chemistry
Stowe, Robert Allen, surface chemistry
Strojny, Edwin Joseph, organic chemistry
Stull, Daniel Richard, thermochemistry
Tabor, Theodore Emmett, organic chemistry
Thurber, William Samuels, organic chemistry
Tobey, Stephen Winter, pharmaceutical chemistry
Tolkmith, Henry, organic chemistry, bio-organic chemistry
Tomalia, Donald Andrew, physical organic chemistry
Torkelson, Theodore Ruben, toxicology
Tou, James Chieh, analytical chemistry, physical chemistry
Turley, June Williams, x-ray crystallography, economic statistics
Turley, Sheldon Gamage, physics
Van Dell, Robert Duane, surface chemistry, colloid chemistry
Vanderkooi, William Nicholas, industrial chemistry
Van Hall, Clayton Edward, analytical chemistry
Vatne, Robert Dahlmeier, parasitology
Vrieland, Gail Edwin, industrial chemistry
Wagner, Eugene Ross, medicinal chemistry
Wahr, John Cannon, physics
Walles, Wilhelm Egbert, organic chemistry, polymer chemistry
Walter, Richard Webb, Jr., microbial biochemistry
Wang, Chun Shan, organic chemistry
Warrick, Earl Leathen, physical chemistry
Watson, Andrew John, weed science
Wehr, Henry William, Jr, organic chemistry
Wendel, Samuel Reece, bioinorganic chemistry
Wessling, Ritchie A, physical chemistry, polymer science
Whalen, Joseph Wilson, microbiology
Wolf, Mark Adam, bacteriology
Wright, Wayne Gordon, weed science, entomology
Wymore, Charles Elmer, inorganic chemistry
Young, David Caldwell, organic chemistry

MILFORD
Fox, Sereck Hall, pharmaceutical chemistry
Marsh, Richard Hayward, analytical chemistry
Prostak, Arnold S, chemical instrumentation

MONTAGUE
Hill, William Francis, Jr, virology, environmental biology

MT CLEMENS
Crevasse, Gary A, food science

MT PLEASANT
Adams, Jack Donald, genetics
Adkins, Julia Elizabeth, mathematics
Benson, Edmund Walter, inorganic chemistry

MICHIGAN

Bland, Roger Gladwin, entomology
Briggs, Edgar Van, physics
Brockman, Ellis R, microbiology
Burlington, Roy Frederick, environmental physiology
Caldwell, Larry D, ecology
Cook, Theodore Warren, chemistry
Cratin, Paul David, physical chemistry
Current, David Harlan, experimental solid state physics, nuclear magnetic resonance
Curry, La Verne Leon, insect taxonomy
Cuthbert, Nicholas Le Huray, ornithology
Delia, Thomas J, organic chemistry, medicinal chemistry
Dietrich, Richard Vincent, petrology
Domanik, Richard Anthony, biochemistry, biophysics
Fairbanks, Michael Bruce, comparative physiology
Filson, Malcolm Harold, chemistry
Fisk, Donald, mathematics
Gump, J R, inorganic chemistry
Hampton, Raymond Earl, plant pathology
Hohn, Matthew Henry, economic botany, limnology
Jacobson, Baruch S, biophysics
Kabe, Frederick Carl, organic chemistry
Kiefer, Wayne Eugene, geography
Koehler, Lawrence D, developmental biology
Krull, John Norman, ecology, wildlife biology
Kyser, Forrest DeWayne, physical geography, cultural geography
Lampky, James Robert, microbiology
Lenon, Herbert Lee, fish biology
Levich, Calman, biophysics
Lieberman, Leonard, anthropology, sociology
Lindfors, Karl Russell, physical chemistry
Lorand, John Peter, physical organic chemistry
Magnell, Kenneth Robert, audiology
Marple, Robert P, geography
Maxwell, Keith L, speech science
McDermott, Leon Anson, chemistry
McGaugh, Maurice Edron, geography
Miller, William Anton, mathematics
Moore, Wayne Elden, petroleum geology
Petersen, Quentin Richard, organic chemistry
Phelps, Frederick Martin, III, atomic spectroscopy
Powell, Richard L, speech pathology, audiology
Scheel, Carl Alfred, entomology
Shonk, Carl Ellsworth, biochemistry
Slocum, Robert Richard, solid state physics, radiological physics
Swart, William Lee, mathematics
Waggoner, Wilbur J, mathematics
Whitmore, Edward Hugh, geometry
Whitney, Marion Isabelle, geology
Wright, Kenneth Arthur, nuclear physics, electronics
Wujek, Daniel Everett, botany
Young, Janice Edith, zoology
Yuill, Robert Stanley, urban geography, comparative physiology, economic geography
Zeoli, Harold Wilson, mathematics, physics

NORTHPORT
Vander Brook, Milton John, pharmacology

NORTH MUSKEGON
Dean, Gordon Spencer, pharmaceutical chemistry
Heitmeier, Donald Elmer, organic chemistry

NAZARETH
Allen, Cheryl, biochemistry
Zeleznik, Pauline, inorganic chemistry, physical chemistry

MUSKEGON
Carlson, James C, radiological physics
De Wall, Gordon, organic chemistry
Jackson, William Gordon, chemistry
Sandri, Joseph Mario, organic chemistry
Schroeder, William, Jr, organic chemistry
Shirley, Robert Louis, organic chemistry

OAK PARK
Ben, Manuel, physical chemistry, physics
Johnson, Leonard Gustave, applied statistics
Orloff, Harold David, organic chemistry

OKEMOS
Brubaker, Carl H, Jr, inorganic chemistry, organometallic chemistry
Henderson, Hugh E, animal husbandry, industrial microbiology
Linnell, Albert Paul, astrophysics
Solomon, Marvin David, biology
Su, George Chung-Chi, organic chemistry
Su, Ruth Wolf, analytical mathematics
Vinge, Clarence L, economic geography, historical geography
Wright, Jonathan William, forest genetics

OXFORD
Coleman, Neil Lloyd, geology, fluid mechanics
Scott, Robert Blackburn, Jr, organic chemistry

OLIVET
Fleming, Richard Cornwell, zoology
Gruen, Fred Martin, entomology, mycology
Roberts, John Maurice, chemistry
Speare, Edward Phelps, zoology

PARMA
Leatherman, Anna D, plant ecology

PENTWATER
Huffman, John William, obstetrics & gynecology

PETOSKY
Hogan, Bartholomew William, medicine

PLYMOUTH
Atchley, Ralph Warren, organic chemistry
Domke, Charles J, air pollution
Penney, Robert Vincent, theoretical physics
Plautz, Donald Melvin, forensic science
Slifkin, Sam Charles, organic chemistry

PONTIAC
Baginski, Eugene S, biochemistry, physiology
Benham, Ross Stephen, microbiology
Dodson, Helen Walter (Mrs Edmond L Prince), solar physics
Kantrowitz, Adrian, surgery
Truant, Joseph Paul, immunology

PORTAGE
Buller, Robert Henry, pharmacology

PORT HURON
Dawe, Harold Joseph, chemistry

RAPID CITY
Hough, Jack Luin, geology
Ingle, Dwight Joyce, physiology

ROCHESTER
Arnold, Harvey James, mathematical statistics
Ayres, Paul E, industrial hygiene
Beardmore, William Boone, microbiology
Bragg, Louis Richard, mathematics
Brieger, Gottfried, organic chemistry
Butterworth, Francis M, genetics
Dettman, John Warren, mathematical analysis
Doherty, Paul Michael, solid state physics
Ebert, William Robley, pharmaceutical chemistry
Evans, David Hunden, mathematics
Feeman, George Franklin, organic chemistry
Friedman, Thomas Baer, human genetics, developmental genetics
Froemke, Jon, mathematics
Giblin, Frank Joseph, biochemistry
Goudsmit, Esther Marianne, invertebrate embryology, biochemistry
Hammerle, William Gordon, physics
Harman, Kenneth Millard, organic chemistry
Henry, Egbert Winston, plant physiology
Hightower, Kenneth Ralph, biophysics
Hoffman, William Charles, applied mathematics
Hunter, Robert Douglas, physiological ecology
Johnson, George Philip, mathematical analysis
Joswick, Harry Loren, microbiology
Ketcham, Paul Abbott, microbial physiology
Kinsey, Victor Everett, chemistry
Lindemann, Charles Benard, cell physiology, biophysics
Livermore, Brian Paul, medical microbiology
Mallow, Jeffry Victor, atomic physics
Malm, Donald E G, topology
McKay, James Harold, mathematics
McKinley, John McKeen, theoretical physics
Miller, Steven Ralph, physical chemistry
Mobley, Ralph Claude, nuclear physics
Obear, Frederick W, inorganic chemistry
Pak, Moon Jae, medical physiology
Pino, Lewis Nicholas, organic chemistry
Reddan, John R, cell physiology
Reddy, Venkat N, biochemistry
Riley, Michael Verity, biochemistry, ophthalmology
Romeo, John Thomas, plant chemistry, systematic botany
Roy, Arun K, biochemistry, endocrinology
Russell, Joel W, physical chemistry
Schmidt, Parbury Pollen, chemical physics
Sevilla, Michael Douglas, physical chemistry, biophysical chemistry
Singer, Philip, behavioral anthropology
Smith, Harvey Alvin, mathematics, physics
Stern, Robert Louis, organic chemistry
Taylor, Robert Craig, inorganic chemistry
Tepley, Norman, medical physics
Tipler, Paul A, nuclear physics
Tomboulian, Paul, organic chemistry
Torch, Reuben, protozoology
Unakar, Nalin J, cell biology
Wallace, William Donald, solid state physics
Williamson, Robert Marshall, physics
Wilson, Walter LeRoy, physiology
Winkler, Barry Steven, physiology
Young, John Davis, organic chemistry

ROYAL OAK
Al-Saadi, Abdul A, cell biology, microscopic anatomy
Bernstein, Jay, pathology
Bloor, Robert John, radiotherapy
Brackenridge, David Ross, organic chemistry
Epstein, Emanuel, clinical chemistry
Harrold, Gordon Coleson, chemistry
Horwitz, Norman H, radiological physics
Jackisch, Philip Frederick, organic chemistry
Mader, Ivan John, medicine
Morita, Yoshikazu, nephrology, internal medicine
Poulik, Dave, immunology, immunochemistry

ROGERS CITY
Applegate, Vernon Calver, fisheries

SAGINAW
Bissell, Grosvenor Willse, internal medicine

ST JOSEPH
Cutler, Warren Gale, fluid physics
Roth, Norman Gilbert, microbiology

ST CLAIR SHORES
Kuiler, Charles Peter, synthetic organic chemistry

SAULT STE MARIE
Anderson, Melvin Lee, inorganic chemistry
Behmer, David J, fisheries
Chandra, Purna, agricultural microbiology
Duwe, Arthur Edward, immunology, embryology

SOUTHGATE
Aepli, Otto Theodore, research administration

SPRING ARBOR
Hawkins, Charles Edward, plasma physics, astrophysics
Johnson, David Alfred, physical inorganic chemistry
Whiteman, Eldon Eugene, zoology

STANTON
Cole, Clarence Lorraine, animal husbandry

STERLING HEIGHTS
Hiebrink, Earl Henry, research administration
Murie, Richard A, analytical chemistry

STURGIS
Benne, Erwin John, biochemistry

SUTTONS BAY
Solem, Anson Donald, physics

THREE RIVERS
Heustis, Albert Edward, public health administration

TIPTON
Freeman, Fred W, forestry, horticulture

TRAVERSE CITY
Glarum, Sivert N, physical chemistry

SOUTHFIELD
Bettman, Max, physical chemistry
Bode, James Daniel, physical chemistry
Braunlich, Peter Fritz, experimental solid state physics, atomic physics
Buell, Wayne H, physical chemistry
Burlant, William Jack, organic chemistry, chemical engineering
Cutler, Sidney Joshua, medical statistics, epidemiology
Gleason, Gale R, Jr, environmental science
Hakala, Reino William, chemical physics, applied mathematics
Halberstadt, Marcel Leon, physical chemistry
Holmes, Larry A, polymer chemistry
Jacko, Michael George, physics, chemical engineering
Kalt, Melvin Barry, environmental management
Keating, Patrick Norman, applied physics
Knowles, David M, geology
Knudson, Vernie Anton, limnology
Marburger, Richard Eugene, technological forecasting
McGlinn, Edward James, technological forecasting
McNair, Ruth Davis, clinical chemistry
Michel, Richard Edwin, solid state physics
Mosher, Robert Eugene, analytical chemistry
Mullin, Charles R, physical chemistry
Person, Steven John, mammalian physiology
Sawatari, Takeo, optics
Smith, Bryce Everton, ecology
Spurgeon, William Marion, physical chemistry
Vaughn, Clarence Benjamin, oncology
Weatherby, Gerald Duncan, biochemistry

SOUTH HAVEN
Marshall, Charles Wheeler, organic chemistry

SCOTTVILLE
Sutton, Dale Dinkins, botany

SHINGLETON
Verne, Louis Joseph, wildlife biology, ecology

AFTON
Bird, Charles Norman, organic chemistry

AITKIN
Upadhyaya, Rajarama Belle, genetics

ALBERT LEA
Speers, George M, poultry nutrition

ARDEN HILLS
Stephens, Dale Nelson, organic chemistry

AUSTIN
Baumann, Wolfgang Josef, biochemistry, lipid chemistry
Chipault, Jacques Robert, biochemistry
Hill, Eldon G, nutritional biochemistry
Holman, Ralph Theodore, biochemistry, nutrition
Jenkin, Howard M, microbiology
Lin, Jiann-Tsyh, biochemistry
Lundberg, Walter Oscar Paul, lipid chemistry, biochemistry
Schlenk, Hermann, organic chemistry, biochemistry
Schmid, Harald Heinrich Otto, biochemistry
Stearns, Eugene Marion, Jr, biochemistry
Wedmid, George Yuri, biological chemistry

BAUDETTE
Greenwell, Benjamin Elmer, biochemistry

BEMIDJI
Avelsgaard, Roger A, algebra
Baker, Robert Charles, ecology, science education
Beitzel, Richard Earl, organic chemistry, science education
Borchers, Harold Allison, entomology
Britton, William Giering, physical chemistry, inorganic chemistry
Hazard, Evan Brandao, vertebrate zoology
Knodel, Raymond Willard, mathematics education
Kraft, Donald J, plant physiology
Ludwig, James Pinson, ecology, vertebrate biology
Peters, Harold Truman, entomology
Saccoman, Frank (Michael), anatomy, zoology

BLOOMINGTON
Aagard, Roger L, physics, mathematics
Bernal G, Enrique, optical physics, lasers
Crawford, Thomas Michael, electronic technology
Heist, Herbert Ernest, zoology
Heuschele, Ann, aquatic biology
Koepke, Barry George, materials science
Kruse, Paul Walters, Jr, solid state physics, electrooptics
Lutes, Olin Silas, electromagnetism
McClure, Benjamin Thompson, electronic physics
Mikhail, Adel Ayad, medicinal chemistry, biomedical engineering
Petracek, Francis James, medicinal chemistry
Ready, John Fetsch, lasers
Schuldt, Spencer Burt, applied mathematics
Scott, Troy Alexander, Jr, physical chemistry
Stallard, Richard E, anatomy, periodontology
Tufte, Obert Norman, solid state physics

BURNSVILLE
Baker, James Dennard, mathematical analysis

CANNON FALLS
Hartzell, Thomas H, food science

CHASKA
Lee, William Glen, food science
Nelson, John Howard, food science
Rosevear, John William, biochemistry
Snyder, Leon Carleton, horticulture

YPSILANTI
Anderson, Charles Thomas, physical inorganic chemistry
Belcher, Robert Orange, botany
Breedlove, Charles B, science education
Brewer, Stephen Wiley, Jr, analytical chemistry
Brown, Donald Frederick Mackenzie, biological sciences
Buckeye, Donald Andrew, mathematics
Carter, Giles Frederick, metallurgical chemistry, archaeological chemistry
Caswell, Herbert Hall, Jr, ecology, ornithology
Collins, Ronald William, inorganic chemistry
Compere, Edward L, Jr, organic chemistry
Ehrlich, Allen S, anthropology
Fennel, William Edward, invertebrate zoology
Gessert, Walter Louis, physics
Graves, Bruce Bannister,
Hee, Christopher Edward, mathematics, chemical engineering
Hicks, Kenneth Ward, physical inorganic chemistry
Howe, George Marvel, climatology, geography
Hurst, Edith Marie Maclennan, neuroanatomy
Johnson, Fred Tulloch, physics
Koo, Delia Wei, mathematics, statistics
Liu, Stephen C Y, microbiology, immunology
Loeber, Adolph Paul, physics
MacMahan, Horace Arthur, Jr, earth science
McDonald, James Robert, geography of France, population geography
Moore, John Ward, inorganic chemistry, physical chemistry
Moses, Gerald Robert, speech pathology
Ogden, Lawrence, geology
Parsons, Karl Alfred, physics
Pearson, Ross Norton, cultural geography, geography of Latin America
Porter, James Colegrove, plasma physics, statistical mechanics
Powell, Ralph Robert, physical chemistry
Ramsay, Ogden Bertrand, organic chemistry, history of chemistry
Raphael, C Nicholas, geography, geomorphology
Rengan, Krishnaswamy, nuclear chemistry, analytical chemistry
Richards, Lawrence Phillips, mammalogy, vertebrate paleontology
Robbins, Omer Ellsworth, Jr, physical chemistry
Roth, Richard Francis, physics
Schullery, Stephen Edmund, physical biochemistry
Scott, Ronald McLean, biochemistry
Silver, Robert, physics
Spike, Clark Ghael, inorganic chemistry
Sullivan, John M, organic chemistry
Thomas, Clinton Edward, physics, astronomy
Tirtha, Ranjit, geography
Turner, Daniel Stoughton, petroleum, geology
Volz, Paul Albert, mycology, botany
Waffie, Elizabeth Lenora, parasitology, invertebrate zoology
West, Bruce David, biochemistry
Williamson, Jerry Robert, organic chemistry, polymer chemistry
Work, Stewart D, organic chemistry
Yamauchi, Masanobu, inorganic chemistry
Yu, Shih-An, botany

ZEELAND
Hoepfinger, Lynn Morris, food chemistry

Niu, Joseph H Y, inorganic chemistry, organic chemistry
Panek, Edward John, organometallic chemistry, organic polymer chemistry
Patton, John Thomas, Jr, organic polymer chemistry
Pizzini, Louis Celeste, organic chemistry
Schaaf, Robert Lester, organic chemistry
Schnuda, Nasr Danial, pathology, electron microscopy
Stryker, Walter Albert, pathology
Vogel, Philip Christian, physical organic chemistry
Vogt, Herwart Curt, polymer chemistry

WEST BLOOMFIELD
Chessin, Hyman, physical chemistry

WYANDOTTE
Bissing, Donald Eugene, organic chemistry
Burkard, Perle Netius, inorganic chemistry, analytical chemistry
Cramer, John Joseph, textile chemistry
Davis Pauls, organic chemistry
Floutz, William Vaughn, analytical chemistry
Gallagher, James A, polymer chemistry
Garvin, Donald Frank, industrial microbiology, clinical microbiology
Hartman, Robert John, industrial organic chemistry
Howe, Norman Elton, Jr, physical organic chemistry
Kan, Peter Tai Yuen, organic chemistry
Levis, William Walter, Jr, organic chemistry
Login, Robert Bernard, organic chemistry, polymer chemistry
Lundsted, Lester Gordon, industrial organic chemistry
Mericola, Francis Carl, inorganic chemistry
Moore, Richard Anthony, organic chemistry
Narayan, Tv Lakshmi, organic chemistry, polymer chemistry
Netherton, Lowell Edwin, inorganic chemistry, physical chemistry

Lewis, Lynn Loraine, analytical chemistry
Majkowski, Richard Francis, plasma physics, spectroscopy
Marick, Louis, physics, engineering
Mayer, William John, physical chemistry
McBreen, James, electrochemistry
McEwen, David John, analytical chemistry
McGrath, James Joseph, inhalation toxicology
Meibuhr, Stuart Gene, electrochemistry
Montgomery, George Paul, Jr, lasers, semiconductors
Muench, Nils Lilienberg, physics
Nash, David Henry George, applied mathematics
Nine, Harmon D, metal physics
Ottaviani, Robert Augustine, organic polymer chemistry
Perkins, Walton A, III, computer science
Reck, Ruth Annette, atmospheric physics, atmospheric chemistry
Reising, Richard F, physical inorganic chemistry
Riley, Bernard Jerome, radiochemistry, physical chemistry
Roessler, David Martyn, physics, spectroscopy
Rohde, Steve Mark, applied mathematics, tribology
Rouze, Stanley Rupie, metal physics, electron microscopy
Rowe, Leonard C, corrosion
Saur, Roger Leo, physical chemistry
Schreiber, Thomas Paul, electron microscopy
Sloane, Thompson Milton, physical chemistry
Smith, George Wolfram, chemical physics
Smith, John Robert, theoretical solid state physics
Snyder, Dexter Dean, physical chemistry, solid state chemistry
Steele, Martin Carl, solid state electronics
Stephenson, David Allen, molecular physics
Swets, Don Eugene, solid state physics, chemical physics
Tallman, Richard Louis, chemistry, corrosion
Taylor, Kathleen C, physical chemistry
Thacker, Raymond, electrochemistry, physical chemistry
Thomas, Donald Henry, applied mathematics, applied statistics
Tibbetts, Gary George, physics
Tracy, Joseph Charles, Jr, solid state physics, surface physics
Tuesday, Charles Sheffield, research administration
Wasielewski, Paul Francis, physics, operations research
Wedel, Richard Glenn, electrochemistry
Weiss, Philip, organic chemistry
Williams, Ronald Lloyde, physical chemistry, environmental chemistry
Wims, Andrew Montgomery, physical chemistry, polymer chemistry

TRENTON
Deck, Charles Francis, inorganic chemistry, radiochemistry
Francis, Cecil Vernon, analytical chemistry
Leaf, Clyde William, organic chemistry

TROY
Adinoff, Bernard, chemistry
Beck, John R, vertebrate ecology, economic biology
Epel, Joseph Norman, physical chemistry
Lipschutz, Louis Sanderson, psychiatry, psychoanalysis
Smart, James Blair, physical inorganic chemistry
Wong, Chun-Ming, organic polymer chemistry

UNIVERSITY CENTER
Catacosinos, Paul Anthony, stratigraphy, geology
Eastland, George Warren, Jr, inorganic chemistry, physical chemistry
Gorden, Berner J, organic chemistry
Kumler, Philip L, organic chemistry, polymer chemistry
Levine, Samuel, physical chemistry
Luce, Everett N, pharmacy
Mock, Richard Armitage, chemistry
Moehs, Peter John, chemistry
Owsley, William Burr, entomology
Pelzer, Charles Francis, human genetics, zoology
Plaush, Albert Charles, physical chemistry
Ts'o, Timothy On To, pharmacology, pathology

WARREN
Abu-Isa, Ismat Ali, physical chemistry, polymer chemistry
Albers, Walter Anthony, Jr, research administration, solid state physics
Baxter, William John, metal physics
Beacom, Seward Elmer, electrochemistry
Bender, Howard Sanford, polymer chemistry, organic chemistry
Bleil, Carl Edward, physics
Brown, William Bernard, polymer chemistry
Buchholz, Jeffrey Carl, surface physics
Burkstrand, James Michael, surface physics
Cadle, Stephen Howard, analytical chemistry, environmental chemistry
Cairns, Elton James, electrochemistry, physical chemistry
Chance, Robert L, analytical chemistry, inorganic chemistry
Charbonneau, Larry Francis, organic polymer chemistry
Cooper, Maurice D, inorganic chemistry, analytical chemistry
Dearlove, Thomas John, polymer chemistry
Donohue, Robert J, spectroscopy, physical optics
Evans, Leonard, operations research, human factors
Fry, David Lloyd George, physics
Furey, Robert Lawrence, organic chemistry
Gara, Aaron Delano, optics
Gardlund, Zachariah Gust, polymer chemistry, organic chemistry
Gay, Jackson Gilbert, theoretical solid state physics
Gifford, Fay Evan, metal physics
Gland, John Louis, surface chemistry
Gordon, William John, applied mathematics
Griffin, John Leander, physical chemistry
Hart, Donald John, physical organic chemistry, polymer chemistry
Hays, Donald R, organic chemistry
Herman, Robert, theoretical physics
Hill, John Campbell, solid state physics
Hoare, James Patrick, electrochemistry
Jamerson, Frank Edward, physics
Johnson, Jack (Lamar), analytical chemistry
Joseph, Bernard William, energy conversion
Klimisch, Richard L, organic chemistry
Koistinen, Donald Peter, metal physics
Krebs, William H, industrial hygiene
Laukonis, Joseph Vainys, physics
Lavery, John Joseph, organic chemistry
Lee, Robert W, experimental physics
Lee, William Richard, analytical chemistry

517

CLOQUET
Alm, Alvin Arthur, forestry
Brown, Bruce Antone, forest ecology
Cornell, Richard Henry, paper chemistry
Gullion, Gordon W., wildlife management
Hallgren, Alvin Roland, forestry, forest management
White, Edwin Henry, forestry, soil science

COLERAINE
Johnson, Tegner Albin, inorganic chemistry, analytical chemistry

COLLEGEVILLE
Ford, Norman Lee, ornithology, ecology
Hughes, Mark, organic chemistry
Kalinowski, Walbert C., mathematics
Michel, Bede Eugene, chemistry
Theisen, Wilfred Robert, history of science, physics
Valley, Leonard Maurice, molecular physics

COON RAPIDS
Esmay, Donald Levern, organic chemistry

COURTLAND
Blaylock, Lynn Gail, animal nutrition
Stockland, Wayne Luvern, nutrition

CROOKSTON
Lofgren, James R., plant breeding, genetics
Marx, George Donald, agriculture, animal physiology

DULUTH
Adams, John Edward, cultural geography, geography of the Caribbean
Ahlgren, Clifford Elmer, forestry
Ahlgren, George E., plant physiology
Ahlgren, Isabel Fulton, botany
Alich, Agnes Amelia, organometallic chemistry
Anderson, Paul M., biochemistry
Beck, Lloyd, pharmacology
Bethuis, Lyda Carol, genetics
Brungs, William Aloysius, Jr., pollution biology
Burgstahler, Sylvan, mathematics
Bydalek, Thomas Joseph, analytical chemistry, inorganic chemistry
Cahoon, Mary Odile, cell physiology
Caple, Ronald, organic chemistry
Carlson, John Bernard, plant anatomy
Carlson, Robert M., organic chemistry
Carter, Robert Eldred, pediatrics, chemistry
Casserberg, Bo R., physics
Chamberlin, Thomas Wilson, Sr., geography of the Soviet Union
Christensen, Glenn Marvin, biology
Collins, Hollie L., zoology
Cowles, Edward J., chemistry
Darby, David G., invertebrate paleontology
Darland, Raymond Winston, ecology
Davidson, Donald Miner, Jr., geology, tectonics
Drewes, Lester Richard, biochemistry
Duval, Anna Marie, biochemistry
Forbes, Donna Jean, neuroscience
Glick, Francis James, organic chemistry
Grant, James Alexander, petrology
Green, John Chandler, geology
Guergiuan, John Leo, pharmacology, biochemistry
Haller, Edwin Wolfgang, physiology, neuroendocrinology
Hamilton, Thomas Reid, microbiology, pathology
Hanson, Howard Grant, physics
Harriss, Donald K., physical chemistry
Hedman, Stephen Clifford, genetics, molecular biology
Heller, Robert Leo, stratigraphy, paleontology
Heller, Lois Jane, medical physiology
Hilding, Anderson C., medical research
Hoag, Leverett Paddock, geography of Scandinavia
Hofslund, Pershing Benard, zoology
Jones, Bernard R., fish biology
Jordan, Thomas Fredrick, physics theoretical
Kallio, Arvo, horticulture, genetics
Kaups, Matti, cultural geography
Knabe, George W., Jr., pathology
Kroening, John Leo, physics

Krogstad, Blanchard Orlando, insect ecology, invertebrate ecology
Leppi, Theodore John, anatomy
Lukasewycz, Omelan Alexander, immunobiology
Magnuson, Vincent Richard, chemistry, crystallography, inorganic chemistry
Marsden, Ralph Walter, geology
Matsch, Charles Leo, geology, glacial geology
McEwen, William Robert, mathematics
McLaughlin, James L., plant pathology
Monson, Paul Herman, plant taxonomy
Moore, Francis Bertram, physical chemistry
Mount, Donald I., fish biology
Nichol, James Charles, physical chemistry
Odlaug, Theron Oswald, parasitology
Poe, Donald Patrick, analytical chemistry
Pozos, Robert Steven, neurophysiology, biophysics
Rapp, George Robert, Jr., geology
Riehl, Mary Agatha, organic chemistry
Routs, Timothy Gerald, social anthropology, ethnology
Salo, Wilmar Lawrence, biochemistry
Severson, Arlen Raynold, biological structure, cell physiology
Spoor, William Arthur, zoology, physiology
Stauffer, Edward Keith, medical physiology
Swain, Wayland Roger, public health, environmental biology
Sydor, Michael, environmental physics
Thompson, Larry Clark, inorganic chemistry
Witzig, Frederick Theodore, geography
Young, Steven Wilford, sedimentary petrology
Ziegler, Richard James, virology

EDEN PRAIRIE
MacDonald, Noel Charles, surface physics
Moore, Glenn D., entomology, plant breeding
Romig, Robert William McClelland, plant pathology

EDEN PROVINCE
Porter, Frederic Edwin, bacteriology

EDINA
Luck, John Virgil, microbiology
Smith, Robert Elijah, computer science, statistics

ELY
Schuldt, Marcus Dale, computer science

FERGUS FALLS
Lindley, Stanley Bryan, psychiatry, psychology

GLYNDON
Johnson, Freeman Keith, genetics, plant breeding

GRAND RAPIDS
Boelter, Don Howard, soil physics, plant physiology
Rust, Joseph William, animal husbandry
Zasada, Zigmond Anthony, forestry

HOPKINS
Engh, Robert Oswald, physical optics
Johnson, Robert Glenn, electrophysics

INTERNATIONAL FALLS
McPherson, William Hakes, paper chemistry

LAMBERTON
Ford, John Harlan, agronomy
Nelson, Wallace Warren, soils

LAUDERDALE
Anderson, Jean Hackett, astronomy

LE SUEUR
Evans, Marshall Pierson, crop breeding
Curme, John Henry, plant genetics
Kaukis, Karl, genetics

Purdy, James Lealon, plant genetics
Reiling, Theodore Paul, horticulture
Stoll, William Francis, food science

MAHTOMEDI
Kuehn, Jerome H., fish biology

MANKATO
Anderson, Lawrence Conrad, geography of the Polar Lands
Balcziak, Louis William, educational statistics
Ballard, Neil Brian, zoology, parasitology
Beech, Ellsworth Benjamin, physical chemistry
Burns, Bert E., geography
Burton, Daniel Frederick, botany
Burton, Verona Devine, botany
Choe, Hyung Tae, plant physiology
Coonce, Harry B., mathematics
Ehrle, Elwood Bernhard, plant taxonomy
Engh, Helmer A., Jr., genetics
Flinner, Jack L., nuclear chemistry
Frydendall, Merrill J., vertebrate ecology, animal behavior
Gebel, Robert Eugene, inorganic chemistry
Glick, Forrest Irving, low temperature physics
Gordon, Donald, plant taxonomy
Graham, Robert Leslie, inorganic chemistry, physical chemistry
Grove, Arthur M., geography
Grundmeier, Ernest Winston, physical chemistry
Hartzler, Harrod Harold, physics
Henderson, Donald Lee, computer sciences
Henney, Robert Charles, analytical chemistry
Herickhoff, Robert John, physics
Holden, John B., Jr., analytical chemistry, environmental sciences
Inman, Fred Winston, nuclear physics
Jewsbury, Wilbur, chemistry
Johnson, Theodore Reynold, microbiology, immunology
Krabbenhoft, Kenneth Louis, microbiology
Lund, Arnold Jerome, bacteriology
McCarty, John Edward, organic chemistry
Miller, Carl Elmer, molecular physics
Mordue, Dale Lewis, physics, mathematics
Morley, Gayle L., theoretical solid state physics
Perisho, Clarence R., biochemistry
Ralston, Douglas Edmund, biochemistry
Schuster, Sanford Lee, solid state physics
Sehe, Charles Theodore, developmental biology
Standeford, Leo Vern, astronomy
Turner, Veras D., mathematics
Weber, Berton Charles, organic chemistry
Wheaton, Burdette Carl, algebra
Zell, LaRoy W., cytology, plant taxonomy

MAPLEWOOD
Kropp, James Edward, organic chemistry

MARSHALL
Barker, Laren Dee Stacy, vertebrate physiology
Bowman, Leo Henry, analytical chemistry
Carberry, Edward Andrew, inorganic chemistry, organometallic chemistry
Frazier, Ralph Paul, zoology
Halgren, Lee A., entomology, invertebrate ecology
Hsi, Eugene Yu-Tseng, systematic botany
Sparling, Dale R., geology
Spencer, Richard L., biochemistry
Surdy, Ted E., bacteriology, biochemistry
Thomas, John Paul, physical inorganic chemistry

MENDOTA HEIGHTS
Oberle, Thomas M., inorganic chemistry

MINNEAPOLIS
Abdel-Monem, Mahmoud Mohamed, medicinal chemistry
Abrahamson, Dean Edwin, forensic science, environmental sciences
Abul-Hajj, Jusuf J., biochemistry, organic chemistry

Abuzzahab, Faruk S., Sr., psychiatry, pharmacology
Ackerman, Eugene, computer sciences
Adams, John Stephen, urban geography, urban economics
Adams, Russell Blair, geography, statistics
Adiarte, Arthur Lardizabal, biophysical chemistry
Aeppli, Alfred, mathematics
Agre, Courtland LeVerne, chemistry
Ahmed, Khalil, biochemistry
Aines, Philip Deane, nutrition
Alexander, Carl Stuart, medicine
Alexander, Emmit Calvin, Jr., geochronology
Alter, Milton, neurology, epidemiology
Alton, Earl Robert, inorganic chemistry
Amplatz, Kurt, radiology
Anders, Marion Walter, toxicology
Anderson, Dwight Lyman, microbiology
Anderson, Joseph Tomlinson, nutrition
Anderson, Ray Carl, medicine
Anderson, Ray Harold, food science
Anderson, Thomas Page, physical medicine & rehabilitation
Andersen, Brian Herbert, analytical chemistry
Andrews, Fred Albert, biochemistry
Aronowitz, Frederick, quantum physics
Aronson, Donald Gary, mathematics
Athelstan, Gary Thomas, physical medicine & rehabilitation, psychology
Awad, Essam A., physical medicine & rehabilitation
Azar, Miguel M., immunology, pathology
Bacaner, Marvin Bernard, physiology
Baker, Abe Bert, neurology
Baker, Griffin Jonathan, entomology
Baker, Michael Harry, applied chemistry
Baldwin, Arthur Richard, chemistry
Banerjee, Subir Kumar, geophysics, paleomagnetism
Barber, Donald E., health physics
Barker, Norval Glen, nutritional biochemistry
Barks, Paul Allan, organic chemistry
Barnett, Ronald E., physical organic chemistry
Barnwell, Franklin Hershel, comparative physiology
Bartsch, Glenn Emil, biostatistics
Bauer, Gustav Eric, anatomy
Bauer, Henry, bacteriology
Bauman, Howard Eugene, food science
Bayman, Benjamin, nuclear physics
Bearinger, Van W., food science
Beggs, William H., medical microbiology
Behnke, James Ralph, food science
Behrens, Richard, weed science
Belmont, Arthur David, meteorology
Benson, Ellis Starbrand, pathology
Berg, Marie Hirsch, biochemistry
Berman, Reuben, cardiology
Bernstein, William Carl, cancer
Berry, James Frederick, biochemistry
Beyer, William W., food science
Birney, Elmer Clea, mammalogy
Bixby, John N., food chemistry, biochemistry
Blackard, Clyde Erhardt, surgery, urology
Blackburn, Henry Webster, Jr., epidemiology
Blackshear, Gertrude Liebl, physiology
Blair, John Morris, experimental nuclear physics
Blake, Rolland Laws, geology
Blazevic, Donna Jean, medical microbiology
Bloedel, James R., neurophysiology
Blumenfeld, Martin, cell biology, molecular biology
Bodley, James William, biochemistry
Boen, James Robert, biostatistics
Bond, Richard Guy, environmental health
Borch, Richard Frederic, organic chemistry
Borchert, John Robert, geography
Braden, Charles McMurray, mathematics
Bradford, David S., orthopedic surgery
Brand, Karl Gerhard, microbiology
Bransted, William, dentistry
Brasted, Robert Crocker, inorganic chemistry
Brauer, George Ulrich, mathematics
Bright, Robert C., geology
Britton, Doyle, structural chemistry
Bronson, David Lee, virology
Brook, Alan J., phycology
Brown, William Fuller, Jr., physics
Bruels, Mark Charles, radiological physics, medical physics

Brunning, Richard Dale, hematology, pathology
Bryant, Robert George, biophysical chemistry, inorganic chemistry
Buchwald, Henry, surgery
Buckley, Joseph J, anesthesiology
Buehler, Robert Joseph, mathematical statistics
Burchell, Howard Bertram, medicine
Buttrill, Sidney Eugene, Jr, analytical chemistry
Cahill, Laurence James, Jr, space physics
Caldwell, Alfred Craig, soil chemistry
Cameron, Robert Horton, mathematics
Capps, Mary Jayne, physiological psychology, sensory physiology
Carpenter, Anna-Mary, anatomy
Carr, Charles William, biochemistry
Carr, Robert Wilson, Jr, chemical kinetics, chemical engineering
Carter, John Vernon, biophysical chemistry
Cavert, Henry Mead, physiology, medical school administration
Cervenka, Jaroslav, medical genetics, cancer
Chanin, Lorne Maxwell, plasma physics
Chern, Ming-Fen Myra, biostatistics, population genetics
Chi, Che, neuroscience
Chou, Shelley Nien-Chun, neurosurgery
Christianson, George, food science
Ciriacy, Edward W, family medicine
Cleary, Paul Patrick, molecular biology, medical microbiology
Clement, Jacob James, radiobiology
Click, Robert Edward, immunobiology
Coe, John Ira, pathology
Cohen, Harold P, biological chemistry
Cohn, Jay Norman, cardiovascular diseases
Collins, Amy L Tsui, radiotherapy, immunology
Collins, Robert Joseph, physics
Conroy, Lawrence Edward, inorganic chemistry
Corbin, Kendall Wallace, evolutionary biology, population ecology
Coulter, Herbert David, Jr, anatomy
Courant, Hans Wolfgang Julius, high energy physics
Coury, Arthur Joseph, organic chemistry, polymer chemistry
Cowan, Donald William, medicine
Crawford, Bryce (Low), Jr, physical chemistry
Cronk, Ted Clifford, food science
Crowley, Leonard Vincent, medicine, pathology
Cunningham, David Kenneth, agricultural biochemistry
Dahler, John S, physical chemistry
Dalmasso, Agustin Pascual, medicine, immunology
Daravingas, George Vasilios, food science, biochemistry
Das Gupta, Somesh, mathematical statistics
Davidson, Kris, astrophysics
Davis, Howard Ted, physical chemistry, chemical physics
Dearden, Douglas Morey, zoology, genetics
Decelles, George Arthur, Jr, food science
Dehnhard, Dietrich, nuclear physics
Delaney, John P, surgery, physiology
De Master, Eugene Glenn, biochemistry
Dempsey, John Nicholas, physical chemistry
Dempsey, Mary Elizabeth, biochemistry
Derr, Robert Frederick, biochemistry
Di Gangi, Frank Edward, chemistry
Dixit, Padmaker Kashinath, anatomy, nutritional biochemistry
Dodson, Raymond Monroe, organic chemistry
Doe, Richard P, endocrinology, internal medicine
Donker, John D, animal science
Doughman, Donald James, ophthalmological surgery
Drew, Bruce Arthur, applied mathematics
Dunshee, Bryant R, food science
Durst, Jack Rowland, biochemistry, food science
Duvall, Arndt John, III, otology, otolaryngology
Dworkin, Martin, microbiology
Eagon, John Alonzo, mathematics
Eaton, John Wallace, hematology, physiology
Eaton, Morris Leroy, mathematical statistics
Ebert, Richard Vincent, clinical medicine
Edstrom, Ronald Dwight, biochemistry
Egberg, David Curtis, organic chemistry
Eifrig, David Eric, ophthalmology
Eisenberg, M Michael, surgery, gastroenterology
Elliott, Arthur York, microbiology, virology
Ellwood, Paul M, Jr, rehabilitation medicine
Emery, Donald F, food chemistry
Erickson, Kenneth Neil, elementary particle physics, space physics
Erlandsen, Stanley L, histology, microscopic anatomy
Ettinger, Milton G, neurology
Fabes, Eugene Barry, mathematical analysis
Falk, Abraham, clinical medicine
Faras, Anthony James, virology
Farnham, Rouse Smith, soil science
Fenton, Stuart William, organic chemistry
Fester, Keith Edward, electrochemistry
Fitzgerald, Thomas James, microbiology
Foreman, Harry, radiological health
Forester, Ralph H, polymer chemistry
Foster, Donald Myers, biochemistry, biochemical pharmacology
Fox, Irwin J, cardiovascular physiology
Fraley, Elwin E, urology, pathology
Frank, David Lewis, topology
Frantz, Ivan DeRay, Jr, biochemistry
Freier, Esther Fay, physiological chemistry
Freier, George David, atmospheric physics
Freier, Phyllis S, physics
French, Lyle Albert, neurosurgery
Fritsch, Carl Walter, biochemistry
Fuhrken, Gebhard, mathematics
Fuller, Benjamin Franklin, Jr, medicine
Gaal, Ilse Lisl Novak, symbolic logic, algebra
Gaal, Steven Alexander, applied mathematics
Gage, Kenneth Seaver, geophysics, meteorology
Gallagher, John Sill, astrophysics
Gasiorowicz, Stephen G, particle physics
Gassman, Paul G, organic chemistry
Gatewood, Lael Cranmer, computer science, biometry
Gault, Neal L, Jr, medicine
Geisser, Seymour, mathematical statistics
Gentry, William Ronald, physical chemistry, chemical physics
Gerlach, Luther Paul, cultural anthropology, social anthropology
Gershenson, Hillel Halkin, mathematics
Gersmehl, Philip J, geography
Giese, Clayton, chemical physics
Giese, David Lyle, applied statistics, academic administration
Gilbertsen, Victor Adolph, surgery, cancer
Gilbertson, Donald Edmund, parasitology, malacology
Gilboe, Daniel Pierre, biochemistry
Gil de Lamadrid, Jesus, mathematics
Gisvold, Ole, pharmaceutical chemistry
Gobel, Fredarick L, cardiovascular diseases
Goetz, Frederick Charles, internal medicine
Goldberg, Nelson D, pharmacology, biochemistry
Goldman, Allen Marshall, low temperature physics
Goldman, Anne Ipsen, biometrics
Goldstein, Stuart Frederick, cell physiology
Goltz, Robert W, dermatology, histopathology
Goodkind, Richard Jerry, prosthodontics
Goon, David James Wong, physical organic chemistry
Gorlin, Robert James, oral pathology
Gougoutas, Jack Zanos, x-ray crystallography, solid state chemistry
Grage, Theodor B, surgery
Grande, Francisco, physiology
Graubard, Mark Aaron, physiology
Gray, Ernest David, biochemistry
Gray, Gary Ronald, biochemistry
Green, Leon William, mathematics
Greenberg, Leonard Jason, biochemistry, immunology
Greene, Velvl William, microbiology
Grim, Eugene, physiology
Gulliver, Robert David, II, geometry, analytical mathematics
Gutmann, Helmut Rudolph, biochemistry
Gyberg, Arlin Enoch, analytical chemistry, physical chemistry
Haaland, John Edward, science administration, biophysics
Haddad, Louis Charles, biophysical chemistry
Hall, Henry Thompson, geochemistry, mineralogy
Hall, Wendell Howard, internal medicine
Halley, James Woods, (Jr), theoretical physics
Halpern, Daniel, medicine
Hamermesh, Morton, physics
Hanna, Patrick E, medicinal chemistry
Hanson, Russell Floyd, internal medicine, biochemistry
Hargrove, James Lee, reproductive endocrinology
Harper, Laurence Raymond, Jr, mathematics
Harris, John Edward, ophthalmology, biochemistry
Harris, Morton E, algebra
Harrison, Stuart Amos, chemistry
Hart, John Fraser, cultural geography, geography of North America
Harvey, Charles Arthur, mathematics
Hastings, Donald Wilson, medicine
Hatcher, John Burton, chemistry, physics
Hausman, William, psychiatry, academic administration
Hebbel, Robert, pathology
Heggestad, Carl B, anatomy
Hegre, Orion Donald, anatomy, tissue culture
Heinisch, Roger Paul, optics, physics
Henkel, James Gregory, medicinal chemistry, organic chemistry
Henrikson, Ernest Hilmer, speech pathology
Herforth, Robert S, genetics
Herman, William S, zoology
Heston, Leonard L, psychiatry, genetics
Hexter, Robert Maurice, physical chemistry
Hildebrand, Frank Childs, biochemistry, food science
Hintz, Norton Mark, nuclear physics
Hitchcock, Claude Raymond, surgery
Hobbie, Russell Klyver, physics
Holahan, John L, Sr, food science
Holtzman, Jordan L, pharmacology
Holum, John Robert, organic chemistry
Hooke, Robert Lebaron, geology
Hopkins, Carl Douglas, animal behavior
Hovde, Ruth Frances, medical technology
Howard, James Bryant, biochemistry
Howard, Robert Bruce, medicine
Huang, Jack Shih Ta, applied physics
Humphrey, Edward William, surgery, physiology
Humphreys, Roberta Marie, astronomy
Isaacson, Robert John, orthodontics, anatomy
Jacob, Harry S, internal medicine, hematology
Jacobs, David R, Jr, biostatistics
Jenkins, Howard Bryner, mathematics
Jepson, William W, medicine, psychiatry
Jezeski, James John, microbiology, food science
Jindrich, Vladimir, economic geology, exploration geology
Jodeit, Max A, Jr, mathematics
Johnson, Donovan Albert, mathematics
Johnson, Eugene A, biostatistics
Johnson, Hildegard Binder, historical geography, political geography
Johnson, John Alexander, physiology
Johnson, Joseph Richard, medicine
Johnson, Ross Glenn, cell biology
Johnson, Russell Clarence, microbiology
Johnson, Walter Heinrick, Jr, physics, mass spectrometry
Jones, Duane Arnold, carbohydrate chemistry
Jones, Roger Stanley, high energy physics
Jorgenson, Gordon Victor, surface physics
Kabat, Hugh F, pharmacy administration
Kahn, Alan Richard, bioengineering, physiology
Kahn, Donald W, mathematics
Kallestad, Steven Bix, microbiology, immunochemistry
Kallianpur, Gopinath, mathematics
Kaplan, Manuel E, internal medicine, hematology
Karl, Curtis Lee, organic chemistry, polymer chemistry
Katz, Morris Howard, food science, biochemistry
Kaufmann, Henry Hans, food technology
Keenan, Kathleen Margaret, biostatistics
Kellogg, Paul Jesse, plasma physics
Kelly, William Daniel, surgery
Kennedy, Byrl James, internal medicine
Kennedy, William Robert, neurology
Kerr, Norman Story, developmental biology
Kettelkamp, Ben H, physiology
Keynes, Harvey Bayard, mathematics
Keys, Ancel (Benjamin), physiology, nutrition
Khan, Faiz Mohammad, biophysics, radiological physics
Kiang, David Teh-Ming, internal medicine, oncology
Kinderlehrer, David (Samuel), mathematics
King, Charles McDonald, Jr, pharmacy
Kinsey, James Humphreys, applied physics
Kiste, Robert Carl, anthropology
Kjelsberg, Marcus Olaf, biostatistics
Knox, Charles Kenneth, neurophysiology, engineering
Koelsch, Charles Frederick, chemistry
Koerner, James Frederick, biochemistry
Koffler, Henry, microbiol
Kolthoff, Izaak Maurits, analytical chemistry
Kottke, Frederic James, physical medicine
Krbechek, Leroy O, organic chemistry
Kreevoy, Maurice M, organic chemistry, chemical kinetics
Krivit, William, pediatrics, hematology
Kronenberg, Richard Samuel, medicine, physiology
Kubicek, William George, physiology
Kuder, Robert Clarence, plastics chemistry
Kurzepa, Henryka Janina, microbiology, cancer
La Bonte, Anton Edward, computer science
Lamprech, Earl Duwain, food science
Lawton, Wallace Clayton, dairy bacteriology
Layton, Edwin Thomas, Jr, history of technology
Lee, Chiung Puh, physiology
Lee, Jui Shuan, physiology
Lee, Sylvan Burton, microbiology, biochemistry
Leete, Edward, organic chemistry
Leininger, Harold Vernon, microbiology
Leon, Arthur Sol, medical research, clinical pharmacology
Leonard, Arnold S, surgery
Levitt, David George, physiology
Levitt, Michael D, gastroenterology
Levitt, Seymour H, radiotherapy
Liao, Ji-Chia, anesthesiology, pharmacology
Lieberman, Gerald Alan, ecology
Liedtke, Claus-Eberhard, health sciences, computer sciences
Lifson, Nathan, physiology
Lillehei, Richard Carlton, surgery
Lin, Sping, neurochemistry, physiology
Linck, Albert John, plant physiology
Lindall, Arnold Walfred, endocrinology
Lindgren, Bernard William, mathematics
Link, Bernard Alvin, meat science
Lipsky, Sanford, physical chemistry
Littman, Walter, mathematical analysis
Lober, Paul Hallam, pathology
Loewenson, Ruth Brandenburger, biometrics
Loken, Merle Kenneth, nuclear medicine, biophysics
Lorber, Victor, physiology
Loud, Warren Simms, mathematics
Lovald, Roger Allen, organic polymer chemistry
Lucas, Russell Vail, Jr, pediatric cardiology
Lumry, Rufus Worth, physical chemistry
Lund, Curtis Joseph, obstetrics & gynecology
Luyten, Willem Jacob, astronomy
Lyon, Richard Hale, microbiology, biochemistry
Lyons, Walter Andrew, meteorology, air pollution
Machne, Xenia, physiology
MacKay, Kenneth Donald, organic chemistry
Mackenzie, Charles Westlake, III, biochemistry
MacWherter, John Baird, mathematics
Malmquist, Carl Phillip, psychiatry
Mantis, Homer Theodore, physics, meteorology
Markland, Alan Colin, surgery, urology
Markus, Lawrence, mathematics
Marsh, Frederick Leon, analytical chemistry, physical chemistry
Martens, Leslie Vernon, dentistry
Mather, Eugene Cotton, cultural geography
Mauersberger, Konrad, aeronomy
Mayer, William Joseph, information science
McCarthy, Charles Alan, mathematical analysis
McCollister, Robert John, internal medicine
McCrossen, Garner, mathematics

MINNESOTA

McDermott, Richard P, speech pathology, audiology
McGehee, Richard Paul, mathematics
McHugh, Richard B, biometrics, biostatistics
McKenna, Robert Wilson, pathology
McKhann, Charles Fremont, surgery, microbiology
McKinnell, Robert Gilmore, developmental biology
McKinney, Frank, zoology
McQuarrie, Donald G, surgery, computer sciences
Mead, Chester Alden, theoretical chemistry
Meehan, Edward Joseph, chemistry
Meier, Manfred John, neuropsychology
Merrell, David John, genetics
Meskin, Lawrence Henry, dental epidemiology
Messer, Harold Henry, dental research
Meyer, Maurice Wesley, physiology, dentistry
Meyers, Norman George, mathematics
Michael, Alfred Frederick, Jr, pediatrics, nephrology
Miller, Byron Sloane, cereal chemistry
Miller, Frank Charles, anthropology
Miller, Jack W, pharmacology
Miller, Kenneth Wayne, pharmacology, biopharmaceutics
Miller, Willard, Jr, applied mathematics, mathematical physics
Miller, Willie, plant science, soil science
Miller, Wilmer Glenn, polymer chemistry
Miracle, Chester Lee, mathematics
Mirkin, Bernard Leo, pediatrics, clinical pharmacology
Mizuno, Nobuko Shimotori, biochemistry
Moe, John, orthopedic surgery
Moller, Karlind Theodore, speech pathology
Mooney, Harold Morton, geophysics
Moore, Jerry Lamar, nutrition
Moore, Joseph B, economic entomology
Moore, Richard, medical physics, biomedical engineering
Moscowitz, Albert, physical chemistry
Mossotti, Victor Giovoni, physical chemistry, analytical chemistry
Mueller, Rolf Karl, applied physics, underwater acoustics
Mullen, Joseph David, biochemistry, food science
Munro, William Delmar, numercial analysis
Murray, Murray J, medicine
Murrill, Rupert Ivan, anthropology
Murthy, Varanasi Rama, geochemistry
Mykleby, Ray W, food technology
Nagasawa, Herbert Tsukasa, medicinal chemistry
Najarian, John Sarkis, surgery
Nazy, John Robert, industrial organic chemistry
Nelson, Kenneth Gordon, pharmaceutics
Ney, Edward Purdy, physics
Nicoloff, Demetre M, cardiovascular surgery, bioengineering
Nier, Alfred Otto Carl, physics
Noland, Wayland Evan, organic chemistry
Nomura, Kaworu Carl, solid state physics
Notation, Albert David, endocrinology, biochemistry
Nussbaum, Allen, solid state physics
Nuttall, Frank Q, internal medicine
Olsen, Edmund Severn, Jr, dentistry, anatomy
Olson, Magnus, zoology
Olson, Melvin Martin, organic chemistry
Opie, Joseph Wendell, organic chemistry, food technology
Orey, Steven, mathematics
Overdahl, Curtis J, soils
Overend, John, physical chemistry
Owens, Boone Bailey, electrochemistry
Page, Arthur R, immunology, pediatrics
Parmelee, David Freeland, ornithology
Paulus, Harold John, environmental health, air pollution
Pedoe, Daniel, mathematics
Peerman, Dwight Ellsworth, polymer chemistry
Pepin, Robert Osborne, physics, space sciences
Perrozzi, Joseph Richard, food technology, bacteriology
Petersen, Robert J, organic chemistry
Peterson, Lowell E, physical chemistry
Pfannkuch, Hans Olaf, hydrogeology
Pflug, Irving John, food science, microbiology
Pignolet, Louis H, inorganic chemistry
Plagemann, Peter Guenter Wilhelm, virology, biochemistry
Pohl, William Francis, mathematics
Poppe, Carl Hugo, physics
Poppele, Richard E, neurophysiology
Porter, Philip Wayland, geography
Portoghese, Philip S, medicinal chemistry
Pour-El, Marian Boykan, mathematics
Prager, Stephen, physical chemistry
Pratt, Douglas Charles, plant physiology
Prem, Konald Arthur, obstetrics & gynecology
Prikry, Karel Libor, mathematics
Prince, James T, virology, medical microbiology
Prineas, Ronald James, cardiovascular diseases
Pruitt, William Edwin, probability
Puleston, Dennis Edward, anthropology
Purple, Richard L, physiology, neurophysiology
Quebbemann, Aloysius John, pharmacology
Quie, Paul Gerhardt, pediatrics
Rassweiler, Merrill (Paul), physics
Rathbun, William B, biochemistry
Reed, Elizabeth Wagner, human genetics
Reed, Sheldon Clark, genetics
Regal, Philip Joe, behavioral biology, evolution
Reich, Edgar, mathematics
Reilly, Bernard Edward, microbial genetics, virology
Rejto, Peter A, mathematics
Remley, Frank Morris, crop breeding, plant breeding
Resch, Joseph Anthony, neurology
Reynolds, John Weston, pediatrics
Reynolds, Warren Lind, inorganic chemistry
Richards, Jonathan Ian, mathematics
Richter, Wayne H, mathematics
Rippie, Edward Grant, pharmaceutics
Roach, J Robert, organic chemistry, food chemistry
Roberts, Alan H, psychophysiology
Roberts, Joel Laurence, mathematics
Robinson, Edwin James, Jr, parasitology
Rodgers, Nelson Earl, bacteriology, biochemistry
Rogers, Palmer, Jr, biochemistry, microbiology
Roll, Peter Guy, physics
Roon, Robert Jack, biochemistry, microbiology
Rosen, Judah Ben, applied mathematics
Rosner, Jonathan Lincoln, physics
Rottmann, Warren Leonard, cell biology
Rowe, William Leal, cultural anthropology
Ruggero, Mario Alfredo, neurophysiology
Ruschmeyer, Orlando R, public health biology, limnology
Ryder, John C, Jr, plant physiology, economic entomology
Sabath, Leon David, medicine, microbiology
Sako, Yoshio, surgery
Sandell, Ernest Burger, chemistry
Sarles, Harvey B, ethology, anthropology
Sawchuk, Ronald John, biopharmaceutics, medical research
Schacht, Lee Eastman, human genetics
Schachtele, Charles Francis, dental research, microbiology
Schaffer, Erwin Michael, periodontology
Schiele, Burtrum Clarence, psychiatry
Schmid, William Dale, animal ecology, comparative animal physiology
Schmidtke, Jon Robert, immunology
Schmitt, Otto Herbert, biophysics, biomedical engineering
Schugel, LaVerne, veterinary medicine, animal nutrition
Schultz, Alvin Leroy, internal medicine, endocrinology
Schuman, Leonard Michael, epidemiology
Schwartz, Samuel, medical research
Schwartzberg, Joseph Emanuel, geography
Sell, George Roger, mathematics
Serrin, James B, mathematics
Shapiro, Alan Elihu, history of science
Shapiro, Burton Leonard, oral biology, genetics
Shapiro, Joseph, limnology
Shapiro, Leonard David, mathematics
Sheppard, John Richard, cell biology, biochemistry
Sherck, Charles Keith, food science
Sheridan, Judson Dean, cell physiology, cell biology
Shideman, Frederick Earl, pharmacology
Sibuya, Yasutaka, mathematics
Siem, Robert Arthur, microbiology
Simmons, Richard Lawrence, surgery, immunology
Simon, Geza, cardiovascular diseases, internal medicine
Singer, Leon, biochemistry
Sinha, Akhouri Achyutanand, zoology, anatomy
Siniff, Donald Blair, ecology, biometry
Sladek, Norman Elmer, pharmacology
Sloan, Robert Evan, stratigraphy, paleontology
Smith, Donald Edgar, biochemistry
Smith, Franklin Chapin, Sr, mathematics
Smith, Quenton Terrill, biochemistry, oral biology
Smithberg, Morris, embryology, neuroanatomy
Snow, John Elbridge, organic chemistr
Soechting, John F, neurophysiology
Soine, Taito Olaf, medicinal chemistry
Somers, Perrie Daniel, biochemistry
Song, Chang Won, radiobiology, immunology
Sorenson, Robert Lowell, anatomy
Speidel, Edna W, biochemistry
Speidel, Thomas Michael, orthodontics
Spencer, Robert Francis, anthropology
Sperber, William Henry, microbiology, biochemistry
Spink, Wesley William, internal medicine
Spliethoff, William Ludwig, organic chemistry
Staba, Emil John, pharmacognosy
Stein, Marvin L, mathematics, computer science
Steinhauser, Fredric R, geography
Storvick, David A, mathematics
Stuewer, Roger Harry, history of physics
Sudderth, William David, mathematics, mathematical statistics
Sudo, Sara Zeece, microbiology, biochemistry
Sulerud, Ralph L, genetics, zoology
Sullivan, Betty, biochemistry
Sullivan, William Albert, Jr, surgery
Sundberg, Ruth Dorothy, anatomy
Sung, Joo Ho, pathology, neuropathology
Swaiman, Kenneth F, neurochemistry, pediatric neurology
Swain, Frederick Morrill, Jr, geology
Swofford, Harold S, Jr, chemistry
Taft, David Dakin, polymer chemistry, organic chemistry
Takemori, Akira Eddie, pharmacology
Tang, Yau-Chien, physics
Taylor, Henry Longstreet, physiology
Templeman, Gareth J, physical chemistry, analytical chemistry
Terzuolo, Carlo A, physiology
Theologides, Athanasios, internal medicine, oncology
Thompson, Fay Morgen, occupational health, environmental chemistry
Thorpe, Neal Owen, biochemistry
Till, Michael John, pedodontics
Timm, Gerald Wayne, biomedical engineering, electrical engineering
Tobian, Louis, internal medicine
Tordoff, Harrison Bruce, zoology
Torres, Fernando, neurology, neurophysiology
Touba, Ali R, food technology
Treloar, Alan Edward, medical anthropology, biometrics
Truhlar, Donald Gene, physical chemistry, molecular physics
Tsien, Hsienchyang, microbiology
Tsuchiya, Henry Mitsumasa, bacteriology
Tuan, Yi-Fu, cultural geography
Tuna, Naip, internal medicine, cardiovascular diseases
Turnipseed, Marvin Roy, reproductive physiology, biochemistry
Turrittin, Hugh Lonsdale, mathematics
Ulstrom, Robert, medicine
Ungar, Frank, biochemistry, endocrinology
Van Bergen, Frederick Hall, anesthesiology
van der Ziel, Aldert, physics
Van Hulle, Glenn Joseph, food science
Van Pilsum, John Franklin, biochemistry
Varco, Richard Lynn, surgery
Vennes, Jack A, gastroenterology, internal medicine
Verbrugge, Frank, academic administration
Verby, John E, family medicine, community health
Vernier, Robert L, pediatrics
Vesley, Donald, environmental health
Villars, Charles Earl, organic polymer chemistry, industrial organic chemistry
Vince, Robert, medicinal chemistry
Visscher, Maurice B, physiology
Waddington, Cecil Jacob, physics, astrophysics
Wagenaar, Raphael Omer, dairy bacteriology
Wagenknecht, Austin Clayton, biochemistry
Wahba, Isaac Jack, food chemistry
Waite, Daniel Elmer, oral surgery
Walker, Raymond John, space physics
Wallace, Franklin Gerhard, parasitology
Wang, Yang, cardiology
Wangensteen, Ove Douglas, physiology
Wangensteen, Owen Harding, surgery
Wannamaker, Lewis William, pediatrics, microbiology
Ward, Wallace Dixon, psychoacoustics
Warner, Dwain Willard, zoology
Warner, John Ward, astronomy
Warner, Raymond M, Jr, physics
Warner, William Hamer, applied mathematics
Warwick, Warren J, pediatrics
Wathen, Ronald Larry, medical research
Watson, Cecil James, medicine
Watson, Dennis Wallace, microbiology
Wattenberg, Lee Wolff, pathology
Weaver, Lawrence Clayton, pharmacology
Webb, John W, geography
Webster, David Dyer, medicine, neurology
Weckwerth, Vernon Ervin, biostatistics, hospital administration
Wehner, Gottfried Karl, physics
Weiblen, Paul Willard, geology, petrology
Weinberger, Hans Felix, applied mathematics
Werntz, James Herbert, Jr, physics
Wertheimer, Albert I, public health administration
Wertz, John Edward, chemistry
Weyhmann, Walter Victor, solid state physics, low temperature physics
White, Wallace Fletcher, pharmacology
Wilder, Russell Morse, medicine, psychiatry
Williams, Lawrence Ernest, nuclear medicine, biophysics
Willits, Richard Ellis, dairy microbiology, biochemistry
Wilson, Archie Spencer, chemistry
Wilson, Leonard Gilchrist, history of medicine
Wilson, Michael John, endocrinology, cell biology
Winchell, C Paul, medicine
Winckler, John Randolph, physics
Witkop, Carl Jacob, Jr, human genetics, oral pathology
Wittcoff, Harold, organic chemistry
Wright, Francis Stuart, pediatric neurology
Wright, Herbert Edgar, Jr, paleoecology
Yablonski, Michael Eugene, medicine, physiology
Yasmineh, Walid Gabriel, biochemistry
Yellin, Absalom Moses, psychophysiology, child psychiatry
Yunis, Edmond, medicine, pathology
Yunis, Jorge J, genetics, pathology
Zieve, Leslie, medicine
Zimmerman, Ben George, pharmacology
Zimmermann, William, Jr, physics
Zinneman, Horace Helmut, internal medicine
Zoltai, Tibor, mineralogy, crystallography

MINNETONKA

Jensen, James Le Roy, chemistry
Robertsen, John Alan, medical microbiology

MONTICELLO

Hokanson, Kenneth Eric Fabian, fish biology

MOORHEAD

Bailey, Carl Leonard, physics
Bartel, Monroe H, biology
Behforooz, Ali, computer science, experimental statistics
Brummond, Dewey Otto, biochemistry
Condell, Yvonne C, human genetics
Dahlberg, Duane Arlen, physics, environmental sciences
Dierenfeldt, Karl Emil, physical chemistry, nuclear chemistry
Dinga, Gustav Paul, inorganic chemistry
Hamburg, James F, historical geography, geography of the Great Plains
Harrison, Wilks Douglas, geomorphology, meteorology
Heuer, Charles Vernon, mathematics
Heuer, Gerald Arthur, mathematics
Holoien, Martin Olaf, computer science
Homann, H Robert, biochemistry
Johnson, Ivan M, zoology, physiology
Johnson, Oscar Walter, zoology, histology
Klug, Harold Philip, physical chemistry
Kowanko, Nicholas, organic chemistry
Lin, Benjamin Ming-Ren, computer sciences

Loeffler, Robert J., botany
MacKellar, William John, analytical
  chemistry
Mathiason, Dennis R., inorganic
  chemistry
McCashland, Benjamin William, cell
  physiology
Meeks, Benjamin Spencer, Jr, organic
  chemistry
Ostercamp, Daryl Lee, organic chemistry
Parsons, Jesse Leroy, bacteriology
Paulson, Carlton, parasitology
Pemble, Richard Hope, ecology,
  phytogeography
Sautter, Chester A, experimental atomic
  physics, environmental physics
Shaw, James Edward, organic chemistry
Shimabukuro, Mary Abrahamsen, plant
  physiology
Sipson, Roger Fredrick, physics
Skjegstad, Kenneth, botany
Strong, Judith Ann, physical chemistry
Thomsen, Warren Jessen, mathematics
Tolbert, Robert John, plant anatomy
Treumann, William Borgen, physical
  chemistry
VanAmburg, Gerald Leroy, plant
  ecology, botany
Weibust, Robert Smith, genetics, zoology
Werth, Richard George, organic
  chemistry
Wesley, Walter Glen, theoretical physics
Yeh, Hsin-Yang, mathematical physics,
  particle physics

MORRIS

Abbott, Robinson Shewell, ecology
Abbott, Rose Marie Savelkoul, plant
  morphology
Agarwal, Som Prakash, cosmic ray
  physics, nuclear physics
Hirsh, Merle Norman, atomic physics,
  molecular physics
Holt, Robert F, soil conservation
Hoppe, David Matthew, vertebrate
  biology
Latterell, Joseph J, analytical chemistry
Olson, Tamlin Curtis, soil physics
Ordway, Ellen, entomology
Radke, Jerry Kieth, micrometeorology,
  soil physics
Roquitte, Bimal C, physical chemistry
Roshal, Jay Yehudie, botany
Smith, Ralph Emerson, agricultural
  economics
Straw, Thomas Eugene, aquatic biology
Voorhees, Ward Byron, soil conservation
Warnes, Dennis Daniel, agronomy, weed
  science

NAVARRE

Crawford, Ronald Lyle, microbiology,
  ecology
Wood, John Martin, biochemistry,
  organic chemistry

NEW BRIGHTON

Mattison, Phillip LeRoy, organic
  chemistry

NEW HOPE

Hurtey, William Charles, food science,
  biochemistry
Pyne, Alvan Wesley, food science,
  biochemistry

NORTHFIELD

Boardman, Shelby Jett, geology
Bowers, Delores Maureen, analytical
  chemistry, water chemistry
Buchwald, Caryl Edward, environmental
  geology
Burton, Alice Jean, biochemistry,
  microbiology
Carlin, Charles Herrick, organic
  chemistry
Casper, Barry Michael, theoretical
  physics
Cederberg, James W, molecular physics,
  physics
Child, William Clark, Jr, physical
  chemistry
Christensen, Fritjof Ernest, physics
Dyer-Bennet, John, algebra
Fick, Herbert John, plastics chemistry,
  polymer chemistry
Finholt, Albert Edward, chemistry
Finholt, James E, physical chemistry,
  inorganic chemistry
Fisher, James F, anthropology, social
  anthropology
Grimsrud, David T, low temperature
  physics

Gropen, Arthur Louis, topology
Hansen, Harold Westberg, plant
  morphology
Hanson, Allen Louis, physical chemistry
Hardgrove, George Lind, Jr, physical
  chemistry
Hendrickson, Herman Stewart, II,
  biochemistry
Henrickson, Eiler Leonard, economic
  geology
Jensen, Paul, population ecology
Keller, Harry Bert, III, physics
Larson, Nora Leona, bacteriology
Lofthus, Orin Merwin, zoology
Marshall, John Clifford, analytical
  chemistry
Mathews, Robert Thomas, astronomy
Mohrig, Jerry R, organic chemistry
Muir, William Howard, plant pathology,
  physiology
Nau, Richard William, mathematics
Noer, Richard Juul, physics
Orr, Howard Dennis, zoology
Palm, John Daniel, human genetics,
  biology
Pearson, Wesley A, organic chemistry
Petersen, Arnold Jerome, genetics,
  ornithology
Ramette, Richard Wales, physical
  chemistry, analytical chemistry
Reitz, Robert Alan, physics
Riesman, Paul Hastings, anthropology
Schuster, Seymour, mathematics
Seebach, J Arthur, Jr, mathematics
Shoger, Ross L, zoology, developmental
  biology
Spessard, Gary Oliver, synthetic organic
  chemistry
Stanaitis, Otonas Edmundas, analytical
  mathematics
Steen, Lynn Arthur, mathematics
Tarr, Donald Arthur, inorganic chemistry
Thomas, Bruce Robert, experimental
  physics
Titus, William James, low temperature
  physics, statistical mechanics
Vessey, Theodore Alan, mathematical
  analysis
Wagenbach, Gary Edward, parasitology,
  zoology
Wegner, Kenneth Warren, mathematics
Wolf, Frank Louis, mathematics
Zischke, James Albert, parasitology,
  invertebrate zoology

PAYNESVILLE

Hansen, Mike, veterinary medicine

PLYMOUTH

Amata, Charles David, physical
  chemistry

ROBBINSDALE

Floyd, Don Edgar, organic polymer
  chemistry

ROCHESTER

Albert, Alexander, endocrinology
Anderson, Howard Arne, medicine
Anderson, Milton Winfield, internal
  medicine, cardiology
Atassi, Zouhair, organic chemistry,
  biochemistry
Baggenstoss, Archie Herbert, pathology
Bahn, Robert Carlton, pathology
Baldus, William Phillip, internal
  medicine, gastroenterology
Banner, Edward Arther, obstetrics &
  gynecology
Bartholomew, Lloyd Gibson, internal
  medicine, gastroenterology
Bayrd, Edwin Dorrance, medicine
Beeler, George W, Jr, biomedical
  engineering
Berggren, Michael J, radiological physics
Bisel, Harry Ferree, cancer
Blinks, John Rogers, pharmacology
Bollman, Jesse Louis, pathology
Brandenburg, Robert O, medicine
Brimijoin, William Stephen,
  pharmacology, neurobiology
Brown, Arnold Lanehart, Jr, pathology
Brown, Joe Robert, neurology
Burke, Edmund C, pediatrics
Butt, Hugh Roland, internal medicine
Cain, James Clarence, internal medicine,
  gastroenterology
Calvanico, Nickolas Joseph, biochemistry
Carr, David Turner, internal medicine,
  oncology
Carryer, Haddon McCutchen, internal
  medicine, allergy
Carter, Earl Thomas, physiology

Chevalier, Peter Andrew,
  cardiopulmonary physiology,
  physiology
Childs, Donald Smythe, Jr, nuclear
  medicine
Clagett, Oscar Theron, thoracic surgery
Cody, D Thane, otolaryngology,
  physiology
Coventry, Mark Bingham, orthopedic
  surgery
Cowlishaw, John David, biophysics
Culp, Ormond S, urology
Dahlin, David Carl, pathology
Danielson, Gordon Kenneth,
  cardiovascular surgery
Darley, Frederic Loudon, speech
  pathology
Daugherty, Guy Wilson, clinical
  medicine
Decker, David Garrison, obstetrics &
  gynecology
Didisheim, Paul, hematology,
  cardiovascular diseases
Dockery, Malcolm Birt, pathology
Dousa, Thomas Patrick, internal
  medicine, physiology
Du Shane, James William, pediatrics,
  cardiology
Ellefson, Ralph Donald, organic
  chemistry
Elveback, Lillian Rose, biostatistics
Engel, Andrew G, neuropathology,
  biochemistry
Erich, John B, plastic surgery
Erickson, Donald Johan, medicine
Fairbairn, John F, II, medicine
Fairbanks, Virgil, internal medicine,
  hematology
Faulconer, Albert, Jr, anesthesiology
Ferris, Deward Olmsted, surgery
Fleisher, Gerard Adalbert, chemistry
Fowler, Ward Scott, physiology
Gastineau, Clifford Felix, medicine
Geraci, Joseph E, internal medicine
Gibilisco, Joseph, dentistry
Gleich, Gerald J, immunology, internal
  medicine
Goldstein, Norman Philip, neurology,
  neurochemistry
Good, Clarence Allen, radiology
Gordon, Hymie, medical genetics,
  internal medicine
Gores, Robert James, oral surgery
Gorman, Colum A, endocrinology,
  internal medicine
Greene, Laurence Francis, urology
Gross, John Burgess, medicine
Hagedorn, Albert Berner, medicine
Hallberg, Olav Erik, otolaryngology
Hayles, Alvin Beasley, pediatrics
Hedgecock, Le Roy Darien, audiology,
  speech pathology
Henderson, John Warren, ophthalmology
Hermans, Paul E, internal medicine,
  infectious diseases
Hill, John Roger, proctology
Hofmann, Alan Frederick,
  gastroenterology
Hollenhorst, Robert William, medicine,
  ophthalmology
Hyatt, Robert Eliot, physiology, internal
  medicine
Jiang, Nai-Siang, biochemistry
Johnson, Einer Wesley, Jr, orthopedic
  surgery
Jones, James Donald, biochemistry
Jordon, Robert Earl, dermatology,
  immunology
Jowsey, Jenifer, physiology, orthopedics
Juergens, John Louis, internal medicine
Karlson, Alfred Gustav, medical
  microbiology, pathology
Kaye, Michael Peter, physiology, surgery
Kearns, Thomas P, ophthalmology
Kelly, Patrick Joseph, orthopedic surgery
Kerr, Frederick William Lawson,
  neurosurgery, neuroanatomy
Keys, Thomas Edward, history of
  medicine
Kline, Bruce Clayton, molecular biology,
  medical microbiology
Koelsche, Giles Alexander, internal
  medicine
Kottke, Bruce Allen, biochemistry,
  internal medicine
Kurland, Leonard T, medicine,
  epidemiology
Labarthe, Darwin Raymond,
  epidemiology
Lambert, Edward Howard, medical
  physiology
Lofgren, Karl Adolph, surgery
Logan, George Bryan, pediatrics
Lovestedt, Stanley Almer, oral surgery
Lucas, Alexander Ralph, child psychiatry
Lufkin, Edward Gwynne, endocrinology,
  internal medicine
Lynn, Hugh Bailey, surgery

MacCarty, Collin Stewart, neurosurgery
Maher, Frank Thomas, medicine
Maldonado, Jorge Eusebio, hematology,
  electron microscopy
Mann, Kenneth Gerard, biochemistry
Markowitz, Harold, immunochemistry
Martin, Gordon Mather, physical
  medicine
Martin, Harold Roland, medicine
Mathieson, Don Romuald, immunology
Mattox, Vernon Ross, organic chemistry
Mayberry, William Eugene, endocrinology
McCall, John Temple, biochemistry
McConahey, William McConnell, Jr,
  internal medicine, endocrinology
McCullough, Edwin Charles, medical
  physics
McDuffie, Frederic Clement, medicine
McGill, Douglas B, gastroenterology
Meyers, Paul, microbiology, virology
Miller, Roland Drew, medicine
Millikan, Clark Harold, medicine,
  neurology
Mizuno, Shigeki, molecular biology
Moersch, Herman John, clinical medicine
Moertel, Charles George, medicine
Morlock, Carl G, medicine
Mulder, Donald William, clinical
  neurology
Needham, Gerald Morton, bacteriology
Nelson, Ralph A, nutrition, physiology
Okazaki, Haruo, pathology,
  neuropathology
Olsen, Arthur Martin, internal medicine
Orvis, Alan Leroy, medical physics
Owen, Charles Archibald, Jr,
  biochemistry
Paradise, Norman Francis, physiology
Perry, Harold Otto, dermatology
Phillips, Sidney Frederick,
  gastroenterology
Pierre, Robert V, internal medicine
Polley, Howard Freeman, medicine
Priestley, James Taggart, surgery
Pruitt, Raymond Donald, medicine
Randall, Raymond Victor, medicine
Rehder, Kai, anesthesiology, physiology
ReMine, William Hervey, surgery
Riggs, Byron Lawrence, internal
  medicine
Ritts, Roy Eliot, Jr, microbiology,
  immunology
Roberts, Glenn Dale, clinical
  microbiology, medical mycology
Robertson, James Sydnor, medical
  physics, nuclear medicine
Roland, Charles Gordon, communication
  science, history of medicine
Rome, Howard Phillips, psychiatry
Rovelstad, Randolph Andrew, medicine
Ryan, Robert J, medicine, endocrinology
Salassa, Robert Maurice, medicine
Sauer, William George, medicine
Savedoff, Malcolm Paul, astrophysics
Sayre, George Pomeroy, pathology
Shepherd, John Thompson, physiology
Shorter, Roy Gerrard, experimental
  medicine
Siekert, Robert George, neurology
Slocumb, Charles Henry, internal
  medicine
Smith, Lucian Anderson, medicine
Spelsberg, Thomas Coonan, genetics,
  biochemistry
Spittell, John A, Jr, internal medicine,
  cardiovascular diseases
Sprague, Randall George, internal
  medicine
Stickney, James Minott, medicine
Stillwell, George Keith, physical
  medicine
Stobo, John David, immunology
Strong, Cameron Gordon, internal
  medicine, nephrology
Summerskill, William Hedley John,
  gastroenterology
Swanson, David Wendell, psychiatry
Swift, George Herbert, Jr, mathematics,
  computer science
Taylor, William F, biostatistics
Thompson, John Harold, Jr, parasitology
Tomasi, Thomas B, Jr, immunology,
  biochemistry
Tyce, Francis Anthony, psychiatry
Tyce, Gertrude Mary, biochemistry
Van Dyke, Russell Austin, biochemistry,
  microbiology
Veneziale, Carlo Marcello, biochemistry,
  dermatology
Ward, Louis Emmerson, medicine
Washington, John A, II, clinical
  microbiology, clinical pathology
Whisnant, Jack Page, medicine,
  neurology
Winkelmann, Richard Knisely,
  dermatology
Wollaeger, Eric Edwin, clinical medicine
Wood, Earl Howard, physiology

Wyatt, Andy Jack, animal breeding
Yoss, Robert Eugene, neurology, neuroanatomy

ROSEMOUNT
Wilcox, Clifford LaVar, dairy husbandry

ROSEVILLE
Carstens, Allan Matlock, operations research
Cunningham, John L., mycology, plant pathology
Klein, William Arthur, organic chemistry

ST CLOUD
Anderson, Rowland C., mathematics
Barker, Shirley Hugh, zoology
Bruton, Charles William, biology, animal ecology
Carpenter, John Harold, high temperature chemistry
Coppock, Henry Aaron, historical geography, cultural geography
Eckroth, John M., physical chemistry
Erickson, Wayland Lee, systematic botany, evolutionary biology
Ezell, Michael K., biophysics
Garrity, Alfred H Jr., zoology, botany
Grewe, Evelyn Payne, cultural anthropology
Hatcher, Harold Hoffman, botany
Hopkins, Joan, audiology, speech pathology
Jacobson, Louise H., mathematics
Johnson, Vincent Arnold, zoology, physiology
Kammermeier, Martin A., speech pathology
Lewis, Standley Eugene, entomology, paleontology
McCue, John Francis, parasitology
McMullen, James Clinton, chemistry
Mork, David Peter Sogn, biology
Peck, John Hubert, medical entomology, zoology
Pou, Wendell Morse, physics
Rehwaldt, Charles A., genetics, plant physiology
Sorensen, David T., inorganic chemistry
Tideman, Philip Lundsten, resource geography, agricultural geography
Watkins, Ivan Warren, physics
Williams, Steven Frank, fisheries, ichthyology
Younger, Philip Genevus, physics

ST JOSEPH
Anderson, Ingrid, nutrition
Grell, Mary, cytogenetics
Muggli, Joanne, mathematics
Westkaemper, Remberta, botany

ST PAUL
Abbe, Ernst Cleveland, botany
Abere, Joseph Francis, polymer chemistry
Adams, Roger James, physical chemistry
Adams, Russell S Jr., soil chemistry
Addis, Paul Bradley, food science
Albrecht, Norman Edward, mathematics
Allen, Charles Eugene, animal science, biochemistry
Allen, Martin, physical chemistry
Allen, Neil Keith, poultry nutrition
Anderson, Victor Elving, human genetics, behavioral genetics
Anderson, Robert Neils, plant physiology
Anderson, Gaylord West, public health
Anderson, Gerald S., solid state physics
Anderson, John Seymour, biochemistry
Anderson, Neil Albert, plant pathology, mycology
Andrus, Milton Henry, Jr., organic polymer chemistry, organic chemistry
Antonius (Kennelly), Mary, organic chemistry
Arneman, Harold Frederick, soils
Arnold, John P., veterinary medicine
Arthaud, Raymond Louis, animal science
Baker, Donald Gardner, microclimatology
Bakuzis, Egolfs Voldemars, forest ecology
Banit, Eldon Harris, organic chemistry
Banttari, Ernest E., plant pathology
Barnes, Donald Kay, genetics, plant breeding
Barnes, Donald McLeod, veterinary pathology, microbiology

Barnes, Raymond D., reproductive biology, laboratory animal science
Batzer, Harold Otto, entomology
Beebe, George Warren, photographic chemistry
Bennick, William Joseph, parasitology
Berg, John Richard, physical chemistry, inorganic chemistry
Berg, Robert W., poultry genetics
Bingham, Christopher, statistics
Birkeland, Stephen P., solid state chemistry
Birnbaum, David, elementary particle physics
Bissonnette, Howard Louis, plant pathology
Blake, George Rowland, soil physics
Blomquist, Charles Howard, biochemistry
Blomquist, Mary M Osborn, statistics, mathematics
Bloomfield, Victor Alfred, biophysical chemistry
Boerwinkle, Fred P., organic chemistry, physical chemistry, inorganic chemistry
Bolles, Theodore Frederick, inorganic chemistry
Borders, Alvin Marshall, quantitative genetics
Borgeson, Carl, agronomy
Boylan, William J., animal breeding
Breene, William Michael, food science, biochemistry
Brenner, Mark, horticulture, plant physiology
Brooks, Marion Alice, entomology
Brun, William Alexander, crop physiology
Brunelle, Thomas E., biochemistry
Bryce, Hugh Glendinning, organic chemistry
Buchholz, Allan C., physical organic chemistry, polymer chemistry
Bushnell, William J., phytopathology
Bushong, Jerold Ward, plant pathology
Busta, Francis Frederick, food microbiology
Button, Allan Clifford, organic chemistry
Cadotte, John Edward, organic chemistry
Caldecott, Richard S., genetics
Caldwell, Elwood F., food science
Cardwell, Vernon Bruce, agronomy; crop physiology
Carlson, Philip R., mathematics
Carlson, Robert Leonard, photographic chemistry
Carney, James Joseph, organic chemistry
Carlson, Roy David, poultry husbandry
Carnean, Willard Handy, forest soils
Carruth, Betty Ruth, nutrition
Case, Marvin Theodore, veterinary pathology, toxicology
Castro, Anthony Edward, virology; immunology
Chaffin, Tommy L., organic chemistry
Chang, Shaw Fai, biochemical
Chevone, Boris Ivan, insect physiology
Chiang, Huai C., entomology
Chow, Christopher N., electrooptics
Christensen, Clyde Martin, plant pathology
Christian, Curtis Gilbert, organic chemistry
Christian, Paul Jackson, systematic entomology
Christians, Charles J., animal breeding
Christmann, Marvin Henry, physics, mathematics
Clapp, Charles Edward, soil biochemistry
Clemens, Lawrence Martin, organic chemistry
Coad, L Keith, operations research, statistics
Comstock, Ralph Ernest, genetics, animal breeding
Comstock, Verne Edward, plant breeding, plant genetics
Conard, Gordon Joseph, biochemical pharmacology
Conlin, Bernard Joseph, dairy science, animal breeding
Conway, Alvin Charles, pharmacology
Cook, Edwin Francis, entomology
Cook, Jack E., organic polymer chemistry
Coulter, Samuel Todd, dairy manufacturing
Croft, Thomas Stone, organic chemistry
Csallany, Agnes Saari, organic chemistry
Cushing, Edward John, ecology
Cutkomp, Laurence Kremer, insect toxicology; economic entomology
Czarnecki, Caroline Mary Anne, veterinary anatomy
Dagley, Stanley, biochemistry
Davis, David Warren, plant breeding, vegetable crops

Davis, Horace Raymond, organic chemistry
DeGrande, Gary Gaston, organic chemistry; medicinal chemistry
Desborough, Sharon Lee, plant genetics
Deviny, Edward John, industrial organic chemistry, radiation chemistry
Dickson, Arthur Donald, physical chemistry; physics
Diedrich, James Loren, polymer chemistry
Diesch, Stanley L., veterinary public health, veterinary microbiology
Doomes, Earl, organic chemistry
Dowdy, Robert H., soil science
Downing, William Lawrence, protozoology
Doyle, Margaret John, nutrition
Duke, Gary Earl, avian physiology, ecology
Duncan, Archibald, inorganic chemistry, analytical chemistry
Dybvig, Douglas Howard, organic chemistry
Dziuk, Harold Edmund, animal physiology
Edwards, Jesse Efrem, pathology
Eian, Gilbert Lee, organic chemistry
Eide, Carl John, plant pathology
Elder, James Tait, physics
Elling, Laddie Joe, plant breeding
Enfield, Franklin D., genetics
Ensign, Thomas Charles, solid state physics
Epley, Richard Jess, meat science, food science
Erickson, Edward Herbert, medicinal chemistry
Erickson, John Gerhard, organic chemistry
Erickson, Randall L., physical chemistry, polymer chemistry
Erickson, Robert W., forest products
Eubank, William Roderick, physical chemistry
Evans, Roger Lynwood, inorganic chemistry
Evensen, Thomas James, organic chemistry
Ewing, Sidney Alton, veterinary parasitology
Fan, David P., biology
Farnham, Paul Rex, structural geology; geophysics
Fenster, William E., soil chemistry, geophysics
Fienberg, Stephen Elliott, statistics
Filipovich, George, solid state physics
Fleming, Peter B., inorganic chemistry
Fleming, Walter, mathematics
Fletcher, Thomas Francis, veterinary anatomy
Fogelson, David Eugene, geophysics
Forro, Frederick Jr., biophysics
Fox, Neil Stewart, industrial organic chemistry
Frank, William Charles, organic chemistry; photographic chemistry
Freier, Herbert Edward, analytical chemistry
French, David Weston, forest pathology
Frenkel, Albert W., plant physiology
Frenzel, Louis Daniel, Jr., biology, ecology
Fridinger, Thomas Lee, pesticide chemistry, organic chemistry
Friedlander, William Sheffield, organic chemistry
Fritts, Robert Washburn, physics
Froshiser, Fred Imanuel, plant pathology
Fuchs, James Allen, biochemistry
Gabor, Thomas, physical chemistry
Gagnon, John Gregory, analytical chemistry
Galbraith, William, biochemical pharmacology
Gander, John E., biochemistry
Gast, Robert Gale, soil chemistry
Geadelmann, Jon Lee, plant breeding
Geiger, Robert Warren, organic chemistry; aquatic biology
Gengenbach, Burle Gene, plant genetics
Germann, Paul Julian, biology; botany
Gerjejansen, Roland O., forest products
Gibbons, M Seraphim, mathematics
Glass, Robert Louis, biochemistry
Goblirsch, Richard Paul, mathematics
Gobran, Ramsis, organic chemistry
Goken, Garold Lee, polymer chemistry; organic chemistry
Good, A. L., veterinary physiology
Gooding, John Alan, agronomy
Goodrich, Richard Douglas, animal nutrition
Gordon, Joan, food science, nutrition
Gorham, Eville, limnology; ecology

Graham, Edmund F., physiology; cryobiology
Grant, Edwin Allen, Jr., organic chemistry
Grava, Janis (John), soil science
Gray, Grace Warner, pharmacology
Green, Charles Darwin, analytical chemistry; environmental sciences
Green, Charles E. Jr., plant physiology; plant genetics
Gregersen, Hans Miller, forest economics, resource management
Griffiths, Henry Joseph, veterinary parasitology
Grigal, David Francis, soil science, forest ecology
Groth, James Vernon, plant pathology
Gruenhagen, Richard Dale, weed science
Grundner, Matthias James, electromagnetics, engineering management
Guehler, Paul Frederick, organic chemistry
Gurr, Graham Edward, physics, crystallography
Hagemark, Kjell Ingvar, solid state chemistry
Hall, John Walton, botany
Hanson, Lester Eugene, animal science
Hapke, Bern, nuclear medicine
Ham, George Eldon, soil microbiology
Hard, Cecil Gustav, horticulture
Harein, Phillip Keith, entomology
Harriman, Benjamin Ramage, organic chemistry
Hamill, Dennis W., solid state physics
Hamm, Franklin Albert, solid state science
Hammer, Walton James, medicinal chemistry
Hamre, Melvin L., food science, poultry science
Hanlon, Griselda Frances, veterinary radiology
Hansen, Henry Leo, forestry
Hanson, Patrick Vincent, nutrition
Heiner, Robert E., genetics, plant breeding
Hart, Una Lynch, applied chemistry, cosmetic chemistry
Hatch, Robert Atchin, mineralogy
Hatcher, Herbert John, microbiology
Hauser, Edward Russell, physical biochemistry
Haygreen, John G., wood technology
Hegarty, Patrick Vincent, nutrition
Herk, Leonard Frank, Jr., applied chemistry
Hertz, Leonard B., horticulture
Hewitt, Frederick George, physics
Hicks, Dale R., plant breeding, statistics
Hill, Brian Kellogg, nuclear medicine
Hill, Eddie P., botany, mycology
Hochman, Harry, marine chemistry
Hodson, Alexander Carlton, entomology
Hoebel, Edward Adamson, anthropology; ethnology
Hogberg, Rudolph Karl, geology
Hogle, Donald Hugh, physical chemistry
Holler, Albert Cochran, chemistry
Hooper, Alan Bacon, microbial physiology
Hossfeld, Ralph (Lowell), wood chemistry
Hovin, Arne William, genetics, plant breeding
Howell, Peter Adam, physical chemistry
Huege, William Frederick, Jr., agronomy
Hunter, Alan Graham, animal physiology
Hunter, Samuel W., thoracic surgery; research
Huston, Keith Arthur, administration
Hutjens, Michael Francis, animal science; nutrition
Irving, Frank Dunham, forestry
Ithakissios, Dionyssis Spiros, radiochemistry
Janes, Donald Lucian, solid state chemistry
Jankus, Edward Francis, physiology
Jansen, Bernard Joseph, mathematics
Jenness, Robert, biochemistry
Jessen Carl Roger, radiology; veterinary medicine
Johnson, Donald W., optics
Johnson, Edgar Gustav, plant pathology
Johnson, Herbert Gordon, plant breeding
Johnson, Herbert Windal, genetics, plant pathology
Johnson, Kenneth Harvey, veterinary pathology

Johnston, Manley Roderick, organic polymer chemistry
Jonas, Herbert, plant physiology
Jones, Lester Tyler, Jr, physical chemistry
Joos, Richard W, biochemistry
Jordan, Peter Albion, ecology, wildlife management
Jordan, Robert Manseau, animal husbandry
Judd, Teresita, anatomy
Kasper, Eugene Mitchell, obstetrics & gynecology
Kaufert, Frank Henry, forest products
Keane, David Donagh, chemistry
Keffer, Charles Joseph, solid state physics, crystallography
Kemp, Arne K, forest products
Kennedy, Bill Wade, plant pathology
Kerley, Troy Lamar, pharmacology
Kernkamp, Milton F, plant pathology
Kerr, Sylvia Joann, developmental biology, microbiology
Kessler, Kenneth J, Jr, plant pathology
Kim, Sung Kyu, theoretical physics
King, Thomas Henry, plant pathology
Kirkwood, Samuel, biochemistry
Knutson, Charles Dwaine, computer sciences
Kommedahl, Thor, plant pathology
Konhauser, Joseph Daniel Edward, mathematics
Kotz, Arthur Rudolph, solid state electronics
Kovacic, Joseph Edward, organic chemistry
Krogh, Lester Christensen, organic chemistry
Krupa, Sagar, phytopathology, air pollution
Kulman, Herbert Marvin, entomology
Kumar, Mahesh C, veterinary microbiology, public health
Kurmis, Vilis, forestry
Kurtti, Timothy John, insect pathology, insect physiology
Kurtz, Harold John, veterinary pathology
Kvam, Donald Clarence, pharmacology
Labuza, Theodore Peter, food science, physical chemistry
Lambert, Jean William, agronomy
Lanegran, David Andrew, urban geography
Langager, Bruce Allen, polymer chemistry
Larntz, Kinley, applied statistics, mathematical statistics
Larson, Gerald Willis, organic chemistry
Larson, Vaughn Leroy, veterinary medicine
Larson, William Day, physical chemistry, analytical chemistry
Larson, William Earl, soil science
Lauer, Florian Isidore, horticulture
Lawrence, Donald Buermann, ethnobotany, plant ecology
Leahy, Sidney Marcus, organic chemistry
Leary, Rolfe Albert, forest mensuration
Lebsock, Kenneth L, agronomy
Lee, Pui Kum, optics
Leman, Allen Duane, veterinary medicine
Lepp, Henry, geology
Levitt, Jacob, plant physiology
Li, Paul H, horticulture, plant physiology
Libby, William Harris, organic chemistry
Liener, Irvin Ernest, biochemistry, nutrition
Lindeberg, George Kline, solid state physics
Lindorfer, Robert Karl, bacteriology
Lockwood, Robert Greening organic polymer chemistry, industrial organic chemistry
Loken, Keith I, veterinary microbiology
Long, James Earl, toxicology, industrial hygiene
Lorimer, Nancy L, genetics
Lovrien, Rex Eugene, physical biochemistry
Lowrey, Robert Dean, photochemistry
Lynch, Francis Watson, dermatology
MacDonald, David Howard, plant nematology, pomology
Mace, Arnett C, Jr, forest hydrology
MacGregor, John Malcolm, soil science
Mader, Erich Otto, plant pathology
Mainen, Eugene Louis, organic chemistry
Mannering, Gilbert James, pharmacology, biochemistry
Marshall, William Hampton, wildlife management
Marten, Gordon C, agronomy
Martin, Frank Burke, mathematics
Martin, James Tillison, behavioral physiology
Martin, Margaret Pearl, forest biometry
Martin, William Paxman, soil microbiology
Mather, Edward Chantry, veterinary medicine
Mather, George Wells, veterinary medicine
Matteson, John Warren, economic entomology
McBrady, John J, physical chemistry
McBrayer, James Franklin, ecology
McCann, Lester J, vertebrate zoology
McCown, Joseph Dana, industrial organic chemistry, research administration
McCurdy, David Whitwell, cultural anthropology
McDonald, William John, physics, mathematics
McKay, Larry Lee, microbiology
McKelvey, John Leyland, obstetrics, gynecology
McKinney, James T, surface physics
McMillan, John Frank, vertebrate ecology
Meade, Robert J, animal science, nutrition
Mech, Lucyan David, wildlife ecology
Megard, Robert O, zoology
Meiske, Jay C, animal husbandry
Mendel, Arthur, organic chemistry
Mentone, Pat Francis, inorganic chemistry, electrochemistry
Meredith, Harvey L, soil physics
Merriam, Lawrence Campbell, Jr, forest management
Meyer, Franklin Vincent, number theory
Meyer, Merle P, forestry
Mikkelson, Raymond Charles, solid state physics
Miles, William Raymond, forestry
Miller, Carl Stinson, physical chemistry
Miller, Fletcher A, surgery
Miller, Gerald R, weed science
Miller, Matthew William, photographic chemistry
Miller, William Eldon, forest entomology
Mirocha, Chester Joseph, plant pathology
Mishmash, Harold Edward, analytical chemistry
Mitchell, Clifford L, pharmacology
Mitchell, William Cobbey, solid state physics, energy conversion
Mizuno, William George, bacteriology
Montgomery, Peter Williams, physical chemistry
Moore, Perry Alldredge, organic chemistry, physical chemistry
Morath, Richard Joseph, organic chemistry
Morey, Glenn Bernhardt, geology
Morley, Thomas, systematic botany
Morris, Howard Arthur, food science
Morse, Lura Myra, nutrition, biochemistry
Moss, Dale Nelson, agronomy, plant physiology
Mudge, Joseph William, animal breeding, genetics
Mularie, William Mack, solid state physics
Mulhausen, Robert Oscar, internal medicine, medical administration
Munson, Robert Dean, soil fertility, plant nutrition
Mutsch, Edward L, organic chemistry
Nelsestuen, Gary Lee, biochemistry
Nelson, Robert A, veterinary medicine, toxicology
Neuvar, Erwin W, inorganic chemistry
Newman, Norman, organic chemistry
Newmark, Richard Alan, analytical chemistry
Nicholson, Geoffrey Charles, inorganic chemistry, ceramics
Nippoldt, Bertwin W, analytical chemistry
Noetzel, David Martin, entomology, zoology
Norris, Carol Lee, chemical physics
Nowlin, Duane Dale, water chemistry
Nylund, Robert Einar, horticulture
Ober, Robert Elwood, biochemistry, pharmacology
Oehmke, Richard Wallace, applied chemistry, research administration
Oelke, Ervin Albert, agronomy, plant physiology
Ogburn, Phillip Nash, cardiovascular physiology
Ohman, John Hamilton, research administration, forestry
Olsen, Rodney L, analytical chemistry
Olson, Paul B, analytical chemistry
Osborne, Carl Andrew, veterinary medicine, pathology
Otterby, Donald Eugene, nutrition
Otto, Harley John, agronomy
Owens, Kenneth Eugene, physical chemistry
Ownbey, Gerald Bruce, plant taxonomy
Packard, Vernal Sidney, Jr, food science
Parham, Walter Edward, geology, mineralogy
Parsons, Lawrence Reed, environmental physiology, horticulture
Patel, Kalyanji U, organic chemistry, chemical engineering
Paterson, William Gordon, chemistry
Pearlson, Wilbur H, chemistry
Pellett, Harold M, horticulture
Perlich, Robert Willard, analytical chemistry
Perman, Victor, veterinary pathology
Perry, John Francis, Jr, surgery
Peterson, Allan George, entomology
Peterson, Harold Oscar, medicine, radiology
Petrellis, Panayotis C, organic chemistry, photochemistry
Phillips, Richard Edward, animal behavior, avian physiology
Phillips, Ronald Lewis, cytogenetics, plant breeding
Pitt, Donald Alfred, physical chemistry
Pletcher, Wayne Albert, organic chemistry
Plovnick, Ross Harris, chemistry
Pocius, Alphonsus Vytautas, physical chemistry
Polson, David Ernest, plant breeding, plant genetics
Pomeroy, Benjamin Sherwood, veterinary microbiology
Pontinen, Richard Ernest, physics
Prager, Julianne Heller, organic chemistry
Price, Roger DeForrest, entomology
Prokop, Robert A, fluorine chemistry, electrochemistry
Puerckhauer, Gerhard Wilhelm Richard, organic chemistry
Radcliffe, Edward B, entomology
Randen, Neil Allen, organic chemistry, cosmetic chemistry
Rapatz, Gabriel Louis, cryobiology
Rasmusson, Donald C, plant genetics
Read, Paul Eugene, horticulture
Ree, Buren Russel, organic chemistry
Reich, Charles, research administration, organic chemistry
Reid, Thomas S, organic chemistry
Reineccius, Gary (Aubrey), food science
Rempel, William Ewert, animal breeding
Reynolds, Allan Eastman, physical chemistry
Rice, David E, organic chemistry, polymer chemistry
Rice, Thomas Kenneth, immunobiology, microbiology
Richards, Albert Glenn, insect physiology
Roberts, A Wayne, mathematics
Roberts, James Herbert, experimental nuclear physics
Robertson, Jerry Earl, organic chemistry, medicinal chemistry
Robins, Janis, physical chemistry, organic chemistry
Robinson, Glen Moore, III, physical chemistry
Robinson, Robert George, agronomy
Roelfs, Alan Paul, plant pathology
Rohlfing, Stephen Roy, microbiology, infectious diseases
Rosenberg, Murray David, cell biology, medicine
Rothman, Paul George, agronomy
Rowe, William A, analytical chemistry
Rowell, John Bartlett, plant pathology
Rubens, Sidney Michel, physics
Rubenstein, Irwin, molecular biology
Rudolf, Paul Otto, silviculture, genetics
Runquist, Olaf A, organic chemistry, chemical education
Rust, Richard Henry, soil science
Rutledge, Robert L, physical chemistry
Ryan, James Arthur, inorganic chemistry
Ryan, John Peter, physical chemistry
Rynkiewich, Michael Allen, anthropology
Sahyun, Melville Richard Valde, photographic chemistry
Salmon, Oliver Norton, physical chemistry
Sandberg, Carl Lorens, polymer chemistry
Sanders, Richard Mark, data processing
Satterthwaite, Ridgway, geography
Saunders, Donald Roy, toxicology
Sautter, Jay Howard, veterinary pathology
Savereide, Thomas J, industrial organic chemistry
Scherrer, Robert Allan, organic chemistry
Schipper, Arthur Louis, Jr, plant pathology, plant physiology
Schlotthauer, John Carl, veterinary parasitology
Schmelzle, Ambrose Francis, physical chemistry
Schmid, Alois Rudolph, agronomy
Schmid, Jack Robert, pharmacology, physiology
Schue, John R, algebra
Schultz, Donald Raymond, inorganic chemistry
Schwartz, Albert Truman, biophysical chemistry
Seffl, Raymond James, organic chemistry
Seibold, Carol Duke, photographic chemistry
Senkus, Raymond, physical chemistry
Shepard, Joseph William, physical chemistry
Sherman, Patsy O'Connell, polymer chemistry, applied chemistry
Shoffner, Robert Nurman, poultry genetics
Shope, Richard Edwin, Jr, virology, immunology
Short, Everett C, Jr, biochemistry
Siefken, Mark William, organic polymer chemistry
Silverman, William Bernard, plant pathology
Simkins, Charles Abraham, soil fertility, agronomy
Sjolander, John Rogers, organic chemistry
Skilling, Darroll Dean, plant pathology, forestry
Skok, Richard Arnold, forest economics
Skoog, Ivan Hooglund, organic chemistry
Slowinski, Emil J, Jr, physical chemistry
Smail, James Richard, embryology, marine biology
Smith, David Philip, solid state physics
Smith, George Henry, organic chemistry
Smith, James Knox, histopathology
Smith, John Douglas, animal nutrition
Smith, Lawrence Hubert, plant physiology
Smith, Lloyd Lyman, fisheries, pollution biology
Smith, Samuel, polymer chemistry
Snell, John B, plastics chemistry
Snustad, Donald Peter, genetics
Snyder, Leon Allen, genetics
Sobieski, James Fulton, chemistry
Sorensen, Dale Kenwood, veterinary science
Soulen, Thomas Kay, biochemistry
Southwick, David Leroy, geology
Sparrow, Gene Rodell, mass spectroscopy, surface chemistry
Spradley, James Phillip, anthropology, ethnography
Spurrell, Francis Arthur, veterinary medicine, radiation biology
Stadelmann, Eduard Joseph, plant physiology, cell physiology
Stakman, Elvin Charles, plant pathology, mycology
Stebbings, William Lee, structural chemistry
Stepan, Alfred Henry, organic chemistry
Stevens, Jerry Bruce, veterinary pathology, biochemistry
Stewart, Elwin Lynn, mycology
Stienstra, Ward Curtis, plant pathology
Stocker, Fred Butler, organic chemistry
Stowe, Clarence M, pharmacology, veterinary medicine
Strand, Oliver Eric, agronomy, plant physiology
Strehlow, Wolfgang Hans, physical chemistry, research administration
Stricklin, Buck, organic chemistry
Stucker, Robert Evan, plant breeding, statistics
Stushnoff, Cecil, horticulture
Stuthman, Deon Dean, plant genetics, plant breeding
Stutzman, Jacob William, pharmacology
Sucoff, Edward Ira, plant physiology
Swan, James Byron, soil science
Swan, Patricia B, nutrition, biochemistry
Swedish, Frank, organic chemistry
Swingle, Karl F, pharmacology
Talbott, Richard Lloyd, organic chemistry
Tamsky, Morgan Jerome, physical chemistry
Tarleton, Raymond Joseph, biochemistry
Tatini, Sita Ramayya, microbiology, food science
Taylor, Charles William, organic chemistry
Taylor, Philip Seyfang, entomology
Taylor, Robert Franklin, resource management, research administration
Taylor, Robert Joe, population ecology
Tester, John Robert, ecology
Thalacker, Victor Paul, organic chemistry

Thatcher, Walter Eugene, analytical chemistry
Thomas, Elmer Lawrence, food science
Thompson, Mary E., physical chemistry
Thompson, Philip Gerhard, inorganic chemistry
Thompson, Roy Lloyd, agronomy
Throckmorton, James Rodney, pesticide chemistry
Tiers, George Van Dyke, physical chemistry, organic chemistry
Tingstad, James Edward, pharmacy
Todd, Jerry William, analytical chemistry
Toren, George Anthony, organic chemistry
Toren, Paul Edward, analytical chemistry
Torp, Bruce Alan, inorganic chemistry
Touchberry, Robert Walton, animal breeding
Trepka, Robert Dale, organic chemistry
Turnquist, Orrin Clinton, horticulture
Usenik, Edward A., veterinary medicine
Underhill, James Campbell, zoology
Van Valkenburg, Jeptha Wade, Jr., colloid chemistry, surface chemistry
Varberg, Dale Elithon, mathematics
Vaughan, Alfred Leland, molecular physics, mass spectroscopy
Vickers, Zata Marie, food science
Wade, James Joseph, medicinal chemistry
Waibel, Paul Edward, poultry nutrition
Walczak, Hubert R., mathematics
Warner, Huber Richard, biochemistry
Waters, Thomas Frank, aquatic ecology, fisheries
Weber, Alvin Francis, cytology
Webers, Gerald F., paleontology
Weisberg, Sanford, applied statistics
Welch, Claude Alton, zoology
Weller, Milton Webster, animal ecology
Wetter, Alphonse Nicholas, physiology, anatomy
Wen, Richard Yutze, organic polymer
Wetmore, Clifford Major, lichenology
Wheaton, Jonathan Edward, neuroendocrinology
White, Donald Benjamin, ornamental horticulture, genetics
White, Joe Wade, physical organic chemistry
Whitemore, Howard Lloyd, veterinary medicine
Widmer, Richard Ernest, horticulture
Witman, William F., biology
Witry, Esperance, biology
Wold, Finn, biochemistry
Wilcoxson, Roy Dell, plant pathology
Wilkins, Harold, horticulture, plant physiology
Willard, Paul W., physiology,
Willard, Robert Jackson, geology
Williams, Jesse Bascom, animal science
Williams, Robert Calvin, analytical chemistry, physical chemistry
Williams, Todd Robertson, medicinal chemistry
Wolsey, Wayne C., inorganic chemistry
Wollner, Thomas Edward, organic chemistry, polymer chemistry
Wood, Francis A., air pollution, phytopathology
Woodward, Clare K., biochemistry
Woodward, Val Waddoups, genetics
Workman, Wesley Ray, organic chemistry
Wyatt, Ellis Junior, parasitology, invertebrate zoology
Yapel, Anthony Francis, Jr., physical chemistry, biophysical chemistry
Yarian, Dean Robert, organic chemistry
Young, Charles Wesley, dairy husbandry
Zaske, Darwin Erhard, clinical pharmacology
Zemjanis, Raimunds, reproductive physiology
Zeyen, Richard John, plant pathology
Zollinger, Joseph LaMar, organic chemistry
Zottola, Edmund Anthony, food science

ST PETER
Carlson, Keith J., geology
Fuller, Richard M, solid state physics
Glass, Arthur Warren, genetics
Hamrum, Charles Lowell, entomology
Hilding, Stephen R., analytical chemistry
Jensen, Richard Erling, analytical chemistry
Langsjoen, Arne Nels, organic chemistry
Mason, Charles Perry, botany
Ryan, Peter Michael, mathematics

Tanner, Ward Dean, Jr., wildlife management, ecology
Vande Vusse, Frederick John, parasitology

SHAKOPEE
Olsen, Douglas Alfred, physical chemistry

STANTON
Kleese, Roger Allen, plant genetics

STILLWATER
Meade, Reginald Eson, food science, food technology
Tait, William Charles, theoretical solid state physics

TWIN CITIES
Atchison, Thomas Calvin, Jr., physics
McLain, Albertson Lamson, fish biology
Savanick, George Adrian, mining, solid state science

VIRGINIA
Broderick, Alan Thomas, economic geology
Mauston, Glenn Warren, entomology

WASECA
Frederick, Edward C., animal science, physiology
Lueschen, William Everett, agronomy
Miller, Kenneth Philip, animal breeding
Randall, Gyles Wade, soil science

WAYZATA
Geiger, James Woodrow, industrial chemistry

WEST ST PAUL
Wear, Robert Lee, organic polymer chemistry

WHITE BEAR LAKE
Bohon, Robert Lynn, environmental sciences, chemistry
Heltemes, Eugene Casmir, physics, magnetism
Holmen, Reynold Emanuel, polymer chemistry, organic chemistry
McKeown, James John, physical chemistry
Sorensen, David Perry, organic chemistry
Wright, Charles Dean, polymer chemistry

WINONA
Allen, James Durwood, organic polymer chemistry
Alsum, Donald James, reproductive physiology
Anderson, Cyrus Vincent, animal nutrition, parasitology
Bayer, Thomas Norton, geology, paleontology
Doerr, Robert George, analytical chemistry
Donovan, John Francis, economic geology, earth sciences
Foss, Frederick William, Jr., inorganic chemistry
Fremling, Calvin R., limnology, entomology
Houtz, Ray Clyde, chemistry
Jarvinen, Richard Dalvin, mathematics
Kowles, Richard Vincent, genetics
McConville, David Raymond, limnology, fisheries
Morgan, Donald R., solid state physics
O'Rourke, Richard Clair, philosophy of science, botany
Pahl, George Leo, radiobiology
Raymond, Marion Robert, zoology
Rislove, David Joel, organic chemistry
Schwendewin, Gerd (Anton), physical geography, geography
Severin, Charles Hilarion, limnology
Thomas a Kempis, Mary, botany
Trusk, Ambrose, mathematics
Wald, Wilbur J., polymer chemistry
Wilson, Dan Leroy, biology

CRYSTAL SPRINGS
Windham, Steve Lee, horticulture

FOREST
Sadler, Clarence Reagan, physiology, parasitology

ABERDEEN
Scott, Franklin James, polymer chemistry

BAY ST LOUIS
Baker, Robert Andrew, environmental sciences
Ekberg, Donald Roy, environmental physiology
Koopman, Francis Christian, hydrology
Smathers, Garrett Arthur, plant ecology, science administration
Tai, Han, analytical chemistry
Trampus, Anthony, mathematics
Wade, Richard Archer, biological oceanography
Wolverton, Billy Charles, chemistry

BILOXI
Watkins, Charles, psychiatry

BRANDON
Landers, Holbrook, meteorology

BROOKHAVEN
Johnson, Carl Erick, organic chemistry
Peacock, Milton O., biochemistry
Ward, James Wellington, neuroanatomy, parasitology

CARTHAGE
Ajemian, Martin, anatomy

CLEVELAND
Millican, Troy Bea, limnology
Myers, Richard Showse, physical chemistry
Outlaw, Henry Earl, biochemistry, molecular biology
Raspet, Mabel Wilson, agronomy, botany
Steen, James Southworth, biology, immunology
Stewart, Robert Archie, II, palynology, plant ecology
Walters, Eleanor Boyd, mathematics
White, Jesse Steven, parasitology

CLINTON
Cannon, Jerry Wayne, biochemistry
Carlock, Henry Arthur, nuclear physics
Cox, Prentiss Gwendolyn, developmental biology
Deer, George Wendell, mathematics
Germany, Archie Herman, organic chemistry
Legg, John Wallis, physical chemistry
Nobles, William Lewis, pharmaceutical chemistry
Ohme, Paul Adolph, mathematics
Patterson, Joseph Gilbert, geology
Sadler, William Otho, limnology
Strange, William Ernest, experimental nuclear physics, spectroscopy
Whitlock, Lapsley Craig, experimental chemistry
Winton, Raymond Sheridan, molecular spectroscopy

COLUMBUS
Eckstein, Eleanor Foley, nutrition
Feland, Sarah Elizabeth, nutritional biochemistry
Fulton, Macdonald, medical microbiology
Gildea, Ray Yeakle, resource geography
Marsalis, Sula Johnson, chemistry
Ottinger, Carol Blanche, vertebrate ecology
Parker, William Skinker, mathematics
Payne, Dwight Arthur, inorganic chemistry
Scheer, Eleanor Ruth, microbiology
Sherman, Harry Logan, botany, ecology
Van Dusen, Clarence Raymond, speech pathology
Wierengo, Cyril John, Jr., organic chemistry

FOREST
Sadler, Clarence Reagan, physiology, parasitology

GREENVILLE
Kincade, Robert Tyrus, entomology, weed science

GRENADA
Moody, Julius Reynard, entomology

GULFPORT
Amburgey, Terry L., forest pathology
Carter, Fairie Lyn, analytical chemistry
DeGroot, Rodney Charles, plant pathology, mycology
Dinus, Ronald John, forest genetics
Haverty, Michael Irving, insect ecology
Henry, Berch Waldo, forest pathology
Loe, Robert Wayne, plant physiology
Smith, Virgil Kirkland, Jr., entomology
Sperka, Christina Kanschat, entomology
Walkinshaw, Charles Howard, Jr., phytopathology

HATTIESBURG
Anderson, Gary, physiological ecology, parasitology
Bedenbaugh, Angela Lea Owen, chemistry
Bedenbaugh, John Holcombe, organic chemistry
Blomquist, Gary James, biochemistry
Bowen, Richard Lee, physical geology
Brent, Charles Ray, physical chemistry, organic chemistry
Brundage, William Gregory, medical microbiology, immunochemistry
Cade, Ruth Ann, applied mathematics
Cliburn, Joseph William, zoology, botany
Cross, Ralph Donald, climatology, water resources geography
Elakovich, Stella Daisy, organic chemistry
Felder, Virginia Isabelle, mathematics
Fish, Arthur Geoffrey, invertebrate zoology
Folse, Raymond Francis, Jr., fluid dynamics
Grantham, Billy Joe, aquatic biology
Griffin, Anselm Clyde, III, physical organic chemistry
Gruchy, David Francis, zoology
Hampton, Kenneth Gerald, organic chemistry
Herzog, Richard (Franz Karl), physics
Howard-Peebles, Patricia Nell, human genetics
Howell, John Emory, inorganic chemistry
Hughes, William E., solid state physics
Irby, Bobby Newell, science education
King, Robert William, mathematics
Krubsack, Arnold J., organic chemistry
Larsen, James Bouton, comparative physiology, biochemistry
Martin, Billy Joe, cell biology,
Morrell, Joseph Salvador, mathematics
Norris, Donald Earl, Jr., parasitology
Owen, Donald Robertson, Jr., polymer
Paulson, Oscar Lawrence, petroleum geology, environmental geology
Pierce, Richard Harry, Jr., environmental chemistry
Pinkerton, Frank Henry, chemistry
Pinson, James Wesley, physical chemistry
Rayborn, Grayson Hanks, atomic physics
Smith, Byron Colman, comparative anatomy
Smith, Gaston, mathematics
Thames, Shelby Freland, polymer chemistry, organic chemistry
Toom, Paul Marvin, biochemistry
van Aller, Robert Thomas, organic chemistry, biochemistry
Walker, James Frederick, physiology, histology
Webster, Porter Grigsby, mathematics
Wertz, David Lee, inorganic chemistry, physical chemistry
West, Rose Gayle, physical chemistry
Wildman, Gary Cecil, polymer chemistry
Wooten, Jean W., botany
Yarbrough, Karen Marguerite, genetics, microbiology

INDIANOLA
Black, Emmet Russell, Jr., entomology

ITTA BENA
Balam, Baxish Singh, soil chemistry
Boykins, Ernest Aloysius, Jr., zoology
Khan, Sharif Ahmad, experimental solid state physics

Pillai, Thanumalaya Perumal, solid state physics
Singh, Jaswant, plant pathology

JACKSON

Allen, George Perry, microbiology
Andy, Orlando Joseph, neurosurgery
Arceneaux, Joseph Lincoln, microbiology, bacterial physiology
Ashburn, Allen David, anatomy, physiology
Ball, Carroll Raybourne, anatomy
Barnett, William Oscar, surgery
Batson, Blair Everett, pediatrics
Batson, Margaret Bailly, psychiatry, pediatrics
Bell, Oliver E, Jr, biochemistry
Bell, Rondal E, cell physiology
Bell, Warren Napier, hematology
Berdon, John Kenneth, periodontics
Berndt, William O, pharmacology, toxicology
Berry, Roy Alfred, Jr, organic chemistry
Bishop, Allen David, Jr, inorganic chemistry, physical chemistry
Blake, Thomas Mathews, cardiology
Brace, Robert Allen, cardiovascular physiology, chemical engineering
Brooks, Thomas Joseph, Jr, preventive medicine
Butchko, Gregory Michael, immunology
Byers, Benjamin Rowe, microbiology
Cain, Charles Eugene, organic chemistry
Clark, Carroll Thomas, organic chemistry
Cole, Wilfred Q, pediatrics
Cowley, Allen Wilson, Jr, physiology, cardiovascular physiology
Crowell, Jack Wesley, physiology, biophysics
Currier, Robert David, neurology
Davidson, William Martin, orthodontics
Dawson, James Robertson, Jr, pathology, bacteriology
Dodgen, Charles Lee, biochemistry
Douglas, Ben Harold, physiology
Evers, Carl Gustav, medicine, pathology
Gatipon, Glenn Blaise, psychopharmacology, neurophysiology
Gentry, Glenn Aden, virology, biochemical pharmacology
Goetz, Catherine Gertrude, pathology
Greenfield, Wilbert, physiology
Grogan, James Bigbee, microbiology
Guyton, Arthur Clifton, physiology
Haerer, Armin Friedrich, neurology
Haining, Joseph Leo, biochemistry
Hardy, James D, surgery, biochemistry
Hayes, Andrew Wallace, biochemistry, toxicology
Hellems, Harper Keith, cardiology
Ho, Ing Kang, neurochemistry, pharmacology
Hogue, Raymond Ellsworth, physical medicine
Hume, Arthur Scott, pharmacology, toxicology
Hutchison, William Forrest, parasitology, preventive medicine
Jackson, John Fenwick, internal medicine, genetics
Johnson, Ben Butler, internal medicine
Johnson, Samuel Britton, ophthalmology
Keegan, Hugh Lawrence, medical entomology
Klein, Richard Lester, pharmacology, cell physiology
Langford, Herbert Gaines, internal medicine, physiology
Lewis, Jesse C, mathematics, computer sciences
Lockwood, William Rutledge, internal medicine, experimental pathology
Mann, Wallace Vernon, Jr, dentistry
McCaa, Connie Smith, biochemistry
McDonald, Janet, mathematics
McKeown, James Preston, ethology
Mehendale, Harihara Mahadeva, insect toxicology
Mellen, Frederic Francis, geology
Montalvo, Jose Miguel, medicine, pediatrics
Mosey, Lois Margot, obstetrics & gynecology
Nelson, Norman Crooks, surgery, endocrinology
Norman, Roger Atkinson, Jr, physiology, chemical engineering
O'Callaghan, Dennis John, animal virology, medical microbiology
Oelshlegel, Frederick James, Jr, biochemistry, genetics
Olsen, Kenneth Wayne, protein chemistry, biochemistry
Pearce, David Harry, biomedical engineering, respiratory physiology
Peeler, Dudley Flavius, Jr, neuropsychology
Perkins, James, chemistry
Pfaffman, Madge Anna, pharmacology
Phillips, Samuel, medicine, hospital administration
Raj, Baldev, botany
Randall, Charles Chandler, microbiology
Read, Virginia Hall, biochemistry, endocrinology
Reid, Milton Roy, entomology, botany
Riggs, Schultz, mathematics
Rovelstad, Gordon Henry, dentistry
Shive, Robert Allen, Jr, mathematics
Shore, Fred L, organic chemistry
Shreve, Darrell Rhea, mathematics
Sloan, Robert Dye, medicine
Smith, Thomas Marion, biochemistry, parasitology
Stephens, Charlene Barr, otolaryngology
Straka, William Charles, astrophysics
Suess, James Francis, psychiatry
Sulya, Louis Leon, biochemistry
Taylor, Aubrey Elmo, physiology, biophysics
Thiede, Henry A, obstetrics & gynecology
Thureson-Klein, Asa Kristina, biology, pharmacology
Turner, Manson Don, physiology
Uzodinma, John E, preventive medicine, microbiology
Walley, Willis Wayne, zoology, ecology
Watson, David Goulding, pediatrics
Watson, Robert Lee, epidemiology, public health
White, Harold Birts, Jr, biochemistry
Williams, William Lane, anatomy
Woodley, Charles Leon, biochemistry
Yensen, Arthur Eric, ecology
Young, David Bruce, physiology

LELAND

Broadfoot, Walter Marion, soils

LONG BEACH

Boudreau, Robert Donald, meteorology

LORMAN

Brady, Ruth Mary, physical chemistry
Edney, Norris Allen, biology
Lawson, Verna Rebecca, plant physiology
Parker, Henry Louis, zoology
Russell, Joseph Louis, fluorine chemistry
Vadhwa, Om Parkash, agronomy

MATHERVILLE

Mathers, Alexander Pickens, organic chemistry

MERIDIAN

Coleman, Otto Harvey, agronomy
Freeman, Kelly Carey, soils, field crops
Zummo, Natale, plant pathology

MISSISSIPPI STATE

Alley, Earl Gifford, pesticide chemistry, photochemistry
Anderson, Edmund Hughes, mathematics
Andrews, Cecil Hunter, agronomy
Andrews, Gordon Louis, entomology
Arner, Dale H, wildlife management
Arthur, Robert David, biochemistry
Atchison, Thomas Andrew, numerical analysis
Baker, Bryan, Jr, animal husbandry, animal physiology
Barker, Marvin Windel, organic chemistry
Batson, William Edward, Jr, plant pathology
Bearden, Henry Joe, reproductive physiology
Behr, Lyell Christian, organic chemistry
Blakeman, Crawford Harris, Jr, anthropology
Boyd, Leroy Houston, animal science
Brown, Lewis Raymond, bacteriology
Bunch, Harry Dean, agronomy
Butler, Charles Morgan, statistics, operations research
Cardwell, Joe Thomas, dairy chemistry
Cole, Avean Wayne, weed science
Combs, Leon Lamar, III, chemical physics
Combs, Robert L, Jr, entomology
Cook, Robert Lee, molecular spectroscopy
Crawford, Crayton McCants, physical chemistry
Creech, Roy G, genetics, plant breeding
Cross, William Henley, entomology, ecology
Cummings, Kenneth Ross, animal nutrition
Davich, Theodore Bert, economic entomology
Day, Elbert Jackson, poultry nutrition
Deaton, James Washington, poultry science
Dilworth, Benjamin Conroy, poultry nutrition, biochemistry
Douglas, Alvin Gene, agronomy, plant breeding
Duffey, Donald Creagh, physical chemistry, organic chemistry
Elam, William Warren, forest physiology
Essig, Henry Werner, animal nutrition
Fitzpatrick, George, environmental biology, entomology
Foil, Robert Rodney, forestry
Frazier, James Lewis, entomology, behavioral physiology
Fuquay, John Wade, animal physiology, animal nutrition
Glick, Bruce, physiology, genetics
Graves, Clinton Hannibal, Jr, plant pathology
Green, Henry Burwell, entomology
Hammett, Harrell Lee, horticulture, plant breeding
Hare, Mary Louise Eckles, cytology, botany
Hare, Woodrow Wilson, plant pathology
Harris, Frank Aubrey, entomology
Hedin, Paul A, biological chemistry
Hegwood, Donald Augustine, horticulture
Heitz, James Robert, biochemistry
Hepner, Leon Wilburne, entomology
Hesketh, J D, plant physiology, plant ecology
Hickok, Leslie George, plant cytogenetics
Hodges, Harry Franklin, plant physiology
Hodges, John Deavours, plant physiology, silviculture
Jackson, Jerome Alan, ornithology, ecology
Jenkins, Johnie Norton, plant genetics, agronomy
Jones, Gordon Ervin, physics
Kellogg, Thomas Floyd, biochemistry
Kennedy, Maurice Venson, biochemistry
Kilgore, Lois Taylor, nutrition
Knight, William Eric, plant breeding
Koch, Robert B, biochemistry
Kubena, Leon Franklin, nutrition, biochemistry
Lancaster, James D, agronomy
Lane, Harry Cleburne, plant physiology
Lashomb, James Harold, entomology
Laswell, Troy James, geology
Lindley, Charles Edward, animal husbandry
Lyon, Duane Edgar, forest products
Malloy, Thomas Bernard, Jr, molecular spectroscopy
Malone, Linda Catron, statistics
Marshall, James Tilden, Jr, food science, dairy science
Mastin, Charles Wayne, mathematical analysis
May, James David, poultry physiology
McGinnis, Gary David, carbohydrate chemistry
McLaughlin, Roy Earl, entomology, protozoology
Merwine, Norman Charles, agronomy, crop science
Minyard, James Patrick, organic chemistry
Mitchell, Earl Bruce, entomology
Mitchell, Henry Cooper, entomology
Moak, James Emanuel, forest economics
Moore, Evon Lamar, horticulture
Myers, Merle Wentworth, geography
Nash, Clyde E, soil chemistry, mineralogy
Nebeker, Thomas Evan, forest entomology
Nelson, Lyle Engnar, soils
Nickerson, John Charles, III, physics
Norment, Beverly Ray, entomology
Overcash, Jean Parks, horticulture
Parrott, William Lamar, entomology
Peterson, Harold LeRoy, soil microbiology
Peterson, Jon Holbrook, Jr, anthropology
Pettry, David Emory, soil morphology
Pitre, Henry Nolle, Jr, entomology, plant pathology
Pluenneke, Ricks Henry, plant physiology
Porter, Walter Kenneth, Jr, plant physiology
Potts, Howard Calvin, plant breeding
Ramsey, Dero Saunders, dairy science
Riggs, Karl A, mineralogy, petrology
Roberts, Edward Guernsey, silviculture
Rogers, Robert Wayne, animal science
Rohde, Florence Virginia, applied mathematics
Rosenkranz, Eugen Emil, plant pathology
Russell, Ernest Everett, geology
Schmittle, Samuel Conrad, veterinary pathology
Schuster, Michael Frank, economic entomology
Scott, Gene E, plant breeding
Sheely, Clyde Quitman, chemistry
Short, Paul Henry, forest products, wood chemistry
Singletary, Clyde C, horticulture
Solomon, Jimmy Lloyd, mathematics
Spencer, James Alphus, plant pathology
Spikes, Paul Wenton, mathematical analysis
Stojanovic, Borislav Jovan, microbiology, biochemistry
Sullivan, Alfred Dewitt, forestry, biometrics
Switzer, George Lester, forest ecology, forest soils
Taylor, Clayborne D, electromagnetics
Taylor, Fred William, wood science, wood technology
Thomas, Charles Hill, poultry husbandry, genetics
Thompson, Alonzo Crawford, organic chemistry, pharmaceutical chemistry
Thompson, Emmett Frank, forest economics
Thompson, Warren Slater, wood science & technology
Tilley, John Leonard, mathematics
Tischer, Robert George, microbiology, food technology
Toole, Eben Richard, forest pathology
Turner, Barbara Holman, ecology
Turner, James William, animal breeding, statistics
Vanderford, Harvey Birch, agronomy
Vaughan, Charles Edwin, agronomy
Watson, James Ray, Jr, plant taxonomy, paleobotany
Watson, Vance H, agronomy
Whisler, Frank Duane, soil physics
Wiles, Alfred Barksdale, plant pathology
Wilson, Clifton Arlie, entomology
Wilson, Robert Paul, biochemistry
Wilson, Wilbur William, physical chemistry
Wise, Louis Neal, agronomy
Wolfe, James Leonard, vertebrate zoology, animal behavior
Wynn, Jack Thomas, anthropology
Yao, Joe, wood science, wood chemistry
Young, Poh-Shien, particle physics

MORTON

Ousterhout, Lawrence Elwyn, nutrition, biochemistry

OCEAN SPRINGS

Cook, David Wilson, microbiology
Dawson, Charles Eric, ichthyology
Eleuterius, Lionel Numa, marine botany
Gunter, Gordon, zoology
Howse, Harold Darrow, cell biology
Otvos, Ervin George, sedimentology, geomorphology
Overstreet, Robin Miles, parasitology, marine biology
Venkataramiah, Amaraneni, comparative physiology, marine zoology
Woodmansee, Robert Asbury, marine ecology

OXFORD

Dodge, Austin Anderson, pharmacy
Grissinger, Earl H, soil science, physics
Hammond, Elmer Lionel, pharmacy, pharmacy administration
Happ, Stafford Coleman, geology, engineering
McHenry, John Roger, soil science
Meyer, Lawrence Donald, soil conservation
Ritchie, Jerry Carlyle, ecology
Sheffield, Roy Dexter, mathematics
Shields, Fletcher Douglas, physics
Ursic, Stanley John, watershed management, forest hydrology

PASCAGOULA

Ford, Robert Sedgwick, food science
Hull, Edgar, internal medicine
Juhl, Rolf, fish biology

PASS CHRISTIAN

Arceneaux, George, plant breeding, agronomy

525

MISSISSIPPI

PICAYUNE
Bauer, Stewart Thomas, organic chemistry
Sugathan, Kanneth Kochappan, organic chemistry

POPLARVILLE
Spiers, James Monroe, plant physiology, agronomy

STARKVILLE
Brook, Ted Stephens, entomology
Emerich, Donald Warren, analytical chemistry
Hardee, Dicky Dan, entomology, ecology
Heller, Robert, mathematics
Hoover, Clifford Dale, agronomy
Lei, Kai Yui, nutrition
Lusk, John William, dairy science
Murmann, Richard P., physical chemistry, soil science
Pitts, Gerald Nelson, plant physiology
Ranney, Carleton David, plant pathology
Sartor, Clyde Flake, entomology
Thompson, Wilfred Roland, Jr., agronomy

STATE COLLEGE
Chambers, Howard Wayne, toxicology
Croft, Walter Lawrence, physics
Crow, Terry Tom, electromagnetism
Delouche, James Curtis, botany
Ferguson, Joseph Luther, Jr., plasma physics
Fisher, Thomas Henry, organic chemistry, physical chemistry
Futrell, Mary Feltner, biochemistry, human nutrition
Futrell, Maurice Chilton, plant pathology
Grimley, Eugene Burhans, III, inorganic chemistry
Howell, Everette Irl, physics
Johnston, George Washington, structural geology
Lins, Thomas Wesley, marine geology, cytology
Lloyd, Edwin Phillips, entomology
Locke, John Flowers, botany
Luter, William Dean, analytical chemistry, chemical engineering
McMahan, William H, inorganic chemistry
Mickelson, John Clair, microbiology
Miles, Delbert Howard, chemistry
Neel, William Wallace, entomology
Pegram, Joe D., applied statistics, computer science
Scarborough, Charles T, Jr., mathematics
Wang, Augustine Weisheng, microbiology

STONEVILLE
Bagga, Harmahinder Singh, plant pathology
Bailey, Jack Clinton, entomology
Baker, James Bert, forest soils
Baker, Ralph Stanley, silviculture
Beland, Gary LaVern, economic entomology
Blackmon, Bobby Glenn, forest soils
Bowman, Donald Houts, agronomy
Brewer, Franklin Douglas, entomology
Davis, Robert Gene, plant pathology
Duke, Stephen Oscar, plant physiology
Elmore, Carroll Dennis, plant physiology
Filer, Theodore H. Jr., plant pathology, soils
Frick, Kenneth Eugene, entomology
Keeling, Bobbie Lee, plant pathology
Kennedy, Harvey Ellis, Jr., forest soils
Klien, Thomas Clarence, plant genetics
Kreasky, Joseph Bernard, plant nutrition
Laster, Marion Logan, entomology
Manning, Cleo Willard, plant breeding
Martin, Dial Franklin, entomology
McCracken, Francis Irvin, forest pathology
McWhorter, Chester Gray, plant physiology, weed science
Merk, Marvin Eugene, entomology
Pfrimmer, Theodore Roscoe, entomology
Solomon, James Doyle, forest entomology
Vogt, George Britton, biosystematics

TOUGALOO
Mehrotra, Bam Deo, biochemistry
Sra, Kewal Singh, mathematics

UNIVERSITY
Ahmed, Ismail Yousef, inorganic chemistry
Baker, John Keith, medicinal chemistry, physical chemistry
Barden, Ned Thorson, microbiology
Bickerstaff, Thomas Alton, mathematics
Bolen, Lee Napier, Jr., nuclear physics, acoustics
Borne, Ronald Francis, medicinal chemistry, organic chemistry
Britton, Otha Leon, numerical analysis
Causey, William McLain, mathematical analysis
Childress, Noel A., geometry
Cook, David Edwin, topology
Cullen, Abbey Boyd, Jr., applied physics
Davis, Wilbur Marvin, pharmacology
Doorenbos, Norman John, pharmacognosy, medicinal chemistry
Elsohly, Mahmoud Ahmed, pharmacognosy
Gilmore, William Franklin, medicinal chemistry
Guess, Wallace Louis, pharmacy
Heimer, Norman Eugene, organic chemistry
Hickenbottom, John Powell, biochemistry
Hufford, Charles David, pharmacy, pharmacognosy
Huneycutt, Maeburn Bruce, mycology
Kelly, Robert Emmett, physics
Kitchin, Irwin Clark, embryology
Klingen, Theodore James, physical inorganic chemistry
Knight, Luther Augustus, Jr., aquatic biology
Longest, William Douglas, invertebrate zoology
Magee, Lyman Abbott, microbiology
McClurkin, Iola Taylor, biology
McDonald, Francis Guy, biochemistry
McGaha, Young John, zoology
McLaughlin, Kenneth Phelps, geology
Metzger, Robert Melville, chemistry, chemical physics
Morrison, Marcus Eugene, bacteriology, ophthalmology
Norman, William Harvey, cell physiology, science education
Pace, Henry Buford, physiology
Panetta, Charles Anthony, organic chemistry
Peters, Randall Douglas, acoustics, biophysics
Pullen, Thomas Marion, botany
Reidy, James Joseph, physics
Reynolds, William Roger, geology
Riley, Thomas Nolan, medicinal chemistry
St Amand, Wilbrod, cytogenetics
Sam, Joseph, pharmaceutical chemistry, organic chemistry
Stefani, Andrew Peter, physical chemistry
Stokes, Russell Aubrey, mathematics
Truax, Robert Lloyd, chemistry
Turner, Carlton Edgar, pharmacognosy
Waller, Coy Webster, pharmaceutical chemistry
Waters, Irving Wade, pharmacology
Watson, Edna Sue, immunology
Wilson, Marvin Cracraft, pharmacology
Yun, Kwang-Sik, physical chemistry

UTICA
Gupta, Gian Chand, colloid chemistry, soil chemistry

VICKSBURG
Barko, John William, aquatic ecology
Kirby, Conrad Joseph, Jr., ecology
Kolb, Charles Rudolph, geology
Lee, Charles Richard, soil chemistry, plant nutrition
Mather, Katharine Kniskern, geology
Peekna, Andres, physics, mechanical engineering
Saucier, Roger Thomas, physical geography, environmental geology
Thornton, Kent W., aquatic ecology, systems science
Weissman, William, surface chemistry

MISSOURI

ANDERSON
Barger, John Walter, physical chemistry

BALLWIN
Gash, Virgil Walter, organic chemistry
Mottus, Edward Hugo, organic chemistry
Redmore, Derek, organic chemistry

BOLIVAR
Clark, Jasper Arnold, botany
Harris, Beverly Howard, mathematical analysis
Hogue, Ralph Stewart, bacteriology
Kort, Margaret Alexander, histology, cell biology

CANTON
Brodmann, John Milton, organic chemistry
Harris, Henson, mathematics, academic administration
Wiltshire, Charles Thomas, aquatic ecology

CAPE GIRARDEAU
Bahn, Emil Lawrence, Jr., inorganic chemistry, nuclear chemistry
Bieber, Gene Lawrence, agronomy, plant physiology
Bolen, Homer Roscoe, zoology
Braasch, Norman L., entomology, botany
Diehl, Stanley Gregg, botany, microscopy
Dudgeon, Edna, genetics
Francis, Richard L., organic chemistry
Froemsdorf, Donald Hope, organic chemistry
Gersbacher, Willard Marion, zoology, ecology
Hodges, Sidney Edward, physics
Huckabay, John Porter, botany, biosystematics
Knox, Burnal Ray, geology
Kullberg, Russell Gordon, botany
Lowell, Gary Richard, geology
Meyer, William Ellis, animal science
Pinnick, Herbert Robert, Jr., physical chemistry
Readnour, Jerry Michael, inorganic chemistry
Rutledge, Harley Dean, solid state physics
Train, Carl T, parasitology, invertebrate zoology
Williams, Charles Edwin, geography

CHESTERFIELD
Linder, Solomon Leon, optics

CLARENCE
Wallace, Doris David, speech pathology, audiology

CLAYTON
Black, Wayne Edward, public health
Lam, Robert Lee, neurology
Oppenheimer, Henry Ernest, medicine
Redmond, John Charles, geophysics, engineering management
Soule, Samuel David, obstetrics & gynecology

COLUMBIA
Acedo, Gregoria N, plant genetics, plant pathology
Adair, Kent Thomas, forestry, management sciences
Addinger, Hans Karl, veterinary virology, cancer
Aldrich, Calvin Dale, mathematics
Aldrich, Richard John, agronomy
Alexander, Charles William, agronomy
Anast, Constantine Spiro, pediatrics
Anderson, Helen Lester, nutrition
Anderson, Laurel Ethan, field crops, weed science
Anderson, Philip Carr, dermatology
Anderson, Ralph Robert, dairy husbandry
Anderson, Richard Orr, aquatic ecology
Arendt, Billy Dean, algebra
Arnold, Richard C., public health, preventive medicine
Asay, Kay Harris, crop breeding, agronomy
Bailey, Milton (Edward), food chemistry
Barbero, Giulio J., pediatrics, gastroenterology
Barnes, Asa, Jr., pathology, hematology
Barnes, Ronald Edward, medical physiology
Barrett, James Thomas, immunology, microbiology
Barry, Billy Dean, entomology
Barsky, Constance Kay, geochemistry
Baskett, Thomas Sebree, wildlife biology
Basu, Asit Prakas, statistics
Bauman, John E, Jr., inorganic chemistry
Beckett, Jack Brown, plant genetics
Beem, John Kelly, geometry
Benfer, Robert Alfred, anthropology
Benson, Norman G., biology, limnology
Berrier, Harry Hilbourn, veterinary pathology
Biehler, Harold Victor, physiology
Biever, Kenneth Duane, entomology
Black, Samuel P W., medicine
Blackwell, Paul K, II, computer science, mathematics
Blanchar, Robert W., soil chemistry, biochemistry
Blenden, Donald, veterinary public health
Blumenthal, Leonard Mascot, mathematics
Braaten, Melvin Ole, statistics, genetics
Brawner, Thomas Allan, microbiology, virology
Brazille, James E., neurophysiology, neuroanatomy
Breitenbach, Robert Peter, zoology
Brooke, Clement Eustace, pediatrics
Brown, Esther Marie, anatomy
Brown, Gregory Neil, tree physiology, cryobiology
Brown, Herbert Ensign, anatomy
Brown, James Milton, plant ecology, resource management
Brown, James Richard, soil fertility
Brown, Merton F., mycology, forest pathology
Brown, Olen Ray, microbiology, biochemistry
Brugger, Robert Melvin, nuclear physics
Buening, Gerald Matthew, immunology, veterinary virology
Burcham, Paul Baker, mathematics
Burdick, Allan Bernard, genetics
Burns, Thomas Wade, internal medicine, endocrinology
Byington, Keith H., biochemical pharmacology
Carpenter, Alden B, geochemistry
Case, Arthur Adam, veterinary medicine
Cavanah, Lloyd (Earl), agronomy
Chapel, James L., child psychiatry
Chapman, Carl Haley, anthropology, archaeology
Chase, Gerald Roy, biostatistics
Clark, Jack L, animal husbandry
Coe, Edward Harold, Jr., genetics
Colwill, Jack M, internal medicine
Cornell, Creighton N, physiology
Courtney, Gladys A, mammalian physiology, endocrinology
Cowan, David Lawrence, solid state physics
Cox, Gene Spracher, physiology
Craig, Wilfred Stuart, entomology
Crownover, Richard McCranie, mathematical analysis
Crumley, Carole Linda, anthropology
Cumbie, Billy Glenn, botany
Cutts, James Henry, histology
Dale, Homer Eldon, veterinary physiology
Darkow, Grant Lyle, meteorology
Davies, David K., geology
Davis, David R., psychiatry
Davis, James Othello, physiology
Day, Billy Neil, animal husbandry
Dean, Anthony Marion, physical chemistry
Decker, John D, anatomy, embryology
Decker, Wayne Leroy, meteorology
DeFacio, W Brian, theoretical high energy physics
deRoos, Roger McLean, zoology
Dewhirst, Leonard Wesley, veterinary parasitology
Dibb, David Walter, soil fertility
Dickhaus, Donald William, internal medicine, cardiology
Dillard, Joe G, zoology
Dolphin, Robert Earl, entomology
Donnell, Henry Denny, Jr., epidemiology
Dropkin, Victor Harry, zoology
Duncan, Donald Pendleton, forestry
Dunn, David Baxter, botany
Dykstra, Richard Lynn, statistics
Edwards, Terry Winslow, astronomy
Eggers, George W Nordholtz, Jr., medicine, anesthesiology

Eisenbrandt, Leslie Lee, pharmacology
Eisenstark, Abraham, bacteriology
Eklund, Darrel Lee, applied statistics
Elder, William Hanna, wildlife conservation
Engley, Frank B, Jr, microbiology
Enns, Wilbur Ronald, entomology
Ethinton, Raymond Lindsay, geology, paleontology
Eyestone, Willard Halsey, pathology
Fahim, Mostafa Safwat, reproductive biology
Fairchild, Mahlon Lowell, entomology
Fales, William Harold, veterinary bacteriology
Farish, Donald James, ethology, population genetics
Farmer, John Neville, parasitology
Feather, Milton S, biochemistry
Fields, Marion Lee, food science, microbiology
Fisher, Theodore Roosevelt, agriculture, soil chemistry
Fleming, Warren R, physiology
Fletchall, Oscar Hale, weed science
Flynn, Margaret A, nutrition
Folks, Homer Clifton, soils
Forte, Leonard Ralph, pharmacology
Franck, Wallace Edmundt, mathematical statistics
Franz, John Matthias, biochemistry
Freeman, Thomas J, geology
Funk, John Leon, ecology, fish biology
Gardner, Peter Michael, anthropology
Garner, George Bernard, agricultural biochemistry
Garner, Harold E, cardiovascular disease, anesthesiology
Gaus, Arthur Edward, horticulture
Gavan, James Anderson, anthropology, anatomy
Gehrke, Charles William, chemistry
George, Melvin Douglas, mathematics
Gerhardt, Klaus Otto, organic chemistry, analytical chemistry
Ghiron, Camillo A, biophysics
Gingrich, Newell Shiffer, physics
Goldberg, Herbert Sam, microbiology, medical education
Goodge, William Russell, human anatomy, vertebrate morphology
Goodman, Robert Norman, plant pathology
Gowans, Charles Shields, genetics, phycology
Graham, Ellis Ray, soils
Green, James Arnold, anatomy
Griffin, William Thomas, obstetrics & gynecology
Griggs, Douglas M, Jr, physiology, medicine
Grossman, Robert Bruce, agronomy
Hahn, Allen W, veterinary medicine, biomedical engineering
Hall, David Goodsell, III, obstetrics & gynecology
Hansen, Harry Louis, entomology, biology
Harrington, James Bishop, Jr, meteorology
Harris, Patrick Donald, physiology, electrical engineering
Harrison, Arthur Pennoyer, Jr, microbiology
Hart, William Milton, ophthalmology
Hazelwood, Donald Hill, zoology
Hedrick, Harold Burdette, animal husbandry
Heimberg, Murray, biochemistry, pharmacology
Hemphill, Delbert Dean, horticulture, environmental health
Hensley, Eugene Benjamin, solid state physics
Hentges, David John, microbiology, biochemistry
Herman, Harry August, dairy husbandry, animal breeding
Hertzler, Ann Atherton, nutrition
Hibbard, Aubrey D, horticulture
Himmelberg, Glen Ray, petrology
Hoffman, Jarett David, entomology
Holroyd, Louis Vincent, physics
Hoover, Loretta White, dietetics
Hopps, Howard Carl, pathology
Horrocks, Rodney Dwain, agronomy
Hostetler, Paul B, geochemistry, geology
Huang, Justin C, theoretical physics
Huckaba, James Albert, algebra
Huggans, James Lee, entomology
Hultsch, Roland Arthur, physics
Hurst, Robert Rowe, nuclear physics
Hutcheson, David Paul, animal nutrition, biostatistics
Ide, Carl Heinz, ophthalmology
Ignoffo, Carlo Michael, entomology, invertebrate pathology
Jacobs, Marc Quillen, mathematics
Jensen, Harlan Ellsworth, ophthalmology
Johannsen, Christian Jakob, agronomy
Johns, William Davis, geochemistry
Johnson, Clayton Henry, Jr, petrology
Johnson, David Robert, agronomy, crop physiology
Johnson, Harold David, physiology
Johnson, Joseph Alan, physiology
Jones, Allan W, biology, physiology
Jones, John Richard, limnology
Joseph, Donald J, otolaryngology
Kaiser, Edwin Michael, organic chemistry
Katti, Shriniwas Keshav, analytical statistics, applied statistics
Kay, Michael Aaron, nuclear chemistry
Kearby, William H, forest entomology, insect ecology
Keaster, Armon Joseph, entomology
Keller, Walter David, geology, ceramics
Keown, Kenneth K, anesthesiology
Kikudome, Gary Yoshinori, cytogenetics
Kilburn, Kaye Hatch, internal medicine
Kim, Hyunyong, physical chemistry, quantum chemistry
Kimber, Gordon, cytogenetics
Knowles, Charles Otis, entomology
Koeppe, Owen John, biochemistry
Koirtyohann, Samuel Roy, analytical chemistry
Kostbade, J Trenton, geography
Krause, Gary F, statistics
Krause, William John, anatomy, histology
Kung, Ernest Chen-Tsun, meteorology
Kuntz, Robert Roy, physical chemistry
Kurtz, William Boyce, forest resource economics
Lam, Ping-Fun, mathematics
Lamberti, Joseph W, psychiatry
Lambeth, Victor Neal, vegetable crops
Lange, Leo Jerome, mathematics
Lapping, Mark Barry, resource management, forestry
Larson, Russell L, biochemistry
Lasley, John Foster, animal breeding
Leeson, Charles Roland, anatomy
Leighton, Walter (Woods), mathematics
Leonhard, Frederick Wilhelm, physics
Lindberg, Donald Allan Bror, pathology, computer science
Loan, Raymond Wallace, immunobiology
Lobeck, Charles Champlin, pediatrics
Lodwick, Gwilym Savage, radiology, bioengineering
Loeppky, Richard N, physical organic chemistry
Lower, William Russell, genetics, environmental health
Lowrance, Edward Walton, anatomy
Lucas, Fred Vance, pathology
Luckey, Thomas Donnell, biochemistry, nutrition
Lussier, Roger Jean, inorganic chemistry
Maaseidvaag, Frode, bioengineering, electrophysiology
Malin, John Michael, inorganic chemistry, photochemistry
Mangel, Margaret, food science, nutrition
Manson, Donald Joseph, physics, electronics
Marco, Philip Joseph, psychiatry
Marienfeld, Carl J, pediatrics, preventive medicine
Marquardt, William Harrison, anthropology
Marshall, Charles Edmund, colloid chemistry, soil science
Marshall, Robert T, microbiology, food science
Marston, Norman Lee, entomology
Martin, Arlene Patricia, biochemistry
Martin, Charles Everett, veterinary physiology
Martin, Richard Harvey, medicine, physiology
Martz, Fredric A, animal nutrition, dairy science
Matches, Arthur Gerald, agronomy
Mayer, Foster Lee, Jr, toxicology, aquatic ecology
Mayer, William Dixon, medicine
McClure, Robert Charles, veterinary anatomy
McCormick, John Pauling, bio-organic chemistry
McCune, Emmett L, veterinary microbiology, avian pathology
McGinnes, Edgar Allan, wood chemistry, wood technology
McQuade, Henry Alonzo, cytology, cytogenetics
Mehrle, Paul Martin, Jr, biochemistry, physiology
Mengel, Charles E, medicine, hematology
Merilan, Charles Preston, dairy husbandry
Mertz, Dan, plant physiology
Metter, Dean Edward, herpetology
Meyer, Dallas Kremer, physiology
Meyer, Hermann, veterinary anatomy
Middleton, Charles Cheavens, experimental pathology
Miles, Charles Donald, plant physiology
Millikan, Daniel Franklin, Jr, botany
Millikan, Larry Edward, dermatology, immunology
Moore, Aimee N, nutrition
Morehouse, Lawrence G, veterinary pathology, veterinary microbiology
Moscatelli, Ezio Anthony, biochemistry, neurochemistry
Muhrer, Merle E, biochemistry
Murmann, Robert Kent, inorganic chemistry
Murphy, Leslie Carlton, microbiology
Musacchia, Xavier Joseph, physiology, zoology
Naumann, Hugh Donald, food microbiology
Nelson, Curtis Jerome, agronomy
Neuffer, Myron Gerald, genetics
Niemeyer, Kenneth H, veterinary medicine
Nightingale, Dorothy Virginia, organic chemistry
Noble, William Allister, cultural geography, anthropology
Noteboom, William Duane, biochemistry
Novak, Alfred, health sciences
O'Dell, Boyd Lee, biochemistry, nutrition
Oestreich, Alan Emil, radiology
Olson, LeRoy David, veterinary pathology, veterinary virology
Ortwerth, Beryl John, biochemistry, ophthalmology
Osweiler, Gary D, veterinary toxicology
Otto, David Arthur, marine biology, paleobotany
Overholser, Milton David, anatomy
Pape, Brian Eugene, toxicology, pathology
Parisi, Joseph Thomas, microbiology
Patton, John Franklin, medicine
Peck, Raymond Elliott, geology
Peters, Elroy John, agronomy
Petty, Clinton Myers, geometry
Pfander, William Harvey, nutrition
Pflieger, William Leo, ichthyology, fisheries
Pickett, Edward Ernest, analytical chemistry
Pierce, James Otto, II, environmental health, industrial health
Pierce, Keith Robert, mathematics
Platner, Wesley Stanley, physiology, environmental physiology
Poehlman, John Milton, plant breeding, agriculture
Politis, Demetrios John, plant pathology, electron microscopy
Puttler, Benjamin, entomology
Quissell, David Olin, biochemistry
Rabjohn, Norman, organic chemistry
Redei, Gyorgy Pal, genetics, plant physiology
Reeder, John Hamilton, mathematics
Reynolds, James Blair, fish biology, limnology
Riggs, Hammond Greenwald, Jr, microbiology
Robbins, Michael C, anthropology
Rodabaugh, David Joseph, algebra, numerical analysis
Roetman, Ernest Levane, applied mathematics
Rogers, Marlin Norbert, floriculture
Rosenquist, Bruce David, veterinary microbiology
Rowlett, Ralph Morgan, anthropology, archaeology
Russell, Robert Lee, physiology, pharmacology
Russell, Thomas Randall, fisheries management
St Omer, Vincent Victor, neuroscience, veterinary pharmacology
Savage, Jimmie Euel, poultry nutrition
Scherr, David DeLano, orthopedic surgery, microbiology
Schlemper, Elmer Otto, inorganic chemistry
Schmidt, Donald Arthur, veterinary pathology, clinical pathology
Schmidt, Paul Woodward, physics
Schoettger, Richard A, fish biology, environmental biology
Schowengerdt, George Carl, agriculture
Schrader, Keith William, mathematics
Schuder, John Claude, bioengineering, biophysics
Schupp, Guy, nuclear physics
Searles, Scott, Jr, organic chemistry
Sears, Ernest Robert, plant cytogenetics
Sears, Lotti Maria Steinitz, cytology
Sechler, Dale Truman, plant breeding
Sehgal, Om Parkash, plant pathology, virology
Selby, Lloyd A, veterinary medicine, epidemiology
Sentilles, F Dennis, Jr, mathematics
Settergren, Carl David, forestry, watershed management
Sewell, Homer B, animal nutrition
Sharp, John Malcolm, Jr, hydrogeology
Shear, David Ben, biophysics
Siegel, Edward, biophysics
Siegel, Elsie P, cell biology, endocrinology
Silver, Donald, cardiovascular surgery, thoracic surgery
Sleper, David Allen, plant breeding
Smith, James Elmo, Jr, floriculture, horticulture
Smith, Richard Chandler, forest economics
Solorzano, Robert Francis, virology, medical microbiology
Sorenson, Marion W, ethology, zoology
South, Frank E, physiology
Spier, Robert Forest Gayton, anthropology
Spradling, Stuart Leslie, chemistry, agricultural engineering
Spratt, John Stricklin, Jr, surgery
Srivastava, Probodh K, cytogenetics, reproductive physiology
Stalling, David Laurence, analytical chemistry, organic chemistry
Stephenson, Alfred Benjamin, poultry husbandry
Stephenson, Hugh Edward, Jr, thoracic surgery, cardiovascular surgery
Stewart, Wellington Buel, pathology
Stitt, James Harry, geology, paleontology
Stuth, Charles James, algebra
Su, Kwei Lee, lipid biochemistry
Sun, Albert Yung-Kwang, biochemistry, neurochemistry
Thomas, Gustave Daniel, entomology
Thomas, Lloyd Brewster, chemistry
Thompson, Richard Claude, inorganic chemistry
Tompson, Clifford Ware, physics
Townsend, John Ford, pathology
Tritschler, Louis George, veterinary medicine, surgery
Trombly, Thelma Woodhouse, speech pathology, audiology
Troutner, David Elliott, nuclear chemistry
Tsutakawa, Robert K, statistics
Tumbleson, Myron Eugene, biochemistry, nutrition
Twente, Janet, physiology, zoology
Twente, John W, zoology
Unklesbay, Athel Glyde, geology
Unklesbay, Nan F, food science, nutrition
Utz, Winfield Roy, Jr, mathematical analysis
Veum, Trygve Lauritz, animal nutrition, veterinary physiology
Viele, George Washington, geology, tectonics
Volkert, Wynn Arthur, radiochemistry, radiobiology
Vorbeck, Marie L, biochemistry, microbiology
Wallin, Jack Robb, plant pathology
Wang, Richard J, cell biology, genetics
Weide, Kenneth Duane, pathology, veterinary medicine
Weinstein, Ira, endocrinology, biochemical pharmacology
Weiss, James Moses Aaron, psychiatry
Werner, Samuel Alfred, solid state physics, nuclear engineering
Westfall, Bertis Alfred, pharmacology, physiology
Wheeler, Jesse Harrison, Jr, geography, historical geography
Wheeler, Robert Lee, mathematics
White, Arnold Allen, biochemistry
White, Harry Houston, neurology
White, Henry W, physics
White, Joseph Mallie, anesthesiology
Whitley, James R, biochemistry, nutrition
Willett, Joseph Erwin, plasma physics
Williams, Charles Herbert, biochemistry
Wingo, Curtis W, entomology
Winthrop, Joel Albert, applied mathematics
Wixom, Robert Llewellyn, biochemistry
Wolfram, Thomas, solid state physics
Wood, Joseph M, botany, paleobotany
Wood, Randall Dudley, biochemistry, organic chemistry
Woodruff, Calvin Watts, medicine
Woodruff, Clarence Merrill, agronomy
Wosilait, Walter Daniel, pharmacology, biochemistry
Wright, Dennis Charles, neurosciences
Wyllie, Thomas Dean, plant pathology

Yanders, Armon Frederick, genetics
Yelon, William B., experimental solid state physics
Yonke, Thomas Richard, entomology, systematics
York, Donald Harold, neuroscience
Zahler, Warren Leigh, biological chemistry
Zatzman, Marvin Leon, physiology
Zemmer, Joseph Lawrence, Jr., algebra
Zuber, Marcus Stanley, agronomy

CREVE COEUR
Henis, Jay Myls Stuart, physical chemistry
Myers, Richard Lee, analytical chemistry, environmental systems & technology
Williams, Byron Lee, Jr., organic chemistry

EUREKA
Coles, Richard Warren, physiological ecology, animal behavior
Hartroff, Phyllis Merrit, experimental pathology

FAYETTE
Freidline, Charles Eugene, inorganic chemistry, analytical chemistry
Li, Edward Hsien-Chi, plant physiology, cell biology
Long, Norman Oliver, inorganic chemistry, organic chemistry
McIntosh, William David, mathematics
Momberg, Harold Leslie, zoology, histology
Peery, Larry Joe, physics
Smith, Vincent Francis, Jr., physical organic chemistry

FERGUSON
Hertenstein, Harold (Nelson), chemistry

FLORISSANT
King, Perry, Jr., radiochemistry

FORSYTH
Allen, Fred William, physiology

FULTON
Fickess, Douglas Ricardo, zoology, physiology
Hinde, Howard Parrish, plant ecology, cytology
McClary, Joseph Edward, biochemistry
McCreigh, Robert Willis, organic chemistry
McVeigh, Ilda, bacteriology, botany
Schaefer, Harold Franklin, chemical microscopy, physical chemistry
Sloss, Frank Brooke, mathematics
Schultz, John E., organic chemistry
Williams, Henry Warrington, zoology, animal behavior
Winters, Roger, physics

GALLATIN
Shawver, Murl Charles, biology

GLENDALE
Craver, Clara Diddle (Smith), spectrochemistry, petroleum chemistry

HALLSVILLE
Hanson, Willis Dale, fish biology

INDEPENDENCE
Ferm, Robert James, petroleum chemistry

JEFFERSON CITY
Burchard, Jeanette, microbiology, serology
Byrd, Willis Edward, physical chemistry
Chowdhury, Ikbalur Rashid, soil fertility, agronomy
Cook, Nathan Howard, cytology, zoology
Domke, Herbert Reuben, public health
Dowdy, William Wallace, biology
Eigner, Joseph, microbiology, environmental management
Finley, David Emanuel, biology, botany

Gnaedinger, Richard H., food science, animal husbandry
Krishna, Gopal X., nutrition, biochemistry
Lewis, Paul Edward, reproductive physiology
Miller, Herman T., biochemistry, immunology
Paul, Kamalendu Bikash, plant physiology, horticulture
Polowy, Henry, applied mathematics
Purkett, Charles A., Jr., fish biology
Schwartz, Elizabeth Reeder, zoology
Spurrier, Elmer R., microbiology, public health
Talley, Spurgeon Morris, animal nutrition
Warner, David Charles, mathematical physics, plasma physics
Wilson, Edward Matthew, dairy science

KAISER
Wachtel, Louis William, biochemistry

JOPLIN
Baiamonte, Vernon D., physical chemistry
Boehning, Rochelle Lloyd, geometry
Mosher, Melvin Wayne, physical organic chemistry
Orr, Orry Edwin, vertebrate zoology
Phillips, Russell Allan, physics
Whittle, Philip Rodger, organic chemistry, forensic chemistry

KANSAS CITY
Afghani, Hisham T., analytical chemistry
Alms, Thomas H., microbiology, immunology
Anderson, Robert Gordon, botany
Arnold, John D., chemotherapy, clinical pharmacology
Baldauf, Richard John, environmental education
Barnelow, Russell George, Jr., bacteriology, bacterial physiology
Beadle, Buell Wesley, chemistry
Beckloff, Gerald Lee, medicine
Bellet, Eugene Marshall, pesticide chemistry
Berenbom, Max, biochemistry
Berman, Elizabeth Alexandra, algebra
Blatz, Paul E., biochemistry
Boyer, Philip A., Jr., medicine
Breed, Laurence Woods, organic polymer chemistry
Brown, Elwyn S., anesthesiology
Bryant, Paul James, solid state physics
Burdick, Harold Charles, biology, physiology
Burger, Henry G., cultural anthropology
Burkholder, David Frederick, pharmacy
Butterworth, Bernard Bert, zoology
Calkins, William Graham, internal medicine, gastroenterology
Campbell, Robert Calvin, mathematical analysis
Carlson, Roger, mathematical statistics
Castles, Thomas R., pharmacology
Chappelow, Cecil Clendis, Jr., organic chemistry
Cheng, Chia-Chung, organic chemistry
Cheng, Kuang Lu, analytical chemistry
Chien, Ping-Lu, organic chemistry
Christensen, John Bert, anatomy
Chu, Luke Lo-Hwa, biochemistry, endocrinology
Clare, Stewart, zoology, biochemistry
Cobick, A Doyle, Jr., economic entomology, information science
Cohn, David Valor, biochemistry, endocrinology
Cole, Wilbur Vose, neuroanatomy
Colen, Alan Hugh, biophysical chemistry
Connolly, John W., inorganic chemistry, biological chemistry
Coveney, Raymond Martin, Jr., geology
Dahlgren, George, physical chemistry
Dale, Wesley John, organic chemistry
Dias, Jerry Ray, physical organic chemistry, chemical engineering
Dimond, Edmunds Grey, cardiology
Downs, James Joseph, physical chemistry, organic chemistry
Doyle, William Cletus, geometry
Droll, Henry Andrew, inorganic chemistry
Duncan, William Perry, synthetic organic chemistry, radiochemistry
Eisenmann, Kurt, applied mathematics
ElAttar, Tawfik Mohammed Ali, biochemistry
Elliott, Joseph Robert, clinical chemistry
Engel, James Francis, synthetic organic chemistry
Field, Marvin Frederick, microbiology
Fisher, Harvey Franklin, biochemistry

Fuhlhage, Donald Wayne, organic chemistry
Garrison, Robert Gene, medical microbiology
Gentile, Richard J., stratigraphy
Gier, Ronald E., dentistry
Gillian, James Horace, mathematics
Gillingham, James Morris, pharmaceutical chemistry
Givier, Robert L., medicine, pathology
Goebel, Edwin DeWayne, environmental geology
Goetz, Kenneth Lee, physiology
Goldberg, Merill B., mathematics
Goodson, Louis Hoffman, organic chemistry
Graham, Walter Donald, biochemistry
Green, Vernon Albert, biochemical pharmacology
Gross, Gordon E., applied physics
Grosskreutz, Joseph Charles, energy conversion
Grunt, Jerome Alvin, pediatrics, endocrinology
Gwynne-Thomas, Eric Hubert, geography, anthropology
Haggerty, William Joseph, Jr., analytical chemistry
Hall, Lynn Raymond, surgery
Hamilton, James Wilburn, biochemistry
Haney, William Garland, Jr., pharmaceutical chemistry
Harrington, Glenn William, parasitology, microbial genetics, biophysics
Harriman, Philip Darling, microbial genetics
Harris, Michael Joseph, analytical chemistry
Hartman, John Leo, Jr., medical physiology
Hava, Milos, clinical pharmacol
Hellerstein, Stanley, pediatrics, clinical chemistry
Hemberger, Judith Ann, perinatal biology
Herndon, Betty Larue, pharmacology
High, Marathon Eby, physics
Hignite, Charles E., pharmacology
Hill, John Joseph, physics
Hill, Shirley Ann, mathematics education
Hipman, Paul Lorenz, geology
Holder, Thomas M., pediatric surgery
House, William Burner, biochemistry, nutrition
Hubbard, Forest Craven, organic chemistry
Hubbard, Harold Mead, research administration
Hughes, Lawrence Ambrose, meteorology
Hylton, Alvin Roy, environmental sciences
Johnson, Richard Dean, pharmaceutical chemistry
Jones, Daniel L., microbiology, clinical pathology
Jones, Ronald Dale, computer science
Justesen, Don Robert, neuropsychology
Kennedy, James A., enzymology
Kezlan, Thomas Phillip, mathematics
King, William Roy, Jr., geology
Knapp, Theodore Martin, neuropsychology
Kos, Edward Stanley, microbiology
Krum, Jack Kern, food technology
Lanman, Robert Charles, biochemical pharmacology
Lapp, Thomas William, environmental chemistry, industrial chemistry
Lawless, Edward William, science policy, physical inorganic chemistry
Lee, Cheng-Chun, pharmacology, toxicology
Levy, Edward Robert, organic chemistry
Liao, Tsung-Kai, pharmaceutical chemistry
Liebnitz, Paul W., mathematics
Linn, Carl Barnes, organic chemistry
Liu, Frank Tsung Yuan, physiology
Lott, Peter F., physical chemistry, analytical chemistry
Luke, Yudell Leo, mathematics
MacDonald, Carolyn Trott, theoretical chemistry, applied mathematics
MacFarlane, John O'Donnell, medical microbiology, oncology
MacGregor, Ronal Roy, biochemistry
Major, Schwab Samuel, Jr., physics
Malinowski, Henry John, health sciences
Marvin, James Osbert, food technology
McCoy, Layton Leslie, organic chemistry
McCoy, Melnykovych, George, cell biology
Melvin, Robert Burrow, biochemistry
Metz, Florence Irene, physical chemistry, inorganic chemistry
Miles, Charles David, invertebrate zoology
Miller, Lowell D., biochemistry, pharmacology

Millich, Frank, polymer chemistry, photochemistry
Mitchell, Henry Andrew, mammalian physiology
Moffat, James, chemistry
Mohamed, Aly Hamed, genetics
Murphy, John Joseph, biochemistry, plant physiology
Myers, Richard F., vertebrate zoology, academic administration
Neogen, John William, physical chemistry, inorganic chemistry
Nelson, Gayle Herbert, anatomy
Niebergall, Paul J., pharmaceutics
Noback, Richardson K., internal medicine
Nuessle, Noel Oliver, pharmacy
Panzek, Eldon Joseph, geology
Pearson, Bennie Jake, mathematics
Peters, Paul James, behavioral biology
Phillips, James M., physics
Podrebarac, Eugene George, organic chemistry, medicinal chemistry
Puhl, Richard James, pesticide chemistry
Query, Marvin Richard, physics
Ringle, David Allan, physiology, immunology
Rommel, Frederick Allen, medical microbiology, immunopathology
Rost, William Joseph, pharmaceutical chemistry
Roth, Genevieve D., dentistry
Royall, Norman Norris, Jr., mathematics
Russell, Robert Julian, Jr., mammalogy
Salmon, Shirley Joan, speech pathology
Savage, George Roland, microbiology
Schanker, Lewis Stanley, pharmacology
Scherrer, Joseph Henry, polymer chemistry
Schmitz, Kenneth Stanley, biophysical chemistry
Schuler, Martin N., biochemistry
Schwartz, Norman Louis, dentistry, biology
Scott, Robert Crawford, chemistry
Selner, Ronald George, physiology, endocrinology
Sengupta, Sailes Kumar, mathematical statistics, operations research
Servoss, Reva R., physical chemistry
Shortridge, Robert William, information science, chemistry
Sice, Jean, pharmacology
Singh, Shri Krishna, mathematics
Sirridge, Marjorie Spurrier, hematology, laboratory medicine
Skapataon, Joseph Bjorn, agricultural chemistry
Slosky, Myron Norton, industrial chemistry
Smith, Ivan C., environmental chemistry, physical inorganic chemistry
Smith, John Chandler, anatomy, pharmacology
Spangler, William J., microbiology
Sprinz, Helmuth, pathology
Stanley, Charles William, analytical chemistry, pesticide chemistry
Stauffer, Truman Parker, Sr., physical geography
Stern, Daniel Henry, limnology, ecology
Stern, Michele Suchard, plant physiology, biochemistry
Stewart, Jack Lauren, dentistry
Strickland, William Alexander, Jr., pharmacy
Summicht, Russell William, dentistry
Talbot, Ted Delwyn, analytical chemistry
Templeton, Arch W., medicine, radiology
Thomas, Timothy Farragut, physical chemistry
Timberlake, Joseph William, clinical biochemistry
Tuttle, Warren Wilson, neuropharmacology
Waggoner, Terry Bill, organic chemistry, biochemistry
Wagner, Myron L., chemistry
Waring, Richard C., physics
Warren, Halleck Burkett, Jr., bacteriology
Welch, Quintin B., genetics, applied statistics
Wells, Frank Edward, microbiology, food science
Wesner, Bruce Richard, topology
Werner, Herbert Allen, medicine
Wheeler, James Donlan, pharmaceutical chemistry, biochemistry
White, Thomas George, immunopathology
Whitten, Elmer Hammond, medical physiology
Wilkinson, Charles Brock, medicine, psychiatry
Wilkinson, Ralph Russell, analytical chemistry, environmental management
Willis, Lyle Wilmot, pharmacy
Winkler, Bruce Conrad, biochemistry

Woodhouse, Edward John, chemistry, toxicology
Wright, Charles V, Jr, sanitary engineering, public health
Young, George Robert, biological chemistry
Zech, Arthur Conrad, agronomy, biochemistry
Zee-Cheng, Kwang Yuen, organic chemistry, medicinal chemistry

KIRKSVILLE
Bell, Max Ewart, plant morphology
Black, John David, vertebrate zoology
Bond, Robert Franklin, physiology
Chornock, Francis William, biochemistry, human anatomy
Crummy, Pressley Lee, histology, human anatomy
Denslow, John Stedman, osteopathy
Freeland, Max, analytical chemistry
Hall, William Francis, speech pathology
Hanks, David L, mycology
Hix, Elliott Lee, pharmacology
Hudgins, Patricia Montague, pharmacology
Jones, Everet Clyde, marine biology
Julyan, Frederick John, anatomy, histology
Magruder, Willis Jackson, chemistry, science education
McMurry, Earl William, physics, geophysics
Mock, Orin Bailey, reproductive physiology, mammalogy
Nothdurft, Robert Ray, physics
Ohler, Edwin Allen, physiology
Pandya, Krishnakant Hariprasad, pharmacology, pharmacognosy
Peavler, Robert Jean, inorganic chemistry
Peissner, Lorraine C, physiology
Rechtien, James Joseph, biomaterials
Rosebery, Dean Arlo, fisheries
Sells, Gary Donnell, plant physiology
Selser, Will Lindsey, science education, biology
Shaddy, James Henry, ecology, entomology
Snyder, George Edward, anatomy, embryology
Stilwell, Kenneth James, mathematics
Umanzio, Carl Beeman, microbiology
Woods, Dale, mathematics

KIRKWOOD
Brown, Harold Probert, organic polymer chemistry
Hutton, William Elmer, microbiology
Johnson, John Harpster, organic chemistry
Orban, Edward, applied chemistry
Patrick, Tracy Minard, Jr, organic chemistry
Salivar, Charles Joseph, organic chemistry
Swisher, Robert Donald, organic chemistry, environmental chemistry
Waldron, Harold Francis, applied chemistry
Wildi, Bernard Sylvester, organic chemistry
Wilson, Clyde Livingston, soil physics

LIBERTY
Dixon, Marvin Porter, physical chemistry, physical organic chemistry
Geiker, Charles Don, astronomy, physics
Hilton, Wallace Atwood, physics
Holloway, Thomas Thornton, chemical physics
Philpot, John Lee, physics
Wagenknecht, Burdette Lewis, botany

LOUISIANA
Bukowick, Peter Anthony, organic chemistry, biochemistry

MANCHESTER
Konneker, Wilfred R, medical physics, nuclear physics
Myers, Howard, chemical physics

MARSHALL
Bearce, Winfield Hutchinson, organic chemistry
Freeark, Clayton Wayne, chemistry

MARYLAND HEIGHTS
Smith, Terry Douglas, medicinal chemistry

MARYVILLE
Beeks, John Charles, soils, field crops
Carpenter, Sammy, organic chemistry
Hart, Richard Allen, entomology, biochemistry
Kenner, Morton Roy, mathematics, geology
Mallory, Bob Franklin, geology, paleontology
Maxwell, Dwight Thomas, mineralogy
Moss, Ronnie Lee, statistics, operations research
Mueller, Irene Marian, plant ecology
Oomens, Frederick Walter, agriculture
Padgitt, Dennis Darrell, dairy husbandry
Rosenburg, Dale Weaver, organic chemistry

MEXICO
Danser, James Weart, economic geology, engineering geology

MOUNTAIN GROVE
Brown, Gerald Richard, pomology, viticulture
Hanson, Kenneth Warren, horticulture
Moore, James Frederick, Jr, plant pathology
Townsend, Howard Garfield, Jr, entomology

MT VERNON
Justus, Norman Edward, agronomy

NORTH KANSAS CITY
Bechtle, Gerald Francis, organic chemistry
Brizio-Molteni, Loredana, reconstructive surgery

ODESSA
Carlyle, David Wesley, inorganic chemistry

OLIVETTE
Jaworski, Ernest George, biological chemistry, molecular biology
Raffelson, Harold, pharmaceutical chemistry

OVERLAND
Hall, Frank Foy, biochemistry

PACIFIC
Ruehle, Archie Edwin, pharmacognosy

PARKVILLE
Hamilton, John Meacham, biology
Pivonka, William, organic chemistry

PERRYVILLE
Powers, William Thomas, biology

POINT LOOKOUT
Davis, D Wayne, environmental biology
De Jong, Marvin Lee, radio astronomy
White, Winifred Sharlene, biology

PORTAGEVILLE
Duclos, Leo Albert, plant breeding
Kerr, Harold Delbert, weed science
Sappenfield, William Paul, agronomy, plant breeding
Shannon, James Grover, plant breeding

PUXICO
Fredrickson, Leigh H, wildlife ecology

RAYTOWN
Pritham, Gordon Herman, physiological chemistry

ROLLA
Adawi, Ibrahim (Hasan), theoretical solid state physics
Alexander, Ralph William, Jr, solid state physics
Alldredge, Gerald Palmer, physics, materials science
Anderson, Richard Alan, atomic physics, molecular physics
Bain, Lee J, mathematical statistics

Bartley, David Lauren, solid state physics, cloud physics
Beistel, Donald W, physical chemistry
Bell, Robert John, spectroscopy, solid state physics
Bertrand, Gary Lane, physical chemistry
Beveridge, Thomas Robinson, geology
Biolsi, Louis, Jr, theoretical chemistry
Bolter, Ernst A, geochemistry, geology
Brown, Harry Allen, magnetism
Carstens, John C, physics
Cochran, Andrew Aaron, biophysics, quantum physics
Cuthbertson, George Raymond, physical chemistry
Daane, Adrian Hill, inorganic chemistry
Erkiletian, Dickran Hagop, Jr, mathematics
Eyer, Jerome Arlan, exploration geology, exploration geophysics
Fellows, Larry Dean, stratigraphy, environmental geology
Fuller, Harold Q, physics
Gale, Nord Loran, microbiology, biochemistry
Gerson, Robert, physics
Grant, Sheldon Kerry, mineralogy, petrology
Grimm, Louis John, applied mathematics
Haddock, Aubra Glen, topology
Hagni, Richard D, economic geology, petrology
Hale, Barbara Nelson, physics
Hale, Edward Boyd, solid state physics
Hamblen, John Wesley, computer science
Hardtke, Fred Charles, Jr, chemistry
Hatfield, Charles, Jr, mathematics
Hicks, Troy L, mathematics
Hill, Otto Herman, physics
Howe, Wallace Brady, geology
Hufham, James Birk, bacteriology, genetics
James, William Joseph, solid state chemistry, electrochemistry
Johnson, Charles Andrew, mathematics
Kassner, James Lyle, Jr, cloud physics
Kisvarsanyi, Geza, economic geology
Koenig, John Waldo, paleontology
Lee, Ralph Edward, computer science
Levenson, Leonard L, physics
Long, Gary John, physical inorganic chemistry
Lund, Louis Harold, chemical physics
Manuel, Oliver K, nuclear chemistry, geochemistry
Maxwell, James Christie, geology, geophysics
McDonald, Hector O, physical inorganic chemistry
McFarland, Charles Elwood, physics
McFarland, Robert Harold, physics
Nicholson, Larry Michael, biochemistry
Nygaard, Kaare Johann, atomic physics, quantum optics
Park, John Thornton, atomic physics
Parks, William Frank, theoretical physics
Parry, Myron Gene, theoretical mechanics, applied mechanics
Pauls, Franklin Benjamin, nuclear physics
Penico, Anthony Joseph, mathematical physics
Plummer, Otho Raymond, applied mathematics, mathematical physics
Plummer, Patricia Lynne Moore, cloud physics, theoretical physics
Podzimek, Josef, cloud physics, meteorology
Proctor, Paul Dean, economic geology, structural geology
Pursell, Lyle Eugene, mathematical analysis
Rakestraw, Roy Martin, mathematics
Rechtien, Richard Douglas, geophysics
Rigler, A Kellam, numerical analysis, optics
Roach, Donald Vincent, physical chemistry
Robertson, Bobby Ken, physical chemistry
Rupert, Gerald Bruce, geophysics
Russell, Robert Raymond, chemistry
Schearer, Laird D, atomic physics
Schmitt, John Leigh, astronomy
Siehr, Donald Joseph, biochemistry
Snow, William Rosebrook, atomic physics, molecular physics
Spartin, Don Merle, solid state physics
Spreng, Alfred Carl, stratigraphy
Stampfer, Joseph Frederick, Jr, chemistry
Stanojevic, Caslav V, mathematics
Stoffer, James Osber, organic chemistry
Tappmeyer, Wilbur Paul, inorganic chemistry
Thompson, Thomas Luther, geology
Warner, Don Lee, hydrogeology, geological engineering

Wulfman, David Swinton, synthetic organic chemistry, physical organic chemistry
Zung, Joseph T, physical chemistry, quantum chemistry

ST CHARLES
Bornmann, John Arthur, physical chemistry
Brescia, Vincent Thomas, microbial genetics
Delaney, Patrick Francis, Jr, metabolism, academic administration
Holt, John Melvin, mathematics
Talbot, Mary, ecology

ST JOSEPH
Davis, Larry E, animal science, animal nutrition
Ewart, Ralph Bradley, science writing, botany
Hendrickson, Adolph Alexander, bacteriology
Holt, Wendell Levern, chemistry
Musser, Samuel John, biochemistry
Nelson, Robert John, mathematics, statistics
Peniston, Francis L, bacteriology

ST LOUIS
Ackermann, Philip Gulick, biochemistry
Adams, Kenneth Howard, organic chemistry
Alderfer, Ronald Godshall, ecology
Alkjaersig, Norma Kirstine (Mrs A P Fletcher), biochemistry
Allen, Garland Edward, III, history of science, history of biology
Allen, William E, Jr, medicine, radiology
Alt, Gerald Horst, organic chemistry
Althaus, Ralph Elwood, plant pathology
Ames, Donald Paul, physical chemistry
Anagnostopoulos, Constantine E, organic chemistry, polymer chemistry
Andalafte, Edward Ziegler, mathematics
Andrews, John Jacob, mathematics
Anthony, E James, child psychiatry
Arens, Max Quirin, virology
Armbruster, Charles William, organic chemistry
Arneson, Axel Norman, obstetrics & gynecology
Averett, John E, systematic botany, phytochemistry
Avioli, Louis, nuclear medicine, endocrinology
Bachman, Gerald Lee, organic chemistry
Bader, Robert Smith, evolutionary biology
Baernstein, Albert, II, mathematical analysis
Bailey, Paul Townsend, physics
Baizer, Manuel M, organic chemistry
Baker, Joseph Willard, organic chemistry
Balbes, Raymond, mathematics
Ballinger, Walter F, II, surgery
Balthazor, Terrell Mack, organic chemistry
Baltz, Howard Burl, applied statistics, management sciences
Banaszak, Leonard Jerome, crystallography, biochemistry
Barkate, John Albert, microbiology, food science
Barner, Hendrick Boyer, surgery
Barnes, Byron Ashwood, pharmacology
Barnett, Kenneth Wayne, inorganic chemistry, organic chemistry
Barnstorff, Henry Dreses, industrial analysis
Barton, Lawrence, inorganic chemistry
Bauermeister, Herman Otto, industrial organic chemistry, chemical engineering
Baumann, Thiema Marie Wolf, health sciences, human genetics
Bazzano, Gaetano, internal medicine, biochemistry
Beestman, George Bernard, agricultural chemistry, weed science
Bellone, Clifford John, immunobiology
Belt, Charles Banks, Jr, economic geology
Bennet, Velma, geography
Bennett, John William, cultural anthropology, ecology
Bensinger, David August, dentistry, periodontology
Berndt, Alan Fredric, physical chemistry, dental research
Bettonville, Paul John, pharmacology
Betz, Norman Leo, behavioral physiology, food biochemistry
Bhagat, Budh Dev, physiology, pharmacology
Bingenheimer, Levi Edwin, Jr,

pharmaceutical chemistry, analytical chemistry
Binns, Walter Robert, lasers, cosmic ray physics
Bischoff, Eric Richard, developmental biology, cell biology
Blanton, Madison Van, biochemistry, biology
Block, Eric, organic chemistry
Bloomfield, Jordan Jay, organic chemistry
Blum, Victor Joseph, geophysics
Blumenthal, Herman T., pathology
Boedeker, Roy Vincent, obstetrics & gynecology
Boime, Irving, molecular pharmacology
Bolef, Dan Isadore, solid state physics
Boothby, William Munger, mathematics
Bose, Subir Kumar, virology, molecular biology
Boyarsky, Saul, urology, physiology
Boyd, Donald Mitchell, biochemistry, microbiology
Bradshaw, Ralph Alden, biochemistry
Brew, William Barnard, chemistry
Brice, James Coble, geomorphology
Brill Kenneth Gray, Jr., geology
Brodeur, Armand Edward, radiobiology
Brodman, Estelle, history of medicine, information science
Brot, Frederick Elliot, biochemical genetics
Brown, Barbara Illingsworth, physiological chemistry
Brown, David Henry, biological chemistry
Brown, Elmer Burrell, internal medicine, hematology
Brown, Ian McLaren, magnetic resonance
Brunges, Robert Anthony, solid state physics
Brunngraber, Eric Gustav, biochemistry
Bruns Lester George, pharmacy, genetics
Bsharah, Lewis, chemistry
Buchdahl, Rolf, physics
Bunge, Mary Bartlett, cytology
Bunge, Richard Paul, anatomy, cell biology
Burbage, Joseph James, inorganic chemistry
Burch, Helen Bulbrook, biochemistry
Burgess, James Harland, physics
Burleson, James C., organic chemistry
Burns, Francis John, surgery
Burton, Harold, neurophysiology
Burton, Robert Main, biochemistry
Bushkovitch, Alexander Viatcheslav, theoretical physics
Butcher, Harvey Raymond, Jr., surgery
Calandra, Alexander, physics, chemistry
Callis, Clayton Fowler, inorganic chemistry
Cantwell, John Christopher, mathematics
Carey, Paul L., biochemistry
Carpenter, Will Dockery, plant physiology
Caughlan, John Arthur, organic chemistry
Caulder, Jerry Dale, weed science
Cavanagh, Denis, obstetrics & gynecology
Celesia, Gastone G., neurology.
Chickos, James S., organic chemistry.
Choi, Sung Chil, biostatistics
Chraplyvy, Zenobius Volodymyr, physical organic chemistry
Chaplin, Hugh H., Jr., medicine
Chapman, Douglas Wilfred, organic chemistry
Churchill, Ralph John, water chemistry, environmental chemistry
Cicero, Theodore James, biochemical pharmacology
Clark, Frank S., organic chemistry
Clark, John Walter, theoretical nuclear physics, theoretical astrophysics
Clark, Sidney Gilbert, organic chemistry
Clayton, Paula Jean, psychiatry
Clayton, Robert Allen, biochemistry
Cleaves, Arthur Bailey, geology
Cockrell, Ronald Spencer, biochemistry
Codd, John Edward, transplantation immunology, thoracic surgery
Coerver, Helen Joseph, inorganic chemistry
Cohen, Adolph Irvin, neurobiology
Cole, Basil Chambrus, bacteriology
Collard, William David, oncology, virology
Collins, Lloyd Raymond, anthropology
Commoner, Barry, biology

Conklin, Robert A., analytical chemistry, organic chemistry
Conn, James Frederick, cereal chemistry
Connell, Rosemary, physiology
Connett, William C., mathematical analysis
Connors, Natalie Ann, histology
Conoyer, John Weedon, geography
Converse, Jimmy G., chemical instrumentation
Coret, Irving Allen, pharmacology
Corey, Joyce Yagla, inorganic chemistry
Corey, Eugene R., inorganic chemistry, crystallography
Coscia, Carmine James, biochemistry
Cover, Morris Seifert, veterinary pathology
Craddock, John Harvey, inorganic chemistry
Craver, John Kenneth, chemistry, technological forecasting
Crumrine, Ann Louise, synthetic organic chemistry
Crutchfield, Marvin Mack, physical chemistry
Csapo, Arpad Istvan, physiology
Curphey, Thomas John, organic chemistry
Cutler, Hugh Carson, botany
D'Agrosa, Louis Salvatore, medical physiology
Dahlgren, Robert R., veterinary pathology
Dahms, Donald J., structural chemistry
Dahms, Thomas Edward, physiology
Daly, John Francis, algebra
D'Amico, John J., organic chemistry
Danforth, William H., medicine
Danis, Peter Godfrey, pediatrics
Daughaday, William Hamilton, medicine
Davidse, Gerrit, taxonomic botany
Davie, Joseph Myrten, immunology
Davis, James Wendell, biochemistry, statistics
Daw, Nigel Warwick, neurophysiology, psychophysics
Dean, Walter Keith, inorganic chemistry
Delaney, Robert Michael, theoretical physics
Deming, John Miley, agronomy
Derby, Albert, developmental biology, endocrinology
Dewar, Norman Ellison, microbiology, biochemistry
De Weer, Paul Joseph, biophysics, physiology
Dietrich, Martin Walter, environmental chemistry, analytical chemistry
DiFate, Victor George, organic chemistry, pesticide chemistry
Dill, Dale Robert, organic chemistry, pharmaceutical chemistry
Dina, Stephen James, plant ecology
Diuguid, Lincoln Isaiah, organic chemistry
Dodge, Philip Rogers, pediatrics
Doisy, Edward Adelbert, Sr., biochemistry
Doisy, Edward Adelbert, physiology, biochemistry
Donati, Robert M., internal medicine, nuclear medicine
Dombush, Rhea L., health sciences
Dorner, Robert Wilhelm, biochemistry
Dub, Michael, organometallic chemistry, organic chemistry
Durbin, Marshall Elza, anthropology
Durbin, Mridula Adenwala, anthropology
Dus, Karl M., biochemistry, biochemical pharmacology
Dutra, Gerard Anthony, organic chemistry, agricultural chemistry
Dwyer, John Duncan, systematic botany
Dyroff, David Ray, industrial chemistry
Ebner, Jerry Rudolph, chemistry
Echols, Dorothy Jung, geology
Edwards, Carol Abe, applied mathematics
Ehrenthal, Irving, biochemistry
Eibert, John Jr., pharmaceutical chemistry
Eichling, John O., radiation biophysics
Eigel, Edwin George, Jr., number theory
Eime, Lester Oscar, industrial chemistry
Eldredge, Donald Herbert, physiology, biophysics
Eliasson, Sven Gustav, neurology
Eliceiri, George Louis, cell biology, biochemistry
Eller, Charles Howe, epidemiology, public health administration
Elliott, Martha Schwandt, physiology
Elliott, William H., biochemistry, organic chemistry
Ellison, John Vogelsanger, electronics, underwater acoustics
Elvin-Lewis, Memory P F., virology, bacteriology

Emery, Edward Mortimer, physical chemistry
Etheredge, Edward Ezekiel, surgery, transplantation immunology
Everett, James Peck, Jr., animal nutrition
Fabian, Leonard William, anesthesiology
Farrar, Martin Wilbur, organic chemistry, polymer chemistry
Fedders, Peter Alan, physics
Feder, Joseph, biochemistry
Feenberg, Eugene, physics
Feigin, Ralph David, pediatrics, infectious diseases
Field, Byron Dustin, spectroscopy
Feiler, William A Jr., industrial chemistry
Feir, Dorothy Jean, insect physiology
Feldman, Hyman Morris, mathematics
Felix, Robert Hanna, psychiatry
Ferrendelli, James Anthony, neurology, neuropharmacology
Finger, Thomas Emanuel, neuroanatomy
Finn, James Bernard, histology
Fishman, Marvin Allen, pediatric neurology, rehabilitation
Fitch, Coy Dean, medicine, biochemistry
Fleischman, Julian B., microbiology
Fleming, Theodore Harris, zoology, mammalian ecology
Fletcher, Anthony Phillips, internal medicine
Fore, Paul Lewis, fish biology
Forsberg, John Herbert, inorganic chemistry
Forster, Denis, inorganic chemistry
Fowler, Lewis, physical chemistry, analytical chemistry
Fox, Michael Wilson, ethology
Francis, Faith Ellen, chemistry
Franz, John Edward, bio-organic chemistry
Frawley, Thomas Francis, medicine
Freeman, John Jerome, physical chemistry
Freeman, Robert Clarence, organic chemistry
Freerks, Marshall Cornelius, atomic physics
Freese, Raymond William, geometry
Freiberg, George William, plant pathology, plant physiology
Freiwald, Ronald Charles, topology
French, Charles Leroy, chemistry
Frieden, Carl, biochemistry
Friedman, Harvey Paul, embryology, immunology
Friedman, Lawrence David, genetics
Fryer, Minot Packer, surgery
Gaertner, Van Russell, organic chemistry
Gallagher, Neil Ignatius, internal medicine, hematology
Gammel, John Ledel, theoretical physics
Gandhi, Sham Sunder, pharmacology
Gantner, George E., Jr., medicine, pathology
Garey, Carroll Laverne, chemistry
Garin, David L., organic chemistry
Gaspar, Peter Paul, physical organic chemistry
Geisman, Raymond August, Sr., analytical chemistry, radiochemistry
Geller, David Melville, biochemistry
Gentry, Alwyn Howard, taxonomic botany
Gerard, Gary Floyd, molecular biology, virology
Gerfen, Charles Otto, physical chemistry
Germuth, Frederick George, Jr., pathology
Glaser, Luis, biochemistry
Gleaves, John Thompson, physical chemistry
Gledhill, William Emerson, microbiology, biochemistry
Godt, Henry Charles, Jr., organic chemistry
Gold, Alvin Hirsh, biochemistry, pharmacology
Goldring, Sidney, neurosurgery
Goldstein, Milton Norman, cell biology
Graham, Paul Roger, organic chemistry, environmental management
Green, Maurice, biochemistry
Greene, Daryle E., nutrition, biochemistry
Greenley, Robert Z., polymer chemistry
Greider, Marie Helen, cytology
Griffith, Edward Jackson, physical chemistry
Griffiths, David Warren, physical organic chemistry, bio-organic chemistry
Grobman, Arnold Brams, zoology, academic administration
Groenweghe, Leo Carl Denis, inorganic chemistry, computer sciences
Grogan, Donald E., cell biology
Groves, Warren Olley, physical chemistry, semiconductors

Guerry, Davenport, Jr., analytical chemistry
Gutsche, Carl David, organic chemistry, bio-organic chemistry
Guze, Samuel Barry, psychiatry
Haberman, Warren Otto, entomology
Haimo, Deborah Tepper, mathematical analysis
Haimo, Franklin, mathematics
Hallett, Floyd Prentice, pharmaceutical chemistry
Hamburger, Viktor, zoology
Hamm, Philip Curtis, plant physiology
Hammann, William Curl, organic chemistry
Hansen, Gary Ralph, synthetic organic chemistry, mineralogy
Harvey, Joseph Claude, biochemistry, anatomy
Hansen, Hobart Raymond, medical research
Hardwick, William Aubrey, Jr., research administration, food chemistry
Harford, Carl Gayler, medicine
Harris, Harold H., physical chemistry
Hartman, Boyd Kent, neurochemistry
Harvey, Cecil Claude, biochemistry
Hays, Thomas Hamilton, otolaryngology
Haynes, William Miller, analytical chemistry
Hedgecock, Loyd Wilson, bacteriology
Hedrick, Ross Melvin, organic chemistry
Heider, Rudolph Louis, organic chemistry
Heiner, Mary Grace, organic chemistry
Heiniger, Samuel Allen, organic chemistry
Herelendy, Frank, physiological chemistry
Herber, John Frederick, organic chemistry
Helmholz, Lindsay, chemistry
Henson, Bob Londes, low temperature physics
Herrmann, Robert Bernard, seismology
Hershey, Falls Bacon, surgery
Hicks, Patricia Fain, biological chemistry, communication science
Hight, Robert, solid state physics
Hill, Dale Eugene, physics
Hill, Helene Zimmermann, biology, genetics
Hill, Jim Tom, metabolism, toxicology
Hillman, Richard Ephraim, medical genetics
Heinrich, Ross Raymond, geophysics
Heithaus, Joseph John, Jr., petroleum chemistry
Heitsch, Charles Weyand, inorganic chemistry
Hinchey, Norman Shreve, geology
Hinkebein, John Arnold, physical chemistry
Hirsch, Richard Henry, organic chemistry
Hirschberg, Carlos Benjamin, biochemistry
Hirschman, Isidore Isaac, Jr., mathematics
Hirsh, Ira Jean, psychophysics
Hirzy, John William, organic polymer chemistry, plastics chemistry
Hochwalt, Carroll Alonzo, organic chemistry
Hoffman, Joyce Bennett, botany
Hoffman, Michael K., physical organic chemistry, mass spectrometry
Hofling, Charles Kreimer, psychiatry
Hohenberg, Charles Morris, physics
Hohnstedt, Leo Frank, inorganic chemistry
Holloezy, John O., physiology
Holm, Myron James, organic chemistry, food chemistry
Holmes, William Farrar, biochemistry
Holtzer, Alfred Melvin, chemistry
Homeyer, August Henry, chemistry
Hopkins, Daniel T., nutritional biochemistry
Hopkins, Johns Wilson, molecular biology
Horenstein, Simon, neurology
Hortmann, Alfred Guenther, organic chemistry
Horwitz, Max Kenneth, nutrition
Howe, Robert Kenneth, organic chemistry
Hudgens, Richard Watts, psychiatry
Hunsley, James Ray, biochemistry
Hunt, Carlton Cuyler, neurophysiology
Hunt, James Howell, ecology
Hunter, Francis Edmund, Jr., biochemistry, pharmacology
Husted, Robert Forest, agronomy
Hwang, Ung Kee, anatomy, embryology

Hyndman, Harry Lester, organic chemistry
Israel, Martin Henry, physics
Ivey, Francis James, food science
Jackson, Craig Merton, biochemistry
Jackson, Henry Woodrow, theoretical physics
Jacobsmeyer, Vincent Paul, solid state physics
James, Philip Benjamin, particle physics
Jamieson, Norman Clark, organic chemistry
Janney, James G, Jr, cardiology
Jarett, Leonard, clinical pathology, biochemistry
Jaynes, Edwin Thompson, theoretical physics
Jeffrey, John J, biological chemistry, endocrinology
Jenkins, James Allister, mathematics
Johnson, Mary Frances, inorganic chemistry
Johnson, William K, organic chemistry
Johnston, Marilyn Frances Meyers, biochemistry, immunology
Jones, Edward George, anatomy
Jones, Marvin Thomas, physical chemistry
Jonsson, Valgard, public health, medical microbiology
Kaelble, Emmett Frank, analytical chemistry, spectroscopy
Kaiser, George C, thoracic surgery
Kalota, Dennis Jerome, organic chemistry
Kaminski, Donald Leon, surgery
Katz, Israel Norman, mathematics, statistics
Katzman, Philip Aaron, biochemistry
Keane, John Francis, Jr, biophysics, physiology
Keay, Leonard, biochemistry, microbiology
Keller, Robert Ellis, analytical chemistry
Kelley, John Daniel, chemical physics
Kelly, R Emmet, medicine
Kenyon, Allen Stewart, polymer science
Kern, Roland James, chemistry
Kernaghan, Marie, physics
Kertz, Alois Francis, animal nutrition
Kessler, Gerald, biochemistry
Kester, William Lee, environmental physics
Kidwell, Roger Lynn, synthetic organic chemistry
Kim, Yee Sik, biochemistry, pharmacology
King, Barry Frederick, microscopic anatomy, cytology
King, Morris Kenton, internal medicine
King, Thomas Morgan, physical inorganic chemistry
Kinsella, Ralph A, Jr, internal medicine
Kinsky, Stephen Charles, biochemistry
Kipnis, David Morris, internal medicine, endocrinology
Kirk, David Livingstone, developmental biology
Kirk, Marilyn Chaloupka, developmental biology
Kissane, John M, medicine
Klarmann, Joseph, physics
Klein, William McKinley, Jr, plant taxonomy
Klotz, John William, genetics, philosophy of science
Knight, William Allen, Jr, gastroenterology
Knox, Walter Robert, organic chemistry
Kobayashi, George S, mycology, biochemistry
Koerner, William Elmer, physical chemistry
Kohl, Daniel Howard, plant physiology
Kohl, Seena B, cultural anthropology
Kornfeld, Stuart Arthur, hematology
Korol, Bernard, psychopharmacology
Krueger, Paul A, organic chemistry
Krukowski, Marilyn, physiology, developmental biology
Ksycki, Mary Joecile, electrochemistry
Kuhlman, Robert Eugene, orthopedic surgery
Kuhn, Charles, III, pathology
Kuo, Tseng-Tong, pathology
Kurz, Joseph Louis, physical organic chemistry
Lacy, Paul Eston, pathology
Lake, Lorraine Frances, anatomy, rehabilitation medicine
Laks, Hillel, thoracic surgery
Landau, William M, neurology, neurophysiology
Lang, Gerhard Paul, inorganic chemistry
Lannert, Kent Philip, industrial organic chemistry
Lanson, Herman Jay, organic chemistry
Larsen, David W, physical chemistry
Larson, Wilbur John, analytical chemistry
Laskowski, Leonard Francis, Jr, medical microbiology, clinical microbiology
Layloff, Thomas, analytical chemistry
Lee, Charlotte Elizabeth Outland, biochemistry
Lee, Emerson Howard, physical chemistry
Lee, Howard Augustus, biochemistry
Lee, Ralph Hewitt, inorganic chemistry, academic administration
Lee, Sook, solid state physics
Leventhal, Jacob J, atomic physics, molecular physics
Levi-Montalcini, Rita, neurology
Levin, Harold Leonard, geology, paleontology
Levinskas, George Joseph, toxicology, pharmacology
Levy, Irwin, neurology, psychiatry
Levy, Ronald Fred, topology
Lewis, Walter Hepworth, botany
Li, Tao Ping, organic chemistry
Lin, Yeong-Jer, meteorology, atmospheric sciences
Lind, Arthur Charles, applied physics, space physics
Lindenblad, Gordon Eric, biochemistry
Lindhorst, Taylor Erwin, mycology
Lipkin, David, organic chemistry
Lish, Paul Merrill, pharmacology
Lissant, Ellen Kern, phycology
Little, John Russell, Jr, immunology
Loeb, Virgil, Jr, oncology, hematology
Loewy, Arthur DeCosta, neuroanatomy
Loncrini, Donald Francis, organic chemistry
Long, Lawrence William, biochemistry
Longmore, William Joseph, biochemistry
Losee, Michael Leonard, organic chemistry
Lovett, Eva G, organic chemistry
Lowry, Oliver Howe, pharmacology, biochemistry
Ludwig, Frederick John, Sr, analytical chemistry, organic chemistry
Macias, Edward S, nuclear chemistry, air pollution
Maddy, Kenneth Hilton, biochemistry
Magee, Steve Carl, biochemistry
Magill, Robert Earle, bryology, taxonomy
Majerus, Philip W, hematology, medicine
Malone, Leo Jackson, Jr, inorganic chemistry
Mange, Franklin Edwin, organic chemistry
Maniotis, James, mycology
Markowski, Henry Joseph, organic polymer chemistry
Marr, James Joseph, infectious diseases, internal medicine
Marrs, Barry Lee, microbiology, biochemistry
Marshall, Fred Taylor, organic chemistry
Marshall, Garland Ross, biochemistry
Marshall, Kenneth Chenery, orthodontics
Martin, Donald Effon, meteorology
Martin, Edward Eugene, agricultural chemistry
Martonosi, Anthony, biochemistry
Marvel, John Thomas, organic chemistry, biochemistry
Massie, Edward, cardiology
Massopust, Leo Carl, Jr, anatomy
Masters, William Howell, obstetrics & gynecology
Mathews, Francis Scott, biochemistry
Mattson, Leland Neil, chemistry
Matzner, Edwin Arthur, organic chemistry
May, Walter Ruch, physical inorganic chemistry
McAlister, William H, medicine, radiology
McCarthy, Albert Joseph Patrick, historical geography, geography of Europe
McClure, James Nathaniel, Jr, psychiatry
McConaghy, John Stead, Jr, physical organic chemistry
McDonald, David William, organic chemistry
McDougal, David Blean, Jr, pharmacology
McDowell, Robert Hull, mathematics
McElfresh, Arthur Edward, Jr, medicine
McElheny, George Clark, organic chemistry
McGee, James Francis, atomic physics
McHugh, Kenneth Laurence, organic chemistry
Medoff, Gerald, microbiology
Melechen, Norman Edward, genetics
Melick, William F, urology
Menne, Thomas Joseph, theoretical physics
Menton, David Norman, anatomy, histology
Menz, Leo Joseph, biophysics, cryobiology
Mesmer, Gustav, applied mechanics
Metzner, Wendell Phillips, chemistry
Meyer, Ferdinand Clark, organic chemistry
Middelkamp, John Neal, pediatrics
Mieure, James Philip, chemistry
Miller, James Gegan, ultrasound, medical biophysics
Miller, Jane Alsobrook, chemistry
Miller, Joyce Mary, biochemistry
Minard, Edwin Lincoln, bacteriology
Minnich, Virginia, hematology, nutrition
Mitchell, Brian James, geophysics
Moedritzer, Kurt, organometallic chemistry
Mohrman, Harold W, chemistry
Molnar, Charles Edwin, neurophysiology, computer science
Montgomery, Arthur Vernon, physiology
Montgomery, George Edward, anthropology
Moog, Florence, developmental biology
Moore, Blake William, biochemistry
Moore, Edward Lee, inorganic chemistry
Moore, Morris, mycology, dermatology
Moore, Richard Newton, organic chemistry
Morgan, Wyman, inorganic chemistry
Morr, Charles Vernon, food chemistry
Morris, Daniel W, molecular biology, biochemistry
Morris, Donald Eugene, organometallic chemistry
Morrison, George Robert, biochemistry, internal medicine
Morrison, Richard Donald, dentistry
Morse, Ronald Loyd, industrial organic chemistry
Mount, Ramon Albert, organic chemistry
Moyer, Frank H, cytology, environmental science
Mudd, J Gerard, cardiology
Mueller, Joseph Robert, biochemistry, endocrinology
Muller, Marcel Wettstein, physics
Mulligan, James Anthony, animal behavior
Mulligan, Leo Virgil, surgery
Munch, John Howard, organic chemistry
Munch, Ralph Howard, physical chemistry
Munns, Theodore Willard, biochemistry
Munson, Arvid W, agricultural statistics
Murino, Clifford John, meteorology
Murphy, George Earl, psychiatry
Murray, Robert Wallace, physical organic chemistry
Nadel, Eli Maurice, physiology, pathology
Nadel, Marvin Keith, chemistry
Naeger, Leonard L, pharmacology
Nason, Howard King, chemistry
Nelson, James S, neuropathology
Neubert, Ralph Lewis, research administration
Newell, Jon Albert, biochemistry
Nicholas, Harold Joseph, biochemistry
Nichols, Herbert Wayne, biology, phycology
Nicholson, Eugene Haines, mathematics, physics
Nielsen, Lawrence Ernie, polymer science
Nikolai, Robert Joseph, engineering mechanics, orthodontics
Norberg, Richard Edwin, physics
North, Gerald R, theoretical physics
Nuetzel, John Arlington, internal medicine
Nussbaum, Adolf Edward, mathematics
Nutter, James Douglas, physical chemistry
Nuttli, Otto William, geophysics
Oftedahl, Marvin Loren, organic chemistry
Ogilvie, James Louis, analytical chemistry
Ogura, Joseph H, otolaryngology
O'Leary, James Lee, anatomy
Oliver, G Charles, cardiovascular disease
Olney, John William, psychiatry, neuropathology
Olson, Robert Eugene, biochemistry, medicine
O'Neal, Patricia L, psychiatry
Ornstein, Wilhelm, applied mathematics, mathematical analysis
Osterland, C Kirk, immunology, internal medicine
Overman, Ralph Theodore, medical education
Owsley, Dennis Clark, synthetic organic chemistry
Pace, Caroline S, physiology
Padberg, Harriet A, mathematics
Paine, Robert, physiology
Pallmann, Albert J, atmospheric physics, environmental sciences
Parker, Brent M, internal medicine, cardiology
Parker, Charles W, internal medicine, immunology
Parker, Mary Langston, medicine, endorincology
Paulik, Frank Edward, organometallic chemistry
Pearlman, Alan Lee, neurobiology, neurology
Peck, William Arno, endocrinology
Perez, Carlos A, radiotherapy
Perkins, R Marlin, zoology
Perkoff, Gerald Thomas, internal medicine
Perry, Horace Mitchell, Jr, medicine
Perry, Randolph, Jr, industrial chemistry
Peterson, Gerald E, mathematical analysis, computer sciences
Peterson, Roy Reed, anatomy
Phillips, Bruce Edwin, organic chemistry
Pickard, Barbara Gillespie, plant physiology
Pickard, William Freeman, biophysics, biomedical engineering
Pierce, John Albert, internal medicine
Piper, Roger D, physical chemistry
Platt, William Rady, pathology
Plattner, Stuart Mark, economic anthropology
Podosek, Frank A, mass spectrometry, meteoritics
Pogell, Burton M, biochemistry
Porter, Clark Alfred, plant pathology
Premachandra, Bhartur N, endocrinology, immunology
Price, Joseph Levering, anatomy
Pullon, Peter Akins, oral pathology
Purkerson, Mabel Louise, medical science, nephrology
Quay, Paul Michael, thermal physics, philosophy of science
Quinn, John Frederick, industrial chemistry, engineering management
Radke, Rodney Owen, soil science
Raichle, Marcus Edward, neurology
Raizman, Paula, organic chemistry
Ramsey, Penny Tina, medical microbiology
Ramsey, Robert Bruce, neurochemistry
Rana, Mohammed Waheeduz-Zaman, human anatomy
Rao, Gandikota V, meteorology
Rao, Papineni Seethapathi, reproductive physiology, cardiovascular physiology
Raskas, Heschel Joshua, biochemistry
Ratts, Kenneth Wayne, organic chemistry
Raven, Peter Hamilton, botany, population biology
Raw, Cecil John Gough, physical chemistry
Reid, Stanley Lyle, organic chemistry
Reinhard, Edward Humphrey, medicine
Reis, Raymond Hugo, zoology
Reutzel, Lawrence Frederick, biometrics, computer science
Reynolds, Fred C, orthopedic surgery
Rice, Bernard, physical chemistry
Richard, William Ralph, Jr, industrial organic chemistry, paper chemistry
Richards, Charles Norman, physical organic chemistry
Richards, Graydon Edward, soil fertility
Richardson, Ralph J, solid state physics
Richins, Calvin Alexander, anatomy
Rigden, John Saxby, acoustics
Riles, James Byrum, algebra
Rinehart, Keith Edward, nutrition, biochemistry
Ring, John Robert, anatomy
Robbins, Ernest Aleck, biochemistry
Robins, Eli, biochemistry, psychiatry
Robinson, Donald Alonzo, organic chemistry
Rochberg, Richard Howard, mathematics
Rodin Ervin Y, applied mathematics
Rodriguez, Eugene, immunology
Rolf, Doris Barbara, physiology
Roodman, Stanford Trent, biochemistry, immunology
Roos, Albert, physiology
Rosan, Robert Carl, pathology
Rosenthal, Harold Leslie, biochemistry
Roth, James Frank, physical chemistry
Rothrock, Thomas Stephenson, organic chemistry
Rouse, Robert Arthur, theoretical chemistry
Rovainen, Carl (Marx), neurophysiology
Rudman, Sanford Winton, biomathematics, biophysics
Ruh, Mary Frances, physiology, endocrinology
Russo, Michael Eugene, chemistry
Sabacky, M Jerome, organic chemistry
Sabharwal, Chaman Lal, mathematics

MISSOURI

Saeger, Victor William, environmental chemistry, analytical chemistry
Sage, Martin, zoology, physiology
Sandel, Thomas Theodore, psychophysiology
Sansone-Bazzano, Gail, biochemistry, nutrition
Sarantites, Demetrios George, nuclear chemistry, nuclear physics
Satkowski, William Briscoe, organic chemistry
Scallet, Barrett Lerner, chemistry
Scandrett, John Harvey, physics
Schaefer, Jacob Franklin, physical chemistry
Schaefer, Henry Maximilian, mathematics
Scharpf, Lewis George, Jr., environmental sciences
Schatz, Robert James, polymer chemistry
Schenkenberg, Philip Rawson, inorganic chemistry, cellulose chemistry
Schisla, Robert M., organic chemistry
Schleppnik, Alfred Adolf, synthetic organic chemistry, chemoreception
Schlesinger, Milton J., biochemistry
Schlesinger, Sondra, virology, microbiology
Schlessinger, David, molecular biology, genetics
Schneider, Julius Edward, biochemical genetics
Schober, Charles Coleman, psychiatry, psychoanalysis
Schuck, James Michael, bio-organic chemistry, protein chemistry
Schuldt, Erich Henry, microbiology, food technology
Schulte, Irwin, medicine, virology
Schultz, Robert George, organic chemistry
Schulze, Irene Theresa, virology, biochemistry
Schumacher, Ignatius, organic chemistry, medicinal chemistry
Schwall, Robert Ely, medicine
Schwartz, Alan Lee, mathematics
Schwartz, Henry Gerard, neurosurgery
Schweiss, John Francis, medicine
Sears, J Kern, plastics chemistry
Seeley, Robert Dudley, biochemistry
Seltzer, Jo Louise, biochemistry
Senay, Leo Charles, Jr., physiology
Senturia, Ben Harlan, otolaryngology
Sewell, Raymond F., animal nutrition
Sexton, Owen J., ecology, vertebrate zoology
Shank, Robert Ely, medicine
Shannugam, Govindaswamy, molecular biology, virology
Sharp, Dexter Brian, chemistry
Shaver, Kenneth John, food chemistry
Shaw, Roger Walz, solid state physics
Shelton, Damon Charles, organic chemistry
Sherman, William Reese, biochemistry
Shieh, Kenneth Kuang-Zen, industrial microbiology
Shourd, Melvin Lee, physiology, paleoecology
Shrauner, Barbara Abraham, plasma physics, biophysics
Shrauner, James Ely, physics
Shreffler, Donald Cecil, genetics
Shull, Franklin Buckley, physics
Shuter, Eli Ronald, neurology, neurochemistry
Siegel, Barry Alan, nuclear medicine, radiology
Silberberg, Ruth, pathology
Silbert, David Frederick, biochemistry
Silver, Simon David, molecular biology, microbial physiology
Silverman, Sol Richard, communications science
Simmons, David J., physiology, endocrinology
Slavin, Raymond Granam, internal medicine, allergy
Slocombe, Robert Jackson, organic polymer chemistry
Sly, William S., medical genetics
Smith, David Warren, analytical chemistry
Smith, Ellen Marian Evans, experimental embryology, biochemistry
Smith, John Russell, medicine
Smith, Kathleen, psychiatry
Smith, Kenneth Rupert, Jr., neurosurgery
Smith, Lowell R., organic chemistry
Snetsinger, David Clarence, poultry nutrition
Snyder, John Marshall, animal science
Snyder, Joseph Quincy, chemistry
Solberg, Lawrence Arthur, Jr., internal medicine, physiology
Solodar, Arthur John, organic chemistry

Sonnenwirth, Alexander Coleman, medical microbiology
Speziale, Angelo John, organic chemistry
Spitznagel, Edward Lawrence, Jr., algebra
Stacey, Larry Milton, physics
Stahl, Philip Damien, physiology, cell biology
Stalker, Harrison Dailey, genetics, evolutionary biology
Staniforth, Robert Arthur, inorganic chemistry
Staple, Tom Weinberg, medicine, radiology
Starcher, Barry Chapin, nutrition, biochemistry
Starkloff, Gene B., surgery
Stauder, M Francis Borgia, surgery
Stauder, William, geophysics, seismology
Stearns, Robert Inman, inorganic chemistry
Steers, Arthur Walter, pharmaceutical chemistry
Steger, Theodore Roosevelt, Jr., polymer physics
Steinke, Frederick H., poultry nutrition
Stejskal, Edward Otto, physical chemistry
Stenseth, Raymond Eugene, organic chemistry, pharmacy
Stephen, Charles Ronald, anesthesiology
Stephens, John Arnold, organic chemistry
Stevens, Sue Cassell, biochemistry
Stewart, Carleton C., immunology, biophysics
Strickberger, Monroe Wolf, evolution, genetics
Stroud, Malcolm Herbert, medicine
Suga, Nobuo, physiology
Suhovecky, Albert J., plant pathology
Sullivan, James Michael, neuroanatomy
Summer, Robert Jocelyn, natural products chemistry
Sundfors, Ronald Kent, solid state physics
Sweet, Frederick, biochemistry, organic chemistry
Sweet, Herbert C., internal medicine
Symington, Janey, virology, plant physiology
Taibleson, Mitchell H., mathematics
Takano, Masaharu, physical chemistry, applied physics
Tallman, Ralph Colton, organic chemistry
Tan, Charles Hua-Min, internal medicine, biochemistry
Taylor, David Lawrence, paper chemistry, chemical engineering
Taylor, Herbert Bradley, surgical pathology
Taylor, John Joseph, anatomy
Taylor, Keith Mar, organic chemistry
Teaford, Margaret Elaine, biochemistry, clinical chemistry
Teicher, Harry, colloid chemistry
Teng, James, chemistry
Teresine (Lewis), Mary, mathematics
Ternberg, Jessie L., surgery
Ter-Pogossian, Michel Mathew, medical physics
Thale, Thomas Richard, psychiatry
Thoma, George Edward, internal medicine, nuclear medicine
Thomas, Charles Allen, chemistry
Thomas, Lewis Jones, Jr., anesthesiology
Thompson, Quentin Elwyn, industrial organic chemistry
Thomson, Gordon Merle, biostatistics
Tipton, George Murtha, analytical chemistry
Titus, Dudley Seymour, food technology
Tolmach, Leonard Joseph, radiobiology
Tomkins, David Francis, analytical chemistry
Torack, Richard M., pathology
Totel, Gregory Lee, physiology
Touchette, Norman Walter, chemistry
Townsend, Jonathan, physics
Tusing, Thomas William, pharmacology
Tuthill, Samuel Miller, analytical chemistry
Ulett, George Andrew, psychiatry
Unland, Mark Leroy, physical chemistry
Valentine, Bob Leon, medical microbiology, biochemistry
Valeriote, Frederick Augustus, biophysics
Van Beaumont, Karel William, physiology
Van Deripe, Donald R., pharmacology
Van Kley, Harold, biochemistry, protein chemistry
Varner, Joseph Elmer, plant physiology

Vietti, Teresa Jane, pediatrics, hematology
Vincent, Stanislaw Aleksander, geophysics
Vineyard, Billy Dale, organic chemistry
Von Rohr, Beatrice Louise, applied mathematics
Wagenknecht, John Henry, organic chemistry, electrochemistry
Waggle, Doyle H., cereal chemistry
Wagner, Robert G., solid state physics
Wahl, Arthur Charles, nuclear chemistry
Walbot, Virginia Elizabeth, developmental biochemistry
Waldo, Willis Henry, inorganic chemistry
Walker, Robert Mowbray, space physics
Walsh, Raymond Robert, physiology
Wang, Maw Shiu, agronomy, spectrochemistry
Warren, James C., endocrinology, biochemistry
Wasson, Richard Lee, organic chemistry
Wax, Rosalie H., anthropology
Weber, Karl Keen, dentistry
Weber, Morton M., microbiology, physiology
Weeks, Paul Martin, surgery
Wegria, Rene, medicine
Wehunt, W. Ralph Lee, soil fertility, plant physiology
Weichselbaum, Theodore Edwin, biochemistry
Weiss, Guido Leopold, mathematics
Weiss, Virgil Wayne, polymer chemistry
Weissler, Harold Edward, food chemistry
Weissman, Samuel Isaac, physical chemistry
Welch, Michael John, radiochemistry, nuclear medicine
Weldon, Virginia V., pediatric endocrinology, medical administration
Welland, Grant Vincent, mathematics
Westura, Edwin Eugene, internal medicine, cardiovascular diseases
Wette, Reimut, biostatistics, biomathematics
Wharton, Ferdinand Decatur, Jr., animal nutrition, poultry nutrition
Wheeler, Elmer Perley, industrial hygiene
Whyte, Michael Peter, internal medicine
Wiest, Walter Gibson, biochemistry
Wilke, Frederick Walter, mathematics
Wilkens, Lon Allan, neurobiology
Williamson, Frank Shaver, Jr., inorganic chemistry
Williamson, James Lawrence, science, animal nutrition
Willman, Vallee L., surgery
Wilson, Frank B., speech, audiology
Wilson, James Dennis, chemistry
Wilson, Richard Hansel, plant physiology, biochemistry
Winter, Henry Frank, Jr., physiology
Winter, Rudolph Ernst Karl, organic chemistry
Woodbrey, James Calvin, polymer chemistry, polymer science
Woodruff, Robert Arnold, Jr., psychiatry
Woolsey, Robert S., medicine, neurology
Woolsey, Thomas Allen, neuroanatomy
Wright, David Lee, mathematics
Wygant, James Calvin, organic chemistry
Yadav, Kamaleshwari Prasad, analytical chemistry
Yaris, Robert, physical chemistry
Yeager, Vernon LeRoy, anatomy
Yohe, Cleon Russell, algebra
Young, Paul Andrew, anatomy
Yu, Shiu Yeh, biochemistry, organic chemistry
Zacher, Albert Richard, high energy physics
Zetlmeisl, Michael Joseph, electrochemistry
Zienty, Ferdinand B., chemistry
Zimmer, Arthur James, pharmaceutical chemistry
Zuzack, John W., organic chemistry

SALEM

Davis, James Howell, economic geology, mining geology
Stone, Bobbie Dean, solid state chemistry, inorganic chemistry

SPRINGFIELD

Aley, Thomas John, hydrology
Bond, Lora, botany
Carlson, Douglas W., physical chemistry
Cooley, Maxwell Louis, nutrition

ST PETERS

Gutsche, Henry W., solid state chemistry
Rogers, Nelson Floyd, forestry

SALEM

Davis, James Howell, economic geology, mining geology
Stone, Bobbie Dean, solid state chemistry, inorganic chemistry

WARRENSBURG

Bell, Marvin Drake, solid state physics
Belshe, John Francis, zoology, ecology
Berkland, Terrill Raymond, science education
Bowling, David Ivan, physics
Brown, Claude Harold, mathematics
Castaner, David, systematic botany
Elliott, Alice, zoology, parasitology
Emerson, John Wilford, geology
Fauk, Dennis Derwin, organic chemistry
Hawksley, Oscar, ecology
Hess, John Berger, ecology, genetics
Hewitt, Sam Parker, plant physiology
Hinerman, Charles Ovalee, science education
Holmes, Neal Jay, science education
Johnson, Larry K., mathematics
Killingbeck, Stanley, physical chemistry
Martin, Frank Elbert, solid state physics
Oshima, Eugene Akio, science education
Peck, William B., arachnology
Davidson, Steve Edwin, agronomy
Fronabarger, Carl Valentine, mathematics
Gibbons, James Joseph, analytical chemistry, inorganic chemistry
Gordon, Albert Raye, cell biology, electron microscopy
Greene, Joseph S., mathematics, operations research
Grindstaff, Wyman Keith, inorganic chemistry
Hayes, William Clifton, Jr., geology
Ingersol, Robert Harding, mammalogy, ecology
Irgens, Roar L., microbiology
Kurtz, Vincent E., paleontology, stratigraphy
Larson, Sager Daryl, physical chemistry
Madsen, Fred Christian, animal nutrition
Mantel, Erwin Joseph, geochemistry
Marshall, Richard, biochemistry
Miller, James Frederick, micropaleontology, invertebrate paleontology
Monfore, Gervaise Edwin, instrumentation
Mortensen, Harley Eugene, organic biochemistry
Northrip, John Willard, biophysics
O'Brien, James Francis, physical chemistry, inorganic chemistry
Padron, Jorge Louis, physics
Philibert, Robert Lawrence, endocrinology
Redfearn, Paul Leslie, Jr., bryology
Riley, James A., physics
Roy, Rabindra (Nath), physical chemistry, analytical chemistry
Schmidt, Bruno (Francis), economic geology, petrography
Sheets, Ralph Waldo, environmental chemistry
Shiflett, Lilburn Thomas, mathematics
Starks, Paul (Breckenridge), animal botany
Stevenson, Robert Thomas, biology
Stombaugh, Tom Atkins, biology
Stuffebeam, Charles Edward, animal science
Thielmann, Vernon James, chemistry
Thompson, Clifton C., physical chemistry
Thompson, Kenneth Clair, economic chemistry
Weber, Wallace Rudolph, systematic botany
Wilbur, James Myers, Jr., organic chemistry
Wolf, George William, astronomy

TARKIO

McIntosh, Harold Leroy, physics

UNIVERSITY CITY

Coffey, Joseph Francis, chemistry, chemical engineering
Davis, Hallowell, neurophysiology, audiology
Kennell, David Epperson, molecular biology

VAN BUREN

Feltz, Elmer T., virology, bacteriology

VIDA

Harmon, Sidney M., applied mathematics

VIENNA

Oviatt, Charles Dixon, applied statistics

Pfeifer, Gerard David, biochemistry
Savery, Harry P, physiology, endocrinology
Seeley, James Lee, biology
Smith, Howard Keith, mathematics
Stumpff, Robert D, physical chemistry
Waite, Albert B, reproductive physiology, genetics
Welch, Lin, speech pathology
Zey, Robert L, organic chemistry

WEBSTER GROVES
Gray, Raymond Francis, microbiology
Guthrie, David Burrell, organic chemistry
Lissant, Kenneth Jordan, colloid chemistry
Raether, Louis Otto, organic chemistry

MONTANA

BIGFORK
Prescott, Gerald Webber, phycology
Tibbs, John Francisco, protozoology

BILLINGS
Balster, Clifford Arthur, geology
Barnhart, David M, physical chemistry
Bayliss, Berenice, bacteriology
Bintz, Gary Luther, comparative physiology, environmental physiology
Clark, Wilson Farnsworth, environmental sciences, conservation
Elliott, Eugene Willis, inorganic chemistry, botany
Fanshawe, John Richardson, II,
Fisher, Frederick Stephen, geology, petroleum geology
Gloege, George Herman, chemistry
Graham, Raymond, physical chemistry, molecular spectroscopy
Jamison, William H, mathematics
Kirkpatrick, Jay Franklin, animal physiology
Layman, Wilbur A, analytical chemistry, physical chemistry
McRae, Robert James, science education
Milstead, Wayne Lavine, systematic botany
Ramsey, Richard Harold, II, phytopathology, mycology
Rollins, Myron B, soil conservation, watershed management
Rouse, John Thomas, geology
Sinclair, Elizabeth Faye, pathology
Stannard, William A, mathematics

Dorgan, William Joseph, developmental biology, cell biology
Drumheller, John Earl, magnetic resonance
Drummond, James, animal science
Dunn, Darrel Eugene, hydrogeology
Dynes, J Robert, animal science
Edie, Milton James, geography
El-Negoumy, Abdul Monem, agricultural biochemistry
Emerson, Kenneth, physical chemistry, inorganic chemistry
Eslick, Robert Freeman, plant breeding, agronomy
Evans, George Edward, ornamental horticulture
Feltner, Kurt C, plant physiology
Ferguson, Albert Hayden, soil physics
Firchammer, Burton Deforest, veterinary microbiology, veterinary immunology
Fiscus, Alvin G, virology, microbiology
Frahm, Elmer Edward, chemistry
Gerry, Henry, organic chemistry
Goering, Kenneth Justin, chemistry
Gould, William Robert, III, zoology
Hansen, Rodney Thor, mathematics, statistics
Hapner, Kenneth D, biochemistry
Hastings, Ellsworth (Bernard), entomology
Hehn, Erhardt Richard, agronomy
Henry, Myron S, applied mathematics
Hermanson, John Carl, solid state physics
Hess, Adrien LeRoy, mathematics education
Hewitt, George Berlyn, entomology
Howald, Reed Anderson, physical chemistry
Hull, Maurice Walter, veterinary physiology
Jackson, Larry Lee, biochemistry
Jennings, Paul W, organometallic chemistry
Julian, Gordon Ray, chemistry
Jutila, John W, immunology, medical microbiology
Keyes, Everett A, animal husbandry
Kinnersley, William Morris, theoretical physics
Kirkpatrick, Larry Dale, physics
Lapeyre, Gerald J, solid state physics
Lee, Floyd Denman, nuclear physics
Lozano, Edgardo A, bacteriology
Mackie, Richard John, wildlife management, ecology
Mathre, Donald Eugene, plant pathology
McAllister, Byron Leon, mathematics
McBee, Richard Harding, microbiology
McFeters, Gordon Alwyn, microbial physiology, biochemistry
McGuire, Charles Francis, agronomy, cereal chemistry
McMillan, James Alexander, neurophysiology
McNeal, Francis H, plant breeding, plant genetics
Metcalf, Homer Noble, horticulture
Mills, Ira Kelly, plant physiology
Montagne, John M, physical geology, environmental geology
Moore, Robert Emmett, vertebrate ecology
Moss, Buelon Rexford, animal nutrition, dairy science
Mundy, Bradford Philip, chemistry
Myers, Lyle Leslie, biochemistry, immunochemistry
Nelson, Nels M, microbiology
Newman, Clarence Walter, animal nutrition
Newman, Franklin Scott, microbiology, virology
Nielsen, Gerald Alan, soil science, ecology
Nordtvedt, Kenneth L, theoretical physics
Olsen, Ralph A, soil chemistry
Onsager, Jerome Andrew, entomology
Pagenkopf, Andrea L, nutrition
Pagenkopf, Gordon K, physical inorganic chemistry
Payne, Gene F, range management
Pepper, James Hubert, zoology, entomology
Perry, David Anthony, ecology
Phillips, Dwight Edward, anatomy
Pickett, James M, plant physiology
Picton, Harold D, zoology
Reed, Norman D, immunology
Robbins, John Edward, biochemistry
Roemhild, George R, entomology
Rogers, Samuel John, biochemistry
Rosa, Richard John, engineering physics
Rugheimer, Norman MacGregor, physics
Rumely, John Hamilton, plant ecology, plant taxonomy
Ryder, Gerald H, mathematical analysis

Schaeffer, Jurgen Richard, plant breeding, cytogenetics
Scharen, Albert Lois, plant pathology
Scharff, Donald Kenneth, entomology, ecology
Schmidt, Victor Hugo, solid state physics
Schmitroth, Louis Anthony, mathematics
Sharp, Eugene Lester, plant pathology
Shepard, James F, plant pathology, plant virology
Skaar, Palmer David, genetics
Skogley, Earl O, soil fertility, plant nutrition
Smith, Ervin Paul, animal science
Smith, Richard James, surface physics, magnetohydrodynamics
Strobel, Gary A, plant pathology
Stuart, David Gordon, microbiology
Super, Arlin B, meteorology
Swartz, William John, mathematics
Taylor, Donald Curtis, mathematics
Taylor, George Allan, plant breeding, genetics
Temple, Kenneth Loren, geoenvironmental science
Thomas, Oscar Otto, animal nutrition
Visscher, Saralee Neumann, entomology
Walter, William Goff, microbiology
Watling, Harold, zoology
Wend, David Van Vranken, mathematics
Whitesitt, John Eldon, mathematics
Wiegand, Roy Vernon, atomic spectroscopy
Wiesner, Loren Elwood, seed physiology
Woodriff, Ray Alan, chemistry, spectroscopy
Woodward, Ray R, genetics
Worley, David Eugene, parasitology
Wright, John Clifford, limnology

BUTTE
Berg, Richard Blake, geology
Devaul, Guillaume Pierre, physics
Dresser, Hugh W, geology, paleontology
Earll, Fred Nelson, geology
Gless, Elmer E, acarology
Goebel, Jack Bruce, mathematics
Groff, Sidney Lavern, economic geology
Johns, Willis Merle, economic geology, geochemistry
King, Ralph Hughes, economic geology
Koch, Edwin George, chemistry
McCaslin, John Garfield, physical chemistry
McLeod, Kenneth Neil, physical chemistry
Murray, Joseph, organic chemistry
Sawyer, Paul Thompson, plant taxonomy
Wideman, Charles James, geophysics

HELENA
Harrington, Joseph D, zoology, physiology
Lackman, David Buell, bacteriology, public health
Manion, James J, ecology
Smith, Jean E, animal ecology

KALISPELL
Britton, Michael Paul, plant pathology
Schumacher, Robert E, aquatic biology, fisheries management

MILES CITY
Pahnish, Otto Floyd, animal breeding

MISSOULA
Allendorf, Frederick William, population genetics
Alt, David D, geology, geochemistry
Arno, Stephen Francis, forest ecology
Banaugh, Robert Peter, applied mathematics, geophysics
Behan, Mark Joseph, physiological ecology
Bessac, Frank Bagnall, anthropology
Blake, George Marston, forestry
Bolle, Arnold William, forestry
Browman, Ludvig Gustav, zoology
Brunson, Royal Bruce, zoology
Bryan, Charles A, mathematics
Bryan, Gordon Henry, pharmacology
Catalfomo, Philip, pharmacognosy
Chaney, Robert Bruce, Jr, audiology
Chessin, Meyer, plant physiology, environmental management
Craighead, John J, ecology
Crosby, Gary Wayne, structural geology, geophysics
Crowley, John Max, geography
Curry, Robert Rodney, geomorphology, ecology
Derrick, William Richard, mathematics
Diettert, Reuben Arthur, botany
Eddleman, Lee E, range ecology
Erickson, Ronald E, environmental chemistry
Faust, Richard Ahlvers, microbiology
Fellin, David Gene, forest entomology
Fessenden, Ralph James, chemistry
Fevold, Harry Richard, biochemistry, endocrinology
Field, Chris, cultural geography
Field, Richard Jeffrey, chemical kinetics
Fields, Robert William, paleontology, sedimentology
Gordon, Clarence Conrad, mycology
Grossman, Stanley I, applied mathematics
Habeck, James Robert, ecology
Harris, John Tom, wildlife management
Harvey, Alan Eric, plant pathology
Harvey, LeRoy Hatfield, botany
Hayden, Richard John, physics
Hill, Walter Ensign, biophysics,
Hyndman, Donald William, petrology
Jakobson, Mark John, nuclear physics
Jenni, Donald Alison, ethology
Juday, Richard Evans, organic chemistry
Kamego, Albert Amil, physical organic chemistry
Kelsey, Rick Guy, plant chemistry
Kilgore, Delbert Lyle, Jr, environmental physiology
Kinsella, John Michael, parasitology, invertebrate zoology
Konizeski, Richard L, vertebrate paleontology
Koostra, Walter L, microbiology
Krier, John Peter, wood science & technology
Lange, Ian M, geology, geochemistry
Lange, Robert Walter, forest management
Larson, Carl Leonard, bacteriology
Loftsgaarden, Don Owen, mathematical statistics
Lory, Earl Christian, physical chemistry
Lotan, James E, forestry, ecology
Lowe, James Harry, Jr, entomology
Lyon, Leonard Jack, wildlife ecology, forest ecology
Mallory, William R, theoretical physics, optics
Malouf, Carling I, anthropology
Manis, Merle E, mathematics
Margrave, Thomas Ewing, Jr, astronomy
McKelvey, Robert William, mathematics
McRae, Daniel George, mathematics
Medora, Rustem Sohrab, pharmacognosy, biology
Mell, Galen P, biochemistry
Melton, William Grover, Jr, vertebrate paleontology

FT BENTON
Brown, Paul Lawson, soil fertility

GREAT FALLS
Chamberlain, Virgil Ralph, geology, engineering
Long, James William, biochemistry
Stimpfling, Jack Herman, genetics

HAMILTON
Bell, John Frederick, bacteriology
Bergman, Robert Kaye, immunology
Brinton, Elias (Lyle) Patterson, developmental biology
Coe, John Emmons, immunology
Cox, Herald Rea, virology
Eklund, Carl Milton, epidemiology
Gerloff, Robert Kay, microbiology, public health
Hadlow, William John, veterinary pathology
Keirans, James Edward, medical entomology, parasitology
Lodmell, Donald Louis, virology
Milner, Kelsey Charles, medical microbiology
Munoz, John Joaquin, immunology, microbiology
Ormsbee, Richard Armstrong, biochemistry
Owen, Cora Rust, bacteriology
Philip, Cornelius B, epidemiology
Ribi, Edgar, biophysics
Stoenner, Herbert George, bacteriology, virology
Thomas, Leo Alvon, parasitology, medical microbiology
Yunker, Conrad Erhardt, zoology, parasitology

HAVRE
Holmes, Charles Henry, academic administration

MONTANA

Metzgar, Lee Hollis, ecology
Morris, Melvin Solomon, range science
Nakamura, Mitsuru J., microbiology
Nimlos, Thomas John, forest soils, ecology
O'Gara, Bartholomew Willis, zoology, fish & game management
Osterheld, Robert Keith, physical inorganic chemistry
Parker, Charles D., audiology
Patent, Gregory Joseph, comparative endocrinology
Peterson, James Algert, geology
Peterson, John Alvin, mathematics, computer science
Pettinato, Frank Anthony, pharmaceutical chemistry
Pfeiffer, Egbert Wheeler, vertebrate zoology
Pfister, Robert Dean, silviculture, plant ecology
Pierce, William R., forest management
Porter, Leonard Edgar, nuclear physics
Preece, Sherman Joy, Jr., plant taxonomy
Ream, Robert Ray, botany, ecology
Reinhardt, Howard Earl, mathematical statistics
Rudbach, Jon Anthony, microbiology, immunology
Schaefer, James Michael, anthropology
Schmidt, Wyman Carl, silviculture
Shafizadeh, Fred, organic chemistry
Sharrock, Floyd Wayne, anthropology
Shearer, Raymond Charles, forestry
Sheldon, Andrew Lee, zoology, ecology
Sheridan, Richard P., plant physiology, biology
Slocum, Sally Virginia, anthropology
Smith, Charlene Galloway, physical anthropology
Solberg, Richard Allen, plant pathology
Stark, Nellie May, plant ecology
Steele, Robert Wilbur, forest management
Stevens, Robert R., mathematics
Stewart, John Mathews, organic chemistry
Talbot, James Lawrence, geology
Taylor, Dee C., anthropology, archaeology
Taylor, John Jacob, microbiology
Thomas, Forrest Dean, II, inorganic chemistry
Thornton, Melvin LeRoy, botany
Ushijima, Richard N., microbiology, virology
Van Horne, Robert Loren, pharmacognosy, pharmacy
Van Meter, Wayne Paul, inorganic chemistry
Wailes, John Leonard, pharmacy
Wambach, Robert F., forest economics, resource administration
Waters, William Lincoln, organic chemistry, organometallic chemistry
Weber, William Mark, geology
Wehrenberg, John P., mineralogy
Weidman, Robert McMaster, geology
Weist, George Ferdinand, Jr., zoology
Weist, Katherine Morret, ethnology
Winston, Donald, geology
Woodbury, George Wallis, Jr., physical chemistry
Wright, Philip Lincoln, zoology
Yale, Irl Keith, mathematics
Yates, Leland Marshall, physical chemistry

PROCTOR
Sahinen, Uuno Mathias, economic geology

SIDNEY
Aase, Jan Kristian, soil science
Halvorson, Ardell David, soil science
Siddoway, Francis H., soil science
White, Larry Melvin, range science
Wight, Jerald Ross, range science

WEST GLACIER
Kessell, Stephen Robert, resource management

NEBRASKA

CHADRON
Agenbroad, Larry Delmar, geology

BLAIR
Grube, George Edward, ecology

Struempler, Arthur W., analytical chemistry
Swanson, Jack Lee, physical chemistry, biochemistry

CLAY CENTER
Cundiff, Larry Verl, animal breeding, population genetics
Doupnik, Ben Lee, Jr., phytopathology
Gregory, Keith Edward, animal breeding
Koch, Robert Milton, animal breeding
Kohlmeier, Ronald Harold, veterinary physiology
Peters, Leroy Lynn, entomology
Prior, Ronald Leon, animal nutrition, biochemistry
Teague, Howard Stanley, nutrition

CONCORD
Rehm, George W., soil fertility, agronomy
Witkowski, John Frederick, entomology

DONIPHAN
Gross, Delmer Ferd, plant breeding

FREMONT
Johnson, John Ronald, parasitology
Johnston, Richard Milton, mathematics
Lueninghoener, Gilbert Carl, geology

GLENVIL
Robison, Norman Glenn, genetics, statistics

GRANT
Eastin, John A., agronomy

HARRISON
Meade, Grayson Eichelberger, geology

HASTINGS
Aiken, John M., veterinary medicine
Alberts, Arnold A., organic chemistry
Moulton, John Maxim, zoology
Paradise, Michael Emmanuel, academic administration
Sachtleben, Clyde Clinton, physics, science education
Warner, Charles D., organic chemistry

HOLDREGE
Dragoun, Frank J., hydrology, civil engineering

KEARNEY
Blickensderfer, Peter W., analytical chemistry
Bliese, John Carl William, zoology
Clark, Ronald David, organic chemistry, academic administration
Cochran, John Rodney, speech pathology
Cole, Harvey E., ecology
Fougeron, Myron George, endocrinology
Fox, Donald E., organic chemistry
Ickenberry, Richard W., plant pathology
Kuecker, John Frank, physical chemistry
Larsen, Leland Malvern, algebra
Lund, Douglas E., genetics
Nelson, Theodora S., mathematics
Pickens, Charles Glenn, mathematics
Searcy, Nelson Donald, geography of the Great Plains, resource geography
Swanson, James A., physical geography
Underhill, Glenn, theoretical physics, astronomy

LEWISTON
Reierson, James (Dutton), nuclear physics, operations research

LINCOLN
Adams, Alfred Birk, biochemistry, microbiology
Adams, Charles Henry, animal science, meat science
Adams, John Lester, poultry husbandry
Ahlschwede, William T., animal breeding
Allington, William B., plant pathology
Arnold, Roy Gary, food science
Arthaud, Vincent Henry, animal science
Bader, Kenneth L., agronomy, academic administration
Bagley, Walter Thaine, forestry, plant ecology
Ball, Harold James, insect physiology
Barnawell, Earl B., comparative endocrinology
Basoco, Miguel Antonio, mathematics
Bass, Edmund P., veterinary virology
Batten, Charles Francis, physical chemistry
Baumgarten, Henry Ernest, synthetic organic chemistry
Bentall, Ray, geology
Berdanier, Charles Reese, Jr., soil morphology, mineralogy
Blouet, Brian Walter, geography
Blunn, Cecil Thomas, genetics
Boellstorff, John David, stratigraphy, geochronology
Boohar, Richard Kenneth, developmental biology
Boosalis, Michael Gus, plant pathology
Borchers, Raymond (Lester), biochemistry
Boswell, Charles Leland, organic chemistry
Bowman, Robert Goldthwait, geography
Bradley, Richard E., dentistry, periodontics
Brakke, Myron Kendall, plant virology
Brockemeyer, Eugene William, pharmacy chemistry
Broman, Robert Fabel, electroanalytical chemistry
Brown, Albert Loren, veterinary microbiology
Brumbaugh, John (Albert), developmental genetics
Buenker, Robert J., physical chemistry
Bullerman, Lloyd Bernard, food microbiology, food science
Burns, Donal Joseph, atomic physics
Burnside, Orvin C., weed science, agronomy
Byerly, Paul Robertson, Jr., physics
Calabrese, Francis Anthony, anthropology, archaeology
Caldwell, Warren W., anthropology
Carlson, Marvin Paul, geology
Carr, James David, analytical chemistry
Case, Ronald Mark, wildlife ecology
Charnicki, Walter Francis, pharmaceutical chemistry
Chesnin, Leon, soil chemistry, plant nutrition
Chivukula, Ramamohana Rao, mathematics
Clark, Ralph B., plant physiology, plant biochemistry
Clough, John Wendell, geophysics
Coleman, George Hunt, nuclear chemistry
Compton, William A., genetics, statistics
Conard, Elvene Clyde, agronomy
Costello, Donald F., statistics, computer science
Cox, Henry Miot, mathematics, psychometrics
Coyne, Dermot P., plant breeding
Cromwell, Norman Henry, organic chemistry, chemotherapy
Cunningham, Peter John, animal breeding
Daly, Joseph Michael, plant biochemistry, plant physiology
Dam, Richard, biochemistry, nutrition
Davidson, John Fraser, plant taxonomy
Davies, Eric, plant physiology
Davis, Eldon Vernon, virology
Demuth, John Robert, organic chemistry
Dewey, Kenneth Frederic, meteorology
Dickason, Elvis Arnie, entomology
Dillon, Roy Dean, agriculture
Doane, Ted H., animal breeding
Dockum, Norman Leslie, histology
Domb, Ellen Ruth (Colmer), experimental solid state physics
Dreeszen, Vincent Harold, geology, hydrology
Drew, James Van, soil science
Dunkle, Larry D., phytopathology
Dyer, John Kaye, microbiology, immunology
Eastin, Jerry Dean, agronomy
Eckhardt, Craig Jon, physical chemistry
Eldridge, Franklin Elmer, animal genetics, cytogenetics
Ellington, Earl Franklin, animal physiology
Fagerstrom, John Alfred, invertebrate paleontology
Farlin, Stanley Dean, animal nutrition
Ferguson, Donald Leon, veterinary parasitology
Ferrill, Mitchell, forest ecology, forest mensuration
Finkler, Paul, particle physics, nuclear medicine
Fuller, Robert Gohl, solid state physics
Furrer, John D., agronomy
Gardner, Charles Olda, quantitative genetics
Gauger, Wendell Lee, mycology
George, T Adrian, organometallic chemistry, inorganic chemistry
Georgi, Carl Edward, microbial biochemistry
Gugler, Carl Wesley, malacology
Gunderson, Harvey Lorraine, zoology, mammalogy
Gunnerson, James Howard, anthropology, natural history
Gupta, Naba K., biochemistry
Guyer, Paul Quentin, animal husbandry
Haderlie, Lloyd Conn, weed science
Halfar, Edwin, topology
Hall, Leon Morris, Jr., mathematical analysis
Hansen, Bernard Lyle, physics
Hanway, Donald Grant, agronomy
Hardt, Alfred Black, physiology
Hardy, Robert J., physics
Harlan, Phillip Walker, soil morphology
Harm, Stanton Douglas, anatomy
Harris, Lewis Eldon, pharmaceutical chemistry
Harris, Robert Hutchison, inorganic chemistry
Hartung, Theodore Eugene, food science
Haskins, Francis Arthur, plant genetics
Helms, Thomas Joseph, insect morphology, insect pathology
Hergenrader, Gary Lee, limnology, fish biology
Hewes, Leslie, geography
Hill, Robert Mathew, biochemistry
Hill, Roscoe Earle, entomology
Holtzclaw, Henry Fuller, Jr., inorganic chemistry
Holzhey, Charles Steven, soil morphology, soil classification
Howell, Daniel Bunce, physical chemistry
Jackson, Lloyd K., atomic physics
Jaecks, Duane H., atomic physics
James, Garth A., endodontics
Janovy, John, Jr., zoology
Jaswal, Sitaram Singh, physics
Johnsgard, Paul Austin, zoology
Johnson, Virgil Allen, agronomy
Johnston, Robert Benjamin, biochemistry
Jones, Emerson, physics
Joseph, David Winram, elementary particle physics
Kaul, Robert, physics
Kehr, William R., agronomy
Keith, David Lee, entomology
Kies, Constance, nutrition
Kinbacher, Edward John, plant physiology
Kindler, Sharon Dean, entomology
Kingsbury, Charles Alvin, organic chemistry
Kirby, Roger D., physics

Klucas, Robert Vernon, microbial biochemistry, plant biochemistry
Knoche, Herman William, agricultural biochemistry
Kramer, Earl Sidney, mathematics
Kramer, William S, dentistry
Krutak, Paul Russell, geology, micropaleontology
Landolt, Paul Albert, physiology
Lane, Leslie Carl, plant virology
Langenberg, Willem G, plant pathology, plant virology
Larsen, Max Dean, algebra
Larson, Larry Lee, reproductive physiology
Laursen, Paul Herbert, organic chemistry
Lavy, Terry Lee, weed science, soil chemistry
Lawson, Merlin Paul, climatology
Leavitt, William Grenfell, algebra
Lee, Jean Chor-Yin, biochemistry
Leonhardt, Earl A, mathematics
Leopold, Aldo Carl, plant physiology
Leung, Kam-Ching, astronomy, astrophysics
Lewis, David Thomas, agronomy, soil morphology
Lewis, William James, mathematics
Lipson, Joseph, physics
Lommasson, Robert Curtis, plant morphology
Lonsdale, Richard Ellis, economic geography, geography of the Soviet Union
Looker, James Howard, organic chemistry
Luke, Stanley D, mathematics
Lynch, John Douglas, zoology, herpetology
Lynn, Warren Clark, soil science
Macek, Joseph, physics
Manglitz, George Rudolph, entomology
Maranville, Jerry Wesley, agronomy
Marianelli, Robert Silvio, inorganic chemistry
Marsh, Connell Leroy, biochemistry
Mattern, Paul Joseph, analytical chemistry
Maxcy, Ruthford Burt, food science, microbiology
Mayo, Z B, entomology
Mazurak, Andrew Peter, soil physics
McCalla, Thomas Mark, soil microbiology
McCarty, Melvin Knight, weed science
McClendon, John Haddaway, plant physiology
McConnell, Henry Elden, microbiology
McGill, David Park, agronomy
McGill, Lawrence David, veterinary pathology
McGlone, Robert Ernest, speech sciences, hearing sciences
McIntosh, Charles Barron, geography
Mebus, Charles Albert, veterinary pathology
Meints, Russel H, developmental biology
Meisels, Gerhard George, physical chemistry, analytical chemistry
Meisters, Gary Hosler, mathematics
Mientka, Walter Eugene, mathematics
Militzer, Walter Ernest, biochemistry
Miller, Donald Wright, mathematics
Miller, Dwight Dana, genetics
Miller, Gary A, nutrition, food science
Misheloff, Michael Norman, physics
Mitchum, Ronald Kem, mass spectrometry
Modrak, John Bruce, pharmacology
Moline, Waldemar John, agronomy
Morgan, Thomas Anthony, physics
Morris, Mary Rosalind, plant cytogenetics
Moser, Lowell E, agronomy
Mourer, Kermit L, pharmaceutical chemistry
Mumm, Robert Franklin, biometrics, genetics
Murray, Wallace Jasper, medicinal chemistry
Neild, Ralph E, horticulture, agricultural climatology
Nelson, Robert B, structural geology
Newell, Laurence Cutler, agronomy
Nickol, Brent Bonner, parasitology
Nielsen, Merlyn Keith, animal breeding
Ogden, Edwin Burman, mathematical physics
O'Keefe, Robert Bernard, horticulture, forestry
Olson, Robert August, soils
Omtvedt, Irvin T, animal science
Owen, Foster Gamble, dairy nutrition
Page, Walter, zoology
Parrish, David Joe, plant pphysiology
Payne, Loyal Cobb, veterinary physiology
Pearlstein, Edgar Aaron, solid state physics
Pederson, Darryll Thoralf, hydrogeology
Peo, Ernest Ramy, Jr, animal nutrition
Peterson, Gary A, soil science, statistics
Peterson, Glenn Walter, plant pathology
Petri, Lester Reinhold, hydrology, water chemistry
Pritchard, Mary (Louise) Hanson, parasitology
Pruess, Kenneth Paul, entomology
Rack, Edward Paul, chemistry, radiochemistry
Rapp, William Frederick, Jr, entomology
Raun, Earle Spangler, entomology
Reed, Eugene Clifton, hydrogeology, petroleum geology
Rinne, Vernon Wilmer, dentistry
Robertson, Robert James, exercise physiology
Roche, Edward Browining, medicinal chemistry, chemical pharmacology
Roeth, Frederick Warren, weed science
Rosenberg, Harry, pharmacognosy, biochemistry
Rosenberg, Norman J, micrometeorology, soil physics
Ross, Sam Jones, Jr, soil morphology
Ross, William Max, agronomy
Rowland, Neil Wilson, botany
Rudd, Millard Eugene, atomic physics
Rugg, Dean Sprague, geography
Ryan, Charles Futrell, pharmacology, toxicology
Samson, James Alexander Ross, atomic physics
Sander, Donald Henry, agronomy
Sandsted, Rudolph Marion, cereal chemistry
Sartori, Leo, theoretical astrophysics
Saski, Witold, pharmaceutics
Satterlee, Lowell Duggan, food chemistry, biochemistry
Schiess, Alvin V, oral pathology
Schlitt, Dan Webb, theoretical high energy physics, mathematical physics
Schmidt, Edward George, astronomy
Schmidt, John Wesley, agronomy
Schneider, Hubert H, mathematical logic
Scholz, John Joseph, Jr, physical chemistry
Schultz, Charles Bertrand, geology, vertebrate paleontology
Schutz, Wilfred M, genetics, statistics
Sellmyer, David Julian, solid state physics
Shahani, Khem Motumal, food technology, biochemistry
Shaw, David Harold, physiology, pharmacology
Shores, Thomas Stephen, mathematics
Simon, Norman Robert, astrophysics
Skoug, David L, mathematical analysis
Smith, Gary Lee, virology
Smith, Russell, geology
Sorensen, Robert Carl, soil chemistry
Specht, James Eugene, plant physiology, plant breeding
Splettstoesser, John Frederick, geology
Srb, Joseph Emil, Jr, computer sciences
Stafford, George Ewing, pediatrics, public health
Stanley, Kenneth Oliver, geology
Staples, Robert, entomology
Starace, Anthony Francis, atomic physics
Staudinger, Wilbur Leonard, plant pathology, microbiology
Steadman, James Robert, plant pathology
Stemm, Robert Marvin, orthodontics
Stoddard, Robert Hugh, geography
Stohs, Sidney John, pharmacognosy, biochemistry
Stout, Thompson Mylan, geology, vertebrate paleontology
Stratbucker, Robert A, physiology, bioengineering
Sturgeon, George Dennis, solid state chemistry, high temperature chemistry
Sullivan, Robert Emmett, dentistry, pedodontics
Sullivan, Thomas Wesley, poultry nutrition, biochemistry
Surkan, Alvin John, physics, applied mathematics
Sutherland, William Neil, soil fertility, agricultural economics
Swift, Lloyd Harrison, plant morphology
Tanner, Lloyd George, geology
Taylor, Donald James, astronomy
Tharp, Gerald D, vertebrate physiology
Thompson, Thomas Leo, bacterial physiology
Thornton, Melvin Chandler, mathematics
Thorp, Eldon Marion, soil conservation
Thorson, Thomas Bertel, zoology
Tong, Yung Liang, statistics, mathematics
Treves, Samuel Blain, volcanology, petrology
Tsai, Bilin Paula, chemical physics
Ullman, Frank Gordon, solid state physics
Underdahl, Norman Russell, bacteriology, virology
Van Den Bos, Jan, computer science, atomic physics
Vanderzee, Cecil Edward, physical chemistry
Van Etten, James L, microbial physiology
Van Haverbeke, David F, taxonomy, silviculture
Vidaver, Anne Marie Kopecky, bacteriology
Vidaver, George Alexander, biochemistry
Voorhies, Michael Reginald, vertebrate paleontology
Waggener, Donald Todd, dentistry
Wagner, Frederick William, biochemistry
Walsh, Gary Lynn, air pollution, environmental sciences
Wampler, Joe Forrest, mathematics
Ward, John K, animal nutrition
Wayne, William John, geology
Webster, William Wallace, oral surgery
Weinberg, Daniela, cultural anthropology
Weymouth, John Walter, solid state physics, archaeometry
Wheeler, Desmond Michael Sherlock, organic chemistry
Wiegand, Sylvia Margaret, algebra
Wiese, Richard Anton, soil fertility
Wilder, Violet Myrtle, biochemistry
Wilkins, Charles Lee, analytical chemistry
Williams, James Henry, Jr, agronomy
Wishart, David John, cultural geography
Wotton, Robert Moore, anatomy
Wysong, David Serge, plant pathology
Young, Joseph Oran, botany
Zechmann, Albert W, mathematics
Zimmerman, Edward John, theoretical physics, philosophy of science

NORTH PLATTE

Campbell, John Bryan, entomology
Clanton, Donald Cather, animal nutrition
Danielson, David Murray, animal nutrition
Hibbs, Clair Maurice, veterinary pathology, microbiology
Nichols, James T, range management, agronomy
White, Raymond Gene, veterinary science

OMAHA

Aita, John Andrew, neurology, psychiatry
Andrews, Richard Vincent, physiology
Angle, William Dodge, internal medicine
Badeer, Henry Sarkis, physiology
Baker, Robert Norton, medicine
Baldwin, Lynne Juedeman, computer sciences
Barak, Anthony Joseph, biochemistry
Bardawil, Wadi Antonia, pathology
Barker, Kenneth Leroy, biochemistry, endocrinology
Bartholow, George William, psychiatry
Barton, John M, anatomy
Baumann, Donald Joseph, inorganic chemistry
Baumel, Julian Joseph, anatomy
Baumstark, John Spann, biochemistry, microbiology
Becker, Donald A, botany, plant ecology
Belknap, Robert Wayne, physiology
Bennett, Arthur Lawrence, physiology
Berdanier, Carolyn Dawson, nutritional biochemistry
Berton, William Morris, pathology
Booth, Richard W, medicine
Brody, Alfred Walter, physiology, pharmacology
Brooks, Merle Eugene, biology, botany
Burch, Robert Emmett, internal medicine, biochemistry
Busch, Karl Heinrich Daniel, zoology
Campbell, Gary Thomas, neuroendocrinology
Carver, Michael Joseph, biochemistry
Castater,Robert Dewitt, operations analysis
Chaperon, Edward Alfred, immunology, microbiology
Cipolla, Sam J, atomic physics, nuclear physics
Clark, Francis John, neurophysiology
Clifford, George O, internal medicine, hematology
Connolly, John Francis, orthopedics
Conrad, Robert Dean, laboratory animal medicine
Cook, David Edgar, biochemistry
Cooper, Raymond B, weed science
Cordes, Davies Marcia A, molecular spectroscopy
Crampton, James Mylan, pharmacology
Creek, Robert Omer, endocrinology
Curtin, Charles Byron, biology
Curtis, Gary Lynn, biochemistry, immunology
Czerwinski, Ann Langley, pharmacology
Dalley, Arthur Frederick, II, gross anatomy, morphology
Dalske, Howard Frederick, pharmacology
Danke, Richard John, animal nutrition
Davies, Kennard Michael, physics
Davis, Richard Bradley, internal medicine, hematology
Davis, William Clayton, surgery
Dossel, William Edward, histology, embryology
Dubes, George Richard, genetics
Earle, Alvin Mathews, anatomy
Eaton, Merrill Thomas, Jr, psychiatry
Ebadi, Manuchair, neuropharmacology, neurochemistry
Eddy, Dennis Eugene, biochemistry
Eisen, James David, human genetics, cytogenetics
Elder, John Thompson, Jr, pharmacology
Eliot, Robert S, cardiology, chemistry
Ellingson, Robert James, psychobiology, electroencephalography
Feldhaus, Richard Joseph, surgery
Feleppa, Alfred E, Jr, microscopic anatomy, hematology
Fierer, Joshua A, immunopathology
Fishkin, Arthur Frederic, biochemistry
Flocken, John W, theoretical solid state physics
Foley, John F, internal medicine
Fried, Rainer, biochemistry
Friedlander, Walter Jay, neurology
Fuller, Derek Joseph Haggard, hematical analysis
Fullilove, Susan Louise, developmental biology
Fusaro, Ramon Michael, dermatology
Gainsforth, Burdette Livingston, orthodontics
Gallagher, Thomas Francis, biochemistry
Gambal, David, biochemistry
Gardner, Paul Jay, anatomy
Gedgoud, John Leo, pediatrics
German, John Dee, surgery
Gerraughty, Robert Joseph, pharmacy
Gessaman, Margaret Palmer, mathematical statistics
Gessert, Carl F, pharmacology
Ghosh, Chitta Ranjan, immunology, microbiology
Gibbs, Gordon Everett, pediatrics
Gifford, Harold, ophthalmology
Gilmore, Joseph Patrick, physiology
Gingell, Ralph, biochemistry, drug metabolism
Gloor, Walter Thomas, Jr, pharmacy
Goldsmith, Dale Preston Joel, biochemistry
Goodfellow, Elsie F, neuroanatomy, histology
Grandjean, Carter Jules, biochemistry, cancer
Greco, Salvatore Joseph, pharmacy
Grinnell, Edward Hoepfner, pharmacology
Grissom, Robert Leslie, internal medicine
Guenther, Raymond A, low temperature physics, solid state physics
Guirgis, Hoda A, epidemiology, biostatistics
Haddix, George Franklin, mathematics, computer science
Haeder, Paul Albert, applied mathematics
Hamsa, William Rudolph, orthopedic surgery
Hard, Walter Leon, neuroanatomy
Harman, Denham, biochemistry
Haskell, Albert Russell, pharmacology, biochemistry
Hatch, Robert Harold, physiology
Hay, Charles Alfred, microbiology, virology
Heaney, Robert Proulx, internal medicine
Heidrick, Margaret Louise, biochemistry, immunology
Heilenday, Frank W, operations research
Henn, Mary Josephine, internal medicine
Hexum, Terry Donald, pharmacology
Higgs, Gary Kent, economic geography, cartography
Hill, Marvin Francis, anatomy, physiology
Hill, Mary Mechtilde, inorganic chemistry, radiation chemistry
Himwich, Williamina (Elizabeth) Armstrong, physiology
Hodgson, Paul Edmund, surgery
Hofert, John Frederick, biochemistry
Hoffman, Donald Richard, immunology
Holcslaw, Terry Lee, pharmacology

Holyoke, Edward Augustus, anatomy
Huffman, Max Niel, biochemistry
Hunt, Howard Beeman, radiology
Hupka, Arthur Lee, pharmacology
Issenberg, Phillip, biochemistry, toxicology
Javel, Eric, neurophysiology
Jensen, Richard Harvey, anatomy, immunology
Johnson, John Arnold, biochemistry, immunology
Johnson, John Raymond, physiology, pharmacology
Jones, Ernest Olin, radiological physics
Joyner, William Lyman, physiology
Kass, Irving, medicine
Keller, John George, toxicology, physiology
Kennedy, Robert E., theoretical physics, physical oceanography
Keppel, Charles Robert, inorganic chemistry
Klein, Francis Michael, organic chemistry
Ladwig, Harold Allen, neurology
Lee, Kyu Yawp, biochemistry
Lee, Leroy William, urology, surgery
Lehnhoff, Henry John, Jr., internal medicine
Lemon, Henry Martyn, medicine
Lim, Thomas Pyung Kee, thoracic diseases
Linstromberg, Walter William, organic chemistry
Long, Mary Jean, medical technology, biochemistry
Luby, Robert James, obstetrics & gynecology
Lunt, Steele Ray, entomology, genetics
Lynch, Benjamin Leo, oral surgery
Lynch, Henry T., medical genetics, internal medicine
Macaluso, Mary Christelle, anatomy
Maclean, Donald Isadore, physical chemistry
Magee, Donal Francis, physiology, pharmacology
Mahovald, Theodore Augustus, biochemistry
Malashock, Edward Marvin, urology
Maloney, John P., mathematical analysis
Mann, Michael David, neurophysiology
Marquardt, Dawn Nilan, organic chemistry
Maser, Morton D., cell biology
Matschiner, John Thomas, biochemistry
McCarthy, Robert Elmer, immunology, microbiology
McClurg, James Edward, biochemistry, reproductive biology
McFadden, Harry Webber, Jr., medical microbiology, infectious diseases
McIntire, Matilda S., pediatrics
McLaughlin, Charles William Jr., surgery
McWhorter, Clarence Austin, medicine
Meader, Roland Darrell, anatomy
Meier, Gilbert W., physiological psychology
Metcalf, William Kenneth, human anatomy
Miller, Norman Gustav, microbiology
Mirvish, Sidney Solomon, cancer
Mohiuddin, Syed M., cardiology, internal medicine
Montague, Patricia Tucker, mathematics
Montague, Stephen, mathematics
Mooring, Paul K., pediatric cardiology, pediatrics
Mordeson, John N., mathematics
Moriarty, C Michael, physiology, biophysics
Murthy, Veeraraghavan Krishna, biochemistry, physiology
Musselman, Merle McNeil, surgery
Nagel, Donald Lewis, cancer, organic chemistry
Nayak, Ramnath V., medicine, endocrinology
Paustian, Frederick Franz, internal medicine, gastroenterology
Pearson, Paul (Hammond), public health
Phalen, William Edmund, biochemistry
Phares, Cleveland Kirk, biochemistry, parasitology
Phillips, Hugh Jefferson, physiology
Piepho, Robert Walter, pharmacology, neurochemistry
Platt, Peyton Thomas, medicine
Platz, James Ernest, evolutionary biology
Pratt, Robert Walter, plant physiology
Prior, Paul Verdayne, plant morphology
Prioreschi, Plinio, pharmacology, medicine
Quigley, Herbert Joseph, Jr., pathology, chemistry

Raha, Chitta Ranjan, organic chemistry
Ramaley, Judith Aitken, reproductive biology
Ramaley, Robert Folk, biochemistry, microbiology
Rasmussen, Edith Svoboda, animal nutrition, biochemistry
Rigby, Perry G., internal medicine, hematology
Rogan, Eleanor Groeniger, biochemistry
Rongone, Edward Laurel, biochemistry
Ross, Oscar Burr, animal nutrition
Ruegamer, William Raymond, biochemistry
Ryan, Wayne L., biochemistry, immunology
Salhany, Jimmy Mitchell, biophysics
Sanders, Christine Culp, medical microbiology, infectious diseases
Sanders, W Eugene, Jr., medicine, microbiology
Sanger, Warren Glenn, medical genetics
Schaefer, Arnold Edward, nutrition
Scheerer, Anne Elizabeth, mathematics
Schenken, John Rudolph, pathology, bacteriology
Schieferstein, Robert Harold, plant physiology
Schiltz, Gordon B., geography
Schlesinger, Allen Brian, embryology, environmental biology
Schmehl, Francis Lawrence, organic chemistry, research administration
Scholar, Eric M., pharmacology
Scholes, Norman W., pharmacology, physiology
Sellers, Robert Douglas, thoracic surgery
Seshachalam, Dutta, biochemistry, microbiology
Severn, Charles B., anatomy, embryology
Severn, Matthew Joseph, medical microbiology
Sharp, Edward A., mathematics
Sharpe, Roger Stanley, zoology
Shearer, Greg Otis, physical organic chemistry
Sheehan, John Francis, cytopathology
Shipp, Joseph Calvin, medicine
Shroder, John Ford, Jr., geology, geomorphology
Shubik, Philippe, pathology, oncology
Sisson, Joseph A., pathology
Skutley, Francis Miles, medicine
Small, LaVerne Doreyn, medicinal chemistry
Smith, Edward Russell, biochemistry
Smith, Jack Louis, biochemistry, nutrition
Snipp, Robert Leo, physical chemistry
Sorrell, Michael Floyd, gastroenterology
Sparks, Robert D., internal medicine, gastroenterology
Staff, Charles Hubert, food science, organic chemistry
Stageman, Paul Jerome, biochemistry
Starr, Phillip Henry, psychiatry
Steenburg, Richard Wesley, surgery
Stone, Daniel Boxall, internal medicine, endocrinology
Sullivan, James F., medicine
Sutherland, David M., plant taxonomy
Takemura, Kaz Horace, organic chemistry
Thomas, John Martin, pediatrics
Thomas, Robert Joseph, cell biology, epidemiology
Thurman, Richard Gary, chemistry
Tobin, Richard Bruce, physiology
Todd, Gordon Livingston, electron microscopy
Toth, Bela, pathology, oncology
Tremaine, Mary M., microbiology
Tuma, Dean J., biological chemistry
Urban, Theodore Joseph, physiology
Van Leeuwen, Gerard, pediatrics
Vogel, Philip E., geography
Von Riesen Victor Lyle, medical microbiology
Waggener, Ronald E., radiology
Walker, Charles Eugene, cereal chemistry
Wallace, Lawrence, biochemistry
Ware, Frederick, physiology, internal medicine
Watt, Dean Day, biochemistry
Weber, Allen Thomas, microbiology
Wells, Ibert Clifton, biochemistry
White, Roberta Jean, virology
Wigton, Robert Spencer, medicine
Wikening, Marvin C., animal nutrition, biochemistry
Willingham, Allan King, biochemistry
Wilson, Richard Barr, pathology
Witson, Cecil L., psychiatry
Wolf, Gerald Lee, radiology

Wood, James Kenneth, synthetic organic chemistry
Zebolsky, Donald Michael, physical chemistry
Zepf, Thomas Herman, physics
Zucker, Irving H., physiology

PERU
Christ, John Conrad, biology
Long, Daryl Clyde, soil science

SCOTTSBLUFF
Daigger, Louis A., soil fertility
Kerr, Eric Donald, phytopathology
Nelson, Lenis Alton, agronomy
Weihing, John Lawson, plant pathology

SEWARD
Adams, Clark Edward, mammalian ecology
Brandhorst, Carl Theodore, zoology, botany
Meyer, Herbert A., inorganic chemistry, analytical chemistry

SILVER CREEK
Starostka, Raymond Walter, agronomy

WAYNE
Hirt, Bethold Joseph, embryology, anatomy
Johar, Jogindar Singh, environmental chemistry, fluorine chemistry
Maier, Charles Robert, botany
Rasmussen, Russell Lee, inorganic chemistry, bio-organic chemistry
Sutherland, Robert Carver, plant physiology, plant morphology

NEVADA

BOULDER CITY
Dill, David Bruce, physiology
Felts, Wayne Moore, geology
Miles, Maurice Jarvis, analytical chemistry, environmental chemistry
O'Farrell, Thomas Paul, mammalian ecology

CARSON CITY
Dehne, Edward James, occupational medicine, public health
Miller, Richard Gordon, vertebrate zoology
Worts, George Frank, Jr., geology

ELKO
Hunt, Charles Warren, III, geology

EUREKA
Kitchen, Joseph Henry, watershed management

HENDERSON
Rhees, Raymond Charles, inorganic chemistry, analytical chemistry

LAS VEGAS
Aizley, Paul, mathematics
Babero, Bert Bell, parasitology
Baepler, Donald H., ornithology
Barger, James Daniel, pathology
Bell, Katherine Lapsley, plant ecology
Billingham, Edward J., Jr., analytical chemistry
Brooks, Sheilagh Thompson, physical anthropology
Chamberlain, Charles Kent, geology, invertebrate paleontology
Coffer, Henry Ford, colloid chemistry
Deacon, James Everett, vertebrate zoology
Douglas, Charles Leigh, mammalian ecology
Douthett, Elwood Moser, nuclear chemistry
Dunham, Linda Thompson, biological chemistry
Fiero, George William, Jr., geology, environmental sciences
Gentile, Arthur Christopher, plant physiology
Goldman, Aaron Sampson, statistics
Graham, Malcolm, mathematics
Grenda, Stanley C., inorganic chemistry

Jobst, Joel Edward, nuclear physics
Kaufmann, Robert Frank, geology, hydrogeology
Kepper, John C., sedimentary petrology, stratigraphy
Knutson, Carroll Field, geoenvironmental science
Mann, Bruce Jameson, health physics, nuclear engineering
McClellan, William Alan, environmental geology
McMillion, Leslie Glen, hydrology
Murvosh, Chad M., aquatic ecology
Niles, Wesley E., botany
O'Farrell, Michael John, mammalian ecology, behavioral biology
Peckham, Alan Embree, hydrology
Potter, Gilbert David, physiology
Schreiweis, Donald Otto, co-mparative anatomy, vertebrate embryology
Scott, Donald Ray, physical chemistry, analytical chemistry
Smith, Robert Bruce, organic chemistry
Snaper, Alvin Allyn, research administration
Tarsey, Alexandre Rolf, applied physics
Tew, Richard Wilcox, microbiology
Titus, Richard Lee, organic chemistry
Warren, Claude Nelson, archaeology, cultural anthropology
Werth, John Sr Clair, Jr., mathematics, petrology
Wilbanks, John Randall, structural geology
Williams, Nelson Noel, human ecology
Yousef, Mohamed Khalil, environmental physiology

RENO
Altick, Philip Lewis, atomic physics
Anderson, Bernard A., speech pathology, audiology
Arnett, William Harold, economic entomology
Bach, L Matthew N., neurophysiology
Bailey, Curtiss Merkel, animal breeding
Barnes, George, electronics
Bechtel, Robert Christy, entomology
Beesley, Edward Maurice, mathematics
Bettler, Philip C., physics
Bingler, Edward Charles, geology
Blackadar, Bruce Evan, mathematics
Blincoe, Clifton (Robert), physical biochemistry
Bohman, Verle Rudolph, animal nutrition
Bohmont, Dale W., plant science, academic administration
Bonham, Harold F., Jr., geology
Brady, Allen H., mathematics, computer sciences
Brown, Russell Wilfrid, medical microbiology
Burkhart, Richard Delmar, physical chemistry
Carnahan, Chalon Lucius, hydrology
Case, Clinton Meredith, surface physics, hydrology
Cathey, William Newton, solid state physics
Christensen, Glen C., wildlife management
Cooley, Richard Lewis, hydrogeology
Cooney, Donald George, microbiology
Cords, Howard Paul, agronomy, weed science
Craner, John Wesley, biochemistry
Davis, Robert Dabney, mathematics
d'Azevedo, Warren Leonard, anthropology
Dreiling, Charles Ernest, neurochemistry
Eckert, Richard Edgar, Jr., range science
Eichborn, Bartane R., solid state physics, ceramic engineering
Evans, Raymond Arthur, range science
Fenske, Paul Roderick, hydrogeology
Firby, James R., invertebrate paleontology, stratigraphy
Flint, Delos Edward, economic geology
Foote, Wilford Darrell, animal physiology
Fowler, Don D., anthropology, research
Frazier, Thomas Vernon, physics
Fuson, Reynold Clayton, organic chemistry
Gifford, Richard Oliver, soil physics
Goudsmit, Samuel Abraham, spectroscopy
Guenthner, Harold Reinhold, agronomy
Haas, Otto Henry, paleontology, geology
Haber, Meryl H., pathology
Hallett, John, cloud physics

Hardesty, Donald Lynn, anthropology
Harrington, Rodney E, biophysics, physical chemistry
Heisler, Charles Rankin, biochemistry
Hibbard, Malcolm Jackman, geology
Hoffer, Thomas Edward, physical meteorology
Horton, Robert Carlton, geology
Houghton, John G, climatology, physical geography
Hsu, Liang-Chi, mineralogy, petrology
Jensen, Edwin Harry, agronomy
Johnson, Bruce Paul, solid state physics
Kemp, Kenneth Courtney, physical organic chemistry
Kendall, Michael Welt, gross anatomy, microscopic anatomy
Kersten, Earl William, geography
Kimble, Gerald Wayne, numerical analysis
Klieforth, Harold Ernest, meteorology
Kliwer, James Karl, physics
Knoll, Jack, animal physiology
Koh, Young O, statistics, computer science
Kozel, Thomas Randall, medical mycology, medical bacteriology
Lamb, Dennis, cloud physics
La Rivers, Ira John, insect taxonomy
Larson, Edward Richard, geology
Larson, Lawrence T, economic geology, mineralogy
Lawrence, Edmond Francis, exploration geology
LeMay, Harold E, Jr, inorganic chemistry
Lewis, Roger Allen, biochemistry
Lightner, David A, organic chemistry
Lindner, Luther Edward, pathology
Linkletter, George Onderdonk, quaternary geology, geochemistry
Lintz, Joseph, Jr, geology
Lupan, David Martin, medical mycology
Lynn, Edward Joseph, psychiatry
MacDonald, David J, physical chemistry, inorganic chemistry
Marsh, David Paul, nuclear physics
Maxey, George Burke, hydrology
McMinn, Trevor James, mathematics
Mead, Robert Warren, animal parasitology, invertebrate zoology
Meeuwig, Richard O'Bannon, forest ecology
Miller, Watkins Wilford, soil fertility, water pollution
Moore, Edwin Neal, atomic physics
Morris, Robert James, chemistry
Mozingo, Hugh Nelson, botany
Nelson, John Henry, inorganic chemistry, organometallic chemistry
Nork, William Edward, hydrogeology
O'Brien, Thomas Doran, inorganic chemistry
O'Harra, John Lewis, veterinary medicine
Papke, Keith George, geology
Pardini, Ronald Shields, biochemistry, pharmacology
Payne, Anthony Luke, economic geology, mineral economics
Peterson, Frederick Forney, soil classification, soil genesis
Pfaff, Donald Chesley, mathematics
Pough, Frederick Harvey, mineralogy
Reitz, Ronald Charles, biochemistry
Risser, Arthur Crane, Jr, zoology, geophysics
Robertson, Joseph Henry, range conservation
Rose, Charles Buckley, organic chemistry
Ruf, Robert Henry, Jr, horticulture
Ryall, Alan S, Jr, seismology
Ryser, Fred A, Jr, zoology
Scheid, Vernon Edward, economic geology, mineral economics
Scott, Lawrence Tressler, synthetic organic chemistry
Scott, William Taussig, atmospheric physics, philosophy of science
Sharp, John Van Alstyne, hydrogeology, water resources
Sheps, Lillian, plant physiology, desert ecology
Shin, Hyung Kyu, physical chemistry, theoretical chemistry
Sill, Richard Clements, physics
Slemmons, David Burton, geology, geophysics
Smiley, Vern Newton, optics, atmospheric physics
Smith, George Thomas, pathology
Squires, Patrick, meteorology
Steele, Fred A, Jr, air pollution
Telford, James Wardrop, atmospheric physics, computer science
Thyr, Billy Dale, plant pathology, mycology
Tibbitts, Forrest Donald, embryology
Tompson, Robert Norman, mathematical analysis; applied mathematics
Tueller, Paul T, plant ecology, range management
Van Remoortere, Emile C, cardiovascular physiology, pharmacology
Vig, Baldev K, cytogenetics
Voskuil, Walter Henry, geology
Ward, John M, physiology
Welch, William Henry, Jr, biochemistry
Went, Frits Warmolt, botany
Wheeler, George Carlos, entomology
Wheeler, Jeanette Norris, entomology
Winterberg, Friedwardt, theoretical physics
Winzeler, Robert Lee, cultural anthropology, ethnography
Woodyard, Jack Ramon, atomic physics
Young, Ralph Alden, soil fertility

SPARKS

Williams, Loring Rider, inorganic chemistry

NEW HAMPSHIRE

ALTON BAY

Waters, Nelson Fenn, genetics

ANTRIM

Turner, Harry Jackson, Jr, water pollution

ATKINSON

Lacaillade, Charles William, medical entomology

BEDFORD

Normandeau, Donald Arthur, aquatic ecology
Shevenell, Thomas Cortland, oceanography, sedimentology

BERLIN

Danforth, Raymond Hewes, pulp chemistry

CENTER HARBOR

Rose, Harry Melvin, clinical medicine, microbiology

CHOCORUA

Dunham, Theodore, Jr, astrophysics, biophysics

CONCORD

Frost, Terrence Parker, water pollution
Zumbrunnen, Charles Edward, dentistry

CONTOOCOOK

English, Albert Charles, physical chemistry

DOVER

Masuelli, Frank John, organic chemistry

DUBLIN

Kraichnan, Robert Harry, statistical mechanics

DURHAM

Allen, Fred Ernest, veterinary medicine
Amell, Alexander Renton, physical chemistry
Andersen, Kenneth K, organic chemistry
Anderson, Franz Elmer, oceanography, marine geology
Arnoldy, Roger L, space physics
Barrett, James Passmore, forestry
Batho, Edward Hubert, mathematics
Bickle, Robert Louis, entomology
Borror, Arthur Charles, protozoology
Bothner, Wallace Arthur, petrology
Bowman, James Sheppard, entomology
Brown, Wendell Stimpson, physical oceanography
Bruns, Paul Eric, forestry
Bullock, Wilbur Lewis, parasitology
Chapman, Donald Harding, geomorphology, glacial geology
Chasteen, Norman Dennis, bioinorganic chemistry
Chesbro, William Ronald, microbiology
Chupp, Edward Lowell, physics
Clark, David Gordon, physics
Collins, Walter Marshall, animal genetics
Copeland, Arthur Herbert, Jr, mathematics
Croker, Robert Arthur, marine ecology
Crow, Garrett Eugene, systematic botany
Daggett, Albert Frederick, chemistry
Dawson, John Frederick, physics
Dingman, Jane Van Zandt, evolutionary biology
Dingman, Stanley Lawrence, hydrology, resource management
Downs, Richard Erskine, anthropology
Dunlop, William Robert, avian pathology, cell biology
Dunn, Gerald Marvin, plant breeding, genetics
Dunn, Stuart, botany
Ellis, David Wertz, analytical chemistry
Federer, C Anthony, forest meteorology, forest hydrology
Foret, John Emil, developmental biology
Francq, Edward Nathaniel Lloyd, vertebrate ecology, animal behavior
Garrett, Peter Wayne, forest genetics
Gaudette, Henri Eugene, geology, geochemistry
Grant, Clarence Lewis, analytical chemistry
Haendler, Helmut Max, inorganic chemistry
Hall, Francis Ramey, hydrology
Haney, James Filmore, limnology
Harter, Robert Duane, soil chemistry
Herbst, Edward John, biochemistry
Hill, John Ledyard, wood science & technology
Hodgdon, Albion Reed, botany
Holter, James Burgess, dairy science
Hoornbeek, Frank Kent, genetics
Houston, Robert Edgar, Jr, physics
Hoyle, Merrill Cassius, biochemistry, plant physiology
Hubbard, Colin D, physical chemistry, biological chemistry
Ikawa, Miyoshi, chemistry
Jackson, Herbert William, pollution biology
Jacoby, Alexander Robb, mathematics
Johnson, Richard Edward, mathematics
Jones, Galen Everts, marine microbiology
Jones, Paul Raymond, organic chemistry
Kaufmann, Richard L, physics
Keener, Harry Allan, animal nutrition
Kiang, Yun-Tzu, population genetics, plant breeding
Klippenstein, Gerald Lee, protein chemistry
Koch, David William, agronomy
Kuo, Shan Sun, applied mathematics, computer science
Lambert, Robert Henry, atomic physics
Lavoie, Marcel Elphege, zoology
LeBlanc, Robert George, geography
Lockwood, John Alexander, physics
Loy, James Brent, plant science, developmental genetics
Lyle, Gloria Gilbert, organic chemistry
Lyle, Robert Edward, organic chemistry
Mathieson, Arthur C, botany
Mayewski, Paul Andrew, glacial geology, geomorphology
Metcalf, Theodore Gordon, microbiology
Meyers, Theodore Ralph, geology
Milne, Lorus Johnson, biology
Milne, Margery (Joan Greene), biology
Moore, Berrien, III, mathematics
Morrison, James Daniel, organic chemistry
Mott, David Gordon, ecology, forest entomology
Mower, Lyman, physics
Mulhern, John E, Jr, physics
Norton, Robert James, economic entomology, phytopathology
Owens, Charles Wesley, physical chemistry
Peirce, Lincoln Carret, plant breeding
Pierce, Edward Ronald, health sciences, medical genetics
Pilar, Frank Louis, physical chemistry
Pistole, Thomas Gordon, microbiology, immunology
Pollard, James Edward, physiology, horticulture
Porter, Clarence A, parasitology
Poulton, Bruce R, animal science
Prince, Allan Bixby, soil chemistry
Radlow, James, applied mathematics
Reeves, Roger Marcel, entomology
Rich, Avery Edmund, plant pathology
Ringrose, Richard Caig, poultry nutrition
Rogers, Owen Maurice, genetics, plant breeding
Ross, Shepley Littlefield, mathematics
Routley, Douglas George, horticulture
Safford, Lawrence Oliver, forestry, ecology
St Onge, Richard Norbert, nuclear physics, medical physics
Sasner, John Joseph, Jr, comparative physiology, invertebrate zoology
Sawyer, Philip John, zoology
Schneer, Cecil Jack, geology
Schreiber, Richard William, botany, cell biology
Schreiner, Ernst Jefferson, forest genetics
Shepard, Harvey Kenneth, theoretical high energy physics
Shigo, Alex Lloyd, plant pathology
Shore, Samuel David, mathematics
Shortle, Walter Charles, plant pathology
Skoglund, Winthrop Charles, poultry husbandry
Slanetz, Lawrence William, microbiology
Smith, Roderick MacDowell, fisheries
Smith, Samuel Cooper, biochemistry
Stewart, Glenn William, geology
Stewart, James Anthony, biochemistry
Strout, Richard Goold, zoology, parasitology
Swan, Emery Frederick, invertebrate zoology
Teeri, Arthur Eino, biochemistry
Tischler, Herbert, invertebrate paleontology
Tokay, F Harry, audiology, speech & hearing sciences
Uebel, Jacob John, organic chemistry
Urban, Willard Edward, Jr, biometrics, animal breeding
Waldron, Stephen, physics
Wallace, Oliver P, Sr, forestry
Wallace, William Huston, economic geography, geography of new england
Webber, William R, physics, astrophysics
Weber, James Harold, inorganic chemistry
Wells, Otho Sylvester, horticulture
Wheeler, Charles Mervyn, Jr, physical chemistry
Wheeler, Ellsworth Haines, Jr, biological oceanography
Wilkinson, Ronald Craig, forest genetics, physiology
Wright, John Jay, atomic physics
Wright, Paul Albert, reproductive endocrinology
Zsigray, Robert Michael, microbiology

ENFIELD

Cummings, Charles Sumner, II, physics

EXETER

Hoffman, Richard Albert, physics

GILMANTON IRON WORKS

Strout, George M, applied physics

GORHAM

Morse, Erwin Emerson, physical chemistry

HANCOCK

Kerwin, Richard Martin, bacteriology

HANOVER

Allen, Robert Day, cell biology, biophysics
Almy, Thomas Pattison, internal medicine
Alverson, Hoyt Sutliff, anthropology
Anderson, Duwayne Marlo, earth sciences, soil physics
Arkowitz, Martin Arthur, mathematics
Assur, Andrew, geophysics
Bagne, Farideh, medical physics
Ballard, William Whitney, vertebrate embryology, experimental morphology
Bartlett, Donald, Jr, physiology
Berger, Edward Michael, genetics
Bickel, Thomas Fulcher, algebra
Bilello, Michael Anthony, meteorology, climatology
Birnie, Richard Williams, geology
Black, Craig Patrick, animal physiology, animal ecology
Bogart, Kenneth Paul, mathematics
Boley, Forrest Irving, plasma physics, astrophysics
Borison, Herbert Leon, pharmacology
Bowen, Douglas Malcomson, organic chemistry
Braun, Charles Louis, chemical physics, experimental solid state physics

NEW HAMPSHIRE

Brinck-Johnson, Truls, endocrinology, biochemistry
Brown, Edward Martin, mathematics
Brown, Jerry, soil science
Brown, Marianne, topology
Brown, Stanley Alfred, biomaterials
Carleton, Richard Allyn, cardiology
Carpenter, Stanley John, cytology
Chaffee, Robert Gibson, paleontology
Chambers, Wilbert Franklin, neuroanatomy
Christy, Robert Wentworth, physics
Cleland, Robert Lindbergh, physical chemistry
Clendenning, William Edmund, dermatology
Copenhaver, John Harrison, Jr., biochemistry
Croasdale, Thomas, biochemistry
Crowell, Richard Henry, mathematics
Cushman, Samuel Wright, cell physiology, biochemistry
Daubenspeck, John Andrew, biomedical engineering, physiology
Davis, William Potter, Jr., physics
Decker, Robert Wayne, geophysics, academic administration
DeMaggio, Augustus Edward, botany, physiology
Dennison, David Severin, biophysics
Detar, Reed L., physiology
Doyle, William Thomas, physics
Dupree, Louis Benjamin, ethnology, archaeology
Edwards, Brian Ronald, medical physics
English, Jackson Pollard, organic chemistry
English, Van Harvey, geography
Ferm, Vergil Harkness, biology, embryology
Folley, Jarrett Harter, internal medicine, gastroenterology
Forster, Roy Philip, physiology
Friedman, Matthew Joel, psychiatry, pharmacology
Gale, Thomas Francis, teratology, anatomy
Galton, Valerie Anne, endocrinology
Gilbert, John Jouett, invertebrate zoology, freshwater biology
Gosselin, Robert Edmond, pharmacology
Gray, Clarke Thomas, microbiology, biochemistry
Gribble, Gordon W., organic chemistry
Ham, Thomas Hale, hematology
Hanlon, David Paul, biochemistry, physiology
Harbury, Henry Alexander, biochemistry
Harris, Edward Day, Jr., medicine, rheumatology
Haugen, Richard Kenneth, physical geography
Hoefnagel, Dick, pediatrics, genetics
Holmes, Richard Turner, animal ecology, animal behavior
Hornig, James Frederick, physical chemistry
Horowitz, Paul Martin, biochemistry, physical chemistry
Huggins, Elisha R., theoretical physics
Huke, Robert Edward, geography
Jackson, William Thomas, plant physiology
Jacobs, Nicholas Joseph, bacteriology
Jagels, Richard H., plant morphology
Johnson, Gary Dean, geology
Johnson, Noye Monroe, geochronology, hydrogeology
Karl, Richard C., surgery
Kelley, Maurice Leslie, Jr., medicine
Kemeny, John George, mathematics, academic administration
Kidder, John Newell, physics, physiological optics
Kilham, Lawrence, virology
King, Allen Lewis, physics
Kreider, Donald Lester, mathematical logic
Kurtz, Thomas Eugene, computer systems
Lahr, Charles Dwight, mathematical analysis
Lamperti, John Williams, mathematics
Lang, Gerald Edward, plant ecology
Langway, Chester Charles, Jr., geology
Lawrence, Walter Edward, theoretical solid state physics
Layton, William Malloy, Jr., pathology, marine biology
Leiby, Robert William, organic chemistry
Lemal, David M., organic chemistry
Lienhard, Gustav E., biochemistry
Lindgren, David Treadwell, geography
Lipowski, Zbigniew J., psychiatry
Longnecker, Daniel Sidney, pathology
Lubin, Martin, cell biology

Luehrmann, Arthur Willett, Jr., solid state physics, computer science
Lyons, John Bartholomew, geology
MacDonald, Gordon James Fraser, geophysics
Marin-Padilla, Miguel, pathology
Matsuo, Keizo, polymer chemistry
McCalley, Roderick Canfield, magnetic resonance
McCann, Frances Veronica, physiology, electrophysiology
McIntyre, Oswald Ross, hematology
McKim, Harlan L., geology, soil science
McNair, Andrew Hamilton, geology
Mellor, Malcolm, applied physics, engineering
Merritt, Katharine, immunology
Mook, Delo Emerson, II, astronomy
Mudge, Gilbert Horton, physiology, medicine
Munck, Allan Ulf, endocrinology
Naitove, Arthur, surgery, physiology
Nattie, Eugene Edward, pulmonary physiology
Nice, Philip Oliver, pathology, microbiology
Noda, Lafayette Hachiro, biochemistry
Norman, Robert Zane, mathematics
Nutt, David Clark, oceanography, geography
Nye, Robert Eugene, Jr., physiology
Olson, Maynard Victor, inorganic chemistry
Ou, Lo-Chang, physiology, biochemistry
Payson, Henry Edwards, psychiatry
Pfefferkorn, Elmer Roy, Jr., microbiology, parasitology
Pipes, Paul Bruce, low temperature physics
Plakun, Geraldine Taiani, mathematical analysis
Poole, James Plummer, botany
Prosser, Reese Trego, mathematics
Pyte, Agnar, theoretical physics
Rawnsley, Howard Melody, pathology
Reiners, William A., ecology
Reynolds, Robert Coltart, Jr., petrology, geochemistry
Rieser, Leonard M. Jr., atomic physics, nuclear physics
Roos, Thomas Bloom, zoology
Rosen, Leonard Craig, environmental physics, energy conversion
St. John, Walter McCoy, neurophysiology
Salans, Lester Barry, medicine, metabolism
Schultz, John Charles, insect ecology, vertebrate ecology
Shafer, Paul Richard, organic chemistry
Sharp, John Joseph, biochemistry
Shepard, June Smith, cytogenetics
Simon, Eliot Morton, physics, chemistry
Simpson, Robert Bonebrake, geography
Slesnick, William Ellis, mathematics
Smith, Edith Lucile, biochemistry, microbiology
Smith, Roger Powell, toxicology
Snapper, Ernst, mathematical logic
Snell, James Laurie, mathematical logic
Sobel, Raymond, child psychiatry
Soderberg, Roger Hamilton, inorganic chemistry
Sokol, Hilda Weyl, neuroendocrinology
Sonnerup, Bengt Ulf Osten, space physics, fluid mechanics
Spencer, Thomas A., organic chemistry
Spiegel, Evelyn Schafer, developmental biology
Spiegel, Melvin, biology
Sterncik, Edward Selby, medical physics
Sterrett, Kay Fife, geophysics, physical chemistry
Stibitz, George Robert, medical research
Stockmayer, Walter Hugo, physical chemistry
Stoiber, Richard Edwin, geology, volcanology
Stolk, Jon Martin, psychiatry, pharmacology
Strauss, Bella S., medicine
Strohbehn, John Walter, biomedical engineering, radiophysics
Swinzow, George K., physics
Takagi, Shunsuke, mechanics
Tanzer, Radford Chapple, medicine
Tenney, Stephen Marsh, physiology
Trumpower, Bernard Lee, biochemistry
Tucker, Gary Jay, psychiatry
Valtin, Heinz, physiology, internal medicine
Velez, Samuel Jose, neurophysiology
Walsh, John Edmond, plasma physics
Watson, Thomas Richard, Jr., surgery
Weeks, Wilford Frank, glaciology, hydrology
Weiner, John, chemical physics
Whybrow, Peter Charles, psychiatry

Wilenski, Gerald, physical chemistry
Williamson, Richard Edmund, mathematics
Wurster-Hill, Doris Hadley, cytogenetics
Yager, James Donald, Jr., cell biology, chemical carcinogenesis

HILLSBORO
Baldwin, Henry Ives, forestry

HENNIKER
Brown, Sanborn Conner, physics
Holton, Adolphus, geology
Puglia, Charles Raymond, invertebrate zoology

JAFFREY
Evans, Llewellyn Thomas, ethology
MacCready, Robert Alvan, public health

KEENE
Bayr, Klaus J., geography of Africa, cultural geography
Goder, Harold Arthur, plant ecology
Havill, Thomas Lampert, geography, anthropology
Neil, Thomas C., organic chemistry, analytical chemistry
Papadopoulos, Alex Spero, applied statistics, operations research
Quirk, James Denis, chemical physics, theoretical physics
Stepenuck, Stephen Joseph, Jr., environmental chemistry
Tourgee, Ronald Alan, mathematical statistics

LACONIA
Andrews, Henry Nathaniel, Jr., paleobotany
Evvard, John Cooper, aerodynamics

LEBANON
Stettenheim, Peter, ornithology

LITTLETON
Hatt, Robert Torrens, mammalogy, ethnography

LONDONDERRY
Buss, Keen, zoology
Hebert, Normand Claude, electroanalytical chemistry

LYME CENTER
Margolis, George, pathology

MANCHESTER
Bibeau, Armand A., zoology
Custer, Michael, chemistry
Damour, Paul Lawrence, physical chemistry
Fozdar, Birendra Singh, horticulture
Grenier, Marie-Anne Cecile, education
Lavigne, Andre Andre, organic chemistry
Normandin, Robert F., radiation biology
Plantenberg, Gonzaga, physics
Stahl, Barbara Jaffe, comparative anatomy, evolution
Swift, Robinson Marden, physical chemistry
Upham, Roy Herbert, organic chemistry
Wilks, Phillip Howard, high temperature chemistry

MELVIN VILLAGE
Page, Lincoln Ridler, geology

MEREDITH
Widger, William Knowlton, Jr., meteorology

MERRIMACK
Fricke, Edwin Francis, nuclear physics

NASHUA
Bileau, Claire of the Savior, physiology
Folweiler, Robert Cooper, materials science
Kuppenheimer, John D., Jr., optics
Schwartz, Jack, engineering physics

NEW BOSTON
Bunting, Mary Ingraham, microbiology

NEW LONDON
Bent, Donald Frederick, microbiology
Dodge, William G., chemistry, chemical engineering

NEWMARKET
Barton, William R., geology

OSSIPEE
Borg, Robert Munson, ecology

PETERBOROUGH
Clarkson, Merton Robert, veterinary medicine
Morison, Robert Swain, neurophysiology

PLYMOUTH
Dow, Maynard Weston, geography of North America
Fralick, Richard Allston, marine phycology
Long, Joseph K., anthropology
Salmons, George Beverly, physics
Spencer, Larry T., aquatic ecology, invertebrate zoology
Taffe, William John, atmospheric physics
Wixson, Eldwin A., Jr., mathematics

RINDGE
Briggs, Lloyd Cabot, anthropology
Dethlefsen, Edwin Stewart, biological anthropology, demography

STRATHAM
Gowen, Frederick Arthur, horticulture

SUNAPEE
MacNeill, Arthur Edson, physiology, biomedical engineering

SUNCOOK
Rice, Dale Wilson, physical chemistry

WEARE
Dearborn, Roland Balch, horticulture

NEW JERSEY

ALLENTOWN
Campbell, Donald Gray, reproductive physiology
Okarma, Theodore Joseph, husbandry, dairy science
Panitz, Eric, parasitology, protozoology

ALPINE
Bedrosian, Karakian, food technology
Schiffmann, Robert F., food science, microwave engineering

ASBURY
Palermo, Felice Charles, organic polymer chemistry

ASBURY PARK
Both, Eberhard, physical chemistry
Jacobson, Avrohm, psychiatry, psychoanalysis

ATLANTIC HIGHLANDS
Lasthuysen, Willem, organic chemistry

AUDUBON
Kozanowski, Henry Nikodem, physics

BASKING RIDGE
Fisk, James Brown, physics
Forster, Warren Schumann, organic chemistry
Hoff, Dale Richard, medicinal chemistry
Lien, Arthur Philip, organic chemistry, resource management
Lorenz, Donald H., organic chemistry, polymer chemistry

Rigtrink, Merle Dale, colloid chemistry
Smith, Philip T., physical chemistry, analytical chemistry
Van Hook, James Paul, physical chemistry, engineering management

BAYONNE
Conbere, John Philip, industrial chemistry
De Cicco, Henry, mathematics, operations research

BEACH HAVEN
Lyman, Frank Lewis, toxicology

BELLE MEAD
Slusarchyk, William Allen, organic chemistry
Spitzer, Penn Fulton, Jr., organic chemistry

BELVIDERE
Hemmendinger, Henry, physics

BERKELEY HEIGHTS
Adams, Phillip, organic chemistry
Babson, Robert Daniel, organic chemistry
Belohlav, Leo Rudolf, research administration
Brown, John Angus, organic chemistry, physical chemistry
Brown, Walter Lyons, solid state physics
Chen, William Kwo-Wei, polymer chemistry, chemical engineering
Daoust, Donald Roger, microbiology
Goliob, Fred, physical chemistry
LeSher, Dean Allen, clinical pharmacology, nephrology
Meyer, Victor Bernard, organic polymer chemistry
Munsell, Monroe Wallwork, organic chemistry, petroleum technology
Rubino, Andrew M, inorganic chemistry, physical chemistry
Smith, Harlan Millard, physical chemistry
Somkaite, Rozalija, pharmaceutical chemistry, analytical chemistry
Sorkin, Howard, organic chemistry
Van de Castle, John F., polymer chemistry, petroleum chemistry
Whiter, Paul Francis, research administration
Zoss, Abraham Oscar, industrial chemistry
Kreutner, William, pharmacology, biochemistry
Larson, Daniel A., atomic physics
Lawrason, F Douglas, internal medicine, pediatrics
Leer, John Addison, Jr., medicine, pediatrics
Leitz, Frederick Henry, pharmacology
Lin, Chin-Chung, biochemical pharmacology
Long, James Frantz, physiology
Mallams, Alan Keith, natural products chemistry
Marquez, Joseph A., biochemistry
McAllister, William Albert, physical chemistry
Munroe, Joscelyn Spencer, cancer
Nafissi-Varchei, Mohammad Mehdi, organic chemistry
Neary, Edward R., medicine
Neustadt, Bernard Ray, medicinal chemistry
Paustian, John Earle, organic chemistry
Peets, Edwin Arnold, molecular biology, biochemistry
Perlman, Preston Leonard, physiology
Popper, Thomas Leslie, organic chemistry
Reimann, Hans, organic chemistry
Roebke, Heide, organic chemistry
Rogers, Dexter, science education
Russell, Harvey R., synthetic organic chemistry, high temperature chemistry, engineering management
Sausville, Joseph Winston, biochemistry
Schafer, Thomas Wayne, microbiology, biochemistry
Schimmel, Nelson Hirsch, medicine
Schrogie, John Joseph, internal medicine, clinical pharmacology
Shattes, Walter John, experimental physics
Siegel, Marshall Mayer, physical chemistry, spectroscopy
Smith, Elizabeth Melva, steroid chemistry, synthetic organic chemistry
Smith, Sidney R, Jr, zoology, biochemistry
Sperber, Nathan, organic chemistry
Steinman, Martin, medicinal chemistry, organic chemistry
Symchowicz, Samson, drug metabolism, biochemical pharmacology
Tabachnick, Irving I A, pharmacology
Taber, Robert Irving, psychopharmacology, neuropharmacology
Testa, Raymond Thomas, microbiology, biochemistry
Thompson, Robert Edward, pharmacology
Thornton, William Andrus, Jr., physics
Topliss, John G., medicinal chemistry
Villani, Frank Conrad, pediatrics
Vogt, Frank Conrad, chemistry
Wachtel, Anselm, solid state chemistry
Wagman, Gerald Howard, microbial biochemistry
Waitz, Jay Allan, chemotherapy
Watjick, Arthur Saul, endocrinology
Weinstein, Marvin Joseph, physiology, microbiology
Wolkoff, Hal Norman, pharmaceutics
Yakubik, John, pharmaceutical chemistry
Youngstrom, Richard Earl, organic chemistry, radiochemistry

BERNARDSVILLE
Deem, Mary Lease, synthetic organic chemistry
Fuls, Ellis, semiconductors

BIRMINGHAM
Klein, Richard M, inorganic chemistry, organic chemistry

BLACKWOOD
Halpern, Myron Herbert, medical science

BLOOMFIELD
Barnett, Allen, neuropharmacology
Bhalla, Ranbir J R Singh, solid state science
Capeci, Nicholas Ernest, internal medicine
Carr, Raymond Niel, mathematical statistics, biostatistics
Chiu, Peter Jiunn-Shyong, pharmacology
Collins, Elliott Joel, endocrinology
Corth, Richard, physical chemistry, analytical chemistry
Daniels, Peter John Lovell, organic chemistry
Draper, Richard William, steroid chemistry
Ehrreich, Stewart Joel, pharmacology
Ewart, Roswell Horr, polymer science
Fahrenholtz, Kenneth Earl, organic chemistry
Gold, Elijah Herman, organic chemistry
Gourzis, James Theophile, pharmacology
Green, Michael John, organic chemistry
Heimlich, Ernest Maurice, pediatrics
Iezzoni, Domenic G, medicine
Johnson, Gordon Lee, pharmacology, physiology
Katchen, Bernard, biochemistry
Klem, Edward Benson, molecular biology, cell biology

BOONTON
Kononenko, Oleg K, organic chemistry, pharmaceutical chemistry
Rabel, Fredric M, analytical chemistry

BORDENTOWN
Weber, Paul Van Vranken, plant pathology, plant breeding

BOUND BROOK
Ackart, Watson Boudinot, industrial microbiology
Amick, Chester Albert, physical chemistry, colloid chemistry
Auerbach, Victor, polymer chemistry
Baker, Leonard Morton, organic chemistry, polymer chemistry
Balsley, Richard Benjamin, organic chemistry
Banick, William Michael, Jr, analytical chemistry
Behrens, Rudolf Adolf, rubber chemistry
Brady, Thomas E., synthetic organic chemistry, textile chemistry
Burhans, Allison Stilwell, plastics chemistry
Burley, David Richard, physical chemistry
Carrick, Wayne Lee, organic chemistry, polymer chemistry
Chow, Sui-Wu, organic chemistry
Cotter, Robert James, polymer chemistry
Cowles, Craig Schuyler, industrial organic chemistry
Davidson, Daniel Lee, polymer chemistry
Flynn, Kenneth G, organic chemistry
Foster, Walter H, Jr, physical chemistry, analytical chemistry
Fry, John Sedgwick, organic chemistry
Garty, Kenneth Thomas, organic chemistry
Gerber, Samuel Michael, organic chemistry
Goeke, George Leonard, organic chemistry, organometallic chemistry
Gorham, William Franklin, organic chemistry, polymer chemistry
Hale, Warren Frederick, polymer chemistry
Hardy, William Baptist, organic chemistry
Heffron, Peter John, industrial organic chemistry, synthetic organic chemistry
Horton, Robert Louis, organic chemistry
Innes, John Edwin, industrial organic chemistry
Kaizerman, Samuel, organic chemistry
Karol, Frederick J, polymer chemistry
Keogh, Michael John, organic chemistry
Kirshenbaum, Gerald Steven, polymer chemistry
Klingsberg, Erwin, organic chemistry
Konzelman, Leroy Michael, organic chemistry
Kopf, Peter W, physical chemistry, polymer chemistry
Leavitt, Julian Jacob, chemistry
Loffelman, Frank Fred, organic chemistry
Longfield, James Edgar, physical chemistry
Mackenzie, Neil Mitchill, organic chemistry
Maulding, Donald Roy, synthetic organic chemistry
McCarthy, Neil Justin, Jr, organic polymer chemistry
McCleary, Harold Russell, physical chemistry
McComas, Wilbur Harrison, Jr, analytical chemistry
McKay, Donald Edward, organic chemistry
Megson, Frederic Houghton, organic chemistry
Miller, John Clark, physical chemistry
Misner, Robert E, organic chemistry
Mosby, William Lindsay, organic chemistry
Oppelt, John Christian, organic chemistry
Orloff, Malcolm Kenneth, quantum chemistry
Parsons, James Sidney, analytical chemistry
Pinto, Frank G, organic chemistry
Potts, James Edward, polymer chemistry, radiation chemistry
Rauhut, Michael McKay, organic chemistry
Reichle, Walter Thomas, chemistry
Reinking, Norman Herbert, organic polymer chemistry
Roth, Philip B, organic chemistry
Ruby, Philip Randolph, organic chemistry
Russell, Henry Franklin, organic chemistry
Santacana-Nuet, Francisco, analytical chemistry
Sarakwash, Michael, environmental management
Saxon, Robert, polymer chemistry
Schober, Donald Lincoln, organic polymer chemistry
Shaw, Montgomery Throop, polymer science
Shaw, Richard Gregg, polymer chemistry
Shepherd, Floyd, polymer chemistry
Sherr, Allan Ellis, physical organic chemistry
Singh, Ajaib, physical organic chemistry, polymer chemistry
Slagan, Peter Michael, industrial organic chemistry
Soffer, Irving Herbert, physical chemistry
Spurr, Orson Kirk, Jr, physical chemistry
Stueben, Kenneth Charles, organic chemistry, polymer chemistry
Susi, Peter Vincent, organic chemistry
Swift, Abbott Montague, organic chemistry
Thompson, Harold G, organic chemistry
Toothill, Richard B, organic chemistry
Tsang, Sien Moo, photochemistry
Vial Theodore Merriam, rubber chemistry
Whelan, John Michael, polymer chemistry
Willeboordse, Friso, analytical chemistry
Wynstra, John, organic polymer chemistry
Young, Robert Hayward, photochemistry, organic polymer chemistry

BRADLEY BEACH
Tashlick, Irving, organic chemistry, polymer chemistry

BRIDGETON
Springer, John Kenneth, plant pathology, nematology

BRIDGEWATER
Aspelin, Gary Bertil, organic chemistry
Bann, Robert (Francis), organic chemistry
Borysko, Emil, biology
Caldwell, Carlyle Gordon, plant chemistry
Dess, Howard Melvin, inorganic chemistry
Knapp, Malcolm Hammond, industrial organic chemistry
Lieberman, Emanuel Roy, bioengineering
Linn, Bruce Oscar, organic chemistry
Mayer, Warren Clifford, food science, engineering
Merritt, Richard Howard, horticulture, academic administration
Ray-Chaudhuri, Dilip K, polymer chemistry, organic chemistry
Rutenberg, Morton Wolf, organic chemistry
Searle, Norma Zizmer, physical chemistry
Starkovsky, Nicolas Alexis, organic chemistry
Watrel, Warren George, microbiology, biochemistry
Wurzburg, Otto Bernard, carbohydrate chemistry
Wykoff, Matthew Henry, experimental surgery, physiology

BRIGANTINE
O'Connor, Charles Timothy, medical entomology

BROWNS MILLS
Downing, Daniel Francis, cardiology

BUDD LAKE
Unrau, David George, biochemistry

CALDWELL
Adler, Beatriz Raquel, biology
Breslin, Maureen Elizabeth, analytical chemistry
Kurnick, Allen Abraham, chemistry, nutrition
Proudfoot, Bernadette Agnes, biology
Scannell, James Parnell, biochemistry
Sherman, Seymour, systems analysis
Tracy, M Joanna, experimental medicine

CAMDEN
Bacha, William Joseph, Jr, parasitology
Berner, David Leo, biochemistry
Bidwell, Leonard Nathan, mathematics
Black, Otis Deitz, industrial chemistry
Bluebond-Langner, Myra Honore, anthropology
Briskey, Ernest J, food science, biochemistry
Charney, Jesse, biochemistry
Conrad, Louis Johnson, chemistry
Coriell, Lewis L, bacteriology
Dahle, Leland Kenneth, cereal chemistry
Davis, Stanley Gannaway, physical chemistry
Denton, Arnold Eugene, biochemistry
Dion, Arnold Silva, biochemistry
Durand, James Blanchard, biology
Feldman, William, clinical chemistry
Gagliardi, L John, chemical physics, biophysics
Good, Ralph Edward, plant ecology
Greene, Arthur E, cell biology
Hartwick, Robert Frank, animal behavior, biological oceanography
Jackson, Richard Field, food technology
Jacobson, Glen Arthur, food chemistry

Katz, Sidney A., radiochemistry, analytical chemistry
Lasfargues, Etienne Yves, microbiology, oncology
Lee, Hsin-Yi, developmental biology
Loercher, Lars, plant physiology
Lohr, Lester Jay, analytical chemistry, organic chemistry
McGarrity, Gerard John, microbiology, infectious diseases
Moore, Dan Houston, biophysics
Nichols, Warren Wesley, cytogenetics, pediatrics
Owens, Clifford, physical inorganic chemistry
Perkins, William Edward, food microbiology, food science
Peterson, Arthur Carl, bacteriology, immunology
Phillips, Wendell Francis, analytical chemistry
Reeve, Eldrow, plant physiology, soils
Scheiner, Donald M., biochemistry, microbiology
Sheffield, Joel Benson, cell biology, virology
Stempen, Henry, microbiology
Stone, Edward John, organic chemistry, biochemistry
Sullivan, Andrew Jackson, biochemistry, food chemistry
Toji, Lorraine Hellenga, biochemistry
Vollmer, James, physics
Weissman, Gerard Selwyn, plant physiology
Willits, Charles Haines, organic chemistry
Witte, Michael, organic chemistry, pollution chemistry
Younce, Gordon Baldwin, structural geology, physical oceanography
Younkin, Stuart G., plant pathology

CAPE MAY COURT HOUSE
Wood, Albert Elmer, vertebrate paleontology

CARLSTADT
Burniston, George Kissam, analytical chemistry
Hotelling, Eric Bell, organic chemistry
Sulzberg, Theodore, organic polymer chemistry
Vill, John Joseph, physical organic chemistry

CARTERET
Clark, Raymond George, chemistry

CEDAR GROVE
Henry, Arnold William, polymer science
Kraiman, Eugene Alfred, polymer chemistry
Ram, Gerson Louis, biochemistry
Sello, Stephen Balthazar, textile chemistry
Williams, Thomas Henry, structural chemistry

CEDAR KNOLLS
Brody, Gerald, research administration, chemotherapy
Goble, Frans Cleon, microbiology, animal pathology
Gold, William, dental research
Guernsey, Richard Montgomery, acoustics
Jacob, James Thecattil, pharmaceutics
Salem, Harry, pharmacology, toxicology
Theimer, Edgar E., analytical chemistry, pharmaceutical chemistry

CHATHAM
Conciatori, Anthony Bernard, organic chemistry
Druding, Leonard F., inorganic chemistry
Glover, Alan Marsh, physics
Granadesikan, Mrudulla, mathematical statistics
Greze, John Paul, Jr., chemistry
King, Cecil Victor, chemistry
Moran, William Joseph, organic chemistry
Paterson, Arthur Renwick, analytical chemistry
Payne, Elmer Curry, chemistry
Prochaska, Robert Joseph, organic chemistry
Robinson, Edwin Allin, chemistry
Struve, William Scott, organic chemistry
Thomas, Jess William, applied physics
Thompson, Henry Theron, organic chemistry

CHERRY HILL
Boyd, Glenn D., food science
Burgess, Kenneth Alexander, physical chemistry
Chibnik, Sheldon, organic chemistry
Davis, Frank, biochemistry, food science
Diamantis, William, pharmacology
Fulde, Roland Charles, food science
Glassman, Jerome Martin, pharmacology
Gonshery, Marvin E., optics
Gopalakrishnan, Pungampoondi Velamur, immunochemistry
Granito, Charles Edward, chemistry, information science
Green, Joseph, polymer chemistry
Harun, Joseph Stanley, clinical research
Hendler, Edwin, human physiology
Hicks, Arthur Earl, rubber chemistry
High, LeRoy Bertolet, organic chemistry, electrooptics
Hills, Stanley, electrochemistry
Jones, Daniel Gethen, organic chemistry
Keane, Kenneth William, nutrition, biochemistry
Kornstein, Edward, engineering management
Ling, Alfred Soy Chou, neurochemistry, endocrinology
Lu, Gordon Go, pharmacology, medicine
McGehee, Charles Leroy, analytical chemistry
Munger, Robert Shoop, analytical chemistry
Narin, Francis, research administration, computer sciences
Newcombe, Jack, organic chemistry
Papazian, Louis Arthur, physical chemistry
Parisse, Anthony John, cosmetic chemistry
Pflug, Gerald Ralph, pharmaceutics
Pinski, Gabriel, applied mathematics, information science
Plechner, Sophie L., biochemistry
Rashkin, Jay Arthur, physical chemistry
Reinhart, Richard Joseph, biomedical engineering, physiology
Rosen, Bernard H., chemistry, chemical engineering
Roth, Shirley H., chemistry
Schlichting, David Arthur, clinical medicine
Seagers, William James, organic chemistry
Smiles, Kenneth Albert, environmental physiology
Sofia, R Duane, pharmacology
Soldati, Gianluigi, organic chemistry
Sorkin, Marshall, cosmetic chemistry
Stoy, William S., chemistry
Tudor, David Cyrus, poultry pathology
Wooding, William Minor, experimental statistics, chemistry
Wulf, Ronald James, pharmacology
Zimmerman, Henry B., chemistry
Zimmerman, Stanley Dean, mathematics
Zimmerman, Henry B., chemistry, electron microscopy

CHESTER
Kratochvil, Joseph Howard, cosmetic chemistry

CINNAMINSON
Hager, Onslow Bonner, pollution chemistry
Hikida, Hideaki Robert, plant breeding, plant genetics
Nickeson, Richard L., horticulture, plant pathology
Raymer, William Bruce, plant physiology
Reynard, George Bergin, botany

CLARK
Barton, Preston Nichols, medicine
Hanley, Arnold V., analytical chemistry
Lieberman, Samuel Victor, chemistry, medicinal chemistry
Wu, Mu Tsu, organic chemistry

CLIFTON
Ackerman, Joseph Francis, organic chemistry
Brandman, Harold A., organic chemistry
Burrell, Harry, polymer chemistry
Coscarelli, Waldimero, microbiology, biochemistry
Demos, Christopher Harry, chemistry
Dorsky, Julian, chemistry
Genet, Rene P H, organic chemistry
Hirsch, Arthur, organic chemistry
Hochstetler, Alan Ray, synthetic organic chemistry
Keaveney, William Patrick, organic chemistry
Marton, Oliver L., cosmetic chemistry
Mowitz, Arnold Martin, analytical chemistry
Mullen, Patricia Ann, cosmetic chemistry
Saad, Hosny Younes A., physical chemistry, physical pharmacy
Shaffer, Gary W., organic chemistry, photochemistry
Shur, E Gustave, chemistry
Sommers, David A., geology, hydrology
Williams, Neal Thomas, engineering physics

COLONIA
Rutkowski, Alfred John, organic chemistry

COLTS NECK
McGough, William Edward, psychiatry, physiological psychology
Thurston, Robert Norton, physics

CONVENT STATION
Arrow, Leslie Earle, physiological chemistry, research administration
Boulton, Mary, analytical chemistry
Lawlor, Anna Catherine, zoology
Levi, Barbara Goss, high energy physics
Smith, Marian Jose, biochemistry

CRANBURY
Banerjee, Bhola Nath, toxicology, teratology
Barker, James Emory, inorganic chemistry
Bodin, Jerome Irwin, pharmaceutical chemistry
Bond, Elizabeth Dux, analytical chemistry

CRANFORD
Beck, John Louis, analytical chemistry
Bradstreet, Raymond Bradford, chemistry, organic chemistry, analytical chemistry
Farrar, Thomas C., physical chemistry
Fay, Roger Richard, biological oceanography, marine ecology
Fordice, Michael W., clinical chemistry
Kaback, Stuart Mark, organic chemistry
Rinaldi, Leonard Daniel, mathematics
Sechrist, Warren Doyle, plastics
Shane, Norman Abraham, analytical chemistry
Sowa, Frank Joseph, chemistry
Swackhamer, Farris Saphar, organic chemistry
Wilson, Robert Curtis, chemistry
Yeh, Kuo-Chen, organic chemistry

CRESSKILL
Eilberg, Ralph G., biochemistry
Teeters, Wilbur Oldroyd, physical chemistry

DEEPWATER
Alekman, Stanley L., physical organic chemistry
Brooks, Robert Alan, organic chemistry
Crouse, Dale McClish, organic chemistry
Day, Herman O'Neal, Jr., physical chemistry
DeGeiso, Richard Charles, analytical chemistry
Foley, Henry Grant, organic chemistry
Franklin, Richard Crawford, organic chemistry
Garland, Charles E., organic chemistry, color science
Griffin, Dale Miller, Jr., organic chemistry
Jensen, Philip Wright, analytical chemistry
Krahler, Stanley Earl, organic chemistry
Mallonee, James Edgar, industrial organic chemistry
Manos, Philip, organic polymer chemistry
McCarthy, John Randolph, organic chemistry
Moores, Mead Stephen, chemistry
Murray, William Singler, chemistry
Payne, Donald Hughel, physical chemistry
Petersen, Wallace Christian, chemistry
Pohland, Hermann W., organic chemistry
Purzig, Donald Edward, organic chemistry
Rimmer, Robert W., organic chemistry
Steiner, Werner Douglas, organic chemistry
Twelves, Robert Ralph, organic chemistry
Varon, Albert, analytical chemistry
Watson, Richard Noble, organic chemistry
Williams, William Michael, organic chemistry
Witterholt, Vincent Gerard, organic chemistry
Zeftel, Leo, organic chemistry

DEMAREST
Crosman, Arthur Marston, zoology

DENVILLE
Hansen, Holger Victor, organic chemistry, cell biology
Rose, Patricia McGovern, chemistry
Westerdahl, Raymond P., physical chemistry

DOVER
Abel, James Edward, x-ray crystallography, explosives
Avrami, Louis, physics
Daly, Robert E., analytical chemistry
Downs, David S., solid state physics
Fair, Harry David, Jr., solid state physics
Gentner, Robert F., physical chemistry
Girard, Rene Jean, atomic physics
Griffin, William Dallas, pathology
Heim, Louise Marie, anatomy, physiology
Kirshenbaum, Abraham David, physical chemistry, explosives
Lanzerotti, Mary Yvonne DeWolf, physical chemistry
Matsuguma, Harold Joseph, inorganic chemistry
Morrow, Scott, microscopy, inorganic chemistry
Owens, Frank James, solid state physics
Prask, Henry Joseph, solid state physics
Readdy, Arthur F., Jr., materials engineering
Sandus, Oscar, physical chemistry
Sharkoff, Eugene Gibb, physics
Sheffield, Oliver Epravel, physical chemistry, organic chemistry
Torok, Andrew, Jr., chemistry
Veal, Dean Johnson, analytical chemistry
Vezzoli, Gary Christopher, solid state physics, chemical physics
Wiegand, Donald Arthur, solid state physics

DUNELLEN
Kleiner, Walter Bernhard, electrochemistry
Perkel, Robert Jules, polymer chemistry, organic chemistry

EAST BRUNSWICK
Beede, Charles Herbert, organic chemistry
Chen, James L., pharmaceutical chemistry
Deinet, Adolph Joseph, organic chemistry
Feldman, Martin Louis, organic chemistry
Grande, John Anthony, weed science
Grattan, Jerome Francis, biochemistry
Kimmel, Paul Issac, physical chemistry
LaVia, Anthony L., pharmaceutical chemistry
Marchisotto, Robert, pharmaceutical chemistry
Marcus, David, pharmacy
Noshay, Allen, organic chemistry, polymer chemistry
Woodruff, Edythe Parker, topology
Woodruff, Robert Wilson, polymer physics

EAST HANOVER
Arcese, Paul LeRoy, organic chemistry
Bagdon, Robert Edward, medical research
Barrett, Walter Edward, pharmacology
Bond, Robert Wallace, chemistry

Brown, Richard Emery, organic chemistry
Burns, Robert B P., medicine, pharmacology
Burrell, Craig Donald, medicine, endocrinology
Coombs, Renate Bangert, organic chemistry
Coombs, Robert Victor, organic chemistry
Dain, Jeremy George, organic chemistry, drug metabolism
Dugger, Harry A., organic chemistry
Eberle, Marcel Karl, organic chemistry
Frey, Albert Joseph, organic chemistry, medicinal chemistry
Gogerty, John Harry, pharmacology
Hardtmann, Goetz E, organic chemistry, medicinal chemistry
Harrington, Francis Eugene, physiology, genetics
Hay, Peter Marsland, industrial chemistry
Houlihan, William J., organic chemistry, medicinal chemistry
Iorio, Louis Carmen, pharmacology
Jacobs, Allen Leon, pharmaceutical chemistry
Kirby, Ben Harrison, organic chemistry
Kornblum, Saul S, physical pharmacy, physical chemistry
Laufer, Robert J., organic chemistry
Luthy, Jakob Wilhelm, chemistry
Maldacker, Thomas Anton, analytical chemistry
Maulding, Hawkins Valliant, Jr, physical chemistry, physical pharmacy
Michaelis, Arthur Frederick, pharmaceutics, physical pharmacy
Perrine, John W, Jr, biology
Schwartz, Benjamin Sam, microbiology
Schwarz, Hans Jakob, organic chemistry, biochemistry
Shellenberger, Carl, H, physiology, pharmacology
Solomon, Thomas Allan, pharmacology, cardiovascular physiology
Trapold, Joseph Hugh, pharmacology
Turi, Paul George, industrial pharmacy, analytical chemistry

EAST MILLSTONE
Rapp, William Rodger, pathology
Russell, Thomas J, Jr, biochemistry

EAST ORANGE
Auerbach, Oscar, medicine
Ayvazian, L Fred, internal medicine
Boehme, Diethelm Hartmut, pathology, neuropathology
Cvijanovich, George, physics
Ertel, Norman H, internal medicine, endocrinology
Fanale, Louisa P., zoology, immunology
Gallo, Michael Anthony, toxicology, biochemical pharmacology
Goodloe, Paul Miller, II, chemistry
Hein, Rosemary Ruth, developmental anatomy, developmental biology
Hobby, Gladys Lounsbury, microbiology
Johnson, Eben Lennart, mineralogy, petrology
Landau, Edward Frederick, organic chemistry, textile chemistry
Luisada-Opper, Anita Victoria, biochemistry, clinical chemistry
McGuire, David Kelty, science policy
Most, Joseph Morris, physical chemistry
Muller, Thomas C, organic chemistry
Sellmer, George Park, zoology
Serlin, Oscar, surgery
Simmons, Jean Elizabeth Margaret, mathematics
Snygg, John Morrow, applied mathematics
Sobel, Harold John, pathology, histochemistry
Trubowitz, Sidney, hematology, cell physiology

EAST RUTHERFORD
Dunkel, Morris, organic chemistry
Sparks, Allen Kay, organic chemistry

EDGEWATER
Cahn, Arno, organic chemistry
Healey, Frank Henry, physical chemistry
Pader, Morton, industrial chemistry
Peterson, Dennis Randall, biochemistry
Prince, Martin Irwin, organometallic chemistry
Princen, Henricus Mattheus, physical chemistry, surface chemistry
Richberg, Carl George, food science

Ritchey, Thomas William, microbiology, dental research

EDISON
Abramovici, Miron, organic chemistry, physical organic chemistry
Aykan, Kamran, inorganic chemistry
Brezenski, Francis T, microbiology, biochemistry
Brown, Stanley Monty, petroleum chemistry, surface chemistry
Brugger, John Edward, physical chemistry
Butter, Stephen Allan, inorganic chemistry
Campbell, Larry Edwin, physical inorganic chemistry
Cohn, Johann Gunther Ernst, chemistry
Conrad, Lester I, cosmetic chemistry
Eckert, Joseph Nicolaus, analytical chemistry
Eiseman, Fred S, organic chemistry
Falk, Charles David, physical inorganic chemistry
Gladstone, Harold Maurice, organic chemistry
Heiba, El-Ahmadi Ibrahim, organic chemistry, physical organic chemistry
Hemstock, Glen Alton, colloid chemistry
Hindin, Saul Gerald, surface chemistry
Hoeschele, James David, bioinorganic chemistry
Isaac, Peter Ashley Hammond, surface chemistry, pulp & paper technology
Jurewicz, Anthony Theodore, organic chemistry
Kaufman, Harold Alexander, organic chemistry
Kennedy, James Vern, chemistry, research administration
Lafornara, Joseph Philip, environmental sciences
Lomanitz, Rachel, medical microbiology, medical mycology
Lui, Yiu-Kwan, physical chemistry
Maso, Henry Frank, cosmetic chemistry
Matzner, Markus, organic chemistry
McCullough, John Price, physical chemistry
McGowan, H Christopher, ceramics, ceramic engineering
Milkowski, John David, medicinal chemistry
Murray, James Gordon, organic polymer chemistry
Murray, John Joseph, organic chemistry
Napier, Roger Paul, organic chemistry
Pappas, James John, organic chemistry
Pearce, David Archibald, agricultural chemistry
Phillips, Donald David, organic chemistry
Powell, Leo S, organic chemistry
Prapas, Aristotle George, organic polymer chemistry, petroleum chemistry
Pugliese, Michael, organic chemistry
Schwab, Frederick Charles, polymer chemistry
Seger, Francis Michael, organic chemistry
Simpson, Frank Martin, Jr, analytical chemistry, physical chemistry
Spano, Francis A, organic chemistry, biochemistry
Stern, Eric Wolfgang, chemistry
Storfer, Stanley J, organic chemistry
Strong, Jerry Glenn, pesticide chemistry
Woltermann, Gerald M, inorganic chemistry
Young, Lewis Brewster, organic chemistry

ELIZABETH
Bundy, Wayne Miley, clay mineralogy
Harrison, Jack Lamar, geology
Moll, William Francis, Jr, geochemistry, mineralogy
Rudel, Harry William, organic chemistry
Wohl, Arnold J, pharmacology

ELIZABETHPORT
Myles, William John, organic chemistry

EMERSON
Ishler, Norman Hamilton, food science

ENGLEWOOD
Dellenback, Robert Joseph, medical physiology, academic administration
Garber, John Douglas, organic chemistry
Toralballa, Leopoldo Vasquez, mathematics

ENGLEWOOD CLIFFS
Bernetti, Raffaele, organic chemistry, analytical chemistry
Eiszner, James Richard, chemistry
Feliciotti, Enio, food chemistry
Franceschini, Remo, food science, food technology
Freeman, Jere Evans, technological forecasting
Graham, Harold Nathaniel, organic chemistry
Kascic, Michael Joseph, Jr, mathematics
Manley, Charles Howland, food chemistry
Opdyke, Donald Lloyd J, pharmacology
Ranadive, Arvind S, food science
Rupp, Frank Adolph, microbiology
Saiger, George Lewis, epidemiology, medical statistics
Sanderson, Gary Warner, food chemistry
Supran, Michael Kenneth, food science

ESSEX FELLS
Kirchoff, William F, microbiology
Solmssen, Ulrich Volckmar, organic chemistry
Sturznegger, August, organic chemistry, chemical engineering

FAIR HAVEN
De Sessa, Michael Anthony, research administration
Scheffler, Edward Reinhard, analytical chemistry

FAIR LAWN
Bednarcyk, Norman Earle, food science
Gaj, Bernard Joseph, organic chemistry
Geller, Milton, analytical chemistry, pharmaceutical chemistry
McAnelly, John Kitchel, food microbiology
Rusoff, Irving Isadore, nutrition
Rutkin, Philip, organic chemistry
Siracusano, Vincent C, microbiology, biochemistry
Sodano, Charles Stanley, natural products chemistry
Sosis, Paul, organic chemistry
Stillings, Bruce Robert, nutrition

FAIRFIELD
Blank, Zvi, solid state chemistry
Kline, Charles Howard, physical chemistry
Prince, Herbert N, microbiology, toxicology
Waugh, John Blake-Steele, electronics, engineering physics

FANWOOD
Atkinson, Russell H, physical chemistry, engineering management
Whittingham, Michael Stanley, solid state chemistry

FARMINGDALE
Clark, Arthur Randolph, chemistry
Navarra, John Gabriel, earth sciences, science education

FLANDERS
Campbell, Clement, Jr, chemistry
Tremain, Henry Earl, chemistry

FLEMINGTON
Baughn, Charles (Otto), (Jr), biology
Falkiewicz, Michael Joseph, physical chemistry
Frear, George Lewis, applied chemistry
Hoge, William Anthony, paper chemistry
Ketterer, Paul Anthony, analytical chemistry
Olmstead, Edwin Vincent, pathology

FLORHAM PARK
Ariemma, Sidney, polymer chemistry, organic chemistry
Eisner, Mark Joseph, operations research
Hakala, Neil Victor, chemistry
Kiernan, William John, physical chemistry
Schoenholz, Daniel, organic chemistry
Segers, Richard George, mathematics, operations research
Smith, Paul Vergon, Jr, chemistry

FRANKLIN LAKES
Pizzarello, Roy Aloysius, textile chemistry

FREEHOLD
Bondar, Richard Jay Laurent, clinical biochemistry, clinical chemistry
Bungay, Henry Robert, III, microbial bioengineering
Flora, Robert Montgomery, biochemistry, enzymology
Gill, James Wallace, biochemistry

FT LEE
Thompson, Laura, cultural anthropology, applied anthropology

FT MONMOUTH
Behl, Wishvender K, physical chemistry, electrochemistry
Berkowitz, Harry Leo, radiation physics, theoretical physics
Bomke, Hans Alexander, physics
Creedon, John E, plasma physics
Epstein, Seymour, physics, engineering
Hafner, Erich, solid state physics
Helbert, John N, radiation chemistry, polymer chemistry
Hunger, Herbert Ferdinand, electrochemistry
Kedesdy, Horst H, physics
Kohn, Jack Arnold, crystallography
Kotin, Leon, mathematics
Kronenberg, Stanley, physics
McAfee, Walter Samuel, theoretical physics
Mette, Herbert L, solid state physics
Reder, Friedrich H, physics
Salomon, Mark, physical chemistry
Savage, Robert O, Jr, inorganic chemistry
Schneider, Sol, physics, electronics
Streever, Ralph L, solid state physics
Tauber, Arthur, chemistry, magnetism
Tebo, Edith Janssen, physics, astrophysics

GARFIELD
Ceprini, Mario Q, organic chemistry
Chase, Vernon Lindsay, textile chemistry
Hollihan, John Philip, colloid chemistry
Latta, Bruce McKee, physical organic chemistry
Riccobono, Paul Xavier, organic chemistry, polymer chemistry
Ross, Stanley Elijah, organic polymer chemistry, textiles
Stoner, George Green, organic chemistry

GILLETTE
Sacks, William, polymer chemistry, polymer engineering
Townley, Robert William, chemistry
Vanderspurt, Thomas Henry, physical inorganic chemistry

GLADSTONE
Blackburn, Benjamin (Coleman), botany, dendrology
Henderson, Norman Leo, physical pharmacy

GLASSBORO
Dike, Paul Alexander, geology
Gershenowitz, Harry, history of science, philosophy of science
Husain, Syed M, plant pathology
Jenkins, Alfred Martin, organic chemistry
Renlund, Robert N, phycology
Stansfield, Charles Arthur, resource geography
Stoudt, Harry Nathaniel, botany
Timon, William Edward, Jr, mathematical statistics

GLEN RIDGE
Herzog, Hershel Leon, chemistry
Papa, Domenick, chemistry

GLEN ROCK
Barnes, Douglas, economic entomology, ecology
Morck, Roland Anton, biochemistry
Vinson, Leonard J, biochemistry
Weaver, John Martin, biochemistry, microbiology

NEW JERSEY

GREAT MEADOWS
Kwartler, Charles Edward, organic chemistry
Robin, Michael, organic chemistry

GREEN BROOK
Woske, Harry Max, internal medicine, cardiology

HACKENSACK
Alfano, Michael Charles, periodontology, nutrition
Baden, Ernest, anatomic pathology, oral pathology
Boucher, Louis Jack, dentistry, anatomy
Chasens, Abram I., periodontics, dentistry
Di Paolo, Rocco John, orthodontics
Ehrenfeld, Robert Louis, organic chemistry
Ferstandig, Louis Lloyd, organic chemistry
Freund, Thomas Steven, biochemistry
Kaslick, Ralph Sidney, periodontics, oral medicine
Katz, Albert Barry, biochemistry
Ramazzotto, Louis John, dentistry, physiology
Rivetti, Henry Conrad, dentistry, prosthodontics
Ross, Norton Morris, clinical pharmacology
Rowley, George Richard, biochemistry, physiology
Smith, David Joseph, biochemistry

HACKETTSTOWN
Browder, Eli Jefferson, neurosurgery
Perry, Donald Dunham, organic chemistry, polymer chemistry
Stone, Charles Dean, food science, biochemistry
Tecozky, Melvin, inorganic chemistry
Thomas, Alan, food science

HADDONFIELD
Barber, George Winston, biochemistry
Greer, Albert H., organic chemistry
Miller, Ralph Albert, food science

HADDON HEIGHTS
Kaiser, Carl, medicinal chemistry

HAMBURG
Glidden, Richard Mills, rubber chemistry

HAMMONTON
Jaffe, Jonah, pharmaceutical chemistry
Ordile, Carol Maria, biostatistics
Tindall, Charles Gordon, Jr, forensic science

HARRINGTON PARK
Regna, Peter P., organic biochemistry

HARRISON
Schaaf, Kurt Herbert, organic chemistry

HASBROUCK HEIGHTS
Viscelli, Thomas Alfonse, biochemistry

HAWTHORNE
Lerner, Lawrence Robert, industrial organic chemistry
Neri, Rudolph Orazio, endocrinology
Puckna, John Godfrey, organic chemistry

HIGH BRIDGE
Davis, Benjamin Harold, plant pathology

HIGHLAND PARK
Bernstein, Eugene H., biochemistry, virology
Cordon, Martin, organic chemistry
Cronin, Jane Smiley (Mrs Joseph C Scanlon), biomathematics
Schipper, Edgar, organic chemistry
Stangel, Harvey J., agronomy, plant physiology

HIGHLANDS
MacKenzie, Clyde Leonard, Jr, marine biology, marine ecology
Olla, Bori Liborio, animal behavior, fish biology
Pacheco, Anthony Louis, fish biology, fisheries
Pearce, Jack B., marine ecology, zoology
Sinderman, Carl James, marine biology, parasitology
Stolen, Joanne Siu, immunology
Thomas, James Postes, zoology, ecology
Walford, Lionel Albert, fish biology

HIGHTSTOWN
Carpenter, William Graham, polymer chemistry
Herman, Daniel Francis, organic chemistry
Kruh, Daniel, organic chemistry
Lee, Min-Shiu, physical chemistry, polymer
Lyding, Arthur R., polymer chemistry, organic chemistry
Lynd, Langtry Emmett, chemical metallurgy, mineralogy
Malone, William Maxton, organic chemistry
Schmidt, Kurt, experimental physics
Sie, Hsien-Gieh, biochemistry
Weintritt, Donald J., colloid chemistry, petroleum chemistry
White, Robert Winslow, organic chemistry, polymer chemistry

HILLSDALE
Brescia, Frank, physical chemistry
Krey, Philip W., radiochemistry, meteorology
Landrum, Billy Frank, organic chemistry, polymer chemistry

HILLSIDE
Chavkin, Leonard Theodore, pharmacy chemistry
Cohen, George Lester, physical chemistry
Frey, William Carl, statistics, chemistry
Henry, Sydney Mark, microbiology, physiology
Kaellis, Eugene, dentistry
Letterman, Herbert, analytical chemistry
Marcus, Arnold David, pharmacy, science education
Piala, Joseph Joseph, pharmacology
Rosenberg, Allan (Herbert), physical chemistry
Sterbenz, Francis Joseph, biochemical pharmacology
Teneza, Thomas Michael, organic chemistry
Weintraub, Leonard, organic chemistry, pharmaceutical chemistry

HO HO KUS
Hofmann, Corris Mabelle, pharmaceutical chemistry

HOBOKEN
Allen, Paul, Jr, organic chemistry
Anderson, James Leroy, theoretical physics
Andrews, Rodney Denlinger, Jr, physical chemistry
Arase, Elizabeth Martha, physics
Arase, Tetsuo, physics
Bahary, William S., biophysical chemistry
Barnes, Robert Lee, organic chemistry
Bernstein, Jeremy, theoretical physics
Bose, Ajay Kumar, organic chemistry
Bostick, Winston Harper, physics
Brucker, Edward Byerly, physics
Clough, Francis Bowman, inorganic chemistry
Crabtree, James Bruce, mathematics
Daunt, John Gilbert, physics
Fajans, Jack, physics
Friedman, Edward Alan, solid state physics, academic administration
Hires, Richard Ives, physical oceanography
Jones, Francis Thomas, physical chemistry
Koller, Earl Leonard, physics
Lapidus, Ivan Richard, molecular biophysics, theoretical physics
Layzer, Arthur James, physics
Levine, Lawrence Elliott, applied mathematics
Malinowski, Edmund R., physical chemistry
Manhas, Maghar Singh, organic chemistry
Meissner, Hans Walter, experimental physics
Pollara, Luigi Zunmo, chemical physics
Pollock, Franklin, physics
Robb, Ernest Willard, organic chemistry
Rogers, Kenneth Cannicott, physics
Rosen, Bernard, plasma physics, solid state physics
Rothberg, Gerald Morris, solid state physics
Salwen, Harold, theoretical physics
Schiller, Ralph, theoretical physics
Schmidt, George, plasma physics
Seidl, Milos, plasma physics
Stivala, Salvatore Silvio, physical chemistry
Strumpf, Albert, applied mathematics
Taylor, Snowden, high energy physics, particle physics
Tindell, Ralph S., mathematics
van der Veen, James Morris, organic chemistry, x-ray crystallography
Volpe, Angelo Anthony, organic chemistry, polymer chemistry
White, Myron Edward, mathematics, computer science
Wright, George Buford, solid state physics
Yevick, George Johannus, physics

HOLMDEL
Arnaud, Jacques, electromagnetics
Ashkin, Arthur, experimental physics
Bergman, John George, Jr, inorganic chemistry
Boyle, Willard Sterling, physics
Brown, Alfred Bruce, Jr, electron physics
Buchsbaum, Solomon Jan, physics
Burke, Paul J., operations research
Burrus, Charles Andrew, Jr, optics
Chao, Min-Te, statistics
Cohen, Leonard George, communications, plasma physics
Courtney-Pratt, Jeofry Stuart, physics, engineering
Curtis, Thomas Hasbrook, applied physics, acoustics
Cutler, Cassius Chapin, physics
Dacey, George Clement, solid state physics, telecommunications
Dohrnanyi, Julius S., theoretical physics
Fork, Richard Lynn, solid state physics, quantum optics
Foschini, Gerard Joseph, applied mathematics, systems engineering
Foster, Norman Francis, physical chemistry, science education
Froehlich, Fritz Edgar, physics
Gillette, Dean, mathematics
Gordon, James Power, physics
Hagan, James Joyce, biochemistry
Headley, Arthur Bruce, statistics
Hogg, David Clarence, radio physics
Hooke, John Allen, operations research
Hou, Tien Fang, mathematical statistics
Hubbard, William Marshall, optical physics, communication engineering
Jackel, Lawrence David, experimental solid state physics, low temperature physics
Jacobs, Ira, physics, telecommunications
Jain, Aridaman Kumar, applied statistics
James, Dennis Bryan, physics
Jarvis, John Frederick, computer science
Jones, Howard, organic chemistry
Kaminow, Ivan Paul, applied physics, electrooptics
Knudsen, John Roland, applied mathematics
Kogelnik, H W., physics, electronics
Kompfner, Rudolf, physics, electronics
Mercer, Robert Allen, physics
Mitchell, Olga Mary Mracek, physics
Mollenauer, Linn F., physics
Murphy, Ray Bradford, mathematical statistics, applied statistics
McCallum, Charles John, Jr, operations research
Nahory, Robert Edward, physics
Neal, Scotty Ray, applied mathematics
Newsome, Ross Whitted, physics
Orrok, George Timothy, physics
Ortel, William Charles Gormley, electronics, communications
Penzias, Arno A., radio astronomy, physics
Poindexter, Edward Haviland, chemical physics
Polonsky, Ivan Paul, mathematics, computer science
Reudink, Douglas O., mathematics, physics
Rigrod, William Walter, electronics
Roberts, Charles Sheldon, computer science, plasma physics
Russell, Charles Addison, polymer chemistry, analytical chemistry
Schulte, Harry John, Jr, experimental physics
Sell, Darrell, solid state physics, systems engineering
Shah, Jagdeep C., solid state physics
Shay, Joseph Leo, solid state physics
Shipley, Edward Nicholas, physics
Silfvast, William Thomas, physics
Sinclair, James Douglas, analytical chemistry, corrosion
Smith, Peter William, physics, electrical engineering
Smith, Thomas Harry Francis, toxicology
Stolen, Rogers Hall, solid state physics
Stone, Julian, physics
Swann, Dale William, applied mathematics, operations research
Tell, Benjamin, physics
Thomas, Ronald Emerson, mathematical statistics, operations research
Tischendorf, John Allen, applied statistics
Tomlinson, Walter John, III, physics
Trambarulo, Ralph, physics, physical chemistry
Turner, Edward Harrison, magnetism, microwave electronics
Warters, William Dennis, physics
Weber, Heinz Paul, physics, optics
Weschler, Charles John, physical inorganic chemistry
Wilson, Robert Woodrow, radio astronomy
Woerner, Robert Leo, low temperature physics
Wolman, Eric, operations research, systems engineering
Wolontis, Vidar Michael, mathematics
Worlock, John M., solid state physics
Young, James Arthur, Jr., physics

IRONIA
Pollock, Bernard David, physical chemistry

JAMESBURG
Sears, Raymond W., physics

JERSEY CITY
Bacon, Charles Vincent, chemistry
Baker, Herman, metabolism, nutrition
Belmonte, Rocco George, biology
Bergen, Catherine Mary, science education
Brous, Jack, inorganic chemistry, physical chemistry
Cinotti, William Ralph, prosthodontics, periodontics
Clark, Mary Jane, biochemistry
Dreyfuss, Robert George, inorganic chemistry, glass technology
Dyer, Judith Gretchen, entomology
Fingerhut, Marilyn Ann, cell biology
Frances, Saul, bacteriology
Friedrich, Benjamin C., science education, human biology
Garone, John Edward, human biology
Goldstein, Philip, applied physics
Grant, James J., Jr, theoretical physics
Gruszczyk, Jerome Henry, animal physiology
Jansons, Vilma Karina, microbiology
Kelly, Robert P., cell physiology
Lada, Arnold, biochemistry, organic chemistry
Laughlin, Alice, biochemistry, analytical chemistry
Madrazo, Alfonso A., pathology
Makar, Boshra Halim, pure mathematics
Mehta, Mahesh J., physical chemistry
Merianos, John James, medicinal chemistry
Miller, Carl Henry, Jr, physiological chemistry, electron microscopy
Morneweck, Samuel, organic biochemistry
Mulcahy, Gabriel Michael, pathology
Pamer, Treva Louise, biochemistry
Parker, Leroy A, Jr, dentistry, natural science
Pegolotti, James Alfred, organic chemistry
Peterson, Arthur F., microbiology, physical chemistry
Petriello, Richard P., nematology
Petit, Barbara Jane, physiology
Poiani, Eileen Louise, mathematics
Raines, Thaddeus Joseph, physical chemistry
Rapuano, Joseph A., dentistry
Redden, Patricia Ann, physical chemistry

Riggs, Richard, mathematics education
Rosenberg, Evelyn Kivy, biology
Rosenberg, Herman, mathematics
Rosenthal, Murray William, pharmaceutical chemistry
Rubin, Ephraim Leo, theoretical chemistry
Russell, Helen Ross, biology
Schuh, Joseph Edward, cytology
Sein, John Jaan, theoretical physics
Seuging, Earl William, Jr, pharmacy, chemical engineering
Shovlin, Francis Edward, endodontics, microbiology
Thomas, Larry Emerson, applied mathematics
Treitler, Theodore Leo, physical chemistry
Troiano, Marlin Frank, oral surgery
Turner Matthew X, cytology, immunology
Vinton, Paul Wesley, dentistry
Weinrich, Marcel, physics
Weisberg, Joseph Simpson, oceanography, geology

KEARNY
Jadmicek, Bohumil Robert, polymer chemistry
Petry, Robert Kendrick, biology

KENDALL PARK
Underhill, Donald Kraft, zoology

KENILWORTH
Albu, Evelyn D, medical education
Baldini, James Thomas, nutrition, dermatology
Berkman, Robert Nelson, veterinary pathology
Coan, Stephen B, organic chemistry
Cohen, Lester Allan, physical organic pharmacology
Coppola, John Anthony, endocrinology, pharmacology
Harris, Jerome Jones, medicine, communications
Howard, John Campbell, Jr, medicine
Miller, Ronald Kent, immunobiology, allergy
Scrafani, Joseph Thomas, pharmacology, medical education
Wachs, Gerald N, dermatology

KINNELON
Babcock, George, Jr, pharmacology

KIRKWOOD
McDonnell, William Vincent, pathology

LAFAYETTE
Gray, William David, pharmacology
Lighty, Paul Elliott, chemical physics

LAKE HOPATCONG
Smith, Chester Lional, physics, telemetry

LAKEHURST
Stein, Gustav Albert, pharmaceutical chemistry

LAKEWOOD
Bernstein, Isidor Mayer, chemistry
Coakley, Mary Peter, physical chemistry, inorganic chemistry
Cook, Marie Mildred, zoology, developmental biology
Goedertier, Kuei-Ling Li, physics
Hussey, Kathleen Louise, biology
O'Hara, Elizabeth Mary, organic chemistry
Olin, Arthur David, organic chemistry
Otte, Herman Frederick, economic geography
Shen, Nai-Hsuan Chang, biochemistry
Shive, John Northrup, electronics
Sloyan, Mary Stephanie, mathematics

LAWRENCEVILLE
Balinski, Michel Louis, applied mathematics
Ferraro, Charles Frank, analytical chemistry
Funke, Philip T, mass spectrometry, organic chemistry
Harendza-Harinxma, Alfred Josef, chemical physics
Harrison, Ross Arthur, mathematics

Lurie, Joan B, theoretical solid state physics
McCarroll, William Henry, inorganic chemistry
Olechowski, Jerome Robert, physical chemistry, organometallic chemistry
Sheats, John Eugene, organometallic chemistry, physical organic chemistry
Spink, Walter John, stratigraphy, structural geology
Sugerman, Abraham Arthur, psychiatry
Venable, Patricia Lengel, botany

LEONIA
Boorse, Henry Abraham, low temperature physics
Farinholt, Larkin Hundley, chemistry

LIBERTY CORNER
Sawyer, Baldwin, physics
Wheeler, Ora Leon, chemistry

LINCOLN PARK
Mitchell, William Alexander, food science, carbohydrate chemistry
Tracy, David J, organic chemistry

LINDEN
Alpert, Norman, petroleum technology, petroleum chemistry
Angier, Derek John, polymer chemistry
Bachman, Kenneth Charles, physical chemistry, fuel science
Berger, Martin, physics
Bertrand, Rene Robert, physical chemistry, chemical engineering
Black, James Francis, physical chemistry
Brenner, Douglas, polymer science
Canter, Nathan H, polymer physics
Chianelli, Russell Robert, physical chemistry
Dines, Martin Benjamin, organometallic chemistry, inorganic chemistry
Eckart, Robert E, biochemistry, toxicology
Eckert, John Andrew, electrochemistry
Feldman, Nicholas, petroleum chemistry, analytical chemistry
Fischer, Traugott Erwin, surface physics
Frankenfeld, John William, organic chemistry
Fuller, Everett J, physical chemistry
Gamble, Fred Ridley, Jr, solid state science
Gardiner, John Brooke, organic chemistry
Geissler, Paul Robert, chemical kinetics
Gessler, Albert Murray, rubber chemistry
Ghosh, Amal Kumar, solid state physics
Goldblatt, Irwin Leonard, tribology
Gould, Kenneth Alan, organic chemistry, fuel science
Grafstein, Daniel, chemistry, nuclear science
Grandolfo, Marian Carmela, inorganic chemistry
Griffith, Martin G, organic chemistry, physical chemistry
Hall, Homer James, information science
Harrison, John William, petroleum chemistry
Hartgerink, Ronald Lee, organic chemistry
Holt, Eugene Lawrence, physical chemistry
Holzwarth, George Michael, biophysics, physical chemistry
Hou, Ching-Tsang, biochemistry
Hsu, Edward Ching-Sheng, physical chemistry
Huang, John S, chemical physics, surface physics
Jaruzelski, John Janusz, industrial organic chemistry
Keirns, Mary Hull, chemistry, automotive engineering
Kirshenbaum, Isidor, chemistry
Klemann, Lawrence Paul, organometallic chemistry
Konecky, Milton Stuart, organic chemistry
Krafer, Elise, organic chemistry
Kramer, George Mortimer, physical chemistry
Kresge, Edward Nathan, polymer organic chemistry
Kuntz, Irving, polymer chemistry
Laskin, Allen I, microbiology
Leary, Ralph John, organic chemistry
Longo, John M, inorganic chemistry
Looney, Ralph William, physical chemistry, polymer chemistry
Lovett, John Robert, petroleum chemistry, lubrication engineering

Lundberg, Robert Dean, polymer chemistry
Lyon, Richard Kenneth, physical chemistry
Magee, Ellington McFall, physical chemistry
Manuel, Thomas Asbury, rubber chemistry
Michael, James Richard, organic chemistry
Milliman, George Elmer, analytical chemistry
Mohan, Raam Ramanuja, microbiology
Myerson, Albert Leon, physical chemistry
Newill, Vaun Archie, epidemiology, internal medicine
O'Farrell, Charles Patrick, rubber chemistry
Panzer, Jerome, organic chemistry
Polizzotti, Richard Samuel, surface chemistry
Pollak, Kurt, organic chemistry
Rebick, Charles, physical chemistry
Roper, Robert, polymer chemistry
Rosenfeld, Daniel, petroleum chemistry
Salvesen, Robert H, organic chemistry, polymer chemistry
Scala, Robert Andrew, toxicology
Schriesheim, Alan, organic chemistry
Schwartz, Robert David, petroleum microbiology
Schwarz, James Alan, surface chemistry
Shaw, Robert Frank, solid state physics
Silbernagel, Bernard George, physics
Smith, Richard Pearson, physical chemistry
Song, Won-Ryul, polymer chemistry
Spenadel, Lawrence, physical chemistry, polymer chemistry
Spilker, Clarence William, organic chemistry
Sprague, Robert Hicks, organic chemistry
Starnes, Paul Kiser, organic chemistry, analytical chemistry
Thaler, Warren Alan, organic chemistry
Thompson, Arthur Howard, solid state physics
Ver Strate, Gary William, polymer chemistry, polymer physics
Weisgerber, George Austin, petroleum chemistry
Wellman, William Edward, organic chemistry
Whitney, Thomas Allen, organic chemistry
Wilchinsky, Zigmund Walter, polymer physics
Young, Archie Richard, II, inorganic chemistry

LITTLE FALLS
Boucher, Bertrand Philip, economic geography, geography of the Middle East
Gordon, Harry William, biochemical pharmacology
Greenwood, Ivan Anderson, atomic physics, instrumentation
Postelnek, William, organic chemistry, materials science
Simpson, James Henry, Jr, physics

LITTLE FERRY
Harvey, Robert Joseph, organic chemistry
Naglieri, Anthony N, industrial organic chemistry

LITTLE SILVER
Sincius, Joseph Anthony, photographic chemistry

LIVINGSTON
Breen, James Langhorne, obstetrics & gynecology
Del Guercio, Louis Richard M, surgery, pharmacology
Dien, Chi-Kang, organic chemistry
Hutter, Robert V P, pathology, oncology
Lipowski, Stanley Arthur, organic polymer chemistry
Manning, John Paul, histochemistry, anatomy
Rubin, Herbert, physical chemistry
Schiffers, Justus J, science writing
Shabica, Anthony Charles, Jr, organic chemistry
Skeist, Irving, polymer chemistry
Weintraub, Lester, organic polymer chemistry
Yen, Lewis C, physical chemistry, chemical engineering
Yeung, Shiu Fong, applied mathematics

LODI
Hoffmann, Sandor Alexander, organic chemistry
Molnar, Nicholas M, organic chemistry

LONG BRANCH
Vaun, William Stratin, medicine

LONG VALLEY
Mizzoni, Renat Herbert, organic chemistry
Olsen, Robert Thorvald, synthetic organic chemistry
Yost, William Lassiter, organic chemistry

LYNDHURST
Cohen, Sidney, organic chemistry
Owen, Louis John, organic chemistry

LYONS
Feller, Ralph Paul, dentistry, prosthodontics
Freeman, Leslie Sherwood, neuropsychiatry

MADISON
Aguiar, Adam Martin, organic chemistry
Baker, Edgar Gates Stanley, biology
Baylouny, Raymond Anthony, organic chemistry
Bush, Louise Fulton, marine biology, taxonomy
Crowley, Thomas Henry, mathematics
Fenstermacher, Robert Lane, atmospheric physics, radio astronomy
Francoeur, Robert Thomas, embryology, evolution
Greenspan, Bernard, mathematics
Griffo, James Vincent, Jr, natural history, parasitology
Hardy, George Fisk, polymer physics
Hay, Alden Wendell, physical chemistry
Huber, Ivan, entomology
Justus, Philip Stanley, geology, tectonics
Keenan, Edward Milton, mathematics
Lang, Frank Theodore, physical chemistry
Lenz, Paul Heins, physiology, endocrinology
Martin, Robert Allen, vertebrate paleontology
Mascio, Afework Asghedom, immunology
Midlige, Frederick Horstmann, Jr, virology, microbiology
Miller, James Monroe, analytical chemistry
Muller, Peter S, biological anthropology
Multer, H Gray, geology
Nagle, James John, genetics, evolution
Nelson, Gregory Victor, inorganic chemistry
Ollom, John Frederick, physics
Phillips, Joy Burcham, physiology
Pollock, Leland Wells, ecology, invertebrate zoology
Rohrs, Harold Clark, cell physiology
Scott, Donald Albert, organic chemistry
Siebert, Donald Robert, chemical physics
Smith, William Edgar, experimental pathology
Strange, Ronald Stephen, inorganic chemistry
Wagner, Richard Carl, mathematics
Weissman, Paul Morton, organometallic chemistry, electrochemistry
Zuck, Robert Karl, botany

MAHWAH
Borowitz, Grace Burchman, organic chemistry
Halpern, Teodoro, experimental solid state physics, instrumentation
Jacobs, Marian Beckmann, clay mineralogy, oceanography
Kaufman, Herman S, polymer science, science education
Kramer, Henry Herman, nuclear medicine, medical bacteriology
Roth, Rodney J, mathematics
Sall, Theodore, microbiology
Shine, Robert John, organic chemistry

MAPLEWOOD
Crowe, Bernard Francis, organic chemistry
Hoffmann, Paul Otto, physics
Joffe, Joseph, physical chemistry
Leal, Joseph Rogers, organic chemistry
Talbot, Ashley Frederick, geography

NEW JERSEY

MARLTON
Lotkin, Mark Max, mathematics
Troisi, Raphael Angelo, physical chemistry

MARTINSVILLE
Tenzer, Rudolf Kurt, physics

MATAWAN
Mrozik, Helmut, organic chemistry

MAYS LANDING
Brown, Thomas Edward, plant physiology

MAYWOOD
Harewood, Ken Rupert, cancer
Jensen, Keith Edwin, cancer
Larson, David L., virology; biochemistry
Mayyasi, Sami Ali, virology
McBride, Tom Joseph, bacteriology
Traul, Karl Arthur, immunology; virology
Wolff, John Shearer, III, biochemistry, virology

MEDFORD
Maitlen, Eldon Gene, physiology
Marr, Eleanor B., organic chemistry

MENDHAM
Desjardins, Raoul, clinical pharmacology, experimental medicine
Elkholy, Hussein A., physics
Kaprelian, Edward Karnig, optics, photography
Moolten, Sylvan E., internal medicine; pathology
Pribor, Hugo Casimer, pathology

MENLO PARK
Albert, Charles Gerald, chemistry
Haden, Walter Linwood, Jr., physical inorganic chemistry
Tedeschi, Robert James, organic chemistry

METUCHEN
Brickman, Leo, organic chemistry
Kahn, Donald Jay, organic chemistry
Lofstrom, John Gustave, analytical chemistry

MIDDLESEX
Herrick, Guy Scott, organic chemistry

MILBURN
Wells, Edward Henry, mathematics

MILFORD
Bald, Kenneth Charles, physical chemistry
Richardson, Henry Howe, entomology

MILLINGTON
Kalnin, Ilmar L., materials science
Langley, Robert Charles, organic chemistry
Taylor, Alfred Henry, Jr., physical chemistry

MILLTOWN
Abbey, Anthony Alfred, bacteriology
Barnard, William Sprague, physical chemistry
Chatterjee, Pronoy Kumar, polymer chemistry; physical chemistry
Clark, Harold Eugene, plant physiology
Hynek, Walter Joseph, physics, engineering
Kraskin, Kenneth Stanford, microbiology
Morbey, Graham Kenneth, textile chemistry; chemical engineering
Schwenker, Robert Frederick, Jr., cellulose chemistry; polymer chemistry
Smith, Delmont K., organic chemistry
Steiger, Fred Harold, applied chemistry, cosmetic chemistry
Toth, William James, polymer chemistry

MILLVILLE
Smith, Dudley Cozby, physical chemistry; glass technology

MONMOUTH JUNCTION
Luttinger, Lionel, physical chemistry, colloid chemistry
Smith, Richard Neilson, physical chemistry; electrochemistry

MONTCLAIR
Bierenbaum, Marvin L., cardiology
Charney, William, microbiology; nutrition
Crowe, George A., physical chemistry
Daniels, Gert L., zoology, anthropology
Fleischman, Alan Isadore, biochemistry, microbiology
Gourary, Barry Sholom, applied physics
Gourary, Mina Haskind, applied statistics
Leingruber, Willy, organic chemistry
Lyndrup, Mark Leroy, physical biochemistry
Metzer, Carl Martin, physics
Paisley, Nancy Sandelin, biochemistry
Shrier, Adam Louis, resource management, chemical engineering
Watson, Frank Yandle, pathology
Wilson, Thomas Lee, physical chemistry

MONTVALE
Arquete, Gordon James, physical chemistry
Bach, Frederick Louis, organic chemistry
Cohen, Murray Samuel, organic chemistry
Prindle, Robert Franklin, chemistry
Rector, Robert Wayman, computer science, mathematics

MONTVILLE
Duerg, William Henry, physics

MOORESTOWN
Allen, Harry Willis, entomology
Fordyce, David Buchanan, analytical chemistry
Frey, Sheldon Ellsworth, organic chemistry
Gerber, Charles Edwin, entomology
Haake, James Walter, operations research
Heidrick, Lee E., plant pathology
Hurt, H David, nutrition
King, William Connor, physics
Kohl, John C., Jr., cytogenetics, environmental sciences
Lichy, Charles Thorne, agricultural chemistry
LoCicero, Joseph Castelli, organic chemistry
Matlack, Louis Rogers, polymer chemistry; physical chemistry
Patterson, Omar Leroy, physics
Reboul, Theo Todd, III, solid state physics
Souder, Philip Walburn, food chemistry

MORRIS PLAINS
Babson, Arthur Lawrence, clinical chemistry
Bapatla, Krishna M., pharmaceutical chemistry
Belt, Roger Francis, crystallography
Black, Jack, pharmacology
Blum, Stanley Walter, biochemistry
Brizarelli, Giuliano, pathology
Brogle, Richard Charles, food science
Bulusu, Suryanarayana, organic chemistry
Carroll, James Joseph, biochemistry
Commarato, Michael A., pharmacology
Connor, David Thomas, organic chemistry
Crane, Laura Jane, clinical biochemistry
Crew, Malcolm Charles, drug metabolism
Di Carlo, Frederick Joseph, pharmacology
DiPasquale, Gene, biology, physiology
Dubinsky, Barry, pharmacology
Dubnick, Bernard, pharmacology
Eades, Charles Hubert, Jr., biochemistry
Emele, Jane Frances, pharmacology, dermatology
Fand, Theodore Ira, organic chemistry, pharmaceutical chemistry
Fuhrmann, Robert, physical organic chemistry

MORRISTOWN
Aharoni, Shaul Moshe, polymer chemistry
Anderson, Lowell Ray, inorganic chemistry
Armor, John N., physical inorganic chemistry, organometallic chemistry
Barrett, Joseph John, molecular spectroscopy
Bartnoff, Shepard, nuclear physics
Bauer, Frederick William, organic chemistry, research administration
Baughman, Ray Henry, chemical physics, polymer science
Berenbaum, Morris Benjamin, polymer chemistry, organic chemistry
Bernholz, William Francis, lipid chemistry, surface chemistry
Bhatia, Kishan, organic chemistry
Chance, Ronald Richard, solid state chemistry
Cheema, Zafrullah K., organic chemistry, pharmaceutical chemistry
Cipriani, Cipriano, industrial chemistry
DeFelice, Eugene Anthony, internal medicine
Degginger, Edward R., organic chemistry
Dehn, Joseph William, Jr., organic chemistry
Denkewalter, Robert George, organic chemistry
DiGiuseppe, Michael Anthony, inorganic chemistry
Doczi, John, pharmaceutical chemistry
Erkkila, Armas Victor, organic chemistry, physical chemistry
Evans, Francis Eugene, organic chemistry
Farone, William Anthony, physical chemistry
Genzer, Jerome Daniel, organic chemistry
Giles, Ralph E., pharmacology
Gilleo, Matthias Alten, electrophysics
Gilman, John Joseph, physics, metallurgy
Gingold, James Lehman, biochemistry
Glick, Aaron, clinical pharmacology
Green, David Francis, physiological chemistry
Greenough, Ralph Clive, analytical chemistry
Hammond, Willis Burdette, organic chemistry
Harris, Robert Laurence, physical chemistry
Hasegawa, Ryusuke, solid state physics, metal physics
Hoff, Gail Richards, immunology
Hoppens, Harold Albert, organic chemistry
Hoskins, Walter Hugh, pharmaceutical chemistry
Huang, Yung-Chen, biochemistry
Johnston, Gene Woods, pharmacy management
Kapadia, Yash M., pharmacy
Kaplan, Harvey Robert, pharmacology
Katsampes, Chris Peter, pediatrics
Kessler, Henry A., pharmacy
Kesten, Yall, polymer science
Klainer, Albert S., internal medicine, infectious diseases
Klucko, Sylvester, synthetic organic chemistry
Kokta, Milan Rastislav, solid state chemistry
Kotlar, Abraham Morris, physical chemistry; polymer physics
Kronish, Donald Paul, microbiology, microbial biochemistry
Kushner, George Samuel, analytical chemistry
Largman, Theodore, organic chemistry
Lee, Lester Tsung-Cheng, polymer chemistry; organic chemistry
Leinweber, Franz Josef, drug metabolism
Lewis, Arnold D., analytical chemistry
Libowitz, George Gotthart, solid state chemistry
Lichtman, Irwin A., physical chemistry
Lieberman, Herbert A., pharmaceutical chemistry
Little, Edwin Demetrius, organic chemistry
Logan, Ralph Andre, solid state physics
MacFarlane, Robert, Jr., physical chemistry
Martin, James Franklin, microbiology
Maxwell, Donald Robert, pharmacology
Meltzer, Robert Israel, organic chemistry
Miller, Floyd Laverne, physical chemistry
Minmack, William Edward, optics
Miskel, John Joseph, Jr., polymer chemistry; organic chemistry
Morrison, Glenn C., organic chemistry
Mueller, Max Best, organic chemistry, physical chemistry
Ninger, Fred Constant, pharmaceutical chemistry
Norik, Arthur Jack, energy conversion
Nunn, Leslie Grey, Jr., organic chemistry
Penner, Melvin H., analytical chemistry
Perry, Kenneth W., cell physiology
Pianotti, Roland Salvatore, microbiology
Pietrusza, Edward Walter, organic chemistry
Pitts, Robert Gary, dental research, microbial physiology
Pollack, Maxwell Aaron, organic chemistry
Prevorsek, Dusan Ciril, polymer physics, polymer chemistry
Ratcliffe, Charles Thomas, inorganic chemistry
Rieger, Martin Max, physical organic chemistry
Ringel, Samuel Morris, microbiology
Robichaud, Roger Charles, psychopharmacology
Robinson, Harry John, medicine
Rogic, Milorad Mihailo, organic chemistry
Roldan, Luis Gonzalez, crystallography, chemistry
Ropp, Richard C., solid state chemistry
Rose, Ira Marvin, organic chemistry, analytical chemistry
Ruddy, Arlo Wayne, organic chemistry
Schick, Martin J., colloid chemistry
Schmitt, George Joseph, polymer chemistry
Schule, Elmer Christian, organic chemistry
Schultze, Helmuth W., chemistry
Schwartz, Edward, toxicology
Schwender, Charles Frederick, medicinal chemistry
Sepkoski, Joseph John, organic chemistry
Sheehan, Desmond, organic physics
Sibilia, John Philip, chemical physics, inorganic chemistry
Siegel, Sheldon, pharmaceutical chemistry
Simon, Thomas H., pharmacy
Sircar, Jagadish Chandra, organic chemistry
Skovronek, Herbert Samuel, organic chemistry
Skrypa, Michael John, organic chemistry
Stein, Jerome D., Jr., physiology
Steinberg, Eliot, organic chemistry
Stewart, Sheila Frances, reproductive endocrinology
Stone, Edward, organic chemistry, polymer chemistry
Tannenbaum, Stanley, inorganic chemistry
Tetenbaum, Marvin Theodore, organic chemistry
Tuccio, Sam Anthony, lasers
Turner, Frank Joseph, microbiology
Turner, James E., biochemistry
Unangst, Paul Charles, medicinal chemistry
Vanderkooi, Nicholas, Jr., chemistry
Verburg, Robert Martin, industrial chemistry
von Strandtmann, Max, medicinal chemistry
Weir, James Henry, III, medical research
Weisberg, Jerry, clinical pharmacology
Winbury, Martin M., pharmacology; physiology
Wise, Lawrence David, organic chemistry
Wittekind, Raymond Richard, organic chemistry
Wohlers, Herbert C., organic chemistry, research administration
Wolin, Alan George, food science
Young, William Donald, Jr., bacteriology, systematics
Zinnes, Harold, organic chemistry

MOUNTAIN LAKES
Bellet, Richard Joseph, polymer chemistry
Buyske, Donald Albert, biochemistry, organic chemistry
Gora, Thaddeus F., Jr., theoretical solid state physics
Punnam, Robert Conrad, organic chemistry
Schell, Emil Daniel, mathematics

MOUNTAINSIDE

Auston, David H., lasers
Burton, Gilbert W., organic chemistry, polymer chemistry
Conley, Robert F., inorganic chemistry
Nelson, Aaron Louis, organic chemistry
Terry, Milton Everett, mathematical statistics
Watson, Richard White, Jr., microbiology

MURRAY HILL

Abrahams, Sidney Cyril, crystallography
Adams, Arthur Curtis, inorganic chemistry
Ahlers, Guenter, physical chemistry, solid state physics
Allara, David Lawrence, physical organic chemistry
Alles, Harold Gene, instrumentation, communications engineering
Anderson, Philip Warren, theoretical physics
Andres, Klaus, physics
Appelbaum, Joel A., solid state physics
Arthur, John Read, Jr, surface physics
Aspnes, David E., physics
Bagley, Brian G., solid state physics
Baker, Brenda Sue, computer science
Baker, William Oliver, physical chemistry, polymer science
Baraff, Gene Allen, solid state physics
Barker, A S, Jr., solid state physics
Barmatz, Martin Bruce, acoustics
Barns, Robert L., crystallography
Bayer, Douglas Leslie, computer science
Becker, Gordon Edward, surface physics
Benes, Vaclav Edvard, mathematics
Bergh, Arpad A., physical chemistry
Berkley, David A., acoustics
Berreman, Dwight Winton, physics
Biondi, Frank Joseph, chemistry, electronics
Blount, Leo Seth, physical organic chemistry
Blount, Eugene Irving, theoretical solid state physics
Blue, James Lawrence, applied mathematics, mathematical physics
Blumberg, William Emil, physics, biophysics
Blyler, Lee Landis, Jr, rheology, polymer engineering
Boddy, Philip J., physical chemistry
Bovey, Frank Alden, polymer chemistry
Bowden, Murrae John Stanley, polymer chemistry
Brinkman, William F., solid state physics
Brown, William Stanley, computer science
Brus, Louis Eugene, chemical physics
Buck, Thomas M, surface physics
Buckley, Reginald R, inorganic chemistry, physical chemistry
Burbank, Robinson Derry, x-ray crystallography
Burckbuchler, Frederick V, solid state electronics
Burnett, Joseph Ashby, solid state physics
Cardillo, Mark J, chemical physics
Carruthers, John Robert, materials science
Chambers, John McKinley, statistics, computer science
Chan, Maureen Gillen, polymer chemistry
Chandross, Edwin A, organic chemistry, physical chemistry
Chang, Chuan Chung, physics
Chang, Kuang-Chou, physical organic chemistry
Chen, Ho Sou, solid state physics
Cho, Alfred Y., surface physics, semiconductors
Cohen, Richard Lewis, solid state science
Collier, Robert Jacob, electron optics
Copeland, John Alexander, solid state physics
Curran, Robert Kyran, physics
Dahlberg, Susan Clardy, surface chemistry
Daly, Daniel Francis, Jr., solid state physics
D'Asaro, Lucian Arthur, physics
DeLoach, Bernard Collins, Jr., physics
DiDomenico, Mauro, Jr., solid state physics

Dietz, Robert E, solid state physics
Dillon, Joseph Francis, Jr, magnetism
DiLorenzo, James V., physical chemistry, inorganic chemistry
Dingle, Raymond, solid state physics, chemical physics
Di Salvo, Francis Joseph, solid state physics, chemistry
Dixon, Richard Wayne, solid state physics
Dynes, Robert Carr, solid state physics
Edelson, David, environmental chemistry
Eisinger, Josef, biophysics
Falconer, Warren Edgar, physical chemistry
Farrow, Leonilda Altman, chemical physics
Feit, Eugene David, organic polymer chemistry
Feldman, Martin, physics
Fletcher, Robert Chipman, physics
Fleury, Paul A, solid state physics, spectroscopy
Fowlkes, Edward B, II, statistics
Fox, Phyllis, mathematics, computer science
Frankenthal, Robert Peter, electrochemistry, corrosion
Freed, Donald Joseph, analytical chemistry, photochemistry
French, William George, physical chemistry, glass technology
Freund, Robert Stanley, chemical physics
Gajewski, Fred John, chemistry
Gamota, George, low temperature physics
Garey, Michael Randolph, mathematics, computer science
Garrett, Charles Geoffrey Blythe, physics
Geschwind, Stanley, physics
Gibbs, Hyatt McDonald, atomic physics, quantum optics
Gilbert, Edgar Nelson, mathematics
Gilmer, George Hudson, statistical mechanics
Ginsberg, Alvin Paul, inorganic chemistry
Giordmaine, Joseph Anthony, optical physics
Glarum, Sivert Herth, physical chemistry
Gnanadesikan, Ramanathan, statistical analysis
Golding, Brage, Jr., experimental solid state physics, physical acoustics
Gordon, Eugene Irving, physics
Gossard, Arthur Charles, physics
Graedel, Thomas Eldon, atmospheric chemistry, air pollution
Graham, Ronald Lewis, mathematics
Greywall, Dennis Stanley, low temperature physics
Gyorgy, Ernst Michael, experimental solid state physics
Hagedorn, Fred Bassett, magnetism
Hagelbarger, David William, communications science
Hagstrum, Homer Dupre, surface physics
Halpern, Donald F, organometallic chemistry
Hamming, Richard W., computer science
Hannay, Norman Bruce, physical chemistry
Hansen, Ralph Holm, plastics chemistry
Harbus, Fredric Ira, operations research
Hauser, Joachim J, solid state physics
Hawkins, Walter Lincoln, polymer chemistry
Hebard, Arthur Foster, low temperature physics
Helfand, Eugene, polymer science, statistical mechanics
Heller, Adam, physical chemistry
Henry, Charles H., solid state physics, semiconductor physics
Hensel, John Charles, solid state physics
Herbst, Robert Taylor, mathematics
Herring, Conyers, theoretical solid state physics
Hohenberg, Pierre Claude, theoretical physics, solid state physics
Hornbeck, John Austin, solid state physics
Hutson, Andrew Rhodes, solid state physics
Hwang, Frank K, statistics, discrete mathematics
Irvine, Merle M, physics
Jayaraman, Aiyasami, high pressure physics

Johnson, David Stifler, computer science, mathematics
Johnson, Leo Francis, lasers
Johnson, Stephen Curtis, computer science
Josenhans, James Gross, solid state electronics
Joyce, William Baxter, theoretical physics
Julesz, Bela, vision, physiological optics
Kadota, T Theodore, mathematics, communications
Kahng, Dawon, physics, semiconductors
Kane, Evan O, solid state physics
Kaplan, Martin L, physical organic chemistry
Kaufman, Linda Carol, numerical analysis, computer science
Kenney, Gary Dale, industrial hygiene
Kerwin, Robert Eugene, photochemistry
Kettenring, Jon Roberts, applied statistics, mathematical statistics
Kimerling, Lionel Cooper, solid state science
Kinsel, Tracy Stewart, solid state physics, quantum electronics
Klauder, John Rider, theoretical physics
Kleiner, Beat, applied statistics
Knight, Stephen, experimental physics
Krauskopf, John, psychophysiology
Krueger, Paul Carlton, inorganic chemistry, physical chemistry
Kruskal, Joseph Bernard, mathematics
Kunzler, John Eugene, physical chemistry
Kushner, Richard Allan, physical chemistry
Kwei, Ti-Kang, polymer chemistry, physical chemistry
Lamola, Angelo Anthony, photobiology, photochemistry
Landau, Henry Jacob, mathematics
Lang, David (Vern), semiconductors, physics
Lanzerotti, Louis John, geophysics, space physics
Laudise, Robert Alfred, solid state chemistry
Lazay, Paul Duane, solid state physics
Leeds, Morton W, organic chemistry
Leventhal, Marvin, astrophysics, atomic physics
Levine, Barry Franklin, lasers
Lewis, John Allen, applied mathematics
Ligenza, Joseph Raymond, solid state chemistry
Lin, Shen, computer science, mathematics
Lines, Malcolm Ellis, theoretical solid state physics
Link, Gordon Littlepage, physical chemistry
Liwshitz, Mordehai, space physics
Lloyd, Stuart Phinney, mathematics
Loan, Leonard Donald, polymer chemistry
Loomis, Thomas Clement, analytical chemistry
Lycklama, Heinz, nuclear physics, computer science
Lyons, Kenneth Brent, solid state physics, atmospheric chemistry
MacChesney, John Burnette, chemistry
Mac Rae, Alfred Urquhart, physics
MacWilliams, Florence Jessie, mathematics
Magerlein, John Harold, experimental solid state physics
Maldonado, Juan Ramon, applied physics
Mallows, Colin Lingwood, mathematical statistics
Marcus, Robert Boris, physical chemistry
Maydan, Dan, applied physics, electrooptics
Mazo, James Emery, applied physics
McAfee, Kenneth Bailey, Jr, chemical physics
McCall, David Warren, physical chemistry
McIntyre, James Douglass Edmonson, physical chemistry
McKay, Kenneth Gardiner, solid state electronics, telecommunications
McKenna, James, mathematics, physics
McKnight, Lee Graves, chemical physics
McRae, Eion Grant, physical chemistry, physics
McWhan, Denis B, solid state physics
Meiboom, Saul, chemical physics
Merz, James L, solid state physics
Miller, Barry, electrochemistry
Miller, Gabriel Lorimer, physics
Miller, Robert Charles, optical physics, solid state physics
Miller, Terry Alan, chemical physics
Mills, Allen Paine, Jr., atomic physics
Milner, Paul Chambers, physical chemistry

Mitchell, James Winfield, analytical chemistry
Mollenauer, James Frederick, nuclear physics, computer science
Morgan, Samuel Pope, mathematical physics, computer science
Morrison, John Allan, applied mathematics
Murphy, Bernard T, physics
Narayanamurti, Venkatesh, experimental solid state physics
Nassau, Kurt, solid state chemistry, physics
Nelson, Donald Frederick, physics
Nelson, Larry Dean, mathematics
Nelson, Terence John, theoretical physics
Newton, Carolyn McCrory, electroanalytical chemistry
Nielsen, James Willard, solid state chemistry
Nilsen, Walter Grahn, chemical physics
North, James Clayton, solid state physics
Okinaka, Yutaka, electrochemistry, analytical chemistry
Olive, Joseph P, communications science
Padden, Frank Joseph, Jr., polymer physics
Page, Robert Clinton, clinical pharmacology, industrial medicine
Paoli, Thomas Lee, applied physics, quantum electronics
Pappalardo, Leonard Thomas, organic chemistry, polymer chemistry
Parisi, George I, physical chemistry
Patterson, Gary David, polymer physics, chemical physics
Pearson, Arthur David, chemistry
Persky, George, solid state physics, electrical engineering
Peterson, George Earl, physics
Petroff, Pierre Marc, materials science, physics
Pfeiffer, Loren Neil, experimental physics
Phillips, James Charles, solid state physics
Pierce, Russell Dale, magnetism
Platzman, Philip M, solid state physics
Pollak, Henry Otto, mathematics
Prim, Robert Clay, mathematics
Reed, Joseph, organometallic chemistry
Reed, William Alfred, experimental solid state physics
Rentzepis, Peter M, chemical physics
Rice, Thomas Maurice, theoretical solid state physics
Rider, Don Keith, organic polymer chemistry, materials engineering
Robbins, Murray, inorganic chemistry
Roberts, Ronald Frederick, surface chemistry
Rode, Daniel Leon, solid state physics, plasma physics
Roe, Ryong-Joon, physical chemistry
Rose, Carl Martin, Jr, physics
Rosenblum, Martin Jacob, mathematics
Rosencwaig, Allan, physics
Rosler, Lawrence, computer sciences
Rossol, Frederick Carl, magnetism
Rousseau, Denis Lawrence, physical chemistry, solid state physics
Rowe, John Edward, surface physics, experimental solid state physics
Rowell, John Martin, physics
Schneider, Martin V, physics
Schoenberg, Leonard Norman, electrochemistry
Schonhorn, Harold, physical chemistry, surface chemistry
Schreibeis, William J, industrial hygiene
Schryer, Norman Loren, applied mathematics
Schwartz, Gary Paul, surface chemistry
Schwartz, Newton, solid state chemistry
Schweikert, Daniel George, computer sciences, applied mathematics
Scovil, Henry Evelyn Derrick, solid state physics
Seidel, Thomas Edward, physics, solid state electronics
Sevick, Jerry, applied physics
Sharpe, Louis Haughton, physical chemistry
Shen, Lawrence Y L, physical chemistry
Shepp, Lawrence Alan, mathematics
Shulman, Robert Gerson, chemical physics
Shumate, Paul William Jr, solid state physics
Sinclair, William Robert, physical inorganic chemistry
Sinden, Frank William, mathematics
Singh, Shobha, lasers, solid state physics
Sittig, Erhard Karl, physics
Slichter, William Pence, physical chemistry
Sloane, Neil James Alexander, mathematics, electrical engineering
Slusher, Richart Elliott, physics

Smith, George Elwood, solid state electronics
Smith, Neville Vincent, physics
Smolinsky, Gerald, organic chemistry,
Snoke, Lloyd Randolph, forestry
Snyder, Lawrence Clement, chemical physics
Speers, Louise (Mrs Henry Croix), organic chemistry
Sterlight, Himan, physical chemistry
Strauss, Walter, physics, electrical engineering
Stuck, Barton W, applied mathematics
Sturge, Michael Dudley, experimental solid state physics
Sullivan, Miles Vincent, physical chemistry
Surko, Clifford Michael, physics
Tabor, William Joseph, magnetism
Taylor, Gary N, organic chemistry
Testardi, Louis Richard, solid state physics
Thomas, David Gilbert, physical inorganic chemistry
Thomas, Gordon Albert, experimental solid state physics
Thompson, Larry Flack, polymer chemistry
Thompson, William Baldwin, geophysics
Thomson, Michael George Robert, electron optics
Thornber, Karvel Kuhn, solid state physics
Tomblin, Fred Fitch, space science, nuclear physics
Tonelli, Alan Edward, polymer physics
Torza, Sergio, surface chemistry, hydrodynamics
Tsui, Daniel Chee, solid state physics
Tukey, John Wilder, statistics, statistical analysis
Tully, John Charles, chemical physics
Tyson, J Anthony, astrophysics
van der Ziel, Jan Peter, solid state physics
van Roosbroeck, Willy Werner, semiconductors
Vasile, Michael Joseph, physical chemistry
Vella-Coleiro, George, physics
Vernon, Russel, physics, nuclear engineering
Vitcha, James F, organic chemistry
Voorhoeve, Rudolf Johannes Herman, surface chemistry, physical inorganic chemistry
Walsh, Walter Michael, Jr, solid state physics
Walstedt, Russell E, solid state physics
Waltz, Maynard Carleton, semiconductors
Warner, Daniel Douglas, numerical analysis
Warren, William Willard, Jr, nuclear magnetic resonance
Watson, Hugh Alexander, physics
Weber, Thomas Andrew, theoretical chemistry
Weeks, John David, chemical physics
Wertheim, Gunther Klaus, solid state physics
Wilkening, George Martin, environmental health
Williams, William Howard, mathematics, statistics
Wilson, John Anthony, solid state science
Wilson, Lynn Olson, applied mathematics
Winslow, Field Howard, organic chemistry
Wisenhausen, Hans S, applied mathematics
Wolfe, Raymond, solid state physics
Wood, Darwin Lewis, physics, chemistry
Yafet, Yako, solid state physics
Yanini, Michael, physical chemistry, computer sciences
Yarnell, Charles Frederick, physical chemistry
Ziptel, Christie Lewis, low temperature physics

NESHANIC STATION
Johnston, Christian William, physical chemistry

NEPTUNE
Kornfeil, Fred, physical chemistry, electrochemistry
Swingle, Donald Morgan, applied physics, systems engineering

Kersta, Lawrence George, physics, electrical engineering

NEW BRUNSWICK
Abou-Sabe, Morad A, molecular biology
Abrahams, Elihu, physics
Adair, Suzanne Frank, biopharmaceutics
Adams, William Henry, theoretical chemistry
Alderfer, Russell Brunner, soils
Andrews, Horace Porter, experimental statistics
Appelgate, Alan, radiation chemistry
Applegate, James Edward, wildlife biology
Asano, Akira, pharmaceutics
Auslander, David E, pharmaceutics
Avers, Charlotte Jo, cytogenetics
Axelrod, David E, microbiology; genetics
Bazin, Maurice Jacques, physics
Babcock, MacLean Jack, nutrition administration
Babcock, Philip Arnold, pharmacognosy; botany
Bagatell, Fillmore Kenneth, dermatology; biochemistry
Bailey, Catherine Hayes, pomology
Bailey, Leonard Charles, pharmaceutical chemistry
Bamberger, Curt, industrial organic chemistry
Barlaz, Joshua, mathematics
Bartha, Richard, agricultural ecology, microbial ecology
Battle, Warren Rich, agronomy
Bauman, Robert Andrew, organic chemistry
Beck, John Edwin, medical administration
Berg, Jeffrey Howard, organic chemistry
Bikales, Norbert M, organic chemistry, polymer chemistry
Bingham, Carleton Dille, analytical chemistry; radiochemistry
Bird, George Richmond, photographic chemistry
Bird, John William Clyde, physiology; biochemistry
Boikess, Robert S, organic chemistry
Bonner, Daniel Patrick, microbiology, chemotherapy
Bouis, Paul Andre, organic chemistry
Bowers, Roy Anderson, pharmacy
Boyden, Alan Arthur, systematic zoology
Bredon, Glen E, mathematics
Brewer, Glenn A, Jr, chemistry
Bronzan, John Brayton, high energy physics
Brown, Harry Darrow, biochemistry
Brown, Robert Lee, bacterial physiology; biochemistry
Brownlee, Paula Pimlott, organic chemistry
Bruno, Charles Frank, biochemistry, bacteriology
Bruno, Gerald A, bionucleonics, biochemistry
Brush, John Edwin, population geography; urban geography
Brush, Miriam Kelly, nutrition
Bryan, Wilbur Lowell, pharmaceutical chemistry
Bryson, Vernon, microbiology; genetics
Bumby, Richard Thomas, number theory
Burger, Joanna, ethology; ecology
Burns, David Jerome, agricultural economics; resource economics
Bush, Karen Jean, biochemistry, analytical chemistry
Butler, Terence, mathematics
Cappellini, Raymond Adolph, plant pathology
Carr, Herman Yaggi, condensed state physics
Carr, Michael John, geology
Ceponis, Michael John, plant pathology
Champe, Sewell Preston, molecular biology; genetics
Chandler, Louis, biophysics, radiation physics
Chang, Stephen Szu Shiang, food chemistry
Chase, Theodore, Jr, enzymology
Chen, James Che Wen, developmental biology; botany
Chen, Tseh-An, botany
Childers, Norman Franklin, horticulture
Christman, Edward Arthur, radiation chemistry
Christofferson, Eric, geological oceanography
Cino, Paul Michael, microbiology; industrial
Clark, Frank Eugene, mathematics
Clement, William H, organic chemistry

Cohen, Arthur, mathematical statistics
Cohen, Kenneth Joel, experimental high energy physics
Cohn, Yehudi Aryeh, anthropology
Cohn, Richard Moses, algebra
Cohn, Robert M, pharmaceutical chemistry
Compton, Charles Chalmer, entomology
Cook, Elizabeth Anne, microbiology
Cousins, Robert John, nutritional biochemistry; nutrition
Cross, John Milton, pharmaceutical chemistry
Cyr, Gilman Norman, pharmacy
Daines, Robert Henry, plant pathology
Dalal, Siddhartha Ramanlal, statistics
Daun, Henryk, food science
Davis, Bill David, developmental biology
Davis, Spencer Harwood, Jr, plant pathology
Dekker, Jacob Christoph Edmond, mathematics
Denney, Donald Berend, organic chemistry
Dennis, Emmet Adolphus, parasitology
DiFazio, Louis T, drug metabolism, pharmaceutical chemistry
Doleck, Elwyn Haydn, radiological physics
Douglas, Lowell Arthur, clay mineralogy, soil genesis
Dreyfuss, Jacques, biochemistry
Drinkwater, William Otho, horticulture
Duell, Robert William, agronomy; crops
Dunbar, Phyllis Marguerite, physical chemistry
Dunham, John Malcolm, analytical chemistry
Durkin, Dominic J, horticulture, plant physiology
Dursch, Friedrich, organic chemistry
Eck, Paul, horticulture, soils
Efraty, Avi, organometallic chemistry
Ehrenfeld, David W, ecology, conservation
Ellentuck, Erik, mathematical logic
Elliott, Joanne, mathematics
Eltz, Robert Walter, microbiology
Erickson, Charles Edward, inorganic chemistry
Evans, Joseph Liston, nutrition, animal nutrition
Eveleigh, Douglas Edward, microbiology
Fairbrothers, David Earl, botany
Faith, Carl Clifton, mathematics
Farmanfarmaian, Allahverdi, physiology
Finkelstein, Paul, biochemistry
Finstein, Melvin S, microbiology
Fisher, Hans, nutrition
Fisher, Kenneth Walter, molecular genetics
Florey, Klaus, organic chemistry
Forgash, Andrew John, entomology
Forman, Richard T T, plant ecology
Foster, Siddhartha Burnham, parasitology, invertebrate zoology
Fox, Stephen Knowlton, Jr, geology, paleontology
Frankel, Harry Meyer, physiology
Frantz, Beryl May, analytical biochemistry
Freeman, Ira M, physics
Fuller, Robert Arthur, physical chemistry
Funk, Cyril Reed, Jr, plant breeding
Gaffney, Barbara Jean, organic chemistry; biochemistry
Gaughran, Eugene Robert Lawrence, microbiology
Geczik, Ronald Joseph, physiology; pharmacology
Gerber, Nancy Nichols, organic chemistry
Getis, Arthur, urban geography
Gewirtzman, Leonard, mathematics
Gilbert, Seymour George, food science
Ginsburg, Michael, mathematics
Glaberson, William I, low temperature physics
Glaser, Charles, biochemistry
Glashausser, Charles Michael, nuclear physics
Gonshor, Harry, pure mathematics
Goode, Philip Ranson, nuclear physics
Goodman, Lionel, chemical physics, photobiology
Goodman, Roe William, mathematical analysis
Gordon, Ruth Evelyn, bacteriology
Gradeff, Peter S, industrial organic chemistry; synthetic organic chemistry
Granett, Philip, economic entomology
Green, James Weston, physiology
Grozier, Michael Lawrence, clinical pharmacology

Gunckel, James Eugene, botany
Gupta, Ayodhya P, entomology
Gusmano, Ernest Ambrose, biology,
Haard, Norman F, plant physiology; nuclear medicine
Haddon, Virginia Rae, organic chemistry
Hagmann, Lyle Everest, entomology
Haines, William Joseph, pharmaceutical chemistry
Halisky, Philip Michael, phytopathology
Hamil, Martha M, structural geology,
Hamilton, Leonard W, psychophysiology
Hamilton, Willard Charlson, surface chemistry; physical chemistry
Hansen, Elton J, entomology
Harrington, David Rogers, physics
Hart, Nathan Hoult, vertebrate anatomy
Harvey, Richard Alexander, biochemistry
Haskin, Harold H, physiology
Hasser, George Lee, Jr, physiology;
Hess, Sidney Marvin, pharmacology;
Heuser, Leon John, organic chemistry
Hewins, Roger Herbert, mineralogy,
Hayakawa, Kan-ichi, food science
Havens, Abram Vaughn, meteorology,
Haughey, Francis James, health physics
Hazard, Katharine Elizabeth,
Hedrick, Charles Edward, Jr, analytical chemistry
Helfrich, Kenneth, pesticide chemistry
Herber, Rolfe H, chemical physics, solid state physics
Herman, Robert, parasitology,
Hidick-Smith, Gavin, pediatrics
Hill, George Carver, parasitology
Hopper, Arthur Frederick, embryology
Hordon, Robert Marshall, geography; water resources
Hoyt, William Lind, mathematics
Huber, Raymond C, pharmacy
Hudrik, Anne Marie, synthetic organic chemistry
Hudrik, Paul Frederick, synthetic organic chemistry
Hutchins, Hastings Harold, Sr, pharmaceutical chemistry
Iannarone, Michael, bacteriology
Iinicki, Richard Demetry, weed science
Imlay, Richard Larry, high energy physics
Indyk, Henry Walter, agronomy
Isied, Stephan Saleh, inorganic chemistry
Ivashkiv, Eugene, pharmaceutics, chemistry
Jacobson, Harold, physical chemistry,
Jenifer, Franklyn G, microbiology,
Jenkins, William Robert, nematology
Johnson, Joseph Lemuel, Jr, mathematics
Johnson, William Bradford, horticulture
Justin, James Robert, agronomy; field crops
Kabadi, Balachandra N, physical pharmacy; pharmaceutical chemistry
Kade, Charles Frederick, Jr, chemistry
Kadin, Harold, analytical chemistry;
Kaplan, Murray Lee, nutrition, metabolism
Karmas, Endel, food science
Katz, Irwin Alan, pharmaceutics,
Kenney, Edward Joseph, organic
King, Louis Delwin, pharmacy
Kirschbaum, Joel Jerome, biochemistry, analytical chemistry
Kleyn, Dick Henry, dairy science,
Knill, James Reginald, clinical pharmacology; gastroenterology
Koft, Bernard Waldemar, bacteriology
Kojima, Haruo, low temperature physics
Koller, Noemie, nuclear physics, solid state physics
Kosinski, Antoni A, mathematics
Krauthamer, George Michael, neurophysiology
Krenos, John Robert, chemical physics
Kripalani, Kishin J, drug metabolism, biopharmaceutics
Kuchler, Robert Joseph, microbiology
Kugel, Henry W, experimental atomic physics
Laity, Richard Warren, electrochemistry, physical chemistry
Lampen, J Oliver, microbiology

Lan, Shih-Jung, drug metabolism
Landweber, Peter Steven, mathematics
Langreth, David Chapman, physics
Last, Jerold Alan, biochemistry
Leader, Solomon, mathematics
Leath, Paul Larry, solid state physics
Leathem, James Hain, reproductive endocrinology
Lechevalier, Hubert Arthur, microbiology
Lechevalier, Mary P, microbiology
Leck, Charles Frederick, ornithology, ecology
Lee, Robert John, medical physiology
Leeder, Joseph Gordon, dairy chemistry
Lester, David, biochemistry, pharmacology
Levin, Joseph David, industrial microbiology
Levitt, Norman Jay, mathematics
Lewis, Gwynne David, plant pathology
Li, Jane Chiao, applied statistics
Lindenfeld, Peter, low temperature physics
Litchfield, Carol Darline, marine microbiology, microbial physiology
Litchfield, Charles Carter, biochemistry
Lord, Geoffrey Haverton, pathology
Lordi, Nicholas George, pharmacy
Lovelace, Claud William Venton, theoretical physics
Loveland, Robert Edward, biology
Luthi, Bruno, solid state physics
MacKay, Vivian Louise, molecular biology, biochemical genetics
MacKenzie, James W, thoracic surgery
Maglich, Bogdan, experimental nuclear physics, particle physics
McLean, William L, physics
Meade, John Arthur, agronomy
Mechlinski, Witold, medicinal chemistry, analytical chemistry
Medwick, Thomas, analytical chemistry
Mekjian, Aram Zareh, nuclear physics
Merrill, Leland (Gilbert), Jr, environmental science
Meyers, Edward, microbiology
Michel, Gerd Wilhelm, natural products chemistry
Milnor, Tilla Savanuck Klotz, mathematics
Mitchell, Barry Miller, mathematics
Mitchell, Ralph Gerald, dairy husbandry, genetics
Mixner, John Paulding, animal physiology
Mohammed, Kasheed, nutritional biochemistry, human physiology
Morris, N Ronald, pharmacology
Morrow, Darrell Roy, solid state physics, polymer physics
Morse, Roy Earl, food science
Morss, Lester Robert, physical inorganic chemistry
Moss, Robert Allen, organic chemistry
Muckenhoupt, Benjamin, mathematical analysis
Murray, Bertram George, Jr, zoology
Murray, Raymond Carl, sedimentary petrology
Murty, Dasika Radha Krishna, organic chemistry
Murty, Hari Sriram, biochemistry
Myers, Ronald Fenner, nematology
Nashed, Wilson, pharmacy
Nickerson, Walter John, microbiology, biochemistry
Niederman, Robert Aaron, molecular biology, biochemistry
Niedermeyer, Alfred O, analytical chemistry
Nimeck, Maxwell W, microbiology
Nussbaum, Roger David, mathematical analysis
O'Farrell, Helen Krogull, biochemistry
Oldemar, Gerald M, physical chemistry
O'Leary, Robert Kent, materials science, toxicology
Olson, Wilma King, biophysical chemistry, polymer chemistry
Olsson, Richard Keith, geology
O'Nan, Michael Ernest, algebra
Osofsky, Barbara Langer, algebra
Padhi, Sally Bulpitt, microbiology, insect pathology

Page, Charles Henry, neurophysiology, comparative physiology
Palczuk, Nicholas Charles, immunology
Palser, Barbara Frances, botany
Pan, Bonnie Sun, food science
Papastephanou, Constantin, biochemistry, analytical chemistry
Parreira, Helio Correa, surface chemistry
Passmore, Howard Clinton, immunogenetics
Pearson, Paul Guy, ecology
Peterson, Joseph Louis, plant pathology, mycology
Petryshyn, Walter Volodymyr, mathematics
Pheasant, Richard, organic chemistry, research administration
Piburn, Michael D, geology
Pieczenik, George, molecular biology
Pietri, Charles Edward, analytical chemistry, nuclear chemistry
Pietruszko, Regina, biochemistry
Plano, Richard James, physics
Platt, Thomas Boyne, bacteriology
Poet, Raymond B, biochemistry
Pollack, Bernard Leonard, plant breeding
Potenza, Joseph Anthony, physical chemistry, inorganic chemistry
Pramer, David, microbial ecology
Prockop, Darwin J, biochemistry, medicine
Psuty, Norbert Phillip, geography
Quinn, James Amos, botany, ecology
Race, Stuart Rice, entomology
Rafter, John Arthur, biostatistics
Ramage, Carroll Herbert, dairy nutrition
Ramsden, Hugh Edwin, organic chemistry
Reece, Ralph Parlette, animal physiology
Reilly, Hilda Christine, microbiology
Richardson, Bonham Churchill, cultural geography, geography of the Caribbean
Riemer, Donald Neil, aquatic biology
Rifino, Carl Biaggio, pharmaceutical chemistry
Ritter, Klaus Guenter, applied mathematics, operations research
Robbins, Allen Bishop, physics
Roberts, Fred Stephen, combinatorics, mathematical psychology
Rockmore, Ronald Marshall, theoretical high energy physics
Rodman, Morton Joseph, pharmacology
Rosen, Joseph David, food chemistry, pesticide chemistry
Rosenstein, Joseph Geoffrey, mathematical logic
Rothenberg, David, computer sciences, applied mathematics
Rouslin, William, biochemistry
Rovee, David Thomas, biology
Safa, Helen Icken, applied anthropology
St John, Ann Carlson, microbial physiology
Sak, Joseph, statistical mechanics
San Filippo, Joseph, Jr, organic chemistry, organometallic chemistry
Sannes, Felix Rudolph, elementary particle physics
Santamarina, Enrique, pathology, endocrinology
Sauers, Carl Kilbourne, organic chemistry
Sauers, Ronald Raymond, organic chemistry
Schaefer, John F, Jr, pharmacology, biochemistry
Schaffner, Carl Paul, biochemistry
Schlesinger, Robert Walter, microbiology
Schofer, Jerry Parker, population geography, geography
Schreiber, Eric Christian, pharmacology, organic chemistry
Schueler, Paul Edgar, physical organic chemistry
Schugar, Harvey, inorganic chemistry, bioinorganic chemistry
Sciarrone, Bartley John, pharmaceutics
Segelman, Alvin Burton, pharmacognosy, phytochemistry
Segelman, Florence Pettler, pharmacognosy, phytochemistry
Shank, Brenda Mae Buckhold, cell physics
Shapiro, Joel Alan, theoretical high energy physics
Shapiro, Ralph, nutrition, biochemistry
Sheehan, Brian Talbot, biochemical engineering
Shinkai, John H, pharmaceutical chemistry
Shombert, Donald James, physical chemistry
Simpson, Robert Wayne, virology
Singhvi, Sampat Manakchand, biopharmaceutics
Sipos, Frank, peptide chemistry
Sipos, Tibor, biochemistry, microbiology

Siverston, John Neilos, analytical chemistry, biometrics
Smith, Carter Riley, horticulture, pomology
Snyder, William Enoch, plant physiology
Solberg, Myron, food science, food microbiology
Solotorovsky, Morris, bacteriology
Sommers, Jay Richard, organic chemistry
Spoerlein, Marie Teresa, pharmacology, physiology
Sprague, Milton Alan, agronomy
Stephen, Michael John, physics
Stier, Elizabeth Fleming, biochemistry
Stockel, Richard F, organic chemistry
Stout, Benjamin Boreman, forest ecology
Strauss, George, biophysical chemistry
Strauss, Ulrich Paul, physical chemistry
Streu, Herbert Thomas, entomology, zoology
Strumeyer, David Hyman, biochemistry
Sturkie, Paul David, physiology
Sundberg, Robert Lee, organic chemistry, research administration
Sutherland, Donald James, physiology, biochemistry
Swallow, William Hutchinson, biostatistics
Sweeney, William John, mathematics
Swift, Fred Calvin, entomology, ecology
Szarka, Laszlo Joseph, bioengineering
Taft, Earl J, algebra
Tedrow, John Charles Fremont, soils
Teicher, Henry, mathematical statistics
Temmer, Georges Maxime, nuclear physics
Thoma, Richard William, biochemistry
Tierney, Myles, algebra, mathematical logic
Tobias, Irwin, physical chemistry
Toby, Sidney, physical chemistry
Torrey, Henry Cutler, magnetic resonance
Toth, Stephen John, soil chemistry
Trama, Francesco Biagio, limnology
Treves, Jean Francois, pure mathematics
Umbreit, Wayne William, bacteriology, biochemistry
Valentine, Joan Selverstone, bioinorganic chemistry
Vander Noot, George Ward, animal nutrition
Vander Wende, Christina, biochemistry, pharmacology
Varney, Eugene Harvey, plant pathology
Vayda, Andrew Peter, human ecology, anthropology
Vazakas, Aristotle John, medicinal chemistry
Verlangieri, Anthony Joseph, biochemistry, toxicology
Viebrock, Frederick William, biochemistry
Vrijenhoek, Robert Charles, evolutionary biology
Wacker, Peter Oscar, cultural geography, historical geography
Wadke, Deodatt Anant, pharmacy
Wainio, Walter W, enzymology, biochemistry
Walsh, Bertram (John), mathematical analysis
Walton, Grant Fontain, soil science, resource administration
Watts, Terence Leslie, physics
Weidner, Richard Tilghman, physics, magnetic resonance
Weigend, Guido Gustav, geography
Weiss, Mitchell Joseph, invertebrate zoology
Weiss, Peter Reifer, theoretical physics
Weissman, Sigmund, crystallography
Welker, William V, Jr, weed science, horticulture
West, Richard Fussell, forestry, wood technology
Westman, James Ross, biology
Wheeden, Richard Lee, mathematics
White-Stevens, Robert Henry, environmental biology
Wildnauer, Richard Harry, dermatology, polymer chemistry
Willemsen, Roger Wayne, physiological ecology, botany
Willey, Ann Morris, human genetics
Willson, John Ellis, veterinary medicine, toxicology
Wilson, Armin Guschel, organic chemistry, medicinal chemistry
Wilson, Billy Ray, entomology
Wilson, Evelyn H, organic chemistry
Wilson, Robert Lee, algebra
Winnett, George, environmental sciences, analytical chemistry
Witkin, Evelyn Maisel, microbial genetics
Wolfe, Peter Edward, geology
Wolfson, Kenneth Graham, mathematics

Wong, Keith Kam-Kin, biochemistry, physiology
Wong, Tang-Fong Frank, theoretical high energy physics
Worthington, Thomas Kimber, low temperature physics
Wynveen, Robert Allen, health physics, medical physics
Yamin, Samuel Peter, particle physics
Yoshida, Shiro, nuclear physics
Zahler, Raphael, mathematics
Zamick, Larry, nuclear physics
Zapolsky, Harold Saul, theoretical physics
Zenchelsky, Seymour Theodore, analytical chemistry
Zilinskas, Barbara Ann, biochemistry
Zimmerberg, Hyman Joseph, mathematics

NEW LISBON
Marucci, Philip Edward, entomology

NEW MILFORD
Nash, Nat H, food science

NEW MONMOUTH
Beispiel, Myron, organic chemistry

NEW PROVIDENCE
Biagetti, Richard Victor, electrochemistry, inorganic chemistry
Clendinning, Robert Andrew, organic chemistry, polymer chemistry
DeCastro, Arthur, analytical chemistry
Gaylord, Norman Grant, polymer chemistry
Morrison, Barbara Ann, applied mathematics
O'Sullivan, Thomas Denis, physical chemistry, electrochemistry

NEW SHREWSBURY
Goldstein, Albert, synthetic organic chemistry, organic polymer chemistry
Roos, Leo, physical organic chemistry, photochemistry

NEWARK
Agron, Sam Lazrus, geology
Aissen, Michael Israel, mathematics
Alger, Elizabeth A, internal medicine, anatomy
Anderson, James Donald, herpetology
Andruskiw, Roman Ihor, mathematical analysis
Ballinger, Carter M, anesthesiology
Bartell, Pasquale, medical microbiology
Bassett, Emmett W, microbiology, biochemistry
Bauman, John W, Jr, physiology, biochemistry
Behrle, Franklin C, pediatrics
Bennet, Ian Cecil, dentistry
Bennun, Alfredo, biochemistry, biophysics
Berendsen, Peter Barney, anatomy, cell biology
Bergen, Stanley S, Jr, internal medicine
Boccabella, Anthony Vincent, anatomy
Boylan, Edward, mathematics
Brezenoff, Henry Evans, neuropharmacology
Briody, Bernard Aloysius, medical bacteriology
Bullock, John, endocrinology
Carey, George Warren, urban geography, analytical statistics
Cherlin, George (Yale), mathematics
Chinard, Francis Pierre, physiological chemistry, internal medicine
Cinotti, Alfonse A, medicine, ophthalmology
Cohen, Marion Deutsche, mathematical analysis
Condouris, George Anthony, pharmacology
Cook, Stuart D, neurology, neurosciences
Crow, John H, plant ecology
Davis, Robert Foster, Jr, plant physiology
Deasy, Michael Joseph, periodontics
DeFouw, David O, anatomy, physiology
Diecke, Friedrich Paul Julius, physiology
Ditman, John Gordon, physical chemistry
Dumm, Mary Elizabeth, nutrition
Eckhardt, Eileen Theresa, pharmacology
Edinger, Henry Milton, physiology, neurophysiology
Ehrlich, Julian, biophysical chemistry, biochemistry

El Guindy, Mahmoud Ismail, chemical metallurgy
Ende, Norman, pathology
Farber, Elliot, physics
Fasano, Anthony Vincent, anatomy, neuroendocrinology
Feder, Harvey Herman, anatomy, reproductive physiology
Feldman, Harold Samuel, psychiatry, psychopharmacology
Feldman, Lawrence A., microbiology, virology
Ferrante, Frank L., physiology
Fitzgerald, Patrick Henry, physical organic chemistry
Flynn, Edward Joseph, pharmacology
Fono, Andrew, organic chemistry
Frank, Oscar, biochemistry, nutrition
Fraser, Donald Boyd, physical chemistry
Freeman, James R., optics, solid state physics
Friedman, Kenneth Joseph, physiology, biophysics
Frisell, Wilhelm Richard, enzymology, metabolism
Fu, Shou-Cheng Joseph, biochemistry
Furness, Geoffrey, microbiology, venereal diseases
Gardiner, Lion Frederick, marine zoology
Garner, Hessie Filmore, geomorphology
Garrison, Vivian Eva, anthropology
Gautreau, Ronald, physics
Gertner, Sheldon Bernard, pharmacology
Gilani, Shamshad H., teratology, cardiology
Gilliland, William Nathan, geology
Gillis, Robert Edward, microbiology
Gilmont, Ernest Rich, organic chemistry
Giuffrida, Robert Eugene, organic chemistry
Gona, Amos G., endocrinology, developmental anatomy
Grady, Hugh Gerard, pathology
Greenfield, Sydney Stanley, botany
Hall, Stan Stanley, organic chemistry, bio-organic chemistry
Hemmes, Paul Richard, Jr., physical chemistry
Higgins, James Francis, industrial chemistry
Hilding, David Anderson, otolaryngology
Hirschberg, Erich, biochemistry
Hollinshead, May B., anatomy
Horn, Leif, biophysics
House, Earl Lawrence, anatomy
Howland, Richard David, pharmacology
Hutcheon, Duncan Elliot, pharmacology
Imaeda, Tamotsu, bacteriology, leprology
Jonnard, Raymond, physical chemistry
Jordan, Frank, bio-organic chemistry, biophysical chemistry
Joselow, Morris M., public health
Kahn, Arthur Jole, physiology
Kaminetzky, Harold Alexander, obstetrics & gynecology
Kaminski, Zsigmund Charles, medical microbiology, infectious diseases
Kaplan, Harry Arthur, neurosurgery
Kasper, Andrew E., Jr., paleobotany
Kimmel, Howard S., physical chemistry
Kingery, Bernard Troy, physics
Kirschner, Marvin Abraham, internal medicine, endocrinology
Kluiber, Rudolph W., inorganic chemistry
Komisaruk, Barry Richard, psychobiology, neurophysiology
Koren, Charles, mathematical analysis
Kozam, George, anatomy, pathology
Krey, Phoebe Regina, rheumatology
Kumar, Surender, organic chemistry, biochemistry
Kushnick, Theodore, pediatrics
Lalancette, Roger A., analytical chemistry, crystallography
Lanzoni, Vincent, pharmacology, clinical medicine
Layman, William Arthur, psychiatry
Lazaro, Eric Joseph, surgery
Lea, Michael Anthony, biochemistry
Leevy, Carroll M., medicine, nutrition
Levine, Robert, pediatric cardiology
Levinson, Gilbert E., medicine, cardiology
Levy, Leon Sholom, computer science
Loeser, Eugene William, medicine
LoMonaco, Carmine Joseph, endodontics
Louria, Donald Bruce, internal medicine, microbiology
Maiello, John Michael, mycology, environmental biology
Mallams, John Thomas, medicine, radiology
Manhold, John Henry, Jr., pathology
Manspeizer, Warren, geology
Maraspin, Lyno Evelino, human anatomy, endocrinology
Marks, Richard Henry Lee, biochemistry

McArdle, Joseph John, neuropharmacology, neurophysiology
McDowell, Sam Booker, herpetology
Meyer, Conrad Frederick, developmental plant anatomy
Miller, Tony Jasper, developmental biology
Monse, Ernst Ulrich, physical chemistry
Mooney, Richard Warren, physical chemistry, inorganic chemistry
Moskowitz, Harvey D., biomedical engineering, dental materials
Mycek, Mary J., biochemistry, pharmacology
Nenno, Robert Peter, psychiatry
Neville, William E., surgery
Nolasco, Jesus Bautista, physiology
Nussbaum, Murray, medicine, hematology
Ocken, Paul Robert, biochemistry, organic chemistry
Opdyke, David Franklin, physiology
Palmer, Eddy Davis, gastroenterology
Panson, Gilbert Stephen, physical chemistry, organic chemistry
Parker, Richard C., physical chemistry
Perkins, Ben Harrison, chemistry
Perlmutter, Howard D., organic chemistry
Perryman, James Harvey, neurophysiology, biology
Petryshyn, Walter A., otolaryngology
Peyser, Gideon, mathematics
Poetz, Robert George, inorganic chemistry
Puffer, John H., petrology, geochemistry
Quinones, Mark A., preventive medicine, public health
Raab, Jacob Lee, physiology
Redwood, William Raymond, biochemistry, biophysics
Regan, Timothy Joseph, internal medicine, cardiology
Riva, Humbert Lewis, obstetrics & gynecology
Rosenkrantz, Jacob Alvin, hospital administration
Rush, Benjamin Franklin, Jr., surgery
Sacher, Alex, immunology
Saha, Anil, immunology
Salgado, Ernesto D., pathology
Salzarulo, Leonard Michael, physics, chemical engineering
Sather, Bryant Thomas, physiology, ecology
Schlegel, James M., physical chemistry
Schramm, Charles H., organic chemistry
Schrier, Melvin Henry, chemistry
Schuback, Philip, periodontics, oral pathology
Schwalb, Marvin N., mycology, microbiology
Shafter, Morris Frank, medical
Shapiro, Herman Simon, biochemistry, genetics
Sherr, Stanley I., biochemistry
Shilman, Avner, analytical chemistry, physical chemistry
Shloming, Robert, applied statistics
Siegel, Allan, neurobiology, neuroanatomy
Silver, Herbert Graham, physical chemistry
Snyder, William H., organic chemistry, polymer chemistry
Somberg, Ethel Weiss, biochemistry
Sonnenblick, Benjamin Paul, environmental chemistry
Spruch, Grace Marmor, physics
Stefancsik, Ernest Anton, organic chemistry
Strausser, Helen R., physiology, zoology
Suchow, Lawrence, solid state chemistry, inorganic chemistry
Swan, Kenneth G., surgery
Tamburro, Carlo Horace, internal medicine, hepatology
Tassoni, Joseph Paul, entomology
Theokritoff, George, paleontology
Thomas, Claudewell Sidney, psychiatry, public health
Thompson, Hugh Walter, organic chemistry
Turoff, Murray, computer science, operations research
Vassiliou, Andreas H., mineralogy
Veldhuis, Benjamin, organic chemistry
Vincent, Gordon Ross, dentistry, dental materials
Von Hagen, D Stanley, pharmacology
Walters, Thomas Richard, pediatrics, hematology
Weill, Carol Edwin, organic chemistry
Weis, Judith Shulman, developmental biology, aquatic biology
Weis, Peddrick, anatomy, embryology
Weisse, Allen B., cardiology, medicine

Weissman, William Kent, neurosurgery, biomedical engineering
Wenger, Franz, physical chemistry
West, Charles P., organic chemistry
Wies, William Walter, geology, immunochemistry
Wilhoft, Daniel C., zoology
Wilson, Mabel Florey, analytical chemistry
Wilson, Robert G., chemistry
Wolfson, Edward A., preventive medicine, medicine
Yamane, George M., oral medicine, oral pathology
Yang, Chung Shu, biochemistry
Yu, Mang Chung, anatomy, neuroanatomy
Zajcew, Mykola, chemistry

NEWFOUNDLAND
Steiger, Leonard William, chemistry

NEWTON
Mehring, Arnon Lewis, Jr., nutrition
Spooner, Alfred Brent, organic chemistry

NORTH BRUNSWICK
Bolton, Laura Lee, experimental biology
Gander, Robert Johns, polymer chemistry
Kaplan, Leonard Louis, pharmaceutics
Linegar, Charles Ramon, pharmacology
Linsky, Cary Bruce, biological chemistry
Lynch, Peter J., physical chemistry
Ludwig, Bernard John, organic chemistry
Marshall, James R., pharmaceutics, health sciences

NORTH PLAINFIELD
Angel, Henry Seymour, organic chemistry
Beattie, Thomas Robert, organic chemistry
Blicher, Adolph, solid state science
Burns, Mary Grace, biochemistry
Lucchesi, Peter J., physical chemistry

NUTLEY
Antoshkiw, Thomas, pharmaceutical chemistry
Banziger, Ralph Frederick, pharmacology administration
Bartl, Paul, biophysics, science administration
Batcho, Andrew David, organic chemistry
Bauernfeind, Jacob Christopher, nutrition, biochemistry
Benz, Wolfgang, mass spectrometry
Berger, Julius, biochemistry
Berger, Leo, medicinal chemistry
Berkowitz, Barry Alan, pharmacology
Beskid, George, bacteriology
Bhattacharyya, Pranab K., physical chemistry, analytical chemistry
Blaivas, Murray A., clinical chemistry
Blessel, Kenneth Wayne, analytical chemistry
Blume, Arthur Joel, neurochemistry
Bontempo, John A., microbiology
Borenstein, Benjamin, food chemistry, organic chemistry
Brin, Myron, biochemistry
Brooks, Marvin Alan, analytical biochemistry
Brot, Nathan, biochemistry
Burns, John J., biochemistry
Cardinale, George Joseph, biochemistry
Chassan, Jacob Bernard, psychiatry, statistics
Coffey, John William, biochemistry
Cohen, Noal, synthetic organic chemistry, natural products chemistry
Conney, Allan Howard, biochemistry, pharmacology
Cook, Alan, chemistry
Cook, Leonard, pharmacology
Cordts, Richard Henry, Jr., nutrition
Coutinho, Claude Bernard, experimental medicine
Dairman, Wallace, biochemistry, pharmacology

NORTHVALE
Ruderman, Irving Warren, electrooptics
Wiener, Benjamin, biochemistry

Dalton, Colin, biochemical pharmacology
Davidson, Arnold B., psychopharmacology
Davis, Raymond Vincent, biochemistry, immunochemistry
DeLorenzo, William F., bacteriology, chemotherapy
Derege, Michael E., organic chemistry
De Silva, John Arthur F., analytical chemistry, pharmaceutical chemistry
Ennis, Herbert Leo, microbiology
Faust, Richard Edward, pharmaceutical chemistry
Felix, Arthur M., bio-organic chemistry
Field, George Francis, organic chemistry
Finkelstein, Jacob, organic chemistry, medicinal chemistry
Flexner, Leo Aaron, chemistry
Foenzler, Ernest Carl, pharmaceutical chemistry
Forrester, Frank Robert, analytical chemistry
Foulke, Donald Gardner, organic chemistry, electrochemistry
Free, John Marshall, applied physics, electrooptics
Fryer, Rodney Ian, medicinal chemistry
Gamzu, Elkan, psychopharmacology, psychology
Garcia, Edward Ernest, organic chemistry
Garland, William Arthur, medicinal chemistry
Gaut, Zane Noel, clinical pharmacology
Gibson, Kenneth David, molecular biology
Givens, Samuel Virtue, biostatistics
Goldberg, Arthur H., physical pharmacy, biopharmaceutics
Green, Erika Ana, medical microbiology, science writing
Grethe, Guenter, organic chemistry
Grimmell, William C., mathematical chemistry
Grunberg, Emanuel, chemotherapy
Gruter, Frederick Herbert, microbiology
Gurien, Harvey, organic chemistry, analytical chemistry
Hagel, Robert B., analytical chemistry
Hallman, Harold Frank, endocrinology
Hamilton, James Guthrie, biochemistry
Harrison, Yvonne E., pharmacology, drug metabolism
Hayes, Terence James, veterinary parasitology
Herzfeld, Charles Maria, physics
Heveran, John Edward, analytical chemistry
Hines, Leonard Russell, biochemistry
Hoar, Richard Morgan, developmental anatomy, reproductive biology
Horecker, Bernard Leonard, biochemistry
Huber, William George, veterinary medicine
Humiec, Frank S., Jr., inorganic chemistry
Idson, Bernard, chemistry
Infeld, Martin Howard, pharmaceutics
Jacobson, Martin Michael, biochemical pharmacology
Jampolsky, Lester Mischa, organic chemistry
Jernow, Jane L., organic chemistry
Kaback, Howard Ronald, biochemistry
Kamm, Jerome Jr., biochemistry
Kaplan, Stanley Albert, pharmaceutics
Kasparek, Stanislav Vaclav, organic chemistry
Keith, Dennis Dalton, organic chemistry
Kent, Robert Eugene, organic chemistry
Kierstead, Richard Wightman, organic chemistry
Klein, Bernard, organic chemistry
Koechlin, Bernard Alphons, biochemistry
Konikoff, John Jacob, biomedical engineering, electrophysiology
Kuntzman, Ronald Grover, biochemistry, pharmacology
Lauer, Rudolph Frank, pharmaceutical chemistry
Laurencot, Henry Jules, Jr., drug metabolism
Levy, Alan C., physiology, toxicology
Liebman, Arnold Alvin, organic chemistry, radiochemistry
Lu, Anthony Y H., biochemistry
MacDonald, Alexander, Jr., analytical chemistry
Machlin, Lawrence Judah, nutritional biochemistry
Maehr, Hubert, natural products chemistry
Maestrone, Gianpaolo, veterinary medicine, microbiology
Magid, Louis, pharmaceutical chemistry

Margolis, Frank L., neurochemistry
Maricq, John, organic chemistry
Mauer, Irving, genetics
Maynard, Donald Earle, biochemistry
McKinley, Raymond Earl, veterinary medicine, animal nutrition
McSharry, William Owen, analytical chemistry
Mergens, William Joseph, analytical chemistry
Messersmith, Robert E, veterinary medicine
Meyers, Kenneth Purcell, cell biology, endocrinology
Micheli, Robert Angelo, organic chemistry
Miller, David Lee, physical organic chemistry
Miller, Oscar Neal, nutrition, biochemistry
Mitrovic, Milan, veterinary medicine
Mlodozeniec, Arthur Roman, physical chemistry, pharmacy
Moros, Stephen Andrew, analytical chemistry
Mowles, Thomas Francis, endocrinology, biochemistry
Newmark, Harold Leon, organic chemistry, biochemistry
Ning, Robert Ye Fong, medicinal chemistry, organic chemistry
Nishikawa, Alfred Hirotoshi, biochemistry
Ochoa, Severo, medicine
O'Connell, James Anthony, physical chemistry
Ofengand, Edward James, biochemistry, molecular biology
Oliveto, Eugene Paul, organic chemistry
Osborne, Melville, pharmacology, physiology
Partridge, John Joseph, Jr, organic chemistry
Paulsrud, John Reynold, biochemistry
Perry, Clark William, synthetic organic chemistry
Pestka, Sidney, biochemistry, medicine
Pifer, Charles William, pharmaceutical chemistry
Pool, William Robert, toxicology, pharmacology
Randall, Lowell Orlando, pharmacology
Rosen, Perry, organic chemistry
Rubin, Saul Howard, biochemistry
Runser, Richard Henry, physiology, medicine
Sachs, Howard, biochemistry
Sadek, Salah Eldine, pathology, toxicology
Salvador, Richard Anthony, pharmacology
Savard, Edward Victor, microbiology, parasitology
Schallek, William Barrett, pharmacology
Schiffrin, Milton Julius, physiology
Schlosser, Walter Jr, pharmacology
Schwartz, Morton Allen, drug metabolism
Scott, William Edwin, chemistry
Senkowski, Bernard Zigmund, pharmaceutical chemistry
Sepinwall, Jerry, psychopharmacology, physiological psychology
Shapiro, Stanley Seymour, biochemistry
Shatkin, Aaron Jeffrey, biochemistry, virology
Sheridan, Jane Connor, analytical chemistry
Sherman, Michael Ian, developmental biology
Sigg, Ernest Beat, physiology
Skalka, Anna Marie, molecular biology, virology
Sorensen, Lloyd J, microbiology
Sorter, Peter F, organic chemistry, information science
Spector, Sydney, pharmacology
Spiegel, Herbert Eli, biochemistry, clinical chemistry
Stefko, Paul Lowell, experimental surgery, pharmacology
Stempel, Arthur, microchemistry
Steyermark, Al, microchemistry
Sullivan, Ann Clare, biochemistry
Sullivan, William Richard, chemistry
Swarm, Richard Lee, pathology
Tabenkin, Benjamin, chemistry
Teitel, Sidney, medicinal chemistry
Theilheimer, William, organic chemistry
Toome, Voldemar, physical chemistry
Tsolas, Orestes, molecular biology, biochemistry
Udenfriend, Sidney, biochemistry
Unowsky, Joel, industrial microbiology, microbial genetics
Valentine, Donald H, Jr, organometallic chemistry, photochemistry

Vandevoorde, Jacques Pierre, biochemistry
Vane, Floie Marie, drug metabolism
Van Pelt, Wesley Richard, health physics, environmental health
Walser, Armin, medicinal chemistry
Wang, Su-Sun, biochemistry
Watkins, Paul Donald, microbiology, biochemistry
Weber, Joseph, organic chemistry
Wehrli, Pius Anton, chemistry
Wei, Chung-Chen, organic chemistry, photochemistry
Weissbach, Arthur, biochemistry
Weissbach, Herbert, biochemistry, molecular biology
Westley, John William, organic chemistry, biochemistry
White, Lawrence S, physics
Whitman, Erwin N, research administration, clinical pharmacology
Wittman, James Smythe, III, biochemistry, nutrition
Yard, Allan Stanley, pharmacology
Young, Richard Lawrence, analytical biochemistry, neurochemistry
Yu, George Chinshih, statistics
Ziering, Albert, organic chemistry
Zomzely-Neurath, Claire Eleanore, neurobiology

OAKLAND
Bluestein, Bernard Richard, organic chemistry
Bluestein, Claire, industrial organic chemistry
Friedman, Seymour K, organic chemistry, colloid chemistry
Nelson, Lawrence Barclay, chemistry
Spitsbergen, James Clifford, physical polymer chemistry

OCEAN
DiMasi, Gabriel Joseph, electrochemistry, chemical engineering

OCEANPORT
Henrich, Christopher John, pure mathematics

OLD BRIDGE
Sisco, William E, industrial organic chemistry

ORADELL
Choman, Bohdan Russell, microbiology
Gilbert, Allan Henry, synthetic organic chemistry
Riso, R Richard, organic chemistry
Tong, Mary Powderly, mathematics
Van Dam, M, photography, organic chemistry

ORANGE
Brown, Bernard Beau, organic chemistry
Gabriel, Richard Francis, mathematics

PALISADES PARK
Greenberg, Arthur, physics
Kelley, Maurice Joseph, organic chemistry
Sykes, Donald Joseph, photographic chemistry, statistical analysis

PARAMUS
Carrock, Frederick E, physical chemistry, organic chemistry
Chandler, Horace W, physical chemistry, chemical engineering
Charen, George, physics, chemistry
Eisner, Philip Nathan, physics
Esser, Aristide Henri, psychiatry, human ecology
Fein, Marvin Michael, organic chemistry
Harrison, James Beckman, organic chemistry
Hulyalkar, Ramchandra K, polymer chemistry
Lewis, Seymour, microbiology, botany
Marder, Herman Lowell, organic polymer chemistry, research administration
Mercier, Philip Laurent, physical chemistry
Rudner, Bernard, organic chemistry
Shinohara, Makoto, polymer chemistry, polymer science
Sidi, Henri, organic chemistry, polymer chemistry
Vouras, Paul Peter, geography

PARK RIDGE
Blagg, John Creighton Lee, chemistry
Conger, Robert Perrigo, plastics chemistry
Kosel, George Eugene, electrophotography
Moyer, Arden Wesley, physiological chemistry

PARLIN
Behrens, Herbert Ernest, physics, photography
Botjer, William George, inorganic chemistry, engineering
Cohen, Abraham Bernard, organic chemistry, photochemistry
Evans, Warren William, physical chemistry
Flannagan, Gordon Neel, physical chemistry
Foy, Walter Lawrence, physical chemistry
Gervay, Joseph Edmund, organic chemistry
Gray, Russell Houston, color science, photographic chemistry
Halfon, Marc, photographic chemistry, organic chemistry, polymer chemistry
Hauser, William P, photographic chemistry
Held, Robert Paul, physical chemistry
Hendricks, Robert William, physical chemistry
Lohner, Donald J, organic chemistry
Manger, Charles Walter, physical chemistry
Pilato, Jack Carmen, photographic chemistry
Schoenthaler, Arnold Charles, organic chemistry
Valiaveedan, George Devasia, organic chemistry
Warfield, Peter Foster, organic chemistry
Weiss, James Paul, optics

PARSIPPANY
Aaronoff, Burton Robert, organic chemistry
Antognini, Joe, plant science
Dean, Donald E, cosmetic chemistry, pharmaceutical chemistry
Dietz, George Robert, nuclear science
Goodman, Louis P, food science
Jenkins, James William, organic chemistry
Kranz, Alan Zelig, physics, research administration
Martin, Robert Lewis, information science, research administration
Overberg, Richard Joseph, polymer chemistry
Roach, Paul G, analytical chemistry, organic chemistry
Sexsmith, David Randal, organic chemistry
Shrawder, Elsie June, clinical chemistry
Stevenson, David, organic chemistry, biochemistry
Theile, Fred Charles, organic chemistry, biochemistry
Yoder, Donald Maurice, plant pathology

PASSAIC
Bolla, Robert Irving, zoology, molecular biology
Feltman, Reuben, oral medicine

PATERSON
Fytelson, Milton, chemistry
Heyman, Karl, organic chemistry
Moyerman, Robert Max, organic chemistry, analytical chemistry

PAULSBORO
Brennan, James A, organic chemistry, petroleum chemistry
Caesar, Philip D, petroleum chemistry
Chester, Arthur Warren, inorganic chemistry
Ciric, Julius, physical chemistry, chemical engineering
Garwood, William Everett, organic chemistry
Glass, John Richard, analytical chemistry
Grey, Peter, analytical chemistry
Kirklin, Perry William, petroleum chemistry
Kobrin, Robert Jay, chemical instrumentation
Koehl, William John, Jr, fuel science, air pollution
Kokotailo, George T, solid state physics
Kuhl, Gunter Hinrich, inorganic chemistry
Landis, Philip Sherwood, organic chemistry
Lawton, Stephen Latham, structural chemistry
Leaman, Wilbur Kauffman, petroleum chemistry
Manley, Leo William, chemistry, mathematics
Myers, Claude Grenville, physical chemistry, engineering
Nace, Donald Miller, physical chemistry
Neiswender, David Daniel, petroleum chemistry
Otto, Ferdinand Philip, organic chemistry
Plank, Charles Joseph, petroleum chemistry
Raich, Henry, physical chemistry
Schick, John William, petroleum chemistry
Sedlak, Michael, organic chemistry, analytical chemistry
Sherry, Howard S, physical inorganic chemistry, chemical engineering
Socolofsky, John Frederick, chemistry
Voltz, Sterling Ernest, physical chemistry
Wilson, John William, Jr, petroleum chemistry, microscopy
Wu, Ellen Lem, physical chemistry

PEMBERTON
Ludin, Roger Louis, nuclear physics
Moyer, Samuel Edward, population genetics

PENNINGTON
Little, Silas, Jr, forestry

PENNS GROVE
Pagano, Alfred Horton, organic chemistry
Railing, Wilford Edward, industrial organic chemistry
Thomas, Robert Joseph, chemistry

PENNSAUKEN
Hagberg, Elroy Carl, food science
Labbee, Marcel D, agricultural chemistry, food technology
Silleck, Clarence Frederick, chemistry
Stein, Sidney J, physical chemistry, microelectronics

PERTH AMBOY
Dahill, Robert T, Jr, organic chemistry
Resnik, Robert Kenneth, inorganic chemistry
Ross, Charles Augustus, surgery
Zief, Morris, industrial chemistry

PHILLIPSBURG
Barnard, Alfred James, Jr, chemistry, analytical chemistry
Campbell, Bruce Henry, analytical chemistry
Bohrer, James Calvin, organic chemistry
Bond, Arthur Chalmer, physical chemistry

PINE BEACH
Elliot, John, textile chemistry

PISCATAWAY
Avigad, Gad, biological chemistry
Batzar, Kenneth, pollution chemistry, inorganic chemistry
Becker, Joseph Gerald, microbiology
Blades, Charles Ernest, organic chemistry
Breckenridge, Bruce (McLain), pharmacology, physiology
Brostrom, Charles Otto, pharmacology, biochemistry
Brostrom, Margaret Ann, biochemistry, pharmacology
Browning, Edward T, pharmacology, neurochemistry
Clark, Irwin, biochemistry
Clarke, David Bruce, industrial organic chemistry
Costello, Christopher Hollet, pharmacology
Crane, Robert Kellogg, biological chemistry, physiology
Cross, Richard James, internal medicine, community health
Danzig, Meyer Hillel, chemistry

549

De Salva, Salvatore Joseph, neurology, pharmacology
DiBella, Eugene Peter, pharmacology
Dubin, Donald T., biochemistry, cell biology
Ederberg, Robert, psychophysiology
Egger, Maurice David, neurophysiology
Eichholz, Alexander, biochemistry
Eider, Norman George, analytical chemistry
Eigen, Edward, chemistry, microbiology
Fath, Joseph, organic chemistry
Gaffar, Abdul, immunochemistry
Geller, Herbert M., neurophysiology, neuropharmacology
Gerecht, John Fred, chemistry
Gilman, Martin Robert, toxicology
Goldstein, Leonide, neuropharmacology
Goodman, Donald, polymer chemistry
Gottesman, Roy Tully, organic chemistry
Grand, Paul Sheldon, organic chemistry
Hearney, Elaine Schmidt, neurosciences
Hess, Arthur, anatomy
Hoffmann, Hans-Peter Gerhard, cell biology, biochemistry
Jacobson, Howard Newman, physiology
Johnson, Hilding Reynold, analytical chemistry
Kestenbaum, Richard Charles, bacteriology
Khachadurian, Avedis K., medicine
Kohn, Herbert Myron, neuropsychology, electroencephalography
Kuo, Peter Te, internal medicine
Landor, John Henry, surgery
Lautsch, Elizabeth Virginia, pathology
Lenard, John, biochemistry
Macdonald, Gordon J., physiology, anatomy
Malamed, Sasha, cell biology
McAllister, William Turner, molecular biology
McCoy, John Roger, chemotherapy, comparative pathology
Meuly, Walter C., organic chemistry
Miller, Robert Stephen, polymer chemistry, polymer engineering
Morrison, Ashton Byron, metabolism, pathology
Mueller, Peter Sterling, psychiatry
Murphree, Henry Bernard Scott, clinical pharmacology
Murray, Leo Thomas, inorganic chemistry, organic chemistry
Page, Ernest Winslow, obstetrics & gynecology
Perr, Irwin Norman, psychiatry, forensic medicine
Pierce, Robert Charles, physical chemistry
Pierson, William Grant, organic chemistry, physical chemistry
Pollack, Irwin W., psychiatry
Raska, Karel Frantisek, Jr., pathology, microbiology
Raskova, Jana D., pathology
Salkind, Alvin J., bioengineering
Sanders, Marilyn Magdanz, biochemistry
Schlissel, Harvey Joel, microbiology
Schurman, Jack Vair, research administration
Sircar, Ila, synthetic organic chemistry, medical research
Solis-Gaffar, Maria Corazon, biochemistry, medicinal chemistry
Spiro, Herzl Robert, psychiatry, social psychology
Stanaback, Robert John, organic chemistry
Seiden, James Philip, organic polymer chemistry
Shea, Stephen Michael, morphology, cell biology
Shelden, Robert Merten, pathology
Shimamura, Tetsuo, reproductive physiology, embryology
Sims, Victor A., organic chemistry
Sinha, Arabinda Kumar, physiology, neurobiology
Scullin, James Philip, organic polymer chemistry
Stevens, Thomas McConnell, virology, entomology
Stevenson, Nancy Roberta, physiology, nutrition
Stollar, Victor, microbiology, virology
Strohl, William Allen, virology
Takahashi, Mark T., physical biochemistry
Thomas, Virginia Lee, microscopy
Trowbridge, James Rutherford, surface chemistry
Tsapogas, Makis Joakim, surgery, biomedical sciences
Turse, Richard S., analytical chemistry, spectrochemistry
Uitto, Jouni Jorma, dermatology, biochemistry
Wakeman, Irving B., analytical chemistry

Weiss, Harvey Richard, physiology, pharmacology
Weiss, Sidney, pharmacology
Wilson, Frank Joseph, anatomy, cell biology
Wolf, Thomas, analytical chemistry
Wolff, Donald John, biochemistry, pharmacology
Yacowitz, Harold, nutrition

PITMAN

Huber, Melvin Lefever, organic chemistry, information science
Rope, Barton Whitefield, petroleum chemistry

PLAINFIELD

Boucher, Raymond Edward, organic chemistry
Drelich, Arthur (Herbert), textile chemistry, microscopy
Fisher, Robert George, neurosurgery
Hyun, Bong Hak, pathology
Kendall, David Nelson, physical chemistry, spectroscopy
Lowenheim, Frederick Adolph, electrochemistry
Oroshnik, William, organic chemistry
Riley, David Waegar, theology
Sor, Kamil, soil science
Starke, Emory Potter, mathematics
Szymanski, Chester Dominic, biochemistry
Tessler, Martin Melvyn, organic chemistry
Von, Isaiah, industrial organic chemistry

PLAINSBORO

Fitzgerald, Cornelius Gilbert, organic chemistry

POMONA

Bean, Ralph J., mathematics
Cromartie, William James, Jr., ecology, natural history
Hunt, Kenneth Whitten, botany, conservation
Kirch, Murray R., mathematics
Nanzetta, Philip Newcomb, academic administration
Plank, Donald Leroy, mathematics
Sharon (Schwadron), Yitzhak Yaakov, theoretical nuclear physics
Taylor, Harold Evans, plasma physics, space physics

PORT MURRAY

Messerly, George Henry, physical chemistry

POMPTON LAKES

Kaniecki, Thaddeus John, organic chemistry
Malesky, Evan M., mathematics

POMPTON PLAINS

PRINCETON

Abeles, Benjamin, solid state physics
Adler, Stephen L., theoretical high energy physics
Ager, John Winfrid, Jr., chemistry
Akselrad, Aline, experimental solid state physics, magnetism
Alberts, Bruce M., molecular biology
Alge, Roger Casanova, solid state physics
Allard, William Kenneth, pure mathematics
Allen, Leland Cullen, chemistry
Almgren, Frederick Justin, Jr., mathematics
Alyea, Hubert Newcombe, physical chemistry
Amick, James Albert, solid state electronics, semiconductors
Anderson, Charles Hammond, solid state physics
Arendt, Volker Dietrich, polymer chemistry, rubber chemistry
Arunasalam, Vickramasingam, physics
Asato, Goro, organic chemistry
Ashley, Samuel Edward Quatrough, analytical chemistry
Axtmann, Robert Clark, chemical engineering
Bahcall, John Norris, astrophysics
Bahcall, Neta Assaf, astrophysics
Baird, Donald, vertebrate paleontology
Baltzer, Philip Keene, physics
Ban, Vladimir Sinisa, solid state science

Barach, Richard L., medicine
Bargmann, Valentine, mathematical physics
Barker, Richard Gordon, organic chemistry
Barr, William J., energy conversion
Barringer, Donald F. Jr., metabolism
Bartolini, Robert Alfred, electrooptics
Barton, Lucian Anthony, inorganic chemistry, organic chemistry
Bassett, Alton Herman, textile chemistry
Baum, Bruton Murry, organic chemistry, microbiology
Beaton, Albert E., statistics
Bennett, James Hallam, mathematics, biology
Bennett, Ralph Edgar, botany
Berger, Harold, parasitology
Berk, Bernard, chemistry
Berkelhammer, Gerald, organic chemistry
Berkowitz, Sidney, industrial organic chemistry
Bernabei, Stefano, plasma physics
Bernstein, Jack, medicinal chemistry
Bertin, Eugene P., analytical chemistry, inorganic chemistry
Blanc, Joseph, solid state science, physical chemistry
Blinn, Roger Lee, pesticide chemistry
Bloom, Allen, organic chemistry, photochemistry
Bloom, Stanley, electronics
Boczkowski, Ronald James, analytical chemistry
Bodamer, George Willoughby, chemistry
Bol, Kees, plasma physics
Bonini, William Emory, geology, geophysics
Bonner, John Tyler, developmental biology
Borel, Armand, mathematics
Borman, Aleck, biochemistry
Boyd, John Edward, agricultural biochemistry, drug metabolism
Bradway, Keith E., pulp & paper technology, organic chemistry
Bridger, Robert Frederick, organic chemistry
Briggs, George Roland, physics
Browder, William, topology
Brown, Lawrence Milton, petroleum chemistry
Brown, William Everett, biochemistry
Bruce, Victor Gardiner, biological rhythms
Bryan, Kirk, (Jr.), meteorology
Bulbenko, George Fedir, organic chemistry
Burnett, Bruce Burton, analytical chemistry, polymer physics
Butcher, Earl Orlo, biology
Calcote, Hartwell Forrest, physical chemistry
Carlson, David Emil, solid state physics
Carlson, Ronald H., industrial chemistry
Carnes, James Edward, solid state electronics
Carroll, Marcus Newman, Jr., physiology, pharmacology
Carver, Thomas Ripley, physics
Cecchi, Joseph Leonard, plasma physics, atomic physics
Cevasco, Albert Anthony, organic chemistry
Chang, Kern Ko Nan, electronic physics
Chen, Catherine S H., polymer chemistry
Cherry, William Henry, physics
Chiang, Yuen-Sheng, physical chemistry
Cho, Kon Ho, physical chemistry
Church, John Armistead, cellulose chemistry, physical chemistry
Clewe, Thomas Hailey, internal medicine, medical research
Cluff, Leighton Eggertsen, infectious diseases, allergy
Cody, George Dewey, solid state physics
Conley, Jack Michael, analytical chemistry
Coty, Vernon Frank, microbiology
Cox, Edward Charles, genetics
Crabb, David Wendell, anthropology
Crane, Roger L., mathematics
Crowell, Edward Hirsch, cell biology
Crowell, Samuel Alan, plasma physics
Cruickshank, P A., pesticide chemistry
Cruz, Mamerto Manahan, Jr., chemistry

Cullen, Daniel Edward, operations research
Cullen, Glenn Wherry, solid state chemistry
Cullison, David Arthur, medicinal chemistry
Cummins, Richard Williamson, organic chemistry, synthetic organic chemistry
Cushman, David Wayne, biochemical pharmacology
Dahlen, Francis Anthony, Jr., geophysics
Dalrymple, Ronald Howell, meat science
Danielson, Robert E., astrophysics
Dawson, Ray Fields, natural products chemistry
Deckert, Cheryl A., surface chemistry
Deffeyes, Kenneth Stover, oceanography
De Lay, Roger Lee, ruminant nutrition
De Wolf, David Alter, theoretical physics
Diassi, Patrick Andrew, organic chemistry
Dicke, Robert Henry, astrophysics
Dimock, Dirck L., physics
Donley, Hugh Lancelot, geology, paleobotany
Dorf, Erling, geology, paleobotany
Dowty, Eric, mineralogy
Dreeben, Arthur B., inorganic chemistry
Dreskin, Sanford A., physics
Dresner, Joseph, surface physics, solid state physics
Drfica, Karl, molecular biology
Dyson, Freeman John, mathematical physics, astrophysics
Eachus, Spencer William, chemistry
Easton, William Bigelow, computer science
Edelman, Franz, applied mathematics
Egan, Michael Eugene, behavioral science
Eggert, Robert Glenn, animal science
Ekstrom, Lincoln, physical chemistry
Elden, Richard Edward, industrial chemistry
Ellis, Robert Anderson, Jr., physics
Endler, John Arthur, evolutionary biology, population biology
Engel, Stanford Lowell, pharmacology
Eringen, Ahmed Cemal, applied mechanics
Eubank, Harold Porter, physics
Fagan, Ernest Brad, entomology
Fankhauser, Gerhard, developmental biology
Farrar, George Elbert, Jr., internal medicine
Faughnan, Brian Wilfred, solid state physics
Fefferman, Charles Louis, mathematical analysis
Felder, William, physical chemistry
Feldstein, Nathan, physical chemistry
Fels, Stephen Brook, dynamic meteorology, planetary atmospheres
Fernandez, James William, anthropology
Feyerherm, Marvin Paul, physics
Firman, Melvin Curtis, bacteriology
Fischer, Alfred George, geology
Fishburne, Edward Stokes, III, molecular physics, aeronautical engineering
Fishman, Morris, organic chemistry
Fitch, Val Logsdon, physics
Fletcher, Martin J., biochemistry
Fogleman, Ralph William, veterinary medicine
Fonger, William Hamilton, physics
Fontijn, Arthur, physical chemistry
Fornaess, John Erik, mathematical analysis
Fox, Leonard P., electrochemistry
Franko-Filipasic, Borivoj Richard Simon, organic chemistry
Frazier, William Robert, microbiology
Free, Charles Alfred, biochemistry
Fresco, Jacques Robert, biochemistry, molecular biology
Friedman, Louis David, chemistry
Frieman, Edward Allan, plasma physics
Frilette, Vincent Joseph, organic chemistry
Frost, David, embryology
Fukui, George Masaaki, microbiology
Furth, Harold Paul, physics
Gadebusch, Hans Henning, medical microbiology
Gale, George Osborne, microbiology, science administration
Galivan, Robert Milo, Jr., synthetic organic chemistry, organic polymer chemistry
Garvey, Gerald Thomas, nuclear physics
Gatterdam, Paul Esch, chemistry
Geertz, Hildred, cultural anthropology
Gelfand, Jack Jacob, chemical physics
Gelperin, Alan, neurophysiology, behavioral biology

Giarrusso, Frederick Frank, organic chemistry
Gilbert, Richard Lapham, Jr., chemistry
Gillham, John K., polymer science
Gilvarg, Charles, biochemistry
Godel, Kurt, mathematical logic, philosophy of science
Goedertier, Peter V., mathematics, physics
Goldberg, Morton Edward, pharmacology
Goldberger, Marvin Leonard, theoretical high energy physics
Goldhust, Marvin Bertram, immunology
Goldmacher, Joel E., organic chemistry
Goldstein, Bernard, surface physics
Goldstein, Theodore Philip, organic chemistry, geochemistry
Golin, Stuart, solid state physics, computer science
Gordon, Eric Michael, natural products chemistry, medicinal chemistry
Gorog, Istvan, electrical engineering, physics
Gottlieb, Melvin Burt, physics
Gould, James L., ethology
Greene, John M., plasma physics
Grisham, Larry Richard, plasma physics
Groel, John Trueman, medicine
Gross, David (Jonathan), theoretical high energy physics
Groth, Edward John, III, astrophysics
Grove, Donald Jones, physics
Guiducci, Mariano A., organic chemistry
Gunning, Robert Clifford, mathematics
Gurk, Herbert Morton, mathematics
Haines, Bernard A, Jr., pharmaceutical chemistry
Hall, Kimball Parker, organic chemistry
Hall, Richard Eugene, inorganic
Hammer, Jacob Meyer, solid state physics, quantum physics
Hanak, Joseph J., solid state science
Harding, Maurice James Charles, organic chemistry
Hargraves, Robert Bero, geology
Harris, Don Navarro, biochemistry, physiology
Haubrich, Dean Robert, pharmacology, neurochemistry
Hawryluk, Richard Janusz, plasma physics
Hayashi, Yoshikazu, dynamic meteorology
Hedberg, Hollis Dow, geology
Heiberger, Charles Adam, chemistry
Heilwell, Israel Joel, physics, biophysics
Heinemann, Heinz, physical chemistry
Hendel, Hans William, plasma physics
Hendrickson, Arlo Dennis, mathematical statistics
Hernqvist, Karl Gerhard, quantum electronics, lasers
Hill, Kenneth Lee, plant physiology, entomology
Hillier, James physics, research administration
Hinnov, Einar, physics
Hockings, Eric Francis, solid state
Hoffman, Henry Tice, Jr., analytical chemistry
Holland, Paul William, statistics
Hollister, Lincoln Steffens, geology, petrology
Holloway, John Leith, Jr, dynamic meteorology
Honig, Richard Edward, mass spectrometry
Hopfield, John Joseph, solid state physics, biophysics
Horn, Henry Stainken, population ecology, animal behavior
Iwasawa, Kenkichi, mathematics
Jacobs, William Paul, plant development
Jacobus, David Penman, medicine, radiobiology
Jaffe, Walter Joseph, astronomy
Janeff, Donka Grigorova, bacteriology, serology
Janeff, Jan Dimitroff, serology, bacteriology
Jass, Herman Earl, biochemistry
Javick, Richard Anthony, analytical chemistry

Jenkins, Edward Beynon, astrophysics
Jenney, Elizabeth Holden, pharmacology
Jensen, Douglas Andrew, elementary particle physics
Jobes, Forrest Crossett, Jr., physics
Johns, Don Herbert, analytical chemistry
Johnson, Frank Harris, molecular biology
Johnson, George Chrysler, chemistry
Johnson, John Lowell, plasma physics
Johnson, Larry Claud, physics
Johnson, William Pitner, veterinary medicine
Jones, Alan Richard, physical chemistry, chemical engineering
Jones, Maitland, Jr, organic chemistry
Judson, Sheldon, (Jr), geology
Kamath, Yashavanth Katapady, physical chemistry, polymer chemistry
Kantor, Sidney, parasitology, protozoology
Kaplan, Michael, radiation chemistry
Kapoor, Inder Prakash, metabolism, insect toxicology
Karpel, Richard Leslie, biochemistry
Kaskey, Gilbert, statistics
Kastrinos, William, Jr., biology, zoology
Kauffman, Ellwood, mathematics
Kauzmann, Walter (Joseph), physical chemistry, protein chemistry
Kemp, Gordon Arthur, microbiology, research administration
Kenney, John T, physical chemistry
Kern, Werner, solid state chemistry, inorganic chemistry
Kerr, George Thomson, surface chemistry
King, John Albert, pharmacology, research administration
Kinsman, David John James, geology, geochemistry
Kirschner, Marc Wallace, biochemistry, cell biology
Kiser, Jackson Sebree, medical microbiology
Kleinberg, William, physiology
Klopfenstein, Ralph Walter, applied mathematics
Kochen, Simon Bernard, mathematics
Kohn, Joseph John, mathematics
Komishi, Masakazu, biology
Koster, William Henry, synthetic organic chemistry
Kressel, Henry, solid state physics
Kruskal, Martin David, mathematical physics, applied mathematics
Kucherlapati, Raju Suryanarayana, human genetics
Kuehn, Christa Gisela, industrial chemistry
Kuhn, Harold William, mathematics
Kulsrud, Helene E, applied mathematics, system analysis
Kulsrud, Russell Marion, plasma physics
Kurihara, Yoshio, meteorology
Kurshan, Jerome, physics
Lachance, Paul Albert, nutrition, food science
Lang, William Harry, petroleum chemistry, fuel science
Langlands, Robert P., mathematics
Langridge, Robert, molecular biology
Larach, Simon, physical inorganic chemistry, solid state chemistry
Latourette, Harold Kenneth, organic chemistry
Laurie, Victor William, physical chemistry
Law, Harold Bell, physics
Lawrence, Philip Linwood, geophysics
Layton, Herbert Wallace, bacteriology
Leavitt, Richard Irwin, microbiology
Leech, John G, pulp chemistry, paper chemistry
Leibler, Richard Arthur, mathematics
Lemonick, Aaron, high energy physics
Leverenz, Humboldt Walter, solid state science
Levin, Edwin Roy, solid state science
Levine, Aaron William, organic chemistry
Levine, Arnold Jay, virology
Levy, Hiram, II, atmospheric chemistry, atomic physics
Liao, Hsiang Peng, organic chemistry
Lieb, Elliott Hershel, mathematical physics
Lies, Thomas Andrew, inorganic chemistry
Lipp, Steven Alan, inorganic chemistry
Lisk, Robert Douglas, physiology, endocrinology
Little, Robert, plasma physics
Liu, Wen Chih, biochemistry
Lo, Elizabeth Shen, organic chemistry
Lobunez, Walter, industrial chemistry
Los, Marinus, chemistry
Lovell, James Byron, entomology
Lozier, Gerald Scott, physical chemistry

Lutz, Albert William, agricultural chemistry
Lutz, Charles William, physical chemistry, inorganic chemistry
MacKellar, Donald Gordon, toxicology, organic chemistry
Magder, Jules, physical chemistry, inorganic chemistry
Magee, Richard Joseph, organic chemistry
Maglio, Vincent Joseph, paleobiology, vertebrate paleontology
Margolin, Solomon, pharmacology, endocrinology
Mark, Peter Herman, surface physics, materials science
Mason, John Frederick, petroleum geology
Matsuda, Yoshiyuki, plasma physics
Maxim, Leslie Daniel, operations research
May, Robert McCredie, ecology
McClure, Donald Stuart, physical chemistry
McDermott, Walsh, medicine
McDowell, Wilbur Benedict, organic chemistry, pharmaceutical chemistry
McGuinness, Aims Chamberlain, pediatrics
McKinstry, Doris Naomi, clinical pharmacology
McMahon, David Harold, analytical chemistry
McPherson, Charles Allen, organic chemistry
McRipley, Ronald James, medical microbiology
Meyerhofer, Dietrich, solid state physics
Miale, Joseph Nicolas, petroleum chemistry
Miller, Arthur, physical chemistry, inorganic chemistry
Miller, Bernard, polymer science, textiles
Miller, Charles Frederick, III, mathematics
Miller, Philip, organic chemistry, pesticide chemistry
Miller, William John, physical chemistry
Mills, Robert Gail, nuclear physics
Mills, William Harold, mathematics
Milnor, John Willard, mathematics
Miner, Robert Scott, Jr, organic chemistry
Miraglia, Gennaro J., medical microbiology, physiology
Mischler, Terrence Wynn, endocrinology, clinical pharmacology
Mislow, Kurt Martin, chemistry
Mitchell, Thomas Owen, organic chemistry, ceramics
Motley, Robert W, plasma physics
Mucenieks, Paul Raymond, physical chemistry, electrochemistry
Mueller, George L, veterinary medicine
Montagu, Ashley, physical anthropology, cultural anthropology
Montgomery, Deane, topology
Moore, Arnold Robert, solid state physics
Moore, John Coleman, mathematics
Morgan, William Jason, geophysics
Morse, Marston, mathematics
Morton, Donald Charles, astrophysics
Moss, Herbert Irwin, inorganic chemistry
Murthy, Vishnubhakta Shrinivas, pharmacology
Nager, Urs Felix, organic chemistry, biochemistry
Naro, Paul Anthony, organic chemistry, physical chemistry
Naumann, Robert Alexander, physical chemistry
Neidleman, Saul L, biochemistry
Nelson, Joseph Edward, mathematics
Newton, (William) Austin, developmental genetics
Nichols, Joseph, organic chemistry
Nickerson, Helen Kelsall, mathematics
North, Dwight Olcott, theoretical physics
O'Connor, Robert Barnard, Jr, physics
Offenhauer, Robert Dwight, organic chemistry
Okabayashi, Michio, plasma physics
Olsen, Gregory Hammond, solid state physics
Olson, David Harold, physical chemistry
Olson, Harry Ferdinand, electroacoustics
Ondetti, Miguel Angel, organic chemistry
O'Neal, Thomas Denny, plant physiology, agronomy
O'Neill, Gerard Kitchen, physics
Oort, Abraham H, dynamic meteorology
Orlanski, Isidoro, geophysics
Osmond, Humphry Fortescue, medicine, psychiatry
Ostriker, Jeremiah P, astrophysics

Palmere, Raymond M, agricultural biochemistry
Pan, Samuel Cheng, biochemistry
Pankavich, John Anthony, parasitology
Pankove, Jacques I, solid state physics
Paoloni, Frank John, plasma physics
Papaioannaou, Christos George, organic chemistry
Papakyriakopoulos, Christos Dimitriou, mathematics
Parker, William Lawrence, organic chemistry
Pensack, Joseph Michael, animal nutrition, animal physiology
Perkins, Harolyn King, physical chemistry
Petrillo, Edward William, synthetic organic chemistry
Pfeiffer, Carl Curt, pharmacology
Philander, Samuel George, physical oceanography
Phinney, Robert A, geophysics
Pickard, Porter Louis, Jr, organic chemistry, research administration
Pinch, Harry Louis, physical inorganic chemistry
Pinder, George Francis, groundwater hydrology
Piroue, Pierre Adrien, physics
Pluscec, Josip, organic chemistry
Pobiner, Harvey, analytical chemistry, physical chemistry
Podolak, Esther, mathematical analysis
Porkolab, Miklos, plasma physics
Post, Douglass Edmund, Jr, plasma physics
Post, Theodore B, endocrine physiology
Powell, Richard James, solid state physics
Prichard, Benjamin Arnold, Jr, plasma physics
Puar, Mohindar S, physical organic chemistry, analytical chemistry
Quarles, Richard Wingfield, polymer chemistry
Rabitz, Herschel Albert, physical chemistry
Radimer, Kenneth John, inorganic chemistry
Rau, Eric, industrial chemistry
Reagan, William Joseph, inorganic chemistry
Rebenfeld, Ludwig, organic chemistry
Rechester, Alexander, plasma physics
Redfield, David, solid state physics
Reich, Murray H, polymer chemistry
Reisner, John Henry, physics
Reynolds, George Thomas, biophysics, high energy physics
Rivkin, Israel, immunology
Roberts, Floyd Edward, Jr, organic chemistry
Robertson, James Allen, inorganic chemistry
Rose, Albert, physics
Rosenblum, Charles, physical chemistry
Rosenbluth, Marshall N, theoretical physics
Robinson, Bruce B, plasma physics
Rodewald, Paul Gerhard Jr, organic chemistry
Rogers, Charles Henry, ornithology
Rogers, David Elliott, internal medicine
Rogers, Eric Malcolm, physics
Rogerson, John Bernard, Jr, astrophysics
Rollmann, Louis Deane, inorganic chemistry
Rome, Herbert Ronald, mathematics
Romei, Samuel Hanan, biomedical engineering
Rouse, George Charles, medicinal chemistry, organic chemistry
Rovnyak, George Charles, medicinal chemistry, organic chemistry
Rowe, Carleton Norwood, physical chemistry
Royce, Barrie Saunders Hart, solid state physics
Rubin, Bernard, pharmacology
Rubin, Donald Bruce, statistics
Rutherford, Paul Harding, plasma physics
Ross, Bruce Brian, dynamic meteorology
Ross, Daniel Louis, organic chemistry
Ross, Donald Alexander, physics
Ross, Lawrence James, physical organic chemistry
Sabisky, Edward Stephen, solid state physics
Saffer, Alfred, physical chemistry
Sagal, Matthew Warren, physical chemistry
Saldick, Jerome, physical chemistry
Sams, Burnett Henry, III, mathematics
Sanders, Gary Hilton, experimental high energy physics
Sanders, Thomas Garrison, biochemistry, genetics
Schenkel, Robert H, parasitology, immunobiology

Schivell, John Francis, plasma physics
Schleyer, Paul Von Rague, physical organic chemistry
Schmidt, John Allen, plasma physics
Schnable, George Luther, inorganic chemistry
Schnatterly, Stephen E, microelectronics
Schnatterly, Stephen E, solid state physics
Schwartz, Jeffrey, organometallic chemistry
Schwartz, Jeffrey Lee, microbial biochemistry
Schwarzschild, Martin, astronomy, astrophysics
Schweizer, Albert Edward, inorganic chemistry
Scott, Eric James Young, chemistry
Seidl, Frederick Gabriel Paul, mathematical physics, nuclear physics
Selberg, Atle, mathematics
Semenik, Nick Sarden, organic chemistry, information science
Shanefield, Daniel J, physical chemistry
Sheehan, John Timothy, organic chemistry, scientific bibliography
Shenstone, Allen Goodrich, atomic spectroscopy
Sherr, Rubby, nuclear physics
Shidlovsky, Igal, chemistry
Shimura, Goro, mathematics
Shirk, Richard Jay, bacteriology, data processing
Shoemaker, Frank Crawford, experimental high energy physics
Shoemaker, John Daniel, Jr, chemistry
Shor, Aaron Louis, veterinary medicine, animal nutrition
Sichel, Enid Keil, physics
Silvestri, Anthony John, chemistry
Simkins, Karl LeRoy, Jr, animal science
Simmons, Glen Raymond, chemistry
Simon, Barry Martin, mathematical physics
Sinnis, James Constantine, plasma physics
Smolochowski, Roman, solid state physics, planetary physics
Smyth, Charles Phelps, physical chemistry
Sohler, Arthur, biochemistry, microbiology
Smagorinsky, Joseph, dynamic meteorology
Smith, Arthur John Stewart, experimental high energy physics
Smith, William Hayden, astronomy
Soper, Davison Eugene, theoretical high energy physics
Southgate, Peter David, physics
Spencer, Donald Clayton, mathematics
Spiro, Thomas George, chemistry
Spitzer, Lyman, Jr, astronomy, astrophysics
Staebler, David Lloyd, physics
Stange, Hugo, organic chemistry, optics, solid state physics
Sommers, Henry Stern, Jr, physics
Soos, Zoltan Geza, physical chemistry
Stein, Elias M, mathematics
Steinberg, Malcolm Saul, developmental biology, embryology
Stephenson, Robert Saul, mathematics
Sterzer, Fred, physics
Stewart, John Conyngham, geology
Stickle, Gene P, chemical engineering, biochemistry
Stigliani, William Michael, physical chemistry
Stix, Thomas Howard, plasma physics
Stoner, John Clark, veterinary medicine
Summerfield, Martin, physics
Sumner, Richard Lawrence, high energy physics
Sundeen, Joseph Edward, organic chemistry
Suppe, John Edward, geology, geophysics
Surko, Pamela Toni, elementary particle physics
Swartz, George Allan, physics
Sweet, Melvin Millard, number theory
Takahashi, Hironori, plasma physics
Tang, James Jub-Ling, applied mechanics, heat transfer
Tanner, Earl C, physics
Taylor, Edward Curtis, organic chemistry
Taylor, Theodore Brewster, applied physics
Taylor, Wendell Hertig, chemistry
Terborgh, John J, population biology, plant physiology
Thaler, Jon Jacob, experimental high energy physics
Thompson, Ida, paleobiology
Thurman, Melburn D, anthropology
Tietjen, James Joseph, physical chemistry
Toeplitz, Barbara Keeler, spectroscopy

Tompkins, Robert, developmental genetics
Treiman, Sam Bard, physics
Treves, Gino Robert, agricultural chemistry
Trotter, Hale Freeman, mathematics
Tucker, Albert William, mathematics
Turkevich, John, physical chemistry
Vandeputte, John, organic chemistry
Van Houten, Franklyn Bosworth, geology
van Raalte, John A, research administration
Varma, Ravi Kannadikovilakom, medicinal chemistry
Venuto, Paul B, petroleum chemistry
Verkhovsky, Boris Samuel, systems analysis, resource management
Violante, Michael Robert, inorganic chemistry, physical chemistry
Volpp, Gert Paul Justus, organic chemistry
Von Hippel, Frank, theoretical physics
Vossen, John Louis, physics
Wade, Peter Cawthorn, medicinal chemistry, organic chemistry
Wagner, Frank A, Jr, organic chemistry
Wang, Chih Chun, materials science, physical chemistry
Wang, Guang Tsan, veterinary medicine, parasitology
Watson, Geoffrey Stuart, mathematical statistics
Watt, William Russell, organic chemistry
Watts, Daniel Jay, chemistry, botany
Weakliem, Herbert Alfred, Jr, physical chemistry
Webb, Robert Carroll, high energy physics
Webster, William Merle, physics
Weil, Andre, mathematics
Weil, Nicholas A, mechanics
White, James Russell, physical chemistry
Weiler, Edward John, astronomy
Weimer, Katherine E, physics
Weimer, Paul Kessler, physics
Weisenborn, Frank L, organic chemistry
Wigner, Eugene Paul, mathematical physics
Weisz, Paul Burg, chemistry, chemical engineering
Welky, Irving, clinical pharmacology
Wheeler, John Archibald, theoretical physics
White, Milton Grandison, experimental high energy physics
White, Roscoe Beryl, high energy physics
Whitehurst, Darrell Duayne, organic chemistry
Whitney, Hassler, mathematics
Whitney, Wendell Keith, entomology, cosmology
Wickes, William Castles, cosmology
Wightman, Arthur Strong, mathematical physics
Wibur, Robert Daniel, animal nutrition, animal physiology
Wilkinson, David Todd, physics
Willard, Paul Edwin, organic chemistry
Williams, Albert Lloyd, organic chemistry, agriculture
Williams, Brown F, solid state physics
Williams, Dale Gordon, physical chemistry
Williams, Richard, physical chemistry
Williams, Robert Hackney, organic chemistry
Winkley, Donald Charles, inorganic chemistry
Witherell, Michael Stewart, elementary particle physics
Witman, George Bodo, III, cell biology
Wittke, James Pleister, optical physics
Wojnar, Robert John, immunobiology, biochemistry
Wojtowicz, Peter Joseph, theoretical physics
Woodbridge, Joseph Eliot, clinical chemistry, physical chemistry
Woodward, J Guy, physics
Wright, Christine Gerda, molecular biology, structural chemistry
Wronski, Christopher Roman, solid state physics
Wu, Chung Pao, solid state electronics
Yale, Harry Louis, organic chemistry
Yamada, Masaaki, plasma physics
Ycom, Perry Niel, inorganic chemistry
York, Donald Gilbert, astrophysics
Yost, John Franklin, organic chemistry
Zado, Franjo M, analytical chemistry
Zanzucchi, Peter John, analytical chemistry
Zee, Anthony, theoretical high energy physics
Zierler, Neal, mathematics
Zulalian, Jack, metabolism

Zworykin, Vladimir Kosma, physics

PRINCETON JUNCTION
Hudyn, Donald Edward, organic chemistry
Shallcross, Frank Van (Loon), physical chemistry

RAHWAY
Albers-Schonberg, Georg, organic chemistry
Andoe, Joseph David, physical organic chemistry, computer science
Andrews, Alice E (Mrs Theodore B Hunt), mathematics
Birnbaum, Jerome, microbiology, biochemistry
Boos, Richard Newton, analytical chemistry
Brink, Norman George, health sciences
Brooks, Jerry R, reproductive endocrinology
Brown, Horace Dean, medicinal chemistry
Burg, Richard William, microbial biochemistry
Campbell, William Cecil, zoology
Chemerda, John Martin, organic chemistry
Cox, James Lee, animal nutrition
Denny, George Hutcheson, organic chemistry
Devlin, Richard Gerald, Jr, zoology, immunology
Dorn, Conrad Peter, Jr, pharmaceutical chemistry
Dulaney, Eugene Lambert, microbiology
Durette, Philippe Lionel, organic chemistry
Gittler, Melvin Hyman, inorganic chemistry
Egerton, John Richard, parasitology
Fink, David Warren, analytical chemistry
Firestone, Raymond A, organic chemistry
Foster, Alvin Garfield, veterinary medicine, microbiology
Gal, George, organic chemistry
Gaudino, Mario, physiology, medicine
Gilbert, Jack Pittard, analytical chemistry
Gitterman, Charles Oscar, microbiology
Gould, David Huntington, organic chemistry
Graesslie, Otto Edward, chemotherapy
Grier, Nathaniel, medicinal chemistry
Guthrie, Roger Thackston, organic chemistry
Gwatkin, Ralph Buchanan Lloyd, reproductive biology, cell biology
Hannah, John, organic chemistry
Hendin, David, microbiology
Heneghan, Leo Francis, industrial chemistry
Hensens, Otto Derk, natural products chemistry, physical biochemistry
Hoffman, Roger Allen, organic chemistry
Hoffman, Theodore F, organic chemistry
Hoogsteen, Karst, physical chemistry
Huff, Jesse William, biochemistry
Humes, John Leroy, endocrinology
Inamine, Edward Seiyu, biochemistry
Jacob, Theodore August, organic chemistry
Jahn, Ernesto, animal science, ruminant nutrition
Jones, William Howry, organic chemistry
Joshua, Henry, organic chemistry, separation science
Kaczka, Edward Anthony, chemistry
Kaplan, Louis, microbiology
Karady, Sanor, organic chemistry
Katzen, Howard M, biochemistry
Kenney, James Franklin, polymer chemistry
Kropp, Helmut, medical research
Krukoff, Boris Alexander, economic botany
Kuehl, Frederick Albert, Jr, organic chemistry
Kuna, Samuel, pharmacology
Kuo, Chan-Hwa, organic chemistry
Lago, Barbara (Drake), genetics
Leaning, William Henry Dickens, research administration
Leone, Ida Alba, pollution biology
Lin, Tsau-Yen, organic chemistry
Love, George M, organic chemistry
MacConnell, John Griffith, natural products chemistry
Malkin, Martin F, research administration
Mandel, Lewis Richard, biochemistry, pharmacology
Marburg, Stephen, organic chemistry
Mayernik, John Joseph, microbiology

McCauley, James A, physical chemistry
McManus, Edward Clayton, parasitology
Mellin, Theodore Nelson, physiology, pharmacology
Mezey, Kalman C, medicine, pharmacology
Michael, Laurent, nutrition, veterinary medicine
Miller, Anna Kathrine, bacteriology
Miller, Brinton Marshall, microbiology, protozoology
Miller, Ian McKenzie, microbiology
Miller, Robert Frederick, animal nutrition
Miller, Thomas William, natural products chemistry
Morse, Lewis David, biochemistry
Muir, Larry Allen, animal nutrition & physiology
Mullen, Robert Terrence, physical chemistry
Mulvey, Dennis Michael, organic chemistry
Muniz, Raul A, poultry pathology
Mushett, Charles Wilbur, pathology
Nessel, Robert J, pharmacy, pharmaceutical chemistry
Oien, Helen Grossbeck, biochemistry
Ostlind, Dan A, parasitology, entomology
Page, Lynn Merton, microbiology
Patchet, Arthur Allan, organic chemistry
Perper, Robert J, immunopathology
Peretta, Armond Thomas, molecular spectroscopy, analytical chemistry
Phillips, Richard Fifield, organic chemistry
Pines, Seemon H, organic chemistry
Pingeon, Rene, medicine
Poe, Martin Turner, biophysics, physical biochemistry
Putter, Irving, natural products chemistry, organic chemistry
Radlick, Philip Chris, organic chemistry
Rasmusson, Gary Henry, medicinal chemistry
Rickes, Edward Lawrence, natural products chemistry
Rogers, Edward Franklin, organic chemistry
Rogers, Thomas Olin, microbiology
Rosegay, Avery, radiochemistry, organic chemistry
Rothrock, John William, biochemistry
Saperstein, David Dorn, physical chemistry
Saret, Lewis Hastings, organic chemistry, analytical chemistry
Schoenewaldt, Erwin Frederick, synthetic chemistry, research administration
Shen, Tsung Ying, organic chemistry
Siegmund, Otto Hanns, veterinary medicine
Simmons, Ivor Lawrence, physical chemistry
Singleton, Bert, analytical chemistry
Slater, Robert Lee, veterinary parasitology
Sletzinger, Meyer, organic chemistry
Smith, Gary E, animal nutrition
Smith, George Byron, physical chemistry
Stapley, Edward Onley, microbiology
Steelman, Sanford Lewis, clinical pharmacology
Steffens, James Jeffrey, bio-organic chemistry, drug metabolism
Stoudt, Thomas Henry, microbiology
Taub, David, organic chemistry
Thiele, Elizabeth Henriette, biochemistry
Tull, Roger James, organic chemistry
Vagelos, P Roy, biochemistry
Vanden Heuvel, William John Adrian, III, drug metabolism
Wagner, Arthur Franklin, organic chemistry
Walton, Edward, organic chemistry
Walton, Robert Bruce, microbiology
Wang, Ching Chung, biochemistry, parasitology
Washko, Floyd Victor, veterinary pathology
Wax, Richard Gerald, microbial genetics
Weinstock, Leonard M, organic chemistry
Wendler, Norman Lord, organic chemistry
Westlake, Harry Edward, Jr, organic chemistry
Williams, Myra Nicol, molecular biophysics
Windholz, Thomas Bela, organic chemistry
Wittick, James John, analytical chemistry
Wolf, Donald Edwin, organic chemistry
Wolf, Frank James, organic chemistry
Woodruff, Harold Boyd, microbiology

Zambito, Arthur Joseph, organic chemistry
Zimmerman, Morris, biochemistry
Zimmerman, Sheldon Bernard, microbiology

RAMSEY
Lorey, Frank William, paper chemistry
Maleny, Robert Timothy, chemistry

RANDOLPH
Neidhardt, Walter Jim, low temperature physics, quantum physics

RARITAN
Allen, George Otis, physiology
Baker, William J, immunohematology
Baughman, Dwight Joe, biophysics
Bishop, David C, immunology, immunochemistry
Chasin, Mark, biochemistry, enzymology
Cronk, Gary Arnold, pharmacology
Graham, Henry Alexander, Jr, immunohematology
Greenslade, Forrest C, pharmacology
Hahn, DeWon, endocrinology, reproductive physiology
Hajos, Zoltan George, organic chemistry
Hartop, William Lionel, Jr, research administration
Heilman, Richard Dean, pharmacology
Hirsch, Allen Frederick, medicinal chemistry
Karmas, George, organic chemistry
Killinger, Joanne Marie, biochemical pharmacology
Kupferberg, Alfred Ballen, microbiology
Kupperman, Herbert Spencer, endocrinology
Lehrer, Harris Irving, biochemistry, immunochemistry
Levine, Philip, immunology, cancer biochemistry
Marx, Joseph Vincent, clinical biochemistry
McGuire, John L, pharmacology
McKee, Albert Preston, bacteriology
McShefferty, John, pharmaceutical chemistry
Morigi, Eugene Mario Edmund, bacteriology, immunology
Neumann, Norbert Paul, biochemistry
Persico, Francis J, immunology
Pollack, William, immunology, immunochemistry
Reckel, Rudolph P, immunology, biochemistry
Reiss, Alice Margaret, immunology
Sawardeker, Jawahar Sazro, physical chemistry, analytical chemistry
Shaw, Eugene, microbiology, virology
Struck, Jacob, Jr, biochemistry
Teipel, John William, biochemistry
Wagstaff, Paul Arlen, microbiology
Weintraub, Howard Steven, biopharmaceutics
Wellerson, Ralph, Jr, microbiology
Zolton, Raymond Peter, hematology
Zuckerman, Leo, laboratory medicine

RED BANK
Tynes, Arthur Richard, optics

RIDGEFIELD
Cobb, R M Karapetoff, paper chemistry

RIDGEFIELD PARK
Schwing, Karl Josef, chemistry

RIDGEWOOD
Chertkoff, Marvin Joseph, pharmaceutical chemistry, physical pharmacy
Cooney, Robert Clair, organic chemistry
Erchak, Michael, Jr, polymer chemistry
Gunberg, Paul F, organic chemistry
Rigler, Neil Edward, environmental chemistry
Schwarz, John Samuel Paul, industrial organic chemistry
Williams, James Horace, organic chemistry
Yasso, Warren E, geology, geomorphology

RIVER EDGE
Polye, William Ronald, physics
Teeters, Wilber Otis, chemistry

RIVERTON
Cunningham, Charles Everett, plant genetics, agronomy
Johnson, Robert Shepard, numerical analysis
Miller, Robert Ernest, plant pathology, soil microbiology

ROBBINSVILLE
Buono, Frederick J, microbiology, biochemistry

ROCKAWAY
Schlossman, Mitchell Lloyd, cosmetic chemistry
Welt, Martin A, nucleonics

ROCKY HILL
Cacella, Arthur Ferreira, polymer chemistry
Peters, Timothy Victor, polymer chemistry

ROSELLE
Rosen, William Edward, organic chemistry

ROSELLE PARK
Levenstein, Irving, endocrinology, toxicology

RUTHERFORD
Bender, Max, physical chemistry, colloid chemistry
Bird, Harvey Harold, plasma physics, geophysics
Citron, Irvin Meyer, analytical chemistry, science education
De Korte, Aart, physical chemistry
Hundert, Murray Bernard, polymer chemistry, organic chemistry
Klosek, Richard C, microbiology
Kronenwett, Frederick Rudolph, bacteriology
Luedemann, Lois W, geochemistry, materials science
Picozzi, Anthony, clinical pharmacology, preventive dentistry
Pollack, Jerome Marvin, geology
Szebenyi, Emil, comparative anatomy, embryology
Trifan, Daniel Siegfried, physical chemistry, organic chemistry

SADDLE BROOK
Sheeran, Stanley Robert, organic chemistry

SADDLE RIVER
Caspers, Horst J, physical chemistry, analytical chemistry
Hilton, Clifford L, chemistry
Howell, Charles Frederick, medicinal chemistry
Jensen, Otto Gerhard, chemistry

SALEM
Eby, John Martin, organic chemistry

SAYREVILLE
Sullivan, William Francis, physical inorganic chemistry

SCOTCH PLAINS
Brown, George Lincoln, chemistry
Christensen, Burton Grant, bio-organic chemistry, medicinal chemistry
Forster, Eric Otto, physical organic chemistry
Hancock, James William, organic polymer chemistry
Hoffman, Dorothea Heyl, organic chemistry
Knudsen, George Andrew, Jr, physical organic chemistry
Merkle, F Henry, pharmaceutical chemistry
Ruyle, William Vance, organic chemistry, medicinal chemistry
Scott, George Vane, colloid chemistry
Shewmaker, James Edward, physical chemistry
Tornqvist, Erik Gustav Markus, polymer chemistry, biochemistry

Weppelman, Roger Michael, biochemistry, parasitology

SHIP BOTTOM
Stockton, John Richard, research management

SHORT HILLS
Baron, Frank A, organic chemistry
Bogaty, Herman, research administration, technology
Brown, John Rowland, Jr, research administration
Buhs, Rudolf Paul, drug metabolism
Cary, Boyd Balford, Jr, acoustics, fluid physics
Le Duc, J Adrien Maher, inorganic chemistry, physical chemistry
San Soucie, Robert Louis, mathematics, research administration
Schwartz, Harold, pharmacology, physiology
Wittenberg, Albert M, physical electronics

SHREWSBURY
Field, Norman J, physics

SKILLMAN
Brill, William Franklin, organic chemistry
Sinclair, John C, neurophysiology, biomedical engineering

SOMERSET
Beachem, Michael Thomas, organic chemistry
Bloch, Alfred, biochemistry
Dzierzanowski, Frank John, inorganic chemistry
Howe, Eugene Everett, pharmaceutical chemistry
Jung, John Andrew, Jr, organic chemistry
Milionis, Jerry Peter, organic chemistry
Skoultchi, Martin Milton, organic chemistry, polymer chemistry

SOMERVILLE
Abodeely, Robert Assad, microbiology
Bernady, Karel Francis, pharmaceutical chemistry
Buday, Paul Vincent, pharmacology
Casella, John Francis, polymer chemistry
Chao, Tsai Hsiang, organic chemistry
Cheniceck, Albert George, polymer chemistry
Cobb, Walter R, agriculture, science education
Considine, Richard George, virology, immunology
Craig, Peter Harry, pathology
Dondershine, Frank Haskin, microbiology
Dreisbach, Paul Franklin, organic chemistry
Fagan, Paul V, biochemistry, organic chemistry
Frascella, Daniel W, physiology, biochemistry
Fredericks, Robert Joseph, x-ray crystallography
Geshner, Robert Andrew, physics, mathematics
Gordon, Saul, analytical chemistry
Greene, Elias Louis, immunology, virology
Harris, Barton A, internal medicine, research administration
Hoffman, Allan Jordan, pharmaceutical chemistry, biochemistry
Jones, David Stevens, immunopathology, immunology
Jones, John Paul, Jr, medicine
Kanojia, Ramesh Maganlal, medicinal chemistry
Kereluk, Karl, microbiology, public health
Koesterer, Martin George, microbiology
Koh, Won Young, medical microbiology
Kronenthal, Richard Leonard, polymer chemistry
Lefar, Morton Saul, organic chemistry
Levy, Alan, polymer chemistry
Lewis, Charles Edward, organic chemistry
Melveger, Alvin Joseph, chemistry
Mohan, Arthur G, organic chemistry
Novick, William Joseph, Jr, pharmacology, biochemistry
Policoff, Leonard David, medicine
Ray, James Alton, veterinary pathology, veterinary microbiology

Reissmann, Thomas Lincoln, analytical chemistry
Romano, Salvatore James, analytical chemistry
Salthouse, Thomas N, histology, cytochemistry
Smith, Gail Bevington, veterinary microbiology, veterinary pharmacology
Thelin, Jack Horstmann, chemistry
Tsuzuki, Toshio, physical chemistry
Vieira, Ernest Charles, physical chemistry
Wasserman, David, polymer chemistry
Whitlock, Leigh Stuart, information science, computer systems
Woodward, David Lee, pharmacology
Zell, Howard Charles, organic chemistry

SOUTH AMBOY
Herrington, Kermit (Dale), chemistry
Sanderson, Benjamin S, physical chemistry
Slepetys, Richard Algimantas, physical chemistry, inorganic chemistry

SOUTH HACKENSACK
Leung, Albert Yuk-Sing, pharmacognosy, microbiology

SOUTH ORANGE
Ander, Paul, physical chemistry
Andrushkiw, Joseph Wasyl, mathematics
Auer, Edward Everett, polymer chemistry
Augustine, Robert Leo, organic chemistry
Bulas, Romuald, physical chemistry
Celiano, Alfred, chemical kinetics, physical inorganic chemistry
DeProspo, Nicholas Dominick, anatomy, academic administration
Doman, Elvira (Mrs John H Holder), biochemistry, enzymology
Ewing, Galen Wood, physical chemistry, analytical chemistry
Franke, Charles H, algebra
Freed, Ruth Shelley, history of anthropology, ethnography
Hirsch, Jerry Allan, organic chemistry
Hirsch, Roland Felix, analytical chemistry, environmental chemistry
Huchital, Daniel H, inorganic chemistry
Intemann, Gerald William, particle physics
Kana, Alfred Jan, applied statistics, analytical statistics
Katz, Frank Fred, parasitology
Keller, John Randall, plant pathology
Kim, Moon W, mathematical analysis
Kraft, Herbert Clemens, anthropology
Kramer, Stanley Zachary, neuropharmacology
Krause, Eliot, population genetics
Love, Linda J Cline, analytical chemistry
McGuinness, Eugene T, biochemistry
Orsi, Ernest Vinicio, cell biology, virology
Saccoman, John Joseph, mathematical analysis
Singer, Arnold Jack, bacteriology
Sonnessa, Anthony J, physical chemistry
Stamer, Peter Eric, physics
Sternberg, David, physics, oceanography

SOUTH PLAINFIELD
Anderson, Louis Weston, analytical chemistry
Goffman, Martin, electrochemistry, environmental chemistry
Hollander, Max Leo, chemistry
Kenkare, Divaker B, biological chemistry, physical chemistry
Kimmel, Elias, physical chemistry
Sattur, Theodore W, analytical chemistry
Youssef, Mary Naguib, statistics

SOUTH VINELAND
Inglessis, Criton George S, organic chemistry

SPARTA
Fabry, Andras, toxicology, veterinary science
Tyler, Stanley Warren, physiological chemistry

SPRINGFIELD
Birdsall, Henry Alfred, physical chemistry
Frost, Bettina Mary, microbiology
Feig, Gerald, organic chemistry
Leiff, Morris, organic chemistry

NEW JERSEY

STANTON
Goger, Pauline Rohn, internal medicine

SUCCASUNNA
Bluestein, Allen Channing, organic chemistry
Levi, David Winterton, polymer chemistry

SUMMIT
Adair, Dennis Wilton, biopharmaceutics
Adair, Frank William, microbiology
Anschel, Joachim, pharmaceutics
Ball, Ralph Henry, plastics chemistry, polymer chemistry
Berardinelli, Frank Michael, polymer chemistry
Bernard, Patrick Spitaletta, physiology, biochemistry
Besso, Michael M., organic chemistry
Bird, Joseph Gordon, clinical pharmacology
Bowman, Robert Mathews, synthetic organic chemistry
Braunstein, David Michael, organic pathology
Brofazi, Fred Robert, analytical chemistry, polymer chemistry
Brown, Alfred Edward, organic chemistry, pharmaceutical chemistry
Carney, Richard William James, organic chemistry
Ciotti, Paul Peter, magnetism
de Stevens, George, organic chemistry
Diener, Robert Max, toxicology
Edelman, Robert, organic chemistry, polymer chemistry
Finch, Neville, organic chemistry
Firschein, Hilliard E., biochemistry
Fitch, Howard Montgomery, chemistry
Fitchett, Gilmer Trower, organic chemistry
Gaunt, Robert, endocrinology
Goldberg, Albert Isaac, chemistry
Gray, Frederick William, organic chemistry
Gray, James Robert, agronomy
Groeger, Theodore Oskar, organic chemistry
Gschwend, Heinz W., organic chemistry, medicinal chemistry
Hannon, Martin J., polymer science
Hanson, Harry Thomas, polymer science, organic chemistry
Harris, Luke, clinical pharmacology
Hass, Henry Bohn, petroleum chemistry
Hay, Eleanor (Louise) Clarke, endocrinology
Hay, Ian Leslie, polymer physics, electron microscopy
Heymann, Hans, biochemistry
Huebner, Charles Ferdinand, pharmaceutical chemistry
Hughes, O Richard, industrial chemistry
Jaffe, Michael, polymer physics, physical chemistry
Jones, Granville Lillard, industrial psychiatry
Kayan, Sabih, medical research
Keith, Harvey Douglas, physics
Kitchen, Leland Joseph, chemistry
Knoppers, Antonie Theodoor, pharmacology
Konopka, Edward Alexander, bacteriology
Layng, Edwin Tower, chemistry
Leeson, Lewis Joseph, pharmacy
Luberoff, Benjamin Joseph, industrial chemistry
Lukas, George, organic chemistry
Maggio-Cavaliere, Mary, clinical pharmacology
Margenson, Richard Bennett, organic chemistry
Marra, Michael Dominick, biochemistry, bioanalysis
Marsh, John Lee, information science
May, Ernest Max, organic chemistry
McMahon, Paul E., physical chemistry
McMillan, Brockway, mathematics
Model, Frank Steven, chemistry, polymer science
Mohacsi, Erno, organic chemistry, physical chemistry
Naimark, George Modell, biochemistry
Nelson, John Archibald, chemistry
Plummer, Albert J., pharmacology
Poist, John Edward, organometallic chemistry
Polestak, Walter John S., physical chemistry

Powers, Edward James, organic chemistry, textile technology
Quinn, Gertrude Patricia, pharmacology
Rhum, David, polymer chemistry, organic chemistry
Richardson, Peter Charles, physical chemistry
Rinehart, Roy K., biochemistry
Roberts, Richard Hillman, medicine
Robson, Ronald D., pharmacology
Rosenthal, Arnold Joseph, physical chemistry
Saelens, Jeffrey K., pharmacology
Schwarz, Maurice Jacob, organic chemistry
Sprague, Basil Sheldon, textiles, physics
Stackman, Robert W., organic chemistry, polymer chemistry
Sullivan, Peter Kevin, polymer physics
Taylor, Paul Duane, chemistry
Thompson, Samuel Wesley, II, veterinary pathology, histochemistry
Tishler, Frederick, analytical chemistry
Traina, Vincent Michael, toxicology
Trombore, Forrest Allen, physical chemistry
Ulshafer, Paul R., organic chemistry
Vinick, Fredric James, synthetic organic chemistry
Vona, Joseph Albert, organic chemistry
Wagner, William Edward, Jr., clinical pharmacology
Wasserman, Edel, theoretical chemistry, organic chemistry
Werner, Lincoln Harvey, organic chemistry
Wilson, Donald Richard, organic chemistry, polymer chemistry
Winter, Roland Arthur Edwin, organic chemistry
Wissbrun, Kurt Falke, polymer science, rheology
Ziegler, John Benjamin, synthetic organic chemistry
Zitomer, Fred, analytical chemistry

TEANECK
Abend, Phillip Gary, organic chemistry
Arthur, Wallace, nuclear physics, space physics
Bein, Donald, mathematics
Benedict, Joseph T., inorganic chemistry, academic administration
Bieber, Samuel, developmental biology
Catanzaro, Edward John, geochemistry
Cooperman, Philip, mathematics
Davis, Morton David, mathematics
Ehrenstorfer, Sieglinde K M, inorganic chemistry, physical chemistry
Fram, Harvey, food science
Gershon, Sol D., chemistry
Gildart, Lee William, physics
Gimelli, Salvatore Paul, pharmaceutical chemistry
Goldstein, Byron Bernard, biophysics
Haase, Oswald, solid state physics
Hobart, Everett W., analytical chemistry
Isquith, Irwin R., protozoology, ecology
Kallmann, Silve, analytical chemistry
Lapidus, Arnold, mathematics
Lapin, David Marvin, biology
Liberles, Arno, organic chemistry
Meister, Charles William, plant pathology, plant physiology
Moeller, Karl Dieter, spectroscopy, optics
Naiman, Barnet, analytical chemistry
Rhodes, Rondell H., developmental biology, histology
Saxena, Anjali, neuroendocrinology, animal behavior
Senter, Harvey, mathematics
Shulman, George, industrial organic chemistry
Spagnoli, Harriet, zoology, morphology
Steiner, Gilbert, mathematics
Tavolga, Margaret Cordsen, endocrinology
Tomasetti, Vincent Paul, physics
Tracht, Myron Edward, pathology
Walsh, Peter, physics
Weinberger, Harold, organic chemistry
Wieder, Sol, physics
Winters, Harvey, microbiology, biochemistry
Wolsson, Kenneth, mathematics
Zemansky, Mark Waldo, physics

TENAFLY
Bradley, William Robinson, industrial hygiene, toxicology
Eastwood, Judith Janssen, ecology
Turro, Nicholas John, organic chemistry
Wolfe, Hugh Campbell, physics

THREE BRIDGES
Cruthers, Larry Randall, veterinary parasitology
Guerrero, Raul Jaime, veterinary medicine
Linkenheimer, Wayne Henry, physiology
Maplesden, Douglas Cecil, veterinary medicine
O'Connor, Jeremiah Joseph, animal nutrition
Reynolds, William Aden, veterinary medicine
Szanto, Joseph, veterinary parasitology

TITUSVILLE
Dermody, William Joseph, organic chemistry, science administration

TOMS RIVER
Adamo, Joseph Albert, nematology
Ballina, Rudolph August, organic chemistry
Deutsch, Dennis Leslie, industrial organic chemistry
Grosz, Oliver, chemistry
Harris, Charles, pathology
Lang, Philip Charles, organic chemistry
Lee, Do-Jae, organic chemistry
Lohmann, Karl H, organic chemistry
Mansfield, Kevin Thomas, organic chemistry
O'Neal, Grady Malcolm, organic chemistry
Stingl, Hans Alfred, industrial organic chemistry
Szap, Peter Charles, analytical chemistry
Tamburin, Henry John, industrial organic chemistry

TOTOWA
Brandau, Robert Paul, cosmetic chemistry, colloid chemistry
Strianse, Sabat John, pharmacy

TRENTON
Adler, Seymour Jacob, analytical chemistry
Aronovic, Sanford Maxwell, analytical chemistry
Barclay, Robert, Jr., polymer chemistry, industrial organic chemistry
Carlson, James H, genetics
Casper, Berenice Margaret, urban geography
Deems, Robert Eugene, plant pathology
Ferrigno, Thomas Howard, organic chemistry
Flato, Jud B., physical chemistry
Gobran, Riad Hilmy, polymer chemistry
Goldfield, Martin, public health, epidemiology
Goodin, Jerome, physical chemistry, electrochemistry
Goodkind, Morton Jay, internal medicine
Hanisch, Siegfried, mathematics
Hausdoerffer, William H., mathematics
House, Edward Holcombe, physical chemistry
Johnson, Clarence Albert, physical chemistry
Kang, Chia-Chen Chu, fuel science, petroleum science
Klug, William Stephen, biology
Kontrovitz, Mervin, micropaleontology, paleoecology
Kuehn, Harold Herman, mycology, biology
Lazarus, Allan Kenneth, organic chemistry
Lecher, David Wayne, meteorology
Mayer, Thomas C., developmental biology
Meisler, Harold, hydrogeology
Moses, Herbert A., atomic physics
Moury, Daniel Norman, biochemistry
Nechamkin, Howard, inorganic chemistry
Oddis, Leroy, endocrinology
Pfeiffer, Raymond John, astronomy
Prebluda, Harry Jacob, agricultural biochemistry, microbiology
Pregger, Fred Titus, physics, science education
Rothman, Milton A., plasma physics
Santaniello, Anthony Frank, physical chemistry, organic chemistry

Slezak, Frank Bier, organic chemistry
Smith, Eileen Patricia, physical organic chemistry
Srinhongse, Sunthorn, virology, tropical medicine
Star, Aura E., botany
Turk, Jessie Rose, geography of Latin America, political geography
Valentine, Frank Rossiter, organic chemistry
Vena, Joseph Augustus, cytology
Widmer, Kemble, geology
Zucker, Melvin Joseph, solid state physics

TUCKERTON
Greubel, Paul William, chemistry

UNION
Abeles, Francine, mathematics
Akerboom, Jack, food technology
Arnold, Frederic G., biology, human physiology
Bardell, David, virology
Bayan, Aris Paul, microbiology
Bouboulis, Constantine Joseph, organic chemistry
Cullmann, Ralph E., inorganic chemistry
Galperin, Irving, physical chemistry
Getzin, Paula Mayer, theoretical chemistry
Goodenough, Eugene Ross, food science
Gooding, Chester Martin (Briggs), organic chemistry
Grauman, Joseph Uri, high energy physics
Hayat, M A., biology
Hennings, George, environmental biology
Hlavacek, Robert John, food chemistry
Ilavsky, Jan, bacteriology
Irwin, John (Henry) Barrows, astronomy
Kelch, Walter L., astrophysics
Kirsch, Nathan Carl, pharmaceutical chemistry
Linden, Duane B., plant genetics, cell biology
Madison, Caroline Rabb, zoology, genetics
Malbrock, Jane C., mathematical analysis
Maul, Stephen Bailey, industrial microbiology
Metz, Robert, stratigraphy
Meyerson, Arthur Lee, marine geology, marine geochemistry
O'Connor, Joseph Michael, organic chemistry
Osborne, Frank Harold, microbiology
Rathman, Dorothy Marie, nutrition
Reid, Hay Bruce, Jr., plant physiology, botany
Rosenberg, Morton Murray, poultry husbandry
Rosenthal, Judith Wolder, endocrinology
Salisbury, Lynn, organic chemistry
Sexton, Edwin Leon, biochemistry, nutrition
Smith, Amelia Lillian, physiology
Smittle, Richard Baird, food microbiology
Virkar, Raghunath Atmaram, physiology; invertebrate zoology
Walking, Arthur Ernest, analytical chemistry, food chemistry
Wells, Phillip Richard, food science, dairy science
Zimmerman, R Erik, astrophysics

UNION BEACH
Easterday, Otho Dunreath, pharmacology, radiobiology
Evers, William John, organic chemistry
Fearns, Edward Cranshaw, chemistry
Freeman, Stanley Knoel, organic chemistry
Hall, John B., organic chemistry
Jacobs, Morton Howard, analytical chemistry
Katz, Ira, food science
Light, Kenneth Karl, organic chemistry
Morris, James Albert, enzymology
Mussinan, Cynthia June, analytical chemistry
Schreiber, William Lewis, organic chemistry
Seitz, Eugene W., microbial biochemistry
Steinbach, Leonard, organic chemistry
Taylor, William Irving, organic chemistry
Theimer, Ernst Theodore, chemistry
Vopicka, Edward, industrial organic chemistry
Walrad, John Pierce, analytical chemistry
Warren, Craig Bishop, physical organic chemistry

UPPER MONTCLAIR
Becker, Joseph F, chemistry, biochemistry
Chai, Winchung A, mathematical analysis, applied mathematics
Clifford, Paul Clement, statistics, quality control
Cribben, Larry Dean, plant ecology
Gallopo, Andrew Robert, organic chemistry,
Garik, Valdemar L, physical chemistry
Gawley, Irwin H, Jr, chemistry
Goldstein, Robert Martin, aquatic ecology
Hamilton, Charles Leroy, geology
Hoadley, Alfred Damon, plant pathology, biology
Kalmanson, Kenneth, mathematics
Klein, David Xavier, organic chemistry
Klein, Harriet Esther Manelis, anthropology
Koditschek, Leah K, microbial ecology
Koepp, Stephen John, zoology, histopathology
Kowalski, Ludwik, nuclear physics, nuclear chemistry
Kowalski, Stephen Wesley, inorganic chemistry, science education
Kuhnen, Sybil Marie, botany
Lehr, Hanns H, pharmaceutical chemistry
Lynde, Richard Arthur, inorganic chemistry
McCormick, Jon Michael, marine ecology
McGee, Edward Arthur, operations research
Meyer, Albert William, physical chemistry
Nussbaum, Alexander Leopold, organic chemistry, biochemistry
Oppel, Edwin Irving, economic geography
Pai, Anna Chao, genetics, embryology
Quintana, Bertha Beatrice, anthropology
Ramsdell, Robert Cole, geology
Rehn, John William Holman, entomology
Roberts-Marcus, Helen Miriam, biostatistics
Sawits, Marie Schlam, nutrition, physiology
Shubeck, Paul Peter, ecology, entomology
Sternbach, Leo H, medicinal chemistry
Sternberg, Rolf, urban geography, geography of Latin America
Stewart, Ruth Carol, mathematics
Stoddard, James H, mathematics
Thiruvathukal, John Varkey, geophysics, oceanography

VERONA
Sherman, Clarence Steiner, chemistry
Turer, Jack, organic chemistry, physical chemistry

VINCENTOWN
Burrows, Robert Beck, parasitology
Hall, Robert Henry, physiology, pharmacology

VINELAND
Hamby, Robert Jay, physiological ecology, marine biology

WARETOWN
Vaughn, Thomas Hunt, organic chemistry

WARREN
Barnes, Lucien, Jr, analytical chemistry
Boyce, William Martin, mathematics
Downing, George V, Jr, physical chemistry
Glamkowski, Edward Joseph, organic chemistry
Landon, Donald Omar, physics, chemistry
Rosen, Marvin, organic chemistry, polymer chemistry
Waggoner, William Charles, physiology, toxicology
Zukas, Danute, organic chemistry

WASHINGTON CROSSING
Bittle, James Long, veterinary microbiology, virology
Emery, Jerrell Bemis, bacteriology, virology
Glick, Phillip Ray, resource management, laboratory animal medicine
Hambly, Lavern R, veterinary medicine
McCurdy, Dennis, veterinary pharmacology

WATCHUNG
Barabas, Eugene S, polymer chemistry, organic chemistry
Gurry, Robert Wilton, physical chemistry
Jacobs, Sanford, inorganic chemistry
Langer, Arthur Walter, Jr, organic chemistry
Minckler, Leon Sherwood, Jr, organic chemistry
Sheinaus, Harold, pharmacy
Whitcomb, Gordon Putnam, organic chemistry

WAYNE
Bell, John Barr, Jr, organic chemistry
Berry, William Lee, organic chemistry
Boyle, Richard James, industrial organic chemistry
Burstyn, Harold Lewis, history of science, history of technology
Burwasser, Herman, physical chemistry
Clyne, Robert Martin, industrial medicine
DeAcetis, William, organic chemistry
DeBaun, Robert Matthew, information science, statistics
DiBenedetto, Frank Edward, animal physiology
Drake, Rosemarie Sheila, neurophysiology
Fatora, Frank Charles, Jr, organic chemistry
Fitzsimmons, James G, geography, science education
Flanders, Clifford Auten, analytical chemistry
Gamertsfelder, George Royce, optics
Gilman, Norman Washburn, organic chemistry
Graham, David E, organic chemistry
Greyson, Jerome C, organic chemistry
Grosser, Frederick, organic chemistry
Hall, Robert Lindsay, photobiology
Hen, John, surface chemistry
Hort, Eugene Victor, organic chemistry
Hutter, Edwin Christian, electrooptics
Isaacson, Allen, physiology, biophysics
Keilgren, John, organic chemistry
Koehler, Truman L, Jr, statistics
Levine, Donald Martin, zoology, parasitology
McSweeney, Ellsworth Edward, chemistry
Mino, Guido, polymer chemistry
Moskowitz, Mark Lewis, photographic chemistry
Paul, Albert P, organic chemistry
Poeschel, Gordon Paul, parasitology
Priesing, Charles Paul, organic chemistry
Quo, Sih-Gwan, organic chemistry, biochemistry
Rivela, Louis John, inorganic chemistry
Rosengren, John, freshwater biology
Salsbury, Jason Melvin, industrial organic chemistry
Scalera, Mario, organic chemistry
Scalfarotto, Robert Emil, petroleum chemistry, paper chemistry
Shaffer, Charles Boyd, toxicology
Shinn, Alvin Fleetwood, entomology, ecology
Slagg, Norman, physical chemistry, chemical kinetics
Smith, Terry Edward, polymer chemistry
Speert, Arnold, physical organic chemistry
Spivak, Monroe Leon, biology, conservation
Trice, William Henry, research administration
Van Order, Robert Bruce, organic chemistry
Voos, Jane Rhein, microbiology
Weisbrot, David R, population genetics
Werth, Jean Marie, microbial physiology
White, Robert Keller, psychophysiology, radiobiology
Withers, Edward Donald, analytical chemistry
Wolstenholme, William Ernest, mathematics
Wood, Irwin Boyden, parasitology

WEST CALDWELL
Afonso, Adriano, organic chemistry
Surmatis, Joseph D, organic chemistry

WEST END
Balton, Isidore Alfred, geophysics

WEST LONG BRANCH
Barnes, Derek A, plasma physics, fluid dynamics
Canavan, Robert I, mathematics
Dorfman, Donald, fresh water & marine biology
Garner, William Vaughn, entomology
Guilfoyle, Richard Howard, mathematics
Holt, Everett William, mathematics
Hunter, Edwin Thomas, nuclear physics
Kijewski, Louis Joseph, theoretical physics
Kuntz, Richard A, mathematics
Mack, James Patrick, cell biology
Richlin, Jack, physical chemistry
Rouse, Irwin Louis, organic chemistry
Shapiro, Irwin Louis, biochemistry
Smith, Robert Owens, experimental solid state physics, environmental physics
Spiegel, Leonard Emile, ecology

WEST NEW YORK
Prussak, Philip Morris, cosmetic chemistry

WEST ORANGE
Hershberg, Emmanuel Benjamin, organic chemistry
Jaffe, Sol Samson, electrochemistry
Konigsberg, Moses, organic chemistry
Mampe, Charles Douglass, entomology
Mason, Warren Perry, acoustics
Moody, Joseph E, Jr, pharmaceutical chemistry
Nobile, Arthur, microbiology
Reich, Leo, polymer chemistry
Scott, Peter John, textile chemistry
Sheppard, Herbert, biochemistry
Weinstein, David, cytogenetics
Zatzkis, Henry, theoretical physics, applied mathematics

WEST PATERSON
Harris, Murray T, physical chemistry

WEST TRENTON
Saferstein, Richard, forensic science

WESTFIELD
Addinall, Carl Rupert, organic chemistry
Bartok, William, physical chemistry, chemical engineering
Beach, Leland Kenneth, chemistry
Bollinger, Frederick William, organic chemistry
Bouchal, Alexander Wayne, cosmetic chemistry
Brown, Ralph Andres, analytical chemistry
Chamberlin, Earl Martin, organic chemistry
Chandler, Roger Eugene, organic chemistry
Cotty, Val Francis, pharmacology, toxicology
Cuckler, Ashton Clinton, animal parasitology
Curran, John Phineas, pediatrics, medical genetics
Curry, Michael Joseph, plastics chemistry
Cutler, Frank Allen, Jr, chemistry, research administration
Detweiler, William Kenneth, organic chemistry
Frolich, Per Keyser, petroleum chemistry, biochemistry
Grow, George Copernicus, Jr, geology
Hazen, George Gustave, organic chemistry
Kaeding, Warren William, organic chemistry
Kelley, Joseph Matthew, Jr, polymer chemistry
Luckmann, Frederick H, food technology, analytical chemistry
Lynn, John Wendell, organic chemistry
Maravetz, Lester L, agricultural chemistry
McGroarty, Joseph A, industrial chemistry
Mertel, Holly Edgar, organic chemistry
Mulinos, Michael George, clinical pharmacology
Ott, Walther Henry, poultry science, biometrics
Prince, Leon Maximilian, surface chemistry
Rehner, John, Jr, physical chemistry
Schwartz, Bertram, surface chemistry
Scott, Robert Wallace, physics
Shacklett, Comer Drake, physical organic chemistry
Slates, Harry Lovell, organic chemistry
Stein, Charles W C, organic chemistry
Sweet, Ronald Lancelot, organic chemistry

WESTWOOD
Carey, Benjamin Watson, pediatrics
Kulp, John Laurence, geochemistry
Martin, John David, nuclear physics, atmospheric physics
Roach, Kenneth Alphonsa, physics
Schutz, Donald Frank, geochemistry

WHIPPANY
Bennett, Robert Putnam, organic chemistry
Bogert, Bruce Plympton, physics
Bronzo, Joseph Alexander, applied mathematics, electrical engineering
Carter, Ashley, electrophysics, acoustics
Chasalow, Ivan G, operations research
D'Adamo, Anthony, pharmaceutical chemistry
Daul, George Cecil, organic chemistry
Davies, Richard Edgar, organic chemistry, plastics
Deem, Gary Spencer, fluid dynamics, applied mathematics
Easley, James W, physics
Eckler, Albert Ross, statistics, operations research
Epstein, Marvin Phelps, mathematics
Farkass, Imre, thermal physics, telecommunications
Francis, Samuel Hopkins, ionospheric physics
Fretwell, Lyman Jefferson, Jr, acoustics
Gianola, Umberto Ferdinando, physics
Hammer, Richard Benjamin, cellulose chemistry, organic chemistry
Hinkle, John Marion, physics
Holford, Richard L, applied mathematics
Holt, Vernon Emerson, heat transfer, environmental control
Jagerman, David Lewis, mathematics, electrical engineering
Kukin, Ira, environmental chemistry
Labianca, Frank Michael, underwater acoustics
Looney, Duncan Hutchings, physics
McCumber, Dean Everett, physics
Mitchell, John Peter, physics
Nichols, Rudolph Henry, Jr, acoustics
Philips, Thomas O, theoretical physics, acoustics
Portnoy, Norman Abbye, organic chemistry, cellulose chemistry
Riordan, William J, mathematics
Ritger, Paul David, applied mathematics
Robbins, Naomi Bograd, statistics
Rothleder, Stephen David, underwater acoustics
Scales, William Webb, underwater acoustics
Svokos, Steve George, biological chemistry
Thelman, John Patrick, cellulose chemistry
Turbak, Albin Frank, organic chemistry
Walker, Richard Alden, underwater acoustics, ocean engineering
Wintermitz, Thomas W, electromagnetics, communications engineering
Worley, Robert Dunkle, physics
Zabor, John William, physical chemistry
Zipfel, George G, Jr, acoustics

WILLINGBORO
Glickman, Samuel Arthur, organic chemistry
Owens, Frederick Hammann, organic chemistry

WINDSOR
Capwell, Robert J, physical chemistry

WINSLOW
Martin, George Lloyd, inorganic chemistry
Willson, Donald Bruce, physical inorganic chemistry, extractive metallurgy

WOODBRIDGE
Clark, Robert Long, medicinal chemistry

WOODBURY
Badertscher, Darwin Earl, organic chemistry; information science
Dahlstrom, Bertil Philip, Jr., physical chemistry
Feasley, Charles Frederick, environmental health
Lufkin, James E., organic chemistry

WOODCLIFF LAKE
Hall, Luther Axtell Richard, science writing, organic chemistry
Keller, Dolores Elaine, reproductive physiology, microbiology
Martin, Eric Wentworth, medical communication
Wright, William Blythe, Jr., pharmaceutical chemistry

WOOD RIDGE
Mathe, Clarence Eugene, Jr., chemistry

WOODSTOWN
Gross, Peter Fredrick, organic chemistry
Klinke, David J., organic chemistry

WYCKOFF
Kroll, Emanuel, biochemistry
Louis, Kwok Toy, textile chemistry

NEW MEXICO

ALAMOGORDO
Jones, Mark Wallon, physics

ALBUQUERQUE
Aberle, Sophie Bledsoe, nutrition
Acosta, Phyllis Brown, foods, nutrition
Ahluwalia, Harjit Singh, cosmic ray physics, high energy astrophysics
Alden, Earl F., forestry
Allen, Richard Crenshaw, Jr., applied mathematics
Alpert, Seymour Samuel, physics
Amos, Donald E., mathematics, chemical engineering
Anderson, David Hamel, physical chemistry, chemical physics
Anderson, Robert Alan, experimental physics
Anderson, Robert Edwin, pathology
Anderson, Roger Yates, geology
Appenzeller, Otto, neurology, medicine
Arnold, Charles, Jr., polymer chemistry
Arnold, George W., solid state physics
Assink, Roger Alyn, polymer chemistry
Atencio, Alonzo C., biochemistry
Atterbom, Hemming A., physiology
Auerbach, Irving, physical chemistry, organic chemistry
Babb, David Daniel, electromagnetism
Bailey, Paul Bernard, mathematics
Baker, Thomas Irving, microbiology, biochemistry
Banister, John Robert, physics
Barnaby, Bruce E., physics, electrical engineering
Barr, George E., applied mathematics
Barrett, Richard Allan, anthropology, ethnology
Bartel, Lewis Clark, theoretical solid state physics, geophysics
Baschart, Harry Wetherald, anthropology, ethnology
Beattie, Alan Gilbert, solid state physics
Beckel, Charles Leroy, theoretical physics
Beckerman, Stephen Joel, anthropology
Beckner, Everet Hess, plasma physics
Bell, Stoughton, mathematics, computer science
Bennett, Iven, climatology; geography
Bergeron, Kenneth Donald, plasma science
Berry, Louis Milton, chemistry, statistical mechanics
Bice, David Earl, immunology
Biggs, Frank, mathematical physics
Binford, Lewis R., anthropology, archaeology
Bingham, Felton Wells, experimental atomic physics
Black, Bruce Allen, geology
Boade, Rodney Russett, physics, acoustics
Bock, Philip Karl, anthropology
Becker, Bruce Bernard, radiobiology
Borders, James Alan, solid state physics

Bourne, Earl Whitfield, histology, cytology
Bowers, Klaus Dieter, physics
Bradford, William Henry, applied mathematics
Brannen, Joseph P., mathematics
Brannon, Paul J., spectroscopy
Breiland, John Gustavson, meteorology
Bruce, David Kenneth, solid state physics
Brogdon, Byron Gilliam, medicine
Brookins, Douglas Gridley, geochemistry
Brooks, Antone L., cytogenetics
Brower, Keith Lamar, physics
Broyles, Carter D., physics, engineering science
Bryant, Howard Carnes, physics
Burford, Thomas Maynard, mathematics
Butler, Michael Alfred, solid state physics
Buyers, Archie Girard, environmental chemistry
Cabrera, Edelberto Jose, immunology; malariology
Calkins, Myron Eugene, analytical chemistry
Callender, Jonathan Ferris, geology, tectonics
Calvert, Ralph Lowell, mathematics
Campbell, Howard, wildlife biology, herpetology
Campbell, John Martin, anthropology
Campbell, Robert Dale, geography
Cano, Gilbert Lucero, atomic physics
Carasso, Alfred Samuel, mathematics
Carlson, Gary Alden, physical chemistry
Carpenter, Robert Leland, biophysics
Caton, Roy Dudley, Jr., electroanalytical chemistry
Chabai, Albert John, physics
Chandler, Colston, theoretical physics
Chick, Thomas Wesley, pulmonary physiology
Cipriano, Leonard Francis, physiology
Claassen, Richard Strong, solid state physics
Clark, Sophie Paul, analytical chemistry; physical chemistry
Clauser, Milton John, plasma physics
Coats, Richard Lee, nuclear physics, reactor physics
Cochran, Paul Terry, cardiology; medical education
Coleman, Curtis Burger, environmental chemistry
Coleman, William Fletcher, physical inorganic chemistry, molecular spectroscopy
Conant, Roger, herpetology
Cords, Carl Ernest, Jr., virology; medical physics
Curro, John Gillette, polymer physics
Cowan, Maynard, Jr., microbiology
Cramer, James D., nuclear physics
Crawford, Clifford Smeed, invertebrate ecology, invertebrate physiology
Creps, Elaine Sue, neuroanatomy, neuroendocrinology
Cuderman, Jerry Ferdinand, experimental atomic physics
Cuthrell, Robert Eugene, surface chemistry; surface physics
Damerow, Richard Aasen, physics
Damon, Edward George, environmental biology, pathobiology
Daub, Guido Herman, organic chemistry
Davis, Herbert Thaddeus, III, statistics, mathematics
Davis, James Avery, mathematical statistics
Davis, Jeffrey Robert, mathematics
Davison, Lee Walker, applied mechanics
Dawes, William Redin, Jr., high energy physics, solid state physics
Dean, Christopher, physics
De Boer, Jelle, physiology; radiobiology
Degenhardt, William George, vertebrate zoology
Divett, Robert Thomas, information science, history of medicine
DeMarr, Ralph Elgin, mathematics
Demski, Leo Stanley, neuroanatomy, animal behavior
Depatie, David A., physics
Dieter, Scott Edward, anatomy; pathology
Dillon, Richard Thomas, radiation genetics, systems analysis
Dittmer, Howard James, botany, morphology

Duszynski, Donald Walter, parasitology, ecology
Eagle, Donald Frohlichstein, medical physics
Edwards, Leon Roger, solid state physics
Edwards, W Sterling, surgery
Elston, Wolfgang Eugene, volcanology; economic geology
Elwell, Albert R., physics
Emin, David, solid state physics
Entringer, Roger Charles, number theory
Epstein, Bernard, mathematics
Ewing, Rodney Charles, mineralogy, geochemistry
Ewing, Ronald Ira, physics
Feibelman, Peter Julian, theoretical solid state physics
Findley, James Smith, vertebrate zoology
Finley, James Dale, Jr., internal medicine
Fitzgerrel, William Wright, physiology
Fitzsimmons, John Paul, high energy physics
Flanagan, Robert Joseph, system analysis
Fletcher, Edward Royce, biodynamics
Fox, Ronald Lee, fluid mechanics, atomic physics
Franzak, Edmund George, theoretical physics
Freyer, Gustav John, space physics
Galey, William Raleigh, physiology, biophysics
Galt, John Kirtland, solid state science
Gardner, Kenneth Drake, Jr., internal medicine
Gauster, Wilhelm Belrupt, solid state physics
George, Raymond S., nuclear chemistry
Gerardo, James Bernard, lasers, atomic physics
Griego, Richard Jerome, mathematics
Grissom, John Thomas, atomic physics, microelectronics
Guinn, Theodore, mathematics
Gurbaxani, Shyam Hassomal, solid state electronics
Haaland, David Michael, physical chemistry
Hackett, Colin Edwin, fluid physics
Hackett, Nora Reed, cell physiology
Hadley, George Ronald, plasma physics
Hahn, Liang-Shin, mathematics
Halpin, Yu Hak, quantum optics
Halpin, Walter J., physics
Hansmann, Eugene William, aquatic ecology; algology
Hargis, Philip Joseph, Jr., quantum optics
Harpending, Henry Cosad, biological anthropology
Heckman, Richard Cooper, physics
Heinberg, Milton, physics, operations research
Henderson, Rogene Faulkner, biochemistry
Henderson, Thomas Richard, biochemistry
Hendrickson, Morris S., mathematics
Hendryson, Irvin Edward, orthopedic surgery
Hentel, William, medicine, pathology
Hereford, Joseph Pierce, geography; geology
Hersh, Reuben, mathematical analysis
Hessel, Kenneth Ray, optics
Hibben, Frank Cummings, anthropology, archaeology
Hickey, Wayne C., Jr., range management, animal husbandry
Hicks, Darrell Lee, applied mathematics
Hill, Ronald Ames, physics
Hillman, Abraham P., mathematics
Hirsch, Frederic Geake, neurosciences
Hoeft, Lothar Otto, applied physics
Hoff, Clarence Clayton, systematic zoology
Hollstein, Ulrich, organic chemistry
Houston, Jack E., surface physics
Howard, Robert Eugene, pathology, biochemistry
Howarth, John Lee, medical physics, biophysics

Hudson, Frank Peter, atmospheric chemistry, air pollution
Hughes, Robert Clark, chemical physics
Hulme, Bernie Lee, applied mathematics
Hyder, Charles Latif, environmental physics, solar physics
Jacobsen, Lynn C., geology
Jennings, Charles Warren, electrochemistry
Jennings, Daniel Thomas, forest entomology; arachnology
Johnson, Daniel Everett, medical microbiology
Johnson, Gordon Verle, plant physiology
Johnson, Ralph T., Jr., physics
Johnson, William Wayne, genetics
Jones, Eric Daniel, solid state physics
Jones, Orval Elmer, applied mechanics
Jones, Thomas Evan, analytical chemistry; inorganic chemistry
Jones, Vernon Douglas, pharmacology
Jordan, Scott Wilson, pathology
Kastella, Kenneth George, physiology
Kahn, Milton, chemistry
Keil, Klaus, meteorites, petrology
Keller, Donald V., experimental physics
Kelley, Robert Otis, biology
Kelley, Vincent Cooper, geology
Kenna, Bernard Thomas, nuclear chemistry; analytical chemistry
Kennedy, Jerry Dean, physics
Kennedy, Lynn Wean, data processing
Kepler, Raymond Glen, experimental solid state physics
Key, Charles R., pathology
Kidd, David Eugene, phycology
King, David Solomon, astrophysics
King, James Claude, low temperature physics, solid state physics
Kjeldgaard, Edwin Andreas, applied physics
Kligerman, Morton M., radiotherapy
Knotek, Michael Louis, solid state physics
Koopmans, Lambert Herman, mathematical statistics
Kornfeld, Mario O., neuropathology
Kraus, Alfred Andrew, Jr., physics
Kudo, Albert Masakiyo, petrology, volcanology
Kugel, Robert Benjamin, pediatrics, high temperature chemistry
Ladman, Aaron Julius, anatomy
Langley, Robert Archie, solid state physics
Law, David H., internal medicine, gastroenterology
Lawson, Katheryn Emanuel, microscopy
Lay, John Charles, veterinary medicine
Leach, John Kline, cardiology; physiology
Leavitt, Christopher Pratt, physics
LeBaron, Francis Newton, biochemistry
Lehrman, George Philip, pharmacy administration
Levine, Herman Saul, physical chemistry, high temperature chemistry
Levy, Samuel C., electrochemistry, electroanalytical chemistry
Lewis, James Vernon, mathematics
Lieberman, Morton Leonard, physical chemistry
Ligon, James David, zoology
Lincoln, Richard Criddle, applied physics
Ling, Donald Percy, mathematics
Lister, Robert Hill, cultural anthropology
Lockwood, Grant John, experimental atomic physics, electron physics
Loftfield, Robert Berner, organic chemistry; biochemistry
Lubash, Glenn David, internal medicine, nephrology
Luft, Ulrich Cameron, human physiology
Lundgren, David Lee, microbiology
Lustgarten, Catherine Sue, veterinary medicine
Lynch, Richard Wallace, chemical physics, chemical engineering
Lynn, Peter C., applied physics
MacCallum, Crawford John, radiation physics
Manz, Bruno Julius, theoretical physics, mathematics
Martin, William Clarence, plant taxonomy
Mason, William Van Horn, aerospace medicine
Mattox, Donald Moss, physics, materials science
Mauderly, Joe L., pulmonary physiology
Mauro, Jack Anthony, physiological optics
McClellan, Roger Orville, inhalation toxicology; veterinary toxicology
McClure, Gordon Wallace, physics

McGuire, Eugene J, atomic physics
McLaren, Leroy Clarence, microbiology, virology
McLaughlin, Donald Reed, chemical physics
Merewether, David Evan, electromagnetics
Merritt, Melvin Leroy, geophysics
Metzler, Richard Clyde, pure mathematics
Miller, Glenn Houston, physics
Mitchell, Merle, mathematics
Modreski, Peter John, geochemistry
Mogford, James A., physics
Mokler, Brian Victor, inhalation toxicology, air pollution
Moler, Cleve B., mathematics
Molles, Manuel Carl, Jr., ecology, ichthyology
Morosin, Bruno, physical chemistry
Morris, Glen Jeffs, plasma physics, electrooptics
Morrison, Donald Ross, mathematics, computer science
Moseley, Robert David, Jr, radiology
Mourant, Walter Arthur, veterinary physiology
Muggenburg, Bruce Al, veterinary physiology
Murphy, Richard Ernest, physical geography
Myers, Samuel Maxwell, Jr, solid state physics
Napolitano, Leonard Michael, anatomy
Narath, Albert, solid state physics
Neal, James Thomas, geology
Nelson, Gerald Clifford, nuclear physics
Nevison, Thomas Oliver, Jr., physiology, aerospace medicine
Newman, Stanley Stewart, anthropology
Nordstrom, Terry Victor, solid state electronics
Northrop, David A, geochemistry
Northrup, Clyde John Marshall, Jr, thermodynamics, geochemistry
Norwood, Frederick Reyes, solid mechanics, applied mathematics
Oliver, Carl Edward, mathematics
Omer, George Elbert, Jr, orthopedic surgery
Owen, George Murdock, pediatrics, nutrition
Paine, Robert Treat, Jr, inorganic chemistry
Palmer, Darwin L, medicine, infectious diseases
Palmer, Eugene Charles, neurochemistry, pharmacology
Palmer, Richard Everett, physics
Papadopoulos, Eleftherios Paul, organic chemistry
Park, Su-Moon, electroanalytical chemistry, photochemistry
Pastuszyn, Andrzej, biochemistry
Peek, James Mack, theoretical chemistry
Peercy, Paul S, solid state physics
Pena, Hugo Gabriel, radiobiology, nuclear medicine
Perkins, Walter George, physical chemistry
Perret, William Riker, geophysics
Peterson, Alan W, physics, astronomy
Peterson, Donald Palmer, mathematics
Pettit, Richard Bolton, solid state physics
Pfleger, Raymond C, biochemistry
Pickrell, John A, biochemistry
Pierson, Hugh Ottho, chemical metallurgy, high temperature chemistry
Pollay, Michael, neurosurgery, neurophysiology
Potter, Loren David, botany
Powers, Dana Auburn, inorganic chemistry, high temperature chemistry
Powis, Raymond Leslie, physiology, biomedical engineering
Prinz, Martin, petrology, mineralogy
Priola, Donald Victor, physiology, pharmacology
Proper, Robert, internal medicine
Pruess, Steven Arthur, numerical analysis
Qualls, Clifford Ray, mathematical statistics
Quinn, Rod King, physical inorganic chemistry
Ratner, Albert, endocrinology, physiology
Reed, Jack Wilson, meteorology
Regener, Victor H., physics
Renken, James Howard, mathematical physics
Reyes, Philip, biochemistry
Rhodes, John Marshall, neuropsychology
Rice, James Kinsey, chemical physics, lasers
Rich, John Charles, astrophysics, lasers
Richards, Peter Michael, magnetism
Riedesel, Marvin LeRoy, physiology
Rigsby, Bruce Joseph, anthropology
Roberts, Irwin Herbert, parasitology

Robertson, Merton M, experimental physics
Rypka, Eugene Weston, bacteriology
Saland, Linda C, anatomy, cytology
Samara, George Albert, physics, chemical engineering
Sasmor, Daniel Joseph, physical chemistry, materials science
Sattler, Allan R, atomic physics, nuclear physics
Scaletti, Joseph Victor, bacteriology
Scallen, Terence, biochemistry
Schaefer, Dale Wesley, chemical physics
Scharn, Herman Otto Friedrich, mathematics, physics
Schirber, James E, solid state physics
Schoenfeld, Robert George, biochemistry
Schulman, John, Jr., internal medicine
Schultze, Max Otto, biochemistry
Schwerin, Karl H, anthropology
Schwoebel, Richard Lynn, physics
Scott, Hugh Logan, III, experimental nuclear physics
Scott, Norman Jackson, Jr, herpetology, vertebrate ecology
Seager, Carleton Hoover, physics
Searle, John Randolph, biomedical engineering
Shafi, Mohammad, information science
Silverman, Paul Hyman, parasitology, immunology
Simmons, Gustavus James, combinatorial mathematics
Slater, Peter John, mathematics
Snead, Rodman E, geography
Snipes, Morris Burton, radiation biology
Solomon, Sidney, physiology
Somntag, Arch Christian, pharmacology, clinical chemistry
Sopher, Roger Louis, pathology
Southward, Harold Dean, nuclear physics, solid state
Sparks, Morgan, chemistry
Spencer, William J, solid state physics
Springfield, Harry Wayne, range ecology
Spuhler, James Norman, anthropology, biology
Standefer, Jimmy Clayton, biochemistry, clinical chemistry
Steinberg, Stanly, applied mathematics
Stevens, Alfred Lyman, energy physics
Stewart, Joseph Letie, audiology, speech pathology
Stone, Alexander Paul, mathematics
Stuart, David Edward, anthropology, research administration
Stubbs, Morris Frank, chemistry
Stuetzer, Otmar Michael, physics
Sun, James Ming-Shan, mineralogy, geophysics
Sweeney, Mary Ann, plasma physics, astrophysics
Swinson, Derek Bertram, physics
Switendick, Alfred Carl, solid state physics
Tapp, Charles Millard, physics
Tapscott, Robert Edwin, chemistry
Taylor, G Jeffrey, geochemistry, petrology
Taylor, Warren Egbert, metrology
Thacher, Philip Duryea, solid state physics
Theis, Charles Vernon, geology
Thomas, Roy, physics
Thompson, James James, mathematics
Thompson, Samuel Lee, theoretical physics
Tisone, Gary C, atmospheric physics
Toepfer, Alan James, physics
Tokuda, Sei, immunology, microbiology
Tollefsrud, Philip Bjorn, experimental nuclear physics
Trauth, Charles Arthur, Jr, mathematics
Truby, Frank Keeler, chemical physics
Trujillo, Ralph Eusebio, biochemistry
Ulrich, John August, microbiology, bacteriology
Vander Jagt, David Lee, biochemistry
Van Domelen, Bruce Harold, physics
Van Epps, Dennis Eugene, immunology
Velez, William Yslas, mathematics
Vittitoe, Charles Norman, electrodynamics
Vook, Frederick Ludwig, physics
Vorher, Helmult Wilhelm, obstetrics & gynecology
Walker, Arthur Earl, medicine
Wall, Francis Joseph, biostatistics, applied statistics
Walters, Edward Albert, physical organic chemistry
Warren, William Ernest, applied mathematics
Wayland, James Robert, Jr, astrophysics, agricultural physics
Weart, Wendell D, geophysics

Weaver, Harry Talmadge, solid state physics
Wenger, Sherman Alexander, geology
Werkema, George Jan, physical chemistry, health physics
Weston, James T., medicine, pathology
Whan, Ruth Elaine, physical chemistry
Wild, Gaynor (Clarke), biochemistry
Williams, David C, physics
Williams, David Cary, nuclear chemistry, nuclear physics
Williams, Ralph C, Jr, internal medicine, immunology
Wilson, Grant Ivins, animal parasitology
Wimer, Bruce Meade, internal medicine, hematology
Wise, David Haynes, population ecology
Wolfe, David M, high energy physics
Wood, Stephen Craig, physiology
Woodfin, Beulah Marie, biochemistry
Woodward, Lee Albert, structural geology
Workman, Peter L, genetics, anthropology
Wright, Thomas Payne, physics, plasma physics
Yeh, Hsu-Chi, environmental sciences, mechanical engineering
Yuhas, John M., radiobiology, cancer
Yu-Sun, Clara Chuan Chang, mycology, genetics

ALCALDE
Trujillo, Philip M, agronomy

ARTESIA
Barnes, Carl Eldon, agronomy, field crops

CAMPUS STATION
Reiter, Marshall Allan, geophysics

CARLSBAD
Glassbrook, Clarence I, physical chemistry
Underkofler, Leland Alfred, biophysical chemistry

CHAMA
Cataline, Elmon Lamont, pharmacy

CLOUDCROFT
Levy, Robert Aaron, solid state physics

CLOVIS
Finkner, Ralph Eugene, plant breeding, agronomy
Hsi, David Ching Heng, plant pathology

DEMING
Walton, George, physical chemistry, forensic science

EAGLE NEST
Allen, Rhesa McCoy, Jr, geology

ESPANOLA
Froman, Darol Kenneth, nuclear physics
Manley, John Henry, nuclear physics

GALLUP
Ordway, Nelson Kneeland, pediatrics
Tempest, Bruce Dean, infectious diseases

GLENWOOD
Sumner, Lowell, biology

GRANTS
Rapaport, Irving, mining geology

HOLLOMAN AFB
Griffin, Travis Barton, biochemistry
Iatropoulos, Michael John, comparative pathology, experimental pathology
Rosenblum, Ira, pharmacology, toxicology

KIRTLAND AFB
Avizonis, Petras V, lasers, physical optics
Bettis, Jerry Ray, optical physics
Brabson, George Dana, Jr, physical chemistry

Burns, Erskine John Thomas, plasma physics
Glowienka, John Clement, plasma physics
Guenther, Arthur Henry, chemistry, physics
Hadley, Steven George, physical chemistry
Jungling, Kenneth Corneal, optics, lasers
Payton, Daniel N, III, physics
Ristvet, Byron Leo, marine geochemistry

LAS CRUCES
Adams, J Mack, computer science
Alexander, Martin Dale, inorganic chemistry
Allen, Rex Wayne, animal parasitology
Ames, Lynford Lenhart, physical chemistry
Anderson, Marlowe George, zoology
Baltensperger, Arden Albert, plant breeding
Barrera, Cecilio Richard, microbial physiology
Bernstein, Marvin Harry, comparative physiology
Birnbaum, Edward Robert, inorganic chemistry, bioinorganic chemistry
Boes, Eldon C, topology
Booth, John Austin, plant pathology
Botsford, James L, microbial physiology, biochemistry
Burleson, George Robert, experimental nuclear physics, experimental high energy physics
Burr, Alexander Fuller, atomic physics, solid state physics
Burris, Albert, physics
Cardenas, Manuel, experimental statistics
Casillas, Edmund Rene, biochemistry
Chen, Tuan Wu, theoretical physics
Clemons, Russell Edward, economic geology
Cleveland, Ernest Lynn, physics
Coburn, Horace Hunter, physics
Cotter, Donald James, horticulture, plant physiology
Cuffey, James, astronomy
Darnall, Dennis W, biochemistry
Daugherty, LeRoy Arthur, agronomy, soil morphology
Davis, Charles A, wildlife ecology
Davis, Dennis Duval, organic chemistry, organometallic chemistry
Davis, Dick D, agronomy
Daw, Harold Albert, physics
Day, Mary-Lou, nutrition
DePree, John Deryck, mathematics
Donart, Gary B, range science
Dressel, Ralph William, physics
Dunford, Max Patterson, plant cytogenetics, biosystematics
Enzie, Joseph Vincent, horticulture
Ewing, Gordon J, physical chemistry
Finkner, Morris Dale, experimental statistics
Fowler, James Lowell, crop physiology
Francis, David Wesson, poultry husbandry
Garrett, Edgar Ray, speech pathology
Gibbens, Robert Parker, range management
Giever, John Bertram, mathematics
Glaze, Richard Michael, computer science, data processing
Good, William Breneman, physics
Gould, Walter Leonard, weed science, natural resources
Hageman, James Howard, biochemistry, bacterial physiology
Hanson, Fred Sumner, (III), physics, aerospace sciences
Heaton, Maria Malachowski, quantum chemistry
Herbel, Carlton Homer, range science
Hoffman, Robert Vernon, organic chemistry
Horney, Amos Grant, chemistry
Howard, Volney Ward, Jr, wildlife research
Hunt, Charles Butler, geology
Ingraham, Richard Lee, theoretical physics
Jester, Douglas Brewer, fisheries management
Johannes, Robert, physics
Kellogg, David Wayne, nutrition, animal science
Kinzer, H Grant, entomology
Kist, Joseph Edmund, mathematics
Kuehn, Glenn Dean, biochemistry
Kunz, Kaiser Schoen, physics
La Pointe, Joseph L, zoology
Leyendecker, Philip Jordon, phytopathology
Liebert, Wolfgang, algebra

NEW MEXICO

Lindsey, Donald Leroy, plant pathology
Loustaunau, Joaquin, mathematics
Lwowski, Walter Wilhelm Gustav, organic chemistry
Maim, Norman R., plant breeding, agronomy
Mandelker, Mark, mathematics
McCaslin, Bobby Duane, soil fertility
Melgaard, Kenneth Gilbert, analytical chemistry
Melton, Billy Alexander, Jr., plant breeding
Miller, August, atmospheric physics
Miyamoto, Seichi, soil physics, soil chemistry
Monagle, John Joseph, Jr., organic chemistry
Mott, David Lowe, physics
Nakayama, Roy Minoru, plant breeding, horticulture
Nelson, Arnold Bernard, animal nutrition
Nunemaker, John Coleman, internal medicine
O'Brien, Robert Thomas, bacterial physiology, biochemistry
O'Connor, George Albert, soil chemistry
Ortiz, Melchor, Jr., applied statistics
Phillips, Keith L., mathematical analysis
Pieper, Rex Delane, range science, plant ecology
Quinones, Ferdinand Antonio, plant breeding
Raitt, Ralph James, Jr., vertebrate zoology
Ram, Budh, physics
Randhawa, Jagir Singh, physics
Ray, Earl Elmer, genetics
Reitmeyer, William L., astronomy, aerospace engineering
Richardson, Albert Edward, physical chemistry, radiochemistry
Rogers, Gerald Stanley, mathematical statistics
Rowan, Robert, Jr., analytical chemistry
Schemitz, Sanford David, wildlife research, wildlife ecology
Sheck, Ronald Calvin, cultural geography, geography of Latin America
Sluyter, Marshall M., applied mathematics, theoretical mechanics
Smith, Garmond Stanley, animal nutrition
Southward, Glen Morris, fish biology, statistics
Spellenberg, Richard (William), plant taxonomy
Staffeldt, Eugene Edward, plant pathology
Stark, Richard Harlan, computer science
Stromberg, Thorsten Frederick, low temperature physics
Sullivan, Darrell Thornton, horticulture
Summers, Donald Balch, chemistry
Thaeler, Charles Schropp, Jr., zoology
Theiner, Otto, physics
Thomas, Gerald Waylett, range ecology
Throneberry, Glyn Ogle, plant physiology
Todsen, Thomas Kamp, science administration, botany
Tromble, John M., hydrology, soil science
Turner, Ralph B., biochemistry
Urquhart, N Scott, statistics
Vance, Irvin Elmer, mathematics education
Van Heuvelen, Alan, physics
Villa, Vicente Domingo, microbial physiology
Walker, Carol L., mathematics
Walker, Elbert Abner, mathematics
Watts, John Gordon, entomology
Weeks, Owen Bayard, chemistry
Whitford, Walter George, ecology
Widmoyer, Fred Bixler, horticulture
Wierenga, Peter J., soil physics, agronomy
Wilkins, Ralph G., inorganic chemistry
Wisner, Robert Joel, algebra
Wisner, John Philip, anthropology
Zartman, David Lester, genetics
Zimmerman, James Roscoe, zoology, entomology
Zund, Joseph David, mathematics, mathematical physics

LAS VEGAS

Amai, Robert Lin Sung, organic chemistry
Anselmo, Vincent C., radio chemistry, physical chemistry
Bejnar, Waldemere, geology
Clark, Ronald Duane, organic chemistry, environmental chemistry
Lindeborg, Robert G., bioecology, mammalogy
Llamas, Vicente Jose, solid state physics

LOS ALAMOS

Aamodt, R. L., physics
Agnew, Harold Melvin, physics
Agnew, Lewis Edgar, Jr., physics
Alire, Richard Marvin, chemistry
Ames, Susan, astrophysics
Amols, Howard Ira, radiological physics, radiobiology
Amsden, Anthony Avery, fluid dynamics
Anderson, Ernest Carl, biophysics
Andrew, James E., experimental solid state physics
Apt, Kenneth E., nuclear chemistry
Argo, Harold Virgil, astronomy
Armstrong, Dale Dean, nuclear physics
Asbridge, John Robert, space physics
Asprey, Larned Brown, physical chemistry, inorganic chemistry
Auchampaugh, George Fredrick, experimental nuclear physics
Baggett, Lester Marchant, nuclear physics, hydrodynamics
Baker, Floyd B., physical chemistry
Baker, George Allen, Jr., statistical mechanics, applied mathematics
Balagna, John Paul, analytical chemistry, radiochemistry
Balog, George, lasers
Bame, Samuel Jarvis, Jr., space physics
Barasch, Guy Errol, atmospheric physics
Barfield, Walter David, physics
Barnes, John Fayette, atomic physics
Barnhart, Benjamin J., genetics
Bartlett, Roger James, solid state physics
Baxman, Horace Roy, chemistry
Bayhurst, Barbara P., nuclear chemistry
Beattie, Willard Horatio, physical chemistry
Bell, George Irving, reactor physics, theoretical biology
Bendt, Philip Joseph, cryogenics, experimental nuclear physics
Benjamin, Robert Fredric, experimental physics
Bennett, Elbert White, experimental nuclear physics
Bennett, Michael J., nuclear physics
Bentley, Richard Foster, nuclear physics
Bergen, Delmar Wesley, nuclear physics
Best, George Harold, physics
Beyer, William A., mathematics
Birely, John H., chemical physics
Blair, Allen G., nuclear physics
Blais, Normand C., chemical physics
Blake, Richard L., astrophysics
Bohachevsky, Ihor O., fluid mechanics, applied mathematics
Bohannon, George Edmond, nuclear physics, particle physics
Bolsterli, Mark, theoretical physics
Bonner, Billy Edward, nuclear physics
Bowersox, David F., physical chemistry, analytical chemistry
Bowman, Allen Lee, physical chemistry
Bowman, James David, nuclear physics
Bradbury, James Norris, experimental high energy physics
Bradbury, James Norris, experimental physics
Brandow, Baird H., theoretical physics
Breedlove, James Robby, Jr., physics, computer science
Breshears, Wilbert Dale, physical chemistry
Britt, Harold Curran, nuclear physics
Brock, Ernest George, electrophysics
Brolley, John Edward, Jr., nuclear physics
Brown, Leon Joseph, chemistry
Brown, Robert Theodore, physics
Browne, Charles Idol, radiochemistry
Brownlee, Robert Rex, astrophysics
Brunfield, Philip Lincoln, hydrodynamics
Bruckner, Lawrence Adam, statistical analysis
Bunker, Merle E., nuclear physics
Burck, Anthony John, solar physics, x-ray astronomy
Burman, Robert L., nuclear physics
Burton, Blendin L., physics
Butler, Gilbert W., physics
Butler, Harold S., computer science
Butler, Thomas Daniel, fluid dynamics

McGahan, Merritt Wilson, plant morphology
Schlosser, Jon A., mathematics, physics
Schufle, Joseph Albert, physical chemistry, history of science
Searcy, Charles Jackson, mathematics
Shields, Lora Mangum, biology
Verma, Sadanand, algebra, topology
Yarger, Frederick Lynn, physics

Cady, Howard Hamilton, physical chemistry
Caldwell, John Thomas, physics
Campbell, David Kelly, theoretical high energy physics
Campbell, Evan Edgar, industrial hygiene
Campbell, George Melvin, physical chemistry, analytical chemistry
Campillo, Anthony Joseph, lasers, biophysics
Canada, Robert, nuclear physics
Cantrell, Cyrus D. III, physics
Cantwell, Robert Murray, theoretical physics
Carbone, Robert James, physics
Carruthers, Peter A., theoretical physics
Carter, Leland LaVelle, reactor physics, nuclear engineering
Cartwright, David Chapman, chemical physics, molecular physics
Cashwell, Edmond Darrell, applied mathematics
Catlett, Duane Stewart, physical chemistry, chemical metallurgy
Chambers, William Hyland, nuclear science
Conner, Jerry Power, nuclear physics
Conner, James Huntington, space physics
Cooper, Frederick Michael, elementary particle physics
Cooper, Ralph Sherman, theoretical physics
Cowan, George A., radiochemistry
Cowan, Robert Duane, atomic physics
Cox, Arthur Nelson, astrophysics, hydrodynamics
Cox, Lawrence Edward, spectrochemistry
Cram, Leighton Scott, biophysics
Critchfield, Charles Louis, physics
Cross, Jon Byron, chemical physics, physical chemistry
Crowell, John Marshall, biophysics
Daly, Bartholomew Joseph, fluid dynamics
Daniels, William Richard, radiochemistry
Davis, Alvin Herbert, theoretical physics
Davis, William Chester, physics, explosives
Deal, William E., Jr., hydrodynamics, high pressure physics
Deaven, Larry Lee, cell biology, cytogenetics
Deinken, Herman Porter, physics
Dendy, Joel Eugene, Jr., numerical analysis
Devaney, Joseph James, lasers, mathematical physics
Dicello, John Francis, Jr., medical biophysics, nuclear physics
Dietz, Rudolph John, inorganic chemistry
Dinegar, Robert Hudson, physical chemistry
Diven, Benjamin Clinton, nuclear physics
Dougherty, John E., physics
Dowdy, Edward Joseph, nuclear science
Doyle, Thomas Carlson, mathematics
Dreicer, Harry, plasma physics
Dropesky, Bruce Joseph, nuclear chemistry
Du Bois, Donald Frank, theoretical physics
DuBois, Frederick Williamson, inorganic chemistry
Dudziak, Donald John, nucleonics
Dunning, Thomas Harold, Jr., theoretical chemistry
Edwards, David Franklin, solid state physics
Ehler, Arthur Wayne, plasma physics
Eiler, Philip Gary, chemistry
Ellis, Reed Hobart, Jr., physics
Ellis, Walton P., surface chemistry
Emigh, Charles Robert, physics
Engelhardt, Albert George, physics
Enger, Merlin Duane, biochemistry
Engleman, Rolf, Jr., physical chemistry
Erdal, Bruce Robert, nuclear chemistry
Erickson, Dennis John, solid state physics
Erpenbeck, Jerome John, theoretical chemistry
Essington, Edward Herbert, soil science
Ettinger, Harry Joseph, industrial hygiene

Evans, Albert Edwin, Jr., nuclear physics
Evans, Foster, theoretical physics
Evans, Winifred Doyle, physics
Everett, Cornelius Joseph, mathematics
Eyster, Eugene Henderson, physical chemistry
Familaro, Kendall Ferris, nuclear physics
Farrell, John A., nuclear physics
Feldman, Barry Joel, quantum optics, lasers
Fenstermacher, Charles Alvin, experimental physics
Ficket, Wildon, theoretical physics
Figueira, Joseph Franklin, lasers
Fisher, Robert Alan, lasers, optics
Flicker, Herbert, lasers, molecular spectroscopy
Fluharty, Rex Gilbert, nuclear physics
Flynn, David Robert, physics
Ford, George Pratt, chemistry
Forslund, David Wallace, plasma physics
Foster, Duncan Graham, Jr., nuclear physics
Fowler, Clarence Maxwell, physics
Fowler, Eric Beaumont, soil chemistry
Frank, Robert Morris, theoretical physics
Freiberg, Jeffrey Philip, plasma physics
Fries, Ralph Jay, physical chemistry
Fritz, Joseph N., solid state physics
Fu, Jerry Hui Ming, plasma physics, space physics
Fukushima, Eiichi, solid state physics, chemical physics
Fultyn, Robert Victor, computer sciences
Gerstl, Siegfried Adolf Wilhelm, reactor physics
Gibbs, William Royal, physics
Gibson, Benjamin Franklin, V., physics
Gillespie, Claude Milton, nuclear physics
Giorgi, Angelo Louis, radiochemistry
Glasgow, Dale William, nuclear physics
Glass, George, high energy physics
Glass, Neel Warren, nuclear physics
Goad, Walter Benson, Jr., biophysics
Godfrey, Brendan Berry, plasma physics, numerical analysis
Godfrey, Thomas Nigel King, physics
Goldstein, John Cecil, lasers
Goldstein, Louis, theoretical physics, low temperature physics
Gosling, John Thomas, space physics, solar physics
Grant, Patrick Michael, nuclear chemistry, radiochemistry
Green, Walter Verney, metal physics
Gregg, Charles Thornton, biochemistry
Greiner, Norman Roy, chemical kinetics, lasers
Grilly, Edward Rogers, thermodynamics
Grisham, Genevieve Dwyer, chemistry
Grover, George Maurice, nuclear physics
Gubernatis, James Edward, solid state physics
Gula, William Peter, plasma physics
Gurley, Lawrence Ray, biochemistry, chemical engineering
Gursky, Martin Lewis, theoretical physics
Haberstich, Albert, plasma physics
Hagerman, Donald Charles, physics
Hakkila, Eero Arnold, analytical chemistry
Hammel, Edward Frederic, physical chemistry
Han, Ki Sup, electrooptics
Hanson, Wayne Carlyle, wildlife management, radiation ecology
Hardekopf, Robert Allen, nuclear physics
Harlow, Francis Harvey, Jr., theoretical physics, fluid dynamics
Hatcher, Charles Richard, physics
Hayes, Francis Newton, organic chemistry
Hayward, Thomas Doyle, experimental chemistry
Helland, Jerome A., biophysics
Heller, Leon, nuclear physics
Helmick, Herbert Hanna, instrumentation, nuclear physics
Hemmendinger, Arthur, nuclear physics
Henderson, Dale Barlow, plasma physics
Henins, Ivars, physics
Henkel, Richard Luther, nuclear physics
Hensley, John Coleman, II, experimental chemistry
Hill, Henry Hunter, physics
Hirt, Cyril William, Jr., theoretical physics
Hoard, Donald Ellsworth, biochemistry
Hoerlin, Herman William, physics
Hoffman, Cyrus Miller, particle physics

Hoffman, Darleane Christian, nuclear chemistry
Hoffman, Marvin Morrison, physics
Holland, Redus Foy, molecular spectroscopy
Holley, Charles Elmer, Jr, geochemistry, thermochemistry
Holm, Dale M, biophysics
Hopkins, John Chapman, nuclear physics
Howell, Jo Ann Shaw, computer sciences
Hoyt, Harry Charles, physics
Huang, Chao-Yuan, solid state physics
Huebner, Walter F, atomic physics, astrophysics
Hunter, Raymond Eugene, nucleonics
Hutson, Richard Lee, elementary particle physics
Hwang, Chester F, physics
Ingraham, John Charles, plasma physics
Ingram, Marylou, clinical medicine
Jackson, Darryl Dean, chemical instrumentation
Jackson, Jasper Andrew, Jr, experimental physics
Jackson, Sydney Vern, nuclear chemistry
Jahoda, Franz Carl, experimental physics
Janney, Donald Herbert, physics
Jarboe, Thomas Richard, plasma physics
Jarmie, Nelson, nuclear physics
Jeffries, Robert Alan, optical physics
Jones, Eric Manning, hydrodynamics, astrophysics
Jones, Llewellyn Hosford, physical chemistry
Judd, O'Dean P, lasers, plasma physics
Karr, Hugh James, nuclear physics
Keaton, Posey W, Jr, nuclear physics
Keenan, Thomas K, inorganic chemistry
Keepin, George Robert, Jr, nuclear physics, instrumentation
Keller, William Edward, quantum physics
Kerley, Gerald Irwin, physical chemistry
Kerr, Donald M, Jr, research administration, ionospheric physics
Kerr, Eugene Charles, physical chemistry
Keyworth, George A, II, experimental nuclear physics
Kindel, Joseph Martin, plasma physics
Kmetko, Edward Andrew, physics
Knapp, Edward Alan, physics
Knight, Jere Donald, physical chemistry
Kraemer, Paul Michael, cell biology
Krohn, Burton Jay, molecular physics
Krupka, Milton Clifford, high temperature chemistry
Kubas, Gregory Joseph, inorganic chemistry
Kunz, Walter Ernest, radiation physics
Laquer, Henry L, cryogenics
Larson, Thomas E, physical chemistry, organic chemistry
Lathrop, Kaye Don, reactor physics
Laughlin, Alexander William, geochemistry, economic geology
Lawrence, James Neville Peed, health physics, physics
Lazarus, Roger Ben, theoretical physics
Lee, Clarence Edgar, mathematical physics
Lee, David Mallin, experimental nuclear physics
Leland, Wallace Thompson, lasers
Leon, Melvin, theoretical physics
Lewis, Harold Ralph, physics
Lewis, Warren Burton, chemistry
Liberman, David Arthur, theoretical physics
Liberman, Irving, lasers, optical physics
Liebenberg, Donald Henry, plasma physics, lasers
Lilley, John Richard, plasma physics
Lindman, Erick Leroy, Jr, plasma physics
Lindstrom, Ivar E, Jr, physics
Lohrding, Ronald Keith, statistical analysis, research administration
London, Robert Elliot, biophysical chemistry
Loree, Thomas Robert, solid state physics, lasers
Louck, James Donald, mathematical physics
Loughran, Edward Dan, analytical chemistry
Lyon, Luther Lawrence, Jr, physical inorganic chemistry
Lyons, Peter Bruce, plasma physics
MacDougall, Duncan Peck, physical chemistry
Macek, Robert James, elementary particle physics, experimental nuclear physics
Mader, Charles Lavern, physical oceanography
Malanify, John Joseph, nuclear physics
Malik, John Stanley, experimental physics
Malone, Robert Charles, theoretical physics
Mann, Joseph Bird, (Jr), chemical physics
Mark, J Carson, mathematics, mathematical physics
Mason, Caroline Faith Vibert, inorganic chemistry
Mason, Rodney Jackson, plasma physics
Mather, Joseph Walter, physics
Matlack, George Miller, radiochemistry
Mausner, Leonard Franklin, nuclear chemistry
McCrory, Robert Lee, Jr, plasma physics, hydrodynamics
McDowell, Robin Scott, molecular spectroscopy
McGetchin, Thomas R, geology
McGuire, Austin Dole, particle physics
McKay, Michael Darrell, statistics
McNaughton, Michael Walford, experimental nuclear physics
Merts, Athel Lavelle, atomic physics, astrophysics
Metropolis, Nicholas Constantine, applied mathematics, theoretical physics
Metzger, Daniel Schaffer, physics
Miller, Leston Wayne, electrodynamics
Mills, Robert Leroy, physical chemistry
Minerbo, Gerald N, theoretical physics
Miranda, Gilbert A, cryogenics, nuclear magnetic resonance
Mischke, Richard E, particle physics, nuclear physics
Mjolsness, Raymond C, fluid dynamics, atomic physics
Moody, David Coit, III, inorganic chemistry
Moore, Benjamin LaBree, nuclear physics
Moore, Michael Stanley, nuclear physics
Morales, Raul, analytical chemistry
Morris, Charles Edward, solid state physics
Mosley, John Ross, physical chemistry
Moss, Calvin E, experimental nuclear physics
Motz, Henry Thomas, nuclear physics
Mueller, Karl Hugo, Jr, low temperature physics
Mueller, Marvin Martin, physics
Muir, Douglas William, nuclear science
Mulford, Robert Neal Ramsay, physical chemistry, physical metallurgy
Mullaney, Paul F, physics, biophysics
Mullins, Lawrence J, Jr, physical chemistry
Nagle, Darragh (Edmund), physics
Neeper, Donald Andrew, physics
Neher, Leland K, nuclear physics
Nereson, Norris (George), lasers
Newton, Thomas William, inorganic chemistry, physical chemistry
Nicholson, Nicholas, experimental nuclear physics
Nickols, Norris Allan, nuclear physics
Nicolaenko, Basil, mathematics
Nielson, Clair W, physics
Nieto, Michael Martin, theoretical physics
Nix, James Rayford, theoretical nuclear physics
Norman, Jay Harold, nuclear physics
Norris, Andrew Edward, nuclear chemistry
Norton, John Leslie, mathematical physics
Noth, Paul Henry, clinical medicine
O'Brien, Harold Aloysious, Jr, nuclear chemistry, nuclear medicine
Ogard, Allen E, nuclear chemistry
Ohlsen, Gerald G, nuclear physics
Olinger, Bart, high pressure physics
Olsen, Clayton Edward, solid state physics
Olsen, Kenneth Harold, astrophysics, geophysics
Olson, William Marvin, high temperature chemistry
Onstott, Edward Irvin, inorganic chemistry, physical chemistry
Orth, Charles Joseph, nuclear chemistry
Osborne, Robert Kidder, physics
Ott, Donald George, chemistry, synthetic organic chemistry
Overton, William Calvin, Jr, physics
Pack, Russell T, quantum mechanics, statistical mechanics
Parker, Jack Lindsay, nuclear physics, cosmic ray physics
Patterson, James Howard, chemistry
Paxton, Hugh Campbell, nuclear physics
Peaslee, Alfred Tredway, Jr, theoretical physics
Peek, Harry Milton, physical chemistry, atmospheric physics
Penneman, Robert Allen, inorganic chemistry
Perelson, Alan Stuart, biophysics, immunology
Perkins, Roger Bruce, nuclear physics
Perry, Dennis Gordon, nuclear chemistry
Perry, Joseph Earl, Jr, physics
Petersen, Donald Francis, pharmacology
Peterson, Charles Leslie, physical chemistry, corrosion
Peterson, Dean Everett, high temperature chemistry
Peterson, Robert W, nuclear physics
Peterson, Rolf Eugene, nuclear physics
Phillips, Donald Davis, experimental physics
Phillips, James A, nuclear physics, plasma physics
Piltch, Martin Stanley, lasers
Pimbley, George Herbert, Jr, mathematics
Plassmann, Elizabeth Hebb, nuclear physics
Plassmann, Eugene Adolph, nuclear physics
Pongratz, Morris Bernard, space physics
Poore, Emery Ray Vaughn, nuclear physics
Prestwood, Rene Jesse, radiochemistry
Putnam, Thomas Milton, Jr, experimental nuclear physics
Quinn, Warren Eugene, physics
Rabideau, Sherman Webber, physical chemistry
Radziemski, Leon Joseph, Jr, atomic spectroscopy
Ragan, Charles Ellis, III, nuclear physics
Raju, Mudundi Ramakrishna, radiation biophysics, radiotherapy
Rall, Stanley Carlton, Jr, biochemistry
Ramshaw, John David, chemical physics
Ranken, William Allison, energy conversion
Rappaport, Stephen Morris, industrial hygiene
Ratliff, Robert L, biochemistry
Reavis, James Gene, nuclear chemistry
Reedy, Robert Challenger, nuclear chemistry, cosmochemistry
Reichelt, Walter Herbert, nuclear physics, plasma physics
Rein, James Earl, analytical chemistry
Ribe, Fred Linden, physics
Richter, John Lewis, nuclear physics
Rinker, George Albert, Jr, theoretical nuclear physics, atomic physics
Rivard, William Charles, fluid physics
Robb, William Derek, atomic physics
Robinson, C Paul, lasers, chemical physics
Rockwood, Stephen Dell, lasers
Rogers, Raymond N, analytical chemistry
Roof, Raymond Bradley, Jr, crystallography
Rose, Donald Glenn, physical chemistry
Rosen, Louis, physics
Rosenblum, Stephen Saul, experimental solid state physics, experimental nuclear physics
Rouse, Prince Earl, Jr, physical chemistry
Ruppel, Hans Max, theoretical physics
Salmi, Ernest William, nuclear physics
Salzman, Gary Clyde, biophysics
Sampson, Thomas Edward, nuclear science
Sandenaw, Thomas Arthur, metal physics
Sandmeier, Henry Armin, nuclear physics
Saponara, Arthur G, biochemistry
Sattizahn, James Edward, Jr, physical chemistry
Saunders, George Cherdron, immunology
Sawyer, George Alanson, plasma physics
Sayer, Arthur Robert, physics
Schelberg, Arthur Daniel, physics
Schermer, Robert Ira, nuclear physics, cryogenics
Schillaci, Mario Edward, radiation physics, particle physics
Schneider, Barry I, molecular physics, theoretical chemistry
Schneider, Jacob David, experimental atomic physics, engineering design
Schott, Garry Lee, physical chemistry
Schreiber, Raemer Edgar, nuclear physics
Schulte, Harry Frank, industrial hygiene
Schultz, Rodney Brian, applied physics, optics
Scolman, Theodore Thomas, nuclear physics
Seagrave, John Dorrington, nuclear physics, optical physics
Seamon, Robert Edward, nuclear physics
Sedlacek, William Adam, atmospheric chemistry
Seeger, Philip Anthony, nuclear physics
Shapiro, Stanley Leland, physics
Shera, E Brooks, nuclear physics
Sherman, Robert Howard, physical chemistry, cryogenics
Sherwood, Arthur Robert, plasma physics
Silbar, Richard Robert, physics
Silver, Richard N, theoretical solid state physics
Simmons, James E, experimental nuclear physics
Simmons, Leonard Micajah, Jr, theoretical high energy physics
Slansky, Richard Cyril, theoretical high energy physics
Smith, David Allen, molecular biology, radiobiology
Smith, Gordon Meade, physical chemistry
Smith, James Lawrence, solid state physics
Smith, Louis Charles, physical organic chemistry
Smith, Maynard E, analytical chemistry
Sollid, Jon Erik, optics
Spalding, John F, radiobiology
Spall, Walter Dale, analytical chemistry
Spaulding, Robert Lee, Jr, chemistry
Spillman, George Raymond, physics
Stein, Nelson, nuclear physics
Stein, William Earl, electron physics, nuclear physics
Steinhaus, David Walter, atomic physics, spectrochemistry
Stephenson, Gerard J, Jr, theoretical nuclear physics
Steyert, William Albert, physics
Stokes, Richard Hivling, experimental physics
Stone, Sidney Norman, optical physics, astrophysics
Stratton, William R, physics
Streetman, John Robert, physical chemistry
Strniste, Gary F, molecular biology, biological chemistry
Strong, Ian B, physics, astronomy
Stroud, F Agnes Naranjo Schmink, radiobiology
Sullivan, John Henry, physical chemistry
Sunier, Jules Willy, experimental nuclear physics
Swartz, Blair Kinch, numerical analysis
Swenson, Donald Adolph, physics, mathematics
Sydoriak, Stephen George, physics
Talley, Thurman Lamar, physics
Taschek, Richard Ferdinand, physics
Taylor, Raymond Dean, low temperature physics
Taylor, Thomas Newton, surface physics
Telle, John Martin, lasers
Terrell, James, (Jr), physics
Tesmer, Joseph Ransdell, experimental nuclear physics
Theobald, J Karl, physics
Thieme, Melvin T, physics, mathematics
Thiessen, Henry Archer, physics
Thomas, Keith Skelton, physics
Thomas, Robert Glenn, radiobiology, biophysics
Thomassen, Keith I, plasma physics
Thompson, Donald Leo, physical chemistry
Thorn, Robert Nicol, physics
Tisinger, Richard Martin, Jr, data processing
Tobey, Robert Allen, cell biology, cancer
Todd, Jay, Jr, physics
Travis, James Roland, physics, explosives
Travis, John Richard, physics, fluid dynamics
Tuck, James Leslie, physics
Turner, Leaf, theoretical physics
Veeser, Lynn Raymond, nuclear physics
Venable, Douglas, physics
Vier, Dwayne Trowbridge, physical chemistry
Visscher, William M, theoretical physics
Voelz, George Leo, occupational medicine
Wagner, Paul, physical chemistry
Walker, James Joseph, theoretical physics
Wallace, Terry Charles, physical chemistry
Waller, Ray Albert, mathematical statistics
Walsh, John M, physics
Walters, Ronald Arlen, biochemistry, radiobiology
Walton, Roddy Burke, nuclear physics
Wampler, Fred Benny, physical chemistry, photochemistry
Ward, John William, high temperature chemistry
Warren, John Lucius, solid state physics, research administration
Waterbury, Glenn Raymond, analytical chemistry
Waterman, Michael S, mathematics, statistics
Watt, Bob E, nuclear physics
Wendroff, Burton, mathematics
Westervelt, Donald Ramsey, physics

NEW MEXICO

Wewerka, Eugene Michael, physical chemistry
Whaley, Thomas Williams, organic chemistry
White, George Nichols, Jr., applied mathematics
White, Paul Chapin, physics
Whitley, Jerry Barnard, nuclear chemistry, nuclear physics
Williams, Joel Mann, Jr., fuel science, physical organic chemistry
Williams, Robert Allen, nuclear chemistry
Williamson, Ralph Elmore, astronomy
Wimett, Thomas Frederick, reactor physics
Wing, George Milton, applied mathematics
Wolfberg, Kurt, radiochemistry, nuclear chemistry
Wollan, John Jerome, low temperature physics
Wood, Gerry Odell, industrial hygiene
Wood, John Herbert, solid state physics
Wood, William Wayne, statistical mechanics
Wright, Bradford Lawrence, physics
Yarnell, John Leonard, physics
Yasuda, Stanley K., analytical chemistry
York, George William, lasers
Young, Frederick, physics
Young, Phillip Gaffney, nuclear physics
Zeigler, Royal Keith, mathematics, applied statistics
Zelezny, William Francis, physical chemistry
Zeltmann, Alfred Howard, physical chemistry
Zinn, John, applied physics

MESILLA PARK
Culpepper, Gideon Alston, mathematics, statistics

PORTALES
Agogino, George Allan, anthropology, social psychology
Bohrer, Vorsila Laurene, ethnobotany
Buscemi, Philip Augustus, limnology, environmental management
Frost, Everett Lloyd, anthropology, archaeology
Genaro, Antonio Louis, vertebrate ecology
Irwin-Williams, Cynthia Cora, anthropology, archaeology
Mettenet, William Joseph, geography, cartography
Neill, Warren Joseph, physical chemistry, radiochemistry
Pitt, William Daniel, geology
Propes, Ernest (A), algebra
Russell, Thomas Webb, environmental chemistry, organic chemistry
Sae, Andy S W., biochemistry
Secor, Jack Behren, physiological ecology
Sharma, Ram Krishan, organic chemistry
Sittler, Orvind Dayle, physics, biophysics
Spalding, Dan Wesley, physics
Sublette, James Edward, zoology
Taylor, Robert Gay, microbiology
Thomas, William G., physical chemistry
Wilson, Carroll Klepper, mathematics
Yos, David Albert, botany, history of biology

ROSWELL
Chapman, Randolph Wallace, geology
Havenor, Kay Charles, geology
Roudebush, William Campbell, physics

SANTA FE
Dutton, Bertha Pauline, anthropology, archaeology
Ellis, Florence Hawley, ethnology
Harris, Reece Thomas, conservation, ornithology
Hubbard, John Patrick, ornithology
Huey, William S., wildlife management
King, L D Percival, physics
Lambert, Marjorie Ferguson, anthropology, archaeology
Lange, Robert Echlin, Jr., wildlife diseases
Livingston, Milton Stanley, nuclear physics
Overhage, Carl F J., experimental physics
Overstreet, Elizabeth Claire (Fisher), geology
Peterson, Roger Shipp, botany
Redman, Leslie Merrill, chemistry

Reed, Erik Kellerman, anthropology, archaeology
Rinehart, John Sargent, physics
Schwartz, Douglas Wright, anthropology, academic administration
Stanislawski, Michael Barr, cultural anthropology, archaeology
Taylor, Walter Willard, anthropology
Walsh, Joseph Matthew, biochemistry, physical chemistry
Wauer, Roland H., ecology, ornithology
Willmon, Thomas L., medicine, physiology
Wolff, Theodore Albert, entomology

SOCORRO
Arterburn, David Roe, mathematical analysis
Austin, George Stephen, geology
Ball, Ralph Wayne, mathematics
Beane, Richard Edward, geochemistry, economic geology
Bieberman, Larry, physics
Bodine, Marc Williams, Jr., geochemistry, petroleum geology
Brandvold, Donald Keith, biochemistry, physical chemistry
Brierley, James Alan, microbiology
Brook, Marx, physics
Brower, Kay Robert, organic chemistry
Budding, Antonius Jacob, geology
Chapin, Charles Edward, geology
Clark, Barry Gillespie, astronomy
Colgate, Stirling Auchincloss, physics
Condie, Kent Carl, petrology, geochemistry
Fallon, Leslie Dodds, physics
Flower, Rousseau Hayner, geology, paleontology
Ford, Kenneth William, theoretical physics
Friberg, Martin Samuel, analytical mathematics
Garwick, Jan Vaumund, computer science
Gross, Gerardo Wolfgang, geophysics
Gutjahr, Allan L., mathematical statistics, operations research
Hatch, Melvin (Jay), organic chemistry
Holmes, Charles Robert, geophysics
Hume, William, physics, research administration
Keizer, Clifford Richard, physical chemistry
Klett, James Dean, cloud physics
Kottlowski, Frank Edward, economic geology
Kuelmer, Frederick John, petrology
LeFebre, Vernon Glen, thermodynamics
Lochman-Balk, Christina, invertebrate paleontology, stratigraphy
Lomanitz, Ross, theoretical physics
McGehee, Ralph Marshall, applied mathematics, numerical analysis
Moore, Charles Bachman, Jr., atmospheric physics
O'Donnell, Brian Desmond, cosmic ray physics
Ohline, Robert Wayne, analytical chemistry
Petschek, Albert George, physics
Popp, Carl John, inorganic chemistry
Raymond, David James, physics, meteorology
Renault, Jacques Roland, geology
Robertson, James Magruder, economic geology
Sanchez, Gilbert, microbiology, parasitology
Sanford, Allan Robert, geophysics
Sharples, Alan, mathematics
Shortess, David Keen, genetics, plant physiology
Smith, Clay Taylor, geology
Smoake, James Alvin, physiology, biochemistry
Sturgil, John Roman, geophysics, engineering science
Thompson, Anthony Richard, radio astronomy
Thompson, Samuel, III, petroleum geology, stratigraphy
Weber, Robert Harrison, geology
Wilkening, Marvin Hubert, atmospheric physics

SILVER CITY
Cunningham, John E., geology
Hayward, Bruce Jolliffe, zoology
Heiner, Terry Charles, botany, forestry
Morton, John West, Jr., organic chemistry
Muna, Martin Hammond, arachnology, desert ecology
Zimmerman, Dale A., ornithology, ecology

SUNSPOT
Altrock, Richard Charles, solar physics
Beckers, Jacques Maurice, astrophysics
Canfield, Richard Charles, solar physics
Dunn, Richard B., astronomy, mechanical engineering
Evans, John Wainwright, Jr., astrophysics
Fisher, Richard R., astrophysics, solar astronomy
Keil, Stephen Lesley, solar physics
Nye, Alan Hall, solar physics
Simon, George Warren, astrophysics
Wagner, William John, astronomy
Worden, Simon Peter, solar physics

TAOS
Allen, James Sircom, nuclear physics

UNIVERSITY PARK
Ader, Olin Blair, applied mathematics
Ellington, Joe J., entomology
Evans, Latimer Richard, chemistry
Holland, Lewis, animal breeding, genetics
Kruse, Arthur Herman, mathematics
Roberson, Robert H., poultry nutrition, biochemistry

WHITE ROCK
Bryant, Ernest Atherton, radiochemistry

WHITE SANDS MISSILE RANGE
Ballard, Harold Noble, atmospheric physics
Paul, Robert Hugh, physics, electrical engineering
Webb, Willis Lee, meteorology

NEW YORK

ALBANY
Able, Kenneth Paul, behavioral & community ecology
Alexander, Robert Spence, physiology
Allaway, Norman C., biostatistics
Allen, Eric Raymond, physical chemistry
Allen, Sydney Henry George, biochemistry
Amiraian, Kenneth, biochemistry
Andrews, Charles Luther, physics
Aronson, John Noel, biochemistry
Augustyn, Joan Mary, biochemistry
Baglioni, Corrado, biochemical genetics, immunology
Bailey, William Charles, nuclear magnetic resonance
Bakhru, Hassaram, nuclear physics
Balint, John Alexander, medicine
Bank, Shelton, physical organic chemistry
Barlow, James Lawrence, medical microbiology, infectious diseases
Barron, Kevin D., medicine, neurology
Baum, Werner Christian, biology
Baxter, Donald Henry, radiotherapy
Beckman, Lewis David, virology, molecular biology
Beebe, Richard Townsend, internal medicine
Becker, Donald A., biochemistry
Bell, Bruce McConnell, paleobiology
Bell, James Milton, psychiatry
Benedict, Peter Carl, geology
Benenson, Raymond Elliott, nuclear physics
Benson, Lois Mary, virology
Berns, Donald Sheldon, physical chemistry, biophysics
Blanchard, Duncan Cromwell, marine sciences
Blount, Stanley Freeman, resource management
Bocchieri, Samuel Francis, microbiology, biochemistry
Bonney, Robert John, biochemistry
Boomsliter, Paul Colgan, speech pathology, audiology
Bosart, Lance Frank, dynamic meteorology
Browe, John Harold, nutrition
Brown, David Frederick, medicine
Brown, Harold Hubley, clinical chemistry
Brown, Stephen Clawson, histochemistry, invertebrate zoology
Bryan, Ashley Monroe, plant biochemistry
Buerle, David E., water resources
Buhac, Ivo, gastroenterology, internal medicine
Bulloff, Jack John, physical inorganic chemistry

Burke, Kevin Charles, geology
Caliguiri, Lawrence Anthony, pediatrics
Carmack, Robert..., anthropology
Chen, Chang-Hwei, physical chemistry
Cheng, Li-Jen, experimental solid state physics
Chessin, Henry, physics
Chi, Benjamin, physics
Childs, Lindsay Nathan, mathematics
Chodos, Robert Bruno, medicine
Chui, Siu-Tat, theoretical solid state physics
Cicchinelli, Alexander L., biostatistics, academic administration
Closson, William Deane, organic chemistry
Connola, Donald Pascal, forest entomology, forest genetics
Corbett, James William, solid state physics
Coulston, Frederick, pathology, parasitology
Covert, Scott Veasey, medical microbiology
Cowger, Marilyn L., pediatrics, biochemistry
Cowling, Vincent Frederick, mathematics
Crabill, Edward Vaughn, academic administration
Cue, Nelson, experimental nuclear physics
Cullen, James Henry, meteorology
Czapski, Ulrich Hans, geophysics
Daoud, Assaad S., pathology
Das, Tara Prasad, atomic physics, molecular physics
Davies, Jack Neville Phillips, epidemiology
Davis, Edward Dewey, mathematics
Deibel, Rudolf, pediatrics, virology
Dennis, Dermot Joseph, dermatology, pharmacology
Dennis, Emery Westervelt, bacteriology, parasitology
Dewey, John Frederick, structural geology
Dickerman, Herbert W., biochemistry, internal medicine
Doudney, Charles Owen, biochemistry, genetics
Doyle, Joseph Theobald, cardiology
Dulak, Norman Charles, cell biology
Dutton, Cynthia Baldwin, internal medicine
Dutton, Robert Edward, Jr., physiology, biomedical engineering
Eadon, George Albert, organic chemistry
Eckert, Charles, surgery
Edmonds, Richard H., anatomy
Edwards, Charles, physiology, biophysics
Ewart, Mervyn H., biochemistry
Fakundiny, Robert Harry, environmental geology, regional geology
Farmer, Walter Ashford, science education
Farrell, Margaret Alice, mathematics education
Federighi, Francis D., computer science
Fenton, John William, II., biochemistry, immunology
Fenton, William Nelson, anthropology, ethnology
Ferguson, Frank Currier, Jr., pharmacology
Fescyan, Sezar, statistical mechanics
Fox, Sally Ingersoll, microbiology
Freitag, Julia Louise, preventive medicine, epidemiology
Frisch, Harry Lloyd, theoretical chemistry, statistical mechanics
Fritz, Katherine Elizabeth, pathology, immunology
Froelich, Ernest, microbiology, chemotherapy
Frost, Robert Edwin, inorganic chemistry
Fuhs, George Wolfgang, microbiology, limnology
Furst, Peter, anthropology
Gabrielsen, Ann Emily, immunology
Galvan, John H., biochemistry
Garg, Jagadish Behari, nuclear physics
George, Elmer, Jr., microbiology, biochemistry
Gershman, Lewis C., biophysics
Gibson, Walter Maxwell, experimental physics
Gillette, Charles Edgar, archaeology
Glenn, Joseph Leonard, physical anthropology
Gokhale, Narayan Ramchandra, atmospheric physics
Goldfarb, Roy David, cardiovascular physiology

Goldstein, Jerome Charles, otolaryngology
Gordon, Hugh, mathematical analysis
Gordon, Morris Aaron, medical mycology
Griffel, Maurice, physical chemistry
Grimley, Philip M, medicine, pathology
Grimwood, Brian Gene, biochemical genetics
Grinnell, Robert S, Jr, marine geology
Groblewski, Gerald Eugene, pharmacology, toxicology
Grumbach, Leonard, physiology
Hacker, Bruce, biochemical pharmacology, virology
Haidak, Gerald Lewis, anatomy
Haines, John Haldor, mycology
Hall, Charles A, medicine
Hamilton, Harry Lemuel, Jr, micrometeorology
Han, Jaok, cardiology, physiology
Hanks, Jane Richardson, ethnology
Harris, Raymond, medical sciences, health sciences
Harrison, Frederick Williams, anatomy
Hauser, Richard Scott, ecology
Hechemy, Karim E, microbial physiology
Heinig, Katherine H, botany
Heiser, Willard Wayne, geography of Europe
Hemenway, Curtis Leland, astronomy, physics
Henshaw, Robert Eugene, comparative physiology, environmental physiology
Herdman, Roger Cole, public health, pediatrics
Hicks, Robert Eugene, psychophysics
Hirsch, Helmut V B, neurobiology
Hof, Liselotte Bertha, biochemistry
Hook, Ernest Benjamin, epidemiology, teratology
Horn, Eugene Harold, endocrinology, anatomy
Hotchin, John Elton, virology
Hunsberger, Isaac Moyer, organic chemistry
Hurwitz, Charles, bacteriology
Husain, Liaquat, nuclear chemistry
Imai, Hideshige, pathology, anatomy
Ingenito, Alphonse J, pharmacology
Ingraham, Hollis Steadman, epidemiology
Inomata, Akira, theoretical physics
Isachsen, Yngvar William, geology
Jacklet, Jon Willis, neurophysiology, animal behavior
Jacobs, Richard L, orthopedic surgery, biochemistry
Jacobson, Herbert (Irving), reproductive endocrinology, biochemistry
Jaeger, Robert Gordon, population ecology
Jamnback, Hugo Andrew, Jr, medical entomology
Jenkins, Joe Wiley, mathematics
Jensen, Wallace Norup, hematology
Jiusto, James E, meteorology, atmospheric physics
Jones, David Hartley, biochemistry
Kaftan-Kassim, May A, radio astronomy
Kaminsky, Laurence Samuel, biochemistry
Kanter, Gerald Sidney, physiology
Kao, Oranda Hai-Wen, physical chemistry, biochemistry
Katz, Julius, medicine
Kaufmann, William, pathology
Kaye, Gordon I, anatomy, pathology
Kaye, Nancy Weber, embryology, endocrinology
Keeshan, Margaret M, biology, protozoology
Kelly, Richard Delmer, biology, science education
Kelly, Sally Marie, clinical pathology
Kiley, John Edmund, medicine
Kim, Carl Stephen, biophysics
Kim, Jai Soo, atmospheric physics
Kimelberg, Harold Keith, biochemistry
King, Tsoo E, biochemistry
Knobloch, Hilda, pediatrics
Kolb, Lawrence Coleman, psychiatry
Korns, Robert Fulton, epidemiology
Kraft, Alan M, psychiatry
Kuivila, Henry Gabriel, organic chemistry
Laffin, Robert James, microbiology
Lam, Kwok-Wai, biochemistry
Landmesser, Charles Monroe, anesthesiology
Larney, Violet Hachmeister, mathematics
Laurenzi, Bernard John, chemical physics
Lawrence, David A, immunology
Lawson, William Burrows, biochemistry
Lee, Kyu Taik, medicine, pathology
Lee, Young-Hoon, solid state physics
Lentini, Eugene Anthony, physiology
Levitas, Alfred Dave, physics
Lininger, Lloyd Lesley, mathematics
Li-Scholz, Angela, experimental physics
Lobo, Angelo Peter, bio-organic chemistry
Loegering, Daniel John, physiology
Lomonaco, Samuel James, Jr, mathematics
Long, Arthur Owen, physical chemistry
Loose, Leland David, physiology, immunology
Luduena, Froilan Pindaro, pharmacology
Lumpkin, Lee Roy, dermatology, pathology
MacColl, Robert Joseph, physical chemistry, biochemistry
MacGregor, Thomas Harold, mathematics
Mackiewicz, John Stanley, parasitology
Maley, Frank, biochemistry
Maley, Gladys Feldott, biochemistry
Malik, Asrar Bari, cardiovascular physiology
Marfey, Sviatopolk Peter, organic chemistry, biochemistry
Marois, Robert Leo, pharmacology
Marr, Paul Donald, urban geography, regional planning
Marsh, Bruce Burton, nuclear physics
Martin, George Edward, geometry
Mascarenhas, Joseph Peter, developmental biology
Mason, Larry Gordon, population biology
Matuszek, John Michael, Jr, radiological health, radiochemistry
McChesney, Evan William, biochemistry
McGee-Russell, Samuel M, biology, electron microscopy
McGrath, John F, physical chemistry, inorganic chemistry
McKinley, Daniel Lawson, ecology, natural history
McLaren, Eugene Herbert, atmospheric chemistry
McManus, Elizabeth Catherine, physics, mathematics
McNaught, Donald Curtis, limnology
McSharry, James John, microbiology, virology
Megirian, Robert, pharmacology
Miller, Richard Avery, endocrinology
Millis, Albert Jason Taylor, cell biology
Mitchell, Richard Sheppard, plant taxonomy, plant morphology
Miyashiro, Akiho, petrology
Mizejewski, Gerald Jude, embryology, immunology
Mohnen, Volker A, atmospheric sciences, physics
Mou, Thomas William, medicine, preventive medicine
Murphey, Rodney Keith, neurobiology
Myer, Yash Paul, physical biochemistry
Narahara, Hiromichi Tsuda, biochemistry, metabolism
Nelson, William Pierrepont, III, medicine
Newman, Stuart Alan, developmental biology
O'Brien, Francis Joseph, pharmacy
Oesterreich, Roger Edward, physiological psychology
Ogawa, Hajimu, mathematics
Olafsson, Patrick Gordon, organic chemistry
Oliver, Anne Rebecca, physics
Orville, Richard Edmonds, atmospheric physics, spectroscopy
Ozarow, Vernon, physical chemistry
Palmer, Ralph Simon, vertebrate zoology
Pasamanick, Benjamin, psychiatry
Paul, Boris Jerome, physical medicine
Peabody, Richard Arthur, biochemistry
Pemrick, Suzanne Marie, muscular physiology, protein chemistry
Peng, Shi-Kaung, pathology
Philip, A G Davis, astronomy
Pickering, Richard Joseph, pediatrics, immunology
Pittman, Kenneth Arthur, drug metabolism, toxicology
Plummer, Thomas H, Jr, biochemistry
Pollara, Bernard, pediatrics, immunology
Poulos, Dennis A, neurophysiology
Powers, Samuel Ralph, Jr, surgery
Procita, Leonard, pharmacology
Pryor, Marvin J, physics
Putman, George Wendell, geology, geochemistry
Ratcliff, Keith Frederick, nuclear physics, theoretical physics
Ray, Surendra Nath, atomic physics
Rehfuss, Mary, organic chemistry
Reilly, Edwin David, Jr, computer science, physics
Reilly, Marguerite, protozoology, microbiology
Reinecke, Robert Dale, ophthalmology
Reiner, John Maximilian, mathematical biology
Renzema, Theodore Samuel, experimental solid state physics
Reynolds, George William, Jr, atmospheric physics
Rickard, Lawrence Vroman, paleontology, stratigraphy
Rikmenspoel, Robert, biophysics, instrumentation
Roach, John Faunce, radiology
Rodgers, James Earl, atomic physics, molecular physics
Rodgers, John Barclay, Jr, internal medicine, gastroenterology
Rome, Doris Spector, cytology, pathology
Rosenzweig, Norbert, theoretical physics
Roth, Laura Maurer, solid state physics
Royce, George James, anatomy
Roznowski, Donald Martin, urban geography, geography of the Northeastern United States
Saba, Thomas Maron, medical physiology, biophysics
Salmoiraghi, Gian Carlo, physiology
Sarma, Ramaswamy Harihara, biochemistry
Saturno, Antony Fidelas, chemical physics
Saunders, John Warren, Jr, embryology
Schaefer, Vincent Joseph, chemistry, meteorology
Schmalberger, Donald C, astrophysics
Scholes, Charles Patterson, biophysics
Scholz, Wilfried, nuclear physics
Schwartz, Sidney A, research administration, physiology
Scovill, William Albert, surgery
Serrone, David M, pharmacology, toxicology
Sherman, Malcolm J, mathematics
Sica, Albert Joseph, pharmacy
Simpson, Karl William, fresh water biology, entomology
Singer, Walter, pharmaceutical chemistry
Sipe, Harry Craig, science education
Skalko, Richard Gallant, anatomy
Smiley, Malcolm Finlay, mathematics
Snow, Dean Richard, anthropology, archaeology
Snow, John Thomas, organic chemistry
Soike, Kenneth F, virology
Spoor, Ryk Peter, pharmacology, physiology
Squires, Donald Fleming, marine biology, paleontology
Stahl, Walter Bernard, parasitology
Stein, Arthur A, medicine, pathology
Stevens, Roy White, medical microbiology, immunology
Stewart, Margaret McBride, vertebrate ecology, herpetology
Stoll, William Russell, pharmacology, chemistry
Story, Harold S, solid state physics, nuclear magnetic resonance
Strominger, Norman Lewis, neuroanatomy
Stross, Raymond George, ecology
Sturman, Lawrence Stuart, virology
Sun, Chih-Ree, high energy physics, nuclear physics
Surrey, Alexander Robert, organic chemistry
Swartz, Donald Percy, obstetrics & gynecology
Tappa, Donald W, ecology
Taylor, Charles Bruce, pathology
Tedeschi, Henry, physiology
Thomas, Edward Sandusky, Jr, mathematics
Thomas, Wilbur Addison, pathology
Tompkins, Victor Norman, pathology
Treble, Donald Harold, biochemistry
Trimble, Robert Bogue, biochemistry, genetics
Truscott, Frederick Herbert, plant physiology
Turinsky, Jiri, physiology, biochemistry
Turner, Nura Dorothea, mathematics
Tutas, Daniel Joseph, chemistry, biochemistry
Urban, Paul, cell biology
Van Camp, Harlan Larue, biophysics
Vanderlinde, Raymond E, biochemistry
Vanko, Michael, biochemistry
Vianna, Nicholas Joseph, epidemiology
Vonnegut, Bernard, physical chemistry
Wallace, Dwight Tousch, anthropology, archaeology
Wallace, Robert Henry, medical physics
Wallach, Stanley, medicine
Waller, Roger Milton, groundwater geology
Wang, Chung Shan, physics
Wasley, William Lingel, textile chemistry, polymer chemistry
Weber, Peter B, biochemistry
Weigel, Christoph, solid state physics
Weinberg, Jerry L, astrophysics, atmospheric physics
Welch, Charles Stuart, surgery
Welch, Harold Francis, surgery
Wharton, James Dumont, public health, preventive medicine
White, Albert M, clinical pharmacy
Whitney, Philip Roy, geochemistry, petrology
Wiggers, Harold Carl, physiology
Wilken, Donald Rayl, pure mathematics
Wingate, Martin Bernard, obstetrics & gynecology
Winn, Hudson Sumner, zoology
Wolfe, Jack Morris, anatomy
Wolin, Meyer Jerome, microbial ecology, microbial physiology
Wong, Patrick Yu-Pei, gastroenterology
Wright, Arthur William, pathology
Wysolmerski, Theresa, zoology, ecology
Yencha, Andrew Joseph, physical chemistry, atmospheric chemistry
Yong, Fook Choy, biological chemistry
Zambernard, Joseph, cytology
Zenner, Walter P, anthropology
Ziegler, Frederick Dixon, biochemistry, physiology
Zuckerman, Jerold J, inorganic chemistry, organometallic chemistry
Zwarich, Ronald James, physical chemistry

ALFRED
Barton, James Don, Jr, plant ecology
Butler, Lewis Clark, mathematics
Carr, Roger Byington, astronomy, physics
Close, Richard Thomas, computer science, electromagnetic theory
Condrate, Robert Adam, solid state chemistry
Crayton, Philip Hastings, inorganic chemistry
Douglass, Roger Thackrey, mathematics
Finlay, Peter Stevenson, zoology
Huntington, David Hans, agriculture
Lamprey, Headlee, chemistry
Love, Robert Lyman, health sciences
Monroe, Eugene Alan, mineralogy, crystallography
Moritz, Roger Homer, mathematics, engineering statistics
Ogden, Elston Gordon, botany
Rausch, James Peter, marine zoology, physiology
Rossington, David Ralph, physical chemistry
Rough, Gaylord Earl, zoology
Rulon, Richard M, physical chemistry, ceramics
Sands, Richard Dayton, organic chemistry
Sass, Daniel B, paleontology, environmental sciences
Scholes, Samuel Ray, Jr, chemistry
Shively, Carl E, bacteriology
Sloan, Robert W, mathematics
Smith, Vincent C, floriculture, marketing
Stopper, William W, poultry husbandry, agricultural economics
Stull, John Leete, physics, ceramics
Taylor, James Addison, geography
Towe, George Coffin, physics, science education

ALPLAUS
Bundy, Francis Pettit, physics

AMHERST
Addelman, Sidney, experimental statistics
Benenson, David Maurice, plasma physics
Bialas, Wayne Francis, operations research, statistics
Conkling, Edgar Clark, geography
Cusick, Thomas William, number theory, combinatorics
Desu, Manavala Mahamunulu, statistics
Dickey, James Mills, statistics
Eberlein, Patricia James, numerical analysis, computer science
Fadell, Albert George, mathematics
Findler, Nicholas Victor, computer science, applied mathematics
Frieder, Gideon, computer sciences
Gelber, Richard David, biostatistics
Giese, Rossman Frederick, Jr, crystallography
Greizerstein, Walter, organic chemistry
Hodge, Dennis, geophysics, petrology
Isbell, John Rolfe, mathematics
Kazarinoff, Nicholas D, mathematics
King, John Stuart, geology
Lagakos, Stephen William, biostatistics
Lavin, Philip Todd, biostatistics

NEW YORK

Lentnek, Barry, economic geography, geography of Latin America
Lisk, George F., organic chemistry
Marble, Duane F., geography
McConnell, James E., economic geography
Mears, Whitney Harris, chemistry
Mielgovski, William Leonard, biostatistics
Mitchell, Josephine Margaret, mathematics
Naroll, Raoul, cultural anthropology, social anthropology, statistical analysis
Newton, Howard Joseph, statistical analysis
Otterbein, Keith Frederick, anthropology
Pagano, Marcello, statistics
Parzen, Emanuel, statistics
Piech, Margaret Ann, mathematics
Ralston, Anthony, computer science
Rein, Robert, quantum chemistry, biophysics
Schoenfeld, David Alan, biostatistics
Sedransk, Joseph Henry, statistics
Stanley, Kenneth Earl, statistics
Steegmann, Albert Theodore, Jr., biological anthropology, physical anthropology
Townsend, Patricia Kathryn, cultural anthropology
Vaidhyanathan, V. S., biophysics
Wan, Yieh-Hei, topology
Williams, Scott Warner, mathematics
Wittie, Larry Dawson, computer science
Zielezny, Zbigniew Henryk, mathematical analysis

AMSTERDAM

Vullo, William Joseph, organic chemistry

ANNANDALE-ON-HUDSON

Josephson, Betty Louise, biology
Kato, Karen Friedman, molecular biology
Rosenthal, Michael R., physical inorganic chemistry

APO NEW YORK

Clement, Duncan, science administration
Coogan, John Michael, research administration, science policy
Ely, Ray E., animal nutrition
Hadley, Donald G., sedimentology, stratigraphy
Hobbs, Jesse H., medical entomology
Killsgaard, Thor H., economic geology
Lacher, Heinrich, pure mathematics
Lee, Vernon Harold, medical entomology
Moore, Michael Cabot, inorganic chemistry
Pryce, Aubrey William, physics
Radke, Myron Glen, medical parasitology, therapeutics
Reichard, Douglas Warren, organic chemistry
Wallace, Craig Kesting, medicine
Worl, Ronald Grant, economic geology
Zimmerman, John Harvey, medical entomology

ARDSLEY

Barclay, Ralph Kinney, biochemistry
Bateman, John Hugh, polymer chemistry, organic chemistry
Brunings, Karl John, organic chemistry, medicinal chemistry
Cargill, David Innes, biochemistry
Cash, William Davis, biological chemistry
Catsiff, Ephraim Herman, physical chemistry, polymer chemistry
Chart, Jerome James, endocrinology
Durrell, William S., organic chemistry, polymer chemistry
Francis, John Elsworth, organic chemistry
Grad, Arthur, mathematics
Kariner, Jerrold, structural chemistry
Kerwar, Suresh, molecular biology
Klenchuk, Peter Paul, organic chemistry
Knaak, James Bruce, biochemistry, toxicology
Knell, Martin, organic chemistry
Ku, Edmond Chiu-Choon, biochemistry
Luders, Richard Christian, analytical chemistry
Malicki, Carol Ann, biochemistry
Maragoudakis, Michael E., biochemical pharmacology
Mungle, Jo Anne, anatomy
Nicolson, Paul Clement, chemistry
Oronsky, Arnold Lewis, biochemistry, physiology
Peterson, Janet Brooks, organic chemistry
Petrack, Barbara Kepes, biochemistry
Psychoyos, Stacy, biochemistry
Ramey, Chester Eugene, organic chemistry
Renfroe, Harris Burt, organic chemistry, surface chemistry
Seltzer, Raymond, polymer chemistry, organic chemistry
Steinberg, David H., organic chemistry
Steinetz, Bernard George, Jr., endocrinology
Ullman, Montague, psychiatry, parapsychology
Watthey, Jeffrey William Herbert, organic chemistry, medicinal chemistry
Weiss, Jonas, polymer chemistry

ARGYLE

Wiegert, Philip E., organic chemistry, physical chemistry

ARMONK

Branscomb, Lewis McAdory, atomic physics, science policy
Clarke, Frank Henderson, organic chemistry, medicinal chemistry
Horton, Thomas Roscoe, mathematics
Joenk, Rudolph John, Jr., solid state physics
Kaufman, Samuel, information science
Schechman, Barry H., solid state physics
Smith, Edgar Clarence, Jr., computer science

AURORA

Berman, Barry L., low temperature physics
Burch, Charles, plant morphology
Byrne, Barbara Jean McManamy, animal genetics
Byrne, Bruce Campbell, animal genetics
Delaney, Charles MacGregor, physical chemistry
Perry, John Murray, mathematics
Shiflett, Ray Calvin, mathematics
Shilepsky, Arnold Charles, mathematics
Sullivan, Patricia Ann Nagengast, biology
Van de Poel, Josephus, organic chemistry, analytical chemistry

BAINBRIDGE

Winslow, Alfred Edwards, organic chemistry

BALLSTON SPA

Johnston, Don Richard, chemical physics

BAY SHORE

Wolpert, Arthur, psychiatry, child psychiatry

BAYSIDE

Anderson, Allan George, mathematics
Arnowich, Beatrice, physical organic chemistry
Berleant-Schiller, Riva, anthropology
Cline, Sylvia Good, cell physiology
Federico, Olga Marie, animal physiology, cytology
Gross, Leo, biophysics
Johansson, Mildred P., biology
Levine, Leo Meyer, entomology
Lowe, Kurt, petrology
Pleasants, Elsie W., inorganic chemistry, organic chemistry
Rance, Hugh, structural geology
Reiner-Deutsch, William, microbiology

BEACON

Arkell, Alfred, organic chemistry
Becker, Harry Carroll, analytical chemistry
Christensen, Edward Richards, petroleum chemistry
Coppoc, William Joseph, chemistry
Davis, Marshall Earl, fuel technology
Dille, Kenneth Leroy, organic chemistry, petroleum chemistry
Foucher, Walter David, Jr., petroleum chemistry
Francis, Stanley Arthur, physical chemistry
Grina, Larry Dale, organic chemistry, petroleum chemistry
Helmuth, Walter Wilhelm, organic chemistry
Herbstman, Sheldon, organic chemistry
Hileman, Robert E., organic chemistry
Holder, Charles Burt, Jr., chemistry
Kablaoui, Mahnoud Shafiq, chemistry
Kerr, Edwin Robert, physical chemistry
Kolaian, Jack H., colloid chemistry, surface chemistry
Konyathy, Joseph Charles, inorganic chemistry, analytical chemistry
Larkin, John Michael, organic chemistry
Lasday, Albert Henry, environmental management
Levine, Stephen Alan, organic chemistry
Ludlum, Kenneth Hills, physical chemistry
Lyons, Joseph F., petroleum chemistry
Mitchell, John Jacob, physical chemistry
Moser, Charles Edwin, chemistry
Paterson, John Arthur, petroleum chemistry
Powell, Justin Christopher, physical organic chemistry
Rand, Salvatore John, physical chemistry
Riordan, Michael Davitt, petroleum chemistry
Rubin, Isaac D., polymer chemistry, organic chemistry
Schallenberg, Elmer Edward, petroleum chemistry, research administration
Siegart, William Raymond, organic chemistry
Stehouwer, David Mark, organic chemistry
Sweeney, William Mortimer, petroleum chemistry
Tessier, John Edward, research administration
Ward, Frank Kernan, organic chemistry, polymer chemistry
Webb, Allen Nystrom, physical chemistry
Weetman, David G., organic chemistry
Wisner, Jackson Ward, Jr., organic chemistry
Yaffe, Roberta, organic chemistry

BEAR MOUNTAIN

Stewart, Donald Borden, environmental management

BEDFORD

Mintz, Esther Uress, physics

BEECHHURST

Eaton, James Edmonds, algebra

BELLEROSE

Sankar, D V Siva, biochemistry
Stenzel, Wolfram G., physics

BELLMORE

Kross, Robert David, polymer chemistry, analytical chemistry

BELLPORT

Damask, Arthur Constantine, physics

BERLIN

Burdick, George Edgar, zoology

BETHPAGE

Armen, Harry A., Jr., applied mechanics
Baukinght, Charles Wesley, physical chemistry
Chang, Joseph Yung, physical chemistry
Egan, Walter George, solid state physics, electrical engineering
Favale, Anthony John, nuclear physics
Garnet, Hyman, applied mechanics, mathematics
Grinoch, Paul, nuclear science
Kivilghn, Herbert Daniel, Jr., physical chemistry, inorganic chemistry
Lisa, Joseph Daniel, physiology, biomedical engineering
Loeffler, Albert L. Jr., magnetohydrodynamics
Popkin, George Lionel, dermatology
Schmied, Edward Joseph, nuclear physics
Tripp, Ralph Harry, mathematics
Yagi, Fumio, mathematics
Yen, Elizabeth Hsi, mathematical statistics

BIG FLATS

Lynn, Merrill, polymer chemistry

BINGHAMTON

Adams, Andrew Borden, anatomy, radiology
Ahluwalia, Gurjit Singh, biochemistry
Barr, Donald Eugene, chemistry
Bartle, Glenn Gardner, geology
Battin, James Edmund, photographic chemistry
Battin, William T., zoology
Beard, Helen Pearl, mathematical analysis
Beerbower, James Richard, paleontology
Bland, Robert Gary, operations research, applied mathematics
Butler, Joseph Herbert, economic geography
Christian, John Jermyn, endocrinology, pathobiology
Clark, Charles Austin, organic chemistry
Coates, Donald Robert, geology
Colligan, John Joseph, electromagnetics
Conant, Robert Henry, photographic chemistry
Constable, James Harris, solid state physics, low temperature physics
Craft, George Arthur, mathematics
Doetschman, David Charles, chemical physics
Donnelly, Thomas Wallace, geology
Donovick, Peter Joseph, psychophysiology
Eisch, John Joseph, organic chemistry
Enos, Paul (Portenier), geology
Fischthal, Jacob Henry, parasitology
Foster, Brian Lee, anthropology
Germann, Donald Pitt, chemistry
Gray, Jeffrey W., applied physics
Greenberg, Newton Isaac, theoretical physics
Grierson, James Douglas, paleobotany, plant morphology
Haber, Alan Howard, radiation botany, plant physiology
Hall, Dick Wick, pure mathematics
Hall, Judy Dale, cell biology
Hanson, David Lee, statistics
Hart, Robert John, physics
Haugh, John Richard, ecology, vertebrate zoology
Hoffmann, Hans, cultural anthropology
Houghton, Charles Joseph, topology
Hull, Carl Max, organic chemistry
Hunter, Hugh Edwards, geology
Innes, Kenneth Keith, chemical physics, molecular physics
Janauer, Gilbert E., physical analytical chemistry
Jennings, Carl Anthony, organic chemistry, photographic chemistry
Jensen, Roy A., microbiology
Jones, Gerald Walter, photographic chemistry
Karl, Robert Raymond, Jr., chemical physics
Kaufman, Donald Wayne, mammalian ecology, evolutionary biology
Kent, James Ronald Fraser, mathematical analysis
Kissling, Don Lester, paleoecology, sedimentology
Klimko, Eugene M., mathematics
Konowalow, Daniel Dimitri, theoretical chemistry
Kronk, Hudson V., mathematics
Kull, Fredrick J., biochemistry
Landry, Stuart Omer, Jr., zoology
Lazaroff, Norman, microbiology
Lercher, Bruce L., mathematical logic
Little, Michael Alan, physical anthropology
Loew, Leslie Max, physical organic chemistry
Lopresi, Frank James, physical chemistry
MacDonald, William David, structural geology
Mackey, Karen Ethel, computer science
Madan, Stanley Krishen, inorganic chemistry
McAuley, Louis Floyd, mathematics
McAuley, Patricia Tulley, topology
McDuffie, Bruce, chemistry
Michelson, Karen L., ethnology
Moore, George Edward, physics
Morisawa, Marie, geology, geomorphology
Mourning, Michael Charles, chemistry
Mueller, August P., immunology, serology
Myers, Clifford Earl, inorganic chemistry, physical chemistry
Nelson, Charles Arnold, high energy & theoretical physics
Norcross, Bruce Edward, organic chemistry
Pattee, Howard Hunt, Jr., theoretical biology, systems theory
Penfield, Robert Harrison, physics
Posner, Herbert Bernard, plant physiology
Raboy, Sol, nuclear physics

Rauch, Emil Bruno, organic chemistry
Redman, Charles Lincoln, anthropology
Rightmire, George Philip, physical anthropology, biological anthropology
Roberson, Herman Ellis, geology, mineralogy
Salisbury, Matthew Harold, geophysics
Schrier, Eugene Edwin, physical chemistry
Schumacher, George John, phycology
Shepherd, Julian Granville, invertebrate physiology
Sheridan, Peter Sterling, inorganic chemistry
Shrift, Alex, plant physiology
Sorauf, James E, geology
Sprung, Joseph Asher, organic chemistry
Starzak, Michael Edward, biophysical chemistry
Sterling, Nicholas J, mathematics
Timofeeff, Nicolay P, environmental science, geography
Transue, William Reagle, mathematical analysis
Trucker, Donald Edward, organic chemistry
Trumbore, Roger H, physiology
Vander Velde, Edward Jay, cultural geography
Van Riper, Joseph Edwards, geography
Verbit, Lawrence, organic chemistry
Vidale, Rosemary J, geology
Wilmoth, James Herdman, biology
Wu, Francis Taming, geophysics
Wu, Tsu Ming, physics
Yao, Jerry Shi Kuang, photographic science
Yeh, Noel Kuei-Eng, particle physics
Ziebur, Allen Douglas, mathematics

BLAUVELT
Pochapsky, Theodore Elias, physical oceanography, ocean engineering

BOHEMIA
Cole, James A, experimental physics, systems engineering
Simpson, Wilburn Dwain, nuclear physics

BREWSTER
Bacon, Ralph Hoyt, engineering physics
Hsu, Tsong-Han, organic polymer chemistry

BRIARCLIFF MANOR
Arnold, Emil, solid state physics
Ault, Wayne Urban, geochemistry
Berger, Jay Manton, theoretical physics, computer science
Bhargava, Rameshwar Nath, physics, solid state physics
Birdsall, Clair Mallery, analytical chemistry
Booth, Eugene Theodore, physics
Chizinsky, Walter, reproductive physiology, vertebrate biology
Corfield, Peter William Reginald, inorganic chemistry, x-ray crystallography
DeBitetto, Dominick John, experimental physics, physical optics
Dougherty, Joseph Patrick, electrooptics
Friar, Wayne, systematic zoology
Gustafson, Carl Gustaf, Jr, organic chemistry
Killion, Philip James, physical biochemistry
Kurtz, Stewart K, electrooptics
Neumark, Gertrude Fanny, solid state physics
Stupp, Edward Henry, solid state physics
Sullivan, Jerry Stephen, solid state electronics
Thomas, Tudor Lloyd, physical chemistry
Tuckerman, Bryant, mathematics
Watts, William Wilbur, academic administration, institutional research
Zernike, Frits, physics
Zwicker, Walter Karl, solid state chemistry

BROCKPORT
Adams, Robert W, geology
Barr, Charles E, cell physiology
Bixler, John Wilson, analytical chemistry
Bobear, Jean B, systematic botany
Cassie, Robert MacGregor, geology
Cloutier, Elmer Joseph, entomology
Clune, Francis Joseph, Jr, archaeology, anthropology
Damann, Kenneth Eugene, botany
Finley, Kay Thomas, physical organic chemistry

Geer, Ira W, science education, meteorology
Gehris, Clarence Winfred, botany, plant ecology
Grunwald, Hubert Peter, solid state physics
Haines, Terry Alan, aquatic ecology
Hammond, H David, botany, plant physiology
Hewitt, Philip Cooper, geology, paleontology
Hill, Derek Leonard, physical chemistry, electrochemistry
Hubbard, John Edward, hydrology, climatology
Kallen, Thomas William, inorganic chemistry
Kline, Larry Keith, molecular biology
Liebe, Richard Milton, geology
Makarewicz, Joseph Chester, limnology
Mancuso, Richard Vincent, nuclear physics
McLean, Robert J, phycology
Miller, Sanford Stuart, mathematical analysis
Morris, John Emory, oncology, biochemistry
Mosher, John Ivan, human ecology
Mumpton, Frederick Albert, mineralogy
Noonan, Thomas Wyatt, astronomy, physics
Petersen, Ingo Hans, organic synthetic chemistry
Pribil, Stephen, thermodynamics, low temperature physics
Pritchard, Parmely Herbert, microbial ecology
Rockhill, Theron D, mathematics
Rumage, Kennard Walter, geography
Schmidt, Victor Edward, geology
Smith, M Estellie, cultural anthropology
Starr, Theodore Jack, microbiology
Stephany, Edward O, mathematics, statistics
Thomas, Charles S, mammalogy, science education
Thompson, Robert Poole, developmental biology
Tillson, David Stanley, anthropology
Vesting, Martha Meredith, organic chemistry

BRONX
Abramson, Morris Barnet, physical biochemistry, neurochemistry
Aiello, Edward Lawrence, pharmacology
Aisen, Philip, biochemistry, medicine
Alms, Gregory Russell, chemical physics
Anchel, Marjorie Wolff (Mrs Herbert Rackow), organic chemistry
Badding, Victor George, organic chemistry
Baker-Cohen, Katherine France, zoology
Bank, Norman, nephrology
Bard, Gily Epstein, plant ecology
Barile, Raymond Conrad, physical chemistry, inorganic chemistry
Barksdale, Alma Whiffen, mycology
Barnett, Henry Lewis, pediatrics
Basile, Dominick V, morphology, bryology
Batt, Conrad William, biochemistry
Baum, Stephen Graham, internal medicine, infectious diseases
Baumgarten, Reuben Lawrence, physical organic chemistry
Becker, Herman Frederick, paleobotany
Becker, Norwin Howard, pathology, histochemistry
Beckerman, Barry Lee, ophthalmology
Beller, Barry M, cardiology, internal medicine
Bellhorn, Margaret Burns, biochemistry
Belserene, Emilia Pisani, astronomy
Bennett, Michael Vander Laan, neurophysiology
Bernabei, Austin M, nuclear physics
Bhatnagar, Anil Kumar, solid state physics, low temperature physics
Biempica, Luis, gastroenterology, pathology
Black, William James, biochemistry, enzymology
Blaufox, Morton D, internal medicine, nuclear medicine
Bloch, Eric, biochemistry
Bloom, Barry R, immunology
Blumenfeld, Olga O, biochemistry
Boissevain, Ethel (Mrs Arthur Lesser, Jr), anthropology, archaeology
Borgese, Thomas A, physiology, biochemistry
Bowman, James Floyd, II, sedimentology, oceanography
Boyar, Robert Martin, endocrinology
Bradlow, Herbert Leon, biological chemistry

Brande, Edward Woodrow, number theory, algebra
Bray, Norman Francis, physical chemistry
Bricker, Neal S, internal medicine
Briehl, Robin Walt, molecular biology, internal medicine
Bruning, Donald Francis, ornithology
Buckley, Nancy Margaret, physiology
Budick, Burton, nuclear physics, atomic physics
Buschke, Herman, neurology, psychology
Campbell, Edwin Stewart, chemical physics
Cardone, Vincent J, meteorology
Carroll, Robert William, physical chemistry
Chase, John William, biochemical genetics, molecular biology
Chen, Yuh-Ching, pure mathematics
Chung, Kuk Pyo, astrophysics, particle physics
Chute, John Lawrence, Jr, geology, oceanography
Clay, John Paul, physical chemistry
Cloney, Robert Dennis, physical chemistry
Cohen, Burton D, internal medicine
Costantino-Ceccarini, Elvira, neurochemistry
Crain, Stanley M, neurophysiology
Cronquist, Arthur John, systematic botany
Crowe, George Joseph, solid state physics
Daly, Marie Maynard, biochemistry
Delson, Eric, anthropology, primatology
Dembitzer, Herbert, cell biology, experimental pathology
Dillemuth, Frederick Joseph, physical chemistry
Dougherty, Charles Michael, organic chemistry
Downs, Frederick Jon, biochemistry
Drosness, Daniel Leed, community health, hospital administration
Dudzinski, Diane Marie, biology
Duncalf, Deryck, medicine, anesthesiology
Dyer, Joan L, mathematics
Eagle, Harry, medicine
Edelmann, Chester M, Jr, pediatrics
Edward, Cosmas, genetics
Eisenstadt, Maurice, physics
Elkin, Milton, radiology
Elliott, Robert Hare Egerton, Jr, physiology, surgery
Engel, Jerome, Jr, neurophysiology, neurology
Engelke, Charles Edward, nuclear physics, atomic physics
Epling, Gary Arnold, physical organic chemistry, photochemistry
Escher, Doris Jane Wolf, cardiology
Etkin, William, zoology
Fabry, Mary E Riepe, biophysical chemistry
Falke, Ernest Victor, molecular genetics
Finberg, Laurence, pediatrics, physiology
Fischel, Edward Elliot, internal medicine
Fishman, Jack, organic chemistry, biochemistry
Forbes, James, entomology
Fraad, Lewis M, pediatrics
Frank, Charles Warren, cardiology, internal medicine
Fredrick, Jerome Frederick, biochemistry
French, Joseph H, pediatrics, neurology
Friedman, Alan Herbert, ophthalmology
Friedman, Ephraim, ophthalmology
Friedman, Max Martin, clinical chemistry
Fujimoto, George Iwao, biochemistry
Fukushima, David Kenzo, biochemistry
Fulop, Milford, internal medicine
Furman, Seymour, surgery
Garcia, Mariano, mathematics
Gaskin, Felicia, physical biochemistry, neurochemistry
Geehern, Margaret Kennedy, atmospheric physics
Geever, Erving Francis, pathology
Gentile, Philip, inorganic chemistry
Giblin, Denis Richard, neurology
Gidez, Lewis Irwin, biochemistry
Gizis, Evangelos John, biochemistry, food science
Gliedman, Marvin, surgery
Goldfischer, Sidney L, pathology, cytochemistry
Goldring, Irene P, experimental zoology, cell physiology
Gollub, Seymour, nuclear medicine
Golubow, Julius, biochemistry
Goodfellow, Robert David, biochemistry, cell physiology
Goodwin, Paul Newcomb, radiological physics

Gordon, Harry Haskin, pediatrics
Greenhill, Maurice H, psychiatry
Greenwald, Bernard William, agricultural chemistry, biochemistry
Gross, Ludwik, cancer
Gueft, Boris, pathology, cytochemistry
Gussin, Arnold E S, plant physiology
Guthwin, Hyman, cell physiology, biochemistry
Hamerman, David Jay, internal medicine
Held, Abraham Albert, mycology
Hellman, Leon, cancer research
Henkind, Paul, ophthalmology
Herbert, Victor, medicine
Hershey, Solomon George, anesthesiology
Hervey, Annette, botany
Herz, Fritz, biochemistry, cell biology
Hilton, James Garrett, internal medicine
Hirano, Asao, neuropathology
Hogan, William, physics
Holmgren, Noel Herman, plant taxonomy
Holmgren, Patricia Kern, plant taxonomy
Horwitz, Susan Band, molecular pharmacology, biochemistry
Hurwitz, Jerard, microbiology
Irwin, Howard Samuel, Jr, taxonomy, botany
Isaac, Richard Eugene, mathematics
Isaacs, Godfrey Leonard, mathematical analysis
Jaffe, Ernst Richard, medicine
Jensen, Thomas E, cell biology
Johnston, Helen, physical chemistry
Jones, C Robert, cell physiology
Kaiser, Irwin Herbert, obstetrics & gynecology
Kane, Conrad Gabriel, physics
Kaplan, Barry Hubert, oncology, biochemistry
Karmen, Arthur, medicine, clinical pathology
Keen, Linda, mathematics
Kern, Harold L, biochemistry
Kern, Michael Don, avian physiology
Kinney, Alvin Edgar, mathematics
Kirchner, Richard Martin, inorganic chemistry, structural chemistry
Klinger, Harold P, genetics
Knutson, Donald Ivar, mathematics
Kochen, Joseph Abraham, pediatrics, hematology
Koerner, Diona Heather, biochemistry, endocrinology
Koss, Leopold George, pathology, cytology
Kowalsky, Arthur, biophysical chemistry, systematic botany
Koyama, Tetsuo, systematic botany
Kuperman, Albert Sanford, pharmacology
Kurczynski, Thaddeus Walter, human genetics, neurology
Lacy, Patricia, microbiology, genetics
Laderman, Jack, mathematics
Lalezari, Parviz, medicine, physiology
Larson, Daniel Lewis, medicine
LaRuffa, Anthony Louis, cultural anthropology
Laufman, Harold, surgery
Lauson, Henry Dumke, physiology
Lawyer, Tiffany, Jr, neurology
Lazarus, Marc Samuel, physical inorganic chemistry
Ledeen, Robert, biochemistry, neurobiology
Lee, Teh Hsun, biochemistry
Leff, Judith, plant physiology
Leung, Irene Sheung-Ying, mineralogy
Levenson, Stanley Melvin, surgery
Levi, Howard, mathematics
Levine, Isidore, internal medicine
Levine, Walter (Gerald), pharmacology
Levy, Susanna Agnes, clinical chemistry, toxicology
Lieber, Charles Saul, internal medicine, nutrition
Lilly, Frank, genetics, oncology
Lisman, Henry, mathematics
Listowsky, Irving, biochemistry
Liverhant, Solomon Elieser, nuclear physics, mathematics
Lowy, Bertram Alan, biochemistry
Ludwig, Armin K, urban geography, geography of Brazil
Luteyn, James Leonard, systematic botany
Maguire, Bassett, systematic botany
Mahony, John Daniel, nuclear chemistry
Maio, Joseph James, microbiology, biochemistry
Maitra, Umadas, biochemistry, molecular biology
Makman, Maynard Harlan, pharmacology, biochemistry
Malhotra, Ashwani, enzymology, cardiovascular physiology
Manner, Georg Karl, biochemistry, pharmacology

Marcus, Donald M., internal medicine, immunochemistry
Mark, Herbert, medicine, cardiology
Marx, Gertie F., anesthesiology
Matz, Robert, internal medicine
McConn, Rita, biochemistry, physiology
McLaughlin, John J (Anthony), microbiology
Mencher, Joan Phyllis, applied anthropology, economic anthropology
Meyer, Paul Richard, anthropology
Michel, John Thomas, plant taxonomy
Mickel, John Thomas, plant morphology
Middleton, John F'M, anthropology
Miller, Herbert Kenneth, biochemistry
Mirone, Leonora, biochemistry
Moore, Cyril A., biochemistry, biophysics
Morell, Pierre, biochemistry, biophysics
Mukerji, Ambuj, nuclear physics
Mukherjee, Asit B., cytogenetics, developmental genetics
Murphy, Daniel Barker, organic chemistry
Nair, Ramachandran Mukundalayam Sivarama, natural product chemistry
Nathan, Helmuth M, surgery, history of medicine
Nathenson, Stanley G, immunobiology, immunochemistry
Newman, Harry, urology
Niemetz, Julian, hematology
Niklas, Karl Joseph, biomathematics, geochemistry
Nitowsky, Harold Martin, medicine, genetics
Norin, Allen Joseph, cell biology, transplantation immunology
Novikoff, Alex Benjamin, biochemistry
O'Connor, Cecilian Leonard, physics
O'Connor, John Francis, biochemistry, neurology, pediatric neurology
Oppenheimer, Jack Hans, endocrinology, internal medicine
Orkin, Louis R, anesthesiology
Pappas, George Demetrios, neurobiology
Pe, Maung Hla, physics
Peisach, Jack, biochemistry
Perlmutter, Frank, horticulture
Pesetsky, Irwin, anatomy, neuroembryology
Phillips, Esther Rodlitz, mathematics, history of science
Poiak, Aaron, obstetrics & gynecology
Polowczyk, Carl John, photochemistry, electrochemistry
Posmentier, Eric S, geophysics
Prance, Ghillean T, botany
Prestwidge, Kathleen Joyce, biology
Prince, Jack, physics
Purpura, Dominick Paul, neurophysiology
Quinn, Cosmas Edward, biology
Rachlin, Joseph Wolfe, aquatic biology
Ragins, Herzl, surgery
Raine, Cedric Stuart, neuropathology
Ralston, Elizabeth Wall, algebra
Rapin, Isabelle (Mrs Harold Oaklander), neurology, pediatric neurology
Reiner, Leopold, medicine
Rickett, Harold William, botany
Ritter, Irving Frederick, mathematics
Ritter, Walter Paul, neuropsychology
Robinson, Edward J., physics
Rodriguez-Trias, Helen, medicine, pediatrics
Rofwarg, Howard Philip, psychiatry
Rogerson, Clark Thomas, mycology
Roheim, Paul Samuel, physiology, medicine
Ronney, Seymour L., obstetrics & gynecology
Rose, Arthur L., neurology, pediatrics
Rose, Israel Harold, mathematics
Rosen, Ora Mendelsohn, enzymology, endocrinology
Rosenbaum, Robert Morris, biology
Rosenfeld, Robert Samson, biochemistry
Rotstein, Jerome, medicine, biochemistry
Rowan, A James, neurology, electroencephalography
Rozett, Richard Walter, analytical chemistry
Ruben, Robert Joel, otolaryngology, cell biology
Rubin, Charles Stuart, biochemistry
Ryall, Ronald W., neuropharmacology
Salzman, Leon, psychiatry
Scharff, Matthew Daniel, cell biology, immunology
Scharrer, Berta Vogel, anatomy

Schaumberger, Norman, mathematics
Schechter, Mannie M, radiology
Scheflen, Albert E., psychiatry, behavioral anthropology
Schein, Clarence Jacob, surgery
Scheinberg, Israel Herbert, medicine
Scheinberg, Labe Charles, neurology
Schildkraut, Carl Louis, biological chemistry
Schimmel, Herbert, biomathematics, biophysics
Schnare, Paul Stewart, mathematics
Schorr, Julian, pediatrics, hematology
Schulman, Harold, obstetrics & gynecology
Schulman, LaDonne Heaton, molecular biology, biochemistry
Schwartz, Harold Leon, endocrinology, biochemistry
Searles, Arthur Langley, organic chemistry
Seifter, Eli, nutrition
Sensenig, Chester, mathematics
Shafritz, David Andrew, molecular biology, medicine
Shapiro, Lucille, molecular biology, biochemistry
Sharpless, Nansie Sue, biochemistry
Shaw, Frederick Carleton, invertebrate paleontology
Shechter, Yaakov, systematics, medical mycology
Shinn, Robert A., internal medicine
Shinn, Seung-il, genetics, cell biology
Shookhoff, Howard Benedict, tropical medicine
Shulman, Harold, mathematics
Sidel, Victor William, community health, preventive medicine
Silverman, Philip Michael, biochemistry, molecular biology
Singer, Jacques Mauriciu, microbiology
Singer, William Merrill, topology
Singh, Pauline Iirik, neuropsychology
Skalski, Stanislaus, solid state physics
Skoultchi, Arthur, cell biology
Smith, Gary Lane, botany
Smith, Gerrit Joseph, theoretical physics, philosophy of science
Smith, Joseph Jay, obstetrics & gynecology
Smith, Norman Obed, physical chemistry
Smoller, Sylvia Wassertheil, biostatistics, epidemiology
Sobel, Edna H., medicine, pediatrics
Soeiro, Ruy, cell biology, biochemistry
Soffer, Richard Luber, biochemistry
Spielholtz, Gerald I., analytical chemistry
Spray, David Conover, neurophysiology
Stanken, Dennis M, physiological ecology, marine zoology
Stanley, Evan Richard, cell biology
Steere, William Campbell, botany
Steeves, Richard Allison, virology
Stefano, George Bogdon, neurobiology
Stein, John Michael, neurophysiology
Sternlieb, Irmin, gastroenterology
Stone, Daniel Joseph, medicine
Sullivan, Daniel Joseph, entomology, animal behavior
Suzuki, Kinuko, pathology, neuropathology
Suzuki, Kunihiko, neurochemistry, neuropathology
Sweeny, Arthur, Jr., organic chemistry
Taylor, Francis B., mathematics
Tenenbaum, Saul, microbiology
Terry, Robert Davis, neuropathology
Thysen, Benjamin, obstetrics & gynecology
Toraballa, Gloria C., chemistry, biochemistry
Ullman, Jack Donald, nuclear physics
Villa, Juan Francisco, inorganic chemistry
Waelsch, Salome Glueckohn, genetics, developmental biology
Walker, Ruth Angelina, organic chemistry
Wallace, Susan Scholes, molecular biology, biophysics
Wallach, Jacques Burton, medicine, pathology
Walton, Cyprian James, biology, entomology
Warner, Jonathan Robert, molecular biology, cell biology
Wasacz, John Peter, organic chemistry
Weber, Alfons, physics
Weil, Peter H., thoracic surgery
Weinreb, Steven Martin, organic chemistry
Weitzman, Elliot D., neurology, neurophysiology

Weitzman, Mary C., histology, endocrinology
White, John Greville, chemistry
Wilber, Laura Ann, audiology, speech pathology
Wise, Harold B., social medicine, internal medicine
Wisniewski, Henryk Miroslaw, neuropathology, pathology
Wittenberg, Beatrice A., biochemistry, physiology
Wittenberg, Jonathan B., biochemistry, physiology
Wittner, Murray, physiology, parasitology
Wolf, Julius, medicine
Wolinsky, Harvey, medicine, pathology
Wu, Cheng-Wen, biophysics, biochemistry
Wu, Felicia Ying-Hsiueh, biophysics, biochemistry
Yalow, Rosalyn Sussman, medical physics
Yellin, Edward L., bioengineering, cardiovascular physiology
Yuen, Po Sang, polymer chemistry, organic chemistry
Zingesser, Lawrence H, radiology
Zinnes, Irving Isadore, theoretical physics, molecular physics
Zunoff, Barnett, medicine

BRONXVILLE

Altschul, Rolf, physical organic chemistry
Cogan, Edward J., mathematical logic
De Carlo, Charles R., mathematics, academic administration
Gardner, William Howlett, polymer chemistry
Goldman, Irving, cultural anthropology, social anthropology
Guggenbuhl, Laura, mathematics
McNatt, Eugene Melton, biophysics
Ortner, Sherry B., cultural anthropology
Riegelhaupt, Joyce Firstenberg, anthropology
Ventriglia, Anthony E., algebra, applied mathematics

BROOKHAVEN

Drew, Ruth Miriam, bacteriology
Kouts, Herbert John Cecil, physics

BROOKLYN

Ackerman, Bruce David, pediatrics
Albaum, Harry Gregory, biochemistry
Alexander, Leslie Luther, medicine, radiology
Alfieri, Gaetano T., physical chemistry
Allen, Arthur Charles, pathology
Allen, Emma Gates, microbiology, virology
Altura, Bella T., physiology
Altura, Burton Myron, physiology, pharmacology
Amassian, Vahe Eugene, neurophysiology
Anderson, Harrison Clarke, pathology
Arlow, Jacob, psychiatry
Aronowitz, Leonard, physics
Aronson, Seymour, physical chemistry
Assadourian, Fred, mathematics
Babaian, Rostom, virology, biology
Bachman, George, mathematics
Backeland, Frederick, psychiatry
Banks, Ephraim, inorganic chemistry
Beam, Carl Adams, microbial genetics, radiology
Beatty, John Joseph, anthropology
Becker, Joshua A., medicine, radiology
Beckman, Frank Samuel, mathematics, computer science
Begletier, Henri, psychophysiology, neurophysiology
Belford, Julius, pharmacology
Bellin, Judith Schryver, physical chemistry, photobiology
Belsky, Melvin Myron, biology
Benton, Joseph George, medicine
Beringer, Frederick Marshall, organic chemistry
Bernheimer, Harriet P., microbial genetics
Bernstein, David, otorhinolaryngology
Bertoni, Henry Louis, electrophysics
Bhattacharji, Somdev, structural geology
Bielski, Benon H J., radiation chemistry
Blau, Harold, mathematics
Bleicher, Sheldon Joseph, internal medicine, metabolism
Bloodstein, Oliver, speech pathology
Bloomfield, Dennis Alexander, cardiology, internal medicine
Boardman, John, theoretical physics
Bollet, Alfred Jay, medicine
Bond, Albert Haskell, Jr., nuclear physics
Bond, Blanche Ann, biochemistry

Borofsky, Samuel, mathematics
Borowsky, Harry Herbert, chemistry
Botino, Michael Louis, geochemistry
Boyer, Carl Benjamin, mathematics
Boyer, Howard, physics
Bramwell, Fitzgerald Burton, physical chemistry
Brase, Peter Charles, chemistry, academic administration
Bregman, Judith, chemical physics
Brooks, Chandler McCuskey, physiology
Brown, Audrey Kathleen, pediatrics, hematology
Brust, Manfred, physiology
Burstein, Samuel Z., applied mathematics
Bushnell, John Hempstead, anthropology
Calabretta, Peter Joseph, organometallic chemistry
Carr, Julius Jay, biochemistry
Carriere, Rita Margaret, histology
Carroll, Harvey Franklin, physical chemistry
Carson, Steven, pharmacology
Casper, Anne Cohen, endocrinology
Celeste, Vincent, mathematics
Chan, Phillip C., chemistry
Charton, Marvin, physical organic chemistry
Chesley, Leon Carey, biochemistry
Cheung, Paul James, marine biology
Choudhury, Deo C., nuclear physics
Chung, Kyung Won, reproductive endocrinology
Clark, Duncan William, medicine
Clark, Edward Aloysius, theoretical physics
Clemetson, Charles Alan Blake, obstetrics & gynecology
Cohen, Irwin A., inorganic chemistry
Cole, Milton Walter, solid state physics, low temperature physics
Collins, Frank Charles, physical chemistry
DeBolt, Lawrence Clifford, polymer chemistry
Debons, Albert Frank, physiology
Detwiler, Thomas C., biochemistry
Dichter, Michael, petroleum chemistry, polymer chemistry
Dickes, Robert, psychiatry
Dilgen, St Francis, organic chemistry
Dillard, Clyde Ruffin, inorganic chemistry
Cook, Albert William, neurosurgery
Cook, Charles Davenport, medicine
Copley, Alfred Lewin, physiology
Craig, John Philip, microbiology
Cramer, Eva Brown, infectious diseases
Crump, Jesse Franklin, biomedical engineering, internal medicine
Damadian, Raymond, biochemistry, biophysics
Dreizen, Paul, biophysics, medicine
Drenick, Rudolf F., applied mathematics
Dropkin, John Joseph, solid state physics
Duncan, Donald Stuart, physics
Eaton, Norman Ray, microbiology
Eckhardt, Ronald A., cell biology
Eichna, Ludwig Waldemar, medicine
Eirich, Frederick Roland, chemistry
Enerman, Sidney, organic chemistry
Emert, Jack Isaac, organic chemistry
Emmanuel, George, cardiopulmonary physiology
Engelhardt, David Meyer, psychiatry
Enquist, Irving Fritiof, surgery
Evans, Hugh E., pediatrics, infectious disease
Fabricand, Burton Paul, physics
Farber, Julius, pharmacology
Feder, Walter, medicine
Feinman, Richard David, biochemistry
Feldman, Joseph Gerald, biostatistics, epidemiology
Felton, Stephen M., organic chemistry
Ferentz, Melvin, physics
Ferraro, John J., organic chemistry
Fialkow, Aaron David, mathematics
Filandro, Anthony Salvatore, organic chemistry
Fink, Austin Ira, ophthalmology
Finkelstein, Abraham Bernard, applied mathematics
Finston, Harmon Leo, nuclear chemistry
Firriolo, Domenic, physiology, anatomy
Ford, Donald Herbert, anatomy, physiology
Forman, William, mathematics
Forster, Sigmund, physical medicine

Franco, Victor, theoretical nuclear physics, atomic physics
Frank, Lawrence, medicine
Freilich, Gerald, mathematics
Fried, George H, physiology
Fried, Vojtech, physical chemistry
Friedman, Eli A, internal medicine, immunology
Friedman, Paul, organic chemistry
Friedman, Sigmund L, medicine
Furchgott, Robert Francis, pharmacology, biochemistry
Gabriel, Mordecai Lionel, biology, academic administration
Galdston, Iago, medicine
Gannon, William E, radiology
Gardner, Bernard, surgery
Garret, Marta, medicine, pathology
Gerber, Bernard Robert, physical chemistry, biology
Gerber, Donald Albert, internal medicine
Gerber, Leon E, geometry
Gewanter, Herman Louis, organic chemistry
Gewirtz, Allan, mathematics
Ghadimi, Hossein, medicine, pediatrics
Gibson, George, chemistry
Gidari, Anthony Salvatore, hematology, physiology
Ginell, Robert, chemical physics
Glass, Leonard, medicine, neonatology
Glickman, Walter A, physics
Glickstein, Joseph, analytical chemistry
Godino, Charles F, mathematics
Goldberg, David Elliott, inorganic chemistry
Goldberg, Martin A, applied mechanics
Goldman, James Allan, academic administration, philosophy of science
Goldner, Martin Gerhard, internal medicine
Goldstein, Solomon, mycology
Gootman, Phyllis Myrna Adler, neurophysiology, physiology
Gortler, Leon Bernard, organic chemistry
Gould, Lawrence A, internal medicine, cardiology
Green, Edward H, electronics
Greenblatt, Irving Jules, biochemistry
Gringauz, Alex, medicinal chemistry
Grob, David, internal medicine
Gross, Milton Michael, psychiatry
Grossman, Edward Joseph, pharmaceutical chemistry
Grosso, Leonard, histology, reproductive physiology
Guggenheimer, Heinrich Walter, mathematics
Gugig, William, organic chemistry
Gustav, Bonnie Lee, physical anthropology, anthropometrics
Guttman, Rita, biophysics, neurophysiology
Haberfield, Paul, organic chemistry
Hamilton, James Bruce, anatomy
Hammerman, David Lewis, physiology
Hankoff, Leon Dudley, psychiatry
Harrison, Nancy Evelyn, algebra
Hayes, Carol J, biological rhythms
Helly, Walter S, operations research
Henneman, Dorothy, medicine, physiology
Herbach, Leon Howard, statistics, engineering statistics
Hillman, Robert Wright, medicine
Himes, Marion, biology, cytology
Hirschberg, Albert I, organic chemistry
Hirschman, Albert, biochemistry, histology
Hirshon, Jordon Barry, algology, biological rhythms
Hirshon, Ronald, mathematics
Hochberg, Murray, mathematics
Hochstadt, Harry, applied mathematics, mathematical analysis
Hoggard, Patrick Earle, physical inorganic chemistry
Horowitz, Carl, polymer chemistry, textile chemistry
Howe, Michael Luray, immunology
Howell, James MacGregor, quantum chemistry
Howery, Darryl Gilmer, physical chemistry
Hsu, Howard Huai Ta, biochemistry, endocrinology
Huang, Suei-Rong, chemistry
Hucke, Dorothy Marie, biostatistics
Hurst, Donald D, genetics
Hurwitz, Solomon, mathematics
Indictor, Norman, organic chemistry, polymer chemistry
Ingalls, James Warren, Jr, pharmacology
Iovino, Anthony Joseph, physiology
Ishida, Takanobu, physical chemistry, nuclear engineering
Jablow, Joseph, anthropology
Jacobson, Homer, physical biochemistry
Jakway, Jacqueline Sinks, anatomy
Jindrak, Karel, pathology
Jirgensons, Arnold, polymer chemistry
Jochsberger, Theodore, physical organic chemistry
Johnson, Robert Hall, organic chemistry
Joseph, Solomon, industrial chemistry
Josephson, Alan S, medicine, immunology
Joshi, Madhusudan Shankarrao, reproductive physiology, endocrinology
Juretschke, Hellmut Joseph, solid state physics
Kallman, Klaus D, genetics, ichthyology
Kanof, Abram, medicine
Kao, Chien Yuan, physiology, pharmacology
Kao, Frederick Fengtien, physiology
Kardonsky, Stanley, nuclear chemistry, analytical chemistry
Kauffman, Shirley Louise, pathology
Kaufman, Albert Irving, medical physiology, neurophysiology
Kawatra, Mahendra P, physics
Kaye, Irving Allan, organic chemistry
Keitel, Hans George Emil, pediatrics
Kenney, Michael, clinical pathology, parasitology
Kesner, Leo, biochemistry, analytical chemistry
Khan, Mozzam Ali, neurochemistry, clinical chemistry
Kim, Giho, clinical biochemistry
King, Benton Davis, medicine
Kirpekar, Sadashiv M, pharmacology
Kirschenbaum, Donald (Monroe), biochemistry
Kissin, Benjamin, internal medicine
Kiszenick, Walter, physics, electrical engineering
Kjeldaas, Terje, Jr, theoretical physics
Klein, Herbert Alan, nuclear medicine
Kleinman, Chemia Jacob, physics
Kleinman, Daniel J, dentistry
Kline, Morris, applied mathematics
Koerting, Lola Elisabeth, genetics, cytology
Kohl, Schuyler G, medicine
Koizumi, Kiyomi, physiology, biochemistry
Kottmeier, Peter Klaus, surgery
Kovacic, Jerald J, algebra
Kraus, Shirley Ruth, physiology, pharmacology
Krieger, Harvey, surgery
Krieger, Joseph Bernard, theoretical solid state physics, atomic physics
Kuan, Hsin Min, physics
Kuchinskas, Edward Joseph, biochemistry
Lake, Robert Edgar, physics
Lander, Patricia Slade, cultural anthropology, social anthropology
Landers, Aubrey Wilfred, mathematics
Laster, Leonard, medicine, science policy
Lavine, Leroy S, orthopedic surgery
Lazarus, Sydney Simon, pathology
Lebowitz, Jacob Mordecai, nuclear physics
Lee, Ching-Tse, animal behavior
Lee, Kwang Soo, pharmacology
Lee, Stanley L, internal medicine, hematology
Lee, Wei-Li S, biochemistry
Lerner, Leon Maurice, biochemistry
LeVeen, Harry Henry, surgery
Levere, Richard David, internal medicine, hematology
Levey, Harold Abram, endocrine physiology
Levi, Enrico, energy conversion, plasma physics
Levin, Gerson, mathematics
Levin, Norman Lewis, zoology, parasitology
Levine, Ira Noel, physical chemistry
Levy, Norman B, psychiatry, psychosomatic medicine
Lewin, Anita Hana, physical organic chemistry
Lieberman, Burton Barnet, mathematics
Liguori, Vincent Robert, marine microbiology
Lippe, Robert Lloyd, chemistry
Lipsey, Sally Irene, mathematics education
Loebl, Ernest Moshe, chemical physics, physical ch
emistry
Lofgren, Ruth, biology, science education
London, Morris, biochemistry
Loscalzo, Anne Grace, microchemistry, analytical chemistry
Lyons, Harold Aloysius, medicine
Ma, Tsu Sheng, microchemistry, organic chemistry
Macomber, Richard Wiltz, paleontology, stratigraphy
Maderson, Paul F A, developmental anatomy
Maier, Mary Louise, inorganic chemistry, physical chemistry
Margolis, Renee Kleinmann, pharmacology, neurochemistry
Mark, Herman Francis, physical chemistry
Marsh, Walton Howard, biochemistry
Marshall, Clifford Wallace, applied mathematics, applied statistics
Mary, Nouri Y, pharmacognosy
Maslow, Philip Herman, organic chemistry
Mazumdar, Purabi, experimental solid state physics
McCormick, Patrick Gary, analytical chemistry
Melnick, Ronald L, cell physiology
Mendelsohn, Lawrence Barry, physics
Menes, Meir, physics
Mennitt, Philip Gary, physical chemistry
Mesnikoff, Alvin Murray, psychiatry, psychoanalysis
Meyer, Leo Martin, hematology
Milman, Doris H, medicine, psychiatry
Mindlin, Rowland L, pediatrics, community health
Minkowitz, Stanley, pathology
Mishkin, Eli Absalom, applied physics
Mittier, James Carlton, physiology, endocrinology
Morawetz, Herbert, physical chemistry
Moriber, Louis G, cell biology
Morse, Stephen Ivor, microbiology, medicine
Mosesson, Michael W, biochemistry, internal medicine
Motzkin, Shirley M, anatomy, developmental biology
Muensterberger, Werner, psychiatry, psychoanalysis
Mule, Salvatore Joseph, pharmacology, biochemistry
Murphy, Donald Henry, bioengineering
Murray, Irwin MacKay, anatomy
Namba, Tatsuji, neurology, pharmacology
Neimark, Harold Carl, microbiology, immunology
Nelson, James H, Jr, obstetrics & gynecology
Nelson, Peter K, botany
Newstein, Maurice, theoretical physics
Nigrelli, Ross Franco, protozoology, parasitology
O'Gorman, John Michael, chemistry
Orenstein, Henry, anthropology
Palocz, Istvan, electrophysics
Park, Samuel, mathematical analysis
Parnell, Jerome Patrick, anatomy
Person, Philip, biochemistry, dentistry
Peterson, Norman Cornelius, physical chemistry
Petrucci, Sergio, physical chemistry
Pinck, Robert Lloyd, radiology
Pincus, Joseph B, biochemistry
Pinkston, James Oliver, physiology
Plakogiannis, Fotios M, pharmaceutics
Plimpton, Calvin Hastings, medicine
Plotz, Charles M, internal medicine, rheumatology
Pokorny, Kathryn Stein, protozoology, cell biology
Post, Benjamin, crystallography
Post, Bernard Saul, rehabilitation medicine, biomedical engineering
Pradhan, Suresh B, pharmacy administration
Prener, Robert, mathematics education
Pryles, Charles Victor, infectious diseases, nephrology
Qazi, Qutubuddin H, pediatrics, genetics
Raeder, Arthur O, orthodontics
Reich, Nathaniel Edwin, medicine
Reichsman, Franz Karl, psychosomatic medicine
Reidlinger, Anthony A, organic chemistry
Rieder, Ronald Frederic, medicine, hematology
Riss, Walter, neuroanatomy, psychology
Rizzuto, Anthony B, microbiology, biochemistry
Rogers, Donald Warren, physical chemistry, analytical chemistry
Rollino, John A, physical chemistry
Romain, Charles B, organic chemistry
Rosen, Milton Jacques, surface chemistry, applied chemistry
Rosenblum, Leonard Allen, ethology
Roth, Benjamin, elementary particle physics
Rothwell, Norman Vincent, cytogenetics
Rubenfeld, Sidney, medicine, radiology
Ruggieri, George D, marine biology
Sahni, Viraht, atomic physics, solid state physics
Saifer, Abraham, biochemistry
Salthe, Stanley Norman, evolutionary biology
Savage, E Lynn, sedimentology
Sawyer, Philip Nicholas, thoracic surgery
Sayres, Alden R, nuclear physics
Scalia, Frank, neurobiology
Scheman, Paul, dentistry, oral medicine
Schmidling, David (Gilbert), organometallic chemistry, theoretical chemistry
Schneck, Larry, pediatric neurology, neurochemistry
Schneider, Joseph, chemistry
Schreibman, Martin Paul, zoology, comparative endocrinology
Schuel, Herbert, cell biology, biochemistry
Schuster, Frederick Lee, protozoology, electron microscopy
Sciarra, John J, industrial pharmacy
Sechzer, Philip Haim, anesthesiology
Seligman, Stephen Jacob, infectious diseases
Selsky, Melvyn Ira, botany
Senitzky, Benjamin, physics
Shafiq, Saiyid Ahmad, cell biology
Shaftan, Gerald Wittes, surgery
Shakin, Carl M, nuclear physics
Shalita, Alan Remi, dermatology, biochemistry
Sharefkin, David Michael, organic chemistry
Sharefkin, Jacob George, organic chemistry
Shaw, Leonard G, systems analysis, electrical engineering
Sheid, Bertrum, biochemistry
Shen, Samuel Yi-Wen, chemistry, physics
Sherman, Burton Stuart, anatomy
Shine, William Morton, organic chemistry
Silverman, Morris, biochemistry, bacteriology
Silverstein, Emanuel, biochemistry, genetics
Simmons, Harry Dady, Jr, medicinal chemistry, internal medicine
Sinder, Leon, anthropology
Singer, James, mathematics
Skorinko, George, physics
Snow, Wolfe, mathematics
Sobel, Michael I, theoretical physics
Solish, George Irving, obstetrics & gynecology, human genetics
Solomon, Nathan A, nuclear medicine
Spain, David M, pathology
Spiegel, Allen David, public health, communications
Stanley, Nathaniel Richard, mathematics
Stein, Philip, medical physics, physics
Steiner, Morris, pediatrics
Stempak, Jerome G, anatomy
Stempel, Edward, pharmacy
Stempien, Martin F, Jr, biochemistry
Stern, Marvin, psychiatry
Stern, Richard M, solid state physics
Stoler, David, elementary particle physics, quantum optics
Stolfi, Julius E, internal medicine
Stracher, Alfred, biochemistry
Strauss, Steven, pharmacy
Struzynski, Raymond Edward, low temperature physics
Stuckey, Jackson H, surgery
Sturm, Edward, geology, mineralogy
Sultzer, Barnet Martin, microbiology
Talbert, George Brayton, endocrinology, anatomy
Tamir, Theodor, electrophysics, electrical engineering
Teiger, Martin, physics, astronomy
Terzuoli, Andrew Joseph, mathematics
Tesser, Herbert, physics
Tittler, Irving Albert, zoology
Tolles, Walter Edwin, biophysics, physiology
Tooney, Nancy Marion, biochemistry, biophysics
Trail, Carroll C, nuclear physics
Tricomi, Vincent, obstetrics & gynecology
Trombetta, Louis David, histology
Vacirca, Salvatore John, radiation physics, health physics
Valenti, Carlo, medicine
Valentine, Wilbur Goodrich, petrology, mineralogy
Van Den Bosch, Frank Joseph Gerard, biophysics, bioengineering
Vasicka, Alois, obstetrics & gynecology
Vassalle, Mario, physiology
Vastola, Edward Francis, medicine, neurology
Vitrogan, David, science education
Volpert, Eugene M, biochemistry, endocrinology
Vroman, Leo, physiology

NEW YORK

Wakade, Arun Ramchandra, pharmacology
Waldinger, Hermann V., mathematics
Wang, Kia K., geology; stratigraphy
Ward, Robert T., zoology; cytology
Waterhouse, Keith R., urology; surgery
Webb, Edmund Leslie, anesthesiology
Weil, Georges Gustave, mathematical analysis; applied mathematics
Weitzner, Stanley Wallace, anesthesiology
Wellmann, Klaus Frederich, anatomic pathology; clinical pathology
Weltman, A Stanley, physiology; endocrinology
Werthamer, Seymour, pathology; computer science
West, William T., histology; anatomy
Wheeler, George Edward, plant anatomy
Wieder, Carl Leslie, Jr, plant morphogenetics
Wiener, Alexander S., immunohematology; medicine science
Williams, Evan Thomas, nuclear chemistry
Wilson, Gustavus Edwin, Jr, bio-organic chemistry
Wiseman, George Edward, inorganic chemistry; organic chemistry
Withner, Carl Leslie, Jr, plant biology; cancer
Wolf, Ira Kenneth, mathematics, information science
Wolfe, Jack, chemistry
Wolin, Harold Leonard, microbiology
Yablonsky, Harvey Allen, physical chemistry
Yang, Dorothy Chuan-Ying, pediatrics, neurology
Yuska, Henry, organic chemistry
Zak, Frederick Gerard, pathology
Zavitsas, Andreas Athanasios, physical organic chemistry
Zeldin, Martel, inorganic chemistry
Zelnick, Ernest, audiology
Zieger, Herman Ernst, organic chemistry
Zuckerberg, Hyam L., occupational medicine
Zuckerman, Israel, mathematics
Zupko, Arthur George, pharmacology

BUFFALO

Abbott, Robert Classic, computer sciences, statistical analysis
Abel, Ernest Lawrence, psychopharmacology
Adler, Richard, surgery
Albert, Oscar J., bacteriology, food
Alexander, A Allan, vertebrate anatomy, taxonomy
Allen, Gary Irving, neurophysiology
Al-Nakeeb, Shaheen Mustafa, veterinary pathology; veterinary surgery
Alvis, Harry J., occupational medicine
Ambrus, Clara Maria, hematology, pharmacology
Ambrus, Julian Lawrence, hematology, oncology
Anderson, Lloyd James, radiophysics
Anderson, Wayne Keith, medicinal chemistry, organic chemistry
Andric, Robert Francis, biogeography, ornithology
Annino, Raymond, analytical chemistry, physical chemistry
Aquilina, Joseph Thomas, medicine
Back, Nathan, pharmacology
Bahl, Om Parkash, chemistry
Baier, Robert Edward, surface chemistry, biophysics
Bailey, William T., mathematics
Bakay, Louis, neurosurgery
Baker, John Cummins, plasma physics
Banker, Richard Burton, immunology
Bannerman, Robin Mowat, genetics, hematology
Bannon, Robert Edward, optometry
Barbeck, Joseph, mathematics
Bardos, Thomas Joseph, medicinal chemistry
Barker, Kenneth Ray, zoology; cell biology
Barnard, Eric A., biochemistry, biochemical pharmacology
Beachley, Orville Theodore, inorganic chemistry

BROOKVILLE

Cahn, Phyllis Hofstein, fish biology; animal behavior
Garretson, Craig Martin, electrophysics
Gordon, Florence S., mathematical statistics

Bealmear, Patricia Maria, experimental pathology; transplantation immunology
Beiter, Marion, mathematics
Belinger, Larry Lee, animal physiology
Bellino, Francis Leonard, steroid biochemistry, molecular biology
Bello, Jake, physical biochemistry
Bender, Merrill Arthur, medicine
Bereman, Robert Deane, inorganic chemistry, bioinorganic chemistry
Berezney, Ronald, cell biology
Bernardis, Lee L., physiology
Bertelli, Rosalie, biomathematics
Beth, Eric Walter, physics
Beuther, Ernst Herman, microbiology
Bieron, Joseph F., biochemistry
Birkett, Frank Elliot, chemistry
Bishop, Beverly Petterson, physiology
Blaisdell, Robert Ferris, systematic botany; plant morphology
Blau, Monte, nuclear medicine
Bloch, Alexander, biochemical pharmacology; cancer
Blumenson, Leslie Eli, mathematical biology; cancer
Bock, Fred G., biochemistry
Boon, Donald Arthur, clinical chemistry
Borst, Lyle Benjamin, physics
Box, Harold C., physics
Boyer, Donald Wayne, physics, aerophysics
Boyian, John W., physiology; medicine
Brady, William Gordon, applied mathematics
Breen, Gail Anne Marie, neurosciences, molecular biology
Breit, Gregory, physics
Brink, Gilbert Oscar, experimental atomic physics
Brody, Harold, anatomy
Brownlee, Alexander C., radiobiology
Bruce, Alan Kenneth, radiobiology
Bruce, Erika, pediatrics
Bruckenstein, Stanley, analytical chemistry, electrochemistry
Buchan, Ronald Forbes, medicine
Buffett, Rita Frances, biology; physiology
Bunnell, Ivan Lee, medicine
Busenberg, Euryblades, geochemistry
Cadenhead, David Allan, physical chemistry
Calkin, Parker, geology
Caputo, Joseph Anthony, physical organic chemistry
Chen, Freeman Philip, chemical physics
Chen, Joseph Ke-Chou, medical microbiology
Cheng, Yung-Chi, biochemical pharmacology
Chheda, Girish B., medicinal chemistry, biochemistry
Chicote, Max Eli, clinical chemistry, laboratory medicine
Chu, Tsann Ming, biochemistry
Churchill, Melvyn Rowen, inorganic chemistry; crystallography
Ciancio, Sebastian Gene, periodontology
Claebaux, Marie Striegel, physical anthropology
Clemency, Charles V., geochemistry, clay mineralogy
Coburn, Robert A., medicinal chemistry, physical organic chemistry
Cody, Vivian, crystallography
Cohen, Elias, immunology
Cohen, Sidney, pharmacy
Collins, June McCormick, anthropology; ethnology
Collord, James, pedodontics
Constantine, Anthony Benedict, pathology
Conway, Walter Donald, organic chemistry
Cook, Gerhard Albert, physical chemistry
Cooper, Robert Michael, pharmacy
Coppens, Philip, crystallography; chemistry
Creaven, Patrick Joseph, clinical pharmacology
Cropp, Gerd J A., physiology; pediatrics
Cudkowicz, Gustavo, immunobiology; transplantation immunology
Dannhauser, Walter, physical chemistry
Dao, Thomas Ling Yuan, medicine
Davis, Paul Joseph, endocrinology
De, Nimai C., bio-organic chemistry

Dean, David Campbell, internal medicine, cardiology
De Luca, Chester, biochemistry
Dentan, Robert Knox, anthropology; American studies
Dickson, Stanley, speech pathology; audiology
Dinan, Frank J., organic chemistry, analytical chemistry
Dobson, Richard Lawrence, dermatology
Donovan, Thomas Arnold, inorganic chemistry
Dorset, Douglas Lewis, crystallography; biological structure
Doubleday, Charles E., Jr, physical organic chemistry
Dougherty, Thomas John, radiobiology
Doyle, Darrell Joseph, biochemistry
Drinnan, Alan John, oral pathology
Duax, William Leo, x-ray crystallography; molecular biophysics
Dutta, Shib Prasad, organic chemistry, medicinal chemistry
Dziak, Rose Mary, cell physiology
Earing, Mason Humphry, organic polymer chemistry
Ebert, Charles H V., physical geography
Edwards, John Anthony, medical genetics
Edwards, Leila, clinical chemistry
Ehrlich, Paul, physical chemistry
Elliott, Rosemary Estridge, bacterial genetics, biochemistry
Elliott, Willard Buford, biochemistry
Ellis, Elliot F., pediatrics
Ellison, Rose Ruth, oncology
Ellison, Solon Arthur, microbiology
English, James Andrew, dentistry
Erasmus, Beth De Wet, physiology
Eschner, Edward George, radiology
Evans, Mary Jo, microbiology, cancer immunology
Evans, Richard Todd, microbiology, immunology
Ewell, Raymond (Henry), chemistry, economics
Faber, Donald S., neurobiology
Farhi, Leon Elie, cardiopulmonary physiology; environmental physiology
Farnsworth, Marjorie Whyte, genetics
Farnsworth, Wells Eugene, biochemistry
Feagans, William Marion, anatomy
Feder, Leo Richard, organic chemistry
Ferber, Kelvin Halket, industrial hygiene, occupational health
Ferencz, Charlotte, pediatric cardiology
Finnegan, Richard Allen, organic chemistry
Fischman, Stuart L., oral pathology
Flanagan, Thomas Donald, virology; immunology
Fleysher, Maurice Henry, medicinal chemistry
Flickinger, Reed Adams, developmental biology
Fopeano, John Vincent, Jr, biochemistry
Fournier, Charles Russell, invertebrate physiology
Frantz, Charles, ethnology; social anthropology
Freeman, Arnold I., pediatrics
Friedman, Harvey Martin, mathematical logic, mathematical analysis
Friedman, Irwin, medicine
Fuda, Michael George, theoretical nuclear physics
Fung, Ho-Leung, pharmaceutics
Gabrieli, Elemer Rudolph, pathology
Gailani, Salman, medicine, cancer
Garrick, Laura Morris, molecular biology
Garrick, Michael D., biochemistry
Gasparini, Francis Marino, low temperature physics
Geisler, Fred Harden, medical physiology
Gelbaum, Bernard Russell, mathematics
Genco, Robert J., immunochemistry, microbiology
Gessner, Peter K., pharmacology
Gibaldi, Milo, pharmacology
Glomski, Chester Anthony, anatomy, hematology
Gnau, Donald Vaughn, electronics
Good, Robert James, surface chemistry, chemical engineering
Gordon, Mildred, cell biology
Grafton, Thurman Stanford, laboratory animal medicine
Green, Larry J., orthodontics
Green, David Gorham, cardiology
Greene, George W., Jr, oral physiology
Greizerstein, Hebe Beatriz, pharmacology
Grossberg, Allan Louis, immunology
Grushka, Eli, physical chemistry, analytical chemistry
Gulati, Subhash Chander, molecular biology
Gurney, Ramsdell, internal medicine

Gurwara, Sweet K., medicinal chemistry, cancer
Guthrie, Robert, medical genetics, microbiology
Guttuso, James, endodontics
Hadley, Wayne Franklin, animal behavior, ecology
Hallpern, Raoul, mathematics
Hakala, Maire Tellervo, biochemical pharmacology
Han, Tin, immunology, cancer
Hanson, Perry Oliver, geography
Hanson, Susan Easton, geography
Harker, David, crystallography
Harris, Gordon McLeod, physical chemistry
Harris, Richard Allen, applied mathematics, system analysis
Hartman, James Keith, surface physics, electronic engineering
Hastings, Stuart Pendleton, mathematics
Haupert, John Selby, geography
Hauptman, Herbert Aaron, crystallography; physiological ecology
Hausmann, Ernest, biochemistry
Hayes, Everett Russell, anatomy
Hebeler, Peter, biochemical pharmacology
Heberle, Juergen, electrodynamics, experimental physics
Helmstetter, Charles E., biophysics
Henderson, Edward S., internal medicine, oncology
Herreid, Clyde F. II, comparative physiology, physiological ecology
Hreshchyshyn, Myroslaw M., obstetrics & gynecology
Hichar, Joseph Kenneth, physiology,
Hill, William Joseph, statistics
Hiller, Lejaren Arthur, acoustics
Ho, John Ting-Sum, experimental solid state physics
Hohmann, Philip George, biochemistry
Holtermann, Ole A., microbiology, biology
Hong, Chung H., medicinal chemistry
Hong, Suk Ki, physiology
Howard, Richard John, physics
Howell, Barbara Jane, physiology
Hoyer, Fridolin Alfonse, organic chemistry
Huang, Chester Chen-Chiu, biology; cytogenetics
Hubbard, John Casterman, pathology
Huebsch, William M., topology
Hull, McAllister Hobart, Jr, theoretical physics, nuclear physics
Hullar, Theodore Lee, organic chemistry, medicinal chemistry
Iijima, Herbert, biochemistry; molecular biology
Ilmet, Ivar, physical chemistry, analytical chemistry
Ingall, John, cancer, surgery
Ip, Clement Cheung-Ying, biochemistry
Ip, Margot Morris, biochemistry; biology
Isseroff, Hadar, physiology; biology
Jacobson, Kenneth Allan, biophysics
Jager, Blair Valdemar, internal medicine
Jain, Pyare Lal, physics
Janik, Daniel Robert, lasers
Jarvis, Richard Stanley, hydrology
Jones, Edward Stephen, organic chemistry
Jones, Oliver Perry, anatomy
Jupnik, Helen, optics
Jusko, William Joseph, pharmacy
Kallen, Frank Clements, vertebrate anatomy
Kalman, Thomas Ivan, biochemical pharmacology; medicinal chemistry, neurology
Kalyanaraman, Krishnaswamy, polymer chemistry
Kamath, Vasanth Rathnakar, polymer chemistry
Kaplan, Melvin, organic chemistry
Kartha, Gopinath, physics
Katz, Jack, audiology
Katz, Ulrich, cloud physics
Kaufman, Sol, operations research, mathematical logic
Koenig, Edward, neurochemistry
Kino, Akiko, mathematical logic
Kinzly, Robert Edward, electrooptics, optical engineering
Kirdani, Rashad Y., organic chemistry; biochemistry
Kite, Joseph Hiram, Jr, immunology
Klein, Edmund, medicine, dermatology
Klingman, Gerda Isolde, pharmacology
Klingman, Jack Dennis, biochemistry
Klocke, Robert Albert, pulmonary diseases; pulmonary physiology
Koester, Charles John, optics
Kohler, Robert Henry, physics
Korynyk, Walter, medicinal chemistry, biochemistry
Kosman, Daniel Jacob, biochemistry

Kowalski, David Francis, biochemistry
Krasner, Joseph, biochemistry
Krasney, John Andrew, cardiovascular physiology
Kress, Lawrence Francis, biochemistry
Kresse, Jerome Thomas, organic chemistry
Kristal, Mark Bennett, behavioral biology
Kroeker, Warren Dean, biochemistry
Kumar, Vijay, laboratory medicine
Kundu, Nakuleswar, organic chemistry, biochemistry
Kurland, Robert John, nuclear magnetic resonance, biophysical chemistry
Lambert, Reginald Max, bacteriology, immunology
Lanigan, M Regina, bacteriology, cell physiology
Lanphier, Edward Howell, medicine, physiology
Larson, Donald Alfred, botany
Laskowski, Michael, biochemistry
Laug, George Milton, biology
Lawani, Samuel Adetunji, physical chemistry
Lawvere, Francis William, algebra
Lee, James B, endocrinology, metabolism
Lee, Joseph Ching-Yuen, anatomy
Lee, Lih-Syng, biochemistry, biophysics
Lees, Helen, biochemistry
Lehotay, Judith Mona, forensic medicine, pathology
Leibovic, K Nicholas, neurosciences, biophysics
Leone, James A, physical chemistry, instrumentation
Levitt, Abel, internal medicine
Levy, Gerhard, pharmacology
Lewis, Elmer James, organic chemistry, physics
Libby, Paul Robert, biochemistry
Lin, Duo-Liang, physics
Linzer, Rosemary, oral microbiology
Lippschutz, Eugene J, cardiology
Lorch, Joan, cell biology, protozoology
Lore, John M, Jr, otolaryngology, surgery
Lusis, Aldons Jekabs, molecular biology
Lyons, Joseph Paul, operations research, public health administration
MacGillivray, Archibald Dean, applied mathematics
MacGillivray, Margaret Hilda, endocrinology, immunology
MacKnight, Franklin Collester, petroleum geology
MacLeay, Ronald E, synthetic organic chemistry
Magill, Kenneth Derwood, Jr, mathematics
Magorian, Thomas R, geophysics
Magoss, Imre V, urology
Manly, Kenneth Fred, virology
Marchetta, Frank Carmelo, medicine
Markus, Gabor, biochemistry
Marra, Edward Francis, preventive medicine
Mashimo, Paul Akira, oral microbiology
Maslow, David E, embryology, cell biology
Massaro, Edward Joseph, toxicology, biochemistry
Mayers, George Louis, bio-organic chemistry, immunochemistry
Mayhew, Eric George, cell biology
McCarthy, Paul James, physical inorganic chemistry
McIsaac, Robert James, pharmacology
McIver, James W, Jr, quantum chemistry
McLimans, William Fletcher, virology, microbiology
McMenamy, Rapier Hayden, biochemistry, physical chemistry
McPartland, Richard Paul, biochemistry
Meisler, Miriam Horowitz, biochemistry
Meloon, Daniel Thomas, Jr, analytical chemistry
Merrick, Joseph M, biochemistry, microbiology
Mihich, Enrico, pharmacology
Milano, Michael John, analytical chemistry
Miles, Philip Giltner, botany
Milgrom, Felix, medical bacteriology, immunology
Mindell, Eugene R, orthopedic surgery
Mink, Irving Bernard, hematology
Minowada, Jun, pathology, virology
Mirand, Edwin Albert, biology
Mittelman, Arnold, surgery, medicine
Mohn, James Frederic, immunology, medical bacteriology
Montgomery, Mabel D, mathematics
Mott, Thomas, mathematics
Mukherjee, Anil B, cell biology, genetics
Munson, Benjamin Ray, biochemistry, oncology
Murphey, William Howard, biochemistry, microbiology
Myhill, John, mathematics
Nancollas, George H, physical chemistry, inorganic chemistry
Narotsky, Saul, veterinary medicine, poultry pathology
Naughton, John Patrick, internal medicine, cardiology
Neeman, Moshe, organic chemistry, biochemistry
Neter, Erwin, bacteriology, immunology
Newberger, Edward, mathematics
Nickerson, Peter Ayers, pathology, endocrinology
Noble, Robert Warren, Jr, biophysical chemistry
Noell, Werner K, physiology
Nolan, James P, medicine
Noller, David Conrad, organic chemistry
Northcott, Jean, physical chemistry, organic chemistry
Novotny, Jaroslav, medicinal chemistry
Noyes, Wilbur Fiske, virology
O'Hare, George Alfred, chemistry
Ohki, Shinpei, biophysics
Ohr, Eleonore A, cell physiology
Opler, Marvin Kaufmann, anthropology, psychiatry
Ortman, Harold R, dentistry
Osawa, Yoshio, organic chemistry, endocrinology
Padwa, Albert, organic chemistry
Paganelli, Charles Victor, biophysics, physiology
Paigen, Kenneth, biochemical genetics
Pandit, Hemchandra M, animal physiology
Pannill, Fitzhugh Carter, Jr, internal medicine
Papahadjopoulos, Panayotis Demetrios, biochemistry, biophysics
Parsons, Donald Frederick, biophysics, molecular biology
Parthasarathy, Rengachary, biophysics, crystallography
Payne, S Howard, prosthodontics
Pentney, Roberta Pierson, biology, neuroanatomy
Pepkowitz, Leonard Paul, chemistry
Phelan, John T, medicine
Phillips, Paul John, polymer science
Pickren, John Warren, pathology
Piech, Kenneth Robert, optical physics
Pierucci, Olga, molecular biology, physics
Pine, Martin J, bacteriology, biochemistry
Plager, John Everett, biochemistry, internal medicine
Poste, George Henry, experimental pathology
Powell, Richard Anthony, dentistry
Pragay, Desider Alexander, clinical biochemistry, clinical chemistry
Preisler, Harvey D, oncology, cell biology
Prentice, Theodore C, medicine
Pressman, David, immunochemistry
Price, Frederick William, molecular biology, biochemistry
Privitera, Carmelo Anthony, biology
Rahn, Hermann, physiology
Ram, Michael, theoretical physics
Randall, Eric A, bryology
Rattazzi, Mario Cristiano, human genetics, biochemical genetics
Raymonda, John Warren, physical chemistry
Rechnitz, Garry Arthur, analytical biochemistry
Reeves, Robert Blake, physiology
Reichert, Jonathon F, experimental solid state physics, magnetic resonance
Reichlin, Morris, medicine
Reitan, Paul Hartman, geochemistry, petrology
Rennick, Barbara Ruth, physiology
Rennie, Donald Wesley, physiology
Reuss, Ronald Merl, human biology, science education
Reynard, Alan Mark, pharmacology
Riggs, Douglas Shepard, pharmacology
Ritchie, Calvin Donald, organic chemistry
Roalsvig, Jan Per, experimental nuclear physics, high energy physics
Robinson, Martin Alvin, inorganic chemistry
Roholt, Oliver A, Jr, immunology
Rohrer, Douglas C, x-ray crystallography
Rosen, Fred, biochemistry, nutrition
Rosen, Robert, mathematical biology
Ross, Alexander, organic chemistry
Rothstein, Morton, biochemistry
Ruddick, James John, physics
Rustgi, Moti Lal, nuclear physics
Sachs, Frederick, biophysics
Sachs, Mendel, theoretical physics
Sako, Kumao, medicine, surgery
Salomone, Ramon Angelo, organic chemistry
Saltarelli, Cora G, mycology, genetics
Sandberg, Avery Aba, internal medicine
Sanders, Benjamin Elbert, biochemistry
Sansone, Frances Marie, human anatomy
Sarcione, Edward James, biochemistry
Saroff, Jack, biochemistry, endocrinology
Schaaf, Norman George, maxillofacial prosthetics, prosthodontics
Schanuel, Stephen Hoel, mathematics
Schenk, Worthington G, Jr, surgery
Schimert, George, cardiovascular surgery
Schlagenhauff, Reinhold Eugene, neurology
Schmidt, Richard Nicholas, statistics, mathematics
Schneckenburger, Edith Ruth, mathematics
Schwartz, Michael Averill, pharmaceutical chemistry
Scott, Peter Douglas, biomedical engineering
Segal, Harold Lewis, biochemistry
Sengbusch, Howard George, parasitology
Seon, Byeong Kuk, immunology, cancer
Seyfert, Carl K, Jr, structural geology, petrology
Sharma, Ram Ashrey, medicinal chemistry
Shedd, Donald Pomroy, surgery
Shefter, Eli, medicinal chemistry, pharmaceutics
Sheppard, Chester Stephen, organic chemistry
Shimaoka, Katsutaro, medicine, oncology
Siegel, John H, surgery, physiology
Siemankowski, Francis Theodore, geology, science education
Siemens, Albert John, pharmacology
Simmons, Noel, biochemistry
Singh, Surjit, physical chemistry, analytical chemistry
Sinha, Dilip, endocrinology, zoology
Sinks, Lucius Frederick, biophysics, pediatrics
Sirianni, Joyce E, physical anthropology, dental anthropology
Slack, Nelson Hosking, biostatistics
Slaunwhite, Wilson Roy, Jr, biochemistry
Small, Saul Mouchly, psychiatry
Smith, Alden Ernest, science education, aquatic biology
Smith, Bernard H, neurology
Smith, Cedric Martin, pharmacology
Smith, Charles James, neuropsychology
Smith, David August, geography
Smith, George David, physical chemistry, x-ray crystallography
Snell, Fred Manget, biophysics
Sokal, Joseph Emanuel, internal medicine, oncology
Solo, Alan Jere, medicinal chemistry, organic chemistry
Spangler, Robert Alan, biophysics
Srebro, Richard, physiology
Srivastava, Bejai Inder Sahai, biochemistry, molecular biology
Stanton, Richard Edmund, theoretical chemistry
Staple, Peter Hugh, physiology, dentistry
Staubitz, William Joseph, medicine, urology
Stein, Arthur, mathematical statistics
Stein, Robert Carrington, ornithology, bioacoustics
Steinmeier, Robert C, biochemistry
Stern, Samuel T, mathematics
Stewart, Kenton M, zoology, limnology
Storr, John Frederick, aquatic ecology
Stouter, Vincent Paul, zoology, neuroendocrinology
Stusnick, Eric, acoustics
Stutzman, Leon, internal medicine, oncology
Sulkowski, Eugene, biochemistry
Surgalla, Michael Joseph, medical microbiology
Surgenor, Douglas MacNevin, biochemistry
Swamy, Vijay Chinnaswamy, pharmacology
Swank, Richard Tilghman, biochemistry
Swartz, Gordon Elmer, zoology, embryology
Sweeney, Robert Anderson, phycology, limnology
Szuchet, Sara, physical biochemistry
Tamari, Dov, mathematics
Targowski, Stanislaw P, immunology, veterinary microbiology
Tesmer, Irving Howard, stratigraphy, paleontology
Thielking, David H, physics
Thomas, Charles Carlisle, Jr, academic administration, nucleonics
Tieckelmann, Howard, bio-organic chemistry
Treanor, Charles Edward, physics, aerodynamics
Treanor, Katherine P, genetics
Triggle, David J, pharmacology, medicinal chemistry
Tritsch, George Leopold, biochemistry
Tufariello, Joseph James, organic chemistry
Tunis, Marvin, biochemistry
Tweto, John Halvor, biological chemistry
Updegraff, William Edward, nuclear physics, computer science
Urban, John, science education, biology
Urbscheit, Nancy Lee, respiratory physiology, physical medicine & rehabilitation
Uschold, Richard L, mathematics
Van Hattum, Rolland James, speech pathology, audiology
Van Liew, Hugh Davenport, medical physiology
Van Liew, Judith Bradford, physiology
van Oss, Carel J, immunochemistry, physical biochemistry
Van Verth, James Edward, organic chemistry
Vlad, Peter, cardiology
Voorhess, Mary Louise, pediatrics, endocrinology
Wagner, Charles Roe, organic chemistry
Wagner, Gerald Roy, organic chemistry, geology
Wang, Jui Hsin, biochemistry
Ware, Carolyn Bogardus, neuroanatomy, physiological psychology
Warfel, John Hiatt, anatomy
Warner, Paul Longstreet, Jr, medicinal chemistry
Warner, Robert, pediatrics
Waters, Donald Hilton, neuropharmacology
Webber, Richard Harry, anatomy
Weeks, Charles Merritt, biophysics
Weinfeld, Herbert, biochemistry
Weiss, Leonard, pathology, cell biology
Wenner, Charles Earl, biochemistry
Wiles, Robert Allan, industrial organic chemistry
Williams, Timothy Cheney, animal behavior
Wilson, Richard J, physics, optics
Winkler, Norman Walter, biochemistry, dermatology
Winkler, Sheldon, dentistry
Wobschall, Darold C, biophysics, electrical engineering
Wurster, Walter Herman, physics
Yalkovsky, Ralph, oceanography
Yang, John Yun-Wen, environmental chemistry
Yang, Man-chiu, biochemistry
Zakrzewski, Sigmund Felix, biochemistry
Zapisek, William Francis, biochemistry, developmental biology
Zeigel, Robert Francis, virology, cytology
Zeschke, Richard Herman, family medicine, immunology
Zielezny, Maria Anna, biostatistics
Zingaro, Joseph S, science education, chemistry
Zobel, Carl Richard, molecular biophysics, biophysical chemistry
Zottoli, Steven Jaynes, neurobiology
Zusman, Jack, psychiatry, public health

BURNT HILLS
Launer, Philip Jules, analytical chemistry

BURT
Kolczynski, James Robert, organic polymer chemistry

CANANDAIGUA
Mariner, Allen Shan, psychiatry
Mooney, Charles Frank, physics

CANTON
Ash, William James, animal genetics, developmental genetics
Baker, David Kenneth, physics
Bloomer, Robert Oliver, petrology
Bowers, Landon Emanual, bacteriology
Budd, Thomas Wayne, plant physiology, molecular biology
Crowell, Kenneth L, vertebrate ecology, zoogeography
Crowell, Robert Merrill, zoology, entomology
Elberty, William Turner, Jr, geology
Erickson, John Mark, invertebrate paleontology
Finch, Thomas Lassfolk, physics
Gage, Clarke Lyman, organic chemistry
Green, John Irving, ecology, botany

NEW YORK

Hornung, David Eugene, physiology
Lufburrow, Robert Allen, physics
Oesper, Peter, biological chemistry
Parker, Francis Dunbar, applied mathematics
Peckham, Donald Charles, science education
Robinson, William Kirley, physics
Romer, Alfred, physics, science education
Romey, William Dowden, geology, science education
Rupp, John Jay, inorganic chemistry
Stradling, Samuel Stuart, organic chemistry
Street, James Stewart, geology
Strodt, Walter Charles, mathematics
Turcheck, Joseph Edward, applied mathematics
Warner, Edward Nelson, ichthyology
Warner, Frederic Cooper, geometry
Wells, Russell Frederick, biology
West, Kenneth Calvin, analytical chemistry
Yourtee, Lawrence Karn, organic chemistry

CAYUGA
Rowoth, Olin Arthur, poultry nutrition, poultry husbandry

CENTRAL ISLIP
Blume, Sheila Bierman, psychiatry
Merlis, Sidney, psychiatry, neurology

CENTRAL VALLEY
Sharma, Mahesh Chandra, mathematics

CHAPPAQUA
Agresta, Joseph, physics, applied mathematics
Clum, Harold Haydn, plant physiology
Dayan, Jason Edward, organic chemistry
Goldsby Arthur Raymond, organic chemistry
Willoughby, Ralph Arthur, mathematics

CHAZY
Czarnetzky, Edward John, agriculture
Earl, Alfred Ellsworth, toxicology
Pollock, John Joseph, toxicology, pharmacology

CHEEKTOWAGA
Wolinski, Leon Edward, organic polymer chemistry

CHENANGO FORKS
Moore, Ralph Gower Davies, organic chemistry

CHURCHVILLE
Aradine, Paul William, inorganic chemistry, analytical chemistry

CLARENCE
Greatbatch, Wilson, biomedical engineering
Hopkins, George C., polymer chemistry
Mason, John Hugh, polymer chemistry

CLIFTON PARK
Nystrom, Richard Alan, physiology, medical education

CLINTON
Boggs, Norman Towar, III, organic chemistry
Cameron, George Harvey, physics
Chiquoine, A Duncan, cell biology
Cratty, Leland Earl, Jr., physical chemistry, surface chemistry
Denney, Donald John, physical chemistry
Ellis, John Francis, biology
Gerold, Nicolas John, histochemistry
Gulick, Walter Lawrence, psychophysics, psychophysiology
Hawley, David, structural geology
Jones, Grant Drummond, stratigraphy
Kinnel, Robin Bryan, organic chemistry, ethnology
McManus, Lawrence Robert, ecology
Pearle, Philip Mark, theoretical physics
Potter, Donald N., geology
Putala, Eugene Charles, botany
Ring, James Walter, chemical physics
Rogers, Philip Virgilius, endocrinology

COBLESKILL
Smalley, Ralph Ray, agronomy
Southwick, Richard Arthur, plant breeding

COLD SPRING HARBOR
Albrecht-Buehler, Guenter Wilhelm, cell biology
Broker, Thomas Richard, molecular genetics
Bukhari, Ahmad Iqbal, molecular biology, genetics
Chow, Louise Tsi, molecular biology
Gesteland, Raymond Frederick, biochemistry, molecular biology
Grodzicker, Terri Irene, molecular genetics
Hershey, Alfred Day, virology
Lewis, James Bryan, molecular biology
McClintock, Barbara, genetics, cytology
Roberts, Richard John, molecular biology
Rothermel, Barbara, chemistry
Topp, William Carl, cell biology
Watson, James Dewey, molecular biology
Zipser, David, molecular genetics

COLONIE
Hoguet, Robert Gerard, polymer chemistry

CONGERS
Kramer, Alfred William, Jr., morphological biology
Stasiw, Roman Orest, clinical chemistry, physics

COOPER SQUARE
Forman, Stanley Maurice, physics

COOPERSTOWN
Ashley, Charles Allen, pathology
Blumenstock, David A., medicine
Bordley, James, III, medicine
Kydd, David Mitchell, medicine, metabolism
Reed, Roberta Gable, bio-organic chemistry
Sauer, Leonard A., cell biology, biochemistry
Vaules, David Wilson, cardiology
Warren, William A., biochemistry

CORNING
Adams, P B, surface chemistry
Ambrosine, Joseph Paul, chemistry
Armistead, William Houston, Jr., chemistry, glass technology
Baum, George, organic biochemistry
Beall, George Halsey, geochemistry, ceramics
Borrelli, Nicholas Francis, optical physics
Boyd, David Charles, glass technology
Brill, Robert H., physical chemistry
Britton, Marvin Gale, mineralogy, ceramics
Carpenter, T J, organic chemistry, analytical chemistry
Dalton, Robert Hennah, chemistry
Davis, James K., chemistry, research administration
Dumbaugh, William Henry, Jr., inorganic chemistry, physical chemistry
Evans, Doris L., crystallography, physics
Fehlner, Francis Paul, physical chemistry
Filbert, Augustus Myers, physical chemistry
Fischer, David John, physical chemistry
Garfinkel, Harmon M., physical chemistry, bio-organic chemistry
Gustin, Vaughn Kenneth, analytical chemistry
Guyer, Edwin Michael, physics
Hares, George Bigelow, inorganic chemistry
Herczog, Andrew, physical chemistry
Hersh, Leroy S., physical biochemistry
Hertl, William, physical chemistry
Hoekstra, Karl Egmond, mineralogy, glass technology
Holland, Hans J., inorganic chemistry, x-ray crystallography
Hood, Harrison Porter, physical chemistry
Kane, William Theodore, x-ray crystallography
Keck, Donald Bruce, physics
Kozlowski, Theodore R., physical inorganic chemistry
Lawless, William N., solid state physics
Lepp, Cyrus Andrew, clinical biochemistry, clinical chemistry
Lichtenstein, Ivan Edgar, analytical chemistry, inorganic chemistry
MacAvoy, Thomas Coleman, organic chemistry
MacDowell, John Fraser, geology, chemistry
Maurer, Robert Distler, applied physics
Medrud, Ronald Curtis, x-ray crystallography, ceramics
Messing, Ralph Allan, enzymology
Munier, John Hammond, physics
Olshansky, Robert, theoretical physics
Oppermann, Robert Arthur, physical chemistry
Parker, Charles Jeremiah, physics
Plummer, William Allan, physical chemistry
Ramsey, William Scott, microbiology
Reade, Richard Francis, physical chemistry
Rothermel, Daphne Land, inorganic chemistry
Rothermel, Joseph Jackson, chemistry, glass technology
Schreurs, Jan Willem Herman, physical chemistry, magnetic resonance
Schucker, Gerald D., analytical chemistry
Segatto, Peter Richard, analytical chemistry, physical chemistry
Seward, Thomas Philip, III, physics
Shaver, William Walker, chemistry, glass technology
Sheldon, John Lewis, chemistry, glass technology
Shoup, Robert D., physical inorganic chemistry
Smith, David Preston, applied physics
Spremulli, Paul Francis, physics
Stookey, Stanley Donald, physical chemistry
Su, Yao Sin, analytical chemistry
Swinehart, Bruce Arden, analytical chemistry
Van Cott, Harrison Corbin, mineralogy
Voss, Raymond Olson, chemistry, microscopy
Weedall, Howard H., immunochemistry, enzymology
Wexell, Dale Richard, inorganic chemistry
Williams, Jean Paul, analytical chemistry
Yaverbaum, Sidney, medical microbiology, immunochemistry

CORONA
Feinberg, Donald Lester, neurology, pediatrics

CORTLAND
Allen, Ross Lorraine, public health
Batzing, Barry Lewis, microbiology
Brennan, Daniel Joseph, stratigraphy, sedimentology
Brownell, Joseph William, geography
Chaturvedi, Ram Prakash, atomic physics
Clemens, William Bryson, bacteriology
Fisk, George Raymond, ornithology
Gustafson, John Alfred, animal nutrition
Harsh, John F., groundwater hydrology
Hawkins, William Max, economic geology, mineralogy
Hay, Robert E., economic geology
Heaslip, William Graham, invertebrate paleontology
Horak, Karel, anatomy, physiology
Jones, Richard Conrad, botany
Leaf, Boris, statistical mechanics
Leninger, Charles W., mathematics
Leon, Kenneth Allen, fish biology
Mason, Elliot Bernard, physiology
McConnell, James Francis, organic chemistry
Miller, Richard J., physical chemistry
Nasrallah, Mikhail Elia, biology, genetics
Podoliak, Henry Andrew, analytical chemistry
Poston, Hugh Arthur, reproductive physiology, animal nutrition
Reynolds, Harlan Kendall, experimental atomic physics
Reynolds, Norman Bruce, physiology
Schick, Robert Dean, phycology
Silberman, Robert G., organic chemistry
Spence, Alexander Perkins, comparative anatomy, embryology
Spink, Charles Harlan, analytical chemistry
Swinehart, James Stephen, organic chemistry
Waldbauer, Eugene Charles, natural history

CROSS RIVER
Cole, Henderson, physics

CROTON-ON-HUDSON
Brennan, Donald George, mathematics, national security
Halitsky, James, air pollution, meteorology

DEER PARK
Packard, Karle Sanborn, Jr., physics

DELANSON
Cofrancesco, Anthony J., industrial organic chemistry

DELHI
Hepple, Harold Rhine, chemistry
Richards, James Austin, Jr., physics
Stone, Winfield S., veterinary medicine

DELMAR
Archer, Sydney, medicinal chemistry, organic chemistry
Bennett, James Gordy, Jr., organic chemistry
Brady, George W., physical chemistry
Hawthorne, Frank Sylvester, mathematics
Hulme, William Arthur, pharmacology
Lawrence, William Mason, fisheries
Muntz, John Adolph, biochemistry
Ogden, Eugene Cecil, botany
Ritchie, John Augustus, anthropology
Schulenberg, John William, chemistry
Stanley, Lester Nelson, organic chemistry
Thompson, William Rae, mathematical statistics
Weinstein, Abbott Samson, biostatistics, public health administration

DEWITT
Crenshaw, Ronnie Ray, medicinal chemistry

DOBBS FERRY
Adler, Stephen Fred, industrial chemistry
Altscher, Siegfried, polymer chemistry, organic chemistry
Ang, Catharina Yung-Kang Wang, food science
Anzenberger, Joseph F, Sr, organic chemistry
Brokke, Mervin Edward, organic chemistry
Casey, Adria Catala, organic chemistry, medicinal chemistry
Durnick, Thomas Jackson, analytical chemistry, spectroscopy
Fearing, Ralph Burton, organic chemistry, textile chemistry
Harris, Bernard, applied physics
Hartman, Sven Richard, organic chemistry
Honig, Milton Leslie, organic chemistry
Jaffe, Fred, organic chemistry
Kim, Ki-Soo, polymer chemistry
Kirkpatrick, William John, physical chemistry
Kiss, Klara, physical chemistry, polymer chemistry
Liu, Chong Tan, inorganic chemistry
Liu, Sophia Yan, organic chemistry
Livingston, G E., food science
Lurio, Allen, physics
McClure, Judson P., organic chemistry
Melachouris, Nicholas, food science, biochemistry
Mirviss, Stanley Burton, synthetic organic chemistry
Moss, Frank Anthony James, inorganic chemistry, physical chemistry
Muntz, Ronald Lee, organometallic chemistry, organic chemistry
Postman, Robert Derek, mathematics
Strother, Connelle Osburn, industrial chemistry
Tesoro, Giuliana C., organic polymer chemistry
Timony, Peter Edward, organic chemistry
Toy, Arthur Dock Fon, industrial chemistry
Via, Francis Anthony, organic chemistry
Vopicka, Ellen Vandersee, biology, microbiology
Walsh, Edward Nelson, organic chemistry
Zipp, Arden Peter, physical inorganic chemistry

Weil, Edward David, organic chemistry
Williams, Ross Edward, physics
Yu, Arthur J, polymer chemistry
Ziegler, William Arthur, analytical chemistry

DRYDEN
Gould, Robert Henderson, physics

DUNKIRK
Snader, Daniel Webster, mathematics

DURHAM
Munroe, Marshall Evans, mathematics

EAST CONCORD
Mumbach, Norbert R, applied chemistry

EAST FISHKILL
Anderson, Frank Wallace, physical chemistry, electron microscopy
Heller, William R, solid state physics, applied mathematics
O'Brien, Redmond R, mathematics

EAST GREENBUSH
Gruett, Monte Deane, synthetic organic chemistry
Neumann, Helmut Carl, pharmaceutical chemistry
Traver, Janet Hope, biochemistry

EAST HAMPTON
Magill, Thomas Pleines, microbiology, immunology

EAST HILLS
Levine, Samuel W, physical chemistry

EAST MEADOW
Archambeau, John Orin, radiology
Collipp, Platon Jack, pediatrics
Friedman, Mark Hirsch, veterinary medicine
Klein, Sidney Wayne, medicine
Maddaiah, Vaddanahally Thimmaiah, biochemistry
Praissman, Melvin, biochemistry, physical chemistry
Ritterman, Murray B, applied mathematics
Weinstock, Irwin Morton, biochemistry

EAST ROCHESTER
Buonocore, Michael, organic chemistry
Monahan, Alan Richard, physical chemistry, polymer chemistry

EAST SETAUKET
Garber, Donald I, nuclear physics
Habicht, Ernst Rollemann, Jr, environmental sciences, energy conversion

EAST SYRACUSE
Mays, David Lee, analytical chemistry

EASTPORT
Price, Jessie Isabel, veterinary microbiology

ELMA
Sweeney, Richard F, organic chemistry

ELMHURST
Li, Koibong, microbiology
Pochazevsky, Rubem, radiology

ELMIRA
Foster, Donald Bartley, developmental biology, plant morphology
Lindsay, William Germer, Jr, physiology
Molloy, Andrew R, chemistry
Potter, Louise Frances, microbiology
Ruffer, David G, vertebrate zoology
Shabanowitz, Harry, mathematics
Spremulli, Gertrude H, biochemistry
Stephens, Lawrence James, organic chemistry, science education
Whitney, Rae, biology

ELMONT
Pierson, Willard James, Jr, meteorology, oceanography

ELMSFORD
Blurton, Keith F, physical chemistry
Holt, William Robert, mathematical statistics, medical statistics
Menke, John Roger, physics
Mittelman, Phillip Sidney, nuclear physics
Munies, Robert, industrial pharmacy, research administration
Steinberg, Herbert Aaron, mathematics
Stetter, Joseph Robert, physical chemistry, surface chemistry
Troubetzkoy, Eugene Serge, theoretical physics
Vinocur, Myron, medicine, pathology

ELNORA
Shell, Donald Lewis, mathematics

ENDICOTT
Albrecht, William Melvin, physical chemistry, materials engineering
Dreikorn, Russell E, physical chemistry
Elmore, Glenn Van Ness, electrophotography
Macur, George J, physical chemistry
Pedroza, Gregorio Cruz, industrial organic chemistry
Sacher, Edward, physical chemistry
Underkofler, William Leland, analytical chemistry
Yasko, Richard N, nuclear physics

FAIRPORT
Cunningham, Geoffrey Everett, physical chemistry
Dehm, Richard Lavern, electrophotography
Feder, Donald Perry, optics
Holman, James Lawson, physical chemistry
Hubsch, Harold Lawrence, pharmaceutical chemistry
Milne, Gordon Gladstone, optics
Parent, Richard Alfred, organic chemistry
Trevoy, Donald James, solid state chemistry, energy conversion
Webb, Julian Pierce, experimental physics

FARMINGDALE
Abbatiello, Michael James, zoology, biology
Balamuth, Lewis, physics
Barke, Harvey Ellis, economic entomology, botany
Christensen, Eric, biology
DiLiello, Leo Ralph, microbiology
Frishman, Austin Michael, entomology
Giannotti, Ralph Alfred, organic chemistry
Gross, Stanley H, space physics, planetary atmospheres
Hessel, Alexander, physics
Highland, Harold Joseph, applied statistics
Hilbert, Kenneth Franklin, poultry pathology
Hopfer, Samuel, physics
Kramer, Paul Robert, physics
Lamberg, Stanley Lawrence, biochemistry, histology
La Tourrette, James Thomas, physics
Marcuwitz, Nathan, mathematical physics
Mark, Robert Vincent, organic chemistry
Pellegrini, Frank C, organic chemistry
Pyenson, Louis L, entomology, plant pathology
Rothstein, Robert, medical administration
Scarl, Donald B, quantum optics
Shmoys, Jerry, electrophysics
Stahlman, Clarence L, dairy industry
Stockbridge, Robert R, animal husbandry, poultry husbandry
Ursino, Joseph Anthony, organic chemistry
Vinciguerra, Michael Joseph, physical chemistry
Walter, William Trump, electrophysics
Williamson, Charles Edward, plant pathology

FAYETTEVILLE
Cheney, Lee Cannon, medicinal chemistry

Gottstein, William J, pharmaceutical chemistry
Lein, Joseph, microbiology
Menotti, Amel Romeo, chemistry

FISHERS
Milton, Kirby Mitchell, chemistry

FISHKILL
Bergmann, Eric Arnold, analytical chemistry, polymer chemistry
Lessor, Arthur Eugene, Jr, analytical chemistry, crystallography
Newman, Stanley Ray, organic chemistry
Patmore, Edwin Lee, organic chemistry, inorganic chemistry
Schlicht, Raymond Charles, synthetic organic chemistry

FLORAL PARK
Sebrell, William Henry, nutrition

FLUSHING
Aaronson, Sheldon, microbial biochemistry
Adams, Elizabeth, organic chemistry
Alsop, David W, cytology, morphology
Archibald, Ralph George, mathematics
Axelrad, George, organic chemistry
Baum, Paul M, experimental physics
Bittman, Robert, bio-organic chemistry
Borenstein, Samuel R, physics, biology
Boylan, Elizabeth Shippee, vertebrate embryology, oncology
Braun, Martin, applied mathematics
Brueckner, Hannes Kurt, geochemistry, structural geology
Calhoun, Robert Ellsworth, quantitative genetics
Chako, Nicholas, mathematics
Coch, Nicholas Kyros, sedimentology, marine geology
Cotter, Maurice Joseph, physics
Czerniakiewicz, Anastasia Juana, mathematics
Diesendruck, Leo, physics
Disch, Raymond L, physical chemistry
Eastham, James Norman, mathematics
Eidinoff, Maxwell Leigh, physical chemistry, biochemistry
Engel, Robert Ralph, synthetic organic chemistry
Essman, Walter Bernard, psychophysiology
Fass, Arnold Lionel, mathematics
Feely, Herbert William, geochemistry
Ferrari, Lawrence A, invertebrate physiology
Finks, Robert Melvin, invertebrate paleontology, geology
Fischer, C Rutherford, molecular physics, solid state physics
Fremont, Herbert Irwin, mathematics
Frumkes, Thomas Eugene, neurosciences, physiological optics
Glasse, Robert Marshall, social anthropology, cultural anthropology
Goldman, Norman L, chemistry
Goodman, Seymour, computer science, physical chemistry
Gordon, Benjamin Solomon, pathology
Gregersen, Edgar Alstrup, anthropology, linguistics
Habib, Daniel, palynology, micropaleontology
Hansen, Edward Charles, anthropology
Harvey, Alexander Louis, physics
Hatcher, Robert Douglas, solid state physics
Hechler, Stephen Herman, mathematical logic, topology
Hecht, Max Knobler, vertebrate biology
Herrey, Erna Miranda Julia, physics
Hershenov, Joseph, mathematics
Hoffmann, Banesh, mathematical physics
Hogg, James Felter, biochemistry
Itzkowitz, Gerald Lee, mathematics
Johansson, Tage Sigvard Kjell, zoology
Kanter, Louis Harold, mathematics
Kaplan, Martin L, comparative physiology, insect pathology
Kashin, Philip, neurophysiology, insect physiology
Kirby, Robert Emmet, analytical chemistry
Klarfeld, Joseph, theoretical physics
Koehler, Dale Roland, nuclear physics
Koeppl, Gerald Walter, chemical physics
Krinsley, David, sedimentology
Kupferberg, Kenneth Maurice, physics, electronics
Landau, Ronald Wolf, plasma physics
Lilly, Daniel McQuillan, protozoology
Locke, David Creighton, chemistry

Longsworth, Lewis Gibson, physical chemistry
Ludman, Allan, geology, petrology
MacIntyre, Giles T, vertebrate zoology, vertebrate paleontology
Magnien, Ernest, organic chemistry
Mansfield, Larry Everett, mathematics
Marcus, Leslie F, biometry, paleontology
Marien, Daniel, genetics, zoology
Marion, Alexander Peter, physical chemistry
Mattson, Peter Humphrey, geology
McIntyre, Andrew, marine geology, marine biology
Mendelson, Elliott, mathematical logic
Moise, Edwin Evariste, mathematics
Neuberger, Jacob, solid state physics
Newman, Walter S, quaternary geology
Orenstein, Albert, physics
Owen, Roger C, cultural anthropology
Paskin, Arthur, physics
Paulson, Edward, mathematical statistics
Rafanelli, Kenneth R, theoretical physics
Robinson, Myron, air pollution, environmental science
Rothenberg, Ronald Isaac, mathematics, statistics
Roze, Uldis, biochemistry
Sard, Arthur, mathematics
Schreiber, Edward, ceramics, geophysics
Schulman, Jerome M, chemical physics
Shore, Ferdinand John, nuclear physics
Shub, Michael I, mathematics
Slater, Mariam Kreiselman, cultural anthropology
Speidel, David H, geochemistry, petrology
Stark, Joel, speech pathology
Stemple, Joel G, mathematics
Strait, Peggy, mathematics
Szalay, Jeanne, cell biology
Thurber, David Lawrence, geochemistry
Tilson, Bret Ransom, mathematics
Tropp, Burton E, biochemistry
Washton, Nathan Seymour, science education, environmental science
Wasserman, Marvin, evolutionary biology
Weaver, John Scott, geophysics
Weintraub, Sol, mathematics
Weiss, Norman Jay, mathematical analysis
Williamson, Robert Samuel, physics
Woods, Philip Sargent, cell biology, molecular biology
Youdin, Myron, biomedical engineering

FOREST HILLS
Dodes, Irving Allen, numerical analysis
Jean, George Noel, organic chemistry
Konig, Otto, chemistry, information science
Oser, Bernard Levussove, nutrition, toxicology

FPO NEW YORK
Ferguson, Carl E, agriculture
Husk, George Ronald, organometallic chemistry
Schenck, Harry Allen, acoustics
Schulman, James Herbert, solid state physics, optical physics

FRANKLIN SQUARE
Labianca, Dominick A, organic chemistry, polymer chemistry

FREDONIA
Arroe, Hack, atomic physics
Barnard, Walther M, mineralogy, geochemistry
Benton, Allen Haydon, zoology
Bernstein, Kenneth, molecular genetics, virology
Boenig, Robert William, biology
Chilton, Bruce L, mathematics
Chimenti, Frank A, mathematical analysis
Connelly, John Joseph, Jr, experimental solid state physics
Danese, Arthur E, mathematics
Davey, Paul Oliver, physics
Dingledy, David Peter, physical chemistry, photochemistry
Dowds, Richard E, mathematics
Dunham, Valgene Loren, plant physiology, plant biochemistry
Fahnestock, Robert Kendall, geology
Fox, Kevin A, behavioral physiology, reproductive biology
Gates, Olcott, geology
Gilman, Richard Atwood, geology
Keller, Roy Alan, analytical chemistry
Lincoln, Charles Albert, theoretical physics

NEW YORK

Luntz, Myron, radiation physics
Mantai, Kenneth Edward, plant physiology
Mayer, Julian Richard, management, environmental physiology
Metzger, William Richard, environmental chemistry
Moos, Gilbert Ellsworth, geology
Neven, Maurice C., physical chemistry, organic chemistry
Olson, Frank R., mathematics
Peterson, Donald Neil, geophysics
Polimeni, Albert D., mathematics
Schulze, Lothar Walter, science education
Stanley, Willard Francis, zoology
Supple, Jerome Henry, organic chemistry
Thumm, Byron Ashley, chemistry
Wood, Kenneth George, limnology
Yunghans, Wayne N., cytology
Zimmer, George P., biology, science education

FREEPORT
Yamins, Jacob Louis, food science

FT EDWARD
Sackoff, Martin M., organic polymer chemistry

FULTON
Merrifield, Paul Elliot, colloid chemistry

GARDEN CITY
Berger, Joel Gilbert, chemistry
Bettelheim, Frederick A., physical chemistry
Brenowitz, A Harry, biology
Brodie, Edmund Darrell, Jr., herpetology, behavioral biology
Burge, Robert Ernest, Jr., organic chemistry
Burke, Edward Aloysius, theoretical physics
Canfield, William H., speech pathology, audiology
Cassin, Joseph M., microbial ecology
Churchill, Algernon Coolidge, phycology, plant ecology
Cohen, Marvin, pharmacology, psychopharmacology
Conklin, Marie Eckhardt, botany, genetics
Diakow, Carol, animal behavior
Dooley, James Keith, systematic ichthyology
Eickelberg, W Warren B, biomechanics
Evert, Henry Earl, organic chemistry
Fliedner, Leonard John, Jr., medicinal chemistry
Ganti, Venkat Rao, organic chemistry
Garrell, Martin Henry, physics
Gochenaur, Sally Elizabeth, mycology
Goldberg, Stephen Zalmund, inorganic chemistry, structural chemistry
Goldschmidt, Eric Nathan, organic chemistry
Grillo, Ramon S., biology
Grob, Howard Shea, reproductive physiology, histophysiology
Halliday, Robert William, inorganic chemistry
Hampton, John Kyle, Jr., physiology
Hecht, Eugene, physics
Heston, William May, Jr., physical chemistry
Hodgson, Jonathan Peter Edward, mathematics
Jaffe, Bernard Mordecai, physics
Jainchill, Jerome, genetics, biochemistry
Jervis, Herbert Hunter, genetics
Johnsen, Roger Craig, genetics
Kalicki, Henrietta, genetics
Kranzer, Herbert C., applied mathematics
Kumbar, Mahadevappa M., physical chemistry, biophysics
Kuss, Herbert, mathematics
Lachman, Leon, pharmacy
Lakritz, Julian, organic chemistry
Landesberg, Joseph Marvin, organic chemistry
Lemos, Anthony M., theoretical physics, solid state physics
Lener, Walter, entomology
Lubell, David, mathematics
Lund, Richard, vertebrate paleontology
March, Jerry, organic chemistry
Meyer, Walter Joseph, mathematics
Moon, Sung, organic chemistry
Morrone, Terry, plasma physics
Napolitano, Joseph J., biology, protozoology
Newman, Theodore Joseph, physics, electronics
Payton, Robert Gilbert, applied mathematics
Pohle, Frederick V., applied mathematics
Rajagopalan, Parthasarathi, organic chemistry, medicinal chemistry
Rasero, Lawrence J., pharmacy
Robinson, Howard Addison, physics
Roellig, Harold Frederick, paleontology
Rubin, Alan A., pharmacology
Rudman, Reuben, x-ray crystallography, structural chemistry
Russell, George Keith, plant physiology, biochemistry
Ryan, Simeon P., biology
Schor, Joseph Martin, biochemistry
Sirkin, Leslie A., geology, palynology
Sisti, Anthony Joseph, organic chemistry
Smart, Kathryn Marilyn, microbiology, virology
Smeriglio, Alfred John, biology, comparative anatomy
Steinmetz, William John, applied mathematics
Weinstein, Stephen Henry, drug metabolism
Wenig, Jeffrey, pharmacology, physiology
Yanowitch, Michael, applied mathematics
Zajac, Alfred, physics

GARNERVILLE
Redalieu, Elliot, medicinal chemistry
Zugbe, Frederick T., forensic medicine, cardiovascular diseases

GENESEO
Alley, Phillip Wayne, solid state physics
Battersby, Harold Ronald Eric, anthropology, anthropological linguistics
Chen, James Ralph, particle physics, atomic physics
Clark, George Richmond, III, paleoecology, marine biology
Deutsch, John Ludwig, physical chemistry, spectroscopy
Forest, Herman Silva, environmental biology
Graham, William Joseph, behavioral ecology
Harke, Douglas J., physics, science education
Hathaway, Richard Brackett, geology, petrology
Huff, G Bradley, surface physics, low temperature physics
Innis, Donald Quayle, geography
Jackson, John Edward, agricultural biochemistry, organic chemistry
Joshi, Bhairav Datt, physical chemistry, quantum chemistry
Judkins, Russell Alan, anthropology
King, Lawrence J., plant physiology
Kinsey, Kenneth F., physics
Latorella, A Henry, genetics, algology
Leela, Srinivasa (G), mathematics
Levine, Jeffrey, mathematics
Lougeay, Ray Leonard, physical geography, remote sensing
Meisel, David Dering, astronomy, astrophysics
Meyer, Daniel L., bacteriology
Nahabedian, Kevork Vartan, chemistry
Reber, Jerry D., nuclear physics
Reid, Allen Francis, biophysics
Ristow, Bruce W., physical chemistry
Ritter, Edward, protozoology
Roecker, Robert Maar, vertebrate zoology
Scatterday, James Ware, geology
Schaefer, Paul Theodore, mathematics
Sells, Robert Lee, physics
Small, William Andrew, mathematics
Smith, Richard Alan, biochemistry
Smith, Richard Frederick, organic chemistry
Thompson, Charles Frederick, population ecology, ornithology
Ulmer, David Heading Bartine, Jr., physiology, microbiology
Young, Richard A., geology

GENEVA
Acree, Terry Edward, food chemistry
Aten, Carl Faust, Jr., physical chemistry
Barton, Donald Wilber, plant breeding
Blackburn, Thomas Roy, analytical chemistry, physical chemistry
Bourke, John Butts, biochemistry
Bourne, Malcom Cornelius, food science
Bowers, William Sigmond, invertebrate physiology, biochemistry
Braun, Alvin Joseph, plant pathology
Braverman, Samuel William, plant pathology
Cain, John Carlton, plant physiology
Campbell, Larry Enoch, solid state physics
Carle, Kenneth Roberts, physical organic chemistry
Chapman, Paul Jones, entomology
Clark, Benjamin Edward, agronomy
Cummins, James Nelson, pomology
Curtis, Otis Freeman, Jr., plant morphogenesis research
Davis, Alexander Cochran, economic entomology
Dickson, Michael Hugh, plant breeding
Dolan, Desmond Daniel, plant pathology
Downing, Donald Leonard, food science
Einset, John, pomology
Gambrell, Lydia Jahn, biology
Gilmer, Robert McCullough, phytopathology
Gilpatrick, John Daniel, plant pathology
Glass, Edward Hadley, entomology
Hackler, Lonnie Ross, biochemistry, nutrition
Hang, Yong Deng, food science, water pollution
Harman, Gary Elvan, plant pathology
Hill, Ada Sinz, organic chemistry
Hill, John Hamon Massey, organic chemistry
Hrazdina, Geza, biochemistry
Hunter, James Edward, plant pathology
Kender, Walter John, pomology
Kerlan, Joel Thomas, endocrinology
Khan, Anwar Ahmad, plant physiology, plant biochemistry
Klein, John Sharpless, applied mathematics
Kuhr, Ronald John, agricultural chemistry, insect toxicology
LaBelle, Robert Lawrence, food science
Lakso, Alan Neil, pomology, plant physiology
Lamb, Robert Consay, pomology
Lee, Chang Yong, food science
Lienk, Siegfried Eric, entomology
Marx, Gerald Alvin, agronomy, plant pathology
Massey, Louis Melville, Jr., plant pathology
Mattick, Leonard Robert, biochemistry
Moyer, James Charles, food technology
Nellis, Lois Fonda, microbiology
Nelson, Thomas Robert, organic chemistry
Nittler, LeRoy Walter, agriculture
Oberle, George David, horticulture, plant genetics
Ourecky, Donald K., genetics, cytology
Peck, Nathan Hiram, horticulture
Robinson, Richard Warren, horticulture
Robinson, Willard Bancroft, food science
Roelofs, Wendell Lee, organic chemistry
Russell, Allan Melvin, physics
Ryan, Allan Alexander, chemistry
Schaefers, George Albert, entomology
Shallenberger, Robert Sands, food chemistry, food technology
Shannon, Stanton, soil chemistry
Shaulis, Nelson Jacob, viticulture
Sleeper, David Allanbrook, biology
Splittstoesser, Clara Quinnell, entomology
Splittstoesser, Don Frederick, bacteriology
Stamer, John Richard, microbiology
Stearns, Brenton Fisk, energy conversion, surface physics
Steinkraus, Keith Hartley, microbiology
Stoewsand, Gilbert Saari, nutrition, toxicology
Szkolnik, Michael, plant pathology
Tao, Kar-Ling James, plant physiology
Taschenberg, Emil Frederick, entomology
Tashiro, Haruo, economic entomology
Trammel, Kenneth, entomology
Uyemoto, Jerry Kazumitsu, plant pathology, plant virology
Van Buren, Jerome Paul, biochemistry
Vittum, Morrill Thayer, agronomy
Walter, Reginald Henry, food chemistry
Way, Roger Darlington, pomology
Wolff, Emily Tower, botany
Woodrow, Donald L., stratigraphy, sedimentology

GHENT
Davis, David, plant pathology

GLEN COVE
Apellaniz, Joseph E P, photographic chemistry
Carlson, Arthur Stephen, pathology
Mayberger, Harold Woodrow, obstetrics & gynecology
Pall, David B., physical chemistry
Pomerance, Janet Bellcourt, mathematics
Van Slyke, Richard M., operations research
Whitacre, Francis Marion, biology, chemistry

GLEN HEAD
Halpert, Bernard, bacteriology, biochemistry
Mikes, John Andrew, chemistry

GLENS FALLS
Burrell, Harold Paul Charles, organic chemistry
Green, Floyd Wilson, chemistry
Huckle, William George, inorganic chemistry
Inman, Charles Gordon, organic chemistry
Jones, George Francis, organic chemistry

GLEN OAKS
Klein, Donald Franklin, psychiatry, psychopharmacology

GLENWOOD
Kreiter, Victor Peter, Jr., chemistry

GOSHEN
Balassa, Leslie Ladislaus, pharmaceutical chemistry

GRAND ISLAND
Carr, Russell L K, organic chemistry
Coats, Alma Winifred, polymer chemistry
Dorfman, Edwin, organic chemistry, polymer chemistry
Geering, Emil John, organic chemistry
Hirsch, Stephen Simeon, organic chemistry
Mackay, John Kelvin, surface chemistry
Moore, Donald R., organic chemistry, polymer chemistry
Rosenfeld, Jerold Charles, organic chemistry, polymer chemistry
Weinberg, Norman L., organic chemistry

GRANITE SPRINGS
Kaikow, Julius, geography, geology

GREAT NECK
Harrington, John Vincent, colloid chemistry
Hiskey, Clarence Francis, analytical chemistry
Patra, Sushant Kumar, polymer chemistry
Smith, Bernard, physics
Solomon, Frank I, electrochemistry
Sparberg, Esther Braun, history of science, chemistry
Zitrin, Arthur, psychiatry

GREENPORT
Bachrach, Howard Lloyd, biochemistry
Breese, Sydney Salisbury, Jr., veterinary virology
Butterfield, Walter K., veterinary virology
Callis, Jerry Jackson, veterinary medicine
Campbell, Charles Haywood, microbiology
Cowan, Keith Morris, immunochemistry
Cunliffe, Harry R., veterinary immunology, animal virology
Dardiri, Ahmed Hamed, veterinary virology
DeBoer, Carl John, virology, immunology
Ferris, Dean Hunter, veterinary microbiology
Graves, John Henry, veterinary medicine
Hamdy, Farouk Mohamed, animal medicine, virology
McVicar, John West, veterinary medicine, virology
Pan, In-Chang, immunopathology, veterinary immunology

Richmond, Jonathan Young, virology, genetics
Schloer, Gertrude M, microbiology, genetics
Trautman, Rodes, biophysics
Yedloutschnig, Ronald John, veterinary medicine

GREENVALE
Biondo, Frank X, microbiology, mycology
Bonvicino, Guido Eros, organic chemistry
Dwyer, James Michael, physical chemistry, forensic science
Feit, Irving N, organic chemistry
Flemings, Milton Baker, insect physiology
Garner, James Gregory, cytogenetics, microbiology
Greenlaw, Jon Stanley, ornithology, population ecology
Karp, Stewart, analytical chemistry
Kelly, Martin Joseph, physics
LoGerfo, John J, zoology, medical technology
Lowell, Seymour, physical chemistry
Marengo, Norman Payson, botany
Maturo, Joseph Martin, III, biochemistry, physiology
Meiselman, Newton, botany, biology
Meskill, Victor Peter, microbiology, cytology
Newman, Bernard, chemistry
Nicholson, Isadore, organic chemistry
Reissig, Jose Luis, genetics, molecular biology
Rocks, Lawrence, chemistry
Schaffrath, Robert Eben, organic chemistry
Seckler, Bernard David, mathematics
Shields, Joan Esther, organic chemistry
Shodell, Michael J, cell biology
Stevenson, John Crabtree, mathematics
Tralli, Nunzio, physics
Wallach, Sylvan, mathematics

GUILDERLAND
Cullen, Marion Permilla, biochemistry, nutrition
Engster, Maryann Sandra, invertebrate zoology, electron microscopy
Fietz, William Adolf, low temperature physics
Stevenson, Isaac Glenn, chemistry, bacteriology
Walker, Michael Stephen, engineering physics, experimental solid state physics

HAMBURG
Booth, Robert Edwin, organic chemistry
Gorzynski, Eugene Arthur, infectious diseases

HAMILTON
Aveni, Anthony, astronomy
Bryan, Philip Steven, inorganic chemistry
Cochran, John Charles, organic chemistry
Crook, Philip George, microbiology
DeNoyer, Linda Kay, astrophysics
Dodd, Jack Gordon, (Jr), atomic physics
Gibbons, David Louis, physical chemistry, biophysics
Gibbons, Katherine Bond, hematology
Goodwin, Robert Earle, vertebrate zoology
Henshaw, Clement Long, physics, science education
Herman, Theodore, geography
Hoffman, Roger Alan, comparative endocrinology, reproductive biology
Hoham, Ronald William, phycology, algology
Holbrow, Charles H, experimental nuclear physics, information science
Kastens, Merritt Louis, industrial chemistry
Lewis, David Kenneth, physical chemistry
Linsley, Robert Martin, invertebrate paleontology
Lloyd, James Newell, physics
Longyear, John Munro, III, archaeology, physical anthropology
Malin, Shimon, theoretical physics
Mansfield, Victor Neil, astrophysics
Mitchell, John Taylor, developmental biology, experimental embryology
Myers, Raymond J, zoology
Pownall, Malcolm Wilmor, mathematics
Shatoff, Larry David, mathematics

Thurner, Joseph John, inorganic chemistry
Todd, Robert Emerson, zoology, embryology
Trumbull, Elmer Roy, Jr, organic chemistry
Wardwell, James Fletcher, mathematics
Weyter, Frederick William, biochemical genetics
Wolf, Walter Alan, biochemistry, organic chemistry

HAMLIN
Sawdey, George Washington, organic chemistry

HAMMONDSPORT
Rice, Andrew Cyrus, food bacteriology

HARRISON
Gunther-Mohr, Gerard Robert, physics
Hess, Henry M, organic chemistry

HASTINGS
Nethercot, Arthur Hobart, Jr, physics

HASTINGS-ON-HUDSON
Gans, Eugene Howard, pharmacy, chemistry
Gaylin, Willard, psychiatry

HAUPPAUGE
Nuzzi, Robert, marine biology
Wheatley, George Milholland, preventive medicine
Zaki, Mahfouz H, environmental medicine, public health

HEMPSTEAD
Abrahams, Irving, microbiology, immunology
Agnello, Eugene Joseph, organic chemistry
Aronson, Irving, theoretical physics
Brunschwig, Bruce Samuel, physical inorganic chemistry, chemical kinetics
Bumcrot, Robert J, mathematics
Eklund, Karl E, applied physics
Erb, Kenneth, microbiology
Friedman, Helen Lowenthal, physical optics, health physics
Galinsky, Irving, cytology, genetics
Glaser, Herman, physics
Grimes, Gary Wayne, developmental genetics
Gruber, Gary Richard, physics, mathematics
Holmes, Richard Remsen, organic chemistry
Horowitz, Esther, speech pathology, psychology
Inaba, Masaharu George, economic geography, regional economy
Johnson, Robert Walter, ecology
Kaplan, Eugene Herbert, parasitology, science education
Knee, David Isaac, mathematics
Lesser, Alexander, cultural anthropology
McCourt, Robert Perry, vertebrate zoology, radiation biology
Miller, Solomon, cultural anthropology
Phillips, Carleton Jaffrey, mammalogy, oral biology
Prover, Corinne Bea, plant morphogenesis, plant physiology
Reisinger, Joseph G, physics
Rosman, Howard, physical chemistry
Schneiweiss, Jeannette W, physiology, biometrics
Sterrett, Frances Susan, environmental chemistry
Walcher, Azelle Brown, mathematics
Wemyss, Courtney Titus, Jr, zoology
Wolff, Manfred Paul, geology, sedimentology

HIGHLAND FALLS
Cardillo, Frances M, botany

HOBART
Graham, Dean McKinley, biophysics

HOLLISWOOD
Kleiman, Howard, algebra, number theory

HONEOYE FALLS
Ames, Wendell Russell, epidemiology

HOPEWELL JUNCTION
Agnihotri, Ram K, chemistry
Chu, Wei-Kan, experimental atomic physics, engineering physics
Genin, Dennis Joseph, solid state physics, engineering physics
Kaplan, Leon H, physical chemistry
Klein, Donald Lee, inorganic chemistry
Lever, Reginald Frank, materials science
Nadas, Arthur Joseph, mathematical statistics
Needham, Charles D, polymer chemistry
Pliskin, William Aaron, semiconductors, chemical physics
Rupprecht, Hans S, physics
Schick, Jerome David, chemistry, physics
Schwartz, Geraldine Cogin, physical chemistry
Vromen, Benjamin H, physical chemistry
Weeks, William Thomas, computer sciences

HORSEHEADS
Laponsky, Alfred Baer, physical electronics

HOUGHTON
Calhoon, Stephen Wallace, Jr, analytical chemistry
Christensen, Larry Wayne, organic chemistry
Luckey, Robert Ruel Raphael, mathematics
Munro, Donald W, Jr, physiology, genetics
Paine, Samuel Hugh, physics, earth sciences
Piersma, Bernard J, physical chemistry
Shannon, Frederick Dole, organic chemistry, polymer chemistry
Trexler, Frederick David, physics
Whiting, Anne Margaret, vertebrate anatomy, embryology

HUDSON FALLS
George, Philip Donald, chemistry

HULETTS LANDING
Emerson, Alfred Edwards, zoology, ecology

HUNTINGTON
Gross, George Lloyd, applied mathematics
Harmon, John Baltzell, medicinal chemistry, science administration
Stein, Samuel H, photographic chemistry, physical organic chemistry
Von Keszycki, Carl Heinrich, physics

HUNTINGTON STATION
Weinberg, Sidney B, pathology

IRVINGTON
Wijangco, Antonio Robles, high energy physics

ISLIP
Rakieten, Morris Louis, clinical pathology
Rakieten, Nathan, pharmacology, toxicology

ITHACA
Adelmann, Howard Bernhardt, zoology
Adkins, Elizabeth Kocher, behavioral biology, reproductive endocrinology
Adler, Kraig (Kerr), animal behavior, evolution

NEW YORK

Adorno, David Samuel, mathematics, statistics
Ainslie, Harry Robert, animal nutrition
Aist, James Robert, plant pathology, plant cytology
Albrecht, Andreas Christopher, quantum chemistry, solid state chemistry
Alexander, Martin, soil microbiology, microbial ecology
Alexander, Ralph William, medicine
Alexander, Renee R, molecular biology, biochemistry
Anderson, John Maxwell, zoology
Anderson, Ronald Eugene, genetics, plant breeding
Apgar, Barbara Jean, nutrition
Appel, Max J, veterinary virology
Arion, William Joseph, biochemistry
Armbruster, Gertrude D, food science
Arneson, Phil Alan, plant pathology
Arnold, Richard Warren, soil morphology
Aronson, A L, pharmacology, toxicology
Artusio, Joseph F, Jr, anesthesiology
Ascher, Marcia, numerical analysis, applied mathematics
Ascher, Robert, anthropology, archaeology
Auer, Peter Louis, plasma physics, theoretical physics
Austic, Richard Edward, nutrition
Baker, Robert Carl, food science
Banks, Harlan Parker, plant morphology, anatomy
Barlow, John Peleg, oceanography
Barnes, Richard Henry, nutrition
Bateman, Durward F, plant pathology
Bates, David Martin, taxonomic botany
Batterman, Boris William, x-ray crystallography, experimental solid state physics
Bauer, Simon Harvey, chemical kinetics, structural chemistry
Beachy, Roger Neil, plant virology
Beer, Steven Vincent, plant pathology
Bensadoun, Andre, nutrition, physiology
Berg, Clifford Osburn, fresh water biology
Berger, Franklin Gordon, biochemistry, molecular biology
Bergman, Emmett Norlin, physiology
Bergmark, William R, organic chemistry
Berkelman, Karl, experimental high energy physics
Bernard, John Milford, plant ecology
Berstein, Israel, mathematics
Bethe, Hans Albrecht, theoretical physics
Billera, Louis Joseph, mathematics, operations research
Bird, John Malcomb, geology
Blackler, Antonie W C, developmental biology
Blanpied, George David, pomology, physiology
Block, Henry David, applied mathematics
Blomquist, Alfred Theodore, organic chemistry
Bloom, Arthur Leroy, geology
Bloom, Stephen Earl, cytology, cytogenetics
Blumen, Isadore, statistics
Bonnichsen, Bill, petrology, economic geology
Boothroyd, Carl William, plant pathology
Bouldin, David Ritchey, soil fertility
Bowering, Jean, nutrition
Bowers, Raymond, physics
Bramble, James H, applied mathematics, numerical analysis
Brann, James Lewis, Jr, entomology
Brattsten, Lena B, biochemistry
Brecht, Patrick Ernest, plant physiology, vegetable crops
Brodie, Bill Burl, nematology
Brown, William Louis, biology
Bruner, Dorsey William, microbiology
Brussaert, Peter Frans, population biology
Brutsaert, Wilfried, hydrology
Buchanan, David Royal, textile physics
Buhrman, Robert Alan, solid state physics
Burns, John A, planetary science, celestial mechanics
Buskirk, Ruth Elizabeth, animal behavior, animal ecology
Butler, Karl Douglas, Sr, agriculture
Cady, Thomas Joseph, ornithology
Calnek, Bruce Wixson, veterinary virology
Camhi, Jeffrey Martin, neurophysiology, animal behavior
Campbell, Samuel Gordon, immunology
Campbell, Thomas Colin, biochemistry, nutrition
Capranica, Robert R, neurobiology, electrical engineering
Carmichael, Leland E, veterinary microbiology

Nickum, John Gerald, zoology, conservation
Norcross, Neil Linwood, immunology
Noronha, Fernando M Oliveira, virology
Novak, Joseph Donald, biology
Obendorf, Ralph Louis, plant physiology, agronomy
Oberly, Gene Herman, pomology
O'Brien, Richard Desmond, neurochemistry
Oglesby, Ray Thurmond, aquatic biology
Oliver, Jack Ertle, geophysics
Olson, Gerald Walter, soil morphology, agronomy
Oltenacu, Elizabeth Allison Branford, animal science
Orear, Jay, particle physics
Ostrander, Charles Evans, poultry science
Ozbun, Jim L, plant physiology
Pack, Albert Boyd, agricultural meteorology
Palm, Charles Edmund, entomology
Palmer, Katherine Van Winkle, paleontology, stratigraphy
Paolillo, Dominick Joseph, botany
Pardee, William Durley, plant breeding, agronomy
Parker, Kenneth Gardner, plant pathology
Parratt, Lyman George, physics
Pasternack, Robert Francis, physical inorganic chemistry, bioinorganic chemistry
Patton, Robert Lee, insect physiology
Payne, Lawrence Edward, applied mathematics
Pechuman, Laverne Leroy, entomology
Peverly, John Howard, plant biochemistry, soil chemistry
Philbrick, Shailer Shaw, geology
Pierce, Ellis A, animal science
Pimentel, David, ecology, entomology
Plaisted, Robert Leroy, plant breeding
Platek, Richard Alan, mathematical logic
Podleski, Thomas Roger, physiology, biochemistry
Pohl, Robert O, solid state physics
Pond, Wilson Gideon, animal nutrition
Poppensiek, George Charles, veterinary medicine
Porter, Richard Francis, physical inorganic chemistry
Post, John E, veterinary pathology, veterinary oncology
Postle, Donald Sloan, veterinary science
Potter, Norman N, food science, food microbiology
Pough, Frederick Harvey, environmental physiology, herpetology
Powell, Loyd Earl, Jr, pomology, plant physiology
Prabhu, Narahari Umanath, mathematics, statistics
Pratt, Arthur John, vegetable crops
Purchase, Mary Elizabeth, human ecology, surface chemistry
Racker, Efraim, biochemistry
Raffensperger, Edgar M, entomology
Raney, Edward Cowden, ichthyology
Rankin, John Metcalf, radio astronomy
Rawlins, William Arthur, economic entomology
Regenstein, Joe Mac, food science
Reid, John Thomas, animal nutrition
Reid, William Shaw, agronomy, soils
Reppy, John David, physics
Rhodin, Thor Nathaniel, Jr, chemical physics
Richardson, Robert Coleman, low temperature physics
Richmond, Milo Eugene, vertebrate zoology, reproductive biology
Rivers, Jerry Margaret, nutrition
Robson, Douglas Sherman, biometrics
Rochow, William Frantz, plant pathology
Rockcastle, Verne Norton, ecology
Roe, Daphne A, nutrition, medicine
Romanoff, Anastasia, biology
Root, Richard Bruce, ecology
Rosenberg, Alex, mathematics
Rosenthal, John William, mathematics
Rothaus, Oscar Seymour, mathematics
Ruoff, Arthur Louis, physical chemistry
Sack, Wolfgang Otto, anatomy, veterinary medicine
Sagan, Carl, planetary sciences
Salpeter, Edwin Ernest, astrophysics
Salpeter, Miriam Mirl, neurobiology, cytology
Salton, Gerard, applied mathematics
Saunders, Joseph Lloyd, entomology, nematology
Schaaf, Herbert Martin, agronomy
Schaffner, William Robert, aquatic ecology
Schano, Edward Arthur, poultry husbandry
Scheraga, Harold Abraham, biophysical chemistry
Schmidt, Willard Carl, microbiology, medicine
Scholer, Frederick Richard, organometallic chemistry
Schryver, Herbert Francis, veterinary physiology, veterinary pathology
Schultz, Arthur George, organic chemistry
Schultz, Otto Ernst, plant pathology
Schultz, Ronald David, immunology, veterinary virology
Schwark, Wayne Stanley, pharmacology, toxicology
Schwartz, Ruth, nutrition
Scott, Fredric Winthrop, veterinary virology
Scott, Milton Leonard, nutrition
Scott, Thomas Walter, soil fertility
Searle, Shayle Robert, statistics
Seeley, Harry Wilbur, bacteriology
Seeley, John George, floriculture
Sellers, Alvin Ferner, veterinary physiology
Semmelhack, Martin F, organic chemistry
Shapiro, Stuart Louis, astrophysics
Sharp, Lauriston, anthropology
Sheffy, Ben Edward, nutrition, microbiology
Sheldrake, Raymond, Jr, horticulture
Sherald, Allen Franklin, developmental genetics
Sherbon, John Walter, dairy chemistry
Sherf, Arden Frederick, plant pathology
Shipe, William Franklin, food science
Shively, James Nelson, veterinary pathology
Siegel, Benjamin Morton, biophysics
Siegel, James T, anthropology
Sienko, Michell Joseph, physical inorganic chemistry
Sievers, Albert John, III, solid state physics
Sievers, Sally Riedel, mathematical statistics
Silcox, John, physics
Silsbee, Robert Herman, physics
Silverman, Albert, nuclear physics
Sinclair, Thomas Russell, agronomy, micrometeorology
Sinclair, Wayne A, plant pathology
Slack, Samuel Thomas, nutrition, physiology
Slate, Floyd Owen, applied chemistry
Smiley, Richard Wayne, plant pathology, soil microbiology
Smith, Edward Holman, economic entomology
Smith, Laura Lee Weisbrodt, food chemistry
Smith, Robert John, cultural anthropology
Smith, Sedgwick Edward, animal nutrition
Smock, Robert Mumford, pomology
Solomon, Daniel Lester, biometrics, mathematical statistics
Spalding, Robert Wilber, animal husbandry
Spanswick, Roger Morgan, biophysics, plant physiology
Spitzer, Frank L, mathematics
Spofford, Sally Hoyt, ornithology
Spofford, Walter Richardson, zoology, anatomy
Sprague, Gale Clifford, physics
Srb, Adrian Morris, genetics
Steponkus, Peter Leo, plant physiology, horticulture
Sternstein, Martin, mathematics
Stevens, Charles Edward, veterinary physiology
Stillions, Merle C, laboratory animal science, nutrition
Stinson, Harry Theodore, Jr, genetics
Stone, Earl Lewis, Jr, forest soils
Stone, Margaret Hodgman, plant taxonomy
Stouffer, James Ray, animal science
Strichartz, Robert Stephen, mathematics
Sudan, Ravindra Nath, plasma physics
Sundelin, Ronald M, elementary particle physics
Swader, Fred Nicholas, soil science
Sweedler, Moss Eisenberg, mathematics
Sweet, Robert Dean, vegetable crops
Talman, Richard Michael, physics
Tang, Chung Liang, physics
Tapper, Daniel Naptali, neurophysiology, radiation biology
Tasker, John B, veterinary medicine, clinical pathology
Tauber, Maurice Jesse, entomology
Taylor, Howard Milton, III, applied mathematics, statistics
Tennant, Bud C, veterinary medicine, gastroenterology
Terzian, Yervant, astronomy
Teukolsky, Saul Arno, theoretical astrophysics
Thomison, Joel Douglas, mathematics
Thompson, Daniel Quale, wildlife ecology, conservation
Thompson, John C, Jr, physical biology
Thompson, John Fanning, plant biochemistry
Thompson, Steven Risley, zoology, genetics
Thurston, Herbert David, plant pathology
Tigner, Maury, high energy physics
Tingey, Ward M, entomology
Tomkins, John Preston, horticulture
Topoleski, Leonard Daniel, plant breeding, vegetable crops
Travers, William Brailsford, geology
Travis, Bernard Valentine, medical entomology, parasitology
Travis, Hugh Farrant, animal nutrition, physiology
Trimberger, George William, dairy science
Tukey, Harold Bradford, Jr, horticulture
Turcotte, Donald Lawson, fluid mechanics, geophysics
Turk, Kenneth Leroy, dairy husbandry
Tyler, Winfield Warren, solid state physics
Uhl, Charles Harrison, botany
Uhler, Lowell Dohner, insect ecology
Usher, David Anthony, bio-organic chemistry
Utermohlen, Virginia, immunology
Van Campen, Darrell R, nutrition, biochemistry
Vandemark, Noland Leroy, physiology
Van Etten, Hans D, plant pathology
VanSoest, Peter John, animal nutrition
Van Tienhoven, Ari, animal physiology
Van Vleck, Lloyd Dale, genetics, animal science
Venkataraghavan, R, analytical chemistry
Veverka, Joseph, astronomy, planetary sciences
Vogel, Glenn Charles, inorganic chemistry
Wadell, Lyle H, animal breeding
Wagner, David Kendall, experimental solid state physics
Wallace, Bruce, genetics
Wallace, Donald Howard, plant genetics
Wang, Hsien Chung, mathematics
Warner, Richard G, animal nutrition
Wasserman, Robert Harold, physiology, biochemistry
Weaver, Robert Michael, clay mineralogy, soil science
Weaver, Terry L, microbiology
Webb, Watt Wetmore, physics
Webster, Dwight Albert, biology
Weiss, Lionel Ira, mathematical statistics
Welch, Ross Maynard, plant nutrition, plant physiology
Wellington, George Harvey, animal science
Wells, John West, paleontology, marine zoology
West, James Edward, mathematics
Wharton, Charles Benjamin, plasma physics
White, David Hywel, experimental high energy physics
White, James Carrick, food microbiology
Whitlock, John Hendrick, veterinary parasitology, parasitology
Whitlock, Robert Henry, veterinary medicine, pathology
Whittaker, Robert Harding, ecology
Widom, Benjamin, physical chemistry
Wiesenfeld, John Richard, chemical kinetics, photochemistry
Wilcox, Charles Frederick, Jr, physical chemisty, organic chemistry
Wilkins, Bruce Tabor, resource management
Wilkinson, Christopher Foster, entomology, organic chemistry
Wilkinson, Robert Elzworth, plant pathology
Williams, Harold Henderson, biochemistry
Wilson, David Buckingham, biochemistry
Wilson, Jack Belmont, plant pathology
Wilson, Kenneth Geddes, physics
Wimsatt, William Abell, histophysiology, reproductive biology
Winch, Fred Everett, Jr, forestry, silviculture
Winter, Alexander J, immunobiology
Wolga, George Jacob, physics
Woodward, William Mooney, nuclear physics
Wootton, John Francis, biochemistry
Wright, Lemuel Dary, biochemistry
Wright, Madison Johnston, agronomy
Wu, Ray J, biochemistry
Yamamoto, Tomoko, biophysics
Yan, Tung-Mow, high energy physics
Yennie, Donald Robert, theoretical high energy physics
Yoder, Olen Curtis, plant pathology
Young, Charlotte Marie, nutrition
Young, Robert John, animal nutrition
Young, Roger Grierson, comparative biochemistry
Zahler, Stanley Arnold, microbial genetics
Zaitlin, Milton, plant virology
Zall, Robert Rouben, food science
Zilversmit, Donald Berthold, physiological chemistry
Zwerman, Paul Joseph, conservation

JACKSON HEIGHTS

Funicelli, Nicholas Anthony, fisheries, marine ecology
Storm, Leo Eugene, computer science, statistics

JAMAICA

Bartilucci, Andrew J, pharmacy
Benjaminson, Morris Aaron, microbiology, developmental genetics
Birnbaum, Ernest Rodman, inorganic chemistry
Bodi, Lewis Joseph, physical chemistry
Brady, James Edward, inorganic chemistry
Brody, Selma Blazer, physics
Brooks, Hugh Campbell, cultural geography
Brown, Arthur Barton, mathematical analysis
Brunner, Michael, molecular biology
Butkov, Eugene, theoretical physics
Butler, John Joseph, hematology
Callahan, Willie Russell, applied mathematics
Capobianco, Michael F, mathematics, statistics
Concannon, Joseph N, parasitology, radiobiology
Cover, Richard Edward, analytical chemistry
D'Adamo, Amedeo Filiberto, Jr, molecular biology, neurobiology
Eckroth, David Raymond, organic chemistry
Efthymiou, Constantine John, medical microbiology, bacteriology
Eisen, Henry, pharmaceutical chemistry
Feldman, Lawrence, nuclear physics
Fong, Conrad Tuck Onn, bio-organic chemistry
Fugmann, Ruth Adele, immunology, chemotherapy
Fuller, Charles E, ethnology, applied anthropology
Gabai, Hyman, mathematics
Greco, Claude Vincent, organic chemistry
Haldar, Dipak, biochemistry
Hampton, Suzanne Harvey, developmental biology, genetics
Hartig, William John, biology, protozoology
Holleran, Eugene Martin, physical chemistry
Horan, Harold Arthur, analytical chemistry
Jack, Robert Cecil Milton, biochemistry
Jain, Duli Chandra, chemical physics, molecular spectroscopy
Joseph, James, physics
Kapoor, Amrit Lal, medicinal chemistry, pharmaceutical chemistry
King, Reatha Clark, inorganic chemistry, physical chemistry
Kovacs, Joseph, bio-organic chemistry
Kupchik, Eugene John, organometallic chemistry
Lau-Cam, Cesar A, pharmacognosy, phytochemistry
Lengyel, Istvan, organic chemistry
Levin, Eugene (Manuel), physics
Lewis, Leslie Arthur, genetics, microbiology
Liberti, Alfred Vincent, vertebrate anatomy, vertebrate physiology
Lockshin, Richard Ansel, physiology, developmental biology
Loring, Arthur Paul, geology, oceanography
Lurye, Jerome Robert, mathematics
Lynch, Vincent De Paul, pharmacology
Machover, Maurice, mathematical analysis
Malkevitch, Joseph, mathematics
Manche, Emanuel Peter, chemistry
Mancuso, Vincent J, mathematics

NEW YORK

McCarthy, Donald John, algebra
McGee, Thomas Howard, chemical kinetics, photochemistry
Morgan, Richard C., chemical
Narici, Lawrence Robert, mathematical analysis
Oppo, Giuseppe, elementary particle physics, theoretical physics
Pasfield, William Horton, physical chemistry
Pisano, Michael A., microbiology
Pomilla, Frank Rocco, atomic physics
Posinsky, Sollie Henry, anthropology
Roistacher, Seymour Lester, dentistry
Scheiner, Peter, organic chemistry
Schwartz, Melvin J., theoretical physics
Sherman, Lawrence, endocrinology, internal medicine
Silver, Lawrence, medicine
Skarulis, John Anthony, physical chemistry
Spergel, Martin Samuel, high energy physics, astrophysics
Stalter, Richard, botany, plant ecology
Stern, Martin, oral surgery
Sun, Siao Fang, physical chemistry
Taschdjian, Claire Louise, medicine, mycology
Testa, Anthony Carmine, physical chemistry, photochemistry
Tuller, Annita, mathematics
Wagner, Vaughn Edwin, environmental health, medical entomology
Ward, Charles O., pharmacology, toxicology
Wirth, John Christian, biochemistry, organic chemistry
Yaspan, Arthur, mathematics, operations research
Young, Stephen Dean, invertebrate physiology, marine biology
Zimmerman, Jay Alan, mammalian physiology

JAMESTOWN
DJang, Arthur H K., pathology, nuclear medicine
Lucey, Carol Ann, theoretical physics, philosophy of science

JERICHO
Wolf, Henry, applied mathematics, celestial mechanics

JOHNSON CITY
Reynoso, Gustavo D., medicine, pathology

JOHNSONVILLE
Moore, Kenneth Howard, physics

JOHNSTOWN
Brown, Robert Getman, industrial

KATONAH
Ventura, William Paul, pharmacology
Weil, Thomas P., public health, hospital administration

KENMORE
Bray, Joseph Moyer, mining geology
Montague, Harriet Frances, mathematics
Parnes, Milton N., mathematics
Starks, Fred W., organic chemistry

KEUKA PARK
Prest, Dorothy B., microbiology
Stimson, Miriam Michael, organic chemistry
Webber, Edgar Ernest, botany
White, James Edwin, biology, ecology

KEW GARDENS
Baczewski, Alexander, colloidal chemistry

KINDERHOOK
Fisher, Donald William, paleontology

KINGS PARK
Blumenthal, Irving Jack, electroencephalography, psychiatry

KINGSTON
Benz, George William, organic chemistry

Lorenzen, Jerry Alan, surface chemistry

LAGRANGEVILLE
Miller, Lewis F., polymer chemistry, colloid chemistry

LAKE PLACID
Dougall, Donald K., biochemistry, experimental morphology
Verma, Devi C., agricultural biochemistry

LANSING
Sandsted, Roger France, vegetable crops

LARCHMONT
Brown, Edward Allan, physics, mathematics
Modell, Walter, pharmacology

LATHAM
Brown, Severn Parker, economic geology, structural geology

LAURELTON
Greenberg, Jacob, biochemistry
Strats, Gerald, inorganic chemistry

LAWRENCE
Klahr, Carl Nathan, applied physics

LE ROY
Wheeler, Edward Stubbs, research administration

LEWISTON
Bronk, Theodore Tobias, pathology
Conn, Paul Kohler, materials science
Cook, Edward Hoopes, Jr., physical chemistry, inorganic
Dexter, Theodore Henry, inorganic chemistry
Geiger, Marion Braxton, industrial
Wadlinger, Robert Louis Peter, chemistry

LINDENHURST
Ellis, Samuel Benjamin, electroanalytical chemistry

LITTLE NECK
Liban, Eric, applied mathematics
Swenson, Hugo Nathanael, physics

LIVERPOOL
Lane, Keith Aldrich, analytical chemistry
Thurber, Walter Arthur, ornithology

LIVINGSTON
Clarkson, Vernon A., horticulture

LOCKPORT
Alfen, Franz Juergen, chemistry

LONG EDDY
Hope, Henry Bell, analytical chemistry

LONG ISLAND CITY
Benumof, Reuben, engineering physics
Cantor, Abraham, biochemistry
Levy, Milton, biochemistry
Marov, Gaspar J., sugar chemistry
Palochak, Muriel E., industrial microbiology
Tibbetts, Merrick Sawyer, organic

LOUDONVILLE
Arnett, Ross Harold, Jr., entomology
Bevak, Joseph Perry, physical chemistry
Brown, Patricia Stocking, zoology, endocrinology
Fraser, Douglas Fyfe, ecology
Hanhauser, Martin A., mathematics, mathematical physics
Kreuzer, James Leon, organic chemistry
LaRow, Edward J., aquatic ecology, invertebrate zoology
Quinan, James Roger, chemistry
Sullivan, Seraphin A., physics
Waldrop, Ann Lyneve, crystallography

Wittig, Kenneth Paul, invertebrate physiology

LYNBROOK
Forray, Marvin Julian, mathematics

MACEDON
Clayton, William Joseph, chemistry
Weiner, Milton Lawrence, polymer chemistry, plastics chemistry

MAHOPAC
Joseph, Alexander, forensic science, physics
Rupert, John Paul, polymer science

MALVERNE
Beckmann, Albert Jules, pediatrics, public health
Freireich, Abraham Walter, internal medicine
Vogel, Alfred Morris, chemistry, instrumentation

MAMARONECK
Beuzeville, Carlos F., pediatrics
King, Robert Evans, geology
Ogilvy, C Stanley, mathematics

MANHASSET
Anderson, Lucia Lewis, microbiology
Bechtold, Edwin William, applied optics
Bowie, Robert McNeil, physics
Gulotta, Stephen Joseph, cardiology
Loring, Marvin F., radiology
Oka, Masamichi, pathology
Scherr, Lawrence, internal medicine, cardiology
Smith, John Kelly, immunology

MANLIUS
Hovey, Charles Louis, entomology

MASSAPEQUA
Lezer, Leon Robert, medicine

MASSAPEQUA PARK
Feit, Julius, space physics, solar physics
Rosin, Seymour, physics

MECHANICVILLE
Hardman, Bruce Bertolette, organic chemistry

MELVILLE
Freudenthal, Hugo David, environmental sciences
Wallace, John Longstreet, solid state physics

MERRICK
Molomut, Norman, oncology
Zadoff, Leon Nathan, physics

MEXICO
Brunson, John Taylor, physiology

MIDDLE VILLAGE
Schubert, Walter John, organic chemistry, biochemistry

MIDDLEPORT
Cook, Ronald Frank, analytical chemistry
Di Sanzo, Carmine Pasqualino, nematology
Drummond, Paul Edward, organic chemistry
Engel, John Francis, organic chemistry
Folts, Dwight David, entomology
Gates, Robert Leroy, biochemistry
Gibbons, Loren Kenneth, organic chemistry
Graham, Jack Raymond, organic chemistry
Harnish, Wayne Nelson, plant pathology
Krog, Norman Eiler, plant physiology, research administration
Montgomery, Ronald Eugene, chemistry, pesticide chemistry
Peake, Clinton J., organic chemistry, biological chemistry

Ramsey, Arthur Albert, pesticide chemistry
Sticker, Robert Earl, organic chemistry
Towns, Donald Lionel, physical organic chemistry, chemical engineering
Willard, Joe Raymond, organic chemistry
Young, Sanford Tyler, organic chemistry, analytical chemistry

MIDDLETOWN
Knowlton, Carroll Babbidge, Jr., entomology
Lauer, Gerald J., aquatic ecology, limnology
Wood, Raymond Arthur, zoology, parasitology

MILLBROOK
Donohue, John J., marine geology, geoenvironmental science
Elias, Thomas S., systematic botany, morphology
Goodland, Robert James, ecology, environmental management
Karnosky, David Frank, forest genetics
Setliff, Edson Carmack, forest pathology, mycology

MILLER PLACE
Miller, William Alfonso, physics

MINEOLA
Li, Min Chiu, oncology

MOHEGAN LAKE
Petro, Anthony James, physical chemistry

MONROE
Martus, Joseph Armand, history of chemistry

MONSEY
Gutcho, Sidney Jack, biochemistry
Morduchowitz, Abraham, organic polymer chemistry
Naumann, Alfred Wayne, physical chemistry
Szabo, Steve Stanley, agronomy

MONTAUK
Staker, Robert Dale, phycology
Wourms, John Barton, cell biology, developmental biology

MORRISVILLE
Greene, Kingsley L., ecology
Whipple, Royson Newton, food technology

MT VERNON
Butensky, Irwin, cosmetic chemistry, pharmaceutical chemistry
Cooper, Aaron David, analytical chemistry
Garizio, John Ernest, pharmaceutical chemistry
Haas, David Jean, instrumentation, biophysics
Hanus, Edward Joseph, pharmacy
Heyd, Allen, pharmaceutics
Hoy, James Benjamin, entomology
Kaufman, Raymond, experimental physics
Kelton, Gilbert, physics
London, S J., medicine, clinical pharmacology
Nielsen, Peter Adams, bacteriology
O'Brian, Dennis Martin, physiology, toxicology
Paolini, Francis Rudolph, x-ray spectrometry
Poetsch, Chester E., pharmaceutical chemistry
Smith, Douglas Stewart, organic chemistry, research administration
Summa, A Francis, pharmaceutical chemistry

NANUET
Albright, Jay Donald, organic chemistry
Bogard, Terry L., organic chemistry

NEW CITY

Fantini, Amedeo Alexander, microbial genetics
Krueger, James Elwood, organic chemistry
Kuzmak, Joseph Milton, physical chemistry
Travers, John Joseph, biochemistry

NEW HYDE PARK

Abramson, Allan Lewis, otolaryngology
Berkman, James Israel, medicine, pathology
Gootman, Norman Lerner, cardiovascular diseases, cardiovascular physiology
Grand, Stanley, physical chemistry
Hoory, Shlome, medical physics
Isenberg, Henry David, medical microbiology
Lipsitz, Philip Joseph, medicine, pediatrics
Rosenthal, Alexander Herman, obstetrics & gynecology
Rosenthal, Arthur Frederick, biochemistry
Sawitsky, Arthur, hematology, oncology
Scheib, Richard, Jr., physics
Wiener, Stanley L., internal medicine, experimental pathology

NEW PALTZ

Anastasio, Salvatore, mathematics
Bernstein, Burton, lasers, optical physics
Brain, James Lewton, anthropology
Bregman, Allyn Aaron, cytogenetics
Brenner, Gilbert J., palynology
Campion, James J., analytical chemistry, physical chemistry
Erhlich, Robert, physics
Gawer, Albert Henry, physical chemistry
Gray, James Clarke, developmental anatomy
Harkavy, Allan Abraham, magnetic resonance, visual physiology
Hendel-Sebestyen, Giselle, anthropology
Heyl, George Richard, geology, economic geology
Ho, Hon Hing, mycology, phytopathology
Knapp, Ronald Gary, geography of China, cultural geography
Konigsberg, Alvin Stuart, atmospheric physics, biometeorology
Krauss, Peter H., mathematics
Krieg, David Charles, animal behavior, vertebrate zoology
Kudzin, Stanley Francis, organic chemistry
Manos, Constantine T., sedimentology
Meng, Heinz Karl, biology
Nunes, Thomas Lester, chemical instrumentation
Nydegger, LeRoy B., parasitology
Ordway, Richard John, geology
Patsis, Angelo Vlasios, physical chemistry, chemical engineering
Pease, Robert Louis, physics
Redmond, Billy Lee, anatomy, entomology
Robison, Gerson B, mathematics
Rutstein, Martin S, mineralogy, geochemistry
Santoro, Thomas, microbiology
Schnell, George Adam, population geography, economic geography
Senechalle, David Albert, mathematics
Shaw, Joe William, Jr., mathematics
Shelton, Austin Jesse, ethnology, cultural anthropology
Slater, John Vernon, radiobiology, comparative physiology
Spencer, Selden J., biological sciences
Stein, Philip, physiology
Straus, David Bradley, biochemistry
Ullman, Arthur William James, mathematics

NEW ROCHELLE

Acerbo, Samuel Nicholas, organic chemistry
Blackwood, Carlton E, biochemistry
Bullen, Thomas Gerrard, experimental physics
Campisi, Louis Sebastian, inorganic chemistry
Driscoll, John G, mathematics
Ghidoni, Estelle, biology
Levkov, Jerome Stephen, physical chemistry
Machlis, Samuel, textile chemistry
Morrison, Nathan, mathematical statistics

Murphy, James Joseph, experimental solid state physics
Orna, Mary Virginia, physical chemistry, analytical chemistry
Pappas, George Stephen, physiology
Power, Richard B, biology
Rousseau, Viateur, organic chemistry
Ruben, Samuel, electrochemistry, medicinal chemistry
Soloway, Harold, organic chemistry
Soloway, Saul, organic chemistry
Stanionis, Victor Adam, applied mathematics

NEW YORK

Abbott, Edwin Hunt, inorganic chemistry
Abbott, Joan, developmental biology
Abell, Liese Lewis, biochemistry
Abrahamson, Adolf Avraham, theoretical physics
Abramsky, Tessa, biochemistry
Abramson, Arthur Simon, rehabilitation medicine
Ackerman, Roy Alan, bioengineering
Acs, George, biochemistry
Adams, Stuart Lyle, bacteriology
Adisman, I Kenneth, prosthodontics
Adler, Alexandra, neurology, psychiatry
Adler, Frank Leo, bacteriology, immunology
Adler, Kurt Alfred, psychiatry
Agosta, William Carleton, organic chemistry
Ahluwalia, Daljit Singh, applied mathematics
Ahrens, Edward Hamblin, Jr., medicine
Al-Askari, Salah, immunology, urology
Albert, Roy Ernest, environmental medicine, pulmonary physiology
Alderman, Michael Harris, public health
Alexander, Alexandre Emil, mineralogy, ceramics
Alexander, Benjamin, medicine
Alexander, George Jay, biological taxonomy, ethnobotany
Alexander-Jackson, Eleanor Gertrude, bacteriology
Alland, Alexander, Jr, anthropology, cultural anthropology
Allen, Fred Harold, Jr, immunohematology
Alt, Franz Leopold, mathematics
Altschul, Siri von Reis, botanical taxonomy, ethnobotany
Altshuler, Bernard, biomathematics
Altszuler, Norman, pharmacology, endocrinology
Alvares, Alvito Peter, pharmacology, biochemistry
Amadon, Dean, ornithology
Ammirato, Philip Vincent, plant physiology
Anderson, Albert Douglas, physical medicine, rehabilitation medicine
Anderson, Ethel Irene, biochemistry
Anderson, Orval Roger, cell biology
Anderson, Paul J, neuropathology, neurology
Anderson, Sydney, vertebrate zoology
Anshel, Michael, mathematics, computer science
Ante, Robert, economic geography, geography of Central Asia
Antonsen, Donald Hans, industrial chemistry
Apfelberg, Benjamin, psychiatry
Appelbaum, Emanuel, internal medicine
Applebaum, Edmund, dentistry
April, Ernest W, anatomy, muscular physiology
Aranow, Henry, internal medicine
Archibald, Reginald MacGregor, biochemistry, pediatric endocrinology
Arden, Sheldon Bruce, medical microbiology
Arents, John (Stephen), physical chemistry
Arias, Irwin Monroe, medicine, gastroenterology
Arieti, Silvano, psychiatry
Arkin, Arthur Malcolm, psychiatry, physiological psychology
Arkin, Herbert, applied statistics, analytical statistics
Arluk, David Jay, dermatology, anatomy
Armstrong, Robert Thexton, chemistry
Arons, Michael Gene, physics
Aronson, Lester Ralph, animal behavior, neurosciences
Artzi, Karen, developmental genetics
Arvesen, James Norman, mathematical statistics, applied statistics
Arzt, Sholom, mathematics
Asanuma, Hiroshi, neurophysiology
Aschner, Joseph Felix, solid state physics, semiconductors

Asimov, Isaac, science writing, biochemistry
Askari, Amir, pharmacology, biochemistry
Atchison, Joseph Edward, pulp chemistry, paper chemistry
Auerbach, Morris Baline, endodontics
Augenlicht, Leonard Harold, cell biology, biochemistry
Auld, Peter A McF., medicine, pediatrics
Aull, Felice, physiology
Axel, Richard, molecular biology
Axelrod, Norman Nathan, optics, optical engineering
Axenrod, Theodore, organic chemistry
Aylesworth, Thomas Gibbons, zoology
Bacchi, Cyrus Joseph, parasitology, biological chemistry
Bacharach, Martin Max, biology, genetics
Baer, Rudolf L, dermatology, immunology
Baez, Silvio, physiology
Bailin, Gary, biochemistry
Baker, David H, pediatrics, radiology
Balagura, Saul, neurosurgery, physiological psychology
Balazs, Endre Alexander, biochemistry
Balboni, Edward Raymond, insect physiology
Balis, Moses Earl, biochemistry
Ballantyne, Donald Lindsay, Jr, transplantation biology
Baltay, Charles, nuclear physics
Banay-Schwartz, Miriam, biochemistry, organic chemistry
Banfield, Armine Frederick, economic geology
Banks, Clarence Kenneth, organic chemistry, graphics
Bar, Asher, speech pathology, audiology
Barach, Alvan Leroy, internal medicine
Barany, Kate, physical chemistry, biophysics
Barber, Sherburne Frederick, mathematics
Barfield, Mary Ashton, reproductive physiology
Barka, Tibor, experimental medicine
Barker, Harold Grant, surgery
Barker, June Northrop, physiology
Barker, Louis Allen, pharmacology, neurochemistry
Barksdale, Lane, microbiology
Barnes, Allan Campbell, medicine
Barnes, William Alexander, medicine
Barrett, Edward Joseph, organic chemistry
Barron, Bruce Albrecht, obstetrics & gynecology
Barrow, Thomas D, petroleum geology
Barry, John Young, mathematics
Barsky, James, information science
Bartalos, Mihaly, medical genetics, pharmacology
Bassett, Charles Andrew Loockerman, orthopedic surgery, cell physiology
Bastos, Milton Lessa, toxicology
Batt, Ellen Rae, physiology
Batten, Roger Lyman, invertebrate paleontology
Battista, Arthur Francis, neurosurgery, neurophysiology
Batton, Robert Ralph, neuroanatomy
Bauer, Frances Brand, applied mathematics
Baumel, Philip, high energy physics
Baumslag, Gilbert, mathematics
Bavley, Abraham, chemistry
Bayes, Alfred Lee, inorganic chemistry
Bazer, Jack, mathematics, physics
Beach, Eliot Frederick, biochemistry
Beardsley, Robert Eugene, microbiology, genetics
Bearn, Alexander Gordon, medicine
Beattie, Diana Scott, biochemistry
Beattie, Edward J, surgery
Beck, Donald Richardson, quantum chemistry, atomic physics
Becker, Carl George, pathology
Becker, David Victor, medicine
Becker, Ernest Lovell, internal medicine
Beckhorn, Frederick F, pathology, oncology
Bederson, Benjamin, experimental physics

Bedford, John Michael, anatomy, physiology
Beech, John Alan, chemistry, pharmacology
Beg, Mirza Abdul Baqi, theoretical physics
Begg, Charles Frederic, pathology, hematology
Behre, Charles Henry, Jr, geology
Behrends, Ralph Eugene, elementary particle physics
Behrman, Richard Elliot, pediatrics, physiology
Beinfest, Sidney, organic chemistry
Bell, Alfred Lee Loomis, Jr., medicine
Bellak, Leopold, psychiatry, psychoanalysis
Bellino, Vito Victor, organic chemistry
Belman, Sidney, biochemistry
Bender, Morris Boris, neurology
Bendich, Aaron, biochemistry
Benesch, Reinhold, biochemistry
Benesch, Ruth Erica, biochemistry
Benet, Sula, cultural anthropology
Bennet, Dorothea, genetics
Bennett, Burton George, environmental sciences
Bennett, Ivan Loveridge, Jr, pathology
Bentley, Peter John, physiology, pharmacology
Benua, Richard Squier, nuclear medicine
Benuck, Myron, neurochemistry
Beranbaum, Samuel Louis, radiology
Beredjick, Nicky, organic chemistry, polymer chemistry
Berger, Eugene Y, medicine
Berger, Lawrence, medicine
Berger, Melvyn Stuart, mathematics
Berger, Selman A, analytical chemistry, environmental chemistry
Bergmann, Louis Lawrence, anatomy
Berkowitz, Jerome, mathematics
Bernardi, Salvatore Dante, mathematics
Berl, Soll, biochemistry, neurochemistry
Berliner, Kurt, cardiology
Berman, Paul Ronald, atomic physics, lasers
Berman, Rose Louise, biology, natural sciences
Berman, Simeon Moses, mathematical statistics
Bermon, Stuart, solid state physics
Berne, Bruce J, chemical physics, physical chemistry
Bernheimer, Alan Weyl, microbiology, toxicology
Bernkopf, Michael, mathematics
Bernstein, Irwin S, mathematics
Berry, Gail Wruble, psychiatry
Bers, Lipman, mathematics
Bertani, Laura Marie, biochemistry, organic chemistry
Bertles, John F, internal medicine, hematology
Bertsch, Walter Frank, biophysics, plant physiology
Bethel, Joseph Jay, biochemistry
Bevans, Margaret, pathology
Beveridge, David L, physical chemistry
Beychok, Sherman, biochemistry, immunobiology
Bickerman, Hylan A, respiratory physiology, clinical pharmacology
Bieber, Irving, medicine, psychiatry
Bierman, Arthur, theoretical physics
Bigger, J Thomas, Jr, cardiology, pharmacology
Bigler, Rodney Errol, nuclear physics, biophysics
Bird, Junius Bouton, anthropology
Birke, Ronald Lewis, electrochemistry, analytical chemistry
Birman, Joan Sylvia, mathematics
Birman, Joseph Leon, theoretical physics
Black, Virginia H, cell biology, developmental biology
Blackman, Samuel William, organic chemistry
Blakeslee, Alton Lauren, science writing
Blanc, William Andre, pathology
Blanck, Martin, physiology, biophysics
Blanton, Richard Edward, anthropology
Blatt, Sylvia, clinical chemistry, science administration
Biau, Lawrence Martin, medical physics
Blechman, Harry, microbiology
Bleyman, Lea Kanner, genetics, cell biology
Bliss, Dorothy Elizabeth, invertebrate physiology
Blizard, David Arthur, behavioral biology
Bloch, Arthur David, genetics, pediatrics
Blu, Karen Isobell, anthropology
Bluestone, E Michael, medical administration, public health
Bock, Walter Joseph, vertebrate morphology, evolution
Bodansky, Oscar, medicine, biochemistry

Bogart, Bruce Ian, cell biology
Boley, Scott Jason, pediatric surgery
Bond, George Clement, anthropology
Bonnes, Roy Walter, chemistry
Booman, Keith Albert, physical organic chemistry
Borek, Carmia Ganz, cell biology
Borenfreund, Ellen, biochemistry
Borland, John Raymond, communication science
Bornstein, Lawrence A., physics
Borowitz, Irving Julius, bio-organic chemistry
Borowitz, Sidney, physics
Borowska, Zofia Kurylo, biochemistry
Borowsky, Richard Lewis, evolution, genetics
Boskey, Adele Ludin, structural chemistry
Bosniak, Morton A., radiology
Botstein, Charles, radiology
Bottone, Edward Joseph, microbiology
Bowman, William Henry, clinical biochemistry
Boyd, Robert Neilson, organic chemistry
Boyer, Timothy Howard, theoretical physics
Boykin, Lorraine Stith, nutrition
Brachfeld, Norman, biochemistry, cardiovascular physiology
Brackenbury, Robert William, developmental biology
Bradley, Francis J., health physics
Bradley, Stanley Edward, medicine
Brady, Robert Townsend, petroleum geology
Braestrup, Carl Bjorn, radiological physics
Bragg, John Kendal, chemical physics
Brand, Leonard, biochemistry
Brandaleone, Harold, medicine, cardiology
Brandt, Philip Williams, anatomy, physiology
Brandt, Werner, physics
Brandwein, Paul Franz, botany
Branley, Franklyn M., astronomy
Brasel, Jo Anne, pediatric endocrinology, nutrition
Braun, Armin Charles, cell biology, plant biology
Brazeau, Paul, pharmacology
Breed, Ernest Spencer, surgery, physiology
Breidenbach, Harold, medicine, anesthesiology
Brennan, Goodwin Milton, ophthalmology
Brenner, Henry Clifton, solid state chemistry
Breslow, Esther M.G., biophysical chemistry
Breslow, Ronald, organic chemistry
Bressler, Robert S., cell biology, developmental anatomy
Briggs, Donald K., medicine, hematology
Brightman, I. Jay, medicine, public health
Brink, Frank, Jr., biophysics
Briscoe, Anne M., biochemistry, physiology
Briscoe, William Alexander, physiology, medicine
Brodman, Keeve, clinical medicine
Brodsky, William Aaron, physiology
Brody, Marcia, photobiology, biochemistry
Brody, Seymour Steven, biophysics
Broecker, Wallace, geochemistry
Broeg, Charles Burton, chemistry
Brohn, Frederick Herman, biochemistry
Bromberg, Eleazer, applied mathematics
Bronstein, Eugene L., medicine
Brooks, Lester Allen, rubber chemistry
Brown, Maurice Vertner, acoustics, geophysics
Brucker, Harvey Jerome, biophysics, electronics
Brunlik, Joseph V., medicine
Brunngraber, Elinor Flora, biochemistry
Bryant, John Harland, public health
Budner, Stanley, public health
Budzilovich, Gleb Nicholas, pathology, neuropathology
Bucker, Elmer Daniel, embryology, neuroanatomy
Buettner-Janusch, John, physical anthropology, primatology
Bulkin, Bernard Joseph, physical chemistry, inorganic chemistry
Bull, John, ornithology, ecology
Burchenal, Joseph Holland, medicine
Burger, Richard Melton, molecular biology
Burke, Gail De Planque, radiation physics
Burlington, Harold, physiology
Burman, Louis Robert, physics
Burnette, Llewellyn Wilson, organic chemistry

Burney, Leroy Edgar, public health
Burns, Fredric Jay, oncology
Burrill, Claude Wesley, mathematics
Burstein, Shlomo, biochemistry
Bush, Ian, physiology, biochemistry
Butler, Vincent Paul, Jr., internal medicine, immunochemistry
Butterworth, Julian Scott, medicine
Byrnes, Francis Clair, communication science, academic administration
Cabeen, Samuel Kirkland, chemistry, information science
Cadbury, William Edward, Jr., physical chemistry
Cahill, Kevin M., medicine, tropical medicine
Calhoun, David H., microbiology, biochemistry
Camerini-Davalos, Rafael, medicine
Candia, Oscar A., physiology, biophysics
Canfield, Robert E., medicine, endocrinology
Canizares, Orlando, dermatology
Cantor, Charles Robert, biophysical chemistry
Cappell, Sylvain Edward, topology
Carbone, Gabriel, environmental medicine
Carlson, Eric Theodore, psychiatry
Carlson, Robert Leonard, anthropology
Carneiro, Robert Leonard, anthropology
Carol, Bernard, mathematical statistics, computers
Carpenter, Malcolm Breckenridge, anatomy, physiology
Carr, Malcolm Wallace, anatomy, surgical pathology
Carr, Ronald E., ophthalmology, visual physiology
Carroll, Edward Major, mathematics education
Carter, Sidney, neurology
Case, Kenneth Myron, physics
Case, Robert B., cardiology, physiology
Caspe, Saul, biochemistry
Catell, McKeen, pharmacology
Cavallito, Chester John, organic chemistry, physiological chemistry
Cefola, Michael, nuclear chemistry, inorganic chemistry
Celis, Teodoro F.R., molecular biology, microbiology
Chaganti, Raju Sreerama Kamalasana, genetics
Chaikin, Lawrence, oral surgery
Chalmers, Thomas Clark, internal medicine, gastroenterology
Chambers, Robert Warner, biochemistry
Chan, Wah Yip, pharmacology, reproductive physiology
Chang, Chu Huai, radiology
Chang, Kwang-Poo, parasitology
Chang, Ngee Pong, theoretical high energy physics
Chang, Shih-Yung, quantum chemistry
Chang, William Wei-Lien, histology, pathology
Chanley, Jacob David, biochemistry
Chapman, Carleton Burke, medicine
Chappell, Richard Lee, neurobiology
Chargaff, Erwin, biochemistry
Charmatz, Richard, micropaleontology
Charmatz, Mark Ray, III, astronomy
Chase, Norman E., medicine
Chase, Randolph Montieth, Jr., medicine, immunology
Chasin, Lawrence Allen, biochemical genetics
Chasis, Herbert, medicine
Chatterjee, Sampriti, statistics, operations research
Chavel, Isaac, geometry
Chen, John Heng, biochemistry
Chen, Shu, physiology
Childress, William Stephen, fluid mechanics, applied mathematics
Chitwood, Henry Cady, organic chemistry
Chiu, Hong-Yee, astrophysics
Chodroff, Saul, chemistry
Choppin, Purnell Whittington, internal medicine, virology
Chou, Ting-Chao, pharmacology, chemotherapy
Chow, Yuan Shih, mathematics
Christ, Adolph Ervin, child psychiatry
Christensen, Robert Lee, atomic physics, computer science
Christian, Charles L., immunology
Christman, Judith Kershaw, biochemistry, molecular biology
Christy, Nicholas Pierson, medicine
Chrysanthou, Chrysanthos, experimental pathology
Chu, Chia-Kun, applied mathematics
Chu, Florence Chien-Hwa, radiology
Chuckrow, Vicki G., mathematics
Chung, Victor, physics

Church, Shepard Earll, (Jr), physical chemistry
Churg, Jacob, pathology
Chusid, Joseph George, neurology
Cizek, Louis Joseph, physiology
Clark, Salem Thomas, physical chemistry
Clarke, Donald Dudley, biochemistry
Clarkson, Allen Boykin, Jr., parasitology
Clarkson, Bayard D., hematology, oncology
Claus, George, microbiology, marine biology
Clauss, Roland Charles, ornithology
Clement, Dayton Harris, physics
Clewell, Eugene Everett, surgery, enzymology
Clifton, Raymond Otto, medicinal chemistry
Clinton, Howard Edward, parasitology, neurology
Cmelja, David H., prosthodontics
Coelho, Sidney Quex, pediatrics
Cohan, Harvey, mathematics
Colby, Clarence, biochemical genetics, virology
Cole, Harold S., medicine
Cole, Jerome F., environmental health, pharmacy
Coleman, Morton, hematology, oncology
Coleman, Peter Stephen, biochemistry, biophysics
Coles, James Stacy, physical chemistry
Collier, Jack Reed, embryology
Collins, Richard Cornelius, parasitology, entomology
Conan, Neal Joseph, Jr., medicine
Conant, Francis Paine, cultural anthropology, ethnology
Conant, James Bryant, chemistry
Condon, Francis Edward, chemistry
Connell, Elizabeth Bishop, obstetrics & gynecology
Connelly, Clarence Morley, biophysics
Converse, John Marquis, plastic surgery, transplantation biology
Cooke, Lloyd Miller, chemistry
Cool, Rodney Lee, physics
Cooper, George Wallace, Jr., reproductive biology
Cooper, George William, physiology, hematology
Cooper, Irving S., neurosurgery
Cooper, Louis Zucker, pediatrics, diseases, infectious
Cooper, Norman S., pathology
Cooper, Owen, food technology
Cooperman, Jack M., nutritional biochemistry
Copulsky, William, industrial chemistry
Cornell, Alan, food science
Cornell, James S., biological chemistry, endocrinology
Cosmatos, Alexandros, organic chemistry
Cote, Lucien Joseph, biochemistry, neurology
Cotter, Mary Virginia, genetics
Cotzias, George Constantin, medicine
Courtnand, Andre Frederic, medicine
Cousminer, Harold L., micropaleontology, geology
Cowell, James Leo, microbiology
Cowen, David, neuropathology
Cox, Dudley, microbiology, cell physiology
Cox, Rody Powell, medical genetics
Craddock, Elysse Margaret, biology, cytogenetics
Craig, Stanley Harold, radiology
Cramer, Joseph Benjamin, psychiatry, psychoanalysis
Cranefield, Paul Frederic, physiology
Crapanzano, Vincent Bernard, anthropology
Craviato, Humberto, neuropathology, neurology
Crawford, Oakley H., physical chemistry
Crikelair, George F., surgery
Crotty, William Joseph, plant morphogenesis
Crouch, Billy G., physiology, radiation biology
Cucci, Cesare Eleuterio, medicine
Cummin, Alfred Samuel, physical chemistry

Cummings, Ralph Waldo, Jr., agricultural economics
Cummins, Herman Z., quantum optics, solid state physics
Cunningham, Bruce Arthur, biochemistry
Cunningham, Dorothy J., physiology
Cunningham, Nicholas, pediatrics, public health
Curnen, Edward Charles, Jr., pediatrics, microbiology
Curnen, Mary G McCrea, epidemiology, pediatrics
Currie, Julia Ruth, neurosciences
Cushman, Paul, Jr., endocrinology, drug abuse
Cutita, Janice Ann, physical chemistry
Cutler, Janice Ann, physical chemistry, dentistry
Dack, Simon, medicine
Dailey, Benjamin Peter, chemistry
D'Alesandro, Philip Anthony, parasitology, immunology
Dalton, Philip Benjamin, organic chemistry
Daly, John F., otolaryngology
Dancis, Joseph, medicine, biochemistry
D'Angio, Giulio J., radiotherapy
Daniels, Farrington, Jr., physiology
Dannenberg, Joseph, physical organic chemistry, theoretical chemistry
Darling, Robert Croly, physical medicine, rehabilitation
Darlington, Gretchen Ann Jolly, genetics
Darnell, James Edwin, Jr., cell biology
Darzynkiewicz, Zbigniew Dzierzykraj, cell biology, cytochemistry
Dash, Barry Harold, pharmacy
Datta, Ranajit Kumar, neurochemistry, neuropharmacology
Davidow, Bernard, toxicology, clinical biochemistry
Davis, Elizabeth B., psychiatry
Davis, Harold Larue, nuclear physics
Davis, Martin (David), mathematical logic
Davis, Thomas Wilders, physical chemistry
Davis, William Thompson, veterinary medicine
Dawson, Charles Reginald, organic chemistry, biochemistry
Day, Jack Calvin, organic chemistry
Day, Noorbibi Kassam, immunobiology
Day, Stacey Biswas, medicine, biochemistry
Deane, Norman, internal medicine, experimental surgery
de Duve, Christian Rene, biochemistry, cytology
Defendi, Vittorio, pathology
DeForest, Peter Rupert, forensic science, criminalistics
Degani, Meir Hershenkorn, meteorology, physics
DeGraff, Arthur Christian, medicine
De Harven, Etienne, cytology, biological structure
Deitch, Arline D., cell biology, cytopathology
de la Chapelle, Clarence Ewald, medicine
Delagi, Edward F., medicine
Delahay, Paul, chemical physics
De Lemos, Carmen Loretta, cell biology
De Lillo, Nicholas Joseph, mathematical logic, computer science
Delmonte, Lilian, hematology
Deming, Quentin Burritt, medicine
Demopoulos, Harry Byron, experimental pathology
Demopoulos, James Thomas, physical medicine & rehabilitation
Derby, Bennett Marsh, neuropathology, neurology
Deresiewicz, Herbert, theoretical mechanics, applied mechanics
Derman, Cyrus, mathematical statistics, operations research
de Salegui, Miriam, biochemistry, parasitology
Despommier, Dickson, parasitology, immunology
Deuschle, Kurt W., medicine
Deutsch, Murray Lewis, computer sciences
Dev, Vaithilingam Gangathara, genetics, cytogenetics
DeVoe, Arthur Gerard, ophthalmology
Devons, Samuel, physics
Diacumakos, Elaine G., cell biology
Dickerman, Stuart Carlton, organic chemistry
Dietz, George William, Jr., biochemistry, molecular biology
Dilley, William G., cell biology, oncology
Dinning, James Smith, nutrition
Di Salvo, Nicholas Armand, orthodontics, physiology
Djordjevic, Bozidar, radiobiology, biophysics
Dockstader, Frederick J., anthropology

Dole, Gertrude Evelyn, cultural anthropology, ethnology
Dole, Vincent Paul, medicine
Dolgin, Martin, cardiology, internal medicine
Donn, William L, climatology, aeronomy
Donohoo, Horrie Van Waldo, geophysics
Donsker, Monroe David, mathematics
Dorfman, Howard David, pathology
Dorsey, Thomas Edward, biochemistry
Douglas, Gordon Watkins, obstetrics & gynecology
Do-Van-Quy, Dominic, cancer
Dowling, Herndon Glenn, (Jr), herpetology
Downey, John A, medicine, physiology
Doyle, Eugenie F, pediatric cardiology
Dreiling, David A, surgery
Drusin, Lewis Martin, preventive medicine, infectious diseases
Dubnau, David, molecular biology
Dubos, Rene Jules, pathology
Dudley, Patricia, invertebrate zoology
Duffy, Philip, neurology, neuropathology
Duffy, Thomas Edward, neurochemistry
Dumont, Allan E, surgery, physiology
Dunbar, Howard Stanford, neurosurgery
Dutka, Jacques, mathematics
Duvoisin, Roger C, neurology
Dwyer, Dennis Michael, parasitology, protozoology
Dyer, Eldon, mathematics
Eagle, John Frederick, pediatrics
Eakins, Kenneth E, pharmacology, ophthalmology
Eames, Edwin, anthropology, ethnology
Eastwood, Abraham Bagot, cytopathology, electron microscopy
Ebel, Alfred, rehabilitation medicine, electromyography
Eberstein, Arthur, biophysics
Ebert, Paul Allen, cardiovascular surgery, physiology
Ebihara, May Mayko, cultural anthropology, ethnology
Edberg, Stephen Charles, microbiology, immunology
Edelman, Gerald Maurice, biochemistry
Eder, Howard Abram, medical research
Edmonds, Peter Derek, biomedical engineering
Edmonds, Sylvan Milton, chemistry
Eggena, Patrick, physiology
Ehrenfeld, Sylvain, mathematical statistics
Ehrenpreis, Leon, mathematics
Ehrenpreis, Seymour, biochemistry, pharmacology
Ehrenreich, Theodore, pathology
Eilenberg, Samuel, mathematics
Eisele, Carolyn, mathematics
Eisenberg, Max A, biochemistry, microbiology
Eisenbud, Merril, environmental health
Eisenman, Joseph Sol, physiology, neurophysiology
Eisenmann, Eugene, ornithology
Eisenstein, Albert Bernard, internal medicine, endocrinology
Ekholm, Gordon Frederick, anthropology
Eldredge, Niles, paleobiology
Elizan, Teresita S, neurology, virology
Elliott, Ralph Francis, animal nutrition
Ellis, John Taylor, pathology
Ellner, Paul Daniel, medical microbiology
Elsbach, Peter, medicine, physiology
Elwyn, David Hunter, biochemistry
Ely, Charles A, anatomy
Ember, Carol Ruchlis, cultural anthropology
Emberson, Richard Maury, physics
Emerson, William Keith, malacology
Emmers, Raimond, medical physiology, neurophysiology
Engelhardt, Dean Lee, cell biology, molecular biology
Englard, Sasha, biochemistry
Engle, Mary Allen English, pediatrics, cardiology
Engle, Ralph Landis, Jr, medicine, hematology
Entner, Nathan, biochemistry
Epp, Edward Rudolph, medical physics, radiation physics
Epstein, Isadore, astronomy, astrophysics
Epstein, Jeanne Alice, medicine
Erickson, Bruce Wayne, organic chemistry, organic biochemistry
Erlanger, Bernard Ferdinand, biochemistry
Erlbach, Erich, physics
Erlenmeyer-Kimling, L, medical genetics
Ernst, John L, organic polymer chemistry
Evans, James William, agricultural biochemistry
Evenson, Donald Paul, cell biology, virology
Everhard, Martin Edward, physical chemistry, surgery
Everhart, Donald Lee, immunology, biochemistry
Evers, William L, polymer chemistry
Extermann, Richard C, physics
Fabianek, John, biochemistry, physiology
Fabry, Thomas Lester, biophysical chemistry, gastroenterology
Fagan, John J, geology
Fahn, Stanley, neurology, neuropharmacology
Fairbridge, Rhodes Whitmore, geology
Falk, Catherine T, population genetics
Falk, Harold, statistical mechanics
Farber, Saul Joseph, medicine
Farnsworth, Marie, chemistry
Farnsworth, Patricia Nordstrom, physiology, biochemistry
Farrow, Joseph Helms, medicine
Feigelson, Muriel, biochemistry
Feigelson, Philip, biochemistry
Feigin, Irwin Harris, neuropathology
Feinberg, Gerald, theoretical high energy physics
Feinberg, Robert Samuel, bio-organic chemistry
Feiner, Rose Resnick, bacteriology
Feinman, Max L, surgery, anatomy
Feit, Carl, immunology
Feldman, Edgar A, geometry
Fell, Colin, physiology
Ferrer, Jose M, surgery
Ferry, Andrew P, ophthalmology, pathology
Fertig, John William, biostatistics, public health
Feuer, Irving, physical chemistry
Ficken, Frederick Arthur, mathematics
Field, Frank Henry, physical chemistry
Fieve, Ronald Robert, psychiatry
Finby, Nathaniel, radiology
Fine, Albert Samuel, biochemistry
Finfrock, Dwight Curtis, agronomy
Fink, Frederick Charles, microbiology
Finkbeiner, John Aris, medicine
Finkelstein, David, theoretical physics
Finkenstaedt, John Turner, medicine, biochemistry
Finster, Mieczyslaw, physiology, pharmacology
Fischbarg, Jorge, physiology, ophthalmology
Fischer, Arthur H, physical chemistry
Fischman, Harlow Kenneth, medical genetics, cytogenetics
Fisher, Leonard V, medicine
Fisher, Saul Harrison, psychiatry
Fishman, Louis, biochemistry
Fishman, Marvin, immunology
Fishman, Myer M, biochemistry
Fitzgerald, Patrick James, pathology
Flach, Frederic Francis, psychiatry
Flanagan, Joseph Edward, mathematics
Flaschen, Stewart Samuel, science policy, resource management
Flatto, Leopold, mathematics
Fleisher, Martin, clinical chemistry, immunochemistry
Florman, Alfred Leonard, pediatrics
Flynn, George William, chemical physics
Foght, James Loren, textile chemistry
Foldes, Francis Ferenc, anesthesiology
Foley, David Allen, pathobiology, parasitology
Foley, Henry Michael, physics
Foley, William Thomas, medicine
Fontana, Vincent J, immunology, medicine
Foreman, Bruce Milburn, Jr, information science
Fortner, Joseph Gerald, surgery, biology
Fox, Arthur Charles, medicine
Fox, Charles Lewis, Jr, microbiology, surgery
Fraenkel, George Kessler, chemical physics
Franck, Richard W, organic chemistry
Franklin, Edward Claus, medicine
Franklin, Kenneth Linn, astronomy
Franson, Richard Carl, biochemistry
Fraser, Blair Allen, immunogenetics
Frater, Robert William Mayo, thoracic surgery, cardiovascular surgery
Freed, Stanley Arthur, anthropology, ethnology
Freedman, Aaron David, internal medicine, biochemistry
Freedman, Alfred Mordecai, psychiatry
Freedman, Lewis Simon, neurochemistry, neuropharmacology
Freeman, Donald Chester, Jr, biomedical engineering
Freiberg, Samuel Robert, plant physiology, biology
Freudenthal, Peter, environmental health, meteorology
Freyermuth, Harlan Benjamin, organic chemistry
Fried, Jerrold, biophysics
Fried, Morton H, cultural anthropology
Friedberg, Richard Michael, theoretical physics
Friedenberg, Richard M, radiology
Friedhoff, Arnold Jerome, psychiatry
Friedland, Beatrice L, biology
Friedland, Joan Martha, biochemistry, neurochemistry
Friedman, Selwyn Marvin, biochemical genetics
Friedrichs, Kurt Otto, mathematics
Friend, Charlotte, medical microbiology
Fritz, Henry Edward, organic chemistry
Frosch, William Arthur, psychiatry, psychoanalysis
Fruhman, George Joshua, anatomy
Frye, Graham Eugene, physics
Fu, Lorraine Shao-Yen, mathematics
Fuchs, Anna-Riitta, reproductive physiology
Fuchs, Fritz Friedrich, obstetrics & gynecology
Furst, Milton, physics
Furth, Jacob, cancer, experimental pathology
Gabelman, Norman, cell biology, virology
Galdston, Morton, medicine
Galin, Miles A, ophthalmology
Gall, William Einar, cell biology, biochemistry
Gallagher, Patrick Ximenes, mathematics
Galton, Suzanne A, organic chemistry, biochemistry
Gambino, Salvatore Raymond, clinical pathology, clinical chemistry
Gang, Henry, biochemistry, clinical chemistry
Gans, David, mathematics
Gans, Henry, surgery, physiological chemistry
Gans, Paul Jonathan, chemical physics
Garabedian, Paul Roesel, mathematics
Gardner, Esther Polinsky, neurophysiology
Gardner, William Ullman, anatomy
Garfinkel, Lawrence, epidemiology, biostatistics
Garret, Rudolph, pathology
Garro, Anthony Joseph, microbiology, microbial genetics
Gartner, Lawrence Mitchel, medicine, physiology
Gavasci, Anna Teresa, petrology
Geacintov, Nicholas, chemical physics, biophysics
Geffen, Abraham, radiology
Gelbard, Alan Stewart, enzymology, nuclear medicine
Gelbart, Abe, mathematics
Gellhorn, Alfred, cancer
Generales, Constantine D John, internal medicine, aerospace medicine
Genghof, Dorothy Schaefer, microbiology
Gennaro, Joseph Francis, anatomy
Gentry, John Tilmon, public health, hospital administration
German, James Lafayette, III, human genetics
Gershberg, Herbert, internal medicine
Gershinowitz, Harold, physical chemistry
Gerson, Raymond E, biomedical engineering, environmental engineering
Gerst, Paul Howard, surgery, physiology
Gersten, Joel Irwin, theoretical solid state physics, surface physics
Gerstner, Robert, human anatomy, neuroanatomy
Gertler, Menard M, internal medicine
Gettler, Joseph Daniel, physical organic chemistry
Ghosh, Nimai Kumar, experimental medicine
Gibian, Thomas George, chemistry
Gigli, Irma, dermatology, immunology
Gilardi, Gerald Leland, medical bacteriology
Gilbert, Harriet S, hematology
Gilder, Helena, biochemistry
Gilgore, Sheldon G, internal medicine, endocrinology
Gilman, John Richard, Jr, astrophysics
Gilman, Sid, clinical neurology, neurophysiology
Ginsberg, Harold Samuel, virology, microbiology
Ginsberg, Jonathan I, mathematical analysis
Girolamo, Rita Frances, radiology, nuclear medicine
Gisin, Balthasar Friedrich, organic chemistry, synthetic organic chemistry
Glabman, Sheldon, internal medicine, nephrology
Glaeser, John Douglas, geology
Glass, George B Jerzy, gastroenterology
Glassgold, Alfred Emanuel, theoretical physics
Glick, Stanley Dennis, neuropharmacology
Glimm, James Gilbert, mathematics
Glover, Roland Leigh, electrochemistry
Glusman, Murray, neuropsychiatry
Godman, Gabriel C, cell biology, pathology
Goebel, Walther Frederick, biochemistry
Goetchius, George Richard, medical microbiology
Goetz, Robert Hans, medicine
Golbey, Robert (Bruce), chemotheraphy
Gold, Albert, solid state physics
Gold, Allen Morton, biochemistry, organic chemistry
Goldberg, Burton David, pathology, cell biology
Goldberg, Henry Peter, cardiology
Goldberg, Joseph Louis, organic chemistry
Golden, Robert K, computer science, communications science
Goldensohn, Eli Samuel, neurology
Goldfeder, Anna, experimental pathology
Goldman, Malcolm, mathematics
Goldschmidt, Bernard Morton, cancer
Goldsmith, Edward I, surgery
Goldsmith, Eli David, developmental biology, research administration
Goldstein, David, environmental medicine
Goldstein, Eli, medicine
Goldstein, Gideon, immunology, internal medicine
Goldstein, Gilbert, biochemistry
Goldstein, Herbert, physics
Goldstein, Inge F, epidemiology, biophysics
Goldstein, Jack, biochemistry
Goldstein, Martin, physical chemistry
Goldstein, Max, mathematics
Goldstein, Menek, biochemistry
Goldstein, Robert, medicine
Golomb, Frederick M, surgery
Gomatos, Peter John, biochemistry, virology
Gomez, Daniel Guillermo, neuroanatomy, neurophysiology
Good, Robert Alan, pediatrics
Goode, Robert P, biology
Goodgold, Joseph, medicine
Goodhart, Robert Stanley, medicine
Goodman, DeWitt Stetten, internal medicine, biochemistry
Goodman, Fred, molecular genetics
Goodman, Irving, biochemistry
Goodman, Jerome, physical chemistry
Gordon, Albert Saul, physiology
Gordon, Arnold J, organic chemistry
Gordon, Arnold L, physical oceanography
Gordon, Edwin Earl, medicine
Gorlin, Richard, medicine
Gottfried, Eugene Leslie, hematology
Gottsegen, Robert, dentistry
Goulian, Dicran, plastic surgery
Goulianos, Konstantin, high energy physics
Gourevitch, George, psychophysiology
Grad, Harold, applied mathematics, plasma physics
Graef, Irving Philip, pathology, internal medicine
Graf, Liselotte, pathology, immunology
Graff, Samuel M, applied mathematics
Grafstein, Bernice (Mrs Howard Shanet), physiology
Grafton, Robert Bruce, applied mathematics
Granick, Sam, plant physiology, biochemistry
Grant, Neil George, plant physiology
Grauer, Amelie L, biochemistry
Gray, Clarence Cornelius, III, agronomy
Green, Jack Peter, pharmacology
Green, Marvin, neonatology, pediatrics
Green, Michael Enoch, physical chemistry
Green, Morris, radiation biology
Green, Saul, biochemistry
Greenbaum, Lowell Marvin, pharmacology, biochemistry
Greenberg, Howard, theoretical physics
Greenberger, Daniel Mordecai, theoretical physics
Greenebaum, Michael, optical physics, atmospheric physics
Greenleaf, Frederick P, mathematical analysis
Greenspan, Joseph, instrumentation, physical chemistry
Greenwald, Isidor, biochemistry
Greenwood, Mary Rita Cooke, physiology, nutrition
Gregor, Harry Paul, applied chemistry

Gregory, Carter H., geophysics
Gregory, James McKanna, operations research, computer science
Gregory, John Delafield, biochemistry
Greif, Roger Louis, physiology
Gresik, Edward Louis, anatomy
Griffin, Donald Redfield, biology
Griffiths, Raymond Bert, biology, medicine
Grimm, Charles Henry, organic chemistry
Griswold, Joseph Garland, behavioral biology
Groeneveld Meijer, Willem Otto Jan, economic geology
Gromisch, Donald S., medicine, pediatrics
Gross, Daniel Russell, anthropology
Gross, Jonathan Light, mathematics
Gross, Robert Alfred, plasma physics
Grossfeld, Joseph, behavioral genetics
Grossi, Carlo E., surgery
Grossman, George, mathematics
Grossman, Jacob, medicine
Grossman, William Elderkin Leffingwell, analytical chemistry
Grossman, William, plasma physics
Gruenewald, Ruben, immunology, bacteriology
Grundfest, Harry, physiology
Grynbaum, Bruce B., medicine
Guenther, Harry Wilbert, textile chemistry
Guertin, Donald Lucius, analytical chemistry
Guha, Arabinda, enzymology
Gump, Frank E., surgery
Gumport, Stephen Lawrence, surgery
Gund, Tamara H Mladineo, biochemistry
Gunnison, Albert Farrington, environmental medicine, toxicology
Gupta, Rajendra, atomic physics
Gupta, Sudhir, immunology
Gurel, Okan, applied mathematics, applied mechanics
Gurpide, Erlio, biochemistry, biophysics
Gusberg, Saul Bernard, medicine
Gwilt, John Ruff, science administration, applied chemistry
Haas, Albert B., medicine
Hachigian, Jack, mathematics, statistics
Haddad, Jamil Raouf, pathology
Hagstrom, Jack Walter Carl, pathology
Haimovici, Henry, physiology, surgery
Haines, Thomas Henry, lipid chemistry, cell biology
Haleed Safia Arifi, molecular spectroscopy
Hadley, Susan Jane, pulmonary diseases, medicine, immunology
Hafner, Alden Norman, optometry
Hafner, Theodore, microwave electronics
Haft, David Edward, medicine, biochemistry
Halford, Ralph Stanley, physical chemistry
Hamilton, Mary Jane Gill, physical biochemistry
Hammer, Henry Felix, pharmaceutical chemistry
Hammett, Louis Plack, chemistry
Hance, William Adams, economic geography, geography of Africa
Handler, Evelyn Erika, physiology, hematology
Hanks, Edgar C., anesthesiology
Hanks, Jess Paul, ecology, botany
Hanson, Ronald Lee, biochemistry
Happer, William, Jr., atomic physics
Harber, Leonard C., dermatology
Hardy, Edward Peirson, Jr., analytical chemistry
Harley, John Henry, analytical chemistry, environmental health
Harner, Michael James, anthropology
Harpold, Michael Alan, bio-organic chemistry
Harrington, Charles Christopher, anthropology
Harris, Cyril Manton, acoustics, electroacoustics
Harris, Edgar C., anesthesiology
Harris, Henry William, medicine
Harris, Marvin, cultural anthropology
Harris, Matthew N., surgery, oncology
Harris, Ruth Cameron, pediatrics
Harrison, Malcolm Charles, mathematics, computer science
Harrison, Shirley Wanda, chemical physics
Hart, Hiram, physics
Hart, William James, Jr., food science

Hartline, Haldan Keffer, physiology
Hartman, Lawton Mervale, III, science policy
Harvey, Rajane M., medicine
Haschemeyer, Audrey Elizabeth Veazie, biological chemistry, environmental physiology
Hashim, Sami A., internal medicine
Hass, William K., medicine, neurology
Hassan, Ikram Ul., biology
Hasson, Alvin Stanley, mathematics
Hausner, Melvin, geometry
Havens, William Westerfield, Jr., nuclear physics
Havran, Robert Thomas, biochemistry
Hayes, Guy Scull, medicine, public health administration
Haymovits, Asher, medicine, biochemistry
Hays, James D., geology
Hecht, Charles Edward, statistical mechanics
Hehre, Edward James, microbiology, immunology
Heidelberger, Michael, immunochemistry
Heikkila, Richard Elmer, biochemistry, neurobiology
Heimann, Harry, environmental medicine, occupational medicine
Hendley, Charles Daniel, pharmacology, physiology
Hendrickson, William George, chemistry
Henig, Philip Edward, internal medicine, preventive medicine
Henley, Walter L., pediatrics
Henneman, Philip Harry, medicine
Hennessy, Douglas John, chemistry
Henrikson, Ray Charles, anatomy, cell biology
Henry, Timothy James, molecular biology, electron microscopy
Henze, Carlo, pharmacology, physiology
Herget, Herbert L., organic chemistry
Herr, Harry Wallace, urology
Herskovits, Theodore Tibor, physical chemistry
Herskowitz, Irwin Herman, genetics
Herter, Frederic P., surgery
Herz, Marvin Ira, psychiatry
Hetrick, David Adam Werner, otolaryngology
Hetzel, Donald Stanford, organic chemistry
Hiatt, Robert Burritt, surgery, physiology
Hiebert, John Mark, pharmaceutics
Higgins, George Kendall, pathology
Hilal, Sadek K., radiology
Hiller, Jacob Moses, neuropharmacology
Hillman, Dean Elof, neurobiology
Hinkle, Lawrence Earl, Jr., medicine
Hinterbuchner, Catherine Nicolaides, physical medicine & rehabilitation
Hinterbuchner, Ladislav Paul, neurology
Hirschfeld, Henry Israel, invertebrate zoology, cell biology
Hirst, George Keble, virology
Hisey, Robert Warren, organic chemistry, chemical engineering
Hix, Homer Bennett, organic chemistry, personnel administration
Hoberman, Henry Don, biochemistry
Hochwald, Gerald Martin, neurology, immunology
Hockett, Robert Casad, biochemistry
Hodes, Horace Louis, bacteriology, pediatrics
Hoffman, Donald Bertrand, toxicology
Hoffman, Heiner, oral microbiology
Hoffman, Joseph, medicine
Hoffman, Philip, biochemistry
Hoffman, Frederick Gustave, endocrinology
Holland, Albert Harold, Jr., medicine
Holland, James Frederick, medicine

Hollander, Vincent Paul, internal medicine
Holloway, Ralph L., Jr., physical anthropology, neurosciences
Holman, Cranston William, surgery
Holmes, Ralph Jerome, geology, mineralogy
Holtzman, Eric, cell biology, neuroscience
Holub, Donald Arthur, internal medicine
Holzman, Robert Stephen, immunology, infectious diseases
Hoogstraal, Harry, biology
Hopper, Richard H., geology
Horiuchi, Kensuke, molecular biology
Horn, Christian Friedrich, chemistry
Hornick, Edward J., psychiatry
Horosko, Roger N., nuclear physics
Horowitz, Sidney Lester, orthodontics
Horvath, Fred Ernest, neurophysiology
Horvitz, Gerald, theoretical physics, solid state physics
Horwitz, Marshall Sydney, microbiology, pediatrics
Hotchkiss, Robert S., surgery, urology
Hotchkiss, Rollin Douglas, biochemistry, gerontology
Houde, Raymond Wilfred, pharmacology, therapeutics
Houle, Joseph E., mathematics, academic administration
Housepian, Edgar M., neurosurgery, neurophysiology
Howland, William Stapleton, medicine, anesthesiology
Hoyer, Horst Walter, physical chemistry
Hruza, Zdenek, physiology, gerontology
Hsieh, Yu-Nian, magnetic resonance
Hsu, Konrad Chang, immunobiology
Hsu, Ming-Ta, molecular biology
Hu, Pung Nien, physics
Huang, Sylvia Lee, biophysics, biochemistry
Hutchison, Dorris Jeannette, developmental pharmacology, microbiology
Hunter, Seymour Herbert, microbiology
Hutterer, Ferenc, clinical biochemistry, clinical pathology
Hyman, Arthur Bernard, medicine
Hyman, Milton, dentistry
Hurvitz, Arthur Isaac, comparative pathology, experimental pathology
Hutchings, Donald Edward, developmental psychobiology
Hughes, Edward Francis Xavier, community health
Hughes, G M K, biomedical engineering
Hunt, Robert Elton, petroleum geology
Hunt, Roy Edward, chemistry
Hunt, Beryl Eleanor, medicine
Hunt, Howard Francis, medical psychology
Imparato, Anthony Michael, medicine
Imperato, Pascal James, tropical medicine, preventive medicine
Insull, William, Jr., biochemistry, cardiovascular diseases
Inturrisi, Charles E., pharmacology
Ioachim, Harry L., pathology, oncology
Isaac, Erich, cultural geography, biogeography
Isaacs, Leslie Laszlo, low temperature physics, experimental solid state physics
Isaacson, Eugene, mathematics
Iyengar, V K Sundararaja, histochemistry, cytochemistry
Izli, Turan M, psychiatry
Jabbur, Ramzi Jibrail, industrial
Jackanicz, Theodore Michael, reproductive endocrinology
Jacobs, Joseph H, science education
Jacobs, Myron Samuel, pathology
Jacobson, Harold Gordon, radiology
Jacobson, Willard James, science education
Jaffe, Israeli Aaron, internal medicine
Jaffe, Jerome Herbert, psychopharmacology
Jaffe, Marvin Richard, analytical chemistry, inorganic chemistry
Jaffe, Morry, experimental chemistry
Jagiello, Georgiana Mary, genetics
Jahiel, Rene, microbiology
Jain, Anrudh Kumar, biostatistics, population studies
Jakowska, Sophie (Mrs C L Jeannopoulos), pathobiology
James, Leonard Stanley, medicine, pediatrics
Jameson, Arthur Gregory, medicine
Janowitz, Henry David, gastroenterology
Jarowski, Charles I., pharmaceutical chemistry
Jastrow, Robert, atmospheric physics

Javitt, Norman B., medicine, physiology
Jehle, Leon Paul, physical chemistry
Jelinek, Maurice H., chemistry
Jesaitis, Margeris Adomas, molecular biology, biochemistry
Jicha, Henry Louis, Jr., economic geology
Jirku, Helmut, endocrinology, biochemistry
John, E Roy, neurophysiology, psychophysiology
John, Fritz, mathematics
Johnson, Alan J., hematology
Johnson, Carl Lynn, pharmacology, biochemistry
Johnson, Dewey, Jr., biochemistry, nutrition
Johnson, Howard Claus Edmund, chemistry
Johnson, Kenneth Gerald, internal medicine
Johnson, Philip Martyn, radiology
Johnston, Barbara Jane, medicine
Jolly, Clifford J., physical anthropology, primatology
Jones, Delmos Jehu, cultural anthropology, social anthropology
Joseph, Edward David, psychiatry, psychoanalysis
Joseph, Peter Maron, medical physics
Josephson, Ben, Jr., solid state physics
Kabat, Elvin Abraham, biochemistry
Kac, Mark, mathematics
Kahn, Norman, neuropharmacology, medical education
Kakari, Sophia, biochemistry
Kakari, Sophia, biochemistry
Kaku, Michio, theoretical high energy physics
Kaley, Gabor, physiology, experimental pathology
Kalikstein, Kalman, physics
Kalos, Malvin Howard, theoretical physics
Kambysellis, Michael Panagiotis, developmental genetics
Kamner, Mildred Elsie (Mrs Edward M Tolman), medicine, chemistry
Kamrin, Benjamin Barnett, anatomy
Kandel, Eric Richard, neurobiology, psychiatry
Kang, Sungzong, biophysics, molecular spectroscopy
Kanick, Virginia, radiology
Kanig, Joseph Louis, pharmaceutics
Kaplan, Harold Irwin, psychiatry, psychoanalysis
Kaplan, Helen Singer, psychiatry
Kaplan, Raymond, solid state physics
Kaplan, Morton Fischel, physics
Kappas, Attallah, metabolism, pharmacology
Karal, Frank Charles, Jr., applied mathematics
Karkas, John D., biochemistry
Karlin, Arthur, neurobiology
Karp, Samuel Noah, applied mathematics
Karpatkin, Simon, biochemistry
Karpiak, Stephen Edward, neuroscience
Kaslow, Haven DeLoss, biochemistry
Kasses, Kenneth George, pharmacology
Katsoyannis, Panayotis G., biochemistry
Katz, Arnold Martin, medicine, physiology
Katz, George Maxim, engineering
Katz, Jacob Feuer, orthopedic surgery
Katz, Louis, molecular biology, computer science
Katz, Michael, pediatrics, virology
Katz, Thomas Joseph, organic chemistry
Katzman, Robert, neurology
Kauder, Otto Samuel, organic chemistry
Kaufman, Edward Godfrey, dentistry
Kaufman, Mavis Anderson, neuropathology
Kavaler, Frederic, physiology
Kavanagh, Kevin Enda, organic chemistry
Kay, Kingsley, toxicology, occupational health
Kayden, Herbert J., medicine
Kean, Benjamin Harrison, tropical medicine
Keller, Joseph Bishop, mathematics
Keller, Kenneth Frank, geology
Kellner, Aaron, pathology
Kelly, Dennis D., neuropsychology
Kennedy, Michael Craig, neurobiology
Kent, Stephen Brian Henry, biological chemistry
Kernberg, Otto F., psychiatry
Keusch, Gerald Tilden, infectious diseases

Khanna, Shyam Mohan, audiology, physiology
Khoobiar, Sargis, petroleum chemistry, chemical engineering
Khuri, Nicola Najib, theoretical physics
Kidd, John Graydon, pathology, bacteriology
Kieras, Fred John, biochemistry
Kilbourne, Edwin Dennis, microbiology, virology
Kim, Harry Hi-Soo, pathology
Kim, Jae Ho, radiobiology, radiotherapy
Kim, Kwang Shin, medical microbiology, horticulture
Kim, Young Tai, biochemistry, immunology
Kindle, Cecil Haldane, geology, paleontology
Kindt, Thomas James, biochemistry, immunogenetics
King, Donald West, Jr., pathology
King, James Clement, genetics
King, Kendall Willard, nutrition
King, Marvin, physical optics, electrical engineering
King, Te Piao, organic chemistry
King, Thomas Creighton, medicine, thoracic surgery
King, Thomas K C, pulmonary physiology, pulmonary diseases
Kingston, Charles Richard, forensic science
Kinney, John Martin, surgery
Kinzey, Warren Glenford, anatomy, physical anthropology
Kirchberger, Madeleine, physiology
Kirkham, Frederic Theodore, Jr., medicine
Kirsten, Edward Bruce, neuropharmacology, cell physiology
Kirsteuer, Ernst, invertebrate zoology
Kiyasu, John Yutaka, biochemistry
Klaas, Nicholas Paul, organic chemistry
Klass, Morton, cultural anthropology, ethnology
Klein, Morton, operations research
Klerer, Julius, physical chemistry, solid state chemistry
Kliegman, Jonathan Morris, organic chemistry
Klivington, Kenneth Albert, academic administration, neuroscience
Klumpp, Theodore George, internal medicine
Kneip, Theodore Joseph, analytical chemistry, environmental chemistry
Knowles, John Hilton, internal medicine, pulmonary diseases
Knox, Charles Emery, plastics chemistry
Koch, Herman William, physics
Kochwa, Shaul, biochemistry, immunochemistry
Koffler, David, immunopathology
Kohl, Richard Niemes, psychiatry
Koide, Samuel Saburo, biochemistry
Kolchin, Ellis Robert, mathematics
Kolesar, Peter John, operations research, statistical analysis
Komar, Arthur Baraway, theoretical physics
Konde, Anthony Joseph, physical chemistry
Koopman, Karl Friedrich, mammalogy
Kopac, Milan James, physiology, zoology
Kopeloff, Lenore Moolten, microbiology
Kopf, Alfred Walter, medicine, oncology
Kopperman, Ralph David, mathematical logic
Koranyi, Adam, mathematics
Korein, Julius, neurology
Korff, Serge Alexander, physics
Kraft, David Werner, physics
Krakoff, Irwin Harold, medicine
Krakow, Joseph S, biochemistry, molecular biology
Kramer, Bernard, solid state physics
Kramer, Elmer E, medicine
Kramer, Fred Russell, molecular biology, biochemistry
Kramer-Lassar, Edna Ernestine, mathematics
Krasna, Alvin Isaac, biochemistry
Krasnow, Frances, biochemistry
Kream, Jacob, bio-chemistry, clinical chemistry

Kreibich, Gert, cell biology
Kremer, Chester B, chemistry
Kremzner, Leon T, biochemistry
Krey, Lewis Charles, neuroendocrinology
Krieger, Dorothy T, physiology
Krieger, Howard Paul, neurology
Krim, Mathilde, cytogenetics, virology
Krishna, Kumar, zoology
Kritz, Arnold H, plasma physics
Krooth, Robert S, genetics
Krugman, Saul, pathology
Krulwich, Terry Ann, biochemistry, microbial physiology
Krumbein, Simeon Joseph, electrochemistry, surface physics
Krumerman, Martin Saul, anatomic pathology, clinical pathology
Krummel, Edward William, organic chemistry
Krupa, Paul L, parasitology, electron microscopy
Ku, Peh Suen, environmental physics
Kubler, William Frank, Jr., chemistry
Kuhn, Leslie A, cardiology
Kuhns, William Joseph, clinical pathology, immunology
Kumbaraci, Turkan Emine, medical statistics
Kunkel, Henry George, medicine
Kuo, John Tsung (Fen), geophysics
Kupfer, Sherman, internal medicine, physiology
Kupfermann, Irving, neuropsychology
Kuramoto, Simpey, food science
Kutscher, Austin Harrison, dentistry
Lackner, Henriette, hematology
LaDue, John Samuel, internal medicine
Lagomarsino, Raymond J, radiochemistry
Lai, Tze Leung, mathematics, statistics
Lajtha, Abel, biochemistry
Lake, James Albert, biology
Lamb, Albert R, Jr., medicine
Lamdin, Ezra, medicine, physiology
Lamm, Michael Emanuel, immunology, pathology
Lancefield, Rebecca Craighill, bacteriology
Landers, Mary Kenny, mathematics
Landesman, Perry Robert, medicine
Landovitz, Leon Fred, solid state physics
Landsberger, Frank Robbert, physical biochemistry, virology
Lang, Enid Asher, psychiatry
Lang, Robert Eugene, physical chemistry
Langer, Kurt, internal medicine
Langer, Arthur M, mineralogy, environmental sciences
Langford, William Siddon, child psychiatry
Langner, Thomas S, psychiatry, epidemiology
Lanks, Karl William, pathology, molecular biology
Lanman, Jonathan T, pediatrics
Lanyon, Wesley Edwin, ornithology
Lapham, Roger Fulmer, medicine
Lapin, Evelyn P, neurochemistry, enzymology
Laragh, John Henry, physiology, medicine
Lasker, Sigmund E, physical chemistry
Lasley, Betty Jean, microbiology
Lathem, Willoughby, medicine
Lattes, Raffaele, medicine
Lattimer, John Kingsley, urology
Laubach, Gerald David, organic chemistry
Lauer, Dolor John, industrial medicine, toxicology
Laughlin, John Seth, medical physics
Laurence, Kenneth Allen, microbiology
Lavietes, Beverly Blatt, developmental biology, oncology
Lawless, Robert Dale, anthropology
Lawrence, Henry Sherwood, medicine
Lax, Anneli, mathematics
Lax, Melvin, theoretical solid state physics, quantum optics
Lax, Peter David, mathematics
Leacock, Eleanor Burke, applied anthropology
Lebow, Arnold, mathematics
Lebowitz, Joel Louis, physics
Lederer, Ludwig George, physiology
Lederman, Leon Max, nuclear physics
Lee, John Joseph, marine microbiology, protozoology
Lee, Mathew Hung Mun, physical medicine & rehabilitation
Lee, Sung Gue, biochemistry
Lee, Tsung Dao, theoretical physics
Lee, Wonyong, high energy physics
Lee-Huang, Sylvia, biochemistry, molecular biology
Leeper, Robert Dwight, nuclear medicine, endocrinology
Lefkowitz, Issai, solid state physics

Lefkowitz, Ruth Samson, mathematics
Lefkowitz, Stanley A, physical inorganic chemistry
Lefton, Phyllis, mathematics, number theory
Lehne, Richard Karl, organic chemistry
Lehrer, Gerard Michael, neurology
Lehrer, Harold Z, radiology
Lehrman, Leon, organic chemistry
Leibman, Lawrence Fred, organic chemistry
Leibowitz, Sarah Fryer, psychopharmacology
Leis, Donald George, organic chemistry
Leiter, Elliot, medicine, urology
Leiter, Louis, internal medicine
Lerman, Stephen Paul, immunology
Lerner, Rita Guggenheim, information science
Leslie, Charles Miller, medical anthropology
Leslie, Stephen Howard, medicine
Lesser, Elliott, medical parasitology, protozoology
Lesser, Gerson Theodore, internal medicine, physiology
Levandowsky, Michael, marine ecology, mathematical biology
Levene, Howard, mathematical statistics
Levenson, Morris E, mathematics
Levy, Samuel, hospital administration, health administration
Levi, Roberto, pharmacology
Levin, Aaron R, pediatrics, cardiology
Levine, Bernard Benjamin, immunology, medicine
Levine, Eli Morris, chemistry
Levine, Louis, genetics, animal behavior
Levine, Milton Isra, pediatrics
Levinson, Sidney Bernard, chemistry, chemical engineering
Levinthal, Cyrus, biophysics
Levitan, Max, genetics, anatomy
Levitt, Barrie, pharmacology, internal medicine
Levitt, Marvin Frederick, nephrology
Levitz, Mortimer, biochemistry, endocrinology
Levy, Arthur Louis, analytical chemistry, physical chemistry
Levy, David Edward, neurology
Levy, Ezra, inorganic chemistry, physical chemistry
Levy, Peter Michael, solid state physics
Levy, Seymour Z, physical chemistry, analytical chemistry
Lewin, Seymour M, oncology
Lewis, Frank George, botany, ecology
Lewis, Irving James, community health
Lewis, John L, Jr., obstetrics & gynecology
Li, Lu Ku, biochemistry, protein chemistry
Li, Steven Shoei-Lung, genetics, biochemistry
Lichter, Robert (Louis), organic chemistry
Lieberman, Seymour, biochemistry
Liebeskind, Herbert, physical chemistry
Liebling, Richard Stephen, mineralogy
Liebman, Frederick Melvin, physiology
Liegner, Leonard M, oncology
Lier, Frank George, surgery
Lin, Leu-Fen Hou, biochemistry
Linder, Regina, microbiology
Link, Richard Forest, mathematical statistics
Linke, Harald Arthur Bruno, microbiology
Linkie, Daniel Michael, endocrinology, physiology
Linksz, Arthur, ophthalmology
Linn, Kurt O, geology
Lipkin, George, cell biology, dermatology
Lipkin, Mack, internal medicine, psychiatry
Lipkin, Martin, gastroenterology, oncology
Lipmann, Fritz (Albert), biochemistry
Lippard, Stephen J, inorganic chemistry, biophysical chemistry
Lippmann, Bernard Abram, quantum mechanics
Lippmann, Heinz Israel, medicine
Lippmann, Morton, environmental health
Lipsicas, Max, chemistry, physics
Litman, Gary William, immunology, biochemistry
Litwak, Robert Seymour, surgery
Liu, Chi Tan Chang, immunochemistry
Liu, Si-Kwang, veterinary pathology
Lizardi, Paul Modesto, cell biology, biochemistry
Lloyd, Kenneth Oliver, biochemistry, immunochemistry
LoBue, Joseph, physiology, hematology
Localio, S Arthur, surgery
Lockshin, Michael Dan, rheumatology, immunology

Lockwood, Arthur H, molecular biology, cell biology
Long, Margaret Eleanor, cytochemistry
Lopez, Carlos, immunology, virology
Lopez, Rafael, pediatrics, hematology
Lorch, Edgar Raymond, mathematics
Lord, Jere Williams, Jr., surgery
Lorente De No, Rafael, physiology
Loucas, Spiro P, organic chemistry, biochemistry
Lovejoy, Derek R, solid state physics
Low, Barbara Wharton, biochemistry
Low, Manfred Josef Dominik, physical chemistry
Low, Niels Leo, medicine
Lowman, Edward Wynne, physical medicine
Lubkin, Gloria Becker, physics
Lucca, John J, dentistry
Luck, David Jonathan Lewis, cytology
Luckett, Winter Patrick, anatomy, embryology
Luckey, Egbert Hugh, internal medicine
Ludlam, William Myrton, optometry, physiological optics
Luhby, Adrian Leonard, hematology, pediatrics
Luine, Victoria Nall, neurochemistry
Lukas, Daniel Stanley, cardiology
Lustig, Harry, physics
Luttinger, Joaquin Mazdak, theoretical physics
Lutzker, Edythe, history of science, history of medicine
Lyne, Everett, physiology
Lyser, Katherine May (Mrs E Shouby), neuroembryology, oncology
Maas, Werner Karl, molecular genetics
Maben, Jerrold William, science education, science writing
Mackles, Leonard, cosmetic chemistry, pharmaceutical chemistry
MacLeod, John, physiology
Madden, Robert E, thoracic surgery, cardiovascular surgery
Magiulo, Anthony Rudolph, microbiology, biochemistry
Mahoney, Colette, biology
Mahoney, Richard Theodore, reproductive biology, chemical physics
Maier, John, public health
Maines, Mahin D, pharmacology
Maker, Howard Smith, neurochemistry, neurology
Mais, Leonard I, neurosurgery
Malitz, Sidney, psychiatry, psychopharmacology
Mancinelli, Alberto L, plant physiology
Mandel, Benjamin, virology
Mandel, Edward H, dermatology
Mandel, Irwin D, preventive dentistry, oral biology
Mandl, Ines, biochemistry
Manger, William Muir, medicine
Maniatis, George Marinos, biochemistry, developmental biology
Manning, James Joseph, physical chemistry, spectrochemistry
Manning, James Matthew, biochemistry
Manski, Wladyslaw J, microbiology, immunochemistry
Mantel, Linda Habas, comparative physiology
Marcus, Aaron Jacob, internal medicine, hematology
Marcy, Willard, organic chemistry, chemical engineering
Margolin, Paul, genetics, microbiology
Margolis, Richard Urdangen, pharmacology, biochemistry
Marica, Hildegard Rand, psychiatry
Marino, Robert Anthony, physics
Mark, Lester Charles, medicine
Marks, Morton, neurology, psychiatry
Marks, Neville, neurobiology
Marks, Paul A, internal medicine, biochemistry
Marler, Peter, zoology
Marmur, Julius, molecular biology, biochemistry
Marovitz, William F, anatomy, otolaryngology
Marquardt, Hans Wilhelm Joe, pharmacology, cancer
Marra, Dorothea Catherine, surface chemistry, colloid chemistry
Marshak, Robert Eugene, theoretical physics, astrophysics
Marshall, Carter Lee, preventive medicine
Marshall, Thomas C, physics
Marshall, Victor Fray, surgery
Martin, Constance Rigler, endocrinology
Martin, Daniel S, surgery
Mashburn, Louise Tull, biochemistry
Mashburn, Thompson Arthur, Jr., biochemistry

Masland, Richard Lambert, psychiatry, neurology
Mason, Michael E., biochemistry, food science
Massa, Louis, chemical physics
Mather, William B., Jr., electrochemistry
Mattikow, Morris, chemistry
Mattis, Daniel Charles, solid state physics
Mauro, Alexander, biophysics
Maurzrall, David Charles, biophysics
Mawr, Klaus, internal medicine, hematology
Mazunder, Rajarshi, biochemistry
Mazur, Abraham, biochemistry
McCabe, John Patrick, mathematics
McCarthy, Eugene Gregory, public health
McCarty, Maclyn, medical bacteriology
McCauley, John Corran, Jr., orthopedic surgery
McClelland, Charles Paul, organic chemistry
McClement, John Henry, pulmonary diseases
McClung, Andrew Colin, soil fertility
McCollester, Duncan L., cancer
McCoy, Oliver Rufus, parasitology
McCrory, Wallace Willard, pediatrics
McDivitt, James Frederick, mineral economics
McDowell, Fletcher Hughes, neurology
McEwen, Bruce Sherman, neurobiology
McFall, Elizabeth, biochemistry
McGary, Charles Wesley, Jr., polymer chemistry
McKelvey, John Jay, Jr., entomology
McKelvie, Neil, organic chemistry
McKenna, Francis Eugene, information science
McKenna, Malcolm Carnegie, vertebrate paleontology
McLaughlin, James E., physics
McManis, Douglas R., historical geography
McNeils, Edward Joseph, organic chemistry
McNitt, James R., geology
McSherry, Charles K., surgery
Mead, Margaret, cultural anthropology
Meisel, Seymour Lionel, organic chemistry, research administration
Meislich, Herbert, organic chemistry
Meister, Alton, biochemistry
Melamed, Myron Roy, medicine, pathology
Melamid, Alexander, economic geography, regional economics
Melezin, Abraham, geography
Melkonian, Edward, physics
Mellet, James Silvan, vertebrate paleontology
Mellins, Robert B., pediatrics
Mellors, Robert Charles, cardiopulmonary physiology, pathology
Melnick, Edward Lawrence, mathematics, statistics
Meltzer, Herbert Lewis, biochemistry
Mencher, Ely, geology
Menczel, Jehuda H., environmental chemistry
Mendell, Rosalind B., cosmic ray physics
Mendlowitz, Milton, internal medicine
Menon, Maya Devi, insect endocrinology
Merchant, Roland Samuel, biostatistics, economic statistics
Merriam, George Rennell, Jr., medicine
Merrifield, Robert Bruce, biochemistry
Mersheimer, Walter Lyon, surgery
Merskey, Clarence, physiology, medicine
Metcalf, William, surgery
Metraux, Rhoda, cultural anthropology, applied anthropology
Mettler, Frederick Albert, neurology, anatomy
Metzner, Jerome, botany, cytology
Meyer, Karl, biochemistry
Meyers, Philip Robert, topology, biomathematics
Michele, Arthur A., orthopedic surgery
Michels, Robert, psychiatry, psychoanalysis
Middleton, David, applied physics, applied mathematics
Milgrom, Harry, science education, science writing
Milligan, Jack Walter, internal medicine
Miller, David Clair, systematic entomology, invertebrate zoology
Miller, Dorothy Anne Smith, genetics, cytogenetics
Miller, Franklin Stuart, economic geology
Miller, James E., meteorology

Miller, James Milton, computer science, statistics
Miller, Julian Malcolm, chemistry, physics
Miller, Lawrence Peter, plant chemistry, toxicology
Miller, Maria G., animal behavior
Miller, Neal Elgar, psychophysiology
Miller, Orlando Jack, human genetics, obstetrics & gynecology
Miller, Walter E., physical chemistry
Miller, Wilbur Hobart, organic chemistry, biochemistry
Miller, William, experimental solid state physics
Millman, Stephen Jerry, virology
Millman, Robert Barnet, public health
Millman, Sidney, nuclear magnetic resonance, microwave electronics
Milstoc, Mayer, medicine, pathology
Milvy, Paul, biophysics, radiation biology
Mindich, Leonard Eugene, microbial physiology
Minick, Charles Richard, pathology
Minkoff, John, communication science, physics
Mitchel, Peter, ethology
Monahan, Wayne Gordon, biophysics, nuclear medicine
Monder, Carl, biochemistry
Mones, Robert J., neurology
Mitchell, Alexander Rebar, biochemistry
Mitchell, Ormond Glenn, human anatomy, histology
Mitrag, Thomas Waldemar, pharmacology, biochemistry
Moore, Charles Henkel, chemistry
Moore, James A., medicine
Moore, Maurice Lee, medicinal chemistry
Moore, Stanford, biochemistry
Moor-Jankowski, Jan K., immunogenetics, primatology
Moran, Juliette May, organic chemistry
Morawetz, Cathleen Synge, applied mathematics
Morgan, Councilman, microbiology, virology
Mori, Ken, pathology
Morishima, Akira, pediatrics, cytogenetics
Morishima, Hisayo Oda, medicine
Moroson, Harold, radiation immunology, radiobiology
Morrill, Gene A., biochemistry, developmental biology
Morris, Edward Craig, anthropology
Morris, Melvin Lewis, dentistry
Morrow, Jack I., physical inorganic chemistry
Morse, Jane H., immunology
Mosbach, Erwin Heinz, biochemistry
Moscovitz, Howard, cardiovascular physiology
Moser, Jurgen (Kurt), mathematics
Moses, Campbell J., medicine
Moses, William, comparative biochemistry
Moesson, Zehnan I., mathematics
Moskowitz, Jules Warren, physical chemistry
Moskowitz, Martin A., mathematics
Moss, Melvin Lane, biochemistry, science administration
Moss, Melvin Lionel, anatomy
Moss-Salentijn, Letty, anatomy, dental research
Most, Harry, tropical medicine
Motz, Lloyd, astrophysics, nuclear physics
Motz, Robin Owen, plasma physics, internal medicine
Mount, Lester Adran, neurosurgery
Mountain, Isabel Morgan, immunology, virology
Mudgett, Meredith, immunochemistry
Muller, Miklos, biological chemistry, parasitology
Munnell, Equinn W., medicine
Murphy, Edward Joseph, biophysics
Murphy, George Edward, experimental pathology
Murphy, James Slater, microbiology
Murphy, Martin Joseph, Jr., immunohematology
Murphy, Mary Lois, medicine
Murphy, Robert Francis, anthropology
Murphy, Terence W., bioengineering, public health
Muschel, Louis Henry, immunology

Myers, Charles William, herpetology
Myers, George Henry, electrical engineering, biomedical engineering
Myers, Sarah Kerr, cultural geography, geography of Latin America
Myers, Warren Powers Laird, internal medicine, oncology
Myint, Than, biochemistry
Nachbar, Martin Stephen, medical microbiology
Nachman, Ralph Louis, medicine
Nachmansohn, David, neurology
Nag, Moni, anthropology, population studies
Nahas, Gabriel Georges, pharmacology
Naidorf, Irving Joseph, dentistry
Nakanishi, Koji, organic chemistry
Narragon, Ernest Ashley, physical chemistry, organic chemistry
Nash, Harold Anthony, biochemistry
Nastuk, William Leo, physiology
Nathan, Henry C., biology
Nathan, Marc A., neurophysiology, hygiene
Nelson, Norton, environmental medicine, biochemistry
Nemir, Rosa Lee, pediatrics
Neu, Harold Conrad, medicine, pharmacology
Neumann, Paul Gerhard, oceanography, meteorology
Neurath, Alexander Robert, virology
Neuwirth, Robert Samuel, obstetrics & gynecology
New, Maria Iandolo, pediatrics
Newell, Norman Dennis, geology
Newling, Bruce Edgar, urban geography, economic geography
Newman, James Martin, mathematics
Ngai, Shih Hsun, anesthesiology
Nicholas, James A., surgery
Nicholson, Thomas Dominic, astronomy, navigation
Nicholson, William Jamieson, environmental health
Nigen, Alan Mark, protein chemistry
Nightingale, Eugene Richard, Jr., physical chemistry
Niman, John, applied mathematics
Nirenberg, Louis, mathematics
Nisselbaum, Jerome Seymour, neurochemistry
Noback, Charles Robert, anatomy
Nocenti, Mero Raymond, physiology
Norman, Alex, radiology
Norton, William Thompson, biochemistry
Nossel, Hymie L., hematology
Novick, Richard P., microbial genetics
Novick, Robert, physics
Ober, William B., pathology
Oberstar, Helen Elizabeth, cosmetic chemistry
O'Brien, Keran, radiation physics
Ockman, Nathan, molecular biophysics
O'Connell, Robert James, neurophysiology
Oettgen, Herbert Friedrich, medicine, immunology
Ohnuma, Takao, internal medicine, cancer
Oja, Tonis, physics, chemistry
Okamoto, Michiko, pharmacology
Old, Lloyd John, cancer, immunology
Orahovats, Peter Dimiter, pharmacology, physiology
O'Leary, J. Austin John, physics
O'Leary, William Michael, microbiology, biochemistry
O'Malley, Edward Paul, psychiatry, pharmacology
Omura, Yoshiaki, cardiology, electrophysiology
Ong, Eng-Bee, biochemistry
Oyama, Katsuyuki, meteorology
Oppenheim, Elliot, chemistry, physiology
Oratz, Murray, biochemistry
Oreskes, Irwin, biochemistry, immunology
Organ, James Albert, zoology
Orkin, Lazarus Allerton, medicine, surgery
Ornstein, Leonard, cell biology, biophysics
Orr, William Frederick, phychiatry
Orris, Leo, environmental medicine
Ortman, Robert A., zoology

Osinchak, Joseph, anatomy
Osipow, Lloyd Irving, physical chemistry
Osler, Abraham George, immunology
Osserman, Elliott Frederick, medicine
Oster, Gerald, biophysics, physical chemistry
Osterriter, John Ferdinand, occupational medicine
Ovary, Zoltan, immunology
Owen, Joel, statistics, operations research
Pacella, Bernard Leonard, psychiatry
Padawer, Jacques, anatomy, cell biology
Padnos, Norman, physical chemistry
Pagels, Heinz Rudolf, theoretical physics
Pais, Abraham, theoretical physics
Paisley, David M., organic chemistry
Palmer, Richard Bradbury, petroleum geology
Palmer, Robert Howard, internal medicine, gastroenterology
Pancella, Edward Dannelly, biochemistry
Panessa, Barbara Jean, cell biology
Pang, Peter Kai To, comparative physiology, endocrinology
Pantuck, Eugene Joel, anesthesiology, biochemical pharmacology
Papanicolaou, George Constantine, mathematics
Paronetto, Fiorenzo, pathology
Parshley, Mary Stearns, anatomy
Parsons, Robert W., Jr., organic chemistry
Pasik, Pedro, neurophysiology, neuroanatomy
Pasik, Tauba, neurophysiology, neuroanatomy
Passo, Stanley Samuel, neuroendocrinology
Pasternack, Bernard Samuel, epidemiology, biostatistics
Pastel, Dhun Burjor, anthropology
Patt, Charles Richard, mathematics
Patterson, Eugene B., animal nutrition
Patterson, Richard Westcott, physiology, anesthesiology
Patterson, Russel Hugo, Jr., neurosurgery
Pechukas, Philip, theoretical chemistry
Peirce, Edmund Converse, II, physiology, surgery
Peitz, Betsy, reproductive physiology
Peluso, Ada, mathematics
Penicnak, Adrian John, physiology, clinical chemistry
Pentel, Leon, radiology, radiology
Percus, Jerome K., theoretical physics
Perdue, Henry Stafford, biochemistry
Pereira, Gerard P., anatomy, histology
Pereira, Michael Alan, biochemical pharmacology
Perel, James Maurice, clinical pharmacology, psychopharmacology
Perlmutter, Alfred, ichthyology
Perrotta, James, chemistry
Peters, Theodore, Jr., biochemistry
Peterson, Barry Wayne, neurophysiology
Peterson, Ralph Edward, endocrinology
Pfaff, Donald Wells, physical pharmacology
Pfeiffer, Raymond L., ophthalmology
Phansalkar, Sadashiv Vinayak, animal behavior
Phillips, Frederick Stanley, pharmacology
Phillips, David Mann, cell biology
Phillips, George Wygant, Jr., physical chemistry
Phillips, Gerald B., medicine
Phillips, Louise Lang, hematology
Phillips, Mildren E., pathology
Phillips, Stephanie Gordon, cell biology
Phillips-Quagliata, Julia Molyneux, immunology, zoology
Pickering, Miles Gilbert, nuclear chemistry
Pierce, James Clarence, surgery, immunology
Pigman, Ward, biochemistry, health sciences
Piliero, Sam Joseph, endocrinology, hematology
Pilkington, Lou Ann, physiology
Pines, Kermit L., internal medicine
Pino, John Anthony, animal sciences
Pinsky, Carl Muni, oncology
Pinto, John, biological chemistry, nutritional biochemistry
Piomelli, Sergio, hematology
Piore, Emanuel Ruben, physics
Pirani, Conrad Levi, pathology
Pirone, Pascal Pompey, plant pathology
Pi-Sunyer, F Xavier, endocrinology, nutrition
Pitts, Robert Franklin, physiology
Pizzarello, Donald Joseph, radiology
Plum, Fred, neurology
Pohorecky, Larissa Alexandra, pharmacology, neuroendocrinology

Poindexter, Charles Aden, medicine
Pokorny, Frank Joseph, botany
Polatin, Phillip, psychiatry
Pollack, Burton Robert, medical administration
Pollack, Richard M, mathematics
Pollak, Fred Hugo, solid state physics
Pollart, Dale Flavian, polymer chemistry
Polley, Margaret J, immunochemistry
Pons, Marcel William, virology
Pontius, Dieter J J, organic chemistry, analytical chemistry
Ponzio, Nicholas Michael, immunology
Poole, Brian Howard, cell biology, biochemistry
Pope, Martin, solid state physics
Popper, George H P, geology
Popper, Hans, pathology
Poppers, Paul Jules, anesthesiology
Porter, Milton Reeves, surgery
Posey, Clayton Eugene, forest genetics
Posner, Aaron Sidney, physical chemistry, biochemistry
Posner, Gerald Seymour, oceanography
Posner, Jerome B, neurology
Posner, Philip, physiology
Post, Joseph, medicine
Potter, Guy Dill, radiology
Potter, Jacobus Louw, internal medicine, experimental pathology
Potter, Roy Frank, solid state physics
Poutsiaka, John William, physiology
Powers, Jack W, organic chemistry
Powers, John Clancey, Jr, organic chemistry, biochemistry
Prasad, Ishwari, microbial genetics, microbiology
Preiser, Stanley, numerical analysis
Prendergast, Kevin Henry, astronomy
Prensky, Wolf, genetics, cell biology
Primack, Marshall Philip, endocrinology
Prose, Philip H, pathology
Proskauer, Eric S, physical chemistry
Prudden, John Fletcher, surgery
Prutkin, Lawrence, anatomy
Pullman, Maynard Edward, biochemistry
Purohit, Surendra Nath, reactor physics, nuclear physics
Puszkin, Saul, cell biology, neurobiology
Puttlitz, Donald Herbert, medical microbiology
Pye, Orrea Florence, nutrition
Quartararo, Ignatius Nicholas, dentistry
Rabi, Isidor Isaac, physics
Rabinovitch, Michel Pinkus, cell biology, experimental medicine
Rabinowitz, James Robert, biophysics
Rachele, Julian Richard, biochemistry
Rackow, Herbert, anesthesiology
Radel, Stanley Robert, theoretical chemistry
Ragan, Charles (Alexander), Jr, medicine
Rainer, John David, psychiatry, medical genetics
Rainwater, Leo James, physics
Ramaley, James Francis, mathematics
Ramstad, Paul Ellertson, food science
Randol, Burton Swank, mathematics
Randt, Clark Thorp, neurology
Ransohoff, Joseph, neurosurgery
Rapaport, Felix Theodosius, surgery, transplantation immunology
Rapport, Maurice M, biochemistry
Rasweiler, John Jacob, IV, reproductive physiology, anatomy
Ratcliffe, Nicholas Morley, geology
Rathnam, Premila, biochemistry, endocrinology
Ratliff, Floyd, neurophysiology
Ratner, Sarah, biochemistry
Rauch, Harry Ernest, mathematics
Rausen, Aaron Reuben, pediatrics, hematology
Ray, Bronson Sands, surgery
Reader, George Gordon, medicine, public health
Reasenberg, Julian Robert, synthetic organic chemistry, mineralogy
Redi, Olav, experimental atomic physics
Redisch, Walter, physiology
Redman, Colvin Manuel, biochemistry
Redo, Saverio Frank, surgery
Reed, George Elliott, cardiovascular surgery, surgery
Reeke, George Norman, Jr, crystallography
Reemtsma, Keith, surgery
Rees, Mina Spiegel, mathematics
Reese, Robert Trafton, immunology
Reidenberg, Marcus Milton, pharmacology, medicine
Reinmuth, William Henry, analytical chemistry
Reis, Donald J, neurology, neurophysiology
Reiss, Frederick, dermatology, mycology
Remes, Nathaniel L, organic chemistry
Rendell-Baker, Leslie, anesthesiology
Rennert, Joseph, photochemistry
Resnik, Frank Edward, analytical chemistry
Reuben, John Philip, biophysics
Reuben, Richard N, medicine
Reynolds, William Elliot, medicine
Rhodes, Yorke Edward, physical organic chemistry
Ribble, John C, pediatrics
Ricci, John Ettore, physical chemistry
Rich, Herbert, medical statistics
Richardson, Ralph William, Jr, horticulture
Richardson, Robert William, statistical mechanics, theoretical nuclear physics
Richardson, Ronald John, physical chemistry, information science
Richart, Ralph M, pathology, obstetrics & gynecology
Richman, Alex, psychiatry, epidemiology
Richter, Donald, statistics
Richter, Frederick Paul, petroleum chemistry
Rickard, James Alexander, physics
Riesenfeld, Alphonse, anthropology
Rifkind, Arleen B, endocrinology, pharmacology
Rifkind, Richard A, medicine, hematology
Riker, Walter Franklyn, Jr, pharmacology
Rinaldi, Robert Arthur, cell physiology
Ripley, Suzanne, anthropology, primatology
Ripps, Harris, physiology
Ritchie, Donald Dirk, mycology
Rizack, Martin A, biochemistry, pharmacology
Robbins, Herbert Ellis, mathematical statistics
Robbins, William Jacob, plant physiology
Roberts, Norbert Joseph, medicine
Roberts, Robert Bryan, anesthesiology, obstetrics & gynecology
Robertson, Dale Norman, organic chemistry
Robinson, David Zav, physics
Robinson, Harry L, pathology
Roboz, John, physical chemistry
Rochovansky, Olga Maria, biochemistry, virology
Rodman, Toby C, cytogenetics
Rodriguez, Joaquin, geology
Rogoff, Joseph Bernard, medicine
Roizin, Leon, neuropathology
Rolle, Gloria Katharine, histology, embryology
Ronn, Avigdor Meir, chemical physics
Rosalsky, Maurice B, geomorphology
Rosano, Henri Louis, chemistry
Rosen, Carol Zwick, low temperature physics
Rosen, Donn Eric, ichthyology
Rosen, Harry Mark, medical administration
Rosen, John Friesner, biochemistry, metabolism
Rosen, Paul, physics, electronics
Rosen, Stephen, science writing, science policy
Rosenberg, Leonard, physics
Rosenberg, Robert, physics
Rosenberry, Terrone Lee, biochemistry, neurochemistry
Rosenbloom, Paul Charles, mathematics
Rosenbluth, Jack, cytology
Rosenfeld, Louis, biochemistry
Rosenfeld, Norman Samuel, mathematics
Rosenfield, Richard Ernest, hematology, immunology
Rosenstein, Solomon Nathan, dentistry
Rosenthal, Stanley Arthur, medical mycology, microbiology
Rosenthal, William S, gastroenterology, physiology
Rosoff, Betty, endocrinology
Rosoff, Morton, physical chemistry, surface chemistry
Rossi, Harald Herman, radiation biophysics
Rossman, Isadore, geriatrics
Rossman, Toby Gale, microbiology
Rosso, Pedro, perinatal biology
Roth, Daniel, medicine, pathology
Roth, Howard, food chemistry
Roth, Sidney Goldwater, mathematics
Rothballer, Alan Burns, neuroanatomy
Rothen, Alexandre, chemistry
Rothenberg, Lawrence Neil, medical physics
Rothenberg, Sheldon Philip, hematology, internal medicine
Rothermel, Robert E, medicine, public health
Rothschild, Edmund Otto, internal medicine, endocrinology
Rothschild, Marcus Adolphus, internal medicine
Rovit, Richard Lee, neurosurgery
Rowe, Arthur Wilson, biochemistry, cryobiology
Rowe, William Bruce, biochemistry
Rowland, Lewis Phillip, neurology
Roy, Debdutta, biochemistry
Roze, Janis Arnold, human ecology
Rozen, Jerome George, Jr, systematic entomology, academic administration
Rubel, Paula G, cultural anthropology
Rubin, Albert Louis, internal medicine
Rubin, Arnold David, hematology, cell biology
Rubin, Emanuel, pathology
Rubin, Kenneth, physics
Rubin, Samuel H, internal medicine
Rubin, Vera, anthropology
Rubinow, Sol Isaac, biomechanics, biomathematics
Rubinson, Kalman, neuroanatomy
Ruderman, Malvin Avram, theoretical physics
Rudner, Rivka, genetics
Rudzinska, Maria Anna, zoology, protozoology
Rusk, Howard A, medicine
Ruskin, Richard A, obstetrics & gynecology
Russ, Gerald A, nuclear medicine
Russell, Charlotte Sananes, organic chemistry, biochemistry
Russell, Martin, physical geology
Rutherford, Floyd James, science education
Rutishauser, Urs Stephen, biochemistry, immunology
Sabatini, David Domingo, cell biology, biochemistry
Sachs, Allan Maxwell, physics
Sackler, Arthur M, psychiatry
Sacks, Martin, invertebrate zoology
Sacksteder, Richard Carl, mathematics
Sadock, Benjamin, medicine, psychiatry
Sadock, Virginia A, psychiatry
Saffer, Charles Martin, Jr, industrial chemistry
Safir, Aran, medicine, ophthalmology
Sage, Harold Hubert, medicine
Sager, Clifford J, psychiatry
Sakita, Bunji, theoretical high energy physics
Saks, Norman Martin, cell physiology, physiological ecology
Salanitre, Ernest, medicine, anesthesiology
Salant, Abner, food technology
Sall, Sanford, obstetrics & gynecology
Salser, Josephine See, microbiology, biochemistry
Salton, Milton Robert James, microbiology
Salwen, Bert, anthropology, archaeology
Salzberg, Hugh William, physical chemistry
Sampath, Angus C, medical microbiology
Sampson, Phyllis Marie, biochemistry
Samuels, Stanley, neurochemistry
Sanders, Frank Kingsley, biology
Sanders, John Essington, geology
Sandow, Alexander, biophysics
Sank, Diane, biological anthropology, medical genetics
Santos-Buch, Charles A, experimental pathology
Santulli, Thomas V, surgery
Sarachik, Myriam Paula, solid state physics
Sarkar, Nurul Haque, biophysics, oncology
Sarmousakis, James Nicholas, physical chemistry
Sassa, Shigeru, hematology, biochemistry
Sawyer, Stanley Arthur, mathematics
Sawyer, Wilbur Henderson, physiology, pharmacology
Saxena, Brij B, biochemistry, endocrinology
Saxena, Vishv Prakash, physical biochemistry, medical research
Sayers, Ross, clinical pharmacology
Scaglione, Peter Robert, pediatrics
Scalora, Frank Salvatore, mathematics
Scarpelli, Emile Michael, pediatrics, cardiopulmonary physiology
Schachter, David, medicine
Schaefer, George, obstetrics & gynecology
Schaeffer, Bobb, vertebrate paleontology
Schaefler, Sam, microbiology
Schaffel, Simon, geology, oceanography
Schaffer, Abraham Isaac, electrophysiology
Schaffner, Fenton, medicine
Scharf, Walter, gemology
Schatten, Robert, mathematics
Schaye, Alvin Albert, medicine
Schearer, Sherwood Bruce, population studies, science administration
Schechter, Martin, mathematical analysis
Schenkein, Isaac, biological chemistry
Scherer, William Franklin, medicine, microbiology
Scherf, David, cardiovascular diseases
Scheuer, James, physiology, biochemistry
Schiavi, Raul Constante, psychiatry
Schiessler, Robert Walter, chemistry, technological forecasting
Schildkrout, Enid, anthropology, ethnography
Schindler, Hans, petroleum chemistry
Schindler, Susan, mathematics
Schlaeger, Ralph, radiology
Schlesinger, Edward Bruce, neurosurgery
Schlissel, Arthur, mathematics
Schmid, Wilfried, mathematics
Schneckloth, Roland Edmunds, internal medicine
Schneider, Alfred Marcel, mathematical statistics, operations research
Schneider, Allan Stanford, physical chemistry, biochemistry
Schneierson, S Stanley, microbiology, infectious diseases
Schoenfeld, Robert Louis, electronics, computer science
Scholte, Bob, anthropology, sociology
Schreiber, Sidney S, physiology, nuclear medicine
Schubert, Edward Thomas, biochemistry
Schuh, Randall Tobias, systematic entomology
Schuster, David Israel, organic chemistry
Schwartz, Abraham, mathematics
Schwartz, Arthur Harold, psychiatry, academic administration
Schwartz, Gerald Peter, biochemistry
Schwartz, Irving Leon, physiology, medicine
Schwartz, Jacob Theodore, mathematics
Schwartz, Leonard H, chemistry
Schwartz, Morton K, biochemistry, clinical chemistry
Schwartz, Norman Martin, biology
Schwartz, Sheldon, medicine
Schwarz, Gerhart Steven, radiology
Schweitzer, Morton David, epidemiology
Sciarra, Daniel, medicine
Scopp, Irwin Walter, oral medicine, periodontics
Scott, Virgil Cole, internal medicine
Scott, Walter Neil, medicine, physiology
Scott, William Addison, III, biochemistry, genetics
Seaman, William B, radiology
Sechzer, Jeri Altneu, physiological psychology, neurobiology
Sedlis, Alexander, obstetrics & gynecology
Seegal, Beatrice Carrier, immunology, cancer
Seelig, Mildred Sylvia, medicine, nutrition
Segal, Alvin, biochemistry, cancer
Segal, Bernice G, physical chemistry
Segal, Sheldon Jerome, endocrinology, embryology
Seidman, Irving, pathology
Seifter, Joseph, pharmacology
Seifter, Sam, biochemistry
Seitz, Frederick, physics
Selby, Henry M, medicine
Selikoff, Irving John, environmental medicine, public health
Sellers, Peter Hoadley, mathematics
Seneca, Harry, internal medicine, bacteriology
Senescu, Robert A, psychiatry
Senterfit, Laurence Benfred, microbiology, immunology
Serber, Robert, physics
Sessa, Grazia L, biochemistry
Severiens, Johannes Coenraad, nuclear physics
Severn, David Jones, information science
Seybolt, John Francis, pathology
Shafland, James L, human anatomy
Shahidi, Syed Abdus-Salam, microbiology
Shahn, Ezra, molecular biology, biophysics
Shamoian, Charles Anthony, psychiatry, biochemical pharmacology
Shamos, Morris Herbert, biophysics, laboratory medicine
Shands, Harley Cecil, psychiatry
Shane, Harold D, mathematical statistics
Shannon, James Augustine, physiology
Shapiro, Harry Lionel, biological anthropology
Shapiro, Jack Sol, mathematics
Shapiro, Rebecca Lillian, industrial bacteriology
Shapiro, Robert, biochemistry, organic chemistry
Shapiro, William Richard, neurology
Sharkey, John Bernard, inorganic chemistry, instrumentation
Sharkey, Margaret Mary, cell biology

Sharma, Ram Karan, chemistry
Sharma, Sansar C., neurobiology
Sharp, Lewis Inman, medicine
Sharpe, Lawrence, medicine
Sharpe, William D., pathology, history of medicine
Sharpless, Thomas Kite, cytology, biomedical engineering
Shatin, Harry, dermatology
Shechter, Leon, organic chemistry, polymer chemistry
Shedlovsky, Leo, colloid chemistry, electrochemistry
Shedlovsky, Theodore, physical chemistry
Sheer, Charles, physics
Shefer, Sarah, biochemistry
Shetupsky, David I., physics
Sherlock, Paul, internal medicine, gastroenterology
Sherman, Robert S., medicine
Shields, Robert James, medicine
Shils, Maurice Edward, medicine, nutrition
Shimer, John Asa, geomorphology
Short, Lester Le Roy, Jr., ornithology
Shpiz, Joseph M., elementary particle physics
Shriver, Joyce Elizabeth, anatomy
Shulman, Herbert Byron, topology
Shulman, Sidney, immunology
Siderowitz, Joshua, physics
Sidran, Miriam, solid state physics
Siegal, Frederick Paul, immunology, internal medicine
Siegel, Charles David, biology
Siegel, Irwin Michael, physiology, genetics
Siegel, Laurane Geary, neurochemistry, enzymology
Siegel, Maurice L., organic chemistry
Siegel, Morris, preventive medicine, epidemiology
Siegmund, David O., mathematical statistics
Siekevitz, Philip, biochemistry
Siffert, Robert S., orthopedic surgery
Silagi, Selma, genetics, cancer
Silber, Robert, internal medicine, hematology
Silva-Hutner, Margarita, mycology, microbiology
Silver, Archie Aaron, psychiatry
Silver, Rae, animal behavior
Silver, Richard Tobias, medicine
Silverman, Sydel, anthropology
Simon, Eric Jacob, neurochemistry, pharmacology
Sinclair, David, applied physics
Singer, Alan Granger, bio-organic chemistry
Singer, Burton Herbert, statistics
Singh, Harbhajan, biochemistry
Singh, Inder Jit, anatomy, dentistry
Sinnette, Calvin Herman, pediatrics
Siskind, Gregory William, immunology
Sit, William Yu, algebra
Skinner, Elliott Percival, anthropology
Skogerson, Lawrence Eugene, biochemistry
Slack, Lewis, nuclear physics
Slattery, Louis R., surgery
Slipp, Samuel, psychiatry, psychoanalysis
Sloan, Donald Leroy, Jr., biochemistry
Slobody, Lawrence Boris, pediatrics
Smith, Archie Lee, biochemistry
Smith, Austin Edwards, pharmacology, therapeutics
Smith, Carl Arthur, surgery
Smith, Cassandra Lynn, molecular biology
Smith, Clarence Lavett, zoology
Smith, Frederick William, experimental solid state physics
Smith, Issar, molecular biology
Smith, John Edgar, nutritional biochemistry
Smith, Nicole Schupf, neuropsychology
Smith, Paul Althaus, mathematics
Smithwick, Elizabeth Mary, immunology
Snyderman, Selma Eleanore, pediatrics, medicine
Socha, Wladyslaw Wojciech, immunology, pathology
Sohmer, Bernard, mathematics
Sohn, David, pathology, toxicology
Sokol, Herman, organic chemistry
Solecki, Ralph Stefan, anthropology, archaeology
Sommers, Sheldon Charles, pathology
Sonenberg, Martin, endocrinology, biochemistry
Song, Sun Kyu, neuropathology

Sonneborn, Henry, III, chemistry
Soren, Arnold, orthopedic surgery
Southam, Anna L., medicine
Southren, A Louis, internal medicine, endocrinology
Southworth, Hamilton, internal medicine
Spaet, Theodore H., medicine
Spar, Jerome, meteorology
Speck, Paul William, meteorology
Speck, William T., pediatrics
Spector, Abraham, biochemistry
Spector, Leonard B., biochemistry
Spencer, Frank Cole, medicine
Spencer, William Alden, physiology, neurophysiology
Speth, John David, anthropology
Spiegel, Allen J., pharmacy, pharmaceutical chemistry
Spiegel, Edward A., astrophysics
Spiegelman, Martha, embryology
Spiegelman, Solomon, biochemistry
Spielman, Harold S., electronics, science education
Spies, Gunther Otto, plasma physics
Spilman, Edra Lavergene, biochemistry
Spindel, William, chemistry
Spitaley, George Leonard, immunology, parasitology
Spitzer, Joseph Maurice, cardiology
Spivak, Joseph, geology
Sprinson, David Benjamin, biochemistry
Spritz, Norton, biochemistry
Spruch, Larry, theoretical physics
Squires, Arthur Morton, physical chemistry
Srinivasan, Parthbychery, biochemistry
Stahl, S Sigmund, periodontology, oral pathology
Stahl, William J., biochemistry
Stamatelatos, Michael G., nuclear science
Stammelman, Mortimer Jacob, physical chemistry
Stanton, Nancy Kahn, mathematics
Stark, Dennis Michael, immunology
Stark, Richard B., plastic surgery
Starnes, Ordway, entomology
Stecher, Emma Dietz, organic chemistry
Stecher, Milton, physics
Stein, Thomas B, Jr., computer science
Steel, Charles Melvin, obstetrics & gynecology
Steinberg, Leonard, medicine
Steinmetz, Charles Robert, rheumatology
Steigman, Alex J., pediatrics
Stein, Marvin, psychiatry
Stein, William Howard, biochemistry
Stein, Zena, epidemiology
Steinmetz, Charles Henry, occupational health, medical administration
Stenger, Richard J., pathology, electron microscopy
Stenzel, Kurt Hodgson, nephrology
Sternfeld, Leon, medical administration, research administration
Stevens, John Joseph, cancer
Stevens, Nelson Pierce, geochemistry
Stevens, Richard S., marine geology, physical oceanography
Stevenson, Neil Elizabeth, topology
Stevenson, Stuart Shelton, pediatrics
Stewart, Albert Clifton, radiation chemistry
Stewart, Richard Willis, atmospheric physics
Stiller, Richard L., biochemistry
Stinson, Stephen Charles, organic chemistry
Stitchfield, Frank E., orthopedic surgery
Stock, Charles Chester, chemotherapy
Stock, Herbert Carl, pathology
Stoker, James Johnston, mathematics, mechanics
Stokes, Peter E., medicine, psychiatry
Stolov, Harold L., physics, meteorology
Stone, David, biochemistry, genetics
Stone, Gilbert C H, biochemistry
Stone, Martin L., obstetrics & gynecology
Stoneham, Richard George, mathematics
Stork, Gilbert Jesse, synthetic organic chemistry
Stotzky, Guenther, microbial ecology
Strand, Fleur Lillian, biology
Straus, Bernard, internal medicine
Straus, Lotte, pathology
Strobos, Robert Julius, neurology
Stroke, Hinko Henry, atomic spectroscopy, nuclear physics
Stultz, Robert Lee, Jr., physical chemistry
Stutman, Leonard Jay, cardiology
Stutman, Osias, immunology; pathology
Suciu-Foca, Nicole M., immunology; pathology
Suggitt, Robert Murray, physical chemistry

Sullivan, Walter James, physiology, biophysics
Sullivan, Walter Seager, science writing, journalism
Sun, Alexander Shikuang, biochemistry, cell biology
Sunderlin, Charles Eugene, organic chemistry
Susdorf, Dieter Hans, immunology
Susser, Mervyn W, epidemiology, social medicine
Sutton, Constance Rita, anthropology
Svikolos, Nikola, mathematics, system analysis
Swan, Roy Craig, Jr., anatomy
Swartout, John Arthur, research administration, reactor engineering
Sweeney, William Victor, biophysics
Swenberg, Charles Edward, biophysics
Swick, Kenneth Eugene, mathematics
Swinney, Harry Leonard, fluid physics
Swisolod, Norbert Ira, biochemistry, endocrinology
Sy, Jose, biochemistry
Sykes, Marguerite Prince, cancer
Szer, Wlodzimierz, biochemistry
Taggart, John Victor, physiology, medicine
Tainter, Melvin, applied mathematics
Takahashi, Taro, geochemistry, geophysics
Taketomo, Yasuhiko, psychiatry
Tam, Andrew Ching, atomic physics, molecular physics
Tamir, Hadassah, neuroscience, biochemistry
Tamm, Igor, virology, medicine
Tan, Charlotte, cancer
Tannenbaum, Harold E., science education
Tannenbaum, Michael J., high energy physics
Tannenbaum, Robert S., computer science
Tantravahi, Ramana V., human genetics, cytogenetics
Tanzer, Charles, science education, medical microbiology
Tapley, Donald Fraser, internal medicine
Tapper, Frederick Drach, mathematical physics
Taranta, Angelo, internal medicine
Tarantino, Laura Mary, biochemistry
Tattersall, Ian Michael, physical anthropology; primatology
Tatum, Howard James, obstetrics & gynecology
Taub, Abraham, pharmaceutical chemistry
Taub, Robert Norman, immunology; hematology
Taussig, Robert Trimble, plasma physics
Tavolga, William Nicolai, animal behavior
Taylor, Anna Newman, physiology, anatomy
Tchertkoff, Victor, pathology
Tedford, Richard Hall, vertebrate paleontology, stratigraphy
Teichner, Victor Jerome, psychiatry, psychoanalysis
Teitelbaum, Charles Leonard, analytical chemistry
Tender, Moses David, microbiology
Tenenbaum, Leon Edward, chemistry
Tennant, Judith R., virology
Teply, Lester Joseph, biochemistry
Terres, John Kenneth, environmental biology
Terris, Milton, epidemiology
Tessler, Arthur Ned, urology
Texon, Meyer, medicine
Thaddeus, Patrick, physics, astrophysics
Thau, Rosemarie B Zischka, biochemistry, endocrinology
Thomas, Alexander, psychiatry
Thomas, Arthur L., chemistry
Thomas, David Hurst, anthropology, archaeology
Thomas, Lewis, internal medicine, pathology
Thompson, David Duvall, internal medicine
Thompson, Samuel Alcott, surgery
Thompson, Gerald Edmund, internal medicine, nephrology
Thorbecke, Geertruida Jeanette, immunology, experimental pathology
Tice, David Anthony, thoracic surgery, cardiovascular surgery
Tichauer, Erwin Rudolph, biomedical engineering, occupational health
Tidwell, Troy Haskell, Jr., electrochemistry
Tiersten, Martin Stuart, physics
Tietjen, John H, biological oceanography, invertebrate zoology

Tiger, Lionel, biological anthropology, social structure
Ting, Er Yi, physiology, internal medicine
Ting, Lu, applied mathematics
Tobach, Ethel, animal behavior
Tobin, Margaret Edna, physiology
Todd, Daniel F., dentistry
Toennessen, Gary Herbert, environmental sciences
Tokumaru, Tadasu, virology
Tomashefsky, Philip, experimental pathology
Tomasz, Alexander, biochemistry, cell biology
Tomasz, Maria, organic chemistry, biochemistry
Tonna, Edgar Anthony, cell physiology, cell chemistry
Tonndorf, Juergen, physiology
Topoff, Howard Ronald, biology, animal behavior
Tourin, Richard Harold, energy conversion
Trager, William, parasitology
Trakatellis, Anthony C., biochemistry, molecular biology
Travis, Larry Dean, astrophysics
Treat, Asher Eugene, entomology
Troll, Walter, biochemistry, organic chemistry
Troutman, Richard Charles, ophthalmology
Trubek, Max, medicine
Trussell, Ray, medicine, epidemiology
Tryon, Edward Polk, theoretical high energy physics, cosmology
Tsong, Yun Yen, biochemistry, organic chemistry
Turell, Robert, surgery
Turino, Gerard Michael, medicine
Twombly, Gray Huntington, obstetrics & gynecology
Udem, Stephen Alexander, virology, infectious diseases
Uhlenbeck, George Eugene, physics
Urdang, Arnold, pharmaceutical chemistry, pharmacy
Uretsky, Myron, computer science, data processing
Usami, Shunichi, physiology
Valentine, Fred Townsend, immunology, infectious diseases
Vanamee, Parker, physiology
Van Burkalow, Anastasia, geomorphology
Vanderberg, Jerome Philip, medical entomology; cell physiology
Vandewiele, Raymond Laurent, obstetrics & gynecology
van Duuren, Benjamin Louis, organic chemistry
Van Gelder, Richard George, mammalogy
Van Italie, Theodore Bertus, medicine
Van Norton, Roger Norman, applied mathematics, computer science
Van Woert, Melvin H., internal medicine
Varicchio, Frederick, biochemistry, developmental biology
Varro, Stephen, Jr., ecology, plant nutrition
Vasquez, Alphonse Thomas, mathematics
Veenema, Ralph J., urology
Veith, Frank James, surgery, transplantation biology
Vick, Gerald Kieth, organic chemistry
Vilcek, Jan Tomas, microbiology, virology
Vincent, George Paul, chemistry
Vivona, Stefano, preventive medicine, public health
Vnek, John, protein chemistry, enzymology
Vogel, Henry, microbiology, serology
Vogel, Bruce R., mathematics
Volavka, Jan, psychiatry
Volchok, Herbert Lee, geochemistry
Vratsanos, Spyros M., biochemistry
Wachtel, Jonathan Mark, plasma physics
Wagner, Bernard Meyer, pathology
Wagner, Richard H., plant ecology
Wagreich, Harry, biochemistry
Wahleri, John Howard, vertebrate paleontology
Wainfan, Elsie, biochemistry
Wajda, Isabel, pharmacology, neurochemistry
Wald, Francine Joy, science education
Wald, Samuel Stanley, roentgenology
Walker, Warren Elliott, operations research, urban research and development
Walker, William Comstock, paper chemistry, physical chemistry
Waller, Hardress Jocelyn, physiology

Wallerstein, Harry, medicine, hematology
Waltcher, Irving, polymer chemistry
Wang, Dalton Ta Tung, plant biochemistry
Wang, Hao, mathematics
Wang, Hsueh-Hwa, pharmacology, physiology
Wang, Shih Chun, physiology, pharmacology
Warburton, Dorothy, human genetics
Warburton, Frederick E, evolution, population genetics
Wasserman, Aaron Osias, vertebrate zoology
Wasserman, Edward, pediatrics
Wasserman, Louis Robert, medicine
Wasserman, Robert Fletcher, medicine
Watson, Robert Joseph, geophysics
Wearn, Richard Benjamin, organic chemistry
Weaver, Sylvia Short, marine biology
Webb, Kempton Evans, geography
Weber, Julius, cytology, photography
Wecker, Stanley C, vertebrate ecology
Weinberg, Myron Simon, toxicology, research administration
Weiner, Herbert, psychiatry, neurology
Weinstein, Bernard Ira, biochemistry
Weinstein, Edwin Alexander, neuropsychiatry
Weinstein, Harel, biophysical chemistry, quantum chemistry
Weinstein, I Bernard, medicine
Weisberg, Herbert, medicine
Weiss, Benjamin, biochemistry, organic chemistry
Weiss, Dennis, micropaleontology, environmental geology
Weiss, Gerson, reproductive endocrinology
Weiss, Harvey Jerome, internal medicine
Weiss, Irma Tuck, biochemistry, organic chemistry
Weiss, Jay M, psychophysiology
Weiss, Marvin, pharmaceutical chemistry, science administration
Weiss, Paul Alfred, biology
Weissmann, Gerald, cell biology, internal medicine
Weitzner, Harold, applied mathematics, plasma physics
Weksler, Marc Edward, medicine, immunology
Wellner, Daniel, biochemistry
Wellner, Vaira Pamiljans, biochemistry
Welner, Erna Alture, virology
Werner, Sidney Charles, medicine
Wernick, William, mathematics
Wertheim, Arthur Robert, medicine
Wescoe, W Clarke, pharmacology, experimental medicine
Wessler, Stanford, medicine
Wetmur, James Gerard, biophysical chemistry
Weyl, F Joachim, mathematics
Wham, George Sims, textile chemistry
Wheatley, Victor Richard, biochemistry
Wheeler, Edward Norwood, organic chemistry
Whitby, Owen, statistics
White, Russell Alan, geography
Whitfield, Graham Frank, surgery
Whitsell, John Crawford, II, surgery
Wick, Gian Carlo, physics
Widmark, Rudolph M, immunology, medical microbiology
Wigand, Jeffrey Stephen, endocrinology, biochemistry
Wigger, H Joachim, medicine, pathology
Wijnen, Joseph M H, physical chemistry, photochemistry
Wilen, Samuel Henry, organic chemistry
Wiley, Richard Haven, pharmacology
Wilk, Sherwin, biochemistry, pharmacology
Williams, Marshall Henry, Jr, physiology, internal medicine
Williams, Robert Bruce, physical oceanography
Williams, Roger Wright, medical entomology, parasitology
Williamson, Samuel Johns, low temperature physics
Willis, Carl Raeburn, Jr, pharmacy, pharmaceutical chemistry
Willis, William Shedrick, Jr, anthropology
Willoughby, Stephen Schuyler, mathematics education
Wilner, George Dubar, hematology, pathology
Wilson, Marion Evans, medical microbiology
Wilson, Victor Sidney J, neurophysiology
Winawer, Sidney J, internal medicine, gastroenterology
Windhager, Erich E, physiology, biophysics

Winick, Myron, pediatrics, nutrition
Winicov, Edith, organic chemistry
Winikoff, Beverly, public health, nutrition
Winkler, Marvin Howard, biophysical chemistry
Winokur, Morris, biology
Winston, James J, food chemistry
Winter, Jeanette, microbiology
Winter, Joseph Wolfgang, microbiology
Winters, Robert Wayne, pediatrics, physiology
Wischnitzer, Saul, biology, anatomy
Wishnick, Marcia M, biochemistry, pediatrics
Wit, Andrew Lewis, cardiovascular physiology, pharmacology
Witkin, George Joseph, dentistry
Witkin, Steven S, molecular biology, virology
Witkus, Eleanor Ruth, biology
Wites, Janet Turk, statistics, biostatistics
Witz, Gisela, cancer
Woessner, Ronald Arthur, forestry, genetics
Wohlgelernter, Devora Kasachkoff, mathematics
Wolf, Abner, neuropathology
Wolf, Bernard Saul, radiology
Wolf, Robert Lawrence, physiology, biochemistry
Wolff, James A, pediatrics, hematology
Wolff, Marianne, pathology, surgical pathology
Wolff, Steven, organic chemistry
Wolff, William I, surgery
Wolman, Sandra R, pathology, cytogenetics
Wolsky, Alexander Albert, developmental biology
Woltjer, Lodewyk, astrophysics
Wood, Henry Nelson, biochemistry
Woodard, Helen Quincy, biochemistry, radiobiology
Woodruff, Marvin Wayne, urology
Woodward, Arthur Eugene, physical chemistry
Worgul, Basil Vladimir, cell biology
Wortman, Leo Sterling, Jr, plant breeding
Wotherspoon, Neil, physical chemistry, biophysics
Wotman, Stephen, dentistry
Wrenn, McDonald Edward, environmental health, radiological health
Wright, Jane Cooke, medicine
Wright, Bill C, soil science
Wu, Cheng-Tsu, geography of China, geography
Wu, Chien-Shiung, physics
Wu, Daisy Yen, nutrition
Wynder, Ernst Ludwig, preventive medicine, epidemiology
Wyssbrod, Herman Robert, physiology, biophysics
Yahr, Melvin David, neurology
Yalow, Abraham Aaron, physics, medical biophysics
Yarmush, David Leon, applied mathematics
Yasumura, Seiichi, endocrinology
Yeh, Samuel D J, biochemistry
Yehle, Clifford Omer, genetics, microbiology
Yip, Lily Chung, biochemistry, parasitology
Yoeli, Meir, tropical medicine, hematology
Yoho, James Gibson, forest economics
Young, Charles William, internal medicine, cancer
Young, Morris Nathan, ophthalmology, surgery
Young, Ronald Jerome, biochemistry, molecular biology
Yu, Ts'ai-Fan, medicine, metabolism
Yucoglu, Yusuf Ziya, internal medicine, cardiology
Zablow, Leonard, biophysics, neurophysiology
Zadeh, Norman, operations research
Zadunaisky, Jose Atilio, physiology, biophysics
Zalusky, Ralph, internal medicine, hematology
Zanjani, Esmail Dabaghchian, hematology, physiology
Zare, Richard Neil, chemical physics
Zatuchni, Gerald Irving, obstetrics & gynecology
Zedeck, Morris Samuel, pharmacology, biochemistry
Zeek, William Charles, chemistry
Zegarelli, Edward Victor, dentistry
Zeitz, Louis, biophysics, physics
Zimmerman, Barry, organic chemistry
Zimmerman, Harry Martin, pathology
Zinder, Norton David, molecular genetics
Zizza, Frank, pathology

Zlot, William Leonard, mathematics
Zobler, Leonard, geography, hydrology
Zolla-Pazner, Susan Beth, immunology
Zubay, Geoffrey, biometrics
Zubin, Joseph, biometrics, psychology
Zucker, Marjorie Bass, physiology, hematology
Zucker-Franklin, Dorothea, cell biology
Zuckerman, Martin Michael, mathematical logic
Zuckerman, Samuel, organic chemistry
Zumbrunn, John Robert, mathematics
Zuzolo, Ralph C, cell physiology
Zwanziger, Daniel, theoretical physics
Zweifel, Richard George, herpetology

NEWBURGH
Bartleson, Christian James, psychophysics
DeLuca, Patrick John, microbiology, cytology
Hornibrook, Walter John, plastics chemistry
Lessor, Edith Dora, analytical chemistry
Peckham, Richard Stark, zoology

NEWCOMB
Brocke, Rainer H, mammalian ecology
Tierson, William Cornelius, forest ecology, wildlife ecology

NIAGARA FALLS
Allenbach, Charles Robert, inorganic chemistry, physical chemistry
Batha, Howard Dean, physical chemistry
Bean, C Thomas, Jr, polymer chemistry
Bharadwaj, Prem Datta, nuclear physics
Chadwick, George F, plastics chemistry, solid state physics
Chamberlin, Howard Allen, polymer chemistry, textile chemistry
Collins, Franklyn, solid state physics
Davis, Abram, spectroscopy, analytical chemistry
Fair, Frank Vernon, analytical chemistry
Felton, John James, organometallic chemistry
Forsyth, Paul Francis, industrial chemistry
Goddard, John Burnham, inorganic chemistry
Gordon, Irving, organic chemistry, information science
Grotheer, Morris Paul, electrochemistry, physical chemistry
Hansen, Donald Joseph, physical chemistry
Heintz, Edward Allein, analytical chemistry
Hicks, William Thomas, physical chemistry
Hindersinn, Raymond Richard, organic polymer chemistry
Klingele, Harold Otto, medicinal chemistry, organic chemistry
Larry, John Robert, physical chemistry
Lee, Sung Ki, organic chemistry, polymer chemistry
Lemper, Anthony Louis, organic chemistry
Matkovich, Vlado Ivan, inorganic chemistry
Maul, James Joseph, organic chemistry
Parker, William Edward, electronics
Pernert, John Carl, organic chemistry, inorganic chemistry, research administration
Rosenberg, Richard Martin, inorganic chemistry, research administration
Slagowski, Eugene Louis, polymer physics
Sojka, Stanley Anthony, physical organic chemistry
Takahashi, Akio, polymer chemistry
Weltman, Clarence A, physical chemistry, organic chemistry
Witschard, Gilbert, organic chemistry

NIAGARA UNIVERSITY
Britten, Bryan Terrence, biology, zoology
Dineen, Eugene Joseph, organic chemistry, analytical chemistry
Gilman, John Frances, mathematics
Hubbard, Richard Alexander, II, physical inorganic chemistry
Kiely, Lawrence J, anatomy, physiology
Krause, Josef Gerald, organic chemistry
Morton, Thomas Harlow, biology
Reedy, John Joseph, biology

NISKAYUNA
Krastins, Gunar, nuclear physics

NORTH CHATHAM
Potts, Gordon Oliver, endocrinology

NORTH TONAWANDA
Blumenthal, Warren Barnett, industrial chemistry
Klaiber, George Stanley, physics
Lauffenburger, James C, solid state physics
Rasch, Carl Henry, chemistry

NORTHPORT
Kosbab, Frederic Paul Gustav, psychiatry, internal medicine
Sherman, Jacques Lawrence, Jr, internal medicine
Shreeve, Walton Wallace, nuclear medicine, medicine

NORWICH
Alaimo, Robert J, organic chemistry
Andersen, Kenneth J, research administration, microbiology
Bickerton, Robert Keith, pharmacology
Burns, Richard Henry, pharmacology
Burrous, Stanley Emerson, chemotherapy
Castellion, Alan William, pharmacology
Chamberlain, Robert English, microbiology
Chang, Yi-Chi, pharmacology
Conkin, John Douglas, biopharmaceutics
Denning, George Smith, Jr, bio-organic chemistry
Ebetino, Frank Frederick, organic chemistry
Ells, Victor Raymond, physical chemistry
Gagnon, Leo Paul, pharmacy
Gever, Gabriel, organic chemistry, research administration
Goldenberg, Marvin M, pharmacology, physiology
Hewitt, William Francis, Jr, physiology
Hill, Richard A, pharmacy, chemical engineering
Holt, Leroy Henry, animal husbandry
Honkomp, Leroy J, pharmacology
Jacks, Thomas Mauro, medical microbiology
Johnson, Roland Norman, medicinal chemistry, organic chemistry
Kohls, Robert E, veterinary parasitology, entomology
Leathem, William Dolars, zoology
Levin, Robert Aaron, clinical chemistry
Liao, Hsueh-Liang, analytical chemistry
McKenzie, Walter Lawrence, industrial pharmacy, research administration
Michels, Julian Getz, organic chemistry
Morrison, Joseph Louis, biochemistry, drug metabolism
Neill, Alexander Bold, organic chemistry
Pelosi, Stanford Salvatore, Jr, organic chemistry
Perry, Samuel Cassius, animal nutrition
Pogue, John Parker, veterinary medicine
Pong, Schwe Fang, pharmacology
Porter, David Bruce, veterinary medicine
Ratto, Peter Angelo, pharmaceutical chemistry
Schwan, Thomas James, medicinal chemistry
Snyder, Harry Raymond, Jr, organic chemistry, medicinal chemistry
Spencer, Claude Franklin, organic chemistry
Stark, John Frederick, pharmaceutical chemistry
Taylor, Terry Mac, ruminant nutrition
White, Ralph Lawrence, Jr, synthetic organic chemistry
Windsor, Donald Arthur, information science, invertebrate zoology
Wright, George Carlin, synthetic organic chemistry
Yu, Chia-Nien, organic chemistry

NYACK
Arnold, Jeffrey, biochemistry, data processing
Fulmer, William, organic chemistry, spectroscopy
Ninkovich, Dragoslav, geology

OAKDALE
Brown, Robert Zanes, ecology, animal behavior
Haske, Bernard Joseph, organic chemistry
Kamran, Mervyn Arthur, entomology, ecology
Moeller, Henry William, marine biology
Spingola, Frank, chemistry

OAKFIELD
Denk, Ronald H, organic chemistry, lasers

OCEANSIDE
Kaplan, William, organic chemistry

OLD CHATHAM
Reilly, Edgar Milton, Jr., ornithology

OLD FORGE
Simmons, Frederick Charles, forest products

OLD WESTBURY
Chester, Clive Ronald, mathematics
Drossman, Melvyn Miles, biomedical engineering, computer sciences
Goldberg, Conrad Stewart, physics, computer science
Hoyte, Robert Mikell, organic chemistry
Levy, Paul, applied mathematics
Lulla, Kotusingh, atomic physics
Meade, Linda Celida, biochemical genetics
Mitacek, Eugene Jaroslav, biochemistry
Pierre, Leon L., bacteriology, biochemistry
Shapiro, Samuel S., applied statistics
Shigeura, Harold Takeo, biochemistry
Spector, Bertram, environmental health
Von Winbush, Samuel, inorganic chemistry, physical chemistry

OLEAN
Biggs, Homer Gates, biochemistry
Ciadella, Cataldo, organic chemistry
Fetscher, Charles Arthur, organic polymer chemistry
McElroy, Wilbur Renfrew, polymer chemistry, polymer engineering
Moss, Leo (David), pathology
Pande, Gyan Shanker, chemistry

ONEONTA
Acholonu, Alexander Dozie, zoology
Anderson, Adolph (Gustof), physical chemistry, academic administration
Ang, Jan Kee, horticulture, plant physiology
Armstrong, William Lawrence, organic chemistry
Bak, David Arthur, physical chemistry, organic chemistry
Bowers, Spotswood D Jr., physical chemistry, nuclear chemistry
Brzeik, Ronald Michael, number theory
Bukovsan, Laura A., genetics
Bukovsan, William, endocrinology
Butts, William Lester, entomology
Chiang, Joseph Fei, physical chemistry, applied physics
Corry, Martha Lucille, geography
Dayton, Bruce R., plant ecology
Deitz, William Harris, microbiology, bacteriology
Deubler, Earl Edward, Jr., ichthyology
Dixon, William Brightman, physical chemistry
Dubins, Mortimer Ira, science education, geology
Egan, Francis P., mathematics
Fielder, Douglas Stratton, experimental nuclear physics
Fleisher, Penrod Jay, geomorphology, glaciology
Harman, Willard Nelson, freshwater ecology, malacology
Hartley, Charles LeRoy, fluid physics
Hickey, Roger, experimental physics
Holway, James Gary, human ecology
Hutchison, David M., geology
Knauer, Bruce Richard, organic chemistry
Koehn, Paul V., biochemistry
Kotz, John Carl, inorganic chemistry, organometallic chemistry
Lawrence, James Lester, invertebrate zoology
Lawson, Kent DeLance, parasitology
Lutz, John George, organic physics, experimental physics
Miller, James Robert, organic chemistry
New, John G, environmental sciences, vertebrate zoology
Palmer, Arthur N., hydrogeology
Pence, Harry Edmond, inorganic chemistry
Phillips, Robert Rhodes, ethology, ecology
Raemsch, Bruce Ellenwood, anthropology
Read, Albert James, physics
Rhee, Haewun, mathematics
Sales, John Keith, structural geology
Sanik, John Jr., analytical chemistry
Settle, Wilbur Jewell, botany

Shannon, Jerry A. Jr., science education, biology
Simmonds, Robert T., paleobiology
Singh, Madho, genetics, biometrics
Sohacki, Leonard Paul, limnology, zoology
Stagg, Ronald M., physiology, endocrinology
Titus, Robert Charles, invertebrate paleontology
Van Tassell, Morgan Howard, microbiology
Wang, Charles T P., physics
Wilson, Philo Calhoun, marine geology, stratigraphy
Wilson, William D., parasitology
Wohlford, Duane Dennis, geology, petrology
Yang, Chao-Hui, mathematics

ORANGEBURG
Bagchi, Sakti Prasad, biochemistry
Bergner, Per-Erik Emil, theoretical biology
Camboy, Leslie Alan, physics
Coggon, Philip, natural products chemistry
Gradijan, Jack R., biomedical engineering
Grant, Frederick Warren, Jr., organic chemistry
Kohn, Michael, biomedical engineering, neurophysiology
Laska, Eugene, mathematics, statistics
May, Paul S., microbiology
Mowat, John Halley, organic chemistry
Sacks, William, biochemistry
Schayer, Richard William, biochemical pharmacology
Shah, Bhupendra K., statistics, biometrics
Siegel, Carole Ethel, biostatistics
Simpson, George M., psychiatry
Vestergaard, Per B., psychiatry

ORCHARD PARK
Carruthers, Christopher, cancer chemistry
Dempsey, Daniel Francis, chemistry
Eibeck, Richard Elmer, inorganic chemistry
Sand, Seaward Alwyn, genetics

OSSINING
Armanini, Louis Anthony, physical chemistry
Dahms, Harald, physical chemistry, electrochemistry
Greenstein, Leon M., physical chemistry
Hibben, Craig Rittenhouse, plant pathology
Kuritzkes, Alexander Mark, organic chemistry
Miller, Harold A., physical chemistry
Watkins, George Raymond, physical chemistry
Walker, Edward John, solid state physics

OSWEGO
Bishop, Charles Aldrich, ethnology, anthropology
Brown, Ronald Alan, theoretical physics, solid state physics
Burling, James P., mathematics
Cassens, Patrick, mathematical analysis
Chase, Sherret Spaulding, genetics, botany
Chermack, Eugene E A., physical meteorology
Cox, Donald David, plant science
Cutler, Richard Oscar, geography
Deming, Robert W., mathematics
Dristy, Forrest E., topology
Fisher, Edward, mathematical physics
Ghobrial, Girgis Bakhoum, geography
Gillette, Norman John, paleobotany
Gregory, Stephen Albert, astrophysics
Hammill, Terrence Michael, mycology, cytology
Harrison, John Robert, embryology
Hyde, Kenneth E., inorganic chemistry
Lackey, James Alden, mammalogy
Liebenauer, Paul (Henry), experimental nuclear physics
Lipsig, Joseph, physical chemistry
Magliveras, Spyros Simos, mathematics
Mahajan, Kishan Paul, biochemistry, organic chemistry
Marsh, Leland C., botany
Martin, Kathryn Helen, histophysiology
Maurer, Robert Eugene, chemistry
Maxwell, George Ralph, II, ecology, ornithology
Monroe, Pearle Arvel, organic chemistry, biochemistry
Moore, Richard Byron, microbiology, marine ecology

Moore, Rufus Adolphus, physics
Morgan, Thomas Edward, astrophysics
Nappi, Anthony Joseph, insect physiology, pathology
Nelson, Sigurd Oscar, Jr., arachnology
Nugent, Robert Charles, geology
O'Donnell, Raymond Thomas, analytical chemistry, physical chemistry
O'Dwyer, John J., theoretical solid state physics
Orr, Richard Clayton, mathematics
Pituga, George D., astronomy
Powers, Harold O., human genetics, cytogenetics
Shaver, Paul Merl, science education
Shineman, Richard Shubert, inorganic chemistry
Silveira, Augustine, Jr., organic chemistry
Stopher, Emmet Carson, mathematics
Van Geet, Anthony Leendert, physical chemistry, analytical chemistry
Wernick, Robert J., mathematics
Wise, Ernest George, radiation biology, microbiology

OVID
Boynton, Damon, tropical agriculture

OWEGO
Standish, Charles Junior, mathematics

OXFORD
Currie, Gustavus Noel, II, pharmacology
Halliday, Robert Parker, pharmacology

OYSTER BAY
Prytz, Bo, biochemistry

OZONE PARK
Libackyj, Anfir, physical chemistry

PAINTED POST
Bartholomew, Roger Frank, physical chemistry
Martin, Francis W., physical chemistry
Murphy, James A., physical chemistry, surface chemistry
Randels, Robert Basil, physics

PALISADES
Alsop, Leonard E., geophysics
Alvarez, Walter, volcanology, tectonics
Be, Allan Wie Hwa, geology, biology
Biscaye, Pierre Eginton, marine geochemistry
Dalziel, Ian William Drummond, structural geology
Ewing, John I., geophysics
Heezen, Bruce Charles, geology
Herron, Thomas J, geophysics
Horai, Ki-iti, geophysics, lunar science
Hunkins, Kenneth, oceanography
Langseth, Marcus G, Jr., geophysics, oceanography
Ludwig, William Jackson, marine geophysics
Nafe, John Elliott, physics, geophysics
Olson, Walter Sigfrid, petroleum geology, tectonics
Opdyke, Neil, geology, geophysics
Roels, Oswald A., biochemistry, marine biology
Ryan, William B F., oceanography
Saito, Tsunemasa, micropaleontology
Scholz, Christopher Henry, geophysics
Schweickert, Richard Alan, geology, tectonics
Sharma, Gurdial Mal, organic chemistry
Sykes, Lynn Ray, geophysics
Talwani, Manik, geophysics
Thorndike, Edward Moulton, physics
Wollin, Goesta, oceanography

PAUL SMITHS
Hamilton, Chester Eugene, organic chemistry
Rutherford, William, Jr, forestry

PEARL RIVER
Aiston, Stewart Samuel, bacteriology
Allen, George Rodger, Jr., medicinal chemistry
Anderson, George Washington, organic chemistry
Angier, Robert Bruce, organic chemistry
Ball, Edwin Lawrence, pharmaceutics

Barringer, William Charles, pharmaceutical chemistry
Bauman, Norman, immunology, medicine
Bell, Paul Hadley, biochemistry
Bernstein, Seymour, organic chemistry
Blank, Robert H, biology
Bolte, Henry Frederick, veterinary pathology
Bone, Donald Robert, organic chemistry
Boothe, James Howard, organic chemistry
Borders, Donald B, organic chemistry
Boshart, Charles Ralph, pharmacology
Brabander, Herbert Joseph, organic chemistry
Brancone, Louis Maria, chemistry
Breuer, Charles B., chemistry
Brockman, John A, Jr., immunology
Carrier, Steven Theodore, medical statistics, biostatistics
Cerni, Costantino Peter, virology, immunology
Chan, Peter Sinchun, pharmacology, biochemistry
Chau, Raymond Ying Pui, toxicology, pharmacology
Child, Ralph Grassing, medicinal chemistry
Church, Robert Fitz (Randolph), organic chemistry
Cohen, Elliot, medicinal chemistry
Cooper, Murray Sam, microbiology
Cosulich, Donna Bernice, drug metabolism
Curran, William Vincent, organic chemistry
Davis, Selby Brinker, research administration
Day, Ivana Podvalova, pharmacology, biology
Denton, John Joseph, medicinal chemistry
De Renzo, Edward Clarence, biochemistry
Diermeier, Harold Frederick, physiology
Durr, Friedrich (E), infectious diseases, therapeutics
Ellenbogen, Leon, biochemical pharmacology, hematology
Ellestad, George A., organic chemistry
Esse, Robert Carlyle, organic chemistry
Fedrick, James Love, medicinal chemistry
Fields, Thomas Lynn, organic chemistry
Foley, Dennis Joseph, pharmacology
Forbes, Martin, microbiology
Frazza, Everett Joseph, organic chemistry
Gallagher, James Daniel, medicine
Goodman, Joseph Jacob, plant pathology
Gordon, Gloria, electron microscopy
Gordon, Samuel, toxicology
Gore, William Earl, organic chemistry
Govier, William Charles, research administration
Greenblatt, Eugene Newton, pharmacology
Gura, Louis D., statistics
Halwer, Murray, physical chemistry
Hardy, Elizabeth MacGregor, organic chemistry
Heller, Milton David, medical research
Hlavka, Joseph John, organic chemistry
Hodgson, William Gordon, analytical chemistry
Houser, Vincent Paul, psychopharmacology
Howell, Mary Gertrude, information science, research administration
Izzo, Patrick Thomas, chemistry
Jackson, Benjamin, toxicology
Jarolmen, Howard, medical microbiology, genetics
Kanegis, Leon Abbott, experimental pathology, toxicology
Keirns, James Jeffery, biochemistry
Kele, Roger Alan, industrial microbiology
Kissel, John Walter, pharmacology
Kohlbrenner, Philip John, organic chemistry
Kroll, Henry Michael, bacteriology
Kushner, Samuel, pharmaceutical chemistry
Lang, Stanley Albert, Jr, organic chemistry
Latimer, Clinton Narath, neurophysiology
Liberman, Daniel Franklin, clinical microbiology, industrial microbiology
Lindsay, Harry Lee, virology
Lovell, Frederick Maurice, x-ray crystallography
Martell, Michael Joseph, Jr, industrial chemistry, pharmacy
McClintock, David K, biochemistry

McCoy, Donald W., microbiology, biochemistry
McEvoy, Francis Joseph, chemistry
McGahren, William James, organic chemistry
Meyer, Walter Edward, pharmaceutical chemistry
Morrison, John Agnew, chemistry
Murdock, Keith Chadwick, synthetic organic chemistry, medicinal chemistry
Nash, Robert Arnold, pharmaceutical chemistry, pharmacy
Noble, John F., pharmacology
Osterberg, Arnold Curtis, pharmacology
Pangides, John., biochemistry
Parham, Margaret Regina, information science
Patterson, Ernest Leonard, biochemistry
Paul, Rolf, organic chemistry
Pearl, William, physiology, pharmacology
Personeus, Gordon Rowland, organic chemistry
Phelps, Allen Spencer, bacteriology
Poorvin, David Walter, cardiovascular physiology
Porter, John Norman, mycology
Press, Jeffery Bruce, organic chemistry
Roepke, Raymond Rollin, biochemistry
Safir, Sidney Robert, medicinal chemistry
Saldarini, Ronald John, physiology, biochemistry
Sax, Karl Jolivette, chemistry
Scarpone, Anthony John, pharmacy
Schaffer, Sheldon Arthur, biological chemistry
Schroer, Richard Allen, biochemical pharmacology
Shepherd, Robert Gordon, medicinal chemistry, organic chemistry
Shull, Gilbert Malcolm, biochemistry
Shultz, Walter, biopharmaceutics
Sjolander, Newell Oscar, industrial microbiology
Sloboda, Adolph Edward, cell physiology, pharmacology
Sparano, Benjamin Michael, animal pathology, toxicology
Sutherland, George Leslie, organic chemistry
Swanzey, Eugene Harry, medicine
Sweeney, William Michael, internal medicine
Szumski, Stephen Aloysius, medical administration
Tedeschi, David Henry, pharmacology
Thiessen, Reinhardt, Jr., biochemistry
Thomson, Alexander, medicine
Tobkes, Martin, organic chemistry
Tolman, Edward Laurie, pharmacology
Tomcufcik, Andrew Stephen, organic chemistry
Tonelli, George, veterinary medicine
Traitor, Charles Eugene, pharmacology, toxicology
Tresner, Homer David, mycology
Upeslacis, Janis, pharmaceutical chemistry
Vasington, Paul John, microbiology, virology
Wagie, Gilmour Lawrence, pharmacology
Waitman, Reuben Homer, food chemistry
Warren, James Donald, medicinal chemistry, synthetic organic chemistry
Webb, John Schurr, analytical chemistry
Weiss, Martin Joseph, organic chemistry
Wilkinson, Raymond George, organic chemistry
Williams, Susan Catherine Frary, biochemistry
Wissner, Allan, organic chemistry
Woodworth, Curtis Wilmer, organic chemistry

PEEKSKILL
Pomper, Seymour, microbiology

PELHAM
Baylor, Curtis Horton, internal medicine, occupational medicine
Carlozzi, Michael, clinical pharmacology
Pough, Richard Hooper, ornithology, ecology

PELHAM MANOR
Barsa-Newton, Mary Claire, insect physiology
De Nisco, Stanley Gabriel, nutrition

PENFIELD
Chisholm, James Joseph, optics, spectroscopy
Gering, Robert Lee, zoology, science education

Kermisch, Dorian, optics, electromagnetics
Schein, Lawrence Brian, experimental solid state physics
Thompson, Robert Deane, instrumentation

PERU
Kraemer, J Hugo, forestry

PILOT KNOB
Suits, Chauncey Guy, physics

PITTSFORD
Boudakian, Max Minas, fluorine chemistry
Crossmon, Germain Charles, microscopy chemistry
Dessauer, John Hans, chemistry, chemical engineering
Hursh, John Bachman, physiology
Insalaco, Michael Anthony, organic physiology
Jensen, David Edward, mineralogy
Mott, Frederick Dodge, preventive medicine
O'Reilly, James Michael, polymer physics, polymer chemistry
Smith-Lewis, Margaret J., pharmacology
Thomas, Telfer Lawson, organic chemistry

PLAINVIEW
Baiardi, John Charles, hematology, physiology
Beckerley, James Gwavas, physics
D'Amato, Richard Frank, clinical microbiology, infectious diseases
Held, Kalman M., physical chemistry
Nerken, Albert, chemistry
Sadowski, Henry, electronics, physics

PLATTSBURGH
Barnett, Stockton Gordon, III, geology
Baum, Stuart J., inorganic chemistry, bioorganic chemistry
Blood, Charles Allen, Jr., organic chemistry
Bobka, Rudolph J., physical chemistry
Clark, Ralph M., zoology
Dawson, James Clifford, sedimentology, sedimentary petrology
Donaldson, Robert Rymal, science education, physics
Ellsworth, Robert King, plant biochemistry
Gillett, Lawrence B., geology
Harris, Charles Leon, neurobiology
Harris, Robert L., organic polymer chemistry
Hartnett, William Edward, mathematics
Klein, Harold George, vertebrate zoology
Kokoszka, Gerald Francis, physical inorganic chemistry
Krueger, William E., organic chemistry
Liu, Houng-Zung, biochemical genetics, microbial genetics
Loach, Kenneth William, analytical chemistry
McGraw, James Carmichael, zoology, parasitology
Moore, Richard Davis, biophysics
Munk, Vladimir, microbiology, biochemistry
Myer, Glenn Evans, oceanography, meteorology
Nevin, Floyd Reese, acarology
Nolan, James Robert, botany
Perkins, Harold Jackson, biochemistry
Reuter, Gerald Louis, veterinary pharmacy, agricultural pharmacy
Rheingold, Arnold L., organometallic chemistry
Riley, John Astwood, applied mathematics, operations research
Ryan, Donald F., physical chemistry
Sheats, George Frederic, physical chemistry
Stancliff, Merton Wesley, anthropology
Sudds, Richard Huyette, Jr., parasitology
Szydlik, Paul Peter, physics
Walker, Philip Caleb, biology, palynology
Waterhouse, Joseph Stallard, human anatomy, human physiology
Worrall, Winfield Scott, organic chemistry

Robinson, Albert Dean, genetics, mycology
Rosenthal, Donald, analytical chemistry, physical chemistry
Serway, Raymond A., solid state physics
Simone, Leo Daniel, plant morphology, bryology
Spencer, Armond E., mathematics
Stephens, Clarence Francis, mathematics
Stombler, Milton Philip, experimental solid state physics
Thickstun, William Russell, Jr., mathematics
Thygesen, Kenneth Helmer, experimental solid state physics
Vandiver, Bradford B., geology
Vaska, Lauri, inorganic chemistry
Zevos, Nicholas, physical chemistry
Zuman, Petr, electrochemistry

POINT LOOKOUT
Krewer, Semyon E., physics

POMONA
Misek, Bernard, pharmaceutical chemistry, cosmetic chemistry

PORT CHESTER
Beam, John E., food science, dairy science
Halsey, John Joseph, chemistry, operations research
Kachikian, Rouben, microbiology
Miller, Kenneth Sielke, mathematics

PORT JEFFERSON
Bateman, John Laurens, radiobiology, oncology
Haworth, Leland John, physics
Wheeler, George William, experimental physics

PORT JEFFERSON STATION
Reichenthal, Jules, medicinal chemistry

PORT WASHINGTON
Braids, Olin Capron, soil chemistry, water chemistry
Luttrell, George Howard, analytical chemistry
Miller, David W., groundwater geology
Osborg, Hans, physical chemistry
Rowe, Irving, physics, electronics
Schneider, Frank L., analytical chemistry

POTSDAM
Anderson, Elmer E., solid state physics
Arais, Sigurds, solid state physics
Biegen, Joseph Robert, surface physics
Brunauer, Stephen, colloid chemistry, surface chemistry
Campbell, Bruce (Nelson), Jr., biochemistry, organic chemistry
Carl, James Dudley, geochemistry
Cerwonka, Robert Henry, marine ecology
Chin, Der-Tau, electrochemistry, chemical engineering
Cooke, Derry Douglas, physical chemistry
Czanderna, Alvin Warren, surface physics, surface chemistry
DeGhett, Victor John, ethology
Dubinsky, Edward Leonard, mathematics
Foisy, Hector B., mathematics
Frederic, Paul Burgess, geography
Goodrich, Frank Chauncey, surface chemistry
Gulick, Luther Halsey, Jr., geography
Hafer, Paul Egan, ecology, science education
Helbig, Herbert Frederick, atomic physics
Horst, G Roy, zoology, histology
Hughes, Harold Kenneth, physics, cybernetics
Isenberg, George Raymond, Jr., biology
Jekeli, Walter, physics
Jellinek, Hans Helmut Gunter, physical chemistry
Jerri, Abdul J., applied mathematics
Jones, George Lett, Jr., physical chemistry
Jordan, David M., organic chemistry
Juo, Pei-Show, virology, biochemistry
Kerker, Milton, physical chemistry
Kratohvil, Josip, colloid chemistry
LaHaye, Philip Arthur, botany
Lovass-Nagy, Victor, applied mathematics
Martin, Martin Claude, experimental solid state physics
Matijevic, Egon, physical chemistry, colloid chemistry
McNulty, Peter J., biophysics, high energy physics
Meites, Louis, physical chemistry, analytical chemistry
Merritt, Paul Eugene, analytical chemistry
Moore, Frank Leslie, Jr., physics
O'Brien, Neal Ray, geology
Partch, Richard Earl, organic chemistry, medicinal chemistry
Partenheimer, Walter, inorganic chemistry
Plane, Robert Allen, bioinorganic chemistry, spectroscopy
Popp, Frank Donald, organic chemistry
Powers, David Leusch, applied mathematics
Rabson, Gustave, mathematics
Revetta, Frank Alexander, geophysics

POUGHKEEPSIE
Albers, Henry, astronomy
Asprey, Winifred Alice, mathematics, computer sciences
Ballard, Robert Wilson, medicine
Beck, Curt Werner, organic chemistry
Bettencourt, Joseph S, Jr., parasitology
Blaker, John Warren, optics
Braun, Robert Denton, analytical chemistry
Cecil, Thomas E., geometry
Chafetz, Harry, organic chemistry
Clark, Patricia Ann, physical chemistry
Danziger, Lawrence, statistics
Fleisher, Harold, physics, mathematics
Gounaris, Anne Demetra, biochemistry
Herwitz, Paul Stanley, mathematics
Hillis, Mary Olive, analytical chemistry
Hooper, George Bates, biology
Hopper, Steven Phillip, organometallic chemistry
Htoo, Maung Shwe, physical chemistry
Johnsen, John Herbert, geology
Johnson, Corinne Lessig, microbiology, biochemistry
Johnson, Patricia R., mammalian physiology, nutrition
Lang, William Warner, acoustics
LaPietra, Richard Andrew, physical chemistry, thermodynamics
Levine, Solomon Leon, analytical chemistry
Lewis, Paul Herbert, physical chemistry
Link, Vernon Bennett, public health
Linner, Edward Robert, physical chemistry
Lumb, Ethel Sue, embryology, cytology
Maissel, Leon I, physics, computer sciences
Maling, George Croswell, Jr., physics
Masters, Burton Joseph, solid state science
Menapace, Lawrence William, organic chemistry
Merriell, David McCray, mathematics
Michelson, Malvin J., organic chemistry
Mikhail, Wadie F., statistics
Mittal, Kashmiri Lal, physical chemistry
Monaco, Lawrence Henry, zoology
Mooney, Patricia May, solid state physics
Mucci, Joseph Francis, physical chemistry
Muschio, Henry M., human genetics
Mutter, Walter Edward, physics
Newton, Abba Verbeck, geometry
Pierce, Madelene Evans, ecology
Polivka, Raymond Peter, mathematics
Ranzoni, Francis Verne, mycology
Rebhuhn, Deborah, applied mathematics
Regnier, Jerome, geology
Rehwoldt, Robert E., analytical chemistry
Schmeckenbecher, Arnold F., inorganic chemistry
Spicer, Donald Z., mathematics
Stearns, Robert L., physics
Tavel, Morton, theoretical physics
Tokay, Elbert, biology
Tremelling, Michael, Jr., physical organic chemistry
Turley, Hugh Patrick, cell physiology
Wajda, Edward Stanley, physics
Warthin, Aldred Scott, Jr., paleontology
White, William Wallace, operations research
Williams, Donald Benjamin, biology
Wright, Margaret Ruth, zoology
Zorzoli, Anita, biochemistry, physiology

POUND RIDGE
McMahon, Rita Mary, cytology

PURCHASE
Alscher, Ruth Paula, cell physiology

NEW YORK

Applewhite, Philip Boatman, behavioral physiology
Davis, Paul Lawrence, mathematics
Ehrman, Lee, population genetics
Foner, Nancy, anthropology
Hotchkiss, Frederick Hatfield Clark, invertebrate zoology, paleontology
Hyde, Robert Wallace, organic chemistry, food chemistry
Morehouse, Sheila McInness, inorganic chemistry
Pavlos, John, organic chemistry
Tenenbaum, Joel, dynamic meteorology
Williams, Curtis Alvin, Jr, biochemistry
Wolsky, Maria de Issekutz, cell biology

QUEENS VILLAGE
Goldschmidt, Leontine, psychobiology
Korenyi, Charles, psychiatry, dermatology
Orr, Alfonso, neurophysiology, experimental neurology
Squire, Richard Douglas, genetics, radiation biology
Whittier, John Rensselaer, neurology, psychiatry

RANSOMVILLE
Jones, Ralph, Jr, internal medicine

RAVENA
Hager, Richard Arnold, plant pathology, microbiology

RENSSELAER
Ackerman, James Howard, medicinal chemistry
Albertson, Noel Frederick, medicinal chemistry
Albro, Lewis Pearson, analytical chemistry
Alexander, Ernest John, organic chemistry
Bailey, Dennis Mahlon, organic chemistry
Bell, Malcolm Rice, organic chemistry
Benson, Robert Franklin, organic chemistry
Beyler, Arthur Lewis, endocrinology
Billmeyer, Fred Wallace, Jr, chemistry
Blackmore, William Peter, pharmacology
Borisenok, Walter A, pharmacy
Bradford, James Carrow, gastroenterology
Cafruny, Edward Joseph, pharmacology
Came, Paul E, virology, oncology
Carabateas, Philip M, organic chemistry, medicinal chemistry
Casler, David Robert, pharmaceutics
Christianson, Robert George, organic chemistry
Clarke, Robert La Grone, pharmaceutical chemistry
Clemans, Stephen D, structural chemistry
Collins, Joseph Charles, Jr, organic chemistry
Creange, John Ellyson, endocrinology
Cutler, Royal Anzly, Jr, organic chemistry
Daum, Sol Jacob, organic chemistry, medicinal chemistry
Davison, Clarke, drug metabolism
Dickinson, William Borden, medicinal chemistry
Donikian, Marc Roupen, pharmacology
Drobeck, Hans Peter, toxicology, experimental pathology
Dupont, Paul Emile, organic chemistry
Edelson, Jerome, biological chemistry, drug metabolism
Farah, Alfred Emil, medical research, pharmacology
Ferrari, Richard Alan, biochemistry
Fleischer, Thomas B, paper chemistry
Freele, Hugh W, parasitology, entomology
Goss, William Albert, microbiology
Harding, Homer Robert, endocrinology
Johnson, Robert Ed, medicinal chemistry
Johnson, Thomas Lynn, organic chemistry
Kelly, Clark Andrew, analytical chemistry, pharmaceutical chemistry
Kirchner, Frederick Karl, organic chemistry, information science
Kuhng, Rudolph K, physical organic chemistry
Lesher, George Yohe, pharmaceutical chemistry, synthetic organic chemistry
Lorenz, Roman R, organic chemistry
Lubitz, Betty Baum, analytical chemistry, organic chemistry

chemistry
Martini, Catherine Marie, physical chemistry
Michaud, Ronald Normand, microbiology
Miller, Theodore Charles, pharmaceutical chemistry, research administration
Minatoya, Hiroaki, pharmacology, physiology
Nachod, Frederick Constantine, physical chemistry
Paikoff, Myron, pharmaceutical chemistry
Pazienza, Joseph Peter, pharmacy
Plue, Arnold Frederick, organic chemistry
Portmann, Glenn Arthur, pharmaceutics
Rosenberg, Franklin J, pharmacology
Rossi, David, biology, organic chemistry
Shepardson, John U, analytical chemistry
Shriver, Ellsworth Harold, research administration
Siggins, James Ernest, medicinal chemistry
Skulan, Thomas William, pharmacology, physiology
Steinberg, Bernard Albert, microbiology
Terminiello, Louis, enzymology
Tkacheff, Joseph, Jr, pharmacy
Walker, Bernard Forestier, physical chemistry
Wallace, William Eldred, chemistry
Webb, William Gatewood, medicinal chemistry
Wentland, Mark Philip, organic chemistry
Wood, David, analytical chemistry
Wyatt, Benjamin Woodrow, chemistry
Yarinsky, Allen, parasitology, chemotherapy
Zalay, Ethel Suzanne, organic chemistry
Zenitz, Bernard Leon, medicinal chemistry

RENSSELAERVILLE
Daigleish, Robert Campbell, systematic entomology

REXFORD
White, Donald Robertson, optical physics

RIDGE
Zahnd, Hugo, biochemistry

RIVERDALE
Caso, Marguerite Miriam, chemistry
Evans, Lance Saylor, plant physiology, histochemistry
Fask, Alan S, statistics, operations research
Iannuzzi, Melanie Mary, physical inorganic chemistry
Kronstein, Max, polymer chemistry
Zielinski, Theresa Julia, theoretical chemistry

RIVERHEAD
Bing, Arthur, plant physiology
Cetas, Robert Charles, plant pathology
Selleck, George Wilbur, plant ecology, weed science
Semel, Maurie, entomology

ROCHESTER
Abbott, John Richards, organic chemistry
Abood, Leo George, biochemistry
Adduci, Jerry M, organic chemistry
Adelstein, Peter Z, physical chemistry
Adin, Anthony, physical inorganic chemistry
Ahrenkiel, Richard K, solid state physics
Albrecht, Frederick Xavier, physical organic chemistry
Aldridge, William Gordon, anatomy
Allen, Charles Francis Hitchcock, organic chemistry
Allen, Gary William, physical organic chemistry
Allen, Lewis Edwin, organic chemistry
Allen, Peter Zachary, immunology
Allentoff, Norman, photographic chemistry
Alling, Norman Larrabee, mathematics
Aitland, Henry Wolf, physics
Altman, Joseph Henry, physics
Altman, Kurt Ison, biochemistry, radiobiology
Ambrose, Robert T, analytical chemistry, environmental chemistry
Ames, Stanley Richard, nutritional biochemistry, animal nutrition
Amsel, Lewis Paul, pharmaceutical chemistry

Anagnostopoulos, Constantine N, solid state electronics
Anderson, Albert Edward, organic chemistry
Anderson, Donald Grigg, medicine
Anderson, Donald Hervin, analytical chemistry
Arces, Joseph A, organic chemistry, polymer chemistry
Archie, William C, Jr, physical organic chemistry
Armour, Eugene Arthur, organic chemistry, photographic chemistry
Armstrong, Clay M, physiology, biophysics
Astill, Bernard Douglas, biochemistry
Atkins, Robert W, psychoanalysis, psychiatry
Atwater, Edward Congdon, internal medicine, history of medicine
Augustine, (Small), Marie, chemistry
Babineau, G Raymond, psychiatry
Bacon, Robert Elwin, photographic science
Baden, Harry Christian, analytical chemical physics
Baetzold, Roger C, physical chemistry
Bagchi, Pranab, colloid chemistry, surface chemistry
Balduzzi, Piero, microbiology, virology
Bale, William Freer, biophysics
Bannister, Thomas Turpin, biology
Barber, Eugene Douglas, biochemistry, clinical chemistry
Bard, Charleton Cordery, photographic chemistry
Barkey, Kenneth Thomas, organic chemistry
Barr, Charles (Robert), photographic chemistry
Bartlett, James Williams, Jr, psychiatry
Bartlett, John Richard, neuroscience
Bass, Jon Dolf, photographic chemistry
Bassett, William Akers, mineralogy, geophysics
Bauer, Eldon Eugene, physical chemistry, chemical engineering
Baum, John, rheumatology, immunology
Baum, Martin David, photographic chemistry
Baumeister, Philip Werner, physics
Beavers, Dorothy (Anne) Johnson, chemistry
Becker, Richard William, photographic chemistry
Beavers, Leo Earice, organic chemistry
Belly, Robert T, microbial ecology
Bennett, John M, internal medicine, hematology
Bent, Richard Lincoln, organic chemistry
Berg, George G, toxicology
Berg, Olga Aronowitz, cytochemistry
Berg, Richard Allen, endocrinology, environmental sciences
Bernhard, William Allen, biophysics
Beyer, George Leidy, polymer chemistry
Bickmore, John Tarry, applied physics, engineering physics
Bigelsen, Jacob, physical chemistry
Bigler, William Norman, biochemistry
Bignall, Keith E, neurophysiology
Bishop, Charles Anthony, organic chemistry
Black, Donald Lee, physical chemistry, electron microscopy
Blinn, H Marshall, nuclear chemistry
Blum, Peter, algebra, geometry
Bogdanowicz, Mitchell Joseph, materials science
Bohrod, Milton George, pathology, clinical pathology
Booms, Robert Edward, photographic chemistry
Borgstedt, Harold Heinrich, pharmacology, toxicology
Borsenberger, Paul Michael, materials science
Bosmann, Harold Bruce, biophysics, pharmacology
Botger, Gary Lee, physical chemistry
Bouyoucos, John Vinton, applied physics
Bowen, Zeddie Paul, geology, paleontology
Bowie, Edward John Walter, hematology
Boyd, Eleanor H, neurophysiology, neuropharmacology
Boyno, John S, photographic chemistry
Boyd, Eugene Stanley, pharmacology
Bradford, William L, pediatrics, bacteriology
Bramlet, Roland C, radiological physics
Brand, John S, biophysics
Brandriss, Michael W, immunology, infectious diseases
Brault, Albert Thomas, physical inorganic chemistry

Brayer, Franklin T, medicine
Breckenridge, Robert T, internal medicine, hematology
Breneman, Edwin Jay, psychophysics
Brewer, Allen A, dentistry
Brody, Bernard B, medicine, chemistry
Breneman, Eric Richard, analytical chemistry
Brown, James, speech pathology
Brown, James Wallace, III, photographic chemistry
Brust, David Philip, photographic chemistry
Burnham, Dwight Comber, solid state physics, information science
Buff, Frank Paul, physical chemistry
Bunch, Phillip Carter, medical physics
Burgener, Francis Andre, radiology
Burgmaier, George John, organic chemistry, photography
Burkey, Bruce Curtiss, solid state physics
Burness, Donald Mac Arthur, organic chemistry
Brown, Jerram L, ethology
Brown, John Lott, psychophysiology
Brumley, Corwin Hoyt, electronics, optics
Cappel, Carl Robert, organic chemistry, photographic chemistry
Capretta, Umberto, analytical chemistry
Catto, Peter James, plasma physics
Chaffee, Eleanor, physical inorganic chemistry
Cardillo, Thomas E, medicine
Carroll, Burt Harrng, chemistry
Casaretti, George William, pathology
Caspari, Ernst Wolfgang, developmental genetics, behavioral genetics
Castner, Theodore Grant, Jr, solid state
Cathcart, John Almon, organic chemistry
Campbell, Gerald Allan, organic polymer chemistry
Chamberlain, Phyllis Ione, inorganic chemistry
Chang, Jack Che-Man, analytical chemistry
Chapas, Richard Bernard, photographic chemistry
Chapman, Derek D, organic chemistry
Chen, Chin Hsin, synthetic organic chemistry
Chen, Chung Wei, applied statistics
Chen, Inan, solid state physics
Chen, Stephen P K, chemistry
Chen, Tsang Jan, polymer chemistry
Childers, Robert Lee, photographic chemistry
Cleare, Henry Murray, physics
Clemens, Carl Frederick, chemistry
Cline, Douglas, nuclear physics
Ching, Melvin Chung Hing, anatomy
Choi, Byung Ho, pathology
Christensen, James Roger, virology
Ciccarelli, Roger N, polymer chemistry
Clark, Alfred, Jr, applied mathematics
Clark, Patricia Ann Andre, astrophysics
Clark, Robert A, organic chemistry
Clark, Roger William, molecular biology, genetics
Clark, Walter, chemistry
Clarke, John Ross, solid state electronics
Clarkson, Thomas William, toxicology
Cockett, Abraham Timothy K, urology, physiology
Cohen, Hyman L, organic chemistry
Cohen, Jacob Isaac, photographic chemistry
Cohen, Jules, internal medicine, cardiology
Cohen, Julius Jay, physiology
Cohen, Nicholas, immunology
Coble, Roger M, inorganic chemistry, botany
Cole, David Le Roy, physical chemistry, chemical kinetics
Cole, Russell Cleven, operations research
Coll, Hans, chemistry
Coleman, Babette Brown, botany
Coleman, James R, cell physiology
Coleman, Paul David, neurobiology
Collier, Susan S, photochemistry
Condit, Paul Brainard, organic chemistry
Conners, Gary Hamilton, applied physics
Connolly, Lewis Timothy, mechanics
Conwell, Esther Marly, solid state physics
Coombs, William, Jr, biomedical engineering

Cooper, Robert Arthur, Jr., medicine, oncology
Cooper, Walter, physical chemistry
Costa, Lorenzo F., electronic spectroscopy
Craig, Albert Burchfield, Jr., physiology, medicine
Crane, Edward Mastin, photographic science
Craven, Robert Lee, organic chemistry
Cravitz, Leo, bacteriology, immunology
Cunningham, Michael Paul, photochemistry, photographic chemistry
Cunningham, Robert Gail, surface physics, solid state physics
Cunov, Carl Henry, medicinal chemistry, organic chemistry
Curme, Henry Garrett, physical chemistry
Cupery, Kenneth N., optics
Dahneke, Barton Eugene, environmental health, particle physics
Dallon, Dale Sherman, research administration
Dalton, James Christopher, organic chemistry, photochemistry
Damouth, David Earl, physics
Daniel, Daniel S., physical chemistry, organic chemistry
Dann, John Robert, organic chemistry
Dappen, Glen Marshall, organic chemistry
Darlak, Robert, organic chemistry
Davidson, Theodore, materials science, physical chemistry
Davis, Marvin Lester, synthetic organic chemistry
Dearing, William Hill, internal medicine
Deboer, Charles D., photochemistry
Dedinas, Jonas, physical chemistry, photochemistry
Delano, Erwin, optics
del Cerro, Manuel (Perez), neurosciences, developmental biology
de Mauriac, Richard Arthur, organic chemistry
DePalma, James John, optics, mathematics
De Selms, Roy Charles, organic chemistry
DeVries, Ralph Milton, nuclear physics
DeWeese, James A., surgery
Dexter, David Lawrence, theoretical solid state physics
Dickerson, Dorsey Glenn, photographic chemistry
Dieterich, David Allan, photographic science
Douglass, Pritchard Calkins, analytical chemistry
Dounce, Alexander Latham, biochemistry
Drews, Reinhold Eldor, physics
Drexhage, Karl Heinz, physical chemistry, polymer chemistry
Duke, Charles Bryan, theoretical solid state physics, research administration
Dulmage, William James, physical chemistry
Dunham, Kenneth Royal, applied chemistry
Dunn, J Stanley, inorganic chemistry
Dutton, David B., physics
Dykstra, Thomas Karl, organic chemistry, polymer chemistry
Eachus, Raymond Stanley, physical inorganic chemistry
Eberle, Helen I, molecular biology, biophysics
Eberlein, William Frederick, mathematics
Eberly, Joseph Henry, theoretical physics
Eddy, Hubert Allen, radiobiology, pathology
Edgerton, Richard Oliver, organic chemistry
Edgerton, Robert Flint, chemistry
Edsberg, Robert Leslie, analytical chemistry
Ehrlich, Sanford Howard, chemical physics
Eikenberry, Jon Nathan, physical organic chemistry
Eisenberg, Richard, inorganic chemistry
Elder, Fred Kingsley, Jr., physics

Elins, Herbert Samuel, photographic chemistry
Elwood, James Kenneth, organic chemistry
Emch, Gerard G., mathematical physics
Emerson, Ernest Benjamin, Jr., surgery
Emmel, Victor Meyer, histology
Engebrecht, Ronald Henry, photochemistry
Engel, George Libman, medicine, psychiatry
Erdmann, Duane John, photographic chemistry
Erickson, Wayne Francis, organic chemistry
Esders, Theodore Walter, biochemistry
Evans, Hugh Lloyd
Ewing, Joan Rose, electrophysics
Fabish, Thomas John, solid state physics, physical chemistry
Faloon, William Wassell, gastroenterology
Farley, Eugene Shedden, Jr., medicine, public health
Farran, Charles Frederick, physical chemistry, photography
Faul, William Henry, organic chemistry, photographic chemistry
Feather, David Hoover, materials science
Feldman, Isaac, chemistry, biophysics
Feldman, Larry Howard, photographic chemistry
Feldman, William, physics, research administration
Ferbel, Thomas, particle physics
Ferin, Juraj, environmental medicine, inhalation toxicology
Fernandez, Jose Martin, organic chemistry
Fields, Alfred E., polymer physics
Fields, Donald Lee, organic chemistry
Figueras, John, organic chemistry
Figueras, Patricia Ann McVeigh, organic chemistry
Fischer, Harry William, radiology
Fischer, Leewellyn C, photographic chemistry
Fix, Delbert Dale, photographic chemistry
Flannery, John B, Jr., physical chemistry
Fleming, James Charles, photographic chemistry
Foos, Barbara Ann, mathematics
Forbes, Gilbert Burnett, pediatrics
Ford, John Albert, Jr, organic chemistry
Forest, Edward, physical chemistry
Forrest, John Wilson, optics
Forsyth, James M., optics
Fowler, Edward Herbert, pathology
Fowler, William Frank, Jr., chemistry
Fox, Charles Junius, organic chemistry
Frank, David Stanley, biochemistry
Frazer, John P., otolaryngology
Freeman, John Paul, photographic chemistry
Freeman, Richard B., internal medicine, nephrology
Freimer, Marshall Leonard, mathematics
Frickey, Paul Henry, microbiology, virology
Friedrich, Louis Elbert, organic chemistry
Fritz, Garold Frederic, solid state physics
Froix, Michael Francis, physical chemistry, polymer physics
Fulbright, Harry Wilks, nuclear physics
Gallo, Charles Francis, Jr., engineering physics
Gans, Roger Frederick, fluid mechanics, physics
Garay, Gustav John, entomology, geophysics
Garber, Calvin Samuel, physical chemistry
Garber, John Warburton, Jr., organic chemistry
Gardner, Sylvia Alice, organometallic chemistry
Gardner, William Lee, physical chemistry, inorganic chemistry
Garfield, Lawrence James, physical chemistry
Gates, Allen Hazen, Jr., reproductive biology, genetics
Gates, John Warburton, Jr., organic chemistry
Gates, Marshall DeMotte, Jr., organic chemistry
Gaudioso, Stephen Lawrence, physical chemistry
Geddes, Amos Leslie, physical chemistry
Geisler, Grace, analytical chemistry, inorganic chemistry
George, Thomas Frederick, theoretical chemistry
Gerace, Paul Louis, inorganic chemistry, electrophotography
Gerard, Jesse Thomas, analytical chemistry, nuclear chemistry

Giannini, Donald Dominic, nuclear magnetic resonance
Gibbs, Finley P., physiology, anatomy
Giering, John Edgar, pharmacology, physiology
Gill, James Edward, cytochemistry
Gillies, Alastair J., anesthesiology, pharmacology
Gilman, Paul Brewster, Jr., physical chemistry
Gilman, Robert Edward, organic chemistry
Gilmour, Hugh Stewart Allen, physical chemistry, photography
Gilmour, Marion Nyholm H., microbiology, dental research
Givens, Miles Parker, optics
Gledhill, Ronald James, chemistry, information science
Glegg, Ronald Edward, chemistry
Gluck, Ronald Monroe, solid state chemistry, solid state electronics
Goddard, Murray Cowdery, color science
Goh, Kong-Oo, hematology, cytogenetics
Gohlke, Roland Schulz, mass spectrometry
Goldman, Leonard Manuel, plasma physics
Goldstein, Louis Arnold, surgery
Goldstein, Paul, computer science
Goodhue, Charles Thomas, biochemistry, microbiology
Goodrich, D Wells, psychiatry
Gore, Ira, pathology
Gorovsky, Martin A, biology, cell biology
Gosling, John William, organic chemistry, photography
Gove, Harry Edmund, physics
Gramiak, Raymond, medicine, radiology
Granberry, Julian, anthropology
Greene, William Allan, psychosomatic medicine
Groner, Carl Fred, electrophotography
Gross, Paul Randolph, developmental biology, molecular biology
Gross, Susan C, polymer chemistry
Grota, Lee J, neuroendocrinology
Grover, Paul L, Jr., medical education, communication science
Gruenbaum, William Tod, synthetic organic chemistry
Grum, Franc, optics, photometry
Guevara, Alfredo Ruben, photographic chemistry, photography
Gundersen, Kare, internal medicine
Gundersen, Norman Gustav, mathematics
Gunter, Karlene Klages, biophysics, radiological physics
Gunter, Thomas E, Jr, solid state physics, biophysics
Gunther, Wolfgang Hans Heinrich, organic chemistry
Gysling, Henry J, inorganic chemistry, organometallic chemistry
Haase, Jan Raymond, photographic chemistry
Hadidian, Zareh, pharmacology
Hagen, Carl Richard, physics
Haines, Charles Wills, applied mathematics
Haist, Grant Milford, chemistry
Hall, Robert Burnett, Jr., economic geography, geography of Asia
Hall, William Jackson, mathematical statistics
Hamb, Fredrick Lynn, organic polymer chemistry
Hamblen, David Philip, chemical physics, biophysics
Hamilton, John Frederick, solid state physics
Hamilton, Lewis R., organic chemistry
Hanshaw, James Barry, pediatrics, virology
Hanson, Wesley Turnell, Jr., photographic chemistry
Harden, Philip Howard, entomology
Hare, John Donald, virology
Harnish, Donald Philip, organic chemistry
Harrison, Howard N., physiology
Hattman, Stanley, microbiology, virology
Hawkins, Gilbert Allan, solid state physics
Hawks, George H, III, organic chemistry
Hayles, William Joseph, physical chemistry, inorganic chemistry
Haynie, William Howard, physics
Hays, Dan Andrew, solid state physics
Hayward, John Standish, solid state electronics
Head, William Francis, Jr, pharmaceutical chemistry, analytical chemistry
Heggeness, Franklin, physiology

Heininger, Clarence George, Jr., physical chemistry
Heifer, Herman Lawrence, astrophysics
Hellmann, Richard Jason, organic polymer chemistry
Hempelmann, Louis Henry, Jr., radiology
Hempfling, Walter Pahl, microbiology
Hendess, Raymond William, photographic chemistry
Henion, Richard S., organic chemistry
Henrichs, Paul Mark, physical organic chemistry, structural chemistry
Hensler, Joseph Raymond, chemistry
Henzel, Richard Paul, organic chemistry
Herdle, Lloyd Emerson, cellulose chemistry
Herkstroeter, William G., photochemistry
Herting, David Clair, nutritional biochemistry
Herz, Arthur H, surface chemistry
Hickman, Kenneth Claude Devereux, water chemistry
Hicks, Harry Frank, Jr., physics
Higgins, George Clinton, physics
Hilborn, David Alan, biochemistry
Hilf, Russell, biochemistry
Hill, Cliff Otis, physical chemistry, photochemistry
Hill, Doyle Eugene, biochemistry
Hinkle, David Currier, biochemistry
Hinshaw, J Raymond, surgery
Hinshaw, Jerald Clyde, organic chemistry
Hoffmeister, Elaine Helen, organic chemistry, photochemistry
Hollander, Joshua, neurology, biochemistry
Holler, Jacob William, medicine
Holtfreter, Johannes Friedrich Karl, embryology
Holtz, Carl Frederick, photographic chemistry
Honig, Carl Robert, physiology
Hopkins, Betty Jo Henderson, radiation biology, experimental embryology
Hopkins, Robert Earl, physics
Horak, Jerry Robert, electronics, physics
Hormats, Ellis Irving, electron microscopy, applied chemistry
Horowicz, Paul, physiology, biophysics
Hoss, Wayne Paul, neurochemistry
Houde, Robert A, communications science
Howe, Dennis George, optics
Hoyen, Harry Alexander, Jr., solid state chemistry, surface chemistry
Huizenga, John Robert, nuclear physics
Hunter, Lloyd Philip, physics
Hyde, Richard Witherington, respiratory physiology
Iker, Howard Paul, psychosomatic medicine
Inglis, James, statistics, mathematical statistics
Irani, N F, physical chemistry
Isaacson, Henry Verschay, organic polymer chemistry
Isett, Lawrence C, experimental solid state physics, surface physics
Istock, Conrad Alan, animal ecology
Jackson, James Edward, mathematical statistics
Jacobsen, Edward Hastings, physics, electron optics
Jacox, Ralph Franklin, internal medicine
Jaenike, John Robert, medicine
James, Thomas Howard, photographic chemistry
Jarvis, James Gordon, physics
Jenkins, Philip Winder, organic chemistry
Johansen, Erling, dental research, oral pathology
Johnson, Arthur Lee, chemistry
Johnson, Hollister, Jr., chemistry
Johnson, Myrle F., physical chemistry
Jones, Evan Thomas, organic chemistry
Jones, Jean Elmore, organic chemistry
Joynt, Robert James, neurology
Julian, Donald Benjamin, photographic chemistry, chemical microscopy
Kalenda, Norman Wayne, organic chemistry
Kamm, Donald E, nephrology, physiology
Kampmeier, Jack A, organic chemistry
Kapecki, Jon Alfred, physical chemistry, x-ray crystallography
Kaplan, Mark Steven, photographic chemistry
Karlson, Richard Warren, photographic chemistry
Kauffmann, Alvern Walter, mathematics
Kaye, Jerome Sidney, cell biology
Keevert, John Edward, Jr, photographic chemistry
Kellson, Julian, mathematical statistics, operations research

Kellogg, Lillian Marie, solid state chemistry
Kelly, Thomas Michael, physics, solid state science
Kemperman, Johannes Henricus Bernardus, mathematics
Kende, Andrew S., organic chemistry
Kenkre, Vasudev Mangesh, theoretical solid state physics, statistical mechanics
Kennard, Kenneth Clayton, chemistry
Kent, Frank William, organic chemistry
Kestner, Melvin Michael, organic chemistry
Ketchum, Donald Frederick, analytical chemistry
Khanna, Ravi, polymer chemistry, chemical engineering
Khosla, Rajinder Paul, solid state physics
Klipper, Robert William, biomathematics
Kimberg, Daniel Victor, internal medicine, gastroenterology
Kimmich, George Arthur, biochemistry
King, James Reber, chemistry
Kingslake, Rudolf, optics
Kingsley, Harry Durwood, surgery
Kinsella, John J., physics
Kissel, Thomas Robert, analytical chemistry
Klanderman, Bruce Holmes, organic chemistry
Klein, Gerald Wayne, polymer chemistry
Klemperer, Martin R., hematology, oncology
Klijanowicz, James Edward, organic chemistry
Klingensmith, Merle Joseph, plant physiology
Klipstein, Frederick August, medicine
Klose, Thomas Richard, chemistry
Knechel, William Franklin, organic chemistry
Knigge, Karl Max, neuroendocrinology
Knoll, Henry Albert, vision
Knox, Robert Seiple, solid state physics, biophysics
Koeng, Fred R., organic chemistry
Kofron, James Thomas, Jr., photographic chemistry
Koller, James Edward, photographic chemistry
Koltun, Daniel S., theoretical nuclear physics
Krelick, Robert W., physical chemistry
Kriss, Michael Allen, physics, photography
Krueger, Spencer M., organic chemistry
Krugh, Thomas Richard, biophysical chemistry
Kuan, Teh Soong, chemistry, physics
Kuipers, George Albertus, physics
Kurz, Clark N., optics
Kurz, Richard Karl, photographic
Lasagna, Louis (Cesare), pharmacology
Laties, Victor Gregory, psychopharmacology
LaCelle, Paul (Louis), hematology,
Ladd, John Herbert, physical chemistry, physics
Lamberts, Robert L., physics
Lambeth, David N., magnetism
Landry, Edward F., molecular genetics
Lange, Christopher Stephen, radiobiology, biophysics
Langworthy, Harold Frederick, mathematics, optics
Lapham, Lowell Winship, neuropathology
Large, Robert F., analytical chemistry, photographic science
Lau, Philip T S., organic chemistry
Lavine, James Philip, theoretical physics
Lavine, Richard Bengt, mathematical analysis
Law, Paul Arthur, photographic chemistry
Lawrence, Christopher William, genetics, radiobiology
Lawrence, Robert Marshall, anesthesiology
Lawton, William Harvey, applied statistics
Leach, Leonard Joseph, toxicology, environmental health
Leaders, Floyd Edwin, Jr., pharmacology
LeBlanc, Jerald Thomas, photographic science, organic chemistry
Leddy, John Plunkett, immunology
Leder, Lewis Beebe, experimental physics
Lee, Joseph Chuen Kwun, pathology, cell biology
Lee, Teh-Hsuang, solid state physics
Leone, Ronald Edmund, organic chemistry
Leubner, Gerhard Walter, organic chemistry

Leubner, Ingo Herwig, physical chemistry
Levinson, Steven R., analytical chemistry, photography
Liang, Kai, physical chemistry
Lichtman, Marshall A., hematology, biophysics
Liebman, Alan Joel, physics, electrophotography
Liebman, Susan Weiss, molecular genetics
Lin, Shu-Ren, radiology
Lincoln, Lewis Lauren, photographic chemistry
Lindholm, Robert D., physical chemistry
Lindsay, Jacque K., organic chemistry
Lobkowicz, Frederick, elementary particle physics
Lochart, Haines Boots, Jr., environmental chemistry, biochemistry
Lok, Roger, organic chemistry
Long, Michael Edgar, physical chemistry
Looker, Jerome J., organic chemistry
Loose, David Lawrence, solid state physics, semiconductors
Losee, Fred Lester, biochemistry
Lovecchio, Frank Vito, analytical chemistry
Lovecchio, Karen K., electrochemistry
Lowry, Karl, neurophysiology
Lubbers, Gerrit, solid state electronics
Lubin, Moshe J., plasma physics, aerodynamics
Luckey, George William, physical chemistry
Lum, Kin K., photographic chemistry
Lundgren, Lawrence William, Jr., environmental geology
Lurie, Arnold Paul, organic chemistry
Lyman, William Chester, Jr., organic chemistry
Lynch, Eugene Joseph Michael, physics
MacAdam, David Lewis, physics
Machele, Delwyn Earl, organic chemistry
MacWilliam, Edgar Alexander, physical chemistry
Maier, Thomas O., inorganic chemistry
Maille, Hugh David, radiology, health physics
Malan, Rodwick Lapur, organic chemistry
Maniloff, Jack, biophysics, microbiology
Manning, James Arthur, pediatrics, pediatric cardiology
Marchand, Erich Watkinson, optics
Marchetti, Alfred Paul, physical chemistry
Marinetti, Guido V., biochemistry
Marquis, Robert E., microbial physiology
Martellock, Arthur Carl, polymer chemistry, organic chemistry
Martinez, Alberto Magin, organic chemistry, photography
Mason, Max Garrett, surface physics
Massa, Dennis Jon, physical chemistry, polymer physics
Masurekar, Prakash Sharatchandra, industrial microbiology, biochemical engineering
Matulic, Ljubomir Francisco, quantum optics
Mauer, Paul Bernard, optical physics
Mavis, Richard David, biochemistry
May, John Elliott, Jr., physics
May, John Walter, surface physics, surface chemistry
McCormack, Grace, microbiology
McCormack, Robert Morris, plastic surgery
McCusker, Jane, epidemiology
McDugle, Woodrow Gordon, Jr., inorganic chemistry, physical chemistry
McKelvey, John Murray, theoretical chemistry, organic polymer chemistry
McLean, Donald Francis, organic chemistry, photographic chemistry
McNally, James Green, Jr., organic chemistry
Meisler, Arnold Irwin, oncology
Melissinos, Adrian Constantin, particle physics
Mench, John Warren, organic chemistry
Mendel, John Richard, polymer chemistry
Mendel, Maryann Madeliene, photographic chemistry, synthetic organic chemistry
Menguy, Rene, medicine
Mercer, Thomas T., industrial hygiene, health physics
Merin, Robert Gillespie, anesthesiology, pharmacology
Merkel, Paul Barrett, photochemistry, photographic chemistry
Mermelstein, Robert, polymer chemistry

Merrigan, Joseph A., photographic chemistry
Merrill, Ronald Eugene, synthetic organic chemistry
Merrill, Stewart Henry, organic polymer chemistry
Metakides, George, mathematical logic
Meyer, Garson, industrial chemistry, medicine
Meyerowitz, Sanford, medicine, psychiatry
Michaelson, Solomon M, radiation biology, physiology
Mihajlov, Vsevolod S, physical chemistry, organic chemistry
Miller, Donald Smith, mathematics
Miller, Gerald, medicine
Miller, Howard Anthony, pharmacy, organic chemistry
Miller, Joel Steven, inorganic chemistry
Miller, Leon Lee, biochemistry
Miller, Morton W, radiobiology, cytogenetics
Miller, Raymond Sumner, analytical chemistry
Miller, Warren James, physical chemistry
Millikan, Allan G, photography, astronomy
Milton, Rene, anthropology, archaeology
Miner, Clifford E., physical chemistry
Mitacek, Paul, Jr., physical chemistry
Mitchell, Gary F., photographic chemistry
Montean, John J., science education
Montroll, Elliott Waters, mathematics, physics
Mooberry, Jared Ben, synthetic organic chemistry
Moon, Neil Sennett, synthetic organic chemistry
Mooney, Robert Arthur, applied chemistry
Moore, Carol Wood, genetics
Moore, Duncan Thomas, optics, optical engineering
Moore, Robert Stephens, polymer science
Morgan, Herbert Roy, microbiology, medicine
Morgan, William L., Jr., internal medicine, cardiovascular diseases
Morken, Donald A., biophysics
Morokuma, Keiji, quantum chemistry
Morrill, Terence Clark, organic chemistry
Morrow, Paul Edward, toxicology, pharmacology
Morton, John Henderson, surgery
Mosak, Richard David, mathematical analysis
Moser, Frank, solid state physics
Mott, George Robson, physics
Motter, Robert Franklin, photography, arachnology
Muchmore, William Breuleux, statistics
Mudholkar, Govind S, mathematical statistics
Muench, Donald Leo, mathematics
Muenter, Annabel Adams, photographic chemistry, physical chemistry
Muenter, John Stuart, physical chemistry, spectroscopy
Murray, Thomas J., organic chemistry
Myers, Drewfus Young, Jr., organic polymer chemistry, colloid chemistry
Nachlin, Leopoldo, mathematics
Nahas, Aly, biochemical pharmacology
Nedderman, Howard Charles, physics
Neddermeyer, Peter Arthur, analytical chemistry
Nelson, Clarence Norman, physics
Nelson, Kyler Fischer, electrooptics
Nelson, Robert Andrew, polymer chemistry, cellulose chemistry
Neuberger, Dan, photographic chemistry
Neuman, Margaret Wrightington, pharmacology
Neuman, William Frederick, biochemistry, biophysics
Newburg, Edward A., applied mathematics
Nichols, Donald Richardson, clinical medicine
Nieh, Marjorie T., organic chemistry
Nielsen, Paul Livingstone, polymer chemistry
Notides, Angelo C., endocrinology, biochemistry
Nur, Uzi, cytogenetics
O'Brien, David F., biophysical chemistry
Odoroff, Charles Lazar, biostatistics
Ogden, Philip Myron, physics
Ohlsson-Wilhelm, Betsy Mae, genetics, molecular biology
Oliver, Gene Leech, photographic chemistry, organic chemistry
Olmsted, Joanna Belle, cell biology
O'Loane, James Kenneth, chemical physics

Olsen, Ronald G, photography, organic chemistry
O'Mara, Robert E., radiology, nuclear medicine
Oppenheimer, Larry Eric, colloid chemistry
Orbison, James Lowell, pathology
Orem, Michael William, physical chemistry
Orzechowski, Adam, physical chemistry, microscopy
Osborn, Harland James, synthetic organic chemistry, photographic chemistry
Oster, Carl Frederick, electron microscopy, photochemistry
Osterhoudt, Hans Walter, physical chemistry
Owen, John Reindel, organic chemistry
Owen, James Carl, optics
Pabico, Rufino C., medicine, nephrology
Pacansky, Thomas John, polymer chemistry
Paine, Robert H., organic chemistry, inorganic chemistry
Panner, Bernard J., pathology
Parks, Ronald Dee, experimental physics
Parsons, Timothy F., organic chemistry
Patten, Stanley Fletcher, Jr., pathology
Patterson, William Bradford, surgery
Patton, Elizabeth VanDyke, physical chemistry, oncology
Patton, James Edward, photographic chemistry, chemical engineering
Paxton, K Bradley, applied mathematics, electrical engineering
Pearlman, Donald, solid state chemistry
Pearson, George John, applied physics, photographic science
Peek, Frank Willard, animal behavior, neurobiology
Pell, Erik Mauritz, solid state physics
Penney, David P., histology, electron microscopy
Perez-Albuerne, Evelio A., physical chemistry
Perlstein, Jerome Howard, solid state chemistry
Perry, Edmond S., physical chemistry
Perry, Ernest John, physical chemistry
Peskin, James Charles, physiology
Petropoulos, Constantine Chris, organic polymer chemistry
Phelps, Daniel James, solid state physics
Pietrzykowski, Anthony D., analytical chemistry
Pinney, Jack Erwin, applied physics, instrumentation
Piper, Douglas Edward, organic chemistry
Platzer, Richard France, internal medicine
Pochan, John Michael, polymer physics, physical chemistry
Ponticello, Ignazio Salvatore, organic chemistry
Popp, Gerhard, inorganic chemistry
Poslusny, Jerrold Neal, synthetic organic chemistry, electrochemistry
Prakash, Louise, molecular genetics
Prakash, Satya, genetics
Prest, William Marchant, Jr., polymer physics
Price, Harry James, inorganic chemistry
Priddle, Osgood Daniel, Jr., clinical pharmacology, toxicology
Priest, William James, chemistry
Puskin, Jerome Sanford, biophysics
Racy, John Cecil, psychiatry
Raimi, Ralph Alexis, mathematical analysis
Raman, Varadaraja Venkata, theoretical physics, history of science
Randolph, John Fitz, mathematics
Rao, Poduri S R S., statistics
Raphael, Harold James, wood science, wood technology
Rasch, Arthur Allyn, photographic chemistry
Raup, David Malcolm, invertebrate paleontology
Ravindran, Nair Narayanan, organic chemistry
Rebel, William J., polymer chemistry
Reckhow, Warren Addison, organic chemistry, information science
Reed, Kenneth Joseph, physical chemistry
Rees, William Wendell, organic chemistry
Regan, Thomas Hartin, organic chemistry
Reid, Charles David, physics
Reiffer, Clifford Bruce, medical administration
Reynolds, George Arthur, organic chemistry
Rezanka, Ivan, physics

NEW YORK

Richards, Jack Lester, organic chemistry, photographic technology
Richter, Goetz Wilfried, pathology, cell biology
Rickmers, Albert D, applied statistics
Riecke, Edgar Erick, organic polymer chemistry
Rifkin, Barry Richard, experimental pathology, oral pathology
Riggi, Stephen Joseph, physiology, pharmacology
Ritterson, Albert L, medical parasitology
Rob, Charles G, surgery
Roberts, Harry Edward, photographic chemistry, organic chemistry
Rogoff, Stanley Myron, radiology
Romano, John, psychiatry
Rose, Philip I, physical chemistry
Ross, Robert Edward, organic chemistry
Rossi, Louis J, organic chemistry
Rossiter, Bryant William, organic chemistry
Roth, Frieda, microbiology, virology
Rothstein, Rodney Joel, molecular genetics
Roudabush, Robert Lee, zoology
Rowley, Peter Templeton, human genetics, internal medicine
Rubin, Bruce Joel, electrophotography
Rubin, Philip, radiology
Rubin, Sidney, psychiatry
Rubulis, Albert, biochemistry
Rudolph, Jerome Howard, obstetrics & gynecology
Rush, Kent Rodney, organic chemistry
Saeva, Franklin Donald, chemistry
Sagura, John Joseph, photographic chemistry
Salesin, Eugene Dennis, photochemistry
Salminen, Ilmari Fritiof, organic chemistry
Sauter, Frederick Joseph, organic chemistry
Saltsburg, Howard Mortimer, physical chemistry
Sandhu, Mohammad Akram, organic polymer chemistry
Sandifer, James Roy, electrochemistry
Sanford, Karl John, clinical biochemistry
Satran, Richard, neurology
Saunders, Vernon Irving, solid state chemistry
Saunders, William Hundley, Jr, physical organic chemistry
Savage, Dennis Jeffrey, organic chemistry
Saward, Ernest Welton, internal medicine
Schaefer, James Robert, organic chemistry, microbiology
Schallhorn, Charles H, photographic chemistry
Schleigh, William Robert, organic chemistry
Schlessinger, Richard H, organic chemistry
Schmale, Arthur H, Jr, medicine, psychiatry
Schneider, Robert L, polymer chemistry
Schmipelsky, Paul Nicholas, analytical chemistry
Schottmiller, John Charles, physical chemistry
Schultz, William Clinton, synthetic organic chemistry
Schupp, Orion Edwin, III, analytical chemistry
Schuster, Daniel Bradley, psychiatry
Schwartz, Allan James, psychiatry, transactional analysis
Schwartz, Seymour I, thoracic surgery
Schwartz, Frederick Anton, anatomy, physiology
Scott, David Evans, anatomy, neuroendocrinology
Scott, Francis Leslie, medicinal chemistry, organic chemistry
Seanor, Donald A, chemistry
Searle, Roger, organic chemistry
Segal, Harry L, gastroenterology
Segal, Sanford Leonard, mathematics
Sessler, John Charles, physics, operations research
Seus, Edward J, organic chemistry, photographic chemistry
Shah, Pravin Mangaldas, cardiology
Shahin, Michael Mahmood, physical chemistry
Shamoo, Adil E, physiology, biophysics
Shapiro, Sidney, solid state physics
Sharp, Richard Lee, organic chemistry
Sharpless, Stewart Lane, astronomy
Sheridan, Michael N, anatomy, electron microscopy
Sherman, Fred, genetics, biophysics
Sherman, George Charles, physical optics
Shiao, Daniel Da-Fong, physical chemistry, physical biochemistry
Shrager, Peter George, physiology, biophysics
Shuck, John Winfield, mathematics
Sieg, Albert Louis, organic chemistry
Sigai, Andrew Gary, high temperature chemistry
Silha, Robert Emmett, dentistry
Silverman, Robert Andrew, photographic chemistry
Simon, Albert, plasma physics
Simon, Robert David, microbial physiology, developmental biology
Simonsen, David Raymond, physical chemistry
Simson, Joseph Michael, photographic chemistry
Sinclair, Douglas C, optics
Singh, Arjun, genetics
Skillman, David Corwin, photography
Skucas, Jovitas, radiology
Sladek, Celia Davis, medical microbiology
Slattery, Paul Francis, high energy physics
Sliva, Philip Oscar, solid state physics
Smith, Donald Arthur, polymer chemistry
Smith, Donald Luke, mathematics
Smith, Douglas Lee, x-ray crystallography
Smith, Frank Ackroyd, physiological chemistry
Smith, Gale Eugene, photographic chemistry
Smith, Harold Ladd, Jr, organic chemistry
Smith, Howard Michael, optics
Smith, Michael, physical chemistry
Smith, Thomas Woods, organic polymer chemistry
Smith, Wendell Franklyn, Jr, physical organic chemistry
Snavely, Benjamin Breneman, quantum electronics, solid state physics
Snider, Ray S, anatomy, physiology
Snoke, Roy Eugene, enzymology
Solodar, Warren E, organic chemistry
Sorem, Allan Louis, physics
Sorrentino, Sandy, Jr, neuroendocrinology
Sowinski, Raymond, biochemistry
Spangler, Fred Walter, chemistry
Spar, Irving Leo, immunology
Spaulding, Richard Alan, electrooptics
Spayd, Richard W, organic chemistry
Spencer, Harry Edwin, physical chemistry
Sproull, Robert Lamb, physics
Stannard, James Newell, radiobiology, toxicology
Staples, Jon T, organic chemistry, polymer chemistry
Stark, Egon, microbiology
Starr, John Edward, photography
Stauffer, Robert Eliot, physical chemistry
Steel, R Knight, internal medicine
Stein, Norman, mathematics
Stephen, Keith H, inorganic chemistry
Sterman, Melvin David, colloid chemistry, polymer chemistry
Stern, Max Herman, organic chemistry
Stiles, John I, (Jr), molecular genetics
Stockman, David Lyle, physical chemistry
Stoeber, Werner, physical chemistry, biophysics
Stone, Arthur Harold, pure mathematics
Stone, Dorothy Maharam, mathematics
Stone, Joe Thomas, physical organic chemistry
Stotz, Elmer Henry, biochemistry
Stradford, H Todd, orthopedic surgery, pathology
Strauss, John Steaven, psychiatry
Strome, Forrest C, Jr, solid state physics, lasers
Stroud, Carlos Ray, quantum optics
Stroud, Jackson Swavely, experimental solid state physics, engineering
Sturmer, David Michael, organic chemistry, photographic chemistry
Sumberg, David Allan, experimental physics
Sundberg, Michael William, physical chemistry
Sutherland, Judith Elliott, polymer chemistry
Sutton, Robert George, geology
Swicklik, Leonard Joseph, organic chemistry
Swingley, Charles Stephen, physical chemistry
Swoyer, Vincent Harry, academic administration, computer science
Taber, Harry Warren, microbiology, molecular biology
Tan, Julia S, polymer science, physical chemistry
Tan, Yen T, solid state physics, surface physics
Tan, Zoilo Cheng-Ho, photographic chemistry
Tanaka, Ryo, neurochemistry, biochemistry
Taves, Donald R, medicine, radiobiology
Taylor, Jack Eldon, physics
Taylor, Stuart Robert, physiology, biophysics
Teasdale, William Brooks, organic chemistry
Teegarden, David Morrison, organic chemistry
Teegarden, Kenneth James, physics
Terepka, Anthony Raymond, internal medicine
Terhaar, Clarence James, toxicology
Thaler, Otto Felix, psychiatry, psychoanalysis
Thirtle, John Robson, organic chemistry
Thomas, Garth Johnson, neuropsychology
Thomas, Harold Todd, photochemistry
Thompson, Brian J, optics
Thompson, Crayton Beville, organic chemistry
Thorndike, Edward Harmon, elementary particle physics
Tinker, John Frank, chemistry, information science
Tischer, Thomas Norman, photographic chemistry
Todd, Hollis N, photography, physics
Tollefson, Charles Ivar, biochemistry
Tometsko, Andrew M, biochemistry, organic chemistry
Tong, Lee Karl Jan, physical chemistry
Toribara, Taft Yutaka, biophysics, chemistry
Townes, Philip Leonard, genetics
Transue, Laurence Frederick, physical chemistry
Treat, Donald Fackler, medical education
Tregillus, Leonard Warren, physical chemistry
Tremmel, Carl George, analytical chemistry
Trotter, Philip James, physical chemistry
Tuite, Robert Joseph, research administration, photographic chemistry
Tuites, Richard Clarence, organic chemistry, polymer chemistry
Tulagin, Vsevolod, chemistry
Turnblom, Ernest Wayne, organic chemistry
Turner, Andrew B, organic chemistry
Turner, Douglas Hugh, biophysical chemistry
Turner, Michael D, medicine, biochemistry
Tutihasi, Simpei, solid state physics
Tweet, Arthur Glenn, solid state physics
Ureles, Alvin L, medicine
Urry, Ronald Lee, reproductive physiology, urology
Vaala, Allen Richard, experimental physics
Van Der Voorn, Peter C, electrophotography
Van Heyningen, Roger, solid state physics
Van Horn, Hugh Moody, astrophysics
VanKerkhove, Alan Paul, optical physics
Van Meter, James P, organic chemistry
Van Norman, Gilden Ramon, photographic chemistry, polymer chemistry
Vanas, Don Woodruff, physical chemistry
Van Den Berghe, John, organic chemistry, polymer chemistry
Vander Valk, Paul David, organic chemistry
Vasicek, Oldrich Alfonso, mathematical statistics
Veverbrants, Egils, medicine
Vinal, Richard S, inorganic chemistry
Vishniac, Helen Simpson, microbial ecology, mycology
Voglesong, William Frederick, photometry, photogrammetry
Von Bacho, Paul Stephan, Jr, photographic chemistry
Vostal, Jaroslav Joseph, pharmacology, toxicology
Walker, Michael Sidney, photophysics
Wallace, Robert Bruce, thoracic surgery, cardiovascular surgery
Wallace, Thomas Patrick, physical chemistry, polymer science
Walsh, Edward Joseph, organic chemistry, photographic science
Wardell, William Michael, clinical pharmacology
Waterhouse, Christine, medicine
Watts, Charles Edward, mathematics
Weber, David Alexander, medical physics
Weber, Frank E, food science
Weed, Robert I, hematology
Weisler, Leonard, organic chemistry
Weiss, Bernard, psychopharmacology, toxicology
Wellman, Russel Elmer, physical chemistry
Weymann, Helmut Dietrich, rheology
Wheeless, Leon Lum, Jr, electrical engineering, biomedical engineering
Whitaker, Harry Allen, neuropsychology
Whiteley, Thomas Edward, organic chemistry
Whitlock, L Ronald, analytical chemistry
Wiberg, John Samuel, biochemistry, genetics
Wien, Richard W, Jr, physical chemistry
Wilder, Donald Richard, mathematics, optics
Wilhelm, James Maurice, microbiology, molecular biology
Wilkes, Glenn Richard, inorganic chemistry
Williams, Carl James, Jr, photographic chemistry
Williams, David James, physical chemistry
Williams, Jack L R, organic chemistry
Williams, Thomas Franklin, internal medicine
Willis, Roland George, physical organic chemistry
Wilson, Burton David, organic chemistry
Wilson, James, Jr, organic chemistry
Wilson, John Charles, organic polymer chemistry
Wilson, Robert Hallowell, environmental health
Wittliff, James Lamar, biochemistry
Wolf, Emil, mathematical physics
Wolfarth, Eugene F, physical organic chemistry
Wolfe, Robert Norton, physics
Wolfe, Nikolaus Emanuel, chemistry, electronics
Woodall, David Monroe, plasma physics
Woodward, Donald Jay, physiology
Wrathall, Donald Prior, physical chemistry
Wright, Charles Joseph, organic chemistry
Wright, John Fowler, physical chemistry
Wrobel, Joseph Jude, chemical physics
Wu, Tai Wing, biochemistry, biochemical engineering
Wynne, Lyman Carroll, psychiatry, psychology
Yamamoto, Yasushi Stephen, organic chemistry, photographic chemistry
Young, Frank E, pathology, microbiology
Young, Gail Sellers, Jr, mathematics
Young, Lawrence Eugene, medicine
Young, Ralph Howard, chemical physics, solid state physics
Youngquist, Mary Josephine, photographic chemistry
Yudelson, Joseph Samuel, chemistry, photochemistry
Zander, Helmut A, dentistry
Ziegler, Hrolfe Read, surgery
Zigman, Seymour, biochemistry
Zimmer, James Griffith, preventive medicine, community health
Zuehlke, Carl William, analytical chemistry
Zwick, Daan Marsh, photographic chemistry, physics

ROCKLAND LAKE
Schneider, Ross Nelson, mathematics

ROCKVILLE CENTRE
Jones, Justine H, protozoology
McDonnell, Joseph Francis, Jr, pharmaceutical chemistry, nutrition
Ragone, Stephen Edward, geochemistry, water chemistry
Ross, Stewart Hamilton, geology
Smith, Frank Engelbert, mathematics
Wesolowski, S Adam, surgery

ROCKY POINT
Tucker, Robert H, theoretical physics

ROME
McNamara, James Henry, organic chemistry
Nelson, Mark Radford, astrophysics

ROOSEVELT ISLAND
Levine, Seymour, pathology, neuropathology

ROSENDALE
Archimovich, Alexander S., biology, agronomy

ROSLYN HEIGHTS
Perlman, Ely, allergy, immunology
Rogatz, Peter, public health
Saletan, Leonard Timothy, chemistry

ROUSES POINT
Beall, Desmond, biochemistry
Fox, Chester David, pharmaceutical chemistry
Johnson, Raymond Nils, analytical chemistry
Kho, Boen Tong, analytical chemistry
Krol, George J., physical chemistry, analytical chemistry
Lin, Song-Ling, physical pharmacy, biopharmacy
Orzech, Chester Eugene, Jr., analytical chemistry
Park, Moo Kwang, pharmaceutics
Pramoda, Maturu Krishna, pharmaceutics
Smith, Alan Jerrard, analytical chemistry
Surpuriya, Vijay B., pharmaceutical chemistry, physical pharmacy
Tsuk, Andrew George, physical chemistry, polymer chemistry

RYE
Albrecht, Alberta Marie, microbial physiology, cancer
Anderson, Lucy Macdonald, cancer
Archibald, Francis Magoun, lipid chemistry
Baldwin, Robert Russel, biochemistry
Biedler, June Lee, cell genetics
Brown, George Bosworth, bio-organic chemistry
Burness, Alfred Thomas Henry, virology, molecular biology
Cavalieri, Liebe Frank, physical chemistry
Chang, Tien-ding, genetics
Chello, Paul Larson, biochemical pharmacology
Choi, Yong Chun, biochemistry, organic chemistry
Coffey, Ronald Gibson, biochemistry, pharmacology
Colmey, John C., food science
Cotton, Robert Henry, food science, nutrition
Falco, Elvira Allegra (Mrs Bass), medicinal chemistry
Fischbach, Eugene, microbiology, food science
Fogh, Jorgen (Engell), medicine
Fox, Jack Jay, biochemistry
Giner-Sorolla, Alfreo, organic chemistry
Horan, Paul Karl, biophysics, virology
Kassel, Robert Lawrence, cell physiology, immunology
Kennedy, Edwin Russell, environmental chemistry, marine ecology
Kenyon, Alan J., biochemistry, immunology
Khan, Paul, food science
Kim, Yoon Bern, immunology
Kiszkiss, David F., immunology
Kreis, Willi, clinical pharmacology, chemotherapy
Lam, Fuk Luen, chemistry
Lee, Tzoong-Chyh, organic chemistry, bio-organic chemistry
Mehta, Bipinchandra Mohanlal, microbial genetics, microbial physiology
Meiera, Peter William, cell biology
Parham, James Crowder, II, bio-organic chemistry
Peterson, Robert H F, microbiology
Roberts, Joseph, biochemistry
Rolfes, Thomas J., poultry science, food technology
Rosenberg, Barbara Hatch, biochemistry
Schmid, Franz Anton, cancer pharmacology, genetics
Sirotnak, Francis Michael, molecular pharmacology
Skipski, Vladimir Pavlovich, biochemistry
Stohrer, Gerhard, organic chemistry, biochemistry
Tarnowski, George Serge, cancer
Teller, Morris N., cancer
Turner, Earl Wilbert, biochemistry, protein chemistry
Watanabe, Kyoichi A., organic chemistry, biochemistry

ST BONAVENTURE
Anderson, Kenneth Ellsworth, bacteriology
Bothner, Richard Charles, ecology, anatomy
Budzinski, Walter Valerian, experimental solid state physics
Eaton, Stephen Woodman, ornithology
Geiger, Joseph M., theoretical physics
Gritmon, Timothy F., inorganic chemistry
Hach, Edwin E., Jr., physical chemistry
Hartman, Ronald Earl, microbiology, biochemistry
Jacques, Felix Anthony, biology
Neeson, John Francis, nuclear physics
Turek, William Norbert, synthetic organic chemistry
White, Albert George, Jr., microbial physiology
White, James Patrick, microbial physiology
Worden, John Lorimer, Jr., public health administration

SANBORN
Kwitowski, Paul Thomas, inorganic chemistry

SARANAC LAKE
Hospelhorn, Verne D., biochemistry
Kubica, George P., medical bacteriology
Mackaness, George Bellamy, immunology
McGregor, Douglas D., experimental pathology
Miller, Thomas Edward, immunochemistry
Volkman, Alvin, pathology, immunology

SARATOGA SPRINGS
Baum, Parker Bryant, physical chemistry, inorganic chemistry
Crocker, Denton Winslow, invertebrate zoology
Fahey, Charlotte Wieghard, organic chemistry
Gallagher, Orvoell Roger, anthropology
Golden, Ben Roy, genetics
Howard, Harold Henry, botany, limnology
Johnson, Kenneth George, geomorphology
Mahoney, Robert Patrick, microbial physiology, electron microscopy
McKibben, John Joseph, mathematics
Samworth, Eleanor A., physical chemistry
Thomas, John Jenks, geology
Valency, Vivian Briones, chemistry
Walter, Paul Hermann Lawrence, inorganic chemistry
Williams, Mary Elizabeth, mathematics

SCARSDALE
Benson, Wilbur Maxwell, pharmacology, toxicology
Gunsberg, Ephraim, biochemistry, analytical chemistry
Harrar, Jacob George, plant pathology
Leerburger, Benedict Alan, Jr., science writing, journalism
Meyer, Werner Franz, mathematics
Mooney, Richard T., radiological health physics
Renner, Sheldon Samson, biological chemistry

SCHENECTADY
Afterput, Siegfried, organic chemistry
Alpher, Ralph Asher, physics
Anthony, Donald Joseph, physics
Arendt, Ronald H., physical chemistry
Aroian, Leo Avedis, mathematical statistics
Aven, Manuel, solid state physics
Bacon, Egbert King, chemistry
Bails, Earl William, chemistry
Banucci, Eugene George, organic polymer chemistry
Bartram, Stanley F., x-ray crystallography
Bean, Charles Palmer, physics, biophysics
Bellamy, Winthrop Dexter, microbiology, biochemistry
Ben Daniel, David J., physics, research administration
Bennett, Alan Jerome, solid state physics
Bergeron, John Albert, cytology
Berkowitz, Ami Emanuel, magnetism
Bick, Theodore A., chemistry
Birecka, Helena M., plant physiology, biochemistry
Bolebuck, Edith Maude, polymer chemistry
Bolon, Donald A., organic chemistry
Boyer, Barbara Conta, biology
Boyer, John Frederick, population biology
Bradshaw, John Alden, electrophysics
Braidwood, Clinton Alexander, organic chemistry
Bray, James William, theoretical solid state physics
Breiter, Manfred Wolfgang, electrochemistry, physical chemistry
Brown, John Francis, Jr., physical organic chemistry
Bueche, Arthur Maynard, polymer chemistry, physical chemistry
Butzel, Henry M, Jr., genetics
Carlson, Richard Oscar, physics
Carson, Chester Carrol, analytical chemistry
Chakraborty, Ananda Mohan, microbial genetics
Chrenko, Richard Michael, spectroscopy, solid state physics
Coggeshall, A Darling, polymer chemistry
Comly, James B., applied physics
Comstock, George Milton, physics
Cook, Newell Choice, organic chemistry
Crivello, James V., organic chemistry
Daniels, John Maynard, water chemistry
Davis, William Donald, physical chemistry
DeBlois, Ralph Walter, physics
DeLuca, John Anthony, solid state chemistry
DeVries, Robert Charles, mineralogy
Dietz, Leonard Allan, physics
Doigan, Paul, physical chemistry
Edelich, Richard, physical chemistry
Ehrlich, Lewis S., solid state physics
Engeler, William E., solid state physics
Eshbach, John Robert, reactor physics
Ewaraye, Andrew Otieku, solid state physics
Factor, Arnold, physical organic chemistry
Fairchild, William Warren, mathematics
Farrington, Gregory Charles, chemistry, electrochemistry
Fenner, Gunther Erwin, solid state physics
Finkbeiner, Herman Lawrence, physical chemistry, organic chemistry
Fiske, Milan Derbyshire, solid state physics
Fleischer, Robert Louis, physics
Flom, Donald Gordon, physical chemistry
Fox, Augustus Henry, mathematics
Francis, Norman, theoretical physics
Friedman, Jack P., computer science, physics
Fromageot, Henri Pierre-Marcel, biochemistry
Frosch, Robert Peter, physical chemistry, spectroscopy
Gaines, George Loweree, Jr., physical chemistry
Galginaitis, Simeon Vitis, physics
Gallagher, Charles Joseph, physics
George, Carl Joseph Winder, marine ecology
Giaver, Ivar, immunology, physics
Glascock, Homer Hopson, Jr., physics
Goble, Alfred Theodore, optics
Golbersuch, David Clarence, physics
Gray, Peter Vance, solid state physics
Greskovich, Charles David, ceramics
Groot, Cornelius, chemistry
Grubb, Willard Thomas, physical chemistry
Haas, Walter Oskar, Jr., inorganic chemistry
Hall, Robert Noel, physics
Ham, Frank Slagle, physics
Hannan, Edward Lees, operations research
Hart, Edward Walter, theoretical physics, materials science
Hart, Howard Roscoe, Jr., physics
Hay, Allan Stuart, organic chemistry, polymer chemistry
Hebb, Malcolm Hayden, physics
Hanley, Stanley Young, polymer physics
Holub, Fred F., chemistry
Houston, John Mapes, physics
Huggins, Charles Marion, physical chemistry
Hurwitz, Henry, Jr., nucleonics
Interrante, Leonard V., inorganic chemistry
Johnson, Peter Dexter, applied physics
Johnston, William George, solid state physics
Jones, Stanley Leslie, analytical chemistry
Jordan, Manuel Albert, organic chemistry
Joynson, Reuben Edwin, Jr., physics
Julinac, Peter C., polymer physics
Kambour, Roger Peabody, physical chemistry, polymer physics
Kaplan, Joel Howard, immunology, cell biology
Kashnow, Richard Allen, physics
Kasper, John Simon, physical chemistry
Kennicott, Philip Ray, physical chemistry
Kingsley, Jack Dean, solid state physics
Kinsinger, Richard Estyn, plasma physics
Krantz, Karl Walter, organic chemistry
Kurey, Thomas John, nuclear physics, reactor physics
Lambert, Francis Lincoln, physiology
Larsson, Robert Dustin, chemistry
Lawson, James Lewellyn, physics
Lawson, Richard Woodruff, physiology
LeBlanc, Oliver Harris, Jr., physical chemistry
LeGrand, Donald George, physical chemistry
Levinson, Lionel Monty, solid state physics
Lewis, Edwin Augustus Stevens, thermal physics
Lichtenstein, Roland Max, physics
Lichtenstein, Pearl Rubenstein, astrophysics
Loughlin, Timothy Arthur, applied mathematics
Lubitz, Cecil Robert, nuclear physics
Luborsky, Fred Everett, physical chemistry
Ludwig, Gerald W., semiconductors, solid state physics
Lynk, Edgar Thomas, lasers
MacLeary, Michael Risley, organometallic chemistry
Markovitz, Mark, organic chemistry, polymer chemistry
Marshall, Walter Lincoln, physical chemistry
Martin, Frederick Johnson, physical chemistry
Martin, William Butler, Jr., physical chemistry
Martin, William Butler, Jr., organic chemistry, spectroscopy
McDonald, Robert Skillings, physical chemistry, spectroscopy
McElligott, Peter Edward, surface chemistry, tribology
McHugh, James Anthony, Jr., physical chemistry, nuclear chemistry
McKee, Douglas William, physical chemistry
Messmer, Richard Paul, chemical physics
Mihran, Theodore Gregory, microwave electronics
Miller, Alexander Andrew, physical chemistry
Miller, Dudley Grant, nuclear chemistry, corrosion chemistry
Mogro-Campero, Antonio, space physics, geophysics
Moody, Leroy Stephen, chemistry
Nagamatsu, Henry T., fluid physics, plasma physics
Neugebauer, Constantine Aloysius, physical chemistry
Neumann, George Joseph, biochemistry
Niedrach, Leonard William, chemistry
Norton, Francis James, physical chemistry
Oliver, David W., solid state physics
Orlando, Charles M., organic chemistry
Overmire, Thomas Gordon, biology
Palm, John Andrew, ceramics
Pasher, Peter Edward, electronics
Peak, David, theoretical physics
Pemrick, Raymond Edward, textile chemistry
Petersen, Kenneth C., organic polymer chemistry
Philipp, Herbert Reynold, experimental physics
Pilcher, Valter Ennis, physics
Piper, William Weidman, solid state physics
Plumley, Harold Jamison, applied physics, solar energy
Pollock, Herbert Chermside, physics
Possin, George Edward, experimental solid state physics
Povich, Michael Jean, colloid chemistry, corrosion
Powers, Robert William, physical chemistry
Prener, Jerome Sidney, chemistry
Pritchard, Robert Leslie, physical chemistry
Prochazka, Svante, ceramics

Quinn, Clayton Byerley, organic polymer chemistry
Quisenberry, Karl Spangler, Jr., physics
Rappaport, Raymond, Jr., embryology
Rautenberg, Theodore Herman, physics
Raymond, Richard Collyer, physics
Real, Cathleen Clare, algebra
Redington, Rowland Wells, solid state physics
Rich, Joseph Anthony, plasma physics
Roberts, Benjamin Washington, Jr., solid state physics
Roe, Glenn Magnus, applied mathematics
Roedel, George Frederick, polymer chemistry
Rohr, Robert Charles, physics
Rosenberg, Steven Loren, microbiology
Rosolowski, Joseph Henry, solid state physics
Roth, Walter Lester, solid state chemistry
Pzad, Stefan Jacek, physical chemistry, theoretical
Salahub, Dennis Russell, theoretical chemistry, theoretical solid state physics
Scaife, Charles Walter John, inorganic chemistry
Schaefer, Robert William, analytical chemistry
Schick, Kenneth Leonard, solid state physics, biophysics
Schmee, Josef, statistics
Schmitt, Roland Walter, solid state physics
Schroeter, Siegfried Hermann, chemistry
Schultheis, James J., physics
Schwarz, Winfred Max, physics
Sears, Alan Roy, physical chemistry
Seiken, Arnold, mathematics
Sharbaugh, Amandus Harry, chemistry, physics
Sheffer, Howard Eugene, organic chemistry
Shores, David Aurth, metallurgical chemistry
Shultz, Allan R., physical chemistry
Simons, David Stuart, mass spectrometry
Simons, Edward Louis, environmental chemistry
Slack, Glen Alfred, solid state physics
Smith, Jack Howard, theoretical physics
Smith, William Edward, industrial organic chemistry
Sowa, John Robert, organic chemistry
Stearns, Richard Edwin, mathematics, computer science
Stein, Richard James, polymer chemistry
Stout, Virgil L., physics
Stricos, David Peter, analytical chemistry
Strong, Herbert Maxwell, high pressure physics, physical optics
Stuart, Robert Lee, radiochemistry
Styles, Twitty Junius, parasitology, biology
Su, Tah-Mun, organic chemistry, bioengineering
Swartz, Charles Dana, physics
Takekoshi, Tohru, polymer chemistry, organic chemistry
Tantaporn, Wirojana, experimental solid state physics
Taylor, Dale Frederick, electrochemistry
Thompson, Charles Denison, corrosion
Thornton, Roy Fred, chemical engineering, electrochemistry
Tiemann, Jerome J, solid state physics
Titus, Walter Franklin, physics
Tobiessen, Peter Laws, plant ecology
Treu, Jesse Isaiah, biophysics, optics
Tucker, Charles Winfred, Jr., physical chemistry
Uzgiris, Egidijus E, physics
Vedder, Willem, chemistry
Vosburgh, Kirby Gannett, applied physics, experimental physics
Vought, Robert Howard, physics
Walton, Warren Lewis, organic chemistry
Watkins, George Daniels, physics
Webster, Harold Frank, physics
Weick, Charles Frederick, inorganic chemistry
Wentorf, Robert H, Jr., physical chemistry
Werner, Thomas Clyde, analytical chemistry
Whetten, Nathan Rey, physics
White, Dwain Montgomery, organic chemistry, polymer chemistry
Will, Fritz Gustav, electrochemistry
Wilson, Ronald Harvey, solid state physics, energy conversion
Witting, Harald Ludwig, plasma physics
Wolfe, John Kavanaugh, organic chemistry
Wolny, Friedrich Franz, organic polymer chemistry
Wong, Joe, physical chemistry, solid state science
Woodbury, Henry Hugh, physics

Wright, Archibald Nelson, physical chemistry
Yang, Kei-Hsiung, quantum optics, medical physics
Young, James Roger, physics
Yunick, Robert P, organic polymer chemistry
Zalewski, Edmund Joseph, organic polymer chemistry
Zeltmann, Eugene W, physical chemistry

SCOTIA
Barreto, Ernesto, electrodynamics, fluid dynamics
Dunn, Cecil Gordon, physics
Esch, Louis James, reactor physics
Lupinski, John Henry, polymer chemistry
Moore, George Edmund, physical chemistry

SCOTTSVILLE
Cohen, Arthur Isaac, physiology, biochemistry

SEAFORD
Wolk, Robert George, ornithology

SELDEN
Gordon, Sheldon P., applied mathematics, mathematical statistics
Kirchner, Carl Edward John, bacteriology
Lewis, Carmie Perrotta, histology
Mecklosky, Morton, mathematics

SELKIRK
Cooper, Glenn Dale, organic chemistry
Katchman, Arthur, organic chemistry
Shank, Charles Philip, polymer chemistry

SENECA FALLS
Allen, Donald Stewart, chemistry
Brown, Richard Irwin, environmental sciences, physics
Curtis, Paul Robinson, microbiology
Hosley, Edward, anthropology
Janssen, Jerry Frederick, organic chemistry
Kuempel, John Rickey, analytical chemistry, inorganic chemistry
McAuley, Auley Anderson, zoology
Miller, Gary L, ecology, botany
Paige, Richard Joseph, geography, geography of the Middle East

SETAUKET
Sokoloff, Leon, pathology
Wolf, Alfred Peter, organic chemistry, physical chemistry

SHOREHAM
Friedes, Joseph Leonard, nuclear physics, high energy physics

SIDNEY
Morris, Wilford Ernst, physics

SINCLAIRVILLE
Kochersberger, Robert Charles, zoology

SKANEATELES
Schaedle, Michail, plant physiology

SLINGERLANDS
Harris, Albert Hall, medicine

SMITHTOWN
Cohen, Martin Gilbert, lasers
Luther, Herbert George, chemistry

SNYDER
Fitzpatrick, Robert Charles, geophysics
Pine, Ellen Kann, bacteriology, biochemistry

SOLVAY
Follows, Alan Greaves, chemistry
Gancy, Alan Brian, physical chemistry, inorganic chemistry
Keeler, Robert Adolph, inorganic chemistry
Poncha, Rustom Pestonji, inorganic chemistry, physical chemistry
Smalley, Edmund Walter, inorganic chemistry

SOUTH OZONE PARK
Taschdjian, Edgar, genetics

SOUTH SALEM
Gore, Robert Cummins, physical chemistry

SOUTHAMPTON
Berkebile, Charles Alan, mineralogy, crystallography
Briles, George Herbert, organic chemistry
Burke, William Thomas, Jr., biology, biochemistry
Haresign, Thomas, zoology
Hehre, Edward James, Jr, phycology, plant taxonomy
Melter, Robert Alan, mathematics
Reisman, Howard Maurice, ichthyology
Siegel, Alvin, chemical oceanography
Sipes, Richard Grey, medical anthropology, population studies
Tiedke, Kenneth Earle Riordan, applied anthropology, ethnology
Wood, Walter Abbott, geography

SOUTHOLD
Polatnick, Jerome, biochemistry

SPARKHILL
Rachinsky, Michael Richard, science education, biochemistry

SPRING VALLEY
Church, Gilbert, zoology, embryology
Forgacs, Joseph, medical microbiology
Hewitt, Redginal Irving, parasitology
Lombardo, J Robert D, biochemistry
McCormick, Michael E, biochemistry
Schneider, Paul, clinical chemistry
Witzel, Frank, organic chemistry

SPRINGVILLE
Cairns, John Mackay, embryology

STAMFORD
Montgomery, James Douglas, botany, taxonomy

STATEN ISLAND
Addy, John Keith, physical organic chemistry
Annan, Murvel Eugene, genetics
Auerbach, Andrew Bernard, polymer chemistry
Beckenstein, Edward, mathematical analysis
Blei, Ira, biophysical chemistry
Brockerhoff, Hans, biochemistry, neurochemistry
Carp, Richard Irvin, microbiology, virology
Christianson, John Dean, plant ecology
Clitheroe, H John, endocrinology, gynecology
Colosi, Natale, bacteriology
Connally, George Gordon, geology, petrology
Deshmukh, Diwakar Shankar, biochemistry
Donahue, Sheila, pathology, neuropathology
Eger, F Martin, theoretical physics
Erlichson, Herman, physics
Fox, William, physical chemistry, colloid chemistry
Gabrielsen, Bjarne, organic chemistry
Gaull, Gerald E, pediatrics
Gerber, James Norman, analytical chemistry
Goldberg, Philip A, physics
Gould, Robert Michael, neurochemistry
Henkel, Elmer Thomas, physics, science education
Jensen, Jens Trygve, radiochemistry
Kaarsberg, Ernest Andersen, geophysics
Kanzler, Walter Wilhekm, animal behavior, bioethics
Karl, Seung Chul, immunology
Kershaw, Edythe Marie, bacteriology
Kiley, Charles Walter, ichthyology
Kim, Kwang Soo, virology, microbiology
Lin, Fu Hai, bacteriology, biochemistry
Loo, Yen Hoong, biochemistry
Macri, Alfred Roger, chemistry, geology
Nace, Paul Foley, marine biology, endocrinology

Nankivell, John (Elbert), physics
Odian, George G, polymer chemistry
Pascal, Theresa A, immunology
Priddy, Ralph Banta, vertebrate zoology, insect taxonomy
Rabe, Ausma, neuropsychology
Raju, Pullarkat Krishnan, biochemistry
Rapaport, Ionel, psychiatry, genetics
Rassin, David Keith, neurochemistry
Robusto, C Carl, mathematics, physics
Ruark, Annette, physiology
Schain, Philip, microbiology, clinical pathology
Schulz, Johann Christoph Friedrich, organic chemistry
Schwartz, Joseph A, high energy physics, biophysics
Shapiro, Leonard, organic chemistry
Silberstein, Richard M, psychiatry
Soifer, David, cell biology
Stern, Adolph John, chemistry
Sturman, John Andrew, biochemistry, nutrition
Tallan, Harris H, biochemistry
Thormar, Halldor, virology
Wolf, Pierre L, microbiology
Yarns, Dale A, animal physiology

STERLING
Shaver, Richard Cornell, inorganic chemistry

STONY BROOK
Ackerman, Lauren Vedder, pathology
Adler, Alfred, mathematics
Alexander, John Macmillan, Jr, physical chemistry, nuclear science
Altman, Lawrence Jay, organic chemistry
Arnheim, Norman, molecular biology, evolution
Baer, Paul Nathan, periodontology
Balazs, Nandor Laszlo, theoretical physics
Barcus, William Dickson, Jr., mathematics
Battley, Edwin Hall, microbiology, biochemistry
Baum, Lloyd, dentistry
Baylor, Edward Randall, marine ecology
Beltrami, Edward J, mathematics
Bergofsky, Edward Harold, physiology
Berlad, Abraham Leon, chemical physics
Bonner, Francis Truesdale, chemistry
Bowman, Malcolm James, physical oceanography, electronic engineering
Bretsky, Peter William, invertebrate paleontology
Bretsky, Sara (Su) Stewart, invertebrate paleontology
Brown, Gerald Edward, theoretical physics
Brown, Joel Edward, neurophysiology
Brown, Paula, anthropology
Brynes, Paul Jeffrey, molecular pharmacology
Burnett, William Craig, marine geochemistry
Carleton, Herbert Ruck, opical physics, solid state science
Carlson, Albert Dewayne, Jr., invertebrate physiology
Carlson, Elof Axel, genetics
Carpenter, Edward J, biological oceanography
Carrasco, Pedro, anthropology
Carter, Neville Louis, geology, geophysics
Charlap, Leonard Stanton, mathematics
Chen, Yung Ming, applied mathematics
Chu, Benjamin Peng-Nien, physical chemistry
Cirillo, Vincent Paul, biochemistry
Dayal, Ramesh, marine geochemistry
Debiak, Ted Walker, nuclear chemistry
Dresden, Max, theoretical physics
Dudock, Bernard S, biochemistry
Duedall, Iver Warren, oceanography, physical chemistry
Ebin, David G, mathematics
Edmunds, Leland Nicholas, Jr., biological rhythms, cell biology
Eisenbud, Leonard, theoretical physics
Erk, Frank Chris, genetics
Esaias, Wayne Evor, biological oceanography, photobiology
Farkas, Hershel M, mathematics

Faron, Louis Charles, anthropology, history of anthropology
Feingold, Arnold Moses, nuclear physics
Finch, Stephen Joseph, statistics
Fink, Max, neuropsychiatry
Flessa, Karl Walter, paleontology
Fogg, George Garrett, plant ecology, systematic botany
Fossan, David B., physics
Fowler, Frank Wilson, organic chemistry
Fowler, James A., vertebrate embryology
Fox, David, theoretical physics
Fox, William Cassidy, mathematics
Freedman, Daniel Z., theoretical physics
Freundlich, Martin, biochemistry, molecular biology
Friedman, Harold Leo, physical chemistry, inorganic chemistry
Fritts, Harry Washington, Jr., medicine
Fusco, Madeline M., physiology
Futuyma, Douglas Joel, population biology
Gazzaniga, Michael Saunders, psychobiology
Geiger, H Jack, epidemiology, community medicine
Gelernter, Herbert Leo, physics, computer science
Gerst, Irving, mathematics
Glass, Hiram Bentley, genetics
Godfrey, Henry Philip, immunology
Goldfarb, Theodore D., physical chemistry, science policy
Goldhaber, Alfred Scharff, theoretical physics
Good, Myron Lindsay, physics
Gough, Michael, microbial genetics
Graf, Erlend Haakon, low temperature physics
Grollman, Arthur Patrick, pharmacology, medicine
Gwinnett, A John, oral microanatomy, restorative dentistry
Habicht, Gail Sorem, immunology
Haim, Albert, inorganic chemistry
Hanson, David M., chemical physics
Hanson, Gilbert N., petrology
Hardorp, Johannes Christfried, astrophysics
Harris, Stewart, statistical mechanics
Hartung, Jack Burdair, planetary sciences, geochronology
Jackson, Andrew D., Jr., theoretical nuclear physics
Janoff, Aaron, experimental pathology
Jen, Shen, applied physics
Johnson, Francis, organic chemistry
Johnson, Philip M., physical chemistry, molecular spectroscopy
Jona, Franco Paul, surface physics
Heley, Patrick James, solid state chemistry, organometallic chemistry
Herley, Patrick James, solid state
Hirota, Noboru, chemistry
Iden, Charles R., physical chemistry
Inke, Gabor, anatomy, physical anthropology
Jones, Raymond F., plant physiology
Kahn, Peter B., theoretical physics
Kao, Yi-Han, physics
Karten, Harvey J., neuroanatomy
Kelly, James P., neurobiology
Kerber, Robert Charles, organic chemistry
Kim, Charles Wesley, immunology, parasitology
Kim, Woo Jong, mathematics
Kirkorian, Abraham D., plant physiology
Kirz, Janos, experimental high energy physics, optical physics
Klavins, Janis Vilberts, medicine
Kleinberg, Israel, oral biology
Knacke, Roger Fritz, astronomy
Koehn, Richard Karl, population genetics
Kosower, Edward Malcolm, organic chemistry
Kra, Irwin, mathematics
Krantz, Allen, organic chemistry
Kumpel, Paul Gremminger, Jr., topology
Kuo, Thomas Tzu Szu, theoretical physics
Kuschner, Marvin, pathology
Kwan, John Ying-Kuen, astrophysics
Lambe, Edward Dixon, physics
Lane, Bernard Paul, pathology
Lanning, Edward P., anthropology, archaeology
Lankowsky, Philip, pediatrics, hematology
Lauterbur, Paul Christian, physical chemistry

Lee, Linwood Lawrence, Jr., nuclear physics
Lee-Franzini, Juliet, experimental physics
Le Fevre, Paul Green, cell physiology
Leibowitz, Martin Albert, applied mathematics, operations research
le Noble, William Jacobus, organic chemistry
Lesiewicz, Jeanne Lee, molecular biology
Levine, Sumner Norton, physical chemistry
Levinton, Jeffrey Sheldon, ecology, paleontology
Levy, Alan B, organometallic chemistry
Levy, Harvey Merrill, biochemistry, physiology
Lutz, Barry Lafean, astrophysics, molecular spectroscopy
Lyman, Harvard, plant physiology, molecular biology
Maskin, Bernard, mathematics
McCoy, Barry, theoretical physics
McGrath, Robert L., nuclear physics
McHugh, John Laurence, fisheries management
McTernan, Edmund J., public health, health administration
Meisetas, Leonard E., internal medicine
Mendelson, Martin, physiology
Merriam, Robert William, developmental biology
Metcalf, Harold, physics
Moos, Carl, biochemistry, cell physiology
Mould, Richard A., theoretical physics
Muether, Herbert Robert, nuclear physics
Nathans, Robert, solid state physics, urban studies
Nemerson, Yale, hematology
Oaks, J Howard, dentistry
O'Connor, Joel Sturgens, ecology, fisheries
Okaya, Yoshi Haru, crystallography
Okubo, Akira, physical oceanography, mathematical biology
Osher, Stanley Joel, mathematics
Owen, Tobias Chant, astrophysics, planetary sciences
Paldy, Lester George, science education, science policy
Palevitz, Barry Allan, cell biology, plant physiology
Palmer, Allison Ralph, invertebrate paleontology
Papike, James Joseph, crystallography, mineralogy
Paul, Peter, nuclear physics
Peress, Nancy E, neuropathology
Pollack, Robert Elliot, cell biology, oncology
Pond, Thomas Alexander, nuclear physics
Porter, Richard Needham, physical chemistry, theoretical chemistry
Prewitt, Charles Thompson, earth science
Ramirez, Fausto, organic chemistry
Riley, Monica, molecular genetics
Robinson, Charles Vernon, radiobiology
Rohf, F James, biometry, population biology
Rosenfeld, Martin Herbert, microbiology, clinical biochemistry
Sachs, John Richard, hematology, physiology
Sah, Chih-Han, mathematics
Sarma, Raghupathy, molecular biology, biochemistry
Schaeffer, Oliver Adam, cosmochemistry
Schechter, Nisson, biochemistry
Schneider, Robert Fournier, chemical physics
Schoen, Max H, public health
Schubel, Jerry Robert, marine geology
Shea, Dion Warren Joseph, theoretical physics, statistical mechanics
Shevchik, Nigel John, solid state physics
Silsbee, Henry Briggs, physics
Simer, Gary Albert, statistics, applied statistics
Simon, Michal, astronomy
Simon, Sanford Ralph, biochemistry
Simpson, Melvin Vernon, biochemistry
Slobodkin, Lawrence Basil, ecology
Smith, Douglas Graham, animal behavior
Smith, John, physics
Smith, Roger Alan, theoretical physics
Smolker, Robert Eliot, zoology
Sokal, Robert Reuven, population biology, taxonomy
Solomon, Philip M., astrophysics, molecular physics
Soroff, Harry S., surgery
Springer, Charles S Jr., inorganic chemistry
Sprouse, Gene Denson, physics
Sreebny, Leo Morris, pathology, biochemistry

Srivastav, Ram Prasad, applied mathematics
Stell, George Roger, physics,
Stern, Jack Tuteur, Jr., biomechanics
Sternglanz, Rolf, molecular biology, biochemistry
Stevenson, Robert Findlay, anthropology
Stillerman, Maxwell, pediatrics
Strasser, Elvira Rapaport, mathematics
Stroke, George W., physical optics
Strotman, Daniel, nuclear physics
Sujishi, Sei, inorganic chemistry
Swanson, Robert Lawrence, physical oceanography, civil engineering
Swartz, Clifford Edward, experimental high energy physics
Szasz, Peter, mathematics
Szusz, James, mechanics
Tasi, James, mechanics
Terry, Orville Whitfield, marine ecology
Tewarson, Reginald P., applied mathematics, computer science
Theys, John C., astrophysics
Thorpe, John Alden, geometry, cosmology
Toll, John Sampson, theoretical physics
Tu, Shu-i, biophysical chemistry
Tucker, Alan Curtiss, mathematics
Tunik, Bernard D., physiology
Turner, Clarence Marshall, physics
Twarog, Betty Mack, physiology
Tycko, Daniel H, computer sciences, experimental high energy physics
Upton, Arthur Canfield, experimental pathology
Van der Kloot, William George, physiology
Varanasi, Prasad, planetary atmospheres, spectroscopy
Walcott, Benjamin, comparative physiology
Walcott, Charles, behavioral physiology
Weisberger, William I., theoretical high energy physics
Weisbroth, Steven H., laboratory animal medicine, comparative pathology
Weiser, David M., inorganic chemistry
Weyl, Peter K., environmental management
Wheeler, Margaret Cameron, anthropology
Whitten, Jerry Lynn, theoretical chemistry
Williams, George Christopher, zoology
Williamson, David Lee, genetics
Wimmer, Eckard, molecular biology, virology
Wingate, Catharine L., radiological physics, biophysics
Witkovsky, Paul, sensory physiology
Yang, Chen Ning, physics
Yoles, Stanley Fausst, medicine,
Yuan, William Jen Chun, mathematical statistics, medical statistics
Zaustinsky, Eugene Michael, geometry
Zemanian, Armen Humpartsoum, applied mathematics

STONY POINT

Erickson, Raymond Leroy, organic chemistry
Kesslin, George, organic chemistry

SUFFERN

Ahuja, Satinder, pharmaceutical chemistry
Anderson, Jon C., cosmetic chemistry, pharmaceutical chemistry
Copson, Harry Rollason, corrosion
Dill, Aloys John, electrochemistry
Illian, Carl Richard, analytical chemistry, pharmacy
Karg, Gerhart, physical chemistry, cosmetic chemistry
Kender, Donald Nicholas, analytical chemistry
Kline, Berry James, pharmaceutical chemistry
Mollica, Joseph Anthony, pharmaceutical chemistry, analytical chemistry
Padmanabhan, G R., analytical chemistry, physical chemistry
Pisano, Frank D., physical pharmacy
Rhoda, Richard Noble, chemistry
Rich, Arthur Gilbert, physical pharmacy, pharmaceutical chemistry
Rowland, Charles Sherman, organic chemistry
Shay, Edward Griffin, industrial chemistry
Stober, Henry Carl, analytical chemistry, pharmaceutical chemistry
Zeitin, Benjamin Raphael, toxicology, pharmacology

SYOSSET

Pereira, Martin Rodrigues, environmental physiology
Zinman, Walter George, chemistry

SYRACUSE

Ackerman, Norman Bernard, surgery, cancer
Akers, Sheldon Buckingham, Jr., mathematics
Alexander, Maurice Myron, vertebrate ecology
Allen, Douglas Charles, insect ecology
Ames, Ira Harold, cell biology
Argyris, Bertie, zoology
Argyris, Thomas Stephen, experimental pathology
Auchincloss, Joseph Howland, Jr., internal medicine
Balbinder, Elias, genetics
Barlow, Robert Brown, Jr., neurophysiology
Barth, Karl Frederick, mathematical analysis
Bartos, Henry R., medicine
Becker, Robert O., orthopedic surgery
Beer, Sylvan Zavi, physical chemistry
Behrend, Donald Fraser, wildlife management, environmental sciences
Berglund, John Verne, silviculture, forest ecology
Bergmann, Peter Gabriel, theoretical physics
Bergstrom, William H., pediatrics
Berry, Herbert Weaver, physics
Beverung, Warren Neil, medicinal chemistry
Bharati, Agehananda, / cultural anthropology, South Asian studies
Biermann, Wendell J., physical chemistry
Bierwagen, Max Eugene, pharmacology
Billotti, Joseph Eugene, mathematics
Black, Peter Elliot, watershed management
Blackman, Jerome, mathematics
Blaha, Stephen, elementary particle physics
Boone, Gary M., petrology, mineralogy
Bortoff, Alexander, physiology
Boteron, Donald George, organic chemistry
Bowen, Kenneth Alan, mathematical logic
Bowles, Gordon Townsend, anthropology
Bowman, Roger Holmes, biochemistry
Bradner, William Turnbull, bacteriology, oncology
Brennan, Robert Owings, theoretical physics
Brezner, Jerome, entomology
Brouillette, Walter, solid state physics
Brower, James Clinton, invertebrate paleontology
Brumberger, Harry, physical chemistry
Bryan, John Kent, plant physiology, biochemistry
Buchanan, Ronald Leslie, medicinal chemistry
Bunkfeldt, Rudolf, biochemistry
Burdick, Daniel, surgery
Burger, Walter, radiobiology, virology
Bursnall, John Trehane, structural geology
Burtt, Benjamin Pickering, physical chemistry, radiation chemistry
Byles, Peter Henry, anesthesiology
Cargo, Gerald Thomas, mathematical analysis
Caruso, Frank San Carlo, pharmacology
Chamberlain, Charles Craig, medical physics, radiobiology
Chan, Samuel H P, biochemistry, molecular biology
Church, Philip Throop, topology
Claridge, Charles Alfred, microbiology
Cole, Nancy, mathematics
Collette, Alfred Thomas, botany, human genetics
Collins, George H., pathology, neurology
Colton, Raymond H., otolaryngology
Conan, Robert James, Jr., physical chemistry
Coppola, Patrick Paul, physical chemistry, electronics
Cote, Wilfred Arthur, Jr., electron microscopy, wood science
Cubitt, John Malcolm, geology, statistical analysis
Cunia, Tiberius, statistics
Dabrowiak, James Chester, chemistry
Daly, Robert Ward, psychiatry, psychoanalysis

Davidson, Robert W, wood science, wood technology
De Gennaro, Louis D, zoology
de Laubenfels, David John, geography, botany
De Waard, Dirk, geology
De Zeeuw, Carl Henri, wood technology, mechanics
Dibble, Marjorie Veit, nutrition, food science
Dickison, Harry Leo, pharmacology
Dindal, Daniel Lee, ecology
Di Stefano, Henry Saverio, anatomy
Dittmer, Donald Charles, organic chemistry
Dobkin, Allen Benjamin, anesthesiology
Doisy, Richard Joseph, biochemistry
Dougherty, Robert Malvin, microbiology, virology
Druger, Marvin, genetics
Dunham, Philip Bigelow, physiology
Eich, Robert, medicine
Elander, Richard Paul, microbiology
Elwood, John Clint, biochemistry
Eschner, Arthur Richard, forestry, hydrology
Essery, John M, organic chemistry, medicinal chemistry
Farrell, James Kenneth, organic chemistry
Feldman, Alan Sidney, audiology, speech pathology
Feldman, Harry Alfred, medicine
Ficalora, Peter, physical chemistry, solid state chemistry
Flaherty, Edward John, Jr, theoretical physics
Florini, James Ralph, biochemistry
Fondy, Thomas Paul, biochemistry, chemotherapy
Frank, Thomas Stolley, mathematics, computer science
Furth, Frank Willard, internal medicine, hematology
Gabrusewycz-Garcia, Natalia, cell biology, cytology
Garcia, Alfredo Mariano, biology, anatomy
Gardner, Lytt Irvine, pediatrics
Garvey, Justine Spring, immunochemistry
Gaver, Robert Calvin, drug metabolism
Gilbert, Paul Wilner, mathematics
Ginsburg, Nathan, physics
Godfrey, John Carl, organic chemistry
Goings, Richard Lewis, dairy nutrition, dairy husbandry
Goldberg, Joshua Norman, physics
Goldberg, Marvin, high energy physics
Goodisman, Jerry, chemical physics
Goodman, Donald Charles, neuroanatomy, comparative neurology
Gorbatsevich, Serge N, pulp & paper technology
Gordon, Gerald Bernard, pathology
Gordon, Maxwell, medicinal chemistry
Gottlieb, Arlan J, internal medicine, hematology
Granatek, Alphonse Peter, pharmaceutical chemistry
Granatek, Edmund Stanley, pharmacy, pharmaceutical chemistry
Graver, Jack Edward, mathematics
Graves, William Earl, endocrinology, physiology
Griffin, David H, mycology
Grillot, Gerald Francis, organic chemistry
Gwaltney, John L, anthropology, ethnography
Gylys, Jonas Antanas, pharmacology
Hahn, Roger C, organic chemistry
Hainsworth, Fenwick Reed, zoology
Hammond, Warner Smith, anatomy
Hand, Bryce Moyer, sedimentology
Harlow, William Morehouse, forestry
Harris, Carl Matthew, operations research, industrial engineering
Hartenstein, Roy, physiology, biochemistry
Harth, Erich Martin, neurosciences, mathematical biophysics
Heinemann, Bernard, bacteriology
Hemmingsen, Erik, mathematics
Henderson, Donald, sensory psychology, sensory physiology
Hennigan, Robert Dwyer, environmental management
Henry, John Bernard, medicine
Herrington, Lee Pierce, forest meteorology, micrometeorology
Hill, Rolla B, Jr, pathology, developmental biology
Hirth, Robert Stephen, veterinary pathology
Hoekstra, Justin Bernard, pharmacology
Holz, George Gilbert, Jr, protozoology, physiology

Honig, Arnold, semiconductors, low temperature physics
Hooper, Irving R, medicinal chemistry
Horel, James Alan, physiological psychology, neuroanatomy
Horn, Allen Frederick, Jr, forest management
Horwitz, Nahmin, physics
Hottendorf, Girard Harold, toxicology, comparative pathology
Howard, Philip Hall, environmental sciences
Hoyt, Ernest Basil, physical chemistry
Hsu, Robert Ying, biochemistry
Hunt, George Albert, microbiology
Ise, Charles Masao, biochemistry
Jackson, Daniel Francis, limnology
Jacobs, Ross, biochemistry
Jahn, Edwin Cornelius, organic chemistry
Jensen, Robert Granville, economic geography, geography of the Soviet Union
Johnson, David Aaron, organic chemistry
Johnson, Guy, Jr, mathematical analysis
Johnson, John William, silviculture
Jones, David B, pathology
Jordan, Edward Hill, plant morphogenesis
Juby, Peter Frederick, medicinal chemistry
Jurkat, Wolfgang Bernhard, mathematical analysis
Kalogeropoulos, Theodore E, high energy physics
Kanda, Frank Albert, chemistry, metallurgy
Kaplan, Harvey, theoretical physics
Keller, Douglas Vern, Jr, physical chemistry
Kerr, Marilyn Sue, developmental biology
Kerridge, Kenneth A, medicinal chemistry, organic chemistry
Ketchledge, Edwin Herbert, forest ecology, bryology
Kibbey, Donald Eugene, mathematics
Kieffer, Stephen A, radiology
King, Robert (Bainton), surgery
Klemperer, Friedrich Wilhelm, medicine
Kohls, Carl William, mathematics
Kopel, William Otis, pediatrics
Kriebel, Mahlon E, physiology, comparative physiology
Kronman, Martin Jesse, physical chemistry
Kuehnert, Charles Carroll, plant morphogenesis
Kurczewski, Frank E, entomology
Lalonde, Robert Thomas, organic chemistry
Landaw, Stephen Arthur, internal medicine, hematology
Lande, Sheldon Sidney, environmental chemistry
Lane, Alexander Z, biochemistry, medicine
Lanier, Gerald Norman, forest entomology
Lardy, Lawrence James, numerical analysis
Larson, Charles Conrad, forestry administration
Leaf, Albert Lazarus, forest soils
Leitner, Felix, microbiology, chemotherapy
Leopold, Bengt, paper chemistry, pulp & paper technology
Levine, Robert Alan, medicine, pharmacology
Levinstein, Henry, physics
Levy, Hans Richard, biochemistry
Lilien, Otto Michael, genitourinary surgery
Lindberg, John Albert, Jr, mathematical analysis
Lissner, David, mathematics
LoSurdo, Antonio, physical chemistry, water chemistry
Lourie, Herbert, neurosurgery
Lowe, Josiah Lincoln, mycology
Lozner, Eugene Leonard, clinical medicine
Lugthart, Garrit John, Jr, entomology, genetics
Lundgren, Donald George, microbiology
Luner, Philip, physical chemistry
Macero, Daniel Joseph, analytical chemistry
MacNintch, John Edwin, biochemistry
Madissoo, Harry, veterinary toxicology
Mallov, Samuel, pharmacology
Marino, Andrew Anthony, biophysics
Marsden, David Henry, agricultural microbiology
Marton, Renata, chemistry
Marucci, Americo Alvin, immunochemistry
Marzolf, John George, solid state physics

Matthews, David Livingstone, agronomy, plant breeding
Mattson, Harold F, Jr, applied mathematics
McAfee, John Gilmour, radiology
McDowell, Larry Leon, plant pathology
McGinn, Clifford, organic chemistry
McGraw, James Lorenz, ophthalmology
McGregor, Donald Neil, organic chemistry
McNaughton, Samuel J, plant ecology
Mehrotra, Kishan Gopal, mathematical statistics
Meinig, Donald William, cultural geography, geography of the United States
Merriam, Daniel Francis, geology
Meyer, Franz, biochemistry
Meyer, Haruko, biochemistry, microbiology
Meyer, John Austin, analytical chemistry
Michalek, Joel Edmund, statistics
Miller, Allen H, physics
Miller, Howard Charles, entomology
Miller, John Henry, plant physiology
Miller, Myron, internal medicine, endocrinology
Miller, Pauline Monz, botany, information science
Millman, George Harold, ionospheric physics
Minckler, Leon Sherwood, forestry, ecology
Misiek, Martin, microbiology
Mitchell, Myron James, ecology
Moneti, Giancarlo, elementary particle physics
Morgan, John Clifford, II, mathematical statistics, pure mathematics
Morgan, Kathryn A, mathematics
Moyer, Robert (Findley), radiation physics
Mozell, Maxwell Mark, sensory physiology, psychophysiology
Mozley, James Marshall, Jr, biomedical engineering
Mueller, Justus Frederick, zoology
Muller, Ernest Hathaway, geology
Muller, Otto Heinrich, physiology
Muller-Schwarze, Dietland, animal behavior
Murray, Thomas Edmund, solid state physics
Nafie, Laurence Allen, physical chemistry
Nakatsugawa, Tsutomu, insect toxicology, insecticide toxicology
Negishi, Ei-Ichi, organic chemistry
Nelson, Douglas A, clinical pathology, hematology
Nesbitt, Robert Edward Lee, Jr, obstetrics & gynecology
Nettleton, Donald Edward, Jr, organic chemistry, biochemistry
Neville, John F, Jr, medicine
Newman, James L, geography
Nye, Osborne Barr, Jr, invertebrate paleontology
Oates, Richard Patrick, biostatistics
O'Brien, James Francis, cytology, microbiology
Oken, Donald, psychiatry
Oski, Frank, medicine, pediatrics
Pachter, Irwin Jacob, organic chemistry
Pardee, Otway O'Meara, computer science, mathematics
Parkins, Charles Warren, otorhinolaryngology, speech & hearing science
Payne, Harrison H, zoology, wildlife conservation
Pearse, George Ancell, Jr, analytical chemistry
Pentoney, Richard Ellis, forestry
Petriceks, Janis, forest economics
Pfeffer, Morris, biochemistry
Pfluger, Clarence Eugene, analytical chemistry, x-ray crystallography
Phillips, Arthur William, Jr, microbiology, environmental sciences
Phillips, Richard Hart, psychiatry
Plate, Henry, agronomy
Porter, Gilbert Harris, animal nutrition
Prestayko, Archie William, biochemistry
Preston, James Benson, physiology
Price, Kenneth Elbert, chemotherapy
Prior, John Thompson, pathology
Prucha, John James, structural geology
Pusch, Allen Lewis, clinical pathology
Raynal, Dudley Jones, plant ecology
Redl, George, medicinal chemistry
Reed, George Farrell, otolaryngology
Reichmanis, Maria, experimental biophysics
Reynolds, John C, computer sciences
Ringler, Neil Harrison, aquatic ecology
Roberts, Richard Norman, biochemistry

Robertson, Douglas Reed, anatomy, endocrinology
Robinson, Beatrice Barbara, phycology
Robinson, David Bancroft, psychiatry
Robinson, Joseph Douglass, pharmacology
Rogers, Lloyd Sloan, surgery
Rohrlich, Fritz, theoretical physics
Ross, Gilbert Stuart, neurology
Rotheim, Minna B, microbial genetics
Rowntree, Rowan Allen, resource geography
Russell, Virginia Ann, analytical chemistry, inorganic chemistry
Sage, Gloria W, physical chemistry, clinical chemistry
Sage, Martin Lee, chemical physics
Sagerman, Robert H, medicine, radiology
Samson, Donald C, internal medicine
Sandell, Dewey Jay, physical chemistry
Sapino, Chester, Jr, organic chemistry
Sarko, Anatole, polymer chemistry, polymer physics
Schechter, Joseph M, physics
Schechter, Marshall David, child psychiatry
Schmidt, Richard Penrose, medicine, neurology
Schmitz, Henry, organic chemistry
Schroder, Klaus, solid state physics, materials science
Schuerch, Conrad, organic chemistry
Schwartz, Nathan, applied mathematics
Semon, Warren Lloyd, mathematics
Serfozo, Richard Frank, operations research
Sheehe, Paul Robert, epidemiology
Sherman, Frederick George, biology
Shirley, Hardy Lomax, forest management, forest economics
Sibert, Elbert Ernest, computer science, mathematics
Silverborb, Savel Benhard, forestry
Silverstein, Robert Milton, natural products chemistry
Simeone, John Babtista, forest entomology
Siminoff, Paul, virology, immunobiology
Skaar, Christen, wood technology
Sleezer, Paul David, organic chemistry
Slepecky, Ralph Andrew, microbiology
Smid, Johannes, physical chemistry, polymer chemistry
Smith, James Eldon, microbiology
Smith, James F, mathematical analysis
Smith, Kenneth Judson, Jr, physical chemistry, polymer physics
Smulyan, Harold, internal medicine, cardiology
Sopher, David E, geography
Spitzer, Roger Earl, pediatrics, immunology
Staab, Frank William, organic chemistry
Steele, Robert L, dairy science
Steinschneider, Alfred, pediatrics
Stern, Silviu Alexander, physical chemistry, chemical engineering
Stewart, William Andrew, medicine
Stiteler, William Merle III, statistics, forestry
Streeten, David Henry Palmer, internal medicine
Sturr, Joseph Francis, psychophysiology
Subramanian, Gopal, nuclear medicine, chemical engineering
Sullivan, David Thomas, developmental genetics
Szasz, Thomas Stephen, psychiatry
Szwarc, Michael, physical chemistry
Talham, Robert J, applied mathematics, acoustics
Tanenbaum, Stuart William, biochemistry, microbiology
Tepper, Herbert Bernard, plant anatomy
Tepperman, Helen Murphy, physiological chemistry
Tepperman, Jay, medicine
Thiele, Victoria Florence, nutrition
Timell, Tore Erik, organic chemistry
Townsend, John Marshall, medical anthropology
Trimble, Mary Ellen, physiology, biology
Trischka, John Wilson, physics
Tsai, Yuan-Hwang, microbiology
Turkki, Pirkko Reetta, nutrition, food science
Valentine, Fredrick Arthur, forest genetics
VanDruff, Larry Wayne, wildlife biology
Verrillo, Ronald Thomas, psychophysics, neurosciences
Vida, Julius Adalbert, medicinal chemistry
Vincow, Gershon, physical chemistry
Vook, Richard Werner, solid state physics
Vournakis, John Nicholas, biophysics, molecular biology

Wali, Kameshwar C., theoretical physics
Walnut, Thomas Henry, Jr., quantum chemistry
Walton, Daniel C., plant biochemistry, plant physiology
Wamser, Christian Albert, inorganic chemistry
Wang, Chun-Juan Kao, mycology
Waterman, Daniel, mathematics
Watkins, Mark E., mathematics
Wayner, Matthew John, neurosciences, psychopharmacology
Webb, Watts Rankin, surgery
Webb, William Leonard, vertebrate ecology
Weiner, Irwin M., pharmacology
Wellner, Marcel, physics
Werner, Robert George, zoology
Wessel, Gunter Kurt, physics
Westerfeld, Wilfred Wiedey, biochemistry
Westlake, Robert Elmer, internal medicine, cardiology
Wheatley, William Bacon, organic
Wilcox, Hugh Edward, plant physiology
Williams, Donald Errol, physical chemistry
Williams, William Joseph, medicine, biochemistry
Willner, David, organic chemistry
Wirth, Henry Edgar, chemistry
Wolf, Larry Louis, zoology
Woodin, William Graves, allergy
Wright, Herbert N., psychoacoustics, audiology
Yavorsky, John Michael, wood technology
Ycas, Martynas, biology
Yntema, Chester Loomis, anatomy
Yourno, Joseph Dominic, molecular biology; microbial genetics
Zabel, Robert Alger, forest pathology
Zwislocki, Jozef John, psychophysics, neurosciences

TARRYTOWN

Angell, Charles Leslie, physical chemistry
Aspinall, Samuel Rusnisell, organic chemistry; research administration
Atwood, Gilbert Richard, physical chemistry
Becker, Barbara, biochemistry
Beretsky, Irwin, biomedical engineering, bioacoustics
Bezman, Richard David, physical chemistry
Blake, Robert J., physical chemistry
Bolton, Anthony Peter, physical chemistry
Breck, Donald Wesley, inorganic chemistry
Cummerow, Robert Leggett, organic chemistry
Cupper, Robert Alton, organic chemistry; petroleum chemistry
Doebler, Gerald Francis, biochemistry
Duisman, Jack Arnold, physical chemistry
Dunks, Gary Burr, organometallic chemistry
Durante, Anthony Joseph, organic chemistry
Dwyer, Robert Francis, analytical chemistry; physical chemistry
Eagar, Robert Gouldman, Jr., bio-organic chemistry
Eisenhardt, William Anthony, Jr., organic chemistry
Farina, Peter R., bio-organic chemistry
Faucher, Joseph A., cosmetics chemistry
Flanigen, Edith Marie, inorganic chemistry
Flank, William H., physical chemistry, catalysis
Friedlander, Henry Z., organic chemistry
Friedmann, Herman H., physical chemistry, spectroscopy
Gortsema, Frank Peter, inorganic chemistry
Grattan, James Alex, bio-organic chemistry
Green, Gerald, immunochemistry
Grossman, Steven Harris, biochemistry
Haas, Gerhard Julius, microbiology, biochemistry
Hamilton, Robert William, Jr., environmental medicine
Hargitay, Bartholomew, physical chemistry, colloid chemistry
Haynes, Harry Leonard, entomology
Hickenell, Gary L., food chemistry
Jex, Victor Bird, physical organic chemistry
Kassi, Paul Haruo, physical chemistry, food chemistry,
Kolor, Michael Garrett, mass spectrometry

Lawrence, Robert Howard, Jr., plant physiology; biochemistry
Lee, Martin Jerome, biochemistry
Leung, Pak Sang, colloid chemistry
Livengood, Samuel Miller, organic chemistry
Lo, Donald Hung-Tak, clinical chemistry
Macek, Thomas Joseph, pharmaceutical chemistry
Mackay, Lottie Elizabeth Bohm, science writing, science education
Margoshes, Marvin, analytical chemistry
Marsden, James G., chemistry
Matteo, Robert Allan, organic chemistry
Matteo, Martha R., biochemistry
Mays, Rolland Lee, analytical chemistry
McGuinness, Michael Joseph, Jr., physical chemistry
McKeon, James Edward, organic chemistry
McMullen, Charles Henry, organic chemistry
Molme, Sheldon Walter, biochemistry
Mulhaupt, Joseph Timothy, physical chemistry
Neale, Robert S., organic chemistry
Oberhadt, Bruce J., biophysics
Pace, Salvatore Joseph, chemistry
Patton, Robert Lyle, inorganic chemistry
Phillips, Benjamin, organic chemistry
Pike, John Nazarian, optical physics
Prescott, Henry Emil, Jr., botany; food science
Projetto, Lillian Jacqueline, biology
Prokai, Bela, organometallic chemistry
Ranney, Maurice William, physical chemistry
Rinfret, Arthur Piers, biophysics
Rosemund, Walter Richard, polymer chemistry
Saleeb, Fouad Zaki, surface chemistry; physical chemistry
Schilling, Curtis Louis, Jr., polymer chemistry; organic chemistry
Schreiner, Heinz Rupert, biochemistry; physical chemistry
Scott, William James, electrochemistry
Shapiro, Stephen Irving, biomedical engineering
Snyder, Lloyd Robert, clinical chemistry
Solomon, Jack, physical chemistry
Spokes, Gilbert Neil, research administration
Sterman, Samuel, physical chemistry
Sweet, Richard Clark, analytical chemistry; physical chemistry
Szekely, Andrew Geza, physical chemistry
Tapp, William Jouette, organic chemistry
Tinsley, Samuel Weaver, organic chemistry
Wang, Shu Lung, data processing
Weeks, Thomas Joseph, Jr., physical inorganic chemistry
Weidlein, Edward Ray, Jr., organic chemistry
Welch, Thomas Harris, organic chemistry
Wolf, Philip Frank, physical organic chemistry
Woodcock, Charles Martin, food science

THIELLS

Goldstein, Fred Bernard, biochemistry

THORNWOOD

Tucciarone, John Peter, mathematics, law

TOMKINS COVE

Fahrenbach, Marvin Jay, clinical pharmacology

TONAWANDA

Benzinger, James Robert, plastics
Berger, S Edmund, organic chemistry
Chang, Ching Ming, fluid physics, engineering science
Sauer, Charles William, organic chemistry
Siconolfi, Carmine Anthony, organic chemistry

TROY

Abdali, Syed Kamal, computer sciences
Aikens, David Andrew, electroanalytical chemistry
Albrecht, Robert Michael, epidemiology
Altwicker, Elmar Robert, organic chemistry
Amadigo, John C., applied mechanics
Auclair, Walter, cell biology; developmental biology

Baechler, Raymond Dallas, organic chemistry
Bailey, Ronald Albert, inorganic chemistry
Baldwin, George C., physics
Bassett, Lewis Gordon, analytical chemistry
Bauer, Walter Hermann, physical chemistry
Bayly, M Brian, geology
Bing, Kurt, mathematics
Birnbaum, Meyer Harold, physics, rheology
Block, Robert Charles, nuclear physics, nuclear engineering
Boyce, William Edward, applied mathematics
Breed, Helen Illick, zoology
Breed, Henry Eltinge, physics
Brown, Edmond, theoretical solid state physics
Bunce, Stanley Chalmers, organic chemistry
Campbell, James Dow, mathematics
Carlson, Mildred V., biochemistry
Carter, Richard Leston, mathematical statistics
Casabella, Philip A., physics
Clark, Herbert Mottram, radiochemistry
Clesceri, Lenore Stanke, biochemistry; microbiology
Cluxton, David H., experimental physics
Corelli, John Charles, radiation biophysics, solid state science
Current, Michael Ira, surface physics
Diaz, Joaquin Basilio, applied mathematics
Di Prima, Richard Clyde, applied mathematics
Diwan, Joyce Johnson, cell biology
Dondes, Seymour, physical chemistry
Doremus, Robert Howard, physical chemistry
Downey, Joseph Robert, Jr., physical inorganic chemistry
Dunn, James Robert, geology
Ehrlich, Henry Lutz, microbiology
Eppenstein, Walter, experimental physics
Ferris, James Peter, biochemistry, organic chemistry
Fleishman, Bernard Abraham, applied mathematics
Friedman, Gerald Manfred, sedimentology; sedimentary petrology
Gehrig, Robert Frank, microbiology
Gladstone, Matthew Theodore, chemistry
Glaros, George Raymond, organic chemistry
Gutmann, Ronald Jay, microwave
Habetler, George Joseph, mathematical
Handelman, George Herman, applied mathematics
Harper, Richard Allan, biophysics, physics
Harteck, Paul, physical chemistry
Heidenthal, Gertrude Antoinette, zoology
Hepfinger, Norbert Francis, organic chemistry; polymer chemistry
Herbrandson, Harry Fred, organic chemistry
Hickok, Robert Lyman, Jr., plasma physics
Hollinger, Henry Boughton, physical chemistry
Hollingsworth, Jack W., mathematics
Horn, Edward Gustav, ecology; population biology
Hsieh, Ryine Tsu-Shou, biophysics
Hubbard, James K., organic chemistry
Huntington, Hillard Bell, solid state physics
Huston, Antoinette Killen, mathematics
Jacobson, Melvin Joseph, applied mathematics, underwater acoustics
Janz, George John, physical chemistry
Johnson, William Harding, physiology; biophysics
Jorgensen, Helmut Erik Milo, physical chemistry
Katz, J Lawrence, biophysics, biomedical engineering
Katz, Samuel, geophysics
Krause, Sonja, physical chemistry
LaFleur, Robert George, geomorphology
Landau, Joseph Victor, molecular biology
Leitner, Alfred, mathematical physics
Lemke, Carlton Edward, mathematics
Levinger, Joseph S., physics
Levy, Roland Albert, solid state physics
Luchins, Edith Hirsch, mathematics
Maley, Martin Paul, physics
Marcelli, Joseph F., organic chemistry
McGloin, Paul Arthur, mathematics, computer science

McKinley, William Albert, physics
McNaughton, Robert, mathematics, computer sciences
Medicus, Heinrich Adolf, nuclear physics
Meltzer, Alan Sidney, astronomy
Miller, Donald Spencer, geochemistry
Miller, Kenneth John, theoretical chemistry
Min, Kongki, nuclear physics
Moore, James Alfred, organic chemistry, polymer chemistry
Moss, Gerald, biomedical engineering, surgery
Muckenfuss, Charles, theoretical chemistry
Nettel, Stephen J E, solid state physics
Newell, Jonathan Clark, physiology; biomedical engineering
Newton, Thomas Allen, organic chemistry
Ohanian, Hans C., physics
Ollmann, Loyal Taylor, mathematics
Park, Richard Avery, IV, environmental sciences
Parsons, Robert Hathaway, comparative physiology
Pfau, Charles Julius, virology
Potts, Kevin T., organic chemistry
Preiss, Kevin Louis, nuclear chemistry
Resnick, Robert, physics
Richtol, Herbert H., analytical chemistry, photochemistry
Rockwood, William Philip, endocrinology, physiology
Rogers, Edwin Henry, applied mathematics
Ross, Sydney, colloid chemistry
Roy, Harry, plant biochemistry
Rubenfeld, Lester A., mathematical
Salinger, Gerhard Ludwig, low temperature physics
Schmiedeshoff, Frederick William, research administration, applied
Slack, Nancy G., plant ecology
Smith, Lois C., organic chemistry
Smith, Richard Lloyd, astrophysics
Somerscales, Euan Francis Cuthbert, heat transfer; fluid mechanics
Sperber, Daniel, physics
Sternstein, Sanford Samuel, polymer physics, polymer engineering
Stoler, Paul, experimental nuclear physics
Strong, Robert Lyman, physical chemistry
Surapaneni, Chalapathi Rao, synthetic organic chemistry
Tataczuk, Joseph Richard, nuclear physics; computer engineering
Tiersten, Harry Frank, applied mechanics
Wait, Samuel Charles, Jr., physical chemistry
Walker, Roland, fish pathology
Warden, Joseph Tallman, biophysical chemistry
Warnock, Walter George, mathematics
Wedler, Frederick Charles Oliver, chemistry
White, Frederick Andrew, physics
Wiberley, Stephen Edward, analytical chemistry
Wiedemeier, Herbert, inorganic chemistry
Wilkinson, John Wesley, mathematical statistics
Wilson, Dwight Elliott, Jr., virology
Winhold, Edward John, physics
Wisdom, Norvell Edwin, Jr., physical chemistry
Wunderlich, Bernhard, polymer chemistry
Yergin, Paul Flohr, nuclear physics
Youker, John, physical chemistry
Zelman, Allen, biophysics, bioengineering

TRUMANSBURG

Hodge, Walter Henricks, economic botany

TUCKAHOE

Aboul-Enein, Hassan Youssef, medicinal chemistry, pharmacology
Ceroni, Peter, pharmacology
Grebow, Peter Eric, biopharmaceutics
Johnston, Martha M., biochemistry
Kamath, Burde Laxminarayan, biopharmaceutics
Loev, Bernard, organic chemistry; medicinal chemistry
McGrath, William Robert, pharmacology
Mutson, Daniel, pharmaceutical chemistry
Smith, Charles Giles, biochemistry
Weinryb, Ira, biochemistry

TUXEDO PARK
Benning, Calvin James, organic chemistry
Cohen, Norman, radiological physics, radiobiology
Cubberley, Adrian H, research administration, organic chemistry
Eudy, William Wayne, microbiology, biochemistry
Frankle, William Ernest, paper chemistry
George, Kenneth Dudley, nuclear medicine
Godsay, Madhu, cellulose chemistry
Hall, Frederick Keith, cellulose chemistry
Henderson, David Rippey, organic chemistry
Heusser, Calvin John, botany
Heusser, Linda Olga, palynology
Hung, John Hui-Hsiung, physical chemistry
Leong, Basil K U, pharmocology, toxicology
Love, James Allan, polymer chemistry
Luce, James Edward, paper chemistry
MacDonald, Donald MacKenzie, cellulose chemistry
Manning, James Harvey, paper chemistry
May, Edwin Anthony, biomedical engineering
McIntosh, Douglas Carl, botany, wood technology
Molinski, Victor Joseph, nuclear chemistry
Murphy, John Joseph, organic polymer chemistry, photochemistry
Nass, Harold William, nuclear chemistry, analytical chemistry
Nelson, Kurt Herbert, analytical chemistry
Nesty, Glenn Albert, organic chemistry
Noel, Dale Leon, analytical chemistry
Parisek, Charles Bruce, organic chemistry, polymer chemistry
Parkash, Ved, veterinary medicine, veterinary surgery
Smeltzer, Richard Homer, plant morphogenesis
Solomon, Jerome Jay, physical chemistry, massspectrometry
Stark, John Howard, carbohydrate chemistry
Stier, Paul Max, physics
Taylor, R Perry, physical chemistry
Tuthill, Harlan Lloyd, pharmaceutics
Ward, James Andrew, physical chemistry
Wright, Robert W, physical organic chemistry, polymer chemistry

UNIVERSITY HEIGHTS
Hellman, Henry Martin, chemistry

UPTON
Adams, Peter David, solid state physics
Adler, George, solid state chemistry, physical organic chemistry
Alburger, David Elmer, physics
Allen, Augustine Oliver, physical chemistry, radiation chemistry
Anderson, Carl William, molecular biology, animal virology
Anderson, Robert Christian, organic chemistry
Aronson, Robert Bernard, biochemistry
Atkins, Harold Lewis, nuclear medicine, radiology
Auerbach, Clemens, chemistry
Auerbach, Elliot H, nuclear physics
Axe, John Donald, chemical physics
Baker, Charles Parker, physics
Baltz, Anthony John, theoretical nuclear physics
Barton, Mark Q, physics
Barvenik, Frank W, marine microbiology, microbial ecology
Bender, Michael A, cytogenetics
Bennett, Gerald William, radiation physics, medical physics
Bernstein, Herbert Jacob, computing
Beuhler, Robert James, Jr, physical chemistry
Bhat, Mulki Radhakrishna, nuclear physics
Bird, Richard Putnam, biophysics
Bittner, John William, nuclear physics
Blewett, John Paul, physics
Blumberg, Leroy Norman, physics
Blume, Martin, physics
Bond, Peter Danford, experimental nuclear physics
Bond, Victor Potter, radiation biophysics, experimental medicine
Borg, Donald Cecil, medical research
Brown, Hugh Needham, nuclear physics
Brown, Stanley Gordon, elementary particle physics
Buslik, Arthur J, reactor physics, solid state physics
Campbell, Graham Hays, computer sciences
Carsten, Arland L, radiobiology, health physics
Casten, Richard Francis, nuclear physics
Chanana, Arjun Dev, experimental pathology, surgery
Chasman, Chellis, nuclear physics
Chrien, Robert Edward, nuclear physics
Christman, David R, organic chemistry, pharmaceutical chemistry
Chu, Yung Yee, nuclear chemistry
Chung, Suh Urk, high energy physics
Cleveland, Bruce Taylor, nuclear physics
Cohn, Stanton Harry, physiology, radiobiology
Commerford, Spencer Lewis, biochemistry
Conard, Robert Allen, medical research
Connolly, Philip Louis, physics
Corliss, Lester Myron, solid state physics
Courant, Ernest David, theoretical physics
Cox, David Ernest, crystallography, solid state chemistry
Creutz, Carol Ann, inorganic chemistry
Creutz, Michael John, theoretical physics
Cronkite, Eugene Pitcher, medicine
Cumming, James Burton, nuclear chemistry
Dahl, Per Fridtjof, nuclear physics
Danby, Gordon Thompson, physics
Davis, Raymond, Jr, chemistry
der Mateosian, Edward, nuclear physics
Dienes, George Julian, solid state physics
Divadeenam, Mundrathi, nuclear physics
Dodson, Richard Wolford, physical chemistry
Doniger, Jay, molecular biology
Dover, Carl Bellman, theoretical nuclear physics
Dreiss, Gerard Julius, theoretical physics
Dunn, John Patrick James, molecular biology
Egan, James John, physical chemistry
Ehrenson, Stanton Jay, theoretical chemistry, physical organic chemistry
Engstrom, Herbert Leonard, experimental solid state physics
Evans, John Charles, Jr, radiochemistry
Fairchild, Ralph Grandison, radiological physics
Fajer, Jack, physical chemistry
Feldberg, Stephen William, electrochemistry
Fischer, George J, physics
Flaum, Charles, nuclear physics
Foelsche, Horst Wilhelm Julius, high energy physics
Fowler, Joanna S, organic chemistry
Frazer, Benjamin Chalmers, solid state physics
Freed, Simon, neurophysiological chemistry
Friedlander, Gerhart, nuclear chemistry
Fuller, Richard Clair, theoretical nuclear physics
Gangwer, Thomas E, physical chemistry
Garber, Meyer, physics
Goland, Allen Nathan, solid state physics
Goldhaber, Gertrude Scharff, physics
Goldhaber, Maurice, physics
Goldstein, Charles Irwin, applied mathematics
Goodman, Leon Judias, radiological physics, dosimetry
Gordon, Barry Maxwell, physical chemistry, nuclear chemistry
Greene, Lewis Joel, biochemistry
Grover, James Robb, chemical physics, nuclear science
Growcock, Frederick Bruce, physical chemistry, fuel science
Hamilton, Leonard Derwent, experimental medicine
Hankes, Lawrence Valentine, biochemistry
Harbottle, Garman, nuclear chemistry
Hastings, Julius Mitchell, solid state chemistry
Haustein, Peter Eugene, nuclear chemistry
Hendrie, Joseph Mallam, applied physics, nuclear engineering
Higinbotham, William Alfred, physics
Hillman, Manny, nuclear chemistry, radiochemistry
Hillman, William Sermolino, plant physiology
Hind, Geoffrey, biochemistry, plant physiology
Holroyd, Richard Allan, radiation chemistry
Holtzman, Seymour, comparative endocrinology
Hough, Paul Van Campen, molecular biology
Hudis, Jerome, nuclear chemistry, inorganic chemistry
Hull, Andrew P, health physics
Humphrey, John William, experimental high energy physics
Joel, Darrel Dean, experimental pathology, immunology
Jones, Keith Warlow, experimental atomic physics, applied physics
Jones, William Barclay, physics
Kalbfleisch, George Randolph, particle physics, high energy physics
Kane, Walter Reilly, nuclear physics
Katcoff, Seymour, nuclear chemistry
Kato, Walter Yoneo, reactor physics
Khattak, Chandra Prakash, solid state physics, materials science
Kistner, Ottmar Casper, nuclear physics
Klobuchar, Richard Louis, nuclear chemistry
Knudsen, Knud David, medicine
Koetzle, Thomas F, physical chemistry
Krieger, Theodore Joseph, theoretical physics
Krinsky, Samuel, statistical mechanics, theoretical solid state physics
Ku, Thomas Hsiu-Heng, nuclear chemistry, radiochemistry
Kuper, J B Horner, physics, electronics
Kycia, Thaddeus F, high energy physics
Lacks, Sanford, genetics, biochemistry
Lai, Kwan Wu, high energy physics
Lambrecht, Richard Merle, physical chemistry, nuclear medicine
Lazareth, Otto William, Jr, solid state physics, reactor physics
Ledbetter, Myron C, cell biology, botany
LeFevre, Marian E Willis, physiology
Leipuner, Lawrence Bernard, high energy physics, elementary particle physics
Levine, Melvin Mordecai, engineering physics
LeVine, Micheal Joseph, nuclear physics
Levy, Paul Warren, solid state physics
Li, Kelvin K, physics
Lindenbaum, Seymour Joseph, experimental high energy physics
Liou, Horng Ing, physics
Louttit, Robert Irving, experimental high energy physics
Love, Robert Alexander, radiological health, industrial medicine
Love, William Alfred, physics
Lynn, Kelvin Gideon, solid state physics, metallurgy
MacKenzie, Donald Robertson, physical chemistry
Majumdar, Debaprasad, physics
Marr, Robert B, physics
McInturff, Alfred D, high energy physics, cryogenics
Meinhold, Charles Boyd, health physics
Metz, Donald J, physical chemistry
Michael, Paul Andrew, physics
Miles, Francis Turquand, inorganic chemistry, nuclear engineering
Millener, David John, theoretical nuclear physics
Morris, Thomas Wendell, physics
Mughabghab, Said F, nuclear physics
Muzinich, Ivan J, theoretical high energy physics
Nakadomari, Hisamitsu, electrochemistry
Nathan, Alan Marc, experimental nuclear physics
Nauman, Charles Hartley, biology, genetics
Newman, Leonard, atmospheric chemistry, analytical chemistry
Newton, Marshall Dickinson, theoretical chemistry
Norton, Elinor Frances, analytical chemistry
Olness, John William, nuclear physics
Olson, John Melvin, photobiology
Paige, Frank Eaton, Jr, elementary particle physics
Palmedo, Philip F, nuclear physics, systems analysis
Papavasiliou, Paul S, internal medicine
Parzen, George, theoretical physics
Passell, Laurence, solid state physics
Pasternack, Simon, mathematical physics
Pearlstein, Sol, nuclear physics
Peierls, Ronald F, theoretical high energy physics
Perlman, Morris Leonard, solid state chemistry
Phillips, Robert Hastings, nuclear physics
Piel, William Frederick, nuclear physics
Pisano, Daniel Joseph, Jr, nuclear physics
Popenoe, Edwin Alonzo, biochemistry
Price, Glenn Albert, academic administration, reactor physics
Prodell, Albert Gerald, physics
Protopopescu, Serban Dan, experimental high energy physics
Rahm, David Charles, physics
Raka, Eugene C, physics
Rau, R Ronald, physics
Remsberg, Louis Philip, Jr, nuclear chemistry
Rohrig, Norman, radiological physics
Rubinson, William, nuclear chemistry
Sailor, Vance Lewis, nuclear science
Sakitt, Mark, elementary particle physics
Samios, Nicholas Peter, physics
Sampson, William B, physics
Sayre, Edward Vale, physical chemistry
Schoenborn, Benno P, molecular biology, biophysics
Schwartz, Stephen Eugene, physical chemistry
Schwarz, Harold A, physical chemistry
Schwarzschild, Arthur Zeiger, nuclear physics
Schweitzer, Donald Gerald, reactor physics
Seltzer, Stanley, physical chemistry, organic chemistry
SethuRaman, S, meteorology
Setlow, Jane Kellock, biophysics
Setlow, Richard Burton, biophysics
Sevian, Walter Andrew, applied mathematics, data processing
Shaw, Elliott Nathan, biochemistry
Shellabarger, Claire J, radiobiology, endocrinology
Shimamoto, Yoshio, theoretical physics, mathematics
Shirane, Gen, solid state physics
Siegelman, Harold William, plant biochemistry
Slatkin, Daniel Nathan, pathology
Smith, Harold Hill, genetics
Smith, Lyle W, physics
Snead, Clarence Lewis, Jr, solid state physics
Sparrow, Arnold Hicks, genetics, cytology
Sparrow, Rhoda Cornish, cytology, radiobiology
Srinivasan, Supramaniam, electrochemistry, bioelectrochemistry
Stang, Louis George, Jr, radiochemistry
Stelts, Marion Lee, nuclear physics
Sternheimer, Rudolph Max, atomic physics
Stone, John Patrick, endocrinology, radiobiology
Stoner, Richard Dean, immunology
Strongin, Myron, low temperature physics
Strozier, John Allen, Jr, surface physics
Studier, Frederick William, biophysics
Sunyar, Andrew William, nuclear physics
Sutin, Norman, physical inorganic chemistry
Thieberger, Peter, nuclear physics
Thomas, Robert, crystallography, physical chemistry
Thorndike, Alan Moulton, physics
Toraskar, Jayashree Ravalnath, x-ray astronomy, particle physics
Trigg, George Lockwood, theoretical physics
Trueman, Thomas Laurence, physics
Tuhy, Peter Mirko, organic biochemistry
Underbrink, Alan George, plant cytology, radiobiology
Van Steenbergen, Arie, physics
Van't Hof, Jack, cell biology, radiobiology
Vineyard, George Hoagland, physics
Wagner, George Joseph, plant physiology
Wall, Joseph S, biophysics
Wanderer, Peter John, Jr, high energy physics
Wang, Ling-Lie, high energy physics
Warburton, Ernest Keeling, nuclear physics
Wegner, Harvey E, nuclear physics
Weiss, Jerome, physical chemistry
Weneser, Joseph, theoretical physics
Weston, Ralph E, Jr, chemical kinetics
Williams, Graheme John Bramald, structural chemistry, crystallography
Wiswall, Richard H, Jr, physical chemistry
Yuan, Luke Chia Liu, physics
Zucker, Martin Samuel, plasma physics, radiation physics

UTICA
Baird, Malcolm Barry, physiology, biochemical pharmacology
Behrens, Mildred Esther, zoology
Birnbaum, Linda Silber, molecular biology
Cerny, Laurence Charles, biophysical chemistry
Chamberlain, James Luther, zoology
Connor, Robert Sherman, parasitology
Cutler, Edward Bayler, marine biology
Gotwald, William Harrison, Jr, systematic entomology
Iodice, Arthur Alfonso, biochemistry

NEW YORK

Jesaitis, Raymond G., physical organic chemistry
Macfarlane, Sidley Kerr, geography
Massie, Harold Raymond, molecular biology
Meijer, Robert Randal, physics
Mendez-Bauer, Carlos, obstetrics & gynecology, physiology
Moe, Gordon Kenneth, physiology
Nicholson, Clara K., anthropology
Pier, Harold William, organic chemistry
Stovens, Daniel, pathology
Tuttle, Richard Suneson, physiology, pharmacology
Wu, Chi-Huei, medical microbiology
Wulff, Verner John, neurophysiology

VALHALLA
Benfield, William Harvey, internal medicine, pharmacology
Blumberg, Harold, pharmacology
Browner, Robert Herman, neurobiology, comparative neurology
Chan, Po Chuen, cell biology
Cohen, Leonard A., cell biology
Cottrell, Thomas S., pathology
Danishefsky, Isidore, biochemistry, organic chemistry
Fisher, Clark Alan, microbiology, immunology
Frank, Morton Howard, physiology
Freund, Matthew J., physiology, pharmacology
Giacomelli, Filiberto, experimental pathology
Giannini, Margaret Joan, pediatrics
Guideri, Giancarlo, pharmacology
Gustein, William H., pathology, biophysics
Hecht, Stephen Samuel, organic chemistry, environmental chemistry
Herman, Lawrence, cell biology, electron microscopy
Hoffmann, Dietrich, organic chemistry
Horowitz, Martin I., biochemistry
Hui, Ferdinand W., pharmacology, neuropharmacology
Inchiosa, Mario Anthony, Jr., pharmacology, biochemistry
Lehr, David, pharmacology, medicine
Loud, Alden Vickery, cell biology, biophysics
McBride, Raymond Andrew, immunobiology, pathology
Messina, Edward Joseph, cardiovascular physiology
Milmore, John Edward, neuroendocrinology, pharmacology
Mohos, Steven Charles, pathology, immunology
Moschera, John Anthony, biochemistry
Ortic, Donald, cell biology
Parker, Frank S., biochemistry, spectrochemistry
Peterson, Rudolph Nicholas, biochemistry, pharmacology
Pike, Eileen Halsey, medical parasitology, medical microbiology
Rappaport, Irving, immunology
Reddy, Bandaru Sivarama, biochemistry, nutrition
Rohman, Michael, cardiovascular surgery, thoracic surgery
Rook, George David, pediatrics
Rosenkranz, Herbert S., microbiology, oncology
Schierman, Louis W., immunogenetics
Schiller, Sara, biochemistry
Schmeltz, Irwin, bio-organic chemistry
Settles, Harry Emerson, anatomy
Shapiro, Donald M., applied mathematics, computer science
Shehadi, William Henry, radiology
Shenker, Martin, optics
Siegel, Henry, pathology
Sobol, Bruce J., medicine, physiology
Tabachnick, Milton, biochemistry
Theis, Gail Ann, immunology
Trubach, Janet, neurophysiology, biophysics
Wasserman, Felix Emil, virology, microbial genetics
Weiner, Richard, physiology
Weisburger, John Hans, biochemical pharmacology, oncology
Wenk, Eugene J., anatomy
Wiener, Joseph, pathology, cell biology
Williams, Gary Murray, pathology

VERONA
Lasky, Jack Samuel, organic chemistry, polymer chemistry

VESTAL
Pimbley, Walter Thornton, physics

VOORHEESVILLE
Severinghaus, Charles William, wildlife management

WADING RIVER
Kogel, Marcus David, public health, academic administration

WANAKENA
Castagnozzi, Daniel M., forestry

WANTAGH
Gottesman, Elihu, pharmaceutical chemistry, safety engineering
Gregg, John Henry, physics

WAPPINGERS FALLS
Allred, Harry Milburn, physics
Corbells, Roger, fuel science, high temperature chemistry
Estes, John H., organic chemistry, metallurgy
McCoy, David Ross, organic chemistry
Nolan, John Thomas, Jr., petroleum chemistry, research administration
Oberender, Frederick G., organic chemistry, petroleum chemistry
Raniseski, John Walter, photochemistry, physical organic chemistry

WARWICK
Osterholtz, Frederick David, organic chemistry, radiation chemistry

WATERFORD
Ashby, Bruce Allan, organic chemistry, organometallic chemistry
Bluestein, Ben Alfred, chemistry, polymer chemistry
Bobear, William Joseph, polymer chemistry
Brewer, Stuart Dexter, organometallic chemistry
Modic, Frank Joseph, physical chemistry, organic chemistry
Pfeifer, Charles William, rubber chemistry
Torkelson, Arnold, organometallic chemistry
Wirth, Joseph Glenn, organic chemistry

WATERTOWN
Rauscher, Grant K., physical chemistry

WATERVLIET
Ahmad, Iqbal, physical chemistry, metallurgy
Gray, Alma Marcus, theoretical solid state physics
Gray, Donald M., physics
Weigle, Robert Edward, applied mechanics, applied mathematics

WAVERLY
Cox, George Elton, experimental pathology

WEBSTER
Abkowitz, Martin Arnold, experimental solid state physics
Beatty, Charles Lee, polymer science
Berkes, John Stephan, surface science
Brillson, Leonard Jack, surface physics
Carter, David L., solid state science
Ceasar, Gerald P., chemical physics
Cherin, Paul, solid state science
Chow, Che Chung, physical chemistry
Chow, Tsu-Sen, applied mathematics, mechanics
Donoian, Haig Cadmus, physical chemistry, colloid chemistry
Elder, Fred A., physical chemistry
Elter, John Frederick, electrophotography
Epstein, Arthur Joseph, physics
Erhardt, Peter Franklin, polymer physics
Farley, Thomas Albert, heat transfer
Felty, Evan J., physical chemistry, electrophotography
Ganguly, Bishwa Nath, physics
Gibson, Harry William, physical organic chemistry, organic polymer chemistry
Griffiths, Clifford H., experimental solid state physics, polymer physics
Grushkin, Bernard, physical chemistry, organic chemistry
Gundlach, Robert William, physics
Hunt, Herman Dowd, photographic chemistry
Imperial, George Romero, surface chemistry
Kittelberger, John Stephen, physical chemistry
Kuder, James Edgar, organic chemistry
Lee, Lieng-Huang, organic chemistry
Levy, Mortimer, research administration
Liang, Keng-San, solid state science
Limburg, William W., organic polymer
Lipari, Nunzio Ottavio, solid state physics, molecular physics
Muller, Olaf, inorganic chemistry, solid state chemistry
Murphy, Cornelius Bernard, chemistry
Nash, Robert Joseph, surface chemistry
Nealey, Richard H., organic chemistry
Nielsen, Paul Herron, solid state physics
O'Malley, James Joseph, polymer chemistry, polymer physics
Pearson, James Earl, physical chemistry
Pilato, Philip Anthony, chemistry
Pinsler, Heinz Willi, physics, physical chemistry
Przybylowicz, Edwin P., analytical chemistry
Risko, John James, experimental solid state physics
Roetling, Paul G., physics, optics
Salaneck, William R., solid state physics
Schmidlin, Frederick W., solid state physics
Schwarz, William Merlin, Jr., electrochemistry, physical chemistry
Shaw, Rodney, optics
Slowik, John Henry, solid state physics
Springett, Brian E., physics
Stover, Raymond Webster, physics
Sullivan, Michael Francis, photographic chemistry
Taylor, James Earl, research administration
Tokoli, Emery G., organic chemistry
Trommel, Jan, physical chemistry
Waehner, Kenneth Arthur, analytical chemistry
Ward, Anthony Thomas, physical chemistry
Wegl, John Wolfgang, physical chemistry
Yip, Kwok Leung, solid state physics
Zallen, Richard, solid state physics
Ziolo, Ronald F., inorganic chemistry, physical chemistry

WEST FALLS
Wilkens, Hans J., biochemistry

WEST HEMPSTEAD
Beckter, William Henry, cytology, histology

WEST NYACK
Adickes, H Wayne, organic chemistry
Alexander, Stuart David, pulp & paper technology
Bernier, Gloria A., mathematical statistics, experimental statistics
Dobbins, Robert Joseph, cellulose chemistry
Helfgott, Cecil, physical chemistry, polymer chemistry
Levit, Lawrence Bruce, physics
Schulz, John Hampshire, paper chemistry
Soltes, Edward John, wood chemistry
Wollwage, Paul Carl, carbohydrate chemistry, pulp chemistry

WEST POINT
Hoff, Wilford J. Jr., organic chemistry
MacWilliams, Donald Gribble, chemistry
Medger, Gerald William, physics, mathematics
Pollin, Jack Murph, mathematics, systems engineering
Saunders, Edward A., solid state electronics, nuclear science
Smith, Frederick Adair, Jr., theoretical mechanics, applied mechanics
Streett, William Bernard, physical chemistry, high pressure physics
Willis, James Stewart, Jr., physics

WEST SAND LAKE
Crain, Alfred V R, analytical chemistry

WESTBURY
Blumenthal, Ralph Herbert, physics
Bradley, Arthur, surface chemistry
Cleland, Marshall Robert, nuclear physics
Finkelstein, Frances, biochemistry
Hahn, Eric Walter, radiation biology
Ingham, Herbert Smith, Jr., physics
Malamud, Herbert, medical physics
Morganstern, Kenard H., physics
Savet, Paul H., electronics, particle physics
Winkler, Bertram Stanley, pharmacology
Winsten, Walter Abbott, biochemistry

WESTHAMPTON
Hoag, Warren George, comparative medicine

WHITE PLAINS
Ajl, Samuel Jacob, microbiology
Albanese, Anthony August, biochemistry
Bergsma, Daniel, preventive medicine
Canavan, Frederick Louis, nuclear physics
Cante, Charles John, physical chemistry, surface chemistry
Chapman, George Herbert, biochemical pharmacology
Cocodrilli, Gus D. Jr., nutrition
Cutler, Louise Marie, chemistry
Ellison, Theodore, toxicology
Flynn, Charles Edward, food chemistry
Fortney, Cecil Garfield, Jr., food chemistry
Frank, Samuel B., dermatology
Gibbs, James Gendron, Jr., psychiatry
Hall, William Myron, Jr., microbiology
Hamilton, Francis Joseph, psychiatry
Hopper, Paul Frederick, organic chemistry, food science
Insalata, Nino F., microbiology, bacteriology
Kirschman, John C., biochemistry
Lapidus, Herbert, pharmacy, pharmacology
Lee, Chi-Hang, natural products chemistry
Levenson, Harold Samuel, food science
Lhamon, William Taylor, psychiatry
Macaluso, Pat, physical chemistry, computer science
Maisky, Stanley Joseph, radiological physics, medical physics
Moser, Marvin, internal medicine, cardiology
Orto, Louise A (Mrs Patrick Famighetti), laboratory medicine
Otis, Herbert Newell, physics
Reussner, George Henry, dental research, nutrition
Robinson, Louis, mathematics, computer sciences
Scala, James, biochemistry
Schenz, Timothy William, chemistry
Sims, Rex J., organic chemistry
Sluder, John Cochran, organic chemistry
Smith, Gerard Peter, physiology
Staub, Herbert Warren, biochemistry
Yeransian, James A., analytical chemistry

WHITESTONE
Katz, Lenard, chemistry

WILLIAMSVILLE
Amborski, Leonard Edward, physical chemistry, polymer chemistry
Claworthy, Willard Hubert, statistics
Fujita, Shigeji, theoretical physics
Light, Rupert Edwin, polymer chemistry
Obenland, Clayton O., industrial chemistry
Rushton, Brian Mandel, chemistry
Shatz, Marvin Mandel, chemistry
Sheard, John Leo, physical chemistry
Thomas, Edward Carl, inorganic chemistry
Woodburn, Henry Milton, organic chemistry
Zelen, Marvin, biometry, mathematical statistics

WOODBURY
Ferris, Philip, oncology
Freeman, Richard Carl, textile chemistry
Hoffman, Robert, dentistry, orthodontics
McDonald, Margaret Ritchie, biochemistry
Mizell, Louis Richard, textile chemistry
Padnos, Morton, virology, oncology

WOODHAVEN
Stolfi, Robert Louis, immunochemistry, immunology

WOODMERE
Brahen, Leonard S., pharmacology

WOODSIDE
Minieri, P Paul, chemistry

YONKERS
Al-Aidroos, Karen Messing, microbial genetics
App, Alva Agee, plant science
Bown, Delos Edward, organic chemistry
Bozarth, Robert F., plant virology
Buckley, Edward Harland, plant physiology
Castillo, Jessica Maguila, insect pathology, nematology
Dovel, William Lawrence, fish biology
Dowling, Joseph Francis, food chemistry
Gershon, Herman, biochemistry
Granados, Robert R., invertebrate pathology, virology
Jacobson, Jay Stanley, plant physiology, analytical chemistry
Katz, Leon, organic chemistry
Kieffer, Robert G., pharmaceutical chemistry
Langridge, William Henry Russell, virology, developmental biology
Limpel, Lawrence Eugene, entomology
MacLean, David Cameron, plant physiology
Manasse, Robert James, microbiology
Mandl, Richard H., biology
McCallan, Samuel Eugene Alan, plant pathology
McCune, Delbert Charles, plant physiology
McNew, George Lee, plant pathology, microbiology
Mirsky, Joseph Herbert, pharmacology
Mukai, Cromwell Daisaku, chemistry, electrophotography
Mussell, Harry W., plant pathology, plant biochemistry
Parmegiani, Raulo, microbial physiology
Roberts, Donald Wilson, insect pathology
Solon, Leonard Raymond, radiological physics
Staples, Richard Cromwell, phytopathology
Titone, Luke Victor, physics
Torgeson, Dewayne Clinton, plant pathology
Weinberger, Lester, organic chemistry, electrophotography
Weinstein, Leonard Harlan, plant physiology, environmental biology
Wellman, Richard Harrison, plant pathology
Wood, Harry Alan, plant virology
Yaniv, Zohara, plant physiology, phytopathology

YORKTOWN HEIGHTS
Adams, Edward Neufville, information science
Adler, Roy Lee, mathematics
Aitken, John Malcolm, solid state physics
Ames, Irving, physics
Anschel, Morris, organic chemistry, computer science
Aviram, Ari, organic chemistry, chemical physics
Bakis, Raimo, physics
Barnes, Earl Russell, mathematics
Barrekette, Euval S., electrooptics, applied mechanics
Bednowitz, Allan Lloyd, x-ray crystallography
Blachman, Arthur Gilbert, physics
Blakeslee, A Eugene, physical chemistry
Blum, Samuel Emil, physical chemistry
Bogardus, Egbert Hal, solid state science
Braslau, Norman, physics
Brodsky, Marc Herbert, solid state physics
Brown, Rodney Duvall, III, solid state physics, biophysics

Burns, Gerald, physics
Chan, Kwing Lam, astrophysics
Chu, Kai-Ching, applied mathematics, system theory
Cohen, Hirsh G, applied mathematics
Cooley, James William, applied mathematics
Craven, Robert Alan, solid state physics
Cronemeyer, Donald Charles, physics
Cullum, Jane Kehoe, applied mathematics
Damerau, Frederick Jacob, information science
Desilets, Brian H., physics
DiStefano, Thomas Herman, solid state physics
Dreyfus, Russell Warren, lasers
Easki, Leo, solid state physics
Elgot, Calvin C, mathematics, computer science
Engler, Edward Martin, physical organic chemistry, solid state chemistry
Feder, Ralph, solid state physics
Fowler, Alan B, solid state physics, surface physics
Freiser, Marvin Joseph, theoretical physics
Garwin, Richard Lawrence, experimental physics
Giess, Edward August, solid state physics
Gilmore, Paul Carl, mathematical logic
Goldberg, Richard, mathematics
Goldstine, Herman Heine, numerical analysis
Gomory, Ralph E, applied mathematics
Griesmer, James Hugo, mathematics
Grischkowsky, Daniel Richard, physics
Grobman, Warren David, solid state physics
Grossman, Edna K, algebra
Gunn, John Battiscombe, physics
Gustavson, Fred Gehrung, applied mathematics, computer sciences
Gutzwiller, Martin Charles, theoretical physics
Hall, John Jay, solid state physics
Haller, Ivan, physical chemistry
Hardman, Karl David, biochemistry
Harris, Erik Preston, solid state physics
Hempstead, Robert Douglas, magnetism
Henkels, Walter Harvey, solid state physics
Hickmott, Thomas Ward, physical chemistry
Hodgson, Rodney, physics
Hoffman, Alan Jerome, mathematics
Holtzberg, Frederick, physical chemistry
Howard, Webster Eugene, Jr., solid state physics
Irene, Eugene Arthur, solid state physics
Janak, James Francis, electrical engineering, physics
Jenks, Seymour Hillel, biophysics
Jepsen, Donald William, surface physics, statistical mechanics
Johnson, Ellis Lane, operations research
Kelisky, Richard Paul, mathematics, data processing
Keller, Seymour Paul, solid state physics
Keyes, Robert William, physics
Kirkpatrick, Edward Scott, solid state physics
Koenig, Seymour Hillel, biophysics
Konheim, Alan G, mathematics
Krongelb, Sol, physics
Kryder, Mark Howard, solid state physics
Laff, Robert Allan, solid state physics
Laibowitz, Robert (Benjamin), applied physics
Landauer, Rolf William, solid state physics, computer science
Lang, Norton David, theoretical solid state physics, surface physics
Lasher, Gordon (Jewett), astrophysics
Lean, Eric Gung-Hwa, acoustics, optics
Levanoni, Menachem, acoustics, fluid physics
Lew, John S., applied mathematics
Liniger, Werner, numerical analysis
Mandelbrot, Benoit, applied mathematics
Marcus, Paul Malcolm, mathematical physics
Martinez, Nilda, organometallic chemistry
Matthews, John Wauchope, physics
McCorkle, Richard Anthony, plasma physics, lasers
McGroddy, James Cleary, solid state physics
Mehran, Farrokh, physics
Melcher, Robert Lee, solid state physics
Miranker, Willard Lee, mathematics
Morehead, Frederick Ferguson, Jr., physical chemistry, experimental solid state physics
Moss, Thomas Henry, biophysics

Myers, Robert Anthony, solid state physics
Nathan, Marshall I, physics
Odeh, Farouk M, applied mathematics
Ors, Jose Alberto, photochemistry
Pantelides, Sokrates Theodore, theoretical solid state physics
Park, Kyu Chang, inorganic chemistry
Pennebaker, William B, Jr., physics
Pennington, Keith Samuel, physics
Petrick, Stanley R., computer sciences
Pomerantz, Melvin, physics
Price, Peter J, theoretical solid state physics
Pugh, Emerson William, solid state electronics
Pytte, Erling, solid state physics
Raider, Stanley Irwin, chemistry
Reich, Haskell, Aaron, applied physics
Reisman, Arnold, physical chemistry, inorganic chemistry
Reuter, Wilhad, analytical chemistry
Rivlin, Theodore J, mathematics
Rosenberg, Arnold Leonard, information science, systems theory
Roth, John Paul, mathematics, computer science
Rutledge, Joseph Dela, mathematics, computer sciences
Sayre, David, x-ray crystallography
Schultz, Theodore David, theoretical physics
Schweitzer, Paul Jerome, operations research, applied mathematics
Scott, Bruce Albert, physical chemistry, inorganic chemistry
Segmuller, Armin Paul, crystallography
Seiden, Philip Edward, solid state physics
Senko, Michael Edward, computer science, information science
Shih, Kwang Kuo, solid state physics, physical chemistry
Shiren, Norman Steven, physics
Silverman, Benjamin David, solid state physics
Slonczewski, John Casimir, solid state physics
Smart, James Samuel, physics
Smith, Archibald William, lasers
Smith, John Ernest, Jr., solid state physics
Sorokin, Peter, solid state physics
Spiller, Eberhard Adolf, x-ray optics
Steinbeck, Herbert D, medicine
Stern, Frank, theoretical solid state physics
Stockmeyer, Larry Joseph, computer science
Teaney, Dale T, solid state physics, biophysics
Thatcher, James W, computer science, mathematics
Thompson, William A, solid state physics
Tomkiewicz, Micha, solid state physics
Torrance, Jerry Badgley, Jr., physics
Toupin, Richard A, mathematical physics, continuum mechanics
Triebwasser, Sol, physics
Turner, William Joseph, solid state physics
Valsamakis, Emmanuel, solid state electronics
Van Vechten, James Alden, theoretical solid state physics, semiconductors
Von Gutfeld, Robert J, solid state physics
von Molnar, Stephan, solid state physics
Wagner, Eric G, mathematics
Welber, Benjamin, high pressure physics
Welch, Peter D, mathematical statistics
Williams, Arthur Robert, solid state physics
Winograd, Shmuel, mathematics
Wolfe, Philip, applied mathematics
Wong, Chak-Kuen, computer science, applied mathematics
Wu, Lilian Shiao-Yen, applied mathematics
Young, Donald Reeder, physics
Yu, Peter Yound, experimental solid state physics, semiconductors
Ziegler, James Francis, nuclear physics
Zory, Peter Stephen, Jr., physics

YOUNGSTOWN
Ueltz, Herbert Frank, high temperature chemistry

NORTH CAROLINA

ABERDEEN
Beesch, Samuel C, chemistry

ALBEMARLE
Leiby, George Martin, epidemiology

ARDEN
Liberatore, Laurence Columbus, chemistry

ASHEVILLE
Boyce, Stephen Gaddy, plant ecology
Butson, Keith D, climatology
Cole, Robert Stephen, solar physics, quantum physics
Cole, Warren Henry, surgery
Cotrufo, Cosimo (Gus), plant physiology
Crutcher, Harold L, meteorology, chemistry
Haggard, William Henry, climatology
Hepting, George Henry, forest pathology
Hudson, Frederick Mitchell, organic chemistry, explosives
Johnston, Harry Henry, microbiology, biochemistry
Krochmal, Arnold, economic botany
McCoy, John J, ornithology
Murray, Linwood Asa, Jr, physical chemistry
Myrick, Alvin Grant, mathematics
Remington, Lloyd Dean, analytical chemistry
Scott, Stewart Melvin, cardiovascular surgery, thoracic surgery
Speers, Charles Frederick, entomology
Squibb, Samuel Dexter, organic chemistry
Stevens, John Gehret, physical chemistry
Vinson, James S, physics
Wills, James E, Jr, physics
Wilson, Jack Charles, algebra

BEAUFORT
Bonaventura, Celia Jean, photobiology
Bonaventura, Joseph, molecular biology
Bookhout, Cazlyn Green, marine zoology
Costlow, John DeForest, marine invertebrate zoology
Forward, Richard Blair, Jr, comparative physiology
Hunter, Wanda Sanborn, zoology
Huntsman, Gene Raymond, fish biology
Kjelson, Martin Anton, III, ecology, fish biology
Lewis, Robert Minturn, fish biology
McCutcheon, Frederick Harold, comparative physiology
Nicholson, William Robert, fisheries
Rice, Theodore Roosevelt, marine ecology
Sullivan, James Bolling, biochemistry, zoology
Sutherland, John Patrick, marine ecology
Sykes, James Enoch, fisheries
Thayer, Gordon Wallace, ecology

BELMONT
Stuart, Jeanne Jones, zoology

BESSEMER CITY
Bach, Ricardo O, inorganic chemistry, physical chemistry
Gillespie, Arthur Samuel, Jr, industrial chemistry
Hoffman, Doyt K, Jr, industrial chemistry
Kamienski, Conrad William, organic chemistry
Stubblefield, Charles Bryan, analytical chemistry

BOILING SPRINGS
Harrelson, Michael Asbury, botany, ecology

BOONE
Bowkley, Herbert Louis, inorganic chemistry
Buckland, Golden Thaddeus, mathematics
Carpenter, Irvin Watson, Jr., plant taxonomy
Connolly, Walter Curtis, physics
Curd, Rudy Leroy, mathematics
Derrick, Finnis Ray, zoology
Durham, Harvey Ralph, topology
Eargle, George Marvin, applied mathematics
Epperson, Terry Elmer, Jr., geography
Glover, Sandra Jean, entomology
Graham, Ray Logan, mathematics
Henson, Richard Nelson, parasitology
Hubbard, William Ralph, biochemistry, endocrinology
Johnson, James Edwin, physical chemistry
Lane, Ernest Paul, topology

Lindsay, James Gordon, Jr., nuclear physics
Mamola, Karl Charles, solid state physics
McKinney, Frank Kenneth, invertebrate paleontology, biostratigraphy
Miles, George Benjamin, organic chemistry
Nicklin, Robert Clair, solid state physics
Olander, Donald Paul, analytical chemistry
Purrington, Burton Lewin, anthropology, archaeology
Randall, John Frank, biology
Raymond, Loren Arthur, structural geology, petrology
Reiman, Robert Ellis, geography
Rhyne, Thomas Crowell, physical inorganic chemistry
Robinson, Kent, biology, science education
Rokoske, Thomas Leo, solid state physics
Sanders, Oliver Paul, mathematics
Sink, Donald Woodrin, inorganic chemistry
Smith, James Reaves, algebra
Soeder, James W, organic chemistry
Stillwell, Harold Daniel, physical geography, biogeography
Stubblefield, Beauregard, topology
Vergeer, Teunis, biology
Webb, Earl, Jr., geology
Yoder, Julian Clifton, geography

BREVARD
Brown, Raymond Arthur, biochemistry
Cartledge, Groves Howard, chemistry
Cohen, Edward David, physical chemistry, chemical engineering
Eastes, John Wesley, organic chemistry
Hembree, George Hunt, physical chemistry
Koob, Derry Delos, phycology, limnology

BUIES CREEK
Beard, Luther Stanford, systematic botany
Hovis, Louis Samuel, chemistry, physics
Howard, Clarence Edward, mineralogy, crystallography
Jung, James Moser, organic chemistry
Martin, Richard Harold, geology
McIntyre, Robert Allen, Jr, zoology
Taylor, Jerry Duncan, mathematics
Yarbrough, Charles Gerald, zoology

BURLINGTON
Arnold, Luther Bishop, Jr., textile chemistry
Flagg, Raymond Osbourn, botany, science administration
Powell, Thomas Edward, biology

CARY
Bursu, James Edward, medicinal chemistry
Heagle, Allen Streeter, plant pathology

CAMP LEJEUNE
Carroll, Harold Wilson, nutrition, biochemistry
Grothaus, Roger Harry, medical entomology
Shepard, Maurice Charles, medical bacteriology

CEDAR MOUNTAIN
Mendel, Clifford William, mathematics
Van Wagtendonk, Willem Johan, biochemistry

CHAPEL HILL
Abernathy, James Ralph, biostatistics
Adams, Joseph Edison, botany
Anderson, Carl Elmore, biochemistry
Anderson, John Joseph Baxter, physiology, nutrition
Bachenheimer, Steven Larry, virology
Baer, Tomas, chemical physics
Baker, Charles Ray, applied mathematics, engineering
Balfour, Marshall Coulter, public health
Bane, John McGuire, Jr., physical oceanography
Barnett, Thomas Buchanan, medicine
Barrow, Emily Mildred Stacy, physiology
Barry, Edward Gail, genetics
Baseman, Joel Barry, microbiology
Basile, David Giovanni, geography
Battigelli, Mario C., industrial medicine

Bawden, James Wyatt, physiology, dentistry
Bear, Richard Scott, molecular biology, biophysics
Bell, Clyde Ritchie, botany
Bell, Fred E, biochemistry
Bennett, Henry Stanley, anatomy, cell biology
Benson, Walter Russell, pathology
Berkut, Michael Kalen, biological chemistry
Berryhill, Walter Reece, medicine
Bevin, A Griswold, plastic reconstructive surgery
Blaug, Seymour Morton, pharmacy
Bleyman, Michael Alan, molecular genetics
Blythe, William Brevard, internal medicine
Boatman, Ralph Henry, Jr., public health
Bott, Kenneth F., microbiology, genetics
Bowers, Wayne Alexander, physics
Breese, George Richard, pharmacology, neurobiology
Breeman, Robert Alan, medicine
Brinkhous, Kenneth Merle, pathology
Briscoe, Charles Victor, physics
Brockington, Donald Leslie, anthropology
Brookhart, Maurice S, organic chemistry
Brown, Richard Malcolm, Jr., botany
Buck, Richard Pierson, physical chemistry, analytical chemistry
Buckwalter, Joseph Addison, surgery
Bunce, Paul Leslie, medicine
Burford, Hugh Jonathan, pharmacology, medical education
Bursey, Maurice Moyer, analytical chemistry
Butler, James Robert, geology, petrology
Butler, Thomas Cullom, pharmacology
Cameron, Edward Alexander
Campbell, Suzann Kay, physical therapy
Caplow, Michael, biochemistry, enzymology
Carson, Johnny Lee, environmental biology, cytopathology
Cassel, John Charles, epidemiology
Chaffe, Elmer Fenn, immunology
Chamberlin, Harrie Rogers, pediatrics
Chandross, Ronald Jay, biophysics
Chaney, Stephen Gifford, biochemistry
Choe, Jae Yne, pharmacology
Choi, Sang-Il, solid state physics
Christiansen, Wayne Arthur, theoretical chemistry
Christman, Russell Fabrique, chemistry
Chuang, Hanson Yit-Kuan, biochemistry, pathology
Clegg, Thomas Boykin, experimental nuclear physics
Clem, Judy Roberta, parasitology
Clyde, Wallace Alexander, Jr., pediatrics, infectious diseases
Cocolas, George Harry, medicinal chemistry
Coe, Joffre Lanning, anthropology
Coffey, James Cecil, Jr., endocrinology, parasitology
Coke, James Logan, organic chemistry
Collier, Francis Nash, Jr., inorganic chemistry
Cook, Warren Ayer, industrial hygiene
Cooper, Cary Wayne, biology, pharmacology
Cooper, Herbert Asel, experimental pathology, pediatrics
Costello, Donald Paul, zoology, cell biology
Couch, John Nathaniel, botany
Coulter, Norman Arthur, Jr., physiology, biophysics
Craige, Ernest, medicine
Crandell, Clifton E., dentistry
Crane, Julia Gorham, cultural anthropology
Crawford, James Homer, Jr., solid state physics
Crawford, James Joseph, L., microbiology
Crenshaw, Miles Aubrey, physiology
Cromartie, William James, infectious diseases
Crounse, Robert Griffith, dermatology
Curtis, Thomas Edwin, psychiatry
Dalldorf, Frederic Gilbert, pathology
Daniels, Robert Edward, anthropology
Davidian, Nancy McConnell, biochemistry, drug metabolism
Davis, Henry Mauzee, physical chemistry
Davis, Morris Schuyler, astronomy

Davis, Robert Lloyd, mathematics
Daw, John Charles, medical education
Dearman, Henry Hursell, physical chemistry
Dennison, John Manley, geology
Denny, Floyd Wolfe, Jr., pediatrics
D'eustachio, Dominic, physics
Dickison, William Campbell, plant morphology, plant anatomy
Dillard, Margaret Bleick, radiological physics, mathematical physics
Dodd, Arthur V, geography, meteorology
Downie, David Ernest, bioengineering, cardiovascular physiology
Dudley, Kenneth Harrison, pharmacology, organic chemistry
Dugger, Gordon Shelton, neurosurgery
Dunn, David Evan, rock mechanics, structural geology
Dunphy, Donal, pediatrics
Dy, Kian Seng, solid state physics
Easterling, William Ewart, Jr., obstetrics & gynecology
Eckel, Frederick Monroe, pharmacy
Edgell, Marshall Hall, biochemistry, genetics
Elandt-Johnson, Regina C., mathematics, statistics
Eldridge, Frederic L, respiratory physiology
Eliel, Ernest Ludwig, organic chemistry
Ellis, Fred Wilson, pharmacology, biochemistry
Engels, William Louis, zoology
Ennis, Ella Gray Wilson, physiology
Eyre, John Alexander, psychiatry
Eyre, John D., geography
Farel, Paul Bertrand, psychobiology, neurophysiology
Farmer, Thomas Wohlsen, neurology
Faust, Robert Gilbert, cell physiology, biophysics
Feduccia, John Alan, zoology
Feiss, Paul Geoffrey, economic geology, geochemistry
Ferguson, John Howard, physiology
Filley, John Paton, psychiatry
Finn, Arthur Leonard, physiology, biophysics
Fischer, Janet Jordan, internal medicine
Fishman, George Samuel, operations research, applied statistics
Fleming, William Leroy, preventive medicine
Fogleman, Wavell Wainwright, environmental chemistry, inorganic chemistry
Folds, James Donald, microbiology, immunology
Fordham, Christopher Columbus, III., medicine
Fox, Donald Lee, air pollution, atmospheric chemistry
Franceschetti, Donald Ralph, physics, solid state physics
Frankenberg, Dirk, biological oceanography
Fraser, David Allison, occupational health, electron microscopy
Frazier, Patricia Dianne (Murphy), biochemistry
Freymann, Moye Wicks, public health
Fulghum, Robert Schmidt, bacteriology
Fullagar, Paul David, geology
Gatzy, John T Jr., pharmacology
Gardner, Robert B, geometry
Geissinger, Ladnor Dale, mathematics
Gensel, Patricia Gabbey, botany, paleobotany
Geratz, Joachim Dieter, pathology
Glasser, Richard Lee, neurobiology
Glassman, Edward, neurobiology, genetics
Giezen, William Paul, medicine
Goldman, Leonard Jay, infectious diseases
Gooder, Harry, microbiology, genetics, bioacoustics
Gottschalk, Carl William, physiology, nephrology
Goulson, Hilton Thomas, parasitology
Goz, Barry, pharmacology, biochemistry
Graham, John Borden, pathology
Graves, William Howard, mathematical analysis, algebra
Greenberg, Bernard George, biostatistics, public health
Gregg, John Marshall, oral surgery
Greulach, Victor August, plant physiology
Grisham, Joe Wheeler, medicine, pathology
Grizzle, James Ennis, biostatistics
Gross, Kenneth Irwin, mathematical analysis

Gulick, John, anthropology
Hagadorn, Irvine R., comparative physiology; endo crinology
Hager, George Philip, Jr., medicinal chemistry
Hairston, Nelson George, zoology
Haisley, Waldo Emerson, physics
Hall, Iris Beryl Haddon, cell physiology, biochemistry
Hall, Thomas Livingston, public health
Halleck, Seymour Leon, psychiatry
Hanker, Jacob S, histochemistry, cell biology
Hanst, Philip Lincoln, physical chemistry
Harned, Herbert Spencer, Jr., medicine
Harper, Curtis, pharmacology
Harris, Albert Kenneth, Jr., embryology, cell biology
Harrison, John Kennel, IV., biochemistry
Hatfield, William E, inorganic chemistry
Haughton, Geoffrey, immunogenetics, immunology
Hendricks, Charles Henning, obstetrics & gynecology
Hendricks, James Richard, parasitology
Henson, O'Dell Williams, Jr., anatomy
Henson, Anna Miriam (Morgan), neurobiology
Hermans, Jan Joseph, physical chemistry
Hernandez, John Peter, solid state physics
Herring, William Benjamin, internal medicine
Higley, Lester Bodine, orthodontics
Hirsch, Philip Francis, pharmacology
Hiskey, Richard Grant, synthetic organic chemistry, protein chemistry
Hodgson, Derek John, structural chemistry, inorganic chemistry
Hoefling, Wassily, mathematical statistics
Holbrook, David James, Jr., biochemistry
Holcomb, George Ruble, physical anthropology
Hollander, Walter, Jr., internal medicine, nephrology
Hollinshead, William Henry, anatomy
Hollister, William Gray, psychiatry, public health
Hommersand, Max Hoyt, botany
Honigmann, John Joseph, anthropology
Howland, Joe Wiseman, internal medicine, radiobiology
Hubbard, Paul Stancyl, Jr., magnetic resonance, molecular physics
Hudson, R Page, Jr., pathology; forensic medicine
Huffines, William Davis, medicine, pathology
Hulka, Jaroslav Fabian, obstetrics & gynecology
Humm, Douglas George, physiology
Hutchison, Clyde Allen, III., molecular biology, genetics
Johnson, Charles Sidney, Jr., physical chemistry
Ibrahim, Michel A., epidemiology
Ikenberry, Lynn David, psychiatry
Ingram, Roy Lee, geology
Irvin, Joseph Logan, biochemistry
Isenhour, Thomas Lee, analytical chemistry
Ishaq, Khalid Sulaiman, medicinal chemistry
Jarnagin, Richard Calvin, physical chemistry
Jenner, Charles Edwin, biology
Jenner, William Elliott, mathematics
Jenzano, Anthony Francis, astronomy
Johnson, George, Jr., medicine, surgery
Johnson, James Donald, statistics
Johnson, Norman Lloyd, statistics
Jones, Claiborne Stribling, zoology
Kafer, Enid Rosemary, anesthesiology
Karel, Martin Lewis, mathematics
Kaylor, Cornelius Timpson, anatomy
Kiffney, Gustin Thomas, Jr., physics
Killingsworth, Lawrence Madison, clinical chemistry, pathology
King, Jerome Stovall, neurophysiology
Kingdon, Henry Shannon, biochemistry, hematology
Kirkman, Henry Neil, medicine, pediatrics
Kizer, John Stephen, neuroendocrinology
Knight, Samuel Bradley, chemistry
Knobeloch, F X Calvin, pathology; audiology
Koch, William Edward, anatomy
Koch, William Julian, mycology
Koeppe, John K, insect physiology, endocrinology

Krigman, Martin Ross, neuropathology, neurobiology
Kropp, Paul Joseph, organic chemistry
Kuebler, Roy Raymond Jr, mathematical statistics
Kuenzler, Edward Julian, ecology
Kumaroo, Kuziyilethu Krishnan, biochemistry
Kuno, Motoy, physiology
Kusy, Robert Peter, dental research, polymer science
Kusyk, Christine Johanna, genetics, biochemistry
Lachenbruch, Peter Anthony, biostatistics
Langdell, Robert Dana, pathology
Larsh, John E, Jr, medical parasitology
Lassiter, William Edmund, physiology, nephrology
Lay, Douglas M, anatomy, zoology
Leander, John David, psychopharmacology
Lee, Kuo-Hsiung, medicinal chemistry, natural products chemistry
Lehman, Harvey Eugene, developmental biology, embryology
Lehman, Lillian Margot Youngs, embryology
Leighton, Dorothea Cross, medicine, psychiatry
Lemberg, Howard Lee, chemical physics
LeVeau, Barney Francis, physical medicine
Lieth, Helmut H F, botany, ecology
Lindahl, Roy Lawrence, dentistry
Linder, Forrest Edward, statistics
Lipton, Morris Abraham, experimental psychiatry, medicine
Little, Linda West, environmental biology
Little, William Frederick, organic chemistry
Loeffler, Larry James, organic chemistry
Lucchesi, John Charles, developmental genetics
Ludwig, Edward James, nuclear physics
Lundblad, Roger Lauren, biochemistry, hematology
Lyman, John, chemical oceanography
Macdonald, James Ross, solid state physics
MacRae, Edith Krugelis, anatomy
Mailman, Richard Bernard, toxicology, neurophysiology
Malindzak, George Steve, Jr, medical physiology, environmental health
Mangelsdorf, Paul Christoph, economic botany, genetics
Manire, George Philip, microbiology
Mann, William Robert, applied mathematics
Maroni, Gustavo Primo, developmental genetics, biochemical genetics
Martens, Christopher Sargent, marine chemistry
Massey, Jimmy R, botany
Mattocks, Albert McLean, pharmaceutical chemistry
Mayer, Eugene Stephen, medical education
Mazade, Noel Andre, public health administration
McBay, Arthur John, toxicology, chemistry
McCormick, J Frank, ecology
McDonagh, Jan M, biochemistry, pathology
McDonagh, Richard Patrick, Jr, physiology, experimental pathology
McGinnis, Michael Randy, mycology
McIlwain, David Lee, neurochemistry
McKee, Robert Lambert, chemistry
McLendon, William Woodard, pathology, laboratory medicine
McMahan, Elizabeth Anne, zoology
McMillan, Campbell White, pediatrics, hematology
McMillan, Donald Edgar, pharmacology, psychology
McNamara, Mary Colleen, neurobiology
McNitt, Rand Edwin, mycology
Mechanic, Gerald, organic chemistry, biochemistry
Merzbacher, Eugen, theoretical physics
Mewborn, Ancel Clyde, algebra
Meyer, Thomas J, inorganic chemistry
Miller, Augustus Taylor, Jr, physiology
Miller, C Arden, pediatrics, public health
Miller, Milton Leonard, psychiatry
Misch, Donald William, cell biology
Mitchell, Earl Nelson, physics
Montgomery, Royce Lee, gross anatomy, neuroanatomy
Morris, Charles Elliot, neurology
Morrow, John Charles, III, physical chemistry
Mueller, Helmut Charles, zoology, animal behavior
Mueller, Robert Arthur, anesthesiology, pharmacology
Munson, Paul Lewis, pharmacology
Murray, Royce Wilton, analytical chemistry
Mushak, Paul, biochemistry, chemistry
Nayfeh, Shihadeh Nasri, biochemistry, endocrinology
Nelson, Robert Mellinger, orthodontics
Neumann, A Conrad, oceanography
Newbold, John Edward, molecular biology, virology
Newsome, James Frederick, surgery
Ney, Robert Leo, medicine, endocrinology
Nopanitaya, Waykin, pathology, electron microscopy
Nordsiek, Frederic William, nutrition
Noyes, Claudia Margaret, analytical biochemistry
Oldenburg, Theodore Richard, pedodontics
Olive, Lindsay Shepherd, botany
Olsen, James Leroy, industrial pharmacy
Omran, Abdel Rahim, epidemiology
Ontjes, David A, endocrinology
Palmatier, Everett Dyson, physics
Palmer, Jeffress Gary, internal medicine, hematology
Parr, Robert Ghormley, theoretical chemistry
Patterson, Hubert C, surgery
Patterson, Joseph Flanner, Jr, anesthesiology
Peach, Roy, histology
Peacock, James Lowe, anthropology
Pearlman, William Henry, biochemistry
Pedersen, Lee G, physical chemistry, chemical physics
Peet, Robert Krug, ecology
Peng, Tai-Chan, pharmacology
Penniall, Ralph, biochemistry
Penniston, John Thomas, biological chemistry
Perl, Edward Roy, physiology
Perlmutt, Joseph Hertz, physiology
Pettis, Billy James, mathematics
Pfaender, Frederic Karl, microbial ecology
Pfaltzgraff, John Andrew, mathematics
Piantadosi, Claude, medicinal chemistry
Pick, James Raymond, laboratory animal science
Pollitzer, William Sprott, physical anthropology, genetics
Poole, Doris Theodore, biochemistry, pharmacology
Powell, Don Watson, gastroenterology
Prange, Arthur Jergen, Jr, psychiatry
Proffit, William R, orthodontics, physiology
Quade, Dana Edward Anthony, statistics
Radford, Albert Ernest, botany
Ragland, Paul C, geochemistry, petrology
Ramp, Warren Kibby, physiology, biochemistry
Raney, Richard Beverly, orthopedic surgery
Reichler, Robert Jay, child psychiatry
Reilley, Charles Norwood, analytical chemistry
Reist, Parker Cramer, industrial hygiene
Rice, Oscar Knefler, physical chemistry
Richardson, William Perry, preventive medicine
Rieke, Reuben Dennis, organic chemistry
Roberts, Harold R, internal medicine, hematology
Roberts, Louis Douglas, physics
Rogers, John James William, geology, geochemistry
Romanovicz, Dwight Keith, cell biology
Rosenfeld, Leonard Sidney, medicine, public health
St Jean, Joseph, Jr, micropaleontology, invertebrate paleontology
Sapp, Oscar LeMay, III, internal medicine, gastroenterology
Sar, Madhabananda, veterinary medicine, physiology
Savory, John, clinical chemistry, pathology
Scatliff, James Howard, radiology
Schlessinger, Michael, algebra
Schneider, Howard Albert, biochemistry, nutrition
Schroeer, Dietrich, physics
Schwab, John Harris, bacteriology, immunology
Scott, Tom Keck, plant physiology
Semeniuk, Fred Theodor, pharmaceutical chemistry
Sen, Pranab Kumar, biostatistics
Sessions, John Turner, Jr, medicine
Shafroth, Stephen Morrison, atomic physics
Shankle, Robert Jack, dentistry
Sharp, David Gordon, biophysics
Shearin, Paul Edmondson, physics
Sheps, Cecil George, preventive medicine
Shiffman, Morris A, environmental health, veterinary medicine
Shinkman, Paul G, physiological psychology
Shuman, Mark S, environmental chemistry, electroanalytical chemistry
Silver, Marvin, solid state physics
Silverman, Myron Simeon, bacteriology, immunology
Singleton, Mary Clyde, anatomy, physical medicine & rehabilitation
Slifkin, Lawrence, solid state physics
Small, Ernest William, oral surgery, maxillofacial surgery
Smiley, Gary Ray, orthodontics
Smith, Charles E, psychiatry
Smith, Charles Sydney, Jr, physics
Smith, Harold Carter, biochemistry
Smith, Walter Laws, mathematical statistics
Smith, William Walker, mathematics
Sobel, Eugene L, statistics
Sobsey, Mark David, microbiology, environmental health
Sonner, Johann, mathematics
Spitznagel, John Keith, microbiology, medicine
Srivastava, Kunwar Krishna, microbiology
Stafford, Darrel Wayne, zoology, molecular biology
Stiven, Alan Ernest, ecology, population biology
Stover, Betsy Jones, radiobiology, pharmacology
Straley, Joseph Ward, spectroscopy
Straughn, William Ringgold, Jr, bacteriology
Stumpf, Walter Erich, neuroendocrinology, pharmacology
Sugioka, Kenneth, anesthesiology
Summer, George Kendrick, biochemistry, nutrition
Swanton, Margaret Catherine, pathology
Switzer, Boyd Ray, nutrition, biochemistry
Talbert, Luther M, obstetrics & gynecology
Talmage, Roy Van Neste, physiology
Taylor, Duane Francis, dental materials
Taylor, Isaac Montrose, medicine
Taylor, William West, pharmacy
Textoris, Daniel Andrew, geology
Thomas, Colin Gordon, Jr, surgery
Thomas, Henry Carrison, physical chemistry
Thomas, Jon Charles, astronomy
Thomas, William Grady, audiology
Thompson, Herman O, pharmacy
Tolle, Jon Wright, mathematics, operations research
Toverud, Svein Utheim, pharmacology, endocrinology
Trevino, Daniel Louis, neurophysiology
Turner, Derek T, physical chemistry, polymer chemistry
Twarog, Robert, microbial biochemistry
Tyan, Marvin L, internal medicine, experimental biology
Underwood, Louis Edwin, pediatrics, endocrinology
Underwood, Newton, physics
Vakilzadeh, Javad, public health, epidemiology
Van Cleave, Charles Durward, anatomy
Van Ert, Mark Dewayne, industrial hygiene
Van Wyk, Judson John, pediatrics, endocrinology
Via, William Fredrick, Jr, pedodontics
Wagner, Robert H, biochemistry
Wahl, Jonathan Michael, mathematics
Walker, Richard Isley, medicine
Warren, Donald W, dentistry
Watson, Robert Briggs, medicine
Weatherly, Norman F, medical parasitology
Webster, William Phillip, dentistry
Weiss, Charles Manuel, environmental biology
Wells, Henry Bradley, biostatistics
Wertz, Gail T Williams, virology
Wheeler, Clayton Eugene, Jr, dermatology
Wheeler, Walter Hall, vertebrate paleontology
White, James Rushton, biochemistry
White, Raymond Petrie, Jr, anatomy, oral surgery
White, William Alexander, geology
Whitsel, Barry L, neurophysiology, neuropharmacology
Whitten, David G, physical organic chemistry, biophysical chemistry
Whittinghill, Maurice, zoology
Wier, Jack Knight, pharmacognosy
Wilcox, Benson Reid, cardiovascular surgery, thoracic surgery
Wiley, Richard Haven, Jr, ethology, ecology
Wilkins, Homer Clifton, physics
Willhoit, Donald Gillmor, radiation health
Wilson, Frank Crane, medicine, orthopedic surgery
Wilson, John Eric, biochemistry
Wogen, Warren Ronald, mathematics
Wolfenden, Richard Vance, biochemistry
Wood, William Bainster, medicine
Woods, James Watson, medicine
Workman, Erwin Franklin, Jr, biochemistry
Wright, Fred Boyer, mathematics
Wyrick, Priscilla Blakeney, bacteriology
Yarnell, Richard Asa, anthropology, ethnobotany
Yonce, Lloyd Robert, physiology
York, James Wesley, Jr, theoretical physics
Young, Daniel Test, internal medicine

CHARLOTTE
Archer, David Anderson, numerical analysis
Arvey, Martin Dale, zoology, ornithology
Aspland, John Richard, textile chemistry
Baker, Elizabeth McIntosh, invertebrate zoology
Barach, Joseph Leonard, polymer science
Barnes, James Crowell, textile physics
Basavappa, Parannara, mathematics
Berry, George William, chemistry
Brown, Richard Dean, animal behavior, bioacoustics
Browne, Colin Lanfear, textile chemistry
Burson, Sherman Leroy, Jr, organic chemistry
Busby, Hubbard Taylor, Jr, organic chemistry, industrial organic chemistry
Bush, Stewart Fowler, physical chemistry
Busse, Robert Franklyn, biochemistry, organic chemistry
Chopra, Baldeo K, plant pathology, microbiology
Clay, James William, geography, cartography
Coffey, Janice Carlton, systematic botany
Colvard, Dean Wallace, animal science, animal economics
Cote, Philip Norman, organic chemistry
Doerr, Marvin LeRoy, organic chemistry
DuBois, Thomas David, inorganic chemistry
Duncan, Gordon Duke, biochemistry
Edwards, Nancy Claire, embryology, developmental biology
Embry, Mary Rodriguez, mathematical analysis
Evins, Charles Victor, analytical chemistry
Fehon, Jack Harold, zoology
Fisher, Harold M, inorganic chemistry
Gibson, Robert Harry, analytical chemistry
Greenspan, Frank Philip, organic chemistry
Guerrant, Ralph Eugene, clinical biochemistry
Haines, Harry Caum, forest management
Hamit, Harold F, surgery
Hartford, Winslow Hopper, physical inorganic chemistry
Haviland, James Roger, plant physiology
Hedstrom, John Richard, mathematics
Herdklotz, John Key, chemistry
Hildreth, Philip Elwin, genetics
Hill, Eric Stanley, physical chemistry
Hogan, Gene Richard, radiobiology, animal physiology
Hoyle, Hughes Bayne, Jr, mathematics
Huber, Wilson Frederick, organic chemistry
Huffman, Allan Murray, organic chemistry
Johnson, Phillip Eugene, mathematics
Keith, Charles Herbert, physical chemistry
Kuppers, James Richard, physical chemistry
Lee, Thomas Alan, astronomy
Little, John Stanley, organic chemistry
Lucas, Thomas Ramsey, mathematics
Martin, Virginia Lorelle, biology, parasitology
Matthews, James Francis, cytology, plant taxonomy
Mayes, Terrill W, plasma physics, atomic spectroscopy
McEwen, Mildred Morse, chemistry
Menhinick, Edward Fulton, ecology, physiology
Miller, Bradford, clinical biochemistry
Niccolai, Nilo Anthony, mathematics, computer science

Nunnally, Nelson Rudolph, geography, physical
Oates, Jimmie C., physics
Oberhofer, Edward Samuel, nuclear physics
Orr, Douglas Milton, Jr., geography
Orwoll, Richard David, physical chemistry
Ostrowski, Ronald Stephen, genetics
Pickett, John Harold, analytical chemistry
Pollak, Victor Louis, physics
Reiter, Harold Braun, mathematics
Roth, Walter John, operations research
Schell, Joseph Francis, geometry
Schuie, Arlo Willard, mathematics
Steele, Richard, chemistry
Stuart, Alfred Wright, urban geography
Swanson, Samuel Edward, geochemistry, petrology
Theuer, William John, organic chemistry
Vermilion, Robert Everett, physics
Vickroy, David Gill, analytical chemistry
Walsh, Thomas David, chemistry
Watts, Judith Elizabeth Culbreth, clinical chemistry

CULLOWHEE
Bassett, Joseph Yarnall, Jr., organic chemistry
Chapman, John Judson, geology
Clark, Louis Watts, physical chemistry
Coyle, Frederick Alexander, biology
Dorwin, John T., anthropology, archaeology
Eller, Dean James Gerald, ecology
Gilman, Albert F. III, mathematics
Horsman, Arden William, geology
Horton, James Heathman, taxonomy, botany
Lumb, Roger H., animal physiology
Martin, Charles John, applied mathematics
Mathews, David A., organic chemistry
McGowan, William Courtney, Jr., solid state physics
Moore, Allen Murdoch, ecology, zoology
Morris, Gene Franklin, organic chemistry
Phillips, Yvonne, geography, anthropology
Pittillo, Jack Daniel, plant ecology
Robinson, Harold Frank, quantitative genetics, plant breeding
Teague, David Boyce, applied mathematics
Terango, Larry, speech pathology, audiology
Thomas, John Pelham, mathematics
Wallace, James William, Jr., plant biochemistry, plant physiology
West, Jerry Lee, fish biology, zoology
Woosley, Royce Stanley, organic chemistry
Wright, Clarence Paul, genetics
Youmans, Hubert Lafay, analytical chemistry

DAVIDSON
Bernard, Richard Ryerson, mathematics
Bryan, Horace Alden, inorganic chemistry
Burnett, John Nicholas, analytical chemistry
Daggy, Tom, biology
Fredericksen, James Monroe, organic chemistry
Frey, William Francis, experimental physics
Gable, Ralph William, physical chemistry
Gallent, John Bryant, chemistry
Hopkins, John Isaac, nuclear physics
Jackson, Robert Bruce, Jr., developmental biology
Kimmel, Donald Loraine, Jr., analysis
Klein, Benjamin Garrett, mathematical analysis
McGavock, William Gillespie, mathematics
Miller, Verna Jean, ethology, animal behavior
Putnam, Jerry L., comparative anatomy, histology
Roberts, Jerry Allan, mathematics
Stroud, Junius Brutus, algebra
White, Locke, Jr., chemical physics
Wolf, Albert Allen, elementary particle physics

DURHAM
Adams, Dolph Oliver, pathology, immunology
Amos, Dennis Bernard, immunogenetics

Anderson, Lewis Edward, botany
Anderson, Nels Carl, Jr., endocrinology, biophysics
Anderson, Roger Fabian, forest entomology
Anlyan, William B., surgery
Anderson, William George, neurology
Appel, Stanley Hersh, psychopharmacology
Arena, Jay M., pediatrics
Bailey, Joseph Randle, zoology
Baker, Lenox Dial, orthopaedic surgery
Ballard, Lewis Franklin, solid state physics, electrical engineering
Barnes, Robert Lloyd, forest physiology
Barr, Roger Coke, biomedical engineering, computer science
Bates, William Wannamaker, Jr., physical chemistry
Baylin, George Jay, radiology
Bell, Robert Maurice, biochemistry, molecular biology
Bennett, Peter Brian, anesthesiology, physiology
Bernheim, Frederick, pharmacology
Bernheim, Mary Lilias Christian, biochemistry
Biedenharn, Lawrence Christian, Jr., nuclear physics
Biermann, Alan Wales, computer science
Billings, William Dwight, botany, ecology
Bilpuch, Edward George, physics
Blum, Jacob Joseph, physiology, biophysics
Bolognesi, Dani Paul, virology, immunology
Bonar, Robert Addison, biochemistry
Bonk, James F., chemistry
Boon, James Alexander, ethnology
Boynton, John E., genetics
Bradford, William Dalton, pathology, biochemistry
Bradsher, Charles Kilgo, organic chemistry
Bressler, Bernard, medicine
Brodie, Harlow Keith Hammond, psychiatry
Brown, Frances Campbell, organic chemistry
Bryan, Virginia Schmitt, botany, academic administration
Bryant, Herman Grey, Jr., physical chemistry
Buckley, Charles Edward, internal medicine, allergy
Buckley, Rebecca Hatcher, pediatric allergy, pediatric immunology
Bunce, James Arthur, physiological ecology
Burdick, Donald Smiley, statistics
Busse, Ewald William, psychiatry
Callaway, Jasper Lamar, dermatology
Carlitz, Leonard, mathematics
Carlsen, Richard Chester, neurophysiology
Cartmill, Matt, physical anthropology
Casseday, John Herbert, neurosciences, psychology
Chan, Moses Hung-Wai, low temperature physics
Chandler, Arthur Cecil, Jr., medicine, ophthalmology
Chesnut, Donald Blair, physical chemistry
Christensen, Norman Leroy, Jr., plant ecology
Chuang, Ronald Yan-Li, biochemistry, oncology
Clapp, James R., internal medicine
Clark, Howard Garmany, biomedical engineering
Clippinger, Frank Warren, Jr., orthopedic surgery
Conant, Norman Francis, mycology, bacteriology
Corless, Joseph Michael James, biophysics, vision
Counce, Sheila Jean, genetics, embryology
Culberson, Chicita Frances, organic chemistry
Culberson, William Louis, botany
Currie, William Deems, biochemistry
Cusson, Ronald Yvon, theoretical nuclear physics
Davidson, Jack Dougan, nuclear medicine
Davis, David A., medicine
Dawson, Jeffrey Robert, immunology
Dawson, John William, organic chemistry
Day, Eugene Davis, immunology
Dees, Susan Coons, pediatrics
De Lucia, Frank Charles, molecular physics, quantum electronics
Demaria, William John Amsterdam, pediatrics
Dent, Sara Jamison, anesthesiology
Dressel, Francis George, mathematics

Duke, Kenneth Lindsay, anatomy
Eichlepp, Jane G., pathology
Ellinwood, Everett Hews, Jr., psychiatry, neuropharmacology
Erickson, Harold Paul, biophysics, electron microscopy
Erickson, Robert Porter, psychophysiology
Estes, Edward Harvey, Jr., medicine
Evans, Lawrence Eugene, physics
Evans, Ralph Aiken, physics
Everett, John Wendell, anatomy
Fairbank, Henry Alan, low temperature physics
Fetter, Bernard Frank, pathology
Fiscus, Edwin Lawson, plant physiology
Fletcher, William Henry, endocrinology, cell biology
Fletcher, John Thomas, mathematics
Fluke, Donald John, radiation biophysics
Forster, Jean Lois, genetics, cell biology
Fortney, Lloyd Ray, high energy physics
Fowler, John Alvis, psychoanalysis
Fox, Richard G, cultural anthropology
Fridovich, Irwin, biochemistry
Friedl, Ernestine, cultural anthropology, enzymology
Frothingham, Thomas Eliot, tropical medicine, public health
Gallie, Thomas Muir, computer science
Garvey, Robert Michael, molecular spectroscopy
George, Charles Redgenal, entomology, parasitology
Georgiade, Nicholas George, plastic surgery
Gillham, Nicholas Wright, genetics
Githens, Sherwood, Jr., physics
Glenn, James Francis, urology
Goldner, Joseph Leonard, orthopedic surgery
Goldwater, Leonard John, occupational medicine
Gooding, Linda R., immunology
Goodrich, Jack K., radiology
Gordy, Walter, physics
Goshaw, Alfred Thomas, experimental high energy physics
Graham, Doyle Gene, neuropathology, anatomic pathology
Green, Donald MacDonald, genetics
Green, Robert Lee, Jr., psychiatry
Greene, Ronald C., biochemistry
Greenfield, Joseph C. Jr., cardiovascular physiology
Gregg, John Richard, zoology
Grimson, Keith Sanford, surgery
Gross, Paul Magnus, physical chemistry
Grossman, Herman, radiology, pediatrics
Guild, Walter Rufus, molecular biology, biophysics
Gutknecht, John William, physiology, marine biology
Habig, Robert L., clinical chemistry
Hackel, Donald Benjamin, pathology
Hall, Dwight Hubert, molecular genetics
Hall, James Ewbank, biophysics
Hall, Kenneth Daland, anesthesiology
Hall, William Charles, neuropsychology, neuroanatomy
Hammond, William Edward, biomedical engineering, computer sciences
Han, Moo-Young, particle physics
Harmel, Merel H., anesthesiology
Harris, Cecil Craig, nuclear medicine
Harris, Harold Joseph, psychiatry, psychoanalysis
Harris, Jerome Sylvan, pediatrics
Harrison, Lura Ann, physiology
Hellmers, Henry, plant physiology, forest physiology
Hempel, Franklin Glenn, physiology, zoology
Hendrix, James Paisley, medicine
Henkens, Robert William, biophysical chemistry
Herget, William F., physics
Heron, Duncan, geology
Heyman, Albert, internal medicine, neurology
Hill, Gale Bartholomew, microbiology, infectious diseases
Hill, Robert Lee, biochemistry
Hilliard, Roy C., chemistry
Hine, Frederick Roy, psychiatry
Hirschberg, Nell, microbiology
Hobbs, Marcus Edwin, physical chemistry
Hodel, Margaret Jones, combinatorial analysis, number theory
Hodel, Richard Earl, mathematics
Holmes, Edward Warren, biochemical genetics

Hudson, William Rucker, surgery, otolaryngology
Hughes-Schrader, Sally, cytology
Hylander, William Leroy, primatology, dental research
Izydore, Robert Andrew, organic chemistry
Jeffs, Peter W., organic chemistry
Jennings, Robert Burgess, pathology
Jobsis, Frans Frederik, physiology, experimental pathology
Johnson, Armead H., immunogenetics
Johnson, Kurt Edward, developmental biology
Johnson, Terry Walter, Jr., mycology
Johnston, William Webb, pathology, cytology
Joklik, Wolfgang Karl, biochemistry, virology
Jones, James David, pediatrics, psychiatry
Kalianos, Andrew George, organic chemistry
Kamin, Henry, biochemistry
Katz, Samuel Lawrence, virology, pediatrics
Kaufman, Bernard, biochemistry
Kempner, Walter, internal medicine, cell physiology
Kerby, Grace Partridge, medicine
Kersey, Robert Lee, Jr., analytical chemistry
Kim, Ki-Hyon, medical physics, nuclear medicine
Kinney, Thomas DeArman, pathology
Kirshner, Norman, biochemistry
Klein, Dolph, clinical microbiology
Klintworth, Gordon K., pathology, anatomy
Klopfer, Peter Hubert, zoology
Knoerr, Kenneth Richard, forestry
Kootsey, Joseph Mailen, physiology
Kraines, David Paul, topology
Kramer, Paul Jackson, plant physiology
Kredich, Nicholas M, internal medicine, biochemistry
Krigbaum, William Richard, physical chemistry
Kylstra, Johannes Arnold, medicine, physiology
La Barre, Weston, anthropology
Lack, Leon, biochemistry
LaManna, Joseph Charles, neurosciences
Laszlo, John, medicine, biochemistry
Lawson, Dewey Tull, low temperature physics
Lazarus, Gerald Sylvan, dermatology
Lefkowitz, Robert Joseph, molecular pharmacology, medical science
Lester, Richard Garrison, radiology
Levy, Nelson Louis, immunology
Lewis, Harold Walter, nuclear physics
Lieberman, Melvyn, physiology
Lin, Stephen Fang-Maw, physical chemistry
Livingstone, Daniel Archibald, ecology
Llewellyn, Charles Elroy, Jr., psychiatry
Lochmuller, Charles Howard, analytical chemistry
Long, Ernest Croft, physiology
Loveland, Donald William, mathematics, computer science
Lowenbach, Hans, psychiatry
Luken, William Louis, theoretical chemistry
Lynn, William Sanford, medicine
Lynts, George Willard, geology
Mandel, Lazaro J., physiology, biophysics
Manson, Earle Lowry, Jr., molecular physics
Massengill, Raymond, speech pathology, audiology
McBryde, Angus Murdoch, pediatrics
McCarty, Kenneth Scott, biochemistry
McCollum, Donald E., orthopedic surgery
McKee, Patrick Allen, medical research, hematology
McLees, Byron D., biochemistry
McManus, Thomas (Joseph), cell physiology, hematology
McPhail, Andrew Tennent, chemistry
Mendell, Lorne Michael, neurophysiology
Menzel, Daniel B., pharmacology
Metzgar, Richard Stanley, immunology
Meyer, Johannes Horst Max, physics
Mills, Elliott, pharmacology, physiology
Mold, James Davis, bio-organic chemistry
Moore, John Wilson, neurophysiology
Morris, James Joseph, Jr., internal medicine, cardiology
Moses, Montrose James, cell biology, electron microscopy

Anderson, Laura Gaddes, experimental embryology, human cytogenetics
Arthur, B Wayne, entomology, toxicology
Aycock, Benjamin Franklin, textile chemistry
Ballantine, Larry Gene, agricultural chemistry
Bates, William K, biochemistry
Beaumont, Ralph Harrison, Jr, chemistry
Benfey, Otto Theodor, physical organic chemistry, history of science
Bennett, David Gordon, population geography
Bixler, Dean A, textile chemistry
Boehle, John, Jr, agronomy, weed science
Bryden, Robert Richmond, animal ecology
Byrd, Kenneth Alfred, algebra
Callahan, Kemper Leroy, plant pathology
Cassidy, James Edward, metabolism, analytical chemistry
Cherry, Edward Taylor, entomology
Chopra, Naiter Mohan, organic chemistry
Church, Charles Alexander, Jr, mathematics
Clark, Clifton Bob, solid state physics
Cleveland, Gregor George, biophysics, magnetic resonance
Collins, Henry A, agronomy, plant physiology
Cowett, Everett R, agronomy, plant physiology
Croft, Wilma Janice, chemistry, information science
Cutter, Lois Jotter, systematic botany, plant morphology
Davis, Myrtis, mathematics
Dilts, Joseph Alstyne, inorganic chemistry
Dovenmuehle, Robert Henry, psychiatry
Dozier, Craig Lanier, geography
Dye, William Thomson, Jr, organic chemistry
Eason, Robert Gaston, neuropsychology, psychophysiology
Eberhart, Bruce Maclean, biochemical genetics
Edwards, Gerald Alonzo, physical chemistry
Ellis, John Fletcher, weed science
Findley, William Robert, applied chemistry
Fischer, William Carl, Jr, agricultural chemistry
Forrester, Sherri Rhoda, organic chemistry
Gall, Lorraine Sibley, bacteriology
Gentry, Karl Ray, mathematics
Gordh, George Rudolph, Jr, topology
Graves, Artis P, zoology, embryology
Graves, John Lowell, quantum chemistry, biochemistry
Gray, Richard C, animal breeding, animal physiology
Greer, James Edward, textile chemistry, bioacoustics
Hageseth, Gaylord Terrence, physics
Hall, Seymour Gerald, textile chemistry
Hays, Donald Brooks, entomology
Hendrickson, Herbert T, vertebrate zoology
Herman, Harvey Bruce, analytical chemistry
Hermanson, Harvey Philip, soil chemistry, mathematical biology
Herr, David Guy, statistics, mathematics
Hildebrandt, Theodore Ware, computer science
Hill, Alfred, Jr, medical entomology
Hudson, Albert Berry, natural products chemistry
Irvine, James Bosworth, physical chemistry
Jezorek, John Robert, analytical chemistry, inorganic chemistry
Jones, Stephen Thomas, organic chemistry
Kane, Gordon Philo, pesticide chemistry
Kennedy, Wadaran Latamore, dairy husbandry
Kirk, Philip Moore, organic chemistry
Knight, David Bates, organic chemistry
Kupferer, Harriet Jane, anthropology
LeBaron, Homer McKay, pesticide chemistry
Lewis, Claude Irenius, analytical chemistry
Ljung, Harvey Albert, analytical chemistry
Logan, Cheryl Ann, animal behavior, neuropsychology
Long, Andrew Fleming, Jr, mathematics
Love, William P, mathematics education
Lutz, Paul E, invertebrate zoology, ecology
MacInnes, David Fenton, Jr, inorganic chemistry
Mueller, Nancy Schneider, developmental biology, reproductive biology
Murdock, Gordon Robert, invertebrate zoology
Murray, Francis John Albert, mathematics
Myers, John Albert, organic chemistry
Narashashi, Toshio, neurophysiology, neuropharmacology
Nashold, Blaine S, neurosurgery, neurophysiology
Naumann, Dorothy Ethel, medicine
Naylor, Aubrey Willard, plant physiology
Nelson, A Carl, Jr, mathematical statistics, mathematics
Nichols, Jack Loran, biochemistry
Nicklas, Robert Bruce, cell biology
Norden, Jeanette Jean, neurobiology
Nozaki, Yasuhiko, biophysical chemistry
O'Barr, William McAlston, anthropology
Obrist, Walter Dorn, electroencephalography, experimental neurology
Odom, Guy Leary, neurosurgery
O'Fallon, William M, biostatistics, mathematical statistics
O'Foghludha, Fearghus Tadhg, medical physics
Osterhout, Suydam, virology
Ottolenghi, Athos, pharmacology
Padilla, George M, cell physiology, biophysics
Palethorpe, George, organic chemistry
Palmer, Richard Alan, inorganic chemistry
Parham, William Eugene, organic chemistry
Parker, Joseph B, Jr, psychiatry
Patrick, Merrell Lee, applied mathematics, computer science
Patterson, David Thomas, physiological ecology, weed science
Pattillo, Walter Hugh, Jr, parasitology
Paulson, James Carsten, biochemistry
Peele, Talmage Lee, neuroanatomy, neurology
Peet, Mary Monnig, plant physiology
Peete, Charles Henry, Jr, medicine
Peete, William P J, surgery
Perkins, Ronald Dee, petrology, marine geology
Peterson, David West, applied mathematics
Pfeiffer, Eric A, psychiatry
Pfeiffer, John B, Jr, medicine
Philpot, Jane, botany
Philpott, Philip Elliott, pure mathematics
Pickrell, Kenneth LeRoy, surgery
Pilkey, Orrin H, marine geology
Poirier, Jacques Charles, theoretical chemistry, physical chemistry
Porter, Frederick Stanley, Jr, pediatrics
Porter, Ned Allen, organic chemistry
Postlethwait, Raymond Woodrow, surgery
Pratt, Philip Chase, pathology
Preston, Jack, organic chemistry, polymer chemistry
Protter, Philip Elliott, pure mathematics
Purser, Fred O, nuclear physics, particle physics
Quin, Louis Dubose, organic chemistry
Quinn, Galen Warren, orthodontics
Rajagopalan, K V, biochemistry
Ralston, Charles William, forest soils
Ramm, Dietolf, computer science
Reckless, John B, psychiatry, psychosomatic medicine
Reed, Michael Charles, mathematical physics
Reedy, Michael K, cell biology
Renuart, Adhemar William, pediatric neurology, neurochemistry
Reynolds, Jacqueline Ann, biochemistry
Rhoads, John McFarlane, psychiatry
Roberson, Nathan Russell, Jr, nuclear physics
Roberts, John Henderson, mathematics
Robertson, James David, anatomy
Robinson, Hugh Gettys, physics
Robinson, Roscoe Ross, internal medicine, nephrology
Rosen, Lawrence, anthropology
Rosse, Wendell Franklyn, medicine, immunology
Ruffin, Julian Meade, gastroenterology
Rundles, Ralph Wayne, internal medicine
Sabiston, David Coston, Jr, surgery
Sage, Harvey J, immunochemistry, biochemistry
Salber, Eva Juliet, public health, community medicine
Salzano, John Victor, physiology
Sanders, Aaron Perry, cell physiology, radiobiology
Sandhu, Shabeg Singh, genetics, agriculture
Sarneski, Joseph Edward, analytical chemistry
Schanberg, Saul M, neuropharmacology
Schiebel, Herman Max, surgery
Schmalz, Alfred Chandler, organic chemistry
Schmidt-Koenig, Klaus, zoology
Schmidt-Nielsen, Knut, physiology
Schooler, James M, Jr, biochemistry, physiology
Scott, David William, immunology
Scoville, Richard Arthur, mathematics
Sealy, Will Camp, thoracic surgery
Searles, Richard Brownlee, phycology
Seigler, Hilliard Foster, surgery, immunology
Sessoms, Stuart McGuire, internal medicine
Shauck, Maxwell Eustace, Jr, mathematics
Shaw, Barbara Ramsay, biophysical chemistry
Shelburne, John Daniel, pathology
Shipp-Watson, Mary Elizabeth, bacteriology
Shoenfeld, Joseph Robert, mathematics
Siegel, Lewis Melvin, biochemistry
Sieker, Herbert Otto, medicine
Simon, Sidney Arthur, physical biology
Slotkin, Theodore Alan, pharmacology
Smith, David Alexander, mathematics
Smith, Peter, physical chemistry
Smith, Ralph Earl, virology, oncology
Smith, Wirt Wilsey, medicine
Snow, Thomas Russell, cardiovascular physiology
Snyderman, Ralph, internal medicine, immunology
Somjen, George G, physiology, pharmacology
Sommer, Joachim Rainer, pathology
Southwick, Everett West, organic chemistry
Spach, Madison Stockton, pediatric cardiology
Spock, Alexander, pediatrics, allergy
Spooner, George Hansford, clinical chemistry, physical chemistry
Stackelberg, Olaf Patrick, mathematics
Stambaugh, William James, forest pathology
Starmer, C Frank, computer science, medicine
Stead, Eugene Anson, Jr, medicine
Stickel, Delford LeFew, surgery
Stone, Donald Eugene, botany, genetics
Stone, Erika Mares, mathematics
Stopford, Woodhall, internal medicine, clinical toxicology
Strain, Boyd Ray, physiological ecology
Strobel, Howard Austin, analytical chemistry, physical chemistry
Styron, Charles Woodrow, medicine
Sweeny, Hale Caterson, mathematical statistics
Sylvia, Avis Latham, cell physiology
Tanford, Charles, biochemistry
Thurstone, Frederick Louis, biomedical engineering, electrical engineering
Ting-Beall, Hie Ping, cell physiology
Torre-Bueno, Jose Rollin, physiology
Totton, Ezra Lester, organic chemistry, biochemistry
Tourian, Ara Yervant, biochemical genetics
Townes, Mary McLean, cell physiology
Trivedi, Kishor Shridharbhai, computer sciences
Tucker, Vance Alan, comparative physiology
Tuthill, Richard Lovejoy, geography of Africa, economic geography
Tyczkowski, Edward Albert, organic chemistry
Tyor, Malcolm Paul, medicine
Tyrey, Lee, neuroendocrinology, physiology
U, Raymond, genetics, radiobiology
Vanaman, Thomas Clark, biochemistry, microbiology
Verghese, Margrith Wehrli, genetics, immunogenetics
Verwoerdt, Adrian, psychiatry, psychoanalysis
Vogel, Francis Stephen, pathology
Vogel, Steven, zoology, physiology
Wachtel, Howard, biomedical engineering, physiology
Wadsworth, Joseph Allison Cannon, ophthalmology, medicine
Wainwright, Stephen Andrew, invertebrate zoology
Walker, William Delany, nuclear physics
Wallace, Andrew Grover, internal medicine, cardiovascular physiology
Walter, Richard L, nuclear physics
Ward, Calvin Lucian, genetics
Ward, Frances Ellen, immunogenetics
Warfield, Albert Harry, organic chemistry
Warner, Seth L, algebra
Way, Katharine, physics
Webster, Robert Edward, molecular biology, molecular genetics
Wescott, William B, oral pathology
Wheat, Robert Wayne, microbiology, biochemistry
White, Richard Alan, developmental anatomy, morphology
Widmann, Frances King, pathology
Wilbur, Henry Miles, zoology, ecology
Wilbur, Karl Milton, physiology
Wilbur, Robert Lynch, plant taxonomy
Wilder, Pelham, Jr, organic chemistry
Willett, Hilda Pope, microbiology
Williams, John Frederick, chemistry
Wilson, William Preston, psychiatry
Wise, Dwayne Allison, cytogenetics
Wittels, Benjamin, pathology
Wolbarsht, Myron Lee, biophysics, biomedical engineering
Woodbury, Max Atkin, biomathematics, computer science
Woodhall, Barnes, neurosurgery
Wyngaarden, James Barnes, metabolism
Young, William Glenn, Jr, thoracic surgery
Zung, William Wen-Kwai, psychiatry, psychopharmacology
Zwadyk, Peter, Jr, microbiology

ELIZABETH CITY
Alam, Mohammed Ashraful, analytical chemistry, physical chemistry
Choudhury, Abdul Latif, theoretical high energy physics
Cooke, Herman Glenn, zoology, entomology
Garrigus, Woodford McDowell, geography
Husain, Delawar, mathematics
Jenkins, Jimmy Raymond, biological structure
Jordan, Wade Hampton, Jr, electrochemistry, physical chemistry
Khan, Sekender Ali, plant pathology
Sutton, Louise Nixon, mathematics

ELON COLLEGE
Danieley, Earl, organic chemistry
Perkins, Kenneth Warren, parasitology
Rao, Raghavendra D, plant cytology
Reddish, Paul Sigman, biology
Ryals, George Lynwood, Jr, invertebrate ecology

ENKA
Meierhoefer, Alan William, organic chemistry
Muldrow, Charles Norment, Jr, polymer chemistry
Parker, James Philips, physics

ETOWAH
Bever, Enid L, biochemistry

FAYETTEVILLE
Chao, Tyng Tsair, chemistry
Horner, William Wesley, inorganic chemistry
Jawa, Manjit S, applied mathematics, continuum mechanics
Knuckles, Joseph Lewis, parasitology

FLETCHER
Konsler, Thomas Rhinehart, horticulture
Shelton, James Edward, soil chemistry, soil fertility

FRANKLIN
Douglass, James Edward, forestry
Smith, Walton Ramsey, forest products
Swift, Lloyd Wesley, Jr, micrometeorology, forest hydrology

GASTONIA
Ellestad, Reuben B, chemistry

GOLDSBORO
Herrett, Richard Allison, biochemistry, plant physiology

GREENSBORO
Adelberger, Rexford E, nuclear physics

Madan, Rabinder Nath, theoretical physics
Magee, Aden Combs, III, nutrition
Mandurey, Ilma Morell, mathematics
Marco, Gino Joseph, biochemistry, organic chemistry
Mason, Dorothy Stafford, geography
McCormack, Francis Joseph, plasma physics
McCrady, Edward, III, developmental biology
McIvin, Ronald Ray, cultural anthropology
McPherson, Joe Wayne, solid state physics
Meisner, Gerald Warren, elementary particle physics
Miller, Robert L., physical chemistry
Minnemeyer, Harry Joseph, bio-organic chemistry
Morrison, Ralph M., plant physiology, microbiology
Morse, Richard Kenneth, developmental physiology
Mountjoy, Joseph Bode, anthropology, archeology
Murphy, Robert T., analytical chemistry
Nile, Terence Anthony, organometallic chemistry
Patterson, Ronald Brinton, organic chemistry
Posey, Eldon Eugene, mathematics
Potthoff, Richard Frederick, medical chemistry
Powers, William Allen, III, statistics
Purdom, Emil Garness, optics
Puterbaugh, Walter Henry, organic chemistry
Reardon, Anna Joyce, physics
Reed, William Edward, soil chemistry
Reid, Jack Richard, organic chemistry, research
Richards, Russell Fayette, economic entomology
Robbins, Louise Marie, physical anthropology
Rogers, Hollis Jetton, botany
Salvin, Victor S., chemistry
Sandin, Thomas Robert, solid state physics
Schauer, Richard C., physiology
Schickedantz, Paul David, organic chemistry, analytical biochemistry
Schroeder, Juel Pierre, organic chemistry
Schultz, Frederick John, organic chemistry
Sen, Lalita, economic geography, transportation geography
Sher, Richard B., topology
Soderquist, David Richard, psychoacoustics
Sookne, Arnold Maurice, chemistry
Spears, Alexander White, III, organic chemistry, physical chemistry
Stavn, Robert Hans, ecology
Sumner, Darrell Dean, medicinal chemistry, metabolism
Thekkekandan, Joseph Thomas, organic chemistry
Tucker, Charles Leroy, Jr., chemistry
Van Geluwe, John David, entomology
Van Pelt, Arnold Francis, Jr., ecology
Vanselow, Clarence Hugo, physical chemistry
Vaughan, Jerry Eugene, mathematics, topology
Vaughan, Theresa Phillips, algebra
Vick, Alphonso Roscoe, botany, zoology
Weaver, John Arthur, inorganic chemistry
Webb, Alfreda Johnson, veterinary medicine
Webb, Burleigh C., agronomy, plant physiology
Wehner, Philip, industrial organic chemistry
White, Booker Taliaferro, biochemistry
Whitlock, Richard T., theoretical physics
Willett, Richard Michael, mathematics
Wilson, James Franklin, microbiology, genetics

GREENVILLE
Adler, Carl George, theoretical physics
Allen, Wendall E., microbiology, microbial genetics
Ayers, Caroline LeRoy, physical chemistry
Ayers, Paul Wayne, chemistry
Bailey, Donald Etheridge, physical organic chemistry
Beckman, David Lee, pulmonary physiology

HENDERSONVILLE
Herman, Carlton Martin, parasitology

Bellis, Vincent J, Jr., aquatic ecology
Birchard, Ralph Edwin, urban geography, geography of africa, europe
Bishop, B A., geology
Blackwell, Floyd Orris, environmental health
Bland, Charles E., mycology
Boyette, Joseph Greene, vertebrate ecology, academic administration
Brinn, Jack Elliott, Jr., endocrinology, electron microscopy
Brinson, Mark McClellan, plant ecology, limnology
Brown, Charles Quentin, geology
Burden, Hubert White, reproductive endocrinology
Byrd, James William, physics
Coulter, Byron Leonard, theoretical physics
Cramer, Robert Eli, cartography, geography of north america
Daugherty, Patricia A., genetics
Davis, Graham Johnson, aquatic ecology
Davis, Kenneth Joseph, mathematics
Debnath, Lokenath, applied mathematics, pure mathematics
Debnath, Sadhana, analytical chemistry, physical chemistry
Dough, Robert Lyle, science education
Everett, Grover Woodrow, analytical chemistry
Haggard, Paul Wintzel, mathematics
Hall, William Earl, pharmacy, physical chemistry
Hampton, Carolyn Hutchins, zoology, parasitology
Hayek, Dean Harrison, physiology
Heckel, Edgar, physical chemistry
Heckrotte, Carlton, zoology
Helms, Rufus Marshall, experimental physics
Ito, Takeru, biochemistry
Jeffreys, Donald Bearss, plant physiology, microbiology
Jennings, Albert Ray, mathematics
Johnson, Frankford Milam, mathematics
Joyce, James Martin, nuclear physics
Knight, Clifford Burnham, insect ecology
Lamb, Robert Charles, organic chemistry
Laurie, John Sewall, experimental biology
Lawrence, Irvin E, Jr., embryology, histology
Leahy, Edward Prior, geography
Li, Chia-Yu, electroanalytical chemistry
Lowry, Jean, geology
Lunney, David Clyde, chemical instrumentation
Martin, George Carlyle, geology
Matthes, Floyd E., science education
Mauger, Richard L., geology
McAllister, Warren Alexander, inorganic chemistry
McDaniel, James Scott, parasitology
McDaniel, Susan Griffith, zoology
McEnally, Terence Ernest, Jr., physics, electronic engineering
Nye, Sylvanus William, pathology
Paul, William Larry, pharmacy
Pennington, Sammy Noel, analytical biochemistry
Phelps, David Sutton, archaeology, physical anthropology
Pignani, Tullio Joseph, mathematics
Putnam, Serpas Jerome, physiology
Read, Floyd M., physics
Riggs, Stanley R., geology
Ryan, Edward Parsons, invertebrate zoology, developmental biology
Saunders, Frank Wendell, mathematics
Sayetta, Thomas C., magnetic resonance
Schweisthal, Michael R., anatomy
Sehgal, Prem P., plant physiology
Sheppard, Moses Maurice, science education
Simpson, Everett Coy, zoology, endocrinology
Smith, Susan T., biochemistry, clinical chemistry
Sowell, Katye Marie Oliver, mathematics
Spickerman, William Reed, applied mathematics
Stella, Donald, climatology
Stephenson, Richard Allen, geomorphology, marine sciences
Thurber, Robert Eugene, physiology, biophysics
Variashkin, Paul, solid state physics
Watrous, Blanche Greene, anthropology
Waugh, William Howard, physiology, internal medicine
Webber, Carroll A, Jr., mathematics
Winters, Loren Mel, atomic physics
Wootes, Wallace Ralph, pharmacology

LAURINBURG
Anand, Satish Chandra, plant breeding
Barnes, Donald George, physical chemistry, chemical physics
Clausz, John Clay, mycology

King, Willis, fisheries
Koppenhoefer, Robert Mack, petroleum chemistry
Sanders, Rosaltha Hagan, physiology

HICKORY
Blackburn, Thomas Henry, mathematics
Boliek, Irene, zoology
Chou, David Yuan Pin, physical chemistry
Garrett, James Richard, mathematics
Lugn, Alvin Leonard, geology
Spuller, Robert L., protozoology, parasitology
Unglaube, James M., organic chemistry
Wells, Charles Van, botany

HIGH POINT
Epperson, Edward Roy, inorganic chemistry
Leach, James Moore, organic chemistry
Page, Nelson Franklin, mathematics
Ward, John Everett, Jr., mycology, ecology
Weeks, Leo, genetics
Wilson, Christopher Lumley, chemistry
Yeats, Frederick Tinsley, botany

HIGHLANDS
Crumpler, Thomas Bigelow, analytical chemistry
Sanford, John Theron, geology

JAMESTOWN
Tweedy, Billy Gene, plant pathology

JONESVILLE
Keever, Catherine, ecology

KINSTON
Abashian, Steven, organic chemistry
Billica, Harry Robert, chemistry
Bosley, David Emerson, physical chemistry
Burrus, Robert Tilden, inorganic chemistry, physical chemistry
Daniels, William Ward, physical chemistry, research administration
Davis, Hawthorne Antoine, solid state physics
Ericson, Corey William, theoretical chemistry, physical chemistry
Evans, Evan Franklin, polymer chemistry
Grindstaff, Teddy Hodge, physical chemistry
Hartzler, Jon David, organic polymer chemistry
Hartzog, James Victor, physical chemistry
Haseley, Edward Albert, physical chemistry, polymer science
Hodge, James Dwight, polymer chemistry
Hogsed, Milton Jones, organic chemistry
Jones, William Jonas, Jr., organic chemistry
Kobsa, Henry, physical chemistry
Kupstas, Edward Eugene, organic chemistry
Latschar, Carl Ernest, physical chemistry
McKay, Jerry Bruce, organic chemistry, polymer chemistry
Meelheim, Richard Young, physical chemistry
Olsen, John Sylvester, textile chemistry
Pacofsky, Edward Anthony, organic chemistry
Profitt, Thomas Jefferson, Jr., organic chemistry
Reese, Cecil Everett, polymer chemistry
Reese, Millard Griffin, Jr., organic chemistry
Sargeant, Peter Barry, textile chemistry, research administration
Shaffner, Thomas Jackson, textile physics
Van Veld, Robert Dale, textile physics
Watson, William Harrison, polymer chemistry
Whitehouse, Bruce Alan, polymer chemistry, textile chemistry

LAKE TOXAWAY
Marston, Alfred Lawrence, chemistry

Helm, James Leroy, plant breeding, genetics
Marks, Stuart A., anthropology, wildlife research
Miller, George Tyler, Jr., chemistry, human ecology
Newton, H Calvin, Jr., plant pathology, plant breeding
Rolland, William Woody, nuclear physics
Smith, Thomas Earle, plant pathology
Styron, Clarence Edward, Jr., ecology
Wetmore, David Eugene, organic chemistry, petroleum chemistry

LENOIR
Edwards, Ben E., industrial organic chemistry
Wiesler, Donald Paul, organic chemistry

LEXINGTON
Derrick, Mildred Elizabeth, physical chemistry
Sink, Woodford Grady, physical chemistry

LITTLE SWITZERLAND
Walker, Lillie Cutlar, preventive medicine, pediatrics

MARS HILL
Devries, David J., mathematics
Diercks, Fred Herman, medical microbiology, epidemiology
Holtkamp, Freddy Henry, organic chemistry
Outen, Lora Milton, zoology
Sams, Emmett Sprinkle, mathematics

MISENHEIMER
Crowl, Robert Harold, animal genetics
Echols, Joseph Todd, Jr., physical chemistry
Jackman, Donald Coe, inorganic chemistry, analytical chemistry
Macey, Wade Thomas, mathematics
Manly, Jethro Oates, phycology
Mobley, Jean Bellingrath, mathematics
Riemann, James Michael, organic chemistry
Robertson, Clyde Henry, entomology
Stephenson, Harold Patty, molecular spectroscopy

MONTREAT
Snyder, Karl Doughty, zoology

MOREHEAD CITY
Baker, Edward George, mathematics
Chestnut, Alphonse F., marine ecology, zoology
Fahy, William Earl, ichthyology
Kohlmeyer, Jan Justus, mycology
Schwartz, Frank Joseph, ichthyology
Voris, Aaron LeRoy, nutrition

MT AIRY
Bradley, Harris Walton, chemistry

MURFREESBORO
Warren, Richard, physics

NORTH WILKESBORO
Smith, Wendell Phillips, ornithology

OTEEN
Joyner, John T, III, internal medicine, nuclear medicine

OXFORD
Baumhover, Alfred Henry, economic entomology
Burk, Lawrence G., genetics
Chaplin, James Ferris, genetics
DeJong, Donald Warren, plant cytochemistry
Elsey, Kent D., entomology
Sievert, Richard Carl, animal pathology
Spurr, Harvey Wesley, Jr., plant pathology
Stewart, Paul Alva, ecology, ornithology
Williamson, Ralph Edward, plant physiology

PEMBROKE
Teague, Harold Junior, organic chemistry
Wallingford, John Stuart, physics

PENROSE
Rehberg, Chessie Elmer, organic chemistry

PINEHURST
Steffee, Clyde Harold, pathology

PISGAH FOREST
Cline, Warren Kent, organic chemistry
Kirk, David Clark, Jr, physical chemistry
Lea, David Chester, paper chemistry
Martin, Richard Hugo, physical chemistry
Rapoport, Lorence, organic polymer chemistry

RALEIGH
Anderson, Charles Eugene, morphology
Andrews, Walter Glenn, poultry science
Apple, Jay Lawrence, phytopathology
Armstrong, Frank Bradley, Jr, biochemistry
Aurand, Leonard William, biochemistry
Averre, Charles Wilson, III, plant pathology
Axtell, Richard Charles, medical entomology
Aycock, Robert, phytopathology
Baird, Jack Vernon, soil science
Baitinger, William F, Jr, textile chemistry
Baker, Simon, geography
Baldwin, James Gordon, plant nematology, plant pathology
Ballinger, Walter Elmer, horticulture
Barefoot, Aldos Cortez, Jr, wood science & technology
Barkalow, Frederick Schenck, Jr, zoology
Barker, Kenneth Reece, plant pathology
Barrick, Elliott Roy, animal science
Bartholomew, William Victor, soils
Batte, Edward G, veterinary parasitology
Beeler, Joe R, Jr, solid state physics
Behlow, Robert Frank, animal parasitology
Bennett, Willard Harrison, plasma physics
Bent, Henry Albert, physical chemistry
Beute, Marvin Kenneth, plant pathology
Bewley, Glenn Carl, developmental genetics
Bhattacharyya, Bibhuti Bhushan, mathematical statistics
Bishir, John William, applied mathematics
Blake, Carl Thomas, agronomy
Blalock, Thomas Jacks, inorganic chemistry
Blum, Udo, botany, ecology
Bordner, Jon D B, organic chemistry, biochemistry
Bostian, Carey Hoyt, genetics
Bowen, Lawrence Hoffman, physical chemistry
Bradbury, Phyllis Clarke, protozoology
Bradley, Julius Roscoe, Jr, entomology
Brett, Charles H, entomology
Brim, Charles A, plant breeding
Brooks, Wayne Maurice, insect pathology, protozoology
Brown, Henry Seawell, economic geology
Bryan, Carl Eddington, chemistry
Bryan, Frederick Allen, Jr, system analysis, applied physics
Bryant, Ralph Clement, forest management
Bullock, Roberts Cozart, mathematics
Bumgardner, Carl Lee, organic chemistry
Buol, Stanley Walter, soil science
Burniston, Ernest Edmund, applied mathematics
Burns, George Robert, soil fertility
Burns, Joseph Charles, plant physiology, animal nutrition
Busbice, Thaddeus H, genetics, plant breeding
Campbell, William Vernon, entomology
Carmichael, Halbert Hart, chemical kinetics
Carson, Robert James, III, geology, geomorphology
Carter, Roy Merwin, wood technology
Caruolo, Edward Vitangelo, animal physiology, medical physiology
Caruthers, Leo Thomas, Jr, health physics
Cassel, D Keith, soil physics
Cates, David Marshall, polymer chemistry, textile chemistry
Caves, Thomas Courtney, theoretical chemistry
Chamblee, Douglas Scales, agronomy
Chandler, Richard Edward, mathematics
Chaney, David Webb, organic chemistry
Chang, Hou-Min, wood chemistry
Chang, Irene Ching Lai, nutrition
Charlton, Harvey Johnson, mathematics
Clawson, Albert J, animal nutrition
Clayton, Carlyle Newton, plant pathology
Cobb, Grover Cleveland, Jr, nuclear physics
Cochran, Fred Derward, horticulture
Cockerham, Columbus Clark, population genetics, quantitative genetics
Collins, William Kerr, plant breeding, genetics
Colwell, William Maxwell, veterinary microbiology
Cook, Maurice Gayle, soil science, agronomy
Cook, Robert Edward, genetics
Cooke, Henry Charles, mathematics
Cooper, Arthur Wells, ecology
Coots, Alonzo Freeman, physical chemistry
Cope, Will Allen, genetics, plant breeding
Copeland, Billy Joe, ecology
Corbin, Frederick Thomas, weed science, plant physiology
Cowling, Ellis Brevier, plant pathology
Cox, Frederick Russell, soil science
Cummings, George August, soil science
Curtin, Terrence M, veterinary medicine, physiology
Dahiya, Raghunath S, food microbiology
Danby, John Michael Anthony, celestial mechanics
Daniels, Raymond Bryant, soil genesis, geomorphology
Dauterman, Walter Carl, insect toxicology, biochemistry
Davey, Charles Bingham, soil microbiology
Davis, Charles Alfred, applied mathematics
Davis, David Edward, ecology
Davis, William Robert, physics
DeArmond, M Keith, physical chemistry, inorganic chemistry
Dillard, Emmett Urcey, animal genetics
Dillman, Richard Carl, veterinary pathology
Doak, George Osmore, pharmaceutical chemistry
Dobrogosz, Walter Jerome, microbiology, biochemistry
Doerr, Phillip David, wildlife ecology
Doggett, Wesley Osborne, plasma physics
Donaldson, William Emmert, poultry nutrition
Downs, Robert Jack, botany
Droessler, Earl George, meteorology
Duffield, John Warren, forestry
Duncan, Harry Ernest, plant pathology
Dunn, Joseph Charles, applied mathematics, cybernetics
Echandi, Eddie, plant pathology
Edens, Frank Wesley, poultry physiology, avian physiology
Eisen, Eugene J, genetics, statistics
Elkan, Gerald Hugh, bacteriology
Elleman, Thomas Smith, physical chemistry
Elliott, William Whitefield, mathematics
Ellis, Don Edwin, plant pathology
Ellwood, Eric Louis, forest products
Emery, Donald Allen, plant breeding
Etchells, John Lincoln, bacteriology
Evans, James Brainerd, bacteriology
Fabacher, David Lawrence, toxicology
Falter, John Max, entomology
Farrier, Maurice Hugh, acarology, forest entomology
Fike, William Thomas, Jr, agronomy
Fitts, James Walter, soil fertility
Fleming, Henry Pridgen, food science, microbiology
Fornes, Raymond Earl, polymer physics, textile physics
Freedman, Leon David, organometallic chemistry
Fulp, Ronald Owen, algebra
Funderburg, John Broadus, Jr, animal ecology, vertebrate zoology
Galletta, Gene John, genetics
Gardner, Marianne Lepp, combinatorics
Garlich, Jimmy Dale, nutrition, biochemistry
Gehle, Marvin Harlan, institutional research
Gerig, Thomas Michael, mathematical statistics
Gerstel, Dan Ulrich, genetics
Getzen, Forrest William, physical chemistry
Ghosh, Kalyan K, polymer chemistry, physical chemistry
Gilbert, Richard Dean, polymer chemistry
Gilbert, William Best, agronomy
Gilliam, James Wendell, soil chemistry
Gilliland, Stanley Eugene, food microbiology
Glazener, Edward Walker, poultry husbandry
Gleit, Chester Eugene, analytical chemistry, physical chemistry
Gold, Harvey Joseph, biomathematics
Goldfinger, George, color science, polymer science
Goldstein, Irving Solomon, wood chemistry, paper chemistry
Goldston, Eugene Frizzelle, soils
Goldthwait, Charles Francis, textile chemistry
Goode, Lemuel, physiology
Gooding, Guy V, Jr, phytopathology
Goodman, Major M, genetics, evolutionary biology
Gottlieb, Gilbert, psychobiology
Grandage, Arnold Herbert Edward, experimental statistics
Grau, Craig Robert, phytopathology
Gregory, Max Edwin, food science
Gregory, Walton Carlyle, plant breeding
Grosch, Daniel Swartwood, genetics, zoology
Gross, Harry Douglass, agronomy
Grube, Geraldine Joyce Terenzoni, theoretical nuclear physics
Guion, Thomas Hyman, textile chemistry
Gupta, Bhupender Singh, textile physics
Guthrie, Frank Edwin, entomology
Haase, David Glen, low temperature physics
Hader, Robert John, mathematical statistics
Hafley, William LeRoy, forestry, experimental statistics
Hain, Fred Paul, forest entomology
Halaby, Sami Assad, physical chemistry, solid state chemistry
Hall, George Lincoln, theoretical physics, solid state physics
Hamilton, Pat Brooks, microbiology, biochemistry
Hanck, Kenneth William, analytical chemistry
Hansen, Donald Joseph, mathematics
Hanson, Warren Durward, plant genetics, quantitative genetics
Hardin, James Walker, botany
Harkema, Reinard, zoology, parasitology
Harrington, Walter Joel, mathematics
Hart, Clarence Arthur, forestry
Hartwig, Robert Eduard, applied mathematics
Hayne, Don William, biometrics
Haynes, Frank Lloyd, Jr, plant breeding, genetics
Heath, Ralph Carr, geology
Hebert, Teddy T, plant pathology
Heck, Walter Webb, plant physiology, air pollution
Henderson, Nannette Smith, plant pathology
Henderson, Warren Robert, horticulture
Hentz, Forrest Clyde, Jr, inorganic chemistry, physical chemistry
Hersh, Solomon Philip, polymer chemistry
Hill, Charles Horace, Jr, nutrition, biochemistry
Hines, Martin Patterson, veterinary medicine
Hirschmann, Hedwig, plant nematology
Hodgson, Ernest, comparative biochemistry, toxicology
Hoover, Maurice Wilbur, horticulture
Horie, Yasuyuki, physics
Horner, Sally Melvin, inorganic chemistry
Horton, Horace Robert, biochemistry
Horvitz, Daniel Goodman, statistics
Huang, Jeng-Sheng, plant pathology
Huisingh, Donald, plant pathology, biochemistry
Huneycutt, James Ernest, Jr, mathematical analysis
Hunter, Arvel Hatch, soil fertility
Hunter, William James, microbiology
Hyatt, George, Jr, dairy husbandry
Ingram, William Prentiss, Jr, chemistry, nutrition
Jackson, William Addison, plant nutrition, soil fertility
Jenkins, Alvin Wilkins, Jr, plasma physics, astrophysics
Jenkins, Samuel Forest, Jr, plant pathology
Johnson, Bobby Ray, biochemistry, food science
Johnson, Charles Edward, atomic physics
Johnson, William Lawrence, animal nutrition
Jones, Evan Earl, biochemistry
Jones, Guy Langston, plant breeding
Jones, Ivan Dunlavy, food science
Jones, James Robert, animal husbandry, nutrition
Kahn, Joseph Stephan, plant biochemistry
Kamprath, Eugene John, soil fertility
Katzin, Gerald Howard, physics
Keller, Kenneth Raymond, plant breeding
Kellison, Robert Clay, forest genetics
Kelly, Harry Charles, experimental physics
Kennedy, George Grady, economic entomology
Kepler, Carol R, lipid chemistry
Kimbrough, Everett Lamar, agronomy
Knight, Kenneth Lee, entomology
Knopp, James A, biophysical chemistry
Knott, Fred Nelson, animal nutrition
Knowles, Charles Ernest, physical oceanography
Koh, Kwangil, mathematics
Kunz, Hans Joseph, physics, physical chemistry
Lado, Fred, physics
Lammi, Joe Oscar, forestry, international economics
Larson, Roy Axel, horticulture
Latch, Dana May, algebra, topology
Leatherwood, James M, animal nutrition
Lecce, James Giacomo, microbiology
Lee, Joshua Alexander, genetics
Legates, James Edward, animal genetics
LeGrand, Harry E, hydrogeology
Leidy, Ross Bennett, pesticide chemistry
Leith, Carlton James, geology
Levi, Michael Phillip, forest products
Levine, Jack, mathematics
Levine, Samuel Gale, organic chemistry
Levings, Charles Sandford, III, genetics
Lewis, Paul Edwin, mathematics
Lewis, William Mason, weed science
Linnerud, Ardell Chester, experimental statistics
Linton, Thomas LaRue, fisheries
Lipscomb, Elizabeth Lois, biochemistry
Loeppert, Richard Henry, organic chemistry
Long, George Gilbert, inorganic chemistry
Long, Raymond Carl, plant physiology
Longmuir, Ian Stewart, biochemistry, physiology
Lucas, George Blanchard, plant pathology
Lucas, Henry Laurence, Jr, statistics
Lucas, Leon Thomas, plant pathology, microbiology
Luh, Jiang, algebra
Lytle, Charles Franklin, zoology
Main, Alexander Russell, biochemistry, enzymology
Main, Charles Edward, plant pathology, plant physiology
Maki, Tenho Ewald, forest soils, forest hydrology
Mann, Thurston (Jefferson), genetics
Manring, Edward Raymond, physics
Marlin, Joe Alton, applied mathematics
Martin, LeRoy Brown, Jr, mathematics, operations research
Martof, Bernard Stephen, zoology
Mason, David Dickenson, applied statistics
Matzinger, Dale Frederick, quantitative genetics
McCants, Charles Bernard, soil science
McCollum, Robert Edmund, soils, plant biochemistry
McDaniel, Benjamin Thomas, animal genetics, animal breeding
McKenzie, Wendell Herbert, human genetics
McLaughlin, Foil William, agronomy
McNeill, John J, microbiology
McPeters, Arnold Lawrence, polymer science
McVay, Francis Edward, applied statistics
Memory, Jasper Durham, physics
Menius, Arthur Clayton, Jr, physics
Mercando, Neil Aldo, aquatic ecology
Mettler, Lawrence Eugene, biology
Miles, Marion Lawrence, organic chemistry
Miller, Conrad Henry, plant physiology
Miller, Grover Cleveland, zoology
Miller, Philip Arthur, plant breeding
Miller, Robert James, II, plant physiology, ecology
Mills, William Clearon, Jr, poultry science, marketing
Miner, Gordon Stanley, soil fertility, agronomy
Mistric, Walter J, Jr, entomology

NORTH CAROLINA

Mitchell, Gary Earl, physics
Mochrie, Richard Douglas, dairy science
Moll, Robert Harry, genetics
Monroe, Robert James, statistics
Moore, Harry Ballard, Jr., entomology
Moore, Robert Parker, agronomy
Moreland, Charles Glen, physical chemistry
Moreland, Donald Edwin, plant physiology; weed science
Morgan, George Wallace, poultry physiology
Moss, Marvin Kent, physics, theoretical mechanics
Murphy, Charles Franklin, plant breeding
Nace, Raymond Lee, hydrology
Nahikian, Howard Movess, mathematics
Nair, Sreekantan S., statistics, operations research
Nankoong, Gene, forest genetics, evolution
Nelson, Paul Victor, horticulture, plant nutrition
Nesbitt, William Belton, horticulture, plant breeding
Neunzig, Herbert Henry, entomology
Nicholaides, John J., soil fertility
Nickerson, Gifford Spruce, ethnology; applied anthropology
Nielsen, Lowell Wendell, plant pathology
Noggle, Glenn Ray, plant physiology
Nobstad, Arnold Ragnvald, mathematics
Nusbaum, Charles Joseph, plant pathology
Oliver, George Joseph, health physics
Oppenheim, Robert Preston, crop physiology; crop production
Patty, Richard Roland, molecular spectroscopy
Perry, Astor, agronomy
Perry, Jerome John, microbiology
Perry, Thomas Oliver, forest genetics
Philips, Lyle Llewellyn, cytogenetics
Phipps, Patrick Michael, plant pathology, mycology
Pope, Daniel Townsend, horticulture
Porterfield, Ira Deward, dairy husbandry
Powell, Nathaniel Thomas, plant pathology
Preston, Richard Joseph, Jr., forestry
Prince, Jerome John, microbiology
Purcell, Albert Ernest, biochemistry
Purrington, Suzanne T., organic chemistry
Quay, Thomas Lavelle, animal ecology
Querry, John William, mathematics
Quesenbery, Charles P., statistics
Rabb, Robert Lamar, entomology
Rakes, Allen Huff, dairy nutrition
Ramsey, Harold Arch, animal nutrition
Rawlings, John Oren, biometrics
Reagan, Thomas Eugene, entomology
Reinert, Richard Allyn, plant pathology, horticulture
Richmond, Rollin Charles, population genetics
Rickard, David Alan, plant nematology
Rigney, Jackson Ashcraft, statistics, plant breeding
Roberts, John Fredrick, zoology; cell biology
Roberts, William Milner, food science
Robertson, Robert L., entomology
Robison, Odis Wayne, genetics, reproductive physiology
Rochow, Theodore George, chemical microscopy
Rock, George Calvert, entomology
Rogerson, Asa Benjamin, plant physiology; agronomy
Rollins, Robert Leroy, Jr., psychiatry
Rose, Nicholas John, mathematics
Ross, John Paul, plant pathology
Roulier, John Arthur, mathematics
Rutherford, Henry Ames, organic chemistry
Sagan, Hans, mathematics
Sanchez, Pedro Antonio, agronomy, soil fertility
Sargent, Frank Dorrance, quantitative genetics, reproductive physiology
Sasser, Joseph Neal, nematology
Saucier, Walter Joseph, meteorology
Sawin, Stephen Sanford, inorganic chemistry
Saylor, LeRoy C., forestry genetics
Scandalios, John George, genetics

Schaffer, Henry Elkin, genetics
Schetzina, Jan Frederick, physics
Schlichting, Harold Eugene, Jr., physiology
Schmidt, Donald Peter, plant nematology; plant pathology
Schrader, John William, agronomy
Schreiner, Anton Franz, inorganic chemistry
Scofield, Herbert Temple, plant physiology
Scott, Harry Eldon, entomology
Seagondollar, Lewis Worth, physics
Seltmann, Heinz, plant physiology
Sheets, Thomas Jackson, plant pathology
Sherwood, Jesse Eugene, atomic physics
Shreffler, Carol Kauffman, chemistry, biochemistry
Silber, Robert, mathematics
Sisler, Edward C., plant physiology
Skroch, Walter Arthur, weed science
Smith, Benjamin Warfield, plant cytology, genetics
Smith, Clyde F., entomology
Smith, Donald Eugene, cell physiology; endocrinology
Smith, Frank Houston, analytical chemistry, biochemistry
Smith, Leonard, organic chemistry
Sorensen, Kenneth Alan, entomology
Speck, Marvin Luther, microbiology
Spence, Herbert E., mathematics
Sprenkel, Richard Keiser, entomology
Stamm, Alfred Joaquin, physical chemistry; wood chemistry
Stannett, Vivian Thomas, polymer chemistry
Steel, Robert George Douglas, statistical analysis
Steensen, Donald H.J., forest economics
Stephens, Stanley George, evolutionary biology
Stinner, Ronald Edwin, population ecology
Strider, David Lewis, plant pathology
Strobel, James Walter, plant pathology
Struble, Raimond Aldrich, mathematics
Stuber, Charles William, genetics
Sutton, Paul Porter, chemistry, physics
Sutton, Turner Bond, plant pathology
Swaisgood, Harold Everett, protein biochemistry
Tabor, Elbert Cecil, biochemistry
Tarver, Fred Russell, Jr., food technology; bacteriology
Terry, David Lee, soil science
Thaxton, James Paul, avian physiology
Theil, Elizabeth, biochemistry
Theil, Michael Herbert, polymer chemistry; developmental biology
Thomas, Frank Bancroft, horticulture, food technology
Thomas, Joab Langston, botany
Thomas, Richard Joseph, wood technology
Thompson, Donald Loraine, agronomy
Tilley, David Ronald, nuclear physics
Timothy, David Harry, plant genetics, evolution
Tove, Samuel B., biochemistry, nutrition
Tove, Shirley Ruth, bacteriology
Triantaphylou, Anastasios Christos, cytogenetics, nematodes
Troyer, James Richard, plant physiology
Tucker, Paul Arthur, textiles, microscopy
Tucker, William Preston, organic chemistry
Ulberg, Lester Curtiss, reproductive physiology
Ullrich, David Frederick, mathematics
Underwood, Herbert Arthur, Jr., biological rhythms
Unrath, Claude Richard, horticulture
Vance, Miles Elliott, medical technology
Vandenbergh, John Garry, animal behavior, endocrinology
van der Vaart, Hubertus Robert, mathematics, statistics
Van Dyke, Cecil Gerald, plant pathology
Voelker, Robert Allen, genetics
Volk, Richard James, plant nutrition
Wahl, George Henry, Jr., structural chemistry
Walsh, William K., textile chemistry, chemistry engineering
Walter, William Mood, Jr., food science
Walter, Arthur, nuclear physics
Ward, James B., poultry nutrition
Ward, Thomas Marsh, physical chemistry
Webb, Neil Broyles, biochemistry, microbiology
Weber, Allen Howard, micrometeorology

Weber, Jerome Bernard, soil chemistry; weed science
Weed, Sterling Barg, soil chemistry
Weekman, Gerald Thomas, entomology
Welby, Charles William, hydrogeology; stratigraphy
Welch, Aaron Waddington, botany
Welty, Ronald Earle, plant pathology
Wensman, Earl Allen, crop breeding
Wertz, Dennis William, physical chemistry
Wesler, Oscar, mathematics, statistics
Wessling, Wolfgang Heinrich, genetics, physics
Weybrew, Joseph Arthur, plant chemistry
Whalery, Wilson Monroe, textile chemistry
Whitford, Larry Alston, botany
Wilson, James Blake, applied mathematics
Wilson, Lorenzo George, horticulture, vegetable crops
Winstead, Nash Nichls, plant pathology
Winton, Lowell Sheridan, animal nutrition
Wise, George Herman, animal physiology
Witt, Peter Nikolaus, pharmacology
Wollum, Arthur George, II, soils, microbiology
Woltz, Willie Garland, agronomy, soil fertility
Woodhouse, William Walton, Jr., soils
Work, Robert Wyllie, polymer chemistry
Worsham, Arch Douglas, weed science, plant physiology
Wright, Charles Gerald, entomology
Wu, Shirley Shao-ning, biochemistry
Wynne, Johnny Calvin, plant breeding, plant genetics
Yamamoto, Robert Takaichi, entomology
Yarbrough, John Alonzo, botany; biology
Yarbrough, Mary Elizabeth, nutrition
Young, David Allan, insect taxonomy
Young, Mahlon Gilbert, chemistry
Zobel, Bruce John, forest genetics
Zumwalt, Lloyd Robert, physical chemistry

REIDSVILLE

Ammondson, Clayton John, physical chemistry

RESEARCH TRIANGLE PARK

Albro, Philip William, biochemistry
Allis, John W., physical chemistry
Alpert, Leo, environmental sciences, meteorology
Anderson, Richmond Karl, physiological chemistry, medicine
Baccanari, David Patrick, enzymology
Baron, Ronald L., biochemistry, entomology
Bayless, David Lee, statistics
Blackman, Carl F., Jr., molecular biology, physical chemistry
Boghosian, Charles, low temperature physics, solid state physics
Brieaddy, Lawrence Edward, pharmaceutical chemistry; organic chemistry
Burch, George Nelson Blair, polymer chemistry
Burchall, James J., bacterial physiology
Burger, Robert M., solid state physics, electronics
Burke, James Joseph, Jr., polymer physics
Carroll, F Ivy, organic chemistry
Chae, Kun, medicinal chemistry, biochemistry
Chandra, Jagdish, mathematics
Chenery, Peter Jaspersen, science administration, information science
Chhabra, Rajendra S., pharmacology, toxicology
Ching, Jason Kwock Sung, environmental science
Cifian, Mikael, theoretical physics
Clark, Edgar William, insect physiology, forest entomology
Cohn, Naomi Kenda, cell biology; cancer
Cook, Clarence Edgar, organic chemistry
Courtney, Katherine Diane, pharmacology
Cuatrecasas, Pedro, biochemistry, medicine
Davis, Richard Cecil, textile chemistry, science policy
DeBrunner, Ralph Edward, organic chemistry
De Miranda, Paulo, pharmacology, biochemistry
de Serres, Frederick Joseph, environmental health, microbial genetics
DiAugustine, Richard Patrick, molecular pharmacology

Dixon, Robert Louis, pharmacology, biochemistry
Drechsel, Paul David, polymer chemistry; textile chemistry
Drew, Robert Taylor, health physics
Drooz, Arnold T., forest entomology
Durham, William Fay, biochemistry
Edelman, David Anthony, biostatistics
Eling, Thomas Edward, pharmacology, biochemistry
Elion, Gertrude Belle, biochemistry, pharmacology
Ellis, Charles Herbert, physiology
Ellison, Alfred Harris, surface chemistry, air pollution
Elmer, Curtis, applied chemistry
Ely, Ralph Lawrence, Jr., nuclear physics
Ferone, Robert, microbiology, biochemistry
Ferris, Robert Monsour, biochemistry, neurochemistry
Fishbein, Lawrence, organic chemistry
Fouts, James Ralph, pharmacology
Fowler, Bruce Andrew, toxicology
Friedlander, Herbert Norman, physical organic chemistry, research administration
Fyfe, James Arthur, enzymology
Gardner, Edward, Jr., immunology
Garmon, Ronald Gene, analytical chemistry
Garner, Jasper Henry Barksdoll, mycology; plant pathology
Gay, Bruce Wallace, Jr., atmospheric chemistry
Ghirardelli, Robert George, organic chemistry
Glenn, William Alexander, statistics
Golberg, Leon, toxicology; experimental pathology
Goldstein, Joyce Allene, pharmacology
Goodall, McChesney, physiology
Griffing, George Warren, theoretical physics
Grivsky, Eugene Michael, organic chemistry
Hamilton, Harry L., Jr., meteorology
Hammer, William Frederick, polymer chemistry
Harfenist, Morton, organic chemistry
Hart, Larry Glen, pharmacology
Hass, James Ronald, analytical chemistry, environmental health
Hastings, Felton Leo, insect toxicology
Henderson, Thomas Rannoy, organic chemistry
Hill, John Benjamin, pharmacology
Hitchings, George Herbert, biochemistry
Hodges, Charles Sasnette, Jr., plant pathology
Hoel, David Gerhard, biostatistics
Holland, Virgil Fortune, polymer science
Holstius, Elvin Albert, pharmacy
Holzworth, George Charles, meteorology
Horton, Robert John Munro, epidemiology
Hosler, Charles R., meteorology
Hueter, Francis Gordon, animal toxicology
Hurlbert, Bernard Stuart, chemistry
Ingram, Peter, physics
Jorgensen, Franklin M., genetics
Johnson, Jacques R., soil science
Jurgelsky, William, Jr., experimental pathology; pharmacology
Kelley, James LeRoy, medicinal chemistry
Kelsey, John Edward, medicinal chemistry
Kimmel, Carole Anne, teratology
Kimmel, Gary Lewis, reproductive endocrinology
King, Henry Lee, organic chemistry
Kirby, James Ray, analytical chemistry
Knapp, Kenneth T., environmental chemistry
Knudsen, John Peter, physics
Knudsen, Thomas Paul, mass spectrometry, physical chemistry
Kolman, Wilfred Aaron, clinical pharmacology; biometrics
Koop, John C., experimental statistics
Koster, Rudolf, neuropharmacology
Krenitsky, Thomas Anthony, biochemistry
Kuhlman, Elmer George, plant pathology
Kull, Frederick Charles, bacteriology
Lawless, Philip Austin, engineering physics
Lawson, Julian Keith, Jr., organic chemistry

Lee, Robert E, Jr., physical chemistry, air pollution
LeMunyan, Cobert Duane, environmental health
Levy, Louis A, organic chemistry
Lewis, Robert Glenn, organic chemistry
Liddle, Charles George, veterinary medicine, radiation biology
Liepins, Raimond, polymer chemistry, organic chemistry
Lilyquist, Marvin Russell, organic chemistry
Lontz, Robert Ian, physics
Malling, Heinrich Valdemar, genetics, microbiology
Martin, William Royall, Jr, organic chemistry, business administration
Mason, Robert Edward, statistics, ecology
Matthews, Hazel Benton, Jr, biochemical pharmacology
Maurer, Ralph Rudolf, reproductive physiology, reproductive endocrinology
Maxwell, Robert Arthur, pharmacology
McKinney, James David, environmental chemistry
Mehta, Nariman Bomanshaw, drug metabolism, pharmaceutical chemistry
Miller, Richard Lee, biochemistry
Moody, Max Dale, medical bacteriology
Morosoff, Nicholas, physical chemistry, polymer chemistry
Morrison, Robert W, Jr, organic chemistry
Moseman, Robert Fredrick, pesticide chemistry
Moustafa, Laila Ahmed, developmental biology, genetics
Murayama, Takayuki, polymer science
Namm, Donald H, pharmacology
Nelson, Donald J, biochemical
Nichol, Charles Adam, pharmacology, biochemistry
Nielsen, Lawrence Arthur, organic chemistry
Niemeyer, Lawrence E, meteorology
Olf, Heinz Gunther, polymer physics
Oswald, Edward Odell, biochemistry
Pellizzari, Edo Domenico, analytical chemistry, analytical biochemistry
Phillips, Arthur Page, organic chemistry
Phillips, Grace Briggs, microbiology
Philpot, Richard Michael, biochemistry
Pitt, Colin Geoffrey, organometallic chemistry
Poole, William Kenneth, statistics
Pooler, Francis, Jr, meteorology
Posner, Herbert S, biochemistry, pharmacology
Preiss, Donald Merle, organic chemistry
Quinn, Richard Paul, immunochemistry
Rakestraw, Lawrence Frederick, inorganic chemistry
Rakita, Philip Erwin, organometallic chemistry
Rall, David Platt, pharmacology
Ray, Paul H., microbiology, biochemistry
Richter, Harold Gene, inorganic chemistry
Rideout, Janet Lister, chemistry
Ringwald, Eugene Lee, organic polymer chemistry
Ripperton, Lyman Alonzo, biochemistry
Robl, Hermann R, theoretical physics
Rosenthal, David (Walter), analytical chemistry
Roth, Barbara, organic chemistry
Scaringelli, Frank Philip, environmental chemistry, mass spectrometry
Scharver, Jeffrey Douglas, organic chemistry
Schroeder, David Henry, biochemical pharmacology
Shah, Babubhai Vadilal, statistics
Sheridan, William, genetics
Shy, Carl Michael, medicine, epidemiology
Sigel, Carl William, organic chemistry
Silverman, Bernard, organic chemistry, polymer chemistry
Soares, Eugene Robbins, mammalian genetics
Sovocool, George Wayne, physical organic chemistry
Sparacino, Charles Morgan, organic chemistry
Spector, Thomas, biochemistry
Spielvogel, Bernard Franklin, inorganic chemistry, nuclear magnetic resonance
Squire, David R, polymer chemistry, physical chemistry
Staples, Robert Edward, teratology
Suttle, Jimmie Ray, physics, electrical engineering
Swan, Algernon Gordon, physiology, biophysics

Swaringen, Roy Archibald, Jr, organic chemistry
Taylor, Hugh Alan, analytical chemistry
Thomas, Hollis Allen, entomology
Thompson, Richard John, analytical chemistry, air pollution
Tsai, Tsui Hsien, pharmacology
Tucker, Walter Eugene, Jr, veterinary pathology
Tulis, Jerry John, microbiology, immunology
Van Stee, Erhard Wendel, pharmacology, toxicology
Vinegar, Ralph, pharmacology
Vukovich, Fred Matthew, dynamic meteorology, physical oceanography
Wakeley, Jay Townsend, operations research, applied statistics
Wall, Monroe Eliot, biochemistry
Waters, Michael Dee, biochemistry, cell biology
Welch, Richard Martin, biochemical pharmacology
Wentworth, Gary, organic chemistry, polymer chemistry
Werner, Richard Allen, entomology
White, Helen Lyng, biochemistry
Williams, Joel Lawson, polymer chemistry
Williams, Leland Hendry, mathematics, computer science
Wilson, William Enoch, Jr, atmospheric chemistry
Wilson, William Ewing, molecular pharmacology
Wise, Edmund Merriman, Jr, microbial biochemistry
Wolberg, Gerald, immunology, microbiology
Wooten, Frank Thomas, biomedical engineering
Worf, Douglas Lowell, environmental sciences
Worth, James Judson Blackley, geophysics, meteorology
Wright, James Francis, comparative pathology, toxicology
Wyman, George Martin, organic chemistry
Yasuda, Hirotsugu, polymer chemistry, physical chemistry
Yeowell, David Arthur, pharmaceutical chemistry
Zimmerman, Thomas Paul, biochemical pharmacology

ROCKY MOUNT
Sharer, Archibald Wilson, zoology

RUTHERFORDTON
Brown, Robert Calvin, pathology, medicine
Kerr, Richard John, organic chemistry

SALISBURY
Baranski, Michael Joseph, plant taxonomy
Boyd, Robert Edward, Sr, medicinal chemistry
Buxton, Jay A, entomology
Deal, Glenn W, Jr, chemistry
Detty, Wendell Eugene, organic chemistry
Horrington, Emily Mae, nutrition chemistry
Kirk, Daniel Eddins, biology
Langebeck, Mary Therese, astronomy
Munavalli, Somashekhar, organic chemistry, biochemistry

SOUTHERN PINES
Vilbrandt, Charles Frank, physical chemistry

SOUTHPORT
Green, Melvin William, pharmaceutical chemistry
Tipton, Isabel Hanson, physics
Tipton, Samuel Ridley, physiology

SWANNANOA
Eggler, Willis Alexander, botany
Penfound, William Theodore, plant ecology

TRYON
Bucy, Paul Clancy, neurosurgery, neurology
Karpov, Boris George, astronomy

WAYNESVILLE
Wickham, William Terry, Jr, polymer chemistry, polymer physics

WEAVERVILLE
Serota, Cornelia Ann Roach, bacteriology, cytogenetics

WILKESBORO
May, Kenneth Nathaniel, food technology

WILMINGTON
Ackart, Richard Jenks, pathology
Adams, David A, resource management
Adcock, Louis Henry, analytical chemistry
Bane, Gilbert Winfield, marine ecology, biological oceanography
Biggs, Walter Clark, Jr, vertebrate zoology, animal behavior
Brauer, Ralph Werner, pharmacology
Cahill, Charles L, physical chemistry, biochemistry
Clator, Irvin Garrett, nuclear physics, explosives
Cleary, William James, marine geology, sedimentary petrology
Dankel, Thaddeus George, Jr, mathematical physics
De Loach, Will Scott, chemistry
Greim, Barbara Ann, algebra
Hart, Haskell Vincent, physical chemistry, inorganic chemistry
Hornack, Frederick Mathew, physical chemistry
Kaplan, Arthur Lewis, health physics, nuclear engineering
Kapraun, Donald Frederick, phycology
Levy, Jack Benjamin, organic chemistry
Lindquist, David Gregory, ichthyology
Lundeen, Carl Victor, Jr, biochemistry
Mann, Diana Witherspoon, neurophysiology
Martin, Ned Harold, bio-organic chemistry
McCrary, Anne Bowden, invertebrate zoology, marine biology
Norris, Fletcher R, mathematics
Parnell, James Franklin, ornithology, ecology
Randall, Duncan Peter, physical geography
Reynolds, Joshua Paul, zoology
Sabella, James C, ethnography
Sieren, David Joseph, taxonomic botany
Thayer, Paul Arthur, marine geology, sedimentology
Tiedeman, John Albert, physics
Toney, Fred, Jr, mathematics
Wilhoit, Eugene Dennis, physical chemistry
Young, Davis Alan, petrology, mineralogy
Zullo, Victor August, invertebrate paleontology, zoology

WILSON
Dickerman, John Melville, medical microbiology
Frazier, Robert Carl, mathematics
Tyndall, Jesse Parker, biology, science education

WINSTON-SALEM
Alexander, Eben, Jr, neurosurgery
Alexander, Paul Marion, plant pathology
Allen, Charles Marvin, cytology
Amen, Ralph DuWayne, physiological ecology
Ashburn, Gilbert, organic chemistry
Ausband, John R, otolaryngology
Baird, Herbert Wallace, structural chemistry
Banks, Eugene Pendleton, anthropology
Becker, Veryl E, plant physiology
Blankespoor, Ronald Lee, organic chemistry
Bo, Walter John, anatomy, endocrinology
Boyce, William Henry, urology
Brehme, Robert W, theoretical physics
Buchanan, James Wesley, physical chemistry, inorganic chemistry
Burt, Richard Lafayette, obstetrics & gynecology
Carmichael, Richard Dudley, mathematical analysis
Cayer, David, gastroenterology
Clapp, William Lee, analytical chemistry

Clark, Glenn R, human anatomy, vertebrate morphology
Colby, Frank Gerhardt, chemistry
Cook, Lawrence C, organic chemistry
Cooper, Miles Robert, oncology, hematology
Cowgill, Robert Warren, biochemistry
Cundiff, Robert Hall, research administration, analytical chemistry
Davidson, Ivan William Frederick, physiology, pharmacology
Davis, Courtland Harwell, Jr, neurology, surgery
Dickerson, James Perry, organic chemistry
Dimmick, John Frederick, animal physiology
Dimock, Ronald Vilroy, Jr, invertebrate zoology, physiological ecology
Dixon, Robert Leland, radiological physics
Dobbins, James Talmage, Jr, analytical chemistry
Dodge, William Howard, oncology
Drexler, Henry, microbiology
Edwards, James Wesley, zoology, genetics
Ekstrand, Kenneth Eric, medical physics
Esch, Gerald Wisler, animal parasitology, ecology
Eure, Herman Edward, parasitology
Evans, David Kenneth, anthropology
Felts, John Harvey, internal medicine
Finch, Robert Allen, anatomy
Flory, Walter S, Jr, botany, genetics
Foushee, J Henry Smith, Jr, pathology
Gentry, Ivey Clenton, mathematics
Giles, Jesse Albion, III, organic chemistry
Goodman, Harold Orbeck, genetics
Green, Harold David, physiology
Greiss, Frank C, Jr, obstetrics & gynecology
Gross, Paul Magnus, Jr, physical chemistry
Groves, David Lynn, immunology, microbiology
Gusdon, John Paul, obstetrics & gynecology, immunology
Harrell, T Gibson, analytical chemistry
Harrill, James Albert, otolaryngology
Harris, Geraldine Bender, microbiology
Hayes, Donald M, internal medicine, hematology
Hayes, John Terrence, orthopedic surgery
Headley, Robert N, cardiology
Heckman, Robert Arthur, organic chemistry
Hegstrom, Roger Allen, theoretical chemistry, atomic physics
Heise, Eugene Royce, immunology, microbiology
Hempel, Judith Cato, inorganic chemistry, theoretical chemistry
Herndon, Claude Nash, human genetics
Hightower, Felda, surgery
Hinman, Alanson, pediatrics
Hinze, Willie Lee, analytical chemistry
Howard, Fredric Timothy, number theory
Howell, Charles Maitland, dermatology, allergy
Hulcher, Frank H, biochemistry
Huntley, Carolyn Coker, pediatrics
Hutaff, Lucile W, medicine
Hutchins, Phillip Michael, physiology, biomedical engineering
Janeway, Richard, medicine, neurology
Johnson, Delwin Phelps, analytical chemistry
Johnson, Joseph Eggleston, III, internal medicine, infectious disease
Jones, Don Carl, biochemistry
Jones, Samuel O'Brien, chemistry, chemical engineering
Kalnins, Zelma A Grinfelds, cytopathology
Kaufmann, John Simpson, clinical pharmacology, internal medicine
Kerr, Sandria Neidus, mathematical analysis
King, Jonathan Stanton, biochemistry
Kremkau, Frederick William, bioacoustics
Kucera, Louis S, virology, oncology
Kuhn, Raymond Eugene, immunobiology, parasitology
Leffingwell, John C, organic chemistry
Levitt, Melvin, neurobiology
Lieberman, Edward Marvin, physiology
Little, James Maxwell, pharmacology, physiology
Love, Samuel Harris, microbiology
Lyerly, Larry Alexander, analytical chemistry
Marshall, Richard Blair, pathology
Martin, James Franklin, medicine, radiology
Maynard, Charles Douglas, nuclear medicine

McCall, Charles Emory, infectious diseases
McCreight, Charles Edward, anatomy
McDonald, James Clifton, mycology
McGraw, Charles Patrick, biomedical engineering, neurophysiology
McKinney, William Markley, neurology
McWilliams, Kenneth Richard, physical anthropology
Meads, Manson, medicine
Meredith, Jesse Hedgepeth, medicine
Meschan, Isadore, radiology
Miller, Harry Brown, organic chemistry
Miller, Inglis, Jr., physiology, anatomy
Moates, Robert Franklin, organic chemistry
Morehead, Robert P., pathology
Myers, Richard Thomas, surgery
Myrvik, Quentin Newell, microbiology, immunology
Neumann, Calvin Lee, physiology
Newell, Marjorie Pauline, organic chemistry
Noftle, Ronald Edward, inorganic chemistry, fluorine chemistry
Nowell, John William, chemistry
Olive, Aulsey Thomas, entomology
Parker, Roger Edwin, pharmacology, physiology
Pennell, Timothy Clinard, surgery
Pfiel, Charles George, thoracic surgery
Piehl, Donald Herbert, physical inorganic chemistry
Powers, Leland Earle, preventive medicine
Prichard, Robert Williams, pathology
Proctor, James Thornton, psychiatry
Rapela, Carlos Enrique, medical physiology, medical research
Remy, Charles Nicholas, biochemistry
Reynolds, John Hughes, IV, physical chemistry
Rhyne, A. Leonard, physiology
Richardson, Stephen H., microbiology, statistics
Rodgman, Alan, organic chemistry
Rowland, Ralph L., organic chemistry
St. Clair, Richard William, biochemistry, pathology
Sawyer, C Glenn, cardiology
Sawyer, John Wesley, geometry
Schumacher, Joseph Nicholas, organic chemistry
Senkus, Murray, chemistry
Shepperson, Jacqueline Ruth, parasitology
Shields, Howard William, solid state physics
Sidhu, Bhag Singh, plant genetics, plant science
Simms, Nathan Frank, Jr., pure mathematics
Simon, Jimmy L., medicine
Skinner, Newman Sheldon, Jr., physiology
Spurr, Charles Lewis, medicine
Squires, William Campbell, microbiology, electron microscopy
Strittmatter, Cornelius Frederick, biochemistry
Sullivan, Robert Little, genetics
Teague, Claude Edward, Jr., chemistry
Tefft, Stanton Knight, cultural anthropology, ethnology
Thomas, Mary Beth, developmental biology, cell biology
Thorne, Frederick A., analytical chemistry
Toole, James Francis, medicine
Truscott, Basil Lionel, biology, anatomy
Turner, James Eldridge, neurobiology, electron microscopy
Turner, Thomas Jenkins, solid state physics
Waite, Moseley, biochemistry, organic chemistry
Weigl, Peter Douglas, ecology, animal behavior
Whitley, Joseph Efird, radiology
Whitley, Nancy O'Neil, radiology
Witcofski, Richard Lou, medical biophysics, nuclear medicine
Wyatt, Raymond L., plant taxonomy
Youn, Ernest H., medicine
Zeya, Hasan Ismail, immunology, bacteriology

NORTH DAKOTA

BISMARCK
Johnson, Alyn William, organic chemistry

DICKINSON
Freeman, Myron L., plant taxonomy

FARGO
Adams, Terrance Sturgis, insect physiology
Anderson, Melvern K., agronomy
Andrews, Myron Floyd, veterinary parasitology
Aschbacher, Peter William, animal physiology
Bakke, Jerome E., biochemistry
Banasik, Orville James, biochemistry
Barker, William T., systematic botany
Bauer, Armand, soil science
Beryhill, David Lee, bacteriology
Bleier, William Joseph, reproductive biology
Bolin, Fonsoe M., veterinary science
Bristol, Douglas Walter, organic chemistry
Broberg, Joel Wilbur, inorganic chemistry
Bromel, Mary Cook, microbial ecology
Brophy, John Allen, geology
Buchanan, Marion Lynn, animal husbandry
Busch, Robert Henry, genetics, statistics
Callenbach, John Anton, entomology
Campbell, Edward Charles, physics
Carlson, Robert Bruce, entomology
Carter, Jack Franklin, agronomy
Cassel, Joseph Franklin, ornithology, ecology
Clambey, Gary Kenneth, plant ecology
Comita, Gabriel William, zoology
Cook, Benjamin Jacob, insect physiology, biochemistry
D'Appolonia, Bert Luigi, cereal chemistry
Davis, David G., plant physiology, tissue culture
Davison, Kenneth Lewis, nutrition
Deckard, Edward Lee, crop physiology
Dinusson, William Erling, animal nutrition, animal husbandry
Duysen, Murray E., plant physiology
Edgerly, Charles George Morgan, dairy husbandry
Erickson, Duane Otto, animal nutrition
Erickson, John Robert, plant breeding
Esslinger, Theodore Lee, lichenology
Fick, Gerhard Nelson, plant breeding, genetics
Fisher, Doris M., mathematical statistics
Fisher, George Robert, dairy husbandry
Fleeker, James R., biochemistry
Fleetwood, Charles Wesley, chemistry
Flor, Harold Henry, plant pathology
Foster, Albert Earl, plant breeding
Frear, Donald Stuart, agricultural biochemistry
Freeman, Thomas Patrick, plant anatomy, plant morphology
Galitz, Donald S., plant physiology
Garvey, Roy George, inorganic chemistry
Gerst, Jeffery William, zoology, physiology
Gilles, Kenneth Albert, agricultural biochemistry
Glass, James Clifford, nuclear physics
Goetz, Harold, plant ecology, range ecology
Gordon, Mark Stephen, quantum chemistry
Grier, James William, zoology
Gruber, John B., solid state physics, optical physics
Hammond, James Jacob, agronomy
Hare, Robert Ritzinger, Jr., applied mathematics, operations research
Harrold, Robert Lee, animal nutrition
Hassoun, Ghazi Qasim, theoretical physics
Hill, Loren Wallace, chemistry
Hodgson, Richard Holmes, botany
Holland, Neal Stewart, horticulture
Hosford, Robert Morgan, Jr., plant pathology
Huguelet, Joseph Edward, plant pathology, virology
Jacobsen, Neil Soren, academic administration
Joachim, Frank G., entomology
Johansen, Robert H., horticulture
Johnson, LaDon Jerome, animal husbandry
Joppa, Leonard Robert, genetics

BOTTINEAU
Tinus, Richard Willard, plant physiology
Van Deusen, James Lowell, forestry

Kelling, Clayton Lynn, veterinary virology
Khalil, Shoukry Khalil Wahba, pharmacognosy
Kiesling, Richard Lorin, plant pathology
Klosterman, Harold Joseph, biochemistry
Knob, Robert Duane, physical chemistry
Koob, Warren Donald, geography
Kuang, Huan Pao, mathematics, statistics
La Chance, Leo Emery, genetics
Lamoureux, Gerald Lee, biochemistry
Lana, Edward Peter, horticulture
Leopold, Roger Allen, entomology
Littlefield, Larry James, plant pathology
Lucken, Karl Allen, crop breeding
Lund, Hartvig Roald, agronomy
Maan, Shivcharan Singh, genetics, plant breeding
Magaran, Edward O., pharmaceutical chemistry
Mathsen, Ronald M., mathematics
McDonald, Clarence Eugene, cereal chemistry
McDonald, Ian Cameron Crawford, entomology, genetics
McMahon, Kenneth James, bacteriology
Meyer, Dwain Wilber, agronomy
Miller, Clarence Ashton, mammalian physiology, animal behavior
Miller, Stephen Douglas, weed science
Moore, Vaughn Clayton, physics, biophysics
Morris, Melvin L., inorganic chemistry
Mulkern, Gregory Benedict, entomology
Nalewaja, John Dennis, weed science
Nelson, Dennis Raymond, biochemistry
Nelson, Donald Carl, horticulture, plant physiology
Norum, Enoch Betuel, soils
Pappas, Betty Colleen, organic chemistry
Pappas, Socrates Peter, photochemistry
Paulson, Gaylord D., biochemistry
Pekas, Jerome Charles, animal physiology, animal nutrition
Pomonis, James George, organic chemistry
Post, Richard Lewis, entomology
Proshold, Fredrick Irving, entomology, radiation biology
Puyear, Robert Louis, comparative physiology
Quick, James S., plant breeding, genetics
Quraishi, Mohammed Sayeed, insect toxicology
Rao, Kanneganti Nageswara, mathematics
Rao, Nutakki Gouri Sankara, toxicology
Rathmann, Franz Heinrich, chemistry
Reinecke, John Philip, entomology
Riemann, John G., entomology
Roath, William Wesley, plant breeding
Rowe, Paul Preston, mathematics
Rudesill, James Turner, organic chemistry
Schipper, Ithel Arie, veterinary virology, veterinary pharmacology
Schmidt, Claude Henri, entomology
Scholz, Earl Walter, horticulture, plant physiology
Schooler, Arnold B., agronomy
Schulz, John Theodore, entomology
Scoby, Donald Ray, environmental biology
Sell, Jerry Lee, animal nutrition
Shelver, William H., medicinal chemistry, organic chemistry
Shimabukuro, Richard Hideo, plant physiology
Shuey, William Carpenter, cereal chemistry
Sinha, Mahendra Kumar, surface physics
Sleeper, Bayard Paul, microbial physiology
Smith, Glenn Sanborn, plant breeding
Staples, George Emmett, veterinary medicine, animal nutrition
Sill, Gerald G., biochemistry, organic chemistry
Stiver, James Frederick, medicinal chemistry, bionucleonics
Stolzenberg, Gary Eric, pesticide chemistry
Sugihara, James Masanobu, organic chemistry, biophysical chemistry
Tallman, Dennis Earl, analytical chemistry
Tanaka, Fred Shigeru, chemistry
Tanner, Noall Stevan, pharmacognosy, pharmacology
Terranova, Andrew Charles, toxicology
Thacker, Edward Jesse, nutrition
Tidd, Robert Frederick, mathematics

Tilton, James Earl, animal physiology
Timian, Roland Gustav, plant virology
Vacik, Dorothy Nobles, organic chemistry, pharmaceutical chemistry
Vacik, James P., pharmaceutical chemistry, bionucleonics
Vasey, Edfred H., soil fertility
Vincent, Muriel C., pharmacy, pharmaceutical chemistry
Volborth, Alexis, geochemistry, analytical chemistry
Walsh, David Ervin, cereal chemistry
Watson, Clifford Andrew, cereal chemistry
Whited, Dean Allen, genetics, agronomy
Whitman, Warren Charles, range science
Wicks, Zeno W. Jr., polymer chemistry
Williams, Norman Dale, plant genetics
Williams, Phleus P., microbiology, biochemistry
Worden, David Gilbert, solid state physics, plasma physics
Zimmer, David E., plant pathology, plant breeding
Zimmerman, Don Charles, biological chemistry, plant physiology
Zubriski, Joseph Cazimer, soil science

GRAND FORKS
Akers, Thomas Kenny, comparative physiology, environmental physiology
Albright, Bruce Calvin, neurosciences
Auyong, Theodore Koon-Hook, pharmacology
Bale, Harold David, x-ray crystallography
Beckering, Willis, physical inorganic chemistry
Brown, Ralph Clarence, geography
Brumleve, Stanley John, physiology
Bzoch, Ronald Charles, mathematics
Christensen, Odin Dale, mineralogy
Connelly, Jerald Leonard, biochemistry
Cornatzer, William Eugene, biochemistry
Crawford, Richard Dwight, wildlife ecology
Cross, Timothy Aureal, sedimentology, tectonics
Cvancara, Alan Milton, paleobiology
Dando, William Arthur, geography, microclimatology
DeBoer, Benjamin, pharmacology
Duer, Frederick G., zoology, comparative physiology
Duerre, John A., microbial physiology, biochemistry
Dyson, John Edgar, biochemistry
Ederstrom, Helge Ellis, physiology
Evans, Gary William, biochemistry, nutrition
Facey, Vera L., plant ecology, taxonomic botany
Filippi, Gordon Michael, microbiology, medical technology
Fischer, Robert George, microbiology
Frank, Richard Ernst, analytical chemistry
Gerhard, Lee C., stratigraphy, petrology
Givand, Samuel Harold, analytical chemistry
Gregory, Michael Baird, mathematical analysis
Hamre, Christopher John, zoology, anatomy
Harrell, James W., Jr., nuclear magnetic resonance
Hauntz, Edgar Alfred, endocrinology
Holland, Frank Delno, Jr., paleontology
Holloway, Harry Lee, Jr., parasitology
Jacob, Robert Allen, nutrition, biochemistry
Jacobs, Francis Albin, biochemistry
Jalal, Syed M., cytogenetics
Johnson, Gary Edwin, geography, remote sensing
Joos, Howard Arthur, pediatrics, pediatric cardiology
Kannowski, Paul Bruno, zoology
Karner, Frank Richard, geology
Kelleher, James Joseph, immunology
Keller, Reed Theodore, medical microbiology
Kemper, Gene Allen, internal medicine
Klabunde, Kenneth John, organic chemistry, inorganic chemistry
Klevay, Leslie Michael, nutrition, internal medicine
Kraus, Olen, physics
Kulevsky, Norman, physical chemistry
Landry, Richard Georges, applied statistics
Larson, Omer R., parasitology
Lee, Ya Pin, biochemistry
Low, Frank Norman, anatomy
Lykken, Glenn Irven, physics
Marvin, Richard Martin, medical microbiology, medical mycology

McBride, Woodrow H, mathematics
Miller, Roy G, Jr, organic chemistry
Moore, Walter Leroy, geology
Muraskin, Murray, physics
Myron, Duane R, biochemistry, nutrition
Neel, Joe Kendall, Sr, limnology
Nicolls, Ken E, anatomy, endocrinology
Nielsen, Forrest Harold, nutrition, biochemistry
Noble, Edwin Austin, geology
Nordlie, Robert Conrad, biochemistry
Ollerich, Dwayne A, anatomy, neuroanatomy
Olmstead, Edwin Guy, medicine
Oring, Lewis Warren, ecology, ethology
Owen, Alice Koning, embryology
Pekarek, Robert Sidney, microbiology, biochemistry
Radonovich, Lewis Joseph, structural chemistry
Rao, B Seshagiri, spectroscopy, optics
Ray, Paul Dean, biochemistry
Reid, John Reynolds, Jr, glaciology, geology
Robinson, Thomas John, mathematics
Rue, James Sandvik, mathematics
Sandstead, Harold Hilton, internal medicine, nutrition
Schneider, Frederick Ewing, anthropology
Schobert, Harold Harris, high temperature chemistry
Seabloom, Robert W, mammalogy
Severson, Roland George, organic chemistry
Sheridan, William Francis, genetics, cell biology
Shubert, Lester Elliot, phycology
Snook, Theodore, histology, embryology
Soonpaa, Henn H, solid state physics
Stenberg, Virgil Irvin, organic chemistry
Stewart, James Allen, physical chemistry
Summers, Lawrence, organic chemistry
Tupa, Dianna Lou Dowden, phycology, morphology
Uherka, David Jerome, numerical analysis
Vennes, John Wesley, bacteriology
Wali, Mohan Kishen, plant ecology, environmental biology
Waller, James R, microbiology, biochemistry
Walsh, Scott Wesley, endocrinology, reproductive physiology
Wardner, Carl Arthur, organic chemistry
Wills, Bernt Lloyd, geography
Winger, Milton Eugene, mathematics, statistics
Woolsey, Neil Franklin, organic chemistry
Zogg, Carl A, medical physiology, nutrition

JAMESTOWN
Adomaitis, Vytautas Albin, water science
Morrow, Larry Alan, range science, weed science
Power, James Francis, soil science
Ries, Ronald Edward, range science, range ecology
Willis, Wayne Owen, soil physics

MAYVILLE
Carlson, Kenneth Theodore, chemistry
Ralston, Robert D, plant ecology

MINOT
Berkey, Gordon Bruce, astrophysics, physics
Bickel, Edwin David, malacology, environmental geology
Clausen, Eric Neil, geomorphology
Farnum, Bruce Wayne, organic chemistry
Johnson, Arnold Richard, Jr, analytical chemistry
Ladendorf, Agnes J, mathematics
Martin, DeWayne, petrology
Thompson, Michael Bruce, embryology

OHIO

ADA
Awad, Albert T, natural products chemistry, pharmacognosy
Beltz, LeRoy Duane, pharmaceutical chemistry
Berton, John Andrew, geometry
Bettinger, Donald John, inorganic chemistry
Bhattacharya, Amar Nath, pharmacology, physiology
Dawson, John E, reproductive physiology
Edin, Albert Irving, pharmacology
Gangemi, Francis A, molecular physics, quantum physics
Gossel, Thomas Alvin, pharmacology
Haight, Howard Lewis, inorganic chemistry
Hawbecker, Byron L, organic chemistry
Laing, Charles Corbett, ecology
Mallin, Morton Lewis, microbiology, biochemistry
Meyer, Samuel Lewis, plant morphology
Moore, Nelson Jay, ethology, ornithology
Nelson, Eric V, entomology, apiculture
Pillai, Raman Narayana, mathematical statistics, mathematics
Stansloski, Donald Wayne, medicinal chemistry, clinical pharmacy
Stright, I Leonard, mathematics
Stuart, David Marshall, pharmacy, chemistry
Theodore, Joseph M, Jr, pharmacy
Vandor, Sandor Laszlo, biochemical pharmacology
Weimer, David, gas dynamics
Wilhelm, Dale Leroy, inorganic chemistry

AKRON
Aggarwal, Sundar Lal, polymer chemistry
Alliger, Glen, organic chemistry
Altenau, Alan Giles, analytical chemistry
Ambelang, Joseph Carlyle, organic chemistry, rubber chemistry
Ambler, Michael Ray, polymer science
Ambrose, Richard Joseph, polymer chemistry
Anhorn, Victor John, physical chemistry
Auerbach, Melvin, organic chemistry, organometallic chemistry
Averill, Seward Junior, rubber chemistry
Avgeropoulos, George N, physical chemistry
Bachmann, John Henry, physical chemistry
Bain, Roger J, geology
Bauer, Richard G, polymer chemistry, organic chemistry
Beckman, Joseph Alfred, polymer chemistry
Bergomi, Angelo, organic chemistry
Beyer, William Hyman, mathematical statistics, mathematics
Blackwood, Unabelle Boggs, nutrition
Boustany, Kamel, rubber chemistry
Bouton, Thomas Chester, polymer science
Burford, Arthur Edgar, structural geology
Calderon, Nissim, polymer chemistry, organometallic chemistry
Cameron, Douglas Ewan, topology
Campbell, Donald R, analytical chemistry, radiochemistry
Campbell, Robert Henry, analytical chemistry, organic chemistry
Carlson, Earl John, polymer chemistry
Carson, Robert Cleland, mathematics
Causa, Alfredo G, polymer science, textiles
Cheng, Tai Chun, organic polymer chemistry, organometallic chemistry
Claxton, William Eugene, mathematics, statistics
Cole, John Oliver, chemistry
Coleman, John Franklin, organic chemistry, polymer chemistry
Conant, Floyd Sanford, polymer physics
Cook, Wendell Sherwood, organic chemistry
Coran, Aubert Y, pharmacy, organic chemistry
Corbett, Robert G, economic geology, geochemistry
Costanza, Albert James, polymer chemistry, rubber chemistry
Crane, Grant, organic chemistry
Cunningham, Robert Elwin, polymer chemistry
Darling, Stephen Deziel, organic chemistry
Dial, William Richard, organic chemistry
D'Ianni, James Donato, polymer chemistry

Dollwet, Helmar Hermann Adolf, plant physiology, plant biochemistry
Dove, Ray Allen, chemistry
Dreyfuss, Patricia, polymer chemistry
Dubravcic, Milan Frane, analytical chemistry, food science
Duddey, James E, organic chemistry, polymer chemistry
Dudek, Thomas Joseph, polymer science
Dutt, Ashok Kumar, urban geography, geography of Asia
Eaker, Charles Mayfield, chemistry
Elmer, Otto Charles, organic chemistry
Erickson, David Edward, physical chemistry, colloid chemistry
Fall, Harry H, physical organic chemistry
Farona, Michael F, inorganic chemistry
Fetters, Lewis, physical chemistry, polymer chemistry
Fieldhouse, John W, synthetic organic chemistry
Fielding-Russell, George Samuel, polymer science
Finelli, Anthony Francis, organic chemistry
Foley, Howard Kenneth, polymer chemistry
Folk, Theodore Lamson, analytical chemistry
Forster, Michael Jay, polymer physics
Franks, Paul C, geology
Frederick, John Edgar, physical chemistry
Freeman, Aaron Eliot, pathology, cancer research
Friedman, Emil Martin, rubber chemistry, polymer physics
Garn, Paul Donald, analytical chemistry, physical chemistry
Gent, Alan Neville, physics, mechanics
Gippin, Morris, organic chemistry
Gleim, Clyde Edgar, organic chemistry
Gloth, Richard Edward, organic chemistry, organometallic chemistry
Glymph, Eakin Milton, rubber chemistry
Gregg, Earl Charles, Jr, organic chemistry
Griffin, Charles Frank, physics
Griffin, Claibourne Eugene, Jr, organic chemistry
Gruber, Elbert Egidius, polymer chemistry
Gurley, Thomas Gordon, organic polymer chemistry
Gwinn, John Fredrick, developmental physiology
Halasa, Adel Farhan, polymer chemistry
Hanlon, Thomas Lee, physical chemistry
Hanten, Edward W, urban geography
Hargis, I Glen, physical chemistry
Hausch, Walter Richard, rubber chemistry
Hayes, Robert Arthur, organic chemistry
Hergenrother, William Lee, organic polymer chemistry
Herold, Robert Johnston, organic chemistry
Hillegass, Donald Victor, biochemical engineering
Hilton, Ashley Stewart, analytical chemistry
Hirst, Robert Charles, physical chemistry
Hively, Robert Arland, analytical chemistry
Hoekje, Howard Hail, physical chemistry
Homer, George Mohn, biochemistry
Houser, John J, photochemistry
Hyndman, John Robert, physical chemistry
Igel, Howard Eugene, pathology, virology
Jackson, Dale Latham, zoology
Johnson, Edward Lee, analytical chemistry
Kanakkanatt, Sebastian Varghese, polymer physics
Kang, Jung Wong, organic chemistry
Kay, Edward Leo, organic chemistry
Keck, Max Hans, polymer chemistry
Keller, Roger F, Jr, zoology
Kelly, Walter James, organic chemistry, polymer chemistry
Kennedy, Joseph Paul, polymer chemistry
Kerr, Andrew, Jr, medicine
Kime, Joseph Martin, physics
Kline, Richard Henry, organic chemistry
Kofron, William G, organic chemistry
Kollar, William L, polymer chemistry
Korsok, Albert Joseph, geography
Krishen, Anoop, analytical chemistry
Krivis, Alan Frederick, analytical chemistry
Kuan, Tiong H, polymer science
Kunze, Adolf Wilhelm Gerhard, geophysics
Kurz, James Eckhardt, physical chemistry, polymer chemistry
Kuska, Henry (Anton), physical chemistry

Kyker, Gary Stephen, inorganic chemistry, polymer chemistry
Lake, Robert Samuel, animal virology, cell biology
Lal, Joginder, polymer chemistry
Lawson, David Francis, organic chemistry
Ledinko, Nada, virology
Lehmicke, David John, textile physics
Leshin, Richard, organic chemistry
Lewis, Thomas Brinley, polymer science
Livigni, Russell Anthony, polymer chemistry
Livingston, Daniel Isadore, physical chemistry
Lohr, Delmar Frederick, Jr, polymer chemistry
Loughborough, Dwight Logan, physics
Lucas, Kenneth Ross, analytical chemistry, electrochemistry
Luckenbach, Thomas Alexander, physical chemistry, polymer chemistry
Lyon, William D, theoretical chemistry
Ma, Laurence Jun-Chao, geography
MacGregor, Ian Robertson, botany
Macior, Lazarus Walter, botany
Maender, Otto William, chemistry
Mallay, James Francis, nuclear physics
Marker, Leon, physical chemistry
Marshall, Richard Allen, polymer chemistry
Marwitt, John Paul, ethnology, archaeology
Mast, William Carlton, polymer chemistry
Mayor, Rowland Herbert, rubber chemistry
McIntyre, Donald, polymer chemistry
Merten, Helmut Ludwig, organic chemistry
Meyer, Glen Ernest, polymer chemistry
Milone, Charles Robert, organic polymer chemistry
Mochel, Virgil Dale, physical chemistry, physics
Morita, Eiichi, rubber chemistry
Morris, Marion Clyde, polymer chemistry
Morton, Maurice, polymer chemistry
Mosser, John Snavely, physical chemistry, environmental management
Mostardi, Richard Albert, physiology
Muse, Joel, Jr, industrial organic chemistry
Naples, Felix John, organic chemistry
Nelson, Charles Jay, polymer chemistry
Nelson, Wayne Franklin, inorganic chemistry
Nicholson, David William, mechanics
Noble, Allen George, geography
Nokes, Richard Francis, veterinary medicine, anatomy
Nunn, Dorothy Mae, microbiology
Oberster, Arthur Eugene, organic chemistry
Oetjen, Robert Adrian, molecular spectroscopy
Ofstead, Eilert A, polymer chemistry
Olive, John H, cell physiology, aquatic biology
Orcutt, Frederic Scott, Jr, ethology, ornithology
Oziomek, James, organic chemistry, polymer chemistry
Pace, Henry Alexander, organic chemistry
Patel, Siddharth Manilal, analytical chemistry, pollution chemistry
Paxton, Thomas Rice, physical chemistry
Piirma, Aleksander, polymer chemistry
Piirma, Irja, polymer chemistry
Pinnick, Harry Thomas, solid state physics
Powell, Grace L, economic geography
Prettyman, Clinton Elmer, computer sciences
Prettyman, Irven Bernhard, physics
Prudence, Robert Thomas, organic chemistry
Purdon, James Ralph, Jr, polymer chemistry
Pyle, Gerald Fredric, urban geography
Rader, Charles Phillip, organic chemistry, rubber chemistry
Reed, Thomas Freeman, polymer chemistry, rubber chemistry
Reilly, Charles Bernard, organic chemistry, polymer chemistry
Reilly, Patrick J, polymer chemistry
Richardson, Barry Lovell, radiobiology
Ritter, Hartien Sharp, physical chemistry
Roberts, Durward Thomas, Jr, organic polymer chemistry
Rogers, Thomas Henry, Jr, chemistry, chemical engineering
Ross, Louis, mathematics, statistics
Rowland, George Peabody, Jr, physical chemistry

OHIO

Saltman, William Mose, rubber chemistry
Schirmer, Joseph P., Jr., organic chemistry
Schneider, Ronald E., theoretical physics, nuclear physics
Schoenberg, Emanuel, industrial chemistry
Schulz, Donald Norman, organic chemistry, polymer chemistry
Scott, Kenneth Walter, polymer chemistry
Shaver, Forrest Wheeler, rubber chemistry
Smith, Ward Arden, chemistry
Snyder, Carl Edward, chemistry
Spanninger, Philip Andrew, organic chemistry
Stambaugh, Richard Bulla, physics
Steichen, Richard John, analytical chemistry
Stephens, Howard L., polymer chemistry
Stoutamire, Warren Petrie, plant taxonomy, evolution
Stux, Paul, organic chemistry
Suich, Ronald Charles, statistics
Sumner, Thomas, organic chemistry
Sweet, Leonard, statistics
Tate, David Paul, organic chemistry
Taucci, Enes Barbara, mathematics, biometry
Tazuma, James Junkichi, chemistry
Teeter, James Wallis, geology, paleontology
Thackeray, Ernest Russel, physics
Throckmorton, Morford Church, chemistry
Thurman, George Raymond, mathematics, physics
Urbanic, Anthony Joseph, colloid chemistry, polymer chemistry
Volk, Murray Edward, organic chemistry, radiochemistry
Von Meerwall, Ernst Dieter, physics
Wadelin, Coe William, chemistry
Waisbrot, Samuel William, organic chemistry
Wang, Jin-Liang, polymer chemistry
Warner, Walter Charles, polymer chemistry
Wideman, Lawson Gibson, organic chemistry
Williger, Ervin John, plastics chemistry, rubber chemistry
Wilson, Charles Woodson, III, physics
Wilson, Matthew Woodrow, physical chemistry
Wingard, Paul Sidney, physical geology
Wise, Paul Henry, organic chemistry
Wise, Raleigh Warren, research administration, rubber chemistry
Wright, Robert L., organic chemistry
Young, Evan Johnson, rubber chemistry
Zwicker, Benjamin M G., chemistry

ASHLAND

Anderson, Marlin Dean, animal nutrition, animal physiology
Carpenter, Dorothy Irene, mathematics
Cole, Jerry Joe, analytical chemistry
Craine, Elliott Maurice, agricultural biochemistry
Guldenzopf, Emil Charles, zoology
Hennigan, Bernard Robert, sedimentology, stratigraphy
Kiggins, Edward M., microbiology
Kouba, Rudolph Frank, biochemistry, physical organic chemistry
Kriens, Richard Duane, organic chemistry
McGinnis, Charles Henry, Jr., avian physiology
Meredith, William Edward, microbiology
Poorman, Alan Gene, mathematics
Rhoades, Rendell, zoology

ALLIANCE

Blaser, Robert U., research administration, resource management
Brueske, Charles H., plant physiology
Cels, Robert, physical chemistry, electrochemistry
Clark, William Glenn, algebra
Epp, Leonard G., developmental biology
Gherng, Walter L., acoustics
Golestaneh, Ahmad Ali, theoretical physics
Markley, William A., Jr., mathematics
Murdoch, Arthur, organic chemistry
Osterman, George B., biology
Rice, William Abbott, physical geology
Rodman, James Purcell, administration
Turnquist, Truman Dale, analytical chemistry
Wise, Robert George, Jr., geology

ASHVILLE

Flierl, Donald William, physical chemistry

ATHENS

Adams, Jerry L., nuclear physics
Ahman, Mold Uddin, groundwater hydrology, geophysics
Atkins, Charles E., mathematics
Atkins, Charles Gilmore, microbial genetics
Barnhart, Clyde Sterling, Sr., entomology
Blair, Robert Louie, mathematics
Blazyk, Jack, biochemistry
Breitenbacher, Ernst, physics
Brient, Charles E., nuclear physics
Cappelletti, Ronald Louis, experimental solid state physics, metal physics
Cavender, James C., mycology
Chen, Charles Chin-Tse, physics
Cohn, Norman Stanley, cytology, cytochemistry
Day, Jesse Harold, physical chemistry
Denbow, Carl (Herbert), mathematics
Dilley, James Paul, theoretical physics
Downey, Ronald J., cell physiology
Dunlap, Paul R., statistics
Eblin, Lawrence Powell, colloid chemistry, history of chemistry
Edwards, John Elza, physics
Eldridge, Klaus Emil, mathematics
Elliott, Rush, zoology
Finlay, Roger W., nuclear physics
Fisher, Stanley Parkins, petroleum geology
Goedicke, Victor Alfred, astronomy
Graffius, James Herbert, botany
Heck, Oscar Benjamin, parasitology
Hendricker, David George, inorganic chemistry
Hikida, Robert Seichi, experimental morphology
Houk, Clifford C., inorganic chemistry
Hummon, William Dale, aquatic ecology, population biology
Hunt, Earle Raymond, physics
Huntsman, William Duane, organic chemistry
Huwe, Darrell O., particle physics
Ingham, Robert Kelly, organic chemistry
Jaffe, Mordecai J., plant physiology
Jasper, Samuel Jacob, mathematics
Jewett, John Gibson, physics
Jones Witters, Patricia H., reproductive physiology
Kline, Robert Joseph, chemistry
Koshel, Richard Donald, nuclear physics
Lane, Raymond Oscar, nuclear physics
Larson, Laurence Arthur, plant physiology
Latz, Howard N., analytical chemistry
Lawrence, James Vantine, bacteriology
Lloyd, Robert Michael, systematic botany
Lustfield, Charles Davenport, mathematical analysis
Maier, Siegfried, microbiology
McQuate, John Truman, genetics
Mehr, Cyrus B., mathematics, electrical engineering
Miller, Charles Edward, mycology
Onley, David S., theoretical physics
Peterson, Wesley John, zoology
Randall, Charles Addison, Jr., physics
Rapaport, Jacobo, experimental nuclear physics
Redlich, Robert Walter, electromagnetism
Rollins, Roger William, experimental solid state physics, low temperature physics
Romoser, William Sherburne, entomology
Rovner, Jerome Sylvan, arachnology, animal behavior
Sanford, Edward Richard, physics
Seibert, Henri Cleret, vertebrate ecology
Smith, Geoffrey W., geology
Spring, Ray Frederick, mathematics
Stumpf, Folden Burt, physics
Sturgeon, Myron Thomas, paleontology, stratigraphy
Svendsen, Gerald Eugene, ethology
Sympson, Robert F., analytical chemistry, electrochemistry
Segelken, Warren George, magnetic resonance
Tennent, David Maddux, biochemistry
Van Osdall, Thomas Clark, inorganic chemistry
Weidenhamer, Harry E., geology

Taylor, John Langdon, Jr., medical education
Tong, James Ying-Peh, physical inorganic chemistry
Ungar, Irwin A., plant ecology
Vancko, Robert Michael, nuclear physics
Wagner, Thomas Edwards, biochemistry, endocrinology
Walker, Richard V., medical bacteriology, immunology
Westenbarger, Gene Arlan, physical chemistry
Wicke, Howard Henry, topology
Wilson, James Albert, physiology, biochemistry
Winkler, Robert Randolph, organic chemistry
Wistendahl, Warren Arthur, plant ecology
Witters, Weldon L., reproductive physiology, biology
Worrell, John Mays, Jr., mathematics, medicine
Wright, Louis Edgar, theoretical physics, nuclear physics
Yeats, Robert Sheppard, geology
Yun, Seung Soo, acoustics

AURORA

Curran, Daniel R., physics, mathematics

AVON LAKE

Collins, Edward A., physical chemistry
Essig, Henry J., organic polymer chemistry
Nakajima, Nobuyuki, polymer physics
O'Mara, Michael Martin, analytical chemistry

BARBERTON

Baughman, Glenn Laverne, organic polymer chemistry
Carlson, Gordon Andrew, inorganic chemistry
Crano, John Carl, organic chemistry
DeWitt, Bernard James, chemistry
Dietz, Paul Luther, Jr., physical chemistry, electrochemistry
Ecke, George Graff, organic chemistry
Ewald, Fred Peterson, Jr., analytical chemistry
Gard, Leavitt Nelson, analytical chemistry
Johnson, Harlan Bruce, physical chemistry
Jones, Frederick Mason, III, organic chemistry
Krass, Dennis Keith, synthetic organic chemistry, agricultural chemistry
Krivak, Thomas Gerald, inorganic chemistry
Krochta, William G., analytical chemistry
Laning, Stephen Henry, analytical chemistry
Lavanish, Jerome Michael, organic chemistry
Marinsons, Aleksandrs, electrochemistry
Moore, John Williamson, physical chemistry
Reich, Donald Arthur, organic chemistry
Rinehart, Jay Kent, chemistry
Steiger, Roger Arthur, high temperature chemistry
Stevens, Henry Conrad, organic chemistry
Strong, Walker Albert, chemistry
Wagner, Melvin Peter, polymer chemistry, organic chemistry
Welch, Cletus Norman, physical chemistry

BEACHWOOD

Gans, David Manus, chemistry

BEDFORD

Morgan, Stanley L., pharmaceutics, chemical engineering
Uebele, Curtis Eugene, polymer chemistry
Wickes, Glenn French, chemistry
Zaremsky, Baruch, organometallic chemistry, plastics chemistry

BEREA

Annear, Paul Richard, astronomy
Baker, Charles Edward, physical chemistry
Brokaw, Richard Spohn, physical chemistry

BEXLEY

Duga, Jules Joseph, research administration, science policy

BLANCHESTER

Vallee, Richard Earl, physical chemistry, inorganic chemistry

BLUFFTON

Kaufmann, Maurice John, plant pathology
Pannabecker, Richard Floyd, zoology
Schirch, Laverne Gene, biochemistry, organic chemistry
Suter, Robert Winford, chemistry
Weaver, John Richard, physical chemistry

BOWLING GREEN

Anderson, Thomas Dale, geography
Applebaum, Charles H., mathematical logic
Baxter, William D., immunobiology
Beck, Doris Jean, microbial genetics
Blinn, Elliott L., inorganic chemistry
Bowman, Donald Whitney, physics
Brecher, Arthur Seymour, biochemistry
Brent, Morgan McKenzie, microbiology
Clemans, George Burtis, organic chemistry, biochemistry
Cobb, Thomas Berry, physics, chemistry
Cossaboom, Robert T., geography, geology
Crandall, Arthur Jared, physics
Crang, Richard Earl, plant cytology, electron microscopy
Crawford, Paul Vincent, cartography, physical geography
Dafforn, Geoffrey Alan, bio-organic chemistry
Dean, Donald Stewart, botany
Den Besten, Ivan Eugene, physical chemistry
Duquet, Robert Theodore, meteorology
Eakin, Richard R., mathematics
Emmitty, Davy A., phytopathology
Endres, Paul Frank, physical chemistry
Fisher, T Richard, botany
Forsyth, Jane Louise, glacial geology, geology
Frank, Ralph William, geography of Asia
Graham, James Douglas, developmental genetics, experimental hematology
Graves, Louis Charles, mathematics
Graves, Robert Charles, entomology
Hall, William Heinlen, chemistry
Hallberg, Carl William, parasitology
Hamilton, Ernest Scovell, plant ecology
Hammer, Averill John, inorganic chemistry
Hanre, Harold Thomas, biology
Heberlein, Gary T., microbiology, plant physiology
Herman, George, speech pathology
Hern, Thomas Albert, mathematics
Hilliard, Stephen Dale, embryology, human anatomy
Hiltner, John, Jr., geography
Hoare, Richard David, invertebrate paleontology
Horvath, Raymond S., microbial physiology
Howe, John A., vertebrate paleontology
Hudson, William Nathaniel, mathematical analysis, mathematical statistics
Hurst, Peggy Morison, inorganic chemistry
Hyman, Melvin, speech pathology
Jackson, William Bruce, animal ecology
Jensen, Adolph Robert, analytical chemistry
Johnson, Carlos Sigfrid, Jr., mathematics
Kinstle, Charles F., geology
Krabill, David Milton, mathematics
Laha, Radha Govinda, analytical mathematics, pure mathematics
Leetch, James Frederick, mathematics
Miller, John Wesley, Jr., marine zoology, entomology
Peterjohn, Glenn William, zoology, physiology
Procior, David George, physiology
Riggle, Timothy A., mathematics
Schochet, Melvin Leo, physical chemistry
Smith, Calvin Albert, botany
Stansfield, Roger Ellis, organic chemistry
Swartz, Karl D., solid state physics
Wallis, Robert L., physics

Leone, Charles Abner, immunology, radiation biology
Limbird, Arthur George, resource geography
Long, Clifford A., mathematics
Lougheed, Milford Seymour, geology
Lowe, Rex Loren, phycology
Lukacs, Eugene, mathematics
Mancuso, Joseph J., economic geology
Martin, Elden William, animal physiology, ecology
McMorris, Fred Raymond, mathematics, biomathematics
Meyer, Norman James, physical chemistry
Newman, David S., physical chemistry
Noble, Reginald Duston, plant physiology
O'Brien, Thomas V., topology
Oster, Irwin Isaac, genetics
Owen, Donald Edward, stratigraphy, geology
Pawlowicz, Edmund F., geophysics
Ptak, Roger Leon, astrophysics
Rabalais, Francis Cleo, parasitology
Rendina, George, biochemistry
Rich, Charles Clayton, geology
Rohatgi, Vijay, mathematical statistics
Romans, Robert Charles, botany
Rothe, Kenneth Warren, elementary particle physics
Satyanarayana, Motupalli, algebra
Schurr, Karl M., entomology
Scott, John Paul, zoology, psychology
Scovell, William Martin, bioinorganic chemistry
Shrestha, Mohan Narayan, geography
Singleton, Edgar Bryson, molecular physics
Smith, Bruce Wayne, economic geography, urban geography
Srinivasan, Vakula S., electrochemistry, analytical chemistry
Statz, Joyce Ann, computer sciences
Steiner, Ray Phillip, number theory
Steinker, Don Cooper, paleontology
Stoner, Ronald Edward, solid state physics
Thibault, Roger Edward, ecology
Townsend, Ralph N., mathematics
Vessey, Stephen H., animal behavior, ecology
Vincze, Lajos, anthropology
Walters, Lester James, Jr., geochemistry
Weber, Joseph Elliott, geochemistry
Weber, Waldemar Carl, mathematics
Williams, James Garner, mathematics

BRECKSVILLE
Baranwal, Krishna Chandra, polymer chemistry, physical chemistry
Beatty, James Roger, physics
Berens, Alan Robert, polymer science
Clark, Milton B., physical chemistry
Dannis, Mark Libman, polymer chemistry
DeWitt, Elmer John, organic polymer chemistry
Dickens, Elmer Douglas, Jr, polymer science
Diem, Hugh Egbert, spectroscopy
Dinbergs, Kornelius, organic polymer
Dreyfuss, Max Peter, polymer chemistry
Frederick, Marvin Ray, organic chemistry
George, Paul John, organic chemistry
Gross, Malcolm Edmund, polymer chemistry
Harmon, Dale Joseph, polymer chemistry
Horne, Samuel Emmett, Jr., polymer chemistry, organic chemistry
Huddleston, George Richmond, Jr, physical chemistry, chemical engineering
Kelley, Philip Carlos, polymer chemistry
Kroenke, William Joseph, chemistry
Layer, Robert Wesley, organic chemistry
Lehr, Marvin Harold, polymer physics, polymer chemistry
McRowe, Arthur Watkins, organic chemistry
Parker, Richard Ghrist, organic chemistry
Roha, Max Eugene, organic chemistry
Schollenberger, Charles Sundy, polymer chemistry, organic chemistry
Siebert, Alan Roger, physical chemistry, polymer chemistry
Stevens, Harry Nelson, rubber chemistry
Traynor, Lee, organic chemistry, polymer chemistry
Tucker, Harold, polymer chemistry
Tyler, Willard Philip, analytical chemistry
Westfahl, Jerome Clarence, organic chemistry
Wilkes, Charles Eugene, chemistry

Yanko, John Alexis, polymer science, rubber chemistry
Zakriski, Paul Michael, analytical chemistry

BROADVIEW HEIGHTS
Nicholas, Paul Peter, organic chemistry

BURTON
Smith, Franklin Danford, chemistry

CANTON
Howland, Willard J., radiology
Johnson, Ronald Gene, radiation biophysics, radiological health
Pandya, Mahendra Kodarlal, medicinal chemistry, biochemistry
Ritter, Ronald Dale, inorganic chemistry
Starchman, Dale Edward, medical physics
Stephens, Marvin Wayne, biochemistry

CARROLL
Melvin, John Harper, geology

CARROLLTON
Hall, Judd Lewis, polymer chemistry

CEDARVILLE
Helmick, Larry Scott, organic chemistry

CELINA
Norris, Bill Eugene, biology

CENTERVILLE
De Sando, Richard John, physical chemistry, x-ray crystallography
Hobrock, Don Leroy, physical chemistry, inorganic chemistry
Sellers, Douglas Edwin, analytical chemistry

CHAGRIN FALLS
Duke, June Temple, organic chemistry, polymer chemistry
Mattson, Raymond Harding, organic chemistry
Morris, William Collins, chemistry

CHARDON
Rohner, Mary Christopher, experimental embryology, neurophysiology

CHESTERLAND
Hsu, Roger Yun Kung, organic chemistry
Lick, Wilbert James, environmental sciences, applied mathematics
Veraguth, Arnold John, analytical chemistry

CHILLICOTHE
Austin, Robert Andrae, paper chemistry
Davis, Gerald Titus, paper chemistry
Green, Robert Patrick, pulp & paper technology
Lee, Yu-Sun, organic chemistry
Redd, John Coleman, pulp & paper technology
Robinson, James Vance, paper chemistry
Rutledge, Wyman Coe, applied physics
Shackle, Dale Richard, chemistry
Spatz, Sydney Martin, organic chemistry
Young, Ainslie Thomas, Jr, polymer chemistry, photochemistry

CINCINNATI
Adolph, Robert J, internal medicine
Albrecht, William Lind, medicinal chemistry
Alexander, James Wesley, surgery, immunology
Alexander, John J, inorganic chemistry, organometallic chemistry
Ampulski, Robert Stanley, mass spectrometry
Anderson, Robert L, biochemistry, nutrition
Anfinsen, Jon Robert, analytical chemistry
Aring, Charles Dair, neurology
Artman, Neil Ross, organic chemistry
Austern, Barry M, analytical chemistry, biochemistry
Badger, Donald W, physiology

Bahr, Gustave Karl, radiological physics, nuclear medicine
Baker, Frank Weir, organic chemistry
Bambury, Ronald Edward, medicinal chemistry
Banerjee, Dilip Kumar, analytical chemistry
Bannan, Elmer Alexander, microbiology
Barlow, Anthony, polymer chemistry
Barrett, Fred Oliver, chemistry
Basom, Charles Ray, anatomy
Baur, Fredric John, Jr, physiological chemistry, organic chemistry
Beatey, Janice Carson, plant ecology
Beck, Lloyd Willard, chemistry
Behnke, William David, biological chemistry, biophysics
Behrmann, Eleanor Mitts, chemistry, organic biochemistry
Bell, Harold, mathematics
Bell, Mary, cell biology, human anatomy
Bender, Daniel Frank, physical organic chemistry, analytical chemistry
Bendure, Raymond Lee, physical chemistry, surface chemistry
Benedict, James Harold, biochemistry
Benjamin, L, surface chemistry
Benzing, George, III, pediatric cardiology, physiology
Berg, Gerald, virology, microbiology
Berman, Alex, history of pharmacy
Berman, Donald, virology, water pollution
Berman, Jerome Richard, internal medicine
Biehl, Joseph Park, neurology
Binhammer, Robert T, anatomy
Black, Hugh Elias, veterinary pathology
Blaha, Gordon C, anatomy, histology
Blanchard, Richard Lee, radiochemistry, ecology
Blaney, Donald John, biochemistry
Blejer, Hector P, occupational medicine, toxicology
Blewett, Charles William, organic chemistry
Blizzard, Richard Reese, Sr, organic chemistry, mathematics
Block, Stanley L, psychiatry
Blohm, Thomas Robert, biochemistry
Bobst, Albert M, biophysical chemistry, organic biochemistry
Bonventre, Peter Frank, microbiology
Boolchand, Punit, experimental solid state physics, experimental nuclear physics
Booth, Gary Edwin, organic chemistry
Boswell, Donald Eugene, organic chemistry
Bozian, Richard C, biochemistry, internal medicine
Brain, Devin King, industrial organic chemistry
Brause, Allan R, food chemistry
Briner, William Watson, microbiology
Broaddus, Charles D, organic chemistry
Brod, John Sydney, organic chemistry
Brodish, Alvin, physiology
Broge, Robert Walter, physical chemistry
Brooks, Stuart Merrill, pulmonary disease, occupational medicine
Brown, Ronald David, developmental biology
Brownscheidle, Carol Mary, histochemistry
Bryant, Shirley Hills, pharmacology, physiology
Bubel, Hans Curt, medical bacteriology
Budde, William L, analytical chemistry
Budde, Clifford Charles, analytical chemistry
Buehler, Edwin Vernon, toxicology
Buncher, Charles Ralph, biostatistics, epidemiology
Bunde, Carl Albert, clinical pharmacology
Burg, William Robert, industrial hygiene, analytical chemistry
Butz, Andrew, insect physiology
Callen, Joseph Edward, analytical chemistry, research administration
Campbell, Jeptha Edward, Jr, food science
Campbell, Kirby I, environmental health
Caplan, Paul E, chemical engineering, industrial hygiene
Carlson, Gustav Gunnar, anthropology
Carpenter, Lawrence Edward, biochemistry
Carr, Albert A, organic chemistry, medicinal chemistry
Carter, Owen, Jr, physical chemistry
Caster, Kenneth Edward, geology, paleontology
Cawein, Madison Julius, hematology, clinical pharmacology
Centner, Rosemary Louise, organic chemistry

Chalkley, Roger, mathematics, algebra
Chambers, Cecil William, bacteriology
Chambers, Lee Mason, analytical chemistry, electrochemistry
Chang, Shih Lu, microbiology, public health
Charters, Elaine Mary, zoology, endocrinology
Chen, I-Wen, radiobiology, biochemistry
Cholak, Jacob, analytical chemistry
Christian, Robert Thomas, virology
Clark, Corodon Scott, public health, environmental health engineering
Clark, Leland Charles, Jr, biochemistry
Clarke, Norman Arthur, microbiology
Clowers, Churby Conrad, Jr, analytical chemistry
Cody, Terence Edward, environmental health
Collins, Gary Brent, phycology, aquatic ecology
Connell, Alastair McCrae, gastroenterology
Connor, Daniel S, synthetic organic chemistry, applied chemistry
Conway, Gene Farris, internal medicine, cardiovascular diseases
Cook, Elton Straus, medicinal chemistry
Cooke, William Bridge, mycology
Cooley, William Edward, dental research
Cooper, Gary Pettus, neurophysiology
Coots, Robert Herman, biochemistry
Cotton, Wyatt Daniel, organic chemistry
Courchene, William Leon, physical chemistry
Crafts, Roger Conant, anatomy
Crandall, Dana Irving, biochemistry
Crouse, Nathan Norman, organic chemistry
Culbertson, William Richardson, medicine
Curry, John D, inorganic chemistry
Dage, Richard Cyrus, pharmacology
Danian, Michael S, pharmacy
Daniels, Robert (Sanford), medicine, psychiatry
Daoud, Georges, cardiology
Darragh, Richard T, food science, biochemistry
Davis, Edward Melvin, organic chemistry
Davis, Neil Clifton, biochemistry
Davis, Richard Arnold, geology, paleontology
Day, Richard Allen, biochemistry
Dean, Robert Berridge, colloid chemistry, pollution control
DeJong, Diederik Cornelis Dignus, plant taxonomy
Delcamp, Robert Mitchell, organic chemistry
Dement, John McCray, industrial hygiene
Di Salvo, Joseph, cardiovascular physiology
DeSesso, John Michael, human anatomy, teratology
Dizenhuz, Israel Michael, psychiatry
Dolfini, Joseph E, organic chemistry, medicinal chemistry
Deuschle, Frederick Marion, anatomy, orthodontics
Deutsch, Edward Allen, inorganic chemistry
DeYoung, Joyce Lewis, pharmaceutics
Diller, Violet Marion, biophysics
Dimond, Harold Lloyd, organic chemistry
Dirks, Brinton Mario, agricultural biochemistry
Donaldson, Virginia Henrietta, medicine
Drew, Howard Felshaw, organic chemistry
Dreyer, William Albert, zoology, ecology
Dube, Harvey Albert, chemistry
Dukes-Dobos, Francis N, exercise physiology, occupational health
Dumbleton, John Herbert, biophysics, orthopedics
Dunbar, John Scott, radiology
Dunham, David Waring, astronomy, celestial mechanics
Edwards, Brenda Kay, biostatistics
Elder, Richard Charles, inorganic chemistry
Elia, Victor John, environmental health, analytical biochemistry
Ellis, Larry Edward, organic chemistry, pharmaceutical chemistry
Emmett, Edward Anthony, environmental health, occupational medicine
Erickson, Raymond C, microbiology
Erman, William F, organic chemistry
Erway, Lawrence Clifton, Jr, developmental genetics
Esposito, F Paul, theoretical physics
Etges, Frank Joseph, parasitology, malacology

OHIO

Evans, Arthur T, urology, surgery
Fairchild, Edward Joseph, II, pharmacology, physiology
Falkner, Frank Tardrew, pediatrics
Fallat, Ronald Walter, medical research, metabolism
Farrier, Noel John, bioinorganic chemistry, nutritional biochemistry
Fayter, Richard George, Jr, organic chemistry
Fedder, Richard Charles, physics
Feldman, Julian, organic chemistry
Felson, Benjamin, radiology
Fenichel, Henry, low temperature physics
Field, Michael, physics
Fike, Winston, analytical chemistry
Finelli, Vincent Nicola, biochemistry, environmental health
Flautt, Thomas Joseph, Jr, physical chemistry
Foote, Gordon Lee, organic chemistry
Foster, Joseph Frederick, cultural anthropology, anthropological linguistics
Foulkes, Ernest Charles, physiology
Fowler, Noble Owen, internal medicine, cardiovascular disease
Francis, Marion David, biochemistry, medical research
Frank, Charles Edward, organic chemistry
Frank, Donald Joseph, pediatrics
Franke, Ernst Karl, biophysics
Freeberg, Fred E, analytical chemistry
Freedman, Jules, medicinal chemistry
Freisheim, James Harold, biochemistry
Fryxell, Robert Edward, physical inorganic chemistry
Fujii, Akira, medicinal chemistry, biochemistry
Fulford, Margaret Hannah, botany
Gall, Edward Alfred, pathology
Ganschow, Roger Elmer, biochemistry, genetics
Garascia, Richard Joseph, chemistry
Garner, Reuben John, environmental health
Gartside, Peter Stuart, biostatistics
Geldreich, Edwin Emery, microbiology
Ghering, Mary Virgil, enzymology, biochemistry
Gibson, John Phillips, pathology, toxicology
Gibson, Thomas William, organic chemistry
Gilbert, Nathan, physical chemistry
Gilbert, Theodore William, Jr, analytical chemistry
Glasser, Arthur Charles, pharmaceutical chemistry
Glueck, Helen Iglauer, medicine
Goebel, Charles Gale, organic chemistry
Goetz, Richard W, industrial organic chemistry
Goldman, Stephen Allen, physical chemistry, magnetic resonance
Goldsmith, Richard E, medicine
Goldstein, Sidney, pharmacology
Gonzalez, Luis L, surgery
Goodman, Bernard, theoretical solid state physics
Gosselink, Eugene Paul, organic chemistry
Gottschang, Jack Louis, mammalogy, herpetology
Gough, Robert George, organic chemistry
Gould, Harry J, III, comparative neurology
Grab, Eugene Granville, Jr, food science
Grabenstetter, Robert John, physical chemistry
Gray, John Augustus, III, physical chemistry
Green, Floyd J, organic & analytical chemistry
Greenberg, Daniel, physical chemistry
Greene, Hoke Smith, physical chemistry, organic chemistry
Greider, Harold William, chemistry
Griffin, Claude Lane, pharmacology
Grisar, Johann Martin, organic chemistry, medicinal chemistry
Gruenstein, Eric Ian, biochemistry
Grupp, Gunter, experimental medicine, physiology
Gunther, Marian W J, theoretical physics
Gustafson, David Harold, organic chemistry
Haberman, Jon Phillip, analytical chemistry
Hagan, Charles Patrick, synthetic organic chemistry
Hagenauer, Fedor, child psychiatry, psychoanalysis
Hall, James Lawrence, anatomy
Halsall, Hallen Brian, biochemistry

Hamblet, Frederick Edwards, epidemiology
Hammond, Paul B, toxicology
Hanenson, Irwin Boris, medicine
Hannah, Sidney Allison, environmental chemistry
Hansen, Harold Louis, chemistry
Harris, Elliott Stanley, research administration, environmental health
Hart, John Birdsall, physics
Hatzenbuhler, Douglas Albert, physical chemistry
Heckert, David Clinton, physical organic chemistry, process chemistry
Helm, Robert Albert, internal medicine
Helmsworth, James Alexander, thoracic surgery
Henschel, Austin Ferdinand, physiology
Herget, Paul, astronomy
Herrmann, Kenneth Walter, physical chemistry, surface chemistry
Hess, Evelyn V, internal medicine, immunology
Heuring, Vincent Paul, physical chemistry
Hiatt, Harold, psychiatry
Higgins, Daniel Joseph, botany
Hiles, Richard Allen, biochemistry, toxicology
Hirtle, Donald Stephen, chemistry
Hoekenga, Mark T, internal medicine, research administration
Hoff, John, microbiology
Hoffman, David J, developmental physiology, biochemistry
Holder, Ian Alan, microbiology
Holtkamp, Dorsey Emil, endocrinology, medical science
Horgan, Stephen William, medicinal chemistry
Horowitz, Myer George, biochemistry
Horwitz, Harry, oncology
Hosler, John Frederick, organic chemistry
Hoyt, John Manson, polymer chemistry
Huber, Harold E, pharmaceutical chemistry
Hudak, William John, pharmacology
Huether, Carl Albert, Jr, population genetics
Huff, Warren D, geology
Hug, George, pediatrics, biochemistry
Hummel, Robert P, surgery
Hung, William Mo-Wei, organic chemistry
Hunter, John Earl, bacteriology
Hutchison, Robert B, organic chemistry
Isaac, Barry LaMont, cultural anthropology
Isbrandt, Lester Reinhardt, physical chemistry, nuclear magnetic resonance
Isenecker, Lawrence Elmer, mathematics
Issitt, Peter David, immunohematology
Jackson, Earl Roger, food chemistry
Jaffe, Hans, physical chemistry, theoretical chemistry
Jarroid, Thomas, internal medicine
Jebsen, Robert H, physical medicine & rehabilitation
Jenks, William Furness, economic geology
Jha, Shacheenatha, experimental nuclear physics
Joffe, Frederick M, biochemistry, food technology
Johnson, Barry Lee, electrical engineering, biomedical engineering
Johnson, Gary R, comparative pathology
Johnson, Howard M, immunology, microbiology
Johnson, Robert Gudwin, organic chemistry
Johnson, William H, organic chemistry
Johnston, John O'Neal, reproductive endocrinology, neuropharmacology
Johnston, John O'Neal, reproductive endocrinology, neuropharmacology
Joiner, William Cornelius Henry, solid state physics
Jones, Dale Robert, physics
Jones, Winton D, Jr, medicinal chemistry
Judd, Claude Ivan, medicinal chemistry
Judge, Leo Francis, Jr, microbiology
Juneja, Prem S, bio-organic chemistry
Kalter, Harold, genetics, teratology
Kandel, Alexander, pharmacology
Kaneshiro, Edna Sayomi, cell biology, biochemistry
Kaplan, Fred, organic chemistry
Kaplan, Phyllis Deen, chemistry, biochemistry
Kaplan, Samuel, cardiology
Kaplan, Stanley Meisel, psychiatry
Kapp, Frederic T, psychiatry
Kariya, Takashi, pharmacology
Karp, Richard Dale, immunology
Kashyap, Moti Lal, internal medicine

Kassner, James Edward, industrial organic chemistry
Kawahara, Fred Katsumi, petroleum pollution
Keily, Hubert Joseph, analytical chemistry
Keller, Jeffrey Thomas, neuroanatomy
Keller, Stephen Jay, molecular biology, biochemistry
Kennedy, Richard J, organic chemistry
Kenner, Bernard Alexander, bacteriology
Kereiakes, James Gus, physics
Kilinc, Attila Ishak, geochemistry
Kirkland, Jerry J, microbiology
Kleinman, Leonard I, physiology, neonatology
Klemm, Donald J, limnology, freshwater biology
Kline, Daniel Louis, physiology
Klingenberg, Joseph John, analytical chemistry
Knoblaugh, Armand Frank, physics
Knowles, Harvey C, Jr, internal medicine
Kopfler, Frederick Charles, environmental chemistry
Kopp, John F, analytical chemistry
Kramer, Milton, psychiatry
Kreishman, George Paul, biophysical chemistry
Krueger, Robert Carl, biochemistry
Krueger, Robert George, immunology, virology
Krueger, Russell Francis, microbiology
Kuemmel, Donald Francis, analytical chemistry
Kuhn, William Lloyd, pharmacology
Kupel, Richard E, organic chemistry, inorganic chemistry
Kwiatek, Jack, industrial organic chemistry
Lamperti, Albert A, neuroendocrinology
Lang, Dennis Robert, cell physiology
Lang, James Frederick, drug metabolism
Lange, Willy, microbial ecology
Larkin, Edward P, food microbiology, virology
Laughlin, Robert Gene, physical organic chemistry
Lawson, Kenneth Dare, electron microscopy
Leake, Lowell, Jr, mathematics
Leavitt, Wendell W, endocrinology, reproductive physiology
Lee, Si Duk, environmental sciences, biological chemistry
Lemmerman, Karl Edward, physical chemistry
Lerner, Sidney Isaac, occupational medicine
Lessard, James Louis, biochemistry
Levin, Robert Harold, organic chemistry
Levine, Maita Faye, mathematics
Lewis, Trent R, occupational health
Leyland, Harry Mours, organic chemistry
Lichstein, Herman Carlton, medical microbiology
Lichtin, J Leon, pharmaceutical chemistry, cosmetic chemistry
Light, Irwin Joseph, pediatrics
Lindquist, John Raymond, chemistry
Lingrel, Jerry B, biochemistry
Lipicky, Raymond John, internal medicine, pharmacology
Lipsich, H David, mathematics
Loder, Edwin Robert, analytical chemistry
Logan, Ted Joe, industrial chemistry
Lollar, Robert Miller, leather chemistry, environmental chemistry
Loomans, Maurice Edward, dermatology
Loper, John C, microbiology, genetics
Loudon, Robert G, internal medicine
Lovett, Joseph, environmental health, public health
Lubin, Clarence Isaac, mathematical analysis
Lucke, William E, analytical chemistry
Lundberg, Charles Andrew, Jr, synthetic organic chemistry
Lutton, Edwin Scott, physical chemistry
Lydy, David Lee, surface chemistry
Mabis, Alton John, physical chemistry
MacGee, Joseph, biochemistry, analytical chemistry
Macht, Martin Benzyl, physiology
Mack, Harry Patterson, anatomy
MacKenzie, Robert Douglas, biochemistry
MacMillan, Bruce Gregg, surgery
Macpherson, Colin Robertson, pathology
Manning, Herbert Lee, microbiology, water pollution
Manudhane, Krishna Shankar, industrial pharmacy
Margolin, Esar Gordon, internal medicine
Mark, Harry Berst, Jr, electrochemistry, analytical chemistry

Martin, Daniel William, physics
Martin, Lester W, pediatric surgery
Marvin, Philip Roger, solid state science
Mast, Roy Clark, physical chemistry, inorganic chemistry
Mattheis, Eula Bingham, environmental health
Mattingly, Steele F, animal husbandry, veterinary medicine
Mattson, Fred Hugh, biochemistry
Mauer, Alvin Marx, medicine
Mayer, Gerald Douglas, microbiology
McAdams, Arthur James, pediatrics, pathology
McCarty, Frederick Joseph, organic chemistry, medicinal chemistry
McCune, Homer Wallace, physical chemistry
McCuskey, Robert Scott, anatomy
McFarren, Earl Francis, analytical chemistry
McLaurin, Robert L, neurosurgery
McMaster, Robert H, medicine
McNee, Robert Bruce, economic geography, urban geography
McVean, Duncan Edward, pharmaceutical chemistry
Meadows, Brian T, high energy physics
Meal, Larie L, physical chemistry, analytical chemistry
Meckel, Alfred Hans, dentistry
Meeks, Frank Robert, physical chemistry
Megel, Herbert, immunology
Meineke, Howard Albert, anatomy
Meinert, Walter Theodore, organic chemistry
Menefee, Max Gene, pathology
Merkes, Edward Peter, mathematics
Meyer, David Lachlan, invertebrate paleontology
Meyer, Ralph Roger, cell biology, biochemistry
Meyer, Walter H, nutrition
Michael, Jacob Gabriel, immunology
Michael, Leslie William, chemistry
Michael, William R, pharmacology, biochemistry
Michaels, John Edward, cell biology
Michaelson, I Arthur, pharmacology
Middleton, Francis Marvin, sanitary chemistry
Miller, Franklyn David, physical chemistry, organic chemistry
Miller, Raymond Edwin, physics
Milliken, Spencer Rankin, physical chemistry
Minda, Carl David, mathematical analysis
Mohlenkamp, Marvin Joseph, Jr, food chemistry
Montgomery, Daniel Michael, radiation ecology, radiochemistry
Moore, Wellington, Jr, veterinary toxicology, inhalation toxicology
Moulton, Bruce Carl, endocrinology, biochemistry
Murthy, Gopala Krishna, dairying
Myhre, David V, food chemistry
Nathan, Paul, physiology
Newberne, James Wilson, veterinary pathology
Nicholson, D Allan, organometallic chemistry, analytical chemistry
Niemeier, Richard William, environmental health
Niewenhuis, Robert James, anatomy, electron microscopy
Nobis, John Francis, organic chemistry
Nolen, Granville Abraham, toxicology
Norris, Paul Edmund, pharmaceutical chemistry
Nussbaum, Mirko, experimental high energy physics
O'Brien, James Edward, astronomy, solar physics
O'Connor, David Evans, food chemistry
O'Dell, Thomas Beniah, pharmacology
Odioso, Raymond C, organic chemistry
Oehlschlaeger, Herman Fred, organic chemistry
Oertel, Richard Paul, spectrochemistry
Ogle, James D, biochemistry
O'Neill, Richard Thomas, physical inorganic chemistry
Orchin, Milton, organic chemistry
Orr, Charles Henry, computer science
Orthoefer, John George, veterinary medicine
Palmer, Alan, epidemiology
Palopoli, Frank Patrick, organic chemistry
Parker, Roger A, medicinal chemistry
Partin, John Calvin, medicine, cell biology
Perry, Robert Leonard, applied statistics
Petering, Harold George, biochemistry, biochemical pharmacology
Peters, Joseph John, biology

Peterson, Donald J, organic chemistry
Petke, Frederick Edward, industrial chemistry
Pickrum, Harvey Marvin, microbiology
Piker, Philip Edward, psychiatry
Pinzka, Charles Frederick, mathematics
Place, Gerald Alan, agronomy, soil chemistry
Pollak, Victor Eugene, internal medicine, nephrology
Potter, Paul Edwin, geology
Prairie, Richard Lane, biochemistry
Pratt, Edward Lowell, pediatrics
Price, Robert Harper, physical chemistry
Pritchett, Ervin Garrison, organic chemistry
Pryor, Wayne Arthur, geology
Purcell, Thomas Charles, environmental chemistry, industrial hygiene
Raychaudhuri, Anilbaran, biology, immunology
Reilly, Frank Daniel, human anatomy, medical physiology
Reller, Herbert Henry, biochemistry, dermatology
Reynolds, David Stephen, applied statistics, applied mathematics
Richardson, Alfred, Jr, organic chemistry
Ritschel, Wolfgang Adolf, biopharmaceutics
Ritter, Edmond Jean, biochemistry
Ritter, Harry Woodward, infectious diseases
Ritterhoff, Robert J, medicine
Rizzi, George Peter, organic chemistry
Roddy, William Thomas, chemistry
Roder, Wolf, economic geography
Rolwing, Raymond H, mathematics
Rosen, Aaron A, organic chemistry, analytical chemistry
Rosevear, Francis Burt, physical chemistry
Ross, William Donald, psychiatry
Rudney, Harry, biochemistry
Russell, James Edward, physics
Russell, Paul Telford, biochemistry
Ryan, James Anthony, soil chemistry, soil biochemistry
Ryan, Kenneth Bruce, geography
Saenger, Eugene L, radiology, nuclear medicine
Safferman, Robert S, microbiology
Sallee, Eugene Merriditl, analytical chemistry
Saltzman, Bernard Edwin, environmental health, analytical chemistry
Santner, Joseph Frank, mathematical statistics, operations research
Scarpino, Pasquale Valentine, microbiology
Schafer, Mary Louise, food chemistry
Schieve, James Ferdinand, internal medicine
Schiff, Gilbert Martin, infectious diseases, virology
Schlossman, Irwin S, organic chemistry
Schmidt, Paul J, organic chemistry
Schnettler, Richard Anselm, medicinal chemistry, organic chemistry
Schreiner, Albert William, internal medicine, hematology
Schroll, Gene E, organic chemistry
Schubert, William K, pediatrics
Schulz, Peter, geography of europe, political geography
Scott, Ralph Carmen, internal medicine, cardiology
Scott, William James, Jr, teratology, developmental biology
Selevan, Sherry Gail, epidemiology, occupational health
Sevenants, Michael R, food chemistry
Sharp, George Oscar, analytical chemistry
Shawhan, Gerald L, mathematics, operations research
Shaya, Steven Alan, physical chemistry
Shelton, Robert Schember, chemotheraphy
Shemano, Irving, pharmacology
Shields, George Seamon, internal medicine
Shine, Daniel Phillip, cosmetic chemistry
Shumrick, Donald A, otorhinolaryngology
Sigell, Leonard, pharmacology
Silberstein, Edward B, nuclear medicine
Sill, Arthur DeWitt, organic chemistry
Silverman, Frederic Noah, pediatrics, radiology
Sjoerdsma, Albert, experimental medicine, clinical pharmacology
Skorcz, Joseph Anthony, industrial organic chemistry
Smith, Carl Clinton, biochemistry
Smith, Robert, family medicine
Smith, Roger Dean, pathology, virology
Snider, Jerry Allen, bryology, cytology
Snyder, Fred Hugh, biochemistry, toxicology
Sodd, Vincent J, nuclear chemistry, nuclear medicine
Somoza, Cesar, pathology
Sorenson, John R J, pharmacy
Soukup, Victor Gerald, organic chemistry
Sperti, George Speri, biophysics
Sprague, Estel Dean, physical chemistry
Sprang, Cornelius Austin, organic chemistry
Srivastava, Laxmi Shanker, endocrinology
Stadler, Louis Benjamin, pharmaceutical chemistry, analytical chemistry
Stafford, Howard A, economic geography
Staker, Donald David, organic chemistry
Stander, Richard Wright, obstetrics & gynecology
Stauffer, Clyde E, biochemistry
Sterken, Gordon Jay, organic chemistry
Stevenson, Jean Moorhead, surgery
Stewart, Robert Lewis, psychiatry
Stokinger, Herbert Ellsworth, toxicology
Stormont, Robert Tulloch, pharmacology
Streng, William Harold, physical chemistry
Strobel, Rudolf G K, biochemistry
Suranyi, Peter, high energy physics
Suskind, Raymond Robert, medical science, health sciences
Sutherland, James McKenzie, pediatrics
Tafuri, John Francis, entomology
Tan, Henry S I, pharmaceutical chemistry
Temple, Robert Dwight, organic chemistry
Terrell, Ross Clark, organic chemistry
Thayer, John Stearns, inorganic chemistry, organometallic chemistry
Thompson, James Edwin, organic chemistry
Titchener, James Lampton, psychiatry, psychoanalysis
Tofe, Andrew John, nuclear medicine
Townsend, Samuel Franklin, biology, anatomy
Troller, John Arthur, microbiology
Troup, Stanley Burton, internal medicine, hematology
Trufant, Samuel Adams, medicine, neurology
Tsang, Reginald C, neonatology, nutrition
Tsay, Jia-Yeong, biostatistics, mathematical statistics
Tsuei, Yeong Ging, applied mechanics
Tuan, Tai-Fu, theoretical physics
Twedt, Robert Madsen, microbiology
Uchtman, Vernon Albert, physical inorganic chemistry
Umminger, Bruce Lynn, comparative physiology, environmental physiology
Ungar, Gerald S, mathematics
Ursillo, Richard Carmen, pharmacology
VanLandingham, Samuel Leighton, botany, phycology
Van Maanen, Evert Florus, pharmacology
Vernardakis, Theodore Galaction, physical chemistry, materials science
Vestal, James Robie, microbiology
Vester, John William, biochemistry, internal medicine
Vilter, Richard William, medicine
Vonderbrink, Sally Ann, analytical chemistry
Voss, Jack Goddard, microbiology
Wakefield, Shirley Lorraine, solid state chemistry
Warkany, Josef, pediatrics
Watanakunakorn, Chatrchai, internal medicine, infectious diseases
Weaver, James Edmund, pharmacology
Webb, Norval Ellsworth, Jr, pharmacy, pharmaceutical chemistry
Weber, George Russell, bacteriology
Wee, Elizabeth Liu, biophysical chemistry, polymer chemistry
Weiner, Murray, medicine
Weintraub, Philip Marvin, organic chemistry
Weiser, Herman Joshua, Jr, analytical chemistry
Weisfeld, Lewis Bernard, physical organic chemistry, polymer chemistry
Werner, Raymond Edmund, organic chemistry
West, Clark Darwin, immunology, nephrology
Wexler, Bernard Carl, experimental medicine
Wharton, Harry Whitney, analytical chemistry
Whitehead, William Earl, psychophysiology
Whitman, Roy Milton, psychiatry
Widder, James Stone, immunology, biochemistry
Wiech, Norbert Leonard, biochemistry, nutrition
Will, John Junior, hematology
Williams, Josephine Louise, physical chemistry, surface chemistry
Wilson, James Graves, embryology, teratology
Winget, Gary Douglas, plant biochemistry, plant physiology
Witman, Robert Charles, organic chemistry
Witten, Louis, physics
Wolf, Laurence Grambow, cultural geography, urban geography
Wolfe, Roger Thomas, organic chemistry, research administration
Woodward, James Kenneth, pharmacology
Wozencraft, Paul, pathology
Wright, George Joseph, drug metabolism, biochemistry
Wright, William Robert, theoretical physics
Xintaras, Charles, industrial health
Yates, Albert Carl, theoretical physical chemistry
Yellin, Wilbur, spectroscopy, physical inorganic chemistry
Yoshimura, Sei, chemotherapy, virology
Youngquist, Rudolph William, food biochemistry
Zeffren, Eugene, bio-organic chemistry
Zilch, Karl T, organic chemistry
Ziller, Stephen A, Jr, nutrition, toxicology
Zimmer, Hans, organic chemistry
Zimmerman, Ernest Frederick, pharmacology
Zoglio, Michael Anthony, pharmaceutical chemistry

CIRCLEVILLE

Burton, Robert Louis, polymer chemistry
Edman, James Richard, organic chemistry, research administration
England, Richard Jay, polymer chemistry, analytical chemistry
Heacock, James Flaviel, organic chemistry
Heffelfinger, Carl John, polymer physics
Hovermale, Ralph Allen, polymer science
Katz, Morton, organic chemistry, polymer chemistry
Kreuz, John Anthony, organic polymer chemistry
McIntyre, William Ernest, Jr, chemistry
Parish, Darrell Joe, organic polymer chemistry
Theis, Richard James, organic polymer chemistry
Walker, Charles Carey, organic polymer chemistry
Woolsey, Gerald Bruce, physical chemistry

CLEVELAND

Adelson, Lester, pathology
Albats, Paul, cosmic ray physics
Alexander, John William, analytical chemistry
Alford, Harvey Edwin, organic chemistry
Alicino, Nicholas J, animal science
Alldridge, Norman Alfred, botany
Alley, Keith Edward, neurobiology
Andeen, Carl Gustav, solid state physics
Andrist, Anson Harry, physical organic chemistry
Ankeney, Jay Lloyd, medicine
Anton, Aaron Harold, pharmacology
Antonucci, Frank Ralph, organic polymer chemistry
Arnoff, E Leonard, mathematics
Arnold, Allen Parker, organic chemistry
Ashby, William, physics
Astle, Melvin Jensen, organic chemistry
Astrachan, Lazarus, biochemistry
Aziz, Philip Michael, physical chemistry
Bachman, Paul Lauren, organic chemistry
Bacon, Roger, materials science
Baer, Eric, polymer chemistry, plastics engineering
Baer, Helmut W, nuclear physics
Baker, Peter C, developmental biology, neurobiology
Baker, Saul Phillip, geriatrics, cardiology
Ball, David Ralph, industrial organic chemistry, fuel science
Ball, Lawrence Ernest, polymer chemistry, organic chemistry
Baltazzi, Evan S, organic chemistry
Banks, Philip Oren, geology, geochemistry
Barbour, Stephen D, genetics, biochemistry
Bartlett, Noel Sloane, applied statistics
Battisto, Jack Richard, immunology, microbiology
Batza, Eugene M, audiology, speech pathology
Baum, Gerald L, medicine
Bauman, Richard Gilbert, physical chemistry
Beard, William Clarence, mineralogy
Beck, Harris Graybill, organic chemistry
Belles, Frank Edward, fuel science
Benade, Arthur Henry, acoustics
Ben-Dor, Shmuel, anthropology
Benjamin, Philip Palamoottil, nuclear chemistry, nuclear medicine
Bensusan, Howard Bernard, biological chemistry
Benton, Kenneth Curtis, polymer chemistry
Bettice, John Allen, mammalian physiology
Bevington, Philip Raymond, nuclear physics
Biaglow, John E, biochemistry
Bikerman, Jacob Joseph, physical chemistry
Billiar, Reinhart Billie, reproductive endocrinology, biochemistry
Binkley, Roger Wendell, organic chemistry, photochemistry
Black, Richard H, applied mathematics
Blackwell, John, biophysics, polymer science
Blades, John Dieterle, physics
Bland, John Edward, archaeology, biological anthropology
Blank, Robert Eugene, organic chemistry
Bliss, Myron, Jr, entomology
Bloch, Edward Henry, anatomy
Blomgren, George Earl, physical chemistry
Boaz, Willard Denton, medicine
Bockhoff, Frank James, physical chemistry
Bodanszky, Agnes Adrienne, organic chemistry
Bodanszky, Miklos, bio-organic chemistry
Bohinski, Robert Clement, biochemistry
Bond, Douglas Danford, psychiatry
Bordenca, Carl, organic chemistry
Bost, Robert Orion, toxicology
Bowerfind, Edgar Sihler, Jr, internal medicine
Bradley, Martin Patrick Timothy, analytical chemistry
Brenner, Lorry Jack, immunology
Brodd, Ralph James, physical chemistry
Brodkey, Jerald Steven, neurosurgery, biomedical engineering
Brown, Charles Albert, inorganic chemistry
Brown, Helen Bennett, nutrition, cardiovascular diseases
Brown, Robert William, physics
Brugger, Thomas C, child psychiatry
Buck, Charles (Carpenter), mathematics
Bumpus, Francis Merlin, organic chemistry, biochemistry
Burgyan, Aladar, inorganic chemistry
Burke, John James, paleontology
Burlage, Stanley R, applied physics
Butkus, Antanas, analytical chemistry, cardiovascular diseases
Butler, Daniel Knowles, computer science, reactor physics
Callahan, James Louis, inorganic chemistry
Camiener, Gerald Walter, biochemistry, microbiology
Caplan, Arnold I, developmental biology, biochemistry
Carlson, Keith Douglas, physical chemistry
Carome, Edward F, physics
Carpenter, Charles C J, internal medicine, infectious disease
Carrabine, John Anthony, x-ray crystallography, inorganic chemistry
Carter, Charles Edward, biochemistry
Carter, John Robert, pathology
Casper, Karl Joseph, physics
Caston, J Douglas, biochemistry, embryology
Caughey, John Lyon, Jr, medical education
Chambers, William Edward, analytical chemistry
Chan, Paula Pui-Ying, numerical analysis
Chandrasekhar, Bellur S, physics
Charms, Bernard, cardiopulmonary physiology
Chester, Edward M, internal medicine
Chu, Chi Hsuin Ulli, anatomy
Chu, Ching-Wu, solid state physics
Clapham, Wentworth B, Jr, ecology, environmental management
Clark, Robert Arthur, applied mathematics
Clise, Ronald Leo, genetics, zoology
Cole, Monroe, neurology
Collins, Clifford B, physics

Compton, Eli Dee, organic chemistry,
environmental management
Conomy, John Paul, neurology
Cook, Robert Thomas, pathology,
biochemistry
Cook, William R., Jr., solid state
chemistry, mineralogy
Cooper, Cecil, biochemistry
Croft, George Thomas, physics
Crouch, Marshall Fox, physics
Cupp, Lloyd Anthony, virology, cell
biology
Cupas, Chris Angelo, organic chemistry
Dahm, Arnold J., physics
Damley, Ralph Lawrence, organic
chemistry
Darling, Samuel Mills, chemistry
Dayton, James Anthony, Jr., electron
optics
Dean, Burton Victor, operations research
Del Duca, Betty Spahr, physical
chemistry
DiCarlo, James Anthony, solid state
physics
Del Villano, Bert Charles, immunology,
virology
De Marco, Thomas Joseph,
pharmacology, periodontology
DeMarinis, Frank, genetics
Denko, Charles W., biochemistry
Deodhar, Sharad Dinkar, pathology,
biochemistry
Depalma, Ralph G., surgery
Giamati, Charles C., Jr., nuclear physics
Dickerman, Richard Curtis, genetics,
physics
Dietrich, Harry Joseph, physical
Dworken, Harvey J., internal medicine,
gastroenterology
Eastwood, Douglas William,
anesthesiology
Eck, Thomas G., physics
Eckel, Robert Edward, medicine
Eckstein, Richard Waldo, medicine
Egan, Joseph Michael, mathematics
Egar, Margaret Wells, anatomy, electron
microscopy
Eiben, Robert Michael, medicine
Eichner, Eduard, obstetrics, gynecology
Eisler, Milton, toxicology, pharmacology
Eisner, Robert Lawrence, experimental
high energy physics
Ells, Frederick Richard, polymer
chemistry
Elzam, O E., soils, plant nutrition
Emmons, Hamilton, operations research
Epstein, Samuel Stanley, pathology,
environmental sciences
Evans, Helen Harrington, biochemistry
Evans, Robert Morton, polymer
chemistry
Evans, Thomas Edward, molecular
biology
Everson, Howard E., inorganic chemistry,
physical chemistry
Fanger, Michael Walter, biochemistry,
immunology
Farrell, David E., physics
Fay, Philip S., petroleum
Ferencz, Nicholas, dentistry
Ferrario, Carlos Maria, cardiovascular
physiology, cardiovascular diseases
Fickinger, William Joseph, high energy
physics
Foley, Joseph Michael, neurology
Foos, Raymond Anthony, physical
chemistry
Ford, William Frank, theoretical physics,
applied mathematics
Fordyce, James Stuart, physical
chemistry, electrochemistry
Foreman, Darhl Lois, zoology,
endocrinology

De Marco, Thomas Joseph,
Deodhar, Sharad Dinkar, pathology,
Dietz, David (Henry), science writing
Dobyns, Brown M., surgery
Donovan, Sandra Steranka, physical
chemistry, electrochemistry
Dorer, Frederic Edmund, biochemistry
Dornette, William Henry Lueders,
anesthesiology
Doss, Nagib A., organic chemistry
Duff, Robert Hodge, electron
microscopy, physical
Dunbar, Robert Copeland, physical
chemistry
Dunkle, David Hosbrook, vertebrate
paleontology
Duperuis, Clarence Welsey,
anthropology

Forman, Ralph, physics
Frankel, Victor H., orthopedic surgery,
bioengineering
Friede, Reinhard L., neuropathology,
histochemistry
Friedell, Hymer Louis, radiology
Friedman, Jean Small, embryology
Froelich, George Crumback, metallurgical
chemistry
Fryberg, George Crumback, metallurgical
chemistry
Frye, Glenn McKinley, Jr., physics
Fuller, Richard Kenneth, biomedical
engineering
Gardner, Ralph Alexander, physical
chemistry
Garner, Harry Richard, electrochemistry
Genuth, Saul M., endocrinology
Gerrity, Ross Gordon, experimental
pathology, cardiovascular diseases
Ghose, Hirendra M., physical chemistry,
inorganic chemistry
Gibbons, Donald Frank, biomedical
engineering
Giffen, William Martin, Jr., polymer
population ecology
Goldblatt, Harry, pathology
Goldstein, Melvin C., anthropology,
population ecology
Goldwait, David Atwater, biochemistry
Goodrich, Cecile Ann, physiology
Gordon, Joseph R., organic chemistry,
polymer chemistry
Gordon, William Livingston, solid state
physics
Goss, Leonard Joyce, animal pathology
Grasselli, Jeanette Gecsy, spectroscopy
Grasselli, Robert Karl, physical chemistry
Gravenstein, Joachim Stefan, medicine,
anesthesiology
Greenbaum, Sheldon Boris, organic
chemistry
Greene, Arthur Frederick, Jr., analytical
chemistry, inorganic chemistry
Greene, Janice L., organic chemistry
Gregg, Earle Covington, physics
Greinke, Ronald Alfred, analytical
chemistry
Griggs, Robert C., medicine
Grimes, Hubert Henry, solid state
physics
Grodsky, Irvin T., theoretical physics
Gronski, Jan Maksymilian, mathematics,
control theory
Guenther, Paul Ernest, mathematics
Gutmann, Andrew Titus, organic
chemistry, physical organic chemistry
Haas, Erwin, biochemistry
Hajek, Otomar, mathematics
Halbedel, Harold Seibert, industrial
chemistry
Hall, John Frederick, geology
Hall, Philip Wells, III, medicine,
physiology
Hambourger, Paul David, experimental
solid state physics
Hamermesh, Bernard, physics
Harper, Edward O'Neil, medicine,
psychiatry
Harpst, Jerry Adams, biophysical
chemistry
Harris, John William, clinical medicine
Hart, Kathleen Therese, biochemistry
Hatfield, Garry Kent, toxicology
Haybron, Ronald M., physics
Healy, James C., rubber chemistry,
polymer engineering
Hedstrom, Gerald Walter, mathematics
Hellerstein, Herman Kopel, medicine
Hendricks, Walter James, physics
Henry, Charles Eric,
electroencephalography
Herndon, Charles Harbison, orthopedic
surgery
Hewes, Ralph Allan, solid state physics
Heymann, Walter, pediatric nephrology
Hill, Mary Ann Gertrude, inorganic
chemistry, physical chemistry
Hilliker, David Lee, number theory
Hilton, Peter John, mathematics
Hinko, Edward N., medicine, psychiatry
Hirschmann, Hans, biological chemistry
Hoffman, Richard Wagner, solid state
physics
Hogg, Robert W., microbiology, genetics
Hoke, Donald I., polymer chemistry,
organic chemistry
Holden, William Douglas, surgery
Holubec, Zenovie Michael, petroleum
chemistry
Hopfer, Ulrich, cell biology
Hopfinger, Anton J., polymer science,
biological chemistry

Hopkins, Amos Lawrence, physiology
Hoshiko, Tomuo, physiology
Houser, Harold Byron, epidemiology
Hower, John, Jr., geochemistry
Hubay, Charles Alfred, surgery
Hull, Bradley Zangerle, operations
research
Hunscher, Helen Alvina (Mrs H P
Wilkinson), nutrition
Hunt, Dominic Joseph, inorganic
chemistry
Hunter, Joseph Lawrence, physics
Hurwitz, David Allan, pharmacology
Husni, Elias A., cardiovascular surgery
Hutchison, William Marwick, analytical
chemistry
Idol, James Daniel, Jr., industrial
chemistry, organic polymer chemistry
Ignatoski, Joseph Adam, plant pathology,
weed science
Igou, Donald Kaye, clinical biochemistry
Inkley, Scott Russell, internal medicine
Izant, Robert James, Jr., pediatric surgery
Jacob, Chempthra Varughese,
immunology
Jamieson, Alexander MacRae, chemical
physics, polymer science
Joyce, Blaine R., physical chemistry
Kainen, Paul Chester, mathematics
Kaiserman-Abramof, Ita Rebeca,
neurobiology, neurocytology
Johnson, Thomas Raymond, cell biology
Jones, Arthur Letcher, petroleum
chemistry
Jones, Joie, Pierce, medical research,
ultrasound
Jones, Paul Kenneth, biostatistics
Kantor, Paul B., physics
Kaplan, Arnold Raymond, human
microscopy
Kean, Edward Louis, biochemistry
Keck, Max Johann, solid state physics
Kelly, Raymond Crain, toxicology
Kelsall, Margaret Aston, biochemistry,
histology
Kennell, John Hawks, pediatrics,
behavioral science
Kenney, Malcolm Edward, inorganic
chemistry
Kent, Barbara Wynne, anatomy
Keplinger, Orin Clawson, polymer
chemistry, explosives
Kerkay, Julius, clinical chemistry
Khairallah, Philip Asad, physiology,
pharmacology
King, Katherine Chung-Ho, pediatrics,
metabolism
Kirby, Albert Charles, physiology
Klein, Howard Clarason, organic
chemistry
Klein, LeRoy, biochemistry, medical
research
Klein, Robert Herbert, physics
Kleinerman, Jerome, pathology,
physiology
Klopman, Gilles, chemistry
Knoke, James Dean, biostatistics
Knox, Kerro, inorganic chemistry
Ko, Wen Hsiung, bionics
Koenig, Jack L., polymer chemistry,
physical chemistry
Koga, Rokutaro, astrophysics
Kohl, Fred John, physical chemistry
Kohn, Robert Rothenberg, experimental
pathology
Koletsky, Simon, pathology
Kosmahl, Henry G., electron physics
Kowal, Jerome, internal medicine,
biochemistry
Kowalski, Kenneth L., theoretical nuclear
physics
Kozawa, Akiya, electrochemistry
Krampitz, Lester Orville, bacteriology,
microbiology
Kreider, Eric Russell, ceramics,
inorganic chemistry
Kretchmer, Henry Edmund,
anesthesiology
Kushner, Irving, rheumatology, medicine
Kushner, Arthur Simon, organic
chemistry, photochemistry
Kumar, Manjula Satyendra, immunology,
endocrinology
Kumins, Charles Arthur, surface
chemistry
Lad, Robert Augustin, chemistry
Lake, Robin Benjamin, biometrics,
biomedical engineering
Lalli, Anthony, roentgenology
Landau, Bernard Robert, medicine,
biochemistry
Lando, Jerome B., polymer science
Larson, Myra Joan, genetics

Lasek, Raymond J., neurobiology
Lauver, Milton Renick, plasma physics
Lauver, Richard William, physical
chemistry
Lauzau, Wilbur R., inorganic chemistry,
physical chemistry
Lavik, Paul Sophius, biochemistry
Leach, Ernest Bronson, mathematics
Lee, Leavie Edgar, Jr., pathology, medical
administration
Leitman, Marshall J., applied
mathematics, continuum physics
Lenkoski, L Douglas, medicine,
psychiatry
Lentz, Kenneth Eugene, biochemistry
Leonards, Jack Ralph, biochemistry
Lesh-Laurie, Georgia Elizabeth,
developmental biology
LeSuer, William Monroe, petroleum
chemistry
Leutner, Frederick Stanley, clinical
biochemistry
Levy, Matthew Nathan, physiology
Lewis, John Erwin, microbiology
Lewis, Lena Armstrong, physiology
Lewis, Sally, botany, horticulture
Lindley, Barry Drew, physiology,
biophysics
Lipset, Mortimer Broadwin, internal
medicine
Litt, Morton Herbert, polymer chemistry
Litt, A Brian, obstetrics & gynecology
Lock, James Albert, theoretical nuclear
physics
Lohwater, A J., mathematics
Long, Edward B., aquatic ecology
Love, David S., zoology
Luoma, John Robert Vincent, physical
chemistry
Lynch, William C., mathematics,
computer science
Machlup, Stefan, physics
Macintyre, Malcolm Neil, cytogenetics
Macintyre, William James, nuclear
medicine, medical physics
Macklin, Martin, physical biology
Mahajan, Damodar K., biochemistry,
endocrinology
Mahan, Harold Dean, ornithology
Makowski, Mieczyslaw Paul,
electrochemistry, physical chemistry
Malhotra, Om Parkash, physiology
Maron, Samuel Herbert, physical
chemistry, polymer science
Marsters, Roger Westcott, biochemistry,
immunohematology
Marty, Roger Henry, topology
Masson, Georges Marie Charles,
endocrinology
Mawardi, Osman Kamel, plasma physics,
acoustics
Mayo, Charles Edward, electrochemistry
Mayo, Joseph William, biochemistry
McCann, Roger C., mathematics
McCarty, Lewis Vernon, chemistry
McCloskey, Teresemarie, mathematical
logic
McCorkle, Lois Pake, epidemiology,
biostatistics
McEvoy, Donald, biochemistry
McGervey, John Donald, experimental
solid state physics
McKenzie, Douglas Hugh, anthropology
McLean, Edward Bruce, ornithology,
ecology
McLemore, Benjamin Henry, Jr.,
mathematical statistics
McMillan, Garnett Ramsay,
photochemistry
Megargle, Robert G., analytical chemistry
Meinhardt, Norman Anthony, organic
polymer chemistry
Messineo, Luigi, biochemistry, biophysics
Metcalfe, Joseph Edward, III, physical
chemistry, fuel technology
Metz, David A., audiology, speech
pathology
Milberger, Ernest Carl, chemistry
Mieyal, John Joseph, pharmacology,
biochemistry
Miller, Charles G., microbiology
Miller, Leonard Edward, organic
chemistry
Miller, Max, medicine
Millis, John Schoff, medical education
Miraldi, Floro D., bioengineering, nuclear
medicine
Mitchell, George Redmond, Jr., organic
polymer chemistry
Moore, Allen Charlton, chemistry,
research administration
Moore, Lee E., neurophysiology,
biophysics

Zubler, Edward George, physical chemistry
Zull, James E, biochemistry, cellular biology

CLEVELAND HEIGHTS
Blank, John Moulton, physics
Donner, Henry Frederick, geology
Fritsch, Klaus, electronics, fluid physics
Jaffe, Hans, electrooptics
Kotnik, Louis John, organic chemistry, chemical engineering
Nelson, Raymond John, mathematics
Parkinson, David B, physics
Price, Jacob Waide, clinical chemistry

CLEVES
Brant, Russell Alan, geology

COLUMBUS
Abel, Alan Wilson, physical chemistry
Ackerman, Gustave Adolph, Jr, anatomy, medicine
Adams, Dale W, agricultural economics
Addanki, Somasundaram, clinical chemistry, biochemistry
Adelman, Albert H, physical chemistry
Alban, Evan Kenneth, horticulture
Alben, James O, biochemistry
Allaire, Francis Raymond, dairy science
Allen, Alfred Ells, physical inorganic chemistry, solid state science
Allen, Harry Prince, mathematics
Allen, Norman, neurology
Allred, John B, biochemistry, nutrition
Anderson, Larry Bernard, analytical chemistry
Anderson, Richard Jasper, geology
Andrus, Paul Grier, electrophotography
Angerer, Clifford Ackerman, physiology
Antler, Morton, applied chemistry
Arewa, E Ojo, anthropology
Arnfield, Anthony John, physical geography, climatology
Arns, Robert George, nuclear physics
Atwell, Robert James, medicine
Austin, Alfred Ells, physical inorganic chemistry, solid state science
Baba, Nobuhisa, pathology, electron microscopy
Baggot, John Desmond, veterinary pharmacology
Bailey, Cecil Dewitt, dynamics
Balcerzak, Stanley Paul, internal medicine, hematology
Baldwin, Eldon Dean, agricultural economics
Baldwin, Maynard Martin, organic chemistry
Banwart, George J, food microbiology
Barber, George Alfred, biochemistry
Barnes, Herbert M, animal science
Baroody, Eugene Michael, physics
Barr, Harry L, reproductive physiology, genetics
Bashe, Winslow Jerome, Jr, pediatrics, preventive medicine
Bates, Robert Latimer, economic geology
Batley, Frank, radiology
Beach, Paul L, nuclear physics
Beal, Jack Lewis, pharmacognosy
Bearse, Arthur Everett, organic chemistry
Beatty, Glenn Hurst, statistics
Beer, Albert Carl, physics
Behn, Robert Collins, meteorology
Behrman, Edward Joseph, biochemistry
Bell, John Clarence, mathematics
Bendixen, Leo E, plant physiology
Benton, Duane Allen, biochemistry
Berggren, Ronald B, surgery
Bergstrom, Stig Magnus, geology, invertebrate paleontology
Berliner, Lawrence J, biophysical chemistry
Bernays, Peter Michael, physical inorganic chemistry, information science
Bernhagen, Ralph John, geology
Berry, David A, organic chemistry
Berry, Richard Lee, medical entomology, insect taxonomy
Bianchine, Joseph Raymond, pharmacology
Biersdorf, William Richard, visual physiology, ophthalmology
Birkeland, Jorgen Maurice, bacteriology
Birky, Carl William, Jr, genetics
Black, John Wilson, speech & hearing science
Blackwell, Harold Richard, vision
Blair, Billie D, entomology
Blatt, S Leslie, nuclear physics
Blau, Henry Hess, physical inorganic chemistry
Blocher, John Milton, Jr, physical chemistry
Blozis, George G, pathology

Morgan, Harold Eugene, physics
Moritz, Alan Richards, pathology
Mortensen, Earl Miller, physical chemistry
Mortimer, Edward Albert, Jr, medicine
Mudrak, Anton, organic chemistry
Muir, William A, human genetics, anatomy
Munger, George Donald, plant pathology
Murrish, David Earl, comparative physiology
Myers, Ira Thomas, physics
Nankervis, George Arthur, pediatrics, microbiology
Nash, Harry Charles, solid state physics, optics
Neet, Kenneth Edward, biochemistry
Neiderhiser, Dewey Harold, biochemistry
Nelson, Paul Andrew, medicine
Neville, Janice Nelson, nutrition
Nichols, William Herbert, statistical mechanics
Nordlander, John Eric, organic chemistry
Nygaard, Oddvar Frithjof, radiobiology
Ogden, William Frederick, mathematics, computer science
Olah, George Andrew, organic chemistry
Olah, Judith Agnes, organic chemistry
Oleinick, Nancy Landy, biochemistry
Olson, Walter T, chemistry
Olynk, Paul, chemistry
Ott, Teunis Jan, operations research, mathematical statistics
Ouzts, Johnny Drew, medical entomology
Owen, James Emmet, inorganic chemistry
Pace, Eugene Leonard, chemistry
Page, Irvine Heinly, cardiovascular diseases
Paine, Robert Madison, metallurgical chemistry
Palik, Emil Samuel, physical chemistry
Pao, Yoh-Han, physics
Parker, David J, physical chemistry, analytical chemistry
Parker, Robert Frederic, medicine
Parkinson, Truman David, applied mechanics
Paynter, John, Jr, applied chemistry
Pearson, Olof Hjalmar, medicine
Pensky, Jack, biochemistry
Peppler, Richard Bond, industrial chemistry
Perlmutter, Henry Irwin, radiology
Persky, Lester, surgery, urology
Pesch, Peter, astronomy
Petti, Richard James, geometry, numerical analysis
Plonsey, Robert, biomedical engineering
Pollack, John L, optics
Pories, Walter J, surgery, nutrition
Post, Robert Elliott, electrochemistry
Potter, Ralph Miles, physical chemistry
Pratt, Richard J, organic chemistry
Primiano, Frank Paul, Jr, biomedical engineering
Pritchard, Walter Herbert, medicine
Pruitt, Ralph L, mathematics, mathematics education
Przybylski, Ronald J, zoology, cytology
Rakita, Louis, cardiovascular diseases, internal medicine
Ram, Madhira Dasaradhi, surgery, experimental pathology
Rammelkamp, Charles Henry, internal medicine
Ratnoff, Oscar Davis, internal medicine
Recknagel, Richard Otto, physiology, toxicology
Reagan, James W, pathology
Reed, George Benson, pathology, pediatrics
Reiling, Gilbert Henry, engineering physics
Reisman, Arnold, operations research, industrial engineering
Rejali, Abbas Mostafavi, radiology, nuclear medicine
Reynard, Kennard Anthony, polymer chemistry, inorganic chemistry
Rightmire, Robert, nuclear chemistry
Rippon, William Barton, biological structure
Ritchey, William Michael, physical chemistry
Robbins, Frederick Chapman, pediatrics
Robbins, Norman, neurophysiology
Robertson, Abel Alfred Lazzarini, Jr, pathology
Robins, Richard Dean, organic chemistry
Robinson, Donald Keith, particle physics
Robinson, Stewart Marshall, mathematics
Rogers, Charles Edwin, physical chemistry
Rose, Gene Fuerst, mathematics
Rosen, Irving, polymer chemistry

Rosen, Mortimer Gilbert, obstetrics & gynecology
Rosenblatt, Judah Isser, mathematical statistics
Ross, John Franklin, chemistry
Rossi, Edward P, oral pathology, dentistry
Rothchild, Irving, reproductive endocrinology
Rowland, Vernon, psychiatry
Ruchelman, Maryon Waldman (Mrs Roderick A Kratoville), petroleum chemistry, system analysis
Rushforth, Norman B, animal behavior, epidemiology
Rustad, Ronald Cameron, physiology
Rynbrandt, Donald Jay, biochemistry
Sable, Henry Zodoc, biochemistry
Saby, John Sanford, physics
Saidel, Gerald Maxwell, biomedical engineering
Sakami, Warwick, biochemistry
Sara, Raymond Vincent, materials science
Sarbach, Donald Victor, organic chemistry
Savin, Samuel Marvin, geochemistry, geology
Sayers, George, endocrinology
Schacter, Bernice Zeldin, immunogenetics
Schafer, Irwin Arnold, pediatrics, genetics
Schilling, Edward George, statistics
Schindler, Stephen Michael, astrophysics
Schlosser, Herbert, solid state physics, molecular biophysics
Schubert, Daniel Sven Paul, psychiatry, psychology
Schuele, Donald Edward, solid state physics
Schwartz, Colin John, anatomic pathology, experimental pathology
Scott, David Bytovetzski, biology
Segall, Benjamin, theoretical physics
Seldin, Emanuel Judah, engineering physics
Selim, Mostafa Ahmed, obstetrics & gynecology, gynecologic oncology
Sena, Elissa Purnell, molecular biology
Senderoff, Seymour, electrochemistry
Senturia, Jerome Basil, comparative physiology, environmental physiology
Shainoff, John Rieden, biophysics
Shamberger, Raymond J, biochemistry
Shankland, Robert Sherwood, physics
Shaw, Wilfrid Garside, inorganic chemistry
Sheibley, Fred Easly, organic chemistry
Shelton, James Reid, polymer science
Shipley, Reginald Alden, medicine
Sholander, Marlow, mathematics
Sibley, Willis Elbridge, anthropology, academic administration
Silberger, Allan Joseph, mathematics
Simha, Robert, physics, biophysics
Singer, Joseph, electrochemistry
Singer, Leonard Sidney, physical chemistry
Sporzynski, Adam Przemyslaw, organic chemistry
Standish, Norman Weston, organic chemistry
Stavitsky, Abram Benjamin, immunology
Steckler, Harold Arthur, inorganic chemistry
Stehli, Francis Greenough, geology
Steinberg, Arthur Gerald, medical genetics, human genetics
Stephenson, Betty Ann, bio-organic chemistry
Stephenson, Charles Bruce, astronomy
Stephenson, Francis Creighton, physics
Stevenson, Frank Robert, solid state physics
Stewart, Charles Neil, surface physics
Stoll, John Roth, II, mathematics
Stone, John Grover, II, geology
Stonehouse, Albert James, inorganic chemistry, physical chemistry
Storaasli, John Phillip, medicine
Stratton, Ralph Atwood, medicine, urology
Strain, William Henry, chemistry
Stuebe, Carl, organic chemistry
Stuehr, John Edward, biophysical chemistry

Sullivan, Charles Raymond, physics
Sullivan, Julia Christine, microbiology
Sung, Chien-Bor, research administration, interdisciplinary sciences
Sunshine, Irving, clinical chemistry
Surbey, Donald Lee, organic chemistry
Swift, Terrence James, physical chemistry, biochemistry
Swinehart, Carl Francis, inorganic chemistry
Tandler, Bernard, cytology, electron microscopy
Taslitz, Norman, neuroanatomy
Tavill, Anthony Sydney, gastroenterology
Taylor, Philip Liddon, solid state physics
Thaler, Raphael Morton, theoretical physics
Thomas, Charles I, ophthalmology
Thompson, Eric Douglas, solid state physics
Thompson, Wilmer Leigh, Jr, clinical pharmacology, medicine
Tobocman, William, theoretical nuclear physics
Tomashefski, Joseph Francis, physiology
Towns, Robert Lee Roy, chemistry
Trautman, William Dean, computer sciences
Travis, Randall Howard, physiology
Trivisonno, Joseph, Jr, solid state physics
Tucker, Arthur Smith, medicine, radiology
Tucker, Harvey Michael, otolaryngology, surgery
Turnbull, Bruce Felton, physical chemistry
Urbach, Frederick Lewis, inorganic chemistry
Utter, Merton Franklin, biochemistry
Valenzuela, Rafael, immunopathology, pathology
Van Horn, David Downing, metal physics, mathematics
Vertes, Victor, internal medicine
Vignos, Paul Joseph, Jr, biochemistry, rheumatology
Vogt, Lester Herbert, Jr, inorganic chemistry
Voneida, Theodore J, neurobiology
Wagoner, Glen, physics
Wainer, Eugene, inorganic chemistry
Walker, Jearl Dalton, optics
Walmsley, Mildred Marie, geography
Walsh, James Aloysius, physical organic chemistry
Walter, Edward Joseph, geophysics
Walters, Virginia F, physics
Walton, Alan George, biophysical chemistry
Warren, Kenneth S, tropical medicine
Way, Frederick, III, computer science
Weast, John Calvin, chemistry
Weaver, John Calvin, chemistry
Weaver, William Michael, organic chemistry
Webb, Thomas Howard, organic chemistry, lubrication engineering
Weinberg, Irving, physics
Weis, Dale Stern, microbiology, protozoology
Weisman, Russell, medicine, hematology
Wells, Charles Frederick, algebra
Wentz, William Budd, obstetrics & gynecology, oncology
Weschler, Joseph Robert, organic chemistry
Whalen, William James, physiology
White, Robert J, surgery, neurophysiology
Whitman, Charles Inkley, physical chemistry
Whitman, Donald Ray, theoretical chemistry, physical chemistry
Willard, Harvey Bradford, nuclear physics
Willard, James Matthew, biochemistry
Williams, Michael Eugene, vertebrate paleontology
Williams, Robert Frank, medicine
Willson, Karl Stuart, electrochemistry, corrosion
Wilson, Richard Mac, electrochemistry, physical chemistry
Wisotzky, Joel, dentistry, dental research
Wolin, Lee Roy, psychology, neurophysiology
Wolinsky, Emanuel, infectious diseases
Wood, Galen Theodore, nuclear physics
Wood, Harland G, biochemistry, microbiology
Yeager, Ernest Bill, electrochemistry
Yeager, John Frederick, chemistry
Yen-Watson, Belinda R S, immunology
Zacks, Shelemyahu, mathematical statistics, statistical analysis
Zilsel, Paul Rudolph, theoretical physics
Zirkes, Al, radiological health

Boatman, Joseph Brasher, physiology
Bohning, Richard Howard, botany
Bojanic, Ranko, mathematical analysis
Bond, Charles Clayton, geology
Bookhout, Theodore Arnold, wildlife research
Bope, Frank Willis, pharmaceutical chemistry
Borror, Donald Joyce, entomology, ornithology
Bourguignon, Erika Eichhorn, anthropology
Bowerman, Ernest William, chemistry
Bowers, John Dalton, agriculture
Bowman, Bernard Ulysses, Jr., medical microbiology
Bowman, Mary Lynne, environmental management
Bozler, Emil, physiology
Brand, Benson Glenn, chemistry
Branson, Dorothy Swingle, clinical microbiology
Bridge, John Robert, structural chemistry, information science
Brierley, Gerald Philip, biochemistry
Briggs, John Dorian, insect pathology
Britt, N Wilson, limnology, entomology
Brockman, Harold W., mathematics
Brodhag, Alex Edgar, Jr., organic chemistry
Brog, Kenneth Clair, physical chemistry, photochemistry
Brooman, Eric William, electrochemistry
Brown, L Carlton, physics
Brown, Lawrence Alan, urban geography, population geography
Brown, Patricia Lynn, information science
Brown, Robert Bruce, mathematics
Brownell, Katharine Anna, physiology
Browning, Robert Hamilton, medicine
Bull, Colin Bruce Bradley, geophysics
Bulman, Warren Eugene, physics
Burch, Joseph Eugene, organic chemistry
Burkman, Allan Maurice, pharmacology
Burt, James Kay, veterinary radiology
Busch, Daryle Hadley, inorganic chemistry
Byrd, Paul L., astronomy, astrophysics
Byers, Thomas Jones, biology
Cahill, Vern Richard, meat science
Cahoon, Garth Arthur, horticulture
Caley, Earle Radcliffe, analytical chemistry
Calvert, Jack George, photochemistry, physical chemistry
Capen, Charles Chabert, veterinary pathology
Capriotti, Eugene Raymond, physics
Carey, Larry Campbell, surgery
Carlton, Richard Walter, sedimentary petrology
Carr, Francis W., mathematical analysis
Carroll, Katharine Dean, dermatology
Casetti, Emilio, geography
Castle, Richard Thomas, animal nutrition
Chase, Dan L., analytical chemistry
Chipley, John Raymond, microbiology
Chism, Grady William, III., food science
Chorpenning, Frank Winslow, microbiology, immunobiology
Clark, Clarence Floyd, zoology
Clark, David Lee, animal behavior, anatomy
Clarke, William James, veterinary medicine
Clay, Mary Ellen, entomology, biology
Clendenon, Nancy Ruth, neurochemistry
Cline, Jack Henry, animal nutrition
Cline, Morris George, plant physiology, ecology
Coe, Kenneth Loren, chemistry
Cole, Clarence Russell, veterinary pathology, medical administration
Colinvaux, Paul Alfred, ecology, paleoecology
Collings, Edward William, experimental solid state physics
Collins, George W. II, astronomy
Collins, Horace Rutter, geology
Collins, William John, entomology
Collinson, James W., geology
Connor, Lawrence John, entomology, apiculture
Conroy, Charles William, periodontics, oral pathology
Cook, Harry Lee, agronomy
Copeland, James Clinton, microbial genetics, molecular genetics
Corbato, Charles Edward, geology, geophysics
Cornwell, David George, biochemistry
Corson, Samuel Abraham, pharmacology, physiology
Cox, Kevin Robert, geography
Craig, Richard Anderson, theoretical physics, solid state physics

Cramblett, Henry G., pediatrics, virology
Crites, John L., zoology
Croft, Charles Clayton, microbiology, public health
Crosswhite, F Joe, mathematics education
Croxton, Frank Cushaw, chemistry, research administration
Culbertson, Billy Muriel, organic chemistry, physical chemistry
Curry, John Joseph, III, neuroendocrinology
Czyzak, Stanley Joachim, physics
Dabah, Roger, microbiology, biochemistry
Darby, Ralph Lewis, information science, chemical engineering
Davidson, Ralph Howard, entomology
Davidson, Richard Shoots, plant pathology
Davis, Bruce Wilson, applied mathematics
Dean, David W., mathematics
Dean, Donald Harry, molecular genetics
Deatherage, Fred E., food biochemistry, nutrition
Deep, Ira Washington, plant pathology
de Fiebre, Conrad William, microbiology, biochemistry
Dehne, George Clark, analytical chemistry, inorganic chemistry
DeLor, Camille Joseph, medicine
Delphia, John Maurice, vertebrate morphology
Demko, George Joseph, geography
Deviney, Marvin Lee, Jr., physical chemistry, rubber chemistry
Devor, Arthur William, biochemistry
Dew, William Calland, dentistry
Dewein, Louis F., physiology
DeWet, Pieter D., neuroscience
Dickey, Frederick Pius, physics
Dickman, John Theodore, biochemistry
Diesem, Charles D., veterinary anatomy
Dingee, David Aaron, plasma physics
Disinger, John Franklin, conservation
Divis, Bohuslav B., mathematics, number theory
Dodd, Matthew Charles, bacteriology
Donohue, Timothy R., nuclear physics
Donovan, Edward Francis, veterinary medicine
Dorfman, Leon Monte, chemical kinetics, radiation chemistry
Dorn, Charles Richard, veterinary public health, epidemiology
Doskotch, Raymond Walter, pharmacognosy, biochemistry
Draughs, Edmund, surface chemistry
Drees, David T., veterinary pathology
Drennan, James Elliott, physics
Drobot, Stefan, mathematics
Droege, John Walter, fuel science
Drozda, William, mathematics
Dudewicz, Edward John, statistics
Dugan, Patrick R., microbiology, biochemistry
Dukes, John R., engineering
Ebner, Charles Arthur, low temperature physics
Eckert, George Frank, physical chemistry
Edwards, David Olaf, physics
Egitis, Irma, anatomy, dermatology
Egilts, John Arnold, anatomy, otolaryngology
Ehlers, Ernest George, mineralogy
Eilett, Clayton Wayne, plant pathology
Ellingson, Harold Victor, aerospace medicine, preventive medicine
Elliot, David Hawksley, geology, petrology
Erb, John Hoffman, dairy science
Erickson, Richard Ames, physics
Erven, Bernard Lee, agricultural economics
Euwema, Robert Noel, theoretical solid state physics
Evans, Michael Leigh, botany
Everett, Kaye, geology
Fairand, Barry Philip, lasers
Falb, Richard D., biochemistry
Farley, James Gunter, geochemistry
Faure, Gunter, geochemistry
Fawcett, Sherwood Luther, research administration
Fechheimer, Nathan S., animal genetics
Feigenbaum, Abraham Samuel, biochemistry, nutrition
Ferrar, Joseph C., mathematics
Ferris, Thomas Francis, internal medicine

Fetter, Arthur Williams, veterinary pathology
Firestone, Richard Francis, physical chemistry
Fisk, Frank Wilbur, insect physiology
Foltz, Rodger L., organic chemistry
Foreman, Dennis Walden, Jr., mineralogy, physical chemistry
Forster, D Lynn, agricultural economics
Foster, Wilfrid Raymond, mineralogy
Fox, Edward L., exercise physiology
Fraenkel, Gideon, chemistry
Frajola, Walter Joseph, biochemistry
Frank, Sylvan Gerald, surface chemistry, pharmaceutics
Fratianne, Douglas G., plant physiology
Frea, James Irving, bacteriology
Freimanis, Atis K., radiology
Fretz, Thomas Alvin, ornamental horticulture
Freud, Geza, mathematical analysis
Freudenthal, Ralph Ira, biochemical pharmacology
Frey, Perry Allen, biochemistry
Fried, John, cultural anthropology
Fry, Glenn Ansel, physiological optics
Fullmer, June Zimmerman, history of science
Gardner, David L., biomedical engineering, physiology
Gabel, Albert A., veterinary surgery
Gaines, Gordon Bradford, surface physics, solid state physics
Gaines, James R., solid state physics
Gardier, Robert Woodward, pharmacology
Garraway, Michael Oliver, plant pathology
Garret, Alfred Benjamin, chemistry
Garrow, Robert Joseph, mathematics, physics
Gatewood, Buford Echols, mathematics
Gatherum, Gordon Elwood, forest physiology, soils
Gaudette, Leo Eward, biochemistry
Gaughran, George Richard Lawrence, anatomy
Gaunt, Abbot Stott, zoology
Gealy, William James, geology
Geisman, Jean Richard, horticulture
George, William Leo, Jr., plant breeding
Gerald, Michael Charles, pharmacology
Gibson, William Miles, pediatrics
Gideon, Donald Nason, physics
Giesy, Robert, plant morphology
Gifford, David Stevens, organic chemistry
Gilbert, Gareth E., botany
Giltz, Maurice Leroy, biology, zoology
Gist, George Reinecker, agronomy, academic administration
Globe, Samuel, applied physics
Gold, Robert, number theory
Goldthwait, Richard Parker, geology
Goleman, Denzil Lyle, economic entomology
Golledge, Reginald George, urban geography
Gomes, Wayne Reginald, reproductive biology
Good, Ernest Eugene, wildlife management
Goodson, Leslie Alan, organic chemistry
Gordon, Donald Theile, plant pathology, virology
Gould, Ira A. (Jr), dairy industry
Gould, Wilbur Alphonso, food science, horticulture
Graef, Walter L., organic chemistry, pharmaceutical chemistry
Graffeo, Anthony Philip, analytical chemistry
Graham, Bruce Douglas, pediatrics
Grannis, George Franklin, clinical biochemistry, gerontology
Gray, Eoin Wedderburn, plasma physics, high temperature chemistry
Green, Mary Eloise, nutrition
Greenberg, Stanley, cardiovascular diseases
Greenwald, Lewis, comparative physiology
Gregory, Ian (Walter De Grave), psychiatry
Grieser, Daniel R., optical physics
Griffing, J Bruce, genetics
Griswold, Bernard Lee, fish biology
Grossie, James Allen, food technology
Grotta, Henry Monroe, organic chemistry
Grubbs, Robert Custis, physiology

Gruber, H Thomas, physics engineering
Gruemer, Hanns-Dieter, clinical chemistry, biochemistry
Hall, Elton Harold, physical chemistry
Hall, Geroge Frederick, agronomy
Hameberg, William, medicine, anesthesiology
Hamlin, Robert Louis, cardiovascular physiology
Hamman, Donald Jay, electronics
Hamparian, Vincent, virology
Hanks, Richard Donald, chemistry
Hansen, Paul M T, food science, dairy technology
Hanson, Kenneth Marvin, physiology
Harder, John Dwight, wildlife biology, reproductive physiology
Harper, Willis James, dairying
Harris, Preston Mayne, physical chemistry
Hart, Ronald Wilson, radiobiology, molecular biophysics
Hartman, Fred Oscar, pomology
Harvey, Walter Robert, animal breeding
Hassell, John Allen, physical chemistry
Hausman, Hershel J., nuclear physics
Havener, William H., ophthalmology
Hayes, Thomas G., histology
Haynes, Ralph Edwards, pediatrics, medical microbiology
Hebbard, Frederick Worthman, optometry, physiological optics
Heer, Clifford V., lasers, quantum optics
Henke, Clarence Henry, mathematics
Henney, Jeannette Hillman, cultural anthropology
Henderlong, Paul Robert, agronomy, plant physiology
Henderson, Dennis Roger, agricultural economics
Herdendorf, Charles Edward, III, geology, limnology
Herr, Donald Edward, agronomy
Hiatt, Edwin Peele, physiology
Hicks, John Frederick Gross, (Jr), physical chemistry
Hill, Richard M, physiological optics
Hillis-Colinvaux, Llewellya Williams, marine biology
Hines, Frank Lawrence, entomology
Hine, Jack, chemistry
Hines, Harold C., dairy science
Hink, Walter Fredric, entomology, cell biology
Hitchcock, Fred Andrews, physiology
Hock, Arthur George, agronomy, soil morphology
Hoff, Donald Jerome, soil science
Hohn, Ronald Bruce, veterinary surgery
Holaday, William J., medicine, surgical pathology
Holdsworth, Robert Powell, Jr., entomology
Hollander, Philip B., pharmacology, biophysics
Hollenbeck, Zeph John Reid, obstetrics & gynecology
Holowaychuk, Nicholas, soil science
Holzaepfel, John Houston, obstetrics & gynecology
Holzinger, Thomas Walter, food science
Hopper, Norman Wayne, crop physiology, plant physiology
Horn, David Jacobs, ecology, entomology
Horner, James William, Jr, information science, organic chemistry
Horowitz, Aaron, veterinary anatomy
Horrocks, Lloyd Allen, neurochemistry
Horton, Derek, organic chemistry
Hosansky, Norman Leon, organic chemistry
Hostetter, Jeptha Ray, anatomy, electron microscopy
Hostetler, Heber P. III, plant morphology, phycology
House, Verl Lee, genetics
Howard, William Henry Richard, radiology
Howes, James E. Jr., pollution chemistry
Hsia, John S., number theory, algebra
Hsu, Hsiung, applied physics
Huffman, Richard William, dentistry
Hughes, Daniel Thomas, cultural anthropology, social anthropology
Humbertson, Albert O, Jr, anatomy, neuroanatomy
Huneke, John Philip, mathematics
Hunker, Henry L., geography
Hunt, Fern Ensminger, foods, microbiology
Hunt, William Edward, neurosurgery

Hurt, Paul Victor, agricultural statistics, agricultural economics
Hushak, Leroy J., agricultural economics
Ihas, Gary Gene, low temperature physics
Ingalls, William Lisle, veterinary pathology
Isenberg, Allen (Charles), pharmaceutical chemistry, information science
Ish, Carl Jackson, chemistry
Isler, Gene A., animal science
Ives, David Homer, enzymology, biochemistry
Jaap, Robert George, poultry husbandry
Jakobsen, Robert John, physical chemistry
Jalil, Mazhar, acarology
Janson, Blair F., plant pathology
Jastram, Philip Sheldon, physics
Jaynes, John Alva, food science
Jentgen, Richard Louis, research management
Johnson, Carl Sand, resource management
Johnson, George Robert, animal husbandry
Johnson, Jay Wolbert, agronomy
Johnson, Lee Frederick, molecular biology
Johnson, Robert Oscar, applied mathematics
Johnson, Tillman Joseph, plant morphology, anatomy
Johnson, William Buhlmann, mathematical analysis
Johnston, Herbert Norris, chemistry
Jorns, Marilyn Schuman, enzymology
Josephson, Ronald Victor, food chemistry, biochemistry
Jossem, Edmund Leonard, physics
Kaelbling, Rudolf, psychiatry
Kanakkanatt, Antony, polymer chemistry
Kapral, Frank Albert, medical microbiology, immunology
Kartha, Mukund K., radiology, biophysics
Keller, Geoffrey, astronomy
Keller, Martin David, epidemiology
Kenan, Richard P., solid state physics
Kern, Charles William, theoretical chemistry
Kerr, Douglas S., computer sciences
Kerr, Kirklyn M., veterinary pathology
Kiehl, Samuel Jacob, Jr, analytical chemistry, biochemistry
Kilman, James William, cardiovascular surgery, thoracic surgery
Kim, Young Sik, high energy physics
Kimball, Paul Clark, molecular biology, virology
King, Charles C., ecology, entomology
King, James S., neuroanatomy
King, John Edward, anatomy, histology
Kiplinger, Donald Carl, horticulture
Kircher, John Frederick, physical chemistry
Klapper, Michael H., biochemistry
Klassen, Karl Peter, surgery
Klingensmith, Raymond W., nuclear physics
Knies, Phillip Thomas, medicine
Knopp, Walter, psychiatry
Koestner, Adalbert, veterinary pathology
Kolodziej, Bruno J., microbiology, microbial physiology
Kontras, Stella B., pediatrics, genetics
Kormacker, Karl, theoretical physics, biophysics
Korringa, Jan, theoretical physics
Kottman, Roy Milton, animal breeding
Kozel, Philip C., horticulture, plant physiology
Kraus, John Daniel, electrical engineering, astronomy
Krause, Horatio Henry, inorganic chemistry
Kreier, Julius Peter, microbiology, veterinary medicine
Krill, Walter Roland, veterinary medicine
Kristoffersen, Thorvald, food science
Kroenberg, Bernd, food science, nutrition
Kruger, Fred Albert, biochemistry, endocrinology
Kunz, Albert L., physiology
Kuwana, Theodore, analytical chemistry, electrochemistry
Kyung, Jai Ho, organic chemistry
Landin, Joseph, mathematics
Lang, Raymond W., bacteriology, immunology
La Pidus, Jules Benjamin, medicinal chemistry
LaRocque, Joseph Alfred Aurele, geology
Larsen, Philip O., plant pathology
Larson, Donald W., agricultural economics
Lebrie, Stephen Joseph, medical physiology
Lee, Warren Ford, agricultural economics
Leighty, Edith Gardner, biochemistry

Leininger, Robert Irvin, biomedical engineering, polymer chemistry
Leitzel, James Robert C., mathematics
Leitzel, Joan Phillips, mathematics
Lessler, Milton A., physiology
Leussing, Daniel, Jr, analytical chemistry
Levine, Raphael David, theoretical chemistry
Levy, Arthur, atmospheric chemistry, fuel science
Lewis, James Edward, physical chemistry, organic chemistry
Lewis, Neil Jeffrey, medicinal chemistry, biochemistry
Liberator, Frederick Anthony, biochemistry
Lim, David J., otolaryngology, electron microscopy
Lin, Denis Chung Kam, analytical chemistry, mass spectrometry
Lipetz, Leo Elijah, biophysics, neurosciences
Lipsky, Joseph Albin, physiology
Liss, Leopold, neuropathology
Litchfield, John Hyland, food science, industrial microbiology
Lobuglio, Albert Francis, hematology, immunology
Loening, Kurt L., physical chemistry, organic chemistry
Logan, Terry James, soil chemistry
Long, John Frederick, veterinary pathology
Long, Terrill Jewett, botany
Long, Thomas Ross, solid state physics
Losekamp, Bernard Francis, polymer chemistry, organic chemistry
Lott, John Alfred, analytical chemistry
Lougher, Edwin Henry, physical chemistry, research administration
Lowney, Edmund Dillahunty, dermatology
Lowther, Gerald Eugene, optometry, visual physiology
Ludwick, Thomas Murrell, dairy husbandry
Lustick, Sheldon Irving, environmental physiology, vertebrate zoology
Lyon, William Francis, economic entomology
MacWood, George Eugene, chemistry
Mainland, Gordon Bruce, theoretical high energy physics
Makley, Torrence Aloysius, Jr, ophthalmology
Mallozzi, Philip James, physics
Mamrak, Sandra Ann, computer sciences
Manner, Richard John, chemistry, molecular spectroscopy
Markworth, Alan John, physics
Martin, David P., agronomy
Martin, George Franklin, Jr, neuroanatomy
Marzluf, George A., genetics
Matesich, Mary Andrew, physical chemistry
Mathews, Collis Weldon, physical chemistry, molecular spectroscopy
Mathis, Robert Fletcher, computer science, mathematics
McIntyre, Russell Theodore, biochemistry
McKenzie, Garry Donald, geology
McLachlan, Dan, Jr, chemical physics
McLean, Eugene Otis, soil chemistry
Means, Gary Edward, biochemistry
Mayer, Victor James, earth sciences, science education
McConnell, Duncan, dental research, materials science
McCormick, Francis B., agricultural economics
McDowell, Margaret Ann, bacteriology
McGinnis, John Thurlow, ecology, biostatistics
Meek, Devon Walter, inorganic chemistry
Meiling, Richard L., obstetrics & gynecology
Meites, Samuel, clinical chemistry
Meleca, Cosmo Benjamin, medical education
Melton, Carl Wesley, zoology, electron microscopy
Mendenhall, George David, physical organic chemistry
Merola, A John, biochemistry, microbiology
Merritt, Robert Edward, leather chemistry, information science
Messenger, John Cowan, cultural anthropology
Messier, Robert Louis, biochemistry
Metanomski, Wladyslaw Val, polymer chemistry, information science
Meyer, Bernard Sandler, botany

Meyer, Richard Lee, agricultural economics
Meyers, Leroy Frederick, mathematics
Mezey, Eugene Julius, inorganic chemistry
Michal, Edwin Keith, physiology
Mickle, Earl John, mathematics
Middleton, Arthur Everts, solid state physics, electronics
Mikolajcik, Emil Michael, dairy microbiology
Milford, Frederick John, physics
Miller, Duane Douglas, medicinal chemistry
Miller, James Frederick, fuel science
Miller, Joseph Nelson, parasitology
Miller, Richard Edward, physical chemistry, chemical physics
Miller, Richard Lloyd, entomology, horticulture
Miller, Robert Harold, soil science, microbiology
Miller, Robert W., agronomy
Mills, Robert Laurence, theoretical physics
Milo, George Edward, virology, biochemistry
Miskimen, Mildred, zoology
Mitchell, Rodger (David), population biology
Mitchell, Walter Edmund, Jr, astronomy
Moore, George Emerson, Jr, geology
Moore, Patricia Ann, biochemistry, physiology
Moore, Richard Owen, biochemistry
Morgan, James Frederick, nuclear physics
Morrison, David Lee, nuclear chemistry, physical chemistry
Mouk, Robert Watts, polymer chemistry, synthetic organic chemistry
Mourad, A George, geodesy, petroleum engineering
Mueller, Ivan I., geodesy, geophysics
Mulhausen, Hedy Ann, biochemistry, information science
Mulligan, Bernard, theoretical physics, nuclear physics
Murdick, Philip W., veterinary medicine
Murphy, Samuel G., cancer, oncology
Musgrave, Orlo Lynn, soils
Myers, William Graydon, medical biophysics, chemistry
Myser, Willard C., zoology
Naber, Edward Carl, poultry nutrition
Nagode, Larry Allen, veterinary pathology
Nathan, Richard Arnold, organic chemistry
Neher, Maynard Bruce, chemistry
Nelson, John White, pharmacology
Nelson, Reginald David, organic chemistry
Newsom, Gerald Higley, astronomy, atomic spectroscopy
Newton, William Allen, Jr, pediatrics, pathology
Nicholson, Richard Benjamin, nuclear physics
Nishikawara, Margaret T., physiology
Noltimier, Hallan Costello, geophysics
Noyes, David Holbrook, physiology
Nuenke, Richard Harold, biochemistry
Nutter, William Ermal, biochemistry
Ockerman, Herbert W., food chemistry, statistics
Ohta, Masao, physical chemistry
Olsen, Richard George, virology
Olson, Carter LeRoy, analytical chemistry
Otolenghi, Abramo Cesare, medical microbiology
Ouellette, Robert J., organic chemistry
Ouimet, Alfred J, Jr, information science, solid state science
Pace, William Greenville, surgery
Paddock, Elton Farnham, plant cytogenetics
Palmer, Dwight Miller, anatomy, neuropsychiatry
Pappas, Peter William, parasitology
Paquette, Leo Armand, organic chemistry
Parikh, Sarvabhaum Sohanlal, nuclear chemistry, analytical chemistry
Parker, Milton Marvin, psychiatry
Parks, Lloyd McClain, pharmaceutical chemistry

Parrett, Ned Albert, meat science
Parrish, Wayne, zoology
Parson, Louise Alayne, mathematics
Parsons, John Lawrence, agronomy
Partyka, Robert Edward, plant pathology
Patil, Popat N., pharmacology, physiology
Patten, George Phillip, geography
Paul, Lawrence Thomas, physiology
Paul, Ronald Stanley, physics
Peng, Andrew Chung Yen, food chemistry
Pepper, Paul Milton, mathematics
Perkins, Robert Louis, infectious diseases, internal medicine
Perlman, Philip Stewart, genetics, cell biology
Permar, Dorothy, dentistry
Peterle, Tony J., biology, ecology
Petrarca, Anthony Edward, organic chemistry, information science
Pettyjohn, Wayne A., hydrology
Pfister, Robert M., microbiology
Phillis, William Avery, III, entomology
Pieper, Heinz Paul, cardiovascular physiology
Pitzer, Russell Mosher, theoretical chemistry
Plaine, Henry Leroy, genetics
Platau, Gerard Oscar, organic chemistry
Platt, Robert Swanton, Jr, plant physiology
Plimpton, Rodney F., Jr, animal science, meat science
Ploughe, William D., nuclear physics
Pobereskin, Meyer, physical chemistry
Poirier, Frank Eugene, biological anthropology, behavioral anthropology
Pollack, J Dennis, microbiology, medical mycology
Ponomarev, Paul, mathematics
Popham, Richard Allen, botany
Powell, Charles Carleton, Jr, plant pathology
Powell, Warren Howard, organic chemistry
Powers, Jean Hensel, biostatistics
Powers, Thomas E, veterinary pharmacology
Price, John Worthington, vertebrate zoology
Pride, Douglas Elbridge, economic geology
Prior, John Alan, medicine
Protheroe, William Mansel, astronomy
Putnam, Loren Smith, ornithology
Raciszewski, Zbigniew, physical organic chemistry
Raghavan, Valayamghat, plant morphogenesis
Rahwan, Ralf George, pharmacology, toxicology
Rall, Jack Alan, medical physiology
Ralley, Thomas G., mathematics
Randall, John Reed, geography
Randels, James Bennett, computer science
Randles, Chester, microbiology
Rao, Kandarpa Narahari, physics, astrophysics
Rapp, Richard Henry, geodesy
Rask, Norman, agricultural economics
Ravely, Melville Fuller, organic chemistry
Ray, Dale Allen, plant genetics
Ray, Richard Schell, veterinary pharmacology
Ray-Chaudhuri, Dwijendra Kumar, applied mathematics, pure mathematics
Rayner, John Norman, geography, climatology
Reed, Randall R, animal physiology
Reed, William Robert, physical inorganic chemistry
Reeves, Roy Franklin, mathematics
Reibel, Kurt, physics
Reiner, Charles Brailove, pathology, pediatrics
Reisch, Kenneth William, horticulture
Renoll, Mary Wilhelmine, chemistry
Reuning, Richard Henry, pharmacology
Reynolds, Nancy Miller, dentistry
Rheins, Melvin S., microbiology, immunology
Richardson, Kathleen Schueller, organic chemistry
Richardson, Keith Erwin, physiological chemistry
Riedl, John Orth, Jr, mathematics
Rieske, John Samuel, biochemistry
Riley, William Robert, physics
Riner, John William, mathematics
Ritchie, Austin E., agriculture, agricultural education
Roark, Terry P., astronomy, astrophysics
Robinson, Kenneth Robert, organic chemistry
Rollins, Howard A, Jr, horticulture
Rose, Lawrence Lyon, computer science

Rosen, Samuel, microbiology
Rosenfeld, Irene, toxicology; pathological physiology
Ross, Arnold Ephraim, mathematics
Ross, Robert Talman, biophysical chemistry
Roth, Robert Earl, science education, natural resources
Rothenbuhler, Walter Christopher, zoology
Rothstein, Jerome, physics
Rowlett, Russell Johnston, Jr., information science
Ryerson, George Douglas, organic chemistry
Saidudin, Syed, endocrinology; reproductive physiology
St. Pierre, Ronald Leslie, histology, physiology
Saladin, Jimmie James, optometry, visual physiology
Salisbury, Rupert, pharmacy
Saltzer, Charles, mathematics
Sams, Richard Alvin, analytical chemistry
Saunders, William H., otolaryngology
Scanlan, Mary Ellen, organic chemistry
Schafer, George Miles, soils
Schick, Harold, natural resources
Schmidt, Berlie Louis, agronomy, soil conservation
Schmidt, Glen Henry, dairy husbandry
Schmidt, John Arvid, Jr., medical mycology
Schoessler, John Paul, optometry, visual physiology
Schopf, James Morton, paleobotany
Schram, Eugene P., inorganic chemistry
Schultz, James Edward, mathematics
Schumm, Dorothy Elaine, biochemistry
Schwab, George Lewis, bacteriology
Schwartz, Charles Marvin, physical chemistry
Schwartz, Daniel Manning, physics
Schwerzel, Robert Edward, photochemistry; physical organic chemistry
Sciulli, Paul William, biological anthropology; dental anthropology
Scott, Roy Albert, III, physical chemistry; molecular biology
Sedor, Edward Andrew, organic chemistry
Sehgal, Surinder K., algebra
Semple, Robert Keith, economic geography
Senhauser, Donald Albert, pathology
Serif, George Samuel, biochemistry
Seyler, Richard G., nuclear physics
Seymour, Roland Lee, mycology
Shafer, Thomas Howard, plant physiology
Shewell, John Robert, synthetic organic chemistry
Shilliday, Theodore Smith, experimental physics
Shore Sheldon Gerald, inorganic chemistry
Shaudys, Edgar T., agricultural economics
Shavit, Isaiah, theoretical chemistry
Shaw, John H., physics
Shechter, Harold, organic chemistry
Sheppard, William James, fuel science
Sherwood, Bob Edwin, chemistry
Siever, Carl Frank, chemistry
Simon, Ralph, physics
Sinclair, Richard Glenn, II, polymer chemistry
Skavaril, Russell Vincent, genetics
Skillman, Thomas G., medicine
Slatter, Walter LeClare, dairy technology
Sietebak, Arne, astrophysics
Sliemers, Francis Anthony, Jr., polymer chemistry
Smeck, Neil Edward, agronomy
Smith, Charles Roger, veterinary physiology
Smith, Charles Weistead, physiology
Snell, Junius Fielding, biochemistry
Snyder, Milton Jack, chemistry
Soifer, Morton Marshall, pediatrics
Sokoloski, Theodore Daniel, pharmacy, physical chemistry
Somerson, Norman L., medical microbiology

Sondheimer, Norman Keith, computer sciences
Sotos, Juan Fernandez, pediatrics
Speiser, Rudolph, physical chemistry
Spitzer, Jeffrey Chandler, organic chemistry; information science
Sprecher, Howard W., biochemistry
Srivastava, Ramesh C., statistics
Staby, George Lester, horticulture, plant physiology
Stairs, Gordon R., insect pathology
Stansbery, David Honor, zoology
Stanton, Noel Russell, high energy physics
Staubus, John Reginald, dairy science
Stephens, James Fred, poultry science
Stevens, Vernon Cecil, physiology
Stevenson, Thomas Dickson, medicine
Stickel, Philip Rice, meteorology
Stieglitz, Ronald Dennis, geology
Stobaugh, Robert Earl, organic chemistry
Stockwell, Charles Warren, neurosciences
Stoner, Elaine Carol Blatt, physical chemistry
Stow, Richard W., biophysics
Stroube, Edward W., agronomy
Stuckey, Ronald Lewis, botany
Stuessy, Tod Falor, systematic botany
Sucheston, Martha Elaine, developmental anatomy
Suie, Ted, microbiology; immunology
Summerson, Charles Henry, geology
Sunderman, Duane Neuman, research administration
Sutter, John Frederick, geology, geochronology
Sutton, Paul, soil fertility
Swanson, Carroll Arthur, plant physiology
Swartzentruber, Paul Edwin, organic chemistry
Sweet, Thomas Richard, chemistry
Sweet, Walter Clarence, geology; invertebrate paleontology
Swenton, John Stephen, physical organic chemistry
Swiger, Louis Andre, animal genetics
Swinehart, Philip Ross, solid state electronics
Sydnor, Thomas Davis, ornamental horticulture
Taaffe, Edward James, geography
Taft, Clarence Egbert, botany
Tanaka, Katsumi, physics
Tassava, Roy A., developmental biology; comparative endocrinology
Tate, Fred Alonzo, organic chemistry, information science
Tayama, Harry K., horticulture, floriculture
Taylor, George Stanley, soil physics
Taylor, Jack Neel, urology
Taylor, Thomas Norwood, paleobotany
Taylor, William Johnson, physical chemistry, theoretical chemistry
Teater, Robert Woodson, agronomy
Teteris, Nicholas John, obstetrics & gynecology
Tettenhorst, Rodney Tampa, mineralogy
Tharp, Vernon Lance, veterinary medicine
Theisen, Cynthia Theres, organic chemistry
Thomas, Donald C., virology
Thomas, Ralph Edward, operations research, statistics
Thompson, Victor Carl, apiculture, entomology
Thomson, William Alexander Brown, biochemistry, analytical chemistry
Throckmorton, Peter E., organic chemistry
Tikson, Michael, mathematics
Tjoe, Sarah Archambault, pharmacology
Tough, James Thomas, low temperature physics
Trautman, Milton Bernhard, zoology
Trimble, Harold Callander, mathematics
Triplehorn, Charles A., entomology
Trout, Dennis Alan, air pollution, meteorology
Troxel, Allen Wendell, plant pathology
Tull, Jack Phillip, mathematics
Turkel, Rickey Martin, organic chemistry
Turner, Edward V., pediatrics
Tynik, William John, animal nutrition
Tzagournis, Manuel, medicine, endocrinology
Underusch, William Charles, organic chemistry
Uniacke, Charles Allyn, visual physiology, optometry
Uotila, Urho Antti, geodesy
Utgard, Russell Oliver, geology, science education

Vahey, David William, optical physics, geometry
Valentine, Barry Dean, entomology; herpetology
Vander Stouw, Gerald Gordon, organic chemistry, information science
Van Winkle, Quentin, chemistry
Vasko, John Stephen, cardiovascular & thoracic surgery
Venard, Carl Ernest, zoology
Venzke, Walter George, veterinary anatomy
Verber, Carl Michael, optical physics
Verhoek, Frank Henry, physical chemistry
Vertrees, Robert Layman, economics, resource management
Visconti, James Andrew, pharmacy
Vivian, Virginia Ann, nutrition
Vleck, Donald Henry, electronics, research administration
Vogt, Charles Frederick, organic chemistry
Volk, Garth William, soil chemistry
von Haam, Emmerich, pathology
Wada, Walter W., physics
Wagner, Alan R., pathology
Watts, Bert Kerr, mathematics
Waldron, Acie Chandler, agronomy, entomology
Walker, Francis Edwin, agricultural economics
Walker, Joanne Gillespie, physiology
Wall, Robert Leroy, medicine
Walters, Craig Thompson, plasma physics, lasers
Walters, Martha I., clinical chemistry
Walum, Herbert, chemistry
Warner, John Scott, organic chemistry
Warren, James Vaughn, internal medicine
Watters, James I., chemistry
Weddleton, Richard Francis, polymer chemistry
Webb, Thomas Evan, biochemistry
Weiss, Harold Samuel, physiology
Wenden, Henry Edward, mineralogy
Westfall, Arthur Oscar, hydrology; agricultural engineering
Wharton, George Willard, Jr., zoology
White, John Joseph, III, solid state physics, applied physics
White, Sidney Edward, geology
Whitney, Donald Ransom, statistics, mathematical statistics
Whittingham, David James, organic chemistry
Whittle, Betty Ann, nutrition
Wigen, Philip E., solid state physics, magnetism
Wikoff, Helen Landman, biochemistry
Wilding, Lawrence Paul, soil science
Wilkinson, James Freeman, agronomy
Williams, Benjamin Hayden, anatomy, orthodontics
Williams, James Hutchison, obstetrics & gynecology
Williams, Mary Bearden, evolutionary biology, philosophy of science
Williams, Thomas Rhys, cultural anthropology; ethnography
Willis, Judith Ione, immunobiology; human anatomy
Wilke, Thomas Aloys, mathematical statistics
Wilson, Charles Marshall, food bacteriology
Wilson, George Porter, III, veterinary surgery; immunology
Wilson, George Rodger, agriculture, animal science
Wilson, Henry E., internal medicine, hematology
Wilson, Richard Ferrin, animal science education, agricultural economics
Wilson, Richard Michael, mathematics
Wing, Robert Farquhar, astronomy
Winter, Alden Raymond, poultry science
Winter, Chester Caldwell, medicine
Wismar, Beth Louise, anatomy, medical education
Witiak, Donald T., organic chemistry, medicinal chemistry
Woelfel, Julian Bradford, dentistry
Wojicki, Andrew V., inorganic chemistry, organometallic chemistry
Wolf, Harold Herbert, pharmacology
Wolff, David A., cell biology, virology
Wolken, George, Jr., chemical physics, theoretical chemistry
Wood, Van Earl, physics
Wukelic, George Edward, atmospheric physics, space physics
Wyman, Milton, mathematics
Wyman, Boswick Frampton, mathematics
Wyman, Milton, veterinary medicine, ophthalmology

Yaqub, Jill Courtaney Donaldson Spencer, geometry
Yashon, David, neurosurgery
Yeary, Roger A., veterinary pharmacology
Yohn, David Stewart, microbiology
Young, Sydney Sze Yih, quantitative genetics, population genetics
Yovits, Marshall Clinton, computer sciences
Yu, Greta Y., physics, operations research
Zaye, David Francis, information science, analytical chemistry
Zimmering, Shimshon, mathematical analysis
Zimpfer, Paul (Ellsworth), bacteriology
Zollinger, Robert Milton, surgery
Zuspan, Frederick Paul, obstetrics & gynecology
Zwilling, Bruce Stephen, immunobiology

COSHOCTON

Boyd, John Robert, analytical chemistry, spectroscopy
Chichester, Frederick Wesley, soil chemistry, plant physiology
Fausey, Norman Ray, soil physics
McGuinness, James L., hydrology
Mikesell, Sharell Lee, polymer chemistry
Pyle, James Johnston, plastics chemistry
Trewiler, Carl Edward, polymer chemistry, polymer agriculture

CUYAHOGA FALLS

Gates, Raymond Dee, polymer chemistry
Hamill, James Junior, chemical microscopy
Krueger, Robert A., organic chemistry
Murphy, Walter Thomas, organic chemistry, polymer chemistry

DAYTON

Adamczak, Robert L., physical chemistry
Alexander, C Alex, medical administration
Alf, Carol Jean, statistics, pure mathematics
Andrews, Merrill Leroy, plasma physics
Bajpai, Praphulla K, physiology; immunology
Bakan, Joseph A., chemistry administration
Bales, Howard E., spectroscopy, research
Barbour, Clyde D., ichthyology
Barbour, Prem Parkash, biochemistry
Batino, Rubin, physical chemistry
Beljan, John Richard, aerospace medicine, biomedical engineering
Berens, Alan Paul, mathematical statistics
Bertelson, Robert Calvin, synthetic organic chemistry
Besancon, Robert Martin, optical physics
Bigley, Nancy Jane, bacteriology
Birden, John Harlan, inorganic chemistry; immunology
Blackwell, Barry M., psychopharmacology
Brams, Stewart L., rubber chemistry
Broderick, Lynne Sechrist, microbiology, genetics
Bueche, Frederick Joseph, physics
Burky, Albert John, physiological ecology
Butler, John Mann, organic chemistry, polymer chemistry
Callan, Edwin Joseph, atomic physics
Canfield, James Howard, organic chemistry, information science
Carson, Albert B., mathematics
Chantell, Charles J., vertebrate anatomy
Chudd, Cletus Charles, organic chemistry
Conley, Robert T., organic chemistry, polymer chemistry
Cooney, Joseph Jude, microbiology
Coppage, William Eugene, algebra
Cothern, Charles Richard, nuclear physics
Cummings, Sue Carol, inorganic chemistry, bioinorganic chemistry
Dahm, Donald B., physical chemistry
Davis, Henry Werner, computer science
Detrio, John A., solid state physics
DeWall, Richard A., thoracic surgery, cardiovascular surgery
Dimopoullos, George Takis, microbiology
Dombrowski, Joanne Marie, mathematical analysis
Duff, Willard Moyle, physiology
Ericksen, Wilhelm Skjetstad, mathematics
Evers, Robert C., polymer chemistry

Eveslage, Sylvester Lee, organic chemistry
Fairheller, William Russell, Jr, environmental chemistry
Ferguson, Harry, mathematics, mechanics
Fernelius, Nils Conard, surface physics, experimental solid state physics
Fortman, John Joseph, inorganic chemistry
Fox, Bernard Lawrence, organic chemistry
Frey, Mary Anne Bassett, medical physiology
Fritz, Herbert Ira, nutrition, biochemistry
Furtado, Victor Cunha, health physics, environmental health engineering
Gehatia, Matatiahu T, physical chemistry, applied mathematics
Geiger, Donald R, plant physiology
Glaser, Roger Michael, exercise physiology
Glasgow, David Gerald, polymer chemistry
Goldman, Charles C, mathematics
Gotshall, Robert William, physiology, endocrinology
Gregor, Clunie Bryan, geology
Guderley, Karl Gottfried, mathematics, aerodynamics
Gustafson, Steven Carl, molecular spectroscopy, ultrasound
Haas, Trice Walter, physical chemistry
Haber, Robert Morton, mathematics
Haddox, Charles Hugh, Jr, clinical chemistry
Hagee, George Richard, nuclear physics, radiological physics
Hall, John Felix, Jr, physiology
Hanson, Harvey Myron, theoretical physics
Hardy, Edgar Erwin, organic chemistry
Harris, Richard Jacob, optical physics
Hay, Russell Earl, Jr, agronomy
Helminiak, Thaddeus Edmund, chemistry
Hemsky, Joseph William, nuclear physics
Hengehold, Robert Leo, solid state physics
Hess, George G, organometallic chemistry, analytical chemistry
Hewes, Cecil Gordon, anatomy
Heyd, Josef William, chemistry
Honda, Shigeru Irwin, cell physiology
Hou, Shou L, applied physics
Hubschman, Jerry Henry, invertebrate physiology
Hutchings, Brian Lamar, microbial biochemistry
Jaffee, Oscar Charles, experimental embryology
Jehn, Lawrence A, applied mathematics
Jensen, Richard Jorg, systematic botany, plant ecology
Johnston, George Taylor, optical physics
Jones, Edward Grant, chemical physics
Jones, John, Jr, mathematics
Kane, James Joseph, organic chemistry
Kantor, George Joseph, biophysics, molecular biology
Karl, David Joseph, physical chemistry
Keil, Robert Gerald, physical chemistry
Keller, Harold Willard, mycology
Kepes, Joseph John, nuclear & reactor physics
Kezdi, Paul, cardiology
Klein, Sherwin Jared, psychophysiology
Kmetec, Emil Philip, biological chemistry
Kolmen, Samuel Norman, physiology
Krishnaiah, Paruchuri Rama, statistics
Land, Peter L, solid state physics, ceramics
Laufersweiler, Joseph Daniel, ecology, botany
Lucier, John J, organic chemistry
MacEwen, James Douglas, toxicology
Malone, Philip Garcin, geochemistry
Maneri, Carl C, mathematics
Mann, Leonard Andrew, physics
Martino, Joseph Paul, operations research, science administration
McCloskey, John W, mathematical statistics
McDonald, Larry William, pathology, neuropathology
McDougall, Kenneth J, microbial genetics
McFarland, Charles R, microbiology
McMillin, Carl Richard, medical research

Medicus, Gustav Konrad, electron physics
Meese, Jon Michael, solid state physics
Michaelis, Carl I, organic chemistry
Mueller, Sabina Gertrude, botany
Nelson, Gilbert Harry, clinical chemistry
Nemergut, Paul Joseph, Jr, applied mechanics
Nixon, Charles William, acoustics, audiology
Noel, James A, geology
Noland, George Bryan, academic administration
Nussbaum, Noel Sidney, physiology, endocrinology
O'Hare, John Michael, theoretical solid state physics, experimental solid state physics
Park, Won Joon, mathematics, statistics
Pedrotti, Leno Stephano, lasers, optical physics
Phillips, Chandler Allen, biomedical engineering
Potoczny, Henry Basil, mathematics
Poynter, James William, physical chemistry
Pryor, Paul L, optical physics, military systems
Pushkar, Paul, geology, geochemistry
Quinn, Dennis Wayne, applied mathematics
Rab, Paul Alexis, zoology
Rake, Adrian Vaughan, biochemistry, behavioral biology
Ramsey, James Marvin, environmental physiology
Ray, John Robert, resource geography, remote sensing
Richard, Benjamin H, structural geology, geophysics
Rodin, Alvin E, pathology
Ross, Charles Burton, physics
Rossmiller, John David, biochemistry
Rutner, Emile, chemical physics
Sachs, David, mathematics
Schaefer, Donald John, computer science, mathematics
Schmidt, Ronald Grover, hydrogeology
Schneider, James Roy, physics
Schraut, Kenneth Charles, mathematics
Schwelitz, Faye Dorothy, cell biology
Seiger, Marvin Barr, genetics
Serve, Munson Paul, organic chemistry, biochemistry
Sherwin, Jo-Ann Major, geology
Singer, Sanford Sandy, biochemistry
Skees, Hugh Benedict, applied chemistry
Skinner, Gordon Bannatyne, physical chemistry
Smith, Michael James, analytical chemistry, environmental chemistry
Sonstein, Stephen Allen, microbiology
Spanier, Edward J, inorganic chemistry
Spencer, Walter William, clinical chemistry
Spjalter, Leonard, organic chemistry
Springer, George Henry, geology
Stahl, Saul, mathematics
Stander, Joseph W, algebra
Stansbrey, John Joseph, physical chemistry
Steinlage, Ralph Cletus, mathematics
Stuhlman, Robert August, laboratory animal medicine, medical research
Tamborski, Christ, organic chemistry, fluorine chemistry
Taylor, Michael Lee, analytical chemistry, medicinal chemistry
Taylor, Paul John, analytical chemistry
Thomas, Joseph Francis, Jr, solid state physics, materials science
Tiernan, Thomas Orville, chemical physics, analytical chemistry
Toman, Karel, crystallography
Van Deusen, Richard L, polymer chemistry
Von Gierke, Henning Edgar, bioacoustics, biomechanics
Webster, James Albert, organic chemistry
Williams, Patrick Kelly, ecology
Winslow, Leon E, mathematical analysis
Wolfe, Paul Jay, nuclear physics, exploration geophysics
Wood, David Roy, spectroscopy
Wood, Timothy Smedley, ecology
Wrist, Peter Ellis, physics, mathematics
Wu, Richard Li-Chuan, physical chemistry
Yaney, Perry Pappas, physics
Yanko, William Harry, radiochemistry, organic chemistry
Zechiel, Leon Norris, astrophysics, optics

DEFIANCE
Buccino, Raymond, Jr, clinical chemistry
deRoth, Gerardus Cabble, zoology
Frey, James R, bacteriology, immunology

Mikula, Bernard C, genetics, taxonomy
Miller, Harry Galen, physics

DELAWARE
Berry, Frederick Hamer, forest pathology
Bossert, Roy Garner, organic chemistry
Burns, George W, genetics
Burnside, Phillips Brooks, physics
Cannon, William Nelson, Jr, entomology
Crowl, George Henry, glacial geology, geomorphology
Decker, Jane M, plant anatomy
Dillman, Lowell Thomas, nuclear physics
Dochinger, Leon S, plant pathology
Donley, David Edward, entomology, zoology
Freed, James Melvin, cell physiology
Fry, Anne Evans, zoology, developmental biology
Fusch, Richard Dennis, urban geography
Ganis, Sam Eugene, mathematics
Gatz, Arthur John, Jr, zoology
Gregory, Garold Fay, phytopathology
Hahnert, William Franklin, invertebrate zoology
Harris, William N, geography
Hock, Winand Karl, plant pathology
Hull, David Lee, applied mathematics, statistics
Hull, John Winter, cytogenetics, plant breeding
Ichida, Allan A, mycology, bacteriology
Jensen, Keith Frank, forestry
Keenan, Philip Childs, astronomy
Maxwell, Howard Nicholas, physics
McQuigg, Robert Duncan, physical chemistry
Meek, Violet Imhof (Mrs Deveon W), inorganic chemistry
Mendenhall, Robert Vernon, mathematical logic
Patton, Wendell Keeler, invertebrate zoology
Peacock, John William, entomology
Peterson, Carl, theoretical chemistry
Radabaugh, Dennis Charles, animal behavior
Roberts, Bruce R, plant physiology
Ross, Eldon Wayne, forest pathology
Russell, Leonard Nelson, nuclear physics
Sanger, Jon Edward, aquatic ecology
Schreiber, Lawrence, plant pathology
Selsikar, Carl Edward, plant pathology
Shirling, Elwood Brent, bacteriology, microbiology
Staley, David H, mathematics
Stanger, Philip DeMott, astronomy
Stull, William Charles, plant genetics
Townsend, Alden Miller, tree physiology
Tuchinsky, Philip Martin, mathematics
Wick, Lawrence Bernard, organic chemistry
Wilcox, Harold Edwin, chemistry
Wilson, Lauren R, inorganic chemistry, environmental chemistry
Wilson, Robert Lee, mathematics

DELPHOS
Peltier, Leslie Copus, astronomy

DUBLIN
Nielsen, Carl Eby, physics

EAST CLEVELAND
Bidelman, William Pendry, astronomy
Datta, Ranajit K, solid state chemistry
Gross, Peter George, astrophysics
Grummitt, Oliver Joseph, organic chemistry
Hickok, Robert Lee, physical chemistry, materials science
Krieger, Irvin Mitchell, physical chemistry
MacLennan, Donald Allan, atomic physics
McCuskey, Sidney Wilcox, astronomy
Moyer, James Ward, inorganic chemistry
Wolff, Gunther Arthur, physical chemistry

EAST LIVERPOOL
Shreve, Gregory Monroe, anthropology, ethnography

ELMORE
Walsh, Kenneth Albert, chemical metallurgy, ceramics

ELYRIA
Schales, Otto, chemistry

EUCLID
Ricksecker, Ralph E, chemical metallurgy, metallurgical engineering

FAIRBORN
Chattoraj, Shib Charan, inorganic chemistry, analytical chemistry
Oestreicher, Hans Laurenz, mathematics
Weishaupt, Clara Gertrude, botany

FAIRVIEW PARK
Cahoon, Nelson Corey, chemistry
Delvigs, Peter, organic polymer chemistry

FAYETTEVILLE
Klee, Albert Joseph, statistics, chemical engineering

FINDLAY
Haller, Charles Regis, petroleum geology
Hewitt, Charles Hayden, geology, mineralogy
Schoonmaker, George Russell, geology

FREMONT
Schmidt, Walter Harold, agronomy

GAHANNA
Yantis, Richard P, mathematics, operations research

GAMBIER
Batt, Russell Howard, physical chemistry
Burns, Robert David, zoology
Finkbeiner, Daniel Talbot, II, mathematics
Greenslade, Thomas Boardman, Jr, physics, history of physics
Jegla, Thomas Cyril, comparative physiology, comparative endocrinology
Johnson, John Alan, nuclear physics
Lindstrom, Wendell Don, mathematics
McLeod, Robert Melvin, mathematical analysis
Miller, Franklin, Jr, physics, science education
Pappenhagen, James Meredith, analytical chemistry
Wohlpart, Alfred, plant physiology, molecular biology
York, Owen, Jr, organic chemistry
Yow, Francis Wagoner, embryology

GARRETTSVILLE
Spoehr, Albert Frederick, chemistry

GATES MILLS
Smalheer, Calvin Van Laaten, organic chemistry

GRANVILLE
Alrutz, Robert Willard, ecology
Archibald, Kalman Dale, ichthyology, embryology
Berg, James Irving, solid state science
Biefeld, Paul Franklin, physical chemistry
Bonar, Daniel Donald, mathematics
Bork, Kennard Baker, geology, paleontology
Brown, John Boyer, phycial chemistry, polymer chemistry
Chang, Robert Chi-Heng, polymer chemistry
Day, Frank, Jr, chemistry
De Gray, Ronald Willoughby, chemistry
Dewhurst, Harold Ainslie, physical chemistry
Doyle, Richard Robert, organic chemistry, biochemistry
Galloway, Gordon Lynn, inorganic chemistry
Gamble, Francis Trevor, physics
Gerdy, James Robert, developmental biology, genetics
Gilbert, George Lewis, inorganic chemistry
Goodman, Felicitas Daniels, anthropological linguistics, anthropology
Graham, Charles Edward, geology
Grant, Roderick M, Jr, solid state physics, medical physics
Grimm, George Walter, physics
Haubrich, Robert Rice, biology
Hoffman, William Andrew, Jr, analytical chemistry
Jalbert, Jeffrey Scott, nuclear physics

Johnson, Carl Arnold, organic chemistry
Johnston, Norman Wilson, polymer chemistry
Larson, Lee Edward, physics
Mahard, Richard Harold, geomorphology
Malcuit, Robert Joseph, petrology, astrogeology
McKinnis, Charles Leslie, chemistry
Mickelson, Michael Eugene, spectroscopy, planetary atmospheres
Norris, Gail Royal, radiation biophysics
Pettgrew, Raleigh K., mammalian physiology
Prentice, Wilbert Neil, numerical analysis
Prusaczyk, Joseph Edward, physical chemistry
Siefert, August Carl, inorganic chemistry
Spessard, Dwight Rinehart, organic chemistry
Sterrett, Andrew, mathematics
Strauss, Carl Richard, polymer chemistry
Stukus, Philip Eugene, microbiology, molecular biology
Trusty, Josann Watkins, surface physics
Wheeler, Samuel Crane, Jr., physics
White, Jerry Eugene, organic chemistry
Windle, William Frederick, anatomy
Winters, Ronald Ross, nuclear physics, astrophysics
Wismer, Robert Kingsley, x-ray crystallography

GROVE CITY
Smithson, George Raymond, Jr., inorganic chemistry

HAMILTON
Betten, Cornelius, Jr., chemistry
Clark, Robert Kenley, physics

HARTVILLE
Korver, Gailerd Lee, polymer chemistry

HILLIARD
Carver, Eugene Arthur, meteoritics

HIRAM
Andrews, John Timothy Sawford, physical chemistry, thermodynamics
Barrow, James Howell, Jr., protozoology, parasitology
Becker, Lawrence Charles, nuclear physics
Berg, Dwight Hillis, plant morphology
Denham, Joseph Milton, organic chemistry
Johnson, Wendell Gilbert, mathematics
Knight, Walter Rea, psychology
MacDowell, Robert W., mathematics
Pickford, Grace Evelyn, endocrinology, taxonomy
Rosser, Edward Barry, inorganic chemistry
Scalzi, Francis Vincent, organic chemistry
Slotterbeck, Oberta Ann, mathematics

HUDSON
Corbett, Gail Rushford, taxonomic botany
Dowell, Michael Brendan, high temperature chemistry
Hein, Richard William, industrial organic chemistry
Miller, Allan Stephen, solid state physics
Semegen, Stephen Thomas, chemistry
Weinstein, Arthur Howard, polymer chemistry, rubber chemistry

HURON
Hille, Kenneth R., limnology, aquatic toxicology

INDEPENDENCE
Baskin, Yehuda, geology, mineralogy
Harrington, Roy Victor, inorganic chemistry
Leckie, Donald Stewart, operations research, statistics
Sherman, Anthony Michael, polymer chemistry
Spindler, Donald Charles, analytical chemistry

KENT
Adams, Walter Church, endocrinology, physiology
Allender, David William, theoretical solid state physics
Anderson, Bryon Don, nuclear physics
Anderson, John Jerome, geology
Baker, John Warren, mathematical analysis, topology
Berger, Kenneth Walter, audiology
Bethel, Edward Lee, mathematics
Bhargava, Trilok Nath, applied statistics
Bordne, Erich Fred, geography
Brooks, Foster (Lindsey), chemistry
Brown, Glenn Halstead, chemistry
Brown, Richard Kettel, mathematics
Bush, Laurens Earle, mathematics
Buttlar, Rudolph O., inorganic chemistry
Byrne, John Maxwell, plant anatomy
Carlson, Ernest Howard, geology, mineralogy
Christensen, Stanley Howard, solid state physics
Cibula, Adam Burt, biology, entomology
Cleaver, Charles E., mathematics
Codding, Edward George, analytical chemistry
Coogan, Alan H., paleontology; geology
Cooke, George Dennis, aquatic ecology
Cooperrider, Tom Smith, plant taxonomy
Cummins, Kenneth Burdette, mathematics
De La Fuente, Rollo K., plant physiology
de Vries, Adriaan, X-ray crystallography, physical chemistry
Dexter, Ralph Warren, ecology; history of biology
Diestel, Joseph, mathematics
Doane, Joseph William, nuclear magnetic resonance
Dressler, Byron Brown, organic chemistry
Duffy, Norman Vincent, Jr., inorganic chemistry
Dutta, Hiram Moyee, anatomy, biology
Easterling, George Riley, zoology
Epstein, Bart Jacob, urban geography
Feinberg, Melvin Joel, chemical physics
Feldmann, Rodney Mansfield, invertebrate paleontology
Ferguson, Marion Lee, physiology
Fernelius, Willis Conard, chemistry
Fishel, Derry Lee, organic chemistry
Foote, Benjamin Archer, entomology
Fort, Raymond Cornelius, Jr., physical organic chemistry
Frank, Glenn William, geology
Franklin, Wilbur Mitchell, physics, metallurgy
Frederickson, Edward Arthur, geology
Fridy, John Albert, mathematics
Frieden, Edward Hirsch, biochemistry, endocrinology
Gallicchio, Vincent, zoology
Gelernter, Edward, chemical physics
Gesinski, Raymond Marion, cell biology
Gordon, John Edward, organic chemistry, physical chemistry
Gould, Edwin Sheldon, inorganic chemistry, organic chemistry
Graham, Alan Keith, plant morphology, paleobotany
Graham, Shirley Ann, botany, taxonomy
Greenberg, Mark Shiel, analytical chemistry
Harshbarger, Frances, mathematics
Heath, Robert Thornton, biophysics
Hemlich, Richard Allen, geology
Hobbs, Clinton Howard, botany
Hodgkins, Jordan Atwood, geography
Hubin, Wilbert N., solid state physics
Irmiter, Theodore Ferer, food chemistry
Jenkins, Emerson D., algebra
Knotts, Glenn Richard, public health
Liebelt, Annabel Glockler, cancer
Liebelt, Robert Arthur, anatomy, experimental pathology
Lovejoy, Owen, human biology, biomechanics
Mack, Richard Norton, plant ecology
Madey, Richard, nuclear physics
Manes, Milton, physical chemistry
Maxwell, Glenn, mathematics
McCandless, Byron Howard, mathematics
McComas, Murray Ratcliffe, geology
McGrath, James Williamson, physics
Moroi, David S., theoretical physics
Movius, William Gust, inorganic chemistry
Myers, Ralph Thomas, physical inorganic chemistry
Myers, Raymond Reever, chemistry
Neff, Vernon Duane, physical chemistry
Neuzil, John Paul, mathematics
Nicholson, Victor Alvin, mathematics
Olson, Stanley William, medicine
Orr, Lowell Preston, vertebrate ecology, herpetology
Powell, Robert Ellis, mathematics
Pruter, Olaf, anthropology
Pynadath, Thomas I., biochemistry, organic chemistry
Raup, Hallock Floyd, geography
Rhodes, Russell G., physiology
Riley, Charles Victor, biology; ecology
Ruch, Richard Julius, physical chemistry
Saupe, Alfred (Otto), physics, physical chemistry
Scheepfle, George Kern, physics
Shane, Orrin Clifton, anthropology
Silvidi, Anthony Alfred, physics
Smith, Frank A., mathematics
Spielberg, Nathan, physics
Stevenson, Joseph Ross, developmental biology, endocrinology
Stokes, Robert Mitchell, comparative physiology
Szmuc, Eugene Joseph, geology
Taylor, Jay Eugene, chemistry
Tuan, Debbie Fu-Tai, chemical physics
Ulrich, David Lee, physics
Varga, Richard S., mathematics
Walz, Frederick George, biochemistry
Warren, Kenneth Lyle, physics
Witten, Thomas Riner, experimental nuclear physics
Zobel, Herbert Lawrence, geography of the United States & Canada

KETTERING
Barr, David Ross, mathematical statistics
Emrick, Donald Day, organic chemistry
McGraw, Hugo Richard, organic chemistry

LAKEWOOD
Dragt, Gerrit, analytical chemistry
Enger, Carl Christian, biomedical engineering
James, Milton, electrochemistry
Johannes, Karl A., mathematics, statistics
Kordesch, Karl Victor, electrochemistry
Theusch, Colleen Joan, number theory, operations research

LANCASTER
Fox, Robert Kriegbaum, physical inorganic chemistry
Gray, William Dudley, mycology
Wilson, Larry Eugene, analytical chemistry

LIMA
Colombini, Victor Domenic, geology

LOVELAND
Brunzie, Gerald Franklin, analytical chemistry

LOWELL
Moynihan, Robert Edward, physical chemistry

LYNDHURST
Rynasiewicz, Joseph, analytical chemistry

MANSFIELD
Allen, Raymond Clayton, floriculture
Bart, George James, plant pathology
Berra, Tim Martin, ichthyology, zoogeography
Bowling, Arthur Lee, Jr., theoretical high energy physics
Rathmanna, Dasara V., physical chemistry

MARIETTA
Anderson, Robert Lester, physics
Brown, William Paul, genetics, animal physiology
Capps, Raymond Haul, chemistry
Chey, Tong Chull, physics
Davis, William Howard, physics
Dyar, Robert Matthew, organic chemistry
Fraser, Robert B., mathematics
Gilde, Hans-Georg, organic chemistry
Grose, Herschel Gene, organic chemistry
Hohman, William H., inorganic chemistry
Knecht, Laurance A., analytical chemistry
Porter, Galen, chemistry
Punderson, John Oliver, organic chemistry; polymer science
Seyler, Paul Jacob, zoology
Steel, Warren G., geology

MARION
Albernaz, Jose Geraldo, neurosurgery, neuroanatomy
Aubrecht, Gordon James, II, high energy physics
Pettijohn, Terry Frank, animal behavior
Tarvin, Donald, chemistry
Thompson, Arthur Carsten, physical chemistry
Walp, Russell Lee, botany
Wright, Robert Paul, paleoecology, invertebrate paleontology

MARYSVILLE
Husaini, Saeed A., food science
Schery, Robert Walter, botany

MASSILLON
Wehrle, Louis, Jr., food microbiology

MAUMEE
Berning, Peter H., mathematics
Powell, Homer Eugene, physics

MENTOR
Emrich, William Oscar, organic chemistry
Lewandowski, Thaddeus, bacteriology

MIAMISBURG
Attala, Albert, physical chemistry, analytical chemistry
Barnett, Albert Gerald, environmental physics, health physics
Blanke, Bertram Charles, physical chemistry
Bowman, Robert Clark, Jr., chemical physics
Carfagno, Daniel Gaetano, physical chemistry, inorganic chemistry
Cave, William Thompson, physical chemistry
Chong, Clyde Hok Heen, analytical chemistry
Essig, Gustave Alfred, physics
Fushimi, Fred Chikashi, computer science
Haas, Francis Xavier, Jr., nuclear physics
Hartzel, Lawrence Woodring, organic chemistry
Hastings, James Donald, nuclear chemistry, physical chemistry
Ivey, Jerry Lee, physical chemistry
Jaeger, Ralph R., inorganic chemistry
Johnston, Charles Paul, inorganic chemistry
Kershner, Carl John, physical chemistry, radiochemistry
Kokenge, Bernard Russell, inorganic chemistry
Lemming, John Frederick, nuclear physics
Lonadier, Frank Dalton, physical chemistry, inorganic chemistry
Love, Calvin Miles, inorganic chemistry
McConville, George T., physics
Otto, George W., thermal physics
Rembold, Eugene Albert, physical chemistry
Rogers, Donald Richard, inorganic chemistry
Seabaugh, Pyrtle W., analytical chemistry, applied statistics
Smith, Warren Harvey, physical chemistry
Sullenger, Don Bruce, solid state chemistry
Wiedenheft, Charles John, chemistry
Wilkes, William Roy, cryogenics
Wittenberg, Layton Junior, inorganic chemistry

MIAMIVILLE
Adams, Philip Delmar, biochemistry
Majors, Paul Alexander, bacteriology
Qusno, George L., microbiology, biochemistry
Thomas, McCalip Joseph, chemistry

MIDDLEBURG
Thomas, Alexander Edward, III, organic chemistry; analytical chemistry

MIDDLETOWN
Bendure, Robert J., inorganic chemistry, pollution chemistry

Bergstrom, David Wallace, zoology
Ikenberry, Luther Curtis, analytical chemistry
Koch, Ronald Joseph, physics
Mark, Earl Larry, physical chemistry
Molnar, Stephen P., organic chemistry, inorganic chemistry
Vian, Richard W., geology
Woodruff, Joseph Franklin, instrumentation, materials science

MILAN
Tiedjens, Victor Alphons, horticulture

MOGADORE
Cooper, Jack Loring, polymer chemistry
Cox, William Lester, physical chemistry, organic chemistry

MT ST JOSEPH
Deasy, Clara Louise, biochemistry
Deiters, Rose Mary, inorganic chemistry
Gonzalez, Paula, science education, human anatomy
Sferra, Pasquale Richard, entomology

MT VERNON
Lashley, Gerald Ernest, numerical analyses

NEW CONCORD
Bradford, James McClellan, physics
Crandell, Merrell Edward, physics
Dasch, Clement Eugene, entomology, taxonomy
Frye, Charles Isaac, sedimentary petrology, stratigraphy
Griffith, Gordon Lamar, physics
Knight, Lyman Coleman, mathematics
Kovach, Jack, geology
Landolt, Robert George, organic chemistry
Meyers, James Harlan, geology
Nieman, George Carroll, physical chemistry, spectroscopy
Quinn, David Lee, neuroendocrinology, reproductive physiology
Saksena, Vishnu P., zoology, aquatic biology
Smith, James L., mathematics
Wallace, William J., inorganic chemistry
Zettel, Larry Joseph, mathematics

NEWARK
Czepyha, Chester George Reinhold, atmospheric physics
St John, Fraze Lee, zoology

NORTH CANTON
Bebb, Robert Lloyd, organic chemistry
Thomson, Dale S., developmental biology

NORTH ROYALTON
Eckstein, Bernard Hans, physical chemistry

NORTHFIELD
Badger, George Franklin, biostatistics
Semon, Waldo Lonsbury, chemistry

NORTHWOOD
Gorski, Theodore William, biochemistry, microbiology
Greenberg, William Michael, applied physics

NORTON
Doran, Thomas J, Jr, organic chemistry

NORWOOD
Schumacher, Roy Joseph, organic chemistry

OBERLIN
Ackermann, Martin Nicholas, inorganic chemistry
Anderson, David Leonard, physics
Andrews, George Harold, mathematics, numerical analysis
Baum, John Daniel, mathematics
Benzing, David H., biology
Bromund, Werner Hermann, analytical chemistry, organic chemistry
Brummett, Anna Ruth, embryology, cytology

Carlton, Terry Scott, theoretical chemistry
Craig, Norman Castleman, physical chemistry
Egloff, David Allen, ecology
Foreman, Helen Pulver, micropaleontology
Fuchsman, William Harvey, bioinorganic chemistry
Goldberg, Samuel, mathematics
Hawkins, Peter Jack, physical chemistry
High, Lee Rawdon, Jr, geology
Hilborn, Robert Clarence, atomic physics, lasers
Levin, Richard Alexander, microbial physiology, genetics
Luck, Dennis Noel, molecular biology
Mittleman, Don, mathematics, computer science
Palmieri, Joseph Nicholas, nuclear physics
Powell, James Lawrence, geochemistry, petrology
Renfrow, William Burns, Jr, organic chemistry
Richards, Richard Peter, paleoecology
Richards, Walter Bruce, nuclear physics
Rosenweig, Abraham, crystallography, mineralogy
Schoonmaker, Richard Clinton, physical chemistry
Scott, George Taylor, physiology
Sherman, Thomas Fairchild, biology
Skinner, William Robert, petrology, structural geology
Snider, Joseph Lyons, atomic physics, astrophysics
Vance, Elbridge Putnam, mathematics
Walker, Warren Franklin, Jr, zoology
Warner, Robert Edson, physics
Weinstock, Robert, mathematical physics
Yinger, John Milton, anthropology

OXFORD
Allenspach, Allan Leroy, developmental biology
Anderson, Thomas Brown, speech pathology, audiology
Arfken, George Brown, Jr, physics
Baldwin, Arthur Dwight, Jr, geology
Barnhart, Donald Delbert, medical microbiology
Barrett, Gary Wayne, ecology
Bever, James Edward, geology
Bhattacharjee, Jnanendra K, microbial genetics
Boesel, Marion Waterman, entomology, zoology
Bohn, Sherman Elwood, mathematics
Bolger, Edward M, mathematics
Brady, Robert James, microbiology
Breidenbach, Gearold Peter, virology
Buckingham, John Herbert, physical chemistry
Bullock, Robert M, III, mathematics
Burke, Dennis Keith, topology
Cantrell, Joseph Sires, physical chemistry, solar physics
Capel, Charles Edward, mathematics
Chang, Luke Li-Yu, mineralogy
Chung-Phillips, Alice, theoretical chemistry, quantum chemistry
Claussen, Dennis Lee, physiological ecology
Cobbe, Thomas James, forest ecology
Cox, Milton D, algebra, geometry
Daniel, Paul Mason, zoology
Deonier, D L., entomology, ecology
DeVillez, Edward Joseph, comparative physiology, comparative biochemistry
De Vita, Joseph Michael, ecology
Eicher, John Harold, organic chemistry
Eshbaugh, William Hardy, plant taxonomy, ethnobotany
Fiehler, Harlan Edward, analytical chemistry
Gass, Frederick Stuart, mathematics
Gordon, Gilbert, inorganic chemistry
Grassmick, Robert Alan, protozoology, parasitology
Gregg, Thomas G, genetics
Griffin, Charles Campbell, biochemistry
Griffing, David Francis, solid state physics
Guttman, Sheldon, zoology
Hayes, Robert Edward, protozoology
Hefner, Robert Arthur, zoology
Heimsch, Charles W, plant anatomy, morphology
Houk, Thomas William, biophysics
Ingersoll, Edwin Marvin, zoology
James, Floyd Laem, organic chemistry
Jantzen, Carl Raymond, cultural anthropology
Jaworski, Jan Guy, biochemistry
Julian, Glenn Marcenia, nuclear physics

Karipides, Anastas, inorganic chemistry, physical chemistry
Katon, John Edward, physical chemistry
Kelly, Donald C, theoretical physics
Kochan, Ivan, immunology
Koehler, Anne Bramble, algebra
Koehler, Donald Otto, pure mathematics
Krause, Paul Frederick, physical chemistry
Kullman, David Elmer, mathematics
Laatsch, Richard G, mathematical analysis
Lang, Andrew George, paper chemistry
Limper, Karl Esslinger, micropaleontology
Lloyd, Howell Clevenger, geography
Macklin, Philip Alan, quantum mechanics, astronomy
MacMasters, William Joseph, plant physiology
Martin, Wayne Dudley, geology
Mattox, Karl, phycology
McClure, Jerry Weldon, plant chemistry
McWilliams, Robert Gene, stratigraphy
Mundell, Percy Meldrum, organic chemistry
Newman, David William, plant physiology
Nicholson, Nancy Lynne, natural sciences, botany
Nielson, Read R., physiology
Norman, John William, chemistry
Palmer, Robert Alexander, reproductive physiology
Park, Chull, mathematics, aeronautical engineering
Payne, Stanley E, mathematics
Peck, Lyman Colt, mathematics
Perino, Janice Vinyard, plant ecology, physiological ecology
Peterson, Robert C, pulp chemistry, paper chemistry
Pfohl, Ronald John, developmental biology
Phillips, David Berry, physical chemistry
Plybon, Benjamin Francis, mathematical physics
Pope, John Keyler, paleontology, geology
Poth, James Edward, nuclear physics
Priest, Joseph Roger, nuclear physics
Pyle, James L, organic chemistry, environmental chemistry
Rayle, Richard Eugene, genetics
Reinhart, Roy Herbert, geology, vertebrate paleontology
Rockwood, Susan Williams, microbiology
Schuurmann, Frederick James, mathematics
Scotford, David Matteson, geochemistry, petrology
Sebastian, John Francis, physical organic chemistry, biochemistry
Simmel, Edward Clemens, behavioral genetics, animal behavior
Thompson, John Leslie, environmental sciences
Smith, Joseph Donald, biochemistry
Smith, Richard Vergon, urban geography, population geography
Smith, Robert Sefton, mathematics
Snider, John William, physics
Stewart, David Perry, geomorphology
Stueber, Alan Michael, geochemistry
Taylor, Douglas Hiram, ethology
Travis, Dennis Michael, botany, genetics
Treick, Ronald Walter, microbial genetics
Tung, John Shih-Hsiung, mathematics
Vaughn, Charles Melvin, parasitology, protozoology
Vaughn, Jack C, biochemical cytology
Weidner, Bruce Vanscoyoc, chemistry
Weller, Harry, zoology
Whitten-Wolfe, Barbara L, mathematical physics
Wiley, Ronald Lee, respiratory physiology, pulmonary physiology
Williamson, Clarence Kelly, microbiology
Wilson, David Franklin, neurophysiology
Wilson, Thomas Kendrick, plant morphology
Wilson, William Ernest, botany, plant pathology
Winner, Robert William, ecology
Wissing, Thomas Edward, fresh water ecology

PAINESVILLE
Battershell, Robert Dean, pesticide chemistry
Bimber, Russell Morrow, organic chemistry
Bluestone, Henry, organic chemistry
Bodson, Herman, physical chemistry, physics
Chin, Wei Tsung, agricultural chemistry

Cryberg, Richard Lee, organic chemistry
Dietrich, Joseph Jacob, organic chemistry
Fordham, James Lynn, physical chemistry, research administration
Gibbins, Betty Jane, chemistry, science education
Harris, Richard Lee, organic polymer chemistry
Holm, Robert Eric, plant physiology, biochemistry
Kluth, Fred Carl, public health
Ku, Han San, plant physiology, biochemistry
Magee, Thomas Alexander, organic chemistry
Malkin, Irving, inorganic chemistry, electrochemistry
Marks, Alfred Finlay, agricultural chemistry
McCain, George Howard, organic chemistry
Metz, Fred Lewis, organic chemistry
Moser, Robert E, bio-organic chemistry
O'Leary, Kevin Joseph, polymer chemistry
Pollack, Norman Mark, bio-organic chemistry, medicinal chemistry
Powers, Larry James, medicinal chemistry
Richner, Donald Roosevelt, geology
Scozzie, James Anthony, organic chemistry
Sieglaff, Charles Lewis, polymer chemistry, colloid chemistry
Taylor, Kirman, analytical chemistry, clinical chemistry
Wise, Benjamin Nathan, cell biology, developmental biology
Worthington, James Brian, analytical chemistry

PARMA
Barr, John Baldwin, physical chemistry
Bright, Arthur Aaron, experimental solid state physics
Cessna, John Curtis, organic chemistry
Coulter, Paul (David Todd), analytical chemistry, applied chemistry
Jones, Creighton Clinton, physics
Laughlin, Ethelreda R, biochemistry, science education
Lewis, Richard Thomas, physical chemistry
Meers, Joseph Tinsley, research administration
Powers, Robert Allen, research administration
Schaefer, Hugh Ferdinand, chemistry
Smith, Robert Emery, solid state physics
Van Lier, Jan Antonius, physical chemistry

PERRYSBURG
Howald, Jeremiah Mark, organic chemistry

PIKETON
Kaplan, Roy Irving, physical inorganic chemistry
Saraceno, Anthony Joseph, inorganic chemistry
Seufzer, Paul Richard, inorganic chemistry, research administration
Wohlfort, Sam Willis, analytical chemistry, materials science

POLAND
Foldvary, Elmer, organic chemistry
MacLean, Bonnie Kuseske, entomology

REYNOLDSBURG
Roach, William Kenney, entomology

RICHFIELD
Ramp, Floyd Lester, organic chemistry

ROCKY RIVER
Dull, Raymond Broadwell, physics
Jucaitis, Pranas Francis, chemistry
Weitz, John Hills, mineralogy, economic geology

ROSS
Bissett, Donald Lynn, biochemistry

ST BERNARD
Schneider, Charles Aloysius, organic chemistry

SANDUSKY
Franks, John Anthony, Jr., organic chemistry
Ross, John Edward, industrial hygiene, health physics

SHAKER HEIGHTS
Burt, Gerald Dennis, organic chemistry
Chace, Frederic Mason, mining geology
Garst, Josephine Burgis, chemistry
Kline, Irene Tabitha, biochemistry
Ludwig, Jerome Howard, organic chemistry
Selker, Milton Leonard, physical organic chemistry
Selover, Theodore Briton, Jr., physical chemistry

SOLON
Durand, Edward Allen, inorganic chemistry, science writing
Knipple, Warren Russell, industrial chemistry, petroleum chemistry
Moss, Robert Henry, physical inorganic
Nestor, Ontario Horia, physics

SOUTH EUCLID
Guth, Sylvester Karl, physics, illumination

SPRINGFIELD
Bolls, Nathan J, Jr., vertebrate physiology
Bush, Everett Homer, geography
Byers, Walter Hayden, physics
Curry, Howard Millard, organic morphology
deLanglade, Ronald Allan, plant morphology
Dice, Stanley Frost, mathematics
Gerrard, Thomas Aquinas, geology
Glasoe, Paul Kirkwold, chemistry
Hageboeck, Myron Paul, experimental physics
Hahn, Samuel Wilfred, mathematics
Hitt, John Burton, cell biology
Houchin, Ollie Boyd, biochemistry
Lutz, Arthur Leroy, nuclear physics
Mason, David Lamont, botany
Morris, Robert William, geology, invertebrate paleontology
Nave, Floyd Roger, geomorphology, paleontology
Pickering, Ed Richard, plant physiology
Powelson, Elizabeth Eugenie, genetics
Rendon, Leandro, physiology, biochemistry
Sartoris, Nelson Edward, organic chemistry
Shaffer, Charles Franklin, immunology, transplantation biology
Westneat, David French, analytical chemistry
Wilson, Eric LeRoy, mathematics
Woodard, Ralph Emerson, research administration, reactor physics

STEUBENVILLE
Campbell, Thomas Hodgen, mycology
Cerroni, Rose E., biology, physiology
Convery, Robert James, organic chemistry
Madson, Willard Hegland, chemistry
Slater, James Louis, inorganic chemistry

STOW
Harwood, Harold James, polymer chemistry
Hauenstein, Jack David, polymer chemistry, physical chemistry
Nielsen, Stuart Dee, organic chemistry, polymer chemistry
Simpson, Billy Doyle, organic chemistry
Studebaker, Merton Leland, rubber chemistry, colloid chemistry
Svetlik, Joseph Frank, rubber chemistry

STRONGSVILLE
Dunning, John Walcott, organic chemistry
Hoert, Charles William, physical chemistry
Jacobs, Harvey, physical chemistry, organic chemistry
Koch, Stanley D., organic chemistry
Kuo, Cheng-Yih, polymer science
McGinniss, Vincent Daniel, polymer chemistry
Nieman, Theodore Frank, polymer chemistry, computer science
Norman, George Russel, organic chemistry
O'Neill, Richard Delos, bacteriology
Provder, Theodore, physical chemistry
Woo, James T K, organic chemistry

SUNBURY
Buckingham, William Thomas, communications science, plant breeding

SYLVANIA
Carlson, William Samuel, geology, meteorology
Hibbits, James Oliver, analytical chemistry
Muchow, Gordon Mark, physical chemistry

TALLMADGE
Anderson, John Norton, polymer chemistry

TIFFIN
Baker, David Bruce, plant physiology
Barlow, George, physiology
Cordell, Richard William, analytical chemistry
Groce, John Wesley, biochemistry
Hintz, Howard William, zoology, entomology
Lilly, Percy Lane, plant taxonomy
Reno, Martin A., chemical physics
Steele, William F., mathematics

TOLEDO
Anderson, Marion C., surgery
Bachman, Kenneth Allen, pharmacology
Bacon, Frank Rider, physical inorganic chemistry
Bailey, James L., applied mathematics
Basile, Robert Manlius, physical geography
Bass, Robert Eugene, physics
Bennett, Richard Thomas, plastics chemistry
Bentley, Herschel Lamar, chemistry
Berning, Jean Ackerman, computer sciences
Bishop, David Wakefield, physiology
Black, Arthur Herman, analytical chemistry
Blakemore, William Stephen, surgery
Block, Paul, Jr., organic chemistry
Bohn, Randy G., low temperature physics, cryogenics
Brady, Leonard Everett, organic chemistry
Brundage, Donald Keith, organic chemistry
Budd, Geoffrey Colin, cell biology, physiology
Burnham, Jeffrey C., microbiology
Burow, Duane Frueh, structural chemistry, physical chemistry
Charlesworth, Lloyd James, Jr., sedimentology
Chen, Shui-Chin, clinical chemistry, toxicology
Chidambaraswamy, Jayanthi, mathematics
Chrysochoos, John, physical chemistry
Claybrook, James Russell, biological chemistry
Clifford, Donald H., veterinary surgery
Conrad, Malcolm Alvin, mineralogy
Crisp, Michael Dennis, quantum optics, lasers
Curtis, Lorenzo Jan, atomic physics
Davis, Wilford Lavern, research administration, applied statistics
Deck, Robert Thomas, quantum optics
Delsemme, Armand Hubert, astrophysics
DiDio, Liberato John Alphonse, anatomy, electron microscopy
Diehn, Bodo, biophysical chemistry, radiochemistry
DuBrul, Ernest, biochemistry, developmental biology
Dunipace, Donald William, physics
Edwards, Jimmie Garvin, high temperature chemistry
Ellis, David Greenhill, theoretical physics
Emery, Byron Elwyn, geography
Faber, Lee Edward, endocrinology, biochemistry
Fechner, Robert Bernard, organic chemistry
Ferguson, Shirley Martha, psychiatry
Foster, Alfred Field, physical chemistry
Foster, Edward Stanford, Jr., applied physics
Foster, Harold Marvin, industrial chemistry
Fraleigh, Peter Charles, aquatic ecology
Francel, Josef, chemistry
Freimer, Earl Howard, microbiology, infectious diseases
Fry, James Leslie, organic chemistry
Gano, James Edward, organic chemistry, photochemistry
Garg, Mohan Lal, medical statistics
Garlid, Keith David, pharmacology
Gatten, Robert Edward, Jr., environmental physiology, comparative physiology
Glatzer, Louis, microbial genetics, molecular biology
Godar, Edith Marie, organic chemistry, analytical chemistry
Golden, Alfred, pathology
Goldman, Stephen L., genetics
Gray, Don Norman, organic polymer chemistry
Hageage, George John, Jr., medical microbiology
Hall, Ronald Henry, spectrochemistry, analytical chemistry
Hanson, Daniel James, pathology
Hatfield, Craig, geology
Herliczek, Siegfried H, organic polymer chemistry
Heyman, Laurel Elaine, inorganic chemistry
Hider, Shibley A., organic chemistry
Hoffman, Lawrence Arnes, geography
Howard, John Malone, surgery
Huff, Norman Thomas, theoretical chemistry
Jabarin, Saleh Abd El Karim, polymer chemistry
Jackson, Robert Franklin, mathematical physics
Jacobs, Richard Lee, organic chemistry
Judis, Joseph, biochemical pharmacology
Jyung, Woon Heng, plant physiology
Kemph, John Patterson, psychiatry, physiology
Kertz, George J., mathematics
Kneller, William Arthur, engineering geology, economic geology
Kollen, Wendell James, surface physics
Krohn, Albertine, chemistry
Kruse, Ferdinand Hobert, physical chemistry
Kummer, Martin, mathematical physics, applied mathematics
Kunkle, George Robert, geology, hydrology
Lee, Harold Hon-Kwong, biology, virology
Lee, Haynes A., lasers, glass technology
Levin, Jerome Allen, biochemical pharmacology
Lewis, Donald W., economic geography
Li, Chi-Tang, physical chemistry, crystallography
Long, Calvin Lee, biochemistry
Manning, Maurice, biochemistry, chemistry
Mayers, Richard Ralph, nuclear physics
Mayor, Stephen Joseph, neurophysiology
McCorquodale, Donald James, biochemistry
McGrady, Angele Vial, physiology
McGraw, Delford Armstrong, physics
McNamara, Michael Joseph, community health, preventive medicine
Montgomery, Charles Gray, theoretical physics
Morris, Lucien Ellis, anesthesiology
Morse, Dennis Ervin, human anatomy
Moss, L Howard, III, microbiology, virology
Mulrow, Patrick J., medicine
Muraco, William Anthony, urban geography
Nasatir, Maimon, cell biology
Nelson, George Francis, Jr., physical chemistry
Nelson, Leonard, physiology, reproductive physiology
Nikodem, Robert Bruce, physical chemistry, inorganic chemistry
Nolan, James Francis, physics
Ogg, Frank Chappel, Jr., applied mathematics
Page, Robert Griffith, medicine
Pansky, Ben, anatomy
Parker, Gordon Arthur, analytical chemistry
Patrick, James R., pediatrics, pathology
Pearlmutter, Anne Frances, biochemistry
Peterson, James Oliver, organic chemistry, fluorine chemistry
Pinheiro, Marilyn Lays, neurosciences, audiology
Poplawsky, Robert P., physics
Pribor, Donald B., cell physiology, cryobiology
Rawat, Arun Kumar, biochemistry
Rayport, Mark, neurosurgery
Reimann, Erwin M., biochemistry
Ritzert, Roger William, biochemistry
Ross, James Neil, Jr., cardiovascular physiology, veterinary medicine
Ruedisili, Lon Chester, hydrogeology, environmental geology
Ryan, Joseph Dennis, organic chemistry
Saffran, Judith, biochemistry
Saffran, Murray, biochemistry
Santelli, Thomas Robert, organic chemistry
Saul, Frank Philip, biological anthropology, anatomy
Schectman, Richard Milton, atomic physics
Schlender, Keith K., biochemistry, pharmacology
Schradie, Joseph, pharmacognosy
Senitzer, David, immunology
Shade, John William, geochemistry, mineralogy
Shields, Paul Calvin, pure mathematics
Shoemaker, Richard W., mathematics
Shore, Richard Eugene, developmental biology
Simon, Henry John, solid state physics
Singh, Iqbal, orthopedic surgery
Smith, Clifford James, animal physiology
Soloff, Melvyn Stanley, endocrinology
Spielberg, Stephen E., mathematics
Sporek, Karel Frantisek, analytical chemistry, organic chemistry
Stein, Junior, mathematics
Steinberg, Stuart Alvin, mathematics
Taylor, Lynn Johnston, chemistry, polymer chemistry
Thomas, Lazarus Daniel, physical chemistry
Thompson, Herbert Bradford, structural chemistry
Thompson, Lancelot Churchill Adalbert, inorganic chemistry
Tidrick, Robert Thompson, surgery
Trexler, Bryson Douglas, Jr., hydrogeology
Vayo, Harris Westcott, applied mathematics
Walmsley, Frank, inorganic chemistry
Walmsley, Judith Abrams, inorganic chemistry
Walters, Jack Henry, obstetrics, gynecology
Watson, Barry, biophysical chemistry
Weinberg, David Samuel, polymer chemistry
Weinberg, Michael C., statistical mechanics
Wente, Henry Christian, mathematics
Wenzel, Richard Louis, public health administration, preventive medicine
White, Peter, hematology
Wiband, John Truax, petrology
Williams, Elmer Lee, physical chemistry
Williamson, William, Jr., atomic physics, astrophysics
Witt, Adolf Nicolaus, astrophysics
Wu, Ting-Chi, biochemistry
Yanof, Howard Merar, biophysics, physiology

TWINSBURG
Uy, Oscar Manuel, physical chemistry, high temperature chemistry

UNION
Watrous, Ralph Melvin, physics

UNIONTOWN
Wise, Richard Melvin, organic chemistry

UNIVERSITY HEIGHTS
Cummings, Jean Marie, biology
Gaul, Richard Joseph, organic chemistry
Spitler, Ernest George, chemistry

UPPER ARLINGTON
Fuller, James Osborn, geology
Kistheimer, John Robert, agricultural chemistry

URBANA
Klinck, Ross Edward, physical chemistry

VIENNA
Parker, Bernard, microscopy

WADSWORTH
Holland, William Frederick, industrial chemistry
Kreider, Leonard Cale, chemistry

WATERVILLE
Mayfield, Harold Ford, ornithology

WAVERLY
Woltz, Frank Earl, physical chemistry

WEST ALEXANDRIA
Gage, Frederick Worthington, chemistry

WESTERVILLE
Deever, David Livingstone, pure mathematics
Herschler, Michael Saul, embryology, cytogenetics
Ogle, Pearl Rexford, Jr, inorganic chemistry, physical chemistry
Phinney, George Jay, vertebrate ecology
Place, Robert Daniel, inorganic chemistry, physical chemistry
Tegenkamp, Thomas Richard, biology, genetics
Willis, Jeanne Eleanor, paleobotany

WESTLAKE
Brannen, William Thomas, Jr, organic chemistry
Fox, Thomas Allen, reactor physics
Hatfield, Marcus Rankin, electrochemistry

WHITEHOUSE
Martin, Glenn Ellis, physical chemistry, polymer
Meier, James Archibald, polymer chemistry

WICKLIFFE
Grieshammer, Lawrence Louis, analytical chemistry, physical chemistry
Vogel, Paul William, organic chemistry
Zalar, Frank Victor, organic chemistry, polymer chemistry

WILLOUGHBY
Coleman, Lester Earl, (Jr), organic chemistry, polymer chemistry

WILMINGTON
Bayless, Philip Leighton, chemistry
Parker, James Willard, ornithology, ecology

WOOSTER
Barta, Allan Lee, agronomy, crop physiology
Beane, Donald Gene, mathematics
Berglund, Donna Lou, inorganic chemistry
Berry, Stanley Z, plant breeding
Blake, Roland Charles, horticulture
Bohl, Edward Homer, veterinary virology
Borders, Charles LaMonte, Jr, biochemistry
Bromund, Richard Hayden, analytical chemistry
Brown, James Harold, forest ecology
Brown, Keith Irwin, poultry science, reproductive physiology
Byers, Floyd Michael, animal nutrition
Clements, Robert Lawrence, food biochemistry
Conrad, Harry Russell, nutrition
Coplin, David Louis, plant pathology
Coyle, Elizabeth Eleanor, biology, botany
Cropp, Frederick William, III, geology, palynology
Cross, Robert Franklin, veterinary pathology
Davis, Richard Richardson, agronomy
Dehority, Burk Allyn, ruminant nutrition, microbiology
Dollinger, Elwood Johnson, cytogenetics, plant breeding
Donoho, Clive Wellington, Jr, horticulture, plant physiology
Eby, Robert L, agriculture
Elwell, David Leslie, low temperature physics
Ferguson, LLoyd C, microbiology
Ferree, David C, horticulture, pomology
Findley, William Ray, Jr, plant breeding
Fobes, Melcher Prince, mathematical analysis
Gallander, James Francis, food science
Gingery, Roy Evans, biochemistry, plant physiology
Haghiri, Faz, agronomy
Hall, Franklin Robert, entomology
Hampton, Charles Robert, algebra
Haynes, Leroy Wilbur, organic chemistry
Herr, Leonard Jay, plant pathology
Hibbs, John William, animal nutrition
Hill, Robert George, Jr, horticulture
Hinton, Claude Willey, genetics
Hotink, Harry A J, plant pathology, bacteriology
Jeffers, Daniel L, agronomy, plant physiology
Jones, Brian McCoy, plant pathology
Jones, James Edward, veterinary medicine
Kawase, Makoto, plant physiology, horticulture
Kieffer, William Franklin, physical chemistry
Klein, Michael Gardner, entomology
Klosterman, Earle Wayne, animal husbandry
Knoke, John Keith, entomology
Kohler, Erwin Miller, veterinary microbiology
Koucky, Frank Louis, Jr, mineralogy, geochemistry
Kretchman, Dale Warren, horticulture
Kriebel, Howard Burtt, forest genetics
Ladd, Thyril Leone, Jr, entomology
Lafever, Howard N, plant breeding, genetics
Larson, Merlyn Milfred, forest physiology
Leben, Curt (Charles), plant pathology
Lindquist, Richard Kenneth, entomology
Louie, Raymond, plant pathology
Mahan, Donald Clarence, animal nutrition
McClenahen, James Richard, environmental biology, forest ecology
McDowell, Theodore C, floriculture, horticulture
Mederski, Henry John, plant physiology
Moke, Charles Burdette, geology
Moore, Earl Neil, veterinary medicine
Moorhead, Philip Darwin, veterinary pathology
Moxon, Alvin Lloyd, biochemistry, nutrition
Murray, Finnie Ardrey, Jr, reproductive physiology
Musick, Gerald Joe, entomology
Nault, Lowell Raymond, entomology
Nestor, Karl Elwood, genetics
Niehaus, Merle Hinson, agronomy
Nielsen, David Gary, entomology
Niemczyk, Harry D, entomology
Palmquist, Donald Leonard, animal nutrition
Perley, James E, plant physiology
Powell, David Lee, physical chemistry
Redman, Donald Roger, veterinary medicine, veterinary surgery
Reinheimer, John David, organic chemistry
Rings, Roy Wilson, entomology
Robertson, Jack Alex, mycology, phycology
Rogers, Charles Fletcher, agricultural biochemistry
Rowe, Randall Charles, plant pathology
Saif, Yehia Mohamed, veterinary microbiology
Schanbacher, Floyd Leon, biochemistry
Schmittenner, August Fredrick, plant pathology
Shambaugh, George Franklin, entomology
Smith, Clyde Konrad, veterinary microbiology
Smith, Kenneth Larry, dairy science, immunobiology
Spotts, John Allen, plant physiology
Streeter, John Gemmil, plant physiology, agronomy
Treece, Robert Eugene, economic entomology
Triplett, Glover Brown, Jr, agronomy
Van Doren, David Miller, Jr, agronomy, soil physics
Van Keuren, Robert W, agronomy
Vimmerstedt, John P, soil science, forestry
Vogt, Albert R, plant physiology
Warner, John Ward, Jr, mathematics
Weaver, Andrew Albert, entomology
Weaver, Clyde Richard, biometry
Weidensaul, T Craig, plant pathology
Whitmore, Frank William, forest physiology
Willett, Lynn Brunson, animal physiology, dairy science
Williams, Lansing Earl, plant pathology
Williams, Theodore Roosevelt, analytical chemistry
Wilson, Charles L, cell physiology
Wise, Donald L, cell physiology
Yamazaki, William Toshi, cereal chemistry
Zimmerman, Tommy Lynn, soil conservation

WORTHINGTON
Brown, William Anderson, inorganic chemistry
Chapman, Floyd Barton, horticulture
Clark, Burr, Jr, agricultural biochemistry
Crenin, Howard Lee, food science
Hickman, Howard Minor, organic chemistry
Irwin, William Elliot, biochemistry, food technology
Lehr, Jay H, hydrology, groundwater geology
Link, William Edward, analytical chemistry
Robinson, Radcliffe Franklin, biology, food technology
Stephen, William Archibald, apiculture
Toeniskoetter, Richard Henry, industrial chemistry
Zassenhaus, Hans J, mathematics

WRIGHT-PATTERSON AFB
Anderson, Wayne Jay, optical physics
Back, Kenneth Charles, pharmacology
Braun, Wolfgang G, physics
Buell, Glen R, organic chemistry, analytical chemistry
Case, Carl Tyler, plasma physics
de Treville, Robert T P, occupational medicine
Dorko, Ernest A, organic chemistry, physical chemistry
Downing, Reginald Harton, mathematics
Dunlap, Duane Sherbert, operations research
Ehlers, Gerhard Friedrich Louis, polymer physics, polymer science
Eisentraut, Kent James, analytical chemistry
Harter, Harman Leon, mathematical statistics
Hartrum, Thomas Charles, bioengineering
Holt, James Franklin, magneto hydrodynamics
Johnson, Daniel Leon, bioacoustics
Kaplan, Bernard, mathematical physics
Kararian, Leon Edward, biomechanics
Kellerstrass, Ernst Junior, geophysics, civil engineering
Kissell, Kenneth Eugene, astronomy, space physics
Kissen, Abbott Theodore, physiology
Langer, Dietrich Wilhelm Josef, solid state physics
Loughran, Girard Andrew, Sr, organic chemistry
Mavco, George Edward, astronomy, optics
Mundie, J Ryland, medicine
Nielsen, Philip Edward, plasma physics, solid state physics
Nikolai, Paul John, mathematics
Park, Yoon Soo, solid state physics
Pickett, David Franklin, Jr, electrochemistry, electrochemical engineering
Pinson, Ernest Alexander, biophysics
Poirier, Charles Philip, nuclear physics
Replogle, Clyde R, bioengineering
Sanderson, Richard Blodgett, physics
Slonim, Arnold Robert, biochemistry, physiology
Spicer, John A, mathematics
Spry, Robert James, solid state physics
Tamburino, Louis Anthony, mathematical physics
Warren, Richard Hawks, mathematics
Weichel, Hugo, physics

XENIA
Carpenter, Robert Francis, molecular chemistry
Pershing, William Raymond, optics
Prince, Alton Ernest, biology
Shaw, Elwood R, analytical chemistry

YELLOW SPRINGS
Armstrong, Marvin Douglas, biological chemistry
Asakawa, George, physical chemistry
Bernstein, Stanley Carl, physical organic chemistry
Bieri, Robert, oceanography, ecology
Blau, Julian Herman, mathematics
Burling, Richard Lancaster, physics
Churchill, Edmund, mathematics
Corbin, James Lee, organic chemistry
Darrow, Robert A, enzymology
Fleischman, Darrell Eugene, biophysics
Garner, Albert Y, organic chemistry
Hertzberg, Hans Theodore Edward, physical anthropology, anthropometry
Houston, William Bernard Jr, geometry
Israel, Harry, III, dentistry, anatomy
Keister, Donald Lee, microbial biochemistry
Kleineberg, Gerd A, analytical chemistry
Knecht, Walter Ludwig, physics
Lacey, Beatrice Cates, psychophysiology
Lacey, John Irving, psychophysiology, neurophysiology
Lamborg, Marvin, biochemistry
Loud, Oliver Schule, history & philosophy of science
Mayne, Berger C, plant physiology
McConville, John Theodore, physical anthropology
Murie, Martin L, vertebrate zoology
Newton, William Edward, inorganic chemistry
Peters, Gerald Alan, plant physiology
Pueppke, Steven Glenn, phytopathology, plant biochemistry
Raved, Dan, plant physiology,
Reporter, Minocher C, developmental biology, biochemistry
Robertson, Richard Thomas, neurobiology
Roche, Alexander F, child growth, anthropometrics
Samuel, Edmund William, developmental biology
Seely, Gilbert Randall, physical chemistry
Silverman, Robert, mathematics
Sokolowsky, Daniel, mathematics
Stewart, Albert Burns, experimental atomic physics
Taylor, Charles Emory, nuclear magnetic resonance
Tulecke, Walter, botany
Turoff, Robert David, solid state physics
Ucko, David A, inorganic chemistry, bioinorganic chemistry
Varandani, Partab T, biochemistry, endocrinology
Webb, Paul, medical physiology, environmental medicine
White, John Francis, geology
Whitesell, William James, theoretical physics
Williams, Francis Trueman, organic chemistry
Williams, Robert Ellis, biology
Yalman, Richard George, inorganic chemistry

YOUNGSTOWN
Anton, John Ralph, urban geography
Bishop, Edwin Vandewater, physics
Bridgham, Catherine Mitchell, biochemistry
Buoni, John J, mathematical analysis
Cochran, William Ronald, physics
Cohen, Irwin, organic chemistry, structural chemistry
Dalbec, Paul Euclide, solid state physics
Del Bene, Janet Elaine, quantum chemistry
Dobbelstein, Thomas Norman, analytical chemistry
Ducey, Paul Richard, anthropology, history of anthropology
Faires, John Douglas, mathematical analysis
Gebelein, Charles G, polymer chemistry, organic chemistry

Hanzely, Stephen, physics
Harris, Ann Graetsch, geology
Harris, C Earl, Jr., geology
Karas, James Glynn., zoology
Kelley, George W, Jr., zoology
Khawaja, Ikram Ullah, economic geology
Kiniazis, James William, anthropology, ethnography
Klein, Albert Jonathan, topology
Koknat, Friedrich Wilhelm, inorganic chemistry, x-ray crystallography
Kreutzer, Richard D., cytogenetics
Krill, Karl Emil, academic administration
Lateef, Abdul Bari, forensic science, chemistry
Lukin, Marvin, organic chemistry
MacLean, David Belmont, insect ecology; insect taxonomy
Mahadeviah, Inally, nuclear chemistry, inorganic chemistry
McLennan, Donald Elmore, electrodynamics
Mettee, Howard Dawson, spectrochemistry
Moorhead, William Dean, theoretical physics
Peterson, Paul Constant, acarology; systematics
Rand, Leon, organic chemistry
Santos, Eugene (Sy), mathematics, computer science
Schidcrout, Steven Michael, physical chemistry
Schroeder, Lauren Alfred, ecology
Smith, Francis White, physical chemistry, analytical chemistry
Smith, Robert Kingston, inorganic chemistry
Sobota, Anthony E., plant physiology
Van Norman, John Donald, analytical chemistry, clinical chemistry
Van Zandt, Paul Doyle, parasitology
Von Ostwalden, Peter Weber, organic chemistry
Whipkey, Kenneth Lee, mathematics, statistics
Yemma, John Joseph, cytochemistry
Yingst, Arthur Grady, inorganic chemistry
Young, Warren Melvin, astronomy
Yozwiak, Bernard James, mathematics

ZANESVILLE
Brumbaugh, Richard J., chemistry
Marshall Anne (Corinne), biological structure

OKLAHOMA

ADA
Carter, William Alfred, zoology, ornithology
Dunlap, William Joe, organic biochemistry, environmental chemistry
Eddington, Carl Lee, biochemistry
Enfield, Carl George, soil physics
Hornsby, Arthur Grady, soil physics, soil chemistry
Law, James Pierce, Jr., water pollution
Love, Harry Schroeder, Jr., botany, ecology
McKnight, Thomas John, parasitology
Moyer, James Earl, microbiology
Shew, Delbert Craig, mass spectrometry
Weems, Malcolm Lee Bruce, physics

ALVA
Nighswonger, Paul Floyd, plant ecology
Phelps, Jack, mathematics
Rogers, Stearns Walter, biochemistry, organic chemistry
Shorter, Daniel Albert, entomology, zoology

ARDMORE
Bates, Richard Pierce, agronomy, plant breeding
Dell'Orco, Robert T., cell biology
Kampschmidt, Ralph Fred, biochemistry
Kizer, Donald Earl, biochemistry
Leu, Richard William, microbiology, immunology
Merriman, Charles Richard, biochemistry
Patterson, Manford Kenneth, Jr., biochemistry, cancer

BARNSDALL
Coffman, Harold H., chemistry

BARTLESVILLE
Anderson, Paul Dean, analytical chemistry
Axe, William Nelson, organic chemistry
Bailey, F Wallace, physical chemistry
Bailey, Grant Carter, petroleum chemistry
Baldwin, Bernard Arthur, physical chemistry, molecular spectroscopy
Ball, John Sigler, petroleum chemistry
Borst, Roger Lee, mineralogy
Bost, Howard William, organic chemistry
Bresson, Clarence Richard, applied chemistry
Campbell, Robert Wayne, polymer chemistry
Carr, Donald Eaton, chemistry
Childs, William Ves, fluorine chemistry, electrochemistry
Cleary, James William, organic chemistry, polymer chemistry
Cobb, Raymond Lynn, organic chemistry
Collins, Arlee Gene, geochemistry
Cook, Charles Falk, nuclear physics, physics
Cruzan, Charles Grant, engineering
Doss, Richard Courtland, applied chemistry
Dreiling, Mark Jerome, physics
Dwiggins, Claudius William, Jr., physical chemistry
Eastman, Alan Dan, inorganic chemistry
Eccleston, Barton Henry, air pollution, fuel science
Eilerts, Charles Kenneth, numerical analysis, petroleum engineering
Elliott, Sheldon Ellwood, exploration geophysics, applied mathematics
Erdman, John Gordon, petroleum, organic geochemistry
Fahey, Darryl Richard, organometallic chemistry
Farrar, Ralph Coleman, organic chemistry
Fenstermaker, Roger William, chemistry
Finch, Jack Norman, physical chemistry
Findlay, Robert Artemas, petroleum chemistry
Fodor, Lawrence Martin, polymer chemistry
Fozzard, George Broward, organic chemistry
Furrow, Clarence Lee, organic chemistry
Gabriel, Henry, inorganic chemistry, physical chemistry
Gall, James William, physical chemistry
Gardner, John Omen, electron microscopy
Gregg, Robert Quinly, spectroscopy
Guillory, Jack Paul, physical chemistry
Gunnell, Thomas Jefferson, physical chemistry
Hardage, Bob Adrian, earth sciences
Harris, Jesse Ray, physical chemistry, ceramic engineering
Hartzfeld, Howard Alexander, organic chemistry
Heckelsberg, Louis Fred, physical chemistry
Hill, Harold Wayne, Jr., polymer chemistry
Hitzman, Donald Oliver, bacteriology
Hogan, John Paul, organic chemistry
Holmes, Clifford Newton, geology
Holtz, Hans Dietrich, physical organic chemistry
Hoover, Gary McClellan, physics
Horton, Robert Louis, physical chemistry
Hsieh, Henry Lien, polymer chemistry
Hughes, William Bond, inorganic chemistry
Iorns, Terry Vern, analytical chemistry
Janzen, Jay, physical chemistry
Jayaraman, H, organic chemistry
Johansen, Robert Torolf, physical chemistry
Johnson, Wallace Delmar, organic chemistry
Johnston, Harlin Dee, chemistry
Jones, Faber Benjamin, polymer chemistry
Kleinschmidt, Roger Frederick, organic chemistry, petroleum technology
Kraus, Gerard, physical chemistry
Lanning, William Clarence, physical chemistry
Lauffer, Donald Eugene, physics
Lawson, Bob Leroy, chemistry
Lorenz, Philip Boalt, physical chemistry
Mahan, John Elmer, organic chemistry
Mark, Harold Wayne, physical organic chemistry
McCoy, Raymond Duncan, physical chemistry, instrumentation

McDaniel, Max Paul, surface chemistry
Michalowski, Joseph Thomas, geochemistry
Miller, John Walcott, analytical chemistry
Mills, King Louis, Jr., petroleum chemistry
Moczygemba, George A., polymer chemistry, inorganic chemistry
Morgan, Thomas David, physics
Morris, David Albert, exploration geology
Mosher, Loren Cameron, petroleum geology, paleontology
Mowery, Richard Allen, Jr., analytical chemistry
Naylor, Floyd Edmond, polymer chemistry
Narell, John Reynolds, organic chemistry
O'Shaughnessy, Marion Thomas, Jr., physical chemistry, polymer chemistry
Owen, James Robert, physical chemistry
Paxson, John Ralph, analytical chemistry
Pitchford, Armin Cloyst, inorganic chemistry
Pritchard, James Edward, chemistry
Randall, James Carlton, Jr., physical chemistry
Rascoe, Bailey, Jr., geology
Reid, James Alexander, physical chemistry
Ripley, Dennis Leon, chemistry
Runnels, John Hugh, chemistry
Rycheck, Mark Rule, inorganic chemistry
Schiff, Sidney, organic chemistry
Schmidt, Thomas William, chemical physics
Smith, Donald Charles, physics
Smith, James Edward, geochemistry, geophysics
Sonnenfeld, Richard John, organic chemistry
Scott, Donald William, physical chemistry
Schorno, Karl Stanley, organic geochemistry
Selman, Charles Melvin, organic polymer chemistry
Shell, Francis Joseph, physical chemistry
Short, James N., polymer chemistry, research administration
Shotton, James Arthur, organic chemistry
Shue, Robert Sidney, polymer chemistry, organometallic chemistry
Souder, Wallace William, nuclear physics, geophysics
Stacy, Carl J., polymer science
Tanner, Joseph Jarratt, petroleum geology
Todd, Samuel Spaulding, physical chemistry, thermochemistry
Tomaja, David Louis, organometallic chemistry
Trepka, William James, polymer chemistry, rubber chemistry
Tucker, Paul William, petroleum, natural resources
Udipi, Kishore, polymer chemistry
Waldrop, Morgan A, solid state physics, atomic physics
Wegner, Gene H, petroleum microbiology
Welch, Melvin Bruce, inorganic chemistry, polymer chemistry
Wilp, Elmar Konstantin, organic chemistry
Winter, William Kenneth, physics
Witt, Donald Reinhold, chemistry
Zelinski, Robert Paul, polymer chemistry, rubber chemistry
Zuech, Ernest A., organic chemistry

BETHANY
Beaver, W Don, chemistry
Greer, Earl Vincent, mathematics
Heasley, Gene, chemistry
Reinbold, Paul Earl, analytical chemistry, inorganic chemistry
Walker, Keith Gerald, molecular physics
Young, Sharon Clairene, zoology

BROKEN ARROW
Johnson, James Franklin, geophysics

CHICKASHA
Frere, Maurice Herbert, watershed management, water pollution

DEL CITY
Broadie, Larry Lewis, neuropharmacology

DEWEY
Clark, Bill Paz, semiconductors, solid state physics
Stratton, Charles Abner, colloid chemistry; science education

DUNCAN
Banks, William Patrick, electrochemistry
Frost, Jackie Gene, chemistry
Holtmyer, Martin Dean, petroleum chemistry; polymer chemistry

DURANT
Davis, Robert Jaquette, Jr., soil microbiology
Dwight, Leslie Alfred, mathematics
Hazell, Don Bliss, range management, plant ecology
Hibbs, Leon, mathematics, academic administration
Kilpatrick, Earl Buddy, fish biology
Krattiger, John Trubert, mathematics
Menzel, Ronald George, soil chemistry
Olness, Alan, water pollution, soil science
Polson, William Jerry, nuclear physics
Robinson, Jack Landry, analytical chemistry
Taylor, Raymond John, vertebrate zoology, ichthyology
Taylor, Constance Elaine Southern, ecology; systematic botany
Wade, William Frank, analytical chemistry
Wright, John Ricken, chemistry

EDMOND
Anderson, Roger Clark, botany, ecology
Bogenschutz, Robert Parks, physiology; biology
Boyce, Donald Joe, mathematics
Carlstone, Darry Scott, particle physics
Cox, Beverley Lenore, zoology, physiology
Derrick, Grace Ethel, embryology
Fosberg, Mary Dee Harris, computer science
Frazier, Floyd Wendell, crop breeding
Frey, Delton Ruben, chemistry
Hamm, Thomas Edward, organic chemistry; biochemistry
Harden, Virginia Pauline, bacteriology
Hocker, Reginald Orson, microbiology
Hornuff, Lothar Edward, Jr., entomology
King, John Paul, solid state physics
Loman, Laverne, mathematics
Marks, Luther Whitfield, III, physics
Rachlin, Carol King, anthropology, ethnology
Rice, Earl Clifton, mathematics
Richardson, Vertin Homer, chemistry; science education
Russell, Norman Hudson, Jr., biology
Trout, Verdine Elza, science education, ecology
Yoesting, Clarence C., physics, science education

ENID
Everly, Charles Ray, organic chemistry
Horton, Philip Bish, solid state physics
Lewis, B Kenneth, inorganic chemistry
Mason, Lysle C., mathematics
Murphy, Marjory Beth, cell physiology; biochemistry
Williams, Cecil R., biology, chemistry

EL RENO
Furry, Benjamin K., rubber chemistry

GOODWELL
Cutter, Paul Ramey, physical chemistry
England, Milton (W), animal breeding
Gardner, Richard Lynn, biochemistry
Martin, Jerry Junior, animal nutrition, animal physiology
Peck, Raymond A., field crops
Ramon, Serafin, cytogenetics
Reeves, Homer Eugene, agronomy

GUTHRIE
Farmer, James, applied mathematics
Thomas, Sarah Nell, physiology, radiation biology

GUYMON
Shults, Mayo Glenwood, mathematics

LANGSTON
Brinkerhoff, Lloyd Allen, plant pathology
Hawxby, Keith William, plant physiology, aquatic biology
Jones, Walter Larue, entomology
Venere, Ralph Joseph, Sr, plant pathology
Wall, Charles Ephraim, chemistry

LAWTON
Harwood, William H, physical chemistry
McKellips, Terral Lane, mathematics
Tyler, Jack D, ornithology, ecology
Wagner, Harry Mahlon, mathematics

MIAMI
Cunningham, Frank W, chemistry
Lux, Carl Ray, nuclear chemistry

NORMAN
Alberty, Ronnie Lee, meteorology
Amsden, Thomas William, invertebrate paleontology
Andree, Richard Vernon, mathematics
Atkinson, Gordon, physical chemistry, inorganic chemistry
Babb, Stanley Ernest, Jr, physics
Bastian, Joseph, neurobiology
Beevers, Leonard, plant physiology
Bell, Robert Eugene, archaeology, physical anthropology
Bernhart, Arthur, mathematics
Bittle, William Elmer, ethnology, anthropological linguistics
Blatt, Harvey, petrology
Bissitt, Charles W, pharmacy
Boke, Norman Hill, plant anatomy, morphology
Bourassa, Ronald Ray, solid state physics
Branch, David Reed, astrophysics
Branson, Carl Colton, stratigraphy
Braver, Gerald, genetics
Breen, Marilyn, geometry
Brixey, John Clark, mathematics
Broersma, Sybrand, molecular physics
Brown, Harley Procter, invertebrate zoology, entomology
Brown, Rodger Alan, meteorology
Brown, Vivia Jean, pharmacy
Burr, John Green, photochemistry, radiation chemistry
Burwell, James Robert, elementary particle physics, high energy physics
Carpenter, Charles Congden, zoology
Carpenter, Mary Pitynski, biochemistry
Chartock, Michael Andrew, ecology, science policy
Christian, Sherril Duane, physical chemistry
Cieszko, Leon Stanley, biochemistry
Clark, James Bennett, microbial physiology
Clemens, Howard Paul, zoology
Cohn, Jack, theoretical physics, electrodynamics
Cox, Donald Cody, virology, molecular biology
Cozad, George Carmon, medical mycology, immunology
David, Paul Rembert, zoology
Davies-Jones, Robert Peter, meteorology
Donahue, Hayden Hackney, psychiatry
Doviak, Richard J, electromagnetism, meteorology
Du Bois, Robert Lee, geophysics
Eddy, George Amos, meteorology
Eliason, Stanley B, mathematical analysis
Estes, James Russell, systematic botany
Ewing, George McNaught, mathematics
Feyock, Stefan, computer sciences
Fiorica, Vincent, physiology
Fischbeck, Helmut J, physics
Fletcher, John Samuel, plant physiology, cell biology
Fogel, Norman, physical inorganic chemistry
Fowler, Richard Gildart, physics
Frech, Roger, physical chemistry
Friedman, Samuel Arthur, geology
Frings, Hubert William, comparative physiology
Frings, Mable Ruth, animal behavior
Gilliland, Martha Winters, environmental sciences
Goff, Richard Allen, experimental embryology
Golden, David E, atomic physics, spectroscopy

Gonzalez-Arce, Teofilo Francisco, computer science
Goodman, George Jones, plant taxonomy
Goodman, James Marion, geography
Greer, John Keever, mammalogy, ecology
Hagen, Arnulf Peder, inorganic chemistry
Haines, Howard Bodley, zoology, physiology
Harper, Charles Wood, Jr, invertebrate paleontology
Harris, Loyd Ervin, pharmaceutical chemistry
Harrison, William Earl, organic geochemistry
Hellack, Jenna Jo, zoology
Hill, Loren Gilbert, ichthyology, ecology
Hopla, Cluff Earl, medical entomology
Howard, Robert Adrian, physics
Hoy, Harry Eugene, geography
Huber, Wolfgang Karl, psychiatry
Huff, William Nathan, mathematical analysis
Huffaker, James Neal, physics
Huffman, George Garrett, geology
Huneke, Harold Vernon, mathematics
Hutchison, Victor Hobbs, physiological ecology
Johnson, Kenneth Sutherland, geology
Kaul, Pushkar Nath, pharmacology, clinical pharmacology
Kay, David Clifford, geometry
Kessler, Edwin, III, meteorology
Kitts, David Burlingame, vertebrate paleontology, geology
Klehr, Edwin Henry, water chemistry
Kraynak, Matthew Edward, nutrition
Lancaster, John, microbial genetics
Larsh, Howard William, medical mycology
Lee, Jean T, geophysics, meteorology
Lehr, Roland E, organic chemistry
Loewen, Kenneth Leroy, mathematics
Magarian, Robert Armen, medicinal chemistry, organic chemistry
Magid, Andy Roy, mathematics
Mankin, Charles John, geology
Marchand, Alan Philip, physical organic chemistry
McCarthy, John, meteorology
McDonald, Bernard Robert, mathematics
Merrill, James Allen, obstetrics & gynecology, pathology
Morris, John Wesley, geography
Murphy, George Washington, physical chemistry
Murphy, Juneann Wadsworth, medical microbiology
Myers, Arthur John, geomorphology
Neely, Stanley Carrell, physical chemistry
Norden, John Alexander, geology, geophysics
Nostrand, Richard Lee, geography
Olson, Ralph Eugene, geography
Opler, Morris Edward, anthropology
Peavey, Jerris Hinkins, plant physiology
Pento, Joseph Thomas, endocrinology, pharmacology
Petry, Robert Franklin, nuclear physics
Ray, Peter Sawin, meteorology
Reid, William Thomas, mathematical analysis
Renner, John Wilson, science education, physics
Rice, Elroy Leon, botany
Ritzman, Carl Harry, / speech pathology
Rohrbaugh, Lawrence Milburn, plant physiology
Roller, Duane Henry DuBose, history of science
Rose, William Dake, petroleum geology
Rubin, Leonard Roy, topology
St John, Robert Mahard, experimental atomic physics
Sasaki, Yoshi Kazu, dynamic meteorology
Schaefer, Joseph Thomas, meteorology
Schiel, Joseph Bernard, Jr, biogeography
Schindler, Charles Alvin, microbiology, biochemistry
Schmitz, Francis John, natural products chemistry, marine chemistry
Schnell, Gary Dean, zoology
Self, Frank, zoology
Seto, Frank, zoology
Shay, Dennis (John), physics
Smith, Eddie Carol, biochemistry
Smith, Kirby Campbell, mathematics
Sommers, Ella Blanche, pharmacy
Sorenson, William George, mycology
Springer, Charles Eugene, mathematics
Stoever, Edward Carl, Jr, structural geology, science education
Sutherland, Patrick Kennedy, paleobiology, stratigraphy
Sutton, George Miksch, ornithology
Thies, Roger E, psychosomatic medicine

Thompson, Gary Lynn, economic geography, geography of the Soviet Union
Thompson, James Neal, Jr, genetics
Turner, Billie Lee, II, geography, anthropology
van der Helm, Dick, physical chemistry
Wender, Simon Harold, biochemistry
Whitmore, Mary (Elizabeth) Rowe, anatomy, zoology
Whitmore, Stephen Carr, physics
Wilbanks, Thomas John, geography
Wilson, Leonard Richard, geology, palynology
Zallen, Eugenia Malone, food science

OKLAHOMA CITY
Adams, Gail Dayton, Jr, radiological physics
Alaupovic, Petar, biochemistry, organic chemistry
Allison, John Everett, anatomy
Altmiller, Dale Henry, clinical chemistry, medical genetics
Amick, Lawrence Douglas, neurology, psychiatry
Anderson, David Walter, medical physics, nuclear physics
Anderson, Paul Sigfried, Jr, biometry, bacteriology
Anglin, J Hill, Jr, organic biochemistry
Asal, Nabih Rafia, epidemiology, biostatistics
Baker, Louis Reed, anesthesiology
Baker, Mary Rebecca, anesthesiology
Baldwin, Roger Allan, organic chemistry, fuel science
Barczak, Virgil J, mineralogy, geology
Bell, Richard Dennis, physiology
Bills, John Lawrence, chemistry
Bodine, Charles David, obstetrics & gynecology
Bogardus, Carl Robert, Jr, medicine, radiology
Bottomley, Richard H, internal medicine, oncology
Bottomley, Sylvia Stakle, internal medicine, hematology
Bradford, Reagan Howard, biochemistry
Branch, John Curtis, parasitology
Braun, Winfred Quentin, chemistry
Brecher, Gerhard Adolf, physiology
Briggs, Thomas, biochemistry
Brooks, Margaret Hoover, cytogenetics, biochemical genetics
Brown, William Ernest, dentistry
Bulmer, Glenn Stuart, medical mycology, microbiology
Burke, Richard Michael, clinical medicine
Byrd, Daniel Madison, III, pharmacology
Cain, William Aaron, immunology, microbiology
Carubelli, Raoul, biochemistry
Chandler, Albert Morrell, biochemistry
Chowdhury, Tushar Kumar, biophysics, physiology
Christensen, Howard Dix, neuropharmacology
Chrysant, Steven George, cardiovascular diseases
Clapper, Thomas Wayne, organic chemistry
Clark, Mervin Leslie, medicine
Coalson, Jacqueline Jones, pathology
Coalson, Robert Ellis, anatomy
Coffman, Moody Lee, theoretical physics, acoustics
Coleman, Ronald Leon, biochemistry
Coston, Tullos Oswell, ophthalmology
Cox, Andrew Chadwick, biochemistry, biophysics
Craig, Louis Elwood, organic chemistry
Crane, Charles Russell, biochemistry
Crosby, Warren Melville, obstetrics & gynecology
Czerwinski, Anthony William, nephrology, pharmacology
Daron, Garman Harlow, anatomy
Davis, Robert Elliott, inorganic chemistry
Deckert, Gordon Harmon, psychiatry, academic administration
Delaney, Robert, biochemistry
Deutsch, Stanley, anesthesiology
Dille, John Robert, aerospace medicine
Dubowski, Kurt Max, clinical biochemistry, toxicology
Dugan, Kimiko Hatta, microscopic anatomy
Erickson, John William, exploration geology
Everett, Mark Allen, medicine, dermatology
Everett, Mark Reuben, biochemistry

Fahmy, Aly, surgical pathology, clinical pathology
Faulkner, Kenneth Keith, anatomy
Felton, Frances Grace, medical microbiology
Felts, William Joseph Lawrence, anatomy
Ferretti, Joseph Jerome, biochemistry, microbiology
Floyd, Robert A, biophysics
Fong, Kuo-Lan, biochemistry
Friedberg, Wallace, radiobiology
Froelich, Robert Earl, psychiatry
Frohlich, Edward David, internal medicine
Gable, James Jackson, Jr, internal medicine
Ganesan, Devaki, internal medicine, pharmacology
Gleaton, Harriet Elizabeth, anesthesiology
Gray, Peter Norman, biomedical sciences
Griffin, Martin John, biochemistry
Groom, Dale, medicine, cardiology
Grubb, Alan S, community health
Grubb, Randall Barth, developmental biology
Gumbreck, Laurence Gable, endocrinology
Gunn, Chesterfield Garvin, Jr, internal medicine, neurophysiology
Hall, William Harvey, gastroenterology
Hammarsten, James Francis, internal medicine
Hampton, James Wilburn, hematology, oncology
Harkins, Rosemary Knighton, gross anatomy
Harris, Ray Edgar, analytical chemistry
Hartsuck, Jean Ann, biochemistry, x-ray crystallography
Haug, Norman L, community health
Hermann, John Alexander, applied chemistry
Hinshaw, Lerner Brady, physiology
Hodgins, Daniel Stephen, biochemistry
Hole, John J, mathematics
Holloway, Frank A, psychology, neurosciences
Hornbrook, K Roger, pharmacology
Howard, Robert Palmer, internal medicine, history of medicine
Hurst, Thomas Leonard, ceramics
Hyde, Richard Moorehead, microbiology, immunology
Ivey, Michael Hamilton, medical parasitology
Johnson, B Connor, biochemistry, nutrition
Jones, Ben Morgan, psychopharmacology
Keyl, Milton Jack, physiology
Killion, Jerald Jay, immunobiology, cancer
King, Mary Margaret, cell physiology, drug metabolism
Kirkham, William R, pathology
Kleen, Harold J, geology
King, Ozro Ray, reproductive endocrinology
Kollmorgen, G Mark, cell biology, immunology
Koss, Michael Campbell, pharmacology
Kraikitpanitch, Sompong, internal medicine, nephrology
Kremers, Howard Earl, industrial chemistry
Larsen, Earl George, biochemistry
Latimer, Steve B, biochemistry, organic chemistry
Lerner, Michael Paul, virology, cell biology
Lhotka, John Francis, Jr, histochemistry, human anatomy
Lindeman, Robert D, internal medicine, physiology
Long, Calvin H, analytical chemistry
Lucid, Michael Francis, inorganic chemistry
Lynn, Thomas Neil, Jr, medicine, preventive medicine
Mandal, Anil Kumar, cardiovascular diseases
Manoharan, A Chelvanayakam, physics, mathematics
Massion, Walter Herbert, anesthesiology, physiology
Mayes, Jary S, biochemistry, human genetics
McCallum, Roderick Eugene, microbiology, biochemistry
McCay, Paul Baker, biochemistry
McClellan, Betty Jane, ophthalmology
McClure, Coye Willard, ophthalmology
McClure, Theodore Dean, neuroanatomy
McConathy, Walter James, lipid chemistry, protein chemistry
McFadden, Ernest B, physiology
McKenzie, Jess Mack, physiology

623

OKLAHOMA

Melton, Carlton Earl, Jr., physiology
Metcoff, Jack, physiology
Miller, Leonard Robert, pathology, cell biology
Miller, Marion Paul, human anatomy
Mock, David Clinton, Jr., internal medicine
Moore, Joanne Iweta, pharmacology
Muchmore, Harold Gordon, internal medicine
Murray, Edward Conley, inorganic chemistry
Nanninga, Harold Eugene, ecology, limnology
Nordquist, Robert Ersel, pathology, virology
Olson, Robert Leroy, physiology,
O'Neal, Robert Munger, pathology
Oniko, Joseph Andrew, biochemistry
Papper, Solomon, internal medicine, metabolism
Paredes, Alfonso, psychiatry
Parker, Don Earl, applied statistics
Parrish, R Gibson, anesthesiology
Parry, William Lockhart, urology
Patnode, Robert Arthur, immunology
Pennock, Bernard Eugene, biomedical engineering, pulmonary physiology
Pinto, P Vincent C., clinical chemistry,
Revzin, Alvin Morton, pharmacology, neurophysiology
Rhoades, Everett Ronald, internal medicine, microbiology
Richter, Kenneth Murrel, histology, embryology
Riley, Harris D, Jr., pediatrics
Robertson, Wilbert Joseph, Jr., inorganic chemistry
Schneeberger, Charles Michael, mathematics
Schneider, Robert Arnold, physiology
Scott, Lawrence Vernon, virology
Seely, J Rodman, pediatrics
Shapiro, Stewart, dentistry; public health
Shurley, Jay Talmadge, human ecology
Silberg, Stanley Louis, epidemiology,
Smith, Carl Walter, Jr., nuclear medicine, endocrinology
Smith, Joseph Darrel, pediatrics
Smith, Paul Winston, pharmacology, toxicology
Smith, Philip Edward, parasitology
Smith, Vivian Sweibel, health sciences
Smith, William Ogg, medicine
Snow, Clyde Collins, physical anthropology
Sohler, Katherine Berridge, epidemiology
Sokatch, John Robert, bacteriology
Stanley, Allan John, endocrinology, embryology
Staples, Albert Franklin, oral surgery,
Steen, Wilson D., preventive medicine, physiology
Stith, Rex David, public health
Tang, Jordan J N., biochemistry
Tapia, Fernando, psychiatry
Taylor, Fletcher Brandon, Jr., internal medicine
Thurman, William Gentry, pediatrics, oncology
Tobias, Jerry Vernon, psychoacoustics
Torres-Pinedo, Ramon, pediatrics, gastroenterology
Toussieng, Povl Winning, psychiatry
Trachewsky, Daniel, endocrinology, molecular biology
Van De Steeg, Garet Edward, radiochemistry, physical chemistry
Vestal, Bedford Mather, animal behavior, vertebrate zoology
Walloch, Richard Arthur, physiology
Walsh, Gerald Michael, pharmacology
Wang, Chi-Sun, biochemistry
Weiss, Adolph Kurt, physiology
West, Kelly M., medicine
Whitcomb, Walter Henry, internal medicine, nuclear medicine
White, Clayton Samuel, medicine, physiology
Whitsett, Thomas L., internal medicine, clinical pharmacology
Wicke, Charles Robinson, anthropology, archaeology
Williams, George Rainey, medicine
Woods, James W., medical education
Woodward, James D., dentistry, prosthodontics
Wurth, Michael John, organic chemistry, photography

Yunice, Andy Aniece, biochemistry, environmental science
Zimmer, Louis George, geology

PONCA CITY

Albright, James Curtice, physics, mathematics
Allen, Marvin Carrol, analytical chemistry
Allred, Raymond Charles, petroleum microbiology, environmental sciences
Byerly, Perry Edward, geophysics
Cabbiness, Dale Keith, analytical chemistry
Conley, Francis Raymond, inorganic chemistry
Cowley, Thomas Gladman, analytical chemistry, spectrochemistry
Durr, Albert Matthew, Jr., organic chemistry
Eby, Harold Hildenbrandt, organic chemistry
Evens, Floyd Monte, analytical chemistry, spectroscopy
Fertl, Walter Hans, geophysics
Fonseca, Anthony Gutierre, inorganic chemistry
Gregory, M Duane, physical chemistry
Hanning, Mynard C., spectrochemistry
Ho, Thomas Tong-Yun, organic geochemistry
Hopkins, John Raymond, geophysics
Kennedy, Flynn, organic chemistry
Kirk, James Curtis, organic chemistry
Leslie, Wallace Dean, analytical chemistry
Libbey, William Jerry, organic chemistry
Linder, Donald Ernst, analytical chemistry
Lunden, Allan Jay, organic chemistry
Matson, Ted P, applied chemistry, surface chemistry
McGuire, Stephen Edward, industrial chemistry
Miller, Edsel Leo, organic chemistry
Minor, John Threecivelous, organic chemistry, information science
Monn, Donald Edgar, analytical chemistry
Motz, Kaye La Marr, industrial organic chemistry
Nielsen, Allen Madsen, microbial physiology
Payton, Charles Ellis, petrology
Perkins, Gerald, Jr., analytical chemistry, physical chemistry
Schwab, Peter Austin, polymer chemistry
Shook, D'Arcy Adriance, physical chemistry
Steffey, Oran Dean, industrial hygiene
Tanis, James Iran, exploration geophysics
Thomason, William Hugh, physical chemistry, corrosion
Tillman, Richard Milton, organic chemistry
Washecheck, Paul Howard, industrial organic chemistry
Waters, Kenneth Harold, geophysics
Whitfill, Donald Lee, physical inorganic chemistry
Woods, Warren Whitney, petroleum chemistry

POTEAU

Chessmore, Roy A., plant breeding

PRYOR

Weakley, Martin LeRoy, research administration, agricultural chemistry

ROFF

Dutfer, William Riley, aquatic biology

SAND SPRINGS

Maguire, Keith Dean, inorganic chemistry

SHAWNEE

Black, Jeffrey Howard, herpetology, ecology
Canham, Richard Gordon, physical chemistry
Heinze, John Edward, microbiology
Hurley, James Edgar, microbiology
Mills, John Norman, biochemistry
Neptune, William Everett, physical chemistry
Purdie, Jack Olen, chemistry
Whitington, Melvin Othal, Jr., microbiology

STILLWATER

Abbott, Donald Clayton, biochemistry, cereal chemistry
Agnew, Jeanne Le Caine, mathematics
Ahmad, Shair, mathematics
Ahring, Robert M., agronomy, soils
Baker, Frank Hamon, animal science
Banks, Donald Jack, botany
Barbour, Helen F., human nutrition
Barnes, George Lewis, plant pathology
Basler, Eddie, Jr., plant physiology
Beames, Calvin G Jr., physiology
Berlin, Kenneth Darrell, organic chemistry
Berthoff, Dennis E., mathematics
Brock, William Elihu, veterinary pathology
Broemeling, Lyle D., statistics
Bruneau, Leslie Herbert, genetics
Buck, Richard F., atmospheric physics
Buckner, Ralph Gupton, veterinary medicine
Burchard, Hermann Georg, mathematics
Burks, Sterling Leon, ecology, limnology
Bush, Linville John, dairy nutrition
Butler, Charles Thomas, biophysics
Campbell, Raymond Earl, horticulture
Carney, George Olney, historical geography, cultural geography
Carraway, Kermit Lee, biochemistry
Carroll, Arthur George, plant physiology
Constvet, Richard E., veterinary microbiology, veterinary pathology
Crockett, Jerry J., plant ecology
Croy, Lavoy I., crop physiology, plant biochemistry
Cunningham, Clarence Marion, physical chemistry
Curd, Milton Rayburn, zoology
Dermer, Otis Clifford, industrial organic chemistry
Devlin, Joseph Paul, physical chemistry, physics, magnetism
Dixon, George Sumter, Jr., solid state physics
Dodd, David Cedric, veterinary pathology
Dorris, Troy Clyde, limnology, water pollution
Drew, William Arthur, arachnology
Durham, Norman Nevill, bacteriology
Duvall, Paul Frazier, Jr., topology,
Edmison, Marvin Tipton, academic administration, organic chemistry
Edwards, Lewis Hiram, agronomy, genetics
Eikenbary, Raymond Darrell, entomology, forestry
Eisenbraun, Edmund Julius, organic chemistry
Essenberg, Margaret Kottke, biochemistry, plant pathology
Essenberg, Richard Charles, biochemistry
Eubanks, Isaac Dwaine, inorganic chemistry
Fisher, Donald D., mathematics, computer science
Fite, Robert Carl, geography
Folks, John Leroy, analytical statistics,
Frahm, Richard R., population genetics, animal breeding
Freeman, Robert David, physical chemistry
Frey, Merwin Lester, veterinary microbiology
Friend, Jonathon D., veterinary medicine
Garner, Duane LeRoy, reproductive physiology, animal science
Gee, Lynn LaMarr, microbiology
Gholson, Robert Karl, biochemistry
Glass, Bryan Pettigrew, mammalogy
Glenn, Bertis Lamon, veterinary pathology, comparative pathology
Goff, Gerald K., mathematics
Gorin, George, physical biochemistry
Gough, Francis Jacob, plant pathology
Gray, Fenton, soil science
Greer, Howard A L., weed science
Gries, George Alexander, plant physiology
Grula, Edward Alan, microbiology
Grula, Mary Muedeking, microbial ecology
Guenther, John James, meat science
Hair, Jakie Alexander, entomology
Halliburton, Larry Eugene, solid state physics
Harrison, Aix B., exercise physiology
Hecock, Richard Douglas, geography
Henrickson, Robert Lee, animal science
Hillier, James Calvin, animal nutrition
Hodnett, Ernest Matelle, organic chemistry
Holbert, Donald, applied statistics
Howard, James Henri, ethnology

Howell, Daniel Elza, entomology
Huffine, Wayne Winfield, agronomy
Hurst, Jerry G., physiology
Jennings, David Phipps, physiology,
Jewett, John William, mathematical analysis
Jobe, John M., topology
Johnson, Becky Beard, physiology
Jones, Eric Wynn, veterinary surgery
Jones, Randall Jefferies, soil chemistry
Jones, Roy Winfield, experimental embryology
Jordan, Helen Elaine, veterinary parasitology
Keener, Marvin Stanford, mathematical analysis
Kent, Douglas Charles, geology
Koeppe, Roger Erdman, biochemistry
Kohnke, Elton Everett, physics
Kottarski, Ignacy Icchak, mathematics
Lafon, Earl Edward, solid state physics
Lange, James Neil, Jr., physics
Langwig, John Edward, forest products
Leach, Franklin Rollin, biochemistry
LeGrand, Frank Edward, genetics,
Leivo, William John, physics
Lewis, James Chester, wildlife ecology
Lindsey, Dortha Ruth, health science
Luce, William Glenn, plant ecology
Mains, Gilbert Joseph, physical chemistry
Martin, Joel Jerome, solid state physics
Mattock, Ralph S., agronomy
Mayer, George Pat, endocrinology, physiology
McLachlan, Eugene Kay, mathematics
McMurphy, Wilfred E., agronomy
McPherson, James King, plant ecology
Meicher, Ulrich Karl, molecular biology,
Merkle, Owen George, plant breeding
Miller, Helen Carter, vertebrate zoology,
Mitchell, Earl Douglass Jr., biochemistry
Monfus, Andrew W., veterinary pathology
Monnett, Victor Brown, geology
Moore, Thomas Edwin, inorganic chemistry
Morrill, Lawrence George, soil chemistry, fertility
Morrison, Robert Dean, biostatistics
Mottola, Horacio Antonio, analytical chemistry
Murray, Jay Clarence, genetics,
Naff, John Davis, geology
Nelson, Eldon Carl, biochemistry
Newcomer, Wilbur Stanley, physiology
Newell, George Watts, poultry husbandry
Noble, Robert Lee, animal nutrition
Olson, Harold Cecil, dairy bacteriology
Owens, Fredric Newell, animal nutrition
Ownby, Charlotte Ledbetter, veterinary anatomy
Ownby, James Donald, plant physiology
Panciera, Roger J., veterinary pathology
Payne, Richard N., horticulture
Peters, Don Clayton, entomology
Plaxico, James Samuel, agricultural economics
Pohl, Herbert Ackland, chemical physics
Powell, Jeff, range science
Powell, Richard Conger, solid state physics
Price, Richard Graydon, entomology
Purdie, Neil, inorganic chemistry,
Raff, Lionel M., chemical physics
Reed, Lester W., soil chemistry, mineralogy
Rich, Travis Dean, reproductive physiology
Richardson, Lavon Preston, physiology
Richardson, Paul Ernest, plant anatomy, plant morphology
Rolf, Lester Leo, Jr., pharmacology,
Rooney, John Francis, geography
Ross, Alex R., physical geology, geomorphology
Roszel, Jeffie Fisher, cytopathology, veterinary pathology
Russell, Charles Clayton, nematology
Rutledge, Delbert Leroy, entomology
Samuel, Mark Aaron, theoretical high energy physics
Sanford, William Corbin, developmental biology

Santelmann, Paul William, weed science, agronomy
Sauer, John Robert, entomology
Schroeder, Leon William, astrophysics, astronomy education
Scott, Hugh Lawrence, Jr, biophysics
Shaw, James Harlan, wildlife ecology, wildlife research
Shelton, John Wayne, geology
Sibley, William Arthur, solid state physics
Silker, Theodore Henry, forest ecology, soils
Smith, Edward Lee, plant genetics, field crops
Spivey, Howard Olin, biophysical chemistry
Staley, Theodore Earnest Leon, veterinary anatomy, veterinary physiology
Starks, Kenneth James, entomology
Stewart, Gary Franklin, geology
Stocking, Hobart Ebey, geology
Stone, John Elmer, geology
Stone, John Floyd, soil physics
Stritzke, Jimmy Franklin, agronomy
Sturgeon, Edward Earl, forestry
Sturgeon, Roy V, Jr, plant pathology
Summerfelt, Robert C, fish biology
Summers, Geoffrey P, solid state physics
Taliaferro, Charles M, plant breeding, plant genetics
Teate, James Lamar, forest ecology
Tennille, Newton Bridgewater, veterinary radiology
Thayer, Rollin Harold, poultry nutrition
Thomas, John Eugene, phytopathology
Thornton, John William, zoology, cytology
Todd, Glenn William, plant physiology
Toetz, Dale W., fish biology, limnology
Totusek, Robert, animal nutrition
Tucker, Billy Bob, agronomy
Turman, Elbert Jerome, animal science
Tweedie, Stephen William, cultural geography
Tyrl, Ronald Jay, plant taxonomy
Uehara, Hiroshi, mathematics
Usher, William Mack, mathematical statistics
Varga, Louis P., analytical chemistry, radiochemistry
Verhalen, Laval Mathias, plant breeding, plant genetics
Wadsworth, Dallas Fremont, plant pathology
Walker, Nathaniel, forest management, forest economics
Walter, George Rozier, Jr, biochemistry
Walters, Lowell Eugene, animal science
Weeks, David Lee, experimental statistics
Weibel, Dale Eldon, agronomy
Wells, Milton Ernest, animal breeding
Westhaus, Paul Anthony, atomic physics, molecular physics
Wettemann, Robert Paul, reproductive physiology
Whatley, James Arnold, animal breeding
Whitcomb, Carl Erwin, horticulture, plant ecology
Whiteman, Joe V., animal breeding
Wilhm, Jerry L., limnology, ecology
Wilson, Timothy M., solid state physics
Young, Harry Curtis, Jr, plant pathology
Young, Jerry H, entomology

TAHLEQUAH
Brooks, Nathan Cyrus, geography
Collier, Robert Eugene, microbiology
Nolan, George Junior, physical chemistry
Smith, Charles Clinton, Jr, zoology, physiology

TULSA
Ames, Roger Lyman, geochemistry
Barclay, Harriett G., botany
Barker, Colin G, organic geochemistry
Basan, Paul Bradley, sedimentology, paleoecology
Bednar, Jonnie Bee, mathematics, computer science
Bennison, Allan P., geology
Bingham, Robert J., biochemistry, food chemistry
Blair, Albert Patrick, zoology
Bradley, John Samuel, geology
Broin, Thayne Leo, geology
Bryden, Elmer Louis, geology
Buck, Paul, plant ecology
Burkhart, Sarah Maybelle, mathematics
Busch, Daniel Adolph, geology
Butler, Edward Byron, physical chemistry
Cairns, Thomas W., mathematics
Carter, Harry Nelson, geology
Chenoweth, Philip Andrew, geology, stratigraphy

Clement, William Glenn, geophysics, electrical engineering
Coffey, Mitchael Dewayne, organic chemistry
Cornish, John Henry, hydrology
Couch, Richard Wesley, plant physiology
Devonshire, Leonard Norton, inorganic chemistry, analytical chemistry
Dickey, Parke Atherton, petroleum geology
Disney, Ralph Willard, geology
Dott, Robert E., physics, astronomy
Elmore, Robert E., physics, astronomy
Ferris, Craig, geophysics
Foster, Gerald Lawrence, organic chemistry
French, William Stanley, exploration geophysics
Frisillo, Albert Lawrence, geophysics
Garst, Arthur Wilhelm, chemistry
Goldstein, August, Jr, petroleum geology
Graves, Roy William, Jr, geology
Grayson, John Francis, palynology
Green, Thom Henning, petroleum geology
Harrison, Hugh Thomas, industrial geology
Hartman, Roger Duane, solid state physics, medical physics
Hawkins, James Edward, geophysics
Hedlund, Richard Warren, geology, palynology
Hobson, John Peter, Jr, geology
Horn, Myron Kay, geology, geochemistry
Horton, John, oncology
Houlihan, Rodney T., physiology, endocrinology
Hyne, Norman John, marine geology, physical oceanography
Jaques, William Everett, pathology
Jennemann, Vincent Francis, computer science, exploration geophysics
Jensen, Clyde B., pharmacology
Jinks, Douglas David, economic geology, exploration geology
Jones, James Homer, chemistry
Jones, Katherine Maurice, science education
Justice, James Horace, applied mathematics
Keith, Brian Duncan, sedimentary petrology
Kemp, Louis Franklin, Jr, mathematics
Kemp, Marwin K., physical chemistry
Konkel, Philip M., petroleum geology
Kucera, Clare H., organic chemistry
Kuenhold, Kenneth Alan, atomic physics, engineering physics
Lane, Harold Richard, paleontology
Laster, Stanley Jerral, geophysics
Letcher, John Henry, physics, computer science
Levetin-Avery, Estelle, mycology, botany
Lindsay, Hague Leland, Jr, vertebrate zoology
Link, Peter K., geology, meteorology
Maddin, Charles Milford, analytical chemistry
Martner, Samuel (Theodore), geophysics
Massey, Linda Kathleen Locke, cell biology
Masters, Bruce Allen, micropaleontology
Mattis, Allen Francis, petroleum geology
Matuszak, David Robert, geology
May, Hubert Eugene, biochemistry
McCollough, Edward Heron, geology
McCoy, Jerome Dean, physics
Meadors, Victor Gerald, physical chemistry
Merrill, Robert Kimball, exploration geology
Merriman, John Edward, internal medicine
Meshri, Dayaldas Tanumal, inorganic chemistry, physical chemistry
Meyerhoff, Arthur Augustus, geology
Mooney, Paul David, medicinal chemistry, biochemistry
Moritz, Carl Albert, geology
Mossman, Reyel Wallace, geophysics
Murray, Frederick Nelson, structural geology, stratigraphy
Naymik, Daniel Allan, mathematics, geology
Nelson, Eldon Lane, Jr, medical physiology
Nelson, John Marvin, Jr, entomology
Nesbitt, Stuart Stoner, chemistry
Nettles, John Barnwell, obstetrics & gynecology
Oliphant, Charles Winfield, geology
Parks, Kenneth Lee, industrial chemistry
Pittman, Edward D., petrology, sedimentology
Pope, Paul Terrell, statistics, mathematics
Reno, Harley, ichthyology, environmental management

Robertson, William G, medical physiology
Rogers, Steffen Harold, cell biology, parasitology
Sanderson, George Albert, geology, paleontology
Schemel, Mart Philip, geology
Scott, Robert W., geology, paleontology
Shirley, Barbara Anne, physiology, endocrinology
Shotts, Adolph Calveran, organic chemistry
Smith, Norman Cutler, geology, biology
Soday, Frank John, organic chemistry, anthropology
Soodsma, James Franklin, biochemistry, enzymology
Steinmetz, Richard, geology
Steward, James Gordon, mathematics, computer science
Sturbaum, Barbara Ann, physiology
Terriere, Robert T., geology
Thapar, Mangat Rai, seismology, geophysics
Thomas, Clarence Delmar, atomic physics
Thompson, Robert Richard, organic geochemistry
Thompson, Thomas Luman, geology
Thurman, Lloy Duane, plant ecology
Treitel, Sven, geophysics
Trygvason, Eysteinn, geophysics
Veatch, Ralph Wilson, mathematics
Verville, George Julius, geology
Visher, Glenn S., geology
Wiegel, William Edward, geology, population ecology
Williams, Philip Sidney, physics, geophysics
Winter, Thomas Greeley, physics
Woolsey, Marion Elmer, microbiology

WEATHERFORD
Armoudian, Garabed, electrodynamics, theoretical physics
Castleberry, George E, inorganic chemistry
Decker, Rolan Van, physical biochemistry
Dickison, Walter Lee, pharmacy, chemistry
Frame, Harlan D., inorganic chemistry, physical chemistry
Hamm, Donald Ivan, organic chemistry
Hertzler, Donald Vincent, organic chemistry
Hill, Benny Joe, physics
Keller, Bernard Gerard, Jr, pharmacy
Kriesel, Douglas Clare, medicinal chemistry
Lovell, James F., environmental biology, science administration
Lynn, Robert Thomas, animal behavior, ecology
McGurk, Donald J, organic chemistry
Messmer, Dennis A., microbiology
Nithman, Charles Joseph, pharmacy
Reichmann, Keith Wilford, pharmacy
Ulrich, Floyd Seymour, medicinal chemistry, analytical chemistry
Von Wicklen, Frederick Charles, biochemistry
White, Harold McCoy, organic chemistry
Wolgamott, Gary, microbiology, molecular biology

OREGON

ALBANY
Bennington, Kenneth Oliver, geochemistry
Elger, Gerald William, chemistry
Ferrante, Michael John, physical chemistry, thermodynamics
Henry, Jack Leland, chemistry
Ko, Hon-Chung, physical chemistry
Mickelberry, William Charles, food science
Nafziger, Ralph Hamilton, chemical metallurgy
Scott, Peter Carlton, plant physiology
Wood, Floyd William, physical metallurgy

ARCH CAPE
Richmond, James Frank, geology

ASHLAND
Badger, Rodney Allan, organic chemistry

Bartlett, James Kenneth, chemistry
Battaile, Julian, biochemistry
Bowman, Eugene W., mathematics
Coffey, Marvin Dale, entomology, zoology
Couch, Jack Gary, nuclear physics
Cross, Stephen P., mammalogy
Flower, Michael Joe, developmental biology, molecular biology
Fowler, Gregory L., genetics
Gardner, Murray Curtis, geology
Hollenbeck, Irene, botany
Lang, Frank Alexander, systematic botany
Linn, DeVon Wayne, limnology
Lloyd, Prescott Rees, bacteriology
MacCracken, Elliott, science education, mathematics
MacGraw, Frank Moss, geography
McNeal, Roy Wilson, geography
Montgomery, Richard Glee, mathematics
Nitsos, Ronald Eugene, plant physiology
Pennington, Lloyd Drew, analytical chemistry, academic administration
Purdom, William Berlin, economic geology
Rio, Sheldon T., mathematics
Seevers, Robert Edward, physical chemistry
Wolfe, Gordon A., solid state physics

ASTORIA
Babbitt, Jerry, food science, biochemistry
Crawford, David Lee, food science, food biochemistry

AURORA
Ticknor, Robert Lewis, ornamental horticulture

BEAVERTON
Alexander, Nancy J., reproductive physiology, immunology
Barofsky, Douglas Fred, mass spectrometry
Beatty, Clarissa Hager, physiology
Bocek, Rose Mary, biochemistry, clinical chemistry
Brenner, Robert Murray, cytology
Daves, Glenn Doyle, Jr, organic chemistry, bio-organic chemistry
Eaton, Gordon Gray, animal behavior
Elliott, Richard Amos, theoretical physics
Eror, Nicholas George, Jr, solid state chemistry, materials science
Fahrenbach, Wolf Henrich, histology, cytology
Grand, Theodore I, gross anatomy
Halko, David Joseph, inorganic chemistry
Hill, John Donald, veterinary medicine, cardiology
Hoskins, Dale Douglas, biochemistry, reproductive physiology
Howard, Charles Frank, Jr, biochemistry
Huntzicker, James John, air pollution, atmospheric chemistry
Hurst, James Kendall, bioinorganic chemistry
Kittinger, George William, biochemistry
Kontaxis, Nicholas E., biochemistry
LaSalle, Marjorie, immunology, immunohematology
Loehr, Thomas Michael, chemistry
Malinow, Manuel R., cardiology
Malley, Arthur, immunology
Massey, Gail Austin, lasers
Montagna, William, cytology
Norman, Reid Lynn, neuroendocrinology
Palotay, James Lajos, comparative pathology
Perry, Lorin Edward, fisheries
Phoenix, Charles Henry, behavioral physiology, reproductive biology
Pitter, Richard Leon, cloud physics, air pollution
Quadri, Syed Kaleemullah, neuroendocrinology
Resko, John A., reproductive physiology
Roper, John Gordon, experimental solid state physics
Spies, Harold Glen, neuroendocrinology, reproductive biology
Swanson, Lynwood Walter, physical chemistry
Van Horn, Richard Norman, primatology, reproductive biology
Verhoeven, Leon A., fish biology
West, Edward Staunton, biochemistry
Zingeser, Maurice Roy, anatomy, orthodontics

BEND
Baker, Richard William, physical chemistry; polymer chemistry
Cochran, Patrick Holmes, forest soils
Glading, Ben, wildlife conservation
Lonsdale, Harold Kenneth, physical chemistry

BURNS
Gomm, Fred Bryant, range management

CENTRAL POINT
DeTray, Donald Ervin, veterinary medicine

CHARLESTON
Rudy, Paul Passmore, Jr., comparative physiology

CLACKAMAS
Pratt, Carol Bert, physiological optics

COOS BAY
Burde, Donald Eugene, mathematics
Hower, Charles Oliver, nuclear chemistry; physical chemistry
Kelley, Raymond H., nuclear physics

CORBETT
Campbell, Colin Arthur, food science

CORVALLIS
Adams, Darius Mainard, forest economics
Adams, John Allen, animal nutrition
Adams, Holyoke Purinton, animal nutrition
Aho, Paul E., plant pathology
Allen, Thomas Cort, Jr., plant pathology
Anderson, Arthur W., bacteriology
Anderson, Carl Leonard, hygiene, public health
Anderson, Norman Herbert, aquatic ecology; entomology
Anglemier, Allen Francis, food science
Anselone, Philip Marshall, mathematical analysis
Apple, Spencer Butler, Jr., horticulture
Appleby, Arnold Pierce, weed science
Armstrong, Donald James, plant physiology
Arnold, Bradford Henry, pure mathematics
Arscott, George Henry, poultry nutrition
Ayres, James Walter, biopharmaceutics
Baggett, James Ronald, horticulture
Baisted, Derek John, organic chemistry; biochemistry
Ballantine, C. S., mathematics
Bard, Robert Charles, physical geography; cartography
Baross, John Allen, marine microbiology
Bartsch, Alfred Frank, aquatic biology
Bayne, Christopher Jeffrey, physiology
Beaudreau, George Stanley, biochemistry
Becker, Robert Richard, biochemistry
Bell, John Frederick, forest mensuration, forest management
Bernier, Paul Emile, poultry genetics
Berry, Ralph Eugene, entomology
Beuter, John H., forest economics
Bills, Donald Duane, food science
Bishop, Norman Ivan, plant physiology
Block, John Harvey, chemistry
Bodvarsson, Gunnar, applied mathematics, geophysics
Boedtker, Olaf A., solid state physics
Bogart, Ralph, animal science
Bond, Carl Eldon, ichthyology
Bone, Jesse Franklin, veterinary medicine
Bostwick, David Arthur, paleontology
Bouck, Gerald R., aquatic ecology; fisheries physiology
Boucot, Arthur James, geology, paleontology
Brady, James Joseph, surface physics
Brandt, William Henry, plant physiology
Brookes, Victor Jack, entomology
Brown, George Wallace, forest hydrology
Brown, James Russell, mathematical analysis
Brunk, Hugh Daniel, statistics, mathematics
Bryden, Harry Leonard, physical oceanography

Bublitz, Walter John, Jr., pulp chemistry, paper chemistry
Buhler, Donald Raymond, biochemical pharmacology
Burch, David Stewart, physics
Burt, Wayne Vincent, oceanography
Byrne, John Vincent, oceanography
Cain, Robert Farmer, food science
Caldwell, Douglas Ray, oceanography
Calhoun, Wheeler, Jr., agronomy
Callaway, Richard Joseph, physical oceanography
Calvin, Lyle D., statistics
Cameron, H Ronald, plant pathology
Cardenas, Mary Janet M, protein chemistry; enzymology
Carey, Andrew Galbraith, Jr., biological oceanography
Carlson, David Hilding, mathematics
Carter, David Southard, mathematics
Chambers, Kenton Lee, botany
Ching, Te May, plant physiology
Chilgren, John Douglas, physiology
Christensen, Bert Einar, synthetic organic chemistry
Church, David Calvin, animal nutrition
Compton, Oliver Cecil, pomology
Constantine, George Harmon, Jr., pharmacy
Conte, Frank Philip, comparative physiology; cell physiology
Converse, Richard Hugo, phytopathology
Corden, Malcolm Ernest, plant pathology
Couch, Richard W., geophysics, seismology
Cowan, John Ritchie, plant breeding
Coyier, Duane L., horticulture
Crabtree, Garvin (Dudley), horticulture, weed science
Craig, Albert Morrison, biophysics
Crowell, Hamblin Howes, entomology
Cutler, Melvin, solid state physics
Cutshall, Norman Hollis, chemical oceanography
Daniels, Malcolm, physical chemistry
Daterman, Gary Edward, forest entomology; insect ecology
Davis, Howard Fred, physics
Davis, Wilbur Arthur, anthropology
Dawson, Murray Drayton, soils, plant physiology
Dawson, Peter Sanford, population biology; genetics
Decius, John Courtney, physical chemistry
Decker, Fred William, forestry
Deeney, Anne O'Connell, microbiology, biochemistry
Denison, William Clark, botany
Dickinson, Ernest Milton, veterinary medicine
Dilworth, John Richard, forestry
Donaldson, John Russell, limnology; fish biology
Dornfeld, Ernst John, cell biology
Dost, Frank Norman, physiology; toxicology
Doudoroff, Peter, zoology
Drake, Charles Whitney, atomic physics
Dunn, John Asher, anthropology
Dyrness, Christen Theodore, forest soils
Dyson, Robert Duane, cell biology, enzymology
Easterday, Harry Tyson, nuclear physics
Edgren, James W., forest ecology
Elliker, Paul R., microbiology
England, David Charles, animal breeding
Enlows, Harold Eugene, petrography
Evans, James, Harold J., plant nutrition, plant biochemistry
Fairchild, Clifford Eugene, atomic physics
Fang, Sheng Chung, agricultural chemistry
Faulkenberry, Gerald David, statistics
Feldman, Milton H., physical chemistry
Fendall, Roger K., agronomy
Ferguson, George Ray, entomology
Ferrell, William Kreiter, forestry
Field, Cyrus West, economic geology, geochemistry
Fincke, Margaret Louise, nutrition
Fink, Gregory Burnell, pharmacology
Firey, William James, mathematics
Flaherty, Francis Joseph, geometry
Fontana, Peter R., theoretical physics
Foote, Wilson Hoover, agronomy

Frakes, Rodney Vance, agronomy
Franklin, Jerry Forest, ecology; forestry
Frazier, William Allen, olericulture
Fredericks, William John, physical chemistry
Fredriksen, Richard L., soils, forestry
Freed, Virgil Haven, biochemistry
Freeman, Peter Kent, organic chemistry
Frenkel, Robert Edgar, geography
Freund, Harry, analytical chemistry
Frolander, Herbert Farley, biological oceanography
Fryer, John Louis, microbiology; fisheries
Fuchigami, Leslie H., horticulture, plant physiology
Fullerton, Dwight Story, medicinal chemistry
Gamble, Wilbert, biochemistry
Gardner, E Hugh, soil fertility
Garren, Ralph, Jr., horticulture
Garton, Ronald Ray, aquatic biology
Gillett, James Warren, biochemistry
Gleicher, Gerald Jay, organic chemistry
Goetze, Norman Richard, agronomy
Goheen, Harry Earl, mathematics
Gordon, Jacqueline Irene, optics
Gordon, Louis Irwin, chemical oceanography
Goulding, Robert Lee, Jr., entomology
Griffiths, David John, physics
Guenther, Ronald Bernard, mathematical physics
Hagedorn, Charles, soil microbiology, forest ecology
Hall, James Dane, fish biology
Halter, Albert Nelson, agricultural economics, systems analysis
Hampton, Richard Owen, / plant pathology
Han, Youn Woo, microbiology
Hansen, Henry Paul, palynology
Hardison, John Robert, plant pathology
Harper, James Arthur, poultry science
Harr, Robert Dennis, hydrology
Harward, Moyle E., soil mineralogy
Haunold, Alfred, plant genetics
Hawkes, Stephen J., analytical chemistry, polymer chemistry
Hawthorne, Betty Eileen, nutrition
Heath, G Ross, oceanography, marine geology
Hedberg, Kenneth Wayne, physical chemistry
Heinrichs, Donald Frederick, geophysics
Heinzelman, Oliver Harry, geography
Henderson, Robert Wesley, research administration
Hermann, Richard Karl, plant ecology
Hewson, Edgar Wendell, meteorology
Highsmith, Richard Morgan, Jr., resource geography
Hillemann, Howard Herbert, zoology
Hisaw, Frederick Lee, Jr., zoology,
Hogg, Thomas Clark, anthropology
Hohenboken, William Daniel, animal breeding, animal science
Holton, Robert Lawrence, marine biology; pollution biology
Hopkins, Theodore Emo, physical chemistry
Horner, Chester Ellsworth, plant pathology
Horton, Howard Franklin, fish biology
Huddleston, James Herbert, soil morphology
Hughes, Kenneth Marion, insect pathology; electron microscopy
Hunter, Larry Clifton, mathematics
Huyer, Adriana (Jane), physical oceanography
Isenberg, Irvin, biophysics, biochemistry
Jackson, Thomas Lloyd, soils, agronomy
Jemison, George Meredith, forestry
Jensen, Harold James, nematology
Jensen, James Herbert, plant pathology
Jensen, John Granville, resource geography
Jensen, Varon, plant physiology
Johnson, Elizabeth Cox, parasitology
Johnson, Walter Curtis, biophysical chemistry
Jones, Daniel Patrick, history of science
Kamm, James A., economic entomology
Kaplan, Edward Lynn, mathematics
Kas, Arnold, mathematics
Kelley, John Paul, electrical engineering, radiological physics
Kennick, Walter Herbert, meat science
Kifer, Paul Edgar, food science
Kimeldorf, Donald Jerome, radiation biology
Kling, Gerald Fairchild, soil science
Knapp, Stuart Edward, veterinary parasitology
Knittel, Martin Dean, microbiology
Koehler, Carlton Smith, economic entomology

Kohlhepp, Sue Joanne, analytical biochemistry
Komar, Paul D., oceanography, marine geology
Kottman, Clifford Alfons, mathematics
Krahmer, Robert Lee, forest products
Krane, Kenneth Saul, experimental nuclear physics
Krantz, Gerald William, entomology
Krauss, Robert Wallfar, plant physiology
Kronstad, Warren Ervind, genetics, agronomy
Krueger, Hugo Martin, pharmacology, physiology
Krueger, William Clement, inorganic chemistry
Krueger, James Harry, inorganic chemistry
Kuhn, Laverne Duane, geological oceanography
Lagerstedt, Harry Bert, horticulture
Lahey, James Frederick, physical geography; meteorology
Landers, John Herbert, Jr., animal nutrition
Larson, Robert Elof, pharmacology; toxicology
Lattin, John D., entomology
Lavender, Denis Peter, plant physiology
Laver, Murray Lane, organic chemistry
Lawrence, Francis Joseph, plant breeding
Lawrence, Robert D., geology
Leach, Charles Morley, plant pathology
Lee, Jong Sun, microbiology
Lee, William Orvid, field crops
Leklem, James Erling, nutrition
Leong, Jo-Ann Ching, virology
Libbey, Leonard Morton, food science
Linderman, Robert G., plant pathology
Lindstrom, Fredrick Thomas, applied mathematics
Lohr, Dennis Evan, physical chemistry
Lombard, Porter Bronson, horticulture
Lonseth, Arvid Turner, mathematical analysis; numerical analysis
Loomis, Walter David, biochemistry
Loveland, Walter (David), nuclear chemistry
Lowry, William Prescott, meteorology
Lu, Kuo Chin, soil microbiology; plant pathology
Lyford, John H. Jr., ecology
MacDonald, Donald Laurie, biochemistry
Mack, Harry John, horticulture
MacSwan, Iain Christie, plant pathology
MacVicar, Robert William, chemistry
Madsen, Victor Arviel, theoretical nuclear physics
Maksymiuk, Bohdan, entomology, forestry
Maloney, Thomas Edward, environmental sciences
Malueg, Kenneth Wilbur, limnology
Martignoni, Mauro Emilio, virology; invertebrate pathology
Marvell, Elliot Nelson, organic chemistry
Mason, Richard Randolph, forestry; entomology
Mattson, Donald Eugene, veterinary virology
McCauley, James Elias, biological oceanography
McCustion, Willis Lloyd, genetics, plant breeding
McGill, Lois Sather, food science
McGuire, William Saxon, agronomy
McIntire, Charles David, aquatic ecology
McKenzie, Frederick Francis, animal science
McKinny, Milford D., wood technology
Meehan, William Robert, fish biology
Mehlig, Joseph Parke, analytical chemistry
Mesecar, Roderick Smit, physical oceanography; electrical engineering
Meslow, E Charles, ecology; wildlife research
Millemann, Raymond Eagan, zoology
Miller, Charles Benedict, biological oceanography
Miller, Lorraine Theresa, nutrition, biochemistry
Miller, Paul William, plant pathology
Miller, Terry Lee, biochemistry
Minore, Don, ecology
Mitchell, Russel Gene, forest entomology
Mix, Michael Cary, invertebrate pathology; radiation biology
Montgomery, Morris William, food science, biochemistry
Moore, Duane Grey, soil science
Moore, Frank Ludwig, endocrinology
Moore, Larry Wallace, plant pathology
Moore, Thomas Carroll, plant physiology
Morgan, Max Eugene, dairy bacteriology
Morita, Richard Yukio, microbiology; oceanography

Morita, Toshiko N, food microbiology
Morris, John Edward, developmental biology
Morris, Roy Owen, biochemistry
Naiman, Robert Joseph, aquatic ecology
Narasimhan, Mysore N L, applied mathematics, engineering science
Neal, Victor Thomas, physical oceanography
Nebeker, Alan V, entomology, aquatic ecology
Nelson, A Gene, agricultural economics
Nelson, Earl Edward, plant pathology
Neshyba, Steve, physical oceanography
Newberger, Stuart Marshall, mathematics
Newburgh, Robert Warren, biochemistry
Newton, Michael, forest ecology, weed science
Nibler, Joseph William, physical chemistry
Nicodemus, David Bowman, physics
Niem, Alan Randolph, geology
Nixon, Joseph Eugene, toxicology
Norris, Logan Allen, plant physiology, pesticide chemistry
Norris, Thomas Hughes, physical inorganic chemistry, chemical kinetics
Northam, Ray Mervyn, urban geography
Nowotny, Kurt A, industrial chemistry, inorganic chemistry
Oberhettinger, Fritz, mathematics
Oldfield, James Edmund, animal nutrition
Oles, Keith Floyd, geology
Oman, Paul Wilson, entomology
Owczarzak, Alfred, cell biology, biological structure
Owston, Peyton Wood, forest physiology
Padfield, Harland Irvine, applied anthropology, behavioral anthropology
Paine, David Philip, forest mensuration
Park, Paul Kilho, oceanography
Parker, Jesse Elmer, poultry husbandry
Parks, Leo Wilburn, microbial physiology
Parsons, Theran Duane, inorganic chemistry
Paulson, Clayton Arvid, oceanography, meteorology
Pawlowski, Norman E, organic chemistry
Pearcy, William Gordon, animal ecology, biological oceanography
Pearson, George Denton, animal virology
Pease, James Robert, resource management
Petersen, Bent Edvard, mathematics
Petersen, Roger Gene, biometrics
Peterson, Ernest W, meteorology
Peterson, Kermit Joseph, veterinary medicine
Peterson, Spencer Alan, limnology
Petzel, Florence E, textiles
Phinney, Harry Kenyon, botany
Piepmeier, Edward Harman, analytical chemistry
Pilcher, K Stephen, virology, medical microbiology
Pillsbury, Dale Ronald, physical oceanography
Poole, Albert Roberts, mathematics
Powelson, Robert Loran, plant pathology
Powers, Charles F, limnology
Preston, Eric Miles, ecology
Pritchard, Austin Wyatt, comparative physiology
Pyott, William Tucker, range ecology, systems analysis
Pytkowicz, Ricardo Marcos, physical chemistry, oceanography
Quatrano, Ralph Stephen, botany
Quinn, William Hewes, physical oceanography, meteorology
Ralston, Allen Thurman, animal nutrition
Ramsey, Fred Lawrence, mathematical statistics
Reed, Donald James, biochemistry
Reese, Hamit Darwin, biochemistry
Resch, Helmuth, forest products
Richert, Anton Stuart, particle physics, nuclear physics
Ritcher, Paul Osborn, entomology
Roberts, Alfred Nathan, horticulture
Roberts, Paul Alfred, genetics
Robinson, Dan D, forestry
Romancier, Robert Marshall, forest ecology
Roth, Lewis Franklin, forest pathology, mycology
Rowe, Kenneth Eugene, experimental statistics, biometry
Ruben, John Alex, morphology
Rudinsky, Julius Alexander, forest entomology
Rutherford, James Charles, invertebrate ecology
Ryan, Roger Baker, insect ecology
Ryker, Lee Chester, bioacoustics, entomology
Sager, Robert William, pharmacy
Sandine, William Ewald, bacteriology, microbiology
Saunders, Roy Bly, mathematics
Scanlan, Richard Anthony, food science
Schecter, Larry, nuclear physics
Scheffer, Theodore Comstock, forest products
Schmitt, Roman A, physical chemistry
Schmitz, John Albert, veterinary pathology & microbiology
Schrumpf, Barry James, range ecology, remote sensing
Schultz, Harold William, food science
Schultz, Harry Wayne, pharmaceutical chemistry
Scott, Allen Brewster, physical chemistry
Seely, Justus Frandsen, statistics, mathematics
Seidler, Ramon John, microbiology
Seyb, Leslie Philip, environmental chemistry
Shabica, Stephen Vale, biological oceanography
Shay, Junior Ralph, plant pathology
Shoemaker, Clara Brink, stuctural chemistry, crystallography
Shoemaker, David Powell, solid state chemistry, structural chemistry
Simons, William Haddock, mathematics
Simonson, Gerald Herman, soil science, agronomy
Sinnhuber, Russell Otto, food science, toxicology
Sisson, Harriet E, pharmacy
Slabaugh, Wendell Hartman, colloid chemistry
Small, Lawrence Frederick, biological oceanography
Smith, Courtland Lester, cultural anthropology, applied anthropology
Smith, Dean Harley, veterinary medicine, clinical pathology
Smith, Frank Herschel, botany
Smith, John Wolfgang, mathematics
Smith, Kennan Taylor, mathematics
Smith, Robert Lloyd, physical oceanography
Spencer, James Brookes, history of science
Spinrad, Bernard Israel, physics, nuclear engineering
Stalley, Robert Delmer, mathematics
Stein, William Ivo, forestry
Stephen, William Procuronoff, entomology
Stevenson, Elmer Clark, horticulture
Stoltenberg, Carl H, forest economics
Stone, William Matthewson, mathematics
Storm, Robert MacLeod, zoology
Stormshak, Fredrick, reproductive endocrinology
Storvick, Clara A, nutrition
Sutherland, Charles F, forestry economics
Swanson, Lloyd Vernon, reproductive endocrinology
Swenson, Knud George, entomology
Swenson, Leonard Wayne, physics
Taubeneck, William Harris, petrology
Taylor, Edward Morgan, geology, petrology
Terriere, Leon C, biochemsitry
Thiede, Jorn, geological oceanography
Thies, Richard William, organic chemistry
Thomas, Thomas Darrah, nuclear chemistry, physical chemistry
Thompson, Clarence Garrison, entomology
Thompson, Maxine Marie, genetics, horticulture
Tingey, David Thomas, plant physiology, air pollution
Tinsley, Ian James, biochemistry
Torley, Robert Edward, chemistry
Trappe, James Martin, mycology, forest pathology
Trione, Edward John, biochemistry
Triska, Frank John, aquatic ecology, microbial ecology
Trout, Edrie Dale, physics
Tubb, Richard Arnold, limnology, fisheries
Tullock, Robert Johns, soil chemistry
Ullery, Charles Howard, soil science, agronomy
Van Andel, Tjeerd Hendrik, marine geology
Van Dyke, Henry, microbiology
Van Holde, Kensal Edward, physical chemistry
Van Vliet, Antone Cornelis, wood science, communications
Vaughan, Edward Kemp, plant pathology
Volk, Veril Van, soil chemistry
Vomocil, James Arthur, soil science, agronomy
Wagner, Harry Henry, fish biology, ecology
Wang, Chih Hsing, radiochemistry
Waring, Richard H, plant ecology
Warren, Charles Edward, fish biology
Wasserman, Allen Lowell, solid state physics
Weber, Lavern J, pharmacology, academic administration
Wedman, Elwood Edward, veterinary microbiology
Wellons, Jesse Davis, III, wood technology, polymer science
West, William Irvin, forestry
Westwood, Melvin (Neil), pomology, plant physiology
Weswig, Paul Henry, animal nutrition
Whanger, Philip Daniel, nutritional biochemistry
White, James David, organic chemistry
Wickman, Herbert Hollis, physical chemistry, biophysics
Wiens, John Anthony, animal ecology, animal behavior
Wight, Howard Morgan, wildlife ecology
Williams, Max Bullock, analytical chemistry
Williamson, Stanley Ellsworth, science education
Willis, David Lee, biology, radiation biology
Wilson, Charles Owens, pharmaceutical chemistry
Wilson, Howard Le Roy, mathematics
Winters, Ronald Howard, pharmacology
Woodburn, Margy Jeanette, food science, microbiology
Worrest, Robert Charles, photobiology, marine ecology
Wrolstad, Ronald Earl, food science, agricultural chemistry
Wu, Szu Hsiao Arthur, physiology
Yang, Hoya Y, food technology
Yearick, Elisabeth Stelle, nutrition, biochemistry
Yoke, John Thomas, inorganic chemistry
Young, J Lowell, soil biochemistry
Young, Roy Alton, plant pathology
Youngberg, Chester Theodore, forest soils
Zaerr, Joe Benjamin, forest physiology
Zak, Bratislav, forest pathology
Zaneveld, Jacques Ronald Victor, oceanography
Zobel, Donald Bruce, plant ecology

DALLAS
Bedell, Thomas Erwin, animal husbandry, range management

DUNDEE
Brewster, John La Due, physics

EUGENE
Aikens, Clyde Melvin, anthropology, archaeology
Anderson, Frank Wylie, mathematics
Andrews, Fred Charles, mathematical statistics
Bajer, Andrew, cell biology
Baldwin, Ewart Merlin, geology
Baldwin, John E, physical organic chemistry
Barker, David Lowell, neurobiology
Barnett, Homer Garner, anthropology, applied anthropology
Barrar, Richard Blaine, mathematics
Beyer, Wendell T, computer science
Blank, Horace Richard, Jr, regional geology
Boekelheide, Virgil Carl, organic chemistry
Boggs, Sam, Jr, geology
Bradshaw, William Emmons, animal physiology, animal ecology
Carroll, George C, mycology
Carter, Elizabeth Francis, anthropology
Castenholz, Richard William, botany, microbiology
Ch'en, Shang-Yi, spectroscopy
Christensen, Ned Jay, audiology
Civin, Paul, mathematics
Clancy, Clarence William, genetics
Cook, Stanton Arnold, ecology, evolution
Crasemann, Bernd, atomic physics
Crasemann, Jean M, molecular genetics
Csonka, Paul L, elementary particle physics
Curtis, Charles Whittlesey, mathematics
Dahlquist, Frederick Willis, biophysical chemistry
Daly, James Edward, mathematical analysis
Dart, Francis Eliot, physics
Deshpande, Nilendra Ganesh, elementary particle physics
Dolby, Lloyd Jay, organic chemistry, natural products chemistry
Donley, Michael William, geography, geomorphology
Donnelly, Russell James, physics
Dorjahn, Vernon Robert, anthropology
Dumond, Don Edward, anthropology, archaeology
Dyke, Thomas Robert, physical chemistry
Ebbighausen, Edwin G, astronomy
Evonuk, Eugene, physiology
Fenna, Roger Edward, molecular biology, x-ray crystallography
Frank, Peter Wolfgang, ecology
Freeman, Robert, mathematics
Gaber, Bruce Paul, physical biochemistry
Ghent, Kenneth Smith, mathematics
Girardeau, Marvin Denham, Jr, statistical mechanics
Girsch, Stephen John, photobiology
Goles, Gordon George, geochemistry
Goswami, Amit, theoretical physics
Grant, Philip, developmental biology, neuroembryology
Gray, Jane, palynology, paleoecology
Griffith, O Hayes, biophysical chemistry
Harris, Patricia J, cell biology, electron microscopy
Harrison, David Kent, mathematics
Herbert, Edward, biochemistry, molecular biology
Herrick, David Rawls, theoretical chemistry
Herskowitz, Ira, molecular genetics
Higgins, Richard J, solid state physics
Hoffer, Alan R, mathematics
Holser, William Thomas, mineralogy
Holzafel, Christina Marie, biogeography
Hoyle, Graham, zoology, physiology
Hwa, Rudolph Chia-Chao, theoretical high energy physics
Johannessen, Carl L, biogeography, cultural geography
Jost, Patricia Cowan, biophysical chemistry, molecular biology
Kays, M Allan, geology, petrology
Keana, John F W, organic chemistry
Kemp, James Chalmers, astrophysics
Kezer, James, zoology
Kimmel, Charles Brown, developmental biology
Kittleman, Laurence Roy, Jr, geology
Klemm, LeRoy Henry, organic chemistry
Klopfenstein, Charles E, physical organic chemistry, chemical instrumentation
Koch, Richard Moncrief, geometry
Koenig, Thomas W, organic chemistry
Krueger, Eugene Rex, mathematics, computer sciences
Lefevre, Harlan W, nuclear physics
Lowndes, Douglas H, Jr, low temperature physics
Loy, William George, geography, cartography
Lund, Ernest Howard, geology
Maier, Eugene Alfred, mathematics
Matthews, Brian Wesley, molecular biology, x-ray crystallography
Maynard, Edith Adele, neurobiology
Mazo, Robert Marc, theoretical chemistry
McBirney, Alexander Robert, geology, petrology
McClure, Joel William, Jr, theoretical physics, solid state physics
McConnaughey, Bayard Harlow, marine biology
McDaniels, David K, nuclear physics, solar energy
McFee, Malcolm, anthropology
Mooney, Larry Albert, clinical chemistry
Moravcsik, Michael Julius, high energy physics, science policy
Moreno-Black, Geraldine S, physical anthropology
Morris, Robert Wharton, ichthyology
Moursund, David G, computer science, mathematics education
Munz, Frederick Wolf, visual physiology
Niven, Ivan (Morton), mathematics
Novick, Aaron, biophysics
Novitski, Edward, genetics
Noyes, Richard Macy, physical chemistry, chemical kinetics
Orr, William N, geology
Overley, Jack Castle, nuclear physics
Palmer, Theodore W, mathematics
Park, Kwangjai, solid state physics, optics
Patton, Clyde Perry, geography
Peticolas, Warner Leland, physical biochemistry
Pond, Judson Samuel, inorganic chemistry
Postlethwait, John Harvey, developmental biology, genetics

OREGON

Powell, John Leonard, physics
Price, Edward Thomas, geography
Rayfield, George W., physics
Reithel, Francis Joseph, biochemistry
Ross, Kenneth Allen, mathematics
Salmon, Theodora Nussmann,
Savage, Norman Michael, paleontology,
  evolutionary biology
Scheer, Bradley Titus, physiology
Schellman, John Anthony, physical
  chemistry
Simonds, Paul Emery, anthropology
Simpson, William Tracy, chemistry
Sistrom, William R., microbiology
Smith, Gerald Ralph, molecular biology
Soderwall, Arnold Larson, endocrinology
Stahl, Franklin William, genetics
Staples, Lloyd William, geology
Stern, Theodore, anthropology
Streisinger, George, genetics
Struble, George W., computer science
Swinehart, Donald Fought, chemical
  kinetics
Tate, Robert Flemming, mathematical
  statistics
Tepfer, Sanford Samuel, plant
  morphology
Trent, Walter Russell, chemistry
Truax, Donald R., mathematical statistics
Udovic, Joseph Daniel, population
  biology
Van Schaack, George Booth,
  mathematics, plant taxonomy
von Hippel, Peter Hans, biophysical
  chemistry
Wanner, Gregory Hugh, theoretical
  physics
Ward, Lewis Edes, Jr., mathematics
Weill, Daniel Francis, petrology,
  geochemistry
Weston, James A., developmental biology,
  cell biology
Witanen, Wayne Alfred, bionics,
  computer sciences
Wimber, Donald Edward, botany,
  cytology
Wolfe, Raymond Grover, Jr.,
  biochemistry
Wright, Charles R. B., algebra
Young, Phillip D., ethnology
Youngquist, Walter, geology
Zimmerman, Robert Lyman, physics

FOREST GROVE

Bhattacharya, Ramendra Kumar,
  mathematics
Carter, Richard Thomas, parasitology,
  invertebrate zoology
Gerke, John Royal, microbiology
Gilbert, Margaret Shea, biology
Griffith, William Thomas, physics
Levine, Leonard, neurophysiology
Malcolm, David Robert, entomology
Meyer-Arendt, Jurgen Richard, optics
Richards, Oscar White, environmental
  physiology
Roth, Niles, physiological optics,
  optometry
Thorn, Frank, psychophysiology

HILLSBORO

Fry, Louis Rummel, orthopedic surgery

HOOD RIVER

Mellenthin, Walter M., horticulture
Zwick, Robert Ward, entomology

KLAMATH FALLS

Cade, Stephen C., forestry
Fisk, LeRoy (Henry), parasitology
Wilson, Frank MacDonnell, medical
  administration

LA GRANDE

Anderson, Ernest Clifford, biology
Catlin, Seth, mathematical logic
Geist, Jon Michael, soil chemistry, plant
  nutrition
Gilbert, David Erwin, atomic
  spectroscopy
Hermens, Richard Anthony, physical
  chemistry
Skovlin, Jon Matthew, range science,
  wildlife biology
Stephas, Paul, physics
Thomas, Jack Ward, wildlife biology

LAKE OSWEGO

Carter, Charles Conrad, neurology
Emmett, Paul Hugh, physical chemistry

Gabler, Walter Louis, biochemistry
Hughes, Paul Warren, geology
Lapidus, Leo, mathematics

MARYLHURST

Gfeller, Barbara, genetics

MCMINNVILLE

Alin, John Suemper, algebra
Anderson, Carl Martin, chemistry
Bell, Anthony E., surface physics
Boling, John Landrum, zoology
Buckley, Patricia M., microbiology,
  biochemistry
Charbonnier, Francis Marcel, physics
Crook, James Richard, parasitology,
  epidemiology
Day, John Arthur, meteorology
Dell, Roger Marcus, mathematical
  analysis
Dirks-Edmunds, Jane Claire, ecology
Dolan, Winthrop Wiggin, mathematics
Dyke, Walter Payne, physics
Farris, Richard Austin, marine ecology
Hamby, Drannan Carson,
  electrochemistry
Hinrichs, Clarence H., physics
Johansen, Herman Andrew, inorganic
  chemistry
Jones, Robert Edward, physics
Springer, Charles Haviec, organic
  chemistry

MEDFORD

Aldrich, Willard Walker, horticulture

MONMOUTH

Bandick, Neal Raymond, human
  physiology; medical education
Cummins, Ernie Lee, science education
Evett, Jay Fredrick, physics, biophysics
Gallagher, James Weldon, climatology,
  geography
Greco, Peter V., geography
Griffin, Paul F., economic geography
Jennings, Charles David, oceanography
Johnson, John Morris, botany, cytology
Liedtke, James Dale, chemistry
Nelson, Norman Neibuhr, mathematics
Novak, Robert Otto, mycology, plant
  pathology
Penk, Anna Michaelides, mathematical
  education
Postl, Anton, organic chemistry
Walker, Kenneth Merriam, aquatic biology
White, Donald Harvey, nuclear physics,
  particle physics

NEWBERG

Chittick, Donald Ernest, physical
  chemistry
Martin, Gordon Wyatt, parasitology,
  developmental biology

NEWPORT

Caldwell, Richard Stanley, physiological
  ecology
Gonor, Jefferson John, invertebrate
  zoology; marine biology
Hedgpeth, Joel Walker, aquatic biology
Olson, Robert Eldon, zoology;
  parasitology
Tyler, Albert Vincent, fisheries

NYSSA

Hoff, John Conrad, plant breeding
Oldemeyer, Donald LeRoy, plant
  breeding
Simantel, Gerald M., plant breeding
Trupp, Clyde Rulon, plant breeding

PENDLETON

Almaraz, Raymond Richard, soil science
Lund, Steve, agronomy
Ramig, Robert E., soil conservation, soil
  fertility
Rickman, Ronald Wayne, soil
  conservation
Rohde, Charles Raymond, plant breeding

PORTLAND

Adams, Thomas C., forest economics,
  international economics
Ahuja, Jagdish C., statistics, mathematics
Allen, John Eliot, geology
Anderson, James G., chemistry
Arch, Stephen William, neurobiology
Ashbaugh, James G., geography
Atherton, John Harvey, cultural
  anthropology, archaeology
Autrey, Robert Luis, organic chemistry,
  science education
Bacon, Robert Lewis, anatomy
Balogh, Charles B., mathematics
Bardana, Emil John, Jr., allergy,
  immunology
Barnack, Neal Herbert, neurobiology,
  neuropsychology
Barnum, Dennis W., chemistry
Bartley, Murray Hill, Jr., anatomy, oral
  pathology
Baros, Dagmar, clinical biochemistry
Baros, Frantisek, clinical biochemistry
Beals, Margaret K., orthopedic surgery
Beard, Margaret Elzada, cell biology
Beaulieu, John David, stratigraphy,
  environmental geology
Benson, Gilbert Thomas, geology
Benson, John Alexander, Jr.,
  gastroenterology
Benson, Ralph Criswell, obstetrics &
  gynecology
Bergman, Norman, anesthesiology
Bigley, Robert Harry, internal medicine,
  genetics
Bilbao, Marcia Kepler, radiology
Bissonnette, John Maurice, obstetrics &
  gynecology
Blachly, Paul H., psychiatry
Black, John Alexander, biochemistry
Bluemle, Lewis W., Jr., internal medicine
Boddy, Dennis Warren, entomology,
  zoology
Bonhorst, Carl W., biochemistry
Brehm, Bertram George, Jr., botany
Briegleb, Philip Anthes, forestry
Bristow, John David, medicine
Brodie, Laird Charles, physics
Brooke, Clarke Harding, geography
Brookhart, John Mills, physiology
Brooks, Robert E., experimental
  pathology
Brown, Bruce Willard, analytical
  chemistry, inorganic chemistry
Bruce, David, forest mensuration
Bruckner, Robert Joseph, dentistry
Brummett, Robert E., pharmacology,
  otolaryngology
Buchan, George Colin, medicine,
  neuropathology
Buck, Douglas L., dentistry
Butler, John Ben, Jr., mathematics
Byrne, John Richard, mathematics
Calvin, Clyde Lacey, plant anatomy
Campbell, Charles J., fish biology,
  fisheries management
Campbell, John Richard, pediatric
  surgery
Campbell, Robert A., pediatrics
Carleton, Blondel Henry, physiology
Carolin, Valentine Mott, Jr., insect
  ecology
Cater, Frank Sydney, mathematics
Chakraborty, Prabir Kumar, physiology;
  neuroendocrinology
Charlton, David Berry, chemistry;
  bacteriology
Chester, Clarence Lucian, pathology;
  anatomy
Christensson, Hubert Edwin, mathematics
Christensen, Leonard, medicine, surgery
Church, Larry B., nuclear chemistry,
  radiochemistry
Clark, David Thurmond, parasitology
Clark, William Melvin, Jr., medical
  education administration
Clarkson, Quentin Deane, biostatistics
Clayomb, Cecil Keith, biochemistry
Cleaver, Frederick Charles, fish biology
Coleman, Ralph Orval, Jr., speech
  pathology
Collins, Ernest Hobart, experimental
  physics
Connor, William Elliott, internal
  medicine
Corbin, Ludlow, phycology
Courtney, Dale Elliott, geography
Cowan, Frederick Fletcher, Jr.,
  pharmacology
Critchlow, Burtis Vaughn, anatomy
Cronyn, Marshall William, organic
  chemistry
Curtis, Robert Orin, forest mensuration
D'Agostino, Anthony N., pathology;
  neuropathology
Darrow, Thomas D., herpetology
Dart, John Olney, geography
David, Norman Austin, pharmacology
Davis, Kenneth Edward, physics
Dawson, Peter J., pathology
De Maria, F. John, obstetrics &
  gynecology
Denney, Donald Duane, psychiatry
Dennis, Daniel L., surgery
De Vries, Dale Byron, physical
  chemistry, physics
DeWeese, David D., otolaryngology,
  surgery
Dittmer, Karl, chemistry
Dow, Robert Stone, anatomy, physiology
Drews, Robin Arthur, anthropology
Duncan, James Alan, physical organic
  chemistry
Dunne, Thomas Gregory, physical
  inorganic chemistry
Durfee, Raphael B., obstetrics &
  gynecology
Edwards, Benjamin Frank, developmental
  biology
Edwards, Miles John, medicine
Eicher, George J., fish biology
Elwell, Leonard Hubert, physiology
Engel, Rudolf, medicine
Enneking, Eugene A., mathematical
  statistics
Enneking, Marjorie, mathematics
Everett, Frank G., periodontology
Faber, Jan Job, biophysics, biophysics
Faust, Charles Harry, Jr., molecular
  biology, immunology
Fay, Warren Henry, speech pathology,
  audiology
Feeney, Mary Lynette, cell biology,
  ophthalmology
Feldman, Jacob Harold, biochemistry
Ferguson, Denzel Edward, vertebrate
  zoology
Ferguson, James Williams, organic
  chemistry
Fixott, Henry Cline, roentgenology
Fletcher, William Sigourney, surgery
Forbes, Richard Bryan, vertebrate
  zoology, ecology
Fox, Kaye Edward, pharmacology
French, Kathrine Story, anthropology
Frisch, Arthur Wain, bacteriology
Gabourel, John Dustan, pharmacology
Gallo, Anthony Edward, Jr., neurosurgery
Garcia, Gary Lee, fluorine chemistry
Garrison, George Alfred, forest ecology,
  range ecology
Gassaway, Alexander Ramsey, geography
Gatewood, Dean Charles, biochemistry
Gatz, Carole R., physical chemistry
Greenleaf, Merwyn Ronald, public health
Greer, Monte Arnold, endocrinology
Grette, Donald Pomeroy, organic
  chemistry
Griswold, Herbert Edward, internal
  medicine
Grover, M Roberts, Jr., medicine, medical
  administration
Gurevitch, Mark, physics
Gwilliam, Gilbert Franklin, neurobiology,
  invertebrate zoology
Hall, Frederick Columbus, plant ecology
Hall, Joseph Alfred, chemistry
Hallum, Jules Verne, virology
Hamilton, James Arthur Roy, fish
  biology
Hann, Robert A., forest products, wood
  technology
Hammett, James Roy, genetics
Hammond, Paul Ellsworth, geology
Hancock, John Edward Herbert, organic
Haney, Hance Francis, medicine
Haney, Harvey John William, geology
Harris, John Kenneth, topology
Harris, Nellie Robbins, human
  physiology
Harville, John Patrick, fisheries
  management
Hawley, John William, geology,
  geomorphology
Hearon, William Montgomery, organic
  chemistry
Hecht, Frederick, genetics, pediatrics
Higgins, Paul Daniel, physiological
  ecology, botany
Holman, Charles Nixon, medicine
Horton, Aaron Wesley, biophysical
  chemistry
Howard, Donald Grant, solid state
  physics
Hsu, Kwan, biophysics
Hu, Funan, dermatology
Hutchens, Tyra Thornton, clinical
  pathology
Iglewski, Wallace, virology, molecular
  biology
Irwin, James Wesley, biophysics
Irwin, Samuel, psychopharmacology
Isom, John B., pediatrics, neurology
Jacob, Stanley W., medicine, surgery
Jastak, J Theodore, oral surgery,
  anesthesia
Jensen, Bruce A., mathematics
Jones, Richard Theodore, medicine,
  biochemistry
Jones, Russell Stine, pathology
Jump, Ellis Burnett, anatomy
Kabat, David, biochemistry, genetics

Keedy, Curtis Russell, physical chemistry, radiochemistry
Kendall, John Walker, Jr, endocrinology
Kessel, Edward Luther, zoology
Keyes, Jack Lynn, physiology
Kilbourn, Joan Priscilla Payne, medical microbiology
Kilgour, Gordon Leslie, biochemistry
Killian, Thomas Joseph, engineering physics, acoustics
Kim, Kenneth, microbiology
Kinersly, Thorn, dentistry, dental research
Kleinholz, Lewis Hermann, physiology
Koler, Robert Donald, medical genetics, hematology
Kramer, Fritz Louis, geography
Kripaehne, William W, surgery
Labby, Daniel Harvey, medicine, psychiatry
Laffitte, Herbert Bonell, dentistry, periodontology
Lambie, Margaret B McClements, cosmic ray physics, power engineering
Larsen, Martin Lee, clinical chemistry
Lea, Malcolm Sinclair, biology
Leadley, John David, mathematics
Lees, Martin H, pediatrics, cardiology
Leslie, Gerrie Allen, immunology, immunochemistry
Leung, Benjamin Shuet-Kin, endocrinology, oncology
Levinson, Alfred Stanley, organic chemistry
Lewis, Howard Phelps, internal medicine
Limpert, Frederick Arthur, hydrology
Linman, James William, medicine
Lippert, Byron E, phycology
Lis, Adam W, biochemistry
Lis, Elaine Walker, nutrition, biochemistry
Litt, Michael, biochemistry
Liu-Ger, Tsu-Huei, theoretical physics
Low, Robert James, geophysics
Lu, Kuo Hwa, biostatistics, genetics
Lucas, Oscar Nestor, hematology, physiology
Lutz, Raymond Paul, physical organic chemistry
Lycan, D Richard, geography
Macpherson, Cullen H, biophysics
Macy, Ralph William, parasitology
Marriage, Lowell Dean, wildlife conservation
Marshall, Frederick James, histology
Martin, Robert Leonard, physics
Meike, Mary B, auditory physiology
Menashe, Victor D, pediatrics, cardiology
Metcalfe, James, medicine
Meyer, Ernest Alan, microbiology
McCawley, Elton Leeman, pharmacology
Mickelsen, John Raymond, physical chemistry
Moore, Richard Donald, pathology
Morgan, Bruce Henry, food microbiology
Morgenstern, Alan Lawrence, psychiatry
Morris, James F, medicine
Morse, Stephen Allen, microbiology
Morton, William Edwards, epidemiology
Mullooly, John P, biostatistics, biomathematics
Neill, William Alexander, medicine, cardiology
Nelsen, Roger Bain, mathematics
Newman, Lester Joseph, cytogenetics
Newman, Thomas McClellan, anthropology
Niles, Nelson Robinson, pathology, anatomy
Nussbaum, Rudi Hans, solid state physics
Olsen, George Duane, clinical pharmacology
Osterud, Harold T, public health, preventive medicine
Palmer, Leonard A, geology
Parker, Richard Bennett, bacteriology
Parker, William Lockwood, physics
Parrott, Marshall Ward, radiological health
Parsons, Roger Bruce, soil genesis, morphology
Pearson, Anthony Augustus, anatomy
Perdue, Edward Michael, physical chemistry
Petersen, Richard Randolph, limnology
Peterson, Clare Gray, surgery, physiology
Phillips, David, psychophysiology
Pierce, Joe Eugene, / anthropology

Pirofsky, Bernard, immunohematology
Porter, George A, medicine
Portman, Oscar William, medicine
Poulsen, Thomas Martin, geography
Prescott, Gerald H, medical genetics, oral medicine
Quinton-Cox, Robert, histology, cytology
Rabiner, Saul Frederick, hematology, internal medicine
Rampone, Alfred Joseph, physiology
Rasmussen, Lester Paul, pediatrics
Reeves, Melvin Mitchell, cardiovascular surgery
Rempfer, Gertrude Fleming, physics
Rempfer, Robert Weir, mathematics
Reynolds, Robert Eugene, physics
Richards, Thomas Charles, biological structure
Rickert, David A, environmental sciences
Rickles, Norman Harold, oral pathology
Rigas, Demetrios A, biochemistry, biophysics
Riker, William Kay, pharmacology
Rittenberg, Marvin Barry, immunology, microbiology
Ritzmann, Leonard W, internal medicine, cardiology
Roberts, Joseph Buffington, mathematics
Roe, David Kelmer, electrochemistry, analytical chemistry
Rose, Norman Carl, organic chemistry
Rosenwinkel, Earl Richard, plant ecology
Ruben, Laurens Norman, immunology
Russell, Peter James, molecular genetics
Sarles, Lynn Redmon, physics, research administration
Saslow, George, psychiatry
Sato, Makoto, neurophysiology, physiology
Savara, Bhim Sen, dentistry
Scheans, Daniel Joseph, anthropology
Schink, Chester Albert, organic chemistry
Schmidt, Harvey John, Jr, mathematics
Scott, Arthur Ferdinand, inorganic chemistry
Seaman, Geoffrey Vincent F, physical biochemistry, biomaterials
Seshu, Lilly Hannah, mathematics
Shearer, Thomas Robert, biochemistry
Sherwin, Duane O, psychiatry, neurology
Shore, James Henry, psychiatry
Siegel, Benjamin Vincent, virology
Silverman, Morris Bernard, inorganic chemistry
Simpson, Leonard, invertebrate zoology
Skinner, Richard Emery, physics, mathematics
Smith, Catherine Agnes, otology, anatomy
Snell, William E, orthopedic surgery
Snow, Michael Dennis, phytopathology, microbial ecology
Sorenson, Fred M, dentistry
Spencer, Elaine, biochemistry
Stafford, Helen Adele, plant physiology
Stanley, Robert Lauren, mathematics
Starr, Albert, thoracic surgery
Starr, Merle Arthur, electronics
Starr, Patricia Rae, microbiology
Stauffer, James, genetics
Steen, John Carl, physical anthropology
Stevens, Janice R, neurology, psychiatry
Suttles, Wayne Prescott, anthropology
Swan, Kenneth Carl, ophthalmology, pharmacology
Swank, Roy Laver, neurology
Swanson, John Robert, biochemistry
Swanson, Robert E, medical physiology
Takeo, Makoto, physics
Talbott, Richard Evans, neurophysiology
Tam, Kwok-Wai, mathematical analysis, operations research
Tanz, Ralph, pharmacology, physiology
Tauber, Selmo, applied mathematics
Taubman, Robert Edward, psychiatry, psychology
Taylor, Eugene Emerson, psychiatry, public health
Taylor, Mary Lowell Branson, microbial physiology
Taylor, Samuel Edwin, pharmacology
Taylor, Walter Herman, Jr, microbial physiology
Tedford, Myron Duncan, anatomy, human genetics
Terkla, Louis Gabriel, dentistry
Thayer, Lewis A, analytical chemistry
Thompson, Charles Calvin, oral pathology
Thoms, Richard Edwin, paleontology
Tinnin, Robert Owen, plant ecology
Tocher, Richard Dana, plant physiology
Todd, William Remington, biochemistry
Tracy, Wilbert E, pedodontics
Trainer, Joseph B, medical physiology
Tunturi, Archie Robert, anatomy
Turner, Larry Webster, vertebrate biology

Underwood, Rex J, medicine
Van Bruggen, John Timothy, biochemistry
Vernon, Jack Allen, psychophysiology
Volland, Leonard Allan, plant ecology
Wack, Paul Edward, nuclear physics
Walsh, John Richard, internal medicine
Walton, Richard Bruce, physics
Weaver, Charles R, biology, chemistry
Weaver, David Dawson, medicine, human genetics
Weaver, Morris Eugene, anatomy, zoology
Weaver, William Judson, cryobiology
Weber, Vinson M, dentistry
Weimar, Virginia Lee, physiology
Weir, William David, physical chemistry
Weitlauf, Harry, reproductive physiology
Wendel, Herbert A, pharmacology, internal medicine
Wetzel, Karl Joseph, experimental nuclear physics
Wheeler, Nicholas Allan, mathematical physics
White, Horace Frederick, physical chemistry
Williams, Christopher P S, pediatrics, medicine
Williams, Lloyd Bayard, mathematical analysis
Wirtz, John Harold, natural history, vertebrate zoology
Woolley, LeGrand H, oral pathology, dentistry
Wright, Kenneth Harold, entomology
Wright, Wellesley Horton, dentistry, periodontology
Wuepper, Kirk Dean, dermatology
Wyandt, Herman Edwin, Jr, cytogenetics, medical genetics
Yatsu, Frank Michio, neurology
Young, Norton Bruce, speech pathology, audiology

REDMOND
Johnson, Malcolm Julius, agronomy

ROGUE RIVER
Burkig, Jack Whipple, nuclear physics

ROSEBURG
Dejmal, Roger Kent, physiology

SALEM
Bitz, Miriam L, geochemistry
Breakey, Donald Ray, biology
Dade, Philip Eugene, agronomy
Duell, Paul Merwyn, inorganic chemistry
Hawke, Scott Dransfield, physiological ecology
Hudak, Norman John, organic chemistry
Iltis, Donald Richard, mathematics
Jacoby, Lawrence John, organic chemistry
Luther, Chester Francis, mathematics
Mattes, Frederick Henry, analytical chemistry
McQuate, Robert Samuel, bio-inorganic chemistry
Montague, Daniel Grover, physics
Mortimore, Donald Merton, analytical chemistry
Payton, Arthur David, electrochemistry
Purbrick, Robert Lamburn, physics
Springer, Martha Edith, biology
Thorsett, Grant Orel, molecular biology

SPRINGFIELD
Hine, John Maynard, organic polymer chemistry

TIGARD
Hedrick, Leslie Ray, microbiology

PENNSYLVANIA

ABINGTON
Groner, Miriam Georgia, biology, genetics
Hellriegel, John Curtis, Jr, biochemistry
Moore, Gordon George, organic chemistry
O'Neal, Harry Roger, physical chemistry
Schuster, Ingeborg I M, physical organic chemistry
Tulsky, Emanuel Goodel, radiology
Wilson, Frederick Sutphen, family medicine

ALCOA CENTER
Becker, Aaron Jay, physical chemistry, high temperature chemistry
Bonewitz, Robert Allen, electrochemistry, corrosion
Bonsignore, Patrick Vincent, polymer chemistry
Carpenter, Lee Graydon, industrial chemistry
Cochran, Charles Norman, physical chemistry
Danchik, Richard S, analytical chemistry
Eiland, Ehrlich Mayo, industrial chemistry
Farrah, George Henry, industrial hygiene
Frank, William Benson, industrial chemistry, information science
Guthrie, Joseph D, analytical chemistry
Hartman, Harold Beers, analytical chemistry
Hollingsworth, Ernest Howard, physical chemistry
Kondis, Thomas John, industrial chemistry
Kramer, Raymond Arthur, analytical chemistry
Manhart, Joseph Heritage, organic polymer chemistry, physical chemistry
Matocha, Charles K, spectrochemistry
Milz, Wendell Collins, physical chemistry
Minford, James Dean, metallurgical chemistry
Obbink, Russell C, analytical chemistry, organic chemistry
Rolles, Rolf, surface chemistry
Tingle, William Herbert, instrumentation, spectroscopy
Wahnsiedler, Walter Edward, chemical physics
Wallace, Paul Francis, surface chemistry
Wefers, Karl, crystallography, chemistry
Williams, James Earl, Jr, physical chemistry
Wohleber, David Alan, chemistry
Zeley, Walter Gaunt, electrochemistry

ALLENTOWN
Adda, Lionel Paul, solid state science
Barr, John Tilman, organic chemistry
Basseches, Harold, physical chemistry
Berry, Robert Walter, inorganic chemistry, physical chemistry
Boyer, Robert Allen, physics
Clement, Gerald Edwin, chemistry
Cohen, Howard Melvin, solid state chemistry
DeLaMater, George (Bearse), organic chemistry
Drum, Charles Monroe, solid state physics
Forster, John Heslop, physics
Garbarini, Victor C, electrochemistry
Gerstenberg, Dieter, solid state physics
Gittler, Franz Ludwig, physical chemistry
Goldey, James Mearns, physics
Hall, Peter M, physics
Hart, William Ardley, industrial chemistry
Haruta, Kyoichi, physics
Hatch, Richard C, physical chemistry
Hensler, Donald H, solid state electronics
Jacodine, R J, solid state physics
Kayhart, Marion, genetics
Koehler, Truman Lester, mathematics
Latshaw, David Rodney, analytical chemistry
Mandel, John Herbert, medical microbiology
Manly, Donald G, organic chemistry
Mantell, Gerald Jerome, organic polymer chemistry
McEvoy, James Edward, petroleum chemistry
Meeker, Thrygve Richard, physical chemistry
Miller, Paul, solid state physics
Morabito, Joseph Michael, electronic spectroscopy
Mortimer, Charles Edgar, organic chemistry, history of science
Pearman, G Timothy, acoustics
Perry, Mary Hertzog, analytical chemistry, organic chemistry
Pfahnl, Arnold, physics
Pfeiffer, Heinz Gerhard, energy conversion
Rand, Myron Joel, physical chemistry
Raub, Harry Lyman, III, physics
Recktenwald, Gerald William, physical chemistry, research administration
Rice, Eugene Worthington, clinical biochemistry, clinical chemistry

PENNSYLVANIA

Robertson, Nat Clifton, physical chemistry
Rosenzweig, Walter, physics
Schaeffer, Robert L., Jr., plant taxonomy
Schmauch, George Edward, environmental chemistry
Schmidt, Paul F., semiconductors, surface physics
Shive, Donald Wayne, analytical chemistry
Smart, G N Russell, organic chemistry
Smith, William Mayo, Jr., organic polymer chemistry
Smith, William Novis, Jr., physical organic chemistry
Smits, Friedolf M., physics
Stehly, David Norvin, inorganic chemistry
Trainer, John Ezra, Sr., ornithology
Vaughan, James Roland, microbiology, biochemistry
Weston, John Colby, histology, embryology
White, Malcolm Lunt, physical chemistry
White, Thaddeus Everett, Jr., physical chemistry

AMBLER
Amar, Henri, theoretical physics
Barth-Wehrenalp, Gerhard, industrial chemistry
Bishop, John Russell, agricultural chemistry
Cooke, Anson Richard, plant physiology
Craig, Paul Norman, medicinal chemistry
Deger, Thomas Edward, organic chemistry
Drach, John Edward, industrial organic chemistry
Eckfeldt, Edgar Lawrence, chemistry
Fertig, Stanford Newton, weed science
Harrison, Stanley L., analytical chemistry, environmental sciences
Leeper, Robert Walz, organic chemistry
Leister, Harry M., physical chemistry
McLane, Stanley Rex, Jr., horticulture, plant physiology
Osborne, William Wesley, bacteriology, public health
Precopio, Frank Mario, organic chemistry
Quimby, Daniel Joseph, physical organic chemistry, plant chemistry
Reeves, Richard Franklin, environmental chemistry
Schultz, Everett Maynard, medicinal chemistry
Segal, Hirsh Sholom, analytical chemistry
Steinbrecher, Lester, inorganic chemistry
Strohm, Paul F., agricultural chemistry, organic chemistry

ANNVILLE
Argol, Jeanne, microbiology
Bailey, David Newton, analytical chemistry
Hearsey, Bryan Vandiver, mathematics
Lockwood, Karl Lee, organic chemistry
Mayer, Joerg Werner Peter, mathematics
Moe, Owen Arnold, chemistry
Neidig, Howard Anthony, physical chemistry, organic chemistry
Rhodes, Jacob Lester, nuclear physics
Spencer, James Nelson, physical chemistry
Thompson, Phillip Eugene, solid state physics
Williams, Stephen Edward, plant physiology
Wolfe, Allan Frederick, invertebrate physiology, histology

ALLISON PARK
Chang, Wen-Hsuan, organic chemistry
Dowbenko, Rostyslaw, organic chemistry
Frick, Neil Huntington, physical chemistry, polymer chemistry
Hauser, Mary Martin, biology
Parker, Earl Elmer, organic polymer chemistry
Pierce, Percy Everett, physical chemistry
Wismer, Marco, polymer chemistry

ALTOONA
Carney, Albert Stricker, organic chemistry
Chan, Cheung-King, botany, horticulture
Miskovsky, Nicholas Matthew, physics
Oh, Yoon Yong, mathematics
Russo, Thomas Joseph, organic chemistry
Sheridan, Laurence Ward, mathematics, meteorology
Singh, Sukhjit, topology

BERWYN
Blake, John Wilson, aquatic biology, pollution ecology
Dohany, Julius Eugene, polymer chemistry
Fisher, Sallie Ann, water chemistry, pollution chemistry
Robinette, Hillary, Jr., chemistry
Schmid, James Addison, biogeography, environmental management
Wing, Merle Wesley, entomology

BERWICK
Gale, William F., aquatic ecology

BETHEL PARK
Doub, William Blake, nuclear physics
Fette, Clarence William, mathematical physics

BETHLEHEM
Achey, Frederick Augustus, analytical chemistry
Adenstedt, Rolf Karl, mathematics, mathematical statistics
Allen, Eugene (Murray), physical chemistry, color science
Amstutz, Edward Delbert, organic chemistry
Anderson, David Martin, environmental health
Assmus, Edward Ferdinand, Jr., algebra
Barber, Saul Benjamin, zoology
Benson, Brent W., biophysics, radiation chemistry
Benz, Edward John, clinical pathology, microbiology
Bergmann, Ernest Eisenhardt, lasers
Bergstresser, Kenneth A., biology, zoology
Blythe, Philip Anthony, applied mathematics, fluid mechanics
Borse, Garold Joseph, theoretical physics
Cheng, Thomas Clement, biology
Chu, Vincent Hao Kwong, inorganic chemistry, physical chemistry
Collier, Herman Edward, Jr., analytical chemistry, inorganic chemistry
Crelling, John Crawford, geology
Curtis, Cassius W., physics
Davis, Donald Miller, topology
Edelen, Dominic Gardiner Bowling, applied mathematics
Emrich, Raymond Jay, fluid dynamics, fluid mechanics
Erdogan, Fazil, applied mechanics
Erickson, Edwin Sylvester, Jr., mineralogy, petrography
Evenson, Edward B., glacial geology, geomorphology
Feigl, Frank Joseph, solid state physics
Fetsko, Jacqueline Marie, physical chemistry
Folk, Robert Thomas, nuclear physics
Follweiler, Douglas MacArthur, analytical chemistry
Fowkes, Frederick Mayhew, chemistry
Fowler, Wyman Beall, solid state physics
Gaumer, Albert Edwin Hellick, ecology
Ghosh, Bhaskar Kumar, mathematical statistics
Grunder, Fred Irwin, environmental administration
Hartford, Bradford C., analytical chemistry
Halperin, Theodore, mathematical logic
Hall, Wade Eckes, physical chemistry
Heindel, Ned Duane, organic chemistry, medicinal chemistry
Herman, Sidney Samuel, biological oceanography
Hoagland, Elaine, ecology
Hole, Gilbert Lee, geology
Hsing, Chuan Chih, mathematics
Hughes, Michael Charles, analytical chemistry
Hulbert, Matthew H., analytical chemistry
Jenkins, George Robert, geology
Kazaka, Jacob Yakovos, applied mechanics, continuum mechanics
Kendig, Martin William, physical chemistry
Khabbaz, Samir Anton, mathematics
Kim, Yong Wook, atomic physics, fluid dynamics
King, Jerry Porter, mathematics
Klier, Kamil, physical chemistry
Kohls, Donald W., exploration geology
Kanofsky, Alvin Sheldon, elementary particle physics
Kraihanzel, Charles S., inorganic chemistry
Krawiec, Steven Stack, molecular biology
Kulp, Stuart S., organic chemistry
Kydonieus, Anastasios D., applied mechanics
Leidheiser, Henry, Jr., physical chemistry
Lewis, Willard Deming, physics
Lovejoy, Roland William, physical chemistry
Main, Frederic Hall, economic geology
Maisberger, Richard Griffith, biology
Manson, John Alexander, physical chemistry, polymer science
McAllister, Gregory Thomas, Jr., mathematics
McAllister, Marialuisa N., computer sciences
McCluskey, George E., Jr., astrophysics
McLennan, James Alan, Jr., statistical mechanics
Merkel, Joseph Robert, microbial biochemistry, marine microbiology
Micale, Fortunato Joseph, physical chemistry, colloid chemistry
Nagell, Raymond H., geology
Nigh, Harold Eugene, solid state physics
Ohnesorge, William Edward, analytical chemistry
Owen, Bradford Breckenridge, chemistry
Pitcher, Arthur Everett, mathematics
Pritchard, Hayden N., botany, histochemistry
Radin, Sheldon Henry, plasma physics
Rauch, Stewart Emmart, Jr., organic chemistry
Richards, Adrian Frank, oceanography, ocean engineering
Rivlin, Ronald Samuel, applied mathematics
Rodgers, Robert Stanleigh, analytical chemistry
Roeder, Edward A., solid state physics
Ryan, John Donald, geology
Sands, Jeffrey Alan, biophysics
Sawyers, Kenneth Norman, applied mathematics
Scaffer, Stephen Ward, biochemistry
Schechter, Murray, mathematics
Schiff, Robert, medical research
Schray, Keith James, bio-organic chemistry
Solar, Charles Bertram, geology
Shaffer, Russell Allen, theoretical physics
Shoemaker, Carlyle Edward, inorganic chemistry
Simmons, Gary Wayne, physical chemistry, surface chemistry
Simpson, Dale R., petrology, mineralogy
Simpson, Samuel John, economic geology
Smith, Gerald Francis, applied mathematics, mechanics
Smith, Wesley R., fluid physics, molecular physics
Smyth, Donald Morgan, solid state chemistry
Snyder, Andrew Kagey, mathematics
Snyder, John M., surgery
Spatz, Wilber De Villa Bernhart, physics
Sperling, Leslie Howard, physical science
Spriggs, Richard Moore, academic administration, ceramic engineering
Summer, John Randolph, geophysics
Thwaite, Robert David, geology, physical science
Trutt, David, physics
Vanderhoff, John W., polymer chemistry, colloid chemistry
Van Sciver, Wesley J., physics
Watterson, Kenneth Franklin, inorganic chemistry
Wheeler, Donald Bingham, Jr., physics
Wilansky, Albert, mathematics
Yoshino, Timothy Phillip, parasitology, immunobiology
Young, Thomas Edwin, organic chemistry
Zeroka, Daniel, theoretical chemistry
Zettlemoyer, Albert Charles, colloid chemistry, surface chemistry

ARDMORE
Koch, Robert Harry, astronomy
Schweiker, George Christian, chemistry, research administration
Solomons, Cyril, physical chemistry

AVONDALE
Bott, Thomas Lee, microbiology
Howell, Edward Tillson, organic chemistry
Larson, Richard Allen, bio-organic chemistry
Merner, Richard Raymond, industrial organic chemistry, science administration
Rogers, Emery Herman, physics
Rowland, Fred W., instrumentation
Vannote, Robin L., aquatic ecology

BANGOR
Hallet, Lawrence Trenery, analytical chemistry
Stolten, Hans Joseph, physical chemistry

BALA-CYNWYD
Kravitz, Edward, science writing
Rubin, Benjamin Arnold, microbiology

BEAVER
Rudolph, Robert Lewis, analytical chemistry

BEAVER FALLS
Adams, Roy Melville, chemistry
Bruce, Harold Asa, biology
Cruzan, John, ecology
Hartman, Kenneth Eugene, organic chemistry, medicinal chemistry
McMillion, Theodore Miller, zoology
Pinkerton, John Edward, experimental nuclear physics, electronic engineering

BLUE BELL
Baltz, Alfred, physics, physical chemistry
Delaney, Frank Michael, mathematics
Doyle, William David, solid state physics
Goldberg, Norman, physics
Greiter, Aaron Philip, physical inorganic chemistry, magnetism
Jones, James Holden, organic chemistry
Seitchik, Jerold Alan, solid state physics
Smith, Josephine Reist, medical microbiology
Stein, Barry Fred, solid state physics

BOALSBURG
Hyre, Russell A., plant pathology

BOOTHWYN
Bechara, Ibrahim, organic chemistry
Driscoll, Gary Lee, organic chemistry

BOYERTOWN
Cerbulis, Janis, biochemistry, agriculture
Gaines, Richard Venable, economic geology, mineralogy
Gustison, Robert Abdon, inorganic chemistry

BRACKENRIDGE
Burk, David Lawrence, solid state physics

BIGLERVILLE
Asquith, Dean, economic entomology
Bode, William Morris, entomology
Lewis, Frank Herbert, plant pathology
Stouffer, Richard Franklin, plant pathology, virology

BLOOMSBURG
Adams, Bruce Edward, physical geography, geography of Europe
Anderson, Wayne Philpot, inorganic chemistry
Benson, Barrett Wendell, organic chemistry
Cole, James Edward, ethology, vertebrate zoology
Ennam, John A., geography
Farber, Phillip Andrew, cell biology, cytogenetics
Frantz, Wendelin R., petroleum geology, stratigraphy
Gates, Halbert Frederick, physics
Herbert, Michael, bacteriology, biochemistry
Klenner, Jerome James, physiology
Kroschewsky, Julius Richard, botany, phytochemistry
Lanterman, Harold H., chemistry
Mack, Lawrence Lloyd, physical chemistry, biochemistry
Mingrone, Louis V., botany
Taebel, Wilbert August, chemistry
Taylor, Merlin Gene, physics
Taylor, Norman Edward, physical chemistry
White, Norman Marshall, chemistry

BRADFORD
Boutros, Osiris Wahba, plant physiology
Boutros, Susan Noblit, cytogenetics
Gray, June Pfister, physical chemistry, chemical engineering
Heck, Edward Timmel, geology
Keder, Wilbert Eugene, physical inorganic chemistry

BRIDGEVILLE
de Torok, Denes Gabor, biology

BRISTOL
Clarke, Duane Grockett, chemistry
De Jong, Gary Joel, analytical chemistry
Drake, Billy Blandin, biochemistry, enzymology
Fellmann, Robert Paul, chemistry
Goldman, Theodore Daniel, organic polymer chemistry
Graham, Roger Kenneth, polymer chemistry
Gutbezahl, Boris, polymer chemistry
Harrop, William Henry, rubber chemistry, plastics chemistry
Kittle, Paul Alvin, physical chemistry, organic chemistry
Kopchik, Richard Michael, organic polymer chemistry
Lane, Constance A, polymer chemistry, plastics chemistry
Larkin, Robert Hayden, analytical chemistry
Li, Wu-Shyong, physical organic chemistry
Mueller, Donald Scott, plastics chemistry, polymer chemistry
Neubeck, Clifford Edward, chemistry
Opie, Thomas Ranson, organic chemistry
Hwang, Bruce You-Huei, organic chemistry, medicinal chemistry
Pipenberg, Kenneth James, analytical chemistry
Schindler, Frederick James, polymer chemistry
Sheasley, William David, physical chemistry
Singh, Udai Pratap, chemistry
Weese, Richard Henry, organic polymer chemistry
Yanai, Hideyasu Steve, analytical chemistry

BROWNSVILLE
Duda, John J, medical microbiology, electron microscopy

BRYN ATHYN
Allen, Edward Franklin, physics

BRYN MAWR
Anderson, Jay Martin, physical chemistry, environmental sciences
Berliner, Ernst, organic chemistry
Berliner, Frances (Bondhus), organic chemistry
Conner, Robert Louis, zoology
Crawford, Maria Luisa Buse Petrology, mineralogy
Crawford, William Arthur, geochemistry
Cunningham, Frederic, Jr, mathematics
De Laguna, (Lopez de Leo), Frederica (Annis), cultural anthropology, archaeology
Goodale, Jane Carter, anthropology
Hoyt, Rosalie Chase, physics
Kaney, Anthony Rolland, genetics
Koroly, Mary Jo, biochemistry, cell biology
Lucy, Frank Allen, physics
Mallory, Clelia Wood, organic chemistry
Mallory, Frank Bryant, organic chemistry
Oppenheimer, Jane Marion, biology
Oxtoby, John Corning, mathematics
Platt, Lucian B, structural geology
Prescott, David Julius, protein chemistry, neurochemistry
Pruett, John Robert, physics, computer science

Pruett, Patricia Onderdonk, cell physiology, biophysics
Rogerson, Allen Collingwood, molecular biology
Scheidy, Samuel F, veterinary medicine
Smith, Stephen Roger, physics
Varimbi, Joseph, electrochemistry
Young, Jay Maitland, physical biochemistry, enzymology
Zimmerman, George Landis, physical chemistry

BUTLER
Mould, Richard Everett, solid state physics, glass technology
Southwick, Russell Duty, physics

CALIFORNIA
Agrawal, Jagdish Chandra, applied mathematics
Balling, Jan Walter, animal behavior
Bausor, Sydney Charles, botany, microbiology
Betz, Gabriel Pohl, geography
Bitonti, John, speech pathology, audiology
Cignetti, Jess A, science education
Fusco, Gabriel Carmine, organic chemistry
Hood, Samuel Lowry, biochemistry
Hunter, Barry B, mycology, plant pathology
Kells, Milton Carlisle, physical chemistry
Leavy, Thomas A, earth science
Machusko, Andrew Joseph, Jr, mathematics
Minnick, Robert Fletcher, cultural geography
Moon, Thomas Charles, science education, ecology
Mullins, Jeanette Somerville, microbial ecology, plant taxonomy
Riggle, John H, mathematics
Tomikel, John, earth science
Vajk, Raoul, geophysics

CALLERY
Hough, William Vernon, inorganic chemistry

CAMBRIDGE SPRINGS
Jenkins, Charles Rivington, archeology, anthropology
Szymanski, Herman A, physical chemistry

CAMP HILL
Herpel, Coleman, mathematics
Witkoski, Francis Clement, organic chemistry

CARLISLE
Baric, Lee Wilmer, mathematics
Benson, John Edward, surface chemistry
Benson, William Howard, mathematics
Biebel, Paul Joseph, phycology
Crist, Ray Henry, chemistry
Hanson, Henry W A, III, sedimentology
Herber, Elmer Charles, zoology
Jeffries, William Bowman, invertebrate zoology
Laws, Kenneth Lee, solid state physics, meteorology
Laws, Priscilla Watson, nuclear physics
Leyon, Robert Edward, analytical chemistry
Light, John Henry, mathematics
Long, Howard Charles, physics
Luetzelschwab, John William, health physics
McDonald, Barbara Brown, cytology
McDonald, Daniel James, genetics
Potter, Noel, Jr, geology
Rogers, Horace Elton, physical chemistry, analytical chemistry
Roper, Gerald C, physical chemistry
Scharer, William Richard, organic chemistry
Seaford, Henry Wade, Jr, anthropology
Sheeley, Richard Moats, medicinal chemistry, organic chemistry
Sia, Richard Mae, physics
Vernon, William W, geology
Wolf, Neil Steven, plasma physics
Wolgemuth, Kenneth Mark, marine geochemistry
Ziegler, George William, Jr, physical chemistry

CARNEGIE
Briggs, Reginald Peter, geology

CENTER VALLEY
Dooling, John Stuart, physical chemistry, academic administration
Gadek, Frank Joseph, organic chemistry
Talbott, Francis Leo, nuclear physics

CHADDS FORD
Arnold, Harold Wilfred, chemistry
Boardman, Harold, industrial chemistry
Kliman, Harvey Louis, physical chemistry, polymer physics
Middleton, William Joseph, organic chemistry
Smith, Elgene Arthur, plastics chemistry

CHALFONT
Carty, Daniel T, organic chemistry, polymer chemistry
Woltersdorf, Otto William, Jr, medicinal chemistry

CHAMBERSBURG
Dropp, John Jerome, biology, histology
Grove, Davison Greenawalt, botany
MacDonald, Eve Lapeyrouse, developmental biology, cell biology
Monack, Louise Charlotte, chemistry
Portmann, Walter Oddo, mathematics
Scheele, Harry George, anthropology, archaeology
Waggoner, Margaret Ann, nuclear physics, history of science
Yingst, Harvey Austin, physical inorganic chemistry

CHELTENHAM
Conlon, Daniel Rupert, chemistry, instrumentation
Rabiger, Dorothy June, organic chemistry

CHESTER
Arnold, David Brown, physical organic chemistry
Clark, James Edward, internal medicine
Conroy, James Strickler, fuel chemistry
Czepel, Thomas P, paper chemistry
DeCaro, Thomas F, cellular physiology
Gottlieb, Irvin M, inorganic chemistry, analytical chemistry
Kamel, Hyman, mathematics
O'Tanyi, Theodore John, Jr, neurosciences
Rao, H V Ramakrishna, developmental biology
Smith, Allen Anderson, developmental biology, histochemistry
Smith, Harold Edmond, physics
Storlazzi, Joseph Jordan, biology, bacteriology
Thornton, Elizabeth K, physical organic chemistry
Wolfe, Dorothy Wexler, mathematics

CHESTER SPRINGS
Peacock, Samuel Moore, Jr, neurophysiology

CHESWICK
Myers, Earl Eugene, petroleum chemistry

CHEYNEY
Stevenson, Robert William, organic chemistry
Taylor, Donald Fulton, pathobiology, biochemistry
Taylor, John H, organic chemistry
Vinton, William Howells, organic chemistry, inorganic chemistry

CLAIRTON
Irwin, Philip George, physical organic chemistry
Wilson, Dean George, physical chemistry, inorganic chemistry

CLARION
Beck, Paul Edward, organic chemistry
Buckwalter, Tracy Vere, Jr, geology
Dinsmore, Bruce Heasley, ecology, aquatic biology
Harmon, George Andrew, biochemistry, microbiology
Hart, William James, physical chemistry, inorganic chemistry
Kodrich, William Ralph, ecology
Konitzky, Gustav Adolf, anthropology, archaeology

Linton, Kenneth Jack, aquatic ecology, biometry
Moore, John Robert, plant ecology
Morrow, Terry Oran, microbial genetics
Oakes, Lester Charles, geography
Ossesia, Michel Germain, mathematics
Shirey, George S, geography
Shontz, Charles Jack, animal ecology, human ecology
Snedegar, William H, Jr, nuclear physics
Totten, Don Edward, cultural geography
Twiest, Gilbert Lee, ornithology, science education

CLARKS SUMMIT
Appleton, Martin David, biochemistry

CLEARFIELD
Landy, Richard Allen, mineralogy

COATESVILLE
Bell, Robert Lloyd, medicine, neurosurgery
Franzl, Robert E, immunobiology
Lame, Edwin Lever, radiology
Sprince, Herbert, biochemistry, pharmacology

COLLEGEVILLE
Anderson, Robert Sven, lasers
Dennis, Foster Leroy, geometry
Di Cuollo, C John, biochemistry, microbiology
Hager, Robert B, organic chemistry
Hess, Ronald Eugene, organic chemistry
Howard, Robert Stearns, biology
Kruse, Conrad Edward, bacteriology
Levesque, Charles Louis, organic chemistry
Lewis, Everett Vernon, statistics
Schultz, Blanche Beatrice, mathematics
Schultz, Ray Karl, chemistry
Snyder, Evan Samuel, nuclear physics
Staiger, Roger Powell, organic chemistry

CONSHOHOCKEN
Francis, Peter Schuyler, physical organic chemistry
Lange, Klaus Robert, physical chemistry, surface chemistry
Shearer, Charles M, analytical organic chemistry
Tucker, Charles R, textile chemistry
Varga, Charles E, physical organic chemistry

COOPERSBURG
Lam, Gow Thue, microbiology

CORNWELLS HEIGHTS
Lever, Cyril, Jr, organic chemistry
Rosenstock, Paul Daniel, organic chemistry

COUDERSPORT
Reding, Georges Rene, psychiatry

DANVILLE
Fleetwood, Mildred Kaiser, immunopathology
Wells, George Sherman, biochemistry

DEVON
Levin, Michael Howard, ecology, resource management
McCormick, Jack Sovern, environmental management, ecology
Young, John Albion, Jr, petroleum geology

DINGMANS FERRY
Phillips, Charles John, physics

DOWNINGTOWN
Adrounie, V Harry, environmental management

DOYLESTOWN
Adelson, Lionel Morton, marine ecology, histology
Allison, William Hugh, mycology
Balke, Claire Coddington, chemistry
Berthold, Robert, Jr, apiculture, entomology
Blackmon, Clinton Ralph, plant breeding

PENNSYLVANIA

Blumenfield, David, horticulture, plant physiology
Coyne, Veronica E., internal medicine, immunology
Elson, Jesse, physical chemistry
French, Ellery Walter, entomology, ecology
Garrett, Michael Benjamin, physical chemistry, medicinal chemistry
Kahan, I Howard, avian pathology, microbiology
Orr, Robert S., organic chemistry
Weber, Charles Walton, analytical chemistry, inorganic chemistry

DREXEL HILL
Langner, Paul Harry, Jr., cardiology
Morton, Harry E., bacteriology
Rosen, Harry, pharmacology
Rosenbaum, Eugene Joseph, physical chemistry
Turck, Joseph Abraham Valentine, Jr., organic chemistry
Van Meter, Clarence Taylor, medicinal chemistry, pharmaceutical chemistry

DRUMORE
Mathur, Dilip, fish biology

DUNMORE
Baildon, John David, geometry, topology
Weiss, James Allyn, organic chemistry

EAST PITTSBURGH
Johnson, Woodrow Eldred, physics

EAST STROUDSBURG
Anderson, Neil Owen, cell physiology
Baxevanis, John James, geography
Cheng, Cheng-Yin, soil chemistry, analytical chemistry
Cheng, Chiang-Shuei, plasma physics
Fremount, Henry Neil, biology, medical parasitology
Grainger, Thomas Hutcheson, Jr., medical bacteriology
Haase, Bruce Lee, aquatic ecology, human ecology
Kelsey, Ruben Clifford, comparative endocrinology
Kicska, Paul A., physics
Maclay, Charles Wylie, mathematics
Murphy, Clarence John, organic chemistry, inorganic chemistry
Palaitis, Waldemar, physical organic chemistry, biophysical chemistry
Pruser, Etha Marie, geography
Rao, Balakrishna Raghavendra, insect physiology
Rymon, Larry Maring, animal ecology, conservation
Saich, Richard K., plant pathology
Schramm, Robert Frederick, chemistry
Shwe, Hla, high energy physics, nuclear physics
Wagner, Timothy Knight, solid state physics
Wilkinson, Arthur, mathematics, computer science

EASTON
Allison, Jerry Robert, organic chemistry
Ansbacher, Theodore Henry, solid state physics
Barr, James K., physical chemistry, research administration
Bruun, Johannes Hadeln, chemistry, chemical engineering
Chase, Robert Silmon, Jr., vertebrate zoology
Cook, Robert Crossland, chemical physics
Copes, Joseph Paul, physical chemistry, organic chemistry
Crockett, David Scott, inorganic chemistry
Daniels, Wiley Edgar, organic chemistry
Erich, Lester Charles, physics
Faas, Richard William, geology, paleontology
Fried, Bernard, parasitology
Hart, William Forris, organic chemistry
Hicks-Bruun, Mildred M., physical chemistry
Hogenboom, David L., high pressure physics, nuclear magnetic resonance
Homberg, Otto Albert, organic chemistry
Hoskin, George Perry, physiology, invertebrate pathology
James, Laylin Knox, Jr., surface chemistry
Jeffers, William Allen, Jr., low temperature physics
Keck, Winfield, physics
Kugler, Blanca Louise, surface chemistry
Locke, Harold Ogden, physical chemistry, analytical chemistry
Miller, Thomas Gore, organic chemistry
Novaco, Anthony Dominic, theoretical solid state physics, surface physics
Pauli, George H., photochemistry
Peace, George Earl, Jr., analytical chemistry
Relkin, Richard, endocrinology
Roper, Paul James, structural geology, petrology
Saalfrank, Charles W., analytical chemistry
Sherma, Joseph A., analytical chemistry
Stableford, Louis Tranter, biology
Walters, Lee Rudyard, organic chemistry

EDINBORO
Bowne, Samuel Winter, Jr., chemistry, genetics
Brand, Richard Robert, geography
Brandt, Donald Paul, urban geography, population geography
Butler, Ronald G., mathematics
Carls, Ralph A., bacteriology
Come, Thomas V., science education
DeFigio, Francis V., mycology
Gatzy, John Thomas, biology
Gilinan, David Anthony, physics
Legge, Thomas Steiner, zoology, physiology
Long, Harriet Ruth, geography
Lowenhaupt, Benjamin, biophysics, plant physiology
Lucas, Stephen Bernard, science education, health sciences
Lukert, Michael T., geology, geochemistry
Masing, Ulv, geography
McKay, James Brian, analytical chemistry
McKinley, James Ernest, mathematics
Miller, Gerson Harry, mathematics, medical statistics
Nair, K Aiyappan, mathematical statistics
Newby, Neal Dow, Jr., theoretical physics
Olsen, Glenn W., mathematics education
Robinson, Curtis, plant physiology
Sarma, Pramod Lal, analytical chemistry
Schneider, Michael Charles, geology
Snyder, Donald Benjamin, biology, ecology
Thomas, Paul Milton, immunobiology, ichthyology
Wagner, David Loren, low temperature physics, solid state physics
Wainer, Arthur, biochemistry
Walkiewicz, Thomas Adam, nuclear physics, solid state physics
Wegweiser, Arthur E., geology
Weller, Richard Irwin, physics, academic administration
Wheeler, Donald Alsop, genetics
Witthuhn, Burton Orrin, geography

EIGHTY FOUR
Guild, Lloyd V., chemistry

ELIZABETHTOWN
Custer, Hubert Minter, physics
Lehr, Raymond Bruce, anthropology, archaeology
Pepper, Rollin E., microbiology
Ranck, John Philip, physical chemistry
Reeder, Ray Robert, physical chemistry, inorganic chemistry
Spangler, Martin Ord Lee, organic chemistry, biochemistry
Stambaugh, Oscar Frank, chemistry

ELKINS PARK
Baum, Harry, analytical chemistry

ELWYN
Clark, Gerald Robert, psychiatry

EMMAUS
Musser, David Muselman, organic chemistry

ERIE
Alstadt, Don Martin, physics
Balmer, Louis Whiteside, organic chemistry
Becker, Robert Hugh, organic chemistry
Bollinger, Richard Coleman, applied mathematics
Burkhard, Charles (Austin), organic chemistry
Cole, Quintin Perry, chemistry
Cunningham, Harry N Jr, chemistry
Eckroat, Larry Raymond, genetics, fish biology
Fasemyer, Mary Celine, mathematics
Gammon, Richard Anthony, microbiology
Gruenwald, Geza, plastics chemistry
Gunther, Donald Albert, analytical chemistry, biochemistry
Guthrie, Donald Arthur, organic chemistry, radiation chemistry
Halleck, Frank Eugene, biomedical engineering
Hostetler, Robert Paul, mathematics education
Kueber, John Ralph, Jr., industrial chemistry
Larson, Roland Edwin, mathematics
Leffield, Robert Francis, inorganic chemistry
Lewis, Armand Francis, physical chemistry
Masteller, Edwin C., biology, entomology
McDowell, John Robert, physical chemistry
McKinstry, Donald Michael, animal physiology, herpetology
Obermanns, Henry Ernst, chemistry
Pelczar, Francis A., organic chemistry
Poydock, Mary Eymard, experimental medicine
Sexsmith, Frederick Hamilton, chemistry
Shim, Benjamin Kin Chong, physical chemistry
Spector, Calvin, plant physiology, genetics
Tongren, John Corbin, chemistry
Weschler, Mary Charles, environmental chemistry

ESSINGTON
McConnell, Albert Lawrence, chemistry

EVANS CITY
Mausteller, John Wilson, physical chemistry
Rodgers, Sheridan Joseph, analytical chemistry

EXPORT
Buzzell, Edward S., physics, electrochemistry
Richter, Helen Wilkinson, physical chemistry

EXTON
Barrett, Wayne Thomas, physical chemistry
Brown, Patrick Michael, physical chemistry
Grady, Harold Roy, physical chemistry
Kunasz, Ihor Andrew, economic geology

FLOURTOWN
Lipsitz, Paul, organic chemistry

FOGELSVILLE
Warmkessel, Carl Andrew, geology

FT WASHINGTON
Almond, Harold Russell, Jr., analytical chemistry, organic chemistry
Appino, James B., industrial pharmacy
Auyang, King, organic chemistry
Benica, William Steinhart, pharmacy
Brenner, Ronald John, pharmaceutical chemistry
Buehler, John David, pharmaceutical chemistry
Cain, Cornelius Kennady, medicinal chemistry
Ciccone, Patrick Edwin, psychopharmacology, psychiatry
Cressman, William Arthur, pharmacology
Cullumbine, Harry, pharmacology
Diamond, Julius, medicinal chemistry
Erlich, Ronald Harvey, analytical chemistry
Furgivele, Angelo Ralph, pharmacology
Gardocki, Joseph F., pharmacology
Gilpin, Roger Keith, analytical chemistry
Greenberg, Samuel Mendel, biochemistry
Gussin, Robert Z., pharmacology
Hageman, William E., pharmacology, physiology
Madison, William Leon, pharmaceutical chemistry
Maryanoff, Bruce Eliot, synthetic organic chemistry, medicinal chemistry
Migdalof, Bruce Howard, drug metabolism
Millstein, Lloyd Gilbert, medical physiology, information science
Muschek, Lawrence David, biochemical pharmacology
Peters, Lawrence, pharmacology
Piperno, Elliot, veterinary medicine, pharmacology
Plostnieks, Janis, organic chemistry
Poos, George Ireland, organic chemistry
Pruss, Thaddeus P., pharmacology
Rasmussen, Chris Royce, medicinal chemistry
Reavey-Cantwell, Nelson Henry, medicine
Reynolds, Brian Edgar, medicinal chemistry, organic chemistry
Rudolph, Jeffrey Stewart, pharmacy, chemistry
Seay, Patrick Herbert, pharmacology
Slotnick, Victor Bernard, medicine
Smyth, Robert Daniel, biochemistry, immunochemistry
Studt, William Lyon, medicinal chemistry
Tuwiler, Gene Floyd, biochemistry
Umen, Michael Jay, endocrinology
Uscavage, Joseph Peter, bacteriology, mycology
Walker, Robert Winn, physical chemistry
Walking, Walter Douglas, pharmacy
Wong, Stewart, pharmacology
Zalipsky, Jerome Jaroslaw, analytical chemistry

FURLONG
Davis, Philip Seals, physical chemistry

GETTYSBURG
Barnes, Robert Drane, invertebrate zoology
Beach, Neil William, zoology
Cavaliere, Alphonse Ralph, botany
Cowan, David J., physics
Darrah, William Culp, paleobotany
Fortnum, Donald Holly, physical chemistry
Fryling, Robert Howard, mathematics
Haskins, Joseph Richard, nuclear physics
Hendrickson, Thomas James, physics
Holder, Leonard Irvin, mathematics
Logan, Rowland Elizabeth, physiology
Mara, Richard Thomas, physics
Mikesell, Jan Erwin, plant anatomy
Parker, William Evans, inorganic chemistry
Rowland, Alex Thomas, steroid chemistry
Schildknecht, Calvin Everett, polymer chemistry
Schroeder, Allen C., animal physiology
Weiland, Glenn Statler, biochemistry
Winkelmann, John Roland, vertebrate zoology, mammalogy

GIBSONIA
Buechler, Peter Robert, chemistry
Erikson, J Alden, organic chemistry
McBane, Bruce Newton, industrial chemistry
Robinson, John W., Jr., organic chemistry
Swift, Harold Eugene, physical inorganic chemistry

GLADWYNE
Gilstein, Jacob Burrill, physics, research administration
Montgomery, Hugh, medicine

GLEN MILLS
Meade, Alston Bancroft, entomology
Robinson, John N. W., Jr., leather chemistry

GLENOLDEN
Capaldi, Eugene Carmen, organic chemistry
Connor, James Edward, Jr., physical organic chemistry
Leonard, John Joseph, chemistry
Roseinthal, Rudolph, organic chemistry

Ryan, Patrick Walter, organic polymer chemistry
Shalit, Harold, organic chemistry
Sheng, Ming Nan, organic chemistry
Zajacek, John George, organic chemistry
Zehner, Lee Randall, organic chemistry

GLEN RIDDLE-LIMA
Arnold, Mary Tryson, forensic science, mycology

GLENSHAW
Walsh, William Louis, organic chemistry
Wuenschel, Paul Clarence, geophysics

GLENSIDE
Breyer, Arthur Charles, physical chemistry, inorganic chemistry
Cartier, Peter G, physical chemistry
Cooper, Murray Irving, entomology
Lang, Edgar Reed, polymer chemistry
Machlowitz, Roy Alan, analytical biochemistry, virology
McGonigal, Paul J, physical chemistry, chemical engineering
Mikulski, Chester Mark, inorganic chemistry
Rose, Raymond Wesley, Jr, molecular genetics
Rothman, Edward Samuel, organic chemistry

GRANTHAM
Cassel, David Wayne, algebra
Hoover, Kenneth Bert, ecology
Paine, Dwight Milton, mathematics

GREENSBURG
Bettwy, Mary Leon, inorganic chemistry
Flamman, M Muriel, botany
Infanger, Ann Martin, genetics
Mann, Jacinta, applied statistics
Ozarda, Ahsen T, radiotherapy, nuclear medicine
Solomon, Miriam Grace, chemistry
Walker, William Howard, aquatic ecology
Winters, Mary Ann, biochemistry

GREENVILLE
Bennett, Richard Bond, organic chemistry
Heald, Emerson Francis, physical chemistry, geochemistry
Ingersoll, Ronald John, radiation biology, molecular biology
Kolossvary, Bela Gabriel, physics
Mason, Walter Harry, physiology, pharmacology
Nichols, John C, mathematics
Ode, Philip E, population biology, entomology
Pronay, Andrew C, chemistry
Safford, Edward LaPorte, inorganic chemistry
Sheppard, David W, physics
Stallwood, Robert Antony, nuclear physics
Starks, Aubrie Neal, Jr, analytical chemistry
Yeardley, Nelson Paul, mathematics

GROVE CITY
Brenner, Frederic J, ecology, behavioral biology
Conder, Harold Lee, organometallic chemistry
Fabian, Michael William, vertebrate zoology, embryology
Jeffers, Edmund E, biology, microbiology
Naegele, Edward Wister, Jr, organic chemistry
Shaw, John Thomas, organic chemistry
Swezey, William Weekley, biology

GWYNEDD VALLEY
Beyer, Karl Henry, Jr, pharmacology, physiology
Boyle, Mary Maurice, physics, biophysics

HANOVER
Elehwany, Nazmy Elhamy, horticulture

HARRISBURG
Berkheiser, Samuel William, clinical pathology
Botdorf, Ruth G, analytical chemistry, inorganic chemistry
Boyer, Lee Emerson, mathematics
Cutt, Roger Alan, physiological psychology, audiology
Fall, Rodger Tanner, structural geology
Geyer, Alan Raymond, environmental geology
Gilmore, Hugh Richmond, Jr, pathology
Griswold, Robert Edward, analytical chemistry, clinical chemistry
Hoskins, Donald Martin, geology
Inners, Jon David, geology
Kahn, David, solid state physics
Kwalwasser, William David, organic chemistry
Laughlin, Robert David, physics
Loughry, Frank Glade, soil conservation
McKee, Ruth Stauffer, algebra
Oliver, Morris Albert, engineering statistics
Quickel, Kenneth Elwood, medicine
Redmond, John Peter, plastics chemistry
Reed, Allan Hubert, electrochemistry, physical chemistry
Root, Samuel I, geology
Schrack, William Dunton, Jr, public health
Sevon, William David, III, geology
Shumway, Clare Nelson, (Jr), pediatrics, hematology
Slysh, Roman Stephan, polymer chemistry
Socolow, Arthur A, economic geology
Sullivan, George Allen, information science, statistics
Tokuhata, George K, epidemiology, public health
Tristan, Theodore A, radiology
Valley, Karl Roy, insect taxonomy
Wheeler, Alfred George, Jr, entomology

HATBORO
Vernon, John Ashbridge, organic chemistry

HAVERFORD
Chesick, John Polk, physical chemistry
Davidon, William Cooper, mathematical physics, numerical analysis
Dunathan, Harmon Craig, organic chemistry
Finger, Irving, genetics
Gavin, Robert M, Jr, physical chemistry
Goff, Christopher Godfrey, molecular biology, biochemical genetics
Green, Elizabeth Ufford, cytology
Green, Louis Craig, astrophysics
Kessler, Dietrich, cell biology
Lerman, Charles Lew, bio-organic chemistry
Loewy, Ariel Gideon, physiology
MacKay, Colin Francis, physical chemistry
Matacic, Slavica Smit, biochemistry
Meiss, Alfred Nelson, biology
Mihram, George Arthur, mathematical statistics, systemic sciences
Miller, Douglas Gordon, physics
Newirth, Terry L, organic biochemistry
Partridge, Robert Bruce, astronomy
Santer, Melvin, microbiology
Showe, Michael Kent, biochemistry, molecular biology
Wintner, Claude Edward, organic chemistry

HAVERTOWN
Flitter, David, organic chemistry
Gorby, Charles K, experimental medicine, pharmacology
Koch, Walter Theodore, organic chemistry
Miehle, William, mathematics
Morgan, James Frederick, industrial hygiene
Turner, William Russel, petroleum chemistry, polymer chemistry

HAZLETON
Montjar, Monty Jack, physical chemistry
Morana, Simon Joseph, organic chemistry
Orbin, David Paul, nematology, botany
Romberger, Karl Arthur, analytical chemistry, physical chemistry
Smith, David, physical chemistry

HELLERTOWN
Gerhard, Sherman Leidich, applied physics
Landis, Eugene Markley, physiology, medicine

HERSHEY
Arthur, Alan Thorne, reproductive biology
Baird, Irwin Lewis, anatomy
Brodie, Bernard Beryl, pharmacology
Bullock, Leslie Patricia, endocrinology
Connor, John D, pharmacology
Daniels-Severs, Anne Elizabeth, pharmacology
Davidson, Eugene Abraham, biochemistry
Demers, Laurence Maurice, clinical pathology, endocrinology
Dvorchik, Barry Howard, pharmacology
Flaim, Kathryn Erskine, physiology
Flaim, Stephen Frederick, physiology
Fritz, Paul John, biochemistry
Glaser, Ronald, virology, cytology
Gluntz, Martin Lucius, organic chemistry, food technology
Harrison, Timothy Stone, surgery
Hass, Louis F, biochemistry
Hernandez, Milton John, cardiovascular physiology
Hill, Charles Whitacre, biochemistry, genetics
Hill, Richard Norman, genetics, cell biology
Hyman, Richard W, microbiology, biochemistry
Idell-Wenger, Jane Arlene, biochemistry, physiology
Jefferson, Leonard Shelton, physiology
Jeffries, Graham Harry, internal medicine, gastroenterology
Kennedy, Donald Alexander, medical anthropology
Krieg, Arthur F, pathology
La Noue, Kathryn F, biochemistry, cell physiology
Lausch, Robert Nagle, immunobiology
Li, Jeanne B, biochemistry, physiology
Lipton, Allan, internal medicine, oncology
Mann, Elton W, microbiology, plant pathology
Marquez, Ernest Domingo, biological chemistry
McCallister, Lawrence P, microscopic anatomy
McPherson, Alexander, biological structure
Meyers, Elwood William, chemistry
Morgan, Howard E, physiology, biochemistry
Mortel, Rodrigue, obstetrics & gynecology
Mortimore, Glenn Edward, physiology, cell biology
Munger, Bryce Leon, human anatomy, cell biology
Naeye, Richard L, pathology
Nahrwold, David Lange, surgery, gastroenterology
Nahrwold, Michael Lange, anesthesiology, neurosciences
Neely, James Robert, cardiovascular physiology
Oliver, George Davis, Jr, medical physics, radiation physics
Passananti, Gaetano Thomas, enzymology
Prystowsky, Harry, medicine
Pubols, Benjamin Henry, Jr, neurophysiology
Purnell, Dallas Michael, experimental pathology, cancer
Rannels, Donald Eugene, Jr, physiology
Rapp, Fred, virology
Roark, Dennis Edward, biochemistry, biophysics
Rohner, Thomas John, urology
Rose, Richard Carrol, physiology
Rosenberg, Abraham, biochemistry
Schengrund, Cara-Lynne, biochemistry
Severs, Walter Bruce, pharmacology
Shiman, Ross, biochemistry
Smith, Steven Joel, biochemistry
Vargo, Steven William, audiology
Vesell, Elliot S, molecular pharmacology, biochemical pharmacology
Waldhausen, John Anton, surgery
Ward, Walter Frederick, physiology, endocrinology
Weisz, Judith, reproductive physiology, neuroendocrinology
Whitfield, Carol Faye, physiology, biochemistry
Williams, Geneva Hyland, developmental biology
Woodside, Kenneth Hall, physiology, biochemistry
Yeakel, Allen Egger, anesthesiology
Zagon, Ian Stuart, neurobiology, protozoology

Zoumas, Barry Lee, nutrition, food science

HOMESTEAD
Wang, Yen, radiology, nuclear medicine

HORSHAM
Carter, Orwin Lee, physical chemistry

HUNTINGDON
Blaisdell, Baalis Edwin, applied mathematics
Engle, Irene May, theoretical physics
Fagot, Wilfred Clark, applied mathematics, mathematical physics
Fisher, Robert L, animal ecology, mammalogy
Hartzler, Eva Ruth, biochemistry
Mitchell, Donald J, physical chemistry, crystallography
Norris, Wilfred Glen, chemical physics
Rockwell, Donald Mason, organic chemistry
Rockwell, Kenneth H, zoology
Russey, William Edward, organic chemistry
Schettler, Paul Davis, Jr, physical chemistry
Senft, Joseph Philip, physiology
Trexler, John Peter, geology, stratigraphy
Wampler, Dale Lee, structural chemistry, inorganic chemistry
Washburn, Robert Henry, stratigraphy, structural geology
Zimmerer, Robert P, plant physiology, microbiology

HUNTINGDON VALLEY
Bauer, William, Jr, organic chemistry, chemical engineering
Baumgarten, Werner, biochemistry
Boettner, Fred Easterday, pharmaceutical chemistry
Dickstein, Jack, organic chemistry, polymer chemistry
Emmons, William David, organic chemistry
Karo, Wolf, industrial organic chemistry, organic polymer chemistry

IMMACULATA
Feighan, Maria Josita, physical chemistry, organic chemistry
Malter, Margaret Quinn, organic chemistry
Suter, M St Agatha, biology

INDIANA
Berry, Richard Emerson, solid state physics
Buckwalter, Gary Lee, physics
Chambers, Jack Virgil, physical chemistry
Eddy, Jerry Kenneth, nuclear physics
Fuget, Charles Robert, physical chemistry
Gallati, Walter William, zoology
Gault, Thomas Gower, geography
Gillis, Bernard Thomas, organic chemistry
Granata, Walter Harold, Jr, petrology, stratigraphy
Hartline, Richard, biochemistry, organic chemistry
Hoffmaster, Donald Edeburn, botany, conservation
Hoyt, John Paul, mathematical statistics
Humphreys, Jan Gordon, entomology
Kukarni, Gopal Shrinivas, urban geography, geography of South Asia
Lanham, Betty Bailey, ethnology
Liegey, Francis William, microbiology
Marks, Ronald Lee, inorganic chemistry
Matolyak, John, magnetism
McCoy, Ronald Eugene, mathematics
McKelvey, Donald Richard, physical chemistry
Miller, Vincent Paul, Jr, geography
Noz, Marilyn E, theoretical physics
Patsiga, Robert A, organic chemistry, polymer chemistry
Pickering, Jerry L, systematic botany
Purdy, David Lawrence, biomedical engineering
Riban, David Michael, science education
Schrock, Gould Frederick, botany
Strawcutter, Richard, vertebrate zoology
Syty, Augusta, chemistry
Tackett, Stanford F, analytical chemistry
Vallowe, Henry Howard, endocrinology
Winslow, David Clinton, geography

633

Woodard, Robert Louis, science education, astronomy
Wunz, Paul Richard, Jr., organic chemistry
Zenisek, Cyril James, chemistry
Zimmerman, Donald Nathan, physical inorganic chemistry

IRVONA
Gabrysh, Andrew Francis, physics, metallurgy

IRWIN
Lindsay, William Tenney, Jr., physical chemistry, organic chemistry

JEANNETTE
Masciantonio, Philip (X), physical chemistry, organic chemistry

JENKINTOWN
Bidlack, Verne Claude, Jr., organic chemistry
Bowen, Paul Ross, botany
Nemec, Joseph William, organic chemistry
Riddle, Edward Hollister, chemistry
Spaulding, Earle Henry, microbiology
Vogel, Martin, organic chemistry, polymer chemistry
Willett, Norman P., microbiology
Wilson, Harold Frederick, organic chemistry

JOHNSTOWN
Idzkowsky, Henry Joseph, endocrinology, embryology
Moyer, John Henry, medicine
Toigo, Angelo, medicine
Van Ausdal, Ray Garrison, atomic physics, musical acoustics
Zets, John Stephen, physics

KENNETT SQUARE
Allam, Mark Whittier, surgery
Armstrong, Robert John, horticulture, genetics
Baile, Clifton Augustus, III, nutrition, physiology
Bartovics, Albert, polymer chemistry
Bender, Reinhold, organic chemistry
Brackett, Benjamin Gaylord, veterinary medicine, biochemistry
Dalton, Richard Lee, inorganic chemistry
Donawick, William Joseph, surgery
Fackelman, Gustave Edward, veterinary surgery
Froning, Joseph Fendall, organic chemistry
Fulton, Robert Burwell, III, economic geology
Gorton, Bert SoRelle, organic chemistry
Huttleston, Donald Grunert, plant taxonomy
Hwang, Jen, avian pathology
King, Charles Glen, biochemistry
Kronfeld, David Schultz, nutrition, veterinary physiology
Martin, Aaron Jay, analytical chemistry
McClellan, William Robert, organometallic chemistry
McFeely, Richard Aubrey, veterinary medicine, cytogenetics
Meadows, Geoffrey Walsh, industrial chemistry
Merritt, Alfred M. II, gastroenterology, veterinary medicine
Morse, Guy Emery, veterinary bacteriology
Raker, Charles W., veterinary surgery
Ralston, Robert Henry, polymer science
Reid, Charles Feder, veterinary radiology
Seibert, Russell Jacob, botany
Umbreit, Gerald Ross, analytical chemistry
Woodward, David Willcox, organic chemistry

KING OF PRUSSIA
Advani, Shyam Bhojraj, medicinal chemistry, agricultural chemistry

KIMBERTON
Ullyot, Glenn Edgar, organic chemistry

KERSEY
Bolstad, Luther, plastics chemistry

Benzinger, William Donald, industrial chemistry
Block, Burton Peter, inorganic chemistry
Bohen, Joseph Michael, organic chemistry
Bohm, Howard Allan, medicinal chemistry
Bricks, Bernard Gerard, lasers
Burgess, Thomas Edward, clinical chemistry
Cheung, Peter Pak Lun, dental materials
Clavan, Walter, analytical chemistry
Dimmig, Daniel Ashton, organic polymer chemistry, industrial organic chemistry
Emrich, Grover Harry, environmental geology
Fainberg, Arnold Harold, physical chemistry
Ferren, Richard Anthony, physical organic chemistry, polymer chemistry
Fox, Adrian Samuel, organic chemistry, hematology
Gardner, David Milton, physical chemistry
Gillman, Hyman David, inorganic chemistry
Haines, Paul Gordon, organic chemistry
Hauptschein, Murray, organic chemistry
Jurecic, Anton, dental research
Kashatus, William C., pathology, hematology
King, James P., physical inorganic chemistry
Koester, Robert Charles, pesticide chemistry
Mahecke, Henry Elmore, industrial chemistry
Miller, Harold James, plant pathology
Nannelli, Piero, polymer chemistry
Popoff, Ivan Christoff, organic chemistry
Reifenberg, Gerald H, organometallic chemistry
Robinson, Donald Nellis, organic chemistry, polymer chemistry
Robinson, Douglas Walter, analytical chemistry
Signorino, Charles Anthony, organic chemistry
Smith, James Graham, Jr., analytical chemistry
Smith, Perrin Gary, organic chemistry
Stefanou, Harry, polymer physics

KINTNERSVILLE
Bertsch, Charles Rudolph, inorganic chemistry
Calkins, Charles Richard, chemistry

KUTZTOWN
Agoos, William Bailey, physics
Green, William Asa, bacteriology, zoology
Hamel, Coleman Rodney, physiology
Kaiser, Russell Florentine, chemistry
Pirnot, Thomas Leonard, algebra
Piscitelli, Joseph, physiology
Slick, Max Harrell, geology
Tinsman, James Herbert, Jr., physical anthropology
Webb, Glenn R., malacology

Christie, Peter Allan, organic polymer chemistry
Connors, William Matthew, biochemistry, industrial chemistry
Coursin, David Baird, pediatrics
Darlington, James McCown, zoology
Dieck, Ronald Lee, inorganic chemistry, polymer chemistry
Duck, William N., Jr., biochemistry, physical chemistry
Dunlap, Lawrence Hallowell, polymer chemistry
Ehrhart, Wendel A., polymer chemistry
Eisenberg, Rita B., audiology
Emmons, Larrimore Brunneler, optics
Engstrom, Ralph Warren, environmental medicine
Feit, Ira (Nathan), developmental biology, microbiology
Fink, Collin Ethelbert, inorganic chemistry
Fishel, John B., organic chemistry
Ford, Milton David, chemistry
Freedman, Jacob, geology
Garrett, Thomas Boyd, theoretical polymer chemistry
Gavan, Francis Michael, physics
Gerstell, Richard, ecology
Goodyear, William Frederick, Jr., wood science, wood technology
Hadley, Charles Peleg, physics
Hager, Nathaniel Ellmaker, Jr., thermal physics, optical physics
Hartranft, George Robert, organic polymer chemistry
Hazeltine, James Ezra, Jr., physical chemistry
Heller, Hugh Andrews, chemistry
Hess, Earl Hollinger, agricultural chemistry, environmental chemistry
Hipple, John Alfred, physics
Hofferth, Burt Frederick, organic chemistry
Hoffman, Richard Bruce, physics
Holzinger, Charles Henry, behavioral anthropology
Holzinger, Joseph Rose, mathematics
Hood, Richard Fred, physics
Irwin, William Edward, plastics
Jacobson, Bernard, mathematics
John, Kenneth Rydal, limnology
Judge, Joseph Malachi, organic chemistry
Kauffman, Marvin Earl, geology
King, David Beeman, biology
Kinsey, W Fred, III, anthropology, archaeology
Koehn, George Willis, chemistry
Krogman, Wilton Marion, physical anthropology
McDermott, John Joseph, parasitology, marine biology
Miller, William Robert, Jr., electron physics, solid state physics
Moggio, William Aldo, applied chemistry, environmental science
Moscony, John Joseph, chemistry
Moss, John Hall, geology
Newton, Robert Collier, chemistry
Optiz, Herman Ernest, inorganic chemistry
Perlis, Irwin Bernard, plant chemistry
Pike, Carl Stephen, plant physiology
Poshkus, Algirdas C., organic chemistry
Quinn, Edwin John, organic chemistry, polymer chemistry
Reehling, Harold Arthur, chemistry
Remer, Joseph Francis, organic chemistry
Reuwer, Joseph Francis, Jr., physical organic chemistry, polymer chemistry
Reynolds, Samuel R M., anatomy, physiology
Richardson, Jonathan L., limnology
Rosenstein, George Morris, Jr., mathematics
Sanders, Charles Irvine, chemistry
Scott, George William, Jr., physics
Seeds, Michael August, astronomy
Simon, Ralph Emanuel, physics
Smith, Arthur Leo, organic chemistry
Snavely, Fred Allen, inorganic chemistry
Spalding, George Robert, physics
Sundaresan, Peruvemba Ramnathan, nutritional biochemistry
Sutter, Philip Henry, physics
Suydam, Frederick Henry, organic chemistry
Van Horn, Ruth Warner, organic chemistry
Western, Donald Ward, mathematics
White, Howard Sorrel, organic chemistry
Wiebe, Robert A., petrology
Work, James Leroy, polymer chemistry
Yoder, Claude H, inorganic chemistry
Zentmyer, David Taylor, organic chemistry

LAFAYETTE HILL
Barnhart, William Siddall, organic polymer chemistry
Burns, Allan Fielding, research administration, soil chemistry
Campbell, Thomas Cooper, physical organic chemistry
Hertzenberg, Elliot Paul, inorganic chemistry
Katsanis, Eleftherios P., colloid chemistry, surface chemistry
Kress, Bernard Hiram, organic chemistry
Schleyer, Walter Leo, chemistry
Scholnick, Frank, organic chemistry

LANCASTER
Albright, Fred Ronald, biochemistry
Bagley, George Everett, polymer chemistry
Bedient, Philip E., mathematics
Behrens, Ernst Wilhelm, polymer physics
Bellaire, Frank Roland, meteorology
Bhagavan, Hemmige, nutrition, biochemistry
Bolgiano, Nicholas Charles, polymer chemistry
Browning, Daniel Dwight, organic chemistry
Bruns, Charles Alan, physics
Cherry, Leonard Victor, solid state physics

LANSDALE
Bicking, John Beeh, organic chemistry
De Tommaso, Gabriel Louis, organic chemistry, polymer chemistry
Kunitz, Moses, biochemistry
Ponticello, Gerald S., organic chemistry
Tate, Charles Luther, pathology, veterinary medicine
Wohl, Bernard G, organic chemistry, medicinal chemistry
Zwickey, Robert Earl, veterinary medicine

LANSDOWNE
Hanes, Robert Bruce, physiology

LATROBE
Duman, Maximilian George, systematic botany
Dzombak, William Charles, physical chemistry
Emling, Bertin Leo, organic chemistry
Heid, Roland Leo, physics
Roth, Owen, zoology, histology
Taubler, James H., microbiology

LEECHBURG
Blair, McClellan Gordon, solid state physics

LEVITTOWN
James, Herbert I., physical chemistry
Neely, James W., physical inorganic chemistry
Young, Irving Gustav, analytical chemistry, physical chemistry

LEWISBURG
Abrahamson, Warren Gene, II, plant ecology, population biology
Allen, Jack C, Jr., geology
Anderson, Owen Thomas, physics
Auten, John Thompson, soil science
Baclawski, Leona Marie, organic chemistry
Becker, Stephen Fraley, theoretical physics
Cooper, John Neale, physical chemistry, inorganic chemistry
Cotter, Edward, sedimentology
Edmonds, James D J., physics
Ellis, Richard John, plant physiology
Harclerode, Jack E., zoology
Heine, Harold Warren, organic chemistry
Henry, Richard Warfield, physics
Hoffman, Daniel Lewis, endocrinology, developmental biology
Kaminsky, Kenneth S., statistics
Kieff, Lester, analytical chemistry
Klimko, Lawrence Andrew, mathematical statistics
Leshner, Alan Irvin, physiological psychology
Lonski, Joseph, developmental biology
Magalhaes, Hulda, zoology
McDiffett, Wayne Francis, ecology
Nicholsen, Richard Peter, structural geology
Nyquist, Sally Elizabeth, cell biology
Oesterling, James Frederick, research administration
Ohl, Donald Gordon, mathematics
Pearson, David D., immunobiology, biochemistry
Pinter, Charles Claude, mathematics
Rackoff, Jerome S., vertebrate paleontology
Ray, David Scott, mathematics
Root, Charles Arthur, inorganic chemistry
Schweinsberg, Allen Ross, mathematical analysis
Scouten, William Henry, enzymology, protein chemistry
Sigler, Laurence Edward, mathematics
Smith, Manning Amison, genetics
Tonzetich, John, genetics
Veening, Hans, analytical chemistry
Willeford, Bennett Rufus, Jr., physical chemistry
Winstead, Meldrum Barnett, organic chemistry

LIBRARY
Gorin, Everett, physical chemistry
Sudbury, John Dean, physical chemistry

LIMERICK
Carriker, Roy C., physics

LINCOLN UNIVERSITY
Branson, Herman Russell, physics
Christensen, Sabinus Hoegsbro, physics
Gunn, Harold Dale, cultural anthropology
Johnson, Leroy Dennis, organic chemistry
Johnson, William Thomas Mitchell, physical chemistry, cardiovascular physiology
Rudd, DeForest Porter, inorganic chemistry

LINESVILLE
Tryon, Clarence Archer, zoology

LINWOOD
Mitchell, Maurice McClellan, Jr., physical chemistry

LITITZ
Welch, Rodney Channing. food science

LOCK HAVEN
Botros, Raouf, chemistry
Crosby, Alan Hubert, organic chemistry
Klens, Paul Frank, microbiology
Phelps, Dean G., mathematics
Pursell, Mary Helen, genetics
Renfrew, Edgar Earl, industrial organic chemistry
Scherer, Robert C, animal ecology
Schwalbe, Paul Wayman, population ecology, ornithology
Settlemyer, Kenneth Theodore, taxonomic botany
Thomas, Joseph Charles, mathematics
Williamson, Hugh A., organic chemistry, science education
Yoho, Timothy Price, developmental biology, entomology

LORETTO
Driesch, Albert John, organic chemistry
Duryea, William R., environmental biology, soil microbiology
Dickson, Paul Wesley, Jr., physics
Taylor, George Russell, physical chemistry

MALVERN
Harrison, George Conrad, Jr., surface chemistry
Hunter, James Bruce, chemistry
Ladisch, Rolf Karl, biochemistry, electrochemistry
Paulson, Stuart R., endocrinology
Richardson, Henry Russell, mathematics
Stearns, Richard S., physical chemistry
Tucci, Edmond Raymond, physical inorganic chemistry
Tweedie, Adelbert Thomas, polymer chemistry

MANHEIM
Wronski, Joseph Peter, physics

MANSFIELD
Dowling, John, Jr., physics
Gassner, Edward, plant physiology
Hartman, John Alan, organic chemistry
Mullen, George Henry, theoretical physics
Powell, Manly Joy, physical chemistry
Sidler, Jack D., organic chemistry
Trindell, Roger Thomas, cultural geography, historical geography

MAPLE GLEN
Bagdon, Walter Joseph, toxicology, pharmacology

Schrum, Robert Wallace, petroleum chemistry

MARCUS HOOK
Alexander, Frank Creighton, Jr., physics
Ambs, William Frederick, surface chemistry
Daughenbaugh, Randall Jay, physical organic chemistry
Davis, Brian Clifton, physical organic chemistry
Dickason, Alan Frederick, organic chemistry
Dux, James Philip, physical chemistry
Dyer, John, polymer chemistry
Ellis, Sidney Glenn, microscopy, textile physics
Hersh, Sylvan David, chemistry
Hoegberg, Erick Ingvar, chemistry, information science
Hosler, Peter, petroleum chemistry
Huang, Der-Shing, organic chemistry
Kennedy, Robert Michael, organic chemistry
King, Richard Warren, analytical chemistry, physical chemistry
Ladenheim, Harry, physical organic chemistry
Lothrop, Everett Winfred, Jr., textile physics, polymer physics
Lyons, James Edward, organic chemistry, organometallic chemistry
MacFadden, Kenneth Orville, analytical chemistry
Mascioli, Rocco Lawrence, organic chemistry
Mayo, Ralph Elliott, physical chemistry
Meichiore, John J., petroleum chemistry
Milligan, Barton, organic chemistry
Norman, Oscar Loris, organic chemistry
Norton, Richard Vail, organic chemistry
Oishi, Masayoshi, organic chemistry
Pearson, Frank Gardiner, chemistry
Phifer, Lyle Hamilton, analytical chemistry
Swann, William B., analytical chemistry
Toney, Frank Morgan, physical chemistry
Vanderwerff, William D., organic chemistry
Wells, James Edward, physical inorganic chemistry
Wynkoop, Raymond, chemical engineering, chemistry

MARIETTA
Bernstein, Alan, virology
McCarthy, Frank John, microbiology

MARTINSBURG
Coursen, David Linn, rock mechanics

McKEESPORT
Fussell, Catharine Pugh, cell biology, botany
Gordon, William Edwin, physical chemistry
Maricondi, Chris, inorganic chemistry
Wells, Jacqueline Gaye, algebra
Zavodni, John J., physiology

McMURRAY
Barthauer, Gerald Lee, environmental chemistry, resource management
Sharpe, Andrew Jackson, Jr., organic chemistry, polymer chemistry

MEADOWBROOK
Beavers, Ellington McHenry, organic chemistry
Clymer, Harold Arthur, research administration
McKeever, Charles H., organic chemistry
Schneider, Henry C., medicine

MEADVILLE
Bivens, Richard Lowell, physical chemistry
Brown, Richard Leland, physics
Bugbee, Robert Earl, taxonomy
Cable, Charles Allen, algebra, number theory
Harrell, Ronald Earl, pure mathematics
Harrison, Samuel S, geomorphology
Howard, James F, stratigraphy, paleoecology
Klikoff, Lionel G, botany, ecology
Lundgren, J Richard, algebra
Lutton, Lewis Montfort, vertebrate zoology
Monson, Arvid Monroe, microbiology, genetics

Parsons, William Howard, geology
Reisner, Gerald Seymour, microbiology
Rhinesmith, Herbert Silas, organic chemistry
Rodgers, Glen Ernest, inorganic chemistry
Steen, Frederick Henry, mathematics
Swank, Rolland Laverne, mathematics
Walsh, Edward Joseph, Jr., organic chemistry
Wurst, Glen Gilbert, genetics
Yellen, Jay, algebra

MECHANICSBURG
Bragg, Lincoln Ellsworth, mathematical physics

MEDIA
Barmby, David Stanley, physics
Black, Robert Corl, botany, plant physiology
Cooper, Jane Elizabeth, genetics
Demos, Miltiades Stavros, mathematics
Engle, Robert Fry, Jr., chemistry
Gallagher, George Arthur, organic chemistry
Georgopulos, Peter Demetrios, nuclear physics
Morris, Albert Gregory, inorganic chemistry
Nonemaker, Larry Franklin, organic chemistry, polymer chemistry
Saul, Leon Joseph, psychiatry
Tomezsko, Edward Stephen John, physical chemistry

MENDENHALL
Balthis, Joseph Hendrickson, Jr., inorganic chemistry

MERION STATION
Deischer, Claude Knauss, chemistry
Fogg, John Milton, Jr., botany

MIDDLETOWN
Bissinger, Barnard Hinkle, mathematics, engineering
Fusco, Robert Angelo, entomology
McDermott, Robert Emmet, forestry
Nichols, James Otis, forest entomology
Richards, Winston Ashton, mathematical statistics
Simko, Robert Alexander, geography of Africa, political geography
Towers, Barry, forest pathology, mycology

MILLERSVILLE
Casselberry, Samuel Emerson, anthropology, archaeology
Henderson, Alex, vertebrate morphology, biology
Jordan, Esther May, geology
Kilheffer, Edward Malcolm, geography
Kohr, Charles Byron, physics
Koken, James E, chemistry
Oostdam, Bernard Lodewijk, oceanography, marine geology
Ostrovsky, David Saul, zoology
Parks, James C, plant systematics
Ratzlaff, Willis, limnology
Sasin, Richard, organic chemistry
Stauffer, George Franklin, astronomy
Stuecek, Guy Linsley, plant physiology
Ting, Shih-Fan, chemistry
Van Horn, John A., physics
Weiss, Gerald S., inorganic chemistry
Wolf, Charles Frostle, mathematics
Yeager, Sandra Ann, organic chemistry
Yurkiewicz, William J., insect physiology, biochemistry

MILTON
Schlimme, Donald Vincent, food science

MONACA
Derby, James Victor, analytical chemistry
Johnston, Gordon Robert, organic chemistry
Kanzelmeyer, James Herbert, analytical chemistry
Kogut, Leonard S, inorganic chemistry
Thanigasalam, Kandiah, number theory

MONROEVILLE
Altares, Timothy, Jr., physical chemistry, polymer chemistry
Andria, George D, mathematical analysis, energy conversion

Angeloni, Francis M, analytical chemistry
Bandi, William R, analytical chemistry
Bi, Le-Khac, polymer chemistry
Birch, Homer James, analytical chemistry
Blommers, Elizabeth Ann, organic polymer chemistry
Boggs, William Emerson, physical chemistry, analytical chemistry
Boodman, Norman S, fuel science
Booth, William Thomas, Jr, organic chemistry
Brownell, George L, organic chemistry
Bullock, Paul David, research administration, applied statistics
Carter, Paul Richard, surface chemistry
Chaudhary, Sohan Singh, physical organic chemistry
Creagan, Robert Joseph, applied physics
Davie, William Raymond, organic chemistry
Detrick, Robert Sherman, chemistry
Doak, Kenneth Worley, polymer chemistry
Dressler, Hans, organic chemistry, biochemistry
Farquhar, Mary Janet, analytical chemistry
Fugger, Joseph, industrial chemistry
Gleason, Edward Hinsdale, Jr, organic chemistry
Glick, Charles Frey, analytical chemistry
Gormley, William Thomas, organic chemistry
Gray, Ralph J, petrography, geology
Harju, Philip Herman, physical chemistry
Harris, James Joseph, inorganic chemistry
Helwig, Lawrence E, physical chemistry, corrosion
Hudson, Robert McKim, physical chemistry, inorganic chemistry
Huffman, Gerald P, solid state physics
Hurwitz, Jan Krossi, analytical chemistry
Igoe, John Waite, research administration
Ingram, Alvin Richard, organic polymer chemistry
Kersting, Raymond James, polymer chemistry
Kifer, Edward W, inorganic chemistry, physical chemistry
Kitazawa, George, wood chemistry
Klapproth, William Jacob, Jr, polymer chemistry
Krouskop, Ned Carter, physical chemistry
Lake, Robert D, polymer chemistry
Laubscher, Aner Nearhood, organic chemistry
Leach, Charles Willard, forestry
Lee, Richard J, solid state physics
MacDonald, Hubert C. Jr., physical chemistry, analytical chemistry
Maclay, William Nevin, physical chemistry, research administration
Mainier, Robert, analytical chemistry
Martin, John F, analytical chemistry
Melnick, Laben Morton, analytical chemistry
Milkovich, Ralph, polymer chemistry
Miller, William Reynolds, Jr, polymer physics, physical chemistry
Mirabella, Francis Michael, Jr, polymer chemistry
Moult, Roy Hepworth, organic polymer chemistry
Nagel, Roger Miles, polymer chemistry
Obrycki, Richard, analytical chemistry, organic chemistry
Onopchenko, Anatoli T, organic chemistry
Oriani, Richard Anthony, physical chemistry
Perry, Paul Eberline, surface chemistry
Peters, John Thomas, organic chemistry
Pignocco, Arthur John, physical chemistry
Pillar, Walter Oscar, polymer chemistry, forest products
Podgurski, Harry Howard, physical chemistry
Pollock, Dorothy Jean, polymer physics
Rall, Waldo, physics, research administration
Rhee, Kee Hyun, physical chemistry
Robbins, Thomas Ennis, Jr, organic chemistry
Sands, Daniel Edward, research administration
Schowalter, Kenneth Arthur, fuel science
Schwerer, Frederick Carl, applied physics, materials science
Selwitz, Charles Myron, organic chemistry
Snyder, Paul Edwin, physical chemistry

PENNSYLVANIA

Sorensen, Frederick Allen, mathematical statistics
Straub, William Albert, analytical chemistry
Sweeney, Harold A., analytical chemistry, polymer chemistry
Welsh, David Albert, organic chemistry, polymer chemistry
Wicker, Everett E., nuclear science
Wilson, Leon William, Jr., polymer chemistry
Wried, Henry Anderson, physical chemistry

MONT ALTO
Keiper, Ronald R., animal behavior
Sanders, Mary Elizabeth, genetics

MONTOURSVILLE
Fineman, Morton A., physical chemistry
Sonntag, Norman Oscar Victor, organic chemistry

MONTROSE
Millard, Frederick William, photographic chemistry

MURRYSVILLE
Berry, William Francis, geology

MT JOY
Mariner, Thomas, physics

MT POCONO
Bolyn, Anthony Edward, medical microbiology

NANTICOKE
Miller, Eugene D., physical chemistry
Minsavage, Edward Joseph, medical microbiology

NARBERTH
English, Oliver Spurgeon, psychiatry
Gail, John Frederick, physical chemistry
Soulen, John Richard, physical chemistry

NEW BRITAIN
Brown, W Ray, veterinary pathology

NEW CUMBERLAND
Abramson, Edward, optics
Peiffer, Howard R., metallurgy, physics

NEW HOPE
Goff, Sidney, biological chemistry
Hollander, Leonore, biochemistry, clinical chemistry
Reich, Irving, colloid chemistry
Zeltner, Carl Naeher, organic chemistry

NEW KENSINGTON
Englehart, Edwin Thomas, Jr., chemical metallurgy
Foster, Perry Alanson, Jr., physical chemistry
Grubbs, Donald Keeble, geology, geochemistry
Hinderliter, Hilton Fay, nuclear physics
Hoops, Stephen C., organic chemistry
Kinosz, Donald Lee, industrial chemistry
Kolka, Alfred Jerome, organic chemistry
Leftault, Charles Joseph, Jr., analytical chemistry, mechanical engineering
Russell, Allen Stevenson, engineering

NEW WILMINGTON
DeSieno, Robert P., physical chemistry
De Witt, Hobson Dewey, organic chemistry
Fawley, John Philip, physiology
Harms, Clarence Eugene, parasitology
Hendry, Richard Allan, biochemistry
Johnson, William Lewis, solid state physics
Lewis, Phillip Albert, analytical chemistry
Long, Kenneth Maynard, inorganic chemistry
Travis, Robert Victor, entomology
Warrick, Percy, Jr., physical organic chemistry
Zehr, Floyd Joseph, experimental physics

NEWTOWN
Grey, James Tracy, Jr., physical chemistry
Smith, Norman Hankele, mathematics

NEWTOWN SQUARE
Condo, Albert Carman, Jr, petroleum chemistry
Cosby, John Norman, organic chemistry, engineering
Robinson, Corinne (Hogden), nutrition

NORRISTOWN
Abrams, Ellis, industrial chemistry
Bacher, Frederick Addison, physical chemistry
Bergnes, Manuel, medicine
Boger, William Pierce, medicine
Cohen, Edward Morton, pharmaceutical chemistry, analytical chemistry
Conte, John Salvatore, organic chemistry
Janicki, Casimir A., analytical chemistry
Monaco, Anthony L., pharmacy
Monsimer, Harold Gene, organic chemistry
Thampi, Nagendran Sankaranarayanan, biochemistry, pharmacology

NORTH VERSAILLES
Barnett, Herald Alva, physical chemistry

NORTH EAST
Haeseler, Carl W., pomology, horticulture
Jubb, Gerald Lombard, Jr., entomology
McConaughy, David Lester, chemical physics

NORTH WALES
Cohen, Stuart Colin, organic polymer chemistry
Erenrich, Evelyn Schwartz, enzymology, physical biochemistry
Field, Arthur Kirk, immunobiology, virology
Freud, Paul J, solid state physics
Gillen, Raymond Daniel, physical chemistry, chemical engineering
Grooms, Thomas Albin, biological chemistry, enzymology
Levin, Herman Westley, biochemistry, enzymology
Mattis, Paul Alvin, toxicology
Nelson, Roger Edwin, physical chemistry
Plantz, Philip Edward, food science, biochemistry
Taylor, Robert Morgan, analytical chemistry, electrochemistry
Weiss, Edward Leonhardt, physics, metrology
Wertheimer, Alan Lee, optics
Wooldridge, David Paul, entomology

NORTHUMBERLAND
Grafius, Melba A., biochemistry

NORWOOD
Rogers, Ralph Loucks, industrial organic chemistry

OAKMONT
Lemmon, Donald H., spectroscopy
Mears, Robert Bruce, metallurgical chemistry

ORELAND
Allison, Patricia (Lee Van Burgh), plant pathology

ORRTANNA
Parker, Frank Wilson, agronomy

PALMERTON
Brand, John Robert, physical inorganic chemistry
Lloyd, Thomas Blair, industrial chemistry, colloid chemistry
Tennant, Charles Beard, inorganic chemistry

PALMYRA
Johnson, Ogden Carl, food technology, research administration

PAOLI
Arnold, Leslie K., mathematics
Barker, William Hamblin, II, operations research, pure mathematics
Belkin, Barry, operations research
Bolmarcich, Joseph John, operations research, applied mathematics
Bossard, David Charles, operations research
Boxill, Gale Clark, pharmacology
Corwin, Thomas Lewis, mathematical statistics
Gore, Edward Michael, pharmacology
Kipp, Egbert Mason, physical chemistry
Music, Jack Farris, physical chemistry, physics
Scranton, Bruce Edward, mathematical analysis, operations research
Stone, Lawrence David, operations research, mathematics
Wagner, Daniel Hobson, mathematics

PAUPACK
Steers, Edward, microbiology, chemistry

PEACH GLEN
Oyler, James Russell, biochemistry

PENNSYLVANIA FURNACE
Blankenhorn, Paul Richard, wood science & technology; materials engineering

PETROLIA
Phillips, Joseph, organic chemistry

PHILADELPHIA
Abaidoo, Kodwo-James R., hematology, physiology
Abruzzo, John L., medicine
Ackerman, James L., orthodontics
Ackerman, Guenter Rolf, organic chemistry, pharmaceutical chemistry
Actor, Paul, chemotherapy
Adams, John Kendal, geology
Adelman, Richard Charles, gerontology, biochemistry
Aden, David Paul, immunobiology
Adler, Irving Larry, organic chemistry, analytical chemistry
Adler, John Henry, biological chemistry
Adler, Martin William, pharmacology
Aegerter, Ernest E., medicine
Agersborg, Helmer Pareli Kjerschow, Jr., physiology, toxicology
Ajzenberg-Selove, Fay, nuclear physics
Aker, Franklin David, neuroanatomy, anatomy
Albright, Robert Lee, organic chemistry
Alburn, Harvey Eugene, biochemistry
Aleo, Joseph John, pathology
Alexander, Fred, internal medicine
Alexander, James King, microbiology, biochemistry
Allen, Arthur, biochemistry
Alper, Carl, clinical biochemistry
Alper, Robert, biochemistry
Alperin, Richard Junius, cytochemistry
Alteveer, Robert Jan George, developmental biology, biophysics
Amado, Ralph, theoretical physics
Ambrovage, Anne Marie, physiology
Amenta, Peter Sebastian, anatomy
Ancona, Umberto, industrial chemistry, chemical engineering
Anderson, Carl Einar, biochemistry
Anderson, Edwin J., paleoecology
Anderson, Herbert Hale, inorganic chemistry
Anderson, Melvin Albert, epidemiology, public health
Anderson, Theodore Gustave, bacteriology
Anderson, Thomas Foxen, biophysics
Andrew, Barbara Jean, research administration
Andros, George James, obstetrics & gynecology
Angelakos, Evangelos Theodorou, physiology
Angstadt, Carol Newborg, biochemistry
Anton, Howard, mathematics
Aponte, Gonzalo Enrique, medicine
Argabright, Loren N., mathematics
Armbruster, David Charles, polymer chemistry, plastics chemistry
Aronson, Carl Edward, pharmacology, toxicology
Aronson, John Ferguson, physiology
Arrington, Wendell S., science administration, biostatistics
Artzy, Rafael, geometry, algebra
Arvan, Dean Andrew, clinical pathology, biochemistry
Asakura, Toshio, biochemistry, hematology
Ascinsky, Eugene, physiology
Auerbach, Arthur Henry, psychiatry
Auerbach, Leonard B., physics
Austrian, Robert, internal medicine, bacterial genetics
Avadhani, Narayan G., biochemistry, molecular genetics
Aviado, Domingo Mariano, pharmacology, physiology
Axler, David Allan, virology
Aycock, Nancy Rae, human anatomy
Bacon, Harry Elliot, gastroenterology
Bacon, Norman Ira, information science
Badler, Ian Matheson, internal medicine, pharmacy
Bagdassarian, Andranik, biochemistry
Baggott, James Patrick, biochemistry
Bahl, Surendra Mohan, physical pharmacy
Bahar, Leon Y., applied mechanics
Bailey, David George, biochemistry, enzymology
Baird, Henry W., III, pediatrics, neurology
Bajcsy, Ruzena K., computer science
Baker, Alan Paul, biochemistry
Baker, Harold Weldon, analytical chemistry
Baker, Walter Wolf, neuropharmacology
Balamuth, David, nuclear physics
Baldridge, Robert Crary, biochemistry
Baldwin, Paul Clay, biochemistry
Balin, Howard, obstetrics & gynecology
Ballard, Ian Matheson, internal medicine, clinical pharmacology
Bandy, Alan Ray, atmospheric chemistry
Barba, William P. II, pediatrics
Barker, Clyde Frederick, surgery
Barker, Earl Stephens, internal medicine
Barriga, Omar Oscar, parasitology, immunology
Barry, William Eugene, medicine
Barth, Max, physical chemistry
Bartram, John Bowman, pediatrics
Baserga, Renato, pathology
Bashey, Reza Ismail, biochemistry
Bates, M Noble, histology, embryology
Baum, Joseph Herman, pathology
Baum, O Eugene, psychiatry, psychoanalysis
Baum, Robert Harold, biochemistry, microbiology
Baurer, Stanley, radiology
Bayer, Theodore, physical chemistry
Bayer, Margret Helene Janssen, plant physiology
Bearman, Toni Carbo, information science
Beasley, Andrew Bowie, genetics
Beauchamp, Gary Keith, animal behavior
Beck, Aaron Temkin, psychiatry
Beckman, Alexander Lynn, neurophysiology
Beem, John Raymond, medicine
Beerman, Herman, dermatology
Begany, Albert John, pharmacology
Begun, Marjam Gojchiemer, analytical chemistry
Behar, Marjam Gojchiemer, analytical chemistry
Beicht, George John, physical chemistry, inorganic chemistry
Beizer, Lawrence H., internal medicine, hematology
Beller, Martin Leonard, orthopedic surgery
Bello, Carmen T., internal medicine, pharmacology
Bello, Leonard John, biochemistry
Belmont, Herman S., psychiatry, psychoanalysis
Benarde, Melvin Albert, public health
Bender, Paul Elliot, synthetic organic chemistry
Bennett, Hugh Deveraux, medicine
Benson, Charles Everett, biochemical genetics, medical microbiology
Bergen, John Vanderveer, pharmacy, medicinal chemistry
Berges, David Alan, organic chemistry, medicinal chemistry
Bergon, Lloyd, biochemistry
Berkoff, Charles Edward, organic chemistry, medicinal chemistry
Berkowitz, David B., biochemistry, molecular biology
Berman, Helen Miriam, biological structure
Bernardo, Peter D., physical chemistry
Berney, Steven, rheumatology, immunology
Berry, Richard G., neuropathology

Besarab, Anatole, nephrology, physiology
Bethke, George William, Jr, applied physics
Bianchi, Carmine Paul, cell physiology
Bickel, Robert John, mathematics
Bierly, James N, Jr, physics
Bierly, Mahlon Zwingli, Jr, medicine
Binkley, Sue Ann, biological rhythms, endocrinology
Binnendijk, Leendert, astronomy
Bitcover, Ezra Harold, leather chemistry
Blackburn, Dale Warren, pharmaceutical chemistry
Blain, Daniel, psychiatry
Blank, Benjamin, organic chemistry, medicinal chemistry
Blank, Fritz, medical mycology
Blasie, J Kent, biophysics, physics
Bleecker, Margit, neurochemistry, neuroendocrinology
Blitzstein, William, astronomy
Bloomer, James L, organic chemistry
Blough, Herbert Allen, virology
Bludman, Sidney Arnold, theoretical physics
Blum, Harold Francis, physiology, biophysics
Blum, Haywood, biophysics, solid state physics
Blumberg, Baruch Samuel, medicine, medical anthropology
Blumenthal, Reuben R, marine biology
Blumstein, George I, medicine
Bock, Charles Walter, quantum chemistry, nuclear physics
Bogacz, John, bacteriology
Bohlke, James Erwin, ichthyology
Boland, James P, cardiovascular surgery, thoracic surgery
Bonakdarpour, Akbar, radiology
Bond, James, taxonomic ornithology
Bondi, Amedeo, microbiology
Bongiovanni, Alfred Marius, biology, medical science
Bonner, Walter Daniel, Jr, cell biology
Borei, Hans Georg, invertebrate physiology, ecology
Borkowski, Walter Leonard, organic chemistry
Borun, Thaddeus M, biochemistry
Bose, Shyamalendu M, solid state physics, metal physics
Bosee, Roland Andrew, biochemistry
Botelho, Stella Yates, physiology
Bowers, Paul Applegate, obstetrics & gynecology
Bradford, Spencer Graves, academic administration, physiology
Bradley, James Henry Stobart, dynamic meteorology, numerical analysis
Brady, John Paul, psychiatry
Brady, J Luther Weldon, Jr, medicine, radiology
Braverman, Jerome David, applied statistics, statistics
Braxton, Wilbert Leo, physics
Breazeale, Robert David, organic chemistry
Breedis, Charles, pathology
Brennan, James Thomas, radiotherapy, radiobiology
Brent, Robert Leonard, embryology
Brest, Albert N, medicine
Brigham, M Prince, surgery, biochemistry
Brightman, Vernon, dentistry, oral medicine
Brighton, Carl T, orthopedic sugery
Brinigar, William Seymour, Jr, biochemistry
Brinster, Ralph Lawrence, embryology, reproductive physiology
Brobeck, John Raymond, physiology
Brodey, Robert S, veterinary surgery, oncology
Brody, Howard, physics
Brody, Jerome Ira, internal medicine, hematology
Bromels, Edward, physical chemistry, environmental chemistry
Brooks, Frank Pickering, physiology
Brown, Allan Harvey, plant physiology
Brown, Darrell Quentin, radiation biophysics
Brown, Eleanor Moore, physical biochemistry
Brownell, Arnold S, biophysics
Brownstein, Barbara L, cell biology
Brucker, Paul Charles, family medicine
Brutcher, Frederick Vincent, Jr, organic chemistry
Buchanan, Robert Lester, Jr, food science
Buchin, Irving D, orthodontics
Buck, Clayton Arthur, biochemistry, cancer
Budney, Mary Lillian, microbiology
Burg, Fredric David, medical education, pediatrics

Burgess, Benjamin Franklin, Jr, physiology, biochemistry
Burka, Edward Richard, hematology, biochemistry
Burke, James David, organic chemistry
Burley, J William Atkinson, plant physiology, radiation biology
Burstein, Elias, solid state physics
Busby, Robert Clark, mathematics
Byler, David Michael, physical inorganic chemistry
Cagan, Robert Howard, biochemistry
Cahill, Jerry Edward, chemical physics
Calabi, Eugenio, mathematics
Caldwell, Henry Cecil, Jr, pharmaceutical chemistry, organic chemistry
Calesnick, Benjamin, pharmacology
Calhoun, George Milton, organic chemistry
Callen, Herbert Bernard, statistical mechanics
Cammarata, Arthur, physical organic chemistry
Campbell, Kenneth Bruce, cardiovascular physiology
Campo, Robert D, biochemistry, biology
Cander, Leon, internal medicine, clinical physiology
Carpenter, Gary Grant, pediatrics
Carr, John Weber, III, mathematics
Carrington, Elsie Reid, obstetrics & gynecology
Caso, Louis Victor, immunology, histology
Caspari, Max Edward, solid state physics
Castor, LaRoy Northrop, cell biology
Cava, Michael Patrick, organic chemistry
Ceron, Gabriel, anatomy, teratology
Cevallos, William Hernan, biochemistry, metabolism
Chacko, George Kutty, biochemistry
Chait, Arnold, medicine
Chakrin, Lawrence William, pharmacology
Chambers, Richard, neurology
Chambers, Robert Rood, organic chemistry
Chance, Britton, biophysics, biochemistry
Chang, Sak Chui, pathology, cytology
Chang, Ted Teh-Liang, physical chemistry, analytical chemistry
Charalampous, Frixos C, biochemistry
Charkes, N David, nuclear medicine
Chase, Grafton D, physical chemistry
Chase, Harold Frederick, anesthesiology
Chase, Robert A, reconstructive surgery
Chein, Orin Nathaniel, mathematics
Chemburkar, Pramod Bhaurao, pharmaceutical sciences, chemical engineering
Chen, Robert Chia-Hua, computer science, electrical engineering
Chen, Sow-Yeh, oral pathology
Chepenik, Kenneth Paul, developmental biology
Chernack, Neil S, internal medicine
Ch'ih, John Juwei, biological chemistry
Child, Proctor Louis, pathology
Childress, Scott Julius, organic chemistry
Chilton, Neal Warwick, oral medicine
Christensen, Albert Kent, anatomy, cell biology
Christian, Howard Harris, biology, data processing
Christoph, Francis Theodore, Jr, topology, nuclear physics, optical physics
Ciaccio, Edward I, biochemistry
Cilley, Jonathan Hubbard, biochemistry
Civan, Mortimer M, physiology
Clancy, Carl Francis, bacteriology
Clark, Charles Christopher, biochemistry
Clark, John Kapp, physiology, pharmacology
Clark, Junius Manson, immunology, microbiology
Clark, David Walter, dentistry
Cohen, Gary H, microbiology, virology
Cohen, Jeffrey M, astrophysics, theoretical physics
Clark, Robert Alfred, psychiatry
Clark, Wallace Henderson, Jr, medicine, pathology
Cleland, Richard Cook, applied statistics
Coburn, Ronald F, physiology
Cohen, Beverly Shapiro, medical biophysics

Connelly, Damian, mathematics
Conover, Thomas Ellsworth, biochemistry
Conrad, Bruce, topology
Conyne, Richard Francis, polymer chemistry
Codley, Eugene Leon, internal medicine, cardiology
Coon, Julius Mosher, pharmacology
Cooney, John Anthony, atmospheric physics
Cooper, David Young, medicine
Cooper, Donald Russell, surgery
Cooperman, Barry S, physical organic chemistry, biochemistry
Corbin, Alan, neuroendocrinology, reproductive physiology
Corman, Lew Andre, clinical pharmacology, internal medicine
Corn, Herman, periodontology
Cornelison, Floyd S, Jr, psychiatry
Corner, George Washington, anatomy
Cornfeld, David, pediatrics
Cortner, Jean A, biochemical genetics
Coughlin, Raymond Francis, algebra
Cox, Robert Harold, physiology, bioengineering
Cramer, Richard David, III, medicinal chemistry, physical organic chemistry
Crandall, Edward D, pulmonary physiology
Crane, August Reynolds, medicine, pathology
Creech, Hugh John, immunochemistry, cancer
Cristofalo, Vincent Joseph, physiology, biochemistry
Cropper, Walter V, instrumentation, research administration
Crouch, Ralph Boyett, mathematics
Crouse, Gail, anatomy
Crutchfield, Floy Love, endocrinology
Cuff, David J, cartography, physical geography
Cundy, Kenneth Raymond, medical microbiology, clinical microbiology
Cunningham, Howard Charles, organic chemistry
Custer, Richard Philip, oncology
Dahl, Anthony Orville, botany
Dalal, Fram Ruston, clinical chemistry, biochemistry
D'Angelo, Savino Albert, endocrinology
Davies, Helen Jean Conrad, biochemistry, microbiology
Davies, Richard Oelbaum, veterinary physiology
Davies, Robert Ernest, biochemistry, physiology
Davies, Ronald Edgar, biochemistry
Davis, Franklin A, organic chemistry
Davis, George Morgan, malacology
Davis, John K, optics
Davis, Neil Monas, pharmacy
Davis, Richard, neurobiology, neuropharmacology
Davis, Richard A, neurosurgery
Davis, Robert Harry, endocrinology
Day, Harvey James, internal medicine, hematology
De Alvarez, Russell Ramon, obstetrics & gynecology
Deas, Thomas C, anesthesiology
DeBias, Domenic Anthony, physiology
De Cani, John Stapley, applied statistics
De Courcy, Samuel Joseph, Jr, medical microbiology
Dees, Bowen Causey, physics, science administration
Deforest, Adamadia, virology
de la Haba, Gabriel Luis, biological chemistry
Delluva, Adelaide Marie, biochemistry
DeMarinis, Robert Michael, organic chemistry
DeMartinis, Frederick Daniel, physiology
DeMeio, Romano Humberto, biochemistry
Demitras, Gregory Claude, inorganic chemistry
Denoon, Clarence England, Jr, organic chemistry
De Pace, Dennis Michael, anatomy
De Terra, Noel, zoology
Detweiler, David Kenneth, physiology, pharmacology
Deubler, Mary Josephine, veterinary medicine
DeVault, Don Charles, biophysics, physical chemistry
Devlin, Thomas McKeown, biochemistry
DeWald, Eugene, pharmaceutical chemistry
Dheer, Surendra Kumar, organic chemistry
DiBerardino, Marie Antoinette, developmental biology

DiCarlo, Ernest Nicholas, chemical physics
Dickman, Albert, laboratory medicine
Dickson, James Gillespie, Jr, medicine, physiology
Di George, Angelo Mario, endocrinology, pediatrics
Di Gregorio, Guerino John, pharmacology
Diller, William Frey, Jr, protozoology
DiPalma, Joseph Rupert, pharmacology
Dierassi, Isaac, hematology
Dodson, Peter, paleobiology, vertebrate anatomy
Dolphin, John Michael, pathology
Domenicali, Charles Angelo, solid state physics
Donaldson, James Bowie, medicine
Doner, Landis Willard, carbohydrate chemistry
Donohue, Jerry, crystallography
Dorwart, Bonnie Brice, rheumatology
Doty, Richard Leroy, behavioral biology, psychophysics
Douglas, Bryce, medicinal chemistry
Dreby, Edwin Christian, III, physical chemistry, organic chemistry
Dress, John Allen, physiology, biophysics
Driscoll, Dorothy H, medical biophysics
Drouet, Francis, botany
Duane, Thomas David, ophthalmology
Dubeck, Leroy W, solid state physics
Dubin, Isadore Nathan, pathology
Duling, Irl Noel, organic polymer chemistry, petroleum chemistry
Dunn, George Lawrence, organic chemistry
Durant, Thomas Morton, clinical medicine
Dvonch, William, bio-organic chemistry
Dymicky, Michael, organic chemistry, physical chemistry
Dyson, Robert Harris, Jr, archaeology, anthropology
Eaton, Larry Rodney, atmospheric physics, cloud physics
Ediken, Jack, radiology
Edmonds, Louis Henry, Jr, thoracic surgery
Edwards, McIver Williamson, Jr, physiology, anesthesiology
Effros, Edward George, mathematics
Ehrlich, George Edward, rheumatology, internal medicine
Eichel, Herbert Joseph, biochemistry, enzymology
Eidson, William Whelan, nuclear physics, atomic physics
Eiseley, Loren Corey, anthropology
Eisen, Martin, mathematics
Eisenberg, Roselyn Jane, microbiology, biochemistry
Eisenhardt, Rudolph Hermann, physical biochemistry
Eisenman, Leonard Max, neurosciences
Eisenstein, Toby K, immunology
Elfvin, Myra L, cell biology
Elkins, William L, transplantation immunology
Ellenbogen, William Cromwell, analytical chemistry, research administration
Ellendman, Merrill, industrial chemistry
Elliott, Frank A, neurology
Elsom, Kendall Adams, medicine
Englander, Sol Walter, biophysics
Enterline, Horatio Theodore, pathology
Epple, August Wilhelm, comparative endocrinology
Epstein, Alan Neil, animal behavior, neurophysiology
Erb, Robert Allan, physical chemistry
Erdelyi, Ivan Nicholas, mathematical analysis
Erdman, William James, II, medicine
Erecinska, Maria, medicine, biochemistry
Erickson, James C, III, anesthesiology
Erickson, Ralph O, botany
Erick, Barry J, genetics
Ernst, Audrey Elizabeth, pediatrics
Ernst, Stephen Arnold, cell biology
Ernst, Sue Carlisle, parasitology, cell biology
Ersley, Allan Jacob, medicine
Ertel, Henry Robinson, organic chemistry
Erulkar, Solomon David, neurophysiology
Esfahani, Mojtaba, biochemistry
Esposito, Robert John, solid state physics
Evans, Audrey Elizabeth, pediatrics
Evers, Patricia Weber, pharmacology
Ewens, Warren John, population genetics, information science
Ezekiel, David Hirsch, microbiology
Fabrizio, Angelina Maria, medical microbiology
Fackenthal, Edward, physical chemistry

Faessinger, Robert William, organic chemistry
Faludi, Georgina, medicine, endocrinology
Farber, Paul Alan, microbiology, dentistry
Farley, Belmont Greenlee, biophysics, information science
Farrell, Harold Maron, Jr., biochemistry
Farren, Ann Louise, biochemistry, information science
Faul, Henry, geophysics
Feairheller, Stephen Henry, organic polymer chemistry
Federowicz, Rose Ann, botany, parasitology
Feinberg, Edwin Harold, biology, animal ecology
Fell, James Michael Gardner, pure mathematics
Feiley, Donald Louis, organic chemistry
Ferguson, James Joseph, Jr., biochemistry, medicine
Ficher, Miguel, endocrinology, chemistry
Fiddler, Walter, analytical chemistry
Field, Howard Lawrence, psychiatry
Fields, Harry, obstetrics & gynecology
Fifer, Robert Alan, physical chemistry
Filachione, Edward Mario, chemistry
Fineberg, Charles, thoracic surgery
Finegold, Leonard X., molecular biophysics, polymer physics
Finestone, Albert Justin, medicine
Finkelstein, David, cardiology
Fischer, Grace Mae, physiology
Fischer, John Edward, solid state physics
Fischer, Robert Leigh, clinical chemistry
Fishman, Alfred Paul, physiology
Fitts, Donald Dennis, physical chemistry, chemical physics
Fitts, William Thomas, Jr., surgery
Fitzgerald, Maurice E., analytical chemistry
Fitzpatrick, Thomas Joseph, agricultural chemistry
Flaks, Joel George, biochemistry
Flanagan, Thomas Leo, biochemistry
Foglia, Thomas Anthony, organic chemistry
Forbes, Paul Donald, radiobiology
Forman, Loren Verne, paper chemistry
Forster, Robert E, II, physiology
Fortune, H Terry, nuclear physics
Foster, Neal Robert, ichthyology
Fowler, Elizabeth Haddock, radiology
Fox, Eva Fernandez, bacteriology
Fox, Jay B, Jr., biochemistry
France, Evelyn S (Kalagher), reproductive endocrinology
Flexner, Louis Barkhouse, anatomy
Flick, John A, immunology
Flickinger, George Latimore, Jr., comparative pathology, reproductive physiology
Fluck, Eugene Richards, biochemistry
Flueck, John A, biometrics
Glauser, Elinor Mikelberg, pharmacology, physiology
Glick, Jane Mills, molecular biology
Glusker, Jenny Pickworth, physical biochemistry, x-ray crystallography
Gold, Jerome A, internal medicine, infectious diseases
Gold, Martin, clinical biochemistry
Goldberg, Harry, physiology
Goldberg, Irwin, theoretical physics
Goldberg, Martin, nephrology
Goldberger, Michael Eric, experimental neurology
Golder, Richard Harry, biochemistry
Goldfarb, Alvin F, obstetrics & gynecology, endocrinology
Goldfine, Howard, biochemistry, microbiology
Goldman, Abraham Samuel, chemistry
Goldman, David Eliot, biophysics, physiology
Goldman, Oscar, mathematics
Goldman, Herman, medical microbiology
Goldman, Frederick J, neuropharmacology
Goldsmith, Harry Sawyer, surgery
Goldstein, Franz, gastroenterology, internal medicine
Goldstein, Harold William, air pollution
Goldwein, Manfred Isaac, internal medicine, hematology
Golton, William Charles, analytical chemistry
Good, Norma Frauendorf, plant ecology
Goode, Judith Granich, anthropology
Goodenough, Ward Hunt, cultural anthropology, anthropological linguistics
Goodgal, Sol Howard, genetics
Goodman, David Barry Poliakoff, endocrinology
Goodwin, Peter Warren, stratigraphy
Gordon, Charles Francis, analytical chemistry
Gordon, Paul, applied mathematics
Gorn, Saul, computer sciences, information sciences
Gorson, Robert O, radiology, medical physics
Gosfield, Edward, Jr., internal medicine, cardiology

Gabuzda, Thomas George, hematology
Gaines, Alan McCulloch, geochemistry
Galambos, Janos, statistics
Gambescia, Joseph Marion, internal medicine, gastroenterology
Garcia, Celso-Ramon, obstetrics & gynecology
Garfield, Eugene, chemistry, information science
Garfinkel, David, mathematical biology, bioengineering
Gartto, Anthony Frank, solid state physics
Garrett, Bowman Staples, physical chemistry
Garsky, Victor Michael, organic chemistry
Gasic, Gabriel J, pathology
Gasser, David Lloyd, genetics, immunology
Gautieri, Ronald Francis, pharmacology
Gebhard, Joseph John, inorganic chemistry
Geller, Kenneth N, physics
Geller, Nancy L, mathematical statistics, statistics
Gennaro, Alfonso Robert, medicinal chemistry
Gero, Alexander, chemistry
Gersh, Eileen Susman, cytogenetics
Gersh, Isidore, histochemistry, electron microscopy
Gershenfeld, Louis, bacteriology, hygiene
Gerstein, George Leonard, neurobiology
Gerstenhaber, Murray, mathematics
Ghosh, Amal Kumar, biochemistry
Giannavola, Joseph Anthony, analytical chemistry, chemical instrumentation
Gibley, Charles W, Jr., zoology
Gibson, Robert John, Jr., biophysics
Gilbert, Fred, human genetics
Gilbert, Robert Pettibone, internal medicine
Gilden, Donald Harvey, neurology
Giles, Waldron, physical chemistry
Gill, Frank Bennington, ornithology
Gilpin, Richard William, microbiology,
Ginsberg, Myron David, neurology
Ginsburg, Isadore Wilcher, clinical medicine
George, Philip, biophysical chemistry

Gots, Joseph Simon, medical microbiology
Gottheil, Edward, psychiatry, psychology
Goulden, Clyde Edward, ecology, limnology
Gowdy, Spenser O, mathematics
Graham, William Rendall, physics, molecular biophysics
Grant, Norman Howard, biochemistry
Grappel, Sarah Fay, immunology
Graub, Milton, pediatrics
Gray, Frank Davis, Jr., internal medicine
Gray, Frieda Gersh, medicine
Grebner, Eugene Ernest, biochemistry
Green, Melville Saul, statistical mechanics, theoretical physics
Greene, Leon Charles, physiology
Greenberg, Herman Samuel, science education, organic chemistry
Greenspan, George, microbiology, pharmacology
Greenwald, Harold Leopold, polymer chemistry
Grego, Nicholas John, physiology
Gregory, Francis Joseph, microbiology, biochemistry
Griffin, Richard Norman, analytical chemistry
Griggs, Lee Jackson, medicinal chemistry
Grosse, Aristid Victor, chemistry
Grossman, Louis Irwin, dentistry
Grosswald, Emil, mathematics
Grotzinger, Paul John, surgery
Grove, Daniel Dwight, anesthesiology
Grove, Gary Lee, cell biology
Groves, William G, pharmacology
Gruber, Jacob William, anthropology, history of science
Gruenwald, Peter, pathology
Gruss, Leonard Louis, physical chemistry
Gunton, James D, theoretical physics, statistical mechanics
Gutekunst, Richard Ralph, microbiology
Guthrie, Marshall Beck, medicine, dermatology
Guttman, Samuel Arnold, psychiatry
Guttmann, Mark, nuclear physics
Guy, Eugene James, food science
Haag, Thomas Harry, polymer chemistry
Haase, Gunter R, neurology
Haase, Richard Henry, applied statistics
Haff, Richard Francis, microbiology
Hagis, Peter, Jr., mathematics
Hahn, George Alan, obstetrics & gynecology
Hahn, Richard Allen, pharmacology
Hale, John, radiological physics
Hall, Benedict Mark, genetics
Haltner, Arthur John, physical chemistry
Hameka, Hendrik Frederik, theoretical chemistry
Hamilton, Charles Lewis, physiology, neurosciences
Hamilton, Robert Houston, biochemistry
Hammel, Jay Morris, microbiology
Hammond, Benjamin Franklin, medical microbiology
Hampton, Alexander, organic chemistry, biochemistry
Hand, Peter James, anatomy
Hanson, Richard W, biochemistry
Harakal, Concetta, pharmacology
Harbert, Fred, otolaryngology
Harbold, Mary Leah, physics, acoustics
Hare, William Currie Douglas, veterinary anatomy
Harley, John, radiological physics
Harley, Robison Dooling, ophthalmology
Harnwell, Gaylord Probasco, physics
Harrington, George William, analytical chemistry
Harris, Arthur Brooks, solid state physics
Harris, Susanna, microbiology
Harris, Tzvee N, immunology
Harrison, W Craig, particle physics
Hartman, Mary Ellen, histology
Harvey, William Michael, anatomy
Harvey, William Royal, biology
Haskin, Marvin Edward, radiology
Haskin, Myra Ruth Singer, physical medicine & rehabilitation
Haugaard, Ella Shvartzman, biochemistry
Haugaard, Niels, biochemistry, pharmacology
Hugh, Michael J, chemical physics
Haurani, Farid I, medicine
Hausberger, Franz X, anatomy, physiology
Havas, Helga Francis, microbiology
Havas, Peter, theoretical physics

Havens, Walter Paul, Jr., internal medicine
Hazel, James Frederic, chemistry
Hearney, Elaine Frances, developmental genetics, histology
Heath, Barbara Honeyman, physical anthropology
Heath, Frederick Kriete, internal medicine
Hedges, Thomas Reed, Jr., ophthalmology
Heeger, Alan J, solid state physics
Heiner, Ralph, biochemistry
Helfer, Melvin S, medicine, psychiatry
Helwig, John, Jr., internal medicine, cardiology
Hencken, Frederick Hillens, biochemistry
Herold, Richard Carl, developmental biology
Herzog, Karl Adolph, pharmaceutics, pharmaceutical chemistry
Hess, Marilyn E, pharmacology
Hetzel, Charles Acheson, analytical chemistry
Hewetson, John Francis, virology
Hickey, Richard James, biochemistry, human ecology
Hickman, Don Winston, optometry
Higgins, Joseph John, biophysics
Higgins, Michael Lee, microbiology
Highland, Virgil Lee, physics
Hildebrand, David Kent, statistics
Hilfer, Saul Robert, developmental biology
Hillman, Ralph, biology
Hills, Claude Hibbard, food science
Hiltz, Arnold Aubrey, physical chemistry
Hodges, John Hendricks, medicine
Holbrook, Ruth Robertson, physiology
Holden, Kenneth George, medicinal chemistry
Hollander, Joseph Lee, rheumatology
Holling, Herbert Edward, medicine
Holly, Roy (Groves), obstetrics & gynecology
Holmes, Robert Hicks, pathology
Holmes, William Leighton, biochemistry
Holroyd, Roland, botany
Holroyde, Christopher Peter, oncology
Holst, Gerald Carl, optics
Holtzer, Howard, embryology
Homann, Frederick Anthony, mathematics
Hoober, John Kenneth, biochemistry
Hoover, John Russel Eugene, medicinal chemistry
Horger, Lewis Milton, cardiopulmonary physiology
Horner, George John, medicine
Horwitz, Orville, medicine
Howard, Barbara V, biochemistry, cell culture
Howard, Edgar, Jr., chemistry
Huang, Nancy N, pediatrics
Hubbar, Wayland Michael, actuarial science
Huber, John Franklin, anatomy
Hughes, Joseph F, psychiatry
Hughes, William, textile chemistry
Hummeler, Klaus, virology
Hungerford, David A, cytogenetics
Hurley, Harry James, dermatology
Hussar, Daniel Alexander, pharmacy
Hutchins, Robert Owen, Sr., organic chemistry
Hutchinson, Robert Cranford, anatomy
Hutchinson, Wesley Gillis, microbiology
Huth, Edward J, medicine
Hutton, Thomas Watkins, organic chemistry
Iglewicz, Boris, statistics
Immediata, Tony Michael, organic chemistry, polymer chemistry
Innes, David Lyn, physiology
Inoue, Shinya, biology
Intemann, Robert Louis, physics
Intoccia, Alfred Paul, drug metabolism
Iossifides, Ioulios A, pathology
Ipsen, Johannes Jr., medical statistics, epidemiology
Iralu, Vichazelhu, microbiology
Isard, Harold Joseph, medicine, radiology
Israel, Harold L, pulmonary diseases
Itkin, Irving Herbert, allergy
Iverson, Kenneth Eugene, applied mathematics
Iyengar, Raja M, biochemistry
Jackson, Donald Cargill, physiology
Jackson, Laird G, medicine

638

PENNSYLVANIA

Michael, Henry N, anthropology, dendrochronology
Mihalisin, Ted Warren, low temperature physics
Mikula, James J, physical chemistry
Mildvan, Albert S, biophysics, enzymology
Miller, Arthur Simard, oral pathology
Miller, Eleanor Marie, medical research
Miller, Elizabeth Eshelman, biochemistry, immunology
Miller, Elmer S, anthropology
Miller, Irvin Alexander, theoretical physics
Miller, John George, physical chemistry
Miller, Leonard David, surgery
Miller, Lyle Herbert, neuropsychology, psychophysiology
Miller, Richard Lee, invertebrate zoology, developmental biology
Millman, Irving, immunology, medical microbiology
Mills, Lewis Craig, Jr, internal medicine
Ming, Si-Chun, pathology
Mintz, Beatrice, biology
Miselis, Richard Robert, neurobiology
Misher, Allen, physiology, pharmacology
Mitchell, Albert Haywood, physical chemistry
Mitchell, Kenneth Frank, immunobiology
Mitchell, Theodore, mathematics, operations research
Mitruka, Brij Mohan, microbiology, veterinary medicine
Moat, Albert Groombridge, microbiology
Mode, Charles J, mathematics, population studies
Monson, Frederick Carlton, histology, reproductive biology
Montgomery, Paul Charles, immunology, microbiology
Moore, Earl Neil, physiology, cardiology
Moore, Mary Elizabeth, rheumatology
Moore, Matthew Thibaud, neuropsychiatry
Moorhead, Paul Sidney, cytogenetics
Moran, John J, medicine
Morganroth, Joel, cardiology, internal medicine
Morris, Harold Hollingsworth, Jr, medicine, psychiatry
Morrison, Adrian Russel, neuroanatomy, neurophysiology
Morrison, Donald Franklin, mathematical statistics
Moskowitz, Norman, anatomy
Moss, W Wayne, systematics, entomology
Mote, Michael Isnardi, neurophysiology
Motsavage, Vincent Andrew, pharmacy, physical chemistry
Moulton, David Gillman, physiology
Muldawer, Leonard, solid state physics, medical physics
Murphy, John Joseph, urology, surgery
Murray, Donald Shipley, academic administration, applied statistics
Murtagh, Frederick, Jr, neurosurgery
Myer, George Henry, geology, mineralogy
Myers, David, medicine
Myers, Howard M, pharmacology
Myers, Jacob Martin, psychiatry
Myers, Jeffery, pathology, information science
Myerson, Ralph M, internal medicine
Naide, Meyer, medicine
Narrod, Marian Freedman, pharmacology
Narrod, Stuart Allan, biochemistry
Nash, Carroll Blue, biology, parapsychology
Nath, Amar, physical chemistry, solid state chemistry
Negin, Michael, electrical engineering, biomedical engineering
Nelson, George Leonard, organic chemistry
Nelson, Waldo Emerson, pediatrics
Nemer, Martin Joseph, biochemistry
Nemeth, Andrew Martin, anatomy
Nemir, Paul, Jr, surgery
Nes, William Robert, biochemistry
Neste, Sherman Lester, meteoritics
Neville, Donald Edward, theoretical high energy physics
Nevyas, Jacob, biochemistry
Newman, David John, biochemistry, microbiology
Newman, Pauline, industrial chemistry
Newstein, Herman, meteorology
Nicholas, Gerardus, speleology
Nicholas, Leslie, dermatology, syphilology
Nijenhuis, Albert, mathematics
Nikelly, John G, analytical chemistry
Niu, Mann Chiang, developmental biology, biochemistry
Nixon, Eugene Ray, physical chemistry
Nodiff, Edward Albert, organic chemistry
Noone, Michael John, ceramics
Noordergraaf, Abraham, bioengineering, biophysics
Norcia, Leonard Nicholas, biochemistry, physiology
Noval, Joseph James, biochemistry
Nowak, Thomas, biochemistry
Nowell, Peter Carey, pathology
Nowotny, Alois Henry, immunology, immunochemistry
Noyes, Paul R, organic chemistry, polymer chemistry
Oaks, Wilbur W, internal medicine
O'Brien, Denise (Mrs Jay Ruby), cultural anthropology
O'Brien, Joan A, veterinary medicine, laryngology
Ocone, Luke Ralph, inorganic chemistry
O'Connor, Michael John, neurosurgery
O'Conor, John Stanislaus, physics
Oels, Helen C, pathology, immunology
Oesterling, Myrna Jane, biochemistry
Offenbacher, Elmer Lazard, physics
Ogburn, Clifton Alfred, immunology
Ohnishi, Tsuyoshi, biophysics, biochemistry
O'Kane, Daniel Joseph, bacteriology
O'Keefe, John Joseph, medicine
Oler, Norman, mathematics
Oliet, Seymour, dentistry
O'Neill, John Cornelius, applied mathematics
O'Neill, John Joseph, biochemistry
Oppenheimer, Morton Joseph, physiology
Orkand, Richard K, physiology, neurobiology
Orne, Martin Theodore, psychiatry, psychology
Orzechowski, Raymond Frank, pharmacology
Osofsky, Howard J, obstetrics & gynecology
Osol, Arthur, analytical chemistry
Ostapiak, Mykola, biology, chemistry
Ostrow, Jay Donald, gastroenterology
Otte, Daniel, zoology
Ou, Jonathan Tsien-Hsiong, microbiology, genetics
Outzen, Henry Clair, Jr, anatomy, cancer immunology
Owen, Charles Scott, biophysics
Pachtman, Elliott A, biochemistry
Packman, Elias Wolfe, pharmacology
Pagano, Joseph Frank, microbiology
Paik, Woon Ki, biochemistry
Pairent, Frederick William, biochemistry
Pakman, Leonard Marvin, microbiology
Palmer, Jon (Carl), immunology, molecular biology
Palmer, Larry Alan, neuropsychology
Palumbo, Samuel Anthony, microbiology, food science
Panepinto, Frank William, leather chemistry
Panos, Charles, bacteriology, biochemistry
Papariello, Gerald Joseph, analytical chemistry
Parker, Winfred Evans, physical organic chemistry
Parks, John S, pediatrics, endocrinology
Passow, Eli (Aaron), mathematics
Patel, Mulchand Shambhubhai, biochemistry
Patrick, Ruth (Mrs Charles Hodge IV), botany
Patterson, Donald Floyd, veterinary medicine, medical genetics
Patterson, Elizabeth Knight, biochemistry, cytochemistry
Patterson, Thomas Carl, anthropology, archaeology
Paucker, Kurt, virology
Pauls, John F, applied mathematics, statistics
Payne, Bobby Joe, veterinary pathology
Peachey, Lee DeBorde, cell biology, physiology
Pearlstein, Fred, electrochemistry, corrosion
Pearson, Manuel Malcolm, psychiatry
Peck, Richard Merle, organic chemistry
Pendleton, Robert Grubb, pharmacology
Penneys, Raymond, medicine
Penny, John Sloyan, paleobotany
Penzotti, Stanley Clare, Jr, pharmacy
Pepe, Frank Albert, anatomy
Perchonock, Carl David, organic chemistry, medicinal chemistry
Perlish, Jerome Seymour, biochemistry
Perry, Robert Palese, biophysics
Persky, Harold, biochemistry, endocrinology
Pessen, Helmut, physical chemistry, biochemistry
Peterson, Charles Fillmore, pharmacy
Peterson, Rudolph Price, biology, neurochemistry
Petta, John M, pharmacology, physiology
Pettit, Mary DeWitt, gynecology & obstetrics
Pfeffer, Philip Elliot, physical organic chemistry
Pfeiffer, Francis Richard, organic chemistry
Pfeiffer, Mildred Clara Julia, medicine
Pflugfelder, Hala, mathematics
Phelps, Michael Edward, radiation physics
Phillips, S Michael, immunology
Phillips, Steven Jones, anatomy
Piatt, Jean, anatomy
Piccolini, Richard John, physical chemistry, organic chemistry
Pickands, James, III, statistics
Pieringer, Ronald Arthur, biochemistry
Pierson, Ellery Merwin, statistical analysis
Pietra, Giuseppe G, pathology
Pinkston, John Turner, physical chemistry
Pitkow, Howard Spencer, reproductive endocrinology
Pitts, Paul Miller, Jr, petroleum chemistry
Pizer, Lewis Ivan, microbiology, biochemistry
Plaut, Gerhard Wolfgang Eugen, biochemistry
Plotkin, Stanley Alan, virology, pediatrics
Plummer, E Ward, solid state physics
Polish, Edwin, internal medicine, gastroenterology
Pollack, Robert Leon, biochemistry
Pollikoff, Ralph, virology
Ponessa, Joseph Thomas, medical physiology
Pontarelli, Domenic Joseph, obstetrics & gynecology
Poole, John Anthony, physical chemistry
Poole, John William, pharmaceutical chemistry
Porter, Gerald Joseph, mathematics
Poss, Howard Lionel, astronomy, plasma physics
Pratt, Neal Edwin, anatomy
Prehn, Richmond Talbot, experimental pathology
Preti, George, organic chemistry, mass spectrometry
Price, Charles Coale, organic chemistry
Price, Henry Locher, pharmacology, anesthesiology
Prier, James Eddy, public health
Primakoff, Henry, theoretical physics
Pring, Martin, physiology
Pruzan, Anita M, behavior genetics
Pubols, Lillian Menges, neurophysiology
Puglia, Charles David, pharmacology, environmental physiology
Punnett, Hope Handler, genetics
Punnett, Thomas R, plant biochemistry
Pye, Edward Kendall, biochemistry
Pytlewski, Louis Lawrence, inorganic chemistry
Rabii, Sohrab, theoretical solid state physics
Rabinowitz, Joseph Loshak, biochemistry
Radin, Charles Lewis, mathematical physics
Raffensperger, Edward Cowell, gastroenterology
Rainey, Froelich Gladstone, anthropology
Rakoff, Abraham E, endocrinology
Ralph, Nathan, medicine
Ralston, Edgar Lee, medicine
Randall, Peter, plastic surgery
Rapp, John P, pathology
Rappaport, Harry P, biochemistry
Rashkind, William Jacobson, pediatrics
Rasmussen, Howard, medicine, cell biology
Ratchford, William Paul, physical chemistry, organic chemistry
Ravin, Louis Joseph, pharmacy
Rawson, Arnold Joseph, pathology
Raymon, Louis, mathematics
Raymond, Samuel, medicine, chemistry
Reber, Louis Alexander, physical chemistry, analytical chemistry
Reddy, John Bernard, otology
Reed, Emerson Aloysius, physiology
Reed, George Henry, biophysics, biochemistry
Reed, Sherman Kennedy, chemistry
Reich, Daniel, mathematics
Reichgott, Michael Joel, medicine, clinical pharmacology
Reichle, Frederick Adolph, surgery
Reiff, Harry Elmer, organic chemistry
Reimann, Hobart Ansteth, medicine
Reina, Ruben E, cultural anthropology
Reivich, Martin, neurology
Relman, Arnold Seymour, medicine
Rengert, George Frederick, population geography, geography of latin america
Repper, Charles John, solid state physics
Revesz, George, radiology, electrical engineering
Reynolds, Monica, medical physiology
Rhoads, Jonathan Evans, surgery
Rhodes, William Harker, veterinary medicine
Rich, Michael, mathematics
Richards, Horace Gardiner, paleontology
Richter, George Alvin, Jr, chemistry
Rickels, Karl, psychiatry, psychopharmacology
Ricklefs, Robert Eric, ecology
Ridley, Peter Tone, physiology, pharmacology
Riethof, Thomas Robert, physical chemistry, electrooptics
Rim, Dock Sang, mathematics
Roback, Selwyn, entomology
Robbins, Robert, medicine
Robbins, William S, psychiatry
Roberti, Dominic M, physical chemistry
Roberts, Bryan Wilson, organic chemistry
Roberts, David, biomechanics
Roberts, Dean Winn, medicine, community health
Roberts, Howard Radclyffe, entomology
Roberts, Jay, pharmacology
Roberts, Shepherd (Knapp de Forest), comparative physiology
Roberts, William John, chemistry
Robertson, Robert, malacology
Robinson, Thomas Frank, biophysics, muscular physiology
Rockey, John Henry, ophthalmology, immunochemistry
Rockwell, Harriet Esther, therapeutics
Rodriguez-Peralta, Lorenzo Alberto, anatomy
Rogers, Fred Baker, preventive medicine, public health
Roia, Frank Costa, Jr, economic botany, microbiology
Ronis, Bernard Joseph, otolaryngology
Ronis, Max Lee, otolaryngology
Rooney, James Rowell, veterinary pathology
Roosa, Robert Andrew, medical microbiology, genetics
Rork, Gerald Stephen, pharmaceutical chemistry
Rorke, Lucy Balian, neuropathology
Rorres, Chris, mathematics
Rosan, Burton, medical microbiology
Rose, Irwin Allan, biochemistry
Rosemond, George P, surgery
Rosen, Gerald Harris, theoretical physics
Rosenberg, Philip E, audiology
Rosenblatt, David Bar Macbee, applied physics
Rosenbloom, Joel, biochemistry, biophysics
Rosenfeld, Leonard M, physiology, biochemistry
Rosenman, Sanford Becker, industrial microbiology
Rosenow, Edward Carl, Jr, medicine
Rosenthale, Marvin E, pharmacology
Rosett, Theodore, biochemistry
Ross, Leonard Lester, anatomy, neurobiology
Ross, Morris H, biology
Ross, Oscar Alan, pathology
Ross, Sidney, physics, research administration
Ross, Stephen T, medicinal chemistry
Rossi, George Victor, pharmacology
Roth, James Luther Aumont, gastroenterology
Rothbart, Herbert Lawrence, physical chemistry, analytical chemistry
Rothblat, George H, microbiology, biochemistry
Rothman, Richard Harrison, orthopedic surgery, anatomy
Rovetto, Michael Julien, physiology, biochemistry
Rowlands, David T, Jr, pathology, immunology
Royster, Henry Page, surgery
Rubin, Alan, medicine
Rubin, Leonard Sidney, psychophysiology, psychopharmacology
Rubin, Nathan, organic chemistry
Rubin, Walter, gastroenterology, cell biology
Ruby, Jay W, anthropology
Rudin, Donald Oliver, electrochemistry
Rudkin, George Thomas, genetics
Ruelius, Hans Winfried, drug metabolism
Ruff, George Elson, psychiatry
Rugen, Donald Frederick, organic chemistry
Rump, Ellis Samuel, Jr, physical chemistry
Rush, Alexander, medicine

Russell, Peter Byrom, organic chemistry
Russo, Emanuel Joseph, pharmaceutics
Rusy, Ben F, pharmacology, anesthesiology
Rutgers, Jay G, analytical chemistry
Rutman, Robert Jesse, biochemistry
Ryan, Carolyn J, geography
Ryan, Jon Michael, cell biology
Saaty, Thomas L, mathematics
Sachs, Marvin Leonard, internal medicine
Sacks, David Alan, pharmaceutics
Sadtler, Philip, analytical chemistry
Saegebarth, Klaus Arthur, organic chemistry, polymer chemistry
Saggiomo, Andrew Joseph, medicinal chemistry, research administration
Sakai, Shoichiro, mathematics
Salomon, Robert Ephriam, physical chemistry
Salzberg, Brian Matthew, neurobiology, biophysics
Samitz, M H, dermatology
Sanday, Peggy R, cultural anthropology, behavioral anthropology
Sanger, Joseph William, molecular biology, developmental biology
Santilli, Arthur A, organic chemistry
Sapers, Gerald M, food science
Sapico, Virginia L, biochemistry
Sarma, Akkaraju V N, anthropology, archaeology
Sarma, Dittakavi S R, biochemistry, experimental pathology
Sataloff, Joseph, otology
Sato, Hidemi, cytology, biophysics
Saunders, Harry Link, mammalian physiology
Saunders, Leon Z, veterinary pathology, history of medicine
Scandura, Joseph M, mathematics
Scarpa, Antonio, biochemistry, biophysics
Schad, Gerhard Adam, parasitology
Schaedler, Russell William, medicine, microbiology
Schamberg, Ira Leo, medicine
Schantz, Ilene Sue Cottler, reproductive biology, cytochemistry
Scheie, Harold Glendon, ophthalmology
Scheier, Arthur, optometry
Scheindlin, Stanley, pharmaceutical chemistry
Scheinok, Perry Aaron, research administration, mathematics
Schemm, George Walker, neurosurgery
Schepartz, Bernard, biochemistry
Schick, Paul Kenneth, hematology
Schild, Albert, mathematics
Schiller, John Joseph, mathematics
Schindler, Alan Michael, physics, biophysics
Schleyer, Heinz, biophysics, enzymology
Schlezinger, Nathan Stanley, neurology, psychiatry
Schmidt, Richard Ralph, teratology
Schmuckler, Joseph S, science education
Schnaare, Roger L, pharmaceutics, quantum chemistry
Schnabel, Truman Gross, Jr, internal medicine
Schneider, Henry Joseph, organic chemistry
Schneider, Henry Peter, laboratory animal science, medical research
Schneider, John Joseph, biochemistry
Schneller, George Henry, pharmaceutical chemistry
Schnizer, Arthur Wallace, organic chemistry
Schoen, Kurt L, chemistry
Schor, Stanley, biostatistics
Schott, Hans, physical chemistry
Schramm, Vern Lee, biochemistry
Schrenk, George L, theoretical physics
Schrieffer, John Robert, physics
Schultz, Richard E, psychiatry
Schuyler, Alfred E, botany, taxonomy
Schwan, Herman Paul, biophysics, biomedical engineering
Schwartz, Arthur Gerald, cell biology, cancer
Schwartz, Emanuel Elliot, medicine, radiology
Schwartz, Heinz (Georg), pathology
Schwartz, Irving Robert, medicine, hematology
Schwartzman, Robert M, dermatology, immunology
Schwarz, Gabriel Alexander, medicine
Schwarz, Henry P, biochemistry
Schwarz, Richard Howard, obstetrics & gynecology
Schwegman, Cletus W, surgery
Scidmore, Wright Harwood, optics
Scott, Donald, Jr, neurophysiology
Scott, Dwight Baker McNair, biochemistry
Scott, John Culbertson, physiology
Scott, Michael, neurosurgery
Scrutton, Michael Christopher, biochemistry
Searls, Robert L, biochemistry, embryology
Sedar, Albert William, microscopic anatomy
Segal, Bernard L, internal medicine, cardiology
Segal, Stanton, medicine, biochemistry
Segall, Stanley, organic chemistry, food technology
Segre, Gino C, theoretical physics
Sellers, Alfred Mayer, medicine
Selove, Walter, particle physics
Seltzer, Albert Pincus, surgery
Seltzer, Samuel, dentistry
Senior, John Robert, medicine, gastroenterology
Setler, Paulette Elizabeth, pharmacology, physiology
Settle, Richard Gregg, neuropsychology, psychophysics
Sevy, Roger Warren, pharmacology
Seydel, Horst Gunter, radiotherapy, radiobiology
Shagass, Charles, psychiatry
Shale, David, mathematics
Shapiro, Bernard, nuclear medicine
Shapiro, Herbert, physiology
Shapiro, Irving Meyer, biochemistry
Shapiro, Sandor Solomon, hematology
Sharma, Ran S, biostatistics, mathematical statistics
Sharp, David Howland, theoretical physics
Shatz, Stephen S, mathematics
Shaw, Daniel Leonard, Jr, medicine
Shaw, Ralph Arthur, metabolism, endocrinology
Shayegani, Mehdi, medical microbiology
Shea, John Raymond Michael, Jr, anatomy, histology
Shelley, Walter Brown, dermatology
Shen, Benjamin Shih-Ping, astrophysics, nuclear physics
Shenkin, Henry A, neurosurgery
Sherry, Sol, medicine
Ship, Irwin I, oral medicine
Shockman, Gerald David, microbiology, biochemistry
Showers, Mary Jane C, anatomy
Shuba, Raymond J, analytical chemistry
Shuman, Charles Ross, internal medicine
Siegel, Edward T, veterinary endocrinology
Siegfried, John Barton, visual physiology, electrophysiology
Siegler, Peter Emery, internal medicine
Siegman, Marion Joyce, pharmacology, physiology
Siggia, Eric Dean, statistical mechanics
Sigler, Miles Harold, nephrology
Silberberg, Donald H, neurology
Silbert, Leonard Stanton, physical chemistry, organic chemistry
Silver, Melvin Joel, biochemistry, pharmacology
Silver, Ruth Kunkle, hematology
Silvers, Willys Kent, genetics
Silverstein, Alexander, neurology, psychiatry
Sims, Homer Jennings, organic biochemistry
Singer, Arthur Chester, biostatistics
Singh, Jagbir, statistics
Singh, Rudra Pratap, medicinal chemistry, pharmaceutical chemistry
Sinha, Ramananda, medicine, physiology
Sisenwine, Samuel Fred, drug metabolism, organic chemistry
Sisson, Thomas Randolph Clinton, pediatrics, obstetrics
Sittel, Karl, applied physics
Slavin, Ovid, dentistry, biology
Sloviter, Henry Allan, physiological chemistry
Smith, Allan Laslett, physical chemistry
Smith, Charles Lea, chemistry
Smith, Colleen Mary, biochemistry
Smith, David English, pathology
Smith, David S, pediatrics
Smith, Edward John, analytical chemistry
Smith, Harry Logan, Jr, microbiology
Smith, Hugo Dunlap, pediatrics
Smith, James Lee, microbiology
Smith, Jerry Joseph, physical inorganic chemistry
Smith, Robert Kinsel, petroleum chemistry
Smith, Shaler Gordon, Jr, polymer chemistry, information science
Smith, Susan Trussell, animal behavior
Smith, Theodore Craig, anesthesiology, pharmacology
Smith, W John, biology, animal behavior
Smoliar, Stephen William, information science
Snipes, Charles Andrew, physiology
Snodgrasse, Richard Montgomery, anatomy, physical anthropology
Snow, James Byron, Jr, otolaryngology
Snow, Laurence H, psychiatry
Snyder, Robert, biochemical pharmacology
Snyder, Robert LeRoy, comparative pathology, ecology
Soberman, Robert K, environmental physics
Sodicoff, Marvin, anatomy
Sokol, Frantisek, virology, biochemistry
Sokolowski, Henry Alfred, physics, mathematics
Sollott, Gilbert Paul, organic chemistry
Soloff, Louis Alexander, medicine, cardiology
Solomon, Gene Barry, zoology, parasitology
Solomon, Joseph Alvin, physical chemistry
Solomon, M Michael, organic chemistry, polymer chemistry
Solter, Davor, developmental biology
Soma, Lawrence R, anesthesiology
Somlyo, Andrew Paul, physiology, pathology
Somlyo, Avril Virginia, cell physiology
Sorof, Sam, biochemistry
Soslau, Gerald, biochemistry
Soulen, Renate Leroi, medicine, radiology
Soulsby, Ernest Jackson Lawson, parasitology
South, Mary Ann, pediatrics
Southam, Chester Milton, oncology, virology
Southard, Martha Ellen, radiology
Soven, Paul, physics
Spaeth, George L, ophthalmology
Spann, James Fletcher, (Jr), cardiology
Spear, Joseph Francis, physiology
Spielberg, Kurt, theoretical physics, applied mathematics
Spielman, Richard Saul, population genetics
Spolsky, Christina Maria, cell biology, cancer
Sprague, James Mather, anatomy
Stambaugh, John Edgar, Jr, oncology, clinical pharmacology
Stambaugh, Richard L, biochemistry
Stanley, Edward Livingston, chemistry
Stasheff, James Dillon, mathematics
Staum, Muni M, radiochemistry, pharmaceutical chemistry
Steben, John D, physics
Stedman, Robert John, organic chemistry
Steel, Howard Haldeman, orthopedic surgery
Stein, George Nathan, radiology
Stein, Irvin, orthopedic surgery
Stein, Larry, physiological psychology
Stein, Reinhardt P, organic chemistry
Stein, Samuel C, medicine
Stein, T Peter, biochemistry
Stellar, Eliot, physiological psychology
Stelos, Peter, immunology, immunochemistry
Stephens, William Edwards, physics
Sterling, Peter, neuroanatomy, neurophysiology
Stevens, Lloyd Weakley, surgery
Stewart, George Hamill, biomedical engineering
Stewart, William Charles, statistical analysis, operations research
Stinnett, James LeBaron, psychiatry
Stoloff, Irwin Lester, medicine
Stone, Hrant H, medicine
Storey, Bayard Thayer, cell physiology, physical biochemistry
Storey, Patrick Brendan, medicine
Story, Jon Alan, biochemistry, nutrition
Stoyle, Judith, applied statistics
Straile, William Edwin, biological science, neurosciences
Straumanis, John Janis, Jr, psychiatry
Strauss, Robert R, biochemistry, microbiology
Streng, Alex G, physical chemistry
Studzinski, George P, experimental pathology, cell biology
Stuebing, Edward Willis, chemical physics
Suffet, Irwin Henry, analytical chemistry, environmental chemistry
Sugita, Edwin T, pharmaceutics
Suhadolnik, Robert J, biochemistry
Sullivan, Nicholas, speleology
Sunderman, Frederick William, internal medicine, clinical pathology
Sung, Cheng-Po, biochemistry
Suntharalingam, Naglingam, radiological physics
Suter, Stuart Ross, organic chemistry
Sutman, Frank X, chemistry
Sutnick, Alton Ivan, internal medicine
Sutton, Blaine Mote, medicinal chemistry
Swain, Ansel Parrish, biochemistry, organic chemistry
Swanson, Ernest Allen, Jr, anatomy, histology
Swern, Daniel, organic chemistry
Swisher, Ely Martin, entomology
Szabo, Kalman Tibor, teratology
Szwed, John Francis, anthropology
Tabachnick, Joseph, comparative biochemistry
Tahir-Kheli, Raza Ali, theoretical magnetism
Taichman, Norton Stanley, pathology, immunology
Takashima, Shiro, physical biochemistry, neurophysiology
Takats, Stephen Tibor, cytology, genetics
Talbot, Timothy Ralph, Jr, research administration
Tallarida, Ronald Joseph, biomathematics, pharmacology
Talley, Eugene Alton, agricultural chemistry
Tang, Chung-Muh, meteorology
Tansy, Martin F, physiology
Tappe, John, geology, geochemistry
Tartof, Kenneth D, genetics
Tasman, William S, ophthalmology
Taylor, Donald Rudolph, Jr, electronics, biomedical engineering
Taylor, W J Russell, clinical pharmacology
Tedesco, Thomas Albert, human genetics, biochemical genetics
Tekel, Ralph, organic chemistry
Telfer, William Harrison, reproductive biology, developmental biology
Teller, Daniel Myron, organic chemistry
Templeton, John Y, III, surgery
Teplick, Joseph George, medicine, radiology
Ternes, Joseph Wayne, psychopharmacology
Terracio, Louis, histology
Terry, Luther Leonidas, medicine
Thayer, Charles Walter, invertebrate paleontology, paleoecology
Thelen, Edmund, materials science, industrial chemistry
Thibeault, Jack Claude, inorganic chemistry
Thilo, Edward Rudolf, physics
Thomas, Carmen Christine, dermatology
Thompson, Kenneth David, microbiology, immunology
Thompson, Marvin P, biochemistry
Thompson, Thomas Eaton, solid state physics
Thornton, Edward Ralph, organic chemistry, bio-organic chemistry
Thurman, Ronald Glenn, biochemistry, pharmacology
Tice, Linwood Franklin, pharmaceutical chemistry
Tint, Howard, biochemistry
Tischio, John Patrick, biochemistry
Tislow, Richard Frederick, psychiatry, pharmacology
Titus, Donald Dean, inorganic chemistry
Tobia, Alfonso Joseph, pharmacology
Toglia, Joseph U, neurology
Tomer, Kenneth Beamer, mass spectrometry, organic chemistry
Tomlinson, Hazel M, chemistry
Tooker, Elisabeth (Jane), anthropology
Toporek, Milton, biochemistry
Toton, Edward Thomas, astrophysics
Touchstone, Joseph Cary, organic chemistry, biochemistry
Tourtellotte, Charles Dee, internal medicine, biochemistry
Townend, Robert Edward, physical biochemistry
Trachtman, Mendel, radiation chemistry, photochemistry
Treadway, Robert Holland, physical chemistry
Treadway, William Jack, Jr, immunochemistry
Trench, William Frederick, mathematics
Trice, James Buckner, radiation physics, atomic spectroscopy
Triolo, Anthony J, pharmacology
Trotter, Nancy Louisa, cytology, electron microscopy
Troyer, John Robert, anatomy
Truex, Raymond Carl, anatomy
Tse, Rose (Lou), organic chemistry, medicine
Tsou, Kwan Chung, organic chemistry, biochemistry
Tucci, Anthony Frederick, biochemistry, microbiology
Tucker, Gabriel Frederick, Jr, laryngology

Tuckerman, Murray Moses, pharmaceutical chemistry
Tuddenham, William J., radiology, roentgenology
Tumen, Henry Joseph, medicine
Turchi, Joseph J. internal medicine, hematology
Turco, Salvatore J., pharmacy
Tyson, Ralph Robert, surgery
Uhle, Charles Augustus Woerwag, urology
Ulmer, Gene Carleton, geochemistry
Upton, G Virginia, physiology, biochemistry
Urbach, Frederick, dermatology
Urbach, John Robert, medicine
Uzzell, Thomas Marshall, Jr, systematic biology; vertebrate biology
Valdes-Dapena, Marie A. pediatrics, pathology
Vanderheiden, Bernardo S., biochemistry
Van Dyke, John Howard, anatomy
Van Rossum, George Donald Victor, cell physiology; biochemical pharmacology
Van Scott, Eugene Joseph, dermatology
Varkey, Thanakamma Eapen, organic chemistry
Vars, Harry Morton, physiological chemistry
Vaughan, Victor Clarence, III, pediatrics
Viale, Richard O., biochemistry, biophysics
Viceps-Madore, Dace I., cell biology
Vila, Samuel Campderros, astrophysics
Voet, Donald Herman, crystallography, biochemistry
Vogel, Wolfgang Hellmut, biochemistry, pharmacology
Wachsberger, Phyllis Rachelle, cell biology
Wachtell, George Peter, physics
Wagner, Herman Block, physical
Wahl, Milton Heins, chemistry, science administration
Waldman, Joseph, ophthalmology
Waldron, Ingrid Lore, psychosomatic medicine
Wales, Walter D. physics
Walkenstein, Sidney S., drug metabolism
Walsh, John Paul, organic chemistry
Walsh, Peter Newton, hematology
Walter, Gerald Joseph, organic chemistry
Walz, Donald Thomas, pharmacology
Wampler, Stanley Norman, veterinary medicine, radiobiology
Wallace, Andrew Hugh, medicine
Wallace, Anthony Francis Clarke, cultural anthropology
Wallace, Herbert William, biochemistry, surgery
Wallach, Edward E., obstetrics & gynecology
Wardell, Joe Russell, Jr, pharmacology, physiology
Warner, Francis James, anatomy
Warner, Frank Wilson, III, mathematics
Warren, George Harry, microbiology, chemotherapy
Warren, Richard Joseph, analytical chemistry
Wase, Arthur William, biochemistry
Washburne, Stephen Shepard, organic chemistry
Wasserman, Aaron E., food chemistry
Watrous, James Joseph, physiology
Waxman, Herbert Sumner, medicine, hematology
Wayland, Bradford B. inorganic chemistry
Weaver, Quentin Clifford, paper chemistry
Weber, Annemarie, physiology
Weber, Wilfred T., pathology, immunology
Weeks, Alice Mary Dowse, geology
Weeks, Donald Paul, molecular biology, physiology
Wegner, Marcus Immanuel, biochemistry, nutrition
Weibel, Michael Kent, biochemistry, enzymology
Weidanz, William P., immunology, medical microbiology
Weiler, Eberhardt, immunology, molecular biology
Weiler, Ivan-Jeanne Mayfield, immunology; psychobiology
Weinbaum, George, biochemistry, microbiology
Weinhouse, Sidney, biochemistry, cancer
Weinreb, Eva Lurie, biological structure, cell biology
Weinstock, Joseph, organic chemistry

Weisbach, Jerry Arnold, organic chemistry; medicinal chemistry
Weisberg, Howard Louis, physics
Weisenberg, Richard Charles, cell biology
Weiss, Arthur Jacobs, internal medicine
Weiss, Charles, immunology, medical anthropology
Weiss, William, pulmonary diseases
Wells, Charles Robert Edwin, pediatric cardiology
Werner, Ervin Robert, Jr, organic chemistry
Wesson, Laurence Goddard, Jr, physiology; internal medicine
West, Keith P., zoology, radiation biology
Westlake, Wilfred James, mathematics, statistics
Whaley, Randall McVay, physics
Wheeler, Tamara Stech, anthropology
White, Harry Joseph, organic chemistry
White, Jonathan Winborne, Jr, agricultural chemistry
Whittaker, John Richard, embryology
Wiebelhaus, Virgil D., biochemistry, pharmacology
Wiederman, Mary Purcell, physiology
Wiener, Leslie, cardiology
Wiktor, Tadeusz Jan, virology
Wilcox, Wesley C., microbiology
Wilf, Herbert, mathematics
Wilkerson, Clarence Wendell, Jr, topology
Wilkins, Raymond Leslie, organic chemistry
Williams, Hugh Harrison, experimental high energy physics
Williams, John Roderick, synthetic organic chemistry
Williams, Norman, occupational medicine
Williamson, John Richard, biochemistry, biophysics
Wilpizeski, Chester Robert, otology, speech science
Wilson, Darcy Benoit, immunology
Winchester, Richard Albert, audiology, speech pathology
Winegrad, Albert Irvin, medicine
Winegrad, Saul, physiology
Winicov, Herbert, organic chemistry
Winicov, Ilga Butelis, molecular biology
Winsten, Seymour, biochemistry
Wirts, Charles Wilmer, physiology
Witin, Bernard, bacteriology
Witthoff, John, anthropology
Wittle, John Kenneth, inorganic chemistry, analytical chemistry
Witzleben, Camillus Leo, pathology
Wohl, George T. radiology
Wohlgemuth, John Harold, experimental solid state physics
Wolf, Benjamin, microbiology
Wolf, Don Paul, reproductive biology
Wolf, Leonard Nicholas, biology, zoology
Wolff, Ivan A., organic chemistry; research administration
Wollman, Robert Joseph, otolaryngology
Wollman, Harry, anesthesiology
Woloshin, Henry Jacob, radiology
Wolpe, Joseph, psychiatry, psychology
Wolsky, Alan Martin, physics
Wood, Francis Clark, medicine
Wood, James Edwin, III, medicine
Wood, Margaret Gray, medicine, dermatology
Wood, Thomas Hamil, biophysics
Woodward, Kent Thomas, radiotherapy
Worrell, Wayne L., high temperature chemistry
Woychik, John Henry, biochemistry, physiology
Wrigley, Arthur Nelson, organic chemistry
Yaffe, Sumner J., pediatrics, pharmacology
Yamamoto, Nobuto, microbiology, biophysics
Yang, Chung-Tao, mathematics
Yankell, Samuel L., physiology
Yiannos, Peter N. physical chemistry, engineering
Yonetani, Takashi, biochemistry, biophysics

Young, Charity Louise, cell biology
Young, Harrison Hurst, Jr, physical chemistry
Young, In Min, audiology, psychoacoustics
Young, Irving, medicine, pathology
Yu, Ruey Jiin, clinical pharmacology
Yuan, Edward Lung, physical chemistry
Yushok, Wasley Donald, biochemistry, cancer
Zabara, Jacob, physiology, neurophysiology
Zacharias, David Edward, x-ray crystallography, organic chemistry
Zachariah, Robert Marvin, plant biochemistry
Zacks, Sumner Irwin, pathology
Zaika, Laura Larysa, microbiology
Zajac, Barbara Ann, virology, electron microscopy
Zajac, Ihor, medical microbiology, virology
Zanger, Murray, physical organic chemistry
Zanovak, Paul, pharmaceutics
Zatuchni, Jacob, internal medicine, cardiovascular disease
Zauderer, Bert, magnetohydrodynamics, energy conversion
Zavada, John Michael, Jr, theoretical physics
Zavitsanos, Petros D., physical chemistry
Zawoiski, Eugene Joseph, physiology
Zeidman, Irving, pathology, radiological health, radiological physics
Zelac, Ronald Edward, radiological physics
Zelikoff, Steven Barry, applied statistics, business administration
Zelson, Philip Richard, biochemistry
Zigera, Sumner Root, pediatrics
Zimmerman, Irwin David, neurophysiology, biophysics
Zimmerman, James Joseph, pharmacy, pharmaceutical chemistry
Zimmt, Werner Siegfried, polymer chemistry; pollution chemistry
Zinser, Edward John, analytical chemistry
Zinsser, Harry Frederick, cardiology
Ziskin, Marvin Carl, biomedical engineering
Zitarelli, David Earl, mathematics
Zmijewski, Chester Michael, immunology
Zubrzycki, Leonard Joseph, medical microbiology
Zuccarello, William A. endocrinology
Zurmuhle, Robert W., nuclear physics
Zweiman, Burton, medicine
Zwerling, Israel, psychiatry

PHOENIXVILLE
Rosenthal, Fritz, industrial chemistry

PITTSBURGH
Abraham, Donald James, organic chemistry; medicinal chemistry
Abram, Dinah, microbiology, electron microscopy
Abrams, Richard, biochemistry
Abramson, Stanley L., physics, operations research
Acheson, Willard Phillips, applied mathematics
Adelman, Fred, anthropology
Adibi, Siamak A. gastroenterology, nutrition
Adovasio, James Michael, anthropology
Alarie, Yves, physiology
Alexander, Leroy Elbert, chemistry, crystallography
Allen, Alexander Charles, pediatrics, neonatology
Alo, Richard Anthony, mathematics
Altman, Isidore, applied statistics
Alvino, William Michael, polymer chemistry
Andelman, Julian Barry, environmental chemistry
Anderson, John Howard, chemical physics
Anderson, Julius Horne, Jr, pharmacology, biochemistry
Andrews, Peter Bruce, mathematical logic
Angello, Stephen James, physics, semiconductors
Anspon, Harry Davis, organic chemistry
Anthes, John Allen, organic chemistry
Archer, William Harry, oral surgery
Arnett, Edward McCollin, physical chemistry
Arnold, William H, Jr, experimental physics
Amtzen, Clyde Edward, organic chemistry

Artman, Joseph Oscar, chemical physics, solid state physics
Asenjo, Florencio Gonzalez, mathematical logic
Ashkin, Julius, magnetic resonance
Atchison, Robert Wayne, virology, immunology
Austern, Norman, theoretical nuclear physics
Austin, Janet Evans, chemistry
Axelrod, Abraham Edward, biochemistry
Bache, Charlotte Gertrude, psychiatry
Backus, John King, physical chemistry
Bahill, Andrew Terry, bioengineering, neuroscience
Banerjee, Sushanta Kumar, soil science, botany
Bansal, Krishan Murari, physical chemistry
Baranger, Elizabeth Urey, theoretical physics
Bardsley, James Norman, atomic physics
Bane, Walter Peter, Jr, organic polymer chemistry; industrial organic chemistry
Barnes, Wallace Edward, mathematics
Barnes, Peter David, nuclear physics
Barrett, William A. urology
Barry, Herbert, III, psychopharmacology
Basford, Robert Eugene, biochemistry
Bates, Margaret Westbrook, biochemistry
Baumann, Gert Friedrich, organic polymer chemistry; industrial organic chemistry
Bendet, Irwin (Jack), biophysics
Benedicty, Mario, mathematics
Bennett, Marvin Herbert, neuroanatomy, neurophysiology
Bentley, Ronald, biochemistry
Berger, Luc, solid state physics
Berkey, Edgar, physical chemistry, nuclear science
Berman, David S., vertebrate paleontology
Berry, Guy C., polymer chemistry
Biagas, Wilfred Michael, physical chemistry
Bikerman, Michael, geochronology
Biondi, Manfred Anthony, atomic physics, acronomy
Biordi, Joan Concetta, physical chemistry
Birnbaum, Hermann, organic chemistry
Black, Craig C. vertebrate paleontology
Blank, Albert Abraham, mathematical physics, vision
Blaustein, Bernard Daniel, physical chemistry
Block, Henry William, mathematical statistics
Block, Lawrence Howard, clinical pharmacology
Blumberg, John Otto, mathematics
Blumstein, Alfred, operations research
Boggs, Dane Ruffner, hematology
Borke, Mitchell Louis, pharmaceutical chemistry; organic chemistry
Borie, Andre Bernard, physiology; endocrinology
Boston, John Robert, biomedical engineering, audiology
Bother-By, Aksel Arnold, biophysical chemistry
Boucher, Laurence James, inorganic chemistry
Bowden, Henry James, computer sciences
Bowman, Robert Samuel, physical chemistry; organic chemistry
Brackmann, Richard Theodore, mass spectrometry, atomic physics
Brandt, Gerald Bennett, optical physics
Brandt, James Lewis, physical chemistry
Braun, Daniel Carl, industrial medicine
Brinton, Charles Chester, Jr, microbiology, biophysics
Brocoum, Stephan John, geology, tectonics
Brody, Thomas Peter, solid state physics
Bromberg, J Philip, physical chemistry
Brown, Frederick Ronald, fuel science
Brown, Leonard Keith, anthropology, ethnology
Brown, Paul Edmund, physical chemistry
Brown, Peter Frank, nuclear physics
Brown, Stuart Irwin, ophthalmology
Buchheim, Arno Fritz Gunther, systematic botany
Buck, Robert Edward, food science
Burgess, Jack D. botany
Burgi, Ernest Junior, physical chemistry
Burholt, Dennis Robert, cell biology, cancer

Burt, Robert C, medicine, pathology
Butera, Richard Anthony, solid state chemistry, physical chemistry
Byers, Donald James, organic chemistry
Byrne, Francis Patrick, analytical chemistry
Byrne, George D, numerical analysis
Caretto, Albert A, Jr, nuclear chemistry, physical chemistry
Carlin, Robert Burnell, organic chemistry
Carlson, Gerald Leroy, physical chemistry
Carlson, Kristin Rowe, psychopharmacology
Carpenter, Charles Patten, toxicology, bacteriology
Carr, Walter James, Jr, physics
Carter, James Clyde, inorganic chemistry
Carter, John Lyman, invertebrate paleontology
Cartwright, Thomas Edward, biophysics
Caruso, Sebastian Charles, environmental chemistry
Casassa, Edward Francis, physical chemistry
Cassidy, William Arthur, geochemistry, geology
Castle, Peter Myer, chemical physics, high temperature chemistry
Cauna, Nikolajs, anatomy, cell biology
Chaiken, Robert Francis, chemical physics
Chao, Chong-Yun, mathematics
Chapman, Toby Marshall, bio-organic chemistry, polymer chemistry
Charap, Stanley H, solid state science
Charny, Eugene Joseph, psychoanalysis
Chervenick, Paul A, hematology, internal medicine
Chi, Christina Hadinata, pharmacology
Chi, Donald Nan-Hua, applied mathematics
Chin, Byong Han, biochemistry
Chmura, Norman Walter, microbiology
Choyke, Wolfgang Justus, solid state physics
Chung, Albert Edward, biochemistry
Cleland, Wilfred Earl, elementary particle physics
Clench, Harry Kendon, entomology
Clench, Mary Heimerdinger, ornithology
Coetzee, Johannes Francois, analytical chemistry
Coffman, Charles Vernon, mathematics
Coffman, William Page, ecology, limnology
Coggeshall, Norman David, physics
Cohen, Alvin Jerome, geochemistry
Cohen, Anna Foner, physics
Cohen, Bernard Leonard, nuclear physics
Cohen, Richard Lawrence, child psychiatry
Cohen, Theodore, organic chemistry
Colaizzi, John Louis, pharmaceutics, pharmacy
Coleman, Bernard David, continuum mechanics, mathematical analysis
Coleman, Charles Mosby, clinical chemistry, microbiology
Coltman, John Wesley, musical acoustics, electron optics
Compher, Marvin Keen, Jr, developmental biology, endocrinology
Connamacher, Robert Henle, pharmacology
Conroy, Harold, theoretical chemistry
Corder, Clinton Nicholas, pharmacology, medicine
Corwin, Harry O, genetics
Courtney, Welby Gillette, physical chemistry
Cowgill, Ursula Moser, geochemistry
Cox, Eugene Floyd, organic chemistry, plant physiology
Craig, Raymond S, physical chemistry
Craven, Bryan Maxwell, structural chemistry
Cresswell, Michael William electronics
Crump, Robert Myers, geology
Cullen, Charles G, mathematics
Current, Jerry H, physical chemistry, analytical chemistry
Cutkosky, Richard Edwin, theoretical high energy physics
Cutler, John Charles, public health
Daehnick, Wilfried A W, experimental nuclear physics
Dahlberg, Michael D, zoology
Dakin, Thomas Wendell, physical chemistry
Daniel, Michael Roger, cryogenics, magnetism
Daniels, Gilbert S, botany
Danowski, Thaddeus Stanley, medicine
Darby, Edsel Kenneth, geophysics
Davies, D K, physics
Davies, David Huw, physical chemistry
Davies, Thomas Harrison, biochemistry

DeBenedetti, Sergio, experimental physics
Deegan, Ross Alfred, solid state physics
Degroat, William Chesney, Jr, pharmacology
DeGroot, Morris Herman, mathematical statistics
Deis, Daniel Wayne, solid state physics
De Klerk, John, physics
Del Bel, Elsio, organic chemistry
Dell, Manuel Benjamin, chemistry
Deskins, Wilbur Eugene, mathematics
Detre, Thomas Paul, psychiatry
Detwiler, John Stephen, biomedical engineering
Deuben, Roger R, pharmacology, physiology
Dewey, Edward Russell, statistics
Dinman, Bertram David, occupational medicine, environmental medicine
Diven, Warren Field, biochemistry
Dixit, Balwant N pharmacology
Dixon, George Douglass, polymer chemistry
Dobecki, Thomas Lee, geophysics
Doerder, F Paul, genetics
Doerfler, Leo G, audiology
Donahue, Jack David, stratigraphy
Donahue, Thomas Michael, physics
Douglas, Bodie Eugene, inorganic chemistry
Douglas, Kenneth Thomas, bio-organic chemistry
Dowd, Paul, organic chemistry
Draus, Frank John, biochemistry
Drew, Frances L, public health
Duffin, Richard James, mathematics, physics
Duncan, George Thomas, statistics
Dunkelberger, Tobias Henry, physical chemistry, microchemistry
Dunning, Virginia Alexandria, spectrochemistry
Dwyer, Thomas A, computer science, applied mathematics
Edelman, Leonard Edward, polymer chemistry
Edelstein, Richard Malvin, physics
Edmonds, Mary P, biochemistry
Ehrig, Raymond John, polymer chemistry
Ehrlich, Howard George, cytology
Ehrlich, Mary Ann, botany, plant pathology
Eichenholz, Alfred, internal medicine, metabolism
Eisenstein, Robert Alan, nuclear physics, particle physics
Eisner, Robert Linden, applied physics, forensic science
Elbling, Irving Nelson, organic chemistry
Eldred, Nelson Richards, organic chemistry
Elkins, Thomas Anthony, mathematics
Eller, Eugene Rudolph, paleontology, geology
Elliott, Dan Whitacre, surgery
Ellis, Alan F, organic chemistry
Ellison, Frank Oscar, quantum chemistry
Emmerich, Werner Sigmund, plasma physics
Emtage, Peter Roesch, solid state physics
Engler, Arnold, physics
Epstein, Lawrence Melvin, physical chemistry
Ernsberger, Fred Martin, physical chemistry
Ertel, Robert James, pharmacology, endocrinology
Esposito, John Nicholas, inorganic chemistry
Etter, Lewis Elmer, radiology, roentgenology
Evans, Douglas Fennell, physical chemistry, biophysical chemistry
Fahey, Dennis Martin, organic chemistry
Faires, Barbara Trader, pure mathematics
Fardo, Robert D, applied statistics
Farrauto, Robert Joseph, petroleum chemistry
Faulkner, John Edward, nuclear physics
Federowicz, Alexander John, applied mathematics
Feichtner, John David, quantum electronics
Feingold, David Sidney, biochemistry
Feldman, Donald William, physics
Feldman, Joseph Aaron, pharmaceutical chemistry
Feller, Robert Livingston, physical organic chemistry
Ferguson, Albert Barnett, orthopedic surgery
Ferguson, Herman White, economic geology
Fernald, Herbert Byron, petroleum chemistry

Fessenden, Richard Warren, physical chemistry
Fetkovich, John Gabriel, high energy physics, energy conversion
Field, James Bernard, medicine
Finck, Henry, anatomy, cell biology
Fink, James Paul, applied mathematics
Finn, Frances M, biochemistry
Finseth, Dennis Henry, spectrochemistry
Fireman, Philip, pediatrics, immunology
Fischer, John Lyle, anthropology
Fisher, Bernard, surgery
Fisher, Edwin Ralph, pathology
Fite, Wade Lanford, physics
Fletcher, Ronald D, virology
Flinn, Paul Anthony, physics, metallurgy
Flint, Norman Keith, geology
Foltz, George Edward, synthetic organic chemistry
Fort, Tomlinson, Jr, surface chemistry
Foster, Robert Scott, medical bacteriology
Fox, John Gaston, nuclear physics
Fox, Russell Elwell, atomic physics
Fox, Thomas G, Jr, chemistry, science administration
Franke, Frederick Rahde, medicine
Franzen, James, biophysical chemistry
Freedman, Robert Wagner, organic chemistry
Freeman, James Harrison, organic chemistry, polymer chemistry
Friedberg, Simeon Adlow, physics
Friedman, Sidney, organic chemistry
Fritsch, Arnold Rudolph, nuclear chemistry
Fritz, George Richard, Jr, physiology, endocrinology
Frohliger, John Owen, analytical chemistry
Fromm, Hans, gastroenterology
Frost, Lawrence William, polymer chemistry
Fuchs, Franklin, physiology
Fugassi, James Paul, physical chemistry
Fuller, Michael D, geophysics
Fung, Leslie Wo-Mei, molecular biophysics
Gaffney, Paul Cotter, medicine
Gaik, Geraldine Catherine, health sciences
Gainer, Gordon Clements, organic chemistry
Galinsky, Alvin M, pharmaceutical chemistry
Gall, Donald Alan, biomedical engineering
Galla, Stephen Joseph, medicine, anesthesiology
Garbuny, Max, optical physics, molecular physics
Gardner, Gerald Henry Fraser, mathematics
Garfunkel, Myron Paul, physics
Gatewood, George David, astronomy
Gawron, Oscar, biochemistry
Geisel, Martin Simon, applied statistics, econometrics
Gent, Martin Paul Neville, biophysical chemistry
George, William Arthur, dentistry
Gerhardt, George William, organic chemistry
Gerhart, Howard Leon, polymer chemistry
Gerjuoy, Edward, theoretical physics, atomic physics
Giannetti, Joseph Paul, petroleum chemistry, fuel science
Gibson, Gordon, theoretical physics
Gilbert, William Irwin, petroleum chemistry
Gilbertson, John R, biochemistry
Gill, Thomas James, III, pathology, immunology
Gillery, Frank Howard, physics
Ginsburg, Herbert, experimental statistics
Gitlin, David, pediatrics
Glaid, Andrew Joseph, III, biochemistry
Glickstein, Stanley S, nuclear physics
Gloyer, Stewart Wayne, chemistry
Goldburg, Walter Isaac, solid state physics
Goldman, Robert David, cell biology
Goldstein, E Bruce, psychophysics
Goldstein, Norman Phillip, nuclear physics
Gonter, Clara Ellen, aanalytical chemistry
Gordon, Thomas Pascoe, chemistry
Gottlieb, Frederick Jay, genetics, developmental biology
Gottlieb, Milton, physics
Grannemann, Glenn Niel, energy conversion
Grant, Richard J, physical chemistry
Grauer, Robert Coleman, pathology, endocrinology

Gray, Peter, microscopy, zoology
Green, Lawrence, experimental nuclear physics
Greenshields, John Bryce, physical chemistry
Grieco, Paul Anthony, synthetic organic chemistry
Griffiths, Robert Budington, physics
Grumer, Joseph, fuel science
Grundmann, Christoph Johann, chemistry
Guggenheimer, James, oral medicine
Guilbault, Lawrence James, polymer chemistry
Gulbransen, Earl Alfred, physical chemistry
Gupta, Tapan Kumar, ceramics
Gurtin, Morton Edward, applied mathematics, mechanics
Gwynn, Bernard Henry, organic chemistry
Habermann, Arie Nicolaas, computer science
Hagemann, Ronald Fred, radiation biology
Hall, Charles Allan, numerical analysis
Hall, William Spencer, applied mathematics
Halteman, Eber Kingdon, heat transfer
Hamill, Gilmor Semmes, IV, geology, geophysics
Hanin, Israel, psychopharmacology
Hanna, Norman Edwin, chemistry, research administration
Hansen, J Richard, electrooptics
Hapke, Bruce W, planetary sciences
Harbison, Samuel Pollock, surgery
Hardman, Carl Charles, electrochemistry
Harkins, Thomas Regis, chemistry
Hartman, Richard Thomas, plant ecology
Haun, Robert Dee, Jr, physics, research administration
Hausser, Jack W, organic chemistry
Hayashi, Teruo Terry, medicine, biochemistry
Hayes, Raymond L, Jr, anatomy, embryology
Heath, Edward Charles, biochemistry
Heath, Robert Winship, mathematics
Hedenburg, John Frederick, organic chemistry
Heilman, William Joseph, polymer chemistry
Hein, Richard Earl, chemistry
Hendrix, Roger Walden, molecular biology
Henry, LeRoy Kershaw, botany
Hermann, Theodore S, academic administration
Hertzberg, Martin, physical chemistry
Higman, Henry Booth, neurochemistry, pharmacology
Hill, Robert William, polymer chemistry
Hingson, Robert Andrew, anesthesiology
Hinkel, Robert Dale, chemistry
Hirayama, Chikara, physical chemistry, inorganic chemistry
Ho, Chien, biochemistry, biophysics
Ho, Monto, virology, medicine
Ho, Shih Ming, solid state chemistry
Hobbs, Anson Parker, physical chemistry
Hofer, Lawrence John Edward, physical chemistry
Hoffee, Patricia Anne, biochemistry, microbiology
Hofmann, Klaus Heinrich, biochemistry
Hollingsworth, Charles Alvin, physical chemistry
Homan, Elton Richard, toxicology
Hooke, Robert, mathematics, statistics
Hopkins, Thomas Robert, organic chemistry
Hopper, Sarah Priestly, biochemistry
Horne, William Appler, petroleum chemistry
Hosmer, Henry Liggett, geology
Howell, John N, physiology
Hsieh, Chiao-Min, cultural geography, geography of East Asia
Hsu, In-Ding, applied mathematics
Hull, Harry H, physical chemistry
Hulm, John Kenneth, solid state physics
Iammarino, Richard Michael, pathology
Idowu, Elayne Arrington, algebra, pure mathematics
Ihler, Garret Martin, molecular biology, hematology
Ihler, Karin Ippen, bacterial genetics
Ingham, Albert Irwin, geology
Irving, Philip, analytical chemistry
Isaacs, Thelma Jean, chemistry
Ivey, Henry Franklin, optical physics, solid state physics
Jackovitz, John Franklin, physical inorganic chemistry
Jacob, Leonard, Jr, economic geology
Jacobs, Alan Martin, geology
Jacobson, Lewis A, molecular biology

Jaffe, James Mark, pharmaceutics
Jandhyala, Bhagavan S., pharmacology
Janis, Allen Ira, physics
Jannetta, Peter Joseph, surgery
Jargiello, Patricia, molecular genetics
Jeffrey, George Alan, crystallography
Jen-Jacobson, Linda, biochemistry, biophysics
Jerslow, Robert G., operations research, mathematical logic
Johnson, Edwin Wallace, physical chemistry
Johnson, K Jeffrey, inorganic chemistry
Johnson, Sharon Lejoy, organic chemistry, biochemistry
Johnston, William Dwight, inorganic chemistry
Johnston, Anita Katherine, computer science
Jones, Elizabeth W, microbial genetics
Jordan, Raymond Ellsworth, medicine
Joyce, Ronald Stone, physical chemistry
Joyner, Claude Reuben, medicine
Juselius, Roger Elliott, clinical chemistry
Kadane, Joseph Born, applied statistics, mathematical statistics
Kaplan, Morton, nuclear chemistry, physical chemistry
Kaplan, Sandra Solon, internal medicine, hematology
Kanczak, Norbert M, anatomy
Kansky, Karel Joseph, urban geography, transportation geography
Karol, Meryl Helene, immunochemistry
Karol, Paul Jason, nuclear chemistry, physical chemistry
Kasner, William Henry, physics
Katoh, Arthur, developmental biology
Kaufman, Frederick, physical chemistry
Kay, Robert Leo, physical chemistry
Keller, Frederic, solid state physics
Keisch, Bernard, radiochemistry
Keller, Eldon Lewis, nuclear physics, research administration
Kelley, Dana Robineau, exploration geology
Kelley, William Sheldon, molecular biology
Kellman, Simon, reactor physics
Keyes, Gary Sylvester, elementary particle physics
Khalil, Mohamed Thanaa, biophysics, mathematics
Kibby, Charles Leonard, chemical kinetics, surface chemistry
Kiesewetter, William Burns, medicine
Kiewiet De Jonge, Joost Herman Albert, astronomy
Kiger, Robert William, systematic botany, history of biology
Kimblin, Clive William, physics
King, Arthur Bruce, plasma physics
Kissinger, Leonard Sol, theoretical physics
Kivlat, Fred E, physical chemistry
Kline, Hibbed Van Buren, Jr, geography
Klionsky, Bernard Leon, medicine, pathology
Knipp, John Charles, mathematics
Knobil, Ernst, endocrinology, physiology
Kochhar, Rajindar Kumar, polymer chemistry
Kohman, Truman Paul, nuclear chemistry
Kolodner, Ignace Izaak, mathematics
Korach, Malcolm, organic chemistry
Koros, Aurelia M Carissimo, immunology, cell biology
Kortanek, Kenneth O, operations research
Kozak, Wlodzimierz Maciej, visual physiology, psychophysics
Kozora, Andrew John, physics
Kraemer, Robert Walter, experimental high energy physics
Kraitchman, Jerome, chemical physics
Kreke, Cornelius W, chemistry
Kroon, Paulus Arie, physical biochemistry
Kussmaul, Keith, statistics
LaFountain, Lester James, Jr, geology
Laing, Patrick Gowans, orthopedic surgery
Langdon, Herbert Lincoln, gross anatomy, developmental anatomy
Lange, William James, surface physics
Langer, James Stephen, statistical mechanics
Lansing, Albert Ingram, anatomy
Larsen, Elmer Conrad, physical chemistry
Lassetre, Edwin Nichols, chemical physics
Lautier, Max Augustus, Jr, biophysics
Laush, George, mathematics
Leathen, William Warrick, microbiology
Lee, Chia-Ming, atomic physics
Lehman, Robert Nathan, mathematics, surgery

Lehner, Joseph, mathematical analysis, number theory
Lehoczky, John Paul, statistics
Lemke, Paul Arenz, genetics, microbiology
Lempert, Joseph, magnetohydrodynamics
Leonard, James Joseph, internal medicine
Lesher, Samuel Walter, cell biology, radiobiology
Lester, Roger, medicine
Leston Gerd, organic chemistry
Levine, Robert, organic chemistry
Lewis, Bernard, physical chemistry
Lewis, Charles William, physical chemistry
Lewis, Daniel William, organic chemistry
Lewis, Jessica Helen, medicine
Li, Ching Chun, population genetics, biometrics
Li, Ling-Fong, theoretical high energy physics
Li, Norman Chung, physical chemistry
Liddell, Robert William, Jr, organic chemistry, biochemistry
Lidiak, Edward George, geology
Lieberman, Irving, cell biology
Linn, Jay George, Jr, ophthalmology
Lintner, Anthony Ethelbert, chemistry
Linton, Robert Walter, inorganic chemistry
Lipman, Harry Jerome, food chemistry
Lloyd, Charles Wait, endocrinology
Loeffler, Mary Constance, physical chemistry
Ludwig, Howard C, chemical physics, plasma physics
Lukens, Francis Dring Wetherill, medicine
MacCamy, Richard C, mathematics
MacFarland, Harold Noble, industrial hygiene, toxicology
Madhav, R, organic chemistry
Mackay, Johnstone Sinnott, chemistry
Magovern, George Jerome, thoracic surgery, cardiovascular surgery
Maher, James Vincent, physics
Mahon, John Harold, agricultural chemistry
Mailliot, I Floyd, psychiatry
Malmberg, Paul Rovelstad, solid state electronics
Mandelcorn, Lyon, physical chemistry
Manuck, Barbara Ann, physical biochemistry
Markovitz, Hershel, physical chemistry
Martin, Albert, Jr, microbiology
Martin, Bruce Douglas, pharmaceutical chemistry
Martineau, Perry Cyrus, pathology, pharmacology
Masani, Pesi Rustom, mathematics
Mason, Charles Morgan, physical chemistry
Mateer, Frank Marion, medicine
Mathay, William Lewis, applied physics
Matta, Joseph Edward, environmental physics
Matz, William Howard, physical chemistry
Mayfield, John Eric, molecular biology
Mazelsky, Robert, solid state chemistry
Mazumdar, Mainak, operations research, statistics
McAvoy, Bruce Ronald, physics
McCaslin, Murray Frew, ophthalmology
McConnell, Robert A, parapsychology
McCoy, Clarence John, herpetology
McCoy, Richard Hugh, biochemistry
McCune, Duncan Chalmers, statistical analysis
McGovern, John Joseph, chemistry
McKinney, David Scroggs, physical chemistry
McLain, David Kenneth, mathematical analysis
McLain, Paul Larimer, physiology, pharmacology
McCurry, Patrick Matthew, Jr, chemistry
McDonald, Robert H, Jr, clinical pharmacology
McGandy, Edward Lewis, physical chemistry, public health
McGinnis, Edgar Lee, petroleum chemistry

McNall, John William, applied physics
McPherron, Alan, anthropology, archaeology
Mears, Dana Christopher, chemical metallurgy, medicine
Mechlin, George Francis, Jr, physics
Medearis, Donald N, Jr, pediatrics, microbiology
Meechan, Charles James, solid state physics
Meier, Joseph Francis, polymer chemistry
Melamed, Nathan T, chemistry
Meredith, Ruby Frances, cancer, genetics
Merten, Ulrich, physical chemistry
Metlay, Max, clinical chemistry
Metz, William Clinton, geography
Metzger, Sidney Henry, Jr, applied chemistry
Metzger, Thomas Andrew, pure mathematics
Michael, Joe Victor, physical chemistry
Michael, Norman, inorganic chemistry
Michalik, Edmund Richard, applied mathematics, statistics
Miksch, Edmund Stewart, energy conversion
Miller, Foil Allan, physical chemistry
Miller, Helena Agnes, botany
Minard, David, physiology
Missimer, John Hertel, particle physics, theoretical physics
Moberly, Lawrence Ervin, physical chemistry, inorganic chemistry
Moffet, Eugene Wilkin, organic chemistry
Mokotoff, Michael, medicinal chemistry
Morgan, Paul Vincent, microbiology
Morimoto, Hideo, microbiology
Moskovitz, David, clinical medicine
Moye, Alfred Leon, chemistry
Mullins, William Wilson, physics
Nacatraro, William Frank, biochemistry, lipid chemistry
Nadelhaft, Irving, neurobiology
Nagata, Jun-Iti, topology
Nagin, Daniel Steven, mechanics
Nagle, John F, physics, statistical mechanics
Nakada, Daisuke, biochemistry, microbiology
Narasimhan, Kalatur S V L, solid state chemistry
Nehari, Zeev, mathematics
Neta, Pedatsur, chemistry
Netting, Morris Graham, herpetology
Newell, Allen, computer science, psychology
Newman, Ezra, theoretical physics
Nilan, Thomas George, physics
Noll, Walter, mathematics, continuum mechanics
Noreika, Alexander Joseph, solid state physics
Noren, Gerry Karl, organic polymer
Nutini, Hugo Gino, anthropology
O'Donnell, Thomas John, exploration geophysics
Ohlberg, Stanley Miles, physical chemistry
Okunewick, James Philip, radiation biophysics, cancer
O'Leary, Dennis Patrick, neurophysiology
Oliver, Thomas K, Jr, pediatrics
Osmond, Leslie H, radiology
Ostfield, Howard G, organic polymer
Ove, Peter, cell biology, biochemistry
Owen, David R, applied mathematics
Pack, John Lee, physics
Page, Lorne Albert, experimental physics
Parker, George Anthony, physical chemistry
Parker, James Henry, Jr, cryogenics
Partanen, Carl Richard, ornithology, plant cytology, plant morphogenetics

Pasztor, Laszlo, chemistry
Pattinson, Charles Byron, Jr, petroleum chemistry
Paulson, Mark Clements, organic chemistry
Paviak, Stanley C, organic chemistry
Peckham, William Dieroff, biochemistry, endocrinology
Pederson, Roger Noel, mathematics
Peffer, John Roscoe, polymer chemistry
Pellegrini, John P, Jr, organic chemistry
Pement, Fredric William, chemical engineering
Perkins, William Enfield, geophysics, oceanography
Perrotta, Anthony Joseph, mineralogy, crystallography
Perryman, Charles Richard, radiology
Perzak, Frank John, physical chemistry, explosives
Peterson, Axel Harding, physical chemistry
Peterson, Jack Kenneth, polymer chemistry, physical chemistry
Petrakis, Leonidas, physical chemistry
Phillips, Bruce A, microbiology, virology
Phillips, David Colin, chemistry
Phillips, Otto C, medicine
Phillips, Stephen Lee, molecular biology
Plankey, Walter L, geophysics
Plankey, Francis William, Jr, analytical chemistry
Plant, William J, organic polymer chemistry
Platt, David, immunology
Plazek, Donald John, physical chemistry
Plotnicov, Leonard, anthropology, ethnology
Poel, William Elias, pathology
Pollack, Sidney Solomon, crystallography
Pople, John Anthony, theoretical chemistry
Pratt, Richard Houghton, theoretical physics
Preble, Olivia Toby, virology
Price, Harold M, pathology
Puschett, Jules Bernard, internal medicine, nephrology
Postic, Bosko, medicine, infectious diseases
Foust, Rolland Irvin, chemistry
Powell, Bruce Allan, operations research
Pratt, David W, chemical physics, physical chemistry
Porsching, Thomas August, mathematics
Porter, John Robert, mathematics
Quillin, Charles Robert, cytology
Radford, Kenneth Charles, ceramics, nuclear metallurgy
Raikow, Robert Jay, anatomy, ornithology
Raizen, Carol Eileen, microbial genetics
Rao, Vallabhajosyula U S, solid state
Rapp, Robert, pedodontics, histology
Ravitch, Mark Mitchell, surgery
Raymund, Mahlon, physics
Rayne, John A, physics
Redgate, Edward Stewart, neurophysiology, psychophysiology
Reggell, Leslie, organic chemistry
Reichert, Thomas Andrew, biomedical engineering
Reid, Marlene Barnes, applied anthropology
Reinhart, John Belvin, psychiatry, pediatrics
Reis, Walter Joseph, psychiatry, clinical psychology
Rescher, Nicholas, philosophy of science, symbolic logic
Rice, Robert Vernon, biochemistry
Richey, Willis Dale, physical chemistry
Richmond, James Kenneth, physics
Ricks, Herbert Elias, electrochemistry
Riedel, Ernest Paul, lasers, optical physics
Rike, Paul Miller, cardiology, internal medicine
Riley, Gene Alden, pharmacology
Roberts, John Milton, anthropology
Roche, James Norman, chemistry
Rockette, Howard Earl, Jr, biostatistics
Rodman, Gerald Paul, medicine
Roesmer, Josef, nuclear science
Rogers, Kenneth D, preventive medicine
Rogers, Lewis Henry, analytical chemistry
Roland, George Warren, geology, analysis
Rollins, Harold Bert, invertebrate paleontology, paleoecology
Rosenberg, Jerome Laib, biophysical chemistry
Rosenthal, Theodore Bernard, microscopic anatomy, physiology, theoretical physics
Roskies, Ralph Zvi

Rossow, Alfred George, industrial chemistry
Rostoker, David, geology
Roth, Arthur Jason, mathematical statistics
Rubenstein, Martin, inorganic chemistry, solid state chemistry
Rubin, Herbert, communications, speech pathology
Ruppel, Thomas Conrad, physical chemistry
Russ, James Stewart, high energy physics
Ryan, Frederick Merk, physics
Rycheck, Russell Rule, medicine, epidemiology
Safar, Peter, anesthesiology, critical care medicine
Saladin, Jurg X, nuclear physics
Salazar, Hernando, pathology, public health
Salvin, Samuel Bernard, immunology, mycology
Sanders, Robert Hugh, astrophysics
Sandler, Rivka Black, gerontology
Sandler, Yehuda Ludwig, physical chemistry
Sandmeyer, Esther E, pharmacology, toxicology
Sankar, Suryanarayan G, solid state chemistry
Sansone, Eric Brandfon, industrial hygiene, air pollution
Sargent, Lowrie Barnett, Jr, physical chemistry, tribology
Sashin, Donald, radiological physics, health physics
Sassouni, Viken, dentistry, orthodontics
Sax, Sylvan Maurice, clinical chemistry
Scala, Luciano Carol, organic chemistry
Schachter, Joseph, psychiatry
Schaeffer, William Dwight, colloid chemistry
Schaffer, Juan Jorge, mathematics
Schempp, Ellory, chemical physics
Schiavone, George Joseph, applied mathematics
Schiffer, Lewis Martin, hematology
Schlegel, Alice Elizabeth, cultural anthropology
Schliter, Duane A, mammalogy
Schmid, Victor Adolf, geophysics, paleomagnetism
Schorr, Thomas S, anthropology
Schrayer, Grover J, Jr, organic geochemistry
Schreiber, Kurt Clark, physical organic chemistry
Schubert, Jack, radiation chemistry, bioinorganic chemistry
Schuler, Robert Hugo, physical chemistry
Schultz, Hyman, analytical chemistry
Schultz, Stanley George, physiology
Schumacher, Berthold Walter, experimental physics, physics engineering
Schumacher, Robert Thornton, acoustics
Sebesta, Charles Frederick, mathematics
Seehafer, Marlyn E, food science
Seelbach, Charles William, polymer chemistry, industrial chemistry
Segal, Alan H, dentistry
Seidensticker, Raymond George, applied physics
Seshadri, Kalkunte S, physical chemistry
Shaffer, Douglas Howerth, mathematics
Shalaby, Ragaa Abdel Fattah, biophysics
Shaman, Paul, statistics
Shamos, Michael Ian, computer sciences
Shapiro, Alvin Philip, internal medicine
Shepard, Paul Fenton, particle physics
Shepherd, James Willis, organic chemistry
Sherman, Frank Edward, pathology
Shih, Tsung-Ming, pharmacology
Shiono, Ryonosuke, crystallography
Shostak, Stanley, developmental biology, primatology
Shoupp, William Earl, physics, engineering
Shrivastava, Prakash Narayan, radiological physics
Shull, Kenneth Henry, biochemistry
Siegel, Melvin Walter, molecular physics, mass spectrometry
Siegel, Michael Ian, physical anthropology, primatology
Sillman, Emmanuel I, parasitology, invertebrate zoology
Simon, Joseph Matthew, analytical chemistry
Singer, Joseph Marcus, physical chemistry
Singh, Balwant, microbiology, epidemiology
Singleton, Fred Gray, chemistry
Singleton, Jack Howard, physical chemistry

Siska, Peter Emil, physical chemistry, chemical physics
Skinner, Walter Swart, earth science
Skolnick, Herbert, geology
Slagel, Robert Clayton, organic chemistry
Slaughter, Frank Gill, Jr, mathematics
Slifkin, Malcolm, medical microbiology, virology
Smith, James David Blackhall, polymer chemistry
Smith, Richard Cecil, applied physics
Smudski, James W, pharmacology, dentistry
Smyth, Henry Field, Jr, toxicology
Snider, Albert Monroe, Jr, spectroscopy
Snow, Roland B, mineralogy
Sonis, Meyer, psychiatry
Sorensen, Raymond Andrew, nuclear physics
Southwick, Philip Lee, chemistry
Spatz, Sidney S, oral surgery
Spencer, Ralph Donald, organic chemistry
Spicher, John L, medical technology
Spindt, Roderick Sidney, organic chemistry
Spoehr, Alexander, anthropology
Spooner, Robert Bruce, biomedical engineering
Spritzer, Albert A, medicine
Steffgen, Frederick Williams, fuel science, surface chemistry
Stehle, Philip McLellan, quantum electronics
Sternglass, Ernest Joachim, physics
Stevens, Charles Le Roy, biophysical chemistry
Steward, Omar Waddington, organometallic chemistry, inorganic chemistry
Steward, Glenn Alexander, solid state physics, surface phsyics
Stewart, Kenneth C, acoustics, environmental science
Stewart, Robert F, physical chemistry
Stiff, Robert H, dentistry
Stolc, Viktor, endocrinology, hematology
Strange, John Phillip, physics
Straub, Darel K, inorganic chemistry
Strehler, Allen Frederick, mathematics
Streiff, Anton Joseph, petroleum chemistry
Strick, Ellis, geophysics, oceanography
Stricker, Edward Michael, neuropsychology
Strickler, Herbert Sharpless, chemistry
Strickler, Paul Donovan, organic chemistry
Sullivan, Lloyd John, biochemistry, physical organic chemistry
Summer, Roger D, geophysics
Sun, Kuan-Han, nuclear science, glass technology
Sussman, Maurice, developmental biology, molecular biology
Sussman, Raquel Rotman, microbiology
Sutton, Roger Beatty, experimental high energy physics
Swarts, Elwyn Lowell, physical chemistry
Swawgo, James Lee, anthropology
Swiss, Jack, organic chemistry
Szepesi, Zoltan Paul John, electronics
Tabakin, Frank, theoretical chemistry
Taber, Joseph John, physical chemistry, petroleum engineering
Talley, Charles Peter, analytical chemistry
Tamburino, John, mathematical logic
Taulbee, Orrin Edison, computer science
Taylor, Floyd Heckman, mathematics
Taylor, John Joseph, mathematics
Taylor, Lyle Herman, lasers
Taylor, Paul M, pediatrics, physiology
Theiner, Micha, biochemistry
Thompson, Gerald Luther, applied mathematics
Thompson, Julia Ann, elementary particle physics, high energy physics
Tong, Winton, physiology, biochemistry
Townsend, John Robert, physics
Traub, Joseph Frederick, computer science
Treuting, Waldo Louis, public health
Troen, Philip, medicine
Troy, William Christopher, applied mathematics
Tuden, Arthur, cultural anthropology
Turner, J Howard, human genetics
Twomey, Arthur Cornelius, zoology
Uhlrich, Helen Marie, biology
Uricchio, William Andrew, biology, helminthology
Vagnucci, Anthony Hillary, medicine, physiology
Vanderven, Ned Stuart, physics
Van Dolah, Robert Wayne, chemistry
Van Dyke, Charles H, inorganic chemistry, organometallic chemistry

Vassamillet, Lawrence Francois, solid state physics
Vaux, James Edward, Jr, chemistry, statistics
Vogt, Molly Thomas, biochemistry
Waddell, Walter Harvey, photochemistry
Wagman, Nicholas Emory, astronomy
Wagner, George Richard, solid state physics
Wald, Niel, public health, radiation medicine
Walker, Augustus Chapman, research administration, academic administration
Walker, Robert John, mathematics
Wall, Conrad, III, bioengineering
Wallace, George Egbert, entomology
Wallace, William Edward, solid state chemistry
Wang, Jin Tsai, inorganic chemistry, analytical chemistry
Wang, Ke-Chin, high temperature chemistry, ceramics
Warde, Charles Joseph, electrochemistry
Warfel, David Ross, organic chemistry, polymer chemistry
Warner, John Christian, chemistry
Watson, Joseph Alexander, radiobiology, microbiology
Watson, Richard William, physics
Weaver, Leo James, organic chemistry
Weil, Carrol Solomon, toxicology
Weinberger, Edward Bertram, computer science
Weiner, Robert Allen, solid state physics
Weismann, Theodore James, physical chemistry
Wells, Ralph Gordon, materials science
Wender, Irving, fuel science, organometallic chemistry
Werner, Gerhard, pharmacology
Werner, Harry Jay, geology
Wharton, William Raymond, experimental nuclear physics
White, Douglas Richie, anthropology
Whiteside, Theresa L, immunology, immunopathology
Widnell, Christopher Courtenay, cell biology, biochemistry
Williams, Melvin Donald, anthropology
Williams, William Orville, mathematics, mechanics
Willingham, Charles Baynard, physical chemistry
Windisch, Rita M, clinical chemistry
Winek, Charles L, toxicology, pharmacology
Wingard, Lemuel Bell, Jr, pharmacology
Winicour, Jeffrey, theoretical physics
Witman, Eugene DeWald, agricultural chemistry
Wolfe, Clinton Ray, analytical chemistry
Wolfe, Peter Nord, physics
Wolfenstein, Lincoln, theoretical high energy physics
Wolford, Jack Arlington, psychiatry
Wolfson, Sidney Kenneth, Jr, neurosurgery, biomedical engineering
Wolke, Robert Leslie, nuclear chemistry, radiochemistry
Wolken, Jerome Jay, biophysics
Wootten, Michael John, physical chemistry, water chemistry
Worthington, Charles Roy, biophysics
Wu, Ching-Yong, industrial organic chemistry
Wyler, Oswald, mathematics
Wyllie, Malcolm Robert Jesse, geophysics, engineering
Yao, Shang Jeong, chemical physics, biomedical engineering
Yoldas, Bulent Erturk, ceramics, glass technology
Young, Hugh David, theoretical physics
Young, Lionel Wesley, pediatrics, radiology
Young, Robert Bruce, chemistry
Youngner, Julius Stuart, medical microbiology
Zabusky, Norman J, applied mathematics, computer mathematics
Zaphyr, Peter Anthony, mathematics
Zarrella, William Michael, physical chemistry, organic chemistry
Zener, Clarence Melvin, physics
Zigmond, Michael Jonathan, neuropharmacology, psychopharmacology
Zlochower, Isaac Aaron, physical chemistry
Zollweg, Robert John, experimental physics
Zordan, Thomas Anthony, physical chemistry

PLYMOUTH MEETING
Brink, Robert Harold, Jr, biochemistry

Moss, Jack N, pharmacology, chemotherapy

PORT MATILDA
Boyd, Charles Alexander, chemistry

POTTSTOWN
Feldmeier, Joseph Robert, nuclear physics
Schofield, Richard Alan, medicine
Woods, Frank Robert, hydrodynamics, gas dynamics

QUAKERTOWN
Winters, Earl D, physical chemistry

RADNOR
Bell, Stanley C, organic chemistry, medicinal chemistry
Dalton, David Robert, organic chemistry
Fenichel, Richard Lee, biochemistry, physiology
Hegarty, Charles Paul, bacteriology
Hersberger, Arthur Bucher, organic chemistry
Langeland, William Enberg, organic chemistry
Natt, Michael Philip, information science, communication science
Sellstedt, John H, medicinal chemistry
Stone, Joseph Louis, medical bacteriology
Strike, Donald Peter, pharmaceutical chemistry
Tomarelli, Rudolph Michael, biochemistry, nutrition
Wei, Peter Hsing-Lien, medicinal chemistry
Yurchenco, John Alfonso, medical microbiology

READING
Ambrosiani, Vincent F, physical chemistry
Batdorf, Robert Ludwig, solid state science
Bell, Edwin Lewis, II, herpetology
Birdsall, William John, inorganic chemistry
Chang, Charles Hung, synthetic organic chemistry
Dunn, Charles Nord, solid state physics
Elkind, Michael John, inorganic chemistry
Evans, Edward William, mathematics
Feeman, James Frederic, industrial organic chemistry, research administration
Ghiselin, Jon Brewster, ecology
Goodwin, Charles Arthur, solid state electronics, ceramics
Graybill, Donald Lee, ecology
Green, Marcus Herbert, biology
Gundy, Samuel Charles, biology
Hall, John Sylvester, zoology
Heller, Morgan Silliman, organic chemistry
Hildreth, Eugene Augustus, internal medicine
Horning, Roderick Henry, textile chemistry, environmental chemistry
Hower, Meade M, physics
Jaeger, Charles Wayne, industrial organic chemistry
Khan, Mahmood Ahmed, food science, nutrition
Kremser, Thurman Rodney, physics
Lambert, Joseph Michael, mathematical analysis
Leininger, Paul Miller, physical chemistry
LeLacheur, Robert Murray, physics
Post, Irving Gilbert, solid state electronics
Rabina, Manuel Jose, mathematics
Rapp, Robert Dietrich, organic chemistry
Reider, Malcolm John, organic chemistry
Rowe, Jay Elwood, industrial organic chemistry
Scharfstein, Lawrence Robert, chemical metallurgy
Shapiro, Zalman Mordecai, chemistry
Steiner, Russell Irwin, industrial organic chemistry
Vehse, Robert Chase, solid state physics
Voigt, David Quentin, anthropology, sociology

RICHBORO
Preston, Robert Kreig, organic chemistry

RIDLEY PARK
Nelsen, Olin E, anatomy, embryology

PENNSYLVANIA

RILLTON
Jennings, Burridge, nuclear physics

ROSEMONT
Bryan, Mary Leo, physical organic chemistry, biochemistry
Ward, William Francis, biochemistry
Wolfenberger, Michael Gregg, physiological chemistry

ROSLYN
Santora, Norma Julian, medicinal chemistry

RUSHLAND
Spell, Aldenlee, physical chemistry

RYDAL
Goldman, Robert Barnett, applied physics

SAEGERTOWN
State, Harold M., analytical chemistry, inorganic chemistry
Wohler, James Richard, II, ecology, limnology

ST DAVIDS
Moore, Jay Winston, developmental genetics, poultry genetics
Sheldon, Joseph Kenneth, insect ecology

ST MARYS
Goochee, Herman Francis, inorganic chemistry
Liggett, Lawrence Melvin, analytical chemistry, organic chemistry
Shobert, Erle Irwin, II, physics, materials science

SALTSBURG
Durno, William Henry, chemistry
Knapp, John Samuel, chemistry

SAXONBURG
McCarthy, James Francis, organic chemistry

SAYRE
Beck, William Carl, surgery
Carpender, James Wood Johnson, radiology
Royce, Paul Chadwick, medicine, encocrinology

SCHNECKSVILLE
Gerteis, Robert Louis, inorganic chemistry

SCHUYLKILL HAVEN
Baumann, Jacob Bruce, organic chemistry

SCRANTON
Bartley, Edward Francis, mathematics
Bening, Paul R., microbiology, biochemistry
Burti, Umbay H., organic chemistry
Cann, Michael Charles, organic
Del Vecchio, Vito Gerard, biochemical genetics
Fahey, Paul Farrell, Jr, physics
Haab, Walter, biochemistry, analytical chemistry
Hart, Maurice I, Jr, physical chemistry
Laurence, Maria (Maher), botany
Lavender, Ardis Ray, internal medicine, nephrology
Lobo, Francis X, microbiology
Matthews, Richard John, Jr, pharmacology
McGinnis, Eugene A, physics
Nee, M Coleman, mathematics
Radzikowski, M St Anthony, inorganic chemistry
Sallavanti, Robert Armando, physical chemistry
Shoemaker, Richard Nelson, medical education
Thoman, Charles J, organic chemistry
Vinson, Joe Allen, analytical chemistry

SELINSGROVE
DeMott, Howard Ephraim, botany
Fletcher, Frank William, geology
Goodspeed, Robert Marshall, geology
Grosse, Fred A., physics
Lowright, Richard Henry, sedimentology
McGrath, Thomas Frederick, organic chemistry
Nylund, Robert E., physical chemistry
Potter, Neil H., organic chemistry
Presser, Bruce Douglas, entomology

SEWICKLEY
Gruber, Gerald William, photochemistry
Rogers, Dow Albert, Jr, polymer chemistry

SHARON
Bursey, Charles Robert, invertebrate physiology
Madacsi, David Peter, experimental solid state physics, magnetic resonance
Walsh, Edward John, inorganic chemistry, polymer chemistry

SHARPSVILLE
Kerlick, George David, theoretical physics

SHIPPENSBURG
Baird, Merton Denison, organic chemistry
Barr, Richard Arthur, plant physiology, phytochemistry
Bobonich, Harry Michael, inorganic chemistry
Gould, William Allen, mathematics, computer science
Greenstein, Julius S, zoology, academic administration
Johnston, Jean Vance, organic chemistry
Kelley, William Russell, botany
Kerr, Carl E., mathematics
Kirkland, Gordon Laidlaw, Jr, mammalogy, vertebrate ecology
Laidig, Kermit McClellan, geography
McArthur, William George, mathematics
Morrison, William Joseph, genetics
Peightel, William Edgar, biology
Rae, George Ramsay, geography, geoenvironmental science
Rogers, William Edwin, zoology, parasitology
Schroeder, Thomas Dean, analytical chemistry
Sieber, James Leo, mathematics
Slysh, Anton Roman, botany, microbiology
Spencer, Jack T, botany, agronomy
Wilson, John Randall, physical chemistry

SHIPPINGPORT
Bogar, Louis Charles, radiochemistry, health physics

SLIPPERY ROCK
Archibald, Patricia Ann, algology, phycology
Brady, Wray Grayson, mathematics
Bushnell, Kent O, environmental geology, geophysics
Campaigne, Howard Herbert, mathematics, computer sciences
Chapman, William Frank, glacial geology
Cunkle, Charles Henry, mathematics
Dresden, Carlton F, biochemistry, organic chemistry
Erdman, Kimball S, botany
Gaither, Thomas Walter, botany
Hart, Robert Gerald, reproductive physiology
Holland, Monte W, physics
Hou, Roger Hsiang-Duh, algebra
Hughes, James Charles, urban geography, geography of Latin America
Lindgren, William Frederick, mathematics
Mani, Srinivasa Balasubra, applied anthropology, cultural anthropology
McClure, Clair Wylie, plant ecology
Medve, Richard J, plant ecology
Michel, Kenneth Earl, cytogenetics
Mueller, Charles Frederick, ecology
Papanikolaou, Nicholas E, organic chemistry
Rizza, Paul Frederick, geography
Shultz, Charles H, geology
Smith, Herbert L, inorganic chemistry
Taylor, David Cobb, analytical chemistry
Taylor, Rhoda E., reproductive endocrinology
Urtscheit, Peter, geography
Wilhelm, Eugene J, Jr, biogeography, cultural geography

SOLEBURY
Ewing, Douglas Hancock, physics

SOUTH WILLIAMSPORT
Smith, Willy, nuclear engineering, physics

SOUTHAMPTON
Danco-Moore, Lolita, microbiology
Hsu, Quei-Shiow, biochemistry, physiology
Hurwitz, Melvin David, organic chemistry
Janssen, Richard William, pharmaceutical chemistry, physical pharmacy
Nedwick, John Joseph, industrial organic chemistry, chemical engineering
Ramberg, Edward Granville, physics

SPRING HOUSE
Adler, Harold Ernest, entomology
Bakule, Ronald David, physical organic chemistry
Bauman, Bernard D., organic chemistry
Bayer, Horst Otto, organic chemistry
Belair, Ernest Joseph, physiology
Bortnick, Newman Mayer, synthetic chemistry
Brendley, William H, Jr, physical chemistry, polymer chemistry
Brown, Theodore Gates, Jr, organic chemistry
Burke, Susan Schiff, synthetic organic chemistry, pesticide chemistry
Carley, Harold Edwin, plant chemistry, soil science
Carlson, Glenn Richard, synthetic organic chemistry
Cenci, Harry Joseph, polymer chemistry
Chan, Hak-Foon, agricultural chemistry
Chang, Ching-Jen, physical organic chemistry
Chong, Berni Patricia, synthetic organic chemistry
Chong, Joshua Anthony, synthetic organic chemistry
Clemens, David Henry, organic chemistry
Clovis, James S, physical organic chemistry
Coe, Beresford, organic chemistry
Connor, Stephen R, plant pathology
Cook, Richard Sherrard, organic chemistry
Dajani, Esam Zager/Zafer, pharmacology, toxicology
Decker, Fred W, biochemical pharmacology, drug metabolism
Dyott, Thomas Michael, theoretical chemistry
Fisher, James Delbert, pesticide chemistry
Fitzpatrick, Joseph Michael, synthetic organic chemistry
Ford, Warren Thomas, organic chemistry
Frankel, Lawrence (Stephen), physical chemistry, analytical chemistry
Gaughan, Renata Rysnik, physical chemistry, polymer chemistry
Gehman, David Richard, polymer
Gill, Robert Anthony, physical chemistry
Goode, William Edward, organic
Haggard, Richard Allan, organic chemistry
Haines, Linwood Davis, chemistry
Harren, Richard Edward, chemistry
Honeycutt, Richard Carl, biochemistry
Howell, Thomas James, physical organic chemistry
Ignatowski, Albert J, physical chemistry
Imondi, Anthony Rocco, biochemistry, pharmacology
Johnson, Richard Carl, synthetic organic chemistry
Johnson, Wayne Orrin, agricultural chemistry
Kollman, Gerald Eugene, plant physiology
Kronberger, Karlheinz, polymer chemistry, organic chemistry
Krzeminski, Stephen F, physical chemistry, analytical chemistry
Lashen, Edward S, microbiology
Lewis, Sheldon Noah, physical chemistry, organic chemistry
Lo, Chien-Pen, organic polymer chemistry
Lyman, William Ray, chemistry
Machleder, Warren Harvey, organic chemistry
McCallum, Keith Stuart, analytical chemistry
Melamed, Sidney, polymer chemistry
Merritt, Richard Foster, organic chemistry
Meyer, Thomas Edward, synthetic organic chemistry
Miller, John Joseph, organic chemistry
Naples, John Otto, organic chemistry, polymer chemistry
Niederhauser, Warren Dexter, chemistry
Novak, Ronald William, organic chemistry, polymer chemistry
Nuessle, Albert Christian, textile chemistry
Nyi, Kayson, organic chemistry
Ollinger, Janet, organic chemistry
O'Mara, James Herbert, physical chemistry
Patterson, Dennis Ray, organic chemistry
Peardon, David Lee, veterinary science
Plamondon, Joseph Edward, textile technology
Prentiss, William Case, leather chemistry
Reitman, Larry N, synthetic organic chemistry
Rogerson, Thomas Dean, organic chemistry, agricultural chemistry
Ross, Robert M, organic chemistry, analytical chemistry
Rothman, Alan Michael, radiochemistry
Schaffenburg, Carlos A, medicine
Seidel, Michael Caspar, animal parasitology
Smith, Jerry Morgan, pharmacology, toxicology
Spinner, Joseph F, physical chemistry
Stevens, Travis Edward, organic chemistry
Stroike, James Edward, plant breeding, plant genetics
Swift, Graham, organic polymer chemistry
Thirugnanam, Muthuvelu, plant chemistry
Unger, Victor Herman, nematology, entomology
Van Landuyt, Dennis Clarke, organic chemistry
Warner, Harlow Lester, plant physiology, plant breeding
Weiler, Ernest Dieter, organic chemistry
Wells, Paula Parker, organic chemistry
Wempe, Lawrence Kyran, organic chemistry
Whiteman, John David, physical chemistry
Wiersema, Richard Joseph, inorganic chemistry
Willette, Gordon Louis, petroleum chemistry
Williamson, Martin John, physical chemistry
Wolf, Leslie Raymond, organic chemistry, analytical chemistry
Wuchter, Richard B, organic chemistry, polymer chemistry
Yih, Roy Yangming, agricultural chemistry
Yost, Robert Stanley, organic chemistry

SPRINGDALE
Christenson, Roger Morris, chemistry
Marcus, Robert Toby, color science

SPRINGFIELD
Alber, Herbert Karl, microchemistry
Baker, Lois Van Meter, information science
Beitchman, Burton David, organic chemistry
Hollibaugh, William Calvert, physical chemistry
Kraatz, Charles Parry, pharmacology
Lord, Arthur E, Jr, physics
Spooner, Laurence Whipple, chemistry

STATE COLLEGE
Anderson, Bertil Gottfrid, zoology, pollution biology
Bastuscheck, Clifford Paul, physics
Bauer, Carl August, astronomy
Bell, Ian, industrial organic chemistry
Bissey, Luther Trauger, chemistry
Black, Alex, animal nutrition
Blackadar, Alfred Kimball, meteorology
Bressler, Glenn Otto, agricultural economics
Bryner, Clarence Sheldon, agronomy

Buckalew, John McKinney, dairy production
Buss, Edward George, poultry genetics
Cleveland, Richard Warren, plant breeding
Curry, Haskell Brooks, mathematical logic
Dalton, Howard Clark, embryology, genetics
Downsbrough, George Atha, physics
Farwell, Robert William, physics
Grindall, Emerson Leroy, underwater acoustics
Henisch, Heinz Kurt, solid state physics, history of photography
Herzog, Leonard Frederick, II, geophysics, mass spectrometry
Johnson, John Charles, geophysics
Klevans, Edward Harris, theoretical physics
Lampe, Frederick Walter, physical chemistry
Mount, Lloyd Gordon, industrial organic chemistry
Nelson, Richard Robert, plant pathology
Noll, Clarence Irwin, organic chemistry
Phillips, Bert, inorganic chemistry
Ramachandran, Subramania, organic chemistry, biochemistry
Rose, Arthur, chemistry
Rose, Elizabeth Gates, science writing
Shannon, Jack Corum, plant physiology
Skudrzyk, Eugen J, physics
Smith, Grant Warren, physical chemistry, analytical chemistry
Snowdon, John Colin, applied mechanics, acoustics
Sprague, Howard Bennett, agronomy
Stingelin, Ronald Werner, geology, remote sensing
Stover, James Anderson, Jr, operations research, systems science
Thompson, William, Jr, acoustics
Wilson, Geoffrey Leonard, acoustics
Wilson, Lowell L, animal sciences
Withstandley, Victor DeWyckoff, III, molecular spectroscopy
Zindler, Richard Eugene, mathematics, operations research

STRAFFORD
Feuer, Robert Charles, environmental biology, herpetology

STROUDSBURG
Hankins, William Alfred, virology
Spicer, Daniel Shields, biochemistry

SWARTHMORE
Bender, Allan Douglas, pharmacology
Berrill, Norman John, developmental biology
Bilaniuk, Olexa-Myron, nuclear physics
Duggal, Shakti Prakash, cosmic ray physics, space physics
Elmore, William Cronk, physics
Enders, Robert Kendall, mammalogy
England, James Walton, mathematics
Fehnel, Edward Adam, organic chemistry
Flemister, Launcelot Johnson, physiology
Gaisser, Thomas Korff, particle physics, cosmic ray physics
Hammons, James Hutchinson, physical organic chemistry
Heald, Mark Aiken, plasma physics
Heckscher, Stevens, mathematics
Heintz, Wulff Dieter, astronomy
Hershey, John Landis, astronomy
Hickman, Carole Stentz, paleobiology, malacology
Hickman, James C, plant ecology
Iversen, Gudmund R, applied statistics
Jacobs, Mark, plant physiology
Jenkins, John Bruner, genetics
Keighton, Walter Barker, water chemistry
Kent, Donald Wetherald, Jr, physics
Klotz, Eugene Arthur, mathematics
Lippincott, Sarah Lee, astronomy
Livingston, Luzern Gould, botany
Maass, Alfred Roland, biochemistry
Mangelsdorf, Paul Christoph, Jr, marine geochemistry
Martin, Francis Hall, physical biochemistry
McBride, Duncan Eldridge, experimental solid state physics, metal physics
Meinkoth, Norman August, zoology
Miovic, Margaret Lancefield, microbiology
Mullin, Dermott Joseph, astrophysics
Mullins, Edgar Raymond, Jr, mathematics
Piker, Steven I, anthropology
Pittel, Stuart, theoretical nuclear physics
Pomerantz, Martin Arthur, physics

Rawson, Kenneth Sidney, animal physiology
Rosen, David, mathematics
Rosenberg, Alburt M, biophysics
Rosenthal, Michael David, solid state physics
Savage, Robert E, cell biology
Skeath, J Edward, mathematics
Suplinskas, Raymond Joseph, physical chemistry
Swann, Charles Paul, nuclear physics
Thompson, Peter Trueman, physical chemistry
van de Kamp, Peter, astronomy
Van Patter, Douglas Macpherson, nuclear physics
Weiland, Henry Joseph, physical chemistry, organic chemistry

SWEDELAND
Cooper, David John, organic chemistry, bio-organic chemistry
Sitrin, Robert David, organic chemistry

SWIFTWATER
DeMeio, Joseph Louis, medical microbiology
Fuscaldo, Anthony Alfred, virology, genetics
Lawlis, John Frank, Jr, microbiology
Metzgar, Don P, virology, immunology
Robey, Robert Ellis, microbiology, immunology

TAMAQUA
Bluhm, Harold Frederick, organic chemistry, explosives
Cox, Fred Ward, Jr, industrial chemistry, research administration
Robins, Jack, physical chemistry, polymer chemistry

TELFORD
Land, Anthony Hamilton, organic chemistry

TOWANDA
Alper, Allen Myron, petrology, mineralogy
Boyd, Robert Henry, physical chemistry
Buzzell, John Gibson, polymer chemistry
Chenot, Charles Frederic, solid state chemistry
Chiola, Vincent, inorganic chemistry
Dickens, John Ernest, photographic chemistry
Dunkle, Michael Patrick, physical chemistry, environmental chemistry
Kim, Tai Kyung, physical chemistry
MacInnis, Martin Benedict, chemistry
Mikus, Felix F, physical chemistry
Tulk, Alexander Stuart, applied chemistry

TREVOSE
Albrecht, Steven Harold, physical chemistry
Ball, William, clinical microbiology
Friend, Patric Lee, microbiology, parasitology
Halpern, Ephriam Philip, biochemistry
Hansen, Gerald Delbert, Jr, physical chemistry
Lieberman, Hillel, organic chemistry, microbial biochemistry
Magnusson, Lawrence Bersell, physical inorganic chemistry
Neddenriep, Richard Joe, physical chemistry
Schiesser, Robert H, physical chemistry, surface chemistry
Shema, Bernard Francis, microbiology
Tonkyn, Richard George, organic chemistry, polymer chemistry
Vorchheimer, Norman, polymer chemistry
Zwarun, Andrew Alexander, microbiology, soil science

UNIONTOWN
Anderson, David Robert, zoology, parasitology
Ostrander, Peter Erling, nuclear physics

UNIVERSITY PARK
Abler, Ronald Francis, geography
Abplanalp, Paul LeRoy, neuroanatomy
Adams, Richard Sanford, animal nutrition, dairy husbandry

Alfke, Dorothy, science education
Allcock, Harry Rex, inorganic chemistry, polymer chemistry
Almquist, John Olson, reproductive physiology
Amann, Rupert Preynoessl, reproductive physiology
Andrews, George Eyre, mathematics
Anthes, Richard Allen, meteorology
Anthony, Adam, physiology
Antle, Charles Edward, mathematical statistics
Armentrout, Steve, topology
Arnold, Dean Edward, aquatic ecology
Aronson, Nathan Ned, Jr, biochemistry
Ascah, Ralph Gordon, physical chemistry
Atwater, Harry Albert, solid state physics
Ayers, John E, plant pathology, plant genetics
Ayoub, Christine Williams, mathematics
Ayoub, Raymond G Dimitri, mathematics
Baker, Dale E, agronomy, soil fertility
Baker, Paul Thornell, biological anthropology, environmental health
Barnes, Bruce Herbert, computer science, mathematics
Barnes, Hubert Lloyd, geochemistry
Barsch, Gerhard Richard, solid state physics
Bartoo, James Breese, mathematics
Bates, Thomas Fulcher, mineralogy
Baumgardt, Billy Ray, animal nutrition
Baylor, John E, agronomy
Beattie, James Monroe, horticulture
Beatty, Alice Ferguson, zoology, entomology
Beelman, Robert B, food science, enology
Bell, Maurice Evan, geophysics
Bellis, Edward David, zoology
Benton, Allen William, biochemistry, physiology
Berg, Clyde C, plant breeding, plant genetics
Bergman, Ernest L, horticulture, plant nutrition
Bernheim, Robert A, physical chemistry, chemical physics
Bernlohr, Robert William, biochemistry
Biederman, Edwin Williams, Jr, geology, mineralogy
Bienvenue, Gordon Raymond, audiology, psychoacoustics
Bitler, William Reynolds, solid state physics
Bleuler, Ernst, nuclear physics
Bloom, James R, plant pathology
Boettcher, Arthur Lee, geochemistry, petrology
Bollag, Jean-Marc, soil microbiology
Bortree, Alfred Lee, veterinary science
Boyle, John Samuel, plant pathology
Braune, Maximillian O, virology, immunochemistry
Brenchley, Jean Elnora, molecular biology
Brewer, James Edward, ornamental horticulture
Brickwedde, Ferdinand Graft, low temperature physics
Brindley, George W, mineralogy, solid state physics
Brown, John Lawrence, Jr, applied mathematics, communication science
Brownawell, Woodrow Dale, number theory
Burick, Emil Joseph, chemistry
Burnham, Clifford Wayne, geochemistry
Buskirk, Elsworth Robert, physiology
Butler, Robert Lee, fisheries
Byers, Robert Allan, entomology
Cameron, Edward Alan, forest entomology
Carlson, Toby Nahum, meteorology
Casida, Lester Earl, Jr, microbiology, ecology
Ceglowski, Walter Stanley, microbiology
Chagnon, Napoleon Alphonseau, anthropology
Cheng, Tien-Hsi, zoology, entomology
Clagett, Carl Owen, biochemistry
Cole, Herbert, Jr, plant pathology, agricultural chemistry
Cole, Richard H, agronomy
Coleman, Michael Murray, polymer science
Coon, Beckford Feddersen, entomology
Cooper, Edwin Lavern, fish biology
Cowan, Robert Lee, animal nutrition
Craig, Richard, plant genetics, plant breeding
Cross, Leslie Eric, physics
Cuffey, Roger James, invertebrate paleontology
Cunningham, Robert Lester, soil genesis & morphology
Dachille, Frank, geochemistry

Dachtler, Sally Louise, genetics
Darken, Lawrence Stamper, physical chemistry
Day, James Thomas, mathematics
Deasy, George F, geography
Deering, Reginald Atwell, molecular biophysics
Deines, Peter, geochemistry
De Maine, Paul Alexander Desmond, computer science
Deno, Norman C, physical organic chemistry
de Pena, Rosa G, atmospheric chemistry, cloud physics
DeWalle, David Russell, microclimatology, watershed management
Dimick, Paul Slayton, food science
Dixon, Joseph Ardiff, organic chemistry
Docherty, John Joseph, virology, microbiology
Downs, Roger Michael, geography, environmental psychology
Duich, Joseph M, agronomy
Dunson, William Albert, zoology, physiology
Dutton, John Altnow, meteorology
Dyke, Bennett, biological anthropology, human genetics
Eakin, James (Henry), Jr, agronomy
Eastman, Daniel Robert Peden, physics
Eberhart, Robert J, veterinary medicine
Eckhardt, Robert Barry, biological anthropology, population genetics
Epp, Donald James, agricultural economics, resource economics
Escobar, M Gabriel, cultural anthropology
Fergus, Charles Leonard, mycology
Feuchtwang, Thomas Emanuel, solid state physics, surface physics
Fine, Nathan Jacob, mathematics
Fischer, Patrick Carl, computer science
Fissel, Guy Wilmer, plant chemistry
Fleming, Gordon N, theoretical physics
Fletcher, Peter Whitcomb, forest soils
Flipse, Robert Joseph, dairy science
Fortmann, Henry Raymond, agronomy
Fowler, Horatio Seymour, biology, science education
Frankl, Daniel Richard, surface physics
Fraser, Alistair Bisson, atmospheric physics, optics
Freed, Norman, theoretical nuclear physics
Fritton, Daniel Dale, soil physics
Fritz, James John, physical chemistry
Fung, Daniel Yee Chak, microbiology, food technology
Gaffney, Edwin V, II, cell biology
Gentry, Robert Francis, poultry pathology
Geoffroy, Gregory Lynn, organometallic chemistry
George, John Lothar, biology
Gerhold, Henry Dietrich, forest genetics
Gilmore, Herbert Clarence, dairy science
Given, Peter Hervey, fuel science, organic geochemistry
Glantz, Paul Joseph, medical microbiology
Glasner, Moses, mathematics
Gobble, James Lawrence, animal husbandry
Gokel, George William, organic chemistry
Gold, David Percy, geology, petrology
Gold, Lewis Peter, physical chemistry
Good, Roland Hamilton, Jr, theoretical physics
Goodwin, Kenneth, genetics
Gould, Peter Robin, geography
Graetzer, Reinhard, nuclear physics, biophysics
Grun, Paul, cytology, cytogenetics
Guber, Albert Lee, geology
Gustine, David Lawrence, biochemistry
Graves, Hannon B, ethology
Greenfield, Roy Jay, geophysics
Greig, Joseph Wilson, petrology, physical chemistry
Griffiths, John Cedric, petrology
Grossman, Herbert H, botany
Grotch, Howard, theoretical physics
Grove, Alvin Russell, Jr, anatomy, morphology
Guthrie, Helen A, nutrition
Haas, Charles Gustavus, Jr, inorganic chemistry
Hahn, Kyong T, mathematical analysis
Hagen, John Peter, astronomy
Haight, Frank Avery, applied mathematics
Hale, Edgar Brewer, animal behavior
Hall, Jon K, agronomy, soil chemistry
Hamilton, Gordon Andrew, bio-organic chemistry

647

Hamilton, Robert Hillery, Jr., plant physiology
Hammerstedt, Roy H., biochemistry, reproductive physiology
Haramaki, Chiko, ornamental horticulture, weed science
Hargrove, George Lynn, quantitative genetics, statistics
Harkness, William Leonard, applied statistics
Harrison, Joseph Donald, agronomy
Harrington, Ian Roland, polymer science
Hartsook, Elmer William, nutritional biochemistry
Heicklen, Julian Philip, atmospheric chemistry
Hayes, John Robert, chemistry
Heald, Walter Roland, soil chemistry
Heddleson, Milford Raynord, soil chemistry, conservation
Hershberger, Truman Verne, animal nutrition
Herman, Roger Myers, atomic physics, molecular physics
Hetmansperger, Thomas Philip, statistics
Hicks, Floyd W., poultry sci
Hill, Richard Ray, Jr., plant breeding, genetics
Hitz, Chester W., pomology
Hinish, Wilmer Wayne, agronomy
Hisatsune, Isamu Clarence, physical chemistry
Hollis, Theodore M., physiology
Horrocks, William DeWitt, Jr., inorganic chemistry, bioinorganic chemistry
Hosler, Charles Luther, Jr., meteorology
Hovermale, John Bruce, meteorology
Howell, Benjamin F, Jr., geophysics
Hower, Arthur Aaron, Jr., entomology
Hultquist, Robert Allan, mathematical statistics
Hunt, Edward Eyre, physical anthropology
Hunter, Albert Sinclair, soil fertility
Hunter, Robert P., mathematics
Hutnik, Russell James, forest ecology
Hymer, Wesley C., endocrinology, cell biology
Ignizio, James Paul, operations research
Jackman, Lloyd Miles, organic chemistry
James, Donald Gordon, mathematics
Johnson, Leon Joseph, soil mineralogy
Johnson, Melvin Walter, Jr., agronomy, genetics
Jones, Jennings Hinch, anthropology
Jordan, Joseph, analytical chemistry
Joys, Terence M., medical microbiology
Jung, Gerald Alvin, agronomy, plant physiology
Jurs, Peter Christian, analytical chemistry
Kanwal, Ram Prakash, applied mathematics
Kappenman, Russell Francis, statistics
Kardos, Louis Thomas, soil chemistry
Kazes, Emil, theoretical physics
Keene, Owen David, poultry nutrition
Keener, Carl Samuel, botany
Keeney, Philip G., food science
Keith, MacKenzie Lawrence, geochemistry
Kelly, Bernard Wayne, economics
Kendall, Bruce Reginald Francis, space physics
Kendall, William Anderson, plant physiology
Kerrick, Derrill M., petrology, geochemistry
Kester, Earl Marshall, dairy science
Killian, Gary Joseph, reproductive physiology
Kim, Ke Chung, systematic entomology, medical entomology
King, Thomas B., animal science
Kline, Donald Edgar, physics
Knebone, Leon Russell, mycology
Knievel, Daniel Paul, crop physiology
Kocher, Frank T., mathematics
Kollias, James, physiology
Krall, Allan M., mathematics
Krall, Harry Levern, mathematics
Kroger, Manfred, food science
Kurland, Jeffrey Arnold, primatology
Kurtz, David Allan, pesticide chemistry
Lacasse, Norman L., plant pathology
Lancaster, Otis Ewing, aeronautical engineering, mathematics
Lang, Lawrence George, solid state physics
Langmuir, Donald, geochemistry

Larson, Russell Edward, agricultural education, academic administration
Lavin, Peter Masland, geophysics
Leach, Roland Melville, Jr., animal nutrition
Leath, Kenneth T., plant pathology
Lee, John Denis, meteorology
Levine, Samuel Harold, nuclear physics, reactor physics
Lewis, Peirce Fee, cultural geography
Lieberman, Leslie Sue, biological anthropology
Lindstrom, Eugene Shipman, bacteriology
Lochicel, William A., physics
Long, Theodore Alfred, animal nutrition
Lotz, John Robert, analytical chemistry
Lovell, Harold Lemuel, fuel science
Lowe, John Philip, quantum chemistry
Ludwig, Ernest Harry, microbiology
Lukezic, Felix Leon, plant pathology
MacCluer, Jean Walters, human genetics, population genetics
MacKenzie, David Robert, plant pathology, plant genetics
MacNeil, Joseph H., food science
Mallette, Manney Frank, biochemistry
Mallis, Arnold, entomology
Marriott, Lawrence Frederick, agronomy, soils
Marsolf, J David, agricultural meteorology
Marshall, Harold Gene, plant genetics
Martin, Roy Joseph, Jr., nutrition
Maserick, Peter H., mathematics
Mast, Morris Glen, food science
Mastalerz, John W., horticulture
Matsushima, Satoshi, astronomy, astrophysics
Matthews, Charles Robert, biophysical chemistry
McCubbin, Thomas King, Jr., physics
McKee, Guy William, agronomy
McKinstry, Herbert Alden, solid state physics
McCammon, Mary, mathematics
McCammon, Robert Desmond, solid state physics
Melton, Rex Eugene, forestry
Mendez, Jose De La Vega, physiology
Merrill, William, forest pathology, forest products
Michael, Paul Lee, physics, environmental health
Michels, Joseph William, anthropology
McCarl, Richard Lawrence, biochemistry
McCarthy, Robert David, dairy science
McCarthy, William John, virology,
Miller, E Willard, geography
Miller, Warren Widmer, chemistry
Mills, Wilford Richard, plant pathology
Minard, Robert David, organic chemistry
Mitchell, Robert Bruce, physiology
Moorehead, Thomas J., chemistry
Morrill, Warren Thomas, ethnology,
Muan, Arnulf, geochemistry
Mueller, Erwin W., electron physics
Mueller, Werner Julius, poultry science
Mulay, Laxman Nilakantha, physical chemistry, inorganic chemistry
Mumma, Ralph O., biochemistry
Murphey, Wayne K., wood technology
Neff, William H., zoology, physiology
Nelson, Paul Edward, plant pathology
Newnham, Robert Everett, physics
Noll, Charles Joseph, horticulture, weed science
Norman, John Matthew, micrometeorology, agricultural meteorology
Ohmoto, Hiroshi, geochemistry
Olivero, John Joseph, Jr., aeronomy, meteorology
Olofson, Roy Arne, organic chemistry
O'Mara, Joseph George, genetics
Oswald, John Wieland, plant pathology
Palmer, Howard Benedict, physical chemistry, fuel science
Panofsky, Hans Arnold, micrometeorology
Parizek, Richard Rudolph, environmental geology, hydrogeology,
Parsons, Torrence Douglas, mathematics
Patil, Ganapati P., statistics, mathematics
Patton, Stuart, food biochemistry
Patton, William Henry, veterinary microbiology
Pazur, John Howard, biochemistry
Pearson, David Leander, ecology, ornithology

Pedersen, Svend, biochemistry
Pena, Jorge Augusto, air pollution, atmospheric chemistry
Person, Stanley R., biophysics
Petersen, Donald Harry, plant pathology
Petersen, Gary Walter, soil morphology, science
Petrich, Mario, algebra
Pfahl, Peter Blair, horticulture
Pfeifer, Robert Paul, plant breeding, genetics
Phillips, Allen Thurman, biochemistry
Pigott, Miles Thomas, cryogenics
Pike, Ruth Lillian, nutrition
Pillay, K R Sivasankara, nuclear chemistry, nuclear engineering
Pionke, Harry Bernhard, soil chemistry
Pliva, Josef M., molecular spectroscopy
Polo, Santiago Ramos, molecular physics, solid state physics
Pootjes, Christine Fredricka, microbiology
Pratt, William Winston, nuclear physics
Prout, James Harold, applied physics, acoustics
Pursell, Ronald A., botany
Quinn, Robert George, plasma physics, space sciences
Rank, David Herr, physics
Reed, Robert Willard, solid state physics
Rhoades, Robert A., physiology
Ribeiro, Hugo B., mathematics
Richey, Herman Glenn, Jr., organometallic chemistry
Ridge, John Drew, economic geology
Rigby, Paul Herbert, management sciences, applied statistics
Risby, Terence Humphrey, analytical chemistry
Rosenblatt, Gerd Matthew, physical chemistry
Rothenbacher, Hansjakob, veterinary pathology, veterinary microbiology
Risius, Marvin Leroy, plant breeding
Ritter, Crum Marshall, pomology
Rodgers, Allan L., geography
Roe, Keith Edward, plant taxonomy
Rogowski, Andrew S., soil physics, hydrology
Roy, Rustum, materials science
Rung, Donald Charles, Jr., mathematical analysis
Rutschky, Charles William, entomology
Ryan, Thomas Arthur, Jr., statistics
Sampson, Douglas Howard, astrophysics
Samson, Fred Burton, wildlife ecology
Samuelson, Donald James, mathematics
Sanders, William T., anthropology
Schein, Richard David, plant pathology,
Schempf, John Morey, analytical chemistry
Schisler, Lee Charles, mycology
Schmalz, Robert Fowler, marine geology
Scholten, Robert, geology, tectonics
Scholz, Richard W., nutritional biochemistry, avian physiology
Schraer, Harald, cell biology
Schraer, Rosemary, biochemistry, cell biology
Schwartz, Leland Dwight, poultry pathology
Shamma, Maurice, natural products chemistry
Shellenberger, Paul Robert, dairy science
Shenk, John Stoner, agronomy, plant breeding
Sherritt, Grant Wilson, animal science
Sherwood, Robert Tinsley, plant pathology
Shipman, Robert Dean, forest ecology, silviculture
Shipp, Raymond Francis, agronomy
Shrigley, Robert Leroy, science education
Sibul, Leon Henry, applied mathematics, underwater acoustics
Siegenthaler, Bruce Monroe, speech pathology, audiology
Simkins, Paul Dean, geography
Simonaitis, Romualdas, chemical kinetics, photochemistry
Sink, John Davis, biochemistry, biophysics
Skell, Philip S., organic chemistry
Smilowitz, Zane, entomology
Smith, Cyril Beverley, plant nutrition
Smith, Deane Kingsley, Jr., mineralogy, crystallography
Smith, Samuel H., plant pathology, plant virology
Smyth, Thomas, Jr., insect physiology
Snetsinger, Robert J., economic entomology, arachnology
Snipes, Wallace Clayton, biophysics
Snow, Jean Anthony, plant pathology, mycology

Snyder, Fred Calvin, agriculture
Sopper, William Edward, forestry
Spackman, William, Jr., paleobotany
Specht, Lawrence W., genetics
Squires, Burton Elliott, Jr., computer science
Starling, James Lyne, plant breeding, statistics
Steele, William A., physical chemistry
Steiner, Kim Carlyle, forest genetics
Stere, Athleen Jacobs, biology
Stinson, Richard Floyd, floriculture
Stone, Robert William, microbiology
Strother, Greenville Kash, biophysics
Suhr, Norman Henry, spectroscopy, microbiology
Tammen, James F., plant pathology
Tanabe, Tsuneo Y., physiology
Taylor, William Daniel, biophysics
Tershak, Daniel R., virology
Tetrault, Robert Close, entomology
Therrien, Chester Dale, mycology
Thomas, Walter Ivan, agronomy, genetics
Thomson, Dennis Walter, meteorology
Thornton, Charles Perkins, petrology
Thrower, Peter Albert, physics, materials science
Thwaites, Thomas Turville, nuclear physics
Ting, Annsheng Chien, numerical analysis
Todd, Paul Wilson, biophysics
Traverse, Alfred, palynology, paleobotany
Tsong, Tien Tzou, solid state physics
Tukey, Loren Davenport, horticulture
Underwood, Barbara Ann, nutrition
Usher, Peter Denis, astronomy
Vastola, Francis J., physical chemistry
Vedam, Kuppuswamy, physics, materials science
Villafranca, Joseph John, biochemistry, bio-organic chemistry
Voight, Barry, geology
Waddington, Donald Van Pelt, soil fertility, plant science
Walcher, Dwain N., pediatrics, human development
Wallner, Stephen John, plant physiology
Wangsness, Paul Jerome, animal nutrition
Ward, Wilber W., forestry
Warme, Roger Perry, algebra
Warme, Paul Kenneth, biochemistry
Wartik, Thomas, inorganic chemistry
Washko, John Blasius, agronomy
Waterhouse, William Charles, mathematics
Watrous, George H., Jr., dairy husbandry
Watschke, Thomas Lee, agronomy
Webber, Jon, geochemistry
Weber, Jon, geochemistry
Weber, Robert L., physics
Webster, David Lee, anthropology, archaeology
Wernstedt, Frederick Lage, geography of Southeast Asia
White, Eugene Wilbert, instrumentation, mineralogy
White, John W., horticulture
White, John Blaine, geochemistry, materials science
Whitfield, George Danley, physics
Wickersham, Edward Walker, reproductive physiology, human sexuality
Wiggins, Thomas Arthur, optics
Williams, Anthony Vearncombe, geography, applied statistics
Williams, Eugene G., mineralogy
Williams, Frederick McGee, biology
Wilton, Arthur Charles, cytogenetics, plant breeding
Winkler, Louis, astronomy
Witham, Francis H., plant physiology
Woloshuk, Detlef, computer science, systems science
Wright, James Everett, Jr., genetics
Wright, Lauren Albert, geology
Wuest, Paul Joseph, plant pathology
Yendol, William G., entomology
Yood, Bertram, mathematical analysis
Zabriskie, Franklin Robert, astronomy
Zarkower, Arian, veterinary medicine, immunology
Zelinsky, Wilbur, cultural geography, population geography
Ziegler, John Henry, Jr., meat science
Zimmerman, Leonard Norman, bacteriology
Zook, Harry David, organic chemistry, academic administration

UPPER DARBY
Clark, Floyd Bryan, forestry, research administration
Gordon, William George, protein chemistry
Iberall, Arthur Saul, physics
Lautz, William, plant pathology, air pollution
Shlaifer, Arthur, optometry
Uhlig, Hans Gerd, biology

VALLEY FORGE
Hopkins, George Hallman, rubber chemistry
Muck, Darrel Lee, physical organic chemistry
Verhanovitz, Richard Frank, theoretical nuclear physics, numerical analysis

VERONA
Bryson, Theodore Cornelius, analytical chemistry
Clark, Ralph O, chemistry

VILLANOVA
Armenti, Angelo, Jr., theoretical physics
Baker, Wilber Winston, biochemistry, cell biology
Barlow, Thomas L, mathematics
Beck, Robert Edward, algebra, operations research
Brooks, James O, algebra
Cawley, John Joseph, physical organic chemistry
Clarke, Lilian A, inorganic chemistry
Dollahon, Norman Richard, protozoology, parasitology
Downey, Bernard Joseph, physical chemistry
Doyne, Thomas Harry, biochemistry
Edwards, John R, biochemistry
Ehrmann, Rita Mae, geometry
Grob, Robert Lee, analytical chemistry
Hartmann, Frederick W., mathematics
Hones, Michael J, particle physics
Jenkins, William Bernard, astronomy
Langan, William Bernard, physiology
Leffler, Amos J, inorganic chemistry, physical chemistry
Levitan, Michael Leonard, mathematics
Ludwig, Oliver George, physical chemistry
Lynn, Roger Yen Shen, applied mathematics
Maksymowich, Roman, botany
Malewicz, Thomas Donald, anatomy
Markham, James J, analytical chemistry, oceanography
McClain, John A, zoology
McCook, George Patrick, astronomy
Peterson, Leroy Eric, physics
Riser, Wayne H., veterinary pathology
Sardinas, August A, mathematical analysis
Schauble, J Herman, organic chemistry
Snyder, Martin Avery, applied mathematics
Spritzer, Michael Stephen, analytical chemistry
Ward, Ingeborg L, physiological psychology
Waters, Joseph Hemenway, vertebrate zoology
Wojcik, John F, physical chemistry
Wunderlich, Francis J, physical chemistry
Yeh, George Chiayou, chemical physics
Zajac, Walter William, Jr, organic chemistry

WALLINGFORD
Davis, Francis Kaye, Jr, physical meteorology
Doolittle, Arthur King, fluid physics
Doolittle, Dortha Bailey, organic chemistry
Greenland, Miles Griffith, physics
Mayer, Theodore Jack, petroleum chemistry
Statton, Gary Lewis, polymer chemistry

WALNUTPORT
Parks, James Marshall, Jr., paleoecology, oceanography

WARMINSTER
Berman, David Alvin, inorganic chemistry, corrosion
Cope, Freeman Widener, physical biochemistry, physiology
Feingold, Earl, solid state physics
Krutter, Harry, physics
Kydd, George Herman, cardiopulmonary physiology
Polis, Beryl David, biochemistry
Shmukler, Herman William, biochemistry
Squires, Russell Dill, physiology
Stoll, Alice Mary, medical biophysics
Von Beckh, Harald Johannes, aerospace medicine

WARREN
Horsley, Stephen Braithwaite, plant physiology
Marquis, David Alan, forest ecology
Roach, Benjamin Arthur, forest management
Varma, Rajendra, organic chemistry, biochemistry
Varma, Ranbir S, clinical biochemistry, medical research
Wardi, Ahmad Hassan, biochemistry, analytical chemistry

WARRINGTON
Greenfield, Stanley A, organic chemistry
Halperin, Benjamin David, chemistry, chemical engineering

WARRIORS MARK
Brinkman, Gail Lynn, nutrition, biochemistry

WASHINGTON
Bell, Raymond Martin, physics
Foland, William Douglas, physics
Lawrence, Vinnedge Moore, entomology, ecology
Leake, William Walter, organic chemistry
Michaels, Adlai Eldon, physical chemistry
Mooney, David Samuel, organic chemistry
Parker, Leslie, clinical chemistry
Porter, Homer Clifford, biology
Rabenstein, Albert Louis, mathematics
Schreiner, Ceinwen Ann, genetics
Staskiewicz, Bernard Alexander, physical chemistry
Trelka, Dennis George, animal physiology
Vogel, Norman William, zoology
Wicker, Robert Kirk, physical inorganic chemistry
Wylen, Herbert E, theoretical physics

WAYNE
Adams, Alden Ross, organic chemistry, biochemistry
Glauser, Stanley Charles, pharmacology
Hales, J Vern, atmospheric physics
Hermans, James J, computer science
Leader, Gordon Robert, physical chemistry
Schroyer, James Burnette, geology
Thomas, Carolyn Margaret, analytical chemistry
Walker, David Kenneth, physics
Williams, Russell Raymond, parasitology, invertebrate zoology

WAYNESBURG
Barnett, Leland Bruce, physiological ecology
Bryner, Charles Leslie, botany
LaCount, Robert Bruce, organic chemistry
Maguire, Mildred May, physical chemistry
Wilson, James William, medicinal chemistry

WEST CHESTER
Barclay, Eugene Samuel, biology
Becker, Marshall Joseph, anthropology
Braddock-Rogers, Kenneth, inorganic chemistry
Bravo, Justo Baladjay, inorganic chemistry
Brown, Relis Bastian, zoology
Bugosh, John, physical chemistry, colloid chemistry
Buhle, Emmett Loren, industrial organic chemistry
Chalupa, William Victor, animal nutrition
Chandler, Alfred Bertram, inorganic chemistry, analytical chemistry
Chow, Alfred Wen-Jen, medicinal chemistry, organic chemistry
Cinquina, Carmela Louise, bacteriology

Deist, Robert Paul, pharmaceutical chemistry
Deyrup, Alden Johnson, physical chemistry
Dorchester, John Edmund Carleton, physiology, biochemistry
Durand, Marc L, organic chemistry
Duswalt, Allen Ainsworth, Jr, analytical chemistry
Fabrey, James Douglas, computer science, applied mathematics
Filano, Albert Eugene, mathematics
Garber, Charles A, polymer physics
Gietsos, Constantine, organic chemistry, medicinal chemistry
Greenberg, Seymour Samuel, geology
Herrick, Elbert Charles, organic chemistry, chemical engineering
Kaschenbach, Leo, carbohydrate chemistry
Keller, Joseph Herbert, physical chemistry
Killian, Donald B, chemistry
Langdon, George L J, / geography of Latin America, physical geography
March, Louis Charbonnier, organic chemistry
Martinez, Margaret Yarnall, human anatomy, physiology
Medeiros, Robert Whippen, organic chemistry
Merkel, Timothy Franklin, organic polymer chemistry
Miller, Cecil R, veterinary medicine
Overlease, William R, ecology, botany
Parish, Roger Cook, organic chemistry, medicinal chemistry
Pinkney, Paul Swithin, organic chemistry
Reynolds, Francis Joseph, organic chemistry
Rickert, Russell Kenneth, physics
Robinson, Charles Albert, organic chemistry
Sarantakis, Dimitrios, organic chemistry
Scott, George Clifford, veterinary medicine
Shawhan, Elbert Neil, physics, instrumentation
Shoaf, Mary La Salle, physical chemistry
Sklar, Stanley, pharmaceutics
Theodorides, Vassilios John, veterinary parasitology
Torop, William, inorganic chemistry, science education
Treece, Jack Milan, genetics, biochemistry
Trezise, Willard Joseph, biochemistry, bacteriology
Weiss, Sol, mathematics
Williams, Ardis Mae, physical chemistry
Woodruff, Richard Ira, developmental biology, reproductive biology
Yarosewick, Stanley J, atomic physics, spectroscopy

Bolhofer, William Alfred, organic chemistry
Boyle, John Joseph, virology
Brady, Stephen Francis, organic chemistry
Breault, George Omer, drug metabolism
Brenner, Gerald Stanley, organic chemistry, pharmaceutical chemistry
Castello, Robert Anthony, pharmaceutical chemistry
Ciminera, Joseph Louis, pharmacy, biostatistics
Clineschmidt, Bradley Van, neuropharmacology
Conti, Pierre Andre, laboratory animal medicine
Cook, Margaret Mary, microbiology
Cragoe, Edward Jethro, Jr., medicinal chemistry, organic chemistry
Davies, Richard O, medicine, pharmacology
Dempski, Robert E, pharmaceutics
Duggan, Daniel Edward, biochemical pharmacology
Engelhardt, Edward Louis, organic chemistry
Evans, Ben Edward, synthetic organic chemistry, enzymology
Fanelli, George Marion, Jr, pharmacology
Freidinger, Roger Merlin, organic chemistry
Garfinkle, Barry David, virology, microbiology
Gould, A Lawrence, biometrics
Gray, Alan, virology
Grim, Wayne Martin, pharmaceutical chemistry
Hanson, Hazel Jean, microbiology
Hilleman, Maurice Ralph, virology
Hirschmann, Ralph Franz, organic chemistry
Hite, Mark, toxicology, cytogenetics
Hoffman, Jacob Matthew, Jr, medicinal chemistry, synthetic organic chemistry
Holly, Frederick William, protein chemistry
Hucker, Howard Benjamin, drug metabolism
Ittensohn, Oswald Ludwig, virology, microbiology
Jackson, Gerald James, pharmacy
Jensen, Richard Donald, veterinary pathology
Kwan, King Chiu, pharmaceutical chemistry
Lampson, George Peter, biochemistry
Larson, Vivian M, virology
Lehman, Ernest Dale, biochemistry
Maha, George Edward, medicine, pharmacology
Martin, Christopher Michael, medicine
McAleer, William Joseph, organic chemistry, biochemistry
McClelland, Laurella, virology
McGonigle, Eugene Joseph, analytical chemistry
Mendlowski, Bronislaw, pathology, immunology
Minsker, David Harry, physiology, pharmacology
Neff, Beverly Jean, virology, epidemiology
Nemes, Marjorie M, bacteriology
Novello, Frederick Charles, organic chemistry
Nutt, Ruth Foelsche, medicinal chemistry
Peck, Harold Mitchell, medicine
Peltier, Hubert Conrad, medicine
Porter, Curt Culwell, biochemistry
Powell, Burwell Frederick, organic chemistry
Prasad, Suresh, poultry science, laboratory animal science
Provost, Philip Joseph, microbiology, virology
Prugh, John Drew, medicinal chemistry
Randall, William Carl, physical chemistry
Remy, David Carroll, medicinal chemistry
Restaino, Frederick A, physical pharmacy
Saari, Walfred Spencer, organic chemistry
Salerno, Ronald Anthony, oncology, zoology
Schuchardt, Lee Frank, bacteriology, virology
Schwartz, Joseph Barry, pharmaceutical chemistry
Scriabine, Alexander, pharmacology
Shaffer, James Milton, clinical medicine
Shepard, Kenneth LeRoy, medicinal chemistry
Sinotte, Louis Paul, pharmaceutical chemistry, organic chemistry

WEST MIFFLIN
Abu-Shumays, Ibrahim Khalil, applied mathematics, mathematical physics
Baer, William, nuclear physics
Barker, Franklin Brett, radiochemistry
Barnart, Sidney, electrochemistry, corrosion
Bogard, Andrew Dale, radiochemistry
Clayton, John Charles (Hastings), physical inorganic chemistry
Galey, John Apt, reactor physics
Gordon, Neil E, Jr, chemistry
Gunst, Samuel Burton, experimental physics
Hageman, Louis A, mathematics
Hardy, Judson, Jr, nuclear physics, reactor physics
Hillner, Edward, metallurgical chemistry, metallurgical engineering
Judd, Jane Harter, spectroscopy, analytical chemistry
Markowitz, Joseph Morris, physical chemistry
Milani, Salvatore, nuclear physics
Schultz, Boyd Gilbert, water chemistry
Shure, Kalman, nuclear science
Sussman, Myron Maurice, numerical analysis
Tessler, George, physics
Vogeley, Clyde Eicher, Jr, applied mathematics
Walther, Frank H, mineralogy

WEST POINT
Allen, Henry L, veterinary pathology
Baer, John Elson, pharmacology, drug metabolism
Balant, Charles Paul, medicinal chemistry, research administration
Bayne, Gilbert M, psychopharmacology, therapeutics
Bokelman, Delwin Lee, pathology

PENNSYLVANIA

Smith, Robert Lawrence, organic chemistry; medicinal chemistry
Stone, Clement A., pharmacology
Sweet, Charles Samuel, pharmacology
Toberman, Ralph Owen, analytical chemistry
Tocco, Dominick Joseph, biochemistry, pharmacology
Torchiana, Mary Louise, physiology, pharmacology
Tyrell, Alfred A., biochemistry
Ulm, Edgar H., biochemistry
Van Arman, Clarence Gordon, pharmacology
Veber, Daniel Frank, chemistry
Vella, Philip Peter, medical bacteriology, immunology
Vickers, Stanley, drug metabolism
Watson, Lloyd Sherman, physiology
Wenger, Herbert Charles, pharmacology, physiology
Wolf, George L., veterinary pathology
Woodhour, Allen F., virology
Yarbrough, George Gibbs, neuropharmacology
Zabriskie, John Lansing, Jr., organic chemistry
Zachel, Anthony Gabriel, drug metabolism, mass spectrometry

WILKES-BARRE
Berard, Anthony D. Jr., topology
Bohning, James Joel, photochemistry
Borkowski, Raymond P., physical chemistry
Chock, Jan Sun-Lum, physiology
Donahoe, Frank J, solid state physics
Donahue, William H., plant ecology
Doty, Robert Bruce, bacteriology
Earl, Boyd L., mathematics
El-Ashry, Mohamed T., geology
Faut, Owen Donald, inorganic chemistry
Hayes, Wilbur Frank, zoology
Holden, Stanley J., solid state physics, surface physics
Kimball, Grace Caroline, bacteriology
Kleinsteuber, Tilmann Christoph Werner, physics, physical chemistry
Labows, John Norbert, Jr., organic chemistry
Lee, Chin-Chiu, microbiology; electron microscopy
Merrill, Samuel, III, mathematics
Miller, Theodore Lee, physical chemistry
Mitra, Grihapati, inorganic chemistry
Ogren, Robert Edward, zoology
Paoletti, Robert Anthony, developmental biology
Reif, Charles Braddock, zoology
Reynolds, William Wallace, biology
Rozelle, Ralph B., physical chemistry
Smith, Francis Xavier, organic chemistry
Sours, Richard Eugene, mathematical analysis
Stine, William R, organic chemistry
Swain, Howard Aldred, Jr., physical chemistry
Tillman, Stephen Joel, mathematics
Tobin, Thomas Vincent, biology
Wehman, Henry Joseph, comparative neurology
Wong, Bing Kuen, mathematical analysis

WILLIAMSBURG
Siegel, Robert Ted., particle physics

WILLIAMSPORT
Angstadt, Robert B., animal physiology
Franz, David Alan, analytical chemistry
Hale, Creighton J., medical physiology
Hummer, James Knight, organic chemistry
Kelley, Alden Gerard, plant physiology
Radspinner, John Asa, physical chemistry
Rannels, Herman Wolfe, gynecology; obstetrics & gynecology
Sherbine, K Bruce, zoology; marine biology

WILLOW GROVE
Curtin, Leo Vincent, animal nutrition
Epstein, Samuel David, computer science, systems engineering
Preuss, Albert F., chemistry
Rieders, Fredric, toxicology, chemistry
Schiffman, Louis F., physical chemistry
Schweda, Paul, toxicology
Stokes, Charles Sommers, physical chemistry

WOMELSDORF
Wells, Eugene Hadley, chemistry

WYNCOTE
Connor, Ralph (Alexander), chemistry
Dank, Milton, solid state physics, mathematics
Quinn, Alfred Otto, photogrammetry

WYNDMOOR
Benedict, Robert Curtis, biochemistry
Kimoto, Walter Iwao, food chemistry
Mozersky, Samuel M, biochemistry, enzymology
O'Leary, Virginia Sawyer, food microbiology
Szymanski, Edward Stanley, enzymology

WYNNEWOOD
Krieger, Carl Henry, biochemistry, research administration
Rose, Edward, endocrinology
Smyrk, Charles McCahan, Jr, industrial organic chemistry
Zoelisch, Oscar Cornelius, plant physiology

YARDLEY
Bertozzi, Eugene R, polymer chemistry
Drescher, Robert Frederick, industrial microbiology
Ferrell, D Thomas, Jr., electrochemistry
Hull, Michael Neill, electrochemistry
Konrad, Dusan, electrochemistry
Kugler, George Charles, electrochemistry,
Margalit, Nehemiah, chemistry
Nordblom, George Frederick, physical chemistry
Scott, Charles Edward, chemistry
Venuto, Carmine Joseph, economic geology; mineralogy

YORK
Boas, Charles William, geography
Burkle, Joseph S., nuclear medicine
Forchheimer, Otto Louis, physical chemistry, inorganic chemistry
Harrison, Ernest Augustus, Jr., organic chemistry
Jones, David John, preventive medicine
Khanna, Sardari Lal, physics
Scheirer, Carl Latimer, physical chemistry, solid state chemistry
Smith, Bruce Barton, plant morphology
Tateosian, Louis Hagop, dental materials

RHODE ISLAND

ASHTON
Polevy, John Henry, organic chemistry

BARRINGTON
Leo, Micah Wei-Ming, chemistry
Stein, Alvin, polymer chemistry

BRISTOL
Holstein, Thomas James, developmental genetics
Von Riesen, Daniel Dean, organic chemistry

COVENTRY
Mulvaney, John Francis, organic chemistry
Tien, Rex Yuan, organic chemistry

CRANSTON
Blecharczyk, Walter Joseph, biomedical engineering
Fish, William Arthur, embryology
Martin, Frank Stephen, chemistry
Parker, Edward Arthur, electrochemistry
Velluro, Anthony Francis, applied chemistry; organic chemistry

CUMBERLAND
Walsh, John Thomas, analytical chemistry

DAVISVILLE
Verber, James Leonard, physical oceanography

EAST GREENWICH
Dann, Frank Warren, pharmaceutical chemistry
Lawrence, George Hill Mathewson, systematic botany
LeBlanc, Robert Bruce, inorganic chemistry

EAST PROVIDENCE
Bobe, Ernest Christoph, organic chemistry
Forchielli, Americo Lewis, chemistry
Johnson, Donald Wayne, organic chemistry; photochemistry
Kroll, Harry, organic chemistry

EDGEWOOD
Smith, Homer Pine, physical chemistry

FISKEVILLE
Carlson, Steven Allen, organic chemistry
Habib, David Peter, industrial organic chemistry

GREENVILLE
O'Keefe, J George, physics

HOPE
Russo, William Richard, analytical chemistry

JAMESTOWN
Andrews, Howard Lucius, biophysics

JOHNSTON
Casparian, Sarkis Manoug, organic chemistry; analytical chemistry

KINGSTON
Abell, Paul Irving, organic chemistry
Abushanab, Elie, pharmaceutical chemistry
Albert, Luke Samuel, plant physiology
Alexander, Lewis McElwain, geography
Ballard, Berton Etienne, pharmaceutical chemistry
Bass, Leonard Joel, computer sciences
Beauregard, Raymond A., algebra
Beckman, Carl Harry, plant pathology,
Bell, Robert Gale, biochemistry
Bond, Howard Wissler, chemistry
Brown, Christopher W., physical chemistry
Brown, James Henry, Jr., forest ecology; soils
Brown, Phyllis R., chemistry; biochemistry
Byrne, Robert Howard, oceanography
Cain, James Allan, geology
Carney, Edward J., statistics, computer sciences
Carosella, Nestor Edgar, plant pathology
Carpenter, Philip Lewis, microbiology; immunology
Chang, Pei Wen, veterinary pathology
Cheer, Clair James, organic chemistry
Chichester, Clinton Oscar, food technology
Chipman, Robert K., zoology
Choudry, Amar, nuclear physics
Christopher, Everett Percy, horticulture
Cobble, James William, dairy production
Cohen, Paul Sidney, molecular biology
Constantinides, Spiros Minas, food science, biochemistry
Cosgrove, Clifford James, dairy science,
Costantino, Robert Francis, population biology
Cruickshank, Alexander Middleton, inorganic chemistry; analytical chemistry
Dain, Joel A., biochemistry
Datta, Dilip Kumar, geometry
DeFanti, David R., pharmacology
DeFeo, John Joseph, pharmacology
De Wolf, Robert Abel, zoology

Dietz, Frank Tobias, acoustics, academic administration
Donovan, Gerald Alton, poultry nutrition, biochemistry
Driver, Rodney David, mathematical analysis, electrodynamics
Duce, Robert Arthur, atmospheric chemistry
Duff, Dale Thomas, crop physiology
Durfee, Wayne King, avian physiology, poultry science
Dymsza, Henry A., biochemistry; food science
Etzold, David Frank, pharmaceutical electroacoustics
Fasching, James Le Roy, nuclear chemistry, analytical chemistry
Felbeck, George Theodore, Jr., organic geochemistry
Fish, Marie Poland, biological oceanography
Fisher, Harold Wilbur, biophysics, molecular biology
Fuller, George Charles, biochemical pharmacology
Gieisse, Peter Jacob Maria, mineralogy, crystallography
Goertemiller, Clarence C., Jr., developmental biology, cell biology
Golet, Francis Charles, wildlife ecology; fresh water ecology
Gonzalez, Richard D., physical chemistry
Goodman, Leon, organic chemistry
Gould, Roger Delmon, mycology
Gould, Walter Philip, forestry; wildlife management
Halvorson, William Lee, ecology
Hammen, Carl Schlee, invertebrate physiology; comparative biochemistry
Hanumara, Ramachandra Choudary, statistics
Hargraves, Paul E., biological oceanography; marine botany
Harlin, Marilyn Miler, algology
Harrison, Robert William, zoology,
Hartman, Karl August, Jr., physical chemistry; molecular biology
Hauke, Richard Louis, systematic botany
Havens, James Meryle, meteorology; climatology
Heavers, Barbara Ann, marine zoology
Hemmerle, William J., statistics, computer science
Heppner, Frank Henry, ornithology
Hermes, O Don, petrology; geochemistry
Higbee, Edward, agricultural economics, geography
Hill, Robert Benjamin, comparative physiology
Houston, Chester Warren, bacteriology
Howard, Frank Leslie, plant pathology
Hull, Richard James, plant physiology
Hyland, Kerwin Ellsworth, Jr., entomology; parasitology
Jackson, Noel, plant pathology
Jeffries, Harry Perry, zoology
Kaufman, James Peter, theoretical physics
Kennett, James Peter, micropaleontology; paleoecology
Kerr, Theodore William, Jr., entomology
Kester, Dana R., chemical oceanography
Kirwan, Donald Frazier, nuclear physics
Knauss, John Atkinson, oceanography
Kraus, Douglas Lawrence, physical chemistry
Krause, Dale Curtiss, marine geology;
Ladas, Gerasimos E., mathematical analysis
Lal, Harbans, pharmacology; psychology
Laux, David Charles, immunology, oncology
Lawing, William Dennis, statistics
Lepper, Robert, Jr., cytology; genetics
Letcher, Stephen Vaughan, physics
Levine, Howard Allen, mathematics
Liu, Pan-Tai, applied mathematics
MacKenzie, Scott, Jr., chemistry
Marshall, Nelson, biological oceanography
Mathewson, John Angell, entomology
McGuire, John J., ornamental horticulture
Meade, Thomas Leroy, animal nutrition
Michel, Aloys Arthur, geography
Moore, Theodore Carlton, Jr., oceanography; marine geology
Mottinger, John P., cytogenetics
Mueller, Walter Carl, plant pathology
Napora, Theodore Alexander, biological oceanography
Nelson, Wilfred H., inorganic chemistry
Nixon, Scott West, ecology
Olney, Charles Edward, agricultural chemistry
Osborne, George Edwin, pharmacy

Oviatt, Candace Ann, ecology
Palmatier, Elmer Arthur, botany
Petersen, Harold, Jr, physical inorganic chemistry, information science
Pickart, Stanley Joseph, physics
Pilson, Michael Edward Quinton, marine biology, chemistry
Poggie, John Joseph, Jr, anthropology
Poularikas, Alexander D, plasma physics, electromagnetics
Pratt, David Mariotti, biological oceanography
Purvis, John L, biochemistry
Quinn, James Gerard, biochemistry, organic chemistry
Quirk, Arthur Lincoln, physics
Rand, Arthur Gorham, Jr, food science
Rhodes, Christopher Thomas, pharmacy, chemistry
Roberts, Eliot Collins, plant science
Rockett, Thomas John, mineralogy, ceramics
Rosen, William M, chemistry
Rosenfeld, Stuart Michael, physical organic chemistry
Rosie, Douglas McDonald, analytical chemistry
Rossby, Thomas, physical oceanography
Roxin, Emilio O, mathematics
Sage, Nathaniel McLean; Jr, research administration
Saila, Saul Bernhard, fish biology
Salomon, Milton, agricultural chemistry
Sastry, Akella N, biological oceanography
Schwartzman, Sol, mathematics
Shaw, Richard John, floriculture, ornamental horticulture
Shimizu, Yuzuru, natural products chemistry, pharmacognosy
Shoop, C Robert, ecology
Shutak, Vladimir Gregory, horticulture
Sieburth, John McNeill, marine microbiology
Sigurdsson, Haraldur, petrology, volcanology
Silver, Bernard Euric, biochemistry
Simpson, Kenneth L, food science
Skogley, Conrad Richard, agronomy
Smayda, Theodore John, biological oceanography
Smith, Charles Irvel, medicinal chemistry
Smith, Lewis Turner, animal science, statistics
Stern, Melvin Ernest, hydrodynamics
Stuckey, Irene Hawkins, plant physiology
Surver, William Merle, Jr, developmental genetics
Swift, Dorothy Garrison, biological oceanography
Swift, Elijah, V, biological oceanography
Traxler, Richard Warwick, bacterial physiology
Tremblay, George Charles, biochemistry
Tucker, Ruth Emma, nutrition
Turcotte, Joseph George, organic chemistry, medicinal chemistry
Tynan, Eugene Joseph, palynology
Verma, Ghasi Ram, applied mathematics
Wakefield, Robert Chester, agronomy
Watkins, Norman David, geophysics
Watts, Dennis Randolph, physical
Weisberg, Robert H, physical oceanography
Wilde, Charles Edward, Jr, biology
Winn, Howard Elliott, biological oceanography, bioacoustics
Wolke, Richard Elwood, veterinary pathology
Wood, Norris Philip, bacteriology
Wood, Richard Dawson, botany
Worthen, Leonard Robert, microbiology
Wright, John Ray, soil morphology
Yates, Vance Joseph, veterinary virology
Youngken, Heber Wilkinson, Jr, pharmacognosy

MIDDLETOWN
Childs, Donald Ray, mathematical physics
Formwalt, John McClellan, physics

NARRAGANSETT
Caroselli, Remus Francis, textile chemistry
Gonzalez, Juan Gerardo, biological oceanography
Hegre, Carman Stanford, biochemistry
Ingham, Merton Charles, physical oceanography, biological oceanography
Malcolm, Alexander Russell, biochemical genetics
McKenney, Thomas William, ichthyology, marine biology

Phelps, Donald Kenneth, biological oceanography
Riesenfeld, Peter William, nuclear science
Rogerson, Peter Freeman, analytical chemistry
Scott, Kenneth John, Jr, marine ecology
Seraichekas, Helen Rose, bacteriology, immunology
Sherman, Kenneth, biological oceanography

NEWPORT
Bordelon, Derrill Joseph, mathematics, physics
Di Pippo, Ascanio G, organic chemistry, clinical chemistry
Drummond, Andrew Jamieson, space physics
Fitzgerald, Thomas Michael, acoustics
McKillop, Lucille Mary, mathematics
Morris, George V, physical chemistry, analytical chemistry

NORTH SCITUATE
Dean, John Gilbert, chemistry

PAWTUCKET
Baldini, Mario G, hematology
Mason, Reginald G, Jr, pathology, biochemistry

PORTSMOUTH
Becken, Bradford Albert, acoustics
Bugnolo, Dimitri Spartaco, physics, electrical engineering
Fink, Don Roger, geophysics, electronics

PROVIDENCE
Adamsons, Karlis, Jr, obstetrics & gynecology
Agarwal, Kailash C, biochemistry
Agarwal, Ram Prakash, biochemical pharmacology
Ahlberg, John Harold, applied mathematics, numerical analysis
Anderson, Douglas Dorland, anthropology, archaeology
Anderson, George Albert, mathematics
Arnold, Mary B, pediatric endocrinology
Arnott, Ronald James, geology
Aronson, Stanley Maynard, medicine
Baird, James Clyde, physical chemistry
Barnes, Frederick Walter, Jr, medicine
Baum, Paul Frank, mathematics
Beyer, Robert Thomas, acoustics
Biggins, John, plant physiology, plant biochemistry
Bishopp, Frederic Edward, applied biochemistry
Boyko, Edward Raymond, physical chemistry
Bray, Philip James, solid state physics
Butcher, Fred Ray, biochemistry
Calabresi, Paul, pharmacology,
Callahan, Kenneth Paul, inorganic chemistry
Cane, David Earl, bio-organic chemistry, natural products chemistry
Carpenter, Gene Blakely, physical chemistry, crystallography
Cha, Sungman, biochemistry, pharmacology
Chapman, Kent M, biophysics
Chapple, William Massee, structural geology, tectonics
Chase, Herman Burleigh, animal genetics
Christensen, Burgess Nyles, neurophysiology
Church, George Lyle, botany
Clapp, Leallyn Burr, organic chemistry
Cobb, Sidney, epidemiology
Cole, Robert Hugh, chemical physics
Coleman, Annette Wilbois, cell biology
Coleman, John Russell, developmental biology
Constantine, Herbert Patrick, physiology, medicine
Cooper, Leon N, theoretical physics
Crabtree, Gerald Winston, biochemistry, pharmacology
Cserr, Helen F, physiology
Cutts, David, experimental particle physics
Czech, Michael Paul, biochemistry
Dafermos, Constantine M, applied mathematics
Dahlberg, Albert Edward, biochemistry
Davis, Philip, mathematics
Davis, Robert Paul, biochemistry, medicine

DeFonzo, Alfred Peter, solid state physics
Denhoff, Eric, medicine
Diamond, Israel, pathology
DiLeone, Gilbert Robert, immunology
Dolyak, Frank, physiology
Durst, Lincoln Kearney, number theory
Dyer, Hubert Jerome, botany
Eastman, Joseph Thornton, comparative anatomy, human anatomy
Ebner, Ford Francis, neurosciences
Eckelmann, Frank Donald, geology
Edwards, John Oelhaf, inorganic chemistry
Elbaum, Charles, solid state science, neurosciences
Ellis, Richard Akers, biological structure
Erikson, George Emil, anatomy
Estrup, Faiza Fawaz, biophysics, molecular biology
Estrup, Peder Jan Z, physical chemistry
Fain, John Nicholas, biochemistry
Falb, Peter L, applied mathematics
Fanger, Herbert, pathology
Farnes, Patricia, hematology
Feather, Ben Wayne, psychiatry, psychology
Federer, Herbert, mathematical analysis
Feldman, David, theoretical physics
Fenton, Paul Fredric, biochemistry
Fischer, Glenn Albert, genetics, pharmacology
Fleming, Wendell Helms, mathematics
Franco, Nicholas Benjamin, physical inorganic chemistry, pollution chemistry
Freiberger, Walter Frederick, applied mathematics
Fried, Herbert Martin, theoretical physics
Fruzzetti, Lina Maria, anthropology
Galkowski, Theodore Thaddeus, organic chemistry
Gallagher, Edward Henry, mathematics
Galletti, Pierre Marie, cardiopulmonary physiology, biomedical engineering
Gerbi, Susan Alexandra, cell biology
Gerritsen, Hendrik Jurjen, physics, chemistry
Gibbs, Julian Howard, physical chemistry
Gilbert, Barry Jay, physics
Giletti, Bruno John, geochemistry
Glicksman, Arvin Sigmund, medicine, biology
Glicksman, Maurice, experimental solid state physics, semiconductors
Glickstein, Mitchell, physiological psychology
Goldstein, Leon, physiology
Gonsalves, Neil Ignatius, genetics
Gora, Edwin Karl, theoretical physics
Goss, Richard Johnson, developmental anatomy
Greene, Edward Forbes, physical chemistry
Grenander, Ulf, applied mathematics, mathematical statistics
Griffiths, William C, clinical chemistry
Hagy, Jack Kenneth, mathematics
Hamolsky, Milton William, internal medicine
Hanley, John Joseph, organic chemistry
Harris, Bruno, mathematics
Hartmann, George Charles, mycology
Head, James William, III, stratigraphy, planetary geology
Heath, Dwight Braley, anthropology, ethnology
Heller, Gerald S, solid state physics
Heppner, Gloria Hill, immunology, virology
Hermance, John Francis, geophysics
Hess, Paul C, petrology, geochemistry
Heywood, Peter, cell biology, phycology
Hicks, George Leon, Jr, cultural anthropology, social anthropology
Hillman, Gilbert R, neuropharmacology, biochemistry
Hollos, Marida Clara, anthropology
Holowinsky, Andrew Wolodymyr, plant genetics
Hopkins, Robert West, surgery
Hornig, Donald Frederick, chemistry
Hornig, Lilli Schwenk, organic chemistry, academic administration
Howland, Richard A, mathematics, algebra
Imbrie, John, invertebrate paleontology
Kane, Kyungsik, psychiatry
Karkalas, John, biochemistry
Karlson, Karl Eugene, surgery
Keeffe, Mary Margaret, cytology
Kennedy, Hubert Collings, history of science
Keogh, Richard Neil, cell biology
King, William Travers, physical chemistry

Kolsky, Herbert, physics, mechanics
Krasner, Robert Irving, bacteriology
Laferriere, Arthur L, organic chemistry, inorganic chemistry
Landy, Arthur H, molecular biology, biochemical genetics
Lanou, Robert Eugene, Jr, physics
LaSalle, Joseph Pierre, systems theory
Lawler, Ronald George, physical organic chemistry
Lederberg, Seymour, molecular biology, genetics
Leduc, Elizabeth, cell biology
Legator, Marvin (Seymour),
Leis, Philip Edward, cultural anthropology
Levin, Frank S, nuclear physics
Lewis, Dennis Osborne, organic chemistry
Lichtman, Herbert Charles, internal medicine, clinical pathology
Lindquist, Lawrence Willard, cultural anthropology, applied anthropology
Lindsay, Robert Bruce, physics
Lubin, Jonathan Darby, mathematics
Lusk, Joan Edith, biochemistry
MacKay, Francis Patrick, organic chemistry
Marshall, Jean McElroy, physiology
Martin, Horace F, biochemistry, analytical chemistry
Marzzacco, Charles Joseph, physical chemistry
Mason, Edward Allen, chemical physics
Massey, Walter Eugene, theoretical solid state physics
Massover, William H, cell biology, biophysics
Matthews, Robley Knight, sedimentology
Maxson, Donald Robert, physics
McClure, Donald Ernest, applied mathematics
McGowan, Clement Leo, III, computer science
Mecca, Stephen Joseph, nuclear physics
Meroney, William Hyde, III, internal medicine
Miech, Ralph Patrick, biochemistry, pharmacology
Monchamp, Roch Robert, solid state chemistry
Morris, David Julian, organic chemistry, endocrinology
Morton, Thomas Hellman, organic chemistry
Murtaugh, Walter A, nuclear physics, aeronautics
Mutch, Thomas Andrew, geology
Nace, Harold Russ, organic chemistry
Nagel, Sidney Robert, solid state physics
Oh, William, pediatrics
O'Leary, Gerard Paul, Jr, microbiology, pharmacology
Orton, Colin George, medical physics
Palatt, Paul Jay, biophysics,
Palmer, Keith Henry, biochemical
Parker, Kathlyn Ann, synthetic organic chemistry
Parks, Robert Emmett, Jr, biochemistry, pharmacology
Pearson, Philip Richardson, Jr, plant ecology
Peck, Russell Allen, Jr, physics
Pipkin, Allen Compere, applied mathematics
Poole, Robert Wayne, ecology, statistics
Povar, Morris Leon, veterinary medicine, primatology
Prusch, Robert Daniel, invertebrate physiology
Pueschel, Siegfried M, pediatrics
Pysh, Eugene Stephen, physical chemistry
Quevedo, Walter Cole, Jr, animal genetics
Quinn, Alonzo Wallace, petrology
Quinn, John Joseph, theoretical physics
Randall, Henry Thomas, surgery
Rathcke, Beverly Jean, ecology
Reichart, Charles Valerian, entomology
Reinstein, Lawrence Elliot, medical physics
Rerick, Mark Newton, organic chemistry, systems analysis
Richardson, Peter Damian, biomedical engineering
Rieger, Anne Lloyd, organic chemistry
Rieger, Philip Henri, physical chemistry
Riggs, Lorrin Andrews, physiological psychology
Ripley, Robert Clarence, anatomy, cell biology
Risen, William Maurice, Jr, physical chemistry, inorganic chemistry

Robertshaw, Joseph Earl, nuclear physics, systems analysis
Rosen, Michael Ira, mathematics
Rossner, Lawrence Franklin, astrophysics
Rotman, Boris, microbiology, immunology
Rutherford, Malcolm John, petrology
Saltzman, Martin D. organic chemistry
Schwartz, Robert, pediatrics
Seidel, George Merle, solid state physics
Senft, Alfred Walter, tropical medicine, parasitology
Shapiro, Anatole Morris, physics
Shepherd, Raymond Edward, physiological chemistry
Shipp, William Stanley, biophysics
Shue, Ching-Yann, organic chemistry, medicinal chemistry
Silver, Alene Freudenheim, developmental biology, dermatology
Simeone, Fiorindo Anthony, surgery
Sirovich, Lawrence, fluid dynamics
Soled, Stuart, solid state chemistry
Sommer, Michael Anthony, II, geochemistry
Sparagen, Sanford C. nuclear medicine, internal medicine
Sterling, Anne, developmental genetics
Stern, Leo, perinatal biology; clinical pharmacology
Steward, Robert F. mathematics
Stewart, Frank Moore, mathematics
Stewart, Peter Arthur, biophysics, physiology
Stiles, Phillip John, solid state physics
Stokes, William Moore, organic chemistry
Strauss, Charles Michael, computer science, applied mathematics
Strauss, Elliott William, pathology, anatomy
Strauss, Walter A. mathematics
Su, Chau-Hsing, fluid mechanics, plasma physics
Tauc, Jan, solid state physics
Tefft, Melvin, radiotherapy
Thayer, Walter Raymond, Jr., medicine, gastroenterology
Tullis, Julia Ann, structural geology
Tullis, Terry Edson, structural geology, geophysics
Uhl, Henry Stephen Magraw, internal medicine
Vargas, Lester Lambert, cardiovascular surgery
Viehland, Larry Alan, chemical physics
Vitale, Richard Albert, mathematics, statistics
Walker, Gordon Loftis, mathematics
Wallace-Haagens, Mary Jean, endocrinology; reproductive physiology
Ward, Harold Roy, environmental chemistry
Webb, Thompson, III, palynology, climatology
Wegner, Peter, computer science
Weiner, Jerome Harris, applied mathematics
Weisz, Paul B. embryology
Weitman, Joel Kenneth, immunology, biochemistry
Westervelt, Peter Jocelyn, theoretical physics
Whitman, Philip Martin, mathematics
Widgoff, Mildred, elementary particle physics
Williams, Arthur Olney, Jr., physics
Winner, Irene Portis, cultural anthropology
Wold, Aaron, inorganic chemistry
Wright, Marion Irene, geography
Ying, See Chen, solid state physics
Young, Robert M. biochemistry, biology
Yund, Richard Allen, mineralogy
Zarcaro, Robert Michael, genetics, developmental biology
Zimmering, Stanley, genetics

RIVERSIDE
Laufer, Maurice Walter, psychiatry

RUMFORD
Brillhart, Russell Edward, pharmacology, bacteriology

WAKEFIELD
Tarzwell, Clarence Matthew, water pollution, pollution biology

WARWICK
Bailey, Garland Howard, immunology
Moos, Anthony Manuel, physical chemistry

WEST KINGSTON
Cabelli, Victor Jack, microbiology, environmental health
Eisler, Ronald, marine biology; pollution biology
Hodgkiss, William Searles, chemistry
Larmie, Walter Esmond, floriculture
Levin, Morris A. microbiology
Prager, Jan Clement, marine microbiology
Yevich, Paul Peter, histopathology

WEST WARWICK
Perry, Edward Mahlon, organic chemistry

WOOD RIVER JUNCTION
Pande, Kailash Chandra, organometallic chemistry; polymer chemistry

SOUTH CAROLINA

ABBEVILLE
Burgess, Charles H. geology

AIKEN
Ahlfeld, Charles Edward, reactor physics
Bailey, Charles Edward, physical chemistry
Baumann, Norman Paul, nuclear physics
Baumann, Elizabeth Wilson, analytical chemistry
Benjamin, Richard Walter, nuclear physics
Bibler, Ned Eugene, radiation chemistry
Bowman, Wilfred William, chemistry
Brisbin, I Lehr, Jr., zoology; ecology
Church, John Phillips, reactor physics
Clark, Hugh Kidder, reactor physics
Corey, John Charles, soil physics
Crandall, John Lou, chemistry
Crawford, Todd V. meteorology
Dessauer, Gerhard, nuclear physics
Driggers, Frank Edgar, reactor physics
Durant, W S. reactor physics
Forstner, James Lee, nucleonics
Garten, Charles Thomas, Jr., ecology
Gibbons, J Whitfield, population ecology, herpetology
Giesy, John Paul, Jr., limnology
Graves, William Jean, reactor physics
Gregory, Michael Vladimir, reactor physics
Groh, Harold John, physical chemistry
Hale, William Henry, Jr., inorganic chemistry
Hawkins, Richard Horace, soil
Hayes, David Wayne, oceanography, mineralogy
Hennelly, Edward Joseph, physical chemistry
Hill, Arthur Joseph, Jr., organic chemistry
Hochel, Robert Charles, nuclear chemistry, environmental systems & technology
Holcomb, Herman Perry, analytical chemistry; radiochemistry
Honeck, Henry Charles, reactor physics
Horton, James Henry, Jr., soil chemistry, applied mathematics
Hyder, Monte Lee, nuclear chemistry
Jacober, William John, chemistry
Johnson, David Russell, physical chemistry
Kauffman, George Emmett Clarence, health physics, physics
Kern, Clifford Dalton, meteorology
Long, Paul Eastwood, Jr., meteorology
Longtin, Bruce, thermodynamics
Marine, Ira Wendell, geology; hydrology
McCrosson, F Joseph, reactor physics
McDonell, William Robert, radiation chemistry
McLeod, Kenneth William, plant ecology
Milham, Robert Carr, environmental chemistry
Montenyohl, Victor Irl, chemistry
Mosley, Wilbur Clanton, Jr., solid state physics, materials science
Murbach, Earl Wesley, inorganic chemistry
Nelson, David Herman, vertebrate ecology; aquatic ecology
O'Neill, George Francis, nuclear physics
Orebaugh, Errol Glen, physical chemistry
Owen, John Harding, inorganic chemistry; physical chemistry
Parks, Paul Blair, experimental nuclear physics, reactor physics
Perkins, William Clopton, nuclear chemistry
Peterson, Stephen Frank, analytical chemistry
Phillips, Robert Gibson, mathematics
Pilinger, William Lewis, applied physics
Plodinec, Matthew John, physical chemistry
Price, Vanceton, Jr., geology
Pryor, Ralph J, nuclear physics
Randall, Duncan, physical chemistry
Reinig, William Charles, health physics
Roggenkamp, Paul Leonard, reactor physics, nuclear engineering
Sanders, Samuel Marshall, Jr., health physics, radiobiology
Sharitz, Rebecca Reyburn, ecology, botany
Smith, Paul Kent, physical chemistry
Spooner, John D, zoology; entomology
Stone, John Austin, nuclear chemistry
Suich, John Edward, information science
Thompson, Gary Haughton, physical chemistry, inorganic chemistry
Topp, Stephen V. reactor physics
Wallace, Richard Maitha, physical chemistry
Wheat, John Allen, nuclear chemistry
Wiley, John Robert, nuclear chemistry
Wilson, John Neville, physics, resource management

ANDERSON
Cogswell, George Wallace, organic chemistry

BLACKVILLE
Blackmon, Cyril Wells, plant pathology
Cohoon, Daniel Fred, plant pathology
Granberry, Darbie Merwin, plant breeding
Hughes, Morris Burdette, genetics

CAMDEN
Davis, Merton Louis, physical chemistry
Edwards, John C. analytical chemistry, environmental science
Fulda, Myron Oscar, analytical chemistry

CATAWBA
Yan, Johnson Faa, physical chemistry

CENTRAL
LaBar, Martin, population biology
Wilcox, Floyd Lewis, science education

CHARLESTON
Allen, Robert Carter, pathology
Anderson, Norman Gulack, physiology
Anderson, Wallace Ervin, physics
Andrus, Charles Frederick, horticulture
Ariail-Chaves, Mario Passalaqua, immunobiology
Artz, Curtis Price, surgery
Baggett, Billy, biochemistry
Bagwell, Ervin Eugene, pharmacology
Ballentine, Alva Ray, physical organic chemistry; synthetic organic chemistry
Barrington, Burness Austin, Jr., anatomy, mammalogy
Beard, Percy Morris, Jr., experimental nuclear physics
Bender, Roger Stillman, radiation dosimetry
Bennmann, Joseph David, biopharmaceutics
Berlinghieri, Joel Carl, acoustics, optics
Bills, Alan Morris, carbohydrate chemistry, organic chemistry
Blackburn, John Gill, physiology
Blankenship, James William, pharmacology
Boltjes, Ben Harold, medical microbiology
Bradham, Gilbert Bowman, surgery
Brostoff, Steven Warren, neurochemistry
Buhler, John Embich, dentistry
Burdash, Nicholas Michael, immunology, bacteriology
Burrell, Victor Gregory, Jr., marine sciences
Buse, John Frederick, internal medicine
Buse, Maria F Gordon, medicine
Calder, Dale Ralph, marine science
Cannon, Albert, clinical pathology
Carroll, Robert Leon, physics, mathematics
Chesnut, Clarence, Jr., biology, chemistry
Cleckley, James Jennings, medicine
Colwell, John Amory, internal medicine, physiology
Comer, Stephen Daniel, mathematical logic, algebra
Corcoran, John W, pedodontics
Corley, John Bryson, family medicine
Creighton, Charlie Scattergood, entomology
Crosby, Emory Spear, forest pathology, plant physiology
Cunningham, Earlene Brown, biochemistry, organic chemistry
Curry, Hiram Benjamin, neurology
Daniel, Herman Burch, pharmacology
Debacker, Hilda Spodheim, neuroanatomy, microscopic anatomy
Dodds, Alvin Franklin, biochemistry, pharmaceutical chemistry
Doig, Marion Tilton, III, biochemistry
Donley, Clark Stephen, biomedical engineering
Dougherty, William J. cell biology, histology
Dukes, Philip Duskin, plant pathology
Eldridge, John Charles, medicine
Eurenius, Karl, hematology
Ezell, William Bruce, Jr. medical entomology; environmental health
Farrar, William Edmund, Jr., infectious disease, microbiology
Fery, Richard Lee, genetics, plant breeding
Fingar, Walter Wiggs, dentistry
Fish, Wayne William, biochemistry
Fitts, Charles Thomas, surgery
Fitzharris, Timothy Patrick, cell biology, developmental biology
Frayser, Katherine Regina, physiology
Fredericks, Christopher M. physiology
Freeman, Harry W. vertebrate zoology
Fudenberg, H Hugh, hematology, immunology
Gadsden, Richard Hamilton, biochemistry
Gaffney, Thomas Edward, pharmacology
Gale, Glen Roy, pharmacology
Garlick, Norman Lee, veterinary medicine, comparative pathology
Geraghty, Ronald Mills, pathology
Gibson, Gerald W. organic chemistry
Golod, William Hersh, pharmacology
Goode, Julia Pratt, analytical chemistry
Graber, Charles David, bacteriology; immunology
Green, Rupert L. cardiovascular physiology
Greenfield, Seymour, neurochemistry
Gross, Paul, pathology
Groves, William Ernest, biochemistry, computer sciences
Haefner, Richard Charles, geology
Halsey, John Frederick, physical biochemistry
Halushka, Perry Victor, pharmacology
Hanson, Alvin Walter, physics
Hargest, Thomas Sewell, biomedical engineering
Harrison, Julian R, III, herpetology
Harvin, James Shand, plastic surgery
Haskill, John Stephen, immunopathology
Heere, Leonard J., biochemistry
Hempling, Harold George, physiology
Hennigar, Gordon Ross, Jr., pathology
Hester, Lawrence Lamar, Jr., obstetrics & gynecology
Higerd, Thomas Braden, microbial physiology
Hogan, Edward L., neurology
Hohn, Arno R., pediatrics, cardiology
Horres, Alan Dixon, physiology
Hummers, William Strong, Jr., inorganic chemistry
Hutchinson, Lee Pressley, mathematics
Hynes, John Barry, organic chemistry
Johnson, Henry Stanley, Jr., exploration geology, economic geology
Jones, Alfred, plant genetics
Jonsson, Haldor Turner, Jr., biochemistry
Jumper, Charles Frederick, physical chemistry
Kane, John Joseph, radiology
Karam, Jim Daniel, biochemistry
Katz, Sidney, physiology
Keeler, Martin Harvey, medicine
Keil, Julian Eugene, epidemiology

Kinard, Fredrick William, physiology, physiological chemistry
Kliewer, John Wallace, medical entomology
Knapp, Daniel Roger, pharmacology
Knisely, William Hagerman, anatomy
Krall, Albert Raymond, biochemistry
Ladd, Anthony Thornton, internal medicine
Lam, Chan Fun, biomedical engineering
Larisey, Mary Maxine, taxonomy, plant anatomy
Lawson, Benjamin F, oral medicine, periodontology
Lee, William Hall, Jr, thoracic surgery, cardiovascular surgery
Legerton, Clarence W, Jr, gastroenterology
Leland, Thomas Mikell, physiology, endocrinology
Leopold, Robert Summers, organic chemistry
Levine, Jon Howard, endocrinology
Likes, Carl James, physical chemistry
Lindenmayer, George Earl, biochemical pharmacology
Loadholt, Claude Boyd, biostatistics
Lockard, Isabel, anatomy
Ludwig, Theodore Frederick, prosthodontics
Lurie, Dan, biostatistics
Margolius, Harry Stephen, clinical pharmacology
Mathur, Rajesh Swarup, reproductive endocrinology, steroid chemistry
McCord, William Mellen, biochemistry
McCurdy, Layton, psychiatry
McDonald, John Kennely, biochemistry, endocrinology
McGill, Julian Edward, analytical chemistry
McKee, Kelly Tilson, internal medicine, allergy
McLean, Darrell Marshall, plant pathology
McManus, Joseph Forde Anthony, pathology
Mellette, Russell Ramsey, Jr, child psychiatry
Merrill, Glen Kenton, geology, paleontology
Metcalf, Isaac Stevens Halstead, gross anatomy, comparative anatomy
Miller, Millage Clinton, III, biostatistics
Miller, Ronald Lee, biochemistry
Mithoefer, John Caldwell, medicine
Moncrief, John A, surgery
Moseley, Vince, internal medicine
Newberry, William Marcus, internal medicine
Newlin, Gordon Ermel, microbiology, virology
Newman, Walter Hayes, pharmacology
Nuckles, Douglas Boyd, dentistry
Odom, Homer Clyde, Jr, organic chemistry
Ondo, Jerome G, neuroendocrinology, cancer
Othersen, Henry Biemann, Jr, pediatric surgery
Pannell, Lolita, bacteriology
Patton, Nancy Jane, neurosciences
Pennington, Raymond Carroll, anatomy
Perot, Phanor L, Jr, neurosurgery
Pittman, Fred Estes, gastroenterology, medical research
Prakash, Chandra, immunology, reproductive biology
Pratt-Thomas, Harold Rawling, pathology
Priest, David Gerard, biochemistry, enzymology
Pritchard, John B, physiology
Privitera, Philip Joseph, pharmacology
Pruett, Jack Kenneth, pharmacology
Putney, Blake Fuqua, pharmacy
Putney, Floyd Johnson, medicine, surgery
Redding, Joseph Stafford, anesthesiology
Reed, John K, entomology
Reves, George Everett, mathematics
Richardson, James Albert, pharmacology
Riggs, Benjamin C, psychiatry, psychoanalysis
Rittenbury, Max Sanford, surgery, biochemistry
Robbins, Marion LeRon, horticulture, plant breeding
Robinson, John Frank, economic entomology
Robson, John Robert Keith, nutrition, public health
Roof, Betty Sams, internal medicine
Ross, Joseph C, internal medicine
Rubin, Mitchell Irving, pediatrics, nephrology
Runey, Gerald Luther, physiology
Sabin, Albert Bruce, infectious diseases, virology
Sandifer, Paul Alan, marine zoology
Sandifer, Samuel Hope, medicine
Sawyer, Roy Thomas, invertebrate zoology
Schmidt, Gilbert Carl, biochemistry
Schuman, Stanley Harold, epidemiology, public health
Schwabe, Christian, biological chemistry
Sharry, John Joseph, dentistry
Sherer, Glenn Keith, developmental biology
Sherer, James Pressly, organic chemistry
Silverthorn, Saidee Unglaub, invertebrate physiology
Simson, Jo Anne V, cell biology, cytochemistry
Sitterly, Wayne R, plant pathology
Smiley, James Watson, physiology
Spicer, Samuel Sherman, Jr, pathology
Stillway, Lewis William, biochemistry
Suddick, Richard Phillips, physiology
Sumner, Edward D, pharmacy
Sutton, Charles Samuel, mathematics
Swanson, Arnold Arthur, biochemistry, ophthalmology
Taber, Elsie, embryology, reproductive physiology
Towell, Edward Emerson, chemistry
Vallotton, William Wise, medicine
Waldrep, Alfred Carson, Jr, oral surgery, dentistry
Walter, Wilbert George, pharmacognosy, medicinal chemistry
Wang, An-Chuan, immunogenetics
Wann, Elbert Van, genetics, plant breeding
Weidner, Michael George, Jr, surgery
Wells, John Arthur, plant chemistry
Westphal, Milton C, Jr, medicine, pediatrics
Wheeler, Darrell Deane, physiology
White, Edward, immunology, endodontics
Wilkinson, Joseph Ridley, analytical chemistry
Wise, William Curtis, physiology, biophysics
Wohltmann, Hulda Justine, pediatrics
Worthington, Ward Curtis, Jr, anatomy
Wright, Allen Kent, biophysics, biometry
Young, Franklin Alden, Jr, materials science
Young, Gilbert Flowers, Jr, neurology, pediatrics
Zemp, John Workman, biochemistry

CHESTER
Goldstein, Herman Bernard, textile chemistry, paper chemistry

CLEMSON
Adkins, Theodore Roosevelt, Jr, entomology
Allen, Joe Frank, inorganic chemistry, science education
Allen, Robert Max, forestry
Ashworth, Ralph P, botany
Bailey, Roy Horton, Jr, physical organic chemistry
Barker, Robert Henry, textile chemistry
Barnett, Bobby Dale, poultry nutrition
Barnett, Ortus Webb, Jr, plant virology, plant pathology
Baxter, Ann Webster, microbial physiology
Baxter, Luther Willis, Jr, plant pathology
Beinhart, Ernest George, Jr, plant physiology, agronomy
Birkhead, Paul Kenneth, geology, invertebrate paleontology
Bishop, Muriel Boyd, biochemistry, organic chemistry
Boland, Willard Robert, numerical analysis
Bookmyer, Beverly Brandon, astronomy
Boone, Merritt Anderson, reproductive physiology, poultry physiology
Bose, Anil Kumar, pure mathematics
Brawley, Joel Vincent, Jr, algebra
Bregger, John Taylor, horticulture
Brown, Farrell Blenn, structural chemistry
Burt, Philip Barnes, theoretical physics
Byrd, Wilbert Preston, statistics, genetics
Camper, Nyal Dwight, plant physiology, plant biochemistry
Carter, George Emmitt, Jr, plant physiology
Chaplin, Robert Lee, Jr, physics
Cool, Bingham Mercur, forestry
Craddock, Garnet Roy, agronomy
Dick, John Walter, avian pathology
Dickey, Joseph Freedman, reproductive physiology, electron microscopy
Edwards, Robert Lee, animal nutrition
Epps, William Monroe, plant pathology
Fanning, James Collier, inorganic chemistry
Fennell, Robert E, mathematics
Fox, Richard Charles, forest entomology
Fulton, John David, mathematics
Garrison, Olen Branford, horticulture
Gauthreaux, Sidney Anthony, Jr, vertebrate zoology, animal behavior
Gibson, Pryce Byrd, plant breeding
Gillingham, J T, soils, plant nutrition
Godley, Willie Cecil, genetics
Graben, Henry Willingham, theoretical physics
Graham, William Doyce, Jr, plant genetics, plant breeding
Griffin, Villard Stuart, Jr, geology
Halfacre, Robert Gordon, horticulture
Hare, William Ray, Jr, mathematics
Harlow, Richard Fessenden, wildlife ecology, plant ecology
Haselton, George Montgomery, geology, geomorphology
Hatcher, Robert Dean, Jr, structural geology, geochemistry
Haun, Joseph Rhodes, plant physiology
Hayasaka, Steven S, marine microbiology
Hays, Ruth Lanier, physiology, biochemistry
Hays, Sidney Brooks, entomology, animal physiology
Helms, Carl Wilbert, zoology, physiological ecology
Henningson, Robert Walter, dairy science
Henricks, Donald Maurice, biochemistry, endocrinology
Hind, Alfred Thomas, Jr, mathematics
Hobson, James Harvey, physical chemistry
Holder, David Parker, poultry nutrition
Holleman, Kendrick Alfred, poultry science
Holloway, Rodney Leon, entomology
Holmes, Paul Thayer, statistics
Hudson, Larry Wilson, animal science
Huffman, John William, Jr, organic chemistry
Hughes, Buddy Lee, poultry science, reproductive physiology
Hurst, Victor, dairy husbandry
Jacobus, Otha John, organic chemistry
Janzen, John, dairy science
Jen, Joseph Jwu-Shan, food science, biochemistry
Johnston, Walter Edward, statistics
Jones, Champ McMillian, plant breeding
Jones, Jack Edenfield, poultry science
Jones, Jess Willard, plant breeding
Jones, Ulysses Simpson, Jr, soil fertility
Jutras, Michel Wilfrid, agronomy
Keller, Frederick Jacob, experimental solid state physics
Kenelly, John Willis, Jr, mathematics
King, Edwin Wallace, entomology
King, Willis Alonzo, dairying
Kingsland, Graydon Chapman, plant pathology
Kirkwood, Charles Edward, Jr, computer science
LaFleur, Kermit Stillman, textile chemistry, soil science
Lane, Carl Leaton, forest soils
Laskar, Amulya Lal, physics
Laskar, Renu Chakravarti, mathematics
LaTorre, Donald Rutledge, mathematics
Lazar, James Tarlton, Jr, dairy science
Lewis, Stephen Albert, plant nematology
Lindstrom, Frederick John, analytical chemistry
Ling, Robert Francis, statistics
Loyacano, Harold Anthony, ichthyology, fisheries management
Luedeman, John Keith, algebra
Marullo, Nicasio Philip, organic chemistry
Maxwell, James Donald, plant breeding
McDowell, Harding Keith, chemical physics
McGregor, William Henry Davis, plant physiology, forestry
McKelvey, John Philip, solid state physics
Miller, Donald Piguet, chemical physics, crystallography
Miller, Robert Walker, Jr, plant pathology
Min, Hong Shik, cell physiology, comparative physiology
Mitchell, Jack Harris, Jr, biochemistry
Nettles, William Carl, economic entomology
Noblet, Raymond, insect physiology, invertebrate pathology
O'Dell, Wayne Talmage, dairy science
Ogle, Wayne LeRoy, horticulture
Page, Norwood Rufus, soil chemistry, plant nutrition
Park, George Bennet, analytical chemistry, electroanalytical chemistry
Parks, Clyde Leonard, soil fertility, plant nutrition
Paynter, Malcolm James Benjamin, microbiology, biochemistry
Peele, Thomas Christopher, soil science
Pinder, Albert Reginald, organic chemistry
Porter, John J, physical organic chemistry
Powell, Gary Lee, biochemistry
Proctor, Thomas Gilmer, mathematics
Ray, John Robert, physics
Reneke, James Allen, mathematical analysis
Roberts, Carleton W, organic chemistry
Ruckle, William Henry, mathematics
Rupert, Earlene Atchison, plant genetics, taxonomy
Russell, Charles Bradley, mathematics, statistics
Savitsky, George Boris, physical chemistry
Schoenike, Roland Ernest, forest genetics
Senn, Taze Leonard, horticulture
Shain, William Arthur, forestry
Shepard, Buford Merle, entomology, ecology
Sherrill, Max Douglas, solid state physics
Shively, Jessup MacLean, biochemistry, microbiology
Simon, Frederick Tyler, physical chemistry, color science
Sims, Ernest Theodore, Jr, horticulture, plant physiology
Skelley, George Calvin, Jr, animal science
Skelton, Bobby Joe, horticulture, plant physiology
Skelton, Thomas Eugene, entomology
Skove, Malcolm John, solid state physics
Smith, Bill Ross, soil science
Snipes, David Strange, geology
Sobczyk, Andrew, mathematics
Spencer, Harold Garth, physical chemistry
Stearns, Edwin Ira, physical chemistry
Steiner, Pinckney Alston, solid state physics, magnetic resonance
Stembridge, George Eugene, pomology
Stillwell, Ephraim Posey, Jr, solid state physics
Sturch, Conrad Ray, astronomy
Talbot, Gerald Byron, fish biology
Thompson, Carl Eugene, animal breeding, animal genetics
Tombes, Averett Snead, invertebrate physiology
Turk, Donald Earle, nutrition, biochemistry
Turnipseed, Samuel Guy, entomology
Ulbrich, Carlton Wilbur, low temperature physics
Umphlett, Clyde Jefferson, botany
Van Lear, David Hyde, forestry, soils
Vogel, Henry Elliott, solid state physics
Von Rosenberg, Joseph Leslie, Jr, organic chemistry
Waide, Jack Boid, ecology
Walker, Richard Francis, endocrinology, physiology
Whitney, John Barry, Jr, plant physiology
Wiley, William Henry, genetics, agriculture
Witcher, Wesley, plant pathology
Yardley, Darrell Gene, population genetics
Zehr, Eldon Irvin, plant pathology
Zimmerman, James Kenneth, biochemistry

CLINTON
Carter, Kenneth Nolon, organic chemistry
Womble, Eugene Wilson, mathematics
Yarborough, William Walter, Jr, plasma physics

COLUMBIA
Abbott, William Harold, micropaleontology, oceanography
Abel, Francis Lee, cardiovascular physiology
Adams, James Norman, microbiology, genetics
Adams Smith, William Nelson, medicine, neuroendocrinology
Allen, Donald Orrie, pharmacology
Amma, Elmer Louis, bioinorganic chemistry, biophysical chemistry
Au, Chi-Kwan, theoretical physics
Augustine, James Robert, neuroanatomy
Avignone, Frank Titus, III, nuclear physics
Ayres, William Stanley, anthropology
Batson, Wade Thomas, botany

SOUTH CAROLINA

Bauguess, Carl Thomas, Jr., pharmaceutics
Beamer, Robert Lewis, medicinal chemistry
Beck, Ronald Richard, physiology, endocrinology
Berryhill, Virginia Farmer, cell physiology, chemical embryology
Bideman, Terry Frank, analytical chemistry, environmental chemistry
Bierer, Bert Worman, poultry pathology
Bly, Robert Stewart, physical organic chemistry
Bonner, Oscar Davis, physical chemistry
Bouknight, Joseph Ward, inorganic chemistry
Bryson, Thomas Allan, organic chemistry
Buchanan, W C, geography
Bushong, Allen David, geography
Cargill, Robert Lee, Jr., organic chemistry
Carpenter, John Richard, geochemistry
Caruccio, Frank Thomas, groundwater geology, environmental geology
Cathey, LeConte, radiation physics
Childers, Richard Lee, physics
Cohen, Arthur David, petrology, palynology
Cole, Benjamin Theodore, physiology
Colquhoun, Donald John, geology
Coull, Bruce Charles, ecology, biological oceanography
Cowley, Gerald Taylor, mycology
Darden, Colgate W, III, nuclear physics
Davis, Harry Willard, organic chemistry
Dawson, Wallace Douglas, Jr., genetics, evolution
Dean, John Mark, aquatic ecology
DiSalvo, Arthur F, microbiology
Dunlap, Robert Bruce, enzymology
Durig, James Robert, physical chemistry
Edge, Ronald (Dovaston), nuclear physics, solid state physics
Ehrlich, Robert E, geology, sedimentology
Ely, Berten E, III, microbial genetics
Ermutlu, Ilhan M, psychiatry
Faust, John William, Jr., crystallography, surface physics
Ferguson, Leland Greer, anthropology
Ferm, John Charles, geology
Fincher, Julian H, physical pharmacy
Fisher, Ronald Richard, biochemistry
Freeman, John J, pharmacology
Gilkerson, William Richard, physical chemistry
Gimarc, Benjamin M, theoretical chemistry
Goode, Scott Roy, analytical chemistry
Graef, Philip Edwin, biology
Griffith, Elizabeth Ann Hall, organometallic chemistry, physical chemistry
Haines, Daniel Webster, acoustics, engineering mechanics
Harley, John Barker, biology
Hayes, Miles O, geology, oceanography
Hedberg, Marguerite Zeigel, physiological ecology
Heider, Karl Gustav, anthropology
Herr, John Mervin, Jr., plant embryology
Hodge, Bartow, information science
Husband, David Dwight, botany
Jacobs, Jacqueline E, environmental biology
Jeffrey, Norris Boddie, fisheries management
Jones, Edwin Rudolph, Jr., solid state physics
Kanes, William H, stratigraphy, sedimentology
Kinard, Frank Elird, nuclear science
Kistler, Wilson Stephen, Jr., biochemistry, reproductive biology
Knight, James Milton, theoretical physics
Kosh, Joseph William, neuropharmacology
Kovacik, Charles Frank, geography
Lauter, Felix H, parasitology
Lawrence, David Reed, geology, invertebrate paleontology
Lease, Elmer John, biochemistry
Lee, Talmage Hoyle, mathematics
Lerner, Edward Clarence, theoretical physics
Lewis, Robert Frank, public health
Liu, Edwin Chiap Henn, biochemistry
Locke, Charles Stephen, statistics
Lovingood, Paul Evans, Jr., geography, geology
Lytle, Raymond Alfred, mathematics
Markham, Thomas Lowell, algebra
Matthies, Karl Heinrich, mathematics
McNulty, George Frank, mathematical logic
Mercer, Edward Everett, physical inorganic chemistry

Meredith, Howard Voas, child growth, morphology
Mingin, Julian Vincent, political geography, geography of Europe
Mitchell, Lisle Series, urban geography, economic geography
Morrison, Robert William, pharmacology, pharmacy
Nairn, Alan Eben Mackenzie, geology
Nicol, Charles Albert, mathematics
Norris, Eugene Michael, mathematics
Odom, Jerome David, inorganic chemistry
Padgett, William Jowayne, statistics, mathematics
Peterson, Paul Erik, organic chemistry
Philp, Robert Herron, Jr., analytical chemistry
Poole, Charles Patton, Jr., physics
Poteat, William Louis, anatomy
Freedom, Barry Mason, nuclear physics
Rathbun, Ted Allan, physical anthropology
Reddick-Mitchum, Rhoda Anne, medical microbiology
Reger, Daniel Lewis, organometallic chemistry
Robinson, Robert Earl, organic chemistry
Rodesiler, Paul Frederick, inorganic chemistry
Rohlfing, Duane L, biochemistry, evolution
Russell, Edwin Roberts, chemistry
Sadik, Farid, pharmaceutics
Safko, John Loren, physics
Sargent, Roger Gary, parasitology
Sas, Anthony, geography
Sawyer, Roger Holmes, developmental science
Schuette, Oswald Francis Jr., physics
Schulze, Chris Carl, chemistry, management
Secor, Donald Terry, Jr., geophysics, geology
Silvernail, Richard George, geography
Singh, Ragbhir, agronomy, plant physiology
Snoke, Arthur Wilmot, geology, petrology
Sorell, Henry P, organic chemistry
Stancyk, Stephen Edward, invertebrate zoology, marine ecology
Stephenson, Robert Lloyd, anthropology, archaeology
Stephenson, Robert Moffatt, Jr., mathematics
Stevenson, L Harold, microbiology
Still, Charles Neal, neurology
Stoll, Manfred, mathematical analysis
Strebe, David Diedrich, mathematics
Sutherland, Donald Ralph, anthropology
Taylor, Robert Lee, mathematics, statistics
Teague, Peyton Clark, organic chemistry
Thompson, Eric Fontele, Jr., zoology
Venberg, Frank John, marine biology, physiological ecology
Vernberg, Winona B, environmental physiology
Vodkin, Michael Harold, genetics
Wahab, James Hatton, mathematics
Wall, Charles Robert, mathematics
Watabe, Norimitsu, electron microscopy
Webb, James Woodrow, wildlife management
Weymouth, Richard J, anatomy, endocrinology
Winner, Larry Thomas, chemistry
Wuthier, Roy Edward, biochemistry
Wynn, James Elkanah, medicinal chemistry, analytical chemistry
Yang, Jeong Sheng, topology
Zingmark, Richard G, algology

CONWAY
Freeman, Carl Jackson, Jr., invertebrate zoology

CROSS HILL
Stump, Alexander Bell, protozoology

DENMARK
Guram, Malkiat Singh, zoology; botany
Rawalay, Surjan Singh, chemistry; biochemistry

DUE WEST
Knight, James Aldon, inorganic chemistry
Parkinson, Gilbert Gordon, Jr., organic chemistry

Schmelpfeng, Clarence William, organic chemistry

DUNCAN
Patterson, William Alexander, physical chemistry
Pike, LeRoy, analytical chemistry

ELGIN
Hardwicke, James Ernest, Jr., organic chemistry

FLORENCE
Boyce, John Shaw, Jr., forest pathology
Breazeale, William Horace, physical chemistry
Campbell, Robert Benoni, agronomy
Durant, John Alexander, III, entomology
Johnson, Albert Wayne, entomology
LaPrade, Jesse Cobb, plant pathology
Leggett, Joseph Edwin, entomology
Manwiller, Alfred, plant breeding
Moore, Raymond F, Jr., entomology
Pecka, James Thomas, industrial chemistry
Pitner, John Bruce, soils-agronomy
Reicosky, Donald Charles, agronomy, soil physics
Welbourne, Frank Fitzhugh, botany, zoology

FT MILL
Martin, William Harry, organic chemistry, textiles
Mitchell, Hugh Bertron, radiobiology, aerospace medicine

GAFFNEY
Carpenter, Dwight William, physics
Mehra, Krishna Nandan, parasitology
Pribble, Mary Jo, inorganic chemistry

GEORGETOWN
Wood, Gene Wayne, wildlife ecology

GREENVILLE
Ashby, Peter Jawad, chemical physics, mathematics
Berger, Richard S, organic polymer chemistry
Bien, Paul Beh Nien, physical chemistry
Blevins, Maurice Everett, applied physics, computer sciences
Brantley, William Henry, nuclear physics
Buurman, Clarence Harold, industrial organic chemistry
Clanton, Donald Henry, mathematics
Cope, James Francis, organic polymer chemistry, organometallic polymer
Desai, Vinodrai Ranchhodji, organic polymer chemistry
Dix, James Seward, organic chemistry
Fairbanks, Gilbert Wayne, physiology
Fray, Robert Dutton, mathematics
Goodman, Henry Gaines, synthetic organic chemistry
Hammett, Michael E, mathematics
Harris, William Charles, physical chemistry, spectroscopy
Henson, Joseph Lawrence, entomology, microbiology
Herdklotz, Richard James, chemistry
Hopkins, Allen John, polymer chemistry
Kane-Maguire, Noel Andrew Patrick, inorganic chemistry
Kelly, Robert Withers, zoology, ecology
Kerstetter, Rex E, plant physiology
Knight, Lon Bishop, Jr., physical chemistry
Kubler, Donald Gene, organic chemistry
Ludvigsen, Bernhard (Thoger) Frants (Josef), biochemistry, clinical chemistry
Moses, Ray Napoleon, Jr., astronomy
Motill, Ronald Allen, theoretical physics, cosmology
Nancey, Thomas Ray, physical chemistry, computer science
Piciot, William P, natural science
Poole, John Terry, mathematics
Rodgers, Charles Leland, botany
Scarpellino, Joseph, organic chemistry, polymer chemistry
Scruggs, Jack G, organic chemistry
Sheehan, William C, organic chemistry
Smith, Gary Charles, ecology, vertebrate biology
Soldano, Benny A, physical chemistry
Straton, Lewis Palmer, biochemistry

Taylor, Peter Anthony, physical chemistry, textile engineering
Varin, Roger Robert, physical chemistry
Waites, Alan C, mathematics
Williams, Alan Philip, physical chemistry
Womer, Walter Dale, organic chemistry
Wylie, Clarence Raymond, Jr., geometry

GREENWOOD
Alston, Jimmy Albert, plant breeding
Borick, Paul Michael, bacteriology
Dodson, Norman Elmer, mathematics, science education
Marek, Jerry William, biochemistry
Stewart, Shelton E, botany, zoology
Taylor, Harold Allison, Jr., human genetics, biochemistry
Tolbert, Thomas Warren, chemistry, molecular spectroscopy
Vereen, Larry Edwin, microbiology
Wilson, Jerry Dick, science writing, physics

GREER
Farrar, Grover Louis, organic chemistry
Heberger, John M, physical chemistry
Hopper, Michael James, spectroscopy
Kimmel, Robert Michael, materials engineering, polymer physics

HARTSVILLE
Heald, Alfred Matson, chemistry
Miller, Carol Raymond, plant pathology
Neely, James Winston, organic chemistry
Ropp, Gus Anderson, organic chemistry
Smith, William Edmond, pulp chemistry
Swallow, Richard Louis, zoology, biology
Wise, John Thomas, chemistry, paper

HILTON HEAD ISLAND
Roberts, Rufus Winston, ophthalmology

LYMAN
Holsten, John Robert, organic chemistry

MAULDIN
Knowles, Cecil Martin, organic chemistry
Worsham, Walter Castine, textile chemistry

MT PLEASANT
Force, Carlton Gregory, colloid chemistry
Stryker, Lynden Joel, colloid chemistry, surface chemistry
Witherspoon, Samuel McBridge, anesthesiology

MOUNTVILLE
Badger, Blanche Crisp, mathematics

NEWBERRY
Baumgartner, Luther Leroy, biochemistry, ecology
Beam, Charles Fitzhugh Jr, organic chemistry, polymer chemistry
Fitzpatrick, John Michael, chemical physics
Huthmance, Edward Dennis, Jr., algebra, computer science
Jeremias, Charles George, organic chemistry, inorganic chemistry
Park, Conrad B, chemistry

NORTH CHARLESTON
Adams, Daniel Otis, chemistry, paper technology
Bailey, Carl Williams, III, wood technology
Ball, Frank Jervey, organic chemistry
Doughty, Joseph Bayne, paper chemistry
Falkehag, S Ingemar, organic chemistry, surface chemistry

ORANGEBURG
Abernathy, Robert O, chemistry
Daniel, Victor Wayne, mathematics
Hunter, George William, biochemistry, agricultural chemistry
McAlpin, Cesaria Eugenio, botany

McAlpin, John Harris, mathematics
Payne, James Edward, experimental solid state physics
Payne, Linda Lawson, solid state physics
Powell, Harold, speech pathology, audiology
Roache, Lewie Calvin, zoology
Sandhu, Shingara Singh, soil chemistry
Vroman, Hugh Egmont, biochemistry

PENDLETON
Wood, James C, Jr, solid state physics

ROCK HILL
Brook, Robert B, topology
Davis, Joe Bill, analytical chemistry
Davis, Luckett Vanderford, invertebrate zoology, marine ecology
Ebert, Patricia Dorothy, behavioral genetics
Freeman, John Alderman, zoology
Gamble, Robert Oscar, mathematics
Gustafson, Ralph Alan, mycology
Hodges, Billy Gene, mathematics
Houk, Richard Duncan, botany, taxonomy
Huff, Charles William mathematics
King, Elizabeth Norfleet, cell physiology
King, Joseph Clarence, textile chemistry
Konlande, James Edward, nutrition
Laughlin, Kenneth Clifford, textiles
Olson, John Bernard, ecology, genetics
Sanderfer, Paul Otis, organic chemistry
Tutwiler, Frank Bryan, physical chemistry, organic chemistry
Vail, Charles Brooks, physical chemistry

SENECA
Dillon, John Henry, polymer physics

SPARTANBURG
Bell, Curtis Porter, mathematics
Cavin, William Pinckney, organic chemistry
Cromer, Jerry Haltiwanger, developmental biology, physiology
Dobbs, Harry Donald, biology
Dusenbury, Joseph Hooker, physical chemistry
Earl, Charles Riley, polymer chemistry, physical chemistry
Eastes, Frank Elisha, organic chemistry
Farmer, Larry Bert, organic chemistry
Habib, Emile Edward, chemistry
Harrington, John Wilbur, geology
Highsmith, Philip E, physics
Kuhn, Hans Heinrich, organic chemistry, zoology, physiology
Leonard, Walter Raymond, zoology
Lyons, Harold Dwight, synthetic organic chemistry
Michener, John William, physics
Moore, Lawrence Edward, inorganic chemistry
Olds, Daniel Wayne, physics
Otto, Wolfgang Karl Ferdinand, physical chemistry, textile chemistry
Patton, Ernest Gibbes, plant ecology
Powell, Robert W, Jr, botany
Stephens, Bobby Gene, analytical chemistry
Stewart, William Hogue, Jr, solid state physics
Stokes, David Kershaw, Jr, family medicine
Turner, Jack Allen, ecology
Wagner, William Sherwood, organic chemistry
Weiss, James Owen, organic chemistry, polymer chemistry

SUMTER
Nerbun, Robert Charles, Jr, experimental nuclear physics

WHITE ROCK
Frick, Charles Harold, mathematics

WINNSBORO
Hartz, Roy Eugene, organic polymer chemistry
Kelly, Robert James, organic chemistry

SOUTH DAKOTA

ABERDEEN
Fors, Elton W, mathematics
Fries, James Andrew, physical chemistry, environmental chemistry
Haigh, William E, mathematics
Hein, Warren Walter, nuclear physics
Miller, Gertrude Nevada, plant taxonomy
Saunders, Jack K, Jr, mammalian ecology, wildlife biology

BROOKINGS
Applegate, Richard Lee, entomology, zoology
Bailey, Harold Stevens, pharmaceutical chemistry
Balsbaugh, Edward Ulmont, Jr, entomology
Bergeland, Martin E, veterinary pathology
Brage, Burton L, soil science, geology
Brandwein, Bernard Jay, biochemistry
Branson, Terry Fred, entomology
Brenner, George Marvin, pharmacology
Briggs, Hilton Marshall, animal husbandry
Buchenau, George William, plant pathology
Bush, Leon F, animal nutrition
Carlson, Charles Wendell, animal nutrition
Carson, Paul L Lewellyn, soil fertility
Chen, Chen Ho, plant cytology, plant physiology
Collins, Paul Everett, silviculture
Costello, William James, animal science, meat science
Dracy, Arthur E, dairy husbandry
Duffey, George Henry, theoretical physics
Dybing, Clifford Dean, plant physiology
Embry, Lawrence Bryan, animal husbandry
Emerick, Royce Jasper, animal nutrition
Fine, Lawrence Oliver, soil chemistry, water chemistry
Gardner, Wayne Scott, plant pathology
Gehrke, Henry, inorganic chemistry
Gerloff, Eldean D, agronomy, biochemistry
Glover, Loyd, Jr, agricultural economics
Graetzer, Hans Gunther, physics
Greichus, Algirdas, parasitology, physiology
Greichus, Yvonne A, pesticide chemistry, toxicology
Gross, Guilford C, pharmacology
Grove, John Amos, nutritional biochemistry
Haertel, Lois Steben, aquatic ecology
Hales, Donald Caleb, fish biology, aquatic ecology
Halverson, Andrew Wayne, agricultural chemistry
Hecht, Harry George, physical chemistry
Hietbrink, Bernard E, pharmacology, toxicology
Holden, David Jerome, plant physiology
Horton, Maurice Lee, agronomy, soil ecology
Howard, Ronald M, bacteriology, immunology
Huff, Albert Keith, plant science
Hugghins, Ernest Jay, animal parasitology
Hutcheson, Harvie Leon, Jr, plant ecology
Jensen, Stanley George, plant pathology
Jensen, William Rogers, chemistry
Johnson, Elmer Roger, chemistry
Kantack, Benjamin H, entomology, agronomy
Kieckhefer, Robert William, insect ecology
Kirk, Vernon Miles, economic entomology
Kirkbride, Clyde Arnold, veterinary bacteriology
Kohl, Robert A, soil physics, irrigation
Kohlmeyer, William, poultry science, economics
Kranzler, Albert William, mathematics
Krysan, James Louis, insect physiology
Le Blanc, Floyd Joseph, pharmacy
Lewis, James Kelley, range science
Linder, Raymond, wildlife ecology
Lund, Lillian O, textiles
Lunden, Allyn Oscar, plant breeding, plant genetics
Mankin, Cleon J, plant pathology, mycology
Martin, James Harold, dairy microbiology
McDaniel, Burruss, Jr, entomology
Middaugh, Paul Richard, bacteriology
Miller, Bruce Linn, physics, mathematics
Moore, Raymond A, agronomy, plant physiology
Morgan, Walter Clifford, animal genetics
Muller, Lawrence Dean, dairy husbandry
Musson, Alfred Lyman, animal science
Myers, Gerald Andy, plant anatomy, science education
Nagel, Clatus Martin, plant pathology
Olson, Edwin S, organic chemistry
Olson, Oscar Edward, agricultural biochemistry
Omodt, Gary Wilson, pharmaceutical chemistry
Otta, Jack Duane, plant pathology
Parikh, Gokaldas Chandulal, virology, immunology
Parsons, John G, dairy chemistry
Pengra, Robert Monroe, microbial physiology, soil microbiology
Peterson, Ronald M, horticulture, plant breeding
Prashar, Paul D, horticulture, plant breeding
Price, Philip B, agronomy, horticulture
Redman, Kenneth, pharmacognosy
Reeves, Dale Leslie, plant breeding, agronomy
Roller, Michael Harris, animal physiology
Ross, James George, agriculture
Rue, Rolland R, physical chemistry
Rumbaugh, Melvin Dale, crop breeding
Scalet, Charles George, fish biology, ichthyology
Schingoethe, David John, dairy nutrition
Semeniuk, George, plant pathology
Shank, Daniel Boyd, agronomy
Slyter, Arthur Lowell, reproductive physiology, animal science
Smolik, James Darrell, plant nematology
Spinar, Leo Harold, physical chemistry
Spuhler, Walter S, climatology
Stoner, Warren Norton, virology, entomology
Sutter, Gerald Rodney, entomology
Taylor, Charles Arthur, Jr, plant taxonomy
Thibodeau, Gary Arthur, physiology, pharmacology
Tunheim, Jerald Arden, solid state physics
Vohs, Paul Anthony, Jr, zoology
Wadsworth, William Steele, Jr, organic chemistry
Wahlstrom, Richard Carl, animal science
Walder, Orlin E, mathematics
Walgenbach, David D, entomology, agronomy
Walstrom, Robert John, entomology
Webster, Victor Stuart, organic chemistry
Wells, Darrell Gibson, plant breeding
Westby, Carl A, microbial physiology, biochemistry
Whitehead, Eugene Irving, plant biochemistry
Wilkinson, Thomas Ross, microbiology
Worman, James John, organic chemistry

HURON
LeRoux, Edmund Frank, hydrology
Okonkwo, Augustine Ikechukwuka, biology
Powell, John Edward, hydrology

MADISON
Brashier, Clyde Kenneth, plant taxonomy, plant morphology
Churchill, Constance Louise, organic chemistry

MINOT
Johnson, Arnold Richard, Jr, analytical chemistry

MITCHELL
Hills, C Loran, biochemistry

PIERRE
Diamond, Ben Elkan, medical bacteriology
June, Fred C, fish biology

RAPID CITY
Ashworth, T, solid state physics, thermodynamics
Ballew, David Wayne, number theory
Benson, Dean Clifton, mathematics
Bjork, Philip R, vertebrate paleontology
Bjugstad, Ardell Jerome, range science
Boldt, Charles Eugene, forestry
Chiu, Chin-Shan, atmospheric physics, cloud physics
Davis, Briant LeRoy, geology, geophysics
Gaines, Jack Raymond, organic chemistry
Garske, David Herman, mineralogy
Gilbertson, Lyle Ithiel, inorganic chemistry
Green, Morton, paleontology
Gries, John Paul, geology
Grimm, Carl Albert, mathematics
Hopkins, Don Carlos, physics
Jonte, John Haworth, geochemistry, inorganic chemistry
Looyenga, Robert William, analytical chemistry
Martin, Willard John, physical chemistry
Mickelson, John Chester, geology
Orville, Harold Duvall, meteorology
Patterson, James Deane, physics
Rahn, Perry H, hydrology, geomorphology
Redden, Jack A, geology
Redin, Robert Daniel, solid state physics
Rognlie, Dale Murray, mathematics
Schilz, Carl Edward, chemistry
Schleusener, Richard A, meteorology, engineering
Severson, Kieth Edward, range science, wildlife ecology
Smith, David Reeder, solid state physics
Smith, Paul Letton, Jr, atmospheric physics, electrical engineering
Weyland, Jack Arnold, physics
Willard, John Wesley, chemistry

SIOUX FALLS
Britzman, Darwin Gene, animal nutrition, poultry nutrition
Draeger, William Charles, natural resources, photogrammetry
Froiland, Sven Gordon, biology, ecology
Gaalswyk, Arie, applied mathematics
Gildseth, Wayne, physical chemistry
Gregg, John Bailey, otolaryngology
Hanson, Milton Paul, chemistry
Johnson, Leland Gilbert, comparative physiology, embryology
Kintner, Robert Roy, organic chemistry
Landborg, Richard John, chemistry
Maas, Keith Allan, photographic chemistry
Mobley, Jack Ervin, urology
Nelson, Vernon Ronald, electronics
Prescott, Lansing M, biochemistry, microbiology
Reeves, Robert Grier (Lefevre), economic geology, geophysics
Rogers, Dilwyn John, ecology
Rohde, Wayne G, forestry, remote sensing
Taranik, James Vladimir, exploration geology, photogeology
Thompson, John Darrell, physics, molecular biophysics
Tieszen, Larry L, plant physiology
Vander Lugt, Karel L, solid state physics
Viste, Arlen E, inorganic chemistry
Watkins, Allen Harrison, science administration
Wegner, Karl Heinrich, medicine, pathology

SPEARFISH
Harder, James Otto, geology
Shryock, Gerald Duane, organic chemistry

SPRINGFIELD
Miedema, Eddy, biochemistry, organic chemistry
Regehr, David Lavonne, ecology, botany

UNIVERSITY
Froslie, Harold Milton, physics

VERMILLION
Beaty, Marjorie Heckel, mathematics
Carraher, Charles Eugene, Jr, polymer chemistry, physical chemistry
Coker, Earl Howard, Jr, physical chemistry
Dillon, Raymond Donald, zoology
Dunlap, Donald Gene, vertebrate zoology, animal ecology
Einhellig, Frank Arnold, botany, physiological ecology
Estee, Charles Remington, physical chemistry
Gaush, Charles Richard, virology
Goldman, Max, endocrinology, biology
Gutzman, Wayne Wallace, mathematics, statistics
Harrell, Byron Eugene, zoology
Heisinger, James Fredrick, vertebrate physiology
Hoffman, George R, plant ecology
Johnson, Jeffery Lee, neurophysiology

SOUTH DAKOTA

Jones, Robert William, physics
Kakolewski, Jan Wiktor, psychobiology, psychiatry
Langworthy, Thomas Allan, microbial physiology
Lynn, Raymond J., medical microbiology
Mayberry, William Roy, microbiology, analytical biochemistry
McBroom, Marvin Jack, physiology
McGregor, Duncan J., geology
McVay, Chester Bidwell, surgery
Miller, Norman E., inorganic chemistry
Moller, Gottfried Irving, physics
Moore, Josephine Carroll, neuroanatomy, electromyography
Neuhaus, Otto Wilhelm, biochemistry, enzymology
Peasslee, Margaret H., endocrinology
Padmore, Joel M., analytical chemistry
Parrish, Henry Mack, preventive medicine
Peanasky, Robert Joseph, biochemistry, medicine
Quick, Jacquelin Dunn, medical microbiology
Raab, Wallace Albert, applied mathematics
Raney, Brooks, obstetrics & gynecology
Read, Willard Oliver, physiology
Rinker, George Clark, anatomy
Schmulbach, James C., fish biology
Scott, Earl B, anatomy
Scott, George Prescott, organic chemistry, philosophy of science
Sill, Webster Harrison, Jr., plant pathology, environmental biology
Small, Gary D., biochemistry
Smith, Paul Francis, microbiology
Steele, James Patrick, radiology
Stevenson, Robert Evans, geology
Stoner, Marshall Robert, organic chemistry
Thomas, John Alva, biochemistry
Tipton, Merlin J., geology
Van Bruggen, Theodore, botany
Welty, Joseph D., pharmacology, physiology
Walburg, Charles Herman, fish biology

YANKTON
Catana, Anthony J. Jr., plant ecology, phytopathology
Davidson, Bruce Lloyd, chemistry, philosophy of science
Diggins, Maureen Rita, fresh water ecology, comparative physiology
Duffey, Lowell Myers, experimental embryology
Fasbender, M Veronica, invertebrate zoology, palynology

TENNESSEE

ARNOLD AFB
Krakow, Burton, physical chemistry, spectroscopy
Lennert, Andrew E., physics
Lewis, James W L., molecular physics, acoustics

ATHENS
Bowling, Floyd E., mathematics
Cox, Edmond Rudolph, Jr., phycology, microbiology
Duncan, Budd Lee, physical chemistry
Honaker, Carl Boggess, inorganic chemistry

BRISTOL
Bickford, Charles Allen, forest biometry
Burke, Edward Walter, Jr., physics
Hill, George Neal, veterinary medicine
Keefe, Thomas J., veterinary medicine
MacFadden, Donald Lee, physiology, biology
Mattison, Louis Emil, chemistry
Vann, Robert Lee, pediatrics, pharmaceutics

CARTHAGE
Fischer, Frederick Thomas, economic geology, hydrogeology

CHATTANOOGA
Auerbach, Stewart Hart, pathology
Bergenback, Richard Edward, geology
Brown, Charles A., mathematics
Cain, Carl Jr., analytical chemistry, chemical engineering
Corden, Brian Joseph, medicine, bioinorganic chemistry
Derryberry, Oscar Merton, occupational medicine, public health
Durham, Ross M., neurophysiology, space biology
Fletcher, Lewis Arrowood, biochemistry
Freeman, John Richardson, herpetology
Garth, Richard Edwin, physiology, biology
Glasser, Julian, chemical metallurgy
Gross, Benjamin Harrison, organic chemistry
Hetzler, Morris Clifford, Jr., solid state physics
Hujer, Karel, astronomy, physics
Hutcheson, Joseph William, atomic physics, applied mathematics
Kirk, Joe Eckley, Jr., mathematics
Lane, Eric Trent, theoretical physics
Lents, James Marcelius, physics
Litchford, Robert Gary, parasitology
Massey, Winston Louis, physics
McCay, Myron Stanley, physics
McColl, John Duncan, pharmacology
McDowell, Horace Greeley, geography
McNeely, Robert Lewis, analytical chemistry
Mirhej, Michael Edward, textile technology
Nalley, Samuel Joseph, atomic physics, molecular physics
Nelson, Charles Henry, entomology
Nyman, DeWayne Stanley, mathematics
Peterson, William Roger, organic chemistry
Reeves, Richard Allen, polymer chemistry
Rose, John Logan, Jr., analytical chemistry
Scroggie, Lucy E., analytical chemistry
Spangler, George Wesley, biophysics
Stroud, Robert Wayne, textile chemistry, organic chemistry
Summers, James Thomas, inorganic chemistry
Van Horn, Gene Stanley, systematic botany
Vredeveld, Nicholas Gene, plant pathology
Waddell, Thomas Groth, bio-organic chemistry
Walton, Barbara Ann, vertebrate embryology, histology
Ware, James Gareth, geology
Wilson, Robert Lake, geology
Woods, Maribelle, pharmacology
Wyse, Benjamin Delaney, organic chemistry

CLARKSVILLE
Blanck, Harvey F., Jr., physical chemistry, analytical chemistry
Boehms, Charles Nelson, physiological ecology, freshwater biology
Boercker, Fred D., physics, science education
Ellis, William Haynes, systematic botany, plant ecology
Findley, Diane Ingram, limnology
Ford, Floyd Mallory, zoology
Harris, Durward Smith, bio-organic chemistry
Mayfield, Melburn Ross, physics, science education
Myers, Carol Bruce, algebra
Phillips, Haskell C., bacteriology, lichenology
Provo, Marvin Monroe, animal ecology, zoology
Sears, Robert F., Jr., physics
Snyder, David Hilton, vertebrate biology, ethology
Stokes, William Glenn, mathematics
Stone, Benjamin P., plant physiology
Wibking, Robert Kenton, geography
Wood, Saralue, nuclear physics

CLEVELAND
Dennison, Clifford C., zoology

COLLEGEDALE
Christensen, John, biochemistry
Hefferlin, Ray (Alden), molecular spectroscopy
Peck, Norman Eugene, organic chemistry, analytical chemistry

COLUMBIA
Naddy, Badie Ibrahim, physical chemistry, soil chemistry

CONCORD
Hansard, Samuel Leroy, nutrition

COOKEVILLE
Allen, Vernon R., polymer chemistry, physical chemistry
Ayers, Jerry Bart, science education
Ballal, Srikrishna, botany
Bulow, Frank Joseph, fish biology
Caplenor, Donald, botany
Coburn, Corbett Benjamin, Jr., environmental physiology, physiological ecology
Culp, Frederick Lynn, physics
Dooley, Elmo S., physiology
Farrar, David Turner, physical chemistry
Griffin, Sumner Albert, animal husbandry
Harris, John Wallace, biology
Helton, Walter Lee, economic geology, stratigraphy
Hessley, Rita K., analytical chemistry
Hunter, Gordon Eugene, botany
Knox, Larry William, paleontology
Martin, Robert Eugene, animal ecology
Mazeres, Reginald Merle, algebra
Patil, Surgounda A., mathematics
Skinner, John Taylor, chemistry
Sublett, Robert L., chemistry
Swindell, Robert Thomas, organic chemistry
Ventrice, Carl Alfred, plasma physics, nuclear physics
Wells, John Calhoun, Jr., experimental nuclear physics
Willard, William Kenneth, ecology

DAYTON
Henning, Willard Loren, zoology

ELMWOOD
Hill, William T., geology

ELIZABETHTON
Bryan, Loren Aldro, industrial chemistry
Harvey, Jay Arthur, electrochemistry
Lawrence, Franklin Isaac Latimer, organic chemistry

ERWIN
Ideker, Richard Louis, agriculture

FRANKLIN
Clark, John, petrology

GALLATIN
Miller, Paul Thomas, inorganic chemistry
Thorland, Rodney Harold, solid state physics

GATLINBURG
Bratton, Susan Power, plant ecology

GERMANTOWN
Elowe, Louis N., pharmacy, chemistry
Fenyes, Joseph Gabriel Egon, synthetic organic chemistry

GREENEVILLE
Godfrey, Paul Russell, biochemistry
Gupton, Creighton Lee, genetics
McGavock, Walter Donald, invertebrate zoology, parasitology
Phelan, Earl Walter, chemistry

HARRISON
McPherson, James Louis, polymer chemistry

HARROGATE
Lu, Mary Kwang-Ruey Chao, organic chemistry, mathematics
Ziegler, Robert G., inorganic chemistry

HENDERSONVILLE
Booth, Max Howard, physical chemistry
Holmes, Joseph Charles, chemistry

JACKSON
Barnes, Ronnie C., astrophysics
Beasley, James Gordon, organic chemistry, medicinal chemistry
Carlton, Robert Austin, zoology
Davis, William James, physiology, endocrinology
Keller, Wayne Hicks, physical chemistry
Lentz, Gary Lynn, economic entomology
Mahajan, Satish Chander, physiology, endocrinology

JEFFERSON CITY
Burton, John Williams, experimental nuclear physics
Chapman, Joe Alexander, plant ecology
Childress, Denver Ray, pure mathematics
Chitwood, Howard, mathematics
Dobyns, Roy A., mathematics
Fincher, John Albert, zoology
Herring, Carey Reuben, mathematics education
Myers, Albert Leroy, physical chemistry
Naylor, Gerald Wayne, agronomy, plant physiology
O'Neal, Thomas Norman, experimental chemistry, analytical chemistry
Patrick, Charles Russell, entomology
Patterson, Truett Clifton, inorganic chemistry
Sierk, Herbert Allen, lichenology
Sloan, Albert Russell, mathematics
Sloan, Ben Leroy, zoology
Vananan, Sherman Benton, mathematics
White, June Broussard, inorganic chemistry, analytical chemistry

JOHNSON CITY
Bailey, John H., zoology, science education
Barclay, Frank Hunt, biology
Blevins, Raymond Dean, physiology, biochemistry
Copeland, Thompson Preston, entomology
Cowden, Ronald Reed, cytology, embryology
Cresswell, Arthur, organic chemistry
Franzus, Boris, physical chemistry, organic chemistry
Gaby, William Lawrence, bacteriology
Ginnings, Gerald Keith, mathematics, statistics
Huang, Thomas Tao Shing, physical chemistry
Huskey, Glen E., food microbiology, biochemistry
Iglar, Albert Francis, Jr., environmental health, sanitary engineering
Ikenberry, Roy Dewayne, vertebrate physiology, hematology
Jewett, Robert Elwin, pharmacology
Kinloch, John, algebra
Lawson, James Everett, invertebrate zoology
Lura, Richard Dean, physical organic chemistry
Miller, James L., organic chemistry
Miller, James Robert, solid state physics
Moore, John David, animal physiology
Morgan, Monroe Talton, Sr., environmental health, public health
Nelson, Diane Roddy, invertebrate zoology
Newby, Frank Armon, Jr., physical chemistry
Nicholson, Douglas Gillison, inorganic chemistry
Pafford, William N., science education
Pleasant, James Carroll, mathematics
Powell, Harry Douglas, solid state
Raasch, Lou Reinhart, analytical chemistry
Sakhare, Vishwa M., mathematics
Snell, Robert L., organic chemistry
Suh, Tae-Il, algebra
Tarpley, Wallace Arnell, insect ecology
Wardeska, Jeffrey Gwynn, inorganic chemistry

KINGSPORT
Agett, Albert Henry, chemistry
Ball, Fred, organic chemistry, industrial chemistry
Barton, Kenneth Ray, polymer chemistry, organic chemistry
Bentz, Ralph Wagner, physical chemistry, organic chemistry
Boye, Charles Andrew, Jr., polymer chemistry
Brokaw, George Young, applied chemistry
Browning, Horace Lawrence, Jr., polymer chemistry, physical chemistry
Chambers, Ralph Arnold, polymer chemistry
Chitwood, James Leroy, organic chemistry
Clark, Raymond Donald, organic chemistry
Clark, Richard Bennett, chemistry
Clegg, William Josiah, industrial organic chemistry

Cleveland, James Perry, organic chemistry
Combs, Robert Leonard, physical chemistry, polymer chemistry
Coover, Harry Wesley, Jr, organometallic chemistry, polymer chemistry
Dale, John Irvin, III, organic chemistry
Davis, Burns, polymer chemistry, organic chemistry
Dickason, William Charles, polymer chemistry, organic chemistry
Donaldson, Raymond Edwin, chemistry
Elam, Edward Underwood, organic chemistry
Embree, Norris Dean, applied chemistry
Finch, Gaylord Kirkwood, organic chemistry
Fisher, John Gatewood, organic chemistry
Foster, Charles Howard, organic chemistry
Fuzek, John Frank, physical chemistry
Gerwe, Roderick Daniel, organic chemistry
Gilkey, Russell, polymer chemistry
Glover, Clyde Albert, analytical chemistry
Grant, Peter Malcolm, organic chemistry
Gray, Theodore Flint, Jr, polymer chemistry
Hasek, Robert Hall, organic chemistry
Hedrick, Robert Jerry, industrial organic chemistry
Hicks, Jackson Earl, analytical chemistry
Hill, Hubert Mack, systems analysis, applied statistics
Hoyle, Vinton Asbury, Jr, organic chemistry
Hudnall, Phillip Montgomery, organic chemistry
Hyatt, John Anthony, synthetic organic chemistry
Irick, Gether, Jr, photochemistry
Jackson, Winston Jerome, Jr, polymer chemistry
Jones, Glenn Clark, organic chemistry, polymer chemistry
Kennedy, Robert Wilson, organic chemistry
Kennedy, William Dempsey, physical chemistry
Knowles, M B, organic chemistry
Kreh, Donald Willard, organic chemistry
Krutak, James John, Sr, organic chemistry, physical chemistry
Lappin, Gerald R, chemistry
Larkins, Thomas Hassell, Jr, inorganic chemistry
Lewis, Jack Smith, mass spectrometry
Magoffin, James Edward, physical chemistry
Martin, James Cuthbert, industrial organic chemistry
McCall, Marvin Anthony, organic chemistry
McDaniel, Edgar Lamar, Jr, industrial organic chemistry
McFarlane, Finley Eugene, polymer chemistry
McGraw, Gary Earl, physical chemistry
Meen, Ronald Hugh, synthetic organic chemistry
Miller, Robert Witherspoon, organic chemistry
Moore, Jon Thomas, polymer physics
Moore, Louis Doyle, Jr, polymer chemistry
Morie, Gerald Prescott, analytical chemistry
Nealy, David Lewis, organic chemistry
Nelan, Donald Royce, organic chemistry
Newland, Gordon Clay, polymer chemistry, photochemistry
Nicely, Vincent Alvin, physical chemistry
O'Neill, George Joseph, polymer chemistry
Osborne, Charles Edward, synthetic organic chemistry
Otis, Marshall Voigt, chemistry
Pacifici, James Grady, physical organic chemistry
Patton, Hugh Wilson, physical chemistry
Payne, DeWitt Allen, electrochemistry
Peck, Virgil Glenn, chemistry, physics
Petke, Frederick David, surface chemistry
Poe, James Edgar, organic chemistry
Pond, David Martin, organic chemistry, photochemistry
Reynolds, Jefferson Wayne, physical chemistry
Robbins, Robert, organic chemistry
Russin, Nicholas Charles, physical chemistry
Shearer, Newton Henry, Jr, organic chemistry
Smith, Benjamin Harper, Jr, organic chemistry
Smith, James Luther, organic chemistry
Spadafino, Leonard Peter, organic chemistry
Stanin, Theodore E, organic chemistry
Straley, James Madison, organic chemistry
Sublett, Bobby Jones, organic chemistry, polymer chemistry
Tarpley, Anderson Ray, Jr, physical chemistry, spectroscopy
Thweatt, John G, organic chemistry
Tirman, Alvin, physical chemistry, mathematics
Vachon, Raymond Normand, organic chemistry, polymer chemistry
Van Sickle, Dale Elbert, physical organic chemistry
Watson, Marshall Tredway, physical chemistry, textile chemistry
Whetsel, Kermit Bazil, analytical chemistry
Wicker, Thomas Hamilton, Jr, organic chemistry, polymer chemistry
Wilkin, Louis Alden, organic chemistry
Wooten, Willis Carl, Jr, polymer chemistry
Young, DeWalt Secrist, organic chemistry
Young, Howard Seth, physical chemistry

KINGSTON

Blankenship, Forrest (Farley), physical chemistry
Ketchen, Eugene Earl, physical chemistry
Parkinson, William Walker, Jr, physical chemistry

KNOXVILLE

Adcock, James Luther, fluorine chemistry, inorganic chemistry
Adler, Laszlo, physical acoustics
Aiken, Charles S, cultural geography, economic geography
Aiken, Robert McLean, computer science
Alexeff, Igor, nuclear physics
Allen, Freddie Lewis, agronomy
Amundsen, Clifford C, physiological ecology, microclimatology
Armistead, Willis William, veterinary medicine, medical education
Asp, Carl W, speech & hearing sciences, statistics
Barrett, John William, forestry
Barrett, Lida Kittrell, mathematics
Barth, Karl M, animal nutrition
Bass, Mary Anna, food science, nutrition
Bass, William Marvin, III, physical anthropology, anthropometrics
Beach, Betty Laura, food science, nutrition
Beauchene, Roy E, nutrition, geriatrics
Beck, Raymond Warren, microbiology
Becker, Jeffrey Marvin, microbiology
Bell, Frank F, agronomy
Bell, Marvin Carl, animal nutrition
Bell, Sandra Lucille, botany, cytogenetics
Bingham, Carrol R, nuclear physics
Blass, William Errol, molecular spectroscopy, computer science
Bletner, James Karl, poultry husbandry
Bloor, John E, chemistry
Bowman, Newell Stedman, organic chemistry
Bradley, John Spurgeon, mathematics
Bratton, Gerald Roy, veterinary anatomy
Breazeale, Mack Alfred, physical acoustics, solid state physics
Brent, William B, structural geology, stratigraphy
Briggs, Garrett, sedimentology
Brinkman, Leonard W, Jr, geography
Brown, Arthur, microbiology
Bugg, William Maurice, physics
Bull, William Earnest, inorganic chemistry
Bullock, Austin Larnel, organic chemistry
Bunting, Dewey Lee, zoology
Buntley, George Jule, soil morphology
Burghardt, Gordon Martin, ethology, biopsychology
Burnham, Kenneth Donald, zoology
Byerly, Don Wayne, geology
Callahan, Lloyd Milton, agronomy
Callcott, Thomas Anderson, solid state physics, surface physics
Campbell, Ada Marie, food chemistry
Caponetti, James Dante, botany
Carlson, James Gordon, cytology, radiobiology
Carruth, James Harvey, pure mathematics
Carruth, Kayla Bernard, nutrition
Chamberlain, Charles Calvin, animal nutrition
Chambers, David Smith, applied statistics
Chambers, James Q, analytical chemistry
Chance, Charles Jackson, zoology, fisheries
Chapman, Jefferson, anthropology, archaeology
Chen, James Pai-Fun, biochemistry, immunology
Chernoff, Amoz Immanuel, hematology
Childers, Robert Wayne, theoretical physics
Clark, Edward Shannon, polymer chemistry
Clebsch, Edward Ernst Cooper, plant ecology
Cline, Randall Eugene, applied mathematics
Coggin, Joseph Hiram, microbiology, virology
Cole, Arthur Charles, Jr, entomology
Collins, Jimmie Lee, food science
Condo, George T, high energy physics
Congdon, Charles C, pathology
Core, Harold Addison, forestry
Corrick, James Adam, Jr, animal science
Coulson, Patricia Bunker, reproductive endocrinology, cell physiology
Craven, Claude Jackson, physics
Daniel, Joseph Car, Jr, developmental biology, reproductive physiology
Daverman, Robert Jay, topology
Dean, John Aurie, analytical chemistry
Deeds, William Edward, physics
Demott, Bobby Joe, dairy chemistry
DeSelm, Henry Rawie, plant ecology
Diddle, Albert W, obstetrics & gynecology
Dimmick, Ralph W, animal ecology, wildlife management
Dobbs, David Earl, algebra
Duckett, Kermit Earl, physics
Eastham, Jerome Fields, organic chemistry
Eaves, Edgar Dewey, analytical mathematics
Echternacht, Arthur Charles, vertebrate zoology, ecology
Elston, Stuart B, atomic physics
Etnier, David Allen, ichthyology
Ewing, John Arthur, agronomy, research administration
Eyring, Edward J, orthopedics, physiological chemistry
Farkas, Walter Robert, biochemistry
Faulkner, Charles Herman, anthropology
Fletcher, William Henry, physical chemistry
Fox, Kenneth, molecular spectroscopy, astrophysics
Frandsen, Henry, mathematics
Fraser, Ronald Chester, develomental biology
Fribourg, Henry August, crop ecology
Gailar, Norman Milton, physics
Galbraith, Harry Wilson, organic chemistry
Gerhardt, Reid Richard, entomology
Gibbons, John Howard, nuclear physics, resource management
Girardi, Anthony Joseph, microbiology
Goan, Hugh Charles, poultry husbandry
Goertz, Grayce Edith, food science
Gossett, Dorsey McPeake, agronomy
Gram, Mary Rose, nutrition
Greenawalt, John W, biochemistry
Gregory, Robert Todd, numerical analysis
Gunzburger, Max Donald, applied mathematics
Guthe, Alfred Kidder, anthropology, archaeology
Hall, O Glen, animal nutrition
Hammond, Edwin Hughes, physical geography
Handel, Mary Ann, cell biology
Harris, Edward Grant, theoretical physics
Hart, Edward Leon, high energy physics
Haver, William Emery, topology
Hawkinson, Stuart Winfield, crystallography, biochemistry
Heidel, John Willard, mathematics
Heilman, Alan Smith, botany
Herndon, Walter Roger, botany
Hilty, James Willard, plant pathology
Hinton, Don Barker, mathematics
Hitchcock, John Paul, animal nutrition
Hochman, Benjamin, genetics, zoology
Holtman, Darlington Frank, medical bacteriology
Holton, Raymond William, plant physiology
Horton, Bennett Franklin, anesthesiology
Howell, Joseph Corwin, animal behavior
Hubbard, Daniel Willis, nutrition
Hughes, Karen Woodbury, plant genetics
Hunt, Gordon Ellsworth, botany
Huray, Paul Gordon, solid state physics
Husch, Lawrence S, topology
Ichiki, Albert Tatsuo, immunology
Jacobson, Harry C, physics
James, Jesse, biochemistry
Jamieson, Haley M, animal breeding
Jaynes, Hugh Oliver, food science
Jeon, Kwang Wu, cell biology, developmental biology
Johnson, Leander Floyd, plant pathology
Johnson, Robert W, Jr, geology, geophysics
Johnson, Ronald Roy, biochemistry, animal nutrition
Johnston, Melvin Roscoe, food technology
Jones, Arthur Wynne, parasitology
Jones, Jimmy Barthel, laboratory animal medicine
Jones, Larry W, plant physiology
Jordan, George Samuel, mathematics
Josephson, Leonard Melvin, plant breeding, agronomy
Joshi, Jayant Gopal, biochemistry
Jumper, Sidney Roberts, economic geography, resource geography
Jungreis, Arthur Martin, zoology
Kalbach, Constance, nuclear physics, nuclear chemistry
Kant, Kenneth James, physiology
Keenan, Charles William, inorganic chemistry
Keigher, William Francis, algebra
Kennedy, John Robert, Jr, cytology
Kent, Rosemary (Christine) May, public health education
King, David Thane, physics
Kinstle, James Francis, polymer chemistry
Kitchen, Hyram, biochemistry, hematology
Klaasen, Gene Allen, mathematics
Klein, Edward Lawrence, forestry, forest products
Klepser, Harry John, stratigraphy, geology
Kopp, Otto Charles, geology
Kretchmar, Arthur Lockwood, biochemistry
Krueger, William Arthur, agronomy
Lambdin, Paris Lee, entomology
Lane, Charles A, organic chemistry
Lange, Robert Dale, hematology
Larsen, John W, organic chemistry, physical chemistry
Lasater, Herbert Alan, mathematical statistics, applied statistics
Lessman, Gary M, soil fertility, plant nutrition
Lewis, Russell J, soil chemistry
Lide, Robert Wilson, nuclear physics
Lietzke, Milton Henry, physical chemistry
Liles, James Neil, comparative physiology, insect physiology
Lipscomb, David M, audiology
Long, Robert Grant, geography of Latin America, cartography
Lozzio, Bismarck Berto, hematology, immunology
Lozzio, Carmen Bertucci, medical genetics, cell biology
MacCabe, Jeffrey Allan, developmental biology
Machado, Emilio Alfredo, experimental pathology
Magid, Linda Jenny, physical organic chemistry
Magid, Ronald, organic chemistry
Maher, Stuart Wilder, geology
Mahlman, Harvey Arthur, analytical chemistry, inorganic chemistry
Mamantov, Gleb, inorganic chemistry, analytical chemistry
Mathews, Harry T, mathematics
McArthur, William Henry, zoology
McConnel, Robert Merriman, mathematics
McDonald, Ted Painter, biochemistry
McLaughlin, Robert Everett, paleontology
Milici, Robert Calvin, geology
Miller, Don Dalzell, algebra
Miller, Franics Joseph, analytical chemistry
Miller, Robert Verne, microbial genetics, molecular genetics
Montgomery, Monty J, dairy science
Montie, Thomas C, biochemistry, microbiology
Monty, Kenneth James, biochemistry, cell biology
Mundt, John Orvin, food microbiology
Neil, Hugh Gross, physics
Nielsen, Alvin Herborg, physics
Odland, Lura Mae, nutrition
Oliver, Jack Wallace, veterinary pharmacology
Overcast, Woodrow Webb, dairy bacteriology
Pagni, Richard, organic chemistry

TENNESSEE

Painter, Linda Robinson, radiation physics
Parmalee, Paul Woodburn, zoology
Peacock, Neal Dow, horticulture
Felton, Michael Ramsay, wildlife biology, mammalogy
Penny, Keith, nuclear physics
Perry, Margaret Nutt, academic administration, food science
Peterson, Joseph Richard, physical genetics
Ranke, Thomas Franklin, forestry
Rayside, John Stuart, microbiology
Plemmons, Robert James, mathematics
Present, Richard David, chemical physics
Rainey, Robert Hamric, radiochemistry
Rainey, Harmon Hobson, Jr., plant genetics
Reich, Vernon Henry, agronomy; plant physiology
Reynolds, John Horace, geography
Reynolds, Marjorie Lavers, physiology
Savage, Jane Ramsdell, nutrition
Richardson, Don Orland, animal science, dairy science
Riechert, Susan Elise, zoology
Riedinger, Leo Louis, nuclear physics
Riggsby, William Stuart, molecular biology, biophysics
Ritter, Robert L., physical chemistry
Rowlett, Russell Johnston, III, topology
Rutland, Rufus Burr, horticulture
Schaefer, Philip William, applied mathematics
Schell, Fred Martin, synthetic organic chemistry
Schmitt, Harold William, nuclear physics
Schmudde, Theodore Henry, physical geography; resource geography
Schofield, Frances Armistead, nutrition
Schwarz, Otto John, plant physiology; biochemistry
Schweitzer, George Keene, inorganic chemistry, philosophy of science
Seatz, Lloyd Frank, soil fertility
Sellin, Ivan Armand, atomic physics
Sharp, Aaron John, botany
Sharp, John Buckner, forestry
Sherman, Gordon R., computer sciences, operations research
Shirley, Herschel Vincent, Jr., animal genetics, animal physiology
Shivers, Charles Alex, reproductive biology, developmental biology
Shrode, Robert Ray, animal breeding
Skold, Laurence Nelson, agronomy
Smith, David Kent, botany, bryology
Smith, Hilton Albert, physical chemistry
Smith, John Thurmond, biochemistry
Smith, William Thomas, Jr., physical chemistry
Snyder, Walter Stephen, health physics, mathematics
Solomon, Alan, hematology; oncology
Soni, Kusum, mathematical analysis
Southards, Carroll J., nematology, plant pathology
Springer, Maxwell Elsworth, soil morphology
Stalnmann, Friedemann Wilhelm, mathematics
Stewart, James McDonald, plant physiology
Swanson, Eric Wallace, dairy science
Swingle, Homer Dale, horticulture
Tanner, James Taylor, animal ecology
Taylor, Lawrence August, geochemistry, mineralogy
Thomson, John Oliver, physics
Thor, Eyvind, forest genetics, forest ecology
Turner, Lincoln Hulley, mathematics
Van Hook, William Alexander, physical chemistry
Wade, William Raymond, II, mathematics
Wagner, Carl George, mathematics
Walker, Kenneth Russell, stratigraphy, paleoecology
Wahne, Patricia Lee, botany
Wehry, Earl L., Jr., analytical chemistry
Wells, Garland Ray, forestry, economics
Wert, Jonathan Maxwell, Jr., environmental sciences
Wheeler, Donald Jefferson, applied statistics, mathematical statistics
White, James Wilson, physics
White, Jane Vicknair, nutrition
Whiteside, Melbourne C., aquatic ecology
Wigler, Paul William, biochemistry
Williams, Thomas Franconi, physical chemistry
Wilson, James Larry, fisheries management

Woods, Clifton, III, physical inorganic chemistry
Woods, Frank Wilson, forest ecology, environmental sciences
Woodward, John Morrill, bacteriology
Worley, Smith, Jr., agronomy
Wust, Carl John, immunology
Yoakum, Anna Margaret, analytical chemistry; physical metallurgy

LENOIR CITY
Barnett, Clarence Franklin, physics
Young, Frederick Walter, Jr., physical chemistry

LOOKOUT MOUNTAIN
Lothers, John Edmond, Jr., genetics

LOUISVILLE
Kent, Deane Frederick, geology

LOWLAND
Collins, Jerry Dale, organic chemistry, polymer chemistry

MADISON
Ryden, Fred Ward, microbiology

MARTIN
Airee, Shakti Kumar, physical chemistry, biochemistry
Copeland, David Anthony, chemical physics, inorganic chemistry
Duck, Bobby Neal, cytogenetics
Gagen, James Edwin, organic chemistry
Greman, Laurie M., analytical chemistry
Harding, Charles Enoch, physical organic chemistry
Hathcock, Bobby Ray, agriculture, plant breeding
Henson, James Wesley, plant physiology, cell biology
Jain, Mahendra Kumar, mathematical analysis
James, Ted Ralph, ecology, vertebrate zoology
Kittison, Harold Lee, genetics, botany
Loebkaka, David S., high energy physics
Moore, James Marvin, botany, zoology
Sharma, Gopal Krishan, ecology
Sliger, Wilburn Andrew, fish biology, fresh water biology
Smith, Rufus Albert, Jr., horticulture
Trentham, Jimmy N., microbiology

MARYVILLE
Love, Norman Duane, low temperature physics
Rhodes, William Gale, organic chemistry, biochemistry
Shields, Arthur Randolph, biology
Young, David Paris, science education

MCKENZIE
Black, John Larry, embryology;

MEMPHIS
Adams, Paul Louis, medical physics
Addink, Sylvan, weed science
Aivazian, Garabed Hagpop, psychiatry
Alden, Roland Herrick, anatomy
Albritten, Herbert Graves, soil chemistry; plant biochemistry
Amy, Robert Lewis, zoology
Andersen, Richard Nicolaj, reproductive endocrinology
Anderson, Lewis Daniel, orthopedic surgery
Andrews, James Tucker, dentistry
Argall, Clifford Irving, immunology
Arneson, Dora Williams, chemistry, biochemistry
Arneson, Richard Michael, biochemistry
Atnip, Robert Lee, anatomy
Autian, John, pharmacy
Avis, Kenneth Edward, pharmaceutics
Baker, Clinton Lyle, zoology; limnology
Bancroft, Harold Ramsey, entomology
Bandelin, Fred John, pharmaceutical chemistry
Barber, Melvin Clyde, III, economic geography; urban geography
Barrett, Terence William, biophysics
Bass, Abraham, biochemistry
Baxter, John Edwards, physical chemistry; biochemistry
Beard, James David, medical physiology
Biggers, Charles James, biology, genetics

Bisno, Alan Lester, internal medicine, infectious diseases
Blackwell, Leo Herman, physiology
Blatteis, Clark Martin, physiology
Bowen, Thomas Earle, Jr., physiology; biophysics
Bowns, Beverly Henry, community health, public health
Bozeman, Samuel Richmond, bacteriology
Bradham, Laurence Stobo, biochemistry
Brent, Thomas Peter, biochemistry
Brooks, Sam Raymond, mathematics
Brown, Carl Dee, entomology
Brown, Fountaine Christine, biochemistry
Brown, James Melton, soils
Brown, Michael, industrial pharmacy
Browne, Edward Tankard, Jr., botany
Bruesch, Simon Rulin, anatomy
Bryan, Thornton Emry, family medicine
Bucovaz, Edsel Tony, biochemistry; organic chemistry
Buehler, John A., organic chemistry
Burdick, Frank A., prosthodontics
Butler, Orion Carmichael, earth science
Byers, Lawrence Wallace, biochemistry
Byrne, William Lawrence, biochemistry
Caldwell, Robert William, pharmacology, physiology
Camacho, Alvro Manuel, pediatric endocrinology
Cantrell, William Fletcher, chemotherapy
Cardoso, Sergio Steiner, pharmacology
Carter, James Roland, pharmacology
Chang, Frederic Chewning, organic chemistry
Chen, Huber Jan-Peing, statistics
Cheung, Wai Yiu, biochemistry
Christopher, Robert Paul, physical medicine & rehabilitation
Clark, James William, dentistry
Clark, Larry P., organic chemistry, medicinal chemistry
Claypool, Don Pearson, organic chemistry
Clayton, John Mark, medicinal chemistry; medical research
Cohn, Sidney Arthur, biology
Collins, Bill Martin, speech pathology
Conrad, Jack Randolph, anthropology
Coons, Lewis Bennion, electron microscopy; histology
Corbet, John Harry, geography
Crow, Edwin Lee, anatomy
Cox, Clair Edward, II, surgery, urology
Cox, Frederick Eugene, pediatrics; infectious diseases
Cox, Ray, biochemistry
Cummins, Alvin J., internal medicine
Darlington, Julian Trucheart, invertebrate zoology
Darlington, Robert Wells, virology; electron microscopy
Davies, Dean Fletcher, medicine
Davis, Harry L., internal medicine
Davis, Kenneth Bruce, Jr., vertebrate physiology
Dean, Walter Lee, organic chemistry, inorganic chemistry
Deboo, Phil B., geology, paleontology
Deininger, Robert Wade, geology; petrology
D'Encarnacao, Paul S., psychopharmacology; neurosciences
DeSaussure, Richard Laurens, Jr., neurosurgery
Diggs, Lemuel Whitley, hematology; clinical chemistry
Dillingham, Elwood Oliver, toxicology
Dilts, Preston Vine, Jr., obstetrics & gynecology
Donaldson, Donald Jay, human anatomy, developmental biology
Doody, John Edward, physical chemistry
Duckworth, John Kelly, pathology
Duckworth, William Clifford, endocrinology
Dugdale, Marion, internal medicine, hematology
Edwards, Harold Henry, biophysics
Erickson, Cyrus Conrad, pathology
Etteldorf, James N., pediatrics
Evans, James Spurgeon, anatomy
Farmer, T Albert, Jr., endocrinology
Faudree, Ralph Jasper, Jr., algebra
Fedinec, Alexander, anatomy
Feinstone, Wolffe Harry, bacteriology; science administration
Fitzgerald, Laurence Rockwell, physiology
Fleming, Richard Joseph, mathematics
Folden, Dewey Bray, Jr., animal physiology
Francisco, Jerry Thomas, pathology
Franklin, Stanley Philip, topology
Freeman, Bob A., microbiology

Friedman, Ben I., nuclear medicine, internal medicine
Ganguly, Pankaj, hematology, cell physiology
Garland, Michael McKee, solid state physics
Gates, Ronald Eugene, biochemistry
Geller, Arthur Michael, biochemistry
Gendel, Benjamin Robert, internal medicine
George, Stephen L., biostatistics
Geurin, Stephen D., animal nutrition
Giaroli, John Nello, oral surgery
Gibson, Walter Milton, biology
Gilliom, Richard D., organic chemistry
Glisow, Helmuth Martin, organic chemistry
Givens, James Robert, reproductive endocrinology
Goldner, Karl John, pharmacognosy
Gompertz, Michael L., internal medicine, gastroenterology
Goodwin, James Thomas, aquatic entomology
Granoff, Allan, virology
Gravell, Maneth, virology
Grosvenor, Clark Edward, zoology; endocrinology
Hagen, Arthur Ainsworth, pharmacology
Hammer, Martin, pharmacy
Harvey, Michael Joseph, animal ecology
Hashimoto, Ken, electron microscopy; dermatology
Hendrix, James Harvey, Jr., plastic surgery
Hill, Robert James, biochemistry
Hinds, Nancy Webb, physical chemistry
Hofstetter, Adrian Marie, biology
Hollis, Cecil George, mycology
Houk, Larry Wayne, inorganic chemistry
Howell, Golden Leon, plant physiology
Hughes, James Gilliam, medicine
Hughes, Robert Rule, oncology
Hughes, Walter T., pediatrics; biochemistry
Hutcheson, Eldridge Tillmon, III.
Ijams, Charles Carroll, chemical physics
Incardona, Antonino L., virology;
Ingram, Alvin John, orthopedic surgery
Irving, Charles Clayton, biochemistry
Jabbour, J T., pediatric neurology
Jackson, Carl Wayne, hematology
Jefferson, William Emmett, Jr., pediatrics
Jennings, Billy Ray, microbiology;
Johnson, Warren W., pathology
Jones, Dallas Wayne, physics
Jordan, Robert H., pediatrics
Jordan, Willis Pope, Jr., urology
Jurand, Jerry George, immunology;
Kaltenborn, Howard Scholl, mathematics
Kane, James Francis, microbiology;
Kaplan, Stanley Baruch, medicine,
Karve, Mohan Dattatreya, industrial rheumatology
Kashgarian, Mark, preventive medicine,
Kauker, Michael Lajos, pharmacology
Keith, Eaden Francis, biochemistry;
Kennedy, Michael Lynn, vertebrate zoology
King, Lloyd Elijah, Jr., dermatology;
Kingsbury, David Wilson, virology;
Kirksey, Howard Graden, Jr., science education
Kitiachi, Abbas E., biochemistry; endocrinology
Kossmann, Charles Edward, medicine
Kraus, Alfred Paul, medicine
Kraus, Lorraine Marquardt, biochemistry
Kuiken, Kenneth (Alfred), biochemistry
Lackey, Laurence, geomorphology; quaternary geology
Lakshmanan, Nanja Kada, ecology
Larrabee, Allan Roger, biochemistry
Lassio, Andrew, medicinal chemistry
Lawrence, William Homer, toxicology; pharmacology
Leach, Byron Elwood, biochemistry
Leonard, Edward Charles, Jr., polymer chemistry
Levene, John Reuben, optometry; physiological optics
Longo, Frank Joseph, cell biology
Lounsbury, Richard William, geology
Lumsden, David Norman, geology
Lutey, Richard William, industrial microbiology; plant pathology
Luton, Edgar Frank, internal medicine

TENNESSEE

Lyons, Harold, analytical chemistry, molecular pathology
Manley, Emmett S, pharmacology, physiology
Manthey, Arthur Adolph, physiology
Marcus, Carol Joyce, biochemistry
Margaris, Angelo, mathematics
Marlowe, Edward, pharmaceutical chemistry
Marshall, Robert Herman, inorganic chemistry
Masi, Alfonse Thomas, epidemiology, internal medicine
Mason, James Michael, immunology, experimental pathology
Mathison, Ian William, organic & medicinal chemistry
Matthews, James Swinton, geography of Europe, historical geography
McBride, Elna Browning, applied mathematics
McCall, Charles B, medicine
McDearman, Sara, microbiology, immunology
McGiff, John C, pharmacology, internal medicine
McGowan, Robert William, ornithology
McHenry, Hugh Lansden, biostatistics, mathematical statistics
McKnight, James Pope, dentistry
McLaughlin, Barbara Jean, neurobiology
Mecklenborg, Kenneth Thomas, organic chemistry
Mercer, Gerald Dean, organic chemistry
Meyer, Marvin Chris, clinical pharmacology
Michelson, Israel David, medical bacteriology, pathology
Miller, Neil Austin, forest ecology
Molinary, Samuel Victor, biochemistry, genetics
Morgan, Rebue Marvin, physics
Morrison, John Coulter, obstetrics & gynecology, physiology
Morrison, Martin, biochemistry
Mortimer, Robert George, physical chemistry, theoretical chemistry
Mui, Paul Ting-Kai, biochemistry
Muirhead, Ernest Eric, pathology
Murrell, Leonard Richard, anatomy
Nagel, Fritz John, organic chemistry
Nash, Clinton Brooks, pharmacology
Naumann, Hans Norbert, pathology, pathological chemistry
Neely, Charles Lea, Jr, medicine, hematology
Nemec, Josef, organic chemistry
Nemitz, William Charles, pure mathematics
Nickson, James Joseph, radiology
North, William Charles, anesthesiology, pharmacology
Norton, Virginia Marino, vertebrate zoology, physiology
Novak, Josef Frantisek, cancer biology
Nunez, Loys Joseph, chemistry, toxicology
Overman, Richard Roll, physiology
Packer, Henry, preventive medicine
Partridge, Lloyd Donald, neurophysiology, biomedical engineering
Pate, James Wynford, medicine
Paulson, Jack Charles, pulp chemistry
Pera, John Dominic, organic chemistry
Peterson, Glen Ervin, bacteriology, biochemistry
Phillips, David Richard, biochemistry
Phillips, Jerry Clyde, radiology
Pitner, Samuel Ellis, neurology
Portner, Allen, virology, biochemistry
Pratt, Charles Benton, pediatrics, cancer
Prewitt, Russell Lawrence, Jr, cardiovascular physiology
Price, Robert Allen, pediatrics, pathology
Prudhon, Rolland A, Jr, parasitology
Pulido, Miguel, weed science, plant pathology
Purcell, William Paul, molecular biology
Quintana, Ronald Preston, medicinal chemistry
Raabe, Austin Bauer, organic chemistry, medicinal chemistry
Rabinowitz, Jack Grant, radiology
Ragland, James Benjamin, biochemistry
Rampp, Donald L, speech pathology
Reed, Homer Vernon, oral surgery, dentistry
Reger, James Frederick, cytology
Rendtorff, Robert Carlisle, tropical medicine
Reynolds, Leslie Boush, Jr, physiology, clinical medicine
Riggin, John T, Jr, medicine
Rightsel, Wilton Adair, bacteriology
Roberts, Audrey Nadine, immunology, virology

Roberts, DeWayne, biochemical pharmacology
Robertson, George Gordon, anatomy
Robertson, James Thomas, neurosurgery
Robinson, Charles Nelson, organic chemistry
Robinson, Harry, biostatistics
Rogers, Stanfield, pathology
Rosenbluth, Sidney Alan, pharmacy
Rosensweig, Jacob, thoracic surgery, cardiovascular surgery
Ross, Richard Travis, bacteriology, biochemistry
Runyan, John William, Jr, internal medicine
Rushton, Priscilla Strickland, radiobiology
Russell, Jack Unger, mathematics
Sabesin, Seymour Marshall, internal medicine, gastroenterology
Sanyal, Shyamal Kumar, pediatric cardiology, pulmonary diseases
Sargent, William Quirk, medical physiology
Saxon, James Glenn, anatomy
Scheel, Konrad Wolfgang, cardiovascular physiology
Schneider, Edward Greyer, physiology
Schrank, Gordon Dabney, microbiology
Schwenzer, Kathryn Sarah, plant physiology, weed science
Scringer, Edward Brantly, Jr, mathematics
Seyer, Jerome Michael, biochemistry
Shade, Robert Eugene, physiology, endocrinology
Share, Leonard, physiology
Sharma, Rameshwar Kumar, medicinal chemistry
Sheppard, Charles Wilcox, physics, physiology
Sheth, Bhogilal, physical pharmacy, pharmaceutical chemistry
Shipp, Oliver Elmo, entomology, plant physiology
Shively, John Adrian, pathology, hematology
Simco, Bill Al, ichthyology
Simone, Joseph Vincent, hematology, pediatrics
Simpson, Daniel Martin Henry, soil chemistry, pesticide chemistry
Siskin, Milton, oral medicine, endodontics
Slater, Carl David, organic chemistry
Sloane, Nathan Howard, biochemistry
Smiley, Karl Leroy, Jr, microbiology
Smith, Arlo Irving, biology
Smith, Charles Howard, organic chemistry, medicinal chemistry
Smith, Gerald Patrick, anthropology, archaeology
Smith, Mary Ann Harvey, nutrition
Smith, Omar Ewing, Jr, economic entomology, medical entomology
Smith, Roy Martin, analytical chemistry
Sobol, John Andrew, geography
Solomon, Solomon Sidney, medicine, metabolism
Solomons, William Ebenezer, medicinal chemistry, organic chemistry
Sordinas, Augustus, cultural anthropology, archaeology
Sparer, Phineas Jack, psychiatry, preventive medicine
Spell, William Hux, Jr, biochemistry
Stanford, John Pershing, agronomy
Starr, Jason Leonard, oncology
Staub, Robert J, ecology, botany
Stephens, Harold W, mathematics
Stevenson, Everett E, mathematics
Sticht, Frank Davis, pharmacology
Stiles, Robert Neal, physiology
Stollerman, Gene Howard, medicine
Strauss, Ronald George, pediatrics, hematology
Struve, William George, biochemistry, chemistry
Studebaker, Gerald A, audiology
Sullivan, Jay Michael, cardiovascular diseases
Summit, Robert L, pediatrics, medical genetics
Sutliff, Wheelan Dwight, internal medicine
Swafford, William Bryson, pharmacy
Tan, Wai-Yuan, biostatistics
Tanner, Raymond Lewis, radiological physics, health physics
Taylor, Jack Howard, physics
Taylor, Robert Emerald, Jr, physiology
Thomas, Barbara Smith, topology
Todd, William McClintock, medical microbiology, virology

Townes, Alexander Sloan, internal medicine, immunology
Van Deren, John Medearis, Jr, analytical chemistry
Van Middlesworth, Lester, physiology, medicine
Victor, Leonard Baker, anatomic pathology, clinical pathology
Vogel, Howard H, Jr, zoology
Wakelyn, Phillip Jeffrey, textile chemistry
Walker, David Tutherly, mathematics
Walker, Laurence Graves, geology
Walker, William Stanley, immunology, microbiology
Wander, Joseph Day, neurochemistry
Warren, Charles O, Jr, mycology, plant physiology
Webb, William Logan, psychiatry
Weber, Faustin N, orthodontics
Webster, Robert G, virology, immunology
Welch, Arnold DeMerritt, pharmacology, chemotherapy
Wells, Jack E, pedodontics
Wescott, Lyle DuMond, Jr, organic chemistry
Wessels, Kenneth Edwin, dentistry
Weston, Alan Jay, audiology, speech pathology
White, Richard Paul, pharmacology
Wilcox, Harry Hammond, anatomy
Wilhelm, Walter Eugene, parasitology, protozoology
Williams, Edward Foster, Jr, biochemistry
Williams, George Kenneth, mathematics
Willis, Dawn Butler, microbiology, biochemistry
Wilson, Harwell, surgery
Wilson, Jack Lowery, anatomy
Wilson, James Walter, pathology
Wilson, Robert John, radiological physics, nuclear medicine
Witherspoon, James Donald, animal physiology
Wong, Patrick Yui-Kwong, biochemistry, biochemical pharmacology
Wood, John Grady, neurobiology
Wood, John Lewis, biochemistry
Wood, William Booth, physiology
Woodbury, Robert Arthur, pharmacology, physiology
Woollett, Albert Haines, physics
Young, Joseph Marvin, pathology, anatomy
Zee, Paulus, pediatrics, biochemistry
Zuber, William Henry, Jr, physical chemistry

McGhee, Charles Robert, systematic zoology
McMillion, Ovid Miller, economic geography, geography of North America
Miles, Melvin Henry, electrochemistry, physical chemistry
Murphy, George Graham, herpetology
New, Earl Hiram, ornamental horticulture
Parchment, John Gerald, zoology, ecology
Patten, John A, biology
Rawlins, Nolan Omri, agricultural economics
Ray, John Bernard, marine geography
Reed, Horace Beecher, Jr, medical entomology, insect ecology
Rucker, Ellis Suttle, biology
Scot, Dan Dryden, analytical chemistry, science education
Smith, Helen Leonore, geography of Southeast Asia, economic geography
Smith, Jesse Leo, mathematics
Spraker, Harold Stephen, mathematics
Terrell, Roy Paul, geography
Watts, Exum DeVer, organic chemistry
Wells, Marion Robert, cell physiology
Wiser, Cyrus Wymer, aquatic ecology, physiology
Wiser, James Eldred, analytical chemistry
Woods, Alvin Edwin, food chemistry, biochemistry

MILLIGAN COLLEGE
Gee, Charles William, science education
Leach, Eddie Dillon, microbiology, toxicology
Sisk, Lone L, organic chemistry, physiological chemistry
Wallace, Gary Oren, ecology

MORRISTOWN
Bahner, Carl Tabb, organic chemistry, biomedical engineering

MOUNTAIN HOME
Brindley, Clyde Owens, medicine

MT JULIET
Schulert, Arthur Robert, chemistry

MURFREESBORO
Alexander, Robert Allen, animal husbandry
Anderson, June S, inorganic chemistry, science education
Bigger, Theodore C, soil science
Bordine, Burton W, micropaleontology, paleobiology
Brown, James Walker, Jr, biochemistry
Chandler, Clay Morris, zoology
Cook, James Marion, theoretical physics, mathematical analysis
Dunn, Mary Catherine, zoology
Fletcher, Jesse Lane, animal breeding
Foutch, Harley Wayne, agronomy, plant physiology
Fullerton, Ralph O, geography
Hutcheson, Paul Henry, applied mathematics, computer science
Hutchinson, James Herbert, Jr, organic chemistry
Jamison, King W, Jr, mathematics
Kohland, William Francis, petrology
Lea, James Wesley, Jr, topology, algebra

NASHVILLE
Abernethy, Virginia Deane, medical anthropology
Abram, Harry Shore, psychiatry, psychosomatic medicine
Adams, Robert Walker, Jr, medicine
Adkins, Rutherford Hamlet, nuclear physics
Albridge, Royal, nuclear physics, atomic physics
Allen, Joseph Hunter, radiology
Anderson, Mary Loucile, reproductive endocrinology
Anderson, Robert Spencer, internal medicine
Andrews, John Stevens, Jr, biochemistry
Arenstorf, Richard F, mathematics
Artist, Russell (Charles), botany
Aulsebrook, Lucille Hagan, anatomy
Banks, John Houston, geometry
Barach, John Paul, plasma physics
Bashir, Nasir Ahmad, physiology, zoology
Bass, Allan Delmage, pharmacology
Batson, Oscar Randolph, pediatrics
Bedford, Joel S, radiobiology, chemistry
Bennett, Word Brown, Jr, natural products chemistry
Bolden, Theodore Edward, dentistry, pathology
Bond, Andrew, biochemistry, agronomy
Boone, James Ronald, organic chemistry
Bourne, John Ross, biomedical engineering, electrical engineering
Brady, Robert Nyle, biochemistry
Breeden, Johnnie Elbert, plant ecology
Brigham, Kenneth Larry, pulmonary physiology
Brill, A Bertrand, nuclear medicine
Brode, William Edward, herpetology, vertebrate ecology
Brown, Harold William, parasitology, public health
Bryant, Billy Finney, mathematics
Burkow, Henry, medicine
Burrows, Elizabeth Parker, bio-organic chemistry
Burt, Alvin Miller, III, neurochemistry, neuroanatomy
Bush, Milton Tomlinson, pharmacology
Calhoun, Calvin L, clinical neurology, anatomy
Campbell, James A, biology, ecology
Carter, Clint Earl, animal physiology, parasitology
Celauro, Francis L, mathematics
Chanin, Martin, chemistry

Channell, Robert Bennie, systematic botany
Chapman, John E., pharmacology
Cheatham, William J., pathology
Chiu, Jen-Fu, biochemistry
Chyll, Frank, biochemistry
Claus, Thomas Harrison, medical physiology
Clement, William Madison, Jr., genetics
Cohen, Stanley, biochemistry
Coley, Daniel George, immunology, microbiology
Collins, Mary Jane, psychoacoustics, audiology
Collins, Warren Eugene, nuclear physics
Colowick, Sidney Paul, biochemistry
Coniglio, John Giglio, biochemistry
Coppolino, Henry, psychiatry
Corbin, Jack David, physiology, biochemistry
Crater, Horace William, particle physics, quantum mechanics
Crofford, Oscar Bledsoe, medical research, internal medicine
Crooke, Philip Schuyler, applied mathematics
Crouch, Hubert Branch, zoology
Cunningham, Leon William, biochemistry
Cunningham, William John, dentistry, preventive dentistry
Dale, William Andrew, surgery
Dalton, Larry Raymond, chemical physics
Daniel, Rollin Augustus, Jr., surgery
Darby, William Jefferson, biochemistry, nutrition
Davenport, Guy Rodman, biochemistry
Davies, Jack, anatomy, embryology
De Balbian Verster, Floris, biochemistry, neurochemistry
Deshpande, Krishnanath Bhaskar, physical chemistry, inorganic chemistry
Des Prez, Roger Moister, internal medicine, pulmonary diseases
Dettman, Wolf Dietrich, neurochemistry, neuropharmacology
Dilts, Robert Voorhees, analytical chemistry
Dingell, James V., pharmacology
Di Pietro, David Louis, biochemistry
Di Sabato, Giovanni, immunology
Donald, William David, medicine
Dooley, Wallace T., orthopedic surgery
Driskill, William David, atomic physics, science education
Elam, Lloyd Charles, psychiatry
Elliott, Irvin Wesley, organic chemistry
Elliott, James, H., ophthalmology
Engel, Eric, medicine
Ewig, Carl Stephen, theoretical chemistry
Exton, John Howard, biochemistry
Falk, Leslie Alan, community health, public health
Farrell, Charles Ernest, zoology
Faulkner, Willard Riley, biochemistry
Federspiel, Charles Foster, biostatistics
Fenichel, Gerald M., neurology
Field, Lamar, chemistry
Fleischer, Becca Catherine, biochemistry
Fleischer, Sidney, biochemistry
Flexner, John M., internal medicine, hematology
Foster, John Hoskins, surgery
Fouche, Clarence Estes, Jr., biochemistry
Freeman, John A., neurophysiology, biophysics
Freemon, Frank Reed, neurology
Friaut, James Joseph, invertebrate zoology, entomology
Friesinger, Gottlieb Christian, medicine, physiology
Fuson, Nelson, molecular spectroscopy
Gaity, Joseph Anthony, biochemistry, genetics
Garbers, David Lorm, biological chemistry
Gibbs, Samuel Julian, dental radiology, radiobiology
Ginn, H Earl, internal medicine
Goodwin, Robert Archer, Jr., internal medicine
Goshen, Charles Ernest, medicine
Goss, Donald A., obstetrics & gynecology
Grams, Anne P., mathematics
Green, Louis Douglas, pathology
Griffin, Paul Putnam, orthopedic surgery
Griffith, Michael Grey, textile technology
Grossman, Laurence Abraham, internal medicine, cardiovascular diseases
Hagstrom, Ruth Murray, preventive medicine, epidemiology
Haines, Thomas Walton, medical entomology
Hall, Douglas Scott, astronomy
Hall, Hugh David, oral surgery, physiology

Hall, Larry Cully, analytical chemistry, electrochemistry
Hamilton, Joseph H, Jr., experimental nuclear physics
Hansen, Axel C., ophthalmology
Hara, Saburo, pediatrics, hematology
Harbison, Raymond D., pharmacology, toxicology
Hardie, Robert Howie, physics, astronomy
Hardin, Carolyn Myrick, behavioral physiology, neuroendocrinology
Harford, Earl Raymond, audiology
Harris, Thomas Munson, organic chemistry
Harrison, Robert Edwin, nematology, marine zoology
Harshman, Sidney, biochemistry, microbiology
Hartman, William Herman, pathology, surgical pathology
Hash, John H., biochemistry
Hawkins, Linda Louise, medical microbiology
Hayes, Wayland Jackson, Jr., toxicology
Heiser, Arnold M., astronomy
Hess, Bernard Andes, Jr., organic chemistry
Hollender, Marc Hale, psychiatry, pharmacology
Hollett, Charlotte R., biochemistry, immunology
Hull, George, Jr., entomology
Inagami, Tadashi, biochemistry
Isa, Abdallah Mohammad, microbiology, immunology
James, Alton Everette, Jr., radiology
James, Emory Albarte, Jr., biochemistry, nuclear medicine
Joesten, Melvin D., inorganic chemistry
Johnson, Rother Rodenious, bacteriology, immunology
Johnson, Thomas W., dermatology
Johnston, David Owen, physical chemistry
Jones, Clinton E., applied mathematics
Jones, Ernest Addison, physics
Jones, Mark Martin, inorganic chemistry
Jonsson, Bjarni, mathematics
Kaas, Jon Howard, psychophysiology
Kahlon, Prem Singh, biology, plant genetics
Kaplan, Albert Sydney, virology
Karzon, David T., virology, pediatrics
Kerce, Robert H., mathematics
Key, James Frazier, mathematics
King, Calvin Elijah, biomedical engineering
King, Paul Harvey, biomedical engineering, mechanical engineering
Kono, Tetsuro, physiology, biochemistry
Kral, Robert, plant taxonomy
Krantz, Sanford B., internal medicine, hematology
Kuczenski, Ronald Thomas, psychopharmacology
Lacy, William White, medicine
Lageman, Robert Theodore, physics
Lai, Francis Ming-Hung, physiology, biophysics
Lai, Peter Chengliang, applied mathematics, chemical engineering
Landon, Erwin Jacob, biochemistry
Langford, Paul Brooks, physical organic chemistry
Laskowski, Michael Bernard, neurophysiology, neuroanatomy
Lefkowitz, Lewis Benjamin, Jr., medicine
Lembach, Kenneth James, biochemistry
Lenhert, P Galen, crystallography, physics
LeQuire, Virgil Shields, pathology
Lerman, Leonard Solomon, molecular biology
Liddle, Grant Winder, endocrinology
Linke, Ernest George, clinical biochemistry
Lott, Sam Houston, Jr., physics, health physics
Love, Russell Jacques, speech pathology
Love, Theodore Arceola, mathematics
Lowe, James Urban, Jr., physical organic chemistry
Lukehart, Charles Martin, organometallic chemistry
Lundberg, Gustave Harold, applied mathematics
Luther, Edward Turner, geology
Lynch, John Brown, plastic surgery

Mallette, John M., endocrinology, experimental embryology
Mann, Rama I, organic chemistry
Mann, George Vernon, biochemistry, nutrition
Martin, Thomas Waring, physical chemistry
Martindale, William Earl, biochemistry
Marx, Morris Leon, mathematics
McConnell, Freeman Erton, audiology
McCown, Otis Blakely, surgery
McSwain, Barton, surgery
Meacham, William Feland, medicine, neurosurgery
Meador, Clifton Kirkpatrick, medicine, endocrinology
Megibow, Charles Kimbrough, mathematics
Meng, Raymond Hsien Chang, physiology
Mgbodile, Marcel Ume, biochemistry, nutrition
Mickens, Ronald Elbert, chemical physics
Mitchell, William Marvin, oncology, pathology
Moore, Conrad Taylor, geography, historical geography
Morrison, Edward Joseph, teratology, radiochemistry
Moses, Henry A., biochemistry, physiology
Mosig, Gisela, genetics
Murray, Robert Edward, biochemistry, kinetics
Murrell, James Thomas, Jr., systematic botany
Navalkar, Ram G., microbiology, immunology
Neal, Robert A., toxicology, biochemistry
Neff, Robert Jack, biology physiology
Nelson, Manno Fredrick, Jr., radiochemistry
Nelson, Oscar Tivis, Jr., mathematics
Netsky, Martin George, neurology, neuropathology
Netterville, John T., analytical chemistry
Nichoalds, George Edward, biochemistry, nutrition
Nies, Alan Sheffer, clinical pharmacology
Nunnally, David Ambrose, zoology, biology
Oates, John Alexander, clinical pharmacology, internal medicine
Olson, Erik Joseph, pharmacology
O'Neill, James A, Jr., pediatric surgery
Orgebin-Crist, Marie-Claire, biology
Paine, Thomas Fite, Jr., medicine
Panvini, Robert S., high energy physics
Park, Charles Rawlinson, physiology, biochemistry
Park, Jane Harting, biochemistry
Pearson, Donald Emanual, chemistry
Peatman, William Burling, physical chemistry
Penner, Hellmut Philip, chemistry
Perry, Frank Anthony, surgery
Phelps, Jewell A., geography
Pinkston, William Thomas, nuclear physics, theoretical physics
Pittinger, Charles Bernard, anesthesiology
Plummer, Michael David, mathematics
Pointer, Richard Hamilton, biochemistry
Post, Robert Lickely, physiology
Potter, Thomas Franklin, mathematics
Pratt, Lee Herbert, plant physiology
Puett, John David, biochemistry, endocrinology
Quarterman, Elsie, botany, ecology
Quinn, Robert (William), preventive medicine
Ramayya, Akunuri V., experimental nuclear physics
Ramsey, Lloyd Hamilton, medical administration, internal medicine
Ratner, Lawrence Theodore, mathematics
Reed, Peter William, biochemistry, pharmacology
Reesman, Arthur Lee, geology
Regen, David Marvin, physiology, biochemistry
Reif, Donald John, organic chemistry
Reynolds, Vernon H., surgery, oncology
Rhamy, Robert Keith, urology, physiology
Rhodes, Robert Shaw, hematology
Richardson, Elisha Roscoe, orthodontics, anatomy
Risby, Edward Louis, parasitology, cell biology
Riven, Samuel Saul, internal medicine
Riven, Prince, organic chemistry
Roberson, Jill Sharon, bio-organic chemistry
Robinson, John Price, microbiology
Roos, Charles Edwin, nuclear physics

Rummel, Robert Edwin, physical chemistry
Sachan, Dileep Singh, nutritional biochemistry, microbiology
Sanders, Jay W., audiology
Sanders-Bush, Elaine, pharmacology
Sastry, Bhamidipaty Venkata Rama, pharmacology, medicinal chemistry
Sawyers, John Lazelle, surgery
Schaad, Lawrence Joseph, physical chemistry
Schenker, Steven, internal medicine, gastroenterology
Schmidt, Dennis Earl, neuropharmacology
Scott, Henry William, Jr., surgery
Scott, Mack Tonmie, animal science, veterinary medicine
Sell, Sarah H Wood, pediatrics
Senter, Gilbert Wayman, Jr., organic chemistry
Shanks, Eugene Baylis, mathematics
Shapiro, John Lawton, pathology
Shockley, Dolores Cooper, pharmacology
Shockley, Thomas S., bacteriology
Silberman, Enrique, molecular spectroscopy
Siminoff, Robert, neurophysiology
Singh, Dharmdeo Narayan, cytogenetics
Smith, Charles Bassel, mathematics
Smith, Howard E., organic chemistry
Soderling, Thomas Richard, physiology, biochemistry
Soupart, Pierre, reproductive physiology, biochemistry
Spores, Ronald, anthropology
Springer, John Mervin, chemical physics
Srygley, Fletcher Douglas, solid state physics
Stahlman, Mildred, pediatrics, physiology
Stearns, Richard Gordon, geology
Stephenson, Charles V., solid state physics
Stewart, Lynn Martin, biochemistry, enzymology
Stiles, Raeburn Brackett, human genetics
Still, W Clark, Jr., synthetic organic chemistry
Sulser, Fridolin, pharmacology
Sumartojo, Jojok, sedimentary petrology, economic geology
Sutherland, Claudia Sebeste, pharmacology
Sweetman, Brian Jack, organic chemistry
Tarbell, Dean Stanley, organic chemistry
Tarleton, Gadson Jack, Jr., radiology
Tellinghuisen, Joel Barton, chemistry
Teschan, Paul E., medicine
Thomas, David John, economic anthropology
Thornton, Erly J., poultry husbandry
Todhunter, Elizabeth Neige, nutrition
Tomlinson, Gus, electron microscopy, cell physiology
Torrey, Rubye Prigmore, radiation chemistry, analytical chemistry
Touster, Oscar, molecular biology, biochemistry
Trupin, Joel Sunrise, biochemistry, microbiology
Tuleen, David L., organic chemistry
Vander Zwaag, Roger, biostatistics
Van Wazer, John Robert, chemistry
Vaughn, William King, biostatistics
Venable, John Heinz, Jr., molecular biophysics
Wagner, Conrad, biochemistry
Walden, George Ellis, physical chemistry
Walker, Matthew, surgery
Ward, James William, anatomy
Ward, Robert Porter, vertebrate morphology, vertebrate ecology
Warnock, Laken Guinn, biochemistry
Warren, Mitchum Ellison, Jr., organic chemistry
Watkins, Mary Louise, biochemical pharmacology
Watson, Jack Throck, analytical chemistry, pharmacology
Webb, Glenn Francis, mathematical analysis
Webster, Burnice Hoyle, thoracic diseases
Weil, Jon David, molecular genetics
Wells, Charles Edmon, psychiatry, neurology
Wells, Jack Nulk, medicinal chemistry, pharmacology
Wesson, James Robert, mathematics
West, Harold D., biochemistry
Whitaker, Joe Russell, geography

Whittier, Dean Page, plant morphology
Wiesmeyer, Herbert, microbiology
Wilcox, Henry G, biochemistry, pharmacology
Wilkinson, Grant Robert, pharmacology
Willems, Emilio, anthropology
Williams, Marion Ervin, mycology
Wilson, Benjamin James, microbiology
Wilson, David J, physical chemistry
Wilson, James Lester, zoology
Wilson, John T, pediatrics, pharmacology
Wilson, Vernon Earl, medical administration, family medicine
Wolf, Frederick Taylor, botany
Womack, Frances C, genetics, enzymology
Wood, Henderson Kingsberry, genetics, physiology
Wood, James Lee, inorganic chemistry, thermochemistry
Woosley, Raymond Leon, clinical pharmacology
Yates, Harris Oliver, biology
Zealey, Marion Edward, biochemistry, microbiology
Zarger, Thomas Gordon, forestry, horticulture
Zelenik, John Slowko, obstetrics & gynecology

NEW JOHNSONVILLE
Buchacek, Robert Joseph, inorganic chemistry

NEWPORT
Clark, Stephen Darrough, organic chemistry
Griscom, Richard William, synthetic organic chemistry

NORRIS
Barnett, Paul Edward, forest genetics
Farmer, Robert E, Jr, forestry, plant physiology
Hall, Gordon Earl, zoology, aquatic biology
King, Woodrow Wilson, forest produs
Ripley, Thomas H, biology
Taft, Kingsley Arter, Jr, forestry, genetics

NORTH NASHVILLE
Johnson, Charles William, microbiology

OAK RIDGE
Abraham, Marvin Meyer, physics
Adler, Howard Irving, bacteriology
Akers, Lawrence Keith, physics
Alsmiller, Rufard G, Jr, physics, mathematics
Alton, Gerald Dodd, experimental atomic physics, theoretical atomic physics
Anderson, Richard Louis, physical chemistry
Anderson, Stanley H, ecology, environmental sciences
Andrews, Gould Arthur, internal medicine
Appleton, B R, solid state physics
Arakawa, Edward Takashi, solid state physics
Arnold, William Archibald, physiology
Auerbach, Stanley Irving, radiation ecology
Aull, Luther Bachman, III, science administration, nuclear physics
Auxier, John Alden, health physics, nuclear engineering
Baes, Charles Frederick, Jr, physical inorganic chemistry
Bair, Joe Keagy, nuclear physics
Baker, Philip Schaffner, chemistry
Baldwin, Willis Harford, organic chemistry
Ball, Frances Louise, organic chemistry, electron microscopy
Ball, James Bryan, nuclear chemistry
Ball, Robert P, radiology
Bamberger, Carlos Enrique Leopoldo, physical inorganic chemistry
Barber, Eugene John, physical inorganic chemistry
Barnett, William Edgar, molecular biology
Barrett, John Harold, theoretical solid state physics
Barton, Charles Julian, Sr, environmental health
Barton, James Clyde, chemistry
Bates, John Bryant, chemical physics
Bayne, Charles Kenneth, statistics
Beasley, Cloyd O, Jr, plasma physics

Beauchamp, John J, statistics, mathematics
Becker, Richard Logan, theoretical nuclear physics
Begun, George Murray, physical inorganic chemistry, chemical physics
Bell, Jimmy Todd, physical chemistry
Bell, Persa Raymond, nuclear medicine
Bemis, Curtis Elliot, Jr, nuclear physics, nuclear chemistry
Bernard, Selden Robert, mathematical biology
Bertini, Hugo W, physics
Best, Audrey Nance, bacteriology, biochemistry
Bibb, William Robert, immunology
Biggerstaff, John A, physics
Billen, Daniel, microbiology
Birkhoff, Robert D, experimental solid state physics
Blankenship, James Lynn, experimental solid state physics, semiconductors
Bohlmann, Edward Gustav, geochemistry, corrosion
Bond, Walter D, physical chemistry, chemical engineering
Boston, Charles Ray, physical chemistry
Bowman, Kimiko Osada, mathematical statistics
Bradshaw, Aubrey Swift, fresh water ecology
Braunstein, Helen Mentcher, science writing, information science
Braunstein, Jerry, physical chemistry
Bredderman, Paul John, cell biology, molecular biology
Bredig, Max Albert, physical inorganic chemistry
Brooks, Alfred Austin, Jr, physical chemistry
Brosi, Albert Ralph, chemistry
Brown, David Hazzard, cell biology
Brown, George Marshall, structural chemistry, x-ray crystallography
Brown, Keith Blanchard, applied chemistry
Brown, Lloyd Leonard, physical chemistry
Bruce, Francis Robert, chemistry
Brunton, George Delbert, geology, mineralogy
Bullock, Jonathan S, IV, physical chemistry
Burgess, Robert Lewis, plant ecology, botany
Burns, John Francis, experimental atomic physics
Burns, John Howard, physical chemistry
Burtis, Carl A, Jr, biochemistry, analytical chemistry
Busey, Richard Hoover, physical chemistry, thermodynamics
Busing, William Richard, physical chemistry
Butler, Thomas Arthur, chemistry
Butler, William H, solid state physics
Cable, Joe Wood, solid state physics
Callen, James Donald, plasma physics
Callihan, Alfred Dixon, physics
Cameron, Angus Ewan, mass spectrometry
Campbell, David Owen, radiochemistry
Cantor, Stanley, physical chemistry
Carlson, Roy Douglas, biophysics
Carlson, Thomas Arthur, molecular physics
Carson, Stanley Frederick, microbiology
Carter, H Kennon, experimental nuclear physics
Carter, Harvey Pate, computer science
Carter, Hubert Kennon, experimental physics
Cathcart, John Varn, physical chemistry
Chapman, Thomas Shelby, physical chemistry
Charles, George William, physics
Chen, Yok, physics
Chertok, Robert Joseph, physiology
Child, Harry Ray, solid state physics, magnetism
Chilton, John Morgan, inorganic chemistry
Christie, Warner Howard, mass spectrometry
Christophorou, Loucas Georgiou, atomic physics, molecular physics
Clapp, Neal K, radiobiology, pathology
Clark, Grady Wayne, physics
Clark, Walter Ernest, applied chemistry
Clarke, John F, physics
Cleland, John W, experimental solid state physics
Cloutier, Roger Joseph, health physics
Cohn, Hans Otto, physics
Cohn, Waldo E, biochemistry
Coleman, Charles Franklin, physical chemistry

Collins, Clair Joseph, physical organic chemistry
Coltman, Ralph Read, Jr, solid state physics
Compere, Edgar Lattimore, physical chemistry
Compton, Robert Norman, atomic physics, molecular physics
Condon, James Benton, physical chemistry, surface chemistry
Conger, Bob Vernon, plant genetics
Constantin, Milton J, plant breeding, botany
Cook, John Samuel, cell physiology
Cope, David Franklin, nuclear science, energy conversion
Cosgrove, Gerald Edward, pathology
Crandall, David Hugh, atomic physics, electron optics
Culkowski, Walter Martin, meteorology
Dabbs, John Wilson Thomas, nuclear physics, cryogenics
Darden, Edgar Bascomb, Jr, biophysics
Datz, Sheldon, chemical physics, atomic physics
Davies, Kenneth Thomas Reed, nuclear physics, theoretical physics
Davis, Harold Lloyd, theoretical solid state physics, surface physics
Davis, Wallace, Jr, chemistry
DeAngelis, Donald Lee, ecology
De Saussure, Gerard, nuclear physics
Dickens, Justin Kirk, nuclear physics
Dittner, Peter Fred, atomic physics
Doherty, David George, biochemistry
Donnellan, James Edward, Jr, molecular biophysics
Dorsey, George Francis, polymer chemistry
Dory, Robert Allan, mathematical physics
DuFrain, Russell Jerome, reproductive physiology, radiobiology
Dumont, James Nicholas, zoology, cytology
Dunlap, Julian Lee, physics
Dyer, Frank Falkoner, analytical chemistry, nucleonics
Eatherly, Walter Pasold, physics
Edwards, Kiah, III, developmental biology
Eichler, Eugene, nuclear chemistry, nuclear physics
Einstein, J Ralph, crystallography, biophysics
Ellis, Yurdanur Akovali, nuclear physics
Emerson, Lewis Cotesworth, surface physics
England, Alan Coulter, physics
Epler, James L, genetics, biochemistry
Ewbank, Wesley Bruce, nuclear physics, atomic physics
Farrar, Robert Lynn, Jr, physical chemistry
Faulkner, John Samuel, physics
Ferguson, Robert Lynn, nuclear science
Fields, David Edward, applied physics
Finamore, Frank Joseph, biochemistry, physical chemistry
Fisher, William David, cell physiology
Foard, Donald Andrew, mathematical statistics
Ford, James L C, Jr, physics
Fowler, Joseph Lee, high energy physics
Francis, Chester Wayne, soil science, ecology
Franklin, James Curry, analytical chemistry
Fujimura, Robert, biochemistry
Fulmer, Clyde Benson, physics
Gardiner, Donald Andrew, mathematical statistics
Garrett, William Ray, chemical physics
Gauster, Wilhelm Friedrich, electrophysics
Generoso, Walderico Malinawan, genetics
Gengozian, Nazareth, immunology
Gifford, Franklin Andrew, Jr, meteorology
Glasstone, Samuel, nuclear science
Gleason, Geoffrey Irving, radiochemistry
Goff, Frederick Glenn, ecology
Goldstein, Gerald, analytical chemistry
Goodall, Wilfred Manly, physics
Goodman, Charles David, experimental nuclear physics
Goodman, Joan Wright (Mrs Charles D), physiology
Gossler, David Gilbert, biometrics
Goswitz, Francis Andrew, nuclear medicine, hematology
Goswitz, Helen Vodopick, hematology, nuclear medicine
Gove, Norwood Babcock, nuclear physics
Grell, Ellsworth Herman, genetics
Grell, Rhoda Frank, genetics
Gresky, Alan Tolstoy, nuclear chemistry

Griesemer, Richard Allen, veterinary pathology
Gross, Edward Emanuel, physics
Guerin, Michael Richard, analytical chemistry, cancer
Gwin, Reginald, physics
Haeuslein, Guenter Karl, mathematics
Hahn, Richard Leonard, nuclear chemistry, nuclear physics
Halbert, Edith Conrad, theoretical nuclear physics
Halbert, Melvyn Leonard, nuclear physics
Halperin, Joseph, nuclear chemistry, nuclear physics
Hanna, Steven Rogers, meteorology
Harris, Warren Whitman, chemistry, electron optics
Harris, William Franklin, III, ecology
Hartman, Frederick Cooper, biochemistry, protein chemistry
Harvey, John Arthur, nuclear physics
Hayes, Raymond Leroy, nuclear medicine
Hedrick, Clyde Lewis, Jr, plasma physics
Henderson, Gray Stirling, forest soils, ecology
Henke, Randolph Ray, biochemical genetics
Hibbs, Roger Franklin, chemistry
Hill, Kenneth Wayne, atomic spectroscopy, atomic physics
Hochanadel, Clarence Joseph, physical chemistry, atmospheric chemistry
Hoffman, Everett John, physical chemistry
Holdeman, Jonas Tillman, Jr, physics, computer science
Holleman, James William, biochemistry
Holmes, David Kelley, solid state physics
Holmes, Howard Frank, physical chemistry
Holoway, Clayton Frank, organic chemistry
Horen, Daniel, nuclear physics
Horton, Charles Abell, analytical chemistry
Hsie, Abraham Wuhsiung, genetics, cell biology
Hubbell, Harry Hopkins, Jr, physics, health physics
Humason, Gretchen Lyon, zoology
Hurst, George Sam, radiation physics
Jacobs, Donald, physical chemistry, health physics
Jacobson, Karl Bruce, biochemistry
Jenkins, Leslie Hugh, surface physics
Jenks, Glenn Herbert, physical chemistry
Johnson, Carroll Kenneth, crystallography, biophysics
Johnson, Cleland Howard, nuclear physics
Johnson, Elizabeth Briggs, physics
Johnson, James Steven, Jr, physical chemistry
Johnson, Noah R, nuclear chemistry
Johnson, Philip L, ecology
Jolley, Robert Louis, environmental chemistry
Jones, Alfred Russell, organic chemistry
Jones, Charles Miller, Jr, physics
Jordan, Walter Harrison, physics, nuclear engineering
Katz, Sidney, analytical chemistry
Keilholtz, Gerald Watson, physical chemistry, analytical chemistry
Keim, Christopher Peter, chemistry
Keller, Oswald Lewin, physical chemistry, research administration
Kelley, Myron Truman, analytical chemistry
Kelly, Minton J, high temperature chemistry
Kelman, Bruce Jerry, perinatal biology, environmental sciences
Kelmers, Andrew Donald, inorganic chemistry
Kenney, Francis T, biochemistry
Kerchner, Harold Richard, experimental solid state physics
Ketchel, Melvin M, reproductive physiology
Kim, Hee Joong, nuclear physics
Kim, Jinchoon, plasma physics
Kimball, Richard Fuller, biology
Klatt, Leon Nicholas, electroanalytical chemistry, chemical instrumentation
Koehler, Wallace Conrad, physics
Krause, Herbert Francis, atomic physics, chemical physics
Krause, Manfred Otto, atomic physics, chemical physics
Kyker, Granvil Charles, biochemistry, science administration
Lafferty, Robert Hervey, Jr, analytical chemistry, inorganic chemistry
Larson, Nancy Marie, theoretical physics
Lavelle, George Cartwright, virology

TENNESSEE

Lazar, Norman Henry, physics
Lee, Kai-Lin, biochemistry, endocrinology
Leed, Russell Ernest, physical chemistry
Leibo, Stanley Paul, cryobiology, embryology
Lenhard, Joseph Andrew, health physics, nuclear physics
Lever, William Edwin, statistics
Levy, Henri Arthur, structural chemistry
Lindemer, Terrence Bradford, high temperature chemistry, nuclear chemistry
Littlefield, Gayle, anatomy, cytogenetics
Livingston, Ralph, chemical physics
Livingston, Robert Simpson, physics, research administration
Lombardi, Max H., radiation biology, nuclear medicine
Longworth, James W., biophysics, chemical physics
Lubell, Martin S., solid state physics
Ludeman, Carl Arnold, nuclear physics
Lushbaugh, Clarence Chancelum, pathology
Lyon, William Southern, Jr., radiochemistry
Macklin, Richard Lawrence, nuclear physics
MacLeod, Michael Christopher, molecular biology
MacPherson, Herbert Grenfell, science, nuclear engineering
Maienschein, Fred (Conrad), nuclear science administration
Malinauskas, Anthony Peter, molecular biology
Maya, Leon, inorganic chemistry
Mazur, Peter, cell physiology, cryobiology
McFee, Alfred Frank, cytogenetics, radiobiology
McGill, Robert Mayo, physical inorganic chemistry
McGowan, Francis Keith, physics
McKee, Rodney Allen, solid state kinetics
McLean, Richard Bea, marine biology, animal behavior
McNally, James Rand, Jr., plasma physics, solid state physics
Miller, Arthur Joel, chemistry
Miller, Forest Leonard, Jr., experimental statistics
Miller, James Kincheloe, animal nutrition, physiology
Miller, Philip Dixon, nuclear physics
Mills, Gordon Frederick, physical chemistry, organic chemistry
Minturn, Robert Edward, physical chemistry
Mitra, Sankar, molecular biology, biochemistry
Moak, Charles Dexter, physics
Mook, Herbert Arthur, Jr., physics
Moon, Ralph Marks, Jr., physics
Moore, George Edward, physical chemistry
Moore, Robert Earl, physical inorganic chemistry
Morrow, Roy Wayne, analytical chemistry
Mostoller, Mark Ellsworth, solid state physics
Mraz, Frank Rudolph, nutrition, physiology
Mueller, Theodore Rolf, analytical chemistry, instrumentation
Murphy, Brian Donal, physics
Nehls, James Warwick, inorganic chemistry
Neiler, John Henry, nuclear physics
Nettesheim, Paul, cancer
Neufeld, Jacob, theoretical physics
Newman, Eugene, nuclear physics
Nicklow, Robert Merle, solid state physics
Noonan, Thomas Robert, radiobiology, physiology
Novelli, Guerino David, biochemistry
Nowlin, Charles Henry, applied mathematics, applied physics
Nugent, Leonard James, plasma physics
Oakberg, Eugene Franklin, genetics
Obenshain, Felix Edward, nuclear physics, solid state physics
Odell, Theodore Tellefsen, Jr., cell biology, physiology
Oder, Charles Patchin, radiology
Oen, Ordean Silas, physics
O'Kelley, Grover Davis, nuclear chemistry, cosmochemistry
Olins, Ada Levy, biochemistry, electron microscopy
Olins, Donald Edward, biochemistry
O'Neill, Robert Vincent, ecology
Painter, Gayle Stanford, theoretical solid state physics
Pal, Bimal Chandra, organic chemistry, biochemistry
Papaconstantinou, John, biochemistry
Pare, Victor Kenneth, physics
Payne, Marvin Gay, theoretical physics
Pearlstein, Robert Milton, biophysics, chemical physics
Peelle, Robert W., nuclear physics
Perkins, Eugene Hafen, immunology
Peterson, Sigfred, chemistry
Pigg, Jay Cee, solid state physics
Pinajian, John Joseph, pharmaceutical chemistry, nuclear physics
Plasil, Franz, nuclear physics, nuclear chemistry
Pleasonton, Frances, experimental nuclear physics
Poggenburg, John Kenneth, Jr., nuclear chemistry
Pollard, William Grosvenor, solar physics, energy conversion
Pomato, Nicholas, molecular biology
Pomerance, Herbert (Solomon), physics
Popp, Raymond Arthur, embryology, genetics
Posey, Franz Adrian, electrochemistry, physical chemistry
Postma, Herman, physics
Poston, John Ware, health physics
Powell, George Louis, physical chemistry
Preston, Robert Julian, cytogenetics
Quist, Arvin Sigvard, physical chemistry
Raaen, Helen Parks, analytical chemistry
Raaen, Vernon F., organic chemistry
Ragan, George Leslie, nuclear physics, reactor physics
Rahn, Ronald Otto, biophysics
Rainey, William Thomas, Jr., organic chemistry
Raman, Subramanian, nuclear physics
Randolph, Malcolm Logan, biophysics
Raridon, Richard Jay, physical chemistry
Regan, James Dale, human genetics
Reichle, David Edward, ecology
Reister, David Bryan, mathematics, system analysis
Reynolds, Samuel Allen, analytical chemistry
Ricci, Enzo, applied physics, atomic physics
Rice, Walter Wilburn, analytical chemistry
Richmond, Chester Robert, radiobiology
Ritchie, Rufus Haynes, radiation physics
Robbins, Gordon Daniel, electrochemistry
Robinson, Mark Tabor, physics
Robinson, Russell Lee, nuclear physics
Rona, Elizabeth, radio chemistry, nuclear chemistry
Ronningen, Reginald Martin, nuclear physics
Ross, Harley Harris, analytical chemistry, chemical instrumentation
Rudolph, Philip S., physical chemistry, radiation instrumentation
Rush, Richard Marion, physical chemistry
Russell, Liane Brauch, genetics
Russell, William Lawson, genetics
Rutenberg, Aaron Charles, physical chemistry
Rutkowski, Robert William, physics, information science
Sasser, Lyle Blaine, animal nutrition medicine
Satchler, George Raymond, theoretical physics
Sayer, Royce Orlando, nuclear physics
Schmitt, Charles Rudolph, applied chemistry
Schreyer, James Martin, chemistry
Sears, Mildred Bradley, inorganic chemistry, chemical engineering
Sega, Gary Andrew, molecular genetics
Sekula, Stanley Ted, physics
Sellin, Helen Gill, immunobiology
Serrano, Louis Joseph, laboratory animal medicine
Shaw, Robert Wayne, spectroscopy
Shor, Arthur Joseph, physical chemistry
Shriner, David Sylva, phytopathology, forest ecology
Shugart, Herman Henry, Jr., ecology, zoology
Shugart, Lee Raleigh, biochemistry, microbial physiology
Shults, Wilbur Dorry, II, analytical chemistry
Silva, Robert Joseph, nuclear chemistry
Silver, Ernest Gerard, experimental physics
Silverman, Meyer David, physical inorganic chemistry
Skinner, Dorothy M., molecular biology
Slaga, Thomas Joseph, biochemical pharmacology
Smith, David Huston, chemistry
Smith, George Pedro, physical chemistry
Smith, Harold Glenn, physics
Smith, Lawson Harcourt, radiobiology
Smith, Lester Anah, nuclear physics
Smith, Orville L., nuclear physics
Smith, Wesley Earl, organic chemistry
Snell, Arthur Hawley, nuclear physics, atomic physics
Snyder, Fred Leonard, biochemistry
Sonder, Edward, physics
Souto, Jose, cancer, physiology
Spalding, Gary E., biomathematics
Spejewski, Eugene Henry, nuclear physics
Staats, Percy Anderson, physical chemistry, physics
Steele, Vernon Eugene, radiobiology
Stelson, Paul Hugh, physics
Stevens, Audrey L., biochemistry, microbiology
Stockdale, John Alexander Douglas, physics
Storer, John B., radiobiology
Stoughton, Raymond Woodford, physical chemistry, nuclear chemistry
Strehlow, Richard Alan, chemistry
Stulberg, Melvin Philip, biochemistry
Swartzendruber, Donald Clair, zoology
Sweeton, Frederick Humphrey, physical chemistry
Swenson, Paul Arthur, cell physiology
Sworski, Thomas John, radiation chemistry
Tamura, Tsuneo, soils
Taylor, Ellison Hall, physical chemistry
Tennant, Raymond Wallace, microbiology, virology
Terzaghi, Margaret, cancer
Textor, Robin Edward, mathematics
Thiessen, William Ernest, organic chemistry
Thoe, Robert Steven, atomic physics
Thoma, Roy E., inorganic chemistry, physical chemistry
Thorngate, John Hill, physics
Tompkins, Paul Carter, biochemistry
Toth, Kenneth Stephen, nuclear physics
Totter, John Randolph, biochemistry
Trauger, Donald Byron, physics
Trubey, David Keith, radiation physics
Tsang, Kang Too, plasma physics
Uppuluri, V R Rao, mathematical statistics
Urso, Paul, immunology, zoology
Uziel, Mayo, biological chemistry
Vander Sluis, Kenneth Leroy, physics
Van Rij, Willem Idaniel, computer sciences, plasma physics
Van Winkle, Webster, Jr., environmental sciences
Volkin, Elliot, biochemistry
Waggener, William Cole, inorganic chemistry
Walburg, Harry E., Jr., veterinary medicine
Walker, Raymond Lloyd, analytical chemistry
Wallace, Robin A., reproductive biology
Walker, Fred John, solid state physics, science administration
Ward, Helen Lavina, parasitology
Ward, Robert Cleveland, numerical analysis
Warmack, Robert Joseph, molecular physics
Washburn, Lee Cross, organic chemistry, nuclear medicine
Waters, Larry Charles, biochemistry
Weber, Charles William, analytical chemistry
Wedemeyer, Robert E., physical chemistry
Weeks, Robert A., solid state physics
Wei, Chin Hsuan, x-ray crystallography
Weinberg, Alvin Martin, nuclear physics, theoretical physics
West, John Leslie, pathology, bacteriology
Westbrook, Russell David, solid state physics
Wheaton, John Hobson, plasma physics
White, James Carl, analytical chemistry
Wieland, Bruce Wendell, biomedical engineering
Wilcox, William Jenkins, Jr., engineering physics
Wilkinson, Michael Kennerly, physics, solid state physics
Witherspoon, John Pinkney, Jr., radiation ecology, plant ecology
Wollan, Ernest Omar, physics
Wong, Cheuk-Yin, physics
Wood, Richard Frost, physics
Wright, Harvel Amos, mathematics
Wykle, Raymond Lee, biochemistry
Wymer, Raymond George, research administration
Yakel, Harry L., x-ray crystallography
Yang, Wen-Kuang, biochemistry, medicine
Yarbro, Claude Lee, Jr., ecology
Yeatts, LeRoy Brough, Jr., physical inorganic chemistry
Young, Gale, nuclear physics
Young, Jack Philip, analytical chemistry
Zehner, David Murray, surface physics
Zeldes, Henry, physical chemistry
Zerby, Clayton Donald, physics, mechanical engineering
Zucker, Alexander, nuclear physics

OLD HICKORY
Baxter, James F., polymer chemistry, textile chemistry
Estes, Leland Lloyd, organic polymer chemistry
Lueck, Charles Henry, analytical chemistry
Webb, Charles Alan, physical chemistry

POWELL
Kurzynske, Janet Stickley, nutrition

ROCKWOOD
Schwenker, Harold Constantine, physics

SAVANNAH
Stengle, William Bernard, wood

SEWANEE
Alvarez, Laurence Richards, mathematics
Camp, David Bennett, organic chemistry
Cheston, Charles Edward, forestry
Croom, Frederick Hailey, mathematics
Croom, Henrietta Brown, biochemistry
Ellis, Eric Hans, physics
Foreman, Charles William, zoology
Guenther, William Benton, physical chemistry, inorganic chemistry
Lorenz, Philip Jack, atmospheric physics
Lowe, James N., organic chemistry
McCrady, Edward, embryology, biophysics
Mignery, Arnold Louis, forestry
Owen, Howard Malcolm, zoology
Puckette, Stephen Elliott, mathematics
Ramseur, George Shuford, botany
Ross, Clay Campbell, Jr., mathematics
Smalley, Glendon William, forest soils
Yeatman, Harry Clay, zoology

SIGNAL MOUNTAIN
Hammond, James Alexander, Jr., organic chemistry
Perry, Lloyd Holden, chemistry

TEXAS

ABILENE
Beasley, Clark Wayne, invertebrate zoology
Bottom, Virgil Eldon, physics
Bradford, James C., mathematics
Brewer, John Hanna, medical bacteriology
Brokaw, Bryan Edward, animal breeding education
Craik, Eva Lee, biology, science education
Davis, Alvie Lee, biochemistry

TULLAHOMA
Brewer, LeRoy Earl, Jr., thermal physics
Dicks, John Barber Jr., physics
Huebschman, Eugene Carl, solid state physics
Joseph, Roy D., applied mathematics
Mason, Arthur Allen, molecular spectroscopy
Reddy, Kapuluru Chandrasekhara, applied mathematics, fluid mechanics

DuBose, Leo Edwin, animal science
Dunn, Floyd Warren, biochemistry
Foster, Terry Lynn, microbiology
Hance, Robert Lee, chemical physics, physical chemistry
Harris, Edward Lyndol, analytical chemistry
Harvey, John Frank, geology, engineering
Heidebrecht, A Allen, animal nutrition
Hughes, David Knox, mathematics
Hutchinson, Bennett B, spectrochemistry
Ivey, Robert Charles, chemical physics, structural chemistry
Johnson, William Cone, internal medicine
Jones, William Norton, Jr, organic chemistry
Justice, John Keith, soil fertility
Klassen, David Morris, inorganic chemistry, spectroscopy
Lewis, Donald Everett, biochemistry
McCord, Tommy Joe, biochemistry
Moore, Richard Dana, anatomy
Newman, George Allen, ecology, ornithology
Pilcher, Benjamin Lee, botany
Robinson, Charles Dee, mathematics
Shake, Roy Eugene, plant ecology
Sharp, A C, Jr, magnetism
Sonntag, Roy Windham, organic chemistry
Stevens, William Clark, microbiology, cell physiology
Williams, Kenneth Bock, plant taxonomy, zoology

ALPINE

Abbott, Maxine Langford, paleobotany
Deal, Dwight Edward, environmental geology
Houston, James Grey, organic chemistry, photochemistry
Powell, Albert Michael, botany
Rangra, Avinash K, physical chemistry
Scudday, James Franklin, vertebrate biology, wildlife biology

ALVIN

Houk, Wallace Eugene, entomology

AMARILLO

Barnes, Adele, biology
Denko, John V, pathology
Emerson, David Edwin, analytical chemistry
Faubion, Billy Don, physical chemistry
Gilmore, Earl Howard, mathematics
Kohn, Erwin, physical chemistry, polymer chemistry
Petr, Frank Charles, agronomy
Worley, Richard Dixon, physics

ARLINGTON

Armstrong, Andrew Thurman, physical chemistry, spectroscopy
Arnott, Howard Joseph, botany, cell biology
Baker, Willie Arthur, Jr, inorganic chemistry
Black, Truman D, solid state physics
Blake, Daniel Melvin, inorganic chemistry, organometallic chemistry
Boley, Robert B, bacteriology, immunology
Boon, John Daniel, Jr, geology
Bragg, Louis Hairston, botany
Burkart, Burke, geology, geochemistry
Butler, James Keith, cytology
Cheney, Monroe G, physics, geophysics
Cogdell, Thomas James, organic chemistry
Cooke, James Horton, theoretical physics
Cooper, James Erwin, organic geochemistry
Diana, Leonard M, physics
Dodge, Charles Fremont, geology
Eisenfeld, Jerome, applied mathematics
Ellis, Jason Arundel, physics
Girardot, Peter Raymond, inorganic chemistry
Green, Harold Rugby, number theory
Hall, Clarence Coney, Jr, acarology
Hamilton, William Wingo, mathematics
Heath, Larry Francis, mathematical analysis
Hellier, Thomas Robert, Jr, aquatic biology
Herrmann, Ulrich Otto, physics
Hopkins, Archibald Wilson, phycology
Huggins, Frank Norris, mathematics
Johnson, Grover Leon, physical chemistry

Kennerly, Thomas Everton, Jr, zoology, ecology
Klasky, Sheldon, microbiology
Lakshmikantham, Vangipuram, mathematics
Marquis, Richard Jack, physics
Martin, Donald Ray, inorganic chemistry
McCrady, William B, genetics
McDonald, William Charles, microbiology
McMahon, Robert Francis, III, invertebrate ecology
McNulty, Charles Lee, Jr, paleontology
McNutt, John Dewight, atomic physics, electron physics
Meacham, William Ross, vertebrate zoology
Miner, Norman Allen, molecular biology, virology
Mitchell, A Richard, applied mathematics
Mitchell, Roger W, mathematics
Moore, Marion E, mathematics
Murchison, John Taynton, organic chemistry
Nestell, Merlynd Keith, mathematics
Perkins, Bobby Frank, invertebrate paleontology
Perryman, John Keith, applied mathematics
Pomerantz, Martin, organic chemistry
Pyburn, William F, vertebrate zoology
Rayburn, Louis Alfred, physics
Ricca, Paul Joseph, analytical chemistry
Rubins, Roy Selwyn, magnetic resonance
Sims, Stillman Austin, mathematics
Singh, Harinder, computer science, geophysics
Tennison, Robert L, mathematics
Ternay, Andrew Louis, Jr, organic chemistry
Terrell, Glen Edward, nuclear physics
Thompson, Bonnie Cecil, solid state physics
Whitmore, Donald Herbert, Jr, comparative physiology
Williams, Bennie B, mathematics

AUSTIN

Acosta, Daniel, Jr, pharmacology
Adams, Richard Newbold, social anthropology, cultural anthropology
Alexopoulos, Constantine John, mycology
Altmiller, Henry, physical chemistry, radiation chemistry
Antoniewicz, Peter R, theoretical solid state physics, surface physics
Armendari, Efraim Pacillas, algebra
Arnold, R Keith, forestry
Arumi, Francisco Noe, thermal physics
Aunderson, Aubrey Lee, acoustics, oceanography
Austin, Thomas Howard, organic chemistry
Ayres, Gilbert Haven, analytical chemistry
Backus, Milo M, geophysics
Bailey, Philip Sigmon, organic chemistry
Baker, Dudley Duggan, III, underwater acoustics
Baker, Victor Richard, geomorphology, environmental geology
Baltzer, Otto John, physics
Bard, Allen Joseph, electroanalytical chemistry, physical chemistry
Barker, Daniel Stephen, geology
Barnard, Garland Ray, acoustics
Barnes, Virgil Everett, geology
Barth, Robert Hood, Jr, zoology
Barthel, Frank Ness, radio astronomy
Bash, Nathan Louis, organic chemistry
Bauld, Nathan Louis, organic chemistry
Bebout, Don Gray, sedimentology
Benedict, George Frederick, astronomy
Bengtson, Roger D, plasma physics, atomic physics
Bennett, Richard Henry, petroleum chemistry
Berberian, Sterling Khazag, mathematical analysis
Bernau, Simon J, mathematics
Bernstein, Richard Barry, physical chemistry
Berry, Levette Joe, microbiology
Bicheler, Klaus Richard, mathematics
Biesele, John Julius, cytology
Bing, R H, mathematics
Blackstock, David Theobald, acoustics
Blair, William Franklin, vertebrate biology
Bledsoe, Woodrow Wilson, mathematics
Bloch, David Paul, cytology
Boehme, Hollis Clyde, physics
Boggess, William Randolph, forestry
Boggs, James Ernest, chemistry

Bohm, Arno, high energy physics, theoretical physics
Bold, Harold Charles, botany
Bonner, Hugh Warren, exercise physiology
Bose, Henry Robert, Jr, microbiology
Boyer, Robert Ernst, geology
Brader, Walter Howe, Jr, organic chemistry
Bramblett, Claud Allen, physical anthropology, primatology
Brand, Donald Dilworth, geography, anthropology
Brand, Jerry Jay, plant physiology, photobiology
Breed, Benny Ray, environmental physics, system theory
Breland, Osmond Philip, zoology, entomology
Brenner, Michael Edward, organic chemistry
Bronson, Franklin Herbert, reproductive physiology
Brown, Leonard Franklin, Jr, geology
Brown, Walter Varian, botany
Browne, James Clayton, molecular physics, computer science
Brune, Gunnar Magnus, geology
Buchler, Ira Richard, anthropology
Buckman, Alvin Bruce, electrooptics, optical physics
Bullard, Fred Mason, geology, volcanology
Burk, Creighton, geology
Burlage, Henry Matthew, pharmacy, pharmaceutical chemistry
Bush, Guy L, evolutionary biology
Campbell, Thomas Nolan, anthropology, archaeology
Cannon, John Rozier, mathematics
Cavitt, Stanley Bruce, petroleum chemistry
Chao, Jia-Arng, mathematics
Charnes, Abraham, mathematics, economics
Cheney, Elliott Ward, (Jr), mathematics
Chester, Daniel Leon, computer sciences
Chiu, Charles Bin, elementary particle physics, high energy physics
Choi, Duk-In, plasma physics
Clabaugh, Stephen Edmund, geology
Coker, William Rory, theoretical nuclear physics, experimental nuclear physics
Collins, Francis Allen, solid state physics, chemical physics
Collins, Russell Lewis, chemical physics, plant embryology
Combs, Alan B, pharmacology
Cone, Conrad, organic chemistry, mass spectrometry
Connor, William Keith, acoustics
Cook, Eula Belle Maley, immunology
Cornell, John B, cultural anthropology, social anthropology
Cowley, Alan H, inorganic chemistry
Currier, Vernon Arthur, petroleum chemistry
Curtis, Howard Benton, Jr, mathematics
Cuscurida, Michael, organic chemistry
Daniel, James Wilson, numerical analysis, computer sciences
Dauwalder, Marianne, cell biology
Davies, Christopher Shane, urban geography, urban economics
Davis, Edward Mott, anthropology, archaeology
Davis Raymond E, physical chemistry
Deeming, Terence James, astronomy
DeFord, Exalton Alfonso, Jr, geology
Delco, Exalton Alfonso, Jr, vertebrate geology
Delevoryas, Theodore, paleobotany
Delgado, Jaime Nabor, pharmaceutical chemistry, medicinal chemistry
Desjardins, Claude, physiology
De Vaucouleurs, Gerard Henri, astronomy
Dewar, Michael James Steuart, chemistry
De Wette, Frederik Willem, solid state physics
DeWitt, Bryce Seligman, theoretical physics
DeWitt, Cecile Morette, theoretical physics
Dollard, John D, physics, mathematics
Doluisio, James Thomas, pharmaceutics
Dorman, Henry James, geophysics
Douglas, James Nathaniel, radio astronomy
Drummond, William Eckel, theoretical physics
Duff, Fratis L, medicine
Durbin, John Riley, algebra
Durden, Christopher John, paleontology, systematic entomology
Eakin, Richard Timothy, biophysical chemistry
Eakin, Robert Edward, biochemistry

Earhart, Charles Franklin, Jr, molecular biology, microbial physiology
Eaton, William Thomas, mathematics
Edmonds, Frank Norman, Jr, astrophysics
Edmonson, Don Elton, mathematics
Eifler, Gus Kearney, Jr, geology
Ellis, Glen Edward, acoustics, computer science
Ellison, Samuel Porter, Jr, geology
English, Paul Ward, cultural geography, geography of the Middle East
Eppright, Margaret Anne, biochemistry, nutrition
Epstein, Jeremiah Fain, anthropology, archaeology
Estes, Nelson N, physics
Evans, David Stanley, astronomy
Evans, Neal John, II, astrophysics
Faberge, Alexander Cyril, genetics
Fineg, Jerry, laboratory animal science
Fink, Manfred, atomic physics
Finney, James William, microbiology, biochemistry
Folk, Robert Louis, sedimentary petrology
Folkers, Karl August, chemistry
Fonken, Gerhard Joseph, organic chemistry
Forrest, Hugh Sommerville, molecular biology
Fox, Jack Lawrence, biophysics, biochemistry
Frazee, Jerry D, physical chemistry
Frazer, Marshall Everett, applied physics
Friedman, Charles Nathaniel, mathematics
Frommhold, Lothar Werner, atomic physics, plasma physics
Gardiner, William Cecil, Jr, physical chemistry
Gardner, Clifford S, applied mathematics
Gavenda, John David, experimental solid state physics
Gentle, Kenneth W, plasma physics
Gerth, Frank Emmett, III, algebra, number theory
Gibb, Elizabeth Glenadine, mathematics education
Gilbert, John Carl, physical organic chemistry
Gillman, Leonard, mathematics
Gipson, Robert Malone, industrial organic chemistry
Gjerstad, Gunnar, pharmacy, biochemistry
Gleeson, Austin M, elementary particle physics
Goff, Harold Milton, bioinorganic chemistry
Grant, Verne (Edwin), botany, genetics
Green, Francis Earl, geology, archaeology
Greenberg, Jerome Herbert, preventive medicine
Greenleaf, Newcomb, mathematics
Greenwood, Robert Ewing, mathematics
Griboval, Paul, electron optics
Griffy, Thomas Alan, theoretical physics
Groat, Charles George, economic geology
Guenther, Frederick Oliver, chemistry
Gustavson, Thomas Carl, geology
Guy, William Thomas, Jr, mathematics
Haddad, Eugene, nuclear physics
Haggerty, Michael John, statistical mechanics
Hale, Cecil Harrison, chemistry
Hall, Esther Jane Wood, pharmacy, health care administration
Hamilton, Terrell Hunter, physiology, developmental biochemistry
Hampton, Loyd Donald, physics
Hardesty, Boyd A, biochemistry
Harlow, Richard Leslie, x-ray crystallography
Harmon, Glynn, information science
Haynes, Kingsley Edwin, geography, environmental engineering
Hazeltine, Richard Deimel, plasma physics
Heisler, Joseph Patrick, mathematics
Henry, Christopher Duval, geochemistry, geochronology
Heplar, Joseph Quincy, microbiology
Hillery, Herbert Vincent, acoustics
Hinton, Frederick Lee, plasma physics
Hitchcock, Daniel Augustus, plasma physics, applied mathematics
Hoffman, George W, economic geography, political geography
Hoffmann, Gerald Wayne, nuclear physics
Holcombe, James Andrew, analytical chemistry
Holz, Robert Kenneth, geography of North Africa
Hopkins, George H, Jr, electrical engineering, geophysics

TEXAS

Horton, Claude Wendell, Sr., underwater acoustics
Horton, Claude Wendell, Jr., theoretical physics
Hubbs, Clark, ichthyology
Hudspeth, Emmett Leroy, nuclear physics
Hurd, Ray Merle, electrochemistry
Hutton, Allen Quilin, applied physics
Ingerson, Fred Earl, geology, geochemistry
Ivash, Eugene V., theoretical physics
Jackson, Eugene Bernard, information science
Jacobson, Antone Gardner, developmental biology
Jancarik, Jiri, plasma physics
Jefferys, William H. III, astronomy
Jespersen, Neil David, analytical chemistry
John, Peter William Meredith, mathematical statistics
Johnson, Fred Lowery, Jr., industrial organic chemistry
Johnston, Marshall Conring, systematic botany
Jonas, Edward Charles, clay mineralogy
Judd, Burke Haycock, genetics
Kehle, Ralph Ottmar, geophysics, applied mechanics
Kemmerer, Lorrin Garfield, resource geography
Kinsey, Bernard Bruno, physics
Kitto, George Barrie, biochemistry
Klein, Douglas J, molecular physics, quantum chemistry
Kleinman, Leonard, solid state physics
Klingman, Darwin Dee, operations research, mathematical statistics
Konecci, Eugene B, medical physiology
Koschmieder, Ernst Lothar, fluid dynamics
Kostoff, Morris R, nuclear physics, acoustics
Kwon, Young Koan, mathematics
Kyba, Evan Peter, organic chemistry
Land, Lynton S, geology, geochemistry
Langston, Wann, Jr, vertebrate paleontology
Lankford, Charles Ely, bacteriology
Larimer, James Lynn, neurobiology
Lawrence, James Franklin, mathematics
Lee, Addison Earl, botany, science education
Lee, Chong Sung, biological chemistry
Leffingwell, Thomas Pegg, Jr, cell biology
Leggett, Anne Marie, mathematical logic
LeMaistre, Charles Aubrey, internal medicine, epidemiology
Leslie, Steven Wayne, pharmacology
Lin, Cheng Shan, entomology
Little, Robert Narvaez, Jr, nuclear physics, science education
Lohr, John Michael, plasma physics
Long, Leon Eugene, geochemistry
Long, Walter K, human genetics
Longenecker, John Bender, nutrition, biochemistry
Lorentz, George G, mathematical analysis
Lundelius, Ernest Luther, Jr, vertebrate paleontology
Lynch, Daniel Matthew, plant ecology
Mabry, Tom Joe, organic chemistry, phytochemistry
Maguire, Bassett, Jr, ecology
Maguire, Marjorie Paquette, cytogenetics
Malina, Robert Marion, physical anthropology
Malkemus, John David, organic chemistry
Mandy, William John, immunogenetics
Mann, Ralph Willard, physics
Manners, Ian Robert, geography
Marquis, Edward Thomas, organic chemistry
Martin, Alfred, physical medicinal chemistry
Martin, Frederick N, audiology
Martin, Norman Marshall, mathematics
Martin, Robert Frederick, vertebrate zoology, ecology
Martin, Stephen Frederick, synthetic organic chemistry
Mascheroni, P Leonardo, theoretical physics
Mather, William Bardwell, economic geology
Matsen, Frederick Albert, chemistry, physics

Matzner, Richard Alfred, physics
Maxwell, John Crawford, geology, tectonics
McBride, Earle Francis, sedimentary petrography
McCormick, William Devlin, physics
McCullough, Thomas F, organic chemistry
McDowell, Fred Wallace, geochemistry
McGowen, Joseph Hobbs, geology
McKenna, George Finley, botany
McKinney, Chester Meek, physics
McMillan, Calvin, plant ecology
Mechler, Mark Vincent, physics
Menaker, Michael, comparative physiology
Mikeska, Emory Eugene, physiology
Miksad, Richard Walter, fluid dynamics, meteorology
Miller, Fred John, physics
Miller, Max K, mathematics, geophysics
Millett, Walter Elmer, physics
Monti, Stephen Arion, organic chemistry
Moore, Barry Newton, nuclear physics
Moore, Cornelius Fred, physics
Morgan, Ira Lon, nuclear physics
Morgan, Leon Owen, chemistry
Morris, Fred John, physics
Morton, Robert Alex, sedimentology
Moyer, William C, Jr, underwater acoustics
Muchlberger, William Rudolf, geology
Muir, Thomas Gustave, Jr, acoustics
Mulholland, John Derral, astronomy, celestial mechanics
Munk, Petr, polymer chemistry
Myers, Jack Edgar, photobiology
Nacozy, Paul E, astronomy, celestial mechanics
Nather, Roy Edward, astronomy, physics
Naylor, Carter Graham, petroleum chemistry
Neely, James Alan, anthropology, archaeology
Nematollahi, Jay, medicinal chemistry, microbiology
Newcomb, William Wilmon, Jr, ethnology
Nicol, Joseph Arthur Colin, zoology, marine biology
Nolle, Alfred Wilson, physics
Noyes, William Albert, Jr, physical chemistry
Nunemacher, Jeffrey Lynn, mathematical analysis
Oakes, Melvin Ervin Louis, plasma physics
Oliver, Symmes Chadwick, cultural anthropology
Olum, Paul, mathematics
Osborn, Roger (Cook), chemistry
Owen, Walter Wycliffe, computer sciences
Parsons, Ronald Gene, genetics
Paterson, Raleigh Elwood, chemistry, air pollution
Payne, Jimmie Sturgis, Jr, analytical chemistry
Pearson, Angus George, physics
Pettit, Rowland, organic chemistry
Phillips, Perry Edward, plasma physics
Pianka, Eric R, ecology
Pickett, Herbert McWilliams, physical chemistry
Pixley, Carl Preston, topology
Pledger, Gordon Wayne, mathematics
Poulsen, Lawrence Leroy, biochemistry, statistics
Powers, Edward Lawrence, radiation biology
Pratt, Terrence Wendall, computer science
Prescott, Kenneth Wade, ornithology
Privett, Orville Samuel, agricultural chemistry
Prouse, Ervin Joseph, mathematics
Prud'homme, John Thomas, nuclear physics
Purdy, William Henry, physics
Ramey, Bobbie Joe, organic chemistry
Ravel, Joanne Macow, biochemistry
Reed, Lester James, biochemistry
Reid, Russell Martin, physical anthropology, biological anthropology
Rhodes, John Rathbone, nuclear science
Richardson, Richard Harvey, genetics, ecology
Riffee, William Harvey, pharmacology
Riggs, Austen Fox, II, biochemistry
Riley, Peter Julian, experimental nuclear physics
Robbins, Ralph Robert, astrophysics
Roberts, Royston Murphy, organic chemistry
Robertson, William Woodrow, physics
Rode, Leonard John, bacteriology
Rogers, Lorene Lane, biochemistry

Rosene, Hilda Florence (Mrs E J Lund), electrophysiology
Ross, David Ward, theoretical physics
Rowton, Richard Lee, organic chemistry
Rudmose, H Wayne, physics
Runge, Thomas Marschall, cardiovascular diseases
Sabbagh, Michael E, geography, climatology
Sanchez, Bobby Gene, immunogenetics
Sanders, Charles W, physics
Scherer, Galen Lathrop, mathematics
Schieve, William, theoretical physics, statistical mechanics
Schild, Alfred, mathematical physics
Scholer, Aline Raymond, physiology
Schrank, Auline Raymond, physiology, biophysics
Schubhardt, Vernon Truett, microbiology
Schulze, Heinz, chemistry
Schumaker, Larry L, mathematics
Scott, Alan Johnson, paleontology
Seever, Galen Lathrop, mathematics
Selander, Robert Keith, zoology
Sheffield, William Johnson, pharmacy
Shepley, Lawrence Charles, cosmology
Shipman, Ross Lovelace, geology
Shipsey, Edward Joseph, physical research
Sikiossy, Laurent, computer science
Simonsen, Stanley Harold, chemistry
Smith, Douglas, geology
Smith, Harlan J, astronomy
Smith, Martha Kathleen, mathematics
Smith, Robert Victor, pharmaceutical chemistry
Smith, Spurgeon Eugene, topology
Soland, Richard Martin, operations research
Spear, Irwin, plant physiology, biology
Sprinkle, James (Thomas), invertebrate paleontology
Spurr, Stephen Hopkins, forestry, ecology
Starbird, Michael Peter, topology
Stark, William Richard, mathematical logic
Stavchansky, Salomon Ayzenman, pharmacy, pharmaceutics
Steinfink, Hugo, physical chemistry
Story, Dee Ann, anthropology, archaeology
Stuart, Joe Don, environmental sciences, computer science
Sudarshan, Ennackel Chandy George, theoretical physics
Sullivan, Gerald, pharmacy, pharmacognosy
Sutton, Harry Eldon, human genetics
Swanson, Basil Ian, inorganic chemistry
Swanson, Donald G, plasma physics
Sweet, Charles Edward, bacteriology, mycology
Swift, Jack Bernard, theoretical solid state physics
Szaniszlo, Paul Joseph, microbiology, mycology
Szebehely, Victor, astronomy
Tanner, Alan Roger, industrial organic chemistry, electrochemistry
Taylor, Morris Chapman, nuclear physics, medical physics
Thompson, Guy A Jr, biochemistry
Thompson, James Chilton, physics
Thurston, George Butte, biomedical engineering, biophysics
Truchard, James Joseph, acoustics, electronic engineering
Tull, Robert Gordon, astronomy
Turk, Leland Jan, geology, hydrology
Turner, Billie Lee, systematic botany
Ulrich, Bruce T, low temperature physics
Ulrich, Marie-Helene DeMoulin, astronomy
Urbatsch, Lowell Edward, systematic botany
Urdy, Charles Eugene, x-ray crystallography, inorganic chemistry
Vanden Bout, Paul Adrian, physics
Venema, Gerard Alan, topology
Venneman, Martin Ray, microbiology, immunology
Vick, James Whitfield, mathematics
Vidmar, Paul Joseph, physics
Villarreal, Jesse James, speech pathology
Wade, Charles Gordon, physical chemistry
Wade, William H, physical chemistry
Wagner, Gerald Richard, operations research
Wagner, Norman Keith, meteorology
Wagner, Robert Philip, genetics
Waite, Marilynn Ransom Fairfax, virology
Walker, James Roy, microbiology
Wallace, Reuben Henry, physics
Walston, Dale Edouard, mathematics
Ware, Alan Alfred, plasma physics

Warlick, Charles Henry, mathematics, computer science
Watt, George Willard, chemistry
Weaver, Milo Wesley, mathematics
Webber, Stephen Edward, physical chemistry
Wermund, Edmund Gerald, Jr, geology
Whaley, William Gordon, biology
Wheeler, John Craig, theoretical astrophysics
Wheeler, Marshall Ralph, zoology, genetics
White, Gifford, physics
White, John Michael, chemical physics
Whitesell, James Keller, synthetic organic chemistry
Wiegand, Oscar Fernando, plant physiology
Wilde, Kenneth Alfred, physical chemistry
Williams, Roger John, biochemistry, nutrition
Wilson, John Andrew, geology
Windham, Ronnie Lynn, analytical chemistry
Wittenborn, August Ferdinand, physics
Woodruff, Charles Marsh, Jr, physical geology
Worrell, Lee Frank, pharmaceutical chemistry
Wray, James David, astronomy
Wrotenbery, Paul Taylor, data processing
Wyatt, Robert Eugene, theoretical chemistry
Wynne, Michael, marine phycology
Wyss, Orville, microbiology
Yakatan, Gerald Joseph, pharmacy
Yanchick, Victor A, pharmacy
Yeakey, Ernest Leon, organic chemistry
Yeh, Raymond T, computer science
Yoon, Jong Sik, genetics, evolution
Young, David Monaghan, Jr, mathematics
Young, Keith Preston, environmental geology
Zaidi, Syed Amir Ali, nuclear physics
Ziegler, Daniel, biochemistry

BAYTOWN

Aczel, Thomas, analytical chemistry
Beck, Benny Lee, analytical chemistry
Bosniack, David S, organic chemistry, petroleum chemistry
Chamberlain, Nugent Francis, spectroscopy, electron microscopy
Floyd, Joseph Calvin, organic chemistry
Horoczy, Joseph Thomas, organic chemistry
House, William T, organic chemistry
Johnson, Malcolm Pratt, inorganic chemistry
Karchmer, Jean Herschel, analytical chemistry
Kelso, Edward Albert, petroleum chemistry
Lumpkin, Henry Earl, analytical chemistry
Mossman, Max Abe, physical chemistry
Neavel, Richard Charles, fuel science
Patton, Tad LeMarre, organic chemistry
Plank, Don Allen, organic chemistry, photochemistry
Powers, John Michael, organic chemistry
Reed, John J R, analytical chemistry
Thorn, John Paul, research administration
Trunnell, Jack B, medicine
Vernon, Lonnie William, physical chemistry, fuel science
Williamson, Charles Wesley, plastics chemistry
Wristers, Harry (Jan), organic chemistry

BEAUMONT

Achilles, Robert F, speech pathology, audiology
Aronow, Saul, geology
Baker, Harold Theodore, physical mathematics
Bowling, Clarence C, entomology, agriculture
Broun, Thorowgood Taylor, Jr, industrial chemistry
Cameron, Margaret Davis, organic chemistry
Cogburn, Robert Ray, entomology
Cowan, Russell (Walter), applied mathematics
Craignilee, Julian Pryor, agronomy
Crim, Sterling Cromwell, mathematics
Dorris, Kenneth Lee, physical chemistry
Eads, Erwin Alfred, inorganic chemistry
Eveland, Harmon Edwin, geology
Geddes, David Darwin, biology, physiology
Keller, William John, organic chemistry

Knabeschuh, Louis Henry, rubber chemistry
Lord, Samuel Smith, Jr, analytical chemistry
Matthews, William Henry, III, geology
McCauley, Garry Nathan, soil physics
McGraw, John Leon, Jr, cell biology, parasitology
Ortego, James Dale, inorganic chemistry
Pampe, William R, geology, paleontology
Peebles, Hugh Oscar, Jr, physics
Pizzo, Joseph Francis, physics
Ramsey, Jed Junior, zoology
Reeves, T Joseph, cardiovascular physiology, medicine
Rigney, Carl Jennings, physics
Shepherd, Jimmie George, physics
Sij, John William, plant physiology
Smith, William Russell, mathematics
Stark, Jeremiah Milton, microbiology
Tennissen, Anthony Cornelius, geology
Turco, Charles Paul, parasitology, nematology
Waddell, Henry Thomas, botany
Webb, Bill D, plant biochemistry
Westfall, Dwayne Gene, soil chemistry, agronomy
Wheeler, Robert Reid, petroleum geology
Whittle, John Antony, organic chemistry, biochemistry
Wilson, Mark Ferlin, mass spectrometry
Yerick, Roger Eugene, analytical chemistry
Zunker, Heinz Otto Hermann, pathology

BEDFORD
Searcy, Virgil Shell, agronomy

BEEVILLE
Wiltbank, James N, animal physiology

BELLAIRE
Allison, Jean Batchelor, physical chemistry
Cruser, Stephen Alan, analytical chemistry, physical chemistry
Dahl, Harry Martin, petroleum geology, exploration geology
DeRudder, Ronald Dean, geology
Eisner, Elmer, exploration physics
Hamilton, Daniel Kirk, geology
Huggins, Sara Espe, zoology
Kalfoglou, George, surface chemistry
Luttrell, Eric Martin, petroleum geology
Payne, Thomas Gibson, geology
Shupe, Russell Dwayne, organic chemistry
Soderman, J William, geology
Wilcox, Ronald Erwin, geology

BORGER
Kuper, Donald G, polymer chemistry, polymer engineering
Sircar, Anil Kumer, polymer chemistry

BROOKS AFB
Anderson, Donald Rex, radiobiology
Blouse, Louis E, Jr, bacteriology, microbiology
Burton, Russell Rohan, pathological physiology
Clark, Dale Allen, biochemistry
Douglas, William Kennedy, medicine, aerospace medicine
Douthit, Thomas D Nathan, geophysics
Ellis, James Percy, Jr, biochemistry
Homme, Paul John, microbiology, virology
Hughes, Harry Meachum, mathematical statistics
Ickes, Kenneth G, physical chemistry, biochemistry
Lancaster, Malcolm, aerospace medicine, cardiology
Leverett, Sidney Duncan, Jr, physiology
Richardson, Billy, research administration
Schmidt, Jerome F, medical microbiology
Shaw, Emil Gilbert, biochemistry, nutrition
Troxler, Raymond George, pathology
Wolfe, James Wallace, neurophysiology, psychology
Wolthuis, Roger A, cardiovascular physiology

BROWNSVILLE
Dulmage, Howard Taylor, insect pathology, microbiology
Graham, Harry Morgan, entomology
Hendricks, Donavan Edward, entomology, ecology
Truby, Charles Paul, medical science
Wolfenbarger, Dan A, entomology

BROWNWOOD
Madden, George D, horticulture, plant physiology
Schultz, Linda Dalquest, inorganic chemistry
Stanford, Jack Wayne, plant taxonomy
Stephenson, Danny Lon, organic chemistry, spectroscopy

BRYAN
Dillon, Lawrence Samuel, evolution
Dyksterhuis, Edsko Jerry, range ecology
Lewis, Robert Donald, agronomy
McCulley, William Straight, applied mathematics
Potts, Richard Carmechial, agronomy
Randolph, Henry England, food science
Sarkissian, Igor V, genetics, botany

BUSHLAND
Daniels, Norris Eugene, entomology
Eck, Harold Victor, soil fertility
Goss, Don Woodson, soil science
Mathers, Aubra Clinton, soil science
Porter, Kenneth Boyd, plant breeding
Rogers, Charlie Ellic, entomology
Stewart, Bobby Alton, soil fertility, soil chemistry
Unger, Paul Walter, soil science
Wiese, Allen F, weed science
Winter, Steven Ray, agronomy, plant physiology

CALDWELL
Franceschini, Guy Arthur, meteorology, oceanography

CANYON
Asquith, George Benjamin, petrology, mineralogy
Barieau, Robert (Eugene), physical chemistry
Brooks, Derl, entomology
Burton, Robert Clyde, geology
Busteed, Robert Charles, biology
Caraway, Prentice Alvin, biology
Carlisle, Gene Ozelle, inorganic chemistry
Cook, Hollis Lee, applied mathematics
Cooper, William Anderson, zoology, physiology
Crowder, Gene Autry, physical chemistry
Daugherty, Franklin W, geology, hydrology
Guidry, Marion Antoine, biochemistry
Head, Martha E Moore, atomic spectroscopy, theoretical high energy physics
Higgins, Larry Charles, plant taxonomy, plant morphology
Huffstutler, Ronald, mathematics
Hughes, Jack Thomas, anthroplogy, geology
LaBrie, David Andre, microbiology, biochemical genetics
Malzahn, Ray Andrew, organic chemistry, academic administration
Oliver, Joel Day, structural chemistry
Schram, Alfred C, biochemistry
Schultz, Gerald Edward, vertebrate paleontology
Smallwood, Charles M, animal science
Smith, Charles G, biology, forensic science
Underwood, James Ross, Jr, structural geology, petroleum engineering
Wheeler, David Laurie, geography of the Mediterranean, historical geography
Woodyard, James Douglas, organic chemistry
Wright, Robert Anderson, botany

CEDAR HILL
Stanford, Geoffrey Brian, environmental management

CENTER
Morris, William Lewis, mathematics
Trammell, Jack Harman, Jr, poultry nutrition

COLLEGE STATION
Aberth, Oliver George, mathematics
Adair, Thomas Weymon, III, solid state physics, low temperature physics
Adams, Emory Temple, Jr, biophysical chemistry
Adkisson, Perry Lee, economic entomology
Akins, Ervin Loraine, physiology, endocrinology
Alexander, Robert Benjamin, physical chemistry
Allen, Roland Emery, solid state physics
Amoss, Max St Clair, endocrinology
Anderson, Warren Boyd, soil chemistry
Armstrong, Howard Wayne, parasitology, invertebrate zoology
Arnold, Keith Alan, ornithology
Atkins, Irvin Milburn, plant breeding, agronomy
Atkinson, Robert Leon, poultry nutrition
Axtell, Darrell Dean, inorganic chemistry
Bailey, Everett Murl, Jr, veterinary toxicology
Baker, Robert Donald, forest managment
Bashaw, Elexis Cook, cytogenetics, plant breeding
Bassett, James Wilbur, animal science
Bassichis, William, theoretical nuclear physics
Baur, Joseph Ralph, plant physiology, biochemistry
Bay, Darrell Edward, entomology
Beard, James B, plant physiology
Bell, Alois Adrian, plant pathology
Benedict, Chauncey, plant biochemistry
Berg, Robert R, geology
Berner, Leo Dewitte, Jr, biological oceanography
Bhaskaran, Govindan, insect physiology, developmental biology
Bird, Luther Smith, plant pathology
Black, Samuel Harold, microbiology
Blackburn, Wilbert Howard, watershed management
Blackhurst, Homer Tennyson, horticulture
Blakley, George Robert, mathematics, mathematical biology
Bockholt, Anton John, plant breeding, genetics
Boone, James Robert, mathematics
Borosh, Itshak, mathematics, number theory
Bottino, Nestor Rodolfo, biochemistry
Bouma, Arnold Heiko, marine geology, sedimentology
Bovey, Rodney William, weed science
Bowen, Hollis Hulon, plant breeding, genetics
Brick, Robert Wayne, animal physiology
Bridges, Charles Hubert, veterinary pathology
Bright, Thomas J, biological oceanography
Bronson, Jeff Donaldson, nuclear physics
Brown, Douglas Richard, nuclear physics
Brown, Meta (Louise) Suche, cytogenetics
Brundidge, Kenneth Cloud, meteorology
Bryan, Ronald Arthur, theoretical nuclear physics
Bryant, Vaughn Motley, anthropology, botany
Bryant, William Richards, marine geology, oceanography
Bull, Don Lee, entomology
Burns, Edward Eugene, food science
Butler, Ogbourne Duke, animal science
Calaway, Paul Kenneth, organic chemistry
Caldwell, Jerry, immunogenetics, molecular genetics
Caldwell, Ralph Merrill, phytopathology
Camp, Bennie Joe, biochemistry
Carpenter, Zerle Leon, animal science, food science
Carter, George Francis, cultural geography
Cartwright, Thomas Campbell, animal breeding
Cate, James Richard, Jr, entomology
Cater, Carl Malcom, biochemistry
Chao, Jing, physical chemistry
Chen, Stephen Shiowshiung, physical chemistry
Chiad, Tang, atomic physics
Chui, Charles Kam-Tai, mathematics
Clark, Patrick Joseph, analytical chemistry
Clark, Robert Beck, elementary particle physics
Clark, William Jesse, limnology, aquatic ecology
Clearfield, Abraham, inorganic chemistry
Cobb, Bryant Franklin, III, food science, fisheries
Cochran, Robert Glenn, physics
Collier, Jesse Wilton, agronomy, genetics
Conway, Dwight Colbur, physical chemistry
Cook, Earl Ferguson, geology, geography
Coon, Jesse Bryan, physics
Couch, James Russell, neurology, neuropharmacology
Coulson, Robert N, insect ecology, forest entomology
Creger, Clarence R, biochemistry
Crookshank, Herman Robert, clinical chemistry
Dahm, Karl Heinz, biochemistry
Darnell, Rezneat Milton, ecology
Das, Phanindramohan, atmospheric physics, meteorology
Davis, William B, wildlife management, conservation
Dieckert, Julius Walter, biochemistry
Dill, Charles William, food science, organic chemistry
Dixon, James Ray, herpetology
Dixon, Joe Boris, soil science
Djuric, Dusan, meteorology
Dobson, William Jackson, zoology
Dodd, Jimmie Dale, plant ecology, soils
Dollahite, James Walton, veterinary toxicology
Doran, Edwin, Jr, cultural geography, geography of the Pacific
Drew, Dan Dale, computer science, mathematics
Driscoll, Dennis Michael, meteorology
Dronen, Norman Obert, Jr, parasitology
Duble, Richard Lee, agronomy, plant physiology
Duller, Nelson M, Jr, nuclear physics
Dutson, Thayne R, meat science
Eastin, Emory Ford, botany
Egan, Robert Shaw, lichenology
Eggers, Richard Carl, nuclear chemistry
Ellett, Edwin Willard, veterinary medicine & veterinary surgery
Ellis, William C, animal nutrition
Emery, William Jackson, physical oceanography
Eugster, A Konrad, veterinary virology
Fahlquist, Davis A, geophysics, oceanography
Fanguy, Roy Charles, immunogenetics
Fendler, Eleanor Johnson, physical organic chemistry, bio-organic chemistry
Fendler, Janos Hugo, organic biochemistry
Ferguson, Thomas Morgan, avian physiology
Fife, William Paul, physiology, anatomy
Foster, Billy Glen, microbiology
Frankie, Gordon William, insect ecology
Frederiksen, Richard Allan, plant pathology, plant genetics
Freund, Rudolf Jakob, statistics
Friedman, Melvin, rock mechanics, structural geology
Frisbie, Raymond Edward, entomology
Fryxell, Greta Albrecht, biological oceanography
Fryxell, Paul Arnold, systematic botany
Galvin, Thomas Joseph, parasitology, veterinary medicine
Gangi, Anthony Frank, geophysics
Gardner, Frederick Albert, organic chemistry, microbiology
Gartner, Stefan, Jr, micropaleontology, marine geology
Gates, Charles Edgar, agricultural statistics
Geyer, Richard Adam, oceanography
Giam, Choo-Seng, physical organic chemistry, analytical chemistry
Gilmore, Earl C, genetics
Gilstrap, Franklin Ephriam, entomology
Gingerich, Karl Andreas, inorganic chemistry, physical chemistry
Gladden, James Kelly, physical chemistry
Gleiser, Chester Alexander, veterinary pathology, comparative pathology
Glover, George Irvin, organic chemistry, protein chemistry
Godfrey, Curtis Loveing, soil science
Gold, John Rush, genetics
Gould, Frank Walton, botany
Graves, Robert Gage, experimental nuclear physics
Gravett, Howard L, genetics
Gray, Carl, soil fertility
Green, Phillip Joseph, II, cosmic ray physics
Greenblatt, Gerald A, plant physiology
Griffiths, John Frederick, meteorology
Grigsby, Ronald Davis, biophysical chemistry
Guseman, Lawrence Frank, Jr, mathematics
Haas, Robert Henry, range science, range ecology
Haensly, William Edward, veterinary medicine, gerontology
Halliwell, Robert Stanley, plant virology
Hallion, John McDonell, plant physiology, plant pathology
Ham, Joe Strother, physics, chemistry
Hancock, Charles Kinney, organic chemistry
Handin, John Walter, geophysics
Hanna, Ralph Lynn, entomology
Harding, Kenn E, organic chemistry

665

Harrington, Marion Thomas, inorganic chemistry
Harris, Edward David, biochemistry
Harris, Marvin Kirk, entomology
Harris, Robert Lee, entomology
Harry, Harold William, invertebrate zoology
Hart, Gary Elwood, genetics
Hartfiel, Darald Joe, mathematics
Hartley, Herman Otto, statistics
Harvey, Roger Bruce, poultry pathology
Hassell, Clinton Alton, chemistry, mass spectrometry
Heck, Fred Carl, virology
Hedges, Dorothea Huseby, biochemistry, enzymology
Hedges, Richard Marion, physical chemistry
Henson, Rodger Dale, natural products chemistry
Hidalgo, Richard Jack, veterinary microbiology
Hiebert, John Covell, nuclear physics
Hightower, Dan, veterinary nuclear medicine
Hoeve, Cornelis Abraham Jacob, physical chemistry
Hoffman, Robert A, economic entomology
Hogg, John Leslie, bio-organic chemistry
Holman, Grant Mark, insect physiology
Holmquest, Donald L, nuclear medicine, physiology
Holt, Ethan Cleddy, plant breeding, agronomy
Hoover, William L, inorganic chemistry, soil chemistry
Hopkins, Sewell Hepburn, marine biology
Hossner, Lloyd Richard, soil chemistry, soil fertility
Hurt, John Tom, applied mathematics
Hutta, Paul John, chemical physics
Ichiye, Takashi, physical oceanography
Inglis, Jack Morton, ecology
Irgolic, Kurt Johann, organometallic chemistry, analytical chemistry
Isbell, Arthur Furman, organic chemistry
Jain, Mahavir, experimental nuclear physics
James, Bela Michael, biological oceanography
Joham, Howard Ernest, plant physiology
Jones, Larry Philip, veterinary pathology
Jordan, Wayne Robert, plant physiology, biochemistry
Kattawar, George W, planetary atmospheres
Keeley, Larry Lee, insect physiology
Kelley, John Richard, Jr, fisheries
Kelton, William Henry, microbiology
Kemp, Walter Michael, immunobiology, parasitology
Kenefick, Robert Arthur, nuclear physics
Ketring, Darold L, plant physiology, biochemistry
Kieffer, Nat, genetics
Kim, Hyeong Lak, natural products chemistry
Kimber, Clarissa Therese, geography
King, General Tye, meat science
Kinman, Murray Luther, agronomy
Kirk, Wiley Price, low temperature physics
Kirwan, Albert Dennis, Jr, physical oceanography
Kleerekoper, Herman, animal physiology, animal behavior
Klemm, William Robert, neurosciences
Klipple, Edmund Chester, mathematical analysis
Knight, James Allen, psychiatry
Koenig, Karl Joseph, geology
Kohel, Russell James, plant genetics
Kothmann, Merwyn Mortimer, range science, range management
Kraemer, Duane Carl, reproductive physiology, medicine
Kramer, Paul R, forestry
Kreglewski, Alexander, thermodynamics, petroleum chemistry
Krise, George Martin, physiology
Krueger, Willie Frederick, poultry husbandry, genetics
Kshirsagar, Anant Madhav, statistics, operations research
Kuchnow, Karl Paul, behavioral physiology
Kunkel, Harriott Orren, biochemistry, nutrition
Kunz, Sidney Edmund, entomology, ecology
Kunze, George William, soil mineralogy
Kuttler, Kenneth Latimer, microbiology, protozoology
Laane, Jaan, physical chemistry, spectrochemistry
Landmann, Wendell August, biochemistry

Lane, Gary (Thomas), animal nutrition, biochemistry
Larsen, John Elbert, horticulture
Larsen, Russell D, chemical physics, statistics
Lee, Dean Ralph, organic chemistry
Leighton, Rudolph Elmo, dairy science
Leinweber, Charles Lee, plant physiology
Leribaux, Henri Romain, solid state physics, fluid physics
Lewis, Donald Howard, fish pathology, microbiology
Lewis, Roscoe Warfield, animal nutrition, biochemistry
Liebhafsky, Herman Alfred, chemistry
Lippke, Hagen, animal nutrition
Logan, John Merle, structural geology, tectonophysics
Long, Ernest M, forest genetics
Lopez, Genaro, economic entomology
Loyd, Coleman Monroe, physics
Lunsford, Jack Horner, physical chemistry
Luther, Herbert Adesla, mathematics
Lyda, Stuart D, plant pathology
Ma, Chin Wah, theoretical nuclear physics
Macfarlane, Ronald Duncan, biophysical chemistry
Magill, Clint William, genetics
Magill, Jane Mary (Oakes), biochemistry, genetics
Mariano, Patrick S, chemistry
Martell, Arthur Earl, chemistry
Matis, James Henry , statistics, mathematical statistics
Mattil, Karl Frederick, food technology
Maurer, Fred Dry, veterinary pathology
Maxson, Carlton J, mathematics
Mayer, Richard Thomas, toxicology
McBee, George Gilbert, plant physiology
McConnell, Stewart, virology, immunology
McCrady, James David, veterinary physiology
McDaniel, Milton Edward, plant breeding, genetics
McIntyre, John Armin, nuclear physics
McNiel, Norbert Arthur, genetics
McWilliams, Edward Lacaze, ornamental horticulture
Meinke, Wilmon William, organic chemistry, biochemistry
Mellor, David Bridgwood, poultry science, food technology
Melton, James Ray, analytical chemistry, soil chemistry
Meola, Shirlee May, entomology
Merker, Jerry Wheeler, reproductive physiology
Merrifield, Robert G, silviculture
Meyer, Edgar F, structural chemistry
Meyer, Robert Earl, plant physiology
Milford, Murray Hudson, soil science, soil mineralogy
Miller, Charles Standish, plant physiology
Moehring, David Marion, forest soils, forest ecology
Moore, Bill C, mathematics, operations research
Moore, Richard Wayne, veterinary microbiology
Morgan, Page Wesley, plant physiology, biochemistry
Moyer, Vance Edwards, meteorology
Nagatani, Kunio, nuclear physics
Nagyvary, Joseph, organic chemistry
Nair, Kuttenair Gopinathan, nuclear physics
Namboodiri, Madassery Neelakantan, nuclear chemistry
Natowitz, Joseph Bernard, nuclear chemistry
Naugle, Donald, low temperature physics
Naugle, Norman Wakefield, numerical analysis, applied mathematics
Neff, Richard D, health physics
Neill, William Harold, fish biology
Nickelson, Ranzell, II, food microbiology
Nightingale, Arthur Esten, horticulture
Niles, George Alva, agronomy, plant breeding
Northcliffe, Lee Conrad, nuclear physics
Nowlin, Worth D, Jr, physical oceanography
O'Brien, Daniel H, organic chemistry
O'Connor, Rod, organic biochemistry
O'Donovan, Gerard Anthony, microbiology, biochemistry
Olson, Jimmy Karl, medical entomology
Orts, Frank Ausut, meat science
Osoba, Joseph Schiller, physics
Ott, Aleta Jo Petrik, botany, plant taxonomy
Pace, Carlos Nick, physical biochemistry
Palmer, Rupert Dewitt, weed science
Parker, Travis Jay, geology

Parrish, David Keith, geology
Paterson, Donald Robert, horticulture
Payne, Thomas Lee, entomology
Peck, Merlin Larry, bio-organic chemistry
Pennington, Campbell White, geography
Pequegnat, Linda Lee Haithcock, biological oceanography, taxonomy
Pequegnat, Willis Eugene, oceanography
Pettit, Robert Eugene, plant pathology & physiology
Pierce, Kenneth Ray, veterinary pathology, clinical pathology
Plass, Gilbert Norman, planetary atmospheres
Pooch, Udo Walter, theoretical physics, computer science
Powell, Robert Delafield, plant physiology
Prescott, John Mack, biochemistry
Presley, Bobby Joe, chemical oceanography
Presley, John Thomas, plant pathology
Price, Alvin Audis, veterinary medicine
Price, Jack D, biochemistry, nutrition
Puhalla, John Edward, microbial genetics
Quarles, John Monroe, medical microbiology, oncology
Quisenberry, John Henry, poultry science, animal nutrition
Raulston, James Chester, horticulture
Ray, Lee Edmisten, biochemistry
Reading, John Frank, theoretical physics
Reagor, John Charles, toxicology, biochemistry
Reid, Robert Osborne, oceanography
Reimer, Rollin DeWayne, zoology
Reiser, Raymond, biochemistry
Rezak, Richard, geological oceanography
Rhee, Khee Choon, food science
Richmond, Thomas Rollin, plant breeding
Riggs, John Kamm, animal science, animal nutrition
Ringer, Larry Joel, statistics
Rizzo, Peter Jacob, plant physiology, cell biology
Rodgers, Alan Shortridge, physical chemistry
Roenigk, William J, veterinary medicine
Roller, Herbert Alfred, zoology, biological chemistry
Rooney, Lloyd William, food science
Rosberg, David William, plant pathology
Rose, Timothy Laurence, chemical physics, chemical kinetics
Rowe, Marvin W, nuclear geochemistry
Rowland, Lenton O, Jr, poultry nutrition
Runkles, Jack Ralph, soil physics
Runnels, Robert Clayton, meteorology
Ryan, Cecil Benjamin, poultry husbandry
Sackett, William Malcolm, geochemistry, chemistry
Saslow, Wayne Mark, low temperature physics, solid state science
Schaffner, Joseph Clarence, entomology
Schake, Lowell Martin, animal science
Scheerbaum, Robert R, theoretical physics
Schertz, Keith Francis, cytogenetics
Schink, David R, chemical oceanography
Schmidly, David James, mammalogy, systematics
Scholl, Philip Cornelius, organic chemistry
Schroeder, Harry William, plant pathology
Schroeder, Melvin Carroll, geology
Schroeter, Gilbert Loren, genetics, cytogenetics
Schuessler, Hans Achim, atomic physics
Schuster, Joseph L, range management, ecology
Schwartz, William Lewis, veterinary pathology
Schweikert, Emile Alfred, analytical chemistry
Scifres, Charles Joel, range science, weed science
Scoggins, James R, meteorology
Scott, Martha Richter, marine geochemistry
Scott, Robert Blackburn, geology
Shapiro, Bernard Lyon, organic chemistry, nuclear magnetic resonance
Shapiro, James Norman, geophysics
Shelton, George Calvin, veterinary parasitology
Sicilio, Fred, inorganic chemistry, radiochemistry
Sielken, Robert Lewis, Jr, mathematical statistics, operations research
Simpson, Russell Bruce, veterinary microbiology
Sippel, William Lawrence, veterinary pathology
Sis, Raymond Francis, veterinary anatomy

Smalley, Harry Edwin, veterinary toxicology
Smeins, Fred E, plant ecology
Smith, Dudley Templeton, weed science
Smith, Gary Chester, meat science
Smith, James Douglas, genetics, statistics
Smith, James Willie, Jr, entomology
Smith, Olin Dail, plant breeding
Smith, Roberta Hawkins, plant physiology, plant science
Smith, William Boyce, mathematical statistics
Sonnenfeld, Joseph, cultural geography
Sorensen, Anton Marinus, Jr, animal science, reproductive physiology
Spence, Thomas Wayne, oceanography
Spencer, Terry Warren, geophysics
Sperry, John Jerome, systematic botany
Spraggins, Robert Lee, organic chemistry
Stanton, Robert James, Jr, geology, paleontology
Staten, Raymond Dale, agronomy, botany
Stearns, David Winrod, structural geology
Stephenson, Robert Charles, geology
Sterling, Winfield Lincoln, entomology
Stipanovic, Robert Douglas, natural product chemistry
Storey, James Benton, pomology, plant physiology
Storts, Ralph Woodrow, veterinary pathology
Strawn, Robert Kirk, ichthyology
Sugihara, Thomas Tamotsu, nuclear chemistry
Sweet, Merrill Henry, II, biology
Swoboda, Allen Ray, soil chemistry, agronomy
Szabuniewicz, Michael, veterinary medicine
Taber, Willard Allen, microbiology
Tang, Yi-Noo, physical chemistry, radiochemistry
Thames, Walter Hendrix, Jr, nematology
Thomas, Norman Dwight, anthropology, ethnology
Thompson, Aylmer Henry, meteorology
Thompson, Patrick Haley, entomology
Thurston, Earle Laurence, botany, cell biology
Tieh, Thomas Ta-Pin, mineralogy
Toler, Robert William, virology, plant pathology
Tolmsoff, Walter John, plant pathology, biochemistry
Torgerson, David Franklyn, mass spectrometry
Treybig, Leon Bruce, mathematics
Tribble, Robert Edmond, nuclear physics
Tsutsui, Ethel Ashworth, biochemistry
Tsutsui, Minoru, organometallic chemistry
Unterberger, Robert Ruppe, physics
Van Bavel, Cornelius H M, agronomy, biology
Van Buijtenen, Johannes Petrus, forest genetics
Van Cleave, Horace William, entomology
Vanderzant, Carl, food microbiology
Vanderzant, Erma Schumacher, biochemistry
Van Overbeek, Johannes, plant physiology, biology
Vastano, Andrew Charles, physical oceanography, numerical analysis
Veech, Joseph A, plant pathology, plant physiology
Vinson, S Bradleigh, entomology
Walther, Fritz R, animal behavior
Watkins, Gustav McKee, botany
Watson, Benjamin Franklin, microbiology
Watson, Rand Lewis, nuclear chemistry
Weaver, Richard Wayne, soil microbiology
Weirich, Gunter Friedrich, biological chemistry
Welch, Charles Darrel, agronomy
Weseli, Donald Fenton, immunology
Whealy, Roger Dale, physical chemistry, analytical chemistry
Whitehouse, Ulysses Grant, physical chemistry
Whiteley, Eli Lamar, soil physics
Whitford, Howard Wayne, veterinary microbiology
Whitson, Robert Edd, agricultural economics, range management
Wild, James Robert, molecular biology
Wilhoit, Randolph Carroll, thermodynamics, physical biochemistry
Witzel, Donald Andrew, veterinary physiology
Wolf, Harold William, environmental health
Wright, James Elbert, medical entomology, bacteriology
Young, Andrew Tipton, astronomy

Youngblood, Dave Harper, physics
Zingaro, Ralph Anthony, inorganic chemistry
Zolnowski, Dennis Ronald, nuclear physics
Zwolinski, Bruno John, physical chemistry

COMMERCE
Ashley, Kenneth R, inorganic chemistry
Attrep, Moses, Jr, radiochemistry, geochemistry
Barton, Janice Sweeny, biophysical chemistry
Bedgood, Dale Ray, mathematics, statistics
Bone, Larry Irvin, chemical kinetics, physical chemistry
Chopra, Dev Raj, experimental atomic physics, surface physics
Clements, John Herbert, physics
Clevenger, Richard Lee, organic chemistry, biochemistry
Echols, Joan, vertebrate paleontology
Gale, Douglas Shannon, II, nuclear physics, experimental atomic physics
Ghaly, Tharwat Shahata, geology, petrology
Goddard, Alton R, computer science, mathematics
Gupta, Sujoy, palynology
Hoover, W Farrin, geology, geography
Horton, Otis Howard, dairy husbandry
Ingold, Donald Alfred, ecology, ethology
Jenkins, William Frank, horticulture
Jones, Charles E, molecular physics, atomic physics
Klaus, Ewald Fred, Jr, zoology, entomology
La Prade, Kerby Eugene, geology
McDaniel, Willard Rich, meteorology, geology
McFeeley, James Calvin, botany
McWhirter, Nolan, geology
Min, Kwang-Shik, mathematical physics, nuclear science
Neff, Laurence D, physical chemistry, inorganic chemistry
Norwood, James S, reproductive physiology, genetics
Peterson, Hazel Agnes, geology
Quane, Denis Joseph, inorganic chemistry
Razniak, Stephen L, organic chemistry
Roberts, Evan Paul, botany, biology
Rodriguez, Carlos Eduardo, computer science
Rohrer, Charles Stephen, physical chemistry, inorganic chemistry
Williams, Robert K, entomology, cell biology
Wilson, Bobby Eugene, microbiology, ecology
Zander, Arlen Ray, nuclear physics, atomic physics
Zimmerman, John Richman, physics

CORPUS CHRISTI
Aguilo, Adolfo, industrial chemistry
Blay, Jorge Albert, organic chemistry
Calma, Victor Charles, medical education, pediatrics
Clayton, Neal, seismology, geophysics
Darlington, William Bruce, inorganic chemistry
Edman, Dwight Douglas, industrial organic chemistry
Fisher, Gene Jordan, organic chemistry
Foster, Robbie T, organic chemistry
Garrison, Louis Eldred, oceanography, geology
Greding, Edward J Jr, herpetology
Hobbs, Charles Clifton, Jr, organic chemistry
Holmes, Charles Ward, marine geology, geochemistry
Levy, Leon Bruce, industrial organic chemistry
Markwell, Dick Robert, electrochemistry, analytical chemistry
Milczarek, Chester J, inorganic chemistry
Morrow, Homer Nicholas, Jr, organic chemistry
Nielsen, Donald R, organic chemistry
Oetking, Philip, marine geology
Price, William Armstrong, geology, oceanography
Reilly, William Leo, organic chemistry, analytical chemistry
Robinson, James Vance, entomology
Rogers, Thomas Ralph, industrial chemistry
Scott, Robert Edward, analytical chemistry
Shideler, Gerald Lee, sedimentology, marine geology

Slinkard, William Earl, inorganic chemistry
Starr, Leon; organic chemistry
Stubeman, Robert Frank, physical chemistry
Sugimoto, Roy, organic chemistry
Unruh, Jerry Dean, industrial organic chemistry
Wagner, Frank S, Jr, organic chemistry
Wilkerson, Robert C, physics, mathematics
Witt, Enrique Roberto, organic chemistry

DALLAS
Adams, Henry Richard, veterinary pharmacology
Adcock, Willis Alfred, chemistry
Agatston, Robert Stephen, geology
Albritton, Claude Carroll, Jr, geology
Allen, Linus Scott, applied physics
Anderson, David H, mathematics
Anderson, George Boine, marine geophysics, military systems
Anderson, Richard Gilpin Wood, cell biology, reproductive biology
Andrychuk, Dmetro, spectroscopy
Angona, Frank Anthony, geophysics, acoustics
Antcliffe, Gault Anderson, solid state physics
Armstrong, Robert Plant, anthropology
Aronofsky, Julius S, applied mathematics, statistics
Ashworth, Robert David, anatomy, neurophysiology
Austin, Donald Mac, biological anthropology
Ayres, William Leake, mathematics
Baird, Stephen Sydney, analytical chemistry, electronics
Bashour, Fouad A, biology, medicine
Bate, Robert Thomas, semiconductors
Bebb, Herbert Barrington, solid state physics
Beer, Alan E, obstetrics & gynecology, transplantation biology
Belknap, Herbert John, analytical chemistry
Bell, William Harrison, oral surgery
Berk, Robert Norton, radiology
Berkhout, Aart W J, geophysics, geology
Berry, Paul McClellan, applied mathematics
Bhandarkar, Dileep Pandurang, computer science
Bhat, Uggappakodi Narayan, operations research, statistics
Biehl, Edward Robert, physical chemistry, organic chemistry
Billingham, Rupert Everett, cell biology, immunology
Bishop, Jack Garland, physiology
Bland, Richard P, mathematical statistics
Blanton, Patricia Louise, gross anatomy, periodontics
Blaw, Michael Ervin, pediatric neurology
Blomqvist, Carl Gunnar, cardiology, cardiovascular physiology
Blount, Floyd Eugene, inorganic chemistry, physical chemistry
Bomba, Steven James, applied physics, systems engineering
Bonte, Frederick James, radiology
Borrello, Sebastian Ronald, semiconductors
Boston, James D, biochemistry, enzymology
Bowles, William Howard, biochemistry
Boyd, James Robert, solid state physics
Boyd, William Warren, solid mechanics
Brachman, Malcolm K, theoretical physics
Bracken, Ronald Clay, physical chemistry
Bradley, Virginia, geography
Braun, Hans, molecular biology
Bremer, Lewis Timothy, economic geology
Broodo, Archie, physical chemistry, chemical engineering
Brooks, James Elwood, geology
Brown, Henry Trueheart, physical chemistry
Buehler, Martin Stowell, internal medicine, cardiology
Burch, William Paul, dentistry
Burke, William Henry, nuclear physics
Burris, William Edmon, environmental sciences
Buss, Dennis Darcy, solid state electronics, semiconductors
Butow, Robert A, biochemistry
Byrd, David Lamar, dentistry
Campbell, William Bryson, pharmacology
Capra, J Donald, medicine, immunology
Carbajal, Bernard Gonzales, III, solid state chemistry, semiconductors

Carlson, Donald Eugene, radiobiology, immunobiology
Cates, Vernon E, analytical chemistry, inorganic chemistry
Chandler, Ray James, corrosion
Chapman, John S, medicine
Chapman, Richard Alexander, semiconductors
Chivian, Jay Simon, physics
Chu, Shirley Shan-Chi, solid state science
Claiborne, Lewis T, Jr, acoustics
Clark, Wesley Gleason, pharmacology
Clark, William Kemp, neurosurgery
Clarke, Robert Travis, palynology, paleontology
Clingman, William Herbert, Jr, physical chemistry
Colegrove, Forrest Donald, applied physics
Coleman, Donald James, Jr, physics, solid state electronics
Collings, Charles Kenneth, dentistry
Combes, Burton, internal medicine
Cook, Elvin Lee, physical chemistry
Cooper, Billy Howard, medical mycology, microbiology
Cooper, Howard Gordon, research administration, electrooptics
Cottam, Gene Larry, biochemistry
Cramer, Oneida Morningstar, neuroendocrinology
Crass, Gwendolyn, pathology
Crawford, George Wolf, environmental physics, nuclear physics
Creagh, Linda Truitt, physical organic chemistry
Currarino, Guido, medicine
Custard, Herman Cecil, physical chemistry
Dahm, Cornelius George, seismology
Daly, David DeRouen, neurology
Danhof, Ivan Edward, physiology
Daugherty, Kenneth E, analytical chemistry
Davis, John Barney, microbiology
Davis, Robert Clay, pure mathematics
Davis, Walter Lewis, cell biology
Decker, Robert Scott, developmental biology, cell biology
Dempsey, Walter B, biochemistry
Denison, Rodger Espy, geology
Desai, Kantilal Panachand, exploration geophysics
De Wit, Michiel, lasers
Dexter, Richard John, physics
Dietschy, John Maurice, internal medicine, gastroenterology
Dill, Russell Eugene, physiology, neuroanatomy
Di Maio, Vincent J M, forensic medicine
Dorman, Homer Lee, physiology, biochemistry
Dorn, Gordon L, genetics
Dorris, Roy Lee, pharmacology
Dowben, Robert Morris, biophysics
Downey, Harry Fred, physiology, biophysics
Drane, John Wanzer, statistics, biometry
Dunaway, Donna Kastle, computer science
Dunlap, Henry Francis, geophysics
Dunn, Danny Leroy, organic chemistry, biochemistry
Dyer, Lawrence D, physical chemistry
Eberle, Jon William, biomedical engineering, electromagnetics
Edlin, John Charles, pediatrics
Eichenwald, Heinz Felix, pediatrics, microbiology
Eidels, Leon, biochemistry
Eisenberg, Seymour, medicine
Erdos, Ervin George, pharmacology
Estabrook, Ronald (Winfield), biochemistry
Faris, Sam Russell, physical chemistry, analytical chemistry
Fashena, Gladys Jeannette, pediatrics, pediatric cardiology
Fawcett, Colvin Peter, endocrinology, neuroendocrinology
Felix, Charles Jeffrey, paleobotany
Fink, Chester Walter, pediatrics
Fink, Thomas Robert, biophysical chemistry, surface chemistry
Finkelstein, Richard Alan, microbiology
Fish, John G, polymer chemistry
Foote, Daniel W, internal medicine, metabolism
Foster, Manus R, applied mathematics, geophysics
Foster, William Roderick, physical chemistry
Foulks, Sidney Marshall, exploration geophysics
Frazier, Loy William, Jr, physiology
Freeman, Robert Glen, pathology

French, William Edwin, environmental management, marine geology
Frenkel, Eugene Philip, internal medicine, hematology
Frenkel, Rene, biochemistry
Fry, Edward Irad, biological anthropology
Fry, Peggy Crooke, nutrition
Fugate, Kearby Joe, microbiology, biochemistry
Fultz, R Paul, public health, pedodontics
Gage, Tommy Wilton, pharmacology, physiology
Galbraith, James Nelson, Jr, geophysics
Gardner, William Reavis, physical chemistry
Garriott, James Clark, toxicology, pharmacology
Gaulden, Mary Esther, radiology, radiobiology
George, Raymond Arthur, physical chemistry
Gerken, George Manz, neuropsychology
Gerneth, Dal Charles, mathematics
Gibson, Lee B, micropaleontology
Giesecke, Adolph H, medicine, anesthesiology
Gill, Atticus James, pathology
Ginsburg, Merril Stuart, geophysics
Givens, Wyatt Wendell, nuclear physics
Gleim, Paul Stanley, solid state chemistry, semiconductors
Glen, Robert S, psychiatry
Giorig, Aram, ootology
Goforth, Thomas Tucker, geophysics
Goodman, Frank R, pharmacology
Goth, Andres, pharmacology
Gothelf, Bernard, pharmacology
Graham, Robert Earl, analytical chemistry, steroid chemistry
Griffin, Edmond Eugene, physiology, radiobiology
Grinnell, Frederick, cell biology, biochemistry
Grollman, Arthur, experimental medicine
Guentherman, Robert Henry, physiology, periodontology
Guy, Leona Ruth, immunology
Haas, Herbert, geochronology, geochemistry
Haberecht, Rolf Reinhold, chemistry, physics
Hackenbrock, Charles Robert, cell biology, electron microscopy
Haisty, Robert W, inorganic chemistry, semiconductors
Halpern, Salmon Reclus, anatomy
Halsey, Jonathan Horace, geology
Hanson, William Bert, physics
Harris, Joseph Pollard, Jr, morphology
Hasty, Turner Elilah, physics
Heller, John Philip, physics
Helwig, James A, geology
Henry, Clay Allen, dentistry, microbiology
Heroy, William Bayard, Jr, geology, geophysics
Herrin, Eugene Thornton, Jr, geophysics
Hersh, Louis Barry, biochemistry
Hill, Joseph MacGlashan, pathology
Hilton, Ray, physical chemistry
Hodges, Ralph Richard, Jr, planetary atmospheres
Hoffman, John Harold, atmospheric physics
Holdaway, Michael Jon, petrology, geochemistry
Holmes, Kathryn Voelker, virology, cell biology
Holmes, Randall Kent, internal medicine, microbiology
Holton, William Coffeen, solid state physics
Hubbell, Wayne Charles, magnetism, materials science
Hunting, Alfred Curtis, applied physics
Hurt, William Clarence, periodontics, oral pathology
Jasin, Hugo E, internal medicine, immunology
Jenkins, Marion Thomas, anesthesiology
Jeskey, Harold Alfred, organic chemistry
Johnson, Alice Ruffin, immunology
Johnson, Mary Lynn Miller, fuel science, air pollution
Johnson, Robert Lee, physiology
Johnson, Rowland Edward, solid state chemistry
Johnson, Warren Frederick, physical chemistry
Johnson, William W, geophysics
Johnston, John Marshall, biochemistry
Jones, Eugene Laverne, geology
Jones, Jesse W, chemistry
Jones, Kirkland Lee, population ecology, herpetology

Jones, Morton Edward, physical chemistry
Jordan, Chris Sullivan, biology
Juliussen, J Egil, computer science, information science
Kallus, Frank Theodore, physiology, anesthesiology
Kane, Philip Francis, analytical chemistry
Kang, Chil-Yong, virology
Kaplan, Norman, internal medicine
Karolyi, Elmer Joseph, biology, physiology
Kay, Jacob Lindy, neonatology
Keele, Doman Kent, pediatrics, endocrinology
Keenan, Joseph Aloysius, nuclear chemistry
Kelly, John, Jr, physical chemistry
Kenner, Charles Thomas, analytical chemistry
Kenny, George S, physics
Kettman, John Rutherford, Jr, immunology
Kirkpatrick, Joel Brian, neuropathology
Klein, Jan, biology, immunogenetics
Korfhage, Robert R, mathematics, computer science
Kraus, William Ludwig, cardiology
Kromer, Ralph Eugene, systems analysis
Krusen, Edward Montgomery, medicine
Kuhn, Carl Sellner, Jr, chemistry
Lamb, Robert Edward, analytical chemistry
Lambert, Joseph Parker, dentistry
Land, Geoffrey Allison, medical mycology, medical microbiology
Largent, Max Dale, dentistry
Larrabee, Graydon B, radiochemistry
Lebovitz, Robert Mark, neurophysiology, bioengineering
Linn, John Charles, computer science, systems theory
Linn, Tracy Claud, biochemistry
Lipton, James Matthew, neurosciences
Lospalluto, Joseph John, biochemistry
Lowry, William Thomas, forensic medicine
Lynn, John R, ophthalmology
Macksey, Harry Michael, semiconductors
Madison, Leonard Lincoln, endocrinology, metabolism
Mahler, William Fred, plant taxonomy
Marks, James Frederic, pediatric endocrinology
Martin, Charles Louis, radiology
Martin, Jack, psychiatry
Martin, James Harold, anatomy
Mason, Morton Freeman, biochemistry
Masters, Bettie Sue Siler, biochemistry
Matthews, James Lester, microscopic anatomy
Mayers, Marvin Keene, applied anthropology
Maynard, Robert G, geology
McAlester, Arcie Lee, Jr, paleobiology
McCann, Samuel McDonald, neuroendocrinology, physiology
McCarthy, John Lawrence, Jr, physiological chemistry
McClelland, Robert Nelson, surgery
McCormack, Harold Robert, geophysics
McDonald, Frank Alan, theoretical & nuclear physics
McGarry, John Denis, biochemistry
McKelvy, Jeffrey Forrester, neurobiology
McMurray, Virginia M (Vollmer), physiology
McNaughton, Duncan Anderson, geology
McWilliams, Gerald Vernon, numerical analysis
Mecom, John Oden, zoology, limnology
Medlin, William Louis, solid state physics
Melrose, James C, physical chemistry
Messenger, Joseph Umlah, inorganic chemistry, physical chemistry
Meyer, Willis George, geology
Mia, Abdul Jabbar, plant morphogenesis, electron microscopy
Mikes, Peter, polymer physics
Milewich, Leon, organic chemistry
Miller, Edward Godfrey, Jr, biochemistry, oncology
Miller, Jarrell E, medicine
Miller, William Franklin, medicine
Miller, William Weaver, pediatrics, cardiology
Mills, William Raymond, Jr, nuclear science
Mishelevich, David Jacob, computer sciences, biomedical engineering
Mitchell, Jere Holloway, cardiovascular physiology
Mize, Charles Edward, pediatrics, biochemistry
Mize, Jack Pitts, nuclear physics
Moiola, Richard James, sedimentology
Montgomery, Edward Benjamin, physics
Montgomery, Philip O'Bryan, Jr, pathology
Moore, Donald Vincent, parasitology
Moore, Thomas D, medicine
Moore, Thomas Francis, geochemistry
Moores, Eugene Albert, earth sciences, research administration
Morgan, Raymond Victor, Jr, mathematics
Morkovin, Dimitry, radiology
Morrow, Charles Tabor, acoustics
Mortada, Mohamed, industrial chemistry
Moss, Robert L, neurophysiology, neuroendocrinology
Moticka, Edward James, immunology
Moushegian, George, physiological psychology
Mundt, Philip A, petroleum geology
Musgrave, Albert Wayne, geophysics, engineering
Mut, Stuart Creighton, geophysics
Myers, Ralph, geology
Nations, Claude, cell biology
Neaves, William Barlow, anatomy, cell biology
Nelson, John D, pediatrics
Neureiter, Norman Paul, science policy
Norman, Floyd (Alvin), medicine
North, Richard Ralph, neurology
Nossaman, Norman L, soil fertility
Novak, Ladislav Peter, physiological anthropology
Orr, Wilson Lee, organic chemistry
Ostroff, Anton G, water chemistry
Otto, John B, Jr, chemistry
Owen, Donald Bruce, applied statistics
Padovani, Francois Antoine, physics
Pak, Charles Y, physical chemistry, endocrinology
Pakes, Steven P, pathology, laboratory animal medicine
Palas, Frank Joseph, mathematics
Pan, Poh-Hsi, geophysics
Parker, Janet Lea, cardiovascular physiology, pharmacology
Parker, Sidney G, solid state chemistry
Patton, Bobbie Joe, physics
Pauken, Robert John, petroleum geochemistry
Paulson, Donald Lowell, thoracic surgery
Pavey, George Madison, Jr, underwater acoustics, geophysics
Peattie, Charles Gordon, electrochemistry
Peeples, Wayne Jacobson, geophysics
Penz, P Andrew, electrooptics
Pervin, William Joseph, mathematics, computer science
Peters, Jack Warren, geophysics
Peters, Paul Conrad, surgery, urology
Petersen, Donald H, physical chemistry, materials science
Peterson, Julian Arnold, biochemistry
Petro, Peter Paul, Jr, analytical chemistry
Petty, Charles Sutherland, medicine
Pierce, Alan Kraft, internal medicine
Pike, Robert Merrett, bacteriology
Pipes, Charles Jefferson, Jr, mathematical analysis
Pollack, Gordon (Paul), physical chemistry
Pollock, Michael L, exercise physiology
Porter, John Charles, physiology
Porter, Vernon Ray, inorganic chemistry
Potter, Robert Joseph, optics
Potts, Mark John, geochemistry
Prager, Morton David, biochemistry
Pritchard, Jack Arthur, obstetrics & gynecology
Prough, Russell Allen, biochemistry
Pruett, George Richard, physics
Race, George Justice, pathology, anthropology
Reed, Dale Hardy, exploration geophysics, electrical engineering
Reeves, Perry Clayton, chemistry
Reinberg, Alan R, solid state physics
Rester, David Hampton, nuclear physics
Reynolds, Jack, radiology
Reynolds, Robert Ware, solid state physics
Reynolds, Rolland, pathology
Richards, Arthur, geology
Ridgway, Helen Jane, biochemistry
Ries, Edward Richard, petroleum geology
Roberts, Ammarette, chemistry
Robinson, Ivor, mathematics, theoretical physics
Robinson, Jack Bert, Jr, biochemistry
Robinson, James Everett, solid state physics
Rodriguez, Argelia Velez, mathematics, physics
Roe, Glenn Dana, geochemistry
Rosenblum, Eugene David, microbiology, genetics
Runyan, Walter R, physics
Said, Sami I, internal medicine, physiology
Sallee, Verney Lee, physiology
Sampson, Herschel Wayne, anatomy
Sanders, Ottys E, herpetology
Saunders, Donald Frederick, exploration geology
Schermerhorn, John W, pharmacy, bionucleonics
Schneider, William Aeppli, geophysics
Schroen, Walter, solid state physics
Schucany, William Roger, mathematical statistics, applied statistics
Seay, Charles Frank, Jr, acoustics
Seldin, Donald Wayne, internal medicine
Sexton, Robert Ross, medicine, ophthalmology
Shadduck, John Allen, comparative pathology, virology
Shannon, Wilburn Allen, Jr, medical research, cytochemistry
Shapiro, William, internal medicine, cardiology
Sharp, Lorld Glen, chemistry
Shaw, Don W, physical chemistry
Shiner, Joel Lewis, anthropology
Shore, Parkhurst Alan, pharmacology
Shriner, Ralph Lloyd, organic chemistry
Siddiqui, Wasi Mohammad, plant pathology, microbiology
Skinner, William Carey, physics
Smiley, James Donald, immunology, medicine
Smith, Alice Lorraine, pathology
Smith, Kenneth Grant, geology
Smith, Thomas Elijah, biochemistry
Snavely, Earl Samuel, Jr, physical chemistry
Sogandares-Bernal, Franklin, parasitology
Sohal, Rajinder Singh, biology
Sparkman, Robert Satterfield, surgery
Sprague, Charles Cameron, internal medicine, hematology
Srere, Paul Arnold, biochemistry
Stallcup, William Blackburn, Jr, vertebrate zoology
Starr, David Wright, mathematics
Stembridge, Vernie Albert, pathology
Stokely, Ernest Mitchell, biomedical engineering
Stone, Irving Charles, Jr, forensic science
Stratton, Robert, theoretical physics
Streilein, Jacob Wayne, immunology, genetics
Strieter, Grederick John, physical chemistry
Strom, Edwin Thomas, physical organic chemistry, geochemistry
Sugg, Winfred Lindley, thoracic surgery, cardiovascular surgery
Sumner, George Gardner, physical chemistry, crystallography
Swink, Laurence N, crystallography
Swords, Ruth Riley, dentistry, dental hygiene
Szczesniak, Raymond Albin, medicinal chemistry, nuclear medicine
Tabbert, Robert L, geology
Tasch, Al Felix, Jr, solid state physics
Taurog, Alvin, biochemistry, physiology
Taylor, Alan Neil, physiology, biochemistry
Taylor, Paul Peak, pedodontics
Teal, Gordon Kidd, physical chemistry
Templeton, Gordon Huffine, physiology
Thompson, Jesse Eldon, surgery
Thorstenson, Donald Carl, geochemistry
Tittle, Charles William, physics
Toben, Howard Ray, immunobiology, anatomy
Tompsett, Ralph Raymond, internal medicine
Trachtenberg, Isaac, electrochemistry
Trimble, Russell Harold, chemical physics
Ubelaker, John E, parasitology
Uhr, Jonathan William, medicine
Unger, Roger Harold, internal medicine
Urban, James Bartel, geology, palynology
Uyeda, Kosaku, biochemistry
Vanatta, John Crothers, III, physiology
Vellios, Frank, pathology
Vogt, Thomas Clarence, Jr, physical chemistry
Wagner, Martin James, biochemistry
Wallace, Ben J, ethnology, applied anthropology
Wallick, George Castor, physics
Waterman, Michael Roberts, protein chemistry
Watson, John Thomas, cardiovascular physiology, biomedical engineering
Watts, Roderick Kent, solid state physics
Weart, Richard Claude, stratigraphy, paleontology
Weathersby, Hal Thompson, anatomy
Webster, John Thomas, statistics
Weiser, Daniel, mathematics
Weiss, George B, pharmacology
Wendorf, Fred, anthropology, archaeology
Wetherington, Ronald K, biological anthropology
Wheeler, Alan Clement, statistics, operations research
Wiggans, Donald Sherman, biochemistry
Wildenthal, Kern, physiology, internal medicine
Wilkins, Eugene Morrill, meteorology
Williams, Fred Eugene, physiology
Williams, Richard Kelso, mathematics
Williams, Thomas Ellis, geology
Williamson, Luther Howard, physical chemistry
Williamson, Pierce MacDonald, physical chemistry
Williamson, Thurmond A, organic chemistry
Wilson, Jean Donald, internal medicine
Wilson, Joseph Edward, physical chemistry
Wilson, Peggy Mayfield Dunlap, surface chemistry
Wisseman, William Rowland, physics
Woessner, Donald Edward, physical chemistry, nuclear magnetic resonance
Worthen, Howard George, pediatrics, biochemistry
Wright, Charles Gary, neuroanatomy, otorhinolaryngology
Wrobel, Joseph Stephen, solid state physics
Yarbrough, Henry Floyd, Jr, bacteriology
Yollick, Bernard Lawrence, anatomy, surgery
Zemanek, Joseph, Jr, acoustics
Ziff, Morris, internal medicine
Zimmermann, Eugene Robert Charles, oral pathology, microbiology
Zumberge, James Herbert, quaternary geology

DECATUR

Carlson, Robert Kenneth, inorganic chemistry

DEER PARK

Benson, Herbert Linne, Jr, physical chemistry
Bockstahler, Theodore Edwin, industrial organic chemistry
Coe, Richard Hanson, chemistry
Rabourn, Warren Joseph, industrial organic chemistry
Schissler, Donald Owen, physical chemistry

DENISON

Wester, Elbert Truman, mathematics

DENTON

Aboul-Ela, Mohamed Mohamed, plant physiology
Alford, Betty Bohon, nutrition
Allen, John Ed, mathematics
Anderson, Miles Edward, magnetic resonance
Appling, William David Love, mathematics
Bean, John Lewis, economic geography, urban geography
Bennett, Lloyd M, science education
Besso, Joseph Augustus, Jr, neurobiology
Bilyeu, Russell Gene, mathematical analysis
Brady, William Thomas, physical organic chemistry
Broome, Esther Roberts, textiles
Cantrell, Elroy Taylor, pharmacology, cancer
Carnes, James Edgar, human anatomy
Carrico, James Leon, chemistry
Caswell, Lyman Ray, organic chemistry
Christy, John Harlan, mathematics
Cockerline, Alan Wesley, cytology
Connell, Louis Fred, Jr, physics
Connor, Frank Field, mathematics
Curti, Gabriel Philip, geography
Dawson, David Fleming, mathematical analysis
Deering, William Dougless, physics
Desiderato, Robert, Jr, physical chemistry
Dobson, Gerard Ramsden, organometallic chemistry, chemical kinetics
Duggan, Jerome Lewis, physics
Emmett-Oglesby, Michael Wayne, pharmacology
Erdman, Howard E, genetics, radiation ecology
Escue, Richard Byrd, Jr, chemistry

Fincher, Bobby Lee, algebra
Fitzpatrick, Lloyd Charles, ecology
Foster, Bruce Parks, experimental nuclear physics
Foster, Norman George, physical chemistry, mass spectrometry
Fry, Kenneth Alvin, microbiology
Fuerst, Robert, genetics, microbiology
Garner, Meridon Vestal, mathematics
Gaugl, John F., medical physiology, environmental physiology
Gerdes, Raymond A., genetics
Glaze, William H., environmental chemistry
Gracy, Robert Wayne, enzymology
Hagan, Melvin Roy, mathematics
Hardcastle, James Edward, radiochemistry, biochemistry
Harris, Ben Gerald, enzymology
Hatch, William James, physiology
Hatten, Betty Arlene, medical microbiology, immunology
Hays, Thomas Reese, anthropology, archaeology
Hupp, Eugene Wesley, physiology
Hurdis, Everett Cushing, organic chemistry
Jacobson, Elaine Louise, biochemistry
Jacobson, Myron Kenneth, biochemistry
Johansson, Karl Richard, microbiology
Johnson, James Elver, organic chemistry
Jones, Paul Ronald, organometallic chemistry
Jordan, Terry G, geography
Kaman, Robert Lawrence, biochemistry
Kester, Andrew Stephen, bacteriology
Keyser, Peter D., medical microbiology
King, Edward Frazier, inorganic chemistry
Kobe, Donald Holm, quantum physics
Kutsky, Roman Joseph, biology, biochemistry
Lacko, Andras Gyorgy, biochemistry, microbiology
Langford, Florence, nutrition
Lewis, Paul Weldon, mathematics
Lott, James Robert, physiology, biophysics
Lynn, Joseph Alden, pathology, electron microscopy
Mack, Pauline Beery, chemistry
Mackey, Henry James, solid state physics
Marshall, James Lawrence, organic chemistry
Mecay, William Lloyd, inorganic chemistry
Milner, Alice N., biochemistry
Mohat, John Theodore, mathematics
Norton, Scotty Jim, biochemistry, pesticide chemistry
Parrish, Herbert Charles, mathematical analysis
Purdom, Martha Elda, nutrition, biochemistry
Pyke, Ralph Edward, nutritional biochemistry
Reber, Elwood Frank, nutrition, biochemistry
Redden, David Ray, medical physiology, animal physiology
Redding, Rogers Walker, physical chemistry, physics
Roach, Archibald Wilson Kilbourne, botany
Rozier, Carolyn K., anatomy
Sams, Lewis Calhoun, Jr., spectroscopy, inorganic chemistry
Schlueter, Edgar Albert, zoology
Sears, Raymond Eric John, nuclear magnetic resonance
Seiler, David George, solid state physics
Sherman, Robert Clyde, zoology
Silvey, J K Gwynn, limnology
Skinner, Charles Gordon, organic chemistry, biochemistry
Smith, Don Wiley, agronomy, plant physiology
Smith, William Conrad, physics
Spurlock, James Josiah, chemistry
Stanford, Jack Arthur, limnology
Stewart, George Hudson, physical chemistry
Stewart, Kenneth Wilson, entomology, aquatic ecology
Sybert, James Ray, solid state physics, low temperature physics
Theriot, Leroy James, inorganic chemistry
Truitt, B Price, organic chemistry
Vance, Benjamin Dwain, plant physiology
Vaughan, Nick Hampton, mathematics
Vela, Gerard Roland, microbiology
Vest, Floyd Russell, mathematics education
Vose, George Parlin, biomedical engineering

Wendel, Carlton Tyrus, analytical chemistry
Whitney, William Bernard, organic chemistry
Windham, Pat Morris, physics
Witting, Lloyd Allen, nutrition, lipid chemistry
Zimmerman, Earl Graves, vertebrate biology, population genetics

EAGLE LAKE
Stansel, James Wilbert, plant breeding, plant genetics

EDINBURG
Allison, Terry C., invertebrate zoology
Draeger, Sidney S., mathematics
Elliott, J Lell, physical organic chemistry
Ellis, Fred E., physics, astronomy
Foltz, Virginia C., radiation biology, genetics
Glaser, Frederic M., quantum physics
LeMaster, Edwin William, solid state physics
Lonard, Robert (Irvin), plant taxonomy
Newton, Clarence Jonathan, solid state physics, x-ray crystallography
Ortega, Jacobo, plant pathology, plant genetics
Okten, Charles Clay, physical organic chemistry
Sager, Ray Stuart, inorganic chemistry, physical chemistry
Spelman, John W., mathematics
Wallace, Charles Ray, ichthyology, ecology

EL PASO
Applegate, Howard George, plant physiology
Barnes, Thomas Grogard, physics
Blue, Michael Henry, cosmic ray physics
Bolen, Max Carlton, physics
Bowen, Donald Edgar, solid state physics
Boyer, Delmar Lee, mathematics
Brient, Samuel John, Jr., solid state physics
Bruce, Rufus Elbridge, Jr., physics
Bullard, Edwin Roscoe, Jr., geophysics, geology
Cabanes, William Ralph, Jr., organic chemistry, polymer chemistry
Chrapliwy, Peter Stanley, herpetology
Coleman, Howard S., electrooptics, electrical engineering
Cook, Clarence Sharp, physics
Cornell, William Crownshield, micropaleontology, palynology
Davidson, Charles Nelson, nuclear physics
Davis, Michael I., structural chemistry
Dean, Eugene Alan, mathematics
Duke, Eleanor Lyon, protozoology
Eastman, Michael Paul, physical chemistry
Eklund, Curtis Einar, microbiology
Ellzey, Joanne Tontz, mycology
Ellzey, Marion Lawrence, Jr., chemistry, quantum chemistry
Erke, Keith Howard, medical mycology
Freeman, Charles Edward, Jr., plant ecology
Gates, Joseph Spencer, hydrogeology
Gladman, Charles Herman, mathematics
Hall, Carl Eldridge, mathematics
Hardaway, Robert M, III., surgery
Harris, Arthur Horne, vertebrate zoology
Hatch, Lewis Frederic, organic chemistry
Herndon, William Cecil, physical organic chemistry
Hills, John Moore, geology
Hoffer, Jerry M., geology
Hunter, Jerry Don, cell biology, cell physiology
Klement, Karl Walter, paleontology
Lawson, Juan (Otto), physics
LeMone, David V., invertebrate paleontology, paleobotany
Levitt, Leonard Sidney, inorganic chemistry, physical organic chemistry
Lloyd, Winston Dale, organic chemistry
Lovejoy, Earl Mark Paul, structural geology, geomorphology
Lyerly, Paul Junior, agronomy
Mayberry, Lillian Faye, cell biology, parasitology
McAnulty, William Noel, geology
McIntyre, Robert Gerald, applied mathematics, theoretical physics
Metcalf, Artie Lou, zoology
Misenhimer, Harold Robert, obstetrics & gynecology
Nusynowitz, Martin Lawrence, nuclear medicine, endocrinology

Nymann, James Eugene, mathematics
Pannell, Keith Howard, organometallic chemistry
Parkanyi, Cyril, physical organic chemistry
Prater, Keith Burns, chemistry
Provencio, Jesus Roberto, communications, physics
Randolph, Philip L, physics, research administration
Reid, William Harper, population ecology
Reimann, Bernhard Erwin Ferdinand, electron microscopy
Rivera, William Henry, analytical chemistry
Robertstad, Gordon Wesley, bacteriology
Sass, Neil Leslie, nutritional biochemistry, pharmacology
Schumaker, Robert Louis, physics, mathematics
Shepherd, William Lloyd, mathematics
Strain, William Samuel, vertebrate paleontology
Strauss, Frederick Bodo, mathematics
Taylor, Richard Melvin, agronomy, plant physiology
Wagner, Neal Richard, topology
Webb, Robert G., vertebrate zoology
Whalen, James William, surface chemistry
White, James Edward, geophysics
Worthington, Richard Dane, morphology, herpetology
Young, Evie Fountain, Jr., plant breeding, agronomy

ENNIS
Stout, Walter Clay, dentistry

FORT WORTH
Ehlmann, Arthur J., mineralogy
Fields, Reuben Elbert, nuclear science
Kibler, Kenneth G., geophysics, environmental science
Lysiak, Richard John, physics

FREDERICKSBURG
Springall, Arthur Newton, hospital administration, preventive medicine

FREEPORT
Gum, Wilson Franklin, Jr., organic chemistry, research administration
Hickner, Richard Allan, organic polymer chemistry
Howard, William Lowry, organic chemistry
Lowery, Kirby, Jr., organometallic chemistry, organic polymer chemistry
Martin, Charles William, organic chemistry
Massingill, John Lee, Jr, industrial organic chemistry
May, James Aubrey, Jr., polymer chemistry, organometallic chemistry
McDuff, James Milton, industrial organic chemistry
Miron, Simon, organic chemistry
Oakes, Billy Dean, organic chemistry
Perettie, Donald Joseph, physical organic chemistry
Wilson, John Shirley, inorganic chemistry

FT SAM HOUSTON
Ansbacher, Rudi, obstetrics & gynecology, reproductive biology
Lindberg, Robert Benjamin, medical microbiology
Nelson, Edward Mons, anatomy
Ohlenbusch, Robert Eugene, dentistry, microbiology
Parrish, Rob Gene, biophysics, anesthesiology
Pick, Robert Orville, clinical chemistry
Tillman, Larry Jaubert, histology

FT WORTH
Adamski, Robert J, medicinal chemistry
Babitch, Joseph Aaron, neurochemistry
Barcellona, Wayne J., cell biology
Bartlett, Paul Doughty, chemistry
Battista, Orlando Aloysius, chemistry
Bell, Dorothy Mays, speech pathology
Benison, Betty Bryant, physical education, behavioral physiology
Blanton, William George, biological oceanography
Blount, Charles E., physics
Boltralik, John Joseph, biochemistry
Borgmann, August Russell, toxicology, pathology
Bulbrook, Harry Marshall, chemistry

Bush, Oakleigh Ross, geography
Colquitt, Landon Augustus, mathematics
Couchman, James C., physics
Deaton, Bobby Charles, solid state physics
Deeter, Charles Raymond, mathematical analysis
Dutta-Ahmed, Akhtar, physical inorganic chemistry, theoretical chemistry
Dvorak, Henry Rudolph, nuclear physics
Epker, Bruce Nelson, oral surgery
Faucett, William Munroe, mathematics
Ferguson, Gary Wright, animal behavior, herpetology
Fletcher, Charles Howard, physics, computer science
Forsyth, John Wiley, experimental morphology
Franklin, Ruth Ann, nutrition
Gile, Leland Henry, soil genesis, soil classification
Hamilton, Janet V., physical chemistry
Harris, Elizabeth Forsyth, medical microbiology
Hecht, Gerald, industrial pharmacy
Helbush, Robert Edwin, meteorology, operations research
Hewatt, Willis Gilliland, ecology
Hoefelmeyer, Albert Bernard, cosmetic chemistry
Hoffman, Alexander A J, theoretical physics, computer science
Huckaby, Dale Alan, physical chemistry
Hurley, Neal Lilburn, geophysics, geology
Kagawa, Charles M, medical research
Kelly, Henry Curtis, inorganic chemistry
Koehler, William Henry, inorganic chemistry
Kroh, Glenn Clinton, plant ecology
Lorenzetti, Olfeo J, pharmacology
Lyles, Sanders Truman, bacteriology
Mahendroo, Prem P, solid state physics, chemical physics
McCracken, Michael Dwayne, botany
Miller, Bruce Neil, theoretical physics
Miller, Richard Albert, mathematics
Morgan, Joseph, physics, x-ray crystallography
Moseley, Harrison Miller, physics
Mullins, John Dolan, pharmacy
Murphy, Clifford Elyman, biology
Newland, Leo Winburne, soil chemistry, water chemistry
Parker, Robert Hallett, ecology
Poe, Richard D., analytical chemistry, organic chemistry
Quarles, Carroll Adair, Jr., high energy physics, atomic physics
Raeuchle, Richard Frank, crystallography, physical biochemistry
Raval, Dilip N., physical biochemistry
Reeves, James Blanchette, medical science
Reinecke, Manfred Gordon, organic chemistry
Robb, Charles Arlee, toxicology, pharmacology
Rockower, Edward Brandt, elementary particle physics, operations research
Rowett, Charles Llewellyn, geology
Sanders, Bobby Lee, applied mathematics
Schaeffer, Norman Morris, nuclear science
Schlech, Barry Arthur, microbiology
Secrest, Everett Leigh, research administration, systems science
Smith, William Burton, organic chemistry
Street, James Clark, Jr, biology
Van Zandt, Gertrude, chemistry
Venier, Clifford George, organic chemistry
Vera, Harriette Dryden, medical microbiology
Walper, Jack Louis, geology
Watson, William Harold, Jr, physical chemistry
Webb, Theodore Stratton, Jr., applied physics, aerospace engineering
Wells, Michael Byron, optical physics, radiation physics
Weston, Henry Morgan, chemistry
Wilson, John Human, exploration geophysics
Zeleznick, Lowell D, immunology, allergy

GAINESVILLE
Stewart, Frank Edwin, physics, chemistry

GALVESTON
Abell, Creed Wills, biochemistry, oncology
Aldrich, David Virgil, marine ecology
Allen, Charles Robert, anesthesiology

Aspey, Wayne Peter, animal behavior, ethology
Awasthi, Yogesh C, biological chemistry
Bailey, Byron James, otolaryngology
Baker, Robert David, physiology
Barclay, Lee Armstead (Jr), fisheries, ichthyology
Barnett, Donald Ray, human genetics, biochemical genetics
Barranco, Sam Christopher, radiobiology, cancer
Baur, Paul Schuh, cytology
Bean, William Bennett, medicine
Beckman, Edward Louis, physiology
Blankenship, James Emery, neurophysiology
Blocker, Truman Graves, Jr, surgery
Blount, Raymond Frank, anatomy
Boelsche, ArNell, pediatrics
Bond, Ted P, biology, physiology
Bowman, Barbara Hyde, genetics
Box, Edith Darrow, parasitology
Brandt, Edward Newman, Jr, biostatistics, medicine
Brown, Arthur Morton, physiology, psychology
Bruce, E Ivan, Jr, psychiatry
Bryan, George Thomas, medical education
Burns, Chester Ray, history of medicine
Callliuet, Charles W, fish biology
Callas, Gerald, anatomy, electron microscopy
Calverley, John Robert, neurology
Cannon, Marvin Samuel, human anatomy, cell biology
Chang, Jeffrey Peh-I, cell biology
Clayton, William Howard, physical oceanography
Coggeshall, Richard E, anatomy
Cooley, Robert Nelson, radiology
Coulter, Joe Dan, neurophysiology, psychology
Daeschner, Charles William, Jr, pediatrics
Dahl, Elmer Vernon, pathology, epidemiology
Dawson, Earl B, biochemistry, nutrition
Deiss, William Paul, Jr, internal medicine
Derrick, John Rafter, surgery
Dodge, Warren Francis, pediatrics, preventive medicine
Dufot, Leo Scot, anesthesiology
Duncan, Donald, anatomy
Eaton, Douglas Charles, neurophysiology
Ellis, Sydney, pharmacology, biochemistry
Ewert, Adam, parasitology, medical entomology
Feinstein, Robert, bioengineering
Ferguson, Edward C, III, ophthalmology
Fishman, Harvey Morton, biophysics, physiology
Folse, Dean Sydney, veterinary pathology, veterinary parasitology
Freeman, James Patrick, analytical chemistry
Fuller, Gerald Maxwell, biochemistry
Fuseler, John Burt, chemistry
Gallagher, Joel Peter, neuropharmacology, neurophysiology
Gan, Jose Cajilig, biochemistry
Goldman, Armond Samuel, pediatrics, immunology
Goldstein, Allan L, biochemistry
Gorten, Ralph J, nuclear medicine
Grossman, Robert G, neurosurgery
Guest, Maurice Mason, physiology
Haber, Bernard, biochemistry
Haggard, Mary Ellen, medicine
Hall, Charles Eric, physiology
Hall, Octavia, pathology, physiology
Hancock, Michael B, neurophysiology
Harris, Leonard Crossley, pediatric cardiology
Harris, Nick Steven, immunobiology, cytology
Hejtmancik, Milton R, internal medicine, cardiology
Henry, Billy Wendell, psychiatry
Hild, Walther, anatomy
Hilton, James Gorton, pharmacology
Holoubek, Viktor, biochemistry, molecular biology
Houston, Elsie Washburn, histology, cytology
Houston, Forrest Gish, biochemistry
Hughes, William Stevenson, gastroenterology
Hulet, William Henry, internal medicine
Jenicek, John Andrew, anesthesiology
Jenkins, Vernon Kelly, radiological health, radiology
Jennings, Frank Lamont, pathology
Johnson, Raleigh Francis, Jr, nuclear medicine, radiological physics

Kempen, Rene Richard, pharmacology
Kischer, Clayton Ward, embryology, cell biology
Klebe, Robert John, human genetics, biochemistry
Koenig, Virgil Leroy, biochemistry
Kunze, Diana Lee, medical physiology
Larson, Duane L, plastic surgery, surgery
Latham, Gary V, geophysics
Levin, William Cohn, internal medicine
Levine, Harry, internal medicine
Lewis, Stephen Robert, plastic surgery
Lin, Yong Yeng, bio-organic chemistry
Lockhart, Lillian Hoffman, medicine
Macdonald, Etta Mae, microbiology
Martindale, Robert Warren, medical administration
Matumoto, Tosinato, geophysics
McAdoo, David John, chemistry, physical chemistry
McCormick, William F, neuropathology
McDanald, Eugene Chester, Jr, psychiatry
McDonald, Donald Fiedler, urology
McGanity, William James, obstetrics & gynecology
Mellinger, Melvin Wayne, neuropsychology
Micks, Don Wilfred, preventive medicine, community health
Mills, Gordon Candee, biochemistry
Morris, Harold H, neurology
Mullins, J Fred, dermatology
Nash, Joe Bert, pharmacology
Neal, Richard Allan, fisheries
Nechay, Boldan Roman, pharmacology
Nelson, Edward Blake, biochemistry
Nelson, James Arly, pharmacology
Nghiem, Quang Xuan, medicine, pediatrics
Olson, Leroy Justin, microbiology
Packchanian, Ardzrooy (Arthur), bacteriology, protozoology
Padula, Richard Thomas, surgery
Park, Tai Soo, invertebrate zoology
Patterson, Marcel, medicine
Payton, Patrick Herbert, physical chemistry, nuclear chemistry
Peake, Robert Lee, internal medicine, endocrinology
Pepermaster, Benjamin W, immunology, genetics
Perry, Robert Riley, nuclear physics, computer sciences
Peters, Bruce Harry, neurology
Peterson, Johnny Wayne, microbiology
Pinsker, Harold M, neurophysiology
Pirch, James Herman, pharmacology
Poffenbarger, Phillip Lynn, internal medicine, biochemistry
Poth, Edgar J, surgery
Potter, David Edward, pharmacology
Powell, Leslie Charles, obstetrics & gynecology
Ramanujam, V M Sadagopa, organic chemistry
Ray, Sammy Mehody, marine biology
Rayford, Phillip Leon, physiology
Reynolds, Edward Storrs, Jr, experimental pathology
Ritzmann, Stephan E, immunology, hematology
Root, Jack Glyndon, physical chemistry
Rowe, Edward Barry, surgery
Rudenberg, Frank Hermann, neurophysiology
Russell, Glenn Vinton, neuroanatomy
Russell, John McCandless, physiology
Sauerland, Eberhardt Karl, anatomy
Saunders, Jack Palmer, pharmacology
Schneider, Rose G, medicine
Schottstaedt, Mary Gardner, psychiatry
Schottstaedt, William Walter, medicine, public health
Schow, Carl Emil, Jr, oral surgery, maxillofacial surgery
Schreiber, Melvyn Hirsh, radiology
Scurry, Murphy Townsend, endocrinology
Severin, Charles Matthew, neuroanatomy
Shinnick-Gallagher, Patricia L, neuropharmacology, neurophysiology
Slocum, Harvey Chittenden, medicine
Smith, Leland Leroy, organic chemistry, biochemistry
Sordahl, Louis A, physiology
Sparks, Albert Kirk, marine biology
Srivastava, Satish Kumar, biochemistry, genetics
Stone, Hubert Lowell, physiology
Stout, Landon Clarke, Jr, pathology, internal medicine

Stubbs, Donald William, physiology
Suttle, Andrew Dillard, Jr, radiochemistry, nuclear physics
Swischuk, Leonard Edward, radiology
Szabo, Gabor, physiology, biophysics
Teng, Jon Ie, steroid chemistry, natural products chemistry
Thompson, James Charles, surgery
Thompson, William M, computer sciences
Towler, Martin Lee, neurology, psychiatry
Traber, Daniel Lee, physiology, pharmacology
Travis, Luther Brisendine, pediatrics, nephrology
Trieff, Norman Martin, environmental chemistry, toxicology
Walker, James Richard, physiology
Walker, Joe Aaron, medicine, anesthesiology
Wall, Malcolm Jefferson, Jr, physiology
Wallace, John M, medicine
Watkins, Joel Smith, Jr, geophysics, seismology
Weford, Norman Traviss, biomedical engineering
Wells, Charles Henry, physiology
White, Robert B, psychiatry
Whitney, Dorothy McCartney (Mrs Ludwik Anigstein), microbiology
Williams, Betty Jean, pharmacology
Williams, John F, Jr, internal medicine, cardiology
Willis, William Darrell, Jr, neurophysiology, neuroanatomy
Wilson, McClure, radiology
Wilson, Roy D, anesthesiology, physiology
Wilson, William Buford, marine biology
Wolf, Stewart George, Jr, clinical medicine, internal medicine
Wolma, Fred J, medicine, surgery
Wood, Robert Charles, bacteriology
Worzel, John Lamar, geophysics
Wright, Ann Elizabeth, physics, biophysics
Young, Margaret Claire, anatomy, physiology
Yu, Riley Chaoping, neurobiology, genetics

GARDEN CITY
Larsen, Victor Robinson, Jr, botany

GARLAND
Cook, John Call, geophysics
Gothard, Nicholas, air pollution, physics
Herrington, James Roland, solid state physics, health physics
Sewell, Kenneth Glenn, physics
Skinner, Thomas Junior, optics
Volz, William Beckham, electrooptics

GEORGETOWN
Girvin, Eb Carl, genetics
Lansford, Edwin Myers, Jr, biochemistry
Soulen, Robert Lewis, organic chemistry

GRAPEVINE
Standlee, William Jasper, poultry nutrition

GREENVILLE
James, Gideon T, vertebrate paleontology

HAWKINS
Abney, Cornelio Oyola, mathematics
Clift, Cecil William, agronomy
Malik, Dharam Dev, physiology, genetics
Yaden, Senkalong, zoology

HEMPSTEAD
Berry, Raymond Orvil, animal cytology

HOUSTON
Abrams, Albert, physical chemistry, petroleum engineering
Adams, Charles Rex, petroleum chemistry
Adams, John Allan Stewart, geochemistry
Adams-Mayne, Mabelle Elaine, clinical chemistry, protein chemistry
Agar, Michael Henry, ethnography, anthropological linguistics
Ahearn, Michael John, cell biology
Akutsu, Tetsuzo, medicine

Alexander, James Kermott, internal medicine
Alexandropoulos, Nikos G, solid state physics, x-ray astronomy
Alford, Bobby R, otolaryngology
Alfrey, Clarence P, Jr, hematology, bioengineering
Allen, Herbert Clifton, Jr, internal medicine, nuclear medicine
Allen, Joseph Percival, nuclear physics, radiology
Allen, Julius Cadden, pharmacology, biochemistry
Allen, Robert Henry, organic chemistry
Allen, Patton Tolbert, animal virology
Allred, John Caldwell, physics
Altenburg, Lewis Conrad, cytogenetics
Altshuler, Harold Leon, pharmacology
Anderson, David Eugene, cancer genetics
Anderson, Hugh Riddell, space physics
Anderson, James Howard, physiology, radiology
Anderson, John B, marine geology
Anderson, Robert E, biochemistry
Ansevin, Allen Thornburg, biophysics
Ansevin, Krystyna D, biology
Arbenz, Johann Kaspar, geology
Arculus, Richard J, petrology
Arganbright, Robert Philip, organic chemistry
Aringhaus, Ralph B, biochemistry
Armijo, Larry, mathematics
Armstrong, George Glaucus, Jr, aerospace medicine, physiology
Arnold, George Benjamin, chemistry
Arp, Gerald Kench, biosystematics, remote sensing
Arrighi, Frances Ellen, cell biology
Ashton, Joseph Benjamin, organic chemistry
Atkinson, Gene, academic administration, physics education
Aumann, Glenn D, ecology, ethology
Aune, Janet L, cell biology
Aune, Kirk Carl, physical biochemistry
Ave Lallemant, Hans Gerhard, structural geochemistry, geology
Awapara, Jorge, biochemistry
Bacon, Phillip, geography
Baker, Donald Roy, organic geochemistry
Baker, Harold Nordean, biochemistry, clinical chemistry
Baker, James Haskell, marine zoology
Baker, Lee Edward, engineering
Baker, Stephen Denio, nuclear physics
Baldwin, Joe G, Jr, mathematics
Baldwin, Joseph, information science
Baptist, James (Noel), microbial biochemistry
Barber, Thomas D, geology
Barnes, Charles M, veterinary medicine, medical physics
Barnes, Eugene Miller, Jr, biochemistry
Barr, Frank Theodore, geology
Bar-Sela, Mildred Elwers, anatomy, physiology
Bartel, Allen Hawley, biophysics
Baskir, Emanuel, physics
Bassett, Henry Gordon, stratigraphy
Batten, George Washington, Jr, mathematics
Bauer, Ronald Sherman, organic chemistry
Baumgartner, Frederick Neil, chemistry
Bayhi, Joseph Franklin, geophysics
Beall, Arthur Charles, Jr, medicine
Bean, William Clifton, mathematics
Bear, John L, inorganic chemistry
Becker, Adrian Anthony, geophysics
Becker, Ralph Sherman, molecular spectroscopy
Bedinger, Charles Arthur, Jr, zoology
Beerstecher, Ernest, Jr, biochemistry
Benjamin, Robert Stephen, pharmacology
Bennett, Edward Owen, bacteriology
Benolken, Robert Marshall, cell physiology
Bentley, Kenton Earl, analytical chemistry, remote sensing
Berger, Jerry Eugene, atmospheric chemistry
Bernal, Ivan, chemistry
Berry, Charles A, aerospace medicine
Berry, Paige Keith, biochemical pharmacology
Besozzi, Alfio Joseph, organic chemistry
Billups, W Edward, organic chemistry
Bishop, Stephen Hurst, biochemistry, comparative physiology
Biswal, Nilambar, virology

Black, Homer Selton, cancer
Blackburn, Archie Barnard, psychiatry
Blackwell, Lawrence A., academic administration, safety engineering
Blackwell, Robert Jerry, physics
Blatti, Stanley Parris, biochemistry, animal virology
Blattner, Meera McCuaig, computer sciences
Blattner, Russell John, pediatrics
Blytas, George Constantin, physical chemistry
Bobbit, Jeffrey L., organic chemistry, physical chemistry
Bobitt, Robert LeRoy, medicinal chemistry
Bochner, Salomon, mathematics
Bodey, Gerald Paul, Sr., oncology, infectious diseases
Bogard, Donald Dale, geochemistry
Borax, Eugene, geology
Borda, Robert Paul, neurophysiology
Boudreau, James Charles, neurophysiology
Bourgin, David Gordon, mathematics
Bourland, Charles Thomas, food science
Bowen, James Milton, virology, immunology
Bozarth, Gene Allen, botany, plant pathology
Braden, Patrick O., physics, engineering
Bramlette, William (Allen), geology
Brandenberger, Stanley George, organic chemistry
Brooks, Philip Russell, physical chemistry
Broom, Knox McLeod, Jr., nuclear chemistry
Brown, Barry W., mathematical biology
Brown, Dennison Robert, mathematics
Brown, Harold, medicine
Brown, Lee Roy, Jr., medical microbiology
Brown, Loretta Ann Port, biochemical genetics, endocrinology
Brown, Robert Don, industrial hygiene
Browning, Henry (Charles), anatomy, zoology
Bruch, Hilde, psychiatry, psychoanalysis
Brusie, James Powers, physical chemistry
Buckley, Joseph Paul, pharmacology
Buderer, Melvin Charles, physiology
Burchfiel, Burrell Clark, geology
Burdette, Walter James, surgery
Burdine, John Alton, nuclear medicine
Burks, Thomas F., pharmacology
Burleigh, Bruce Daniel, Jr., biochemistry, protein chemistry
Burst, John Frederick, clay mineralogy
Burzynski, Stanislaw Rajmund, oncology, protein chemistry
Busch, Harris, pharmacology, biochemistry
Bush, Warren Van Ness, petroleum chemistry, chemical engineering
Bushong, Stewart Carlyle, radiological health
Cantrell, William Allen, psychiatry
Cantwell, Thomas, geophysics
Caprioli, Richard Michael, biochemistry, mass spectrometry
Cardus, David, cardiology, biomathematics
Cares, William Ronald, surface chemistry, chemical kinetics
Carman, Max Fleming, Jr., geology
Carroll, Benjamin L., physical chemistry, applied mathematics
Carver, George Evans, Jr, geology
Caskey, Charles Thomas, human genetics
Castano, John Roman, geology
Castro, Gilbert Anthony, physiology, neuroendocrinology
Cate, Thomas Randolph, virology, clinical medicine
Cates, Lindley A., pharmaceutical chemistry, organic chemistry
Catlin, Francis I, otolaryngology
Catterson, Allen Duane, preventive medicine, aerospace medicine
Caudle, Danny Dearl, physical chemistry, corrosion
Cauley, Darrell Lee, wildlife ecology

Cech, Irina, environmental sciences, medical ecology
Chakraborty, Ranajit, population genetics, human genetics
Chalmers, John Harvey, Jr., biochemical genetics, musical acoustics
Chamberlain, Joseph Wyan, aeronomy, astronomy
Chambers, Leslie Addison, biophysics, zoology
Chan, James C., microbiology
Chang, Donald Choy, biophysics, cell physiology
Chappelear, John Emerson, applied mathematics, physics
Chen, Tchaw-Ren, cell biology, cytogenetics
Chimoskey, John Edward, physiology
Chowdhry, Ajit Kumar, reproductive physiology, reproductive endocrinology
Chowdhury, Mridula, reproductive biology
Chuber, Stewart, petroleum geology
Clanton, Uel S, Jr, geochemistry, astrogeology
Clapp, John Garland, Jr., agronomy
Clark, Howard Charles, Jr., geophysics, geoenvironmental science
Clark, James Henry, reproductive physiology, endocrinology
Clark, Randolph Lee, surgery
Clark, Ronald Keith, physical chemistry
Class, Calvin Miller, nuclear physics
Claus-Walker, Jacqueline Lucy, endocrinology
Clay, Michael M., pharmacology
Claycomb, William Creighton, biological chemistry
Clayton, Donald Delbert, astrophysics, nuclear physics
Clee, Thomas Edward, exploration geophysics
Clendening, John Albert, palynology
Closmann, Philip Joseph, physics, chemical physics
Coats, Alfred Cornell, neurophysiology
Cogen, William Maurice, geology
Cogswell, Howard Winwood, analytical chemistry
Collins, Lois Cowan, radiology
Collins, Royal Eugene, theoretical physics
Cominsky, Catherine, zoology, histology
Conn, Paul Joseph, physical chemistry
Cook, Billy Dean, physics
Cook, Ernest Ewart, exploration geophysics
Cook, Howard, topology
Cook, Theodore Davis, geology
Cooley, Denton Arthur, surgery
Cooley, Stone Deavours, physical chemistry
Cooper, Philip Harlan, radiation biology, radiological physics
Copeland, Murray Marcus, surgery
Cornsweet, Tom Norman, vision
Corriere, Joseph N, Jr, urology
Coster, Hendrik Paulus, internal medicine, infectious diseases
Couch, Robert Barnard, medicine, infectious diseases
Courtney, Richard James, virology
Cowles, Joe Richard, plant physiology
Cox, Hollace Lawton, Jr, medical physics, atomic physics
Cox, James Reed, Jr, organic chemistry
Cramer, Michael Brown, pharmacology, toxicology
Cranford, Jerry L., neuropsychology, psychophysiology
Crawford, Morris Lee Jackson, physiological psychology
Crecelius, Robert Lee, chemistry
Crisp, Edward Lee, petroleum geology, geochemistry
Criswell, Bennie Sue, immunology, infectious diseases
Croce, Louis J, physical chemistry, organic chemistry
Cronkright, Walter Allyn Jr, pollution chemistry, analytical chemistry
Curl, Robert Floyd, Jr, physical chemistry
Curry, James Kenneth, geology
Curtis, Doris Malkin, geology
Curtis, Morton Landers, mathematics
Dafny, Nachum, neurophysiology, neuroendocrinology
Davies, Emlyn B., physics
Davies, James Royce, microbial physiology, clinical microbiology
Davis, Joyce S., medicine, pathology
Davis, Virginia Eischen, pharmacology, biochemistry
Davison, Sol, polymer science
Deal, Carl Hosea, Jr, physical chemistry
DeBakey, Lois, information science, medical education
DeBakey, Michael Ellis, surgery

DeBakey, Selma, scientific communications
De Bremaecker, Jean-Claude, geophysics
Decker, Thomas Arno, ophthalmology, psychology
De La Mare, Harold Elison, organic chemistry, polymer chemistry
DeMoss, John Allen, microbiology
Denekas, Milton Oliver, chemistry
Dennis, Edward Wimberly, internal medicine
Derrick, William Sheldon, anesthesiology
Desiderio, Dominic Morse, (Jr), biochemistry, mass spectrometry
Desmond, Murdina Macfarquhar, pediatrics
Dessler, Alexander Jack, space science
Dietlein, Lawrence Frederick, microscopic anatomy, internal medicine
Di Ferrante, Nicola Mario, medicine, biochemistry
Dishart, Kenneth Thomas, organic chemistry
Dmochowski, Leon Ludomir, bacteriology, pathology
Dobrin, Milton Burnett, geophysics
Dobson, Harold Lawrence, biochemistry, internal medicine
Dodd, Edward Elliot, Jr, petroleum chemistry, chemical engineering
Dodd, Gerald Dewey, Jr, radiology
Dodson, Ronald Franklin, electron microscopy, cytology
Donoho, Paul Leighton, physics
Doty, William Earl Neal, geophysics
Downs, Thomas D, biostatistics
Dreesman, Gordon Ronald, immunology, virology
Dreizen, Samuel, nutrition
Dresden, Marc Henri, biochemistry, developmental biology
Driever, Carl William, clinical pharmacology
Driver, Edgar Steward, geophysics
Dubay, George Henry, mathematics
Dubbs, Del Rose M, virology, biochemistry
Duck, Ian Morley, nuclear physics
Dudar, John S., economic geology, environmental geology
Dufour, Reginald James, astronomy
Duke, Michael B., geology
Dunham, Robert Jacob, geology
Dunning, Frank Barrymore, atomic physics
DuPont, Herbert Lancashire, internal medicine, infectious diseases
Duschatko, Robert William, exploration geology
Dyckes, Douglas Franz, biological chemistry
Dzidic, Ismet, physical chemistry, analytical biochemistry
Eandi, Richard D., high energy physics
East, James Lindsay, virology
Eckelman, Walter R., geochemistry
Eckles, Nylene Elvira (Mrs Arthur Kirschbaum), internal medicine
Edmondson, Morris Stephen, organic chemistry
Edwards, George, chemistry
Edwards, John D, petroleum geology
Ehni, George (John), neurosurgery
Eichberg, Joseph, biochemistry
Eisner, Melvin, physics
Elsik, William Clinton, palynology
Engel, Paul Sanford, physical organic chemistry
Ennever, John Joseph, dentistry
Entman, Mark Lawrence, cell physiology
Eshenour, Terry Ray, food science
Estle, Thomas Leo, solid state physics
Etgen, Garret Jay, mathematics
Eugene, Edward Joseph, pharmacy, pharmacology
Euler, Kenneth L., pharmacognosy
Evans, John Edward, bacteriology
Fabre, Louis Fernand, Jr, psychiatry, psychopharmacology
Fahlberg, Willson Joel, immunology
Faillace, Louis A., psychiatry
Falck, Frank James, audiology, speech pathology
Falletta, John Matthew, oncology, immunology
Fan, Joyce Wang, organic chemistry
Fan, Paul Hsiu-Tsu, geology
Farber, Florence Eileen, molecular biology, virology
Farquhar, Gale Burton, organic chemistry
Farr, John B, geophysics, geology
Feagin, Frank J, pharmaceutics, biopharmaceutics
Feldman, Stuart, pharmacy
Fenimore, David Clarke, analytical chemistry, biochemistry

Ferguson, Noel Moore, pharmacognosy, botany
Fernbach, Donald Joseph, pediatrics, hematology
Ferrell, Robert Edward, biochemical genetics
Ferris, Bernard Joe, petroleum geology, organic geochemistry
Fields, William Straus, neurology
Finch, Leiko Hatta, applied mathematics
Finch, Robert David, acoustics
Fisher, Frank M, Jr, parasitology, invertebrate physiology
Fitzgerald, Jerry Mack, analytical chemistry
Fletcher, Evan, cardiology, internal medicine
Fletcher, Gilbert Hungerford, medicine
Forester, Robert Donald, geophysics
Forker, Robert Fencil, organic chemistry, computer science
Fournier, George Richard, palynology, stratigraphy
Fox, Karl Richard, pathology, physiology
Frachtman, Hirsh Julian, clinical medicine
Francis, William Connet, organic chemistry
Francisco, Cecil Jay, Jr, petroleum chemistry, chemical engineering
Franklin, Luther Edward, developmental biology
Franklin, Robert Ray, obstetrics & gynecology
Freebairn, Hugh Taylor, plant biochemistry
Freedman, David Asa, neurology, psychiatry
Freeman, John Clinton, Jr, meteorology, mathematics
Freeman, John Wright, Jr, space physics
Freireich, Emil J, hematology, internal medicine
Friedman, Lawrence Abraham, periodontology, epidemiology
Friedman, Robert Harold, physical chemistry
Friedmann, Naomi, physiology
Fritsche, Herbert Ahart, Jr., clinical chemistry, analytical chemistry
Fuchs, Richard, organic chemistry
Furlong, Norman Burr, Jr., biochemistry
Fuszard, Barbara, rehabilitation medicine
Gale, Laird Housel, physical organic chemistry
Gallagher, Harry Stephen, pathology
Gamble, Jess Franklin, internal medicine, hematology
Garman, William Lee, inorganic chemistry
Garrett, Julius Benjamin, Jr, micropaleontology
Garriott, Owen Kay, space physics
Gatti, Anthony Roger, inorganic chemistry, research administration
Gealy, John Robert, geology
Geanangel, Russell Alan, inorganic chemistry
Gehan, Edmund A, statistics, cancer
Gesell, Thomas Frederick, health physics
Gibson, Joseph Woodward, water chemistry, corrosion
Gibson, Kathleen Rita, physical anthropology, neuroanatomy
Gibson, William Loane, mathematical analysis
Giddings, Lorrain Eugene, Jr, physical chemistry, spectroscopy
Gilbert, Brian E, cell biology
Gildenberg, Philip Leon, neurosurgery, neurophysiology
Gilliland, Robert McMurtry, psychiatry, psychoanalysis
Gillum, Ronald Lee, clinical pathology
Giovanella, Beppino C, cancer
Girard, Louis Joseph, ophthalmology
Glantz, Raymon M, neurophysiology
Glasser, Graham Percy, physical chemistry
Glasser, Jay Howard, biostatistics, community health
Glasser, Ralph Frederick, toxicology, occupational health
Glasser, Stanley Richard, reproductive biology, endocrinology
Going, Dora Henley, medical bacteriology
Goldman, Joseph L, meteorology
Goldschmidt, Millicent, medical microbiology
Goldstein, Margaret Ann, cell biology, muscular physiology
Goodman, Clark (Drouillard), nuclear physics
Gotto, Antonio Marion, Jr., biochemistry, metabolism
Gould, Howard Ross, geology
Grams, Gary Wallace, organic chemistry

Mateker, Emil Joseph, Jr., geophysics
Matney, Thomas Stull, bacteriology
Mattax, Calvin Coolidge, physical chemistry
Matthews, Charles Sedwick, earth sciences
Matthews, Kathleen Shive, biochemistry
Mayes, Billy Woods, II, physics
Mayfield, Ernest Durward, Jr., clinical biochemistry, pathology
Mayne, William Harry, geophysics
Mayor, Heather Donald, virology; molecular biology
McBride, Mollie Elizabeth, medical microbiology
McClure, James Douglas, organometallic chemistry
McClure, Michael Edward, cell biology, biochemistry
McCormick, James E, geology
McCormick, Kenneth James, cancer
McCoy, James Ernest, magnetospheric physics
McCullough, James Douglas, Jr., organic chemistry, physical chemistry
McFarlan, Edward, Jr., petroleum geology, sedimentology
McGavran, Malcolm Howard, pathology
McGookey, Donald Paul, petroleum geology
McGovern, John Phillip, allergy
McGregor, Robert Finley, biochemistry, endocrinology
McIntosh, Henry Deane, internal medicine
McIver, Norman L, geology
McIver, Richard Donald, geochemistry
McKay, David Stewart, volcanology
McKelvey, Eugene Mowry, internal medicine, hematology
McLendon, David Mark, psychopharmacology
McMordie, Warren C, Jr, inorganic chemistry
McMurtrey, Marion John, surgery, molecular biology
McNamara, Dan Goodrich, pediatrics
McSpaden, Jay Byron, audiology
Means, Anthony R, endocrinology
Medina, Emanuel, oncology
Medlin, William Virgil, chemistry
Meerbott, William Keddie, petroleum chemistry
Meistrich, Marvin Lawrence, biophysics, cell biology
Melnick, Joseph Louis, virology
Mendell, David, psychiatry
Merrifield, D Bruce, physical organic chemistry
Merrill, Joseph Melton, medical administration
Metter, Raymond Earl, geology
Middlehurst, Barbara Mary, astronomy
Migliore, Philip Joseph, medicine
Miller, Edward Titus, geophysics
Miller, James Richard, organic chemistry
Miller, Robert DuWayne, physics
Milligan, Winfred Oliver, physical chemistry
Milling, Marcus Eugene, sedimentology, geomorphology
Minear, John W, geophysics
Mintz, A Aaron, medicine
Modisette, Jerry Lee, space physics
Moehlman, Robert Stevens, geology
Moldawer, Marc, medicine, oncology
Monaghan, Patrick Henry, geochemistry, petroleum engineering
Montague, Eleanor D, radiotherapy
Montgomery, Edward Harry, pharmacology, physiology
Moon, Thomas Edward, biometrics
Moore, Charleen Morizot, cytogenetics
Moore, Erin Colleen, biochemistry, oncology
Moreland, Ferrin Bates, forensic toxicology, clinical chemistry
Moreton, Robert Dulaney, radiology
Morgan, Thomas Harlow, astrophysics, planetary sciences
Morris, George Cooper, Jr, surgery
Morrisett, Joel David, biophysics, biochemistry
Moser, James Howard, chemistry, chemical engineering
Mosier, Benjamin, chemistry

Mote, Victor Lee, geography, environmental sciences
Mountain, Clifton Fletcher, surgery
Muhs, Merrill Arthur, organic chemistry, analytical chemistry
Mullineaux, Richard Denison, organic chemistry
Murany, Ernest Elmer, exploration geology
Murphy, Paul Henry, nuclear medicine, medical physics
Murphy, William Harry, geophysics, virology; molecular biology
Murray, Christopher Brock, mathematics
Musgrave, F Story, physiology, surgery
Mutchler, Gordon Sinclair, nuclear physics
Nachtwey, David Stuart, photobiology, radiobiology
Nanz, Robert Hamilton, Jr, geology
Nei, Masatoshi, population genetics, evolution
Neidell, Norman Samson, geophysics
Nelson, Robert S, medicine
Nelson, Thomas Evar, biochemistry
Nestvold, Elwood Olaf, information science, geophysics
Newhouse, Albert, mathematics, computer science
Newton, Berne Loyst, pathology
Newton, William Anthony, bacteriology, microbiology
Nichols, Buford Lee, Jr, medicine, nutrition
Nielsen, Robert Peter, industrial chemistry
Noall, Matthew Wilcox, biochemistry
Nolte, William Anthony, biochemistry, neurophysiology
Norbeck, Edward, cultural anthropology
Nordyke, Ellis Larrimore, biochemistry, neurochemistry
Norman, Carl Edgar, structural geology, rock mechanics
Norris, James Scott, endocrinology
North, William Gordon, stratigraphy
Nozaki, Kenzie, chemistry
Nudelman, Harvey Banet, bioengineering, neurophysiology
Olsen, Rex E, stratigraphy, paleontology
Olson, Danford Harold, organic chemistry
Olson, John Victor, dentistry
Olson, Mark Obed Jerome, biochemistry, pharmacology
O'Malley, Bert W, endocrinology, molecular biology
O'Malley, Matthew Joseph, algebra
O'Neal, Hubert Ronald, organic chemistry
Orange, Arnold, geophysics, electrical engineering
Orengo, Antonio, biochemistry
Oro, Juan, biochemistry
Osborne, Weymar Zack, cosmic ray physics, experimental high energy physics
Otvos, John William, physical chemistry
Overturf, Merrill L, biochemistry
Owen, Guillermo, mathematics
Oxley, Ralph L, organic chemistry
Page, Thornton Leigh, astrophysics, space physics
Palmer, Graham, biochemistry
Papadopoulos, Michael N, physical chemistry
Parker, Robert Allan Ridley, astronomy
Pathak, Sen, cytogenetics
Paton, David, medicine, ophthalmology
Peck, Ernest James, Jr, biochemistry, neurochemistry
Pelley, Ralph L, organic chemistry
Perry, John Harold, anatomy
Peters, Lester John, radiotherapy
Peterson, Lysle Henry, -optometry
Peterson, Richard George, neurobiology
Peurifoy, Paul Vastine, analytical chemistry
Pfaffenberger, Carl Dale, analytical biochemistry, cancer
Pfeiffer, Paul Edwin, mathematics
Pfeifer, Chester Harry, -optometry
Phelps, Richard A, nutrition
Phillips, Gerald C, physics
Phillips, Timothy Dukes, marine chemistry
Philpott, Charles William, cell biology
Phinney, William Charles, geology
Pier, Stanley Morton, environmental health
Pilgeram, Laurence Oscar, biochemistry, physiology
Pinero, Gerald Joseph, histology, cytology
Pitts, Donald Graves, optometry, physiological optics
Pokorny, Alex Daniel, psychiatry

Polinger, Iris Sandra, developmental biology, dermatology
Polking, John C, mathematics
Pope, Alex, soil chemistry
Potter, Andrew Elwin, Jr, planetary atmospheres, remote sensing
Potter, John Fred, atmospheric science
Powell, Benjamin Neff, petrology, mineralogy
Powell, Norborne Berkeley, urology
Pownall, Henry Joseph, biochemistry
Prasad, Naresh, radiobiology, cytogenetics
Prasad, Rupi, medical genetics, enzymology
Prats, Michael, physics
Prescott, Basil Osborne, geology
Preslock, James Peter, endocrinology, reproductive physiology
Price, Harvey Simon, applied mathematics
Pulley, Thomas Edward, marine biology
Purifoy, Dorothy Jane Martin, virology, molecular biology
Pyle, Leonard Duane, computer science, mathematics
Qadri, Syed M Hussain, medical microbiology
Quiocho, Florante A, biochemistry
Rabalais, John Wayne, physical chemistry
Rachford, Henry Herbert, Jr, mathematics, engineering
Raleigh, James Walsh, medicine
Rall, Elizabeth Pretzer, stratigraphy
Rall, Raymond Wallace, exploration geology
Randerath, Erika, biochemical pharmacology
Randerath, Kurt, biochemistry
Rao, Potu Narasimha, cell biology, cytogenetics
Rapean, John C, physical chemistry
Rasco, Marilyn Arnott, biochemistry, genetics
Rauchwerger, Joel M, hematology, anatomy
Rawson, Rulon Wells, medicine
Reaves, Harry Lee, physics, mathematics
Redburn, Dianna Ammons, neurobiology
Reeves, William John, Jr, medicine, biochemistry
Reid, Archibald, McMillan, geology
Reid, Kenneth Ian Gower, industrial chemistry, petroleum
Reilly, Charles Austin, chemical physics
Reisberg, Joseph, chemistry
Remington, Richard Dellerane, biostatistics
Resnikoff, Howard L, mathematics
Reso, Anthony, stratigraphy, petroleum geology
Reynolds, Richard Johnson, organic chemistry
Reynolds, William Walter, physical biochemistry
Rich, Robert Regier, immunobiology
Richardson, Jasper E, engineering physics
Richter, George Holmes, organic chemistry
Riggs, Stuart, medicine, microbiology
Rigor, Benjamin Morales, Sr, anesthesiology
Riley, Charles Marshall, economic geology
Risser, Jacob Rutt, physics
Rix, Cecil Charles, geology
Roach, Francis Aubra, mathematical analysis
Robbins, Donald, physics
Robbins, Enders Anthony, geophysics
Robinson, James H, geophysics, physics
Robinson, Joseph Dewey, physical biochemistry
Robinson, Robert Eugene, analytical chemistry
Robison, George Alan, biochemical pharmacology, endocrinology
Ro-Choi, Tae Suk, biochemistry, pharmacology
Rocklin, Albert Louis, physical chemistry, industrial chemistry
Rodgers, Lawrence Rodney, Sr, medicine
Rodriguez, Dennis Milton, mathematics
Roessler, Robert L, psychiatry
Rogers, James Kenneth, exploration geology
Rogers, Marion Alan, geochemistry, petroleum geology
Rogers, Thomas D, cell biology, biochemistry
Rogers, Thomas Earl, microbiology, immunology
Romsdahl, Marvin Magnus, surgery, biology
Rorschach, Harold Emil, Jr, physics

Rosborough, John Paul, physiology
Rose, George G, medicine
Rosen, Jeffrey Mark, biochemistry, endocrinology
Rosenbaum, Joseph Hans, physics
Rosenquist, Edward P, organic chemistry, physical chemistry
Ross, Doris Laune, biochemistry
Rossen, Roger Downey, immunology, internal medicine
Rossini, Frederick Dominic, physical chemistry
Rowland, Richards Atwell, clay mineralogy
Rudolph, Arnold Jack, pediatrics
Rundel, Robert Dean, physics
Rupert, Joseph Paul, physical chemistry
Russell, Kenneth Lloyd, geology
Russell, William Ogburn, pathology
Rutledge, Felix N, obstetrics & gynecology
Ryder, Elliott Elkington, Jr, organic chemistry
Ryu, Jisoo Vinsky, exploration geophysics
St Cyr, Lewis Alpha, metallurgical chemistry
Saltzberg, Bernard, biomathematics, bioengineering
Samaan, Naguib A, endocrinology
Samorajski, Thaddeus, anatomy
Sample, Thomas Earl, Jr, physical organic chemistry
Sanderson, Alan Nichols, meteorology, chemical engineering
Sangree, John Brewster, Jr, geology, geophysics
Sargent, Frederick, II, human ecology, nutrition
Sarmiento, Roberto, petroleum geology
Sartor, Albin Francis, Jr, petroleum chemistry
Sass, Ronald L, biophysical chemistry
Sato, Clifford Shinichi, biochemistry
Saunders, Grady Franklin, molecular biology
Saunders, Priscilla Prince, biochemical pharmacology
Savage, Howard Edson, cell biology
Savit, Carl Hertz, geophysics
Sawyer, Webster Morrill, Jr, surface chemistry
Scanlon, John Earl, medical entomology
Schaal, Barbara Anna, population biology
Schaffer, Priscilla Ann, virology
Schatz, Joseph Arthur, mathematics
Schindler, William Joseph, endocrinology
Schlamowitz, Max, biochemistry, immunochemistry
Schmidt, Jurgen Volkmar, mathematics
Schneidermann, Nahum, sedimentology, marine geology
Schnur, Sidney, medicine
Schoenberger, Michael, geophysics
Schoolar, Joseph Clayton, psychiatry, pharmacology
Schrader, William Thurber, endocrinology, molecular biology
Schwanecke, Rebecca G Pineda, pediatrics, neonatology
Schwartz, Arnold, pharmacology, biochemistry
Schwarzer, Theresa Flynn, geology, geochemistry
Scott, Philip Dell, ophthalmology
Scott, Russell, Jr, urology
Seely, Donald Randolph, geology
Seibert, Richard Albert, chemistry
Seifert, William Edgar, Jr, analytical chemistry
Seman, Gabriel, virology, cytology
Semar, Cary Lloyd, space physics
Seriff, Aaron Jay, geophysics
Session, John Joe, immunogenetics, cytogenetics
Severs, Richard Keith, environmental health, occupational health
Seybold, William Dempsey, thoracic surgery
Seymour, Raymond Benedict, polymer chemistry
Shaeffer, Joseph Robert, molecular biology, biochemistry
Shalek, Robert James, biophysics
Shanbour, Linda Livingston, physiology, biophysics
Shannon, Ira Lenwood, biology, chemistry
Sharp, James Martin, physics

Sharp, John T., internal medicine, microbiology
Shaw, Charles Raymond, oncology
Shaw, Daniel Bernard, geology
Shaw, Margery Wayne, human genetics, legal genetics
Sheldon, William Robert, physics
Sheng, Hwai-Ping, medical physiology
Sheriff, Robert Edward, geophysics, exploration
Sherwood, John William Charles, geophysics
Shishkevish, Leo J., geology
Shugart, Thomas Reeder, geology
Shum, Wan-Kyng Liu, biochemistry, endocrinology
Sieck, Herman C., geology
Simpson, John Wayne, biochemistry
Singer, Emanuel, operations research, operations analysis
Singh, Jagat, biochemistry
Singleton, David Michael, organic chemistry, inorganic chemistry
Sinha, Anil K., genetics
Sinkhorn, Richard Dennis, mathematics
Sinkovics, Joseph, virology, internal medicine
Smith, Edwin Barkley, Jr., biochemistry, nutrition
Smith, Edwin Lee, physiology
Smith, Frank E., oncology, chemotherapy
Smith, Jan G., geology
Skelley, Dean Sutherland, endocrinology, clinical chemistry
Skinner, James Ernest, neurosciences
Skjonsby, Harold Samuel, histology, anatomy
Smith, Louis C., biochemistry
Smith, Louis Leslie, Jr., medicine
Stolnick, Malcolm Harris, biophysics
Smith, Roy George, bio-organic chemistry, cell biology
Slye, John Marshall, pure mathematics
Smalley, Richard Errett, chemical physics
Smith, Alan Lyle, plant ecology
Smith, Brandes Henry, petroleum chemistry
Spencer, William Albert, rehabilitation, physiology
Spence, Dale William, exercise physiology
Smnutry, Edgar Josef, organic chemistry, mineralogy
Smyth, Joseph Richard, geology
Smolensky, Michael Hale, human physiology
Sperling, Harry George, vision, psychophysics
Smythe, Cheves McCord, medicine, pediatrics
Sneider, Robert Morton, petroleum
Snelson, Sigmund, geology
Snider, Philip Joseph, genetics
Spira, Melvin, medicine, plastic surgery
Spitznogle, Frank Raymond, physics
Spiu, Harlan Jacobson, pathology
Stalones, Reuel Arthur, medicine, epidemiology
Socher, Susan Helen, cell biology
Sommer, Kathleen Ruth, biochemical pharmacology
Stancel, George Michael, biochemistry, endocrinology
Starbuck, Wesley Curtis, pharmacology, physiology
Stebbings, Ronald Frederick, physics
Steele, Grant, stratigraphy, paleontology
Steele, James Harlan, veterinary medicine
Steib, Michael Lee, mathematical analysis
Steinberger, Anna S., cell biology, immunology
Steiner, Marion Rothberg, biochemistry, animal virology
Steiner, Sheldon, virology
Steinhardt, Charles Kendall, organic chemistry
Stenback, Wayne Albert, microbiology, electron microscopy
Sterner, James Hervi, medicine
Stevens, Robert Velman, organic chemistry
Stewart, Charles Ranous, genetics
Stewart, Robert William, geology
Stillwell, Richard Newhall, organic chemistry, computer science
Stockman, Gail Diane, immunology
Stone, Solon Wallingford, geology
Stoner, Graham Alexander, analytical chemistry, agricultural chemistry
Stoops, James King, biochemistry

Storck, Roger Louis, microbiology, biochemistry
Stouffer, John Emerson, biochemistry
Stover, Lewis Eugene, paleontology, palynology
Strada, Samuel Joseph, pharmacology, neurobiology
Stratton, Everett Franklin, geology, geophysics
Streckfuss, Joseph Larry, microbiology, immunology
Strickland, John Willis, geology
Strong, Louise Connally, human genetics, cancer
Stryker, Harry Kane, polymer chemistry, applied statistics
Stubblefield, Travis Elton, cell biology
Stulken, Donald Edward, physiology
Su, Shin-Yi, space physics
Suk, Wadi Nagib, internal medicine, nephrology
Sullivan, James Thomas, Jr., physical chemistry
Sullivan, Margaret P., pediatrics, medicine
Suzuki, Minoru, pathology, sciences
Swanson, Donald Charles, petroleum geology, sedimentology
Swanson, Roger Glenn, geology
Sybers, Harley D., pathology, physiology
Talbot, Prudence, reproductive physiology, cell biology
Talbot, Raymond James, Jr., astrophysics
Tapia, Roland A., numerical analysis
Tapley, Norah duVernet, medicine
Taylor, Fred M., pediatrics
Taylor, Kenneth Orien, computer sciences
Tcholakian, Robert Kevork, biochemistry, reproductive biology
Tebo, Heyl Gremmer, anatomy
Telford, Ruth Jane, anesthesiology
Templeton, Charles Clark, physical chemistry
Templeton, Joe Wayne, immunology, genetics
Teng, Ching Sung, reproductive biology, biochemistry
Teng, Christina Wei-Tien Tu, cell biology, biochemistry
Terry, Robert James, zoology, developmental biology
Tess, Roy William Henry, polymer chemistry
Thallman, Robert H., anatomy
Thames, Howard Davis, Jr., biomathematics
Thoma, George William, pathology
Thompson, Howard K., Jr., computer science, biomathematics
Thompson, James Robert, mathematics, statistics
Thornton, William Edgar, medicine, astronautics
Thrall, Robert McDowell, mathematics
Threinen, David Tronvig, geology
Tilley, Aubra Everett, geophysics, psychology
Tips, Robert Leonard, pediatrics, genetics
Tittman, Jay, nuclear science
Titus, Jack L., pathology
Tixier, Maurice Pierre, geophysics
Todd, Terrence Patrick, geophysics, mining geology
Tonking, William Harry, geology
Touring, Roscoe Manville, geology
Trammell, George Thomas, theoretical physics
Trentin, John Joseph, experimental biology
Trikula, David, biophysics, virology
Tsai, Ming-Jer, biochemistry, molecular biology
Tucker, Charles Thomas, mathematics
Turnrose, Barry Edmund, astronomy
Tuttle, Robert Lewis, microbiology
Tyler, Stephen Albert, anthropology, anthropological linguistics
Ungar, Georges, pharmacology, neurochemistry
Valkovic, Vlado, experimental nuclear physics, applied physics
Valbona, Carlos, pediatrics, physiology
van Dijk, Christiaan Pieter, physical organic chemistry, chemical engineering
Van Eys, Jan, biochemistry
Van Siclen, Dewitt Clinton, geology
Vant-Hull, Lorin Lee, physics
Varsel, Charles John, food chemistry
Veech, William Austin, mathematics
Venketeswaran, S., cell biology, botany
Vick, Robert Lore, physiology, pharmacology
Vinson, David Berwick, psychophysiology

Vizard, Douglas Lincoln, biophysics, molecular biology
Vobach, Arnold R., mathematics
Vogel, James John, biochemistry, nutrition
Vogelfanger, Elliot Aaron, polymer science, organic chemistry
Von Noorden, Gunter Konstantin, ophthalmology
Waggoner, James Arthur, nuclear physics
Wakil, Salih J., biochemistry
Walborg, Earl Fredrick, Jr., biochemistry
Walbrick, Johnny Mac, industrial organic chemistry
Wald, Milton M., chemistry
Walker, Etta Frances, biophysics
Walker, James Benjamin, biochemistry
Walker, Robert Hugh, physics
Wall, Frederick Theodore, physical chemistry
Wallace, Sidney, radiology
Wallick, Earl Taylor, biological chemistry
Walter, Charles Frank, biochemistry, biomathematics
Walters, Geoffrey King, atomic physics
Wang, Taitzer, biochemistry
Wang, Yeu-Ming Alexander, biochemistry, hematology
Ward, Calvin Herbert, plant pathology, physiology
Ward, Darrell N., biochemistry
Warner, Jeffrey, marine ecology
Warner, Marlene Ryan, endocrinology
Webber, Marion George, pharmaceutical chemistry
Wendhal, Ronald, audiology, speech pathology
Wendlandt, Wesley W., inorganic chemistry
Wenkert, Ernest, natural products chemistry
Wentworth, Wayne, analytical chemistry, physical chemistry
Wheatcroft, Merrill Gordon, dental pathology
Wheeler, Harry Ogden, microbiology, immunology
Wheeler, Mary Fanett, numerical analysis
White, David Archer, geology
White, Edgar C., surgery
Whitman, Andrew Peter, mathematics
Whittlesey, John R B., exploration geophysics, mathematical statistics
Widess, Moses B., geophysics
Wiginton, Ralph Ambrose, geophysics
Wilde, Carroll Lamar, mathematics
Widt, David Edwin, reproductive physiology
Wilcott, Mark Robert, III, organic chemistry
Williams, Richard John, geochemistry
Williams, Robert Earl, physical chemistry
Williams, Robert L., neurology
Williams, Robert Pierce, microbiology
Wilmarth, Verl Richard, economic geology
Wilson, James Lee, geology
Wilson, John Howard, biochemistry, genetics
Wilson, Raphael, medical bacteriology
Wilson, Ray Floyd, analytical chemistry, physical chemistry
Winfrey, J C., organic chemistry, analytical chemistry
Withers, Hubert Rodney, radiotherapy
Wolf, Richard Alan, space physics, astrophysics
Wong, Siu Gum, optometry, public health
Wood, Jeanie McMillin, chemistry
Wood, Joe George, anatomy
Wood, Lowell Thomas, physics, science education
Woodruff, Richard L., physical chemistry
Woods, Raymond Douglas, geology
Wray, Granville Wayne, cell biology, biochemistry
Wray, Virginia Lee Pollan, biochemistry

Wright, Anthony Aune, animal behavior, psychophysics
Wright, David Anthony, developmental biology, genetics
Wright, Martin, mathematics
Wu, Changsheng, exploration geophysics
Yamauchi, Toshio, medical genetics
Yeoman, Lynn Chalmers, biochemistry
Young, Sue Ellen, ophthalmology
Young, William Arlen, petroleum geochemistry
Younglove, James Newton, mathematics
Yow, Martha Dukes, pediatrics
Zabel, Carroll Wayne, physics
Zimmerman, Stuart O., mathematical biology
Zingula, Richard Paul, paleontology
Ziatkis, Albert, analytical chemistry & organic chemistry

HUMBLE
Ball, Stanton Mock, geology

HUNTSVILLE
Agan, Raymond John, agriculture
Amato, Vincent Alfred, plant pathology, horticulture
Banta, Marion Calvin, physical chemistry, analytical chemistry
Bounds, John Howard, geography
Brown, Murray Allison, dairy science
Collins, Alva LeRoy, Jr., inorganic chemistry, organometallic chemistry
Cooley, Adrian B. Jr., mathematics
DeShaw, James Richard, biology, environmental biology
DeShazo, Mary Lynn Davison, biochemistry
Dewees, Andre Aaron, population genetics
Guidry, Carlton Levon, organic chemistry
Hall, Hugh Edward, Jr., nuclear physics
Harding, Winfred Mood, biological chemistry
Harrison, Marjorie Hall, astrophysics, astronomy
Hilliard, John Roy, Jr., entomology
Hoage, Terrell Rudolph, entomology, genetics
Humphrey, Ray Eicken, analytical chemistry
Isham, Elmer Rex, plasma physics, solid state physics
Konen, Harry P., numerical analysis
Little, Perry L., nutrition, physiology
Long, James Duncan, zoology
Manka, Charles K., plasma physics, spectroscopy
Mattingly, Glen E., mathematics
McCoy, John Harold, computer science
McDonald, Perry Frank, solid state physics
Meade, Thomas Gerald, parasitology, invertebrate zoology
Moldenhauer, Ralph Roy, ecology
Muecke, Herbert Oscar, mathematics
Nance, John Arthur, nutrition
Padgett, Algie Ross, organic chemistry
Scott, Elton Monroe, geography
Shinkle, Norman Leroy, physics
Stallings, James Cameron, organic chemistry
Stewart, Robert Blaylock, plant pathology
Thomas, Ruth Beatrice, botany
Vick, George R., algebra, geometry
wilson, Everett D., physiology, endocrinology
Wilson, Jack Martin, chemical physics, biophysics

IRVING
Asner, Bernarad A. Jr., applied mathematics
Bice, Claude Wesley, agricultural biochemistry
Coppin, Charles Arthur, mathematical analysis
Cowan, Donald, physics
Doe, Frank Joseph, genetics
Lockett, M Clodovia, biology
Monoston, Benedict Joseph, physics
Schram, Alfred Francis, physical chemistry
Towne, Jack C., biochemistry, radiochemistry

KEENE
Beary, Dexter F., anatomy, biology

KERRVILLE
Drummond, Roger Otto, acarology
Eschle, James Lee, entomology
Gingrich, Richard Earl, medical entomology
Gladney, William Jess, entomology, acarology
Graham, Owen Hugh, medical entomology, veterinary entomology
Palmer, Jack Sidney, veterinary medicine

KILGORE
Whiteside, Charles Hugh, analytical chemistry, environmental sciences

KINGSVILLE
Anderson, Stanley Robert, agronomy
Bailey, Leo L, horticulture, entomology
Bajza, Charles C, geography
Beran, Jo Allan, environmental chemistry
Bogusch, Edwin Robert, ecology
Cecil, David Rolf, mathematics
Chaney, Allan Harold, zoology
Crenshaw, David Brooks, animal breeding
Davis, Richard B, vertebrate ecology, biometry
Elliott, Paul M, physics
Garland, Fred McKee, chemistry
Gillaspy, James Edward, entomology
Grant, Darroll Lee, meat science
Hewett, Lionel Donnell, solar physics
Hildebrand, Henry H, marine biology
Howe, John Wallace, animal breeding
Jolliff, Gary David, agronomy, crop physiology
Kay, Alvin John, mathematical analysis
Kowalik, Virgil C, mathematics
Kruse, Olan Ernest, physics
Marcotte, Ronald Edward, physical chemistry
McCoy, Ralph Hines, wildlife diseases, medical bacteriology
Neher, David Daniel, soils
Nixon, Donald Merwin, agricultural economics
Norwine, James Randolph, climatology
Peacock, John Talmer, plant ecology
Perez, John Carlos, virology
Pratt, David R, physiology, genetics
Ruhnke, Edward Vincent, physical organic chemistry
Suhm, Raymond Walter, stratigraphy
Williges, George Goudie, plant pathology, plant taxonomy
Wood, Carl Eugene, marine ecology, invertebrate zoology

LA PORTE
Morris, Rupert Clarke, organic chemistry
Taylor, Ralph E, geology

LACKLAND AFB
Marraro, Robert V, medical microbiology
McPhaul, John J, Jr, internal medicine, immunology
Myers, Paul Walter, neurosurgery

LAKE JACKSON
Eckman, Michael Kent, animal parasitology
Ham, George Edward, organic chemistry
Kottle, Sherman, physical chemistry
Newton, Robert Andrew, physical organic chemistry
Smith, Grant Newey, biochemistry

LANCASTER
Hurst, Harold Lowery, veterinary pathology

LAREDO
Fulton, John Donaldson, biology

LEAGUE CITY
Chiddix, Max Eugene, organic chemistry

LEVELLAND
Jones, William Grover, veterinary medicine

LONGVIEW
Dawes, John Leslie, organometallic chemistry, organic chemistry
Etter, Raymond Lewis, Jr, organic polymer chemistry
Goodwyn, Jack Ray, polymer chemistry
Gwynn, Donald Eugene, synthetic organic chemistry
Holmes, Jerry Dell, organic chemistry
Hudson, Glenn Vincent, organic chemistry, chemical engineering
McCollum, Anthony Wayne, organic chemistry
Neeley, Charles Mack, physical chemistry, polymer chemistry
Park, Vernon Kee, polymer chemistry
Robinson, Alfred Green, petroleum chemistry
Snapp, Thomas Carter, Jr, organic chemistry, analytical chemistry
Southern, Thomas Martin, analytical chemistry, inorganic chemistry

LUBBOCK
Adamcik, Joe Alfred, organic chemistry
Albin, Robert Custer, animal nutrition
Allen, Archie C, zoology, genetics
Allen, Bonnie L, soil genesis, soil mineralogy
Amir-Moez, Ali R, mathematics
Anderson, John Arthur, biochemistry
Arper, William Burnside, geology
Ashdown, Donald, entomology
Ault, John Willard, mathematics
Barnes, Charles Dee, physiology
Bartsch, Richard Allen, organic chemistry
Bateman, Barry Lynn, computer sciences
Baugh, Clarence L, bacterial physiology
Baumgardner, John Henry, animal nutrition
Behal, Francis Joseph, biochemistry
Bennett, William Frederick, agronomy, plant nutrition
Berlin, Jerry D, cell biology
Berry, Robert Wade, plant pathology
Bertrand, Anson Rabb, soil physics
Bolender, David Leslie, anatomy
Boullion, Thomas L, mathematics, statistics
Brigham, Raymond Dale, plant breeding
Buesseler, John Aure, ophthalmology
Burns, John Mitchell, physiology, endocrinology
Burzlaff, Donald Frederick, range management, agronomy
Camp, Earl D, plant pathology
Cebull, Stanley Edward, structural geology
Chau, Michael Ming-Kee, physical organic chemistry
Coleman, Eugene Alfred, plant physiology
Conover, William Jay, statistics
Cook, Elton Davis, agronomy
Coulter, Murray Whitfield, genetics, plant physiology
Crass, Maurice Frederick, III, physiology, biochemistry
Curl, Samuel Everett, animal physiology, endocrinology
Dahl, Billie Eugene, range management, animal husbandry
Das Gupta, Kamalaksha, physics
Dennis, Joe, biochemistry
Downes, John D, horticulture
Draper, Arthur Lincoln, physical chemistry
Dregne, Harold Ernest, soil science
Duran, Benjamin S, mathematics, statistics
Durham, Ralph Marion, genetics, animal husbandry
Dyson, James Everett, Jr, medical education, medical administration
Elle, George O, horticulture, vegetable crops
Elliot, Arthur McAuley, phytopathology, mycology
Felkner, Ira Cecil, microbial genetics, molecular biology
Fisher, Charles E, weed science
Frigyesi, Tamas L, neurobiology
Genoways, Hugh Howard, mammalogy, zoogeography
George, John Edwin, entomology
Goodin, Joe Ray, plant physiology
Gott, Preston Frazier, physics
Guerrant, William Barnett, Jr, organic chemistry
Gundersen, Martin Adolph, lasers, solid state physics
Guven, Necip, geology, mineralogy
Haller, Walfred Sigmon, biochemistry
Haragan, Donald Robert, meteorology
Harris, Rae Lawrence, Jr, comparative physiology
Hartman, Herman Bernard, comparative physiology
Harvey, Clark, agronomy, plant physiology
Hatfield, Lynn LaMar, atomic physics
Hazlewood, Emmett Allen, mathematics
Hollis, Gilbert Ray, animal nutrition
Holwerda, Robert Alan, bioinorganic chemistry
Howe, David Allen, physics
Huddleston, Ellis Wright, economic entomology
Hudson, Frank Alden, animal science
Hughes, Maysie J H, pharmacology, physiology
Ickes, William K, speech pathology, audiology
Jackson, Raymond Carl, cytogenetics
Johnson, Clinton Charles, dentistry, oral pathology
Jones, J Knox, Jr, vertebrate zoology
Jurica, Gerald Michael, atmospheric physics
Kellogg, Charles Nathaniel, mathematics
Kenny, Mary Alice, nutrition
Kenny, Alexander Donovan, pharmacology, endocrinology
Kice, John Lord, organic chemistry
Kim, Young Nok, theoretical physics
King, Mary Elizabeth, ethnography, archaeology
Kokernot, Robert Hutson, epidemiology, virology
Konkov, Vadim, engineering mechanics, applied mathematics
Krieg, Daniel R, plant physiology
Kuhnley, Lyle Carlton, microbiology
Lamb, Mina Marie Wolf, nutrition
Lamb, Neven P, biological anthropology
Lefkowitz, Stanley S, microbiology, virology
Lodhi, Mohammed Arfin Khan, nuclear physics
Lombardini, John Barry, pharmacology
Marshall, Billy Jack, solid state physics, low temperature physics
Martz, Harry Franklin, Jr, statistics, operations research
Marx, John Norbert, organic chemistry
Mason, Perry Shipley, Jr, organic chemistry
Mattox, Richard Benjamin, geology
Maunder, A Bruce, genetics, plant breeding
Mayer-Oakes, William James, anthropology, archaeology
McKenna, John Morgan, immunology
McPherson, Clara, nutrition, foods
McPherson, Clinton Marsud, chemistry
Mecham, John Stephen, herpetology, evolutionary biology
Meyer, Harold David, numerical analysis
Mills, Jerry Lee, inorganic chemistry
Mires, Raymond William, physics
Mitchell, Robert W, biospeleology, ecology
Mitchell, Roy Ernest, inorganic chemistry, physical chemistry
Morey, Philip Richard, plant anatomy, industrial hygiene
Morrow, Kenneth John, Jr, genetics
Murray, Grover Elmer, geology
Nau, Carl August, preventive medicine
O'Brien, Coleman Art, reproductive physiology
O'Brien, Larry Joe, physiology
O'Brien, Thomas Joseph, physical chemistry, theoretical chemistry
Onken, Arthur Blake, soil chemistry, soil fertility
Orr, Donald Eugene, Jr, animal science
Ott, Billy Joe, soils
Owens, John Charles, economic entomology
Packard, Robert Lewis, zoology
Pettit, Russell Dean, range ecology
Proctor, Vernon Willard, phycology
Quade, Charles Richard, molecular physics
Ramsey, Clovis Boyd, animal husbandry
Redington, Richard Lee, physical chemistry
Reeves, Corwin C, Jr, geology
Reichert, John Douglas, theoretical physics, optical physics
Rekers, Robert George, analytical chemistry
Rich, Patricia Vickers, vertebrate paleontology, biogeography
Rigby, Fred Durnford, mathematics
Riggs, Charles Lathan, plant physiology
Roark, Bruce (Archibald), plant physiology
Rose, Francis L, zoology
Rylander, Michael Kent, ornithology, comparative anatomy
Sanders, Darryl Paul, entomology
Sandlin, Billy Joe, physics
Seliger, William George, anatomy, dentistry
Sevall, Jack Sanders, biochemistry
Shetlar, Marvin Roy, biochemistry
Shine, Henry Joseph, organic chemistry
Shurbet, Deskin Hunt, Jr, seismology
Signor, Donald C, hydrology
Song, Pill-Soon, molecular biophysics
Sosebee, Ronald Eugene, plant physiology, ecology
Steadman, Robert George, textiles
Strandtmann, Russell William, taxonomy, zoology
Strauss, Monty Joseph, mathematics
Tarwater, Jan Dalton, algebra
Tereshkovich, George, horticulture
Thayer, Donald Wayne, microbiology, biochemistry
Thomas, Henry Coffman, physics
Thompson, Leif Harry, reproductive physiology
Tribble, Leland Floyd, animal science
Tyner, George S, surgery, ophthalmology
Ueckert, Darrell Neal, range ecology, insect ecology
Wade, Franklin Alton, exploration geology
Walker, Homer Franklin, mathematics
Walling, Derald Dee, mathematics
Ward, Charles Richard, entomology
Wendt, Charles William, soil physics
White, John Thomas, mathematics
Wilbur, Donald Lee, neuroendocrinology
Wilde, Richard Edward, Jr, solid state chemistry
Wilson, Robert Warren, vertebrate paleontology
Wood, Warren Wilbur, geochemistry, hydrology
Wright, Henry Albert, range management, applied statistics
Yang, Shiang-Ping, nutrition
Young, Arthur Wesley, agronomy, soils
Zinn, Dale Wendel, animal husbandry

MARBLE FALLS
Ott, Ellis Raymond, experimental statistics

MARSHALL
Condray, Ben Rogers, organic chemistry
Handler, Shirley Wolz, biology, genetics
Kersey, William Hewell, biochemistry, analytical chemistry
Moore, Edwin Forrest, economic statistics

MCALLEN
Bushland, Raymond Cecil, entomology
Frederickson, Arman Frederick, geology
Hobbs, Clifford Dean, plant pathology

MCGREGOR
Burnside, Charles H, organic chemistry
Calhoun, Millard Clayton, animal nutrition

MCKINNEY
Laughlin, Harold Emerson, animal ecology, herpetology

MIDLAND
Dally, Jesse LeRoy, geology
Elam, Jack Gordon, geology
Frenzel, Hugh N, petroleum geology
Haseltine, William Lloyd, geology
Holden, Frederick Thompson, petroleum geology
Jones, Theodore Sidney, petroleum geology
Nabi, Hosni Abdel, plant genetics, plant breeding
Schneider, William T, exploration geology
Scobey, Ellis Hurlbut, geology
Thompson, Lucien Orrin, petroleum geology
Toohey, Loren Milton, geology

MISSION
Crystal, Maxwell Melvin, veterinary entomology

NACOGDOCHES
Alexander, Forrest Doyle, mathematics
Bilan, M Victor, forest physiology, plant ecology
Blair, Robert Marks, plant ecology
Burkart, Leonard F, wood chemistry, wood technology
Chang, Mingteh, forest hydrology
Clark, William Dean, mathematical analysis

675

TEXAS

Clayton, Glen Talmadge, physics
Collier, Gerald Loyd, economic geography
Decker, John P., molecular physics, spectroscopy
Faulkner, Russell Conklin, Jr., zoology, experimental morphology
Fish, Stewart Allison, obstetrics & gynecology
Garrett, James M., organic chemistry
Gibson, William Wallace, entomology
Halls, Lowell Keith, range conservation
Hoff, Victor John, cytogenetics
Layton, Gerald Lafayette, geometry
Lowry, Gerald Lafayette, forestry, soil science
Mace, Kenneth Dean, microbiology, plant physiology
Machel, Albert R., analytical chemistry, inorganic chemistry
McCullough, Jack Dennis, limnology
McDonald, Harry Sawyer, comparative physiology; herpetology
McGrath, William Thomas, forest pathology
Miller, Edwin Lynn, zoology
Mims, Charles Wayne, mycology
Myers, Clifford Albert, Jr., forest management, forest mensuration
Naistat, Samuel Solomon, physical chemistry
Nixon, Elray S., plant ecology, plant taxonomy
Price, Kenneth Hugh, mathematics
Proctor, Clarke Wayne, mathematics
Rink, George, forest genetics
Robertson, Walter Volley, vertebrate zoology
Seaton, Jacob Alif, inorganic chemistry
Slagle, Wayne Grey, parasitology
Smith, Hugh Burnice, biology
Steinhoff, Raymond Oakley, geology
Stransky, Jerry Janos, silviculture
Vincent, Jerry William, paleoecology
Walker, Bennie Frank, physical chemistry
Walker, Laurence Colton, silviculture, soils
Watterson, Kenneth Gordon, forest soils, environmental management

NEDERLAND
Bright, Gordon Stanley, chemistry

ODESSA
Cihonski, John Leo, analytical chemistry
Crawford, James Dalton, organic chemistry
Cywinski, Norbert Francis, organic chemistry
Disman, John H., earth sciences, natural resources
Griffin, Rodger W., Jr., organic chemistry
Haupt, Frederic Curt, physical organic chemistry
Kurtz, Edwin Bernard, Jr., plant physiology
McKinney, Charles Oran, population biology
Muts-Duplat, Emilio, geology, petrology
Nash, William Donald, synthetic organic chemistry
Nickel, James Alvin, applied mathematics
O'Callaghan, Robin Kuehler, algebra
Orton, Edward Whitfield, geology
Sward, Edward Lawrence, Jr., physical chemistry
Taylor, Edward Donald, physical chemistry
Toomey, Donald Francis, geology, paleontology

ORANGE
Adkins, John Earl, Jr., analytical chemistry
Baxter, Warren Nesmith, organic chemistry
Bazzelle, William Edward, analytical chemistry, spectroscopy
Denison, Jack Thomas, physical chemistry
Jenkins, Sidney Hartman, Jr., organic polymer chemistry
Lynch, Thomas John, polymer chemistry
Musser, Michael Tuttle, organic chemistry
Nelson, Richard David, physical chemistry
Paton, Leo Wesley, organic chemistry, analytical chemistry
Powell, Richard James, polymer chemistry
Rike, Zeb W. III, organic chemistry
Schueler, Bruno Otto Gottfried, organic chemistry; chemical engineering
Tannahill, Mary Margaret, polymer chemistry
Wells, James Robert, chemistry

OVERTON
Lipe, John Arthur, horticulture, plant physiology
Rouquette, Francis Marion, Jr., agronomy

PAMPA
Bowman, Ferne, foods, nutrition

PANTEX
Sherrod, Lloyd B., animal nutrition

PASADENA
Chesnuwood, Charles Mark, geography
Helms, Boyce Dewayne, industrial chemistry; analytical chemistry
Keyworth, Donald Arthur, analytical chemistry
Ruland, Norman Lee, physical chemistry
Thomas, Benjamin William, physics

PEARLAND
Miller, Emery B., organic chemistry

PECOS
Reese, Nathan Allan, animal nutrition

PITTSBURG
Miner, James Joshua, nutrition,

PLAINVIEW
Cox, James Carl, Jr., physical chemistry, organic chemistry educ: wva wesleyan col, bs, 40; univchemistry educ: wva wesle
Dalton, Lonnie Gene, genetics, agronomy
Davis, William Hatch, plant breeding, genetics
Ellsworth, Robert Lovell, plant breeding, genetics
Hess, Delbert Coy, agronomy
Kidd, Harold J., cytogenetics, plant morphology
Kramer, James Nicholas William, agronomy, genetics
McCoy, Dorothy, mathematics
Reese, Weldon Harold, pollution biology; phycology

PORT ARANSAS
Behrens, Earl William, marine geology
Kamykowski, Daniel, biological oceanography
Miget, Russell John, Marine microbiology
Oppenheimer, Carl Henry, Jr., oceanography; ecology
Parker, Patrick LeGrand, geochemistry, chemistry
Smith, Ned Philip, physical oceanography
Watson, Richard Lee, sedimentology
Wohlschlag, Donald Eugene, marine ecology

PORT ARTHUR
Hoist, Edward Harland, organic chemistry
Macaluso, Anthony, Sr., chemistry
Mertens, Frederick Paul, physical chemistry; corrosion
Odell, Norman Raymond, organic chemistry
Rigdon, Orville Wayne, petroleum chemistry
Stamm, Ralph Eugene, analytical chemistry; petroleum chemistry
Yang, Ovid Y H., pathology; immunology

PORT NECHES
Edwards, Gayle Dameron, petroleum chemistry
Laemmle, George Joseph, organic chemistry

PRAIRIE VIEW
Berry, Jewell Edward, parasitology
Doctor, Vasant Manilal, biochemistry
Humphrey, Ronald DeVere, microbial physiology
Martin, Edward Williford, embryology
Nelson, Ivory Vance, analytical chemistry, biochemistry
Stubblefield, Cedric Taylor, physical chemistry
Thomas, Richard Garland, physics

RANDOLPH AFB
Rowen, Burt, aerospace medicine

REFUGIO
Rodney, Paul Frederick, electromagnetics, mathematical physics

RENNER
Gardenhire, James Homer, plant breeding, genetics
Whitehurst, Sanford Huey, soil conservation
Tompkins, Donald Roy, Jr., physics

RICHARDSON
Allen, Robert Ray, organic chemistry
Allum, Frank Raymond, physics, space physics
Boren, Roger Boatner, entomology
Cale, William Graham, Jr., ecology
Cargo, Douglas Bruce, environmental geology
Christian, Wayne Gillespie, geochemistry
Clowes, Royston Courtenay, microbiology, genetics
Cole, David F., physical chemistry
Collins, Carl Baxter, Jr., lasers
Collister, Carl Harold, plant breeding
Cordell, Robert James, petroleum geology
Creed, David, photochemistry
Crowe, Christopher, geophysics
Cutler, Richard Gail, cell biology
Dobrott, Robert D., crystallography, physical chemistry
Donahoe, Pat, analytical chemistry
Eaker, Charles William, chemical physics
Fenyves, Ervin J., high energy physics, cosmic ray physics
Ferraris, John Patrick, organic chemistry, solid state chemistry
Gray, Carla Winlund, animal virology
Gray, Donald, biophysical chemistry
Gutz, Herbert, microbial genetics
Hafner, Harold C., physical chemistry, inorganic chemistry
Halpern, Martin, geochronology
Harm, Walter, genetics, radiation biology
Heikkila, Walter John, space physics
Helsley, Charles Everett, geology, geophysics
Jagger, John, biophysics, photobiology
Jeanes, Jack Kenneth, chemistry
Jodry, Richard L., exploration geology
Johnson, Francis Severin, space physics, meteorology
Kemper, Jost Hansjosef Karlfried, molecular biology
Kimeldorf, George S., statistics
Klumpar, David Michael, space physics
Kusch, Polykarp, physics
Lang, Dimitrij Adolf, biophysics, molecular biology
Lanza, Guy Robert, aquatic ecology
Lee, George Fred, water chemistry, environmental quality
Maimberg, Earl Winton, chemistry
McCaleb, Stanley B., clay mineralogy
McKean, Joseph Walter, Jr., statistics
McMahon, Beverly Edith, paleomagnetism
Mehal, Edward Walter, inorganic chemistry
Melton, Lynn Ayres, physical chemistry
Midgley, James Eardley, systems theory
Mitterer, Richard Max, geochemistry
Odell, Patrick L., mathematical statistics
Ozsvath, Istvan, mathematical physics
Parr, Christopher Alan, theoretical chemistry; computer science
Patrick, Michael Heath, biophysics
Presnall, Dean C., geochemistry, petrology
Rapp, Donald, physical chemistry
Rindler, Wolfgang, mathematical physics
Rothwell, William Thomas, Jr., micropaleontology; petroleum geology
Rupert, Claud Stanley, biophysics
Sand, Ralph E., physical organic chemistry
Secrest, Bruce Gill, mathematics, aeronautical engineering
Sherry, Allan Dean, bioinorganic chemistry
Slocum, Robert Earle, physics
Smith, Larry, colloid chemistry
Smouse, Thomas Hadley, food science, biochemistry
Sobey, Arthur Edward, Jr., physics, mathematics
Svenson, Royal Jay, crop science, soil science
Taylor, Howard Lawrence, applied mathematics, research administration
Thall, Peter Francis, mathematical statistics
Thompson, William Raymond, biological statistics
Tinsley, Brian Alfred, space physics
Van Ness, John Winslow, statistics
Wakefield, Gene F., physical chemistry, research administration
Walter, William T., aquatic biology, limnology
Werbin, Harold, biological chemistry
Williams, Robert Leroy, physics
Wine, Paul Harris, chemical physics
Winningham, John David, magnetospheric physics
Wiorkowski, John James, statistics

ROBSTOWN
Harris, Venoia M., range management, biology

SAN ANGELO
Archer, Cass L., mathematics education
Creel, Gordon C., genetics
Dawson, Horace Ray, physics
Drake, Edgar Nathaniel, II., physical chemistry; analytical chemistry
Duke, John Walter, mathematics
Evans, Noel Dee, mathematics
Flury, Alvin Godfrey, herpetology
Hademenos, James George, biophysics
Harlan, Horace David, physical science education
Hodge, James Edgar, mathematics
Loyd, David Heron, atomic physics, nuclear physics
Menzies, Carl Stephen, animal husbandry
Parker, Cleofus Varren, Jr., radiation physics
Raun, Gerald George, vertebrate ecology
Rowell, Chester Morrison, Jr., taxonomy
Shelton, James Maurice, animal breeding, genetics
Vincent, Lloyd Drexell, nuclear physics
Welch, Gordon E., microbiology
Young, Bernard Theodore, physics

SAN ANTONIO
Adams, Leon Milton, organic chemistry
Adams, Richard Edward Wood, anthropology; archaeology
Adrian, Erle Keys, Jr., anatomy
Aust, J Bradley, surgery
Ball, M Isabel, bio-organic chemistry
Bartels, Richard Alfred, solid state physics
Bass, Joseph Alonzo, bacteriology; immunology
Beissner, Robert Edward, theoretical solid state physics
Benedict, Irvin J., biology; public health
Bertrand, Helen Anne, biochemistry
Birchak, James Robert, experimental solid state physics
Bishop, Vernon Spilman, physiology; biophysics
Bitter, Harold Louis, physiology; toxicology
Blystone, Robert Vernon, cell biology
Bollinger, James Norman, biochemistry, nutrition
Boyne, Philip John, oral surgery; anatomy
Briggs, Arthur Harold, pharmacology; medicine
Brockett, Royce Merrett, medical microbiology
Brown, Edwin Augustus, economic geology
Brown, Jack Harold Upton, physiology
Brunell, Philip Alfred, pediatrics, virology
Burke, John A., inorganic chemistry
Burmeister, Charles W., physics
Burton, Alexis Lucien, histology; cytology
Cameron, Ivan Lee, zoology; anatomy
Campbell, Paul Andrew, aerospace medicine
Carrier, Oliver, Jr., physiology; pharmacology
Carter, Elmer Buzby, computer science
Carter, Frederick J., mathematics
Celitans, Gerard John, biophysics, nuclear physics

Cobb, Howell Dee, Jr., cell physiology, microbial physiology
Coelho, Anthony Mendes, Jr., physical anthropology, primatology
Cohoon, David Kent, applied mathematics
Collins, James Francis, biochemistry
Coltman, Charles Arthur, Jr., hematology, oncology
Cooper, John C, Jr., audiology
Crawford, Stanley Everett, pediatrics, medical administration
Criscuolo, Dominic, biochemistry
Cummiskey, Charles, inorganic chemistry
Cutler, Paul, internal medicine
Dasgupta, Uttam, biochemistry
Davis, Jefferson C, aerospace medicine
Decker, Walter Johns, biochemistry, toxicology
Dimitroff, Edward, organic chemistry, physical chemistry
Donohoo, John T, zoology, cell physiology
Dupuy, Harstry Joseph, immunology
Earley, Laurence E., internal medicine
Eaton, Jerome F, geophysics
Eddy, Carlton Anthony, reproductive physiology
Elbein, Alan D, microbiology, biochemistry
Engelhardt, Hugo Tristram, internal medicine
Erickson, Howard Hugh, physiology, biomedical engineering
Espey, Lawrence Lee, reproductive physiology, environmental sciences
Faust, Claude Marie, mathematics
Fearing, Olin S, botany
Fetzer, Homer D, atomic physics, nuclear physics
Flawn, Peter Tyrrell, geology, academic administration
Fodor, George Emeric, organic chemistry
Forbis, Orie Lester, Jr, pediatrics, psychiatry
Forland, Marvin, internal medicine
Fremming, Benjamin DeWitt, veterinary physiology
Frimpter, George W, medicine
Gaedke, Rudolph Meggs, nuclear physics
Gardner, Clarence Gerald, applied physics
Ghidoni, John Joseph, pathology
Giffen, Martin Brener, psychiatry, psychology
Goldsmith, Ralph Samuel, medicine, endocrinology
Goldzieher, Joseph William, endocrinology
Goodwin, John Thomas, Jr, organic chemistry
Grant, Arthur E, physical medicine & rehabilitation
Gray, James F, algebra
Greene, Nathan Doyle, immunology, parasitology
Greifenstein, Ferdinand Ernest, anesthesiology
Gruber, George J, bioengineering, epidemiology
Guarino, Armand John, biochemistry
Hale, Henry Bixby, physiology
Hall, Charles William, experimental surgery
Hamm, William Edward, physics
Hanss, Robert Edward, geophysics
Harper, Michael John Kennedy, reproductive physiology, endocrinology
Harris, Norman Oliver, histopathology, public health administration
Harrison, Frank, anatomy
Hartman, Albert William, surgery
Heberling, Richard Leon, virology
Hiatt, Caspar Wistar, III, biophysical chemistry
Hodgson, Barrie John, pharmacology, reproductive physiology
Hoskins, Sam Whitworth, Jr, periodontology
Houston, Marshall Lee, embryology, histology
Howard, Charles, inorganic chemistry, analytical chemistry
Hubscher, Thomas, microbiology, immunology
Hudson, Donald Charles, dental materials, instrumentation
Huffman, Ronald Dean, neuropharmacology, neurophysiology
Hummer, Robert L, veterinary medicine
Hurlbut, Herbert Sumner, medical entomology
Johanson, Waldemar Gustave, Jr, internal medicine, pulmonary diseases
Johnson, Donald Edgar, analytical chemistry
Jones, Malcolm David, radiology
Kalter, Seymour Sanford, virology

Kalu, Dike Ndukwe, physiology
Kaufmann, Anthony J, microbiology, biochemistry
Kniker, William Theodore, allergy, immunology
Koenig, Charles Louis, physical chemistry
Konstam, Aaron Harry, computer sciences
Kuntz, Robert Elroy, parasitology, helminthology
Lamb, Lawrence Edward, medicine
Langland, Olaf Elmer, radiology, dentistry
Lawler, Charles Wesley, chemistry
Lee, George H. II, physical chemistry, analytical chemistry
Lee, John Chung, biochemistry
Lee, Robert Leonard, medicine, psychiatry
Levinson, Charles, cell physiology
Lew, Chel Wing, chemistry
Lloyd, William Reese, pharmacy
Lobasso, Frank Anthony, cosmetic chemistry
Ludden, Thomas Marcellus, biopharmaceutics, drug metabolism
Machia, Bollera Muddappa, plant physiology
Magee, Wayne Edward, virology, biochemistry
Mainster, Martin Aron, ophthalmology, mathematical physics
Masoro, Edward Joseph, physiology
Matthijssen, Charles, microbiology, biochemistry
Mattingly, Stephen Joseph, microbial physiology
Matzkanin, George Andrew, solid state physics
McCown, Malcolm G, algebra
McCracken, Alexander Walker, medical microbiology
McFee, Arthur Storer, surgery
McGannon, Donald E, Jr, geology
McGavock, William Crews, inorganic chemistry
McGehee, Richard Vernon, geology
McGill, Henry Coleman, Jr, pathology
McLain, Donald Davis, Jr, mycology, morphology
McMasters, Robert Earl, neurosciences, clinical neurology
McNutt, Clarence Wallace, genetics
Medina, Miguel Angel, pharmacology, biochemistry
Melville, George S, Jr, organic chemistry
Melville, Marjorie Harris, organic chemistry
Meyer, George G, psychiatry, social psychiatry
Mikiten, Terry Michael, neurophysiology
Millar, John David, analytical chemistry
Modak, Arvind T, neurochemistry, toxicology
Molloy, Marilyn, mathematics
Moody, Eric Edward Marshall, molecular genetics, venereal diseases
Moyer, Mary Pat Sutter, virology, oncology
Moyer, Rex Carlton, cancer, microbiology
Murray, Harold Dixon, malacology
Myers, Betty June, parasitology
Nishimura, Jonathan Sei, biochemistry
Oginsky, Evelyn Lenore, microbial physiology
Olson, Merle Stratte, biochemistry
Oujesky, Helen Matusevich, microbiology, radiation biology
Owen, Edgar Wesley, geology, geography
Paque, Ronald E, microbiology, immunology
Park, Myung Kun, pediatric cardiology, pharmacology
Parsons, Jerry Montgomery, horticulture
Pauerstein, Carl Joseph, obstetrics & gynecology
Persyn, Gilbert A, physics, mathematics
Pestana, Carlos, surgery
Peterson, Donald Frederick, physiology
Peterson, Shailer Alvarey, dentistry
Petty, Scott, Jr, geophysics, petroleum engineering
Pierce, Alexander Webster, Jr, pediatrics
Pish, George, physical chemistry
Plummer, Benjamin Frank, organic chemistry
Prince, John Edward, human genetics
Purdy, Robert H, biochemistry
Radwin, Howard Martin, urology
Rao, Pemmaraju Narasimha, organic chemistry
Rees, Allan W, physical biochemistry
Reiter, Russel Joseph, neuroendocrinology, neuroanatomy

Rennels, Edward Gerald, reproductive biology, endocrinology
Renthal, Robert David, protein chemistry
Robie, Norman William, pharmacology
Rodriguez, Charles F, analytical chemistry
Rollwitz, William Lloyd, magnetic resonance, electronic instrumentation
Root, Harlan D, surgery
Rosenthal, Saul Haskell, psychiatry
Rowlands, John Rhys, biophysical chemistry
Rudolph, Joseph Anthony, optics
Sagik, Bernard Phillip, virology
Samm, Sherwood, mathematics
Sandidge, John Roy, paleontology, geology
Sanford, Barbara Ann, microbiology, immunology
Sank, Victor J, engineering physics
Schilling, Jesse William, crystallography
Schlameus, Herman Wade, organic chemistry
Schmidt, Helmut, physics, parapsychology
Schuetze, Clarke E, organic chemistry
Sears, David Alan, internal medicine, hematology
Sergeant, Tom Paul, cell biology
Shapiro, David M, biochemistry
Shelokov, Alexis Ioann, virology
Shepherd, Albert Pitt, Jr, physiology
Sherrill, William Manning, radiophysics
Shumate, Kenneth McClellan, organic chemistry, science education
Silber, Herbert Bruce, inorganic chemistry
Simmons, Charles Edward, psychiatry
Smith, Charles T, chemistry, dentistry
Smith, Kendall O, microbiology, virology
Smith, Steffen Wesley, clinical chemistry
Smith, Thomas Caldwell, physiology, biophysics
Smith, William Robert, nuclear physics
Sparks, Cecil Ray, acoustics, fluid dynamics
Stavinoha, William Bernard, pharmacology, toxicology
Story, Jim Lewis, neurosurgery
Straus, David Conrad, medical microbiology
Sweet, Richard Franklyn, physics
Synek, Miroslav (Mike), atomic physics, physical chemistry
Taylor, Robert Lee, medical mycology
Taylor, Robert Martin, geography
Thomas, Virginia Lynn, medical microbiology
Thor, Daniel Einar, immunology, microbiology
Thyagarajan, B S, organic chemistry
Townsend, Frank Marion, pathology
Treat, Charles Herbert, numerical analysis, heat transfer
Tulloch, George Sherlock, parasitology
Ulrich, Jacob, zoology
Vance, Joseph Francis, mathematics, statistics
Waggener, Robert Glenn, medical physics, biophysics
Wallace, Jack E, analytical biochemistry
Weichlein, Russell George, microbiology
Weisman, Robert A, biochemistry
Weser, Elliot, medicine, gastroenterology
Westmeyer, Paul, science education
Wharton, David Carrie, biochemistry
Williams, Mary Carr, steroid chemistry
Williams, Vick Franklin, anatomy
Williams, William Thomas, biochemistry, cell physiology
Winborn, William Burt, anatomy
Winters, Wendel Delos, virology, immunology
Woller, William Henry, physical pharmacy, cosmetic chemistry
Wynne, Elmer Staten, medical microbiology
Young, Eleanor Anne, nutrition
Yu, Byung Pal, biochemistry, cell physiology
Zanca, Peter, radiology
Zauder, Howard L, anesthesiology

SAN MARCOS

Alexander, Mary, Louise, genetics
Anderson, Robert E, solid state physics
Born, Alfred Ervin, algebra, topology
Cassidy, Patrick Edward, chemistry
Cude, Willis Augustus, Jr, inorganic chemistry, crystallography
Emery, William Henry Perry, cytology, taxonomy
Fawcett, Newton Creig, electroanalytical chemistry, polymer chemistry
Fitch, John William, III, organometallic chemistry
Gary, Roland Thacher, biology

Gregg, Cecil Manren, agronomy, horticulture
Hannan, Herbert Herrick, zoology, limnology
Hazlewood, Donald Gene, number theory
Hellman, Allen David, cartography, geography
Helm, Raymond E, dairy science, nutrition
Helton, Burrell W, mathematics
Horne, Francis R, comparative physiology
Huffman, David George, parasitology
Jackson, William Roy, Jr, nuclear physics
Lippmann, David Zangwill, physical chemistry
Longley, Glenn, Jr, limnology
Norris, William Elmore, Jr, plant physiology
Northcutt, Robert Allan, mathematics
Parks, Archie Oliver, analytical chemistry
Perry, Reeves Baldwin, physical chemistry
Sissom, Stanley Lewis, invertebrate zoology, aquatic ecology
Tuff, Donald Wray, parasitology, taxonomy
Tulloch, Lynn Hardyn, mathematics
Whitenberg, David Calvin, plant physiology, biochemistry
Whiteside, Bobby Gene, fisheries
Willms, Charles Ronald, biochemistry
Yager, Billy Joe, physical organic chemistry

SEABROOK

Amsbury, David Leonard, geology

SEGUIN

Bischoff, Harry William, zoology, botany
Oestreich, Charles Henry, inorganic chemistry
Perry, Charles Rufus, Jr, mathematics
Reeves, W Preston, organic chemistry
Scheie, Paul Olaf, physics, biophysics

SHERMAN

Barr, Charles Richard, biochemistry
Buscher, Henry Neil, zoology, parasitology
Edwards, Frank C, chemistry
Gourley, Lloyd Eugene, Jr, physics
Imhoff, Michael Andrew, organic chemistry
Kimes, Thomas Fredric, numerical analysis, mathematical analysis
Mackey, John Linn, physical chemistry, inorganic chemistry
McCarley, Wardlow Howard, vertebrate zoology
Pierce, Jack Robert, zoology

SIMONTON

Mercado, Edward J, geophysics, seismology

SINTON

Bolen, Eric George, wildlife ecology
Drawe, D Lynn, range management, wildlife research
Glazener, William Caleb, wildlife management

SLATON

Conselman, Frank Buckley, petroleum geology

SOUTH PADRE ISLAND

Sorensen, Lazern Otto, marine phycology

STEPHENVILLE

Cude, Joe E, topology, algebra
Evans, Raeford G, agronomy, genetics
Fain, Robert C, organometallic chemistry
Garner, Herschel Whitaker, vertebrate zoology, ecology
Gehrmann, William Henry, physiology
Hinkson, Thomas Clifford, physical chemistry
Johanson, Lamar, plant physiology
Keith, Donald Edwards, invertebrate ecology
Kincannon, John Alvin, mathematics, economics
Mason, Tim Robert, animal nutrition, reproductive physiology
McCoy, Jimmy Jewell, mathematical physics, solid state physics

TEXAS

Medlen, Ammon Brown, histology, endocrinology
Morrison, Eston Odell, entomology, parasitology
Tackett, Jesse Lee, soil physics
Trogdon, William Oren, soils

SULPHUR SPRINGS
Morris, Quentin L., organic chemistry

TARLETON
Terry, Ona Joy, organic chemistry

TEMPLE
Adams, Herman Ray, clinical chemistry
Adams, John Edgar, soil physics
Arkin, Gerald Franklin, agricultural engineering
Burnett, Earl, soil science
Burson, Byron Lynn, plant genetics
Determan, Louis Henry, Jr., physics
DeLoach, Culver Jackson, Jr., entomology
Dyck, Walter Peter, internal medicine, gastroenterology
Forscher, Bernard Kronman, biochemistry
Hightower, Nicholas Carr, Jr., physiology
Kissel, David E., soil chemistry, soil fertility
Langsjoen, Per Harald, medicine, cardiology
Leibovitz, Albert, medical microbiology, oncology
Martt, Jack M., internal medicine, cardiology
Ritchie, Joe T., soil physics, physical chemistry
Smaldino, Joseph James, audiology
Tessmer, Carl Frederic, plant breeding
Voigt, Paul Warren, plant breeding
Wallace, Tracy I., internal medicine

TEXARKANA
Pitchford, Leanne Carolyn, atomic physics

TEXAS CITY
Beaver, Earl Richard, physical chemistry
Benson, Royal H., organic chemistry, analytical chemistry
Bishop, Margaret S., geology
Bodre, Robert Joseph, analytical chemistry
Campbell, Dan Norvell, analytical chemistry
Cavender, James Vere, Jr., organic polymer chemistry
Dawes, Gregory W. Jr., polymer chemistry, analytical chemistry
Engle, Damon Lawson, polymer chemistry
Fox, Dale Bennett, petroleum chemistry
Glaspie, Peyron Scott, physical organic chemistry, surface chemistry
Hubisz, John Lawrence, Jr., theoretical physics
Makin, Earle Clement, Jr., chemistry
Maute, Robert Lewis, analytical chemistry
McLeod, Richard Kenneth, physical chemistry
McMullen, Eugene Joseph, chemistry
Nelson, Don B., organic chemistry
Newsom, Raymond A., organic chemistry
Owens, Marvin Lee, Jr., analytical chemistry
Roberts, Grady Leon, physical chemistry
Roth, Robert George, organic chemistry
Singleton, Tommy Clark, organic chemistry
Vickroy, Virgil Vester, Jr., polymer chemistry, physical chemistry

TYLER
Cranford, Robert Henry, mathematics
Drummond, Paul Linwood, geology, mineralogy
Loetterle, Gerald John, geology
Nash, Donald Robert, immunobiology
Scheffer, George Henry, physical chemistry, colloid chemistry
Stewart, James Ray, microbiology

UVALDE
Blankenship, Lytle Houston, wildlife research
Mulkey, James Robert, Jr., agronomy
Varner, Larry Weldon, animal nutrition, wildlife research

WACO
Belew, John Seymour, organic chemistry
Bond, Thomas Jackson, biochemistry
Brown, Bryce Cardigan, vertebrate zoology
Busch, Kenneth Walter, analytical chemistry
Calkins, Harmon Eldred, bacteriology
Cockerell, Leone (Doris), analytical chemistry, inorganic chemistry
Davidson, Floyd Francis, physiology
Dixon, James William, Jr., geology, geography
Dole, Malcolm, physical chemistry
Echelle, Anthony Allan, ichthyology
Eldridge, David Wyatt, microbial physiology
Franklin, Thomas Chester, physical chemistry
Gehlbach, Frederick Renner, ecology
Hardcastle, Donald Lee, atomic physics
Hayward, Oliver Thomas, geology
Hurst, Fannie Mae, botany
Jackson, Sally Womack, microbiology
Johnston, John Spencer, genetics
Krajicek, Dayton Dunbar, dentistry
Lind, Owen Thomas, limnology, biology
McAtee, James Lee, Jr., colloid chemistry
Meadows, Charles Milton, entomology
Morales, Gustavo Adolfo, micropaleontology
Namy, Jerome Nicholas, petrology, stratigraphy
Packard, Robert Gay, acoustics
Park, Shim C., theoretical physics
Pennington, David Eugene, inorganic chemistry
Perry, E L, Jr., mathematics
Pinkus, Albin George, physical organic chemistry
Powers, Darden, nuclear physics
Reeder, Charles Edgar, physical chemistry
Rolf, Howard Leroy, mathematics
Schwetman, Herbert Dewitt, physics
Smith, Cornelia Marshall, morphology
Thompson, William Donald, psychology
Tweedie, Virgil Lee, organic chemistry
Wang, Ken Hsi, nuclear physics
Watkins, Julian F, II, entomology
Widner, William Richard, physiology, bacteriology

VERNON
Allen, Thomas J., range science, weed science
McCully, Wayne Gunter, range science
Rittenhouse, Larry Ronald, animal science
Slosser, Jeffrey Eric, entomology
Stafford, Roy Elmer, plant breeding

VICTORIA
Homberger, Carl Stanley, Jr., organic chemistry
Wood, Craig Adams, mathematics

WICHITA FALLS
Barker, John Grove, zoology
Beyer, Arthur Frederick, paleobotany
Boswell, James Louis, invertebrate zoology, radiation biology
Carpenter, Rose Marie, zoology
Dalquest, Walter Woelber, vertebrate zoology
Epp, Chirold Delain, nuclear physics
Eskew, Cletis Theodore, biology
Gee, David Easton, geology
Holverson, Edwin LeRoy, solid state physics
Horner, Norman V., arachnology
Huffman, Louie Clarence, mathematics
Meux, John Wesley, mathematics
Rogers, Jesse Wallace, physical chemistry
Sund, Eldon H., organic chemistry
Watkins, Jackie Lloyd, geology
Williams, Rickey Jay, physical inorganic chemistry

YOAKUM
Harrison, Arthur Leslie, plant pathology

UTAH

BOUNTIFUL
Bishop, Jay Lyman, metallurgical chemistry, organic chemistry
Borrowman, S Ralph, chemistry

BRIGHAM CITY
Baird, Bruce Lloyd, soils, statistics
Hepworth, John Leonard, inorganic chemistry
Jensen, Wayne Ivan, veterinary microbiology, wildlife diseases
Jones, Leon Lloyd, physical chemistry, theoretical chemistry
Law, Ronald Dee, analytical chemistry
Layton, Lionel H., physical chemistry
Thompson, Grant, organic chemistry

CEDAR CITY
Anderson, Russell D., entomology
Hatch, Conrad V., inorganic chemistry, physical chemistry
Jones, Merrell Robert, solid state physics
Larsen, Wesley P., zoology

DUGWAY
Brauner, Kenneth Martin, analytical chemistry
Orr, Geoffrey F., mycology
Rees, Horace Benner, Jr., virology
Salomon, Lothar I., biochemistry
Spendlove, John Clifton, bacteriology
Stark, Harold Emil, entomology
Thompson, Howard E., entomology
Wallace, Volney, agricultural chemistry

WESLACO
Albach, Roger Fred, organic chemistry
Amador, Jose Manuel, plant pathology
Burleson, Charles Albertis, agronomy
Carter, William Whitney, phytopathology, nematology
Cruse, Robert Ridgely, applied chemistry
Dean, Herbert A., entomology
Dietz, Edward Albert, Jr., analytical chemistry
Fucik, John Edward, horticulture, plant physiology
Gausman, Harold Wesley, agronomy
Gerard, Cleveland Joseph, soil fertility
Griffiths, Francis Priday, food technology, bacteriology
Harding, James Alfred, economic entomology
Hensz, Richard Albert, horticulture
Hipp, Billy Wayne, soil fertility
Leamer, Ross Wilson, soil science
Leyden, Robert Fullerton, agronomy, soils
Lime, Bruce James, food chemistry
Moore, Leonard Oro, research administration, organic chemistry
Rasheed, Khalid, chemistry
Sleeth, Bailey, plant pathology
Smith, Burns Ashby, sugar chemistry
Timmer, Lavern Wayne, plant pathology
Wiegand, Craig Loren, soil physics, remote sensing

HYDE PARK
Shupe, James LeGrande, veterinary medicine

FARMINGTON
Larsen, Wayne Ammon, statistics

LOGAN
Albrechtsen, Rulon S., genetics, plant breeding
Alexander, Richard Raymond, paleoecology
Allred, Keith Reid, crop production, plant biochemistry
Anderson, Jay LaMar, pomology, weed science
Anderson, Jay Oscar, poultry nutrition
Anderson, Melvin Joseph, animal nutrition
Anderson, Richard C., organic chemistry
Arave, Clive W., animal breeding
Austin, Joseph Wells, animal science
Bahler, Thomas Lee, zoology
Baker, Kay Dayne, meteorology, electrical engineering

EPHRAIM
Dobson, Donald C., poultry science
Lohrengel, Carl Frederick, II, geology
Mangelson, Farrin Leon, biochemistry, nutrition
Van Epps, Gordon Alnon, field crops

Balph, David Finley, animal behavior, ecology
Bennett, James Austin, animal breeding
Bennett, William Hunter, agronomy
Blake, Joseph Thomas, veterinary medicine
Bleak, Alvin Thomas, range management, plant ecology
Booker, Elizabeth Anne, biochemistry
Bowman, James Talton, genetics
Box, Thadis Wayne, range management
Boyle, William Sidney, botany
Burnham, Bruce Franklin, biochemistry
Butcher, John Edward, animal production
Caldwell, Martyn Mathews, plant ecology, plant physiology
Campbell, William Frank, radiation botany, plant physiology
Cannon, Lawrence Orson, mathematics
Cannon, Melvin Croxall, chemistry
Carter, Orson Silver, phytopathology
Carter, Paul Bearnson, bacteriology
Chatelain, Jack Ellis, theoretical physics
Clark, C Elmer, physiology, biochemistry
Cochran, George Wilson, plant virology
Crapo, Richley H., anthropological linguistics, cultural anthropology
Daniel, Theodore William, forest management
Davis, Donald Walter, entomology
Davis, Lawrence S., forest economics
DeByle, Norbert V., forestry, watershed management
Dewey, Douglas R., cytogenetics, plant breeding
Dewey, Wade G., plant breeding, genetics
Dixon, Keith Lee, vertebrate zoology
Doney, Devon Lyle, plant genetics
Dwyer, Don D., range management
Edwards, W Farrell, electromagnetics
Elich, Joe, mathematics
Ellis, LeGrande Clark, reproductive endocrinology
Emery, Thomas Fred, biochemistry
Ernstrom, Carl Anthon, dairy chemistry
Fletcher, Joel Eugene, soils
Fonnesbeck, Paul Vance, animal nutrition
Foote, Warren Christopher, animal physiology
Fullerton, Thomas M., crop physiology
Gardner, Eldon John, genetics
Gessaman, James A., physiological ecology, avian physiology
Gillette, Tedford A., food science, meat science
Goodall, David William, botany, statistics
Griffin, Gerald D., plant nematology
Grover, Ben Leo, soil physics
Hammond, Robert Grenfell, mathematics
Hammon, Alvin Russell, vegetable crops
Hanks, Ronald John, soil physics
Hansen, Roger Gaurth, nutritional biochemistry
Hansen, Wilford Nels, chemistry, physics
Hardy, Clyde Thomas, structural geology
Harris, Lorin E., animal nutrition, animal husbandry
Hart, George Emerson, Jr., hydrology, forestry
Hatch, Eastman Nibley, nuclear physics
Haws, Byron Austin, economic entomology
Helm, William Thomas, aquatic ecology
Hill, Kenneth Wilford, agronomy
Hoffmann, James Allen, plant pathology
Holmgren, Arthur Herman, plant taxonomy, ecology
Hsiao, Ting Huan, insect physiology
Hull, Alvin C., Jr., plant ecology
Hunsaker, Lloyd R., dairy science
Hunsaker, Neville Carter, mathematics
Hurst, Rex LeRoy, statistics, computer science
Innis, George Seth, mathematics, system analysis
James, David Winston, agriculture
Johnson, Ralph M., Jr., biological chemistry
Jurinak, Jerome Joseph, soil chemistry
Kadlec, John A., wildlife management, ecology
Keeler, Richard Fairbanks, natural products chemistry
Keller, Gordon N., anthropology
Keller, Wesley, range conservation
Knowlton, George Franklin, entomology
Knowlton, Frederick Frank, wildlife research
Lamb, Robert Cardon, dairy science
Langerman, Neal Richard, biochemistry
Lanner, Ronald Martin, forest genetics, tree physiology
Lee, Garth Loraine, physical chemistry
Lind, Vance Gordon, physics, astrophysics

Low, Jessop Budge, wildlife ecology
MacMahon, James A, ecology, vertebrate zoology
Madsen, Milton Andrew, animal husbandry
Malechek, John Charles, range science, ecology
Matthews, Doyle Jensen, animal breeding
McAdams, Robert Eli, nuclear physics
McAllister, DeVere Richard, agronomy
McDonough, Walter Thomas, physiological ecology
McKell, Cyrus Milo, range management, environmental sciences
Megill, Lawrence Rexford, physics
Mendelhall, Von Thatcher, food science
Miller, Akeley, particle physics
Miller, Gene Walker, plant biochemistry
Miller, Raymond Woodruff, soil fertility, environmental science
Miner, Merthyr Leilani, veterinary pathology
Moore, William Marshall, physical chemistry
Morse, Joseph Grant, inorganic chemistry
Morse, Karen W, inorganic chemistry
Mueggler, Walter Frank, plant ecology
Mumford, David Louis, plant pathology
Neuhold, John Mathew, aquatic ecology
Nye, William Preston, apiculture, entomology
Oaks, Emily Caywood Jordan, vertebrate zoology
Oaks, Robert Quincy, Jr, geology
Olsen, Donald Ray, mineralogy
Olsen, Richard Kenneth, organic chemistry
Otteson, Otto Harry, nuclear physics
Packer, Paul Earl, forestry
Palmblad, Ivan G, ecology, evolution
Parker, Robert Davis Rickard, environmental health, industrial hygiene
Pedersen, Marion Walter, agronomy
Peterson, Edwin Loose, cultural geography, cultural anthropology
Peterson, Howard Boyd, soil chemistry
Pollard, Leonard Heber, olericulture
Pope, Wendell LaVon, numerical analysis
Porcella, Donald Burke, environmental science
Post, Frederick Just, microbial ecology, pollution biology
Rajagopal, P K, fish biology
Richardson, Gary Haight, dairy chemistry, dairy microbiology
Salisbury, Frank Boyer, plant physiology
Salunkhe, Datta K, food technology
Sanders, Raymond Thomas, physiology
Seeley, Schuyler Drannan, plant chemistry
Sharma, Raghubir Prasad, pharmacology, toxicology
Shaw, Richard Joshua, botany
Sigler, William Franklin, fisheries
Simmons, John Robert, biochemistry, genetics
Sisson, Donald Victor, applied statistics
Skujins, John Janis, soil biochemistry, soil microbiology
Smith, Grant Gill, organic chemistry
Smith, R L, soil chemistry
Smith, Winslow Whitney, bacteriology
Southard, Alvin Reid, soil science, geology
Spence, Jack Taylor, analytical chemistry, inorganic chemistry
Spendlove, Rex S, virology, immunology
Spillett, James Juan, animal ecology, wildlife resources
Stalnaker, Clair B, fish biology, genetics
Stanley, Hugh P, electron microscopy, cell biology
Stoddard, George Edward, animal nutrition
Stokes, Allen Woodruff, animal behavior
Street, Joseph Curtis, toxicology, pesticide chemistry
Suprunowicz, Konrad mathematics
Taylor, Robert E, soil conservation
Thatcher, Theodore Ossip, entomology
Theurer, Jessop Clair, plant genetics, plant breeding
Thomas, Don Wylie, veterinary science, animal science
Thomas, James H, plant breeding, genetics
Thorne, David Wynne, soil fertility
Turner, David Lee, statistical analysis
Valentine, Joseph Earl, mathematics
Van Orden, Harris O, organic chemistry
Wadley, Bryce Nephi, plant pathology
Wagner, Frederic Hamilton, biology
Walker, David Rudger, pomology
Warner, Judith Sauve, ecology
Welkie, George William, virology, plant physiology
West, Neil Elliott, ecology
Wiebe, Herman Henry, botany, plant physiology
Wilcox, Ethelwyn Bernice, nutrition
Williams, Miles Coburn, plant physiology
Windham, Michael Parks, mathematics
Wood, John Karl, physics
Wooldridge, Gene Lysle, atmospheric sciences, physics
Workman, John Paul, range management, agricultural economics
Wu, Ming Tsung, chemistry, food science
Wydoski, Richard Stanley, fisheries, zoology
Wyse, Roger Earl, plant physiology, biochemistry
Youssef, Nabil Naguib, morphology, cell biology

MAGNA
Brown, Billings, thermodynamics
Dehm, Henry Christopher, organic chemistry
Dubois, Ronald Joseph, organic chemistry
Elmslie, James Stewart, organic polymer chemistry
Gardner, William H, industrial chemistry
Peterson, Albert H, analytical chemistry
Sauer, Dennis Theodore, inorganic chemistry
Young, Herbert Lewis, industrial organic chemistry

MOAB
Linehan, Urban Joseph, physical geography

OGDEN
Amman, Gene Doyle, forest entomology
Ash, Sidney Roy, geology, paleobotany
Bay, Roger Rudolph, forestry
Beishline, Robert Raymond, physical organic chemistry
Blaisdell, James Pershing, range ecology
Bozniak, Eugene George, aquatic ecology, phycology
Buchanan, Hayle, plant ecology
Buss, Walter Richard, geology, geography
Clarke, Robert Alma, physics, mathematics
Cole, Walter Eckle, entomology, mathematical biology
Copeland, Otis Lee, environmental management, resource management
Croft, Alfred Russell, ecology
Galli, John Ronald, solid state physics
Graff, Darrell Jay, helminthology
Greer, Deon Carr, geography
Gruschow, George F, forest management, silviculture
Guymon, Ervin Park, inorganic chemistry, analytical chemistry
Harrison, H Keith, plant anatomy
Havertz, David S, medical entomology, ecology
Hayes, Sheldon P, immunology, bacteriology
Huish, Howard Paul, physics
Hyde, Kendell Heman, mathematics
James, Helen Jane, analytical chemistry
Jensen, Chester E, biometrics
Jensen, Emron Alfred, parasitology, protozoology
Jensen, John Neil, population genetics, mammalogy
Johanson, Alva Joseph, organic chemistry
Lindmark, Ronald Dorance, forest economics
Miller, Richard Roy, mathematics, chemistry
Miner, Bryant Albert, physical chemistry
Monk, Ralph Warner, plant physiology
Moyle, Richard W, paleontology, geology
Murphy, Don Robison, geography, geology
Neff, Thomas Rodney, geology
Pashley, Emil Frederick, Jr, geology, hydrology
Schallau, Con H, forest economics
Seager, Spencer Lawrence, physical chemistry
Stockland, Alan Eugene, microbiology
Stoker, Howard Stephen, inorganic chemistry, environmental chemistry
Stringham, Reed Millington, Jr, physiology, oral biology
Welch, Garth Larry, academic administration, inorganic chemistry
Young, Orson Whitney, zoology

OREM
Janes, Ray Low, entomology
Merrill, John Jay, physics

PROVO
Allen, Ashael Lester, zoology
Allred, Donald Mervin, entomology, ecology
Allred, Rodney Chase, agronomy
Andersen, Ferron Lee, parasitology
Andersen, William Ralph, biochemical genetics
Anderson, Keith Phillips, physical chemistry
Baer, James L, paleoecology, structural geology
Ballif, Jae R, solar physics, terrestrial physics
Barnes, James Ray, aquatic ecology
Barnett, John Dean, physics
Baumann, Richard William, systematic entomology, aquatic biology
Beck, Jay Vern, biochemistry, microbiology
Bennion, Marion, nutrition, food science
Bills, James LaVar, inorganic chemistry
Bissell, Harold Joseph, geology
Blackham, Angus Udell, physical organic chemistry
Booth, Gary Melvon, entomology
Bradshaw, Jerald Sherwin, organic chemistry
Bradshaw, Willard Henry, microbiology
Bradshaw, William S, biochemistry, developmental biology
Broadbent, Hyrum Smith, chemistry
Brotherson, Jack DeVon, plant ecology
Bryce, Gale Rex, statistics
Bryner, John C, applied physics
Bryner, Loren Conrad, biophysical chemistry
Bullock, Kenneth C, geology
Burton, Sheril Dale, microbiology, biochemistry
Bushman, Jess Richard, geology
Butler, Eliot Andrew, analytical chemistry
Call, Richard A, pathology
Campbell, Douglas Michael, mathematical analysis
Cannon, John Francis, physical chemistry, inorganic chemistry
Carlson, Gary, computer sciences
Carter, Melvin Winsor, applied statistics, biometrics
Castle, Raymond Nielson, chemistry
Chapman, Arthur Owen, histology, pathology
Dean, Charles Edwin, mathematical statistics, computer science
Decker, Daniel Lorenzo, solid state physics
Dibble, William E, physics
Dixon, Dwight R, nuclear physics
Donaldson, David Miller, bacteriology, immunology
Dudley, James Duane, physics
Eastmond, Elbert John, physics
Evenson, William Edwin, theoretical solid state physics
Farmer, James Lee, biochemical genetics
Farnsworth, Raymond Bartlett, agronomy
Faulkner, James Earl, statistics
Fearnley, Lawrence, topology
Ferguson, Helaman Rolfe Pratt, mathematics
Fleming, Donovan Ernest, psychophysiology
Fletcher, Harvey, physics
Fletcher, Harvey Junior, applied mathematics
Freeman, Lawrence Reed, food science
Frost, Herbert Hamilton, vertebrate zoology
Gardner, Andrew Leroy, plasma physics
Gardner, John Hale, magnetic resonance
Gardner, Robert Wayne, animal nutrition
Garner, Lynn E, mathematics
Gill, Gurcharan S, mathematics
Goates, James Rex, physical chemistry
Grey, Alan Hopwood, physical geography
Gubler, Clark Johnson, biochemistry
Hall, Howard Tracy, high pressure physics
Hamblin, William Kenneth, geology
Hansen, George Henry, geology
Hansen, Lee Duane, inorganic chemistry
Harper, Kimball T, botany, soil science
Harrison, Bertrand Fereday, plant physiology, agrostology
Harrison, Bertrand Kent, theoretical physics
Hatch, Dorian Maurice, theoretical physics
Hawkins, Richard Thomas, organic chemistry
Hayward, Charles Lynn, animal ecology
Heckmann, Richard Anderson, parasitology
Heninger, Richard Wilford, physiology, biochemistry
Herrin, Charles Selby, medical entomology, systematic entomology
Hess, Wilford Moser Bill, botany, plant pathology
Higgins, John Clayborn, mathematics
Hill, Armin John, physics
Hill, John Mayes, Jr, biochemistry
Hill, Max W, nuclear physics
Hillam, Kenneth L, mathematics
Hilton, Horace Gill, statistics
Hintze, Lehi Ferdinand, geology
Hoopes, Keith Hale, veterinary medicine
Hoskisson, William A, bacteriology
Izatt, Reed McNeil, inorganic chemistry
Jackson, Richard H, cultural geography
Jaussi, August Wilhelm, physiology
Jeffery, Duane Eldro, genetics, evolutionary biology
Jensen, Gary Lee, experimental nuclear physics
Jensen, Marcus Martin, medical microbiology
Johnson, F Brent, virology
Johnson, John Hal, organic & food chemistry
Jones, Douglas Emron, physics
Jorgensen, Clive D, entomology
Klein, Sigrid Marta, microbiology, biochemistry
Knight, Larry V, physics
Larsen, Don Hyrum, industrial microbiol, medical microbiology
Larsen, Kenneth Martin, applied mathematics
Larson, Everett Gerald, solid state physics, molecular physics
Laws, Wilford Derby, Jr, soil fertility
Layton, Robert L, geography
Mangelson, Nolan Farrin, physical chemistry, nuclear physics
Mangum, John Harvey, biochemistry
Mason, Grant William, cosmic ray physics
McArthur, Eldon Durant, plant genetics
McNamara, Delbert Harold, astrophysics, astronomy
Miller, Wade Elliott, II, vertebrate paleontology, geology
Moore, Glen, plant physiology
Moore, Hal G, mathematics
Murdock, Joseph Richard, plant ecology
Murphy, Joseph Robison, zoology
Nelson, Homer Mark, solid state physics
Nelson, Kay Leroi, physical organic chemistry
Nelson, Sheldon Douglas, soil conservation, soil morphology
Nielson, Howard Curtis, statistics, mathematics
Nordmeyer, Francis R, inorganic chemistry
North, James A, virology, immunology
Ott, J Bevan, physical chemistry
Park, Robert Lynn, animal science, animal breeding
Paul, Edward Gray, organic chemistry
Petersen, Morris Smith, geology
Peterson, C Denis, veterinary medicine, biochemistry
Peterson, John M, mathematics
Phillips, William Revell, mineralogy
Porter, Richard Dee, wildlife ecology
Rasband, S Neil, physics, astrophysics
Reimschussel, Ernest F, horticulture
Rhees, Reuben Ward, neuroendocrinology
Richards, Dale Owen, statistics
Rigby, J Keith, paleontology
Robinson, Donald Wilford, mathematics
Robison, Laren R, plant genetics, weed science
Rushforth, Samuel Roberts, algology
Sagers, Richard Douglas, microbiology
Seegmiller, Robert Earl, embryology
Shumway, Richard Phil, animal physiology
Skarda, Ralph V, Jr, mathematics
Smith, Bruce Nephi, plant physiology
Smith, Howard Duane, mammalogy, ecology
Smith, Nathan McKay, zoology
Snow, Donald Ray, mathematics
Snow, Richard L, physical chemistry
Sorenson, John Leon, anthropology
Stevens, Dale John, physical geography
Strong, William J, acoustics
Stutz, Howard Coombs, genetics
Swensen, Albert Donald, biochemistry
Sykes, Dwane Jay, environmental sciences, range ecology
Tanner, Wilmer W, zoology
Thorne, James Meyers, physical chemistry
Tipton, Vernon John, parasitology, medical entomology
Transtrum, Lloyd G, agriculture, environmental sciences
Vallentine, John Franklin, range science

Van De Graaff, Kent Marshall, gross anatomy, mammalogy
Vanfleet, Howard Bay, solid state physics
Vernon, Leo Preston, biochemistry
Walker, Leroy Harold, mathematics
Walker, Rudger Harper, soil science
Wallentine, Max V., animal science
Weber, Darrell J., biochemistry, plant pathology
Weiss, Jonathan David, experimental solid state physics
Welch, Stanley L., plant taxonomy
White, Clayton M., vertebrate zoology, ecology
White, Fred G., plant biochemistry
Whitehead, Armand T., entomology
Whitton, Leslie, plant cytology, plant genetics
Wickes, Harry E., mathematics
Wilson, Byron J., inorganic chemistry
Wood, Benjamin W., plant science
Wood, Stephen Lane, entomology
Woolley, Earl Madsen, physical chemistry, analytical chemistry
Wright, Donald N., bacteriology
Yearout, Paul Harmon, Jr., mathematics

SALT LAKE CITY

Abildskov, Junior A., medicine
Adler, Robert Garber, analytical chemistry
Ailion, David Charles, solid state physics
Ajax, Ernest Theodore, clinical neurology
Albee, Howard Franklin, geology
Alexander, Guy B., chemistry, metals
Alger, Terry Dean, physical chemistry
Allred, Evan Leigh, organic chemistry
Alvord, Donald C., economic geology
Andersen, Terrell Neils, physical chemistry
Anderson, Robert, cultural anthropology
Andrade, Joseph D., bioengineering, materials science
Archer, Victor Eugene, medicine, epidemiology
Arnow, Theodore, geology
Ash, Kenneth Owen, biochemistry, clinical chemistry
Astling, Elford George, meteorology
Athens, John William, internal medicine, hematology
Bailey, Richard Elmore, medicine
Baldwin, Robert Charles, toxicology, analytical chemistry
Ball, James Stunsman, physics
Bamford, Robert Wendell, economic geology, geochemistry
Barker, Lynn Marshall, applied physics
Barney, Archie Fay, agronomy
Barnhill, Robert E., mathematical analysis
Beck, Edward C., physiological psychology
Behle, William Harroun, ornithology
Bentrude, Wesley George, organic chemistry
Bergeson, Haven Eldred, high energy physics
Biesele, Ferdinand Charles, mathematics
Bliss, Eugene Lawrence, medicine, psychiatry
Bodily, David Martin, physical chemistry
Boyd, Richard Hays, polymer chemistry
Boyd, William Adam, anesthesiology, physiology
Bragg, David Gordon, medicine, radiology
Brandt, Richard Charles, solid state physics
Brooks, Robert M., mathematics
Broom, Arthur Davis, bio-organic chemistry, medicinal chemistry
Brown, Harold Mack, Jr., visual physiology, biophysics
Brown, James Hemphill, ecology
Bruce, Wayne Royal, economic geology
Bryson, Melvin Joseph, clinical biochemistry
Burgess, Cecil Edmund, mathematics
Burgess, Paul Richards, neurobiology
Butterfield, Veloy Hansen, Jr., computer sciences
Cagle, Fredric William, Jr., chemistry
Callaghan, Eugene, economic geology
Camp, Leslie Wilford, geology
Capecchi, Mario Renato, cell biology
Carlson, James Andrew, pure mathematics
Carroll, Dana, molecular biology
Cartwright, George Eastman, internal medicine
Case, James Hughson, topology
Casjens, Sherwood Reid, biochemistry

Castleton, Kenneth Bitner, surgery
Chamberlin, Richard Eliot, mathematics
Choules, George Lew, biochemistry
Christensen, Carl Joseph, physical chemistry
Christiansen, Francis Wyman, geology
Chu, Yaw-En, plant genetics
Clark, Lincoln Dutton, medicine
Clark, Lloyd Allen, economic geology, geochemistry
Clemens, Charles Herbert, mathematics
Cohen, Elaine, mathematical analysis, systems theory
Cole, Nyla J., psychiatry
Coles, William Jeffrey, mathematics
Cook, Kenneth Lorimer, geophysics
Cramer, Harrison Emery, air pollution, meteorology
Creel, Donnell Joseph, neuropsychology
Davey, Gerald Leland, applied mathematics, systems analysis
Davis, Edward Allan, mathematics
DeFord, John W., solid state physics
Dethlefsen, Lyle A., radiobiology, oncology
DeWitt, Charles Wayne, Jr., immunology, bacteriology
Dibble, Charles Elliot, anthropology
Dick, Bertram Gale, physics
Dickinson, William Joseph, developmental genetics
Dickman, Sherman Russell, biochemistry
Dickson, Don Robert, meteorology
Dixon, John Aldous, surgery, physiology
Doelling, Hellmut Hans, economic geology
Durant, Stephen David, zoology
Edmunds, George Francis, Jr., entomology, evolution
Egan, Merritt H., psychiatry
Ehrenfeld, Elvera, cell biology, virology
Eichwald, Ernest J., pathology
Ekdale, Allan Anton, paleoecology
Elbel, Robert E., medical entomology
Englert, Edwin, Jr., internal medicine, gastroenterology
Ensign, Paul Roselle, pediatrics
Epstein, William Warren, bio-organic chemistry
Ernst, Robert R., microbiology, analytical chemistry
Evans, Frederick Read, protozoology
Eyring, Henry, physical chemistry
Eyzaguirre, Carlos, neurophysiology
Fidlar, Marion Moore, geology
Fidone, Salvatore Joseph, physiology, neurochemistry
Fingl, Edward (George), pharmacology
Fowles, Grant Robert, quantum optics
Franz, Donald Norbert, neuropharmacology
Freston, James W., internal medicine, gastroenterology
Furtell, Jean H., physical chemistry
Gaddie, Robert Stanley, sugar chemistry
Gardner, Pete D., organic chemistry
Gardner, Reed McArthur, bioengineering
Gates, Henry Stillman, science education, environmental sciences
Gaufin, Arden Rupert, zoology
Gebhardt, Louis Philipp, Jr., microbiology
Gersten, Stephen M., topology, algebra
Ghandehari, Mohammad Hossein, physical chemistry
Gibb, James Wooley, biochemical pharmacology
Gibbs, Peter (Godbe), physics
Giddings, John Calvin, chemistry
Gin, Thon Too, mineralogy, geology
Glaser, Leslie, topology
Glasgow, Lowell Alan, microbiology, pediatrics
Goates, Wallace Albert, audiology, speech pathology
Golden, Carole Ann, immunology
Goode, Harry Donald, geology
Goodman, Louis Sanford, pharmacology
Goratowski, Melvin Jerome, organic biochemistry, health sciences
Grant, David Morris, physical chemistry
Green, Sherry Merrill, algebra
Greenfield, Harvey Stanley, medical biophysics
Greenlee, Lorance Lisle, biochemistry, science policy
Groom, Donald Eugene, cosmic ray physics
Gross, Fletcher, mathematics
Grosser, Bernard Irving, psychiatry, neuroendocrinology
Grundmann, Albert Wendell, parasitology, medical entomology
Guiliory, William Arnold, physical chemistry

Gurney, Elizabeth Tucker Guice, molecular genetics
Gurney, Theodore, Jr., cell biology, molecular biology
Gustafson, Grant Bernard, mathematical analysis
Hale, Lytle A., structural geology, stratigraphy
Hanly, Edward William, developmental genetics
Harmon, Shirley Ann, molecular biology
Harris, Frank Ephraim, Jr., chemical physics
Harvey, Stewart Clyde, pharmacology
Hathaway, Ralph Robert, experimental biology
Hawkes, H Bowman, physical geography
Heaney, Robert John, atmospheric chemistry, water chemistry
Hearn, Anthony Clem, theoretical physics
Hendricks, Russel Hyer, analytical chemistry
Hill, Archie Clyde, soil chemistry
Hill, Douglas Wayne, medical microbiology
Hill, Harry Raymond, pediatrics, pathology
Hilpert, Lowell Sinclair, economic geology
Hirsh, Harold Frederick, animal ecology
Horch, Kenneth William, neurophysiology, sensory physiology
Horton, Walter James, synthetic organic chemistry
Howell, Elmer Virgil, Jr., microbiology
Huang, Wai Mun, molecular biology
Huber, Robert John, semiconductors
Hughes, Charles Campbell, applied anthropology, medical anthropology
Hutchings, Theron Bird, soil genesis
Jacho, Leonard Wallenstein, neurology
Jee, Webster Shew Shun, anatomy
Jennings, Jesse David, anthropology, archaeology
Jensen, James Norman, solid mechanics
Johnson, Mead Leroy, economic geology
Johnson, LaVell R., biochemistry, organic chemistry
Johnson, Owen W., solid state physics
Jones, Dane Robert, physical chemistry
Jones, Michael Baxter, economic geology
Kadesch, Robert R., physics
Kalka, Morris, mathematical analysis
Kao, Shih-Kung, meteorology, air pollution
Karler, Ralph, pharmacology
Kelly, Michael Thomas, immunology, clinical microbiology
Kemp, John Wilmer, physiological chemistry, pharmacology
Kern, Earl R., medical microbiology
Kim, Sung Wan, biomaterials, biophysical chemistry
King, James Wilhelmsen, geography
Klauber, Melville Roberts, statistics
Kodama, Goji, inorganic chemistry
Koehler, P Ruben, radiology
Kolff, Willem Johan, medicine, clinical medicine
Kuby, Stephen A., biochemistry
Kuehl, LeRoy Robert, biochemistry
Kuida, Hiroshi, internal medicine, physiology
Kwan-Gett, Clifford Stanley, surgery
Lahey, M Eugene, pediatrics
Lark, Carl Gordon, microbiology
Lark, Cynthia Ann, microbiology, virology
Larsen, Austin Ellis, veterinary medicine
Lattman, Laurence Harold, geology
Lau, Brian Richard, mathematics
Lawrence, Paul J., biochemistry
Lee, Glenn Richard, internal medicine, hematology
Legare, Richard J., polymer chemistry, chemical kinetics
Legler, John Marshall, zoology
Leininger, Madeleine Monica, anthropology
Li, Tien-Yien, mathematics
Linker, Alfred, biochemistry
Linn, Thomas Arthur, Jr., analytical chemistry, radiochemistry
Lombardi, Paul Schoenfeld, microbiology, virology
Lords, James Lafayette, plant physiology
Luty, Fritz, solid state physics
Lyman, Donald Joseph, polymer chemistry, biomaterials
Madsen, James Henry, Jr., vertebrate paleontology
Marcus, Stanley, immunology
Mason, Jesse David, mathematics

Mason, Robert C., pharmacology, medicinal chemistry
Matsen, John Martin, medicine, microbiology
Mavor, Huntington, neurology
Maxwell, John Gary, surgery
Maynard, Hugh Bardeen, mathematics
McCloskey, James Augustus, Jr., chemistry
McCoy, Roger Michael, physical geography
McCullough, John Martin, physical anthropology
McDonald, Keith Leon, theoretical physics
McNeil, Charles J r., cytology, anatomy
McNulty, Irving Bazil, plant physiology
Mecham, Merlin J., speech pathology, audiology
Merrill, Jerald Carl, physical chemistry
Michl, Josef, chemistry
Miles, Charles P., pathology, cytology
Millar, Kay, internal medicine, cardiology
Miller, Gerald R., materials science, physics
Millhouse, Oliver Eugene, neurology, anatomy
Moberly, Lawrence Allan, low temperature physics
Moody, Frank Gordon, surgery
Morrison, James Leslie, theoretical physics
Muir, Melvin K., soil chemistry, plant nutrition
Myers, Marcus Norville, analytical chemistry
Nabors, Charles J r., cytology, anatomy
Nackowski, Matthew Peter, geology
Nash, William Purcell, petrology
Negus, Norman Curtiss, zoology
Nelson, Alan R., medical quality assessment
Nelson, Don Harry, medicine
Nelson, Russell Marion, surgery
Nicholes, Paul Scott, bacteriology
Nichols, William Kenneth, pharmacology
Nielsen, Lewis Thomas, entomology
Nielsen, Richard Leroy, economic geology
Nielson, Dennis Lon, economic geology
Nielson, Dianne Ruth Gerber, exploration geology
Noehren, Theodore Henry, internal medicine, pulmonary diseases
Oblad, Alex Golden, petroleum chemistry
Ohie, Ernest Linwood, economic geology
Ohlsen, William David, solid state physics
Okun, Lawrence M., neurobiology
Olsen, Donna Mae, health services research
Olsen, Peter Fredric, ecology
O'Neill, Frank John, microbiology, pathology
Parker, Don Timothy, bacteriology
Parker, Seymour, cultural anthropology
Parkinson, John Stansfield, molecular genetics
Parma, David Hopkins, genetics
Parry, Robert Walter, inorganic chemistry
Parry, William Thomas, mineralogy, geochemistry
Pendleton, Robert Cecil, radiation ecology
Petajan, Jack Hougen, neurology, physiology
Petersen, Robert Virgil, pharmaceutical chemistry
Pettit, L Archer, nutrition
Picard, M Dane, sedimentary petrology
Plummer, Robert Patrick, computer science
Poulter, Charles Dale, chemistry
Prahl, James William, biochemistry
Pratt, Howard Riley, rock mechanics, structural geology
Price, Richard Henry, theoretical physics, astrophysics
Pugmire, Ronald J., physical chemistry
Ragsdale, Ronald O., inorganic chemistry
Reading, James Cardon, biostatistics, mathematical statistics
Ree, Alexius Taikyue, physical chemistry
Reed, Dennis Keith, mathematics
Reed, Donal J., pharmacology
Rees, Don Merrill, entomology
Renzetti, Attilio D., Jr., pulmonary diseases, physiology
Richards, Oliver Christopher, biochemistry
Richards, Ralph Chamberlain, surgery
Ridd, Merrill Kay, geography
Riesenfeld, Richard F., applied mathematics

Rilling, Hans Christopher, biochemistry
Roberts, Theodore S., neurosurgery
Robinette, Martin Smith, audiology
Rogolsky, Marvin, microbiology
Roll, David Byron, medicinal chemistry
Rosenberger, Franz, chemical physics
Ross, Howard Persing, geophysics, economic geology
Roth, Peter Hans, micropaleontology
Rothenberg, Mortimer Abraham, neurochemistry
Roti Roti, Joseph Lee, radiation biophysics, theoretical biology
Ruhling, Robert Otto, exercise physiology
Ruhmann-Wennhold, Ann Gertrude, medicine, endocrinology
Runyon, Ernest Hocking, biology
Rush, Francis Eugene, geology
Rushing, Thomas Benny, topology
Ruttenberg, Herbert David, pediatrics, cardiology
Samuels, Leo Tolstoy, biochemistry
Sandberg, Vernon Dean, astrophysics
Schmitt, Klaus, mathematics
Scott, William Raymond, algebra
Sharp, Byron James, geology
Shininger, Terry Lynn, developmental biology, plant physiology
Shough, Herbert Richard, pharmacognosy, medicinal chemistry
Sill, William Robert, geophysics
Smart, Keith Lorenzo, immunobiology
Smith, David Lee, physical chemistry, analytical chemistry
Smith, Douglas Lee, occupational health, toxicology
Smith, Robert Baer, geophysics, geology
Snellman, Leonard W., meteorology
Snyder, Clifford Charles, plastic surgery
Sorensen, James Alfred, medical physics, nuclear medicine
Sosin, Abraham, solid state physics, materials science
Stern, Ronald John, topology
Stevens, Walter, anatomy, radiobiology
Spikes, John Daniel, biophysics
Spicer, Leonard Dale, physical chemistry
Stokes, William Lee, stratigraphy
Straight, Richard Coleman, photobiology
Sutherland, Bill, physics
Steinmuller, David, transplantation biology
Swanson, John Lee, experimental pathology, microbiology
Sweat, Floyd Marvin, biochemistry
Sweat, Max Leroy, biochemistry
Swigart, John Irvin, experimental physics
Swinyard, Ewart Ainslie, pharmacology
Symko, Orest George, physics
Taylor, Joseph Lawrence, mathematical analysis
Topham, William Sanford, biophysics, bioengineering
Stenger, Frank, applied mathematics
Stensaas, Larry J, neuroanatomy
Stensaas, Suzanne Sperling, neuroanatomy
Townsend, Leroy B, medicinal chemistry
Treshow, Michael, plant pathology
Tsagaris, Theofilos John, internal medicine, cardiology
Tucker, Don Harrell, mathematics
Turkanis, Stuart Allen, pharmacology
Tyler, Frank Hill, medicine
Ure, Roland Walter, Jr, semiconductors
Ursenbach, Wayne Octave, chemistry
Van Kampen, Kent Rigby, veterinary pathology
Van Norman, Richard Wayne, plant physiology
Velick, Sidney Frederick, biochemistry
Viavant, William Joseph, computer science
Vickery, Robert Kingston, Jr, plant evolution
Voorhees, Kent Jay, physical organic chemistry, analytical chemistry
Wahrhaftig, Austin Levy, physical chemistry
Walker, John Lawrence, Jr, physiology
Walling, Cheves, physical organic chemistry
Wang, Chin Hsien, chemistry
Ward, John Robert, medicine
Ward, Stanley Harry, geophysics
Warnock, Robert G, parasitology
Watson, Maxine Amanda, population biology, molecular genetics
Weinstein, Stanley Edwin, mathematics, computer science
Welsh, John Elliott, Sr, geology
Wender, Paul H, psychiatry, child psychiatry
West, Charles Donald, biochemistry

Whelan, James Arthur, economic geology
Wiens, Delbert, systematic biology, evolutionary biology
Wilburn, Richard Lee, physical chemistry
Wilcox, Calvin Hayden, mathematics
Wiley, Bill Beauford, microbiology, immunology
Willden, Charles Ronald, geology
Willett, Douglas W, mathematics
Williams, George Abiah, solid state physics
Williams, Philip, Jr, meteorology
Wilson, Dana E, medicine, metabolism
Wilson, John Coe, geology
Wilson, John F, medicine, pathology
Winn, Grant Saunders, air pollution, biochemistry
Wintrobe, Maxwell Myer, internal medicine
Withrow, Clarence Dean, pharmacology
Wolbach, Robert Albert, physiology
Wolcott, Mark Walton, surgery, thoracic surgery
Wolf, Dieter, theoretical solid state physics, magnetic resonance
Wolfe, James H, mathematics
Wolstenholme, David Robert, molecular biology, cell biology
Wood, Don Clifton, biochemistry
Woodbury, Dixon Miles, pharmacology
Woodbury, John Walter, physiology, biophysics
Woodbury, Lowell Angus, biostatistics
Wullstein, Leroy Hugh, botany, microbiology
Zdunkowski, Wilford G, physical meteorology

SANTA CLARA
Fernelius, Albert Lawrence, microbiology

TOOELE
Crane, George Thomas, microbiology, virology

VERNAL
Godfrey, Andrew Elliott, geomorphology

VERMONT

ARLINGTON
Harwood, Theodore Henry, internal medicine

BENNINGTON
Barker, Jane Ellen, developmental biology, hematology
Coburn, Everett Robert, organic chemistry
Ellem, Kay Adrian Oswald, molecular biology
Flaccus, Edward, plant ecology
Paulsen, Elizabeth Charlotte, biochemistry
Singer, Irwin I, cell biology, virology
Toolan, Helene Wallace, pathology
Van Der Linde, Reinhoud H, mathematics
Wohnus, John Frederick, zoology
Woodworth, Robert Hugo, biology

BRATTLEBORO
Ratte, Charles A, geology

BURLINGTON
Abrams, Jerome Sanford, surgery
Aiken, Robert Bascom, public health
Allen, Christopher Whitney, inorganic chemistry
Alpert, Norman Roland, physiology
Amidon, Ellsworth Lyman, medicine
Atherton, Henry Vernon, dairy manufacturing
Babbott, Frank Lusk, Jr, epidemiology
Balch, Donald James, animal breeding
Barnum, Horace Gardiner, historical geography, urban geography
Barrington, David Stanley, botany
Bartlett, Richmond J, soil chemistry, plant nutrition
Bell, Ross Taylor, systematic zoology
Blakeslee, George M, forest ecology
Bland, John (Hardesty), medicine
Bolton, Wesson Dudley, animal pathology
Boraker, David Kenneth, immunology, immunogenetics
Bouchard, Richard Emile, cardiology, internal medicine

Brammer, Jimmie Duane, developmental biology, neurobiology
Brock, Paul, applied mathematics
Brown, David Basset, inorganic chemistry
Carew, Lyndon Belmont, Jr, animal nutrition, biochemistry
Chamberlain, Erling William, mathematics
Chambers, Alfred Hayes, physiology
Clemmons, Jackson Joshua Walter, biochemistry, pathology
Coffin, Laurence Haines, thoracic surgery, cardiovascular surgery
Cohen, Herbert Daniel, physics
Cook, Philip W, botany
Cooke, Roger Lee, mathematics
Coon, Robert William, pathology
Craighead, John Edward, pathology, virology
Crooks, George Chapman, physiological chemistry
Crowell, Albert Dary, surface physics
Davison, John (Amerpohl), physiology
Detenbeck, Robert Warren, physics
Dodge, Carroll William, botany
Donaghy, Raymond Madiford Peardon, neurosurgery
Dowe, Thomas Whitfield, animal husbandry
Flanagan, Ted Benjamin, physical chemistry
Flanagan, Theodore Ross, agronomy
Folinas, Helen, cell physiology
Foote, Murray Wilbur, biochemistry
Forsyth, Ben Ralph, internal medicine, infectious diseases
Foss, Donald C, nutrition, physiology
Foster, Roger Sherman, Jr, surgery
Freedman, Steven Leslie, neuroanatomy, histology
Gade, Daniel Wayne, cultural geography, biogeography
Gans, Joseph Herbert, pharmacology
Geiger, William Ebling, Jr, analytical chemistry
Gibson, Thomas Chometon, internal medicine, cardiology
Glade, Richard William, embryology
Gladstone, Arthur A, surgery
Gray, Mary Jane, medicine
Gregg, Donald Crowther, organic chemistry
Gump, Dieter W, internal medicine, infectious diseases
Hanson, John Sherwood, medicine, physiology
Haviland, William Arthur, anthropology
Henderson, Donald Cedric, poultry science
Hendley, Edith Di Pasquale, physiology
Henson, Earl Bennette, limnology
Herrlich, Herman Conrad, pharmacology
Hoaglund, Franklin Theodore, orthopedic physics
Houston, Charles Snead, internal medicine
Huessy, Hans Rosenstock, psychiatry
Hyde, Beal Baker, botany
Izzo, Joseph Anthony Jr, mathematics
Jaffe, Julian Joseph, pharmacology
Janney, Clinton Dales, radiological physics
Johnstone, Donald Boyes, microbiology
Jones, Janice Lorraine, biophysics, cell physiology
Juenker, David W, physics
Kelleher, Philip Conboy, biochemistry, internal medicine
Klein, Richard M, plant physiology
Korson, Roy, pathology
Krapcho, Andrew Paul, organic chemistry
Kriebel, John Ernest, theoretical physics
Krizan, John Ernest, theoretical physics
Krupp, Patricia Powers, gross anatomy, experimental morphology
Kuehne, Martin Eric, organic chemistry
Kunin, Arthur Saul, medicine, physiological chemistry
Lachapelle, Rene Charles, medical chemistry
Ladd, William Alexander, electronics, microscopy
Lambert, Lloyd Milton, Jr, solid state physics
Lamden, Merton Philip, biochemistry
Landesman, Richard, developmental biology
Leinbach, Thomas Raymond, geography
Lepeschkin, Eugene, cardiology
Levy, Arthur Maurice, cardiology
Lipson, Richard L, internal medicine
Little, John Ernest, biochemistry
Low, Robert Burnham, physiology
Lucey, Jerold Francis, pediatrics
Luginbuhl, William Hossfeld, pathology

MacCollom, George Butterick, entomology
Mackay, Albert George, surgery
Macmillan, William Hooper, pharmacology
Maeck, John Van Sicklen, obstetrics & gynecology
Magdoff, Frederick Robin, soil chemistry, environmental science
Magnarella, Paul J, cultural anthropology, social anthropology
Martin, Herbert Lloyd, neurology
Marvin, James Wallace, botany
McCormack, John Joseph, Jr, organic chemistry, pharmacology
McCormack, Maxwell Leland, Jr, forestry
McCrorey, Henry Lawrence, physiology
McKay, Robert James, Jr, pediatrics
Meeker, C Irving, medicine
Meeks, Harold Austin, geography
Melville, Donald Burton, biochemistry
Meserve, Bruce Elwyn, mathematics
Meyer, Diane Hutchins, cell biology
Meyer, William Laros, biological chemistry
Miles, Edward Jervis, geography
Mitchell, William Edward, cultural anthropology
Moehring, Joan Marquart, cell biology
Moehring, Thomas John, medical microbiology
Moody, Paul Amos, zoology
Morse, Ellen Hastings, nutrition
Moser, Donald Eugene, mathematics
Mulieri, Berthann Scubon, physiology
Newcombe, David S, rheumatology
Newhall, Chester Albert, anatomy
Newman, Robert Alwin, biochemistry, pharmacology
Nilson, Kay Milligan, dairy science
Novotny, Charles, microbiology
Nyborg, Wesley LeMars, biophysics, acoustics
Parsons, Rodney Lawrence, physiology, biophysics
Pellett, Norman Eugene, ornamental horticulture
Peterson, Oscar Sylvander, Jr, medicine, radiology
Phillips, Charles Alan, virology, internal medicine
Potash, Milton, ecology, limnology
Racusen, David, plant biochemistry
Rae, Stephen, environmental physics
Ravaris, Charles Lewis, psychiatry
Reit, Ernest Marvin I, pharmacology
Rhodes, Dallas D, geomorphology
Ring, B Albert, medicine
Robinson, Donald Stetson, pharmacology, medicine
Rothstein, Howard, cell physiology
Sachs, Thomas Dudley, ultrasonics, biophysics
Sayer, Jane McKinley, bio-organic chemistry
Scarfone, Leonard Michael, physics
Schaeffer, Warren Ira, bacteriology
Schumacher, George Adam, medicine
Simmons, Kenneth Rogers, reproductive physiology
Sims, Ethan Allen Hitchcock, biochemistry, medicine
Sjogren, Robert Erik, microbiology, biochemistry
Smith, Albert Matthews, animal nutrition
Smith, Carol Price, protein chemistry
Sproston, Thomas, Jr, plant pathology
Stanley, Rolfe S, geology
Stevens, Dean Finley, zoology, cell biology
Stinebring, Warren Richard, medical microbiology
Sylwester, David Luther, mathematical statistics, biostatistics
Tabakin, Burton Samuel, medicine
Tampas, John Peter, radiology
Taylor, Fred Herbert, botany
Thanassi, John Walter, biochemistry
Van Buskirk, Frederick William, radiology
Vandermeer, Canute, geography
Vogelmann, Hubert Walter, botany
Waller, Julian Arnold, medicine, public health
Webb, George Dayton, neurophysiology
Welch, James Graham, animal nutrition
Weller, David Lloyd, biochemistry, molecular biology
Wells, Joseph, neuroanatomy, neurobiology
Welsh, George W, III, internal medicine, endocrinology
Whicher, Wendell Jennison, chemistry
White, William North, physical organic chemistry
Whitehorn, David, neurophysiology
Whitmore, Roy Alvin, Jr, forestry

681

VERMONT

Wiggans, Samuel Claude, plant physiology
Williams, Ronald Wendell, physics
Wolf, George Anthony, Jr., medicine
Wood, Glen Meredith, agronomy
Woodworth, Robert Cummings, biochemistry
Woolfson, Arnold Peter, cultural anthropology
Wulff, Claus Adolf, physical chemistry
Young, William Johnson, II, genetics, anatomy

CASTLETON
Freeman, Jeffrey Van Duyne, entomology; forestry
White, Christopher Clarke, mathematics

CHELSEA
Benedict, Donald Lee, physics, research administration

EAST DORSET
Rinse, Jacobus, physical chemistry

EAST THETFORD
Drake, Charles Lum, geophysics

ELY
Watt, James, epidemiology

ESSEX JUNCTION
Adler, Eric, solid state physics
Alberts, Gene S., analytical chemistry
Doll, Charles George, geology
Ferris-Prabhu, Albert Victor Michael, solid state physics, systems analysis
Gardner, Edward Eugene, solid state science
Stapper, Charles Henri, electrical engineering, solid state physics

JOHNSON
Abajian, Paul G., physical chemistry
Hitzeman, Jean Walter, cell biology

MARLBORO
Hayes, John William, biochemistry, solar physics
MacArthur, John Wood, physics

MIDDLEBURY
Andrews, David Henry, anthropology, ethnography
Baldwin, Brewster, geology
Ballou, Donald Henry, mathematics
Bennett, David Arthur, bioinorganic chemistry
Carruth, Philip Wilkinson, mathematics
Cushman, Robert Vittum, hydrogeology
Gleeson, Robert Willard, organic chemistry
Gould, Robert K., acoustics
Harnest, Grant Hopkins, organic chemistry
Hitchcock, Harold Bradford, zoology
Illick, John Rowland, geography of Asia, cartography
Ledlie, David B., organic chemistry
Malmstrom, Vincent Herschel, geography
Moyer, Walter Allen, Jr., organic chemistry
Olinick, Michael, mathematics
Peterson, Bruce Bigelow, mathematics
Pool, Edwin Lewis, analytical chemistry, environmental chemistry
Roberts, Edwin Kirk, physical chemistry
Rosamond, James Donald, organic chemistry
Saul, George Brandon, II, genetics
Schuh, Merlyn Duane, physical chemistry
Terwilliger, Don William, solid state physics
Watters, Christopher Deffner, cell biology
Winkler, Paul Frank, astrophysics

MILTON
Wilm, Harold Gridley, forestry

MONTPELIER
Garbacik, Theodore John, organic chemistry; water resources

NORTHFIELD
Burnham, George Hyndman, physics
Detwyler, Robert, zoology
Heed, Joseph James, Jr., mathematics
Lane, George H., physics
Larsen, Frederick Duane, geomorphology
McIntire, Summer Harmon, physics, academic administration
Piel, Elmar Viking, organic chemistry
Severance, Dean Charles, physics
Weinhous, Martin S., atomic physics

NORWICH
Brown, Robert Goodell, operations research
McKennan, Robert Addison, anthropology

PLAINFIELD
Jervis, Robert Alfred, botany, ecology

PUTNEY
Stewart, Anne Marie, animal physiology
Westing, Arthur Herbert, botany

SHELBURNE
Rowell, Lyman Smith, embryology

SOUTH BURLINGTON
Etherton, Bud, botany
Hill, David Byrne, mathematics, statistics
Keenan, Robert Gregory, environmental chemistry
Schuele, William John, inorganic chemistry

SPRINGFIELD
Jasinski, Jerry Peter, inorganic chemistry

ST MICHAEL'S COLLEGE
Sullivan, Thomas Donald, cytology

TOWNSHEND
Johnson, John Raven, organic chemistry

WATERBURY
Brooks, George Wilson, psychiatry

WHITE RIVER JUNCTION
Barton, Walter E., psychiatry
Crandell, Walter Bain, surgery, metabolism
Rosenstein, Robert, pharmacology

WILLISTON
Stultz, Walter Alva, anatomy

WINOOSKI
Bean, Daniel Joseph, limnology, physiological ecology
Foley, Edward Leo, organic chemistry
Harnett, John (Conrad), biochemistry
Kellner, Stephan Maria Eduard, physical chemistry
Klein, Deana Tarson, mycology
Provost, Ronald Harold, physical chemistry

WOODSTOCK
Plump, Ralph Eugene, chemistry

VIRGINIA

ALBERTA
Bensel, John Philip, solid state physics

ALEXANDRIA
August, Leon Stanley, nuclear science, radiological physics
Beacham, Lowrie Miller, Jr., chemistry
Benson, Loren Allen, physical chemistry
Betz, Daniel Oliver, Jr., nematology, plant pathology
Blandford, Robert Roy, physical oceanography, seismology
Borum, Olin H., research administration, applied chemistry
Bushey, Gordon Lake, physical chemistry
Chase, Helen Christina (Matulic), biostatistics
Crenshaw, Craig Moffett, physics
Eisenman, Richard L., systems analysis
El-Bisi, Hamed Mohamed, food & industrial microbiology
Fitz, Harold Carlton, Jr., space physics, atmospheric physics
Flint, Einar Philip, chemistry
French, David Milton, polymer chemistry
Garman, Willard Hershel, agriculture
Gindhart, Patricia S., anthropology
Hawkins, Richard Albert, physiology, biochemistry
Hendricks, Charles Wendell, microbiology
Hunt, Leon Gibson, applied mathematics, mathematical epidemiology
Kramer, Richard John, plant ecology, environmental management
Krynitsky, John Alexander, chemistry
LaMont, Robert Ellis, geography, military engineering
Leurtizr, John, Jr., chemistry
Martin, James D., organic chemistry
Massell, Paul Barry, number theory
McHale, Edward Thomas, physical chemistry
McIntyre, John Bowie, geology
Mitchell, William Hinckley, physiology
Morrison, Cohn L., physics
Moxham, Robert Morgan, geology
Parmenter, Guy Norris, geography
Peery, Thomas Martin, pathology
Perry, John Stephen, meteorology
Peters, Charles William, computer science
Petrov, Victor P., geography
Poll, Robert Allen, physics
Reichert, Paul F., seismology
Romney, Carl Fredrick, seismology
Roscher, David Moore, photochemistry
Sax, Robert Louis, geophysics
Schelleng, John H., solid state physics
Seddon, John Carl, physics
Shelkin, Barry David, geology
Simmons, Walt R., mathematical statistics
Spears, Joseph Faulconer, biology
Swinnerton, John W., chemistry
Tatro, Peter Richard, physical oceanography
Van Nostrand, Edwin Lewis, physics
Woisard, Edwin Lewis, operations research

AMHERST
Cauwenberg, Winfred Joseph, chemistry

ANNANDALE
Adams, William Hensley, ecology
Alexander, Allen Leander, chemistry
Baker, Paul, Jr., nuclear physics
Baussus-Von Luetzow, Hans Gerhard, mathematics
Burns, Robert Obed, physics
Fregeau, Jerome Heyde, nuclear physics
Galiano, Robert Joseph, physical chemistry
Greinke, Everett D., chemistry, mathematics
Johnson, John Enoch, physical organic chemistry
Nelson, Thomas Charles, forestry
Schneider, Philip Allen David, information science, applied statistics
Strock, Herman, food science, food technology

ARLINGTON
Adams, Caroline Lander, cytology
Alcaraz, Ernest Charles, laser, atmospheric physics
Alverson, Roy Carl, applied mathematics
Babione, Robert William, epidemiology
Badgley, Peter Coles, earth sciences, remote sensing
Bailey, James Stuart, geological oceanography
Barnett, William Arnold, economic statistics, mathematical statistics
Bartis, James Thomas, statistical mechanics, systems analysis
Barton, Alexander James, herpetology, ecology
Berghoefer, Fred G., operations research
Bergmann, Otto, theoretical physics
Berincourt, Ted Gibbs, physics
Biberman, Lucien Morton, physics
Blumenthal, Robert George, physics
Boley, Robert Eugene, algebra
Bonse, Frederick, physics
Bracken, Jerome, operations research
Bram, Joseph, mathematics
Brown, Gerald Leonard, applied mathematics
Bussard, Robert W., engineering physics
Butterworth, Theron Hervey, public health
Chipley, Robert MacNeill, ornithology
Cleek, Given Wood, chemistry, glass technology
Condell, William John, Jr., research administration, optical physics
Cooper, Larry Russell, nuclear physics
Cooper, Paul David, geography
Corcoran, Vincent John, lasers
Dardis, John G., atomic physics, nuclear physics
Davis, Edwin Griffith, organic chemistry
de Haan, Henry John, psychophysiology
de Latour, Christopher, environmental physics
Desiderio, Anthony Michael, systems analysis
Dimmock, John O., physics
Dolezalek, Hans, atmospheric sciences, electrometry
Duryee, William Rankin, experimental pathology, cancer
Easley, Ronald L., physics
Edelsack, Edgar Allen, physics
Emery, Arthur James, Jr., biochemistry
Espenshade, Gilbert Howry, geology
Ewing, Dean Edgar, medical research
Fain, Janice Bloom, physics
Fain, William Wharton, operations research
Featherston, Frank Hunter, physics
Fitzpatrick, Hugh Michael, physics
Fletcher, Harry Huntington, organic chemistry
Galloway, Joseph Homer, veterinary medicine
Garstens, Martin Aaron, solid state physics
Gehrke, Willis Timothy, geography
Giever, Paul Mathew, industrial hygiene
Gonet, Frank, physical chemistry
Grote, Jeffrey Harlow, applied mathematics, operations research
Guzman, Louis Enrique, cultural geography
Haines, Kenneth A., entomology
Hamilton, Thomas Charles, biophysics
Hardy, William Christopher, operations research, systems science
Hargrove, Logan Ezral, acoustics, optics
Harrington, Marshall Cathcart, optics
Hayes, Patrick Louis, applied mathematics, operations research
Helm, Harry Arthur, mathematics
Hendricks, Ernest LeRoy, hydrology
Hicken, John Allen, public health
Howe, John Arthur, mathematics
Hwang, John Dzen, applied mathematics, systems engineering
Jacknow, Joel, physical chemistry
Jackson, Francis J., physics
Jenkins, Robert Ellsworth, Jr., environmental management
Johnson, Charles Nelson, Jr., physics
Johnson, George Leonard, geological oceanography
Jolley, Homer Richard, public health administration, research administration
Jones, Merriam Arthur, chemistry, agricultural chemistry
Kay, Irvin (William), applied mathematics
Kelley, Ralph Edward, atomic physics, molecular physics
Klingsberg, Cyrus, solid state chemistry, ceramics
Langbein, Walter, hydrology
Langevin, Robert Arthur, mathematics
Latter, Richard, theoretical physics
Leisner, Fred Stanley, botany
Libber, Leonard Mitchell, physiology
Lundegard, Robert James, statistics
Malahoff, Alexander, geology, geophysics
Marder, Stanley, information science, systems analysis
Masi, Joseph Francis, physical chemistry
McDonald, Bruce Jerald, mathematical statistics
McGuire, Donald Charles, genetics
Meyers, Earl Lawrence, physical chemistry
Moroni, Eneo C., industrial organic chemistry
Muir, Donald Earl, mathematics
Mullaney, Henry Wendell, electromagnetics, ionospheric physics
Musser, Marc James, (Jr), internal medicine
Nall, Julian Clark, nuclear physics

Narten, Perry Foote, environmental management, horticulture
Neece, George A, physical chemistry
Nelson, David Lynn, physical chemistry
Nicolai, Van Olin, optical physics, lasers
Oberle, Richard Alan, mathematical analysis
Orth, Donald Joseph, geography, toponymy
Ostenso, Ned Allen, geology, marine geophysics
Ostrom, Carl Eric, forestry
Padgett, Doran William, nuclear physics
Parry, Hubert Dean, meteorology
Paskausky, David Frank, physical oceanography, physics
Pewitt, Nelson Douglas, operations research, operations analysis
Phillips, E Alan, optics, nuclear physics
Pilloff, Herschel Sydney, molecular physics, lasers
Pollard, Joseph Page, medicine, biology
Pruitt, Evelyn Lord, geography
Quinn, Thomas Patrick, ionospheric physics
Randall, Royal William, Jr, mathematics
Raney, William Perin, physics, acoustics
Rigby, Malcolm, meteorology
Rosenbaum, David Mark, applied statistics, nucleonics
Rosengren, Jack Whitehead, physics
Rozzell, Thomas Clifton, environmental biology, radiation biology
Ruby, Stanley, physical chemistry, materials science
Ryan, Robert Dean, mathematics
Salkovitz, Edward Isaac, physics, science policy
Scattergood, Leslie Wayne, fish biology
Seifried, Harold Edwin, biochemical pharmacology, occupational health
Selwyn, Philip Alan, chemical physics, fluid mechanics
Shepard, Harold Henry, entomology, pesticide chemistry
Shuster, Carl Nathaniel, Jr, aquatic ecology, invertebrate zoology
Smith, William Lee, exploration geology
Smith, William Vick, lasers
Sooy, Walter Richard, physics
Spies, Joseph Reuben, chemistry, immunology
Stanley, Ronald Alwin, plant physiology
Swindells, Frank Evans, physical chemistry
Sykes, Alan O'Neil, acoustics
Tamarkin, Paul, physics, research management
Taylor, Raymond Leech, zoology, botany
Thomas, Richard Dean, toxicology, medicinal chemistry
Todd, Frank Arnold, veterinary medicine
Tolbert, Gene Edward, economic geology
Trumbull, Richard, psychophysiology
Underhill, Adna Heaton, ichthyology
Van Dersal, William Richard, plant ecology
Vastine, Frederick Davidson, organic chemistry, organometallic chemistry
Warner, Jacob Larue, research administration
Wasylkiwskyj, Wasyl, electromagnetics
Wayland, Russell Gibson, mining geology, engineering
Weesner, William Eldred, organic chemistry
Weinberg, Elliot Hillel, solid state physics
Weiss, Michael David, mathematics
Whitaker, William Armstrong, computer science, natural science
Wolf, Eric W, computer science, systems science
Wolfhard, Hans G, physics, physical chemistry
Woolfolk, Robert William, physical chemistry
Wrenn, Samuel Nathaniel, organic chemistry
Wynne, Kenneth Joseph, inorganic chemistry, polymer chemistry
Zweig, Gunter, pesticide chemistry

ASHLAND
Boldridge, William Franklin, physical chemistry
English, Bruce Vaughan, environmental physics, environmental sciences
McClurkin, John Irving, Jr, biology
Miller, William Schuyler, chemistry
Monroe, Stuart Benton, organic chemistry, polymer chemistry
Temple, Wade Jett, molecular physics
Thompson, Sanford P, theoretical physics

BAILEY'S CROSSROADS
Creager, Joan Guynn, embryology

BLACKSBURG
Ache, Hans Joachim, radiochemistry, radiation chemistry
Ackerman, Clemens John, biochemistry, nutrition
Adkisson, Curtis Samuel, ornithology
Almeida, Silverio Pedro, electrooptics, biophysics
Anderson, Bruce Murray, biochemistry
Appleman, Maria Duarte, bacteriology
Arndt, Richard Allen, physics
Arnold, Jesse Charles, applied statistics, mathematical statistics
Arnold, Jimmy Thomas, mathematics
Aull, Charles Edward, mathematics
Baird, Leemon Claude, mathematics
Ball, Joseph Anthony, pure mathematics
Barden, John Allan, horticulture
Barnett, Lewis Brinkly, physical biochemistry
Bartschmid, Betty Rains, analytical chemistry
Bell, Harold Morton, organic chemistry
Bell, Wilson Bryan, veterinary science
Bingham, Samuel Wayne, plant physiology, weed science
Blaser, Roy Emil, agronomy
Blecher, Marvin, nuclear physics, elementary particle physics
Bloss, Fred Donald, mineralogy
Bollinger, Gilbert A, geophysics, seismology
Booth, Sheldon James, microbiology
Bowden, Robert Lee, Jr, mathematical physics
Bowen, Samuel Philip, theoretical physics, low temperature physics
Boyd, Earl Neal, dairy science
Bragg, Denver Dayton, poultry science
Brice, Luther Kennedy, physical chemistry
Brown, Ross Duncan, Jr, biochemistry
Buikema, Arthur L, Jr, aquatic biology, environmental physiology
Bunce, George Edwin, nutrition, biochemistry
Burkhart, Harold Eugene, forest biometry
Burns, John Allen, applied mathematics
Buss, Glenn Richard, plant breeding, plant genetics
Cairns, John, Jr, limnology
Carr, Scott Bligh, animal nutrition
Carson, Eugene Watson, Jr, agronomy, plant biochemistry
Carter, Robert Clifton, animal breeding, genetics
Chang, In-Kook, plant physiology
Cherry, Donald Stephen, aquatic ecology
Claus, George William, microbiology
Clifford, Alan Frank, inorganic chemistry, fluorine chemistry
Coartney, James S, plant physiology, weed science
Cobb, Whitfield, mathematical statistics
Cochran, Donald Gordon, insect physiology
Cochran, James Alan, applied mathematics
Collins, George Briggs, physics
Collins, William F, food science
Colmano, Germille, veterinary physiology, medical biophysics
Cooler, Frederick William, food technology
Costain, John Kendall, geophysics
Couch, Houston Brown, plant pathology
Cragle, Raymond George, animal physiology, nutrition
Craig, James Roland, geochemistry
Crittenden, Rebecca Slover, algebra
Croxdale, Judith Gerow, botany
Cummins, Cecil Stratford, microbiology
Debney, George Charles, Jr, applied mathematics, theoretical physics
Dessy, Raymond Edwin, chemical instrumentation
Dickman, Raymond F, Jr, topology
Dickson, Kenneth Lynn, aquatic biology, environmental sciences
Dillard, John Gammons, inorganic chemistry, physical chemistry
Domermuth, Charles Henry, Jr, veterinary microbiology
Downing, Robert Lee, wildlife research
Drake, Charles Roy, plant pathology
Driskell, Judy Anne, nutrition
Dubose, Robert Trafton, veterinary medicine, virology
Duerr, William Allen, forest management
Eaton, John LeRoy, insect physiology
Edlund, Milton Carl, physics
Elgert, Klaus Dieter, immunology
Elias, Robert William, physiological ecology
Engel, Ruben William, biochemistry
Essary, Eskel Oren, poultry science
Etgen, William M, dairy science, animal science
Falkinham, Joseph Oliver, III, microbial genetics
Ferguson, Donald Allen, Jr, microbial biochemistry
Ficenec, John Robert, experimental high energy physics
Field, Paul Eugene, physical chemistry, inorganic chemistry
Fletcher, Peter, topology
Flick, George Joseph, food science
Fontenot, Joseph Paul, animal nutrition
Foy, Chester Larrimore, plant physiology
Furr, Aaron Keith, nuclear physics, health physics
Gaines, James Abner, genetics
Genter, Clarence Frederick, agronomy
Gibbs, Gerald V, mineralogy
Gilbert, Murray Charles, petrology
Giles, Robert H, Jr, ecology, wildlife management
Gilmer, Thomas Edward, Jr, solid state physics, semiconductors
Glover, Lynn, III, regional geology
Good, Charles Munder, cultural geography, geography of Africa
Good, Irving John, mathematics, statistics
Gorsline, George William, computer science
Graf, Gottfried Christian, physiology
Graybeal, Jack Daniel, physical chemistry
Grayson, James McDonald, entomology
Green, George G, animal science, animal nutrition
Greenberg, William, mathematical physics
Gregory, Eugene Michael, enzymology
Grender, Gordon Conrad, geology
Griffin, Gary J, biology
Gross, Walter Burnham, avian pathology
Gruenhagen, Richard Hamilton, chemical pesticides
Gum, Ernest Kemp, Jr, microbial biochemistry
Gwazdauskas, Francis Charles, dairy science, reproductive endocrinology
Hackman, Abigail Salyers, bacteriology
Hahn, Thomas Marshall, Jr, physics
Hale, Maynard George, plant physiology
Hall, Philip Layton, organic chemistry
Hannsgen, Kenneth Bruce, mathematical analysis
Harper, Laura Jane, nutrition, human physiology
Harrison, Robert Louis, agronomy
Harshbarger, Boyd, mathematical statistics
Heald, Charles William, dairy science, cytology
Heath, Alan Gard, zoology, comparative physiology
Heikkenen, Herman John, forest entomology
Henderson, Robert Gordon, plant pathology
Herdman, Terry Lee, applied mathematics
Hess, John Lloyd, plant biochemistry, protein chemistry
Hill-Samli, Marqueta, biochemistry
Hinkelmann, Klaus Heinrich, statistics
Holcomb, Carl James, forestry
Holdeman, Lillian Virginia, bacteriology
Holliman, Rhodes Burns, parasitology
Holmes, Clayton Ernest, poultry science
Holt, Perry Cecil, invertebrate zoology, systematics
Holub, James Robert, mathematics
Hosner, John Frank, forestry
Houska, Charles Robert, physics
Howes, Cecil Edgar, genetics, physiology
Hudlicky, Milos, organic chemistry, fluorine chemistry
Hutcheson, Thomas Barksdale, Jr, agronomy
Ifju, Geza, forestry
Ijaz, Mujaddid A, nuclear physics, elementary particle physics
Jacobs, James Albert, physics
Jensen, Donald Ray, mathematical statistics
Jenssen, Thomas Alan, animal behavior, ecology
Johnson, John LeRoy, microbiology
Johnson, Lee W, mathematics
Jones, Gerald Murray, dairy science
Judkins, Wesley Parkhurst, horticulture
Kalison, Seymour Lincoln, veterinary medicine
Kelly, Robert Frank, biochemistry, animal husbandry
Kingston, David George Ian, natural products chemistry
Kirkpatrick, Roy Lee, reproductive physiology
Kok, Loke-Tuck, entomology
Kornegay, Ervin Thaddeus, animal science
Kosztarab, Michael, entomology
Kowalski, John Bernard, microbial genetics
Krieg, Noel Roger, microbiology
Kroontje, Wybe, agronomy
Krutchkoff, Richard Gerald, applied statistics
Lackey, Robert T, fisheries management
Lambe, Robert Carl, plant pathology
Lau, Norman Eugene, environmental management
Lechowich, Richard V, food science, food microbiology
Leighton, Alvah Theodore, Jr, genetics, physiology
Leinhardt, Theodore Edward, solid state physics
Li, Ming Chiang, physics, mathematics
Lightfoot, Donald Richard, biochemical genetics
Lightfoot, Haideh Nezam, molecular biology
Lindstrom, Richard S, horticulture
Litton, George Washington, animal husbandry
Loh, Hung Yu, physics
Long, Dale Donald, experimental physics
Long, Jerome R, physics
Lopez, Anthony, food science
Lowry, Wallace Dean, geology
Marable, Nina Louise, food chemistry
Marlowe, Thomas Johnson, animal genetics, animal physiology
Martens, David Charles, soil science
Mason, John Grove, analytical chemistry
Massey, Peyton Howard, Jr, olericulture, academic administration
McCombs, Clarence Leslie, plant physiology
McCoy, Robert A, topology
McElwee, Robert L, forest management, genetics
McFadden, Leonard, mathematics
McGilliard, Michael Lon, dairy science, animal genetics
McGinnes, Burd Sheldon, wildlife research
McGrath, James Edward, organic polymer chemistry, polymer science
McLaughlin, Gerald Wayne, research management, applied statistics
McNabb, F M Anne, comparative physiology
McNabb, Roger Allen, comparative physiology
McNair, Harold Monroe, analytical chemistry
Miller, John David, plant breeding, plant genetics
Miller, Lawrence Ingram, phytopathology
Miller, Orson K, Jr, mycology
Moore, Laurence Dale, plant pathology
Moore, Walter Edward Cladek, medical microbiology, bacteriology
Mosby, Henry Sackett, wildlife management, forestry
Mullins, Donald Eugene, insect physiology
Murray, John Wolcott, chemistry
Myers, Raymond Harold, mathematical statistics
Nance, Richard E, computer science, operations research
Neer, Keith Lowell, meat science
Niehaus, Walter G, Jr, biochemistry
Ogliaruso, Michael Anthony, physical organic chemistry
Onega, Ronald Joseph, nuclear physics
Orcutt, David Michael, plant physiology
Orcutt, Fred Scott, Sr, biochemistry, bacteriology
Osborne, John Clark, experimental pathology
Pace, Wesley Emory, mathematics
Palmer, James Kenneth, food biochemistry
Parker, Bruce C, phycology
Parkinson, Thomas Franklin, physics
Parry, Charles J, algebra, number theory
Paterson, Robert Andrew, botany
Patty, Clarence Wayne, topology
Pendleton, John Davis, soil chemistry
Phillips, Jean Allen, food science, nutrition
Pienkowski, Robert Louis, entomology
Pierson, Merle Dean, food science
Pirie, Walter Ronald, mathematical statistics, applied statistics
Polan, Carl E, dairy nutrition, biochemistry

Porter, Duncan Macnair, systematic botany, phytogeography
Potter, Lawrence Merle, poultry nutrition
Powell, Andrew Jackson, agronomy
Pristoi, Robert, plant pathology
Read, John Frederick, physical chemistry
Reaves, Paul Marvin, dairy science
Reneau, Raymond B, Jr, soil chemistry
Rhodes, Richard Wesley, plant ecology
Ribbe, Paul Hubert, mineralogy
Riess, Ronald Dean, mathematics
Ritchey, Sanford Jewell, nutrition
Roane, Curtis Woodard, plant pathology
Roane, Martha Kotila, mycology, taxonomic botany
Roberts, James Ernest, Sr, economic entomology
Robertson, Randal McGavock, physics
Robinson, Edwin S, geophysics, geology
Robinson, William H, entomology
Roper, Leon David, theoretical high energy physics, biophysics
Roselle, David Paul, mathematics
Ross, Frederick Keith, structural chemistry, crystallography
Ross, Mary Harvey, entomology
Ross, Robert Donald, ichthyology
Rutherford, Charles, cell biology, biochemistry
Rutland, Leon W, mathematics
Saacke, Richard George, reproductive physiology, cytology
Sanzone, George, chemical physics, chemical kinetics
Scanlon, Patrick Francis, reproductive physiology, wildlife research
Schmidt, Richard Edward, agronomy
Schmidt, Robert Reinhart, biochemistry
Schug, John Charles, physical chemistry
Sears, Charles Edward, economic physiology
Sperry, Jay Franklin, microbial physiology
Starling, Thomas Madison, plant breeding genetics
Steeves, Harrison Ross, III, histochemistry, taxonomy
Stetler, David Albert, plant cytology
Stone, Warren Kenneth, food science
Stout, Ernest Ray, molecular biology, biochemistry
Swader, Jeff Austin, Jr, plant physiology
Taylor, Lincoln Homer, agronomy
Tepitz, Vigdor Louis, high energy physics
Terrill, Thomas Robert, plant breeding, genetics
Tipsword, Ray Fenton, physics
Tolin, Sue Ann, plant virology, phytopathology
Trower, William Peter, elementary particle physics
Turner, Ernest Craig, Jr, entomology
Van Krey, Harry Peter, physiology, agriculture
Vercellotti, John R, organic chemistry, biochemistry
Viers, Jimmy Wayne, physical chemistry
Vinson, William Ellis, population genetics, dairy science
Watson, Douglas F, veterinary science
Webb, Kenneth Emerson, Jr, animal nutrition
Webb, Ryland Edwin, nutritional biochemistry, toxicology
Webster, Jackson Ross, ecology
Weidhaas, James August, Jr, entomology
Wesley, Roy Lewis, food science
West, David Armstrong, genetics
White, John Marvin, genetics, animal breeding
Wightman, James Pinckney, physical chemistry
Wilkins, Tracy Dale, microbiology
Williams, Clayton Drews, theoretical physics, solid state physics
Williams, George Robertson, horticulture
Wills, Wirt Henry, plant pathology
Wilson, Coyt Taylor, plant pathology
Wise, Milton Bee, animal nutrition
Wisman, Everett Lee, poultry science
Wolf, Dale Duane, agronomy
Wolfe, James F, organic chemistry
Yongue, William Henry, protozoology
Yousten, Allan A, microbial physiology
Zelazny, Lucian Walter, soil mineralogy

Zweifel, Paul Frederick, theoretical physics, nuclear science

BLACKSTONE

Semtner, Paul Joseph, economic entomology, insect ecology

BON AIR

Kane, William Paul, organic chemistry
Majewski, Theodore E, organic chemistry
Manzelli, Manlio Arthur, entomology

BRIDGEWATER

Burns, Robert Kyle, zoology, embryology
Cool, Raymond Dean, chemistry
Heisey, Lowell Vernon, organic chemistry
Hill, Lynn Michael, plant sciences
Jopson, Harry Gorgas Michener, zoology
Keihn, Frederick George, solid state chemistry
Martin, John Walter, Jr, pharmaceutical chemistry
Mengebier, William Louis, physiology
Neher, Dean Royce, physics, computer science
Ulrich, Dale V, physics

BRISTOL

Barr, Fred S, medical microbiology

BROADWAY

Singh, Suresh Pratap, poultry science

CENTREVILLE

Thaler, William John, physics

CHARLOTTESVILLE

Abse, David Wilfred, psychiatry
Ackers, Gary Keith, physical biochemistry
Agarwal, Suresh Kumar, radiological physics
Aldrich, Clarence Knight, psychiatry
Allen, Miller Shannon, Jr, pathology, surgery
Andrews, William Lester Self, physical chemistry, spectroscopy
Armstrong, John William, astronomy, space physics
Attinger, Ernst Otto, bioengineering
Ayers, Carlos R, internal medicine
Balmforth, Dennis, chemistry
Barker, Robert Edward, Jr, polymer physics
Bass, Norman Herbert, neurochemistry, neurology
Batson, Alan Percy, computer science
Bauerle, Ronald H, molecular biology, microbiology
Beams, Jesse Wakefield, physics
Beckwith, Julian Ruffin, internal medicine, cardiovascular disease
Benzinger, Rolf Hans, biochemical genetics
Berent, Stanley, clinical psychology, neuroscience
Berkeley, Edmund, botany
Berne, Robert Matthew, medical physiology
Birdsong, McLemore, pediatrics
Blizzard, Robert M, pediatrics
Bobb, Marvin Lester, entomology
Bobbitt, Oliver Bierne, clinical pathology
Bodenstein, Dietrich H F A, zoology
Boll, Thomas Jeffrey, neuropsychology
Boring, John Wayne, physics
Bradbeer, Clive, biochemistry
Brill, Arthur Sylvan, biophysics
Brooker, Gary, pharmacology, biochemistry
Brown, Robert Lamme, astrophysics, atomic physics
Bryan, Robert Finley, crystallography
Burger, Alfred, medicinal chemistry
Burr, Helen Gunderson, speech pathology, audiology
Burton, William Butler, astronomy
Calver, James Lewis, mineralogy
Cardell, Robert Ridley, Jr, cell biology
Carey, Francis Arthur, organic chemistry
Carpenter, Martha Stahr, astronomy
Cawley, Edward Philip, dermatology
Celli, Vittorio, solid state physics
Cohen, David Harris, low temperature physics
Cole, James Webb, Jr, neurophysiology
Coleman, Robert Vincent, physics
Conley, James Franklin, geology
Coopersmith, Michael Henry, theoretical physics

Craig, James William, medicine
Crispell, Kenneth Raymond, medicine
Cromarie, Thomas Houston, biochemistry
Crowell, Thomas Irving, physical organic chemistry
Crowley, Patrick Arthur, nuclear physics, fluid physics
Curme, George Oliver, Jr, chemistry
Daitzman, Reid Joseph, psychiatry
Danielson, Donald Alfred, applied mathematics, applied mechanics
Davis, John Staige, IV, internal medicine
Deaver, Bascom Sine, Jr, experimental solid state physics
Deck, James David, histology
Demas, James Nicholas, photochemistry
Denn, James (Norman), zoology
DeYoung, David Spencer, theoretical astrophysics
Diehl, Fred A, developmental biology
Dreifuss, Fritz Emanuel, medicine
Drucker, William Richard, medicine
Duckworth, Donna Hardy, biochemistry, genetics
Duling, Brian R, cardiovascular physiology
Duncan, Charles Donald, physical organic chemistry, theoretical chemistry
Dunk, Charles Francis, mathematics
Dunn, William Thornton, endocrinology
Duren, William Larkin, Jr, mathematics
Easley, Eliza (Lila) Waller, biochemistry
Ebbesson, Sven O E, neuroanatomy
Economou, Eleftherios Nickolas, solid state physics, surface physics
Edgerton, Milton Thomas, Jr, plastic surgery
Ellison, Robert L, geology
Epstein, Robert Marvin, anesthesiology
Erickson, Edwin E, anthropology, applied statistics
Ern, Ernest Henry, petrology, geology
Findlay, John Wilson, physics
Floyd, Edwin Earl, topology
Foster, Eugene A, pathology
Fowler, Michael, theoretical physics
Fredrick, Laurence William, astronomy
Friedberg, Charles Bruce, experimental solid state physics, low temperature physics
Galloway, James Neville, environmental chemistry
Garnett, Richard Wingfield, Jr, psychiatry
Garrett, Reginald Hooker, biochemistry
Garstang, Michael, meteorology, oceanography
Gear, Adrian R L, biochemistry
Gillenwater, Jay Young, urology
Goldstein, Gerald, microbiology, medicine
Goldstein, Samuel Joseph (Jr), radio astronomy
Goodell, Horace Grant, geology
Greisen, Eric Winslow, radio astronomy
Grimes, Russell Newell, inorganic chemistry, organometallic chemistry
Grisham, Charles Milton, biological chemistry
Guerrant, John Lippincott, internal medicine
Guglielmo, Piet C, physics
Gwaltney, Jack Merrit, Jr, medicine
Hackett, John Taylor, neurophysiology
Hamilton, Howard Laverne, developmental biology
Hamner, Charles Edward, Jr, biochemistry, veterinary medicine
Hampton, Don Allen, magnetic resonance
Hanawalt, Ronald B, plant ecology, pedology
Hanna, George R, neurology
Hanscom, Roger H, geology
Hargis, William Jennings, Jr, biological oceanography, parasitology
Harrison, Willard Wayne, analytical chemistry
Hawkins, David Rollo, psychiatry
Haynes, Robert Clark, Jr, pharmacology
Heeschen, David Sutphin, radio astronomy
Hendley, Joseph Owen, pediatrics
Hereford, Frank Loucks, physics
Hess, George Burns, low temperature physics
Hobbs, Charles Roderick Bruce, Jr, geology
Hogg, David Edward, astronomy
Hook, Edward W, Jr, internal medicine
Hornberger, George Milton, hydrology

Horsley, John Shelton, III, surgery
Howard, Barbara Yoder, nuclear medicine, physical chemistry
Howland, James Secord, mathematics
Huang, Ching-hsien, biochemistry
Huang, Laura Chi, biochemistry
Hunt, Donald F, organic chemistry
Hunt, William B J, Jr, medicine, immunology
Hunter, Thomas Harrison, internal medicine
Huskey, Robert John, genetics
Jacobs, Kenneth Charles, cosmology, theoretical astrophysics
Jane, John Anthony, neurosurgery
Jesser, William Augustus, metal physics, materials science
Johns, Thomas Richards, II, neurology, neurophysiology
Johnson, Lewis Benjamin, Jr, physical chemistry
Johnson, Richard Noring, biomedical engineering
Johnson, Robert Alan, solid state physics, materials science
Jones, Thomas Walter, theoretical astrophysics
Kabir, Prabahan Kemal, theoretical physics
Kadner, Robert Joseph, biochemical physics
Kaut, Charles, anthropology
Keats, Theodore Eliot, radiology
Kelly, Hugh P, physics
Kelly, Mahlon George, Jr, aquatic ecology, biological oceanography
Kent, James Joseph, nuclear physics
Khalifah, Raja Gabriel, chemistry
Kitay, Julian Israel, internal medicine, physiology
Kochhar, Devendra M, anatomy, embryology
Konigsberg, Irwin R, developmental biology
Kreisinger, Robert, molecular biology
Krovetz, L J Jerome, pediatric cardiology, cardiovascular physiology
Kumar, Shiv Sharan, astronomy
Kupchan, S Morris, organic chemistry
Kupke, Donald Walter, biochemistry
Kutchai, Howard C, physiology, biochemistry
Langdon, Robert Godwin, biochemistry
Langman, Jan, anatomy
Lapsley, Alwyn Cowles, physics
Larner, Joseph, biochemistry, pharmacology
Latta, Gordon, mathematics
Lawless, Kenneth Robert, materials science
Leavell, Byrd Stuart, medicine
Lee, Jen-Shih, biomedical engineering
Levine, Jules Ivan, health sciences, medical administration
Limber, David Nelson, astronomy, astrophysics
Litman, Burton Joseph, biochemistry
Longenecker, David Eugene, anesthesiology
Looney, William Boyd, radiobiology, biophysics
MacGregor, Carolyn Harvey, protein chemistry
MacLeod, Robert Meredith, chemistry
Mann, James Edward, Jr, applied mathematics, applied mechanics
Maroney, Samuel Patterson, Jr, zoology
Martin, Nathaniel Frizzel Grafton, mathematics
Martin, Robert Bruce, biophysical chemistry
McCoy, Sue, biochemistry
McGilvery, Robert Warren, biochemistry
McKinsey, Richard Davis, botany
McShane, Richard James, mathematics
Mellon, DeForest, Jr, neurophysiology
Mika, Leonard Aloysius, microbiology
Miller, James Q, neurology, cytogenetics
Miller, Lewis Dudley, theoretical nuclear physics
Miller, Oscar L, Jr, cell biology
Minor, George Ridgway, surgery
Mitchell, John Wesley, physics
Mitchell, Richard Scott, mineralogy, crystallography
Muller, William Henry, Jr, surgery
Murad, Ferid, clinical pharmacology
Murphy, Richard Alan, physiology
Murray, Joseph James, Jr, zoology
Nelson, Bruce Warren, sedimentology, clay mineralogy
Noble, Julian Victor, theoretical physics, mathematical biophysics

Nolan, Stanton Peelle, cardiovascular surgery, cardiovascular physiology
O'Brien, William M, epidemiology, genetics
Odum, William Eugene, ecology
Ogilvie, James William, Jr, biochemistry, organic chemistry
Oliphant, Edward Eugene, reproductive biology, biological chemistry
Ord, John Allyn, physics, mathematics
Osvalds, Valfrids, astronomy
Owen, Frazer Nelson, astronomy
Owen, John Atkinson, Jr, internal medicine, metabolism
Owens, Robert Hunter, mathematics
Paterson, James Lenander, low temperature physics
Paulsen, Elsa Proehl, pediatrics, endocrinology
Peach, Michael Joe, pharmacology, physiology
Perdue, Charles L, Jr, ethnography
Pielke, Roger Alvin, meteorology
Pitt, Loren Dallas, mathematics
Pitts, Grover Cleveland, physiology
Pullen, Edwin Wesley, anatomy
Rall, Theodore William, pharmacology
Ramirez, Donald Edward, mathematics
Rappaport, Jacques, botany
Rebhun, Lionel Israel, zoology, cell biology
Richardson, Frederick S, physical chemistry
Rijke, Arie Marie, biomaterials, polymer science
Riopel, James L, plant morphology
Ritchie, Wallace Parks, Jr, surgery
Ritter, Rogers C, nuclear physics, medical physics
Roberts, Morton Spitz, radio astronomy
Roberts, William Woodruff, Jr, applied mathematics, astronomy
Rodewald, Richard David, cell biology
Rodig, Oscar Rudolf, organic chemistry
Rood, Robert Thomas, astrophysics
Rosenblum, Marvin, mathematics
Rovnyak, James L, mathematics
Rubio, Rafael, cardiovascular physiology
Rudolf, Leslie E, surgery
Runk, Benjamin Franklin Dewees, botany
Rushia, Edwin Louis, anesthesiology
Russell, Catherine Marie, medical microbiology
Ruvalds, John, physics
Ruyle, Eugene Edward, economic anthropology
Sandusky, William Roberts, surgery
Sapir, J David, anthropology
Schatz, Paul Namon, physical chemistry
Schlecht, Richard Guenther, physics
Schnaitman, Carl A, microbiology, mathematics, astronomy
Schuit, Kenneth Edward, anatomy, cell biology
Schwartz, Edith Richmond, biochemistry, orthopedics
Scott, Leonard Lewy, Jr, algebra
Sheppe, William Marco, Jr, psychiatry
Sherman, Samuel Murray, neurosciences, ophthalmology
Simmonds, James Gordon, applied mathematics
Simpson, Joanne, meteorology
Simpson, Robert H, meteorology
Singer, Siegfried Fred, geophysics
Smith, Larry, mathematics
Sobottka, Stanley Earl, physics
Spencer, Hugh Miller, physical chemistry
Sperelakis, Nick, physiology, biophysics
Spirito, Carl Peter, neurobiology, ethology
Spoerl, Edward Schnurr, cell physiology, plant physiology
Steinberg, William, microbiology
Stevenson, Ian, medicine, psychiatry
Steward, Frederick Campion, botany, cell biology
Stewart, John Westcott, physics
Stewart, Lever F, neurology
Stoner, Glenn Earl, electrochemistry
Sturgill, Benjamin Caleb, medicine, pathology
Sundberg, Richard J, organic chemistry
Swanson, Ronald Frederick, biochemistry, developmental biology
Taub, John Marcus, psychophysiology
Teja, Jagdish Singh, psychiatry
Thompson, Thomas Edward, biochemistry
Thornton, Stephen Thomas, experimental nuclear physics
Thornton, William Norman, Jr, obstetrics & gynecology
Thorup, Oscar Andreas, Jr, medicine
Tolbert, Charles Ray, astronomy
Trefil, James S, particle physics
Trinde, Carl Otis, theoretical chemistry
Turner, Barry Earl, radio astronomy

Van Brunt, Richard Joseph, atomic physics, molecular physics
Villar-Palasi, Carlos, pharmacology, biochemistry
Volk, Wesley Aaron, microbiology
Volkan, Vamik, psychiatry
Voss, Gordon D, food technology
Wade, Campbell Marion, astronomy
Wagner, Robert Roderick, virology, microbiology
Wagner, Roy, anthropology
Walker, Thomas Carl, biomaterials, instrumentation
Wallace, Wayne Alexander, resource geography, geography of the Soviet Union
Wangensteen Stephen Lightner, surgery
Ward, Harold Nathaniel, mathematics
Wartman, William Bechmann, pathology
Wawner, Franklin Edward, Jr, materials science
Weary, Peyton Edwin, dermatology
Weber, Hans Jurgen, theoretical nuclear physics
Weinreb, Sander, radio astronomy, microwave engineering
Westfall, Thomas Creed, pharmacology
Wheby, Munsey S, internal medicine
Whitehead, Walter Dexter, Jr, nuclear physics, atomic physics
Whyburn, Lucille Enid, mathematics
Wilkins, Michael Gray, biomedical engineering
Wilsdorf, Doris Kuhlmann, materials science
Wilson, Edward C, internal medicine, gastroenterology
Wilson, Lester A, Jr, medicine, rheumatology
Winfield, John Buckner, immunology, rheumatology
Winkler, Herbert H, microbiology, biochemistry
Winter, Edward H, anthropology, ethnology
Workman, Marcus Orrin, inorganic chemistry
Wright, Theodore Robert Fairbank, developmental genetics
Young, Joseph Spencer, geology
Zieman, Joseph Crowe, Jr, marine ecology, biological oceanography
Ziock, Klaus Otto H, experimental physics

FALLS CHURCH

Aldrich, John Warren, ornithology
Allenby, Richard John, Jr, geophysics
Badin, Elmer John, physical organic chemistry
Barnett, Charles William Henry, petrography
Beliles, Robert Pryor, toxicology
Calkins, James A, geology
Campbell, Alfred Duncan, biochemistry
Davis, John Litchfield, solid state physics
Emlet, Harry Elsworth Jr, systems analysis, health sciences
Englund, John Arthur, operations research
Greenwood, Joseph Albert, mathematics
Hanson, Roy Eugene, geophysics
Hardy, Frank Merril, virology
Harold, LaVerne Collins, veterinary medicine
Hawley, Dorothea Burton, geography
Honkala, Rudolf A, geography
Huddleston, Harold Frank, mathematical statistics
Kiessling, Oscar Edward, mineralogy
Kutschenreuter, Paul Herbert, meteorology, biometeorology
Lawwill, Stanley Joseph, mathematics
Lockhart, Luther Bynum, Jr, organic polymer chemistry
Lumb, Ralph F, physical chemistry, nuclear sciences
Muir, J Lawrence, geology
O'Neill, Thomas Hall Robinson, meteorology
Ordway, Fred (Delancy), physical chemistry
Patel, Appasaheb Raojibhai, pharmaceutical chemistry
Powers, Marcelina Venus, toxicology
Rassam, Ghassan, geology, information science
Shere, Kenneth David, operations research, applied mathematics
Sick, Theodore John, biochemistry, organic chemistry
Wyckoff, Peter Hines, geophysics

Goodson, Louie Aubrey, Jr, textile chemistry
Moran, Thomas James, pathology, medicine
Urbanik, Arthur Ronald, organic chemistry, textile chemistry
Wayland, Rosser Lee, Jr, textile chemistry

DRAKES BRANCH

Davis, Charles Stewart, clinical pharmacology

DUBLIN

Moore, Raymond Kenworthy, geochemistry, mineralogy

EMORY

Bingham, Edgar, geography
Graybeal, Walter Thomas, mathematics
Hancock, Vernon Ray, mathematics
McCoy, Joseph Hamilton, physical science
Nelson, Cecil Morris, physics, chemistry
Pledger, Huey, Jr, organic chemistry, polymer chemistry
Spell, Charles Raymond, analytical chemistry
Treadwell, George Edward, Jr, botany, biochemistry

FAIRFAX

Andrykovitch, George, microbial physiology
Beckler, David (Zander), science administration
Blank, Charles Anthony, physical inorganic chemistry, aeronomy
Bolstein, Arnold Richard, mathematical analysis, operations research
Bradley, Ted Ray, plant taxonomy, ecology
Bratenahl, Charles George, pathology
Brown, Kathleen Older, anthropology
Chambers, Barbara Mae Fromm, mathematics
Cooley, Duane Stuart, meteorology
Cozzens, Robert F, physical chemistry
Crawford, Donald Lee, microbiology
Deanhardt, Marshall Lynn, analytical chemistry
De Los Reyes, B William, space physics
Emsley, Michael Gordon, entomology
Ernst, Carl Henry, herpetology, mammalogy
Feinstein, Hyman Israel, analytical chemistry
Greer, William Louis, chemical physics
Hart, Jayne Thompson, physiology, pharmacology
Hersey, John Brackett, bioacoustics, marine geophysics
Hobbs, Horton Holcombe, III, limnology
Howard, William Grady, mathematical statistics
Johnston, William Cargill, solid state physics
Joyce, Elaine C Elder, environmental physiology
Kelso, Donald Preston, marine ecology
Krug, Robert Charles, organic chemistry
Lankford, William Fleet, nuclear physics
McLean, Max C, physical oceanography
Mielczarek, Eugenie V, solid state physics
Mose, Douglas George, geochemistry, geochronology
Oppelt, John Andrew, mathematics
Packard, Fred Mallery, conservation
Papaconstantopoulos, Dimitrios A, solid state physics
Papp, Zoltan, mathematics
Roth, Ronald John, organic chemistry
Rubin, Alan Barry, physical chemistry
Seidman, Stephen Benjamin, topology
Seipel, John Howard, neurology, biophysics
Shaffer, Jay Charles, systematic entomology
Smith, John Melvin, mathematics education
Smith, Russell K, forest entomology
Stalick, Wayne Myron, organic chemistry
Stanley, Melissa Sue Millam, experimental zoology
Taub, Stephen Robert, genetics
Wall, James Robert, plant genetics
Wilson, John William, III, zoogeography, paleontology
Yonuschot, Gene R, biochemistry

CHESTER

Funkhouser, John William, geology
Twilley, Ian Charles, chemistry

CHRISTIANSBURG

Kramer, Clyde Young, statistics
Webb, Willis Keith, veterinary pathology

CLARKSVILLE

Albrecht, Herbert Richard, agronomy

COLONIAL HEIGHTS

Benepal, Parshotam S, plant genetics, plant physiology

COVINGTON

Sarjeant, Peter Thomson, paper chemistry, engineering

DAHLGREN

Burns, Grover Preston, applied mechanics, theoretical physics
Freiling, Edward Clawson, physical chemistry
Gibson, Luther Ralph, applied physics, space physics
Green, Daniel Thomas, mathematics, physics
Hershey, Allen Vincent, mathematical physics
Hodge, Frederick Allen, radiobiology
Holt, William Henry, solid state physics
Jarnagin, Milton Preston, Jr, mathematics
Knudsen, Dennis Ralph, physical chemistry
LaMonica, Carl J, computer science
Martens, William Stephen, environmental chemistry
Oesterwinter, Claus, astronomy, celestial mechanics
Reid, Lois Jean, mathematics
Ugincius, Peter, geodesy

DANVILLE

Bogart, William Hawkins, Jr, organic chemistry

FARMVILLE

Batts, Billy Stuart, fish biology, ecology
Breil, David A, bryology, plant morphology
Breil, Sandra J, cell physiology, plant histochemistry
Harvill, Alton McCaleb, Jr, phytogeography
Heinemann, Richard Leslie, cytogenetics
Holman, Leta Jane, zoology, acarology
Jackson, Elizabeth Burger, biology, chemistry
Lane, Charles Franklin, physical geography
Lehman, Robert Harold, physiology, ecology
McCombs, Freda Siler, science education
Scott, Marvin Wade, botany, microbiology
Wells, Ouida Carolyn, zoology

FRANKLIN

Farewell, John P, paper chemistry, physical chemistry
George, Fredna Stone, applied statistics, systems analysis
Wilber, Joe Casley, Jr, chemistry, science education

FREDERICKSBURG

Bird, Samuel Oscar, II, vertebrate paleontology
Bowen, Marshall Everett, geography
Cover, Herbert Lee, physical chemistry
Crissman, Judith Anne, inorganic chemistry
Emory, Samuel Thomas, geography
Gratz, Roy Fred, organic chemistry
Insley, Earl Glendon, physical chemistry
Johnson, Rose Mary, zoology
Johnson, Thomas L, biology
Mahoney, Bernard Launcelot, Jr, analytical chemistry, physical chemistry
Nikolic, Nikola M, physics
Pinschmidt, William Conrad, Jr, zoology
Pitts, Leslie Edwin, physics, acoustics
Schwiderski, Ernst Walter, applied mathematics
Sharpley, John Miles, industrial microbiology, industrial biochemistry
Wishner, Lawrence Arndt, biochemistry

FRONT ROYAL

Bovard, Kenly Paul, animal husbandry

Douglas, Jocelyn Fielding, biochemistry

FT BELVOIR
Adams, Martha Lovell, analytical chemistry
Amsutz, Larry Ihrig, solid state physics
Bond, John Walter, Jr., theoretical physics
Cameron, Louis McDuffy, applied physics
Galvin, Cyril Jerome, Jr., oceanography, coastal engineering
Gonano, John Roland, experimental solid state physics
Harris, D Lee, oceanography, meteorology
Hass, Georg, optical physics, solid state physics
Hastings, Andrew Dewey, Jr., physical geography
Kolobielski, Marjan, petroleum chemistry, fuel science
Mengenhauser, James Vernon, physical chemistry, petroleum chemistry
Pohlmann, Juergen Lothar Wolfgang, solid state chemistry
Robison, William Condit, geography
Rosenthal, Jenny Eugenie (Mrs Arthur Branley), physics
Sanfilippo, Francis Anthony, dentistry, biochemistry
Small, Timothy Michael, nuclear physics, applied physics
Spangler, Glenn Edward, chemical physics, environmental technology
Spitzer, Hermann Josef, thermodynamics, low temperature physics
Ullrich, George Werner, theoretical physics
Vuetto, John, Jr., geography, geomorphology

GREAT FALLS
Olin, Jacqueline S., organic chemistry, analytical chemistry

FT MONROE
Christman, Arthur Castner, Jr., physics

GLOUCESTER POINT
Andrews, Jay Donald, zoology
Bender, Michael E., pollution biology
Benner, Blair Richard, chemical metallurgy
Boesch, Donald Friedrich, biological oceanography, systematic zoology
Bowen, Marcia Ann, aquatic ecology
Byrne, Robert John, oceanography
Cardwell, Paul H., colloid chemistry
Davis, William Jackson, aquatic biology
Diaz, Robert James, marine ecology
Evans, David Arthur, statistical analysis, marine science
Fang, Ching Seng, hydromechanics, environmental engineering
Goldsmith, Victor, geological oceanography
Grant, George C., biological oceanography
Haefner, Paul Aloysius, Jr., zoology, biological oceanography
Haven, Dexter Stearns, fish biology
Hoagman, Walter John, fisheries
Huggett, Robert James, marine geochemistry, environmental sciences
Hyer, Paul Vincent, marine science
Kazana, Frederick Y., mycology
Kuo, Albert Yi-Shuong, physical oceanography, hydrodynamics
Loesch, Joseph, marine biology
Lynch, Maurice Patrick, biological oceanography
Merriner, John Vennor, fish biology, marine biology
Munday, John Clingman, Jr., marine sciences, remote sensing
Musick, John A., ichthyology, ecology
Olander, James Alton, inorganic chemistry
Orth, Robert Joseph, biological oceanography
Perkins, Frank Overton, marine biology
Silberhorn, Gene Michael, marine botany
Smith, Craig La Salle, organic chemistry, marine science
Van Engel, Willard Abraham, marine biology
Virnstein, Robert W., marine ecology
Wass, Marvin Leroy, zoology, botany
Wood, John Langille, mycology
Zeigler, John Milton, geology
Zubkoff, Paul Leon, biological oceanography, biochemistry

HARRISONBURG
Atkins, Robert Charles, organic chemistry
Ault, Janet E (Mills), algebra
Brubaker, Kenton Kaylor, biology, horticulture
Casali, Liberty, chemistry
Christiansen, Marjorie Miner, nutrition, biochemistry
Davis, John Edward, Jr., zoology

GREENWOOD
Jackson, Wendell Ford, chemistry

HAMPDEN-SYDNEY
Joyner, Weyland Thomas, Jr., nuclear physics, electronics
Kiess, Edward Marion, physics
Mayo, Thomas Tabb, IV, physics
Porterfield, William Wendell, inorganic chemistry
Selden, Dudley Byrd, mathematics
Shear, William Albert, biology
Sipe, Herbert James, Jr., physical chemistry
Smith, Homer Alvin, Jr. organometallic chemistry
Turney, Tully Hubert, zoology

HAMPTON
Abram, James Baker, Jr., zoology, parasitology
Bell, Cecil Cooper, Jr., surgery
Benda, Stepan Vaclav, physical chemistry
Benson, James Miller, physics
Bonner, Hazel Garrison, botany
Bonner, Robert Dubois, botany
Brooks, Thomas Furman, acoustics
Browell, Edward Vern, environmental systems & technology physics
Byvik, Charles Edward, solid state physics
Chu, William Peter, optical physics
Darden, Geraldine C., algebra
Deepak, Adarsh, air pollution, aeronomy
Esterling, Donald M., theoretical solid state physics, materials science
Fields, Victor Hugo, organic chemistry
Foelsche, Trutz, physics, biophysics
Garrick, Isadore Edward, mathematics, physics
Grant, Frederick Cyril, physics
Harris, Franklin Stewart, Jr., physics
Harvey, Gale Allen, meteor physics
Hohl, Frank, physics, astronomy
Hudson, Roy D., pharmacology
Inge, Frederick Douglass, plant physiology
Johnston, Norman Joseph, organic polymer chemistry
Levine, Joel Stewart, planetary atmospheres
Maestrello, Lucio, acoustics
McCormick, Michael Patrick, atmospheric physics
Melvin, Horace Willis, organic chemistry
Michael, William Herbert, Jr., space sciences
Ortega, James M., numerical analysis
Palmer, Fred Shank, organic chemistry
Raffenetti, Richard Charles, quantum chemistry
St. Clair, Terry Lee, polymer chemistry
Sands, George Dewey, physical chemistry
Satcher, Robert Lee, radiation chemistry
Singh, Jag Jeet, physics, astronomy
Sobieszczanski, Jaroslaw Eugeniusz, applied mechanics, numerical analysis
Stern, Joseph Aaron, microbiology
Straeter, Terry Anthony, applied mathematics
Swissler, Thomas James, environmental physics, mathematical analysis
Voigt, Robert Gary, numerical analysis
Wenzel, Alan Richard, applied mathematics
Wikins, Judd Rice, bacteriology
Wilson, John William, theoretical high energy physics, health physics
Young, Philip Ross, analytical chemistry
Yu, James Chun-Ying, acoustics, fluid dynamics

Rothenberg, Herbert Carl, solid state physics
Thomas, Clayton James, mathematics, operations research
Thornton, Charles De Wane, chemistry, physics

DeGraff, Benjamin Anthony, chemical kinetics
Dendinger, James Elmer, invertebrate physiology
Dua, Prem Nath, veterinary medicine, animal nutrition
Farmer, George Thomas, Jr., geology, paleontology
Fisher, Elwood, parasitology, ecology
Fisher, Gordon McCrea, mathematics, history of science
Grimm, James K., entomology, human anatomy
Ikenberry, Jesse Emmert, mathematics
Jones, William F., genetics, science education
Kauffman, Glenn Monroe, physical chemistry
Kribel, Robert Kendall, plasma physics, magnetohydrodynamics
Lehman, Robert C., biophysics
Mellinger, Clair, plant ecology
Nielsen, Peter Tryon, plant physiology, biophysics
Sanders, William Mack, mathematics
Sellers, Cletus Miller, Jr., physiology
Sherwood, William Cullen, geochemistry
Smith, Harry Francis, algebra
Stucky, Gary Lee, bioinorganic chemistry
Suter, Daniel B., human anatomy, physiology
Taylor, Gerald Reed, Jr., physics
Trelawny, Gilbert Sterling, microbial physiology
Wells, John Clarence, physics
Winstead, Janet, mycology

HERNDON
Gilmore, Charley E., comparative pathology
Horn, Henry Joseph, medicine
Johnston, Carter Dupuy, biochemistry
Stirewalt, Edward Neale, organic chemistry
Woodard, Geoffrey, pharmacology
Woodard, Marie W., biochemistry

HOLLINS
Freitag, Herta Taussig, number theory
Hackman, Roger H., elementary particle physics, theoretical nuclear physics
Morlang, Charles, Jr., plant morphology

HOLLINS COLLEGE
Boatman, Sandra, organic chemistry
Bull, Alice Louise, developmental biology
Gray, Faith Harriet, zoology
Gushee, Beatrice Eleanor, inorganic chemistry
Kurtz, Lawrence Alfred, numerical analysis
Steinhardt, Ralph Gustav, Jr., chemistry, chemical physics
Stewart, Roberta A., organic chemistry
Wine, Russell Lovell, statistics

HOPEWELL
Bertram, Leon Leroy, cellulose chemistry
Leake, Preston Hildebrand, organic chemistry
Moffett, Samuel McKee, organic chemistry
Price, Byron Frederick, analytical chemistry
Rickett, Frederic Lawrence, biochemistry
Schroeder, Walter Arthur, polymer chemistry, organic chemistry
Wartman, William Benjamin, Jr., organic chemistry
Wright, Howard Edwards, Jr., organic chemistry

KEYSVILLE
Barber, Patrick George, physical chemistry, x-ray crystallography

KILMARNOCK
Overholt, John Lough, physical chemistry

KING GEORGE
Hayden, Leonard Octavius, ionospheric physics

LAWRENCEVILLE
Baumbach, Donald Otto, physical chemistry

Samuel, Albert, molecular biology

LEESBURG
Gaskill, Irving E., mathematics
Monroe, Watson Hiner, geology
Rogers, Nancy Graham, medical microbiology

LEXINGTON
Carpenter, Delma Rae, Jr., physics
Deal, Albert Leonard, III, mathematical analysis
Donaghy, James Joseph, physics
Emmons, Lyman Randlett, biology
Gilreath, Esmarch Senn, chemistry
Goehring, John Brown, inorganic chemistry
Goller, Edwin John, organic chemistry
Gupton, Oscar Wilmot, botany
Hickman, Cleveland Pendleton, Jr., zoology
Hundley, Louis Reams, vertebrate zoology, physiology
Johnson, Robert S., geology
Kozak, Samuel J, geology
McGuire, Odell, geology
Minnix, Richard Bryant, physics
Newbolt, William Barlow, physics
Newman, James Blakey, physics
Peters, Philip Boardman, solid state physics
Pickral, George Monroe, Jr., inorganic chemistry, analytical chemistry
Pingree, George, algebra
Roberts, Robert Abram, mathematics
Royston, Robert Winter, mathematics
Sauder, William Conrad, physics
Schwab, Frederic Lyon, sedimentology
Settle, Frank Alexander, Jr., analytical chemistry
Shillington, James Keith, organic chemistry
Smart, Charles William, organic chemistry
Spencer, Edgar Winston, geology
Starling, James Holt, biology, zoology
Swope, Fred C., food science
Turner, Edward Felix, Jr., physics
Watt, William Joseph, inorganic chemistry
Wetmore, Stanley Irwin, Jr., organic chemistry
White, Alan George Castle, bacteriology
Whitney, George Stephen, organic chemistry
Williams, Charles Wiley, mathematics
Williams, Harry Thomas, theoretical physics
Wilson, Robert Lee, Jr., mathematics
Wise, Gene, organic chemistry, physical chemistry
Wise, John Hice, physical chemistry

LORTON
Harvalik, Zaboj Vincent, physics

LOVETTSVILLE
Lemp, John Frederick, Jr., microbiology, bioengineering

LURAY
Caley, David William, family medicine

LYNCHBURG
Ball, Russell Martin, reactor physics, nuclear physics
Banks, David Lee, zoology
Bliss, Dorothy Crandall, ecology
Bliss, Laura, chemistry
Buescher, Brent J., solid state physics
Chulick, Eugene Thomas, nuclear chemistry
Claiborne, Imogene B., inorganic chemistry, analytical chemistry
Dahlgard, Muriel Genevieve, organic chemistry
Engelder, Theodore Carl, nucleonics
Flint, Franklin Ford, biological structure
Gamble, Samuel James Reeves, biochemistry
Garretson, Harold H., chemistry
Hansrote, Charles Johnson, Jr., physical chemistry
Humphreys, Mabel Gweneth, mathematics
Kennedy, Albert Joseph, radio chemistry, corrosion
Marshall, Maryan Lorraine, physical chemistry

McDaniel, Thomas Lee, nuclear chemistry, radiochemistry
Nagler, Benedict, neurology, psychiatry
Osborne, Paul James, invertebrate zoology, physiology
Pettus, William Gower, reactor physics
Phillips, Robert Bass, Jr., mathematics
Pitts, Thomas Griffin, experimental nuclear physics
Ramsey, Gwynn W., plant systematics
Rosser, Shirley Ewart, physics
Roy, Donald H, nuclear physics, reactor physics
Sigler, Julius Alfred, Jr., solid state physics
Stagg, William Ray, physical inorganic chemistry
Sumrall, H Glenn, plant physiology, microbiology
Toops, Edward Chassell, reactor physics
Tulenko, James Stanley, nuclear physics, applied mathematics
Uhl, Dale Lynden, nuclear chemistry
Warren, Holland Douglas, physics
Whidden, Helen Louise, physical chemistry
Worsham, Herbert J, Jr., nuclear physics

MANAKIN SABOT
Hoge, Randolph Harrison, gynecology

MARTINSVILLE
Hazlehurst, David Anthony, physical chemistry, inorganic chemistry
Longhi, Raymond, inorganic chemistry, organic chemistry
Vogelsong, Donald Clair, physical chemistry, polymer chemistry

MCLEAN
Benade, Leonard E, cancer, occupational health
Cooperstein, Raymond, inorganic chemistry
De Schmertzing, Hannibal, analytical chemistry
Eldridge, Francis Reed, Jr., meteorology
Engle, Edison Grove, Jr, forensic science, physical chemistry
Franzosa, Edward Sykes, geology
Fryklund, Verne Charles, Jr, mathematics, geology
Grandy, Charles Creed, mathematics, physics
Greene, David C., physics, acoustics
Hart, Dabney Gardner, science writing, taxonomy
Heverly, John Ross, physics, mathematics
Holz, Betty Weber, operations research, systems analysis
Jacobs, Verne Louis, atomic physics
Jay, George Edgar, Jr, public health administration
Kapos, Ervin, operations research, systems analysis
Keeny, Spurgeon Milton, Jr., physics
Kelsey, David, analytical chemistry
Krugman, Stanley Liebert, forest genetics, forest physiology
Larkin, James Richard, mathematics
Lee, James Stewart, resource geography, cartography
Lowry, Philip Holt, operations research
Malcolm, Janet May, operations research, resource management
Malcolm, John Lowrie, soils
Malkin, William, meteorology
Mansfield, John E, elementary particle physics, theoretical physics
Massel, Gary Alan, plasma physics, operations research
May, Donald Curtis, Jr, operations research
Molloy, Charles Thomas, physics
Molo, William L, meteorology, oceanography
Moritz, Barry Kyler, physics, computer science
Morrow, Leonard Owen, botany
Oehser, Paul Henry, science writing
Ottoson, Harold, mathematics
Pruett, Carl Eugene, medicine
Pugh, George Edgin, operations research, nuclear physics
Rogers, Thomas F, physics, electronics
Schwartz, Benjamin L, operations research
Scoville, Herbert, Jr., physical chemistry
Siebentritt, Carl R, Jr., health physics
Sobol, Stanley Paul, forensic science, medicine
Solomon, Louis Peter, acoustics
Stewart, Thomas Dale, anthropology
Stubblefield, Frank Milton, organic chemistry

Tachmindji, Alexander John, hydrodynamics
Talbot, Lee Merriam, ecology
Tillson, Albert Holmes, histology, crystallography
Tupac, James Daniel, mathematics, physics
Vagina, Livio L, chemistry
Via, Giorgio G, physics
Watson, Bernard Bennett, physics
Watt, Bernice K, nutrition
Wilkinson, William Lyle, operations research
Wilson, Woodrow, Jr, theoretical physics
Zimmerman, Richard E, physics
Zirkind, Ralph, physics

MECHANICSVILLE
Burke, Arthur Wade, Jr., cancer, biophysics

MERRIFIELD
McBride, John Alexander, chemistry

MIDDLEBURG
Bryant, Harry Talbot, agronomy

MIDLOTHIAN
Randall, Russell E, Jr., internal medicine

NEWPORT NEWS
Duval, Addison (McGuire), medicine, psychiatry
Olson, Lee Charles, plant biochemistry, plant physiology
Pugh, Jean Elizabeth, invertebrate zoology
Remsberg, Ellis Edward, atmospheric physics
Sacks, Lawrence J, inorganic chemistry
Wang, Chin San, topology, systems science

NORFOLK
Abernathie, Jan W, horticulture
Adams, Clifford Lowell, physics
Anderson, Samuel, analytical chemistry, inorganic chemistry
Baeumler, Howard William, mathematics, electrical engineering
Barnes, Annie Shaw, anthropology, sociology
Bell, Charles E, Jr., physical organic chemistry
Birdsong, Ray Stuart, biology, ichthyology
Blais, Roger Nathaniel, atmospheric physics, remote sensing
Borchers, Edward Alan, plant breeding
Brown, Paul Lopez, zoology
Chopra, Kuldip P, environmetnal physics, space physics
Clark, Allen Keith, organic chemistry
Clay, Forrest Pierce, Jr., physics
Clendenin, Martha Anne, neurosciences
Conrad, Margaret C, physiology
Copeland, Gary Earl, chemical physics
Cowan, Daniel Francis, pathology
Darby, Dennis Arnold, sedimentary petrology
Dauer, Daniel Martin, marine ecology
Davis, Richard Elden, academic administration, psychiatry
Day, Frank Patterson, Jr, ecology
Delzell, David Edgar, zoology
Duke, Everette Loranza, soil science
Faulconer, Robert Jamieson, pathology
Fleischer, Peter, geological oceanography, sedimentology
Frye, Keith, geochemistry, mineralogy
Gourley, Desmond Robert Hugh, pharmacology
Grosch, Chester Enright, physics
Hanna, William Jefferson, geochemistry
Harries, Wynford John, physics
Helvie, Carl Otto, public health
Hinman, Norman Dean, biochemistry
Holman, Gerald Hall, pediatrics, endocrinology
Holsinger, John Robert, invertebrate zoology, biospeleology
Homsher, Paul John, reproductive biology, acarology
Hotta, Shoichi Steven, biochemistry, medicine
Hsieh, Jen-Shu, health physics, medical physics
Hufstedler, Robert Sloan, inorganic chemistry
Hughes, Hansel Leigh, chemistry
Johnson, James Carl, virology

Johnson, Ronald Ernest, physical oceanography
Jones, Major Boyd, mathematics
Kernell, Robert Lee, nuclear physics
Kiefer, Harold Milton, theoretical physics
Kindle, Earl Clifton, meteorology
Kirk, Paul Wheeler, Jr., mycology
Knight, John Cian, underwater acoustics, operations research
Levy, Gerald Frank, ecology
Lick, Dale W, pure mathematics, applied mathematics
Ludwick, John Calvin, Jr., marine geology
Malbon, Wendell Endicott, mathematics
Mandell, Alan, science education, biology
Manning, Robert Thomas, medicine, biochemistry
Marshall, Harold George, marine biology
McNeil, Philip Eugene, algebra
Merchant, Donald Joseph, microbiology
Michie, David Doss, cardiovascular physiology
Milbocker, Daniel Clement, horticulture
Moore, Theron Langford, organic chemistry
Nugent, Thomas John, plant pathology
Pariser, Harry, dermatology, syphilology
Patek, David Rushton, neuropharmacology
Pfeiffer, George William, mathematics, statistics
Pittman, Melvin Amos, physics
Powell, Allen LaRue, oceanography
Pritchard, Wenton Maurice, solid state physics
Raymond, Anne Frances, marine biology
Richardson, Annie Louise, systematic botany
Rowe, Mark J, biochemistry
Rudd, David, mathematics
Schapiro, Herbert, anatomy, physiology
Schechter, Martin David, psychopharmacology
Schellenberg, Karl A, biochemistry
Scully, Frank E, Jr, chemistry
Sherwood, Calder Smith, III, inorganic chemistry
Sitz, Thomas O, biochemistry
Somers, Kenneth Donald, microbiology
Sonenshine, Daniel E, zoology
Spencer, Randall Scott, paleontology, stratigraphy
Stewart, Franklin Burton, soil fertility
Stillwell, Edgar Feldman, physiology
Swanson, Robert James, endocrinology
Sweits, John Joseph, mathematical analysis
Tweed, John, applied mathematics
Vernick, Sanford H, pathobiology
Walsh, Michael Joseph, biochemistry, pharmacology
Wan, Abraham Tai-Hsin, clinical chemistry, endocrinology
Wei, Diana Yun Dee, mathematics
Whitten, Harrell David, immunology
Williams, Patricia Bell, pharmacology
Williams, Roy Lee, organic chemistry
Woods, Roy Alexander, physics, electronics
Wright, George Leonard, Jr., immunochemistry, microbiology
Young, William Irving, medical genetics
Zaneveld, Jacques Simon, biological oceanography, phycology

NORTH ARLINGTON
Shear, Sidney Kingsbury, physics

OAKTON
Weber, Karl Hansel, organic chemistry

PAEONIAN SPRINGS
Buckardt, Henry Lloyd, agronomy

PAINTER
Baldwin, Robert Edmund, plant pathology
Hofmaster, Richard Namon, entomology
Hohlt, Herman Edward, horticulture

PETERSBURG
Bakshi, Vidya Sagar, applied mathematics
Beck, James Donald, inorganic chemistry
Chakravarti, Kalidas, physical chemistry, surface chemistry
Dardoufas, Kimon C, organic chemistry, chemical engineering
Dunn, Richard Hudson, biology, botany
Foster, Wilfred John Daniel, physical chemistry
Gipson, Mack, Jr, geology

Kubu, Edward Thomas, physical chemistry, polymer physics
Lewis, Cornelius Crawford, agronomy, soil science
Moody, Arnold Ralph, plant pathology
Newkirk, Robert Franklin, physiology
Russi, Simon, anatomic pathology
Saunders, Peter Reginald, physical chemistry, physics
Townes, Charles Henry, medicine, physics
Waters, Charles Emory, chemistry
Weedon, Gene Clyde, polymer chemistry
Wincklhofer, Robert Charles, polymer physics
Woodson, Bernard Robert, algology

PORTSMOUTH
Ellis, Leonard Culberth, organic chemistry
Kelly, Thomas Edward, organic chemistry
Kise, Mearl Alton, organic chemistry

POWHATAN
De La Burde, Roger Z, research administration

PURCELLVILLE
Parker, Pierre E, mining geology

RADFORD
Benson, Katherine Adams, developmental physiology
Boleter, William Theodore, analytical chemistry
Carlile, Clayton George, organic chemistry
Chalgren, Steve Dwayne, microbiology, virology
Chandler, Carl Davis, Jr, analytical chemistry
Dolliver, Claire Vincent, geomorphology
Durrill, Preston Lee, chemistry
Gourley, Eugene Vincent, zoology
Hoffman, Richard Lawrence, systematic zoology
Jarvis, Floyd Eldridge, Jr, genetics
Jones, Franklin M, science education
Lambert, Rogers Franklin, organic chemistry
Lutes, Charlene McClanahan, genetics, developmental biology
Moore, David Jay, animal ecology
Nichols, Joseph Caldwell, mathematics
Scolaro, Reginald Joseph, invertebrate paleontology, paleoecology
Whisonant, Robert Clyde, stratigraphy, sedimentology
Whitaker, Mack Lee, mathematics

RESTON
Albers, John P, geology
Alexander, Robert Houston, geography
Allingham, John Wing, geology, geophysics
Anderson, James Richard, resource geography, agricultural economics
Arth, Joseph George, geochemistry
Austin, Thomas LeRoy, Jr, mathematics
Back, William, hydrogeology
Baedecker, Philip A, nuclear chemistry, geochemistry
Bailey, Roy Alden, geology
Balsley, James Robinson, Jr, geology, geophysics
Becraft, George Earle, geology
Bergin, Marion Joseph, geology
Bergquist, Harlan Richard, geology
Bodenlos, Alfred John, geology
Botbol, Joseph Moses, geology
Bredehoeft, John Dallas, hydrology, geology
Breger, Irving A, organic geochemistry
Brett, Robin, earth sciences
Britton, Maxwell Edwin, botany
Brobst, Donald Albert, economic geology
Brown, Clarence Ervin, geology
Brown, Glen Francis, geology
Callahan, Joseph Thomas, geology
Cannon, William Francis, III, geology
Carter, William Douglas, geology
Chao, Edward Ching-Te, geology
Chidester, Alfred Herman, geology
Clark, Allen LeRoy, economic geology, geochemistry
Clark, Sandra Helen Becker, geology
Cohen, Philip, hydrogeology, groundwater geology
Cox, Dennis Purver, economic geology
Davidson, David Francis, economic geology

Davis, George H., hydrology
Denny, Charles Storrow, geology
DeNoyer, John M, geophysics
Drake, Avery Ala, Jr., geology
Drew, Lawrence James, statistics, geology
Dwornik, Edward John, mineralogy
Eicher, Ralph N., computer sciences
Epstein, Jack Burton, geology
Ericksen, George Edward, economic geology
Evans, Howard Tasker, Jr., mineralogy, geology
Faizi, Salih, mineralogy, geology
Fary, Raymond W. Jr., geology
Faust, George Tobias, mineralogy, geology
Fisher, James Russell, geochemistry, physical inorganic chemistry
Fiske, Richard Sewell, geology
Fleischer, Michael, geochemistry, mineralogy
Foose, Michael Peter, structural geology
Froelich, Albert Joseph, geology
Gair, Jacob Eugene, geology
Gavarecki, Stephen Jerome, geology
Getzen, Rufus Thomas, groundwater hydrology
Goldberg, Jerald Melvin, geology, geography
Goldsmith, Richard, geology
Greeson, Philip Edward, limnology
Grossling, Bernardo Fruedenberg, geophysics, operations research
Guild, Philip White, economic geology
Hack, John Tilton, geology
Hackett, Orwoll Milton, geology
Haering, George, operations research
Hamilton, Robert Morrison, geophysics
Hanna, William F., geophysics
Hanshaw, Bruce Busser, geochemistry
Harwood, David Smith, geology, petrology
Hatch, Norman Lowrie, Jr., structural geology
Hearn, Bernard Carter, Jr., geology
Heindl, Leopold Alexander, geology, hydrology
Hemley, John Julian, geology, chemistry
Higgins, Brenda Baer, geology
Hoover, Linn, geology
Howard, Keith Arthur, geology
Krivoy, Harold Lloyd, geophysics
Klepper, Montis Ruhl, geology
Lang, Solomon Max, hydrology
Langford, Russel Hal, hydrology, geophysics
Lee, Kwang-Yuan, petrology, geophysics
Lesure, Frank Gardner, geology
Lightner, Jerry P., biology
Lindsey, David Allen, geology
Lipin, Bruce Reed, petrology
Ludington, Stephen Dean, geochemistry
Masters, Charles Day, geology
May, Irving, analytical chemistry, geochemistry
McKelvey, Vincent Ellis, geology
Meissner, Charles Roebling, Jr., geology
Mercer, James Wayne, hydrogeology
Meyer, Richard Fastabend, economic geology
Miller, Ralph Leroy, economic geology
Moody, David Wright, hydrology
Murray, Charles Richard, hydrology
Nichols, Donald Ray, environmental geology
Nolan, Thomas Brennan, geology
Nord, Gordon Ludwig, Jr., geology
Oglesby, Gayle Arden, geology
Pavlides, Louis, geology
Pearson, Frederick Joseph, groundwater geology
Peck, Dallas Lynn, geology
Phipps, Richard L., plant ecology
Pickering, Ranard Jackson, geochemistry, hydrology
Pluhowski, Edward John, hydrology
Pomeroy, John S., geology
Prinz, William Charles S., economic geology
Rakestraw, James William, physics
Rankin, Douglas Whiting, volcanology, tectonics

Reed, John Calvin, Jr., geology
Regan, Robert David, geophysics
Rioux, Robert Lester, geology
Robertson, Eugene Corley, geology
Robie, Richard Allen, mineralogy
Roedder, Edwin Woods, geochemistry
Rooney, Lawrence Frederick, economic geology
Root, David Harley, mathematics, energy resources
Roseboom, Eugene Holloway, Jr., geology, geochemistry
Ross, Malcolm, geochemistry, crystallography
Rubin, Meyer, geology
Sato, Motoaki, geochemistry, economic geology
Schmidt, Robert George, geology
Schneider, Robert, groundwater geology
Schnepfe, Marian Moeller, chemistry, analytical chemistry
Schoen, Robert, geology
Scholle, Peter Allen, sedimentology, petrography
Schultz, David Michael, marine geochemistry
Schwenterth, Stanley P., geology
Senfle, Frank Edward, physics
Shaw, Herbert Richard, geology
Sigafoos, Robert Sumner, plant ecology
Singer, Donald Allen, applied statistics, mineral economics
Smedes, Harry Wynn, geology, resource management
Smoot, George Fitzgerald, hydrology
Spall, Henry Roger, geophysics
Stern, Thomas Whital, geochronology
Stewart, David Benjamin, geology, mineralogy
Tanner, Allan Bain, geophysics
Thayer, Thomas Prence, geochemistry
Tooker, Edwin Wilson, economic geology
Toulmin, Priestley, III, geology
Tracey, Joshua Irving, Jr., geology
Trask, Newell Jefferson, Jr., geology
Trescott, Peter Chapin, hydrogeology
Van Alstine, Ralph Erskine, economic geology
Voikmann, Richard Peter, geology
Warshaw, Charlotte Marsh, geochemistry, mineralogy
Weis, Paul Lester, geology
White, Walter Stanley, economic geology
Wier, Karen, geology
Willams, Richard Sugden, Jr., quaternary geology
Witmer, Richard Everett, geography
Wolff, Roger Glen, geology, hydrology
Wones, David R., petrology, geochemistry
Wright, Thomas L., geology, petrology
Zadnik, Valentine Edward, geology, engineering geology
Zen, E-An, geology, petrology
Ziony, Joseph Israel, geology

RICHMOND

Abbott, Lynn De Forrest, Jr., biochemistry
Aceto, Mario Domenico Giulio, pharmacology
Adams, Max David, pharmacology
Allison, Marvin J., microbiology
Alphin, Reevis Stancil, pharmacology, biochemistry
Alsop, John Henry, III, analytical chemistry
Alston, Peter Van, physical organic chemistry, analytical chemistry
Andrako, John, pharmaceutical chemistry, analytical chemistry
Armstrong, Robert G., polymer chemistry, analytical chemistry
Astruc, Juan A., anatomy, neuroanatomy
Atkins, Henry Pearce, mathematics
Bakerman, Seymour, biophysics, pathology
Banks, William Louis, biochemistry, nutrition
Barkman, Erik Fredrik, inorganic chemistry
Bass, Robert Gerald, organic chemistry
Bauer, David Francis, statistics
Bayliss, John Temple, physics
Berg, Gene Arthur, geometry
Bhatnagar, Ajay Sahai, organic chemistry, reproductive endocrinology
Biber, Margaret Clare Boadle, neuropharmacology
Bishop, John Watson, ecology, aquatic biology
Blanke, Robert Vernon, toxicology
Blaylock, W Kenneth, dermatology
Blem, Charles R., vertebrate ecology

Board, John Arnold, obstetrics & gynecology
Bogdanove, Emanuel Mendel, endocrinology
Bond, Judith, biochemistry, physiology
Boots, Marvin Robert, medicinal chemistry
Boots, Sharon G., organic chemistry
Borzelleca, Joseph Francis, pharmacology, toxicology
Bowe, Robert Looby, psychopharmacology
Bowman, Edward Randolph, pharmacology, physiology
Bowman, Faye Johnson, pharmacology, toxicology
Bradley, Sterling Gaylen, microbiology, genetics
Brandt, Richard Bernard, biochemistry
Brehmer, Morris Leroy, aquatic biology
Brierre, Roland Theodore, Jr., polymer physiology
Briggs, Fred Norman, physiology
Brinckerhoff, Harold Guion, chemistry
Brockington, James Wallace, organic chemistry
Brown, Russell Vedder, immunology
Bruce, Robert Black, drug metabolism
Burke, Hanna Suss, organic chemistry
Burke, Jack Denning, anatomy, physiology
Burke, James Oxey, medicine
Burton, Willard Alvin, organic chemistry
Bush, Francis M., zoology
Carpenter, Robert Dean, analytical chemistry
Carter, Walter Hansbrough, Jr., mathematical statistics
Chambers, John William, endocrinology, pharmacology
Childs, Dana Pitt, entomology
Chinnici, Joseph (Frank) Peter, genetics, evolutionary biology
Clabough, Jeanne Whitaker, anatomy
Clamann, H Peter, biomedical engineering
Clarke, Alexander Mallory, biophysics, physiology
Clayton, Charles Curtis, biochemistry
Cleary, Stephen Francis, biophysics, radiobiology
Clough, Oliver Wendell, dentistry
Clough, Stuart Chandler, organic chemistry
Cloyd, Grover David, veterinary medicine
Coleman, Philip Hoxie, virology
Collins, James Malcolm, biochemistry, developmental biology
Cox, Edwin, chemistry, chemical engineering
Creamer, Robert M., analytical chemistry, physical organic chemistry
Creech, Henry Bryant, audiology, speech
Cribbs, Richard Madison, genetics
Crouch, Robert Thomas, plastics
Crump, John C. III, solid state physics
Cunningham, George J., medicine
Dalton, Harry P., biochemistry, pathology
Dannenburg, Warren Nathaniel, biochemical pharmacology
DeBardeleben, John F., organic chemistry, environmental biology
Decker, Robert Dean, botany
Deever, William Ray, inorganic chemistry
DePaola, Dominick Philip, dental research, nutrition
Deszyck, Edward John, biochemistry
Deveney, James Kevin, pure mathematics
De Vries, George Henry, neurochemistry
Dewey, William Leo, pharmacology
Dille, Roger McCormick, chemistry
DuPuis, Robert Newell, chemistry
Dutz, Werner, pharmacology, physiology
Dwyer, Rowland William, Jr., quantum chemistry
Eareckson, William Milton, III, organic polymer chemistry
Edwards, Carolyn Trowbridge, physiology
Edwards, Leslie Erroll, physiology
Edwards, William Brundige, III, organic chemistry, natural product chemistry
Egle, John Lee, Jr., pharmacology, physiology
Eichorn, Paul Anthony, cytology, research administration
Elford, Howard Lee, biochemistry, cancer

Elmore, Stanley Mcdowell, medicine, orthopedic surgery
Elsea, John Robert, toxicology
Elzay, Richard Paul, dentistry, oral pathology
Epstein, Ludwig Ivan, biophysics
Escobar, Mario R., virology, immunology
Estep, Herschel Leonard, medicine, endocrinology
Fagan, Raymond, epidemiology, environmental sciences
Fallon, Harold Joseph, pharmacology
Farago, John, polymer chemistry, applied psychology
Farley, Reuben William, mathematics
Farrar, Lynn McA, clinical pharmacology
Felber, William Henry, biostatistics
Fels, Irving Gordon, biochemistry
Felton, Staley Lee, pharmacology
Fisher, Lyman McA, clinical pathology, hematology
Fletcher, Lowell W., entomology
Flora, Roger E., biostatistics
Ford, George Dudley, physiology
Fore, Harry Waugh, Jr., dentistry
Formica, Joseph Victor, biochemistry
Franko, Bernard Vincent, pharmacology
Freund, Richard John, clinical pharmacology
Friedman, Marvin Alan, toxicology
Funderburk, William Henry, pharmacology
Gager, Forrest Lee, Jr., organic chemistry, natural products chemistry
Gager, Helen McClure, physical chemistry
Gander, George William, biochemistry
Gates, James Edward, phytopathology
Geeraets, Walter J., ophthalmology
Gerow, Clare William, organic polymer chemistry
Gillespie, Jesse Samuel, Jr., organic chemistry
Gladding, Jane B., inorganic chemistry
Glynn, William Allen, mathematics
Goldman, Israel David, biophysics
Goldstein, Lewis Charles, cytology
Goodale, Fairfield, anatomic pathology, clinical pathology
Greenfield, Lazar John, thoracic surgery
Gregory, Daniel Hayes, internal medicine, gastroenterology
Gronholz, LeRoy Frederick, physical chemistry
Guyer, Kenneth Eugene, Jr., biochemistry
Haar, Jack Luther, anatomy, electron microscopy
Hagman, Donald Eric, industrial pharmacy
Hall, Wayne Clark, plant physiology
Ham, William Taylor, Jr., physics
Hardie, Edith L., cardiopulmonary physiology
Harowitz, Charles Lichtenberg, organic chemistry
Harris, Louis Selig, neuropharmacology
Harris, Thomas Mason, embryology, histology
Hartung, Homer Arthur, physical chemistry
Hasegawa, Ichiro, electron microscopy
Haskell, Vernon Charles, physical chemistry
Haynes, Boyd W. Jr., surgery
Hegre, Erling Stanford, anatomy
Hench, Miles Ellsworth, medical microbiology
Henderson, Ulysses Virgil, Jr., organic chemistry
Henry, Neil Wylie, mathematical statistics
Hesch, Elizabeth Beaman, mathematics
Higgins, Edwin Stanley, biochemistry
Hightower, James Anderson, anatomy
Hilsman, Overton Lindner, analytical chemistry
Hoelzel, Charles Bernard, organic chemistry
Hoff, Ebbe Curtis, neurophysiology
Hoffman, Robert Alan, biophysics
Holyoke, Caleb William, Jr., organic polymer chemistry, environmental chemistry
Hornbuckle, Phyllis Ann, psychophysiology
Hossaini, Ali A., clinical pathology
Howell, John Robert, biomathematics
Hsu, Hsiu-Sheng, immunology
Hudgins, Aubrey C. Jr., experimental solid state physics, computer simulation
Huf, Ernst Gustav, medicine

Ikeda, Robert Mitsuru, analytical chemistry
Ingersoll, Everett Harold, anatomy
Inskeep, George Esler, chemistry
Irby, William Robert, medicine
James, George Watson, III, medicine
Jeffrey, Jackson Eugene, biological structure
Jenkins, Herndon, organic chemistry, medicinal chemistry
Jenkins, Robert Walls, Jr, radiochemistry, plant ecology
Jenks, Theodore Eugene, wood science & technology
Jess, Edward Orland, meteorology
Jimenez, Miguel Angel, food science
John, David Thomas, parasitology
Johns, Philip Timothy, biochemistry
Johnson, Miles F, systematic botany
Johnson, William Randolph, Jr, polymer chemistry, theoretical chemistry
Johnston, Charles Louis, Jr, clinical pathology
Johnston, Robert Howard, mathematics
Jollie, William Pucette, anatomy
Jones, John R, medicine, anesthesiology
Jordan, Robert Lawrence, anatomy, teratology
Joshi, Vijay Vinayak, pathology, pediatrics
Kallman, William Michael, psychophysiology
Kapadia, Abhaysingh J, pharmacy
Kaplan, Alan Marc, immunology
Kapp, Mary Eugenia, analytical chemistry
Kay, Saul, surgical pathology
Keefe, William Edward, biophysics, crystallography
Kendig, Edwin Lawrence, Jr, tuberculosis
Kent, Joseph Francis, topology
Kilpatrick, S James, Jr, biostatistics
Kimbrough, Theo Daniel, Jr, physiology, biochemistry
King, E Richard, radiotherapy
King, Lucy Jane, psychiatry
Kirkland, Richard Horace, medicine
Kirn, James Frederick, physical chemistry
Kline, Edward Samuel, biochemistry
Klioze, Oscar, pharmaceutical chemistry
Knappenberger, Paul Henry, Jr, science administration, astronomy
Knighton, Holmes Tutt, bacteriology, dentistry
Kontos, Hermes A, medicine, physiology
Koontz, Warren Woodson, Jr, urology
Kriegman, George, psychiatry, psychoanalysis
Kuhn, William Frederick, analytical chemistry, spectrochemistry
Larson, Paul Stanley, pharmacology
Laszlo, Tibor S, industrial chemistry
Lau, Kenneth W, chemistry
Laupus, William E, medicine
Lawrence, Walter, Jr, surgery
Lawson, Louis Russell, Jr, energy systems & technology, paper chemistry
Lee, Herbert Carl, surgery
Lee, Hyung Mo, medicine, surgery
Leichtert, George Robert, neuroanatomy
Lenhardt, Martin Louis, speech pathology, speech & hearing sciences
Leonard, Charles Arthur, pharmacology
Lilly, Arnys Clifton, Jr, physics
Line, Lloyd Ernest, Jr, physical chemistry
Lippincott, Stuart Wellington, pathology
Lippmann, Irwin, pharmaceutical chemistry
Llewellyn, Gerald Cecil, bionucleonics
Loria, Roger Moshe, virology
Lowenthal, Werner, pharmacy
Lower, Richard Rowland, thoracic surgery, cardiovascular surgery
Lowitz, David Aaron, chemical physics
Lunsford, Carl Dalton, pharmaceutical chemistry
Lurie, Harry I, pathology
MacDonald, Alan Angus, organic chemistry
Major, Robert Wayne, physics
Makhlouf, Gabriel Michel, medicine, physiology
Mardon, David Norman, microbiology
Martin, Albert Edwin, analytical chemistry, pharmaceutical chemistry
Martin, Albert Erskine, Jr, applied physics
Martin, James Henry III, vertebrate physiology
Marvin, Daniel Ezra, Jr, comparative physiology
Mateer, Richard Austin, physical organic chemistry
Mathis, James L, psychiatry
Mauck, Henry Page, Jr, cardiology
Mayer, David Jonathan, neurophysiology

Maynard, William Rose, Jr, analytical chemistry, biochemistry
McCowen, Sara Moss, microbial physiology
McCue, Carolyn M, pediatrics
McDaniel, Terry Wayne, solid state physics
McKennis, Herbert, Jr, biochemistry
Mefford, David Allen, analytical chemistry
Melson, Gordon Anthony, inorganic chemistry
Melton, Thomas Mason, organic chemistry, industrial hygiene
Merz, Timothy, cytogenetics, radiobiology
Meydrech, Edward F, biostatistics, epidemiology
Meyer, Leo Francis, organic polymer chemistry
Michell, Wilson Doe, economic geology
Miller, Larry Gene, pharmaceutics
Minton, Paul Dixon, statistics
Montroy, Leo Dennis, fresh water ecology
Moon, Peter Clayton, dental materials
Moore, Edward Weldon, gastroenterology, electrochemistry
Morahan, Page Smith, virology, immunology
Morgan, Evan, analytical chemistry
Morris, Joseph Richard, topology
Moses, John Herrick, geology
Mosley, Ronald Bruce, solid state physics
Mullinax, Perry Franklin, rheumatology, immunology
Munson, H Randall, Jr, organic chemistry
Murphey, Robert Stafford, medicinal chemistry
Murphey, Wilbur Alford, research administration, polymer chemistry
Myers, William Howard, inorganic chemistry
Nance, Walter Elmore, human genetics, internal medicine
Neal, Marcus Pinson, Jr, radiology
Neale, Claude Linwood, psychiatry, neurology
Nelson, Elvin Clifford, parasitology
Nemuth, Harold I, preventive medicine
Neuroth, Milton L, pharmacy
Newman, Jack Huff, bacteriology, biochemistry
Newman, Richard Holt, radiochemistry
Norvell, John Edmondson, III, neuroanatomy
O'Donohue, Cynthia H, organic chemistry
O'Donohue, Walter John, Jr, pulmonary diseases, internal medicine
Odor, Dorothy Louise, anatomy
O'Neal, Charles Harold, biochemistry
Osborne, J Scott, Jr, physical chemistry
Ottenbrite, Raphael Martin, organic chemistry
Owen, Fletcher Bailey, pharmacology
Owens, Daniel Kenyon, agricultural chemistry
Owens, Kenneth, biochemistry
Owens, Noel Oscar, biology, anatomy
Pakurar, Alice Swope, anatomy
Pang, David Cheloy, muscular physiology
Parcell, Lloyd Jamison, water chemistry
Parchen, Frank Raymond, Jr, physical chemistry
Park, Herbert William, III, rehabilitation medicine
Pastore, Peter Nicholas, otorhinolaryngology
Patterson, John Legerwood, Jr, internal medicine
Payton, Otto D, physical medicine & rehabilitation
Phibbs, Paul Vester, Jr, microbiology
Pittman, Roland Nathan, cardiovascular physiology
Poland, James Leroy, physiology
Popli, Shankar D, pharmaceutics
Porter, Reno Russell, pharmaceutics
Powell, William Allan, analytical chemistry
Price, Donald Dennis, neurophysiology
Proakis, Anthony George, pharmacology
Purchase, Earl Ralph, pollution chemistry
Quattropani, Steven L, anatomy, reproductive biology
Rabung, John Russell, number theory
Rainer, Norman Barry, applied chemistry
Ranck, Ralph Oliver, organic chemistry
Ranniger, Klaus, radiology
Ray, Edward Scott, pulmonary diseases, internal medicine
Raychowdhury, Pratip Nath, applied mathematics, mathematical physics
Reams, Willie Mathews, Jr, developmental biology

Reed, James Robert, Jr, fish biology
Regelson, William, medicine
Reynolds, John Dick, plant embryology
Rice, Nolan Ernest, zoology
Richard, Alfred Joseph, physical chemistry
Richards, Walter L, Jr, zoology
Richardson, David W, cardiology
Ridgway, Ellis Branson, physiology, biophysics
Roberts, Phyllis Silver, biochemistry
Robin, Ronald P, pharmacology
Rogers, Kenneth Scipio, biochemistry
Rosecrans, John A, pharmacology
Rosenblum, William I, neuropathology, pathology
Rosenzweig, Abraham Leon, bacteriology
Rothberg, Simon, biochemistry
Sabet, Sohair Farid, microbiology
Sahli, Brenda Payne, polymer chemistry, pharmaceutical chemistry
Sahli, Muhammad S, organic chemistry, polymer chemistry
Salley, John Jones, oral pathology
Sancilio, Lawrence F, pharmacology
Sansing, Raymond Clayton, statistics, mathematics
Schaeffer, Howard John, medicinal chemistry
Schmeelk, John Frank, applied mathematics
Scott, Robert Bradley, internal medicine, hematology
Sears, Daniel Scott, pesticide chemistry
Segura, Gonzalo, Jr, radiochemistry
Seibel, Hugo Rudolf, anatomy, electron microscopy
Seligman, Robert Bernard, organic chemistry
Shadomy, Helen Jean, mycology, microbiology
Shadomy, Smith, medical microbiology
Shelton, Keith Ray, biochemistry
Shiles, Eugene Joseph, solid state physics, magnetism
Shillady, Donald Douglas, theoretical chemistry
Shull, Don Louis, physical chemistry
Silberman, Henry K, psychiatry
Skarlos, Leonidas, chemistry, mathematics
Sloope, Billy Warren, physics
Smart, Robert Forte, botany
Smith, Harold Linwood, physical pharmacy, biopharmaceutics
Smith, James Doyle, organic chemistry
Smith, Joseph A, agricultural chemistry, chemical engineering
Smith, Leroy, plastic surgery
Smith, Maurice John Vernon, urology
Snodgrass, Michael Jens, microscopic anatomy
Spencer, Frederick J, preventive medicine, public health
Stein, Barry Edward, neurophysiology, developmental physiology
Steinfeld, Jesse Leonard, cancer, medicine
Stepka, William, plant physiology, plant biochemistry
Stickley, Elmer Eugene, radiology, physics
Still, William James Sangster, pathology
Strickland, John Claiborne, biology
Stubbins, James Fiske, medicinal chemistry
Stump, Billy Lee, physical chemistry, organic chemistry
Swell, Leon, biochemistry
Swerlick, Isadore, organic chemistry
Szumski, Alfred John, physiology, neurophysiology
Taggart, George Bruce, theoretical solid state physics
Talley, Claude Parks, physical chemistry
Tang, Terry Chu, microbiology
Tankersley, Robert Walker, Jr, virology, bacteriology
Taylor, Jackson Johnson, physics
Teng, Lina Chen, drug metabolism
Tenney, Wilton R, plant pathology
Tew, John Garn, immunology, microbiology
Thedford, William Andrew, topology, algebra
Thompson, William Taliaferro, Jr, medicine
Tiedemann, Albert William, Jr, analytical chemistry, science administration
Tiller, Calvin Omah, physics
Toney, Marcellus E, Jr, clinical microbiology
Toone, Elam Cooksey, Jr, medicine
Topham, Richard Walton, biochemistry
Towle, David Walter, molecular biology, medical ethics
Townsend, Joel Ives, genetics
Trout, William Edgar, Jr, chemistry

Turnbull, Colin MacMillan, social anthropology
Turnbull, Lemox Birckhead, organic chemistry
Tyson, Bruce Carroll, Jr., analytical chemistry
Van't Riet, Bartholomeus, analytical chemistry
Vennart, George Piercy, pathology
Vilcins, Gunars, analytical chemistry, spectroscopy
Vlahcevic, Zdravko Reno, gastroenterology
Wakeham, Helmut, physical chemistry
Waller, Marion Van Nostrand, serology, immunology
Walton, Philip Wilson, medical physics
Ward, John Wesley, pharmacology
Waroblak, Michael Theodore, organic chemistry
Wasserman, Albert J, internal medicine
Watlington, Charles Oscar, medicine, physiology
Watson, Duane Craig, analytical chemistry
Watts, Daniel Thomas, pharmacology
Weaver, Warren Eldred, chemistry
Welshimer, Herbert Jefferson, bacteriology
Welstead, William John, Jr, organic chemistry
West, Warwick Reed, Jr, invertebrate zoology
Wheeler, Charles Horatio, III, mathematics
Whidby, Jerry Frank, analytical chemistry, physical chemistry
Wickham, James Edgar, Jr, analytical chemistry, inorganic chemistry
Wiebusch, F B, periodontology
Will, Fritz, III, analytical chemistry
Wilson, John Drennan, physiology, radiobiology
Winters, Lawrence Joseph, organic chemistry
Wist, Abund Ottokar, theoretical physics, thermodynamics
Witorsch, Raphael Jay, physiology
Wittemann, Joseph Klaus, dentistry, psychology
Wolf, James Stuart, medicine, surgery
Wood, James Alan, mathematical analysis
Wood, John Henry, pharmaceutics
Woods, Lauren Albert, pharmacology
Woolcott, William Starnold, vertebrate zoology
Worsham, James Essex, Jr., physical chemistry, biomedical engineering
Young, Nelson Forsaith, biochemistry
Young, Reuben B, medicine, pediatrics
Zfass, Alvin Martin, gastroenterology

RIXEYVILLE
Simonpietri, Andre Christophe, earth sciences

ROANOKE
Churchill, Helen Mar, biology
Glanville, James Oliver, inorganic chemistry, analytical chemistry
Haley, Harold Bernard, surgery
Maycock, Jerry Ray, organic chemistry
Morlang, Barbara Louise, nutrition

ROSSLYN
Burns, Russell MacBain, silviculture, plant physiology
Connolly, John Irving, Jr, low temperature physics, optics
Johnson, Charles Minor, physics

SALEM
Anthony, Lee Saunders, physics
Bondurant, Charles W, Jr, polymer chemistry, analytical chemistry
Clarke, Gary Anthony, microbial physiology
Finfgeld, Charles R, physics
Fisher, Charles Harold, organic chemistry
Henson, Paul D, organic chemistry
Huddle, Benjamin Paul, Jr, physical chemistry
Julian, Maureen M, physical chemistry
Lautenschlager, Edward Walter, parasitology, helminthology
Lee, Philip Calvin, mycology
Petersen-Adkisson, Karen, cytogenetics
Plymale, Donald Lee, inorganic chemistry
Thompson, Jesse Clay, Jr, systematics
Walpole, Ronald Edgar, mathematics, statistics

VIRGINIA

SALUDA
Brand, Eugene Dew., psychiatry, psychopharmacology

SEAFORD
Bell, Vernon Lee, Jr., polymer chemistry, organic chemistry

SPRINGFIELD
Berendt, Raymond Donald, acoustics
Drummond, Kenneth Herbert, oceanography
Heilberg, Ernest, applied mathematics
Johnson, Augustus Clark, mathematics
Macek, Andrej, physical chemistry
McMillan, Louis Kelly, Jr., operations research, mathematics
Mulvaney, Thomas Richard, food science
Plutchok, Robert, physics
Sebastian, Richard Lee, acoustics, applied mathematics
Szakal, Andras Kalman, immunobiology, cancer
Thompson, Albert Johnson, Jr., chemistry
Verna, John E., biology
Walter, Charles Robert, Jr., organic chemistry

STANARDSVILLE
Silvette, Herbert, pharmacology

STAUNTON
Chinn, Austin Brockenbrough, medicine
Mehner, John Frederick, ornithology
Patrick, James Burns, organic chemistry
Sawyer, Jane Orrock, mathematics
Spetzman, Lloyd Anthony, botany

STERLING
Augl, Joseph Michael, organic chemistry
Hodge, Mary Wilma, physics
Hoehne, Walter Elmer, meteorology
Martin, James Oran, Jr., operations research, systems analysis
Smith, John Cole, entomology

STRASBURG
Jenkins, Marie Magdalen, invertebrate zoology

SUFFOLK
Alexander, Morris Wilburn, agronomy
Garren, Kenneth Howard, plant pathology
Hallock, Daniel Leroy, agronomy
Porter, Daniel Morris, phytopathology
Reid, Preston Harding, soil chemistry, soil fertility

SWEET BRIAR
Belcher, Jane Colburn, zoology, evolution
Blair, Barbara Ann, biochemistry, plant pathology
Edwards, Ernest Preston, ornithology
Elkins, Judith Molinar, applied mathematics
Lenz, George H., nuclear physics
Long, Thomas Carlyle, genetics
Markle, H Chester, Jr., physical chemistry
McClenon, John R., organic chemistry
Simpson, Margaret, invertebrate zoology
Sprague, Elizabeth F., botany
Wood, Landley Harriss, environmental physiology

TRIANGLE
Thompson, Ronald Halsey, instrumentation, materials science

UNIVERSITY OF RICHMOND
Glassick, Charles Eizweiler, organic chemistry
Seaborn, James Byrd, theoretical nuclear physics

UPPERVILLE
Pollack, Herbert, medicine

VIENNA
Balint, Francis Joseph, mathematics
Ben, Max, pharmacology, physiology
Braddock, Joseph, physics
Burger, Robert Thornton, computer science
Coate, William Bleecker, inhalation toxicology
Friedlander, Zitta Zipora, quantum optics, experimental atomic physics
Gal, Andrew Eugene, organic chemistry
Gargus, James L., toxicology
Good, Robert Campbell, medical microbiology
Gross, Stanley Burton, toxicology, environmental sciences
Herd, George Ronald, mathematical statistics
Herr, Robert Roy, plant pathology
Hodge, Donald Ray, applied physics, statistics
Judkins, Roddie Reagan, chemistry
Kimura, Kazuo Kay, pharmacology, pediatrics
Ramsey, Lessel Leslie, analytical chemistry, food science
Reisler, Donald Laurence, information science, physics
Reno, Frederick Edmund, toxicology, teratology
Rutter, Henry Alouis, Jr., toxicology, biochemistry
Spear, Philip James, entomology
Stanovick, Richard Paul, soil microbiology, biochemistry
Sullivan, Jeremiah B., pharmaceutical chemistry, organic chemistry
Valerio, David Allen, laboratory animal medicine, research administration
Voelker, Richard William, veterinary pathology
Wise, John P., fisheries biology, marine biology

VIRGINIA BEACH
Cox, James Lester, Jr., plasma physics
Jones, Eleanor Green, mathematics
Millhiser, Frederick Roy, industrial chemistry, polymer chemistry
Munyan, Arthur Claude, geology, engineering geology
Ofelt, George Sterling, spectroscopy

WACHAPREAGUE
Kraeuter, John Norman, biological oceanography, marine ecology

WALLOPS ISLAND
Shardanand, atmospheric physics, spectroscopy

WAYNESBORO
Bleasdale, James Lewis, organic chemistry
Bohnfalk, Erwin Frederick, Jr., physical chemistry
Custer, Robert Louis, science education
Dean, David Lee, physical organic chemistry
Dullaghan, Matthew Edward, organic chemistry, analytical chemistry
Euler, Robert Donald, polymer chemistry, textile chemistry
Everett, John Edward, organic chemistry
Fisher, Calvin L., organic chemistry, textiles
Hahn, Walter Leopold, polymer chemistry
Heuberger, Oscar, organic chemistry
Hewett, James Veith, organic chemistry
Hoffman, Henry Allen, Jr., organic chemistry
Jensen, Arnold William, organic chemistry, polymer chemistry
Ketterer, Charles Clifford, organic chemistry
Lodoen, Gary Arthur, polymer chemistry
Milford, George Noel, polymer chemistry
Moyer, Calvin Lyle, organic chemistry
Nix, Sydney Johnston, Jr., organic chemistry
O'Brien, Michael Harvey, organic chemistry
Stiehl, Roy Thomas, Jr., polymer chemistry
Tichenor, Robert Lauren, physical chemistry
Weeks, Gregory Paul, textile chemistry, instrumentation
Wilkinson, William Kenneth, organic chemistry, polymer chemistry

WEEMS
Adkins, John Nathaniel, geophysics

WILLIAMSBURG
Andersen, Carl Marius, theoretical physics
Armstrong, Alfred Ringgold, analytical chemistry
Bick, Kenneth F., geology
Black, Robert Earl Lee, embryology
Brooks, Garnett Ryland, Jr., ecology
Bynum, William Lee, mathematics
Byrd, Mitchell Agee, zoology
Cato, Benjamin Ralph, Jr., mathematics
Coursen, Bradner Wood, physics
Crawford, George William, plasma physics
Crownfield, Frederic Rudolph, Jr., plasma physics
Delos, John Bernard, molecular physics, theoretical physics
Doverspike, Lynn D., atomic & molecular physics
Drew, John H., mathematics
Dupuy, John L., marine science
Durgan, Elford Sturtevant, chemistry
Eichause, Morton, high energy physics
Fashing, Norman James, acarology
Gary, Stephen Peter, plasma physics
Gibbs, Norman Edgar, computer science
Goodwin, Bruce K., geology
Grant, Bruce S., genetics
Griffith, Melvin Eugene, entomology, public health
Gross, Franz Lucretius, theoretical physics
Hall, Gustav Wesley, botany
Hammer, Gary G., physical organic chemistry
Hill, Trevor Bruce, organic chemistry
Hoblit, Louis Douglas, organic chemistry
Hurley, Rupert B., physical chemistry, organic chemistry
Jahn, Alex Karl, organic & polymer chemistry
Kane, John Robert, nuclear physics
Kiefer, Richard Lee, nuclear chemistry
Kossler, William Johns, physics
Low, Emmet Francis, Jr., applied mathematics
Mangum, Charlotte P., invertebrate zoology, comparative physiology
Mathes, Martin Charles, plant physiology
McKnight, John Lacy, theoretical physics, history of science
Moriarty, John Alan, theoretical solid state physics
Nichols, Maynard M., marine geology, oceanography
O'Neil, Peter Vincent, mathematics
Orwoll, Robert Arvid, physical chemistry
Perdrisat, Charles F., nuclear physics, high energy physics
Poole, William George, Jr., numerical analysis
Reinhart, Theodore Russell, archaeology, anthropology
Remler, Edward A., physics
Reynolds, Thomas Lee, mathematics
Schiavelli, Melvyn David, physical organic chemistry
Schone, Harlan Eugene, physics
Scott, Joseph Lee, botany, cytology
Sher, Arden, physics
Soest, Jon Fredrick, physics
Stockmeyer, Paul Kelly, mathematics
Terman, Charles Richard, animal ecology, animal behavior
Thompson, David Wallace, inorganic chemistry, organometallic chemistry
Ticknor, Leland Bruce, physical chemistry
Tyree, Sheppard Young, Jr., inorganic chemistry
Vahala, George Martin, magnetohydrodynamics, plasma physics
Vermeulen, Carl William, microbiology, biochemistry
Von Baeyer, Hans Christian, theoretical physics
Ware, Donna Marie Eggers, plant taxonomy
Ware, Stewart Alexander, plant ecology
Welsh, William Robert, physics
Winter, Rolf Gerhard, nuclear physics
Wiseman, Lawrence Linden, developmental biology

WINCHESTER
Bender, Howard Leonard, plastics
Hickey, Kenneth Dyer, plant pathology
Hill, Clarence Howell, entomology

WINDMILL POINT
Atwood, Wallace Walter, Jr., geomorphology

WISE
Culbertson, George Edward, mathematics
Hooper, William John, Jr., physics, astronomy

WOODBRIDGE
Connors, Philip Irving, academic administration, experimental nuclear physics

WOODSTOCK
Cone, Clarence Donald, Jr., cell biology, oncology
Rothstein, Lewis Robert, chemistry

YORKTOWN
Viohl, Paul, organic chemistry

WASHINGTON

ABERDEEN
Schermer, Eugene DeWayne, environmental chemistry

AMERICAN LAKE
Eliel, Leonard Paul, endocrinology

ANACORTES
Barnes, Ross Owen, earth sciences
Booth, Ernest Sheldon, biology
Stevens, Vernon Lewis, biochemistry, analytical chemistry

BELLEVUE
Bennett, Carl Allen, statistics
Center, Robert E., lasers, chemical physics
Cotton, John Edward, physical chemistry
Dye, David L., physics
Goldkamp, Arthur Harvey, chemistry
Jarvi, Reino A., polymer chemistry
Kratzke, Albert William, applied mathematics
Levy, Richard H., lasers, engineering
Lowenthal, Dennis David, plasma physics, elementary particle physics
Miller, Lewis Samuel, organic polymer chemistry
Sencar, Allen Eugene, analytical chemistry
Vessel, Eugene David, organic polymer chemistry

BELLINGHAM
Abel, William Robert, mathematics
Albers, James Ray, theoretical physics
Anastasio, Angelo, anthropology
Atneosen, Richard Allen, physics
Barrett, William Louis, solid state physics, nuclear engineering
Beck, Myrl Emil, Jr., geophysics
Bender, William, chemistry
Besserman, Marion, physical chemistry
Briggs, William Scott, organic chemistry
Broad, Alfred Carter, invertebrate zoology
Brown, Herbert Allen, vertebrate zoology
Brown, Willard Andrew, science education
Chaney, Robin W., numerical analysis, operations research
Christman, Robert Adam, geology
Clark, James d'Argaville, pulp & paper technology
Craswell, Keith J., mathematics
Critchfield, Howard John, geography, climatology
Crook, Joseph Raymond, inorganic chemistry
Davidson, Melvin G., nuclear physics
Drum, Ryan William, biology
Dube, Maurice Andrew, botany
Easterbrook, Don J., geology
Eddy, Lowell Perry, inorganic chemistry
Ellis, Ross Courtland, geology
Erickson, John (Elmer), cytogenetics
Felicetta, Vincent Frank, organic chemistry
Fonda, Richard Weston, plant ecology
Gerhold, George A., physical chemistry
Grabert, Garland Frederick, anthropology
Harris, Howard Leroy, cultural anthropology
Hashisaki, Joseph, mathematics
Hildebrand, James Leslie, mathematics
King, Donald M., analytical chemistry
King, Ellis Gray, organic chemistry

Knapman, Fred William, organic chemistry
Kraft, Gerald F., entomology
Kriz, George Stanley, Jr., physical organic chemistry
Lampman, Gary Marshall, organic chemistry
Lindsay, Richard H., nuclear physics
Ludwig, Charles Heberle, wood chemistry
Mason, David Thomas, limnology
Monahan, Robert Leonard, geography
Mookherjee, Debnath, urban geography, geography of South Asia
Neal, John Alexander, inorganic chemistry
Neuzil, Edward F., nuclear chemistry
Newman, James Raney, zoology, ecology
Pavia, Donald Lee, organic chemistry, photochemistry
Quigley, Robert James, solid state physics
Rahm, David Allan, geology
Read, Thomas Thornton, mathematical analysis
Reay, John R., mathematics
Riffey, Meribeth M., biology
Ross, Charles Alexander, geology, paleobiology
Ross, June Rosa Pitt, biological oceanography
Rupaal, Ajit S., nuclear physics
Russo, Salvatore Franklin, physical biochemistry
Rygg, Paul Theodore, mathematics
Schneider, David Edwin, physiological ecology, comparative physiology
Schwartz, Maurice Leo, geological oceanography, science education
Scott, James William, historical geography, cultural geography
Senger, Clyde Merle, zoology
Slesnick, Irwin Leonard, science education
Spanel, Leslie Edward, solid state physics
Summers, William Clarke, marine ecology
Swineford, Ada, geology
Taylor, Herbert Cecil, anthropology
Taylor, Ronald, botany, genetics
Tweddell, Colin Ellidge, cultural anthropology, anthropological linguistics
Veit, Jiri Joseph, atomic physics, nuclear physics
Webber, Herbert H., biology
Weyh, John Arthur, analytical chemistry, inorganic chemistry
Whitmer, John Charles, physical chemistry
Wicholas, Mark L., inorganic chemistry
Woll, John William, Jr., mathematics
Yu, Ming-Ho, environmental health, nutrition

BREMERTON
Klett, Hubert Clifford, mycology

BRINNON
Westley, Ronald E., marine biology, fisheries

BUCKLEY
Ruvalcaba, Rogelio H A, pediatric endocrinology

CAMAS
Aspitarte, Thomas (Robert), microbiology
Bennett, Clifton Francis, wood chemistry
Carlson, Lewis John, paper chemistry
Cormack, James Frederick, environmental chemistry
Goheen, David Wade, organic chemistry
Hanby, John Estes, Jr., paper chemistry
Lackey, Homer Baird, applied chemistry
Murdock, Gordon Alfred, physical chemistry
Shilling, Wilbur Leo, organic chemistry, polymer chemistry
Strand, Robert Fenton, forest ecology, forest soils
Vincent, Gerald Glenn, research administration, polymer chemistry
Wither, Ross Plummer, organic chemistry, pulp & paper technology

CARBONADO
Ekvall, Robert Brainerd, anthropology, ethnography

CENTRALIA
Bower, David Roy Eugene, forest biometry
Boyd, Charles Curtis, forestry, agronomy
Daniels, Jess Donald, forest genetics, plant breeding
Drew, T John, forest management, forest genetics
Harper, Warren Charles, forest hydrology
Heninger, Ronald Lee, forest management
Lawrence, William Hobart, forestry, wildlife management
Lewis, David Kent, forest management, forest economics
Long, Alan Jack, forest ecology, forest genetics
Rediske, John Henry, plant physiology
Scott, William, forest soils, soil science
Steinbrenner, Eugene Clarence, forestry, soils
Stonecypher, Roy W, forestry, genetics
Tanaka, Yasuomi, forest ecology
Webster, Stephen Russell, forest soils
Winjum, Jack Keith, forest ecology
Woodman, James Nelson, silviculture, forest physiology

CHEHALIS
Locke, Seth Barton, plant pathology

CHENEY
Anderson, Jeremy, cultural geography, environmental psychology
Bacon, Marion, medical microbiology
Behm, Roy, analytical chemistry
Dalla, Ronald Harold, algebra
Douglas, John Edward, physical chemistry
Forsman, Earl N, atomic physics
Gilmour, Ernest Henry, paleontology
Hall, Wayne Hawkins, mathematics
Hanegan, James L, comparative physiology
Harter, Dana Eugene, chemistry
Horner, Donald Ray, mathematical analysis
Kiver, Eugene P, geomorphology
Lang, Bruce Z, parasitology, immunology
Long, Daniel R, physics
Marshall, Philip Richard, physical chemistry
Mumma, Martin Dale, geology
Ritter, Preston Otto, biochemistry
Schadegg, Francis John, geography
Shin, Suk-Han, geography
Simms, Horace Ridgly, mycology
Sims, Benjamin Turner, mathematics
Snook, James Ronald, geology
Soltero, Raymond Arthur, limnology, biology
Steele, William Kenneth, geology, geophysics
Stevens, Vincent Leroy, biochemistry
Sullivan, Hugh D, mathematics
Swedberg, Kenneth C, plant ecology
Vigfusson, Norman V, genetics, medical genetics
Whelton, Bartlett David, medicinal chemistry
White, Ronald Jerome, zoology, physiology

COLLEGE PLACE
Barnett, Claude C, theoretical physics
Chambers, James Richard, organic chemistry
Chinn, Clarence Edward, chemistry
Clayton, Dale Leonard, animal behavior
Fisk, Lanny Herbert, paleobiology, palynology
Galusha, Joseph G, Jr., animal behavior, ethology
Grable, Albert E, systematic botany, insect ecology
Jones, Carl Trainer, chemistry
McCloskey, Lawrence Richard, zoology
Perry, Alfred Eugene, zoology, ecology
Rigby, Donald W, parasitology

COOK
Zaugg, Waldo S, biochemistry

EDMONDS
Allison, Lowell Edward, soil chemistry

ELLENSBURG
Bennett, Robert Bowen, physics

Brooks, James Eugene, economic geography
Brown, Robert Harrison, vertebrate zoology
Clark, Glen W, parasitology, protozoology
Comstock, Dale Robert, mathematics
Cutlip, William Frederick, mathematics, computer science
Dean, Robert Yost, mathematics
Denman, Clayton Charlton, anthropology
Dumas, Philip Conrad, vertebrate zoology
Duncan, Leonard Clinton, inorganic chemistry
Gaines, Robert D., biochemistry
Hammond, Kenneth Allen, resource geography, resource administration
Harrington, Edward James, vertebrate zoology
Higgins, Ralph Edward, petrology
Johnson, Wilbur Vance, chemical physics
Jones, Jerry Lynn, analytical chemistry, research administration
Kaatz, Martin Richard, physical geography
Kramar, Jeno Louis, pediatrics
Lister, Frederick Monie, mathematics
Lowe, Janet Marie, microbiology, embryology
Lygre, David Gerald, biochemistry
Macinko, George, geography, environmental sciences
Martin, Bernard Loyal, mathematics
Meany, John Eagleton, chemistry
Mitchell, Robert Curtis, astrophysics
Murphy, Elias Smith, Jr., nuclear physics, astronomy
Newschwander, Wilfrid Williams, physical chemistry
Pacha, Robert Edward, microbiology
Ressler, John Quenton, geography
Shrader, John Stanley, science education
Smith, Stamford Dennis, entomology
Smith, William Charles, anthropology, archaeology
Sperry, Willard Charles, nuclear physics
Thelen, Thomas Harvey, genetics

EVERETT
Bhat, Venkatramana Kakekochi, pharmaceutical chemistry, medicinal chemistry

FOX ISLAND
Ray, Dixy Lee, zoology, marine biology

FPO SEATTLE
Nakayama, Takao, phytopathology, horticulture

FRIDAY HARBOR
Burgess, David Ray, developmental biology
Norris, Richard Earl, marine biology
Strathmann, Richard Ray, marine biology, zoology
Willows, Arthur Owen Dennis, neurophysiology

ISSAQUAH
Cole, Dale Warren, forest soils

KENNEWICK
Barton, Gerald Blackett, applied chemistry
Eberhardt, Lester Lee, biology
Leonard, Bowen Raydo, Jr., physics
Rohrmann, George Frederick, microbiology

KENT
Stegen, Gilbert Rolland, fluid mechanics

KIRKLAND
Bupp, Lamar Paul, physical chemistry

LONG BEACH
Clarke, Joy Harold, horticulture

LONGVIEW
Bisson, Peter Andre, aquatic biology
Bothwell, Max Lewis, limnology
Comstock, Gilbert Leroy, forest products
Frankland, Albert Ernest, organic chemistry
Glading, Ralph Edmond, paper chemistry

Marion, Giles Michael, forest soils, soil chemistry
Urling, Gerard Phelps, wood science & technology

MANSON
Ewing, Richard Everett, physical chemistry

MARYSVILLE
Manus, Louis John, dairy science

MERCER ISLAND
Bridgforth, Robert Moore, Jr., physical chemistry
Halpern, David, physical oceanography
Price, James Ferris, mathematics
Riehl, Jerry A., physics, chemistry

MOSES LAKE
Quenelle, Alan Courtland, Jr., biochemistry

MT VERNON
Dursch, Harry Robert, biochemistry, oceanography
Ford, James, biology, zoology
Frank, Andrew Julian, community health
Haglund, William Arthur, plant pathology
Norton, Robert Alan, horticulture
Peabody, Dwight Van Dorn, Jr., weed science

NAHCOTTA
Tartar, Vance, experimental morphology

OAK HARBOR
Mason, Beryl Troxell, neurology, psychiatry
Mason, Herman Charles, public health, immunology

OCEAN PARK
Burr, Irving Wingate, mathematics

OLYMPIA
Anderson, Lee Roy, applied physics, science education
Beug, Michael William, organic chemistry, environmental chemistry
Cellarius, Richard Andrew, botany, biophysics
DeBell, Dean Shaffer, forestry
Desmond, Gerald Raymond, anthropology, ethnology
Dimitroff, George Ernest, algebra
Eickstaedt, Lawrence Lee, ecology, marine biology
Ernsdorff, Bede (Paul), chemistry
Guttman, Burton Samuel, biology
Heebner, Charles Frederick, plant physiology, forestry
Herman, Francis Robert, forest mensuration
Humphrey, Donald Glen, cytogenetics
Kahan, Linda Beryl, neurophysiology
Kelly, Jeffrey John, biochemistry, physiology
Knapp, Robert Hazard, Jr., theoretical physics
Kormondy, Edward John, ecology
Kutter, Elizabeth Martin, molecular biology
Lane, Wallace, preventive medicine
LeMier, Emanuel H, fisheries
Milham, Samuel, Jr., human genetics, epidemiology
Milne, David Hall, entomology
Newman, Paul Harold, information science
Radwan, Mohamed Ahmed, plant physiology
Rau, Weldon Willis, micropaleontology
Reischman, Placidus George, zoology
Sluss, Robert Reginald, entomology, ecology
Taylor, Peter Berkley, marine ecology
Wendler, Henry O, fish biology, fisheries management
Woelke, Charles Edward, fisheries, marine ecology
Youtz, Byron Leroy, nuclear physics

PACIFIC BEACH
Adams, Mark F, analytical chemistry, industrial chemistry

PORT ANGELES

Douglass, John Richmond, organic chemistry; natural history
Grinols, Richard Byron, marine biology
Hart, Edwin James, radiation chemistry

PORT ORCHARD

Doyle, Worthie Lefler, Jr., mathematics

PORT TOWNSEND

James, Harold Lloyd, geology
Ray, Verne Frederick, anthropology

POULSBO

Heller, Carl George, endocrinology, physiology

PROSSER

Aichele, Murit Dean, plant pathology
Allison, Joseph Lewis, plant pathology
Clore, Walter Joseph, horticulture
Cones, Richard Durward, weed science
Cone, Wyatt Wayne, entomology
Dawson, Jean Howard, weed science
Drake, Stephen Ralph, food science
Easton, Gene Douglas, horticulture
Evans, David W., agronomy, plant physiology
Fridlund, Paul Russell, plant pathology
Heinemann, Wilton Walter, animal nutrition
Hoyman, William Greig, plant pathology
Kraft, John M., plant pathology
Martin, Mark Wayne, genetics
Mink, Gaylord Ira, plant virology
Ogg, Alex Grant, Jr., plant physiology, agronomy
Proebsting, Edward Louis, Jr., horticulture
Santo, Gerald Sunao, plant nematology
Silbernagel, Matt Joseph, plant pathology; plant breeding
Skotland, Calvin B., phytopathology
Toyama, Thomas Kazuo, horticulture
Zimmermann, Charles Edward, plant physiology

PULLMAN

Ackerman, Robert Edwin, anthropology; ethnography
Ackley, William Benton, horticulture
Adams, Donald F., analytical chemistry
Adkins, Ronald James, neurophysiology
Akre, Roger David, entomology
Allan, Robert Emerson, agronomy
Anderson, Daniel Craig, animal science
Ashworth, Ural Stephen, food sciences
Batey, Harry Halsted, Jr., inorganic chemistry
Becker, Walter Alvin, genetics
Bender, Paul A., solid state physics
Benson, Robert Leland, entomology
Berryman, Alan Andrew, entomology
Bertramson, Bertram Rodney, agronomy
Bezdicek, David Fred, soil microbiology
Bhatia, Vishnu Narain, pharmacy
Biddulph, Orlin, botany
Bienz, Darrel Rudolph, plant breeding
Bogyo, Thomas P., genetics, statistics
Borchard, Ronald Eugene, veterinary pharmacology
Branen, Alfred Larry, food science
Brewer, Howard Eugene, plant physiology
Brobst, Duane Franklin, veterinary medicine
Brooks, Stanley Nelson, agronomy; plant breeding
Brosemer, Ronald Webster, biochemistry
Bruce, Richard W., forest economics
Bruehl, George William, plant pathology
Bush, Kenneth Arthur, mathematics
Bushaw, Donald (Wayne), mathematics
Bustad, Leo Kenneth, radiobiology
Butler, Alfred Larry, physics
Campbell, Gaylon Sanford, agricultural meteorology; soil physics
Campbell, Malcolm John, space physics
Canode, Chester Lang, agronomy; plant breeding
Carlson, James Roy, biochemistry
Cheng, Hwei-Hsien, soil chemistry; biochemistry
Cho, Byung-Ryul, veterinary medicine, microbiology
Clement, Paul Arnold, mathematics
Cohen, Arthur Leroy, electron microscopy; research administration
Cook, Robert James, plant pathology

Cooke, Manning Patrick, Jr., organic chemistry
Coon, Craig Nelson, nutrition, metabolism
Crawford, Timothy B., veterinary pathology
Cridland, Arthur A., paleobotany; bryology
Crosby, Glenn Arthur, physical chemistry
Croteau, Rodney, biochemistry
Cunnea, William M., mathematics
Daugherty, Rexford Deo, anthropology
Davis, William C., microbiology, immunology
Dickson, John Otis, pharmacology
Dickson, William Morris, veterinary physiology
Dietz, Sherl M., plant pathology
Dingle, Richard William, silviculture
Dodgen, Harold Warren, chemical physics
Donaldson, Edward Enslow, physics
Drake, Charles Hadley, bacteriology
Dresser, Miles Joel, surface physics
Dunker, Alan Keith, biophysics
Dunlap, Jack Sherwin, veterinary parasitology
Duran, Ruben, plant pathology
Duvall, George Evered, physics
Dyer, Irwin Allen, animal science
Eastlick, Herbert Leonard, zoology
Edmonds, Harvey Lee, Jr., neuropharmacology
Ehlers, Melvin H., animal physiology
Elliott, Lloyd Floren, microbiology
Engibous, James Charles, soil
Estergreen, Victor Line, animal physiology; endocrinology
Farrell, Roy Keith, veterinary virology
Filby, Royston Herbert, nuclear chemistry; geochemistry
Fluharty, Dean Milton, veterinary public health, veterinary bacteriology
Foster, Robert Joe, biological chemistry
Fowles, George Richard, high pressure physics
Froseth, John Allen, animal nutrition
Funk, William Henry, limnology; sanitary biology
Galpin, Donald R., medicinal chemistry, pharmacy
Gardner, Walter Hale, soil physics
Garland, John Kenneth, chemistry
George, Donald Wayne, plant breeding
Gibson, Melvin Roy, pharmacognosy
Gilkeson, Raymond Allen, soil science
Goebel, Carl Jerome, environmental sciences, range management
Golinick, Philip D., exercise physiology
Gorham, John Richard, veterinary medicine
Graham, Shirl Orby, phytopathology
Green, Norman Edward, plant pathology
Griswold, Michael David, biochemistry
Hadwiger, Lee A., plant pathology; biochemistry
Hall, Elizabeth Rose, bacteriology
Hamm, Randall Earl, chemistry
Hard, Margaret McGregor, food science, nutrition
Harris, Grant Anderson, range ecology
Harwood, Robert Frederick, entomology
Hausenbuiller, Robert Lee, soils, chemistry
Hecht, Adolph, botany
Hendrix, John Walter, plant pathology
Henson, James Bond, veterinary pathology
Higinbotham, Noe, botany
Hillers, Joe Karl, genetics, animal science
Hindman, Joseph Lee, plant morphology; developmental anatomy
Hinman, George Wheeler, physics
Hosick, Howard Lawrence, cell biology; cancer
Hostetler, Roy Ivan, veterinary medicine
Hunt, John Philip, physical chemistry, inorganic chemistry
Irtani, W. M., horticulture, plant physiology
Jacobson, Marion, nutrition, foods
James, Maurice Theodore, insect taxonomy
Janos, Ludvik, mathematics
Johansen, Carl August, entomology
Johnson, Richard Evan, ornithology
Johnson, Richard James, animal nutrition, biochemistry
Johnson, Roy Andrew, mathematics
Johnson, William Everett, pharmacology
Johnstone, Donald Lee, bacteriology
Jonas, Robert James, animal ecology; wildlife management

Jordan, James Henry, mathematics
Kalin, Elwood Walter, horticulture
Kallaher, Michael Joseph, mathematics
Kenzy, Sam George, veterinary medicine
King, James Roger, physiological ecology
Kirscher, Leonard Burton, physiology
Kittrick, James Allen, soil mineralogy
Klavano, Paul Arthur, pharmacology
Kleinhofs, Andris, genetics, biochemistry
Klostermeyer, Edward Charles, entomology
Knowles, Harold B., physics
Koch, Alan R., physiology
Koehler, Fred Eugene, soil fertility
Koger, Lavon M., veterinary medicine
Konzak, Calvin Francis, genetics, plant breeding
Kosin, Igor Leonid, poultry science, developmental biology
Kramer, John William, clinical pathology
Krakauer, Henry, biophysics
Kromann, Rodney P., animal nutrition
Kunkel, Robert, horticulture
Larsen, Fenton E., horticulture
Larsen, John Herbert, Jr., vertebrate zoology
Law, Alvin George, agronomy
Lawrence, John McCune, biochemistry
Lee, Donald Jack, nutritional biochemistry
Legg, John Ivan, bioinorganic chemistry
Lejune, Andre Joseph, agronomy; genetics
Loescher, Wayne Harold, plant physiology
Loewus, Frank A., biochemistry
Long, Calvin Thomas, mathematics
Lowell, Sherman Cabot, physics, theoretical numerical analysis
Luedecke, Lloyd O., dairy bacteriology
Luther, Norman Y., pure mathematics
Lutz, Julie Haynes, astronomy
Madsen, Louis Linden, animal nutrition
Magnuson, James Andrew, biochemistry, biophysics
Maguire, James Dale, agronomy
Maloy, Otis Cleo, Jr., plant pathology
Martin, Arnold R., pharmaceutical chemistry
Martin, Charles Franklin, pharmaceutical chemistry
Matchett, William H., microbial physiology
Matteson, Donald Stephen, organometallic chemistry
McCurdy, Jon Alan, veterinary anatomy
McFadden, Bruce Alden, biochemistry, microbiology
McGinnis, James, nutrition
McMichael, Kirk Dugald, organic chemistry
McNeal, Brian Lester, soil chemistry
McNeil, Charles Winslow, zoology
Mehringer, Peter Joseph, Jr., ecology, palynology
Merriam, Willis Bungay, geography, anthropology
Mickelsen, W Duane, veterinary medicine
Miles, Maurice Howard, solid state physics
Miller, Dwane Gene, agronomy
Mills, Joseph William, economic geology
Milne, Henry Bayard, organic chemistry
Mitchell, Madeline Enid, nutrition
Moll, Torbjorn, microbiology
Moree, Ray, genetics, zoology
Morrison, Kenneth Jess, agronomy
Moseley, William David, Jr., physical chemistry
Mosher, Milton Monroe, forest management
Muehlbauer, Frederick Joseph, plant breeding
Muzik, Thomas J., plant physiology
Nagel, Charles William, food science
Nakata, Jack Raymond, range ecology, wildlife ecology
Nelson, Tyre Alexander, mathematical analysis
Nilan, Robert Arthur, genetics
Noskowiak, Arthur Fredrick, wood technology
Nyman, Carl John, inorganic chemistry
O'Mary, Clayton Cordice, animal science
Ostrom, Theodore Gleason, mathematics
Ott, Richard L., veterinary medicine
Pack, Merrill Raymond, plant physiology
Padgett, George Arnold, pathology; genetics
Pal, Martin L., biochemistry, genetics
Papendick, Robert I., agronomy
Park, James Lemuel, physics
Parker, Richard Alan, limnology; computer science

Patterson, Max E., plant physiology; horticulture
Peterson, Clarence James, Jr., veterinary pathology
Piper, Richard Carl, veterinary pathology
Pool, Karl, analytical chemistry
Poovaiah, Bachettira Whappa, plant physiology
Poshusta, Ronald D., chemical physics
Preston, Rodney LeRoy, animal nutrition, animal physiology
Pubols, Merton Harold, nutritional biochemistry
Ronald, Robert Charles, organic chemistry
Rasmussen, Lowell W., agronomy
Ray, Billy Roger, colloid chemistry
Rayburn, William Reed, botany
Reeves, Jerry John, endocrinology, reproductive physiology
Ringen, Leif Matt, microbiology
Robertson, Jack M., mathematics
Robocker, Willard Charles, agronomy
Roche, Ben F., Jr., range management
Rogers, Jack David, mycology; plant pathology
Sandstrom, Donald Richard, physics
Satterlund, Donald Robert, forestry
Saunders, Sam Cundiff, mathematical statistics
Schafer, John Francis, plant pathology
Scheibe, Joseph E., plant physiology
Schekel, Kurt Anthony, ornamental horticulture; floriculture
Schroeder, Alice Louise, genetics
Schroeder, Carlton Raymond, physical geography; resource geography
Schroeder, Paul Clemens, experimental zoology
Schultz, Vincent, animal ecology
Schwendiman, John Leo, agronomy; soil conservation
Scott, Willard Frank, geology
Shaw, Charles Gardner, phytopathology; mycology
Sheppard, John Clarence, radiochemistry; inorganic chemistry
Shumway, Lewis Kay, plant genetics
Shutler, Mary Elizabeth, anthropology
Smith, Allan Hathorn, anthropology
Smith, Orrin Ernest, horticulture, plant physiology
Sorem, Ronald Keith, geology
Spence, Kemet Dean, microbiology
Spencer, Guy Roger, veterinary science
Spencer, John Valentine, food technology
Spitzer, Kenneth Dale, spectrochemistry
Stacy, Gardner Wesley, organic chemistry
Stevens, Carl Mantle, II, biochemistry
Stokes, Jacob Leo, microbiology
Stronck, David Richard, science education
Subramanian Ravanasundaram Venkatachalam, polymer science
Swan, Dean George, polymer science
Swanson, Barry Grant, food science
Tripard, Gerald Edward, nuclear physics
Tukey, Ronald Bradford, pomology
Turner, William Jr., entomology
Uribe, Ernest Gilbert, plant biochemistry
Walden, William Earl, computer sciences, mathematics
Wallenius, Roger Wyn, animal nutrition
Warner, Robert Lewis, agronomy; plant physiology
Way, James Leong, pharmacology; toxicology
Webb, William Albert, mathematics
Webster, Gary Dean, geology; paleontology
Went, Hans Adriaan, physiology
Wescott, Richard Breslich, veterinary parasitology
White, Allen Ingolf, pharmaceutical chemistry
Willett, Roger, physical chemistry; chemical physics
Williams, George Jackson, III, plant ecology, physiological ecology
Wilson, Robert Burton, experimental pathology
Wilson, Vernon Eldridge, genetics; phytopathology
Wiser, Horace Clare, mathematics
Woodbridge, Cyril Gordon, horticulture
Worthman, Robert Paul, veterinary anatomy
Young, Francis Allan, psychophysiology
Yount, Ralph Granville, biochemistry

PUYALLUP
Allmendinger, Davis Frederick horticulture
Andrews, Daniel Keller, poultry science
Baker, Aaron Sidney, soil fertility
Bay, Ernest C, medical entomology, economic entomology
Bearse, Gordon Everett, poultry science
Berg, Lawrence Raymond, poultry nutrition
Doughty, Charles Carter, horticulture
Gabrielson, Richard Lewis, plant pathology
Getzin, Louis William, entomology
Goss, Roy Leon, agronomy
Gould, Charles Jay, plant pathology
Heilman, Paul E, forest soils, plant nutrition
Johnson, Folke, plant pathology
Murdock, Fenoi R, animal nutrition
Nordstrom, Jon Owen, physiology
Ryan, George Frisbie, horticulture

REDMOND
Corey, Victor Brewer, physics
Havenstein, Gerald B, animal genetics
Hill, Robert William, physiology
McClary, Cecil Fay, genetics
Poole, Donald Ray, physical chemistry
Schmidt, Eckart W, industrial chemistry, fuel science
Towner, Richard Henry, animal genetics
Zander, Donald Victor, avian pathology

RENTON
Lowry, Betty Jean Ragle, organic chemistry
Mullen, Anthony J, electromagnetism

RICHLAND
Alkezweeny, Abdul Jabbar, cloud physics, air pollution
Ames, Lloyd Leroy, Jr, mineralogy
Andringa, Keimpe, lasers
Babad, Harry, organic chemistry, nuclear chemistry
Bair, William J, radiobiology, radiological health
Baird, Quincey Lamar, reactor physics
Ballou, John Edgerton, inhalation toxicology
Ballou, Nathan Elmer, nuclear chemistry
Barney, Gary Scott, physical inorganic chemistry
Bartlett, Rodney Joseph, quantum chemistry
Becker, Clarence Dale, fisheries
Bierlein, Theo Karl, physical chemistry
Braby, Leslie Alan, radiological physics, radiation biophysics
Brodzinski, Ronald Lee, nuclear science, planetary atmospheres
Brouns, Richard John, chemistry
Brown, Randall Emory, geology
Bunch, Wilbur Lyle, physics
Burger, Leland Leonard, physical inorganic chemistry
Burns, Raymond Edward, inorganic chemistry
Campbell, Milton Hugh, analytical chemistry, nuclear engineering
Carter, John Lemuel, Jr, physics
Colvin, Curtis A, nuclear chemistry, analytical chemistry
Craig, Douglas Kenneth, air pollution, radiobiology
Crouthamel, Carl Eugene, inorganic chemistry
Cushing, Colbert Ellis, Jr, fresh water ecology
Daniel, Jack Leland, electron microscopy, materials science
Deju, Raul A, environmental management, hydrology
Diebel, Robert Norman, analytical chemistry, radiochemistry
Doran, Donald George, solid state physics, metal physics
Droppo, James Garnet, Jr, micrometeorology
Drucker, Harvey, microbiology, biochemistry
Emery, Richard Meyer, limnology, radiation ecology
Evans, Thomas Walter, physical chemistry
Franz, James Alan, physical organic chemistry
Free, Michael John physiology
Fuquay, James Jenkins, meteorology, climatology
Gandolfi, Allen Jay, toxicology
Gibbs, Alan Gregory, applied mathematics
Gillis, Murlin Fern, medicine, physics
Glass, William A, radiological physics
Gold, Raymond, experimental nuclear physics, reactor physics
Gordon, Richard Lee, physicis
Gore, Bryan Frank, nuclear science
Gray, Robert H, aquatic ecology
Greager, Oswald Herman, physical chemistry
Guthrie, George Leslie, solid state physics
Hackett, Patricia Lou, animal nutrition, biochemistry
Hadlock, Ronald K, oceanography, meteorology
Hampton, James C, anatomy, cytology
Harling, Otto Karl, physics
Heestand, Glenn Martin, atomic physics
Hildebrand, Bernard Percy, optics, ultrasound
Hoch, Richmond Joel, aeronomy, space science
Hopkins, Horace H, Jr, physical chemistry
Hungate, Frank Porter, radiobiology
Kalkwarf, Donald Riley, biophysical chemistry
Karagianes, Manuel Tom, experimental surgery, biomaterials
Kaye, James Herbert, radiochemistry
Kissinger, Homer Everett, metal physics
Klepper, Elizabeth Lee (Betty), plant physiology
Lagergren, Carl Robert, physics
Lambert, Maurice C, physical chemistry
Larson, Harold Vincent, health physics
Lee, Richard Norman, atmospheric chemistry
Leitz, Fred John, Jr, physical chemistry
Lewis, Milton, physical chemistry
Liemohn, Harold Benjamin, plasma physics, space physics
Lindenmeier, Charles William, theoretical physics, nuclear physics
Lund, John Edward, veterinary pathology
Mahlum, Daniel Dennis, biochemistry
Matheson, Willard Edward, physics
McClanahan, Beatrice J, radiobiology
McCown, John Joseph, analytical chemistry, radiochemistry
McElroy, William Nordell, reactor physics
Miller, John Howard, radiation physics
Moore, Robert Lee, physical chemistry
Morrey, John Rolph, physical chemistry, inorganic chemistry
Neeley, Victor Isaac, solid state physics
Nelson, Iral Clair, health physics
Newton, Carlos E, Jr, radiation physics
Nicholson, Wesley Lathrop, statistics
Nielsen, Julian Moyes, environmental chemistry
Nightingale, Richard Edwin, physical chemistry, nuclear engineering
Olsen, Larry Carrol, solid state science
Orgill, Montie M, atmospheric physics, air pollution
Page, Thomas Lee, limnology
Palmer, Harvey Earl, radiological physics, medical physics
Palmer, Ray Frederick, radiation biology
Parker, Herbert Myers, health physics
Partridge, Jerry Alvin, inorganic chemistry
Perkins, Richard W, chemistry, physics
Phillips, Richard Dean, environmental physiology, radiation biology
Price, Keith Robinson, plant ecology, plant physiology
Ragan, Harvey Albert, hematology, radiobiology
Reardon, William Albert, physics, operations research
Reeder, Paul Lorenz, nuclear chemistry
Reisenauer, Andrew E, hydrology
Renne, David Smith, meteorology
Richardson, Richard Laurel, applied mathematics, electrical engineering
Rickard, William Howard, Jr, botany
Roake, William Earl, physical chemistry
Roesch, William Carl, physics
Routson, Ronald C, soil chemistry & mineralogy
Russell, James Torrance, applied physics, electronics
Ryan, Jack Lewis, inorganic chemistry
Sanders, Charles Leonard, Jr, radiobiology
Schenter, Robert Earl, nuclear physics
Schmid, Loren Clark, energy conversion
Schneider, Mark Joseph, physiological ecology
Schulz, Wallace Wendell, inorganic chemistry
Scott, Frederick Arthur, analytical chemistry
Shen, Peter Ko-Chun, reactor physics
Sikov, Melvin Richard, radiation biology
Silker, Wyatt Burdette, radiochemistry, physical chemistry
Smith, Francis Marion, nuclear chemistry
Smith, Victor Herbert, organic chemistry, medical research
Soldat, Joseph Kenneth, health physics, radiological physics
Stevenson, Chris G, chemistry
Stewart, Kirkland Bruce, applied statistics
Stitch, Malcolm Lane, physics, engineering
Stokes, Robert Allan, astrophysics
Strand, John A, III, pollution biology
Stromatt, Robert Weldon, analytical chemistry
Styris, David Lee, experimental physics
Sullivan, Maurice Francis, pharmacology
Templeton, William Lees, radiation ecology, marine ecology
Thomas, John M, biometrics, ecology
Thomas, Montcalm Tom, physics
Thompson, Roy Charles, Jr, radiation biology, biochemistry
Tingey, Garth Leroy, physical chemistry
Toburen, Larry Howard, atomic physics, molecular physics
Tombropoulos, Elias George, biochemistry, nutrition
Upson, U Layton, analytical chemistry
Van Tuyl, Harold Hutchison, radiochemistry
Vaughan, Burton Eugene, physiology, biophysics
Vetrano, James Bond, environmental management
Vogel, Richard Clark, physical chemistry
Waters, Elmer Dale, heat transfer, fluid mechanics
Watson, Donald Gordon, radiation ecology
Wedlick, Harold Lee, health physics, environmental health
Wehner, Alfred Peter, medicine
Wheelwright, Earl J, nuclear chemistry
Wildung, Raymond Earl, soil science, environmental chemistry
Wiley, William Rodney, microbiology, biochemistry
Wilson, Walter Ervin, radiological physics
Wogman, Ned Allen, nuclear chemistry, physical chemistry
Wolf, Marvin Abraham, micrometeorology
Woodley, Robert Earl, physical chemistry
Woodruff, Rodger King, meteorology, physical oceanography
Yoshikawa, Herbert Hiroshi, solid state physics
Yunker, Wayne Harry, physical inorganic chemistry

SEATTLE
Aagaard, George Nelson, internal medicine, cardiology
Aagaard, Knut, physical oceanography
Abson, Derek, pulp chemistry, paper chemistry
Adelberger, Eric George, experimental nuclear physics
Adman, Elinor Thomson, biological structure
Adman, Raymond Lance, biochemistry
Ahmed, Saiyed I, microbiology, biochemical genetics
Aldrich, Lewis Eugene, Jr, invertebrate zoology, parasitology
Alexander, Edward Russell, epidemiology
Allan, George Graham organic chemistry
Allee, Brian James, fisheries
Altman, Leonard Charles, immunology
Alverson, Dayton L , marine biology
Alvord, Ellsworth Chapman, Jr, neuropathology
Amend, Donald Ford, fish pathology
Amory, David William, anesthesiology, pharmacology
Andersen, Niels Hjorth, organic chemistry, biochemistry
Anderson, Arthur G, Jr, organic chemistry
Anderson, George Cameron, biology
Anderson, James Jay, chemical oceanography
Anderson, Larry Ernest, biochemistry, neurochemistry
Anderson, Roger Harris, low temperature physics
Ansell, Julian Samuel, urology
Arons, Arnold Boris, physics
Arsove, Maynard Goodwin, mathematics
Arya, Satya Pal Singh, meteorology, fluid dynamics
Atkinson, William Allen, forest management
Avann, Sherwin Parker, mathematics
Badgley, Franklin Ilsley, meteorology
Bailey, Alan James, wood chemistry
Baker, Donald James, Jr, physical oceanography
Bakke, John Langum, medicine
Bakken, Aimee Hayes, cell biology, developmental biology
Bakker, Cornelis Bernardus, psychiatry
Baltzo, Ralph M, health physics
Banse, Karl, biological oceanography
Bare, Barry Bruce, forest management, operations research
Baribo, Lester E, agricultural microbiology
Barnes, Clifford Adrian, oceanography
Barnes, Edwin Ellsworth, analytical chemistry
Barnes, Glover William, immunology, bacteriology
Barron, Edward J, clinical biochemistry
Barrueto, Richard Benigno, biochemistry, organic chemistry
Bassingthwaighte, James B, physiology, biophysics
Bean, Michael Arthur, pathology, immunology
Beaumont, Ross Allen, mathematics
Beckwith, John Bruce, pathology
Beder, Oscar Edward, prosthodontics
Beeson, Paul Bruce, medicine
Bendersky, Martin, topology
Benditt, Earl Philip, pathology
Benedict, Robert Glenn, microbiology
Bennett, Lee Cotton, Jr, marine geophysics
Bergman, Abraham, medicine, pediatrics
Bertin, Ernest Peter, physical chemistry, inorganic chemistry
Best, Edgar Allan, marine biology
Bethel, James Samuel, wood technology
Bevan, Donald Edward, fish biology
Beyers, William Bjorn, economic geography, regional economics
Bichsel, Hans, radiation physics
Bierman, Edwin Lawrence, internal medicine, metabolism
Birnbaum, Zygmunt William, mathematics, statistics
Bishop, Charles Joseph, nuclear chemistry, nuclear physics
Bjornerud, Egil Kristoffer, physics
Black, Richard Glynn, medical science, anesthesiology
Blackmon, John R, internal medicine
Blair, Andrew Dryden, Jr, pharmaceutical chemistry
Blair, John Sanborn, physics
Blandau, Richard Julius, experimental embryology
Blaser, Henry Weston, botany
Blumenthal, Robert McCallum, mathematics
Boatman, Edwin S, microbiology, electron microscopy
Bodansky, David, physics
Bodemer, Charles William, embryology, history of medicine
Boersma, P Dee, ecology
Bolender, Charles L, dentistry
Bolender, Robert P, cell biology
Bonham, Kelshaw, radiobiology
Bonica, John Joseph, anesthesiology
Booker, John Ratcliffe, geophysics, hydrodynamics
Borden, Weston Thatcher, organic chemistry
Bornstein, Paul, biochemistry, medicine
Bostrom, Robert Christian, geophysics
Boulware, David G, theoretical physics
Boyce, Ronald Reed, urban geography, historical geography
Boyden, Edward Allen, anatomy, embryology
Brady, Lynn R, pharmacognosy
Brancato, Frank Paul, medical microbiology
Breslow, Norman Edward, medical statistics
Brewer, William Augustus, environmental management, remote sensing
Brockenbrough, Edwin C, surgery
Brown, Arthur Charles, physiology, biophysics
Brown, George Willard, Jr, biochemistry, environmental chemistry
Brown, Lowell Severt, physics
Brown, Robert Alan, atmospheric physics, geophysics
Brownell, Frank Herbert, III, mathematics
Bruce, David Stewart, animal physiology, environmental physiology
Bruce, Robert Arthur, medicine
Bryant, Ben S, wood technology
Burgner, Robert Louis, fish biology
Burnell, James McIndoe, medicine, physiology
Burnett, Thompson Humphrey, experimental high energy physics

Burns, Robert Earle, oceanography
Businger, Joost Alois, meteorology
Butler, John, medicine
Butts, William Cunningham, laboratory medicine
Cady, George Hamilton, chemistry, fluorine chemistry
Cahn, Robert Nathan, theoretical high energy physics
Calvin, William Howard, neurophysiology, biophysics
Campbell, William Howard, pharmacology
Cannon, Glenn Albert, physical oceanography
Cantrell, James R., surgery
Capp, Grayson L., biochemistry
Carlson, Charles Merton, theoretical chemistry
Carpenter, Roy, marine chemistry, geochemistry
Cattolico, Rose Ann, developmental biology
Chakravarti, Diptiman, radiochemistry, food chemistry
Chambers, Velma Catherine, microbiology
Chandler, Kirby, physical anthropology
Chang, Kuei-Sheng, geography
Chapman, Douglas George, biometrics
Chapman, Warren Howe, urology
Chen, Shi-Han, biochemical genetics, medical genetics
Cheney, Eric Swenson, economic geology
Cheney, Frederick Wyman, anesthesiology
Chew, Kenneth Kendall, marine biology
Chilton, William Scott, organic chemistry
Chrisman, Noel Judson, medical anthropology
Christensen, Gerald M. biochemistry, radiobiology
Christensen, Nikolas Ivan, geophysics, mineralogy
Christian, Gary Dale, analytical chemistry
Church, Phil Edwards, climatology
Clark, Kenneth Courtright, aeronomy
Clark, Robert Amos, infectious diseases
Clawson, David Kay, orthopedic surgery
Cleland, Robert E., plant physiology
Cloney, Richard Alan, developmental biology
Clothier, William Delbert, fish & wildlife management, aquatic biology
Coachman, Lawrence Keyes, oceanography
Codispoti, Louis Anthony, chemical oceanography
Collias, Eugene Evans, oceanography
Congleton, James Lee, ecology, fisheries
Conrad, John Terry, physiology
Cook, Paul Pakes, Jr., ecology, evolution
Cook, Victor, physics
Coombs, Howard Abbott, geology
Cooney, Marion Kathleen, microbiology
Corson, Harry Herbert, mathematics
Cowgill, James Joseph, physical chemistry
Coyle, Marie Bridget, medical microbiology, applied mathematics
Cramer, John Gleason, nuclear physics
Creager, Joe Scott, geological oceanography
Criminale, William Oliver, Jr., physical oceanography, applied mathematics
Crittenden, Alden La Rue, chemistry
Crosson, Robert Scott, geophysics, seismology
Cullen, Bruce F., anesthesiology
Curiel, Caspar Robert, mathematics
D'Aoust, Brian Gilbert, biochemistry
Dash, Jay Gregory, surface physics
Dassow, John Albert, food chemistry
Davidson, Ernest Roy, physical chemistry
Davie, Earl W., biochemistry
Davis, Kathryn Bullock, biostatistics
Day, Robert Winsor, epidemiology
Dehnelt, Hans Georg, experimental physics
Deisher, Robert William, medicine, pediatrics
deJong, Rudolph H., anesthesiology
Dekker, David Bliss, mathematics
Del Moral, Roger, ecology
Deranleau, David A., biochemistry
Despain, Lewis Gail, space physics
DeVito, June Logan, neuroanatomy
Devol, Allan Houston, biological oceanography, chemical oceanography
Deyrup-Olsen, Ingrith Johnson, zoology
Dickson, Lawrence John, mathematical analysis, applied mathematics

Dietzman, Burton D., analytical chemistry
Dillard, David Hugh, surgery
Dille, James Madison, pharmacology
Dodge, Harold T., internal medicine, cardiology
Doerman, August Henry, biology
Donahue, Roger Purtee, medical genetics
Donaldson, James Adrian, otology
Donaldson, Lauren Russell, biology
Douglas, Howard Clark, microbiology
Dreisbach, Robert Hastings, environmental health
Driver, Charles Henry, forest pathology, range management
Duxbury, Alyn Crandall, oceanography
Dwyer, Wendell Arthur, operations research
Eastman, Carol Mary, anthropology
Ebbesmeyer, Curtis Charles, oceanography
Eddy, Edward Mitchell, cell biology, developmental biology
Edmondson, W Thomas, limnology
Edmondson, Yvette Hardman, limnology
Edwards, John S. entomology
Eggers, David Frank, Jr., physical chemistry
Ehlers, Francis Edward applied mechanics, aerodynamics
Eichinger, Bruce Edward, polymer chemistry
Eisdorfer, Carl, psychiatry, psychophysiology
Eklund, Melvin Wesley, microbiology, food microbiology
Ellis, Stephen Dean, elementary particle physics
Emanuel, Irvin, epidemiology
English, Thomas Saunders, biological oceanography
Ensinck, John William, endocrinology
Erdmann, Joachim Christian, solid state physics, electrooptics
Erickson, Harvey D., forest products
Evans, Bernard William, petrology
Evans, Charles Albert, medical microbiology
Everett, Newton B., anatomy
Ewart, Terry E., physics
Fahnestock, George Reeder, forestry
Fain, Samuel Clark, Jr., solid state physics
Fairhall, Arthur William, chemistry
Falkow, Stanley, microbiology
Farner, Donald Sankey, physiology, zoology
Farwell, George Wells, physics
Feigl, Eric O., physiology
Feigl, Polly Catherine, biostatistics
Feldman, Henry Robert, underwater acoustics
Fellner, Carl Heinz, psychiatry
Felsenstein, Joseph, population genetics
Fernald, Robert Leslie, embryology
Falkow, Philip Jack, internal medicine, medical genetics
Fischbach, David Bibb, solid state physics
Fischer, Edmond H. biochemistry
Fischer, Lloyd D. Jr., biostatistics
Fleagle, Robert Guthrie, meteorology
Fleming, Richard Howell, oceanography
Fleming, Richard Seaman, ecology, environmental sciences
Fletcher, Thomas Lloyd, organic chemistry
Folland, Gerald Budge, mathematics
Forrey, Arden W., biochemistry
Fox, John Perigo, virology, epidemiology
Foy, Hjordis M. infectious diseases, preventive medicine
Fredin, Reynold A., fisheries
Fritschen, Leo J., meteorology
Frost, Bruce Wesley, biological oceanography
Fuchs, Albert Frederick, biomedical engineering
Fukuhara, Francis M. fisheries
Fuller, Richard Eugene, geology
Fulmer, Charles V. environmental geology; geological engineering
Funk, Edward C, oral surgery, oral pathology
Futterman, Sidney, biochemistry

Gaddum-Rosse, Penelope, biological structure
Gahler, Arnold Robert, analytical chemistry; pollution chemistry
Gale, Charles C. Jr., physiology
Gale, James Lyman, epidemiology; public health
Galindo, Anibal H. neurophysiology
Gallant, Jonathan A. molecular biology
Gara, Robert, forest management
Gardner, Howard Shafer, paper chemistry; chemical engineering
Garrison, Gerald Ray, applied physics
Gartler, Stanley Michael, genetics
Geballe, Ronald, atomic physics
Gehrig, John D., oral surgery
Gellert, Ronald J., physiology
Gerhart, James Basil, nuclear physics
Gessel, Stanley Paul, forestry, soils
Ghose, Subrata, mineralogy, crystallography
Gibby, David Duane, plant physiology; horticulture
Giblett, Eloise Rosalie, hematology; immunogenetics
Giddens, William Ellis, Jr., veterinary pathology
Gillespie, Robert Gordon, information science; academic administration
Glassley, William Edward, petrology; tectonics
Glickfeld, Barnett W., mathematics
Glicksberg, Irving Leonard, mathematics
Gliomset, John A., biochemistry; physiology
Glude, John Bryce, marine biology
Goldstein, Allen A., applied mathematics
Goldsworthy, Patrick Donovan, biochemistry
Goodner, Charles Joseph, endocrinology
Gorbman, Aubrey, zoology
Gordon, Albert McCague, muscular physiology
Gordon, Milton Paul, biochemistry
Gould, Kenneth Lance, cardiology
Gouterman, Martin (Paul), chemical physics
Graham, C Benjamin, radiology
Graney, Daniel O., anatomy
Grayson, Donald Kenneth, anthropology
Grayson, J Thomas, physical chemistry
Green, Richard Lee, physical chemistry
Green, Thomas Kerr, neuropsychology
Greene, Thomas Frederick, astrophysics
Greengo, Robert Eugene, archaeology; anthropology
Gregg, Michael Charles, physical oceanography
Gregory, Norman Wayne, physical chemistry
Gresens, Randall Lee, petrology
Groman, Neal Benjamin, microbiology
Gruger, Edward H., Jr., bio-organic chemistry
Grunbaum, Branko, geometry
Guntheroth, Warren G., pediatric cardiology; cardiovascular physiology
Guy, Reed Augustus, nuclear physics
Hakomori, Sen-Itiroh, biochemistry; immunochemistry
Hall, Benjamin Downs, biochemical genetics
Hall, Nathan Albert, pharmacology
Halperin, Lawrence Mayer, neuropharmacology
Halperin, Walter, botany
Halpern, Isaac, physics
Halsey, Eric Richard, geometry; mathematics education
Halsey, George Dawson, Jr., physical chemistry
Halver, John Emil, nutrition
Hammarlund, Edwin Roy, pharmacy
Hammermeister, Karl E., cardiology
Hampson, John L., psychiatry
Hansen, James M. anesthesiology
Hansen, Stephen Joseph, number theory
Harlin, Vivian Krause, public health administration
Harrison, Halstead, atmospheric chemistry
Harry, George Yost, mammalogy
Hartmann, John Rudolf, hematology; oncology
Hartwell, Leland Harrison, genetics
Haskins, Edward Frederick, botany
Hatch, Melville Harrison, zoology; entomology
Hatfield, William Charles, agronomy
Hatheway, William Howell, botany
Hauschka, Stephen D., developmental biology; neurobiology
Hawthorne, Donald Clair, genetics
Hayes, Murray Lawrence, mathematics

Hazzard, William Russell, metabolism
Healy, Eugene A. embryology, genetics
Healy, Michael L., oceanography, analytical chemistry
Heinrichs, W LeRoy, biochemistry, obstetrics & gynecology
Hellstrom, Ingegerd Elisabet, immunology
Hellstrom, Karl Erik Lennart, immunology, oncology
Henderson, Maureen McGrath, epidemiology
Henley, Ernest Mark, theoretical nuclear physics
Hennes, John Peter, space physics, optics
Henry, Dora Priaulx, zoology
Henry, Kenneth Albin, fish biology, statistics
Herriott, Jon R., biochemistry, molecular biology
Herrmann, Walter L., obstetrics & gynecology
Hewitt, Edwin, mathematics
Hickey, Barbara Mary, physical oceanography
Hickey, Maurice John, surgery
Hiebert, Paul Gordon, anthropology; applied anthropology
Higgins, Theodore Parker, mathematics
Hildebrandt, Jacob, respiratory physiology
Hildebrandt, Judith R., physiology
Hille, Bertil, physiology; biophysics
Hinshelder, John Joseph, mathematics
Hitchcock, Charles Leo, botany
Hobbs, Peter Victor, atmospheric physics
Hodge, Paul William, astronomy
Hodgins, Harold Osborne, microbiology
Hodson, John M., biochemistry; x-ray crystallography
Hodson, Jean Turnbaugh, restorative dentistry
Hogness, John Rusten, academic administration
Huitric, Alain Corentin, pharmaceutical chemistry
Holmes, Thomas Hall, III, psychiatry
Holton, James R. dynamic meteorology
Horita, Akira, pharmacology
Hornbein, Thomas F., anesthesiology, physiology
Horstman, Sanford W., environmental health, industrial hygiene
Hotson, Hugh Howison, plant pathology
Hruthford, Bjorn F., chemistry
Hudson, Peggy R., phycology
Hueter, Theodor Friedrich, acoustics; marine sciences
Hungerford, Thomas W., mathematics
Illg, Paul Louis, zoology
Ingalls, Robert L., physics
Irish, James David, physical oceanography; forest ecology
Jackson, Kenneth Lee, physiology
Jackson, W A Douglas, geography
Jacobs, Loyd Donald, acoustics
Jacobs, Sue-Ellen, social anthropology
Jans, James Patrick, mathematics
Jayne, Benjamin A., forest products, wood technology
Jensen, Lyle Howard, biophysical chemistry
Jones, Robert F., oncology
Juchau, Mont Rawlings, pharmacology
Jumars, Peter Alfred, biological oceanography
Kakiuchi, Hiroaki George, biochemistry
Kaplan, Alex, physiology; chemistry
Karrick, Neva Louise, chemistry
Kashiwa, Herbert Koro, anatomy; histochemistry
Katsaros, Kristina, atmospheric physics
Katz, Max, fisheries
Keaton, Clark M, physical chemistry
Kehl, Theodore H, computer science; biophysics
Keller, Patricia J., biochemistry
Kelly, Vincent Charles, pediatrics
Kelly, William Albert, neurosurgery
Kennedy, Jesse Ward, medicine; cardiology
Kennedy, Thelma Terry, neurophysiology
Kenney, James Francis, geophysics
Kenney, George Edward, microbiology
Kerlee, Donald De, nuclear physics
Kersetter, James David, chemical physics
Kevorkian, Jirair, applied mathematics
Keyes, Charles Fenton, anthropology
Kingston, John Maurice, medicine
Kirby, William M M, medicine
Klebanoff, Seymour J, infectious diseases, biochemistry
Klee, Victor La Rue, Jr, mathematics
Knafich, Helen B. geophysics

Kocan, Richard M, microbiology, public health
Koehler, James K, biophysics
Kohn, Alan Jacobs, zoology
Kozloff, Eugene Nicholas, zoology
Krauel, Kathryn Kreamer Kopf, biochemistry
Kreibich, Roland, organic chemistry
Krieger, Alex Dony, anthropology
Krienke, Ora Karl, Jr, astronomy
Krier, Carol Alnoth, physical chemistry, research administration
Kronmal, Richard Aaron, biostatistics
Kruckeberg, Arthur Rice, botany
Krumme, Gunter, economic geography, regional economics
Kunstadter, Peter, biological anthropology
Kuo, Cho-Chou, medical microbiology
Kwiram, Alvin L, physical chemistry, chemical physics
Labbe, Robert Ferdinand, biochemistry
LaChapelle, Edward Randle, meteorology
Lagunoff, David, pathology
Laird, Charles David, cell biology
Lam, Ronald Ka-Wei, oceanography
Landau, Barbara Ruth, physiology
Landolt, Marsha LaMerle, fish pathology, wildlife pathology
Lansinger, John Marcus, geophysics, environmental management
Larsen, Lawrence Harold, oceanography, hydrodynamics
Law, David Barclay, dentistry
Laws, E Harold, medicine
Lee, Fang An, hydrodynamics, physical oceanography
Lee, John Alexander Hugh, epidemiology
Lee, Kai Nien, environmental management, science policy
Lee, Wylie In-Wei, polymer physics
LeGore, Richard Stephen, pollution biology, invertebrate pathology
Lehmann, Justus Franz, physical medicine
Lemire, Ronald John, teratology, pediatrics
Leney, Lawrence, wood science, microscopy
Leovy, Conway B, meteorology
Lepse, Paul Arnold, organic chemistry
Levy, Rene Hanania, pharmacodynamics
Lewin, Joyce Chismore, microbiology
Lindenmeyer, Paul Henry, physical chemistry, materials science
Ling, Hsin Yi, micropaleontology
Lingafelter, Edward Clay, Jr, chemistry
Lingren, Wesley Earl, physical chemistry, oceanography
Lister, Clive R B, geophysics
Lockard, Robert Bruce, ethology
Loomis, Ted Albert, pharmacology, toxicology
Loop, John Wickwire, radiology, medicine
Lord, Jere Johns, physics
Low, Loh-Lee, fisheries
Lubatti, Henry Joseph, physics
Luchtel, Daniel Lee, microscopic anatomy
Luft, John Herman, histology
Lumer, Gunter, mathematics
Lytle, Farrel Wayne, solid state physics
Mackler, Bruce, pediatrics
Macklin, John Welton, inorganic chemistry, spectroscopy
Malins, Donald Clive, biochemistry
Mallory, Virgil Standish, geology
Maltzeff, Eugene M, fisheries
Manuwal, David Allen, wildlife ecology
Marchioro, Thomas Louis, surgery
Martin, Arthur Wesley, Jr, physiology
Martin, Carroll James, medicine
Martin, George Monroe, experimental pathology
Martin, Robert Edward, Jr, forestry
Martin, Seelye, oceanography
Marts, Marion Ernest, geography
Masuda, Minoru, psychophysiology
Matches, Jack Ronald, food science, microbiology
Mathisen, Ole Alfred, population studies
Maxfield, Galen Harry, fish biology
McAlister, William Bruce, marine ecology
McCarthy, Walter Charles, medicinal chemistry
McCaughran, Donald Alistair, biometrics, statistics
McClure, George W, Jr, soil science, plant physiology
McCulloh, Thane H, petroleum geology
McDermott, Lillian Christie, physics, science education
McDermott, Mark Nordman, atomic physics
McKee, Bates, regional geology
McManus, Dean Alvis, geology, oceanography
McNeil, Arthur Louis, analytical chemistry
Meeuse, Bastiaan Jacob Dirk, plant biochemistry, plant physiology
Megraw, Robert Arthur, forestry, wood technology
Merendino, K Alvin Aurelius, surgery
Merrill, Ronald Thomas, geophysics
Metz, Robert John Samuel, internal medicine, endocrinology
Meyer, Carl Beat, physical inorganic chemistry
Michael, Ernest Arthur, topology
Miller, Josef Mayer, physiology, psychology
Misch, Peter Hans, geology
Mockett, Paul M, experimental high energy physics, particle physics
Moffett, Benjamin Charles, Jr, dental research
Mohri, Hitoshi, surgery
Montzingo, Lloyd J, Jr, mathematics
Moore, Alton Wallace, dentistry
Moore, Daniel Charles, anesthesiology
Morgan, Beverly Carver, pediatric cardiology
Morison, Ian George, forestry
Moriyasu, Keihachiro, experimental high energy physics
Morrill, Richard Leland, geography, regional economics
Morris, Daniel Luzon, chemistry
Morris, David Robert, biochemistry
Morrison, Kenneth N, dentistry
Morrow, James Allen, Jr, mathematics
Mottet, N Karle, pathology, teratology
Motulsky, Arno Gunther, internal medicine, medical genetics
Murphy, Stanley Reed, physics
Myhre, Richard John, fisheries management
Nakatani, Roy E, fish biology
Nalos, Ervin Joseph, physics
Nameroff, Mark A, developmental biology
Nason, James Duane, cultural anthropology
Neddermeyer, Seth Henry, physics
Nelp, Wil B, internal medicine, nuclear medicine
Nelson, George Richard, paper chemistry, pulp & paper technology
Nelson, Janet Sue Rasey, cancer, radiobiology
Nelson, Wendel Lane, medicinal chemistry
Ness, Linda Ann, mathematics
Nester, Eugene William, microbiology
Neurath, Hans, biochemistry
Newell, Laura, physical anthropology
Newman, Marshall Thornton, anthropology
Nichols, Davis Betz, lasers, nuclear physics
Norman, Andrea Hausman, inorganic chemistry
Northrop, Cedric, preventive medicine
Noyes, John Channing, physics
Nunke, Ronald John, mathematics
Nutley, Hugh, nuclear physics
Odland, George Fisher, anatomy, dermatology
Ohlson, Margaret Alexander, nutrition
Olsen, Sigurd, limnology
Olson, Richard Louis, botany, genetics
Olstad, Roger Gale, science education
Omenn, Gilbert Stanley, medical genetics, human genetics
Oncley, Paul Bennett, acoustics
Opatowski, Izaak, applied mathematics
Ordal, Erling Joseph, microbiology
Orians, Gordon Howell, ecology, evolution
Orr, Jack Edward, pharmacy
Orth, George Otto, Jr, organic chemistry
Osterud, Kenneth Leland, protozoology
Ottenberg, Simon, anthropology, ethnography
Page, Roy Christopher, experimental pathology, periodontology
Paine, Robert T, zoology, ecology
Palka, John Milan, neurophysiology
Palmer, John M, speech pathology, audiology
Parker, Robert G, radiology
Parks, George Kung, geophysics, physics
Parson, William Wood, biochemistry
Patton, Harry Dickson, physiology
Pauley, Gilbert Buckhannan, immunology, fisheries
Paulsen, Charles Alvin, internal medicine, endocrinology
Paulson, Dennis Roy, evolutionary biology, ecology
Pavlin, Edward George, anesthesiology
Penning, John Russell, Jr, applied physics
Pereyra, Walter T, fisheries, ecology
Perrin, Edward Burton, biostatistics
Perry, Mary Jane, biological oceanography
Peters, Philip Carl, physics
Petersdorf, Robert George, medicine
Peterson, Donald Richard, epidemiology
Peterson, James Macon, polymer science, materials science
Phelps, Robert Ralph, mathematics
Phillips, Leon A, medicine, radiology
Phillips, Ronald Carl, marine botany
Pigott, George M, food science
Pilachowski, Catherine Anderson, astronomy, astrophysics
Pilet, Stanford Christian, operations research, astrodynamics
Pina, Eduardo Isidorio, mathematical statistics, operations research
Pious, Donald A, pediatrics, genetics
Plein, Elmer Michael, pharmacy
Plein, Joy Bickmore, pharmacy
Plemons, Terry Dale, acoustics
Pocker, Anna, biochemistry
Pocker, Yeshayau, physical organic chemistry, biochemistry
Podbielancik, Vincent S, chemistry
Pollack, Sylvia Byrne, immunology
Pollock, Helen Mary, microbiology
Pope, Charles Edward, II, gastroenterology
Porte, Daniel, Jr, medicine, metabolism
Porter, Stephen Cummings, quaternary geology
Powell, Michael Robert, biophysics
Prager, Denis Jules, population studies
Proctor, Charles Mahan, biosciences
Prothero, John W, biophysics, biological structure
Pruter, Alonzo Theodore, fisheries
Puff, Robert David, theoretical physics
Pyke, Ronald, mathematical statistics
Quimby, George Irving, ethnology, archaeology
Rabinovitch, Benton Seymour, physical chemistry
Ragozin, David Lawrence, mathematical analysis
Rasmussen, Reinhold Albert, plant physiology, chemical engineering
Rattray, Maurice, Jr, hydrodynamics, physical oceanography
Ray, Charles George, microbiology, pediatrics
Raymond, Charles Forest, geophysics
Read, David Hadley, physical organic chemistry
Reed, Richard John, meteorology
Rehr, John Jacob, solid state physics
Reitan, Ralph Meldahl, neuropsychology
Richards, Francis Asbury, oceanography
Richmond, Virginia, biochemistry
Riddiford, Lynn Moorhead (Mrs James W Truman), insect physiology, developmental biology
Riedel, Eberhard Karl, statistical mechanics, theoretical solid state physics
Riedel, Richard Anthony, dentistry
Riekerk, Hans, forest science
Riley, Vernon Todd, microbiology, virology
Ripley, Herbert Spencer, psychiatry
Rising, Louis Wait, pharmaceutical chemistry
Ritchie, Robert Wells, mathematics, computer science
Ritter, David Moore, inorganic chemistry, physical chemistry
Roberts, Clifford Evans, Jr, medicine, microbiology
Roberts, Norman Hailstone, applied statistics
Robertson, William O, pediatrics, medical administration
Robinovitch, Murray R, oral biology, experimental pathology
Robinson, Rex Julian, analytical chemistry
Rockafellar, Ralph Tyrrell, mathematics
Roden, Gunnar Ivo, physical oceanography, climatology
Rogers, Donald Eugene, fish biology
Roman, Herschel Lewis, genetics
Roosen-Runge, Edward C, microscopic anatomy
Ross, Russell, experimental pathology, cell biology
Rosse, Cornelius, anatomy, hematology
Rothberg, Joseph Eli, physics
Roubal, William Theodore, bio-organic chemistry
Rowell, Loring B, physiology
Royce, William Francis, fisheries, resource management
Rubin, Cyrus E, medicine, gastroenterology
Ruch, Theodore Cedric, physiology, neurophysiology
Rucker, Robert Raymond, fisheries
Rushmer, Robert Frazer, bioengineering
Sabo, Jesse Jerry, Jr, physics
Salo, Ernest Olavi, fisheries, oceanography
Sandler, Laurence Marvin, genetics
Sandstrom, Wayne Mark, underwater acoustics, ordnance
Santisteban, George Anthony, neuroendocrinology, cardiovascular diseases
Sarason, Leonard, mathematics
Sarkanen, Kyosti Vilho, wood chemistry, pulp chemistry
Sauvage, Lester R, surgery
Sayers, Dale Edward, solid state physics
Schaeffer, Walter Howard, forestry
Schell, William R, nuclear chemistry, agricultural chemistry
Scher, Allen Myron, physiology
Schilling, John Albert, surgery
Schluger, Saul, periodontology
Schmidt, Fred Henry, physics
Schoener, Amy, biological oceanography, biogeography
Schoener, Thomas William, ecology
Schomaker, Verner, physical chemistry
Schreuder, Gerard Fritz, forest economics
Schubert, Wolfgang Manfred, organic chemistry
Schurr, John Michael, physical chemistry, molecular biophysics
Schwartz, Stephen Mark, pathology, cardiovascular diseases
Schwarz, M Roy, anatomy, immunology
Scott, David Robert Main, silviculture
Scribner, Belding Hibbard, medicine
Scribner, John David, organic chemistry, oncology
Segal, Jack, topology
Seymour, Allyn H, radiation ecology, fish biology
Shain, Irving, electroanalytical chemistry
Shapiro, Bennett Michaels, biochemistry
Shapiro, Jewel Templeton, environmental management
Shapley, James Louis, audiology, speech pathology
Sharp, Virginia Leah, geography, environmental sciences
Sharpe, Grant William, forestry
Shaw, Cheng-Mei, neuropathology
Shaw, Ross Franklin, zoology
Shepard, Thomas H, pediatrics
Sherman, John Clinton, geography
Sherris, John C, microbiology
Shires, George Thomas, surgery
Shurtleff, David B, pediatrics
Siegel, Ivens Aaron, pharmacology, oral biology
Silliman, Ralph Parks, fish biology
Simpson, John Barclay, behavioral physiology
Skahen, Julia Goodsell, physiology, endocrinology
Skud, Bernard Einar, marine biology, fisheries
Slaughter, John Brooks, computer sciences
Slutsky, Leon Judah, physical chemistry
Smith, Alan Wayne, physical chemistry
Smith, David W, pediatrics
Smith, Elizabeth Knapp, clinical biochemistry, pediatric endocrinology
Smith, Harold Warren, structural chemistry
Smith, James Dungan, physical oceanography, geological oceanography
Smith, Lynwood S, comparative physiology
Smith, Nathan James, pediatrics
Smith, Orville Auverne, Jr, neurophysiology
Smith, Stewart W, geophysics
Smuckler, Edward Aaron, experimental pathology, biochemistry
Snover, Kurt Albert, physics
Sorby, Donald Lloyd, pharmacy, pharmaceutical chemistry
Spackman, Darrel H, biochemistry, cancer
Spadoni, Leon R, obstetrics & gynecology
Spain, David Howard, anthropology
Sparkman, Donal Ross, medicine
Spencer, Merrill Parker, cardiovascular physiology
Spiers, Philip Sackville, epidemiology
Spinelli, John, analytical chemistry, food chemistry
Stadler, David Ross, genetics
Stahl, William Louis, biochemistry, neurochemistry
Staley, James Trotter, bacteriology, microbial ecology

Stamatoyannopoulos, George, medical genetics, hematology
Stamm, Stanley Jerome, medicine
Stansby, Maurice Earl, marine chemistry
Steckler, Bernard Michael, organic chemistry, history of science
Steinberg, Maynard Albert, food science, fisheries
Stenzel, George, forest management, forest engineering
Stern, Edward Abraham, solid state science
Sternberg, Richard Walter, geological oceanography, marine sedimentation
Stetter, Reinhard Friederich, forest genetics
Stewart, Jennifer Keys, endocrinology
Stewart, Richard John, geology
Stibbs, Gerald Denike, dentistry
Stirling, Charles E., physiology, biophysics
Stober, Quentin Jerome, aquatic ecology, water pollution
Storb, Ursula, immunology
Stout, Edgar Lee, mathematics
Stout, George Hubert, natural products chemistry, x-ray crystallography
Stout, Virginia Falk, environmental biochemistry
Strandjord, Paul Edphil, clinical chemistry, laboratory medicine
Strandness, Donald Eugene, Jr., medicine, surgery
Street, Robert Elliot, aerodynamics, numerical analysis
Streib, John Fredrick, physics
Stuiver, Minze, earth science
Stuntz, Daniel Elliot, mycology
Sundsten, John Wallin, anatomy
Swanson, Phillip D., neurology, biochemistry
Swindler, Daris Ray, anthropology
Swope, Charles C., prosthodontics
Szolosi, Daniel Gabriel, embryology, cell biology
Taber, Richard Douglas, wildlife ecology
Taft, Bruce A., physical oceanography
Taub, Frieda B., ecology, nutrition
Taylor, Murray East, analytical chemistry
Teller, David Chambers, physical biochemistry
Teller, Davida Young, vision, psychology
Templeton, Frederic Eastland, radiology
Thiersch, Johannes Bernhard, pathology, cancer
Thomas, Charles Gomer, mathematics, computer sciences
Thomas, David Phillip, forestry
Thomas, Edward Donnall, internal medicine, oncology
Thomas, Morgan D., economic geography
Thompson, Donovan Jerome, biostatistics
Thompson, Richard Baxter, fish biology
Thuline, Horace Crockett, medicine
Tiedeman, George Trent, organic chemistry, polymer chemistry
Toskey, Burnett Roland, mathematics
Towe, Arnold Lester, physiology
Trager, William Frank, pharmaceutical chemistry, organic chemistry
Trefethen, Parker S., fisheries forestry
Truman, James William, insect physiology
Tsukada, Matsuo, ecology, paleoecology
Tucker, Geoffrey Thomas, pharmacology, pharmacy
Turck, Marvin, internal medicine, infectious disease
Turner, Terry Earle, physics
Ueland, Kent, obstetrics & gynecology
Ugolini, Fiorenzo Cesare, soils
Ullman, Edward Louis, geography
Unterstiener, Norbert, glaciology
Utter, Fred Madison, biochemical genetics
Valente, Frank Anthony, reactor physics, nuclear physics
Van Citters, Robert L., cardiovascular physiology
van Belle, Gerald, statistics
Vance, Joseph Alan, geology
Van Cleve, Richard, fisheries
Vandenbosch, Robert, nuclear chemistry
Van Hassel, Henry John, dentistry
Van Hoosier, Gerald L., Jr., laboratory animal science
Varanasi, Usha Suryan, biochemistry, chemistry
Veirs, Carroll Eugene, water pollution
Vilches, Oscar Edgardo, physics
Vincenzi, Frank Foster, pharmacology
Vlases, George Charpentier, plasma physics
Volwiler, Wade, medicine

Waaland, Joseph Robert, algology, cytology
Waggoner, Thomas Runyan, economics
Wahl, Patricia Walker, biostatistics
Waldron, Howard Hamilton, geology
Walker, Richard Battson, botany
Wallace, John Michael, meteorology
Wallerstein, George, astronomy
Walsh, John Joseph, ecology, oceanography
Walsh, Kenneth Andrew, biochemistry
Wang, San-Pin, medical microbiology
Ward, Arthur Allen, Jr., neurosurgery
Ward, Richard John, anesthesiology
Warfield, Robert Breckinridge, Jr., algebra
Warfield, Virginia McShane, mathematical analysis
Washburn, Albert Lincoln, geomorphology, quaternary geology
Wasserman, William Jack, organic chemistry, polymer chemistry
Watson, James Bennett, anthropology
Wearn, Richard Benjamin, Jr., oceanography, fluid dynamics
Webster, Peter John, dynamic meteorology
Wedemeyer, Gary Alvin, fisheries
Wedgwood, Ralph Josiah Patrick, pediatrics
Wehrenberg, Paul James, atomic physics, space physics
Wei, Pax Samuel Pin, physical chemistry
Weinstein, Boris, organic chemistry, bio-organic chemistry
Weinstein, Joseph M., nuclear physics
Weis, Joe H., theoretical high energy physics
Weisbrod, Alan Richard, vertebrate zoology, evolutionary biology
Weiser, Russell Shivley, bacteriology, immunology
Weiss, Gary Bruce, hematology
Weiss, Richard Raymond, meteorology, electrical engineering
Weitkamp, William George, nuclear physics
Welander, Arthur Donovan, fisheries, radiobiology
Wergedal, Jon E., biochemistry
Westrum, Lesnick Edward, neurosurgery, biological structure
Wheeler, Harry Eugene, stratigraphy
Whetten, John T., geology, oceanography
Whisler, Howard Clinton, botany, microbiology
White, Thomas Taylor, surgery
Whiteley, Arthur Henry, zoology
Whiteley, Helen Riaboff, microbial physiology
Whitney, Richard Ralph, fishery biology
Wiebe, Harold T., zoology, biology
Wiederhielm, Curt Arne, cardiovascular physiology
Wilkes, Richard Jeffrey, high energy physics, cosmic ray physics
Williams, Robert Hardin, medicine
Williams, Robert Melvin, physics
Willoughby, William Franklin, pathology, immunology
Wilson, Richard Atwood, gastroenterology
Wilson, John Thomas, Jr., environmental medicine, preventive medicine
Winans, Edgar Vincent, anthropology
Winter, Donald F., applied mathematics
Winter, Peter Michael, anesthesiology
Wintersheid, Loren Covart, thoracic surgery, cardiovascular surgery
Wolf, Norman Sanford, experimental pathology, radiobiology
Wolfe, Dael (Lee), science policy, administrative sciences
Wood, Francis C. Jr., internal medicine, endocrinology
Wood, James W., biology
Wooldridge, David Dilley, forest soils, forest hydrology
Wooton, Peter, medical physics, radiology
Wykhuis, Walter Arnold, dentistry, community health
Yamanaka, William Kiyoshi, nutrition, radiology
Yarington, Charles Thomas, Jr., otorhinolaryngology
Yee, Sinclair Shee-Sing, microelectronics
Young, Allan Charles, physics, physiology
Young, Elton Theodore, molecular biology
Young, Kenneth Kong, high energy physics

Youngmann, Carl Ernst, cartography, geography
Zimmerman, Gary Alan, clinical chemistry

SEQUIM
Raff, Rudolf August Victor, polymer chemistry
Wanek, Alexander Andrew, geology

SHAW ISLAND
Jefferts, Keith Bartlett, atomic physics, fisheries management

SHELTON
Beelik, Andrew, organic chemistry
Briggs, Ben Thoburn, organic chemistry
Conca, Romeo John, organic chemistry
Ewart, Hugh Wallace, Jr., organic chemistry
Goldschmid, Otto, wood chemistry
Gray, Kenneth Russell, chemistry
Hamilton, John Kelvin, agricultural biochemistry
Herrick, Franklin Willard, organic chemistry
Lovell, Edwin Lister, chemistry
Maranville, Lawrence Frank, pulp chemistry
Reintjes, Marten, organic chemistry
Sears, Karl David, chemistry
Selders, Archie Arnold, plant nutrition, forest genetics
Thomas, Berwyn Brainerd, chemistry
Tostevin, James Earle, paper chemistry

SPOKANE
Armstrong, Frank Clarkson, geology
Bocksch, Robert Donald, organic chemistry
Campbell, Neil, geology
Foubert, Edward Louis, Jr., allergy
Gautereaux, Ione, biology
Gillingham, Robert J., solid state physics, electronics
Hansen, Don A., geophysics
Henricksen, Thomas Alva, geology
Hicks, David L., physiological ecology, ornithology
Hurd, Robert Charles, bacteriology
Johnston, Hugh William, organic chemistry
Kelsh, Dennis J., physical chemistry
Liang, Shou Chu, chemistry
Martin, Kenneth Edward, mathematics
Nealen, Joseph Peter, physics
Olson, Edwin Andrew, geology, geochemistry
Page, Harold Alfred, organic chemistry
Shannon, Charles Francis, biochemistry, immunology
Stien, Howard M., zoology, physiology
Willardson, Robert Kent, solid state physics
Winniford, Robert Stanley, physical chemistry
Wyrick, Ronald Earl, biochemistry, allergy

TACOMA
Adams, Harry, nuclear physics
Alcorn, Gordon Dee, botany, ornithology
Anderson, Charles Dean, bio-organic chemistry
Anderson, Norman Roderick, geology, geomorphology
Bender, Walter Louis, forest management
Berry, Keith O., inorganic chemistry, analytical chemistry
Bland, Jeffrey S., bio-organic chemistry
Bohannon, Randolph F., plant physiology, molecular biology
Brown, Bert Elwood, physics
Campbell, William Joseph, meteorology, oceanography
Clifford, Howard James, physical chemistry, molecular spectroscopy
Cordingly, Richard Henry, pulp & paper technology
Danes, Zdenko Frankenberger, geophysics
Davis, Thomas Austin, mathematics
Foxworthy, Bruce L., hydrology
Frost, Thomas Rogers, forest products, chemistry
Gerber, Carl J., psychiatry
Giddings, William Paul, physical organic chemistry
Goman, Edward Gordon, mathematics
Gregory, Arthur Stanley, industrial organic chemistry
Guthrie, Franklin Kirrey, research administration, wood science & technology

WALLA WALLA
Anderson, Edward Frederick, botany, plant taxonomy
Bracher, Katherine, astronomy
Brattain, Walter Houser, biophysics
Drabek, Charles Martin, zoology
Frasco, David Lee, physical chemistry, spectroscopy
Gunsul, Craig J. W., solid state physics, theoretical physics
Howland, Louis Philip, solid state physics
Hutchings, William Lawrence, chemistry
Pengra, James G., nuclear physics
Rempel, Arthur Gustav, inorganic chemistry
Templeton, John Charles, inorganic chemistry
Uhlenhopp, Elliott Lee, biochemistry
Underwood, Douglas Haines, mathematics
Woodward, Glenn Jones, inorganic chemistry, biochemistry

WENATCHEE
Anthon, Edward W., entomology
Benson, Nels R., horticulture, soil science
Burts, Everett C., entomology
Covey, H. Melvin, plant physiology
Covey, Ronald Perrin, Jr., plant pathology
Hoyt, Stanley Charles, entomology
Ketchie, Delmer O., plant physiology, biochemistry
Klock, Glen Orval, forest soils
Larsen, Robert Paul, horticulture
Leeper, John Robert, entomology
Lopushinsky, William, plant physiology
Olsen, Kenneth Laurence, plant physiology
Pierson, Charles Frederick, plant pathology

Hansen, David Henry, physiological ecology, population ecology
Haushild, William Leland, hydrology
Herzog, John Orlando, mathematics
Hodge, Steven McNiven, glaciology
Huestis, Laurence Dean, organic chemistry
Jacobs, Clarence Gilbert, Jr., nuclear physics, solid state physics
Jensen, Creighton Randall, soil physics
Johnson, Murray Leathers, medicine, mammalogy
Johnson, William Lee, mathematics, operations research
Karlstrom, Ernest Leonard, herpetology, ecology
Kilner, Scott Burgoyne, environmental chemistry
Kleyn, John Gerard, microbiology
Knudsen, Jens Werner, Jr., zoology
Leraas, Harold J., biology, dentistry
Lowther, John Stewart, paleontology, paleobotany
Medcalf, Darrell Gerald, carbohydrate chemistry
Mehlhaff, Leon Curtis, analytical chemistry, polymer chemistry
Meier, Mark Frederick, glaciology
Nearn, William Thomas, forest products
Nelson, Martin Emanuel, physics
Nigh, Wesley Gray, physical organic chemistry
Nornes, Sherman Berdeen, surface physics
Ostenson, Burton Thomas, zoology
Pierson, Beverly, Kanda, microbiology
Post, Austin, glaciology
Ritchie, Gary Alan, physiological ecology
Slee, Frederick Watford, nuclear physics
Snell, Robert Isaac, mathematics
Stockmann, Volker Erwin, solid mechanics, polymer physics
Sullivan, John W., cereal chemistry
Tang, Kwong-Tin, physics, physical chemistry
Tobiason, Frederick Lee, chemistry
Trotter, Patrick Casey, pulp chemistry, paper chemistry
Van Enkevort, Ronald Lee, mathematics

VANCOUVER
Crandall, Perry Clarence, horticulture
Lo, Cheng Fan, wood chemistry
Maynard, Russell Milton, pathology
Shanks, Carl Harmon, Jr., entomology
Smith, Sam Corry, nutrition

VAUGHN
Shirley, Frank Connard, forest management, forest economics

Raese, John Thomas, agronomy, plant physiology
Stahly, Edward Arthur, horticulture
Staff, Donald C, pharmaceutical chemistry, bionucleonics
Thompson, James Arthur, environmental management
Williams, Max W, plant physiology, horticulture
Wright, Theodore Richard, plant pathology

WESTPORT
Kimmey, James William, forest pathology

YAKIMA
Butler, Lillian Ida, analytical chemistry, biological chemistry
Davis, Harry Glenwood, entomology
Eustis, William Henry, polymer chemistry
Halfhill, John Eric, entomology
Ingalsbe, David Weeden, food science, natural products chemistry
Millard, George Buente, inorganic chemistry
Moffitt, Harold Roger, acarology
Powell, Donnie Melvin, entomology
Rigby, F Lloyd, agricultural chemistry, biochemistry
Tamaki, George, entomology
Toba, Hachiro Harold, entomology
White, Leland Darrell, entomology, botany

BECKLEY
Zabetakis, Michael George, physical chemistry, safety engineering

BELLE
McGonigal, William E, organic chemistry

WEST VIRGINIA

ATHENS
Bayless, Laurence Emery, biology
Blatt, Jeremiah Lion, molecular biology, science education
Chapman, Carl Joseph, systematic botany
Covey, Winton Guy, Jr, micrometeorology
Fezer, Karl Dietrich, science education, philosophy of science
Graham, Roger Neill, quantum mechanics, computer science
Jones, Wilber Clark, inorganic chemistry
Montgomery, Andrew Harrison, physical chemistry
Rubinstein, Daniel, biophysics

BETHANY
Brown, David T, mathematics
Buckelew, Albert Rhoades, Jr, microbiology
Draper, John Daniel, organic chemistry
Larson, Gary Eugene, botany, biophysics
Sawtell, James Joseph, ecology, aquatic biology
Zwecker, William R, organic chemistry

BLUEFIELD
Elkins, John Rush, organic chemistry
Gilbert, Charles Russell, zoology

BUCKHANNON
Capstack, Ernest, organic chemistry, biochemistry
Holloway, Homer Edward, physical chemistry
Peterson, Frederick Alvin, anthropology, archaeology
Richter, G Paul, inorganic chemistry
Rossbach, George Bowyer, botany
Wolfe, David Francis Zeke, physical chemistry

CHARLESTON
Bailey, Frederick Eugene, Jr, physical chemistry
Bryant, George Macon, polymer chemistry
Cunningham, Newlin Buchanan, industrial chemistry
Davis, Hubert Greenidge, physical chemistry
Dawson, Thomas Larry, physical chemistry, polymer chemistry

Doumaux, Arthur Roy, Jr, organic chemistry
Fitzpatrick, John Thomas, chemistry
Gillespie, William Harry, phytopathology, paleobotany
Islam, Nurul, plant physiology, agronomy
Jacobson, Bernard Howard, organic chemistry
Johnson, George Frederick, organic chemistry
Kurtz, Abraham N, organic chemistry
Marcus, Erich, synthetic organic chemistry, textile chemistry
Matthews, Virgil Edison, organic polymer chemistry
McCain, James Herndon, organic chemistry
Miller, Marvin E, meteorology
Myerly, Richard Crebs, organic chemistry
Nunley, Robert Gray, phycology
Peck, David W, physical organic chemistry
Pianfetti, John Andrew, organic chemistry
Ream, Bernard Claude, organic chemistry
Saville, Paul D, internal medicine
Smith, Glenn Edward, plant pathology
Smith, James Allbee, analytical chemistry
Spencer, Jesse G, physical chemistry, inorganic chemistry
Stansbury, Harry Adams, Jr, chemistry
Thompson, Hartwell Greene, Jr, neurology

ELKINS
Baker, Barton Scofield, agronomy
Elrod, Lloyd Melvin, physiology, zoology
Martin, John Perry, Jr, physical chemistry, analytical chemistry
Peters, John Burl, animal husbandry
Tolstead, William Lawrence, botany

FAIRMONT
Coleman, James Edward, physical chemistry
LaRue, James Arthur, mathematics
Moffa, David Joseph, biochemistry, clinical chemistry
Pritchett, William Henry, developmental biology
Richardson, Rayman Paul, science education
Ruoff, William (David), organic chemistry
Shan, Robert Kuo-Cheng, aquatic ecology
Swiger, Elizabeth Davis, physical chemistry

FALLING WATERS
Coffey, Charles Eugene, physical chemistry

GLENVILLE
Chisler, John Adam, bacteriology, genetics
Deal, Don Robert, botany, science education
Turner, Byron, chemistry

GREEN BANK
Howard, William Eager, III, astronomy
Kellermann, Kenneth Irwin, radio astronomy
Shaffer, David Bruce, radio astronomy

HARPERS FERRY
Webb, Byron Horton, dairy industry
Williams, Clara Hinton, biochemistry

HEDGESVILLE
Biel, David Franklin, physics

HUNTINGTON
Babb, Daniel Paul, inorganic chemistry
Bauserman, Thomas, mathematics
Binder, Franklin Lewis, microbial physiology, mycology
Bonnet, Richard Brian, geology
Chakrabarty, Manoj R, physical inorganic chemistry
Dils, Robert James, science education
Douglass, James Edward, organic chemistry
Dumke, Warren Lloyd, chemical physics, physical chemistry
Fisher, Dorothy A, zoology
Hanrahan, Edward S, physical chemistry

Hardman, Dennis Hunter, mathematics
Hoback, John Holland, physical chemistry
Hogan, John Wesley, mathematics
Lemke, Thomas Franklin, organic chemistry, corrosion
Lepley, Arthur Ray, physical organic chemistry
Manakkil, Thomas Joseph, physics
Mann, Dennis Keith, medical microbiology
Martin, Donald Clayton, experimental physics
Mills, Howard Leonard, plant physiology, plant morphology
Oberly, Ralph Edwin, physics
Parlett, Robert Carleton, microbiology, medicine
Plymale, Edward Lewis, botany
Price, Howard Charles, biochemistry, organic chemistry
Roberts, Joseph Linton, physical chemistry, biochemistry
Rogers, Wiley Samuel, geology
Shanholtzer, Wesley Lee, solid state physics
Shoemaker, Jon Philip, zoology, parasitology
Tarter, Donald Cain, zoology
Warren, John Rush, botany
Weaks, Thomas Elton, plant physiology
Whitley, William Thurmon, mathematics

INSTITUTE
Barnes, Edward B, biophysics, physics
Brimhall, James Elmore, physics
Das Sarma, Basudeb, inorganic chemistry
Jackson, Peter H, organic chemistry, chemistry
Kagen, Herbert Paul, organic chemistry
Krabacher, Bernard, physical organic chemistry
Wallace, William James Lord, physical chemistry

KEARNEYSVILLE
Barrat, Joseph George, plant pathology
Bullock, Graham Lambert, bacteriology, fish pathology
Wolf, Kenneth Edward, microbiology

KEYSER
Davis, Burtron H, physical chemistry, petroleum chemistry
Michael, William Earl, biology, zoology
Murphy, Allen Emerson, geology
Paine, Alan Henry, mathematics
Powell, Maurice Green, physical chemistry

LEWISBURG
Bentley, Patrick E, pharmacology
Blatt, Elizabeth Kempske, medical physiology, neuroendocrinology
Sharp, Roland Paul, osteopathy

MARTINSBURG
Fiala, Silvio Emerich Ivan, physiology, cancer
Harbour, Robert Myron, nuclear chemistry, physical chemistry
Hoch, Hans, physical chemistry
Hoch-Ligeti, Cornelia, pathology
Lewis, Cameron David, organic chemistry
Markiw, Roman Teodor, biochemistry, organic chemistry
Sudweeks, Walter Bentley, industrial chemistry, explosives

MONTGOMERY
Bell, Raymond Frank, applied mathematics, mathematical analysis
Lind, Robert Wayne, theoretical physics
Rana, Abdul R, nuclear physics

MORGANTOWN
Abbott, Okra Jones, poultry nutrition
Abel, William T, fuel science, physical chemistry
Abrahamson, Lila, plant biochemistry
Adams, Robert Evans, plant pathology
Albrink, Margaret Joralemon, internal medicine
Albrink, Wilhelm Stockman, pathology
Amato, R Stephen S, human genetics, pediatrics
Anderson, Gerald Clifton, animal nutrition

Anderson, William Evan, internal medicine
Anderson, William Niles, Jr, applied mathematics
Andrews, Charles Edward, internal medicine
Arkle, Thomas, Jr, geology
Arya, Atam Parkash, nuclear physics
Baer, Charles Henry, plant physiology, ecology
Balasko, John Allan, agronomy, crop physiology
Barnett, Horace Leslie, mycology
Barton, Jay, II, cell physiology
Bennett, Herald Durward, plant morphology, cytology
Beresford, William Anthony, histology, neuroanatomy
Bernard, Harvey Russell, anthropology
Biddington, William Robert, dentistry
Birch, Robert Lee, invertebrate zoology
Blair, James Bryan, biochemistry
Blaydes, David Fairchild, plant physiology, biochemistry
Bradshaw, William Newman, environmental biology
Brooks, James Lee, biochemistry
Buchanan, Hugh, geology, paleontology
Burrell, Robert Guthrie, immunology, microbiology
Butcher, Roy Lovell, reproductive physiology
Butler, Linda, entomology
Canady, William James, biochemistry
Cardwell, Dudley H, petroleum geology, structural geology
Carmichael, Stephen Webb, human anatomy
Carvell, Kenneth Llewellyn, forest ecology
Cech, Franklin Charles, forest genetics
Cenedella, Richard J, biochemistry, pharmacology
Charon, Nyles William, microbiology
Chen, Ping-Fan, geology
Choulis, Nicolas Helias, pharmaceutical chemistry
Clarkson, Roy Burdette, botany
Clovis, Jesse Franklin, systematic botany
Colasanti, Brenda Karen, neuropharmacology
Colby, Howard David, endocrinology
Collins, William Edgar, endocrinology
Cooper, Bernard Richard, solid state physics
Core, Earl Lemley, botany
Covalt-Dunning, Dorothy, animal physiology, animal behavior
Craig, Charles Robert, pharmacology
Culberson, James Lee, anatomy
Cunningham, Allen Byron, mathematics
Deal, Samuel Joseph, medical bacteriology
Dines, Allen I, pharmaceutical chemistry, medicinal chemistry
Dodson, Chester Lee, geology, hydrology
Donaldson, Alan C, geology
Dozsa, Leslie, veterinary medicine
Dunbar, Robert Standish, Jr, animal breeding
Dunning, Dorothy Covalt, animal behavior
Eaves, James Clifton, algebra
Eckert, Herbert L, pediatrics, infectious diseases
Ellington, John S, biochemistry
Elliott, Edward Sumner, plant pathology
Enlow, Donald Hugh, anatomy
Erwin, Robert Bruce, geology
Fleming, William Wright, pharmacology
Flink, Edmund Berney, medicine
Fodor, Gabor, organic chemistry
Franz, Gunter Norbert, physiology, biophysics
Frederickson, Richard Gordon, anatomy, cell biology
Friedman, Morton Henry, human anatomy, microscopic anatomy
Frist, Ramsey Hudson, virology
Fugo, Nicholas William, obstetrics & gynecology
Gabriele, Orlando Frederick, radiology
Gallegly, Mannon Eithu, plant pathology
Gerencser, Mary Ann (Aiken), bacteriology
Gerencser, Vincent Frederic, bacteriology
Gladfelter, Wilbert Eugene, physiology, neurobiology
Gould, Henry Wadsworth, mathematics
Graham, William Lee, astrophysics
Greig, William Elliott, astrophysics
Guthrie, Roland L, biosystematics, dendrology
Gutmann, Ludwig, neurology
Hahon, Nicholas, virology
Haines, Duane Edwin, neuroanatomy, evolutionary biology

WEST VIRGINIA

Hales, Milton Reynolds, medicine, pathology
Hall, George Arthur, (Jr), physical chemistry
Hall, James Lester, physical chemistry
Hamilton, John Edgar, zoology, parasitology
Harris, Charles Lawrence, biochemistry, technology
Heald, Milton Tidd, sedimentary petrology
Hein, Peter Leo, Jr., psychiatry
Hickman, James Blake, physical chemistry
Hiergeist, Franz Xavier, mathematical analysis
Higginbotham, Arlyn Curtis, anatomy, physiology
Higginbotham, Frances Heffrin, anatomy, histochemistry
Hilloowala, Rusi Ardeshir, anatomy, physical anthropology
Horvath, Donald James, animal nutrition
Howard, Stephen Arthur, pharmaceutics
Humphrey, George Louis, physical chemistry
Hurlburt, Henry Winthrop, zoology
Ingle, L Morris, plant physiology
Inskeep, Emmett Keith, reproductive physiology, endocrinology
Jefimenko, Oleg D., physics
Jones, Barbara, pediatrics
Jones, David Smith, anatomy
Jones, John Evan, internal medicine, endocrinology
Kaczmarczyk, Walter J, biochemical genetics
Karr, Clarence, Jr., chemistry
Katz, Sam., biochemistry, biology
Keefer, Robert Faris, soil science
Keller, Edward Clarence, Jr, ecology, biostatistics
Kelley, John Fredric, psychiatry
Kidder, Harold Edward, physiology
Kim, Jin Bai, algebra
Kimmel, Donald Loraine, neuroanatomy
Kirk, Billy Edward, microbiology
Klingberg, William Gene, pediatrics, hematology
Koch, Christian Burdick, wood science, forestry
Koppelman, Ray., biochemistry, biology
Krall, John Morton, biostatistics
Krause, Reginald Frederick, biochemistry
Larrison, Millard Samuel, organic chemistry
Lass, Norman Jay, speech pathology, audiology
Lee, Richard, hydrology
Lessing, Peter, environmental geology microclimatology
Levine, Arnold David, theoretical physics
Lindholm, Dale David, internal medicine, nephrology
Lindsay, Hugh Alexander, physiology
Lospeich, Frederick Jackson, bio-organic chemistry
Love, Betholene Frances, medical technology
Ludlum, John Charles, geology
MacDowell, Denis W H., organic chemistry
Malanga, Carl Joseph, cell physiology, pharmacology
Maloy, Joseph T, analytical chemistry
Marshall, Robert James, animal behavior
Marshall, Joseph Andrew, cardiovascular physiology
Martin, William Gilbert, nutrition, biochemistry
Martinelli, Louis Carl, pharmaceutical chemistry
Martis, Kenneth Charles, geography
Mauger, John William, pharmaceutics
Mawhinney, Michael G., pharmacology
McCafferty, Robert Eugene, anatomy
McClung, Marvin Richard, animal breeding
McCormick, Bailie Jack, inorganic chemistry
McIntyre, Thomas Woodford, computer science
McLaren, George Aiken, agricultural biochemistry
Mengoli, Henry Francis, immunology
Menser, Harry Alvin, Jr, plant physiology
Merow, William Wayne, orthodontics
Michael, Edwin Daryl, wildlife management
Milam, Denver Franklin, medicine
Moe, Paul G., bacteriology
Moore, William Robert, organic chemistry
Moran, Walter Harrison, Jr., surgery, physiology

Morgan, David Zackquill, internal medicine
Morgan, William Keith C., internal medicine
Morgan, Winfield Scott, pathology
Muth, Chester William, organic chemistry
Nakon, Robert Steven, inorganic chemistry
Nath, Joginder, cytology
Neal, Oliver Meader, Jr., horticulture
Norton, Charles Warren, micropaleontology
Nugent, George Robert, neurosurgery
O'Connell, Frank Dennis, pharmacognosy
Olson, Norman O., veterinary medicine
Overberger, James Edwin, dentistry
Overman, Dennis Orton, teratology
Patrick, Homer, nutrition
Pavlovic, Arthur Stephen, solid state physics
Pearson, George Allen, environmental sciences
Peters, Iland Dee, mathematics
Petersen, Jeffrey Lee, physical inorganic chemistry
Peterson, Ronald A., avian physiology
Pinkstaff, Carlin Adam, histology
Pohlman, George Gordon, soils
Popovich, Peter, medical mycology, microbiology
Pore, Robert Scott, microbiology
Price, Paul Holland, geology
Rafter, Gale William, biochemistry
Renton, John Johnston, geochemistry
Resnick, Harold, biochemistry
Reyer, Randall William, developmental biology
Robinson, Robert Leo, pharmacology
Rodman, Nathaniel Fulford, Jr., pathology
Samuel, David Evan, wildlife biology, ornithology
Saxe, Leroy Hallowell, Jr., protozoology
Schabinger, John Robert, dairy husbandry
Schein, Martin Warren, ethology
Schleusner, John William, mathematics
Schubert, Oscar Edmund, horticulture
Scott, Eion George, plant physiology, plant biochemistry
Seehra, Mohindar Singh, solid state physics
Smosna, Richard Allan, stratigraphy
Spradlin, Wilford W., psychiatry
Sprinkle, Philip Martin, otolaryngology
Stewart, Gerald Walter, physical chemistry
Stewart, Joseph Kyle, mathematics
Stewart, Robert Francis, nuclear chemistry, fuel technology
Stickney, John Clifford, physiology, biophysics
Sitzel, Robert Eli, pharmacology
Strohl, John Henry, analytical chemistry
Sutter, Richard P., biochemistry
Thayne, William V., animal breeding, statistics
Thomas, Charles Danser, physics
Thomas, John A., pharmacology
Thomas, Roy Orlando, animal nutrition, dairy husbandry
Ting, Francis Ta-Chuan, geology
Townsend, Edwin C., biometry
Trapp, George E, Jr., applied mathematics
Traynelis, Vincent John, organic chemistry
Trotter, Robert Russell, ophthalmology
Tryban, George P., biochemistry, cancer
Tryon, Earl Haven, silviculture
Turndorf, Herman, anesthesiology
Ulrich, Valentin, genetics
Urquilla, Pedro Ramon, pharmacology
Van Dyke, Knox, pharmacology, biochemistry
Van Eck, Willem Adolph, agronomy, soil science
Van Landingham, Audrey Howard, agricultural chemistry
Veach, Collins, agronomy
Vehse, William E., physics
Veltri, Robert William, microbiology
Vest, Marvin Lewis, mathematical analysis

Voelz, Herbert Gustav, microbiology
Walker, Elizabeth Reed, human anatomy
Wallace, William Edward, Jr., chemical physics, biophysics
Warden, Herbert Edgar, surgery, thoracic surgery
Warshauer, Steven Michael, invertebrate paleontology, paleoecology
Watne, Alvin Lloyd, surgery
Wearden, Stanley, statistics
Weber, Kenneth C., pulmonary physiology
Welch, James Alexander, animal husbandry, animal physiology
Wells, Dana, geology, paleontology
Welton, William Arch, dermatology
Westfall, David Patrick, pharmacology
White, Charles A Jr., obstetrics & gynecology
White, David Evans, forest economics
Wiant, Harry Vernon, Jr., forestry
Williams, Leah Ann, developmental biology
Williams, Thomas Watley, Jr., human anatomy
Williamson, Douglas Bleecker, physics
Wilson, Harold Albert, microbiology
Wilson, Michael Friend, physiology, biophysics
Winston, Anthony, polymer chemistry
Wirtz, George H., immunochemistry, biochemistry
Wright, John Collins, organic chemistry
Yelton, David Baetz, microbiology
Zimmermann, Bernard, surgery
Zinn, Gary William, forest economics

NEW MARTINSVILLE

Britain, J W, polymer chemistry
Chadwick, David Henry, industrial organic chemistry
Cleveland, Thomas Hilburn, organic chemistry
Dean, Warren Edgell, physical chemistry, inorganic chemistry
Peard, William John, inorganic chemistry, analytical chemistry
Rieck, James Nelson, polymer chemistry
Sandridge, Robert Lee, analytical chemistry, organic polymer chemistry

NEWELL

Pittenger, John T., physics

PARKERSBURG

Bond, William Bradford, physical chemistry, organic chemistry
Carlson, Dana Peter, organic chemistry
Cavanaugh, Robert J., organic chemistry
Cook, Elbert Gary, Jr., research administration
Gentzler, Robert E., chemical kinetics
Hussey, Edward Walter, organic chemistry
Jack, John James, analytical chemistry
Knight, Alan Campbell, polymer science
Latham, Roger Alan, polymer chemistry
Lees, Joseph Kolb, solid state physics
Lipp, Hayden Ivan, polymer chemistry
Martin, Wayne Holderness, polymer chemistry
Novak, Ernest Richard, organic chemistry
Putman, Robert Ervin, organic chemistry
Rattenbury, Kenneth Harrison, organic chemistry
Roura, Miguel Jacinto, organic chemistry
Scott, James Alan, analytical chemistry, polymer chemistry
Sperati, Carleton Angelo, polymer chemistry
Von Schritz, Don Morris, chemistry
Woodland, William Charles, physical chemistry

PHILIPPI

Digman, Robert V., organic chemistry
Maruca, Robert Eugene, inorganic chemistry
Myers, Karl Johnson, Sr., radiology
Peters, Franklin Traviss, chemistry

PARSONS

Aubertin, Gerald Martin, watershed management
Patric, James Holton, forest hydrology

PRINCETON

Plass, William T., forest ecology

ST ALBANS

Decker, Quintin William, synthetic organic chemistry
Fisher, John F., analytical chemistry, physical chemistry
Manyik, Robert Michael, petroleum chemistry
Williamson, Kenneth Dale, physical chemistry

SALEM

Bond, Stephon Thomas, inorganic chemistry
England, Wayne H., plant anatomy, mycology

SHEPHERDSTOWN

Bell, Carl F., plant pathology
Diehl, John Edwin, biochemistry
Latterell, Richard L., genetics
Morris, Peter Craig, mathematics
Volker, Eugene Jeno, organic chemistry

SISTERSVILLE

Bailey, Donald Leroy, organic chemistry
Voisiner, Donald Louis, chemistry

SOUTH CHARLESTON

Andrawes, Nathan R., entomology
Atkins, Kenneth Earl, organic chemistry
Backer, Ronald Charles, toxicology
Barnes, Robert Keith, organic chemistry
Bartley, William J., industrial chemistry
Bassett, David R., physical chemistry
Bhasin, Madan M., physical chemistry, surface chemistry
Caflisch, Edward George, organic chemistry
Chang, Yeong-Jen Peter, physical chemistry
Chen, Tsong Meng, plant physiology
Chiasson, Bertrand Arnold, organic chemistry
D'Silva, Themistocles Damasceno Joaquim, synthetic organic chemistry, steroid chemistry
Durden, John Apling, Jr., pharmaceutical chemistry, organic chemistry
Frosick, Frederick Charles, Jr., organic chemistry
Fu, Wallace Yamtak, organic chemistry
Haag, William George, bio-organic chemistry
Hager, Stanley Lee, physical chemistry
Harrison, Arnold Myron, physical chemistry
Henry, Joseph Peter, organic chemistry
Hess, Lawrence George, synthetic organic chemistry
Kaplan, Leonard, organic chemistry
Keller, Frederick Albert, Jr., biochemistry, biochemical engineering
Knopf, Robert John, organic chemistry
Koenig, Harvey Steven, textile chemistry
Koleske, Joseph Victor, polymer chemistry
Kurland, Jonathan Joshua, physical organic chemistry
Kurtz, A Peter, pesticide chemistry
Kuryla, William C., organic chemistry
Lee, Young-Jin, chemistry, organic chemistry
MacFeek, Donald Lester, industrial organic chemistry
Manning, David Treadway, bio-organic chemistry, agricultural chemistry
Marcinkowsky, Arthur Ernest, physical chemistry
Mellen, Gilbert Emery, physics
Miller, Walter Peter, organic polymer chemistry
Moorefield, Herbert Hughes, toxicology
Osborn, Claborn Lee, organic chemistry, radiation chemistry
Pace, William Theodore, organic chemistry
Papa, Anthony Joseph, organic chemistry
Park, Kisoon, physical chemistry
Rife, Robert Seldon, industrial organic chemistry
Robson, John Howard, polymer chemistry, organic chemistry
Sherman, Paul Dwight, Jr., industrial organic chemistry
Smith, Joseph James, administration
Smith, Oliver Wendell, organic chemistry
Smith, Percy Leighton, chemistry

Steinle, Edmund Charles, Jr, organic chemistry
Taller, Robert Arthur, physical organic chemistry
von Dohlen, Werner Claus, surface chemistry
Walker, Wellington Epler, industrial chemistry, organic chemistry
Weiden, Mathias Herman Joseph, insect toxicology
Wheeler, Thomas Neil, organic chemistry
Wilson, Thomas Putnam, chemistry
Young, Frank Glynn, chemistry

VIENNA
Webber, Thomas Gray, organic chemistry

WASHINGTON
Johnson, Richard Lawrence, organic chemistry
Richter, Robert Freeland, industrial organic chemistry

WEIRTON
McGraw, Leslie Daniel, chemistry
Sherockman, Andrew Antolcik, inorganic chemistry

WEST LIBERTY
Campbell, Clyde Del, organic chemistry, biochemistry
Cook, Harold Andrew, agricultural microbiology
Mitra, Rathin, microbiology
Schramm, Robert William, physics
Swan, Frederick Robbins, Jr, ecology
Talley, Lawerence Horace, academic administration, science education

WHEELING
Giza, Chester Anthony, organic chemistry, synthetic organic chemistry
Giza, Yueh-Hua Chen, organic biochemistry
Hanzely, Joseph Bernard, physiology
Knorr, Thomas George, solid state physics
Radford, Loren E, solid state physics

WILLIAMSON
Meyers, H Russell, neurophysiology, neurosurgery

WILLOW ISLAND
Fessler, Robert Glenn, physical chemistry, inorganic chemistry
Moyer, Melvin Isaac, organic chemistry
Noe, James L, organic chemistry
Ulrey, Stephen Scott, industrial organic chemistry

WISCONSIN

APPLETON
Anand, Amarjit S, reproductive physiology, biochemistry
Atala, Rajai Hanna, chemical physics
Baum, Gary Allen, solid state physics
Berry, Andrew Campbell, mathematics
Biesner, William Clark, paper & pulp technology, chemical engineering
Boggs, Lawrence Allen, classical mechanics
Brackenridge, John Bruce, classical mechanics
Brandenberger, John Russell, atomic physics
Buchanan, Marion Alexander, chemistry
Collins, John W, biochemistry, organic chemistry
Cook, David Marsden, theoretical physics, mathematical physics
Cundy, Paul Franklin, inorganic chemistry
Darling, Stephen Foster, organic chemistry
Dixson, Henry Philip, chemistry
Downs, Martin Luther, chemistry
Dugal, Hardev Singh, paper chemistry
Easty, Dwight Buchanan, analytical chemistry
Einspahr, Dean William, forest genetics
Evans, James Stuart, physical chemistry, inorganic chemistry
Green, John Wilson, organic chemistry
Howells, Thomas Alfred, paper chemistry
Isenberg, Irving Harry, wood anatomy
Joel, Cliffe David, biochemistry

Johnson, Donald Curtis, organic chemistry
Johnson, Morris Alfred, plant biochemistry
Kresch, Alan J, physical chemistry, data processing
LaMarca, Michael James, developmental biology
Lauterbach, George Ervin, physical chemistry
Leekley, Robert Mitchell, organic chemistry
Lokensgard, Jerrold Paul, organic chemistry
Long, Richard Gene, mathematics
Maravolo, Nicholas Charles, plant morphogenesis
Mason, Ronald James, anthropology, archaeology
McClenahan, William St Clair, organic chemistry
McKelvey, Ronald Deane, physical organic chemistry, photochemistry
Miller, Arild Justesen, physical chemistry
Nelson, Richard William, physics
Palmquist, John Charles, geology
Park, Robert William, paper chemistry
Pearl, Irwin Albert, wood chemistry
Piper, Carl Victor, analytical chemistry
Radford, David Eugene, mathematics
Read, William Franklin, geology
Richman, Sumner, aquatic ecology
Rosenberg, Robert Melvin, physical biochemistry
Ross, Theodore William, geology
Schroeder, Leland Roy, organic chemistry
Schwab, Helmut, analytical chemistry, physical chemistry
Senge, George H, mathematics
Shibley, Gilbert A, comparative animal physiology
Smith, Thomas Stevenson, solid state physics
Spinner, Theodore, chemistry, physics
Stevens, Michael Fred, industrial chemistry
Stewart, James Collier, mathematics
Stillings, Robert Almon, chemistry
Stratton, Robert Alan, polymer chemistry
Swanson, John William, physical chemistry
Tank, Ronald W, geology
Thompson, Norman Strom, organic chemistry
Van den Akker, Johannes Archibald, thermodynamics, optics
Ward, Kyle, Jr, cellulose chemistry, pulp & paper technology
Winton, Lawson Lowell, forest physiology, tissue culture
Wollwage, John Carl, paper chemistry

ASHLAND
Leugoeb, Rosalia Aloisia, chemistry
Verch, Richard Lee, aquatic biology

BARABOO
Haavik, Arne Goodwin, medicine

BEAVER DAM
Glunz, Paul R, pathology

BELGIUM
Sullivan, Harris Martin, instrumentation, materials science

BELOIT
Alton, Alvin John, food science
Bailey, John Martin, solid state physics
Biester, John Louis, chemistry, academic administration
Brown, William Henry, organic chemistry, biochemistry
Burger, John Allan, geology
Dobson, David A, nuclear physics, atomic physics
Finch, John Vernor, mathematics
Fuller, Edward C, physical chemistry
Garrett, Robert Ogden, physics
Kemler, John Hughes, geography
Kunny, Bartholomew Kenneth, limnology, zoology
Lutz, John Ewald, zoology
Mathews, Frederick John, organic chemistry
Newsome, Richard Duane, plant ecology
Saxe, Bernhard David, organic chemistry
Schroeder, Daniel John, astronomy, optics
Spencer, John Brockett, physical chemistry
Stenstrom, Richard Charles, geology

Stoltzfus, Joseph Christian, nuclear physics
Sweet, Gertrude Evans, zoology
Welch, Donald Ray, plant pathology, biochemistry
Whiteford, Andrew Hunter, ethnology
Wilde, Edwin Frederick, Jr, mathematics
Woodard, Henry Herman, Jr, geology

BIRNAMWOOD
Discher, Clarence August, physical chemistry, inorganic chemistry

BROKAW
Roberts, Richard W, organic chemistry

BROOKFIELD
Hynek, Robert James, analytical chemistry

BURLINGTON
Palm, Elmer Thurman, plant pathology

CEDARBURG
Luetzow, Arthur Edward, chemistry

COTTAGE GROVE
Krueger, John W, organic chemistry

DE FOREST
Elliott, Fred Irvine, reproductive physiology
Larson, Lester Leroy, veterinary medicine
Miller, Paul Dean, animal breeding
Pace, Marvin M, reproductive physiology
Walton, Robert Eugene, animal breeding, animal genetics

DE PERE
Flanigan, Norbert James, anatomy
Hodgson, James Russell, zoology
Klopotek, David L, organic chemistry
Vandehey, Robert C, entomology

DELAFIELD
Helz, Armin Werner, physical chemistry

EAU CLAIRE
Anderson, James Gerard, solid state physics, electronics
Bakken, Arnold, zoology
Beckfield, William John, pathology
Campbell, Donald L, inorganic chemistry
Chess, Karin V T, mathematics
Crowe, David Burns, zoology
Cvancara, Victor Alan, physiology
Dixon, John Charles, entomology
Duerst, Richard William, physical chemistry, analytical chemistry
Fay, Marcus J, plant taxonomy
Font, William Francis, parasitology
Foote, Kenneth Gerald, biology
Fossland, Robert Gerard, reproductive biology, evolutionary biology
Gerberich, John Barnes, microbiology
Gleiter, Melvin Earl, biochemistry, environmental chemistry
Goranson, Leonard D, geography
Jones, Helena Speiser, anatomy
Klink, Joel Richard, organic chemistry
Lewke, Robert Edward, vertebrate ecology
Lim, Johng Ki, genetics
Marking, Ralph H, inorganic chemistry
Naughten, John Charles, zoology
Ochyrmowycz, Leo Arthur, organic chemistry
O'Connell, Kevin Marshall, cell biology, endocrinology
Page, Allen D, physics, science education
Rouse, Thomas C, parasitology,
Saigo, Roy Hirofumi, plant anatomy, plant pathology
St Louis, Robert Vincent, physical chemistry
Schnack, Larry G, organic chemistry
Schultz, Frederick Herman Carl, physics
Seitz, Kerlin (McCullough), geography
Snudden, Birdell Harry, bacteriology, food science
Wahlstrom, Lawrence F, mathematics
Wilcox, Archer Carl, chemistry, biochemistry
Willis, Ronald Porter, geology

ELCHO
Kluchesky, Elmer Francis, analytical chemistry

ELM GROVE
Headlee, Raymond, psychiatry, psychology
Kneen, Eric, agricultural biochemistry, enzymology

FOND DU LAC
Palen, M Imogene, microbiology
Stubbings, Robert Lamb, physical chemistry, biochemistry

FRANKLIN
Alt, Leslie L, chemistry

FRANKSVILLE
Guertin, Jacques P, physical inorganic chemistry

GLENDALE
Wolfson, Leonard Louis, microbiology

GRAFTON
Heyke, Brigitte, veterinary virology
Kolar, Joseph Robert, Jr, veterinary virology
Rude, Theodore Alfred, pathology, microbiology

GREEN BAY
Anderson, Harold J, physical chemistry
Atkisson, Arthur Albert, environmental management
Benham, Graham Harvey, biochemistry
Byrne, Frank Edward, geology
Clifton, James Alfred, applied anthropology, cultural anthropology
Cook, Robert Sewell, vertebrate ecology
Deese, Dawson Charles, biochemistry
Doberenz, Alexander R, nutrition
Fischbach, Fritz Albert, biophysics,
Gandre, Donald Alfred, geography
Girard, Dennis Michael, mathematics, statistics
Guilford, Harry Garrett, zoology, parasitology
Ihrke, Charles Albert, genetics, plant breeding
Jayne, Jack Edgar, chemistry
Jowett, David, statistics, botany
Kaufman, William Carl, Jr, human physiology, biophysics
Kuesel, Donald Charles, food science
Maier, Robert Hawthorne, plant physiology, environmental management
McIntosh, Elaine Nelson, physiological bacteriology, nutrition
McIntosh, Thomas Henry, environmental sciences
Moore, Douglas Houston, applied mathematics
Morgan, Michael Dean, plant ecology
Mowbray, Thomas Bruce, botany, plant ecology
Nair, Gangadharan V M, plant pathology, mycology
Norman, Jack C, nuclear chemistry, radiochemistry
O'Hearn, George Thomas, science education
Petrie, George Whitefield, III, applied mathematics
Reed, John Frederick, botany
Rhyner, Charles R, solid state physics
Schwartz, Leander Joseph, microbiology
Sell, Nancy Jean, chemical physics
Stevens, Richard Joseph, neurosciences
Wiersma, James H, analytical chemistry

HUDSON
Duwell, Ernest John, physical chemistry

JANESVILLE
Miller, Knudt John, horticulture, plant physiology
Rice, Marion McBurney, bacteriology, botany
Rust, Charles Chapin, zoology

JUNEAU
Bernstein, Sheldon, physiological chemistry

KENOSHA

Amin, Omar M., medical entomology, parasitology
Balsano, Joseph Silvio, population biology
Casey, John Addis, physics
Chen, Chong Maw, plant biochemistry
Datta, Surinder P., genetics, immunology
Esser, Robert Emmet, botany, invertebrate zoology
Firebaugh, Morris W., electronics, energy conversion
Fossum, Timothy V., algebra
Fraser, Margaret Shirley, inorganic chemistry
Gasiorkiewicz, Eugene Constantine, plant science, plant pathology
Goodman, Eugene Marvin, cell biology
Greenebaum, Ben, experimental atomic physics, biophysics
Hamm, Kenneth Lee, organic chemistry
Hansen, Paul Vincent, Jr., physical chemistry
Isenberg, Norbert, organic chemistry
Jeanmaire, Robert L., science education
Knight, Homer Talcott, analytical chemistry
Lowenthal, Franklin, mathematics
Marron, Michael Thomas, theoretical chemistry, physical chemistry
Mochon, Marion Johnson, anthropology
Ogren, Herman August, ecology
Pyper, Diane Marie, astronomy
Quass, La Verne Carl, inorganic chemistry
Schneider, Allan Frank, glacial geology
Shea, James H., geology
Shinkle, Michael Paul, entomology, agricultural chemistry
Smith, Eugene Irwin, petrology, astrogeology
Strommen, Dennis Patrick, inorganic chemistry, spectroscopy
Tietel, Ralph Maurice, botany
Vozza, John F., organic chemistry
Weston, Kenneth W., algebra
Williams, Anna Maria, microbiology
Zimmerman, Lorraine May, anthropology

LA CROSSE

Abts, Mary Lavonne, chemistry
Claflin, Tom O., biology
Classen, Harold Arthur, economic geography, geography of Asia
Cowley, Milford A., food chemistry
Egbert, Gary Trent, molecular spectroscopy
Eisbernd, Helen, inorganic chemistry
Fystrom, Dell O., atomic physics
Gray, Edwin R., human anatomy, electromyography
Hartley, Richard Thomas, zoology
Holder, Virgil Harold, geography
Hosler, Charles Frederick, Jr., biochemistry, organic chemistry
Hunn, Joseph Bruce, physiology, fisheries
Joyner, Powell Austin, research management
Kistner, Clifford Richard, inorganic chemistry
Lindner, Kenneth E., radiochemistry
Meyer, Fred Paul, parasitology, fisheries
Nelson, Allen Charles, microbiology
Rausch, Gerald, organic chemistry
Roskos, Roland R., physical chemistry
Senff, Robert E., animal physiology
Tonnis, John A., organic chemistry
Sohmer, Seymour H., systematic botany
Unhebaun, Laraine Marie, plant pathology
Warner, James Howard, botany, plant ecology
Weber, Albert Vincent, botany
Weeks, Thomas F., plant physiology
Winrich, Lonny B., computer science, applied mathematics
Young, Howard Frederick, zoology

LYNDON STATION

Freeman, Smith, biochemistry

MADISON

Abrahamson, Seymour, genetics
Adams, Michael Studebaker, plant ecology
Adler, Julius, biochemistry, genetics
Ahlgren, Henry Lawrence, agronomy
Albright, Edwin C., medicine
Alexander, Samuel Craighead, Jr., anesthesiology, pharmacology
Allen, James R., Jr., pathology
Allen, Oscar Nelson, bacteriology
Allen, Paul James, plant physiology
Allin, Edgar Francis, anatomy

Amundson, Clyde Howard, food science
Andaregg, John William, biophysics
Anderson, Charles Edward, physical meteorology
Anderson, Christopher Marlowe, astronomy
Anderson, John Walberg, anatomy, histology
Anderson, Laurens, biochemistry, bio-organic chemistry
Anderson, Louis Wilton, atomic physics
Anderson, Mary Pikul, groundwater hydrology
Andrew, Robert Harry, agronomy
Ansfield, Fred Joseph, oncology
Apple, James Wilbur, oncology
Amy, Deane Cedric, plant pathology
Arrington, Louis Carroll, poultry science
Askey, Richard Allen, mathematical analysis
Auerbach, Robert, developmental biology, immunology
Backus, Myron Port, mycology
Baechler, Roy Herman, chemistry, forest products
Baerreis, David Albert, anthropology
Bailey, Sturges Williams, clay mineralogy, crystallography
Balish, Edward, microbiology, biochemistry
Bamforth, Betty Jane, anesthesiology
Bard, John C., food science
Barger, Vernon Duane, theoretical high energy physics, elementary particle physics
Baylis, Jeffrey Rowe, ethology
Barnes, Lester E., zoology, physiology
Barschall, Henry Herman, nuclear science
Bartlett, David E., veterinary pathology
Bass, Paul, pharmacology, physiology
Baumann, Carl August, nutritional biochemistry
Beck, Anatole, mathematics
Beck, Gail Edwin, floriculture, plant physiology
Beck, Stanley Dwight, zoology, ichthyology
Beatty, Marvin Theodore, soils, resource management
Beals, Edward Wesley, ecology
Becker, Wayne Marvin, chemistry, environmental management
Beeman, William Waldron, biophysics
Beinert, Helmut, biochemistry, enzymology
Bender, Margaret McLean, organic chemistry
Bender, Paul J., physical chemistry
Benevenga, Norlin Jay, nutrition
Benforado, Joseph Mark, clinical pharmacology
Benjamin, Daniel Marshall, forest entomology
Benjamin, Robert Myles, neurophysiology
Bennett, E Maxine, otolaryngology
Bennett, Kenneth A., biological anthropology, human genetics
Bentley, Charles Raymond, geophysics, glaciology
Berbee, John Gerard, plant pathology, forestry
Bergdoll, Merlin Scott, biochemistry
Berlow, Stanley, pediatrics
Berman, Alvin Leonard, neuroanatomy
Berman, David Theodore, veterinary science
Berry, Michael James, physical chemistry, chemical physics
Bhattacharyya, Gouri Kanta, mathematical statistics, statistics
Bincer, Adam Marian, theoretical physics
Bingham, Edwin Theodore, plant breeding
Binning, Larry Keith, horticulture
Bird, Herbert Roderick, animal nutrition
Bisgard, Gerald Edwin, veterinary physiology
Bittar, Evelyn Edward, physiology
Bjorksten, Johan Augustus, chemistry
Blaedel, Walter John, analytical chemistry
Blair, James Edward, physical chemistry
Blanchard, Converse Herrick, nuclear physics
Blattner, Frederick Russell, biophysics
Blazkovec, Andrew A., immunology
Bleicher, Michael Nathaniel, mathematics
Bliss, Robert Charles, astronomy
Bliss, Fredrick Allen, plant genetics, plant breeding
Blockstein, William Leonard, health sciences, academic administration

Bloodworth, James Morgan Bartow, Jr., pathology
Boake, William Charles, cardiology, internal medicine
Bock, Robert Manley, biophysical chemistry
Booker, Harold E., neurology
Boone, Donald Milford, plant pathology
Borchers, Robert Reece, nuclear physics
Borisy, Gary Guy, molecular biology
Boush, George Mallory, entomology
Boutwell, Roswell Knight, biochemistry
Bownds, M Deric, neurobiology
Bowser, Carl, geochemistry
Box, George Edward Pelham, statistics
Bradley, Robert Lester, Jr., dairy technology, food science
Brandenburg, James H., otolaryngology
Brauer, Fred, mathematics
Bray, Robert Woodbury, meat science
Bremel, Robert Duane, animal physiology, biochemistry
Brickbauer, Elwood Arthur, agronomy
Brill, Winston J., microbiology, biochemical genetics
Brink, Royal Alexander, genetics, botany
Brock, Katherine Middleton, microbiology, microbial ecology
Brock, Thomas Dale, microbiology
Brown, Raymond Russell, biochemistry
Brualdi, Richard Anthony, mathematics
Bruch, Ludwig Walter, theoretical physics
Bruck, Richard Hubert, mathematics
Brugge, John F., physiology
Bryan, George Terrell, medicine, pharmacology
Bryson, Reid Allen, meteorology
Buchanan-Davidson, Dorothy Jean, biochemistry
Buchan, Robert Ray, molecular biology
Buck, Robert Creighton, mathematical analysis
Buongiorno, Joseph, forest economics
Burdsall, Harold Hugh, Jr., mycology
Burger, Warren Clark, plant science
Burgess, Richard Ray, molecular biology
Burkholder, Peter M, pathology, immunology
Burkholder, Wendell Eugene, entomology
Burris, William Chandler, parasitology
Burris, Robert Harza, biochemistry
Cain, John Manford, resource management
Calbert, Harold Edward, biochemistry
Caldwell, William L., radiology
Cameron, Eugene Nathan, economic geology
Cameron, John Roderick, medical physics
Campbell, Harold Alexander, biochemistry
Carbone, Paul P., oncology
Carlson, Stanley David, entomology, physiology
Casey, Charles P., organic chemistry
Casey, Martha L., biochemistry
Casida, Lester Earl, reproductive physiology
Cassens, Robert G., biochemistry
Cassinelli, Joseph Patrick, astrophysics
Caulfield, Daniel Francis, polymer chemistry
Chambliss, Glenn Hilton, microbiology
Chang, Louis Wai-Wah, experimental pathology
Chapman, Arthur Barclay, animal breeding
Chapman, R Keith, entomology
Chard, Chester Stevens, anthropology
Chase, Ivan Dmitri, ethology
Chosy, Julius J., internal medicine, psychosomatic medicine
Chover, Joshua, mathematics
Chu, Fun Sun, biochemistry
Chun, Raymond Wai Mun, pediatrics, neurology
Churchill, Lynn, neurochemistry
Clark, David Leigh, paleontology
Claude, Philippa, cell biology
Clay, Clarence Samuel, geophysics
Cleland, Charles Samuel, neuropsychology, psychology
Cleland, William Wallace, biochemistry
Clifton, Kelly Hardenbrook, experimental biology
Cliver, Dean Otis, virology
Code, Arthur Dodd, astronomy, astrophysics
Cohen, Philip Pacy, physiological chemistry, enzymology
Colas, Antonio E., biochemistry, reproductive physiology
Coleman, Donald George, forestry
Collins, George Edwin, computer science, mathematics
Combs, Ova Beetem, horticulture

Conley, Charles Cameron, mathematics
Conner, Howard Emmett, mathematics
Connors, Kenneth A., pharmaceutics, analytical chemistry
Conrad, John Rudolph, plasma physics
Cooke, Robert E., pediatrics
Cooper, Garrett, dermatology
Cooper, Margaret Moore, textile chemistry
Coppel, Harry Charles, entomology
Corey, Richard Boardman, soil chemistry, soil fertility
Cornwell, Charles Daniel, chemical physics
Cottam, Grant, ecology, academic administration
Craddock, (John) Campbell, structural geology, tectonics
Cramer, Jane Harris, molecular biology
Crummy, Andrew B., radiology
Cryer, Colin Walker, computer science
Cunningham, Gordon Rowe, forestry
Curreri, Anthony Rudolph, thoracic surgery
Curtiss, Charles Francis, theoretical chemistry
Curwen, David, horticulture, food technology
Dahl, Lawrence Frederick, physical inorganic chemistry
Dahlberg, James Eric, biochemistry
Dana, Malcolm Niven, horticulture
Daub, Edward E., history of science
DeLuca, Hector Floyd, biochemistry
DeMars, Robert Ivan, microbiology
Davis, Charles Patrick, microbiology
Davis, Larry Dean, human physiology
de Boor, Carl (Wilhelm) Reinhold, mathematics
Denevan, William Maxfield, cultural geography, geography of Latin America
Dennis, Warren Howard, physiology, biophysics
Desautels, Edouard Joseph, computer sciences
Deutsch, Harold Francis, chemistry
DeWitt, Calvin Boyd, physiological ecology
Dexter, Richard Norman, solid state physics
De Zoeten, Gustaaf A., plant pathology, plant virology
Dick, Elliot C., microbiology, epidemiology
Dicke, Robert Jerome, insect morphology
Dickey, Ronald Wayne, applied mathematics
Dickie, Helen Aird, medicine
Dierschke, Donald Joe, reproductive biology
Dillinger, Joseph Rollen, low temperature physics
Doersch, Ronald Ernest, weed science
Doherty, Lowell Ralph, astrophysics
Donhowe, John M., nuclear physics
Dott, Robert Henry, Jr., sedimentology, tectonics
Dove, William Franklin, genetics, cell biology
Draper, Norman Richard, statistics
Drescher, William James, groundwater hydrology, hydrogeology
Drolsom, Paul Newell, agronomy
Dudley, Alden Woodbury, Jr., neuropathology
Duncan, Charles Lee, bacteriology
Dunn, Stanley Austin, industrial chemistry
Durand, Bernice, theoretical physics
Durand, Loyal, III, theoretical physics
Durand, Ralph Edward, radiobiology
Durbin, Richard Duane, plant pathology
Dury, George H., geomorphology
Easterday, Bernard Carlyle, comparative medicine
Ebel, Marvin Emerson, theoretical high energy physics, solid state physics
Edwards, Gerald Elmo, plant science, physiology
Eggert, Arthur Arnold, analytical chemistry, computer science
Ehrlich, Edward Norman, internal medicine
Eichman, Peter L., neurology, medicine
Ek, Alan R., forestry
Ellarson, Robert Scott, wildlife management

Elmendorf, William Welcome, ethnology, anthropological linguistics
Elson, Charles, nutrition
Emlen, John Thompson, Jr, zoology
Emmert, Gilbert A, plasma physics
Emmons, Richard Conrad, geology
Enders, George Leonhard, Jr, microbiology
Engerman, Ronald Lester, ophthalmology
Epstein, Saul Theodore, theoretical physics
Erdos, Gregory William, mycology
Erickson, Theodore Charles, neurosurgery, neurophysiology
Erwin, Albert R, physics
Esenther, Glenn R, forest entomology
Eslyn, Wallace Eugene, forest pathology
Ethington, Robert Loren, wood science & technology, engineering mechanics
Evans, Dennis Hyde, analytical chemistry, electrochemistry
Evans, George William, animal science
Evenson, Merle Armin, analytical chemistry, toxicology
Evert, Ray Franklin, botany
Fadell, Edward Richard, mathematics
Fahien, Leonard A, pharmacology, biochemistry
Falk, Edward D, instrumentation
Fawcett, Richard Steven, weed science
Feldballe, Jeanette, zoology
Fennema, Owen Richard, food science
Ferry, John Douglass, polymer chemistry
Fetzner, William Norman, bioengineering
Finley, Robert William, geography
First, Neal L, reproductive physiology
Fisher, Ellsworth Henry, entomology
Fitch, Walter M, biochemistry
Fitzgerald, George Patrick, fresh water ecology
Forelli, Frank John, mathematics
Forsberg, Robert Arnold, agronomy, plant genetics
Forster, Francis Michael, clinical neurology
Foster, Edwin Michael, bacteriology
Fox, Allen Sander, genetics, biochemistry
Fraser, Lemuel Anderson, zoology
Fred, Edwin Broun, bacteriology
Freeman, Joan Elizabeth, anthropology, archaeology
Friedman, William Albert, physics
Fry, William Frederick, experimental physics
Fulton, Robert Watt, plant pathology
Gabelman, Warren Henry, plant genetics
Gaines, Donald Frank, inorganic chemistry
Gardner, Wilford Robert, physics
Gates, Robert Maynard, geology
Gaumnitz, Erwin Alfred, applied statistics
Geison, Ronald Leon, neurochemistry, nutrition
Gerloff, Gerald Carl, botany
Gerritsen, Theo, biochemistry
Gibson, John Michael, neurophysiology, bioengineering
Gibson, Mary Morton, neurophysiology
Giese, Ronald Lawrence, forest entomology
Gilbert, Enid May Fischer, pathology
Gilboe, David Dougherty, biochemistry, physiology
Gillespie, Robert Howard, organic chemistry
Gleim, Robert David, bio-organic chemistry
Glover, Benjamin Howell, psychiatry
Glover, Everett D, geochemistry
Goebel, Charles James, theoretical physics
Goepfert, John McDonnell, microbiology, food bacteriology
Goering, Harlan Lowell, organic chemistry
Gojmerac, Walter Louis, entomology
Goldstein, Robert, audiology, psychophysiology
Golubjatnikov, Rjurik, microbiology, epidemiology
Gomez-Ibanez, Daniel Alexander, geography
Goodfriend, Theodore L, internal medicine, biochemistry
Gordon, Edgar Stillwell, biochemistry, medicine
Gormican, Annette, nutrition
Gorski, Jack, biochemistry, endocrinology
Goy, Robert William, psychophysiology
Graham, David Tredway, internal medicine
Graham, Frances Keesler, psychophysiology
Graven, Stanley N, pediatrics, biochemistry
Green, David Ezra, biochemistry
Green, Theodore, III, physical oceanography, fluid mechanics
Greenspan, Donald, mathematics
Greville, Thomas Nall Eden, mathematics
Gritton, Earl Thomas, agronomy
Grummer, Robert Henry, animal husbandry
Guidotti, Charles V, geology, petrology
Guillery, Rainer Walter, neuroanatomy
Gunji, Hiroshi, mathematics
Gurland, John, statistics
Gustafson, David Harold, industrial engineering, preventive medicine
Haeberli, Willy, nuclear physics
Haig, Thomas O, meteorology, space science
Hailman, Jack Parker, ethology
Hall, James Emerson, mathematics
Hall, Robert Everett, veterinary medicine
Hall, Timothy Couzens, plant physiology, plant biochemistry
Halton, John Henry, mathematics, computer science
Hammer, Sigmund Immanuel, geophysics
Hansen, Marc F, pediatrics
Hanson, Earle William, plant pathology, field crops
Hanson, Richard Steven, biochemistry, microbiology
Hanson, Robert Paul, epizootiology
Harkin, John McLay, organic chemistry, biochemistry
Harlow, Harry F, experimental psychology
Harper, Alfred Edwin, biochemistry
Harriman, John E, chemical physics, quantum chemistry
Harris, Bernard, mathematical statistics
Hart, Phillip A, organic chemistry
Hartmann, Henrik Anton, pathology
Hartshorne, Richard, geography, international relations
Harvey, John Grover, mathematics
Harvey, Robert Gordon, Jr, weed science, agronomy
Hasler, Arthur Davis, zoology
Hasselkus, Edward R, ornamental horticulture
Hastenrath, Stefan Ludwig, meteorology, climatology
Hedden, Gregory Dexter, organic chemistry, meteorology
Heidelberger, Charles, biochemistry
Helgeson, John Paul, plant physiology
Herman, Gerald Francis, meteorology
Hickey, Joseph James, wildlife ecology
Hickman, John Marshall, anthropology, applied anthropology
Highley, Terry L, forest pathology
Hildebrandt, Albert Christian, plant pathology
Hilden, Shirley Ann, cell physiology
Hilsenhoff, William Leroy, entomology
Hind, Joseph Edward, neurophysiology
Hine, Ruth Louise, wildlife conservation
Hinze, Harry Clifford, medical microbiology
Hirschfelder, Joseph Oakland, chemistry
Hitchcock, John Thayer, cultural anthropology
Hoar, Donald Wayne, nutrition, biochemistry
Hoekstra, William George, nutritional biochemistry
Hokin, Lowell Edward, biochemistry
Hokin-Neaverson, Mabel, biochemistry
Holden, James Edward, medical physics
Hole, Francis Doan, soil science, geography
Holland, Wilbur Charles, Jr, mathematics
Holm, LeRoy George, plant physiology
Holt, Charles Lee Roy, Jr, geology
Holt, Matthew Leslie, electrochemistry
Hong, Richard, pediatrics, immunology
Horn, Lyle Henry, meteorology
Hougas, Robert Wayne, plant breeding, genetics
Houge, James C, biomedical engineering, biomaterials
Houghton, David Drew, meteorology
Howard, W Terry, animal nutrition, dairy husbandry
Howe, Martha Morgan, molecular genetics, virology
Huber, David Lawrence, theoretical physics
Huston, Norman Earl, nuclear physics
Iha, Thomas H, microbial genetics
Ihde, Aaron John, chemistry, history of science
Iltis, Hugh Hellmut, systematic botany, biogeography
Ingraham, Mark Hoyt, mathematics
Inhorn, Stanley L, pathology
Inman, Ross, molecular biology, physical chemistry
Irwin, Malcolm Robert, zoology, genetics
Isaacs, Irving Martin, mathematics
Jackson, Marion Leroy, agronomy, soil science
Jaeschke, Walter Henry, pathology
Javid, Manucher J, neurosurgery
Johannes, Russell Fredrick, agronomy
Johnson, Bruce McDougall, analytical chemistry, cancer
Johnson, Donald R, meteorology
Johnson, Marvin Joyce, microbial biochemistry
Johnson, Millard Wallace, Jr, mathematics
Johnson, Robert O, surgery, oncology
Johnson, Sture Archie Mansfield, medicine
Jones, Lois Mae, bacteriology
Juhl, John Harold, radiology
Jurist, John Michael, biophysics
Kabler, J D, internal medicine
Kaesberg, Paul Joseph, biophysics, biochemistry
Kahn, Donald R, cardiovascular surgery, thoracic surgery
Karavolas, Harry J, biochemistry
Karreman, Herman Felix, applied mathematics
Kasper, Charles Boyer, biochemistry
Kauffman, Robert Giller, meat science
Keeney, Dennis Raymond, soil fertility, biochemistry
Keesey, Ulker Tulunay, psychophysiology
Keisler, Howard Jerome, mathematical logic
Keith, Lloyd Burrows, wildlife management
Kelman, Arthur, phytopathology
Kelsey, Charles Andrew, medical physics
Kemp, John Daniel, biochemistry
Kendrick, John Edsel, physiology
Kepecs, Joseph Goodman, psychiatry, psychoanalysis
Kermicle, Jerry Lee, genetics
Kerst, Donald William, plasma physics
Kielsmeier, Elwood William, dairy industry
Kleene, Stephen Cole, mathematical logic
Klotz, Jerome Hamilton, biostatistics
Knox, James Clarence, geomorphology, physical geography
Koeppen, Robert Carl, botany
Konrad, John Grey, soil chemistry, water chemistry
Kornguth, Steven E, biochemistry
Korst, Donald Richardson, internal medicine
Kotch, Alex, organic chemistry
Koval, Charles Francis, entomology
Kowal, Robert Raymond, systematic botany, biometry
Kozlowski, Theodore Thomas, plant physiology, forest physiology
Kraushaar, William Lester, physics
Kubinski, Henry A, oncology, molecular biology
Kubler, Hans Jakob, forest products
Kukachka, Bohumil Francis, plant anatomy
Kung, Ching, genetics, neurobiology
Kunin, Calvin Murry, internal medicine, preventive medicine
Kurtz, Thomas Gordon, mathematics
Ladinsky, Judith L, cytology, endocrinology
Laessig, Ronald Harold, clinical chemistry, public health
Lagally, Max Gunter, materials science, surface physics
Lage, Gary Lee, pharmacology
Lalich, Joseph John, pathology
Landucci, Lawrence L, organic chemistry, wood chemistry
Lardy, Henry Arnold, biochemistry
Larsen, Edwin Merritt, inorganic chemistry
Larsen, Eleanor Marie, medical physiology
Larsen, James Arthur, plant ecology
Larson, Frank Clark, medicine
Lattuada, Charles P, bacteriology
Laudon, Lowell Robert, geology
Lawton, Gerald Warren, chemistry
Lea, William Lief, radiological health
Lee, Gerhard Bjarne, soils
Lee, Tee-Ping, biochemistry, pharmacology
Lehmann, William Fredrick, forest products
Lemmer, Kenneth Ellery, surgery
Leonard, Thomas Joseph, developmental biology
Lester, Donald Thomas, forestry
Lettau, Heinz Helmut, geophysics, meteorology
Lettau, Katharina, climatology
Levin, Jacob Joseph, mathematical analysis
Levy, Lawrence, algebra
Lewis, Herbert Samuel, anthropology, ethnography
Lewis, William C, psychiatry
Libby, John Lester, entomology
Lichtenstein, Emanuel Paul, entomology
Lindsay, Robert Clarence, food science, food chemistry
Link, Karl Paul, biochemistry
Linkswiler, Hellen, nutrition
Littlewood, Barbara Shaffer, biochemistry, genetics
Littlewood, Roland Kay, computer science, molecular biology
Locke, Louis Noah, animal pathology
Lodge, Arthur Scott, physics
Longley, B Jack, surgery
Loucks, Orie Lipton, botany, ecology
Lower, Gerald Malcolm, Jr, oncology, biochemistry
Luby, Patrick Joseph, agricultural economics
Lund, Daryl B, food science
Lysenko, Michael George, parasitology
MacCormick, Alasdair John, medical statistics
MacKinney, Archie Allen, Jr, hematology
Madsen, Paul O, urology
Magnuson, John Joseph, hydrobiology
Maher, Louis James, Jr, palynology, quaternary geology
Mangasarian, Olvi Leon, applied mathematics
Manning, Dean David, immunology
March, Robert Herbert, physics
Marlett, Judith Ann, nutrition
Marsh, Benjamin Bruce, meat science, muscular physiology
Marsh, Richard Floyd, veterinary virology, veterinary pathology
Marth, Elmer Herman, food microbiology, dairy microbiology
Martin, David William, meteorology
Mathis, John Samuel, astrophysics
Matthews, Charles George, neuropshychology
Maxwell, Douglas Paul, plant pathology
Mazess, Richard B, biological anthropology
McCabe, Robert Albert, wildlife management
McClellan, Catharine, anthropology
McCown, Brent Howard, physiological ecology, horticulture
McDonald, Malcolm Edwin, parasitology, ecology
McGibbon, William Henry, genetics
McGregor, Sandy, virology
McLellan, Alden, IV, environmental management, resource management
McManus, John Joseph, vertebrate zoology, ecology
McMillan, Daniel Russell, Jr, topology
McShan, William Hartford, endocrinology, biochemistry
McVoy, Kirk Warren, theoretical nuclear physics
Meade, Dale M, plasma physics
Medaris, L Gordon, Jr, petrology
Medler, John Thomas, entomology
Menzel, Wolfgang Paul, theoretical solid state physics
Mertz, Janet Elaine, molecular biology
Metter, Gerald Edward, biostatistics, oncology
Metzenberg, Robert Lee, Jr, biochemistry
Meyer, Richard Ernst, mathematics, geophysics
Meyer, Roland Kenneth, zoology
Miller, Beatrice Diamond, anthropology
Miller, Edward Ernst, applied physics
Miller, Elizabeth Cavert, oncology
Miller, James Alexander, oncology
Miller, Regis Bolden, plant anatomy
Miller, Robert Burnham, applied statistics
Miller, Robert James, anthropology
Millett, Merrill Albert, wood chemistry
Mitchell, John Edwards, plant pathology
Mitchell, Val Leonard, climatology
Mizutani, Satoshi, virology
Moeller, Floyd Edward, apiculture
Moermond, Timothy Creighton, ecology
Moffet, Hugh L, pediatrics, infectious diseases
Mohs, Frederic Edward, surgery
Moore, Edward Forrest, applied mathematics
Moore, John Duain, plant pathology
Moore, Ramon Edgar, mathematics, computer science
Moran, Paul Richard, solid state physics
Morin, Dornis Clinton, plasma physics, electrical engineering
Morris, Sidney Machen, Jr, biological chemistry
Morrissey, John F, medicine, gastroenterology

Morton, Stephen Dana, water chemistry, physical chemistry
Moss, Valentin G., food chemistry
Mueller, Gerald Conrad, biochemistry
Mukerjee, Pasupati, physical chemistry, colloid chemistry
Murdock, John Thomas, soil science
Murdock, Larry Lee, physiology, biochemistry
Murphy, Quillian R., Jr., physiology
Myers, George E., physical chemistry
Nagle, Francis J., exercise physiology, cardiovascular physiology
Nelsen, Stephen Flanders, organic chemistry
Nelson, David Alan, chemistry
Nelson, Oliver Evans, genetics
Nester, F H Max, physical chemistry
Newcomb, Eldon Henry, plant cytology
Ney, Peter E., mathematics
Nichols, Roy Elwyn, veterinary physiology
Nickles, Robert Jerome, medical physics
Niedermeier, Robert Paul, dairy husbandry
Nohel, John Adolph, mathematics
Norris, Dale Melvin, Jr., biology
Nutter, Gene Douglas, physics
O'Leary, Marion Hugh, organic chemistry, biochemistry
Olmsted, Clarence Walter, agricultural geography, geography of the United States & Canada
Olsen, Ward Alan, internal medicine, gastroenterology
Olson, Carl, comparative pathology
Olson, Norman Fredrick, food technology
Opitz, John Marius, pediatrics, medical genetics
Oplinger, Edward Scott, agronomy
Orlik, Peter Paul, mathematics
Orsini, Margaret Ward (Giordano), embryology
Osborn, J Marshall, Jr., mathematics
Osborn, June Elaine, virology, infectious disease
Osborne, Richard Hazelet, human genetics, anthropology
Ostrom, Meredith Eggers, geology
Padgett, Billie Lou, medical microbiology
Palmer, John Gilbert, plant pathology
Parascandola, John Louis, history of science
Passano, Leonard Macruder, III, zoology
Passman, Donald Steven, algebra
Patton, Robert Franklin, forest pathology
Peckham, Ben M., obstetrics & gynecology
Pella, Milton Orville, science education
Pendleton, Johnny Wryas, agronomy
Perlman, David, biochemistry
Perlman, Kato (Katherine) Lenard, organic chemistry
Peruzzotti, George Peter, microbial biochemistry
Peters, Henry A., neurology, psychiatry
Peters, Arthur Edwin, soils
Peterson, Clinton E., horticulture
Peterson, Dallas Odell, geology
Peterson, David Maurice, plant physiology
Peterson, Gary Lee, cell physiology, comparative biochemistry
Petersen, James Clark, anatomy
Pitot, Henry C. III, biochemistry, pathology
Piaut, Walter (Sigmund), cell biology
Pray, Lloyd Charles, geology
Prentice, Neville, agricultural biochemistry
Prepost, Richard, experimental physics
Price, Walter Van Vranken, dairying
Pringle, Dorothy Jutton, nutrition
Puletti, Flavio, neurosurgery
Punwar, Jalamsinh K, chemistry
Quay, Wilbur Brooks, histophysiology, neurobiology
Quirk, John Thomas, tree physiology, forest products
Ragotzkie, Robert Austin, meteorology, oceanography
Rall, Louis Baker, numerical analysis
Ramirez, Guillermo, oncology

Rankin, John, internal medicine, physiology
Rankin, John Horsley Grey, physiology
Rao, Ghanta Nageswara, biochemistry, veterinary science
Raper, Kenneth Bryan, mycology
Record, M Thomas, Jr., biophysical chemistry
Reddan, William Gerald, physiology
Reed, Charles E., internal medicine, allergy
Reeder, Don David, high energy physics
Reeder, William Glase, animal ecology, vertebrate paleontology
Reich, Hans Jurgen, organic chemistry
Reich, Ieva Lazdins, organic chemistry
Reneau, John, audiology, speech pathology
Reynolds, Ernest West, Jr., cardiology
Reznikoff, William Stanton, molecular genetics
Rhode, William Stanley, neurophysiology
Ribelin, William Eugene, veterinary pathology
Rich, Daniel Hulbert, bio-organic chemistry
Richards, Hugh Taylor, nuclear physics
Richardson, Thomas, food chemistry
Ris, Hans, cell biology
Risse, Guenter Bernhard, history of medicine
Robbin, Joel W., mathematics
Roberts, Leigh M., psychiatry
Robinson, Jerry Allen, reproductive physiology, endocrinology
Robinson, Joseph Robert, pharmaceutics
Robinson, Stephen Michael, applied mathematics, operations research
Roesler, Frederick Lewis, atomic physics, optics
Rohweder, Dwayne A., agronomy
Rollefson, Ragnar, physics
Rongstad, Orrin James, wildlife ecology
Rose, David Peter, endocrinology, oncology
Rowe, George G., internal medicine
Rowe, John Westel, molecular biology
Rownd, Robert Harvey, biochemistry, molecular biology
Roussas, George G., mathematical statistics
Ross, Jerzy Edwin, neurophysiology
Ross, Jeffrey, molecular biology
Rosser, John Barkley, mathematics
Rudin, Mary Ellen, mathematics
Rudin, Walter, mathematics
Ruecker, Roland R., virology
Ruff, Robert LaVerne, wildlife ecology
Rusch, Harold Paul, oncology
Russell, David L., mathematics
Rutledge, Jackie Joe, animal breeding
Ruzicka, Francis Frederick, Jr., radiology
Ryan, Allan James, rehabilitation medicine
Ryan, Dale Scott, food science
Sachs, Irving Benjamin, forestry, wood technology
Saeman, Jerome Francis, chemistry, forest products
Sanyer, Necmi, wood chemistry
Sarles, William Bowen, microbiology
Satter, Carol Ann, biological structure
Satter, Larry Dean, animal nutrition
Savage, Blair DeWillis, astronomy
Schantz, Edward Joseph, biochemistry
Scheele, Robert Blain, biophysics
Schenk, Roy Urban, molecular biology
Scherb, Frank, space physics
Schilling, Robert Frederick, medicine
Schmidt, John Richard, agricultural economics, operations research
Schnoes, Heinrich Konstantin, organic chemistry, biochemistry
Schoenberg, Isaac Jacob, mathematics
Scholl, Jesse Myron, agronomy
Schrader, Lawrence Edwin, plant biochemistry, plant physiology
Schrag, John L., polymer chemistry
Schultz, Gwendolyn Monett, geology
Schultz, Loris Henry, dairy science
Schwerdtfeger, Werner, meteorology, geophysics
Segar, William Elias, pediatrics
Sendelbach, Anton G., dairy science, genetics
Senn, Harold Archie, botany
Sequeira, Luis, plant pathology
Seshadri, Sengadu Rangaswamy, applied physics, electrical engineering
Shahidi, Nasrollah Thomas, hematology
Shakhashiri, Bassam Zekin, chemistry
Shands, Hazel Lee, agronomy
Shea, Daniel Francis, mathematics

Shen, Mei-Chang, applied mathematics
Shenefelt, Roy David, entomology
Shipley Meyer, Elva G., endocrinology
Shohet, Juda Leon, plasma physics, electrical engineering
Shrago, Earl, biochemistry, medicine
Sidky, Younan Abdel Malik, immunology, endocrinology
Siebecker, Karl LaFollete, Jr., anesthesiology
Siegel, Frank Leonard, biochemistry
Siegfried, Robert, history of science, chemistry
Sih, Charles John, bio-organic chemistry
Simpson, David Patten, nephrology
Sims, John LeRoy, clinical medicine
Singer, George, acarology
Sinha, Shyamal K., medical microbiology
Siqua, Richard Anthony, astrophysics
Skoog, Folke, plant physiology
Slack, Steven Allen, plant pathology, plant virology
Slautterback, David Buell, cell biology
Smalley, Eugene Byron, plant pathology
Smart, John Roderick, number theory
Smith, Dale, agronomy
Smith, David Clyde, plant breeding
Smith, Donald Ward, microbiology, immunology
Smith, Horace Vernon, Jr., nuclear science, nuclear engineering
Smith, Richard R., plant breeding, plant genetics
Smith, William Kenneth, genetics
Snowdon, Charles Thomas, physiological psychology, animal behavior
Sobkowicz, Hanna Maria, neurology
Solomon, Louis, mathematics
Sonneborn, David R., developmental biology, microbiology
Southall, Aidan William, anthropology, sociology
Spalatin, Josip, virology
Sproll, Julien Clinton, plasma physics
Stahlman, William Duane, history of science
Stahmann, Mark Arnold, biochemistry
Stapf, Robert Joseph, foods
Starkey, Eugene Edward, dairy science
Stauffer, John Frederick, plant physiology
Stauffer, Robert Clinton, biology, history of science
Stearns, Charles R., meteorology
Steinhart, Carol Elder, plant physiology
Steinhart, John Shannon, geophysics, science policy
Stelly, Matthias, soil fertility
Stephenson, David A., hydrology, geology
Sterling, Henry Somers, geography
Stigler, Stephen Mack, statistics
Stolman, James Bernard, anthropology, archaeology
Stone, William Ellis, physiology
Stone, William Harold, genetics, immunology
Stratman, Frederick William, chemistry
Strickon, Arnold, cultural anthropology
Strong, Dorothy Hussemann, food science
Strong, Frank Morgan, biochemistry
Struckmeyer, Burdean Esther, horticulture
Suchman, David, meteorology
Suess, Gene Guy, meat science
Sugiyama, Hiroshi, bacteriology
Sundaralingam, Muttaya, crystallography
Sunde, Milton Lester, poultry nutrition
Sundharadas, Gnanasigamoni, biochemistry, immunology
Suomi, Verner Edward, meteorology
Susman, Millard, genetics
Suttie, John Weston, biochemistry
Swanson, Arthur Martin, biochemistry, food science
Swick, Robert Winfield, biochemistry
Symon, Keith Randolph, physics
Szybalski, Elizabeth Hunter, microbiology
Szybalski, Waclaw, biochemistry, oncology
Takayama, Kuni, biochemistry
Tanner, Champ Bean, microclimatology, soil physics
Tarkow, Harold, physical chemistry
Taylor, James Welch, analytical chemistry
Temin, Howard Martin, oncology, virology
Temin, Rayla Greenberg, genetics
Tews, Jean King, biochemistry
Thompson, Donald Enrique, anthropology, archaeology
Thomson, John Walter, botany

Tiao, George Ching-Hwuan, mathematical statistics, economic statistics
Tibbitts, Theodore William, horticulture, environmental physiology
Toivola, Pertti Toivo Kalevi, medical physiology, neuroendocrinology
Toren, Eric Clifford, Jr., analytical chemistry
Trautman, Jack Carl, dairying
Treichel, Paul Morgan, Jr., inorganic chemistry
Trost, Barry M, organic chemistry
Turner, Edwin Morris, physical chemistry
Turner, Robert E L., mathematics
Uhlenbrock, Dietrich A., applied mathematics, mathematical physics
Uhr, Leonard Merrick, computer science, psychology
Vallee, Richard Bert, biochemistry
Vanderlin, Carl Joseph, Jr., mathematics
Van Duser, Arthur L., preventive medicine, public health
Van Ryzin, John R., mathematical statistics, statistics
Vaughan, Worth E., physical chemistry
Vedejs, Edwin, organic chemistry
Voichick, Michael, mathematics
Wade, Earl Kenneth, plant pathology
Wagner, Raphael Darrel, mathematics
Wahba, Grace, mathematical statistics
Wahl, Eberhard Wilhelm, meteorology, space sciences
Wahlgren, Harold Emil, forest products
Walker, Duard Lee, virology, microbiology
Wallenfeldt, Evert, food science
Walsh, Leo Marcellus, soil fertility, soil science
Walters, John P., analytical chemistry, spectroscopy
Wang, Herbert Fan, geophysics
Ward, David, historical geography, urban geography
Washa, George William, mechanics
Wasow, Wolfgang Richard, mathematics
Wear, John Brewster, Jr., urology
Webb, Maurice Barnett, physics
Weckel, Kenneth Granville, food science
Weinswig, Melvin H., pharmaceutical chemistry
Welker, Carol, neurophysiology
Welker, Wallace I., neurophysiology, neuroanatomy
Wells, Robert Dale, bio-chemistry, molecular biology
Wen, Sung-Feng, medicine
Wendland, Wayne Marcel, meteorology, climatology
Wentworth, Bernard C., avian physiology
Werner, Joan Kathleen, neuroanatomy, gross anatomy
West, Robert, organometallic chemistry
Westman, Jack Conrad, child psychiatry
Whiffen, James Douglass, surgery
Whittingham, William Francis, mycology
Wilds, Alfred Lawrence, organic chemistry
Wilken, David Richard, biochemistry
Will, James Arthur, physiology
Willard, John Ela, physical chemistry
Williams, John Warren, physical chemistry
Williams, Paul Hugh, plant pathology
Wilson, Joe Bransford, bacteriology
Wilson, Perry William, bacteriology
Winder, William Charles, food science, dairy industry
Wipf, Francis Louise, cytology
Wirka, Herman W., orthopedic surgery
Wolberg, William Harvey, surgery, oncology
Wolf, Richard Clarence, physiology, endocrinology
Wolter, Karl Erich, plant physiology
Woods, Robert Claude, physical chemistry
Woolsey, Clinton Nathan, neurophysiology
Worf, Gayle L., plant pathology
Wright, Sewall, genetics, evolution
Yale, Charles E., medicine, surgery
Yatvin, Milton B., radiobiology
Yen, William Mao-Shung, solid state physics
Yohe, James Michael, computer science, mathematics
Youmans, William Barton, physiology
Young, John A., meteorology
Young, Laurence Chisholm, pure mathematics
Young, Raymond A., wood chemistry

Young, William Paul, surgery
Youngs, Robert Leland, forestry
Youngs, Vernon Leroy, organic chemistry
Yu, Hyuk, physical chemistry, polymer chemistry
Yuill, Thomas Mackay, virology, ecology
Zeikus, J Gregory, microbiology
Zile, Maija Helene, biochemistry
Zimmerman, Howard Elliot, chemistry
Zinkel, Duane Forst, organic chemistry
Zografi, George, pharmaceutical chemistry, surface chemistry
Zu Rhein, Gabriele Marie, pathology, neuropathology

MANITOWOC
Fodden, John Henry, pathology
Hartman, Grant Henry, dairy science
Van Denack, Julia Marie, biology
Van Ryzin, Martina, history of science, mathematics

MARINETTE
Johnson, Wendel J, animal ecology, zoogeography
Neuville, Morris Louis, chemistry
Stahl, Neil, mathematical analysis

MARSHFIELD
Larsen, Howard James, dairy nutrition
Marx, James John, Jr, immunopathology
Plotka, Edward Dennis, endocrinology, physiology
Roberts, Ronald C, physical biochemistry
Tewksbury, Duane Allan, biochemistry
Wenzel, Frederick J, biochemistry, biology

MASON
Swanson, Sigurd Arthur, physical chemistry

MAYVILLE
Chandan, Ramesh Chandra, food technology, biochemistry

MEDFORD
Miller, Frank, organic chemistry
Zeit, Walter, anatomy, histology

MENASHA
Minock, Michael Edward, vertebrate ecology
Schenck, Allan, chemistry
Weis, Leonard Walter, geology

MENOMONIE
Anderson, Orfin, botany, microbiology
Bredahl, Edward Arlan, physiology, ecology
Carlson, Oscar Verdell, entomology, invertebrate zoology
Faris, John Jay, physics
Fossum, Steve P, surface physics, vacuum technology
Kainski, John Michael, plant pathology
Kainski, Mercedes H, food science, nutrition
Kleibacker, Wilson McAlarney, organic chemistry
Lowry, Edward MacLean, biology
Nitz, Otto William Julius, organic chemistry
Runnalls, Nelva Earline Gross, nuclear chemistry
Wilson, Richard Howard, zoology, animal behavior

MIDDLETON
Feist, William Charles, wood chemistry, polymer chemistry
Ferry, James A, nuclear physics
Fleischer, Herbert Oswald, forestry
Heizer, Edwin Elbert, dairy husbandry
Herb, Raymond George, physics
Hinkes, Thomas Michael, organic chemistry

MILTON
Van Horn, Lester Milton, zoology, physiology

MILWAUKEE
Abramoff, Peter, zoology, immunology
Adler, Frederick E W, food science
Altshuler, Charles Haskell, pathology

Anderson, Frank David, anatomy
Antholine, William E, physical biochemistry
Anthony, James Douglas, parasitology
Arends, Robert Leander, clinical chemistry
Arnholt, Philip John, botany
Bachhuber, Edward A, surgery, anatomy
Bader, Alfred Robert, organic chemistry
Bahe, Lowell W, physical chemistry
Baier, Joseph George, immunobiology
Baker, George Severt, solid state physics
Bardell, Eunice Bonow, pharmacy
Barnouw, Victor, anthropology, ethnology
Barrett, James Martin, protozoology
Baum, Werne A, meteorology, climatology
Baxter, John Wallace, mycology, plant pathology
Beck, Donald Edward, surface physics
Beeton, Alfred Merle, limnology
Bell, James Henry, algebra
Bellis, Ernest Anthony, organic chemistry
Bender, Philip R, mathematics
Benjamin, Hiram Bernard, physiology, anatomy
Bennett, James Marvin, environmental management
Bernhard, Victor Montwid, medicine
Bernstein, Aleck, microbiology, genetics
Bezdek, James Christian, applied mathematics
Blodgett, Frederic Maurice, pediatrics
Blum, John Leo, phycology
Blum, Julius Rubin, mathematics
Borowiecki, Barbara Zakrzewska, physical geography
Bournique, Raymond August, chemistry
Branch, Garland Marion, Jr, medical technology
Brandt, Werner Wilfried, physical chemistry
Branovan, Leo, mathematics
Braunschweiger, Christian Carl, mathematics
Brebrick, Robert Frank, Jr, solid state chemistry
Bronkowski, Thomas Andrew, applied mathematics
Brookshear, James Glenn, mathematics
Brown, Bruce Elliot, geology, crystallography
Broyles, Robert Herman, biochemistry, developmental biology
Burck, Larry Harold, solid mechanics
Burton, Joe Covington, agricultural chemistry, microbiology
Cadmus, Robert R, preventive medicine, medical administration
Catlin, B Wesley, microbiology
Chen, Shao Lin, biochemistry
Cherayil, George Devassia, biochemistry, chemistry
Chicoye, Eizer, food chemistry
Chow, Yutze, theoretical physics, mathematics
Clarkson, Robert Breck, physical chemistry, surface chemistry
Colton, Ervin, inorganic chemistry
Condon, Robert Edward, surgery, physiology
Constantin, James Michael, leather chemistry
Cook, James Minton, natural products chemistry, medicinal chemistry
Cooper, Elmer James, cereal chemistry
Cremer, Sheldon E, organic chemistry
Crowshaw, Keith, biochemistry, organic chemistry
Curtis, Robin Livingstone, neuroanatomy, physiological psychology
Cybriwsky, Alex, physics
Davis, Starkey D, pediatrics, infectious diseases
DeCosse, Jerome J, medicine
Delfs, Eleanor, obstetrics & gynecology
DeMillo, Richard A, mathematics, computer science
Deshotels, Warren Julius, physics
de Vlaming, Victor Lynn, comparative endocrinology
Dittman, Richard Henry, surface physics
Dodson, Vernon N, toxicology
Doumas, Basil T, clinical chemistry
Downing, Darryl Jon, mathematical statistics
Drucker, William D, internal medicine, endocrinology
Drufenbrock, Diane, mathematics
Dunlop, Douglas Wayne, botany
Duquesnoy, Rene J, immunology
Eckert, Alfred Carl, Jr, analytical chemistry
Edelhauser, Henry F, physiology, ophthalmology

Edwards, Clinton R, geography
Edwards, Robert Bryce, microbiology, immunology
Ehlert, Thomas Clarence, physical chemistry
Eidt, Robert C, geography
Elliott, Bernard Burton, enzymology
Engbring, Norman H, medicine
Erwin, Chesley Para, medicine, pathology
Evans, Silas McAfee, experimental medicine
Feinberg, Benjamin Allen, analytical chemistry, biochemistry
Feng, Paul Yen-Hsiung, nuclear chemistry, radiation chemistry
Ficken, Millicent Sigler, ethology
Ficken, Robert W, ethology
Filip, Donald Joseph, hematology, oncology
Fink, Jordan Norman, allergy, immunology
Fowler, Melvin Leo, anthropology, archaeology
Frackelton, William Hamilton, plastic surgery
Fredricks, Walter William, biochemistry, immunochemistry
Freund, Peter Richard, food science
Friedman, Harris Leonard, organic chemistry
Fujimoto, James Masao, pharmacology
Gaillen, William J, medicine
Gardner, Weston Deuain, gross anatomy
Gingrass, Ruedi Peter, plastic surgery
Girotti, Albert William, biochemistry
Glick, David M, biochemistry, endocrinology
Good, Thomas Arnold, pediatrics
Greene, Jack Bruce, physics
Greenler, Robert George, surface physics
Greiff, Donald, cryobiology
Gross, Garrett John, pharmacology
Grossberg, Sidney Edward, virology, microbiology
Grunewald, Ralph, bionucleonics
Gundersen, Roy Melvin, mathematics
Gutierrez, Peter Luis, biophysics
Haavik, Coryce Ozanne, pharmacology
Hafemann, Dennis Reinhold, neurosciences
Haft, Jay Stuart, anatomy
Hagedorn, Donald James, plant pathology
Hagen, Richard Eugene, biochemistry, food science
Hake, Carl (Louis), environmental health
Hall, Robert Lester, mathematical analysis
Halmbacher, Paul, chemistry
Handrup, Bernarda, physics
Hanneken, Clemens, mathematics
Hardman, Harold Francis, pharmacology
Harmsworth, Rodney V, zoology, limnology
Harris, J Douglas, mathematics
Haworth, Daniel Thomas, inorganic chemistry
Haymaker, Clifford Robert, chemistry
Heim, Lyle Raymond, immunology, microbiology
Helbert, James Raymond, biochemistry
Hellman, Nison Norman, chemistry
Hennen, Sally, developmental biology
Henschel, Ernest O, anesthesiology, cardiopulmonary physiology
Hill, Elgin Alexander, organic chemistry
Himmelfarb, Philip, microbiology, research administration
Hirschboeck, John Stephen, clinical medicine
Hoerl, Bryan G, microbiology, pathology
Hoffman, Norman Edwin, organic chemistry
Holcenberg, John Stanley, clinical pharmacology, oncology
Holzner, Lutz Ernest, geography
Hosko, Michael J Jr, neuropharmacology
Huber, Calvin, chemistry
Hurley, John D, surgery
Hussa, Robert Oscar, biochemistry
Hussey, Clara Veronica, pathology
Huston, John Howard, internal medicine
Hutchinson, George Keating, computer science, management sciences
Hutton, James Robert, molecular biology
Hyde, James Stewart, solid state physics, biophysics
Iorio, Robert John, anatomy
Jache, Albert William, inorganic chemistry
Jackson, Basil Edgar, psychiatry, child psychiatry
James, Bernard Joseph, anthropology
Jameson, Patricia Madoline, microbiology, virology
Janssen, William C, medicine
Johnson, DeWayne Carl, physics

Kaplan, Stanley, teratology, human anatomy
Kapp, Kenneth M, algebra
Kassell, Beatrice, biochemistry
Kaufmann, Fred Henry, botany, history of science
Kehoe, Alice Beck, anthropology, archaeology
Kehoe, Thomas Francis, anthropology, ethnography
Kemmet, Wilfred J, dentistry
Kemp, Robert Grant, biochemistry
Kerrigan, Gerald Austin, pediatrics
Kersting, David William, dermatology
Keulks, George William, physical chemistry
Kittsley, Scott Loren, physical chemistry
Klitgaard, Howard Maynard, physiology
Klundt, Irwin Lee, organic chemistry
Kohler, Elaine Eloise Humphreys, pediatric endocrinology
Korns, Michael Edward, medicine, pathology
Kovacic, Peter, organic chemistry
Krahnke, Harold C, physiological chemistry, pharmacy
Krakow, Gladys, enzymology, immunobiology
Kramer, Elizabeth, biochemistry
Kuffner, Roy Joseph, physical chemistry
Kumaran, A Krishna, developmental biology
Kuzma, Joseph Francis, pathology
Landis, Charles Walter, psychiatry
Larson, Sanford J, neuroanatomy, neurosurgery
Lasca, Norman P, Jr, quaternary geology, geomorphology
Lawrence, Willard Earl, statistics
Layde, Durward Charles, inorganic chemistry
Lech, John James, pharmacology
Lee, Kiuck, nuclear physics
Lehman, Roger H, medicine
Lehmann, Edward Joseph, internal medicine
Leplae, Luc A, theoretical solid state physics
Lepley, Derward, Jr, surgery
Levine, Leonard P, spectroscopy, surface physics
Levy, Moises, solid state physics
Lichtman, David, surface physics
Lochner, Robert Herman, statistics
Long, Sally Yates, embryology, teratology
Losin, Edward Thomas, physical organic chemistry, energy conversion
Lubkin, Eihu, theoretical physics
Lundquist, Marjorie Ann, industrial hygiene
Lurie, Nancy Oestreich, anthropology
Lutz, Donald Alexander, mathematics
Lydolph, Paul E, climatology, geography
MacArthur, Kenneth William, entomology
Magnuson, Eugene Robert, organic chemistry
Makowski, Gary George, mathematics, statistics
Maksud, Michael George, exercise physiology
Marden, Morris, mathematics
Mattingly, Richard Francis, obstetrics & gynecology
Mayer, Harold M, urban geography, urban planning
McCarty, Daniel J, Jr, internal medicine
McDivitt, Maxine Estelle, mineral economics
McDonough, Eugene Stowell, mycopathology, genetics
McFarland, James Thomas, biochemistry
McKenna, James Francis, applied chemistry
McQuistan, Richard Beckett, physics, statistical mechanics
Meetz, Gerald David, immunology, cell biology
Mehler, Alan Haskell, biochemistry
Melvin, John Lewis, physical medicine & rehabilitation, electromyography
Mendelson, Kenneth Samuel, physics
Meyer, Glenn Arthur, neurosurgery
Meyer, Ralph A, Jr, physiology, endocrinology
Miles, Harry McCauley, ecology, physiology
Miller, David Hewitt, climatology
Miller, Paul George, applied mathematics, genetics
Millington, William Frank, plant morphogenesis
Minnich, John Edwin, environmental physiology, vertebrate zoology
Moore, Robert H, mathematics

WISCONSIN

Morehouse, Clarence Koppel, electrochemistry
Morell, Samuel Allen, biochemistry
Mortimer, Clifford Hiley, limnology, physical oceanography
Moyer, John Clarence, mathematics
Mullins, Robert Emmet, mathematical analysis
Mursky, Gregory, geology
Nakamoto, Kazuo, physical chemistry
Nelson, Katherine Greacen, paleontology
Nelson, Thomas Clifford, microbiology
Nemmer, Max, microbiology
Nickerson, Max Allen, vertebrate biology, histochemistry
Norden, Carroll Raymond, zoology
O'Malley, Richard John, mathematical analysis
Papastamatiou, Nicolas, physics
Parker, Leonard Emanuel, theoretical physics, cosmology
Pauli, Richard Allen, geology
Pauly, Ludwig K, entomology
Pawlisch, Paul E, agronomy
Peppler, Henry James, microbiology
Perry, Billy Wayne, biochemistry, clinical chemistry
Petersen, John Robert, medical administration
Peterson, John Cyril, pediatrics
Phillips, Ruth Brosi, genetics
Piacsek, Bela Emery, physiology, endocrinology
Pincus, Howard Jonah, rock mechanics, structural geology
Pinkel, Donald Paul, pediatrics
Pirie, Robert Gordon, oceanography, sedimentology
Pisciotta, Anthony Vito, internal medicine
Potter, David Samuel, physics
Press, Newton, cell biology
Quirke, Terence Thomas, Jr, exploration geology
Rasch, Ellen M, cell biology, cytochemistry
Rasch, Robert, physiology
Rauch, Dolores, cultural geography
Reed, Gerald, chemistry
Remsen, Charles C, III, biology, microbiology
Rieck, Alvin Frank, physiology, embryology
Rieschbach, Richard Edgar, internal medicine
Rimm, Alfred A, genetics, statistics
Ritchie, Betty Caraway, audiology
Roll, Paul Melvin, biochemistry
Rosenzweig, David Yates, internal medicine, pulmonary physiology
Roth, Donald Alfred, medicine
Roth, Harold, solid state physics
Saari, Jack Theodore, physiology
Salamin, Peter Joseph, botany
Sances, Anthony, Jr, biomedical engineering
Sasse, Edward Alexander, clinical chemistry, biochemistry
Sauve, James Willard, mathematics
Savino, Patrick Ronald, medicine
Schaff, Mary Ellen, inorganic chemistry
Schanfield, Moses Samuel, immunogenetics
Schueter, Jean, anatomy
Schmidt, Albert Charles, orthopedic surgery
Schmidt, Donald Paul, internal medicine
Schmige, Glenn Melwood, physics
Schrader, David Martin, physical chemistry
Schulte, William John, Jr, surgery
Schultz, David Harold, computer science, numerical analysis
Schultz, Richard Otto, medicine, ophthalmology
Sether, Lowell Albert, anatomy
Sevilla-Gardiner, Josefina Zialcita, physiological ecology, microbiology
Shanberge, Jacob N, pathology
Shaw, C Frank, III, inorganic chemistry
Shreve, David Carr, mathematics
Shurman, Michael Mendelsohn, astronomy
Siebring, Barteld Richard, chemistry
Siegel, Jack Morton, biochemistry
Siegesmund, Kenneth A, cytology
Sikdar, Dhirendra N, physics, meteorology
Silverberg, James Mark, social anthropology, behavioral anthropology
Silverman, Franklin Harold, speech pathology, applied statistics
Skinner, Lindsay A, applied mathematics
Smith, James John, physiology
Smith, Jerry Howard, organic chemistry, bio-organic chemistry
Smith, Oliver Hugh, molecular genetics
Snider, Dale Reynolds, high energy physics
Soergel, Konrad H, internal medicine, gastroenterology
Solomon, Donald W, analytical chemistry
Sorbello, Richard Salvatore, theoretical solid state physics
Sosnovsky, George, organic chemistry
Spitzbart, Abraham, mathematics
Stearns, Forest, ecology, botany
Steiner, John F, electrochemistry
Steinke, Paul Karl Willi, agricultural microbiology
Stekiel, William John, biophysics
Stern, Robert Malcolm, bacteriology
Stewart, Norman Reginald, cultural geography
Stewart, Richard Donald, internal medicine, toxicology
Stipe, Claude Edwin, anthropology
Straumfjord, Jon Vidalin, Jr, medicine, clinical pathology
Strause, Sterling Franklin, organic chemistry
Suh, John Taiyoung, medicinal chemistry, organic chemistry
Swartz, Harold M, radiobiology, biophysics
Taketa, Fumito, biochemistry
Tani, Smio, theoretical physics
Tappen, Neil Campbell, physical anthropology, primatology
Tautvydas, Kestutis Jonas, biology
Theine, Alice, organic chemistry
Truitt, Robert Lindell, microbiology, transplantation immunology
Turkington, Roger W, medicine, biochemistry
Tuttle, Merlin Devere, population ecology, mammalogy
Unsworth, Brian Russell, biochemistry
Vanselow, Ralf W, physical chemistry, surface chemistry
Van Straten, Mary Petronia, mathematics
Vigil, Eugene Leon, cell biology
Wackman, Peter Husting, physics
Waech, Theodore G, physical chemistry
Walters, William Le Roy, physics
Warner, Eldon Dezelle, endocrinology
Washabaugh, William, anthropology
Wasko, Peter Edmund, meteorology
Watters, Kenneth Lynn, inorganic chemistry, bioinorganic chemistry
Weibel, Armella, mathematics
Weise, Charles Martin, population ecology, ornithology
West, Robert MacLellan, vertebrate paleontology
Whisler, Kenneth Eugene, physiology, clinical chemistry
White, Philip Taylor, neurology
Whitford, Philip Burton, plant ecology
Wilkie, Charles Arthur, inorganic chemistry
Willis, David Edwin, geophysics, geology
Wold, Richard John, marine geophysics
Worthington, Edward Arthur, physical chemistry
Wussow, George C, oral surgery, oral pathology
Youker, James Edward, radiology
Young, Allen Marcus, ecology
Zabransky, Ronald Joseph, clinical microbiology, clinical bacteriology
Zakrzewska, Barbara (Mrs Borowiecki), geography, geology
Zenz, Carl, occupational medicine
Ziegler, Michael Robert, mathematical analysis

MINOCQUA
Pond, Alonzo William, anthropology

MONROE
Scudamore, Harold Hunter, medicine

MUKWONAGO
Hoffmann, Gilbert Frederick, chemistry

NEENAH
Adrian, Alan Patrick, chemistry
Allison, John P, organic chemistry, polymer chemistry
Bauer, Richard M, applied physics
Bernardin, Leo J, paper chemistry
Bletzinger, John Calvin, chemistry
Butler, John Parkman, research administration
Caston, Ralph Henry, physics
Craig, Kenneth Alexander, analytical chemistry
Cushing, Merchant Leroy, carbohydrate chemistry
DeLong, Robert Francis, bacteriology
Didwania, Hanuman Prasad, physical chemistry
Dobbins, Thomas Edward, forest
Eber, Robert Joseph, paper chemistry
Helms, John F, organic chemistry
Hess, Cecil Lawrence, paper chemistry
Hirschy, Harlan W, organic chemistry, research administration
Hoffenberg, Paul Henry, color science
Hossain, Shaff Ul, bacteriology
Lim, Yong Woon, surface chemistry, dairy science
McBride, Landy James, plant physiology
Miller, Fredric N, paper chemistry
O'Connor, James J, organic chemistry
Rogers, Sedgwick Cookerly, wood technology
Schwarz, Eckhard C A, polymer science
Spiegelberg, Harry Lester, paper chemistry
Starshak, Albert Joseph, statistics, data processing
Vogt, Clifford Marshall, organic colloid chemistry
Weber, Robert Emil, polymer chemistry
Whitehead, Howard Allan, bacteriology

OSHKOSH
Bennington, Neville Lynne, zoology
Berge, Douglas G, analytical chemistry
Bhatia, Shyam Sunder, economic geography, geography of South Asia
Bowman, Max I, organic chemistry
Bruyere, Donald Eugene, geography of the Soviet Union, population geography
Crimmins, Timothy Francis, organic chemistry
Drecktrah, Harold Gene, insect morphology
Feng, Kuo Ao, plant physiology
Fetter, Charles Willard, Jr, hydrogeology, hydrology
Gade, Edward Herman Henry, III, mathematics
Gaede, Sandra Ann, physics
Gaede, Herbert Lawrence, geography of Africa, urban geography
Gueths, James E, solid state physics
Hanson, Thomas Lawrence, biochemistry, clinical chemistry
Harriman, Neil Arthur, plant taxonomy
Hodge, William Howard, anthropology
Hoffman, James Irvie, geology
Klicka, John Kenneth, endocrinology
Kuenzi, Norbert James, mathematics
Kurath, Sheldon Frank, polymer chemistry, rheology
LaBerge, Gene L, geology
Larson, Wilbur S, inorganic chemistry, analytical chemistry
Laudon, Thomas S, geology
Mahadeva, Madhu Narayan, zoology
Mahmood, Ibrahim Younis, herpetology
McKee, James W, paleontology
McKenzie, Harvey, mathematics
Neubecker, Robert Duane, pathology
Parker, Dorothy Lundquist, virology
Polcyn, Daniel Stephen, electroanalytical chemistry
Polinow, Gilbert Frederick, physical chemistry
Post, Elroy Wayne, inorganic chemistry, organometallic chemistry
Pritlipp, Robert Walter, mathematics
Propp, Jacob Henry, analytical chemistry
Provinzano, James, economic anthropology
Randerson, Sherman, genetics
Rigney, Mary Margaret, medical microbiology
Rouf, Mohammed Abdur, bacteriology
Schultz, Hilbert Kenneth, operations research
Schwartz, Edward Leo, apiculture
Tews, Leonard L, botany, mycology
Unger, James William, plant morphology
Uitke, Allen R, inorganic chemistry
Yoho, Clayton W, organic chemistry

PLAINFIELD
Hamerstrom, Frederick Nathan, Jr, wildlife ecology, ornithology

PLATTEVILLE
Broughton, William Albert, economic geology
Bulis, George LeRoy, mathematics
Cressman, Harry Keith, soil science, plant physiology
Curtis, Ralph Wendell, metallurgical chemistry
Duewer, Elizabeth Ann, lichenology
Fatzinger, J Dale Roger, geography of Latin America, historical geography
Fenrick, Harold William, physical chemistry
Foulkes, Robert Hugh, experimental zoology
Hansen, Robert Conrad, analytical chemistry, inorganic chemistry
Heidenreich, Charles John, animal husbandry
Higgs, Roger L, crop breeding, genetics
Hoffman, William F, animal science, dairy science
Jahn, J Russell, animal science
Klaassen, Dwight Homer, biochemistry
Lokken, Stanley Jerome, physical chemistry
Trine, Franklin Dawson, mathematics
Tufte, Marilyn Jean, bacteriology
Wagner, Russel Olson, ecology, botany
West, Walter Scott, economic geology, geochemistry
Willis, Harold Lester, entomology

PORT EDWARDS
Norris, Terry Orban, organic chemistry
Rowe, Herbert William, paper chemistry
Schoettler, James Robert, pulp & paper technology

PORT WASHINGTON
Ginsler, Victor William, biochemistry
Svoboda, Glenn Richard, chemistry

POYNETTE
Sullivan, John Joseph, reproductive physiology
Waugh, Donovan Lloyd, soil fertility

RACINE
Beator, Mark Dabney, organic chemistry
Berge, John Williston, polymer chemistry
Beyerlein, Floyd Hilbert, analytical chemistry
Buckman, Alfred Fletcher, colloid chemistry
Conigliaro, Peter James, organic chemistry
Dickerson, Charlesworth Lee, organic chemistry
D'Orazio, Vincent T, organic chemistry
Dwyer, Sean G, physical organic chemistry
Kitzke, Eugene David, environmental health
Lo, Mike Mei-Kuo, physical chemistry
McGray, Robert James, microbiology
Miller, William Knight, analytical chemistry
Oomidhan, Easwaran Sukumaran, entomology, agricultural chemistry
Randall, Francis James, polymer chemistry
Sheppard, Erwin, physical chemistry
Smith, Robert Verne, organic chemistry
Su, Lao-Sou, physical chemistry
Tabet, Georges Elias, chemistry
Tcheurekdjian, Noubar, physical chemistry
Verbrugge, Calvin James, organic chemistry
Vos, Kenneth Dean, physical chemistry
Whyte, Donald Edward, organic chemistry

RHINELANDER
Cecich, Robert Allen, plant anatomy
Clausen, Jewell Johanna, ecology
Clausen, Knud Erik, forest genetics
Dawson, David H, forest genetics
Dickson, Richard Eugene, tree physiology
Godman, Richard M, forest management
Isebrands, Judson G, wood science & technology
Larson, Philip Rodney, forest physiology
Miksche, Jerome Phillip, plant morphology
Nienstaedt, Hans, forest genetics

Pizzolato, Thompson Demetrio, plant anatomy
Ratliff, Francis Tenney, paper technology

RIPON
Beatty, James Wayne, Jr., physical chemistry
Broshar, Wayne Cecil, solid state physics
Carley, David Wilcox, analytical chemistry
Hagquist, Carl Waldemar, zoology
Nichols, Charles, botany
Poole, Dewey Donald, plant breeding
Scamehorn, Richard Guy, organic chemistry
Scott, Earle Stanley, inorganic chemistry
Zei, Dino, experimental physics, history of science

RIVER FALLS
Akins, Virginia, biology
Bostrack, Jack M, botany
Campbell, Warren Adams, astronomy
Delorit, Richard John, weed science
Dollahon, James Clifford, animal breeding
Fahning, Melvyn Luverne, reproductive physiology, veterinary medicine
Goddard, Stephen, ornithology, ecology
Gough, Lillian, mathematics
Greub, Louis John, agronomy
Hall, Lyle Clarence, physical chemistry
Hill, John William, organic chemistry
McLaughlin, James Joseph, mathematics
Michaelson, Merle Edward, plant pathology
Muto, Peter, science education
Scott, Lawrence William, food science, analytical chemistry
Shepherd, John Patrick George, solid state physics
Sukow, Wayne William, molecular biophysics, biophysical chemistry
Walker, Charles Edward, Jr, physics
Wittwer, Leland S, animal nutrition
Zaborowski, Leon Michael, environmental chemistry

ROTHSCHILD
Boye, Frederick C, organic chemistry
Hogan, David James, analytical chemistry, inorganic chemistry

SCHOFIELD
Adams, James William, industrial chemistry

SHAWANO
Sims, John Albert, animal breeding

SHEBOYGAN
Grittinger, Thomas Foster, ecology, botany

STEVENS POINT
Anderson, Raymond Kenneth, wildlife ecology
Andrews, Oliver Augustus, chemistry, science education
Bainter, Monica Evelyn, atomic physics
Baumgartner, Frederick Milton, wildlife conservation
Booke, Henry Edward, ichthyology
Bowers, Frank Dana, bryology, plant taxonomy
Chander, Jagdish, experimental nuclear physics
Chang, Tsuen Kung, geography
Chitharanjan, Dakshinamurthy, organic chemistry
Copes, Frederick Albert, ecology, fisheries
Difford, Winthrop Cecil, geological oceanography
Engelhard, Robert J, forestry
Farnsworth, Carl Leon, organic chemistry
Freckmann, Robert W, plant taxonomy
Geeseman, Gordon E, genetics, botany
Hall, Kent D, reproductive physiology, environmental biology
Harpstead, Milo I, soils
Harris, Joseph Belknap, plant physiology
Hensler, Ronald Fred, soil fertility, soil chemistry
Hillier, Richard David, botany, ecology
Johnson, Charles Henry, mathematical statistics
Lang, Conrad Marvin, physical chemistry
Lee, Chen Hui, forestry
Long, Charles Alan, zoology

McColl, Daniel Clyde, theoretical physics, elementary particle physics
McKinney, William Mark, planetary sciences, physical geography
Milfred, Clarence James, geography, soil science
Miller, Gordon Lee, mathematics
Multhauf, Delmar Charles, geography
Perret, Maurice Edmond, geography
Pierson, Edgar Franklin, biology
Radtke, Douglas Dean, physical inorganic chemistry
Rice, Orville Millard, mathematics
Simpson, Robert E, microbiology
Sommers, Raymond A, analytical chemistry, paper chemistry
Thiesfeld, Virgil Arthur, botany, plant physiology
Thurmaier, Roland Joseph, organic chemistry, polymer chemistry
Trainer, Daniel Olney, wildlife diseases
Trytten, Roland Aaker, chemistry
Walden, Richard Trussell, medicine
Weaver, Robert Hinchman, biochemistry
White, Charley Monroe, ecology, wildlife biology
Wild, Wayne Grant, physics, mathematics

STOUGHTON
Benbow, Ralph Lawrence, solid state physics
Olson, Clifford Gerald, solid state physics
Weaver, John Herbert, solid state physics

STURGEON BAY
Gilbert, Franklin Andrew, Sr, horticulture

SUPERIOR
Anway, Allen R, physics, electronics
Coward, Nathan A, physical chemistry
Davidson, Donald William ecology
Dickas, Albert Binkley, research administration, geology
Hinkkanen, Donald William, physics, statistics
Horton, Joseph William, organic chemistry, environmental chemistry
Koch, Rudy G, science education, plant taxonomy
Lukens, Paul W, Jr, mammalogy
Meyer, Frank Henry, solid state physics, philosophy of science
Oexemann, Stanley William, plant physiology
Roubal, Ronald Keith, inorganic chemistry
Schneiderwent, Myron Otto, physics, science education
Seltzer, Michael Rogers, cultural anthropology, behavioral anthropology
Thomas, Howard Major, physical chemistry
Tychsen, Paul C, geology

TURTLE LAKE
McGuine, Thomas Harry, chemistry

WAUKESHA
Auchter, Harry A, physics
Bayer, Richard Eugene, analytical chemistry, inorganic chemistry
Cayle, Theodore, enzymology, industrial microbiology
Fallgatter, Michael, physical chemistry
Grotz, Leonard Charles, physical chemistry
Herbst, Richard Peter, algology, freshwater ecology
Macintyre, Bruce Alexander, physiology
Michaud, Ted C, zoology
Nelson, John Victor A, food science
Reinders, Victor A, chemistry
Richason, Benjamin Franklin, Jr, geography
Roys, Paul Allen, physics
Shah, Ghulam M, analytical mathematics
Spies, Robert Glenn, organic chemistry
Wendland, Ray Theodore, organic chemistry

WAUPACA
Brynildson, Oscar Marius, zoology
Hunt, Robert L, fish biology, aquatic ecology

WAUSAU
Burmblay, Ray Ulysses, analytical chemistry
Crabtree, Koby Takayashi, microbiology, civil engineering
Grdinic, Marcel Rudolph, organic chemistry
Mallery, Otto Tod, industrial health
Schmitz, William Robert, limnology, fish biology

WAUWATOSA
Boyle, Robert William, physiology
Wrangell, Lewis J, analytical chemistry

WEST BEND
Roth, Marie M, organic chemistry, general chemistry

WEST DE PERE
Poss, Richard Leon, mathematics
Worley, John David, biophysical chemistry

WHITEWATER
Crone, Lawrence John, plant pathology
Cummings, John Albert, radiobiology
Davis, Larry Wallace, mathematics
Dennis, Clifford John, entomology
Drexler, Edward James, analytical chemistry
Engert, Martin, mathematical analysis
Flanagan, Carroll Edward, mathematics
Follmar, Merle Norman, botany
Gollmar, Dorothy May, mathematics
Gorsica, Henry Jan, biochemistry
Klatt, Gary Brandt, mathematics
McCoy, Charles Ralph, physical chemistry
Meyer, Henry, zoology
Najar, Rudolph Michael, mathematics, physics
Nash, Reginald George, parasitology
North, Charles A, zoology, ornithology
Patterson, James Reid, nuclear physics
Romary, John Kirk, physical chemistry
Schlough, James Sherwyn, animal physiology, endocrinology
Seeburger, George Harold, science education, biology
Shinners, Carl W, physics
Smith, Stanley Galen, systematic botany, aquatic ecology
Stekel, Frank D, science education, physics
Stoneman, David McNeel, mathematical statistics
Tiffany, Sharon Weston, anthropology
Tiffany, Walter Warren, anthropology
Underwood, Robert Marshall, economic geography, geography of the Soviet Union
Varney, Charles Broadwell, geography
Wenaas, Paul Emil, chemistry

WILLARD
Leonard, Margaret Ives, chemistry

WILLIAMS BAY
Cudworth, Kyle McCabe, astronomy
Harper, Doyal Alexander, Jr, astronomy
Hobbs, Lewis Mankin, astronomy, astrophysics
Morgan, William Wilson, astronomy

WOOD
Barboriak, Joseph Jan, biochemical pharmacology
Hamilton, Lyle Howard, physiology
Hasegawa, Andrew Takeo, radiobiology, pharmacology
Llaurado, Josep G, biomedical engineering, physiology
Murray, Mary Patricia, kinesiology, medical research
Rose, Harold D, internal medicine
Rosenthal, Waldemar Arthur, clinical chemistry
Spurr, Gerald Baxter, physiology
Theil, George B, internal medicine
Van Horn, Diane Lillian, physiology, electron microscopy
Wang, Richard I H, pharmacology, internal medicine

WYOMING

BEULAH
Simon, Raymond Charles, genetics, cytology

CASPER
Barden, Thomas, geophysics
Barlow, James A, Jr, physical geology
Curry, William Hirst, III, geology
Kelly, Floyd W, Jr, organic chemistry

CHEYENNE
Edwards, William Charles, ecology
Hart, Richard Harold, agronomy

JACKSON
Zeitel, Eugene Paul, hydrology, water resources

LARAMIE
Adams, John Collins, microbial ecology, water pollution
Allbright, Charles Simar, analytical chemistry
Alley, Harold Pugmire, weed science
Anderson, Archie Duane, pharmacology
Archer, Vernon Shelby, analytical chemistry
Arnold, Gordon William, phytopathology
Asplund, Russell Owen, biochemistry
Atherton, Robert W, developmental biology, physiology
Baxter, George T, fish biology, limnology
Bear, Phyllis Dorothy, microbial genetics, molecular biology
Beetle, Alan Ackerman, agronomy
Belden, Everett Lee, veterinary immunology, veterinary microbiology
Bergman, Harold Lee, physiological ecology
Bergstrom, Robert Charles, veterinary parasitology
Bessey, Robert John, physics
Blackstone, Donald Leroy, geology
Bohnenblust, Kenneth E, plant breeding, plant pathology
Bone, Jack Norman, pharmacy
Borgman, Leon Emry, statistics, geology
Bosshardt, David Kirn, analytical biochemistry
Botkin, Merwin P, animal science
Bowerman, Robert Francis, neurophysiology, behavioral biology
Boyd, Donald Wilkin, historical geology
Bridgmon, George Harrison, plant pathology
Brunett, Emery W, medicinal chemistry, pharmacy
Bulgrin, Vernon Carl, physical chemistry
Burkhardt, Christian Carl, entomology
Caldwell, Daniel R, microbiology, biochemistry
Carlson, William Dwight, veterinary radiology, radiation biology
Christensen, Martha, mycology, ecology
Clarke, Lemuel Floyd, zoology
Coates, Geoffrey Edward, organometallic chemistry
Cooke, William Peyton, Jr, mathematics, statistics
Crawford, Daniel John, plant taxonomy
Deane, Darrell Dwight, dairy bacteriology
Decora, Andrew Wayne, physical organic chemistry, spectroscopy
Denison, Arthur B, molecular physics
Denniston, Rollin H II, animal behavior
Diem, Kenneth Lee, zoology
Dinneen, Gerald Uel, chemistry
Dirks, Richard Allen, dynamic meteorology
Dorrence, Samuel Michael, organic chemistry
Dunn, Thomas Guy, reproductive physiology, endocrinology
Duvall, John Joseph, physical organic chemistry
Edmiston, Clyde, physical chemistry
Ellis, William Wesley, nutritional biochemistry
Field, Ray A, meat science
Fisser, Herbert George, range management, ecology
Frerichs, William Edward, micropaleontology
Frison, George Carr, anthropology, archaeology
Gastl, George Clifford, topology, algebra

Gehrz, Robert Douglas, astrophysics
George, John Harold, mathematics
George, Robert Porter, developmental biology, invertebrate zoology
Gill, George Wilhelm, physical anthropology
Grandy, Walter Thomas, Jr., theoretical physics
Guenther, William Charles, statistics
Hackwell, John Arthur, astrophysics
Haines, William Emerson, petroleum chemistry
Haley, Boyd Eugene, biochemistry
Hamilton, John William, agricultural chemistry
Hanna, James Ray, mathematics
Harding, Samuel William, physics
Hilston, Neal William, nutrition
Hofmann, David John, atmospheric physics
Holt, Smith Lewis, Jr., inorganic chemistry
Hough, Hugh Walter, physics
Houston, Robert S., economic geology, petrology
Howatson, John, x-ray crystallography, inorganic chemistry
Hoyt, Gordon Dunwell, physics
Humburg, Neil Edward, agronomy
Hurtubise, Robert John, analytical chemistry
Husain, Syed Alamdar, mathematics
Jacobson, Irven Allan, Jr., physiology
Jaeger, David Allen, organic chemistry
Jenkins, Robert Allan, cell biology
Jenkins, Terry Lloyd, mathematics
Ji, Tae Hwa, biochemistry
Julian, Edward A., pharmacy
Kaiser, Ivan Irvin, biochemistry
Kaltenbach, Carl Colin, reproductive physiology
Kennington, Garth Stanford, animal ecology, animal physiology
Kercher, Conrad J., animal nutrition
Kingston, Newton, zoology, parasitology
Kirk, John David, geology
Kisir, Charles Joseph, psychopharmacology
Kunselman, Arthur Raymond, nuclear physics
Lang, Robert Lee, agronomy
Latham, DeWitt Robert, petroleum chemistry
Lavigne, Robert James, entomology
Lawson, Fred Avery, entomology
Lillegraven, Jason Arthur, paleontology
Lloyd, John Edward, entomology
Love, John David, geology
Magee, Michael Jack, computer sciences
Maki, Leroy Robert, bacteriology
Maurer, John Edward, organic chemistry
May, Morton, range conservation
McAnelly, Charles William, plant pathology
McColloch, Robert James, physical chemistry, biochemistry
McDonald, Francis Raymond, spectroscopy, organic chemistry
McGrew, Paul Orman, geology, paleontology
McMahon, Vern August, plant biochemistry
Mears, Brainerd, Jr., geomorphology
Mena, Roberto Abraham, algebra
Meyer, Edmond Gerald, physical chemistry
Miknis, Francis Paul, nuclear magnetic resonance
Miller, Daniel Newton, Jr., geology
Miller, Glenn Joseph, biochemistry
Morgan, George L., inorganic chemistry
Muller, Burton Harlow, physics
Nelms, George E., anthropology
Nelson, Kenneth Fred, medicinal chemistry
Nelson, Robert B., pharmacology
Netzel, Daniel Anthony, physical chemistry
Noe, Lewis John, chemistry
Northen, Henry Theodore, botany
Pancoe, William Louis, Jr., endocrinology
Paris, Oscar Hall, population ecology, invertebrate ecology
Parker, Michael, zoology, limnology
Parker, Ronald Bruce, mineralogy, structural geology
Paules, Leon H., animal husbandry
Pepin, Theodore John, physics
Petersen, Joseph Claine, petroleum chemistry, organic chemistry
Pfadt, Robert E., entomology
Porter, A Duane, mathematics
Poulson, Richard Edwin, physical chemistry

Radloff, Harold David, dairy science, biochemistry
Raulins, Nancy Rebecca, organic chemistry
Reeder, John Raymond, botany
Reider, Richard Gary, physical geography
Rhoads, Sara Jane, organic chemistry
Rinehart, Edgar A., molecular spectroscopy, lasers
Robinson, Wilbur Eugene, chemistry, fuel science
Rosen, James Martin, physics
Rosenholtz, Ira N., topology
Roth, Ben G., mathematics
Rowland, John H., mathematics
Ryan, Victor Albert, nuclear chemistry
Schaeffer, Riley, inorganic chemistry
Schick, Lee Henry, theoretical nuclear physics
Schoonover, Carroll Owen, animal breeding
Shader, Leslie Elwin, mathematics
Shive, Peter Northrop, geophysics
Simon, William George, nuclear physics
Singleton, Paul C., soil chemistry, soil fertility
Smith, Charles Ray, theoretical physics
Smith, William Norman, mathematical analysis
Smithson, Scott Busby, geophysics, petrology
Smith-Sonneborn, Joan, cell biology
Steadman, John William, electrical engineering, bioengineering
Stoner, Adair, entomology
Stratton, Paul Oswald, animal science
Swift, Brinton L., veterinary medicine
Tabler, Ronald Dwight, hydrology, watershed management
Tucker, James, veterinary science
Vail, Gabor, atmospheric physics
Varineau, Verne John, mathematics
Varnell, Thomas Raymond, biochemistry, physiology
Veal, Donald L., meteorology
Villemez, Clarence Louis, Jr., biochemistry
Ward, Angus Lorin, wildlife research
Wilson, William Thomas, invertebrate pathology, entomology
Winkel, David Edward, physical chemistry, computer science

CANAL ZONE

BALBOA
Campanella, Paul Joseph, II, ecology, environmental management
Glynn, Peter W., marine ecology
Graham, Jeffrey Brent, respiratory physiology
Leigh, Egbert G., Jr., ecology, population genetics
Linares, Olga Frances, anthropology
Lofdn, Horace (Greeley), ecology
Moynihan, Martin Humphrey, animal behavior
Rand, Austin Stanley, evolutionary biology
Rubinoff, Ira, zoology
Smith, Alan Paul, plant ecology
Smith, Neal Griffith, evolutionary biology, ornithology
Windsor, Donald Montgomery, animal behavior

BALBOA HEIGHTS
Baerg, David Carl, medical entomology
Christensen, Howard Anthony, medical entomology, medical parasitology
Galindo, Pedro, entomology
Rossan, Richard Norman, parasitology

COCO SOLO
Hendler, Gordon Lee, marine ecology, invertebrate zoology

MOOSE
Bachman, Charles Herbert, physics

RIVERTON
Day, William W., physics, academic administration
Dobell, Joseph Porter, geology
Jordan, Russell Thomas, virology, immunochemistry
Scott, Richard Walter, plant ecology

CANAL ZONE
Cruz, Carlos, entomology

FT CLAYTON
Portig, Wilfried Helmut, climatology, meteorology

MARGARITA
Boreham, Melvin Murray, medical entomology

PANAMA
Walton, Bryce Calvin, parasitology

GUAM

AGANA
Crane, Sheldon Cyr, environmental sciences
Eldredge, Lucius G., marine biology
Levand, Oscar, organic chemistry
Marsh, James Alexander, Jr., marine ecology
Smith, Douglas Roane, botany
Tsuda, Roy Toshio, phycology

PUERTO RICO

ARECIBO
Campbell, Donald Bruce, planetary sciences
Conklin, Edward Kirkham, radio astronomy
Craft, Harold Dumont, Jr., radio astronomy, radiophysics
Davis, Michael Moore, radio astronomy
Dyce, Rolf Buchanan, planetary sciences
Hagen, Jon Boyd, radiophysics
Hankins, Timothy Hamilton, radio astronomy
Harper, Robert M., atmospheric physics
Meriwether, John Williams, Jr., aeronomy
Sramek, Richard Anthony, radio astronomy
Walker, James Callan Gray, geophysics

BARCELONETA
Carvajal, Fernando, microbiology

CAYEY
De La Sierra, Angell Ortiz, cell biophysics, chemistry

HATO REY
Lamba, Ram Sarup, inorganic chemistry, organic chemistry
Oliver-Padilla, Fernando Luis, science, animal breeding
Viciedo, Eusebio, microbiology, biochemistry

ISABELA
Cruz, Carlos, entomology

MANATI
Hayes, Kenyon (Joseph), organic chemistry
Page, Calvin Ames, microbiology

MAYAGUEZ
Abu-Zeid, Mohyi Eldin, molecular spectroscopy
Aguayo, Carlos G., malacology
Atwood, Donald Keith, chemical oceanography
Bailey, Carroll Edward, solid state physics
Banus, Mario Douglas, marine ecology, environmental chemistry
Biaggi, Virgilio, Jr., ornithology
Clark, David C., mathematical analysis
Copson, David Arthur, biophysics, microbiology
Crabtree, David Melvin, marine biology
Cutress, Charles Ernest, invertebrate zoology
Engstrom, Norman Ardell, marine ecology, zoology
Francis, Eugene A., marine geology
Gonzalez De Alvarez, Genoveva, organic chemistry
Gonzalo, Julio Antonio, solid state physics
Hernandez-Avila, Manuel Luis, physical oceanography
Villella, John Baptist, zoology

RIO PIEDRAS
Adam, Waldemar, physical organic chemistry
Bangdiwala, Ishver Surchand, statistics
Beck, Jonathon Mock, mathematics
Bobonis, Augusto, mathematics
Bruck, David Lewis, genetics, population biology
Cancio, Marta, biochemistry
Candelas, Graciela C., cell biology, molecular biology
Capo, Bernardo Guillermo, agriculture
Curet, Juan Daniel, physical chemistry
Garcia-Morin, Manuel, physical chemistry, analytical chemistry
Garriga-Rodriguez, Francisco, mathematics
Hillyer, George Vanzandt, parasitology, immunology
Kaye, Carmen Jimenez, medicine
King, Nydia Margarita, phytochemistry
Knighten, Robert Lee, algebra
Koss, Joan Dee, anthropology
Liddle, Larry Brook, marine phycology
Lugo, Herminio Lugo, plant physiology
Marcial, Victor A., radiotherapy
Martorell, Luis Felipe, entomology
McDowell, Dawson Clayborn, meteorology
Nunez-Melendez, Esteban, pharmacognosy
Roman, Jesse, plant nematology
Samuels, George, soil fertility, plant physiology
Schubert, Thomas Herman, forestry
Stolberg, Harold Joset, mathematics
Torres-Rodriguez, Victor M., medicine
Vassos, Basil Harilaos, chemical instrumentation, analytical chemistry
Velez, Antonio, horticulture
Virkki, Nilo, insect cytogenetics
Wadsworth, Frank H., forestry

PONCE
Ramos, Lillian, chemistry, science education
Rodriguez, Haydee C., physics, mathematics

OLD SAN JUAN
Del Castillo, Jose, neurophysiology
Ebbs, Jane Cotton, physiology, nutrition

SAN GERMAN
Enriquez, Nitza M., microbiology
Lagally, Ralph Werner, organic chemistry, pharmaceutical chemistry
Lohr de Irizarry, Mildred Tucker, geography
Verter, Herbert Sigmund, organic chemistry

Kay, Mortimer Isaia, crystallography
Koo, Francis Keh Shing, genetics, radiobiology
Lapiaza, Miguel Luis, mathematics
Lee, Rupert Archibald, radiation chemistry
Lugo-Lopez, Miguel Angel, soil physics
Martin, Franklin Wayne, genetics
Martinez, Nadal, Noemi G, organic chemistry, biochemistry
Martinez-Pico, Jose Luis, physical chemistry
Miskimen, Carmen Rivera, plant pathology, virology
Miskimen, George William, neurophysiology, population ecology
Morelock, Jack, oceanography
Moussa, Mounir, Tawfik, sedimentary petrology, stratigraphy
Ortiz-Suarez, Augusto Hermino, mathematics
Pagan-Font, Francisco Alfredo, fisheries management
Pannella, Giorgio, invertebrate paleontology, palynology
Paschal, Eugene Hamer, II, plant breeding
Peinado, Rolando E., algebra, biomathematics
Rivero, Juan Arturo, herpetology
Rodriguez, Anibal, mathematics
Sahai, Hardeo, statistics
Seiglie, George A. micropaleontology
Siegel, Georges Giovanni, physical chemistry
Singh, Rama Shankar, solid state physics
Smith, Alan Lewis, volcanology
Vakili, Nader Gholi, plant pathology
Walker, David Whitman, entomology
Weaver, John Dodsworth, geology

SAN JUAN
Almodovar, Ismael, inorganic chemistry, nuclear chemistry
Arbona, Guillermo, public health
Asenjo, Conrado Federico, chemistry
Bertran, Carlos Enrique, cardiology
Block, Arthur McBride, environmental chemistry, quantum chemistry
Brush, James S, biochemistry
Calabrisi, Paul, anatomy
Cancel, Cruz A, speech pathology, audiology
Castrillon, Jose P A, organic chemistry, radiochemistry
Chinea, Jose Juan, histology, dentistry
Clements, Richard Gerald, soil science, ecology
Clevenger, Ima Fuchs, speech pathology, audiology
Colon, Jose A, meteorology
Colon, Julio Ismael, virology
Corcino, Jose Juan, hematology, gastroenterology
Crosby, Paul Faljean, chemistry
De Mello, W Carlos, physiology
Eberhardt, Manfred Karl, organic chemistry
Fernandez, Frank, anthropology
Fox, Irving, medical entomology
Frontera-Reichard, Jose Guillermo, neuroanatomy
Garcia-Palmieri, Mario R, internal medicine, cardiology
Goyco Daubon, Jose A, nutrition, biochemistry
Haddock, Lillian, internal medicine, endocrinology
Jobin, William Roger, tropical public health
Johnston, John Derland, industrial organic chemistry
Kaye, Sidney, toxicology
Korchin, Leo, oral surgery
Lawhead, James Stout, organic chemistry
Mackay, Ian Francis Stuart, human physiology
Mahgoub, Ahmed, pharmacology
Marcial-Rojas, Raul Armando, pathology
Marinelarena, Rafael, bacteriology
Martinez-Maldonado, Manuel, internal medicine, nephrology
Mendez, Eugenio Fernandez, anthropology
Moreno, Esteban, medicine, pathology
Nigaglioni, Adan, internal medicine, gastroenterology
Ortiz, Antonia, medicine, pediatrics
Pelegrina, Ivan A, obstetrics & gynecology
Pico, Guillermo, ophthalmology
Pomales-Lebron, Americo, bacteriology
Ramirez, J Roberto, bio-organic chemistry
Ramos-Morales, Francisco, internal medicine
Reinecke, Roger Minske, physiology
Rendon, Orlando Roberto, physiology, genetics
Ritchie, Lawrence Starr, medical parasitology
Sandy, Don Glen, speech pathology
Sandza, Joseph Gerard, biochemistry
Santos-Martinez, Jesus, physiology, pharmacology
Sifontes, Jose E, pediatrics
Sifre, Ramon Alberto, internal medicine, gastroenterology
Simpson, George A, photochemistry
Szepsenwol, Josel, anatomy
Toro-Goyco, Efrain, physical chemistry, biochemistry
Torres-Biasini, Gladys, microbiology

SANTURCE
Del Rosario, Leticia, physics
Hanson, Earl Parker, geography
Hilmer, Paul Edward, biochemistry
Medina, Antonio Samuel, medicine, public health
Noya Benitez, Jose Antonio, surgery
Oliver-Gonzales, Jose, parasitology

VIRGIN ISLANDS

CHRISTIANSTED
Donath, Ernest E, physical chemistry

ST CROIX
Bond, Richard Marshall, environmental biology
Canfield, Norton, otology, laryngology
Gladfelter, William Bayard, invertebrate zoology, marine zoology
Moore, Joseph Graessle, anthropology
Ogden, John Conrad, marine ecology
Weinberg, Sidney R, urology
Williams, David Francis, medical entomology

ST THOMAS
Caron, Aimery Pierre, physical chemistry
Dammann, Arthur Eric, fisheries, wildlife management
Lee, Douglas Harry Kedgwin, environmental sciences
MacLean, William Plannette, III, biology
McMillan, Joseph Patrick, zoology, comparative physiology

CANADA

ALBERTA

BEAVERLODGE
Faris, Donald George, genetics, plant physiology
Harris, Robert Ernest, genetics, plant physiology
Rice, Wendell Alfred, soil microbiology

BENALTO
Holsworth, William Norton, wildlife ecology

CALGARY
Ab Iorwerth, Hefin, geophysics
Acheson, Cyrus Harold, geophysics
Adam, Frank Cuthbert, physical chemistry
Aggarwala, Bhagwan D, applied mathematics
Anderson, Paul Knight, vertebrate ecology
Andrichuk, John Michael, petroleum geology, sedimentology
Anger, Clifford D, atmospheric physics
Arai, Hisao Philip, parasitology
Armstrong, David Anthony, physical chemistry, biochemistry
Athar, Mohammed Aqeel, medical microbiology, public health
Bachelor, Frank William, organic chemistry
Baillie, Andrew Dollar, stratigraphy
Baumber, John Scott, physiology
Bayliss, Peter, mineralogy, crystallography
Beaton, James Duncan, soil fertility
Beck, James S, biophysics
Belyea, Helen R, geology
Bewley, John Derek, plant physiology, biochemistry
Bird, Charles Durham, botany
Bland, Brian Herbert, neurobiology, neuropsychology
Bland, Clifford J, cosmic ray physics, radiation physics
Bombardieri, Caurino Cesar, petroleum chemistry
Boorman, Philip Michael, inorganic chemistry
Browder, Leon Wilfred, developmental biology
Buckmaster, Harvey Allen, magnetic resonance
Campbell, Finley Alexander, geology
Challice, Cyril Eugene, physics, biophysics
Chatterjee, Ramananda, solid state physics
Church, Robert Bertram, developmental biology
Clark, Thomas Alan, astronomy
Cochrane, William, pediatrics
Consul, Prem Chandra, statistics, mathematics
Coombes, Charles Allan, physics
Cooper, Keith Edward, physiology
Costerton, J William F, microbiology
Coulson, Michael Robert Cummins, geography, cartography
Cragg, James Birkett, ecology, environmental sciences
Crickmay, Colin Hayter, geology
Davies, Konald Wallace, ecology
Dawson, J W, internal medicine, endocrinology
Dhaliwal, Ranjit S, applied mathematics
Dickson, Arthur David, reproductive biology, medical education
Dixon, Gordon H, molecular biology, developmental biology
Drummond, George I, biochemistry, pharmacology
Duerksen, Jacob Dietrich, molecular biology
Duggan, Hector Ewart, radiology
Enns, Ernest Gerhard, mathematical statistics, operations research
Erdman, Oscar Alvin, geology
Forbis, Richard George, ethnology, archaeology
Foscolos, Anthony E, soil mineralogy, sedimentology
Fox, Frederick Glenn, geology
Furnival, George Mitchell, geology, petroleum
Gaucher, George Maurice, biochemistry, microbiology
Geist, Valerius, zoology, ethology
Ghent, Edward Dale, geology
Giovinetto, Mario Bartolome, climatology, glaciology
Gonzalez, Alfonso, geography of Latin America, population geography
Goren, Howard Joseph, biochemistry, enzymology
Guy, Richard K, mathematics
Hacquebard, Peter Albertus, geology
Hamill, Louis, geography
Hancock, Ronald Lee, biochemistry
Harper, John David, stratigraphy, sedimentology
Harrington, Jonathan W, paleontology, stratigraphy
Harris, Stuart Arthur, geography, soil science
Hartland-Rowe, Richard C B, aquatic ecology
Henderson, Gerald Gordon Lewis, geology
Hills, Leonard Vincent, geology
Hodgson, Gordon Wesley, physical chemistry
Hoekstra, Pieter, geophysics, geotechnical engineering
Hollenberg, Martin James, anatomy
Hopkins, William Stephen, Jr, palynology, geology
Hriskevich, Michael Edward, geology
Huber, Rueben Bissett, biochemistry
Hyne, James Basil, chemistry
Irish, Ernest James Wingett, geology
Jacobson, Ada Leah Hyne, physical chemistry, biochemistry
Johnson, Dudley Paul, mathematics, mathematical statistics
Jones, J P, mathematics
Jones, Melvin D, enzymology
Kapoor, Manju, enzymology
Kariel, Herbert George, cultural geography
Karr, Gerald William, pharmacology, internal medicine
Kerr, James William, geology
Klaassen, Rudolph Waldemar, geology
Klovan, John Edward, geology
Krouse, Howard Roy, physics, chemistry
Krueger, Peter J, physical chemistry, spectrochemistry
Laidlaw, William George, theoretical chemistry
Laurenson, Rae Duncan, anatomy, medical education
LeBlanc, Francis Ernest, neurophysiology, neurosurgery
Lederis, Karl, pharmacology, endocrinology
Leeson, Bruce Frank, ecology, environmental management
Levene, Cyril, anatomy
Levinson, Alfred Abraham, mineralogy
Lewis, David James, psychiatry, medical education
Lin, Chyi-Chyang, genetics
Lorscheider, Fritz Louis, physiology, endocrinology
MacCannell, Keith Leonard, clinical pharmacology
Macdonald, Digby Donald, physical chemistry, electrochemistry
Macqueen, Roger Webb, geology
Majumdar, Samir Ranjan, applied mathematics
Masters, John Alan, petroleum geology
McBee, William, Jr, geology
McCormack, William Charles, anthropology, anthropological linguistics
McCrossan, Robert George, petroleum geology
McGugan, Alan, geology, paleontology
McKay, Alexander Scott, physics
McKay, William Neil, physical chemistry, analytical chemistry
McLeod, Lionel Everett, medical education, medical administration
McMahon, Brian Robert, animal physiology
McMillan, Neil John, geology, soil science
Milner, Eric Charles, mathematics
Milone, Eugene Frank, astronomy, physics
Momsen, Janet Henshall, geography
Momsen, Richard Paul, Jr, geography
Muir, Donald Ridley, physical chemistry, research administration
Myres, Miles Timothy, ecology, ornithology
Nelson, Samuel James, stratigraphy, paleontology
Nielsen, Kenneth Fred, soil fertility, plant physiology
Norford, Brian Seeley, geology, paleontology
Norris, Donald Kring, geology
Ogawa, Junjiro, mathematical statistics
Oliver, Thomas Albert, geology
Ollerenshaw, Neil Campbell, geology
Parkinson, Dennis, botany, microbiology
Parsons, Neville Ronsley, physics
Paul, Reginald, theoretical chemistry
Pearce, Keith Ian, psychiatry, medicine
Pharis, Richard Persons, plant physiology
Pocock, Stanley Albert John, geology, palynology
Pritchard, Gordon, entomology
Pugh, Derek Charles, geology
Raasch, Gilbert O, geology
Rauk, Arvi, chemistry
Raymond, James Scott, anthropology
Read, John Hamilton, pediatrics, preventive medicine
Reid, David Mayne, plant physiology
Robertson, Ross Elmore, physical chemistry, organic chemistry
Roche, Rodney Sylvester, biophysical chemistry, polymer chemistry
Rod, David Lawrence, mathematics
Rosenberg, Herbert Irving, vertebrate morphology
Rosenvall, Lynn Albert, geography
Rowe, Robert Burton, geology
Rowlands, Stanley, medical biophysics, nuclear medicine
Russell, Anthony Patrick, herpetology
Sahney, B N, pure mathematics
Sainsbury, Robert Stephen, neuropsychology
Sanderson, Kenneth Edwin, microbial genetics
Schaer, Jonathan, mathematics
Scheinberg, Eliyahu, population genetics, applied statistics
Schmidt, Volkmar, marine geology
Schofield, Brian, physiology, gastroenterology
Schultz, Gilbert Allan, developmental biology
Sears, Timothy Stephen, photochemistry, physical chemistry
Sharman, Bernard Clout, botany
Sheehan, Bernard Stephen, institutional research
Sirrine, George Keith, petroleum geology
Sorensen, Theodore Strang, physical organic chemistry
Spira, Arthur William, anatomy
Sreenivasan, Sreenivasa Ranga, physics, astrophysics
Staplin, Frank Lyons, geology, paleontology
Stevenson, Kenneth James, protein chemistry
Stone, Michael Gates, mathematics
Stott, Donald Franklin, geology
Sullivan, Herbert J, geology, palynology
Tan, Celine G L, biochemistry
Tan, Yin Hwee, virology, genetics
Tavares, Donald Francis, organic chemistry
Thorpe, Trevor Alleyne, plant physiology
Tollefson, Eric Lars, physical chemistry, chemical engineering
Torrence, Robert James, theoretical physics
Townley, John Lewis, III, petroleum chemistry
Tschuikow-Roux, Eugene, physical chemistry
Tulczyjew, Wlodzimierz Marek, mathematical physics
Van Petten, Garry R, pharmacology
Varadarajan, Kalathoor, mathematics, topology
Veale, Warren Lorne, physiology, neuropsychology
Venkatesan, Doraswamy, space physics, astrophysics
Vroom, Alan Heard, instrumentation, materials science
Wall, John Hallett, micropaleontology
Wallis, Donald Douglas James H, ionospheric physics
Wani, Jagannath K, statistics
Wardlaw, Norman Claude, geology

Watanabe, Mamoru, medicine, biochemistry
Westbrook, David Rex, applied mathematics
Wieser, Helmut, chemistry
Wilkens, Jewel L., insect physiology
Williamson, John Hybert, genetics
Wong, James Chin-Sze, mathematics
Wyder, John Ernest, geophysics, geology
Yandgeni, Raghavendra, physical chemistry, mass spectrometry
Yeager, Howard Lane, analytical chemistry
Young, Frederick Griffin, stratigraphy
Zvengrowski, Peter Daniel, mathematics

EDMONTON

Addicott, John Fredrick, ecology
Adler, John G., solid state physics
Aherne, Francis Xavier, animal nutrition
Ahmed, Asad, molecular genetics
Allen, Willard Finlay, analytical chemistry
Anderson, Arthur James, pharmaceutics
Andrew, William Treleaven, vegetable crops
Antonelli, Peter Louis, topology, geometry
Ayer, William Alfred, organic chemistry
Baadsgaard, Halfdan, geochemistry
Babcock, Elkanah Andrew, structural geology, remote sensing
Bailey, Arthur W., range science
Bain, Gordon Orville, pathology
Bakshi, Trilochan Singh, ecology
Ball, George Eugene, systematics
Bar, Hans-Peter, organic chemistry, pharmacology
Barclay, Harold Barton, cultural anthropology, ethnology
Beatty, David Delmar, comparative physiology
Bell, Harold E., clinical pathology
Bentley, Charles Fred, soil science
Bercov, Ronald David, algebra
Berg, Roy Torguy, animal genetics
Bergmann, John Francis, geography
Bertie, John E., physical chemistry, chemical physics
Betts, Donald Drysdale, theoretical physics
Bhatia, Avadh Behari, theoretical physics, solid state physics
Biggs, David Frederick, pharmacology
Birss, Fraser William, theoretical chemistry
Blackburn, Edward Victor, organic genetics
Blackmore, Robert Valentine, microbiology, biochemistry
Blades, Arthur Taylor, physical chemistry
Bliss, Lawrence Carroll, botany
Boag, David Archibald, animal ecology
Bowen, Peter, internal medicine, human genetics
Bowland, John Patterson, animal nutrition
Bridger, William Aitken, biochemistry
Brown, Robert Stanley, physical organic chemistry
Bruchovsky, Nicholas, endocrinology
Bryan, Alan Lyle, anthropology
Burwash, Ronald Allan, geology
Cameron, Donald Forbes, anesthesiology
Campbell, James Nicoll, microbiology, biochemistry
Campbell, John Duncan, paleobotany
Capri, Anton Zizi, particle physics
Carlson, Lester William, mathematical physics
Carmichael, David James, plant pathology, silviculture
Carmichael, David James, protein chemistry
Cass, Carol E., cell biology, biochemistry
Cass, David D., plant embryology
Cavell, Ronald George, inorganic chemistry
Charnock, John S., pharmacology, biochemistry
Chatten, Leslie George, pharmaceutical chemistry
Chatterton, Brian Douglas Eyre, paleontology
Chen, I-Ngo, computer science
Chia, Fu-Shiang, zoology
Clandinin, Donald Robert, poultry nutrition
Clarke, Bruce Leslie, theoretical chemistry
Clegg, Lawrence Frank Levey, bacteriology, chemistry
Clifford, Hugh Fleming, limnology, invertebrate zoology
Colter, John Sparby, virology
Colter, Herbert Bruce, biochemistry
Cook, Fred D., soil microbiology

Cookson, Francis Bernard, neuroanatomy, histology
Cormack, Robert George Hall, botany
Corns, William George, plant science, weed science
Cossins, Edwin Albert, plant physiology, biochemistry
Cottle, Merva Kathryn Warren, biochemistry
Cottle, Walter Henry, physiology
Coutts, Ronald Thomson, organic chemistry, medicinal chemistry
Craig, Douglas Abercrombie M., insect morphology, invertebrate limnology
Crawford, Robert James, organic chemistry
Creighton, Stephen Mark, organic chemistry, physical chemistry
Davis, Norman Rodger, biochemistry
Davis, Stuart George, physical chemistry
Dawson, Wilfred Kenneth, nuclear physics
Denford, Keith Eugene, phytochemistry
Dick, Henry Marvin, oral pathology
Dixon, John Michael Siddons, medical microbiology
Donald, Elizabeth Ann, nutrition
Dossetor, John Beamish, transplantation immunology, nephrology
Dudas, Marvin Joseph, soil morphology
Dunford, Hugh Brian, physical chemistry
Dunlop, D L., obstetrics & gynecology
Elofson, Richard Macleod, organic chemistry
Evans, William George, insect ecology
Fields, Jerry L., mathematics
Fisher, James Louis, algebra
Fleterick, Robert John, molecular biophysics
Folinsbee, Robert Edward, geology
Fouron, Yves, organic chemistry
Fox, Richard Carr, vertebrate paleontology
Fraga, Serafin, theoretical chemistry
Frank, George Barry, physiology, pharmacology
Fraser, Robert Stewart, internal medicine, cardiology
Freedman, Herbert Irving, mathematical analysis, biomathematics
Freeman, Gordon Russel, physical chemistry, radiation chemistry
Fuller, William Albert, vertebrate ecology
Gallup, Donald Noel, limnology
Garg, Krishna Murari, mathematics
Ghurye, Sudhish G., statistics
Gilbert, James Alan Longmore, internal medicine
Golding, Douglas Lawrence, forest hydrology
Gooding, Ronald Harry, entomology
Gorham, Paul Raymond, plant physiology, phycology
Gough, Denis Ian, geophysics
Gough, Kenneth Henry, audiology, speech pathology
Goyer, Guy Gaston, meteorology
Graham, William Arthur Grover, inorganic chemistry
Greenhill, Stanley E., medicine
Gruhn, Ruth, anthropology, archaeology
Gunning, Harry Emmet, physical chemistry, photochemistry
Hadziyev, Dimitri, food science
Hage, Keith Donald, meteorology
Halferdahl, Laurence Bowes, exploration geology
Hardin, Robert Toombs, biometrics, poultry breeding
Harris, Walter Edgar, analytical chemistry
Haryett, Rowland D., orthodontics
Heath, Charles, physiology
Heming, Bruce Sword, entomology
Henderson, Joseph Franklin, biochemistry
Henderson, Ruth McClintock, physiology, pharmacology
Hickman, Michael, phycology, limnology
Hiratsuka, Yasuyuki, mycology, plant pathology
Hitchon, Brian, geochemistry
Hodges, Robert Stanley, protein chemistry
Hodgetts, Ross Birnie, developmental genetics
Hogg, Alan Mitchell, chemistry
Hohn, Emil Otto, endocrinology
Holmes, John Carl, zoology
Honsaker, John Leonard, atmospheric physics
Hoo, Cheong Seng, mathematics
Hook, Derek John, microbial biochemistry
Hooz, John, organic chemistry
Horricks, Jack Stewart, plant pathology

Hube, Douglas Peter, astronomy
Humphries, Robert Gordon, meteorology
Huston, Mervyn James, pharmacology
Hutchison, Kenneth James, medical physiology
Huzinaga, Sigeru, theoretical chemistry, physics
Igarashi, Satomi J., molecular biology
Ironside, Robert Geoffrey, economic geography
Israel, Werner, theoretical physics
Jackson, Harold, food science
Janke, Wilfred Edwin, soil science
Jeliard, Charles H., medical bacteriology, public health
Johns, Anthony, pharmacology
Jordan, Robert Beatty, inorganic chemistry
Kadis, Vincent William, microbiology, biochemistry
Kalantar, Alfred Husayn, spectroscopy
Kalnins, Abdul Naim, particle physics
Kanasewich, Ernest Raymond, geophysics
Kaneda, Toshi, biochemistry
Kay, Cyril Max, biochemistry
Kebarle, Paul, physical chemistry
Keeping, Eleanor Silver (Dowding), mycology
Keiker, Douglas, statistics
Kennedy, Lorene Louise, botany
Knaus, Edward Elmer, medicinal chemistry
Kopecky, Karl Rudolph, organic chemistry
Kotowycz, George, biophysical chemistry
Kowalewski, Konstany Piotr, endocrinology, experimental surgery
Kratochwil, Byron, analytical chemistry
Kunzle, Hans Peter, mathematical physics
Kuspira, J., cytogenetics
Lakey, William Hall, genito-urinary surgery
Lam, Sheung Tsing, nuclear physics
Lambert, Richard St. John, petrology, geochemistry
Lane, Edwin David, aquatic ecology, fisheries
Lane, Robert K., environmental management
Larke, Robert Peter Bryce, virology
Lauber, Jean Kautz, zoology, physiology
Laycock, Arleigh Howard, geography, hydrology
Leeson, Thomas Sydney, anatomy
Lemieux, Raymond Urgel, organic chemistry
LePage, Gerald Alvin, biochemistry
Lerbekmo, John Franklin, geology
Lesins, Karlis A., cytogenetics, plant breeding
Lewin, Victor, ecology
Lissey, Allan, hydrogeology, engineering geology
Liu, Hsing-Jang, chemistry
Locock, Robert A., pharmaceutical chemistry, pharmacognosy
Longenecker, Bryan Michael, immunology, cell biology
Longley, Richmond Wilberforce, meteorology, climatology
Lovig, Henry Francis Joseph, pure mathematics
Lown, James William, physical organic chemistry
Ludwig, Garry (Gerhard Adolf), mathematics
MacDonald, Ronald Neil Angus, hematology, oncology
Mackay, William Charles, environmental physiology, comparative physiology
MacKenzie, Walter Campbell, medicine
Mackl, Jack W., mathematics
Macpherson, Andrew Hall, ecology, zoogeography
MacRae, Patrick Daniel, dentistry
Madsen, Neil Bernard, biochemistry
Magno, Richard, solid state physics
Mahrt, Jerome L., parasitology, zoology
Malhotra, Sudarshan Kumar, cell biology
Martin, John Scott, physical chemistry
Marusyk, Raymond George, virology
Masamune, Satoru, organic chemistry
McCalla, Arthur Gilbert, plant biochemistry, plant physiology
McCandless, Robert William, instrumentation, materials science
McClung, Ronald Edwin Dawson, chemistry, physics
McCoy, Ernest E., medicine
McDaniel, Robert Stewart, chemical kinetics, data processing
McDonald, W John, nuclear physics
McElhaney, Ronald Nelson, biochemistry, biophysics
McGann, Locksley Earl, cryobiology
McKinney, Richard Leroy, mathematics

McMurchy, Kenneth Allan, oral pathology
McPherson, Harold James, physical geography, resource geography
McPherson, Thomas Alexander, immunology
Mellon, George Barry, geology
Meyer, Walter Carl, restorative dentistry
Micetich, Ronald George, organic chemistry, medicinal chemistry
Millar, Robert Fyfe, applied mathematics
Milligan, Larry Patrick, biochemistry, nutrition
Molnar, George D., endocrinology, internal medicine
Moon, John Wesley, mathematics
Morgan, Antony Richard, molecular biology
Morin, Robert Bennett, organic chemistry
Morton, Roger David, mineralogy
Moschopedis, Speros E., organic chemistry
Moskalyk, Richard Edward, medicinal chemistry
Moss, Gerald Allen, nuclear physics
Mossop, Grant Dilworth, sedimentology, sedimentary petrology
Muldrew, James Archibald, forest entomology
Murdeshwar, Mangesh Ganesh, topology
Mureika, Roman A., mathematics
Murie, Jan O., animal behavior, ecology
Myers, Gordon Edward, microbiology
Narayana, Tadepalli Venkata, mathematics
Nash, Charles William, pharmacology
Nash, David, biochemical genetics, cytogenetics
Neilson, George Croyden, nuclear physics
Nelson, Joseph Schieser, ichthyology
Newbound, Kenneth Bateman, physics
Nihei, Taiichi, developmental biology
Noujaim, Antoine Atil, nuclear pharmacy
Nursall, John Ralph, zoology
Offenberger, Allan Anthony, plasma physics, lasers
Olsen, William Charles, nuclear physics
Olson, Arthur Olaf, biochemistry
Paranchych, William, biochemistry, virology
Paranjape, Bhalachandra Vishwanath, physics
Paterson, Alan Robb Phillips, biochemistry
Paton, David Murray, pharmacology
Pattie, Donald L., biology, ecology
Pawluk, Steve, soil science
Pearson, James Gordon, medicine, radiotherapy
Pearson, Keir Gordon, neurophysiology
Percy, John Smith, medicine
Pearson, Richard Ector, neuroendocrinology, comparative endocrinology
Pinnington, Eric Henry, atomic spectroscopy, astrophysics
Plambeck, James Alan, analytical chemistry, electrochemistry
Pluth, Donald John, soil science
Ponnapalli, Ramachandramurty, mathematics, statistics
Powell, John Martin, biogeography, biometeorology
Poznansky, Mark Joab, physiology, biophysics
Rabenstein, Dallas Leroy, analytical chemistry
Rankin, David, physics, geophysics
Reffenstein, Rhoderic John, pharmacology
Renner, Ruth, nutrition
Riemensnyder, Sherman Delbert, mathematics
Robblee, Alexander (Robinson), poultry nutrition
Robertson, James Alexander, soil fertility, soil chemistry
Robins, Morris Joseph, organic chemistry
Rogers, James Albert, pharmaceutics
Rogers, James Stewart, physical pharmacy
Rogers, James Stewart, experimental physical chemistry
Rostoker, Gordon, space physics
Rowlands, John Alan, metal physics
Russell, James Christopher, physiological chemistry, clinical chemistry
Ruth, Royal Francis, immunology
Rutter, Nathaniel Westlund, quaternary geology
Salmon, Peter Alexander, surgery
Salt, Walter Raymond, histology

Sample, John Thomas, nuclear physics
Samuel, William Morris, parasitology
Sanger, Alan Rodney, inorganic chemistry, organometallic chemistry
Scafe, Donald William, geology, oceanography
Schachter, Melville, physiology
Schiff, Harry, theoretical physics
Schulz, Karlo Francis, physical chemistry
Schultz, John David, forestry
Scotter, George Wilby, ecology
Scraba, Douglas G, biochemistry, virology
Secord, David Cartwright, laboratory animal medicine
Sheppard, Douglas Murray, physics
Shnitka, Theodor Khyam, pathology, cell biology
Shysh, Alec, bionucleonics
Sigal, Richard Frederick, geophysics, theoretical physics
Singh, Teja, forest hydrology, applied statistics
Skoropad, William Peter, plant pathology
Smillie, Lawrence Bruce, biochemistry
Smith, Dorian Glen Whitney, geology, mineralogy
Smith, Peter John, urban geography
Smith, Richard Sidney, physiology
Smith, Samuel Ivan, geography
Smith, Warren Edward, genetics, plant breeding
Smyser, Gerald Stanley, anatomy
Speight, James G, organic chemistry
Spencer, Andrew Nigel, invertebrate physiology
Spencer, Mary Stapleton, biochemistry
Sperber, Geoffrey Hilliard, anatomy, dentistry
Sproule, Brian J, medicine, thoracic diseases
Steele, Clellie Truman, chemistry
Stein, Richard Bernard, neurophysiology, biophysics
Steiner, Andre Louis, animal behavior
Steiner, Johann, geology
Stelck, Charles Richard, geology
Stemke, Gerald W, immunochemistry
Stephens-Newsham, Lloyd G, biophysics
Stephenson, John, theoretical physics
Stewart, Wilson Nichols, botany
Stinson, Glen Monette, experimental nuclear physics
Stinson, Robert Anthony, biochemistry
Stirrat, James Hill, pathology, bacteriology
Strausz, Otto Peter, chemistry
Subbarao, Mathukumalli Venkata, mathematics
Swanson, Robert Harold, forest hydrology, forest physiology
Sykes, Brian Douglas, physical chemistry
Takahashi, Yasushi, theoretical physics
Tamaoki, Taiki, biochemistry, plant pathology
Tanner, Dennis David, organic chemistry
Taylor, Jack Dean, pharmacology
Taylor, William Clyne, pediatrics
Tertzakian, Gerard, organic chemistry
Thomas, Norman Randall, dentistry, physiology
Thorson, Walter Rollier, theoretical chemistry
Timourian, James Gregory, mathematics
Toogood, John Alfred, soils
Toop, Edgar Wesley, ornamental horticulture
Torgerson, Ronald Thomas, high energy physics, theoretical physics
Toshach, Sheila, medical bacteriology
Toth, Jozsef, geophysics
Trost, Walter Raymond, physical chemistry
Umezawa, Hiroomi, theoretical physics
Urtasun, Raul C, oncology, experimental radiobiology
Vaartnou, Herman, plant pathology
Vanden Born, William Henry, weed science, plant physiology
Vitt, Dale Hadley, botany, bryology
Von Borstel, Robert Carsten, genetics
Voorhees, Burton Hamilton, biomathematics
Walton, Peter Dawson, plant breeding
Wang, Lawrence Chia-Huang, physiology, zoology
Webster, Gordon Ritchie, soil chemistry
Weckowicz, Thaddeus Eugene, psychiatry
Wegmann, Thomas George, genetics, immunology
Weichman, Frank Ludwig, experimental solid state physics
Weijer, Jan, genetics
Westlake, Donald William Speck, microbiology
Whitehouse, Ronald Leslie S, electron microscopy, microbiology

Wiebe, Leonard Irving, bionucleonics
Wiggins, Ernest James, physical chemistry, chemical engineering
Willard, Stephen, mathematics
Wilson, Donald Robert, internal medicine
Wilson, Mark Vincent Hardman, vertebrate paleontology
Wolowyk, Michael Walter, pharmacology, cell biology
Wonders, William Clare, geography, geography of northern lands
Wong, Horne Richard, entomology
Woods, Stuart B, solid state physics
Wouk, Arthur, mathematics
Wyman, Max, mathematics
Yamamoto, Tatsuzo, microbiology
Young, Bruce Arthur, agriculture, animal physiology
Zalik, Sara E, developmental biology, cell biology
Zalik, Saul, plant physiology, biochemistry
Zwickel, Fred Charles, wildlife ecology, zoology

LACOMBE
Berkenkamp, Bill Brodie, plant pathology
Fredeen, Howard T, animal genetics, biometry
Jeremiah, Lester Earl, meat sciences
McBeath, Douglas Kay, soil science, plant nutrition

LETHBRIDGE
Allan, John R, plant physiology
Andrews, John Edwin, genetics, plant breeding
Atkinson, Thomas Grisedale, plant pathology
Bailey, Charles Basil Mansfield, animal physiology, nutrition
Bainborough, Arthur Raymond, pathology
Beaty, Chester Broomell, geography
Beckel, William Edwin, zoology
Bell, Robert Graham, agricultural microbiology
Bide, Richard W, pathological chemistry
Blakeley, Philip Earl, entomology
Cheng, Kuo-Joan, agricultural microbiology
Darcel, Colin Le Q, veterinary virology
Daykin, Philip Norman, computer science, mathematics
Dormaar, Johan Frederik, soil chemistry
Fletcher, Roy Jackson, climatology
Freyman, Stanislaw, agronomy, crop physiology
Grant, Marshall Nelson, plant breeding
Hanna, Michael Ross, agronomy
Harper, Alex Maitland, economic entomology
Harper, Frank Richard, phytopathology, mycology
Hawn, Elmer George, phytopathology
Hepler, Loren George, physical chemistry, water chemistry
Hironaka, Robert, animal nutrition
Hobbs, Gordon Andrew, ecology
Holmes, Neil Delvin, economic entomology
Holmes, Owen Gordon, chemistry
Johnston, Alex, range ecology
Kasting, Robert, biochemistry
Kemp, Gavin Arthur, horticulture, plant breeding
Kounosu, Shigeru, high energy physics, theoretical physics
Kuijt, Job, plant anatomy, plant morphology
Langford, Edgar Verden, veterinary bacteriology
Larson, Ruby Ila, cytogenetics
Lebeau, Jack Bertram, plant pathology
MacDonald, Malcolm Duncan, cytogenetics
MacKay, Donald Cyril, soil chemistry, plant chemistry
McCurdy, Keith G, physical chemistry
McDonald, Stuart, biology
McKenzie, Hugh, plant breeding, plant genetics
Meintzer, Rober Bruce, biochemistry
Miller, Elbert Ernest, geography
Nakamura, Kazuo, microbial genetics, phycology
Neal, John Lloyd, Jr, microbiology, soil science
Nelson, Gordon Albert, plant pathology
Nelson, William Arnold, entomology
Papp, Francis Joseph, IV, mathematics
Roberts, David Wilfred Alan, plant physiology, biochemistry
Rood, Joseph Lloyd, physics
Salt, Reginald Wilson, entomology
Slen, Sydney, Bernard, animal nutrition

Smith, Andrew Douglas, agronomy
Smith, Dean Seyward, entomology
Sommerfeldt, Theron G, soil science, physical chemistry
Struble, Dean L, synthetic organic chemistry
Swailes, George Edward, entomology
Torfason, Wilmer Esplin, horticulture
Vesely, John Anthony, animal breeding, genetics
Wagenaar, Emile B, cell biology, genetics
Wells, Stewart Alderson, genetics
Wilkinson, Paul R, entomology
Zieber, George Henry, urban geography

RALSTON
Heggie, Robert Murray, organic chemistry
Holmes, R H Lavergne, organic chemistry
Stewart, William Christopher, physiology, pharmacology

SHERWOOD PARK
Rothwell, Richard Lee, watershed management, forestry

BRITISH COLUMBIA

AGASSIZ
Forrest, Robert J, biochemistry, animal nutrition
Hill, Arthur Thomas, poultry husbandry, poultry physiology
Hunt, John R, poultry nutrition

BURNABY
Albright, Lawrence John, microbiology, oceanography
Aronoff, Samuel, biochemistry
Ballentine, Leslie Edward, physics
Barlow, John Slaney, biochemistry
Bass, David Eli, physiology
Beirne, Bryan Patrick, economic entomology
Benston, Margaret Lowe, physical chemistry
Brown, Robert Charles, economic geography, resource geography
Brown, Thomas Craig, mathematics
Cercone, Nicholas Joseph, computer science
Chow, Yuan Lang, organic chemistry
Clayman, Bruce Philip, solid state physics
Cochran, John Francis, solid state physics
Colbow, Konrad, solid state physics
Crozier, Edgar Daryl, metal physics, semiconductors
Cunningham, Frank Firman, geography
Cushley, Robert John, spectroscopy, organic chemistry
D'Auria, John Michael, nuclear chemistry, radiochemistry
Druehl, Louis D, marine ecology, phycology
Eaves, David Magill, mathematics
Einstein, Frederick W B, inorganic chemistry
Eliot Hurst, Michael Eliot, economic geography, transportation geography
Enns, Richard Harvey, theoretical physics
Evenden, Leonard Jesse, urban geography
Fisher, Francis John Fulton, evolutionary biology, plant physiology
Freedman, Allen Roy, mathematics
Funt, B Lionel, polymer chemistry
Geen, Glen Howard, zoology, aquatic biology
Graham, George Alfred Cecil, applied mathematics, mechanics
Harrop, Ronald, mathematical logic, computer science
Hickerson, Harold, anthropology
Horvath, Ronald J, geography
Irwin, John Charles, physics
Kellman, Martin C, physical geography
Kiehlmann, Eberhard, organic chemistry
Koroscil, Paul Michael, historical geography, cultural geography
Korteling, Ralph Garret, nuclear chemistry
Lardner, Robin Willmott, applied mathematics, solid mechanics
Lower, Stephen K, physical chemistry
Mackauer, Manfred, entomology
MacPherson, A, geography
Malli, Gulzari Lal, quantum chemistry, chemical physics
Mathewes, Rolf Walter, palynology
McClaren, Milton, Jr, botany, mycology

Millard, Kenneth Young, statistical mechanics
Nichol, Christina Janet, biochemistry
Oehlschlager, Allan Cameron, bio-organic chemistry
Oloffs, Peter Christian, pesticide chemistry
Palmer, Leigh Hunt, low temperature physics
Pate, Brian David, chemistry
Peterson, Louis K, inorganic chemistry
Peucker, Thomas K, information science, geography
Rahe, James Edward, plant pathology
Rheumer, George Alfred, geography
Richards, William Reese, biochemistry, bio-organic chemistry
Rieckhoff, Klaus E, chemical physics, solid state physics
Roberts, Michael Charles, geography, geomorphology
Ryeburn, David, topology
Sadleir, Richard Michael Francis Stuart, physiological ecology
Shen, Chung Yi, applied mathematics
Sherwood, A Gilbert, physical chemistry
Shoemaker, Edward Milton, applied mathematics
Slessor, Keith Norman, organic chemistry
Speer, Henry Lee, cell biology, biochemistry
Srivastava, Lalit Mohan, biology
Steed, Guy Percy F, economic geography
Sterling, Theodor David, computer science, biometry
Sutton, Derek, inorganic chemistry
Thomason, Steven Karl, mathematical logic
Unrau, Abraham Martin, biochemistry, plant physiology
Visvanathan, William Elliott, plant physiology
Voigt, Eva-Maria, physical chemistry, statistical mechanics
Wagner, Philip Laurence, cultural geography
Wells, Edward Joseph, physical chemistry
Wilkins, Ebtisam A M Seoudi, bioengineering, chemical engineering
Wilson, Brian Graham, x-ray astronomy
Wong, Shue Tuck, geography, water resources

CASTLEGAR
Mitchell, Donald John, plant physiology

COBBLE HILL
Yonge, Keith A, psychiatry

KAMLOOPS
Granger, Maurice Roy, physical organic chemistry
Hughes, Richard David, geology
McLean, Alastair, range ecology
Van Ryswyk, Albert Leonard, soil fertility, soil morphology
Waldern, Donald E, animal nutrition, biochemistry

NANAIMO
Bell, Gordon Russell, microbiology
Bourne, Neil, invertebrate zoology
Brett, John Roland, fisheries
Healey, Michael Charles, population ecology
Hourston, Alan Stewart, fisheries
Kabata, Zbigniew, parasitology, marine biology
Kennedy, William Alexander, fish biology
Low, Charles James, aquatic ecology
Manzer, James Ivan, marine biology, fisheries
Margolis, Leo, parasitology, fish pathology
Mason, John Christopher, ecology, fisheries biology
Reimer, Don R, forest economics, economic development
Ricker, William Edwin, fish biology
Tully, John Patrick, oceanography
Yamada, Sylvia Behrens, population ecology

NELSON
Kaller, Cecil Louis, mathematics
Thyer, Norman Harold, meteorology, mathematics

NEW WESTMINSTER
Becker, Edward Samuel, pulp technology; pollution control
Roos, John Francis, fisheries management

PENTICTON
Galt, John (Alexander), astrophysics

PORT ALBERNI
Janzen, Wayne Roger, physical chemistry

RICHMOND
Boston, Noel Edward James, physical oceanography; micrometeorology

SIDNEY
Atkinson, Robert George, plant pathology

SUMMERLAND
Denby, Lyall Gordon, pomology
Fisher, Donald Vince, pomology
Hansen, Anton Juergen, plant pathology
Kitson, John Aidan, food science
Lane, William David, plant breeding
Looney, Norman E, pomology; plant physiology
MacGregor, Dugal, food science
Madsen, Harold F., entomology
Mason, John Leslie, plant nutrition
McIntosh, David Livingston, plant pathology
Meheriuk, Michael, plant biochemistry
Porritt, Stanley Wallace, horticulture
Rosher, Ronald Maitland, plant pathology
Russell, Glenn C., soil chemistry
Stevenson, David Stuart, soil physics
Welsh, Maurice Fitzwilliam, horticulture
Wood, Darrell Fenwick, food science

VANCOUVER
Aberle, David Friend, ethnology
Abu-Zahra, Nadia, anthropology
Adams, James Russell, zoology
Addison, Anthony William, bioinorganic chemistry
Ahlborn, Boye, plasma physics
Aho, Aaro E, petrology; economic geology
Aldridge, Keith Douglas, geophysics
Anastasiou, Clifford J, mycology
Anderson, Donald Oliver, epidemiology; medical care administration
Anderson, R F V, mathematics
Applegarth, Derek A, clinical biochemistry
Armstrong, John Edward, quaternary geology; glacial geology
Armstrong, Richard Lee, geochronology; tectonics
Auersperg, Nelly, cancer; cell biology
Auld, Edward George, nuclear physics
Auman, Jason Reid, astrophysics
Axen, David, nuclear physics
Bandoni, Robert Joseph, mycology
Barnard, Adam Johannes, plasma physics
Barnes, Derek, analytical chemistry; research administration
Barnes, William Charles, geology
Barrie, Robert, theoretical physics
Basco, N, photochemistry
Beames, R M, animal nutrition
Beamish, Katherine I, botany; genetics
Beck, Brenda E F, anthropology
Bell, Thomas Norman, physical chemistry
Belluce, Lawrence P, mathematics
Bellward, Gail Dianne, biochemical pharmacology
Belshaw, Cyril Shirley, anthropology
Berger, James Dennis, cell biology
Best, Raymond Victor, geology
Bichard, J W, solid state physics
Biely, Jacob, poultry nutrition
Bisalputra, Thana, botany
Bismanis, Jekabs Edwards, bacteriology
Bloom, Myer, physics
Bohn, Bruce Arthur, botany
Bowmer, Ernest John, pathology; medical microbiology
Bragg, Darrell, nutrition
Bragg, Philip Dell, biochemistry; organic chemistry
Bree, Alan V, physical chemistry

Brion, Christopher Edward, physical chemistry
Bullen, Peter Southcott, mathematics
Bures, Donald John (Charles), mathematics
Burling, Ronald William, oceanography
Burnell, Edwin Elliott, chemical physics
Burridge, Kenelm Oswald Lancelot, history of anthropology; social anthropology
Burton, Albert Frederick, biochemistry; cancer
Campbell, Jack James Ramsay, bacteriology
Carroll, Murray Norman, physical chemistry
Cayford, Afton Herbert, mathematical analysis, number theory
Chacon, Rafael Van Severen, mathematics
Chang, Bomshik, mathematics
Chapman, John Doneric, geography
Chase, Richard L, geology
Chase, William Henry, immunopathology; neuropathology
Ching, Hilda, parasitology
Chitty, Dennis Hubert, population ecology
Chong, Delano Pun, physical chemistry
Christian, Robert Roland, mathematics
Clarke, Garry K C, seismology; glaciology
Clowes, Ronald Martin, seismology
Constantinides, Paris, anatomy; electron microscopy
Coope, John Arthur Robert, molecular physics
Copp, Douglas Harold, physiology
Cowan, Ian McTaggart, mammalogy; wildlife ecology
Cox, Lionel Audley, organic chemistry; physical chemistry
Cram, Carl Frederick, physiology
Creighton, Robert Hervey Jermain, organic chemistry
Crooker, Arthur Mervyn, physics
Crooks, Michael John Chamberlain, low temperature physics
Cullen, William Robert, inorganic chemistry
Dalby, Frederick William, physics
Danner, Wilbert Roosevelt, geology
Darrach, Marvin, biochemistry
Dehnel, Paul Augustus, comparative physiology; invertebrate zoology
De Jong, Sybren Hendrik, geodesy
Desai, Indrajit Dayalji, nutrition, biochemistry
Divinsky, Nathan Joseph, mathematics
Dolmage, Victor, geology
Dolman, Claude Ernest, medical microbiology
Dolphin, David Henry, bio-organic chemistry; bioinorganic chemistry
Douglas, Roy Rene, topology
Dower, Gordon Ewbank, electrocardiography
Drance, S M, ophthalmology
Duff, Wilson, anthropology
Duncan, Douglas Wallace, applied microbiology
Dunell, Basil Anderson, nuclear magnetic resonance
Dunn, Bruce Partridge, cancer, pollution chemistry
Dunn, Henry George, pediatric neurology
Dunn, William Lawrie, pathology
Dutton, Guy Gordon Studdy, organic chemistry
Eaton, George Walter, horticulture
Elliot, Alfred Johnston, ophthalmology
Ellis, Robert Malcolm, geophysics
Evans, Russell Stuart, wood chemistry
Evelyn, Kenneth Austin, medicine
Farley, Albert Leonard, geography
Farmer, James Bernard, physical chemistry
Finlayson, Douglas Gordon, entomology
Finnegan, Cyril Vincent, experimental embryology; developmental biology
Fisher, Harold Dean, zoology
Fishman, Sherold, nuclear medicine, internal medicine
Forbes, Albert Ronald, entomology
Ford, Denys Kensington, medicine, rheumatology
Ford, Peter, embryology, histology
Foreman, Ronald Eugene, marine ecology
Forgacs, Otto Lionel, physical chemistry
Foulks, James Grigsby, pharmacology
Franz, Norman Charles, wood technology
Frazer, Bryan Douglas, insect ecology

Freeman, Roger Dante, child psychiatry
Freeze, Roy Allan, hydrology
Friedman, Constance Livingstone, anatomy
Friedman, Sydney Murray, medicine
Fritz, Carl T, anatomy, biochemistry
Frost, David Cregreen, physical chemistry
Gabrielse, Hubert, geology
Gage, Walter Henry, mathematics
Ganders, Fred Russell, biosystematics, population biology
Gardner, Joseph Arthur Frederick, organic chemistry
Garrett, Frederic Daugherty, neuroanatomy
Gates, Gary Rickey, resource geography
Gerry, Michael Charles Lewis, chemical physics, spectroscopy
Gibson, William Carleton, neurology
Gold, Andrew Vick, physics
Goodeve, Allan McCoy, pharmacognosy
Graham, Kenneth, entomology
Green, Beverley R, biochemistry
Greenwood, Donald Dean, psychoacoustics
Greenwood, Hugh J, geology, petrology
Griffiths, Anthony J F, genetics
Griffiths, George Motley, molecular biology
Gronlund, Audrey Florence, microbiology
Haddock, Philip George, forestry
Haering, Rudolph Roland, experimental and theoretical solid state physics
Hahn, Peter, physiology
Halabisky, Lorne Stanley, applied mathematics
Hall, Thomas Christopher, cancer, oncology
Halsey, Thomas Gordon, fisheries management, limnology
Hamilton, Richard Ian, plant pathology
Hardwick, David Francis, pathology, pediatrics
Hardy, Walter Newbold, solid state physics
Harris, Richard Colebrook, geography
Harrison, Lionel George, physical chemistry
Harrison, Robert Cameron, surgery
Hattori, Toshiaki, neurocytology
Haussmann, Ulrich Gunther, applied mathematics
Hawthorn, Harry Bertram, anthropology
Hayward, Lloyd Douglas, physical organic chemistry
Henderson, Wilson, avian pathology
Higgs, David Archibald, animal physiology, comparative endocrinology
Hinke, Joseph Anthony Michael, anatomy, biophysics
Hoar, William Stewart, comparative physiology
Hochachka, Peter William, comparative biochemistry
Holling, Crawford Stanley, zoology
Holm, David George, genetics
Hooley, Joseph Gilbert, physical chemistry
Hornby, Cedric Albert, plant breeding
Howard, Betty, physics
Howard, Roger, theoretical physics
Hughes, Gilbert C, botany, mycology
Hughes, Maryanne Robinson, avian physiology
Israels, Sydney, pediatrics
Jackson, Anne Louise, immunology
Jacoli, Giulio Guido, molecular biology
James, Brian Robert, inorganic chemistry
James, Douglas Garfield Limbrey, chemistry
Jenkins, Ralph Duncan, mathematics
Jenkins, Leonard Cecil, anesthesiology, pharmacology
Johnson, Roy Ragnar, plasma physics, solid state physics
Johnstone, Frederick Robert Carlyle, surgery
Jones, Garth, nuclear physics
Kaempfer, Frederick Augustus, theoretical physics
Kane, Julius, environmental systems & technology
Kasinsky, Harold Edward, biochemistry, developmental biology
Keeler, Ralph, physiology
Keevil, Norman Bell, chemistry, physics
Kennedy, James M, computer science
Kennedy, Robert William, wood science, technology
Khanna, Shadi Lall, dentistry
Kimmins, James Peter, forest ecology, environmental sciences
Kitts, Warren Dale, animal physiology, biochemistry

Kozak, Antal, forest biometrics
Krainz, Leon, physiology, endocrinology
Krajina, Vladimir Joseph, plant ecology, plant taxonomy
Krishnamurti, Cuddalore Rajagopal, animal physiology, animal biochemistry
Kutney, James Peter, organic chemistry
Laszlo, Charles A, biomedical engineering
Lavkulich, L M, soil science
Lear, Clement S C, orthodontics
LeBlond, Paul Henri, physical oceanography
Ledsome, John R, physiology, medicine
Lee, Melvin, nutrition, biochemistry
Leichter, Joseph, nutrition, food science
Leimanis, Eugene, applied mathematics
Leung, So Wah, dentistry, physiology
Levy, Julia Gerwing, microbiology
Lewis, Alan Graham, biological oceanography, zoology
Liley, Nicholas Robin, zoology
Lioy, Franco, physiology
Loring, William Bacheller, economic geology
Lowe, Morton David, neurophysiology
Lowe, Lawrence E, soil chemistry
Ludwig, Donald A, mathematics
MacCarthy, Hubert Reagh, entomology
MacDonald, John Lauchlin, mathematics
MacDonald, Walter Charlton, gastroenterology
MacFadyen, Donald John, neurology
Mackay, John Ross, geomorphology, cartography
Mackenzie, Cortlandt John Gordon, public health, epidemiology
Mackenzie, George Henry, applied physics
Manville, John Fieve, wood chemistry
Maranda, Eili Kongas, anthropology
Maranda, Pierre, anthropology, ethnology
March, Beryl Elizabeth, poultry nutrition
Marcus, Anthony Martin, psychiatry
Margetts, Edward Lambert, medicine, psychiatry
Marshall, Albert Waldron, mathematics
Martin, Peter Wilson, nuclear physics
Martin, William Henry, geology
Mathur, Vishwa Nath Prasad, forest products
Matthews, Peter Wren, physics
Maze, Jack Reiser, botany
McCreary, John Ferguson, medicine
McCutcheon, William Henry, radio astronomy
McDowell, Charles Alexander, physical chemistry
McElroy, Fred Dee, plant pathology, nematology
McGeer, Edith Graef, organic chemistry
McGeer, Patrick L, biochemistry
McGreer, Donald Edward, organic chemistry
McIntosh, Hamish William, medicine
McIntyre, Alan David, polymer chemistry
McKague, Allan Bruce, organic chemistry
McLean, Donald Millis, medical microbiology, virology
McLennan, Hugh, physiology
McMillan, James Malcolm, theoretical physics
McNeill, John Hugh, pharmacology
McTaggart, Kenneth C, petrology
Micko, Michael M, environmental chemistry
Miller, James Reginald, genetics
Miller, Milton H, psychiatry
Mitchell, John Campbell, dermatology
Miura, Robert Mitsuru, applied mathematics
Miyake, Mikio, oceanography, meteorology
Morrison, Robert Thomas, nuclear medicine
Morrison-Cleator, Iain Goesta, surgery
Mowshowitz, Abbe, computer science
Moyls, Benjamin Nelson, algebra
Murdoch, David Carruthers, regional geology
Murray, Francis S E, physical chemistry, mathematics
Murray, James W, geology
Myall, Robert William T, oral medicine, dentistry
Mysak, Lawrence Alexander, applied mathematics, physical oceanography
Nakai, Shuryo, food chemistry
Nash, Stanley William, mathematical statistics
Newman, Murray Arthur, ichthyology
Nichol, Hamish, psychiatry
Noble, Robert Laing, physiology
Nodwell, Roy, physics
Nordan, Harold Cecil, vertebrate biology

O'Donnell, Vincent Joseph, biochemistry
Ogilvie, Alfred Livingston, periodontics
Ogryzlo, Elmer Alexander, physical chemistry
Oke, Timothy Richard, microclimatology
Okulitch, Vladimir Joseph, geology, invertebrate paleontology
Olmsted, Richard Dale, theoretical physics
Opechowski, Wladyslaw, theoretical physics
Osborne, John Alan, medicine, cardiology
Osmanski, C Paul, anatomy, pathology
Ovalle, William Keith, medical science
Ovenden, Michael W, astronomy, astrophysics
Ozier, Irving, molecular spectroscopy
Palaty, Vladimir, biophysics, physiology
Parfitt, Gilbert J, dentistry, medicine
Parsons, Timothy Richard, oceanography
Patterson, Frank Porter, orthopedic surgery
Patterson, Ralph Francis, organic chemistry, science administration
Pearce, Richard Hugh, clinical chemistry, experimental pathology
Pearson, Richard Joseph, anthropology, archaeology
Pepin, Herbert Spencer, plant pathology
Perks, Anthony Manning, physiology, pharmacology
Pernarowski, Modest, pharmaceutical chemistry
Perry, Thomas Lockwood, biochemistry, pediatrics
Person, Clayton Oscar, genetics
Peterson, Raymond Glen, animal breeding, statistics
Phillips, John Edward, comparative physiology
Pickard, George Lawson, oceanography
Pincock, Richard Earl, organic chemistry
Polglase, William James, chemistry
Pond, George Stephen, physical oceanography
Porter, Gerald Bassett, physical chemistry
Powrie, William Duncan, food science
Press, S James, mathematical statistics
Price, John David Ewart, internal medicine
Procter, Alan Robert, pulp chemistry, paper chemistry
Pryce, Maurice Henry Lecorney, theoretical physics
Quastel, D M J, physiology
Quastel, Juda Hirsch, biochemistry
Raghunathan, Partha, magnetic resonance, chemical physics
Rainwater, James Carlton, statistical mechanics
Ranta, Lawrence Edward, health services administration
Rastall, Peter, theoretical physics
Rau, Jon Llewellyn, geology
Razzell, Wilfred Edwin, microbiology, biochemistry
Reeves, Ogle Raymond, developmental biology, biochemistry
Reid, Cyril, chemical physics
Renney, Arthur James, agronomy
Restrepo, Rodrigo Alvaro, mathematics
Richards, James Frederick, biochemistry
Richer, Harvey Brian, astronomy
Riddell, Ronald Cameron, mathematics
Ridington, William Robin, anthropology, philosophy
Riedel, Bernard Edward, pharmacy
Robinson, John Lewis, geography
Roddick, James Archibald, geology
Rosenthal, Alex, organometallic chemistry
Rotem, Chava Eve, cardiology, experimental medicine
Rouse, Glenn Everett, palynology, paleobotany
Rowles, Charles A, soil morphology
Roydhouse, Richard Helm, dentistry
Runeckles, Victor Charles, plant physiology
Russell, John Robert, geophysics
Sams, John Robert, Jr, chemical physics, physiology
Sanders, Harvey David, pharmacology, physiology
Scheffer, John R, organic chemistry
Schofield, Wilfred Borden, botany
Schwerdtfeger, Charles Frederick, physics
Scudder, Geoffrey George Edgar, entomology
Shaw, Michael, plant pathology, plant physiology
Shuter, William Leslie Hazlewood, radio astronomy
Sinclair, Alastair James, economic geology

Singh, Akhand Pratap, microbial biochemistry
Singh, Vijendra Kumar, neurochemistry, neurosciences
Sion, Maurice, mathematics
Skarsgard, Lloyd Donald, radiation biology, biophysics
Slade, H Clyde, internal medicine, psychiatry
Slawson, William Francis, geochemistry
Slonecker, Charles Edward, anatomy
Smith, John Harry Gilbert, forestry
Smith, Michael, organic chemistry, biochemistry
Snider, Robert Folinsbee, theoretical chemistry
Spitzer, Ralph, chemistry, pathology
Spouge, John Douglas, oral pathology
Spratley, Richard Denis, physical chemistry
Stace-Smith, Richard, plant pathology
Stanley, John, zoology
Stein, Janet Ruth, botany
Stewart, Ross, physical organic chemistry
Stich, Hans F, cell biology, genetics
Stock, John Joseph, microbiology, medical mycology
Strasdine, George Alfred, microbiology, biochemistry
Sutter, Morley Carman, pharmacology
Suzuki, David Takayoshi, genetics
Swanson, Charles Andrew, mathematics
Syeklocha, Delfa, medical bacteriology, immunology
Sziklai, Oscar, biology, forest genetics
Tait, Robert Malcolm, agriculture, animal husbandry
Taylor, Frank John Rupert (Max), marine biology
Taylor, Iain Edgar Park, plant physiology, plant biochemistry
Taylor, Jocelyn Mary, zoology
Taylor, Roy Lewis, plant taxonomy
Tener, Gordon Malcolm, biochemistry
Thirgood, Jack Vincent, forestry
Thurston, Hugh Ansfrid, algebra
Tiffin, Donald Lloyd, marine geophysics, geology
Todd, Mary Elizabeth, anatomy
Ton, Bui An, mathematical analysis
Towers, George Hugh Neil, plant biochemistry
Townsley, Philip McNair, industrial microbiology
Tregunna, E Bruce, plant physiology
Tremaine, Jack H, plant virology
Trotter, James, physical chemistry
Trussell, Paul Chandos, bacteriology
Tung, Marvin Arthur, food science
Turrell, Brian George, physics
Tyhurst, James Stewart, psychiatry
Ulrych, Tadeusz Jan, geophysics
Vassar, Philip Stanley, pathology
Verma, Subhash Chander, pharmacology
Vogt, Erich W, nuclear physics, theoretical physics
Volkoff, George Michael, theoretical physics
Vrba, Rudolf, neurochemistry, immunochemistry
Wada, Juhn A, medicine
Waldichuk, Michael, oceanography
Walker, David Crosby, radiation chemistry
Walker, Gordon Arthur Hunter, astrophysics
Walters, Carl John, systems ecology
Wan, Frederic Yui-Ming, applied mathematics, solid mechanics
Ward, Diana Valiela, ecology, invertebrate zoology
Ward, Lawrence McCue, psychophysics, vision
Warren, Harry Verney, geology, mineralogy
Warren, John Bernard, physics
Watanabe, Tomiya, aeronomy, plasma physics
Webber, William A, histology, physiology
Webster, John Malcolm, biology, parasitology
Weeks, Gerald Stanley, organic chemistry
Weiler, Lawrence Stanley, organic chemistry
Weinberg, Fred, physics
Weintraub, Marvin, plant virology
Wellington, William George, insect ecology, biometeorology
Wellwood, Robert William, wood science & technology
Westwick, Roy, mathematics
Whipple, Francis Oliver, forest products, resource management
White, Bruce Langton, physics
Whitelaw, Donald Mackay, internal medicine
Whittaker, James Victor, mathematics

Wilimovsky, Norman Joseph, ichthyology, fisheries
Williams, David Llewelyn, metal physics
Willington, Robert Peter, forest hydrology
Worrall, John Gatland, dendrology
Wort, Dennis James, plant physiology
Wright, Norman Samuel, plant pathology
Wright, Robert Hamilton, physical chemistry
Wynne-Edwards, Hugh Robert, geology
Young, Maurice Durward, pediatrics, cardiology
Zbarsky, Sidney Howard, biochemistry
Zidek, James Victor, mathematical statistics

VICTORIA

Aikman, George Christopher Lawrence, astrophysics
Algard, Franklin Thomas, embryology
Alkire, William Henry, cultural anthropology, ethnology
Ballantyne, David John, horticulture, plant physiology
Bandy, Percy John, zoology
Barss, Walter Malcomson, physics
Batten, Alan Henry, astronomy
Beer, George Atherley, nuclear physics
Bell, Marcus Arthur Money, plant ecology, ethnobotany
Buchanan, Ronald James, aquatic ecology, phycology
Buckley, James Thomas, biochemistry
Burke, J Anthony, astronomy, astrophysics
Bushnell, Gordon William, structural chemistry, crystallography
Chapman, Robert Pringle, physics
Clark, Malcolm John Roy, environmental chemistry
Clements, Reginald Montgomery, plasma physics
Climenhaga, John Leroy, astrophysics
Cooperstock, Fred Isaac, theoretical physics
Crampton, David, astronomy
Crumrine, Norman Ross, II, anthropology
Dewey, John Marks, physics, fluid dynamics
Dingle, Thomas Walter, theoretical chemistry
Dixon, Keith R, inorganic chemistry
Dosso, Harry William, geophysics
Duffus, Henry John, physics
Ehle, Byron Leonard, numerical analysis, computer science
Ellis, Derek V, biology
Etheridge, David Elliott, forest pathology
Fields, William Gordon, invertebrate zoology
Fontaine, Arthur Robert, marine zoology
Forward, Charles Nelson, urban geography
Foster, Harold Douglas, geomorphology, hydrology
Friedmann, Gerhart B, medical physics, optics
Fyles, James Thomas, geology
Gibbins, Sidney Gore, inorganic chemistry
Glen, Robert, entomology
Graham, Eric Stanley, organic chemistry
Gray, Archibald Fred, social anthropology, cultural anthropology
Grigg, Harold R, physics
Hagmeier, Edwin Moyer, biology
Hart, Josephine Frances Lavinia, marine biology
Hartwick, Frederick David Alfred, astronomy
Hayward, John S, environmental physiology, aerospace medicine
Hewgill, Denton Elwood, mathematics
Hinrichs, Lowell A, mathematics
Horita, Robert Eiji, space physics, geophysics
Horning, William Clarke, physical chemistry, organic chemistry
Hughes, Blyth Alvin, underwater acoustics
Hunt, Richard Stanley, forest pathology
Hurd, Albert Emerson, mathematics
Hutchings, John Barrie, astronomy
Illingworth, Keith, forest genetics
Innes, Morris James Sage, geophysics
Jennings, Stephen Arthur, mathematics
Kidd, Derek John, human physiology
Kirk, Alexander David, physical chemistry
Lai, David Chuen Yan, urban geography, geography of China
Lancaster, George Maurice, mathematics
Leechman, Douglas, anthropology, linguistics
Leeming, David John, mathematics

Lewis, Edward Lyn, oceanography
Littlepage, Jack Leroy, biological oceanography
Lobb, Donald Edward, physics
Lokken, John Erwin, geophysics
Macdonald, Duncan Ross, forest entomology
Mackie, George Owen, zoology
Marko, John Robert, solid state physics
Martin, John Kenneth, pediatrics
Mason, Grenville R, nuclear physics
McCartney, William Douglas, economic geology
McCurdy, Harriet Mace, embryology, endocrinology
McInerney, John Edward, ichthyology
Micklewright, Malcolm A, geography, economics
Miller, Gary Glenn, topology, theoretical physics
Milne, Allen Ritchie, physics, geophysics
Newroth, Peter Russell, marine botany, resource management
O'Brien, Robert Neville, physical chemistry
Odeh, Robert Eugene, mathematical statistics, computer science
Oliver, Brian Malcolm, plasma physics, space physics
O'Reilly, Henry James, plant pathology
Orr-Ewing, Alan Lindsay, forest genetics
Owens, John N, botany
Paden, John Wilburn, mycology
Panda, Rekha, mathematics
Paul, Dorothy Hayman, neurobiology
Paul, Miles Richard, developmental biology
Pearce, Dennis Wiffen, chemistry
Pearce, R Michael, nuclear physics
Petch, Howard Earle, nuclear magnetic resonance
Pfaffenberger, William Elmer, mathematics
Picciotto, Charles Edward, particle physics
Porteous, John Douglas, urban geography
Riddell, James, mathematics
Ring, Richard Alexander, entomology
Robertson, Lyle Purmal, nuclear physics
Rogak, Earl, mathematics
Ross, William Michael, resource management
Ryce, Stephen A, physical chemistry
Safranyik, Laszlo, forest entomology
Sargent, Thomas Edward Hartley, geology
Scarfe, Colin David, astronomy
Sewell, W R Derrick, economic geography
Shinbrot, Marvin, mathematics
Shrimpton, Douglas Malcolm, plant physiology
Smith, Richard Barrie, forest pathology, forest ecology
Srivastava, Hari Mohan, mathematics
Stewart, Robert William, physics
Styles, Ernest Derek, genetics
Styles, Salma Mahmoud, physics
Sullivan, Harry Morton, physics
Sutherland-Brown, Atholl, economic geology
Szczawinski, Adam Franciszek, botany
Tabata, Susumu, physical oceanography
Thomson, David James, acoustics
Thomson, Richard Edward, physical oceanography
Trust, Trevor John, microbiology
Tuller, Stanton Ernest, geography, climatology
Vanterpool, Thomas Clifford, phytopathology
Walker, Edward Robert, meteorology
Warrington, Patrick Douglas, botany
Weaver, John Trevor, geophysics
Whitney, Harvey Stuart, phytopathology, mycology
Wilson, Alex James, biochemistry
Wood, Alex James, biochemistry
Wright, Kenneth Osborne, astrophysics
Ziller, Wolf Gunther, mycology, forest pathology

WEST VANCOUVER

Antia, Naval Jamshedji, organic chemistry, microbiology
Arrott, Anthony, magnetism
Cameron, William Maxwell, physical oceanography
Donaldson, Edward Mossop, comparative endocrinology, fisheries
Robertson, Malcolm Slingsby, mathematics
Stockner, John G, limnology, ecology
Sykes, Paul Jay, Jr, physics, nuclear engineering

Tarr, Hugh Lewis Aubrey, fisheries

MANITOBA

BRANDON
Campbell, Kenneth Wilford, plant breeding, plant cytogenetics
Corrigan, Samuel Walter, applied anthropology, social anthropology
Dyck, Gerald Wayne, reproductive physiology
Garnett, Ian, animal genetics
Letkeman, Peter, chemistry
MacNaughton, William Norman, animal breeding, genetics
McLeod, James Archie, zoology
Miller, Archie Paul, solid state physics
Pepper, Evan Harold, environmental biology
Rounds, Richard Clifford, biogeography, resource geography
Stadel, Christoph, urban geography, political geography
Stewart, David Bradshaw, gynecology, zoology
Strain, John Henry, poultry science
Swierstra, Ernest Emke, animal physiology
Tyman, John Langton, geography
Welsted, John Edward, physical geography

MORDEN
Ali-Khan, Syed Tahir, plant breeding, genetics
Dorrell, Douglas Gordon, plant chemistry
Enns, Henry, plant breeding, genetics
Giesbrecht, John, plant breeding, genetics
Putt, Eric Douglas, plant breeding

PINAWA
Borsa, Joseph, biophysics, virology
Chapman, John Donald, biophysics, radiation biophysics
Dugle, David L, radiation biology, radiation chemistry
Dugle, Janet Mary Rogge, plant taxonomy
Dyne, Peter John, radiation chemistry
Gillespie, Colin J, theoretical biology, radiation biophysics
Greenstock, Clive Lewis, medical biophysics, radiation chemistry
Hart, Robert George, physical chemistry, nuclear science
Iverson, Stuart Leroy, mammalian ecology, radiation ecology
Mehta, Kishor Kalidas, nuclear physics, reactor physics
Petkau, Abram, medical biophysics
Raleigh, James Arthur, organic chemistry
Robertson, Robert Frank Struan, physical chemistry
Runnmery, Terrance Edward, solid state chemistry
Sargent, Frederick Peter, radiation chemistry, radiation physics
Shoesmith, David William, electrochemistry
Singh, Ajit, radiation chemistry, photochemistry
Singh, Harwani, biochemistry
Smith, Roger M, physics, nuclear engineering
Stewart, Reginald Bruce, analytical chemistry
Strathdee, Graeme Gilroy, surface chemistry, inorganic chemistry
Szekely, Joseph George, biophysics
Tewari, Param Hans, surface chemistry, colloid chemistry
Tomlinson, Michael, physical chemistry
Tremaine, Peter Richard, physical chemistry
Vandergraaf, Tjalle T, analytical chemistry, radiochemistry
Weeks, John Leonard, occupational health
Wikjord, Alfred George, nuclear chemistry
Williamson, Arthur Tandy, physical chemistry

ST BONIFACE
Hamonic, Marcel J, pathology

WINNIPEG
Adamson, Ian Young Radcliffe, experimental pathology
Adamson, John Douglas, psychiatry, psychophysiology
Adhikari, P K, internal medicine
Aleksiuk, Michael, environmental physiology
Anderson, Christian Donald, exploration geophysics
Anderson, Kenneth William, mathematics
Armstrong, Donald T, economic geology
Baker, Robert John, plant breeding
Baldwin, W George, psychiatry, electroencephalography
Bankier, Robert John, plant breeding, quantitative genetics
Barber, Robert Charles, physics
Barker, Philip Shaw, entomology, insect toxicology
Beamish, Robert Earl, cardiology
Bendelow, Victor Martin, cereal chemistry
Berczi, Istvan, immunology
Bernier, Claude, phytopathology
Bertalanffy, Felix D, anatomy, histology
Betts, Robert Holladay, biochemistry
Bhakar, Balram Singh, theoretical nuclear physics
Bhuvaneswaran, Chidambaram, biochemistry
Bihler, Ivan, pharmacology, biochemistry
Birt, Arthur Robert, medicine
Biswas, Shib D, chemistry, biochemistry
Blancher, Marcel Corneille, biochemistry
Bock, Ernst, physical chemistry
Bowden, Drummond Hyde, pathology
Brown, William Jeffrey, geography
Brunskill, Gregg J, limnology, geochemistry
Burton, David Norman, biochemistry, microbiology
Bushuk, Walter, cereal chemistry
Campbell, Alan Newton, physical chemistry
Campbell, Allan Barrie, plant breeding
Campbell, Joseph Dempsey, plant nutrition
Campbell, Norman E Ross, microbiology
Carter, Brian Geoffrey, immunobiology
Carter, Stefan A, physiology
Charlton, James Leslie, photochemistry, organic chemistry
Cherniack, Louis, pulmonary diseases
Cherniack, Reuben Mitchell, medicine
Chernick, Victor, pediatrics, physiology
Choi, Nung Won, epidemiology
Chow, Arthur, analytical chemistry
Chow, H Bruce, medicine, genetics
Clayton, James Wallace, biochemistry
Clinch, Norman Frederick, physiology, biophysics
Coish, Harold Roy, theoretical physics
Connor, Robert Dickson, nuclear physics
Cooke, Iain, solid state physics
Cormack, Douglas Villy, biophysics
Crowson, Charles Neville, pathology
Dakshinamurti, Krishnamurti, biochemistry, nutrition
Dandy, James William Trevor, zoology, physiology
De Pena, Joan Finkle, biological anthropology, physical anthropology
Dever, Donald Andrew, plant science
Dhalla, Naranjan Singh, physiology, pharmacology
Dresel, Peter E, pharmacology
Duckworth, Henry Edmison, atomic physics
Dunn, Gerald Emery, physical organic chemistry
Dyck, Peter Leonard, genetics
Eales, John Geoffrey, animal physiology, endocrinology
Earl, Allan Edwin, food science
Eskin, Neason Akiva Michael, food chemistry, biochemistry
Evans, Brian M, geography, history
Evans, Laurie Edward, cytogenetics, plant breeding
Evans, Roger Malcolm, animal behavior
Faiman, Charles, endocrinology, physiology
Falk, Willie Robert, nuclear physics
Feniak, Elizabeth, nutrition
Ferguson, Colin C, surgery
Ferguson, Robert Bury, mineralogy
Findlay, Glen Marshall, toxicology
Finlayson, Henry C, mathematical analysis
Flannagan, John Fullan, freshwater biology, toxicology
Fox, John Gerald, pathology
Friesen, Rhinehart F, obstetrics & gynecology
Froese, Arnold, immunochemistry
Froese, Gerd, biophysics
Gaskell, Peter, physiology
Gaunt, Paul, solid state physics
Gee, John Henry, ecology
Gesser, Hyman Davidson, physical chemistry
Gibson, Maurice Henry Lindsay, anatomy
Gill, Clifford Cressey, plant virology
Goldenberg, Gerald J, internal medicine, oncology
Grant, William Wallace, medicine, pediatrics
Granzberg, Gary Robert, anthropology
Gratzer, George, mathematics
Green, Gordon John, plant pathology
Green, Roger Harrison, ecology
Greenberg, Arnold Harvey, cancer, immunology
Greenway, Clive Victor, physiology, pharmacology
Grewar, David, medicine, pediatrics
Gupta, Rajendra Prasad, mathematical statistics, applied statistics
Gurwith, Marc Joseph, infectious diseases
Hall, Donald Herbert, geophysics
Hamerton, John Laurence, human genetics, cell biology
Hamilton, Ian Robert, microbiology, biochemistry
Hamilton, Robert Duncan, aquatic microbiology
Hanec, William, entomology
Hannan, Charles Kevin, medical physics
Hara, Toshiaki, neurophysiology
Havlicek, Viktor, neurophysiology
Hawirko, Roma Zenovea, microbiology
Haworth, James C, pediatrics
Hecky, Robert Eugene, limnology
Hedin, Robert Arthur, soils
Helgason, Sigurdur Bjorn, plant breeding, cytogenetics
Henderson, James Stuart, pathology
Henry, Bryan Roger, physical chemistry
Hershfield, Earl S, internal medicine
Hildes, John Arthur, physiology
Hill, Robert D, agricultural biochemistry
Hoare, Robert, biochemistry
Hogg, Benjamin Gregory, nuclear physics
Holloway, Arthur F, medical physics
Hoogstraten, Jan, pathology
Hoshino, Kazumasa, oncology, anatomy
Hougen, Frithjof W, plant chemistry
Hryniuk, William, hematology, oncology
Huebner, Erwin, developmental biology, reproductive biology
Hughes, Kenneth Russell, psychophysiology
Hutton, Harold M, physical chemistry
Hyde, John Baskerville, neuroanatomy
Ingalls, Jesse Ray, animal science
Innes, Ian Rome, pharmacology
Irvine, George Norman, physical chemistry
Israels, Lyonel Garry, hematology
Jackson, Togwell Alexander, organic geochemistry, geology
Janzen, Alexander Frank, inorganic chemistry, organometallic chemistry
Johnson, Diane Mary, mathematics
Johnson, Karen Louise, botany, plant ecology
Johnson, Lionel, limnology, fisheries
Jovanovich, Jovan Vojislav, high energy physics, nuclear physics
Kale, Balvant Keshav, mathematical statistics
Kalisikes, Pantouses John, cytogenetics
Kanfer, Julian Norman, biochemistry
Kartzmark, Elinor Mary, physical chemistry
Keleher, James J, fish biology
Kelly, Frederick Miles, medicine
Kerber, Erich Rudolph, cytogenetics
Kerr, Donald Philip, atomic physics, molecular physics
Klass, Alan Arnold, anatomy, surgery
Kondra, Peter Alexander, poultry genetics
Kordova, Nonna, rickettsial diseases
Kosmolak, Frederick Graham, cereal chemistry
Krause, Guenter, algebra
Krolman, Gordon M, ophthalmology
Kwapinski, J B George, microbiology, immunology
LaBella, Frank Sebastian, pharmacology
LaCroix Lucien Joseph, plant physiology
Lambert, William Gordon, psychiatry
Larter, Edward Nathan, genetics, plant breeding
Lawler, George Herbert, fisheries
Lees, Howard, biochemistry
Lindsey, Casimir Charles, ichthyology
Linford, John Herbert, physical chemistry
Lloyd, Lewis Ewan, nutrition
Loly, Peter Douglas, theoretical physics, solid state physics
Loschiavo, Samuel Ralph, insect physiology
Losey, Gerald Otis, algebra
Loudfoon, James Herbert, organic chemistry, biochemistry
MacDougall, John Taylor, surgery
MacEwan, Douglas W, radiology
Macpherson, Roderick Ian, radiology
Maniar, Atish Chandra, microbiology, pediatrics
Marquardt, Ronald Ralph, biochemistry, avian physiology
Martens, John William, plant pathology
Mathewson, Francis Alexander Lavens, internal medicine
Matsuo, Robert R, biochemistry
McAlpine, Phyllis Jean, human genetics, medical genetics
McDonald, Bruce Eugene, nutrition
McDonald, William Craik, plant pathology
McKee, James Stanley Colton, physics
McKinnon, David M, organic chemistry
Medovy, Harry, pediatrics
Mendelsohn, Nathan Saul, mathematics
Metcalfe, David Richard, plant breeding, genetics
Mieczkowski, Zbigniew Ted, geography of the Soviet Union
Mills, John T, plant pathology
Moore, Keith Leon, anatomy
Moorhouse, John A, medicine, endocrinology
Morrish, Allan Henry, physics
Murdoch, Bruce Thomas, experimental nuclear physics
Murray, John Randolph, pharmacology
Nathaniel, Edward J H, anatomy
Nielsen, John Warrington, dentistry
Nielsen, Jens Juergen, entomology
Osgood, Charles Edgar, entomology
Paraskevas, Frixos, immunology
Parker, Robert J, population genetics, animal breeding
Patalas, Kazimierz, limnology, fisheries
Paul, Gilbert Ivan, statistics
Penner, Donald Wills, medicine
Persaud, Trivedi Vidhya Nandan, anatomy, developmental biology
Phillips, George Douglas, veterinary physiology
Preston, William Burton, herpetology, ecology
Pritchard, Ernest Thackeray, biochemistry
Prosen, Harry, psychiatry
Rauch, Josefine Constantia, comparative physiology, zoology
Rayburn, Marion Cecil, Jr, topology
Reed, Howard Newns, ophthalmology
Reyes, Francisco Ismael, reproductive physiology
Richtik, James Morton, cultural geography
Robinson, Arthur Grant, entomology
Rohringer, Roland, plant pathology
Rollo, Ian McIntosh, pharmacology, microbiology
Romcyn, James Augustus, bacteriology, immunology
Ronald, Allan Ross, infectious diseases
Rosenberg, David Michael, aquatic ecology
Roulston, Thomas Mervyn, obstetrics & gynecology
Rudgers, Lawrence Alton, soil science
Ruddini, Edris Rinaldo, immunobiology
Saether, Ole Anton, limnology
Saunders, Michael Graham, medicine
Sawatzky, Harry Leonard, cultural geography, resource geography
Sayed, Hamdy I, infectious diseases, immunology
Schacter, Brent Allan, medical research, hematology
Schaefer, Theodore Peter, physical chemistry
Schindler, David William, limnology
Schoemperlen, Clarence Benjamin, medicine
Schon, Alec, immunology
Scott, David Paul, biophysics, biometrics
Seale, Marvin Ernest, animal breeding
Searle, Clark Wellington, solid state physics
Sen, Sunil Kumar, nuclear physics, atomic physics
Shapiro, Lorin James, ethology, animal behavior
Shaykewich, Carl Francis, soil physics

Sichler, Jiri Jan, mathematics
Sinha, Ranendra, Nath, insect ecology, acarology
Sinha, Snehesh Kumar, applied statistics
Sisler, George C, psychiatry
Sisler, William Wallace, plant breeding
Smith, Lawrie Booth, biology
Smith, Robert Edward, pedology
Soper, Robert Joseph, soil science
Standil, Sidney, physics
Standing, Kenneth Graham, nuclear physics
Staniforth, Richard John, plant ecology
Steele, John Wiseman, pharmaceutical chemistry
Stefansson, Baldur Rosmund, plant breeding
Stephens, Newman Lloyd, physiology, biostatistics
Stevens, Frits Christiaan, biochemistry, protein chemistry
Stobbe, Elmer Henry, agronomy, weed science
Stringam, Elwood Williams, animal science, agriculture
Subrahmaniam, Kathleen, statistics
Subrahmaniam, Kocherlakota, mathematical statistics
Suzuki, Isamu, microbiology, biochemistry
Svenne, Juris Peteris, physics
Sweet, Louise Elizabeth, cultural anthropology
Tabisz, George Conrad, molecular physics
Teller, James Tobias, geology
Thomas, Percy LeRoy, genetics, mycology
Thomas, Robert Spencer David, mathematics
Thomson, Ashley Edwin, medicine, pharmacology
Thorsteinson, Asgeir Jonas, entomology
Tipples, Keith H, cereal chemistry
Tkachuk, Russell, organic chemistry
Townsend, Joan B, anthropology, archaeology
Trick, Gordon Staples, physical chemistry
Trott, John Richard, dentistry, periodontology
Tucker, Frederick Robert, orthopedic surgery
Turnock, William James, population ecology, biological control
Usmani, Riaz Ahmad, numerical analysis
Vail, John Moncrieff, solid state physics
van Oers, Willem Theodorus Hendricus, nuclear physics
Varsamis, Ioannis, psychiatry
Vitti, Trieste Guido, biopharmaceutics
Wallace, Ronald Richard, fresh water chemistry
Wallbank, Alfred Mills, virology
Walton, Charles Hutchinson Acourt, allergy
Walton, Jerry Joseph, medicine
Wang, Jerry Hsueh-Ching, solid state physics
Ward, Fredrick James, limnology, fish biology
Warner, Peter, microbiology
Watters, Frederick Lewis, entomology
Waygood, Ernest Roy, plant physiology, biochemistry
Weber, Shirley Mae, nutrition
Weisman, Harvey, histology, neuropharmacology
Wells, Donald O, nuclear physics
Westmore, John Brian, physical chemistry, inorganic chemistry
Williams, Hugh Cowie, mathematics, computer science
Williams, Joseph John, mathematics
Wilson, Harry David Bruce, economic geology
Winter, John Charles, bacteriology
Winter, Jeremy Stephen Drummond, pediatrics, endocrine physiology
Wiseman, Gordon Marcy, bacteriology
Wong, Chiu Ming, organic chemistry
Wong, Roderick Sue-Cheung, mathematical analysis
Wrogemann, Klaus, biochemistry, medicine
Yamada, Esther V, chemistry
Yurkowski, Michael, biochemistry, nutrition

NEW BRUNSWICK

FREDERICTON
Adamson, Jean Burnham, insect physiology
Anderson, John Murray, biology
Bagnall, Richard Herbert, plant pathology
Baskerville, Gordon Lawson, forestry, ecology
Bonga, Jan Max, plant physiology

Bottomley, Frank, inorganic chemistry
Bradley, Roy Henry Edward, plant virology
Brewer, Douglas G, inorganic chemistry, analytical chemistry
Brooks, Wendell V F, physical chemistry, molecular spectroscopy
Brown, Norman Rae, forest entomology
Burke, Kenneth B S, geophysics
Burt, Michael David Brunskill, biology, parasitology
Collins, William Beck, plant physiology, horticulture
Coulter, Wilson H, microbiology
Cowan, F Brian M, zoology, physiology
Cumming, Bruce Gordon, plant physiology
Edwards, Merrill Arthur, physics, biophysics
Erickson, Vincent Oliver, anthropology
Findlay, John A, organic chemistry
Franklin, Mervyn, microbiology
Govett, Gerald James, geochemistry
Grein, Friedrich, quantum chemistry
Greiner, Hugo R, geology
Hale, William Ernest, geology
Hamilton, Angus Cameron, geodesy
Kaiser, Reinhold, chemical physics
Kayll, Albert James, forestry
Ker, John William, forestry
Kissick, Norman Lennox, forest management
Krause, Helmut, forest soils
Krause, Margarida Oliveira, cell biology
Lajtai, Emery Zoltan, geology
Lees, Ronald Milne, molecular spectroscopy
Little, Charles Harrison Anthony, tree physiology
MacGillivray, M Ellen, entomology
Mahendrappa, Mukkatira Kariappa, forest soils
McAllister, Arnold Lloyd, geology
McKenzie, Joseph Addison, ethology, biochemical systematics
Nicholson, J W G, animal nutrition
Page, Orville T, plant science
Paim, Uno, environmental physiology, fish biology
Pajari, George Edward, petrology, geochemistry
Passmore, Jack, inorganic chemistry, fluorine chemistry
Powell, Graham Reginald, silviculture
Radforth, Norman William, paleobotany
Roller, Kalman Joseph, forest engineering
Saini, Gulshan Rai, soil science
Seabrook, William Davidson, invertebrate neurophysiology
Sebastian, Leslie Paul, wood science
Semeluk, George Peter, physical chemistry
Singh, Rudra Prasad, virology
Sivasubramanian, Pakkirisamy, developmental biology
Sullivan, Donald, mathematics
Taylor, Andrew Ronald Argo, botany
Turnbull, Christopher John, anthropology
Unger, Israel, chemistry
Valenta, Zdenek, organic chemistry
Vanicek, Petr, geodesy, geophysics
Van Slyke, Arthur Lawton, forestry, forest biometry
Varty, Isaac William, forest entomology
Verma, Ram D, spectroscopy
Wall, Ronald Eugene, plant pathology
Weaver, Gerald MacKnight, plant breeding, cytogenetics
Weetman, Gordon Frederick, forestry
Wein, Ross Wallace, plant ecology
Weisner, Louis, mathematics
Whitney, Norman John, mycology, plant pathology
Wiesner, Karel, chemistry
Wiggs, Alfred James, zoology, physiology
Wright, Bruce Stanley, wildlife ecology
Yoo, Bong Yul, plant physiology, cell biology
Young, Donald Alcoe, genetics, plant breeding

MONCTON
Banerjee, R L, solid state physics
Girouard, Fernand E, solid state physics
Israeli, Julius Yigal, physical inorganic chemistry
Jankowski, Christopher K, organic chemistry
Lakshminarayana, J S S, phycology, water pollution
LeBlanc, Leonard Joseph, solid state physics
Mehra, Mool Chand, radiochemistry, inorganic chemistry
Rahman, Matiur, fluid mechanics
Sichel, John Martin, quantum chemistry
Weil, Francis Alphonse, particle physics

SACKVILLE
Adams, Kenneth Allen Harry, organic chemistry
Allen, Clifford Marsden, geology, geochemistry
Barclay, Lawrence Ross Coates, organic chemistry
Crawford, William Stanley Hayes, mathematics
Fensom, David Strathern, plant biophysics, plant physiology
Ferguson, Laing, geology, paleontology
Foster, Kent Ellsworth, mathematical analysis
Grant, Douglas Hope, physical chemistry
Harries, Hinrich, ecology, soil science
MacFarlane, John T, physics
MacLauchlan, Donald Wells, physical chemistry
Matthews, James Horace, nuclear physics
Moore, John Carman Gailey, economic geology
Mosevich, Jack Walter, applied mathematics
Noble, William John, theoretical physics
Reinsborough, Vincent Conrad, physical chemistry
Stallworthy, Wilson Burnett, physiology
Tory, Elmer Melvin, applied mathematics, chemical engineering
Whitla, William Alexander, structural chemistry

ST ANDREWS
Elson, Paul Frederick, fish biology
Hare, Gerard Murdock, fisheries, parasitology
Kohler, Carl, marine biology
Messieh, Shoukry Naseef, fisheries
Needler, Alfred Walker Hollinshead, marine biology
Saunders, Richard Lee, physiology
Scarratt, David Johnson, marine biology
Stasko, Aivars B, aquatic ecology, fisheries
Sutterlin, Arnold M, animal physiology
Wilder, Donald George, fisheries management
Widish, David John, zoology, ecology

ST JOHN
Kelly, Ronald Burger, organic chemistry
Logan, Alan, paleoecology
Macbeth, Robert Alexander, surgery
Stanley, Eric, crystallography
Thomas, Martin Lewis Hall, biology

NEWFOUNDLAND

CORNER BROOK
Rayner-Canham, Geoffrey William, inorganic chemistry, science education

MOUNT PEARL PARK
Nowak, Wieslaw Stanislaw Wladyslaw, resource geography

ST JOHN'S
Aldrich, Frederick Allen, invertebrate zoology
Anderson, Hugh John, synthetic organic chemistry
Andrews, Cater Wilson, zoology, ecology
Bal, Arya Kumar, cell biology
Barnsley, Eric Arthur, microbial physiology
Barrowman, James Adams, gastroenterology
Bennett, Gordon Fraser, entomology, protozoology
Bigelow, Charles C, chemistry
Brackon, Sydney Wilson, physics
Briggs, Jean Louise, anthropology
Brosnan, John Thomas, metabolism
Brueckner, Werner Dietrich, geology
Bullock, Eric, organic chemistry
Burry, John Henry William, mathematics
Campbell, James Stewart, pathology
Cho, Chung Won, molecular physics
Clase, Howard John, inorganic chemistry
Cohen, Kenneth David, petrology, geology
Cowan, Garry Ian McTaggart, systematics
Crossley, David John, geophysics
Davis, Charles (Carroll), limnology, invertebrate zoology
Deutsch, Ernst Robert, geophysics
Dittmer, John Charles, biochemistry
Emara, Yehia Abdelaziz Saleh, genetics
English, Leonard Stanley, immunology
Evans, John William, zoology

Fallis, Alexander Graham, organic chemistry
Feltham, Lewellyn Allister Woodrow, animal physiology
Fletcher, Garth L, animal physiology
Foltz, Nevin D, molecular physics
Geduldig, Donald Stanley, biophysics, electrophysiology
Gogan, Niall Joseph, inorganic chemistry, academic administration
Green, Cloid Darryl, anesthesiology
Green, John M, marine biology
Hall, Barry Gordon, molecular biology, evolution
Hampson, Michael Chisnall, plant pathology
Hew, Choy-Leong, biochemistry
Hoenig, Julius, psychiatry
Hughes, Charles James, petrology
Idler, David Richard, biochemistry
Irfan, Muhammad, nuclear physics
Ivany, J W George, science education, academic administration
Kennedy, Michael John, structural geology, tectonics
Keough, Kevin Michael William, biochemistry
Khalil, Muhammad Ahsan Khan, forest genetics, quantitative genetics
King, Arthur Francis, geology
Laird, Marshall, parasitology, medical entomology
Lal, Mohan, mathematics
Larsen, Bodil Astrid, immunology
Littlefield, James Beaton, medicine
MacDiarmid, William Donald, internal medicine, human genetics
Mannion, John Joseph, geography
May, Arthur William, fish biology
McKillop, J H, geology
Mellor, Clive Sidney, psychiatry
Michalski, Chester James, molecular biology
Middleton, Richard B, microbial genetics, medical law
Miller, Robert Joseph, marine ecology
Mookerjea, Sailen, biochemistry, physiology
Murthy, Gummuluru Satyanarayana, paleomagnetism
Neale, Ernest Richard Ward, geology
Newlands, Michael John, inorganic chemistry
Nolan, Richard Arthur, mycology, physiology
O'Brien, Peter J, biochemistry
Ogilvie, John Franklin, chemical physics
Olsen, Orvil Alva, botany
Orr, James Cameron, biochemistry, organic chemistry
Paine, Robert Patrick Barten, anthropology, sociology
Papezik, Vladimir Stephen, geology
Payne, Jeremiah Frederick, environmental physiology
Payton, Brian Wallace, physiology
Penner, Peter Edwin, biochemistry
Penrose, William Roy, pollution chemistry
Peterson, Stephen Craig, physiology
Pfeiffer, Carl J, pharmacology, gastroenterology
Pottle, Clarence H, psychiatry
Reddy, Satti Paddi, physics, molecular spectroscopy
Rendell, David H, theoretical physics
Rideout, Donald Eric, mathematics
Rochester, Michael Grant, geophysics, astronomy
Scott, John Marshall William, chemistry
Sells, Bruce Howard, biochemistry
Senciall, Ian Robert, biochemistry
Shapiro, Martin, entomology
South, Graham Robin, phycology
Steele, Donald Harold, zoology
Steele, Vladislava Julie, invertebrate physiology, histology
Stein, Allan Rudolph, physical organic chemistry
Summers, William Francis, geography
Templeman, Wilfred, marine biology, fisheries
Thomeier, Siegfried, mathematics, topology
Threlfall, William, parasitology, ornithology
Tipping, Richard H, molecular physics
Tweeddale, Martin George, clinical pharmacology

CANADA

White, Fredric Paul, biochemistry, neurosciences
White, Susan Ruth, neurosciences
Williams, Harold, geology
Wright, James Arthur, geophysics

NOVA SCOTIA

AMHERST
Raymond, Charles Wyatt, geography of the Maritime Provinces

ANTIGONISH
Asadulla, Syed, mathematics
Bunbury, David Leslie, physical organic chemistry
Chiasson, Leo Patrick, genetics
Clarke, Ernest Maurice, physics
Cormier, Randal, geology
English, Philip Stephen, theoretical solid state physics
Ginivan, Francis Joseph, mathematics
Greenidge, Kenneth Norman Haynes, botany
Hunter, Douglas Lyle, theoretical physics
Joshi, Yoginder N., optics, spectroscopy
Liengme, Bernard V. F., physical chemistry
Lynch, Brian Maurice, physical organic chemistry
MacDonald, John James, electrochemistry
McAlduff, Edward J., spectrochemistry
Pink, David Anthony Herbert, theoretical solid state physics, biophysics
Rousell, Gerald, insect physiology
Secco, Etalo Anthony, physical chemistry
Shaw, William S., geology
Steinitz, Michael Otto, solid state physics, materials science

BADDECK
Black, William Francis, ecology, invertebrate zoology

DARTMOUTH
Bailey, William Best, physical oceanography
Blanchard, Jonathan Ewart, geophysics
Brothers, James Alfred, inorganic chemistry
Buckley, Dale Eliot, geochemistry
Day, Lewis Rodman, fisheries management
Dickie, Lloyd Merlin, marine biology
Ford, William Livingstone, oceanography
Grant, Alan Carson, marine geology, geophysics
Haworth, Richard Thomas, marine geophysics
Hodder, Vincent MacKay, geophysics
Keen, Charlotte Elizabeth, marine geophysics
King, Lewis H., marine geology
Loncarevic, Bosko (D), marine geophysics
Loring, Douglas Howard, marine geochemistry
Mann, Barry Sinclair, zoology
Maunsell, Charles Dudley, physics
McMahon, Garfield Walter, acoustics
Mohammed, Auyuab, applied mathematics
Muir, Barry Sinclair, zoology
Pelletier, Bernard Roderick, marine geology
Ross, David I., geophysics
Schofield, Derek, physics
Smith, Harold Duncan, physics
Smith, Stuart D., oceanography
Srivastava, Surat Prasad, geophysics
Sutcliffe, William Humphrey, Jr., zoology
Tan, Francis C., chemical oceanography
Trites, Ronald Wilmot, physical oceanography
Vilks, Gustavs, micropaleontology
Walton, Alan, chemical oceanography, geochemistry

GLEN MARGARET
Noble, J Arnold, surgery

HALIFAX
Ackman, Robert George, organic chemistry
Aldous, John Gray, enzymology
Angelopoulos, Angelos Panayotis, dentistry
Angelopoulos, Edith W., cytology
Archibald, William James, physics
Aterman, Kurt, pathology
Aue, Walter Alois, organic chemistry, analytical chemistry
Blair, Alan Huntley, biochemistry
Boyd, Carl M., biological oceanography
Brewer, Donald, mycology
Bridgeo, William Alphonsus, organic chemistry
Brown, Robert George, microbial biochemistry
Calkin, Melvin Gilbert, physics
Cameron, Malcolm Laurence, physiology
Castell, John Daniel, marine sciences, nutritional biochemistry
Chandler, Reginald Frank, medicinal chemistry
Chapman, David MacLean, anatomy, zoology
Chute, Walter John, organic chemistry
Collins, Janet Valerie, cell biology, insect physiology
Cooke, Herbert Basil Sutton, geology
Cooper, John (Hanwell), human pathology, histochemistry
Davies, Donald Harry, physical chemistry
Dickson, Douglas Howard, microscopic anatomy
Dickson, Robert Clark, internal medicine
Doane, Benjamin Knowles, psychiatry
Doolittle, Warren Ford, III, molecular biology
Doyle, Roger Whitney, ecology, genetics
Dudar, John Douglas, neurophysiology
Dunsworth, Francis Alfred, psychiatry
Dupuy, David Lorraine, astronomy
Dyer, William John, biochemistry
Dykes, Edward Thomas, zoology
Easterbrook, Kenneth Brian, microbiology
Ecobichon, Donald John, biochemical pharmacology
Ellenberger, Herman Albert, occupational health
Falk, Michael, physical chemistry
Farley, John, history of biology
Farmer, Patrick Stewart, medicinal chemistry
Faught, John Brian, inorganic chemistry, x-ray crystallography
Fentress, John Carroll, ethology, neurobiology
Fillmore, Peter Arthur, pure mathematics
Flynn, Patrick, psychopharmacology
Forrest, Thomas Marie, organic chemistry
Fox, Roy Alan, medical research
Fyfe, Forest William, anatomy
Garside, Edward Thomas, zoology
Ghose, Tarunendu, pathology
Gilgan, Michael Wilson, biochemistry
Goble, David Franklin, theoretical physics
Goldbloom, Richard B., medicine
Grantmyre, Edward Bartlett, radiology
Gray, Michael William, biochemistry
Gray, Thomas James, physical chemistry
Grossert, James Stuart, organic chemistry
Haley, Leslie Ernest, genetics
Hall, Brian Keith, developmental biology
Hammerling, James Solomon, medicine, otolaryngology
Hansell, Margaret Mary, anatomy
Harrington, Fred Haddon, ethology
Harvey, Michael John, botany
Hatcher, James Donald, physiology
Hayes, Kenneth Edward, physical chemistry
Helleiner, Christopher Walter, biochemistry
Hirsch, Solomon, medicine, psychiatry
Holmes, Ian, anatomy
Hooper, Donald Lloyd, chemistry, spectroscopy
Hutzinger, Otto, environmental chemistry
Hyndman, Roy D., geophysics, oceanography
Issekutz, Bela, Jr., physiology
Jamieson, William David, physical chemistry
Jimenez-Marin, Daniel, anatomy, genetics
Jones, Robert Orville, psychiatry
Jones, William Ernest, physical chemistry, respiratory diseases
Josenhans, William T., respiratory physiology, respiratory diseases
Kabe, Dattatraya G., mathematical statistics
Kamra, Om Perkash, radiation genetics, cytogenetics
Kapoor, Brij M., plant cytology
Ke, Paul Jenn, analytical biochemistry, food biochemistry
Keen, Michael J., geophysics, oceanography
Kennedy, Frederick James, theoretical physics
Kimmins, Warwick Charles, plant physiology, plant virology
Kind, Leon Saul, microbiology
Knop, Osvald, inorganic chemistry, solid state chemistry
Kruse, Robert Leroy, algebra
Kwak, Jan C T., physical chemistry, electrochemistry
Lall, Santosh Prakash, nutritional biochemistry
Langley, G R., internal medicine, hematology
Langstroth, George Forbes Otty, physics
Lazier, Catherine Beatrice, biochemistry
Lee, Spencer Hon-Sun, medical biology, animal virology
Leflek, Kenneth Thomas, physical organic chemistry
Leighton, Alexander Hamilton, psychiatry, cultural anthropology
Li, Ming Fang, animal nutrition, microbiology
Lilienthal, Bernard, oral biology, microbiology
Longley, William Joseph, reproductive physiology, endocrinology
Maass, Wolfgang Siegfried Gunther, plant physiology, plant biochemistry
MacConnachie, Hugh John, restorative dentistry
Macdonald, Michael Raymond, preventive medicine
MacFarlane, Constance Ida, biology
MacKelvie, Robin Maxwell, bacteriology
MacLeod, Don Putnam, electrophysiology
MacLeod, Stephen Clair, obstetrics & gynecology
Macpherson, Lloyd Bertram, biochemistry
MacSween, Joseph Michael, immunology
Mahony, David Edward, microbiology
Manchester, J Stewart, radiology
Mann, Kenneth H., ecology
March, Robert Henry, physics
Masson, Charles Robb, physical chemistry
Matthews, Frederick White, chemistry, information science
Maxwell, Ian David, pathology
McAllister, Ronald Eric, clinical medicine, electrophysiology
McBride, Richard Phillips, microbiology, ecology
McFarlane, Ellen Sandra, oncology
McLachlan, Jack (Lamont), phycology
McLaren, Ian Alexander, biology
McLean, James Douglas, dentistry
Medioli, Franco, micropaleontology
Mezei, Catherine, biochemistry
Mezei, Michael, pharmaceutics
Milligan, George Clinton, economic geology, structural geology
Mills, Eric Leonard, biological oceanography
Moffitt, Emerson Amos, anesthesiology
Mounib, M Said, reproductive physiology, endocrinology
Murphy, James Wallace, physical chemistry
Murry, Dangety Satyanarayana, physics
Newkirk, Gary Francis, marine ecology, quantitative genetics
Nicholson, John Fraser, medicine, psychiatry
Nowlan, James Parker, mining geology
Odense, Paul Holger, biochemistry
Ogden, James Gordon, III, ecology
Ozere, Rudolph L., pediatrics
Palmer, Frederick B St Clair, biochemistry
Parker, William Arthur, therapeutics
Passey, Chand Arjun, food science, food engineering
Pielou, Douglas Patrick, ecology
Pielou, Evelyn C., ecology, biometrics
Purkis, Ian Edward, anesthesiology
Ramaley, Louis, analytical chemistry
Rautahariu, Pentti M., physiology, biophysics
Ravindra, Ravi, geophysics, cosmology
Regier, Lloyd Wesley, food science, fisheries
Reynolds, Albert Keith, pharmacology
Reynolds, Peter Herbert, geochronology, geochemistry
Richardson, Irvin Whaley, biophysics, biomathematics
Riley, Gordon Arthur, oceanography
Roger, William Alexander, experimental solid state physics
Rojo, Alfonso, ichthyology
Rozee, Kenneth Roy, microbiology, virology
Russell, Douglas William, biochemistry
Rutherford, John Garvey, anatomy
Ryan, Douglas Earl, inorganic chemistry, analytical chemistry
Saunders, Richard L., neuroanatomy, radiology
Schenk, Paul Edward, petrology, stratigraphy
Shaw, Roderick Wallace, meteorology
Shieh, Hang Shan, microbiology
Silvert, William Lawrence, resource management, theoretical solid state physics
Simpson, Allan Angus, physiology, obstetrics & gynecology
Simpson, Antony Michael, physics
Simpson, Frederick James, bacteriology
Smith, Eldon Raymond, cardiology
Steeves, Lea Chapman, medicine
Stevenson, William Denman, medicine
Stewart, Allan Greenwood, biochemistry
Stewart, Chester Bryant, medicine, epidemiology
Stewart, James Edward, bacterial physiology
Stoddard, Carl C., anesthesiology
Swaminathan, Srinivasa, mathematics
Sykora, Oscar P., dentistry
Szerb, John Conrad, physiology
Tan, Kok-Keong, mathematics
Tan, Meng Hee, internal medicine, medical research
Thiebaux, Helen Jean, applied statistics
Thomas, Anthony C., mathematics
Thomas, Kurian K, physiology
Thompson, Anthony C., mathematics
Tingley, Arnold Jackson, mathematics
Totten, James Edward, geometry
Tupper, W R Carl, obstetrics & gynecology
Uthe, John Frederick, fisheries chemistry
Verpoorte, Jacob A., biophysical chemistry
Vethamany, Victor Gladstone, anatomy
Vining, Leo Charles, bio-organic chemistry
von Maltzahn, Kraft Eberhard, botany
Wainwright, Lillian K (Schneider), genetics
Wainwright, Stanley D., chemical embryology
Wangersky, Peter John, oceanography
Welch, J Philip, human genetics
Weld, Charles Beecher, human physiology
Welles, Harry Leslie, pharmaceutics
White, Thomas David, neurochemistry, neuropharmacology
Whiteway, Stirling Giddings, physical chemistry, inorganic chemistry
Wiles, Michael, parasitology, marine biology
Williams, Christopher Noel, gastroenterology
Wong, Alan Yau Kuen, physics, biophysics
Woodbury John F L, medicine
Wright, Jeffrey Lawson Cameron, biological chemistry
Wyman, Harold Robertson, chemistry
Zouros, Eleftherios, population genetics

KENTVILLE
Aalders, Lewis Eldon, cytogenetics
Bishop, Robert Frederick, soil science, plant nutrition
Chipman, E. W., botany, horticulture
Cox, Allan Clayton, nutrition, physiology
Craig, Donald Laird, horticulture
Forsyth, Frank Russell, plant physiology
Hall, Ivan Victor, botany
Harrison, Kenneth Archibald, mycology
MacLellan, Charles Roger, entomology
MacPhee, Albert William, entomology
Proudfoot, F G., poultry genetics, poultry physiology
Ross, Robert Gordon, phytopathology
Specht, Harold Balfour, entomology
Stewart, D K R, ecology
Webster, David Henry, horticulture
Wright, James R, soil chemistry

LUNENBURG
Olson, Rodney Andreen, biology, cell biology

NAPPAN
MacIntyre, Thomas Martin, agricultural chemistry, animal nutrition

PORT WILLIAMS
Curry, George Montgomery, plant physiology

SYDNEY
Arseneau, Donald Francis, cellulose chemistry

TRURO
Bubar, John Stephen, agronomy
Cock, Lorne M., animal nutrition
Handforth, Christopher Peter, pathology
MacRae, Herbert F., biochemistry, animal science
McFadden, Lorne Austin, plant pathology
Roland, Albert Edward, systematic botany

Nriagu, Jerome Okonkwo, geoenvironmental science, geochemistry
Oliver, Barry Gordon, environmental chemistry
Platford, Robert Frederick, physical chemistry
Rukavina, Norman Andrew, sedimentology
Simons, Theodore J., atmospheric sciences, oceanography
Sly, Peter G., marine geology
Thompson, Mary Eleanor, geochemistry
Vollenweider, Richard A., limnology
Weiler, Roland R., geochemistry
Williams, Julian, geoenvironmental science
Wolkoff, Aaron Wilfred, analytical chemistry, environmental chemistry

WOLFVILLE
Basaraba, Joseph, microbiology
Beveridge, James MacDonald Richardson, biochemistry
Bleakney, John Sherman, malacology
Campbell, Alan, entomology, ecology
Dodds, Donald Gilbert, wildlife biology, ecology
Haley, Kenneth David Cann, mathematical statistics
Jeffery, Ralph Lent, mathematics
Linton, Everett Percival, physical chemistry
Lucas, Colin Cameron, biochemistry
MacLatchy, Cyrus Shantz, plasma physics
MacNeill, Rupert Heath, physics, geology
Magarvey, Raymond Halliday, physics, physical meteorology
Moore, Reginald George, invertebrate paleontology
Peach, Michael Edwin, inorganic chemistry
Pearson, Terrance Laverne, pure mathematics
Roscoe, John Miner, chemical kinetics
Smith, Ernest Chalmers, forest ecology
Snow, Douglas Oscar, mathematics
Stevens, George Richard, structural geology, tectonics
Stiles, David A., analytical chemistry, environmental chemistry
Tillotson, James Glen, physics
Toews, Daniel Peter, animal physiology

ONTARIO

AGINCOURT
Clarke, Charles Henry Douglas, biology

AILSA CRAIG
Ho, Keh Ming, crop breeding, cytogenetics

ANCASTER
Laposa, Joseph David, physical chemistry

ARVA
Cole, Randal Hudie, mathematics

BRACEBRIDGE
McTaggart-Cowan, Patrick Duncan, meteorology, science policy

BRESLAU
Pietrzykowski, Tomasz, computer science

BRIGHTON
Mohr, Willard Phillip, food science

BURLINGTON
Afghan, Baderuddin Khan, analytical chemistry, environmental chemistry
Barabas, Silvio, water pollution, environmental health
Burnison, Bryan Kent, microbial ecology
Carey, John Hugh, photochemistry
Chau, Alfred Shun-Yuen, analytical chemistry
Chau, Yiu-Kee, chemical oceanography, limnology
Dutka, Bernard J, microbiology
Glooschenko, Walter Arthur, limnology, biological oceanography
Heatley, A Harold, chemical kinetics
Jones, Philip Arthur, entomology
Kemp, Anthony Lionel, geology, biochemistry
Leppard, Gary Grant, cell biology

CHALK RIVER
Alexander, Thomas Kennedy, nuclear physics
Andrews, Hugh Robert, experimental nuclear physics
Baldwin, William F., radiobiology
Barry, P J S, health physics, micrometeorology
Bigham, Clifford Bruce, nuclear physics
Birnboim, Hyman Chaim, molecular biology
Brown, Fred, solid state science
Brown, Robert Melbourne, environmental chemistry
Buyers, William James Leslie, solid state physics
Carlisle, Alan, ecology
Chidley, Bruce Gordon, physics
Cowper, George, physics
Crocker, Iain Hay, analytical chemistry
Cross, William Gunn, nuclear physics
Davis, Ronald Stuart, reactor physics, nuclear physics
Dolling, Gerald, solid state physics
Duret, Maurice Francis, mathematics, physics
Earle, Eric Davis, nuclear physics
Eastwood, Thomas Alexander, nuclear chemistry, chemical kinetics
Fraser, John Stiles, accelerator physics
Geiger, James Stephen, physics
Glyde, Henry Russell, theoretical solid state physics
Graham, Robert Lockhart, physics
Green, Ralph Ellis, nuclear physics, reactor physics
Gulens, Janis, electroanalytical chemistry
Hammerli, Martin, electrochemistry
Hanna, Geoffrey Chalmers, nuclear physics
Hardy, John Christopher, nuclear physics
Harvey, Malcolm, theoretical nuclear physics
Hauser, Otto Friedrich, nuclear physics
Holford, Richard Moore, health physics
Holtslander, William John, chemistry, analytical chemistry
Ing, Harry, health physics, nuclear physics
Jackson, David Phillip, mathematical physics
Johnson, John Richard, biophysics, health physics
Khanna, Faqir Chand, nuclear physics
Kim, Soo Myung, solid state physics
Lee, Hoong-Chien, physics
Le Surf, Joseph Eric, physical chemistry
Lone, Muhammad Aslam, experimental nuclear physics
Macleod, John Campbell, forestry
Marko, Arthur Myroslaw, medicine, biochemistry
McDonald, Arthur Bruce, physics
McKeown, Joseph, physics
Mitchel, Ronald Edward John, biochemistry
Morgenstern, Erwin Kristian, forest genetics
Newcombe, Howard Borden, radiation genetics, human genetics
Norton, Peter Robert, physical chemistry, surface physics
Ophel, Ivan Lindsay, radiation ecology
Osborne, Richard Vincent, health physics
Paterson, Malcolm Cyril, radiation biophysics, medical biophysics
Pollard, Douglas Frederick William, plant physiology
Pringle, John Peter Scott, solid state chemistry
Santry, D C, nuclear chemistry, solid state science
Seddon, William Arthur, physical chemistry
Selander, William Nils, applied mathematics, engineering
Serdula, Kenneth James, reactor physics

Simpson, Stuart Douglas, medical research
Smith, Donald Reed, physical chemistry
Stevens, William Harmer, analytical chemistry
Stewart, C Gordon, biochemistry, medicine
Svensson, Eric Carl, solid state physics
Swanson, Max Lynn, experimental solid state physics
Taylor, John Gardiner Veitch, physics
Towner, Ian Stuart, theoretical nuclear physics
Tunnicliffe, Philip Robert, physics
Ungrin, James, nuclear physics
Van Wagner, Charles Edward, forestry, chemical engineering
Wang, Ben Shih-pin, forest physiology
Ward, David, experimental nuclear physics
Woods, Alfred David Braine, solid state physics, low temperature physics
Yeatman, Christopher William, forest genetics

CLARKSON
Maycock, Paul Frederick, plant ecology
Pullan, Harry, physics

COBOURG
Murray, Edward Donald, microbial biochemistry

CONCORD
Reffes, Howard Allen, analytical chemistry

COPPER CLIFF
Worsfold, Richard John, geology

CORNWALL
Elsermann, Edi, analytical chemistry

DEEP RIVER
Bartholomew, Gilbert Alfred, nuclear physics
Boyd, Alan William, radiation chemistry
Carmichael, Hugh, cosmic ray physics
Craig, Donald Spence, nuclear physics
Jarvis, Roger George, mathematics
Millar, Charles Howard, reactor physics
Milton, John Charles Douglas, nuclear physics

DELHI
Rosa, Nestor, plant physiology, biochemistry
Zilkey, Bryan Frederick, plant science, biochemistry

DON MILLS
Cunningham, John (Robert), medical physics
Meresz, Otto, organic chemistry, environmental chemistry
Schmidt, Arthur, physical chemistry
Smith, Bennett Lawrence, geology
Volpe, Robert, endocrinology
Wilson, John Tuzo, geophysics, tectonics
Woodward, Hubert Edmund, chemistry

DOWNSVIEW
Abegg, Victor Paul, organic chemistry, history of science
Aspinall, Gerald Oliver, organic chemistry
Bhartendu, atmospheric physics
Boville, Byron Walter, atmospheric science
Clodman, Joseph, meteorology
Colman, Brian, biochemistry
Darewych, Jurij Wasyl', theoretical physics
Davey, Kenneth George, invertebrate physiology, parasitology
Davis, John Tait, economic geography
Denzel, George Eugene, mathematics
Dugan, Charles Hammond, chemical physics
Ezemenari, Fidel Rex Chukwuemeka, atmospheric physics, radiation physics
Fabian, Robert John, mathematical logic
Forer, Arthur H, cell biology
Found, William Charles, economic geography, agricultural economics
Friesen James Donald, molecular biology, microbiology
Gulliver, Philip Hugh, anthropology, ethnography

Heath, Ian Brent, cell biology, mycology
Heidenreich, Conrad Edmund, cultural geography, historical geography
Hume, Valerie Elizabeth, urban geography, geography of Europe
Innanen, Kimmo A, astrophysics
Johnson, Arthur Clark, nuclear physics
Kisman, Kenneth Edwin, biophysics, ultrasound
Kuehn, Lorne Allan, biophysics
Lee, Roy, meteorology
Lee-Ruff, Edward, chemistry
Leith, Thomas Henry, philosophy of science, geophysics
Lever, Alfred B P, inorganic chemistry
Leznoff, Clifford Clark, organic chemistry
Lorch, Lee (Alexander), mathematics
Lundell, O Robert, physical chemistry
MacHattie, Lloyd Elliot, physics
Malcolm, Richard Evelyn Reginald, physiology
Marshall, John U, geography
Martin, Hans Carl, micrometeorology
McArthur, Colin Richard, organic chemistry
McArthur, Neil M, geography
Measures, Raymond Massey, laser physics
Megaw, William James, atmospheric physics
Merrens, Harry Roy, geography
Moens, Peter B, cell biology, cytogenetics
Money, Kenneth Eric, physiology, biology
Munn, Robert Edward, meteorology
Nicholls, Doris Margaret, biochemistry, enzymology
Pelletier, Joan Wick, pure mathematics
Price, John Andrew, cultural anthropology, economic anthropology
Prince, Robert Harry, surface physics
Pritchard, Huw Owen, physical chemistry, theoretical chemistry
Romans, Robert Gordon, protein chemistry, radiation
Sadowski, Chester M, physical chemistry
St Rose, John Elliston, immunology, clinical biochemistry
Saleuddin, Abu S, invertebrate zoology
Schemenauer, Robert Stuart, cloud physics
Schiff, Harold Irvin, physical chemistry
Shah, Govindlal M, atomospheric physics, radiation
Shepherd, Gordon Greeley, aeronomy
Smylie, Douglas Edwin, geophysics
Stauffer, Allan Daniel, atomic physics
Steel, Colin Geoffrey Hendry, invertebrate physiology, comparative endocrinology
Tait, James Simpson, fish biology
Tatham, George, geography of Europe, political geography
Thomas, Morley Keith, meteorology, climatology
Turner, Alice Willard, mathematics
Walker, Ian Munro, inorganic chemistry
Wallis, Anthony, computer science, mathematics
Warkentin, John Henry, geography
Webb, Rodney A, invertebrate physiology, parasitology
Wolfe, Roy Israel, geography
Wood, John David, geography
Young, Robert A, physics, physical chemistry

DUNDAS
Lomas, Harold, surface chemistry

ELMIRA
Costain, Robert Anthony, nutrition

ERINSVILLE
Taylor, Edward Godfrey, physical chemistry

ETOBICOKE
Godson, Warren Lehman, meteorology

FERGUS
Morrison, William D, animal nutrition, animal physiology

GEORGETOWN
Sieveking, William Earl, agronomy, plant breeding

GUELPH

Abrahamson, Edwin William, physical chemistry, molecular biology
Alex, Jack Franklin, plant taxonomy, ecology
Alexander, James Craig, biochemistry
Ambrose, John Daniel, botany
Anderson, Roy Clayton, parasitology
Archibald, James, veterinary medicine
Armstrong, Herbert Stoker, geology
Ashton, Gordon Clemence, statistics
Bailey, Edward D. animal behavior
Banden, John Drummond, agronomy
Barker, Clifford Albert Victor, veterinary medicine
Barker, William George, plant physiology
Barnum, Donald Alfred, veterinary science, microbiology
Bates, Thomas Edward, soil science
Bayley, Henry Shaw, biochemistry, nutrition
Beamish, Frederick William Henry, zoology
Beattie, Robert Walter, chemistry
Beauchamp, Eric G. soil fertility, plant nutrition
Beverley-Burton, Mary, zoology
Bhatnagar, Mahesh Kumar, veterinary anatomy
Bloomfield, Gerald Taylor, economic geography, history of technology
Borr, Mitchell, animal nutrition
Branion, Hugh Douglas, nutrition
Brewer, Arthur David, agricultural chemistry
Briton, Donald MacPhail, cytogenetics
Brooks, Ronald James, ethology
Brown, Donald Murray, agricultural meteorology
Brown, Robert Glenn, nutrition, physiology
Buchanan-Smith, Jock Gordon, animal science, animal nutrition
Bunce, Nigel James, chemistry
Burke, Philip William, organic chemistry, apiculture
Burnside, Edward Blair, population genetics, animal breeding
Burton, John Heslop, animal nutrition
Busch, Lloyd Victor, phytopathology
Chase, Francis Edward, soil microbiology
Christie, Bertram Rodney, crop breeding
Cichocki, Frederick Paul, evolutionary biology, ichthyology
Cocivera, Michael, physical chemistry
Colter, Allan Kennedy, organic chemistry
Corke, Charles Thomas, soil microbiology
Dahms, Fredric Arthur, geography
Dale, Hugh Monro, aquatic ecology
DeMan, John Maria, food chemistry
Derbyshire, John Brian, veterinary virology
Dickinson, William Trevor, hydrology
Dixon, Stuart Edward, entomology
Douglas, Robert John, microbiology
Downey, Ronald Stuart, veterinary medicine
Downie, Harry G. physiology, experimental surgery
Draper, Harold Hugh, nutrition
Ellis, Clifford Roy, economic entomology
Elrick, David Emerson, soil physics
Evans, Edwin Victor, nutrition
Evans, Taylor Herbert, organic chemistry
Eyre, Peter, pharmacology, immunology
Falconer, Allan, physical geography
Fisher, Kenneth Robert Stanley, developmental biology
Fletcher, Ronald Austin, plant physiology
Forsberg, Cecil Wallace, microbiology

Gregory, Kenneth Fowler, microbiology
Gyles, C. L. veterinary microbiology
Haley, Robert Currie, organic chemistry
Hall, Robert, plant pathology
Hallett, Frederick Ross, biophysics
Hardy Fallding, Margaret Hurlstone, developmental biology, histology
Harney, Patricia Marie, genetics, horticulture
Harrison, William Ashley, organic chemistry
Hatch, Roger Conant, pharmacology, toxicology
Henry, Patrick M. organometallic chemistry
Hill, Douglas Calvert, biochemistry, animal nutrition
Hilton, James L. plant physiology
Hofstra, Gerald, plant physiology, plant ecology
Holub, Bruce John, nutritional biochemistry
Horton, Roger Francis, plant physiology
Howell, Dennis George, microbiology, preventive medicine
Hulland, Thomas John, veterinary medicine
Hume, David, crop physiology
Hung, Frederick Fu, economic geography, geography of East Asia
Ingram, Donald George, veterinary immunology
Irvine, Donald McLean, dairy science
Ison, Marietta Maligalig, statistics
Ison, Norberto Tamisin, statistics
Janzen, Edward George, physical organic chemistry
Jeffrey, Kenneth Robert, nuclear magnetic resonance
Jerome, Frederick Nelson, animal genetics
Jordan, David Carlyle, bacteriology
Kannenberg, Lyndon William plant breeding
Karl, Gabriel, theoretical physics
Karstad, Lars, veterinary science
Kasha, Kenneth John, plant cytogenetics
Kaushik, Narinder Kumar, ecology
Khanna, Som Nath, polymer chemistry
King, Gordon James, reproductive physiology
Kokesh, Fritz Carl, biochemistry, bioorganic chemistry
Kulka, Marshall, organic chemistry
Lang, John E, entomology, ecology
Lange, Gordon Lloyd, organic chemistry
Law, Jimmy, theoretical physics
Lotz, Frederick, pharmacology
Lu, Benjamin Chi-Ko, physics
MacKenzie, Innes Keith, physics
MacNaughton, Earl Bruce, physics
Macphee, Kenneth Erskine, organic chemistry
Martini, Ireneo Peter, sedimentology
McDermot, Lawrence Alfred, medicine
McEwen, Freeman Lester, entomology
McPherson, Ross, physics
Mellors, Alan, biochemistry
Middleton, Alex Lewis Aitken, zoology
Miller, Murray Henry, soil fertility
Millman, Barry Mackenzie, biophysics
Milne, Frank James, veterinary medicine
Miniats, Oigerts Pauls, veterinary medicine
Motzok, Ilary, nutrition
Mullen, Kenneth, statistics, mathematics
Musgrave, Anthony John, biology
Newbould, Francis Henry Samuel, veterinary microbiology, veterinary immunology
Newton, Theodore Duddell, theoretical physics
Noakes, David Lloyd George, ethology
Nonnecke, Ib Libner, horticulture
Okashimo, Katsumi, mathematics, academic administration
Ollerhead, Robin Wemp, nuclear physics
Ormrod, Douglas Padraic, plant physiology
Orr, Henry Lloyd, poultry science
Peterson, Robert Lawrence, botany
Platonow, Nicolas W. toxicology, pharmacology
Poll, Jacobus Daniel, theoretical physics
Prokipcak, Joseph Michael, organic chemistry
Protz, Richard, soil science
Raeside, James Inglis, physiology
Raktoe, B Leo, statistics, experimental design

Rauser, Wilfried Ernst, plant physiology, plant biochemistry
Reinberg, Ernest, plant breeding
Richards, Norval Richard, soils
Rickels, Jerald Wayne, plant physiology, horticulture
Robinson, Gerald Arthur, radiobiology
Robinson, John Bertram, microbial ecology
Ronald, Keith, zoology
Safe, Stephen Harvey, chemistry, microbiology
Salvador, Antonio, computer science
Savan, Milton, veterinary medicine, virology
Sears, Markham Karli, insect ecology
Senoff, Caesar V, inorganic chemistry
Sheth, Anil Amratlal, agronomy, computer science
Shuel, Reginald William, plant physiology, insect physiology
Slater, Keith, textiles
Slinger, Stanley James, animal nutrition
Slocombe, Joseph Owen Douglas, parasitology, veterinary medicine
Smith, David William, plant ecology
Smith, Maurice Vernon, apiculture
Sonstegard, Ronald Arlyn, cancer, pathology
Sprague, John Booty, biology
Stanley, David Warwick, food science
Stevens, Ernest Donald, physiology, zoology
Stinson, Robert Henry, biophysics, physics
Stone, John Bruce, animal science
Stoskopf, N C, crop breeding
Suderman, Harold Julius, biochemistry
Sutton, John Clifford, plant pathology
Switzer, Clayton Macfie, agriculture, weed science
Tan, Kok-Chiang, geography
Thomas, Ronald Leslie, agronomy
Thomson, Reginald George, veterinary pathology
Thorsen, Jan, veterinary virology
Tittiger, Franz, meat science,
Tizard, Ian Rodney, immunology
Tomes, Dwight Travis, plant breeding, plant genetics
Tossell, William Elwood, agronomy, crop breeding
Townsend, Gordon Frederick, apiculture
Truscott, Robert Bruce, veterinary microbiology
Usborne, William Ronald, meat sciences
Vail, William Jerald, biophysics, cell biology
Walker, Brian Lawrence, biochemistry,
Wardlaw, Janet Melville, nutrition, home economics
Whiteford, Robert Daniel, veterinary medicine
Willoughby, Russell A. veterinary medicine
Wilson, Jack Harold, biochemistry, microbiology
Wilson, Stephen James, nuclear physics
Windecker, Richard Chase, biophysics, experimental solid state physics
Wright, Russell Emery, medical entomology
Yatsu, Eiju, mineralogy, geology
Zitnak, Ambrose, plant biochemistry, food technology

HAMILTON

Anderson, James Edward, physical anthropology, anatomy
Anderson, Robert Bernard, physical chemistry, catalysis
Bader, Richard Frederick W, theoretical chemistry
Banerjee, Satyendra Nath, cancer
Bayley, Stanley Thomas, biophysics, biochemistry
Behara, Minaketan, mathematics, mathematical statistics
Bell, Russell A, organic chemistry
Billigheimer, Claude Elias, pure mathematics, mathematical analysis
Bourns, Arthur Newcombe, organic chemistry
Branton, Philip Edward, virology
Brash, John Law, physical chemistry
Brockhouse, Bertram Neville, solid state physics
Brown, Ian David, chemical physics
Buchanan, George Dale, science education, anatomy
Bunting, Brian Talbot, physical geography

Burghardt, Andrew Frank, geography
Burke, Dennis Garth, physics
Calvo, Crispin, physical chemistry
Cameron, John Alexander, nuclear physics
Carr, David Harvey, anatomy, cytogenetics
Chernesky, Max Alexander, medical virology
Childs, Ronald Frank, physical organic chemistry
Collins, Malcolm Frank, physics
Cooper, Matthew Owen, anthropology
Corsini, A. analytical chemistry
Crocket, James Harvie, geochemistry, geology
Damas, David John, anthropology, ethnology
Daniel, Edwin Embrey, pharmacology
Datars, William Ross, physics
Davidson, Douglas, cytology
Davidson, Ronald G, pediatrics, medical genetics
Davies, Douglas Mackenzie, entomology
Davison, Thomas Matthew Kerr, mathematics
Dawson, Peter Thomas, surface chemistry
Deml, Peter Boris, immunology
Dingle, Allan Douglas, developmental biology
Dunnett, Charles William, statistics
Eaton, Donald Rex, physical chemistry
Epand, Richard Mayer, biophysical chemistry
Epstein, Nathan Bernic, psychiatry
Evans, Geoffrey, cardiovascular surgery
Ferrier, Barbara May, biochemistry
Fleming, William Herbert, physics
Ford, Derek Clifford, geography, geology
Freeman, Karl Boruch, biochemistry
Freeman, Milton Malcolm Roland, biological anthropology, cultural anthropology
Fritze, Klaus, analytical chemistry
Gadamer, Ernst Oscar, applied mathematics
Gent, Michael, medical statistics
Gillespie, Ronald James, inorganic chemistry, physical chemistry
Goldsmith, Charles Harry, applied statistics, biostatistics
Goodings, David Ambery, solid state physics
Graham, Frank Lawson, virology
Graham, Ronald Powell, analytical chemistry
Greedan, John Edward, solid state chemistry
Hall, Ross Hume, biochemistry
Hannell, Francis George, microclimatology
Harvey, John Wilcox, health physics, environmental management
Heinig, Hans Paul, mathematics
Hileman, Orville Edwin, Jr, analytical chemistry
Hillcoat, Brian Leslie, biochemistry
Hodgins, John Willard, chemistry, chemical engineering
Horsley, Robert James, nuclear physics
Howard-Lock, Helen Elaine, molecular spectroscopy, optics
Husain, Taqdir, pure mathematics
Johns, Martin Wesley, nuclear physics
Kennett, Terence James, nuclear physics
Kershaw, Kenneth Andrew, plant ecology, lichenology
King, Gerald Wilfrid, chemical physics
King, Leslie John, urban geography, regional economics
Kramer, James Richard, geochemistry, geology
Kuehner, John Alan, nuclear physics
Laidlaw, John Coleman, endocrinology
Landes, Ruth, anthropology
Lind, James Forest, surgery
Lock, Colin James Lyne, physical inorganic chemistry, crystallography
Lott, John Norman Arthur, plant anatomy, plant physiology
MacLean, David Bailey, organic chemistry
Mak, Stanley, virology
Marton, John Peter, solid state electronics
McCalla, Dennis Robert, biochemistry, cell biology
McCallion, David John, embryology, teratology
McCallion, William James, mathematics, astronomy
McCandless, Esther Leib, physiology
McCulloch, Peter Blair, cancer
McCullough, John James, organic chemistry

McNutt, Robert Harold, petrology, geochemistry
Middleton, Gerard Viner, geology
Miller, John James, microbiology
Mohanty, Sri Gopal, mathematics, statistics
Morrison, James Alexander, physical chemistry
Morton, Richard Alan, biophysics
Mueller, C Barber, surgery
Mustard, James Fraser, pathology
Nelson, Evelyn Merle, algebra
Nogami, Yukihisa, theoretical physics
Oaks, B Ann, plant physiology
Olsen, Fredric Phillip, chemistry
Pallie, Wazir, human anatomy, medicine
Papageorgiou, George John, urban geography, economic geography
Perey, Daniel Yves Emile, immunology, surgery
Preston, Melvin Alexander, theoretical physics
Prevec, Ludvik Anthony, biology
Pringle, James Scott, botany
Rainbow, Andrew James, medical biophysics
Rawls, William Edgar, virology, epidemiology
Reeds, Lloyd George, agricultural geography
Rice, Peter (Franklin), forest pathology
Richardson, Harold, medical microbiology
Riehm, Carl Richard, algebra
Risk, Michael John, marine ecology, paleoecology
Sackett, David Lawrence, internal medicine, epidemiology
Schwarz, Henry Philip, geochemistry, petrology
Shaw, Denis Martin, geochemistry
Slobodin, Richard, anthropology
Sorger, George Joseph, biochemical genetics, microbiology
Spaulding, William Bray, clinical medicine
Spenser, Ian Daniel, biochemistry
Sprung, Donald Whitfield Loyal, nuclear physics
Stager, Carl Vinton, solid state physics
Stephens, Michael A, mathematical statistics, applied statistics
Stewart, James Drewry, mathematical analysis
Stohr, Walter B, geography
Summers-Gill, Robert George, experimental nuclear physics
Sweeney, George Douglas, pharmacology
Takahashi, Francois Iwao, microbial genetics
Taylor, David Ward, theoretical physics, solid state physics
Thode, Henry George, physical chemistry
Threlkeld, Stephen Francis H, biology, genetics
Timusk, Thomas, physics
Toews, Cornelius J, endocrinology, biochemistry
Tomlinson, Richard Howden, physical chemistry, inorganic chemistry
Uchida, Irene Ayako, genetics
Unruh, William George, theoretical physics
Volkov, Anatole Boris, nuclear physics
Walker, Roger Geoffrey, sedimentology
Walsh, William J, internal medicine
Walton, Derek, physics
Warkentin, John, organic chemistry
Westermann, Gerd Ernst Gerold, geology
Witelson, Sandra Freedman, neuropsychology
Wood, Derick, computer science
Wood, Harold Arthur, geography
Yip, Patrick Cheung-Yum, applied mathematics, nuclear physics
Younglai, Edward Victor, biochemistry, reproductive physiology
Zipursky, Alvin, pediatrics, physiology

HARROW
Buttery, Brian Richard; plant physiology
Buzzell, Richard Irving, plant breeding, plant genetics
Foott, William Henry, entomology
Fulton, James McCullough, soil science, research administration
Haas, Jerry Henry, plant pathology
Jaques, Robert Paul, insect pathology
Johnson, Peter Wade, plant pathology, plant nematology
Layne, Richard C, plant breeding
Marriage, Paul Bernard, plant biochemistry, plant physiology
McClanahan, Robert Joseph, entomology
McKeen, Colin Douglas, plant pathology
Quamme, Harvey Allen, horticulture

Von Stryk, Frederick George organic chemistry
Ward, Gordon Marshall, plant physiology

HAWKESBURY
Gupta, Virendra Nath, paper chemistry
Histed, John Allan, pulp chemistry
Ingruber, Otto Vincent, chemistry, electrochemistry
Kaepner, Werner Martin, microscopy, biochemistry
Kalisch, John Hans, pulp chemistry, paper chemistry
Lemay, Yvan, botany, zoology
Mutton, Donald Barrett, chemistry
Rivington, Donald Erskine, analytical chemistry
Saxton, William Reginald, pulp chemistry
Smith, Richard E, chemistry

ISLINGTON
Bannan, Marvin William, botany
Cale, William Robert, inorganic chemistry, chemical engineering

KANATA
Cameron, Irvine R, polymer chemistry

KARS
Thomas, Barry Holland, biochemistry, pharmacology

KEMPTVILLE
Robertson, George Wilber, agricultural meteorology

KINGSTON
Abrahams, Vivian Cecil, neurophysiology
Allen, James Roy, theoretical physics
Andrew, George McCoubrey, physiology, physical education
Armstrong, Harold Lewis, physics
Bacon, David W, statistics, chemical engineering
Baird, D C, physics
Bartlett, Grant Aulden, geoenvironmental science
Barton, Stuart Samuel, physical chemistry
Beck, Ivan Thomas, internal medicine, gastroenterology
Bencosme, Sergio Arturo, pathology
Berry, Leonard Gascoigne, mineralogy
Bidwell, Roger Grafton Shelford, plant physiology
Bird, Charles Edward, endocrinology, metabolism
Blakeslee, Dennis Lauren, immunobiology
Blyth, Colin Ross, mathematical statistics
Boag, Thomas Johnson, psychiatry, psychoanalysis
Boston, Robert Wesley, pediatrics
Botterell, Edmund Harry, neurosurgery
Breck, Wallace Graham, physical chemistry
Bridle, Alan Henry, astronomy
Brown, Richard Julian Challis, physical developmental biology
Brown, Seward Ralph, geochemistry, ecology
Bryans, Alexander (McKelvey), medicine
Buncel, Erwin, physical organic chemistry
Cairnie, Alan B, radiobiology, cell biology
Campbell, Louis Lorne, mathematics
Canvin, David T, plant physiology, plant biochemistry
Caradus, Selwyn Ross, mathematical analysis
Castner, Henry Walker, geography
Chadwick, June Marie, microbiology
Chapler, Christopher Keith, physiology
Chiong, Miguel Angel, internal medicine, cardiology
Chung, Kwok-Leung, microbiology
Coleman, Albert John, applied mathematics
Colpa, Johannes Pieter, theoretical chemistry, magnetic resonance
Conkie, Walter Ford, physics
Connell, Walter Ford, internal medicine, cardiology
Cooke, Fred, genetics
Corlett, Mabel Isobel, mineralogy
Crowder, Adele A, plant ecology
Crowe, Arlene Joyce, biochemistry
Dacey, John Robert, physical chemistry

Dawes, David Haddon, physical chemistry, polymer science
Day, J H, internal medicine
Dennis, David Thomas, plant biochemistry, plant physiology
Eaton, Bryan Thomas, virology
Edwards, Martin Hassall, physics
Eidinger, David, microbiology, immunology
Ewan, George T, nuclear science
Faulkner, Peter, virology
Fitzpatrick, Michael Morson, geophysics
Flynn, Thomas Geoffrey, biochemistry
Forsdyke, Donald Roy, biochemistry, immunology
Ghent, William Robert, surgery
Giles, Robin, mathematical physics
Good, Harold Marquis, plant pathology
Gordon, Robert Dixon, physical chemistry
Gorman, William Alan, quaternary geology
Greggs, Robert George, paleontology
Gunn, Kenrick Lewis Stuart, cloud physics
Gutelius, John Robert, surgery
Harmsen, Rudolf, systems biology
Harris-Lowe, Rodney Frederic Brandon, low temperature physics
Harrower, George Alexander, physics
Hay, George William, chemistry
Hetherington, R F, neurosurgery
Heyding, Robert Donald, solid state chemistry
Hogarth, Jacke Edwin, applied mathematics, cosmology
Hood, Cornelius Henry, genetics
Hughes, Victor A, physics, astronomy
Hurst, Robert Osmond, biochemistry
Hutchinson, Thomas Sherret, physics
Jellinck, Peter Harry, biochemistry, endocrinology
Jennings, Donald B, medicine, physiology
Jezak, Edward V, physics
Jhamadas, Khem, pharmacology
Johansen, Peter Herman, zoology
Jolliffe, Alfred Walton, geology
Joneja, Madan Gopal, cytogenetics, teratology
Jones, John Kenyon Netherton, organic chemistry
Kaufman, Nathan, pathology
Keast, James Allen, vertebrate ecology
Kelly, Howard Garfield, internal medicine
Kemp, Robert Richard Dingle, mathematics
Kennedy, James Cecil, cell biology, immunology
Kerbel, Robert Stephen, cancer
Kesteven, Michael, astronomy
Kipkie, George Frederick, pathology
Kirby, Bruce John, applied mathematics
Kozub, Raymond Lee, nuclear physics
Kraicer, Jacob, physiology, endocrinology
Kraus, Arthur Samuel, epidemiology
Kummer, Hans Jacob, applied mathematics
Laverty, S G, psychiatry
Law, Cecil E, operations research
Lightstone, Albert Harold, mathematics
Lott, James Stewart, medicine, radiology
Love, Hugh Morrison, physics
Low, Ralph Beverley, surgery
Lynn, Ralph Beverley, surgery
MacArthur, John Duncan, nuclear physics
Mankovitz, Ralph, cell biology
Marks, Gerald Samuel, organic chemistry, pharmacology
McConville, Brian John, psychiatry
McCorriston, James Roland, surgery
McIntosh, Robert Lloyd, physical chemistry
McLay, David Boyd, molecular physics
Milazzo, Francis Henry, microbiology, physiology
Milligan, John Vorley, medical physiology
Milliken, John Andrew, internal medicine
Moir, Robert Young, organic chemistry
Montague, John H, nuclear physics
Moore, Eric G, urban geography, population geography
Morrin, Peter Arthur Francis, medicine
Morris, Gerald Patrick, cell biology, parasitology
Neilson, James Maxwell, geology
Norman, Robert Daniel, mathematics
Norris, A R, inorganic chemistry, physical chemistry
Orzech, Grace Geist, algebra
Orzech, Morris, mathematics
Osborne, Brian S, geography
Page, John Arthur, analytical chemistry
Parker, John Orval, cardiology, cardiovascular physiology
Partington, Michael W, pediatrics

Pearce, Thomas Hulme, petrology, geochemistry
Phibbs, Murray Kenneth, chemistry
Pope, Noel Kynaston, theoretical physics, mathematics
Powles, William Earnest, psychiatry
Price, Raymond Alex, structural geology, tectonics
Puhach, Paul Alexander, theoretical physics
Pullman, Norman J, mathematics
Reid, James Gavin, kinesiology, physical education
Ribenboim, Paulo, mathematics
Rice, Norman Molesworth, mathematics
Riddell, John Barry, geography
Roberts, Denys Thomas, physics
Roberts, Leslie Gordon, mathematics
Robertson, David Murray, neuropathology, pathology
Roeder, Peter Ludwig, geochemistry, petrology
Rogers, Douglas Herbert, metal physics
Romero-Sierra, Cesar Aurelio, anatomy
Rosen, David A, ophthalmology
Ruggles, Richard Irwin, geography
Russell, Kenneth Edwin, polymer chemistry
Sargent, Bernice Weldon, nuclear physics
Sayer, Michael, physics
Schubert, Cedric F, mathematics
Segel, Stanley Lewis, solid state physics
Semple, Robert Evans, physiology
Shurvell, Herbert Francis, spectrochemistry
Simon, Jerome Barnet, gastroenterology
Simpson, Nancy E, human genetics
Sinclair, D G, academic administration
Smith, Barry Thomas Sturt, neonatology, pneumology
Smith, Walter MacFarlane, physical chemistry
Snider, Neil Stanley, theoretical chemistry
Soudek, Dusan Edward, cytogenetics, medical genetics
Sribney, Michael, preventive medicine, epidemiology
Steele, Robert, preventive medicine, epidemiology
Stewart, Alec Thompson, physics
Stewart, Robert Bruce, animal virology
Stroud, Thomas William Felix, statistics
Szarek, Walter Anthony, organic chemistry
Uffen, Robert James, geophysics
Ursell, John Henry, mathematics
Vandewater, Stuart Leslie, anesthesiology
Verner, James Hamilton, mathematics
Wan, Jeffrey Kwok-Sing, physical chemistry
Wasan, Madanlal T, statistics
Watson, Edmond Evelyn, physics
Watts, Donald George, applied statistics
White, Denis Naldrett, medicine
Wilson, Donald Laurence, medicine
Woodside, William, physics, mathematics
Wyatt, Gerard Robert, insect physiology, biochemistry
Yeates, Maurice, urban geography, economic geography

KITCHENER
Chapman, Judith-Anne Williams, medical statistics
McCauley, Robert William, fish biology

LONDON
Alford, William Parker, nuclear physics
Ali, Mir Maswood, mathematical statistics
Allnatt, Alan Richard, theoretical chemistry, physical chemistry
Armstrong, David Thomas, physiology
Arnold, Donald Robert, organic chemistry, photochemistry
Atkinson, Burr Gervais, biochemistry, developmental biology
Baird, Norman Colin, physical chemistry
Baldwin, Howard Wesley, inorganic chemistry
Bancroft, George Michael, physical inorganic chemistry
Barnett, H J M, neurology
Barr, Murray Llewellyn, anatomy
Battle, Helen Irene, developmental biology, human genetics
Beck, Alan Edward, geophysics
Behme, Ronald John, genetics, microbiology
Bidinosti, Dino Ronald, physical chemistry
Bjorklund, Elaine M, geography
Blackwell, John Henry, applied mathematics
Bolton, James R, biophysical chemistry
Bond, Edwin Joshua, zoology

CANADA

Bondy, Donald Clarence, gastroenterology
Borwein, David, mathematics
Bourns, Thomas Kenneth Richard, parasitology
Bowman, Bruce Tamblyn, soil chemistry, physical chemistry
Brand, John C., nuclear physics, quantum electronics
Brannen, Eric, nuclear physics, quantum electronics
Brien, Francis Staples, internal medicine
Brooks, Vernon Bernard, neurophysiology
Brown, Brian Ellman, neurophysiology
Brown, James Douglas, physical chemistry
Brownstone, Yehoshua Shieky, clinical biochemistry
Bryan, Robert Neff, mathematics
Buck, Carol Whitlow, preventive medicine
Buck, Robert Crawforth, anatomy, histology
Burton, Alan Chadburn, biophysics
Cataresu, Franco Romano, physiology
Canham, Peter Bennet, biophysics
Carroll, Kenneth Kitchener, biochemistry
Carroll, Samuel Edwin, cardiovascular surgery, thoracic surgery
Cavers, Paul Brethen, plant ecology, weed science
Creider, Chester Arthur, III, anthropology
Cummins, Joseph E., genetics, cell biology
Cunningham, David A., physiology
Dales, Samuel, virology, cell biology
de Mayo, Paul, organic chemistry, photochemistry
de Veber, Leverett L., hematology, immunology
Donald, Lynda Joan, human genetics
Donisch, Valentine, biochemistry
Dreimanis, Aleksis, quaternary geology
Duncan, Ronald Ian, biophysics
Dunham, Charles Burton, mathematics
Edgar, Alan D., geochemistry, petrology
Ehrman, Joachim Benedict, plasma physics
Fahsel, Dianne, botany
Ferguson, Gary Gilbert, neurosurgery, biophysics
Ferguson, Harry Ian Symons, physics
Fitz-James, Philip Chester, microbial biochemistry
Forsyth, Peter Allan, physics
Fraser, Peter Arthur, physics
Freedman, James M., anthropology
Frei, Jaroslav Vaclav, pathology
Gagen, Walter Leonard, protein chemistry, research administration
Gallaher, Donald Frederick, theoretical physics
Galsworthy, Peter Robert, biochemistry
Gammal, Elias Bichara, obstetrics & gynecology, anatomy
Gardner, David Godfrey, oral pathology
George, John Allen, entomology
Gerson, Donald Franklin, biophysics
Gordon, Myra, organic chemistry
Gowdey, Charles Willis, pharmacology
Goyer, Robert Andrew, pathology
Greyson, Richard Irving, morphology
Groom, Alan Clifford, biophysics, physiology
Guenple, Lee, anthropology, anthropological linguistics
Gunton, Ramsay Willis, medicine
Guthrie, James Peter, bio-organic chemistry
Haines, Roland Arthur, inorganic chemistry
Haq, M Saiful, mathematical statistics
Harding, Paul George Richard, reproductive physiology, perinatology
Harris, Charles Ronald, entomology
Hart, John Francis, computer science, applied mathematics
Haust, Heinz (Heinrich) Leonhard, biochemistry

Haust, M Daria, pathology
Hay, Donald Ross, physics
Heagy, Fred Clark, biochemistry, nuclear medicine
Heimbecker, Raymond Oliver, experimental surgery, cardiovascular surgery
Hickman, Clarence James, mycology, plant pathology
Hill, Martha Adele, analytical chemistry, biochemical engineering
Hinton, George Greenough, pediatric neurology
Hirst, Maurice, chemical pharmacology
Hobbs, George Edgar, psychiatry
Hobkirk, Ronald, biochemistry, endocrinology
Hodder, Robert William, geology
Hopkins, William George, plant physiology
Howell, Walter Colston, organic chemistry
Hunter, Andrew Tate, medicine
Hunter, William Stuart, orthodontics, anthropology
Hutchinson, Richard William, geology
Inch, William Rodger, radiobiology
Judd, William Wallace, entomology
Johnson, Robert H., dentistry
Johnson, Leonard N., dental materials, metallurgical engineering
Jaco, Nicholas Trevenen, pediatrics
Jacobs, Patrick W.M., physical chemistry, solid state chemistry
Kidd, Robert Garth, inorganic chemistry
King, James Frederick, organic chemistry
Krupka, Richard Morley, biochemistry
Landstreet, John Darlington, astrophysics
Lee, Tsung Ting, plant physiology
Lefcoe, Neville, physiology
Lenz, Alfred C., geology
Lewis, John Albert, medicine
Lo, Theodore Ching-Yang, microbial genetics
Locke, Michael, insect physiology
Lorimer, John William, physical chemistry
Loughheed, Thomas Crossley, food science
Lowe, Robert Peter, physics, aeronomy
Lyon, Gordon Frederick, physics
MacDonald, John Campbell Forrester, physics
MacNeill, Ian B., mathematics, statistics
MacPherson, Catherine Frances Conway, immunochemistry, psychiatry
MacRae, Neil D., geology
Magee, Gordon Richey, mathematics
Magee, William Lovel, biochemistry
Manning, George William, medicine
Mansinha, Lalatendu, geophysics, applied mechanics
Marlborough, John Michael, astronomy
McAninch, Lloyd Nealson, urology
McCarter, John Alexander, virology
McCredie, John A., cancer, surgery
McDonald, John William David, hematology, internal medicine
McGowan, James William, atomic physics, molecular physics
McKeen, Wilbert Ezekiel, plant pathology
McLarty, Duncan Archibald, botany
McMillan, Donald Burley, zoology, comparative histology
McMurray, William Colin Campbell, biochemistry
McWhinney, Ian Renwick, medicine
Meath, William John, theoretical chemistry
Medzon, Edward Lionel, virology
Mercer, Paul Frederick, physiology
Mereu, Robert Frank, geophysics, seismology
Miller, David Milroy, physical chemistry
Mills, Donald Grant, pharmacology
Minshall, William Harold, plant physiology
Misra, Rajendra Kumar, statistics
Mitalas, Romas, theoretical physics, astrophysics
Mogenson, Gordon James, psychophysiology
Montemurro, Donald Gilbert, physiology, neuroendocrinology
Moorcroft, Donald Ross, physics
Moore, John Thomas, mathematics
Murray, Robert George Everitt, bacteriology
Murty, Rama Chandra, physics
Nagai, Toshio, physiology, zoology
Naylor, Derek, applied mathematics
Neufeld, Abram Herman, clinical pathology

Nicholls, John Van Vliet, ophthalmology
Nicholson, Norman Leon, geography, academic administration
Nuttall, John, theoretical physics
O'Hea, Eugene Kevin, physiology
Parnell, H Curries, geophysics, geology
Parnell, Anthony George, oral surgery
Payne, Nicholas Charles, chemistry
Philbrick, Allen Kellogg, geography
Philip, Richard Blair, pharmacology
Phipps, James Bird, systematics
Planck, Roy Jonathan, physiology, ecology
Pomerantz, David Kurt, reproductive physiology
Possmayer, Fred, biochemistry, neurobiology
Purko, John, developmental biology
Raud, Heinz Randar, endocrinology
Redinger, Richard Norman, gastroenterology
Reid, Brian Douglas, nuclear medicine
Reynolds-Warnhoff, Patricia, organic chemistry
Richardson, Lloyd Thomas, plant pathology
Ritcey, Leland Frederick Samuel, mathematics
Roach, Margot Ruth, biophysics, medicine
Robinow, Carl Franz, microbiology
Robinson, John, bacteriology
Roslycky, Eugene Bohdan, microbiology
Rosner, Sheldon David, physics
Rossier, Roger James, biochemistry
Roth, Rene Romain, physiology
Sakmar, Ismail Aydin, elementary particle physics, theoretical physics
Scott, Andrew Edington, organic chemistry
Scott, David Maxwell, zoology
Seguin, Jerome Joseph, physiology
Serdarevich, Bogdan, chemistry
Sergovich, Frederick Raymond, cytology, cytogenetics
Shaver, Evelyn Louise, cytogenetics, reproductive biology
Shawyer, Bruce L R., pure mathematics
Sherebrin, Marvin Harold, biophysics
Shivers, Richard Ray, zoology, cell biology
Shute, Evan Vere, obstetrics & gynecology
Sinclair, Nicholas Roderick, immunology, oncology
Singh, Roderick Pataudi, anatomy
Singhal, Sharwan Kumar, immunology
Smith, David Burrard, biochemistry
Soltan, Hubert Constantine, medical genetics
Song, Seh-Hoon, biophysics, cardiopulmonary physiology
Spence, Michael Wishart, anthropology
Spence, Elvins Yuill, organic chemistry
Spoerel, Wolfgang Eberhart G., anesthesiology, pharmacology
Squires, Bruce Paul, physiology
Starkey, John, structural geology
Starratt, Alvin Neil, natural products chemistry
Steele, John Earle, zoology, endocrinology
Stewart, Harold Brown, biochemistry
Stewart, Thomas William Wallace, acoustics, electronics
Stoessl, Albert, natural products chemistry
Stothers, John Bailie, organic chemistry
Stouffer, James L., audiology
Strejan, Gill Henric, immunology
Strickland, Kenneth Percy, biochemistry
Sukava, Armas John, physical chemistry
Sullivan, Paul Joseph, applied mathematics, engineering
Sutherland, Robert Melvin, radiation biology
Talman, James Davis, theoretical physics
Taylor, Charles Patrick Stirling, biophysics
Teteruck, Walter R., prosthodontics, dental materials
Thierrin, Gabriel, mathematics, computer science
Thorn, George Denis, agricultural chemistry
Tong, Bok Yin, solid state physics, biophysics
Trevithick, John Richard, biochemistry
Tu, Chin Ming, microbiology, biochemistry
Tustanoff, Eugene Reno, biochemistry
Valberg, Leslie S., medicine

Vanderwolf, Cornelius Hendrik, neurosciences
Van Huystee, Robert Bernard, biochemistry, botany
Vardanis, Alexander, biochemistry
Vogan, Eric Lloyd, physics
Walden, David Burton, plant genetics, cytogenetics
Walker, Ian Gardner, biochemistry, cell biology
Wallace, Alexander Cameron, medicine, pathology
Ward, Edmund William Beswick, microbiology, plant pathology
Ware, William Romaine, physical chemistry
Warnhoff, Edgar William, organic chemistry
Warntz, William, geography
Warren, Bruce Albert, pathology
Watson, Thomas Alastair, medicine
Watson, William Crawford, internal medicine, gastroenterology
Wehlau, Amelia W., astronomy
Wehlau, William Henry, astronomy
Weilman, Angela Myra, microbial physiology
Whebell, Charles Frederick John, geography
Whippey, Patrick William, solid state physics
White, Gordon Allan, plant physiology
Wiebe, John Peter, physiology, endocrinology
Wilkins, Peter Osborne, microbiology
Willis, Christopher John, inorganic chemistry
Wilson, David George, plant biochemistry
Winder, Charles Gordon, geology
Wolfe, Bernard Martin, medicine
Wooton, Garnet Alexander, physics
Young, Grant McAdam, stratigraphy, sedimentology
Zajic, James Edward, microbiology, law

MAPLE

Burger, Dionys, forest ecology, soils
Henson, Walter Robert, entomology
Ihssen, Peter Edward, genetics, fish biology
Kerr, Stephen Roy, ecology
McCombie, Alen Milne, limnology, fisheries
Raymond, Frank Leroy, resource management
Slankis, Visvaldis, plant physiology, microbiology

MISSISSAUGA

Booth, Kenneth Gordon, cellulose chemistry
Bratina, Woymir John, solid state physics
Collins, Nicholas Clark, population ecology
Cooper, George S., crop science, soil science
De La Iglesia, Felix Alberto, experimental pathology, toxicology
Dorland, Rodger Malone, wood chemistry
Evans, James Eric Lloyd, exploration geology
Golightly, John Paul, geology
Hair, Michael L., physical chemistry, crystallography
Harbour, John Richard, photochemistry
Hopton, Frederick James, environmental chemistry, environmental engineering
Horgen, Paul Arthur, cell biology, molecular biology
Luebbe, Ray Henry, Jr., physical chemistry
Lush, Donald Lawrence, fresh water ecology, limnology
McAdie, Henry George, environmental chemistry, analytical chemistry
Middleton, Hugh William, chemistry, information science
Myers, Mark B., materials science, ceramics
O'Day, Danton Harry, developmental biology
Pretty, Kenneth McAlpine, crop science
Punnam, Donald Fulton, geography
Reed, Juta Kurtis; biochemistry
Reid, Sidney George, organic chemistry
Robinson, Edward Arthur, inorganic chemistry, physical chemistry

Sharp, James H., physical chemistry, chemical physics
Simmons, George Allen, chemistry, glass technology
Sprules, William Gary, aquatic ecology
Staples, Milfred Lawson, textile chemistry, textile technology
Still, Ian William James, organic chemistry
Williams, Michael John, textile chemistry

NIAGARA FALLS
Patchett, Joseph Edmund, mineralogy, ceramics

NIAGARA-ON-THE-LAKE
Ayre, Charles A., biochemistry

OAKVILLE
Alcock, Norman Zinkan, nuclear physics, electrical engineering

ORANGEVILLE
Morrison, Spencer Horton, veterinary medicine, animal nutrition

OTTAWA
Abbott, Frank Sidney, animal physiology
Adams, Gabrielle H M, molecular biology
Adams, William Alfred, glaciology
Ahern, Francis Joseph, applied physics
Ainsworth, Louis, reproductive endocrinology
Allen, Rovelle Harper, clinical chemistry, serology
Alper, Howard, organic chemistry, organometallic chemistry
Amberg, Carl Helmut, chemistry
Anderson, Francis David, geology
Andrew, Bryan Haydn, radio astronomy
Anstey, Thomas Herbert, genetics, plant breeding
Appel, Warren Curtis, pharmacology
Apsimon, John W, organic chemistry
Argus, George William, systematic botany
Armstrong, John Alexander, insect toxicology, forest meteorology
Armstrong, John Buchanan, internal medicine
Armstrong, Robert A, surface physics
Arnold, John Walter, entomology
Ashwin, James Guy, physiology, pharmacology
Auerbach, Lewis Edward, science policy, science writing
Babbit, John David, physics
Back, Margaret Helen, physical chemistry
Ball, William Lee, organic chemistry
Bannard, Robert Alexander Brock, organic chemistry
Barager, William Robert Arthur, geology
Barr, Donald John Stoddart, mycology
Barradas, Remigio Germano, physical chemistry
Barran, Leslie Rohit, agricultural microbiology
Barrington, Ronald Eric, physics
Barron, John Robert, entomology, taxonomy
Barton, Richard Donald, nuclear physics
Bather, Roy, virology
Baum, Bernard R, taxonomy, botany
Beare-Rogers, Joyce Louise, biochemistry, nutrition
Beaulieu, J A E M, bacteriology, immunology
Becker, Edward Coulton, entomology
Becking, George C, biochemistry
Beesack, Paul Richard, mathematical analysis
Behki, Ram M, molecular biology
Belanger, Leonard Francis, histology, histochemistry
Benoiton, Normand Leo, biochemistry
Benson, Cyril Brownlow, physics
Benson, George Campbell, physical chemistry

Bernstein, Harold Joseph, physical chemistry
Berry, Michael John, seismology
Berry, Robert John, solid state physics
Betz, Thomas William, developmental endocrinology
Beznak, Margaret, physiology
Bhatnagar, Dinech C, inorganic chemistry, analytical chemistry
Biefer, Gregory James, physical chemistry
Billingsley, Lawrence Winston, biochemistry
Birnbaum, George I, structural chemistry
Bishop, Charles Johnson, horticulture
Bishop, Claude Titus, immunochemistry, carbohydrate chemistry
Bishop, David Michael, theoretical chemistry
Biswas, Asit Kumar, environmental management
Blackadar, Robert Gordon, geology
Blackwood, Chesley M, biochemistry, food technology
Blake, Weston, Jr, glacial geology
Blevis, Bertram Charles, physics
Boch, Rolf, animal behavior, animal physiology
Bolton, Thomas Elwood, invertebrate paleontology
Boulanger, Paul, serology, veterinary medicine
Bourchier, Robert James, forestry
Bousfield, Edward Lloyd, invertebrate zoology
Bowler, Peter Aldrich, lichenology
Boyle, Robert William, economic geology, geochemistry
Bray, David Frederick, science administration, statistics
Briand, Frederic Jean-Paul, ecology
Bronskill, Joan Frances, insect histology, insect embryology
Broughton, Roger James, neurophysiology, psychobiology
Brousseau, Nicole, optics
Brown, David Lyle, cell biology
Brown, G Malcolm, internal medicine
Brown, Ira Charles, hydrology
Brown, Roger James Evan, geography, geology
Brownstein, Sydney Kenneth, organic chemistry
Brydon, James Emerson, mineralogy, soil science
Burrows, John Ronald, geophysics, space physics
Buchanan, Gerald Wallace, organic chemistry, nuclear magnetic resonance
Buckner, Charles Henry, wildlife ecology
Burba, John Vytautas, radiobiology, pharmacology
Burk, Cornelius Franklin, Jr, geology, information science
Burrows, Vernon Douglas, plant breeding
Butler, Gordon Cecil, biochemistry
Butler, Keith Winston, biophysics
Butler, Harpal Singh, pharmacology, toxicology
Byers, Philip Douglas, optics
Cann, Malcolm Calvin, organic chemistry, biochemistry
Carman, Philip Douglas, optics
Carmody, George R, population genetics
Cartier, Jean Jacques, entomology
Casselman, Warren Gottlieb Bruce, pharmacology, public health administration
Chakrabarti, Chuni Lal, analytical chemistry, inorganic chemistry
Chan, Allan P, research management, plant science
Chang, Fa Yan, weed science, plant physiology
Chapman, John Herbert, food science
Chapman, Ross Alexander, food science
Charlebois, Clarence Thomas, neuroscience, science policy
Chaudhary, Rabindra Kumar, medical microbiology
Chi, Chien Chen, plant pathology
Childers, Walter Robert, plant breeding, plant genetics
Chiykowski, Lloyd Nicholas, entomology
Cipera, John Dominik, organic biochemistry

Clark, David Sedgefield, food microbiology
Clark, John S, soil chemistry
Clark, Robert H, hydrology
Clark, Robert Vernon, plant pathology
Clarke, Arthur Haddleton, Jr, malacology
Clarke, John, historical geography
Clarke, Robert Lee, nuclear physics
Clegg, David John, toxicology, teratology
Clements, John Richard, environmental sciences
Cody, William James, botany
Cohen, Morris, physical chemistry
Coldwell, Blake Burgess, toxicology, forensic science
Collett, Leonard Stanier, geophysics
Collins, William E, urology
Colonnier, Marc, anatomy
Colvin, John Ross, biochemistry, biophysics
Conway, Brian Evans, physical chemistry
Cooch, Frederick Graham, ecology, ornithology
Cook, William Harrison, biochemistry
Copeland, Murray John, micropaleontology, invertebrate paleontology
Corlett, Michael Philip, mycology
Couture, Roger, internal medicine, nephrology
Cove, John James, anthropology, sociology
Cox, Bruce Alden, anthropology, applied anthropology
Craig, Bruce Gordon, glacial geology
Cuddy, Thomas Foster, plant physiology
Cumming, Leslie Merrill, geology
Cunningham, Hugh Meredith, food chemistry
Curren, Thomas, plant pathology
Cvetanovic, Ratimir J, physical chemistry
Daams, Herman, physics
Dale, Douglas Keith, mathematical statistics
Dauphinee, Thomas McCaul, physics
Davidson, Alexander Grant, forest pathology
Davidson, Donald West, physical chemistry
Dawson, Donald Andrew, mathematics, statistics
Dawson, Peter Henry, chemical physics
Day, Gordon Malcolm, anthropology, optics
DeBoo, Robert Ford, economic entomology, forest entomology
de Freitas, Anthony S, biochemistry, physiology
Dence, Michael Robert, geology, meteoritics
Depocas, Florent, physiology
Desai, Raman Lalbhai, physical chemistry
Deslauriers, Roxanne Marie Lorraine, physical biochemistry
Dessureau, Lionel, agronomy, genetics
Diena, Benito B, bacteriology, veterinary medicine
D'Iorio, Antoine, biochemistry
Dixon, John Douglas, pure mathematics
Dlab, Vlastimil, pure mathematics
Dobrowolski, Jerzy Adam, physical optics
Dodson, Edward O, evolutionary biology, genetics
Doe, Learmont Anstice Earlston, oceanography, meteorology
Dolenko, Allan John, organic polymer chemistry
Don, Conway J, radiology
Donaldson, John Allan, geology
Dore, William George, botany
Douglas, Alexander Edgar, molecular spectroscopy
Douglas, Robert John Wilson, geology
Downes, John Antony, entomology
Dunbar, Isobel Moira, geography
Dunn, Andrew Fletcher, experimental physics
Dunn, Donald William, mathematics
Durst, Tony, organic chemistry
Durzan, Donald John, plant physiology, biochemistry
Dyment, John Cameron, solid state physics
Eade, Kenneth Edgar, regional geology
Eagan, Charles J, physiology, biophysics
Eastham, Arthur Middleton, organic chemistry
Edwards, Kenneth Westbrook, elementary particle physics
Edwards, Oliver Edward, organic chemistry
Efford, Ian Ecott, environmental management
Eisenhauer, Hugh Ross, organic chemistry

Elliot, James I, animal nutrition, animal management
Elliott, James Angus, food microbiology
Embleton, Tony Frederick Wallace, acoustics
Emmons, Douglas Byron, agriculture, dairy industry
Emsley, Alan Burns, animal genetics
Emslie, Ronald Frank, geology, petrology
Erfle, James David, ruminant nutrition
Erickson, Lynden Edwin, experimental atomic physics, lasers
Erskine, Anthony J, wildlife biology
Fejer, Stephen Oscar, genetics
Fenton, Edward Warren, solid state physics
Fenton, M Brock, mammalogy
Fenwick, James Clarke, comparative endocrinology
Ferguson, Wilfred Samuel, soils
Fettes, James Joseph, forest entomology
Firstbrook, John Bradshaw, physiology, internal medicine
Fisher, James Wallwin, epidemiology, microbiology
Fisher, John Edwin, plant physiology
Fitt, Peter Stanley, biochemistry
Fitzgerald, Denis Patrick, geography, sociology
Fleischmann, George, plant pathology
Flood, Edward Alison, surface chemistry
Florian, Svatopluk Fred, cytology, genetics
Forman, Sydney Alexander, science policy
Forrester, Warren David, oceanography
Fortier, Yves Oscar, geology
Fortin, Emery, solid state physics
Foster, Charles David Owen, physiological chemistry
Foster, Thomas Salisbury, biochemistry
Fournier, Pierre William, medicine
Franklin, James McWillie, economic geology, stratigraphy
Frankton, Clarence, weed science
Frarey, Murray James, geology
Fraser, John Keith, geography
Fraser, Robert Rowntree, physical organic chemistry
Frebold, Hans (Wilhelm Ludwig August Herman), geology, paleontology
Frederick, George Leonard, pharmacology, toxicology
French, Warren Neil, pharmaceutical chemistry
Friedlaender, Carlo Gotthelf Immanuel, petrography, mineralogy
Fritz, William Harold, paleontology
Fulton, Robert John, geology
Furesz, John, virology
Fyles, John Gladstone, environmental geology, environmental management
Gadd, Nelson Raymond, geology
Garcia, Manuel Mariano, bacteriology
Garner, Cyril Wilbur Luther, pure mathematics
Gattinger, Richard Larry, physics, aeronomy
Gavora, Jan Samuel, poultry breeding
Geiger, Klaus Wilhelm, nuclear physics
George, Albert El Deeb, petroleum chemistry, geochemistry
Gianini, Jacqueline, probability
Gibb, Richard A, geophysics, geology
Gibson, George G, fish pathology
Gillett, John Montague, plant taxonomy
Gillis, Hugh Andrew, radiation chemistry
Gingras, Bernard Arthur, organic chemistry
Gochnauer, Thomas Alexander, microbiology
Godfrey, William Earl, ornithology
Godin, Gabriel, physical oceanography
Gold, Lorne W, engineering physics, glaciology
Goodenough, David George, space physics
Goranson, Edwin Alexander, exploration geology
Gowe, Robb Shelton, poultry breeding, animal breeding
Graham, John Elwood, statistics
Graham, Richard Charles Burwell, pharmacology
Grandchamp, Yvon, mathematics
Grant, Donald Lloyd, toxicology
Grant, Douglas Roderick, environmental geology
Gray, William MacDonald, physics
Greenhalgh, Roy, pesticide chemistry
Greig, Andrew Stephen, virology
Grice, Harold C, pathology, toxicology
Gridgeman, Norman Theodore, biomathematics, history of science
Grunder, Allan Angus, genetics
Gurd, Fraser Newman, surgery

CANADA

Hagen, Paul Beo, biochemistry, pharmacology
Halliday, Ian, meteoritics
Halstead, Ronald Lawrence, soil fertility, soil chemistry
Hamilton, Donald Gregory, cereal crops
Harcourt, Douglas George, insect ecology
Harcsar, Francis George, rubber chemistry, plastics chemistry
Hardy, James Edward, theoretical physics
Hargrove, Clifford Kingston, experimental high energy physics
Harington, Charles Richard, mammalogy
Harker, Peter, stratigraphy, paleontology
Harris, Donald C, mineralogy
Harris, Jules Eli, oncology, immunology
Harrison, James Merritt, geology
Harrison, Peter, economic geography
Hartz, Theodore Robert, magnetospheric physics
Harvey, Ross Buschlen, physical chemistry, information science
Hauschild, Andreas H W, bacteriology, food microbiology
Hawkins, Winthrop Wesley, nutrition
Heacock, Ronald A, organic chemistry, biochemistry
Heaney, David Paul, ruminant nutrition
Heggtveit, Halvor Alexander, pathology
Heroux, Olivier Joseph Paul, environmental physiology
Herzberg, Gerhard, molecular spectroscopy
Hetenyi, Geza Joseph, medical physiology
Hickman, Charles Garner, animal genetics
Hickman, John Roy, environmental health, occupational health
Hidiroglou, Michael, animal nutrition
Higgs, Lloyd Albert, radio astronomy
Hill, Donald P, pathology
Hill, Patrick Arthur, economic geology, environmental geology
Himms-Hagen, Jean, biochemistry
Hird, Brian, atomic physics
Hobson, John Peter, electron physics
Hodgson, Richard John Wesley, theoretical physics
Hoffman, Israel, analytical chemistry
Holland, George Pearson, entomology
Hollbach, Natasha Coffin, radiochemistry
Hollett, Andrew, biological chemistry
Holmes, James Murray, surface chemistry, colloid chemistry
Holmes, John Leonard, physical chemistry
Hopkins, Clarence Yardley, organic chemistry
Hopkins, Nigel John, operations research
Howden, Henry Fuller, entomology
Howland, James Lucien, mathematical analysis
Hrdina, Pavel Dusan, pharmacology
Huber, Carol (Saunderson), crystallography
Hudson, John Leslie, operations research
Hughes, David William, organic chemistry
Hughes, Stanley John, mycology
Hulan, Howard Winston, nutrition
Hurst, Andre, microbiology
Hurst, Donald Geoffrey, physics
Hutchison, William Watt, geology
Ihnat, Milan, analytical chemistry
Illman, William Irwin, mycology
Ingold, Keith Usherwood, physical organic chemistry
Irving, Edward, paleomagnetism
Ivarson, Karl C, soil microbiology
Iyer, Rajul V, microbiology
Jackson, Albert William, biochemistry
Jackson, Charles Ian, geography
Jackson, John Ranicar, pathology
Jackson, Ray Weldon, physics, science policy
Jacobson, Stuart Lee, physiology, biophysics
James, Allen Pinsent, genetics
Jan, Jean-Pierre, solid state physics
Jande, Sohan Singh, anatomy, histology
Jardine, John McNair, analytical chemistry
Jarzen, David MacArthur, palynology, paleobotany
Jenkins, Kenneth James William, biochemistry, nutrition
Jenness, Stuart Edward, geology
Jessop, Alan Michael, geophysics
Johnson, August S, genetics, physiology
Johnson, Byron F, cell biology, microbiology
Johnson, Christine Margaret, virology, molecular biology
Johnson, Frederick Allan, nuclear physics
Johnson, John Peter, physical geography
Johnson, Peter Graham, geomorphology
Johnson, Willard Jesse, biochemistry
Jones, Alister Vallance, aeronomy
Jones, Gareth Hubert Stanley, earth science
Jones, John Dewi, plant biochemistry
Jones, Richard Norman, spectrochemistry
Jorgensen, Erik, environmental management, forestry
Jost, Tadeusz Piotr, geography
Joy, Kenneth Wilfred, plant physiology
Kabayama, Michiomi Abraham, polymer chemistry
Kako, Kyohei Joe, biochemistry
Kalab, Miloslav, biochemistry
Kalsner, Stanley, pharmacology
Kaplan, Harvey, biochemistry
Kaplan, Jacob Gordin, cell biology, molecular biology
Kapron, Felix Paul, optics, solid state physics
Kates, Morris, lipid chemistry
Kell, George Sinclair, chemical physics
Kelly, Francis John, physical chemistry, science policy
Kesarwani, Roop Narain, mathematics
Kessler, Dan, high energy physics
Keys, John David, physics
Keys, Lloyd Kenneth, solid state physics
Khan, Abdul Waheed, biochemistry, microbiology
Khan, Shahamat Ullah, agricultural chemistry
Khera, Kundan Singh, toxicology, teratology
Kindle, Edward Darwin, geology
Kinson, Gordon A, physiology, endocrinology
Kirstjansson, F K, genetics
Klassen, Norman Victor, physical chemistry
Klein, Michael Lawrence, theoretical chemistry
Klotz, Max Otto, pathology
Klug, Dennis Dwayne, physical chemistry
Kodama, Hideomi, mineralogy
Koningstein, Johannes A, chemistry, chemical physics
Korecky, Borivoj, medical physiology
Kovacs, Bela A, pharmacology, allergy
Kovacs, Eve Maria, medicine
Kovar, Jan Bernard, carbohydrate chemistry, physical organic chemistry
Kraay, Gerrit Jacob, genetics
Kramer, John Karl Gerhard, biochemistry, organic chemistry
Kretz, Ralph, petrology, geochemistry
Kruus, Peeter, physical chemistry
Kushner, Donn Jean, microbial biochemistry, microbial physiology
Kutschke, Kenneth Otto, photochemistry
Laham, Quentin Nadime, histology, embryology
Laham, Souheil, toxicology, cancer
Laidler, Keith James, physical chemistry
Lamarche, J L Gilles, physics
Lang, Arthur (Hamilton), economic geology
Langford, Cooper Harold, III, physical inorganic chemistry
Last, John Murray, epidemiology
Laughton, Paul MacDonell, organic chemistry
Lauzier, Louis Marcel, oceanography
Layne, Donald Sainteval, biochemistry
LeBlanc, Fabius, bryology, ecology
Leblanc, Gabriel, geophysics, seismology
Le Blanc, Marcel A R, physics
Lee, Kotik Kai, mathematical physics
Lee, Peter E, plant virology
Lee, Robert K S, marine botany
Leech, Geoffrey Bosdin, geology
Leese, Eric Leslie, mathematics
Lefkovitch, Leonard Philip, entomology, statistics
Legg, Thomas Harry, physics, radio astronomy
Leitch, Leonard Christie, organic chemistry
Lennox, Donald Haughton, hydrology
LeRoy, Donald James, chemical kinetics, photochemistry
Lewis, Trevor John, geophysics
Lewis, Wilfrid Bennett, physics
Lindquist, Evert E, acarology, systematic entomology
Lindsey, George Roy, systems analysis
Ling, George M, pharmacology
Linis, Viktors, mathematics
Lips, Hilaire John, chemistry
Lipsett, Frederick Roy, forensic science
Lister, Earl Edward, ruminant nutrition
Little, Heward Wallace, economic geology
Locke, Jack Lambourne, physics
Logan, Brian Anthony, nuclear physics
Logan, James Edward, clinical chemistry
Loiselle, Roland, plant breeding
Lossing, Frederick Pettit, chemical physics
Love, George Ross, electricity
Lovering, Edward Gilbert, pharmaceutical chemistry
Lucis, Ojars Janis, endocrinology, environmental health
Lucis, Ruta, comparative endocrinology
Ludwig, Ralph Antony, plant pathology
Lusena, Charles V, molecular biology
MacDonald, Stewart Ferguson, organic chemistry
Macdowall, Fergus D H, plant physiology
MacHattie, Leslie Blake, meteorology
Mack, Stanley Zaner, physics
MacLachlan, Donald Stuart, plant pathology
Maclure, Kenneth Cecil, nuclear physics
Macmanus, John Patrick, biochemistry
Macphail, Moray St John, mathematics
Madhosingh, Clarence, mycology, biochemistry
Malaiyandi, Murugan, organic chemistry, analytical chemistry
Maloney, James Eugene, operations research, forest economics
Manchee, Eric Best, geophysics, science administration
Mandl, Paul, applied mathematics
Mannell, William Arnold, toxicology
Manoogian, Armen, physics
Marcus, George Jacob, reproductive biology
Marion, Leo (Edmond), natural products chemistry
Marshall, William Deforrest, pesticide chemistry
Martin, Douglas Leonard, metal physics, thermal physics
Martin, Stanley Morris, biochemistry, microbiology
Martin, William Gerald, physical chemistry, biochemistry
Martin, William Robert, fish biology
Mason, William Richardson Miles, systematic entomology
Matheson, Alastair Taylor, biochemistry, cell biology
Mavrides, Charalampos, biochemistry
Mavroyannis, Constantine, theoretical solid state physics
Maxwell, John Alfred, geochemistry
Maykut, Madelaine Olga, clinical pharmacology
Mazurkiewicz-Kwilecki, Irena Maria, pharmacology
McAllister, Alan Jackson, animal breeding
McAllister, Donald Evan, systematic ichthyology
McAlpine, James Francis, entomology
McCully, Margaret E, biology
McDiarmid, Donald Ralph, ionospheric physics
McDiarmid, Ian Bertrand, space physics
McDonald, Ian Johnson, microbial physiology
McGregor, Duncan Colin, palynology
McGugan, Wesley Alexander, food chemistry
McIntosh, Bruce Andrew, physics, electronics
McKeague, Justin Alexander, soil genesis, soil classification
McKiel, John Albert, microbiology
McKinley, Donald William Robert, aeronomy
McLaren, Digby Johns, geology
McMullen, Robert Michael, sedimentology, information science
McNamara, Allen Garnet, aeronomy
McNeill, John, botany
McPhail, Murchie Kilburn, physiology, biochemistry
Meek, Jack Henry, physics
Mendoza, Celso Enriquez, biochemistry, entomology
Mercer, Malcolm Clarence, fisheries management, systematic zoology
Merriam, Howard Gray, ecology
Merrill, Gordon Clark, geography
Merritt, Edison S, population genetics, poultry breeding
Meunier, Jean-Louis, experimental nuclear physics, experimental solid state physics
Meyboom, Peter, hydrogeology
Michael, Thomas Hugh Glynn, chemistry, science administration
Middleton, Edward James, toxicology
Migicovsky, Bert Baruch, biochemistry
Miller, Charles Douglas F, entomology
Miller, Donald Richard, mathematical biology
Miller, Richard Wilson, biochemistry, enzymology
Milne, John B, inorganic chemistry
Mongeau, J Denis, veterinary medicine, bacteriology
Moon, Thomas William, comparative physiology
Moore, Charles Wayne, economic geography
Moore, John Marshall, Jr, geology, petrology
Moore, Raymond John, cytotaxonomy
Morand, Peter, organic chemistry
Mori, Kanaka Fred, biochemistry, endocrinology
Morita, Hirokazu, organic chemistry
Morley, Harold Victor, pesticide chemistry, environmental chemistry
Morley, Lawrence Whitaker, geophysics
Morris, Derek, metrology, quantum physics
Morrison, Alexander Baillie, toxicology, nutrition
Morrison, Jack William, cytogenetics
Morrow, Barry Albert, physical chemistry
Mortimer, Donald Charles, plant physiology
Morton, Helen Janet, biochemistry, cytology
Mosquin, Theodore, botany, taxonomy
Mountain, William Buckingham, nematology
Mufti, Izhar-Ul Haq, applied mathematics
Mungall, Allan George, experimental atomic physics
Munro, David (Aird), biology
Munro, James, plant pathology
Munroe, Eugene Gordon, entomology, ecology
Murphy, William Frederick, molecular spectroscopy
Murray, Beatrice E, cytogenetics
Murty, Tadepalli Satyanarayana, physical oceanography
Nagai, Jiro, animal breeding
Napke, Edward, medicine, physiology
Narang, Saran A, organic chemistry, molecular biology
Narbaitz, Roberto, embryology
Nasim, Mohammed Anwar, genetics
Neelin, James Michael, biochemistry
Nel, Louis Daniel, topology
Nesbitt, Herbert Hugh John, entomology
Newnham, Robert Montague, forest management, forest mensuration
Niblett, Edward Ronald, geophysics
Nigam, Prakash Chandra, insect toxicology, serology
Nilson, John Anthony, lasers
Norminton, Edward Joseph, applied mathematics
Northover, Francis Henry, applied mathematics
Northwood, Thomas David, acoustics, seismology
Nozzolillo, Constance, plant biochemistry
Offord, David Robert, child psychiatry
O'Shea, Seamus Francis, theoretical chemistry
Paghis, Irvine, radio physics, aeronomy
Panalaks, Thavil, analytical chemistry, food chemistry
Panarella, Emilio, physics
Pandey, Jagdish Narayan, mathematics
Park, Chong Eel, food microbiology
Parmelee, John Aubrey, mycology
Patel, Girishchandra Babubhai, food microbiology
Paynter, Kenneth Jack, anatomy
Peakall, David B, environmental sciences
Peck, Stewart Blaine, entomology, evolutionary biology
Pelletier, Omer, analytical biochemistry, nutritional biochemistry
Perrault, Marcel Joseph, animal physiology, endocrinology
Perry, Malcolm Blythe, biochemistry
Peters, Hobart Frank, animal breeding
Peterson, Bobbie Vern (Robert), entomology
Petruk, William, geology
Pfalzner, Paul Michael, medical physics
Phillips, William Ernest John, animal nutrition
Pigden, Wallace James, animal nutrition
Pirozynski, Krzysztof Andrzej, paleobiology
Pivnick, Hilliard, bacteriology
Plessers, Arthur Gerard, plant genetics
Poirier, Rolland Paul, genetics
Poland, John C, algebra
Polley, John Richard, biochemistry
Polnaszek, Carl Francis, biophysical chemistry
Poole, William Hope, regional geology
Porsild, Alf Erling, botany

Posen, Gerald, internal medicine, nephrology
Potvin, Georges Charles, urban geography
Prasad, Raghubir (Raj), plant physiology
Pressman, Elaine Steingarten, speech pathology
Pressman, Irwin Samuel, mathematics
Prest, Victor Kent, geology
Preston-Thomas, Hugh, experimental physics
Prince, Alan Theodore, chemistry, geology
Pringle, Ross Barton, biochemistry
Pritchard, G Ian, fisheries
Proulx, Pierre R, biochemistry
Przybylska, Maria, x-ray crystallography
Puddington, Ira Edwin, chemistry
Pullan, George Thomas, research administration
Puttaswamaiah, Bannikuppe M, pure mathematics
Quinn, John R, food science
Quon, David Shi Haung, mineralogy
Racine, Michel Louis, algebra
Rajhathy, Tibor, plant genetics & cytology
Rakusan, Karel Josef, physiology
Ramamoorthy, Subramaniam, environmental chemistry
Ramsay, Donald Allan, molecular spectroscopy
Rapport, David Joseph, ecology, theoretical biology
Ray, Ajit Kumar, applied mathematics, aerodynamics
Ray, David Michael, economic geography, geography of Canada
Rayman, Mohamad Khalil, food microbiology
Read, Donald Earle, organic chemistry
Reddoch, Allan Harvey, chemical physics
Redhead, Paul Aveling, physics
Redmond, Douglas Rollen, forestry
Reed, Robert Marshall, plant ecology
Reesor, John Elgin, geology
Rennie, Peter John, soil science
Resnick, Lazer, particle physics, theoretical physics
Ribes, Luis, algebra
Richter, Maxwell, immunology, pathology
Ricour-Singh, Francoise, urban geography
Riel, Rene Rosaire, food science
Ritcey, Gordon M, inorganic chemistry, organic chemistry
Rixon, Raymond Harwood, physiology
Robertson, Hamish Alexander, reproductive endocrinology
Roberts-Pichette, Patricia Ruth, ecology
Robinson, Donald Barker Wellington, organic chemistry
Rogers, Charles Graham, biochemistry, microbiology
Rolfe, John, solid state physics
Rolleston, Francis Stopford, biochemistry
Romo, William Joseph, physics
Roots, Ernest Frederick, geology
Rose, Dyson, microbiology
Rossier, Edmond, medical microbiology
Rothschild, Henri Charles, bionucleonics
Rowsell, Harry Cecil, veterinary pathology
Roxburgh, James Maxwell, biomedical engineering
Ruckerbauer, Gerda Margareta, veterinary immunology
Russell, Dale A, vertebrate paleontology
Russell, Douglas Stewart, analytical chemistry
Rust, Velma Irene, mathematical statistics, econometrics
Ryan, Michael T, biochemistry, polymer chemistry
Sabry, Zakaria I, nutrition, biochemistry
Sadler, Arthur Graham, chemistry, ceramics
Saha, Jadu Gopal, pesticide chemistry
Saidak, Walter John, weed science
Saleh, Wasfy Seleman, surgery, urology
Salkfeld, E Helen, insect physiology, histochemistry
Sampson, Dexter Reid, genetics, plant breeding
Sanders, Corey Leroy, photometry
Sanderson, Edwin S, organic chemistry, polymer chemistry
Sandi, Emil, analytical chemistry, toxicology
Sarwer-Foner, Gerald, psychiatry
Sattar, Syed Abdus, animal virology, water pollution
Savage, Robert Gilmore, agronomy
Savile, Douglas Barton Osborne, botany
Schirmer, Helga H, mathematics
Schneider, Henry, physical chemistry, biochemistry

Schneider, William George, physical chemistry
Schnitzer, Morris, soil chemistry, organic chemistry
Schwartz, Harry, chemistry
Scott, John Stanley, geology
Scott, Peter Michael, organic chemistry
Seaman, William Lloyd, plant pathology
Sellers, Frank Jamieson, pediatric cardiology
Serson, Paul Horne, magnetism
Setterfield, George Ambrose, cytology
Shaw, Edgar Albert George, physics
Shearer, Duncan Allan, analytical chemistry
Sheffer, Harry, physical chemistry
Sherman, Norman K, nuclear physics, radiation physics
Shigeishi, Ronald A, physical chemistry
Shih, Chang-Tai, systematics, biological oceanography
Shoemaker, Robert Alan, mycology
Sibbald, Ian Ramsay, nutrition
Sida, Derek William, applied mathematics, astronomy
Siddiqui, Iqbal Rafat, organic chemistry
Siebrand, Willem, chemical physics, theoretical chemistry
Siminovitch, David, plant physiology
Simmonds, Walter Henry, chemical engineering
Simpson, John Hamilton, theoretical solid state physics, solid state electronics
Sims, Richard Paul Andrew, chemistry
Singh, Surinder Shah, soil science, soil chemistry
Singhal, Radhey Lal, pharmacology
Sinha, Raj P, microbial genetics, biochemical genetics
Sinha, Ramesh Chandra, plant pathology
Siranni, Aurelio Frederick, colloid chemistry, surface chemistry
Sirois, Jean Claude, plant physiology
Sistek, Vladimir, anatomy, surgery
Somers, Emmanuel, chemistry
Somorjai, Rajmund Lewis, theoretical biology, biomathematics
Small, Ernest, biosystematics
Smetana, Ales, entomology, zoogeography
Smith, Charles Haddon, geology
Smith, Derek George, anthropology
Smith, Donald Alan, vertebrate zoology
Smith, Ian Cormack Palmer, biophysics
Smith, Lorraine Catherine, vertebrate biology
Smith, Robert Clinton, theoretical physics
Smith, Walter George, toxicology
Solman, Victor Edward Frick, zoology, limnology
Song, Kong-Sop, solid state physics
Soper, James Herbert, botany
Sowden, Frederick John, soil chemistry
Speller, Stanley Wayne, wildlife biology
Sprout, Gordon Dennis, microbiology
Srivastava, Harishankar Prasad, geography of South Asia, political geography
Stalker, Archibald MacSween, glacial geology
Stavric, Bozidar, toxicology, pharmacology
Stephenson, Norman Robert, biochemistry, science administration
Sternberg, Charles Mortram, vertebrate paleontology
Stevenson, Ian Lawrie, agricultural microbiology, cytology
Stevenson, Ira Morley, geology
Stevenson, James Cameron, marine biology
Stewart, Thomas Henry McKenzie, internal medicine, immunology
Stockton, Gerald William, molecular biophysics, nuclear magnetic resonance
Stockwell, Clifford Howard, geology
Stolow, Nathan, chemistry
Stone, John Ernest, physical chemistry
St-Onge, Denis Alderic, geomorphology
Strang, Robert M, forest ecology
Straus, Jozef, experimental solid state physics
Sutton, Roscoe Murray Davidson, phytopathology, virology
Svejda, Felicitas Julia, ornamental horticulture, plant breeding
Swain, Harry Sheldon, urban geography, research management
Szabo, Alexander, solid state physics
Szabo, Arthur Gustav, organic chemistry
Szyrynski, Victor, psychiatry
Talbot, Bernard, rehabilitation medicine
Tan, Peter Ching-Yao, mathematical statistics, civil engineering
Tanner, James Gordon, geophysics

Tate, Parr Allen, research administration
Taylor, David Ruxton Fraser, geography
Taylor, Gordon, geography
Taylor, Harold Ernest, pathology
Taylor, William Ewart, anthropology, archaeology
Templeton, Ian M, metal physics, low temperature physics
Thiessen, George Jacob, physics
Thomas, Michael David, geophysics
Thompson, Hazen Spencer, plant pathology
Thomson, Keith Patrick Bowmer, physics
Thorpe, Ralph Irving, economic geology
Tickner, Alfred William, physical chemistry, research administration
Todd, Ewen Cameron David, microbiology
Tolnai, Susan, cell biology
Trenholm, Harold Locksley, biochemistry, enzymology
Tross, Ralph G, mathematical physics
Tryphonas, Leander, veterinary pathology, toxicology
Tsai, Chishiun S, biochemistry, enzymology
Tsang, Charles Pak Wai, endocrinology
Tupper, William Macgregor, geology, geochemistry
Turner, Robert Chapman, physical chemistry
Vaartaja, Olli, forest biology
Vaillancourt, Remi Etienne, mathematics
Valadares, Joseph R E, pharmacology, biochemistry
Van Den Berg, L, food science
Van Denheuvel, Franz Aime, organic chemistry, biochemistry
Varshni, Yatendra Pal, astrophysics, solid state physics
Veliky, Ivan Alois, biological chemistry
Vikis, Andreas Charalambous, chemistry
Vladykov, Vadim Dmitrij, taxonomy, ichthyology
Vogelfanger, Isaac Joel, experimental surgery
Wallen, Victor Reid, plant pathology
Warren, Francis Shirley, agronomy
Watanabe, Akira, communications sciences
Watkin, John Emrys, plant biochemistry, environmental chemistry
Watson, Gordon Dulmage, anthropology, archaeology
Watson, Jeffrey, resource management, information science
Watson, Michael Douglas, aeronomy
Waugh, Douglas Oliver William, medical education, pathology
Weber, Jean Robert, geophysics
Weichert, Dieter Horst, seismology
Weinberger, Pearl, plant physiology, environmental biology
Weir, John Robert, environmental biology
Wellar, Barry Sheldon, urban geography, communications
Weresub, Luella Kayla, mycology
Wesche, Rolf Juergen, geography
Westland, Alan Duane, inorganic chemistry
Westwood, William Dickson, physics
Whalley, Edward, physical chemistry
Wheeler, John Oliver, geology
Whitaker, Donald Robert, biochemistry
Whitehead, James Rennie, physics
Whitfield, James F, cell physiology, cancer
Whitham, Kenneth, geophysics
Wiberg, George Stuart, toxicology
Wigfield, Donald Compston, organic chemistry
Wiggin, Norman Jack Bridgman, immunology
Wightman, Frank, plant physiology
Wightman, Robert Harlan, organic chemistry
Wiles, David M, polymer chemistry
Wiles, Donald Roy, inorganic chemistry
Wilkes, Alfred, biology
Wilkinson, Robert Foster, physical chemistry
Wilkinson, Thomas Preston, physical geography
Williams, David Edward, forestry
Williams, David Trevor, environmental chemistry
Williams, Digby Frederick, physical chemistry, solid state chemistry
Williams, Kenneth Stuart, mathematics
Williamson, Denis George, biochemistry
Wilner, Jacob, plant physiology
Wilson, Andrew Hastie, science policy, mechanical engineering
Winthrop, Stanley Oscar, organic chemistry
Wohl, Philip R, mathematics

Wolfson, Joseph Laurence, nuclear physics
Wong, Patrick T T, physical chemistry
Woodward, Harry W, petroleum geology
Woodward, James Crawford, animal nutrition
Wright, James Sherman, theoretical chemistry
Wyszecki, Gunter, physics, mathematics
Yaguchi, Makoto, protein chemistry, agricultural chemistry
Yamazaki, Hiroshi, biochemistry
Yates, Alfred Randolph, food microbiology
Yip, Roderick Wing, physical chemistry, organic chemistry
Yole, Raymond William, stratigraphy, sedimentology
Yoshimoto, Carl Masaru, entomology
Young, Beverley George, physics
Young, James Christopher F, analytical chemistry, entomology
Zuker, Michael, biomathematics

PETERBOROUGH
Adams, William Peter, geography, glaciology
Alfred, Louis Charles Roland, solid state physics, physical metallurgy
Annett, Robert Gordon, biochemistry
Barrett, Peter Fowler, inorganic chemistry
Berrill, Michael, animal behavior
Brown, Stewart Anglin, plant biochemistry
Cheung, Tsun-Sung Happy, marine biology
Earnshaw, John W, electron physics
Edwards, Roy Lawrence, ecology
Guinand, Andrew Paul, mathematics
Harrison, Peter D'Arcy, archaeology, anthropology
Helmuth, Herman Siegfried, physical anthropology, primatology
Henniger, James Perry, mathematics
Hubbell, Linda Jean, anthropology, ethnology
Johnson, Ronald Gordon, nuclear physics, biophysics
Jones, Roger, botany
Kidd, Kenneth E, anthropology
Lewars, Errol George, organic chemistry
Lodge, John 1, physics
March, Raymond Evans, physical chemistry
Nighswander, James Edward, plant pathology
Oldham, Keith Bentley, electrochemistry, physical chemistry
Powles, Percival Mount, marine ecology
Rees, Alun Hywel, organic chemistry
Stairs, Robert Ardagh, physical chemistry
Vastokas, Joan M, anthropology

PORCUPINE
Carlson, Hugh Douglas, economic geology

PORT ARTHUR
Graham, William Muir, ecology, insect behavior

PRESCOTT
Grainger, Robert Moore, dentistry, operations research

REXDALE
Barringer, Anthony R, geology, geophysics
Raheja, Manu Chatrumal, biochemistry, science administration

RICHMOND HILL
Bolton, Charles Thomas, astronomy
Fernie, John Donald, astronomy
Garrison, Robert Frederick, astronomy, astrophysics
Heard, John Frederick, astronomy
Hogg, Helen (Battles) Sawyer, astronomy
MacKae, Donald Alexander, astronomy
Stainer, Dennis William, biochemistry, microbiology

RUTHERGLEN
Lawrence, Louise de Kiriline, ornithology

ST CATHARINES
Auer, Jan Willem, mathematics
Banfield, Alexander William Francis, zoology
Bell, Howard E, mathematics

CANADA

Black, John Earle, solid state physics
Chang, Shao-chien, mathematics
Chermak, Eugene Anthony, physical chemistry
Denton, Trevor, applied anthropology, social anthropology
Finlay, Gordon Roy, inorganic chemistry
Flint, Jean-Jacques, geomorphology, hydrology
Hartman, John Stephen, inorganic chemistry
Haynes, Simon John, economic geology
Headley, Velmer Bentley, mathematics
Hiatt, Richard Rowls, physical organic chemistry
Houston, Arthur Hillier, environmental physiology
Jackson, John Nicholas, urban geography, urban planning
Jolly, Wayne Travis, petrology, volcanology
Koffyberg, Francois Pierre, solid state science
Lepard, David William, molecular physics
Liberty, Bruce Arthur, stratigraphy
Manocha, Manmohan Singh, mycology, plant pathology
Mayberry, John Patterson, operations research
Miller, Jack Martin, inorganic chemistry
Morris, Ralph Dennis, ecology
Moule, David, physical chemistry, molecular spectroscopy
Muller, Eric Rene, mathematics, physics
Peach, Peter Angus, geology
Plint, Colin Arnold, physics
Richardson, Mary Frances, inorganic chemistry, crystallography
Sankey, Charles Alfred, chemistry
Terasmae, Jean, geology, palynology
Tracy, Martin Louis, Jr, genetics
Ursino, Donald Joseph, plant physiology, radiation botany

SARNIA
Berkoff, Robert Bernard, chemistry
Breitman, Leo, polymer chemistry, physical chemistry
Buckler, Ernest Jack, chemistry
Buckley, Bernard Patrick, physical chemistry
Caesar, Cameron Hull, physical chemistry
Chiang, Schumann, chemistry
Dunn, John Robert, physical chemistry
Edwards, Douglas Cameron, rubber chemistry
Favis, Dimitrios Vasilios, physical chemistry, industrial chemistry
Ford, Richard Westaway, industrial chemistry
Gilbert, John Barry, physical chemistry
Glew, David Neville, physical chemistry
Golemba, Frank John, polymer chemistry, photochemistry
Henderson, John Frederick, physical chemistry, polymer chemistry
McCoubrey, James A. environmental chemistry
Rothenbury, Raymond Albert, academic administration
Sheppard, Allen Anson, analytical chemistry, petroleum chemistry
Walker, Jack, chemical engineering
Watts, Harry, physical chemistry
White, William Harold, chemistry, organic chemistry
Young, David Matheson, chemistry

SAULT STE MARIE
Angus, Thomas Anderson, insect pathology
Basham, Jack Tucker, forestry, botany
Cunningham, John Castel, insect pathology
Fast, Paul Gerhardt, entomology
Kondo, Edward Shin-Chi, forest pathology
Retnakaran, Arthur, insect physiology
Sippell, William Lloyd, forest entomology
Tibbles, John James, fisheries
Tyrrell, David, biochemistry
Whitney, Roy Davidson, forest pathology

SCARBOROUGH
Diosy, Andrew, internal medicine
Esack, Ashneed, organic chemistry, computer science
French, Ian Wilfred, biochemistry
Reynolds, John Keith, wildlife management

SHERIDAN PARK
Campbell, Hugh John, organic chemistry
Jones, Maurice Harry, physical chemistry, organic chemistry
Lautenschlaeger, Friedrich Karl, organic chemistry
Manchester, Donald Fraser, organic chemistry
Schwartz, Norman Vincent, rubber chemistry
Sowa, Walter, organic chemistry
Whittier, Angus Charles, physics
Zakaib, Daniel D. petroleum chemistry, analytical chemistry

SIMCOE
Kerr, Ernest Andrew, plant breeding
Loughton, Arthur, horticulture
Proctor, John Thomas Arthur, horticulture, plant physiology

STITTSVILLE
Lee, Hulbert Austin, geology

STRATHROY
Adams, Gordon Albert, microbial biochemistry
MacLulich, Duncan Alexander, wildlife biology

SUDBURY
Alikhan, Muhammad Akhtar, animal physiology, biochemistry
Cameron, Robert Alan, exploration geology
Card, Kenneth D. geology
Copper, Paul, paleontology
Davies, James Frederick, economic geology, petrology
Demers, Serge, astronomy
Goldsack, Douglas E. biophysical chemistry
Hilldrup, David J. mathematics, operations research
Ho, Shung Pun, mathematics
James, Richard Stephen, geochemistry, petrology
Kaye, Brian H. physics
Lakshman, Akarasi Bhojaraj, endocrinology, anatomy
Leclaire, Roger, astronomy, physics
Richardson, David H S. botany
Rousell, Don Herbert, structural geology
Rubin, G A. physics
Sadana, Yoginder Nath, inorganic chemistry
Todd, Leonard, theoretical mechanics
Toni, Youssef Tanious, geography
Westaway, Kenneth C. physical organic chemistry
Ziauddin, Syed, atmospheric physics, space physics

THORNBURY
Chapman, Lyman John, geomorphology, climatology

THORNHILL
Ferguson, James Kenneth Wallace, pharmacology
Ott, Welland Lee, analytical chemistry

THOROLD
Logan, Charles Donald, wood chemistry
Lyne, Leonard Murray, Sr, paper chemistry

THUNDER BAY
Allen, Gordon Ainslie, environmental management, pulp chemistry
Colby, Peter J. aquatic biology, fisheries
Eames, William, mathematical analysis
Harvais, Gaetan Hugues, mycology
Hawton, Larry David, physical chemistry
Kent, Clement F. mathematics
Lindsay, Douglas Rome, plant ecology, taxonomy
Macdonald, Alastair David, botany, morphology
Momot, Walter Thomas, fish biology
Mothersill, John Sydney, geology
Naimpally, Somashekhar Amrith, topology
Ozburn, George W. entomology, biology
Phillips, Brian Antony Morley, physical geography, cartography
Campbell, Ian Maclean, entomology, genetics

TILLSONBURG
Povilaitis, Bronius, genetics, plant breeding

TORONTO
Abrams, John Werner, history of science, operational research
Ackerman, Uwe, mammalian physiology
Ackles, Kenneth Norman, physiology
Adamek, Eduard George, environmental management, air pollution
Alberti, Peter W R M. otolaryngology
Albisser, Anthony Michael, biomedical engineering
Allan, Robert K. biochemistry
Allen, A D. inorganic chemistry
Allin, Elizabeth Josephine, physics
Anderson, Gregor Munro, geochemistry
Andrews, David F. statistics
Anwar, Rashid Ahmad, biochemistry
Armitage, John Briggs, pediatric neurology, biochemistry
Armstrong, Robin L. nuclear magnetic resonance
Ash, Clifford L. radiology
Ashworth, Murray Alexander, human physiology
Atcheson, J D. psychiatry
Atwood, Carl Edmund, entomology
Atwood, Harold Leslie, physiology
Axelrad, Arthur Aaron, hematology, histology
Badenhuizen, Nicolaas Pieter, botany
Baer, Erich, organic chemistry
Baines, A D. physiology, clinical biochemistry
Baird, Ronald James, cardiovascular surgery
Baker, Michael Allen, hematology, cancer
Barbeau, Edward Joseph, Jr, pure mathematics
Barlow, Jon Charles, ornithology
Basu, Prasanta Kumar, ophthalmology
Baxter, Ross M. pharmacy
Bayly, George Henry Uniacke, geology
Beales, Francis William, geology
Beall, Francis Carroll, wood science
Beamish, Fred Earl, analytical chemistry
Beaton, George Hector, nutrition, biochemistry
Bell, Alexander Graham, human genetics
Berger, Jacques, protozoology, invertebrate zoology
Bergsagel, Daniel Egil, internal medicine, oncology
Berris, Barnet, internal medicine
Bersohn, Malcolm, synthetic organic chemistry, computer sciences
Bertram, Ewart George, neuroanatomy
Best, Charles Herbert, physiology
Bharucha, Keki Rustomji, organic chemistry
Bhavnani, Bhagu R. biochemistry, pharmaceutical chemistry
Blair, Alexander Marshall, geography
Bonkalo, Alexander, psychiatry, neurophysiology
Borth, Rudi, endocrinology, biometrics
Boyer, Michael George, plant pathology
Brainerd, Barron, mathematics
Breckenridge, Carl, clinical biochemistry
Brewer, Alan West, atmospheric physics
Broder, Irvin, immunology
Brook, Adrian Gibbs, organic chemistry
Brown, John Rhys, physiology, toxicology
Brown, Milton Herbert, microbiology
Brummer, Johannes J. economic geology, exploration geology
Buckingham, Forrest Morgan, hydrology
Bunting, John William, organic chemistry, biochemistry
Burgess, William Howard, applied chemistry
Burns, George, chemistry
Burrow, Gerard N. internal medicine
Butler, Leonard, genetics
Cain, Roy Franklin, botany
Callahan, John William, biochemistry, neurochemistry
Camerman, Norman, molecular biology, x-ray crystallography
Cameron, Duncan MacLean, Jr, zoology, ecology

Ross, Robert Anderson, surface chemistry, applied chemistry
Vervoort, Gerardus, communications science, mathematics education
Whitfield, John Howard Mervyn, mathematics

Campbell, James, physiology, biochemistry
Campbell, James B. medical microbiology
Caplan, Paula Joan, neuropsychology
Carington, Tucker, physical chemistry
Carswell, Allan Ian, optical physics
Chant, Donald A. entomology
Chisholm, Alexander James, theoretical physics
Cho, Han-Ru, dynamic meteorology
Chou, Chen-Lin, geochemistry
Christensen, Lauritz Royal, pathology
Christie, Alistair D. atmospheric physics, meteorology
Churcher, Charles Stephen, vertebrate paleontology, mammalogy
Cinader, Bernhard, immunochemistry
Clark, A Gavin, microbiology
Clark, Gordon Murray, radiation biology
Clarke, Donald Walter, biochemistry
Clarke, W T W. internal medicine
Clement, Maurice James, theoretical physics
Cobbold, R S C. biomedical engineering
Coceani, Flavio, neurophysiology
Cohen, Saul Louis, biochemistry, endocrinology
Collins, Desmond H. paleontology
Collins-Williams, Cecil, pediatrics, clinical immunology
Conen, Patrick E. pathology
Connell, George Edward, biochemistry
Corbon, Herbert Charles, theoretical physics
Cox, Diane Wilson, human genetics
Coxeter, Harold Scott MacDonald, mathematics
Cradduck, Trevor David, medical physics
Crawford, John S. internal medicine, rehabilitation medicine
Crookston, John Hamill, hematology
Crookston, Marie Cutbush, immunohematology
Cross, Charles Kenneth, analytical chemistry
Crossman, Edwin John, ichthyology
Cruickshank, Bruce, anatomic pathology
Cruise, James E. plant taxonomy
Currie, John Bickell, geology
Curry, Leslie, geography
Daintly, Jack, plant physiology
Daniels, James Maurice, solid state physics, magnetism
Davidson, Charles Mackenzie, food microbiology
Davis, Chandler, mathematics
Davis, Gerald Gordon, polymer chemistry
Dean, William George, geography
Deber, Charles Michael, biological chemistry
Deckers, Jacques (Marie), physical chemistry
DeLury, Daniel Bertrand, mathematics, statistics
Derry, Duncan Ramsay, geology
Derzko, Nicholas Anthony, applied physics
Desai, Rashmi C. mechanics, chemical physics
Desser, Sherwin S. zoology, parasitology
Dignam, Michael John, surface chemistry, electrochemistry
Doane, Frances Whitman, virology, electron microscopy
Donohue, William Leslie, pathology
Donovan, Ross Grant, biochemistry
Dorrington, Keith John, biochemistry, immunology
Dove, John Edward, chemical kinetics
Dubiski, Stanislaw, immunogenetics
Duesing, Constantin Michael, immunochemistry
Duff, George Francis Denton, mathematics
Duncan, Gerald R. pharmacy
Duncan, I B R. medical microbiology
Dunford, Raymond A. analytical chemistry
Dunkley, Colleen Rose, nutrition, biochemistry
Dyer, Alan Edwin, pharmacology
Dymerski, Paul Peter, physical chemistry
Ebbs, J Harry, medicine, pediatrics
Edmund, Alexander Gordon, vertebrate paleontology
Effer, W R. plant physiology
Egan, Thomas J. medicine, pediatrics
Eilers, Erich Werner, pure mathematics
Evans, Nancy Remage, astronomy
Ezrin, Calvin, endocrinology, internal medicine
Fallis, Albert Murray, parasitology
Falls, James Bruce, animal behavior, ecology

Farber, Emmanuel, biochemistry, pathology
Farkas-Himsley, Hannah, bacteriology, microbiology
Farquhar, Ronald McCunn, geochronology
Farrar, John Laird, forestry
Fawcett, Eric, experimental solid state physics, metal physics
Fawcett, James Jeffrey, geology
Feuer, George, biochemical pharmacology, pathological chemistry
Field, Neil Collard, geography
Filseth, Stephen V, physical chemistry
Fisher, Albert Madden, chemistry
Forstner, Janet Ferguson, medical research
Forward, Dorothy Florence, plant physiology
Fowle, Charles David, vertebrate ecology, wildlife management
Franklin, Arthur Edmund, biochemistry
Fraser, Donald, physiology
Fraser, Donald Alexander Stuart, statistics
Fredrickson, John Murray, otolaryngology, neurophysiology
Freedman, Murray H, biochemistry, immunology
Freeman, Reino Samuel, helminthology, parasitology
Fricke, Werner, environmental physics
Friend, William George, insect physiology
Frisken, William Ross, particle physics
Fritz, Irving Bamdas, reproductive biology, biochemistry
Fritz, Madeleine Alberta, paleontology
Fry, Frederick Ernest Joseph, ecology
Gaherty, Geoffrey George, physical anthropology, archaeology
Gait, Robert I, geology, mineralogy
Garland, George David, geophysics
Gittins, John, geology
Gold, Marvin, biochemistry
Goodings, John Martin, physical chemistry
Goodridge, Alan G, biochemistry
Gornall, Allan Godfrey, biochemistry
Graydon, William Frederick, physical chemistry
Grayson, John, physiology
Grear, John Wesley, Jr, botany, taxonomy
Greiner, Peter Charles, mathematics
Griffin, Peter Allan, solid state physics
Guillet, James Edwin, physical chemistry
Gyulai, Eugene Jeno, virology, immunology
Haist, Reginald Evan, physiology
Hallett, Archibald Cameron Hollis, physics
Halperin, Israel, mathematics
Ham, Arthur Worth, histology
Hamilton, John Drennan, pathology
Hare, Frederick Kenneth, geography, meteorology
Harrison, Alexander George, physical chemistry, organic chemistry
Harrison, Joan Elizabeth, medicine
Harvey, Harold H, limnology, fisheries
Hawke, W A, pediatrics, psychiatry
Haynes, Robert Hall, biophysics
Heath, Michele Christine, phytopathology
Hebert, Gerard Rosaire, physics
Heddle, John A M, radiobiology
Heilbronn, Hans Arnold, pure mathematics
Hellebust, Johan Arnvid, algology
Hewitt, David, medical statistics
Hewitt, Donald Francis, geology
Hibberd, John Hunt, restorative dentistry
Higgins, Verna Jessie, plant pathology
Hines, Colin Oswald, aeronomy
Hobson, Robert Marshall, physics
Hockman, Charles Henry, neurophysiology, neuropharmacology
Hoffman, William Howard, cereal chemistry, food technology
Hofmann, Theo, biochemistry, molecular biology
Hollenberg, Charles H, medicine
Holyk, Walter, geology
Horner, Alan Alfred, biochemistry
Hornykiewicz, Oleh, pharmacology, neurochemistry
Howatson, Allan F, biophysics
Hsia, Jen-Chang, molecular pharmacology
Hughes, David Rees, physical anthropology, human biology
Hughes, Francis Norman, pharmacy
Hughes, George M, hydrogeology
Hume, James Nairn Patterson, theoretical physics
Hummel, Brian Christopher W, biochemistry, endocrinology
Hunt, John Wilfred, biophysics, radiation chemistry
Hunter, Geoffrey, quantum chemistry
Hunter, Harold Alexander, oral pathology
Hunter, John, physiology
Hunter, Robin Cyril Adair, psychiatry, psychoanalysis
Hutchinson, Thomas C, botany, environmental science
Hutton, Elaine Myrtle, human genetics
Illis, Alexander, chemical metallurgy
Ingles, Charles James, biochemistry, molecular biology
Iribarne, Julio Victor, physical chemistry
Irving, William Nathaniel, anthropology
Isgur, Nathan Gerald, theoretical high energy physics
Israel, Yedy, pharmacology
Ivey, Donald Glenn, physics, polymer physics
Jackson, Sanford (Hugh), biochemistry
Jacobs, Allan Edward, theoretical solid state physics
Jervis, Robert E, radiochemistry
Johns, Harold E, physics, biophysics
Johnson, Walter Henry, physiology
Jones, John Bryan, organic chemistry, organic biochemistry
Jopling, Alan Victor, geomorphology
Joubin, Franc Renault, chemistry
Juliano, Rudolph Lawrence, cell biology, biophysics
Kadanka, Zdenek Karel, cytogenetics, biochemistry
Kadar, Dezso, pharmacology
Kalant, Harold, pharmacology, cell physiology
Kalow, Werner, pharmacology
Kapral, Raymond Edward, physical chemistry
Kapur, Bhushan M, toxicology, clinical biochemistry
Kay, Ernest Robert MacKenzie, biochemistry
Kenney-Wallace, Geraldine Anne, chemical physics, quantum chemistry
Kerwin, Alfred John, cardiology
Kesler, Stephen Edward, economic geology, exploration geochemistry
Key, Anthony W, particle physics
Khanna, Jatinder Mohan, biochemical pharmacology
Kinsbourne, Marcel, pediatric neurology, experimental psychology
Kluger, Ronald H, organic chemistry, biochemistry
Kouroupis, George Michael, virology
Krishnan, Silvarama S, forensic medicine, biomedical engineering
Krug, John Christian, mycology
Kuksis, Arnis, biochemistry
Laframboise, James Gerald, plasma physics
Lai-Fook, Joan Elsa I-Ling, zoology
Laishes, Brian Anthony, cancer
Lane, Byron George, biochemistry
Langford, Ernest Robert, preventive medicine, public health
Langford, Raymond Robert, limnology
Lawford, George Ross, biochemistry, cell biology
LeBel, Jean Eugene, mathematics
Leers, Wolf-Dietrich, medical microbiology
Lehman, Alfred Baker, mathematics
Leibel, Bernard S, physiology
Lemon, James Thomas, urban geography
le Riche, William Harding, medicine
Letarte-Muirhead, Michelle, biochemistry
Leznoff, Arthur, medicine, immunology
Likuski, Henry John, animal nutrition, poultry nutrition
List, Roland, atmospheric physics
Lister, Maurice Wolfenden, inorganic chemistry
Litherland, Albert Edward, nuclear physics
Little, James Alexander, medicine, metabolism
Liversage, Richard Albert, developmental biology
Llewellyn-Thomas, Edward, medicine, engineering
Logan, David Mackenzie, molecular biology, biochemistry
Logan, Robert Kalman, physics, futurology
Logothetopoulos, J, medicine, physiology
Loughton, Barry G, biology
Love, David Vaughan, forest management
Lovett-Doust, John William, psychiatry
Lowden, J Alexander, neurochemistry
Luck-Allen, Etta Robena, mycology
Lumbers, Sydney Blake, geology, petrology
Macdonald, John Barfoot, microbiology
MacDougall, Edward Bruce, economic geography
MacHattie, Lorne Allister, biophysics, genetics
Machin, J, comparative physiology
Mackay, Rosemary Joan, fresh water ecology
Mackiw, Vladimir Nicolaus, inorganic chemistry, physical chemistry
MacLeod, Donald Richard Eason, medical microbiology
MacPherson, L W, microbiology, public health
Mahaney, William C, physical geography
Mahdy, Mohamed Sabet, virology, immunology
Mahon, William A, clinical pharmacology
Main, James Hamilton Prentice, oral pathology
Mandarino, Joseph Anthony, mineralogy
Manery, Jeanne Forest, physiology, biochemistry
Martin, Julio Mario, physiology
Massiah, Thomas Frederick, organic chemistry
Mastromatteo, Ernest, medicine
Masui, Yoshio, developmental biology
May, Kenneth Ownsworth, mathematics, history of science
Mayhall, John Tarkington, dental anthropology
McAndrews, John Henry, botany, plant ecology
McCulloch, Ernest Armstrong, medicine
McEwan, Ian Hugh, polymer chemistry
McFeat, Tom Farrar Scott, ethnology
McGreal, Douglas Anthony, medicine
McIntyre, Donald Patrick, meteorology, atmospheric sciences
McIver, Susan Bertha, entomology
McLean, Stewart, organic chemistry
McMurchy, Robert Connell, geology
McNeill, Kenneth Gordon, nuclear physics, nuclear medicine
McPherson, Donald Carman, plant ecology
McQueen, Donald James, ecology
McRae, Donald Lane, radiology
Meakin, James William, internal medicine
Meiklejohn, Robert Baikie, obstetrics & gynecology
Meincke, P P M, physics
Melcher, Antony Henry, histology
Mendis, Eustace Francis, solid state physics
Menon, Aravindakshan I, biochemistry
Menzinger, Michael, physical chemistry
Mettrick, David Francis, parasitology, pathological physiology
Michell, Arthur Stephen, forestry
Michener, Charles Edward, geology
Middleton, Peter James, medical virology
Miller, Anthony Bernard, epidemiology
Miller, Richard Graham, biophysics, immunobiology
Minde, Karl Klaus, psychiatry
Minta, Joe Oduro, immunology
Mintz, Sheldon, medical research
Misener, Austin Donald, physics
Monkhouse, Frank C, physiology
Morley, Nina Hope, nutrition
Morley, Thomas Paterson, neurosurgery
Morris, Glenn Karl, entomology, animal behavior
Moscarello, Mario Antonio, medicine, biochemistry
Moskovits, Martin, spectrochemistry, surface chemistry
Movat, Henry Zoltan, pathology
Murasugi, Kunio, mathematics
Murphy, Edward G, medicine, pediatrics
Murphy, John Thomas, medical physiology, neurology
Murray, Robert Kincaid, biochemistry
Mussells, Francis Lloyd, hospital administration
Nairn, John Graham, pharmacy, physical pharmacy
Naldrett, Anthony James, geology
Nautiyal, Jagdish Chandra, forest economics
Nicholls, Ralph William, physics
Nikiforuk, Gordon, biochemistry, dentistry
Nordin, Vidar John, forest pathology
Norris, Geoffrey, geology, paleontology
Norwich, Kenneth Howard, biophysics, physiology
Nuffield, Edward Wilfrid, mineralogy
Nunes, Paul Donald, geochronology
Nyburg, Stanley Cecil, crystallography
Ogilvie, John Charles, applied statistics
Ogryzlo, Metro Alexander, internal medicine
Olin, Philip, mathematical logic
Osoba, David, hematology, immunology
Pace Asciak, Cecil, organic chemistry, biochemistry
Packham, Marian Aitchison, biochemistry
Painter, Robert Hilton, biochemistry
Paloheimo, Jyri Erkki, mathematical biology
Parsons, Margaret Cranston, insect morphology
Parsons, Thomas Sturges, vertebrate zoology
Paterson, Garnet Russell, pharmaceutical chemistry, history of medicine
Patrick, Zenon Alexander, plant pathology
Paul, Derek (Alexander Lever), physics
Paul, William, biophysics
Paul, William Morris, obstetrics & gynecology
Paull, Allan E, statistics
Pearlman, Ronald E, molecular biology
Pearson, Mark Landell, molecular genetics
Peltier, William Richard, dynamic meteorology, geophysics
Percy, John Rees, astronomy
Perrin, Carrol Hollingsworth, analytical chemistry
Peterson, Randolph Lee, mammalogy
Phillips, Alexander James, medical statistics
Phillips, Melville James, pathology
Pimlott, Douglas Humphreys, wildlife ecology
Pinkerton, Peter Harvey, hematology
Poe, Anthony John, inorganic chemistry
Polanyi, John Charles, physical chemistry
Pollard, Jeffrey William, molecular genetics, reproductive endocrinology
Pomeranz, Bruce Herbert, neurophysiology
Popovich, Frank, dentistry
Porter, Charles Jack, clinical chemistry, biochemistry
Pos, Robert, psychiatry, neurophysiology
Poyton, Herbert Guy, radiology
Prentice, James Douglas, physics
Prichard, John Stobo, medicine
Prugovecki, Eduard, mathematical physics
Pruzanski, Waldemar, medicine, immunochemistry
Pugh, Robert E, theoretical physics
Purton, Christopher Roger, astronomy
Pye, Edgar George, economic geology
Quinn, John Phillip, mathematics
Racine, Rene, astronomy
Rae, James Jamieson, organic chemistry
Rae-Grant, Quentin A, psychiatry
Rakoff, Vivian Morris, psychiatry
Ranadive, Narendranath Santuram, immunology, experimental pathology
Ranger, Keith Brian, applied mathematics
Rao, Akkinapally V, food science
Rapoport, Abraham, medicine
Rappaport, Aron M, physiology, surgery
Rauth, Andrew Michael, biophysics
Reed, Thomas Edward, Jr, human genetics, behavioral genetics
Regier, Henry Abraham, ecology, fisheries
Rennie, James Clarence, animal breeding
Rewcastle, Neill Barry, neuropathology
Rhodes, Andrew James, medical microbiology, public health
Ribner, Herbert Spencer, acoustics, aerodynamics
Richardson, John Clifford, neurology
Ridge, Frank Gerald, geography
Rigler, Frank Harold, zoology
Riordan, John Richard, biochemistry
Rising, James David, systematic zoology
Ritchie, Alexander Charles, pathology
Robertson, Elizabeth Chant, nutrition
Robinson, Gilbert De Beauregard, mathematics
Roeder, Robert Charles, astronomy, physics
Rooney, Paul George, mathematics
Roots, Betty Ida, zoology
Roschlau, Walter Hans Ernest, pharmacology
Rosenthal, Peter (Michael), mathematics
Ross, Roderick Alexander, applied mathematics
Ross, Roderick Clendenning, pathology
Rothfels, Klaus Hermann, cytology
Rothman, Arthur I, medical education
Rothstein, Aser, biophysics, cell physiology
Roubicek, Rudolph, microbiology
Rowe, David John, physics
Roy, Dibyendu Nath, organic chemistry
Roy, Theodore Ernest, bacteriology, immunology
Rubin, Leon Julius, food science
Rucklidge, John Christopher, mineralogy, crystallography

CANADA

Russell, Dennis C., mathematics
Russell, Loris Shano, paleontology
Salter, Robert Bruce, orthopedic surgery
Sandham, Herbert James, oral biology, microbiology
Sarkar, Bibudhendra, biochemistry, physical chemistry
Sarkar, Priyabrata, plant cytogenetics
Sass-Kortsak, Andrew, biochemical genetics, gastroenterology
Satterberg, John Arvid, physiological psychology, neurophysiology
Sawyer, W Warwick, mathematics
Schachter, H, biochemistry
Schaefele, Ronald A., mathematical statistics
Scherk, Peter, geometry
Schimmer, Bernard Paul, medical research
Schmid, George Henry, organic chemistry
Scholefield, Peter Gordon, biochemistry
Schwerdtner, Walfried Martin, structural geology
Scott, George David, physics
Scott, John Gerald, hematology, internal medicine
Scott, John Wilson, neurophysiology
Scott, Steven Donald, economic geology, geochemistry
Scott, William Beverley, ichthyology
Scrimgeour, Kenneth Gray, biochemistry
Seaquist, Ernest Raymond, astronomy
Seeman, Philip, neuropharmacology, cell biology
Segal, Harold Jacob, pharmacy administration
Sellers, Edward Alexander, pharmacology
Sellers, Edward Moncrieff, clinical pharmacology
Sen, Amar Kumar, physiology
Sen, Dipak Kumar, mathematical physics
Sessle, Barry John, neurophysiology, oral physics
Shane, Samuel Jacob, internal medicine
Sheinin, Rose, biochemistry, microbiology
Shenitzer, Abe, mathematics
Shephard, Roy Jesse, physiology, medicine
Sherk, Frank Arthur, mathematics
Sifton, Harold Boyd, botany
Silver, Malcolm David, pathology
Sim, Stephen Kahsun, pharmacognosy
Siminovitch, Louis, biophysics, microbiology
Simmons, James William, urban geography
Simon, Gerard Theodor, pathology, electron microscopy
Sokoloff, Jack, theoretical physics
Solandt, Omond McKillop, physiology
Sparling, John H., ecology
Speck, John Edward, periodontology
Spelt, Jacob, geography
Spence, Leslie Percival, microbiology
Spero, Lawrence, pharmacology, biophysics
Srivastava, Muni Shanker, mathematical statistics
Stanacev, Nikola Ziva, biological chemistry
Stancer, Harvey C., psychiatry, neurochemistry
Steiner, George, medical research
Stoicheff, Boris Peter, lasers, molecular spectroscopy
Storey, Arthur Thomas, physiology, orthodontics
Strangway, David W., geophysics
Straus, Neil Alexander, molecular biology
Stryland, Jan Cornelis, molecular physics
Sturgess, Jennifer Mary, microbiology, pathology
Sullivan, Charlotte Murdoch, animal physiology
Summers, Peter William, meteorology
Sunahara, Fred Akira, pharmacology
Swyer, Paul Robert, pediatrics, neonatology
Talesnik, Jaime, pharmacology, physiology
Tallan, Irwin, genetics
Tamsitt, James Ray, zoology
Tan, Bian Djoen, parasitology, radiation biology
Tasker, Ronald Reginald, neurosurgery
Tator, Charles Haskell, neurosurgery
Taylor, Harry William, nuclear physics
Taylor, Robert Mackay, internal medicine
Tayyeb, A, geography
Teare, Frederick Wilson, pharmaceutical chemistry, radiopharmacy
Ten Cate, Arnold Richard, anatomy, dentistry
Thomas, E Llewellyn, medicine, electrical engineering
Thompson, Gordon William, epidemiology
Thompson, James Charlton, inorganic chemistry
Thompson, James Scott, anatomy, genetics
Thompson, Margaret A Wilson, genetics
Till, James Edgar, biophysics, cell biology
Tinker, David Owen, biochemistry, physical chemistry
Tobin, Sidney Morris, medical research, obstetrics & gynecology
Toguri, James M., chemistry
Tovell, Walter Massey, geology
Trainor, Lynne E H., theoretical physics
Trott, S M., mathematics
Underdown, Brian James, immunology
Urquhart, Frederick Albert, zoology
Valdivieso, Dario, microbiology
Valleau, John Philip, statistical mechanics, chemical physics
Van den Bergh, Sidney, astronomy
Van Kranendonk, Jan, theoretical physics
Vanstone, J R., mathematics
Veen-Baigent, Margaret Joan, nutrition
Vlicks, Arnold Evald, experimental nuclear physics, astrophysics
Vosko, Seymour H, theoretical physics
Vranic, Mladen, physiology, endocrinology
Wahl, William George, geology, geophysics
Walker, Alan, inorganic chemistry
Walker, Michael Barry, physics
Warren, Douglas Robson, medicine, occupational medicine, environmental medicine
Watters, Neil Archibald, surgery
Wayman, Morris, organic chemistry, biochemical engineering
Welsh, Harry L., atomic physics, molecular physics
West, Edmund Cary, experimental high energy physics
West, Gordon Fox, geophysics
Westman, Albert Ernest Roberts, physical inorganic chemistry
Whitmore, Gordon Francis, radiobiology, cancer
Whittington, Stuart Gordon, physical chemistry
Wicks, Frederick John, mineralogy
Wiggins, Glenn Blakely, entomology
Wightman, Keith John Roy, internal medicine
Wigle, Ernest Douglas, medicine, cardiology
Williams, George Ronald, biochemistry
Williams, John Peter, plant biochemistry, cytology
Willoughby, Donald S., medical bacteriology, immunology
Wilson, John Cleland, computer science
Winnik, Mitchell Alan, organic chemistry, photochemistry
Wishart, Franklyn Ogilvie, immunology
Witty, Ralph, biochemistry, nutrition
Wong, Jeffrey Tze-Fei, biochemistry
Wong, Samuel Shaw Ming, physics
Woodside, Donald G., orthodontics
Woolf, C R., pulmonary diseases, internal medicine
Wrenshall, Gerald Alfred, medicine
Wright, George F., chemistry
Wright, Kenneth A., zoology, parasitology
Yarranton, George Anthony, ecology, plant morphology
Yates, Keith, physical organic chemistry
Yates, Peter, organic chemistry
Yip, Cecil Cheung-Ching, biochemistry, endocrinology
York, Derek H., geophysics
Ziegler, Peter, organic chemistry, biochemistry
Zimmerman, Arthur Maurice, cell biology
Zimmerman, Selma Blau, embryology, physiology
Zingg, Walter, bioengineering
Zsoter, Thomas, internal medicine
Zubrzycki, Edgar, food technology
Zukotynski, Stefan, solid state physics, electrical engineering

TRENTON

Heeney, Harold Blair, plant physiology
House, Howard Leslie, insect physiology
Miller, Sherwood Robert, pomology

TUNNEY'S PASTURE

Goodman, Tine, electron microscopy, toxicology

VANIER

Boyd, Carl Edmund, pharmacology, toxicology

VINELAND

Adams, Angus Macaulay, food science, enology

VINELAND STATION

Allen, Wayne Robert, plant pathology
Andersen, Emil Thorvald, horticulture, pomology
Bradt, Oliver A., horticulture
Chiba, Mikio, analytical chemistry
Davidson, Thomas Ralph, plant pathology
Fisher, Robert (William), entomology
Fuleki, Tibor, food science, biochemistry
Marks, Charles Francis, nematology
McGinnis, Arthur James, biochemistry, plant pathology
Olthof, Theodorus Hendrikus Antonius, nematology, plant pathology
Phillips, John Henry Howard, pathology
Townshend, John Lynden, plant pathology
Wiebe, John, horticulture

WASHAGO

Ide, Frederick Palmer, fresh water biology

WATERLOO

Abler, Thomas Struthers, ethnology, social anthropology
Aczel, Janos D., mathematical analysis, geometry
Anderson, Anthony, molecular spectroscopy, solid state physics
Appleyard, Edward Clair, geology
Atkinson, George Francis, analytical chemistry, instrumentation
Aziz, Ronald A., chemical physics
Bakos, Gustav Alfons, astronomy
Berman, Gerald, mathematics
Boswell, Frank William Charles, physics
Brisbin, Doreen A., bioinorganic chemistry
Brodie, Don E., solid state physics
Carter, John C H., marine biology, fresh water biology
Carty, Arthur John, inorganic chemistry, organometallic chemistry
Clough, Donald J., operations research
Corbett, James Murray, physics
Cornell, James Morris, animal behavior
Cowan, John Arthur, fluid physics
Crapo, Henry Howland, mathematics
Cross, George Elliot, pure mathematics
Cummings, Larry Jean, algebra
Dagg, Anne Innis, mammalogy
Dagg, Ian Ralph, microwave physics
Danard, Maurice Beverley, meteorology
Davis, Harry Floyd, mathematics
Davison, Sydney George, surface physics
De'ath, Colin Edward, anthropology
Diem, Aubrey, geography
Dixon, Arthur Edward, physics
Dorney, Robert Starbird, environmental sciences
Downer, Roger George Hamill, insect physiology, endocrinology
Dumbroff, Erwin Bernard, plant physiology, biostatistics
Duthie, Hamish, fresh water biology, phycology
Eastman, Philip Clifford, solid state physics
Elsdon, William Lloyd, physical chemistry
Erb, David Kinsey, geology, geography
Farvolden, Robert Norman, hydrogeology
Fernando, Constantine Herbert, freshwater biology, parasitology
Fischer, Charlotte Froese, applied mathematics
Fitzgerald, Maurice Pim, astrophysics
Forbes, William Frederick, biometrics
Fraser-Reid, Bertram Oliver, organic chemistry
Fritz, Peter, geology, geochemistry
Geddes, Keith Oliver, numerical analysis
Gentleman, Jane Forer, statistical analysis
Gentleman, William Morven, mathematics, computer science
Glasser, M Lawrence, solid state physics, applied mathematics
Graham, James W., mathematics, computer science
Grindlay, John, solid state physics
Harrison, Arthur Desmond, fresh water biology
Haruki, Hiroshi, mathematics
Hayashida, Kaye, histology, developmental biology
Head, Clifford Grant, historical geography, cartography
Hill, Herbert Henderson, Jr, analytical chemistry
Huseyin, Koncay, applied mechanics
Hynes, Hugh Bernard Noel, freshwater ecology
Inniss, William Edgar, microbiology
Irish, Donald Edward, physical chemistry
Irving, Robert McCardle, economic geography, geography of Canada
Isenor, Neil R., physics
Kalbfleisch, James G., mathematical statistics
Kannappan, Palaniappan L., mathematics
Karasek, Francis Warren, analytical chemistry
Karrow, Paul Frederick, geology
Kempton, Alan George, microbiology
Kendrick, Bryce, mycology
Kesik, Andrzej B., physical geography
Klamkin, Murray S., mathematics
Kott, Edward, zoology
Krueger, Ralph R., geography
Kruuv, Jack, biophysics, radiobiology
Lawson, David Edward, sedimentology
Lawson, John Douglas, mathematics
Leech, John Watson, solid state physics, theoretical physics
Leslie, James D., solid state physics
Lind, Niels Christian, applied mechanics
Lipshitz, Stanley Paul, applied mathematics
Lyle, William Montgomery, optometry
Malcolm, Michael Alexander, computer sciences, numerical analysis
Manske, Richard Helmuth (Fred), natural products chemistry
Matthews, Burton Clare, soil chemistry
Maynes, Albion Donald, analytical chemistry
McBoyle, Geoffrey Reid, physical geography
McBryde, William Arthur Evelyn, chemistry
McCourt, Frederick Richard Wayne, molecular physics, chemical physics
McKiernan, Michel Amedee, mathematics
McLeod, Henry George, electrochemistry
Moffat, John Blain, physical chemistry
Morris, John Llewellyn, numerical analysis
Morrison, Hugh MacGregor, physics
Morton, John Kenneth, botany
Mullin, Ronald Cleveland, mathematics
Nash, Peter Hugh, Sr, urban geography, academic administration
Nelson, James Gordon, resource geography
O'Brien, Anne T., environmental sciences
Paldus, Josef, theoretical chemistry, theoretical physics
Pathria, Raj Kumar, theoretical physics, statistical mechanics
Peirson, David Robert, plant physiology
Pindera, Jerzy Tadeusz, applied mechanics
Pintar, Milan Mik, nuclear magnetic resonance
Ponzo, Peter James, applied mathematics
Power, Geoffrey, fish biology
Preston, Richard Ellis, geography
Read, Ronald Cedric, mathematics, computer science
Reesor, Glyn Edward, microwave physics
Reeves, Leonard Wallace, physical chemistry
Remole, Arnulf, physiological optics
Rempel, Garry Llewellyn, inorganic chemistry
Roberts, David Llewellyn, astrophysics, physics
Rudin, Alfred, polymer science
Russwurm, Lorne H, geography
Schellenberg, Paul Jacob, mathematics
Shank, Herbert S., mathematics
Sivak, Jacob Gershon, comparative physiology, physiological optics
Slawson, Peter (Robert), air pollution, mechanical engineering
Smith, Howard John Treweek, low temperature physics
Smith, James Graham, organic chemistry
Snieckus, Victor A., organic chemistry
Sprott, David Arthur, mathematical statistics

Steiner, Dieter, geography
Sumner, Donovan Bradshaw, mathematics
Tchir, Morris Frederick, chemistry
Thompson, John Eveleigh, biochemistry, cell biology
Tompa, Frank William, computer sciences
Toogood, Gerald Edward, inorganic science
Torrie, Bruce Harold, solid state physics
Tutte, William Thomas, mathematics
Van der Hoff, Bernard Maria Euphemius, polymer chemistry, polymer physics
Vanderkooy, John, solid state physics, optics
van Emden, Maarten Herman, computer science
Vanstone, Scott Alexander, mathematics
Viswanatha, Thammaiah, biochemistry
Wang, Shao-Fu, theoretical physics, solid state physics
Watson, Wynnfield Young, entomology, invertebrate zoology
Weaver, Sally Mae, anthropology
Wellwood, Arnold Augustus, cytogenetics, plant morphology
Winter, David Arthur, biomedical engineering, electrical engineering
Woo, George Chi Shing, optometry
Woolford, Robert Graham, organic chemistry
Younger, Daniel H, mathematics

WEST HILL
Chan, Lock Lim, polymer chemistry, paper chemistry
Dyer, Charles Chester, cosmology
Greenwood, Brian, geomorphology, sedimentology
King, James Douglas, nuclear physics
Kresge, Alexander Jerry, physical organic chemistry
Nicholson, Thomas Frederick, pathological chemistry
Ritchie, James Cunningham, botany, ecology
Sparrow, Christopher John, physical geography
Tidwell, Thomas Tinsley, organic chemistry

WESTON
Michell, John Humfrey, organic chemistry

WILLOWDALE
Aitken, James Henry, nuclear physics
Ashford, Walter Rutledge, organic chemistry
Bett, Hillyard Dobson, biochemistry
Borensztajn, David Zelman, microbiology
Fenje, Paul, virology, microbiology
Glass, Douglas Gordon, analytical chemistry
Gupta, Krishana Chandara, microbiology, immunochemistry
Healy, George McNeice, biochemistry, virology
Hines, William Grant, physical chemistry
Mohun, William Arthur, chemistry
Hutcheson, Michael Scott, computer science
Hull, Thomas Edward, computer science, criminology
Ing, Wai Kwok, marine biology
Kates, Josef, science policy, systems science
Macmorine, Hilda Mildred Grace, immunology
McNeel, Burdett Harrison, psychiatry, organic chemistry
Perkons, Auseklis Karlis, nuclear science, chemistry
Robb, Leslie Allan, biochemistry, microbiology
Roy, Marie L, organic chemistry
Seefried, Adolf Von, cell physiology, virology
Tarczy, E K, organic chemistry, biological chemistry
Tosoni, Anthony Louis, chemistry
Van Loon, Jon Clement, geology, chemistry
Wilson, Robert James, bacteriology, immunology
Wye, Edwin James, biochemistry
Zamel, Noe, respiratory physiology

WINDSOR
Atkinson, Harold Russell, mathematical analysis
Atkinson, John Brian, atomic physics
Benedict, Winfred Gerald, plant pathology

Blackbourn, Anthony, economic geography
Drake, Gordon William Frederic, atomic physics
Drake, John Edward, inorganic chemistry, spectroscopy
Faught, Donald Thomas, mathematics
Glass, Edward Nathan, theoretical physics
Gravenor, Conrad Percival, geology
Habib, Edwin Emile, physics
Habowsky, Joseph Edmund Johannes, cytology, histology
Hedgecock, Nigel Edward, solid state physics
Helbing, Reinhard Karl Bodo, physics
Hencher, John Lawrence, structural chemistry
Holland, William John, analytical chemistry
Holuj, Frank, solid state physics
Huschilt, John, theoretical physics
Innes, Frank Cecil, geography, historical geography
Jull, Robert Kingsley, geology
Krause, Lucjan, physics
Lall, Amrit, economic geography
Lasker, George Eric, information sciences, psychology
LaValle, Placido Dominick, geomorphology
Lin, Che-Shung, quantum chemistry
McConkey, John William, physics
McCurdy, Howard Douglas, Jr, microbiology
McDonald, James Frederick, mathematical physics
McGarvey, Bruce Ritchie, physical chemistry
McIntosh, John McLennan, synthetic organic chemistry
McKenney, Donald Joseph, physical chemistry
M'Closkey, Robert Thomas, ecology
Ogata, Hisashi, nuclear physics
Okey, Allan Bernhardt, endocrinology, cancer
Petras, Michael Luke, zoology, genetics
Pillay, Dathathry Trichinopoly Natraj, plant physiology
Price, Stanley James Whitworth, physical chemistry
Romsa, Gerald Harry, geography
Rumfeld, Robert Clark, photochemistry, inorganic chemistry
Ruth, Norbert Joseph, physics
Rutherford, Kenneth Gerald, organic chemistry
Sabina, Leslie Robert, virology, microbiology
Sanderson, Marie Elizabeth, physical geography
Scheidegger, Adrian Eugen, geophysics.
Schlesinger, Mordechay, solid state physics
Schmidt, Donald Emil, Jr, bio-organic chemistry
Shklov, Nathan, mathematics
Singh, Ripu Daman, physical anthropology, primatology
Smith, Terence E, petrology, geochemistry
Snyder, Sally, anthropology
Sonnenfeld, Peter, geology
Stebelsky, Ihor, environmental sciences, earth sciences
Symons, David Thorburn Arthur, geophysics, geology
Taylor, Norman Fletcher, biochemistry
Theuws, Jacques Antoine, cultural anthropology
Thibert, Roger Joseph, chemistry
Tracy, Derrick Shannon, mathematical statistics
Trenhaile, Alan Stuart, geomorphology
Tuck, Dennis George, inorganic chemistry, radiochemistry
Turek, Andrew, geochemistry
Van Wijngaarden, Arie, toxicology, reproductive physiology
Virgo, Bruce Barton, ecology, cell physiology
Wallen, Donald George, ecology, cell physiology
Walters, Fred Henry, analytical chemistry
Warner, Alden Howard, developmental biology
Whitt, Darnell Moses, soil physics, field crops
Wigley, Neil Marchand, mathematical analysis
Wong, Chi Song, mathematical statistics, operator theory
Wood, Gordon Walter, organic chemistry, mass spectrometry

YARKER
Riddell, John Evans, economic geology

PRINCE EDWARD ISLAND

CHARLOTTETOWN
Cutcliffe, Jack Alexander, olericulture
Drake, Edward Lawson, invertebrate pathology, entomology
Gupta, Umesh C, soil fertility, plant nutrition
Hanic, Louis A, phycology
Jammu, K S, physics
Lin, Wei-Ching, space physics
Liu, Michael T H, physical chemistry
MacLeod, LLoyd Beck, agronomy
MacQuarrie, Ian Gregor, botany, plant physiology
Manley, Stephen Alexander, forest genetics
Munro, Douglas Carlyle, soil fertility, plant nutrition
Nass, Hans George, agronomy, plant science
Palmer, Glenn Earl, organic chemistry
Rigney, James Arthur, organic chemistry, biochemistry
Suzuki, Michio, plant physiology, biochemistry
White, Ronald Paul, Sr, soil chemistry, soil fertility
Willis, Carl Bertram, plant pathology

QUEBEC

ARVIDA
Girolami, Libero L, analytical chemistry, industrial chemistry
Hollingshead, Ethan Allen, metallurgical chemistry
Lemieux, Paul E, physics
Ostap, Stephen, inorganic chemistry
Southam, Frederick William, physical chemistry

BELOEIL
Coulombe, Louis Joseph, agriculture

CANDIAC
Perron, Yvon G, organic chemistry, medicinal chemistry

CHICOUTIMI
Bouchard, Louis-Marie, urban geography, cartography
Caty, Jean Louis, geology, sedimentology
Chown, Edward Holton, petrology, structural geology
Claveau, Rosario, hematology, internal medicine
Dimroth, Erich, geology
Garneau, Francois Xavier, organic chemistry

DORVAL
Brown, George Bruce, meteorology
Mueller, Walter A, physical chemistry, metallurgy

GRAND'MERE
Ayroud, Abdul-Mejid, chemistry
Goel, Krishan Narain, chemistry

HUDSON
Clayton, David Walton, organic chemistry
Melnychyn, Paul, agricultural biochemistry

HULL
Carlisle, David B, marine biology, fresh water biology
McKay, Kenneth Alexander, bacteriology
Morrissette, Hugues, geography

LACHINE
Sheps, Louis Jack, organic chemistry

LAVAL
Carbonneau, Roch, physiology, endocrinology
Lamoureux, Gilles, medicine, immunology
Portelance, Vincent Damien, microbiology
Potworowski, Edward Francis, immunology

LAVAL DES RAPIDES
Boudreault, Armand, virology, biology
Cantor, Ena D, microbiology, immunology
Dubreuil, Robert, virology, cancer immunology
Frappier, Armand, bacteriology, hygiene
Guerault, Armand, microbiology, immunology
Lemonde, Paul, cancer
Marois, Paul Henri, veterinary medicine
Pavilanis, Vytautas, medicine

LENNOXVILLE
Bernard, Camille Stephen, animal breeding
Brown, Douglas Fletcher, genetics
Fahmy, Mohamed Hamed, genetics, animal husbandry
Haywood-Farmer, John, physical organic chemistry
Hilton, Donald Frederick James, medical entomology, animal parasitology
Hull, James Clark, plant ecology
Langford, Arthur Nicol, botany
Nagpal, Tarlok Singh, nuclear physics
Redding, John Lawford, physics
Ross, William Gillies, geography
Yeats, Ronald Bradshaw, organic chemistry

LONGUEUIL
Stevenson, D Richard, physics

MACDONALD COLLEGE
Meerovitch, Eugene, parasitology
Taper, Charles Daniel, horticulture

MANSONVILLE
MacKay, Donald Douglas, chemistry

MCMASTERVILLE
Gannon, David John, physical chemistry
Holden, Harold William, physical chemistry
Sharp, John Arthur, physical chemistry
Van Zeggeren, Frederik, physical chemistry

MONTREAL
Abshire, Claude James, biochemistry, microbiology
Adamkiewicz, Vincent Witold, physiology, immunology
Addison, John Rundle, low temperature physics
Albert, Paul Joseph, sensory physiology
Ali, Mohamed Ather, zoology
Allenby, Clement W, organic chemistry
Ambrose, Ernest R, dentistry
Anthonisen, Nicholas R, respiratory physiology
Artizzu, Maria, histopathology
Atkinson, Joseph George, organic chemistry
Atwood, John William, computer science
Auclair, Jacques Lucien, entomology
Bagli, Jenanbux Framroz, organic chemistry
Bain, Barbara, experimental medicine
Ban, Thomas Arthur, psychiatry, psychopharmacology
Banik, Upendra K, reproductive physiology, endocrinology
Banville, Bertrand, physics
Barbeau, Andre, neurology
Barcelo, Raymond, physiology, biochemistry
Bariana, Dilbagh Singh, organic chemistry, pharmaceutical chemistry
Barr, Michael, mathematics
Bata, George L, polymer chemistry, polymer physics
Bates, Donald George, history of medicine
Baxter, Donald William, medicine
Beaudin, Jacques, virology, bacterial genetics
Beaudry, Jean Romuald, biosystematics
Beauregard, Ludger, geography

CANADA

Beique, Rene Alexandre, radiation physics
Beland, Jacques (Robert), geology
Bell, Robert Edward, nuclear physics
Belleau, Bernard Roland, organic
Benley, Bruno Georg, pharmacology, therapeutics
Benoist, Jean, physical anthropology
Benoit, Jean Claude, biology, microbiology
Bensley, Edward Horton, history of medicine
Bentley, Kenneth Chessar, oral surgery
Bergeron, Michel, physiology, nephrology
Bilimoria, Minoo Hormasji, microbiology, biochemistry
Bird, John Brian, geography
Birks, Richard Irwin, physiology
Birmingham, Marion Krantz, biochemistry
Blais, Roger A., economic geology
Blostein, Rhoda, biochemistry
Bois, Pierre, anatomy
Bolande, Robert Paul, pathology
Bolker, Henry Irving, wood chemistry
Boll, William George, plant physiology
Bolte, Edouard, reproductive physiology, medicine
Boothroyd, Eric Roger, cytology
Bordeleau, Jean-Marc, psychiatry, pharmacology
Boucher, Roger, organic chemistry
Boulanger, Jean Baptiste, psychoanalysis
Bourgon, Marcel, physical chemistry, inorganic chemistry
Bourne, Frederick Munroe, internal medicine
Bromage, Philip R., anesthesiology
Brown, Alexander Cyril, economic geology, geochemistry
Boyes, John Wallace, genetics
Brandt, J Leonard, internal medicine, nephrology
Brant, Charles Sanford, anthropology
Brawer, James Robin, neurocytology, neuroendocrinology
Brunel, Pierre, marine ecology
Bui, Tien Dai, computer science
Bussey, Arthur Howard, biology
Butas, Constandina, medical microbiology
Butler, Ian Sydney, inorganic chemistry
Caille, Gilles, medicinal chemistry
Cailloux, Marcel Louis, plant physiology
Cameron, Douglas George, medicine
Cantin, Marc, medicine
Carroll, Robert Lynn, vertebrate paleontology
Clark, Thomas Henry, stratigraphy,
Cayen, Mitchell Ness, biochemistry
Chan, Eddie Chin Sun, microbial physiology
Chan, Tak-Hang, organic chemistry
Chang, Thomas Ming Swi, physiology, medical research
Chicoine, Luc, pediatrics
Chretien, Michel, endocrinology
Chubb, Francis Learmonth, organic chemistry
Cleghorn, Robert Allen, psychiatry
Cohen, Monroe W., neurosciences
Cohen, Montague, medical physics
Colebrook, Lawrence David, physical chemistry
Collier, Brian, pharmacology, physiology
Cooke, Patricia M., bacteriology, virology
Cooper, Bernard A., hematology
Cooper, Paul David, pharmacology,
Concougrs, Andreas P., theoretical chemistry
Cormier, Bruno M., psychiatry
Couillard, Pierre, cell physiology
Courville, Jacques, neuroanatomy
Cousineau, Gilles H., molecular biology, biochemistry
Crawhall, John C., biochemistry, medicine
Croll, Neil Argo, parasitology, behavioral physiology
Cronin, Robert Francis Patrick, physiology
Csorgo, Miklos, mathematics
Cummings, John Rhodes, pharmacology
Daessle, Claude, organic chemistry
Daigneault, Aubert, mathematical logic
Dallaire, Louis, medicine, medical genetics
Dansereau, Pierre, ecology
Daoust, Hubert, physical chemistry
D'Aoust, Maurice, plant physiology
Daoust, Roger, histopathology, cancer

David, Jean, biochemistry, food technology
David, Peter P., quaternary geology
Davignon, Jean, medicine
Davis, Martin Arnold, organic chemistry
Deans, Sidney Alfred Vindin, organic chemistry
De Champlain, Jacques, physiology, pharmacology
Degheughi, Romano, organic chemistry, pharmaceutical chemistry
DeJongh, Don C., organic chemistry
de Lamirande, Gaston, biological chemistry
Deland, Andre N., geology
Del Bianco, Walter, nuclear physics
Delorme, Joachim, industrial chemistry
Demers, Jean-Marie, animal nutrition
Demers, Pierre (A E), photography, color science
Demerson, Christopher, medicinal chemistry
Denhardt, David Tilton, molecular biology
Denis, Gustave, nephrology, metabolism
Depommier, Pierre Henri Maurice, nuclear physics
De Repentigny, Jacques, bacteriology, immunology
Derome, Jacques Florian, dynamic meteorology
Descarries, Laurent, neurobiology, neuroanatomy
Desrosiers, Joseph A Jacques, obstetrics & gynecology, physiology
de Takacsy, Nicholas Benedict, nuclear physics
Dixon, John Francis Clemow, industrial chemistry
Dhindsa, K S., cell biology
Dick, James Gardiner, analytical chemistry, electrochemistry
Digby, Peter Saki Bassett, marine biology, physiology
Dirks, John Herbert, physiology, nephrology
Doig, Ronald, geology, geophysics
Donnay, Gabrielle (Hamburger), crystallography
Donnay, Joseph Desire Hubert, crystallography, mineralogy
Donohue, William B., pathology, surgery
Doughty, Mark, organic chemistry
Drummond, Keith N., pediatrics
Drummond, Robert Norman, geography, cartography
Dubuc, Serge, mathematics
Ducharme, Jacques R, pediatrics, biochemistry
Dugas, Hermann, biophysics, organic chemistry
Dunbar, Maxwell John, oceanography
Duncan, Richard Dale, physical organic chemistry
Dupont, Claire Hammel, biochemistry,
Dvornik, Dushan Michael, biochemistry, organic chemistry
Eade, Norman Russell, pharmacology
Eakins, Peter Russell, geology
East, Conrad, meteorology, atmospheric physics
Eddy, Nelson Wallace, nuclear physics
Edward, Deirdre Waldron, biochemistry
Edward, John Thomas, physical organic chemistry
Eisenberg, Adi, physical chemistry
Eker, Kurt, physical chemistry
Elkin, Eugene Mitchell, chemical engineering
Elliott, Kenneth Allan Caldwell, biochemistry
Elson, John Albert, quaternary geology
Elvidge, Arthur Roland, neurosurgery
Enesco, Hildegard Esper, cell biology
Enesco, Mircea Aaron, biology, endocrinology
Entin, Martin A., plastic surgery, reconstructive surgery
Eu, Byung Chan, theoretical chemistry
Evans, Arvel, mathematics
Fabrikant, Jacob I., radiology
Favre, Henri Albert, organic chemistry
Fazekas, Arpad Gyula, experimental surgery, biochemistry
Feindel, William Howard, neurosurgery
Ferguson, Ronald James, medicine
Firth, David Richard, physics, biophysics
Fisk, Guy Hubert, medicine
Fliszar, Sandor, physical organic chemistry, quantum chemistry
Fong, Jack Sun-Chik, pediatrics, immunology
Fontaine, Gilles Joseph, astrophysics
Fox, Bennett L., mathematics

Frank, Barry, theoretical solid state physics
Fraser, Donald Alexander, biogeography, geography
Fraser, Frank Clarke, medical genetics, teratology
Fraser, Murray Judson, biochemical genetics
Fraser, Robert Gordon, medicine
Freedman, Samuel Orkin, immunology
Fresco, James Martin, analytical chemistry
Frojmovic, Maurice Mony, surface chemistry
Gaertner, Erika Eva, economic botany, forestry
Gagne, Jean-Marie, spectroscopy, optics
Gagnon, Arthur, biochemistry
Gagnon, Marcel, food technology
Gagnon, Real, anatomy
Gardner, Gerard, biology
Gareau, Roger, pathology
Garmaise, David Lyon, organic chemistry
Garnier, Benjamin John, geography
Gascon, Andre L., pharmacology
Gaskell, Robert Weyand, theoretical high energy
Gaudry, Roger, organic chemistry
Genest, Jacques, nephrology, endocrinology
Gianetto, Robert, biochemistry
Gill, James Edward, geology
Giri, Narayan C., mathematical statistics
Givner, Morris Lincoln, biochemistry
Goresky, Carl A., internal medicine
Gotz, Manfred, organic chemistry
Goulard, Bernard, theoretical physics,
Gloor, Pierre, neurophysiology
Goyer, Robert G., pharmacology
Grad, Bernard, experimental biology
Graham, Aloysius, organic chemistry
Graham, Angus Frederick, microbiology
Grant, Peter Raymond, zoology
Gravel, Denis Fernand, photochemistry
Gregoire, Fernand, respiratory physiology, allergy
Grice, Reginald Hugh, geology
Grosser, Arthur Edward, physical chemistry
Gubersky, Victor R, dermatology
Gunn, Bernard M, geochemistry
Gurudata, Neville, physical organic chemistry
Gutkind, Peter C W., anthropology
Guttmann, Ronald D., immunology
Hach, Vladimir, organic chemistry, medicinal chemistry
Hakka, Leo Ernest, physical organic chemistry
Hamlet, Zacharias, organic chemistry
Hanessian, Stephen, organic chemistry
Harper, Pierre Paul, fresh water ecology
Harpp, David Noble, chemistry
Harrod, John Frank, physical chemistry
Hawkins, David Geoffrey, immunology
Heaps, Harold Stanley, computer science
Hedgecock, Frederick Thomas, physics
Heller, Irving Henry, medicine, neurology
Herr, Ferenc, pharmacology
Herschorn, Michael, mathematics
Herz, Carl Samuel, mathematics
Hesse, Reinhard, sedimentology, marine geology
Hillman, Elizabeth S., pediatrics,
Hills, Theodore Lewis, agricultural geography
Hitschfeld, Walter, meteorology, physics
Hogan, James Joseph, nuclear chemistry
Horrobin, David Frederick, endocrinology, cardiovascular physiology
Hossain, Esau Abbas, biochemistry
Hossain, Afzal, biochemistry
Humber, Leslie George, organic chemistry
Ibrahim, Ragai Kamel, plant biochemistry, plant tissue culture
Ikawa-Smith, Fumiko, anthropology
Ingram, Richard Grant, physical oceanography
Isler, Henri Gustave, histology, biochemistry
Itiaba, Kibe, biochemistry, physiology
Jarochowski, Maria Anna, economic geography, population geography
Jasmin, Gaetan, pathology
Jasper, Herbert Henry, neurophysiology

Jirkovsky, Ivo, organic chemistry, medicinal chemistry
Joffe, Anatole, mathematics
Johnstone, Rose M., biochemistry
Jolicoeur, Pierre, biomathematics, biometrics
Joncas, Jean Harry, virology, infectious diseases
Jones, Geoffrey Melvill, physiology, aerospace medicine
Jonsson, Wilbur Jacob, mathematics
Just, George, organic chemistry
Kafer, Etta (Mrs E R Boothroyd), genetics
Kahlenberg, Arthur, biochemistry
Kalant, Norman, endocrinology, metabolism
Kalman, Calvin Shea, high energy physics, mathematical physics
Kapoor, Narinder N., zoology, physiology
Karasaki, Shuichi, developmental biology, cancer
Kaufman, Hyman, mathematics
Kinch, Robert Arthur Hugh, obstetrics & gynecology
Kipling, Arlin Lloyd, solid state physics
Kunos, George, pharmacology
Lalli, Carol Marie, marine biology
Klemola, Tapio, mathematics
L'Abbe, Maurice, mathematics
Kluepfel, Dieter, microbiology, biochemistry
Kopriwa, Beatrix Markus, histology
Kowalik, Janusz Szczesny, numerical analysis
Kramil, Michael Joseph Anthony, biochemistry, organic chemistry
Kravitz, Henry, psychiatry
Krnjevic, Kresimir, neurophysiology
Kuchel, Otto George, nephrology
Lachance, John Paul, biology
Lamonde, Andre M., pharmacy
Langleben, Manuel Phillip, glaciology,
Lajoie, Jean, geology
Lal, Samarthji, neuropsychiatry
Lala, Peeyush Kanti, cancer, cell biology
Lam, Harry Chi-Sing, theoretical high energy physics
Lam, Vinh-Te, physical chemistry
Lambek, Joachim, mathematics
Lanthier, Andre, medicine, biochemistry
Latour, Roger, pharmacy
Lawrence, Donald Gilbert, neurology
Lawson, Norman C., plant breeding
Leblond, Charles Philippe, anatomy
Lesperance, Pierre J., invertebrate paleontology
Letarte, Jacques, pediatrics, biochemistry
Leduc, Gerard, fisheries, biochemistry
Lee, Jonathan K P., nuclear physics
Lefebvre, Rene, medicine
Lefebvre, Yvon, organic chemistry,
LeTourneux, Jean, theoretical nuclear physics
Lehmann, Heinz Edgar, psychiatry
Lehnert, Shirley Margaret, radiobiology
Leighton, Henry George, meteorology
Lemieux, Guy, internal medicine
Leonard, John Alex, industrial chemistry
L'Ecuyer, Jacques, nuclear physics
Leung, Kin-Vinh, numerical analysis
Levesque, Rene J A., nuclear physics
Levi, Irving, medicinal chemistry
Levy, Samuel Wolfe, clinical chemistry
Lewis, John Bradley, marine biology
Lewis, Martin Gwent, cancer, pathology
Limerick, Jack McKenzie, Sr., chemistry
Lindsay, James Gordon, physical chemistry, ceramics
Ling, Daniel, audiology, communications
Lippmann, Wilbur, biochemical pharmacology
Lipsett, Solomon George, physical inorganic chemistry
Lis, Martin, experimental medicine
Lloyd, Trevor, geography
Lorrain, Paul, electromagnetism
Lowenthal, Julius, biochemistry
Lukosevicius, Petras Povilas, agriculture, plant breeding
Lussier, Jean Paul, dentistry
Maag, Urs Richard, statistics
Macintosh, Frank Campbell, physiology
Mackey, Michael Charles, biophysics
Macklem, Peter Tiffany, pulmonary physiology, experimental medicine
Maclachlan, Gordon Alistair, plant biochemistry
MacLean, Lloyd Douglas, surgery

MacLean, Wallace H, geology, geochemistry
MacLeod, Alastair William, psychiatry
MacLeod, Charles Franklyn, ecology
Madison, Dale Martin, animal behavior, vertebrate biology
Magnin, Etienne Nicolas, ichthyology
Maly, Edward J, ecology, evolution
Marner, Orval Albert, chemistry
Mamet, Bernard Leon, geology
Mangat, Bhupinder Singh, plant biochemistry & plant physiology
Mankiewicz, Edith Marion, microbiology
Manley, Rockliffe St John, polymer chemistry
Marc-Aurele, Julien, physiology
Marchant, Cosmo, industrial chemistry
Marchessault, Robert Henri, physical chemistry, polymer chemistry
Margolis, Bernard, physics
Marier, Guy, pharmacology
Mark, Shew-Kuey, experimental nuclear physics
Marks, Melvin Issac, infectious diseases, pediatrics
Marsaglia, George, mathematics, computer science
Marshall, David Jonathan, medicinal chemistry, research administration
Marshall, John Stewart, physics, meteorology
Martignole, Jacques, geology
Martin, Robert Francois Churchill, geology
Martin, William Macphail, nuclear physics
Martineau, Bernard, medical bacteriology
Mason, Stanley George, physical chemistry
Mathai, Arakaparampil M, mathematical statistics
Matheson, Ballem Howard, bacteriology
Mathieu, Jean, internal medicine
Mathieu, Leo Gilles, biochemistry, microbiology
Mathieu, Roger Maurice, radiology, chemistry
McGarry, Eleanor E, medicine, endocrinology
Mattingly, Susan Carol, audiology
Maughan, George Burwell, obstetrics & gynecology
McCaughey, T J, anesthesiology
McDonald, Alison Dunstan, epidemiology
McDonald, John Corbett, epidemiology
McElcheran, Donald Elmo, physical chemistry
McGeer, James Peter, physical chemistry
Michaud, Georges Joseph, astrophysics
McIntosh, Alexander Omar, industrial chemistry
McKenzie, John Maxwell, internal medicine, endocrinology
McNeil, Raymond, ornithology
Meighen, Edward Arthur, biochemistry
Menezes, Jose Piedade Caetano Agnelo, microbiology, immunology
Merilees, Philip, dynamic meteorology
Messier, Bernard, experimental pathology
Metrakos, Julius Demetrius, medical genetics
Mezi, Zdenek, dentistry, oral pathology
Milic-Emili, Joseph, physiology
Milner, Brenda (Atkinson), neuropsychology
Misra, Sushil, nuclear physics, solid state physics
Monaro, Sergio, nuclear physics
Mongeau, Maurice, physical medicine
Montreuil, Fernand, medicine, otolaryngology
Moore, Ronald B, nuclear physics
Morais, Rejean, biochemistry
Moore, Robert Hall, pathology
Morgan, James Ebenezer, physical chemistry
Morris, Stanley P, theoretical solid state physics
Moser, William O J, mathematics
Mountjoy, Eric W, geology, stratigraphy
Muir, Wilson Burnett, solid state physics
Mukerjee, Barid, genetics
Mukherji, Kalyan Kumar, geology, geophysics
Murphy, Beverley Pearson, biochemistry, endocrinology
Murphy, Henry Brian Megget, psychiatry, sociology
Myers, Gordon Sharp, medicinal chemistry
Nadeau, Reginald Antoine, cardiovascular physiology, cardiology
Nadler, Norman Jacob, endocrinology, anatomy
Nash, Peter Howard, research administration
Neal, Jack Laurance, biophysics

Negrepontis, Stylianos, mathematics
Neims, Allen Howard, biochemistry, pediatrics
Nelson, Robert Armstrong, Jr, microbiology
Nicholls, Robert Van Vliet, organic chemistry
Nickerson, Mark, pharmacology, therapeutics
Nogrady, George Ladislaus, microbiology, population studies
Nogrady, Thomas, medicinal chemistry
Nowaczynski, Wojciech, endocrinology
O'Connor, Robert Eric, mathematics
Ogilvie, Kelvin Kenneth, bio-organic chemistry
Ogilvie, Richard Ian, clinical pharmacology
O'Meara, Desmond, organic chemistry
Onyszchuk, Mario, inorganic chemistry statistics
Oseasohn, Robert, epidemiology
Oshiro, George, pharmacology
Osmond, Dennis Gordon, anatomy
Outerbridge, John Stuart, physiology, otology
Page, Edouard, biochemistry
Pai, Chik Hyun, medical microbiology, molecular biology
Palaic, Djuro, pharmacology
Pallen, Robert Harris, physical chemistry, polymer chemistry
Pande, Shri Vardhan, biochemistry
Panisset, Jean-Claude, pharmacology
Papineau-Couture, Gilles, physical chemistry
Pappius, Hanna M, neurochemistry
Paquette, Guy, theoretical physics
Parry, John Trevor, geography
Pasztor, Valerie Margaret, zoology
Patel, Nagin K, physical pharmacy
Patel, Popat-Lal Mulji-Bhai, high energy physics
Patterson, Donald Duke, polymer chemistry
Pearson, John Michael, theoretical physics
Pelletier, Real Lucien, plant pathology
Perdue, James F, biochemistry
Perlin, Arthur Saul, organic chemistry
Perlman, Martin Melvin, solid state physics
Phan, Chon-Ton, plant physiology, biochemistry
Phillips, Norman William Frederick, physical chemistry
Piche, Lucien, organic chemistry
Pierard, Jean Arthur, veterinary anatomy, mammology
Pilon, Jean-Guy, entomology
Pinsky, Leonard, genetics
Pirlot, Paul, zoology
Plaa, Gabriel Leon, pharmacology
Plourde, J Rosaire, pharmacy
Polosa, Canio, medicine, physiology
Poole, Ronald John, plant physiology, biophysics
Pounder, Elton Roy, ice physics, physical oceanography
Prasad, Raj Nandan, organic chemistry
Prud Homme, Jacques, physical
Rakhit, Sumanas, organic chemistry
Rasmussen, Theodore Brown, neurology, neurosurgery
Rattray, Basil Andrew, mathematics
Raudorf, Walter Rudolf, electron physics
Redmond, Ninfa Indacochea, pharmacology, toxicology
Revesz, Clara Rona, endocrinology, pediatrics
Richard, Pierre Joseph Herve, palynology, paleoecology
Richer, Claude-Lise, microscopic anatomy, endocrinology
Richer, Jean-Claude, organic chemistry
Rivest, Roland, chemistry
Robb, J Preston, neurology
Roberge, Fernand Adrien, biomedical engineering, neurobiology
Robertson, Alexander Allen, physical chemistry
Robertson, Roderick Francis, physical chemistry
Robillard, Eugene, physiology
Robson, John Michael, nuclear physics
Rochefort, Joseph Guy, biochemistry, endocrinology
Rogers, Roddy R, meteorology
Rona, George, pathology
Rose, Bram, medicine
Rosenberg, Ivo George, mathematics
Rosenthall, Edward, mathematics
Rotenberg, A Daniel, medical physics, nuclear medicine
Rotenberg, Aubey, applied mathematics, computer science

Roth, Charles, applied mathematics, theoretical physics
Rouleau, Ernest, systematic botany
Roux, Jacques F, obstetrics & gynecology
Roy, Claude Charles, pediatrics, gastroenterology
Roy, Guy, biophysics
Rubinstein, David, biochemistry
Ryan, David George, particle physics
Sabidussi, Gert Otto, mathematics
Sainte-Marie, Guy, hematology
St Pierre, Leon Edward, polymer chemistry
Salvador, Romano Leonard, medicinal chemistry
Salzman, Philip Carl, anthropology, ethnography
Sandborn, Bertil Sigvard Edmund, anatomy
Sandiford, Peter Johnston, physics
Sandor, Thomas, biochemistry, endocrinology
Sattler, Rolf, plant morphology
Saull, Vincent Alexander, geophysics
Sayeki, Hidemitsu, mathematics
Schlomiuk, Dana, mathematics
Schlomiuk, Norbert, mathematics
Schreiber, Henry Peter, physical chemistry
Schucher, Reuben, clinical chemistry, biochemistry
Schulman, Herbert Michael, cell biology, biochemistry
Schwerdtfeger, Hans Wilhelm Eduard, mathematics
Scriver, Charles Robert, pediatrics, genetics
Segall, Gordon Hart, organic polymer chemistry
Segall, Harold Nathan, clinical medicine
Seidah, Nabil George, clinical biochemistry, biophysical chemistry
Sekelj, Paul, physiology, biophysics
Selye, Hans, physiology
Seshadri, Vanamamalai, mathematical statistics
Shapiro, Lorne, internal medicine, hematology
Share, Norman N, neuropharmacology
Sharkawi, Mahmoud, pharmacology
Sharma, Ramesh C, nuclear physics, solid state physics
Sharp, Robert Thomas, theoretical physics
Sheldon, Huntington, pathology, medicine
Sherwin, Allan Leonard, neurology, immunochemistry
Shkarofsky, Issie Peter, plasma physics, microwave electronics
Shuster, Joseph, cancer, immunology
Siboo, Russell, immunology, microbiology
Sicotte, Yvon, physical chemistry
Sieniewicz, David James, radiology
Simard, Rene, cell biology, molecular biology
Simard, Therese Gabrielle, anatomy
Simard-Savoie, Solange, pharmacology
Simon, David Zvi, medicinal chemistry
Simon, Morris Arthur, pathology
Sinclair, Ronald, cell biology
Singh, Kartar, microbiology, biochemistry
Skinner, James Stanford, physiology
Skoryna, Stanley C, gastroenterology, experimental surgery
Smith, Adolph E, biophysics
Smith, Philip Edward Lake, anthropology
Solomon, Samuel, biochemistry
Sourkes, Theodore Lionel, biochemistry
Southin, John L, genetics
Spencer, John Hedley, biochemistry
Srinivasacharyulu, Kilambi, mathematics
Srivastava, Tariq Naseer, pure mathematics, statistical analysis
Stansbury, Edward James, physics
Stearn, Colin William, paleontology, stratigraphy
Stevenson, John Sinclair, mineralogy, geology
Stevenson, Louise Stevens, mineralogy
St-Pierre, Jacques, applied statistics
Stratford, Joseph, neurosurgery
Swan, Henry Stewart Drummond, forestry
Takahashi, Shuichi, mathematics
Tam, Kwok Kuen, applied mathematics
Tan, Ah-Ti Chu, chemistry, biochemistry
Taras, Paul, nuclear physics
Tassoul, Jean-Louis, astrophysics
Taurins, Alfred, organic chemistry
Taussig, Andrew, microbiology, biochemistry
Taylor, John Christopher, mathematics
Telford, William Murray, physics, geophysics
Tenenhouse, Alan M, biochemistry, endocrinology

Terroux, Ferdinand Richard, experimental physics
Tiphane, Marcel, geology
Tomas, Francisco, physics, science administration
Tomlinson, George Herbert, chemistry
Tonks, David Bayard, biochemistry, clinical chemistry
Trasler, Daphne Gay, genetics
Tremblay, Gilles, pathology
Trifaro, Jose Maria, pharmacology
Trochu, Louis, biochemistry
Trudel, Gerald Joseph, radiation chemistry, polymer chemistry
Trzcienski, Walter Edward, Jr, geology
Tuck, Norma Gordon Maxwell, pulp chemistry
Turgeon, Jean, mathematics
Vaillancourt, De Guise, internal medicine, rheumatology
Van Eeden, Constance, mathematics
Van Gelder, Nico Michel, biochemistry, physiology
Van Vliet, Carel (Karel) M, physics, electrical engineering
Vas, Stephen Istvan, medical microbiology, immunology
Ventura, Joaquin Calvo, pathology
Verly, Walter G, biochemistry
Verschingel, Roger H C, chemical instrumentation
Vezina, Claude, microbiology
Vieth, Joachim, plant morphology
Viswanathan, Muri A, plant pathology, physiology
Waddell, Eric Wilson, cultural geography, anthropology
Wade, Robert Simson, chemistry, research administration
Wainberg, Mark Arnold, cancer
Wallace, Philip Russell, theoretical physics
Wallace, Raphael Herman, microbiology
Wang, Nai-San, pathology
Warner, Charles, physical chemistry
Warshawsky, Hershey, histology
Wasserman, Aaron Reuben, biochemistry
Watters, Gordon Valentine, neurology, pediatrics
Webber, George Roger, geology
Webster, John H, radiotherapy
Weigensberg, Bernard Irvine, experimental pathology
Weinstock, Melvyn, biological structure
Weisberg, Stanley Herbert, environmental biology
West, Donald Corey, computer science
West, Eric Neil, statistics, computer science
Whitehead, Michael Anthony, theoretical chemistry, quantum chemistry
Widden, Paul Rodney, soil microbiology
Wilson, Charles Maye, mycology
Wilson, Onslow Harus, cell biology
Winkler, Carl Arthur, chemistry
Witschi, Hanspeter Rudolf, toxicology
Wittkower, Eric David, psychiatry
Wolfe, Leonhard Scott, biochemistry, neurochemistry
Wolfson, Nancy Dolly, cell physiology, developmental physiology
Yaffe, Leo, radiochemistry
Yamamoto, Y Lucas, neurosurgery, nuclear medicine
Yaphe, Wilfred, bacteriology
Yelon, Arthur Michael, solid state physics
Yong, Man Sen, pharmacology
Young, Simon Nesbitt, neurosciences
Zabiocka-Esplin, Barbara, neuropharmacology, neurophysiology
Zaborski, Bogdan, geography
Zaidman, Samuel, mathematics
Zienius, Raymond Henry, analytical chemistry
Zlobec, Sanjo, applied mathematics
Zuckermann, Martin Julius, physics

MONTREAL EAST
Wu, Chisung, organic chemistry

OUTREMONT
Cantero, Antonio, biochemistry, pathology

PERKINS
LeRoux, Edgar Joseph, insect ecology

POINTE CLAIRE
Allen, Lawrence Harvey, physical chemistry, colloid chemistry
Atack, Douglas, physical chemistry
Bharucha, Nana R, electrochemistry, surface chemistry

Chappel, Clifford, biology
Claessens, Pierre, electrochemistry
Currah, Jack Ellwood, analytical chemistry
Freter, Kurt Rudolf, organic chemistry
Gendron, Pierre Raoul, physical chemistry
Goring, David Arthur Ingham, wood chemistry
Jurasek, Lubomir, protein chemistry, microbiology
Kubes, George Jiri, pulp chemistry
Laliberte, Laurent Hector, electrochemistry, corrosion
Page, Derek Howard, pulp technology, paper
Possanza, Genus John, pharmacology
Scalian, Anthony Michael, paper chemistry
Stewart, P Brian, immunology, physiology
Trevelyan, Benjamin John, physical chemistry
Vincent, Donald Leslie, organic chemistry
Vroom, Kenneth Edwin, mathematics
Winer, Herbert Isaac, forestry

PREVILLE
Holden, George Wilfrid, chemistry

POINTE CLAIRE-DORVAL
Holme, George, immunology
Stuart, Ronald S, organic chemistry
Wasson, Burton Kendall, medicinal

QUEBEC
Ackermann, Hans Wolfgang, medical biology
Allard, Claude, biochemistry
Anderson, William Alan, molecular biology
Arsenault, Henri H, optics
Beland, Rene, geology, mineralogy
Belanger, Pierre Andre, lasers
Belzile, Rene, animal nutrition
Benoit, Paul, insect taxonomy
Bergeron, Georges Albert, physiology
Bernier, Bernard, soil science
Berthiaume, Laurent, virology
Bertrand, Forest, agriculture
Boivin, Alberic, physics
Boulet, Marcel, food science
Bourbeau, Gerard Auguste, soil morphology
Bourget, Sylvio-J, soil physics, soil conservation
Brassard, Andre, comparative pathology, immunopathology
Brisson, German J, nutrition
Buckland, Roger Basil, poultry genetics, reproductive physiology
Bullen, Miles Rex, genetics, cytogenetics
Burnell, Robert H, organic chemistry
Cailleux, Andre Paul, geography, archaeology
Campagna, Elzear (Alexandre), plant pathology
Cardinal, Andre, phycology
Cescas, Michel Pierre, soil chemistry, soil fertility
Charette, Laurent A, animal breeding
Chevrette, Joseph Edgar, agronomy
Chin, See Leang, lasers
Cloutier, Louis, chemistry
Common, Robert Haddon, agricultural chemistry
Cote, Raymond-Henri, immunochemistry, biophysical chemistry
Cujec, Bibijana Dobovisek, nuclear physics
Daguillard, Fritz, immunology, medicine
Darling, Byron Thorwell, physics
Delisle, Claude, optics
Denariez-Roberge, Marguerite Marie, lasers
Denis, Paul Yves, geography of Latin America, urb
Douglas, Richard Herbert, atmospheric physics
Dufour, Didier, biology, experimental medicine
Engel, Charles Robert, organic chemistry
Filteau, Gabriel, zoology
Finnegan, Raymond Joseph, forest entomology
Forst, Wendell, physical chemistry, theoretical chemistry
Fortier, Claude, physiology
Fortin, J Andre, botany
Fremont, Claude, physics
Gagnon, Andre, physiology
Garneau, Robert (Paul), pathology
Gauthier, Fernand Marcel, plant breeding
Gauvin, Dominique, chemistry
Gauvin, J N Laurie, theoretical physics
Gervais, Paul, agronomy
Giguere, Paul Antoine, physical chemistry
Girouard, Ronald Maurice, horticulture, plant science
Globensky, Yvon Raoul, geology
Godin, Claude, biochemistry
Goodspeed, Frederick Maynard (Cogswell), mathematics
Goulet, Jacques, food science, food microbiology
Goulet, Marcel, forest products
Grandner, Miroslav Marian, forest ecology
Grenier, Fernand, geography
Grenier, Paul Emile, geology
Hatcher, William S, mathematical logic
Heick, Nicole Begin, biochemistry
Herman, Jan Aleksander, physical chemistry
Ho-Kim, Quang, theoretical nuclear physics
Holtmann, Wilfried, animal breeding
Izatt, Jerald Ray, lasers
Jean, Marcel, biochemistry
Julien, Jean-Paul, food chemistry
Keirstead, Karl Freeman, wood chemistry, surface chemistry
Kelly, Paul Alan, medical research
Kerwin, John Larkin, physics
Knowles, Roger, microbiology
Koenig, Paul, physics
Labrie, Fernand, endocrinology, chemical engineering
Lachance, Denis, forest pathology
Lachance, Rene Onesime, plant pathology
Lacroix, Guy, marine ecology
Lacroix, Norbert Hector Joseph, mathematics
Lafond, Andre, forestry
Lafontaine, Jean-Gabriel, cell biology, electron microscopy
Lagueux, Robert, ichthyology, limnology
Landry, Fernand, exercise physiology
Langevin, Raymond J Francois, forestry
Lasalle, Pierre, geomorphology
Laurent, Roger, geology
Laurin, Andre Frederic, petrology, structural geology
LeBel, Ronald Guy, physical chemistry, chemical engineering
Leblanc, Jacques Arthur, physiology
Ledoux, Robert Louis, mineralogy
Lemieux, Jean-Marie, surgery
Lemonde, Andre, biochemistry, physiology
Leonard, Jacques Walter, polymer chemistry, physical chemistry
Lessard, Maurice, chemistry, food technology
Marceau, Gilles, anatomy, surgery
Meisels, Alexander, cytology
Mercier, Ernest, agriculture
Mockle, Jerry Auguste, pharmacognosy, pharmacology
Moldovanu, Graziella, hematology, oncology
Morin, Yves, cardiology
Morrison, Frank Orville, entomology, toxicology
Murthy, Mahadi Raghavandrarao V, biochemistry, neurochemistry
Nadeau, Gerard, theoretical physics
Ouellet, Ludovic, physical chemistry
Ouellet, Gerard Joseph, soils, plant nutrition
Pallotta, Dominick John, biology
Paquin, Roger Joseph Alfred, plant physiology, biochemistry
Paulin, Gaston, dynamic meteorology
Pezolet, Michel, biophysical chemistry
Poirier, Louis, neurology
Potvin, Pierre, physiology
Ramavataram, Kilambi, nuclear physics
Roberge, Andree Groleau, biochemistry
Roy, Jean-Claude, radiochemistry, analytical chemistry
Sabourn, Robert Joseph Edmond, geology
St Arnaud, Gregoire, obstetrics & gynecology
Samson, Euchariste, medicine
Savard, Jean Yves, solid state physics
Savoie, Rodrigue, physical chemistry
Seguin, Louis-Roch, fisheries
Seguin, Maurice Krisholm, geophysics, geology
Simard, Sylvain J, endocrinology
Singh, Pritam, biochemistry, pharmacology
Slobodrian, Rodolfo Jose, physics
Steriade, Mircea, neurophysiology
St-Pierre, Claude, nuclear physics
De Margerie, Jean-Marie, ophthalmology
DeMedicis, E M J A, biochemistry, organic chemistry
Demers, Pierre-Paul, pediatrics, cardiology
Deslongchamps, Pierre, organic chemistry
Desnoyers, Jacques Edouard, physical chemistry
Dilenge, Domenico, radiology
Dunnigan, Jacques, physiology
Dupuis, Gilles, biochemistry
Elhilali, Mostafa M, cancer
Galeano, Cesar, neurophysiology
Hugon, Jean S, electron microscopy, gastroenterology
Juillet, Jacques Andre, ecology, biometry
Kimmerle, Frank, electrochemistry
Konguetsof, Leonidas, mathematics
Lalancette, Jean-Marc, inorganic chemistry, environmental chemistry
Lamarche, Guy, neurophysiology
Lamy, Francois, biochemistry
La Salle, Gerald, medical administration
Lavallee, Marc, biophysics, biomedical engineering
Legault, Albert, systematic botany, phytogeography
Lehoux, Jean-Guy, biochemistry, endocrinology
Lessard, Jean, organic chemistry
Leveque, Theodore Francois, anatomy
Marchessault, Victor Henri, pediatrics, microbiology
Mignault, Jean De L, medicine, cardiology
Munan, Louis, epidemiology
Nawar, Tewfik, nephrology
Nigam, Vijai Nandan, biochemistry, cell biology
O'Neil, Louis-C, entomology
Park, Won Kil, biochemistry, pharmacology
Pelletier, Gerard Eugene, chemistry
Percy, Bernard Jean Francois, surgery
Peticlerc, Claude Jean, biochemistry, clinical biochemistry
Preiss, Benjamin, biological chemistry
Ramon-Moliner, Enrique, neuroanatomy
Schanne, Otto F, biophysics, instrumentation
Seufert, Wolf D, biophysics, molecular biology
Sharma, Madan Lal, entomology, ecology
Tahan, Theodore Wahba, radiotherapy
Tan, Liat, organic chemistry
Thouez, Jean-Pierre Mary, urban geography, economic geography
Tran-Manh, Ngo, nuclear medicine, medical physics

SHERBROOKE
Banville, Marcel, theoretical physics
Bellabarba, Diego, endocrinology
Bessette, France Marie, molecular biophysics
Bounous, Gustavo, surgery, gastroenterology
Bourgaux, Pierre, animal virology, molecular biology
Braiovsky, Carlos Alberto, cell biology, virology
Brown, Gordon Manley, organic chemistry
Cabana, Aldee, physical chemistry, molecular spectroscopy
Cereti, Elena, cardiology, electrophysiology
Cote, Roger Albert, pathology
Cousineau, Leo, hematology, computer science
Swan, Peter Howard, theoretical physics, biochemistry
Turcot, Jacques, surgery
Turrell, George Charles, physical chemistry
Turrell, Sylvia Jones, physical chemistry
Vanier, Jacques, quantum electronics
Villeneuve, Paul Yvon, urban geography
Waithe, William Irwin, cell physiology
Wilson, Cynthia, climatology
Zardecki, Andrzej, quantum optics

SHAWINIGAN
Germain, Leo (Joseph Frederic Marcel), chemistry

SENNEVILLE
Marshall, Harry Borden, chemistry

RIMOUSKI
Ferron, Jean H, ethology

ST HYACINTHE
Bissaillon, Andre, veterinary anatomy, mammalogy
Blouin, Andre, pharmacology
Flipo, Jean, veterinary medicine
Garon, Olivier, comparative anatomy, veterinary medicine
Gelinas, Louis-de-Gonzague, veterinary medicine
Harrison, Robert J, entomology, parasitology
Jabalpurwala, Kaizer E, inorganic chemistry, physical chemistry

ST BRUNO
Hine, Kenneth Ernest, electrochemistry
Roberts, Kenneth David, biochemistry, endocrinology

SOREL
Sweet, Roger George, physical chemistry, inorganic chemistry

SILLERY
Hamelin, Louis-Edmond, geography

STE ANNE DE BELLEVUE
Anastassiadis, Phoebus A, agricultural chemistry
Bachynski, Morrel Paul, plasma physics
Baker, Robert Dale, animal science
Barthakur, Nayana N, physics
Blackwood, Allister Clark, microbiology
Conradi, Jan, solid state physics
Crane, Robert Anthony, laser physics
DeVoe, Irving Woodrow, medical microbiology, electron microscopy
Dion, Henry George, soils
Doneter, Eugene, animal nutrition
Erskine, Gordon John, inorganic chemistry, organometallic chemistry
Estey, Ralph Howard, plant pathology
Farmer, Florence Amelia, nutrition
Grainger, Edward Henry, biological oceanography
Grant, William Frederick, plant breeding
Harpur, Robert Peter, biochemistry
Hill, Stuart Baxter, ecology, entomology
Idziak, Edmund Stefan, food microbiology
Ingram, Jordan Miles, biochemistry
Kevan, Douglas Keith McEwan, entomology
Klinck, Harold Rutherford, agronomy
MacKenzie, Angus Finley, soil chemistry
MacLeod, Robert Angus, microbiology
Mansfield, Arthur Walter, marine biology
McFarlane, John Elwood, insect physiology
McIntyre, Robert John, solid state physics
Millette, Gerard J F, soil chemistry
Moody, Harry John, microwave physics, communications engineering
Nikolaiczuk, Nikolai, poultry nutrition
Percy, Jonathan Arthur, environmental physiology
Sackston, Waldemar Esi, plant pathology
Steppler, Howard Alvey, agronomy
Stewart, Robin Kenny, entomology
Tanner, Charles E, immunology, parasitology
Tanner, Jurate E, microbiology
Ulagaraj, Muniyandy Seydunganallur, entomology
Van Lier, Johannes Ernestinus, biochemistry
Vobecky, Josef, epidemiology
Weber, Joseph, virology

ST LAURENT
Anderson, Donald Arthur, electronic physics
Borella, Luis Enrique, pharmacology
Hirsch, Michael Allen, endocrinology
Martel, Rene R, pharmacology
Sestanj, Kazimir, organic chemistry, physical chemistry

ST JEAN
Chiang, Morgan S, genetics
Favreau, Roger F, physics
Hogue, Eugene J, plant physiology
Marcoux, Jules E, entomology
Paradis, Rodolphe Omer, entomology
Tilley, Donald E, physics

Vickery, Vernon Randolph, taxonomy, entomology
Wacasey, Jervis Winn, biological oceanography
Waksberg, Armand L, mathematics, physics
Warkentin, Benno Peter, environmental management
Alarie, Albert, soil microbiology
Berlinguet, Louis, organic biochemistry
Bolduc, Reginald J, plant physiology, histochemistry
Caron, Wilfrid M, surgery
D'Aoust, Andre Lucien, plant physiology, plant biochemistry
Dionne, Jean-Claude, geology, geography
Doyon, Dominique, plant ecology
Dugal, Louis Paul, physiology
Gasser, Heinz, crop physiology, plant breeding
Hope, Hugh Johnson, agricultural biochemistry
Jurdant, Michel Louis, forest ecology
Koran, Zoltan, forest products, pulp technology
Lavoie, Victorin, plant ecology
Lemay, Jean-Paul, animal physiology, animal breeding
Lepage, Marius, agricultural chemistry, biological chemistry
Lit, John Wai-Yu, optics
Lortie, Marcel, environmental sciences
Marmet, Paul, atomic physics, molecular physics
McLeod, John Malcolm, entomology
McLeod, Lloyd Alexander, physical chemistry, research administration
Moreau, Jean Raymond, food science
O'Grady, Lawrence J, soil fertility
Ouellette, Guillemond Benoit, histopathology
Pelletier, Georges H, endocrinology
Riva, John, paleontology, structural geology
Roberge, Marcien Romeo, microbiology
Willemot, Claude, plant physiology
Winget, Care Henry, forest ecology, tree physiology

TROIS-RIVIERES
Julien, Julien Bernard, phytopathology
Lehman, Eugene H, mathematical statistics
Pelletier, Raymond Marcel, economic geography

VALCARTIER
Giroux, Guy, optics

VARENNES
Cloutier, Gilles Georges, physics
Drouet, Michel Georges, plasma physics, electrical engineering
Gregory, Brian Charles, electronic physics, plasma physics
Jean, Benoit, plasma physics
Johnston, Tudor Wyatt, plasma physics
Parbhakar, Kanwal Jit, lasers, plasma physics
Richard, Claude, plasma physics, optical physics
Stansfield, Barry Lionel, plasma physics
Terreault, Bernard J E J, nuclear science

VERDUN
Yates, Claire Hilliard, analytical chemistry, pharmaceutical chemistry

VILLE DE LAVAL
Chagnon, Andre, virology, tissue culture

VILLE ST-LAURENT
Sehgal, Surendra N, microbiology

WEST MONTREAL
Nisbet, Michael Alan, organic chemistry

WESTMOUNT
Taguchi, Yoshinori, urology, immunology

SASKATCHEWAN

MELFORT
Nuttall, Wesley Ford, soil fertility

MOOSE JAW
Hoag, Gordon Neil, clinical biochemistry

REGINA
Agnew, Robert Morson, immunobiology
Banting, James Daniel, plant physiology
Barton, Richard J, physical chemistry, physical metallurgy
Chandler, William David, physical organic chemistry
Conlan, James, mathematics
Cullimore, Denis Roy, microbial ecology, bacteriology
Dojcsak, Gyozo Victor, geography
Gear, James Richard, organic chemistry, biochemistry
Gordon, William Anthony, paleontology
Grotewold, Andreas, economic geography, international economics
Grover, Rajbans, environmental sciences
Harold, Stephen, biochemistry
Harris, Peter, weed science
Hay, James Robert, weed science
Hontzeas, S, nuclear chemistry
Johnson, Keith Edward, high temperature chemistry
Katz, Leon, science policy
Kaul, S K, pure mathematics
Koh, Eusebio Legarda, applied mathematics
Kybett, Brian David, physical chemistry
Larson, Denis Wayne, physical chemistry
Ledingham, George Filson, systematic botany
Lee, Donald Garry, physical chemistry, organic chemistry
McConnell, Wallace Beverly, biochemistry
Mook, Leonard Jan, animal ecology
Naqvi, Saiyid Ishrat Husain, nuclear physics, astronomy
Quick, William Andrew, botany
Rao, Veldanda Venugopal, pure mathematics
Riddell, William A, biochemistry
Riegert, Paul William, insect physiology
Robertson, Beverly Ellis, crystallography
Robertson, Hugh Elburn, biochemistry, microbiology
Robinson, Robert Reid, physical chemistry
Rummens, F H A, molecular spectroscopy, physical chemistry
Rystephanick, Raymond Gary, theoretical physics
Sato, Daihachiro, mathematics
Secoy, Diane Marie, vertebrate biology
Smith, Allan Edward, agriculture, organic chemistry
Toews, Kornelius Gerhard, mathematics
VanCleave, Allan Bishop, physical chemistry
Vigrass, Laurence William, geology
Walther, Alina, plant physiology, plant ecology
Zacharuk, R Y, insect morphology, insect pathology

SASKATOON
Angel, Joseph Francis, nutrition, biochemistry
Arnold, Ralph Gunther, environmental geology
Ashford, Ross, agronomy
Austenson, Herman Milton, agronomy
Bardwell, John Alexander Eddie, physical chemistry
Begg, Robert William, biochemistry
Bell, John Milton, animal nutrition
Bellamy, Raymond Edward, biology
Blair, Donald George Ralph, biochemistry
Blakeley, Edwin Raymond, microbial biochemistry
Blum, Richard, mathematics
Bone, Robert M, geography
Boulton, Alan Arthur, neurochemistry, psychiatry
Brandell, Bruce Reeves, anatomy, zoology
Braun, Willi Karl, micropaleontology, stratigraphy
Buchan, Douglas John, internal medicine
Burkell, Charles Craig, radiology
Burrage, R H, entomology
Butler, Harry, embryology
Caldwell, William Glen Elliot, geology
Cates, Geoffrey William, clinical pathology
Chakravarti, Aninda Kumar, agricultural geography
Child, Jeffrey James, microbiology
Chinn, Stanley H F, soil microbiology
Christiansen, E A, geology
Church, Norman Stanley, entomology
Coburn, Frank Emerson, psychiatry
Coleman, Leslie Charles, geology

Coupland, Robert Thomas, plant ecology, systems ecology
Craig, Burton Mackay, agricultural biochemistry
Crawford, Roy Douglas, poultry genetics
Currie, Balfour Watson, physics
Dabbs, Donald Henry, horticulture
Davis, Bruce Allan, analytical biochemistry
Davis, Gordon Richard Fuerst, zoology, insect physiology
Dawson, Peter Stephen Shevyn, microbial biochemistry, physiology
Dempster, George, medical microbiology
Dimmock, Jonathan Richard, pharmaceutical chemistry
Downey, Richard Keith, plant breeding
Durley, Richard Charles, plant biochemistry
Dvorak, Jan, cytogenetics
Eager, Richard Livingston, physical chemistry
Emson, Harry Edmund, pathology
Ewen, Alwyn Bradley, insect physiology
Fedoroff, Sergey, embryology, histology
Fowke, Lawrence Carroll, plant cytology
Gamborg, Oluf Lind, plant biochemistry
Gerrard, John Watson, pediatrics
Gibson, Douglas (Lorne), dairy bacteriology
Gilmour, Thomas Henry Johnstone, invertebrate zoology
Goplen, Bernard Peter, genetics, plant breeding
Gorin, Philip Albert James, chemistry
Grant, Donald R, biochemistry, organic chemistry
Greenshields, John Edward Ross, plant breeding
Gregory, John B, aeronomy, meteorology
Grodums, Emma Irene, dentistry, histology
Gruen, Hans Edmund, mycology
Gunn, George Bradford, physical chemistry
Gusta, Lawrence V, plant physiology
Haig, Thomas Harrison Brian, surgery
Hammer, Ulrich Theodore, limnology
Harms, Vernon Lee, systematic botany
Harvey, Bryan Laurence, genetics, plant breeding
Haskins, Reginald Hinton, mycology, microbiology
Hendry, Hugh Edward, geology
Hickie, Robert Allan, pharmacology
Hirose, Akira, plasma physics
Holl, Frederick Brian, plant genetics
Horlick, Louis, medicine
Houston, Clarence Stuart, radiology
Howarth, Ronald Edward, plant chemistry
Howell, William Edwin, animal breeding
Hunt, T E, medicine
Ingledew, William Michael, microbiology, biochemistry
Jaques, Louis Barker, physiology, pharmacology
Jeffrey, James George, pharmaceutical chemistry
Johnson, Gordon E, pharmacology
Jones, Graham Alfred, agricultural microbiology
Juorio, Augusto Victor, neuropharmacology
Kaul, Rudolf, crop physiology
Kavadas, Alexander D, physics
Kennedy, John Edward, physics
Kent, Henry Peter, radiology
King, John, plant physiology
Knight, Arthur Robert, photochemistry
Knott, Douglas Ronald, plant genetics
Knowles, Robert Patrick, plant breeding
Kupsch, Walter Oscar, geology
Kurz, Wolfgang Gebhard Walter, plant microbiology
Langford, Fred F, economic geology
LaRue, Thomas A, biochemistry
Laut, Wilfred Wayne, physiology, biochemistry
Lee, Choi Chuck, organic chemistry
Lehmkuhl, Dennis Merle, entomology, ecology
Llewellyn, Edward John, experimental physics
MacAulay, Wesley Claude, pharmaceutical chemistry
MacDonald, James Cameron, biochemistry
Maginnes, Edward Alexander, horticulture
Maher, William J, vertebrate ecology
Manohar, Rampukar, numerical analysis
Martin, Robert O, biochemistry, organic chemistry
May, Sherry Jan, atmospheric physics
Maybank, John, atmospheric physics
McArthur, Charles Stewart, biochemistry

McCallum, Kenneth James, physical chemistry
McDonald, Ian MacLaren, psychiatry
McEwen, Kathleen Lenore, theoretical chemistry
McLennan, Barry Dean, biochemistry
McLintock, John James Reid, medical entomology
McPhail, Clarence Wilmer Bernard, public health, dentistry
Mills, James Herbert Lawrence, veterinary pathology
Montalbetti, Raymon, physics
Mossman, David John, mineralogy, geology
Munkacsi, Istvan, anatomy
Murphy, Bruce Daniel, reproductive physiology
Naylor, James Maurice, botany
Nelson, Stuart Harper, horticulture
Newstead, James Duncan MacInnes, cell biology, histology
Nicholson, Hugh Hampson, animal nutrition, animal genetics
Nielsen, N Ole, veterinary pathology
O'Shaughnessy, Charles Dennis, statistics
Owen, Bruce Douglas, animal nutrition, animal physiology
Paine, Kenneth William, neurosurgery
Paul, Eldor Alvin, soils
Paulton, Richard John Laurance, microbiology, biochemistry
Pepper, James Morley, organic chemistry
Pepper, Thomas Peter, nuclear physics
Phillis, John Whitfield, neurophysiology, neuropharmacology
Pollak, Victor A, biomedical engineering, information science
Prasad, Kailash, pharmacology
Quail, John Wilson, inorganic chemistry
Rank, Gerald Henry, genetics
Rausch, Robert Lloyd, veterinary parasitology
Redmann, Robert Emanuel, plant ecology
Rennie, Donald Andrews, soil chemistry
Richards, John Howard, geography
Ripley, Earl Allison, biometeorology
Rowe, John Stanley, plant ecology
Rozdilsky, Bohdan, neuropathology
Russell, David Bernard, radiation biology
Saini, Girdhari Lal, mathematics
St Arnaud, Roland Joseph Odilon, soils
Sarjeant, William Antony Swithin, paleontology, paleoecology
Saunders, James Robert, veterinary pathology, veterinary microbiology
Sawhney, Vipen Kumar, developmental biology
Scheltgen, Elmer, biophysics, molecular genetics
Schiefer, H Bruno, veterinary pathology
Senior, John Brian, inorganic chemistry
Shargool, Peter Douglas, plant biochemistry
Shin, Yong-Moo, nuclear physics
Shokeir, Mohamed Hassan Kamel, medical genetics
Simpson, Graham Miller, plant physiology
Singh, Rajinder, mathematical statistics
Sisodia, Chaturbhuj Singh, veterinary pharmacology, toxicology
Skarsgard, Harvey Milton, physics
Skinner, Orville Ray, theoretical physics
Slater, George P, organic chemistry, biochemistry
Slinkard, Alfred Eugene, agronomy
Smith, David Lawrence Thomson, veterinary pathology
Smith, Jeffrey Drew, plant pathology, mycology
Smith, Peter James, physical organic chemistry
Sosulski, Frank Walter, agronomy, nutrition
Spencer, John Francis Theodore, microbiology
Spinks, John William Tranter, chemistry
Stauffer, Mel R, structural geology
Steer, Ronald Paul, physical chemistry
Steeves, Taylor Armstrong, botany
Stewart, John Wray Black, soil science, chemistry
Stringam, Gary Rice, cytogenetics
Sukahe, Prakash Vinayak, physiology
Sutherland, Gerald Bonar, physiology
Sutherland, Ronald George, chemistry
Tinline, Robert Davies, plant pathology
Tomusiak, Edward Lawrence, theoretical nuclear physics
Tourigny, Guy J, physical organic chemistry
Tracie, Carl Joseph, historical geography, geography of Western Canada
Trew, John Allan, biochemistry

Tulloch, Alexander Patrick, organic chemistry
Turel, Franziska Lili Margarete, plant physiology
Tymchatyn, Edward Dmytro, mathematics
Underhill, Edward Wesley, plant biochemistry
Vella, Francis, biochemistry, genetics
Verrall, Ronald Ernest, physical chemistry
Von Rudloff, Ernst Max, organic chemistry
Vose, John Randal, agricultural biochemistry
Waltz, William Lee, photochemistry, radiation chemistry
Watson, Linvill, cultural anthropology
Weil, John A., chemical physics
Wenger, Byron Sylvester, developmental biology
Wenger, Eleanor Lerner, developmental biology
Wetter, Leslie Robert, biochemistry
Whitaker, Sidney Hopkins, geology
Willard, John Royal, entomology, ecology
Williams, Charles Melville, physiology, genetics
Wilson, Michael Robert, geography
Wobeser, Gary Arthur, fish pathology, wildlife pathology
Wood, James Alexander, neurochemistry
Wood, James Douglas, biology
Woodford, Vernon Rich, Jr., biochemistry
Woods, Robert James, radiation chemistry
Wyant, Gordon Michael, anesthesiology
Ying, Kuang Lin, genetics, cytogenetics
Zaleski, Witold Andrew, medicine
Zuck, Donald Anton, pharmaceutical chemistry

SPRINGSIDE
Zilke, Samuel, agronomy, ecology

SWIFT CURRENT
Campbell, Constantine Alberga, soil chemistry
Heinrichs, David Henry, agronomy
Hurd, Edwin Albert, plant breeding
Kilcher, Mark R., agronomy
Lawrence, Thomas, plant breeding
Looman, Jan, botany
Salmon, Raymond Edward, poultry nutrition
Townley-Smith, Thomas Frederick, plant breeding
McLeod, Lloyd Alexander, physical chemistry, research administration

OTHER COUNTRIES

ARGENTINA
Archangelsky, Sergio, paleobotany, palynology
Arguelles, Amilcar Emilio, endocrinology
Cabrera, Angel Lulio, systematic botany, phytogeography
Carrea, Raul, neurosurgery
De Robertis, Eduardo Diego P., histophysiology, cytology
Feinstein, Alejandro, astrophysics
Feldman, Jose M., plant virology
Foglia, Virgilio Gerardo, endocrinology
Gershanik, Simon, seismology
Gracia, Olga, plant virology
Guerrero, Ariel Heriberto, chemistry, biophysical chemistry
Mancini, Robert Eusebio, medicine
Pontis-Videla, Rafael Edmundo, pathology
Sabade, Jorge, astrophysics
Slauciras, Sergeis, astronomy

AUSTRALIA
Assaykeen, Tatiana Anna, endocrinology, pharmacology
Bauer, Francis Harry, geography
Bearman, Richard John, physical chemistry, chemical physics
Bies, David Alan, acoustics
Blakers, Albert Laurence, mathematics
Bockris, John O'Mara, physical chemistry
Bolotin, Herbert Howard, nuclear physics
Britten, Edward James, plant cytogenetics
Clark, William Arthur, bacteriology
Dance, Ian Gordon, inorganic chemistry
Dunlop, Robert Hugh, veterinary pharmacology, veterinary physiology

Durling, Frederick Charles, mathematical statistics, mathematics
Eggen, Olin Jeuck, astrophysics
Frakes, Lawrence Austin, marine geology, sedimentology
Gilmartin, Malvern, biological oceanography
Goncz, John Henry, physics
Haight, John Richard, neurobiology, evolutionary biology
Hales, Anton Linder, geophysics
Hammer, Richard M., physical geography, hydrology, geology
Hase, Donald Henry, zoology
Heatwole, Harold Franklin, zoology
Heinsohn, George E., vertebrate zoology, vertebrate ecology
Johnson, Kenneth Olafur, neurophysiology, biomedical engineering
Kalnajs, Agris Janis, astronomy
Kron, Gerald Edward, astronomy
Kudenov, Jerry David, invertebrate zoology, pollution biology
Libby, Willard Gurnea, geology
Liebermann, Robert C., geophysics
Mahler, Kurt, pure mathematics
Marchalonis, John Jacob, biochemistry, immunology
Milford, Sidney Nevil, environmental physics, physical oceanography
Moore, Walter John, mathematics
Mond, Bertram, mathematics
Nickel, Ernest Henry, mineralogy
Paleg, Leslie G., plant physiology
Parker, Robert Ray, zoology
Pearson, John Carwardine, helminthology
Peaslee, David Chase, theoretical physics,
Robertson, William Archer, paleomagnetism
Rodieck, Robert William, vision genetics
Schwinghamer, Erwin A., microbial genetics
Smith, David Francis, marine ecology, microbial ecology
Vozoff, Keeva, geophysics
Zimmerman, Elwood Curtin, entomology

AUSTRIA
Arledter, Hanns Ferdinand, physical chemistry
Fried, Maurice, plant nutrition, soil fertility
Goresline, Harry Edward, food technology
Hopkins, Leon Lorraine, Jr., nutrition, biochemistry
Jablonski, Eugene, geology; botany
Kickert, Robert Warren, ethnology; applied anthropology
Levien, Roger Eli, systems analysis, information science
Lindquist, Donald Arthur, entomology
Moxham, Robert Lynn, geochemistry
Otway, Harry John, science policy, systems analysis
Szabo, C S Karoly, organic chemistry
Wood, Howard John, astronomy

BANGLADESH
Alam, Ashraf Ul, biochemistry
Jackson, Thad Marshall, immunology, nutrition

BELGIUM
Bowman, Lewis Wilmer, organic chemistry
Conary, Robert Ekvall, physical chemistry
Gerwitz, David L., agricultural chemistry
Hamilton, James Chipman, organic chemistry
Hawkes, Arthur Stanley, physical chemistry
Kovach, Eugene George, organic chemistry, science administration
Kratochvil, Clyde Harding, biochemistry
Laszlo, Pierre, organic chemistry
Sybesma, Christiaan, biophysics
Wickson, Edward James, plastics

BRAZIL
Adler, Ronald John, theoretical physics
Amorim, Dalmo de Souza, cardiovascular diseases, cardiovascular physiology
Barbosa, Octavio, geology
Barnes, Roderick Arthur, organic chemistry
Barry, Cornelius, entomology
Bitancourt, Agesilau Antonio, plant pathology
Brown, Richard Edwin, theoretical chemistry
Bruns, Roy Edward, theoretical chemistry
Coliver, Michael Moore, solid state physics
Contu, Paolo, anatomy, neuroanatomy
Da Cunha, Antonio Brito, population genetics, cytogenetics
D'Ambrosio, Ubiratan, mathematics
De Goes, Paulo, microbiology, immunology
Delavault, Robert Edmund, biochemistry, metallurgy
Dutra Oliveira, Jose Eduardo, nutrition, internal medicine
Foglio, Mario Eusebio, solid state physics, theoretical physics
Freire-Maia, Ademar, human genetics
Freire-Maia, Derita Villalba, human genetics, cytogenetics
Geldart, Lloyd, geophysics
Gibler, John Wesley, plant breeding
Gottlieb, Otto Richard, phytochemistry
Grynszpan, Flavio, biomedical engineering
Guimaraes, Armenio Costa, internal medicine, cardiology
Kerr, Warwick Estevam, genetics
Kiel, Alvin E., solid state physics
Kousky, Vernon E., dynamic meteorology
Krieger, Eduardo Moacyr, medical physiology
Kuznesof, Paul Martin, inorganic chemistry
Laughlin, Charles William, plant nematology
Lewis, Richard Wheatley, Jr., economic geology
Lobo, Luiz Carlos Galvao, endocrinology; physiology
Mendes, Erasmo Garcia, comparative physiology
Moore, James Elton, chemical physics
Moors, Walter B., natural products
Moscovici, Mauricio, anatomy, surgery
Nephme, Amador, parasitology; medicine
Nussenzveig, Herch Moyses, mathematical physics
Palmeira, Ricardo Antonio Ribeiro, cosmic ray physics, space science
Pavan, Crodowaldo, genetics
Peterson, Vern Leroy, astronomy
Rocha e Silvan, Mauricio, pharmacology
Rosenfield, Christine Ann Culp, agricultural chemistry
Salzano, Francisco Mauro, human genetics
Sharvelle, Eric George, plant pathology
Sick, Helmut, zoology
Strong, Frederick Carl, III, analytical chemistry
Westcott, Peter Walter, ecology; zoology
Williams, Calvin Herndon, Jr., physical chemistry, analytical chemistry
Willis, Edwin O'Neill, ethology
Wilson, Eva Donelson, nutrition

BRITISH WEST INDIES
Farr, Thomas Howard, entomology

BURMA
Goossens, Pierre J., economic geology

CHILE
Blanco, Victor Manuel, astronomy
Hesser, James Edward, astrophysics, molecular spectroscopy
Iglesias, Rigoberto, medicine
Kunkel, William Eckart, astronomy
Luco, Joaquin, physiology; neurophysiology
Middleton, Samuel, physiology
Montiel, Francisco, metabolism, microbiology
Osmer, Patrick Stewart, astronomy
Rodriguez-Leiva, Manuel, virology, bacteriology

CHINA, REPUBLIC OF
Beall, Geoffrey, statistics
Chen, Steve Shih-Chieh, agricultural

Clark, Chester William, physics
Gould, Sydney Henry, mathematics
King, Dorothy Wei (Cheng), nutrition
Moomaw, James Curtis, agronomy
Riley, James Joseph, hydrology, plant physiology
Yang, Charles (Yu-Di), plant pathology, physiology

COLOMBIA
Burnell, Louis A, theoretical chemistry
Eberhard, William Granville, behavioral biology, arachnology
Greer, Donald Lee, medical mycology, microbiology
Eskafi, Fred M., entomology
Fleming, Glenn Allen, medical entomology
Francis, Charles Andrew, agronomy, plant breeding
Jennings, Peter Randolph, plant genetics
Mullenax, Charles Howard, animal science, ecology
Pradilla, Alberto Gonzalo, pediatrics, metabolism
Rachie, Kenneth Owen, plant genetics
Ramirez, Jesus Emilio, geophysics
Raun, Ned S., animal nutrition
Thatcher, Vernon Everett, zoology, parasitology
West Eberhard, Mary Jane, animal behavior, evolution

COSTA RICA
Araujo, Jose Emilio Goncalves, soil science
Carballo-Quiros, Alfredo, plant breeding, quantitative genetics
Fernandez, Bernal, microbial physiology
Forsythe, Warren M., soil physics
Greene, George Linden, plant pathology, physiology
Kleesattel, Claus, physics, electrical engineering
Lambert, William M, Jr., mathematics
Mata, Leonardo J., public health, nutrition
Moh, Carl Craig, agronomy, radiation botany
Smith, Susan May, ecology; animal behavior
Tosi, Joseph Andrew, Jr., ecology

CUBA
Joly, Daniel Jose, epidemiology, oncology

DENMARK
Avery, John Scales, quantum chemistry
Berman, Arthur Irwin, science education
Cooke, Hermon Richard, Jr., mining geology
D'Angelo, Nicola, nuclear physics
Deutch, Bernhard Irwin, physics
Garrett, Jerry Dale, experimental nuclear physics
Hansen, Charles M., physical chemistry, chemical engineering
Kahn, Albert, developmental biology
Pape, Leon, biophysics, physics

EAST AFRICA
Wagner, Gerald Gale, immunology

ECUADOR
Feininger, Tomas, geology

EGYPT
Nelson, Cynthia, cultural anthropology
Stewart, Donald Martin, plant pathology

ENGLAND
Blackwell, David (Harold), mathematics, statistics
Booth, Norman E., elementary particle physics
Boullin, David John, clinical pharmacology, physiology
Bromberger, Samuel H., geology
Bronk, John Ramsey, biochemistry, cell physiology
Carney, Gordon C., invertebrate physiology
Conover, Lloyd Hillyard, organic chemistry
Creighton, Thomas Edwin, molecular biology
Dulbecco, Renato, virology
Edsall, Geoffrey, immunology, medicine
Eells, James, Jr., mathematics
Elkinton, Moylan, oseph Russell, medicine
Falk, Gertrude, biophysics
Fermi, Giulio, molecular biology, systems analysis

Fields, Kay Louise, molecular biology, neurosciences
Freeman, Raymond, physical chemistry
Gibbs, R Darnley, botany
Grodzinski, Bernard, plant science
Hall, Theodore (Alvin), polymer science
Hattersley-Smith, Geoffrey Francis, glaciology, geomorphology
Heegaard, Erik Vilhelm, biochemistry
Howe, Stephen Henry, physics
Hughes, Daniel Richard, algebra, geometry
Johnston, David, exploration geophysics
Jolly, Alison Bishop, animal behavior
Jondorf, Werner Robert, biochemistry, pharmacology
Kahn, Kenneth, biochemistry
Kaplan, Martin Mark, public health
Kass, Thomas Lewis, genetics
Koerber, Walter Ludwig, biochemistry, microbiology
Kopal, Zdenek, astronomy
Kordan, Herbert Allen, developmental physiology, plant morphogenetics
Kupperman, Morton, mathematical statistics
Landauer, Walter, experimental embryology
Lane, Nancy Jane, cytochemistry
Leigh, Roger, geography
Lemons, Hoyt, physical geography
Lieb, William Robert, biophysics
Lloyd, David Pierce Caradoc, physiology
Lochhead, John Hutchison, invertebrate zoology
Louis, Thomas Michael, reproductive endocrinology
Mason, Ronald George, geophysics
Nine, Ogden Wells, Jr., petroleum engineering
Novotny, Eva, astronomy
O'riordan, Timothy, resource geography
Parrent, George Burl, Jr., theoretical physics
Purdy, Edward George, geology
Radford, Alan, genetics
Record, Walter Ross, petroleum geology
Resnick, Michael Aaron, biophysics, genetics
Rubin, John Ronald, biophysical chemistry
Schoene, Dwight Lorin, rubber chemistry, agricultural chemistry
Shilliber, Harry Albert, physics, geophysics
Short, Oliver Alton, electronics
Spaght, Monroe Edward, physical chemistry, organic chemistry
Stockell-Hartree, Anne, biochemistry
Stonier, Tom Ted, science policy, cell physiology
Strehlow, Clifford David, environmental health
Swanson-Eartly, Heidi H, endocrinology
Thomasson, Maurice Ray, geology
Thorson, John Wells, neurophysiology, biophysics
Vlitos, August John, plant physiology
Weber, Richard Robert, applied mathematics
Wehinger, Peter Augustus, astronomy
Woodruff, Ronny Clifford, genetics
Wyckoff, Susan, astronomy

ETHIOPIA
Gouin, Pierre Laurier, geophysics
Wolff, Milo Mitchell, physics, electrical engineering

FRANCE
Alvarado, Francisco, biochemistry, physiology
Blumenfeld, Henry A, physics
Bohrer, John Junior, chemistry
Burke, John T, internal medicine
Campbell, Donald Edward, analytical chemistry, ceramics
Capurro, Luis R A, physical oceanography, marine meteorology
Coles, Leslie Stephen, information science
Daifuku, Hiroshi, anthropology, international management
Diner, Bruce Aaron, photobiology
Drea, John James, Jr., entomology
Ertingshausen, Gerhard, clinical chemistry
Focke, Harold Anthony, science education
Gozdan, Walter Joseph, industrial chemistry
Gresser, Ion, virology, experimental pathology
Higginson, John, pathology
Koch-Weser, Jan, internal medicine, clinical pharmacology
Kramish, Arnold, nuclear physics, international relations

Marezio, Massimo, crystallography
Murphy, Preston V, solid state science
Passman, Sidney, physics, research administration
Petrich, Robert Paul, polymer science
Pudles, Julio, biochemistry
Reinstein, Jerome Alan, physical pharmacy, cosmetic chemistry
Renaud, Serge, experimental pathology
Robinson, Berol Lee, nuclear physics
Roderick, Hilliard, nuclear physics
Rose, Harvey Arnold, fluid dynamics
Roth-Schechter, Barbara F, pharmacology
Russell, Charles Daniel, Jr., physical chemistry
Smith, Lee Anderson, micropaleontology, biostratigraphy
Van Ligten, Raoul Fredrik, optics
Voet, Andries, chemistry
Wessel, Richard Deaton, entomology
Widmer, Hans, polymer chemistry
Yarmolinsky, Michael Bezalel, molecular biology
Yarzabal, Luis Alberto, medical microbiology
Zalokar, Marko, developmental genetics

GERMANY, FEDERAL REPUBLIC OF
Anghileri, Leopoldo Jose, radiochemistry, radiopharmacology
Bauer, Ernst Georg, surface physics
Baumann, Winfried, lasers
Becker, Gweneth (Leslie), biology
Beckwith, Newell Pierce, organic chemistry
Blanks, Janet Marie (Clarenbach), neurocytology
Burkhardt, Walter H, computer science
Cavonius, Carl Richard, neuroscience
Chapman, Albert Simonds, geography
Doerfler, Walter, virology
Dulz, Gunther, physical chemistry
Dum, Christian Thomas, plasma physics, theoretical physics
Everling, Friedrich Gustav, nuclear physics
Fahimi, Hossein Dariush, pathology, electron microscopy
Fang, Joong, mathematics, philosophy
Fernald, Russell Dawson, neurobiology
Foeckler, Francis H, Jr., pharmacology
Fox, John Dana, physics
Funkhouser, Edward Allen, plant physiology
Guth, Egbert Karl Anton, inorganic chemistry, physical chemistry
Helmreich, Ernst, biochemistry
Hutchison, John Joseph, organic chemistry, polymer chemistry
Itten, David Frederick, physical chemistry
Keester, Kenneth Lee, solid state chemistry, crystallography
Kober, Ehrenfried H, organic chemistry
Krader, Lawrence, anthropology, ethnology
Legendy, Charles Rudolf, theoretical physics
Leitzmann, Claus, biochemistry, nutrition
Maltese, George J, mathematics
Marvin, Donald Arthur, molecular biology
McClelland, Clyde Lloyd, nuclear physics
Montgomery, Michael Davis, physics
Mota, Ana Celia, low temperature physics
Paproth, Hans Helmut, data processing
Perry, Judith Joanna, astrophysics
Protsch, Reiner Robert Rudolf, physical anthropology, medical anthropology
Reed, Fred DeWitt, Jr., medicinal chemistry
Richert, Hans E, mathematics
Schlag, Edward William, physical chemistry
Schroeder, Manfred Robert, physics
Settles, Ronald Dean, physics, mathematics
Toennies, Jan Peter, molecular physics
Van der Burg, Sjirk, organic chemistry
Vennesland, Birgit, biochemistry, plant biochemistry
von Kaulla, Kurt Nikolaj, experimental medicine
Von Weyssenhoff, Hanns, physical chemistry, physics
Welsh, Richard Stanley, biochemistry
Wiley, John Duncan, solid state physics

GHANA
Lewis, Roger Abbott, pharmacology
Pearson, Robert Edward, organic chemistry, science education

GREECE
Kaffezakis, John George, food science, microbiology

GUATEMALA
Dengo, Gabriel, petrology, structural geology
Guzman Foresti, Miguel Angel, experimental statistics, nutrition
Schieber, Eugenio, plant pathology
Stuart, Laurence Cooper, zoology
Tejada, Carlos, pathology, nutrition

HONDURAS
Davidson, Thomas, Jr., soil chemistry, soil fertility
Myron, Becky Ann, ethology, ecology
Schocken, Klaus, physics
Stover, Robert Harry, plant pathology

HONG KONG
Chan, Yau Wa, physics
Wong, Tin Kin, mathematics

INDIA
Bauer, William Eugene, analytical chemistry
Bullard, Truman Robert, physiology
Eggleston, Forrest Cary, medicine, surgery
Kohn, Gustave K, chemistry
Krantz, Bert Allan, soil science
McGinnis, Robert Cameron, agronomy
Oswalt, Dallas Leon, crop breeding, science education
Royce, Josiah, economic geology, mining engineering
Williams, William Wilson, organic chemistry

INDONESIA
Cleveland, Anne Stack, biochemistry
McIntosh, Jerry Leon, soil chemistry, agronomy
Palmer, Louis Thomas, plant pathology
Salafsky, Bernard P, pharmacology, physiology
Tillman, Allen Douglas, animal husbandry

IRAN
Bush, George Clark, mathematics
Gowing, Donald Proctor, plant physiology
Katzenstein, Jack, physics
Rosen, Norman Charles, petroleum geology, sedimentology
Silverman, Jeremiah Nordau, quantum mechanics
Weiss, Martin George, plant breeding

IRELAND
Klotz, Frederick Succop, theoretical physics
Winslow, John Hathaway, cultural geography, historical geography

ISRAEL
Bekenstein, Jacob David, astrophysics
Beran, Mark Jay, physics, engineering science
Berenson, Lewis Jay, mathematics education
Bergman, Moe, audiology
Bergman, Isadore B, physics
Berfman, Felix, immunochemistry, organic chemistry
Carmeli, Moshe, theoretical physics
Chipman, David Mayer, enzymology, bio-organic chemistry
Cohen, Daniel, public health, epidemiology
Cohen, Maimon Moses, cytogenetics
Czernobilsky, Bernard, pathology
Dorfman, Ben-Zion, genetics
Edelman, Marvin, cell biology, molecular genetics
Eisenberg, Judah Moshe, theoretical physics
Epstein, Benjamin, statistics, operations research
Eschinasi, Emile Haviv, organic chemistry
Flanders, Harley, pure mathematics, applied mathematics
Freed, Myer, biochemistry
Friedland, Stephen Scholom, medical chemistry
Ganchrow, Donald, neuropsychology
Ginsberg, Joseph, physics
Ginsburg, David, organic chemistry
Glaser, Robert, organic chemistry, bio-organic chemistry
Goldman, Dexter Stanley, biological chemistry
Goldman, Morris, parasitology

Goldstein, Marcus Solomon, physical anthropology
Greenblatt, Charles Leonard, microbiology
Greenfield, Arthur Judah, solid state physics
Gressel, Jonathan Ben, developmental biology
Gross, Jack, endocrinology
Hammerman, Ira Saul, biophysics, biological structure
Hirshfeld, Fred Lurie, chemical crystallography
Hirshfeld, Martin Abraham, spectroscopy
Horwitz, Lawrence Paul, theoretical physics
Isaacs, Philip Klein, polymer chemistry
Jakob, Karl Michael, cell biology, plant biochemistry
Katz, Gerald, solid state science, x-ray crystallography
Kaye, Alvin Maurice, molecular biology, reproductive endocrinology
Klein, Michael W, solid state physics, statistical mechanics
Konijn, Hendrik Salomon, statistics, economics
Kraicer, Peretz Freeman, reproductive physiology, endocrinology
Krakower, Gerald W, organic chemistry
Krumbein, Aaron Davis, reactor physics, plasma physics
Landau, Judah, low temperature physics
Lipkin, Harry Jeannot, nuclear physics, particle physics
Low, William, physics
Luban, Marshall, physics
Lutsky, Irving, laboratory animal medicine
Mandelbaum, Hugo, physical oceanography
Metzger, Gershon, chemistry, research administration
Muskat, Joseph Baruch, number theory, computer science
Ohring, George, meteorology
Ophir, Reuven, astrophysics, solid state physics
Otterman, Joseph, space physics
Pekeris, Chaim Leib, geophysics
Perlman, Isadore, nuclear chemistry
Perry, Albert Solomon, insect toxicology, environmental sciences
Quastel, Michael Reuben, immunology, radiobiology
Rabinowitz, Philip, mathematics
Roseman, Joseph Jacob, mathematics
Rosen, Nathan, theoretical physics
Rosenbaum, Milton, psychiatry
Rosenberg, Eugene, biochemistry, bacteriology
Rubin, Mordecai B, organic chemistry
Rudman, Peter S, solid state physics
Segal, Mark, pharmacology
Segel, Lee Aaron, applied mathematics, biomathematics
Shapiro, Jesse Marshall, mathematics
Singer, Arnold, operations research
Snyder, Mitchell, statistics
Solomon, Alan D, applied mathematics, numerical analysis
Sonn, Jack, mathematics
Stern, Alfred, industrial organic chemistry
Stern, Kurt, pathology, cancer
Stolar, Morris Emmanuel, pharmacy, pharmacology

Tauber, Gerald Erich, theoretical physics
Tuler, Floyd Robert, materials science
Vanderhoek, Jack Yehudi, biochemistry, organic chemistry
Weil, Raoul Bloch, physics
Weiss, David Walter, microbiology, immunology
Weiss, Morris J, physical chemistry
Werman, Robert, neurophysiology
Wieder, Irwin, physics, biophysics
Yavin, Avivi I, physics
Ziffer, Jack, microbiology

ITALY
Asselbergs, Edward Anton Maria, plant physiology
Brown, William Malcolm, Jr, plant pathology
Castellani, Maria, mathematics
Chiarappa, Luigi, phytopathology
Feinstein, Myron Elliot, surface chemistry
Ladinsky, Herbert, pharmacology
Lerner, Leonard Joseph, endocrinology
McCarthy, Martin, astronomy
McFarlane, Hugh Murray, chemistry
Meli, Alberto L G, physiology
Papee, Henry Michael, physical chemistry
Pringle, Stanley Leroy, forest economics
Rossi, Sally Wentworth, mineralogy

OTHER COUNTRIES

Shuyler, Harlan R., economic zoology,
medical zoology
Webster, Isabella Margaret, public health
Winn, Edward Barriere, science
administration
Wyman, Jeffries, molecular biology

JAMAICA
Campbell, James Alexander, food
science, nutrition
Lee, James William, geology
Lewis, Charles Bernard, Jr., natural
history

JAPAN
Allen, LeRoy Richard, research
administration, public health
Dickerson, Chester T., Jr.,
olericulture
Heinicke, Ralph Martin, pharmacology
Henick, William Weil, physics
Hildreth, Robert Claire, plant pathology
Loeliger, David A., inorganic chemistry
Marsh, Nat Huyler, organic chemistry
Nishikawa, Osamu, surface physics
Osawa, Eiji, organic chemistry
Pierce, Timothy Ellis, nuclear chemistry,
medical technology
Rich, Ronald Lee, inorganic chemistry
Schor, Robert Hyllel, neurophysiology
Sovers, Ojars Juris, physical chemistry
White, Albert Cornelius, entomology
Worth, Donald Calhoun, nuclear physics

KENYA
Cassard, Daniel Waters, animal science
Cormack, Melville Wallace, plant
pathology
Jachowski, Richard Leo, ethology,
marine biology
Roach, Mary Katherine, biochemistry

KOREA
Kelley, Omer Joseph, soil chemistry
Kim, John Poong-Kil, physical organic
chemistry
Laney, Howard Arthur, plant pathology

LEBANON
Asper, Samuel Phillips, Jr., medical
administration
Avolizi, Robert Joseph, zoology
Azar, Joseph E., infectious diseases,
epidemiology
Basson, Philip Walter, paleobotany, marine
ecology
Bergman, Ronald Arly, anatomy,
physiology
Cowan, James W., nutrition, biochemistry
Fawaz, George, pharmacology,
biochemistry
House, Leland Ralph, plant genetics,
plant breeding
Jabbur, Suhayl Jibra'il, neurophysiology
Kennedy, Edward Stewart, mathematics
Kirkwood, Samuel Brown, neurology
Slade, Landry Thomas, organic chemistry
gynecology
Makenson, John Christopher,
microbiology
McClain, John William, physics
Olmsted, John, III, physical chemistry
Saari, Eugene E., phytopathology
Sabra, Fuad Amin, neurology
Yff, Peter, mathematics

LESOTHO
Hutcheon, Alan Thompson, physical
chemistry

LIBYA
Haeberle, Frederick Roland, exploration
geology
Lawson, Ralph Willard, geology
Lehmann, Elroy Paul, resource
management, petroleum geology
Spencer, Alexander Burke, geology

MALAWI
Holyoke, Thomas Campbell, algebra

MALAYSIA
Green, Jonathan P., zoology, physiology
Riley, James Daniel, pure mathematics
Sastry, Chella Bhaskara Rama, forest
products
Stone, Benjamin Clemens, III, systematic
botany
Tarble, Richard Douglas, hydrology
Young, Frank Evans, physical chemistry

MEXICO
Adem, Julian, meteorology, applied
mathematics
Alvarez-Buylla, Ramon,
neuroendocrinology
Anderson, Robert Glenn, plant genetics,
plant breeding
Balderrama, Francisco E., internal
medicine
Beltran, Enrique, zoology
Bernal-Llanas, Enrique, pharmacology,
microbiology
Bojalil, Luis Felipe, medical microbiology
Borlaug, Norman Ernest, microbiology,
neurophysiology
Brust-Carmona, Hector, physiology,
neurophysiology
Bunge, Marta C., mathematics
Claveran, Ramon A., range science
Costero, Isaac (Tudanca), pathology
de Alba Martinez Jorge, reproductive
physiology, animal husbandry
De Leon, Carlos, phytopathology
Del Pozo, Efren Carlos, physiology
Dominguez, Oscar V., biochemistry,
endocrinology
Dominguez, Xorge Alejandro Sepulveda,
natural products chemistry
Fernandez-Guardiola, Augusto,
electroencephalography
Flores-Gallardo, Hector, organic
chemistry
Garcia-Colin, Leopoldo Scherer, physics,
thermodynamics
Garcia Ramos, Juan, physiology
Garza-Chapa, Raul, human genetics
Genoves, T Santiago, anthropology
Gonzalez-Angulo, Amador,
neuropathology
Graber, Robert Philip, organic chemistry
Grumbles, Jim Bob, range management,
ecology
Gual, Carlos, endocrinology, reproductive
biology
Gutierrez, Luis Garcia, economic
geology, mining
Herz, Josef Edward, steroid chemistry
Hotchkiss, John Calvin, cultural
anthropology, social anthropology
Kaufmann, Isabel Truesdell, organic
chemistry
Kelly, Isabel Truesdell, anthropology,
ethnography
Kingma, Gerhrand, plant breeding
Klatt, Arthur Raymond, plant breeding
Koch, Stephen Douglas, plant taxonomy
Laird, Reggie James, soil science
L'Annunziata, Michael Frank,
agricultural chemistry, soil biochemistry
Lonnitz, Cinna, seismology
Malacara, Daniel, optics
Mallen, Mario Salazar, medicine, history
of medicine
Malo, Salvador Alejandro, surface
physics, experimental atomic
spectroscopy
Mancera, Octavio, organic chemistry
Mann, Charles E., anthropology
Marroquin De La Fuente, Jorge Saul,
plant taxonomy, plant ecology
Matuda, Eizi, plant taxonomy
Morato, Tomas, endocrinology,
reproductive physiology
Muchowski, Joseph Martin, organic
chemistry
Noble, Robert Hamilton, optical physics
Ortega, Alejandro, economic entomology
Osler, Robert Donald, crop breeding,
agronomy
Pardo, Efrain Guillermo, pharmacology
Phillips, Allan Robert, ornithology
Pizarro, Enriqueta, virology
Pietsch, Donald James, entomology,
malariology
Racotta, Radu, physiology
Reyes-Mota, Alfonso, pathology
Rice-Wray, Edris, preventive medicine
Roig, Juan Antonio, neurophysiology
Romo, Jesus, organic chemistry
Rosenbaum, Marcos, mathematical
physics
Rosenkranz, George, organic chemistry
Rudomin, Pedro, neurophysiology
Russek, Mauricio Berman, animal
physiology
Soberon, Guillermo, biochemistry
Swartz, Harry, allergy
Thomas, Prentice Marquet, Jr.,
archaeology, cultural anthropology
Torres-Peimbert, Silvia, astrophysics
Vallarta, Manuel Sandoval, theoretical
physics
Villarreal, Julian Ernesto, pharmacology
Wellhausen, Edwin John, plant breeding
Will, Theodore A., solid state physics
Williams, John Simeon, agriculture, plant
ecology
Winnie, William W., Jr., geography,
sociology
Zilinsky, Francis John, plant breeding

MONACO
Lanni, Yvonne Thery, microbiology

NETHERLANDS
Alanen, Jack David, computer sciences
Bonting, Sjoerd Lieuwe, biochemistry
Etter, Robert Miller, geophysics
Francis, Elliott S., petroleum chemistry
Godefroi, Erik Fred, organic chemistry
Greenberg, Jerome Mayo, astrophysics
Herman, Robert, pathology
Hausman, Bertram, epidemiology
Kellogg, Richard Morrison, organic
chemistry
Knee, Terence Edward Creasey, physical
chemistry, organic chemistry
Lande, Alexander, theoretical nuclear
physics
Lindenmayer, Aristid, theoretical biology
Martin, Kenneth John, clinical chemistry,
analytical chemistry
Mayne, John Winston, organic
statistics
Metzger, W. Wesley James, physics
Myron, Harold William, theoretical solid
state physics
Pottasch, Stuart Robert, astrophysics
Ravesloot, John Lowell, applied
mathematics
Schonbaum, Eduard, pharmacology,
physiology
Schwartz, Alan William, exobiology,
organic geochemistry
Silvera, Isaac F., solid state physics

NEW CALEDONIA
Dahl, Arthur Lyon, marine ecology,
phycology
Reed, Dwayne (Milton), epidemiology

NEW ZEALAND
Bigelow, Robert Sidney, zoology
Bray, John Roger, ecology
Corbet, Philip Steven, zoology,
entomology
Deely, John Joseph, mathematical
statistics
Ellis, Everett Lincoln, wood science,
wood technology
Graves, Nancy Beatrice,
anthropology
Graves, Theodore Dumaine, behavioral
anthropology, social psychology
Green, Roger Curtis, archaeology,
anthropology
Hopkins, Thomas R., biochemistry,
molecular biology
Jensen, Cynthia G., cell biology
Jensen, Lawrence Craig-Winston,
developmental biology
Lewis, Roger Wolcott, biochemistry,
marine biology
Macdonald, John Alan, comparative
physiology
Michalka, Jack, microbial genetics
Olive, Gloria, mathematics
Schatten, Kenneth Howard, solar physics
Schatten, Peter Larry, clinical
biochemistry, medical education
Shaw, Charles Gardner, III, forest
pathology
Stebbens, William Ellis, pathology
Updegraff, David Maule, microbiology

NIGERIA
Arshad, Muhammad Ahmad, soil
conservation
Buddenhagen, Ivan William, plant
pathology
Caveness, Fields Earl, nematology
Detmers, Almut Edel, anatomy,
animal breeding
Heath, Everett, anatomy, reproductive
biology
Nickel, John L., entomology
Rieg, Louis Eugene, geology
Sadik, Sidki, plant physiology
Sanford, William Warren, plant
physiology

NORTHERN IRELAND
Meyers, Martin Bernard, organic
chemistry
Moore, Royall Tyler, mycology
Wilson, Raymond Hiram, Jr., astronomy,
applied mathematics

NORWAY
Klove, Hallgrim, neuropsychology
Rodahl, Kaare, physiology

PAKISTAN
Izuno, Takumi, agronomy

PERU
Arce, Jose Edgar, geophysics, geology
Caceres, Eduardo, oncology
Hay-Roe, Hugh, petroleum geology
Sawyer, Richard Leander, physiology

PHILIPPINES
Brady, Nyle C., agronomy
Feuer, Reeshon, agronomy, soil
morphology
Glover, Francis Nicholas, physics
Harwood, Richard Roland, agronomy
Hennessy, James J., agronomy
Heyden, Francis Joseph, astronomy
Hodgson, John Humphrey, seismology
Ingalls, Ronald Boyd, physical chemistry
Jenks, Robert L., applied physics
Johnson, Donald Haskall, mineralogy,
crystallography
Kremheller, Alfred, physics, physical
chemistry
Khush, Gurdev S., plant breeding
Nemenzo, Francisco, zoology
Schmitt, William Joseph, organic
chemistry
Torio, Joyce Clarke, science
administration, information science
Velasquez, Carmen C., zoology,
parasitology

RWANDA
Massar, Ann Roller, molecular biology

SAUDI ARABIA
Bowsher, Arthur Leroy, Sr., geology
Hoke, John Humphreys, geology,
geophysics
Kang, Tae Wha, microbiology
Maas, Peter, solid state physics,
biophysics
Proudfoot, Vincent Bruce, geography,
archaeology
Stugard, Frederick, geology
Watson, James Wreford, cultural
geography

SCOTLAND
Berg, Arthur R., plant morphology
Colton, David L., mathematics
Friedman, Lionel Robert, solid state
physics
Gibson, James (Benjamin), astronomy
McGarr, Arthur, seismology
Neuse, Eberhard Wilhelm, polymer
chemistry, organometallic
Rogers, William Alan, physics
Shana'a, Joyce M., mathematics
Wolf, Karl Heinz, geology
Wright, Charles Cathbert, chemistry

SINGAPORE
Holwerda, James G., geology
Johnson, Norman Eilden, forest
management, silviculture
Schaetti, Henry Joachim, geology

SOUTH AFRICA, REPUBLIC OF
Cherenack, Paul Francis, mathematics
Simmons, Eugene Lynn, physical
chemistry
Scott, Mary Jean (Mrs E C H Silk),
medical physics
Strickland, Walter Nicholas, genetics
Wolbarst, Anthony Brinton, high
pressure physics, magnetic resonance

SPAIN
Bell, Frank Heaton, plant nematology
Cabrera, Nicolas, solid state physics

SWEDEN
Berglund, Erik, pulmonary diseases
Bertani, Giuseppe, molecular genetics
Bertani, Lillian Elizabeth, microbiology
Carter, Robert Everett, physical organic
chemistry
Koenig, Donald Frederick, biophysics,
crystallography
Lewis, David Harold, cardiology
Wesemeyer, Harald, experimental physics
Wickman, Frans Erik, geochemistry

SWITZERLAND
Baratoff, Alexis, theoretical solid state
physics
Biewer, Myrtle Hildred, physics
Boatner, Lynn Allen, magnetic resonance
Borgnis, Fritz Edward, electronics,
biomedical engineering
Bruck, George, applied physics
Buck, Alfred A., epidemiology, tropical
medicine
Budowski, Gerardo, forestry, plant
ecology
Caro, Lucien G., molecular biology
Daeniker, Hans Ulrich, organic chemistry
Dasmann, Raymond Fredric, ecology
Dierks, Christa, cell biology

Travis, Russell Burton, geology
Wurster, Richard T., horticulture

Duschinsky, Robert Charles, medicinal chemistry
Esteve, Ramon M, Jr, physical organic chemistry
Flanagan, John Vernon, industrial organic chemistry
Franklin, Richard Morris, molecular biophysics, virology
Freedman, Lawrence Raphael, internal medicine
Gelzer, Justus, microbiology, immunology
Golay, Marcel Jules Edouard, physics
Goodman, Howard Charles, immunology
Gratz, Norman G, medical entomology
Henderson, Donald Ainslie, medicine, epidemiology
Henrici, Peter K, mathematics
Holmes, Lewis M, applied physics, solid state physics
Howard, Guy Allen, molecular biology
Hsu, Kenneth Jinghwa, geology
Jacobson, Ann Beatrice, cell biology
Jost, Jean-Pierre, biochemistry, cell biology
Kent, Naim Hassan, immunobiology, parasitology
Kessler, Alexander, medicine, public health
Kiefer, Hansruedi, organic chemistry, bio-organic chemistry
Kitler, Mary Ellen, biostatistics, pharmacology
Koella, Werner Paul, neurophysiology
Konecny, Jan, physical biochemistry
Kuyper, C Keith, entomology
Leuchtenberger, Cecile, cytology
Lu, Frank Chao, pharmacology
Matray, Otto Jack, organic chemistry

Mehler, Ernest Louis, theoretical chemistry
Merz, Walter John, solid state physics
Moholy-Nagy, Hattula, anthropology, archaeology
Moos, Walter Sam, radiology
Mueller, Walter E, solid state physics
Munn, John Irvin, pharmacology
Newell, Kenneth Wyatt, epidemiology
Oertli, Johann Jakob, soil science, plant nutrition
Piroue, Robert Paul, organic chemistry
Prager, William, applied mechanics
Renold, Albert Ernst, experimental medicine, biochemistry
Rubin, Morton Joseph, meteorology
Schnurrenberger, Paul Robert, veterinary public health
Shaw, Joseph Clement, animal nutrition, physiology
Siegel, Herbert, information science, organic chemistry
Sinden, James Whaples, plant pathology
Sobin, Leslie Howard, pathology
Steinberger, Jack, physics
Sternbach, Daniel David, organic chemistry
Tannenbaum, Carl Martin, biochemistry
Tscharner, Christopher J, organic chemistry, physical chemistry
Ward, John Edward, organic chemistry
Wentzel, Gregor, theoretical physics
Zeiss, Harold Hicks, organic chemistry

TANZANIA

Hatfield, Colby Ray, Jr, applied anthropology, ethnology
Wiggin, Henry Carvel, genetics, plant breeding

THAILAND

Chomchalow, Narong, agronomy, plant genetics
Knorr, Louis Carl, plant pathology
Manis, Wallace Eugene, horticulture
McRoberts, Milton R, biochemistry
Olson, Lloyd Clarence, microbiology, pediatrics
Smeltzer, Dale Gardner, agronomy, genetics
Wheeler, Richard Hunting, forest hydrology
Wray, Joe D, pediatrics, public health
Young, William Robert, economic entomology

TONGA

Hitchcock, James Carroll, Jr, medical entomology

TUNISIA

Grigsby, Buford Horace, botany

TURKEY

Crowell, Julian, mathematics
Harrell, Bryant (Eugene, Jr), organic chemistry
Hoffman, Eugene James, cell biology
McMickle, Robert Hawley, physics
Prescott, Jon Michael, plant pathology, plant breeding
Schuman, Robert Paul, nuclear chemistry, physical chemistry

URUGUAY

Buno, Washington Hector, histology
Cernuschi, Felix, physics, astrophysics
Mackinnon, Juan Enrique, medical mycology, insect toxicology

VENEZUELA

Altman, Allen Burchard, geometry, algebra
Andrea, Stephen Alfred, mathematics
Bemski, George, biophysics
Bunting, George Sydney, Jr, plant taxonomy
Chisholm, Roderick G, geophysics
Downing, John Scott, mathematics, topology
Giulianelli, James Louis, physical chemistry
Jaffe, Werner G, nutrition, biochemistry
Kleiss, Ekkehard, embryology, anatomy
Reinosa Fuller, Jose Angel, biochemistry
Roche, Marcel, endocrinology, parasitology
Rodriguez, Gilberto, biological oceanography
Seibert, Peter, mathematics
Steyermark, Julian Alfred, taxonomic botany
Tonn, Robert J, medical entomology
Whittembury, Guillermo, biophysics
Winkler, Virgil Dean, geology

WALES

Bohren, Craig Frederick, applied physics
Haworth, Alfred John, geology
Riffenburgh, Rogert Harry, statistics, biomedical engineering
Schutz, Bernard Frederick, theoretical astrophysics, mathematical physics

ZAMBIA

Brown, Delos D, food science